S0-BWX-117

Atlas of NUTRITIONAL DATA ON UNITED STATES AND CANADIAN FEEDS

Subcommittee on Feed Composition
Committee on Animal Nutrition
Agricultural Board
National Research Council, United States

and

Committee on Feed Composition
Research Branch
Department of Agriculture, Canada

NATIONAL ACADEMY OF SCIENCES

WASHINGTON, D.C. 1971

NOTICE: The study reported herein was undertaken under the aegis of the National Research Council with the approval of its Governing Board. Such approval indicated that the Board considered the problem of national significance, and that the resources of NRC were particularly suitable to the conduct of the project.

The members of the committee were selected for their individual scholarly competence and judgment with due consideration for the balance and breadth of disciplines. Responsibility for all aspects of this report rests with the study committee, to whom we express our sincere appreciation.

Although the reports of committees are not submitted for approval to the Academy membership nor to the Council, each is reviewed according to procedures established and monitored by the Academy's Report Review Committee. The report is distributed only after satisfactory completion of this review process.

This publication was supported in part by the National Institute of Arthritis and Metabolic Diseases (Contract PH 43-64-44, Task Order No. 4). The investigation was supported in part by financial assistance to Utah State University from U.S. Public Health Service (National Center for Urban and Industrial Health) Research Grant UI-00679. Financial assistance was also received from the Utah Agricultural Experiment Station, Utah State University, Logan, Utah; from Brigham Young University, Provo, Utah; and from the Canada Department of Agriculture, Ottawa, Ontario.

and U.S. Department of Agriculture

ISBN 0-309-01919-2

Available from

Printing and Publishing Office
National Academy of Sciences
2101 Constitution Avenue
Washington, D.C. 20418

Library of Congress Catalog Card Number 76-612077

Printed in the United States of America

PREFACE

This publication was prepared by the Subcommittee on Feed Composition: E. W. Crampton, Chairman, and L. E. Harris. It records names of feeds and analytical data on feeds that have been published in the following books, bulletins, and regulations:

Association of American Feed Control Officials. 1969. Official publication. Kentucky Agricultural Experiment Station, University of Kentucky, Lexington.
Canada Feed Act Regulations, as amended; 1967.
Canada Grain Act Regulations, as amended; 1967.
Ewing, W. R. 1963. Poultry nutrition. 5th ed. rev. Ray Ewing Co., Pasadena, Calif.
Morrison, F. B. 1956. Feeds and feeding. 22nd ed. Morrison Publishing Co., formerly at Ithaca, N.Y., now at Clairmont, Alberta, Canada.
National Research Council. 1956. Composition of concentrated by-product feeding stuffs. NAS–NRC Publ. 449. National Academy of Sciences–National Research Council, Washington, D.C.
National Research Council. 1958. Composition of cereal grains and forages. NAS–NRC Publ. 585. National Academy of Sciences–National Research Council, Washington, D.C.
Official Grain Standards of the United States; 1964.
Official Hay and Straw Standards of the United States; 1949.
Schneider, B. H. 1947. Feeds of the world: their digestibility and composition. West Virginia Agricultural Experiment Station, Morgantown.
Scott, M. L., M. C. Nesheim, and R. J. Young. 1969. Nutrition of the chicken. M. L. Scott and Associates. Ithaca, N.Y.
Titus, H. W. 1949. The scientific feeding of chickens. 2nd ed. Interstate Printers & Publishers, Danville, Ill.

Also included are results, to September 1969, of a continuing effort to assemble nutritionally significant compositional data for feeds of Canadian and United States origin that have been published in scientific journals. Unpublished data have been furnished by experiment station laboratories in the United States. Unpublished data have also been provided by the Canada Department of Agriculture Experimental Farms System, the Canadian Fisheries Board, the Board of Grain Commissioners of Canada, members of the Canadian Feed Manufacturers Association, and members of agricultural facilities at provincial universities, colleges, and schools.

In addition, the Board of Grain Commissioners supplied 1,000-g samples of 27 requested kinds and grades of unground grain normally sold through the feed trade for livestock feeding. Calcium and phosphorus data for these grains were available from the Grain Commissioners' laboratory. The Analytical Chemistry Research Service of the Canada Department of Agriculture analyzed the samples for proximate principles, cellulose, and six requested mineral elements; and the Research Branch, Canada Department of Agriculture, financed contract chemical analyses of the samples for their amino acids and for five specified vitamins of the B complex.

Noteworthy features of the tables of feed composition include the following:

- Data on 6,152 feeds are given.
- Data for each feed are presented in a continuous column; the inconvenience of searching in several tables is avoided.
- Data are presented on both an "as fed" and a "dry" basis.
- A coefficient of variation has been calculated for each nutrient for which there are four or more analyses.
- Data on the digestible and metabolizable energy of feeds for cattle, horses, sheep, and swine are given.
- The metabolizable energy of the feeds commonly fed to poultry is stated.
- The net energy of some cattle feeds is stated.

The last three features make it possible to use the calorie system for calculating diets and feed mixtures. Publications in the NRC nutrient requirement series express energy needs and feed composition in calories.[1]

The committees hope that these tables will be useful to feed manufacturers, feed dealers, nutrition consultants, research specialists, government agencies, teachers, students, county agricultural agents, and farmers.

This Atlas differs from NRC Publication 1684, *United States–Canadian Tables of Feed Composition,* in that the latter lists only the feeds (about 400) that are used in the everyday practice of feed manufacturers and feed dealers, whereas the Atlas lists every feed that has been used in North America. The Atlas contains much material that is useful primarily for reference purposes; hence, it is not intended to replace the working tables used by persons who are concerned only with balancing specific diets.

Collection of data on feed composition is a continuing program in the United States and Canada. Thus, it will be possible to revise these tables as new data warrant or as new products of technology appear.

Dr. J. M. Asplund and Professor L. C. Kearl, Animal Science Department, Utah State University, contributed unstintingly to compiling and organizing the data in this publication. The committees are grateful for their assistance.

The committees thank Mr. Bliss H. Crandall, DHI Computing Service, Provo, Utah, and Professor Karl A. Fugal, Computer Center, Utah State University, for helping with the electronic computer programs.

[1]Series: 1, *Nutrient Requirements of Poultry*; 2, *Nutrient Requirements of Swine*; 3, *Nutrient Requirements of Dairy Cattle*; 4, *Nutrient Requirements of Beef Cattle*; 5, *Nutrient Requirements of Sheep*; 6, *Nutrient Requirements of Horses*; 7, *Nutrient Requirements of Mink and Foxes*; 8, *Nutrient Requirements of Dogs*; 9, *Nutrient Requirements of Rabbits*; and 10, *Nutrient Requirements of Laboratory Animals.* Information about these publications is available from the Printing and Publishing Office, National Academy of Sciences, 2101 Constitution Avenue, Washington, D.C. 20418.

COMMITTEE ON ANIMAL NUTRITION
NATIONAL RESEARCH COUNCIL
UNITED STATES

W. M. Beeson, *Chairman*
H. R. Bird
E. W. Crampton
G. K. Davis
R. M. Forbes
L. E. Harris
L. E. Hanson
J. K. Loosli
J. E. Oldfield
A. D. Tillman

SUBCOMMITTEE ON FEED COMPOSITION

E. W. Crampton, *Chairman*
L. E. Harris

COMMITTEE ON FEED COMPOSITION
RESEARCH BRANCH, DEPARTMENT OF AGRICULTURE
CANADA

E. W. Crampton, *Chairman*
J. R. Aitken
J. M. Bell
L. W. McElroy
W. J. Pigden
W. D. Morrison

CONTENTS

GUIDE TO TABLES OF FEED COMPOSITION

NRC FEED NAMES

Analytical data for 6,152 feeds are given in the tables of feed composition that begin on page 3. Each of these feeds has a name based on a system of nomenclature devised by Harris.[1,2] Names based on this system are called NRC (National Research Council) names.

An NRC name consists of several components and gives a qualitative description of a feed. There are eight components:

- Origin (or parent material)
- Species, variety, or kind
- Part eaten
- Process(es) and treatment(s) to which the parent material or the part eaten has been subjected
- Stage of maturity (applicable only to forages)
- Cutting or crop (applicable only to forages)
- Grade, quality designations, and guarantees
- Classification

Some names contain all eight components, but most names contain fewer. Whether a component is included in a name depends on its applicability.

Feeds of the same origin (and of the same species, variety, or kind, if one of these is stated) are grouped into eight classes, each of which is designated in the NRC name by a number in parentheses. This number is the last component in the name. The numbers and the classes they designate are as follows:

(1) Dry forages or dry roughages
(2) Pasture, range plants, and feeds cut and fed green
(3) Silages
(4) Energy feeds
(5) Protein supplements
(6) Minerals
(7) Vitamins
(8) Additives

Feeds are classified in this way because each class has properties that are considered in balancing a diet.

Feeds that in the dry state contain on the average more than 18 percent of crude fiber are classified as forages or roughages. Feeds that contain 20 percent or more of protein are classified as protein supplements. (These guidelines are approximate, and there is some overlapping.) Fruits, nuts, and roots are classified as energy feeds because most of the by-product feeds from these materials furnish, primarily, energy to the animal.

Certain of the terms used in NRC feed names and in discussions of feeds are defined in a glossary (page 761).[3] Stage-of-maturity terms are defined in Table 1. Abbreviations have been devised for some of the terms in the NRC system (Table 2). This was done to make the names fit printing and punch-card space requirements.

The following listings show how three feeds are described:

Components of Name	Feed No. 1	Feed No. 2	Feed No. 3
Origin (or parent material)	Alfalfa	Fish	Oats
Species, variety, or kind	ranger	rockfish	–
Part eaten	aerial part	whole	groats
Process(es) and treatment(s)	dehy	raw	grnd cooked
Stage of maturity	early bloom	–	–
Cutting or crop	cut 1	–	–
Grade, quality, or guarantees	mn 17% protein mx 27% fiber	–	–
Classification	(1)	(5)	(4)

[1] L. E. Harris. 1963. Symposium on feeds and meats terminology. III. A system for naming and describing feedstuffs, energy terminology, and the use of such information in calculating diets. J. Anim. Sci. 22:535–547.

[2] L. E. Harris, E. W. Crampton, A. D. Knight, and A. Denney. 1968. Collection and summarization of feed composition data. I. The National Research Council feed nomenclature system. J. Anim. Sci. 27:1743–1754.

[3] L. E. Harris, J. M. Asplund, and E. W. Crampton. 1968. An international feed nomenclature and methods for summarizing and using feed data to calculate diets. Utah Agr. Exp. Sta. Bull. 479. Utah State University, Logan. 392 p.

The NRC names of these feeds, in normal linear form, are as follows:

No. 1: Alfalfa, ranger, aerial part, dehy, early bloom, cut 1, mn 17% protein mx 27% fiber, (1)
No. 2: Fish, rockfish, whole, raw, (5)
No. 3: Oats, groats, grnd cooked, (4)

The preceding examples illustrate the principle that an NRC feed name consists of components that are applicable to the feed; inapplicable components are omitted.

LOCATING NAMES IN THE TABLES

To locate the NRC name of a feed in the tables of feed composition (see page 3), one must know the name of the parent material (i.e., the origin of the feed) and usually the variety or kind of parent material.

Parent materials are of four types: plant, animal (other than fish and poultry), poultry, and fish. For a feed derived from a plant, the origin term is the name of the plant (e.g., Alfalfa, Barley, Oats), not the word "plant." For a feed derived from animals or poultry, the origin term is the name of the animal or bird (e.g., Cattle, Chicken, Crab, Horse, Sheep, Turkey, Whale). For a feed of fish origin, the origin term is "Fish" followed by the species or variety (e.g., Fish, cod; Fish, salmon).

When the specific origin of a feed derived from animals, poultry, or fish is not known, the origin term is "Animal," "Poultry," or "Fish."

Names having the same origin term are arranged in an order that depends on whether the names include reference to species, variety, or kind. Names that lack such references are arranged under the origin term as follows:

First: numerically, by classes
Second (within a class): alphabetically, by parts eaten, process(es), stage of maturity (in the order in which the stages occur), cutting, and grade

Names that include references to species, variety, or kind are arranged under the origin term as follows:

First: alphabetically, by species, variety, or kind
Second (within species, variety, or kind): numerically, by classes
Third (within a class): alphabetically, by parts eaten, process(es), stage of maturity (in the order in which the stages occur), cutting, and grade

Names that lack references to species, variety, or kind are listed before names that include them.

Following are examples of entries under the origin term "Wheat" that *lack* references to species, variety, or kind:

Wheat, hay, s-c, milk stage, (1)
Wheat, hay, s-c, dough stage, (1)
Wheat, straw, (1)
Wheat, aerial part, fresh, (2)
Wheat, aerial part, ensiled, (3)
Wheat, flour, coarse bolt, feed gr mx 2% fiber, (4)
Wheat, grain, (4)
Wheat, grain, cooked, (4)
Wheat, grain, Canada feed gr mx 13% foreign material, (4)
Wheat, gluten, grnd, (5)
Wheat, germ oil, (7)

Examples of entries under the origin term "Wheat" that *include* references to species, variety, or kind:

Wheat, durum, bran, dry milled, (4)
Wheat, durum, grain, gr sample US, (4)
Wheat, hard red winter, germ, (5)
Wheat, hard spring, grain, (4)
Wheat, hard spring, germ, (5)

A reader who is uncertain about the origin term that introduces an NRC name in the tables of feed composition may find the term by referring to the common name of the feed in which he is interested. Common names appear in their alphabetical places, and with each common name is a reference to the NRC name. For example, the reader will find the following common names and references:

In the "H" listings
Hominy feed (AAFCO) (CFA)–
see Corn, grits by-product, mn 5% fat, (4)
In the "M" listings
Maize–
see Corn
Milk, cattle, fresh–
see Cattle, milk, fresh, (5)
Molasses, cane–
see Sugarcane, molasses, mn 48% invert sugar mn 79.5 degrees brix, (4)

Many feeds have names that were given to them by the Association of American Feed Control Officials (AAFCO), the Canada Feed Act (CFA), or the Canada Grain Act (CGA). In addition, some feeds have regional or local names. These other names are in their alphabetical places and are cross-referenced to the NRC names.

"Linseed meal (CFA)" is an example of an officially adopted name that is cross-referenced to an NRC name. The reader will find this entry in its alphabetical place:

Linseed meal (CFA)–
see Flax, seeds, mech-extd grnd, mx 10% fiber, (5)

Under "Flax" the reader will find:

Flax, seeds, mech-extd grnd, mx 10% fiber, (5)
Linseed meal, mechanical extracted (AAFCO)
Linseed meal (CFA)
Linseed oil meal, expeller extracted

Linseed oil meal, hydraulic extracted
Linseed oil meal, old process
Ref no 5-02-045

The six-digit reference number listed after each name may be used as the "numerical name" of a feed when performing linear programming with electronic computers. The first digit of the number is the classification designation of the feed (1 = dry forages or dry roughages; 2 = pasture, range plants, and feeds cut and fed green; 3 = silages; and so on). See page vii.

INFORMATION CONTAINED IN THE TABLES

The Committee on Feed Composition was established primarily to assemble nutritionally significant information about the composition of the feedstuffs of North America.

Table 3 shows the kinds of information that the committee assembled for each feed. It lists the nutrients and the biological data that are applicable to them. It shows titer, nutritive value index, weight per liter, and other information that helps nutritionists make up diets and feed mixtures for animals. The sequence is the same as in the tables of feed composition. Data for two feedstuffs, sun-cured alfalfa and corn grain, are included. These feedstuffs are appropriate as examples because they are common and because they are among the feedstuffs about which we have the most extensive data. Even for these feedstuffs there are many blank spaces.

For a few nutrients, data obtained by more than one procedure were available and were recorded separately. (For example, two procedures are available for cellulose and three for lignin.) In specific cases, procedures are identified by referring to authors.

In the past, nutritionists balanced diets for a few nutrients. For example, they balanced ruminants' diets for energy, protein, calcium, phosphorus, and carotene (vitamin A). Diets are now balanced for several more minerals and vitamins. To balance the diet for many nutrients, it is necessary to have data in the tables of feed composition for each nutrient that animals need. Where complete information is available about the feeds, nutritionists can use linear programming to balance the diet for a given animal. The data in the tables of feed composition are not complete, and much more work is needed to augment them.

ANALYTICAL DATA

In the tables of feed composition the analytical data are expressed in metric units (see Table 4 for conversion factors), with the exception of the bushel weights of the cereal grains, and are shown on an "as fed" as well as a "dry" basis.

A coefficient of variation (CV) is given for each constituent for which there are four or more analyses. Coefficients of variation are expressed as percentages of the mean and are directly comparable between sets of figures whose means are approximately the same. This information greatly simplifies the task of setting tolerances when linear programming is used to make up feed mixtures or diets.

The tolerance limit to be established depends on the variability of the nutrient in the feed. If the coefficient of variation for a nutrient is more than 20 percent, and if the nutrient is a critical one, the feed should be analyzed. Other things to take into consideration include the stability of the nutrient, weather conditions (for free-choice mineral and vitamin supplements), temperature, and the length of time the feed has been in storage.

Compositional data were gathered and recorded in the steps listed below.[4] Certain values in the tables of feed composition that were not determined experimentally are marked with an asterisk (*); these values were computed from equations that are marked with an asterisk in steps 7 to 12 or were estimated from feeds of similar composition.

1. The original data were collected on source forms designed for this project.[5]
2. The various items in the source forms were coded to permit processing by machine.
3. Data in the source forms were transferred to punch cards.
4. All nutrient data were calculated to the "preferred unit" basis and to a dry-matter basis.
5. Individual values for each nutrient were totaled, means were calculated, and where there were four or more analyses, the coefficient of variation was computed.
6. The mean nitrogen-free extract was calculated with mean data as follows:

$$\text{Mean nitrogen-free extract (\%)} = 100\% - \%\ \text{ash} - \%\ \text{crude fiber} - \%\ \text{ether extract} - \%\ \text{protein}$$

7. Digestible protein was calculated for each kind of animal by the formula

$$\text{Digestible protein} = \frac{\%\ \text{protein} \times \text{protein digestion coefficient}}{100}$$

or by equations* in Table 5 if protein digestion coefficients were not available.

8. Digestible energy for each kind of animal was calculated from:

   a. DE in kcal/kg = GE (kcal/kg) × GE digestion coefficient
   b. The mean digestible energy in kcal/kg
   c. TDN as:

$$\text{*DE in kcal/kg} = \frac{\text{TDN \%}}{100} \times 4{,}409$$

[4] For further information, see the work cited in note 3.

[5] L. E. Harris. 1970. Nutrition research techniques for domestic and wild animals. Vol. I. An international record system and procedures for analyzing samples. Lorin E. Harris, 1408 Highland Drive, Logan, Utah.

9. Metabolizable energy was calculated as follows:

a. *For each kind of animal except chickens and turkeys*
From the average metabolizable energy (ME)

b. *For chickens and turkeys*[6]
From nitrogen-corrected metabolizable energy (ME_n)

c. *For finishing cattle, sheep, and horses*
From DE as follows:

$$*\text{ME in Mcal/kg} = \text{DE (Mcal/kg)} \times 0.82$$

d. *For swine*[7]
From DE as follows:

$$*\text{ME in kcal/kg} = \text{DE (kcal/kg)} \times \frac{96 - (0.202 \times \text{crude protein \%})}{100}$$

10. Net energy (NE) values for some cattle feeds were taken from Lofgreen and Garrett[8] or were computed by their equations* as follows:

$$\text{Log F} = 2.2577 - 0.2213\ \text{ME}$$
$$NE_m \text{ in Mcal/kg} = 77/F$$
$$NE_{gain} \text{ in Mcal/kg} = 2.54 - 0.0314\ F$$

Where these equations apply to cattle feeds, ME is the metabolizable energy in Mcal/kg of dry matter and F is the grams of dry matter per unit of $W^{0.75}$ required to maintain energy equilibrium. In the tables of feed composition, the values NE_m and NE_{gain} are labeled "NE_m Mcal/kg" and "NE_{gain} Mcal/kg" where they apply to finishing or growing cattle.

11. Net energy (NE) values in Mcal/kg of dry matter for some feeds fed to lactating cattle were computed by a formula* developed by Moe and Flatt:[9]

$$\text{NE (Mcal/kg dry matter) for lactating cows} = -0.77 + 0.84\ \text{DE (Mcal/kg dry matter)}$$

The resulting energy values for lactating cows were decreased by 7 percent to allow for variation in feedstuffs and thus ensure adequate energy intake. This decrease amounts to one standard deviation of the prediction equation given above. In the tables of feed composition, this value is designated $NE_{lactating\ cows}$.

Moe and Flatt assume that their formula adjusts the values for maintenance, growth, and pregnancy to utilization for milk production. If this assumption is correct, one value can be used for all these physiological functions of a lactating cow, provided the NRC requirements for lactating cows are used. Moe and Flatt call this value "NE_{milk}." A term that fits better into the NRC terminology for energy is "$NE_{lactating\ cows}$." In the tables of feed composition, the value is labeled "Cattle $NE_{lactating\ cows}$ Mcal/kg."

12. Total digestible nutrients (TDN) were calculated for each kind of animal. In each calculation, one of the five following procedures was used, the procedure depending on the type of animal.

a. From digestion coefficients as follows:

digestible protein in %	× 1
digestible crude fiber in %	× 1
digestible nitrogen-free extract in %	× 1
digestible ether extract in %	× 2.25
	TDN in % = Total

b. From average TDN

c. From formulas* developed by Schneider *et al.*[10]

d. From DE for ruminants as follows:

$$*\text{TDN in \%} = \frac{\text{DE (kcal/kg)}}{4{,}409} \times 100$$

e. From ME as follows:

$$*\text{TDN in \%} = \frac{\text{ME (kcal/kg)}}{3{,}616} \times 100$$

13. Vitamin A and carotene calculations were made in accordance with the relationships described below.

a. The international standards for vitamin A activity as related to vitamin A and beta-carotene are as follows:

One IU of vitamin A = one USP unit
= vitamin A activity of 0.300 μg of crystalline vitamin A alcohol, which corresponds to 0.344 μg of vitamin A acetate or 0.550 μg of vitamin A palmitate

b. Beta-carotene is the standard for provitamin A.
One IU of vitamin A = 0.6 μg of beta-carotene
One mg of beta-carotene = 1,667 IU of vitamin A

International standards for vitamin A are based on the utilization of vitamin A and beta-carotene by the rat. The vitamin A activity for carotene was calculated by assuming that 0.6 μg of beta-carotene equals one IU of vitamin A. Because the various species do not convert carotene to vitamin A in the same ratio as rats, it is suggested that the conversion rates in Table 6 be used.

[6] For methods of calculating ME_n, see: National Research Council. 1966. Biological energy interrelationships and glossary of energy terms. 2nd ed. (Prepared by L. E. Harris.) NAS-NRC Publ. 1411. National Academy of Sciences, Washington, D.C.

[7] For futher information, see: J. M. Asplund and L. E. Harris. 1969. Metabolizable energy values for nutrient requirements for swine. Feedstuffs 41(14):38–39.

[8] G. P. Lofgreen and W. N. Garrett. 1968. A system for expressing net energy requirements and feed values for growing and finishing beef cattle. J. Anim. Sci. 27:793–806.

[9] P. W. Moe and W. P. Flatt. 1969. Net energy value of feedstuffs for lactation. J. Dairy Sci. 52:928. (Abstr.)

[10] B. H. Schneider, H. L. Lucas, M. A. Cipolloni, and H. M. Pavelech. 1952. The prediction of digestibility for feeds for which there are only proximate composition data. J. Anim. Sci. 11:77–83.

TABLE 1 Stage-of-Maturity Terms Used in NRC Feed Names

Preferred Term	Definition	Comparable Terms
For Plants That Bloom		
Germinated	Stage in which the embryo in a seed resumes growth after a dormant period	Sprouted
Early leaf	Stage in which a plant becomes vegetative, before the stems elongate	Fresh new growth, very immature
Immature	Stage in which stems begin to elongate	Before heading out, before inflorescence emergence, prebud stage, young
Prebloom	Stage in which stems are near maximum length; just before blooming	Before bloom, bud stage, budding plants, heading to in bloom, heads just showing, preflowering, stems elongated, tasseling in corn
Early bloom	Stage between initiation of bloom and stage in which 1/10 of the plants are in bloom; some grass heads are in anthesis	Early anthesis, first flower, headed out, in head, up to 1/10 bloom
Mid-bloom	Stage in which 1/10 to 2/3 of the plants are in bloom; most grass heads are in anthesis	Bloom, flowering, flowering plants, half bloom, in bloom, mid-anthesis
Full bloom	Stage in which 2/3 or more of the plants are in bloom	Late anthesis, 3/4 to full bloom
Late bloom	Stage in which blossoms begin to dry and fall and seeds begin to form	Before milk, 15 days after silking, in bloom to early pod, late to past anthesis
Milk stage	Stage in which seeds are well formed but soft and immature	After anthesis, early seed, fruiting, late bloom to early seed, past bloom, pod, post-anthesis, post-bloom, seed developing, seed forming, soft, soft immature
Dough stage	Stage in which the seeds are of doughlike consistency	Nearly mature, seeds well developed, soft dent
Mature	Stage in which plants are normally harvested for seed	Dent, dough to glazing, fruiting, fruiting plants, in seed, kernels ripe, well matured
Overripe	Stage that follows maturity; seeds are ripe and plants have begun to weather (applies mostly to range plants)	Late seed, ripe, very mature, well matured
Dormant	Stage in which plants are cured on the stem; seeds have been cast and weathering has taken place (applies mostly to range plants)	Mature and weathered, seeds cast
Regrowth		
Vegetative	Stage in which regrowth occurs without flowering activity; vegetative crop aftermath; early dry-season regrowth	Vegetative recovery growth
Flowering	Stage in which regrowth occurs with flowering activity; crop aftermath with flowering	Recovery growth; stems elongating
Ripe seed	Stage in which regrowth occurs after frost or after good growing conditions cease; crop aftermath with ripe seed	
For Plants That Do Not Bloom[a]		
4 to 10 days' growth	A specified length of time after plants have started to grow	
11 to 17 days' growth		
18 to 24 days' growth		

[a]These classes are for species that remain vegetative for long periods and apply primarily to the tropics. When the name of a feed is developed, the age classes form part of the name (e.g., Pangolagrass, aerial parts, 18 to 24 days' growth). For plants growing longer than 24 days, the interval is increased by increments of 1 week.

TABLE 2 NRC Feed-Term Abbreviations

AAFCO	Association of American Feed Control Officials	mech	mechanical
Can	Canadian	mech-extd	mechanically extracted, expeller extracted, hydraulic extracted, or old process
CE	Canadian Eastern	µg	microgram
CGA	Canada Grain Act	mg	milligram
CFA	Canada Feeds Act	mm	millimeter
cp	chemically pure	mn	minimum
CW	Canadian Western	mx	maximum
dehy	dehydrated	NRC	National Research Council
extd	extracted	ppm	parts per million
extn	extraction	s-c	sun-cured
extn unspec	extraction unspecified	solv-extd	solvent-extracted
g	gram(s)	spp	species
gr	grade	US	United States
grnd	ground	USP	United States Pharmacopeia
ICU	International Chick Unit	w	with
IU	International Units	wo	without
kcal	kilocalories	wt	weight
kg	kilogram(s)		
lb	pound(s)		

TABLE 3 Sequence in Which Information in the Tables of Feed Composition Is Arranged

			Alfalfa, sun-cured (1)	Corn, grain (4)
Dry matter		%	100.0	100.0
Cattle	dig coef	%	61.	
Goats	dig coef	%		
Horses	dig coef	%		
Rabbits	dig coef	%	52.	
Rats	dig coef	%		
Sheep	dig coef	%	59.	
Swine	dig coef	%		
in vitro	dig coef	%	66.	
Organic matter		%	87.6	
Ash (total)		%	9.9	1.6
Crude fiber (Weende)		%	30.6	2.4
Cattle	dig coef	%	42.	16.
Goats	dig coef	%		
Horses	dig coef	%	38.	17.
Rabbits	dig coef	%		
Rats	dig coef	%		
Sheep	dig coef	%	47.	62.
Swine	dig coef	%	22.	12.
Ether extract		%	1.9	4.7
Cattle	dig coef	%	36.	85.
Goats	dig coef	%		
Horses	dig coef	%	2.	70.
Rabbits	dig coef	%		
Rats	dig coef	%		
Sheep	dig coef	%	42.	86.
Swine	dig coef	%	14.	62.
Free fatty acid		%		
Iodine number				
Fat, total (official)		%		
Fat, Rose Gottlieb		%		
Titer	degrees	C		
Lipid, unsaponifiable material		%		
N-free extract		%	41.5	80.3
Cattle	dig coef	%	70.	90.
Goats	dig coef	%		
Horses	dig coef	%	65.	93.
Rabbits	dig coef	%		
Rats	dig coef	%		
Sheep	dig coef	%	69.	96.
Swine	dig coef	%	49.	80.
Protein (N X 6.25)		%	16.1	10.9
Cattle	dig coef	%	66.	69.
Goats	dig coef	%		
Horses	dig coef	%	72.	78.
Rabbits	dig coef	%	72.	
Rats	dig coef	%		
Sheep	dig coef	%	70.	74.
Swine	dig coef	%	47.	66.
Cattle	dig prot	%	10.6	7.5
Goats	dig prot	%	11.6	7.2
Horses	dig prot	%	11.5	8.5
Rabbits	dig prot	%	11.5	
Rats	dig prot	%		
Sheep	dig prot	%	11.3	8.1
Swine	dig prot	%	7.6	7.2
Cell contents (Van Soest)		%		
Cattle	dig prot	%		
Goats	dig prot	%		
Horses	dig coef	%		
Rabbits	dig coef	%		
Rats	dig coef	%		
Sheep	dig coef	%		
Swine	dig coef	%		
Cell walls (Van Soest)		%		
Cattle	dig coef	%		
Goats	dig coef	%		
Horses	dig coef	%		
Rabbits	dig coef	%		
Rats	dig coef	%		
Sheep	dig coef	%		

TABLE 3 (Continued)

			Alfalfa, sun-cured (1)	Corn, grain (4)
Swine	dig coef	%		
Cellulose (Crampton)		%		
Cattle	dig coef	%		
Goats	dig coef	%		
Horses	dig coef	%		
Rabbits	dig coef	%		
Rats	dig coef	%		
Sheep	dig coef	%		
Swine	dig coef	%		
Cellulose (Matrone)		%	28.8	
Cattle	dig coef	%	54.	
Goats	dig coef	%		
Horses	dig coef	%		
Rabbits	dig coef	%		
Rats	dig coef	%		
Sheep	dig coef	%		
Swine	dig coef	%		
in vitro	dig coef	%		
Fiber, acid detergent (Van Soest)		%		
Cattle	dig coef	%		
Goats	dig coef	%		
Horses	dig coef	%		
Rabbits	dig coef	%		
Rats	dig coef	%		
Sheep	dig coef	%		
Swine	dig coef	%		
Glucosides		%		
Hemicellulose		%		
Cattle	dig coef	%		
Goats	dig coef	%		
Horses	dig coef	%		
Rabbits	dig coef	%		
Rats	dig coef	%		
Sheep	dig coef	%		
Swine	dig coef	%		
Hexosans		%		
Holocellulose		%		
Other carbohydrates		%		
Pectins		%	5.6	
Pectic substances		%		
Pentosans		%		
Soluble carbohydrates		%		
Starch		%	1.6	76.3
Sugars, total		%	7.1	2.0
Lignin (Ellis)		%	9.4	
Cattle	dig coef	%		
Goats	dig coef	%		
Horses	dig coef	%		
Rabbits	dig coef	%		
Rats	dig coef	%		
Sheep	dig coef	%		
Swine	dig coef	%		
Lignin (Van Soest)		%		
Cattle	dig coef	%		
Goats	dig coef	%		
Horses	dig coef	%		
Rabbits	dig coef	%		
Rats	dig coef	%		
Sheep	dig coef	%		
Swine	dig coef	%		

			Alfalfa, sun-cured (1)	Corn, grain (4)
Lignin (Sullivan)		%		
Cattle	dig coef	%		
Goats	dig coef	%		
Horses	dig coef	%		
Rabbits	dig coef	%		
Rats	dig coef	%		
Sheep	dig coef	%		
Swine	dig coef	%		
Fatty acids		%		
Acetic		%		
Arachidic		%		
Arachidonic		%		
Butyric		%		
Capric		%		
Caproic		%		
Caprylic		%		
Docosahexaenoic		%		
Docosapentaenoic		%		
Docosatetraenoic		%		
Eicosadienoic		%		
Eicasahexaenoic		%		
Eicosapentaenoic		%		
Eicosatrienoic		%		
Eicosenoic		%		
Lauric		%		
Linoleic		%	.3	1.9
Linolenic		%		
Myristic		%		
Myristoleic		%		
Octadecatetraenoic		%		
Oleic		%		
Palmitic		%		
Palmitoleic		%		
Propionic		%		
Stearic		%		
Valeric		%		
Lactic acid		%		
Energy	GE Mcal/kg		4.31	4.22
Cattle	GE dig coef	%	60.	
Goats	GE dig coef	%		
Horses	GE dig coef	%		
Rabbits	GE dig coef	%		
Rats	GE dig coef	%		
Sheep	GE dig coef	%	60.	
Swine	GE dig coef	%		
Cattle	DE Mcal/kg		2.37	3.92
Goats	DE Mcal/kg			
Horses	DE Mcal/kg		2.19	4.02
Rabbits	DE Mcal/kg			
Rats	DE Mcal/kg			
Sheep	DE Mcal/kg		2.49	4.22
Swine	DE kcal/kg		1554.	3436.
Cattle	ME Mcal/kg		1.95	3.21
Chickens	ME_n kcal/kg		733.	3811.
Goats	ME Mcal/kg			
Horses	ME Mcal/kg		1.80	3.29
Rabbits	ME kcal/kg			
Rats	ME kcal/kg			
Sheep	ME Mcal/kg		2.04	3.46

TABLE 3 (Continued)

		Alfalfa, sun-cured (1)	Corn, grain (4)
Swine	ME kcal/kg	1441.	3223.
Turkeys	ME_n kcal/kg		
Cattle	NE_m Mcal/kg		
Cattle	NE_{gain} Mcal/kg		
Cattle	$NE_{lactating\ cows}$ Mcal/kg		
Sheep	NE_m Mcal/kg		
Sheep	NE_{gain} Mcal/kg		
Swine	NE_m kcal/kg		
Swine	NE_{gain} kcal/kg		
Cattle	TDN %	53.8	88.8
Goats	TDN %		
Horses	TDN %	49.6	91.1
Rabbits	TDN %		
Rats	TDN %		
Sheep	TDN %	56.4	95.8
Swine	TDN %	35.2	77.9
Nutritive value index (NVI) (Crampton)	%		
Cattle	%		
Goats	%		
Horses	%		
Rabbits	%		
Rats	%		
Sheep	%		
Swine	%		
Aluminum	mg/kg		
Barium	mg/kg		
Boron	mg/kg		
Calcium	%	1.41	.03
Chlorine	%	.31	.06
Chromium	mg/kg		
Cobalt	mg/kg	.225	.02
Copper	mg/kg	20.2	4.5
Fluorine	mg/kg		
Iodine	mg/kg		
Iron	%	.019	.004
Lead	mg/kg		
Magnesium	%	.34	.11
Manganese	mg/kg	61.9	6.5
Molybdenum	mg/kg	8.0	.06
Phosphorus	%	.24	.31
Chickens, chicks	availability %		
Chickens, hens	availability %		
Swine	availability %		
Turkeys, poults	availability %		
Turkeys, hens	availability %		
Chickens, chicks	available %	.22	.11
Chickens, hens	available %		
Swine	available %		
Turkeys, poults	available %		
Turkeys, hens	available %		
Orthophosphate	%		
Phosphorus, citrate soluble	%		
Phosphorus, phytin	%		
Phosphorus, water soluble	%		
Potassium	%	2.18	.31

		Alfalfa, sun-cured (1)	Corn, grain (4)
Salt, sodium chloride	%		
Selenium	mg/kg		
Silicon	%	.70	
Sodium	%	.17	.01
Sulfur	%	.32	.14
Strontium	mg/kg		
Zinc	mg/kg		
Ascorbic acid	mg/kg		
Betaine	%		
Biotin	mg/kg	.18	.07
Carotene	mg/kg	71.4	.4
Choline	mg/kg	1500.	620.
Folic acid	mg/kg	3.40	.36
Niacin	mg/kg	40.	22.0
Pantothenic acid	mg/kg	20.	5.7
Thiamine	mg/kg	3.9	3.0
α-Tocopherol	mg/kg	111.2	
Vitamin B_2	mg/kg	10.6	1.3
Vitamin B_6	mg/kg	9.0	7.0
Vitamin K	mg/kg		
Vitamin B_{12}	mg/kg	0.0	
Vitamin A	IU/g		
Vitamin D_2	IU/g	1.6	
Vitamin D_3	IU/g		
Vitamin D_3	ICU/g		
Alanine	%	.99	
Arginine	%	.70	.40
Chickens, chicks	availability %		
Swine	availability %		
Turkeys, poults	availability %		
Chickens, chicks	available %		
Swine	available %		
Turkeys, poults	available %		
Aspartic acid	%	1.7	
Citrulline	%		
Cysteine	%		
Cystine	%	.3	.10
Glutamic acid	%	1.26	2.70
Glycine	%	.8	.5
Chickens, chicks	availability %		
Swine	availability %		
Turkeys, poults	availability %		
Chickens, chicks	available %		
Swine	available %		
Turkeys, poults	available %		
Histidine	%	.30	.20
Chickens, chicks	availability %		
Swine	availability %		
Turkeys, poults	availability %		
Chickens, chicks	available %		
Swine	available %		
Turkeys, poults	available %		
Hydroxyproline	%		
Isoleucine	%	.80	.50
Chickens, chicks	availability %		
Swine	availability %		
Turkeys, poults	availability %		

TABLE 3 (Continued)

		Alfalfa, sun-cured (1)	Corn, grain (4)
Chickens, chicks	available %		
Swine	available %		
Turkeys, poults	available %		
Leucine	%	1.00	1.20
Chickens, chicks	availability %		
Swine	availability %		
Turkeys, poults	availability %		
Chickens, chicks	available %		
Swine	available %		
Turkeys, poults	available %		
Lysine	%	.60	.30
Chickens, chicks	availability %		
Swine	availability %		
Turkeys, poults	availability %		
Chickens, chicks	available %		
Swine	available %		
Turkeys, poults	available %		
Methionine	%	.10	.20
Chickens, chicks	availability %		
Swine	availability %		
Turkeys, poults	availability %		
Chickens, chicks	available %		
Swine	available %		
Turkeys, poults	available %		
Phenylalanine	%	.60	.50
Chickens, chicks	availability %		
Swine	availability %		
Turkeys, poults	availability %		
Chickens, chicks	available %		
Swine	available %		
Turkeys, poults	available %		
Proline	%	.82	.80
Serine	%	.78	
Threonine	%	.70	.30
Chickens, chicks	availability %		
Swine	availability %		
Turkeys, poults	availability %		
Chickens, chicks	available %		
Swine	available %		
Turkeys, poults	available %		
Tryptophan	%	.10	.10
Chickens, chicks	availability %		
Swine	availability %		
Turkeys, poults	availability %		
Chickens, chicks	available %		
Swine	available %		
Turkeys, poults	available %		
Tyrosine	%	.50	.50
Valine	%	.70	.50
Chickens, chicks	availability %		
Swine	availability %		
Turkeys, poults	availability %		
Chickens, chicks	available %		
Swine	available %		
Turkeys, poults	available %		
Gossypol, total	%		
Gossypol, free	%		
Isothiocyanate	%		
Oxozoladinethione	%		
pH	pH units		
Tannin	%		
Weight per liter	g		

TABLE 4 Weight-Unit Conversion Factors

Units Given	Units Wanted	For Conversion Multiply by
lb	g	453.6
lb	kg	0.4536
oz	g	28.35
kg	lb	2.2046
kg	mg	1,000,000.
kg	g	1,000.
g	mg	1,000.
g	μg	1,000,000.
mg	μg	1,000.
mg/g	mg/lb	453.6
mg/kg	mg/lb	0.4536
μg/kg	μg/lb	0.4536
kcal/kg	kcal/lb	0.4536
kcal/lb	kcal/kg	2.2046
ppm	μg/g	1.
ppm	mg/kg	1.
ppm	mg/lb	0.4536
mg/kg	%	0.0001
ppm	%	0.0001
mg/g	%	0.1
g/kg	%	0.1

TABLE 5 Equations Used to Estimate Digestible Protein (Y) from Protein (X) for Five Animal Kinds and Four Feed Classes

Animal Kind	Feed Class	Regression Equation
Cattle	1	Y = 0.866 X – 3.06
Cattle	2	Y = 0.850 X – 2.11
Cattle	3	Y = 0.908 X – 3.77
Cattle	4	Y = 0.918 X – 3.98
Goats	1 and 2	Y = 0.933 X – 3.44
Goats	3	Y = 0.908 X – 3.77
Goats	4	Y = 0.916 X – 2.76
Horses	1 and 2	Y = 0.849 X – 2.47
Horses	3	Y = 0.908 X – 3.77
Horses	4	Y = 0.916 X – 2.76
Rabbits	1 and 2	Y = 0.772 X – 1.33
Sheep	1	Y = 0.897 X – 3.43
Sheep	2	Y = 0.932 X – 3.01
Sheep	3	Y = 0.908 X – 3.77
Sheep	4	Y = 0.916 X – 2.76

Source: A. D. Knight and L. E. Harris. 1966. Digestible protein estimation for NRC feed composition tables. J. Anim. Sci. 25(2):593.

TABLE 6 Conversion of Beta-Carotene to Vitamin A for Different Species

Species	Conversion of mg of Beta-Carotene to IU of Vitamin A mg IU	IU of Vitamin A Activity (calculated from carotene), %
Standard	1 = 1,667	100.
Beef cattle	1 = 400	24.0
Dairy cattle	1 = 400	24.0
Sheep	1 = 400–500	24.0–30.0
Swine	1 = 500	30.0
Horses		
Growth	1 = 555	33.3
Pregnancy	1 = 333	20.0
Poultry	1 = 1,667	100.
Dogs	1 = 833	50.0
Rats	1 = 1,667	100.
Foxes	1 = 278	16.7
Cat	Carotene not utilized	–
Mink	Carotene not utilized	–
Man	1 = 556	33.3

Source: W. M. Beeson. 1965. Relative potencies of vitamin A and carotene for animals. Fed. Proc. 24:924–926.

TABLES OF FEED COMPOSITION

Persons referring to the tables of feed composition should be aware of the explanations and definitions appearing elsewhere in the Atlas. They will find the following material especially helpful:

ERRATUM

The publisher regrets the following typographical error that appears at the foot of every left-hand page throughout the Tables of Feed Composition.

Footnote (3) should read "silage" instead of "sitage." The complete, correct series of footnotes is reproduced below.

(1) dry forages and roughages
(2) pasture, range plants, and forages fed green
(3) silages
(4) energy feeds
(5) protein supplements
(6) minerals
(7) vitamins
(8) additives

Feed Name or Analyses			Mean As Fed	Mean Dry	C.V. ± %

ABUTILON, INDIANMALLOW. Abutilon incanum

Abutilon, Indianmallow, browse, (2)

Ref No 2 00 004 United States

Analyses			As Fed	Dry	C.V. ± %
Dry matter	%			100.0	
Ash	%			9.9	
Crude fiber	%			22.5	
Ether extract	%			1.8	
N-free extract	%			46.2	
Protein (N x 6.25)	%			19.6	
Cattle	dig prot %	*		14.6	
Goats	dig prot %	*		14.8	
Horses	dig prot %	*		14.2	
Rabbits	dig prot %	*		13.8	
Sheep	dig prot %	*		15.3	

Abutilon, Indianmallow, browse, pre-bloom, (2)

Ref No 2 00 002 United States

Analyses			As Fed	Dry	C.V. ± %
Dry matter	%			100.0	
Calcium	%			2.32	
Magnesium	%			0.58	
Phosphorus	%			0.28	
Potassium	%			2.44	

Abutilon, Indianmallow, browse, mid-bloom, (2)

Ref No 2 00 003 United States

Analyses			As Fed	Dry	C.V. ± %
Dry matter	%			100.0	
Calcium	%			1.99	
Magnesium	%			0.57	
Phosphorus	%			0.26	
Potassium	%			2.96	

ACACIA. Acacia spp

Acacia, pods, s-c, (1)

Ref No 1 00 006 United States

Analyses			As Fed	Dry	C.V. ± %
Dry matter	%		91.4	100.0	
Ash	%		4.2	4.6	
Crude fiber	%		19.7	21.6	
Cattle	dig coef %		16.	16.	
Sheep	dig coef %		32.	32.	
Ether extract	%		1.6	1.8	
Cattle	dig coef %		71.	71.	
Sheep	dig coef %		81.	81.	
N-free extract	%		56.0	61.3	
Cattle	dig coef %		75.	75.	
Sheep	dig coef %		78.	78.	
Protein (N x 6.25)	%		9.8	10.7	
Cattle	dig coef %		51.	51.	
Sheep	dig coef %		49.	49.	
Cattle	dig prot %		5.0	5.5	
Goats	dig prot %	*	6.0	6.5	
Horses	dig prot %	*	6.0	6.6	
Rabbits	dig prot %	*	6.3	6.9	
Sheep	dig prot %		4.8	5.2	
Energy	GE Mcal/kg				
Cattle	DE Mcal/kg	*	2.33	2.55	
Sheep	DE Mcal/kg	*	2.55	2.79	
Cattle	ME Mcal/kg	*	1.91	2.09	
Sheep	ME Mcal/kg	*	2.09	2.29	
Cattle	TDN %		52.8	57.8	
Sheep	TDN %		57.8	63.2	

(1) dry forages and roughages (2) pasture, range plants, and forages fed green (3) silages (4) energy feeds (5) protein supplements (6) minerals (7) vitamins (8) additives

Acacia, pods, s-c, mature, (1)

Ref No 1 00 005 United States

Analyses			As Fed	Dry	C.V. ± %
Dry matter	%		91.5	100.0	
Ash	%		5.1	5.6	
Crude fiber	%		13.9	15.2	
Sheep	dig coef %		21.	21.	
Ether extract	%		2.3	2.5	
Sheep	dig coef %		88.	88.	
N-free extract	%		58.4	63.8	
Sheep	dig coef %		79.	79.	
Protein (N x 6.25)	%		11.8	12.9	
Sheep	dig coef %		52.	52.	
Cattle	dig prot %	*	7.4	8.1	
Goats	dig prot %	*	7.9	8.6	
Horses	dig prot %	*	7.8	8.5	
Rabbits	dig prot %	*	7.9	8.6	
Sheep	dig prot %		6.1	6.7	
Energy	GE Mcal/kg				
Cattle	DE Mcal/kg	*	2.46	2.69	
Sheep	DE Mcal/kg	*	2.63	2.88	
Cattle	ME Mcal/kg	*	2.01	2.20	
Sheep	ME Mcal/kg	*	2.16	2.36	
Cattle	TDN %	*	55.8	61.0	
Sheep	TDN %		59.7	65.3	

Acacia, twigs, s-c, (1)

Ref No 1 00 007 United States

Analyses			As Fed	Dry	C.V. ± %
Dry matter	%		87.6	100.0	
Ash	%		3.6	4.1	
Crude fiber	%		31.5	36.0	
Sheep	dig coef %		21.	21.	
Ether extract	%		1.7	1.9	
Sheep	dig coef %		23.	23.	
N-free extract	%		41.0	46.8	
Sheep	dig coef %		47.	47.	
Protein (N x 6.25)	%		9.8	11.2	
Sheep	dig coef %		56.	56.	
Cattle	dig prot %	*	5.8	6.6	
Goats	dig prot %	*	6.1	7.0	
Horses	dig prot %	*	6.2	7.0	
Rabbits	dig prot %	*	6.4	7.3	
Sheep	dig prot %		5.5	6.3	
Energy	GE Mcal/kg				
Sheep	DE Mcal/kg	*	1.42	1.62	
Sheep	ME Mcal/kg	*	1.17	1.33	
Sheep	TDN %		32.2	36.8	

ACACIA, PRAIRIE. Acacia angustissima

Acacia, prairie, browse, (2)

Ref No 2 00 009 United States

Analyses			As Fed	Dry	C.V. ± %
Dry matter	%			100.0	
Calcium	%			0.97	
Magnesium	%			0.41	
Phosphorus	%			0.29	
Potassium	%			1.28	

Acacia, prairie, browse, immature, (2)

Ref No 2 00 008 United States

Analyses			As Fed	Dry	C.V. ± %
Dry matter	%			100.0	
Ash	%			4.9	
Crude fiber	%			16.9	
Ether extract	%			3.8	
N-free extract	%			47.8	
Protein (N x 6.25)	%			26.6	
Cattle	dig prot %	*		20.5	
Goats	dig prot %	*		21.4	
Horses	dig prot %	*		20.1	
Rabbits	dig prot %	*		19.2	
Sheep	dig prot %	*		21.8	

ACACIA, ROEMER. Acacia roemeriana

Acacia, roemer, browse, (2)

Ref No 2 00 011 United States

Analyses			As Fed	Dry	C.V. ± %
Dry matter	%			100.0	
Calcium	%			1.51	
Magnesium	%			0.28	
Phosphorus	%			0.20	
Potassium	%			1.05	

Acacia, roemer, browse, immature, (2)

Ref No 2 00 010 United States

Analyses			As Fed	Dry	C.V. ± %
Dry matter	%			100.0	
Ash	%			5.3	
Crude fiber	%			20.2	
Ether extract	%			3.6	
N-free extract	%			50.4	
Protein (N x 6.25)	%			20.5	
Cattle	dig prot %	*		15.3	
Goats	dig prot %	*		15.7	
Horses	dig prot %	*		14.9	
Rabbits	dig prot %	*		14.5	
Sheep	dig prot %	*		16.1	

ACACIA, SWEET. Acacia farnesiana

Acacia, sweet, seeds w hulls, (5)

Ref No 5 09 110 United States

Analyses			As Fed	Dry	C.V. ± %
Dry matter	%			100.0	
Protein (N x 6.25)	%			55.0	
Alanine	%			2.37	
Arginine	%			5.06	
Aspartic acid	%			4.84	
Glutamic acid	%			6.93	
Glycine	%			1.87	
Histidine	%			1.27	
Hydroxyproline	%			0.00	
Isoleucine	%			1.93	
Leucine	%			4.13	
Lysine	%			2.59	
Methionine	%			0.50	
Phenylalanine	%			1.93	
Proline	%			2.81	
Threonine	%			1.38	
Tyrosine	%			1.54	
Valine	%			2.15	

ACACIA, WILLARD. Acacia willardiana

Acacia, willard, seeds w hulls, (5)
Ref No 5 09 111 — United States

Feed Name or Analyses		Mean As Fed	Mean Dry	C.V. ± %
Dry matter	%		100.0	
Protein (N x 6.25)	%		35.0	
Alanine	%		1.09	
Arginine	%		1.89	
Aspartic acid	%		2.56	
Glutamic acid	%		5.01	
Glycine	%		1.09	
Histidine	%		1.09	
Hydroxyproline	%		0.11	
Isoleucine	%		1.02	
Leucine	%		2.03	
Lysine	%		1.86	
Methionine	%		0.32	
Phenylalanine	%		1.05	
Proline	%		1.23	
Serine	%		1.26	
Threonine	%		0.88	
Tyrosine	%		0.91	
Valine	%		1.16	

ACLEISANTHES. Acleisanthes spp

Acleisanthes, browse, immature, (2)
Ref No 2 00 013 — United States

Feed Name or Analyses		Mean As Fed	Mean Dry	C.V. ± %
Dry matter	%		100.0	
Ash	%		11.2	
Crude fiber	%		24.7	
Ether extract	%		2.0	
N-free extract	%		37.7	
Protein (N x 6.25)	%		24.4	
Cattle	dig prot % *		18.6	
Goats	dig prot % *		19.3	
Horses	dig prot % *		18.2	
Rabbits	dig prot % *		17.5	
Sheep	dig prot % *		19.7	
Calcium	%		2.65	
Magnesium	%		0.57	
Phosphorus	%		0.15	
Potassium	%		3.37	

ACTINIDIA, BOWER. Actinidia arguta

Actinidia, bower, seeds w hulls, (4)
Ref No 4 09 010 — United States

Feed Name or Analyses		Mean As Fed	Mean Dry	C.V. ± %
Dry matter	%		100.0	
Protein (N x 6.25)	%		16.0	
Cattle	dig prot % *		10.7	
Goats	dig prot % *		11.9	
Horses	dig prot % *		11.9	
Sheep	dig prot % *		11.9	
Alanine	%		0.74	
Arginine	%		1.65	
Aspartic acid	%		1.46	
Glutamic acid	%		3.07	
Glycine	%		0.75	
Histidine	%		0.32	
Hydroxyproline	%		0.00	
Isoleucine	%		0.67	
Leucine	%		1.09	
Lysine	%		0.46	
Methionine	%		0.43	
Phenylalanine	%		0.56	
Proline	%		0.62	
Serine	%		0.61	
Threonine	%		0.53	
Tyrosine	%		0.35	
Valine	%		0.83	

(1) dry forages and roughages (2) pasture, range plants, and forages fed green (3) silages (4) energy feeds (5) protein supplements (6) minerals (7) vitamins (8) additives

ACTINOMERIS, ALTERNIFOLIA. Actinomeris alternifolia

Actinomeris, alternifolia, seeds w hulls, (5)
Ref No 5 09 300 — United States

Feed Name or Analyses		Mean As Fed	Mean Dry	C.V. ± %
Dry matter	%		100.0	
Protein (N x 6.25)	%		43.0	
Alanine	%		1.42	
Arginine	%		2.84	
Aspartic acid	%		3.10	
Glutamic acid	%		7.44	
Glycine	%		1.55	
Histidine	%		0.73	
Hydroxyproline	%		0.09	
Isoleucine	%		1.55	
Leucine	%		2.49	
Lysine	%		1.42	
Methionine	%		0.52	
Phenylalanine	%		0.99	
Proline	%		1.68	
Serine	%		1.46	
Threonine	%		1.16	
Tyrosine	%		0.99	
Valine	%		1.76	

Active dry yeast (AAFCO) –
see Yeast, active, dehy, mn 15 billion live yeast cells per g, (7)

Adlay –
see Jobstears, seeds, (4)

AGAVE. Agave spp

Agave, browse, (2)
Ref No 2 00 014 — United States

Feed Name or Analyses		Mean As Fed	Mean Dry	C.V. ± %
Dry matter	%		100.0	
Ash	%		8.6	
Crude fiber	%		27.3	
Ether extract	%		2.3	
N-free extract	%		55.0	
Protein (N x 6.25)	%		6.8	
Cattle	dig prot % *		3.7	
Goats	dig prot % *		2.9	
Horses	dig prot % *		3.3	
Rabbits	dig prot % *		3.9	
Sheep	dig prot % *		3.3	
Calcium	%		4.31	
Magnesium	%		0.29	
Phosphorus	%		0.11	
Potassium	%		0.74	

AILANTHUS, TREEOFHEAVEN. Ailanthus altissima

Ailanthus, treeofheaven, seeds w hulls, (5)
Ref No 5 09 261 — United States

Feed Name or Analyses		Mean As Fed	Mean Dry	C.V. ± %
Dry matter	%		100.0	
Protein (N x 6.25)	%		28.0	
Alanine	%		0.92	
Arginine	%		2.24	
Aspartic acid	%		1.68	
Glutamic acid	%		6.22	
Glycine	%		1.26	
Histidine	%		0.67	
Hydroxyproline	%		0.17	
Isoleucine	%		0.90	
Leucine	%		1.74	
Lysine	%		0.70	
Methionine	%		0.70	
Phenylalanine	%		1.01	
Proline	%		0.90	
Serine	%		1.15	
Threonine	%		0.67	
Tyrosine	%		0.78	
Valine	%		1.32	

ALDER. Alnus spp

Alder, leaves, s-c, (1)
Ref No 1 00 015 — United States

Feed Name or Analyses		Mean As Fed	Mean Dry	C.V. ± %
Dry matter	%	85.0	100.0	
Ash	%	5.0	5.9	
Crude fiber	%	14.5	17.0	
Sheep	dig coef %	7.	7.	
Ether extract	%	5.1	6.0	
Sheep	dig coef %	24.	24.	
N-free extract	%	41.7	49.1	
Sheep	dig coef %	70.	70.	
Protein (N x 6.25)	%	18.7	22.0	
Sheep	dig coef %	47.	47.	
Cattle	dig prot % *	13.6	16.0	
Goats	dig prot % *	14.5	17.1	
Horses	dig prot % *	13.8	16.2	
Rabbits	dig prot % *	13.3	15.7	
Sheep	dig prot %	8.8	10.3	
Energy	GE Mcal/kg			
Cattle	DE Mcal/kg *	1.79	2.11	
Sheep	DE Mcal/kg *	1.84	2.17	
Cattle	ME Mcal/kg *	1.42	1.73	
Sheep	ME Mcal/kg *	1.51	1.78	
Cattle	TDN % *	40.6	47.8	
Sheep	TDN %	41.8	49.1	

ALFALFA. Medicago sativa

Alfalfa, aerial part, dehy, (1)
Ref No 1 00 047 — United States

Feed Name or Analyses		Mean As Fed	Mean Dry	C.V. ± %
Dry matter	%	89.9	100.0	1
Ash	%	8.2	9.1	16
Crude fiber	%	26.2	29.1	12
Cattle	dig coef %	45.	45.	
Sheep	dig coef %	46.	46.	
Ether extract	%	2.0	2.2	21
Cattle	dig coef %	45.	45.	
Sheep	dig coef %	48.	48.	

Feed Name or Analyses				Mean As Fed	Mean Dry	C.V. ± %
N-free extract		%		37.0	41.2	
Cattle	dig coef	%		72.	72.	
Sheep	dig coef	%		70.	70.	
Protein (N x 6.25)		%		16.6	18.5	17
Cattle	dig coef	%		72.	72.	
Sheep	dig coef	%		72.	72.	
Cattle	dig prot	%		12.0	13.3	
Goats	dig prot	%	*	12.4	13.8	
Horses	dig prot	%	*	11.9	13.2	
Rabbits	dig prot	%	*	11.6	12.9	
Sheep	dig prot	%		12.0	13.3	
Energy	GE	Mcal/kg				
Cattle	DE	Mcal/kg	*	2.29	2.55	
Sheep	DE	Mcal/kg	*	2.29	2.55	
Cattle	ME	Mcal/kg	*	1.88	2.09	
Sheep	ME	Mcal/kg	*	1.88	2.09	
Cattle	TDN	%		52.0	57.9	
Sheep	TDN	%		52.0	57.9	
Calcium		%		1.70	1.89	7
Magnesium		%		0.25	0.28	
Manganese		mg/kg		33.9	37.7	32
Phosphorus		%		0.22	0.25	45
Carotene		mg/kg		132.7	147.7	41
Choline		mg/kg		575.	639.	
Niacin		mg/kg		36.7	40.8	3
Pantothenic acid		mg/kg		27.5	30.6	
Riboflavin		mg/kg		14.3	15.9	1
Vitamin A equivalent		IU/g		221.3	246.2	
Arginine		%		0.54	0.60	
Glutamic acid		%		1.26	1.40	11
Histidine		%		0.36	0.40	
Isoleucine		%		0.90	1.00	
Leucine		%		1.08	1.20	
Lysine		%		0.63	0.70	
Methionine		%		0.18	0.20	
Phenylalanine		%		0.81	0.90	
Serine		%		0.54	0.60	
Threonine		%		0.63	0.70	
Tryptophan		%		0.18	0.20	
Tyrosine		%		0.27	0.30	
Valine		%		0.72	0.80	

Alfalfa, aerial part, dehy, immature, (1)

Ref No 1 00 041 — United States

Feed Name or Analyses				Mean As Fed	Mean Dry	C.V. ± %
Dry matter		%		91.1	100.0	2
Ash		%		10.9	12.0	13
Crude fiber		%		25.6	28.1	17
Cattle	dig coef	%	*	54.	54.	
Sheep	dig coef	%	*	46.	46.	
Ether extract		%		2.1	2.3	19
Cattle	dig coef	%	*	54.	54.	
Sheep	dig coef	%	*	48.	48.	
N-free extract		%		33.3	36.6	
Cattle	dig coef	%	*	75.	75.	
Sheep	dig coef	%	*	70.	70.	
Protein (N x 6.25)		%		19.1	21.0	17
Cattle	dig coef	%	*	70.	70.	
Sheep	dig coef	%	*	72.	72.	
Cattle	dig prot	%		13.4	14.7	
Goats	dig prot	%	*	14.7	16.2	
Horses	dig prot	%	*	14.0	15.4	
Rabbits	dig prot	%	*	13.6	14.9	
Sheep	dig prot	%		13.8	15.1	
Energy	GE	Mcal/kg				
Cattle	DE	Mcal/kg	*	2.41	2.65	
Sheep	DE	Mcal/kg	*	2.26	2.48	
Cattle	ME	Mcal/kg	*	1.98	2.17	
Sheep	ME	Mcal/kg	*	1.85	2.03	
Cattle	TDN	%	*	54.8	60.1	
Sheep	TDN	%	*	51.2	56.2	

Alfalfa, aerial part, dehy, pre-bloom, (1)

Ref No 1 00 042 — United States

Feed Name or Analyses				Mean As Fed	Mean Dry	C.V. ± %
Dry matter		%		89.4	100.0	
Ash		%		9.8	11.0	
Crude fiber		%		26.1	29.2	
Cattle	dig coef	%		54.	54.	
Sheep	dig coef	%		45.	45.	
Swine	dig coef	%		32.	32.	
Ether extract		%		2.3	2.5	
Cattle	dig coef	%		54.	54.	
Sheep	dig coef	%		50.	50.	
Swine	dig coef	%		50.	50.	
N-free extract		%		32.9	36.8	
Cattle	dig coef	%		75.	75.	
Sheep	dig coef	%		69.	69.	
Swine	dig coef	%		71.	71.	
Protein (N x 6.25)		%		18.3	20.5	
Cattle	dig coef	%		70.	70.	
Sheep	dig coef	%		69.	69.	
Swine	dig coef	%		61.	61.	
Cattle	dig prot	%		12.8	14.4	
Goats	dig prot	%	*	14.0	15.7	
Horses	dig prot	%	*	13.4	14.9	
Rabbits	dig prot	%	*	13.0	14.5	
Sheep	dig prot	%		12.6	14.1	
Swine	dig prot	%		11.2	12.5	
Energy	GE	Mcal/kg				
Cattle	DE	Mcal/kg	*	2.40	2.68	
Sheep	DE	Mcal/kg	*	2.19	2.45	
Swine	DE	kcal/kg	*	2003.	2240.	
Cattle	ME	Mcal/kg	*	1.96	2.20	
Sheep	ME	Mcal/kg	*	1.79	2.01	
Swine	ME	kcal/kg	*	1840.	2058.	
Cattle	TDN	%		54.3	60.8	
Sheep	TDN	%		49.6	55.5	
Swine	TDN	%		45.4	50.8	

Alfalfa, aerial part, dehy, early bloom, (1)

Ref No 1 00 044 — United States

Feed Name or Analyses				Mean As Fed	Mean Dry	C.V. ± %
Dry matter		%		89.3	100.0	2
Ash		%		8.8	9.9	6
Crude fiber		%		24.8	27.8	12
Cattle	dig coef	%		49.	49.	
Sheep	dig coef	%		46.	46.	
Ether extract		%		2.3	2.6	23
Cattle	dig coef	%		36.	36.	
Sheep	dig coef	%		28.	28.	
N-free extract		%		37.4	41.9	
Cattle	dig coef	%		72.	72.	
Sheep	dig coef	%		74.	74.	
Protein (N x 6.25)		%		16.0	17.9	9
Cattle	dig coef	%		68.	68.	
Sheep	dig coef	%		71.	71.	
Cattle	dig prot	%		10.9	12.1	
Goats	dig prot	%	*	11.8	13.2	
Horses	dig prot	%	*	11.3	12.7	
Rabbits	dig prot	%	*	11.1	12.5	
Sheep	dig prot	%		11.3	12.7	
Energy	GE	Mcal/kg		3.89	4.35	
Cattle	DE	Mcal/kg	*	2.28	2.56	
Sheep	DE	Mcal/kg	*	2.29	2.56	
Cattle	ME	Mcal/kg	*	1.87	2.10	
Sheep	ME	Mcal/kg	*	1.88	2.10	
Cattle	TDN	%		51.8	58.0	
Sheep	TDN	%		51.9	58.1	

Ref No 1 00 44 — Canada

Feed Name or Analyses				Mean As Fed	Mean Dry	C.V. ± %
Dry matter		%			100.0	
Ash		%			8.6	
Crude fiber		%			32.0	
Cattle	dig coef	%			49.	
Sheep	dig coef	%			46.	
Ether extract		%			1.1	
Cattle	dig coef	%			36.	
Sheep	dig coef	%			28.	
N-free extract		%			42.5	
Cattle	dig coef	%			72.	
Sheep	dig coef	%			74.	
Protein (N x 6.25)		%			16.0	
Cattle	dig coef	%			68.	
Sheep	dig coef	%			71.	
Cattle	dig prot	%			10.9	
Goats	dig prot	%	*		11.5	
Horses	dig prot	%	*		11.1	
Rabbits	dig prot	%	*		11.0	
Sheep	dig prot	%			11.4	
Cellulose (Matrone)		%			35.1	
Energy	GE	Mcal/kg			4.35	
Cattle	DE	Mcal/kg	*		2.56	
Sheep	DE	Mcal/kg	*		2.56	
Cattle	ME	Mcal/kg	*		2.10	
Sheep	ME	Mcal/kg	*		2.10	
Cattle	TDN	%			58.0	
Sheep	TDN	%			58.1	

Alfalfa, aerial part, dehy, early bloom, cut 3, (1)

Ref No 1 00 043 — United States

Feed Name or Analyses				Mean As Fed	Mean Dry	C.V. ± %
Dry matter		%		87.3	100.0	1
Sheep	dig coef	%		67.	67.	
Ash		%		7.8	9.0	4
Crude fiber		%		23.1	26.5	2
Sheep	dig coef	%		50.	50.	
Ether extract		%		2.5	2.9	10
Sheep	dig coef	%		34.	34.	
N-free extract		%		34.9	40.0	
Sheep	dig coef	%		78.	78.	
Protein (N x 6.25)		%		18.9	21.7	9
Sheep	dig coef	%		80.	80.	
Cattle	dig prot	%	*	13.7	15.7	
Goats	dig prot	%	*	14.6	16.8	
Horses	dig prot	%	*	13.9	15.9	
Rabbits	dig prot	%	*	13.4	15.4	
Sheep	dig prot	%		15.1	17.3	
Hemicellulose		%		5.8	6.7	28
Energy	GE	Mcal/kg		3.91	4.48	
Sheep	GE dig coef	%		66.	66.	
Cattle	DE	Mcal/kg	*	2.32	2.66	
Sheep	DE	Mcal/kg		2.58	2.96	
Cattle	ME	Mcal/kg	*	1.91	2.18	
Sheep	ME	Mcal/kg	*	2.12	2.43	
Cattle	TDN	%	*	52.7	60.4	
Sheep	TDN	%		55.9	64.0	

Alfalfa, aerial part, dehy, mid-bloom, (1)

Ref No 1 00 045 — United States

Feed Name or Analyses				Mean As Fed	Mean Dry	C.V. ± %
Dry matter		%		87.4	100.0	
Ash		%		5.5	6.3	

Continued

Feed Name or Analyses			Mean As Fed	Mean Dry	C.V. ± %
Crude fiber	%		25.9	29.7	
Cattle	dig coef %		49.	49.	
Ether extract	%		1.8	2.1	
Cattle	dig coef %		46.	46.	
N-free extract	%		40.2	46.1	
Cattle	dig coef %		73.	73.	
Protein (N x 6.25)	%		13.9	16.0	
Cattle	dig coef %		66.	66.	
Cattle	dig prot %		9.2	10.5	
Goats	dig prot %	*	10.0	11.4	
Horses	dig prot %	*	9.7	11.1	
Rabbits	dig prot %	*	9.6	11.0	
Sheep	dig prot %	*	9.5	10.9	
Energy	GE Mcal/kg				
Cattle	DE Mcal/kg	*	2.34	2.68	
Sheep	DE Mcal/kg	*	2.32	2.66	
Cattle	ME Mcal/kg	*	1.92	2.20	
Sheep	ME Mcal/kg	*	1.91	2.18	
Cattle	TDN %		53.1	60.8	
Sheep	TDN %	*	52.7	60.3	

Alfalfa, aerial part, dehy, gr 1 US, (1)
Ref No 1 00 046 United States

Feed Name or Analyses			Mean As Fed	Mean Dry	C.V. ± %
Dry matter	%		92.2	100.0	
Ash	%		9.5	10.3	15
Crude fiber	%		14.4	15.6	
Ether extract	%		2.0	2.2	
N-free extract	%		43.4	47.1	
Protein (N x 6.25)	%		22.9	24.8	16
Cattle	dig prot %	*	17.0	18.4	
Goats	dig prot %	*	18.2	19.7	
Horses	dig prot %	*	17.1	18.6	
Rabbits	dig prot %	*	16.4	17.8	
Sheep	dig prot %	*	17.3	18.8	
Cellulose (Matrone)	%		18.6	20.2	10
Pectins	%		5.2	5.6	5
Starch	%		1.6	1.7	12
Sugars, total	%		3.0	3.2	13
Lignin (Ellis)	%		8.7	9.4	11
Energy	GE Mcal/kg				
Cattle	DE Mcal/kg	*	2.73	2.96	
Sheep	DE Mcal/kg	*	2.74	2.98	
Cattle	ME Mcal/kg	*	2.24	2.43	
Sheep	ME Mcal/kg	*	2.25	2.44	
Cattle	TDN %	*	62.0	67.2	
Sheep	TDN %	*	62.2	67.5	
Calcium	%		1.60	1.73	
Magnesium	%		0.06	0.06	
Phosphorus	%		0.56	0.61	

Alfalfa, aerial part, dehy chopped, immature, (1)
Ref No 1 07 816 United States

Feed Name or Analyses			Mean As Fed	Mean Dry	C.V. ± %
Dry matter	%			100.0	
Lignin (Ellis)	%				
Sheep	dig coef %			7.	
Energy	GE Mcal/kg			4.51	
Sheep	DE Mcal/kg			2.74	
Sheep	ME Mcal/kg			2.26	
Sheep	TDN %			58.7	

(1) dry forages and roughages (2) pasture, range plants, and forages fed green (3) silages (4) energy feeds (5) protein supplements (6) minerals (7) vitamins (8) additives

Alfalfa, aerial part, dehy chopped, early bloom, (1)
Ref No 1 07 835 United States

Feed Name or Analyses			Mean As Fed	Mean Dry	C.V. ± %
Dry matter	%			100.0	
Ash	%			11.5	
Crude fiber	%			29.3	
Sheep	dig coef %			51.	
Ether extract	%			2.0	
N-free extract	%			38.8	
Protein (N x 6.25)	%			18.4	
Sheep	dig coef %			69.	
Cattle	dig prot %	*		12.9	
Goats	dig prot %	*		13.7	
Horses	dig prot %	*		13.2	
Rabbits	dig prot %	*		12.9	
Sheep	dig prot %			12.7	
Lignin (Ellis)	%			9.2	
Energy	GE Mcal/kg				
Cattle	DE Mcal/kg	*		2.76	
Sheep	DE Mcal/kg	*		2.47	
Cattle	ME Mcal/kg	*		2.27	
Sheep	ME Mcal/kg	*		2.02	
Cattle	TDN %	*		62.7	
Sheep	TDN %			56.0	

Alfalfa, aerial part, dehy chopped, mid-bloom, cut 2, (1)
Ref No 1 07 728 United States

Feed Name or Analyses			Mean As Fed	Mean Dry	C.V. ± %
Dry matter	%			100.0	
Protein (N x 6.25)	%			21.0	
Cattle	dig prot %	*		15.1	
Goats	dig prot %	*		16.2	
Horses	dig prot %	*		15.4	
Rabbits	dig prot %	*		14.9	
Sheep	dig prot %	*		15.4	
Lignin (Ellis)	%			7.4	

Alfalfa, aerial part, dehy chopped, full bloom, cut 2, (1)
Ref No 1 07 727 United States

Feed Name or Analyses			Mean As Fed	Mean Dry	C.V. ± %
Dry matter	%			100.0	
Protein (N x 6.25)	%			20.0	
Cattle	dig prot %	*		14.3	
Goats	dig prot %	*		15.2	
Horses	dig prot %	*		14.5	
Rabbits	dig prot %	*		14.1	
Sheep	dig prot %	*		14.5	
Lignin (Ellis)	%			7.7	

Alfalfa, aerial part, dehy grnd, (1)
Dehydrated alfalfa meal (AAFCO)
Ref No 1 00 025 United States

Feed Name or Analyses			Mean As Fed	Mean Dry	C.V. ± %
Dry matter	%		92.2	100.0	1 1
Sheep	dig coef %		61.	61.	
Ash	%		9.7	10.5	16
Crude fiber	%		22.9	24.8	15 1
Cattle	dig coef %	*	52.	52.	
Sheep	dig coef %		43.	43.	
Swine	dig coef %	*	22.	22.	
Ether extract	%		2.9	3.2	25 1
Cattle	dig coef %	*	61.	61.	
Sheep	dig coef %		44.	44.	
Swine	dig coef %	*	14.	14.	
N-free extract	%		38.8	42.1	7 1
Cattle	dig coef %	*	69.	69.	
Sheep	dig coef %		73.	73.	
Swine	dig coef %	*	49.	49.	
Protein (N x 6.25)	%		17.9	19.4	11
Cattle	dig coef %	*	78.	78.	
Sheep	dig coef %		65.	65.	
Swine	dig coef %	*	47.	47.	
Cattle	dig prot %		14.0	15.1	
Goats	dig prot %	*	13.5	14.7	
Horses	dig prot %	*	12.9	14.0	
Rabbits	dig prot %	*	12.6	13.7	
Sheep	dig prot %		11.7	12.7	
Swine	dig prot %		8.4	9.1	
Cellulose (Matrone)	%		19.4	21.1	
Pentosans	%		12.4	13.4	4
Sugars, total	%		3.7	4.0	16
Lignin (Ellis)	%		8.1	8.8	
Linoleic	%		.400	.434	
Energy	GE Mcal/kg		3.95	4.29	
Sheep	GE dig coef %		61.	61.	
Cattle	DE Mcal/kg	*	2.50	2.71	
Sheep	DE Mcal/kg		2.39	2.59	
Swine	DE kcal/kg	*	1472.	1597.	
Cattle	ME Mcal/kg	*	2.05	2.22	
Chickens	ME_n kcal/kg		1360.	.1475.	
Sheep	ME Mcal/kg	*	1.96	2.13	
Swine	ME kcal/kg	*	1355.	1471.	
Cattle	TDN %		56.7	61.5	
Sheep	TDN %		52.9	57.3	
Swine	TDN %		33.4	36.2	
Calcium	%		1.36	1.48	
Chlorine	%		0.66	0.72	
Cobalt	mg/kg		0.109	0.119	98
Copper	mg/kg		7.4	8.0	
Iron	%		0.021	0.023	
Magnesium	%		0.20	0.22	
Manganese	mg/kg		20.0	21.7	
Molybdenum	mg/kg		0.98	1.07	
Phosphorus	%		0.28	0.31	
Potassium	%		1.34	1.46	
Sodium	%		0.13	0.14	34
Zinc	mg/kg		31.1	33.7	
Zinc	mg/kg		32.3	35.1	49
Carotene	mg/kg		88.0	95.4	
Folic acid	mg/kg		5.04	5.47	53
α-tocopherol	mg/kg		392.8	426.1	
Vitamin A equivalent	IU/g		146.6	159.1	
Arginine	%		0.92	1.00	17
Cystine	%		0.37	0.40	9
Histidine	%		0.28	0.30	26
Isoleucine	%		0.83	0.90	17
Leucine	%		1.29	1.40	11
Lysine	%		1.01	1.10	9
Methionine	%		0.18	0.20	50
Phenylalanine	%		0.83	0.90	11
Threonine	%		0.74	0.80	13
Tryptophan	%		0.28	0.30	16
Tyrosine	%		0.55	0.60	9
Valine	%		0.83	0.90	14

Ref No 1 00 25 Canada

Feed Name or Analyses			Mean As Fed	Mean Dry	C.V. ± %
Dry matter	%		93.0	100.0	
Crude fiber	%				
Sheep	dig coef %		43.	43.	
Ether extract	%				
Sheep	dig coef %		44.	44.	
N-free extract	%				
Sheep	dig coef %		73.	73.	
Protein (N x 6.25)	%		16.0	17.2	
Sheep	dig coef %		65.	65.	

Feed Name or Analyses		Mean As Fed	Mean Dry	C.V. ± %
Cattle	dig prot %	* 11.0	11.8	
Goats	dig prot %	* 11.7	12.6	
Horses	dig prot %	* 11.3	12.1	
Rabbits	dig prot %	* 11.1	12.0	
Sheep	dig prot %	10.4	11.2	
Energy	GE Mcal/kg	3.99	4.29	
Sheep	GE dig coef %	61.	61.	
Cattle	DE Mcal/kg	* 2.52	2.71	
Sheep	DE Mcal/kg	2.41	2.59	
Swine	DE kcal/kg	*1485.	1597.	
Cattle	ME Mcal/kg	* 2.07	2.22	
Sheep	ME Mcal/kg	* 1.98	2.13	
Swine	ME kcal/kg	*1374.	1478.	
Cattle	TDN %	57.1	61.4	
Sheep	TDN %	53.3	57.4	
Swine	TDN %	33.7	36.2	
Vitamin A	IU/g	2223.0	2390.3	

Alfalfa, aerial part, dehy grnd, immature, (1)
Ref No 1 00 017 United States

Feed Name or Analyses		Mean As Fed	Mean Dry	C.V. ± %
Dry matter	%	94.5	100.0	
Carotene	mg/kg	525.8	556.4	27
Vitamin A equivalent	IU/g	876.6	927.6	

Alfalfa, aerial part, dehy grnd, pre-bloom, (1)
Ref No 1 00 018 United States

Feed Name or Analyses		Mean As Fed	Mean Dry	C.V. ± %
Dry matter	%	92.6	100.0	
Ash	%	13.4	14.5	
Crude fiber	%	29.8	32.2	
Sheep	dig coef %	46.	46.	
Ether extract	%	2.0	2.2	
Sheep	dig coef %	53.	53.	
N-free extract	%	31.8	34.3	
Sheep	dig coef %	65.	65.	
Protein (N x 6.25)	%	15.6	16.8	
Sheep	dig coef %	69.	69.	
Cattle	dig prot %	* 10.6	11.5	
Goats	dig prot %	* 11.3	12.2	
Horses	dig prot %	* 10.9	11.8	
Rabbits	dig prot %	* 10.8	11.6	
Sheep	dig prot %	10.7	11.6	
Energy	GE Mcal/kg			
Cattle	DE Mcal/kg	* 2.31	2.49	
Sheep	DE Mcal/kg	* 2.10	2.26	
Cattle	ME Mcal/kg	* 1.89	2.04	
Sheep	ME Mcal/kg	* 1.72	1.86	
Cattle	TDN %	* 52.2	56.4	
Sheep	TDN %	47.5	51.3	

Alfalfa, aerial part, dehy grnd, early bloom, (1)
Ref No 1 00 019 United States

Feed Name or Analyses		Mean As Fed	Mean Dry	C.V. ± %
Dry matter	%	93.2	100.0	
Carotene	mg/kg	246.8	264.8	
Vitamin A equivalent	IU/g	411.4	441.4	

Alfalfa, aerial part, dehy grnd, over ripe, (1)
Ref No 1 00 020 United States

Feed Name or Analyses		Mean As Fed	Mean Dry	C.V. ± %
Dry matter	%	95.9	100.0	
Carotene	mg/kg	2.3	2.4	
Vitamin A equivalent	IU/g	3.9	4.0	

Alfalfa, aerial part, dehy grnd, mn 13% protein, (1)
Ref No 1 00 021 United States

Feed Name or Analyses		Mean As Fed	Mean Dry	C.V. ± %
Dry matter	%	91.3	100.0	1
Ash	%	8.9	9.8	8
Crude fiber	%	29.1	31.9	11
Ether extract	%	1.8	2.0	19
N-free extract	%	37.0	40.6	
Protein (N x 6.25)	%	14.4	15.7	7
Cattle	dig prot %	* 9.7	10.6	
Goats	dig prot %	* 10.3	11.2	
Horses	dig prot %	* 10.0	10.9	
Rabbits	dig prot %	* 9.9	10.8	
Sheep	dig prot %	* 9.8	10.7	
Energy	GE Mcal/kg			
Cattle	DE Mcal/kg	* 2.32	2.54	
Sheep	DE Mcal/kg	* 2.25	2.46	
Cattle	ME Mcal/kg	* 1.90	2.09	
Chickens	ME_n kcal/kg	444.	486.	
Sheep	ME Mcal/kg	* 1.84	2.02	
Cattle	TDN %	* 52.7	57.7	
Sheep	TDN %	* 51.0	55.9	
Calcium	%	1.21	1.32	
Phosphorus	%	0.20	0.22	
Carotene	mg/kg	60.0	65.7	17
Choline	mg/kg	777.	851.	
Niacin	mg/kg	13.3	14.6	
Pantothenic acid	mg/kg	15.5	17.0	
Riboflavin	mg/kg	10.0	11.0	6
Vitamin A	IU/g	22.2	24.3	
Vitamin A equivalent	IU/g	100.0	109.5	
Arginine	%	0.60	0.66	
Cystine	%	0.24	0.26	
Glycine	%	0.70	0.77	
Lysine	%	0.70	0.77	
Methionine	%	0.24	0.26	
Tryptophan	%	0.16	0.18	
Xanthophylls	mg/kg	22.20	24.31	

Alfalfa, aerial part, dehy grnd, mn 15% protein, (1)
Ref No 1 00 022 United States

Feed Name or Analyses		Mean As Fed	Mean Dry	C.V. ± %
Dry matter	%	92.3	100.0	1
Ash	%	9.1	9.8	7
Crude fiber	%	25.5	27.7	5
Cattle	dig coef %	* 52.	52.	
Sheep	dig coef %	* 43.	43.	
Ether extract	%	2.3	2.5	10
Cattle	dig coef %	* 61.	61.	
Sheep	dig coef %	* 31.	31.	
N-free extract	%	39.8	43.1	
Cattle	dig coef %	* 69.	69.	
Sheep	dig coef %	* 68.	68.	
Protein (N x 6.25)	%	15.6	16.9	6
Cattle	dig coef %	* 78.	78.	
Sheep	dig coef %	* 69.	69.	
Cattle	dig prot %	12.2	13.2	
Goats	dig prot %	* 11.4	12.3	
Horses	dig prot %	* 11.0	11.9	
Rabbits	dig prot %	* 10.8	11.7	
Sheep	dig prot %	10.8	11.7	
Energy	GE Mcal/kg			
Cattle	DE Mcal/kg	* 2.47	2.68	
Sheep	DE Mcal/kg	* 2.22	2.41	
Cattle	ME Mcal/kg	* 2.03	2.20	
Chickens	ME_n kcal/kg	666.	722.	
Sheep	ME Mcal/kg	* 1.82	1.97	
Cattle	NE_m Mcal/kg	* 1.21	1.31	
Cattle	NE_{gain} Mcal/kg	* 0.64	0.69	
Cattle	$NE_{lactating\ cows}$ Mcal/kg	* 1.38	1.49	
Cattle	TDN %	56.1	60.8	
Sheep	TDN %	50.4	54.6	
Calcium	%	1.39	1.51	
Phosphorus	%	0.25	0.27	
Potassium	%	1.17	1.26	
Carotene	mg/kg	63.3	68.6	10
Choline	mg/kg	889.	963.	
Niacin	mg/kg	38.7	41.9	
Pantothenic acid	mg/kg	18.0	19.5	
Riboflavin	mg/kg	12.4	13.4	
Vitamin A	IU/g	55.5	60.2	
Vitamin A equivalent	IU/g	105.5	114.3	
Arginine	%	0.71	0.76	
Cystine	%	0.27	0.29	
Glycine	%	0.81	0.87	
Lysine	%	0.81	0.87	
Methionine	%	0.26	0.28	
Tryptophan	%	0.18	0.20	

Ref No 1 00 22 Canada

Feed Name or Analyses		Mean As Fed	Mean Dry	C.V. ± %
Dry matter	%		100.0	
Crude fiber	%		27.7	
Protein (N x 6.25)	%		16.8	
Cattle	dig prot %	*	11.5	
Goats	dig prot %	*	12.2	
Horses	dig prot %	*	11.8	
Rabbits	dig prot %	*	11.6	
Sheep	dig prot %	*	11.6	
Energy	GE Mcal/kg			
Cattle	DE Mcal/kg	*	2.68	
Sheep	DE Mcal/kg	*	2.41	
Cattle	ME Mcal/kg	*	2.20	
Chickens	ME_n kcal/kg		722.	
Sheep	ME Mcal/kg	*	1.97	
Cattle	TDN %		60.8	
Sheep	TDN %		54.6	
Xanthophylls	mg/kg	55.54	60.17	

Alfalfa, aerial part, dehy grnd, mn 17% protein, (1)
Ref No 1 00 023 United States

Feed Name or Analyses		Mean As Fed	Mean Dry	C.V. ± %
Dry matter	%	92.7	100.0	0
Ash	%	9.8	10.5	6
Crude fiber	%	24.3	26.2	1
Cattle	dig coef %	* 52.	52.	
Sheep	dig coef %	* 43.	43.	
Ether extract	%	2.5	2.7	12
Cattle	dig coef %	* 61.	61.	
Sheep	dig coef %	* 31.	31.	
N-free extract	%	38.5	41.6	2
Cattle	dig coef %	* 69.	69.	
Sheep	dig coef %	* 68.	68.	
Protein (N x 6.25)	%	17.6	19.0	1
Cattle	dig coef %	* 78.	78.	
Sheep	dig coef %	* 69.	69.	
Cattle	dig prot %	13.7	14.8	
Goats	dig prot %	* 13.3	14.3	
Horses	dig prot %	* 12.7	13.7	
Rabbits	dig prot %	* 12.4	13.3	
Sheep	dig prot %	12.2	13.1	
Cellulose (Matrone)	%	26.1	28.2	
Pentosans	%	12.5	13.5	
Sugars, total	%	3.8	4.1	
Energy	GE Mcal/kg			
Cattle	DE Mcal/kg	* 2.48	2.68	

Continued

Feed Name or Analyses			Mean As Fed	Mean Dry	C.V. ± %
Sheep	DE Mcal/kg	*	2.23	2.40	
Cattle	ME Mcal/kg	*	2.04	2.20	
Chickens	ME_n kcal/kg		1213.	1309.	
Sheep	ME Mcal/kg	*	1.83	1.97	
Cattle	NE_m Mcal/kg	*	1.21	1.31	
Cattle	NE_{gain} Mcal/kg	*	0.64	0.69	
Cattle	$NE_{lactating\ cows}$ Mcal/kg	*	1.41	1.52	
Cattle	TDN %		56.3	60.8	
Sheep	TDN %		50.5	54.5	
Calcium	%		1.54	1.66	
Chlorine	%		0.60	0.65	
Cobalt	mg/kg		0.272	0.293	
Copper	mg/kg		6.8	7.4	
Iron	%		0.033	0.036	
Manganese	mg/kg		33.0	35.6	
Phosphorus	%		0.23	0.25	
Carotene	mg/kg		110.8	119.5	34
Choline	mg/kg		1058.	1141.	16
Folic acid	mg/kg		6.31	6.81	37
Niacin	mg/kg		22.7	24.5	45
Pantothenic acid	mg/kg		28.4	30.6	19
Riboflavin	mg/kg		15.7	17.0	13
Thiamine	mg/kg		3.9	4.2	17
a-tocopherol	mg/kg		120.0	129.4	
Vitamin A	IU/g		154.0	166.1	
Vitamin A equivalent	IU/g		184.6	199.2	
Arginine	%		0.78	0.84	14
Cystine	%		0.34	0.36	
Glycine	%		0.90	0.97	
Histidine	%		0.37	0.40	14
Isoleucine	%		0.93	1.00	14
Leucine	%		1.30	1.40	6
Lysine	%		0.97	1.05	6
Methionine	%		0.22	0.24	
Phenylalanine	%		0.83	0.90	6
Threonine	%		0.74	0.80	7
Tryptophan	%		0.23	0.25	23
Tyrosine	%		0.56	0.60	10
Valine	%		0.83	0.90	16

Ref No 1 00 023 Canada

Feed Name or Analyses			Mean As Fed	Mean Dry	C.V. ± %
Dry matter	%			100.0	
Crude fiber	%			21.2	
Protein (N x 6.25)	%			19.3	
Cattle	dig prot %	*		13.7	
Goats	dig prot %	*		14.6	
Horses	dig prot %	*		13.9	
Rabbits	dig prot %	*		13.6	
Sheep	dig prot %	*		13.9	
Energy	GE Mcal/kg				
Cattle	DE Mcal/kg	*		2.69	
Sheep	DE Mcal/kg	*		2.41	
Cattle	ME Mcal/kg	*		2.20	
Chickens	ME_n kcal/kg			1309.	
Sheep	ME Mcal/kg	*		1.98	
Cattle	TDN %			60.9	
Sheep	TDN %			54.7	
Xanthophylls	mg/kg		153.98	166.12	

Alfalfa, aerial part, dehy grnd, mn 20% protein, (1)

Ref No 1 00 024 United States

Feed Name or Analyses			Mean As Fed	Mean Dry	C.V. ± %
Dry matter	%		92.1	100.0	0
Ash	%		10.6	11.5	12
Crude fiber	%		20.9	22.7	14
Cattle	dig coef %	*	52.	52.	
Sheep	dig coef %	*	43.	43.	
Ether extract	%		2.7	2.9	16
Cattle	dig coef %	*	61.	61.	
Sheep	dig coef %	*	31.	31.	
N-free extract	%		37.9	41.1	
Cattle	dig coef %	*	69.	69.	
Sheep	dig coef %	*	68.	68.	
Protein (N x 6.25)	%		20.1	21.8	5
Cattle	dig coef %	*	78.	78.	
Sheep	dig coef %	*	69.	69.	
Cattle	dig prot %		15.7	17.0	
Goats	dig prot %	*	15.6	16.9	
Horses	dig prot %	*	14.8	16.0	
Rabbits	dig prot %	*	14.3	15.5	
Sheep	dig prot %		13.9	15.0	
Energy	GE Mcal/kg				
Cattle	DE Mcal/kg	*	2.48	2.70	
Sheep	DE Mcal/kg	*	2.22	2.42	
Cattle	ME Mcal/kg	*	2.04	2.21	
Chickens	ME_n kcal/kg		1585.	1721.	
Sheep	ME Mcal/kg	*	1.82	1.98	
Cattle	TDN %		56.3	61.2	
Sheep	TDN %		50.5	54.8	
Calcium	%		1.65	1.79	11
Copper	mg/kg		15.6	17.0	10
Iron	%		0.039	0.042	12
Manganese	mg/kg		62.8	68.3	17
Phosphorus	%		0.29	0.31	14
Carotene	mg/kg		122.6	133.2	16
Choline	mg/kg		1108.	1203.	
Niacin	mg/kg		37.9	41.2	32
Pantothenic acid	mg/kg		40.6	44.1	11
Riboflavin	mg/kg		15.9	17.2	16
Thiamine	mg/kg		6.7	7.3	15
α-tocopherol	mg/kg		140.0	152.0	
Vitamin A	IU/g		330.3	358.7	
Vitamin A equivalent	IU/g		204.4	222.0	
Arginine	%		0.90	0.98	
Cystine	%		0.36	0.39	
Glycine	%		1.00	1.08	
Lysine	%		1.00	1.08	
Methionine	%		0.32	0.35	
Tryptophan	%		0.24	0.26	
Xanthophylls	mg/kg		330.26	358.67	

Alfalfa, aerial part, dehy grnd pelleted, (1)

Ref No 1 07 848 United States

Feed Name or Analyses			Mean As Fed	Mean Dry	C.V. ± %
Dry matter	%			100.0	
Sheep	dig coef %			61.	
Crude fiber	%			21.6	
Sheep	dig coef %			44.	
Ether extract	%			3.7	
Sheep	dig coef %			59.	
N-free extract	%			38.1	
Sheep	dig coef %			75.	
Protein (N x 6.25)	%			22.2	
Sheep	dig coef %			64.	
Cattle	dig prot %	*		16.2	
Goats	dig prot %	*		17.3	
Horses	dig prot %	*		16.4	
Rabbits	dig prot %	*		15.8	
Sheep	dig prot %			14.2	
Energy	GE Mcal/kg			4.29	
Sheep	GE dig coef %			61.	
Cattle	DE Mcal/kg	*		2.77	
Sheep	DE Mcal/kg			2.62	
Cattle	ME Mcal/kg	*		2.27	
Sheep	ME Mcal/kg	*		2.15	
Cattle	TDN %	*		62.9	
Sheep	TDN %			57.2	

Alfalfa, aerial part, dehy grnd pelleted, immature, (1)

Ref No 1 07 815 United States

Feed Name or Analyses			Mean As Fed	Mean Dry	C.V. ± %
Dry matter	%			100.0	
Lignin (Ellis)	%				
Sheep	dig coef %			3.	
Energy	GE Mcal/kg			4.45	
Sheep	DE Mcal/kg			2.80	
Sheep	ME Mcal/kg			2.30	
Sheep	TDN %			59.5	

Alfalfa, aerial part, dehy grnd pelleted, pre-bloom, cut 2, (1)

Ref No 1 07 734 United States

Feed Name or Analyses			Mean As Fed	Mean Dry	C.V. ± %
Dry matter	%			100.0	
Protein (N x 6.25)	%			26.0	
Cattle	dig prot %	*		19.5	
Goats	dig prot %	*		20.8	
Horses	dig prot %	*		19.6	
Rabbits	dig prot %	*		18.7	
Sheep	dig prot %	*		19.9	
Lignin (Ellis)	%			6.0	

Alfalfa, aerial part, dehy grnd pelleted, early bloom, (1)

Ref No 1 07 847 United States

Feed Name or Analyses			Mean As Fed	Mean Dry	C.V. ± %
Dry matter	%			100.0	
Ash	%			12.4	
Crude fiber	%			25.3	
Sheep	dig coef %			47.	
Ether extract	%			2.3	
N-free extract	%			39.0	
Protein (N x 6.25)	%			21.0	
Sheep	dig coef %			69.	
Cattle	dig prot %	*		15.1	
Goats	dig prot %	*		16.2	
Horses	dig prot %	*		15.4	
Rabbits	dig prot %	*		14.9	
Sheep	dig prot %			14.5	
Lignin (Ellis)	%			9.0	
Energy	GE Mcal/kg				
Cattle	DE Mcal/kg	*		2.67	
Sheep	DE Mcal/kg	*		2.43	
Cattle	ME Mcal/kg	*		2.19	
Sheep	ME Mcal/kg	*		1.99	
Cattle	TDN %	*		60.5	
Sheep	TDN %			55.0	

Alfalfa, aerial part, dehy grnd pelleted, early bloom, cut 2, (1)

Ref No 1 07 733 United States

Feed Name or Analyses			Mean As Fed	Mean Dry	C.V. ± %
Dry matter	%			100.0	
Protein (N x 6.25)	%			26.0	

(1) dry forages and roughages (2) pasture, range plants, and forages fed green (3) silages (4) energy feeds (5) protein supplements (6) minerals (7) vitamins (8) additives

Feed Name or Analyses			As Fed (Mean)	Dry (Mean)	C.V. ± %
Cattle	dig prot %	*		19.5	
Goats	dig prot %	*		20.8	
Horses	dig prot %	*		19.6	
Rabbits	dig prot %	*		18.7	
Sheep	dig prot %	*		19.9	
Lignin (Ellis)	%			6.4	

Alfalfa, aerial part, dehy grnd pelleted, mid-bloom, cut 2, (1)

Ref No 1 07 732 United States

Feed Name or Analyses			As Fed (Mean)	Dry (Mean)	C.V. ± %
Dry matter	%			100.0	
Protein (N x 6.25)	%			21.0	
Cattle	dig prot %	*		15.1	
Goats	dig prot %	*		16.2	
Horses	dig prot %	*		15.4	
Rabbits	dig prot %	*		14.9	
Sheep	dig prot %	*		15.4	
Lignin (Ellis)	%			7.4	

Alfalfa, aerial part, dehy grnd pelleted, full bloom, cut 2, (1)

Ref No 1 07 731 United States

Feed Name or Analyses			As Fed (Mean)	Dry (Mean)	C.V. ± %
Dry matter	%			100.0	
Protein (N x 6.25)	%			20.0	
Cattle	dig prot %	*		14.3	
Goats	dig prot %	*		15.2	
Horses	dig prot %	*		14.5	
Rabbits	dig prot %	*		14.1	
Sheep	dig prot %	*		14.5	
Lignin (Ellis)	%			7.7	

Alfalfa, aerial part, dehy grnd pelleted, mn 15% protein mx 30% fiber, (1)

Ref No 1 00 026 Canada

Feed Name or Analyses			As Fed (Mean)	Dry (Mean)	C.V. ± %
Dry matter	%		92.7	100.0	1
Crude fiber	%		26.1	28.2	6
Ether extract	%		2.8	3.0	14
Protein (N x 6.25)	%		15.8	17.1	9
Cattle	dig prot %	*	10.9	11.7	
Goats	dig prot %	*	11.6	12.5	
Horses	dig prot %	*	11.1	12.0	
Rabbits	dig prot %	*	11.0	11.8	
Sheep	dig prot %	*	11.0	11.9	
Calcium	%		1.51	1.63	7
Phosphorus	%		0.23	0.25	16
Salt (NaCl)	%		0.53	0.57	
Carotene	mg/kg		117.1	126.3	88
Vitamin A	IU/g		138.5	149.5	
Vitamin A equivalent	IU/g		195.2	210.6	

Alfalfa, aerial part, dehy grnd pelleted, mn 17% protein mx 27% fiber mn 220 IU vitamin A equivalent per g, (1)

Ref No 1 00 027 Canada

Feed Name or Analyses			As Fed (Mean)	Dry (Mean)	C.V. ± %
Dry matter	%		93.0	100.0	
Protein (N x 6.25)	%		17.4	18.7	
Cattle	dig prot %	*	12.3	13.2	
Goats	dig prot %	*	13.1	14.0	
Horses	dig prot %	*	12.5	13.4	
Rabbits	dig prot %	*	12.2	13.1	
Sheep	dig prot %	*	12.4	13.4	
Carotene	mg/kg		16.4	17.6	
Vitamin A equivalent	IU/g		27.3	29.3	

Alfalfa, aerial part, dehy grnd pelleted, mn 18% protein mx 25% fiber mn 276 IU vitamin A equivalent per g, (1)

Ref No 1 00 028 Canada

Feed Name or Analyses			As Fed (Mean)	Dry (Mean)	C.V. ± %
Dry matter	%		92.6	100.0	1
Crude fiber	%		18.2	19.7	
Ether extract	%		2.8	3.0	
Protein (N x 6.25)	%		18.8	20.3	4
Cattle	dig prot %	*	13.4	14.5	
Goats	dig prot %	*	14.3	15.5	
Horses	dig prot %	*	13.7	14.8	
Rabbits	dig prot %	*	13.3	14.3	
Sheep	dig prot %	*	13.7	14.8	
Salt (NaCl)	%		1.14	1.23	
Carotene	mg/kg		35.3	38.2	98
Vitamin A equivalent	IU/g		58.9	63.6	

Alfalfa, aerial part, dehy grnd pelleted, mn 20% protein mx 22% fiber mn 331 IU vitamin A equivalent per g, (1)

Ref No 1 00 029 Canada

Feed Name or Analyses			As Fed (Mean)	Dry (Mean)	C.V. ± %
Dry matter	%			100.0	
Crude fiber	%			19.4	
Protein (N x 6.25)	%			21.2	
Cattle	dig prot %	*		15.3	
Goats	dig prot %	*		16.3	
Horses	dig prot %	*		15.5	
Rabbits	dig prot %	*		15.0	
Sheep	dig prot %	*		15.6	

Alfalfa, flowers w leaves, s-c, mid-bloom, cut 1, (1)

Ref No 1 00 155 United States

Feed Name or Analyses			As Fed (Mean)	Dry (Mean)	C.V. ± %
Dry matter	%		86.6	100.0	
Ash	%		8.9	10.3	
Crude fiber	%		17.3	20.0	
Sheep	dig coef %		56.	56.	
Ether extract	%		2.4	2.8	
Sheep	dig coef %		20.	20.	
N-free extract	%		40.4	46.6	
Sheep	dig coef %		72.	72.	
Protein (N x 6.25)	%		17.6	20.3	
Sheep	dig coef %		75.	75.	
Cattle	dig prot %	*	12.6	14.5	
Goats	dig prot %	*	13.4	15.5	
Horses	dig prot %	*	12.8	14.8	
Rabbits	dig prot %	*	12.4	14.3	
Sheep	dig prot %		13.2	15.2	
Energy	GE Mcal/kg				
Cattle	DE Mcal/kg	*	2.51	2.90	
Sheep	DE Mcal/kg	*	2.34	2.70	
Cattle	ME Mcal/kg	*	2.06	2.37	
Sheep	ME Mcal/kg	*	1.92	2.21	
Cattle	TDN %	*	56.9	65.7	
Sheep	TDN %		53.0	61.2	

Alfalfa, hay, fan air dried, (1)

Ref No 1 00 040 United States

Feed Name or Analyses			As Fed (Mean)	Dry (Mean)	C.V. ± %
Dry matter	%		89.5	100.0	2
Ash	%		7.2	8.0	8
Crude fiber	%		26.3	29.4	11
Ether extract	%		1.8	2.1	18
N-free extract	%		36.1	40.3	
Protein (N x 6.25)	%		18.1	20.2	10
Cattle	dig prot %	*	12.9	14.4	
Goats	dig prot %	*	13.8	15.4	
Horses	dig prot %	*	13.1	14.7	
Rabbits	dig prot %	*	12.8	14.3	
Sheep	dig prot %	*	13.2	14.7	
Sugars, total	%		6.4	7.1	
Energy	GE Mcal/kg		3.82	4.27	2
Cattle	DE Mcal/kg	*	2.25	2.51	
Sheep	DE Mcal/kg	*	2.35	2.63	
Cattle	ME Mcal/kg	*	1.85	2.06	
Sheep	ME Mcal/kg	*	1.93	2.16	
Cattle	TDN %	*	51.1	57.0	
Sheep	TDN %	*	53.4	59.6	
Vitamin D_2	IU/g		0.5	0.6	9

Alfalfa, hay, fan air dried, immature, (1)

Ref No 1 00 031 United States

Feed Name or Analyses			As Fed (Mean)	Dry (Mean)	C.V. ± %
Dry matter	%		89.9	100.0	3
Sheep	dig coef %		66.	66.	
Ash	%		7.9	8.8	4
Crude fiber	%		20.4	22.7	9
Ether extract	%		2.2	2.4	16
N-free extract	%		37.6	41.8	
Protein (N x 6.25)	%		21.8	24.3	6
Sheep	dig coef %		79.	79.	
Cattle	dig prot %	*	16.2	18.0	
Goats	dig prot %	*	17.3	19.2	
Horses	dig prot %	*	16.3	18.2	
Rabbits	dig prot %	*	15.7	17.4	
Sheep	dig prot %		17.3	19.2	
Cellulose (Matrone)	%		19.9	22.1	
Sheep	dig coef %		60.	60.	
Energy	GE Mcal/kg		3.93	4.37	
Sheep	GE dig coef %		65.	65.	
Cattle	DE Mcal/kg	*	2.45	2.72	
Sheep	DE Mcal/kg		2.55	2.84	
Cattle	ME Mcal/kg	*	2.01	2.23	
Sheep	ME Mcal/kg	*	2.09	2.33	
Cattle	TDN %	*	55.5	61.7	
Sheep	TDN %	*	54.3	60.4	

Alfalfa, hay, fan air dried, early bloom, (1)

Ref No 1 00 032 United States

Feed Name or Analyses			As Fed (Mean)	Dry (Mean)	C.V. ± %
Dry matter	%		90.6	100.0	2
Sheep	dig coef %		56.	56.	
Ash	%		6.9	7.6	8
Crude fiber	%		29.4	32.5	4
Ether extract	%		1.7	1.9	15
N-free extract	%		36.1	39.8	
Protein (N x 6.25)	%		16.5	18.2	5
Sheep	dig coef %		70.	70.	
Cattle	dig prot %	*	11.5	12.7	
Goats	dig prot %	*	12.3	13.5	
Horses	dig prot %	*	11.8	13.0	
Rabbits	dig prot %	*	11.5	12.7	
Sheep	dig prot %		11.5	12.7	
Cellulose (Matrone)	%		26.6	29.4	
Sheep	dig coef %		52.	52.	
Energy	GE Mcal/kg		4.04	4.46	
Sheep	GE dig coef %		54.	54.	
Cattle	DE Mcal/kg	*	2.19	2.42	

Continued

Feed Name or Analyses			Mean As Fed	Mean Dry	C.V. ± %
Sheep	DE Mcal/kg		2.18	2.41	
Cattle	ME Mcal/kg	*	1.80	1.98	
Sheep	ME Mcal/kg	*	1.79	1.97	
Cattle	TDN %	*	49.7	54.9	
Sheep	TDN %	*	52.3	57.7	
Riboflavin	mg/kg		14.6	16.1	

Alfalfa, hay, fan air dried, early bloom, cut 3, (1)
Ref No 1 07 883 United States

Feed Name or Analyses			Mean As Fed	Mean Dry	C.V. ± %
Dry matter	%		91.5	100.0	
Cattle	dig coef %		60.	60.	
Horses	dig coef %		52.	52.	
Sheep	dig coef %		59.	59.	
Ash	%		7.7	8.4	
Crude fiber	%		30.7	33.6	
Cattle	dig coef %		40.	40.	
Horses	dig coef %		36.	36.	
Sheep	dig coef %		42.	42.	
Ether extract	%		2.1	2.3	
Cattle	dig coef %		35.	35.	
Horses	dig coef %		25.	25.	
Sheep	dig coef %		31.	31.	
N-free extract	%		34.7	37.9	
Cattle	dig coef %		72.	72.	
Horses	dig coef %		67.	67.	
Sheep	dig coef %		71.	71.	
Protein (N x 6.25)	%		16.3	17.8	
Cattle	dig coef %		72.	72.	
Horses	dig coef %		67.	67.	
Sheep	dig coef %		76.	76.	
Cattle	dig prot %		11.7	12.8	
Goats	dig prot %	*	12.0	13.2	
Horses	dig prot %		10.9	11.9	
Rabbits	dig prot %	*	11.4	12.4	
Sheep	dig prot %		12.4	13.5	
Energy	GE Mcal/kg				
Cattle	DE Mcal/kg	*	2.23	2.44	
Horses	DE Mcal/kg	*	1.94	2.12	
Sheep	DE Mcal/kg	*	2.27	2.48	
Cattle	ME Mcal/kg	*	1.83	2.00	
Horses	ME Mcal/kg	*	1.59	1.74	
Sheep	ME Mcal/kg	*	1.86	2.03	
Cattle	TDN %		50.7	55.4	
Horses	TDN %		44.0	48.1	
Sheep	TDN %		51.4	56.2	
Carotene	mg/kg		6.0	6.6	
Vitamin A equivalent	IU/g		10.1	11.0	

Alfalfa, hay, fan air dried, mid-bloom, (1)
Ref No 1 00 033 United States

Feed Name or Analyses			Mean As Fed	Mean Dry	C.V. ± %
Dry matter	%			100.0	
Riboflavin	mg/kg			11.9	

Alfalfa, hay, fan air dried, cut 1, (1)
Ref No 1 00 034 United States

Feed Name or Analyses			Mean As Fed	Mean Dry	C.V. ± %
Dry matter	%		90.8	100.0	
Ash	%		7.2	7.9	
Crude fiber	%		29.9	32.9	
Ether extract	%		1.9	2.1	
N-free extract	%		36.9	40.6	
Protein (N x 6.25)	%		15.0	16.5	
Cattle	dig prot %	*	10.2	11.2	
Goats	dig prot %	*	10.9	12.0	
Horses	dig prot %	*	10.5	11.5	
Rabbits	dig prot %	*	10.4	11.4	
Sheep	dig prot %	*	10.3	11.4	
Energy	GE Mcal/kg				
Cattle	DE Mcal/kg	*	2.21	2.44	
Sheep	DE Mcal/kg	*	2.27	2.50	
Cattle	ME Mcal/kg	*	1.82	2.00	
Sheep	ME Mcal/kg	*	1.87	2.05	
Cattle	TDN %	*	50.2	55.3	
Sheep	TDN %	*	51.6	56.8	
Carotene	mg/kg		38.2	42.1	
Riboflavin	mg/kg		12.8	14.1	
Vitamin A equivalent	IU/g		63.7	70.2	

Alfalfa, hay, fan air dried, cut 2, (1)
Ref No 1 00 035 United States

Feed Name or Analyses			Mean As Fed	Mean Dry	C.V. ± %
Dry matter	%		88.7	100.0	1
Ash	%		7.1	8.0	7
Crude fiber	%		24.8	28.0	8
Ether extract	%		2.0	2.3	21
N-free extract	%		35.5	40.0	
Protein (N x 6.25)	%		19.2	21.7	6
Cattle	dig prot %	*	14.0	15.7	
Goats	dig prot %	*	14.9	16.8	
Horses	dig prot %	*	14.2	16.0	
Rabbits	dig prot %	*	13.7	15.4	
Sheep	dig prot %	*	14.2	16.0	
Energy	GE Mcal/kg				
Cattle	DE Mcal/kg	*	2.27	2.56	
Sheep	DE Mcal/kg	*	2.37	2.67	
Cattle	ME Mcal/kg	*	1.86	2.10	
Sheep	ME Mcal/kg	*	1.94	2.19	
Cattle	TDN %	*	51.4	58.0	
Sheep	TDN %	*	53.7	60.6	
Carotene	mg/kg		35.4	39.9	52
Riboflavin	mg/kg		15.8	17.9	
Vitamin A equivalent	IU/g		59.0	66.5	

Alfalfa, hay, fan air dried, cut 3, (1)
Ref No 1 00 036 United States

Feed Name or Analyses			Mean As Fed	Mean Dry	C.V. ± %
Dry matter	%		88.5	100.0	3
Carotene	mg/kg		44.5	50.3	18
Riboflavin	mg/kg		9.4	10.6	
Vitamin A equivalent	IU/g		74.2	83.8	

Alfalfa, hay, fan air dried, cut 4, (1)
Ref No 1 00 037 United States

Feed Name or Analyses			Mean As Fed	Mean Dry	C.V. ± %
Dry matter	%		90.1	100.0	3
Ash	%		6.8	7.6	5
Crude fiber	%		20.1	22.3	4
Ether extract	%		2.1	2.3	11
N-free extract	%		39.4	43.7	
Protein (N x 6.25)	%		21.7	24.1	3
Cattle	dig prot %	*	16.0	17.8	
Goats	dig prot %	*	17.2	19.0	
Horses	dig prot %	*	16.2	18.0	
Rabbits	dig prot %	*	15.6	17.3	
Sheep	dig prot %	*	16.4	18.2	
Energy	GE Mcal/kg				
Cattle	DE Mcal/kg	*	2.41	2.68	
Sheep	DE Mcal/kg	*	2.59	2.87	
Cattle	ME Mcal/kg	*	1.98	2.20	
Sheep	ME Mcal/kg	*	2.12	2.35	
Cattle	TDN %	*	54.7	60.7	
Sheep	TDN %	*	58.6	65.1	
Sulphur	%		0.23	0.26	10

Alfalfa, hay, fan air dried, gr 1 US, (1)
Ref No 1 00 038 United States

Feed Name or Analyses			Mean As Fed	Mean Dry	C.V. ± %
Dry matter	%		89.6	100.0	2
Ash	%		7.3	8.1	7
Crude fiber	%		26.2	29.3	8
Ether extract	%		1.8	2.0	19
N-free extract	%		36.4	40.6	
Protein (N x 6.25)	%		17.9	20.0	10
Cattle	dig prot %	*	12.8	14.2	
Goats	dig prot %	*	13.6	15.2	
Horses	dig prot %	*	13.0	14.5	
Rabbits	dig prot %	*	12.6	14.1	
Sheep	dig prot %	*	13.0	14.5	
Energy	GE Mcal/kg				
Cattle	DE Mcal/kg	*	2.35	2.63	
Sheep	DE Mcal/kg	*	2.36	2.63	
Cattle	ME Mcal/kg	*	1.93	2.15	
Sheep	ME Mcal/kg	*	1.93	2.16	
Cattle	TDN %		53.3	59.6	
Sheep	TDN %	*	53.4	59.6	
Carotene	mg/kg		40.5	45.2	37
Vitamin A equivalent	IU/g		67.5	75.3	

Alfalfa, hay, fan air dried, gr 2 US, (1)
Ref No 1 00 039 United States

Feed Name or Analyses			Mean As Fed	Mean Dry	C.V. ± %
Dry matter	%		88.3	100.0	
Ash	%		5.9	6.7	
Crude fiber	%		29.6	33.5	4
Ether extract	%		1.3	1.5	
N-free extract	%		34.3	38.9	
Protein (N x 6.25)	%		17.1	19.4	17
Cattle	dig prot %	*	12.1	13.7	
Goats	dig prot %	*	12.9	14.7	
Horses	dig prot %	*	12.4	14.0	
Rabbits	dig prot %	*	12.1	13.6	
Sheep	dig prot %	*	12.3	14.0	
Energy	GE Mcal/kg				
Cattle	DE Mcal/kg	*	2.06	2.33	
Sheep	DE Mcal/kg	*	2.27	2.57	
Cattle	ME Mcal/kg	*	1.69	1.91	
Sheep	ME Mcal/kg	*	1.86	2.11	
Cattle	TDN %	*	46.7	52.9	
Sheep	TDN %	*	51.5	58.3	

Alfalfa, hay, fan air dried grnd pelleted, early bloom, cut 3, (1)
Ref No 1 07 884 United States

Feed Name or Analyses			Mean As Fed	Mean Dry	C.V. ± %
Dry matter	%		91.2	100.0	
Cattle	dig coef %		53.	53.	
Horses	dig coef %		52.	52.	
Sheep	dig coef %		52.	52.	
Ash	%		8.4	9.2	
Crude fiber	%		29.9	32.8	
Cattle	dig coef %		29.	29.	
Horses	dig coef %		30.	30.	
Sheep	dig coef %		31.	31.	
Ether extract	%		1.7	1.9	

(1) dry forages and roughages (2) pasture, range plants, and forages fed green (3) silages (4) energy feeds (5) protein supplements (6) minerals (7) vitamins (8) additives

Feed Name or Analyses		Mean As Fed	Mean Dry	C.V. ± %
Cattle	dig coef %	40.	40.	
Horses	dig coef %	22.	22.	
Sheep	dig coef %	36.	36.	
N-free extract	%	35.4	38.8	
Cattle	dig coef %	64.	64.	
Horses	dig coef %	68.	68.	
Sheep	dig coef %	63.	63.	
Protein (N x 6.25)	%	15.8	17.3	
Cattle	dig coef %	68.	68.	
Horses	dig coef %	68.	68.	
Sheep	dig coef %	68.	68.	
Cattle	dig prot %	10.7	11.8	
Goats	dig prot %	* 11.6	12.7	
Horses	dig prot %	10.7	11.8	
Rabbits	dig prot %	* 11.0	12.0	
Sheep	dig prot %	10.7	11.8	
Energy	GE Mcal/kg			
Cattle	DE Mcal/kg	* 1.92	2.11	
Horses	DE Mcal/kg	* 1.89	2.07	
Sheep	DE Mcal/kg	* 1.93	2.11	
Cattle	ME Mcal/kg	* 1.58	1.73	
Horses	ME Mcal/kg	* 1.55	1.70	
Sheep	ME Mcal/kg	* 1.58	1.73	
Cattle	TDN %	43.6	47.8	
Horses	TDN %	42.9	47.0	
Sheep	TDN %	43.7	47.9	
Calcium	%	1.58	1.73	
Phosphorus	%	1.84	2.02	
Silicon	%	1.15	1.26	
Carotene	mg/kg	6.8	7.5	
Vitamin A equivalent	IU/g	11.4	12.5	

Alfalfa, hay, fan air dried w heat, (1)
Ref No 1 08 660 United States

Feed Name or Analyses		Mean As Fed	Mean Dry	C.V. ± %
Dry matter	%	89.5	100.0	
Ash	%	8.9	9.9	
Crude fiber	%	29.5	33.0	
Ether extract	%	1.6	1.8	
N-free extract	%	31.5	35.2	
Protein (N x 6.25)	%	18.0	20.1	
Cattle	dig prot %	* 12.8	14.4	
Goats	dig prot %	* 13.7	15.3	
Horses	dig prot %	* 13.1	14.6	
Rabbits	dig prot %	* 12.7	14.2	
Sheep	dig prot %	* 13.1	14.6	
Energy	GE Mcal/kg			
Sheep	GE dig coef %	64.	64.	
Cattle	DE Mcal/kg	* 2.31	2.58	
Sheep	DE Mcal/kg	* 2.20	2.46	
Cattle	ME Mcal/kg	* 1.89	2.12	
Sheep	ME Mcal/kg	* 1.81	2.02	
Cattle	TDN %	* 52.4	58.6	
Sheep	TDN %	* 49.9	55.8	

Alfalfa, hay, s-c, (1)
Ref No 1 00 078 United States

Feed Name or Analyses		Mean As Fed	Mean Dry	C.V. ± %
Dry matter	%	91.4	100.0	2
Rabbits	dig coef %	52.	52.	9
Sheep	dig coef %	59.	59.	
Organic matter	%	80.0	87.6	
Ash	%	9.0	9.9	13
Crude fiber	%	28.0	30.6	16
Cattle	dig coef %	42.	42.	
Horses	dig coef %	38.	38.	
Sheep	dig coef %	47.	47.	8
Swine	dig coef %	22.	22.	
Ether extract	%	1.7	1.9	43
Cattle	dig coef %	36.	36.	
Horses	dig coef %	-2.	2.	
Sheep	dig coef %	42.	42.	
Swine	dig coef %	14.	14.	
N-free extract	%	37.1	40.6	25
Cattle	dig coef %	70.	70.	
Horses	dig coef %	65.	65.	
Sheep	dig coef %	69.	69.	9
Swine	dig coef %	49.	49.	
Protein (N x 6.25)	%	15.5	17.0	8
Cattle	dig coef %	66.	66.	
Horses	dig coef %	72.	72.	
Rabbits	dig coef %	72.	72.	5
Sheep	dig coef %	70.	70.	3
Swine	dig coef %	47.	47.	
Cattle	dig prot %	10.2	11.1	
Goats	dig prot %	* 11.4	12.4	
Horses	dig prot %	11.1	12.2	
Rabbits	dig prot %	11.1	12.2	
Sheep	dig prot %	10.9	12.0	
Swine	dig prot %	7.3	8.0	
Cellulose (Matrone)	%	26.3	28.8	9
Starch	%	1.5	1.6	4
Sugars, total	%	6.5	7.1	26
Lignin (Ellis)	%	8.6	9.4	16
Energy	GE Mcal/kg	3.94	4.31	1
Cattle	DE Mcal/kg	* 2.17	2.37	
Horses	DE Mcal/kg	* 2.00	2.19	
Sheep	DE Mcal/kg	* 2.27	2.49	
Swine	DE kcal/kg	*1419.	1553.	
Cattle	ME Mcal/kg	* 1.78	1.94	
Horses	ME Mcal/kg	* 1.64	1.80	
Sheep	ME Mcal/kg	* 1.86	2.04	
Swine	ME kcal/kg	*1313.	1438.	
Cattle	TDN %	49.1	53.8	
Horses	TDN %	45.4	49.7	
Sheep	TDN %	51.6	56.4	11
Swine	TDN %	32.2	35.2	
Calcium	%	1.29	1.41	
Chlorine	%	0.28	0.31	
Cobalt	mg/kg	0.205	0.225	21
Copper	mg/kg	18.5	20.2	
Iron	%	0.017	0.019	
Magnesium	%	0.31	0.34	
Manganese	mg/kg	56.5	61.9	
Phosphorus	%	0.21	0.24	
Potassium	%	1.99	2.18	
Silicon	%	0.64	0.70	25
Sodium	%	0.15	0.17	
Sulphur	%	0.29	0.32	
Biotin	mg/kg	0.16	0.18	15
Carotene	mg/kg	65.2	71.4	33
Folic acid	mg/kg	3.10	3.40	8
α-tocopherol	mg/kg	101.6	111.2	50
Vitamin B_{12}	μg/kg	0.0	0.0	50
Vitamin A equivalent	IU/g	108.7	118.9	
Vitamin D_2	IU/g	1.5	1.6	34
Arginine	%	0.64	0.70	
Histidine	%	0.27	0.30	
Isoleucine	%	0.73	0.80	
Leucine	%	0.91	1.00	
Lysine	%	0.55	0.60	
Methionine	%	0.09	0.10	
Phenylalanine	%	0.55	0.60	
Threonine	%	0.64	0.70	
Tryptophan	%	0.09	0.10	
Valine	%	0.64	0.70	

Ref No 1 00 078 Canada

Feed Name or Analyses		Mean As Fed	Mean Dry	C.V. ± %
Dry matter	%	94.3	100.0	
Organic matter	%	87.2	92.5	
Ash	%	7.1	7.5	
Crude fiber	%	37.2	39.5	
Cattle	dig coef %	42.	42.	
Horses	dig coef %	38.	38.	
Sheep	dig coef %	47.	47.	
Swine	dig coef %	22.	22.	
Ether extract	%	1.1	1.2	
Cattle	dig coef %	36.	36.	
Horses	dig coef %	-2.	-2.	
Sheep	dig coef %	42.	42.	
Swine	dig coef %	14.	14.	
N-free extract	%	37.0	39.2	
Cattle	dig coef %	70.	70.	
Horses	dig coef %	65.	65.	
Sheep	dig coef %	69.	69.	
Swine	dig coef %	49.	49.	
Protein (N x 6.25)	%	11.9	12.6	
Cattle	dig coef %	66.	66.	
Horses	dig coef %	72.	72.	
Rabbits	dig coef %	72.	72.	
Sheep	dig coef %	70.	70.	
Swine	dig coef %	47.	47.	
Cattle	dig prot %	7.8	8.3	
Goats	dig prot %	* 7.8	8.3	
Horses	dig prot %	8.5	9.0	
Rabbits	dig prot %	8.5	9.0	
Sheep	dig prot %	8.4	8.9	
Swine	dig prot %	5.6	5.9	
Energy	GE Mcal/kg	4.07	4.31	
Cattle	DE Mcal/kg	* 2.21	2.34	
Horses	DE Mcal/kg	* 2.04	2.16	
Sheep	DE Mcal/kg	* 2.32	2.47	
Swine	DE kcal/kg	*1422.	1508.	
Cattle	ME Mcal/kg	* 1.81	1.92	
Horses	ME Mcal/kg	* 1.67	1.77	
Sheep	ME Mcal/kg	* 1.91	2.02	
Swine	ME kcal/kg	*1329.	1409.	
Cattle	TDN %	50.0	53.0	
Horses	TDN %	46.2	49.0	
Sheep	TDN %	52.7	55.9	
Swine	TDN %	32.2	34.2	

Alfalfa, hay, s-c, early leaf, cut 1, (1)
Ref No 1 00 048 United States

Feed Name or Analyses		Mean As Fed	Mean Dry	C.V. ± %
Dry matter	%	89.4	100.0	
Ash	%	11.3	12.6	
Crude fiber	%	17.7	19.8	
Sheep	dig coef %	65.	65.	
Ether extract	%	2.9	3.2	
Sheep	dig coef %	46.	46.	
N-free extract	%	35.8	40.0	
Sheep	dig coef %	65.	65.	
Protein (N x 6.25)	%	21.8	24.4	
Sheep	dig coef %	72.	72.	
Cattle	dig prot %	* 16.2	18.1	
Goats	dig prot %	* 17.3	19.3	
Horses	dig prot %	* 16.3	18.2	
Rabbits	dig prot %	* 15.7	17.5	
Sheep	dig prot %	15.7	17.6	
Energy	GE Mcal/kg			
Sheep	DE Mcal/kg	* 2.36	2.63	
Cattle	ME Mcal/kg	* 2.07	2.32	
Sheep	ME Mcal/kg	* 1.93	2.16	

Continued

Feed Name or Analyses			Mean As Fed	Mean Dry	C.V. ± %
Cattle	TDN %	*	53.4	59.7	
Sheep	TDN %		53.4	59.8	

Alfalfa, hay, s-c, immature, (1)
Ref No 1 00 050 United States

Feed Name or Analyses			Mean As Fed	Mean Dry	C.V. ± %
Dry matter	%		90.4	100.0	4
Ash	%		8.7	9.6	
Crude fiber	%		27.5	30.4	
Cattle	dig coef %	*	44.	44.	
Sheep	dig coef %	*	45.	45.	
Ether extract	%		1.5	1.7	
Cattle	dig coef %	*	35.	35.	
Sheep	dig coef %	*	31.	31.	
N-free extract	%		30.3	33.5	
Cattle	dig coef %	*	71.	71.	
Sheep	dig coef %	*	69.	69.	
Protein (N x 6.25)	%		22.4	24.8	
Cattle	dig coef %	*	70.	70.	
Sheep	dig coef %	*	72.	72.	
Cattle	dig prot %		15.7	17.4	
Goats	dig prot %	*	17.8	19.7	
Horses	dig prot %	*	16.8	18.6	
Rabbits	dig prot %	*	16.1	17.8	
Sheep	dig prot %		16.1	17.9	
Energy	GE Mcal/kg				
Cattle	DE Mcal/kg	*	2.23	2.46	
Sheep	DE Mcal/kg	*	2.22	2.46	
Cattle	ME Mcal/kg	*	1.82	2.02	
Sheep	ME Mcal/kg	*	1.82	2.02	
Cattle	NE_m Mcal/kg	*	1.23	1.36	
Cattle	NE_{gain} Mcal/kg	*	0.52	0.57	
Cattle	$NE_{lactating\ cows}$ Mcal/kg	*	1.21	1.34	
Cattle	TDN %		50.5	55.9	
Sheep	TDN %		50.4	55.8	
Calcium	%		1.92	2.12	14
Chlorine	%		0.31	0.34	20
Iron	%		0.023	0.025	30
Magnesium	%		0.23	0.26	15
Manganese	mg/kg		34.9	38.6	36
Phosphorus	%		0.27	0.30	14
Potassium	%		2.04	2.26	16
Sodium	%		0.20	0.22	28
Sulphur	%		0.57	0.63	12
Carotene	mg/kg		452.7	501.1	33
Riboflavin	mg/kg		10.8	11.9	
Vitamin A equivalent	IU/g		754.7	835.3	

Alfalfa, hay, s-c, immature, leafy, (1)
Ref No 1 00 049 United States

Feed Name or Analyses			Mean As Fed	Mean Dry	C.V. ± %
Dry matter	%		88.6	100.0	1
Ash	%		9.0	10.2	5
Crude fiber	%		19.1	21.6	10
Sheep	dig coef %		51.	51.	
Ether extract	%		1.8	2.0	28
N-free extract	%		35.9	40.6	
Sheep	dig coef %		76.	76.	
Protein (N x 6.25)	%		22.8	25.7	8
Sheep	dig coef %		94.	94.	
Cattle	dig prot %	*	17.0	19.2	
Goats	dig prot %	*	18.2	20.5	
Horses	dig prot %	*	17.1	19.3	
Rabbits	dig prot %	*	16.4	18.5	
Sheep	dig prot %		21.4	24.2	
Lignin (Ellis)	%		6.3	7.1	
Energy	GE Mcal/kg				
Cattle	DE Mcal/kg	*	2.47	2.78	
Sheep	DE Mcal/kg	*	2.48	2.80	
Cattle	ME Mcal/kg	*	2.02	2.28	
Sheep	ME Mcal/kg	*	2.03	2.29	
Cattle	TDN %	*	55.9	63.1	
Sheep	TDN %		56.2	63.4	
Carotene	mg/kg		15.1	17.0	
Vitamin A equivalent	IU/g		25.1	28.3	

Alfalfa, hay, s-c, pre-bloom, (1)
Ref No 1 00 054 United States

Feed Name or Analyses			Mean As Fed	Mean Dry	C.V. ± %
Dry matter	%		87.5	100.0	
Ash	%		7.3	8.3	
Crude fiber	%		23.9	27.3	
Cattle	dig coef %		55.	55.	
Ether extract	%		2.7	3.0	
Cattle	dig coef %		58.	58.	
N-free extract	%		36.2	41.3	
Cattle	dig coef %		70.	70.	
Protein (N x 6.25)	%		17.5	20.0	
Cattle	dig coef %		71.	71.	
Cattle	dig prot %		12.4	14.2	
Goats	dig prot %	*	13.3	15.2	
Horses	dig prot %	*	12.7	14.5	
Rabbits	dig prot %	*	12.3	14.1	
Sheep	dig prot %	*	12.7	14.5	
Energy	GE Mcal/kg				
Cattle	DE Mcal/kg	*	2.40	2.74	
Sheep	DE Mcal/kg	*	2.31	2.64	
Cattle	ME Mcal/kg	*	1.97	2.25	
Sheep	ME Mcal/kg	*	1.89	2.17	
Cattle	NE_m Mcal/kg	*	1.19	1.36	
Cattle	NE_{gain} Mcal/kg	*	0.59	0.67	
Cattle	$NE_{lactating\ cows}$ Mcal/kg	*	1.37	1.57	
Cattle	TDN %		54.4	62.1	
Sheep	TDN %	*	52.4	59.9	
Calcium	%		2.15	2.45	
Chlorine	%		0.30	0.34	
Iron	%		0.022	0.025	
Magnesium	%		0.22	0.25	
Manganese	mg/kg		30.1	34.3	
Phosphorus	%		0.32	0.36	
Potassium	%		2.07	2.36	
Sodium	%		0.19	0.22	
Sulphur	%		0.55	0.63	

Alfalfa, hay, s-c, pre-bloom, cut 1, (1)
Ref No 1 00 051 United States

Feed Name or Analyses			Mean As Fed	Mean Dry	C.V. ± %
Dry matter	%		82.6	100.0	
Ash	%		8.5	10.3	
Crude fiber	%		24.1	29.2	
Cattle	dig coef %		43.	43.	
Sheep	dig coef %		51.	51.	
Ether extract	%		1.8	2.2	
Cattle	dig coef %		38.	38.	
Sheep	dig coef %		20.	20.	
N-free extract	%		35.2	42.7	
Cattle	dig coef %		75.	75.	
Sheep	dig coef %		70.	70.	
Protein (N x 6.25)	%		13.1	15.8	
Cattle	dig coef %		62.	62.	
Sheep	dig coef %		78.	78.	
Cattle	dig prot %		8.1	9.8	
Goats	dig prot %	*	9.3	11.3	
Horses	dig prot %	*	9.0	10.9	
Rabbits	dig prot %	*	9.0	10.9	
Sheep	dig prot %		10.2	12.3	
Energy	GE Mcal/kg				
Cattle	DE Mcal/kg	*	2.05	2.48	
Sheep	DE Mcal/kg	*	2.11	2.56	
Cattle	ME Mcal/kg	*	1.68	2.03	
Sheep	ME Mcal/kg	*	1.73	2.10	
Cattle	TDN %		46.4	56.2	
Sheep	TDN %		47.9	58.0	

Ref No 1 00 051 Canada

Feed Name or Analyses			Mean As Fed	Mean Dry	C.V. ± %
Dry matter	%			100.0	
In vitro	dig coef %			66.	6
Crude fiber	%				
Cattle	dig coef %			43.	
Sheep	dig coef %			51.	
Ether extract	%				
Cattle	dig coef %			38.	
Sheep	dig coef %			20.	
N-free extract	%				
Cattle	dig coef %			75.	
Sheep	dig coef %			70.	
Protein (N x 6.25)	%			18.9	3
Cattle	dig coef %			62.	
Sheep	dig coef %			78.	
Cattle	dig prot %			11.7	
Goats	dig prot %	*		14.2	
Horses	dig prot %	*		13.6	
Rabbits	dig prot %	*		13.3	
Sheep	dig prot %			14.8	
Energy	GE Mcal/kg				
Cattle	DE Mcal/kg	*		2.46	
Sheep	DE Mcal/kg	*		2.57	
Cattle	ME Mcal/kg	*		2.02	
Sheep	ME Mcal/kg	*		2.10	
Cattle	TDN %			55.8	
Sheep	TDN %			58.2	
Phosphorus	%			0.29	11
Potassium	%			2.80	19
Sulphur	%			0.25	18

Alfalfa, hay, s-c, pre-bloom, cut 2, (1)
Ref No 1 00 052 United States

Feed Name or Analyses			Mean As Fed	Mean Dry	C.V. ± %
Dry matter	%		86.0	100.0	
Ash	%		7.1	8.3	
Crude fiber	%		28.7	33.3	
Cattle	dig coef %		45.	45.	
Horses	dig coef %		40.	40.	
Sheep	dig coef %		44.	44.	
Ether extract	%		2.2	2.5	
Cattle	dig coef %		38.	38.	
Horses	dig coef %		6.	6.	
Sheep	dig coef %		53.	53.	
N-free extract	%		34.7	40.3	
Cattle	dig coef %		71.	71.	
Horses	dig coef %		70.	70.	
Sheep	dig coef %		71.	71.	
Protein (N x 6.25)	%		13.4	15.6	
Cattle	dig coef %		66.	66.	
Horses	dig coef %		74.	74.	
Sheep	dig coef %		75.	75.	
Cattle	dig prot %		8.8	10.3	
Goats	dig prot %	*	9.5	11.1	
Horses	dig prot %		9.9	11.5	

(1) dry forages and roughages (2) pasture, range plants, and forages fed green (3) silages (4) energy feeds (5) protein supplements (6) minerals (7) vitamins (8) additives

Feed Name or Analyses			Mean As Fed	Mean Dry	C.V. ± %
Rabbits	dig prot %	*	9.2	10.7	
Sheep	dig prot %		10.0	11.7	
Energy	GE Mcal/kg				
Cattle	DE Mcal/kg	*	2.13	2.47	
Horses	DE Mcal/kg	*	2.03	2.35	
Sheep	DE Mcal/kg	*	2.20	2.55	
Cattle	ME Mcal/kg	*	1.74	2.03	
Horses	ME Mcal/kg	*	1.66	1.93	
Sheep	ME Mcal/kg	*	1.80	2.09	
Cattle	TDN %		48.2	56.0	
Horses	TDN %		45.9	53.4	
Sheep	TDN %		49.8	57.9	

Alfalfa, hay, s-c, pre-bloom, cut 3, (1)

Ref No 1 00 053 United States

Feed Name or Analyses			Mean As Fed	Mean Dry	C.V. ± %
Dry matter	%		84.6	100.0	
Ash	%		6.0	7.1	
Crude fiber	%		27.1	32.0	
Horses	dig coef %		38.	38.	
Sheep	dig coef %		47.	47.	
Ether extract	%		2.0	2.4	
Horses	dig coef %		-4.	-4.	
Sheep	dig coef %		31.	31.	
N-free extract	%		32.7	38.7	
Horses	dig coef %		71.	71.	
Sheep	dig coef %		70.	70.	
Protein (N x 6.25)	%		16.7	19.8	
Horses	dig coef %		75.	75.	
Sheep	dig coef %		74.	74.	
Cattle	dig prot %	*	11.9	14.1	
Goats	dig prot %	*	12.7	15.0	
Horses	dig prot %		12.6	14.9	
Rabbits	dig prot %	*	11.8	14.0	
Sheep	dig prot %		12.4	14.7	
Energy	GE Mcal/kg				
Cattle	DE Mcal/kg	*	2.05	2.42	
Horses	DE Mcal/kg	*	2.02	2.39	
Sheep	DE Mcal/kg	*	2.18	2.58	
Cattle	ME Mcal/kg	*	1.68	1.99	
Horses	ME Mcal/kg	*	1.66	1.96	
Sheep	ME Mcal/kg	*	1.79	2.11	
Cattle	TDN %	*	46.5	55.0	
Horses	TDN %		45.8	54.2	
Sheep	TDN %		49.4	58.5	

Alfalfa, hay, s-c, early bloom, (1)

Ref No 1 00 059 United States

Feed Name or Analyses			Mean As Fed	Mean Dry	C.V. ± %
Dry matter	%		90.1	100.0	2
Cattle	dig coef %		61.	61.	
Organic matter	%		81.5	90.5	
Ash	%		7.8	8.7	13
Crude fiber	%		26.9	29.8	13
Cattle	dig coef %		43.	43.	
Sheep	dig coef %		48.	48.	
Ether extract	%		2.0	2.2	28
Cattle	dig coef %		43.	43.	
Sheep	dig coef %		10.	10.	
N-free extract	%		35.7	39.7	8
Cattle	dig coef %		74.	74.	
Sheep	dig coef %		72.	72.	
Protein (N x 6.25)	%		17.7	19.7	9
Cattle	dig coef %		73.	73.	
Sheep	dig coef %		78.	78.	
Cattle	dig prot %		12.9	14.4	
Goats	dig prot %	*	13.4	14.9	
Horses	dig prot %	*	12.8	14.2	
Rabbits	dig prot %	*	12.5	13.9	
Sheep	dig prot %		13.8	15.4	
Cellulose (Matrone)	%		25.3	28.0	6
Cattle	dig coef %		54.	54.	
Pectins	%		5.0	5.6	5
Starch	%		1.6	1.8	31
Sugars, total	%		5.1	5.7	32
Lignin (Ellis)	%		7.7	8.5	10
Energy	GE Mcal/kg		4.00	4.43	
Cattle	GE dig coef %		59.	59.	
Cattle	DE Mcal/kg		2.36	2.62	
Sheep	DE Mcal/kg	*	2.33	2.59	
Cattle	ME Mcal/kg		2.12	2.35	
Sheep	ME Mcal/kg	*	1.91	2.12	
Cattle	NE_m Mcal/kg	*	1.22	1.35	
Cattle	NE_{gain} Mcal/kg	*	0.44	0.49	
Cattle	$NE_{lactating\ cows}$ Mcal/kg	*	1.21	1.34	
Cattle	TDN %		52.7	58.5	
Sheep	TDN %		52.9	58.7	
Calcium	%		1.26	1.40	32
Chlorine	%		0.34	0.38	15
Cobalt	mg/kg		0.085	0.095	
Copper	mg/kg		12.1	13.4	
Iron	%		0.018	0.020	
Magnesium	%		0.27	0.30	16
Manganese	mg/kg		28.4	31.5	28
Phosphorus	%		0.18	0.21	15
Potassium	%		1.87	2.08	11
Sodium	%		0.14	0.15	17
Sulphur	%		0.27	0.30	14
Carotene	mg/kg		161.7	179.5	
Vitamin A equivalent	IU/g		269.6	299.2	

Ref No 1 00 059 Canada

Feed Name or Analyses			Mean As Fed	Mean Dry	C.V. ± %
Dry matter	%			100.0	
Crude fiber	%				
Cattle	dig coef %			43.	
Sheep	dig coef %			48.	
Ether extract	%				
Cattle	dig coef %			43.	
Sheep	dig coef %			10.	
N-free extract	%				
Cattle	dig coef %			74.	
Sheep	dig coef %			72.	
Protein (N x 6.25)	%			17.5	
Cattle	dig coef %			73.	
Sheep	dig coef %			78.	
Cattle	dig prot %			12.8	
Goats	dig prot %	*		12.9	
Horses	dig prot %	*		12.4	
Rabbits	dig prot %	*		12.2	
Sheep	dig prot %			13.7	
Energy	GE Mcal/kg			4.43	
Cattle	GE dig coef %			59.	
Cattle	DE Mcal/kg			2.62	
Sheep	DE Mcal/kg	*		2.59	
Cattle	ME Mcal/kg			2.35	
Sheep	ME Mcal/kg	*		2.12	
Cattle	TDN %			58.5	
Sheep	TDN %			58.7	
Phosphorus	%			0.17	

Alfalfa, hay, s-c, early bloom, cut 1, (1)

Ref No 1 00 055 United States

Feed Name or Analyses			Mean As Fed	Mean Dry	C.V. ± %
Dry matter	%		90.2	100.0	
Ash	%		7.8	8.6	
Crude fiber	%		34.0	37.7	
Sheep	dig coef %		47.	47.	
Ether extract	%		1.3	1.4	
Sheep	dig coef %		13.	13.	
N-free extract	%		33.4	37.0	
Sheep	dig coef %		65.	65.	
Protein (N x 6.25)	%		13.8	15.3	
Sheep	dig coef %		70.	70.	
Cattle	dig prot %	*	9.2	10.2	
Goats	dig prot %	*	9.8	10.8	
Horses	dig prot %	*	9.5	10.5	
Rabbits	dig prot %	*	9.5	10.5	
Sheep	dig prot %		9.7	10.7	
Energy	GE Mcal/kg				
Cattle	DE Mcal/kg	*	2.09	2.32	
Sheep	DE Mcal/kg	*	2.07	2.29	
Cattle	ME Mcal/kg	*	1.71	1.90	
Sheep	ME Mcal/kg	*	1.70	1.88	
Cattle	TDN %	*	47.4	52.5	
Sheep	TDN %		46.9	52.0	

Ref No 1 00 055 Canada

Feed Name or Analyses			Mean As Fed	Mean Dry	C.V. ± %
Dry matter	%			100.0	
Ash	%			8.6	
Crude fiber	%			30.4	
Sheep	dig coef %			47.	
Ether extract	%			1.5	
Sheep	dig coef %			13.	
N-free extract	%			41.5	
Sheep	dig coef %			65.	
Protein (N x 6.25)	%			18.1	10
Sheep	dig coef %			70.	
Cattle	dig prot %	*		12.6	
Goats	dig prot %	*		13.4	
Horses	dig prot %	*		12.9	
Rabbits	dig prot %	*		12.6	
Sheep	dig prot %			12.6	
Energy	GE Mcal/kg				
Cattle	DE Mcal/kg	*		2.50	
Sheep	DE Mcal/kg	*		2.36	
Cattle	ME Mcal/kg	*		2.05	
Sheep	ME Mcal/kg	*		1.93	
Cattle	TDN %	*		56.7	
Sheep	TDN %			53.4	
Calcium	%			1.49	
Copper	mg/kg			30.0	
Magnesium	%			0.33	
Molybdenum	mg/kg			8.00	
Phosphorus	%			0.27	16
Potassium	%			2.49	17
Sulphur	%			0.27	23

Alfalfa, hay, s-c, early bloom, cut 2, (1)

Ref No 1 00 056 United States

Feed Name or Analyses			Mean As Fed	Mean Dry	C.V. ± %
Dry matter	%		87.0	100.0	3
Cattle	dig coef %		62.	62.	
Ash	%		7.9	9.0	8
Crude fiber	%		25.8	29.7	17
Cattle	dig coef %		41.	41.	
Sheep	dig coef %		46.	46.	
Ether extract	%		1.9	2.2	35
Cattle	dig coef %		35.	35.	
N-free extract	%		33.3	38.3	5
Cattle	dig coef %		70.	70.	
Sheep	dig coef %		66.	66.	
Protein (N x 6.25)	%		18.1	20.8	18
Cattle	dig coef %		75.	75.	

Continued

Feed Name or Analyses			Mean As Fed	Mean Dry	C.V. ± %
Sheep	dig coef %		71.	71.	
Cattle	dig prot %		13.5	15.5	
Goats	dig prot %	*	13.9	16.0	
Horses	dig prot %	*	13.2	15.2	
Rabbits	dig prot %	*	12.8	14.7	
Sheep	dig prot %		12.9	14.8	
Cellulose (Matrone)	%		23.9	27.4	
Cattle	dig coef %		55.	55.	
Energy	GE Mcal/kg		3.94	4.53	
Cattle	GE dig coef %		61.	61.	
Cattle	DE Mcal/kg		2.40	2.76	
Sheep	DE Mcal/kg	*	2.04	2.35	
Cattle	ME Mcal/kg	*	1.97	2.27	
Sheep	ME Mcal/kg	*	1.67	1.92	
Cattle	TDN %		48.9	56.2	
Sheep	TDN %		46.3	53.2	

Ref No 1 00 056 Canada

Feed Name or Analyses			Mean As Fed	Mean Dry	C.V. ± %
Dry matter	%			100.0	
Crude fiber	%				
Cattle	dig coef %			41.	
Sheep	dig coef %			46.	
Ether extract	%				
Cattle	dig coef %			35.	
N-free extract	%				
Cattle	dig coef %			70.	
Sheep	dig coef %			66.	
Protein (N x 6.25)	%			17.6	4
Cattle	dig coef %			75.	
Sheep	dig coef %			71.	
Cattle	dig prot %			13.1	
Goats	dig prot %	*		13.0	
Horses	dig prot %	*		12.4	
Rabbits	dig prot %	*		12.2	
Sheep	dig prot %			12.5	
Energy	GE Mcal/kg			4.53	
Cattle	GE dig coef %			61.	
Cattle	DE Mcal/kg			2.76	
Sheep	DE Mcal/kg	*		2.35	
Cattle	ME Mcal/kg	*		2.27	
Sheep	ME Mcal/kg	*		1.92	
Cattle	TDN %			56.1	
Sheep	TDN %			53.2	
Phosphorus	%			0.26	27
Potassium	%			2.64	9
Sulphur	%			0.28	

Alfalfa, hay, s-c, early bloom, cut 3, (1)

Ref No 1 00 057 United States

Feed Name or Analyses			Mean As Fed	Mean Dry	C.V. ± %
Dry matter	%		85.9	100.0	
Cattle	dig coef %		59.	59.	
Ash	%		7.9	9.2	
Crude fiber	%		22.1	25.8	
Sheep	dig coef %		42.	42.	
Ether extract	%		2.4	2.8	
Sheep	dig coef %		30.	30.	
N-free extract	%		35.5	41.3	
Sheep	dig coef %		73.	73.	
Protein (N x 6.25)	%		18.0	21.0	
Cattle	dig coef %		73.	73.	
Sheep	dig coef %		78.	78.	
Cattle	dig prot %		13.1	15.3	
Goats	dig prot %	*	13.8	16.1	
Horses	dig prot %	*	13.1	15.3	
Rabbits	dig prot %	*	12.7	14.8	
Sheep	dig prot %		14.0	16.3	
Cellulose (Matrone)	%		23.6	27.5	
Cattle	dig coef %		53.	53.	
Energy	GE Mcal/kg		3.86	4.50	
Cattle	GE dig coef %		59.	59.	
Cattle	DE Mcal/kg		2.28	2.66	
Sheep	DE Mcal/kg	*	2.24	2.61	
Cattle	ME Mcal/kg	*	1.87	2.18	
Sheep	ME Mcal/kg	*	1.84	2.14	
Cattle	TDN %	*	52.4	61.0	
Sheep	TDN %		50.8	59.2	

Ref No 1 00 057 Canada

Feed Name or Analyses			Mean As Fed	Mean Dry	C.V. ± %
Dry matter	%			100.0	
Crude fiber	%				
Sheep	dig coef %			42.	
Ether extract	%				
Sheep	dig coef %			30.	
N-free extract	%				
Sheep	dig coef %			73.	
Protein (N x 6.25)	%			19.1	
Cattle	dig coef %			73.	
Sheep	dig coef %			78.	
Cattle	dig prot %			14.0	
Goats	dig prot %	*		14.4	
Horses	dig prot %	*		13.8	
Rabbits	dig prot %	*		13.4	
Sheep	dig prot %			14.9	
Energy	GE Mcal/kg			4.50	
Cattle	GE dig coef %			59.	
Cattle	DE Mcal/kg			2.66	
Sheep	DE Mcal/kg	*		2.61	
Cattle	ME Mcal/kg	*		2.18	
Sheep	ME Mcal/kg	*		2.14	
Cattle	TDN %	*		60.3	
Sheep	TDN %			59.1	
Phosphorus	%			0.31	
Potassium	%			2.71	

Alfalfa, hay, s-c, early bloom, leafy, (1)

Ref No 1 00 058 United States

Feed Name or Analyses			Mean As Fed	Mean Dry	C.V. ± %
Dry matter	%			100.0	
Ash	%			7.1	
Crude fiber	%			31.0	
Ether extract	%			2.0	
N-free extract	%			41.1	
Protein (N x 6.25)	%			18.8	
Cattle	dig prot %	*		13.2	
Goats	dig prot %	*		14.1	
Horses	dig prot %	*		13.5	
Rabbits	dig prot %	*		13.2	
Sheep	dig prot %	*		13.4	
Energy	GE Mcal/kg				
Cattle	DE Mcal/kg	*		2.44	
Sheep	DE Mcal/kg	*		2.61	
Cattle	ME Mcal/kg	*		2.00	
Sheep	ME Mcal/kg	*		2.14	
Cattle	TDN %	*		55.3	
Sheep	TDN %	*		59.2	

Alfalfa, hay, s-c, mid-bloom, (1)

Ref No 1 00 063 United States

Feed Name or Analyses			Mean As Fed	Mean Dry	C.V. ± %
Dry matter	%		89.5	100.0	2
Ash	%		7.3	8.2	15
Crude fiber	%		27.4	30.6	5
Cattle	dig coef %		46.	46.	
Sheep	dig coef %		45.	45.	
Ether extract	%		1.4	1.6	21
Cattle	dig coef %		32.	32.	
Sheep	dig coef %		10.	10.	
N-free extract	%		37.7	42.1	5
Cattle	dig coef %		72.	72.	
Sheep	dig coef %		72.	72.	
Protein (N x 6.25)	%		15.7	17.5	6
Cattle	dig coef %		71.	71.	
Sheep	dig coef %		86.	86.	
Cattle	dig prot %		11.1	12.5	
Goats	dig prot %	*	11.6	12.9	
Horses	dig prot %	*	11.1	12.4	
Rabbits	dig prot %	*	10.9	12.2	
Sheep	dig prot %		13.4	15.0	
Cellulose (Matrone)	%		23.8	26.6	
Starch	%		1.6	1.8	4
Sugars, total	%		4.7	5.3	17
Lignin (Ellis)	%		7.8	8.7	
Energy	GE Mcal/kg				
Cattle	DE Mcal/kg	*	2.29	2.56	
Sheep	DE Mcal/kg	*	2.33	2.60	
Cattle	ME Mcal/kg	*	1.88	2.10	
Sheep	ME Mcal/kg	*	1.91	2.14	
Cattle	NE_m Mcal/kg	*	1.11	1.24	
Cattle	NE_{gain} Mcal/kg	*	0.53	0.59	
Cattle	$NE_{lactating\ cows}$ Mcal/kg	*	1.24	1.38	
Cattle	TDN %		51.9	58.0	
Sheep	TDN %		52.9	59.1	
Calcium	%		1.33	1.49	23
Chlorine	%		0.34	0.38	
Copper	mg/kg		13.8	15.4	
Iron	%		0.012	0.013	
Magnesium	%		0.27	0.30	
Manganese	mg/kg		21.3	23.7	
Phosphorus	%		0.22	0.24	28
Potassium	%		1.63	1.82	
Sodium	%		0.14	0.15	
Sulphur	%		0.27	0.30	
Carotene	mg/kg		23.4	26.1	93
Riboflavin	mg/kg		9.5	10.6	
Vitamin A equivalent	IU/g		39.0	43.6	

Ref No 1 00 063 Canada

Feed Name or Analyses			Mean As Fed	Mean Dry	C.V. ± %
Dry matter	%			100.0	
Crude fiber	%				
Cattle	dig coef %			46.	
Sheep	dig coef %			45.	
Ether extract	%				
Cattle	dig coef %			32.	
Sheep	dig coef %			10.	
N-free extract	%				
Cattle	dig coef %			72.	
Sheep	dig coef %			72.	
Protein (N x 6.25)	%			21.4	
Cattle	dig coef %			71.	
Sheep	dig coef %			86.	
Cattle	dig prot %			15.2	
Goats	dig prot %	*		16.5	
Horses	dig prot %	*		15.7	
Rabbits	dig prot %	*		15.2	

(1) dry forages and roughages (2) pasture, range plants, and forages fed green (3) silages (4) energy feeds (5) protein supplements (6) minerals (7) vitamins (8) additives

Feed Name or Analyses			Mean As Fed	Mean Dry	C.V. ± %
Sheep	dig prot %			18.3	
Energy	GE Mcal/kg				
Cattle	DE Mcal/kg	*		2.56	
Sheep	DE Mcal/kg	*		2.61	
Cattle	ME Mcal/kg	*		2.10	
Sheep	ME Mcal/kg	*		2.14	
Cattle	TDN %			58.0	
Sheep	TDN %			59.1	
Calcium	%			2.01	
Phosphorus	%			0.30	
Potassium	%			1.30	

Alfalfa, hay, s-c, mid-bloom, cut 1, (1)

Ref No 1 00 060 United States

Feed Name or Analyses			Mean As Fed	Mean Dry	C.V. ± %
Dry matter	%		87.8	100.0	
Ash	%		7.5	8.6	
Crude fiber	%		29.8	33.9	
Cattle	dig coef %		38.	38.	
Sheep	dig coef %		40.	40.	
Ether extract	%		1.5	1.8	
Cattle	dig coef %		28.	28.	
Sheep	dig coef %		16.	16.	
N-free extract	%		37.8	43.1	
Cattle	dig coef %		73.	73.	
Sheep	dig coef %		68.	68.	
Protein (N x 6.25)	%		11.2	12.7	
Cattle	dig coef %		65.	65.	
Sheep	dig coef %		67.	67.	
Cattle	dig prot %		7.2	8.3	
Goats	dig prot %	*	7.4	8.4	
Horses	dig prot %	*	7.3	8.3	
Rabbits	dig prot %	*	7.4	8.5	
Sheep	dig prot %		7.5	8.5	
Energy	GE Mcal/kg				
Cattle	DE Mcal/kg	*	2.08	2.37	
Sheep	DE Mcal/kg	*	2.01	2.29	
Cattle	ME Mcal/kg	*	1.70	1.94	
Sheep	ME Mcal/kg	*	1.65	1.88	
Cattle	TDN %		47.2	53.7	
Sheep	TDN %		45.7	52.0	

Alfalfa, hay, s-c, mid-bloom, cut 2, (1)

Ref No 1 00 061 United States

Feed Name or Analyses			Mean As Fed	Mean Dry	C.V. ± %
Dry matter	%		90.5	100.0	
Ash	%		7.0	7.7	
Crude fiber	%		31.3	34.6	
Cattle	dig coef %		42.	42.	
Sheep	dig coef %		42.	42.	
Ether extract	%		1.7	1.9	
Cattle	dig coef %		45.	45.	
Sheep	dig coef %		14.	14.	
N-free extract	%		35.9	39.7	
Cattle	dig coef %		73.	73.	
Sheep	dig coef %		71.	71.	
Protein (N x 6.25)	%		14.6	16.1	
Cattle	dig coef %		73.	73.	
Sheep	dig coef %		74.	74.	
Cattle	dig prot %		10.6	11.8	
Goats	dig prot %	*	10.5	11.6	
Horses	dig prot %	*	10.1	11.2	
Rabbits	dig prot %	*	10.0	11.1	
Sheep	dig prot %		10.7	11.8	
Lignin (Ellis)	%		7.7	8.5	
Energy	GE Mcal/kg		4.14	4.58	
Sheep	GE dig coef %		60.	60.	
Cattle	DE Mcal/kg	*	2.28	2.52	
Sheep	DE Mcal/kg		2.48	2.75	
Cattle	ME Mcal/kg	*	1.87	2.07	
Sheep	ME Mcal/kg	*	2.04	2.25	
Cattle	TDN %		51.7	57.1	
Sheep	TDN %		49.7	54.9	

Alfalfa, hay, s-c, mid-bloom, cut 3, (1)

Ref No 1 00 062 United States

Feed Name or Analyses			Mean As Fed	Mean Dry	C.V. ± %
Dry matter	%		83.1	100.0	
Ash	%		7.2	8.7	
Crude fiber	%		23.2	27.9	
Sheep	dig coef %		43.	43.	
Ether extract	%		2.3	2.8	
Sheep	dig coef %		24.	24.	
N-free extract	%		36.2	43.6	
Sheep	dig coef %		76.	76.	
Protein (N x 6.25)	%		14.1	17.0	
Sheep	dig coef %		72.	72.	
Cattle	dig prot %	*	9.7	11.7	
Goats	dig prot %	*	10.3	12.4	
Horses	dig prot %	*	9.9	12.0	
Rabbits	dig prot %	*	9.8	11.8	
Sheep	dig prot %		10.2	12.2	
Energy	GE Mcal/kg				
Cattle	DE Mcal/kg	*	2.19	2.63	
Sheep	DE Mcal/kg	*	2.16	2.60	
Cattle	ME Mcal/kg	*	1.79	2.16	
Sheep	ME Mcal/kg	*	1.77	2.13	
Cattle	TDN %	*	49.6	59.7	
Sheep	TDN %		48.9	58.9	

Alfalfa, hay, s-c, full bloom, (1)

Ref No 1 00 068 United States

Feed Name or Analyses			Mean As Fed	Mean Dry	C.V. ± %
Dry matter	%		87.6	100.0	2
Ash	%		7.8	8.9	
Crude fiber	%		28.2	32.2	
Cattle	dig coef %		45.	45.	
Sheep	dig coef %		38.	38.	
Ether extract	%		1.6	1.9	
Cattle	dig coef %		35.	35.	
Sheep	dig coef %		35.	35.	
N-free extract	%		35.3	40.4	
Cattle	dig coef %		72.	72.	
Sheep	dig coef %		69.	69.	
Protein (N x 6.25)	%		14.6	16.6	
Cattle	dig coef %		72.	72.	
Sheep	dig coef %		75.	75.	
Cattle	dig prot %		10.5	12.0	
Goats	dig prot %	*	10.6	12.1	
Horses	dig prot %	*	10.2	11.6	
Rabbits	dig prot %	*	10.1	11.5	
Sheep	dig prot %		10.9	12.5	
Starch	%		3.6	4.1	
Sugars, total	%		3.5	4.0	15
Lignin (Ellis)	%		8.1	9.3	
Energy	GE Mcal/kg				
Cattle	DE Mcal/kg	*	2.20	2.51	
Sheep	DE Mcal/kg	*	2.09	2.38	
Cattle	ME Mcal/kg	*	1.81	2.06	
Sheep	ME Mcal/kg	*	1.71	1.95	
Cattle	NE_m Mcal/kg	*	1.07	1.22	
Cattle	NE_{gain} Mcal/kg	*	0.48	0.55	
Cattle	$NE_{lactating\ cows}$ Mcal/kg	*	1.17	1.34	
Cattle	TDN %		49.9	57.0	
Sheep	TDN %		47.3	54.0	
Calcium	%		1.11	1.27	6
Cobalt	mg/kg		0.108	0.124	
Copper	mg/kg		11.8	13.4	
Iron	%		0.012	0.014	35
Magnesium	%		0.32	0.36	22
Manganese	mg/kg		27.0	30.9	42
Phosphorus	%		0.20	0.23	19
Potassium	%		1.54	1.76	21
Carotene	mg/kg		21.5	24.5	68
Vitamin A equivalent	IU/g		35.8	40.9	

Alfalfa, hay, s-c, full bloom, cut 1, (1)

Ref No 1 00 064 United States

Feed Name or Analyses			Mean As Fed	Mean Dry	C.V. ± %
Dry matter	%		83.1	100.0	
Ash	%		7.6	9.1	
Crude fiber	%		27.6	33.2	
Cattle	dig coef %		40.	40.	
Sheep	dig coef %		41.	41.	
Ether extract	%		1.8	2.2	
Cattle	dig coef %		39.	39.	
Sheep	dig coef %		29.	29.	
N-free extract	%		34.0	40.9	
Cattle	dig coef %		70.	70.	
Sheep	dig coef %		65.	65.	
Protein (N x 6.25)	%		12.1	14.6	
Cattle	dig coef %		68.	68.	
Sheep	dig coef %		72.	72.	
Cattle	dig prot %		8.3	9.9	
Goats	dig prot %	*	8.5	10.2	
Horses	dig prot %	*	8.2	9.9	
Rabbits	dig prot %	*	8.3	9.9	
Sheep	dig prot %		8.7	10.5	
Energy	GE Mcal/kg				
Cattle	DE Mcal/kg	*	1.97	2.37	
Sheep	DE Mcal/kg	*	1.91	2.30	
Cattle	ME Mcal/kg	*	1.62	1.94	
Sheep	ME Mcal/kg	*	1.57	1.89	
Cattle	TDN %		44.7	53.8	
Sheep	TDN %		43.3	52.1	

Ref No 1 00 064 Canada

Feed Name or Analyses			Mean As Fed	Mean Dry	C.V. ± %
Dry matter	%			100.0	
Sheep	dig coef %			63.	
Crude fiber	%			38.3	
Cattle	dig coef %			40.	
Sheep	dig coef %			56.	
Ether extract	%			1.2	
Cattle	dig coef %			39.	
Sheep	dig coef %			0.	
N-free extract	%				
Cattle	dig coef %			70.	
Sheep	dig coef %			65.	
Protein (N x 6.25)	%			14.4	
Cattle	dig coef %			68.	
Sheep	dig coef %			70.	
Cattle	dig prot %			9.8	
Goats	dig prot %	*		10.0	
Horses	dig prot %	*		9.8	
Rabbits	dig prot %	*		9.8	
Sheep	dig prot %			10.1	
Energy	GE Mcal/kg				
Cattle	DE Mcal/kg	*		2.33	
Sheep	DE Mcal/kg	*		2.36	
Cattle	ME Mcal/kg	*		1.91	
Sheep	ME Mcal/kg	*		1.93	
Cattle	TDN %			52.9	
Sheep	TDN %			53.5	

Continued

Feed Name or Analyses			Mean As Fed	Mean Dry	C.V. ± %
Phosphorus	%			0.22	
Potassium	%			1.99	
Sulphur	%			0.26	

Alfalfa, hay, s-c, full bloom, cut 2, (1)
Ref No 1 00 065 United States

Feed Name or Analyses			Mean As Fed	Mean Dry	C.V. ± %
Dry matter	%		85.6	100.0	
Ash	%		7.9	9.3	
Crude fiber	%		29.0	33.9	
Cattle	dig coef %		46.	46.	
Sheep	dig coef %		46.	46.	
Ether extract	%		1.4	1.7	
Cattle	dig coef %		44.	44.	
Sheep	dig coef %		-0.	-0.	
N-free extract	%		32.9	38.5	
Cattle	dig coef %		71.	71.	
Sheep	dig coef %		63.	63.	
Protein (N x 6.25)	%		14.3	16.8	
Cattle	dig coef %		72.	72.	
Sheep	dig coef %		76.	76.	
Cattle	dig prot %		10.3	12.1	
Goats	dig prot %	*	10.4	12.2	
Horses	dig prot %	*	10.1	11.8	
Rabbits	dig prot %	*	9.9	11.6	
Sheep	dig prot %		10.9	12.7	
Lignin (Ellis)	%		7.1	8.4	
Energy	GE Mcal/kg				
Cattle	DE Mcal/kg	*	2.14	2.50	
Sheep	DE Mcal/kg	*	1.98	2.32	
Cattle	ME Mcal/kg	*	1.75	2.05	
Sheep	ME Mcal/kg	*	1.63	1.90	
Cattle	TDN %		48.4	56.6	
Sheep	TDN %		44.9	52.5	

Alfalfa, hay, s-c, full bloom, cut 3, (1)
Ref No 1 00 066 United States

Feed Name or Analyses			Mean As Fed	Mean Dry	C.V. ± %
Dry matter	%		87.6	100.0	
Ash	%		8.1	9.3	
Crude fiber	%		29.6	33.8	
Cattle	dig coef %		34.	34.	
Sheep	dig coef %		38.	38.	
Ether extract	%		1.4	1.6	
Cattle	dig coef %		42.	42.	
Sheep	dig coef %		0.	0.	
N-free extract	%		38.8	44.3	
Cattle	dig coef %		71.	71.	
Sheep	dig coef %		64.	64.	
Protein (N x 6.25)	%		9.7	11.1	
Cattle	dig coef %		69.	69.	
Sheep	dig coef %		45.	45.	
Cattle	dig prot %		6.7	7.6	
Goats	dig prot %	*	6.0	6.9	
Horses	dig prot %	*	6.1	6.9	
Rabbits	dig prot %	*	6.3	7.2	
Sheep	dig prot %		4.4	5.0	
Energy	GE Mcal/kg				
Cattle	DE Mcal/kg	*	2.01	2.30	
Sheep	DE Mcal/kg	*	1.78	2.03	
Cattle	ME Mcal/kg	*	1.65	1.88	
Sheep	ME Mcal/kg	*	1.46	1.67	
Cattle	TDN %		45.6	52.1	
Sheep	TDN %		40.4	46.1	

(1) dry forages and roughages (2) pasture, range plants, and forages fed green (3) silages (4) energy feeds (5) protein supplements (6) minerals (7) vitamins (8) additives

Alfalfa, hay, s-c, full bloom, stemmy, (1)
Ref No 1 00 067 United States

Feed Name or Analyses			Mean As Fed	Mean Dry	C.V. ± %
Dry matter	%		90.5	100.0	1
Calcium	%		1.22	1.35	6
Iron	%		0.014	0.015	19
Magnesium	%		0.29	0.32	12
Phosphorus	%		0.22	0.24	15
Potassium	%		1.96	2.17	4

Alfalfa, hay, s-c, milk stage, (1)
Ref No 1 00 070 United States

Feed Name or Analyses			Mean As Fed	Mean Dry	C.V. ± %
Dry matter	%		90.2	100.0	6
Ash	%		7.3	8.1	
Crude fiber	%		33.2	36.9	
Sheep	dig coef %		42.	42.	
Ether extract	%		1.9	2.1	
Sheep	dig coef %		28.	28.	
N-free extract	%		34.1	37.8	
Sheep	dig coef %		58.	58.	
Protein (N x 6.25)	%		13.6	15.1	
Sheep	dig coef %		72.	72.	
Cattle	dig prot %	*	9.1	10.0	
Goats	dig prot %	*	9.6	10.7	
Horses	dig prot %	*	9.4	10.4	
Rabbits	dig prot %	*	9.3	10.3	
Sheep	dig prot %		9.8	10.9	
Energy	GE Mcal/kg				
Cattle	DE Mcal/kg	*	2.12	2.35	
Sheep	DE Mcal/kg	*	1.97	2.19	
Cattle	ME Mcal/kg	*	1.74	1.92	
Sheep	ME Mcal/kg	*	1.62	1.79	
Cattle	TDN %	*	48.0	53.2	
Sheep	TDN %		44.7	49.6	
Calcium	%		1.15	1.28	18
Phosphorus	%		0.21	0.23	23
Potassium	%		1.72	1.91	5
Carotene	mg/kg		24.2	26.9	
Pantothenic acid	mg/kg		23.5	26.0	
Vitamin A equivalent	IU/g		40.4	44.8	

Alfalfa, hay, s-c, milk stage, stemmy, (1)
Ref No 1 00 069 United States

Feed Name or Analyses			Mean As Fed	Mean Dry	C.V. ± %
Dry matter	%		90.5	100.0	1
Calcium	%		1.10	1.21	4
Phosphorus	%		0.20	0.22	12
Potassium	%		1.73	1.91	5

Alfalfa, hay, s-c, mature, (1)
Ref No 1 00 071 United States

Feed Name or Analyses			Mean As Fed	Mean Dry	C.V. ± %
Dry matter	%			100.0	
Crude fiber	%				
Cattle	dig coef %	*		44.	
Sheep	dig coef %	*		45.	
Ether extract	%				
Cattle	dig coef %	*		35.	
Sheep	dig coef %	*		31.	
N-free extract	%				
Cattle	dig coef %	*		71.	
Sheep	dig coef %	*		69.	
Protein (N x 6.25)	%				
Cattle	dig coef %	*		70.	
Sheep	dig coef %	*		72.	
Energy	GE Mcal/kg				
Cattle	NE_m Mcal/kg	*		1.17	
Cattle	NE_{gain} Mcal/kg	*		0.47	
Cattle	$NE_{lactating\ cows}$ Mcal/kg	*		1.26	

Ref No 1 00 071 Canada

Feed Name or Analyses			Mean As Fed	Mean Dry	C.V. ± %
Dry matter	%		92.0	100.0	
Crude fiber	%		36.5	39.7	
Ether extract	%		1.0	1.1	
Protein (N x 6.25)	%		12.7	13.9	
Cattle	dig prot %	*	8.2	8.9	
Goats	dig prot %	*	8.7	9.5	
Horses	dig prot %	*	8.5	9.3	
Rabbits	dig prot %	*	8.6	9.4	
Sheep	dig prot %	*	8.3	9.0	
Calcium	%		1.26	1.37	
Phosphorus	%		0.17	0.19	
Carotene	mg/kg		3.9	4.3	
Vitamin A equivalent	IU/g		6.6	7.2	

Alfalfa, hay, s-c, mature, cut 1, (1)
Ref No 1 08 659 Canada

Feed Name or Analyses			Mean As Fed	Mean Dry	C.V. ± %
Dry matter	%			100.0	
Sheep	dig coef %			54.	
Ash	%			7.4	
Crude fiber	%			40.4	
Sheep	dig coef %			53.	
Ether extract	%			1.2	
Sheep	dig coef %			24.	
N-free extract	%			39.0	
Protein (N x 6.25)	%			12.0	
Sheep	dig coef %			57.	
Cattle	dig prot %	*		7.4	
Goats	dig prot %	*		7.8	
Horses	dig prot %	*		7.7	
Rabbits	dig prot %	*		8.0	
Sheep	dig prot %			6.9	
Energy	GE Mcal/kg				
Cattle	DE Mcal/kg	*		1.94	
Sheep	DE Mcal/kg	*		2.03	
Cattle	ME Mcal/kg	*		1.59	
Sheep	ME Mcal/kg	*		1.67	
Cattle	TDN %	*		44.0	
Sheep	TDN %			46.1	
Calcium	%			0.95	
Phosphorus	%			0.15	

Alfalfa, hay, s-c, over ripe, (1)
Ref No 1 00 072 United States

Feed Name or Analyses			Mean As Fed	Mean Dry	C.V. ± %
Dry matter	%		90.8	100.0	
Ash	%		7.5	8.3	
Crude fiber	%		35.8	39.4	
Cattle	dig coef %	*	44.	44.	
Sheep	dig coef %		53.	53.	
Ether extract	%		1.5	1.6	
Cattle	dig coef %	*	35.	35.	
Sheep	dig coef %		28.	28.	
N-free extract	%		34.1	37.5	
Cattle	dig coef %	*	71.	71.	
Sheep	dig coef %		71.	71.	
Protein (N x 6.25)	%		12.0	13.2	
Cattle	dig coef %	*	70.	70.	
Sheep	dig coef %		71.	71.	
Cattle	dig prot %		8.4	9.2	
Goats	dig prot %	*	8.1	8.9	

Feed Name or Analyses	Mean As Fed	Mean Dry	C.V. ± %
Horses dig prot % *	7.9	8.7	
Rabbits dig prot % *	8.0	8.9	
Sheep dig prot %	8.5	9.4	
Energy GE Mcal/kg			
Cattle DE Mcal/kg *	2.18	2.40	
Sheep DE Mcal/kg *	2.32	2.55	
Cattle ME Mcal/kg *	1.79	1.97	
Sheep ME Mcal/kg *	1.90	2.09	
Cattle TDN %	49.5	54.5	
Sheep TDN %	52.6	57.9	
Calcium %	0.37	0.41	
Phosphorus %	0.17	0.19	
Carotene mg/kg	7.2	7.9	
Vitamin A equivalent IU/g	12.0	13.2	

Alfalfa, hay, s-c, cut 1, (1)
Ref No 1 00 073 United States

Feed Name or Analyses	Mean As Fed	Mean Dry	C.V. ± %
Dry matter %	89.4	100.0	3
Ash %	7.8	8.8	16
Crude fiber %	30.1	33.7	10
Cattle dig coef %	40.	40.	
Sheep dig coef %	45.	45.	
Ether extract %	1.7	1.9	14
Cattle dig coef %	35.	35.	
Sheep dig coef %	17.	17.	
N-free extract %	36.9	41.3	
Cattle dig coef %	72.	72.	
Sheep dig coef %	69.	69.	
Protein (N x 6.25) %	12.8	14.4	13
Cattle dig coef %	65.	65.	
Sheep dig coef %	71.	71.	
Cattle dig prot %	8.3	9.3	
Goats dig prot % *	8.9	10.0	
Horses dig prot % *	8.7	9.7	
Rabbits dig prot % *	8.7	9.8	
Sheep dig prot %	9.1	10.2	
Energy GE Mcal/kg	3.99	4.47	
Sheep GE dig coef %	57.	57.	
Cattle DE Mcal/kg *	2.13	2.38	
Sheep DE Mcal/kg	2.28	2.55	
Cattle ME Mcal/kg *	1.75	1.95	
Sheep ME Mcal/kg *	1.87	2.09	
Cattle TDN %	48.3	54.0	
Sheep TDN %	48.8	54.6	
Calcium %	0.68	0.76	
Manganese mg/kg	13.4	15.0	
Phosphorus %	0.44	0.49	
Sulphur %	0.34	0.38	
Carotene mg/kg	26.6	29.8	47
Riboflavin mg/kg	11.0	12.3	
α-tocopherol mg/kg	127.3	142.4	41
Vitamin A equivalent IU/g	44.3	49.6	
Vitamin D2 IU/g	1.2	1.4	

Ref No 1 00 073 Canada

Feed Name or Analyses	Mean As Fed	Mean Dry	C.V. ± %
Dry matter %		100.0	
Ash %		6.6	
Crude fiber %		35.4	
Cattle dig coef %		40.	
Sheep dig coef %		45.	
Ether extract %		1.2	
Cattle dig coef %		35.	
Sheep dig coef %		17.	
N-free extract %		40.2	
Cattle dig coef %		72.	
Sheep dig coef %		69.	
Protein (N x 6.25) %		16.6	

Feed Name or Analyses	Mean As Fed	Mean Dry	C.V. ± %
Cattle dig coef %		65.	
Sheep dig coef %		71.	
Cattle dig prot %		10.8	
Goats dig prot % *		12.0	
Horses dig prot % *		11.6	
Rabbits dig prot % *		11.5	
Sheep dig prot %		11.8	
Energy GE Mcal/kg		4.47	
Sheep GE dig coef %		57.	
Cattle DE Mcal/kg *		2.42	
Sheep DE Mcal/kg		2.55	
Cattle ME Mcal/kg *		1.98	
Sheep ME Mcal/kg *		2.09	
Cattle TDN %		54.8	
Sheep TDN %		55.9	

Alfalfa, hay, s-c, cut 2, (1)
Ref No 1 00 075 United States

Feed Name or Analyses	Mean As Fed	Mean Dry	C.V. ± %
Dry matter %	89.9	100.0	2
Ash %	7.7	8.5	16
Crude fiber %	28.5	31.7	6
Cattle dig coef %	37.	37.	
Sheep dig coef %	45.	45.	
Ether extract %	1.8	2.0	9
Cattle dig coef %	35.	35.	
Sheep dig coef %	36.	36.	
N-free extract %	37.0	41.2	5
Cattle dig coef %	71.	71.	
Sheep dig coef %	71.	71.	
Protein (N x 6.25) %	14.9	16.5	17
Cattle dig coef %	63.	63.	
Sheep dig coef %	75.	75.	
Cattle dig prot %	9.4	10.5	
Goats dig prot % *	10.8	12.0	
Horses dig prot % *	10.4	11.6	
Rabbits dig prot % *	10.3	11.4	
Sheep dig prot %	11.2	12.4	
Starch %	1.6	1.8	4
Sugars, total %	5.3	5.9	16
Lignin (Ellis) %	10.4	11.6	23
Energy GE Mcal/kg	3.89	4.33	2
Cattle GE dig coef %	59.	59.	
Cattle DE Mcal/kg	2.30	2.55	
Sheep DE Mcal/kg *	2.28	2.54	
Cattle ME Mcal/kg *	1.88	2.09	
Sheep ME Mcal/kg *	1.87	2.08	
Cattle TDN %	47.8	53.2	
Sheep TDN %	51.8	57.6	
Calcium %	1.15	1.28	11
Cobalt mg/kg	0.107	0.119	10
Copper mg/kg	12.1	13.4	
Iron %	0.011	0.012	
Magnesium %	0.38	0.42	
Manganese mg/kg	21.2	23.6	26
Phosphorus %	0.31	0.35	34
Potassium %	1.16	1.29	
Carotene mg/kg	18.8	20.9	37
Riboflavin mg/kg	11.3	12.6	10
Vitamin A equivalent IU/g	31.4	34.9	
Vitamin D2 IU/g	0.9	1.1	

Ref No 1 00 075 Canada

Feed Name or Analyses	Mean As Fed	Mean Dry	C.V. ± %
Dry matter %		100.0	
Crude fiber %			
Cattle dig coef %		37.	
Sheep dig coef %		45.	
Ether extract %			

Feed Name or Analyses	Mean As Fed	Mean Dry	C.V. ± %
Cattle dig coef %		35.	
Sheep dig coef %		36.	
N-free extract %			
Cattle dig coef %		71.	
Sheep dig coef %		71.	
Protein (N x 6.25) %		17.8	
Cattle dig coef %		63.	
Sheep dig coef %		75.	
Cattle dig prot %		11.3	
Goats dig prot % *		13.2	
Horses dig prot % *		12.6	
Rabbits dig prot % *		12.4	
Sheep dig prot %		13.4	
Energy GE Mcal/kg		4.33	
Cattle GE dig coef %		59.	
Cattle DE Mcal/kg		2.55	
Sheep DE Mcal/kg *		2.54	
Cattle ME Mcal/kg *		2.09	
Sheep ME Mcal/kg *		2.08	
Cattle TDN %		53.2	
Sheep TDN %		57.6	
Calcium %		1.66	
Phosphorus %		0.28	
Potassium %		1.21	

Alfalfa, hay, s-c, cut 2, leafy, (1)
Ref No 1 00 074 United States

Feed Name or Analyses	Mean As Fed	Mean Dry	C.V. ± %
Dry matter %	89.3	100.0	1
Ash %	7.6	8.5	8
Crude fiber %	19.5	21.9	7
Ether extract %	2.4	2.7	10
N-free extract %	40.6	45.5	
Protein (N x 6.25) %	19.1	21.4	13
Cattle dig prot % *	13.8	15.5	
Goats dig prot % *	14.7	16.5	
Horses dig prot % *	14.0	15.7	
Rabbits dig prot % *	13.6	15.2	
Sheep dig prot % *	14.1	15.8	
Energy GE Mcal/kg			
Cattle DE Mcal/kg *	2.46	2.76	
Sheep DE Mcal/kg *	2.51	2.82	
Cattle ME Mcal/kg *	2.02	2.26	
Sheep ME Mcal/kg *	2.06	2.31	
Cattle TDN % *	55.8	62.6	
Sheep TDN % *	57.0	63.9	
Calcium %	1.71	1.92	15
Phosphorus %	0.27	0.30	12

Alfalfa, hay, s-c, cut 3, (1)
Ref No 1 00 076 United States

Feed Name or Analyses	Mean As Fed	Mean Dry	C.V. ± %
Dry matter %	89.3	100.0	3
Ash %	7.2	8.0	8
Crude fiber %	26.7	30.0	5
Cattle dig coef %	38.	38.	
Sheep dig coef %	44.	44.	
Ether extract %	1.9	2.1	19
Cattle dig coef %	36.	36.	
Sheep dig coef %	26.	26.	
N-free extract %	38.3	42.9	
Cattle dig coef %	69.	69.	
Sheep dig coef %	72.	72.	
Protein (N x 6.25) %	15.2	17.1	6
Cattle dig coef %	70.	70.	
Sheep dig coef %	72.	72.	
Cattle dig prot %	10.7	11.9	

Continued

Feed Name or Analyses			As Fed	Dry	C.V. ± %
Goats	dig prot %	*	11.1	12.5	
Horses	dig prot %	*	10.7	12.0	
Rabbits	dig prot %	*	10.6	11.8	
Sheep	dig prot %		11.0	12.3	
Cellulose (Matrone)	%		18.0	20.2	10
Pectins	%		5.0	5.6	5
Starch	%		1.5	1.7	12
Sugars, total	%		4.1	4.6	48
Lignin (Ellis)	%		9.3	10.4	21
Energy	GE Mcal/kg		3.91	4.38	1
Cattle	GE dig coef %		64.	64.	
Cattle	DE Mcal/kg		2.51	2.81	
Sheep	DE Mcal/kg	*	2.27	2.54	
Cattle	ME Mcal/kg	*	2.05	2.30	
Sheep	ME Mcal/kg	*	1.86	2.08	
Cattle	TDN %		48.8	54.6	
Sheep	TDN %		51.4	57.5	
Phosphorus	%		0.15	0.17	15
Carotene	mg/kg		36.0	40.3	28
Riboflavin	mg/kg		9.1	10.1	
Vitamin A equivalent	IU/g		60.1	67.3	
Vitamin D_2	IU/g		1.2	1.3	

Ref No 1 00 076 Great Britain

Feed Name or Analyses			As Fed	Dry	C.V. ± %
Dry matter	%		89.6	100.0	
Ash	%		9.5	10.6	
Crude fiber	%		29.7	33.1	
Cattle	dig coef %		55.	55.	
Sheep	dig coef %		44.	44.	
Ether extract	%		2.1	2.3	
Cattle	dig coef %		39.	39.	
Sheep	dig coef %		26.	26.	
N-free extract	%		29.3	32.7	
Cattle	dig coef %		75.	75.	
Sheep	dig coef %		72.	72.	
Protein (N x 6.25)	%		19.0	21.2	
Cattle	dig coef %		78.	78.	
Sheep	dig coef %		72.	72.	
Cattle	dig prot %		14.8	16.5	
Goats	dig prot %	*	14.6	16.3	
Horses	dig prot %	*	13.9	15.5	
Rabbits	dig prot %	*	13.5	15.0	
Sheep	dig prot %		13.7	15.3	
Cellulose (Matrone)	%		19.2	21.4	
Cattle	dig coef %		57.	57.	
Pentosans	%		15.3	17.1	
Lignin (Ellis)	%		9.5	10.6	
Cattle	dig coef %		4.	4.	
Energy	GE Mcal/kg		4.24	4.73	
Cattle	GE dig coef %		64.	64.	
Cattle	DE Mcal/kg		2.71	3.03	
Sheep	DE Mcal/kg	*	2.16	2.41	
Cattle	ME Mcal/kg	*	2.22	2.48	
Sheep	ME Mcal/kg	*	1.77	1.98	
Cattle	TDN %		55.0	61.4	
Sheep	TDN %		49.1	54.8	

Alfalfa, hay, s-c, cut 4, (1)
Ref No 1 00 077 United States

Feed Name or Analyses			As Fed	Dry	C.V. ± %
Dry matter	%		90.1	100.0	3
Lignin (Ellis)	%		8.6	9.5	
Sulphur	%		0.30	0.33	14

(1) dry forages and roughages (2) pasture, range plants, and forages fed green (3) silages (4) energy feeds (5) protein supplements (6) minerals (7) vitamins (8) additives

Feed Name or Analyses			As Fed	Dry	C.V. ± %
Carotene	mg/kg		15.7	17.4	98
Vitamin A equivalent	IU/g		26.2	29.0	

Alfalfa, hay, s-c, gr 1 US, (1)
Ref No 1 00 079 United States

Feed Name or Analyses			As Fed	Dry	C.V. ± %
Dry matter	%		87.7	100.0	4
Ash	%		9.9	11.3	16
Crude fiber	%		23.3	26.6	11
Ether extract	%		1.5	1.7	33
N-free extract	%		36.9	42.1	
Protein (N x 6.25)	%		16.0	18.3	22
Cattle	dig prot %	*	11.2	12.8	
Goats	dig prot %	*	11.9	13.6	
Horses	dig prot %	*	11.5	13.1	
Rabbits	dig prot %	*	11.2	12.8	
Sheep	dig prot %	*	11.4	13.0	
Hemicellulose	%		6.1	7.0	27
Energy	GE Mcal/kg				
Cattle	DE Mcal/kg	*	2.39	2.73	
Sheep	DE Mcal/kg	*	2.27	2.59	
Cattle	ME Mcal/kg	*	1.96	2.24	
Sheep	ME Mcal/kg	*	1.86	2.12	
Cattle	TDN %	*	54.2	61.8	
Sheep	TDN %	*	51.5	58.7	
Cobalt	mg/kg		0.099	0.112	22
Silicon	%		0.60	0.68	60
Carotene	mg/kg		41.6	47.4	
Niacin	mg/kg		38.5	43.9	12
Pantothenic acid	mg/kg		19.1	21.8	24
Riboflavin	mg/kg		16.4	18.7	25
Thiamine	mg/kg		2.9	3.3	67
Vitamin A equivalent	IU/g		69.3	79.0	
Tyrosine	%		0.44	0.50	

Alfalfa, hay, s-c, gr 1 US leafy, (1)
Ref No 1 00 080 United States

Feed Name or Analyses			As Fed	Dry	C.V. ± %
Dry matter	%			100.0	
Calcium	%			1.40	
Phosphorus	%			0.19	

Alfalfa, hay, s-c, gr 2 US, (1)
Ref No 1 00 081 United States

Feed Name or Analyses			As Fed	Dry	C.V. ± %
Dry matter	%		90.2	100.0	1
Ash	%		7.8	8.6	11
Crude fiber	%		26.9	29.8	9
Ether extract	%		1.6	1.8	20
N-free extract	%		36.3	40.2	
Protein (N x 6.25)	%		17.7	19.6	6
Cattle	dig prot %	*	12.6	13.9	
Goats	dig prot %	*	13.4	14.8	
Horses	dig prot %	*	12.8	14.2	
Rabbits	dig prot %	*	12.4	13.8	
Sheep	dig prot %	*	12.8	14.2	
Cellulose (Matrone)	%		25.3	28.0	
Hemicellulose	%		13.9	15.4	
Sugars, total	%		9.7	10.8	
Lignin (Ellis)	%		7.6	8.4	
Energy	GE Mcal/kg		4.06	4.50	
Cattle	DE Mcal/kg	*	2.27	2.52	
Sheep	DE Mcal/kg	*	2.35	2.60	
Cattle	ME Mcal/kg	*	1.87	2.07	
Sheep	ME Mcal/kg	*	1.92	2.13	
Cattle	TDN %	*	51.6	57.2	
Sheep	TDN %	*	53.2	59.0	

Feed Name or Analyses			As Fed	Dry	C.V. ± %
Calcium	%		1.24	1.38	33
Iron	%		0.148	0.164	
Magnesium	%		0.35	0.39	
Phosphorus	%		0.22	0.24	17
Potassium	%		1.33	1.47	
Carotene	mg/kg		20.5	22.7	50
Vitamin A equivalent	IU/g		34.1	37.9	
Vitamin D_2	IU/g		1.8	2.0	

Alfalfa, hay, s-c, gr 2 US leafy, (1)
Ref No 1 00 082 United States

Feed Name or Analyses			As Fed	Dry	C.V. ± %
Dry matter	%		87.6	100.0	1
Ash	%		6.3	7.2	14
Crude fiber	%		29.9	34.1	8
Ether extract	%		1.1	1.3	22
N-free extract	%		34.8	39.7	
Protein (N x 6.25)	%		15.5	17.6	21
Cattle	dig prot %	*	10.7	12.2	
Goats	dig prot %	*	11.4	13.0	
Horses	dig prot %	*	11.0	12.5	
Rabbits	dig prot %	*	10.8	12.3	
Sheep	dig prot %	*	10.9	12.4	
Energy	GE Mcal/kg				
Cattle	DE Mcal/kg	*	2.21	2.52	
Sheep	DE Mcal/kg	*	2.22	2.53	
Cattle	ME Mcal/kg	*	1.81	2.07	
Sheep	ME Mcal/kg	*	1.82	2.08	
Cattle	TDN %		50.1	57.2	
Sheep	TDN %	*	50.3	57.4	

Alfalfa, hay, s-c, gr 2 US stemmy, (1)
Ref No 1 00 083 United States

Feed Name or Analyses			As Fed	Dry	C.V. ± %
Dry matter	%		93.7	100.0	3
Ash	%		7.0	7.5	7
Crude fiber	%		34.9	37.2	4
Ether extract	%		1.7	1.8	11
N-free extract	%		36.4	38.8	
Protein (N x 6.25)	%		13.8	14.7	2
Cattle	dig prot %	*	9.1	9.7	
Goats	dig prot %	*	9.6	10.3	
Horses	dig prot %	*	9.4	10.0	
Rabbits	dig prot %	*	9.4	10.0	
Sheep	dig prot %	*	9.1	9.8	
Energy	GE Mcal/kg				
Cattle	DE Mcal/kg	*	2.16	2.30	
Sheep	DE Mcal/kg	*	2.24	2.39	
Cattle	ME Mcal/kg	*	1.77	1.89	
Sheep	ME Mcal/kg	*	1.84	1.96	
Cattle	TDN %	*	48.9	52.2	
Sheep	TDN %	*	50.8	54.3	

Alfalfa, hay, s-c, gr 3 US, (1)
Ref No 1 00 084 United States

Feed Name or Analyses			As Fed	Dry	C.V. ± %
Dry matter	%		90.5	100.0	0
Ash	%		8.5	9.4	13
Crude fiber	%		31.2	34.5	4
Ether extract	%		1.6	1.8	3
N-free extract	%		35.8	39.5	
Protein (N x 6.25)	%		13.4	14.8	9
Cattle	dig prot %	*	8.8	9.8	
Goats	dig prot %	*	9.4	10.4	
Horses	dig prot %	*	9.1	10.1	
Rabbits	dig prot %	*	9.1	10.1	
Sheep	dig prot %	*	8.9	9.8	

Feed Name or Analyses			Mean As Fed	Mean Dry	C.V. ± %
Cellulose (Matrone)	%		28.3	31.3	
Hemicellulose	%		14.0	15.5	
Sugars, total	%		3.9	4.3	
Lignin (Ellis)	%		11.4	12.6	
Energy	GE Mcal/kg				
Cattle	DE Mcal/kg	*	2.22	2.45	
Sheep	DE Mcal/kg	*	2.17	2.40	
Cattle	ME Mcal/kg	*	1.82	2.01	
Sheep	ME Mcal/kg	*	1.78	1.97	
Cattle	TDN %	*	50.4	55.7	
Sheep	TDN %	*	49.3	54.5	
Calcium	%		1.13	1.25	14
Copper	mg/kg		12.2	13.4	
Iron	%		0.037	0.041	98
Magnesium	%		0.36	0.40	23
Manganese	mg/kg		25.3	28.0	
Phosphorus	%		0.21	0.23	30
Potassium	%		1.07	1.18	23
Carotene	mg/kg		18.3	20.2	83
Vitamin A equivalent	IU/g		30.4	33.6	

Alfalfa, hay, s-c, gr 3 US stemmy, (1)
Ref No 1 00 085 United States

Feed Name or Analyses			Mean As Fed	Mean Dry	C.V. ± %
Dry matter	%		88.6	100.0	
Ash	%		6.0	6.8	
Crude fiber	%		35.7	40.3	
Ether extract	%		1.1	1.2	
N-free extract	%		32.0	36.1	
Protein (N x 6.25)	%		13.8	15.6	
Cattle	dig prot %	*	9.3	10.4	
Goats	dig prot %	*	9.8	11.1	
Horses	dig prot %	*	9.5	10.8	
Rabbits	dig prot %	*	9.5	10.7	
Sheep	dig prot %	*	9.4	10.6	
Energy	GE Mcal/kg				
Cattle	DE Mcal/kg	*	1.91	2.16	
Sheep	DE Mcal/kg	*	2.09	2.36	
Cattle	ME Mcal/kg	*	1.57	1.77	
Sheep	ME Mcal/kg	*	1.71	1.93	
Cattle	TDN %	*	43.4	49.0	
Sheep	TDN %	*	47.3	53.4	

Alfalfa, hay, s-c, gr 4 US, (1)
Ref No 1 00 086 United States

Feed Name or Analyses			Mean As Fed	Mean Dry	C.V. ± %
Dry matter	%			100.0	
Cellulose (Matrone)	%			38.2	
Hemicellulose	%			18.3	
Lignin (Ellis)	%			11.7	

Alfalfa, hay, s-c, leafy, (1)
Ref No 1 00 087 United States

Feed Name or Analyses			Mean As Fed	Mean Dry	C.V. ± %
Dry matter	%		90.2	100.0	3
Ash	%		8.6	9.5	8
Crude fiber	%		25.1	27.9	10
Sheep	dig coef %		48.	48.	
Ether extract	%		1.9	2.1	17
Sheep	dig coef %		22.	22.	
N-free extract	%		37.4	41.5	
Sheep	dig coef %		72.	72.	
Protein (N x 6.25)	%		17.2	19.1	14
Sheep	dig coef %		76.	76.	
Cattle	dig prot %	*	12.2	13.5	
Goats	dig prot %	*	13.0	14.4	
Horses	dig prot %	*	12.4	13.7	
Rabbits	dig prot %	*	12.1	13.4	
Sheep	dig prot %		13.1	14.5	
Energy	GE Mcal/kg				
Cattle	DE Mcal/kg	*	2.37	2.63	
Sheep	DE Mcal/kg	*	2.34	2.59	
Cattle	ME Mcal/kg	*	1.94	2.15	
Sheep	ME Mcal/kg	*	1.92	2.12	
Cattle	TDN %	*	53.8	59.6	
Sheep	TDN %		53.0	58.8	
Calcium	%		1.41	1.56	14
Phosphorus	%		0.25	0.28	12
Potassium	%		1.98	2.20	7

Alfalfa, hay, s-c, mn 20% protein, (1)
Ref No 1 00 088 United States

Feed Name or Analyses			Mean As Fed	Mean Dry	C.V. ± %
Dry matter	%		86.2	100.0	
Ash	%		11.3	13.1	
Crude fiber	%		19.3	22.4	
Sheep	dig coef %		52.	52.	
Ether extract	%		2.8	3.3	
Sheep	dig coef %		43.	43.	
N-free extract	%		31.4	36.4	
Sheep	dig coef %		73.	73.	
Protein (N x 6.25)	%		21.4	24.8	
Sheep	dig coef %		77.	77.	
Cattle	dig prot %	*	15.9	18.4	
Goats	dig prot %	*	17.0	19.7	
Horses	dig prot %	*	16.0	18.6	
Rabbits	dig prot %	*	15.4	17.8	
Sheep	dig prot %		16.5	19.1	
Energy	GE Mcal/kg				
Cattle	DE Mcal/kg	*	2.54	2.95	
Sheep	DE Mcal/kg	*	2.30	2.67	
Cattle	ME Mcal/kg	*	2.08	2.42	
Sheep	ME Mcal/kg	*	1.89	2.19	
Cattle	TDN %	*	57.6	66.8	
Sheep	TDN %		52.2	60.5	

Alfalfa, hay, s-c, mn 34% fiber, (1)
Ref No 1 00 089 United States

Feed Name or Analyses			Mean As Fed	Mean Dry	C.V. ± %
Dry matter	%		89.6	100.0	
Ash	%		6.2	6.9	
Crude fiber	%		36.0	40.2	
Cattle	dig coef %		46.	46.	
Sheep	dig coef %		45.	45.	
Ether extract	%		1.4	1.6	
Cattle	dig coef %		33.	33.	
Sheep	dig coef %		16.	16.	
N-free extract	%		33.6	37.5	
Cattle	dig coef %		64.	64.	
Sheep	dig coef %		64.	64.	
Protein (N x 6.25)	%		12.4	13.8	
Cattle	dig coef %		69.	69.	
Sheep	dig coef %		67.	67.	
Cattle	dig prot %		8.5	9.5	
Goats	dig prot %	*	8.5	9.4	
Horses	dig prot %	*	8.3	9.2	
Rabbits	dig prot %	*	8.4	9.3	
Sheep	dig prot %		8.3	9.2	
Energy	GE Mcal/kg				
Cattle	DE Mcal/kg	*	2.10	2.35	
Sheep	DE Mcal/kg	*	2.05	2.29	
Cattle	ME Mcal/kg	*	1.72	1.92	
Sheep	ME Mcal/kg	*	1.68	1.88	
Cattle	TDN %		47.7	53.2	
Sheep	TDN %		46.5	51.9	

Alfalfa, hay, s-c, mx 13% protein, (1)
Ref No 1 00 090 United States

Feed Name or Analyses			Mean As Fed	Mean Dry	C.V. ± %
Dry matter	%		88.5	100.0	
Ash	%		7.5	8.5	
Crude fiber	%		29.1	32.9	
Horses	dig coef %		36.	36.	
Sheep	dig coef %		46.	46.	
Ether extract	%		1.8	2.0	
Horses	dig coef %		24.	24.	
Sheep	dig coef %		25.	25.	
N-free extract	%		38.3	43.3	
Horses	dig coef %		66.	66.	
Sheep	dig coef %		67.	67.	
Protein (N x 6.25)	%		11.9	13.4	
Horses	dig coef %		68.	68.	
Sheep	dig coef %		66.	66.	
Cattle	dig prot %	*	7.6	8.5	
Goats	dig prot %	*	8.0	9.1	
Horses	dig prot %		8.1	9.1	
Rabbits	dig prot %	*	8.0	9.0	
Sheep	dig prot %		7.8	8.8	
Energy	GE Mcal/kg				
Cattle	DE Mcal/kg	*	2.19	2.48	
Horses	DE Mcal/kg	*	1.97	2.23	
Sheep	DE Mcal/kg	*	2.11	2.38	
Cattle	ME Mcal/kg	*	1.80	2.03	
Horses	ME Mcal/kg	*	1.62	1.83	
Sheep	ME Mcal/kg	*	1.73	1.95	
Cattle	TDN %	*	49.7	56.2	
Horses	TDN %		44.7	50.6	
Sheep	TDN %		47.8	54.1	

Ref No 1 00 090 Canada

Feed Name or Analyses			Mean As Fed	Mean Dry	C.V. ± %
Dry matter	%		92.5	100.0	2
Crude fiber	%		32.0	34.6	14
Horses	dig coef %		36.	36.	
Sheep	dig coef %		46.	46.	
Ether extract	%				
Horses	dig coef %		24.	24.	
Sheep	dig coef %		25.	25.	
N-free extract	%				
Horses	dig coef %		66.	66.	
Sheep	dig coef %		67.	67.	
Protein (N x 6.25)	%		11.9	12.9	6
Horses	dig coef %		68.	68.	
Sheep	dig coef %		66.	66.	
Cattle	dig prot %	*	7.5	8.1	
Goats	dig prot %	*	8.0	8.6	
Horses	dig prot %		8.1	8.8	
Rabbits	dig prot %	*	8.0	8.6	
Sheep	dig prot %		7.9	8.5	
Energy	GE Mcal/kg				
Horses	DE Mcal/kg	*	2.04	2.21	
Sheep	DE Mcal/kg	*	2.19	2.37	
Horses	ME Mcal/kg	*	1.67	1.81	
Sheep	ME Mcal/kg	*	1.80	1.94	
Horses	TDN %		46.3	50.1	
Sheep	TDN %		49.7	53.7	
Calcium	%		1.24	1.34	25
Phosphorus	%		0.15	0.16	34
Carotene	mg/kg		43.8	47.4	51
Vitamin A equivalent	IU/g		73.1	79.0	

Feed Name or Analyses			Mean As Fed	Dry	C.V. ± %

Alfalfa, hay, s-c, mx 25% fiber, (1)
Ref No 1 00 091 United States

Feed Name or Analyses			As Fed	Dry	C.V. ± %
Dry matter	%		86.8	100.0	
Ash	%		8.0	9.3	
Crude fiber	%		22.5	25.9	
Cattle	dig coef %		44.	44.	
Sheep	dig coef %		44.	44.	
Ether extract	%		2.3	2.6	
Cattle	dig coef %		40.	40.	
Sheep	dig coef %		29.	29.	
N-free extract	%		38.5	44.4	
Cattle	dig coef %		73.	73.	
Sheep	dig coef %		72.	72.	
Protein (N x 6.25)	%		15.5	17.9	
Cattle	dig coef %		70.	70.	
Sheep	dig coef %		74.	74.	
Cattle	dig prot %		10.8	12.5	
Goats	dig prot %	*	11.5	13.2	
Horses	dig prot %	*	11.0	12.7	
Rabbits	dig prot %	*	10.8	12.5	
Sheep	dig prot %		11.5	13.2	
Energy	GE Mcal/kg				
Cattle	DE Mcal/kg	*	2.24	2.59	
Sheep	DE Mcal/kg	*	2.23	2.57	
Cattle	ME Mcal/kg	*	1.84	2.12	
Sheep	ME Mcal/kg	*	1.83	2.11	
Cattle	TDN %		50.9	58.6	
Sheep	TDN %		50.5	58.3	

Alfalfa, hay, s-c, pure stand, (1)
Ref No 1 00 092 United States

Feed Name or Analyses			As Fed	Dry	C.V. ± %
Dry matter	%			100.0	
Alanine	%			1.20	
Arginine	%			0.70	
Aspartic acid	%			2.50	
Cystine	%			0.30	
Glutamic acid	%			1.60	
Glycine	%			0.70	
Histidine	%			0.30	
Isoleucine	%			0.80	
Leucine	%			1.60	
Lysine	%			0.80	
Methionine	%			0.30	
Phenylalanine	%			0.80	
Proline	%			0.90	
Serine	%			0.80	
Threonine	%			1.00	
Tryptophan	%			0.30	
Tyrosine	%			0.40	
Valine	%			1.00	

Alfalfa, hay, s-c, stemmy, (1)
Ref No 1 00 093 United States

Feed Name or Analyses			As Fed	Dry	C.V. ± %
Dry matter	%		91.1	100.0	2
Ash	%		7.2	7.9	11
Crude fiber	%		33.7	37.0	6
Sheep	dig coef %		45.	45.	
Ether extract	%		1.5	1.7	19
Sheep	dig coef %		35.	35.	
N-free extract	%		35.2	38.7	
Sheep	dig coef %		68.	68.	
Protein (N x 6.25)	%		13.4	14.8	10
Sheep	dig coef %		73.	73.	
Cattle	dig prot %	*	8.9	9.7	
Goats	dig prot %	*	9.4	10.3	
Horses	dig prot %	*	9.2	10.1	
Rabbits	dig prot %	*	9.2	10.1	
Sheep	dig prot %		9.8	10.8	
Energy	GE Mcal/kg				
Cattle	DE Mcal/kg	*	2.11	2.32	
Sheep	DE Mcal/kg	*	2.21	2.43	
Cattle	ME Mcal/kg	*	1.73	1.90	
Sheep	ME Mcal/kg	*	1.81	1.99	
Cattle	TDN %	*	47.9	52.6	
Sheep	TDN %		50.2	55.1	
Calcium	%		0.89	0.98	30
Iron	%		0.014	0.015	19
Magnesium	%		0.29	0.32	12
Phosphorus	%		0.20	0.21	21
Potassium	%		1.80	1.98	7

(1) dry forages and roughages (2) pasture, range plants, and forages fed green (3) silages (4) energy feeds (5) protein supplements (6) minerals (7) vitamins (8) additives

Alfalfa, hay, s-c, very leafy, (1)
Ref No 1 00 094 United States

Feed Name or Analyses			As Fed	Dry	C.V. ± %
Dry matter	%		90.9	100.0	2
Ash	%		8.6	9.5	13
Crude fiber	%		22.4	24.6	9
Ether extract	%		2.5	2.8	16
N-free extract	%		39.2	43.2	
Protein (N x 6.25)	%		18.2	20.0	11
Cattle	dig prot %	*	12.9	14.2	
Goats	dig prot %	*	13.8	15.2	
Horses	dig prot %	*	13.2	14.5	
Rabbits	dig prot %	*	12.8	14.1	
Sheep	dig prot %	*	13.2	14.5	
Energy	GE Mcal/kg				
Cattle	DE Mcal/kg	*	2.49	2.74	
Sheep	DE Mcal/kg	*	2.44	2.69	
Cattle	ME Mcal/kg	*	2.04	2.25	
Sheep	ME Mcal/kg	*	2.00	2.20	
Cattle	TDN %	*	56.5	62.1	
Sheep	TDN %	*	55.4	60.9	
Calcium	%		1.62	1.78	
Phosphorus	%		0.24	0.27	
Potassium	%		1.96	2.15	

Alfalfa, hay, s-c, 13-15% protein, (1)
Ref No 1 00 095 United States

Feed Name or Analyses			As Fed	Dry	C.V. ± %
Dry matter	%		88.7	100.0	
Ash	%		7.2	8.2	
Crude fiber	%		29.5	33.3	
Cattle	dig coef %		42.	42.	
Horses	dig coef %		42.	42.	
Sheep	dig coef %		42.	42.	
Ether extract	%		1.9	2.1	
Cattle	dig coef %		33.	33.	
Horses	dig coef %		18.	18.	
Sheep	dig coef %		24.	24.	
N-free extract	%		36.5	41.2	
Cattle	dig coef %		71.	71.	
Horses	dig coef %		70.	70.	
Sheep	dig coef %		68.	68.	
Protein (N x 6.25)	%		13.5	15.3	
Cattle	dig coef %		69.	69.	
Horses	dig coef %		74.	74.	
Sheep	dig coef %		71.	71.	
Cattle	dig prot %		9.3	10.5	
Goats	dig prot %	*	9.6	10.8	
Horses	dig prot %		10.0	11.3	
Rabbits	dig prot %	*	9.3	10.5	
Sheep	dig prot %		9.6	10.8	
Energy	GE Mcal/kg				
Cattle	DE Mcal/kg	*	2.15	2.42	
Horses	DE Mcal/kg	*	2.15	2.42	
Sheep	DE Mcal/kg	*	2.11	2.38	
Cattle	ME Mcal/kg	*	1.76	1.98	
Horses	ME Mcal/kg	*	1.76	1.99	
Sheep	ME Mcal/kg	*	1.73	1.95	
Cattle	TDN %		48.7	54.9	
Horses	TDN %		48.7	55.0	
Sheep	TDN %		47.9	54.0	

Ref No 1 00 095 Canada

Feed Name or Analyses			As Fed	Dry	C.V. ± %
Dry matter	%		91.3	100.0	2
Crude fiber	%		28.5	31.2	17
Cattle	dig coef %		42.	42.	
Horses	dig coef %		42.	42.	
Sheep	dig coef %		42.	42.	
Ether extract	%				
Cattle	dig coef %		33.	33.	
Horses	dig coef %		18.	18.	
Sheep	dig coef %		24.	24.	
N-free extract	%				
Cattle	dig coef %		71.	71.	
Horses	dig coef %		70.	70.	
Sheep	dig coef %		68.	68.	
Protein (N x 6.25)	%		14.0	15.3	5
Cattle	dig coef %		69.	69.	
Horses	dig coef %		74.	74.	
Sheep	dig coef %		71.	71.	
Cattle	dig prot %		9.6	10.5	
Goats	dig prot %	*	9.9	10.9	
Horses	dig prot %		10.4	11.3	
Rabbits	dig prot %	*	9.6	10.5	
Sheep	dig prot %		9.9	10.9	
Energy	GE Mcal/kg				
Cattle	DE Mcal/kg	*	2.23	2.44	
Horses	DE Mcal/kg	*	2.23	2.45	
Sheep	DE Mcal/kg	*	2.19	2.40	
Cattle	ME Mcal/kg	*	1.83	2.00	
Horses	ME Mcal/kg	*	1.83	2.01	
Sheep	ME Mcal/kg	*	1.80	1.97	
Cattle	TDN %		50.6	55.4	
Horses	TDN %		50.6	55.5	
Sheep	TDN %		49.7	54.5	
Calcium	%		1.47	1.61	29
Phosphorus	%		0.17	0.18	46
Carotene	mg/kg		45.0	49.3	61
Vitamin A equivalent	IU/g		75.0	82.2	

Alfalfa, hay, s-c, 15-17% protein, (1)
Ref No 1 00 096 United States

Feed Name or Analyses			As Fed	Dry	C.V. ± %
Dry matter	%		87.8	100.0	
Ash	%		8.0	9.2	
Crude fiber	%		27.3	31.1	
Cattle	dig coef %		46.	46.	
Horses	dig coef %		41.	41.	
Sheep	dig coef %		46.	46.	
Ether extract	%		1.8	2.1	
Cattle	dig coef %		40.	40.	
Horses	dig coef %		-11.	-11.	
Sheep	dig coef %		34.	34.	
N-free extract	%		34.8	39.7	

Feed Name or Analyses			Mean As Fed	Mean Dry	C.V. ± %
Cattle	dig coef %		72.	72.	
Horses	dig coef %		68.	68.	
Sheep	dig coef %		71.	71.	
Protein (N x 6.25)	%		15.8	18.0	
Cattle	dig coef %		74.	74.	
Horses	dig coef %		75.	75.	
Sheep	dig coef %		73.	73.	
Cattle	dig prot %		11.7	13.3	
Goats	dig prot %	*	11.7	13.3	
Horses	dig prot %		11.8	13.5	
Rabbits	dig prot %	*	11.0	12.5	
Sheep	dig prot %		11.5	13.1	
Energy	GE Mcal/kg				
Cattle	DE Mcal/kg	*	2.25	2.56	
Horses	DE Mcal/kg	*	2.04	2.32	
Sheep	DE Mcal/kg	*	2.21	2.52	
Cattle	ME Mcal/kg	*	1.84	2.10	
Horses	ME Mcal/kg	*	1.67	1.90	
Sheep	ME Mcal/kg	*	1.82	2.07	
Cattle	TDN %		51.0	58.1	
Horses	TDN %		46.2	52.6	
Sheep	TDN %		50.2	57.2	

Ref No 1 00 096 Canada

Feed Name or Analyses			Mean As Fed	Mean Dry	C.V. ± %
Dry matter	%		91.5	100.0	3
Crude fiber	%		27.3	29.8	14
Cattle	dig coef %		46.	46.	
Horses	dig coef %		41.	41.	
Sheep	dig coef %		46.	46.	
Ether extract	%				
Cattle	dig coef %		40.	40.	
Horses	dig coef %		-11.	-11.	
Sheep	dig coef %		34.	34.	
N-free extract	%				
Cattle	dig coef %		72.	72.	
Horses	dig coef %		68.	68.	
Sheep	dig coef %		71.	71.	
Protein (N x 6.25)	%		15.8	17.3	4
Cattle	dig coef %		74.	74.	
Horses	dig coef %		75.	75.	
Sheep	dig coef %		73.	73.	
Cattle	dig prot %		11.7	12.8	
Goats	dig prot %	*	11.6	12.7	
Horses	dig prot %		11.9	13.0	
Rabbits	dig prot %	*	11.0	12.0	
Sheep	dig prot %		11.5	12.6	
Energy	GE Mcal/kg				
Cattle	DE Mcal/kg	*	2.35	2.57	
Horses	DE Mcal/kg	*	2.13	2.33	
Sheep	DE Mcal/kg	*	2.32	2.53	
Cattle	ME Mcal/kg	*	1.93	2.11	
Horses	ME Mcal/kg	*	1.75	1.91	
Sheep	ME Mcal/kg	*	1.90	2.08	
Cattle	TDN %		53.4	58.4	
Horses	TDN %		48.4	52.9	
Sheep	TDN %		52.6	57.5	
Calcium	%		1.48	1.62	28
Copper	mg/kg		8.1	8.8	
Phosphorus	%		0.20	0.22	30
Carotene	mg/kg		51.9	56.7	75
Vitamin A equivalent	IU/g		86.5	94.5	

Alfalfa, hay, s-c, 17-20% protein, (1)

Ref No 1 00 097 United States

Feed Name or Analyses			Mean As Fed	Mean Dry	C.V. ± %
Dry matter	%		85.8	100.0	
Ash	%		8.1	9.5	
Crude fiber	%		23.3	27.2	
Cattle	dig coef %		75.	75.	
Horses	dig coef %		25.	25.	
Sheep	dig coef %		46.	46.	
Ether extract	%		2.0	2.3	
Cattle	dig coef %		53.	53.	
Horses	dig coef %		-36.	-36.	
Sheep	dig coef %		36.	36.	
N-free extract	%		34.9	40.6	
Cattle	dig coef %		72.	72.	
Horses	dig coef %		62.	62.	
Sheep	dig coef %		71.	71.	
Protein (N x 6.25)	%		17.5	20.4	
Cattle	dig coef %		77.	77.	
Horses	dig coef %		75.	75.	
Sheep	dig coef %		77.	77.	
Cattle	dig prot %		13.5	15.7	
Goats	dig prot %	*	13.4	15.6	
Horses	dig prot %		13.1	15.3	
Rabbits	dig prot %	*	12.4	14.4	
Sheep	dig prot %		13.5	15.7	
Energy	GE Mcal/kg				
Cattle	DE Mcal/kg	*	2.58	3.00	
Horses	DE Mcal/kg	*	1.72	2.00	
Sheep	DE Mcal/kg	*	2.23	2.60	
Cattle	ME Mcal/kg	*	2.11	2.46	
Horses	ME Mcal/kg	*	1.41	1.64	
Sheep	ME Mcal/kg	*	1.83	2.13	
Cattle	TDN %		58.4	68.1	
Horses	TDN %		38.9	45.4	
Sheep	TDN %		50.6	58.9	

Ref No 1 00 097 Canada

Feed Name or Analyses			Mean As Fed	Mean Dry	C.V. ± %
Dry matter	%		90.2	100.0	3
Crude fiber	%		24.6	27.3	17
Cattle	dig coef %		75.	75.	
Horses	dig coef %		25.	25.	
Sheep	dig coef %		46.	46.	
Ether extract	%				
Cattle	dig coef %		53.	53.	
Horses	dig coef %		-36.	-36.	
Sheep	dig coef %		36.	36.	
N-free extract	%				
Cattle	dig coef %		72.	72.	
Horses	dig coef %		62.	62.	
Sheep	dig coef %		71.	71.	
Protein (N x 6.25)	%		18.1	20.1	5
Cattle	dig coef %		77.	77.	
Horses	dig coef %		75.	75.	
Sheep	dig coef %		77.	77.	
Cattle	dig prot %		13.9	15.5	
Goats	dig prot %	*	13.8	15.3	
Horses	dig prot %		13.6	15.1	
Rabbits	dig prot %	*	12.8	14.2	
Sheep	dig prot %		13.9	15.5	
Energy	GE Mcal/kg				
Cattle	DE Mcal/kg	*	2.71	3.00	
Horses	DE Mcal/kg	*	1.80	2.00	
Sheep	DE Mcal/kg	*	2.34	2.60	
Cattle	ME Mcal/kg	*	2.22	2.46	
Horses	ME Mcal/kg	*	1.48	1.64	
Sheep	ME Mcal/kg	*	1.92	2.13	
Cattle	TDN %		61.4	68.1	
Horses	TDN %		40.9	45.3	
Sheep	TDN %		53.1	58.9	
Calcium	%		1.72	1.90	26
Phosphorus	%		0.22	0.25	41
Carotene	mg/kg		54.2	60.1	89
Vitamin A equivalent	IU/g		90.3	100.2	

Alfalfa, hay, s-c, 25-28% fiber, (1)

Ref No 1 00 098 United States

Feed Name or Analyses			Mean As Fed	Mean Dry	C.V. ± %
Dry matter	%		89.1	100.0	
Ash	%		8.1	9.1	
Crude fiber	%		26.5	29.8	
Cattle	dig coef %		42.	42.	
Sheep	dig coef %		44.	44.	
Ether extract	%		2.0	2.2	
Cattle	dig coef %		37.	37.	
Sheep	dig coef %		31.	31.	
N-free extract	%		36.9	41.5	
Cattle	dig coef %		71.	71.	
Sheep	dig coef %		72.	72.	
Protein (N x 6.25)	%		15.6	17.5	
Cattle	dig coef %		72.	72.	
Sheep	dig coef %		75.	75.	
Cattle	dig prot %		11.2	12.6	
Goats	dig prot %	*	11.4	12.8	
Horses	dig prot %	*	11.0	12.4	
Rabbits	dig prot %	*	10.8	12.1	
Sheep	dig prot %		11.7	13.1	
Energy	GE Mcal/kg				
Cattle	DE Mcal/kg	*	2.21	2.49	
Sheep	DE Mcal/kg	*	2.26	2.54	
Cattle	ME Mcal/kg	*	1.82	2.04	
Sheep	ME Mcal/kg	*	1.86	2.08	
Cattle	TDN %		50.2	56.4	
Sheep	TDN %		51.3	57.6	
Calcium	%		1.29	1.45	
Phosphorus	%		0.24	0.27	
Potassium	%		1.96	2.20	

Alfalfa, hay, s-c, 28-31% fiber, (1)

Ref No 1 00 099 United States

Feed Name or Analyses			Mean As Fed	Mean Dry	C.V. ± %
Dry matter	%		89.8	100.0	
Ash	%		8.2	9.2	
Crude fiber	%		29.4	32.7	
Cattle	dig coef %		44.	44.	
Sheep	dig coef %		46.	46.	
Ether extract	%		1.8	2.0	
Cattle	dig coef %		36.	36.	
Sheep	dig coef %		36.	36.	
N-free extract	%		35.8	39.9	
Cattle	dig coef %		72.	72.	
Sheep	dig coef %		70.	70.	
Protein (N x 6.25)	%		14.6	16.3	
Cattle	dig coef %		70.	70.	
Sheep	dig coef %		73.	73.	
Cattle	dig prot %		10.2	11.4	
Goats	dig prot %	*	10.5	11.7	
Horses	dig prot %	*	10.2	11.3	
Rabbits	dig prot %	*	10.1	11.2	
Sheep	dig prot %		10.7	11.9	
Energy	GE Mcal/kg				
Cattle	DE Mcal/kg	*	2.22	2.47	
Sheep	DE Mcal/kg	*	2.23	2.49	
Cattle	ME Mcal/kg	*	1.82	2.03	
Sheep	ME Mcal/kg	*	1.83	2.04	
Cattle	TDN %		50.3	56.1	
Sheep	TDN %		50.7	56.4	
Calcium	%		1.21	1.35	
Phosphorus	%		0.22	0.24	
Potassium	%		1.95	2.18	

Feed Name or Analyses		Mean As Fed	Mean Dry	C.V. ± %

Alfalfa, hay, s-c, 31-34% fiber, (1)
Ref No 1 00 100 United States

Feed Name or Analyses		As Fed	Dry	C.V. ± %
Dry matter %		90.3	100.0	
Ash %		7.5	8.3	
Crude fiber %		32.3	35.8	
Cattle dig coef %		44.	44.	
Sheep dig coef %		45.	45.	
Ether extract %		1.6	1.8	
Cattle dig coef %		23.	23.	
Sheep dig coef %		40.	40.	
N-free extract %		35.2	39.0	
Cattle dig coef %		71.	71.	
Sheep dig coef %		68.	68.	
Protein (N x 6.25) %		13.6	15.1	
Cattle dig coef %		70.	70.	
Sheep dig coef %		72.	72.	
Cattle dig prot %		9.5	10.6	
Goats dig prot %	*	9.6	10.7	
Horses dig prot %	*	9.4	10.4	
Rabbits dig prot %	*	9.3	10.3	
Sheep dig prot %		9.8	10.9	
Energy GE Mcal/kg				
Cattle DE Mcal/kg	*	2.19	2.42	
Sheep DE Mcal/kg	*	2.19	2.43	
Cattle ME Mcal/kg	*	1.79	1.99	
Sheep ME Mcal/kg	*	1.80	1.99	
Cattle TDN %		49.6	55.0	
Sheep TDN %		49.8	55.1	
Calcium %		1.13	1.25	
Phosphorus %		0.20	0.22	
Potassium %		1.78	1.97	

Alfalfa, hay, s-c baled, prebloom, cut 1, (1)
Ref No 1 08 875 United States

Feed Name or Analyses		As Fed	Dry	C.V. ± %
Dry matter %		85.7	100.0	
Sheep dig coef %		61.	61.	
Ash %		5.4	6.3	
Sheep dig coef %		15.	18.	
Crude fiber %		28.7	33.6	
Sheep dig coef %		47.	47.	
Ether extract %		1.5	1.8	
Sheep dig coef %		31.	31.	
N-free extract %		33.8	39.5	
Sheep dig coef %		67.	67.	
Protein (N x 6.25) %		16.2	19.0	
Sheep dig coef %		70.	70.	
Cattle dig prot %	*	11.4	13.4	
Goats dig prot %	*	12.2	14.2	
Horses dig prot %	*	11.7	13.6	
Rabbits dig prot %	*	11.4	13.3	
Sheep dig prot %		11.4	13.3	
Cell walls (Van Soest) %		45.4	53.0	
Sheep dig coef %		53.	53.	
Cellulose (Matrone) %		28.5	33.3	
Sheep dig coef %		46.	46.	
Fiber, acid detergent (VS) %		33.9	39.6	
Sheep dig coef %		55.	55.	
Hemicellulose %		11.4	13.4	
Sheep dig coef %		48.	48.	
Pectins %		64.2	74.9	
Pentosans %		0.7	0.8	
Lignin (Ellis) %		5.5	6.5	
Sheep dig coef %		4.	4.	
Energy GE Mcal/kg		3.80	4.44	
Sheep GE dig coef %		55.	55.	
Cattle DE Mcal/kg	*	2.07	2.42	
Sheep DE Mcal/kg		2.09	2.44	
Cattle ME Mcal/kg	*	1.70	1.98	
Sheep ME Mcal/kg	*	1.71	2.00	
Cattle TDN %	*	46.9	54.8	
Sheep TDN %		48.6	56.7	
Dry matter intake g/w^0.75		83.980	98.050	
Gain, cattle kg/day		0.857	1.000	
Gain, sheep kg/day		0.257	0.300	

(1) dry forages and roughages (2) pasture, range plants, and forages fed green (3) silages (4) energy feeds (5) protein supplements (6) minerals (7) vitamins (8) additives

Alfalfa, hay, s–c baled, early bloom, cut 1, (1)
Ref No 1 08 876 United States

Feed Name or Analyses		As Fed	Dry	C.V. ± %
Dry matter %		85.0	100.0	
Sheep dig coef %		63.	63.	
Ash %		7.4	8.7	
Crude fiber %		23.4	27.5	
Ether extract %		2.6	3.0	
N-free extract %		31.9	37.5	
Protein (N x 6.25) %		19.8	23.3	
Cattle dig prot %	*	14.6	17.1	
Goats dig prot %	*	15.6	18.3	
Horses dig prot %	*	14.7	17.3	
Rabbits dig prot %	*	14.2	16.7	
Sheep dig prot %	*	14.8	17.5	
Cell walls (Van Soest) %		35.2	41.4	
Sheep dig coef %		46.	46.	
Fiber, acid detergent (VS) %		28.4	33.4	
Sheep dig coef %		48.	48.	
Hemicellulose %		6.8	8.0	
Lignin (Ellis) %		5.4	6.3	
Sheep dig coef %		1.	1.	
Energy GE Mcal/kg		3.79	4.46	
Sheep GE dig coef %		62.	62.	
Cattle DE Mcal/kg	*	2.30	2.71	
Sheep DE Mcal/kg		2.35	2.77	
Cattle ME Mcal/kg	*	1.89	2.22	
Sheep ME Mcal/kg	*	1.93	2.27	
Cattle TDN %	*	52.2	61.5	
Sheep TDN %	*	50.9	59.9	
Dry matter intake g/w^0.75		112.370	132.200	
Gain, sheep kg/day		0.170	0.200	

Alfalfa, hay, s-c baled, midbloom, cut 1, (1)
Ref No 1 08 877 United States

Feed Name or Analyses		As Fed	Dry	C.V. ± %
Dry matter %		87.1	100.0	
Sheep dig coef %		60.	60.	
Ash %		5.4	6.3	
Sheep dig coef %		21.	24.	
Crude fiber %		29.4	33.7	
Sheep dig coef %		49.	49.	
Ether extract %		1.6	1.8	
Sheep dig coef %		32.	32.	
N-free extract %		38.3	44.0	
Sheep dig coef %		70.	70.	
Protein (N x 6.25) %		12.5	14.3	
Sheep dig coef %		66.	66.	
Cattle dig prot %	*	8.1	9.3	
Goats dig prot %	*	8.6	9.9	
Horses dig prot %	*	8.4	9.7	
Rabbits dig prot %	*	8.5	9.7	
Sheep dig prot %		8.2	9.4	
Cell walls (Van Soest) %		46.1	52.9	
Sheep dig coef %		50.	50.	
Cellulose (Matrone) %		26.7	30.6	
Sheep dig coef %		51.	51.	
Fiber, acid detergent (VS) %		33.9	38.9	
Sheep dig coef %		54.	54.	
Hemicellulose %		12.2	14.1	
Sheep dig coef %		40.	40.	
Lignin (Ellis) %		6.2	7.2	
Sheep dig coef %		7.	7.	
Energy GE Mcal/kg		3.86	4.43	
Sheep GE dig coef %		58.	58.	
Cattle DE Mcal/kg	*	2.12	2.44	
Sheep DE Mcal/kg		2.24	2.57	
Cattle ME Mcal/kg	*	1.74	2.00	
Sheep ME Mcal/kg	*	1.83	2.11	
Cattle TDN %	*	48.1	55.2	
Sheep TDN %		50.5	58.0	
Dry matter intake g/w^0.75		77.200	88.633	
Gain, cattle kg/day		0.439	0.504	
Gain, sheep kg/day		0.153	0.176	

Alfalfa, hay, s-c baled, milk stage, cut 1, (1)
Ref No 1 08 874 United States

Feed Name or Analyses		As Fed	Dry	C.V. ± %
Dry matter %		86.0	100.0	
Sheep dig coef %		51.	51.	
Ash %		5.8	6.8	
Sheep dig coef %		33.	38.	
Crude fiber %		29.1	33.8	
Sheep dig coef %		41.	41.	
Ether extract %		1.9	2.2	
Sheep dig coef %		31.	31.	
N-free extract %		38.7	45.0	
Sheep dig coef %		55.	55.	
Protein (N x 6.25) %		10.5	12.2	
Sheep dig coef %		61.	61.	
Cattle dig prot %	*	6.5	7.5	
Goats dig prot %	*	6.8	7.9	
Horses dig prot %	*	6.8	7.9	
Rabbits dig prot %	*	7.0	8.1	
Sheep dig prot %		6.4	7.4	
Cell walls (Van Soest) %		49.7	57.8	
Cellulose (Matrone) %		31.6	36.7	
Fiber, acid detergent (VS) %		34.1	39.7	
Sheep dig coef %		37.	37.	
Hemicellulose %		15.7	18.2	
Lignin (Ellis) %		6.4	7.4	
Sheep dig coef %		-0.	-0.	
Energy GE Mcal/kg		3.98	4.62	
Sheep GE dig coef %		50.	50.	
Cattle DE Mcal/kg	*	2.10	2.44	
Sheep DE Mcal/kg		1.99	2.31	
Cattle ME Mcal/kg	*	1.72	2.00	
Sheep ME Mcal/kg	*	1.63	1.90	
Cattle TDN %	*	47.5	55.3	
Sheep TDN %		40.9	47.6	
Dry matter intake g/w^0.75		81.872	95.200	
Gain, sheep kg/day		0.058	0.068	

Alfalfa, hay, s-c baled fan-air dried w heat, early bloom, cut 2, (1)
Ref No 1 08 878 United States

Feed Name or Analyses		As Fed	Dry	C.V. ± %
Dry matter %		86.7	100.0	
Sheep dig coef %		55.	55.	
Ash %		5.8	6.7	
Crude fiber %		32.1	37.0	
Ether extract %		2.0	2.3	
N-free extract %		34.2	39.4	

Feed Name or Analyses			Mean As Fed	Mean Dry	C.V. ± %
Protein (N x 6.25)	%		12.7	14.6	
Cattle	dig prot %	*	8.3	9.6	
Goats	dig prot %	*	8.8	10.2	
Horses	dig prot %	*	8.6	9.9	
Rabbits	dig prot %	*	8.6	9.9	
Sheep	dig prot %	*	8.4	9.7	
Cell walls (Van Soest)	%		47.7	55.0	
Sheep	dig coef %		41.	41.	
Fiber, acid detergent (VS)	%		37.7	43.5	
Sheep	dig coef %		42.	42.	
Hemicellulose	%		10.1	11.6	
Lignin (Ellis)	%		8.2	9.5	
Sheep	dig coef %		5.	5.	
Energy	GE Mcal/kg		3.85	4.44	
Sheep	GE dig coef %		53.	53.	
Cattle	DE Mcal/kg	*	2.07	2.39	
Sheep	DE Mcal/kg		2.04	2.35	
Cattle	ME Mcal/kg	*	1.70	1.96	
Sheep	ME Mcal/kg	*	1.67	1.93	
Cattle	TDN %	*	46.9	54.1	
Sheep	TDN %	*	47.4	54.6	
Dry matter intake	$g/W^{.75}$		86.613	99.900	
Gain, sheep	kg/day		0.075	0.086	

Alfalfa, hay, s-c brown, (1)
Ref No 1 00 103 United States

Feed Name or Analyses			Mean As Fed	Mean Dry	C.V. ± %
Dry matter	%		88.7	100.0	2
Ash	%		9.2	10.4	11
Crude fiber	%		25.2	28.4	11
Sheep	dig coef %		68.	68.	
Ether extract	%		1.5	1.7	10
Sheep	dig coef %		-11.	-11.	
N-free extract	%		36.6	41.3	
Sheep	dig coef %		71.	71.	
Protein (N x 6.25)	%		16.1	18.2	10
Sheep	dig coef %		68.	68.	
Cattle	dig prot %	*	11.2	12.7	
Goats	dig prot %	*	12.0	13.5	
Horses	dig prot %	*	11.5	12.9	
Rabbits	dig prot %	*	11.3	12.7	
Sheep	dig prot %		11.0	12.3	
Energy	GE Mcal/kg				
Cattle	DE Mcal/kg	*	2.34	2.64	
Sheep	DE Mcal/kg	*	2.37	2.67	
Cattle	ME Mcal/kg	*	1.92	2.16	
Sheep	ME Mcal/kg	*	1.94	2.19	
Cattle	TDN %	*	53.1	59.9	
Sheep	TDN %		53.7	60.5	
Calcium	%		1.47	1.66	
Phosphorus	%		0.28	0.32	

Alfalfa, hay, s-c brown, gr 3 US, (1)
Ref No 1 00 102 United States

Feed Name or Analyses			Mean As Fed	Mean Dry	C.V. ± %
Dry matter	%		90.4	100.0	
Carotene	mg/kg		4.0	4.4	
Vitamin A equivalent	IU/g		6.6	7.4	

Alfalfa, hay, s-c chopped, (1)
Suncured chopped alfalfa (AAFCO)
Chopped alfalfa hay (AAFCO)
Ref No 1 00 104 United States

Feed Name or Analyses			Mean As Fed	Mean Dry	C.V. ± %
Dry matter	%		92.8	100.0	
Sheep	dig coef %		61.	61.	
Ash	%		9.4	10.1	13
Crude fiber	%		27.3	29.5	20
Cattle	dig coef %		59.	59.	
Sheep	dig coef %		41.	41.	
Ether extract	%		2.4	2.6	47
Cattle	dig coef %		43.	43.	
Sheep	dig coef %		21.	21.	
N-free extract	%		35.8	38.6	
Cattle	dig coef %		72.	72.	
Sheep	dig coef %		71.	71.	
Protein (N x 6.25)	%		17.9	19.3	24
Cattle	dig coef %		78.	78.	
Sheep	dig coef %		79.	79.	12
Cattle	dig prot %		13.9	15.0	
Goats	dig prot %	*	13.5	14.5	
Horses	dig prot %	*	12.9	13.9	
Rabbits	dig prot %	*	12.6	13.5	
Sheep	dig prot %		14.1	15.2	
Lignin (Ellis)	%		7.1	7.6	5
Energy	GE Mcal/kg		4.18	4.51	
Cattle	DE Mcal/kg	*	2.56	2.76	
Sheep	DE Mcal/kg		2.69	2.90	
Cattle	ME Mcal/kg	*	2.10	2.27	
Sheep	ME Mcal/kg		2.18	2.35	
Cattle	TDN %		58.1	62.7	
Sheep	TDN %		51.5	55.5	
Carotene	mg/kg		13.0	14.0	
Vitamin A equivalent	IU/g		21.6	23.3	

Alfalfa, hay, s-c chopped, early bloom, cut 2, (1)
Ref No 1 05 676 United States

Feed Name or Analyses			Mean As Fed	Mean Dry	C.V. ± %
Dry matter	%			100.0	
Sheep	dig coef %			57.	
Ash	%			8.4	
Crude fiber	%			33.8	
Sheep	dig coef %			40.	
Ether extract	%			2.0	
Sheep	dig coef %			11.	
N-free extract	%			34.5	
Sheep	dig coef %			66.	
Protein (N x 6.25)	%			21.3	
Sheep	dig coef %			76.	
Cattle	dig prot %	*		15.4	
Goats	dig prot %	*		16.4	
Horses	dig prot %	*		15.6	
Rabbits	dig prot %	*		15.1	
Sheep	dig prot %			16.2	
Lignin (Ellis)	%			7.6	
Energy	GE Mcal/kg			4.77	
Sheep	GE dig coef %			54.	
Cattle	DE Mcal/kg	*		2.50	
Sheep	DE Mcal/kg			2.58	
Cattle	ME Mcal/kg	*		2.05	
Sheep	ME Mcal/kg	*		2.11	
Cattle	TDN %	*		56.6	
Sheep	TDN %			53.0	

Alfalfa, hay, s-c grnd, (1)
Suncured alfalfa meal (AAFCO)
Ground alfalfa hay (AAFCO)
Ref No 1 00 111 United States

Feed Name or Analyses			Mean As Fed	Mean Dry	C.V. ± %
Dry matter	%		90.7	100.0	2
Ash	%		9.6	10.6	10
Crude fiber	%		27.3	30.1	8
Cattle	dig coef %	*	52.	52.	
Sheep	dig coef %	*	47.	47.	
Swine	dig coef %		22.	22.	
Ether extract	%		2.1	2.3	47
Cattle	dig coef %	*	61.	61.	
Sheep	dig coef %	*	36.	36.	
Swine	dig coef %		11.	11.	
N-free extract	%		34.1	37.6	
Cattle	dig coef %	*	69.	69.	
Sheep	dig coef %	*	72.	72.	
Swine	dig coef %		47.	47.	
Protein (N x 6.25)	%		17.6	19.5	11
Cattle	dig coef %	*	78.	78.	
Sheep	dig coef %	*	72.	72.	
Swine	dig coef %		46.	46.	
Cattle	dig prot %		13.8	15.2	
Goats	dig prot %	*	13.3	14.7	
Horses	dig prot %	*	12.7	14.0	
Rabbits	dig prot %	*	12.4	13.7	
Sheep	dig prot %		12.7	14.0	
Swine	dig prot %		8.1	8.9	
Cellulose (Matrone)	%		21.8	24.0	25
Pentosans	%		12.1	13.4	4
Sugars, total	%		3.6	4.0	16
Lignin (Ellis)	%		7.6	8.4	
Linoleic	%		.300	.330	
Energy	GE Mcal/kg		4.13	4.56	
Cattle	DE Mcal/kg	*	2.39	2.64	
Sheep	DE Mcal/kg	*	2.28	2.52	
Swine	DE kcal/kg	*	1351.	1490.	
Cattle	ME Mcal/kg	*	1.96	2.17	
Chickens	ME_n kcal/kg		929.	1024.	
Sheep	ME Mcal/kg	*	1.87	2.06	
Swine	ME kcal/kg	*	1244.	1372.	
Cattle	TDN %		54.3	59.9	
Sheep	TDN %		51.7	57.1	
Swine	TDN %		30.6	33.8	
Chlorine	%		0.68	0.76	8
Cobalt	mg/kg		0.209	0.230	86
Sodium	%		0.17	0.19	26
Sulphur	%		0.24	0.27	19
Zinc	mg/kg		31.8	35.1	49
Biotin	mg/kg		0.32	0.35	
Carotene	mg/kg		38.2	42.1	72
Choline	mg/kg		1021.	1127.	15
Folic acid	mg/kg		5.72	6.31	47
Niacin	mg/kg		34.4	37.9	18
Pantothenic acid	mg/kg		28.4	31.3	18
Riboflavin	mg/kg		10.2	11.2	22
Thiamine	mg/kg		2.6	2.9	32
α-tocopherol	mg/kg		386.4	426.1	
Vitamin A equivalent	IU/g		63.6	70.2	
Arginine	%		0.91	1.00	19
Cystine	%		0.36	0.40	8
Histidine	%		0.27	0.30	25
Isoleucine	%		0.82	0.90	17
Leucine	%		1.27	1.40	11
Lysine	%		1.00	1.10	9
Methionine	%		0.18	0.20	48
Phenylalanine	%		0.82	0.90	11
Threonine	%		0.73	0.80	13
Tryptophan	%		0.27	0.30	24
Tyrosine	%		0.54	0.60	9
Valine	%		0.82	0.90	14

Ref No 1 00 111 Canada

Feed Name or Analyses			Mean As Fed	Mean Dry	C.V. ± %
Dry matter	%		91.0	100.0	
Crude fiber	%		27.4	30.1	
Swine	dig coef %		22.	22.	
Ether extract	%		2.0	2.1	
Swine	dig coef %		11.	11.	

Continued

Feed Name or Analyses			Mean As Fed	Mean Dry	C.V. ± %
N-free extract	%				
Swine	dig coef %		47.	47.	
Protein (N x 6.25)	%		14.0	15.4	
Swine	dig coef %		46.	46.	
Cattle	dig prot %	*	9.3	10.3	
Goats	dig prot %	*	9.9	10.9	
Horses	dig prot %	*	9.6	10.6	
Rabbits	dig prot %	*	9.6	10.5	
Sheep	dig prot %	*	9.4	10.4	
Swine	dig prot %		6.4	7.1	
Energy	GE Mcal/kg		4.15	4.56	
Cattle	DE Mcal/kg	*	2.40	2.63	
Sheep	DE Mcal/kg	*	2.29	2.52	
Swine	DE kcal/kg	*	1357.	1491.	
Cattle	ME Mcal/kg	*	1.97	2.16	
Chickens	ME_n kcal/kg		1204.	1323.	
Sheep	ME Mcal/kg	*	1.88	2.06	
Swine	ME kcal/kg	*	1261.	1385.	
Cattle	TDN %		54.4	59.7	
Sheep	TDN %		51.9	57.1	
Swine	TDN %		30.8	33.8	
Calcium	%		0.87	0.96	
Phosphorus	%		0.17	0.19	
Salt (NaCl)	%		0.31	0.34	
Pantothenic acid	mg/kg		30.4	33.5	25
Riboflavin	mg/kg		20.2	22.2	22

Alfalfa, hay, s-c grnd, early leaf, cut 1, (1)
Ref No 1 00 106 United States

Analyses			As Fed	Dry	C.V. ± %
Dry matter	%		89.4	100.0	
Ash	%		11.3	12.6	
Crude fiber	%		17.7	19.8	
Sheep	dig coef %		65.	65.	
Ether extract	%		2.9	3.2	
Sheep	dig coef %		46.	46.	
N-free extract	%		35.8	40.0	
Sheep	dig coef %		65.	65.	
Protein (N x 6.25)	%		21.8	24.4	
Sheep	dig coef %		72.	72.	
Cattle	dig prot %	*	16.2	18.1	
Goats	dig prot %	*	17.3	19.3	
Horses	dig prot %	*	16.3	18.2	
Rabbits	dig prot %	*	15.7	17.5	
Sheep	dig prot %		15.7	17.6	
Energy	GE Mcal/kg				
Cattle	DE Mcal/kg	*	2.47	2.76	
Sheep	DE Mcal/kg	*	2.36	2.63	
Cattle	ME Mcal/kg	*	2.02	2.26	
Sheep	ME Mcal/kg	*	1.93	2.16	
Cattle	TDN %	*	56.1	62.7	
Sheep	TDN %		53.4	59.8	

Alfalfa, hay, s-c grnd, immature, (1)
Ref No 1 00 107 United States

Analyses			As Fed	Dry	C.V. ± %
Dry matter	%		91.8	100.0	1
Ash	%		11.8	12.8	9
Crude fiber	%		21.8	23.7	18
Cattle	dig coef %	*	52.	52.	
Sheep	dig coef %	*	47.	47.	
Ether extract	%		2.9	3.2	19
Cattle	dig coef %	*	61.	61.	
Sheep	dig coef %	*	36.	36.	
N-free extract	%		35.2	38.3	
Cattle	dig coef %	*	69.	69.	
Sheep	dig coef %	*	72.	72.	
Protein (N x 6.25)	%		20.2	22.0	13
Cattle	dig coef %	*	78.	78.	
Sheep	dig coef %	*	72.	72.	
Cattle	dig prot %		15.8	17.2	
Goats	dig prot %	*	15.7	17.1	
Horses	dig prot %	*	14.9	16.2	
Rabbits	dig prot %	*	14.4	15.7	
Sheep	dig prot %		14.5	15.8	
Energy	GE Mcal/kg				
Cattle	DE Mcal/kg	*	2.44	2.66	
Sheep	DE Mcal/kg	*	2.31	2.52	
Cattle	ME Mcal/kg	*	2.00	2.18	
Sheep	ME Mcal/kg	*	1.90	2.07	
Cattle	TDN %	*	55.4	60.3	
Sheep	TDN %	*	52.5	57.1	

Alfalfa, hay, s-c grnd, early bloom, (1)
Ref No 1 00 108 United States

Analyses			As Fed	Dry	C.V. ± %
Dry matter	%			100.0	
Crude fiber	%				
Cattle	dig coef %	*		52.	
Sheep	dig coef %	*		47.	
Ether extract	%				
Cattle	dig coef %	*		61.	
Sheep	dig coef %	*		36.	
N-free extract	%				
Cattle	dig coef %	*		69.	
Sheep	dig coef %	*		72.	
Protein (N x 6.25)	%				
Cattle	dig coef %	*		78.	
Sheep	dig coef %	*		72.	

Alfalfa, hay, s-c grnd, full bloom, (1)
Ref No 1 00 109 United States

Analyses			As Fed	Dry	C.V. ± %
Dry matter	%		87.0	100.0	
Ash	%		9.7	11.2	
Crude fiber	%		23.4	26.9	
Sheep	dig coef %		46.	46.	
Ether extract	%		2.3	2.6	
Sheep	dig coef %		28.	28.	
N-free extract	%		36.1	41.5	
Sheep	dig coef %		74.	74.	
Protein (N x 6.25)	%		15.5	17.8	
Sheep	dig coef %		71.	71.	
Cattle	dig prot %	*	10.7	12.4	
Goats	dig prot %	*	11.5	13.2	
Horses	dig prot %	*	11.0	12.6	
Rabbits	dig prot %	*	10.8	12.4	
Sheep	dig prot %		11.0	12.6	
Energy	GE Mcal/kg				
Cattle	DE Mcal/kg	*	2.40	2.75	
Sheep	DE Mcal/kg	*	2.20	2.53	
Cattle	ME Mcal/kg	*	1.96	2.26	
Sheep	ME Mcal/kg	*	1.80	2.07	
Cattle	TDN %	*	54.3	62.4	
Sheep	TDN %		49.9	57.4	

Alfalfa, hay, s-c grnd, cut 2, (1)
Ref No 1 00 110 United States

Analyses			As Fed	Dry	C.V. ± %
Dry matter	%		92.2	100.0	1
Calcium	%		2.07	2.24	
Phosphorus	%		0.27	0.29	
Carotene	mg/kg		126.2	136.9	
Pantothenic acid	mg/kg		64.2	69.7	
Vitamin A equivalent	IU/g		210.4	228.2	

Alfalfa, hay, s-c grnd, mn 13% protein, (1)
Ref No 1 00 112 United States

Analyses			As Fed	Dry	C.V. ± %
Dry matter	%		90.0	100.0	
Carotene	mg/kg		21.0	23.4	22
Riboflavin	mg/kg		9.7	10.8	
Vitamin A equivalent	IU/g		35.1	39.0	

Alfalfa, hay, s-c grnd, mn 15% protein, (1)
Ref No 1 00 113 United States

Analyses			As Fed	Dry	C.V. ± %
Dry matter	%		92.4	100.0	1
Ash	%		9.4	10.2	
Crude fiber	%		27.1	29.3	
Ether extract	%		1.9	2.1	
Energy	GE Mcal/kg		4.00	4.33	
Swine	DE kcal/kg		1497.	1620.	
Swine	ME kcal/kg		1358.	1470.	
Calcium	%		1.38	1.49	8
Phosphorus	%		0.25	0.27	25
Potassium	%		2.13	2.31	4
Carotene	mg/kg		102.9	111.3	12
Riboflavin	mg/kg		11.0	11.9	
Vitamin A equivalent	IU/g		171.5	185.6	

Alfalfa, hay, s-c grnd, mn 17% protein, (1)
Ref No 1 00 114 United States

Analyses			As Fed	Dry	C.V. ± %
Dry matter	%		92.5	100.0	2
Carotene	mg/kg		81.2	87.9	4
Choline	mg/kg		1139.	1232.	15
Niacin	mg/kg		33.2	35.9	28
Pantothenic acid	mg/kg		28.9	31.3	19
Riboflavin	mg/kg		12.1	13.1	22
Thiamine	mg/kg		3.1	3.3	32
Vitamin A equivalent	IU/g		135.4	146.5	

Alfalfa, hay, s-c grnd, mn 20% protein, (1)
Ref No 1 00 116 United States

Analyses			As Fed	Dry	C.V. ± %
Dry matter	%		92.6	100.0	1
Carotene	mg/kg		152.9	165.1	13
Choline	mg/kg		1121.	1210.	
Niacin	mg/kg		38.0	41.0	32
Pantothenic acid	mg/kg		40.6	43.9	11
Riboflavin	mg/kg		15.5	16.8	17
Thiamine	mg/kg		6.7	7.3	15
Vitamin A equivalent	IU/g		254.9	275.3	

(1) dry forages and roughages (2) pasture, range plants, and forages fed green (3) silages (4) energy feeds (5) protein supplements (6) minerals (7) vitamins (8) additives

Alfalfa, hay, s-c grnd, mn 22% protein, (1)

Ref No 1 00 117 United States

Feed Name or Analyses			Mean As Fed	Mean Dry	C.V. ± %
Dry matter	%			100.0	
Carotene	mg/kg			232.4	28
Vitamin A equivalent	IU/g			387.4	

Alfalfa, hay, s-c grnd, stemmy, (1)

Ref No 1 00 118 United States

Feed Name or Analyses			Mean As Fed	Mean Dry	C.V. ± %
Dry matter	%		92.7	100.0	2
Cattle	dig coef %		57.	57.	3
Ash	%		8.3	9.0	5
Crude fiber	%		33.4	36.0	2
Cattle	dig coef %		45.	45.	7
Sheep	dig coef %	*	47.	47.	
Ether extract	%		1.5	1.6	2
Cattle	dig coef %		17.	17.	35
Sheep	dig coef %	*	36.	36.	
N-free extract	%		35.8	38.7	0
Cattle	dig coef %		66.	66.	3
Sheep	dig coef %	*	72.	72.	
Protein (N x 6.25)	%		13.7	14.8	0
Cattle	dig coef %		71.	71.	3
Sheep	dig coef %	*	72.	72.	
Cattle	dig prot %		9.8	10.5	
Goats	dig prot %	*	9.6	10.4	
Horses	dig prot %	*	9.3	10.1	
Rabbits	dig prot %	*	9.4	10.1	
Sheep	dig prot %		9.9	10.6	
Energy	GE Mcal/kg				
Cattle	DE Mcal/kg	*	2.16	2.33	
Sheep	DE Mcal/kg	*	2.32	2.50	
Cattle	ME Mcal/kg	*	1.77	1.91	
Sheep	ME Mcal/kg	*	1.90	2.05	
Cattle	TDN %		48.9	52.7	3
Sheep	TDN %		52.6	56.7	

Alfalfa, hay, s-c grnd pelleted, (1)

Ref No 1 00 124 United States

Feed Name or Analyses			Mean As Fed	Mean Dry	C.V. ± %
Dry matter	%		92.2	100.0	1
Organic matter	%		82.9	89.9	
Ash	%		10.1	10.9	9
Crude fiber	%		23.6	25.6	14
Cattle	dig coef %	*	44.	44.	
Sheep	dig coef %		40.	40.	7
Ether extract	%		2.0	2.2	33
Cattle	dig coef %	*	35.	35.	
Sheep	dig coef %		23.	23.	57
N-free extract	%		39.5	42.9	6
Cattle	dig coef %	*	71.	71.	
Sheep	dig coef %		75.	75.	2
Protein (N x 6.25)	%		17.0	18.4	12
Cattle	dig coef %	*	70.	70.	
Sheep	dig coef %		74.	74.	3
Cattle	dig prot %		11.9	12.9	
Goats	dig prot %	*	12.6	13.7	
Horses	dig prot %	*	12.1	13.1	
Rabbits	dig prot %	*	11.9	12.9	
Sheep	dig prot %		12.6	13.7	
Cellulose (Matrone)	%		23.8	25.8	
Lignin (Ellis)	%		7.3	7.9	20
Energy	GE Mcal/kg				
Cattle	DE Mcal/kg	*	2.29	2.48	
Sheep	DE Mcal/kg	*	2.32	2.52	
Cattle	ME Mcal/kg	*	1.88	2.04	
Sheep	ME Mcal/kg	*	1.90	2.06	
Cattle	TDN %		51.9	56.3	
Sheep	TDN %		52.6	57.1	3
Calcium	%		1.45	1.57	9
Chlorine	%		0.24	0.26	
Cobalt	mg/kg		0.230	0.249	
Copper	mg/kg		31.3	34.0	
Iron	%		0.038	0.041	
Magnesium	%		0.33	0.36	
Manganese	mg/kg		48.0	52.0	
Phosphorus	%		0.22	0.24	30
Potassium	%		2.15	2.33	
Silicon	%		1.13	1.23	
Sodium	%		0.04	0.04	
Sulphur	%		0.25	0.27	25
Carotene	mg/kg		58.7	63.7	69
Vitamin A equivalent	IU/g		97.9	106.2	

Alfalfa, hay, s-c grnd pelleted, immature, (1)

Ref No 1 00 126 United States

Feed Name or Analyses			Mean As Fed	Mean Dry	C.V. ± %
Dry matter	%		93.1	100.0	1
Ash	%		11.0	11.8	5
Crude fiber	%		16.5	17.7	4
Ether extract	%		3.6	3.9	7
N-free extract	%		36.1	38.8	
Protein (N x 6.25)	%		25.9	27.9	6
Sheep	dig coef %		88.	88.	
Cattle	dig prot %	*	19.6	21.1	
Goats	dig prot %	*	21.0	22.5	
Horses	dig prot %	*	19.7	21.2	
Rabbits	dig prot %	*	18.8	20.2	
Sheep	dig prot %		22.8	24.5	
Lignin (Ellis)	%		4.9	5.3	
Energy	GE Mcal/kg				
Cattle	DE Mcal/kg	*	2.82	3.03	
Sheep	DE Mcal/kg	*	2.71	2.91	
Cattle	ME Mcal/kg	*	2.31	2.48	
Sheep	ME Mcal/kg	*	2.22	2.39	
Cattle	TDN %	*	63.9	68.7	
Sheep	TDN %		61.5	66.1	

Alfalfa, hay, s-c grnd pelleted, pre-bloom, (1)

Ref No 1 07 726 United States

Feed Name or Analyses			Mean As Fed	Mean Dry	C.V. ± %
Dry matter	%			100.0	
Crude fiber	%			21.3	
Protein (N x 6.25)	%			26.1	
Sheep	dig coef %			85.	
Cattle	dig prot %	*		19.5	
Goats	dig prot %	*		20.9	
Horses	dig prot %	*		19.6	
Rabbits	dig prot %	*		18.8	
Sheep	dig prot %			22.1	
Lignin (Ellis)	%			6.5	
Energy	GE Mcal/kg				
Sheep	DE Mcal/kg	*		2.66	
Sheep	ME Mcal/kg	*		2.18	
Sheep	TDN %			60.4	

Alfalfa, hay, s-c grnd pelleted, early bloom, (1)

Ref No 1 00 127 United States

Feed Name or Analyses			Mean As Fed	Mean Dry	C.V. ± %
Dry matter	%		90.8	100.0	1
Ash	%		10.1	11.1	4
Crude fiber	%		24.5	27.0	1
Ether extract	%		3.1	3.4	4
N-free extract	%		34.4	37.9	
Protein (N x 6.25)	%		18.8	20.7	2
Sheep	dig coef %		85.	85.	
Cattle	dig prot %	*	13.5	14.9	
Goats	dig prot %	*	14.4	15.9	
Horses	dig prot %	*	13.7	15.1	
Rabbits	dig prot %	*	13.3	14.7	
Sheep	dig prot %		16.0	17.6	
Lignin (Ellis)	%		7.4	8.2	
Energy	GE Mcal/kg				
Cattle	DE Mcal/kg	*	2.51	2.77	
Sheep	DE Mcal/kg	*	2.29	2.52	
Cattle	ME Mcal/kg	*	2.06	2.27	
Sheep	ME Mcal/kg	*	1.88	2.07	
Cattle	TDN %	*	57.0	62.7	
Sheep	TDN %		51.9	57.2	

Alfalfa, hay, s-c grnd pelleted, early bloom, cut 2, (1)

Ref No 1 05 674 United States

Feed Name or Analyses			Mean As Fed	Mean Dry	C.V. ± %
Dry matter	%			100.0	
Protein (N x 6.25)	%			25.0	
Cattle	dig prot %	*		18.6	
Goats	dig prot %	*		19.9	
Horses	dig prot %	*		18.8	
Rabbits	dig prot %	*		18.0	
Sheep	dig prot %	*		19.0	
Lignin (Ellis)	%			7.6	

Alfalfa, hay, s-c grnd pelleted, early bloom, cut 3, (1)

Ref No 1 07 849 United States

Feed Name or Analyses			Mean As Fed	Mean Dry	C.V. ± %
Dry matter	%			100.0	
Organic matter	%			90.2	
Ash	%			9.8	
Crude fiber	%			26.5	
Sheep	dig coef %			32.	
Ether extract	%			1.0	
Sheep	dig coef %			8.	
N-free extract	%			47.6	
Sheep	dig coef %			74.	
Protein (N x 6.25)	%			15.2	
Sheep	dig coef %			71.	
Cattle	dig prot %	*		10.1	
Goats	dig prot %	*		10.7	
Horses	dig prot %	*		10.4	
Rabbits	dig prot %	*		10.4	
Sheep	dig prot %			10.8	
Lignin (Ellis)	%			9.5	
Energy	GE Mcal/kg				
Cattle	DE Mcal/kg	*		2.74	
Sheep	DE Mcal/kg	*		2.41	
Cattle	ME Mcal/kg	*		2.25	
Sheep	ME Mcal/kg	*		1.98	
Cattle	TDN %	*		62.2	
Sheep	TDN %			54.7	
Calcium	%			1.25	
Phosphorus	%			0.20	
Silicon	%			1.03	

Alfalfa, hay, s-c grnd pelleted, mid-bloom, (1)

Ref No 1 07 735 United States

Feed Name or Analyses			Mean As Fed	Mean Dry	C.V. ± %
Dry matter	%		93.5	100.0	
Ash	%		7.8	8.3	
Crude fiber	%		35.8	38.3	

Continued

Feed Name or Analyses			Mean As Fed	Mean Dry	C.V. ± %
Ether extract	%		1.2	1.3	
N-free extract	%		37.3	39.9	
Protein (N x 6.25)	%		11.4	12.1	
Cattle	dig prot %	*	7.0	7.5	
Goats	dig prot %	*	7.4	7.9	
Horses	dig prot %	*	7.3	7.8	
Rabbits	dig prot %	*	7.5	8.0	
Sheep	dig prot %	*	7.0	7.5	
Energy	GE Mcal/kg				
Cattle	DE Mcal/kg	*	1.99	2.13	
Sheep	DE Mcal/kg	*	2.25	2.41	
Cattle	ME Mcal/kg	*	1.64	1.75	
Sheep	ME Mcal/kg	*	1.85	1.98	
Cattle	TDN %	*	45.2	48.4	
Sheep	TDN %	*	51.1	54.7	
Calcium	%		1.34	1.43	
Phosphorus	%		0.19	0.20	

Alfalfa, hay, s-c grnd pelleted, mid-bloom, cut 2, (1)
Ref No 1 05 673 United States

Feed Name or Analyses			Mean As Fed	Mean Dry	C.V. ± %
Dry matter	%			100.0	
Crude fiber	%			28.5	
Protein (N x 6.25)	%			22.7	
Sheep	dig coef %			85.	
Cattle	dig prot %	*		16.6	
Goats	dig prot %	*		17.7	
Horses	dig prot %	*		16.8	
Rabbits	dig prot %	*		16.2	
Sheep	dig prot %			19.3	
Lignin (Ellis)	%			8.0	
Energy	GE Mcal/kg				
Sheep	DE Mcal/kg	*		2.41	
Sheep	ME Mcal/kg	*		1.98	
Sheep	TDN %			54.7	

Alfalfa, hay, s-c grnd pelleted, cut 3, (1)
Ref No 1 00 120 United States

Feed Name or Analyses			Mean As Fed	Mean Dry	C.V. ± %
Dry matter	%			100.0	
Carotene	mg/kg			214.5	
Vitamin A equivalent	IU/g			357.6	

Alfalfa, hay, s-c grnd pelleted, cut 5, (1)
Ref No 1 07 856 United States

Feed Name or Analyses			Mean As Fed	Mean Dry	C.V. ± %
Dry matter	%			100.0	
Crude fiber	%			22.0	
Sheep	dig coef %			39.	
Ether extract	%			1.9	
Sheep	dig coef %			32.	
N-free extract	%			50.0	
Sheep	dig coef %			79.	
Protein (N x 6.25)	%			16.6	
Sheep	dig coef %			74.	
Cattle	dig prot %	*		11.3	
Goats	dig prot %	*		12.1	
Horses	dig prot %	*		11.7	
Rabbits	dig prot %	*		11.5	
Sheep	dig prot %			12.3	
Lignin (Ellis)	%			7.0	
Energy	GE Mcal/kg				

(1) dry forages and roughages (2) pasture, range plants, and forages fed green (3) silages (4) energy feeds (5) protein supplements (6) minerals (7) vitamins (8) additives

Feed Name or Analyses			Mean As Fed	Mean Dry	C.V. ± %
Cattle	DE Mcal/kg	*		2.88	
Sheep	DE Mcal/kg	*		2.73	
Cattle	ME Mcal/kg	*		2.36	
Sheep	ME Mcal/kg	*		2.24	
Cattle	TDN %	*		65.3	
Sheep	TDN %			61.8	
Sulphur	%			0.26	

Alfalfa, hay, s-c grnd pelleted, gr 1 US, (1)
Ref No 1 07 836 United States

Feed Name or Analyses			Mean As Fed	Mean Dry	C.V. ± %
Dry matter	%			100.0	
Organic matter	%			74.6	
Ash	%			25.5	
Crude fiber	%			13.6	
Sheep	dig coef %			44.	
Ether extract	%			1.1	
Sheep	dig coef %			36.	
N-free extract	%			45.5	
Sheep	dig coef %			80.	
Protein (N x 6.25)	%			14.5	
Sheep	dig coef %			73.	
Cattle	dig prot %	*		9.5	
Goats	dig prot %	*		10.0	
Horses	dig prot %	*		9.8	
Rabbits	dig prot %	*		9.8	
Sheep	dig prot %			10.5	
Energy	GE Mcal/kg				
Cattle	DE Mcal/kg	*		2.49	
Sheep	DE Mcal/kg	*		2.37	
Cattle	ME Mcal/kg	*		2.04	
Sheep	ME Mcal/kg	*		1.94	
Cattle	TDN %	*		56.4	
Sheep	TDN %			53.7	
Calcium	%			5.32	
Chlorine	%			0.24	
Magnesium	%			0.42	
Phosphorus	%			0.17	
Potassium	%			1.90	
Silicon	%			7.71	
Sodium	%			0.19	
Sulphur	%			0.24	

Alfalfa, hay, s-c grnd pelleted, gr 2 US, (1)
Ref No 1 07 745 United States

Feed Name or Analyses			Mean As Fed	Mean Dry	C.V. ± %
Dry matter	%		92.9	100.0	
Organic matter	%		82.8	89.1	
Ash	%		10.1	10.9	
Crude fiber	%		20.0	21.5	
Sheep	dig coef %		42.	42.	
Ether extract	%		1.4	1.5	
Sheep	dig coef %		32.	32.	
N-free extract	%		41.4	44.6	
Sheep	dig coef %		74.	74.	
Protein (N x 6.25)	%		20.0	21.5	
Sheep	dig coef %		75.	75.	
Cattle	dig prot %	*	14.5	15.6	
Goats	dig prot %	*	15.4	16.6	
Horses	dig prot %	*	14.7	15.8	
Rabbits	dig prot %	*	14.2	15.3	
Sheep	dig prot %		15.0	16.1	
Energy	GE Mcal/kg				
Cattle	DE Mcal/kg	*	2.70	2.91	
Sheep	DE Mcal/kg	*	2.43	2.61	
Cattle	ME Mcal/kg	*	2.22	2.39	
Sheep	ME Mcal/kg	*	1.99	2.14	
Cattle	TDN %	*	61.3	66.0	

Feed Name or Analyses			Mean As Fed	Mean Dry	C.V. ± %
Sheep	TDN %		55.0	59.2	
Calcium	%		1.46	1.57	
Chlorine	%		1.21	1.30	
Magnesium	%		0.25	0.27	
Phosphorus	%		0.21	0.23	
Potassium	%		2.88	3.10	
Silicon	%		0.61	0.66	
Sodium	%		0.36	0.39	
Sulphur	%		0.31	0.33	
Carotene	mg/kg		28.1	30.2	
Vitamin A equivalent	IU/g		46.8	50.3	

Alfalfa, hay, s-c grnd pelleted, mn 15% protein, (1)
Ref No 1 00 121 United States

Feed Name or Analyses			Mean As Fed	Mean Dry	C.V. ± %
Dry matter	%		90.4	100.0	1
Ash	%		9.6	10.6	
Crude fiber	%		28.2	31.2	2
Ether extract	%		1.8	2.0	2
N-free extract	%		35.3	39.0	
Protein (N x 6.25)	%		15.5	17.2	1
Cattle	dig prot %	*	10.7	11.8	
Goats	dig prot %	*	11.4	12.6	
Horses	dig prot %	*	11.0	12.1	
Rabbits	dig prot %	*	10.8	11.9	
Sheep	dig prot %	*	10.8	12.0	
Energy	GE Mcal/kg				
Cattle	DE Mcal/kg	*	2.34	2.59	
Sheep	DE Mcal/kg	*	2.23	2.46	
Cattle	ME Mcal/kg	*	1.92	2.12	
Sheep	ME Mcal/kg	*	1.82	2.02	
Cattle	TDN %	*	53.1	58.7	
Sheep	TDN %	*	50.5	55.8	

Ref No 1 00 121 Canada

Feed Name or Analyses			Mean As Fed	Mean Dry	C.V. ± %
Dry matter	%		90.0	100.0	
Crude fiber	%		28.9	32.1	
Ether extract	%		2.2	2.5	
Protein (N x 6.25)	%		15.7	17.5	
Cattle	dig prot %	*	10.9	12.1	
Goats	dig prot %	*	11.6	12.9	
Horses	dig prot %	*	11.1	12.4	
Rabbits	dig prot %	*	11.0	12.2	
Sheep	dig prot %	*	11.0	12.3	
Calcium	%		1.55	1.72	
Phosphorus	%		0.18	0.19	
Salt (NaCl)	%		0.73	0.81	

Alfalfa, hay, s-c grnd pelleted, mn 17% protein, (1)
Ref No 1 00 122 United States

Feed Name or Analyses			Mean As Fed	Mean Dry	C.V. ± %
Dry matter	%		91.1	100.0	1
Sheep	dig coef %		61.	61.	
Ash	%		9.2	10.2	2
Crude fiber	%		24.4	26.8	3
Ether extract	%		3.3	3.6	8
N-free extract	%		36.1	39.7	
Protein (N x 6.25)	%		18.0	19.8	1
Sheep	dig coef %		75.	75.	
Cattle	dig prot %	*	12.8	14.1	
Goats	dig prot %	*	13.7	15.0	
Horses	dig prot %	*	13.0	14.3	
Rabbits	dig prot %	*	12.7	13.9	
Sheep	dig prot %		13.5	14.8	
Lignin (Ellis)	%		7.0	7.7	
Energy	GE Mcal/kg		4.11	4.51	
Cattle	DE Mcal/kg	*	2.50	2.74	

Feed Name or Analyses			Mean As Fed	Mean Dry	C.V. ± %
Sheep	DE Mcal/kg		2.51	2.75	
Cattle	ME Mcal/kg	*	2.05	2.25	
Sheep	ME Mcal/kg		2.07	2.27	
Cattle	TDN %	*	56.6	62.2	
Sheep	TDN %	*	53.0	58.2	

Alfalfa, hay, s-c grnd pelleted, mn 20% protein, (1)

Ref No 1 00 123 United States

Feed Name or Analyses			Mean As Fed	Mean Dry	C.V. ± %
Dry matter	%		92.3	100.0	1
Ash	%		11.0	11.9	16
Crude fiber	%		20.2	21.9	13
Ether extract	%		2.9	3.1	13
N-free extract	%		37.4	40.5	
Protein (N x 6.25)	%		20.9	22.6	7
Cattle	dig prot %	*	15.2	16.5	
Goats	dig prot %	*	16.3	17.6	
Horses	dig prot %	*	15.4	16.7	
Rabbits	dig prot %	*	14.9	16.1	
Sheep	dig prot %	*	15.5	16.8	
Energy	GE Mcal/kg				
Cattle	DE Mcal/kg	*	2.69	2.91	
Sheep	DE Mcal/kg	*	2.47	2.68	
Cattle	ME Mcal/kg	*	2.20	2.39	
Sheep	ME Mcal/kg	*	2.03	2.20	
Cattle	TDN %	*	61.0	66.1	
Sheep	TDN %	*	56.1	60.8	

Alfalfa, hay, s-c on riders, (1)

Ref No 1 00 125 United States

Feed Name or Analyses			Mean As Fed	Mean Dry	C.V. ± %
Dry matter	%		86.4	100.0	
Ash	%		7.4	8.6	
Crude fiber	%		29.1	33.7	
Cattle	dig coef %		42.	42.	
Sheep	dig coef %		44.	44.	
Ether extract	%		1.9	2.3	
Cattle	dig coef %		33.	33.	
Sheep	dig coef %		54.	54.	
N-free extract	%		32.1	37.1	
Cattle	dig coef %		70.	70.	
Sheep	dig coef %		64.	64.	
Protein (N x 6.25)	%		15.9	18.4	
Cattle	dig coef %		78.	78.	
Sheep	dig coef %		79.	79.	
Cattle	dig prot %		12.4	14.4	
Goats	dig prot %	*	11.9	13.7	
Horses	dig prot %	*	11.4	13.2	
Rabbits	dig prot %	*	11.1	12.9	
Sheep	dig prot %		12.6	14.5	
Energy	GE Mcal/kg				
Cattle	DE Mcal/kg	*	2.14	2.47	
Sheep	DE Mcal/kg	*	2.13	2.46	
Cattle	ME Mcal/kg	*	1.75	2.03	
Sheep	ME Mcal/kg	*	1.74	2.02	
Cattle	TDN %		48.5	56.1	
Sheep	TDN %		48.2	55.8	

Alfalfa, hay, s-c rained on, (1)

Ref No 1 00 130 United States

Feed Name or Analyses			Mean As Fed	Mean Dry	C.V. ± %
Dry matter	%		86.7	100.0	
Ash	%		6.4	7.4	
Crude fiber	%		32.6	37.6	
Ether extract	%		0.8	0.9	
N-free extract	%		31.1	35.9	
Protein (N x 6.25)	%		15.8	18.2	
Cattle	dig prot %	*	11.0	12.7	
Goats	dig prot %	*	11.8	13.6	
Horses	dig prot %	*	11.3	13.0	
Rabbits	dig prot %	*	11.0	12.7	
Sheep	dig prot %	*	11.2	12.9	
Energy	GE Mcal/kg				
Cattle	DE Mcal/kg	*	2.03	2.34	
Sheep	DE Mcal/kg	*	2.12	2.44	
Cattle	ME Mcal/kg	*	1.66	1.92	
Sheep	ME Mcal/kg	*	1.73	2.00	
Cattle	TDN %		46.0	53.1	
Sheep	TDN %	*	48.0	55.3	
Vitamin D_2	IU/g		0.8	0.9	55

Alfalfa, hay, s-c rained on, mature, cut 2, (1)

Ref No 1 00 129 United States

Feed Name or Analyses			Mean As Fed	Mean Dry	C.V. ± %
Dry matter	%		80.0	100.0	
Carotene	mg/kg		2.3	2.9	
Vitamin A equivalent	IU/g		3.8	4.8	

Alfalfa, hay, s-c rained on baled, early bloom, cut 1, (1)

Ref No 1 08 851 United States

Feed Name or Analyses			Mean As Fed	Mean Dry	C.V. ± %
Dry matter	%		90.6	100.0	
Sheep	dig coef %		58.	58.	
Ash	%		6.6	7.3	
Sheep	dig coef %		25.	28.	
Crude fiber	%		32.6	36.0	
Sheep	dig coef %		53.	53.	
Ether extract	%		1.5	1.7	
Sheep	dig coef %		41.	41.	
N-free extract	%		35.2	38.9	
Sheep	dig coef %		65.	65.	
Protein (N x 6.25)	%		14.6	16.1	
Sheep	dig coef %		67.	67.	
Cattle	dig prot %	*	9.9	10.9	
Goats	dig prot %	*	10.5	11.6	
Horses	dig prot %	*	10.1	11.2	
Rabbits	dig prot %	*	10.1	11.1	
Sheep	dig prot %		9.8	10.8	
Energy	GE Mcal/kg				
Cattle	DE Mcal/kg	*	2.18	2.41	
Sheep	DE Mcal/kg	*	2.27	2.50	
Cattle	ME Mcal/kg	*	1.79	1.97	
Sheep	ME Mcal/kg	*	1.86	2.05	
Cattle	TDN %	*	49.4	54.6	
Sheep	TDN %		51.4	56.7	
Dry matter intake	$g/W^{.75}$		69.309	76.500	
Gain, sheep	kg/day		0.093	0.103	

Alfalfa, hay, s-c sulfur fertilized, (1)

Ref No 1 00 131 United States

Feed Name or Analyses			Mean As Fed	Mean Dry	C.V. ± %
Dry matter	%		88.2	100.0	1
Sulphur	%		0.34	0.39	11

Alfalfa, hay, s-c wafered, (1)

Ref No 1 07 746 United States

Feed Name or Analyses			Mean As Fed	Mean Dry	C.V. ± %
Dry matter	%		91.0	100.0	2
Ash	%		8.8	9.7	6
Crude fiber	%		26.4	29.0	9
Ether extract	%		1.3	1.5	18
N-free extract	%		36.2	39.8	
Protein (N x 6.25)	%		18.2	20.0	
Cattle	dig prot %	*	13.0	14.3	
Goats	dig prot %	*	13.8	15.2	
Horses	dig prot %	*	13.2	14.5	
Rabbits	dig prot %	*	12.8	14.1	
Sheep	dig prot %	*	13.2	14.5	
Lignin (Ellis)	%		6.1	6.8	
Energy	GE Mcal/kg				
Cattle	DE Mcal/kg	*	2.42	2.66	
Sheep	DE Mcal/kg	*	2.36	2.60	
Cattle	ME Mcal/kg	*	1.99	2.19	
Sheep	ME Mcal/kg	*	1.94	2.13	
Cattle	TDN %	*	55.0	60.4	
Sheep	TDN %	*	53.6	58.9	
Silicon	%		0.32	0.35	
Carotene	mg/kg		21.8	24.0	

Alfalfa, leaves, dehy grnd, (1)

Dehydrated alfalfa leaf meal (AAFCO)

Ref No 1 00 137 United States

Feed Name or Analyses			Mean As Fed	Mean Dry	C.V. ± %
Dry matter	%		92.3	100.0	2
Ash	%		11.2	12.2	15
Crude fiber	%		18.5	20.1	20
Ether extract	%		3.1	3.4	13
N-free extract	%		38.6	41.8	
Protein (N x 6.25)	%		20.8	22.6	12
Cattle	dig prot %	*	15.2	16.5	
Goats	dig prot %	*	16.3	17.5	
Horses	dig prot %	*	15.4	16.7	
Rabbits	dig prot %	*	14.9	16.1	
Sheep	dig prot %	*	15.5	16.8	
Linoleic	%		.520	.563	
Energy	GE Mcal/kg				
Cattle	DE Mcal/kg	*	2.63	2.85	
Sheep	DE Mcal/kg	*	2.51	2.71	
Cattle	ME Mcal/kg	*	2.16	2.34	
Chickens	ME_n kcal/kg		1580.	1712.	
Sheep	ME Mcal/kg	*	2.05	2.23	
Cattle	TDN %	*	59.6	64.6	
Sheep	TDN %	*	56.8	61.6	
Chlorine	%		0.31	0.34	
Cobalt	mg/kg		0.200	0.216	8
Sodium	%		0.06	0.07	
Vitamin D_2	IU/g		0.3	0.4	59

Ref No 1 00 137 Canada

Feed Name or Analyses			Mean As Fed	Mean Dry	C.V. ± %
Dry matter	%			100.0	
Crude fiber	%			18.3	
Protein (N x 6.25)	%			19.6	
Cattle	dig prot %	*		13.9	
Goats	dig prot %	*		14.8	
Horses	dig prot %	*		14.2	
Rabbits	dig prot %	*		13.8	
Sheep	dig prot %	*		14.2	

Alfalfa, leaves, fan air dried, (1)

Ref No 1 00 135 United States

Feed Name or Analyses			Mean As Fed	Mean Dry	C.V. ± %
Dry matter	%		87.3	100.0	3
Ash	%		8.4	9.6	11
Crude fiber	%		14.9	17.1	18
Ether extract	%		2.4	2.7	25
N-free extract	%		39.5	45.3	

Continued

Feed Name or Analyses			Mean As Fed	Mean Dry	C.V. ± %
Protein (N x 6.25)	%		22.1	25.3	9
Cattle	dig prot %	*	16.5	18.8	
Goats	dig prot %	*	17.6	20.2	
Horses	dig prot %	*	16.6	19.0	
Rabbits	dig prot %	*	15.9	18.2	
Sheep	dig prot %	*	16.8	19.3	
Energy	GE Mcal/kg		3.75	4.29	2
Cattle	DE Mcal/kg	*	2.54	2.91	
Sheep	DE Mcal/kg	*	2.57	2.94	
Cattle	ME Mcal/kg	*	2.09	2.39	
Sheep	ME Mcal/kg	*	2.11	2.41	
Cattle	TDN %	*	57.7	66.1	
Sheep	TDN %	*	58.2	66.7	
Carotene	mg/kg		68.3	78.3	21
Riboflavin	mg/kg		23.7	27.1	20
Vitamin A equivalent	IU/g		113.9	130.5	

Alfalfa, leaves, fan air dried, cut 1, (1)
Ref No 1 00 132 United States

Feed Name or Analyses			Mean As Fed	Mean Dry	C.V. ± %
Dry matter	%		86.6	100.0	
Ash	%		8.4	9.7	
Crude fiber	%		14.1	16.3	
Ether extract	%		2.3	2.7	
N-free extract	%		40.5	46.8	
Protein (N x 6.25)	%		21.2	24.5	
Cattle	dig prot %	*	15.7	18.2	
Goats	dig prot %	*	16.8	19.4	
Horses	dig prot %	*	15.9	18.3	
Rabbits	dig prot %	*	15.2	17.6	
Sheep	dig prot %	*	16.1	18.5	
Energy	GE Mcal/kg				
Cattle	DE Mcal/kg	*	2.55	2.94	
Sheep	DE Mcal/kg	*	2.56	2.96	
Cattle	ME Mcal/kg	*	2.09	2.41	
Sheep	ME Mcal/kg	*	2.10	2.42	
Cattle	TDN %	*	57.8	66.7	
Sheep	TDN %	*	58.1	67.0	

Alfalfa, leaves, fan air dried, cut 2, (1)
Ref No 1 00 133 United States

Feed Name or Analyses			Mean As Fed	Mean Dry	C.V. ± %
Dry matter	%		87.5	100.0	
Ash	%		8.4	9.6	
Crude fiber	%		14.9	17.0	
Ether extract	%		2.2	2.5	
N-free extract	%		40.0	45.7	
Protein (N x 6.25)	%		22.1	25.2	
Cattle	dig prot %	*	16.4	18.8	
Goats	dig prot %	*	17.6	20.1	
Horses	dig prot %	*	16.6	18.9	
Rabbits	dig prot %	*	15.9	18.1	
Sheep	dig prot %	*	16.8	19.2	
Energy	GE Mcal/kg				
Cattle	DE Mcal/kg	*	2.54	2.91	
Sheep	DE Mcal/kg	*	2.58	2.95	
Cattle	ME Mcal/kg	*	2.09	2.38	
Sheep	ME Mcal/kg	*	2.12	2.42	
Cattle	TDN %	*	57.7	65.9	
Sheep	TDN %	*	58.6	66.9	

(1) dry forages and roughages (2) pasture, range plants, and forages fed green (3) silages (4) energy feeds (5) protein supplements (6) minerals (7) vitamins (8) additives

Alfalfa, leaves, fan air dried, cut 3, (1)
Ref No 1 00 134 United States

Feed Name or Analyses			Mean As Fed	Mean Dry	C.V. ± %
Dry matter	%		85.0	100.0	
Ash	%		8.0	9.4	
Crude fiber	%		15.0	17.6	
Ether extract	%		2.6	3.1	
N-free extract	%		38.8	45.6	
Protein (N x 6.25)	%		20.7	24.3	
Cattle	dig prot %	*	15.3	18.0	
Goats	dig prot %	*	16.3	19.2	
Horses	dig prot %	*	15.4	18.2	
Rabbits	dig prot %	*	14.8	17.4	
Sheep	dig prot %	*	15.6	18.4	
Energy	GE Mcal/kg				
Cattle	DE Mcal/kg	*	2.48	2.91	
Sheep	DE Mcal/kg	*	2.48	2.91	
Cattle	ME Mcal/kg	*	2.03	2.39	
Sheep	ME Mcal/kg	*	2.03	2.39	
Cattle	TDN %	*	56.2	66.1	
Sheep	TDN %	*	56.1	66.1	

Alfalfa, leaves, s-c, (1)
Ref No 1 00 146 United States

Feed Name or Analyses			Mean As Fed	Mean Dry	C.V. ± %
Dry matter	%		89.4	100.0	
Ash	%		8.9	10.0	10
Crude fiber	%		13.6	15.2	10
Cattle	dig coef %	*	44.	44.	
Sheep	dig coef %	*	50.	50.	
Ether extract	%		2.4	2.7	16
Cattle	dig coef %	*	35.	35.	
Sheep	dig coef %	*	0.	0.	
N-free extract	%		42.0	47.0	
Cattle	dig coef %	*	71.	71.	
Sheep	dig coef %	*	78.	78.	
Protein (N x 6.25)	%		22.5	25.1	9
Cattle	dig coef %	*	70.	70.	
Sheep	dig coef %	*	76.	76.	
Cattle	dig prot %		15.7	17.6	
Goats	dig prot %	*	17.9	20.0	
Horses	dig prot %	*	16.9	18.9	
Rabbits	dig prot %	*	16.1	18.1	
Sheep	dig prot %		17.1	19.1	
Energy	GE Mcal/kg		3.81	4.26	2
Cattle	DE Mcal/kg	*	2.36	2.64	
Sheep	DE Mcal/kg	*	2.50	2.79	
Cattle	ME Mcal/kg	*	1.93	2.16	
Sheep	ME Mcal/kg	*	2.05	2.29	
Cattle	TDN %		53.4	59.8	
Sheep	TDN %		56.6	63.3	
Calcium	%		2.22	2.48	
Iron	%		0.034	0.038	
Magnesium	%		0.40	0.45	
Manganese	mg/kg		72.1	80.6	
Phosphorus	%		0.24	0.27	
Potassium	%		2.06	2.30	
Carotene	mg/kg		59.7	66.8	16
Riboflavin	mg/kg		21.3	23.8	11
Vitamin A equivalent	IU/g		99.6	111.4	

Alfalfa, leaves, s-c, mid-bloom, cut 2, (1)
Ref No 1 00 142 United States

Feed Name or Analyses			Mean As Fed	Mean Dry	C.V. ± %
Dry matter	%		87.5	100.0	
Ash	%		9.5	10.9	
Crude fiber	%		15.1	17.3	
Sheep	dig coef %		53.	53.	
Ether extract	%		3.0	3.4	
Sheep	dig coef %		36.	36.	
N-free extract	%		40.4	46.2	
Sheep	dig coef %		78.	78.	
Protein (N x 6.25)	%		19.4	22.2	
Sheep	dig coef %		80.	80.	
Cattle	dig prot %	*	14.1	16.2	
Goats	dig prot %	*	15.1	17.3	
Horses	dig prot %	*	14.3	16.4	
Rabbits	dig prot %	*	13.8	15.8	
Sheep	dig prot %		15.5	17.8	
Energy	GE Mcal/kg				
Cattle	DE Mcal/kg	*	2.63	3.01	
Sheep	DE Mcal/kg	*	2.54	2.90	
Cattle	ME Mcal/kg	*	2.16	2.47	
Sheep	ME Mcal/kg	*	2.08	2.38	
Cattle	TDN %	*	59.7	68.2	
Sheep	TDN %		57.5	65.7	

Alfalfa, leaves, s-c, mid-bloom, cut 3, (1)
Ref No 1 00 143 United States

Feed Name or Analyses			Mean As Fed	Mean Dry	C.V. ± %
Dry matter	%		85.0	100.0	
Ash	%		8.7	10.2	
Crude fiber	%		15.0	17.6	
Sheep	dig coef %		58.	58.	
Ether extract	%		3.2	3.8	
Sheep	dig coef %		24.	24.	
N-free extract	%		40.0	47.0	
Sheep	dig coef %		78.	78.	
Protein (N x 6.25)	%		18.2	21.4	
Sheep	dig coef %		78.	78.	
Cattle	dig prot %	*	13.2	15.5	
Goats	dig prot %	*	14.0	16.5	
Horses	dig prot %	*	13.3	15.7	
Rabbits	dig prot %	*	12.9	15.2	
Sheep	dig prot %		14.2	16.7	
Energy	GE Mcal/kg				
Cattle	DE Mcal/kg	*	2.54	2.99	
Sheep	DE Mcal/kg	*	2.46	2.89	
Cattle	ME Mcal/kg	*	2.09	2.45	
Sheep	ME Mcal/kg	*	2.02	2.37	
Cattle	TDN %	*	57.7	67.9	
Sheep	TDN %		55.8	65.6	

Alfalfa, leaves, s-c, over ripe, stemmy, (1)
Ref No 1 00 144 United States

Feed Name or Analyses			Mean As Fed	Mean Dry	C.V. ± %
Dry matter	%		88.0	100.0	
Ash	%		10.0	11.4	
Crude fiber	%		13.4	15.2	
Horses	dig coef %		52.	52.	
Ether extract	%		1.8	2.0	
Horses	dig coef %		-60.	-60.	
N-free extract	%		41.2	46.8	
Horses	dig coef %		72.	72.	
Protein (N x 6.25)	%		21.6	24.6	
Horses	dig coef %		78.	78.	
Cattle	dig prot %	*	16.1	18.2	
Goats	dig prot %	*	17.2	19.5	
Horses	dig prot %		16.9	19.2	
Rabbits	dig prot %	*	15.5	17.7	
Sheep	dig prot %	*	16.4	18.6	
Energy	GE Mcal/kg				
Cattle	DE Mcal/kg	*	2.65	3.01	
Horses	DE Mcal/kg	*	2.25	2.56	
Sheep	DE Mcal/kg	*	2.60	2.95	

Feed Name or Analyses			Mean As Fed	Mean Dry	C.V. ± %
Cattle	ME Mcal/kg	*	2.17	2.47	
Horses	ME Mcal/kg	*	1.85	2.10	
Sheep	ME Mcal/kg	*	2.13	2.42	
Cattle	TDN %	*	60.2	68.4	
Horses	TDN %		51.1	58.0	
Sheep	TDN %	*	58.9	66.9	

Alfalfa, leaves, s-c, cut 1, (1)
Ref No 1 00 145 United States

Feed Name or Analyses			Mean As Fed	Mean Dry	C.V. ± %
Dry matter	%			100.0	
Ash	%			9.4	
Crude fiber	%			14.0	
Ether extract	%			2.6	
N-free extract	%			47.7	
Protein (N x 6.25)	%			26.3	
Cattle	dig prot %	*		19.7	
Goats	dig prot %	*		21.1	
Horses	dig prot %	*		19.9	
Rabbits	dig prot %	*		19.0	
Sheep	dig prot %	*		20.2	
Energy	GE Mcal/kg				
Cattle	DE Mcal/kg	*		2.98	
Sheep	DE Mcal/kg	*		3.05	
Cattle	ME Mcal/kg	*		2.44	
Sheep	ME Mcal/kg	*		2.50	
Cattle	TDN %	*		67.5	
Sheep	TDN %	*		69.2	

Alfalfa, leaves, s-c grnd, (1)
Ref No 1 00 147 United States

Feed Name or Analyses			Mean As Fed	Mean Dry	C.V. ± %
Dry matter	%		91.4	100.0	3
Ash	%		11.2	12.2	
Crude fiber	%		16.1	17.6	
Sheep	dig coef %		50.	50.	
Ether extract	%		1.9	2.1	
Sheep	dig coef %		0.	0.	
N-free extract	%		40.8	44.6	
Sheep	dig coef %		78.	78.	
Protein (N x 6.25)	%		21.5	23.5	
Sheep	dig coef %		76.	76.	
Cattle	dig prot %	*	15.8	17.3	
Goats	dig prot %	*	16.9	18.5	
Horses	dig prot %	*	16.0	17.5	
Rabbits	dig prot %	*	15.4	16.8	
Sheep	dig prot %		16.3	17.9	
Energy	GE Mcal/kg				
Cattle	DE Mcal/kg	*	2.60	2.84	
Sheep	DE Mcal/kg	*	2.48	2.71	
Cattle	ME Mcal/kg	*	2.13	2.33	
Sheep	ME Mcal/kg	*	2.03	2.22	
Cattle	TDN %	*	59.0	64.5	
Sheep	TDN %		56.2	61.4	
Calcium	%		1.67	1.83	18
Chlorine	%		0.43	0.47	33
Cobalt	mg/kg		0.198	0.216	8
Copper	mg/kg		10.5	11.5	7
Iron	%		0.036	0.039	10
Magnesium	%		0.35	0.38	28
Manganese	mg/kg		40.9	44.8	19
Phosphorus	%		0.24	0.26	25
Potassium	%		2.06	2.25	14
Sodium	%		0.09	0.10	40
Biotin	mg/kg		0.28	0.31	13
Carotene	mg/kg		98.8	108.0	39
Choline	mg/kg		895.	979.	
Folic acid	mg/kg		5.93	6.48	41
Niacin	mg/kg		53.6	58.6	14
Pantothenic acid	mg/kg		28.0	30.6	13
Riboflavin	mg/kg		15.9	17.4	22
Thiamine	mg/kg		4.4	4.9	26
Vitamin A equivalent	IU/g		164.7	180.1	
Vitamin D_2	IU/g		0.3	0.4	59
Arginine	%		1.19	1.30	18
Cystine	%		0.37	0.40	20
Histidine	%		0.37	0.40	22
Isoleucine	%		0.91	1.00	
Leucine	%		1.37	1.50	
Lysine	%		1.01	1.10	13
Methionine	%		0.37	0.40	19
Phenylalanine	%		0.91	1.00	
Threonine	%		0.73	0.80	11
Tryptophan	%		0.46	0.50	15
Valine	%		1.01	1.10	

Alfalfa, leaves, s-c grnd, good quality, (1)
Ref No 1 00 138 United States

Feed Name or Analyses			Mean As Fed	Mean Dry	C.V. ± %
Dry matter	%		92.3	100.0	
Ash	%		12.1	13.1	
Crude fiber	%		16.6	18.0	
Ether extract	%		2.8	3.0	
N-free extract	%		39.7	43.0	
Protein (N x 6.25)	%		21.1	22.9	
Cattle	dig prot %	*	15.4	16.7	
Goats	dig prot %	*	16.5	17.9	
Horses	dig prot %	*	15.6	16.9	
Rabbits	dig prot %	*	15.1	16.3	
Sheep	dig prot %	*	15.8	17.1	
Energy	GE Mcal/kg				
Cattle	DE Mcal/kg	*	2.56	2.77	
Sheep	DE Mcal/kg	*	2.54	2.76	
Cattle	ME Mcal/kg	*	2.10	2.27	
Sheep	ME Mcal/kg	*	2.09	2.26	
Cattle	TDN %	*	58.1	62.9	
Sheep	TDN %	*	57.7	62.5	
Calcium	%		1.69	1.83	
Chlorine	%		0.49	0.53	6
Phosphorus	%		0.25	0.27	
Sodium	%		0.10	0.11	25

Alfalfa, leaves, s-c grnd, gr 2 US, (1)
Ref No 1 00 140 United States

Feed Name or Analyses			Mean As Fed	Mean Dry	C.V. ± %
Dry matter	%			100.0	
Choline	mg/kg			1310.	

Alfalfa, leaves, s-c grnd, mn 18% fiber, (1)
Ref No 1 08 322 United States

Feed Name or Analyses			Mean As Fed	Mean Dry	C.V. ± %
Dry matter	%		91.9	100.0	
Ash	%		10.6	11.5	
Crude fiber	%		20.8	22.6	
Ether extract	%		2.9	3.2	
N-free extract	%		37.8	41.1	
Protein (N x 6.25)	%		19.8	21.5	
Cattle	dig prot %	*	14.3	15.6	
Goats	dig prot %	*	15.3	16.7	
Horses	dig prot %	*	14.5	15.8	
Rabbits	dig prot %	*	14.1	15.3	
Sheep	dig prot %	*	14.6	15.9	
Energy	GE Mcal/kg				
Cattle	DE Mcal/kg	*	2.37	2.58	
Sheep	DE Mcal/kg	*	2.45	2.66	
Cattle	ME Mcal/kg	*	1.95	2.12	
Sheep	ME Mcal/kg	*	2.01	2.18	
Cattle	TDN %	*	53.8	58.5	
Sheep	TDN %	*	55.5	60.4	

Alfalfa, leaves, s-c grnd, mn 20% protein mx 18% fiber, (1)
Suncured alfalfa leaf meal (AAFCO)
Ref No 1 00 246 United States

Feed Name or Analyses			Mean As Fed	Mean Dry	C.V. ± %
Dry matter	%			100.0	
Crude fiber	%				
Cattle	dig coef %	*		39.	
Sheep	dig coef %	*		50.	
Ether extract	%				
Cattle	dig coef %	*		23.	
N-free extract	%				
Cattle	dig coef %	*		72.	
Sheep	dig coef %	*		78.	
Protein (N x 6.25)	%				
Cattle	dig coef %	*		74.	
Sheep	dig coef %	*		76.	

Alfalfa, leaves, immature, (1)
Ref No 1 00 148 United States

Feed Name or Analyses			Mean As Fed	Mean Dry	C.V. ± %
Dry matter	%			100.0	
Calcium	%			2.71	
Iron	%			0.035	
Magnesium	%			0.43	
Manganese	mg/kg			11.0	
Phosphorus	%			0.31	
Potassium	%			1.20	

Alfalfa, leaves, mid-bloom, (1)
Ref No 1 00 149 United States

Feed Name or Analyses			Mean As Fed	Mean Dry	C.V. ± %
Dry matter	%		86.4	100.0	1
Ash	%		9.1	10.5	2
Crude fiber	%		15.8	18.3	4
Ether extract	%		2.9	3.3	8
N-free extract	%		40.3	46.6	
Protein (N x 6.25)	%		18.4	21.3	2
Cattle	dig prot %	*	13.3	15.4	
Goats	dig prot %	*	14.2	16.4	
Horses	dig prot %	*	13.5	15.6	
Rabbits	dig prot %	*	13.1	15.1	
Sheep	dig prot %	*	13.5	15.7	
Energy	GE Mcal/kg				
Cattle	DE Mcal/kg	*	2.56	2.97	
Sheep	DE Mcal/kg	*	2.44	2.82	
Cattle	ME Mcal/kg	*	2.10	2.43	
Sheep	ME Mcal/kg	*	2.00	2.32	
Cattle	TDN %	*	58.1	67.2	
Sheep	TDN %	*	55.3	64.1	

Alfalfa, leaves, full bloom, (1)
Ref No 1 00 150 United States

Feed Name or Analyses			Mean As Fed	Mean Dry	C.V. ± %
Dry matter	%			100.0	
Calcium	%			1.85	
Iron	%			0.031	
Magnesium	%			0.36	
Manganese	mg/kg			10.1	

Continued

Feed Name or Analyses			Mean As Fed	Mean Dry	C.V. ± %
Phosphorus	%			0.31	
Potassium	%			1.51	

Alfalfa, leaves, cut 1, (1)
Ref No 1 00 151 United States

Feed Name or Analyses			Mean As Fed	Mean Dry	C.V. ± %
Dry matter	%		86.6	100.0	1
Ash	%		8.6	9.9	7
Crude fiber	%		15.2	17.6	13
Ether extract	%		2.3	2.7	6
N-free extract	%		40.6	46.9	
Protein (N x 6.25)	%		19.8	22.9	10
Cattle	dig prot %	*	14.5	16.8	
Goats	dig prot %	*	15.5	17.9	
Horses	dig prot %	*	14.7	17.0	
Rabbits	dig prot %	*	14.2	16.3	
Sheep	dig prot %	*	14.8	17.1	
Energy	GE Mcal/kg		3.66	4.23	2
Cattle	DE Mcal/kg	*	2.53	2.92	
Sheep	DE Mcal/kg	*	2.51	2.90	
Cattle	ME Mcal/kg	*	2.08	2.40	
Sheep	ME Mcal/kg	*	2.06	2.38	
Cattle	TDN %	*	57.4	66.3	
Sheep	TDN %	*	57.0	65.8	
Calcium	%		2.94	3.40	
Phosphorus	%		0.19	0.22	
Carotene	mg/kg		70.4	81.3	20
Riboflavin	mg/kg		22.3	25.8	9
Vitamin A equivalent	IU/g		117.4	135.6	

Alfalfa, leaves, cut 2, (1)
Ref No 1 00 152 United States

Feed Name or Analyses			Mean As Fed	Mean Dry	C.V. ± %
Dry matter	%		87.5	100.0	
Energy	GE Mcal/kg		3.78	4.32	1
Calcium	%		2.29	2.62	
Phosphorus	%		0.28	0.32	
Carotene	mg/kg		56.5	64.6	6
Riboflavin	mg/kg		24.7	28.2	21
Vitamin A equivalent	IU/g		94.2	107.7	

Alfalfa, leaves, cut 3, (1)
Ref No 1 00 153 United States

Feed Name or Analyses			Mean As Fed	Mean Dry	C.V. ± %
Dry matter	%		85.0	100.0	1
Ash	%		8.2	9.7	11
Crude fiber	%		14.9	17.5	4
Ether extract	%		2.9	3.4	23
N-free extract	%		39.9	46.9	
Protein (N x 6.25)	%		19.1	22.5	7
Cattle	dig prot %	*	14.0	16.4	
Goats	dig prot %	*	14.9	17.6	
Horses	dig prot %	*	14.1	16.6	
Rabbits	dig prot %	*	13.6	16.0	
Sheep	dig prot %	*	14.2	16.8	
Energy	GE Mcal/kg				
Cattle	DE Mcal/kg	*	2.51	2.95	
Sheep	DE Mcal/kg	*	2.45	2.88	
Cattle	ME Mcal/kg	*	2.06	2.42	
Sheep	ME Mcal/kg	*	2.01	2.36	
Cattle	TDN %	*	56.9	66.9	
Sheep	TDN %	*	55.6	65.4	

(1) dry forages and roughages	(3) silages	(6) minerals
(2) pasture, range plants, and forages fed green	(4) energy feeds	(7) vitamins
	(5) protein supplements	(8) additives

Feed Name or Analyses			Mean As Fed	Mean Dry	C.V. ± %
Calcium	%		3.05	3.59	
Phosphorus	%		0.19	0.22	
Carotene	mg/kg		52.7	61.9	13
Riboflavin	mg/kg		17.6	20.7	2
Vitamin A equivalent	IU/g		87.8	103.3	

Alfalfa, stems, (1)
Ref No 1 00 173 United States

Feed Name or Analyses			Mean As Fed	Mean Dry	C.V. ± %
Dry matter	%		89.9	100.0	2
Sulphur	%		0.17	0.19	
Vitamin D_2	IU/g		1.6	1.8	5

Alfalfa, stems, dehy grnd, (1)
Dehydrated alfalfa stem meal (AAFCO)
Ref No 1 00 157 United States

Feed Name or Analyses			Mean As Fed	Mean Dry	C.V. ± %
Dry matter	%		90.9	100.0	1
Ash	%		7.4	8.2	12
Crude fiber	%		33.9	37.2	8
Ether extract	%		1.7	1.8	10
N-free extract	%		35.5	39.0	
Protein (N x 6.25)	%		12.5	13.7	10
Cattle	dig prot %	*	8.0	8.8	
Goats	dig prot %	*	8.5	9.4	
Horses	dig prot %	*	8.4	9.2	
Rabbits	dig prot %	*	8.4	9.3	
Sheep	dig prot %	*	8.1	8.9	
Energy	GE Mcal/kg				
Cattle	DE Mcal/kg	*	2.12	2.34	
Sheep	DE Mcal/kg	*	2.14	2.36	
Cattle	ME Mcal/kg	*	1.74	1.92	
Sheep	ME Mcal/kg	*	1.76	1.94	
Cattle	TDN %	*	48.2	53.0	
Sheep	TDN %	*	48.6	53.5	
Calcium	%		1.00	1.10	
Phosphorus	%		0.02	0.02	
Vitamin A	IU/g		11.0	12.1	
Xanthophylls	mg/kg		11.01	12.11	

Alfalfa, stems, fan air dried, (1)
Ref No 1 00 156 United States

Feed Name or Analyses			Mean As Fed	Mean Dry	C.V. ± %
Dry matter	%			100.0	
Ash	%			4.6	8
Crude fiber	%			44.6	4
Ether extract	%			1.1	8
N-free extract	%			38.4	
Protein (N x 6.25)	%			11.3	4
Cattle	dig prot %	*		6.7	
Goats	dig prot %	*		7.1	
Horses	dig prot %	*		7.1	
Rabbits	dig prot %	*		7.4	
Sheep	dig prot %	*		6.7	
Carotene	mg/kg			12.1	40
Riboflavin	mg/kg			7.3	26
Vitamin A equivalent	IU/g			20.2	

Alfalfa, stems, s-c, (1)
Ref No 1 00 164 United States

Feed Name or Analyses			Mean As Fed	Mean Dry	C.V. ± %
Dry matter	%		89.1	100.0	
Ash	%		3.7	4.2	9
Crude fiber	%		39.4	44.2	2
Cattle	dig coef %	*	44.	44.	
Sheep	dig coef %	*	37.	37.	

Feed Name or Analyses			Mean As Fed	Mean Dry	C.V. ± %
Ether extract	%		1.1	1.2	13
Cattle	dig coef %	*	35.	35.	
Sheep	dig coef %	*	58.	58.	
N-free extract	%		35.2	39.5	
Cattle	dig coef %	*	71.	71.	
Sheep	dig coef %	*	56.	56.	
Protein (N x 6.25)	%		9.7	10.9	5
Cattle	dig coef %	*	70.	70.	
Sheep	dig coef %	*	51.	51.	
Cattle	dig prot %		6.8	7.6	
Goats	dig prot %	*	6.0	6.7	
Horses	dig prot %	*	6.1	6.8	
Rabbits	dig prot %	*	6.3	7.1	
Sheep	dig prot %		5.0	5.6	
Energy	GE Mcal/kg		3.74	4.20	0
Cattle	DE Mcal/kg	*	2.20	2.47	
Sheep	DE Mcal/kg	*	1.79	2.01	
Cattle	ME Mcal/kg	*	1.81	2.03	
Sheep	ME Mcal/kg	*	1.47	1.65	
Cattle	TDN %		50.0	56.1	
Sheep	TDN %		40.6	45.6	
Calcium	%		0.82	0.92	
Iron	%		0.015	0.017	
Magnesium	%		0.26	0.29	
Manganese	mg/kg		13.2	14.8	
Phosphorus	%		0.17	0.19	
Potassium	%		2.21	2.48	
Carotene	mg/kg		6.1	6.8	52
Riboflavin	mg/kg		5.1	5.7	9
Vitamin A equivalent	IU/g		10.2	11.4	

Alfalfa, stems, s-c, mid-bloom, cut 1, (1)
Ref No 1 00 161 United States

Feed Name or Analyses			Mean As Fed	Mean Dry	C.V. ± %
Dry matter	%		88.2	100.0	
Ash	%		4.8	5.4	
Crude fiber	%		41.4	46.9	
Sheep	dig coef %		37.	37.	
Ether extract	%		1.1	1.2	
Sheep	dig coef %		58.	58.	
N-free extract	%		32.5	36.9	
Sheep	dig coef %		56.	56.	
Protein (N x 6.25)	%		8.5	9.6	
Sheep	dig coef %		51.	51.	
Cattle	dig prot %	*	4.6	5.3	
Goats	dig prot %	*	4.9	5.5	
Horses	dig prot %	*	5.0	5.7	
Rabbits	dig prot %	*	5.4	6.1	
Sheep	dig prot %		4.3	4.9	
Energy	GE Mcal/kg				
Sheep	DE Mcal/kg	*	1.73	1.96	
Sheep	ME Mcal/kg	*	1.42	1.61	
Sheep	TDN %		39.2	44.5	

Alfalfa, stems, s-c, mid-bloom, cut 2, (1)
Ref No 1 00 162 United States

Feed Name or Analyses			Mean As Fed	Mean Dry	C.V. ± %
Dry matter	%		87.3	100.0	
Ash	%		5.6	6.4	
Crude fiber	%		41.6	47.6	
Sheep	dig coef %		41.	41.	
Ether extract	%		1.6	1.8	
Sheep	dig coef %		56.	56.	
N-free extract	%		30.6	35.0	
Sheep	dig coef %		60.	60.	
Protein (N x 6.25)	%		8.0	9.2	
Sheep	dig coef %		51.	51.	
Cattle	dig prot %	*	4.3	4.9	

Feed Name or Analyses			Mean As Fed	Mean Dry	C.V. ± %
Goats	dig prot %	*	4.5	5.1	
Horses	dig prot %	*	4.7	5.3	
Rabbits	dig prot %	*	5.0	5.8	
Sheep	dig prot %		4.1	4.7	
Energy	GE Mcal/kg				
Sheep	DE Mcal/kg	*	1.83	2.09	
Sheep	ME Mcal/kg	*	1.50	1.72	
Sheep	TDN %		41.4	47.5	

Alfalfa, stems, s-c, mid-bloom, cut 3, (1)

Ref No 1 00 163 United States

Feed Name or Analyses			Mean As Fed	Mean Dry	C.V. ± %
Dry matter	%		86.3	100.0	
Ash	%		5.4	6.2	
Crude fiber	%		38.1	44.2	
Sheep	dig coef %		39.	39.	
Ether extract	%		1.2	1.4	
Sheep	dig coef %		29.	29.	
N-free extract	%		34.3	39.7	
Sheep	dig coef %		61.	61.	
Protein (N x 6.25)	%		7.3	8.5	
Sheep	dig coef %		52.	52.	
Cattle	dig prot %	*	3.7	4.3	
Goats	dig prot %	*	3.9	4.5	
Horses	dig prot %	*	4.1	4.7	
Rabbits	dig prot %	*	4.5	5.2	
Sheep	dig prot %		3.8	4.4	
Energy	GE Mcal/kg				
Sheep	DE Mcal/kg	*	1.78	2.06	
Sheep	ME Mcal/kg	*	1.46	1.69	
Sheep	TDN %		40.4	46.8	

Alfalfa, stems, s-c grnd, (1)

Suncured alfalfa stem meal (AAFCO)

Ref No 1 00 165 United States

Feed Name or Analyses			Mean As Fed	Mean Dry	C.V. ± %
Dry matter	%		91.0	100.0	2
Ash	%		7.6	8.4	17
Crude fiber	%		34.6	38.0	16
Ether extract	%		1.5	1.6	27
N-free extract	%		35.3	38.8	
Protein (N x 6.25)	%		12.0	13.2	18
Cattle	dig prot %	*	7.6	8.4	
Goats	dig prot %	*	8.1	8.9	
Horses	dig prot %	*	7.9	8.7	
Rabbits	dig prot %	*	8.1	8.9	
Sheep	dig prot %	*	7.6	8.4	
Energy	GE Mcal/kg				
Cattle	DE Mcal/kg	*	2.11	2.32	
Sheep	DE Mcal/kg	*	2.12	2.33	
Cattle	ME Mcal/kg	*	1.73	1.90	
Sheep	ME Mcal/kg	*	1.74	1.91	
Cattle	TDN %	*	47.8	52.6	
Sheep	TDN %	*	48.2	52.9	
Carotene	mg/kg		13.0	14.3	98
Riboflavin	mg/kg		10.6	11.7	25
Vitamin A equivalent	IU/g		21.7	23.9	

Ref No 1 00 165 Canada

Feed Name or Analyses			Mean As Fed	Mean Dry	C.V. ± %
Dry matter	%		90.9	100.0	
Crude fiber	%		40.5	44.6	
Ether extract	%		1.3	1.4	
Protein (N x 6.25)	%		9.8	10.8	
Cattle	dig prot %	*	5.7	6.3	
Goats	dig prot %	*	6.0	6.6	
Horses	dig prot %	*	6.1	6.7	
Rabbits	dig prot %	*	6.4	7.0	

Feed Name or Analyses			Mean As Fed	Mean Dry	C.V. ± %
Sheep	dig prot %	*	5.7	6.2	
Carotene	mg/kg		49.8	54.8	
Vitamin A equivalent	IU/g		83.1	91.4	

Alfalfa, stems, s-c grnd, immature, (1)

Ref No 1 00 158 United States

Feed Name or Analyses			Mean As Fed	Mean Dry	C.V. ± %
Dry matter	%		93.0	100.0	1
Ash	%		9.4	10.1	5
Crude fiber	%		31.2	33.5	10
Ether extract	%		1.5	1.6	24
N-free extract	%		34.1	36.7	
Protein (N x 6.25)	%		16.8	18.1	2
Cattle	dig prot %	*	11.7	12.6	
Goats	dig prot %	*	12.5	13.4	
Horses	dig prot %	*	12.0	12.9	
Rabbits	dig prot %	*	11.8	12.6	
Sheep	dig prot %	*	11.9	12.8	
Energy	GE Mcal/kg				
Cattle	DE Mcal/kg	*	2.31	2.49	
Sheep	DE Mcal/kg	*	2.26	2.43	
Cattle	ME Mcal/kg	*	1.90	2.04	
Sheep	ME Mcal/kg	*	1.86	2.00	
Cattle	TDN %	*	52.4	56.4	
Sheep	TDN %	*	51.3	55.2	

Alfalfa, stems, s-c grnd, mature, (1)

Ref No 1 00 159 United States

Feed Name or Analyses			Mean As Fed	Mean Dry	C.V. ± %
Dry matter	%		94.4	100.0	
Ash	%		6.3	6.7	
Crude fiber	%		44.7	47.4	
Ether extract	%		1.4	1.5	
N-free extract	%		34.2	36.2	
Protein (N x 6.25)	%		7.7	8.2	
Cattle	dig prot %	*	3.8	4.0	
Goats	dig prot %	*	4.0	4.2	
Horses	dig prot %	*	4.2	4.5	
Rabbits	dig prot %	*	4.7	5.0	
Sheep	dig prot %	*	3.7	3.9	

Alfalfa, stems, immature, (1)

Ref No 1 00 166 United States

Feed Name or Analyses			Mean As Fed	Mean Dry	C.V. ± %
Dry matter	%			100.0	
Calcium	%			1.09	
Iron	%			0.014	
Magnesium	%			0.30	
Manganese	mg/kg			30.9	
Phosphorus	%			0.32	
Potassium	%			1.24	

Alfalfa, stems, mid-bloom, (1)

Ref No 1 00 167 United States

Feed Name or Analyses			Mean As Fed	Mean Dry	C.V. ± %
Dry matter	%		87.3	100.0	1
Ash	%		5.2	6.0	5
Crude fiber	%		40.3	46.2	2
Ether extract	%		1.3	1.5	11
N-free extract	%		32.5	37.2	
Protein (N x 6.25)	%		7.9	9.1	3
Cattle	dig prot %	*	4.2	4.8	
Goats	dig prot %	*	4.4	5.1	
Horses	dig prot %	*	4.6	5.3	

Feed Name or Analyses			Mean As Fed	Mean Dry	C.V. ± %
Rabbits	dig prot %	*	5.0	5.7	
Sheep	dig prot %	*	4.1	4.7	

Alfalfa, stems, full bloom, (1)

Ref No 1 00 168 United States

Feed Name or Analyses			Mean As Fed	Mean Dry	C.V. ± %
Dry matter	%			100.0	
Calcium	%			0.82	
Iron	%			0.016	
Magnesium	%			0.21	
Manganese	mg/kg			30.9	
Phosphorus	%			0.33	
Potassium	%			1.49	

Alfalfa, stems, over ripe, (1)

Ref No 1 00 169 United States

Feed Name or Analyses			Mean As Fed	Mean Dry	C.V. ± %
Dry matter	%		86.6	100.0	
Ash	%		5.9	6.8	
Crude fiber	%		32.9	38.0	
Horses	dig coef %		40.	40.	
Ether extract	%		1.3	1.5	
Horses	dig coef %		38.	38.	
N-free extract	%		33.8	39.0	
Horses	dig coef %		56.	56.	
Protein (N x 6.25)	%		12.7	14.7	
Horses	dig coef %		73.	73.	
Cattle	dig prot %	*	8.4	9.7	
Goats	dig prot %	*	8.9	10.3	
Horses	dig prot %		9.3	10.7	
Rabbits	dig prot %	*	8.7	10.0	
Sheep	dig prot %	*	8.4	9.8	
Energy	GE Mcal/kg				
Cattle	DE Mcal/kg	*	1.94	2.24	
Horses	DE Mcal/kg	*	1.87	2.16	
Sheep	DE Mcal/kg	*	2.08	2.41	
Cattle	ME Mcal/kg	*	1.59	1.84	
Horses	ME Mcal/kg	*	1.54	1.77	
Sheep	ME Mcal/kg	*	1.71	1.97	
Cattle	TDN %	*	44.0	50.8	
Horses	TDN %		42.5	49.1	
Sheep	TDN %	*	47.3	54.6	

Alfalfa, stems, cut 1, (1)

Ref No 1 00 170 United States

Feed Name or Analyses			Mean As Fed	Mean Dry	C.V. ± %
Dry matter	%		88.2	100.0	1
Ash	%		4.1	4.6	10
Crude fiber	%		40.6	46.0	3
Ether extract	%		1.1	1.2	10
N-free extract	%		33.4	37.9	
Protein (N x 6.25)	%		9.1	10.3	6
Cattle	dig prot %	*	5.2	5.9	
Goats	dig prot %	*	5.4	6.2	
Horses	dig prot %	*	5.5	6.3	
Rabbits	dig prot %	*	5.8	6.6	
Sheep	dig prot %	*	5.1	5.8	
Energy	GE Mcal/kg		3.69	4.18	1
Calcium	%		0.76	0.86	
Phosphorus	%		0.11	0.12	
Carotene	mg/kg		4.1	4.6	70
Riboflavin	mg/kg		4.5	5.1	8
Vitamin A equivalent	IU/g		6.8	7.7	

Alfalfa, stems, cut 2, (1)

Ref No 1 00 171 — United States

Feed Name or Analyses				Mean As Fed	Mean Dry	C.V. ± %
Dry matter		%		87.3	100.0	1
Ash		%		5.0	5.7	9
Crude fiber		%		40.9	46.9	3
Ether extract		%		1.3	1.5	16
N-free extract		%		31.3	35.9	
Protein (N x 6.25)		%		8.7	10.0	8
Cattle	dig prot	%	*	4.9	5.6	
Goats	dig prot	%	*	5.1	5.9	
Horses	dig prot	%	*	5.3	6.0	
Rabbits	dig prot	%	*	5.6	6.4	
Sheep	dig prot	%	*	4.8	5.5	
Energy	GE	Mcal/kg		3.69	4.22	0
Calcium		%		0.78	0.89	
Phosphorus		%		0.19	0.22	
Carotene		mg/kg		11.7	13.4	7
Riboflavin		mg/kg		6.4	7.3	8
Vitamin A equivalent		IU/g		19.6	22.4	

Alfalfa, stems, cut 3, (1)

Ref No 1 00 172 — United States

Feed Name or Analyses				Mean As Fed	Mean Dry	C.V. ± %
Dry matter		%		86.9	100.0	3
Ash		%		5.1	5.9	14
Crude fiber		%		39.5	45.5	8
Ether extract		%		1.1	1.3	7
N-free extract		%		32.8	37.7	
Protein (N x 6.25)		%		8.3	9.6	10
Cattle	dig prot	%	*	4.6	5.3	
Goats	dig prot	%	*	4.8	5.5	
Horses	dig prot	%	*	4.9	5.7	
Rabbits	dig prot	%	*	5.3	6.1	
Sheep	dig prot	%	*	4.5	5.2	
Energy	GE	Mcal/kg		3.66	4.21	0
Calcium		%		1.04	1.20	
Phosphorus		%		0.10	0.12	
Carotene		mg/kg		14.4	16.5	24
Riboflavin		mg/kg		5.0	5.7	9
Vitamin A equivalent		IU/g		24.0	27.6	

Alfalfa, stems w molasses added, s-c grnd, (1)

Ref No 1 00 174 — United States

Feed Name or Analyses				Mean As Fed	Mean Dry	C.V. ± %
Dry matter		%			100.0	
Ash		%			8.7	
Crude fiber		%			33.2	
Ether extract		%			1.8	
N-free extract		%			45.5	
Protein (N x 6.25)		%			10.8	
Cattle	dig prot	%	*		6.3	
Goats	dig prot	%	*		6.6	
Horses	dig prot	%	*		6.7	
Rabbits	dig prot	%	*		7.0	
Sheep	dig prot	%	*		6.3	
Energy	GE	Mcal/kg				
Cattle	DE	Mcal/kg	*		2.56	
Sheep	DE	Mcal/kg	*		2.49	
Cattle	ME	Mcal/kg	*		2.10	
Sheep	ME	Mcal/kg	*		2.04	
Cattle	TDN	%	*		58.0	
Sheep	TDN	%	*		56.5	

(1) dry forages and roughages (2) pasture, range plants, and forages fed green (3) silages (4) energy feeds (5) protein supplements (6) minerals (7) vitamins (8) additives

Alfalfa, straw, (1)

Ref No 1 08 323 — United States

Feed Name or Analyses				Mean As Fed	Mean Dry	C.V. ± %
Dry matter		%		92.7	100.0	
Ash		%		6.8	7.3	
Crude fiber		%		40.6	43.8	
Ether extract		%		1.5	1.6	
N-free extract		%		34.6	37.3	
Protein (N x 6.25)		%		9.2	9.9	
Cattle	dig prot	%	*	5.1	5.5	
Goats	dig prot	%	*	5.4	5.8	
Horses	dig prot	%	*	5.5	6.0	
Rabbits	dig prot	%	*	5.9	6.3	
Sheep	dig prot	%	*	5.1	5.5	
Phosphorus		%		0.13	0.14	

Alfalfa, aerial part, fresh, (2)

Ref No 2 00 196 — United States

Feed Name or Analyses				Mean As Fed	Mean Dry	C.V. ± %
Dry matter		%		25.9	100.0	7
Organic matter		%		24.1	92.9	1
Ash		%		2.4	9.1	
Crude fiber		%		5.5	21.1	29
Cattle	dig coef	%	*	44.	44.	
Horses	dig coef	%		35.	35.	
Sheep	dig coef	%		59.	59.	15
Ether extract		%		1.1	4.1	25
Cattle	dig coef	%	*	47.	47.	
Horses	dig coef	%		23.	23.	
Sheep	dig coef	%		42.	42.	39
N-free extract		%		11.3	43.8	10
Cattle	dig coef	%	*	79.	79.	
Horses	dig coef	%		73.	73.	
Sheep	dig coef	%		79.	79.	8
Protein (N x 6.25)		%		5.7	21.9	22
Cattle	dig coef	%	*	78.	78.	
Horses	dig coef	%		79.	79.	
Sheep	dig coef	%		77.	77.	9
Cattle	dig prot	%		4.4	17.0	
Goats	dig prot	%	*	4.4	17.0	
Horses	dig prot	%		4.5	17.3	
Rabbits	dig prot	%	*	4.0	15.6	
Sheep	dig prot	%		4.4	16.8	
Lignin (Ellis)		%		1.4	5.2	20
Energy	GE	Mcal/kg				
Cattle	DE	Mcal/kg	*	0.74	2.87	
Horses	DE	Mcal/kg	*	0.67	2.59	
Sheep	DE	Mcal/kg	*	0.77	2.99	
Cattle	ME	Mcal/kg	*	0.61	2.35	
Horses	ME	Mcal/kg	*	0.55	2.13	
Sheep	ME	Mcal/kg	*	0.64	2.45	
Cattle	NE_m	Mcal/kg	*	0.34	1.32	
Cattle	NE_{gain}	Mcal/kg	*	0.18	0.71	
Cattle	$NE_{lactating\ cows}$	Mcal/kg	*	0.39	1.49	
Cattle	TDN	%		16.8	65.1	
Horses	TDN	%		15.2	58.8	
Sheep	TDN	%		17.6	67.9	
Calcium		%		0.44	1.68	24
Chlorine		%		0.12	0.45	
Copper		mg/kg		2.5	9.5	14
Iron		%		0.008	0.031	27
Magnesium		%		0.08	0.30	27
Manganese		mg/kg		13.0	50.1	60
Phosphorus		%		0.07	0.28	31
Potassium		%		0.54	2.10	24
Sodium		%		0.04	0.16	
Sulphur		%		0.10	0.37	
Biotin		mg/kg		0.13	0.49	13

Feed Name or Analyses				Mean As Fed	Mean Dry	C.V. ± %
Folic acid		mg/kg		0.64	2.47	29
a-tocopherol		mg/kg		157.6	608.5	
Vitamin D_2		IU/g		0.0	0.2	

Alfalfa, aerial part, fresh, early leaf, (2)

Ref No 2 00 175 — United States

Feed Name or Analyses				Mean As Fed	Mean Dry	C.V. ± %
Dry matter		%		14.6	100.0	
Ash		%		1.8	12.1	
Crude fiber		%		3.2	22.1	
Sheep	dig coef	%		64.	64.	
Ether extract		%		0.4	2.4	
Sheep	dig coef	%		38.	38.	
N-free extract		%		5.6	38.1	
Sheep	dig coef	%		80.	80.	
Protein (N x 6.25)		%		3.7	25.3	
Sheep	dig coef	%		84.	84.	
Cattle	dig prot	%	*	2.8	19.4	
Goats	dig prot	%	*	2.9	20.2	
Horses	dig prot	%	*	2.8	19.0	
Rabbits	dig prot	%	*	2.7	18.2	
Sheep	dig prot	%		3.1	21.3	
Energy	GE	Mcal/kg				
Cattle	DE	Mcal/kg	*	0.42	2.88	
Sheep	DE	Mcal/kg	*	0.44	3.00	
Cattle	ME	Mcal/kg	*	0.34	2.36	
Sheep	ME	Mcal/kg	*	0.36	2.46	
Cattle	TDN	%	*	9.5	65.3	
Sheep	TDN	%		9.9	67.9	

Alfalfa, aerial part, fresh, immature, (2)

Ref No 2 00 177 — United States

Feed Name or Analyses				Mean As Fed	Mean Dry	C.V. ± %
Dry matter		%		23.2	100.0	53
Ash		%		2.3	10.0	
Crude fiber		%		4.0	17.2	
Cattle	dig coef	%	*	44.	44.	
Sheep	dig coef	%	*	44.	44.	
Ether extract		%		1.0	4.4	
Cattle	dig coef	%	*	46.	46.	
Sheep	dig coef	%	*	35.	35.	
N-free extract		%		9.8	42.2	
Cattle	dig coef	%	*	74.	74.	
Sheep	dig coef	%	*	72.	72.	
Protein (N x 6.25)		%		6.1	26.1	
Cattle	dig coef	%	*	78.	78.	
Sheep	dig coef	%	*	75.	75.	
Cattle	dig prot	%		4.7	20.4	
Goats	dig prot	%	*	4.9	20.9	
Horses	dig prot	%	*	4.6	19.7	
Rabbits	dig prot	%	*	4.4	18.8	
Sheep	dig prot	%		4.5	19.6	
Energy	GE	Mcal/kg				
Cattle	DE	Mcal/kg	*	0.65	2.81	
Sheep	DE	Mcal/kg	*	0.62	2.69	
Cattle	ME	Mcal/kg	*	0.54	2.31	
Sheep	ME	Mcal/kg	*	0.51	2.21	
Cattle	TDN	%		14.8	63.8	
Sheep	TDN	%		14.2	61.1	
Chlorine		%		0.08	0.36	13
Sodium		%		0.05	0.20	13
Sulphur		%		0.14	0.61	12
Carotene		mg/kg		56.1	241.6	47
Pantothenic acid		mg/kg		9.7	41.7	
Vitamin A equivalent		IU/g		93.4	402.8	

Alfalfa, aerial part, fresh, immature, cut 2, (2)

Ref No 2 00 176 — United States

Feed Name or Analyses			Mean As Fed	Mean Dry	C.V. ± %
Dry matter	%		29.2	100.0	21
Calcium	%		0.69	2.37	
Iron	%		0.053	0.182	
Magnesium	%		0.06	0.19	
Phosphorus	%		0.11	0.37	50
Potassium	%		0.65	2.22	
Sodium	%		0.11	0.39	
Sulphur	%		0.09	0.31	
Zinc	mg/kg		9.0	30.9	
Biotin	mg/kg		0.15	0.51	
Carotene	mg/kg		53.9	184.3	28
Choline	mg/kg		267.	915.	
Folic acid	mg/kg		0.98	3.35	
Niacin	mg/kg		9.3	32.0	
Pantothenic acid	mg/kg		6.1	20.9	
Riboflavin	mg/kg		2.3	7.7	
Thiamine	mg/kg		1.6	5.5	
Vitamin B6	mg/kg		1.55	5.29	
Vitamin A equivalent	IU/g		89.8	307.2	

Alfalfa, aerial part, fresh, pre-bloom, (2)

Ref No 2 00 181 — United States

Feed Name or Analyses			Mean As Fed	Mean Dry	C.V. ± %
Dry matter	%		20.9	100.0	
Ash	%		2.0	9.8	
Crude fiber	%		4.9	23.6	
Sheep	dig coef %		48.	48.	
Swine	dig coef %		43.	43.	
Ether extract	%		0.7	3.2	
Sheep	dig coef %		23.	23.	
Swine	dig coef %		10.	10.	
N-free extract	%		8.8	42.2	
Sheep	dig coef %		75.	75.	
Swine	dig coef %		76.	76.	
Protein (N x 6.25)	%		4.4	21.2	
Sheep	dig coef %		78.	78.	
Swine	dig coef %		71.	71.	
Cattle	dig prot %	*	3.3	15.9	
Goats	dig prot %	*	3.4	16.4	
Horses	dig prot %	*	3.3	15.6	
Rabbits	dig prot %	*	3.2	15.1	
Sheep	dig prot %		3.5	16.6	
Swine	dig prot %		3.2	15.1	
Energy	GE Mcal/kg				
Cattle	DE Mcal/kg	*	0.57	2.74	
Sheep	DE Mcal/kg	*	0.56	2.70	
Swine	DE kcal/kg	*	535.	2558.	
Cattle	ME Mcal/kg	*	0.47	2.25	
Sheep	ME Mcal/kg	*	0.46	2.21	
Swine	ME kcal/kg	*	491.	2346.	
Cattle	TDN %	*	13.0	62.2	
Sheep	TDN %		12.8	61.2	
Swine	TDN %		12.1	58.0	
Calcium	%		0.47	2.26	
Chlorine	%		0.07	0.35	
Magnesium	%		0.05	0.25	
Manganese	mg/kg		5.8	27.7	
Phosphorus	%		0.07	0.35	
Potassium	%		0.49	2.36	
Sodium	%		0.04	0.20	
Sulphur	%		0.13	0.60	

Alfalfa, aerial part, fresh, pre-bloom, cut 1, (2)

Ref No 2 00 178 — United States

Feed Name or Analyses			Mean As Fed	Mean Dry	C.V. ± %
Dry matter	%		17.1	100.0	
Ash	%		1.7	9.7	
Crude fiber	%		5.1	30.0	
Sheep	dig coef %		43.	43.	
Ether extract	%		0.7	3.9	
Sheep	dig coef %		44.	44.	
N-free extract	%		5.9	34.6	
Sheep	dig coef %		69.	69.	
Protein (N x 6.25)	%		3.7	21.8	
Sheep	dig coef %		71.	71.	
Cattle	dig prot %	*	2.8	16.4	
Goats	dig prot %	*	2.9	16.9	
Horses	dig prot %	*	2.7	16.0	
Rabbits	dig prot %	*	2.7	15.5	
Sheep	dig prot %		2.6	15.5	
Energy	GE Mcal/kg		0.78	4.57	
Cattle	DE Mcal/kg	*	0.44	2.57	
Sheep	DE Mcal/kg	*	0.42	2.47	
Cattle	ME Mcal/kg	*	0.36	2.11	
Sheep	ME Mcal/kg	*	0.35	2.03	
Cattle	TDN %	*	10.0	58.2	
Sheep	TDN %		9.6	56.1	

Alfalfa, aerial part, fresh, pre-bloom, cut 2, (2)

Ref No 2 00 179 — United States

Feed Name or Analyses			Mean As Fed	Mean Dry	C.V. ± %
Dry matter	%		23.1	100.0	
Ash	%		2.0	8.7	
Crude fiber	%		5.8	25.1	
Sheep	dig coef %		42.	42.	
Ether extract	%		0.8	3.3	
Sheep	dig coef %		36.	36.	
N-free extract	%		9.8	42.4	
Sheep	dig coef %		75.	75.	
Protein (N x 6.25)	%		4.7	20.5	
Sheep	dig coef %		75.	75.	
Cattle	dig prot %	*	3.5	15.3	
Goats	dig prot %	*	3.6	15.7	
Horses	dig prot %	*	3.4	14.9	
Rabbits	dig prot %	*	3.3	14.5	
Sheep	dig prot %		3.6	15.4	
Energy	GE Mcal/kg				
Cattle	DE Mcal/kg	*	0.64	2.77	
Sheep	DE Mcal/kg	*	0.62	2.66	
Cattle	ME Mcal/kg	*	0.52	2.27	
Sheep	ME Mcal/kg	*	0.50	2.18	
Cattle	TDN %	*	14.5	62.9	
Sheep	TDN %		14.0	60.4	

Ref No 2 00 179 — Canada

Feed Name or Analyses			Mean As Fed	Mean Dry	C.V. ± %
Dry matter	%		25.2	100.0	
Organic matter	%		22.9	90.9	
Ash	%		2.6	10.2	
Crude fiber	%		6.5	25.8	
Sheep	dig coef %		42.	42.	
Ether extract	%		0.4	1.7	
Sheep	dig coef %		36.	36.	
N-free extract	%		10.8	43.1	
Sheep	dig coef %		75.	75.	
Protein (N x 6.25)	%		4.9	19.4	
Sheep	dig coef %		75.	75.	
Cattle	dig prot %	*	3.6	14.4	
Goats	dig prot %	*	3.7	14.7	
Horses	dig prot %	*	3.5	14.0	
Rabbits	dig prot %	*	3.4	13.6	
Sheep	dig prot %		3.7	14.6	
Energy	GE Mcal/kg				
Cattle	DE Mcal/kg	*	0.68	2.69	
Sheep	DE Mcal/kg	*	0.65	2.60	
Cattle	ME Mcal/kg	*	0.56	2.21	
Sheep	ME Mcal/kg	*	0.54	2.13	
Cattle	TDN %	*	15.4	61.1	
Sheep	TDN %		14.8	59.0	

Alfalfa, aerial part, fresh, pre-bloom, cut 3, (2)

Ref No 2 00 180 — United States

Feed Name or Analyses			Mean As Fed	Mean Dry	C.V. ± %
Dry matter	%		23.2	100.0	
Ash	%		2.5	10.8	
Crude fiber	%		5.1	21.9	
Sheep	dig coef %		50.	50.	
Ether extract	%		0.6	2.7	
Sheep	dig coef %		15.	15.	
N-free extract	%		9.3	40.1	
Sheep	dig coef %		67.	67.	
Protein (N x 6.25)	%		5.7	24.5	
Sheep	dig coef %		74.	74.	
Cattle	dig prot %	*	4.3	18.7	
Goats	dig prot %	*	4.5	19.4	
Horses	dig prot %	*	4.3	18.3	
Rabbits	dig prot %	*	4.1	17.6	
Sheep	dig prot %		4.2	18.1	
Energy	GE Mcal/kg				
Cattle	DE Mcal/kg	*	0.59	2.54	
Sheep	DE Mcal/kg	*	0.58	2.51	
Cattle	ME Mcal/kg	*	0.48	2.08	
Sheep	ME Mcal/kg	*	0.48	2.06	
Cattle	TDN %	*	13.4	57.6	
Sheep	TDN %		13.2	56.9	

Alfalfa, aerial part, fresh, early bloom, (2)

Ref No 2 00 184 — United States

Feed Name or Analyses			Mean As Fed	Mean Dry	C.V. ± %
Dry matter	%		23.9	100.0	9
Ash	%		2.4	10.1	11
Crude fiber	%		6.8	28.5	9
Cattle	dig coef %	*	44.	44.	
Sheep	dig coef %		49.	49.	
Ether extract	%		0.7	2.9	13
Cattle	dig coef %	*	46.	46.	
Sheep	dig coef %		38.	38.	
N-free extract	%		9.3	39.1	
Cattle	dig coef %	*	74.	74.	
Sheep	dig coef %		78.	78.	
Protein (N x 6.25)	%		4.6	19.4	7
Cattle	dig coef %	*	78.	78.	
Sheep	dig coef %		79.	79.	
Cattle	dig prot %		3.6	15.1	
Goats	dig prot %	*	3.5	14.6	
Horses	dig prot %	*	3.3	14.0	
Rabbits	dig prot %	*	3.3	13.6	
Sheep	dig prot %		3.7	15.3	
Lignin (Ellis)	%		1.9	7.9	
Energy	GE Mcal/kg				
Cattle	DE Mcal/kg	*	0.63	2.63	
Sheep	DE Mcal/kg	*	0.65	2.75	
Cattle	ME Mcal/kg	*	0.51	2.16	
Sheep	ME Mcal/kg	*	0.54	2.25	
Cattle	TDN %		14.2	59.6	
Sheep	TDN %		14.9	62.3	
Calcium	%		0.56	2.33	9

Continued

Feed Name or Analyses			Mean As Fed	Mean Dry	C.V. ± %
Magnesium	%		0.01	0.03	
Phosphorus	%		0.07	0.31	11
Potassium	%		0.46	1.92	
Carotene	mg/kg		41.6	174.6	42
Vitamin A equivalent	IU/g		69.4	291.1	

Alfalfa, aerial part, fresh, early bloom, cut 1, (2)
Ref No 2 00 182 United States

Feed Name or Analyses			Mean As Fed	Mean Dry	C.V. ± %
Dry matter	%		24.3	100.0	
Ash	%		2.0	8.2	
Crude fiber	%		7.8	31.9	
Sheep	dig coef %		39.	39.	
Ether extract	%		0.8	3.3	
Sheep	dig coef %		47.	47.	
N-free extract	%		9.3	38.4	
Sheep	dig coef %		68.	68.	
Protein (N x 6.25)	%		4.4	18.3	
Sheep	dig coef %		73.	73.	
Cattle	dig prot %	*	3.3	13.4	
Goats	dig prot %	*	3.3	13.6	
Horses	dig prot %	*	3.2	13.0	
Rabbits	dig prot %	*	3.1	12.8	
Sheep	dig prot %		3.2	13.3	
Energy	GE Mcal/kg		1.11	4.55	
Cattle	DE Mcal/kg	*	0.61	2.50	
Sheep	DE Mcal/kg	*	0.59	2.44	
Cattle	ME Mcal/kg	*	0.50	2.05	
Sheep	ME Mcal/kg	*	0.49	2.00	
Cattle	TDN %	*	13.8	56.8	
Sheep	TDN %		13.4	55.3	

Alfalfa, aerial part, fresh, early bloom, cut 2, (2)
Ref No 2 00 183 United States

Feed Name or Analyses			Mean As Fed	Mean Dry	C.V. ± %
Dry matter	%		25.4	100.0	
Ash	%		2.1	8.4	
Crude fiber	%		7.6	30.1	
Sheep	dig coef %		39.	39.	
Ether extract	%		0.8	3.3	
Sheep	dig coef %		36.	36.	
N-free extract	%		10.2	40.3	
Sheep	dig coef %		72.	72.	
Protein (N x 6.25)	%		4.5	17.9	
Sheep	dig coef %		77.	77.	
Cattle	dig prot %	*	3.3	13.1	
Goats	dig prot %	*	3.4	13.3	
Horses	dig prot %	*	3.2	12.7	
Rabbits	dig prot %	*	3.2	12.5	
Sheep	dig prot %		3.5	13.8	
Energy	GE Mcal/kg				
Cattle	DE Mcal/kg	*	0.66	2.58	
Sheep	DE Mcal/kg	*	0.64	2.52	
Cattle	ME Mcal/kg	*	0.59	2.12	
Sheep	ME Mcal/kg	*	0.53	2.07	
Cattle	TDN %	*	14.9	58.5	
Sheep	TDN %		14.5	57.2	

(1) dry forages and roughages (2) pasture, range plants, and forages fed green (3) silages (4) energy feeds (5) protein supplements (6) minerals (7) vitamins (8) additives

Alfalfa, aerial part, fresh, early bloom, cut 3, (2)
Ref No 2 07 905 United States

Feed Name or Analyses			Mean As Fed	Mean Dry	C.V. ± %
Dry matter	%		23.0	100.0	
Ash	%		2.0	8.6	
Crude fiber	%		5.4	23.4	
Ether extract	%		0.7	3.1	
N-free extract	%		9.9	42.9	
Protein (N x 6.25)	%		5.1	22.0	
Cattle	dig prot %	*	3.8	16.6	
Goats	dig prot %	*	3.9	17.1	
Horses	dig prot %	*	3.7	16.2	
Rabbits	dig prot %	*	3.6	15.7	
Sheep	dig prot %	*	4.0	17.5	
Energy	GE Mcal/kg		1.01	4.41	
Cattle	DE Mcal/kg	*	0.65	2.82	
Sheep	DE Mcal/kg	*	0.64	2.76	
Cattle	ME Mcal/kg	*	0.53	2.32	
Sheep	ME Mcal/kg	*	0.52	2.27	
Cattle	TDN %	*	14.8	64.3	
Sheep	TDN %	*	14.4	62.7	

Alfalfa, aerial part, fresh, mid-bloom, (2)
Ref No 2 00 185 United States

Feed Name or Analyses			Mean As Fed	Mean Dry	C.V. ± %
Dry matter	%		24.2	100.0	10
Ash	%		2.1	8.8	4
Crude fiber	%		6.7	27.7	4
Cattle	dig coef %	*	44.	44.	
Sheep	dig coef %		46.	46.	
Ether extract	%		0.7	2.8	29
Cattle	dig coef %	*	46.	46.	
Sheep	dig coef %		36.	36.	
N-free extract	%		9.7	40.3	
Cattle	dig coef %	*	74.	74.	
Sheep	dig coef %		73.	73.	
Protein (N x 6.25)	%		4.9	20.4	7
Cattle	dig coef %	*	78.	78.	
Sheep	dig coef %		74.	74.	
Cattle	dig prot %		3.9	15.9	
Goats	dig prot %	*	3.8	15.6	
Horses	dig prot %	*	3.6	14.9	
Rabbits	dig prot %	*	3.5	14.5	
Sheep	dig prot %		3.7	15.2	
Lignin (Ellis)	%		1.3	5.5	
Energy	GE Mcal/kg				
Cattle	DE Mcal/kg	*	0.65	2.68	
Sheep	DE Mcal/kg	*	0.63	2.63	
Cattle	ME Mcal/kg	*	0.53	2.20	
Sheep	ME Mcal/kg	*	0.52	2.15	
Cattle	TDN %		14.7	60.8	
Sheep	TDN %		14.4	59.6	
Calcium	%		0.49	2.01	
Chlorine	%		0.11	0.45	
Magnesium	%		0.06	0.26	
Phosphorus	%		0.07	0.28	7
Potassium	%		0.50	2.06	29
Sodium	%		0.04	0.16	
Sulphur	%		0.07	0.29	
Zinc	mg/kg		5.7	23.6	

Ref No 2 00 185 Canada

Feed Name or Analyses			Mean As Fed	Mean Dry	C.V. ± %
Dry matter	%		25.2	100.0	
Ash	%		2.2	8.6	
Crude fiber	%		8.7	34.4	
Sheep	dig coef %		46.	46.	
Ether extract	%		0.4	1.5	
Sheep	dig coef %		36.	36.	
N-free extract	%		9.8	39.0	
Sheep	dig coef %		73.	73.	
Protein (N x 6.25)	%		4.2	16.5	
Sheep	dig coef %		74.	74.	
Cattle	dig prot %	*	3.0	11.9	
Goats	dig prot %	*	3.0	12.0	
Horses	dig prot %	*	2.9	11.5	
Rabbits	dig prot %	*	2.9	11.4	
Sheep	dig prot %		3.1	12.2	
Energy	GE Mcal/kg				
Cattle	DE Mcal/kg	*	0.67	2.66	
Sheep	DE Mcal/kg	*	0.64	2.55	
Cattle	ME Mcal/kg	*	0.55	2.18	
Sheep	ME Mcal/kg	*	0.53	2.09	
Cattle	TDN %		15.2	60.2	
Sheep	TDN %		14.6	57.8	

Alfalfa, aerial part, fresh, mid-bloom, cut 1, (2)
Ref No 2 07 838 United States

Feed Name or Analyses			Mean As Fed	Mean Dry	C.V. ± %
Dry matter	%		22.1	100.0	
Protein (N x 6.25)	%		3.4	15.6	
Cattle	dig prot %	*	2.5	11.2	
Goats	dig prot %	*	2.5	11.1	
Horses	dig prot %	*	2.4	10.8	
Rabbits	dig prot %	*	2.4	10.7	
Sheep	dig prot %	*	2.5	11.5	
Energy	GE Mcal/kg		0.99	4.48	

Alfalfa, aerial part, fresh, full bloom, (2)
Ref No 2 00 188 United States

Feed Name or Analyses			Mean As Fed	Mean Dry	C.V. ± %
Dry matter	%		24.7	100.0	5
Ash	%		2.4	9.8	9
Crude fiber	%		7.5	30.4	8
Cattle	dig coef %	*	44.	44.	
Sheep	dig coef %		40.	40.	
Ether extract	%		0.6	2.6	12
Cattle	dig coef %	*	46.	46.	
Sheep	dig coef %		15.	15.	
N-free extract	%		9.8	39.6	
Cattle	dig coef %	*	74.	74.	
Sheep	dig coef %		73.	73.	
Protein (N x 6.25)	%		4.4	17.7	6
Cattle	dig coef %	*	78.	78.	
Sheep	dig coef %		75.	75.	
Cattle	dig prot %		3.4	13.8	
Goats	dig prot %	*	3.2	13.0	
Horses	dig prot %	*	3.1	12.5	
Rabbits	dig prot %	*	3.0	12.3	
Sheep	dig prot %		3.3	13.2	
Energy	GE Mcal/kg				
Cattle	DE Mcal/kg	*	0.64	2.61	
Sheep	DE Mcal/kg	*	0.60	2.43	
Cattle	ME Mcal/kg	*	0.53	2.14	
Sheep	ME Mcal/kg	*	0.49	2.00	
Cattle	TDN %		14.6	59.1	
Sheep	TDN %		13.6	55.2	
Calcium	%		0.38	1.53	14
Chlorine	%		0.11	0.43	9
Iron	%		0.011	0.043	
Magnesium	%		0.07	0.27	16
Manganese	mg/kg		38.4	155.2	
Phosphorus	%		0.07	0.27	13
Potassium	%		0.53	2.13	13
Sodium	%		0.04	0.15	26
Sulphur	%		0.08	0.31	9

Feed Name or Analyses			Mean As Fed	Mean Dry	C.V. ± %
Zinc	mg/kg		3.5	14.1	
Pantothenic acid	mg/kg		7.7	31.3	

Alfalfa, aerial part, fresh, full bloom, cut 1, (2)
Ref No 2 00 186 United States

Feed Name or Analyses			Mean As Fed	Mean Dry	C.V. ± %
Dry matter	%		24.7	100.0	
Ash	%		1.8	7.1	
Crude fiber	%		9.1	36.8	
Sheep	dig coef %		35.	35.	
Ether extract	%		0.8	3.2	
Sheep	dig coef %		62.	62.	
N-free extract	%		9.2	37.1	
Sheep	dig coef %		63.	63.	
Protein (N x 6.25)	%		3.9	15.8	
Sheep	dig coef %		69.	69.	
Cattle	dig prot %	*	2.8	11.3	
Goats	dig prot %	*	2.8	11.3	
Horses	dig prot %	*	2.7	10.9	
Rabbits	dig prot %	*	2.7	10.9	
Sheep	dig prot %		2.7	10.9	
Energy	GE Mcal/kg		1.13	4.58	
Cattle	DE Mcal/kg	*	0.58	2.34	
Sheep	DE Mcal/kg	*	0.56	2.28	
Cattle	ME Mcal/kg	*	0.47	1.92	
Sheep	ME Mcal/kg	*	0.46	1.87	
Cattle	TDN %	*	13.1	53.1	
Sheep	TDN %		12.7	51.6	

Alfalfa, aerial part, fresh, full bloom, cut 2, (2)
Ref No 2 00 187 United States

Feed Name or Analyses			Mean As Fed	Mean Dry	C.V. ± %
Dry matter	%		26.7	100.0	
Ash	%		2.0	7.5	
Crude fiber	%		9.2	34.4	
Sheep	dig coef %		44.	44.	
Ether extract	%		0.7	2.8	
Sheep	dig coef %		24.	24.	
N-free extract	%		10.8	40.6	
Sheep	dig coef %		68.	68.	
Protein (N x 6.25)	%		3.9	14.7	
Sheep	dig coef %		68.	68.	
Cattle	dig prot %	*	2.8	10.4	
Goats	dig prot %	*	2.7	10.3	
Horses	dig prot %	*	2.7	10.0	
Rabbits	dig prot %	*	2.7	10.0	
Sheep	dig prot %		2.7	10.0	
Energy	GE Mcal/kg				
Cattle	DE Mcal/kg	*	0.64	2.41	
Sheep	DE Mcal/kg	*	0.64	2.39	
Cattle	ME Mcal/kg	*	0.52	1.98	
Sheep	ME Mcal/kg	*	0.52	1.96	
Cattle	TDN %	*	14.6	54.6	
Sheep	TDN %		14.5	54.3	

Alfalfa, aerial part, fresh, late bloom, cut 2, (2)
Ref No 2 00 189 United States

Feed Name or Analyses			Mean As Fed	Mean Dry	C.V. ± %
Dry matter	%		28.8	100.0	
Ash	%		1.8	6.2	
Crude fiber	%		11.7	40.5	
Sheep	dig coef %		36.	36.	
Ether extract	%		0.6	2.0	
Sheep	dig coef %		49.	49.	
N-free extract	%		10.8	37.5	
Sheep	dig coef %		62.	62.	
Protein (N x 6.25)	%		4.0	13.8	
Sheep	dig coef %		68.	68.	
Cattle	dig prot %	*	2.8	9.6	
Goats	dig prot %	*	2.7	9.4	
Horses	dig prot %	*	2.7	9.2	
Rabbits	dig prot %	*	2.7	9.3	
Sheep	dig prot %		2.7	9.4	
Energy	GE Mcal/kg				
Cattle	DE Mcal/kg	*	0.63	2.20	
Sheep	DE Mcal/kg	*	0.63	2.18	
Cattle	ME Mcal/kg	*	0.52	1.80	
Sheep	ME Mcal/kg	*	0.51	1.79	
Cattle	TDN %	*	14.4	49.9	
Sheep	TDN %		14.2	49.4	

Alfalfa, aerial part, fresh, milk stage, (2)
Ref No 2 00 190 United States

Feed Name or Analyses			Mean As Fed	Mean Dry	C.V. ± %
Dry matter	%		29.5	100.0	13
Ash	%		2.2	7.5	9
Crude fiber	%		11.9	40.4	10
Ether extract	%		0.7	2.3	10
N-free extract	%		11.1	37.7	
Protein (N x 6.25)	%		3.6	12.1	17
Cattle	dig prot %	*	2.4	8.2	
Goats	dig prot %	*	2.3	7.9	
Horses	dig prot %	*	2.3	7.8	
Rabbits	dig prot %	*	2.4	8.0	
Sheep	dig prot %	*	2.5	8.3	
Energy	GE Mcal/kg				
Cattle	DE Mcal/kg	*	0.64	2.17	
Sheep	DE Mcal/kg	*	0.64	2.16	
Cattle	ME Mcal/kg	*	0.53	1.78	
Sheep	ME Mcal/kg	*	0.52	1.77	
Cattle	TDN %	*	14.5	49.2	
Sheep	TDN %	*	14.4	48.9	
Calcium	%		0.36	1.23	
Phosphorus	%		0.06	0.20	
Potassium	%		0.56	1.91	

Alfalfa, aerial part, fresh, cut 1, (2)
Ref No 2 00 191 United States

Feed Name or Analyses			Mean As Fed	Mean Dry	C.V. ± %
Dry matter	%		22.0	100.0	15
Ash	%		1.8	8.3	
Crude fiber	%		7.2	32.7	
Cattle	dig coef %	*	44.	44.	
Sheep	dig coef %		39.	39.	
Ether extract	%		0.7	3.4	
Cattle	dig coef %	*	46.	46.	
Sheep	dig coef %		50.	50.	
N-free extract	%		8.6	39.2	
Cattle	dig coef %	*	74.	74.	
Sheep	dig coef %		67.	67.	
Protein (N x 6.25)	%		3.6	16.4	
Cattle	dig coef %	*	78.	78.	
Sheep	dig coef %		72.	72.	
Cattle	dig prot %		2.8	12.8	
Goats	dig prot %	*	2.6	11.9	
Horses	dig prot %	*	2.5	11.5	
Rabbits	dig prot %	*	2.5	11.3	
Sheep	dig prot %		2.6	11.8	
Energy	GE Mcal/kg				
Cattle	DE Mcal/kg	*	0.58	2.63	
Sheep	DE Mcal/kg	*	0.53	2.41	
Cattle	ME Mcal/kg	*	0.47	2.16	
Sheep	ME Mcal/kg	*	0.43	1.98	
Cattle	TDN %		13.1	59.7	
Sheep	TDN %		12.0	54.7	
Calcium	%		0.41	1.87	26
Chlorine	%		0.16	0.73	
Iron	%		0.006	0.026	
Magnesium	%		0.01	0.06	
Phosphorus	%		0.08	0.36	38
Potassium	%		0.49	2.21	35
Sodium	%		0.05	0.24	
Sulphur	%		0.08	0.36	
Zinc	mg/kg		4.4	19.8	
Biotin	mg/kg		0.12	0.53	
Carotene	mg/kg		33.8	153.9	
Choline	mg/kg		429.	1951.	
Folic acid	mg/kg		0.60	2.71	
Niacin	mg/kg		9.6	43.7	
Pantothenic acid	mg/kg		7.6	34.6	
Riboflavin	mg/kg		2.4	11.0	
Thiamine	mg/kg		1.3	6.0	
Vitamin B_6	mg/kg		1.65	7.50	
Vitamin A equivalent	IU/g		56.4	256.5	

Alfalfa, aerial part, fresh, cut 2, (2)
Ref No 2 00 193 United States

Feed Name or Analyses			Mean As Fed	Mean Dry	C.V. ± %
Dry matter	%		24.9	100.0	
Ash	%		2.1	8.3	
Crude fiber	%		7.3	29.4	
Cattle	dig coef %	*	44.	44.	
Sheep	dig coef %		41.	41.	
Ether extract	%		0.8	3.1	
Cattle	dig coef %	*	46.	46.	
Sheep	dig coef %		34.	34.	
N-free extract	%		10.2	41.0	
Cattle	dig coef %	*	74.	74.	
Sheep	dig coef %		72.	72.	
Protein (N x 6.25)	%		4.5	18.2	
Cattle	dig coef %	*	78.	78.	
Sheep	dig coef %		74.	74.	
Cattle	dig prot %		3.5	14.2	
Goats	dig prot %	*	3.4	13.5	
Horses	dig prot %	*	3.2	13.0	
Rabbits	dig prot %	*	3.2	12.7	
Sheep	dig prot %		3.4	13.5	
Energy	GE Mcal/kg				
Cattle	DE Mcal/kg	*	0.67	2.68	
Sheep	DE Mcal/kg	*	0.63	2.53	
Cattle	ME Mcal/kg	*	0.55	2.19	
Sheep	ME Mcal/kg	*	0.52	2.08	
Cattle	TDN %		15.1	60.7	
Sheep	TDN %		14.3	57.4	

Alfalfa, aerial part, fresh, cut 2, gr 2 US, (2)
Ref No 2 00 192 United States

Feed Name or Analyses			Mean As Fed	Mean Dry	C.V. ± %
Dry matter	%		87.4	100.0	
Ash	%		7.5	8.6	
Crude fiber	%		27.4	31.3	
Sheep	dig coef %		39.	39.	
Ether extract	%		1.7	2.0	
Sheep	dig coef %		20.	20.	
N-free extract	%		35.2	40.3	
Sheep	dig coef %		66.	66.	
Protein (N x 6.25)	%		15.6	17.8	
Sheep	dig coef %		77.	77.	
Cattle	dig prot %	*	11.4	13.0	
Goats	dig prot %	*	11.5	13.2	
Horses	dig prot %	*	11.0	12.6	
Rabbits	dig prot %	*	10.8	12.4	

Continued

Feed Name or Analyses			Mean As Fed	Mean Dry	C.V. ± %
Sheep	dig prot %		12.0	13.7	
Energy	GE Mcal/kg				
Cattle	DE Mcal/kg	*	2.18	2.50	
Sheep	DE Mcal/kg	*	2.06	2.35	
Cattle	ME Mcal/kg	*	1.79	2.05	
Sheep	ME Mcal/kg	*	1.69	1.93	
Cattle	TDN %	*	49.6	56.7	
Sheep	TDN %		46.7	53.4	

Alfalfa, aerial part, fresh, cut 3, (2)
Ref No 2 00 194 United States

Feed Name or Analyses			As Fed	Dry	C.V. ± %
Dry matter	%		30.6	100.0	39
Ash	%		2.8	9.2	
Crude fiber	%		8.1	26.6	
Cattle	dig coef %		42.	42.	
Ether extract	%		0.7	2.4	
Cattle	dig coef %		38.	38.	
N-free extract	%		13.2	43.2	
Cattle	dig coef %		72.	72.	
Protein (N x 6.25)	%		5.7	18.6	
Cattle	dig coef %		74.	74.	
Cattle	dig prot %		4.2	13.8	
Goats	dig prot %	*	4.3	13.9	
Horses	dig prot %	*	4.1	13.3	
Rabbits	dig prot %	*	4.0	13.0	
Sheep	dig prot %	*	4.4	14.3	
Energy	GE Mcal/kg				
Cattle	DE Mcal/kg	*	0.78	2.56	
Sheep	DE Mcal/kg	*	0.82	2.68	
Cattle	ME Mcal/kg	*	0.64	2.10	
Sheep	ME Mcal/kg	*	0.67	2.20	
Cattle	TDN %		17.8	58.1	
Sheep	TDN %	*	18.6	60.8	
Calcium	%		0.54	1.76	43
Chlorine	%		0.18	0.60	
Iron	%		0.043	0.140	
Magnesium	%		0.03	0.09	
Phosphorus	%		0.11	0.35	50
Potassium	%		0.68	2.21	43
Sodium	%		0.06	0.19	
Sulphur	%		0.10	0.33	
Biotin	mg/kg		0.14	0.46	
Folic acid	mg/kg		0.72	2.36	

Alfalfa, aerial part, fresh, cut 4, (2)
Ref No 2 00 195 United States

Feed Name or Analyses			As Fed	Dry	C.V. ± %
Dry matter	%		63.4	100.0	22
Crude fiber	%				
Cattle	dig coef %	*	44.	44.	
Sheep	dig coef %	*	44.	44.	
Ether extract	%				
Cattle	dig coef %	*	46.	46.	
Sheep	dig coef %	*	35.	35.	
N-free extract	%				
Cattle	dig coef %	*	74.	74.	
Sheep	dig coef %	*	72.	72.	
Protein (N x 6.25)	%				
Cattle	dig coef %	*	78.	78.	
Sheep	dig coef %	*	75.	75.	
Calcium	%		1.58	2.49	
Chlorine	%		0.43	0.68	

(1) dry forages and roughages (2) pasture, range plants, and forages fed green (3) silages (4) energy feeds (5) protein supplements (6) minerals (7) vitamins (8) additives

Feed Name or Analyses			Mean As Fed	Mean Dry	C.V. ± %
Iron	%		0.018	0.029	
Magnesium	%		0.35	0.55	
Phosphorus	%		0.29	0.45	
Potassium	%		1.74	2.75	
Sodium	%		0.14	0.22	
Sulphur	%		0.26	0.41	

Alfalfa, aerial part, fresh sulfur fertilized, (2)
Ref No 2 00 198 United States

Feed Name or Analyses			As Fed	Dry	C.V. ± %
Dry matter	%			100.0	
Choline	mg/kg			1276.	
Niacin	mg/kg			34.4	
Pantothenic acid	mg/kg			39.5	
Riboflavin	mg/kg			7.5	
Thiamine	mg/kg			5.1	
Vitamin B6	mg/kg			6.17	

Alfalfa, aerial part, fresh wilted, (2)
Ref No 2 00 201 United States

Feed Name or Analyses			As Fed	Dry	C.V. ± %
Dry matter	%		69.2	100.0	17
Ash	%		6.2	8.9	10
Crude fiber	%		16.0	23.1	16
Ether extract	%		1.1	1.6	8
N-free extract	%		30.2	43.7	
Protein (N x 6.25)	%		15.7	22.7	7
Cattle	dig prot %	*	11.9	17.2	
Goats	dig prot %	*	12.3	17.7	
Horses	dig prot %	*	11.6	16.8	
Rabbits	dig prot %	*	11.2	16.2	
Sheep	dig prot %	*	12.5	18.1	
Energy	GE Mcal/kg				
Cattle	DE Mcal/kg	*	1.74	2.52	
Sheep	DE Mcal/kg	*	1.72	2.49	
Cattle	ME Mcal/kg	*	1.43	2.07	
Sheep	ME Mcal/kg	*	1.41	2.04	
Cattle	TDN %	*	39.6	57.2	
Sheep	TDN %	*	39.0	56.4	
Carotene	mg/kg		78.7	113.8	20
Vitamin A equivalent	IU/g		131.1	189.6	

Alfalfa, aerial part, fresh wilted, immature, (2)
Ref No 2 00 199 United States

Feed Name or Analyses			As Fed	Dry	C.V. ± %
Dry matter	%		75.5	100.0	1
Ash	%		6.6	8.8	7
Crude fiber	%		16.2	21.4	5
Ether extract	%		1.2	1.6	4
N-free extract	%		33.4	44.3	
Protein (N x 6.25)	%		18.0	23.9	3
Cattle	dig prot %	*	13.7	18.2	
Goats	dig prot %	*	14.2	18.9	
Horses	dig prot %	*	13.5	17.8	
Rabbits	dig prot %	*	12.9	17.1	
Sheep	dig prot %	*	14.5	19.3	
Energy	GE Mcal/kg				
Cattle	DE Mcal/kg	*	2.06	2.72	
Sheep	DE Mcal/kg	*	2.17	2.88	
Cattle	ME Mcal/kg	*	1.69	2.23	
Sheep	ME Mcal/kg	*	1.78	2.36	
Cattle	TDN %	*	46.7	61.8	
Sheep	TDN %	*	49.3	65.2	

Feed Name or Analyses			Mean As Fed	Mean Dry	C.V. ± %

Alfalfa, aerial part, fresh wilted, pre-bloom, cut 1, (2)
Ref No 2 07 898 United States

Feed Name or Analyses			As Fed	Dry	C.V. ± %
Dry matter	%		29.3	100.0	
Protein (N x 6.25)	%		6.5	22.1	
Cattle	dig prot %	*	4.9	16.7	
Goats	dig prot %	*	5.0	17.2	
Horses	dig prot %	*	4.8	16.3	
Rabbits	dig prot %	*	4.6	15.7	
Sheep	dig prot %	*	5.2	17.6	
Energy	GE Mcal/kg		1.32	4.52	

Alfalfa, aerial part, fresh wilted, early bloom, cut 1, (2)
Ref No 2 07 897 United States

Feed Name or Analyses			As Fed	Dry	C.V. ± %
Dry matter	%		35.5	100.0	
Protein (N x 6.25)	%		7.0	19.6	
Cattle	dig prot %	*	5.2	14.6	
Goats	dig prot %	*	5.3	14.8	
Horses	dig prot %	*	5.0	14.2	
Rabbits	dig prot %	*	4.9	13.8	
Sheep	dig prot %	*	5.4	15.3	
Energy	GE Mcal/kg		1.62	4.56	

Alfalfa, aerial part, fresh wilted, full bloom, cut 1, (2)
Ref No 2 07 899 United States

Feed Name or Analyses			As Fed	Dry	C.V. ± %
Dry matter	%		35.7	100.0	
Protein (N x 6.25)	%		6.2	17.3	
Cattle	dig prot %	*	4.5	12.6	
Goats	dig prot %	*	4.5	12.7	
Horses	dig prot %	*	4.4	12.2	
Rabbits	dig prot %	*	4.3	12.0	
Sheep	dig prot %	*	4.7	13.1	
Energy	GE Mcal/kg		1.64	4.59	

Alfalfa, aerial part, fresh wilted, cut 4, (2)
Ref No 2 00 200 United States

Feed Name or Analyses			As Fed	Dry	C.V. ± %
Dry matter	%		73.6	100.0	5
Ash	%		6.6	9.0	8
Crude fiber	%		15.8	21.4	5
Ether extract	%		1.2	1.6	9
N-free extract	%		32.8	44.5	
Protein (N x 6.25)	%		17.3	23.5	3
Cattle	dig prot %	*	13.1	17.9	
Goats	dig prot %	*	13.6	18.5	
Horses	dig prot %	*	12.9	17.5	
Rabbits	dig prot %	*	12.4	16.8	
Sheep	dig prot %	*	13.9	18.9	
Energy	GE Mcal/kg				
Cattle	DE Mcal/kg	*	1.93	2.62	
Sheep	DE Mcal/kg	*	1.87	2.54	
Cattle	ME Mcal/kg	*	1.58	2.15	
Sheep	ME Mcal/kg	*	1.53	2.08	
Cattle	TDN %	*	43.7	59.4	
Sheep	TDN %	*	42.4	57.6	
Carotene	mg/kg		80.5	109.3	16
Vitamin A equivalent	IU/g		134.2	182.3	

Alfalfa, leaves, fresh, (2)

Ref No 2 00 202 United States

Feed Name or Analyses			Mean As Fed	Mean Dry	C.V. ± %
Dry matter	%		88.8	100.0	2
Sulphur	%		0.54	0.61	

Alfalfa, aerial part, ensiled, (3)

Ref No 3 00 212 United States

Feed Name or Analyses			Mean As Fed	Mean Dry	C.V. ± %
Dry matter	%		28.3	100.0	20
Sheep	dig coef %		63.	63.	
Ash	%		2.6	9.3	9
Crude fiber	%		9.1	32.2	14
Cattle	dig coef %	*	52.	52.	
Sheep	dig coef %	*	46.	46.	
Ether extract	%		0.9	3.3	10
Cattle	dig coef %	*	60.	60.	
Sheep	dig coef %	*	52.	52.	
N-free extract	%		10.5	37.2	7
Cattle	dig coef %	*	61.	61.	
Sheep	dig coef %	*	62.	62.	
Protein (N x 6.25)	%		5.1	17.9	10
Cattle	dig coef %	*	67.	67.	
Sheep	dig coef %	*	64.	64.	
Cattle	dig prot %		3.4	12.0	
Goats	dig prot %	*	3.5	12.5	
Horses	dig prot %	*	3.5	12.5	
Sheep	dig prot %		3.2	11.4	
Lignin (Ellis)	%		3.3	11.7	
Energy	GE Mcal/kg		1.33	4.71	
Sheep	GE dig coef %		63.	63.	
Cattle	DE Mcal/kg	*	0.70	2.47	
Sheep	DE Mcal/kg		0.84	2.97	
Cattle	ME Mcal/kg	*	0.57	2.02	
Sheep	ME Mcal/kg	*	0.69	2.43	
Cattle	NE_m Mcal/kg	*	0.34	1.20	
Cattle	NE_{gain} Mcal/kg	*	0.15	0.53	
Cattle	$NE_{lactating\ cows}$ Mcal/kg	*	0.37	1.30	
Cattle	TDN %		15.9	55.9	
Sheep	TDN %		15.1	53.2	
Calcium	%		0.40	1.40	10
Chlorine	%		0.13	0.47	35
Cobalt	mg/kg		0.042	0.150	
Copper	mg/kg		2.8	9.7	7
Iron	%		0.008	0.028	6
Magnesium	%		0.10	0.36	7
Manganese	mg/kg		15.1	53.1	13
Phosphorus	%		0.10	0.36	69
Potassium	%		0.68	2.38	4
Sodium	%		0.05	0.16	24
Sulphur	%		0.10	0.36	18
Carotene	mg/kg		34.4	121.3	46
Vitamin A equivalent	IU/g		57.3	202.1	

Ref No 3 00 212 Canada

Feed Name or Analyses			Mean As Fed	Mean Dry	C.V. ± %
Dry matter	%		29.4	100.0	19
Crude fiber	%		9.3	31.6	8
Protein (N x 6.25)	%		4.6	15.7	22
Cattle	dig prot %	*	3.1	10.5	
Goats	dig prot %	*	3.1	10.5	
Horses	dig prot %	*	3.1	10.5	
Sheep	dig prot %	*	3.1	10.5	
Energy	GE Mcal/kg		1.38	4.71	
Sheep	GE dig coef %		63.	63.	
Cattle	DE Mcal/kg	*	0.72	2.47	
Sheep	DE Mcal/kg		0.87	2.97	
Cattle	ME Mcal/kg	*	0.59	2.02	
Sheep	ME Mcal/kg	*	0.71	2.43	
Cattle	TDN %		16.4	55.9	
Sheep	TDN %		15.6	53.3	
Calcium	%		0.60	2.04	18
Phosphorus	%		0.06	0.20	25
Carotene	mg/kg		22.3	76.0	92
Vitamin A equivalent	IU/g		37.2	126.7	

Alfalfa, aerial part, ensiled, immature, (3)

Ref No 3 00 203 United States

Feed Name or Analyses			Mean As Fed	Mean Dry	C.V. ± %
Dry matter	%		29.1	100.0	0
Ash	%		2.6	8.8	7
Crude fiber	%		8.1	27.8	8
Cattle	dig coef %	*	52.	52.	
Sheep	dig coef %	*	46.	46.	
Ether extract	%		1.4	4.7	12
Cattle	dig coef %	*	60.	60.	
Sheep	dig coef %	*	52.	52.	
N-free extract	%		10.4	35.7	
Cattle	dig coef %	*	61.	61.	
Sheep	dig coef %	*	62.	62.	
Protein (N x 6.25)	%		6.7	23.0	7
Cattle	dig coef %	*	67.	67.	
Sheep	dig coef %	*	64.	64.	
Cattle	dig prot %		4.5	15.4	
Goats	dig prot %	*	5.0	17.1	
Horses	dig prot %	*	5.0	17.1	
Sheep	dig prot %		4.3	14.7	
Energy	GE Mcal/kg				
Cattle	DE Mcal/kg	*	0.74	2.56	
Sheep	DE Mcal/kg	*	0.71	2.43	
Cattle	ME Mcal/kg	*	0.61	2.10	
Sheep	ME Mcal/kg	*	0.58	1.99	
Cattle	TDN %		16.9	58.0	
Sheep	TDN %		16.0	55.1	
Calcium	%		0.51	1.77	14
Phosphorus	%		0.14	0.49	32
Carotene	mg/kg		49.5	170.4	28
Vitamin A equivalent	IU/g		82.6	284.1	

Alfalfa, aerial part, ensiled, pre-bloom, (3)

Ref No 3 00 204 United States

Feed Name or Analyses			Mean As Fed	Mean Dry	C.V. ± %
Dry matter	%		29.9	100.0	
Sheep	dig coef %		66.	66.	
Ash	%		3.0	10.0	
Crude fiber	%		9.3	31.2	
Cattle	dig coef %		50.	50.	
Sheep	dig coef %		53.	53.	
Ether extract	%		1.1	3.7	
Cattle	dig coef %		62.	62.	
Sheep	dig coef %		54.	54.	
N-free extract	%		10.4	34.9	
Cattle	dig coef %		61.	61.	
Sheep	dig coef %		72.	72.	
Protein (N x 6.25)	%		6.1	20.3	
Cattle	dig coef %		70.	70.	
Sheep	dig coef %		77.	77.	
Cattle	dig prot %		4.2	14.2	
Goats	dig prot %	*	4.4	14.7	
Horses	dig prot %	*	4.4	14.7	
Sheep	dig prot %		4.6	15.5	
Fatty acids	%				
Acetic	%		1.673	5.600	
Butyric	%		2.091	7.000	
Lactic acid	%		0.717	2.400	
Energy	GE Mcal/kg		1.44	4.81	
Sheep	GE dig coef %		67.	67.	
Cattle	DE Mcal/kg	*	0.74	2.48	
Sheep	DE Mcal/kg		0.96	3.22	
Cattle	ME Mcal/kg	*	0.61	2.03	
Sheep	ME Mcal/kg	*	0.79	2.64	
Cattle	TDN %		16.8	56.2	
Sheep	TDN %		18.4	61.7	
pH	pH units		5.70	5.70	

Alfalfa, aerial part, ensiled, early bloom, (3)

Ref No 3 00 205 United States

Feed Name or Analyses			Mean As Fed	Mean Dry	C.V. ± %
Dry matter	%		29.5	100.0	12
Ash	%		2.6	8.7	19
Crude fiber	%		9.6	32.6	6
Cattle	dig coef %		51.	51.	
Sheep	dig coef %	*	46.	46.	
Ether extract	%		1.0	3.6	24
Cattle	dig coef %		58.	58.	
Sheep	dig coef %	*	52.	52.	
N-free extract	%		11.4	38.6	
Cattle	dig coef %		58.	58.	
Sheep	dig coef %	*	62.	62.	
Protein (N x 6.25)	%		4.9	16.6	11
Cattle	dig coef %		62.	62.	
Sheep	dig coef %	*	64.	64.	
Cattle	dig prot %		3.0	10.3	
Goats	dig prot %	*	3.3	11.3	
Horses	dig prot %	*	3.3	11.3	
Sheep	dig prot %		3.1	10.6	
Energy	GE Mcal/kg				
Cattle	DE Mcal/kg	*	0.70	2.38	
Sheep	DE Mcal/kg	*	0.70	2.37	
Cattle	ME Mcal/kg	*	0.58	1.95	
Sheep	ME Mcal/kg	*	0.57	1.94	
Cattle	TDN %		15.9	53.9	
Sheep	TDN %		15.9	53.7	
Calcium	%		0.45	1.52	11
Phosphorus	%		0.09	0.30	15
Carotene	mg/kg		41.0	138.7	25
Vitamin A equivalent	IU/g		68.3	231.2	

Alfalfa, aerial part, ensiled, early bloom, cut 1, (3)

Ref No 3 07 844 United States

Feed Name or Analyses			Mean As Fed	Mean Dry	C.V. ± %
Dry matter	%		21.9	100.0	
Sheep	dig coef %		63.	63.	
Protein (N x 6.25)	%		4.8	22.0	
Sheep	dig coef %		78.	78.	
Cattle	dig prot %	*	3.5	16.2	
Goats	dig prot %	*	3.5	16.2	
Horses	dig prot %	*	3.5	16.2	
Sheep	dig prot %		3.8	17.2	
Fatty acids	%				
Acetic	%		1.073	4.900	
Butyric	%		0.986	4.500	
Lactic acid	%		1.402	6.400	
Energy	GE Mcal/kg		1.04	4.76	
Sheep	GE dig coef %		64.	64.	
Sheep	DE Mcal/kg		0.67	3.05	
Sheep	ME Mcal/kg	*	0.55	2.50	
pH	pH units		5.30	5.30	

Feed Name or Analyses				Mean As Fed	Mean Dry	C.V. ± %

Alfalfa, aerial part, ensiled, early bloom, cut 3, (3)
Ref No 3 07 903 United States

Feed Name or Analyses				As Fed	Dry	C.V. ± %
Dry matter		%			100.0	
Sheep	dig coef	%			63.	
Ash		%			9.6	
Crude fiber		%			26.8	
Sheep	dig coef	%			46.	
Ether extract		%			3.5	
Sheep	dig coef	%			48.	
N-free extract		%			37.9	
Sheep	dig coef	%			70.	
Protein (N x 6.25)		%			22.2	
Sheep	dig coef	%			79.	
Cattle	dig prot	%	*		16.4	
Goats	dig prot	%	*		16.4	
Horses	dig prot	%	*		16.4	
Sheep	dig prot	%			17.5	
Energy	GE Mcal/kg				4.57	
Sheep	GE dig coef	%			63.	
Cattle	DE Mcal/kg		*		2.70	
Sheep	DE Mcal/kg				2.88	
Cattle	ME Mcal/kg		*		2.21	
Sheep	ME Mcal/kg		*		2.36	
Cattle	TDN	%	*		61.3	
Sheep	TDN	%			60.2	

Alfalfa, aerial part, ensiled, mid-bloom, (3)
Ref No 3 00 206 United States

Feed Name or Analyses				As Fed	Dry	C.V. ± %
Dry matter		%		34.1	100.0	8
Ash		%		2.9	8.5	10
Crude fiber		%		10.9	31.9	9
Cattle	dig coef	%		57.	57.	
Ether extract		%		1.1	3.3	8
Cattle	dig coef	%		60.	60.	
N-free extract		%		13.1	38.4	
Cattle	dig coef	%		70.	70.	
Protein (N x 6.25)		%		6.1	18.0	12
Cattle	dig coef	%		71.	71.	
Cattle	dig prot	%		4.3	12.7	
Goats	dig prot	%	*	4.3	12.5	
Horses	dig prot	%	*	4.3	12.5	
Sheep	dig prot	%	*	4.3	12.5	
Energy	GE Mcal/kg					
Cattle	DE Mcal/kg		*	0.94	2.74	
Sheep	DE Mcal/kg		*	0.89	2.61	
Cattle	ME Mcal/kg		*	0.77	2.25	
Sheep	ME Mcal/kg		*	0.73	2.14	
Cattle	TDN	%		21.2	62.2	
Sheep	TDN	%	*	20.2	59.1	
Carotene	mg/kg			20.6	60.4	19
Vitamin A equivalent	IU/g			34.3	100.7	

Alfalfa, aerial part, ensiled, mid-bloom, cut 1, (3)
Ref No 3 07 839 United States

Feed Name or Analyses				As Fed	Dry	C.V. ± %
Dry matter		%		23.6	100.0	
Sheep	dig coef	%		59.	59.	
Ash		%		2.0	8.4	
Crude fiber		%		8.0	34.1	
Sheep	dig coef	%		50.	50.	
Ether extract		%		1.0	4.4	
Sheep	dig coef	%		62.	62.	
N-free extract		%		8.8	37.5	
Sheep	dig coef	%		63.	63.	
Protein (N x 6.25)		%		3.7	15.7	
Sheep	dig coef	%		62.	62.	
Cattle	dig prot	%	*	2.5	10.5	
Goats	dig prot	%	*	2.5	10.5	
Horses	dig prot	%	*	2.5	10.5	
Sheep	dig prot	%		2.3	9.7	
Fatty acids		%				
Acetic		%		1.672	7.100	
Butyric		%		0.141	0.600	
Lactic acid		%		1.696	7.200	
Energy	GE Mcal/kg			1.06	4.52	
Sheep	GE dig coef	%		56.	56.	
Cattle	DE Mcal/kg		*	0.56	2.39	
Sheep	DE Mcal/kg			0.60	2.53	
Cattle	ME Mcal/kg		*	0.46	1.96	
Sheep	ME Mcal/kg			0.49	2.08	
Cattle	TDN	%	*	12.8	54.3	
Sheep	TDN	%		13.3	56.4	
pH	pH units			4.70	4.70	

Alfalfa, aerial part, ensiled, full bloom, (3)
Ref No 3 00 207 United States

Feed Name or Analyses				As Fed	Dry	C.V. ± %
Dry matter		%		28.6	100.0	15
Ash		%		2.3	8.1	26
Crude fiber		%		9.5	33.4	6
Ether extract		%		1.0	3.4	10
N-free extract		%		11.1	38.9	
Protein (N x 6.25)		%		4.7	16.3	10
Cattle	dig prot	%	*	3.2	11.0	
Goats	dig prot	%	*	3.2	11.0	
Horses	dig prot	%	*	3.2	11.0	
Sheep	dig prot	%	*	3.2	11.0	
Energy	GE Mcal/kg			1.27	4.43	
Cattle	DE Mcal/kg		*	0.67	2.33	
Sheep	DE Mcal/kg		*	0.70	2.45	
Cattle	ME Mcal/kg		*	0.55	1.91	
Sheep	ME Mcal/kg		*	0.58	2.01	
Cattle	TDN	%	*	15.1	52.8	
Sheep	TDN	%	*	15.9	55.6	
Carotene	mg/kg			11.7	41.0	14
Vitamin A equivalent	IU/g			19.6	68.4	

Alfalfa, aerial part, ensiled, full bloom, cut 1, (3)
Ref No 3 07 841 United States

Feed Name or Analyses				As Fed	Dry	C.V. ± %
Dry matter		%		25.0	100.0	
Sheep	dig coef	%		53.	53.	
Protein (N x 6.25)		%		4.4	17.5	
Sheep	dig coef	%		70.	70.	
Cattle	dig prot	%	*	3.0	12.1	
Goats	dig prot	%	*	3.0	12.1	
Horses	dig prot	%	*	3.0	12.1	
Sheep	dig prot	%		3.1	12.3	
Fatty acids		%				
Acetic		%		1.075	4.300	
Butyric		%		0.975	3.900	
Lactic acid		%		1.150	4.600	
Energy	GE Mcal/kg			1.15	4.59	
Sheep	GE dig coef	%		52.	52.	
Sheep	DE Mcal/kg			0.60	2.38	
Sheep	ME Mcal/kg		*	0.49	1.96	
pH	pH units			5.50	5.50	

Alfalfa, aerial part, ensiled, mature, (3)
Ref No 3 00 208 United States

Feed Name or Analyses				As Fed	Dry	C.V. ± %
Dry matter		%		23.1	100.0	
Ash		%		2.1	8.9	
Crude fiber		%		9.0	39.1	
Ether extract		%		0.7	3.1	
N-free extract		%		7.9	34.0	
Protein (N x 6.25)		%		3.4	14.9	
Cattle	dig prot	%	*	2.3	9.8	
Goats	dig prot	%	*	2.3	9.8	
Horses	dig prot	%	*	2.3	9.8	
Sheep	dig prot	%	*	2.3	9.8	
Energy	GE Mcal/kg					
Cattle	DE Mcal/kg		*	0.50	2.18	
Sheep	DE Mcal/kg		*	0.52	2.24	
Cattle	ME Mcal/kg		*	0.41	1.79	
Sheep	ME Mcal/kg		*	0.43	1.84	
Cattle	TDN	%	*	11.4	49.5	
Sheep	TDN	%	*	11.7	50.7	

Ref No 3 00 208 Canada

Feed Name or Analyses				As Fed	Dry	C.V. ± %
Dry matter		%		37.1	100.0	
Crude fiber		%		12.5	33.7	
Protein (N x 6.25)		%		5.4	14.6	
Cattle	dig prot	%	*	3.5	9.4	
Goats	dig prot	%	*	3.5	9.4	
Horses	dig prot	%	*	3.5	9.4	
Sheep	dig prot	%	*	3.5	9.4	
Calcium		%		0.90	2.43	
Phosphorus		%		0.05	0.13	
Carotene	mg/kg			1.1	3.0	
Vitamin A equivalent	IU/g			1.8	5.0	

Alfalfa, aerial part, ensiled, over ripe, (3)
Ref No 3 00 209 United States

Feed Name or Analyses				As Fed	Dry	C.V. ± %
Dry matter		%		24.0	100.0	
Carotene	mg/kg			1.6	6.6	73
Vitamin A equivalent	IU/g			2.6	11.0	

Alfalfa, aerial part, ensiled, cut 1, (3)
Ref No 3 00 210 United States

Feed Name or Analyses				As Fed	Dry	C.V. ± %
Dry matter		%		24.7	100.0	7
Ash		%		3.1	12.5	
Crude fiber		%		8.1	32.8	
Ether extract		%		1.5	5.9	
N-free extract		%		7.3	29.4	
Protein (N x 6.25)		%		4.8	19.4	12
Cattle	dig prot	%	*	3.4	13.8	
Goats	dig prot	%	*	3.4	13.8	
Horses	dig prot	%	*	3.4	13.8	
Sheep	dig prot	%	*	3.4	13.8	
Energy	GE Mcal/kg					
Cattle	DE Mcal/kg		*	0.64	2.58	
Sheep	DE Mcal/kg		*	0.65	2.62	
Cattle	ME Mcal/kg		*	0.52	2.11	
Sheep	ME Mcal/kg		*	0.53	2.15	
Cattle	TDN	%	*	14.4	58.5	
Sheep	TDN	%	*	14.7	59.4	

(1) dry forages and roughages (2) pasture, range plants, and forages fed green (3) silages (4) energy feeds (5) protein supplements (6) minerals (7) vitamins (8) additives

Feed Name or Analyses			Mean As Fed	Mean Dry	C.V. ± %

Alfalfa, aerial part, ensiled, cut 2, (3)

Ref No 3 00 211 United States

Feed Name or Analyses			As Fed	Dry	C.V. ± %
Dry matter	%		27.6	100.0	6
Ash	%		2.6	9.5	5
Crude fiber	%		10.4	37.7	3
Sheep	dig coef %		39.	39.	
Ether extract	%		1.8	6.4	4
Sheep	dig coef %		80.	80.	
N-free extract	%		8.3	30.3	
Sheep	dig coef %		43.	43.	
Protein (N x 6.25)	%		4.5	16.2	5
Sheep	dig coef %		55.	55.	
Cattle	dig prot %	*	3.0	10.9	
Goats	dig prot %	*	3.0	10.9	
Horses	dig prot %	*	3.0	10.9	
Sheep	dig prot %		2.5	8.9	
Energy	GE Mcal/kg				
Cattle	DE Mcal/kg	*	0.66	2.39	
Sheep	DE Mcal/kg	*	0.59	2.12	
Cattle	ME Mcal/kg	*	0.54	1.96	
Sheep	ME Mcal/kg	*	0.48	1.74	
Cattle	TDN %	*	15.0	54.3	
Sheep	TDN %		13.3	48.1	

Ref No 3 00 211 Canada

Feed Name or Analyses			As Fed	Dry	C.V. ± %
Dry matter	%		18.9	100.0	
Crude fiber	%		6.3	33.3	
Sheep	dig coef %		39.	39.	
Ether extract	%				
Sheep	dig coef %		80.	80.	
N-free extract	%				
Sheep	dig coef %		43.	43.	
Protein (N x 6.25)	%		3.4	18.0	
Sheep	dig coef %		55.	55.	
Cattle	dig prot %	*	2.4	12.6	
Goats	dig prot %	*	2.4	12.6	
Horses	dig prot %	*	2.4	12.6	
Sheep	dig prot %		1.9	9.9	
Energy	GE Mcal/kg				
Sheep	DE Mcal/kg	*	0.40	2.13	
Sheep	ME Mcal/kg	*	0.33	1.74	
Sheep	TDN %		9.1	48.2	
Calcium	%		0.39	2.06	
Phosphorus	%		0.07	0.37	
Carotene	mg/kg		23.8	126.0	
Vitamin A equivalent	IU/g		39.7	210.0	

Alfalfa, aerial part, ensiled direct cut, prebloom, cut 1, (3)

Ref No 3 08 849 United States

Feed Name or Analyses			As Fed	Dry	C.V. ± %
Dry matter	%		23.1	100.0	
Sheep	dig coef %		60.	60.	
Ash	%		1.6	6.9	
Sheep	dig coef %		8.	33.	
Crude fiber	%		8.9	38.5	
Sheep	dig coef %		63.	63.	
Ether extract	%		1.2	5.3	
Sheep	dig coef %		68.	68.	
N-free extract	%		7.5	32.3	
Sheep	dig coef %		55.	55.	
Protein (N x 6.25)	%		3.9	17.0	
Sheep	dig coef %		72.	72.	
Cattle	dig prot %	*	2.7	11.7	
Goats	dig prot %	*	2.7	11.7	
Horses	dig prot %	*	2.7	11.7	
Sheep	dig prot %		2.8	12.2	
Cell walls (Van Soest)	%		12.7	54.8	
Sheep	dig coef %		54.	54.	
Cellulose (Matrone)	%		8.2	35.4	
Sheep	dig coef %		62.	62.	
Fiber, acid detergent (VS)	%		10.8	46.7	
Sheep	dig coef %		54.	54.	
Hemicellulose	%		1.9	8.1	
Sheep	dig coef %		49.	49.	
Pectins	%		13.7	59.1	
Pentosans	%		0.1	0.6	
Lignin (Ellis)	%		1.9	8.3	
Sheep	dig coef %		1.	1.	
Energy	GE Mcal/kg		1.13	4.90	
Sheep	GE dig coef %		62.	62.	
Cattle	DE Mcal/kg	*	0.61	2.63	
Sheep	DE Mcal/kg		0.70	3.04	
Cattle	ME Mcal/kg	*	0.50	2.16	
Sheep	ME Mcal/kg	*	0.58	2.49	
Cattle	TDN %	*	13.8	59.7	
Sheep	TDN %		14.4	62.4	
Dry matter intake	$g/W^{0.75}$		18.388	79.600	

Alfalfa, aerial part, ensiled direct cut, early bloom, cut 1, (3)

Ref No 3 08 880 United States

Feed Name or Analyses			As Fed	Dry	C.V. ± %
Dry matter	%		25.9	100.0	
Sheep	dig coef %		62.	62.	
Ash	%		1.9	7.3	
Sheep	dig coef %		9.	33.	
Crude fiber	%		8.7	33.6	
Sheep	dig coef %		58.	58.	
Ether extract	%		1.4	5.4	
Sheep	dig coef %		70.	70.	
N-free extract	%		9.3	36.0	
Sheep	dig coef %		67.	67.	
Protein (N x 6.25)	%		4.6	17.7	
Sheep	dig coef %		72.	72.	
Cattle	dig prot %	*	3.2	12.3	
Goats	dig prot %	*	3.2	12.3	
Horses	dig prot %	*	3.2	12.3	
Sheep	dig prot %		3.3	12.7	
Energy	GE Mcal/kg				
Cattle	DE Mcal/kg	*	0.73	2.80	
Sheep	DE Mcal/kg	*	0.74	2.86	
Cattle	ME Mcal/kg	*	0.60	2.30	
Sheep	ME Mcal/kg	*	0.61	2.34	
Cattle	TDN %	*	16.4	63.5	
Sheep	TDN %		16.8	64.9	
Dry matter intake	$g/W^{0.75}$		17.042	65.800	
Gain, sheep	kg/day		0.021	0.083	

Alfalfa, aerial part, ensiled direct cut, midbloom, cut 1, (3)

Ref No 3 08 879 United States

Feed Name or Analyses			As Fed	Dry	C.V. ± %
Dry matter	%		27.5	100.0	
Sheep	dig coef %		62.	62.	
Ash	%		1.8	6.4	
Sheep	dig coef %		10.	37.	
Crude fiber	%		9.5	34.4	
Sheep	dig coef %		68.	68.	
Ether extract	%		1.1	4.1	
Sheep	dig coef %		66.	66.	
N-free extract	%		10.6	38.5	
Sheep	dig coef %		53.	53.	
Protein (N x 6.25)	%		4.6	16.6	

Feed Name or Analyses			As Fed	Dry	C.V. ± %
Sheep	dig coef %		75.	75.	
Cattle	dig prot %	*	3.1	11.3	
Goats	dig prot %	*	3.1	11.3	
Horses	dig prot %	*	3.1	11.3	
Sheep	dig prot %		3.4	12.5	
Cell walls (Van Soest)	%		13.7	49.8	
Sheep	dig coef %		52.	52.	
Cellulose (Matrone)	%		10.0	36.3	
Sheep	dig coef %		65.	65.	
Fiber, acid detergent (VS)	%		10.9	39.7	
Sheep	dig coef %		50.	50.	
Hemicellulose	%		2.8	10.1	
Sheep	dig coef %		56.	56.	
Lignin (Ellis)	%		2.1	7.6	
Sheep	dig coef %		6.	6.	
Energy	GE Mcal/kg		1.33	4.82	
Sheep	GE dig coef %		63.	63.	
Cattle	DE Mcal/kg	*	0.76	2.77	
Sheep	DE Mcal/kg		0.84	3.04	
Cattle	ME Mcal/kg	*	0.62	2.27	
Sheep	ME Mcal/kg	*	0.68	2.49	
Cattle	TDN %	*	17.3	62.9	
Sheep	TDN %		17.1	62.3	
Dry matter intake	$g/W^{0.75}$		20.391	74.150	
Gain, cattle	kg/day		0.037	0.136	
Gain, sheep	kg/day		0.037	0.136	

Alfalfa, aerial part, wilted ensiled, (3)

Ref No 3 00 221 United States

Feed Name or Analyses			As Fed	Dry	C.V. ± %
Dry matter	%		38.1	100.0	12
Ash	%		3.3	8.6	9
Crude fiber	%		11.8	30.9	12
Cattle	dig coef %		53.	53.	
Sheep	dig coef %	*	46.	46.	
Ether extract	%		1.4	3.6	14
Cattle	dig coef %		64.	64.	
Sheep	dig coef %	*	52.	52.	
N-free extract	%		14.9	39.1	
Cattle	dig coef %		64.	64.	
Sheep	dig coef %	*	62.	62.	
Protein (N x 6.25)	%		6.8	17.8	9
Cattle	dig coef %		67.	67.	
Sheep	dig coef %	*	64.	64.	
Cattle	dig prot %		4.5	11.9	
Goats	dig prot %	*	4.7	12.4	
Horses	dig prot %	*	4.7	12.4	
Sheep	dig prot %		4.3	11.4	
Energy	GE Mcal/kg				
Cattle	DE Mcal/kg	*	0.98	2.58	
Sheep	DE Mcal/kg	*	0.91	2.38	
Cattle	ME Mcal/kg	*	0.81	2.12	
Sheep	ME Mcal/kg	*	0.74	1.95	
Cattle	NE_m Mcal/kg	*	0.50	1.31	
Cattle	NE_{gain} Mcal/kg	*	0.26	0.69	
Cattle	$NE_{lactating\ cows}$ Mcal/kg	*	0.53	1.38	
Cattle	TDN %		22.3	58.5	
Sheep	TDN %		20.6	54.1	
Calcium	%		0.53	1.40	8
Chlorine	%		0.16	0.41	
Copper	mg/kg		3.5	9.2	14
Iron	%		0.012	0.031	11
Magnesium	%		0.13	0.33	14
Manganese	mg/kg		19.8	51.9	18
Phosphorus	%		0.12	0.33	13
Potassium	%		0.89	2.34	2
Sodium	%		0.05	0.14	
Sulphur	%		0.14	0.36	

Continued

Feed Name or Analyses			Mean As Fed	Mean Dry	C.V. ± %
Carotene	mg/kg		19.6	51.6	34
Vitamin A equivalent	IU/g		32.7	86.0	
Vitamin D2	IU/g		0.2	0.6	

Alfalfa, aerial part, wilted ensiled, immature, (3)
Ref No 3 00 214 United States

Feed Name or Analyses			Mean As Fed	Mean Dry	C.V. ± %
Dry matter	%		37.4	100.0	9
Ash	%		3.3	8.7	6
Crude fiber	%		10.4	27.8	14
Ether extract	%		1.4	3.8	16
N-free extract	%		14.5	38.9	
Protein (N x 6.25)	%		7.8	20.8	6
Cattle	dig prot %	*	5.7	15.1	
Goats	dig prot %	*	5.7	15.1	
Horses	dig prot %	*	5.7	15.1	
Sheep	dig prot %	*	5.7	15.1	
Energy	GE Mcal/kg				
Cattle	DE Mcal/kg	*	0.90	2.40	
Sheep	DE Mcal/kg	*	0.90	2.42	
Cattle	ME Mcal/kg	*	0.73	1.96	
Sheep	ME Mcal/kg	*	0.74	1.98	
Cattle	TDN %	*	20.3	54.3	
Sheep	TDN %	*	20.5	54.8	

Alfalfa, aerial part, wilted ensiled, pre-bloom, (3)
Ref No 3 00 215 United States

Feed Name or Analyses			Mean As Fed	Mean Dry	C.V. ± %
Dry matter	%		29.1	100.0	
Sheep	dig coef %		64.	64.	
Ash	%		2.5	8.6	
Crude fiber	%		8.3	28.4	
Cattle	dig coef %		57.	57.	
Ether extract	%		1.2	4.0	
Cattle	dig coef %		67.	67.	
N-free extract	%		10.8	37.1	
Cattle	dig coef %		68.	68.	
Protein (N x 6.25)	%		6.4	22.0	
Cattle	dig coef %		70.	70.	
Sheep	dig coef %		78.	78.	
Cattle	dig prot %		4.5	15.4	
Goats	dig prot %	*	4.7	16.2	
Horses	dig prot %	*	4.7	16.2	
Sheep	dig prot %		5.0	17.1	
Fatty acids	%				
Acetic	%		1.394	4.800	
Butyric	%		0.755	2.600	
Lactic acid	%		0.697	2.400	
Energy	GE Mcal/kg		1.33	4.56	
Sheep	GE dig coef %		66.	66.	
Cattle	DE Mcal/kg	*	0.80	2.77	
Sheep	DE Mcal/kg		0.87	3.01	
Cattle	ME Mcal/kg	*	0.66	2.27	
Sheep	ME Mcal/kg	*	0.72	2.47	
Cattle	TDN %		18.2	62.8	
Sheep	TDN %	*	19.8	68.2	
pH	pH units		5.30	5.30	

(1) dry forages and roughages
(2) pasture, range plants, and forages fed green
(3) silages
(4) energy feeds
(5) protein supplements
(6) minerals
(7) vitamins
(8) additives

Alfalfa, aerial part, wilted ensiled, early bloom, (3)
Ref No 3 00 216 United States

Feed Name or Analyses			Mean As Fed	Mean Dry	C.V. ± %
Dry matter	%		36.8	100.0	11
Ash	%		3.0	8.2	11
Crude fiber	%		11.7	31.8	5
Cattle	dig coef %		48.	48.	
Ether extract	%		1.2	3.2	12
Cattle	dig coef %		58.	58.	
N-free extract	%		14.9	40.6	
Cattle	dig coef %		57.	57.	
Protein (N x 6.25)	%		6.0	16.3	9
Cattle	dig coef %		60.	60.	
Cattle	dig prot %		3.6	9.8	
Goats	dig prot %	*	4.1	11.0	
Horses	dig prot %	*	4.1	11.0	
Sheep	dig prot %	*	4.1	11.0	
Energy	GE Mcal/kg				
Cattle	DE Mcal/kg	*	0.85	2.30	
Sheep	DE Mcal/kg	*	0.90	2.44	
Cattle	ME Mcal/kg	*	0.69	1.89	
Sheep	ME Mcal/kg	*	0.73	2.00	
Cattle	TDN %		19.2	52.3	
Sheep	TDN %	*	20.3	55.3	

Alfalfa, aerial part, wilted ensiled, early bloom, cut 1, (3)
Ref No 3 07 843 United States

Feed Name or Analyses			Mean As Fed	Mean Dry	C.V. ± %
Dry matter	%		30.3	100.0	
Sheep	dig coef %		63.	63.	
Protein (N x 6.25)	%		6.6	21.8	
Sheep	dig coef %		75.	75.	
Cattle	dig prot %	*	4.9	16.0	
Goats	dig prot %	*	4.9	16.0	
Horses	dig prot %	*	4.9	16.0	
Sheep	dig prot %		5.0	16.4	
Fatty acids	%				
Acetic	%		1.818	6.000	
Butyric	%		0.848	2.800	
Lactic acid	%		0.970	3.200	
Energy	GE Mcal/kg		1.39	4.59	
Sheep	GE dig coef %		63.	63.	
Sheep	DE Mcal/kg		0.88	2.89	
Sheep	ME Mcal/kg	*	0.72	2.37	
pH	pH units		5.20	5.20	

Alfalfa, aerial part, wilted ensiled, mid-bloom, (3)
Ref No 3 00 217 United States

Feed Name or Analyses			Mean As Fed	Mean Dry	C.V. ± %
Dry matter	%		38.1	100.0	10
Ash	%		3.0	7.9	12
Crude fiber	%		11.9	31.2	12
Cattle	dig coef %		52.	52.	
Ether extract	%		1.2	3.1	13
Cattle	dig coef %		68.	68.	
N-free extract	%		15.6	41.1	
Cattle	dig coef %		67.	67.	
Protein (N x 6.25)	%		6.4	16.8	9
Cattle	dig coef %		70.	70.	
Cattle	dig prot %		4.5	11.8	
Goats	dig prot %	*	4.4	11.5	
Horses	dig prot %	*	4.4	11.5	
Sheep	dig prot %	*	4.4	11.5	
Energy	GE Mcal/kg				
Cattle	DE Mcal/kg	*	1.01	2.65	
Sheep	DE Mcal/kg	*	0.99	2.59	

Feed Name or Analyses			Mean As Fed	Mean Dry	C.V. ± %
Cattle	ME Mcal/kg	*	0.83	2.18	
Sheep	ME Mcal/kg	*	0.81	2.12	
Cattle	TDN %		22.9	60.2	
Sheep	TDN %	*	22.4	58.7	

Alfalfa, aerial part, wilted ensiled, full bloom, (3)
Ref No 3 00 218 United States

Feed Name or Analyses			Mean As Fed	Mean Dry	C.V. ± %
Dry matter	%		36.4	100.0	6
Ash	%		2.8	7.7	5
Crude fiber	%		12.1	33.2	4
Ether extract	%		1.0	2.7	9
N-free extract	%		15.1	41.5	
Protein (N x 6.25)	%		5.4	14.9	5
Cattle	dig prot %	*	3.6	9.8	
Goats	dig prot %	*	3.6	9.8	
Horses	dig prot %	*	3.6	9.8	
Sheep	dig prot %	*	3.6	9.8	
Energy	GE Mcal/kg				
Cattle	DE Mcal/kg	*	0.85	2.34	
Sheep	DE Mcal/kg	*	0.84	2.30	
Cattle	ME Mcal/kg	*	0.70	1.92	
Sheep	ME Mcal/kg	*	0.69	1.89	
Cattle	TDN %	*	19.3	53.1	
Sheep	TDN %	*	19.0	52.1	

Alfalfa, aerial part, wilted ensiled, full bloom, cut 1, (3)
Ref No 3 07 840 United States

Feed Name or Analyses			Mean As Fed	Mean Dry	C.V. ± %
Dry matter	%		34.7	100.0	
Sheep	dig coef %		57.	57.	
Fatty acids	%				
Acetic	%		1.249	3.600	
Butyric	%		0.104	0.300	
Lactic acid	%		1.770	5.100	
Energy	GE Mcal/kg		1.58	4.55	
Sheep	GE dig coef %		54.	54.	
Sheep	DE Mcal/kg		0.85	2.46	
Sheep	ME Mcal/kg	*	0.70	2.01	
pH	pH units		5.10	5.10	

Alfalfa, aerial part, wilted ensiled, milk stage, (3)
Ref No 3 00 219 United States

Feed Name or Analyses			Mean As Fed	Mean Dry	C.V. ± %
Dry matter	%		36.3	100.0	
Calcium	%		0.55	1.51	
Phosphorus	%		0.13	0.36	

Alfalfa, aerial part, wilted ensiled, mature, (3)
Ref No 3 00 220 United States

Feed Name or Analyses			Mean As Fed	Mean Dry	C.V. ± %
Dry matter	%		41.1	100.0	
Calcium	%		0.47	1.14	
Phosphorus	%		0.11	0.26	
Carotene	mg/kg		6.0	14.6	19
Vitamin A equivalent	IU/g		10.0	24.3	

Alfalfa, aerial part w AIV preservative added, ensiled, (3)
Ref No 3 00 225 United States

Feed Name or Analyses			Mean As Fed	Mean Dry	C.V. ± %
Dry matter	%			100.0	
Manganese	mg/kg			22.0	

Alfalfa, aerial part w AIV preservative added, ensiled, full bloom, (3)

Ref No 3 00 224 — United States

Feed Name or Analyses			Mean As Fed	Mean Dry	C.V. ± %
Dry matter	%		36.2	100.0	10
Ash	%		2.8	7.8	98
Crude fiber	%		12.0	33.2	9
Cattle	dig coef %		54.	54.	
Ether extract	%		1.0	2.8	8
Cattle	dig coef %		62.	62.	
N-free extract	%		14.9	41.3	
Cattle	dig coef %		60.	60.	
Protein (N x 6.25)	%		5.4	15.1	6
Cattle	dig coef %		66.	66.	
Cattle	dig prot %		3.6	9.9	
Goats	dig prot %	*	3.6	9.9	
Horses	dig prot %	*	3.6	9.9	
Sheep	dig prot %	*	3.6	9.9	
Energy	GE Mcal/kg				
Cattle	DE Mcal/kg	*	0.90	2.49	
Sheep	DE Mcal/kg	*	0.88	2.43	
Cattle	ME Mcal/kg	*	0.74	2.04	
Sheep	ME Mcal/kg	*	0.72	1.99	
Cattle	TDN %		20.4	56.4	
Sheep	TDN %	*	19.9	55.1	

Alfalfa, aerial part w formic acid added, ensiled, early bloom, cut 1, (3)

Ref No 3 08 850 — United States

Feed Name or Analyses			Mean As Fed	Mean Dry	C.V. ± %
Dry matter	%		26.2	100.0	
Sheep	dig coef %		64.	64.	
Ash	%		1.8	7.0	
Sheep	dig coef %		10.	39.	
Crude fiber	%		7.8	29.8	
Sheep	dig coef %		53.	53.	
Ether extract	%		1.3	5.0	
Sheep	dig coef %		72.	72.	
N-free extract	%		10.3	39.2	
Sheep	dig coef %		70.	70.	
Protein (N x 6.25)	%		5.0	19.0	
Sheep	dig coef %		74.	74.	
Cattle	dig prot %	*	3.5	13.5	
Goats	dig prot %	*	3.5	13.5	
Horses	dig prot %	*	3.5	13.5	
Sheep	dig prot %		3.7	14.1	
Energy	GE Mcal/kg				
Cattle	DE Mcal/kg	*	0.62	2.37	
Sheep	DE Mcal/kg	*	0.76	2.88	
Cattle	ME Mcal/kg	*	0.51	1.94	
Sheep	ME Mcal/kg	*	0.62	2.36	
Cattle	TDN %	*	14.1	53.8	
Sheep	TDN %		17.1	65.4	
Dry matter intake	$g/W^{0.75}$		19.729	75.300	
Gain, sheep	kg/day		0.023	0.089	

Alfalfa, aerial part w grain added, ensiled, (3)

Ref No 3 08 324 — United States

Feed Name or Analyses			Mean As Fed	Mean Dry	C.V. ± %
Dry matter	%		25.5	100.0	
Ash	%		2.4	9.4	
Crude fiber	%		7.5	29.4	
Ether extract	%		1.2	4.7	
N-free extract	%		10.6	41.6	
Protein (N x 6.25)	%		3.8	14.9	
Cattle	dig prot %	*	2.5	9.8	
Goats	dig prot %	*	2.5	9.8	
Horses	dig prot %	*	2.5	9.8	
Sheep	dig prot %	*	2.5	9.8	
Energy	GE Mcal/kg				
Cattle	DE Mcal/kg	*	0.69	2.70	
Sheep	DE Mcal/kg	*	0.69	2.72	
Cattle	ME Mcal/kg	*	0.57	2.22	
Sheep	ME Mcal/kg	*	0.57	2.23	
Cattle	TDN %	*	15.6	61.3	
Sheep	TDN %	*	15.7	61.6	

Alfalfa, aerial part w grnd corn ears added, ensiled, (3)

Ref No 3 00 232 — United States

Feed Name or Analyses			Mean As Fed	Mean Dry	C.V. ± %
Dry matter	%		25.9	100.0	
Carotene	mg/kg		37.3	144.0	
Vitamin A equivalent	IU/g		62.2	240.0	

Alfalfa, aerial part w grnd corn grain added, ensiled, (3)

Ref No 3 00 226 — United States

Feed Name or Analyses			Mean As Fed	Mean Dry	C.V. ± %
Dry matter	%		25.5	100.0	
Ash	%		2.0	7.8	12
Crude fiber	%		6.3	24.8	12
Ether extract	%		1.2	4.7	7
N-free extract	%		11.1	43.7	
Protein (N x 6.25)	%		4.8	19.0	14
Cattle	dig prot %	*	3.4	13.5	
Goats	dig prot %	*	3.4	13.5	
Horses	dig prot %	*	3.4	13.5	
Sheep	dig prot %	*	3.4	13.5	
Energy	GE Mcal/kg				
Cattle	DE Mcal/kg	*	0.67	2.61	
Sheep	DE Mcal/kg	*	0.61	2.40	
Cattle	ME Mcal/kg	*	0.55	2.14	
Sheep	ME Mcal/kg	*	0.50	1.97	
Cattle	TDN %	*	15.1	59.2	
Sheep	TDN %	*	13.9	54.4	

Alfalfa, aerial part w sulfuric acid preservative added, ensiled, (3)

Ref No 3 00 227 — United States

Feed Name or Analyses			Mean As Fed	Mean Dry	C.V. ± %
Dry matter	%		24.0	100.0	
Ash	%		2.8	11.6	
Crude fiber	%		7.4	31.0	
Sheep	dig coef %		45.	45.	
Ether extract	%		1.0	4.0	
Sheep	dig coef %		65.	65.	
N-free extract	%		8.3	34.4	
Sheep	dig coef %		60.	60.	
Protein (N x 6.25)	%		4.6	19.0	
Sheep	dig coef %		68.	68.	
Cattle	dig prot %	*	3.2	13.5	
Goats	dig prot %	*	3.2	13.5	
Horses	dig prot %	*	3.2	13.5	
Sheep	dig prot %		3.1	12.9	
Energy	GE Mcal/kg				
Cattle	DE Mcal/kg	*	0.60	2.52	
Sheep	DE Mcal/kg	*	0.56	2.35	
Cattle	ME Mcal/kg	*	0.50	2.07	
Sheep	ME Mcal/kg	*	0.46	1.93	
Cattle	TDN %	*	13.7	57.1	
Sheep	TDN %		12.8	53.4	

Alfalfa, aerial part w phosphoric acid preservative added, ensiled, (3)

Ref No 3 00 231 — United States

Feed Name or Analyses			Mean As Fed	Mean Dry	C.V. ± %
Dry matter	%		28.7	100.0	4
Ash	%		2.7	9.4	12
Crude fiber	%		8.9	31.1	7
Cattle	dig coef %		45.	45.	
Sheep	dig coef %	*	46.	46.	
Ether extract	%		1.1	4.0	17
Cattle	dig coef %		50.	50.	
Sheep	dig coef %	*	52.	52.	
N-free extract	%		10.7	37.2	
Cattle	dig coef %		65.	65.	
Sheep	dig coef %	*	62.	62.	
Protein (N x 6.25)	%		5.2	18.3	9
Cattle	dig coef %		70.	70.	
Sheep	dig coef %	*	64.	64.	
Cattle	dig prot %		3.6	12.7	
Goats	dig prot %	*	3.7	12.8	
Horses	dig prot %	*	3.7	12.8	
Sheep	dig prot %		3.4	11.7	
Energy	GE Mcal/kg				
Cattle	DE Mcal/kg	*	0.70	2.44	
Sheep	DE Mcal/kg	*	0.68	2.37	
Cattle	ME Mcal/kg	*	0.57	2.00	
Sheep	ME Mcal/kg	*	0.56	1.94	
Cattle	TDN %		15.9	55.4	
Sheep	TDN %		15.4	53.7	
Calcium	%		0.40	1.38	
Phosphorus	%		0.32	1.12	
Potassium	%		0.67	2.35	
Carotene	mg/kg		29.7	103.6	72
Vitamin A equivalent	IU/g		49.6	172.7	

Alfalfa, aerial part w phosphoric acid preservative added, ensiled, early bloom, (3)

Ref No 3 00 228 — United States

Feed Name or Analyses			Mean As Fed	Mean Dry	C.V. ± %
Dry matter	%		29.0	100.0	3
Ash	%		2.8	9.8	14
Crude fiber	%		9.6	32.9	4
Cattle	dig coef %		56.	56.	
Ether extract	%		1.1	3.9	24
Cattle	dig coef %		66.	66.	
N-free extract	%		10.7	36.7	
Cattle	dig coef %		64.	64.	
Protein (N x 6.25)	%		4.8	16.7	7
Cattle	dig coef %		69.	69.	
Cattle	dig prot %		3.3	11.5	
Goats	dig prot %	*	3.3	11.4	
Horses	dig prot %	*	3.3	11.4	
Sheep	dig prot %	*	3.3	11.4	
Energy	GE Mcal/kg				
Cattle	DE Mcal/kg	*	0.76	2.60	
Sheep	DE Mcal/kg	*	0.73	2.51	
Cattle	ME Mcal/kg	*	0.62	2.13	
Sheep	ME Mcal/kg	*	0.60	2.05	
Cattle	TDN %		17.1	59.0	
Sheep	TDN %	*	16.5	56.8	

Feed Name or Analyses			Mean As Fed	Mean Dry	C.V. ± %

Alfalfa, aerial part w phosphoric acid preservative added, ensiled, mid-bloom, (3)
Ref No 3 00 229 United States

Analyses			As Fed	Dry	C.V. ± %
Dry matter	%		29.3	100.0	
Ash	%		2.3	8.0	
Crude fiber	%		9.7	33.1	
Cattle	dig coef %		41.	41.	
Ether extract	%		0.9	3.1	
Cattle	dig coef %		37.	37.	
N-free extract	%		11.0	37.6	
Cattle	dig coef %		66.	66.	
Protein (N x 6.25)	%		5.4	18.4	
Cattle	dig coef %		69.	69.	
Cattle	dig prot %		3.7	12.7	
Goats	dig prot %	*	3.8	12.9	
Horses	dig prot %	*	3.8	12.9	
Sheep	dig prot %	*	3.8	12.9	
Energy	GE Mcal/kg				
Cattle	DE Mcal/kg	*	0.69	2.35	
Sheep	DE Mcal/kg	*	0.71	2.43	
Cattle	ME Mcal/kg	*	0.56	1.93	
Sheep	ME Mcal/kg	*	0.58	1.99	
Cattle	TDN %		15.6	53.4	
Sheep	TDN %	*	16.1	55.1	

Alfalfa, aerial part w phosphoric acid preservative added, ensiled, full bloom, (3)
Ref No 3 00 230 United States

Analyses			As Fed	Dry	C.V. ± %
Dry matter	%		29.6	100.0	6
Ash	%		2.3	7.8	6
Crude fiber	%		9.7	32.8	5
Ether extract	%		0.9	3.2	9
N-free extract	%		11.2	37.8	
Protein (N x 6.25)	%		5.4	18.4	7
Cattle	dig prot %	*	3.8	12.9	
Goats	dig prot %	*	3.8	12.9	
Horses	dig prot %	*	3.8	12.9	
Sheep	dig prot %	*	3.8	12.9	
Energy	GE Mcal/kg				
Cattle	DE Mcal/kg	*	0.66	2.22	
Sheep	DE Mcal/kg	*	0.72	2.43	
Cattle	ME Mcal/kg	*	0.54	1.82	
Sheep	ME Mcal/kg	*	0.59	1.99	
Cattle	TDN %	*	14.9	50.4	
Sheep	TDN %	*	16.3	55.1	
Calcium	%		0.41	1.39	7
Phosphorus	%		0.34	1.14	7

Alfalfa, aerial part w molasses added, ensiled, (3)
Ref No 3 00 238 United States

Analyses			As Fed	Dry	C.V. ± %
Dry matter	%		29.9	100.0	8
Ash	%		2.6	8.5	55
Crude fiber	%		8.6	28.8	8
Cattle	dig coef %		47.	47.	
Sheep	dig coef %	*	46.	46.	
Ether extract	%		1.2	4.0	27
Cattle	dig coef %		65.	65.	
Sheep	dig coef %	*	52.	52.	
N-free extract	%		12.4	41.3	
Cattle	dig coef %		70.	70.	
Sheep	dig coef %	*	62.	62.	
Protein (N x 6.25)	%		5.2	17.3	14
Cattle	dig coef %		69.	69.	
Sheep	dig coef %	*	64.	64.	
Cattle	dig prot %		3.6	11.9	
Goats	dig prot %	*	3.6	11.9	
Horses	dig prot %	*	3.6	11.9	
Sheep	dig prot %		3.3	11.1	
Energy	GE Mcal/kg				
Cattle	DE Mcal/kg	*	0.79	2.66	
Sheep	DE Mcal/kg	*	0.72	2.41	
Cattle	ME Mcal/kg	*	0.65	2.18	
Sheep	ME Mcal/kg	*	0.59	1.98	
Cattle	NE_m Mcal/kg	*	0.38	1.27	
Cattle	NE_{gain} Mcal/kg	*	0.19	0.63	
Cattle	$NE_{lactating\ cows}$ Mcal/kg	*	0.42	1.41	
Cattle	TDN %		18.0	60.3	
Sheep	TDN %		16.4	54.7	
Calcium	%		0.46	1.53	
Phosphorus	%		0.09	0.30	
Potassium	%		0.79	2.65	
Carotene	mg/kg		29.1	97.2	56
Vitamin A equivalent	IU/g		48.5	162.1	

(1) dry forages and roughages (3) silages (6) minerals
(2) pasture, range plants, and forages fed green (4) energy feeds (7) vitamins
(5) protein supplements (8) additives

Alfalfa, aerial part w molasses added, ensiled, immature, (3)
Ref No 3 00 233 United States

Analyses			As Fed	Dry	C.V. ± %
Dry matter	%		31.9	100.0	10
Ash	%		2.9	9.0	8
Crude fiber	%		8.2	25.8	5
Ether extract	%		1.3	4.0	8
N-free extract	%		12.8	40.2	
Protein (N x 6.25)	%		6.7	21.0	5
Cattle	dig prot %	*	4.9	15.3	
Goats	dig prot %	*	4.9	15.3	
Horses	dig prot %	*	4.9	15.3	
Sheep	dig prot %	*	4.9	15.3	
Energy	GE Mcal/kg				
Cattle	DE Mcal/kg	*	0.80	2.50	
Sheep	DE Mcal/kg	*	0.77	2.41	
Cattle	ME Mcal/kg	*	0.65	2.05	
Sheep	ME Mcal/kg	*	0.63	1.98	
Cattle	TDN %	*	18.1	56.6	
Sheep	TDN %	*	17.5	54.8	

Alfalfa, aerial part w molasses added, ensiled, pre-bloom, (3)
Ref No 3 00 234 United States

Analyses			As Fed	Dry	C.V. ± %
Dry matter	%		25.3	100.0	
Ash	%		2.2	8.8	
Crude fiber	%		6.5	25.6	
Cattle	dig coef %		59.	59.	
Ether extract	%		1.1	4.3	
Cattle	dig coef %		69.	69.	
N-free extract	%		10.1	40.1	
Cattle	dig coef %		73.	73.	
Protein (N x 6.25)	%		5.4	21.2	
Cattle	dig coef %		74.	74.	
Cattle	dig prot %		4.0	15.7	
Goats	dig prot %	*	3.9	15.5	
Horses	dig prot %	*	3.9	15.5	
Sheep	dig prot %	*	3.9	15.5	
Energy	GE Mcal/kg				
Cattle	DE Mcal/kg	*	0.74	2.94	
Sheep	DE Mcal/kg	*	0.67	2.66	
Cattle	ME Mcal/kg	*	0.61	2.41	
Sheep	ME Mcal/kg	*	0.55	2.18	
Cattle	TDN %		16.9	66.7	
Sheep	TDN %	*	15.3	60.3	

Alfalfa, aerial part w molasses added, ensiled, early bloom, (3)
Ref No 3 00 235 United States

Analyses			As Fed	Dry	C.V. ± %
Dry matter	%		30.1	100.0	14
Ash	%		2.6	8.6	11
Crude fiber	%		9.5	31.4	5
Cattle	dig coef %		50.	50.	
Ether extract	%		1.4	4.7	8
Cattle	dig coef %		72.	72.	
N-free extract	%		12.0	39.9	
Cattle	dig coef %		66.	66.	
Protein (N x 6.25)	%		4.6	15.4	8
Cattle	dig coef %		61.	61.	
Cattle	dig prot %		2.8	9.4	
Goats	dig prot %	*	3.1	10.2	
Horses	dig prot %	*	3.1	10.2	
Sheep	dig prot %	*	3.1	10.2	
Energy	GE Mcal/kg				
Cattle	DE Mcal/kg	*	0.78	2.59	
Sheep	DE Mcal/kg	*	0.75	2.49	
Cattle	ME Mcal/kg	*	0.64	2.13	
Sheep	ME Mcal/kg	*	0.61	2.04	
Cattle	TDN %		17.7	58.8	
Sheep	TDN %	*	17.0	56.4	

Alfalfa, aerial part w molasses added, ensiled, mid-bloom, (3)
Ref No 3 00 236 United States

Analyses			As Fed	Dry	C.V. ± %
Dry matter	%		31.0	100.0	8
Ash	%		2.6	8.5	11
Crude fiber	%		9.4	30.5	5
Cattle	dig coef %		43.	43.	
Ether extract	%		1.0	3.3	40
Cattle	dig coef %		52.	52.	
N-free extract	%		12.7	41.0	
Cattle	dig coef %		65.	65.	
Protein (N x 6.25)	%		5.2	16.8	9
Cattle	dig coef %		65.	65.	
Cattle	dig prot %		3.4	10.9	
Goats	dig prot %	*	3.6	11.5	
Horses	dig prot %	*	3.6	11.5	
Sheep	dig prot %	*	3.6	11.5	
Energy	GE Mcal/kg				
Cattle	DE Mcal/kg	*	0.74	2.39	
Sheep	DE Mcal/kg	*	0.75	2.43	
Cattle	ME Mcal/kg	*	0.61	1.96	
Sheep	ME Mcal/kg	*	0.62	2.00	
Cattle	TDN %		16.8	54.3	
Sheep	TDN %	*	17.1	55.2	

Alfalfa, aerial part w molasses added, ensiled, full bloom, (3)
Ref No 3 00 237 United States

Analyses			As Fed	Dry	C.V. ± %
Dry matter	%		27.8	100.0	
Ash	%		2.6	9.3	
Crude fiber	%		8.1	29.3	
Ether extract	%		0.8	3.0	
N-free extract	%		11.8	42.3	

Feed Name or Analyses			Mean As Fed	Mean Dry	C.V. ± %
Protein (N x 6.25)	%		4.5	16.1	
Cattle	dig prot %	*	3.0	10.8	
Goats	dig prot %	*	3.0	10.8	
Horses	dig prot %	*	3.0	10.8	
Sheep	dig prot %	*	3.0	10.8	
Energy	GE Mcal/kg				
Cattle	DE Mcal/kg	*	0.71	2.55	
Sheep	DE Mcal/kg	*	0.68	2.44	
Cattle	ME Mcal/kg	*	0.58	2.09	
Sheep	ME Mcal/kg	*	0.56	2.00	
Cattle	TDN %	*	16.1	57.8	
Sheep	TDN %	*	15.4	55.4	

Alfalfa, aerial part w molasses added, wilted ensiled, (3)

Ref No 3 00 241 United States

Feed Name or Analyses			Mean As Fed	Mean Dry	C.V. ± %
Dry matter	%		37.1	100.0	
Ash	%		3.2	8.5	
Crude fiber	%		10.9	29.5	
Cattle	dig coef %		52.	52.	
Ether extract	%		1.2	3.1	
Cattle	dig coef %		62.	62.	
N-free extract	%		15.4	41.5	
Cattle	dig coef %		67.	67.	
Protein (N x 6.25)	%		6.4	17.3	
Cattle	dig coef %		67.	67.	
Cattle	dig prot %		4.3	11.6	
Goats	dig prot %	*	4.4	11.9	
Horses	dig prot %	*	4.4	11.9	
Sheep	dig prot %	*	4.4	11.9	
Energy	GE Mcal/kg				
Cattle	DE Mcal/kg	*	0.97	2.61	
Sheep	DE Mcal/kg	*	0.90	2.42	
Cattle	ME Mcal/kg	*	0.79	2.14	
Sheep	ME Mcal/kg	*	0.74	1.99	
Cattle	TDN %		21.9	59.1	
Sheep	TDN %	*	20.4	54.9	
Calcium	%		0.57	1.54	
Phosphorus	%		0.11	0.31	
Potassium	%		0.98	2.65	

Alfalfa, aerial part w molasses added, wilted ensiled, pre-bloom, (3)

Ref No 3 00 239 United States

Feed Name or Analyses			Mean As Fed	Mean Dry	C.V. ± %
Dry matter	%		38.6	100.0	
Ash	%		3.5	9.1	
Crude fiber	%		9.9	25.7	
Cattle	dig coef %		54.	54.	
Ether extract	%		1.4	3.7	
Cattle	dig coef %		65.	65.	
N-free extract	%		15.6	40.5	
Cattle	dig coef %		70.	70.	
Protein (N x 6.25)	%		8.1	21.0	
Cattle	dig coef %		70.	70.	
Cattle	dig prot %		5.7	14.7	
Goats	dig prot %	*	5.9	15.3	
Horses	dig prot %	*	5.9	15.3	
Sheep	dig prot %	*	5.9	15.3	
Energy	GE Mcal/kg				
Cattle	DE Mcal/kg	*	1.06	2.75	
Sheep	DE Mcal/kg	*	0.98	2.55	
Cattle	ME Mcal/kg	*	0.87	2.25	
Sheep	ME Mcal/kg	*	0.81	2.09	
Cattle	TDN %		24.1	62.3	
Sheep	TDN %	*	22.3	57.8	

Alfalfa, aerial part w molasses added, wilted ensiled, early bloom, (3)

Ref No 3 00 240 United States

Feed Name or Analyses			Mean As Fed	Mean Dry	C.V. ± %
Dry matter	%		34.0	100.0	
Ash	%		3.0	8.9	
Crude fiber	%		11.0	32.4	
Cattle	dig coef %		52.	52.	
Ether extract	%		1.0	2.9	
Cattle	dig coef %		63.	63.	
N-free extract	%		13.7	40.4	
Cattle	dig coef %		64.	64.	
Protein (N x 6.25)	%		5.2	15.4	
Cattle	dig coef %		65.	65.	
Cattle	dig prot %		3.4	10.0	
Goats	dig prot %	*	3.5	10.2	
Horses	dig prot %	*	3.5	10.2	
Sheep	dig prot %	*	3.5	10.2	
Energy	GE Mcal/kg				
Cattle	DE Mcal/kg	*	0.85	2.51	
Sheep	DE Mcal/kg	*	0.84	2.46	
Cattle	ME Mcal/kg	*	0.70	2.05	
Sheep	ME Mcal/kg	*	0.68	2.01	
Cattle	TDN %		19.3	56.8	
Sheep	TDN %	*	18.9	55.7	

Alfalfa, aerial part w phosphorus pentachloride preservative added, ensiled, (3)

Ref No 3 00 223 United States

Feed Name or Analyses			Mean As Fed	Mean Dry	C.V. ± %
Dry matter	%		23.3	100.0	34
Ash	%		2.6	11.1	4
Crude fiber	%		8.1	34.7	6
Sheep	dig coef %		51.	51.	
Ether extract	%		0.8	3.3	8
Sheep	dig coef %		62.	62.	
N-free extract	%		7.1	30.6	
Sheep	dig coef %		69.	69.	
Protein (N x 6.25)	%		4.7	20.3	4
Sheep	dig coef %		80.	80.	
Cattle	dig prot %	*	3.4	14.7	
Goats	dig prot %	*	3.4	14.7	
Horses	dig prot %	*	3.4	14.7	
Sheep	dig prot %		3.8	16.2	
Energy	GE Mcal/kg				
Cattle	DE Mcal/kg	*	0.62	2.65	
Sheep	DE Mcal/kg	*	0.61	2.63	
Cattle	ME Mcal/kg	*	0.51	2.17	
Sheep	ME Mcal/kg	*	0.50	2.16	
Cattle	TDN %	*	14.0	60.2	
Sheep	TDN %		13.9	59.7	

Alfalfa, aerial part w phosphorus pentachloride preservative added, ensiled, pre-bloom, (3)

Ref No 3 00 222 United States

Feed Name or Analyses			Mean As Fed	Mean Dry	C.V. ± %
Dry matter	%		38.2	100.0	
Ash	%		3.9	10.2	
Crude fiber	%		11.7	30.6	
Sheep	dig coef %		45.	45.	
Ether extract	%		1.2	3.2	
Sheep	dig coef %		47.	47.	
N-free extract	%		14.3	37.4	
Sheep	dig coef %		75.	75.	
Protein (N x 6.25)	%		7.1	18.6	
Sheep	dig coef %		73.	73.	
Cattle	dig prot %	*	5.0	13.1	
Goats	dig prot %	*	5.0	13.1	
Horses	dig prot %	*	5.0	13.1	
Sheep	dig prot %		5.2	13.6	
Energy	GE Mcal/kg				
Cattle	DE Mcal/kg	*	0.93	2.43	
Sheep	DE Mcal/kg	*	0.99	2.59	
Cattle	ME Mcal/kg	*	0.76	2.00	
Sheep	ME Mcal/kg	*	0.81	2.13	
Cattle	TDN %	*	21.1	55.2	
Sheep	TDN %		22.5	58.8	

Alfalfa, aerial part w salt added, ensiled, (3)

Ref No 3 00 242 United States

Feed Name or Analyses			Mean As Fed	Mean Dry	C.V. ± %
Dry matter	%		26.6	100.0	19
Carotene	mg/kg		31.7	119.0	29
Vitamin A equivalent	IU/g		52.8	198.5	

Alfalfa, aerial part w sodium bisulfite preservative added, ensiled, pre-bloom, cut 1, (3)

Ref No 3 07 846 United States

Feed Name or Analyses			Mean As Fed	Mean Dry	C.V. ± %
Dry matter	%		23.0	100.0	
Sheep	dig coef %		73.	73.	
Protein (N x 6.25)	%		4.7	20.6	
Sheep	dig coef %		80.	80.	
Cattle	dig prot %	*	3.4	14.9	
Goats	dig prot %	*	3.4	14.9	
Horses	dig prot %	*	3.4	14.9	
Sheep	dig prot %		3.8	16.5	
Fatty acids	%				
Acetic	%		0.230	1.000	
Butyric	%		0.069	0.300	
Lactic acid	%		2.185	9.500	
Energy	GE Mcal/kg		1.03	4.48	
Sheep	GE dig coef %		71.	71.	
Sheep	DE Mcal/kg		0.73	3.18	
Sheep	ME Mcal/kg	*	0.60	2.61	
pH	pH units		4.30	4.30	

Alfalfa, aerial part w sodium bisulfite preservative added, ensiled, early bloom, cut 1, (3)

Ref No 3 07 845 United States

Feed Name or Analyses			Mean As Fed	Mean Dry	C.V. ± %
Dry matter	%		22.2	100.0	
Sheep	dig coef %		63.	63.	
Protein (N x 6.25)	%		4.4	19.9	
Sheep	dig coef %		75.	75.	
Cattle	dig prot %	*	3.2	14.3	
Goats	dig prot %	*	3.2	14.3	
Horses	dig prot %	*	3.2	14.3	
Sheep	dig prot %		3.3	14.9	
Fatty acids	%				
Acetic	%		0.777	3.500	
Butyric	%		0.044	0.200	
Lactic acid	%		2.442	11.000	
Energy	GE Mcal/kg		1.01	4.53	
Sheep	GE dig coef %		61.	61.	
Sheep	DE Mcal/kg		0.61	2.76	
Sheep	ME Mcal/kg	*	0.50	2.27	
pH	pH units		4.70	4.70	

Feed Name or Analyses			Mean As Fed	Mean Dry	C.V. ± %

Alfalfa, aerial part w sodium bisulfite preservative added, ensiled, early bloom, cut 3, (3)

Ref No 3 07 904 United States

Analyses			As Fed	Dry	C.V. ± %
Dry matter	%			100.0	
Sheep	dig coef %			65.	
Ash	%			10.3	
Crude fiber	%			25.8	
Sheep	dig coef %			49.	
Ether extract	%			4.2	
Sheep	dig coef %			58.	
N-free extract	%			39.1	
Sheep	dig coef %			72.	
Protein (N x 6.25)	%			20.6	
Sheep	dig coef %			78.	
Cattle	dig prot %	*		14.9	
Goats	dig prot %	*		14.9	
Horses	dig prot %	*		14.9	
Sheep	dig prot %			16.1	
Energy	GE Mcal/kg			4.42	
Sheep	GE dig coef %			63.	
Cattle	DE Mcal/kg	*		2.60	
Sheep	DE Mcal/kg			2.78	
Cattle	ME Mcal/kg	*		2.13	
Sheep	ME Mcal/kg	*		2.28	
Cattle	TDN %	*		58.9	
Sheep	TDN %			62.3	

Alfalfa, aerial part w sodium bisulfite preservative added, ensiled, mid-bloom, cut 1, (3)

Ref No 3 07 837 United States

Analyses			As Fed	Dry	C.V. ± %
Dry matter	%		23.9	100.0	
Sheep	dig coef %		61.	61.	5
Ash	%		1.7	7.2	
Crude fiber	%		8.6	35.8	
Sheep	dig coef %		53.	53.	
Ether extract	%		1.0	4.3	
Sheep	dig coef %		58.	58.	
N-free extract	%		9.1	37.9	
Sheep	dig coef %		63.	63.	
Protein (N x 6.25)	%		3.5	14.8	6
Sheep	dig coef %		68.	68.	4
Cattle	dig prot %	*	2.3	9.7	
Goats	dig prot %	*	2.3	9.7	
Horses	dig prot %	*	2.3	9.7	
Sheep	dig prot %		2.4	10.0	
Fatty acids	%				
Acetic	%		0.825	3.450	
Butyric	%		0.030	0.125	
Lactic acid	%		0.640	2.675	
Energy	GE Mcal/kg		1.06	4.44	0
Sheep	GE dig coef %		57.	57.	5
Cattle	DE Mcal/kg	*	0.63	2.63	
Sheep	DE Mcal/kg		0.61	2.55	
Cattle	ME Mcal/kg	*	0.52	2.16	
Sheep	ME Mcal/kg		0.48	2.01	
Cattle	TDN %	*	14.3	59.7	
Sheep	TDN %		14.0	58.5	
pH	pH units		5.20	5.20	

Alfalfa, aerial part w sodium bisulfite preservative added, ensiled, full bloom, cut 1, (3)

Ref No 3 07 842 United States

Analyses			As Fed	Dry	C.V. ± %
Dry matter	%		26.8	100.0	
Sheep	dig coef %		59.	59.	
Protein (N x 6.25)	%		4.7	17.5	
Sheep	dig coef %		72.	72.	
Cattle	dig prot %	*	3.2	12.1	
Goats	dig prot %	*	3.2	12.1	
Horses	dig prot %	*	3.2	12.1	
Sheep	dig prot %		3.4	12.6	
Fatty acids	%				
Acetic	%		0.724	2.700	
Butyric	%		0.429	1.600	
Lactic acid	%		1.715	6.400	
Energy	GE Mcal/kg		1.19	4.45	
Sheep	GE dig coef %		57.	57.	
Sheep	DE Mcal/kg		0.68	2.54	
Sheep	ME Mcal/kg	*	0.56	2.08	
pH	pH units		4.80	4.80	

Alfalfa, aerial part w sulfur dioxide preservative added, ensiled, immature, (3)

Ref No 3 00 243 United States

Analyses			As Fed	Dry	C.V. ± %
Dry matter	%			100.0	
Ash	%			11.4	8
Crude fiber	%			30.1	6
Ether extract	%			3.8	6
N-free extract	%			32.7	
Protein (N x 6.25)	%			22.0	7
Cattle	dig prot %	*		16.2	
Goats	dig prot %	*		16.2	
Horses	dig prot %	*		16.2	
Sheep	dig prot %	*		16.2	
Energy	GE Mcal/kg				
Cattle	DE Mcal/kg	*		2.39	
Sheep	DE Mcal/kg	*		2.50	
Cattle	ME Mcal/kg	*		1.96	
Sheep	ME Mcal/kg	*		2.05	
Cattle	TDN %	*		54.2	
Sheep	TDN %	*		56.8	

Alfalfa, aerial part w sugar added, ensiled, (3)

Ref No 3 00 244 United States

Analyses			As Fed	Dry	C.V. ± %
Dry matter	%		17.1	100.0	
Ash	%		2.0	11.8	
Crude fiber	%		6.1	35.8	
Sheep	dig coef %		39.	39.	
Ether extract	%		0.4	2.6	
Sheep	dig coef %		44.	44.	
N-free extract	%		5.2	30.5	
Sheep	dig coef %		58.	58.	
Protein (N x 6.25)	%		3.3	19.3	
Sheep	dig coef %		73.	73.	
Cattle	dig prot %	*	2.4	13.8	
Goats	dig prot %	*	2.4	13.8	
Horses	dig prot %	*	2.4	13.8	
Sheep	dig prot %		2.4	14.1	
Energy	GE Mcal/kg				
Cattle	DE Mcal/kg	*	0.38	2.24	
Sheep	DE Mcal/kg	*	0.36	2.13	
Cattle	ME Mcal/kg	*	0.31	1.83	
Sheep	ME Mcal/kg	*	0.30	1.75	
Cattle	TDN %	*	8.7	50.7	
Sheep	TDN %		8.3	48.3	

Alfalfa, aerial part w 8-10% molasses added, wilted ensiled, early bloom, (3)

Ref No 3 00 245 United States

Analyses			As Fed	Dry	C.V. ± %
Dry matter	%		32.7	100.0	
Ash	%		3.0	9.1	
Crude fiber	%		8.6	26.3	
Cattle	dig coef %		51.	51.	
Ether extract	%		0.9	2.8	
Cattle	dig coef %		56.	56.	
N-free extract	%		14.7	45.1	
Cattle	dig coef %		73.	73.	
Protein (N x 6.25)	%		5.5	16.7	
Cattle	dig coef %		67.	67.	
Cattle	dig prot %		3.7	11.2	
Goats	dig prot %	*	3.7	11.4	
Horses	dig prot %	*	3.7	11.4	
Sheep	dig prot %	*	3.7	11.4	
Energy	GE Mcal/kg				
Cattle	DE Mcal/kg	*	0.88	2.69	
Sheep	DE Mcal/kg	*	0.85	2.61	
Cattle	ME Mcal/kg	*	0.72	2.21	
Sheep	ME Mcal/kg	*	0.70	2.14	
Cattle	TDN %		20.0	61.1	
Sheep	TDN %	*	19.3	59.1	

Alfalfa, seeds, (5)

Ref No 5 08 325 United States

Analyses			As Fed	Dry	C.V. ± %
Dry matter	%		88.3	100.0	
Ash	%		4.4	5.0	
Crude fiber	%		8.1	9.2	
Ether extract	%		10.6	12.0	
N-free extract	%		32.0	36.2	
Protein (N x 6.25)	%		33.2	37.6	
Energy	GE Mcal/kg				
Cattle	DE Mcal/kg	*	3.40	3.85	
Sheep	DE Mcal/kg	*	3.37	3.82	
Swine	DE kcal/kg	*	3884.	4399.	
Cattle	ME Mcal/kg	*	2.79	3.16	
Sheep	ME Mcal/kg	*	2.76	3.13	
Swine	ME kcal/kg	*	3434.	3889.	
Cattle	TDN %	*	77.1	87.3	
Sheep	TDN %	*	76.4	86.6	
Swine	TDN %	*	88.1	99.8	

Alfalfa, seed screenings, (5)

Ref No 5 08 326 United States

Analyses			As Fed	Dry	C.V. ± %
Dry matter	%		90.3	100.0	
Ash	%		5.1	5.6	
Crude fiber	%		11.1	12.3	
Ether extract	%		9.9	11.0	
N-free extract	%		33.1	36.7	
Protein (N x 6.25)	%		31.1	34.4	
Energy	GE Mcal/kg				
Cattle	DE Mcal/kg	*	3.28	3.63	
Sheep	DE Mcal/kg	*	3.33	3.68	
Swine	DE kcal/kg	*	3816.	4226.	
Cattle	ME Mcal/kg	*	2.69	2.98	
Sheep	ME Mcal/kg	*	2.73	3.02	
Swine	ME kcal/kg	*	3398.	3763.	
Cattle	TDN %	*	74.4	82.4	

(1) dry forages and roughages (2) pasture, range plants, and forages fed green (3) silages (4) energy feeds (5) protein supplements (6) minerals (7) vitamins (8) additives

Feed Name or Analyses				Mean As Fed	Mean Dry	C.V. ± %
Sheep	TDN	%	*	75.4	83.5	
Swine	TDN	%	*	86.5	95.8	

ALFALFA, ATLANTIC. Medicago sativa

Alfalfa, Atlantic, hay, s-c, (1)

Ref No 1 08 745 United States

Feed Name or Analyses				Mean As Fed	Mean Dry	C.V. ± %
Dry matter		%		89.6	100.0	1
Horses	dig coef	%		59.	59.	4
Organic matter		%		81.9	91.5	1
Ash		%		7.6	8.5	12
Crude fiber		%		27.7	31.0	15
Horses	dig coef	%		36.	36.	19
Ether extract		%		2.3	2.6	9
N-free extract		%		37.5	41.8	5
Horses	dig coef	%		74.	74.	4
Protein (N x 6.25)		%		14.4	16.1	12
Cattle	dig prot	%	*	9.7	10.9	
Goats	dig prot	%	*	10.4	11.6	
Horses	dig prot	%	*	10.0	11.2	
Rabbits	dig prot	%	*	9.9	11.1	
Sheep	dig prot	%	*	9.8	11.0	
Cell contents (Van Soest)		%		45.8	51.1	12
Horses	dig coef	%		76.	76.	4
Cell walls (Van Soest)		%		43.7	48.8	12
Horses	dig coef	%		40.	40.	15
Cellulose (Matrone)		%		26.6	29.7	12
Horses	dig coef	%		50.	50.	10
Fiber, acid detergent (VS)		%		33.2	37.1	12
Horses	dig coef	%		40.	40.	16
Hemicellulose		%		6.7	7.5	19
Horses	dig coef	%		37.	37.	45
Holocellulose		%		33.3	37.1	12
Soluble carbohydrates		%		25.3	28.2	9
Lignin (Ellis)		%		6.7	7.5	12
Horses	dig coef	%		-3.	-3.	98
Energy	GE	Mcal/kg		3.96	4.42	0
Cattle	DE	Mcal/kg	*	2.36	2.63	
Horses	DE	Mcal/kg	*	2.10	2.34	
Sheep	DE	Mcal/kg	*	2.25	2.52	
Cattle	ME	Mcal/kg	*	1.93	2.16	
Horses	ME	Mcal/kg	*	1.72	1.92	
Sheep	ME	Mcal/kg	*	1.85	2.06	
Cattle	TDN	%	*	53.4	59.7	
Sheep	TDN	%	*	51.1	57.1	
Aluminum		mg/kg		55.73	62.22	
Boron		mg/kg		15.924	17.778	
Calcium		%		0.94	1.05	45
Copper		mg/kg		10.9	12.2	
Iron		%		0.007	0.008	
Magnesium		%		0.30	0.33	16
Manganese		mg/kg		77.6	86.7	
Phosphorus		%		0.21	0.24	12
Potassium		%		1.94	2.17	21
Sodium		%		0.01	0.02	
Sulphur		%		0.18	0.20	
Carotene		mg/kg		63.0	70.3	17
Vitamin A equivalent		IU/g		104.9	117.2	

Alfalfa, Atlantic, hay, s-c, immature, (1)

Ref No 1 09 274 United States

Feed Name or Analyses				Mean As Fed	Mean Dry	C.V. ± %
Dry matter		%			100.0	
Sheep	dig coef	%			74.	3
Ash		%			9.5	5
Crude fiber		%			19.3	22
Sheep	dig coef	%			58.	10
Ether extract		%			5.4	11
Sheep	dig coef	%			79.	
N-free extract		%			35.8	5
Sheep	dig coef	%			80.	
Protein (N x 6.25)		%			30.1	6
Sheep	dig coef	%			83.	2
Cattle	dig prot	%	*		23.0	
Goats	dig prot	%	*		24.6	
Horses	dig prot	%	*		23.1	
Rabbits	dig prot	%	*		21.9	
Sheep	dig prot	%			24.9	
Cellulose (Matrone)		%			18.5	
Sheep	dig coef	%			69.	
Lignin (Ellis)		%			5.2	
Sheep	dig coef	%			13.	
Energy	GE	Mcal/kg			4.61	0
Sheep	GE dig coef	%			72.	
Cattle	DE	Mcal/kg	*		3.03	
Sheep	DE	Mcal/kg			3.31	
Cattle	ME	Mcal/kg	*		2.48	
Sheep	ME	Mcal/kg	*		2.71	
Cattle	TDN	%	*		68.7	
Sheep	TDN	%			74.2	

Alfalfa, Atlantic, hay, s-c, prebloom, (1)

Ref No 1 09 154 United States

Feed Name or Analyses				Mean As Fed	Mean Dry	C.V. ± %
Dry matter		%			100.0	
Sheep	dig coef	%			74.	4
Ash		%			9.3	11
Crude fiber		%			19.4	8
Sheep	dig coef	%			55.	5
Ether extract		%			5.4	10
N-free extract		%			39.3	2
Sheep	dig coef	%			79.	4
Protein (N x 6.25)		%			26.6	2
Sheep	dig coef	%			84.	4
Cattle	dig prot	%	*		20.0	
Goats	dig prot	%	*		21.4	
Horses	dig prot	%	*		20.1	
Rabbits	dig prot	%	*		19.2	
Sheep	dig prot	%			22.2	
Cellulose (Matrone)		%			21.8	
Sheep	dig coef	%			66.	
Lignin (Ellis)		%			6.2	
Sheep	dig coef	%			5.	
Energy	GE	Mcal/kg			4.56	0
Sheep	GE dig coef	%			71.	
Cattle	DE	Mcal/kg	*		3.04	
Sheep	DE	Mcal/kg			3.21	
Cattle	ME	Mcal/kg	*		2.49	
Sheep	ME	Mcal/kg	*		2.63	
Cattle	TDN	%	*		68.9	
Sheep	TDN	%	*		63.3	

Alfalfa, Atlantic, hay, s-c, early bloom, (1)

Ref No 1 09 155 United States

Feed Name or Analyses				Mean As Fed	Mean Dry	C.V. ± %
Dry matter		%			100.0	
Sheep	dig coef	%			71.	4
Ash		%			8.3	8
Crude fiber		%			22.6	12
Sheep	dig coef	%			52.	6
Ether extract		%			5.2	8
N-free extract		%			40.0	4
Sheep	dig coef	%			79.	4
Protein (N x 6.25)		%			23.9	6
Sheep	dig coef	%			81.	3
Cattle	dig prot	%	*		17.6	
Goats	dig prot	%	*		18.8	
Horses	dig prot	%	*		17.8	
Rabbits	dig prot	%	*		17.1	
Sheep	dig prot	%			19.3	
Cellulose (Matrone)		%			23.4	5
Sheep	dig coef	%			63.	4
Lignin (Ellis)		%			7.3	11
Sheep	dig coef	%			11.	32
Energy	GE	Mcal/kg			4.57	0
Sheep	GE dig coef	%			68.	3
Cattle	DE	Mcal/kg	*		2.91	
Sheep	DE	Mcal/kg			3.12	
Cattle	ME	Mcal/kg	*		2.39	
Sheep	ME	Mcal/kg	*		2.56	
Cattle	TDN	%	*		66.1	
Sheep	TDN	%	*		61.7	

Alfalfa, Atlantic, hay, s-c, midbloom, (1)

Ref No 1 09 156 United States

Feed Name or Analyses				Mean As Fed	Mean Dry	C.V. ± %
Dry matter		%			100.0	
Sheep	dig coef	%			69.	6
Ash		%			8.2	18
Crude fiber		%			24.4	12
Sheep	dig coef	%			52.	8
Ether extract		%			5.4	10
N-free extract		%			39.4	2
Sheep	dig coef	%			76.	5
Protein (N x 6.25)		%			22.6	6
Sheep	dig coef	%			80.	4
Cattle	dig prot	%	*		16.5	
Goats	dig prot	%	*		17.6	
Horses	dig prot	%	*		16.7	
Rabbits	dig prot	%	*		16.1	
Sheep	dig prot	%			18.0	
Cellulose (Matrone)		%			25.0	8
Sheep	dig coef	%			60.	6
Lignin (Ellis)		%			7.6	8
Sheep	dig coef	%			9.	77
Energy	GE	Mcal/kg			4.62	98
Sheep	GE dig coef	%			65.	7
Cattle	DE	Mcal/kg	*		2.88	
Sheep	DE	Mcal/kg			3.00	
Cattle	ME	Mcal/kg	*		2.36	
Sheep	ME	Mcal/kg	*		2.46	
Cattle	TDN	%	*		65.2	
Sheep	TDN	%	*		60.3	

Alfalfa, Atlantic, hay, s-c, mid-bloom, cut 1, (1)

Ref No 1 08 766 United States

Feed Name or Analyses				Mean As Fed	Mean Dry	C.V. ± %
Dry matter		%		90.9	100.0	1
Horses	dig coef	%		52.	52.	2
Organic matter		%		84.2	92.6	0
Ash		%		6.7	7.4	3
Crude fiber		%		40.0	44.1	2
Horses	dig coef	%		40.	40.	4
Ether extract		%		1.7	1.9	22
Horses	dig coef	%		25.	25.	50
N-free extract		%		32.0	35.3	2
Horses	dig coef	%		72.	72.	5
Protein (N x 6.25)		%		10.4	11.4	7
Horses	dig coef	%		65.	65.	4
Cattle	dig prot	%	*	6.2	6.8	
Goats	dig prot	%	*	6.6	7.2	
Horses	dig prot	%		6.8	7.4	

Continued

Feed Name or Analyses				Mean As Fed	Mean Dry	C.V. ± %
Rabbits	dig prot	%	*	6.8	7.5	
Sheep	dig prot	%	*	6.2	6.8	
Cell contents (Van Soest)		%		36.2	39.8	2
Horses	dig coef	%		76.	76.	3
Cell walls (Van Soest)		%		54.7	60.2	1
Horses	dig coef	%		36.	36.	4
Cellulose (Matrone)		%		34.3	37.7	2
Horses	dig coef	%		45.	45.	3
Fiber, acid detergent (VS)		%		43.7	48.1	2
Horses	dig coef	%		36.	36.	4
Hemicellulose		%		8.6	9.4	11
Horses	dig coef	%		34.	34.	18
Holocellulose		%		39.0	42.9	25
Soluble carbohydrates		%		23.9	26.3	38
Lignin (Ellis)		%		9.4	10.4	1
Horses	dig coef	%		35.	35.	8
Energy	GE Mcal/kg			4.05	4.45	2
Horses	GE dig coef	%		44.	44.	5
Cattle	DE Mcal/kg		*	1.99	2.19	
Horses	DE Mcal/kg			1.77	1.94	
Sheep	DE Mcal/kg		*	2.07	2.28	
Cattle	ME Mcal/kg		*	1.64	1.80	
Horses	ME Mcal/kg		*	1.45	1.59	
Sheep	ME Mcal/kg		*	1.70	1.87	
Cattle	TDN	%	*	45.1	49.6	
Horses	TDN	%		46.6	51.3	
Sheep	TDN	%	*	47.0	51.7	
Calcium		%		0.65	0.71	4
Magnesium		%		0.33	0.36	9
Phosphorus		%		0.15	0.16	6
Potassium		%		0.22	0.24	3
Carotene	mg/kg			13.1	14.4	10
Vitamin A equivalent	IU/g			21.8	24.0	

Alfalfa, Atlantic, hay, s-c, full bloom, (1)
Ref No 1 09 157 United States

Feed Name or Analyses				Mean As Fed	Mean Dry	C.V. ± %
Dry matter		%			100.0	
Sheep	dig coef	%			67.	6
Ash		%			6.8	10
Crude fiber		%			26.5	15
Sheep	dig coef	%			49.	7
Ether extract		%			5.1	2
N-free extract		%			41.4	8
Sheep	dig coef	%			75.	6
Protein (N x 6.25)		%			20.2	3
Sheep	dig coef	%			79.	3
Cattle	dig prot	%	*		14.5	
Goats	dig prot	%	*		15.4	
Horses	dig prot	%	*		14.7	
Rabbits	dig prot	%	*		14.3	
Sheep	dig prot	%			15.9	
Cellulose (Matrone)		%			26.4	
Sheep	dig coef	%			57.	
Lignin (Ellis)		%			8.6	
Sheep	dig coef	%			6.	
Energy	GE Mcal/kg				4.41	2
Sheep	GE dig coef	%			62.	
Cattle	DE Mcal/kg		*		2.76	
Sheep	DE Mcal/kg				2.74	
Cattle	ME Mcal/kg		*		2.26	
Sheep	ME Mcal/kg		*		2.25	
Cattle	TDN	%	*		62.6	
Sheep	TDN	%	*		60.1	

(1) dry forages and roughages (2) pasture, range plants, and forages fed green (3) silages (4) energy feeds (5) protein supplements (6) minerals (7) vitamins (8) additives

Alfalfa, Atlantic, hay, s-c, mature, (1)
Ref No 1 09 158 United States

Feed Name or Analyses				Mean As Fed	Mean Dry	C.V. ± %
Dry matter		%			100.0	
Sheep	dig coef	%			68.	8
Ash		%			7.3	16
Crude fiber		%			26.6	19
Sheep	dig coef	%			49.	6
Ether extract		%			5.4	14
N-free extract		%			41.8	9
Sheep	dig coef	%			76.	6
Protein (N x 6.25)		%			18.9	7
Sheep	dig coef	%			78.	5
Cattle	dig prot	%	*		13.3	
Goats	dig prot	%	*		14.2	
Horses	dig prot	%	*		13.6	
Rabbits	dig prot	%	*		13.3	
Sheep	dig prot	%			14.8	
Cellulose (Matrone)		%			27.3	
Sheep	dig coef	%			59.	
Lignin (Ellis)		%			8.9	
Sheep	dig coef	%			10.	
Energy	GE Mcal/kg				4.67	0
Sheep	GE dig coef	%			63.	
Cattle	DE Mcal/kg		*		2.80	
Sheep	DE Mcal/kg				2.94	
Cattle	ME Mcal/kg		*		2.29	
Sheep	ME Mcal/kg		*		2.41	
Cattle	TDN	%	*		63.5	
Sheep	TDN	%	*		59.1	

Alfalfa, Atlantic, hay, s-c, over ripe, (1)
Ref No 1 09 159 United States

Feed Name or Analyses				Mean As Fed	Mean Dry	C.V. ± %
Dry matter		%			100.0	
Sheep	dig coef	%			62.	8
Ash		%			6.7	23
Crude fiber		%			30.8	18
Sheep	dig coef	%			48.	19
Ether extract		%			4.9	5
N-free extract		%			40.0	8
Sheep	dig coef	%			68.	12
Protein (N x 6.25)		%			17.6	6
Sheep	dig coef	%			73.	5
Cattle	dig prot	%	*		12.1	
Goats	dig prot	%	*		12.9	
Horses	dig prot	%	*		12.4	
Rabbits	dig prot	%	*		12.2	
Sheep	dig prot	%			12.8	
Cellulose (Matrone)		%			30.5	9
Sheep	dig coef	%			54.	7
Lignin (Ellis)		%			10.3	12
Sheep	dig coef	%			7.	73
Energy	GE Mcal/kg				4.34	3
Sheep	GE dig coef	%			55.	6
Cattle	DE Mcal/kg		*		2.65	
Sheep	DE Mcal/kg				2.37	
Cattle	ME Mcal/kg		*		2.17	
Sheep	ME Mcal/kg		*		1.94	
Cattle	TDN	%	*		60.1	
Sheep	TDN	%	*		57.0	

ALFALFA, BUFFALO, Medicago sativa

Alfalfa, buffalo, hay, dehy chopped, cut 3, (1)
Ref No 1 07 895 United States

Feed Name or Analyses				Mean As Fed	Mean Dry	C.V. ± %
Dry matter		%			100.0	
Cattle	dig coef	%			61.	
Sheep	dig coef	%			60.	
Ash		%			10.9	
Crude fiber		%			29.0	
Cattle	dig coef	%			47.	
Sheep	dig coef	%			48.	
Ether extract		%			2.9	
Cattle	dig coef	%			44.	
Sheep	dig coef	%			43.	
N-free extract		%			39.4	
Cattle	dig coef	%			71.	
Sheep	dig coef	%			69.	
Protein (N x 6.25)		%			17.8	
Cattle	dig coef	%			67.	
Sheep	dig coef	%			66.	
Cattle	dig prot	%			11.9	
Goats	dig prot	%	*		13.2	
Horses	dig prot	%	*		12.6	
Rabbits	dig prot	%	*		12.4	
Sheep	dig prot	%			11.7	
Cellulose (Matrone)		%			30.7	
Cattle	dig coef	%			59.	
Sheep	dig coef	%			61.	
Energy	GE Mcal/kg				4.37	
Cattle	GE dig coef	%			58.	
Sheep	GE dig coef	%			57.	
Cattle	DE Mcal/kg				2.53	
Sheep	DE Mcal/kg				2.49	
Cattle	ME Mcal/kg		*		2.08	
Sheep	ME Mcal/kg		*		2.04	
Cattle	TDN	%			56.4	
Sheep	TDN	%			55.7	

Alfalfa, buffalo, hay, s-c, early bloom, (1)
Ref No 1 09 273 United States

Feed Name or Analyses				Mean As Fed	Mean Dry	C.V. ± %
Dry matter		%			100.0	
Cattle	dig coef	%			61.	
Ash		%			8.0	
Crude fiber		%			28.2	
Ether extract		%			2.7	
N-free extract		%			40.3	
Protein (N x 6.25)		%			20.9	
Cattle	dig coef	%			73.	
Cattle	dig prot	%			15.2	
Goats	dig prot	%	*		16.1	
Horses	dig prot	%	*		15.3	
Rabbits	dig prot	%	*		14.8	
Sheep	dig prot	%	*		15.3	
Cellulose (Matrone)		%			29.4	
Cattle	dig coef	%			58.	
Energy	GE Mcal/kg				4.40	0
Cattle	GE dig coef	%			60.	
Cattle	DE Mcal/kg				2.63	
Sheep	DE Mcal/kg		*		2.65	
Cattle	ME Mcal/kg		*		2.16	
Sheep	ME Mcal/kg		*		2.17	
Cattle	TDN	%	*		60.3	
Sheep	TDN	%	*		60.0	

Alfalfa, buffalo, hay, s-c, early bloom, cut 2, (1)
Ref No 1 09 098 United States

Feed Name or Analyses		Mean As Fed	Mean Dry	C.V. ± %
Dry matter	%	87.1	100.0	
Cattle	dig coef %	62.	62.	
Ash	%	7.0	8.0	
Crude fiber	%	23.4	26.9	
Ether extract	%	3.0	3.5	
N-free extract	%	34.8	39.9	
Protein (N x 6.25)	%	18.9	21.7	
Cattle	dig coef %	74.	74.	
Cattle	dig prot %	14.0	16.1	
Goats	dig prot % *	14.6	16.8	
Horses	dig prot % *	13.9	16.0	
Rabbits	dig prot % *	13.4	15.4	
Sheep	dig prot % *	14.0	16.0	
Cell walls (Van Soest)	%	39.5	45.3	
Cellulose (Matrone)	%	21.8	25.0	
Cattle	dig coef %	59.	59.	
Fiber, acid detergent (VS)	%	26.2	30.1	
Energy	GE Mcal/kg	3.82	4.38	
Cattle	GE dig coef %	61.	61.	
Cattle	DE Mcal/kg	2.33	2.67	
Sheep	DE Mcal/kg *	2.32	2.66	
Cattle	ME Mcal/kg *	1.91	2.19	
Sheep	ME Mcal/kg *	1.90	2.18	
Cattle	TDN % *	53.8	61.8	
Sheep	TDN % *	52.5	60.3	

Alfalfa, buffalo, hay, s-c, early bloom, cut 3, (1)
Ref No 1 09 103 United States

Feed Name or Analyses		Mean As Fed	Mean Dry	C.V. ± %
Dry matter	%	87.2	100.0	
Cattle	dig coef %	62.	62.	
Ash	%	8.3	9.5	
Crude fiber	%	24.7	28.3	
Ether extract	%	1.9	2.2	
N-free extract	%	34.0	39.0	
Protein (N x 6.25)	%	18.3	21.0	
Cattle	dig coef %	72.	72.	
Cattle	dig prot %	13.2	15.1	
Goats	dig prot % *	14.1	16.2	
Horses	dig prot % *	13.4	15.4	
Rabbits	dig prot % *	13.0	14.9	
Sheep	dig prot % *	13.4	15.4	
Cell walls (Van Soest)	%	43.3	49.6	
Cellulose (Matrone)	%	27.3	31.3	
Cattle	dig coef %	61.	61.	
Fiber, acid detergent (VS)	%	28.2	32.3	
Energy	GE Mcal/kg	3.81	4.37	
Cattle	GE dig coef %	61.	61.	
Cattle	DE Mcal/kg	2.32	2.67	
Sheep	DE Mcal/kg *	2.27	2.61	
Cattle	ME Mcal/kg *	1.91	2.19	
Sheep	ME Mcal/kg *	1.86	2.14	
Cattle	TDN % *	53.4	61.2	
Sheep	TDN % *	51.5	59.1	

ALFALFA, CULVER. Medicago sativa

Alfalfa, culver, hay, baled fan-air dried w heat, midbloom, cut 2, (1)
Ref No 1 08 869 United States

Feed Name or Analyses		Mean As Fed	Mean Dry	C.V. ± %
Dry matter	%	89.4	100.0	
Sheep	dig coef %	61.	61.	
Organic matter	%	82.4	92.2	
Ash	%	7.0	7.8	
Protein (N x 6.25)	%	17.2	19.2	
Cattle	dig prot % *	12.1	13.6	
Goats	dig prot % *	12.9	14.5	
Horses	dig prot % *	12.4	13.8	
Rabbits	dig prot % *	12.1	13.5	
Sheep	dig prot % *	12.3	13.8	
Cell walls (Van Soest)	%	41.5	46.4	
Fiber, acid detergent (VS)	%	33.1	37.0	
Lignin (Van Soest)	%	7.5	8.4	
Energy	GE Mcal/kg	3.92	4.39	
Nutritive value index (NVI)	%	58.	65.	
Aluminum	mg/kg	236.91	265.00	
Barium	mg/kg	21.456	24.000	
Boron	mg/kg	11.622	13.000	
Calcium	%	0.99	1.11	
Cobalt	mg/kg	0.322	0.360	
Copper	mg/kg	8.0	9.0	
Iron	%	0.028	0.031	
Magnesium	%	0.16	0.18	
Manganese	mg/kg	31.3	35.0	
Molybdenum	mg/kg	0.83	0.93	
Phosphorus	%	0.22	0.25	
Potassium	%	1.48	1.66	
Silicon	%	0.19	0.21	
Sodium	%	0.03	0.03	
Strontium	mg/kg	16.092	18.000	
Zinc	mg/kg	22.4	25.0	
Dry matter intake	g/w^.75	75.454	84.400	

Alfalfa, culver, hay, baled fan-air dried w heat, full bloom, cut 1, (1)
Ref No 1 08 866 United States

Feed Name or Analyses		Mean As Fed	Mean Dry	C.V. ± %
Dry matter	%	91.3	100.0	
Sheep	dig coef %	60.	60.	
Organic matter	%	85.1	93.2	
Ash	%	6.2	6.8	
Protein (N x 6.25)	%	16.3	17.9	
Cattle	dig prot % *	11.4	12.4	
Goats	dig prot % *	12.1	13.3	
Horses	dig prot % *	11.6	12.7	
Rabbits	dig prot % *	11.4	12.5	
Sheep	dig prot % *	11.5	12.6	
Cell walls (Van Soest)	%	44.4	48.6	
Fiber, acid detergent (VS)	%	35.6	39.0	
Lignin (Van Soest)	%	7.9	8.7	
Energy	GE Mcal/kg	3.96	4.34	
Nutritive value index (NVI)	%	62.	68.	
Aluminum	mg/kg	93.13	102.00	
Barium	mg/kg	22.825	25.000	
Boron	mg/kg	12.782	14.000	
Calcium	%	1.02	1.12	
Cobalt	mg/kg	0.502	0.550	
Copper	mg/kg	7.3	8.0	
Iron	%	0.011	0.012	
Magnesium	%	0.16	0.17	
Manganese	mg/kg	21.9	24.0	
Molybdenum	mg/kg	0.55	0.60	
Phosphorus	%	0.20	0.22	
Potassium	%	1.34	1.47	
Silicon	%	0.05	0.06	
Sodium	%	0.05	0.05	
Strontium	mg/kg	15.521	17.000	
Zinc	mg/kg	21.9	24.0	
Dry matter intake	g/w^.75	82.992	90.900	

Alfalfa, culver, hay, s-c, midbloom, cut 2, (1)
Ref No 1 08 868 United States

Feed Name or Analyses		Mean As Fed	Mean Dry	C.V. ± %
Dry matter	%	89.0	100.0	
Sheep	dig coef %	52.	52.	
Organic matter	%	82.2	92.4	
Ash	%	6.8	7.6	
Protein (N x 6.25)	%	17.3	19.4	
Cattle	dig prot % *	12.2	13.7	
Goats	dig prot % *	13.0	14.7	
Horses	dig prot % *	12.5	14.0	
Rabbits	dig prot % *	12.1	13.6	
Sheep	dig prot % *	12.4	14.0	
Cell walls (Van Soest)	%	48.8	54.8	
Fiber, acid detergent (VS)	%	37.2	41.8	
Lignin (Van Soest)	%	7.9	8.9	
Energy	GE Mcal/kg	3.96	4.45	
Nutritive value index (NVI)	%	43.	48.	
Aluminum	mg/kg	153.08	172.00	
Barium	mg/kg	26.700	30.000	
Boron	mg/kg	11.570	13.000	
Calcium	%	0.96	1.08	
Cobalt	mg/kg	0.561	0.630	
Copper	mg/kg	8.9	10.0	
Iron	%	0.019	0.022	
Magnesium	%	0.17	0.19	
Manganese	mg/kg	30.3	34.0	
Molybdenum	mg/kg	1.07	1.20	
Phosphorus	%	0.22	0.25	
Potassium	%	1.34	1.50	
Silicon	%	0.10	0.11	
Sodium	%	0.03	0.03	
Strontium	mg/kg	13.350	15.000	
Zinc	mg/kg	20.5	23.0	
Dry matter intake	g/w^.75	70.577	79.300	

Alfalfa, culver, hay, s-c, full bloom, cut 1, (1)
Ref No 1 08 867 United States

Feed Name or Analyses		Mean As Fed	Mean Dry	C.V. ± %
Dry matter	%	89.0	100.0	
Sheep	dig coef %	60.	60.	
Organic matter	%	81.0	91.0	
Ash	%	8.0	9.0	
Protein (N x 6.25)	%	16.7	18.8	
Cattle	dig prot % *	11.8	13.2	
Goats	dig prot % *	12.5	14.1	
Horses	dig prot % *	12.0	13.5	
Rabbits	dig prot % *	11.7	13.2	
Sheep	dig prot % *	12.0	13.4	
Cell walls (Van Soest)	%	41.3	46.4	
Cellulose (Matrone)	%	29.7	33.4	
Fiber, acid detergent (VS)	%	32.7	36.7	
Lignin (Van Soest)	%	7.2	8.1	
Energy	GE Mcal/kg	3.87	4.35	
Nutritive value index (NVI)	%	52.	59.	
Aluminum	mg/kg	199.36	224.00	
Barium	mg/kg	30.260	34.000	
Boron	mg/kg	14.240	16.000	
Calcium	%	1.15	1.29	
Cobalt	mg/kg	0.498	0.560	
Copper	mg/kg	8.9	10.0	
Iron	%	0.024	0.027	
Magnesium	%	0.17	0.19	
Manganese	mg/kg	40.1	45.0	
Molybdenum	mg/kg	0.93	1.05	
Phosphorus	%	0.20	0.22	
Potassium	%	1.47	1.65	

Continued

Feed Name or Analyses			Mean As Fed	Mean Dry	C.V. ± %
Silicon	%		0.14	0.16	
Sodium	%		0.04	0.04	
Strontium	mg/kg		19.580	22.000	
Zinc	mg/kg		21.4	24.0	
Dry matter intake	g/$w^{0.75}$		68.708	77.200	

ALFALFA, LADAK. Medicago media

Alfalfa, ladak, hay, s-c, (1)

Ref No 1 08 169 Canada

Feed Name or Analyses			Mean As Fed	Mean Dry	C.V. ± %
Dry matter	%			100.0	
Sheep	dig coef %			63.	
Organic matter	%			90.0	
Ether extract	%			1.7	
Sheep	dig coef %			37.	
N-free extract	%			38.7	
Sheep	dig coef %			74.	
Protein (N x 6.25)	%			16.8	
Sheep	dig coef %			76.	
Cattle	dig prot %	*		11.5	
Goats	dig prot %	*		12.2	
Horses	dig prot %	*		11.8	
Rabbits	dig prot %	*		11.6	
Sheep	dig prot %			12.8	
Lignin (Ellis)	%			9.8	
Sheep	dig coef %			9.	
Energy	GE Mcal/kg			4.49	
Sheep	GE dig coef %			61.	
Sheep	DE Mcal/kg			2.74	
Sheep	ME Mcal/kg	*		2.25	
Sheep	TDN %			57.8	

Alfalfa, ladak, hay, s-c, mid-bloom, cut 2, Can 1, (1)

Ref No 1 08 656 Canada

Feed Name or Analyses			Mean As Fed	Mean Dry	C.V. ± %
Dry matter	%			100.0	
Ash	%			6.7	
Crude fiber	%			34.4	
Ether extract	%			1.7	
N-free extract	%			41.1	
Protein (N x 6.25)	%			16.1	
Cattle	dig prot %	*		10.9	
Goats	dig prot %	*		11.6	
Horses	dig prot %	*		11.2	
Rabbits	dig prot %	*		11.1	
Sheep	dig prot %	*		11.0	
Hexosans	%			7.6	
Pentosans	%			4.6	
Lignin (Ellis)	%			10.5	
Energy	GE Mcal/kg			4.35	
Cattle	DE Mcal/kg	*		2.42	
Sheep	DE Mcal/kg	*		2.52	
Cattle	ME Mcal/kg	*		1.99	
Sheep	ME Mcal/kg	*		2.06	
Cattle	TDN %	*		54.9	
Sheep	TDN %	*		57.1	

Alfalfa, ladak, hay, s-c, mid-bloom, cut 3, Can 1, (1)

Ref No 1 08 655 Canada

Feed Name or Analyses			Mean As Fed	Mean Dry	C.V. ± %
Dry matter	%			100.0	
Ash	%			12.3	
Crude fiber	%			24.4	
Ether extract	%			2.9	
N-free extract	%			38.2	
Protein (N x 6.25)	%			22.2	
Cattle	dig prot %	*		16.2	
Goats	dig prot %	*		17.3	
Horses	dig prot %	*		16.4	
Rabbits	dig prot %	*		15.8	
Sheep	dig prot %	*		16.5	
Cellulose (Matrone)	%			27.6	
Hexosans	%			5.4	
Pentosans	%			6.6	
Lignin (Ellis)	%			8.5	
Energy	GE Mcal/kg			4.30	
Cattle	DE Mcal/kg	*		2.54	
Sheep	DE Mcal/kg	*		2.60	
Cattle	ME Mcal/kg	*		2.08	
Sheep	ME Mcal/kg	*		2.13	
Cattle	TDN %	*		57.5	
Sheep	TDN %	*		58.9	

Alfalfa, ladak, hay, s-c, full bloom, cut 1, Can 1, (1)

Ref No 1 08 653 Canada

Feed Name or Analyses			Mean As Fed	Mean Dry	C.V. ± %
Dry matter	%			100.0	
Sheep	dig coef %			58.	
Ash	%			9.5	
Crude fiber	%			37.5	
Ether extract	%			2.3	
N-free extract	%			36.6	
Protein (N x 6.25)	%			14.1	
Cattle	dig prot %	*		9.2	
Goats	dig prot %	*		9.7	
Horses	dig prot %	*		9.5	
Rabbits	dig prot %	*		9.6	
Sheep	dig prot %	*		9.2	
Cellulose (Matrone)	%			36.7	
Hexosans	%			7.1	
Pentosans	%			3.7	
Lignin (Ellis)	%			12.7	
Energy	GE Mcal/kg			4.33	
Cattle	DE Mcal/kg	*		2.49	
Sheep	DE Mcal/kg			2.56	
Cattle	ME Mcal/kg	*		2.05	
Sheep	ME Mcal/kg	*		2.10	
Cattle	TDN %	*		56.6	
Sheep	TDN %	*		51.8	

ALFALFA, LAHONTAN. Medicago sativa

Alfalfa, lahontan, hay, s-c, (1)

Ref No 1 08 638 United States

Feed Name or Analyses			Mean As Fed	Mean Dry	C.V. ± %
Dry matter	%		93.3	100.0	
Ash	%		10.9	11.7	
Crude fiber	%		24.8	26.6	
Ether extract	%		1.1	1.2	
Lignin (Ellis)	%		7.7	8.3	
Silicon	%		1.38	1.48	

Alfalfa, lahontan, hay, s-c, early bloom, (1)

Ref No 1 08 636 United States

Feed Name or Analyses			Mean As Fed	Mean Dry	C.V. ± %
Dry matter	%		92.9	100.0	
Ash	%		10.5	11.3	
Crude fiber	%		23.0	24.7	
Ether extract	%		1.2	1.3	
Lignin (Ellis)	%		7.2	7.8	
Silicon	%		3.77	4.07	

Alfalfa, lahontan, hay, s-c, grnd pelleted, early bloom, (1)

Ref No 1 08 639 United States

Feed Name or Analyses			Mean As Fed	Mean Dry	C.V. ± %
Dry matter	%		93.1	100.0	
Ash	%		10.9	11.7	
Crude fiber	%		20.8	22.3	
Ether extract	%		1.2	1.3	
Lignin (Ellis)	%		6.1	6.5	
Silicon	%		2.28	2.45	

ALFALFA, RAMBLER. Medicago sativa

Alfalfa, rambler, hay, s-c, early bloom, cut 2, (1)

Ref No 1 08 662 Canada

Feed Name or Analyses			Mean As Fed	Mean Dry	C.V. ± %
Dry matter	%			100.0	
Sheep	dig coef %			63.	
Ash	%			6.6	
Crude fiber	%			30.2	
Ether extract	%			2.2	
N-free extract	%			45.1	
Protein (N x 6.25)	%			15.9	
Cattle	dig prot %	*		10.7	
Goats	dig prot %	*		11.4	
Horses	dig prot %	*		11.0	
Rabbits	dig prot %	*		10.9	
Sheep	dig prot %	*		10.8	
Lignin (Ellis)	%			18.1	
Energy	GE Mcal/kg			4.38	
Cattle	DE Mcal/kg	*		2.55	
Sheep	DE Mcal/kg	*		2.63	
Cattle	ME Mcal/kg	*		2.09	
Sheep	ME Mcal/kg	*		2.15	
Cattle	TDN %	*		57.9	
Sheep	TDN %	*		59.6	

ALFALFA, RANGER. Medicago sativa

Alfalfa, ranger, aerial part, dehy baled, early bloom, cut 1, (1)

Ref No 1 08 870 United States

Feed Name or Analyses			Mean As Fed	Mean Dry	C.V. ± %
Dry matter	%		88.7	100.0	
Sheep	dig coef %		62.	62.	
Organic matter	%		81.9	92.3	
Ash	%		8.6	9.7	
Crude fiber	%		24.7	27.8	
Sheep	dig coef %		45.	45.	
Ether extract	%		1.7	1.9	
Sheep	dig coef %		47.	47.	
N-free extract	%		37.3	42.0	
Sheep	dig coef %		74.	74.	
Protein (N x 6.25)	%		16.5	18.6	
Sheep	dig coef %		70.	70.	
Cattle	dig prot %	*	11.6	13.0	

(1) dry forages and roughages (2) pasture, range plants, and forages fed green (3) silages (4) energy feeds (5) protein supplements (6) minerals (7) vitamins (8) additives

Feed Name or Analyses			As Fed	Dry	C.V. ± %
Goats	dig prot %	*	12.3	13.9	
Horses	dig prot %	*	11.8	13.3	
Rabbits	dig prot %	*	11.6	13.0	
Sheep	dig prot %		11.5	13.0	
Cellulose (Crampton)	%				
Sheep	dig coef %		52.	52.	
Energy	GE Mcal/kg				
Cattle	DE Mcal/kg	*	2.41	2.72	
Sheep	DE Mcal/kg	*	2.29	2.58	
Cattle	ME Mcal/kg	*	1.98	2.23	
Sheep	ME Mcal/kg	*	1.88	2.12	
Cattle	TDN %	*	54.8	61.7	
Sheep	TDN %		52.0	58.6	
Dry matter intake	$g/w^{0.75}$		78.145	88.100	

Alfalfa, ranger, aerial part, dehy baled, full bloom, cut 1, (1)

Ref No 1 08 871 United States

Feed Name or Analyses			As Fed	Dry	C.V. ± %
Dry matter	%		88.2	100.0	
Sheep	dig coef %		59.	59.	
Organic matter	%		80.9	91.7	
Ash	%		7.3	8.3	
Crude fiber	%		28.4	32.2	
Sheep	dig coef %		44.	44.	
Ether extract	%		1.9	2.1	
Sheep	dig coef %		48.	48.	
N-free extract	%		36.0	40.8	
Sheep	dig coef %		71.	71.	
Protein (N x 6.25)	%		14.6	16.6	
Sheep	dig coef %		66.	66.	
Cattle	dig prot %	*	10.0	11.3	
Goats	dig prot %	*	10.6	12.0	
Horses	dig prot %	*	10.3	11.6	
Rabbits	dig prot %	*	10.1	11.5	
Sheep	dig prot %		9.7	11.0	
Cellulose (Crampton)	%				
Sheep	dig coef %		52.	52.	
Energy	GE Mcal/kg				
Cattle	DE Mcal/kg	*	2.26	2.56	
Sheep	DE Mcal/kg	*	2.19	2.49	
Cattle	ME Mcal/kg	*	1.85	2.10	
Sheep	ME Mcal/kg	*	1.80	2.04	
Cattle	TDN %	*	51.3	58.1	
Sheep	TDN %		49.7	56.4	
Dry matter intake	$g/w^{0.75}$		70.648	80.100	

ALFALFA, VERNAL. Medicago sativa

Alfalfa, vernal, hay, fan-air dried w heat, mature after frost, cut 4, (1)

Ref No 1 08 897 United States

Feed Name or Analyses			As Fed	Dry	C.V. ± %
Dry matter	%			100.0	
Sheep	dig coef %			43.	
Crude fiber	%			46.5	
Protein (N x 6.25)	%			8.7	
Cattle	dig prot %	*		4.5	
Goats	dig prot %	*		4.7	
Horses	dig prot %	*		4.9	
Rabbits	dig prot %	*		5.4	
Sheep	dig prot %	*		4.4	
Dry matter intake	$g/w^{0.75}$			42.280	

Alfalfa, vernal, hay, s-c, cut 1, (1)

Ref No 1 08 846 United States

Feed Name or Analyses			As Fed	Dry	C.V. ± %
Dry matter	%			100.0	
Sheep	dig coef %			63.	
Ash	%			8.2	
Ether extract	%			2.0	
Protein (N x 6.25)	%			18.7	
Cattle	dig prot %	*		13.1	
Goats	dig prot %	*		14.0	
Horses	dig prot %	*		13.4	
Rabbits	dig prot %	*		13.1	
Sheep	dig prot %	*		13.3	
Fiber, acid detergent (VS)	%			34.9	
Lignin (Ellis)	%			6.9	
Energy	GE Mcal/kg			4.46	
Sheep	DE Mcal/kg	*		2.67	
Sheep	ME Mcal/kg	*		2.19	
Sheep	TDN %			60.7	
Phosphorus	%			0.39	
Sulphur	%			0.29	
Dry matter intake	$g/w^{0.75}$			103.100	
Gain, sheep	kg/day			0.107	

Alfalfa, vernal, hay, s-c, cut 2, (1)

Ref No 1 08 847 United States

Feed Name or Analyses			As Fed	Dry	C.V. ± %
Dry matter	%			100.0	
Sheep	dig coef %			59.	
Ash	%			8.2	
Ether extract	%			2.0	
Protein (N x 6.25)	%			19.5	
Cattle	dig prot %	*		13.8	
Goats	dig prot %	*		14.8	
Horses	dig prot %	*		14.1	
Rabbits	dig prot %	*		13.7	
Sheep	dig prot %	*		14.1	
Fiber, acid detergent (VS)	%			33.8	
Lignin (Ellis)	%			7.5	
Energy	GE Mcal/kg			4.75	
Sheep	DE Mcal/kg	*		2.62	
Sheep	ME Mcal/kg	*		2.15	
Sheep	TDN %			59.4	
Phosphorus	%			0.31	
Sulphur	%			0.34	
Dry matter intake	$g/w^{0.75}$			107.200	
Gain, sheep	kg/day			0.101	

Alfalfa, vernal, hay, s-c, cut 3, (1)

Ref No 1 08 848 United States

Feed Name or Analyses			As Fed	Dry	C.V. ± %
Dry matter	%			100.0	
Sheep	dig coef %			57.	
Ash	%			9.9	
Ether extract	%			1.9	
Protein (N x 6.25)	%			18.8	
Cattle	dig prot %	*		13.2	
Goats	dig prot %	*		14.1	
Horses	dig prot %	*		13.5	
Rabbits	dig prot %	*		13.2	
Sheep	dig prot %	*		13.4	
Fiber, acid detergent (VS)	%			40.4	
Lignin (Ellis)	%			8.4	
Energy	GE Mcal/kg			4.49	
Sheep	DE Mcal/kg	*		2.43	
Sheep	ME Mcal/kg	*		1.99	
Sheep	TDN %			55.1	
Phosphorus	%			0.30	
Sulphur	%			0.27	
Dry matter intake	$g/w^{0.75}$			85.700	
Gain, sheep	kg/day			0.014	

Alfalfa, vernal, hay, s-c, baled, mature, cut 4, (1)

Ref No 1 08 898 United States

Feed Name or Analyses			As Fed	Dry	C.V. ± %
Dry matter	%			100.0	
Sheep	dig coef %			56.	
Crude fiber	%			30.9	
Protein (N x 6.25)	%			11.5	
Cattle	dig prot %	*		6.9	
Goats	dig prot %	*		7.3	
Horses	dig prot %	*		7.3	
Rabbits	dig prot %	*		7.5	
Sheep	dig prot %	*		6.9	
Dry matter intake	$g/w^{0.75}$			69.140	

Alfalfa, vernal, hay, s-c fan-air dried w heat immature, cut 1, (1)

Ref No 1 08 903 United States

Feed Name or Analyses			As Fed	Dry	C.V. ± %
Dry matter	%			100.0	
Sheep	dig coef %			62.	
Crude fiber	%			26.0	
Protein (N x 6.25)	%			24.8	
Cattle	dig prot %	*		18.4	
Goats	dig prot %	*		19.7	
Horses	dig prot %	*		18.5	
Rabbits	dig prot %	*		17.8	
Sheep	dig prot %	*		18.8	
Dry matter intake	$g/w^{0.75}$			83.710	

Alfalfa, vernal, hay, s-c fan-air dried w heat, pre-bloom, cut 1, (1)

Ref No 1 08 908 United States

Feed Name or Analyses			As Fed	Dry	C.V. ± %
Dry matter	%			100.0	
Sheep	dig coef %			60.	
Crude fiber	%			23.3	
Protein (N x 6.25)	%			22.4	
Cattle	dig prot %	*		16.3	
Goats	dig prot %	*		17.5	
Horses	dig prot %	*		16.5	
Rabbits	dig prot %	*		16.0	
Sheep	dig prot %	*		16.7	
Dry matter intake	$g/w^{0.75}$			81.140	

Alfalfa, vernal, hay, s-c fan-air dried w heat, early bloom, cut 1, (1)

Ref No 1 08 904 United States

Feed Name or Analyses			As Fed	Dry	C.V. ± %
Dry matter	%			100.0	
Sheep	dig coef %			58.	
Crude fiber	%			30.7	
Protein (N x 6.25)	%			20.9	
Cattle	dig prot %	*		15.1	
Goats	dig prot %	*		16.1	
Horses	dig prot %	*		15.3	
Rabbits	dig prot %	*		14.8	
Sheep	dig prot %	*		15.3	
Dry matter intake	$g/w^{0.75}$			86.857	

Feed Name or Analyses		Mean As Fed	Mean Dry	C.V. ± %

Alfalfa, vernal, hay, s-c fan-air dried w heat, mid-bloom, cut 1, (1)

Ref No 1 08 905 United States

Analyses		As Fed	Dry	C.V. ± %
Dry matter	%		100.0	
Sheep	dig coef %		57.	
Crude fiber	%		34.2	
Protein (N x 6.25)	%		21.3	
Cattle	dig prot % *		15.4	
Goats	dig prot % *		16.4	
Horses	dig prot % *		15.6	
Rabbits	dig prot % *		15.1	
Sheep	dig prot % *		15.7	
Dry matter intake	$g/W^{.75}$		86.860	

Alfalfa, vernal, hay, s-c fan-air dried w heat, full bloom, cut 1, (1)

Ref No 1 08 906 United States

Analyses		As Fed	Dry	C.V. ± %
Dry matter	%		100.0	
Sheep	dig coef %		55.	
Crude fiber	%		34.6	
Protein (N x 6.25)	%		19.2	
Cattle	dig prot % *		13.5	
Goats	dig prot % *		14.5	
Horses	dig prot % *		13.8	
Rabbits	dig prot % *		13.5	
Sheep	dig prot % *		13.8	
Dry matter intake	$g/W^{.75}$		81.570	

Alfalfa, vernal, hay, s-c fan-air dried w heat, late bloom, cut 1, (1)

Ref No 1 08 907 United States

Analyses		As Fed	Dry	C.V. ± %
Dry matter	%		100.0	
Sheep	dig coef %		53.	
Crude fiber	%		38.4	
Protein (N x 6.25)	%		17.5	
Cattle	dig prot % *		12.1	
Goats	dig prot % *		12.9	
Horses	dig prot % *		12.4	
Rabbits	dig prot % *		12.2	
Sheep	dig prot % *		12.3	
Dry matter intake	$g/W^{.75}$		84.0	

ALFALFA–APPLES. Medicago sativa, Malus spp

Alfalfa-Apples, aerial part and fruit, ensiled, (3)

Ref No 3 00 247 United States

Analyses		As Fed	Dry	C.V. ± %
Dry matter	%	23.6	100.0	4
Ash	%	1.7	7.2	4
Crude fiber	%	7.0	29.6	5
Sheep	dig coef %	49.	49.	
Ether extract	%	0.8	3.2	13
Sheep	dig coef %	47.	47.	
N-free extract	%	11.9	50.4	
Sheep	dig coef %	70.	70.	
Protein (N x 6.25)	%	2.2	9.5	6
Sheep	dig coef %	51.	51.	
Cattle	dig prot % *	1.1	4.9	
Goats	dig prot % *	1.1	4.9	
Horses	dig prot % *	1.1	4.9	
Sheep	dig prot %	1.1	4.8	
Energy	GE Mcal/kg			
Cattle	DE Mcal/kg *	0.65	2.76	
Sheep	DE Mcal/kg *	0.61	2.56	
Cattle	ME Mcal/kg *	0.53	2.26	
Sheep	ME Mcal/kg *	0.50	2.10	
Cattle	TDN % *	14.8	62.5	
Sheep	TDN %	13.7	58.1	

ALFALFA–BROME. Medicago sativa, Bromus spp

Alfalfa-Brome, hay, s-c, (1)

Ref No 1 08 327 United States

Analyses		As Fed	Dry	C.V. ± %
Dry matter	%	89.2	100.0	
Rabbits	dig coef %	58.	58.	
Ash	%	6.2	7.0	
Crude fiber	%	32.5	36.4	
Ether extract	%	2.0	2.2	
N-free extract	%	36.7	41.1	
Protein (N x 6.25)	%	11.8	13.2	
Rabbits	dig coef %	77.	77.	
Cattle	dig prot % *	7.5	8.4	
Goats	dig prot % *	7.9	8.9	
Horses	dig prot % *	7.8	8.8	
Rabbits	dig prot %	9.0	10.1	
Sheep	dig prot % *	7.5	8.4	
Energy	GE Mcal/kg			
Cattle	DE Mcal/kg *	2.16	2.42	
Sheep	DE Mcal/kg *	2.15	2.41	
Cattle	ME Mcal/kg *	1.77	1.98	
Sheep	ME Mcal/kg *	1.76	1.98	
Cattle	TDN % *	48.9	54.8	
Sheep	TDN % *	48.8	54.7	
Calcium	%	0.77	0.86	
Chlorine	%	0.42	0.47	
Copper	mg/kg	11.5	12.9	
Iron	%	0.014	0.016	
Magnesium	%	0.21	0.24	
Manganese	mg/kg	49.2	55.1	
Phosphorus	%	0.20	0.22	
Potassium	%	1.66	1.86	
Sodium	%	0.44	0.49	
Sulphur	%	0.21	0.24	

Alfalfa-Brome, aerial part, fresh, (2)

Ref No 2 08 328 United States

Analyses		As Fed	Dry	C.V. ± %
Dry matter	%	22.5	100.0	
Ash	%	2.2	9.8	
Crude fiber	%	5.3	23.6	
Ether extract	%	0.8	3.6	
N-free extract	%	9.4	41.8	
Protein (N x 6.25)	%	4.8	21.3	
Cattle	dig prot % *	3.6	16.0	
Goats	dig prot % *	3.7	16.5	
Horses	dig prot % *	3.5	15.6	
Rabbits	dig prot % *	3.4	15.1	
Sheep	dig prot % *	3.8	16.9	
Energy	GE Mcal/kg			
Cattle	DE Mcal/kg *	0.62	2.75	
Sheep	DE Mcal/kg *	0.64	2.82	
Cattle	ME Mcal/kg *	0.51	2.26	
Sheep	ME Mcal/kg *	0.52	2.31	
Cattle	TDN % *	14.0	62.3	
Sheep	TDN % *	14.4	64.0	
Calcium	%	0.28	1.24	
Phosphorus	%	0.07	0.31	
Potassium	%	0.63	2.80	

Alfalfa-Brome, aerial part, ensiled, (3)

Ref No 3 08 329 United States

Analyses		As Fed	Dry	C.V. ± %
Dry matter	%	25.0	100.0	
Ash	%	2.1	8.4	
Crude fiber	%	7.7	30.8	
Ether extract	%	1.1	4.4	
N-free extract	%	10.3	41.2	
Protein (N x 6.25)	%	3.8	15.2	
Cattle	dig prot % *	2.5	10.0	
Goats	dig prot % *	2.5	10.0	
Horses	dig prot % *	2.5	10.0	
Sheep	dig prot % *	2.5	10.0	
Energy	GE Mcal/kg			
Cattle	DE Mcal/kg *	0.64	2.56	
Sheep	DE Mcal/kg *	0.62	2.47	
Cattle	ME Mcal/kg *	0.52	2.10	
Sheep	ME Mcal/kg *	0.51	2.03	
Cattle	TDN % *	14.5	58.0	
Sheep	TDN % *	14.0	56.0	
Calcium	%	0.37	1.48	
Phosphorus	%	0.05	0.20	

Alfalfa-Brome, aerial part, wilted ensiled, (3)

Ref No 3 08 330 United States

Analyses		As Fed	Dry	C.V. ± %
Dry matter	%	44.9	100.0	
Ash	%	4.1	9.1	
Crude fiber	%	15.0	33.4	
Ether extract	%	1.0	2.2	
N-free extract	%	17.8	39.6	
Protein (N x 6.25)	%	7.0	15.6	
Cattle	dig prot % *	4.7	10.4	
Goats	dig prot % *	4.7	10.4	
Horses	dig prot % *	4.7	10.4	
Sheep	dig prot % *	4.7	10.4	
Energy	GE Mcal/kg			
Cattle	DE Mcal/kg *	1.06	2.35	
Sheep	DE Mcal/kg *	1.10	2.45	
Cattle	ME Mcal/kg *	0.87	1.93	
Sheep	ME Mcal/kg *	0.90	2.01	
Cattle	TDN % *	23.9	53.3	
Sheep	TDN % *	25.0	55.6	
Calcium	%	0.66	1.47	
Phosphorus	%	0.10	0.22	

ALFALFA–BROME, CHEATGRASS. Medicago sativa, Bromus tectorum

Alfalfa-Brome, cheatgrass, hay, s-c, mature, (1)

Ref No 1 00 248 United States

Analyses		As Fed	Dry	C.V. ± %
Dry matter	%	92.6	100.0	
Ash	%	5.4	5.8	
Crude fiber	%	26.6	28.7	
Ether extract	%	1.3	1.4	
N-free extract	%	52.1	56.3	
Protein (N x 6.25)	%	7.2	7.8	
Cattle	dig prot % *	3.4	3.7	
Goats	dig prot % *	3.6	3.8	
Horses	dig prot % *	3.8	4.2	
Rabbits	dig prot % *	4.3	4.7	
Sheep	dig prot % *	3.3	3.6	

(1) dry forages and roughages (3) silages (6) minerals
(2) pasture, range plants, and forages fed green (4) energy feeds (7) vitamins
(5) protein supplements (8) additives

Feed Name or Analyses			Mean As Fed	Mean Dry	C.V. ± %
Energy	GE Mcal/kg				
Cattle	DE Mcal/kg	*	2.58	2.78	
Sheep	DE Mcal/kg	*	2.37	2.56	
Cattle	ME Mcal/kg	*	2.11	2.28	
Sheep	ME Mcal/kg	*	1.94	2.10	
Cattle	TDN %	*	58.4	63.1	
Sheep	TDN %	*	53.8	58.1	

Alfalfa-Brome, cheatgrass, hay, s-c, gr 2 US, (1)
Ref No 1 00 249 United States

Feed Name or Analyses			Mean As Fed	Mean Dry	C.V. ± %
Dry matter	%		92.6	100.0	
Calcium	%		0.53	0.57	
Phosphorus	%		0.16	0.17	
Carotene	mg/kg		15.7	17.0	
Vitamin A equivalent	IU/g		26.2	28.3	

ALFALFA–BROME, SMOOTH. Medicago sativa, Bromus inermis

Alfalfa-Brome, smooth, hay, fan air dried, (1)
Ref No 1 00 250 United States

Feed Name or Analyses			Mean As Fed	Mean Dry	C.V. ± %
Dry matter	%			100.0	
Carotene	mg/kg			42.8	
Vitamin A equivalent	IU/g			71.3	

Alfalfa-Brome, smooth, hay, s-c, (1)
Ref No 1 00 255 United States

Feed Name or Analyses			Mean As Fed	Mean Dry	C.V. ± %
Dry matter	%		82.5	100.0	28
Ash	%		6.4	7.8	13
Crude fiber	%		27.8	33.7	10
Ether extract	%		2.2	2.7	27
N-free extract	%		32.8	39.7	
Protein (N x 6.25)	%		13.3	16.2	33
Cattle	dig prot %	*	9.0	10.9	
Goats	dig prot %	*	9.6	11.6	
Horses	dig prot %	*	9.3	11.2	
Rabbits	dig prot %	*	9.2	11.1	
Sheep	dig prot %	*	9.1	11.1	
Energy	GE Mcal/kg				
Cattle	DE Mcal/kg	*	2.01	2.44	
Sheep	DE Mcal/kg	*	2.03	2.46	
Cattle	ME Mcal/kg	*	1.65	2.00	
Sheep	ME Mcal/kg	*	1.67	2.02	
Cattle	TDN %	*	45.7	55.4	
Sheep	TDN %	*	46.1	55.8	
Calcium	%		0.85	1.03	15
Chlorine	%		0.39	0.47	
Cobalt	mg/kg		0.071	0.086	
Copper	mg/kg		13.5	16.3	14
Iron	%		0.011	0.013	18
Magnesium	%		0.45	0.54	49
Manganese	mg/kg		35.1	42.5	23
Phosphorus	%		0.25	0.30	30
Potassium	%		1.53	1.85	18
Sodium	%		0.35	0.42	11
Sulphur	%		0.19	0.23	11
Carotene	mg/kg		21.5	26.0	46
Niacin	mg/kg		22.3	27.0	6
Pantothenic acid	mg/kg		19.4	23.5	5
Riboflavin	mg/kg		5.5	6.7	6
Vitamin A equivalent	IU/g		35.8	43.4	

Alfalfa-Brome, smooth, hay, s-c, immature, (1)
Ref No 1 00 251 United States

Feed Name or Analyses			Mean As Fed	Mean Dry	C.V. ± %
Dry matter	%		58.8	100.0	2
Ash	%		4.6	7.9	9
Crude fiber	%		17.8	30.4	
Ether extract	%		1.9	3.2	8
N-free extract	%		22.0	37.5	
Protein (N x 6.25)	%		12.4	21.2	23
Cattle	dig prot %	*	9.0	15.3	
Goats	dig prot %	*	9.6	16.3	
Horses	dig prot %	*	9.1	15.5	
Rabbits	dig prot %	*	8.8	15.0	
Sheep	dig prot %	*	9.1	15.5	
Energy	GE Mcal/kg				
Cattle	DE Mcal/kg	*	1.48	2.53	
Sheep	DE Mcal/kg	*	1.51	2.57	
Cattle	ME Mcal/kg	*	1.22	2.07	
Sheep	ME Mcal/kg	*	1.24	2.11	
Cattle	TDN %	*	33.7	57.3	
Sheep	TDN %	*	34.2	58.3	

Alfalfa-Brome, smooth, hay, s-c, full bloom, (1)
Ref No 1 00 252 United States

Feed Name or Analyses			Mean As Fed	Mean Dry	C.V. ± %
Dry matter	%		92.0	100.0	0
Ash	%		6.2	6.7	13
Crude fiber	%		31.4	34.2	10
Ether extract	%		1.9	2.1	11
N-free extract	%		39.4	42.8	
Protein (N x 6.25)	%		13.2	14.3	21
Cattle	dig prot %	*	8.6	9.3	
Goats	dig prot %	*	9.1	9.9	
Horses	dig prot %	*	8.9	9.7	
Rabbits	dig prot %	*	8.9	9.7	
Sheep	dig prot %	*	8.6	9.4	
Energy	GE Mcal/kg				
Cattle	DE Mcal/kg	*	2.17	2.36	
Sheep	DE Mcal/kg	*	2.30	2.50	
Cattle	ME Mcal/kg	*	1.78	1.94	
Sheep	ME Mcal/kg	*	1.89	2.05	
Cattle	TDN %	*	49.3	53.6	
Sheep	TDN %	*	52.2	56.7	
Calcium	%		1.05	1.14	
Cobalt	mg/kg		0.079	0.086	
Copper	mg/kg		16.4	17.9	
Iron	%		0.011	0.012	
Magnesium	%		0.72	0.78	
Manganese	mg/kg		34.3	37.3	
Phosphorus	%		0.24	0.26	
Potassium	%		1.29	1.40	
Carotene	mg/kg		18.9	20.5	
Vitamin A equivalent	IU/g		31.5	34.2	

Alfalfa-Brome, smooth, hay, s-c, cut 1, (1)
Ref No 1 00 253 United States

Feed Name or Analyses			Mean As Fed	Mean Dry	C.V. ± %
Dry matter	%		89.9	100.0	1
Ash	%		5.6	6.2	3
Crude fiber	%		29.6	32.9	4
Ether extract	%		2.1	2.3	3
N-free extract	%		40.4	44.9	
Protein (N x 6.25)	%		12.3	13.7	20
Cattle	dig prot %	*	7.9	8.8	
Goats	dig prot %	*	8.4	9.3	
Horses	dig prot %	*	8.2	9.2	
Rabbits	dig prot %	*	8.3	9.2	
Sheep	dig prot %	*	8.0	8.9	
Energy	GE Mcal/kg				
Cattle	DE Mcal/kg	*	2.15	2.39	
Sheep	DE Mcal/kg	*	2.29	2.54	
Cattle	ME Mcal/kg	*	1.76	1.96	
Sheep	ME Mcal/kg	*	1.88	2.09	
Cattle	TDN %	*	48.7	54.2	
Sheep	TDN %	*	51.9	57.7	
Calcium	%		1.21	1.35	
Copper	mg/kg		16.1	17.9	
Iron	%		0.012	0.013	
Magnesium	%		0.44	0.49	
Manganese	mg/kg		37.9	42.1	
Phosphorus	%		0.28	0.31	
Potassium	%		1.26	1.40	
Vitamin D_2	IU/g		0.9	1.0	

Alfalfa-Brome, smooth, hay, s-c, cut 2, (1)
Ref No 1 00 254 United States

Feed Name or Analyses			Mean As Fed	Mean Dry	C.V. ± %
Dry matter	%		89.8	100.0	2
Ash	%		5.5	6.1	12
Crude fiber	%		31.3	34.8	0
Ether extract	%		1.9	2.1	14
N-free extract	%		38.9	43.3	
Protein (N x 6.25)	%		12.3	13.7	9
Cattle	dig prot %	*	7.9	8.8	
Goats	dig prot %	*	8.4	9.3	
Horses	dig prot %	*	8.2	9.2	
Rabbits	dig prot %	*	8.3	9.2	
Sheep	dig prot %	*	8.0	8.9	
Energy	GE Mcal/kg				
Cattle	DE Mcal/kg	*	2.09	2.33	
Sheep	DE Mcal/kg	*	2.24	2.50	
Cattle	ME Mcal/kg	*	1.71	1.91	
Sheep	ME Mcal/kg	*	1.84	2.05	
Cattle	TDN %	*	47.4	52.7	
Sheep	TDN %	*	50.9	56.7	
Calcium	%		0.83	0.92	
Copper	mg/kg		16.0	17.9	
Iron	%		0.009	0.010	
Magnesium	%		0.95	1.06	
Manganese	mg/kg		28.9	32.2	
Phosphorus	%		0.19	0.21	

Alfalfa-Brome, smooth, hay, s-c, gr 1 US, (1)
Ref No 1 00 257 United States

Feed Name or Analyses			Mean As Fed	Mean Dry	C.V. ± %
Dry matter	%		89.5	100.0	1
Ash	%		5.7	6.3	18
Crude fiber	%		30.2	33.8	6
Ether extract	%		1.7	1.9	20
N-free extract	%		38.5	43.0	
Protein (N x 6.25)	%		13.4	14.9	8
Cattle	dig prot %	*	8.8	9.9	
Goats	dig prot %	*	9.4	10.5	
Horses	dig prot %	*	9.1	10.2	
Rabbits	dig prot %	*	9.1	10.2	
Sheep	dig prot %	*	8.9	10.0	
Energy	GE Mcal/kg				
Cattle	DE Mcal/kg	*	2.84	3.17	
Cattle	ME Mcal/kg	*	2.33	2.60	
Cattle	TDN %		64.4	71.9	
Calcium	%		1.21	1.35	
Copper	mg/kg		16.0	17.9	
Iron	%		0.012	0.013	
Magnesium	%		0.44	0.49	

Continued

Feed Name or Analyses			Mean As Fed	Mean Dry	C.V. ± %
Manganese	mg/kg		37.7	42.1	
Phosphorus	%		0.28	0.31	
Potassium	%		1.25	1.40	

Alfalfa-Brome, smooth, hay, s-c, gr 2 US, (1)
Ref No 1 00 258 United States

Dry matter	%		90.8	100.0	2
Ash	%		6.1	6.7	10
Crude fiber	%		29.1	32.0	5
Ether extract	%		2.1	2.3	6
N-free extract	%		39.2	43.2	
Protein (N x 6.25)	%		14.3	15.8	17
Cattle	dig prot %	*	9.6	10.6	
Goats	dig prot %	*	10.3	11.3	
Horses	dig prot %	*	9.9	10.9	
Rabbits	dig prot %	*	9.9	10.9	
Sheep	dig prot %	*	9.8	10.7	
Energy	GE Mcal/kg				
Cattle	DE Mcal/kg	*	2.20	2.42	
Sheep	DE Mcal/kg	*	2.33	2.57	
Cattle	ME Mcal/kg	*	1.80	1.99	
Sheep	ME Mcal/kg	*	1.91	2.10	
Cattle	TDN %	*	49.9	55.0	
Sheep	TDN %	*	52.8	58.2	
Calcium	%		0.84	0.92	
Copper	mg/kg		16.2	17.9	
Iron	%		0.009	0.010	
Magnesium	%		0.96	1.06	
Manganese	mg/kg		29.2	32.2	
Phosphorus	%		0.19	0.21	

Alfalfa-Brome, smooth, hay, s-c, gr 3 US, (1)
Ref No 1 00 259 United States

Dry matter	%		90.9	100.0	4
Ash	%		6.4	7.0	11
Crude fiber	%		34.2	37.6	13
Ether extract	%		1.8	2.0	16
N-free extract	%		36.4	40.0	
Protein (N x 6.25)	%		12.2	13.4	24
Cattle	dig prot %	*	7.8	8.5	
Goats	dig prot %	*	8.2	9.1	
Horses	dig prot %	*	8.1	8.9	
Rabbits	dig prot %	*	8.2	9.0	
Sheep	dig prot %	*	7.8	8.6	
Energy	GE Mcal/kg				
Cattle	DE Mcal/kg	*	2.08	2.29	
Sheep	DE Mcal/kg	*	2.17	2.38	
Cattle	ME Mcal/kg	*	1.70	1.87	
Sheep	ME Mcal/kg	*	1.78	1.95	
Cattle	TDN %	*	47.1	51.8	
Sheep	TDN %	*	49.2	54.1	

Alfalfa-Brome, smooth, hay, s-c, gr sample US, (1)
Ref No 1 00 256 United States

Dry matter	%		92.2	100.0	
Ash	%		7.2	7.8	
Crude fiber	%		34.2	37.1	
Ether extract	%		1.8	2.0	
N-free extract	%		35.5	38.5	

(1) dry forages and roughages (2) pasture, range plants, and forages fed green (3) silages (4) energy feeds (5) protein supplements (6) minerals (7) vitamins (8) additives

Feed Name or Analyses			Mean As Fed	Mean Dry	C.V. ± %
Protein (N x 6.25)	%		13.5	14.6	
Cattle	dig prot %	*	8.8	9.6	
Goats	dig prot %	*	9.4	10.2	
Horses	dig prot %	*	9.2	9.9	
Rabbits	dig prot %	*	9.2	9.9	
Sheep	dig prot %	*	8.9	9.7	
Energy	GE Mcal/kg				
Cattle	DE Mcal/kg	*	2.15	2.33	
Sheep	DE Mcal/kg	*	2.19	2.38	
Cattle	ME Mcal/kg	*	1.76	1.91	
Sheep	ME Mcal/kg	*	1.80	1.95	
Cattle	TDN %	*	48.7	52.8	
Sheep	TDN %	*	49.7	53.9	

Alfalfa-Brome, smooth, aerial part, fresh, (2)
Ref No 2 00 262 United States

Dry matter	%		21.6	100.0	
Ash	%		2.1	9.8	
Crude fiber	%		5.5	25.3	
Ether extract	%		0.8	3.6	
N-free extract	%		9.0	41.7	
Protein (N x 6.25)	%		4.2	19.6	
Cattle	dig prot %	*	3.1	14.6	
Goats	dig prot %	*	3.2	14.8	
Horses	dig prot %	*	3.1	14.2	
Rabbits	dig prot %	*	3.0	13.8	
Sheep	dig prot %	*	3.3	15.3	
Energy	GE Mcal/kg				
Cattle	NE_m Mcal/kg	*	0.29	1.35	
Cattle	NE_{gain} Mcal/kg	*	0.16	0.75	
Cattle	$NE_{lactating\ cows}$ Mcal/kg	*	0.34	1.57	

Alfalfa-Brome, smooth, aerial part, fresh, early bloom, (2)
Ref No 2 00 261 United States

Dry matter	%		19.9	100.0	
Energy	GE Mcal/kg				
Cattle	NE_m Mcal/kg	*	0.26	1.32	
Cattle	NE_{gain} Mcal/kg	*	0.14	0.71	
Cattle	$NE_{lactating\ cows}$ Mcal/kg	*	0.30	1.52	
Carotene	mg/kg		20.2	101.4	
Vitamin A equivalent	IU/g		33.6	169.1	

Alfalfa-Brome, smooth, aerial part, ensiled, (3)
Ref No 3 00 268 United States

Dry matter	%		27.5	100.0	26
Ash	%		2.0	7.4	2
Crude fiber	%		9.8	35.6	6
Ether extract	%		1.1	4.1	17
N-free extract	%		10.8	39.3	
Protein (N x 6.25)	%		3.7	13.6	18
Cattle	dig prot %	*	2.4	8.6	
Goats	dig prot %	*	2.4	8.6	
Horses	dig prot %	*	2.4	8.6	
Sheep	dig prot %	*	2.4	8.6	
Energy	GE Mcal/kg				
Cattle	DE Mcal/kg	*	0.65	2.36	
Cattle	ME Mcal/kg	*	0.53	1.94	
Cattle	TDN %	*	14.7	53.6	
Calcium	%		0.18	0.64	
Cobalt	mg/kg		0.031	0.112	
Copper	mg/kg		3.1	11.2	
Iron	%		0.004	0.015	
Magnesium	%		0.05	0.20	

Feed Name or Analyses			Mean As Fed	Mean Dry	C.V. ± %
Manganese	mg/kg		8.3	30.2	
Phosphorus	%		0.05	0.20	
Potassium	%		0.51	1.86	
Carotene	mg/kg		28.5	103.8	69
Vitamin A equivalent	IU/g		47.6	173.1	

Alfalfa-Brome, smooth, aerial part, ensiled, immature, (3)
Ref No 3 00 263 United States

Dry matter	%			100.0	
Carotene	mg/kg			212.5	18
Vitamin A equivalent	IU/g			354.3	

Alfalfa-Brome, smooth, aerial part, ensiled, early bloom, (3)
Ref No 3 00 264 United States

Dry matter	%		32.5	100.0	
Ash	%		3.4	10.5	
Crude fiber	%		9.3	28.6	
Ether extract	%		0.8	2.5	
N-free extract	%		14.1	43.3	
Protein (N x 6.25)	%		4.9	15.1	
Cattle	dig prot %	*	3.2	9.9	
Goats	dig prot %	*	3.2	9.9	
Horses	dig prot %	*	3.2	9.9	
Sheep	dig prot %	*	3.2	9.9	
Energy	GE Mcal/kg				
Cattle	DE Mcal/kg	*	0.87	2.67	
Sheep	DE Mcal/kg	*	0.80	2.46	
Cattle	ME Mcal/kg	*	0.71	2.19	
Sheep	ME Mcal/kg	*	0.66	2.02	
Cattle	TDN %	*	19.7	60.5	
Sheep	TDN %	*	18.1	55.8	

Alfalfa-Brome, smooth, aerial part, ensiled, mid-bloom, (3)
Ref No 3 00 265 United States

Dry matter	%		30.0	100.0	11
Ash	%		2.7	9.1	15
Crude fiber	%		11.2	37.4	7
Ether extract	%		0.8	2.8	12
N-free extract	%		11.2	37.4	
Protein (N x 6.25)	%		4.0	13.3	14
Cattle	dig prot %	*	2.5	8.3	
Goats	dig prot %	*	2.5	8.3	
Horses	dig prot %	*	2.5	8.3	
Sheep	dig prot %	*	2.5	8.3	
Energy	GE Mcal/kg				
Cattle	DE Mcal/kg	*	0.70	2.33	
Sheep	DE Mcal/kg	*	0.68	2.26	
Cattle	ME Mcal/kg	*	0.57	1.91	
Sheep	ME Mcal/kg	*	0.56	1.85	
Cattle	TDN %	*	15.8	52.7	
Sheep	TDN %	*	15.4	51.3	
Calcium	%		0.46	1.53	25
Phosphorus	%		0.16	0.54	28
Carotene	mg/kg		37.1	123.7	
Vitamin A equivalent	IU/g		61.9	206.2	

Alfalfa-Brome, smooth, aerial part, ensiled, mature, (3)
Ref No 3 00 266 United States

Feed Name or Analyses			Mean As Fed	Mean Dry	C.V. ± %
Dry matter	%		24.7	100.0	
Carotene	mg/kg		3.4	13.7	
Vitamin A equivalent	IU/g		5.6	22.8	

Alfalfa-Brome, smooth, aerial part, ensiled, cut 1, (3)
Ref No 3 00 267 United States

Feed Name or Analyses			Mean As Fed	Mean Dry	C.V. ± %
Dry matter	%		22.0	100.0	
Ash	%		1.6	7.3	
Crude fiber	%		8.6	39.2	
Ether extract	%		0.7	3.4	
N-free extract	%		7.9	36.0	
Protein (N x 6.25)	%		3.1	14.1	
Cattle	dig prot %	*	2.0	9.0	
Goats	dig prot %	*	2.0	9.0	
Horses	dig prot %	*	2.0	9.0	
Sheep	dig prot %	*	2.0	9.0	
Energy	GE Mcal/kg				
Cattle	DE Mcal/kg	*	0.47	2.14	
Sheep	DE Mcal/kg	*	0.46	2.08	
Cattle	ME Mcal/kg	*	0.39	1.76	
Sheep	ME Mcal/kg	*	0.38	1.71	
Cattle	TDN %	*	10.7	48.6	
Sheep	TDN %	*	10.4	47.1	

Alfalfa-Brome, smooth, aerial part, wilted ensiled, (3)
Ref No 3 00 269 United States

Feed Name or Analyses			Mean As Fed	Mean Dry	C.V. ± %
Dry matter	%		46.1	100.0	
Ash	%		4.1	8.8	
Crude fiber	%		15.2	33.0	
Ether extract	%		1.1	2.3	
N-free extract	%		18.6	40.4	
Protein (N x 6.25)	%		7.1	15.5	
Cattle	dig prot %	*	4.8	10.3	
Goats	dig prot %	*	4.8	10.3	
Horses	dig prot %	*	4.8	10.3	
Sheep	dig prot %	*	4.8	10.3	
Energy	GE Mcal/kg				
Cattle	DE Mcal/kg	*	1.09	2.36	
Sheep	DE Mcal/kg	*	1.13	2.44	
Cattle	ME Mcal/kg	*	0.89	1.93	
Sheep	ME Mcal/kg	*	0.92	2.00	
Cattle	TDN %	*	24.7	53.5	
Sheep	TDN %	*	25.5	55.4	
Calcium	%		0.72	1.56	
Phosphorus	%		0.12	0.26	

Alfalfa-Brome, smooth, aerial part w molasses added, ensiled, (3)
Ref No 3 00 270 United States

Feed Name or Analyses			Mean As Fed	Mean Dry	C.V. ± %
Dry matter	%		42.7	100.0	
Ash	%		3.8	8.8	11
Crude fiber	%		13.2	31.0	8
Ether extract	%		1.2	2.9	25
N-free extract	%		18.4	43.1	
Protein (N x 6.25)	%		6.1	14.2	12
Cattle	dig prot %	*	3.9	9.1	
Goats	dig prot %	*	3.9	9.1	
Horses	dig prot %	*	3.9	9.1	
Sheep	dig prot %	*	3.9	9.1	
Energy	GE Mcal/kg				
Cattle	DE Mcal/kg	*	1.08	2.54	
Sheep	DE Mcal/kg	*	1.06	2.48	
Cattle	ME Mcal/kg	*	0.89	2.08	
Sheep	ME Mcal/kg	*	0.87	2.03	
Cattle	TDN %	*	24.6	57.6	
Sheep	TDN %	*	24.0	56.3	
Calcium	%		0.71	1.66	
Phosphorus	%		0.15	0.35	

Alfalfa-Brome, smooth, aerial part w sulfur dioxide preservative added, ensiled, (3)
Ref No 3 00 271 United States

Feed Name or Analyses			Mean As Fed	Mean Dry	C.V. ± %
Dry matter	%		22.0	100.0	
Ash	%		1.6	7.3	
Crude fiber	%		8.6	39.2	
Ether extract	%		0.7	3.4	
N-free extract	%		7.9	36.0	
Protein (N x 6.25)	%		3.1	14.1	
Cattle	dig prot %	*	2.0	9.0	
Goats	dig prot %	*	2.0	9.0	
Horses	dig prot %	*	2.0	9.0	
Sheep	dig prot %	*	2.0	9.0	
Energy	GE Mcal/kg				
Cattle	DE Mcal/kg	*	0.47	2.14	
Sheep	DE Mcal/kg	*	0.49	2.24	
Cattle	ME Mcal/kg	*	0.39	1.76	
Sheep	ME Mcal/kg	*	0.40	1.84	
Cattle	TDN %	*	10.7	48.6	
Sheep	TDN %	*	11.2	50.7	

ALFALFA–CLOVER. Medicago sativa, Trifolium spp

Alfalfa-Clover, hay, s-c, (1)
Ref No 1 00 280 United States

Feed Name or Analyses			Mean As Fed	Mean Dry	C.V. ± %
Dry matter	%		89.7	100.0	5
Ash	%		6.9	7.7	15
Crude fiber	%		30.7	34.2	8
Cattle	dig coef %		43.	43.	
Ether extract	%		2.2	2.5	22
Cattle	dig coef %		42.	42.	
N-free extract	%		38.8	43.3	
Cattle	dig coef %		63.	63.	
Protein (N x 6.25)	%		11.1	12.4	13
Cattle	dig coef %		57.	57.	
Cattle	dig prot %		6.3	7.1	
Goats	dig prot %	*	7.3	8.1	
Horses	dig prot %	*	7.2	8.1	
Rabbits	dig prot %	*	7.4	8.3	
Sheep	dig prot %	*	6.9	7.7	
Energy	GE Mcal/kg				
Cattle	DE Mcal/kg	*	2.03	2.26	
Sheep	DE Mcal/kg	*	2.24	2.50	
Cattle	ME Mcal/kg	*	1.67	1.86	
Sheep	ME Mcal/kg	*	1.84	2.05	
Cattle	TDN %		46.1	51.4	
Sheep	TDN %	*	50.8	56.7	

Alfalfa-Clover, hay, s-c, immature, (1)
Ref No 1 00 272 United States

Feed Name or Analyses			Mean As Fed	Mean Dry	C.V. ± %
Dry matter	%		87.2	100.0	
Ash	%		7.2	8.3	
Crude fiber	%		24.2	27.8	
Ether extract	%		1.9	2.2	
N-free extract	%		39.1	44.8	
Protein (N x 6.25)	%		14.7	16.9	
Cattle	dig prot %	*	10.1	11.6	
Goats	dig prot %	*	10.7	12.3	
Horses	dig prot %	*	10.4	11.9	
Rabbits	dig prot %	*	10.2	11.7	
Sheep	dig prot %	*	10.2	11.7	
Energy	GE Mcal/kg				
Cattle	DE Mcal/kg	*	2.26	2.59	
Sheep	DE Mcal/kg	*	2.30	2.64	
Cattle	ME Mcal/kg	*	1.85	2.13	
Sheep	ME Mcal/kg	*	1.89	2.16	
Cattle	TDN %	*	51.3	58.8	
Sheep	TDN %	*	52.2	59.9	

Alfalfa-Clover, hay, s-c, mid-bloom, (1)
Ref No 1 00 273 United States

Feed Name or Analyses			Mean As Fed	Mean Dry	C.V. ± %
Dry matter	%		92.2	100.0	5
Ash	%		8.2	8.9	15
Crude fiber	%		30.6	33.2	6
Ether extract	%		2.1	2.3	20
N-free extract	%		43.6	47.3	
Protein (N x 6.25)	%		7.7	8.3	15
Cattle	dig prot %	*	3.8	4.1	
Goats	dig prot %	*	4.0	4.3	
Horses	dig prot %	*	4.2	4.6	
Rabbits	dig prot %	*	4.7	5.1	
Sheep	dig prot %	*	3.7	4.0	
Energy	GE Mcal/kg				
Cattle	DE Mcal/kg	*	2.41	2.61	
Sheep	DE Mcal/kg	*	2.25	2.44	
Cattle	ME Mcal/kg	*	1.98	2.14	
Sheep	ME Mcal/kg	*	1.85	2.00	
Cattle	TDN %	*	54.6	59.3	
Sheep	TDN %	*	51.1	55.4	

Alfalfa-Clover, hay, s-c, full bloom, (1)
Ref No 1 00 274 United States

Feed Name or Analyses			Mean As Fed	Mean Dry	C.V. ± %
Dry matter	%		95.4	100.0	0
Ash	%		7.2	7.5	23
Crude fiber	%		31.1	32.6	2
Ether extract	%		3.0	3.1	5
N-free extract	%		49.3	51.7	
Protein (N x 6.25)	%		4.9	5.1	6
Cattle	dig prot %	*	1.3	1.4	
Goats	dig prot %	*	1.3	1.3	
Horses	dig prot %	*	1.8	1.9	
Rabbits	dig prot %	*	2.5	2.6	
Sheep	dig prot %	*	1.1	1.1	
Energy	GE Mcal/kg				
Cattle	DE Mcal/kg	*	2.20	2.31	
Sheep	DE Mcal/kg	*	2.30	2.41	
Cattle	ME Mcal/kg	*	1.80	1.89	
Sheep	ME Mcal/kg	*	1.88	1.98	
Cattle	TDN %	*	50.0	52.4	
Sheep	TDN %	*	52.1	54.6	

Alfalfa-Clover, hay, s-c, milk stage, (1)
Ref No 1 00 275 United States

Feed Name or Analyses			Mean As Fed	Mean Dry	C.V. ± %
Dry matter	%		91.5	100.0	5
Ash	%		6.3	6.9	22
Crude fiber	%		33.2	36.3	7
Ether extract	%		2.0	2.2	31

Continued

Feed Name or Analyses			Mean As Fed	Mean Dry	C.V. ± %
N-free extract	%		42.3	46.2	
Protein (N x 6.25)	%		7.7	8.4	39
Cattle	dig prot %	*	3.9	4.2	
Goats	dig prot %	*	4.0	4.4	
Horses	dig prot %	*	4.3	4.7	
Rabbits	dig prot %	*	4.7	5.2	
Sheep	dig prot %	*	3.8	4.1	
Energy	GE Mcal/kg				
Cattle	DE Mcal/kg	*	2.10	2.30	
Sheep	DE Mcal/kg	*	2.19	2.39	
Cattle	ME Mcal/kg	*	1.72	1.88	
Sheep	ME Mcal/kg	*	1.80	1.96	
Cattle	TDN %	*	47.6	52.1	
Sheep	TDN %	*	49.7	54.3	

Alfalfa-Clover, hay, s-c, cut 1, (1)
Ref No 1 00 276 United States

Feed Name or Analyses			Mean As Fed	Mean Dry	C.V. ± %
Dry matter	%		88.0	100.0	
Ash	%		7.0	7.9	10
Crude fiber	%		26.3	29.9	
Ether extract	%		1.9	2.2	
N-free extract	%		38.2	43.4	
Protein (N x 6.25)	%		14.6	16.6	
Cattle	dig prot %	*	10.0	11.3	
Goats	dig prot %	*	10.6	12.0	
Horses	dig prot %	*	10.2	11.6	
Rabbits	dig prot %	*	10.1	11.5	
Sheep	dig prot %	*	10.1	11.5	
Energy	GE Mcal/kg				
Cattle	DE Mcal/kg	*	2.22	2.52	
Sheep	DE Mcal/kg	*	2.28	2.59	
Cattle	ME Mcal/kg	*	1.82	2.07	
Sheep	ME Mcal/kg	*	1.87	2.12	
Cattle	TDN %	*	50.3	57.2	
Sheep	TDN %	*	51.7	58.8	

Alfalfa-Clover, hay, s-c, gr 1 US, (1)
Ref No 1 00 277 United States

Feed Name or Analyses			Mean As Fed	Mean Dry	C.V. ± %
Dry matter	%			100.0	
Calcium	%			1.33	
Cobalt	mg/kg			0.051	
Manganese	mg/kg			24.9	
Phosphorus	%			0.27	
Carotene	mg/kg			35.3	
Vitamin A equivalent	IU/g			58.8	

Alfalfa-Clover, hay, s-c, gr 2 US, (1)
Ref No 1 00 278 United States

Feed Name or Analyses			Mean As Fed	Mean Dry	C.V. ± %
Dry matter	%			100.0	
Carotene	mg/kg			41.7	
Vitamin A equivalent	IU/g			69.5	

Alfalfa-Clover, hay, s-c, gr 3 US, (1)
Ref No 1 00 279 United States

Feed Name or Analyses			Mean As Fed	Mean Dry	C.V. ± %
Dry matter	%		89.4	100.0	
Ash	%		6.7	7.5	
Crude fiber	%		35.5	39.7	

(1) dry forages and roughages (2) pasture, range plants, and forages fed green (3) silages (4) energy feeds (5) protein supplements (6) minerals (7) vitamins (8) additives

Feed Name or Analyses			Mean As Fed	Mean Dry	C.V. ± %
Ether extract	%		0.9	1.0	
N-free extract	%		35.4	39.6	
Protein (N x 6.25)	%		10.9	12.2	
Cattle	dig prot %	*	6.7	7.5	
Goats	dig prot %	*	7.1	7.9	
Horses	dig prot %	*	7.1	7.9	
Rabbits	dig prot %	*	7.2	8.1	
Sheep	dig prot %	*	6.7	7.5	
Energy	GE Mcal/kg				
Cattle	DE Mcal/kg	*	1.77	1.98	
Sheep	DE Mcal/kg	*	1.84	2.06	
Cattle	ME Mcal/kg	*	1.45	1.62	
Sheep	ME Mcal/kg	*	1.51	1.69	
Cattle	TDN %	*	40.2	44.9	
Sheep	TDN %	*	41.7	46.7	

Alfalfa-Clover, aerial part, fresh, (2)
Ref No 2 00 283 United States

Feed Name or Analyses			Mean As Fed	Mean Dry	C.V. ± %
Dry matter	%		23.1	100.0	8
Ash	%		2.3	10.0	12
Crude fiber	%		8.5	36.6	9
Ether extract	%		0.3	1.4	18
N-free extract	%		8.6	37.4	
Protein (N x 6.25)	%		3.4	14.6	22
Cattle	dig prot %	*	2.4	10.3	
Goats	dig prot %	*	2.4	10.2	
Horses	dig prot %	*	2.3	9.9	
Rabbits	dig prot %	*	2.3	9.9	
Sheep	dig prot %	*	2.4	10.6	
Energy	GE Mcal/kg				
Cattle	DE Mcal/kg	*	0.60	2.61	
Sheep	DE Mcal/kg	*	0.60	2.60	
Cattle	ME Mcal/kg	*	0.49	2.14	
Sheep	ME Mcal/kg	*	0.49	2.13	
Cattle	TDN %	*	13.7	59.1	
Sheep	TDN %	*	13.6	58.9	

Alfalfa-Clover, aerial part, fresh, cut 1, (2)
Ref No 2 00 281 United States

Feed Name or Analyses			Mean As Fed	Mean Dry	C.V. ± %
Dry matter	%		24.2	100.0	
Ash	%		2.2	9.2	
Crude fiber	%		9.4	39.0	
Ether extract	%		0.4	1.5	
N-free extract	%		9.3	38.5	
Protein (N x 6.25)	%		2.9	11.8	
Cattle	dig prot %	*	1.9	7.9	
Goats	dig prot %	*	1.8	7.6	
Horses	dig prot %	*	1.8	7.5	
Rabbits	dig prot %	*	1.9	7.8	
Sheep	dig prot %	*	1.9	8.0	
Energy	GE Mcal/kg				
Cattle	DE Mcal/kg	*	0.62	2.58	
Sheep	DE Mcal/kg	*	0.64	2.65	
Cattle	ME Mcal/kg	*	0.51	2.12	
Sheep	ME Mcal/kg	*	0.53	2.18	
Cattle	TDN %	*	14.2	58.6	
Sheep	TDN %	*	14.6	60.2	

Alfalfa-Clover, aerial part, fresh, cut 3, (2)
Ref No 2 00 282 United States

Feed Name or Analyses			Mean As Fed	Mean Dry	C.V. ± %
Dry matter	%		21.9	100.0	
Ash	%		2.1	9.7	
Crude fiber	%		7.2	32.8	
Ether extract	%		0.2	1.1	

Feed Name or Analyses			Mean As Fed	Mean Dry	C.V. ± %
N-free extract	%		7.8	35.6	
Protein (N x 6.25)	%		4.6	20.8	
Cattle	dig prot %	*	3.4	15.6	
Goats	dig prot %	*	3.5	16.0	
Horses	dig prot %	*	3.3	15.2	
Rabbits	dig prot %	*	3.2	14.7	
Sheep	dig prot %	*	3.6	16.4	
Energy	GE Mcal/kg				
Cattle	DE Mcal/kg	*	0.52	2.39	
Sheep	DE Mcal/kg	*	0.52	2.38	
Cattle	ME Mcal/kg	*	0.43	1.96	
Sheep	ME Mcal/kg	*	0.43	1.95	
Cattle	TDN %	*	11.9	54.3	
Sheep	TDN %	*	11.8	53.9	

Alfalfa-Clover, aerial part, ensiled, immature, (3)
Ref No 3 00 284 United States

Feed Name or Analyses			Mean As Fed	Mean Dry	C.V. ± %
Dry matter	%		28.5	100.0	
Carotene	mg/kg		71.1	249.3	
Vitamin A equivalent	IU/g		118.5	415.7	

Alfalfa-Clover, aerial part w AIV preservative added, ensiled, (3)
Ref No 3 00 285 United States

Feed Name or Analyses			Mean As Fed	Mean Dry	C.V. ± %
Dry matter	%		27.5	100.0	
Carotene	mg/kg		12.7	46.1	
Vitamin A equivalent	IU/g		21.1	76.8	

ALFALFA–CLOVER, LADINO. Medicago sativa, Trifolium repens

Alfalfa-Clover, ladino, aerial part, ensiled, (3)
Ref No 3 00 287 United States

Feed Name or Analyses			Mean As Fed	Mean Dry	C.V. ± %
Dry matter	%		27.5	100.0	7
Ash	%		1.8	6.4	18
Crude fiber	%		8.6	31.1	14
Ether extract	%		1.2	4.3	26
N-free extract	%		12.3	44.9	
Protein (N x 6.25)	%		3.7	13.3	13
Cattle	dig prot %	*	2.3	8.3	
Goats	dig prot %	*	2.3	8.3	
Horses	dig prot %	*	2.3	8.3	
Sheep	dig prot %	*	2.3	8.3	
Energy	GE Mcal/kg				
Cattle	DE Mcal/kg	*	0.69	2.52	
Sheep	DE Mcal/kg	*	0.77	2.80	
Cattle	ME Mcal/kg	*	0.57	2.07	
Sheep	ME Mcal/kg	*	0.63	2.30	
Cattle	TDN %	*	15.7	57.2	
Sheep	TDN %	*	17.5	63.6	

Alfalfa-Clover, ladino, aerial part, ensiled, cut 1, (3)
Ref No 3 00 286 United States

Feed Name or Analyses			Mean As Fed	Mean Dry	C.V. ± %
Dry matter	%		27.4	100.0	7
Ash	%		1.6	5.9	10
Crude fiber	%		8.6	31.4	14
Ether extract	%		1.2	4.3	27
N-free extract	%		12.5	45.6	
Protein (N x 6.25)	%		3.5	12.8	13
Cattle	dig prot %	*	2.2	7.9	
Goats	dig prot %	*	2.2	7.9	
Horses	dig prot %	*	2.2	7.9	

Feed Name or Analyses			Mean As Fed	Mean Dry	C.V. ± %
Sheep	dig prot %	*	2.2	7.9	
Energy	GE Mcal/kg				
Cattle	DE Mcal/kg	*	0.69	2.51	
Sheep	DE Mcal/kg	*	0.67	2.43	
Cattle	ME Mcal/kg	*	0.56	2.06	
Sheep	ME Mcal/kg	*	0.55	1.99	
Cattle	TDN %	*	15.6	56.9	
Sheep	TDN %	*	15.1	55.1	

ALFALFA–CLOVER, RED. Medicago sativa, Trifolium pratense

Alfalfa-Clover, red, aerial part, ensiled, (3)

Ref No 3 00 288 United States

Feed Name or Analyses			Mean As Fed	Mean Dry	C.V. ± %
Dry matter	%		25.7	100.0	29
Ash	%		2.2	8.4	2
Crude fiber	%		8.8	34.3	8
Ether extract	%		0.9	3.6	24
N-free extract	%		10.1	39.2	
Protein (N x 6.25)	%		3.7	14.5	21
Cattle	dig prot %	*	2.4	9.4	
Goats	dig prot %	*	2.4	9.4	
Horses	dig prot %	*	2.4	9.4	
Sheep	dig prot %	*	2.4	9.4	
Energy	GE Mcal/kg				
Cattle	DE Mcal/kg	*	0.62	2.40	
Sheep	DE Mcal/kg	*	0.60	2.34	
Cattle	ME Mcal/kg	*	0.51	1.97	
Sheep	ME Mcal/kg	*	0.49	1.92	
Cattle	TDN %	*	14.0	54.5	
Sheep	TDN %	*	13.6	53.1	

Alfalfa-Clover, red, aerial part w sulfur dioxide preservative added, ensiled, full bloom, (3)

Ref No 3 00 289 United States

Feed Name or Analyses			Mean As Fed	Mean Dry	C.V. ± %
Dry matter	%		22.5	100.0	6
Ash	%		1.9	8.5	2
Crude fiber	%		7.9	34.9	9
Ether extract	%		0.9	4.0	5
N-free extract	%		8.8	39.3	
Protein (N x 6.25)	%		3.0	13.3	7
Cattle	dig prot %	*	1.9	8.3	
Goats	dig prot %	*	1.9	8.3	
Horses	dig prot %	*	1.9	8.3	
Sheep	dig prot %	*	1.9	8.3	
Energy	GE Mcal/kg				
Cattle	DE Mcal/kg	*	0.55	2.46	
Sheep	DE Mcal/kg	*	0.54	2.38	
Cattle	ME Mcal/kg	*	0.45	2.02	
Sheep	ME Mcal/kg	*	0.44	1.95	
Cattle	TDN %	*	12.6	55.9	
Sheep	TDN %	*	12.1	53.9	

ALFALFA–CORN. Medicago sativa, Zea mays

Alfalfa-Corn, aerial part, ensiled, good quality, (3)

Ref No 3 00 320 United States

Feed Name or Analyses			Mean As Fed	Mean Dry	C.V. ± %
Dry matter	%		31.3	100.0	
Carotene	mg/kg		24.8	79.1	
Vitamin A equivalent	IU/g		41.3	131.9	

ALFALFA–GRASS. Medicago sativa, Scientific name not used

Alfalfa-Grass, hay, fan air dried, mx 25% grass, (1)

Ref No 1 00 293 United States

Feed Name or Analyses			Mean As Fed	Mean Dry	C.V. ± %
Dry matter	%		89.8	100.0	2
Ash	%		6.4	7.1	5
Crude fiber	%		27.0	30.1	4
Ether extract	%		1.9	2.1	11
N-free extract	%		42.3	47.1	
Protein (N x 6.25)	%		12.2	13.6	2
Cattle	dig prot %	*	7.8	8.7	
Goats	dig prot %	*	8.3	9.2	
Horses	dig prot %	*	8.2	9.1	
Rabbits	dig prot %	*	8.2	9.2	
Sheep	dig prot %	*	7.9	8.8	
Energy	GE Mcal/kg				
Cattle	DE Mcal/kg	*	2.24	2.49	
Sheep	DE Mcal/kg	*	2.34	2.60	
Cattle	ME Mcal/kg	*	1.84	2.05	
Sheep	ME Mcal/kg	*	1.91	2.13	
Cattle	TDN %	*	50.8	56.6	
Sheep	TDN %	*	53.0	59.0	
Carotene	mg/kg		33.5	37.3	
Vitamin A equivalent	IU/g		55.8	62.1	

Alfalfa-Grass, hay, fan air dried, mx 25% grass gr 1 US, (1)

Ref No 1 00 294 United States

Feed Name or Analyses			Mean As Fed	Mean Dry	C.V. ± %
Dry matter	%			100.0	
Vitamin D_2	IU/g			1.6	

Alfalfa-Grass, hay, s-c, (1)

Ref No 1 08 331 United States

Feed Name or Analyses			Mean As Fed	Mean Dry	C.V. ± %
Dry matter	%		89.6	100.0	
Ash	%		6.2	6.9	
Crude fiber	%		30.4	33.9	
Ether extract	%		1.5	1.7	
N-free extract	%		39.5	44.1	
Protein (N x 6.25)	%		12.0	13.4	
Cattle	dig prot %	*	7.7	8.5	
Goats	dig prot %	*	8.1	9.1	
Horses	dig prot %	*	8.0	8.9	
Rabbits	dig prot %	*	8.1	9.0	
Sheep	dig prot %	*	7.7	8.6	
Energy	GE Mcal/kg				
Cattle	DE Mcal/kg	*	2.20	2.46	
Sheep	DE Mcal/kg	*	2.25	2.51	
Cattle	ME Mcal/kg	*	1.81	2.02	
Sheep	ME Mcal/kg	*	1.84	2.06	
Cattle	TDN %	*	49.9	55.7	
Sheep	TDN %	*	51.0	56.9	
Calcium	%		1.18	1.32	
Manganese	mg/kg		30.9	34.4	
Phosphorus	%		0.24	0.27	

Alfalfa-Grass, hay, s-c, immature, mx 25% grass, (1)

Ref No 1 00 295 United States

Feed Name or Analyses			Mean As Fed	Mean Dry	C.V. ± %
Dry matter	%		90.9	100.0	2
Ash	%		7.5	8.3	13
Crude fiber	%		27.4	30.1	8
Ether extract	%		1.9	2.1	11
N-free extract	%		40.7	44.8	
Protein (N x 6.25)	%		13.4	14.7	16
Cattle	dig prot %	*	8.8	9.7	
Goats	dig prot %	*	9.3	10.3	
Horses	dig prot %	*	9.1	10.0	
Rabbits	dig prot %	*	9.1	10.0	
Sheep	dig prot %	*	8.9	9.8	
Energy	GE Mcal/kg				
Cattle	DE Mcal/kg	*	2.31	2.54	
Sheep	DE Mcal/kg	*	2.33	2.56	
Cattle	ME Mcal/kg	*	1.89	2.08	
Sheep	ME Mcal/kg	*	1.91	2.10	
Cattle	TDN %	*	52.4	57.6	
Sheep	TDN %	*	52.8	58.1	

Alfalfa-Grass, hay, s-c, early bloom, mx 25% grass, (1)

Ref No 1 00 296 United States

Feed Name or Analyses			Mean As Fed	Mean Dry	C.V. ± %
Dry matter	%		88.7	100.0	
Calcium	%		1.19	1.34	
Cobalt	mg/kg		0.170	0.192	
Copper	mg/kg		18.0	20.3	
Iron	%		0.017	0.019	
Magnesium	%		0.29	0.33	
Manganese	mg/kg		75.9	85.5	
Phosphorus	%		0.22	0.25	
Potassium	%		1.74	1.96	

Alfalfa-Grass, hay, s-c, midbloom, mx 10% grass, (1)

Ref No 1 09 194 United States

Feed Name or Analyses			Mean As Fed	Mean Dry	C.V. ± %
Dry matter	%			100.0	
Sheep	dig coef %			66.	
Ash	%			9.2	
Crude fiber	%			29.9	
Ether extract	%			1.6	
N-free extract	%			39.9	
Protein (N x 6.25)	%			19.4	
Cattle	dig prot %	*		13.7	
Goats	dig prot %	*		14.7	
Horses	dig prot %	*		14.0	
Rabbits	dig prot %	*		13.6	
Sheep	dig prot %	*		14.0	
Energy	GE Mcal/kg				
Sheep	GE dig coef %			64.	
Cattle	DE Mcal/kg	*		2.63	
Sheep	DE Mcal/kg			2.85	
Cattle	ME Mcal/kg	*		2.15	
Sheep	ME Mcal/kg	*		2.34	
Cattle	TDN %	*		59.6	
Sheep	TDN %	*		58.6	

Alfalfa-Grass, hay, s-c, mid-bloom, mx 25% grass, (1)

Ref No 1 00 297 United States

Feed Name or Analyses			Mean As Fed	Mean Dry	C.V. ± %
Dry matter	%		87.3	100.0	
Calcium	%		0.92	1.06	
Cobalt	mg/kg		0.133	0.152	
Copper	mg/kg		11.4	13.0	
Iron	%		0.035	0.041	
Magnesium	%		0.26	0.30	
Manganese	mg/kg		34.1	39.0	
Phosphorus	%		0.24	0.27	
Potassium	%		1.51	1.73	

Feed Name or Analyses			Mean As Fed	Mean Dry	C.V. ± %

Alfalfa-Grass, hay, s-c, full bloom, mx 25% grass, (1)
Ref No 1 00 298 United States

Analyses	Unit		As Fed	Dry	C.V. ± %
Dry matter	%		91.1	100.0	1
Ash	%		6.2	6.8	8
Crude fiber	%		30.6	33.6	11
Ether extract	%		2.2	2.4	8
N-free extract	%		39.2	43.0	
Protein (N x 6.25)	%		12.9	14.2	13
Cattle	dig prot %	*	8.4	9.2	
Goats	dig prot %	*	8.9	9.8	
Horses	dig prot %	*	8.7	9.6	
Rabbits	dig prot %	*	8.8	9.6	
Sheep	dig prot %	*	8.5	9.3	
Energy	GE Mcal/kg				
Cattle	DE Mcal/kg	*	2.18	2.40	
Sheep	DE Mcal/kg	*	2.28	2.50	
Cattle	ME Mcal/kg	*	1.79	1.97	
Sheep	ME Mcal/kg	*	1.87	2.05	
Cattle	TDN %	*	49.5	54.4	
Sheep	TDN %	*	51.6	56.7	
Carotene	mg/kg		16.9	18.5	
Vitamin A equivalent	IU/g		28.1	30.9	

Alfalfa-Grass, hay, s-c, cut 1, mx 25% grass, (1)
Ref No 1 00 299 United States

Analyses	Unit		As Fed	Dry	C.V. ± %
Dry matter	%		89.3	100.0	2
Ash	%		5.5	6.2	14
Crude fiber	%		34.5	38.6	11
Ether extract	%		1.9	2.1	17
N-free extract	%		36.9	41.3	
Protein (N x 6.25)	%		10.5	11.8	28
Cattle	dig prot %	*	6.4	7.2	
Goats	dig prot %	*	6.8	7.6	
Horses	dig prot %	*	6.7	7.5	
Rabbits	dig prot %	*	6.9	7.8	
Sheep	dig prot %	*	6.4	7.2	
Energy	GE Mcal/kg				
Cattle	DE Mcal/kg	*	1.84	2.06	
Sheep	DE Mcal/kg	*	1.94	2.17	
Cattle	ME Mcal/kg	*	1.51	1.69	
Sheep	ME Mcal/kg	*	1.59	1.78	
Cattle	TDN %	*	41.7	46.7	
Sheep	TDN %	*	43.9	49.2	
Calcium	%		0.90	1.01	25
Cobalt	mg/kg		0.138	0.154	36
Copper	mg/kg		11.6	13.0	43
Iron	%		0.025	0.028	92
Magnesium	%		0.25	0.28	17
Manganese	mg/kg		39.8	44.5	64
Phosphorus	%		0.22	0.25	4
Potassium	%		1.65	1.85	19
Carotene	mg/kg		17.1	19.2	62
Vitamin A equivalent	IU/g		28.6	32.0	

Alfalfa-Grass, hay, s-c, cut 2, mx 25% grass, (1)
Ref No 1 00 300 United States

Analyses	Unit		As Fed	Dry	C.V. ± %
Dry matter	%		89.1	100.0	
Ash	%		5.5	6.2	
Crude fiber	%		33.1	37.1	
Ether extract	%		1.5	1.7	
N-free extract	%		34.9	39.2	
Protein (N x 6.25)	%		14.1	15.8	
Cattle	dig prot %	*	9.5	10.6	
Goats	dig prot %	*	10.1	11.3	
Horses	dig prot %	*	9.8	10.9	
Rabbits	dig prot %	*	9.7	10.9	
Sheep	dig prot %	*	9.6	10.7	
Energy	GE Mcal/kg				
Cattle	DE Mcal/kg	*	2.00	2.24	
Sheep	DE Mcal/kg	*	2.19	2.46	
Cattle	ME Mcal/kg	*	1.64	1.84	
Sheep	ME Mcal/kg	*	1.79	2.01	
Cattle	TDN %	*	45.3	50.8	
Sheep	TDN %	*	49.6	55.7	
Calcium	%		0.94	1.05	
Cobalt	mg/kg		0.126	0.141	
Copper	mg/kg		11.0	12.3	
Iron	%		0.020	0.022	
Magnesium	%		0.26	0.29	
Manganese	mg/kg		31.4	35.3	
Phosphorus	%		0.26	0.29	
Potassium	%		1.54	1.73	
Carotene	mg/kg		12.2	13.7	
Vitamin A equivalent	IU/g		20.3	22.8	

Alfalfa-Grass, hay, s-c, mn 75% grass, (1)
Ref No 1 00 290 United States

Analyses	Unit		As Fed	Dry	C.V. ± %
Dry matter	%			100.0	
Carotene	mg/kg			31.5	38
Vitamin A equivalent	IU/g			52.6	

Alfalfa-Grass, hay, s-c, mn 75% grass gr 2 US, (1)
Ref No 1 00 291 United States

Analyses	Unit		As Fed	Dry	C.V. ± %
Dry matter	%		88.1	100.0	
Calcium	%		0.87	0.99	
Copper	mg/kg		10.7	12.1	
Iron	%		0.029	0.033	
Magnesium	%		0.26	0.29	
Manganese	mg/kg		32.4	36.8	
Phosphorus	%		0.24	0.27	
Potassium	%		1.62	1.84	
Carotene	mg/kg		25.8	29.3	
Vitamin A equivalent	IU/g		43.1	48.9	

Alfalfa-Grass, hay, s-c, mn 75% grass gr 3 US, (1)
Ref No 1 00 292 United States

Analyses	Unit		As Fed	Dry	C.V. ± %
Dry matter	%			100.0	
Carotene	mg/kg			20.3	
Vitamin A equivalent	IU/g			33.8	

Alfalfa-Grass, hay, s-c, mx 25% grass, (1)
Ref No 1 00 301 United States

Analyses	Unit		As Fed	Dry	C.V. ± %
Dry matter	%		90.6	100.0	2
Cobalt	mg/kg		0.136	0.150	30
Carotene	mg/kg		24.2	26.7	44
Vitamin A equivalent	IU/g		40.3	44.5	

Alfalfa-Grass, hay, s-c, mx 25% grass gr sample US, (1)
Ref No 1 00 302 United States

Analyses	Unit		As Fed	Dry	C.V. ± %
Dry matter	%			100.0	
Vitamin D_2	IU/g			2.9	

Alfalfa-Grass, hay, s-c, mx 25% grass gr 1 US, (1)
Ref No 1 00 303 United States

Analyses	Unit		As Fed	Dry	C.V. ± %
Dry matter	%		90.8	100.0	0
Ash	%		7.4	8.1	13
Crude fiber	%		28.9	31.8	8
Ether extract	%		2.1	2.3	7
N-free extract	%		40.9	45.0	
Protein (N x 6.25)	%		11.6	12.8	4
Cattle	dig prot %	*	7.3	8.0	
Goats	dig prot %	*	7.7	8.5	
Horses	dig prot %	*	7.6	8.4	
Rabbits	dig prot %	*	7.8	8.6	
Sheep	dig prot %	*	7.3	8.1	
Energy	GE Mcal/kg				
Cattle	DE Mcal/kg	*	2.39	2.64	
Sheep	DE Mcal/kg	*	2.32	2.55	
Cattle	ME Mcal/kg	*	1.96	2.16	
Sheep	ME Mcal/kg	*	1.90	2.09	
Cattle	TDN %	*	54.3	59.8	
Sheep	TDN %	*	52.6	57.9	
Calcium	%		0.98	1.08	
Phosphorus	%		0.28	0.31	
Carotene	mg/kg		40.6	44.8	28
Vitamin A equivalent	IU/g		67.7	74.6	

Alfalfa-Grass, hay, s-c, mx 25% grass gr 2 US, (1)
Ref No 1 00 304 United States

Analyses	Unit		As Fed	Dry	C.V. ± %
Dry matter	%		90.5	100.0	3
Calcium	%		0.94	1.04	10
Cobalt	mg/kg		0.130	0.143	34
Copper	mg/kg		10.8	11.9	28
Iron	%		0.026	0.029	89
Magnesium	%		0.25	0.28	19
Manganese	mg/kg		31.3	34.6	27
Phosphorus	%		0.21	0.23	21
Potassium	%		1.57	1.73	20
Carotene	mg/kg		18.8	20.7	42
Vitamin A equivalent	IU/g		31.3	34.5	

Alfalfa-Grass, hay, s-c, mx 25% grass gr 3 US, (1)
Ref No 1 00 305 United States

Analyses	Unit		As Fed	Dry	C.V. ± %
Dry matter	%		90.3	100.0	2
Ash	%		7.3	8.1	16
Crude fiber	%		27.7	30.7	17
Ether extract	%		1.9	2.1	33
N-free extract	%		38.7	42.9	
Protein (N x 6.25)	%		14.6	16.2	6
Cattle	dig prot %	*	9.9	11.0	
Goats	dig prot %	*	10.5	11.7	
Horses	dig prot %	*	10.2	11.3	
Rabbits	dig prot %	*	10.1	11.2	
Sheep	dig prot %	*	10.0	11.1	
Energy	GE Mcal/kg				
Cattle	DE Mcal/kg	*	2.26	2.51	
Sheep	DE Mcal/kg	*	2.31	2.56	
Cattle	ME Mcal/kg	*	1.86	2.06	
Sheep	ME Mcal/kg	*	1.90	2.10	

(1) dry forages and roughages (2) pasture, range plants, and forages fed green (3) silages (4) energy feeds (5) protein supplements (6) minerals (7) vitamins (8) additives

Feed Name or Analyses			Mean As Fed	Mean Dry	C.V. ± %
Cattle	TDN %	*	51.3	56.9	
Sheep	TDN %	*	52.4	58.1	
Calcium	%		1.01	1.12	17
Cobalt	mg/kg		0.147	0.163	
Copper	mg/kg		13.3	14.8	
Iron	%		0.017	0.019	
Magnesium	%		0.25	0.28	
Manganese	mg/kg		49.8	55.1	
Phosphorus	%		0.22	0.24	10
Potassium	%		1.76	1.95	
Carotene	mg/kg		8.0	8.8	80
Vitamin A equivalent	IU/g		13.3	14.7	

Alfalfa-Grass, hay, s-c, 15% grass, (1)

Ref No 1 08 736 United States

Feed Name or Analyses			As Fed	Dry	C.V. ± %
Dry matter	%		90.0	100.0	
Ash	%		6.8	7.6	
Crude fiber	%		31.0	34.4	
Ether extract	%		2.4	2.7	
N-free extract	%		34.9	38.8	
Protein (N x 6.25)	%		14.9	16.6	
Cattle	dig prot %	*	10.1	11.3	
Goats	dig prot %	*	10.8	12.0	
Horses	dig prot %	*	10.4	11.6	
Rabbits	dig prot %	*	10.3	11.5	
Sheep	dig prot %	*	10.3	11.4	
Energy	GE Mcal/kg				
Cattle	DE Mcal/kg	*	2.25	2.50	
Sheep	DE Mcal/kg	*	2.21	2.45	
Cattle	ME Mcal/kg	*	1.84	2.05	
Sheep	ME Mcal/kg	*	1.81	2.01	
Cattle	TDN %	*	51.0	56.6	
Sheep	TDN %	*	50.0	55.6	

Alfalfa-Grass, hay, s-c, 40% grass, (1)

Ref No 1 08 744 United States

Feed Name or Analyses			As Fed	Dry	C.V. ± %
Dry matter	%		90.0	100.0	
Ash	%		6.9	7.7	
Crude fiber	%		30.6	34.0	
Ether extract	%		2.3	2.6	
N-free extract	%		36.0	40.0	
Protein (N x 6.25)	%		14.2	15.8	
Cattle	dig prot %	*	9.5	10.6	
Goats	dig prot %	*	10.2	11.3	
Horses	dig prot %	*	9.8	10.9	
Rabbits	dig prot %	*	9.8	10.9	
Sheep	dig prot %	*	9.7	10.7	
Energy	GE Mcal/kg				
Cattle	DE Mcal/kg	*	2.26	2.51	
Sheep	DE Mcal/kg	*	2.21	2.46	
Cattle	ME Mcal/kg	*	1.85	2.06	
Sheep	ME Mcal/kg	*	1.82	2.02	
Cattle	TDN %	*	51.3	57.0	
Sheep	TDN %	*	50.2	55.8	

Alfalfa-Grass, hay, s-c, rained on, mid-bloom, mx 25% grass, (1)

Ref No 1 00 307 United States

Feed Name or Analyses			As Fed	Dry	C.V. ± %
Dry matter	%		88.8	100.0	
Calcium	%		0.84	0.95	
Copper	mg/kg		8.0	9.0	
Iron	%		0.016	0.018	
Magnesium	%		0.20	0.23	
Manganese	mg/kg		21.9	24.7	
Phosphorus	%		0.24	0.27	
Potassium	%		1.72	1.94	

Alfalfa-Grass, hay, s-c, rained on, full bloom, mx 25% grass, (1)

Ref No 1 00 308 United States

Feed Name or Analyses			As Fed	Dry	C.V. ± %
Dry matter	%		88.7	100.0	
Calcium	%		0.78	0.88	
Copper	mg/kg		10.0	11.2	
Iron	%		0.014	0.016	
Magnesium	%		0.20	0.23	
Manganese	mg/kg		25.0	28.2	
Phosphorus	%		0.22	0.25	
Potassium	%		1.26	1.42	

Alfalfa-Grass, hay, s-c, rained on, mx 25% grass, (1)

Ref No 1 00 309 United States

Feed Name or Analyses			As Fed	Dry	C.V. ± %
Dry matter	%		88.8	100.0	
Cobalt	mg/kg		0.135	0.152	
Carotene	mg/kg		7.8	8.8	
Vitamin A equivalent	IU/g		13.1	14.7	

Alfalfa-Grass, aerial part, fresh, immature, mx 25% grass, (2)

Ref No 2 00 310 United States

Feed Name or Analyses			As Fed	Dry	C.V. ± %
Dry matter	%		40.1	100.0	14
Ash	%		3.3	8.3	14
Crude fiber	%		11.3	28.1	5
Ether extract	%		0.9	2.2	10
N-free extract	%		17.6	43.8	
Protein (N x 6.25)	%		7.1	17.6	9
Cattle	dig prot %	*	5.2	12.9	
Goats	dig prot %	*	5.2	13.0	
Horses	dig prot %	*	5.0	12.5	
Rabbits	dig prot %	*	4.9	12.3	
Sheep	dig prot %	*	5.4	13.4	
Energy	GE Mcal/kg				
Cattle	DE Mcal/kg	*	1.04	2.60	
Sheep	DE Mcal/kg	*	1.04	2.60	
Cattle	ME Mcal/kg	*	0.85	2.13	
Sheep	ME Mcal/kg	*	0.85	2.13	
Cattle	TDN %	*	24.0	59.8	
Sheep	TDN %	*	23.6	58.9	

Alfalfa-Grass, aerial part, ensiled, immature, mx 25% grass, (3)

Ref No 3 00 311 United States

Feed Name or Analyses			As Fed	Dry	C.V. ± %
Dry matter	%		26.5	100.0	
Calcium	%		0.22	0.84	
Copper	mg/kg		4.0	15.0	
Iron	%		0.016	0.059	
Magnesium	%		0.11	0.41	
Manganese	mg/kg		7.9	30.0	
Phosphorus	%		0.07	0.27	
Potassium	%		0.43	1.62	
Carotene	mg/kg		64.3	242.5	
Vitamin A equivalent	IU/g		107.1	404.3	

Alfalfa-Grass, aerial part, ensiled, early bloom, mx 25% grass, (3)

Ref No 3 00 312 United States

Feed Name or Analyses			As Fed	Dry	C.V. ± %
Dry matter	%		24.0	100.0	
Crude fiber	%		7.3	30.5	
Protein (N x 6.25)	%		4.0	16.5	
Cattle	dig prot %	*	2.7	11.2	
Goats	dig prot %	*	2.7	11.2	
Horses	dig prot %	*	2.7	11.2	
Sheep	dig prot %	*	2.7	11.2	

Alfalfa-Grass, aerial part, ensiled, midbloom, mx 10% grass, (3)

Ref No 3 09 193 United States

Feed Name or Analyses			As Fed	Dry	C.V. ± %
Dry matter	%			100.0	
Sheep	dig coef %			65.	
Ash	%			9.7	
Crude fiber	%			26.8	
Ether extract	%			3.0	
N-free extract	%			39.5	
Protein (N x 6.25)	%			21.0	
Cattle	dig prot %	*		15.3	
Goats	dig prot %	*		15.3	
Horses	dig prot %	*		15.3	
Sheep	dig prot %	*		15.3	
Energy	GE Mcal/kg				
Sheep	GE dig coef %			64.	
Cattle	DE Mcal/kg	*		2.44	
Sheep	DE Mcal/kg			2.89	
Cattle	ME Mcal/kg	*		2.00	
Sheep	ME Mcal/kg	*		2.37	
Cattle	TDN %	*		55.3	
Sheep	TDN %	*		54.8	

Alfalfa-Grass, aerial part, ensiled, full bloom, mx 25% grass, (3)

Ref No 3 00 313 United States

Feed Name or Analyses			As Fed	Dry	C.V. ± %
Dry matter	%		25.9	100.0	
Crude fiber	%		10.1	39.0	
Protein (N x 6.25)	%		2.7	10.3	
Cattle	dig prot %	*	1.4	5.6	
Goats	dig prot %	*	1.4	5.6	
Horses	dig prot %	*	1.4	5.6	
Sheep	dig prot %	*	1.4	5.6	

Alfalfa-Grass, aerial part, ensiled, mature, mx 25% grass, (3)

Ref No 3 00 314 United States

Feed Name or Analyses			As Fed	Dry	C.V. ± %
Dry matter	%		22.6	100.0	
Crude fiber	%		10.2	45.0	
Protein (N x 6.25)	%		1.8	7.8	
Cattle	dig prot %	*	0.7	3.3	
Goats	dig prot %	*	0.7	3.3	
Horses	dig prot %	*	0.7	3.3	
Sheep	dig prot %	*	0.7	3.3	

Alfalfa-Grass, aerial part, ensiled, cut 1, mx 25% grass, (3)

Ref No 3 00 315 United States

Feed Name or Analyses		Mean As Fed	Mean Dry	C.V. ± %
Dry matter	%	25.5	100.0	
Ash	%	2.5	9.7	
Crude fiber	%	8.1	31.7	
Ether extract	%	0.9	3.7	
N-free extract	%	10.2	40.0	
Protein (N x 6.25)	%	3.8	14.9	
Cattle	dig prot % *	2.5	9.8	
Goats	dig prot % *	2.5	9.8	
Horses	dig prot % *	2.5	9.8	
Sheep	dig prot % *	2.5	9.8	
Energy	GE Mcal/kg			
Cattle	DE Mcal/kg *	0.65	2.57	
Sheep	DE Mcal/kg *	0.69	2.71	
Cattle	ME Mcal/kg *	0.54	2.11	
Sheep	ME Mcal/kg *	0.57	2.23	
Cattle	TDN % *	14.9	58.3	
Sheep	TDN % *	15.7	61.6	

Alfalfa-Grass, aerial part, ensiled, mx 25% grass, (3)

Ref No 3 00 316 United States

Feed Name or Analyses		Mean As Fed	Mean Dry	C.V. ± %
Dry matter	%	32.3	100.0	22
Ash	%	2.8	8.8	13
Crude fiber	%	10.6	33.0	12
Ether extract	%	1.1	3.3	17
N-free extract	%	12.8	39.6	
Protein (N x 6.25)	%	4.9	15.3	22
Cattle	dig prot % *	3.3	10.1	
Goats	dig prot % *	3.3	10.1	
Horses	dig prot % *	3.3	10.1	
Sheep	dig prot % *	3.3	10.1	
Energy	GE Mcal/kg			
Cattle	DE Mcal/kg *	0.78	2.42	
Sheep	DE Mcal/kg *	0.80	2.47	
Cattle	ME Mcal/kg *	0.64	1.99	
Sheep	ME Mcal/kg *	0.65	2.03	
Cattle	TDN % *	17.7	55.0	
Sheep	TDN % *	18.1	56.0	
Cobalt	mg/kg	0.071	0.221	
Carotene	mg/kg	31.7	98.1	66
Vitamin A equivalent	IU/g	52.8	163.5	

Alfalfa-Grass, aerial part, wilted ensiled, (3)

Ref No 3 08 332 United States

Feed Name or Analyses		Mean As Fed	Mean Dry	C.V. ± %
Dry matter	%	36.2	100.0	
Ash	%	3.2	8.8	
Crude fiber	%	11.4	31.5	
Ether extract	%	1.1	3.0	
N-free extract	%	14.4	39.8	
Protein (N x 6.25)	%	6.1	16.9	
Cattle	dig prot % *	4.2	11.5	
Goats	dig prot % *	4.2	11.5	
Horses	dig prot % *	4.2	11.5	
Sheep	dig prot % *	4.2	11.5	
Energy	GE Mcal/kg			
Cattle	DE Mcal/kg *	0.87	2.40	
Sheep	DE Mcal/kg *	0.89	2.45	
Cattle	ME Mcal/kg *	0.71	1.97	
Sheep	ME Mcal/kg *	0.73	2.01	
Cattle	TDN % *	19.7	54.4	
Sheep	TDN % *	20.1	55.5	

Alfalfa-Grass, aerial part, wilted ensiled, mx 25% grass, (3)

Ref No 3 00 317 United States

Feed Name or Analyses		Mean As Fed	Mean Dry	C.V. ± %
Dry matter	%	36.1	100.0	2
Ash	%	3.1	8.6	5
Crude fiber	%	11.2	31.1	5
Ether extract	%	1.1	3.1	6
N-free extract	%	14.5	40.2	
Protein (N x 6.25)	%	6.1	17.0	9
Cattle	dig prot % *	4.2	11.7	
Goats	dig prot % *	4.2	11.7	
Horses	dig prot % *	4.2	11.7	
Sheep	dig prot % *	4.2	11.7	
Energy	GE Mcal/kg			
Cattle	DE Mcal/kg *	0.87	2.40	
Sheep	DE Mcal/kg *	0.88	2.44	
Cattle	ME Mcal/kg *	0.71	1.97	
Sheep	ME Mcal/kg *	0.72	2.00	
Cattle	TDN % *	19.7	54.4	
Sheep	TDN % *	20.0	55.3	
Calcium	%	0.35	0.97	
Phosphorus	%	0.04	0.11	
Carotene	mg/kg	20.1	55.6	33
Vitamin A equivalent	IU/g	33.5	92.6	

Alfalfa-Grass, aerial part w grnd corn grain added, ensiled, mx 25% grass, (3)

Ref No 3 00 318 United States

Feed Name or Analyses		Mean As Fed	Mean Dry	C.V. ± %
Dry matter	%	23.0	100.0	6
Ash	%	2.5	10.9	8
Crude fiber	%	7.7	33.5	3
Ether extract	%	0.8	3.6	7
N-free extract	%	8.9	38.8	
Protein (N x 6.25)	%	3.0	13.2	5
Cattle	dig prot % *	1.9	8.2	
Goats	dig prot % *	1.9	8.2	
Horses	dig prot % *	1.9	8.2	
Sheep	dig prot % *	1.9	8.2	
Energy	GE Mcal/kg			
Cattle	DE Mcal/kg *	0.61	2.64	
Sheep	DE Mcal/kg *	0.62	2.68	
Cattle	ME Mcal/kg *	0.50	2.16	
Sheep	ME Mcal/kg *	0.51	2.20	
Cattle	TDN % *	13.8	59.8	
Sheep	TDN % *	14.0	60.9	
Calcium	%	0.24	1.06	
Phosphorus	%	0.07	0.31	
Carotene	mg/kg	24.1	104.9	
Vitamin A equivalent	IU/g	40.2	174.9	

Alfalfa-Grass, aerial part w phosphoric acid preservative added, ensiled, mx 25% grass, (3)

Ref No 3 00 319 United States

Feed Name or Analyses		Mean As Fed	Mean Dry	C.V. ± %
Dry matter	%	28.6	100.0	
Ash	%	2.8	9.7	
Crude fiber	%	9.7	34.0	
Ether extract	%	1.1	3.7	
N-free extract	%	11.7	40.8	
Protein (N x 6.25)	%	3.4	11.8	
Cattle	dig prot % *	2.0	6.9	
Goats	dig prot % *	2.0	6.9	
Horses	dig prot % *	2.0	6.9	
Sheep	dig prot % *	2.0	6.9	
Energy	GE Mcal/kg			
Cattle	DE Mcal/kg *	0.75	2.62	
Sheep	DE Mcal/kg *	0.78	2.72	
Cattle	ME Mcal/kg *	0.62	2.15	
Sheep	ME Mcal/kg *	0.64	2.23	
Cattle	TDN % *	17.0	59.5	
Sheep	TDN % *	17.6	61.7	

Suncured alfalfa leaf meal (AAFCO) – see Alfalfa, leaves, s-c, grnd, mn 20% protein mx 18% fiber, (1)

ALFALFA–ORCHARDGRASS. Medicago sativa, Dactylis glomerata

Alfalfa-Orchardgrass, hay, s-c, (1)

Ref No 1 00 322 United States

Feed Name or Analyses		Mean As Fed	Mean Dry	C.V. ± %
Dry matter	%	91.2	100.0	4
Ash	%	7.7	8.4	20
Crude fiber	%	29.5	32.4	5
Ether extract	%	2.0	2.2	18
N-free extract	%	37.3	40.9	7
Protein (N x 6.25)	%	14.8	16.2	10
Cattle	dig prot % *	10.0	11.0	
Goats	dig prot % *	10.6	11.7	
Horses	dig prot % *	10.3	11.3	
Rabbits	dig prot % *	10.2	11.2	
Sheep	dig prot % *	10.1	11.1	
Energy	GE Mcal/kg	4.12	4.52	3
Cattle	DE Mcal/kg *	2.26	2.48	
Sheep	DE Mcal/kg *	2.28	2.50	
Cattle	ME Mcal/kg *	1.85	2.03	
Sheep	ME Mcal/kg *	1.87	2.05	
Cattle	TDN % *	51.3	56.2	
Sheep	TDN % *	51.6	56.6	

Alfalfa-Orchardgrass, hay, s-c, early bloom-immature, cut 2, (1)

Ref No 1 09 199 United States

Feed Name or Analyses		Mean As Fed	Mean Dry	C.V. ± %
Dry matter	%		100.0	
Cattle	dig coef %		62.	
Ash	%		8.4	
Crude fiber	%		28.9	
Cattle	dig coef %		64.	
Ether extract	%		1.6	
Cattle	dig coef %		20.	
N-free extract	%		43.8	
Cattle	dig coef %		63.	
Protein (N x 6.25)	%		17.3	
Cattle	dig coef %		69.	
Cattle	dig prot %		12.0	
Goats	dig prot % *		12.7	
Horses	dig prot % *		12.2	
Rabbits	dig prot % *		12.0	
Sheep	dig prot % *		12.1	
Energy	GE Mcal/kg			
Cattle	DE Mcal/kg *		2.59	
Sheep	DE Mcal/kg *		2.63	
Cattle	ME Mcal/kg *		2.12	
Sheep	ME Mcal/kg *		2.15	

(1) dry forages and roughages (2) pasture, range plants, and forages fed green (3) silages (4) energy feeds (5) protein supplements (6) minerals (7) vitamins (8) additives

Feed Name or Analyses			Mean As Fed	Mean Dry	C.V. ± %
Cattle	TDN %			58.7	
Sheep	TDN %	*		59.6	

Alfalfa-Orchardgrass, hay, s-c, cut 1, (1)

Ref No 1 00 321 United States

Feed Name or Analyses			As Fed	Dry	C.V. ± %
Dry matter	%		92.9	100.0	0
Ash	%		9.4	10.1	12
Crude fiber	%		29.4	31.7	3
Ether extract	%		1.5	1.6	24
N-free extract	%		42.7	46.0	
Protein (N x 6.25)	%		9.8	10.6	9
Cattle	dig prot %	*	5.7	6.1	
Goats	dig prot %	*	6.0	6.4	
Horses	dig prot %	*	6.1	6.5	
Rabbits	dig prot %	*	6.4	6.9	
Sheep	dig prot %	*	5.6	6.1	
Energy	GE Mcal/kg				
Cattle	DE Mcal/kg	*	2.53	2.72	
Sheep	DE Mcal/kg	*	2.33	2.51	
Cattle	ME Mcal/kg	*	2.07	2.23	
Sheep	ME Mcal/kg	*	1.91	2.06	
Cattle	TDN %	*	57.3	61.7	
Sheep	TDN %	*	52.9	56.9	

Alfalfa-Orchardgrass, aerial part, fresh, (2)

Ref No 2 00 323 United States

Feed Name or Analyses			As Fed	Dry	C.V. ± %
Dry matter	%		25.0	100.0	
Calcium	%		0.10	0.40	18
Magnesium	%		0.06	0.24	23
Phosphorus	%		0.13	0.52	12

Alfalfa-Orchardgrass, aerial part, ensiled, prebloom-early bloom, cut 1, (3)

Ref No 3 09 195 United States

Feed Name or Analyses			As Fed	Dry	C.V. ± %
Dry matter	%		23.9	100.0	11
Cattle	dig coef %		62.	62.	7
Ash	%		2.1	8.7	6
Crude fiber	%		7.8	32.8	6
Cattle	dig coef %		71.	71.	5
Ether extract	%		1.1	4.5	27
Cattle	dig coef %		58.	58.	23
N-free extract	%		9.4	39.5	8
Cattle	dig coef %		59.	59.	9
Protein (N x 6.25)	%		3.4	14.4	12
Cattle	dig coef %		63.	63.	11
Cattle	dig prot %		2.2	9.1	
Goats	dig prot %	*	2.2	9.3	
Horses	dig prot %	*	2.2	9.3	
Sheep	dig prot %	*	2.2	9.3	
Energy	GE Mcal/kg				
Cattle	DE Mcal/kg	*	0.65	2.72	
Sheep	DE Mcal/kg	*	0.65	2.71	
Cattle	ME Mcal/kg	*	0.53	2.23	
Sheep	ME Mcal/kg	*	0.53	2.22	
Cattle	TDN %		14.8	61.8	6
Sheep	TDN %	*	14.7	61.5	

Alfalfa-Orchardgrass, aerial part, ensiled, prebloom, cut 1, (3)

Ref No 3 09 192 United States

Feed Name or Analyses			As Fed	Dry	C.V. ± %
Dry matter	%		22.0	100.0	
Cattle	dig coef %		58.	58.	
Ash	%		1.8	8.3	
Crude fiber	%		8.5	38.9	
Cattle	dig coef %		69.	69.	
Ether extract	%		1.1	5.0	
Cattle	dig coef %		65.	65.	
N-free extract	%		7.8	35.5	
Cattle	dig coef %		47.	47.	
Protein (N x 6.25)	%		2.7	12.5	
Cattle	dig coef %		56.	56.	
Cattle	dig prot %		1.5	7.0	
Goats	dig prot %	*	1.7	7.5	
Horses	dig prot %	*	1.7	7.5	
Sheep	dig prot %	*	1.7	7.5	
Energy	GE Mcal/kg				
Cattle	DE Mcal/kg	*	0.56	2.53	
Sheep	DE Mcal/kg	*	0.59	2.67	
Cattle	ME Mcal/kg	*	0.46	2.08	
Sheep	ME Mcal/kg	*	0.48	2.19	
Cattle	TDN %		12.6	57.5	
Sheep	TDN %	*	13.3	60.6	

Alfalfa-Orchardgrass, aerial part, ensiled, early bloom-immature, cut 3, (3)

Ref No 3 09 202 United States

Feed Name or Analyses			As Fed	Dry	C.V. ± %
Dry matter	%		32.0	100.0	
Cattle	dig coef %		62.	62.	
Ash	%		3.3	10.4	
Crude fiber	%		7.4	23.2	
Cattle	dig coef %		61.	61.	
Ether extract	%		0.9	2.9	
Cattle	dig coef %		37.	37.	
N-free extract	%		14.0	43.6	
Cattle	dig coef %		63.	63.	
Protein (N x 6.25)	%		6.4	19.9	
Cattle	dig coef %		74.	74.	
Cattle	dig prot %		4.7	14.7	
Goats	dig prot %	*	4.6	14.3	
Horses	dig prot %	*	4.6	14.3	
Sheep	dig prot %	*	4.6	14.3	
Energy	GE Mcal/kg				
Cattle	DE Mcal/kg	*	0.83	2.59	
Sheep	DE Mcal/kg	*	0.77	2.41	
Cattle	ME Mcal/kg	*	0.68	2.12	
Sheep	ME Mcal/kg	*	0.63	1.97	
Cattle	TDN %		18.8	58.8	
Sheep	TDN %	*	17.5	54.6	

Alfalfa-Orchardgrass, aerial part, ensiled, early bloom-immature, cut 4, (3)

Ref No 3 09 203 United States

Feed Name or Analyses			As Fed	Dry	C.V. ± %
Dry matter	%		29.7	100.0	
Cattle	dig coef %		61.	61.	
Ash	%		2.8	9.4	
Crude fiber	%		7.8	26.3	
Cattle	dig coef %		60.	60.	
Ether extract	%		1.1	3.8	
Cattle	dig coef %		54.	54.	
N-free extract	%		12.1	40.7	
Cattle	dig coef %		60.	60.	
Protein (N x 6.25)	%		5.9	19.8	
Cattle	dig coef %		70.	70.	
Cattle	dig prot %		4.1	13.8	
Goats	dig prot %	*	4.2	14.2	
Horses	dig prot %	*	4.2	14.2	
Sheep	dig prot %	*	4.2	14.2	
Energy	GE Mcal/kg				
Cattle	DE Mcal/kg	*	0.77	2.59	
Sheep	DE Mcal/kg	*	0.72	2.43	
Cattle	ME Mcal/kg	*	0.63	2.12	
Sheep	ME Mcal/kg	*	0.59	1.99	
Cattle	TDN %		17.4	58.7	
Sheep	TDN %	*	16.3	55.0	

Alfalfa-Orchardgrass, aerial part, ensiled, full bloom-late bloom, cut 1, (3)

Ref No 3 09 201 United States

Feed Name or Analyses			As Fed	Dry	C.V. ± %
Dry matter	%		29.0	100.0	
Cattle	dig coef %		52.	52.	
Ash	%		2.4	8.3	
Crude fiber	%		9.8	33.9	
Cattle	dig coef %		63.	63.	
Ether extract	%		0.7	2.5	
Cattle	dig coef %		37.	37.	
N-free extract	%		12.8	44.1	
Cattle	dig coef %		51.	51.	
Protein (N x 6.25)	%		3.2	11.2	
Cattle	dig coef %		52.	52.	
Cattle	dig prot %		1.7	5.8	
Goats	dig prot %	*	1.9	6.4	
Horses	dig prot %	*	1.9	6.4	
Sheep	dig prot %	*	1.9	6.4	
Energy	GE Mcal/kg				
Cattle	DE Mcal/kg	*	0.66	2.28	
Sheep	DE Mcal/kg	*	0.64	2.19	
Cattle	ME Mcal/kg	*	0.54	1.87	
Sheep	ME Mcal/kg	*	0.52	1.80	
Cattle	TDN %		15.0	51.7	
Sheep	TDN %	*	14.4	49.6	

Alfalfa-Orchardgrass, aerial part, wilted ensiled, (3)

Ref No 3 00 325 United States

Feed Name or Analyses			As Fed	Dry	C.V. ± %
Dry matter	%		49.0	100.0	22
Ash	%		4.9	10.1	25
Crude fiber	%		14.9	30.5	7
Ether extract	%		2.1	4.3	13
N-free extract	%		19.1	38.9	
Protein (N x 6.25)	%		7.9	16.2	18
Cattle	dig prot %	*	5.4	10.9	
Goats	dig prot %	*	5.4	10.9	
Horses	dig prot %	*	5.4	10.9	
Sheep	dig prot %	*	5.4	10.9	
Energy	GE Mcal/kg				
Cattle	DE Mcal/kg	*	1.28	2.61	
Sheep	DE Mcal/kg	*	1.23	2.51	
Cattle	ME Mcal/kg	*	1.05	2.14	
Sheep	ME Mcal/kg	*	1.01	2.06	
Cattle	TDN %	*	29.0	59.2	
Sheep	TDN %	*	27.9	56.9	
Calcium	%		0.42	0.86	
Phosphorus	%		0.14	0.29	
Carotene	mg/kg		75.5	154.1	
Vitamin A equivalent	IU/g		125.9	256.9	

Alfalfa-Orchardgrass, aerial part, wilted ensiled, prebloom-early bloom, cut 1, (3)

Ref No 3 09 198 United States

Feed Name or Analyses			As Fed	Dry	C.V. ± %
Dry matter	%		37.8	100.0	
Cattle	dig coef %		62.	62.	

Continued

Feed Name or Analyses			Mean As Fed	Mean Dry	C.V. ± %
Ash	%		3.6	9.6	
Crude fiber	%		11.5	30.5	
Cattle	dig coef %		67.	67.	
Ether extract	%		1.4	3.8	
Cattle	dig coef %		59.	59.	
N-free extract	%		15.0	39.8	
Cattle	dig coef %		60.	60.	
Protein (N x 6.25)	%		6.2	16.4	
Cattle	dig coef %		66.	66.	
Cattle	dig prot %		4.1	10.8	
Goats	dig prot %	*	4.2	11.1	
Horses	dig prot %	*	4.2	11.1	
Sheep	dig prot %	*	4.2	11.1	
Energy	GE Mcal/kg				
Cattle	DE Mcal/kg	*	1.00	2.64	
Sheep	DE Mcal/kg	*	0.94	2.48	
Cattle	ME Mcal/kg	*	0.82	2.17	
Sheep	ME Mcal/kg	*	0.77	2.03	
Cattle	TDN %		22.6	60.0	
Sheep	TDN %	*	21.2	56.2	

Alfalfa-Orchardgrass, aerial part, wilted ensiled, prebloom, cut 1, (3)

Ref No 3 09 191 United States

Feed Name or Analyses			Mean As Fed	Mean Dry	C.V. ± %
Dry matter	%		37.5	100.0	
Cattle	dig coef %		60.	60.	
Ash	%		3.8	10.3	
Crude fiber	%		11.5	30.6	
Cattle	dig coef %		68.	68.	
Ether extract	%		1.2	3.1	
Cattle	dig coef %		60.	60.	
N-free extract	%		14.3	38.3	
Cattle	dig coef %		57.	57.	
Protein (N x 6.25)	%		6.7	17.8	
Cattle	dig coef %		66.	66.	
Cattle	dig prot %		4.4	11.7	
Goats	dig prot %	*	4.6	12.4	
Horses	dig prot %	*	4.6	12.4	
Sheep	dig prot %	*	4.6	12.4	
Energy	GE Mcal/kg				
Cattle	DE Mcal/kg	*	0.96	2.57	
Sheep	DE Mcal/kg	*	0.93	2.47	
Cattle	ME Mcal/kg	*	0.79	2.11	
Sheep	ME Mcal/kg	*	0.76	2.03	
Cattle	TDN %		21.8	58.3	
Sheep	TDN %	*	21.0	56.1	

Alfalfa-Orchardgrass, aerial part, wilted ensiled, full bloom, (3)

Ref No 3 00 324 United States

Feed Name or Analyses			Mean As Fed	Mean Dry	C.V. ± %
Dry matter	%		67.0	100.0	5
Ash	%		6.0	9.0	4
Crude fiber	%		21.0	31.4	8
Ether extract	%		2.7	4.1	9
N-free extract	%		25.7	38.4	
Protein (N x 6.25)	%		11.5	17.1	19
Cattle	dig prot %	*	7.9	11.8	
Goats	dig prot %	*	7.9	11.8	
Horses	dig prot %	*	7.9	11.8	
Sheep	dig prot %	*	7.9	11.8	
Energy	GE Mcal/kg				
Cattle	DE Mcal/kg	*	1.65	2.46	
Sheep	DE Mcal/kg	*	1.66	2.48	
Cattle	ME Mcal/kg	*	1.35	2.02	
Sheep	ME Mcal/kg	*	1.36	2.03	
Cattle	TDN %	*	37.3	55.7	
Sheep	TDN %	*	37.6	56.2	

Alfalfa-Orchardgrass, aerial part w 350 lb grnd corn ears added per ton, ensiled, prebloom, cut 1, (3)

Ref No 3 09 200 United States

Feed Name or Analyses			Mean As Fed	Mean Dry	C.V. ± %
Dry matter	%		34.0	100.0	
Cattle	dig coef %		70.	70.	
Ash	%		1.7	4.9	
Crude fiber	%		8.1	23.9	
Cattle	dig coef %		66.	66.	
Ether extract	%		1.7	5.0	
Cattle	dig coef %		75.	75.	
N-free extract	%		17.7	52.1	
Cattle	dig coef %		73.	73.	
Protein (N x 6.25)	%		4.8	14.1	
Cattle	dig coef %		68.	68.	
Cattle	dig prot %		3.3	9.6	
Goats	dig prot %	*	3.1	9.0	
Horses	dig prot %	*	3.1	9.0	
Sheep	dig prot %	*	3.1	9.0	
Energy	GE Mcal/kg				
Cattle	DE Mcal/kg	*	1.08	3.17	
Sheep	DE Mcal/kg	*	0.98	2.89	
Cattle	ME Mcal/kg	*	0.88	2.60	
Sheep	ME Mcal/kg	*	0.80	2.37	
Cattle	TDN %		24.4	71.8	
Sheep	TDN %	*	22.3	65.4	

Alfalfa-Orchardgrass, aerial part w 8 lb sodium metabisulfite preservative added per ton, ensiled, prebloom-early bloom, cut 1, (3)

Ref No 3 09 196 United States

Feed Name or Analyses			Mean As Fed	Mean Dry	C.V. ± %
Dry matter	%		27.0	100.0	
Cattle	dig coef %		62.	62.	
Ash	%		2.5	9.2	
Crude fiber	%		8.1	30.1	
Cattle	dig coef %		71.	71.	
Ether extract	%		1.1	4.0	
Cattle	dig coef %		52.	52.	
N-free extract	%		11.4	42.2	
Cattle	dig coef %		59.	59.	
Protein (N x 6.25)	%		3.9	14.5	
Cattle	dig coef %		64.	64.	
Cattle	dig prot %		2.5	9.3	
Goats	dig prot %	*	2.5	9.4	
Horses	dig prot %	*	2.5	9.4	
Sheep	dig prot %	*	2.5	9.4	
Energy	GE Mcal/kg				
Cattle	DE Mcal/kg	*	0.72	2.66	
Sheep	DE Mcal/kg	*	0.74	2.74	
Cattle	ME Mcal/kg	*	0.59	2.18	
Sheep	ME Mcal/kg	*	0.61	2.25	
Cattle	TDN %		16.3	60.2	
Sheep	TDN %	*	16.8	62.1	

Alfalfa-Orchardgrass, aerial part w 8 lb sodium metabisulfite preservative added per ton, ensiled, full bloom-late bloom, cut 1, (3)

Ref No 3 09 204 United States

Feed Name or Analyses			Mean As Fed	Mean Dry	C.V. ± %
Dry matter	%		31.7	100.0	
Cattle	dig coef %		56.	56.	
Ash	%		3.0	9.5	
Crude fiber	%		9.1	28.9	
Cattle	dig coef %		62.	62.	
Ether extract	%		0.9	2.7	
Cattle	dig coef %		48.	48.	
N-free extract	%		14.0	44.2	
Cattle	dig coef %		54.	54.	
Protein (N x 6.25)	%		4.7	14.7	
Cattle	dig coef %		64.	64.	
Cattle	dig prot %		3.0	9.5	
Goats	dig prot %	*	3.0	9.6	
Horses	dig prot %	*	3.0	9.6	
Sheep	dig prot %	*	3.0	9.6	
Energy	GE Mcal/kg				
Cattle	DE Mcal/kg	*	0.76	2.39	
Sheep	DE Mcal/kg	*	0.74	2.32	
Cattle	ME Mcal/kg	*	0.62	1.96	
Sheep	ME Mcal/kg	*	0.60	1.90	
Cattle	TDN %		17.1	54.1	
Sheep	TDN %	*	16.7	52.6	

ALFALFA–POTATO. Medicago sativa, Solanum tuberosum

Alfalfa-Potato, aerial part and tubers, ensiled, (3)

Ref No 3 03 791 United States

Feed Name or Analyses			Mean As Fed	Mean Dry	C.V. ± %
Dry matter	%		35.9	100.0	
Ash	%		2.2	6.1	
Crude fiber	%		8.6	24.0	
Ether extract	%		0.6	1.7	
N-free extract	%		19.2	53.4	
Protein (N x 6.25)	%		5.3	14.8	
Cattle	dig prot %	*	3.5	9.7	
Goats	dig prot %	*	3.5	9.7	
Horses	dig prot %	*	3.5	9.7	
Sheep	dig prot %	*	3.5	9.7	
Energy	GE Mcal/kg				
Cattle	DE Mcal/kg	*	0.93	2.59	
Sheep	DE Mcal/kg	*	0.87	2.43	
Cattle	ME Mcal/kg	*	0.76	2.12	
Sheep	ME Mcal/kg	*	0.60	1.99	
Cattle	TDN %	*	21.1	58.7	
Sheep	TDN %	*	19.8	55.1	

ALFALFA–RUSSIANTHISTLE. Medicago sativa, Salsola spp

Alfalfa-Russianthistle, aerial part, ensiled, (3)

Ref No 3 00 326 United States

Feed Name or Analyses			Mean As Fed	Mean Dry	C.V. ± %
Dry matter	%		32.3	100.0	
Ash	%		6.9	21.5	
Crude fiber	%		8.1	25.1	
Ether extract	%		0.7	2.3	
N-free extract	%		12.0	37.0	
Protein (N x 6.25)	%		4.6	14.1	
Cattle	dig prot %	*	2.9	9.0	
Goats	dig prot %	*	2.9	9.0	
Horses	dig prot %	*	2.9	9.0	

(1) dry forages and roughages (3) silages (6) minerals
(2) pasture, range plants, and forages fed green (4) energy feeds (7) vitamins
(5) protein supplements (8) additives

Feed Name or Analyses			Mean As Fed	Mean Dry	C.V. ± %
Sheep	dig prot %	*	2.9	9.0	
Energy	GE Mcal/kg				
Cattle	DE Mcal/kg	*	0.79	2.44	
Sheep	DE Mcal/kg	*	0.81	2.52	
Cattle	ME Mcal/kg	*	0.65	2.00	
Sheep	ME Mcal/kg	*	0.67	2.06	
Cattle	TDN %	*	17.9	55.3	
Sheep	TDN %	*	18.4	57.1	
Calcium	%		0.71	2.21	
Magnesium	%		0.36	1.10	
Phosphorus	%		0.05	0.17	

ALFALFA–RYEGRASS, PERENNIAL. Medicago sativa, Lolium perenne

Alfalfa-Ryegrass, perennial, aerial part, fresh, (2)

Ref No 2 00 327 — United States

Feed Name or Analyses			Mean As Fed	Mean Dry	C.V. ± %
Dry matter	%			100.0	
Calcium	%			0.80	
Magnesium	%			0.29	
Potassium	%			3.06	
Sodium	%			0.05	
Sulphur	%			0.03	

ALFALFA–TIMOTHY. Medicago sativa, Phleum pratense

Alfalfa-Timothy, hay, bleached s-c, (1)

Ref No 1 00 330 — United States

Feed Name or Analyses			Mean As Fed	Mean Dry	C.V. ± %
Dry matter	%		92.9	100.0	
Ash	%		5.0	5.4	
Crude fiber	%		30.6	32.9	
Ether extract	%		1.3	1.4	
N-free extract	%		46.3	49.8	
Protein (N x 6.25)	%		9.8	10.5	
Cattle	dig prot %	*	5.6	6.0	
Goats	dig prot %	*	5.9	6.4	
Horses	dig prot %	*	6.0	6.4	
Rabbits	dig prot %	*	6.3	6.8	
Sheep	dig prot %	*	5.6	6.0	
Energy	GE Mcal/kg				
Cattle	DE Mcal/kg	*	2.26	2.43	
Sheep	DE Mcal/kg	*	2.34	2.52	
Cattle	ME Mcal/kg	*	1.85	1.99	
Sheep	ME Mcal/kg	*	1.92	2.06	
Cattle	TDN %	*	51.2	55.1	
Sheep	TDN %	*	53.0	57.1	

Alfalfa-Timothy, hay, fan air dried, (1)

Ref No 1 00 329 — United States

Feed Name or Analyses			Mean As Fed	Mean Dry	C.V. ± %
Dry matter	%		86.1	100.0	
Ash	%		6.2	7.2	
Crude fiber	%		33.1	38.4	
Ether extract	%		1.9	2.2	
N-free extract	%		32.4	37.6	
Protein (N x 6.25)	%		12.6	14.6	
Cattle	dig prot %	*	8.3	9.6	
Goats	dig prot %	*	8.8	10.2	
Horses	dig prot %	*	8.5	9.9	
Rabbits	dig prot %	*	8.6	9.9	
Sheep	dig prot %	*	8.3	9.7	
Energy	GE Mcal/kg				
Cattle	DE Mcal/kg	*	1.96	2.28	
Sheep	DE Mcal/kg	*	2.03	2.35	
Cattle	ME Mcal/kg	*	1.61	1.87	
Sheep	ME Mcal/kg	*	1.66	1.93	
Cattle	TDN %	*	44.4	51.6	
Sheep	TDN %	*	45.9	53.4	
Calcium	%		1.20	1.39	
Phosphorus	%		0.30	0.35	

Alfalfa-Timothy, hay, s-c, (1)

Ref No 1 00 336 — United States

Feed Name or Analyses			Mean As Fed	Mean Dry	C.V. ± %
Dry matter	%		89.0	100.0	
Ash	%		6.4	7.1	
Crude fiber	%		30.9	34.7	
Ether extract	%		2.1	2.3	
N-free extract	%		37.4	42.1	
Protein (N x 6.25)	%		12.2	13.7	
Cattle	dig prot %	*	7.9	8.8	
Goats	dig prot %	*	8.3	9.4	
Horses	dig prot %	*	8.2	9.2	
Rabbits	dig prot %	*	8.2	9.3	
Sheep	dig prot %	*	7.9	8.9	
Energy	GE Mcal/kg				
Cattle	DE Mcal/kg	*	2.12	2.38	
Sheep	DE Mcal/kg	*	2.18	2.45	
Cattle	ME Mcal/kg	*	1.74	1.95	
Sheep	ME Mcal/kg	*	1.79	2.01	
Cattle	TDN %	*	48.1	54.0	
Sheep	TDN %	*	49.5	55.6	
Calcium	%		0.79	0.89	17
Cobalt	mg/kg		0.073	0.082	30
Manganese	mg/kg		32.5	36.5	19
Phosphorus	%		0.18	0.20	35
Carotene	mg/kg		12.9	14.4	65
Vitamin A equivalent	IU/g		21.4	24.1	

Alfalfa-Timothy, hay, s-c, early bloom, (1)

Ref No 1 00 331 — United States

Feed Name or Analyses			Mean As Fed	Mean Dry	C.V. ± %
Dry matter	%			100.0	
Calcium	%			1.04	
Cobalt	mg/kg			0.071	
Manganese	mg/kg			37.9	
Phosphorus	%			0.21	

Alfalfa-Timothy, hay, s-c, mid-bloom, (1)

Ref No 1 00 332 — United States

Feed Name or Analyses			Mean As Fed	Mean Dry	C.V. ± %
Dry matter	%			100.0	
Ash	%			7.7	
Crude fiber	%			33.1	
Ether extract	%			2.4	
N-free extract	%			40.1	
Protein (N x 6.25)	%			16.7	
Cattle	dig prot %	*		11.4	
Goats	dig prot %	*		12.1	
Horses	dig prot %	*		11.7	
Rabbits	dig prot %	*		11.6	
Sheep	dig prot %	*		11.5	
Energy	GE Mcal/kg				
Cattle	DE Mcal/kg	*		2.44	
Sheep	DE Mcal/kg	*		2.50	
Cattle	ME Mcal/kg	*		2.00	
Sheep	ME Mcal/kg	*		2.05	
Cattle	TDN %	*		55.3	
Sheep	TDN %	*		56.6	

Alfalfa-Timothy, hay, s-c, cut 1, (1)

Ref No 1 00 333 — United States

Feed Name or Analyses			Mean As Fed	Mean Dry	C.V. ± %
Dry matter	%		90.2	100.0	
Ash	%		5.7	6.3	
Crude fiber	%		32.1	35.6	
Ether extract	%		1.4	1.6	
N-free extract	%		40.9	45.3	
Protein (N x 6.25)	%		10.1	11.2	
Cattle	dig prot %	*	6.0	6.6	
Goats	dig prot %	*	6.3	7.0	
Horses	dig prot %	*	6.3	7.0	
Rabbits	dig prot %	*	6.6	7.3	
Sheep	dig prot %	*	6.0	6.6	
Energy	GE Mcal/kg				
Cattle	DE Mcal/kg	*	2.05	2.27	
Sheep	DE Mcal/kg	*	2.22	2.46	
Cattle	ME Mcal/kg	*	1.68	1.86	
Sheep	ME Mcal/kg	*	1.82	2.02	
Cattle	TDN %	*	46.4	51.4	
Sheep	TDN %	*	50.4	55.9	
Calcium	%		0.66	0.74	26
Cobalt	mg/kg		0.076	0.084	41
Manganese	mg/kg		33.2	36.8	12
Phosphorus	%		0.13	0.14	9
Vitamin D_2	IU/g		1.3	1.5	27

Alfalfa-Timothy, hay, s-c, cut 2, (1)

Ref No 1 00 334 — United States

Feed Name or Analyses			Mean As Fed	Mean Dry	C.V. ± %
Dry matter	%		91.0	100.0	
Ash	%		7.2	7.9	
Crude fiber	%		27.6	30.3	
Ether extract	%		1.8	2.0	
N-free extract	%		38.5	42.3	
Protein (N x 6.25)	%		15.9	17.5	
Cattle	dig prot %	*	11.0	12.1	
Goats	dig prot %	*	11.7	12.9	
Horses	dig prot %	*	11.3	12.4	
Rabbits	dig prot %	*	11.1	12.2	
Sheep	dig prot %	*	11.2	12.3	
Energy	GE Mcal/kg				
Cattle	DE Mcal/kg	*	2.27	2.50	
Sheep	DE Mcal/kg	*	2.36	2.59	
Cattle	ME Mcal/kg	*	1.86	2.05	
Sheep	ME Mcal/kg	*	1.93	2.13	
Cattle	TDN %	*	51.6	56.7	
Sheep	TDN %	*	53.5	58.8	
Calcium	%		0.92	1.01	12
Cobalt	mg/kg		0.072	0.079	37
Manganese	mg/kg		32.5	35.7	23
Phosphorus	%		0.19	0.21	38

Alfalfa-Timothy, hay, s-c, cut 3, (1)

Ref No 1 00 335 — United States

Feed Name or Analyses			Mean As Fed	Mean Dry	C.V. ± %
Dry matter	%		92.3	100.0	
Calcium	%		0.72	0.78	
Phosphorus	%		0.11	0.12	

Alfalfa-Timothy, hay, s-c, gr 1 US, (1)

Ref No 1 00 338 — United States

Feed Name or Analyses			Mean As Fed	Mean Dry	C.V. ± %
Dry matter	%			100.0	
Ash	%			7.7	

Continued

Feed Name or Analyses		Mean As Fed	Mean Dry	C.V. ± %
Crude fiber	%		33.1	
Ether extract	%		2.4	
N-free extract	%		40.1	
Protein (N x 6.25)	%		16.7	
Cattle	dig prot % *		11.4	
Goats	dig prot % *		12.1	
Horses	dig prot % *		11.7	
Rabbits	dig prot % *		11.6	
Sheep	dig prot % *		11.5	
Energy	GE Mcal/kg			
Cattle	DE Mcal/kg *		2.44	
Sheep	DE Mcal/kg *		2.50	
Cattle	ME Mcal/kg *		2.00	
Sheep	ME Mcal/kg *		2.05	
Cattle	TDN % *		55.3	
Sheep	TDN % *		56.6	
Calcium	%		0.84	19
Cobalt	mg/kg		0.084	31
Manganese	mg/kg		34.8	21
Phosphorus	%		0.17	22
Carotene	mg/kg		13.2	64
Vitamin A equivalent	IU/g		22.1	
Vitamin D2	IU/g		1.7	

Alfalfa-Timothy, hay, s-c, gr 2 US, (1)
Ref No 1 00 339 United States

Feed Name or Analyses		Mean As Fed	Mean Dry	C.V. ± %
Dry matter	%		100.0	
Ash	%		7.8	
Crude fiber	%		32.0	
Ether extract	%		2.4	
N-free extract	%		45.8	
Protein (N x 6.25)	%		12.0	
Cattle	dig prot % *		7.3	
Goats	dig prot % *		7.8	
Horses	dig prot % *		7.7	
Rabbits	dig prot % *		7.9	
Sheep	dig prot % *		7.3	
Energy	GE Mcal/kg			
Cattle	DE Mcal/kg *		2.62	
Sheep	DE Mcal/kg *		2.54	
Cattle	ME Mcal/kg *		2.15	
Sheep	ME Mcal/kg *		2.08	
Cattle	TDN % *		59.4	
Sheep	TDN % *		57.6	
Calcium	%		0.96	36
Cobalt	mg/kg		0.087	64
Manganese	mg/kg		42.1	31
Phosphorus	%		0.25	31
Carotene	mg/kg		14.3	81
Vitamin A equivalent	IU/g		23.9	

Alfalfa-Timothy, hay, s-c, gr 3 US, (1)
Ref No 1 00 340 United States

Feed Name or Analyses		Mean As Fed	Mean Dry	C.V. ± %
Dry matter	%		100.0	
Ash	%		5.6	
Crude fiber	%		42.3	
Ether extract	%		2.4	
N-free extract	%		38.5	
Protein (N x 6.25)	%		11.2	
Cattle	dig prot % *		6.6	
Goats	dig prot % *		7.0	
Horses	dig prot % *		7.0	
Rabbits	dig prot % *		7.3	
Sheep	dig prot % *		6.6	
Energy	GE Mcal/kg			
Cattle	DE Mcal/kg *		1.78	
Sheep	DE Mcal/kg *		2.32	
Cattle	ME Mcal/kg *		1.46	
Sheep	ME Mcal/kg *		1.90	
Cattle	TDN % *		40.4	
Sheep	TDN % *		52.6	
Vitamin D2	IU/g		1.0	

Alfalfa-Timothy, hay, s-c, 20% timothy, (1)
Ref No 1 08 737 United States

Feed Name or Analyses		Mean As Fed	Mean Dry	C.V. ± %
Dry matter	%	90.0	100.0	
Ash	%	6.6	7.3	
Crude fiber	%	32.6	36.2	
Ether extract	%	2.2	2.4	
N-free extract	%	34.9	38.8	
Protein (N x 6.25)	%	13.7	15.2	
Cattle	dig prot % *	9.1	10.1	
Goats	dig prot % *	9.7	10.8	
Horses	dig prot % *	9.4	10.5	
Rabbits	dig prot % *	9.4	10.4	
Sheep	dig prot % *	9.2	10.2	
Energy	GE Mcal/kg			
Cattle	DE Mcal/kg *	2.19	2.44	
Sheep	DE Mcal/kg *	2.17	2.41	
Cattle	ME Mcal/kg *	1.80	2.00	
Sheep	ME Mcal/kg *	1.78	1.97	
Cattle	TDN % *	49.7	55.3	
Sheep	TDN % *	49.1	54.6	

Alfalfa-Timothy, hay, s-c, 45% timothy, (1)
Ref No 1 08 742 United States

Feed Name or Analyses		Mean As Fed	Mean Dry	C.V. ± %
Dry matter	%	90.0	100.0	
Ash	%	5.9	6.6	
Crude fiber	%	33.0	36.7	
Ether extract	%	2.1	2.3	
N-free extract	%	37.8	42.0	
Protein (N x 6.25)	%	11.2	12.4	
Cattle	dig prot % *	6.9	7.7	
Goats	dig prot % *	7.4	8.2	
Horses	dig prot % *	7.3	8.1	
Rabbits	dig prot % *	7.4	8.3	
Sheep	dig prot % *	7.0	7.7	
Energy	GE Mcal/kg			
Cattle	DE Mcal/kg *	2.00	2.22	
Sheep	DE Mcal/kg *	2.21	2.46	
Cattle	ME Mcal/kg *	1.64	1.82	
Sheep	ME Mcal/kg *	1.81	2.01	
Cattle	TDN % *	45.3	50.3	
Sheep	TDN % *	50.1	55.7	

Alfalfa-Timothy, hay, s-c, gr sample US, (1)
Ref No 1 00 337 United States

Feed Name or Analyses		Mean As Fed	Mean Dry	C.V. ± %
Dry matter	%		100.0	
Vitamin D2	IU/g		1.6	

Alfalfa-Timothy, aerial part, fresh, immature, (2)
Ref No 2 00 342 United States

Feed Name or Analyses		Mean As Fed	Mean Dry	C.V. ± %
Dry matter	%	21.8	100.0	
Ash	%	2.3	10.6	
Crude fiber	%	4.6	21.1	
Ether extract	%	0.8	3.7	
N-free extract	%	9.6	44.2	
Protein (N x 6.25)	%	4.4	20.4	
Cattle	dig prot % *	3.3	15.2	
Goats	dig prot % *	3.4	15.6	
Horses	dig prot % *	3.2	14.8	
Rabbits	dig prot % *	3.1	14.4	
Sheep	dig prot % *	3.5	16.0	
Energy	GE Mcal/kg			
Cattle	DE Mcal/kg *	0.67	3.08	
Sheep	DE Mcal/kg *	0.65	2.97	
Cattle	ME Mcal/kg *	0.55	2.53	
Sheep	ME Mcal/kg *	0.53	2.43	
Cattle	TDN % *	15.2	69.9	
Sheep	TDN % *	14.7	67.3	

Alfalfa-Timothy, aerial part, fresh, 50% alfalfa, (2)
Ref No 2 08 574 United States

Feed Name or Analyses		Mean As Fed	Mean Dry	C.V. ± %
Dry matter	%	21.9	100.0	
Ash	%	2.2	10.0	
Crude fiber	%	4.7	21.5	
Ether extract	%	0.8	3.7	
N-free extract	%	9.6	43.8	
Protein (N x 6.25)	%	4.6	21.0	
Cattle	dig prot % *	3.4	15.7	
Goats	dig prot % *	3.5	16.2	
Horses	dig prot % *	3.4	15.4	
Rabbits	dig prot % *	3.3	14.9	
Sheep	dig prot % *	3.6	16.6	
Energy	GE Mcal/kg			
Cattle	DE Mcal/kg *	0.69	3.16	
Sheep	DE Mcal/kg *	0.64	2.92	
Cattle	ME Mcal/kg *	0.57	2.59	
Sheep	ME Mcal/kg *	0.52	2.40	
Cattle	TDN % *	15.7	71.6	
Sheep	TDN % *	14.5	66.3	
Calcium	%	0.30	1.37	
Phosphorus	%	0.08	0.37	
Potassium	%	0.49	2.24	

Alfalfa-Timothy, aerial part, ensiled, (3)
Ref No 3 00 345 United States

Feed Name or Analyses		Mean As Fed	Mean Dry	C.V. ± %
Dry matter	%	27.6	100.0	7
Ash	%	2.2	8.1	16
Crude fiber	%	9.7	35.2	6
Ether extract	%	1.0	3.5	27
N-free extract	%	10.9	39.5	
Protein (N x 6.25)	%	3.8	13.7	15
Cattle	dig prot % *	2.4	8.7	
Goats	dig prot % *	2.4	8.7	
Horses	dig prot % *	2.4	8.7	
Sheep	dig prot % *	2.4	8.7	
Energy	GE Mcal/kg			
Cattle	DE Mcal/kg *	0.66	2.38	
Sheep	DE Mcal/kg *	0.69	2.49	
Cattle	ME Mcal/kg *	0.54	1.95	
Sheep	ME Mcal/kg *	0.56	2.04	

(1) dry forages and roughages (2) pasture, range plants, and forages fed green (3) silages (4) energy feeds (5) protein supplements (6) minerals (7) vitamins (8) additives

Feed Name or Analyses			Mean As Fed	Mean Dry	C.V. ± %
Cattle	TDN %	*	14.9	54.0	
Sheep	TDN %	*	15.6	56.4	

Alfalfa-Timothy, aerial part, ensiled, mid-bloom, (3)
Ref No 3 00 343 United States

Feed Name or Analyses			Mean As Fed	Mean Dry	C.V. ± %
Dry matter	%		24.9	100.0	
Ash	%		1.8	7.1	
Crude fiber	%		9.0	36.1	
Ether extract	%		0.5	2.1	
N-free extract	%		10.4	41.9	
Protein (N x 6.25)	%		3.2	12.8	
Cattle	dig prot %	*	2.0	7.9	
Goats	dig prot %	*	2.0	7.9	
Horses	dig prot %	*	2.0	7.9	
Sheep	dig prot %	*	2.0	7.9	
Energy	GE Mcal/kg				
Cattle	DE Mcal/kg	*	0.56	2.26	
Sheep	DE Mcal/kg	*	0.60	2.41	
Cattle	ME Mcal/kg	*	0.46	1.85	
Sheep	ME Mcal/kg	*	0.49	1.98	
Cattle	TDN %	*	12.7	51.2	
Sheep	TDN %	*	13.6	54.7	

Alfalfa-Timothy, aerial part, ensiled, cut 1, (3)
Ref No 3 00 344 United States

Feed Name or Analyses			Mean As Fed	Mean Dry	C.V. ± %
Dry matter	%		26.8	100.0	6
Ash	%		2.1	7.7	7
Crude fiber	%		9.3	34.7	5
Ether extract	%		1.0	3.9	30
N-free extract	%		10.6	39.5	
Protein (N x 6.25)	%		3.8	14.2	6
Cattle	dig prot %	*	2.4	9.1	
Goats	dig prot %	*	2.4	9.1	
Horses	dig prot %	*	2.4	9.1	
Sheep	dig prot %	*	2.4	9.1	
Energy	GE Mcal/kg				
Cattle	DE Mcal/kg	*	0.64	2.38	
Sheep	DE Mcal/kg	*	0.68	2.52	
Cattle	ME Mcal/kg	*	0.52	1.95	
Sheep	ME Mcal/kg	*	0.55	2.07	
Cattle	TDN %	*	14.5	53.9	
Sheep	TDN %	*	15.3	57.1	

Alfalfa-Timothy, aerial part w phosphoric acid preservative added, ensiled, (3)
Ref No 3 00 346 United States

Feed Name or Analyses			Mean As Fed	Mean Dry	C.V. ± %
Dry matter	%		27.2	100.0	
Ash	%		2.9	10.7	
Crude fiber	%		8.5	31.2	
Ether extract	%		0.9	3.3	
N-free extract	%		10.8	39.7	
Protein (N x 6.25)	%		4.1	15.1	
Cattle	dig prot %	*	2.7	9.9	
Goats	dig prot %	*	2.7	9.9	
Horses	dig prot %	*	2.7	9.9	
Sheep	dig prot %	*	2.7	9.9	
Energy	GE Mcal/kg				
Cattle	DE Mcal/kg	*	0.71	2.61	
Sheep	DE Mcal/kg	*	0.68	2.51	
Cattle	ME Mcal/kg	*	0.58	2.14	
Sheep	ME Mcal/kg	*	0.56	2.06	
Cattle	TDN %	*	16.1	59.3	
Sheep	TDN %	*	15.5	56.8	

Alfalfa-Timothy, aerial part w molasses added, ensiled, (3)
Ref No 3 00 347 United States

Feed Name or Analyses			Mean As Fed	Mean Dry	C.V. ± %
Dry matter	%		27.8	100.0	6
Ash	%		2.3	8.4	11
Crude fiber	%		9.5	34.2	6
Ether extract	%		1.1	4.0	25
N-free extract	%		10.8	39.0	
Protein (N x 6.25)	%		4.0	14.4	15
Cattle	dig prot %	*	2.6	9.3	
Goats	dig prot %	*	2.6	9.3	
Horses	dig prot %	*	2.6	9.3	
Sheep	dig prot %	*	2.6	9.3	
Energy	GE Mcal/kg				
Cattle	DE Mcal/kg	*	0.68	2.44	
Sheep	DE Mcal/kg	*	0.76	2.72	
Cattle	ME Mcal/kg	*	0.56	2.00	
Sheep	ME Mcal/kg	*	0.62	2.23	
Cattle	TDN %	*	15.4	55.3	
Sheep	TDN %	*	17.1	61.7	

ALFALFA, VERNAL-BROME, SMOOTH CANADA COMMON. Medicago sativa, Bromus inermus

Alfalfa, vernal-brome, smooth Canada common, hay, crimped baled fan-air dried w heat, immature, cut 1, (1)
Ref No 1 08 883 United States

Feed Name or Analyses			Mean As Fed	Mean Dry	C.V. ± %
Dry matter	%		90.8	100.0	
Goats	dig coef %		72.	72.	
Protein (N x 6.25)	%		20.2	22.3	
Goats	dig coef %		80.	80.	
Cattle	dig prot %	*	14.7	16.2	
Goats	dig prot %		16.1	17.7	
Horses	dig prot %	*	14.9	16.4	
Rabbits	dig prot %	*	14.4	15.8	
Sheep	dig prot %	*	15.0	16.5	
Cellulose (Crampton)	%				
Goats	dig coef %		73.	73.	
Fiber, acid detergent (VS)	%		26.1	28.8	
Lignin (Van Soest)	%		3.9	4.3	
Energy	GE Mcal/kg		4.00	4.41	
Goats	GE dig coef %		72.	72.	
Goats	DE Mcal/kg		2.86	3.15	
Dry matter intake	$g/w^{0.75}$		62.754	69.150	

Alfalfa, vernal-brome, smooth Canada common, hay, crimped baled fan-air dried w heat, prebloom, cut 1, (1)
Ref No 1 08 884 United States

Feed Name or Analyses			Mean As Fed	Mean Dry	C.V. ± %
Dry matter	%		89.9	100.0	1
Goats	dig coef %		68.	68.	2
Protein (N x 6.25)	%		16.9	18.9	7
Goats	dig coef %		78.	78.	3
Cattle	dig prot %	*	11.9	13.3	
Goats	dig prot %		13.2	14.6	
Horses	dig prot %	*	12.2	13.5	
Rabbits	dig prot %	*	11.9	13.2	
Sheep	dig prot %	*	12.1	13.5	
Cellulose (Crampton)	%				
Goats	dig coef %		70.	70.	5
Fiber, acid detergent (VS)	%		32.4	36.1	3
Lignin (Van Soest)	%		4.9	5.5	12
Energy	GE Mcal/kg		4.00	4.45	0
Goats	GE dig coef %		67.	67.	3
Goats	DE Mcal/kg		2.67	2.97	3
Dry matter intake	$g/w^{0.75}$		56.684	63.080	17

Alfalfa, vernal-brome, smooth Canada common, hay crimped baled fan-air dried w heat, full bloom, cut 1, (1)
Ref No 1 08 886 United States

Feed Name or Analyses			Mean As Fed	Mean Dry	C.V. ± %
Dry matter	%		91.1	100.0	
Goats	dig coef %		59.	59.	
Protein (N x 6.25)	%		13.2	14.5	
Goats	dig coef %		72.	72.	
Cattle	dig prot %	*	8.6	9.5	
Goats	dig prot %		9.5	10.4	
Horses	dig prot %	*	8.9	9.8	
Rabbits	dig prot %	*	8.9	9.8	
Sheep	dig prot %	*	8.7	9.5	
Cellulose (Crampton)	%				
Goats	dig coef %		61.	61.	
Fiber, acid detergent (VS)	%		41.0	45.1	
Lignin (Van Soest)	%		7.1	7.8	
Energy	GE Mcal/kg		4.07	4.47	98
Goats	GE dig coef %		58.	58.	
Goats	DE Mcal/kg		2.34	2.57	
Dry matter intake	$g/w^{0.75}$		58.265	63.975	

Alfalfa, vernal-brome, smooth Canada common, hay, crimped baled fan-air dried w heat, dough stage, cut 1, (1)
Ref No 1 08 887 United States

Feed Name or Analyses			Mean As Fed	Mean Dry	C.V. ± %
Dry matter	%		91.5	100.0	
Goats	dig coef %		56.	56.	
Protein (N x 6.25)	%		12.4	13.6	
Goats	dig coef %		71.	71.	
Cattle	dig prot %	*	8.0	8.7	
Goats	dig prot %		8.8	9.6	
Horses	dig prot %	*	8.3	9.1	
Rabbits	dig prot %	*	8.4	9.2	
Sheep	dig prot %	*	8.0	8.8	
Cellulose (Crampton)	%				
Goats	dig coef %		57.	57.	
Fiber, acid detergent (VS)	%		41.6	45.5	
Lignin (Van Soest)	%		7.6	8.3	
Energy	GE Mcal/kg		4.11	4.49	
Goats	GE dig coef %		55.	55.	
Goats	DE Mcal/kg		2.25	2.46	
Dry matter intake	$g/w^{0.75}$		58.234	63.667	

Alfalfa, vernal-brome, smooth Canada common, hay, crimped baled fan-air dried w heat, midbloom, cut 1, (1)
Ref No 1 08 885 United States

Feed Name or Analyses			Mean As Fed	Mean Dry	C.V. ± %
Dry matter	%		91.9	100.0	
Goats	dig coef %		64.	64.	
Protein (N x 6.25)	%		16.2	17.7	
Goats	dig coef %		76.	76.	
Cattle	dig prot %	*	11.2	12.2	
Goats	dig prot %		12.3	13.4	
Horses	dig prot %	*	11.5	12.5	
Rabbits	dig prot %	*	11.3	12.3	
Sheep	dig prot %	*	11.4	12.4	

Continued

Feed Name or Analyses			Mean As Fed	Mean Dry	C.V. ± %
Cellulose (Crampton)	%				
Goats	dig coef %		67.	67.	
Fiber, acid detergent (VS)	%		38.4	41.9	
Lignin (Van Soest)	%		6.5	7.0	
Energy	GE Mcal/kg		4.12	4.49	
Goats	GE dig coef %		62.	62.	
Goats	DE Mcal/kg		2.53	2.76	
Dry matter intake	$g/W^{0.75}$		63.147	68.750	

ALFALFA–WHEAT. Medicago sativa, Triticum aestivum

Alfalfa-Wheat, aerial part, ensiled, (3)
Ref No 3 00 348 United States

Feed Name or Analyses			Mean As Fed	Mean Dry	C.V. ± %
Dry matter	%			100.0	
Calcium	%			1.32	
Phosphorus	%			0.20	

ALFALFA–WITCHGRASS. Medicago sativa, Panicum spp

Alfalfa-Witchgrass, hay, fan air dried, cut 1, (1)
Ref No 1 00 349 United States

Feed Name or Analyses			Mean As Fed	Mean Dry	C.V. ± %
Dry matter	%		85.5	100.0	
Carotene	mg/kg		4.3	5.1	
Vitamin A equivalent	IU/g		7.2	8.5	

Alfalfa-Witchgrass, hay, s-c, cut 3, (1)
Ref No 1 00 351 United States

Feed Name or Analyses			Mean As Fed	Mean Dry	C.V. ± %
Dry matter	%		94.7	100.0	0
Calcium	%		1.00	1.06	9
Magnesium	%		0.41	0.43	9
Phosphorus	%		0.38	0.40	4

ALFILERIA, REDSTEM. Erodium cicutarium

Alfileria, redstem, hay, s-c, (1)
Ref No 1 08 333 United States

Feed Name or Analyses			Mean As Fed	Mean Dry	C.V. ± %
Dry matter	%		89.2	100.0	
Ash	%		11.8	13.2	
Crude fiber	%		23.4	26.2	
Ether extract	%		2.9	3.3	
N-free extract	%		40.2	45.1	
Protein (N x 6.25)	%		10.9	12.2	
Cattle	dig prot %	*	6.7	7.5	
Goats	dig prot %	*	7.1	8.0	
Horses	dig prot %	*	7.1	7.9	
Rabbits	dig prot %	*	7.2	8.1	
Sheep	dig prot %	*	6.7	7.5	
Energy	GE Mcal/kg				
Cattle	DE Mcal/kg	*	2.13	2.39	
Sheep	DE Mcal/kg	*	2.34	2.62	
Cattle	ME Mcal/kg	*	1.75	1.96	
Sheep	ME Mcal/kg	*	1.92	2.15	
Cattle	TDN %	*	48.4	54.3	
Sheep	TDN %	*	53.0	59.4	
Calcium	%		1.57	1.76	
Phosphorus	%		0.41	0.46	

(1) dry forages and roughages (2) pasture, range plants, and forages fed green (3) silages (4) energy feeds (5) protein supplements (6) minerals (7) vitamins (8) additives

Alfileria, redstem, hay, s-c, mature, (1)
Ref No 1 08 334 United States

Feed Name or Analyses			Mean As Fed	Mean Dry	C.V. ± %
Dry matter	%		89.0	100.0	
Ash	%		8.5	9.6	
Crude fiber	%		31.4	35.3	
Ether extract	%		1.5	1.7	
N-free extract	%		44.1	49.6	
Protein (N x 6.25)	%		3.5	3.9	
Cattle	dig prot %	*	0.3	0.3	
Goats	dig prot %	*	0.2	0.2	
Horses	dig prot %	*	0.8	0.9	
Rabbits	dig prot %	*	1.5	1.7	
Sheep	dig prot %	*	0.1	0.1	
Energy	GE Mcal/kg				
Cattle	DE Mcal/kg	*	1.89	2.12	
Sheep	DE Mcal/kg	*	2.07	2.33	
Cattle	ME Mcal/kg	*	1.55	1.74	
Sheep	ME Mcal/kg	*	1.70	1.91	
Cattle	TDN %	*	42.8	48.1	
Sheep	TDN %	*	47.0	52.8	

Alfileria, redstem, aerial part, fresh, (2)
Ref No 2 00 356 United States

Feed Name or Analyses			Mean As Fed	Mean Dry	C.V. ± %
Dry matter	%		16.4	100.0	
Ash	%		2.7	16.3	18
Crude fiber	%		3.1	18.9	31
Ether extract	%		0.4	2.4	
N-free extract	%		7.5	45.6	
Protein (N x 6.25)	%		2.8	16.8	35
Cattle	dig prot %	*	2.0	12.2	
Goats	dig prot %	*	2.0	12.2	
Horses	dig prot %	*	1.9	11.8	
Rabbits	dig prot %	*	1.9	11.6	
Sheep	dig prot %	*	2.1	12.6	
Energy	GE Mcal/kg				
Cattle	DE Mcal/kg	*	0.43	2.60	
Sheep	DE Mcal/kg	*	0.40	2.45	
Cattle	ME Mcal/kg	*	0.35	2.13	
Sheep	ME Mcal/kg	*	0.33	2.00	
Cattle	TDN %	*	9.7	58.9	
Sheep	TDN %	*	9.1	55.5	
Calcium	%		0.35	2.11	15
Phosphorus	%		0.07	0.42	50
Potassium	%		0.55	3.37	28

Alfileria, redstem, aerial part, fresh, immature, (2)
Ref No 2 00 352 United States

Feed Name or Analyses			Mean As Fed	Mean Dry	C.V. ± %
Dry matter	%			100.0	
Ash	%			13.9	9
Crude fiber	%			12.2	25
Protein (N x 6.25)	%			26.3	14
Cattle	dig prot %	*		20.2	
Goats	dig prot %	*		21.1	
Horses	dig prot %	*		19.9	
Rabbits	dig prot %	*		19.0	
Sheep	dig prot %	*		21.5	
Calcium	%			2.70	11
Phosphorus	%			0.58	41
Potassium	%			3.95	26

Alfileria, redstem, aerial part, fresh, full bloom, (2)
Ref No 2 00 353 United States

Feed Name or Analyses			Mean As Fed	Mean Dry	C.V. ± %
Dry matter	%			100.0	
Ash	%			13.5	
Crude fiber	%			17.8	
Protein (N x 6.25)	%			17.1	
Cattle	dig prot %	*		12.4	
Goats	dig prot %	*		12.5	
Horses	dig prot %	*		12.0	
Rabbits	dig prot %	*		11.9	
Sheep	dig prot %	*		12.9	
Calcium	%			2.54	
Phosphorus	%			0.51	
Potassium	%			3.56	

Alfileria, redstem, aerial part, fresh, milk stage, (2)
Ref No 2 00 354 United States

Feed Name or Analyses			Mean As Fed	Mean Dry	C.V. ± %
Dry matter	%			100.0	
Ash	%			13.6	20
Crude fiber	%			19.9	15
Protein (N x 6.25)	%			11.2	9
Cattle	dig prot %	*		7.4	
Goats	dig prot %	*		7.0	
Horses	dig prot %	*		7.0	
Rabbits	dig prot %	*		7.3	
Sheep	dig prot %	*		7.4	
Calcium	%			2.40	
Phosphorus	%			0.47	
Potassium	%			2.98	

Alfileria, redstem, aerial part, fresh, mature, (2)
Ref No 2 00 355 United States

Feed Name or Analyses			Mean As Fed	Mean Dry	C.V. ± %
Dry matter	%			100.0	
Ash	%			10.9	28
Crude fiber	%			25.8	
Protein (N x 6.25)	%			9.1	8
Cattle	dig prot %	*		5.6	
Goats	dig prot %	*		5.1	
Horses	dig prot %	*		5.3	
Rabbits	dig prot %	*		5.7	
Sheep	dig prot %	*		5.5	
Calcium	%			2.03	
Phosphorus	%			0.18	
Potassium	%			2.28	

ALGAE. Scenedesmus spp

Algae, fan air dried, (5)
Ref No 5 07 749 United States

Feed Name or Analyses			Mean As Fed	Mean Dry	C.V. ± %
Dry matter	%		92.9	100.0	3
Ash	%		16.0	17.3	25
Crude fiber	%		6.7	7.2	29
Ether extract	%		5.4	5.8	40
Cellulose (Matrone)	%		3.3	3.6	
Lignin (Ellis)	%		4.2	4.5	17
Energy	GE Mcal/kg		4.51	4.85	
Calcium	%		2.10	2.26	19
Phosphorus	%		2.22	2.39	
Potassium	%		2.35	2.53	
Silicon	%		0.91	0.98	49

Feed Name or Analyses				Mean As Fed	Mean Dry	C.V. ± %
Carotene		mg/kg		352.1	379.2	
Vitamin A equivalent		IU/g		586.9	632.1	

ALGAE. Scenedesmus quadricauda

Algae, fan air dried, late summer and fall, (5)
Ref No 5 07 750 United States

Feed Name or Analyses				Mean As Fed	Mean Dry	C.V. ± %
Dry matter		%		93.9	100.0	0
Sheep	dig coef	%		54.	54.	
Ash		%		9.9	10.5	30
Crude fiber		%		4.6	4.9	18
Sheep	dig coef	%		-21.	-21.	
Ether extract		%		6.2	6.6	7
Sheep	dig coef	%		46.	46.	
N-free extract		%		25.5	27.1	
Sheep	dig coef	%		67.	67.	
Protein (N x 6.25)		%		47.8	50.9	
Sheep	dig coef	%		72.	72.	
Sheep	dig prot	%		34.4	36.6	
Lignin (Ellis)		%		3.5	3.7	14
Energy		GE Mcal/kg		5.51	5.87	
Sheep	GE dig coef	%		59.	59.	
Cattle		DE Mcal/kg	*	2.64	2.81	
Sheep		DE Mcal/kg		3.25	3.46	
Cattle		ME Mcal/kg	*	2.16	2.30	
Sheep		ME Mcal/kg	*	2.67	2.84	
Cattle		TDN %	*	59.9	63.8	
Sheep		TDN %		56.9	60.5	
Aluminum		mg/kg		1127.28	1200.00	
Calcium		%		1.31	1.39	40
Copper		mg/kg		93.9	100.0	
Iron		%		0.216	0.230	
Lead		mg/kg		93.940	100.000	
Magnesium		%		1.50	1.60	
Manganese		mg/kg		281.8	300.0	
Phosphorus		%		1.99	2.12	14
Potassium		%		0.86	0.92	
Silicon		%		1.04	1.11	47
Sodium		%		0.22	0.23	
Zinc		mg/kg		1690.9	1800.0	
Carotene		mg/kg		208.0	221.4	
Vitamin A equivalent		IU/g		346.7	369.1	

ALGAE. Chlorella spp

Algae, s-c, winter and early spring, (5)
Ref No 5 07 747 United States

Feed Name or Analyses				Mean As Fed	Mean Dry	C.V. ± %
Dry matter		%		91.9	100.0	1
Ash		%		15.0	16.3	17

ALKALI SACATON. Sporobolus airoides

Alkali sacaton, hay, fan air dried, (1)
Ref No 1 04 096 United States

Feed Name or Analyses				Mean As Fed	Mean Dry	C.V. ± %
Dry matter		%			100.0	
Carotene		mg/kg			121.0	
Vitamin A equivalent		IU/g			201.8	

Alkali sacaton, hay, s-c, (1)
Ref No 1 04 097 United States

Feed Name or Analyses				Mean As Fed	Mean Dry	C.V. ± %
Dry matter		%			100.0	
Carotene		mg/kg			102.3	
Vitamin A equivalent		IU/g			170.5	

Alkali sacaton, aerial part, fresh, (2)
Ref No 2 04 102 United States

Feed Name or Analyses				Mean As Fed	Mean Dry	C.V. ± %
Dry matter		%			100.0	
Chlorine		%			0.17	
Sodium		%			0.30	
Sulphur		%			1.02	

Alkali sacaton, aerial part, fresh, immature, (2)
Ref No 2 04 098 United States

Feed Name or Analyses				Mean As Fed	Mean Dry	C.V. ± %
Dry matter		%		35.0	100.0	6
Ash		%		3.3	9.5	2
Crude fiber		%		11.5	32.8	4
Ether extract		%		0.7	2.1	14
N-free extract		%		15.8	45.1	
Protein (N x 6.25)		%		3.7	10.7	11
Cattle	dig prot	%	*	2.4	7.0	
Goats	dig prot	%	*	2.3	6.5	
Horses	dig prot	%	*	2.3	6.6	
Rabbits	dig prot	%	*	2.4	6.9	
Sheep	dig prot	%	*	2.4	6.9	
Energy		GE Mcal/kg				
Cattle		DE Mcal/kg	*	1.01	2.89	
Sheep		DE Mcal/kg	*	0.99	2.82	
Cattle		ME Mcal/kg	*	0.83	2.37	
Sheep		ME Mcal/kg	*	0.81	2.31	
Cattle		TDN %	*	22.9	65.4	
Sheep		TDN %	*	22.4	63.9	
Calcium		%		0.23	0.65	11
Phosphorus		%		0.08	0.23	25
Carotene		mg/kg		67.4	192.7	
Vitamin A equivalent		IU/g		112.4	321.2	

Alkali sacaton, aerial part, fresh, mid-bloom, (2)
Ref No 2 05 601 United States

Feed Name or Analyses				Mean As Fed	Mean Dry	C.V. ± %
Dry matter		%		40.0	100.0	
Ash		%		3.7	9.3	
Crude fiber		%		14.5	36.3	
Ether extract		%		0.7	1.8	
N-free extract		%		18.6	46.5	
Protein (N x 6.25)		%		2.5	6.3	
Cattle	dig prot	%	*	1.3	3.2	
Goats	dig prot	%	*	1.0	2.4	
Horses	dig prot	%	*	1.1	2.8	
Rabbits	dig prot	%	*	1.4	3.5	
Sheep	dig prot	%	*	1.1	2.8	
Energy		GE Mcal/kg				
Cattle		DE Mcal/kg	*	1.09	2.73	
Sheep		DE Mcal/kg	*	0.96	2.40	
Cattle		ME Mcal/kg	*	0.90	2.24	
Sheep		ME Mcal/kg	*	0.79	1.97	
Cattle		TDN %	*	24.8	61.9	
Sheep		TDN %	*	21.8	54.5	

Alkali sacaton, aerial part, fresh, full bloom, (2)
Ref No 2 04 099 United States

Feed Name or Analyses				Mean As Fed	Mean Dry	C.V. ± %
Dry matter		%		40.0	100.0	5
Ash		%		3.8	9.4	11
Crude fiber		%		13.5	33.8	5
Ether extract		%		0.7	1.8	11
N-free extract		%		19.3	48.3	
Protein (N x 6.25)		%		2.7	6.7	7
Cattle	dig prot	%	*	1.4	3.6	
Goats	dig prot	%	*	1.1	2.8	
Horses	dig prot	%	*	1.3	3.2	
Rabbits	dig prot	%	*	1.5	3.8	
Sheep	dig prot	%	*	1.3	3.2	
Energy		GE Mcal/kg				
Cattle		DE Mcal/kg	*	1.09	2.72	
Sheep		DE Mcal/kg	*	0.98	2.46	
Cattle		ME Mcal/kg	*	0.89	2.23	
Sheep		ME Mcal/kg	*	0.81	2.02	
Cattle		TDN %	*	24.7	61.6	
Sheep		TDN %	*	22.4	55.9	
Calcium		%		0.20	0.51	5
Phosphorus		%		0.08	0.20	26

Alkali sacaton, aerial part, fresh, dough stage, (2)
Ref No 2 04 100 United States

Feed Name or Analyses				Mean As Fed	Mean Dry	C.V. ± %
Dry matter		%			100.0	
Calcium		%			0.27	29
Phosphorus		%			0.10	24

Alkali sacaton, aerial part, fresh, mature, (2)
Ref No 2 05 602 United States

Feed Name or Analyses				Mean As Fed	Mean Dry	C.V. ± %
Dry matter		%			100.0	
Ash		%			8.6	27
Crude fiber		%			35.9	2
Ether extract		%			1.7	12
N-free extract		%			49.3	
Protein (N x 6.25)		%			4.5	11
Cattle	dig prot	%	*		1.7	
Goats	dig prot	%	*		0.8	
Horses	dig prot	%	*		1.4	
Rabbits	dig prot	%	*		2.1	
Sheep	dig prot	%	*		1.2	
Energy		GE Mcal/kg				
Cattle		DE Mcal/kg	*		2.69	
Sheep		DE Mcal/kg	*		2.99	
Cattle		ME Mcal/kg	*		2.20	
Sheep		ME Mcal/kg	*		2.45	
Cattle		TDN %	*		60.9	
Sheep		TDN %	*		67.7	
Calcium		%			0.25	21
Phosphorus		%			0.09	37
Carotene		mg/kg			0.7	
Vitamin A equivalent		IU/g			1.1	

Alkali sacaton, aerial part wo lower stems, fresh, early leaf, (2)
Ref No 2 08 690 United States

Feed Name or Analyses				Mean As Fed	Mean Dry	C.V. ± %
Dry matter		%			100.0	
Organic matter		%			88.3	4
Ash		%			11.7	28
Crude fiber		%			29.1	8

Continued

Feed Name or Analyses			Mean As Fed	Mean Dry	C.V. ± %
Ether extract	%			1.8	34
N-free extract	%			47.7	4
Protein (N x 6.25)	%			9.7	29
Cattle	dig prot %	*		6.1	
Goats	dig prot %	*		5.6	
Horses	dig prot %	*		5.7	
Rabbits	dig prot %	*		6.1	
Sheep	dig prot %	*		6.0	
Energy	GE Mcal/kg				
Cattle	DE Mcal/kg	*		2.67	
Sheep	DE Mcal/kg	*		2.48	
Cattle	ME Mcal/kg	*		2.19	
Sheep	ME Mcal/kg	*		2.03	
Cattle	TDN %	*		60.7	
Sheep	TDN %	*		56.3	
Calcium	%			0.43	30
Phosphorus	%			0.12	34
Carotene	mg/kg			107.2	
Vitamin A equivalent	IU/g			178.7	

Alkali sacaton, aerial part wo lower stems, fresh, mid-bloom, (2)

Ref No 2 08 691 United States

Feed Name or Analyses			Mean As Fed	Mean Dry	C.V. ± %
Dry matter	%			100.0	
Organic matter	%			88.6	
Ash	%			11.4	
Crude fiber	%			30.1	
Ether extract	%			1.5	
N-free extract	%			49.6	
Protein (N x 6.25)	%			7.5	
Cattle	dig prot %	*		4.3	
Goats	dig prot %	*		3.6	
Horses	dig prot %	*		3.9	
Rabbits	dig prot %	*		4.5	
Sheep	dig prot %	*		4.0	
Energy	GE Mcal/kg				
Cattle	DE Mcal/kg	*		2.62	
Sheep	DE Mcal/kg	*		2.38	
Cattle	ME Mcal/kg	*		2.15	
Sheep	ME Mcal/kg	*		1.95	
Cattle	TDN %	*		59.4	
Sheep	TDN %	*		53.9	
Calcium	%			0.55	
Phosphorus	%			0.15	

Alkali sacaton, aerial part wo lower stems, fresh, mature, (2)

Ref No 2 08 692 United States

Feed Name or Analyses			Mean As Fed	Mean Dry	C.V. ± %
Dry matter	%			100.0	
Organic matter	%			89.9	
Ash	%			10.1	
Crude fiber	%			29.6	
Ether extract	%			1.7	
N-free extract	%			51.1	
Protein (N x 6.25)	%			7.6	
Cattle	dig prot %	*		4.4	
Goats	dig prot %	*		3.7	
Horses	dig prot %	*		4.0	
Rabbits	dig prot %	*		4.5	
Sheep	dig prot %	*		4.1	
Energy	GE Mcal/kg				
Cattle	DE Mcal/kg	*		2.69	
Sheep	DE Mcal/kg	*		2.52	
Cattle	ME Mcal/kg	*		2.21	
Sheep	ME Mcal/kg	*		2.07	
Cattle	TDN %	*		61.0	
Sheep	TDN %	*		57.1	
Calcium	%			0.26	
Phosphorus	%			0.11	

Alkali sacaton, aerial part wo lower stems, fresh, dormant, (2)

Ref No 2 08 689 United States

Feed Name or Analyses			Mean As Fed	Mean Dry	C.V. ± %
Dry matter	%			100.0	
Organic matter	%			92.7	3
Ash	%			8.3	38
Crude fiber	%			35.0	11
Ether extract	%			1.2	29
N-free extract	%			51.2	4
Protein (N x 6.25)	%			4.2	22
Cattle	dig prot %	*		1.5	
Goats	dig prot %	*		0.5	
Horses	dig prot %	*		1.1	
Rabbits	dig prot %	*		1.9	
Sheep	dig prot %	*		0.9	
Energy	GE Mcal/kg				
Cattle	DE Mcal/kg	*		2.72	
Sheep	DE Mcal/kg	*		2.53	
Cattle	ME Mcal/kg	*		2.23	
Sheep	ME Mcal/kg	*		2.07	
Cattle	TDN %	*		61.6	
Sheep	TDN %	*		57.3	
Calcium	%			0.23	51
Phosphorus	%			0.09	33
Carotene	mg/kg			5.1	
Vitamin A equivalent	IU/g			8.5	

ALMOND. Prunus amygdalus

Almond, hulls, (1)

Ref No 4 00 359 United States

Feed Name or Analyses			Mean As Fed	Mean Dry	C.V. ± %
Dry matter	%		91.0	100.0	2
Ash	%		5.9	6.5	
Crude fiber	%		13.8	15.2	
Ether extract	%		2.2	2.4	
N-free extract	%		65.2	71.7	
Protein (N x 6.25)	%		3.9	4.2	
Cattle	dig prot %	*	0.6	0.6	
Goats	dig prot %	*	0.5	0.5	
Horses	dig prot %	*	1.0	1.1	
Rabbits	dig prot %	*	1.8	2.0	
Sheep	dig prot %	*	0.3	0.4	
Cellulose (Matrone)	%		9.3	10.2	
Hemicellulose	%		8.0	8.8	
Pectins	%		2.2	2.4	
Starch	%		1.5	1.6	
Sugars, total	%		28.0	30.8	14
Lignin (Ellis)	%		5.5	6.0	
Energy	GE Mcal/kg		3.86	4.24	
Cattle	DE Mcal/kg	*	3.11	3.42	
Sheep	DE Mcal/kg	*	2.89	3.18	
Cattle	ME Mcal/kg	*	2.55	2.80	
Sheep	ME Mcal/kg	*	2.38	2.61	
Cattle	TDN %	*	70.6	77.6	
Sheep	TDN %	*	65.6	72.1	

Almond, shells, (1)

Ref No 1 07 754 United States

Feed Name or Analyses			Mean As Fed	Mean Dry	C.V. ± %
Dry matter	%		92.5	100.0	
Ash	%		3.4	3.7	
Crude fiber	%		41.8	45.2	
Ether extract	%		4.8	5.2	
Cellulose (Matrone)	%		27.3	29.5	
Sugars, total	%		3.6	3.9	
Lignin (Ellis)	%		17.0	18.4	

ALTERNANTHERA. Alternanthera spp

Alternanthera, aerial part, fresh, (2)

Ref No 2 00 360 United States

Feed Name or Analyses			Mean As Fed	Mean Dry	C.V. ± %
Dry matter	%			100.0	
Ash	%			21.4	
Crude fiber	%			14.8	
Ether extract	%			1.6	
N-free extract	%			41.1	
Protein (N x 6.25)	%			21.1	
Cattle	dig prot %	*		15.8	
Goats	dig prot %	*		16.2	
Horses	dig prot %	*		15.4	
Rabbits	dig prot %	*		15.0	
Sheep	dig prot %	*		16.7	
Calcium	%			3.06	
Magnesium	%			0.62	
Phosphorus	%			0.30	
Potassium	%			3.75	

ALYCECLOVER. Alysicarpus vaginalis

Alyceclover, hay, s-c, (1)

Ref No 1 00 361 United States

Feed Name or Analyses			Mean As Fed	Mean Dry	C.V. ± %
Dry matter	%		90.0	100.0	
Ash	%		5.8	6.4	3
Crude fiber	%		35.6	39.6	1
Ether extract	%		1.6	1.7	11
N-free extract	%		35.8	39.8	
Protein (N x 6.25)	%		11.2	12.4	4
Cattle	dig prot %	*	6.9	7.7	
Goats	dig prot %	*	7.3	8.2	
Horses	dig prot %	*	7.3	8.1	
Rabbits	dig prot %	*	7.4	8.3	
Sheep	dig prot %	*	6.9	7.7	
Energy	GE Mcal/kg				
Cattle	DE Mcal/kg	*	2.13	2.37	
Sheep	DE Mcal/kg	*	2.16	2.40	
Cattle	ME Mcal/kg	*	1.75	1.94	
Sheep	ME Mcal/kg	*	1.77	1.97	
Cattle	TDN %	*	48.3	53.7	
Sheep	TDN %	*	48.9	54.4	

Alyceclover, aerial part, fresh, (2)

Clover, alyce, fresh

Ref No 2 00 362 United States

Feed Name or Analyses			Mean As Fed	Mean Dry	C.V. ± %
Dry matter	%			100.0	
Cobalt	mg/kg			0.086	
Carotene	mg/kg			104.9	
Vitamin A equivalent	IU/g			174.9	

(1) dry forages and roughages (3) silages (6) minerals
(2) pasture, range plants, and forages fed green (4) energy feeds (7) vitamins
(5) protein supplements (8) additives

Feed Name or Analyses		Mean As Fed	Mean Dry	C.V. ± %

Alyceclover, seeds w hulls, (5)
Ref No 5 09 288 United States

Analyses		As Fed	Dry	C.V. ± %
Dry matter	%		100.0	
Protein (N x 6.25)	%		34.0	
Alanine	%		1.16	
Arginine	%		2.89	
Aspartic acid	%		3.67	
Glutamic acid	%		5.10	
Glycine	%		1.43	
Histidine	%		0.85	
Hydroxyproline	%		0.03	
Isoleucine	%		1.19	
Leucine	%		2.21	
Lysine	%		1.77	
Methionine	%		0.37	
Phenylalanine	%		1.19	
Proline	%		1.39	
Serine	%		1.53	
Threonine	%		1.12	
Tyrosine	%		1.02	
Valine	%		1.29	

AMARANTH. Amaranthus spp

Amaranth, browse, (2)
Ref No 2 00 365 United States

Analyses		As Fed	Dry	C.V. ± %
Dry matter	%		100.0	
Ash	%		19.7	32
Crude fiber	%		13.7	34
Ether extract	%		1.6	19
N-free extract	%		42.1	
Protein (N x 6.25)	%		22.9	19
Cattle	dig prot % *		17.4	
Goats	dig prot % *		17.9	
Horses	dig prot % *		17.0	
Rabbits	dig prot % *		16.3	
Sheep	dig prot % *		18.3	

Amaranth, browse, immature, (2)
Ref No 2 00 363 United States

Analyses		As Fed	Dry	C.V. ± %
Dry matter	%		100.0	
Ash	%		20.2	
Crude fiber	%		11.2	
Ether extract	%		1.5	
N-free extract	%		40.8	
Protein (N x 6.25)	%		26.3	
Cattle	dig prot % *		20.2	
Goats	dig prot % *		21.1	
Horses	dig prot % *		19.9	
Rabbits	dig prot % *		19.0	
Sheep	dig prot % *		21.5	
Calcium	%		4.08	
Magnesium	%		0.92	
Phosphorus	%		0.46	
Potassium	%		4.83	

Amaranth, browse, full bloom, (2)
Ref No 2 00 364 United States

Analyses		As Fed	Dry	C.V. ± %
Dry matter	%		100.0	
Ash	%		19.0	
Crude fiber	%		17.4	
Ether extract	%		1.7	
N-free extract	%		44.1	
Protein (N x 6.25)	%		17.8	
Cattle	dig prot % *		13.0	
Goats	dig prot % *		13.2	
Horses	dig prot % *		12.6	
Rabbits	dig prot % *		12.4	
Sheep	dig prot % *		13.6	
Calcium	%		2.74	
Magnesium	%		0.57	
Phosphorus	%		0.29	
Potassium	%		4.39	

AMARANTH, PROSTRATE. Amaranthus blitoides

Amaranth, prostrate, browse, immature, (2)
Ref No 2 00 366 United States

Analyses		As Fed	Dry	C.V. ± %
Dry matter	%		100.0	
Ash	%		21.2	
Crude fiber	%		11.7	
Ether extract	%		1.4	
N-free extract	%		38.7	
Protein (N x 6.25)	%		27.0	
Cattle	dig prot % *		20.8	
Goats	dig prot % *		21.8	
Horses	dig prot % *		20.5	
Rabbits	dig prot % *		19.5	
Sheep	dig prot % *		22.2	
Calcium	%		4.19	
Magnesium	%		0.91	
Phosphorus	%		0.49	
Potassium	%		5.20	

Amaranth, prostrate, browse, full bloom, (2)
Ref No 2 00 367 United States

Analyses		As Fed	Dry	C.V. ± %
Dry matter	%		100.0	
Ash	%		26.3	
Crude fiber	%		13.1	
Ether extract	%		1.3	
N-free extract	%		40.8	
Protein (N x 6.25)	%		18.5	
Cattle	dig prot % *		13.6	
Goats	dig prot % *		13.8	
Horses	dig prot % *		13.2	
Rabbits	dig prot % *		13.0	
Sheep	dig prot % *		14.2	
Calcium	%		3.20	
Magnesium	%		0.61	
Phosphorus	%		0.18	
Potassium	%		4.56	

AMARANTH, REDROOT. Amaranthus retroflexus

Amaranth, redroot, browse, immature, (2)
Ref No 2 00 368 United States

Analyses		As Fed	Dry	C.V. ± %
Dry matter	%		100.0	
Ash	%		18.8	
Crude fiber	%		10.8	
Ether extract	%		1.5	
N-free extract	%		43.2	
Protein (N x 6.25)	%		25.7	
Cattle	dig prot % *		19.7	
Goats	dig prot % *		20.5	
Horses	dig prot % *		19.3	
Rabbits	dig prot % *		18.5	
Sheep	dig prot % *		20.9	
Calcium	%		3.95	
Magnesium	%		1.04	
Phosphorus	%		0.51	
Potassium	%		5.15	

AMARANTHUS, REDROOT. Amaranthus retroflexus

Amaranthus, redroot, seeds w hulls, (5)
Ref No 5 09 268 United States

Analyses		As Fed	Dry	C.V. ± %
Dry matter	%		100.0	
Protein (N x 6.25)	%		18.0	
Alanine	%		0.27	
Arginine	%		1.33	
Aspartic acid	%		1.40	
Glutamic acid	%		1.84	
Glycine	%		1.39	
Histidine	%		0.41	
Hydroxyproline	%		0.02	
Isoleucine	%		0.59	
Leucine	%		0.92	
Lysine	%		0.83	
Methionine	%		0.32	
Phenylalanine	%		0.63	
Proline	%		0.52	
Serine	%		1.21	
Threonine	%		0.58	
Tyrosine	%		0.59	
Valine	%		0.67	

AMICIA, MEXICAN. Amicia zygomeris

Amicia, Mexican, seeds w hulls, (5)
Ref No 5 09 026 United States

Analyses		As Fed	Dry	C.V. ± %
Dry matter	%		100.0	
Protein (N x 6.25)	%		23.0	
Alanine	%		0.81	
Arginine	%		2.19	
Aspartic acid	%		2.37	
Glutamic acid	%		3.89	
Glycine	%		0.85	
Histidine	%		0.58	
Hydroxyproline	%		0.12	
Isoleucine	%		0.81	
Leucine	%		1.43	
Lysine	%		1.10	
Methionine	%		0.23	
Phenylalanine	%		0.92	
Proline	%		1.13	
Serine	%		1.06	
Threonine	%		0.71	
Tyrosine	%		0.67	
Valine	%		0.90	

AMMONIUM PHOSPHATE, DIBASIC

Ammonium phosphate dibasic, $(NH_4)_2HPO_4$, cp, (6)
Ref No 6 00 371 United States

Analyses		As Fed	Dry	C.V. ± %
Dry matter	%		100.0	
Chlorine	%		0.00	
Iron	%		0.001	
Phosphorus	%		23.47	

Continued

Feed Name or Analyses		Mean As Fed	Mean Dry	C.V. ± %
Potassium	%		0.01	
Sodium	%		0.01	

ANATTOTREE. Bixa orellana

Anattotree, seeds w hulls, (4)

Ref No 4 09 009 United States

Feed Name or Analyses		Mean As Fed	Mean Dry	C.V. ± %
Dry matter	%		100.0	
Protein (N x 6.25)	%		13.0	
Cattle	dig prot %	*	8.0	
Goats	dig prot %	*	9.1	
Horses	dig prot %	*	9.1	
Sheep	dig prot %	*	9.1	
Alanine	%		0.57	
Arginine	%		0.94	
Aspartic acid	%		1.17	
Glutamic acid	%		2.07	
Glycine	%		0.62	
Histidine	%		0.30	
Hydroxyproline	%		0.17	
Isoleucine	%		0.43	
Leucine	%		0.75	
Lysine	%		0.87	
Methionine	%		0.22	
Phenylalanine	%		0.51	
Proline	%		0.70	
Serine	%		0.52	
Threonine	%		0.51	
Tyrosine	%		0.43	
Valine	%		0.57	

ANIMAL. Scientific name not used

Animal, fat, heat rendered, (4)
Animal fat (CFA)

Ref No 4 00 375 Canada

Feed Name or Analyses		Mean As Fed	Mean Dry	C.V. ± %
Dry matter	%	90.0	100.0	
Ether extract	%			
Cattle	dig coef %	90.	90.	
Fatty acids	%	9.9	11.0	98
Energy	GE Mcal/kg	8.40	9.34	1
Chickens	MEn kcal/kg	6861.	7623.	10

Animal, fat, hydrolyzed, feed gr mn 85% fatty acids mx 6% unsaponifiable matter mx 1% insoluble matter, (4)
Hydrolyzed animal fat, feed grade (AAFCO)

Ref No 4 00 376 United States

Feed Name or Analyses		Mean As Fed	Mean Dry	C.V. ± %
Dry matter	%	95.0	100.0	
Ether extract	%	95.0	100.0	
Energy	GE Mcal/kg			
Chickens	MEn kcal/kg	7055.	7426.	

(1) dry forages and roughages (2) pasture, range plants, and forages fed green (3) silages (4) energy feeds (5) protein supplements (6) minerals (7) vitamins (8) additives

Animal, tallow, bleached stabilized, fancy gr, (4)

Ref No 4 07 880 United States

Feed Name or Analyses		Mean As Fed	Mean Dry	C.V. ± %
Dry matter	%		100.0	
Energy	GE Mcal/kg		9.44	
Swine	DE kcal/kg		8130.	
Chickens	MEn kcal/kg	7010.	7010.	
Swine	ME kcal/kg		7900.	

Animal, blood, dehy grnd, (5)
Blood meal (AAFCO)
Blood meal (CFA)

Ref No 5 00 380 United States

Feed Name or Analyses		Mean As Fed	Mean Dry	C.V. ± %
Dry matter	%	89.3	100.0	
Ash	%	4.4	4.9	
Crude fiber	%	0.6	0.7	
Sheep	dig coef %	18.	18.	
Swine	dig coef %	0.	0.	
Ether extract	%	1.4	1.5	
Sheep	dig coef %	38.	38.	
Swine	dig coef %	-115.	-115.	
N-free extract	%	2.7	3.0	
Sheep	dig coef %	25.	25.	
Swine	dig coef %	-103.	-103.	
Protein (N x 6.25)	%	80.2	89.8	6
Sheep	dig coef %	71.	71.	
Swine	dig coef %	78.	78.	
Sheep	dig prot %	56.9	63.7	
Swine	dig prot %	62.5	70.0	
Energy	GE Mcal/kg	5.05	5.66	
Cattle	DE Mcal/kg	* 2.60	2.91	
Sheep	DE Mcal/kg	* 2.60	2.91	
Swine	DE kcal/kg	*2475.	2772.	
Cattle	ME Mcal/kg	* 2.13	2.39	
Chickens	MEn kcal/kg	2845.	3186.	
Sheep	ME Mcal/kg	* 2.13	2.38	
Swine	ME kcal/kg	*1927.	2159.	
Sheep	TDN %	58.9	66.0	
Cattle	TDN %	* 58.8	65.9	
Swine	TDN %	56.1	62.9	
Calcium	%	0.30	0.33	
Chlorine	%	0.26	0.29	
Copper	mg/kg	9.7	10.8	
Iron	%	0.366	0.411	
Magnesium	%	0.21	0.24	
Manganese	mg/kg	5.2	5.8	
Phosphorus	%	0.23	0.26	
Potassium	%	0.09	0.10	
Sodium	%	0.31	0.35	
Sulphur	%	0.32	0.36	
Choline	mg/kg	749.	838.	
Niacin	mg/kg	31.5	35.3	
Pantothenic acid	mg/kg	1.1	1.2	
Riboflavin	mg/kg	1.5	1.7	
α-tocopherol	mg/kg	0.0	0.0	
Arginine	%	3.20	3.59	
Cystine	%	1.40	1.57	
Glycine	%	4.49	5.03	
Histidine	%	3.79	4.25	
Isoleucine	%	0.88	0.99	
Leucine	%	9.91	11.10	
Lysine	%	5.37	6.02	
Methionine	%	1.05	1.17	
Phenylalanine	%	5.18	5.80	
Threonine	%	3.87	4.34	
Tryptophan	%	1.02	1.14	
Valine	%	6.93	7.76	

Ref No 5 00 380 Canada

Feed Name or Analyses		Mean As Fed	Mean Dry	C.V. ± %
Dry matter	%	93.3	100.0	4
Ash	%	6.0	6.4	16
Crude fiber	%	1.7	1.8	
Sheep	dig coef %	18.	18.	
Swine	dig coef %	0.	0.	
Ether extract	%	4.9	5.2	
Sheep	dig coef %	38.	38.	
Swine	dig coef %	-115.	-115.	
N-free extract	%	12.5	13.4	
Sheep	dig coef %	25.	25.	
Swine	dig coef %	-103.	-103.	
Protein (N x 6.25)	%	68.3	73.2	31
Sheep	dig coef %	71.	71.	
Swine	dig coef %	78.	78.	
Sheep	dig prot %	48.5	52.0	
Swine	dig prot %	53.3	57.1	
Energy	GE Mcal/kg	5.28	5.66	
Cattle	DE Mcal/kg	* 2.72	2.91	
Swine	DE kcal/kg	*1219.	1305.	
Cattle	ME Mcal/kg	* 2.23	2.39	
Sheep	ME Mcal/kg	* 2.03	2.17	
Swine	ME kcal/kg	* 990.	1060.	
Cattle	TDN %	* 61.55	65.4	
Sheep	TDN %	56.1	60.1	
Swine	TDN %	27.6	29.6	
Calcium	%	0.09	0.10	
Phosphorus	%	0.28	0.30	
Pantothenic acid	mg/kg	4.3	4.6	

Animal, blood, spray dehy, (5)
Blood flour

Ref No 5 00 381 United States

Feed Name or Analyses		Mean As Fed	Mean Dry	C.V. ± %
Dry matter	%	92.2	100.0	
Ash	%	4.7	5.1	
Crude fiber	%	1.1	1.2	
Sheep	dig coef %	* 18.	18.	
Ether extract	%	1.0	1.1	
Sheep	dig coef %	* 38.	38.	
Swine	dig coef %	*-115.	-115.	
N-free extract	%	0.7	0.8	
Sheep	dig coef %	* 25.	25.	
Swine	dig coef %	*-103.	-103.	
Protein (N x 6.25)	%	84.7	91.9	
Sheep	dig coef %	* 71.	71.	
Swine	dig coef %	* 78.	78.	
Sheep	dig prot %	60.1	65.[illegible]	
Swine	dig prot %	66.1	71.7	
Energy	GE Mcal/kg			
Cattle	DE Mcal/kg	* 3.16	3.43	
Sheep	DE Mcal/kg	* 2.71	2.93	
Swine	DE kcal/kg	*2577.	2795.	
Cattle	ME Mcal/kg	* 2.59	2.81	
Chickens	MEn kcal/kg	2764.	2998.	
Sheep	ME Mcal/kg	* 2.22	2.41	
Swine	ME kcal/kg	*1996.	2165.	
Cattle	TDN %	* 71.7	77.7	
Sheep	TDN %	61.4	66.6	
Swine	TDN %	* 58.5	63.4	
Calcium	%	0.64	0.69	
Chlorine	%	0.25	0.27	
Copper	mg/kg	8.2	8.8	
Iron	%	0.276	0.299	
Magnesium	%	0.04	0.04	

Feed Name or Analyses		Mean As Fed	Mean Dry	C.V. ± %
Manganese	mg/kg	6.4	6.9	
Phosphorus	%	0.48	0.52	
Potassium	%	0.41	0.44	
Sodium	%	0.33	0.36	
Sulphur	%	0.60	0.65	

Ref No 5 00 381 Canada

Feed Name or Analyses		Mean As Fed	Mean Dry	C.V. ± %
Dry matter	%	84.1	100.0	12
Ash	%	7.2	8.6	38
Crude fiber	%	1.6	2.0	
Ether extract	%	6.3	7.4	72
N-free extract	%	1.8	-2.2	
Protein (N x 6.25)	%	71.0	84.3	8
Energy	GE Mcal/kg			
Cattle	DE Mcal/kg	* 2.99	3.56	
Sheep	DE Mcal/kg	* 2.45	2.91	
Swine	DE kcal/kg	*2708.	3219.	
Cattle	ME Mcal/kg	* 2.45	2.92	
Chickens	MEn kcal/kg	2523.	2998.	
Sheep	ME Mcal/kg	* 2.01	2.39	
Swine	ME kcal/kg	*2138.	2542.	
Cattle	TDN %	* 67.8	80.6	
Sheep	TDN %	55.5	66.0	
Swine	TDN %	* 61.4	73.0	
Calcium	%	1.67	1.98	52
Salt (NaCl)	%	1.72	2.04	

Animal, blood albumin, dehy, (5)

Blood albumin, dried

Ref No 5 00 382 United States

Feed Name or Analyses		Mean As Fed	Mean Dry	C.V. ± %
Dry matter	%	88.5	100.0	
Ash	%	4.0	4.5	
Crude fiber	%	0.0	0.0	
Ether extract	%	1.5	1.7	
Protein (N x 6.25)	%	83.0	93.8	
Energy	GE Mcal/kg			
Cattle	DE Mcal/kg	* 3.13	3.53	
Sheep	DE Mcal/kg	* 2.90	3.28	
Swine	DE kcal/kg	*2545.	2875.	
Cattle	ME Mcal/kg	* 2.56	2.90	
Sheep	ME Mcal/kg	* 2.38	2.69	
Swine	ME kcal/kg	*1961.	2216.	
Cattle	TDN %	* 70.9	80.1	
Sheep	TDN %	* 65.8	74.4	
Swine	TDN %	* 57.7	65.2	
Calcium	%	0.54	0.61	
Magnesium	%	0.02	0.02	
Phosphorus	%	0.11	0.12	
Potassium	%	0.23	0.26	
Sulphur	%	0.58	0.66	

Animal, carcass residue, dry rendered dehy grnd, mn 9% indigestible material mx 4.4% phosphorus, (5)

Meat meal (AAFCO)

Meat scrap

Ref No 5 00 385 United States

Feed Name or Analyses		Mean As Fed	Mean Dry	C.V. ± %
Dry matter	%	94.2	100.0	
Ash	%	24.9	26.4	
Crude fiber	%	2.5	2.7	
Ether extract	%	9.4	10.0	
Cattle	dig coef %	* 100.	100.	
Sheep	dig coef %	* 97.	97.	
N-free extract	%	2.5	2.7	
Protein (N x 6.25)	%	54.9	58.3	
Cattle	dig coef %	* 91.	91.	
Sheep	dig coef %	* 82.	82.	
Cattle	dig prot %	50.0	53.0	
Sheep	dig prot %	45.0	47.8	
Energy	GE Mcal/kg			
Cattle	DE Mcal/kg	* 2.73	2.90	
Sheep	DE Mcal/kg	* 2.69	2.86	
Swine	DE kcal/kg	*2957.	3139.	
Cattle	ME Mcal/kg	* 2.24	2.38	
Chickens	MEn kcal/kg	1984.	2106.	
Sheep	ME Mcal/kg	* 2.21	2.35	
Swine	ME kcal/kg	*2490.	2644.	
Cattle	NEm Mcal/kg	* 1.63	1.73	
Cattle	NEgain Mcal/kg	* 1.07	1.14	
Cattle	NE lactating cows Mcal/kg	* 1.92	2.04	
Cattle	TDN %	* 62.0	65.8	
Sheep	TDN %	* 61.1	64.9	
Swine	TDN %	* 67.1	71.2	
Calcium	%	8.49	9.01	
Chlorine	%	1.31	1.39	
Copper	mg/kg	9.7	10.3	
Iron	%	0.044	0.047	
Magnesium	%	0.27	0.29	
Manganese	mg/kg	9.5	10.1	
Phosphorus	%	4.18	4.44	
Potassium	%	0.55	0.58	
Sodium	%	1.68	1.78	
Sulphur	%	0.50	0.53	
Choline	mg/kg	1962.	2083.	
Niacin	mg/kg	56.9	60.4	
Pantothenic acid	mg/kg	4.9	5.1	
Riboflavin	mg/kg	5.3	5.6	
α-tocopherol	mg/kg	1.0	1.1	
Arginine	%	3.90	4.14	
Cystine	%	0.70	0.74	
Glycine	%	8.10	8.60	
Lysine	%	3.80	4.03	
Methionine	%	0.80	0.85	
Tryptophan	%	0.36	0.38	

Animal, carcass residue w blood, dry or wet rendered dehy grnd, mn 9% indigestible material mx 4.4% phosphorus, (5)

Meat meal tankage (AAFCO)

Digester tankage

Ref No 5 00 386 United States

Feed Name or Analyses		Mean As Fed	Mean Dry	C.V. ± %
Dry matter	%	93.7	100.0	2
Ash	%	21.2	22.7	3
Crude fiber	%	1.8	1.9	24
Sheep	dig coef %	* 8.	8.	
Ether extract	%	9.0	9.6	21
Cattle	dig coef %	* 100.	100.	
Sheep	dig coef %	* 78.	78.	
N-free extract	%	2.4	2.6	
Sheep	dig coef %	* 19.	19.	
Protein (N x 6.25)	%	59.2	63.2	9
Cattle	dig coef %	* 91.	91.	
Sheep	dig coef %	* 83.	83.	
Cattle	dig prot %	53.9	57.6	
Sheep	dig prot %	49.2	52.5	
Lignin (Ellis)	%	2.7	2.9	
Energy	GE Mcal/kg	4.61	4.92	
Cattle	DE Mcal/kg	* 2.89	3.08	
Sheep	DE Mcal/kg	* 2.89	3.09	
Swine	DE kcal/kg	*3003.	3206.	
Cattle	ME Mcal/kg	* 2.37	2.53	
Chickens	MEn kcal/kg	2659.	2839.	
Sheep	ME Mcal/kg	* 2.37	2.53	
Swine	ME kcal/kg	2145.	2290.	
Cattle	TDN %	* 65.5	69.9	
Sheep	TDN %	65.6	70.0	
Swine	TDN %	* 68.1	72.7	
Calcium	%	6.43	6.86	31
Chlorine	%	1.77	1.89	72
Cobalt	mg/kg	0.155	0.165	49
Copper	mg/kg	32.8	35.1	36
Iron	%	0.194	0.207	49
Magnesium	%	0.43	0.46	10
Manganese	mg/kg	15.3	16.3	65
Phosphorus	%	3.39	3.62	21
Potassium	%	0.52	0.55	
Sodium	%	1.69	1.81	
Sulphur	%	0.71	0.76	22
Choline	mg/kg	2102.	2245.	30
Folic acid	mg/kg	1.57	1.68	
Niacin	mg/kg	39.7	42.4	19
Pantothenic acid	mg/kg	2.5	2.6	30
Riboflavin	mg/kg	2.5	2.6	57
Thiamine	mg/kg	0.4	0.5	75
Arginine	%	3.64	3.89	21
Cystine	%	0.50	0.54	
Glycine	%	6.63	7.08	
Histidine	%	1.93	2.06	38
Isoleucine	%	1.93	2.06	36
Leucine	%	5.19	5.54	23
Lysine	%	3.84	4.10	17
Methionine	%	0.74	0.79	11
Phenylalanine	%	2.75	2.93	34
Threonine	%	2.44	2.61	23
Tryptophan	%	0.57	0.61	39
Valine	%	4.27	4.56	26

Animal, carcass residue w blood w bone, dry or wet rendered dehy grnd, mn 9% indigestible material mn 4.4% phosphorus, (5)

Meat and bone meal tankage (AAFCO)

Digester tankage with bone

Meat and bone meal digester tankage

Ref No 5 00 387 United States

Feed Name or Analyses		Mean As Fed	Mean Dry	C.V. ± %
Dry matter	%	94.1	100.0	1
Ash	%	28.0	29.8	15
Crude fiber	%	2.4	2.6	19
Ether extract	%	12.4	13.2	23
N-free extract	%	4.2	4.4	
Protein (N x 6.25)	%	47.1	50.0	10
Energy	GE Mcal/kg			
Cattle	DE Mcal/kg	* 2.76	2.94	
Sheep	DE Mcal/kg	* 2.72	2.89	
Swine	DE kcal/kg	*3200.	3399.	
Cattle	ME Mcal/kg	* 2.27	2.41	
Sheep	ME Mcal/kg	* 2.23	2.37	
Swine	ME kcal/kg	*2748.	2920.	
Cattle	TDN %	* 62.7	66.6	
Sheep	TDN %	* 61.6	65.5	
Swine	TDN %	* 72.6	77.1	
Calcium	%	11.47	12.19	
Phosphorus	%	5.25	5.58	

Animal, carcass residue w blood w calcium carbonate added, dry or wet rendered dehy grnd, (5)

Digester tankage w calcium carbonate added

Ref No 5 08 335 United States

Feed Name or Analyses		Mean As Fed	Mean Dry	C.V. ± %
Dry matter	%	94.3	100.0	
Ash	%	20.6	21.8	

Continued

Feed Name or Analyses		Mean As Fed	Mean Dry	C.V. ± %
Crude fiber	%	3.7	3.9	
Ether extract	%	8.1	8.6	
N-free extract	%	4.5	4.8	
Protein (N x 6.25)	%	57.4	60.9	
Energy	GE Mcal/kg			
Cattle	DE Mcal/kg	* 2.82	2.99	
Sheep	DE Mcal/kg	* 2.78	2.94	
Swine	DE kcal/kg	*2954.	3132.	
Cattle	ME Mcal/kg	* 2.31	2.45	
Sheep	ME Mcal/kg	* 2.28	2.41	
Swine	ME kcal/kg	*2473.	2622.	
Cattle	TDN %	* 63.9	67.8	
Sheep	TDN %	* 62.9	66.8	
Swine	TDN %	* 67.0	71.0	

Animal, carcass residue w blood w rumen contents, dry or wet rendered dehy grnd, (5)

Digester tankage w rumen contents

Ref No 5 08 336 United States

Feed Name or Analyses		Mean As Fed	Mean Dry	C.V. ± %
Dry matter	%	94.0	100.0	
Ash	%	16.6	17.7	
Crude fiber	%	5.5	5.9	
Ether extract	%	7.3	7.8	
N-free extract	%	5.3	5.6	
Protein (N x 6.25)	%	59.3	63.1	
Energy	GE Mcal/kg			
Cattle	DE Mcal/kg	* 2.87	3.05	
Sheep	DE Mcal/kg	* 2.82	3.00	
Swine	DE kcal/kg	*2939.	3127.	
Cattle	ME Mcal/kg	* 2.35	2.50	
Sheep	ME Mcal/kg	* 2.31	2.46	
Swine	ME kcal/kg	*2447.	2603.	
Cattle	TDN %	* 65.0	69.1	
Sheep	TDN %	* 63.9	67.9	
Swine	TDN %	* 66.7	70.9	

Animal, carcass residue w bone, dry rendered dehy grnd, mn 9% indigestible material mn 4.4% phosphorus, (5)

Meat and bone meal (AAFCO)

Meat and bone scrap

Ref No 5 00 388 United States

Feed Name or Analyses		Mean As Fed	Mean Dry	C.V. ± %
Dry matter	%	93.6	100.0	1
Ash	%	28.6	30.5	12
Crude fiber	%	1.8	1.9	41
Ether extract	%	11.1	11.9	17
Cattle	dig coef %	* 100.	100.	
Sheep	dig coef %	* 97.	97.	
Swine	dig coef %	100.	100.	
N-free extract	%	2.7	2.9	15
Swine	dig coef %	-45.	-45.	
Protein (N x 6.25)	%	49.5	52.9	8
Cattle	dig coef %	* 91.	91.	
Sheep	dig coef %	* 82.	82.	
Swine	dig coef %	89.	89.	
Cattle	dig prot %	45.0	48.1	
Sheep	dig prot %	40.6	43.4	
Swine	dig prot %	44.0	47.1	
Energy	GE Mcal/kg	3.87	4.14	
Cattle	DE Mcal/kg	* 2.67	2.86	
Sheep	DE Mcal/kg	* 2.64	2.82	
Swine	DE kcal/kg	1974.	2110.	

(1) dry forages and roughages (2) pasture, range plants, and forages fed green (3) silages (4) energy feeds (5) protein supplements (6) minerals (7) vitamins (8) additives

Feed Name or Analyses		Mean As Fed	Mean Dry	C.V. ± %
Cattle	ME Mcal/kg	* 2.19	2.34	
Chickens	ME_n kcal/kg	1924.	2057.	
Sheep	ME Mcal/kg	* 2.17	2.32	
Swine	ME kcal/kg	1647.	1760.	
Cattle	NE_m Mcal/kg	* 1.51	1.61	
Cattle	NE_{gain} Mcal/kg	* 0.96	1.03	
Cattle	$NE_{lactating\ cows}$ Mcal/kg	* 1.78	1.90	
Cattle	TDN %	* 60.6	64.8	
Sheep	TDN %	* 59.9	64.1	
Swine	TDN %	81.6	87.2	
Calcium	%	11.42	12.20	12
Chlorine	%	0.75	0.80	12
Cobalt	mg/kg	0.182	0.195	
Copper	mg/kg	1.5	1.6	
Iron	%	0.050	0.053	21
Magnesium	%	1.13	1.20	18
Manganese	mg/kg	12.3	13.2	47
Phosphorus	%	5.69	6.08	14
Potassium	%	1.46	1.56	
Sodium	%	0.73	0.78	
Choline	mg/kg	1980.	2116.	31
Niacin	mg/kg	46.4	49.6	45
Pantothenic acid	mg/kg	3.6	3.8	30
Riboflavin	mg/kg	4.2	4.5	30
Thiamine	mg/kg	1.1	1.2	56
α-tocopherol	mg/kg	1.0	1.1	
Vitamin B_{12}	µg/kg	98.3	105.1	56
Arginine	%	3.56	3.80	19
Cystine	%	0.60	0.65	33
Glutamic acid	%	10.95	11.70	8
Glycine	%	6.81	7.28	45
Histidine	%	0.90	0.96	33
Isoleucine	%	1.69	1.81	19
Leucine	%	3.09	3.30	11
Lysine	%	3.29	3.52	45
Methionine	%	0.64	0.69	11
Phenylalanine	%	1.79	1.91	18
Threonine	%	1.79	1.91	23
Tryptophan	%	0.29	0.31	18
Valine	%	2.39	2.55	28

Animal, carcass residue w bone, dry rendered solv-extd dehy grnd, (5)

Meat and bone scrap, solvent extracted

Ref No 5 08 337 United States

Feed Name or Analyses		Mean As Fed	Mean Dry	C.V. ± %
Dry matter	%	93.7	100.0	
Ash	%	34.4	36.7	
Crude fiber	%	2.4	2.6	
Ether extract	%	3.7	3.9	
N-free extract	%	3.3	3.5	
Protein (N x 6.25)	%	49.9	53.3	
Energy	GE Mcal/kg			
Cattle	DE Mcal/kg	* 2.01	2.15	
Sheep	DE Mcal/kg	* 2.25	2.40	
Swine	DE kcal/kg	*2220.	2370.	
Cattle	ME Mcal/kg	* 1.65	1.76	
Sheep	ME Mcal/kg	* 1.84	1.97	
Swine	ME kcal/kg	*1893.	2020.	
Cattle	TDN %	* 45.7	48.7	
Sheep	TDN %	* 51.0	54.4	
Swine	TDN %	* 50.4	53.7	

Animal, carcass residue w bone, solv-extd dehy grnd, (5)

Ref No 5 08 320 United States

Feed Name or Analyses		Mean As Fed	Mean Dry	C.V. ± %
Dry matter	%	93.8	100.0	
Ash	%	34.4	36.7	

Feed Name or Analyses		Mean As Fed	Mean Dry	C.V. ± %
Crude fiber	%	2.4	2.6	
Ether extract	%	3.7	3.9	
N-free extract	%	3.3	3.5	
Protein (N x 6.25)	%	50.0	53.3	
Energy	GE Mcal/kg			
Cattle	DE Mcal/kg	* 2.02	2.15	
Sheep	DE Mcal/kg	* 2.25	2.40	
Swine	DE kcal/kg	*2223.	2370.	
Cattle	ME Mcal/kg	* 1.65	1.76	
Sheep	ME Mcal/kg	* 1.85	1.97	
Swine	ME kcal/kg	*1895.	2020.	
Cattle	TDN %	* 45.8	48.8	
Sheep	TDN %	* 51.1	54.5	
Swine	TDN %	* 50.4	53.8	
Calcium	%	12.25	13.06	
Phosphorus	%	6.00	6.40	
Choline	mg/kg	1411.	1504.	
Niacin	mg/kg	47.8	51.0	
Pantothenic acid	mg/kg	3.7	4.0	
Riboflavin	mg/kg	4.4	4.7	
Arginine	%	3.50	3.73	
Cystine	%	0.64	0.68	
Glycine	%	7.30	7.78	
Lysine	%	3.40	3.62	
Methionine	%	0.65	0.69	
Tryptophan	%	0.31	0.33	

Animal, cracklings, dehy grnd, (5)

Cracklings, animal

Ref No 5 08 184 Canada

Feed Name or Analyses		Mean As Fed	Mean Dry	C.V. ± %
Dry matter	%	92.9	100.0	1
Ash	%	31.8	34.2	11
Crude fiber	%	12.1	13.0	
Ether extract	%	11.3	12.2	12
N-free extract	%	-11.3	-12.2	
Protein (N x 6.25)	%	49.1	52.9	11
Energy	GE Mcal/kg			
Cattle	DE Mcal/kg	* 1.99	2.14	
Sheep	DE Mcal/kg	* 2.04	2.20	
Swine	DE kcal/kg	*2421.	2608.	
Cattle	ME Mcal/kg	* 1.63	1.76	
Sheep	ME Mcal/kg	* 1.67	1.80	
Swine	ME kcal/kg	*2066.	2225.	
Cattle	TDN %	* 45.1	48.6	
Sheep	TDN %	* 46.3	49.9	
Swine	TDN %	* 54.9	59.1	
Calcium	%	12.20	13.14	
Phosphorus	%	6.19	6.66	
Sodium	%	1.10	1.19	

Animal, livers, dehy grnd, (5)

Animal liver meal (AAFCO)

Animal liver meal (CFA)

Liver meal

Ref No 5 00 389 United States

Feed Name or Analyses		Mean As Fed	Mean Dry	C.V. ± %
Dry matter	%	92.7	100.0	2
Ash	%	6.3	6.8	24
Crude fiber	%	1.4	1.5	35
Ether extract	%	15.7	17.0	16
N-free extract	%	2.9	3.1	
Protein (N x 6.25)	%	66.5	71.7	6
Energy	GE Mcal/kg			
Cattle	DE Mcal/kg	* 3.87	+.18	
Sheep	DE Mcal/kg	* 3.40	3.67	
Swine	DE kcal/kg	*3956.	4269.	
Cattle	ME Mcal/kg	* 3.18	3.43	

Feed Name or Analyses		Mean As Fed	Mean Dry	C.V. ± %
Sheep	ME Mcal/kg	* 2.79	3.01	
Swine	ME kcal/kg	*3224.	3479.	
Cattle	TDN %	* 87.9	94.8	
Sheep	TDN %	* 77.2	83.3	
Swine	TDN %	* 89.7	96.8	
Calcium	%	0.56	0.61	44
Cobalt	mg/kg	0.135	0.145	29
Copper	mg/kg	89.4	96.5	59
Iron	%	0.063	0.068	30
Manganese	mg/kg	8.8	9.5	57
Phosphorus	%	1.26	1.36	21
Biotin	mg/kg	0.02	0.02	
Folic acid	mg/kg	5.56	6.00	
Niacin	mg/kg	205.0	221.2	41
Pantothenic acid	mg/kg	31.0	33.4	85
Riboflavin	mg/kg	46.3	50.0	29
Thiamine	mg/kg	0.2	0.2	76
Vitamin B_{12}	µg/kg	501.9	541.6	
Arginine	%	4.10	4.43	9
Cystine	%	0.90	0.97	
Glutamic acid	%	8.11	8.75	13
Glycine	%	5.60	6.05	
Histidine	%	1.50	1.62	11
Isoleucine	%	3.40	3.67	
Leucine	%	5.40	5.83	9
Lysine	%	4.80	5.18	19
Methionine	%	1.30	1.40	70
Phenylalanine	%	2.90	3.13	12
Serine	%	2.50	2.70	
Threonine	%	2.60	2.81	12
Tryptophan	%	0.60	0.65	27
Tyrosine	%	1.70	1.84	6
Valine	%	4.20	4.54	13

Ref No 5 00 389 Canada

Feed Name or Analyses		Mean As Fed	Mean Dry	C.V. ± %
Dry matter	%	93.0	100.0	
Pantothenic acid	mg/kg	19.4	20.9	
Riboflavin	mg/kg	26.2	28.2	

Animal, liver w gland tissue, dehy grnd, mn 50% liver, (5)
Animal liver and glandular meal (AAFCO)
Liver and glandular meal
Ref No 5 00 390 United States

Feed Name or Analyses		Mean As Fed	Mean Dry	C.V. ± %
Dry matter	%	93.4	100.0	1
Ash	%	5.7	6.2	22
Crude fiber	%	1.9	2.0	39
Ether extract	%	15.8	16.9	18
N-free extract	%	3.9	4.2	
Protein (N x 6.25)	%	66.0	70.7	4
Energy	GE Mcal/kg			
Cattle	DE Mcal/kg	* 3.91	4.19	
Sheep	DE Mcal/kg	* 3.44	3.69	
Swine	DE kcal/kg	*4006.	4291.	
Cattle	ME Mcal/kg	* 3.21	3.43	
Chickens	ME_n kcal/kg	2860.	3062.	
Sheep	ME Mcal/kg	* 2.82	3.02	
Swine	ME kcal/kg	*3273.	3506.	
Cattle	TDN %	* 88.7	95.0	
Sheep	TDN %	* 78.1	83.7	
Swine	TDN %	* 90.8	97.3	
Calcium	%	0.66	0.71	22
Cobalt	mg/kg	0.202	0.217	
Copper	mg/kg	97.1	104.0	
Iron	%	0.049	0.052	
Manganese	mg/kg	7.3	7.8	55
Phosphorus	%	1.14	1.22	20
Niacin	mg/kg	161.6	173.1	35
Pantothenic acid	mg/kg	106.1	113.6	
Riboflavin	mg/kg	40.7	43.6	28
Thiamine	mg/kg	2.6	2.8	34

Animal, stickwater solubles, vacuum dehy, (5)
Dried meat solubles (AAFCO)
Meat solubles, dried
Ref No 5 00 393 United States

Feed Name or Analyses		Mean As Fed	Mean Dry	C.V. ± %
Dry matter	%	90.0	100.0	
Ash	%	5.7	6.3	
Protein (N x 6.25)	%	80.0	88.9	
Calcium	%	0.45	0.50	
Phosphorus	%	0.67	0.74	
Vitamin B_{12}	µg/kg	881.8	979.8	

Animal, bone, cooked dehy grnd, mn 10% phosphorus, (6)
Feeding bone meal (CFA)
Ref No 6 00 397 United States

Feed Name or Analyses		Mean As Fed	Mean Dry	C.V. ± %
Dry matter	%	93.6	100.0	
Ash	%	59.1	63.1	
Crude fiber	%	1.0	1.1	
Ether extract	%	5.0	5.3	
N-free extract	%	2.5	2.7	
Protein (N x 6.25)	%	26.0	27.8	
Calcium	%	22.96	24.53	
Chlorine	%	0.09	0.10	
Copper	mg/kg	18.7	20.0	
Iron	%	0.044	0.047	
Magnesium	%	0.35	0.37	
Manganese	mg/kg	8.6	9.2	
Phosphorus	%	10.25	10.95	
Potassium	%	0.23	0.25	
Sodium	%	0.74	0.79	
Sulphur	%	0.12	0.13	

Ref No 6 00 397 Canada

Feed Name or Analyses		Mean As Fed	Mean Dry	C.V. ± %
Dry matter	%	94.0	100.0	
Ether extract	%	6.9	7.3	
Protein (N x 6.25)	%	27.4	29.2	
Calcium	%	25.45	27.07	
Phosphorus	%	10.60	11.28	

Animal, bone, cooked solv-extd dehy grnd, (6)
Ref No 6 08 338 United States

Feed Name or Analyses		Mean As Fed	Mean Dry	C.V. ± %
Dry matter	%	93.1	100.0	
Ash	%	63.5	68.2	
Crude fiber	%	1.0	1.1	
Ether extract	%	1.0	1.1	
N-free extract	%	1.9	2.0	
Protein (N x 6.25)	%	25.7	27.6	
Calcium	%	24.02	25.80	
Phosphorus	%	10.65	11.44	

Animal, bone, raw dehy grnd, (6)
Bone raw
Ref No 6 00 398 United States

Feed Name or Analyses		Mean As Fed	Mean Dry	C.V. ± %
Dry matter	%	86.9	100.0	
Ash	%	42.8	49.3	
Crude fiber	%	0.0	0.0	
Ether extract	%	16.2	18.7	
Sheep	dig coef %	-170.	-170.	
N-free extract	%	4.9	5.6	
Protein (N x 6.25)	%	22.9	26.4	13
Sheep	dig coef %	69.	69.	
Sheep	dig prot %	15.8	18.2	
Energy	GE Mcal/kg			
Sheep	DE Mcal/kg	* 0.19	0.22	
Sheep	ME Mcal/kg	* 0.15	0.18	
Sheep	TDN %	4.3	4.9	
Arginine	%	1.79	2.06	9
Histidine	%	0.20	0.23	
Isoleucine	%	0.54	0.62	14
Leucine	%	0.87	1.01	9
Lysine	%	0.95	1.09	8
Methionine	%	0.19	0.22	
Phenylalanine	%	0.53	0.61	15
Threonine	%	0.53	0.62	
Tryptophan	%	0.04	0.05	
Valine	%	0.71	0.82	11

Animal, bone, solv-extd dehy grnd, high protein, (6)
Ref No 6 08 340 United States

Feed Name or Analyses		Mean As Fed	Mean Dry	C.V. ± %
Dry matter	%	92.8	100.0	
Ash	%	58.8	63.4	
Crude fiber	%	1.8	1.9	
Ether extract	%	2.5	2.7	
N-free extract	%	3.8	4.1	
Protein (N x 6.25)	%	25.9	27.9	
Calcium	%	22.20	23.92	
Phosphorus	%	9.65	10.40	

Animal, bone, steamed dehy grnd, (6)
Bone meal, steamed (AAFCO)
Ref No 6 00 400 United States

Feed Name or Analyses		Mean As Fed	Mean Dry	C.V. ± %
Dry matter	%	97.1	100.0	2
Ash	%	77.0	79.3	10
Crude fiber	%	1.4	1.4	
Ether extract	%	11.3	11.6	37
Sheep	dig coef %	* 96.	96.	
N-free extract	%	-5.3	-5.4	
Protein (N x 6.25)	%	12.8	13.2	19
Sheep	dig coef %	* 68.	68.	
Sheep	dig prot %	8.7	9.0	
Aluminum	mg/kg	2811.79	2895.58	
Barium	mg/kg	278.458	286.755	
Calcium	%	29.82	30.71	8
Chlorine	%	0.01	0.01	
Cobalt	mg/kg	0.000	0.000	
Copper	mg/kg	11.1	11.5	98
Fluorine	mg/kg	3466.19	3569.48	
Iodine	mg/kg	33.199	34.188	
Iron	%	2.597	2.674	98
Magnesium	%	0.32	0.33	74
Manganese	mg/kg	22.6	23.3	98
Phosphorus	%	12.49	12.86	6
Potassium	%	0.18	0.19	

Continued

Feed Name or Analyses		Mean As Fed	Mean Dry	C.V. ± %
Sodium	%	5.53	5.69	98
Sulphur	%	2.44	2.51	
Strontium	mg/kg	3184.736	3279.632	
Zinc	mg/kg	125.8	129.6	98
Niacin	mg/kg	4.6	4.8	
Pantothenic acid	mg/kg	2.7	2.8	
Riboflavin	mg/kg	1.0	1.0	

Animal, bone, steamed dehy grnd, mn 12% phosphorus, (6)
Feeding steamed bone meal (CFA)
Bone meal, steamed
Ref No 6 00 399 United States

Dry matter	%	95.7	100.0	2
Ash	%	80.4	84.0	7
Crude fiber	%	1.9	2.0	49
Ether extract	%	1.9	2.0	98
N-free extract	%	4.4	4.6	
Protein (N x 6.25)	%	7.1	7.4	46
Aluminum	mg/kg	1294.14	1351.96	98
Barium	mg/kg	300.768	314.208	18
Calcium	%	30.92	32.30	6
Cobalt	mg/kg	0.062	0.065	28
Copper	mg/kg	13.0	13.6	50
Fluorine	mg/kg	726.75	759.22	57
Iron	%	3.013	3.148	98
Magnesium	%	0.70	0.73	40
Manganese	mg/kg	29.5	30.8	41
Phosphorus	%	14.01	14.63	7
Potassium	%	0.18	0.19	
Sodium	%	20.91	21.85	35
Sulphur	%	0.29	0.30	98
Strontium	mg/kg	1844.326	1926.738	62
Zinc	mg/kg	387.9	405.2	39
Niacin	mg/kg	4.2	4.4	54
Pantothenic acid	mg/kg	2.4	2.5	49
Riboflavin	mg/kg	0.9	0.9	31
Thiamine	mg/kg	0.4	0.5	98

Ref No 6 00 399 Argentina

Dry matter	%	96.9	100.0	
Ash	%	77.4	79.9	
Ether extract	%	0.7	0.7	
Protein (N x 6.25)	%	12.4	12.8	
Aluminum	mg/kg	29.55	30.50	
Barium	mg/kg	259.600	267.905	
Calcium	%	33.35	34.42	
Copper	mg/kg	2.4	2.5	
Fluorine	mg/kg	283.56	292.63	
Iron	%	0.789	0.815	
Magnesium	%	0.37	0.38	
Manganese	mg/kg	15.2	15.7	
Phosphorus	%	12.46	12.86	
Sodium	%	21.68	22.37	
Sulphur	%	0.08	0.08	
Strontium	mg/kg	3706.254	3824.824	
Zinc	mg/kg	127.8	131.9	

Ref No 6 00 399 Belgium

Dry matter	%	96.9	100.0	
Ash	%	77.4	79.9	

(1) dry forages and roughages (2) pasture, range plants, and forages fed green (3) silages (4) energy feeds (5) protein supplements (6) minerals (7) vitamins (8) additives

Feed Name or Analyses		Mean As Fed	Mean Dry	C.V. ± %
Ether extract	%	1.4	1.4	
Protein (N x 6.25)	%	8.6	8.9	
Aluminum	mg/kg	883.43	911.69	
Barium	mg/kg	146.975	151.677	
Calcium	%	31.51	32.52	
Copper	mg/kg	25.6	26.4	
Fluorine	mg/kg	669.36	690.78	
Iron	%	12.059	12.445	
Magnesium	%	0.51	0.53	
Manganese	mg/kg	59.9	61.8	
Phosphorus	%	11.41	11.78	
Sodium	%	21.53	22.22	
Sulphur	%	0.47	0.49	
Strontium	mg/kg	368.231	380.011	
Zinc	mg/kg	191.7	197.8	

Ref No 6 00 399 France

Dry matter	%	94.4	100.0	
Ash	%	87.4	92.6	
Ether extract	%	0.0	0.0	
Protein (N x 6.25)	%	3.2	3.4	
Aluminum	mg/kg	3430.26	3633.76	
Barium	mg/kg	257.384	272.653	
Calcium	%	31.11	32.96	
Copper	mg/kg	20.4	21.6	
Fluorine	mg/kg	317.57	336.41	
Iron	%	6.529	6.916	
Magnesium	%	0.69	0.73	
Manganese	mg/kg	25.9	27.5	
Phosphorus	%	14.89	15.77	
Sodium	%	25.08	26.57	
Sulphur	%	0.34	0.36	
Strontium	mg/kg	772.152	817.958	
Zinc	mg/kg	450.0	476.7	

Ref No 6 00 399 Great Britain

Dry matter	%	97.2	100.0	0
Ash	%	84.4	86.9	4
Ether extract	%	0.7	0.7	97
Protein (N x 6.25)	%	5.2	5.3	57
Aluminum	mg/kg	2745.38	2824.58	98
Barium	mg/kg	265.039	272.685	42
Calcium	%	31.66	32.57	13
Copper	mg/kg	19.5	20.0	98
Fluorine	mg/kg	1801.42	1853.39	98
Iron	%	7.206	7.414	70
Magnesium	%	0.49	0.51	67
Manganese	mg/kg	44.7	45.9	45
Phosphorus	%	15.23	15.67	5
Sodium	%	20.91	21.52	28
Sulphur	%	1.39	1.43	98
Strontium	mg/kg	1364.088	1403.441	98
Zinc	mg/kg	328.2	337.6	69

Ref No 6 00 399 Italy

Dry matter	%	96.7	100.0	
Ash	%	82.4	85.2	
Ether extract	%	0.2	0.2	
Protein (N x 6.25)	%	6.4	6.6	
Aluminum	mg/kg	962.04	994.88	
Barium	mg/kg	160.202	165.669	
Calcium	%	34.49	35.67	
Copper	mg/kg	24.7	25.6	
Fluorine	mg/kg	400.50	414.16	
Iron	%	3.667	3.792	
Magnesium	%	0.65	0.67	
Manganese	mg/kg	43.5	44.9	

Feed Name or Analyses		Mean As Fed	Mean Dry	C.V. ± %
Phosphorus	%	12.31	12.73	
Sodium	%	23.44	24.24	
Sulphur	%	0.88	0.91	
Strontium	mg/kg	1603.694	1658.422	
Zinc	mg/kg	1538.9	1591.4	

Animal, bone, steamed solv-extd dehy grnd, (6)
Ref No 6 08 339 United States

Dry matter	%	95.6	100.0	
Ash	%	85.0	88.9	
Crude fiber	%	1.4	1.5	
Ether extract	%	0.4	0.4	
N-free extract	%	3.3	3.5	
Protein (N x 6.25)	%	5.5	5.8	
Calcium	%	32.11	33.59	
Copper	mg/kg	18.3	19.1	
Iron	%	0.096	0.100	
Magnesium	%	0.64	0.67	
Manganese	mg/kg	57.1	59.7	
Phosphorus	%	14.23	14.88	
Sulphur	%	0.22	0.23	

Animal, bone charcoal, retort-charred grnd, mn 14% phosphorus, (6)
Bone charcoal (AAFCO)
Ref No 6 00 402 United States

Dry matter	%	90.0	100.0	
Protein (N x 6.25)	%	8.5	9.4	
Calcium	%	27.10	30.11	6
Phosphorus	%	12.73	14.14	6

Animal, bone glue residue, dehy grnd, (6)
Ref No 6 00 405 United States

Dry matter	%	96.7	100.0	2
Ash	%	77.6	80.2	6
Crude fiber	%	2.1	2.2	43
Ether extract	%	4.4	4.6	51
N-free extract	%	2.8	2.9	
Protein (N x 6.25)	%	9.8	10.1	35
Calcium	%	28.99	29.98	6
Cobalt	mg/kg	0.736	0.761	18
Copper	mg/kg	19.4	20.1	2
Iron	%	0.085	0.088	11
Magnesium	%	0.99	1.02	
Manganese	mg/kg	11.7	12.1	58
Phosphorus	%	13.54	14.00	7
Potassium	%	0.52	0.54	
Sodium	%	0.07	0.07	
Choline	mg/kg	871.	901.	33
Niacin	mg/kg	4.4	4.6	
Pantothenic acid	mg/kg	1.8	1.8	
Riboflavin	mg/kg	1.1	1.1	34
Thiamine	mg/kg	2.0	2.1	

Animal, bone phosphate, precipitated dehy, mn 17% phosphorus, (6)
Bone phosphate (AAFCO)
Ref No 6 00 406 United States

Dry matter	%	98.7	100.0	
Crude fiber	%	86.4	87.5	
Ether extract	%	0.3	0.3	
Protein (N x 6.25)	%	0.4	0.4	

Feed Name or Analyses			Mean As Fed	Mean Dry	C.V. ± %
Calcium	%		28.00	28.37	
Magnesium	%		0.42	0.43	
Phosphorus	%		11.20	11.35	

Animal, rumen contents, dehy grnd, (7)

Rumen contents, dried

Ref No 7 00 407 United States

Feed Name or Analyses			Mean As Fed	Mean Dry	C.V. ± %
Dry matter	%		87.4	100.0	
Ash	%		9.6	11.0	
Crude fiber	%		24.4	28.0	
Sheep	dig coef %		30.	30.	
Ether extract	%		1.8	2.1	
Sheep	dig coef %		66.	66.	
N-free extract	%		36.2	41.4	
Sheep	dig coef %		74.	74.	
Protein (N x 6.25)	%		15.3	17.6	
Sheep	dig coef %		67.	67.	
Sheep	dig prot %		10.3	11.8	
Energy	GE Mcal/kg				
Cattle	DE Mcal/kg	*	2.14	2.45	
Sheep	DE Mcal/kg	*	2.08	2.37	
Cattle	ME Mcal/kg	*	1.75	2.00	
Sheep	ME Mcal/kg	*	1.70	1.95	
Cattle	TDN %	*	48.6	55.6	
Sheep	TDN %		47.1	53.9	
Calcium	%		0.69	0.79	
Phosphorus	%		0.58	0.67	

Animal, rumen contents w molasses added, dehy, (7)

Ref No 7 00 392 United States

Feed Name or Analyses			Mean As Fed	Mean Dry	C.V. ± %
Dry matter	%		80.2	100.0	
Ash	%		9.8	12.2	
Crude fiber	%		12.8	16.0	
Sheep	dig coef %		14.	14.	
Ether extract	%		2.9	3.6	
Sheep	dig coef %		82.	82.	
N-free extract	%		40.2	50.1	
Sheep	dig coef %		66.	66.	
Protein (N x 6.25)	%		14.5	18.1	
Sheep	dig coef %		60.	60.	
Sheep	dig prot %		8.7	10.9	
Energy	GE Mcal/kg				
Cattle	DE Mcal/kg	*	2.06	2.55	
Sheep	DE Mcal/kg	*	1.87	2.33	
Swine	DE kcal/kg	*	1287.	1604.	
Cattle	ME Mcal/kg	*	1.68	2.09	
Sheep	ME Mcal/kg	*	1.53	1.91	
Swine	ME kcal/kg	*	1188.	1482.	
Cattle	TDN %	*	46.4	57.9	
Sheep	TDN %		42.4	52.8	
Swine	TDN %	*	29.2	36.4	

Animal liver and glandular meal (AAFCO) – see Animal, liver w gland tissue, dehy grnd, mn 50% liver, (5)

Animal liver meal (AAFCO) (CFA) – see Animal, livers, dehy grnd, (5)

ANIMAL–POULTRY. Scientific name not used

Animal-Poultry, carcass residue mx 35% blood, dry or wet rendered dehy grnd, mn 50% protein, (5)

Feeding tankage (CFA)

Ref No 5 00 410 United States

Feed Name or Analyses			Mean As Fed	Mean Dry	C.V. ± %
Dry matter	%			100.0	
Crude fiber	%				
Sheep	dig coef %	*		8.	
Swine	dig coef %	*		-18.	
Ether extract	%				
Cattle	dig coef %	*		100.	
Sheep	dig coef %	*		78.	
Swine	dig coef %	*		96.	
N-free extract	%				
Sheep	dig coef %	*		19.	
Swine	dig coef %	*		45.	
Protein (N x 6.25)	%				
Cattle	dig coef %	*		91.	
Sheep	dig coef %	*		83.	
Swine	dig coef %	*		82.	

Ref No 5 00 410 Canada

Feed Name or Analyses			Mean As Fed	Mean Dry	C.V. ± %
Dry matter	%		94.4	100.0	1
Ash	%		29.8	31.6	24
Ether extract	%		10.2	10.7	34
Protein (N x 6.25)	%		51.2	54.3	6
Calcium	%		11.57	12.26	18
Phosphorus	%		5.50	5.83	18
Sodium	%		0.96	1.02	30

Animal-Poultry, carcass residue mx 5% blood, dry or wet rendered dehy grnd, mn 50% protein, (5)

Meat meal (CFA)

Meat scrap (CFA)

Ref No 5 00 411 Canada

Feed Name or Analyses			Mean As Fed	Mean Dry	C.V. ± %
Dry matter	%		93.7	100.0	1
Ash	%		24.5	26.1	
Crude fiber	%		2.9	3.1	
Ether extract	%		11.7	12.4	26
N-free extract	%		3.8	4.1	
Protein (N x 6.25)	%		50.9	54.3	7
Energy	GE Mcal/kg		2.98	3.18	
Cattle	DE Mcal/kg	*	2.85	3.04	
Sheep	DE Mcal/kg	*	2.77	2.95	
Swine	DE kcal/kg	*	3183.	3395.	
Cattle	ME Mcal/kg	*	2.33	2.49	
Chickens	ME_n kcal/kg		1797.	1918.	
Sheep	ME Mcal/kg	*	2.27	2.42	
Swine	ME kcal/kg	*	2706.	2887.	
Cattle	TDN %	*	64.5	68.8	
Sheep	TDN %	*	62.7	66.9	
Swine	TDN %	*	72.2	77.0	
Calcium	%		10.79	11.51	11
Chlorine	%		4.97	5.31	
Fluorine	mg/kg		88.08	93.97	
Phosphorus	%		5.01	5.35	21
Salt (NaCl)	%		1.26	1.35	
Niacin	mg/kg		45.3	48.3	
Pantothenic acid	mg/kg		7.5	8.0	29
Riboflavin	mg/kg		4.6	4.9	42
Thiamine	mg/kg		0.0	0.0	
Alanine	%		7.78	8.30	
Arginine	%		6.64	7.09	
Aspartic acid	%		7.38	7.88	
Glutamic acid	%		11.71	12.49	
Glycine	%		13.61	14.52	
Histidine	%		1.73	1.85	
Isoleucine	%		2.63	2.81	
Leucine	%		6.20	6.62	
Lysine	%		5.43	5.79	23
Methionine	%		1.10	1.17	16
Phenylalanine	%		3.44	3.67	
Proline	%		8.94	9.54	
Serine	%		4.06	4.34	
Threonine	%		3.13	3.34	
Tryptophan	%		0.30	0.32	
Tyrosine	%		2.01	2.14	
Valine	%		4.32	4.61	

Animal-Poultry, carcass residue mx 5% blood w bone, dry or wet rendered dehy grnd, mn 40% protein, (5)

Meat and bone meal (CFA)

Meat and bone scrap (CFA)

Ref No 5 00 412 Canada

Feed Name or Analyses			Mean As Fed	Mean Dry	C.V. ± %
Dry matter	%		93.5	100.0	1
Crude fiber	%		2.2	2.4	
Ether extract	%		11.2	11.9	13
Protein (N x 6.25)	%		50.1	53.6	4
Calcium	%		10.20	10.91	12
Phosphorus	%		4.83	5.17	11
Salt (NaCl)	%		1.37	1.47	78

Animal-Poultry, carcass residue w bone mx 35% blood, dry or wet rendered dehy grnd, mn 40% protein, (5)

Feeding meat and bone tankage (CFA)

Ref No 5 00 413 United States

Feed Name or Analyses			Mean As Fed	Mean Dry	C.V. ± %
Dry matter	%			100.0	
Crude fiber	%				
Sheep	dig coef %	*		8.	
Swine	dig coef %	*		-18.	
Ether extract	%				
Cattle	dig coef %	*		100.	
Sheep	dig coef %	*		78.	
Swine	dig coef %	*		96.	
N-free extract	%				
Sheep	dig coef %	*		19.	
Swine	dig coef %	*		45.	
Protein (N x 6.25)	%				
Cattle	dig coef %	*		91.	
Sheep	dig coef %	*		83.	
Swine	dig coef %	*		82.	

ANISE. Pimpinella anisum

Anise, seeds, extn unspecified caked, (4)

Ref No 4 00 415 United States

Feed Name or Analyses			Mean As Fed	Mean Dry	C.V. ± %
Dry matter	%		90.8	100.0	
Ash	%		14.5	16.0	
Crude fiber	%		9.7	10.7	
Cattle	dig coef %		3.	3.	
Ether extract	%		16.9	18.6	
Cattle	dig coef %		94.	94.	
N-free extract	%		33.1	36.4	
Cattle	dig coef %		68.	68.	
Protein (N x 6.25)	%		16.6	18.3	
Cattle	dig coef %		54.	54.	

Continued

Feed Name or Analyses			Mean As Fed	Mean Dry	C.V. ± %
Cattle	dig prot %		9.0	9.9	
Goats	dig prot %	*	12.7	14.0	
Horses	dig prot %	*	12.7	14.0	
Sheep	dig prot %	*	12.7	14.0	
Energy	GE Mcal/kg				
Cattle	DE Mcal/kg	*	2.97	3.28	
Sheep	DE Mcal/kg	*	2.92	3.22	
Cattle	ME Mcal/kg	*	2.44	2.69	
Sheep	ME Mcal/kg	*	2.40	2.64	
Cattle	TDN %		67.5	74.3	
Sheep	TDN %	*	66.3	73.1	

APACHEPLUME. Fallugia paradoxa

Apacheplume, seeds w hulls, (5)

Ref No 5 09 027 United States

Feed Name or Analyses			Mean As Fed	Mean Dry	C.V. ± %
Dry matter	%			100.0	
Protein (N x 6.25)	%			31.0	
Alanine	%			0.93	
Arginine	%			2.26	
Aspartic acid	%			2.73	
Glutamic acid	%			6.08	
Glycine	%			1.27	
Histidine	%			0.65	
Hydroxyproline	%			0.03	
Isoleucine	%			1.02	
Leucine	%			1.64	
Lysine	%			0.93	
Methionine	%			0.37	
Phenylalanine	%			1.21	
Proline	%			0.93	
Serine	%			1.09	
Threonine	%			0.81	
Tyrosine	%			0.96	
Valine	%			1.02	

APHANOSTEPHUS. Aphanostephus spp

Aphanostephus, aerial part, fresh, full bloom, (2)

Ref No 2 00 417 United States

Feed Name or Analyses			Mean As Fed	Mean Dry	C.V. ± %
Dry matter	%			100.0	
Ash	%			8.6	
Crude fiber	%			22.6	
Ether extract	%			3.1	
N-free extract	%			51.3	
Protein (N x 6.25)	%			14.4	
Cattle	dig prot %	*		10.1	
Goats	dig prot %	*		10.0	
Horses	dig prot %	*		9.8	
Rabbits	dig prot %	*		9.8	
Sheep	dig prot %	*		10.4	
Energy	GE Mcal/kg				
Cattle	DE Mcal/kg	*		2.97	
Sheep	DE Mcal/kg	*		2.90	
Cattle	ME Mcal/kg	*		2.44	
Sheep	ME Mcal/kg	*		2.38	
Cattle	TDN %	*		67.4	
Sheep	TDN %	*		65.9	
Calcium	%			1.44	
Magnesium	%			0.21	
Phosphorus	%			0.27	
Potassium	%			2.85	

(1) dry forages and roughages (2) pasture, range plants, and forages fed green (3) silages (4) energy feeds (5) protein supplements (6) minerals (7) vitamins (8) additives

Aphanostephus, aerial part, fresh, mature, (2)

Ref No 2 00 418 United States

Feed Name or Analyses			Mean As Fed	Mean Dry	C.V. ± %
Dry matter	%			100.0	
Ash	%			7.0	
Crude fiber	%			28.8	
Ether extract	%			2.8	
N-free extract	%			52.7	
Protein (N x 6.25)	%			8.7	
Cattle	dig prot %	*		5.3	
Goats	dig prot %	*		4.7	
Horses	dig prot %	*		4.9	
Rabbits	dig prot %	*		5.4	
Sheep	dig prot %	*		5.1	
Energy	GE Mcal/kg				
Cattle	DE Mcal/kg	*		2.87	
Sheep	DE Mcal/kg	*		3.00	
Cattle	ME Mcal/kg	*		2.35	
Sheep	ME Mcal/kg	*		2.46	
Cattle	TDN %	*		65.0	
Sheep	TDN %	*		68.0	
Calcium	%			1.05	
Magnesium	%			0.20	
Phosphorus	%			0.20	
Potassium	%			2.54	

Apple pectin pulp, wet –
see Apples, pulp wo pectin, wet, (4)

APPLES. Malus spp

Apples, fruit, ensiled, (3)

Ref No 3 00 419 United States

Feed Name or Analyses			Mean As Fed	Mean Dry	C.V. ± %
Dry matter	%		12.4	100.0	
Ash	%		0.6	4.8	
Crude fiber	%		1.8	14.6	
Ether extract	%		0.7	5.8	
N-free extract	%		8.6	69.3	
Protein (N x 6.25)	%		0.7	5.5	
Cattle	dig prot %	*	0.2	1.2	
Goats	dig prot %	*	0.2	1.2	
Horses	dig prot %	*	0.2	1.2	
Sheep	dig prot %	*	0.2	1.2	
Energy	GE Mcal/kg				
Cattle	DE Mcal/kg	*	0.39	3.16	
Sheep	DE Mcal/kg	*	0.38	3.03	
Cattle	ME Mcal/kg	*	0.32	2.59	
Sheep	ME Mcal/kg	*	0.31	2.48	
Cattle	TDN %	*	8.9	71.6	
Sheep	TDN %	*	8.5	68.7	

Apples, fruit, fresh, (4)

Ref No 2 00 421 United States

Feed Name or Analyses			Mean As Fed	Mean Dry	C.V. ± %
Dry matter	%		17.9	100.0	
Ash	%		0.4	2.2	
Crude fiber	%		1.3	7.3	
Ether extract	%		0.4	2.2	
N-free extract	%		15.3	85.5	
Protein (N x 6.25)	%		0.5	2.8	
Cattle	dig prot %	*	-0.2	-1.3	
Goats	dig prot %	*	-0.0	-0.1	
Horses	dig prot %	*	-0.0	-0.1	
Sheep	dig prot %	*	-0.0	-0.1	
Energy	GE Mcal/kg				
Cattle	DE Mcal/kg	*	0.55	3.07	
Sheep	DE Mcal/kg	*	0.57	3.19	
Swine	DE kcal/kg	*	594.	3321.	
Cattle	ME Mcal/kg	*	0.45	2.52	
Sheep	ME Mcal/kg	*	0.47	2.62	
Swine	ME kcal/kg	*	567.	3169.	
Cattle	TDN %	*	12.5	69.7	
Sheep	TDN %	*	13.0	72.4	
Swine	TDN %	*	13.5	75.3	
Calcium	%		0.01	0.06	
Chlorine	%		0.01	0.06	
Copper	mg/kg		2.2	12.3	
Iron	%		0.001	0.006	
Magnesium	%		0.05	0.28	
Manganese	mg/kg		3.1	17.2	
Phosphorus	%		0.01	0.06	
Potassium	%		0.14	0.78	
Sodium	%		0.01	0.06	
Sulphur	%		0.01	0.06	

Apples, pulp, ensiled, (3)

Ref No 3 00 420 United States

Feed Name or Analyses			Mean As Fed	Mean Dry	C.V. ± %
Dry matter	%		21.4	100.0	14
Ash	%		1.1	4.9	15
Crude fiber	%		4.4	20.6	16
Ether extract	%		1.3	6.3	6
N-free extract	%		12.9	60.4	
Protein (N x 6.25)	%		1.7	7.8	13
Cattle	dig prot %	*	0.7	3.3	
Goats	dig prot %	*	0.7	3.3	
Horses	dig prot %	*	0.7	3.3	
Sheep	dig prot %	*	0.7	3.3	
Energy	GE Mcal/kg				
Cattle	DE Mcal/kg	*	0.70	3.26	
Sheep	DE Mcal/kg	*	0.63	2.94	
Cattle	ME Mcal/kg	*	0.57	2.68	
Sheep	ME Mcal/kg	*	0.51	2.41	
Cattle	TDN %	*	15.8	74.0	
Sheep	TDN %	*	14.2	66.6	
Calcium	%		0.02	0.10	
Phosphorus	%		0.02	0.10	
Potassium	%		0.10	0.48	

Apples, pulp, dehy, (4)

Ref No 4 00 422 United States

Feed Name or Analyses			Mean As Fed	Mean Dry	C.V. ± %
Dry matter	%			100.0	
Ash	%			7.2	
Crude fiber	%			20.8	
Sheep	dig coef %			56.	
Ether extract	%			4.6	
Sheep	dig coef %			38.	
N-free extract	%			61.4	
Sheep	dig coef %			71.	
Protein (N x 6.25)	%			6.0	
Sheep	dig coef %			-7.	
Cattle	dig prot %	*		1.5	
Goats	dig prot %	*		2.7	
Horses	dig prot %	*		2.7	
Sheep	dig prot %			-0.4	
Energy	GE Mcal/kg				
Cattle	DE Mcal/kg	*		2.44	
Sheep	DE Mcal/kg	*		2.59	
Cattle	ME Mcal/kg	*		2.00	
Sheep	ME Mcal/kg	*		2.12	
Cattle	TDN %	*		55.4	
Sheep	TDN %			58.7	

Feed Name or Analyses			Mean As Fed	Mean Dry	C.V. ± %

Apples, pulp, dehy grnd, (4)
Dried apple pomace (AAFCO)
Ref No 4 00 423 United States

Analyses			As Fed	Dry	C.V. ± %
Dry matter	%		89.0	100.0	2
Ash	%		2.3	2.6	20
Crude fiber	%		14.8	16.6	16
Cattle	dig coef %		54.	54.	
Sheep	dig coef %	*	56.	56.	
Ether extract	%		4.8	5.4	13
Cattle	dig coef %		35.	35.	
Sheep	dig coef %	*	38.	38.	
N-free extract	%		62.8	70.5	
Cattle	dig coef %		82.	82.	
Sheep	dig coef %	*	71.	71.	
Protein (N x 6.25)	%		4.4	4.9	21
Cattle	dig coef %		-30.	-30.	
Sheep	dig coef %	*	-7.	-7.	
Cattle	dig prot %		-1.3	-1.4	
Goats	dig prot %	*	1.5	1.7	
Horses	dig prot %	*	1.5	1.7	
Sheep	dig prot %		-0.2	-0.3	
Pectins	%		8.6	9.7	9
Pentosans	%		14.3	16.0	
Energy	GE Mcal/kg		4.08	4.58	
Cattle	DE Mcal/kg	*	2.72	3.06	
Sheep	DE Mcal/kg	*	2.50	2.80	
Swine	DE kcal/kg	*	2114.	2376.	
Cattle	ME Mcal/kg	*	2.23	2.51	
Sheep	ME Mcal/kg	*	2.05	2.30	
Swine	ME kcal/kg	*	2008.	2257.	
Cattle	TDN %		61.7	69.4	
Sheep	TDN %		56.6	63.6	
Swine	TDN %	*	47.9	53.9	
Calcium	%		0.11	0.13	32
Iron	%		0.027	0.030	
Magnesium	%		0.06	0.07	39
Manganese	mg/kg		7.2	8.1	48
Phosphorus	%		0.10	0.12	56
Potassium	%		0.43	0.49	15
Sodium	%		0.12	0.14	
Sulphur	%		0.02	0.02	

Apples, pulp, wet, (4)
Ref No 2 00 424 United States

Analyses			As Fed	Dry	C.V. ± %
Dry matter	%		20.2	100.0	
Ash	%		0.8	3.8	
Crude fiber	%		3.4	16.9	
Sheep	dig coef %		64.	64.	
Ether extract	%		1.2	5.9	
Sheep	dig coef %		45.	45.	
N-free extract	%		13.7	67.8	
Sheep	dig coef %		84.	84.	
Protein (N x 6.25)	%		1.1	5.6	
Sheep	dig coef %		-29.	-29.	
Cattle	dig prot %	*	0.2	1.2	
Goats	dig prot %	*	0.5	2.4	
Horses	dig prot %	*	0.5	2.4	
Sheep	dig prot %		-0.2	-1.6	
Energy	GE Mcal/kg				
Cattle	DE Mcal/kg	*	0.67	3.34	
Sheep	DE Mcal/kg	*	0.64	3.17	
Swine	DE kcal/kg	*	449.	2230.	
Cattle	ME Mcal/kg	*	0.55	2.74	
Sheep	ME Mcal/kg	*	0.52	2.60	
Swine	ME kcal/kg	*	426.	2115.	
Cattle	TDN %	*	15.3	75.8	
Sheep	TDN %		14.5	72.0	
Swine	TDN %	*	10.2	50.6	
Calcium	%		0.02	0.09	
Phosphorus	%		0.02	0.09	
Potassium	%		0.10	0.47	

Apples, pulp wo pectin, dehy, (1)
Dried apple pectin pulp (AAFCO)
Ref No 4 00 425 United States

Analyses			As Fed	Dry	C.V. ± %
Dry matter	%		91.2	100.0	
Ash	%		3.3	3.6	
Crude fiber	%		24.2	26.5	
Ether extract	%		7.3	8.0	
N-free extract	%		49.4	54.2	
Protein (N x 6.25)	%		7.0	7.7	
Cattle	dig prot %	*	2.8	3.1	
Goats	dig prot %	*	3.9	4.3	
Horses	dig prot %	*	3.9	4.3	
Sheep	dig prot %	*	3.9	4.3	

Apples, pulp wo pectin, wet, (4)
Apple pectin pulp, wet (4)
Ref No 2 08 573 United States

Analyses			As Fed	Dry	C.V. ± %
Dry matter	%		16.7	100.0	
Ash	%		0.6	3.6	
Crude fiber	%		5.8	34.7	
Ether extract	%		0.9	5.4	
N-free extract	%		7.9	47.3	
Protein (N x 6.25)	%		1.5	9.0	
Cattle	dig prot %	*	0.7	4.3	
Goats	dig prot %	*	0.9	5.5	
Horses	dig prot %	*	0.9	5.5	
Sheep	dig prot %	*	0.9	5.5	
Energy	GE Mcal/kg				
Cattle	DE Mcal/kg	*	0.49	2.93	
Sheep	DE Mcal/kg	*	0.48	2.87	
Cattle	ME Mcal/kg	*	0.40	2.40	
Sheep	ME Mcal/kg	*	0.39	2.35	
Cattle	TDN %	*	11.1	66.5	
Sheep	TDN %	*	10.9	65.0	
Swine	TDN %	*	1.6	9.4	

ARTICHOKE. Cynara scolymus

Artichoke, aerial part, fresh, (2)
Ref No 2 00 429 United States

Analyses			As Fed	Dry	C.V. ± %
Dry matter	%		27.2	100.0	
Ash	%		2.1	7.7	
Crude fiber	%		4.9	18.0	
Ether extract	%		0.3	1.1	
N-free extract	%		18.5	68.1	
Protein (N x 6.25)	%		1.4	5.1	
Cattle	dig prot %	*	0.6	2.2	
Goats	dig prot %	*	0.4	1.3	
Horses	dig prot %	*	0.5	1.9	
Rabbits	dig prot %	*	0.7	2.6	
Sheep	dig prot %	*	0.5	1.8	
Energy	GE Mcal/kg				
Cattle	DE Mcal/kg	*	0.72	2.66	
Sheep	DE Mcal/kg	*	0.86	3.18	
Cattle	ME Mcal/kg	*	0.59	2.18	
Sheep	ME Mcal/kg	*	0.71	2.61	
Cattle	TDN %	*	16.4	60.4	
Sheep	TDN %	*	15.7	57.6	
Calcium	%		0.44	1.62	
Phosphorus	%		0.03	0.11	
Potassium	%		0.37	1.36	

Artichoke, tubers, ensiled, (3)
Ref No 3 00 430 United States

Analyses			As Fed	Dry	C.V. ± %
Dry matter	%		30.2	100.0	
Ash	%		5.6	18.5	
Crude fiber	%		8.2	27.3	
Sheep	dig coef %		65.	65.	
Ether extract	%		0.5	1.7	
Sheep	dig coef %		34.	34.	
N-free extract	%		13.7	45.4	
Sheep	dig coef %		70.	70.	
Protein (N x 6.25)	%		2.2	7.2	
Sheep	dig coef %		16.	16.	
Cattle	dig prot %	*	0.8	2.8	
Goats	dig prot %	*	0.8	2.8	
Horses	dig prot %	*	0.8	2.8	
Sheep	dig prot %		0.3	1.2	
Energy	GE Mcal/kg				
Cattle	DE Mcal/kg	*	0.64	2.13	
Sheep	DE Mcal/kg	*	0.69	2.29	
Cattle	ME Mcal/kg	*	0.53	1.75	
Sheep	ME Mcal/kg	*	0.57	1.88	
Cattle	TDN %	*	14.6	48.3	
Sheep	TDN %		15.7	51.9	

Artichoke, tubers, fresh, (4)
Ref No 2 08 341 United States

Analyses			As Fed	Dry	C.V. ± %
Dry matter	%		20.5	100.0	
Ash	%		1.7	8.3	
Crude fiber	%		0.8	3.9	
Ether extract	%		0.1	0.5	
N-free extract	%		15.9	77.6	
Protein (N x 6.25)	%		2.0	9.8	
Cattle	dig prot %	*	1.0	5.0	
Goats	dig prot %	*	1.3	6.2	
Horses	dig prot %	*	1.3	6.2	
Sheep	dig prot %	*	1.3	6.2	
Energy	GE Mcal/kg				
Cattle	DE Mcal/kg	*	0.70	3.42	
Sheep	DE Mcal/kg	*	0.72	3.54	
Swine	DE kcal/kg	*	640.	3123.	
Cattle	ME Mcal/kg	*	0.57	2.80	
Sheep	ME Mcal/kg	*	0.59	2.90	
Swine	ME kcal/kg	*	602.	2936.	
Cattle	TDN %	*	15.9	77.6	
Sheep	TDN %	*	16.4	80.2	
Swine	TDN %	*	14.5	70.8	
Phosphorus	%		0.06	0.29	
Potassium	%		0.41	2.00	

ASBESTOSBUSH. Duosperma trachyphyllum

Asbestosbush, leaves, s-c, (1)
Ref No 1 00 432 United States

Analyses			As Fed	Dry	C.V. ± %
Dry matter	%		85.3	100.0	
Ash	%		22.0	25.8	
Crude fiber	%		13.8	16.2	
Sheep	dig coef %		23.	23.	
Ether extract	%		2.0	2.4	
Sheep	dig coef %		35.	35.	

Continued

Feed Name or Analyses				Mean As Fed	Mean Dry	C.V. ± %
N-free extract		%		40.3	47.2	
Sheep	dig coef	%		51.	51.	
Protein (N x 6.25)		%		7.2	8.4	
Sheep	dig coef	%		29.	29.	
Cattle	dig prot	%	*	3.6	4.2	
Goats	dig prot	%	*	3.8	4.4	
Horses	dig prot	%	*	4.0	4.7	
Rabbits	dig prot	%	*	4.4	5.2	
Sheep	dig prot	%		2.1	2.4	

ASH. Fraxinus spp

Ash, leaves, s-c, (1)

Ref No 1 00 434 United States

				As Fed	Dry	C.V.
Dry matter		%		85.0	100.0	
Ash		%		5.7	6.7	
Crude fiber		%		12.7	14.9	
Sheep	dig coef	%		61.	61.	
Ether extract		%		4.7	5.5	
Sheep	dig coef	%		20.	20.	
N-free extract		%		49.6	58.3	
Sheep	dig coef	%		89.	89.	
Protein (N x 6.25)		%		12.4	14.6	
Sheep	dig coef	%		64.	64.	
Cattle	dig prot	%	*	8.1	9.6	
Goats	dig prot	%	*	8.7	10.2	
Horses	dig prot	%	*	8.4	9.9	
Rabbits	dig prot	%	*	8.5	9.9	
Sheep	dig prot	%		7.9	9.3	
Energy	GE Mcal/kg					
Cattle	DE Mcal/kg		*	2.57	3.02	
Sheep	DE Mcal/kg		*	2.73	3.21	
Cattle	ME Mcal/kg		*	2.11	2.48	
Sheep	ME Mcal/kg		*	2.24	2.63	
Cattle	TDN	%	*	58.3	68.6	
Sheep	TDN	%		61.9	72.8	

ASPARAGUS. Asparagus officinalis

Asparagus, stem butts, dehy grnd, (1)

Ref No 1 00 437 United States

				As Fed	Dry	C.V.
Dry matter		%		91.0	100.0	
Ash		%		14.6	16.0	
Crude fiber		%		29.0	31.9	
Sheep	dig coef	%		41.	41.	
Ether extract		%		0.9	1.0	
Sheep	dig coef	%		70.	70.	
N-free extract		%		32.3	35.5	
Sheep	dig coef	%		64.	64.	
Protein (N x 6.25)		%		14.2	15.6	
Sheep	dig coef	%		62.	62.	
Cattle	dig prot	%	*	9.5	10.4	
Goats	dig prot	%	*	10.1	11.1	
Horses	dig prot	%	*	9.8	10.8	
Rabbits	dig prot	%	*	9.7	10.7	
Sheep	dig prot	%		8.8	9.7	
Energy	GE Mcal/kg					
Cattle	DE Mcal/kg		*	1.97	2.17	
Sheep	DE Mcal/kg		*	1.89	2.07	
Cattle	ME Mcal/kg		*	1.62	1.78	
Sheep	ME Mcal/kg		*	1.55	1.70	

(1) dry forages and roughages (2) pasture, range plants, and forages fed green (3) silages (4) energy feeds (5) protein supplements (6) minerals (7) vitamins (8) additives

Feed Name or Analyses				Mean As Fed	Mean Dry	C.V. ± %
Cattle	TDN	%	*	44.9	49.3	
Sheep	TDN	%		42.8	47.0	

Asparagus, aerial part, fresh, immature, (2)

Ref No 2 00 435 United States

				As Fed	Dry	C.V.
Dry matter		%		8.3	100.0	
Ash		%		0.6	7.2	
Crude fiber		%		0.7	8.4	
Ether extract		%		0.2	2.4	
N-free extract		%		4.3	51.8	
Protein (N x 6.25)		%		2.5	30.1	
Cattle	dig prot	%	*	1.9	23.5	
Goats	dig prot	%	*	2.0	24.7	
Horses	dig prot	%	*	1.9	23.1	
Rabbits	dig prot	%	*	1.8	21.9	
Sheep	dig prot	%	*	2.1	25.1	
Energy	GE Mcal/kg			0.26	3.13	
Calcium		%		0.02	0.27	
Iron		%		0.001	0.012	
Phosphorus		%		0.06	0.75	
Potassium		%		0.28	3.35	
Sodium		%		0.00	0.02	
Ascorbic acid		mg/kg		330.0	3975.9	
Niacin		mg/kg		15.0	180.7	
Riboflavin		mg/kg		2.0	24.1	
Thiamine		mg/kg		1.8	21.7	
Vitamin A		IU/g		9.0	108.4	

Asparagus, stem butts, fresh, (2)

Ref No 2 00 436 United States

				As Fed	Dry	C.V.
Dry matter		%		30.2	100.0	
Calcium		%		0.16	0.53	
Phosphorus		%		0.46	1.52	
Carotene		mg/kg		1.0	3.3	
Niacin		mg/kg		1.0	3.3	
Riboflavin		mg/kg		0.1	0.4	
Thiamine		mg/kg		0.1	0.4	
Vitamin A equivalent		IU/g		1.7	5.5	

Asparagus, fruit, s-c, (4)

Ref No 4 00 438 United States

				As Fed	Dry	C.V.
Dry matter		%		88.3	100.0	
Ash		%		5.9	6.7	
Crude fiber		%		13.3	15.1	
Sheep	dig coef	%		47.	47.	
Ether extract		%		8.9	10.1	
Sheep	dig coef	%		88.	88.	
N-free extract		%		47.2	53.4	
Sheep	dig coef	%		89.	89.	
Protein (N x 6.25)		%		13.0	14.7	
Sheep	dig coef	%		65.	65.	
Cattle	dig prot	%	*	8.4	9.5	
Goats	dig prot	%	*	9.5	10.7	
Horses	dig prot	%	*	9.5	10.7	
Sheep	dig prot	%		8.4	9.6	
Energy	GE Mcal/kg					
Cattle	DE Mcal/kg		*	3.32	3.76	
Sheep	DE Mcal/kg		*	3.28	3.71	
Swine	DE kcal/kg		*	2041.	2312.	
Cattle	ME Mcal/kg		*	2.72	3.08	
Sheep	ME Mcal/kg		*	2.69	3.04	
Swine	ME kcal/kg		*	1899.	2150.	
Cattle	TDN	%	*	75.2	85.2	

Feed Name or Analyses				Mean As Fed	Mean Dry	C.V. ± %
Sheep	TDN	%		74.3	84.2	
Swine	TDN	%	*	46.3	52.4	

ASPEN. Populus spp

Aspen, leaves, s-c, (1)

Ref No 1 00 439 United States

				As Fed	Dry	C.V.
Dry matter		%		85.0	100.0	
Ash		%		7.7	9.1	
Crude fiber		%		20.8	24.5	
Sheep	dig coef	%		49.	49.	
Ether extract		%		4.6	5.4	
Sheep	dig coef	%		37.	37.	
N-free extract		%		37.2	43.8	
Sheep	dig coef	%		58.	58.	
Protein (N x 6.25)		%		14.6	17.2	
Sheep	dig coef	%		63.	63.	
Cattle	dig prot	%	*	10.1	11.8	
Goats	dig prot	%	*	10.7	12.6	
Horses	dig prot	%	*	10.3	12.1	
Rabbits	dig prot	%	*	10.2	11.9	
Sheep	dig prot	%		9.2	10.8	
Energy	GE Mcal/kg					
Cattle	DE Mcal/kg		*	1.84	2.16	
Sheep	DE Mcal/kg		*	1.98	2.33	
Cattle	ME Mcal/kg		*	1.50	1.77	
Sheep	ME Mcal/kg		*	1.62	1.91	
Cattle	TDN	%	*	41.6	48.9	
Sheep	TDN	%		44.8	52.7	

ASPEN, QUAKING. Populus tremuloides

Aspen, quaking, browse, (2)

Ref No 2 00 442 United States

				As Fed	Dry	C.V.
Dry matter		%			100.0	
Calcium		%			3.02	
Magnesium		%			0.58	
Phosphorus		%			0.30	
Sulphur		%			0.26	

ASTER. Aster spp

Aster, aerial part, fresh, (2)

Ref No 2 00 443 United States

				As Fed	Dry	C.V.
Dry matter		%			100.0	
Ash		%			7.6	
Crude fiber		%			21.1	
Ether extract		%			4.1	
N-free extract		%			58.8	
Protein (N x 6.25)		%			8.4	
Cattle	dig prot	%	*		5.0	
Goats	dig prot	%	*		4.4	
Horses	dig prot	%	*		4.7	
Rabbits	dig prot	%	*		5.2	
Sheep	dig prot	%	*		4.8	
Calcium		%			1.48	
Magnesium		%			0.29	
Manganese		mg/kg			43.0	
Phosphorus		%			0.31	
Potassium		%			0.94	

Feed Name or Analyses			Mean As Fed	Mean Dry	C.V. ± %

ASTER, ALPINE. Aster alpinus

Aster, alpine, seeds w hulls, (5)
Ref No 5 09 117 United States

Analyses	Unit		As Fed	Dry	C.V.
Dry matter	%			100.0	
Protein (N x 6.25)	%			19.0	
Alanine	%			0.72	
Arginine	%			1.37	
Aspartic acid	%			1.56	
Glutamic acid	%			4.05	
Glycine	%			0.95	
Histidine	%			0.42	
Hydroxyproline	%			0.13	
Isoleucine	%			0.72	
Leucine	%			1.08	
Lysine	%			0.67	
Methionine	%			0.30	
Phenylalanine	%			0.78	
Proline	%			0.76	
Serine	%			0.78	
Threonine	%			0.61	
Tyrosine	%			0.40	
Valine	%			0.84	

ASTER, FREMONT. Aster fremontii

Aster, fremont, aerial part, fresh, (2)
Ref No 2 00 444 United States

Analyses	Unit		As Fed	Dry	C.V.
Dry matter	%			100.0	
Ash	%			8.8	
Ether extract	%			3.1	
Protein (N x 6.25)	%			9.1	
Cattle	dig prot %	*		5.6	
Goats	dig prot %	*		5.1	
Horses	dig prot %	*		5.3	
Rabbits	dig prot %	*		5.7	
Sheep	dig prot %	*		5.5	
Calcium	%			1.26	
Phosphorus	%			0.57	

Aster, fremont, heads, fresh, (2)
Ref No 2 00 445 United States

Analyses	Unit		As Fed	Dry	C.V.
Dry matter	%			100.0	
Calcium	%			0.98	
Phosphorus	%			0.35	

Aster, fremont, leaves, fresh, (2)
Ref No 2 00 446 United States

Analyses	Unit		As Fed	Dry	C.V.
Dry matter	%			100.0	
Ash	%			11.7	
Ether extract	%			4.8	
Protein (N x 6.25)	%			12.2	
Cattle	dig prot %	*		8.3	
Goats	dig prot %	*		7.9	
Horses	dig prot %	*		7.9	
Rabbits	dig prot %	*		8.1	
Sheep	dig prot %	*		8.4	
Calcium	%			1.60	
Phosphorus	%			0.51	

Aster, fremont, stems, fresh, (2)
Ref No 2 00 447 United States

Analyses	Unit		As Fed	Dry	C.V.
Dry matter	%			100.0	
Ash	%			5.8	
Ether extract	%			1.5	
Protein (N x 6.25)	%			5.3	
Cattle	dig prot %	*		2.4	
Goats	dig prot %	*		1.5	
Horses	dig prot %	*		2.0	
Rabbits	dig prot %	*		2.8	
Sheep	dig prot %	*		1.9	
Calcium	%			0.91	
Phosphorus	%			0.73	

ASTER, HEATH. Aster ericoides

Aster, heath, aerial part, fresh, (2)
Ref No 2 00 448 United States

Analyses	Unit		As Fed	Dry	C.V.
Dry matter	%			100.0	
Ash	%			6.4	
Crude fiber	%			21.1	
Ether extract	%			5.0	
N-free extract	%			59.9	
Protein (N x 6.25)	%			7.6	
Cattle	dig prot %	*		4.4	
Goats	dig prot %	*		3.7	
Horses	dig prot %	*		4.0	
Rabbits	dig prot %	*		4.5	
Sheep	dig prot %	*		4.1	
Calcium	%			2.47	
Magnesium	%			0.29	
Phosphorus	%			0.10	
Potassium	%			0.94	

AVENS. Geum spp

Avens, aerial part, (1)
Ref No 1 00 449 United States

Analyses	Unit		As Fed	Dry	C.V.
Dry matter	%		86.6	100.0	
Ash	%		3.4	3.9	
Crude fiber	%		18.4	21.2	
Ether extract	%		15.5	17.9	
N-free extract	%		34.3	39.6	
Protein (N x 6.25)	%		15.1	17.4	
Cattle	dig prot %	*	10.4	12.0	
Goats	dig prot %	*	11.1	12.8	
Horses	dig prot %	*	10.7	12.3	
Rabbits	dig prot %	*	10.5	12.1	
Sheep	dig prot %	*	10.5	12.2	

AVOCADO. Persea spp

Avocado, seeds, extn unspecified grnd, (5)
Ref No 5 00 451 United States

Analyses	Unit		As Fed	Dry	C.V.
Dry matter	%		91.4	100.0	
Ash	%		18.1	19.8	
Crude fiber	%		17.6	19.3	
Sheep	dig coef %		86.	86.	
Ether extract	%		1.1	1.2	
Sheep	dig coef %		80.	80.	
N-free extract	%		36.0	39.3	
Sheep	dig coef %		75.	75.	
Protein (N x 6.25)	%		18.6	20.4	
Sheep	dig coef %		44.	44.	
Sheep	dig prot %		8.2	9.0	
Energy	GE Mcal/kg				
Cattle	DE Mcal/kg	*	2.02	2.22	
Sheep	DE Mcal/kg	*	2.31	2.52	
Cattle	ME Mcal/kg	*	1.66	1.82	
Sheep	ME Mcal/kg	*	1.89	2.07	
Cattle	TDN %	*	45.9	50.2	
Sheep	TDN %		52.3	57.2	

BABASSU. Orbignya spp

Babassu, kernels, extn unspecified grnd, (5)
Ref No 5 00 453 United States

Analyses	Unit		As Fed	Dry	C.V.
Dry matter	%		92.7	100.0	2
Ash	%		5.2	5.6	7
Crude fiber	%		12.4	13.4	12
Sheep	dig coef %		59.	59.	
Ether extract	%		6.8	7.3	20
Sheep	dig coef %		92.	92.	
N-free extract	%		44.4	47.9	
Sheep	dig coef %		82.	82.	
Protein (N x 6.25)	%		23.9	25.8	5
Sheep	dig coef %		86.	86.	
Sheep	dig prot %		20.6	22.2	
Energy	GE Mcal/kg				
Cattle	DE Mcal/kg	*	3.21	3.46	
Sheep	DE Mcal/kg	*	3.45	3.73	
Swine	DE kcal/kg	*	3810.	4110.	
Cattle	ME Mcal/kg	*	2.63	2.84	
Sheep	ME Mcal/kg	*	2.83	3.05	
Swine	ME kcal/kg	*	3459.	3731.	
Cattle	TDN %	*	72.7	78.5	
Sheep	TDN %		78.3	84.5	
Swine	TDN %	*	86.4	93.2	
Calcium	%		0.13	0.14	
Chlorine	%		0.02	0.02	
Copper	mg/kg		41.3	44.5	
Iron	%		0.035	0.038	0
Magnesium	%		0.97	1.04	
Manganese	mg/kg		308.0	332.3	
Phosphorus	%		0.74	0.80	35
Carotene	mg/kg		0.2	0.2	
Riboflavin	mg/kg		1.3	1.4	
Vitamin A equivalent	IU/g		0.4	0.4	
Arginine	%		2.89	3.11	11
Histidine	%		0.40	0.43	
Isoleucine	%		1.00	1.07	17
Leucine	%		1.29	1.40	4
Lysine	%		0.90	0.97	12
Methionine	%		0.30	0.32	18
Phenylalanine	%		0.90	0.97	
Threonine	%		0.60	0.64	9
Tryptophan	%		0.20	0.21	
Tyrosine	%		0.40	0.43	14
Valine	%		1.10	1.18	5

Babassu, kernels, mech-extd grnd, (5)
Babassu oil meal, expeller extracted
Babassu oil meal, hydraulic extracted
Ref No 5 00 454 United States

Analyses	Unit		As Fed	Dry	C.V.
Dry matter	%		91.0	100.0	3
Ash	%		4.6	5.1	9
Crude fiber	%		14.6	16.1	9

Continued

Feed Name or Analyses		Mean As Fed	Mean Dry	C.V. ± %
Ether extract	%	6.0	6.5	23
N-free extract	%	45.9	50.5	
Protein (N x 6.25)	%	19.9	21.8	16
Energy	GE Mcal/kg			
Cattle	DE Mcal/kg	* 3.04	3.34	
Sheep	DE Mcal/kg	* 2.86	3.15	
Swine	DE kcal/kg	*3679.	4046.	
Cattle	ME Mcal/kg	* 2.49	2.74	
Sheep	ME Mcal/kg	* 2.35	2.58	
Swine	ME kcal/kg	*3370.	3705.	
Cattle	TDN %	* 68.9	75.7	
Sheep	TDN %	* 64.9	71.4	
Swine	TDN %	* 83.4	91.8	
Calcium	%	0.18	0.20	
Chlorine	%	0.02	0.02	
Manganese	mg/kg	265.8	292.2	
Phosphorus	%	0.91	1.00	
Choline	mg/kg	719.	791.	
Pantothenic acid	mg/kg	6.2	6.8	
Riboflavin	mg/kg	3.3	3.6	
Vitamin B6	mg/kg	14.56	16.01	

Babassu, kernels, solv-extd grnd, (5)
Babassu oil meal, solvent extracted
Ref No 5 00 455 United States

Feed Name or Analyses		Mean As Fed	Mean Dry	C.V. ± %
Dry matter	%	92.7	100.0	0
Ash	%	6.3	6.8	4
Crude fiber	%	17.6	19.0	5
Ether extract	%	2.3	2.5	7
N-free extract	%	46.8	50.5	
Protein (N x 6.25)	%	19.7	21.3	7
Energy	GE Mcal/kg			
Cattle	DE Mcal/kg	* 2.71	2.93	
Sheep	DE Mcal/kg	* 2.75	2.97	
Swine	DE kcal/kg	*3306.	3566.	
Cattle	ME Mcal/kg	* 2.22	2.40	
Sheep	ME Mcal/kg	* 2.25	2.43	
Swine	ME kcal/kg	*3032.	3270.	
Cattle	TDN %	* 61.5	66.4	
Sheep	TDN %	* 62.4	67.3	
Swine	TDN %	* 75.0	80.9	

Babassu oil meal, expeller extracted –
see Babassu, kernels, mech-extd grnd, (5)

Babassu oil meal, hydraulic extracted –
see Babassu, kernels, mech-extd grnd, (5)

Babassu oil meal, solvent extracted –
see Babassu, kernels, solv-extd grnd, (5)

BABYBONNETS, ROSARY. Coursetia glandulosa

Babybonnets, rosary, seeds w hulls, (5)
Ref No 5 09 023 United States

Feed Name or Analyses		Mean As Fed	Mean Dry	C.V. ± %
Dry matter	%		100.0	
Protein (N x 6.25)	%		39.5	
Alanine	%		1.18	
Arginine	%		2.77	
Aspartic acid	%		2.97	
Glutamic acid	%		5.21	
Glycine	%		1.47	
Histidine	%		0.95	
Hydroxyproline	%		0.01	
Isoleucine	%		1.07	
Leucine	%		1.97	
Lysine	%		1.81	
Methionine	%		0.42	
Phenylalanine	%		1.41	
Proline	%		1.22	
Serine	%		1.43	
Threonine	%		0.98	
Tyrosine	%		1.10	
Valine	%		1.32	

BAHIAGRASS. Paspalum notatum

Bahiagrass, hay, s-c, (1)
Ref No 1 00 462 United States

Feed Name or Analyses		Mean As Fed	Mean Dry	C.V. ± %
Dry matter	%	89.8	100.0	1
Calcium	%	0.45	0.50	15
Iron	%	0.005	0.006	
Magnesium	%	0.17	0.19	50
Phosphorus	%	0.20	0.22	27

Bahiagrass, hay, s-c, mature, (1)
Ref No 1 00 461 United States

Feed Name or Analyses		Mean As Fed	Mean Dry	C.V. ± %
Dry matter	%	90.8	100.0	0
Ash	%	5.4	5.9	8
Crude fiber	%	30.5	33.6	5
Ether extract	%	1.5	1.6	24
N-free extract	%	49.2	54.2	
Protein (N x 6.25)	%	4.3	4.7	8
Cattle	dig prot %	* 0.9	1.0	
Goats	dig prot %	* 0.9	0.9	
Horses	dig prot %	* 1.4	1.5	
Rabbits	dig prot %	* 2.1	2.3	
Sheep	dig prot %	* 0.7	0.8	
Energy	GE Mcal/kg			
Cattle	DE Mcal/kg	* 2.24	2.47	
Sheep	DE Mcal/kg	* 2.18	2.40	
Cattle	ME Mcal/kg	* 1.84	2.02	
Sheep	ME Mcal/kg	* 1.79	1.97	
Cattle	TDN %	* 50.8	55.9	
Sheep	TDN %	* 49.5	54.5	

Bahiagrass, aerial part, fresh, (2)
Ref No 2 00 464 United States

Feed Name or Analyses		Mean As Fed	Mean Dry	C.V. ± %
Dry matter	%	30.3	100.0	6
Ash	%	3.4	11.1	17
Crude fiber	%	9.5	31.5	5
Ether extract	%	0.5	1.6	19
N-free extract	%	14.4	47.6	
Protein (N x 6.25)	%	2.5	8.3	28
Cattle	dig prot %	* 1.5	4.9	
Goats	dig prot %	* 1.3	4.3	
Horses	dig prot %	* 1.4	4.5	
Rabbits	dig prot %	* 1.5	5.0	
Sheep	dig prot %	* 1.4	4.7	
Energy	GE Mcal/kg			
Cattle	DE Mcal/kg	* 0.69	2.29	
Sheep	DE Mcal/kg	* 0.73	2.42	
Cattle	ME Mcal/kg	* 0.57	1.88	
Sheep	ME Mcal/kg	* 0.60	1.98	
Cattle	TDN %	* 15.8	52.0	
Sheep	TDN %	* 16.6	54.9	
Calcium	%	0.13	0.43	
Phosphorus	%	0.04	0.13	

Bahiagrass, aerial part, fresh, mature, (2)
Ref No 2 00 463 United States

Feed Name or Analyses		Mean As Fed	Mean Dry	C.V. ± %
Dry matter	%		100.0	
Ash	%		12.0	9
Crude fiber	%		32.3	4
Ether extract	%		1.8	16
N-free extract	%		46.7	
Protein (N x 6.25)	%		7.2	20
Cattle	dig prot %	*	4.0	
Goats	dig prot %	*	3.3	
Horses	dig prot %	*	3.6	
Rabbits	dig prot %	*	4.2	
Sheep	dig prot %	*	3.7	
Energy	GE Mcal/kg			
Cattle	DE Mcal/kg	*	2.56	
Sheep	DE Mcal/kg	*	2.43	
Cattle	ME Mcal/kg	*	2.10	
Sheep	ME Mcal/kg	*	1.99	
Cattle	TDN %	*	58.1	
Sheep	TDN %	*	55.2	
Calcium	%		0.44	
Phosphorus	%		0.14	

Bahiagrass, aerial part, ensiled, (3)
Ref No 3 00 465 United States

Feed Name or Analyses		Mean As Fed	Mean Dry	C.V. ± %
Dry matter	%	41.7	100.0	
Ash	%	2.6	6.2	
Crude fiber	%	11.3	27.1	
Ether extract	%	0.6	1.5	
N-free extract	%	25.6	61.5	
Protein (N x 6.25)	%	1.5	3.7	
Cattle	dig prot %	* -0.1	-0.3	
Goats	dig prot %	* -0.1	-0.3	
Horses	dig prot %	* -0.1	-0.3	
Sheep	dig prot %	* -0.1	-0.3	
Calcium	%	0.26	0.62	
Magnesium	%	0.01	0.03	
Phosphorus	%	0.09	0.21	

BAKERY. Scientific name not used

Bakery, refuse, dehy, high fat, (4)
Ref No 4 08 342 United States

Feed Name or Analyses		Mean As Fed	Mean Dry	C.V. ± %
Dry matter	%	91.6	100.0	
Ash	%	1.6	1.7	
Crude fiber	%	0.7	0.8	
Ether extract	%	13.7	15.0	
N-free extract	%	64.7	70.6	
Protein (N x 6.25)	%	10.9	11.9	
Cattle	dig prot %	* 6.4	6.9	
Goats	dig prot %	* 7.5	8.1	
Horses	dig prot %	* 7.5	8.1	
Sheep	dig prot %	* 7.5	8.1	
Linoleic	%	.100	.109	
Energy	GE Mcal/kg			
Cattle	DE Mcal/kg	* 3.59	3.92	
Sheep	DE Mcal/kg	* 3.65	3.98	
Swine	DE kcal/kg	*4001.	4368.	
Cattle	ME Mcal/kg	* 2.94	3.21	
Chickens	ME_n kcal/kg	3740.	4083.	

(1) dry forages and roughages (2) pasture, range plants, and forages fed green (3) silages (4) energy feeds (5) protein supplements (6) minerals (7) vitamins (8) additives

Feed Name or Analyses		Mean As Fed	Mean Dry	C.V. ± %
Sheep	ME Mcal/kg	* 2.99	3.27	
Swine	ME kcal/kg	*3745.	4088.	
Cattle	TDN %	* 81.4	88.9	
Sheep	TDN %	* 82.7	90.3	
Swine	TDN %	* 90.7	99.1	

Bakery, refuse, dehy grnd, mx salt declared above 3.5% (4)

Dried bakery product (AAFCO)

Ref No 4 00 466 United States

Feed Name or Analyses		Mean As Fed	Mean Dry	C.V. ± %
Dry matter	%	91.6	100.0	
Ash	%	3.5	3.8	
Crude fiber	%	0.7	0.8	
Ether extract	%	13.7	15.0	
N-free extract	%	62.8	68.6	
Protein (N x 6.25)	%	10.9	11.9	
Cattle	dig prot %	* 6.4	6.9	
Goats	dig prot %	* 7.5	8.1	
Horses	dig prot %	* 7.5	8.1	
Sheep	dig prot %	* 7.5	8.1	
Sugars, total	%	9.2	10.1	
Energy	GE Mcal/kg			
Cattle	DE Mcal/kg	* 3.65	3.98	
Sheep	DE Mcal/kg	* 3.59	3.92	
Swine	DE kcal/kg	*3798.	4148.	
Cattle	ME Mcal/kg	* 2.99	3.27	
Chickens	ME_n kcal/kg	3966.	4332.	
Sheep	ME Mcal/kg	* 2.94	3.22	
Swine	ME kcal/kg	*3555.	3883.	
Cattle	TDN %	* 82.7	90.3	
Sheep	TDN %	* 81.4	89.0	
Swine	TDN %	* 86.1	94.1	
Calcium	%	0.06	0.06	
Cobalt	mg/kg	0.000	0.000	
Copper	mg/kg	0.0	0.0	
Iron	%	0.006	0.007	
Magnesium	%	0.32	0.35	
Manganese	mg/kg	0.0	0.0	
Phosphorus	%	0.47	0.51	
Potassium	%	0.83	0.91	
Sulphur	%	0.02	0.02	
Carotene	mg/kg	4.6	5.0	
Choline	mg/kg	1235.	1349.	
Niacin	mg/kg	19.0	20.8	
Pantothenic acid	mg/kg	14.7	16.0	
Riboflavin	mg/kg	1.9	2.0	
Thiamine	mg/kg	1.5	1.6	
α-tocopherol	mg/kg	410.3	448.1	
Vitamin B_6	mg/kg	26.01	28.42	
Vitamin A	IU/g	3.9	4.2	
Vitamin A equivalent	IU/g	7.7	8.4	
Arginine	%	0.45	0.49	
Cystine	%	0.16	0.17	
Glycine	%	0.90	0.98	
Histidine	%	0.10	0.11	
Leucine	%	0.80	0.87	
Lysine	%	0.30	0.33	
Methionine	%	0.15	0.16	
Threonine	%	0.60	0.66	
Tryptophan	%	0.09	0.10	

Ref No 4 00 466 Canada

Feed Name or Analyses		Mean As Fed	Mean Dry	C.V. ± %
Dry matter	%	94.4	100.0	
Protein (N x 6.25)	%	12.3	13.0	
Cattle	dig prot %	* 7.5	8.0	
Goats	dig prot %	* 8.7	9.2	
Horses	dig prot %	* 8.7	9.2	
Sheep	dig prot %	* 8.7	9.2	
Energy	GE Mcal/kg			
Chickens	ME_n kcal/kg	4089.	4332.	
Calcium	%	0.08	0.08	
Phosphorus	%	0.18	0.19	

BALANITES. Balanites aegyptica

Balanites, seeds w hulls, (5)

Ref No 5 09 034 United States

Feed Name or Analyses		Mean As Fed	Mean Dry	C.V. ± %
Dry matter	%		100.0	
Protein (N x 6.25)	%		22.0	
Alanine	%		0.62	
Arginine	%		2.38	
Aspartic acid	%		1.58	
Glutamic acid	%		3.61	
Glycine	%		1.14	
Histidine	%		0.35	
Hydroxyproline	%		0.09	
Isoleucine	%		0.62	
Leucine	%		1.08	
Lysine	%		0.70	
Methionine	%		0.33	
Phenylalanine	%		0.97	
Proline	%		0.79	
Serine	%		0.64	
Threonine	%		0.53	
Tyrosine	%		0.46	
Valine	%		0.88	

BALSAM, GARDEN. Impatiens balsamina

Balsam, garden, seeds w hulls, (5)

Ref No 5 09 123 United States

Feed Name or Analyses		Mean As Fed	Mean Dry	C.V. ± %
Dry matter	%		100.0	
Protein (N x 6.25)	%		16.0	
Alanine	%		0.53	
Arginine	%		1.28	
Aspartic acid	%		1.25	
Glutamic acid	%		2.50	
Glycine	%		0.82	
Histidine	%		0.34	
Hydroxyproline	%		0.02	
Isoleucine	%		0.59	
Leucine	%		1.02	
Lysine	%		0.74	
Methionine	%		0.27	
Phenylalanine	%		0.82	
Proline	%		0.66	
Serine	%		0.77	
Threonine	%		0.54	
Tyrosine	%		0.48	
Valine	%		0.74	

BALSAMROOT, ARROWLEAF. Balsamorhiza sagittata

Balsamroot, arrowleaf, leaves, mature, (1)

Ref No 1 00 468 United States

Feed Name or Analyses		Mean As Fed	Mean Dry	C.V. ± %
Dry matter	%	91.8	100.0	
Ash	%	13.0	14.2	
Crude fiber	%	15.9	17.3	
Ether extract	%	4.9	5.3	
N-free extract	%	45.5	49.6	
Protein (N x 6.25)	%	12.5	13.6	
Cattle	dig prot %	* 8.0	8.7	
Goats	dig prot %	* 8.5	9.2	
Horses	dig prot %	* 8.3	9.1	
Rabbits	dig prot %	* 8.4	9.2	
Sheep	dig prot %	* 8.1	8.8	

Balsamroot, arrowleaf, leaves, over ripe, (1)

Ref No 1 00 469 United States

Feed Name or Analyses		Mean As Fed	Mean Dry	C.V. ± %
Dry matter	%	92.2	100.0	
Ash	%	12.3	13.3	
Crude fiber	%	18.7	20.3	
Ether extract	%	5.8	6.3	
N-free extract	%	49.8	54.0	
Protein (N x 6.25)	%	5.6	6.1	
Cattle	dig prot %	* 2.0	2.2	
Goats	dig prot %	* 2.1	2.3	
Horses	dig prot %	* 2.5	2.7	
Rabbits	dig prot %	* 3.1	3.4	
Sheep	dig prot %	* 1.9	2.0	

Balsamroot, arrowleaf, stems (1)

Ref No 1 00 471 United States

Feed Name or Analyses		Mean As Fed	Mean Dry	C.V. ± %
Dry matter	%	92.8	100.0	
Ash	%	10.8	11.6	
Crude fiber	%	41.0	44.2	
Ether extract	%	1.2	1.3	
N-free extract	%	35.1	37.8	
Protein (N x 6.25)	%	4.7	5.1	
Cattle	dig prot %	* 1.3	1.4	
Goats	dig prot %	* 1.2	1.3	
Horses	dig prot %	* 1.7	1.9	
Rabbits	dig prot %	* 2.4	2.6	
Sheep	dig prot %	* 1.1	1.1	

Balsamroot, arrowleaf, aerial part, fresh, immature, (2)

Ref No 2 00 472 United States

Feed Name or Analyses		Mean As Fed	Mean Dry	C.V. ± %
Dry matter	%		100.0	
Crude fiber	%		15.2	9
Ether extract	%		5.1	29
Protein (N x 6.25)	%		28.8	12
Cattle	dig prot %	*	22.4	
Goats	dig prot %	*	23.4	
Horses	dig prot %	*	22.0	
Rabbits	dig prot %	*	20.9	
Sheep	dig prot %	*	23.8	
Calcium	%		2.02	14
Phosphorus	%		0.54	21

Balsamroot, arrowleaf, aerial part, fresh, early bloom, (2)

Ref No 2 00 473 United States

Feed Name or Analyses		Mean As Fed	Mean Dry	C.V. ± %
Dry matter	%		100.0	
Crude fiber	%		14.0	
Ether extract	%		5.5	
Protein (N x 6.25)	%		23.9	
Cattle	dig prot %	*	18.2	
Goats	dig prot %	*	18.9	
Horses	dig prot %	*	17.8	

Continued

Feed Name or Analyses				Mean As Fed	Mean Dry	C.V. ± %
Rabbits	dig prot	%	*		17.1	
Sheep	dig prot	%	*		19.3	

Balsamroot, arrowleaf, aerial part, fresh, mid-bloom, (2)
Ref No 2 00 474 United States

Feed Name or Analyses				Mean As Fed	Mean Dry	C.V. ± %
Dry matter		%			100.0	
Crude fiber		%			15.9	
Ether extract		%			4.3	
Protein (N x 6.25)		%			18.7	
Cattle	dig prot	%	*		13.8	
Goats	dig prot	%	*		14.0	
Horses	dig prot	%	*		13.4	
Rabbits	dig prot	%	*		13.1	
Sheep	dig prot	%	*		14.4	
Phosphorus		%			0.31	

Balsamroot, arrowleaf, aerial part, fresh, full bloom, (2)
Ref No 2 00 475 United States

Feed Name or Analyses				Mean As Fed	Mean Dry	C.V. ± %
Dry matter		%			100.0	
Crude fiber		%			17.5	
Ether extract		%			3.8	
Protein (N x 6.25)		%			17.0	
Cattle	dig prot	%	*		12.3	
Goats	dig prot	%	*		12.4	
Horses	dig prot	%	*		12.0	
Rabbits	dig prot	%	*		11.8	
Sheep	dig prot	%	*		12.8	
Calcium		%			1.66	
Phosphorus		%			0.26	

Balsamroot, arrowleaf, aerial part, fresh, milk stage, (2)
Ref No 2 00 476 United States

Feed Name or Analyses				Mean As Fed	Mean Dry	C.V. ± %
Dry matter		%			100.0	
Crude fiber		%			19.1	
Ether extract		%			2.4	
Protein (N x 6.25)		%			13.9	
Cattle	dig prot	%	*		9.7	
Goats	dig prot	%	*		9.5	
Horses	dig prot	%	*		9.3	
Rabbits	dig prot	%	*		9.4	
Sheep	dig prot	%	*		9.9	
Calcium		%			1.40	
Phosphorus		%			0.17	

Balsamroot, arrowleaf, aerial part, fresh, mature, (2)
Ref No 2 00 477 United States

Feed Name or Analyses				Mean As Fed	Mean Dry	C.V. ± %
Dry matter		%			100.0	
Crude fiber		%			21.6	15
Ether extract		%			2.9	8
Protein (N x 6.25)		%			10.0	17
Cattle	dig prot	%	*		6.4	
Goats	dig prot	%	*		5.9	
Horses	dig prot	%	*		6.0	
Rabbits	dig prot	%	*		6.4	
Sheep	dig prot	%	*		6.3	
Calcium		%			1.02	
Phosphorus		%			0.14	

(1) dry forages and roughages (2) pasture, range plants, and forages fed green (3) silages (4) energy feeds (5) protein supplements (6) minerals (7) vitamins (8) additives

Balsamroot, arrowleaf, aerial part, fresh, over ripe, (2)
Ref No 2 00 467 United States

Feed Name or Analyses				Mean As Fed	Mean Dry	C.V. ± %
Dry matter		%		92.9	100.0	
Ash		%		11.8	12.7	
Crude fiber		%		24.2	26.0	
Sheep	dig coef	%		59.	59.	
Ether extract		%		5.8	6.2	
Sheep	dig coef	%		74.	74.	
N-free extract		%		35.8	38.5	
Sheep	dig coef	%		75.	75.	
Protein (N x 6.25)		%		15.4	16.6	
Sheep	dig coef	%		77.	77.	
Cattle	dig prot	%	*	11.1	12.0	
Goats	dig prot	%	*	11.2	12.0	
Horses	dig prot	%	*	10.8	11.6	
Rabbits	dig prot	%	*	10.7	11.5	
Sheep	dig prot	%		11.9	12.8	
Energy	GE Mcal/kg					
Cattle	DE Mcal/kg		*	2.79	3.00	
Sheep	DE Mcal/kg		*	2.76	2.97	
Cattle	ME Mcal/kg		*	2.28	2.46	
Sheep	ME Mcal/kg		*	2.26	2.43	
Cattle	TDN	%	*	63.2	68.0	
Sheep	TDN	%		62.5	67.3	

BALSAMSCALE, PANAMERICAN. Elyonurus tripsacoides

Balsamscale, Panamerican, browse, mature, (1)
Ref No 1 00 480 United States

Feed Name or Analyses				Mean As Fed	Mean Dry	C.V. ± %
Dry matter		%		92.5	100.0	
Ash		%		5.0	5.4	
Crude fiber		%		36.2	39.1	
Ether extract		%		1.6	1.7	
N-free extract		%		44.9	48.5	
Protein (N x 6.25)		%		4.9	5.3	
Cattle	dig prot	%	*	1.4	1.5	
Goats	dig prot	%	*	1.4	1.5	
Horses	dig prot	%	*	1.9	2.0	
Rabbits	dig prot	%	*	2.6	2.8	
Sheep	dig prot	%	*	1.2	1.3	
Energy	GE Mcal/kg					
Cattle	DE Mcal/kg		*	1.89	2.04	
Sheep	DE Mcal/kg		*	2.13	2.30	
Cattle	ME Mcal/kg		*	1.55	1.67	
Sheep	ME Mcal/kg		*	1.74	1.88	
Cattle	TDN	%	*	42.8	46.2	
Sheep	TDN	%	*	48.2	52.1	
Calcium		%		0.23	0.25	
Phosphorus		%		0.09	0.10	

BANANA. Musa spp

Banana, leaves, deribbed dehy grnd, (1)
Ref No 1 00 482 United States

Feed Name or Analyses				Mean As Fed	Mean Dry	C.V. ± %
Dry matter		%			100.0	
Ash		%			12.5	
Crude fiber		%			19.2	
Ether extract		%			4.3	
N-free extract		%			47.4	
Protein (N x 6.25)		%			16.6	
Cattle	dig prot	%	*		11.3	
Goats	dig prot	%	*		12.0	
Horses	dig prot	%	*		11.6	
Rabbits	dig prot	%	*		11.5	
Sheep	dig prot	%	*		11.5	
Energy	GE Mcal/kg					
Sheep	DE Mcal/kg		*		2.66	
Sheep	ME Mcal/kg		*		2.18	
Sheep	TDN	%	*		60.2	
Calcium		%			0.59	
Iron		%			0.017	
Phosphorus		%			0.22	
Carotene		mg/kg			160.3	
Niacin		mg/kg			58.4	
Riboflavin		mg/kg			47.2	
Thiamine		mg/kg			1.3	
Vitamin A equivalent		IU/g			267.2	

Banana, aerial part, fresh, (2)
Ref No 2 00 483 United States

Feed Name or Analyses				Mean As Fed	Mean Dry	C.V. ± %
Dry matter		%		16.0	100.0	
Ash		%		2.1	13.1	
Crude fiber		%		3.8	23.7	
Sheep	dig coef	%		54.	54.	
Ether extract		%		0.1	0.8	
Sheep	dig coef	%		58.	58.	
N-free extract		%		9.0	56.0	
Sheep	dig coef	%		85.	85.	
Protein (N x 6.25)		%		1.0	6.4	
Sheep	dig coef	%		55.	55.	
Cattle	dig prot	%	*	0.5	3.3	
Goats	dig prot	%	*	0.4	2.5	
Horses	dig prot	%	*	0.5	3.0	
Rabbits	dig prot	%	*	0.6	3.6	
Sheep	dig prot	%		0.6	3.5	
Energy	GE Mcal/kg					
Cattle	DE Mcal/kg		*	0.39	2.44	
Sheep	DE Mcal/kg		*	0.46	2.86	
Cattle	ME Mcal/kg		*	0.32	2.00	
Sheep	ME Mcal/kg		*	0.38	2.35	
Cattle	TDN	%	*	8.9	55.3	
Sheep	TDN	%		10.4	65.0	

Banana, fruit, dehy grnd, (4)
Ref No 4 00 484 United States

Feed Name or Analyses				Mean As Fed	Mean Dry	C.V. ± %
Dry matter		%		86.3	100.0	
Ash		%		2.6	3.0	
Crude fiber		%		1.0	1.2	
Sheep	dig coef	%		-323.	-323.	
Ether extract		%		0.5	0.6	
Sheep	dig coef	%		106.	106.	
N-free extract		%		78.6	91.1	
Sheep	dig coef	%		85.	85.	
Protein (N x 6.25)		%		3.5	4.1	
Sheep	dig coef	%		-11.	-11.	
Cattle	dig prot	%	*	-0.1	-0.1	
Goats	dig prot	%	*	0.9	1.0	
Horses	dig prot	%	*	0.9	1.0	
Sheep	dig prot	%		-0.3	-0.4	
Energy	GE Mcal/kg					
Cattle	DE Mcal/kg		*	2.80	3.25	
Sheep	DE Mcal/kg		*	2.83	3.28	
Swine	DE kcal/kg		*	3348.	3879.	
Cattle	ME Mcal/kg		*	2.30	2.67	
Sheep	ME Mcal/kg		*	2.32	2.69	
Swine	ME kcal/kg		*	3186.	3692.	
Cattle	TDN	%	*	63.5	73.6	

Feed Name or Analyses			Mean As Fed	Mean Dry	C.V. ± %
Sheep	TDN %		64.3	74.5	
Swine	TDN %	*	75.9	88.0	

Banana, fruit, fresh, (4)
Ref No 2 00 485 United States

Feed Name or Analyses			Mean As Fed	Mean Dry	C.V. ± %
Dry matter	%		24.3	100.0	
Ash	%		0.8	3.3	
Crude fiber	%		0.5	2.1	
Ether extract	%		0.2	0.8	
N-free extract	%		21.7	89.3	
Protein (N x 6.25)	%		1.1	4.5	
Cattle	dig prot %	*	0.0	0.2	
Goats	dig prot %	*	0.3	1.4	
Horses	dig prot %	*	0.3	1.4	
Sheep	dig prot %	*	0.3	1.4	
Energy	GE Mcal/kg		0.85	3.50	
Cattle	DE Mcal/kg	*	0.90	3.71	
Sheep	DE Mcal/kg	*	0.93	3.83	
Swine	DE kcal/kg	*	916.	3768.	
Cattle	ME Mcal/kg	*	0.74	3.04	
Sheep	ME Mcal/kg	*	0.76	3.14	
Swine	ME kcal/kg	*	871.	3582.	
Cattle	TDN %	*	20.4	84.1	
Sheep	TDN %	*	21.1	86.9	
Swine	TDN %	*	20.8	85.4	
Calcium	%		0.01	0.03	
Iron	%		0.001	0.003	
Phosphorus	%		0.03	0.11	
Potassium	%		0.37	1.52	
Sodium	%		0.00	0.00	
Ascorbic acid	mg/kg		100.0	411.5	
Niacin	mg/kg		7.0	28.8	
Riboflavin	mg/kg		0.6	2.5	
Thiamine	mg/kg		0.5	2.1	
Vitamin A	IU/g		1.9	7.8	

Banana, skins, dehy grnd, (4)
Ref No 4 00 486 United States

Feed Name or Analyses			Mean As Fed	Mean Dry	C.V. ± %
Dry matter	%		88.0	100.0	
Ash	%		9.2	10.5	
Crude fiber	%		7.6	8.6	
Sheep	dig coef %		22.	22.	
Ether extract	%		7.1	8.1	
Sheep	dig coef %		40.	40.	
N-free extract	%		57.3	65.1	
Sheep	dig coef %		80.	80.	
Protein (N x 6.25)	%		6.8	7.7	
Sheep	dig coef %		34.	34.	
Cattle	dig prot %	*	2.7	3.1	
Goats	dig prot %	*	3.8	4.3	
Horses	dig prot %	*	3.8	4.3	
Sheep	dig prot %		2.3	2.6	
Energy	GE Mcal/kg				
Cattle	DE Mcal/kg	*	2.30	2.61	
Sheep	DE Mcal/kg	*	2.48	2.82	
Swine	DE kcal/kg	*	2165.	2460.	
Cattle	ME Mcal/kg	*	1.88	2.14	
Sheep	ME Mcal/kg	*	2.03	2.31	
Swine	ME kcal/kg	*	2045.	2324.	
Cattle	TDN %	*	52.2	59.3	
Sheep	TDN %		56.2	63.9	
Swine	TDN %	*	49.1	55.8	

BANANA, COMMON. Musa paradisiaca sapientum

Banana, common, skins, fresh, (4)
Ref No 2 00 487 United States

Feed Name or Analyses			Mean As Fed	Mean Dry	C.V. ± %
Dry matter	%		16.2	100.0	
Ash	%		1.9	11.7	
Crude fiber	%		1.6	10.0	
Ether extract	%		1.4	8.7	
N-free extract	%		10.3	63.5	
Protein (N x 6.25)	%		1.0	6.1	
Cattle	dig prot %	*	0.3	1.6	
Goats	dig prot %	*	0.5	2.8	
Horses	dig prot %	*	0.5	2.8	
Sheep	dig prot %	*	0.5	2.8	
Calcium	%		0.06	0.35	
Chlorine	%		0.10	0.64	
Magnesium	%		0.04	0.23	
Phosphorus	%		0.05	0.32	
Potassium	%		0.93	5.72	
Carotene	mg/kg		9.7	60.0	
Vitamin A equivalent	IU/g		16.2	100.0	

BARLEY. Hordeum vulgare

Barley, bran, (1)
Ref No 1 00 515 United States

Feed Name or Analyses			Mean As Fed	Mean Dry	C.V. ± %
Dry matter	%		91.0	100.0	
Ash	%		6.4	7.0	
Crude fiber	%		19.3	21.3	
Sheep	dig coef %		52.	52.	
Ether extract	%		3.9	4.3	
Sheep	dig coef %		126.	126.	
N-free extract	%		50.0	54.9	
Sheep	dig coef %		53.	53.	
Protein (N x 6.25)	%		11.4	12.5	
Sheep	dig coef %		56.	56.	
Cattle	dig prot %	*	7.1	7.8	
Goats	dig prot %	*	7.5	8.3	
Horses	dig prot %	*	7.4	8.2	
Rabbits	dig prot %	*	7.6	8.3	
Sheep	dig prot %		6.4	7.0	
Energy	GE Mcal/kg				
Cattle	DE Mcal/kg	*	2.42	2.66	
Sheep	DE Mcal/kg	*	2.38	2.62	
Cattle	ME Mcal/kg	*	1.98	2.18	
Sheep	ME Mcal/kg	*	1.95	2.15	
Cattle	TDN %	*	54.9	60.3	
Sheep	TDN %		54.0	59.4	

Barley, hay, s-c, (1)
Ref No 1 00 495 United States

Feed Name or Analyses			Mean As Fed	Mean Dry	C.V. ± %
Dry matter	%		87.7	100.0	2
Ash	%		6.6	7.6	9
Crude fiber	%		23.7	27.0	10
Sheep	dig coef %		50.	50.	
Ether extract	%		1.9	2.2	17
Sheep	dig coef %		47.	47.	
N-free extract	%		47.8	54.5	
Sheep	dig coef %		67.	67.	
Protein (N x 6.25)	%		7.7	8.7	19
Sheep	dig coef %		56.	56.	
Cattle	dig prot %	*	4.0	4.5	
Goats	dig prot %	*	4.1	4.7	
Horses	dig prot %	*	4.3	5.0	
Rabbits	dig prot %	*	4.8	5.4	
Sheep	dig prot %		4.3	4.9	
Energy	GE Mcal/kg				
Cattle	DE Mcal/kg	*	2.16	2.46	
Sheep	DE Mcal/kg	*	2.21	2.52	
Cattle	ME Mcal/kg	*	1.77	2.02	
Sheep	ME Mcal/kg	*	1.82	2.07	
Cattle	NE_m Mcal/kg	*	1.08	1.23	
Cattle	NE_{gain} Mcal/kg	*	0.50	0.57	
Cattle	$NE_{lactating\ cows}$ Mcal/kg	*	1.18	1.34	
Cattle	TDN %	*	48.8	55.7	
Sheep	TDN %		50.2	57.2	
Calcium	%		0.25	0.29	
Phosphorus	%		0.22	0.25	
Potassium	%		1.30	1.49	

Barley, hay, s-c, immature, (1)
Ref No 1 00 488 United States

Feed Name or Analyses			Mean As Fed	Mean Dry	C.V. ± %
Dry matter	%		88.0	100.0	
Cobalt	mg/kg		0.054	0.062	
Phosphorus	%		0.36	0.41	

Barley, hay, s-c, full bloom, (1)
Ref No 1 00 489 United States

Feed Name or Analyses			Mean As Fed	Mean Dry	C.V. ± %
Dry matter	%		86.9	100.0	2
Ash	%		7.1	8.2	7
Crude fiber	%		26.2	30.1	9
Ether extract	%		1.8	2.1	6
N-free extract	%		44.1	50.7	
Protein (N x 6.25)	%		7.7	8.9	7
Cattle	dig prot %	*	4.0	4.6	
Goats	dig prot %	*	4.2	4.9	
Horses	dig prot %	*	4.4	5.1	
Rabbits	dig prot %	*	4.8	5.5	
Sheep	dig prot %	*	4.0	4.6	
Calcium	%		0.20	0.23	7
Phosphorus	%		0.27	0.31	22

Barley, hay, s-c, milk stage, (1)
Ref No 1 00 490 United States

Feed Name or Analyses			Mean As Fed	Mean Dry	C.V. ± %
Dry matter	%		84.3	100.0	
Ash	%		6.6	7.8	
Crude fiber	%		27.3	32.4	
Sheep	dig coef %		52.	52.	
Ether extract	%		1.6	1.9	
Sheep	dig coef %		44.	44.	
N-free extract	%		41.7	49.5	
Sheep	dig coef %		57.	57.	
Protein (N x 6.25)	%		7.1	8.4	
Sheep	dig coef %		49.	49.	
Cattle	dig prot %	*	3.6	4.2	
Goats	dig prot %	*	3.7	4.4	
Horses	dig prot %	*	3.9	4.7	
Rabbits	dig prot %	*	4.3	5.2	
Sheep	dig prot %		3.5	4.1	
Energy	GE Mcal/kg				
Cattle	DE Mcal/kg	*	1.97	2.34	
Sheep	DE Mcal/kg	*	1.90	2.25	
Cattle	ME Mcal/kg	*	1.62	1.92	
Sheep	ME Mcal/kg	*	1.56	1.85	
Cattle	TDN %	*	44.7	53.0	
Sheep	TDN %		43.0	51.1	

Feed Name or Analyses			Mean As Fed	Mean Dry	C.V. ± %
Barley, hay, s-c, dough stage, (1)					
Ref No 1 00 491			United States		
Dry matter	%		85.8	100.0	
Ash	%		6.2	7.2	
Crude fiber	%		20.8	24.2	
Sheep	dig coef %		47.	47.	
Ether extract	%		2.0	2.3	
Sheep	dig coef %		58.	58.	
N-free extract	%		50.0	58.3	
Sheep	dig coef %		71.	71.	
Protein (N x 6.25)	%		6.9	8.0	
Sheep	dig coef %		55.	55.	
Cattle	dig prot %	*	3.3	3.9	
Goats	dig prot %	*	3.5	4.0	
Horses	dig prot %	*	3.7	4.3	
Rabbits	dig prot %	*	4.2	4.8	
Sheep	dig prot %		3.8	4.4	
Energy	GE Mcal/kg				
Cattle	DE Mcal/kg	*	2.20	2.56	
Sheep	DE Mcal/kg	*	2.28	2.65	
Cattle	ME Mcal/kg	*	1.80	2.10	
Sheep	ME Mcal/kg	*	1.87	2.18	
Cattle	TDN %	*	49.8	58.1	
Sheep	TDN %		51.6	60.2	
Calcium	%		0.16	0.19	
Phosphorus	%		0.25	0.29	

Feed Name or Analyses			Mean As Fed	Mean Dry	C.V. ± %
Barley, hay, s-c, mature, (1)					
Ref No 1 00 492			United States		
Dry matter	%		85.0	100.0	
Ash	%		6.4	7.5	
Crude fiber	%		16.4	19.3	
Sheep	dig coef %		41.	41.	
Ether extract	%		1.3	1.5	
Sheep	dig coef %		46.	46.	
N-free extract	%		54.7	64.4	
Sheep	dig coef %		75.	75.	
Protein (N x 6.25)	%		6.2	7.3	
Sheep	dig coef %		65.	65.	
Cattle	dig prot %	*	2.8	3.3	
Goats	dig prot %	*	2.9	3.4	
Horses	dig prot %	*	3.2	3.7	
Rabbits	dig prot %	*	3.7	4.3	
Sheep	dig prot %		4.0	4.7	
Energy	GE Mcal/kg				
Cattle	DE Mcal/kg	*	2.24	2.63	
Sheep	DE Mcal/kg	*	2.34	2.76	
Cattle	ME Mcal/kg	*	1.84	2.16	
Sheep	ME Mcal/kg	*	1.92	2.26	
Cattle	TDN %	*	50.7	59.7	
Sheep	TDN %		53.1	62.5	

Feed Name or Analyses			Mean As Fed	Mean Dry	C.V. ± %
Barley, hay, s-c, over ripe, (1)					
Ref No 1 00 493			United States		
Dry matter	%		85.0	100.0	
Calcium	%		0.12	0.14	
Phosphorus	%		0.26	0.30	

(1) dry forages and roughages (2) pasture, range plants, and forages fed green (3) silages (4) energy feeds (5) protein supplements (6) minerals (7) vitamins (8) additives

Feed Name or Analyses			Mean As Fed	Mean Dry	C.V. ± %
Barley, hay, s-c, cut 2, (1)					
Ref No 1 00 494			United States		
Dry matter	%		88.0	100.0	
Phosphorus	%		0.25	0.28	

Feed Name or Analyses			Mean As Fed	Mean Dry	C.V. ± %
Barley, hulls, (1)					
Barley hulls (AAFCO)					
Ref No 1 00 496			United States		
Dry matter	%		93.2	100.0	
Ash	%		6.3	6.8	
Crude fiber	%		22.1	23.7	
Ether extract	%		2.1	2.2	
N-free extract	%		55.4	59.4	
Protein (N x 6.25)	%		7.4	7.9	
Cattle	dig prot %	*	3.5	3.8	
Goats	dig prot %	*	3.7	3.9	
Horses	dig prot %	*	3.9	4.2	
Rabbits	dig prot %	*	4.4	4.8	
Sheep	dig prot %	*	3.4	3.7	
Energy	GE Mcal/kg				
Cattle	DE Mcal/kg	*	2.40	2.57	
Sheep	DE Mcal/kg	*	2.47	2.65	
Cattle	ME Mcal/kg	*	1.97	2.11	
Chickens	ME_n kcal/kg		800.	858.	
Sheep	ME Mcal/kg	*	2.03	2.18	
Cattle	TDN %	*	54.2	58.2	
Sheep	TDN %	*	56.1	60.2	

Feed Name or Analyses			Mean As Fed	Mean Dry	C.V. ± %
Barley, malt hulls, (1)					
Malt hulls (AAFCO)					
Ref No 1 00 497			United States		
Dry matter	%		92.9	100.0	3
Ash	%		4.8	5.2	29
Crude fiber	%		21.2	22.8	15
Ether extract	%		1.3	1.4	31
N-free extract	%		55.2	59.4	
Protein (N x 6.25)	%		10.4	11.2	38
Cattle	dig prot %	*	6.2	6.6	
Goats	dig prot %	*	6.5	7.0	
Horses	dig prot %	*	6.5	7.0	
Rabbits	dig prot %	*	6.8	7.3	
Sheep	dig prot %	*	6.1	6.6	
Energy	GE Mcal/kg				
Cattle	DE Mcal/kg	*	2.46	2.65	
Sheep	DE Mcal/kg	*	2.55	2.74	
Cattle	ME Mcal/kg	*	2.02	2.17	
Sheep	ME Mcal/kg	*	2.09	2.25	
Cattle	TDN %	*	55.8	60.1	
Sheep	TDN %	*	57.8	62.2	

Feed Name or Analyses			Mean As Fed	Mean Dry	C.V. ± %
Barley, straw, (1)					
Ref No 1 00 498			United States		
Dry matter	%		86.9	100.0	3
Organic matter	%		79.8	91.8	
Ash	%		6.0	6.9	11
Crude fiber	%		36.2	41.6	10
Sheep	dig coef %		57.	57.	
Ether extract	%		1.7	1.9	11
Sheep	dig coef %		43.	43.	
N-free extract	%		39.5	45.4	
Sheep	dig coef %		47.	47.	
Protein (N x 6.25)	%		3.6	4.1	12
Sheep	dig coef %		19.	19.	
Cattle	dig prot %	*	0.4	0.5	
Goats	dig prot %	*	0.3	0.4	
Horses	dig prot %	*	0.9	1.0	
Rabbits	dig prot %	*	1.6	1.8	
Sheep	dig prot %		0.7	0.8	
Lignin (Ellis)	%		8.1	9.3	
Energy	GE Mcal/kg				
Cattle	DE Mcal/kg	*	1.68	1.94	
Sheep	DE Mcal/kg	*	1.82	2.09	
Cattle	ME Mcal/kg	*	1.38	1.59	
Sheep	ME Mcal/kg	*	1.49	1.72	
Cattle	NE_m Mcal/kg	*	0.88	1.01	
Cattle	NE_{gain} Mcal/kg	*	0.12	0.14	
Cattle	$NE_{lactating\ cows}$ Mcal/kg	*	0.65	0.75	
Cattle	TDN %	*	38.2	43.9	
Sheep	TDN %		41.3	47.5	
Calcium	%		0.31	0.35	18
Chlorine	%		0.59	0.68	
Iron	%		0.029	0.033	21
Magnesium	%		0.12	0.13	24
Manganese	mg/kg		14.4	16.6	53
Phosphorus	%		0.09	0.10	31
Potassium	%		1.63	1.88	12
Sodium	%		0.12	0.14	
Sulphur	%		0.15	0.17	20

Feed Name or Analyses			Mean As Fed	Mean Dry	C.V. ± %
Ref No 1 00 498			Canada		
Dry matter	%		93.4	100.0	1
Crude fiber	%		38.7	41.4	11
Sheep	dig coef %		57.	57.	
Ether extract	%				
Sheep	dig coef %		43.	43.	
N-free extract	%				
Sheep	dig coef %		47.	47.	
Protein (N x 6.25)	%		3.6	3.8	27
Sheep	dig coef %		19.	19.	
Cattle	dig prot %	*	0.3	0.3	
Goats	dig prot %	*	0.1	0.1	
Horses	dig prot %	*	0.7	0.8	
Rabbits	dig prot %	*	1.5	1.6	
Sheep	dig prot %		0.7	0.7	
Energy	GE Mcal/kg				
Sheep	DE Mcal/kg	*	1.96	2.10	
Sheep	ME Mcal/kg	*	1.61	1.72	
Sheep	TDN %		44.4	47.5	
Calcium	%		0.21	0.22	35
Phosphorus	%		0.04	0.05	70
Carotene	mg/kg		2.2	2.3	
Vitamin A equivalent	IU/g		3.6	3.9	

Feed Name or Analyses			Mean As Fed	Mean Dry	C.V. ± %
Barley, aerial part, fresh, (2)					
Ref No 2 00 511			United States		
Dry matter	%		22.0	100.0	
Ash	%		2.5	11.4	
Crude fiber	%		5.1	23.1	
Sheep	dig coef %		59.	59.	
Ether extract	%		0.8	3.6	
Sheep	dig coef %		56.	56.	
N-free extract	%		9.6	43.5	
Sheep	dig coef %		72.	72.	
Protein (N x 6.25)	%		4.0	18.4	
Sheep	dig coef %		71.	71.	
Cattle	dig prot %	*	3.0	13.5	
Goats	dig prot %	*	3.0	13.7	
Horses	dig prot %	*	2.9	13.1	

Feed Name or Analyses				Mean As Fed	Mean Dry	C.V. ± %
Rabbits	dig prot	%	*	2.8	12.9	
Sheep	dig prot	%		2.9	13.0	
Energy		GE Mcal/kg				
Cattle		DE Mcal/kg	*	0.65	2.96	
Sheep		DE Mcal/kg	*	0.61	2.76	
Cattle		ME Mcal/kg	*	0.53	2.42	
Sheep		ME Mcal/kg	*	0.50	2.26	
Cattle		TDN %	*	14.8	67.0	
Sheep		TDN %		13.8	62.5	
Calcium		%		0.10	0.46	
Phosphorus		%		0.08	0.38	
Potassium		%		0.35	1.58	

Barley, aerial part, fresh, immature, (2)

Ref No 2 00 499 United States

Feed Name or Analyses				Mean As Fed	Mean Dry	C.V. ± %
Dry matter		%		22.7	100.0	14
Crude fiber		%				
Sheep	dig coef	%	*	59.	59.	
Ether extract		%				
Sheep	dig coef	%	*	56.	56.	
N-free extract		%				
Sheep	dig coef	%	*	72.	72.	
Protein (N x 6.25)		%				
Sheep	dig coef	%	*	71.	71.	
Calcium		%		0.14	0.61	35
Phosphorus		%		0.11	0.47	24
Potassium		%		0.69	3.06	32

Barley, aerial part, fresh, early bloom, (2)

Ref No 2 00 500 United States

Feed Name or Analyses				Mean As Fed	Mean Dry	C.V. ± %
Dry matter		%		11.7	100.0	
Ash		%		1.9	16.5	9
Crude fiber		%		2.3	19.4	5
Sheep	dig coef	%	*	59.	59.	
Ether extract		%		0.5	4.0	22
Sheep	dig coef	%	*	56.	56.	
N-free extract		%		4.3	36.4	
Sheep	dig coef	%	*	72.	72.	
Protein (N x 6.25)		%		2.8	23.7	18
Sheep	dig coef	%	*	71.	71.	
Cattle	dig prot	%	*	2.1	18.0	
Goats	dig prot	%	*	2.2	18.7	
Horses	dig prot	%	*	2.1	17.7	
Rabbits	dig prot	%	*	2.0	17.0	
Sheep	dig prot	%		2.0	16.8	
Energy		GE Mcal/kg				
Cattle		DE Mcal/kg	*	0.33	2.83	
Sheep		DE Mcal/kg	*	0.31	2.62	
Cattle		ME Mcal/kg	*	0.27	2.32	
Sheep		ME Mcal/kg	*	0.25	2.15	
Cattle		TDN %	*	7.5	64.1	
Sheep		TDN %	*	7.0	59.5	

Barley, aerial part, fresh, mid-bloom, (2)

Ref No 2 00 501 United States

Feed Name or Analyses				Mean As Fed	Mean Dry	C.V. ± %
Dry matter		%		18.6	100.0	5
Ash		%		1.7	9.4	9
Crude fiber		%		4.3	23.0	5
Sheep	dig coef	%		65.	65.	
Ether extract		%		0.8	4.1	13
Sheep	dig coef	%		58.	58.	
N-free extract		%		8.6	46.5	
Sheep	dig coef	%		70.	70.	
Protein (N x 6.25)		%		3.2	17.1	9
Sheep	dig coef	%		73.	73.	
Cattle	dig prot	%	*	2.3	12.4	
Goats	dig prot	%	*	2.3	12.5	
Horses	dig prot	%	*	2.2	12.0	
Rabbits	dig prot	%	*	2.2	11.9	
Sheep	dig prot	%		2.3	12.5	
Energy		GE Mcal/kg				
Cattle		DE Mcal/kg	*	0.56	2.99	
Sheep		DE Mcal/kg	*	0.53	2.88	
Cattle		ME Mcal/kg	*	0.46	2.45	
Sheep		ME Mcal/kg	*	0.44	2.36	
Cattle		TDN %	*	12.6	67.9	
Sheep		TDN %		12.1	65.3	

Barley, aerial part, fresh, full bloom, (2)

Ref No 2 00 502 United States

Feed Name or Analyses				Mean As Fed	Mean Dry	C.V. ± %
Dry matter		%		20.5	100.0	
Calcium		%		0.05	0.23	
Phosphorus		%		0.07	0.33	

Barley, aerial part, fresh, late bloom, (2)

Ref No 2 00 503 United States

Feed Name or Analyses				Mean As Fed	Mean Dry	C.V. ± %
Dry matter		%		23.4	100.0	
Ash		%		2.2	9.4	
Crude fiber		%		7.0	30.0	
Sheep	dig coef	%		56.	56.	
Ether extract		%		0.9	3.8	
Sheep	dig coef	%		62.	62.	
N-free extract		%		9.7	41.4	
Sheep	dig coef	%		73.	73.	
Protein (N x 6.25)		%		3.6	15.4	
Sheep	dig coef	%		70.	70.	
Cattle	dig prot	%	*	2.6	11.0	
Goats	dig prot	%	*	2.6	10.9	
Horses	dig prot	%	*	2.5	10.6	
Rabbits	dig prot	%	*	2.5	10.6	
Sheep	dig prot	%		2.5	10.8	
Energy		GE Mcal/kg				
Cattle		DE Mcal/kg	*	0.70	2.98	
Sheep		DE Mcal/kg	*	0.65	2.78	
Cattle		ME Mcal/kg	*	0.57	2.45	
Sheep		ME Mcal/kg	*	0.53	2.28	
Cattle		TDN %	*	15.8	67.7	
Sheep		TDN %		14.8	63.1	

Barley, aerial part, fresh, milk stage, (2)

Ref No 2 00 504 United States

Feed Name or Analyses				Mean As Fed	Mean Dry	C.V. ± %
Dry matter		%		27.8	100.0	
Ash		%		2.0	7.2	
Crude fiber		%		6.3	22.7	
Sheep	dig coef	%		56.	56.	
Ether extract		%		0.8	2.9	
Sheep	dig coef	%		49.	49.	
N-free extract		%		15.5	55.7	
Sheep	dig coef	%		74.	74.	
Protein (N x 6.25)		%		3.2	11.5	
Sheep	dig coef	%		69.	69.	
Cattle	dig prot	%	*	2.1	7.7	
Goats	dig prot	%	*	2.0	7.3	
Horses	dig prot	%	*	2.0	7.3	
Rabbits	dig prot	%	*	2.1	7.5	
Sheep	dig prot	%		2.2	7.9	
Energy		GE Mcal/kg				
Cattle		DE Mcal/kg	*	0.82	2.93	
Sheep		DE Mcal/kg	*	0.80	2.87	
Cattle		ME Mcal/kg	*	0.67	2.40	
Sheep		ME Mcal/kg	*	0.65	2.35	
Cattle		TDN %	*	18.5	66.5	
Sheep		TDN %		18.1	65.1	

Barley, aerial part, fresh, dough stage, (2)

Ref No 2 00 505 United States

Feed Name or Analyses				Mean As Fed	Mean Dry	C.V. ± %
Dry matter		%		24.1	100.0	
Ash		%		1.5	6.4	5
Crude fiber		%		8.2	34.2	17
Ether extract		%		0.7	3.0	3
N-free extract		%		11.4	47.5	
Protein (N x 6.25)		%		2.1	8.9	13
Cattle	dig prot	%	*	1.3	5.5	
Goats	dig prot	%	*	1.2	4.9	
Horses	dig prot	%	*	1.2	5.1	
Rabbits	dig prot	%	*	1.3	5.5	
Sheep	dig prot	%	*	1.3	5.3	
Energy		GE Mcal/kg				
Cattle		DE Mcal/kg	*	0.71	2.95	
Sheep		DE Mcal/kg	*	0.69	2.88	
Cattle		ME Mcal/kg	*	0.58	2.42	
Sheep		ME Mcal/kg	*	0.57	2.36	
Cattle		TDN %	*	16.1	66.9	
Sheep		TDN %	*	15.7	65.3	
Calcium		%		0.05	0.22	
Phosphorus		%		0.07	0.28	

Barley, aerial part, fresh, mature, (2)

Ref No 2 00 5 6 Canada

Feed Name or Analyses				Mean As Fed	Mean Dry	C.V. ± %
Dry matter		%		89.2	100.0	
Crude fiber		%		24.8	27.8	
Protein (N x 6.25)		%		7.4	8.3	
Cattle	dig prot	%	*	4.4	4.9	
Goats	dig prot	%	*	3.8	4.3	
Horses	dig prot	%	*	4.1	4.6	
Rabbits	dig prot	%	*	4.5	5.1	
Sheep	dig prot	%	*	4.2	4.7	
Calcium		%		0.25	0.28	
Phosphorus		%		0.19	0.21	
Carotene		mg/kg		47.0	52.7	
Vitamin A equivalent		IU/g		78.3	87.8	

Barley, aerial part, fresh, cut 1, (2)

Ref No 2 00 508 United States

Feed Name or Analyses				Mean As Fed	Mean Dry	C.V. ± %
Dry matter		%		25.6	100.0	19
Calcium		%		0.18	0.70	37
Phosphorus		%		0.14	0.56	20
Potassium		%		0.83	3.24	37

Barley, aerial part, fresh, cut 2, (2)

Ref No 2 00 509 United States

Feed Name or Analyses				Mean As Fed	Mean Dry	C.V. ± %
Dry matter		%		22.3	100.0	
Calcium		%		0.21	0.94	
Phosphorus		%		0.13	0.58	
Potassium		%		0.62	2.78	

Feed Name or Analyses			Mean As Fed	Mean Dry	C.V. ± %
Barley, aerial part, fresh, cut 3, (2)					
Ref No 2 00 510			United States		
Dry matter	%		20.7	100.0	
Calcium	%		0.14	0.68	
Phosphorus	%		0.11	0.53	
Potassium	%		0.55	2.66	
Barley, aerial part, ensiled, (3)					
Ref No 3 00 512			United States		
Dry matter	%		25.0	100.0	8
Ash	%		2.6	10.2	13
Crude fiber	%		9.2	37.0	6
Ether extract	%		1.0	4.0	12
N-free extract	%		9.6	38.5	
Protein (N x 6.25)	%		2.6	10.4	11
Cattle	dig prot %	*	1.4	5.7	
Goats	dig prot %	*	1.4	5.7	
Horses	dig prot %	*	1.4	5.7	
Sheep	dig prot %	*	1.4	5.7	
Energy	GE Mcal/kg				
Calcium	%		0.08	0.32	14
Phosphorus	%		0.09	0.34	50
Potassium	%		0.39	1.56	
Ref No 3 00 512			Canada		
Dry matter	%		32.0	100.0	20
Crude fiber	%		9.1	28.3	14
Protein (N x 6.25)	%		3.4	10.5	26
Cattle	dig prot %	*	1.8	5.8	
Goats	dig prot %	*	1.8	5.8	
Horses	dig prot %	*	1.8	5.8	
Sheep	dig prot %	*	1.8	5.8	
Cattle	DE Mcal/kg	*	0.54	2.15	
Sheep	DE Mcal/kg	*	0.56	2.24	
Cattle	ME Mcal/kg	*	0.44	1.76	
Sheep	ME Mcal/kg	*	0.46	1.84	
Cattle	TDN %	*	12.2	48.7	
Sheep	TDN %	*	12.7	50.9	
Calcium	%		0.13	0.41	43
Phosphorus	%		0.08	0.26	23
Carotene	mg/kg		9.5	29.5	48
Vitamin A equivalent	IU/g		15.8	49.2	
Barley, aerial part, ensiled, full bloom, (3)					
Ref No 3 07 896			United States		
Dry matter	%		18.1	100.0	
Ash	%		1.2	6.7	
Crude fiber	%		6.1	33.6	
Ether extract	%		0.7	3.7	
N-free extract	%		8.2	45.4	
Protein (N x 6.25)	%		1.9	10.6	
Cattle	dig prot %	*	1.1	5.9	
Goats	dig prot %	*	1.1	5.9	
Horses	dig prot %	*	1.1	5.9	
Sheep	dig prot %	*	1.1	5.9	
Energy	GE Mcal/kg		0.80	4.42	
Cattle	DE Mcal/kg	*	0.39	2.14	
Sheep	DE Mcal/kg	*	0.41	2.25	
Cattle	ME Mcal/kg	*	0.32	1.75	
Sheep	ME Mcal/kg	*	0.33	1.84	
Cattle	TDN %	*	8.8	48.6	
Sheep	TDN %	*	9.3	51.2	
Barley, aerial part w phosphoric acid preservative added, ensiled, (3)					
Ref No 3 00 513			United States		
Dry matter	%			100.0	
Calcium	%			0.32	
Phosphorus	%			0.74	
Barley, aerial part w molasses added, ensiled, (3)					
Ref No 3 00 514			United States		
Dry matter	%			100.0	
Calcium	%			0.30	
Phosphorus	%			0.28	
Barley, elevator chaff and dust, (4)					
Ref No 4 00 521			United States		
Dry matter	%		90.5	100.0	
Ash	%		5.7	6.3	
Crude fiber	%		11.0	12.2	
Cattle	dig coef %		35.	35.	
Ether extract	%		1.8	2.0	
Cattle	dig coef %		67.	67.	
N-free extract	%		58.2	64.3	
Cattle	dig coef %		56.	56.	
Protein (N x 6.25)	%		13.8	15.2	
Cattle	dig coef %		60.	60.	
Cattle	dig prot %		8.3	9.1	
Goats	dig prot %	*	10.1	11.2	
Horses	dig prot %	*	10.1	11.2	
Sheep	dig prot %	*	10.1	11.2	
Energy	GE Mcal/kg				
Cattle	DE Mcal/kg	*	2.09	2.31	
Swine	DE kcal/kg	*	2319.	2562.	
Sheep	DE Mcal/kg	*	1.98	2.19	
Cattle	ME Mcal/kg	*	1.71	1.90	
Sheep	ME Mcal/kg	*	1.63	1.80	
Swine	ME kcal/kg	*	2155.	2381.	
Cattle	TDN %		47.4	52.4	
Sheep	TDN %	*	45.0	49.7	
Swine	TDN %	*	52.6	58.1	
Barley, flour, coarse bolt, (4)					
Ref No 4 00 522			United States		
Dry matter	%		89.2	100.0	
Ash	%		1.8	2.0	
Crude fiber	%		1.5	1.7	
Sheep	dig coef %		-243.	-243.	
Ether extract	%		2.5	2.8	
Sheep	dig coef %		36.	36.	
N-free extract	%		70.4	78.9	
Sheep	dig coef %		97.	97.	
Protein (N x 6.25)	%		13.1	14.7	
Sheep	dig coef %		70.	70.	
Cattle	dig prot %	*	8.5	9.5	
Goats	dig prot %	*	9.5	10.7	
Horses	dig prot %	*	9.5	10.7	
Sheep	dig prot %		9.2	10.3	
Energy	GE Mcal/kg				
Cattle	DE Mcal/kg	*	2.51	2.81	
Sheep	DE Mcal/kg	*	3.34	3.74	
Swine	DE kcal/kg	*	3684.	4128.	
Cattle	ME Mcal/kg	*	2.06	2.30	
Sheep	ME Mcal/kg	*	2.74	3.07	
Swine	ME kcal/kg	*	3427.	3841.	
Cattle	TDN %	*	56.9	63.7	
Sheep	TDN %		75.8	84.9	
Swine	TDN %	*	83.5	93.6	
Barley, flour by-product mill run, (4)					
Barley mill by-product (AAFCO)					
Barley mixed feed (CFA)					
Ref No 4 00 523			United States		
Dry matter	%		90.0	100.0	
Ash	%		4.1	4.6	
Crude fiber	%		14.1	15.7	
Ether extract	%		2.5	2.8	
N-free extract	%		58.8	65.3	
Protein (N x 6.25)	%		10.5	11.7	
Cattle	dig prot %	*	6.1	6.7	
Goats	dig prot %	*	7.1	7.9	
Horses	dig prot %	*	7.1	7.9	
Sheep	dig prot %	*	7.1	7.9	
Energy	GE Mcal/kg				
Cattle	DE Mcal/kg	*	2.77	3.08	
Sheep	DE Mcal/kg	*	2.97	3.30	
Swine	DE kcal/kg	*	2101.	2335.	
Cattle	ME Mcal/kg	*	2.28	2.53	
Sheep	ME Mcal/kg	*	2.44	2.71	
Swine	ME kcal/kg	*	1968.	2186.	
Cattle	TDN %	*	62.8	69.8	
Sheep	TDN %	*	67.4	74.9	
Swine	TDN %	*	47.7	53.0	
Barley, flour by-product wo hulls, (4)					
Barley middlings					
Ref No 4 00 547			United States		
Dry matter	%		89.3	100.0	
Ash	%		4.2	4.8	
Crude fiber	%		9.2	10.4	
Horses	dig coef %		22.	22.	
Sheep	dig coef %		39.	39.	
Ether extract	%		4.4	5.0	
Horses	dig coef %		47.	47.	
Sheep	dig coef %		97.	97.	
N-free extract	%		56.9	63.7	
Horses	dig coef %		70.	70.	
Sheep	dig coef %		80.	80.	
Protein (N x 6.25)	%		14.5	16.3	
Horses	dig coef %		68.	68.	
Sheep	dig coef %		82.	82.	
Cattle	dig prot %	*	9.8	10.9	
Goats	dig prot %	*	10.8	12.1	
Horses	dig prot %		9.9	11.1	
Sheep	dig prot %		11.9	13.3	
Energy	GE Mcal/kg				
Cattle	DE Mcal/kg	*	2.69	3.01	
Horses	DE Mcal/kg	*	2.49	2.78	
Sheep	DE Mcal/kg	*	3.12	3.49	
Swine	DE kcal/kg	*	2658.	2976.	
Cattle	ME Mcal/kg	*	2.20	2.47	
Horses	ME Mcal/kg	*	2.04	2.28	
Sheep	ME Mcal/kg	*	2.55	2.86	
Swine	ME kcal/kg	*	2464.	2760.	

(1) dry forages and roughages (2) pasture, range plants, and forages fed green (3) silages (4) energy feeds (5) protein supplements (6) minerals (7) vitamins (8) additives

Feed Name or Analyses		Mean As Fed	Mean Dry	C.V. ± %
Cattle	TDN %	* 60.9	68.2	
Horses	TDN %	56.4	63.2	
Sheep	TDN %	70.7	79.1	
Swine	TDN %	* 60.3	67.5	

Barley, grain, (4)
Ref No 4 00 549 United States

Feed Name or Analyses		Mean As Fed	Mean Dry	C.V. ± %
Dry matter	%	89.0	100.0	3
Ash	%	3.0	3.4	71
Crude fiber	%	5.3	6.0	16
Cattle	dig coef %	* 9.	9.	
Horses	dig coef %	46.	46.	
Sheep	dig coef %	44.	44.	
Swine	dig coef %	8.	8.	
Ether extract	%	1.7	1.9	19
Cattle	dig coef %	* 61.	61.	
Horses	dig coef %	34.	34.	
Sheep	dig coef %	75.	75.	
Swine	dig coef %	44.	44.	
N-free extract	%	67.4	75.7	4
Cattle	dig coef %	* 90.	90.	
Horses	dig coef %	88.	88.	
Sheep	dig coef %	89.	89.	
Swine	dig coef %	89.	89.	
Protein (N x 6.25)	%	11.6	13.0	8
Cattle	dig coef %	* 73.	73.	
Horses	dig coef %	82.	82.	
Sheep	dig coef %	77.	77.	
Swine	dig coef %	75.	75.	
Cattle	dig prot %	8.5	9.6	
Goats	dig prot %	* 8.2	9.2	
Horses	dig prot %	9.5	10.7	
Sheep	dig prot %	8.9	10.0	
Swine	dig prot %	8.7	9.8	
Starch	%	64.1	72.0	
Lignin (Ellis)	%	2.8	3.1	
Linoleic	%	.850	.944	
Energy	GE Mcal/kg			
Cattle	DE Mcal/kg	* 3.17	3.56	
Horses	DE Mcal/kg	* 3.20	3.59	
Sheep	DE Mcal/kg	* 3.26	3.66	
Swine	DE kcal/kg	*3122.	3506.	
Cattle	ME Mcal/kg	* 2.60	2.92	
Chickens	ME_n kcal/kg	2635.	2960.	
Horses	ME Mcal/kg	* 2.62	2.95	
Sheep	ME Mcal/kg	* 2.67	3.00	
Swine	ME kcal/kg	*2915.	3274.	
Cattle	NE_m Mcal/kg	* 1.90	2.13	
Cattle	NE_{gain} Mcal/kg	* 1.25	1.40	
Cattle	$NE_{lactating\ cows}$ Mcal/kg	* 2.05	2.30	
Cattle	TDN %	71.9	80.8	
Horses	TDN %	72.5	81.5	
Sheep	TDN %	73.9	83.0	
Swine	TDN %	70.8	79.5	
Calcium	%	0.07	0.08	
Chlorine	%	0.15	0.17	
Copper	mg/kg	5.8	6.5	
Iron	%	0.008	0.009	
Magnesium	%	0.13	0.15	
Manganese	mg/kg	8.0	8.9	
Phosphorus	%	0.40	0.45	
Potassium	%	0.49	0.55	
Sodium	%	0.06	0.07	
Sulphur	%	0.15	0.17	
Choline	mg/kg	988.	1110.	
Niacin	mg/kg	52.9	59.4	
Pantothenic acid	mg/kg	8.1	9.1	
Riboflavin	mg/kg	1.8	2.0	
α-tocopherol	mg/kg	36.0	40.4	
Arginine	%	0.57	0.64	12
Cystine	%	0.19	0.21	15
Glutamic acid	%	3.03	3.40	
Glycine	%	0.40	0.45	
Histidine	%	0.27	0.30	19
Isoleucine	%	0.53	0.60	19
Leucine	%	0.80	0.90	19
Lysine	%	0.47	0.52	40
Methionine	%	0.17	0.20	38
Phenylalanine	%	0.62	0.70	16
Serine	%	0.45	0.50	
Threonine	%	0.36	0.40	14
Tryptophan	%	0.16	0.18	30
Tyrosine	%	0.36	0.40	37
Valine	%	0.62	0.70	12

Ref No 4 00 549 Canada

Feed Name or Analyses		Mean As Fed	Mean Dry	C.V. ± %
Dry matter	%	90.0	100.0	
Crude fiber	%			
Horses	dig coef %	46.	46.	
Sheep	dig coef %	44.	44.	
Swine	dig coef %	8.	8.	
Ether extract	%			
Horses	dig coef %	34.	34.	
Sheep	dig coef %	75.	75.	
Swine	dig coef %	44.	44.	
N-free extract	%			
Horses	dig coef %	88.	88.	
Sheep	dig coef %	89.	89.	
Swine	dig coef %	89.	89.	
Protein (N x 6.25)	%	11.9	13.2	
Horses	dig coef %	82.	82.	
Sheep	dig coef %	77.	77.	
Swine	dig coef %	75.	75.	
Cattle	dig prot %	* 7.3	8.2	
Goats	dig prot %	* 8.4	9.4	
Horses	dig prot %	9.6	10.8	
Sheep	dig prot %	9.1	10.1	
Swine	dig prot %	8.9	9.9	
Energy	GE Mcal/kg			
Cattle	DE Mcal/kg	* 3.21	3.56	
Horses	DE Mcal/kg	* 3.23	3.59	
Sheep	DE Mcal/kg	* 3.29	3.66	
Swine	DE kcal/kg	*3156.	3506.	
Cattle	ME Mcal/kg	* 2.63	2.92	
Chickens	ME_n kcal/kg	2664.	2960.	
Horses	ME Mcal/kg	* 2.65	2.95	
Sheep	ME Mcal/kg	* 2.70	3.00	
Swine	ME kcal/kg	*2945.	3272.	
Cattle	TDN %	72.7	80.8	
Horses	TDN %	73.3	81.5	
Sheep	TDN %	74.7	83.0	
Swine	TDN %	71.6	79.5	

Barley, grain, cooked, (4)
Ref No 4 00 524 United States

Feed Name or Analyses		Mean As Fed	Mean Dry	C.V. ± %
Dry matter	%	85.3	100.0	
Ash	%	2.2	2.6	
Crude fiber	%	6.7	7.9	
Swine	dig coef %	29.	29.	
Ether extract	%	0.8	0.9	
Swine	dig coef %	42.	42.	
N-free extract	%	66.0	77.4	
Swine	dig coef %	87.	87.	
Protein (N x 6.25)	%	9.6	11.2	
Swine	dig coef %	67.	67.	
Cattle	dig prot %	* 5.4	6.3	
Goats	dig prot %	* 6.4	7.5	
Horses	dig prot %	* 6.4	7.5	
Sheep	dig prot %	* 6.4	7.5	
Swine	dig prot %	6.4	7.5	
Energy	GE Mcal/kg			
Cattle	DE Mcal/kg	* 2.98	3.50	
Sheep	DE Mcal/kg	* 3.04	3.57	
Swine	DE kcal/kg	*2933.	3438.	
Cattle	ME Mcal/kg	* 2.45	2.87	
Sheep	ME Mcal/kg	* 2.49	2.92	
Swine	ME kcal/kg	*2749.	3223.	
Cattle	TDN %	* 67.6	79.3	
Sheep	TDN %	* 69.0	80.9	
Swine	TDN %	66.5	78.0	

Barley, grain, cracked, (4)
Ref No 4 05 637 United States

Feed Name or Analyses		Mean As Fed	Mean Dry	C.V. ± %
Dry matter	%	91.8	100.0	
Ash	%	2.6	2.8	
Crude fiber	%	5.0	5.5	
Ether extract	%	1.2	1.3	
Cellulose (Matrone)	%	5.3	5.8	
Lignin (Ellis)	%	1.0	1.1	
Silicon	%	3.00	3.27	

Barley, grain, grnd, (4)
Ref No 4 00 526 United States

Feed Name or Analyses		Mean As Fed	Mean Dry	C.V. ± %
Dry matter	%	91.9	100.0	
Ash	%	3.0	3.3	
Crude fiber	%	5.9	6.4	
Cattle	dig coef %	* 10.	10.	
Sheep	dig coef %	* 45.	45.	
Swine	dig coef %	* 11.	11.	
Ether extract	%	1.3	1.4	
Cattle	dig coef %	* 60.	60.	
Sheep	dig coef %	* 80.	80.	
Swine	dig coef %	* 44.	44.	
N-free extract	%			
Cattle	dig coef %	* 91.	91.	
Sheep	dig coef %	* 91.	91.	
Swine	dig coef %	* 89.	89.	
Protein (N x 6.25)	%			
Cattle	dig coef %	* 75.	75.	
Sheep	dig coef %	* 79.	79.	
Swine	dig coef %	* 77.	77.	
Lignin (Ellis)	%	1.3	1.5	
Energy	GE Mcal/kg	4.14	4.51	
Cattle	GE dig coef %	76.	76.	
Cattle	DE Mcal/kg	3.15	3.43	
Sheep	DE Mcal/kg	* 3.43	3.73	
Swine	DE kcal/kg	*3209.	3492.	
Cattle	ME Mcal/kg	* 2.58	2.81	
Sheep	ME Mcal/kg	* 2.81	3.06	
Cattle	TDN %	74.4	81.0	
Sheep	TDN %	77.8	84.7	
Swine	TDN %	72.8	79.2	
Silicon	%	0.32	0.35	

Ref No 4 00 526 Canada

Feed Name or Analyses		Mean As Fed	Mean Dry	C.V. ± %
Dry matter	%	88.1	100.0	2
Cattle	dig coef %	76.	76.	
Ash	%	3.2	3.6	
Crude fiber	%	5.9	6.7	31

Continued

Feed Name or Analyses		Mean As Fed	Mean Dry	C.V. ± %
Ether extract	%	1.7	1.9	14
N-free extract	%	64.4	73.1	
Protein (N x 6.25)	%	12.9	14.6	17
Cattle	dig prot %	* 8.3	9.4	
Goats	dig prot %	* 9.4	10.6	
Horses	dig prot %	* 9.4	10.6	
Sheep	dig prot %	* 9.4	10.6	
Cellulose (Matrone)	%	6.4	7.3	
Energy	GE Mcal/kg	3.97	4.51	
Cattle	GE dig coef %	76.	76.	
Cattle	DE Mcal/kg	3.02	3.43	
Sheep	DE Mcal/kg	* 3.29	3.73	
Swine	DE kcal/kg	*3075.	3492.	
Cattle	ME Mcal/kg	* 2.47	2.81	
Sheep	ME Mcal/kg	* 2.70	3.06	
Swine	ME kcal/kg	*2861.	3249.	
Cattle	TDN %	71.3	81.0	
Sheep	TDN %	74.6	84.7	
Swine	TDN %	69.7	79.2	
Calcium	%	0.07	0.08	56
Chlorine	%	0.31	0.35	
Phosphorus	%	0.34	0.38	23
Salt (NaCl)	%	0.19	0.21	17
Niacin	mg/kg	53.5	60.8	
Pantothenic acid	mg/kg	6.0	6.8	
Riboflavin	mg/kg	1.4	1.6	
Thiamine	mg/kg	3.9	4.5	
Tryptophan	%	0.15	0.17	

Barley, grain, grnd, feed gr, (4)

Ref No 4 08 343 United States

Feed Name or Analyses		Mean As Fed	Mean Dry	C.V. ± %
Dry matter	%	89.3	100.0	
Ash	%	3.2	3.6	
Crude fiber	%	6.2	6.9	
Ether extract	%	1.9	2.1	
N-free extract	%	66.2	74.1	
Protein (N x 6.25)	%	11.8	13.2	
Cattle	dig prot %	* 7.3	8.2	
Goats	dig prot %	* 8.3	9.3	
Horses	dig prot %	* 8.3	9.3	
Sheep	dig prot %	* 8.3	9.3	
Energy	GE Mcal/kg			
Cattle	DE Mcal/kg	* 2.88	3.23	
Sheep	DE Mcal/kg	* 3.17	3.55	
Swine	DE kcal/kg	*3014.	3375.	
Cattle	ME Mcal/kg	* 2.36	2.65	
Sheep	ME Mcal/kg	* 2.60	2.91	
Swine	ME kcal/kg	*2813.	3150.	
Cattle	TDN %	* 65.3	73.2	
Sheep	TDN %	* 71.8	80.4	
Swine	TDN %	* 68.3	76.5	

Barley, grain, malted, (4)

Ref No 4 08 344 United States

Feed Name or Analyses		Mean As Fed	Mean Dry	C.V. ± %
Dry matter	%	92.4	100.0	2
Ash	%	2.5	2.7	13
Crude fiber	%	6.1	6.6	11
Ether extract	%	2.0	2.2	18
N-free extract	%	69.7	75.5	
Protein (N x 6.25)	%	12.1	13.1	7
Cattle	dig prot %	* 7.4	8.0	
Goats	dig prot %	* 8.5	9.2	
Horses	dig prot %	* 8.5	9.2	
Sheep	dig prot %	* 8.5	9.2	
Energy	GE Mcal/kg			
Cattle	DE Mcal/kg	* 3.39	3.67	
Sheep	DE Mcal/kg	* 3.31	3.58	
Swine	DE kcal/kg	*3240.	3506.	
Cattle	ME Mcal/kg	* 2.78	3.01	
Sheep	ME Mcal/kg	* 2.72	2.94	
Swine	ME kcal/kg	*3025.	3273.	
Cattle	TDN %	* 77.0	83.3	
Sheep	TDN %	* 75.1	81.3	
Swine	TDN %	* 73.5	79.5	
Calcium	%	0.06	0.06	
Phosphorus	%	0.42	0.45	
Potassium	%	0.37	0.40	
Arginine	%	0.65	0.70	10
Histidine	%	0.28	0.30	23
Isoleucine	%	0.55	0.60	17
Leucine	%	0.83	0.90	12
Lysine	%	0.74	0.80	19
Methionine	%	0.18	0.20	19
Phenylalanine	%	0.55	0.60	12
Threonine	%	0.46	0.50	7
Tryptophan	%	0.18	0.20	
Tyrosine	%	0.46	0.50	10
Valine	%	0.65	0.70	15

(1) dry forages and roughages (2) pasture, range plants, and forages fed green (3) silages (4) energy feeds (5) protein supplements (6) minerals (7) vitamins (8) additives

Barley, grain, pearled, (4)

Ref No 4 00 527 United States

Feed Name or Analyses		Mean As Fed	Mean Dry	C.V. ± %
Dry matter	%	88.9	100.0	
Ash	%	0.9	1.0	
Crude fiber	%	0.5	0.6	
Ether extract	%	1.0	1.1	
N-free extract	%	78.3	88.1	
Protein (N x 6.25)	%	8.2	9.2	
Cattle	dig prot %	* 4.0	4.5	
Goats	dig prot %	* 5.1	5.7	
Horses	dig prot %	* 5.1	5.7	
Sheep	dig prot %	* 5.1	5.7	
Energy	GE Mcal/kg	3.49	3.93	
Cattle	DE Mcal/kg	* 3.18	3.57	
Sheep	DE Mcal/kg	* 3.43	3.86	
Swine	DE kcal/kg	*3773.	4244.	
Cattle	ME Mcal/kg	* 2.61	2.93	
Chickens	ME_n kcal/kg	2445.	2750.	
Sheep	ME Mcal/kg	* 2.81	3.17	
Swine	ME kcal/kg	*3552.	3995.	
Cattle	TDN %	* 72.1	81.1	
Sheep	TDN %	* 77.9	87.6	
Swine	TDN %	* 85.6	96.3	
Calcium	%	0.02	0.02	
Iron	%	0.002	0.002	
Phosphorus	%	0.19	0.21	
Potassium	%	0.16	0.18	
Sodium	%	0.00	0.00	
Niacin	mg/kg	27.4	30.8	18
Riboflavin	mg/kg	0.5	0.6	
Thiamine	mg/kg	1.2	1.3	
Arginine	%	0.44	0.50	
Histidine	%	0.18	0.20	
Lysine	%	0.27	0.30	
Methionine	%	0.09	0.10	
Phenylalanine	%	0.62	0.70	
Threonine	%	0.36	0.40	
Valine	%	0.53	0.60	

Barley, grain, rolled, (4)

Ref No 4 00 528 United States

Feed Name or Analyses		Mean As Fed	Mean Dry	C.V. ± %
Dry matter	%	91.4	100.0	1
Ash	%	2.6	2.8	19
Crude fiber	%	5.3	5.8	11
Ether extract	%	1.4	1.6	23
N-free extract	%	70.5	77.2	
Protein (N x 6.25)	%	11.5	12.6	
Sheep	dig coef %	85.	85.	
Cattle	dig prot %	* 6.9	7.6	
Goats	dig prot %	* 8.0	8.8	
Horses	dig prot %	* 8.0	8.8	
Sheep	dig prot %	9.7	10.6	
Lignin (Ellis)	%	1.4	1.6	
Energy	GE Mcal/kg	4.01	4.39	
Cattle	GE dig coef %	74.	74.	
Sheep	GE dig coef %	92.	92.	
Cattle	DE Mcal/kg	2.97	3.25	
Sheep	DE Mcal/kg	3.69	4.04	
Swine	DE kcal/kg	*3251.	3556.	
Cattle	ME Mcal/kg	* 2.43	2.66	
Sheep	ME Mcal/kg	* 3.03	3.31	
Swine	ME kcal/kg	*3038.	3324.	
Cattle	NE_{gain} Mcal/kg	1.06	1.16	
Cattle	TDN %	* 67.2	73.5	
Sheep	TDN %	* 74.7	81.7	
Swine	TDN %	* 73.7	80.7	
Phosphorus	%	0.37	0.41	
Silicon	%	0.32	0.35	

Ref No 4 00 528 Canada

Feed Name or Analyses		Mean As Fed	Mean Dry	C.V. ± %
Dry matter	%	87.4	100.0	
Cattle	dig coef %	75.	75.	
Sheep	dig coef %	91.	91.	
Crude fiber	%	4.9	5.6	
Ether extract	%	1.8	2.1	
Protein (N x 6.25)	%	13.1	15.0	6
Sheep	dig coef %	85.	85.	
Cattle	dig prot %	* 8.5	9.8	
Goats	dig prot %	* 9.6	11.0	
Horses	dig prot %	* 9.6	11.0	
Sheep	dig prot %	11.1	12.7	
Energy	GE Mcal/kg	3.82	4.37	
Cattle	GE dig coef %	74.	74.	
Sheep	GE dig coef %	92.	92.	
Cattle	DE Mcal/kg	2.83	3.23	
Sheep	DE Mcal/kg	3.51	4.02	
Cattle	ME Mcal/kg	* 2.32	2.65	
Sheep	ME Mcal/kg	* 2.88	3.30	
Cattle	NE_{gain} Mcal/kg	1.01	1.16	
Cattle	TDN %	* 70.0	73.2	
Sheep	TDN %	* 79.6	91.1	
Calcium	%	0.05	0.06	
Phosphorus	%	0.40	0.45	

Barley, grain, thresher-run, mn wt 48 lb per bushel mn 10% mx 20% foreign material, (4)

Ref No 4 08 159 Canada

Feed Name or Analyses		Mean As Fed	Mean Dry	C.V. ± %
Dry matter	%	90.0	100.0	
Ash	%	2.6	2.9	
Crude fiber	%	4.5	5.0	
Ether extract	%	0.9	1.0	
N-free extract	%	70.2	78.0	
Protein (N x 6.25)	%	11.8	13.1	
Cattle	dig prot %	* 7.3	8.1	

Feed Name or Analyses		As Fed	Dry	C.V. ± %
Goats	dig prot %	* 8.3	9.3	
Horses	dig prot %	* 8.3	9.3	
Sheep	dig prot %	* 8.3	9.3	
Energy	GE Mcal/kg	4.21	4.68	0
Cattle	DE Mcal/kg	* 2.79	3.10	
Sheep	DE Mcal/kg	* 3.25	3.61	
Swine	DE kcal/kg	*3275.	3639.	
Cattle	ME Mcal/kg	* 2.29	2.54	
Sheep	ME Mcal/kg	* 2.66	2.96	
Swine	ME kcal/kg	*3057.	3397.	
Cattle	TDN %	* 63.2	70.3	
Sheep	TDN %	* 73.6	81.8	
Swine	TDN %	* 74.3	82.5	

Barley, grain, thresher-run, mn wt 48 lb per bushel mx 10% foreign material, (4)

Ref No 4 08 158 Canada

Feed Name or Analyses		As Fed	Dry	C.V. ± %
Dry matter	%	90.0	100.0	0
Ash	%	2.8	3.1	3
Crude fiber	%	4.5	5.0	11
Ether extract	%	1.0	1.1	15
N-free extract	%	70.5	78.3	1
Protein (N x 6.25)	%	11.3	12.6	7
Cattle	dig prot %	* 6.8	7.6	
Goats	dig prot %	* 7.9	8.8	
Horses	dig prot %	* 7.9	8.8	
Sheep	dig prot %	* 7.9	8.8	
Energy	GE Mcal/kg	4.19	4.66	0
Cattle	DE Mcal/kg	* 3.11	3.46	
Sheep	DE Mcal/kg	* 3.25	3.61	
Swine	DE kcal/kg	*3252.	3613.	
Cattle	ME Mcal/kg	* 2.56	2.84	
Sheep	ME Mcal/kg	* 2.67	2.96	
Swine	ME kcal/kg	*3039.	3377.	
Cattle	TDN %	* 70.6	78.5	
Sheep	TDN %	* 73.8	82.0	
Swine	TDN %	* 73.7	81.9	

Barley, grain, thresher-run, mx wt 48 lb per bushel mn 10% mx 20% foreign material, (4)

Ref No 4 08 156 Canada

Feed Name or Analyses		As Fed	Dry	C.V. ± %
Dry matter	%	90.0	100.0	0
Ash	%	2.8	3.1	5
Crude fiber	%	5.9	6.6	6
Ether extract	%	1.0	1.2	20
N-free extract	%	68.5	76.1	1
Protein (N x 6.25)	%	11.7	13.0	5
Cattle	dig prot %	* 7.2	8.0	
Goats	dig prot %	* 8.2	9.1	
Horses	dig prot %	* 8.2	9.1	
Sheep	dig prot %	* 8.2	9.1	
Energy	GE Mcal/kg	4.02	4.46	11
Cattle	DE Mcal/kg	* 3.07	3.41	
Sheep	DE Mcal/kg	* 3.20	3.56	
Swine	DE kcal/kg	*3098.	3442.	
Cattle	ME Mcal/kg	* 2.52	2.80	
Sheep	ME Mcal/kg	* 2.63	2.92	
Swine	ME kcal/kg	*2892.	3214.	
Cattle	TDN %	* 69.7	77.4	
Sheep	TDN %	* 72.6	80.7	
Swine	TDN %	* 70.3	78.1	

Barley, grain, thresher-run, mx wt 48 lb per bushel mx 10% foreign material, (4)

Ref No 4 08 155 Canada

Feed Name or Analyses		As Fed	Dry	C.V. ± %
Dry matter	%	90.0	100.0	
Ash	%	2.8	3.2	
Crude fiber	%	5.1	5.7	
Ether extract	%	0.8	0.9	
N-free extract	%	69.5	77.2	
Protein (N x 6.25)	%	11.8	13.1	
Cattle	dig prot %	* 7.2	8.0	
Goats	dig prot %	* 8.3	9.2	
Horses	dig prot %	* 8.3	9.2	
Sheep	dig prot %	* 8.3	9.2	
Energy	GE Mcal/kg	4.17	4.63	
Cattle	DE Mcal/kg	* 3.05	3.39	
Sheep	DE Mcal/kg	* 3.22	3.58	
Swine	DE kcal/kg	*3178.	3531.	
Cattle	ME Mcal/kg	* 2.50	2.78	
Sheep	ME Mcal/kg	* 2.64	2.93	
Swine	ME kcal/kg	*2967.	3297.	
Cattle	TDN %	* 69.2	76.9	
Sheep	TDN %	* 73.0	81.2	
Swine	TDN %	* 72.1	80.1	

Barley, grain, germinated, (4)

Ref No 4 00 529 United States

Feed Name or Analyses		As Fed	Dry	C.V. ± %
Dry matter	%		100.0	
Biotin	mg/kg		0.90	
Niacin	mg/kg		115.1	
Pantothenic acid	mg/kg		9.9	
Riboflavin	mg/kg		7.3	
Thiamine	mg/kg		9.0	
Vitamin B6	mg/kg		0.44	

Barley, grain, Can 1 feed mn wt 46 lb per bushel mx 4% foreign material, (4)

Ref No 4 00 531 United States

Feed Name or Analyses			As Fed	Dry	C.V. ± %
Dry matter	%			100.0	
Crude fiber	%				
Cattle	dig coef %	*		10.	
Sheep	dig coef %	*		45.	
Ether extract	%				
Cattle	dig coef %	*		60.	
Sheep	dig coef %	*		80.	
N-free extract	%				
Cattle	dig coef %	*		91.	
Sheep	dig coef %	*		91.	
Protein (N x 6.25)	%				
Cattle	dig coef %	*		75.	
Sheep	dig coef %	*		79.	
Energy	GE Mcal/kg	*		3.21	
Cattle	DE Mcal/kg	*		3.66	
Sheep	DE Mcal/kg	*		3.81	
Cattle	ME Mcal/kg	*		3.00	
Chickens	ME_n kcal/kg	*		2758.	
Sheep	ME Mcal/kg	*		3.13	
Cattle	TDN %	*		82.9	
Sheep	TDN %	*		86.4	

Ref No 4 00 531 Canada

Feed Name or Analyses		As Fed	Dry	C.V. ± %
Dry matter	%	88.0	100.0	2
Ash	%	2.3	2.6	6
Crude fiber	%	5.0	5.7	15
Ether extract	%	1.9	2.2	14
N-free extract	%	67.2	76.4	
Protein (N x 6.25)	%	11.6	13.2	12
Cattle	dig prot %	* 7.2	8.2	
Goats	dig prot %	* 8.2	9.3	
Horses	dig prot %	* 8.2	9.3	
Sheep	dig prot %	* 8.2	9.3	
Cellulose (Matrone)	%	4.1	4.7	
Energy	GE Mcal/kg	2.82	3.21	
Cattle	DE Mcal/kg	* 3.22	3.66	
Sheep	DE Mcal/kg	* 3.35	3.81	
Swine	DE kcal/kg	*3185.	3619.	
Cattle	ME Mcal/kg	* 2.64	3.00	
Chickens	ME_n kcal/kg	2427.	2758.	8
Sheep	ME Mcal/kg	* 2.75	3.13	
Swine	ME kcal/kg	*2973.	3378.	
Cattle	TDN %	73.0	82.9	
Sheep	TDN %	76.1	86.4	
Swine	TDN %	* 72.2	82.1	
Calcium	%	0.05	0.06	29
Copper	mg/kg	5.6	6.3	
Magnesium	%	0.10	0.11	
Manganese	mg/kg	13.5	15.3	
Phosphorus	%	0.35	0.40	18
Potassium	%	0.38	0.43	
Selenium	mg/kg	0.365	0.415	
Sodium	%	0.01	0.01	
Zinc	mg/kg	35.1	39.9	
Choline	mg/kg	1.	1.	
Niacin	mg/kg	62.6	71.1	
Pantothenic acid	mg/kg	9.2	10.4	
Riboflavin	mg/kg	1.2	1.4	
Thiamine	mg/kg	0.9	1.0	
Vitamin B6	mg/kg	3.09	3.51	
Alanine	%	0.45	0.52	
Arginine	%	0.54	0.62	
Aspartic acid	%	0.66	0.75	
Glutamic acid	%	2.73	3.10	
Glycine	%	0.48	0.55	
Histidine	%	0.24	0.27	
Isoleucine	%	0.42	0.47	
Leucine	%	0.78	0.89	
Lysine	%	0.39	0.44	
Methionine	%	0.14	0.15	
Phenylalanine	%	0.56	0.64	
Proline	%	1.09	1.24	
Serine	%	0.44	0.51	
Threonine	%	0.37	0.42	
Tyrosine	%	0.29	0.33	
Valine	%	0.56	0.64	

Barley, grain, Can 2 feed mn wt 43 lb per bushel mx 10% foreign material, (4)

Ref No 4 00 532 United States

Feed Name or Analyses			As Fed	Dry	C.V. ± %
Dry matter	%			100.0	
Crude fiber	%				
Cattle	dig coef %	*		10.	
Sheep	dig coef %	*		45.	
Ether extract	%				
Cattle	dig coef %	*		60.	
Sheep	dig coef %	*		80.	
N-free extract	%				
Cattle	dig coef %	*		91.	
Sheep	dig coef %	*		91.	
Protein (N x 6.25)	%				
Cattle	dig coef %	*		75.	
Sheep	dig coef %	*		79.	

Continued

Feed Name or Analyses		Mean As Fed	Mean Dry	C.V. ± %
Energy	GE Mcal/kg			
Cattle	DE Mcal/kg	*	3.66	
Sheep	DE Mcal/kg	*	3.81	
Cattle	ME Mcal/kg	*	3.00	
Sheep	ME Mcal/kg	*	3.12	
Cattle	TDN %	*	82.9	
Sheep	TDN %	*	86.4	

Ref No 4 00 532 Canada

Feed Name or Analyses		Mean As Fed	Mean Dry	C.V. ± %
Dry matter	%	87.4	100.0	2
Ash	%	2.3	2.6	7
Crude fiber	%	5.0	5.7	15
Ether extract	%	1.9	2.2	18
N-free extract	%	66.7	76.4	
Protein (N x 6.25)	%	11.5	13.2	14
Cattle	dig prot %	*7.1	8.1	
Goats	dig prot %	*8.1	9.3	
Horses	dig prot %	*8.1	9.3	
Sheep	dig prot %	*8.1	9.3	
Cellulose (Matrone)	%	4.9	5.6	
Energy	GE Mcal/kg			
Cattle	DE Mcal/kg	*3.19	3.66	
Sheep	DE Mcal/kg	*3.33	3.81	
Swine	DE kcal/kg	*3157.	3614.	
Cattle	ME Mcal/kg	*2.62	3.00	
Sheep	ME Mcal/kg	*2.73	3.12	
Swine	ME kcal/kg	*2947.	3373.	
Cattle	TDN %	72.4	82.9	
Sheep	TDN %	75.5	86.4	
Swine	TDN %	*71.6	82.0	
Calcium	%	0.05	0.06	28
Copper	mg/kg	5.6	6.4	
Iron	%	0.005	0.006	
Magnesium	%	0.14	0.16	
Manganese	mg/kg	14.5	16.6	
Phosphorus	%	0.36	0.41	18
Selenium	mg/kg	0.387	0.443	
Zinc	mg/kg	39.4	45.1	
Niacin	mg/kg	100.5	115.0	
Pantothenic acid	mg/kg	9.6	11.0	
Riboflavin	mg/kg	1.6	1.8	
Thiamine	mg/kg	0.8	0.9	
Vitamin B6	mg/kg	2.99	3.42	
Alanine	%	0.45	0.51	
Arginine	%	0.51	0.59	
Aspartic acid	%	0.72	0.82	
Glutamic acid	%	2.70	3.09	
Glycine	%	0.45	0.51	
Histidine	%	0.21	0.24	
Isoleucine	%	0.41	0.47	
Leucine	%	0.76	0.87	
Lysine	%	0.42	0.48	
Methionine	%	0.12	0.13	
Phenylalanine	%	0.56	0.64	
Proline	%	1.17	1.34	
Serine	%	0.45	0.51	
Threonine	%	0.38	0.43	
Tyrosine	%	0.27	0.31	
Valine	%	0.56	0.64	

(1) dry forages and roughages (2) pasture, range plants, and forages fed green (3) silages (4) energy feeds (5) protein supplements (6) minerals (7) vitamins (8) additives

Barley, grain, Can 3 feed mx 20% foreign material, (4)

Ref No 4 00 533 United States

Feed Name or Analyses		Mean As Fed	Mean Dry	C.V. ± %
Dry matter	%		100.0	
Crude fiber	%			
Cattle	dig coef %	*	10.	
Sheep	dig coef %	*	45.	
Swine	dig coef %	*	11.	
Ether extract	%			
Cattle	dig coef %	*	60.	
Sheep	dig coef %	*	80.	
Swine	dig coef %	*	44.	
N-free extract	%			
Cattle	dig coef %	*	91.	
Sheep	dig coef %	*	91.	
Swine	dig coef %	*	89.	
Protein (N x 6.25)	%			
Cattle	dig coef %	*	75.	
Sheep	dig coef %	*	79.	
Swine	dig coef %	*	77.	
Energy	GE Mcal/kg			
Cattle	DE Mcal/kg	*	3.64	
Sheep	DE Mcal/kg	*	3.80	
Swine	DE kcal/kg	*	3547.	
Cattle	ME Mcal/kg	*	2.98	
Sheep	ME Mcal/kg	*	3.12	
Cattle	TDN %	*	82.5	
Sheep	TDN %	*	86.2	
Swine	TDN %	*	80.4	

Ref No 4 00 533 Canada

Feed Name or Analyses		Mean As Fed	Mean Dry	C.V. ± %
Dry matter	%	87.1	100.0	2
Ash	%	2.3	2.7	7
Crude fiber	%	5.4	6.2	16
Ether extract	%	2.0	2.3	15
N-free extract	%	66.1	75.9	
Protein (N x 6.25)	%	11.2	12.9	9
Cattle	dig prot %	*6.8	7.8	
Goats	dig prot %	*7.8	9.0	
Horses	dig prot %	*7.8	9.0	
Sheep	dig prot %	*7.8	9.0	
Cellulose (Matrone)	%	5.0	5.8	
Energy	GE Mcal/kg			
Cattle	DE Mcal/kg	*3.17	3.64	
Sheep	DE Mcal/kg	*3.31	3.80	
Swine	DE kcal/kg	*3089.	3547.	
Cattle	ME Mcal/kg	*2.60	2.98	
Sheep	ME Mcal/kg	*2.71	3.12	
Swine	ME kcal/kg	*2886.	3313.	
Cattle	TDN %	71.8	82.5	
Sheep	TDN %	75.1	86.2	
Swine	TDN %	70.1	80.4	
Calcium	%	0.08	0.09	75
Copper	mg/kg	5.6	6.4	
Iron	%	0.005	0.006	
Magnesium	%	0.14	0.16	
Manganese	mg/kg	16.1	18.5	
Phosphorus	%	0.34	0.39	21
Selenium	mg/kg	0.482	0.553	
Zinc	mg/kg	33.0	37.9	
Niacin	mg/kg	90.6	104.0	
Pantothenic acid	mg/kg	13.4	15.4	
Riboflavin	mg/kg	1.4	1.6	
Thiamine	mg/kg	0.9	1.0	
Vitamin B6	mg/kg	2.76	3.16	
Alanine	%	0.45	0.51	
Arginine	%	0.50	0.58	
Aspartic acid	%	0.71	0.81	
Glutamic acid	%	2.74	3.14	
Glycine	%	0.45	0.52	
Histidine	%	0.23	0.27	
Isoleucine	%	0.45	0.51	
Leucine	%	0.80	0.92	
Lysine	%	0.42	0.48	
Methionine	%	0.10	0.11	
Phenylalanine	%	0.62	0.71	
Proline	%	1.17	1.34	
Serine	%	0.46	0.53	
Threonine	%	0.39	0.44	
Tyrosine	%	0.23	0.27	
Valine	%	0.57	0.66	

Barley, grain, gr 1 US mn wt 47 lb per bushel mx 1% foreign material, (4)

Ref No 4 00 535 United States

Feed Name or Analyses		Mean As Fed	Mean Dry	C.V. ± %
Dry matter	%	88.4	100.0	2
Ash	%	2.6	2.9	16
Crude fiber	%	5.0	5.7	8
Cattle	dig coef %	*10.	10.	
Sheep	dig coef %	*45.	45.	
Ether extract	%	1.8	2.1	11
Cattle	dig coef %	*60.	60.	
Sheep	dig coef %	*80.	80.	
N-free extract	%	67.6	76.4	
Cattle	dig coef %	*91.	91.	
Sheep	dig coef %	*91.	91.	
Protein (N x 6.25)	%	11.5	13.0	8
Cattle	dig coef %	*75.	75.	
Sheep	dig coef %	*79.	79.	
Cattle	dig prot %	8.6	9.7	
Goats	dig prot %	*8.1	9.1	
Horses	dig prot %	*8.1	9.1	
Sheep	dig prot %	9.1	10.2	
Energy	GE Mcal/kg			
Cattle	DE Mcal/kg	*3.22	3.64	
Sheep	DE Mcal/kg	*3.35	3.79	
Swine	DE kcal/kg	*3163.	3577.	
Cattle	ME Mcal/kg	*2.64	2.99	
Sheep	ME Mcal/kg	*2.75	3.11	
Swine	ME kcal/kg	*2954.	3340.	
Cattle	TDN %	73.1	82.6	
Sheep	TDN %	76.1	86.0	
Swine	TDN %	*71.7	81.1	
Calcium	%	0.24	0.27	
Phosphorus	%	0.36	0.41	
Thiamine	mg/kg	4.3	4.9	11

Barley, grain, gr 2 US mn wt 45 lb per bushel mx 2% foreign material, (4)

Ref No 4 00 536 United States

Feed Name or Analyses		Mean As Fed	Mean Dry	C.V. ± %
Dry matter	%	88.4	100.0	2
Crude fiber	%			
Cattle	dig coef %	*10.	10.	
Sheep	dig coef %	*45.	45.	
Ether extract	%			
Cattle	dig coef %	*60.	60.	
Sheep	dig coef %	*80.	80.	
N-free extract	%			
Cattle	dig coef %	*91.	91.	
Sheep	dig coef %	*91.	91.	
Protein (N x 6.25)	%			
Cattle	dig coef %	*75.	75.	
Sheep	dig coef %	*79.	79.	
Starch	%	53.4	60.4	3

Feed Name or Analyses		Mean As Fed	Mean Dry	C.V. ± %
Calcium	%	0.05	0.06	
Phosphorus	%	0.31	0.35	
Riboflavin	mg/kg	1.4	1.5	
Thiamine	mg/kg	4.7	5.3	11

Barley, grain, gr 3 US mn wt 43 lb per bushel mx 3% foreign material, (4)

Ref No 4 00 537 United States

Feed Name or Analyses		Mean As Fed	Mean Dry	C.V. ± %
Dry matter	%	88.1	100.0	
Ash	%	2.6	3.0	
Crude fiber	%	6.1	6.9	
Cattle	dig coef %	* 10.	10.	
Sheep	dig coef %	* 45.	45.	
Ether extract	%	1.7	2.0	
Cattle	dig coef %	* 60.	60.	
Sheep	dig coef %	* 80.	80.	
N-free extract	%	66.6	75.6	
Cattle	dig coef %	* 91.	91.	
Sheep	dig coef %	* 91.	91.	
Protein (N x 6.25)	%	11.1	12.7	
Cattle	dig coef %	* 75.	75.	
Sheep	dig coef %	* 79.	79.	
Cattle	dig prot %	8.4	9.5	
Goats	dig prot %	* 7.8	8.8	
Horses	dig prot %	* 7.8	8.8	
Sheep	dig prot %	8.8	10.0	
Energy	GE Mcal/kg			
Cattle	DE Mcal/kg	* 3.17	3.60	
Sheep	DE Mcal/kg	* 3.32	3.76	
Swine	DE kcal/kg	*3026.	3435.	
Cattle	ME Mcal/kg	* 2.60	2.95	
Sheep	ME Mcal/kg	* 2.72	3.09	
Swine	ME kcal/kg	*2828.	3210.	
Cattle	TDN %	71.9	81.6	
Sheep	TDN %	75.2	85.4	
Swine	TDN %	* 68.6	77.9	
Calcium	%	0.05	0.06	
Phosphorus	%	0.34	0.39	

Barley, grain, gr 5 US mn wt 36 lb per bushel mx 6% foreign material, (4)

Barley grain, light

Ref No 4 00 540 United States

Feed Name or Analyses		Mean As Fed	Mean Dry	C.V. ± %
Dry matter	%	88.5	100.0	1
Ash	%	3.3	3.7	10
Crude fiber	%	8.2	9.3	8
Cattle	dig coef %	* 10.	10.	
Sheep	dig coef %	* 45.	45.	
Ether extract	%	2.0	2.3	7
Cattle	dig coef %	* 60.	60.	
Sheep	dig coef %	* 80.	80.	
N-free extract	%	63.1	71.3	
Cattle	dig coef %	* 91.	91.	
Sheep	dig coef %	* 91.	91.	
Protein (N x 6.25)	%	11.9	13.4	8
Cattle	dig coef %	* 75.	75.	
Sheep	dig coef %	* 79.	79.	
Cattle	dig prot %	8.9	10.1	
Goats	dig prot %	* 8.4	9.5	
Horses	dig prot %	* 8.4	9.5	
Sheep	dig prot %	9.4	10.6	
Energy	GE Mcal/kg	3.94	4.45	
Cattle	DE Mcal/kg	* 3.08	3.48	
Sheep	DE Mcal/kg	* 3.27	3.70	
Swine	DE kcal/kg	*2761.	3122.	
Cattle	ME Mcal/kg	* 2.53	2.86	

Feed Name or Analyses		Mean As Fed	Mean Dry	C.V. ± %
Sheep	ME Mcal/kg	* 2.68	3.03	
Swine	ME kcal/kg	*2576.	2912.	
Cattle	TDN %	69.9	79.0	
Sheep	TDN %	74.1	83.8	
Swine	TDN %	* 62.6	70.8	

Barley, grain, Pacific coast, (4)

Ref No 4 07 939 United States

Feed Name or Analyses		Mean As Fed	Mean Dry	C.V. ± %
Dry matter	%	88.8	100.0	
Ash	%	2.4	2.7	
Crude fiber	%	6.0	6.7	
Ether extract	%	1.9	2.1	
N-free extract	%	69.8	78.7	
Protein (N x 6.25)	%	8.7	9.8	
Cattle	dig prot %	* 4.5	5.0	
Goats	dig prot %	* 5.5	6.2	
Horses	dig prot %	* 5.5	6.2	
Sheep	dig prot %	* 5.5	6.2	
Linoleic	%	.850	.957	
Energy	GE Mcal/kg			
Cattle	DE Mcal/kg	* 3.05	3.43	
Sheep	DE Mcal/kg	* 3.23	3.64	
Swine	DE kcal/kg	*3054.	3439.	
Cattle	ME Mcal/kg	* 2.50	2.81	
Chickens	ME_n kcal/kg	2616.	2946.	
Sheep	ME Mcal/kg	* 2.65	2.98	
Swine	ME kcal/kg	*2872.	3234.	
Cattle	NE_m Mcal/kg	* 1.71	1.93	
Cattle	NE_{gain} Mcal/kg	* 1.15	1.29	
Cattle	$NE_{lactating\ cows}$ Mcal/kg	* 2.02	2.27	
Cattle	TDN %	* 69.1	77.8	
Sheep	TDN %	* 73.3	82.5	
Swine	TDN %	* 69.3	78.0	
Calcium	%	0.06	0.07	
Phosphorus	%	0.33	0.37	
Choline	mg/kg	1005.	1131.	
Niacin	mg/kg	44.6	50.3	
Pantothenic acid	mg/kg	7.4	8.3	
Riboflavin	mg/kg	1.3	1.5	
α-tocopherol	mg/kg	36.0	40.5	
Arginine	%	0.51	0.57	
Cystine	%	0.15	0.17	
Glycine	%	0.41	0.46	
Lysine	%	0.30	0.34	
Methionine	%	0.13	0.15	
Tryptophan	%	0.11	0.13	

Barley, grain, gr sample US, (4)

Ref No 4 00 534 United States

Feed Name or Analyses		Mean As Fed	Mean Dry	C.V. ± %
Dry matter	%	83.9	100.0	1
Ash	%	2.4	2.9	12
Crude fiber	%	4.6	5.5	12
Ether extract	%	1.6	1.9	16
N-free extract	%	64.6	77.0	
Protein (N x 6.25)	%	10.7	12.7	10
Cattle	dig prot %	* 6.4	7.7	
Goats	dig prot %	* 7.4	8.9	
Horses	dig prot %	* 7.4	8.9	
Sheep	dig prot %	* 7.4	8.9	
Energy	GE Mcal/kg			
Cattle	DE Mcal/kg	* 2.72	3.24	
Sheep	DE Mcal/kg	* 3.03	3.61	
Swine	DE kcal/kg	*3010.	3588.	
Cattle	ME Mcal/kg	* 2.23	2.65	
Sheep	ME Mcal/kg	* 2.49	2.96	
Swine	ME kcal/kg	*2813.	3353.	

Feed Name or Analyses		Mean As Fed	Mean Dry	C.V. ± %
Cattle	TDN %	* 61.6	73.4	
Sheep	TDN %	* 68.7	81.9	
Swine	TDN %	* 68.3	81.4	
Thiamine	mg/kg	5.0	6.0	8

Barley, grain screenings, (4)

Ref No 4 00 542 United States

Feed Name or Analyses		Mean As Fed	Mean Dry	C.V. ± %
Dry matter	%	88.6	100.0	
Ash	%	3.1	3.5	
Crude fiber	%	6.5	7.3	
Cattle	dig coef %	* 10.	10.	
Sheep	dig coef %	-22.	-22.	
Swine	dig coef %	* 11.	11.	
Ether extract	%	2.3	2.6	
Cattle	dig coef %	* 60.	60.	
Sheep	dig coef %	83.	83.	
Swine	dig coef %	* 44.	44.	
N-free extract	%	65.2	73.6	
Cattle	dig coef %	* 91.	91.	
Sheep	dig coef %	93.	93.	
Swine	dig coef %	* 89.	89.	
Protein (N x 6.25)	%	11.5	13.0	
Cattle	dig coef %	* 75.	75.	
Sheep	dig coef %	80.	80.	
Swine	dig coef %	* 77.	77.	
Cattle	dig prot %	8.6	9.7	
Goats	dig prot %	* 8.1	9.1	
Horses	dig prot %	* 8.1	9.1	
Sheep	dig prot %	9.2	10.4	
Swine	dig prot %	8.9	10.0	
Energy	GE Mcal/kg			
Cattle	DE Mcal/kg	* 3.16	3.57	
Sheep	DE Mcal/kg	* 3.20	3.62	
Swine	DE kcal/kg	*3082.	3479.	
Cattle	ME Mcal/kg	* 2.59	2.93	
Sheep	ME Mcal/kg	* 2.63	2.97	
Swine	ME kcal/kg	*2878.	3248.	
Cattle	NE_m Mcal/kg	* 1.64	1.85	
Cattle	NE_{gain} Mcal/kg	* 1.09	1.23	
Cattle	$NE_{lactating\ cows}$ Mcal/kg	* 1.95	2.20	
Cattle	TDN %	71.7	81.0	
Sheep	TDN %	72.7	82.0	
Swine	TDN %	69.9	78.9	

Ref No 4 00 542 Canada

Feed Name or Analyses		Mean As Fed	Mean Dry	C.V. ± %
Dry matter	%	89.9	100.0	
Crude fiber	%	18.0	20.0	
Sheep	dig coef %	-22.	-22.	
Ether extract	%			
Sheep	dig coef %	83.	83.	
N-free extract	%			
Sheep	dig coef %	93.	93.	
Protein (N x 6.25)	%	11.6	12.9	
Sheep	dig coef %	80.	80.	
Cattle	dig prot %	* 7.1	7.9	
Goats	dig prot %	* 8.1	9.1	
Horses	dig prot %	* 8.1	9.1	
Sheep	dig prot %	9.3	10.3	
Energy	GE Mcal/kg			
Cattle	DE Mcal/kg	* 3.11	3.46	
Sheep	DE Mcal/kg	* 3.10	3.45	
Swine	DE kcal/kg	*3029.	3370.	
Cattle	ME Mcal/kg	* 2.55	2.83	
Sheep	ME Mcal/kg	* 2.55	2.83	
Swine	ME kcal/kg	*2829.	3147.	
Cattle	TDN %	70.5	78.4	

Continued

Feed Name or Analyses	Unit		Mean As Fed	Mean Dry	C.V. ± %
Sheep	TDN %		70.4	78.3	
Swine	TDN %		68.7	76.4	
Calcium	%		0.41	0.46	
Phosphorus	%		0.29	0.32	

Barley, groats, (4)
Ref No 4 00 543 United States

Feed Name or Analyses	Unit		Mean As Fed	Mean Dry	C.V. ± %
Dry matter	%		89.6	100.0	2
Ash	%		2.1	2.3	15
Crude fiber	%		2.1	2.4	21
Cattle	dig coef %	*	10.	10.	
Sheep	dig coef %	*	45.	45.	
Ether extract	%		1.9	2.2	17
Cattle	dig coef %	*	60.	60.	
Sheep	dig coef %	*	80.	80.	
N-free extract	%		71.2	79.5	
Cattle	dig coef %	*	91.	91.	
Sheep	dig coef %	*	91.	91.	
Protein (N x 6.25)	%		12.3	13.7	17
Cattle	dig coef %	*	75.	75.	
Sheep	dig coef %	*	79.	79.	
Cattle	dig prot %		9.2	10.3	
Goats	dig prot %	*	8.8	9.8	
Horses	dig prot %	*	8.8	9.8	
Sheep	dig prot %		9.7	10.8	
Pentosans	%		9.2	10.3	10
Starch	%		55.5	62.0	15
Energy	GE Mcal/kg				
Cattle	DE Mcal/kg	*	3.39	3.78	
Sheep	DE Mcal/kg	*	3.48	3.88	
Swine	DE kcal/kg	*	3581.	3998.	
Cattle	ME Mcal/kg	*	2.78	3.10	
Sheep	ME Mcal/kg	*	2.85	3.18	
Swine	ME kcal/kg	*	3339.	3728.	
Cattle	TDN %		76.8	85.7	
Sheep	TDN %		78.9	88.1	
Swine	TDN %	*	81.2	90.7	
Calcium	%		0.07	0.08	36
Magnesium	%		0.12	0.13	
Phosphorus	%		0.36	0.40	23
Potassium	%		0.54	0.60	

Barley, malt, (4)
Ref No 4 08 345 United States

Feed Name or Analyses	Unit		Mean As Fed	Mean Dry	C.V. ± %
Dry matter	%		90.6	100.0	
Ash	%		2.3	2.5	
Crude fiber	%		1.8	2.0	
Ether extract	%		1.6	1.8	
N-free extract	%		70.6	77.9	
Protein (N x 6.25)	%		14.3	15.8	
Cattle	dig prot %	*	9.5	10.5	
Goats	dig prot %	*	10.6	11.7	
Horses	dig prot %	*	10.6	11.7	
Sheep	dig prot %	*	10.6	11.7	
Energy	GE Mcal/kg				
Cattle	DE Mcal/kg	*	3.30	3.64	
Sheep	DE Mcal/kg	*	3.32	3.67	
Swine	DE kcal/kg	*	3664.	4044.	
Cattle	ME Mcal/kg	*	2.70	2.98	
Sheep	ME Mcal/kg	*	2.72	3.01	
Swine	ME kcal/kg	*	3401.	3753.	
Cattle	TDN %	*	74.8	82.6	
Sheep	TDN %	*	75.4	83.2	
Swine	TDN %	*	83.1	91.7	
Calcium	%		0.08	0.09	
Copper	mg/kg		5.5	6.1	
Iron	%		0.006	0.007	
Magnesium	%		0.18	0.20	
Manganese	mg/kg		18.5	20.4	
Phosphorus	%		0.47	0.52	
Potassium	%		0.43	0.47	
Sodium	%		0.08	0.09	

Barley, malt, dehy, (4)
Ref No 4 00 578 United States

Feed Name or Analyses	Unit		Mean As Fed	Mean Dry	C.V. ± %
Dry matter	%		91.0	100.0	2
Ash	%		2.2	2.4	53
Crude fiber	%		3.3	3.6	49
Ether extract	%		1.9	2.1	24
N-free extract	%		69.9	76.8	
Protein (N x 6.25)	%		13.7	15.1	7
Cattle	dig prot %	*	9.0	9.8	
Goats	dig prot %	*	10.0	11.0	
Horses	dig prot %	*	10.0	11.0	
Sheep	dig prot %	*	10.0	11.0	
Pentosans	%		10.5	11.5	1
Energy	GE Mcal/kg				
Cattle	DE Mcal/kg	*	3.18	3.50	
Sheep	DE Mcal/kg	*	3.31	3.64	
Swine	DE kcal/kg	*	3528.	3876.	
Cattle	ME Mcal/kg	*	2.61	2.87	
Sheep	ME Mcal/kg	*	2.72	2.99	
Swine	ME kcal/kg	*	3279.	3604.	
Cattle	TDN %	*	72.2	79.3	
Sheep	TDN %	*	75.2	82.6	
Swine	TDN %	*	80.0	87.9	
Calcium	%		0.06	0.07	36
Cobalt	mg/kg		0.053	0.058	34
Copper	mg/kg		5.5	6.1	38
Iron	%		0.006	0.007	29
Magnesium	%		0.18	0.20	18
Manganese	mg/kg		18.5	20.4	78
Phosphorus	%		0.46	0.51	27
Potassium	%		0.43	0.47	11
Sodium	%		0.08	0.09	60
Choline	mg/kg		895.	984.	65
Niacin	mg/kg		56.7	62.3	29
Pantothenic acid	mg/kg		7.9	8.7	37
Riboflavin	mg/kg		2.9	3.1	33
Thiamine	mg/kg		4.0	4.4	24
Vitamin B_6	mg/kg		7.05	7.75	15
Arginine	%		0.40	0.44	20
Glutamic acid	%		2.60	2.86	19
Histidine	%		0.30	0.33	
Isoleucine	%		0.60	0.66	
Leucine	%		0.70	0.77	
Lysine	%		0.50	0.55	32
Methionine	%		0.20	0.22	
Phenylalanine	%		0.60	0.66	13
Serine	%		0.50	0.55	
Threonine	%		0.40	0.44	
Tryptophan	%		0.20	0.22	
Tyrosine	%		0.10	0.11	80
Valine	%		0.70	0.77	11

Barley, pearl by-product, (4)
Pearl barley by-product (AAFCO)
Barley feed (CFA)
Ref No 4 00 548 United States

Feed Name or Analyses	Unit		Mean As Fed	Mean Dry	C.V. ± %
Dry matter	%		89.6	100.0	2
Ash	%		5.1	5.7	30
Crude fiber	%		10.7	11.9	30
Cattle	dig coef %	*	10.	10.	
Sheep	dig coef %		18.	18.	
Swine	dig coef %		21.	21.	
Ether extract	%		3.5	3.9	12
Cattle	dig coef %	*	60.	60.	
Sheep	dig coef %		90.	90.	
Swine	dig coef %		77.	77.	
N-free extract	%		57.1	63.8	8
Cattle	dig coef %	*	91.	91.	
Sheep	dig coef %		74.	74.	
Swine	dig coef %		84.	84.	
Protein (N x 6.25)	%		13.2	14.7	10
Cattle	dig coef %	*	75.	75.	
Sheep	dig coef %		81.	81.	
Swine	dig coef %		79.	79.	
Cattle	dig prot %		9.9	11.0	
Goats	dig prot %	*	9.6	10.7	
Horses	dig prot %	*	9.6	10.7	
Sheep	dig prot %		10.7	11.9	
Swine	dig prot %		10.4	11.6	
Cellulose (Matrone)	%		4.6	5.1	
Pentosans	%		11.9	13.3	6
Lignin (Ellis)	%		2.7	3.0	
Energy	GE Mcal/kg				
Cattle	DE Mcal/kg	*	2.98	3.33	
Sheep	DE Mcal/kg	*	2.73	3.05	
Swine	DE kcal/kg	*	2939.	3280.	
Cattle	ME Mcal/kg	*	2.44	2.73	
Sheep	ME Mcal/kg	*	2.24	2.50	
Swine	ME kcal/kg	*	2734.	3051.	
Cattle	TDN %		67.6	75.5	
Sheep	TDN %		61.9	69.1	
Swine	TDN %		66.6	74.4	
Calcium	%		0.04	0.05	59
Iron	%		0.010	0.011	21
Magnesium	%		0.16	0.18	
Manganese	mg/kg		30.4	33.9	14
Phosphorus	%		0.41	0.46	22
Potassium	%		0.60	0.67	
Sulphur	%		0.05	0.06	
Carotene	mg/kg		0.4	0.5	
Choline	mg/kg		1194.	1333.	21
Folic acid	mg/kg		0.77	0.85	21
Niacin	mg/kg		63.2	70.6	47
Pantothenic acid	mg/kg		7.7	8.5	46
Riboflavin	mg/kg		2.2	2.4	27
Thiamine	mg/kg		5.9	6.6	37
Vitamin A equivalent	IU/g		0.7	0.8	

Ref No 4 00 548 Canada

Feed Name or Analyses	Unit		Mean As Fed	Mean Dry	C.V. ± %
Dry matter	%		89.7	100.0	
Crude fiber	%				
Sheep	dig coef %		18.	18.	
Swine	dig coef %		21.	21.	
Ether extract	%				
Sheep	dig coef %		90.	90.	
Swine	dig coef %		77.	77.	
N-free extract	%				
Sheep	dig coef %		74.	74.	
Swine	dig coef %		84.	84.	

(1) dry forages and roughages (2) pasture, range plants, and forages fed green (3) silages (4) energy feeds (5) protein supplements (6) minerals (7) vitamins (8) additives

Feed Name or Analyses			Mean As Fed	Mean Dry	C.V. ± %
Protein (N x 6.25)		%	12.5	13.9	
Sheep	dig coef	%	81.	81.	
Swine	dig coef	%	79.	79.	
Cattle	dig prot	%	* 7.9	8.8	
Goats	dig prot	%	* 9.0	10.0	
Horses	dig prot	%	* 9.0	10.0	
Sheep	dig prot	%	10.1	11.3	
Swine	dig prot	%	9.9	11.0	
Energy	GE Mcal/kg				
Cattle	DE Mcal/kg		* 2.99	3.33	
Sheep	DE Mcal/kg		* 2.73	3.05	
Swine	DE kcal/kg		*2942.	3280.	
Cattle	ME Mcal/kg		* 2.45	2.73	
Sheep	ME Mcal/kg		* 2.24	2.50	
Swine	ME kcal/kg		*2742.	3057.	
Cattle	TDN	%	67.7	75.5	
Sheep	TDN	%	62.0	69.1	
Swine	TDN	%	66.7	74.4	
Calcium		%	0.07	0.08	
Phosphorus		%	0.43	0.48	

Barley, brewers grains, dehy, (5)
Barley brewers dried grains
Ref No 5 00 516 United States

Feed Name or Analyses			Mean As Fed	Mean Dry	C.V. ± %
Dry matter		%	91.1	100.0	
Ash		%	3.7	4.1	
Crude fiber		%	18.1	19.9	
Ether extract		%	5.7	6.3	
N-free extract		%	43.6	47.9	
Protein (N x 6.25)		%	20.0	22.0	
Energy	GE Mcal/kg				
Cattle	DE Mcal/kg		* 2.61	2.86	
Sheep	DE Mcal/kg		* 2.68	2.94	
Swine	DE kcal/kg		*2788.	3060.	
Cattle	ME Mcal/kg		* 2.14	2.35	
Sheep	ME Mcal/kg		* 2.20	2.41	
Swine	ME kcal/kg		*2554.	2803.	
Cattle	TDN	%	* 59.0	64.8	
Sheep	TDN	%	* 60.8	66.7	
Swine	TDN	%	* 63.2	69.4	

Barley, distillers grains, dehy, (5)
Barley distillers dried grains (AAFCO)
Ref No 5 00 518 United States

Feed Name or Analyses			Mean As Fed	Mean Dry	C.V. ± %
Dry matter		%	92.0	100.0	
Ash		%	1.8	2.0	
Crude fiber		%	10.1	11.0	
Ether extract		%	11.6	12.6	
N-free extract		%	40.8	44.3	
Protein (N x 6.25)		%	27.7	30.1	
Energy	GE Mcal/kg				
Cattle	DE Mcal/kg		* 2.81	3.05	
Sheep	DE Mcal/kg		* 2.87	3.12	
Swine	DE kcal/kg		*2985.	3245.	
Cattle	ME Mcal/kg		* 2.30	2.50	
Sheep	ME Mcal/kg		* 2.36	2.56	
Swine	ME kcal/kg		*2684.	2917.	
Cattle	TDN	%	* 63.6	69.1	
Sheep	TDN	%	* 65.1	70.8	
Swine	TDN	%	* 67.7	73.6	

Barley, distillers grains, wet, (5)
Ref No 2 00 519 United States

Feed Name or Analyses			Mean As Fed	Mean Dry	C.V. ± %
Dry matter		%	26.0	100.0	
Ash		%	0.8	3.1	
Crude fiber		%	3.6	13.8	
Ether extract		%	3.0	11.5	
N-free extract		%	10.3	39.6	
Protein (N x 6.25)		%	8.3	31.9	
Energy	GE Mcal/kg				
Cattle	DE Mcal/kg		* 0.78	3.01	
Sheep	DE Mcal/kg		* 0.80	3.08	
Swine	DE kcal/kg		* 818.	3148.	
Cattle	ME Mcal/kg		* 0.64	2.47	
Sheep	ME Mcal/kg		* 0.66	2.53	
Swine	ME kcal/kg		* 774.	2978.	
Cattle	TDN	%	* 17.7	68.2	
Sheep	TDN	%	* 18.1	69.8	
Swine	TDN	%	* 18.6	71.4	

Barley, malt sprout cleanings w hulls, dehy, mx 24% protein, (5)
Malt cleanings (AAFCO)
Ref No 5 00 544 United States

Feed Name or Analyses			Mean As Fed	Mean Dry	C.V. ± %
Dry matter		%	92.1	100.0	1
Ash		%	5.3	5.8	12
Crude fiber		%	15.1	16.4	16
Ether extract		%	1.7	1.8	19
N-free extract		%	51.1	55.5	
Protein (N x 6.25)		%	18.9	20.5	15
Energy	GE Mcal/kg				
Cattle	DE Mcal/kg		* 2.84	3.08	
Sheep	DE Mcal/kg		* 2.83	3.07	
Swine	DE kcal/kg		*2972.	3227.	
Cattle	ME Mcal/kg		* 2.33	2.53	
Sheep	ME Mcal/kg		* 2.32	2.52	
Swine	ME kcal/kg		*2732.	2966.	
Cattle	TDN	%	* 64.3	69.8	
Sheep	TDN	%	* 64.2	69.7	
Swine	TDN	%	* 67.4	73.2	

Barley, malt sprouts w hulls, dehy, (5)
Malt sprouts (CFA)
Ref No 5 00 546 Canada

Feed Name or Analyses			Mean As Fed	Mean Dry	C.V. ± %
Dry matter		%	91.5	100.0	
Protein (N x 6.25)		%	25.8	28.2	

Barley, malt sprouts w hulls, dehy, mn 24% protein, (5)
Malt sprouts (AAFCO)
Ref No 5 00 545 United States

Feed Name or Analyses			Mean As Fed	Mean Dry	C.V. ± %
Dry matter		%	91.9	100.0	2
Ash		%	6.2	6.7	8
Crude fiber		%	14.4	15.6	10
Sheep	dig coef	%	64.	64.	
Ether extract		%	1.4	1.6	25
Sheep	dig coef	%	68.	68.	
N-free extract		%	44.2	48.1	
Sheep	dig coef	%	64.	64.	
Protein (N x 6.25)		%	25.7	28.0	6
Sheep	dig coef	%	91.	91.	
Sheep	dig prot	%	23.4	25.5	

Feed Name or Analyses			Mean As Fed	Mean Dry	C.V. ± %
Energy	GE Mcal/kg				
Cattle	DE Mcal/kg		* 2.77	3.01	
Sheep	DE Mcal/kg		* 2.78	3.03	
Swine	DE kcal/kg		*3222.	3507.	
Cattle	ME Mcal/kg		* 2.27	2.47	
Chickens	ME_n kcal/kg		1428.	1554.	
Sheep	ME Mcal/kg		* 2.28	2.48	
Swine	ME kcal/kg		*2911.	3168.	
Cattle	TDN	%	* 62.8	68.4	
Sheep	TDN	%	63.1	68.6	
Swine	TDN	%	* 73.1	79.5	
Calcium		%	0.24	0.26	18
Chlorine		%	0.36	0.39	
Magnesium		%	0.18	0.19	
Manganese		mg/kg	31.4	34.2	33
Phosphorus		%	0.77	0.84	8
Potassium		%	0.21	0.23	
Sodium		%	1.34	1.46	
Sulphur		%	0.79	0.86	
Niacin		mg/kg	57.7	62.8	2
Riboflavin		mg/kg	10.3	11.3	8
Thiamine		mg/kg	9.1	9.9	
Arginine		%	1.21	1.32	
Cystine		%	0.25	0.28	
Lysine		%	1.32	1.43	
Methionine		%	0.30	0.33	
Tryptophan		%	0.41	0.45	

BARLEY, ATLAS 46. Hordeum vulgare

Barley, atlas 46, grain, (4)
Ref No 4 07 861 United States

Feed Name or Analyses			Mean As Fed	Mean Dry	C.V. ± %
Dry matter		%	91.6	100.0	
Ash		%	2.9	3.1	
Crude fiber		%	6.3	6.9	
Ether extract		%	1.9	2.1	
N-free extract		%	72.1	78.7	
Protein (N x 6.25)		%	8.4	9.1	
Cattle	dig prot	%	* 4.0	4.4	
Goats	dig prot	%	* 5.1	5.6	
Horses	dig prot	%	* 5.1	5.6	
Sheep	dig prot	%	* 5.1	5.6	
Energy	GE Mcal/kg				
Cattle	DE Mcal/kg		* 3.24	3.54	
Sheep	DE Mcal/kg		* 3.32	3.63	
Swine	DE kcal/kg		*3413.	3726.	
Cattle	ME Mcal/kg		* 2.66	2.90	
Sheep	ME Mcal/kg		* 2.73	2.98	
Swine	ME kcal/kg		*3058.	3338.	
Cattle	TDN	%	* 73.6	80.3	
Sheep	TDN	%	* 75.4	82.4	
Swine	TDN	%	* 77.4	84.5	

BARLEY, CALIFORNIA MARIOUT. Hordeum vulgare

Barley, California mariout, grain, grnd, (4)
Ref No 4 05 640 United States

Feed Name or Analyses			Mean As Fed	Mean Dry	C.V. ± %
Dry matter		%	92.6	100.0	
Ash		%	5.0	5.5	
Crude fiber		%	9.4	10.2	
Ether extract		%	1.8	1.9	
Lignin (Ellis)		%	2.9	3.2	

Feed Name or Analyses		Mean As Fed	Mean Dry	C.V. ± %

Barley, California mariout, grain, rolled, (4)
Ref No 4 05 639 United States

Feed Name or Analyses		As Fed	Dry	C.V. ± %
Dry matter	%	92.2	100.0	
Ash	%	3.2	3.5	
Crude fiber	%	5.4	5.9	
Ether extract	%	1.1	1.2	
Lignin (Ellis)	%	1.6	1.7	

BARLEY, FOXTAIL. Hordeum jubatum

Barley, foxtail, aerial part, fresh, over ripe, (2)
Ref No 2 00 550 United States

Feed Name or Analyses		As Fed	Dry	C.V. ± %
Dry matter	%		100.0	
Ash	%		16.9	
Crude fiber	%		39.0	
Ether extract	%		0.9	
N-free extract	%		39.4	
Protein (N x 6.25)	%		3.8	
Cattle	dig prot %	*	1.1	
Goats	dig prot %	*	0.1	
Horses	dig prot %	*	0.8	
Rabbits	dig prot %	*	1.6	
Sheep	dig prot %	*	0.5	

BARLEY, HULL-LESS. Hordeum vulgare

Barley, hull-less, grain, (4)
Ref No 4 00 552 United States

Feed Name or Analyses		As Fed	Dry	C.V. ± %
Dry matter	%	85.9	100.0	
Ash	%	2.0	2.3	
Crude fiber	%	1.6	1.9	
Swine	dig coef %	28.	28.	
Ether extract	%	2.0	2.4	
Swine	dig coef %	38.	38.	
N-free extract	%	67.7	78.9	
Swine	dig coef %	95.	95.	
Protein (N x 6.25)	%	12.5	14.6	
Swine	dig coef %	82.	82.	
Cattle	dig prot %	* 8.1	9.4	
Goats	dig prot %	* 9.1	10.6	
Horses	dig prot %	* 9.1	10.6	
Sheep	dig prot %	* 9.1	10.6	
Swine	dig prot %	10.2	11.9	
Energy	GE Mcal/kg			
Sheep	DE Mcal/kg	* 3.18	3.71	
Swine	DE kcal/kg	*3367.	3922.	
Sheep	ME Mcal/kg	* 2.61	3.04	
Swine	ME kcal/kg	*3133.	3650.	
Sheep	TDN %	* 72.2	84.1	
Swine	TDN %	76.4	89.0	

Ref No 4 00 552 Canada

Feed Name or Analyses		As Fed	Dry	C.V. ± %
Dry matter	%	89.3	100.0	
Crude fiber	%	1.6	1.8	
Swine	dig coef %	28.	28.	
Ether extract	%	1.7	1.9	
Swine	dig coef %	38.	38.	

(1) dry forages and roughages (2) pasture, range plants, and forages fed green (3) silages (4) energy feeds (5) protein supplements (6) minerals (7) vitamins (8) additives

Feed Name or Analyses		As Fed	Dry	C.V. ± %
N-free extract	%			
Swine	dig coef %	95.	95.	
Protein (N x 6.25)	%	16.8	18.8	
Swine	dig coef %	82.	82.	
Cattle	dig prot %	* 11.9	13.3	
Goats	dig prot %	* 12.9	14.5	
Horses	dig prot %	* 12.9	14.5	
Sheep	dig prot %	* 12.9	14.5	
Swine	dig prot %	13.7	15.3	
Energy	GE Mcal/kg			
Cattle	DE Mcal/kg	* 3.02	3.51	
Swine	DE kcal/kg	*3497.	3916.	
Cattle	ME Mcal/kg	* 2.47	2.88	
Swine	ME kcal/kg	*3224.	3610.	
Cattle	TDN %	* 68.5	79.7	
Swine	TDN %	79.3	88.8	
Calcium	%	0.02	0.02	
Phosphorus	%	0.40	0.45	

Barley, hull-less, grain, nitrogen fertilized, (4)
Ref No 4 00 551 United States

Feed Name or Analyses		As Fed	Dry	C.V. ± %
Dry matter	%	84.9	100.0	
Ash	%	2.4	2.8	
Crude fiber	%	2.1	2.5	
Swine	dig coef %	31.	31.	
Ether extract	%	2.2	2.6	
Swine	dig coef %	50.	50.	
N-free extract	%	61.9	72.9	
Swine	dig coef %	91.	91.	
Protein (N x 6.25)	%	16.3	19.2	
Swine	dig coef %	74.	74.	
Cattle	dig prot %	* 11.6	13.6	
Goats	dig prot %	* 12.6	14.8	
Horses	dig prot %	* 12.6	14.8	
Sheep	dig prot %	* 12.6	14.8	
Swine	dig prot %	12.1	14.2	
Energy	GE Mcal/kg			
Cattle	DE Mcal/kg	* 3.01	3.54	
Sheep	DE Mcal/kg	* 3.05	3.60	
Swine	DE kcal/kg	*3154.	3715.	
Cattle	ME Mcal/kg	* 2.46	2.90	
Sheep	ME Mcal/kg	* 2.50	2.95	
Swine	ME kcal/kg	*2905.	3422.	
Cattle	TDN %	* 68.3	80.4	
Sheep	TDN %	* 69.2	81.6	
Swine	TDN %	71.5	84.2	

BARLEY, LITTLE. Hordeum pusillum

Barley, little, aerial part, fresh, immature, (2)
Ref No 2 00 553 United States

Feed Name or Analyses		As Fed	Dry	C.V. ± %
Dry matter	%		100.0	
Calcium	%		0.50	23
Magnesium	%		0.15	
Phosphorus	%		0.26	21
Potassium	%		1.44	

BARLEY, MOUSE. Hordeum murinum

Barley, mouse, aerial part, fresh, immature, (2)
Ref No 2 00 560 United States

Feed Name or Analyses		As Fed	Dry	C.V. ± %
Dry matter	%		100.0	
Calcium	%		0.50	
Phosphorus	%		0.23	

BARLEY, WESTERN, Hordeum vulgare

Barley, western, grain, gr 5 US mn wt 36 lb per bushel mx 4% foreign material, (4)
Barley, western, grain, light
Ref No 4 00 566 United States

Feed Name or Analyses		As Fed	Dry	C.V. ± %
Dry matter	%		100.0	
Crude fiber	%			
Cattle	dig coef %	*	10.	
Sheep	dig coef %	*	45.	
Ether extract	%			
Cattle	dig coef %	*	60.	
Sheep	dig coef %	*	80.	
N-free extract	%			
Cattle	dig coef %	*	91.	
Sheep	dig coef %	*	91.	
Protein (N x 6.25)	%			
Cattle	dig coef %	*	75.	
Sheep	dig coef %	*	79.	

BARLEY, WINTER. Hordeum vulgare

Barley, winter, straw, (1)
Ref No 1 00 567 United States

Feed Name or Analyses		As Fed	Dry	C.V. ± %
Dry matter	%	84.1	100.0	
Ash	%	4.9	5.8	
Crude fiber	%	36.7	43.6	
Sheep	dig coef %	56.	56.	
Ether extract	%	1.3	1.5	
Sheep	dig coef %	54.	54.	
N-free extract	%	37.6	44.7	
Sheep	dig coef %	46.	46.	
Protein (N x 6.25)	%	3.7	4.4	
Sheep	dig coef %	18.	18.	
Cattle	dig prot %	* 0.6	0.8	
Goats	dig prot %	* 0.6	0.7	
Horses	dig prot %	* 1.1	1.3	
Rabbits	dig prot %	* 1.7	2.1	
Sheep	dig prot %	0.7	0.8	
Energy	GE Mcal/kg			
Cattle	DE Mcal/kg	* 1.60	1.90	
Sheep	DE Mcal/kg	* 1.76	2.10	
Cattle	ME Mcal/kg	* 1.31	1.56	
Sheep	ME Mcal/kg	* 1.45	1.72	
Cattle	TDN %	* 36.2	43.1	
Sheep	TDN %	40.0	47.6	

Barley, winter, straw, treated w sodium hydroxide wet, (1)
Ref No 2 00 568 United States

Feed Name or Analyses		As Fed	Dry	C.V. ± %
Dry matter	%	24.0	100.0	
Ash	%	1.4	5.8	
Crude fiber	%	13.7	57.1	
Sheep	dig coef %	80.	80.	

Feed Name or Analyses				Mean As Fed	Mean Dry	C.V. ± %
Ether extract		%		0.4	1.7	
Sheep	dig coef	%		35.	35.	
N-free extract		%		7.9	33.0	
Sheep	dig coef	%		44.	44.	
Protein (N x 6.25)		%		0.6	2.4	
Sheep	dig coef	%		-110.	-110.	
Cattle	dig prot	%	*	-0.1	-0.9	
Goats	dig prot	%	*	-0.2	-1.1	
Horses	dig prot	%	*	-0.0	-0.3	
Rabbits	dig prot	%	*	0.1	0.5	
Sheep	dig prot	%		-0.5	-2.6	
Energy	GE	Mcal/kg				
Cattle	DE	Mcal/kg	*	0.54	2.23	
Sheep	DE	Mcal/kg	*	0.62	2.60	
Cattle	ME	Mcal/kg	*	0.44	1.83	
Sheep	ME	Mcal/kg	*	0.51	2.13	
Cattle	TDN	%	*	12.1	50.6	
Sheep	TDN	%		14.1	58.9	

Barley, winter, grain, (4)
Ref No 4 00 569 United States

Feed Name or Analyses				Mean As Fed	Mean Dry	C.V. ± %
Dry matter		%		87.5	100.0	
Ash		%		2.5	2.9	
Crude fiber		%		5.0	5.7	
Swine	dig coef	%		25.	25.	
Ether extract		%		1.8	2.0	
Swine	dig coef	%		1.	1.	
N-free extract		%		69.0	78.8	
Swine	dig coef	%		90.	90.	
Protein (N x 6.25)		%		9.3	10.6	
Swine	dig coef	%		81.	81.	
Cattle	dig prot	%	*	5.0	5.8	
Goats	dig prot	%	*	6.1	6.9	
Horses	dig prot	%	*	6.1	6.9	
Sheep	dig prot	%	*	6.1	6.9	
Swine	dig prot	%		7.5	8.6	
Energy	GE	Mcal/kg				
Cattle	DE	Mcal/kg	*	3.15	3.60	
Sheep	DE	Mcal/kg	*	3.19	3.65	
Swine	DE	kcal/kg	*	3124.	3570.	
Cattle	ME	Mcal/kg	*	2.58	2.95	
Sheep	ME	Mcal/kg	*	2.62	2.99	
Swine	ME	kcal/kg	*	2932.	3351.	
Cattle	TDN	%	*	71.5	81.7	
Sheep	TDN	%	*	72.4	82.7	
Swine	TDN	%		70.9	81.0	

BARLEY, YELLOW. Hordeum vulgare

Barley, yellow, grain, Can 2 CW, (4)
Ref No 4 00 570 United States

Feed Name or Analyses				Mean As Fed	Mean Dry	C.V. ± %
Dry matter		%			100.0	
Crude fiber		%				
Cattle	dig coef	%	*		10.	
Sheep	dig coef	%	*		45.	
Swine	dig coef	%	*		11.	
Ether extract		%				
Cattle	dig coef	%	*		60.	
Sheep	dig coef	%	*		80.	
Swine	dig coef	%	*		44.	
N-free extract		%				
Cattle	dig coef	%	*		91.	
Sheep	dig coef	%	*		91.	
Swine	dig coef	%	*		89.	
Protein (N x 6.25)		%				
Cattle	dig coef	%	*		75.	
Sheep	dig coef	%	*		79.	
Swine	dig coef	%	*		77.	

Barley, yellow, grain, Can 3 CW, (4)
Ref No 4 00 571 United States

Feed Name or Analyses				Mean As Fed	Mean Dry	C.V. ± %
Dry matter		%			100.0	
Crude fiber		%				
Cattle	dig coef	%	*		10.	
Sheep	dig coef	%	*		45.	
Ether extract		%				
Cattle	dig coef	%	*		60.	
Sheep	dig coef	%	*		80.	
N-free extract		%				
Cattle	dig coef	%	*		91.	
Sheep	dig coef	%	*		91.	
Protein (N x 6.25)		%				
Cattle	dig coef	%	*		75.	
Sheep	dig coef	%	*		79.	
Energy	GE	Mcal/kg				
Cattle	DE	Mcal/kg	*		3.59	
Sheep	DE	Mcal/kg	*		3.76	
Cattle	ME	Mcal/kg	*		2.94	
Sheep	ME	Mcal/kg	*		3.08	
Cattle	TDN	%	*		81.4	
Sheep	TDN	%	*		85.3	

Ref No 4 00 571 Canada

Feed Name or Analyses				Mean As Fed	Mean Dry	C.V. ± %
Dry matter		%		86.5	100.0	
Ash		%		2.2	2.5	
Crude fiber		%		6.0	6.9	
Ether extract		%		1.6	1.8	
N-free extract		%		63.5	73.4	
Protein (N x 6.25)		%		13.2	15.3	
Cattle	dig prot	%	*	8.7	10.0	
Goats	dig prot	%	*	9.7	11.2	
Horses	dig prot	%	*	9.7	11.2	
Sheep	dig prot	%	*	9.7	11.2	
Energy	GE	Mcal/kg				
Cattle	DE	Mcal/kg	*	3.11	3.59	
Sheep	DE	Mcal/kg	*	3.25	3.76	
Swine	DE	kcal/kg	*	3039.	3514.	
Cattle	ME	Mcal/kg	*	2.55	2.94	
Sheep	ME	Mcal/kg	*	2.67	3.08	
Swine	ME	kcal/kg	*	2824.	3265.	
Cattle	TDN	%		70.4	81.4	
Sheep	TDN	%		73.8	85.3	
Swine	TDN	%	*	68.9	79.7	

BARLEY, 2-ROW, Hordeum distichon

Barley, 2-row, grain, Can 2 CW mn wt 49 lb per bushel mn 90% purity mx 1.5% foreign material, (4)
Ref No 4 00 572 United States

Feed Name or Analyses				Mean As Fed	Mean Dry	C.V. ± %
Dry matter		%			100.0	
Crude fiber		%				
Cattle	dig coef	%	*		10.	
Sheep	dig coef	%	*		45.	
Swine	dig coef	%	*		11.	
Ether extract		%				
Cattle	dig coef	%	*		60.	
Sheep	dig coef	%	*		80.	
Swine	dig coef	%	*		44.	
N-free extract		%				
Cattle	dig coef	%	*		91.	
Sheep	dig coef	%	*		91.	
Swine	dig coef	%	*		89.	
Protein (N x 6.25)		%				
Cattle	dig coef	%	*		75.	
Sheep	dig coef	%	*		79.	
Swine	dig coef	%	*		77.	
Energy	GE	Mcal/kg				
Cattle	DE	Mcal/kg	*		3.70	
Sheep	DE	Mcal/kg	*		3.84	
Swine	DE	kcal/kg	*		3609.	
Cattle	ME	Mcal/kg	*		3.03	
Sheep	ME	Mcal/kg	*		3.15	
Cattle	TDN	%	*		83.9	
Sheep	TDN	%	*		87.1	
Swine	TDN	%	*		81.9	

Ref No 4 00 572 Canada

Feed Name or Analyses				Mean As Fed	Mean Dry	C.V. ± %
Dry matter		%		86.8	100.0	1
Ash		%		2.1	2.4	9
Crude fiber		%		4.0	4.6	20
Ether extract		%		2.0	2.3	17
N-free extract		%		66.9	77.0	
Protein (N x 6.25)		%		11.8	13.6	8
Cattle	dig prot	%	*	7.4	8.5	
Goats	dig prot	%	*	8.4	9.7	
Horses	dig prot	%	*	8.4	9.7	
Sheep	dig prot	%	*	8.4	9.7	
Energy	GE	Mcal/kg				
Cattle	DE	Mcal/kg	*	3.21	3.70	
Sheep	DE	Mcal/kg	*	3.33	3.84	
Swine	DE	kcal/kg	*	3133.	3609.	
Cattle	ME	Mcal/kg	*	2.63	3.03	
Sheep	ME	Mcal/kg	*	2.73	3.15	
Swine	ME	kcal/kg	*	2921.	3365.	
Cattle	TDN	%		72.8	83.9	
Sheep	TDN	%		75.6	87.1	
Swine	TDN	%		71.0	81.9	
Calcium		%		0.03	0.03	
Phosphorus		%		0.27	0.31	
Alanine		%		0.49	0.56	
Arginine		%		0.55	0.64	
Aspartic acid		%		0.68	0.78	
Glutamic acid		%		3.18	3.66	
Glycine		%		0.47	0.54	
Histidine		%		0.26	0.30	
Hydroxyproline		%		0.38	0.44	
Leucine		%		0.79	0.91	
Lysine		%		0.41	0.47	
Methionine		%		0.14	0.16	
Phenylalanine		%		0.61	0.70	
Proline		%		1.16	1.34	
Serine		%		0.52	0.59	
Threonine		%		0.39	0.45	
Tyrosine		%		0.33	0.38	
Valine		%		0.52	0.59	

Barley, 2-row, grain, Can 3 CW, (4)
Ref No 4 00 573 United States

Feed Name or Analyses				Mean As Fed	Mean Dry	C.V. ± %
Dry matter		%			100.0	
Crude fiber		%				
Cattle	dig coef	%	*		10.	
Sheep	dig coef	%	*		45.	
Swine	dig coef	%	*		11.	
Ether extract		%				
Cattle	dig coef	%	*		60.	
Sheep	dig coef	%	*		80.	

Continued

Feed Name or Analyses		As Fed	Dry	C.V. ± %
Swine	dig coef %	*	44.	
N-free extract	%			
Cattle	dig coef %	*	91.	
Sheep	dig coef %	*	91.	
Swine	dig coef %	*	89.	
Protein (N x 6.25)	%			
Cattle	dig coef %	*	75.	
Sheep	dig coef %	*	79.	
Swine	dig coef %	*	77.	
Energy	GE Mcal/kg			
Cattle	DE Mcal/kg	*	3.70	
Sheep	DE Mcal/kg	*	3.84	
Swine	DE kcal/kg	*	3608.	
Cattle	ME Mcal/kg	*	3.03	
Sheep	ME Mcal/kg	*	3.15	
Cattle	TDN %	*	83.9	
Sheep	TDN %	*	87.1	
Swine	TDN %	*	81.8	

Ref No 4 00 573 Canada

Feed Name or Analyses		As Fed	Dry	C.V. ± %
Dry matter	%	86.5	100.0	0
Ash	%	2.1	2.4	9
Crude fiber	%	4.0	4.6	25
Ether extract	%	2.0	2.3	11
N-free extract	%	66.8	77.2	
Protein (N x 6.25)	%	11.7	13.5	5
Cattle	dig prot %	* 7.3	8.4	
Goats	dig prot %	* 8.3	9.6	
Horses	dig prot %	* 8.3	9.6	
Sheep	dig prot %	* 8.3	9.6	
Energy	GE Mcal/kg			
Cattle	DE Mcal/kg	* 3.20	3.70	
Sheep	DE Mcal/kg	* 3.32	3.84	
Swine	DE kcal/kg	*3121.	3608.	
Cattle	ME Mcal/kg	* 2.62	3.03	
Sheep	ME Mcal/kg	* 2.72	3.15	
Swine	ME kcal/kg	*2911.	3365.	
Cattle	TDN %	72.6	83.9	
Sheep	TDN %	75.3	87.1	
Swine	TDN %	70.8	81.8	
Calcium	%	0.03	0.03	
Phosphorus	%	0.28	0.32	

BARLEY, 6-ROW. Hordeum vulgare

Barley, 6-row, grain, Can 1 CW mn wt 50 lb per bushel mn 95% purity mx 1% foreign material, (4)

Ref No 4 00 574 Canada

Feed Name or Analyses		As Fed	Dry	C.V. ± %
Dry matter	%	86.5	100.0	
Ash	%	2.2	2.5	
Crude fiber	%	4.6	5.3	
Ether extract	%	2.0	2.3	
N-free extract	%	66.9	77.3	
Protein (N x 6.25)	%	10.8	12.5	
Cattle	dig prot %	* 6.5	7.5	
Goats	dig prot %	* 7.5	8.7	
Horses	dig prot %	* 7.5	8.7	
Sheep	dig prot %	* 7.5	8.7	
Energy	GE Mcal/kg			
Cattle	DE Mcal/kg	* 2.84	3.28	
Sheep	DE Mcal/kg	* 3.15	3.64	

(1) dry forages and roughages (2) pasture, range plants, and forages fed green (3) silages (4) energy feeds (5) protein supplements (6) minerals (7) vitamins (8) additives

Feed Name or Analyses		As Fed	Dry	C.V. ± %
Swine	DE kcal/kg	*3155.	3647.	
Cattle	ME Mcal/kg	* 2.33	2.69	
Sheep	ME Mcal/kg	* 2.58	2.98	
Swine	ME kcal/kg	*2949.	3409.	
Cattle	TDN %	* 64.3	74.4	
Sheep	TDN %	* 71.4	82.5	
Swine	TDN %	* 71.5	82.7	

Barley, 6-row, grain, Can 2 CW mn wt 48 lb per bushel mn 90% purity mx 1.5% foreign material, (4)

Ref No 4 00 575 United States

Feed Name or Analyses		As Fed	Dry	C.V. ± %
Dry matter	%		100.0	
Crude fiber	%			
Cattle	dig coef %	*	10.	
Sheep	dig coef %	*	45.	
Swine	dig coef %	*	11.	
Ether extract	%			
Cattle	dig coef %	*	60.	
Sheep	dig coef %	*	80.	
Swine	dig coef %	*	44.	
N-free extract	%			
Cattle	dig coef %	*	91.	
Sheep	dig coef %	*	91.	
Swine	dig coef %	*	89.	
Protein (N x 6.25)	%			
Cattle	dig coef %	*	75.	
Sheep	dig coef %	*	79.	
Swine	dig coef %	*	77.	
Energy	GE Mcal/kg	*	2.22	
Cattle	DE Mcal/kg	*	3.66	
Sheep	DE Mcal/kg	*	3.81	
Swine	DE kcal/kg	*	3573.	
Cattle	ME Mcal/kg	*	3.00	
Sheep	ME Mcal/kg	*	3.13	
Cattle	TDN %	*	83.1	
Sheep	TDN %	*	86.5	
Swine	TDN %	*	81.0	

Ref No 4 00 575 Canada

Feed Name or Analyses		As Fed	Dry	C.V. ± %
Dry matter	%	87.2	100.0	2
Ash	%	2.2	2.6	13
Crude fiber	%	4.9	5.6	26
Ether extract	%	1.9	2.2	16
N-free extract	%	66.9	76.8	
Protein (N x 6.25)	%	11.2	12.8	9
Cattle	dig prot %	* 6.8	7.8	
Goats	dig prot %	* 7.8	9.0	
Horses	dig prot %	* 7.8	9.0	
Sheep	dig prot %	* 7.8	9.0	
Cellulose (Matrone)	%	2.6	3.0	
Energy	GE Mcal/kg	1.94	2.22	
Cattle	DE Mcal/kg	* 3.19	3.66	
Sheep	DE Mcal/kg	* 3.33	3.81	
Swine	DE kcal/kg	*3114.	3573.	
Cattle	ME Mcal/kg	* 2.62	3.00	
Sheep	ME Mcal/kg	* 2.73	3.13	
Swine	ME kcal/kg	*2909.	3337.	
Cattle	TDN %	72.4	83.1	
Sheep	TDN %	75.4	86.5	
Swine	TDN %	70.6	81.0	
Calcium	%	0.03	0.04	
Copper	mg/kg	4.7	5.4	
Iron	%	0.003	0.004	
Magnesium	%	0.16	0.19	
Manganese	mg/kg	26.3	30.2	
Phosphorus	%	0.31	0.36	

Feed Name or Analyses		As Fed	Dry	C.V. ± %
Selenium	mg/kg	0.400	0.459	
Zinc	mg/kg	39.2	45.0	
Niacin	mg/kg	78.1	89.6	
Pantothenic acid	mg/kg	5.3	6.0	
Riboflavin	mg/kg	1.0	1.2	
Thiamine	mg/kg	0.8	0.9	
Vitamin B_6	mg/kg	3.28	3.76	
Alanine	%	0.51	0.59	
Arginine	%	0.58	0.66	
Aspartic acid	%	0.73	0.84	
Glutamic acid	%	2.94	3.37	
Glycine	%	0.50	0.58	
Histidine	%	0.26	0.29	
Isoleucine	%	0.45	0.51	
Leucine	%	0.83	0.96	
Lysine	%	0.44	0.50	
Methionine	%	0.14	0.16	
Phenylalanine	%	0.63	0.73	
Proline	%	1.16	1.33	
Serine	%	0.45	0.52	
Threonine	%	0.40	0.46	
Tyrosine	%	0.33	0.38	
Valine	%	0.62	0.71	

Barley, 6-row, grain, Can 3 CW mn wt 46 lb per bushel mn 85% purity mx 4% foreign material, (4)

Ref No 4 00 576 United States

Feed Name or Analyses		As Fed	Dry	C.V. ± %
Dry matter	%		100.0	
Crude fiber	%			
Cattle	dig coef %	*	10.	
Sheep	dig coef %	*	45.	
Swine	dig coef %	*	11.	
Ether extract	%			
Cattle	dig coef %	*	60.	
Sheep	dig coef %	*	80.	
Swine	dig coef %	*	44.	
N-free extract	%			
Cattle	dig coef %	*	91.	
Sheep	dig coef %	*	91.	
Swine	dig coef %	*	89.	
Protein (N x 6.25)	%			
Cattle	dig coef %	*	75.	
Sheep	dig coef %	*	79.	
Swine	dig coef %	*	77.	
Energy	GE Mcal/kg			
Cattle	DE Mcal/kg	*	3.66	
Sheep	DE Mcal/kg	*	3.81	
Swine	DE kcal/kg	*	3569.	
Cattle	ME Mcal/kg	*	3.00	
Sheep	ME Mcal/kg	*	3.12	
Cattle	TDN %	*	82.9	
Sheep	TDN %	*	86.3	
Swine	TDN %	*	80.9	

Ref No 4 00 576 Canada

Feed Name or Analyses		As Fed	Dry	C.V. ± %
Dry matter	%	86.8	100.0	1
Ash	%	2.3	2.7	7
Crude fiber	%	4.9	5.6	18
Ether extract	%	1.8	2.1	16
N-free extract	%	66.7	76.8	
Protein (N x 6.25)	%	11.1	12.8	5
Cattle	dig prot %	* 6.8	7.8	
Goats	dig prot %	* 7.8	9.0	
Horses	dig prot %	* 7.8	9.0	
Sheep	dig prot %	* 7.8	9.0	
Cellulose (Matrone)	%	3.4	4.0	

Feed Name or Analyses		Mean As Fed	Mean Dry	C.V. ± %
Energy	GE Mcal/kg			
Cattle	DE Mcal/kg	* 3.17	3.66	
Sheep	DE Mcal/kg	* 3.30	3.81	
Swine	DE kcal/kg	*3097.	3569.	
Cattle	ME Mcal/kg	* 2.60	3.00	
Sheep	ME Mcal/kg	* 2.71	3.12	
Swine	ME kcal/kg	*2893.	3334.	
Cattle	TDN %	72.0	82.9	
Sheep	TDN %	74.9	86.3	
Swine	TDN %	70.2	80.9	
Calcium	%	0.04	0.05	
Copper	mg/kg	5.5	6.4	
Iron	%	0.004	0.005	
Magnesium	%	0.14	0.16	
Manganese	mg/kg	15.1	17.4	
Phosphorus	%	0.28	0.32	
Selenium	mg/kg	0.878	1.012	
Zinc	mg/kg	32.7	37.7	
Niacin	mg/kg	80.2	92.4	
Pantothenic acid	mg/kg	7.1	8.1	
Riboflavin	mg/kg	1.2	1.4	
Thiamine	mg/kg	0.8	0.9	
Vitamin B6	mg/kg	2.54	2.93	

Barley, 6-row, grain, Can 4 CW, (4)

Ref No 4 00 577 United States

Feed Name or Analyses		Mean As Fed	Mean Dry	C.V. ± %
Dry matter	%		100.0	
Crude fiber	%			
Cattle	dig coef %	*	10.	
Sheep	dig coef %	*	45.	
Swine	dig coef %	*	11.	
Ether extract	%			
Cattle	dig coef %	*	60.	
Sheep	dig coef %	*	80.	
Swine	dig coef %	*	44.	
N-free extract	%			
Cattle	dig coef %	*	91.	
Sheep	dig coef %	*	91.	
Swine	dig coef %	*	89.	
Protein (N x 6.25)	%			
Cattle	dig coef %	*	75.	
Sheep	dig coef %	*	79.	
Swine	dig coef %	*	77.	
Energy	GE Mcal/kg			
Cattle	DE Mcal/kg	*	3.64	
Sheep	DE Mcal/kg	*	3.80	
Swine	DE kcal/kg	*	3557.	
Cattle	ME Mcal/kg	*	2.99	
Sheep	ME Mcal/kg	*	3.12	
Cattle	TDN %	*	82.6	
Sheep	TDN %	*	86.2	
Swine	TDN %	*	80.7	

Ref No 4 00 577 Canada

Feed Name or Analyses		Mean As Fed	Mean Dry	C.V. ± %
Dry matter	%	86.5	100.0	0
Ash	%	2.2	2.6	8
Crude fiber	%	5.2	6.0	17
Ether extract	%	1.8	2.1	18
N-free extract	%	66.1	76.4	
Protein (N x 6.25)	%	11.2	12.9	6
Cattle	dig prot %	* 6.8	7.9	
Goats	dig prot %	* 7.9	9.1	
Horses	dig prot %	* 7.9	9.1	
Sheep	dig prot %	* 7.9	9.1	
Energy	GE Mcal/kg			
Cattle	DE Mcal/kg	* 3.15	3.64	
Sheep	DE Mcal/kg	* 3.29	3.80	
Swine	DE kcal/kg	*3077.	3557.	
Cattle	ME Mcal/kg	* 2.58	2.99	
Sheep	ME Mcal/kg	* 2.69	3.12	
Swine	ME kcal/kg	*2873.	3322.	
Cattle	TDN %	71.5	82.6	
Sheep	TDN %	74.5	86.2	
Swine	TDN %	69.8	80.7	

Barley brewers dried grains –
see Barley, brewers grains, dehy, (5)

Barley distillers dried grains (AAFCO) –
see Barley, distillers grains, dehy, (5)

Barley feed (CFA) –
see Barley, pearl by-product, (4)

Barley grain, light –
see Barley, grain, gr 5 US mn wt 36 lb per bushel mx 6% foreign material, (4)

BARLEYGRASS. Hordeum leporinum

Barleygrass, aerial part, fresh, (2)

Ref No 1 07 758 United States

Feed Name or Analyses		Mean As Fed	Mean Dry	C.V. ± %
Dry matter	%		100.0	
Ash	%		6.6	
Crude fiber	%			
Sheep	dig coef %		17.	
Ether extract	%		1.5	
Sheep	dig coef %		89.	
N-free extract	%			
Sheep	dig coef %		63.	
Protein (N x 6.25)	%		4.8	
Sheep	dig coef %		48.	
Cattle	dig prot %	*	2.0	
Goats	dig prot %	*	1.0	
Horses	dig prot %	*	1.6	
Rabbits	dig prot %	*	2.4	
Sheep	dig prot %		2.3	
Cellulose (Matrone)	%		37.6	
Other carbohydrates	%		40.5	
Lignin (Ellis)	%		8.9	
Energy	GE Mcal/kg		5.50	
Sheep	GE dig coef %		46.	
Sheep	DE Mcal/kg		2.53	
Sheep	ME Mcal/kg	*	2.07	
Sheep	TDN %		54.0	
Silicon	%		1.40	

BARLEYGRASS, FOXTAIL. Hordeum jubatum

Barleygrass, foxtail, aerial part, dehy grnd, (1)

Ref No 1 08 234 United States

Feed Name or Analyses		Mean As Fed	Mean Dry	C.V. ± %
Dry matter	%	92.7	100.0	
Organic matter	%	86.8	93.7	
Ash	%	5.9	6.4	
Crude fiber	%	31.3	33.8	
Ether extract	%	1.6	1.7	
N-free extract	%	50.5	54.5	
Protein (N x 6.25)	%	3.4	3.6	
Cattle	dig prot %	* 0.1	0.1	
Goats	dig prot %	* 0.0	0.0	
Horses	dig prot %	* 0.6	0.6	
Rabbits	dig prot %	* 1.4	1.5	
Sheep	dig prot %	* -0.1	-0.1	
Energy	GE Mcal/kg			
Cattle	DE Mcal/kg	* 2.30	2.48	
Sheep	DE Mcal/kg	* 2.20	2.38	
Cattle	ME Mcal/kg	* 1.89	2.04	
Sheep	ME Mcal/kg	* 1.81	1.95	
Cattle	TDN %	* 52.2	56.4	
Sheep	TDN %	* 49.9	53.9	
Silicon	%	3.35	3.61	

Barleygrass, foxtail, aerial part, dehy grnd, dormant, (1)

Ref No 1 08 235 United States

Feed Name or Analyses		Mean As Fed	Mean Dry	C.V. ± %
Dry matter	%	92.5	100.0	
Organic matter	%	86.5	93.6	
Ash	%	6.0	6.5	
Crude fiber	%	33.7	36.5	
Ether extract	%	1.1	1.2	
N-free extract	%	48.7	52.7	
Protein (N x 6.25)	%	3.0	3.2	
Cattle	dig prot %	* -0.1	-0.2	
Goats	dig prot %	* -0.3	-0.3	
Horses	dig prot %	* 0.2	0.3	
Rabbits	dig prot %	* 1.1	1.2	
Sheep	dig prot %	* -0.4	-0.4	
Energy	GE Mcal/kg			
Cattle	DE Mcal/kg	* 2.10	2.27	
Sheep	DE Mcal/kg	* 2.14	2.32	
Cattle	ME Mcal/kg	* 1.72	1.86	
Sheep	ME Mcal/kg	* 1.76	1.90	
Cattle	TDN %	* 47.7	51.6	
Sheep	TDN %	* 48.6	52.5	
Silicon	%	3.13	3.38	

Barleygrass, foxtail, leaves, dehy grnd, immature, (1)

Ref No 1 08 236 United States

Feed Name or Analyses		Mean As Fed	Mean Dry	C.V. ± %
Dry matter	%	92.7	100.0	
Organic matter	%	83.6	90.2	
Ash	%	9.1	9.8	
Crude fiber	%	18.8	20.3	
Ether extract	%	4.0	4.3	
N-free extract	%	46.2	49.8	
Protein (N x 6.25)	%	14.6	15.8	
Cattle	dig prot %	* 9.8	10.6	
Goats	dig prot %	* 10.5	11.3	
Horses	dig prot %	* 10.1	10.9	
Rabbits	dig prot %	* 10.1	10.9	
Sheep	dig prot %	* 10.0	10.7	
Silicon	%	1.35	1.46	

Barleygrass, foxtail, leaves, dehy grnd, pre-bloom, (1)

Ref No 1 08 237 United States

Feed Name or Analyses		Mean As Fed	Mean Dry	C.V. ± %
Dry matter	%	93.9	100.0	
Organic matter	%	84.5	90.0	
Ash	%	9.4	10.0	
Crude fiber	%	15.3	16.3	
Ether extract	%	3.9	4.2	
N-free extract	%	47.9	51.0	
Protein (N x 6.25)	%	17.4	18.5	
Cattle	dig prot %	* 12.2	13.0	
Goats	dig prot %	* 13.0	13.8	
Horses	dig prot %	* 12.4	13.2	
Rabbits	dig prot %	* 12.2	13.0	

Continued

Feed Name or Analyses			Mean As Fed	Mean Dry	C.V. ± %
Sheep	dig prot %	*	12.4	13.2	
Silicon	%		2.41	2.57	

Barleygrass, foxtail, leaves, dehy grnd, early bloom, (1)
Ref No 1 08 238 United States

Feed Name or Analyses			Mean As Fed	Mean Dry	C.V. ± %
Dry matter	%		93.0	100.0	
Crude fiber	%		19.6	21.1	
Ether extract	%		5.7	6.1	
Protein (N x 6.25)	%		15.3	16.4	
Cattle	dig prot %	*	10.4	11.1	
Goats	dig prot %	*	11.0	11.9	
Horses	dig prot %	*	10.7	11.5	
Rabbits	dig prot %	*	10.5	11.3	
Sheep	dig prot %	*	10.5	11.3	

Barleygrass, foxtail, leaves, dehy grnd, mid-bloom, (1)
Ref No 1 08 239 United States

Feed Name or Analyses			Mean As Fed	Mean Dry	C.V. ± %
Dry matter	%		92.8	100.0	
Organic matter	%		83.2	89.7	
Ash	%		9.6	10.3	
Crude fiber	%		14.8	16.0	
Ether extract	%		3.8	4.1	
N-free extract	%		46.7	50.3	
Protein (N x 6.25)	%		17.9	19.3	
Cattle	dig prot %	*	12.7	13.7	
Goats	dig prot %	*	13.5	14.6	
Horses	dig prot %	*	12.9	13.9	
Rabbits	dig prot %	*	12.6	13.6	
Sheep	dig prot %	*	12.9	13.9	
Silicon	%		1.80	1.94	

Barleygrass, foxtail, leaves, dehy grnd, dormant, (1)
Ref No 1 08 240 United States

Feed Name or Analyses			Mean As Fed	Mean Dry	C.V. ± %
Dry matter	%			100.0	
Protein (N x 6.25)	%			4.2	
Cattle	dig prot %	*		0.6	
Goats	dig prot %	*		0.5	
Horses	dig prot %	*		1.1	
Rabbits	dig prot %	*		1.9	
Sheep	dig prot %	*		0.3	

Barleygrass, foxtail, leaves w some stems, dehy grnd, dough stage, (1)
Ref No 1 08 241 United States

Feed Name or Analyses			Mean As Fed	Mean Dry	C.V. ± %
Dry matter	%		93.4	100.0	
Organic matter	%		84.4	90.4	
Ash	%		9.0	9.6	
Protein (N x 6.25)	%		16.1	17.2	
Cattle	dig prot %	*	11.1	11.8	
Goats	dig prot %	*	11.8	12.6	
Horses	dig prot %	*	11.3	12.1	
Rabbits	dig prot %	*	11.2	11.9	
Sheep	dig prot %	*	11.2	12.0	
Silicon	%		3.20	3.43	

(1) dry forages and roughages (2) pasture, range plants, and forages fed green (3) silages (4) energy feeds (5) protein supplements (6) minerals (7) vitamins (8) additives

Barley middlings –
see Barley, flour by-product wo hulls, (4)

Barley mill by-product (AAFCO) –
see Barley, flour by-product mill run, (4)

Barley mixed feed (CFA) –
see Barley, flour by-product mill run, (4)

Barley, western, grain, light –
see Barley, western, grain, gr 5 US mn wt 36 lb per bushel mx 4% foreign material, (4)

BARNYARDGRASS. Ecinochloa crusgalli

Barnyardgrass, hay, s-c, milk stage, (1)
Ref No 1 00 579 United States

Feed Name or Analyses			Mean As Fed	Mean Dry	C.V. ± %
Dry matter	%		84.2	100.0	
Ash	%		7.7	9.1	
Crude fiber	%		31.0	36.8	
Sheep	dig coef %		61.	61.	
Ether extract	%		1.8	2.1	
Sheep	dig coef %		60.	60.	
N-free extract	%		34.0	40.4	
Sheep	dig coef %		53.	53.	
Protein (N x 6.25)	%		9.8	11.6	
Sheep	dig coef %		57.	57.	
Cattle	dig prot %	*	5.9	7.0	
Goats	dig prot %	*	6.2	7.4	
Horses	dig prot %	*	6.2	7.4	
Rabbits	dig prot %	*	6.4	7.6	
Sheep	dig prot %		5.6	6.6	
Energy	GE Mcal/kg				
Cattle	DE Mcal/kg	*	1.95	2.31	
Sheep	DE Mcal/kg	*	1.98	2.35	
Cattle	ME Mcal/kg	*	1.60	1.90	
Sheep	ME Mcal/kg	*	1.62	1.93	
Cattle	TDN %	*	44.2	52.5	
Sheep	TDN %		44.9	53.3	

Barnyardgrass, aerial part, fresh, immature, (2)
Ref No 2 00 580 United States

Feed Name or Analyses			Mean As Fed	Mean Dry	C.V. ± %
Dry matter	%			100.0	
Ash	%			11.2	
Crude fiber	%			22.0	
Ether extract	%			2.9	
N-free extract	%			50.2	
Protein (N x 6.25)	%			13.7	
Cattle	dig prot %	*		9.5	
Goats	dig prot %	*		9.3	
Horses	dig prot %	*		9.2	
Rabbits	dig prot %	*		9.2	
Sheep	dig prot %	*		9.8	
Energy	GE Mcal/kg				
Cattle	DE Mcal/kg	*		2.77	
Sheep	DE Mcal/kg	*		2.91	
Cattle	ME Mcal/kg	*		2.27	
Sheep	ME Mcal/kg	*		2.39	
Cattle	TDN %	*		62.9	
Sheep	TDN %	*		66.0	

BAYBERRY. Myrica spp

Bayberry, browse, immature, (2)
Ref No 2 00 581 United States

Feed Name or Analyses			Mean As Fed	Mean Dry	C.V. ± %
Dry matter	%			100.0	
Ash	%			12.7	
Crude fiber	%			29.8	
Ether extract	%			2.2	
N-free extract	%			44.5	
Protein (N x 6.25)	%			10.8	
Cattle	dig prot %	*		7.1	
Goats	dig prot %	*		6.6	
Horses	dig prot %	*		6.7	
Rabbits	dig prot %	*		7.0	
Sheep	dig prot %	*		7.1	

BEAN. Phaseolus spp

Bean, aerial part wo pods, dehy, (1)
Ref No 1 00 586 United States

Feed Name or Analyses			Mean As Fed	Mean Dry	C.V. ± %
Dry matter	%		89.3	100.0	
Ash	%		12.1	13.5	
Crude fiber	%		23.0	25.8	
Ether extract	%		1.5	1.7	
N-free extract	%		36.8	41.2	
Protein (N x 6.25)	%		15.9	17.8	
Cattle	dig prot %	*	11.0	12.4	
Goats	dig prot %	*	11.8	13.2	
Horses	dig prot %	*	11.3	12.6	
Rabbits	dig prot %	*	11.1	12.4	
Sheep	dig prot %	*	11.2	12.5	
Energy	GE Mcal/kg				
Cattle	DE Mcal/kg	*	2.56	2.86	
Sheep	DE Mcal/kg	*	2.26	2.53	
Cattle	ME Mcal/kg	*	2.10	2.35	
Sheep	ME Mcal/kg	*	1.85	2.07	
Cattle	TDN %	*	58.0	65.0	
Sheep	TDN %	*	51.2	57.3	
Calcium	%		0.13	0.14	
Phosphorus	%		0.24	0.27	

Bean, hay, dehy grnd, (1)
Ref No 1 00 582 United States

Feed Name or Analyses			Mean As Fed	Mean Dry	C.V. ± %
Dry matter	%		90.3	100.0	
Ash	%		8.7	9.6	
Crude fiber	%		22.3	24.7	
Sheep	dig coef %		48.	48.	
Ether extract	%		3.7	4.1	
Sheep	dig coef %		85.	85.	
N-free extract	%		40.8	45.2	
Sheep	dig coef %		77.	77.	
Protein (N x 6.25)	%		14.8	16.4	
Sheep	dig coef %		70.	70.	
Cattle	dig prot %	*	10.1	11.1	
Goats	dig prot %	*	10.7	11.9	
Horses	dig prot %	*	10.3	11.5	
Rabbits	dig prot %	*	10.2	11.3	
Sheep	dig prot %		10.4	11.5	
Energy	GE Mcal/kg				
Cattle	DE Mcal/kg	*	2.56	2.84	
Sheep	DE Mcal/kg	*	2.63	2.91	
Cattle	ME Mcal/kg	*	2.10	2.33	
Sheep	ME Mcal/kg	*	2.15	2.39	

Feed Name or Analyses			Mean As Fed	Mean Dry	C.V. ± %
Cattle	TDN %	*	58.1	64.4	
Sheep	TDN %		59.6	66.0	

Bean, hay, s-c, (1)
Ref No 1 00 583 United States

Analyses			As Fed	Dry	C.V. ± %
Dry matter	%		89.6	100.0	2
Ash	%		9.4	10.5	16
Crude fiber	%		24.2	27.0	8
Ether extract	%		2.6	2.9	25
N-free extract	%		39.3	43.9	
Protein (N x 6.25)	%		14.1	15.7	20
Cattle	dig prot %	*	9.4	10.5	
Goats	dig prot %	*	10.0	11.2	
Horses	dig prot %	*	9.7	10.9	
Rabbits	dig prot %	*	9.7	10.8	
Sheep	dig prot %	*	9.5	10.7	
Energy	GE Mcal/kg				
Cattle	DE Mcal/kg	*	2.48	2.77	
Sheep	DE Mcal/kg	*	2.28	2.55	
Cattle	ME Mcal/kg	*	2.03	2.27	
Sheep	ME Mcal/kg	*	1.87	2.09	
Cattle	TDN %	*	56.2	62.8	
Sheep	TDN %	*	51.8	57.8	

Bean, stems, (1)
Ref No 1 00 584 United States

Analyses			As Fed	Dry	C.V. ± %
Dry matter	%			100.0	
Riboflavin	mg/kg			1.8	
Thiamine	mg/kg			1.1	

Bean, straw, (1)
Ref No 1 00 585 United States

Analyses			As Fed	Dry	C.V. ± %
Dry matter	%		88.4	100.0	2
Ash	%		10.0	11.3	20
Crude fiber	%		33.7	38.1	13
Ether extract	%		1.3	1.5	19
N-free extract	%		35.3	39.9	
Protein (N x 6.25)	%		8.1	9.2	20
Cattle	dig prot %	*	4.3	4.9	
Goats	dig prot %	*	4.5	5.1	
Horses	dig prot %	*	4.7	5.3	
Rabbits	dig prot %	*	5.1	5.8	
Sheep	dig prot %	*	4.3	4.8	
Energy	GE Mcal/kg				
Cattle	DE Mcal/kg	*	2.04	2.31	
Sheep	DE Mcal/kg	*	2.07	2.34	
Cattle	ME Mcal/kg	*	1.67	1.89	
Sheep	ME Mcal/kg	*	1.70	1.92	
Cattle	TDN %	*	46.3	52.4	
Sheep	TDN %	*	47.0	53.2	

Bean, cannery residue, fresh, (2)
Ref No 2 00 587 United States

Analyses			As Fed	Dry	C.V. ± %
Dry matter	%		9.4	100.0	
Crude fiber	%		1.3	13.5	
Ether extract	%		0.3	3.0	
Protein (N x 6.25)	%		2.2	23.5	
Cattle	dig prot %	*	1.7	17.9	
Goats	dig prot %	*	1.7	18.5	
Horses	dig prot %	*	1.6	17.5	
Rabbits	dig prot %	*	1.6	16.8	
Sheep	dig prot %	*	1.8	18.9	

Analyses			As Fed	Dry	C.V. ± %
Carotene	mg/kg		2.3	24.9	
Riboflavin	mg/kg		1.7	18.1	
Vitamin A equivalent	IU/g		3.9	41.5	

Bean, leaves, fresh, (2)
Ref No 2 00 588 United States

Analyses			As Fed	Dry	C.V. ± %
Dry matter	%		28.8	100.0	12
Ash	%		3.4	11.8	50
Crude fiber	%		3.3	11.3	30
Ether extract	%		1.4	4.8	28
N-free extract	%		15.6	54.1	
Protein (N x 6.25)	%		5.2	18.0	24
Cattle	dig prot %	*	3.8	13.2	
Goats	dig prot %	*	3.8	13.4	
Horses	dig prot %	*	3.7	12.8	
Rabbits	dig prot %	*	3.6	12.6	
Sheep	dig prot %	*	4.0	13.8	
Energy	GE Mcal/kg				
Cattle	DE Mcal/kg	*	0.79	2.73	
Sheep	DE Mcal/kg	*	0.74	2.57	
Cattle	ME Mcal/kg	*	0.64	2.24	
Sheep	ME Mcal/kg	*	0.61	2.11	
Cattle	TDN %	*	17.8	61.9	
Sheep	TDN %	*	16.8	58.3	

Bean, pods, fresh, (2)
Ref No 2 00 589 United States

Analyses			As Fed	Dry	C.V. ± %
Dry matter	%		28.2	100.0	
Ash	%		1.2	4.2	
Crude fiber	%		10.8	38.3	10
Ether extract	%		0.4	1.5	49
N-free extract	%		13.7	48.6	
Protein (N x 6.25)	%		2.1	7.4	25
Cattle	dig prot %	*	1.2	4.2	
Goats	dig prot %	*	1.0	3.5	
Horses	dig prot %	*	1.1	3.8	
Rabbits	dig prot %	*	1.2	4.4	
Sheep	dig prot %	*	1.1	3.9	
Energy	GE Mcal/kg				
Cattle	DE Mcal/kg	*	0.77	2.74	
Sheep	DE Mcal/kg	*	0.81	2.89	
Cattle	ME Mcal/kg	*	0.63	2.25	
Sheep	ME Mcal/kg	*	0.67	2.37	
Cattle	TDN %	*	17.5	62.1	
Sheep	TDN %	*	18.5	65.5	

Bean, aerial part, ensiled, (3)
Ref No 3 00 590 United States

Analyses			As Fed	Dry	C.V. ± %
Dry matter	%		18.9	100.0	17
Ash	%		2.8	14.8	33
Crude fiber	%		5.4	28.6	12
Swine	dig coef %		46.	46.	
Ether extract	%		0.7	3.6	21
Swine	dig coef %		37.	37.	
N-free extract	%		7.0	37.3	
Swine	dig coef %		77.	77.	
Protein (N x 6.25)	%		3.0	15.8	10
Swine	dig coef %		78.	78.	
Cattle	dig prot %	*	2.0	10.6	
Goats	dig prot %	*	2.0	10.6	
Horses	dig prot %	*	2.0	10.6	
Sheep	dig prot %	*	2.0	10.6	
Swine	dig prot %		2.3	12.3	

Analyses			As Fed	Dry	C.V. ± %
Energy	GE Mcal/kg				
Cattle	DE Mcal/kg	*	0.50	2.66	
Sheep	DE Mcal/kg	*	0.49	2.60	
Swine	DE kcal/kg	*	476.	2520.	
Cattle	ME Mcal/kg	*	0.41	2.18	
Sheep	ME Mcal/kg	*	0.40	2.13	
Swine	ME kcal/kg	*	442.	2339.	
Cattle	TDN %	*	11.4	60.3	
Sheep	TDN %	*	11.1	58.9	
Swine	TDN %		10.8	57.2	

Bean, aerial part w hydrochloric acid preservative added, ensiled, (3)
Ref No 3 00 592 United States

Analyses			As Fed	Dry	C.V. ± %
Dry matter	%		20.0	100.0	13
Ash	%		3.9	19.6	17
Crude fiber	%		5.6	27.8	7
Sheep	dig coef %		37.	37.	
Ether extract	%		0.6	3.0	20
Sheep	dig coef %		57.	57.	
N-free extract	%		7.2	36.0	
Sheep	dig coef %		75.	75.	
Protein (N x 6.25)	%		2.7	13.6	14
Sheep	dig coef %		66.	66.	
Cattle	dig prot %	*	1.7	8.6	
Goats	dig prot %	*	1.7	8.6	
Horses	dig prot %	*	1.7	8.6	
Sheep	dig prot %		1.8	9.0	
Energy	GE Mcal/kg				
Cattle	DE Mcal/kg	*	0.44	2.19	
Sheep	DE Mcal/kg	*	0.44	2.21	
Cattle	ME Mcal/kg	*	0.36	1.80	
Sheep	ME Mcal/kg	*	0.36	1.81	
Cattle	TDN %	*	9.9	49.6	
Sheep	TDN %		10.0	50.1	

Bean, bran, (1)
Ref No 1 00 593 United States

Analyses			As Fed	Dry	C.V. ± %
Dry matter	%		89.2	100.0	
Ash	%		3.7	4.2	
Crude fiber	%		42.5	47.6	
Sheep	dig coef %		85.	85.	
Ether extract	%		0.4	0.5	
Sheep	dig coef %		100.	100.	
N-free extract	%		38.8	43.5	
Sheep	dig coef %		65.	65.	
Protein (N x 6.25)	%		3.7	4.2	
Sheep	dig coef %		0.	0.	
Cattle	dig prot %	*	0.0	0.0	
Goats	dig prot %	*	1.0	1.1	
Horses	dig prot %	*	1.0	1.1	
Energy	GE Mcal/kg				
Cattle	DE Mcal/kg	*	2.59	2.90	
Sheep	DE Mcal/kg	*	2.75	3.08	
Cattle	ME Mcal/kg	*	2.12	2.38	
Sheep	ME Mcal/kg	*	2.25	2.53	
Cattle	TDN %	*	61.3	68.7	
Sheep	TDN %		52.3	69.9	

Bean, seeds, (5)
Ref No 5 00 594 United States

Analyses			As Fed	Dry	C.V. ± %
Dry matter	%		84.3	100.0	
Ash	%		3.1	3.7	

Continued

Feed Name or Analyses			Mean As Fed	Mean Dry	C.V. ± %
Crude fiber	%		6.6	7.8	
Horses	dig coef %		76.	76.	
Sheep	dig coef %		74.	74.	
Ether extract	%		1.3	1.5	
Horses	dig coef %		15.	15.	
Sheep	dig coef %		85.	85.	
N-free extract	%		47.1	55.9	
Horses	dig coef %		94.	94.	
Sheep	dig coef %		92.	92.	
Protein (N x 6.25)	%		26.2	31.1	
Horses	dig coef %		88.	88.	
Sheep	dig coef %		88.	88.	
Horses	dig prot %		22.9	27.2	
Sheep	dig prot %		23.1	27.3	
Energy	GE Mcal/kg				
Cattle	DE Mcal/kg	*	2.98	3.53	
Horses	DE Mcal/kg	*	3.19	3.79	
Sheep	DE Mcal/kg	*	3.25	3.86	
Swine	DE kcal/kg	*	3307.	3922.	
Cattle	ME Mcal/kg	*	2.44	2.90	
Horses	ME Mcal/kg	*	2.62	3.11	
Sheep	ME Mcal/kg	*	2.67	3.16	
Swine	ME kcal/kg	*	2968.	3519.	
Cattle	TDN %	*	67.5	80.1	
Horses	TDN %		72.4	85.9	
Sheep	TDN %		73.8	87.5	
Swine	TDN %	*	75.0	88.9	

BEAN, BROAD. Vicia faba

Bean, broad, pods, dehy grnd, (1)
Ref No 1 00 595 United States

Feed Name or Analyses			Mean As Fed	Mean Dry	C.V. ± %
Dry matter	%		92.5	100.0	
Ash	%		6.8	7.3	
Crude fiber	%		16.5	17.8	
Sheep	dig coef %		58.	58.	
Ether extract	%		1.0	1.1	
Sheep	dig coef %		55.	55.	
N-free extract	%		52.9	57.2	
Sheep	dig coef %		78.	78.	
Protein (N x 6.25)	%		15.4	16.6	
Sheep	dig coef %		32.	32.	
Cattle	dig prot %	*	10.5	11.3	
Goats	dig prot %	*	11.1	12.0	
Horses	dig prot %	*	10.8	11.6	
Rabbits	dig prot %	*	10.6	11.5	
Sheep	dig prot %		4.9	5.3	
Energy	GE Mcal/kg				
Cattle	DE Mcal/kg	*	2.59	2.80	
Sheep	DE Mcal/kg	*	2.51	2.72	
Cattle	ME Mcal/kg	*	2.12	2.29	
Sheep	ME Mcal/kg	*	2.06	2.23	
Cattle	TDN %	*	58.7	63.4	
Sheep	TDN %		57.0	61.6	

BEAN, FIELD. Phaseolus spp

Bean, field, pods, s-c, (1)
Ref No 1 08 346 United States

Feed Name or Analyses			Mean As Fed	Mean Dry	C.V. ± %
Dry matter	%		91.8	100.0	
Ash	%		3.9	4.2	
Crude fiber	%		34.8	37.9	
Ether extract	%		1.0	1.1	
N-free extract	%		45.0	49.0	
Protein (N x 6.25)	%		7.1	7.7	
Cattle	dig prot %	*	3.3	3.6	
Goats	dig prot %	*	3.5	3.8	
Horses	dig prot %	*	3.8	4.1	
Rabbits	dig prot %	*	4.3	4.6	
Sheep	dig prot %	*	3.2	3.5	
Energy	GE Mcal/kg				
Cattle	DE Mcal/kg	*	2.10	2.29	
Sheep	DE Mcal/kg	*	2.18	2.38	
Cattle	ME Mcal/kg	*	1.73	1.88	
Sheep	ME Mcal/kg	*	1.79	1.95	
Cattle	TDN %	*	47.7	52.0	
Sheep	TDN %	*	49.5	53.9	
Calcium	%		0.78	0.85	
Phosphorus	%		0.10	0.11	
Potassium	%		2.02	2.20	

(1) dry forages and roughages (2) pasture, range plants, and forages fed green (3) silages (4) energy feeds (5) protein supplements (6) minerals (7) vitamins (8) additives

Bean, field, straw, (1)
Ref No 1 00 596 United States

Feed Name or Analyses			Mean As Fed	Mean Dry	C.V. ± %
Dry matter	%		89.1	100.0	
Ash	%		7.4	8.3	
Crude fiber	%		40.1	45.0	
Ether extract	%		1.4	1.6	
N-free extract	%		34.1	38.3	
Protein (N x 6.25)	%		6.1	6.8	
Cattle	dig prot %	*	2.6	2.9	
Goats	dig prot %	*	2.6	2.9	
Horses	dig prot %	*	3.0	3.3	
Rabbits	dig prot %	*	3.5	4.0	
Sheep	dig prot %	*	2.4	2.7	
Fatty acids	%		1.4	1.6	
Energy	GE Mcal/kg				
Cattle	DE Mcal/kg	*	1.92	2.15	
Sheep	DE Mcal/kg	*	1.94	2.18	
Cattle	ME Mcal/kg	*	1.57	1.76	
Sheep	ME Mcal/kg	*	1.59	1.79	
Cattle	TDN %	*	43.4	48.7	
Sheep	TDN %	*	44.0	49.4	
Calcium	%		1.67	1.87	
Magnesium	%		0.12	0.13	
Phosphorus	%		0.13	0.15	
Potassium	%		1.02	1.14	
Sulphur	%		0.10	0.11	

BEAN, GREEN. Phaseolus spp

Bean, green, aerial part wo pods, dehy, (1)
Bean, snap, vines
Ref No 1 08 347 United States

Feed Name or Analyses			Mean As Fed	Mean Dry	C.V. ± %
Dry matter	%		89.3	100.0	
Ash	%		12.9	14.4	
Crude fiber	%		21.4	24.0	
Ether extract	%		1.5	1.7	
N-free extract	%		35.2	39.4	
Protein (N x 6.25)	%		18.3	20.5	
Cattle	dig prot %	*	13.1	14.7	
Goats	dig prot %	*	14.0	15.7	
Horses	dig prot %	*	13.3	14.9	
Rabbits	dig prot %	*	12.9	14.5	
Sheep	dig prot %	*	13.4	15.0	
Energy	GE Mcal/kg				
Cattle	DE Mcal/kg	*	2.00	2.24	
Sheep	DE Mcal/kg	*	2.06	2.31	
Cattle	ME Mcal/kg	*	1.64	1.84	
Sheep	ME Mcal/kg	*	1.69	1.89	
Cattle	TDN %	*	45.3	50.7	
Sheep	TDN %	*	46.8	52.4	
Calcium	%		1.29	1.44	
Phosphorus	%		0.24	0.27	

Bean, green, dehy, (1)
Ref No 1 00 597 United States

Feed Name or Analyses			Mean As Fed	Mean Dry	C.V. ± %
Dry matter	%		88.5	100.0	
Ash	%		8.0	9.0	
Crude fiber	%		22.4	25.3	
Sheep	dig coef %		46.	46.	
Ether extract	%		3.4	3.8	
Sheep	dig coef %		82.	82.	
N-free extract	%		39.8	45.0	
Sheep	dig coef %		74.	74.	
Protein (N x 6.25)	%		15.0	16.9	
Sheep	dig coef %		69.	69.	
Cattle	dig prot %	*	10.2	11.6	
Goats	dig prot %	*	10.9	12.3	
Horses	dig prot %	*	10.5	11.9	
Rabbits	dig prot %	*	10.4	11.7	
Sheep	dig prot %		10.3	11.7	
Energy	GE Mcal/kg				
Cattle	DE Mcal/kg	*	2.46	2.78	
Sheep	DE Mcal/kg	*	2.48	2.80	
Cattle	ME Mcal/kg	*	2.02	2.28	
Sheep	ME Mcal/kg	*	2.04	2.30	
Cattle	TDN %	*	55.8	63.1	
Sheep	TDN %		56.3	63.6	

BEAN, KIDNEY. Phaseolus vulgaris

Bean, kidney, straw, (1)
Ref No 1 00 598 United States

Feed Name or Analyses			Mean As Fed	Mean Dry	C.V. ± %
Dry matter	%		85.7	100.0	
Ash	%		8.9	10.4	
Crude fiber	%		29.7	34.7	
Cattle	dig coef %		52.	52.	
Sheep	dig coef %		51.	51.	
Ether extract	%		1.4	1.6	
Cattle	dig coef %		30.	30.	
Sheep	dig coef %		53.	53.	
N-free extract	%		37.2	43.4	
Cattle	dig coef %		67.	67.	
Sheep	dig coef %		72.	72.	
Protein (N x 6.25)	%		8.5	9.9	
Cattle	dig coef %		67.	67.	
Sheep	dig coef %		54.	54.	
Cattle	dig prot %		5.7	6.6	
Goats	dig prot %	*	4.9	5.8	
Horses	dig prot %	*	5.1	5.9	
Rabbits	dig prot %	*	5.4	6.3	
Sheep	dig prot %		4.6	5.3	
Energy	GE Mcal/kg				
Cattle	DE Mcal/kg	*	2.07	2.42	
Sheep	DE Mcal/kg	*	2.12	2.48	
Cattle	ME Mcal/kg	*	1.70	1.98	
Sheep	ME Mcal/kg	*	1.74	2.03	
Cattle	TDN %		47.0	54.8	
Sheep	TDN %		48.2	56.2	

Bean, kidney, leaves, fresh, (2)

Ref No 2 00 599 United States

Feed Name or Analyses			Mean As Fed	Mean Dry	C.V. ± %
Dry matter	%		20.0	100.0	
Ash	%		3.3	16.7	
Crude fiber	%		2.8	14.0	
Ether extract	%		1.1	5.4	
N-free extract	%		6.8	34.2	
Protein (N x 6.25)	%		5.9	29.7	
Cattle	dig prot %	*	4.6	23.1	
Goats	dig prot %	*	4.9	24.3	
Horses	dig prot %	*	4.5	22.7	
Rabbits	dig prot %	*	4.3	21.6	
Sheep	dig prot %	*	4.9	24.7	
Carotene	mg/kg		50.0	250.0	
Riboflavin	mg/kg		2.2	11.0	
Vitamin A equivalent	IU/g		83.4	416.8	
Arginine	%		0.34	1.70	
Histidine	%		0.16	0.80	
Isoleucine	%		0.32	1.60	
Leucine	%		0.44	2.20	
Lysine	%		0.28	1.40	
Methionine	%		0.04	0.20	
Phenylalanine	%		0.28	1.40	
Threonine	%		0.28	1.40	
Tryptophan	%		0.08	0.40	
Valine	%		0.36	1.80	

Bean, kidney, seeds, (5)

Ref No 5 00 600 United States

Feed Name or Analyses			Mean As Fed	Mean Dry	C.V. ± %
Dry matter	%		88.8	100.0	
Ash	%		3.7	4.2	
Crude fiber	%		4.2	4.7	
Sheep	dig coef %		49.	49.	
Ether extract	%		1.3	1.5	
Sheep	dig coef %		35.	35.	
N-free extract	%		57.7	64.9	
Sheep	dig coef %		88.	88.	
Protein (N x 6.25)	%		21.9	24.7	
Sheep	dig coef %		67.	67.	
Sheep	dig prot %		14.7	16.5	
Energy	GE Mcal/kg		3.40	3.83	
Cattle	DE Mcal/kg	*	3.28	3.69	
Sheep	DE Mcal/kg	*	3.42	3.85	
Swine	DE kcal/kg	*	3705.	4173.	
Cattle	ME Mcal/kg	*	2.69	3.03	
Sheep	ME Mcal/kg	*	2.81	3.16	
Swine	ME kcal/kg	*	3372.	3798.	
Cattle	TDN %	*	74.4	83.8	
Sheep	TDN %	*	77.6	87.4	
Swine	TDN %	*	84.0	94.6	
Calcium	%		0.11	0.12	
Iron	%		0.007	0.008	
Phosphorus	%		0.40	0.45	
Potassium	%		0.98	1.10	
Sodium	%		0.01	0.01	
Niacin	mg/kg		22.8	25.7	
Riboflavin	mg/kg		2.0	2.2	
Thiamine	mg/kg		5.1	5.7	
Vitamin A	IU/g		0.2	0.2	

BEAN, LIMA. Phaseolus limensis

Bean, lima, aerial part wo pods, dehy, (1)

Ref No 1 00 605 United States

Feed Name or Analyses			Mean As Fed	Mean Dry	C.V. ± %
Dry matter	%			100.0	
Ash	%			11.6	
Crude fiber	%			29.3	
Ether extract	%			1.6	
N-free extract	%			45.0	
Protein (N x 6.25)	%			12.5	
Cattle	dig prot %	*		7.8	
Goats	dig prot %	*		8.2	
Horses	dig prot %	*		8.1	
Rabbits	dig prot %	*		8.3	
Sheep	dig prot %	*		7.8	

Bean, lima, hay, s-c, (1)

Ref No 1 00 601 United States

Feed Name or Analyses			Mean As Fed	Mean Dry	C.V. ± %
Dry matter	%			100.0	
Cellulose (Matrone)	%			30.4	

Bean, lima, leaves, s-c grnd, (1)

Ref No 1 00 602 United States

Feed Name or Analyses			Mean As Fed	Mean Dry	C.V. ± %
Dry matter	%		29.4	100.0	8
Ash	%		3.1	10.5	63
Crude fiber	%		3.3	11.1	31
Ether extract	%		1.4	4.8	21
N-free extract	%		16.6	56.3	
Protein (N x 6.25)	%		5.1	17.3	21
Cattle	dig prot %	*	3.5	11.9	
Goats	dig prot %	*	3.7	12.7	
Horses	dig prot %	*	3.6	12.2	
Rabbits	dig prot %	*	3.5	12.0	
Sheep	dig prot %	*	3.6	12.1	
α-tocopherol	mg/kg		207.7	706.6	
Arginine	%		0.24	0.80	
Histidine	%		0.06	0.20	
Isoleucine	%		0.18	0.60	
Leucine	%		0.35	1.20	
Lysine	%		0.18	0.60	
Phenylalanine	%		0.38	1.30	
Threonine	%		0.21	0.70	

Bean, lima, straw, (1)

Ref No 1 00 603 United States

Feed Name or Analyses			Mean As Fed	Mean Dry	C.V. ± %
Dry matter	%			100.0	
Ash	%			8.2	
Crude fiber	%			31.0	
Ether extract	%			1.8	
N-free extract	%			51.4	
Protein (N x 6.25)	%			7.6	
Cattle	dig prot %	*		3.5	
Goats	dig prot %	*		3.7	
Horses	dig prot %	*		4.0	
Rabbits	dig prot %	*		4.5	
Sheep	dig prot %	*		3.4	

Bean, lima, straw w molasses added, (1)

Ref No 1 00 604 United States

Feed Name or Analyses			Mean As Fed	Mean Dry	C.V. ± %
Dry matter	%			100.0	
Ash	%			9.4	
Crude fiber	%			24.8	
Ether extract	%			1.4	
N-free extract	%			57.5	
Protein (N x 6.25)	%			6.9	
Cattle	dig prot %	*		2.9	
Goats	dig prot %	*		3.0	
Horses	dig prot %	*		3.4	
Rabbits	dig prot %	*		4.0	
Sheep	dig prot %	*		2.8	

Bean, lima, aerial part wo pods, fresh, (2)

Ref No 2 00 609 United States

Feed Name or Analyses			Mean As Fed	Mean Dry	C.V. ± %
Dry matter	%			100.0	
Riboflavin	mg/kg			1.5	

Ref No 2 00 6 9 Canada

Feed Name or Analyses			Mean As Fed	Mean Dry	C.V. ± %
Dry matter	%		91.0	100.0	
Crude fiber	%		17.2	18.9	
Protein (N x 6.25)	%		10.6	11.6	
Cattle	dig prot %	*	7.1	7.8	
Goats	dig prot %	*	6.8	7.4	
Horses	dig prot %	*	6.8	7.4	
Rabbits	dig prot %	*	7.0	7.7	
Sheep	dig prot %	*	7.1	7.8	

Bean, lima, pods, fresh, (2)

Ref No 2 00 607 United States

Feed Name or Analyses			Mean As Fed	Mean Dry	C.V. ± %
Dry matter	%		28.2	100.0	
Crude fiber	%		10.9	38.6	13
Ether extract	%		0.5	1.9	51
Protein (N x 6.25)	%		2.0	7.2	34
Cattle	dig prot %	*	1.1	4.0	
Goats	dig prot %	*	0.9	3.3	
Horses	dig prot %	*	1.0	3.6	
Rabbits	dig prot %	*	1.2	4.2	
Sheep	dig prot %	*	1.0	3.7	
Carotene	mg/kg		6.6	23.4	54
Riboflavin	mg/kg		1.1	4.0	11
Vitamin A equivalent	IU/g		11.0	39.0	

Bean, lima, stems, fresh, (2)

Ref No 2 00 608 United States

Feed Name or Analyses			Mean As Fed	Mean Dry	C.V. ± %
Dry matter	%		26.6	100.0	
Crude fiber	%		10.4	39.2	29
Ether extract	%		0.6	2.4	
Protein (N x 6.25)	%		2.4	9.2	26
Cattle	dig prot %	*	1.5	5.7	
Goats	dig prot %	*	1.4	5.1	
Horses	dig prot %	*	1.4	5.3	
Rabbits	dig prot %	*	1.5	5.8	
Sheep	dig prot %	*	1.5	5.6	
Carotene	mg/kg		12.2	45.9	38
Riboflavin	mg/kg		1.3	5.1	
Vitamin A equivalent	IU/g		20.3	76.4	

Feed Name or Analyses			Mean As Fed	Mean Dry	C.V. ± %

Bean, lima, aerial part, ensiled, (3)
Ref No 3 00 610 United States

Feed Name or Analyses			As Fed	Dry	C.V. ± %
Dry matter	%		25.0	100.0	
Ash	%		2.4	9.6	
Crude fiber	%		7.7	30.8	
Ether extract	%		0.5	2.0	
N-free extract	%		11.5	46.0	
Protein (N x 6.25)	%		2.9	11.6	
Cattle	dig prot %	*	1.7	6.8	
Goats	dig prot %	*	1.7	6.8	
Horses	dig prot %	*	1.7	6.8	
Sheep	dig prot %	*	1.7	6.8	
Energy	GE Mcal/kg				
Cattle	DE Mcal/kg	*	0.56	2.23	
Sheep	DE Mcal/kg	*	0.59	2.35	
Cattle	ME Mcal/kg	*	0.46	1.83	
Sheep	ME Mcal/kg	*	0.48	1.93	
Cattle	TDN %	*	12.6	50.5	
Sheep	TDN %	*	13.4	53.4	

Bean, lima, aerial part wo seeds, ensiled, (3)
Ref No 3 00 611 United States

Feed Name or Analyses			As Fed	Dry	C.V. ± %
Dry matter	%		27.3	100.0	
Ash	%		4.0	14.7	
Crude fiber	%		8.1	29.7	
Ether extract	%		0.8	2.9	
N-free extract	%		11.1	40.6	
Protein (N x 6.25)	%		3.3	12.1	
Cattle	dig prot %	*	2.0	7.2	
Goats	dig prot %	*	2.0	7.2	
Horses	dig prot %	*	2.0	7.2	
Sheep	dig prot %	*	2.0	7.2	
Energy	GE Mcal/kg				
Cattle	DE Mcal/kg	*	0.62	2.26	
Sheep	DE Mcal/kg	*	0.64	2.34	
Cattle	ME Mcal/kg	*	0.51	1.85	
Sheep	ME Mcal/kg	*	0.52	1.92	
Cattle	TDN %	*	14.0	51.3	
Sheep	TDN %	*	14.5	53.1	

Bean, lima, seeds, mature, (5)
Ref No 5 00 613 United States

Feed Name or Analyses			As Fed	Dry	C.V. ± %
Dry matter	%		89.6	100.0	
Ash	%		4.1	4.6	
Crude fiber	%		4.6	5.1	
Ether extract	%		1.3	1.5	
N-free extract	%		58.9	65.7	
Protein (N x 6.25)	%		20.7	23.1	
Energy	GE Mcal/kg		3.45	3.85	
Cattle	DE Mcal/kg	*	3.28	3.67	
Sheep	DE Mcal/kg	*	3.42	3.82	
Swine	DE kcal/kg	*	3742.	4176.	
Cattle	ME Mcal/kg	*	2.69	3.01	
Sheep	ME Mcal/kg	*	2.80	3.13	
Swine	ME kcal/kg	*	3417.	3814.	
Cattle	TDN %	*	74.5	83.1	
Sheep	TDN %	*	77.7	86.7	
Swine	TDN %	*	84.9	94.7	
Calcium	%		0.08	0.09	
Chlorine	%		0.03	0.03	
Copper	mg/kg		8.2	9.1	
Iron	%		0.009	0.010	
Magnesium	%		0.18	0.20	
Manganese	mg/kg		16.1	18.0	
Phosphorus	%		0.38	0.42	
Potassium	%		1.61	1.80	
Sodium	%		0.02	0.02	
Sulphur	%		0.20	0.22	
Niacin	mg/kg		19.0	21.2	
Riboflavin	mg/kg		1.7	1.9	
Thiamine	mg/kg		4.8	5.4	

(1) dry forages and roughages (2) pasture, range plants, and forages fed green (3) silages (4) energy feeds (5) protein supplements (6) minerals (7) vitamins (8) additives

BEAN, MOTH. Phaseolus aconitifolius

Bean, moth, hay, s-c, (1)
Ref No 1 00 614 United States

Feed Name or Analyses			As Fed	Dry	C.V. ± %
Dry matter	%		86.2	100.0	
Ash	%		10.3	12.0	
Crude fiber	%		25.3	29.4	
Sheep	dig coef %		52.	52.	
Ether extract	%		1.5	1.7	
Sheep	dig coef %		10.	10.	
N-free extract	%		34.2	39.7	
Sheep	dig coef %		65.	65.	
Protein (N x 6.25)	%		14.8	17.2	
Sheep	dig coef %		67.	67.	
Cattle	dig prot %	*	10.2	11.8	
Goats	dig prot %	*	10.9	12.6	
Horses	dig prot %	*	10.5	12.1	
Rabbits	dig prot %	*	10.3	11.9	
Sheep	dig prot %		9.9	11.5	
Energy	GE Mcal/kg				
Cattle	DE Mcal/kg	*	1.97	2.27	
Sheep	DE Mcal/kg	*	2.01	2.34	
Cattle	ME Mcal/kg	*	1.60	1.86	
Sheep	ME Mcal/kg	*	1.65	1.92	
Cattle	TDN %	*	44.3	51.4	
Sheep	TDN %		45.7	53.0	

Bean, moth, pods w seeds, (5)
Ref No 5 00 615 United States

Feed Name or Analyses			As Fed	Dry	C.V. ± %
Dry matter	%		91.7	100.0	
Ash	%		5.0	5.5	
Crude fiber	%		11.9	13.0	
Ether extract	%		0.3	0.3	
N-free extract	%		55.7	60.7	
Protein (N x 6.25)	%		18.8	20.5	
Energy	GE Mcal/kg				
Cattle	DE Mcal/kg	*	2.92	3.19	
Sheep	DE Mcal/kg	*	2.90	3.16	
Swine	DE kcal/kg	*	2810.	3064.	
Cattle	ME Mcal/kg	*	2.40	2.61	
Sheep	ME Mcal/kg	*	2.38	2.59	
Swine	ME kcal/kg	*	2582.	2816.	
Cattle	TDN %	*	66.3	72.3	
Sheep	TDN %	*	65.7	71.6	
Swine	TDN %	*	63.7	69.5	

BEAN, MUNG. Phaseolus aureus

Bean, mung, hay, s-c, (1)
Ref No 1 00 616 United States

Feed Name or Analyses			As Fed	Dry	C.V. ± %
Dry matter	%		90.3	100.0	1
Ash	%		7.9	8.8	23
Crude fiber	%		23.8	26.4	9
Ether extract	%		2.2	2.4	14
N-free extract	%		46.5	51.6	
Protein (N x 6.25)	%		9.8	10.9	9
Cattle	dig prot %	*	5.7	6.4	
Goats	dig prot %	*	6.0	6.7	
Horses	dig prot %	*	6.1	6.8	
Rabbits	dig prot %	*	6.4	7.1	
Sheep	dig prot %	*	5.7	6.3	
Energy	GE Mcal/kg				
Cattle	DE Mcal/kg	*	2.25	2.49	
Sheep	DE Mcal/kg	*	2.16	2.39	
Cattle	ME Mcal/kg	*	1.84	2.04	
Sheep	ME Mcal/kg	*	1.77	1.96	
Cattle	TDN %	*	51.0	56.5	
Sheep	TDN %	*	48.9	54.2	
Carotene	mg/kg		99.9	110.7	39
Vitamin A equivalent	IU/g		166.5	184.5	

Bean, mung, aerial part, fresh, (2)
Ref No 2 00 619 United States

Feed Name or Analyses			As Fed	Dry	C.V. ± %
Dry matter	%		16.0	100.0	
Ash	%		2.1	13.1	
Crude fiber	%		4.1	25.6	
Ether extract	%		0.4	2.3	
N-free extract	%		6.6	41.4	
Protein (N x 6.25)	%		2.8	17.6	
Cattle	dig prot %	*	2.1	12.9	
Goats	dig prot %	*	2.1	13.0	
Horses	dig prot %	*	2.0	12.5	
Rabbits	dig prot %	*	2.0	12.3	
Sheep	dig prot %	*	2.1	13.4	
Energy	GE Mcal/kg				
Cattle	DE Mcal/kg	*	0.40	2.47	
Sheep	DE Mcal/kg	*	0.39	2.42	
Cattle	ME Mcal/kg	*	0.32	2.03	
Sheep	ME Mcal/kg	*	0.32	1.98	
Cattle	TDN %	*	9.0	56.0	
Sheep	TDN %	*	8.8	54.9	

Bean, mung, aerial part, fresh, immature, (2)
Ref No 2 00 617 United States

Feed Name or Analyses			As Fed	Dry	C.V. ± %
Dry matter	%			100.0	
Carotene	mg/kg			221.1	
Vitamin A equivalent	IU/g			368.6	

Bean, mung, aerial part, fresh, milk stage, (2)
Ref No 2 00 618 United States

Feed Name or Analyses			As Fed	Dry	C.V. ± %
Dry matter	%			100.0	
Carotene	mg/kg			98.8	
Vitamin A equivalent	IU/g			164.6	

Bean, mung, aerial part, ensiled, (3)

Ref No 3 00 620 — United States

Feed Name or Analyses			Mean As Fed	Mean Dry	C.V. ± %
Dry matter	%		27.3	100.0	
Ash	%		7.2	26.4	
Crude fiber	%		5.2	19.1	
Ether extract	%		1.2	4.5	
N-free extract	%		9.9	36.1	
Protein (N x 6.25)	%		3.8	13.9	
Cattle	dig prot %	*	2.4	8.9	
Goats	dig prot %	*	2.4	8.9	
Horses	dig prot %	*	2.4	8.9	
Sheep	dig prot %	*	2.4	8.9	
Energy	GE Mcal/kg				
Cattle	DE Mcal/kg	*	0.64	2.34	
Sheep	DE Mcal/kg	*	0.65	2.39	
Cattle	ME Mcal/kg	*	0.52	1.92	
Sheep	ME Mcal/kg	*	0.54	1.96	
Cattle	TDN %	*	14.5	53.1	
Sheep	TDN %	*	14.8	54.3	

Bean, mung, seeds, (5)

Ref No 5 08 185 — United States

Feed Name or Analyses			Mean As Fed	Mean Dry	C.V. ± %
Dry matter	%		89.8	100.0	
Ash	%		3.7	4.2	
Crude fiber	%		3.9	4.3	
Ether extract	%		1.3	1.4	
N-free extract	%		57.0	63.5	
Protein (N x 6.25)	%		23.9	26.6	
Energy	GE Mcal/kg		3.42	3.81	
Cattle	DE Mcal/kg	*	3.32	3.70	
Sheep	DE Mcal/kg	*	3.41	3.80	
Swine	DE kcal/kg	*	3722.	4144.	
Cattle	ME Mcal/kg	*	2.72	3.03	
Sheep	ME Mcal/kg	*	2.80	3.12	
Swine	ME kcal/kg	*	3373.	3756.	
Cattle	TDN %	*	75.4	83.9	
Sheep	TDN %	*	77.3	86.1	
Swine	TDN %	*	84.4	94.0	
Calcium	%		0.13	0.14	
Iron	%		0.008	0.009	
Phosphorus	%		0.34	0.38	
Potassium	%		1.04	1.15	
Sodium	%		0.01	0.01	
Niacin	mg/kg		26.1	29.1	
Riboflavin	mg/kg		2.1	2.4	
Thiamine	mg/kg		3.8	4.3	
Vitamin A	IU/g		0.8	0.9	

BEAN, MUNGO. Phaseolus mungo

Bean, mungo, aerial part, fresh, (2)

Ref No 2 00 621 — United States

Feed Name or Analyses			Mean As Fed	Mean Dry	C.V. ± %
Dry matter	%		16.0	100.0	
Ash	%		2.6	16.0	
Crude fiber	%		4.3	26.8	
Cattle	dig coef %		72.	72.	
Ether extract	%		0.4	2.5	
Cattle	dig coef %		51.	51.	
N-free extract	%		5.6	35.3	
Cattle	dig coef %		71.	71.	
Protein (N x 6.25)	%		3.1	19.4	
Cattle	dig coef %		82.	82.	
Cattle	dig prot %		2.5	15.9	
Goats	dig prot %	*	2.3	14.7	
Horses	dig prot %	*	2.2	14.0	
Rabbits	dig prot %	*	2.2	13.6	
Sheep	dig prot %	*	2.4	15.1	
Energy	GE Mcal/kg				
Cattle	DE Mcal/kg	*	0.45	2.78	
Sheep	DE Mcal/kg	*	0.46	2.90	
Cattle	ME Mcal/kg	*	0.37	2.28	
Sheep	ME Mcal/kg	*	0.38	2.38	
Cattle	TDN %		10.1	63.1	
Sheep	TDN %	*	10.5	65.7	

Bean, mungo, aerial part, ensiled, dough stage, (3)

Ref No 3 00 622 — United States

Feed Name or Analyses			Mean As Fed	Mean Dry	C.V. ± %
Dry matter	%		27.3	100.0	
Ash	%		7.2	26.4	
Crude fiber	%		5.2	19.1	
Cattle	dig coef %		48.	48.	
Ether extract	%		1.3	4.6	
Cattle	dig coef %		72.	72.	
N-free extract	%		9.8	36.0	
Cattle	dig coef %		58.	58.	
Protein (N x 6.25)	%		3.8	13.9	
Cattle	dig coef %		54.	54.	
Cattle	dig prot %		2.0	7.5	
Goats	dig prot %	*	2.4	8.9	
Horses	dig prot %	*	2.4	8.9	
Sheep	dig prot %	*	2.4	8.9	
Energy	GE Mcal/kg				
Cattle	DE Mcal/kg	*	0.54	1.98	
Sheep	DE Mcal/kg	*	0.53	1.95	
Cattle	ME Mcal/kg	*	0.44	1.63	
Sheep	ME Mcal/kg	*	0.44	1.60	
Cattle	TDN %		12.3	45.0	
Sheep	TDN %	*	12.1	44.2	

BEAN, NAVY. Phaseolus vulgaris

Bean, navy, seeds, (5)

Ref No 5 00 623 — United States

Feed Name or Analyses			Mean As Fed	Mean Dry	C.V. ± %
Dry matter	%		89.7	100.0	
Ash	%		4.1	4.6	
Crude fiber	%		4.2	4.7	
Ether extract	%		1.5	1.6	
N-free extract	%		57.2	63.8	
Protein (N x 6.25)	%		22.7	25.3	
Energy	GE Mcal/kg		3.42	3.82	
Cattle	DE Mcal/kg	*	3.30	3.68	
Sheep	DE Mcal/kg	*	3.41	3.80	
Swine	DE kcal/kg	*	3725.	4153.	
Cattle	ME Mcal/kg	*	2.71	3.02	
Swine	ME kcal/kg	*	3386.	3775.	
Chickens	ME_n kcal/kg		2330.	2598.	
Sheep	ME Mcal/kg	*	2.80	3.12	
Cattle	NE_m Mcal/kg	*	1.76	1.96	
Cattle	NE_{gain} Mcal/kg	*	1.18	1.31	
Cattle	$NE_{lactating\ cows}$ Mcal/kg	*	2.06	2.30	
Cattle	TDN %	*	74.8	83.4	
Sheep	TDN %	*	77.2	86.1	
Swine	TDN %	*	84.5	94.2	
Calcium	%		0.15	0.17	
Chlorine	%		0.04	0.04	
Copper	mg/kg		9.9	11.0	
Iron	%		0.010	0.011	
Magnesium	%		0.17	0.19	
Manganese	mg/kg		18.5	20.6	
Phosphorus	%		0.52	0.58	
Potassium	%		1.23	1.38	
Sodium	%		0.05	0.06	
Sulphur	%		0.23	0.26	
Choline	mg/kg		1758.	1960.	
Niacin	mg/kg		24.4	27.2	
Pantothenic acid	mg/kg		3.1	3.4	
Riboflavin	mg/kg		2.0	2.2	
Thiamine	mg/kg		6.5	7.3	
Arginine	%		1.50	1.67	
Cystine	%		0.20	0.22	
Lysine	%		1.59	1.78	
Methionine	%		0.25	0.28	
Tryptophan	%		0.25	0.28	

BEAN, PINTO. Phaseolus spp

Bean, pinto, seeds, (5)

Ref No 5 00 624 — United States

Feed Name or Analyses			Mean As Fed	Mean Dry	C.V. ± %
Dry matter	%		90.3	100.0	
Ash	%		4.3	4.8	
Crude fiber	%		4.0	4.5	
Sheep	dig coef %		62.	62.	
Ether extract	%		1.3	1.4	
Sheep	dig coef %		65.	65.	
N-free extract	%		57.9	64.1	
Sheep	dig coef %		96.	96.	
Protein (N x 6.25)	%		22.8	25.2	
Sheep	dig coef %		87.	87.	
Sheep	dig prot %		19.8	21.9	
Energy	GE Mcal/kg				
Cattle	DE Mcal/kg	*	3.31	3.67	
Sheep	DE Mcal/kg	*	3.52	3.89	
Swine	DE kcal/kg	*	3739.	4140.	
Cattle	ME Mcal/kg	*	2.72	3.01	
Sheep	ME Mcal/kg	*	2.88	3.19	
Swine	ME kcal/kg	*	3399.	3764.	
Cattle	TDN %	*	75.1	83.2	
Sheep	TDN %		79.7	88.3	
Swine	TDN %	*	84.8	93.9	
Calcium	%		0.14	0.16	
Phosphorus	%		0.35	0.39	

BEAN, TEPARY. Phaseolus acutifolius

Bean, tepary, hay, s-c, (1)

Ref No 1 08 348 — United States

Feed Name or Analyses			Mean As Fed	Mean Dry	C.V. ± %
Dry matter	%		90.0	100.0	
Ash	%		10.5	11.7	
Crude fiber	%		24.8	27.6	
Ether extract	%		2.9	3.2	
N-free extract	%		34.7	38.6	
Protein (N x 6.25)	%		17.1	19.0	
Cattle	dig prot %	*	12.1	13.4	
Goats	dig prot %	*	12.9	14.3	
Horses	dig prot %	*	12.3	13.7	
Rabbits	dig prot %	*	12.0	13.3	
Sheep	dig prot %	*	12.3	13.6	
Energy	GE Mcal/kg				
Cattle	DE Mcal/kg	*	2.12	2.35	
Sheep	DE Mcal/kg	*	2.25	2.49	
Cattle	ME Mcal/kg	*	1.74	1.93	
Sheep	ME Mcal/kg	*	1.84	2.05	
Cattle	TDN %	*	48.1	53.4	
Sheep	TDN %	*	50.9	56.6	

Feed Name or Analyses		Mean As Fed	Mean Dry	C.V. ± %

Bean, tepary, hay, s-c, immature, (1)
Ref No 1 00 625 — United States

Analyses		As Fed	Dry	C.V. ± %
Dry matter	%	89.2	100.0	1
Ash	%	10.4	11.7	8
Crude fiber	%	25.1	28.1	6
Ether extract	%	2.9	3.2	9
N-free extract	%	34.3	38.5	
Protein (N x 6.25)	%	16.5	18.5	9
Cattle	dig prot %	* 11.6	13.0	
Goats	dig prot %	* 12.3	13.8	
Horses	dig prot %	* 11.8	13.2	
Rabbits	dig prot %	* 11.6	13.0	
Sheep	dig prot %	* 11.7	13.2	
Energy	GE Mcal/kg			
Cattle	DE Mcal/kg	* 2.10	2.35	
Sheep	DE Mcal/kg	* 2.21	2.48	
Cattle	ME Mcal/kg	* 1.72	1.93	
Sheep	ME Mcal/kg	* 1.81	2.03	
Cattle	TDN %	* 0.47	53.0	
Sheep	TDN %	* 50.1	56.1	

Bean, tepary, seeds, (5)
Ref No 5 08 349 — United States

Analyses		As Fed	Dry	C.V. ± %
Dry matter	%	90.5	100.0	
Ash	%	4.2	4.6	
Crude fiber	%	3.4	3.8	
Ether extract	%	1.4	1.5	
N-free extract	%	59.3	65.5	
Protein (N x 6.25)	%	22.2	24.5	
Energy	GE Mcal/kg			
Cattle	DE Mcal/kg	* 2.97	3.28	
Sheep	DE Mcal/kg	* 3.07	3.40	
Swine	DE kcal/kg	*3124.	3452.	
Cattle	ME Mcal/kg	* 2.43	2.69	
Sheep	ME Mcal/kg	* 2.52	2.78	
Swine	ME kcal/kg	*2846.	3145.	
Cattle	TDN %	* 67.3	74.4	
Sheep	TDN %	* 69.7	77.0	
Swine	TDN %	* 70.9	78.3	

Bean, snap, vines –
see Bean, green, aerial part wo pods, dehy, (1)

BEARGRASS. Xerophyllum spp

Beargrass, browse, (2)
Ref No 2 00 626 — United States

Analyses		As Fed	Dry	C.V. ± %
Dry matter	%	49.4	100.0	
Ash	%	3.7	7.4	
Crude fiber	%	20.7	42.0	
Ether extract	%	1.1	2.3	
N-free extract	%	20.3	41.0	
Protein (N x 6.25)	%	3.6	7.3	
Cattle	dig prot %	* 2.0	4.1	
Goats	dig prot %	* 1.7	3.4	
Horses	dig prot %	* 1.8	3.7	
Rabbits	dig prot %	* 2.1	4.3	
Sheep	dig prot %	* 1.9	3.8	

(1) dry forages and roughages (2) pasture, range plants, and forages fed green (3) silages (4) energy feeds (5) protein supplements (6) minerals (7) vitamins (8) additives

Beargrass, aerial part, ensiled, (3)
Ref No 3 00 627 — United States

Analyses		As Fed	Dry	C.V. ± %
Dry matter	%	35.3	100.0	
Ash	%	3.4	9.6	
Crude fiber	%	12.5	35.5	
Ether extract	%	1.6	4.5	
N-free extract	%	15.7	44.5	
Protein (N x 6.25)	%	2.1	5.9	
Cattle	dig prot %	* 0.6	1.6	
Goats	dig prot %	* 0.6	1.6	
Horses	dig prot %	* 0.6	1.6	
Sheep	dig prot %	* 0.6	1.6	

Beargrass, yucca, dried –
see Yucca, beargrass, aerial part, s-c, (1)

Beargrass, yucca, fresh –
see Yucca, beargrass, aerial part, fresh, (2)

BEECH. Fagus spp

Beech, leaves, s-c, (1)
Ref No 1 00 628 — United States

Analyses		As Fed	Dry	C.V. ± %
Dry matter	%	86.0	100.0	
Ash	%	4.1	4.8	
Crude fiber	%	21.6	25.1	
Sheep	dig coef %	26.	26.	
Ether extract	%	2.2	2.5	
Sheep	dig coef %	33.	33.	
N-free extract	%	47.6	55.4	
Sheep	dig coef %	42.	42.	
Protein (N x 6.25)	%	10.5	12.2	
Sheep	dig coef %	6.	6.	
Cattle	dig prot %	* 6.5	7.5	
Goats	dig prot %	* 6.8	7.9	
Horses	dig prot %	* 6.8	7.9	
Rabbits	dig prot %	* 7.0	8.1	
Sheep	dig prot %	0.6	0.7	
Energy	GE Mcal/kg			
Cattle	DE Mcal/kg	* 1.34	1.56	
Sheep	DE Mcal/kg	* 1.23	1.43	
Cattle	ME Mcal/kg	* 1.10	1.28	
Sheep	ME Mcal/kg	* 1.01	1.17	
Cattle	TDN %	* 30.4	35.4	
Sheep	TDN %	27.8	32.4	

Beech, twigs, s-c, (1)
Ref No 1 00 629 — United States

Analyses		As Fed	Dry	C.V. ± %
Dry matter	%	85.4	100.0	
Ash	%	2.0	2.4	
Crude fiber	%	36.6	42.8	
Sheep	dig coef %	12.	12.	
Ether extract	%	1.4	1.6	
Sheep	dig coef %	10.	10.	
N-free extract	%	41.5	48.6	
Sheep	dig coef %	34.	34.	
Protein (N x 6.25)	%	3.9	4.6	
Sheep	dig coef %	7.	7.	
Cattle	dig prot %	* 0.8	0.9	
Goats	dig prot %	* 0.7	0.9	
Horses	dig prot %	* 1.2	1.4	
Rabbits	dig prot %	* 1.9	2.2	
Sheep	dig prot %	0.3	0.3	
Energy	GE Mcal/kg			
Cattle	DE Mcal/kg	* 0.96	1.12	
Sheep	DE Mcal/kg	* 0.84	0.99	
Cattle	ME Mcal/kg	* 0.79	0.92	
Sheep	ME Mcal/kg	* 0.69	0.81	
Cattle	TDN %	* 21.6	25.3	
Sheep	TDN %	19.1	22.3	

Beech, nuts w hulls, mech-extd grnd, (4)
Ref No 4 00 630 — United States

Analyses		As Fed	Dry	C.V. ± %
Dry matter	%	89.2	100.0	
Ash	%	5.3	5.9	
Crude fiber	%	19.3	21.7	
Sheep	dig coef %	21.	21.	
Ether extract	%	17.1	19.2	
Sheep	dig coef %	96.	96.	
N-free extract	%	31.5	35.3	
Sheep	dig coef %	31.	31.	
Protein (N x 6.25)	%	16.0	18.0	
Sheep	dig coef %	70.	70.	
Cattle	dig prot %	* 11.2	12.5	
Goats	dig prot %	* 12.2	13.7	
Horses	dig prot %	* 12.2	13.7	
Sheep	dig prot %	11.2	12.6	
Energy	GE Mcal/kg			
Cattle	DE Mcal/kg	* 3.47	3.89	
Sheep	DE Mcal/kg	* 2.73	3.06	
Cattle	ME Mcal/kg	* 2.84	3.19	
Sheep	ME Mcal/kg	* 2.24	2.51	
Cattle	TDN %	* 78.6	88.1	
Sheep	TDN %	62.0	69.5	
Calcium	%	0.57	0.63	
Phosphorus	%	0.29	0.33	
Potassium	%	0.61	0.68	

Beef liver –
see Cattle, livers, raw, (5)

Beef scrap –
see Cattle, carcass residue, dry rendered dehy grnd, (5)

BEET, COMMON. Beta vulgaris

Beet, common, aerial part, fresh, (2)
Ref No 2 00 631 — United States

Analyses		As Fed	Dry	C.V. ± %
Dry matter	%	9.1	100.0	
Ash	%	2.0	22.0	
Crude fiber	%	1.3	14.3	
Ether extract	%	0.3	3.3	
N-free extract	%	3.3	36.3	
Protein (N x 6.25)	%	2.2	24.2	
Cattle	dig prot %	* 1.7	18.4	
Goats	dig prot %	* 1.7	19.1	
Horses	dig prot %	* 1.6	18.1	
Rabbits	dig prot %	* 1.6	17.3	
Sheep	dig prot %	* 1.8	19.5	
Energy	GE Mcal/kg	0.24	2.64	
Cattle	DE Mcal/kg	* 0.23	2.49	
Sheep	DE Mcal/kg	* 0.21	2.35	
Cattle	ME Mcal/kg	* 0.19	2.05	
Sheep	ME Mcal/kg	* 0.18	1.93	
Cattle	TDN %	* 5.1	56.6	
Sheep	TDN %	* 4.9	53.4	
Calcium	%	0.12	1.31	
Iron	%	0.003	0.036	

Feed Name or Analyses			Mean As Fed	Mean Dry	C.V. ± %
Manganese	mg/kg		6.9	75.6	
Phosphorus	%		0.04	0.44	
Potassium	%		0.57	6.26	
Sodium	%		0.13	1.43	
Ascorbic acid	mg/kg		300.0	3296.7	
Niacin	mg/kg		4.0	44.0	
Riboflavin	mg/kg		2.2	24.2	
Thiamine	mg/kg		1.0	11.0	
Vitamin A	IU/g		61.0	670.3	

Beet, common, roots, fresh, (4)

Ref No 2 08 350 United States

Feed Name or Analyses			Mean As Fed	Mean Dry	C.V. ± %
Dry matter	%		13.0	100.0	
Ash	%		1.5	11.5	
Crude fiber	%		0.9	6.9	
Ether extract	%		0.1	0.8	
N-free extract	%		8.9	68.5	
Protein (N x 6.25)	%		1.6	12.3	
Cattle	dig prot %	*	1.0	7.3	
Goats	dig prot %	*	1.1	8.5	
Horses	dig prot %	*	1.1	8.5	
Sheep	dig prot %	*	1.1	8.5	
Energy	GE Mcal/kg				
Cattle	DE Mcal/kg	*	0.46	3.54	
Sheep	DE Mcal/kg	*	0.43	3.31	
Swine	DE kcal/kg	*	326.	2505.	
Cattle	ME Mcal/kg	*	0.38	2.90	
Sheep	ME Mcal/kg	*	0.35	2.72	
Swine	ME kcal/kg	*	304.	2342.	
Cattle	TDN %	*	10.4	80.3	
Sheep	TDN %	*	9.8	75.1	
Swine	TDN %	*	7.4	56.8	
Calcium	%		0.03	0.23	
Chlorine	%		0.03	0.23	
Copper	mg/kg		1.3	10.2	
Iron	%		0.002	0.015	
Magnesium	%		0.02	0.15	
Manganese	mg/kg		7.1	54.3	
Phosphorus	%		0.04	0.31	
Potassium	%		0.28	2.15	
Sodium	%		0.10	0.77	
Sulphur	%		0.02	0.15	

BEET, MANGELS. Beta spp

Beet, mangels, aerial part w crowns, fresh, (2)

Ref No 2 00 632 United States

Feed Name or Analyses			Mean As Fed	Mean Dry	C.V. ± %
Dry matter	%		12.6	100.0	
Ash	%		2.4	19.2	
Crude fiber	%		1.4	11.4	
Sheep	dig coef %		77.	77.	
Ether extract	%		0.5	4.2	
Sheep	dig coef %		47.	47.	
N-free extract	%		6.1	48.2	
Sheep	dig coef %		79.	79.	
Protein (N x 6.25)	%		2.1	17.0	
Sheep	dig coef %		82.	82.	
Cattle	dig prot %	*	1.6	12.3	
Goats	dig prot %	*	1.6	12.4	
Horses	dig prot %	*	1.5	12.0	
Rabbits	dig prot %	*	1.5	11.8	
Sheep	dig prot %		1.8	13.9	
Energy	GE Mcal/kg				
Cattle	DE Mcal/kg	*	0.34	2.68	
Sheep	DE Mcal/kg	*	0.36	2.88	
Cattle	ME Mcal/kg	*	0.28	2.20	
Sheep	ME Mcal/kg	*	0.30	2.36	
Cattle	TDN %	*	7.6	60.7	
Sheep	TDN %		8.2	65.2	

Beet, mangels, aerial part w crowns, ensiled, (3)

Ref No 3 00 635 United States

Feed Name or Analyses			Mean As Fed	Mean Dry	C.V. ± %
Dry matter	%		12.4	100.0	
Ash	%		2.7	21.5	
Crude fiber	%		1.6	13.3	
Sheep	dig coef %		74.	74.	
Ether extract	%		0.5	3.7	
Sheep	dig coef %		39.	39.	
N-free extract	%		5.8	47.0	
Sheep	dig coef %		76.	76.	
Protein (N x 6.25)	%		1.8	14.6	
Sheep	dig coef %		71.	71.	
Cattle	dig prot %	*	1.2	9.4	
Goats	dig prot %	*	1.2	9.4	
Horses	dig prot %	*	1.2	9.4	
Sheep	dig prot %		1.3	10.3	
Energy	GE Mcal/kg				
Cattle	DE Mcal/kg	*	0.33	2.66	
Sheep	DE Mcal/kg	*	0.32	2.59	
Cattle	ME Mcal/kg	*	0.27	2.18	
Sheep	ME Mcal/kg	*	0.26	2.13	
Cattle	TDN %	*	7.5	60.3	
Sheep	TDN %		7.3	58.8	

Beet, mangels, aerial part w crowns, wilted ensiled, (3)

Ref No 3 00 634 United States

Feed Name or Analyses			Mean As Fed	Mean Dry	C.V. ± %
Dry matter	%		13.6	100.0	
Ash	%		2.8	20.7	
Crude fiber	%		2.1	15.4	
Sheep	dig coef %		72.	72.	
Ether extract	%		0.7	5.0	
Sheep	dig coef %		45.	45.	
N-free extract	%		5.9	43.1	
Sheep	dig coef %		67.	67.	
Protein (N x 6.25)	%		2.1	15.8	
Sheep	dig coef %		69.	69.	
Cattle	dig prot %	*	1.4	10.6	
Goats	dig prot %	*	1.4	10.6	
Horses	dig prot %	*	1.4	10.6	
Sheep	dig prot %		1.5	10.9	
Energy	GE Mcal/kg				
Cattle	DE Mcal/kg	*	0.35	2.56	
Sheep	DE Mcal/kg	*	0.34	2.47	
Cattle	ME Mcal/kg	*	0.29	2.10	
Sheep	ME Mcal/kg	*	0.28	2.02	
Cattle	TDN %	*	7.9	58.1	
Sheep	TDN %		7.6	55.9	

Beet, mangels, roots, ensiled, (3)

Ref No 3 00 636 United States

Feed Name or Analyses			Mean As Fed	Mean Dry	C.V. ± %
Dry matter	%		10.3	100.0	
Ash	%		1.9	18.9	
Crude fiber	%		0.8	7.6	
Sheep	dig coef %		39.	39.	
Swine	dig coef %		77.	77.	
Ether extract	%		0.1	1.2	
Sheep	dig coef %		40.	40.	
Swine	dig coef %		-67.	-67.	
N-free extract	%		6.4	62.5	
Sheep	dig coef %		79.	79.	
Swine	dig coef %		96.	96.	
Protein (N x 6.25)	%		1.0	10.0	
Sheep	dig coef %		45.	45.	
Swine	dig coef %		75.	75.	
Cattle	dig prot %	*	0.5	5.3	
Goats	dig prot %	*	0.5	5.3	
Horses	dig prot %	*	0.5	5.3	
Sheep	dig prot %		0.5	4.5	
Swine	dig prot %		0.8	7.5	
Energy	GE Mcal/kg				
Cattle	DE Mcal/kg	*	0.27	2.65	
Sheep	DE Mcal/kg	*	0.26	2.55	
Swine	DE kcal/kg	*	325.	3153.	
Cattle	ME Mcal/kg	*	0.22	2.17	
Sheep	ME Mcal/kg	*	0.22	2.09	
Swine	ME kcal/kg	*	305.	2963.	
Cattle	TDN %	*	6.2	60.1	
Sheep	TDN %		6.0	57.8	
Swine	TDN %		7.4	71.5	

Beet, mangels, roots, (4)

Mangel, roots

Ref No 2 00 637 United States

Feed Name or Analyses			Mean As Fed	Mean Dry	C.V. ± %
Dry matter	%		13.8	100.0	54
Ash	%		1.3	9.7	12
Crude fiber	%		1.0	7.5	11
Cattle	dig coef %		59.	59.	
Sheep	dig coef %		79.	79.	
Swine	dig coef %		64.	64.	
Ether extract	%		0.1	0.6	39
Cattle	dig coef %		-10.	-10.	
Sheep	dig coef %		-16.	-16.	
Swine	dig coef %		-67.	-67.	
N-free extract	%		9.8	71.0	6
Cattle	dig coef %		95.	95.	
Sheep	dig coef %		97.	97.	
Swine	dig coef %		94.	94.	
Protein (N x 6.25)	%		1.6	11.3	24
Cattle	dig coef %		70.	70.	
Sheep	dig coef %		69.	69.	
Swine	dig coef %		67.	67.	
Cattle	dig prot %		1.1	7.9	
Goats	dig prot %	*	1.0	7.6	
Horses	dig prot %	*	1.0	7.6	
Sheep	dig prot %		1.1	7.7	
Swine	dig prot %		1.0	7.5	
Energy	GE Mcal/kg				
Cattle	DE Mcal/kg	*	0.48	3.49	
Sheep	DE Mcal/kg	*	0.50	3.61	
Swine	DE kcal/kg	*	475.	3440.	
Cattle	ME Mcal/kg	*	0.40	2.86	
Sheep	ME Mcal/kg	*	0.41	2.96	
Swine	ME kcal/kg	*	445.	3224.	
Cattle	NE_m Mcal/kg	*	0.25	1.80	
Cattle	NE_{gain} Mcal/kg	*	0.17	1.20	
Cattle	$NE_{lactating\ cows}$ Mcal/kg	*	0.29	2.12	
Cattle	TDN %		10.9	79.2	
Sheep	TDN %		11.3	81.8	
Swine	TDN %		10.8	78.0	
Calcium	%		0.03	0.22	
Chlorine	%		0.19	1.41	
Iron	%		0.003	0.022	
Magnesium	%		0.03	0.22	
Phosphorus	%		0.03	0.22	
Potassium	%		0.31	2.28	
Sodium	%		0.11	0.76	
Sulphur	%		0.03	0.22	

BEET, RED. Beta vulgaris

Beet, red, roots, fresh, (4)

Ref No 2 00 638 United States

Feed Name or Analyses		Mean As Fed	Mean Dry	C.V. ± %
Dry matter	%	12.7	100.0	
Ash	%	1.1	8.7	
Crude fiber	%	0.8	6.3	
Ether extract	%	0.1	0.8	
N-free extract	%	9.1	71.7	
Protein (N x 6.25)	%	1.6	12.6	
Cattle	dig prot %	* 1.0	7.6	
Goats	dig prot %	* 1.1	8.8	
Horses	dig prot %	* 1.1	8.8	
Sheep	dig prot %	* 1.1	8.8	
Energy	GE Mcal/kg	0.43	3.39	
Cattle	DE Mcal/kg	* 0.43	3.39	
Sheep	DE Mcal/kg	* 0.43	3.41	
Swine	DE kcal/kg	* 366.	2879.	
Cattle	ME Mcal/kg	* 0.35	2.78	
Sheep	ME Mcal/kg	* 0.36	2.80	
Swine	ME kcal/kg	* 342.	2691.	
Cattle	TDN %	* 9.8	76.8	
Sheep	TDN %	* 9.8	77.3	
Swine	TDN %	* 8.3	65.3	
Calcium	%	0.02	0.13	
Iron	%	0.001	0.006	
Phosphorus	%	0.03	0.26	
Potassium	%	0.34	2.64	
Sodium	%	0.06	0.47	
Ascorbic acid	mg/kg	100.0	787.4	
Niacin	mg/kg	4.0	31.5	
Riboflavin	mg/kg	0.5	3.9	
Thiamine	mg/kg	0.3	2.4	
Vitamin A	IU/g	0.2	1.6	

BEET, SUGAR. Beta saccharifera

Beet, sugar, aerial part w crowns, dehy, (1)

Ref No 1 00 640 United States

Feed Name or Analyses		Mean As Fed	Mean Dry	C.V. ± %
Dry matter	%	84.5	100.0	3
Ash	%	15.9	18.9	16
Crude fiber	%	11.5	13.6	23
Sheep	dig coef %	66.	66.	
Swine	dig coef %	49.	49.	
Ether extract	%	1.0	1.2	10
Sheep	dig coef %	9.	9.	
Swine	dig coef %	0.	0.	
N-free extract	%	45.5	53.9	9
Sheep	dig coef %	83.	83.	
Swine	dig coef %	80.	80.	
Protein (N x 6.25)	%	10.5	12.4	9
Sheep	dig coef %	61.	61.	
Swine	dig coef %	44.	44.	
Cattle	dig prot %	* 6.5	7.7	
Goats	dig prot %	* 6.9	8.2	
Horses	dig prot %	* 6.8	8.1	
Rabbits	dig prot %	* 7.0	8.3	
Sheep	dig prot %	6.4	7.6	
Swine	dig prot %	4.6	5.5	
Energy	GE Mcal/kg			
Cattle	DE Mcal/kg	* 2.24	2.65	
Sheep	DE Mcal/kg	* 2.29	2.71	
Swine	DE kcal/kg	* 2048.	2424.	
Cattle	ME Mcal/kg	* 1.83	2.17	
Sheep	ME Mcal/kg	* 1.87	2.22	
Swine	ME kcal/kg	* 1914.	2266.	
Cattle	TDN %	* 50.8	60.1	
Sheep	TDN %	51.8	61.4	
Swine	TDN %	46.4	55.0	

(1) dry forages and roughages (2) pasture, range plants, and forages fed green (3) silages (4) energy feeds (5) protein supplements (6) minerals (7) vitamins (8) additives

Beet, sugar, leaves, s-c grnd, (1)

Ref No 1 00 641 United States

Feed Name or Analyses		Mean As Fed	Mean Dry	C.V. ± %
Dry matter	%	8.6	100.0	
Ash	%	2.8	32.8	
Crude fiber	%	0.6	6.7	14
Ether extract	%	0.6	7.2	16
N-free extract	%	2.4	27.8	
Protein (N x 6.25)	%	2.2	25.5	10
Cattle	dig prot %	* 1.6	19.0	
Goats	dig prot %	* 1.7	20.4	
Horses	dig prot %	* 1.6	19.2	
Rabbits	dig prot %	* 1.6	18.4	
Sheep	dig prot %	* 1.7	19.4	
Energy	GE Mcal/kg			
Cattle	DE Mcal/kg	* 0.19	2.20	
Sheep	DE Mcal/kg	* 0.19	2.21	
Cattle	ME Mcal/kg	* 0.15	1.80	
Sheep	ME Mcal/kg	* 0.15	1.81	
Cattle	TDN %	* 4.3	49.8	
Sheep	TDN %	* 4.3	50.1	
α-tocopherol	mg/kg	32.6	381.2	
Arginine	%	0.09	1.10	
Histidine	%	0.03	0.30	
Isoleucine	%	0.09	1.10	
Leucine	%	0.15	1.70	
Lysine	%	0.12	1.40	
Methionine	%	0.03	0.40	
Phenylalanine	%	0.09	1.00	
Threonine	%	0.09	1.00	
Tryptophan	%	0.03	0.30	
Valine	%	0.11	1.30	

Beet, sugar, seed hulls, (1)

Ref No 1 00 643 United States

Feed Name or Analyses		Mean As Fed	Mean Dry	C.V. ± %
Dry matter	%	85.3	100.0	
Ash	%	8.4	9.9	
Crude fiber	%	28.2	33.1	
Sheep	dig coef %	11.	11.	
Ether extract	%	2.1	2.5	
Sheep	dig coef %	70.	70.	
N-free extract	%	35.4	41.5	
Sheep	dig coef %	46.	46.	
Protein (N x 6.25)	%	11.1	13.0	
Sheep	dig coef %	56.	56.	
Cattle	dig prot %	* 7.0	8.2	
Goats	dig prot %	* 7.4	8.7	
Horses	dig prot %	* 7.3	8.6	
Rabbits	dig prot %	* 7.4	8.7	
Sheep	dig prot %	6.2	7.3	
Energy	GE Mcal/kg			
Cattle	DE Mcal/kg	* 1.13	1.33	
Sheep	DE Mcal/kg	* 1.28	1.50	
Cattle	ME Mcal/kg	* 0.93	1.09	
Sheep	ME Mcal/kg	* 1.05	1.23	
Cattle	TDN %	* 25.7	30.1	
Sheep	TDN %	29.0	33.9	

Beet, sugar, straw, (1)

Ref No 1 00 644 United States

Feed Name or Analyses		Mean As Fed	Mean Dry	C.V. ± %
Dry matter	%	81.6	100.0	
Ash	%	8.3	10.2	
Crude fiber	%	35.2	43.1	
Sheep	dig coef %	25.	25.	
Ether extract	%	0.8	1.0	
Sheep	dig coef %	36.	36.	
N-free extract	%	31.9	39.1	
Sheep	dig coef %	41.	41.	
Protein (N x 6.25)	%	5.4	6.6	
Sheep	dig coef %	41.	41.	
Cattle	dig prot %	* 2.2	2.7	
Goats	dig prot %	* 2.2	2.7	
Horses	dig prot %	* 2.6	3.1	
Rabbits	dig prot %	* 3.1	3.8	
Sheep	dig prot %	2.2	2.7	
Energy	GE Mcal/kg			
Cattle	DE Mcal/kg	* 1.05	1.29	
Sheep	DE Mcal/kg	* 1.09	1.34	
Cattle	ME Mcal/kg	* 0.86	1.06	
Sheep	ME Mcal/kg	* 0.89	1.10	
Cattle	TDN %	* 23.9	29.3	
Sheep	TDN %	24.7	30.3	

Beet, sugar, aerial part w crowns, fresh, (2)

Ref No 2 00 649 United States

Feed Name or Analyses		Mean As Fed	Mean Dry	C.V. ± %
Dry matter	%	15.9	100.0	20
Ash	%	3.3	20.6	19
Crude fiber	%	1.7	10.7	6
Sheep	dig coef %	90.	90.	
Ether extract	%	0.4	2.5	48
Sheep	dig coef %	56.	56.	
N-free extract	%	7.7	48.7	14
Sheep	dig coef %	87.	87.	
Protein (N x 6.25)	%	2.8	17.5	24
Sheep	dig coef %	81.	81.	
Cattle	dig prot %	* 2.0	12.7	
Goats	dig prot %	* 2.0	12.8	
Horses	dig prot %	* 2.0	12.3	
Rabbits	dig prot %	* 1.9	12.1	
Sheep	dig prot %	2.2	14.1	
Energy	GE Mcal/kg			
Cattle	DE Mcal/kg	* 0.41	2.57	
Sheep	DE Mcal/kg	* 0.48	3.05	
Cattle	ME Mcal/kg	* 0.33	2.11	
Sheep	ME Mcal/kg	* 0.40	2.50	
Cattle	TDN %	* 9.3	58.4	
Sheep	TDN %	11.0	69.1	
Calcium	%	0.16	1.01	
Chlorine	%	0.09	0.56	58
Copper	mg/kg	2.2	13.6	
Iron	%	0.003	0.017	
Magnesium	%	0.17	1.07	
Manganese	mg/kg	7.7	48.4	
Phosphorus	%	0.04	0.22	
Potassium	%	0.92	5.79	
Sodium	%	0.09	0.54	19
Sulphur	%	0.09	0.57	13
Carotene	mg/kg	5.5	34.8	
Riboflavin	mg/kg	1.1	6.6	
Vitamin A equivalent	IU/g	9.2	58.1	

Feed Name or Analyses			Mean As Fed	Mean Dry	C.V. ± %

Beet, sugar, aerial part w crowns, fresh, pre-bloom, (2)

Ref No 2 00 645 United States

Feed Name or Analyses			As Fed	Dry	C.V. ± %
Dry matter	%		13.4	100.0	
Ash	%		3.0	22.4	
Crude fiber	%		1.3	9.5	
Sheep	dig coef %		82.	82.	
Ether extract	%		0.2	1.4	
Sheep	dig coef %		54.	54.	
N-free extract	%		7.3	54.3	
Sheep	dig coef %		73.	73.	
Protein (N x 6.25)	%		1.7	12.4	
Sheep	dig coef %		79.	79.	
Cattle	dig prot %	*	1.1	8.4	
Goats	dig prot %	*	1.1	8.1	
Horses	dig prot %	*	1.1	8.1	
Rabbits	dig prot %	*	1.1	8.2	
Sheep	dig prot %		1.3	9.8	
Energy	GE Mcal/kg				
Cattle	DE Mcal/kg	*	0.33	2.46	
Sheep	DE Mcal/kg	*	0.35	2.60	
Cattle	ME Mcal/kg	*	0.27	2.02	
Sheep	ME Mcal/kg	*	0.29	2.13	
Cattle	TDN %	*	7.5	55.7	
Sheep	TDN %		7.9	58.9	

Beet, sugar, aerial part w crowns, fresh, full bloom, (2)

Ref No 2 00 646 United States

Feed Name or Analyses			As Fed	Dry	C.V. ± %
Dry matter	%		22.2	100.0	32
Chlorine	%		0.05	0.22	
Sodium	%		0.12	0.56	
Sulphur	%		0.12	0.56	
Carotene	mg/kg		7.9	35.5	61
Vitamin A equivalent	IU/g		13.1	59.2	

Beet, sugar, crowns, fresh, (2)

Ref No 2 00 648 United States

Feed Name or Analyses			As Fed	Dry	C.V. ± %
Dry matter	%		18.0	100.0	
Ash	%		3.6	19.9	
Crude fiber	%		1.9	10.4	
Sheep	dig coef %		61.	61.	
Ether extract	%		0.3	1.8	
Sheep	dig coef %		38.	38.	
N-free extract	%		9.2	51.1	
Sheep	dig coef %		76.	76.	
Protein (N x 6.25)	%		3.0	16.8	
Sheep	dig coef %		79.	79.	
Cattle	dig prot %	*	2.2	12.2	
Goats	dig prot %	*	2.2	12.2	
Horses	dig prot %	*	2.1	11.8	
Rabbits	dig prot %	*	2.1	11.6	
Sheep	dig prot %		2.4	13.3	
Energy	GE Mcal/kg				
Cattle	DE Mcal/kg	*	0.46	2.57	
Sheep	DE Mcal/kg	*	0.47	2.65	
Cattle	ME Mcal/kg	*	0.38	2.11	
Sheep	ME Mcal/kg	*	0.39	2.17	
Cattle	TDN %	*	10.5	58.3	
Sheep	TDN %		10.8	60.0	
Biotin	mg/kg		0.00	0.00	
Folic acid	mg/kg		0.04	0.24	
Niacin	mg/kg		8.5	47.4	
Pantothenic acid	mg/kg		5.0	28.0	
Riboflavin	mg/kg		1.1	6.4	
Thiamine	mg/kg		1.1	6.0	
Vitamin B_6	mg/kg		1.62	9.04	

Beet, sugar, leaves, fresh, (2)

Ref No 2 00 650 United States

Feed Name or Analyses			As Fed	Dry	C.V. ± %
Dry matter	%			100.0	
Calcium	%			1.26	
Chlorine	%			2.86	
Copper	mg/kg			6.0	
Iron	%			0.018	
Magnesium	%			0.81	
Manganese	mg/kg			371.3	
Phosphorus	%			0.22	
Potassium	%			6.00	
Sulphur	%			0.48	
Biotin	mg/kg			0.02	
Folic acid	mg/kg			0.31	
Niacin	mg/kg			10.8	
Pantothenic acid	mg/kg			5.7	
Riboflavin	mg/kg			5.5	
Thiamine	mg/kg			2.4	
a-tocopherol	mg/kg			460.8	
Vitamin B_6	mg/kg			4.41	

Beet, sugar, stems, fresh, (2)

Ref No 2 00 652 United States

Feed Name or Analyses			As Fed	Dry	C.V. ± %
Dry matter	%		9.0	100.0	
Ash	%		2.1	23.4	
Crude fiber	%		1.2	13.5	
Protein (N x 6.25)	%		1.4	15.6	
Cattle	dig prot %	*	1.0	11.2	
Goats	dig prot %	*	1.0	11.1	
Horses	dig prot %	*	1.0	10.8	
Rabbits	dig prot %	*	1.0	10.7	
Sheep	dig prot %	*	1.0	11.5	
Calcium	%		0.09	1.02	
Magnesium	%		0.10	1.07	
Phosphorus	%		0.02	0.22	
Potassium	%		0.24	2.63	
Sulphur	%		0.05	0.58	
Biotin	mg/kg		0.00	0.02	
Carotene	mg/kg		3.1	34.4	27
Folic acid	mg/kg		0.01	0.11	
Niacin	mg/kg		0.3	3.5	
Pantothenic acid	mg/kg		0.2	2.6	
Riboflavin	mg/kg		0.3	3.5	
Thiamine	mg/kg		0.0	0.4	
Vitamin B_6	mg/kg		0.16	1.76	
Vitamin A equivalent	IU/g		5.2	57.3	

Beet, sugar, aerial part, ensiled, (3)

Ref No 3 00 654 United States

Feed Name or Analyses			As Fed	Dry	C.V. ± %
Dry matter	%		15.4	100.0	
Ash	%		4.1	26.7	
Crude fiber	%		1.7	11.3	
Sheep	dig coef %		80.	80.	
Ether extract	%		0.5	3.0	
Sheep	dig coef %		45.	45.	
N-free extract	%		7.3	47.7	
Sheep	dig coef %		79.	79.	
Protein (N x 6.25)	%		1.8	11.5	
Sheep	dig coef %		68.	68.	
Cattle	dig prot %	*	1.0	6.6	
Goats	dig prot %	*	1.0	6.6	
Horses	dig prot %	*	1.0	6.6	
Sheep	dig prot %		1.2	7.7	
Energy	GE Mcal/kg				
Sheep	DE Mcal/kg	*	0.39	2.53	
Sheep	ME Mcal/kg	*	0.32	2.07	
Sheep	TDN %		8.8	57.4	

Ref No 3 00 654 Canada

Feed Name or Analyses			As Fed	Dry	C.V. ± %
Dry matter	%			100.0	
Ash	%			33.3	
Crude fiber	%			13.3	
Sheep	dig coef %			80.	
Ether extract	%			3.5	
Sheep	dig coef %			45.	
N-free extract	%			34.7	
Sheep	dig coef %			79.	
Protein (N x 6.25)	%			15.2	
Sheep	dig coef %			68.	
Cattle	dig prot %	*		10.1	
Goats	dig prot %	*		10.1	
Horses	dig prot %	*		10.1	
Sheep	dig prot %			10.3	
Energy	GE Mcal/kg				
Cattle	DE Mcal/kg	*		2.41	
Sheep	DE Mcal/kg	*		2.28	
Cattle	ME Mcal/kg	*		1.98	
Sheep	ME Mcal/kg	*		1.87	
Cattle	TDN %	*		54.6	
Sheep	TDN %			51.8	
Calcium	%			1.47	
Phosphorus	%			0.36	

Beet, sugar, aerial part w AIV preservative added, ensiled, (3)

Ref No 3 00 657 United States

Feed Name or Analyses			As Fed	Dry	C.V. ± %
Dry matter	%		20.5	100.0	
Ash	%		7.5	36.5	
Crude fiber	%		2.3	11.4	
Ether extract	%		0.3	1.7	
N-free extract	%		8.0	38.9	
Protein (N x 6.25)	%		2.4	11.5	
Cattle	dig prot %	*	1.4	6.7	
Goats	dig prot %	*	1.4	6.7	
Horses	dig prot %	*	1.4	6.7	
Sheep	dig prot %	*	1.4	6.7	
Energy	GE Mcal/kg				
Sheep	DE Mcal/kg	*	0.47	2.30	
Sheep	ME Mcal/kg	*	0.39	1.88	
Sheep	TDN %	*	10.7	52.1	

Beet, sugar, aerial part w AIV preservative added, ensiled, pre-bloom, (3)

Ref No 3 00 656 United States

Feed Name or Analyses			As Fed	Dry	C.V. ± %
Dry matter	%		20.5	100.0	
Ash	%		7.5	36.5	
Crude fiber	%		2.3	11.4	
Sheep	dig coef %		85.	85.	
Ether extract	%		0.3	1.7	
Sheep	dig coef %		35.	35.	
N-free extract	%		8.0	38.9	
Sheep	dig coef %		75.	75.	
Protein (N x 6.25)	%		2.4	11.5	
Sheep	dig coef %		62.	62.	
Cattle	dig prot %	*	1.4	6.7	

Continued

Feed Name or Analyses			Mean As Fed	Mean Dry	C.V. ± %
Goats	dig prot %	*	1.4	6.7	
Horses	dig prot %	*	1.4	6.7	
Sheep	dig prot %		1.5	7.1	
Energy	GE Mcal/kg				
Sheep	DE Mcal/kg	*	0.43	2.09	
Sheep	ME Mcal/kg	*	0.35	1.71	
Sheep	TDN %		9.7	47.3	

Beet, sugar, aerial part w crowns, ensiled, (3)

Ref No 3 00 660 United States

Feed Name or Analyses			Mean As Fed	Mean Dry	C.V. ± %
Dry matter	%		22.5	100.0	
Ash	%		7.4	33.0	
Crude fiber	%		2.8	12.6	
Sheep	dig coef %		80.	80.	
Ether extract	%		0.6	2.6	
Sheep	dig coef %		45.	45.	
N-free extract	%		8.8	38.9	
Sheep	dig coef %		80.	80.	
Protein (N x 6.25)	%		2.9	13.0	
Sheep	dig coef %		71.	71.	
Cattle	dig prot %	*	1.8	8.0	
Goats	dig prot %	*	1.8	8.0	
Horses	dig prot %	*	1.8	8.0	
Sheep	dig prot %		2.1	9.1	
Energy	GE Mcal/kg				
Sheep	DE Mcal/kg	*	0.53	2.33	
Sheep	ME Mcal/kg	*	0.43	1.91	
Cattle	NE_m Mcal/kg	*	0.26	1.14	
Cattle	NE_{gain} Mcal/kg	*	0.09	0.42	
Sheep	TDN %		11.9	52.9	
Calcium	%		0.22	0.98	
Phosphorus	%		0.05	0.22	
Potassium	%		1.28	5.70	

Beet, sugar, aerial part w crowns, washed ensiled, (3)

Ref No 3 00 661 United States

Feed Name or Analyses			Mean As Fed	Mean Dry	C.V. ± %
Dry matter	%		13.3	100.0	
Ash	%		3.1	23.0	
Crude fiber	%		2.0	15.3	
Sheep	dig coef %		77.	77.	
Ether extract	%		0.3	2.3	
Sheep	dig coef %		38.	38.	
N-free extract	%		6.2	46.8	
Sheep	dig coef %		79.	79.	
Protein (N x 6.25)	%		1.7	12.6	
Sheep	dig coef %		62.	62.	
Cattle	dig prot %	*	1.0	7.7	
Goats	dig prot %	*	1.0	7.7	
Horses	dig prot %	*	1.0	7.7	
Sheep	dig prot %		1.0	7.8	
Energy	GE Mcal/kg				
Sheep	DE Mcal/kg	*	0.34	2.58	
Sheep	ME Mcal/kg	*	0.28	2.11	
Sheep	TDN %		7.8	58.4	

(1) dry forages and roughages (2) pasture, range plants, and forages fed green (3) silages (4) energy feeds (5) protein supplements (6) minerals (7) vitamins (8) additives

Beet, sugar, aerial part w phosphorus pentachloride preservative added, ensiled, (3)

Ref No 3 00 658 United States

Feed Name or Analyses			Mean As Fed	Mean Dry	C.V. ± %
Dry matter	%		16.9	100.0	
Ash	%		3.1	18.4	
Crude fiber	%		1.7	10.1	
Sheep	dig coef %		79.	79.	
Ether extract	%		0.3	1.5	
Sheep	dig coef %		13.	13.	
N-free extract	%		9.5	56.5	
Sheep	dig coef %		87.	87.	
Protein (N x 6.25)	%		2.3	13.5	
Sheep	dig coef %		77.	77.	
Cattle	dig prot %	*	1.4	8.5	
Goats	dig prot %	*	1.4	8.5	
Horses	dig prot %	*	1.4	8.5	
Sheep	dig prot %		1.8	10.4	
Energy	GE Mcal/kg				
Cattle	DE Mcal/kg	*	0.49	2.91	
Sheep	DE Mcal/kg	*	0.51	3.00	
Cattle	ME Mcal/kg	*	0.40	2.39	
Sheep	ME Mcal/kg	*	0.42	2.46	
Cattle	TDN %	*	11.2	66.1	
Sheep	TDN %		11.5	68.0	

Beet, sugar, aerial part w potato flakes added, ensiled, (3)

Ref No 3 00 659 United States

Feed Name or Analyses			Mean As Fed	Mean Dry	C.V. ± %
Dry matter	%		20.4	100.0	
Ash	%		7.6	37.1	
Crude fiber	%		1.8	8.8	
Sheep	dig coef %		75.	75.	
Ether extract	%		0.2	1.0	
Sheep	dig coef %		29.	29.	
N-free extract	%		8.9	43.6	
Sheep	dig coef %		85.	85.	
Protein (N x 6.25)	%		1.9	9.5	
Sheep	dig coef %		60.	60.	
Cattle	dig prot %	*	1.0	4.9	
Goats	dig prot %	*	1.0	4.9	
Horses	dig prot %	*	1.0	4.9	
Sheep	dig prot %		1.2	5.7	
Energy	GE Mcal/kg				
Sheep	DE Mcal/kg	*	0.45	2.21	
Sheep	ME Mcal/kg	*	0.37	1.81	
Sheep	TDN %		10.2	50.0	

Beet, sugar, aerial part w pulp, ensiled, (3)

Ref No 3 00 663 United States

Feed Name or Analyses			Mean As Fed	Mean Dry	C.V. ± %
Dry matter	%		15.9	100.0	
Ash	%		2.1	13.0	
Crude fiber	%		2.1	13.1	
Sheep	dig coef %		82.	82.	
Ether extract	%		0.6	3.8	
Sheep	dig coef %		65.	65.	
N-free extract	%		8.8	55.2	
Sheep	dig coef %		85.	85.	
Protein (N x 6.25)	%		2.4	14.9	
Sheep	dig coef %		68.	68.	
Cattle	dig prot %	*	1.6	9.8	
Goats	dig prot %	*	1.6	9.8	
Horses	dig prot %	*	1.6	9.8	
Sheep	dig prot %		1.6	10.1	
Energy	GE Mcal/kg				
Sheep	DE Mcal/kg	*	0.51	3.23	

Feed Name or Analyses			Mean As Fed	Mean Dry	C.V. ± %
Sheep	ME Mcal/kg	*	0.42	2.65	
Sheep	TDN %		11.7	73.4	

Beet, sugar, pulp, ensiled, (3)

Ref No 3 00 662 United States

Feed Name or Analyses			Mean As Fed	Mean Dry	C.V. ± %
Dry matter	%		12.2	100.0	11
Ash	%		0.6	5.1	25
Crude fiber	%		3.9	32.0	20
Sheep	dig coef %		68.	68.	
Ether extract	%		0.2	1.9	42
Sheep	dig coef %		-65.	-65.	
N-free extract	%		5.9	48.5	
Sheep	dig coef %		83.	83.	
Protein (N x 6.25)	%		1.5	12.5	9
Sheep	dig coef %		64.	64.	
Cattle	dig prot %	*	0.9	7.6	
Goats	dig prot %	*	0.9	7.6	
Horses	dig prot %	*	0.9	7.6	
Sheep	dig prot %		1.0	8.0	
Energy	GE Mcal/kg				
Cattle	DE Mcal/kg	*	0.37	3.04	
Sheep	DE Mcal/kg	*	0.36	2.96	
Cattle	ME Mcal/kg	*	0.30	2.49	
Sheep	ME Mcal/kg	*	0.30	2.43	
Cattle	TDN %	*	8.4	68.9	
Sheep	TDN %		8.2	67.2	

Beet, sugar, pulp w potato flakes added, ensiled, (3)

Ref No 3 00 664 United States

Feed Name or Analyses			Mean As Fed	Mean Dry	C.V. ± %
Dry matter	%		17.9	100.0	
Ash	%		0.9	5.1	
Crude fiber	%		3.0	16.5	
Sheep	dig coef %		78.	78.	
Ether extract	%		0.1	0.4	
Sheep	dig coef %		140.	140.	
N-free extract	%		12.1	67.8	
Sheep	dig coef %		90.	90.	
Protein (N x 6.25)	%		1.8	10.2	
Sheep	dig coef %		71.	71.	
Cattle	dig prot %	*	1.0	5.5	
Goats	dig prot %	*	1.0	5.5	
Horses	dig prot %	*	1.0	5.5	
Sheep	dig prot %		1.3	7.2	
Energy	GE Mcal/kg				
Cattle	DE Mcal/kg	*	0.63	3.51	
Sheep	DE Mcal/kg	*	0.65	3.63	
Cattle	ME Mcal/kg	*	0.52	2.88	
Sheep	ME Mcal/kg	*	0.53	2.98	
Cattle	TDN %	*	14.2	79.6	
Sheep	TDN %		14.7	82.4	

Beet, sugar, roots, ensiled, (3)

Ref No 3 00 665 United States

Feed Name or Analyses			Mean As Fed	Mean Dry	C.V. ± %
Dry matter	%		17.5	100.0	
Ash	%		0.9	5.3	
Crude fiber	%		2.1	12.2	
Sheep	dig coef %		62.	62.	
Swine	dig coef %		58.	58.	
Ether extract	%		0.4	2.4	
Sheep	dig coef %		32.	32.	
Swine	dig coef %		0.	0.	
N-free extract	%		12.5	71.5	
Sheep	dig coef %		71.	71.	
Swine	dig coef %		93.	93.	

Feed Name or Analyses		Mean As Fed	Mean Dry	C.V. ± %
Protein (N x 6.25)	%	1.5	8.8	
Sheep	dig coef %	39.	39.	
Swine	dig coef %	0.	0.	
Cattle	dig prot %	* 0.7	4.2	
Goats	dig prot %	* 0.7	4.2	
Horses	dig prot %	* 0.7	4.2	
Sheep	dig prot %	0.6	3.4	
Energy	GE Mcal/kg			
Cattle	DE Mcal/kg	* 0.52	2.97	
Sheep	DE Mcal/kg	* 0.49	2.80	
Swine	DE kcal/kg	* 567.	3241.	
Cattle	ME Mcal/kg	* 0.43	2.44	
Sheep	ME Mcal/kg	* 0.40	2.29	
Swine	ME kcal/kg	* 534.	3054.	
Cattle	TDN %	* 11.8	67.4	
Sheep	TDN %	11.1	63.4	
Swine	TDN %	12.9	73.5	

Beet, sugar, aerial part w crowns, s-c, (1)

Ref No 1 00 666 United States

Feed Name or Analyses		Mean As Fed	Mean Dry	C.V. ± %
Dry matter	%	89.3	100.0	
Ash	%	17.7	19.8	
Crude fiber	%	13.2	14.8	
Sheep	dig coef %	73.	73.	
Swine	dig coef %	45.	45.	
Ether extract	%	1.2	1.4	
Sheep	dig coef %	28.	28.	
Swine	dig coef %	-18.	-18.	
N-free extract	%	45.6	51.0	
Sheep	dig coef %	84.	84.	
Swine	dig coef %	82.	82.	
Protein (N x 6.25)	%	11.6	13.0	
Sheep	dig coef %	68.	68.	
Swine	dig coef %	48.	48.	
Cattle	dig prot %	* 7.1	8.0	
Goats	dig prot %	* 8.2	9.2	
Horses	dig prot %	* 8.2	9.2	
Sheep	dig prot %	7.9	8.9	
Swine	dig prot %	5.6	6.3	
Energy	GE Mcal/kg			
Cattle	DE Mcal/kg	* 2.38	2.66	
Sheep	DE Mcal/kg	* 2.48	2.78	
Swine	DE kcal/kg	*2122.	2377.	
Cattle	ME Mcal/kg	* 1.95	2.18	
Sheep	ME Mcal/kg	* 2.03	2.28	
Swine	ME kcal/kg	*1981.	2219.	
Cattle	TDN %	* 53.9	60.4	
Sheep	TDN %	56.3	63.0	
Swine	TDN %	48.1	53.9	

Beet, sugar, aerial part w crowns, washed chopped dehy, (1)

Ref No 1 00 667 United States

Feed Name or Analyses		Mean As Fed	Mean Dry	C.V. ± %
Dry matter	%	90.8	100.0	
Ash	%	15.2	16.7	
Crude fiber	%	10.3	11.3	
Sheep	dig coef %	59.	59.	
Ether extract	%	1.0	1.1	
Sheep	dig coef %	25.	25.	
N-free extract	%	53.4	58.8	
Sheep	dig coef %	79.	79.	
Protein (N x 6.25)	%	11.0	12.1	
Sheep	dig coef %	62.	62.	
Cattle	dig prot %	* 6.5	7.1	
Goats	dig prot %	* 7.6	8.3	
Horses	dig prot %	* 7.6	8.3	
Sheep	dig prot %	6.8	7.5	
Energy	GE Mcal/kg			
Cattle	DE Mcal/kg	* 2.32	2.56	
Sheep	DE Mcal/kg	* 2.45	2.70	
Cattle	ME Mcal/kg	* 1.91	2.10	
Sheep	ME Mcal/kg	* 2.01	2.21	
Cattle	TDN %	* 52.7	58.0	
Sheep	TDN %	55.6	61.2	

Beet, sugar, aerial part w roots, dehy, (4)

Ref No 1 00 678 United States

Feed Name or Analyses		Mean As Fed	Mean Dry	C.V. ± %
Dry matter	%	86.9	100.0	
Ash	%	9.3	10.7	
Crude fiber	%	15.0	17.3	
Sheep	dig coef %	85.	85.	
Ether extract	%	0.5	0.6	
Sheep	dig coef %	-188.	-188.	
N-free extract	%	53.7	61.8	
Sheep	dig coef %	88.	88.	
Protein (N x 6.25)	%	8.3	9.6	
Sheep	dig coef %	65.	65.	
Cattle	dig prot %	* 4.2	4.8	
Goats	dig prot %	* 5.2	6.0	
Horses	dig prot %	* 5.2	6.0	
Sheep	dig prot %	5.4	6.2	
Energy	GE Mcal/kg			
Cattle	DE Mcal/kg	* 2.67	3.07	
Sheep	DE Mcal/kg	* 2.79	3.21	
Cattle	ME Mcal/kg	* 2.19	2.52	
Sheep	ME Mcal/kg	* 2.29	2.63	
Cattle	TDN %	* 60.9	69.7	
Sheep	TDN %	63.2	72.8	

Beet, sugar, molasses, Steffens, (4)

Ref No 4 08 351 United States

Feed Name or Analyses		Mean As Fed	Mean Dry	C.V. ± %
Dry matter	%	78.7	100.0	
Ash	%	8.8	11.2	
Crude fiber	%	0.0	0.0	
Ether extract	%	0.0	0.0	
N-free extract	%	62.1	78.9	
Protein (N x 6.25)	%	7.8	9.9	
Cattle	dig prot %	* 4.0	5.1	
Goats	dig prot %	* 5.0	6.3	
Horses	dig prot %	* 5.0	6.3	
Sheep	dig prot %	* 5.0	6.3	
Energy	GE Mcal/kg			
Cattle	DE Mcal/kg	* 2.90	3.68	
Sheep	DE Mcal/kg	* 2.80	3.56	
Swine	DE kcal/kg	*2539.	3226.	
Cattle	ME Mcal/kg	* 2.38	3.02	
Sheep	ME Mcal/kg	* 2.29	2.92	
Swine	ME kcal/kg	*2386.	3032.	
Cattle	TDN %	* 65.7	83.5	
Sheep	TDN %	* 63.5	80.6	
Swine	TDN %	* 57.6	73.2	
Calcium	%	0.11	0.14	
Chlorine	%	1.99	2.53	
Magnesium	%	0.01	0.01	
Phosphorus	%	0.02	0.03	
Potassium	%	4.66	5.92	
Sodium	%	0.92	1.17	
Sulphur	%	0.44	0.56	

Beet, sugar, molasses, mn 48% invert sugar mn 79.5 degrees brix, (4)

Beet molasses (AAFCO)

Molasses (CFA)

Ref No 4 00 668 United States

Feed Name or Analyses		Mean As Fed	Mean Dry	C.V. ± %
Dry matter	%	77.5	100.0	7
Ash	%	8.9	11.4	13
Crude fiber	%	0.0	0.0	13
Ether extract	%	0.1	0.1	98
N-free extract	%	61.9	79.9	
Cattle	dig coef %	* 105.	105.	
Sheep	dig coef %	* 94.	94.	
Protein (N x 6.25)	%	6.6	8.5	27
Cattle	dig coef %	* 57.	57.	
Sheep	dig coef %	* -2.	-2.	
Cattle	dig prot %	3.8	4.9	
Goats	dig prot %	* 3.9	5.1	
Horses	dig prot %	* 3.9	5.1	
Sheep	dig prot %	-0.1	-0.2	
Energy	GE Mcal/kg			
Cattle	DE Mcal/kg	* 2.69	3.47	
Sheep	DE Mcal/kg	* 2.77	3.57	
Swine	DE kcal/kg	*2463.	3178.	
Cattle	ME Mcal/kg	* 2.20	2.84	
Chickens	ME_n kcal/kg	1930.	2490.	
Sheep	ME Mcal/kg	* 2.27	2.93	
Swine	ME kcal/kg	*2322.	2996.	
Cattle	NE_m Mcal/kg	* 1.58	2.04	
Cattle	NE_{gain} Mcal/kg	* 1.05	1.36	
Cattle	$NE_{lactating\ cows}$ Mcal/kg	* 1.95	2.52	
Cattle	TDN %	* 61.1	78.8	
Sheep	TDN %	* 62.8	81.1	
Swine	TDN %	* 55.9	72.1	
Calcium	%	0.12	0.16	80
Chlorine	%	1.25	1.62	
Cobalt	mg/kg	0.381	0.492	
Copper	mg/kg	17.4	22.5	
Iron	%	0.005	0.006	
Magnesium	%	0.23	0.29	86
Manganese	mg/kg	4.6	5.9	
Phosphorus	%	0.03	0.03	76
Potassium	%	4.71	6.07	
Sodium	%	1.15	1.49	
Sulphur	%	0.47	0.61	
Choline	mg/kg	867.	1119.	
Niacin	mg/kg	42.2	54.5	12
Pantothenic acid	mg/kg	4.6	6.0	
Riboflavin	mg/kg	2.4	3.1	

Beet, sugar, pulp, dehy, (4)

Dried beet pulp (AAFCO)

Dried beet pulp (CFA)

Ref No 4 00 669 United States

Feed Name or Analyses		Mean As Fed	Mean Dry	C.V. ± %
Dry matter	%	90.6	100.0	3
Ash	%	4.8	5.3	27
Crude fiber	%	18.2	20.1	10
Cattle	dig coef %	69.	69.	
Horses	dig coef %	79.	79.	
Sheep	dig coef %	79.	79.	
Swine	dig coef %	83.	83.	
Ether extract	%	0.5	0.6	45
Cattle	dig coef %	-138.	-138.	
Horses	dig coef %	34.	34.	
Sheep	dig coef %	-212.	-212.	
Swine	dig coef %	-233.	-233.	

Continued

Feed Name or Analyses			Mean As Fed	Mean Dry	C.V. ± %
N-free extract		%	58.4	64.5	2
Cattle	dig coef	%	86.	86.	
Horses	dig coef	%	91.	91.	
Sheep	dig coef	%	86.	86.	
Swine	dig coef	%	89.	89.	
Protein (N x 6.25)		%	8.7	9.6	8
Cattle	dig coef	%	45.	45.	
Horses	dig coef	%	69.	69.	
Sheep	dig coef	%	54.	54.	
Swine	dig coef	%	41.	41.	
Cattle	dig prot	%	3.9	4.3	
Goats	dig prot	%	* 5.5	6.0	
Horses	dig prot	%	6.0	6.6	
Sheep	dig prot	%	4.7	5.2	
Swine	dig prot	%	3.6	3.9	
Pectins		%	11.0	12.1	62
Pentosans		%	24.0	26.4	
Lignin (Ellis)		%	5.0	5.5	
Energy	GE	Mcal/kg	3.84	4.24	2
Cattle	DE	Mcal/kg	* 2.87	3.17	
Horses	DE	Mcal/kg	* 3.26	3.60	
Sheep	DE	Mcal/kg	* 2.95	3.25	
Swine	DE	kcal/kg	*3000.	3310.	
Cattle	ME	Mcal/kg	* 2.36	2.60	
Chickens	ME_n	kcal/kg	611.	674.	
Horses	ME	Mcal/kg	* 2.67	2.95	
Sheep	ME	Mcal/kg	* 2.42	2.67	
Swine	ME	kcal/kg	*2822.	3113.	
Cattle	NE_m	Mcal/kg	* 1.44	1.60	
Cattle	NE_{gain}	Mcal/kg	* 0.93	1.03	
Cattle	$NE_{lactating\ cows}$	Mcal/kg	* 1.71	1.90	
Cattle	TDN	%	65.2	71.9	
Horses	TDN	%	74.0	81.6	
Sheep	TDN	%	66.8	73.7	
Swine	TDN	%	68.0	75.1	
Calcium		%	0.68	0.75	19
Chlorine		%	0.04	0.04	
Cobalt		mg/kg	0.101	0.112	46
Copper		mg/kg	12.5	13.8	25
Iron		%	0.030	0.033	40
Magnesium		%	0.27	0.30	98
Manganese		mg/kg	34.9	38.5	40
Phosphorus		%	0.09	0.10	19
Potassium		%	0.19	0.21	39
Sodium		%	0.17	0.19	25
Sulphur		%	0.20	0.22	40
Zinc		mg/kg	0.7	0.7	
Carotene		mg/kg	0.2	0.2	
Choline		mg/kg	824.	909.	21
Niacin		mg/kg	16.2	17.9	48
Pantothenic acid		mg/kg	1.5	1.7	34
Riboflavin		mg/kg	0.7	0.7	46
Thiamine		mg/kg	0.4	0.5	34
Vitamin A equivalent		IU/g	0.4	0.4	
Vitamin D_2		IU/g	0.6	0.7	
Arginine		%	0.30	0.33	15
Cystine		%	0.01	0.01	
Histidine		%	0.20	0.22	
Isoleucine		%	0.30	0.33	29
Leucine		%	0.60	0.66	22
Lysine		%	0.65	0.71	14
Methionine		%	0.00	0.01	
Phenylalanine		%	0.30	0.33	
Threonine		%	0.40	0.44	
Tryptophan		%	0.09	0.10	

(1) dry forages and roughages (2) pasture, range plants, and forages fed green (3) silages (4) energy feeds (5) protein supplements (6) minerals (7) vitamins (8) additives

Feed Name or Analyses			Mean As Fed	Mean Dry	C.V. ± %
Tyrosine		%	0.40	0.44	22
Valine		%	0.40	0.44	11

Ref No 4 00 669 — Canada

Feed Name or Analyses			Mean As Fed	Mean Dry	C.V. ± %
Dry matter		%	91.4	100.0	
Ash		%	5.9	6.5	
Crude fiber		%	12.1	13.2	
Cattle	dig coef	%	69.	69.	
Horses	dig coef	%	79.	79.	
Sheep	dig coef	%	79.	79.	
Swine	dig coef	%	83.	83.	
Ether extract		%			
Cattle	dig coef	%	-138.	-138.	
Horses	dig coef	%	34.	34.	
Sheep	dig coef	%	-212.	-212.	
Swine	dig coef	%	-233.	-233.	
N-free extract		%			
Cattle	dig coef	%	86.	86.	
Horses	dig coef	%	91.	91.	
Sheep	dig coef	%	86.	86.	
Swine	dig coef	%	89.	89.	
Protein (N x 6.25)		%			
Cattle	dig coef	%	45.	45.	
Horses	dig coef	%	69.	69.	
Sheep	dig coef	%	54.	54.	
Swine	dig coef	%	41.	41.	
Energy	GE	Mcal/kg	3.87	4.24	
Cattle	DE	Mcal/kg	* 2.90	3.17	
Horses	DE	Mcal/kg	* 3.29	3.60	
Sheep	DE	Mcal/kg	* 2.97	3.25	
Swine	DE	kcal/kg	*3023.	3307.	
Cattle	ME	Mcal/kg	* 2.38	2.60	
Chickens	ME_n	kcal/kg	616.	674.	
Horses	ME	Mcal/kg	* 2.70	2.95	
Sheep	ME	Mcal/kg	* 2.44	2.66	
Cattle	TDN	%	65.7	71.9	
Horses	TDN	%	74.5	81.6	
Sheep	TDN	%	67.4	73.7	
Swine	TDN	%	68.6	75.0	
Silicon		%	0.17	0.19	

Beet, sugar, pulp, steamed, (4)
Ref No 4 00 670 — United States

Feed Name or Analyses			Mean As Fed	Mean Dry	C.V. ± %
Dry matter		%	90.2	100.0	
Ash		%	5.1	5.6	
Crude fiber		%	14.2	15.7	
Swine	dig coef	%	91.	91.	
Ether extract		%	0.4	0.4	
Swine	dig coef	%	26.	26.	
N-free extract		%	61.0	67.6	
Swine	dig coef	%	92.	92.	
Protein (N x 6.25)		%	9.7	10.7	
Swine	dig coef	%	63.	63.	
Cattle	dig prot	%	* 5.3	5.8	
Goats	dig prot	%	* 6.4	7.0	
Horses	dig prot	%	* 6.4	7.0	
Sheep	dig prot	%	* 6.4	7.0	
Swine	dig prot	%	6.1	6.7	
Energy	GE	Mcal/kg			
Cattle	DE	Mcal/kg	* 3.50	3.88	
Sheep	DE	Mcal/kg	* 2.94	3.25	
Swine	DE	kcal/kg	*3319.	3680.	
Cattle	ME	Mcal/kg	* 2.87	3.18	
Sheep	ME	Mcal/kg	* 2.41	2.67	
Swine	ME	kcal/kg	*3115.	3453.	
Cattle	TDN	%	* 79.3	87.9	

Feed Name or Analyses			Mean As Fed	Mean Dry	C.V. ± %
Sheep	TDN	%	* 66.6	73.8	
Swine	TDN	%	75.3	83.5	

Beet, sugar, pulp, wet, (4)
Ref No 2 00 671 — United States

Feed Name or Analyses			Mean As Fed	Mean Dry	C.V. ± %
Dry matter		%	11.3	100.0	
Ash		%	0.5	4.7	
Crude fiber		%	3.4	30.1	
Cattle	dig coef	%	* 69.	69.	
Sheep	dig coef	%	82.	82.	
Ether extract		%	0.2	2.1	
Cattle	dig coef	%	*-138.	-138.	
Sheep	dig coef	%	-71.	-71.	
N-free extract		%	5.8	51.4	
Cattle	dig coef	%	* 86.	86.	
Sheep	dig coef	%	88.	88.	
Protein (N x 6.25)		%	1.3	11.7	
Cattle	dig coef	%	* 45.	45.	
Sheep	dig coef	%	55.	55.	
Cattle	dig prot	%	0.6	5.3	
Goats	dig prot	%	* 0.9	8.0	
Horses	dig prot	%	* 0.9	8.0	
Sheep	dig prot	%	0.7	6.4	
Energy	GE	Mcal/kg			
Cattle	DE	Mcal/kg	* 0.32	2.81	
Sheep	DE	Mcal/kg	* 0.36	3.22	
Cattle	ME	Mcal/kg	* 0.26	2.31	
Sheep	ME	Mcal/kg	* 0.30	2.64	
Cattle	NE_m	Mcal/kg	* 0.17	1.52	
Cattle	NE_{gain}	Mcal/kg	* 0.11	0.95	
Cattle	$NE_{lactating\ cows}$	Mcal/kg	* 0.20	1.75	
Cattle	TDN	%	7.2	63.8	
Sheep	TDN	%	8.3	73.0	
Calcium		%	0.10	0.86	
Phosphorus		%	0.01	0.10	
Potassium		%	0.02	0.19	

Beet, sugar, pulp, wet pressed, (4)
Ref No 2 08 582 — United States

Feed Name or Analyses			Mean As Fed	Mean Dry	C.V. ± %
Dry matter		%	14.2	100.0	
Ash		%	0.7	4.9	
Crude fiber		%	4.6	32.4	
Ether extract		%	0.4	2.8	
N-free extract		%	7.1	50.0	
Protein (N x 6.25)		%	1.4	9.9	
Cattle	dig prot	%	* 0.7	5.1	
Goats	dig prot	%	* 0.9	6.3	
Horses	dig prot	%	* 0.9	6.3	
Sheep	dig prot	%	* 0.9	6.3	
Energy	GE	Mcal/kg			
Cattle	DE	Mcal/kg	* 0.48	3.37	
Sheep	DE	Mcal/kg	* 0.45	3.20	
Cattle	ME	Mcal/kg	* 0.39	2.76	
Sheep	ME	Mcal/kg	* 0.37	2.62	
Cattle	TDN	%	* 10.8	76.4	
Sheep	TDN	%	* 10.3	72.5	

Beet, sugar, pulp w molasses, dehy, (4)
Ref No 4 00 672 — United States

Feed Name or Analyses			Mean As Fed	Mean Dry	C.V. ± %
Dry matter		%	92.3	100.0	3
Ash		%	5.7	6.2	7
Crude fiber		%	13.0	14.1	44
Cattle	dig coef	%	86.	86.	
Sheep	dig coef	%	78.	78.	

Feed Name or Analyses			As Fed	Dry	C.V. ± %
Swine	dig coef	%	84.	84.	
Ether extract		%	0.7	0.7	44
Cattle	dig coef	%	9.	9.	
Sheep	dig coef	%	-127.	-127.	
Swine	dig coef	%	-17.	-17.	
N-free extract		%	63.8	69.2	9
Cattle	dig coef	%	79.	79.	
Sheep	dig coef	%	91.	91.	
Swine	dig coef	%	89.	89.	
Protein (N x 6.25)		%	9.0	9.8	5
Cattle	dig coef	%	66.	66.	
Sheep	dig coef	%	66.	66.	
Swine	dig coef	%	25.	25.	
Cattle	dig prot	%	6.0	6.5	
Goats	dig prot	%	* 5.7	6.2	
Horses	dig prot	%	* 5.7	6.2	
Sheep	dig prot	%	6.0	6.5	
Swine	dig prot	%	2.3	2.4	
Energy	GE	Mcal/kg			
Cattle	DE	Mcal/kg	* 2.99	3.24	
Sheep	DE	Mcal/kg	* 3.19	3.45	
Swine	DE	kcal/kg	*3074.	3331.	
Cattle	ME	Mcal/kg	* 2.45	2.65	
Chickens	ME_n	kcal/kg	660.	715.	
Sheep	ME	Mcal/kg	* 2.61	2.83	
Swine	ME	kcal/kg	*2891.	3132.	
Cattle	NE_m	Mcal/kg	* 1.86	2.03	
Cattle	NE_{gain}	Mcal/kg	* 1.24	1.34	
Cattle	$NE_{lactating\ cows}$	Mcal/kg	* 1.82	1.97	
Cattle	TDN	%	67.7	73.4	
Sheep	TDN	%	72.2	78.3	
Swine	TDN	%	69.7	75.6	
Calcium		%	0.56	0.61	18
Magnesium		%	0.15	0.16	
Phosphorus		%	0.08	0.08	29
Potassium		%	1.64	1.78	
Sulphur		%	0.39	0.42	
Carotene		mg/kg	0.2	0.2	
Choline		mg/kg	826.	895.	
Niacin		mg/kg	16.3	17.7	
Pantothenic acid		mg/kg	1.5	1.7	
Riboflavin		mg/kg	0.7	0.7	
Vitamin A equivalent		IU/g	0.4	0.4	

Beet, sugar, pulp w Steffens filtrate, dehy, (4)
Dried beet product (AAFCO)
Ref No 4 00 675 United States

Feed Name or Analyses			As Fed	Dry	C.V. ± %
Dry matter		%	91.9	100.0	
Ash		%	6.1	6.6	
Crude fiber		%	15.7	17.1	
Ether extract		%	0.3	0.4	
N-free extract		%	58.9	64.1	
Protein (N x 6.25)		%	10.9	11.9	
Cattle	dig prot	%	* 6.3	6.9	
Goats	dig prot	%	* 7.4	8.1	
Horses	dig prot	%	* 7.4	8.1	
Sheep	dig prot	%	* 7.4	8.1	
Energy	GE	Mcal/kg			
Cattle	DE	Mcal/kg	* 3.45	3.75	
Sheep	DE	Mcal/kg	* 2.91	3.16	
Swine	DE	kcal/kg	*1780.	1937.	
Cattle	ME	Mcal/kg	* 2.83	3.08	
Sheep	ME	Mcal/kg	* 2.38	2.59	
Swine	ME	kcal/kg	*1666.	1813.	
Cattle	TDN	%	* 78.2	85.1	
Sheep	TDN	%	* 65.9	71.8	
Swine	TDN	%	* 40.4	43.9	

Beet, sugar, roots, (4)
Ref No 2 00 677 United States

Feed Name or Analyses			As Fed	Dry	C.V. ± %
Dry matter		%	20.1	100.0	32
Ash		%	1.1	5.5	41
Crude fiber		%	1.2	5.9	15
Cattle	dig coef	%	29.	29.	
Horses	dig coef	%	53.	53.	
Sheep	dig coef	%	89.	89.	
Swine	dig coef	%	88.	88.	
Ether extract		%	0.1	0.6	33
Cattle	dig coef	%	20.	20.	
Horses	dig coef	%	0.	0.	
Sheep	dig coef	%	50.	50.	
Swine	dig coef	%	-45.	-45.	
N-free extract		%	16.3	81.2	6
Cattle	dig coef	%	93.	93.	
Horses	dig coef	%	90.	90.	
Sheep	dig coef	%	100.	100.	
Swine	dig coef	%	97.	97.	
Protein (N x 6.25)		%	1.4	6.8	35
Cattle	dig coef	%	-34.	-34.	
Horses	dig coef	%	70.	70.	
Sheep	dig coef	%	82.	82.	
Swine	dig coef	%	36.	36.	
Cattle	dig prot	%	-0.4	-2.3	
Goats	dig prot	%	* 0.7	3.5	
Horses	dig prot	%	1.0	4.8	
Sheep	dig prot	%	1.1	5.6	
Swine	dig prot	%	0.5	2.5	
Energy	GE	Mcal/kg			
Cattle	DE	Mcal/kg	* 0.67	3.31	
Horses	DE	Mcal/kg	* 0.72	3.57	
Sheep	DE	Mcal/kg	* 0.82	4.07	
Swine	DE	kcal/kg	* 761.	3782.	
Cattle	ME	Mcal/kg	* 0.55	2.71	
Horses	ME	Mcal/kg	* 0.59	2.93	
Sheep	ME	Mcal/kg	* 0.67	3.34	
Swine	ME	kcal/kg	* 720.	3578.	
Cattle	TDN	%	15.1	75.1	
Horses	TDN	%	16.3	81.0	
Sheep	TDN	%	18.6	92.2	
Swine	TDN	%	17.3	85.8	
Calcium		%	0.05	0.24	
Chlorine		%	0.10	0.49	
Copper		mg/kg	1.6	8.1	
Iron		%	0.001	0.006	
Magnesium		%	0.04	0.18	
Manganese		mg/kg	41.4	205.7	
Phosphorus		%	0.05	0.24	
Potassium		%	0.31	1.52	
Sodium		%	0.10	0.49	
Sulphur		%	0.01	0.06	

Beet, sugar, roots, s-c, (4)
Ref No 4 00 676 United States

Feed Name or Analyses			As Fed	Dry	C.V. ± %
Dry matter		%	93.8	100.0	
Ash		%	3.7	3.9	
Crude fiber		%	4.5	4.8	
Cattle	dig coef	%	14.	14.	
Horses	dig coef	%	32.	32.	
Swine	dig coef	%	66.	66.	
Ether extract		%	0.2	0.2	
Cattle	dig coef	%	-392.	-392.	
Horses	dig coef	%	42.	42.	
Swine	dig coef	%	9.	9.	
N-free extract		%	81.4	86.7	
Cattle	dig coef	%	92.	92.	
Horses	dig coef	%	91.	91.	
Swine	dig coef	%	97.	97.	
Protein (N x 6.25)		%	4.1	4.3	
Cattle	dig coef	%	-38.	-38.	
Horses	dig coef	%	63.	63.	
Swine	dig coef	%	64.	64.	
Cattle	dig prot	%	-1.5	-1.6	
Goats	dig prot	%	* 1.1	1.2	
Horses	dig prot	%	2.6	2.7	
Sheep	dig prot	%	* 1.1	1.2	
Swine	dig prot	%	2.6	2.8	
Energy	GE	Mcal/kg			
Cattle	DE	Mcal/kg	* 3.17	3.38	
Horses	DE	Mcal/kg	* 3.45	3.68	
Sheep	DE	Mcal/kg	* 3.50	3.73	
Swine	DE	kcal/kg	*3728.	3973.	
Cattle	ME	Mcal/kg	* 2.60	2.77	
Horses	ME	Mcal/kg	* 2.83	3.02	
Sheep	ME	Mcal/kg	* 2.87	3.06	
Swine	ME	kcal/kg	*3546.	3779.	
Cattle	TDN	%	72.0	76.7	
Horses	TDN	%	78.3	83.4	
Sheep	TDN	%	* 79.4	84.6	
Swine	TDN	%	84.6	90.1	

Beet, sugar, solubles w low potassium salts and glutamic acid, condensed, (5)
Condensed beet solubles product (AAFCO)
Ref No 5 00 679 United States

Feed Name or Analyses			As Fed	Dry	C.V. ± %
Dry matter		%	60.5	100.0	25
Ash		%	12.4	20.6	25
N-free extract		%	24.5	40.5	
Protein (N x 6.25)		%	18.3	30.3	20
Iron		%	0.034	0.056	
Potassium		%	3.10	5.14	
Sodium		%	4.27	7.06	
Choline		mg/kg	126479.	209057.	
Alanine		%	0.58	0.95	
Aspartic acid		%	0.48	0.79	
Glutamic acid		%	2.21	3.66	
Glycine		%	0.38	0.64	
Isoleucine		%	0.38	0.64	
Leucine		%	0.58	0.95	
Phenylalanine		%	0.10	0.16	
Serine		%	0.10	0.16	
Tryptophan		%	0.19	0.32	
Tyrosine		%	0.29	0.48	
Valine		%	0.29	0.48	

Beet molasses (AAFCO) –
see Beet, sugar, molasses, mn 48% invert sugar mn 79.5 degrees brix, (4)

BEETS. Beta spp

Beets, aerial part, dehy grnd, (1)
Ref No 1 08 845 Canada

Feed Name or Analyses			As Fed	Dry	C.V. ± %
Dry matter		%	90.4	100.0	
Crude fiber		%	8.5	9.4	
Ether extract		%	2.8	3.1	
Protein (N x 6.25)		%	19.1	21.1	
Cattle	dig prot	%	* 13.8	15.2	
Goats	dig prot	%	* 14.7	16.3	

Continued

Feed Name or Analyses			Mean As Fed	Mean Dry	C.V. ± %
Horses	dig prot %	*	14.0	15.5	
Rabbits	dig prot %	*	13.5	15.0	
Sheep	dig prot %	*	14.0	15.5	

Beggarweed –
see Tickclover, cherokee, aerial part, fresh, (2)

Beggarweed hay –
see Tickclover, cherokee, hay, s-c, (1)

BENT. Agrostis spp

Bent, hay, s-c, (1)
Ref No 1 00 681 United States

Feed Name or Analyses			Mean As Fed	Mean Dry	C.V. ± %
Dry matter	%		87.4	100.0	3
Ash	%		6.4	7.3	14
Crude fiber	%		27.8	31.8	5
Ether extract	%		2.0	2.3	28
N-free extract	%		45.1	51.6	
Protein (N x 6.25)	%		6.1	7.0	17
Cattle	dig prot %	*	2.6	3.0	
Goats	dig prot %	*	2.7	3.1	
Horses	dig prot %	*	3.0	3.5	
Rabbits	dig prot %	*	3.6	4.1	
Sheep	dig prot %	*	2.5	2.8	
Energy	GE Mcal/kg				
Cattle	DE Mcal/kg	*	1.99	2.28	
Sheep	DE Mcal/kg	*	2.15	2.46	
Cattle	ME Mcal/kg	*	1.63	1.87	
Sheep	ME Mcal/kg	*	1.77	2.02	
Cattle	TDN %	*	45.1	51.6	
Sheep	TDN %	*	48.8	55.9	

Bent, hay, s-c, over ripe, (1)
Ref No 1 00 680 United States

Feed Name or Analyses			Mean As Fed	Mean Dry	C.V. ± %
Dry matter	%		87.0	100.0	
Ash	%		5.7	6.5	
Crude fiber	%		28.1	32.3	
Sheep	dig coef %		57.	57.	
Ether extract	%		1.6	1.8	
Sheep	dig coef %		49.	49.	
N-free extract	%		44.9	51.6	
Sheep	dig coef %		46.	46.	
Protein (N x 6.25)	%		6.8	7.8	
Sheep	dig coef %		37.	37.	
Cattle	dig prot %	*	3.2	3.7	
Goats	dig prot %	*	3.3	3.8	
Horses	dig prot %	*	3.6	4.2	
Rabbits	dig prot %	*	4.1	4.7	
Sheep	dig prot %		2.5	2.9	
Energy	GE Mcal/kg				
Cattle	DE Mcal/kg	*	1.89	2.17	
Sheep	DE Mcal/kg	*	1.80	2.07	
Cattle	ME Mcal/kg	*	1.55	1.78	
Sheep	ME Mcal/kg	*	1.48	1.70	
Cattle	TDN %	*	42.9	49.3	
Sheep	TDN %		40.9	47.0	

(1) dry forages and roughages (2) pasture, range plants, and forages fed green (3) silages (4) energy feeds (5) protein supplements (6) minerals (7) vitamins (8) additives

Bent, hay, gr 1 US, (1)
Ref No 1 00 682 United States

Feed Name or Analyses			Mean As Fed	Mean Dry	C.V. ± %
Dry matter	%		92.8	100.0	
Ash	%		8.3	8.9	
Crude fiber	%		29.6	31.9	
Ether extract	%		1.8	1.9	
N-free extract	%		46.0	49.6	
Protein (N x 6.25)	%		7.1	7.7	
Cattle	dig prot %	*	3.3	3.6	
Goats	dig prot %	*	3.5	3.7	
Horses	dig prot %	*	3.8	4.1	
Rabbits	dig prot %	*	4.3	4.6	
Sheep	dig prot %	*	3.2	3.5	
Energy	GE Mcal/kg				
Cattle	DE Mcal/kg	*	2.17	2.34	
Sheep	DE Mcal/kg	*	2.29	2.46	
Cattle	ME Mcal/kg	*	1.78	1.92	
Sheep	ME Mcal/kg	*	1.88	2.02	
Cattle	TDN %	*	49.4	53.2	
Sheep	TDN %	*	51.9	55.9	
Calcium	%		0.33	0.36	
Phosphorus	%		0.22	0.24	
Carotene	mg/kg		11.3	12.1	
Vitamin A equivalent	IU/g		18.8	20.2	

Bent, aerial part, fresh, immature, (2)
Ref No 2 00 683 United States

Feed Name or Analyses			Mean As Fed	Mean Dry	C.V. ± %
Dry matter	%		29.8	100.0	8
Ash	%		3.1	10.3	12
Crude fiber	%		6.7	22.6	13
Ether extract	%		1.1	3.6	11
N-free extract	%		13.6	45.5	
Protein (N x 6.25)	%		5.4	18.0	16
Cattle	dig prot %	*	3.9	13.2	
Goats	dig prot %	*	4.0	13.4	
Horses	dig prot %	*	3.8	12.8	
Rabbits	dig prot %	*	3.7	12.6	
Sheep	dig prot %	*	4.1	13.8	
Energy	GE Mcal/kg				
Cattle	DE Mcal/kg	*	0.90	3.00	
Sheep	DE Mcal/kg	*	0.87	2.92	
Cattle	ME Mcal/kg	*	0.73	2.46	
Sheep	ME Mcal/kg	*	0.71	2.40	
Cattle	TDN %	*	20.3	68.1	
Sheep	TDN %	*	19.8	66.3	
Calcium	%		0.19	0.65	12
Copper	mg/kg		6.0	20.1	
Iron	%		0.013	0.045	
Magnesium	%		0.06	0.21	16
Manganese	mg/kg		52.8	177.0	
Phosphorus	%		0.12	0.39	18
Potassium	%		0.68	2.27	7

BENT, COLONIAL. Agrostis tenuis

Bent, Colonial, hay, s-c, (1)
Ref No 1 00 684 United States

Feed Name or Analyses			Mean As Fed	Mean Dry	C.V. ± %
Dry matter	%		86.9	100.0	1
Ash	%		6.2	7.2	5
Crude fiber	%		27.5	31.7	4
Sheep	dig coef %		63.	63.	
Ether extract	%		2.2	2.5	23
Sheep	dig coef %		30.	30.	
N-free extract	%		44.7	51.5	
Sheep	dig coef %		60.	60.	
Protein (N x 6.25)	%		6.2	7.1	5
Sheep	dig coef %		42.	42.	
Cattle	dig prot %	*	2.7	3.1	
Goats	dig prot %	*	2.8	3.2	
Horses	dig prot %	*	3.1	3.6	
Rabbits	dig prot %	*	3.6	4.2	
Sheep	dig prot %		2.6	3.0	
Fatty acids	%		2.9	3.4	
Energy	GE Mcal/kg				
Cattle	DE Mcal/kg	*	2.32	2.67	
Sheep	DE Mcal/kg	*	2.13	2.45	
Cattle	ME Mcal/kg	*	1.90	2.19	
Sheep	ME Mcal/kg	*	1.74	2.01	
Cattle	TDN %	*	52.7	60.7	
Sheep	TDN %		48.3	55.5	
Phosphorus	%		0.18	0.20	
Potassium	%		1.39	1.60	

Bent, Colonial, aerial part, fresh, (2)
Ref No 2 08 352 United States

Feed Name or Analyses			Mean As Fed	Mean Dry	C.V. ± %
Dry matter	%		29.4	100.0	
Ash	%		3.0	10.2	
Crude fiber	%		6.4	21.8	
Ether extract	%		1.1	3.7	
N-free extract	%		13.2	44.9	
Protein (N x 6.25)	%		5.7	19.4	
Cattle	dig prot %	*	4.2	14.4	
Goats	dig prot %	*	4.3	14.6	
Horses	dig prot %	*	4.1	14.0	
Rabbits	dig prot %	*	4.0	13.6	
Sheep	dig prot %	*	4.4	15.1	
Energy	GE Mcal/kg				
Cattle	DE Mcal/kg	*	0.90	3.06	
Sheep	DE Mcal/kg	*	0.86	2.94	
Cattle	ME Mcal/kg	*	0.74	2.51	
Sheep	ME Mcal/kg	*	0.71	2.41	
Cattle	TDN %	*	20.4	69.5	
Sheep	TDN %	*	19.6	66.7	
Calcium	%		0.19	0.65	
Phosphorus	%		0.12	0.41	
Potassium	%		0.65	2.21	

Bent, Colonial, aerial part, fresh, immature, (2)
Ref No 2 00 686 United States

Feed Name or Analyses			Mean As Fed	Mean Dry	C.V. ± %
Dry matter	%		29.8	100.0	15
Calcium	%		0.20	0.66	7
Magnesium	%		0.06	0.21	16
Phosphorus	%		0.12	0.39	9
Potassium	%		0.68	2.27	7
Carotene	mg/kg		39.3	132.1	17
Vitamin A equivalent	IU/g		65.5	220.1	

Bent, Colonial, aerial part, fresh, immature, pure stand, (2)
Ref No 2 00 685 United States

Feed Name or Analyses			Mean As Fed	Mean Dry	C.V. ± %
Dry matter	%		28.8	100.0	3
Calcium	%		0.17	0.60	10
Magnesium	%		0.05	0.19	31
Phosphorus	%		0.11	0.39	15
Potassium	%		0.71	2.45	

Feed Name or Analyses			Mean As Fed	Mean Dry	C.V. ± %

Bent, Colonial, aerial part, fresh, pure stand, (2)
Ref No 2 00 687 United States

Feed Name or Analyses			As Fed	Dry	C.V. ± %
Dry matter	%		28.8	100.0	2
Ash	%		2.9	10.2	5
Crude fiber	%		6.8	23.5	3
Ether extract	%		1.0	3.5	8
N-free extract	%		13.0	45.1	
Protein (N x 6.25)	%		5.1	17.7	8
Cattle	dig prot %	*	3.7	12.9	
Goats	dig prot %	*	3.8	13.1	
Horses	dig prot %	*	3.6	12.6	
Rabbits	dig prot %	*	3.6	12.3	
Sheep	dig prot %	*	3.9	13.5	
Energy	GE Mcal/kg				
Cattle	DE Mcal/kg	*	0.87	3.01	
Sheep	DE Mcal/kg	*	0.83	2.89	
Cattle	ME Mcal/kg	*	0.71	2.47	
Sheep	ME Mcal/kg	*	0.68	2.37	
Cattle	TDN %	*	19.7	68.3	
Sheep	TDN %	*	18.8	65.4	

BENT, CREEPING. Agrostis palustris

Bent, creeping, hay, s-c, over ripe, (1)
Ref No 1 00 688 United States

Feed Name or Analyses			As Fed	Dry	C.V. ± %
Dry matter	%			100.0	
Ash	%			5.8	
Crude fiber	%			30.8	
Ether extract	%			1.4	
N-free extract	%			57.7	
Protein (N x 6.25)	%			4.3	
Cattle	dig prot %	*		0.7	
Goats	dig prot %	*		0.6	
Horses	dig prot %	*		1.2	
Rabbits	dig prot %	*		2.0	
Sheep	dig prot %	*		0.4	
Energy	GE Mcal/kg				
Cattle	DE Mcal/kg	*		2.66	
Sheep	DE Mcal/kg	*		2.46	
Cattle	ME Mcal/kg	*		2.18	
Sheep	ME Mcal/kg	*		2.02	
Cattle	TDN %	*		60.4	
Sheep	TDN %	*		55.7	

BENT, VELVET. Agrostis canina

Bent, velvet, aerial part, fresh, immature, (2)
Ref No 2 00 689 United States

Feed Name or Analyses			As Fed	Dry	C.V. ± %
Dry matter	%			100.0	
Ash	%			10.2	
Crude fiber	%			20.6	
Ether extract	%			3.0	
N-free extract	%			46.9	
Protein (N x 6.25)	%			19.3	
Cattle	dig prot %	*		14.3	
Goats	dig prot %	*		14.6	
Horses	dig prot %	*		13.9	
Rabbits	dig prot %	*		13.6	
Sheep	dig prot %	*		15.0	
Energy	GE Mcal/kg				
Cattle	DE Mcal/kg	*		2.75	
Sheep	DE Mcal/kg	*		2.96	
Cattle	ME Mcal/kg	*		2.26	
Sheep	ME Mcal/kg	*		2.43	
Cattle	TDN %	*		62.4	
Sheep	TDN %	*		67.2	
Copper	mg/kg			20.1	
Iron	%			0.045	
Manganese	mg/kg			177.0	
Phosphorus	%			0.49	

BENT, WATER. Agrostis verticillata

Bent, water, aerial part, fresh, immature, (2)
Ref No 2 00 690 United States

Feed Name or Analyses			As Fed	Dry	C.V. ± %
Dry matter	%			100.0	
Calcium	%			0.53	
Phosphorus	%			0.18	

Bent, water, aerial part, fresh, milk stage, (2)
Ref No 2 00 691 United States

Feed Name or Analyses			As Fed	Dry	C.V. ± %
Dry matter	%			100.0	
Ash	%			14.5	
Crude fiber	%			29.7	
Ether extract	%			2.4	
N-free extract	%			44.1	
Protein (N x 6.25)	%			9.3	
Cattle	dig prot %	*		5.8	
Goats	dig prot %	*		5.2	
Horses	dig prot %	*		5.4	
Rabbits	dig prot %	*		5.8	
Sheep	dig prot %	*		5.7	
Energy	GE Mcal/kg				
Cattle	DE Mcal/kg	*		2.44	
Sheep	DE Mcal/kg	*		2.87	
Cattle	ME Mcal/kg	*		2.00	
Sheep	ME Mcal/kg	*		2.35	
Cattle	TDN %	*		55.3	
Sheep	TDN %	*		65.0	

BENT, WINTER. Agrostis hiemalis

Bent, winter, hay, s-c, milk stage, (1)
Ref No 1 00 692 United States

Feed Name or Analyses			As Fed	Dry	C.V. ± %
Dry matter	%			100.0	
Ash	%			8.9	
Crude fiber	%			35.2	
Ether extract	%			3.3	
N-free extract	%			44.6	
Protein (N x 6.25)	%			8.0	
Cattle	dig prot %	*		3.9	
Goats	dig prot %	*		4.0	
Horses	dig prot %	*		4.3	
Rabbits	dig prot %	*		4.8	
Sheep	dig prot %	*		3.7	
Energy	GE Mcal/kg				
Cattle	DE Mcal/kg	*		2.51	
Sheep	DE Mcal/kg	*		2.39	
Cattle	ME Mcal/kg	*		2.06	
Sheep	ME Mcal/kg	*		1.96	
Cattle	TDN %	*		56.9	
Sheep	TDN %	*		54.2	

Bent, winter, aerial part, fresh, immature, (2)
Ref No 2 00 693 United States

Feed Name or Analyses			As Fed	Dry	C.V. ± %
Dry matter	%			100.0	
Calcium	%			0.39	
Phosphorus	%			0.26	

Bent, winter, aerial part, fresh, milk stage, (2)
Ref No 2 00 694 United States

Feed Name or Analyses			As Fed	Dry	C.V. ± %
Dry matter	%			100.0	
Ash	%			8.8	
Crude fiber	%			34.4	
Ether extract	%			2.8	
N-free extract	%			44.4	
Protein (N x 6.25)	%			9.6	
Cattle	dig prot %	*		6.1	
Goats	dig prot %	*		5.5	
Horses	dig prot %	*		5.7	
Rabbits	dig prot %	*		6.1	
Sheep	dig prot %	*		5.9	
Energy	GE Mcal/kg				
Cattle	DE Mcal/kg	*		2.84	
Sheep	DE Mcal/kg	*		2.82	
Cattle	ME Mcal/kg	*		2.33	
Sheep	ME Mcal/kg	*		2.31	
Cattle	TDN %	*		64.4	
Sheep	TDN %	*		64.0	

BERMUDAGRASS. Cynodon dactylon

Bermudagrass, aerial part, dehy, immature, (1)
Ref No 1 00 697 United States

Feed Name or Analyses			As Fed	Dry	C.V. ± %
Dry matter	%		91.3	100.0	
Ash	%		11.4	12.5	
Crude fiber	%		16.8	18.4	
Cattle	dig coef %		53.	53.	
Ether extract	%		1.3	1.4	
Cattle	dig coef %		27.	27.	
N-free extract	%		51.7	56.6	
Cattle	dig coef %		46.	46.	
Protein (N x 6.25)	%		10.1	11.1	
Cattle	dig coef %		54.	54.	
Cattle	dig prot %		5.5	6.0	
Goats	dig prot %	*	6.3	6.9	
Horses	dig prot %	*	6.3	7.0	
Rabbits	dig prot %	*	6.6	7.2	
Sheep	dig prot %	*	6.0	6.5	
Energy	GE Mcal/kg				
Cattle	DE Mcal/kg	*	1.72	1.88	
Sheep	DE Mcal/kg	*	1.92	2.10	
Cattle	ME Mcal/kg	*	1.41	1.54	
Sheep	ME Mcal/kg	*	1.57	1.72	
Cattle	TDN %		38.9	42.6	
Sheep	TDN %	*	43.4	47.6	

Bermudagrass, aerial part, dormant, (1)
Ref No 1 00 696 United States

Feed Name or Analyses			As Fed	Dry	C.V. ± %
Dry matter	%			100.0	
Carotene	mg/kg			2.0	
Vitamin A equivalent	IU/g			3.3	

Bermudagrass, hay, fan air dired, (1)
Ref No 1 00 698 United States

Feed Name or Analyses				Mean As Fed	Mean Dry	C.V. ± %
Dry matter		%			100.0	
Carotene		mg/kg			147.5	19
Vitamin A equivalent		IU/g			245.9	

Bermudagrass, hay, s-c, (1)
Ref No 1 00 703 United States

Feed Name or Analyses				Mean As Fed	Mean Dry	C.V. ± %
Dry matter		%		91.2	100.0	2
Ash		%		7.5	8.2	
Crude fiber		%		26.8	29.4	
Sheep	dig coef	%	*	52.	52.	
Ether extract		%		1.7	1.8	
Sheep	dig coef	%	*	44.	44.	
N-free extract		%		48.1	52.7	
Sheep	dig coef	%	*	52.	52.	
Protein (N x 6.25)		%		7.2	7.9	
Sheep	dig coef	%	*	51.	51.	
Cattle	dig prot	%	*	3.4	3.8	
Goats	dig prot	%	*	3.6	3.9	
Horses	dig prot	%	*	3.8	4.2	
Rabbits	dig prot	%	*	4.3	4.7	
Sheep	dig prot	%		3.7	4.0	
Fatty acids		%		1.8	2.0	
Energy	GE Mcal/kg					
Cattle	DE Mcal/kg		*	2.01	2.20	
Sheep	DE Mcal/kg		*	1.95	2.14	
Cattle	ME Mcal/kg		*	1.64	1.80	
Sheep	ME Mcal/kg		*	1.60	1.75	
Cattle	NE_m Mcal/kg		*	0.97	1.06	
Cattle	NE_{gain} Mcal/kg		*	0.23	0.25	
Cattle	TDN	%	*	45.5	49.9	
Sheep	TDN	%		44.2	48.5	
Calcium		%		0.37	0.41	
Iodine		mg/kg		0.104	0.115	
Magnesium		%		0.15	0.17	
Phosphorus		%		0.19	0.21	
Potassium		%		1.43	1.57	
Carotene		mg/kg		88.3	96.9	23
Vitamin A equivalent		IU/g		147.3	161.5	

Bermudagrass, hay, s-c, immature, (1)
Ref No 1 00 699 United States

Feed Name or Analyses				Mean As Fed	Mean Dry	C.V. ± %
Dry matter		%		90.6	100.0	1
Ash		%		8.2	9.0	39
Crude fiber		%		22.5	24.8	13
Sheep	dig coef	%	*	52.	52.	
Ether extract		%		1.6	1.8	18
Sheep	dig coef	%	*	44.	44.	
N-free extract		%		45.8	50.6	
Sheep	dig coef	%	*	52.	52.	
Protein (N x 6.25)		%		12.5	13.8	13
Sheep	dig coef	%	*	51.	51.	
Cattle	dig prot	%	*	8.1	8.9	
Goats	dig prot	%	*	8.5	9.4	
Horses	dig prot	%	*	8.4	9.2	
Rabbits	dig prot	%	*	8.4	9.3	
Sheep	dig prot	%		6.4	7.0	
Energy	GE Mcal/kg					
Cattle	DE Mcal/kg		*	1.85	2.04	
Sheep	DE Mcal/kg		*	1.92	2.12	
Cattle	ME Mcal/kg		*	1.51	1.67	
Sheep	ME Mcal/kg		*	1.57	1.74	
Cattle	TDN	%	*	41.9	46.3	
Sheep	TDN	%	*	43.5	48.0	

Bermudagrass, hay, s-c, mid-bloom, (1)
Ref No 1 00 700 United States

Feed Name or Analyses				Mean As Fed	Mean Dry	C.V. ± %
Dry matter		%		92.9	100.0	
Ash		%		9.5	10.2	
Crude fiber		%		25.9	27.9	
Sheep	dig coef	%	*	52.	52.	
Ether extract		%		2.0	2.2	
Sheep	dig coef	%	*	44.	44.	
N-free extract		%		46.6	50.2	
Sheep	dig coef	%	*	52.	52.	
Protein (N x 6.25)		%		8.8	9.5	
Sheep	dig coef	%	*	51.	51.	
Cattle	dig prot	%	*	4.8	5.2	
Goats	dig prot	%	*	5.0	5.4	
Horses	dig prot	%	*	5.2	5.6	
Rabbits	dig prot	%	*	5.6	6.0	
Sheep	dig prot	%		4.5	4.8	
Energy	GE Mcal/kg					
Cattle	DE Mcal/kg		*	2.01	2.16	
Sheep	DE Mcal/kg		*	1.95	2.10	
Cattle	ME Mcal/kg		*	1.64	1.77	
Sheep	ME Mcal/kg		*	1.60	1.72	
Cattle	TDN	%	*	45.5	49.0	
Sheep	TDN	%	*	44.3	47.6	

Bermudagrass, hay, s-c, full bloom, (1)
Ref No 1 00 701 United States

Feed Name or Analyses				Mean As Fed	Mean Dry	C.V. ± %
Dry matter		%		92.2	100.0	
Ash		%		9.9	10.7	
Crude fiber		%		26.3	28.5	
Sheep	dig coef	%	*	52.	52.	
Ether extract		%		1.7	1.8	
Sheep	dig coef	%	*	44.	44.	
N-free extract		%		46.9	50.9	
Sheep	dig coef	%	*	52.	52.	
Protein (N x 6.25)		%		7.5	8.1	
Sheep	dig coef	%	*	51.	51.	
Cattle	dig prot	%	*	3.6	4.0	
Goats	dig prot	%	*	3.8	4.1	
Horses	dig prot	%	*	4.1	4.4	
Rabbits	dig prot	%	*	4.5	4.9	
Sheep	dig prot	%		3.8	4.1	
Energy	GE Mcal/kg					
Cattle	DE Mcal/kg		*	1.98	2.15	
Sheep	DE Mcal/kg		*	1.92	2.08	
Cattle	ME Mcal/kg		*	1.62	1.76	
Sheep	ME Mcal/kg		*	1.57	1.71	
Cattle	TDN	%	*	45.0	48.8	
Sheep	TDN	%	*	43.5	47.2	

Bermudagrass, hay, s-c, mature, (1)
Ref No 1 00 702 United States

Feed Name or Analyses				Mean As Fed	Mean Dry	C.V. ± %
Dry matter		%		92.2	100.0	3
Ash		%		6.5	7.0	35
Crude fiber		%		27.6	29.9	6
Sheep	dig coef	%	*	52.	52.	
Ether extract		%		1.5	1.6	25
Sheep	dig coef	%	*	44.	44.	
N-free extract		%		51.2	55.5	
Sheep	dig coef	%	*	52.	52.	
Protein (N x 6.25)		%		5.5	6.0	6
Sheep	dig coef	%	*	51.	51.	
Cattle	dig prot	%	*	2.0	2.1	
Goats	dig prot	%	*	2.0	2.2	
Horses	dig prot	%	*	2.4	2.6	
Rabbits	dig prot	%	*	3.0	3.3	
Sheep	dig prot	%		2.8	3.1	
Energy	GE Mcal/kg					
Cattle	DE Mcal/kg		*	1.93	2.09	
Sheep	DE Mcal/kg		*	1.99	2.16	
Cattle	ME Mcal/kg		*	1.58	1.71	
Sheep	ME Mcal/kg		*	1.64	1.77	
Cattle	TDN	%	*	43.8	47.5	
Sheep	TDN	%	*	45.2	49.1	

Bermudagrass, hay, s-c, gr 1 US, (1)
Ref No 1 00 704 United States

Feed Name or Analyses				Mean As Fed	Mean Dry	C.V. ± %
Dry matter		%		89.0	100.0	
Ash		%		3.8	4.3	
Crude fiber		%		25.3	28.4	
Ether extract		%		2.1	2.4	
N-free extract		%		44.7	50.2	
Protein (N x 6.25)		%		13.1	14.7	
Cattle	dig prot	%	*	8.6	9.7	
Goats	dig prot	%	*	9.1	10.3	
Horses	dig prot	%	*	8.9	10.0	
Rabbits	dig prot	%	*	8.9	10.0	
Sheep	dig prot	%	*	8.7	9.8	
Energy	GE Mcal/kg					
Cattle	DE Mcal/kg		*	1.97	2.21	
Sheep	DE Mcal/kg		*	1.90	2.13	
Cattle	ME Mcal/kg		*	1.61	1.81	
Sheep	ME Mcal/kg		*	1.56	1.75	
Cattle	TDN	%	*	44.6	50.1	
Sheep	TDN	%	*	43.0	48.3	

Bermudagrass, hay, s-c, weathered, mature, (1)
Ref No 1 08 353 United States

Feed Name or Analyses				Mean As Fed	Mean Dry	C.V. ± %
Dry matter		%		90.0	100.0	
Ash		%		6.8	7.6	
Crude fiber		%		38.8	43.1	
Ether extract		%		0.9	1.0	
N-free extract		%		37.7	41.9	
Protein (N x 6.25)		%		5.8	6.4	
Cattle	dig prot	%	*	2.3	2.5	
Goats	dig prot	%	*	2.3	2.6	
Horses	dig prot	%	*	2.7	3.0	
Rabbits	dig prot	%	*	3.3	3.6	
Sheep	dig prot	%	*	2.1	2.4	
Energy	GE Mcal/kg					
Cattle	DE Mcal/kg		*	1.69	1.88	
Sheep	DE Mcal/kg		*	1.86	2.07	
Cattle	ME Mcal/kg		*	1.39	1.54	
Sheep	ME Mcal/kg		*	1.53	1.70	
Cattle	TDN	%	*	38.4	42.7	
Sheep	TDN	%	*	42.3	47.0	

(1) dry forages and roughages (2) pasture, range plants, and forages fed green (3) silages (4) energy feeds (5) protein supplements (6) minerals (7) vitamins (8) additives

Bermudagrass, aerial part, fresh, (2)
Ref No 2 00 712 United States

Feed Name or Analyses			Mean As Fed	Mean Dry	C.V. ± %
Dry matter	%		30.9	100.0	39
Ash	%		3.8	12.4	
Crude fiber	%		7.9	25.6	
Ether extract	%		0.6	2.0	
N-free extract	%		15.1	48.8	
Protein (N x 6.25)	%		3.5	11.2	
Cattle	dig prot %	*	2.3	7.4	
Goats	dig prot %	*	2.2	7.0	
Horses	dig prot %	*	2.2	7.0	
Rabbits	dig prot %	*	2.3	7.3	
Sheep	dig prot %	*	2.3	7.4	
Energy	GE Mcal/kg				
Cattle	DE Mcal/kg	*	0.82	2.66	
Sheep	DE Mcal/kg	*	0.90	2.92	
Cattle	ME Mcal/kg	*	0.67	2.18	
Sheep	ME Mcal/kg	*	0.74	2.39	
Cattle	TDN %	*	18.6	60.3	
Sheep	TDN %	*	20.4	66.2	
Calcium	%		0.17	0.56	
Magnesium	%		0.07	0.24	
Phosphorus	%		0.06	0.20	
Potassium	%		0.68	2.20	
Carotene	mg/kg		86.7	281.1	30
Vitamin A equivalent	IU/g		144.6	468.6	

Bermudagrass, aerial part, fresh, early bloom, (2)
Ref No 2 00 707 United States

Feed Name or Analyses			Mean As Fed	Mean Dry	C.V. ± %
Dry matter	%			100.0	
Calcium	%			0.58	
Magnesium	%			0.14	
Phosphorus	%			0.23	
Potassium	%			2.13	

Bermudagrass, aerial part, fresh, mid-bloom, (2)
Ref No 2 08 354 United States

Feed Name or Analyses			Mean As Fed	Mean Dry	C.V. ± %
Dry matter	%		35.0	100.0	
Ash	%		3.5	10.0	
Crude fiber	%		9.8	28.0	
Ether extract	%		0.7	2.0	
N-free extract	%		17.4	49.7	
Protein (N x 6.25)	%		3.6	10.3	
Cattle	dig prot %	*	2.3	6.6	
Goats	dig prot %	*	2.2	6.2	
Horses	dig prot %	*	2.2	6.3	
Rabbits	dig prot %	*	2.3	6.6	
Sheep	dig prot %	*	2.3	6.6	
Energy	GE Mcal/kg				
Cattle	DE Mcal/kg	*	0.98	2.79	
Sheep	DE Mcal/kg	*	0.90	2.58	
Cattle	ME Mcal/kg	*	0.80	2.29	
Sheep	ME Mcal/kg	*	0.74	2.12	
Cattle	TDN %	*	22.2	63.4	
Sheep	TDN %	*	20.5	58.6	
Calcium	%		0.14	0.40	
Magnesium	%		0.06	0.17	
Phosphorus	%		0.07	0.20	
Potassium	%		0.55	1.57	

Bermudagrass, aerial part, fresh, full bloom, (2)
Ref No 2 00 708 United States

Feed Name or Analyses			Mean As Fed	Mean Dry	C.V. ± %
Dry matter	%		35.0	100.0	
Ash	%		3.5	10.0	5
Crude fiber	%		9.1	26.1	5
Ether extract	%		0.7	2.0	8
N-free extract	%		18.1	51.6	
Protein (N x 6.25)	%		3.6	10.3	12
Cattle	dig prot %	*	2.3	6.6	
Goats	dig prot %	*	2.2	6.2	
Horses	dig prot %	*	2.2	6.3	
Rabbits	dig prot %	*	2.3	6.6	
Sheep	dig prot %	*	2.3	6.6	
Energy	GE Mcal/kg				
Cattle	DE Mcal/kg	*	0.97	2.78	
Sheep	DE Mcal/kg	*	0.95	2.70	
Cattle	ME Mcal/kg	*	0.80	2.28	
Sheep	ME Mcal/kg	*	0.77	2.21	
Cattle	TDN %	*	22.1	63.0	
Sheep	TDN %	*	21.4	61.2	
Calcium	%		0.19	0.54	8
Magnesium	%		0.06	0.17	38
Phosphorus	%		0.07	0.20	21
Potassium	%		0.55	1.57	11

Bermudagrass, aerial part, fresh, milk stage, (2)
Ref No 2 00 709 United States

Feed Name or Analyses			Mean As Fed	Mean Dry	C.V. ± %
Dry matter	%		49.3	100.0	3
Ash	%		5.2	10.5	10
Crude fiber	%		12.5	25.3	6
Ether extract	%		0.8	1.6	16
N-free extract	%		27.1	55.0	
Protein (N x 6.25)	%		3.7	7.6	27
Cattle	dig prot %	*	2.1	4.4	
Goats	dig prot %	*	1.8	3.7	
Horses	dig prot %	*	2.0	4.0	
Rabbits	dig prot %	*	2.2	4.5	
Sheep	dig prot %	*	2.0	4.1	
Energy	GE Mcal/kg				
Cattle	DE Mcal/kg	*	1.30	2.63	
Sheep	DE Mcal/kg	*	1.33	2.69	
Cattle	ME Mcal/kg	*	1.06	2.16	
Sheep	ME Mcal/kg	*	1.09	2.21	
Cattle	TDN %	*	29.4	59.7	
Sheep	TDN %	*	30.1	61.0	
Calcium	%		0.26	0.53	
Magnesium	%		0.08	0.16	
Phosphorus	%		0.11	0.22	
Potassium	%		0.54	1.09	

Bermudagrass, aerial part, fresh, dough stage, (2)
Ref No 2 00 710 United States

Feed Name or Analyses			Mean As Fed	Mean Dry	C.V. ± %
Dry matter	%		58.9	100.0	
Ash	%		5.1	8.6	
Crude fiber	%		16.4	27.8	
Ether extract	%		1.1	1.9	
N-free extract	%		32.3	54.9	
Protein (N x 6.25)	%		4.0	6.8	
Cattle	dig prot %	*	2.2	3.7	
Goats	dig prot %	*	1.7	2.9	
Horses	dig prot %	*	1.9	3.3	
Rabbits	dig prot %	*	2.3	3.9	
Sheep	dig prot %	*	2.0	3.3	
Energy	GE Mcal/kg				
Cattle	DE Mcal/kg	*	1.60	2.72	
Sheep	DE Mcal/kg	*	1.55	2.64	
Cattle	ME Mcal/kg	*	1.31	2.23	
Sheep	ME Mcal/kg	*	1.27	2.16	
Cattle	TDN %	*	36.3	61.6	
Sheep	TDN %	*	35.2	59.8	
Calcium	%		0.31	0.52	
Magnesium	%		0.04	0.07	
Phosphorus	%		0.11	0.19	

Bermudagrass, aerial part, fresh, mature, (2)
Ref No 2 00 711 United States

Feed Name or Analyses			Mean As Fed	Mean Dry	C.V. ± %
Dry matter	%			100.0	
Ash	%			7.8	24
Crude fiber	%			28.5	6
Ether extract	%			2.0	22
N-free extract	%			55.9	
Protein (N x 6.25)	%			5.8	17
Cattle	dig prot %	*		2.8	
Goats	dig prot %	*		2.0	
Horses	dig prot %	*		2.5	
Rabbits	dig prot %	*		3.1	
Sheep	dig prot %	*		2.4	
Energy	GE Mcal/kg				
Cattle	DE Mcal/kg	*		2.72	
Sheep	DE Mcal/kg	*		2.67	
Cattle	ME Mcal/kg	*		2.23	
Sheep	ME Mcal/kg	*		2.19	
Cattle	TDN %	*		61.7	
Sheep	TDN %	*		60.5	
Calcium	%			0.40	25
Magnesium	%			0.15	
Phosphorus	%			0.18	25
Potassium	%			1.01	

Bermudagrass, aerial part, ensiled, (3)
Ref No 3 07 751 United States

Feed Name or Analyses			Mean As Fed	Mean Dry	C.V. ± %
Dry matter	%		25.7	100.0	
Cattle	dig coef %		57.	57.	
Ash	%		1.6	6.1	
Crude fiber	%		8.5	33.1	
Cattle	dig coef %		67.	67.	
Ether extract	%		1.3	5.0	
Cattle	dig coef %		50.	50.	
N-free extract	%		11.4	44.3	
Cattle	dig coef %		52.	52.	
Protein (N x 6.25)	%		3.0	11.6	
Cattle	dig coef %		58.	58.	
Cattle	dig prot %		1.7	6.7	
Goats	dig prot %	*	1.7	6.7	
Horses	dig prot %	*	1.7	6.7	
Sheep	dig prot %	*	1.7	6.7	
Lignin (Ellis)	%		3.3	12.9	
Energy	GE Mcal/kg		1.23	4.79	
Cattle	GE dig coef %		51.	51.	
Cattle	DE Mcal/kg		0.63	2.44	
Sheep	DE Mcal/kg	*	0.72	2.79	
Cattle	ME Mcal/kg	*	0.51	2.00	
Sheep	ME Mcal/kg	*	0.59	2.29	
Cattle	TDN %		14.7	57.3	
Sheep	TDN %	*	16.2	63.3	
Calcium	%		0.13	0.50	
Phosphorus	%		0.08	0.31	
Carotene	mg/kg		50.1	195.0	

Continued

Feed Name or Analyses			Mean As Fed	Mean Dry	C.V. ± %
Vitamin A equivalent	IU/g		83.5	325.1	
pH	pH units		4.85	4.85	

Bermudagrass, aerial part w bacitracin preservative added, ensiled, (3)

Ref No 3 07 753 — United States

Feed Name or Analyses			Mean As Fed	Mean Dry	C.V. ± %
Dry matter	%		27.8	100.0	11
Cattle	dig coef %		60.	60.	
Ash	%		1.8	6.3	5
Crude fiber	%		8.9	32.0	5
Cattle	dig coef %		68.	68.	
Ether extract	%		1.4	5.2	33
Cattle	dig coef %		63.	63.	
N-free extract	%		11.7	42.1	7
Cattle	dig coef %		53.	53.	
Protein (N x 6.25)	%		4.0	14.4	8
Cattle	dig coef %		63.	63.	
Cattle	dig prot %		2.5	9.0	
Goats	dig prot %	*	2.6	9.3	
Horses	dig prot %	*	2.6	9.3	
Sheep	dig prot %	*	2.6	9.3	
Lignin (Ellis)	%		2.7	9.9	
Energy	GE Mcal/kg		1.32	4.75	
Cattle	GE dig coef %		57.	57.	
Cattle	DE Mcal/kg		0.76	2.72	
Sheep	DE Mcal/kg	*	0.77	2.76	
Cattle	ME Mcal/kg	*	0.62	2.23	
Sheep	ME Mcal/kg	*	0.63	2.27	
Cattle	TDN %		16.9	60.7	
Sheep	TDN %	*	17.4	62.7	
Calcium	%		0.13	0.47	9
Phosphorus	%		0.08	0.27	13
Carotene	mg/kg		34.2	123.0	
Vitamin A equivalent	IU/g		57.0	205.0	
pH	pH units		4.75	4.75	

Bermudagrass, aerial part w grnd corn added, ensiled, (3)

Ref No 3 07 752 — United States

Feed Name or Analyses			Mean As Fed	Mean Dry	C.V. ± %
Dry matter	%		28.0	100.0	
Cattle	dig coef %		59.	59.	
Ash	%		1.7	6.0	
Crude fiber	%		8.5	30.4	
Cattle	dig coef %		63.	63.	
Ether extract	%		1.2	4.4	
Cattle	dig coef %		54.	54.	
N-free extract	%		13.4	48.1	
Cattle	dig coef %		58.	58.	
Protein (N x 6.25)	%		3.1	11.1	
Cattle	dig coef %		60.	60.	
Cattle	dig prot %		1.9	6.7	
Goats	dig prot %	*	1.8	6.3	
Horses	dig prot %	*	1.8	6.3	
Sheep	dig prot %	*	1.8	6.3	
Lignin (Ellis)	%		3.3	11.8	
Energy	GE Mcal/kg		1.35	4.84	
Cattle	GE dig coef %		56.	56.	
Cattle	DE Mcal/kg		0.76	2.71	
Sheep	DE Mcal/kg	*	0.79	2.84	
Cattle	ME Mcal/kg	*	0.62	2.22	
Sheep	ME Mcal/kg	*	0.65	2.33	
Cattle	TDN %		16.5	59.0	
Sheep	TDN %	*	18.0	64.3	
Calcium	%		0.13	0.46	
Phosphorus	%		0.09	0.31	
Carotene	mg/kg		20.1	72.0	
Vitamin A equivalent	IU/g		33.6	120.0	
pH	pH units		4.75	4.75	

(1) dry forages and roughages (2) pasture, range plants, and forages fed green (3) silages (4) energy feeds (5) protein supplements (6) minerals (7) vitamins (8) additives

BERMUDAGRASS, COASTAL. Cynodon dactylon

Bermudagrass, coastal, aerial part, dehy grnd, cut 1, (1)

Ref No 1 09 212 — United States

Feed Name or Analyses			Mean As Fed	Mean Dry	C.V. ± %
Dry matter	%		87.0	100.0	
Cattle	dig coef %		64.	64.	
Ash	%		6.1	7.0	
Crude fiber	%		24.3	27.9	
Cattle	dig coef %		69.	69.	
Ether extract	%		3.3	3.8	
N-free extract	%		38.2	43.9	
Cattle	dig coef %		59.	59.	
Protein (N x 6.25)	%		15.1	17.4	
Cattle	dig coef %		72.	72.	
Cattle	dig prot %		10.9	12.5	
Goats	dig prot %	*	11.1	12.8	
Horses	dig prot %	*	10.7	12.3	
Rabbits	dig prot %	*	10.5	12.1	
Sheep	dig prot %	*	10.6	12.2	
Energy	GE Mcal/kg		4.00	4.60	
Cattle	GE dig coef %		61.	61.	
Cattle	DE Mcal/kg		2.44	2.80	
Sheep	DE Mcal/kg	*	2.29	2.63	
Cattle	ME Mcal/kg	*	2.00	2.30	
Sheep	ME Mcal/kg	*	1.88	2.16	
Cattle	TDN %		54.6	62.8	
Sheep	TDN %	*	51.9	59.6	
Calcium	%		0.29	0.33	
Phosphorus	%		0.23	0.26	

Bermudagrass, coastal, aerial part, dehy grnd pelleted, cut 1, (1)

Ref No 1 09 213 — United States

Feed Name or Analyses			Mean As Fed	Mean Dry	C.V. ± %
Dry matter	%		89.9	100.0	
Cattle	dig coef %		62.	62.	
Ash	%		6.1	6.7	
Crude fiber	%		23.7	26.3	
Cattle	dig coef %		62.	62.	
Ether extract	%		3.3	3.7	
N-free extract	%		43.2	48.0	
Cattle	dig coef %		60.	60.	
Protein (N x 6.25)	%		13.7	15.2	
Cattle	dig coef %		68.	68.	
Cattle	dig prot %		9.4	10.4	
Goats	dig prot %	*	9.7	10.8	
Horses	dig prot %	*	9.4	10.5	
Rabbits	dig prot %	*	9.4	10.4	
Sheep	dig prot %	*	9.2	10.2	
Energy	GE Mcal/kg		4.12	4.58	
Cattle	GE dig coef %		58.	58.	
Cattle	DE Mcal/kg		2.39	2.66	
Sheep	DE Mcal/kg	*	2.41	2.68	
Cattle	ME Mcal/kg	*	1.96	2.18	
Sheep	ME Mcal/kg	*	1.97	2.20	
Cattle	TDN %		54.6	60.7	
Sheep	TDN %	*	54.6	60.7	
Calcium	%		0.30	0.34	
Phosphorus	%		0.22	0.25	

Bermudagrass, coastal, aerial part, dehy grnd pelleted, cut 4, (1)

Ref No 1 09 222 — United States

Feed Name or Analyses			Mean As Fed	Mean Dry	C.V. ± %
Dry matter	%		89.3	100.0	
Cattle	dig coef %		59.	59.	
Ash	%		5.4	6.1	
Crude fiber	%		22.5	25.2	
Cattle	dig coef %		56.	56.	
Ether extract	%		3.9	4.4	
N-free extract	%		41.4	46.4	
Cattle	dig coef %		57.	57.	
Protein (N x 6.25)	%		16.0	18.0	
Cattle	dig coef %		72.	72.	
Cattle	dig prot %		11.5	12.9	
Goats	dig prot %	*	11.9	13.3	
Horses	dig prot %	*	11.4	12.8	
Rabbits	dig prot %	*	11.2	12.5	
Sheep	dig prot %	*	11.3	12.7	
Energy	GE Mcal/kg		4.09	4.58	
Cattle	GE dig coef %		57.	57.	
Cattle	DE Mcal/kg		2.31	2.59	
Sheep	DE Mcal/kg	*	2.43	2.73	
Cattle	ME Mcal/kg	*	1.90	2.12	
Sheep	ME Mcal/kg	*	2.00	2.23	
Cattle	TDN %		52.7	59.1	
Sheep	TDN %	*	55.2	61.8	
Calcium	%		0.27	0.31	
Phosphorus	%		0.21	0.24	

Bermudagrass, coastal, hay, fan air dried w heat, immature, (1)

Ref No 1 09 206 — United States

Feed Name or Analyses			Mean As Fed	Mean Dry	C.V. ± %
Dry matter	%			100.0	
Cattle	dig coef %			58.	
Ash	%			6.5	
Crude fiber	%			32.1	
Cattle	dig coef %			62.	
Ether extract	%			1.5	
Cattle	dig coef %			31.	
N-free extract	%			49.3	
Cattle	dig coef %			56.	
Protein (N x 6.25)	%			10.6	
Cattle	dig coef %			62.	
Cattle	dig prot %			6.6	
Goats	dig prot %	*		6.4	
Horses	dig prot %	*		6.5	
Rabbits	dig prot %	*		6.9	
Sheep	dig prot %	*		6.1	
Energy	GE Mcal/kg				
Cattle	DE Mcal/kg	*		2.43	
Sheep	DE Mcal/kg	*		2.53	
Cattle	ME Mcal/kg	*		1.99	
Sheep	ME Mcal/kg	*		2.07	
Cattle	TDN %			55.1	
Sheep	TDN %	*		57.3	

Bermudagrass, coastal, hay, fan air dried w heat grnd pelleted, cut 4, (1)

Ref No 1 08 743 — United States

Feed Name or Analyses			Mean As Fed	Mean Dry	C.V. ± %
Dry matter	%		92.8	100.0	
Ash	%		5.4	5.8	
Crude fiber	%		24.5	26.4	
Ether extract	%		1.4	1.5	
N-free extract	%		47.4	51.1	

Feed Name or Analyses			Mean As Fed	Mean Dry	C.V. ± %
Protein (N x 6.25)	%		14.1	15.2	
Cattle	dig prot %	*	9.4	10.1	
Goats	dig prot %	*	10.0	10.7	
Horses	dig prot %	*	9.7	10.4	
Rabbits	dig prot %	*	9.7	10.4	
Sheep	dig prot %	*	9.5	10.2	
Cellulose (Matrone)	%		20.3	21.9	
Lignin (Ellis)	%		5.8	6.2	
Energy	GE Mcal/kg				
Cattle	DE Mcal/kg	*	2.41	2.59	
Sheep	DE Mcal/kg	*	2.59	2.79	
Cattle	ME Mcal/kg	*	1.97	2.13	
Sheep	ME Mcal/kg	*	2.12	2.29	
Cattle	TDN %	*	54.6	58.9	
Sheep	TDN %	*	58.8	63.3	
Calcium	%		0.98	1.06	
Phosphorus	%		0.85	0.92	
Niacin	mg/kg		23.3	25.1	
Pantothenic acid	mg/kg		7.3	7.8	
Riboflavin	mg/kg		2.4	2.6	
a-tocopherol	mg/kg		146.9	158.3	
Vitamin A equivalent	IU/g		41.5	44.7	
Cystine	%		0.04	0.05	
Glycine	%		0.49	0.53	
Lysine	%		0.69	0.75	
Methionine	%		0.17	0.18	
Tryptophan	%		0.21	0.22	

Bermudagrass, coastal, hay, fan air dried w heat grnd pelleted, cut 5, (1)

Ref No 1 08 773 — United States

Feed Name or Analyses			Mean As Fed	Mean Dry	C.V. ± %
Dry matter	%		91.6	100.0	
Ash	%		3.8	4.2	
Crude fiber	%		25.6	28.0	
Ether extract	%		1.0	1.1	
N-free extract	%		53.4	58.3	
Protein (N x 6.25)	%		7.7	8.4	
Cattle	dig prot %	*	3.9	4.2	
Goats	dig prot %	*	4.0	4.4	
Horses	dig prot %	*	4.3	4.7	
Rabbits	dig prot %	*	4.7	5.2	
Sheep	dig prot %	*	3.8	4.1	
Cellulose (Matrone)	%		21.9	23.9	
Lignin (Ellis)	%		7.3	8.0	
Energy	GE Mcal/kg				
Cattle	DE Mcal/kg	*	2.52	2.75	
Sheep	DE Mcal/kg	*	2.38	2.60	
Cattle	ME Mcal/kg	*	2.06	2.25	
Sheep	ME Mcal/kg	*	1.95	2.13	
Cattle	TDN %	*	57.1	62.3	
Sheep	TDN %	*	54.0	58.9	
Calcium	%		1.02	1.11	
Phosphorus	%		0.82	0.89	
Niacin	mg/kg		25.6	28.0	
Pantothenic acid	mg/kg		7.6	8.3	
Riboflavin	mg/kg		2.7	3.0	
a-tocopherol	mg/kg		204.8	223.5	
Vitamin A equivalent	IU/g		39.5	43.2	
Cystine	%		0.03	0.04	
Glycine	%		0.22	0.25	
Lysine	%		0.37	0.40	
Methionine	%		0.09	0.10	
Tryptophan	%		0.10	0.11	

Bermudagrass, coastal, hay, s-c, (1)

Ref No 1 00 716 — United States

Feed Name or Analyses			Mean As Fed	Mean Dry	C.V. ± %
Dry matter	%		91.0	100.0	1
Cattle	dig coef %		56.	56.	4
Sheep	dig coef %		52.	52.	5
Ash	%		4.5	4.9	41
Crude fiber	%		29.6	32.5	10
Cattle	dig coef %		55.	55.	6
Sheep	dig coef %		57.	57.	10
Ether extract	%		1.5	1.7	39
Cattle	dig coef %		45.	45.	22
Sheep	dig coef %		51.	51.	29
N-free extract	%		46.5	51.1	7
Cattle	dig coef %		56.	56.	4
Sheep	dig coef %		56.	56.	8
Protein (N x 6.25)	%		9.0	9.9	26
Cattle	dig coef %		55.	55.	17
Sheep	dig coef %		58.	58.	14
Cattle	dig prot %		4.9	5.4	
Goats	dig prot %	*	5.2	5.8	
Horses	dig prot %	*	5.4	5.9	
Rabbits	dig prot %	*	5.7	6.3	
Sheep	dig prot %		5.2	5.7	
Cellulose (Matrone)	%		28.1	30.9	10
Cattle	dig coef %		69.	69.	
Lignin (Ellis)	%		7.5	8.3	23
Energy	GE Mcal/kg		3.97	4.37	21
Cattle	GE dig coef %		54.	54.	4
Sheep	GE dig coef %		54.	54.	7
Cattle	DE Mcal/kg		2.15	2.36	5
Sheep	DE Mcal/kg		2.16	2.37	8
Cattle	ME Mcal/kg	*	1.76	1.94	
Sheep	ME Mcal/kg	*	1.77	1.94	
Cattle	NE_m Mcal/kg	*	0.98	1.08	
Cattle	NE_{gain} Mcal/kg	*	0.24	0.26	
Cattle	$NE_{lactating\ cows}$ Mcal/kg	*	0.78	0.86	
Cattle	TDN %		49.1	53.9	3
Sheep	TDN %		49.9	54.8	6
Calcium	%		0.31	0.34	29
Magnesium	%		0.15	0.16	40
Phosphorus	%		0.14	0.16	18
Carotene	mg/kg		74.4	81.7	
Vitamin A equivalent	IU/g		124.0	136.2	

Bermudagrass, coastal, hay, s-c, 25 to 31 days growth, (1)

Ref No 1 09 207 — United States

Feed Name or Analyses			Mean As Fed	Mean Dry	C.V. ± %
Dry matter	%			100.0	
Sheep	dig coef %			58.	
Ash	%			8.3	
Crude fiber	%			27.1	
Ether extract	%			2.4	
N-free extract	%			51.3	
Protein (N x 6.25)	%			11.0	
Cattle	dig prot %	*		6.4	
Goats	dig prot %	*		6.8	
Horses	dig prot %	*		6.8	
Rabbits	dig prot %	*		7.1	
Sheep	dig prot %	*		6.4	
Cellulose (Matrone)	%			29.0	
Lignin (Ellis)	%			5.7	
Energy	GE Mcal/kg				
Cattle	DE Mcal/kg	*		2.61	
Sheep	DE Mcal/kg			2.54	
Cattle	ME Mcal/kg	*		2.14	
Sheep	ME Mcal/kg	*		2.09	
Cattle	TDN %	*		59.1	
Sheep	TDN %	*		59.4	

Bermudagrass, coastal, hay, s-c, 32 to 38 days growth, (1)

Ref No 1 09 208 — United States

Feed Name or Analyses			Mean As Fed	Mean Dry	C.V. ± %
Dry matter	%			100.0	
Sheep	dig coef %			55.	
Ash	%			7.4	
Crude fiber	%			27.0	
Ether extract	%			3.0	
N-free extract	%			48.5	
Protein (N x 6.25)	%			14.3	
Cattle	dig prot %	*		9.3	
Goats	dig prot %	*		9.9	
Horses	dig prot %	*		9.6	
Rabbits	dig prot %	*		9.7	
Sheep	dig prot %	*		9.4	
Cellulose (Matrone)	%			28.7	
Lignin (Ellis)	%			7.1	
Energy	GE Mcal/kg				
Cattle	DE Mcal/kg	*		2.71	
Sheep	DE Mcal/kg			2.54	
Cattle	ME Mcal/kg	*		2.22	
Sheep	ME Mcal/kg	*		2.08	
Cattle	TDN %	*		61.5	
Sheep	TDN %	*		60.2	

Bermudagrass, coastal, hay, s-c, 60 to 66 days growth, (1)

Ref No 1 09 211 — United States

Feed Name or Analyses			Mean As Fed	Mean Dry	C.V. ± %
Dry matter	%			100.0	
Sheep	dig coef %			61.	
Ash	%			7.4	
Crude fiber	%			28.5	
Ether extract	%			3.5	
N-free extract	%			45.2	
Protein (N x 6.25)	%			15.4	
Cattle	dig prot %	*		10.3	
Goats	dig prot %	*		10.9	
Horses	dig prot %	*		10.6	
Rabbits	dig prot %	*		10.6	
Sheep	dig prot %	*		10.4	
Cellulose (Matrone)	%			30.0	
Lignin (Ellis)	%			6.5	
Energy	GE Mcal/kg				
Cattle	DE Mcal/kg	*		2.69	
Sheep	DE Mcal/kg			2.81	
Cattle	ME Mcal/kg	*		2.21	
Sheep	ME Mcal/kg	*		2.31	
Cattle	TDN %	*		61.0	
Sheep	TDN %	*		58.9	

Bermudagrass, coastal, hay, s-c, 53 to 59 days growth, (1)

Ref No 1 09 210 — United States

Feed Name or Analyses			Mean As Fed	Mean Dry	C.V. ± %
Dry matter	%			100.0	
Sheep	dig coef %			47.	
Ash	%			7.1	
Crude fiber	%			29.3	
Ether extract	%			4.0	
N-free extract	%			51.7	

Continued

Feed Name or Analyses			Mean As Fed	Mean Dry	C.V. ± %
Protein (N x 6.25)	%			7.9	
Cattle	dig prot %	*		3.8	
Goats	dig prot %	*		3.9	
Horses	dig prot %	*		4.2	
Rabbits	dig prot %	*		4.8	
Sheep	dig prot %	*		3.7	
Cellulose (Matrone)	%			30.9	
Lignin (Ellis)	%			7.7	
Energy	GE Mcal/kg				
Cattle	DE Mcal/kg	*		2.59	
Sheep	DE Mcal/kg			2.12	
Cattle	ME Mcal/kg	*		2.12	
Sheep	ME Mcal/kg	*		1.74	
Cattle	TDN %	*		58.7	
Sheep	TDN %	*		57.2	

Bermudagrass, coastal, hay, s-c, 39 to 45 days growth, (1)

Ref No 1 09 209 United States

Feed Name or Analyses			Mean As Fed	Mean Dry	C.V. ± %
Dry matter	%			100.0	
Sheep	dig coef %			53.	
Ash	%			7.7	
Crude fiber	%			29.8	
Ether extract	%			2.7	
N-free extract	%			47.7	
Protein (N x 6.25)	%			12.2	
Cattle	dig prot %	*		7.5	
Goats	dig prot %	*		7.9	
Horses	dig prot %	*		7.9	
Rabbits	dig prot %	*		8.1	
Sheep	dig prot %	*		7.5	
Cellulose (Matrone)	%			31.3	
Lignin (Ellis)	%			7.2	
Energy	GE Mcal/kg				
Cattle	DE Mcal/kg	*		2.53	
Sheep	DE Mcal/kg			2.34	
Cattle	ME Mcal/kg	*		2.07	
Sheep	ME Mcal/kg	*		1.92	
Cattle	TDN %	*		57.3	
Sheep	TDN %	*		58.7	

Bermudagrass, coastal, hay, s-c, 18 to 24 days growth, (1)

Ref No 1 09 205 United States

Feed Name or Analyses			Mean As Fed	Mean Dry	C.V. ± %
Dry matter	%			100.0	
Sheep	dig coef %			57.	
Ash	%			8.8	
Crude fiber	%			28.5	
Ether extract	%			3.3	
N-free extract	%			47.9	
Protein (N x 6.25)	%			11.5	
Cattle	dig prot %	*		6.9	
Goats	dig prot %	*		7.3	
Horses	dig prot %	*		7.3	
Rabbits	dig prot %	*		7.5	
Sheep	dig prot %	*		6.9	
Cellulose (Matrone)	%			31.4	
Energy	GE Mcal/kg				
Cattle	DE Mcal/kg	*		2.47	
Sheep	DE Mcal/kg			2.57	
Cattle	ME Mcal/kg	*		2.02	

(1) dry forages and roughages (2) pasture, range plants, and forages fed green (3) silages (4) energy feeds (5) protein supplements (6) minerals (7) vitamins (8) additives

Feed Name or Analyses			Mean As Fed	Mean Dry	C.V. ± %
Sheep	ME Mcal/kg	*		2.11	
Cattle	TDN %	*		56.0	
Sheep	TDN %	*		58.7	

Bermudagrass, coastal, hay, s-c, 25 to 31 days growth, cut 4, (1)

Ref No 1 09 223 United States

Feed Name or Analyses			Mean As Fed	Mean Dry	C.V. ± %
Dry matter	%		88.1	100.0	
Cattle	dig coef %		59.	59.	
Ash	%		6.2	7.0	
Crude fiber	%		27.9	31.7	
Ether extract	%		2.0	2.3	
N-free extract	%		41.2	46.8	
Protein (N x 6.25)	%		10.7	12.2	
Cattle	dig coef %		63.	63.	
Cattle	dig prot %		6.8	7.7	
Goats	dig prot %	*	7.0	7.9	
Horses	dig prot %	*	6.9	7.9	
Rabbits	dig prot %	*	7.1	8.1	
Sheep	dig prot %	*	6.6	7.5	
Cellulose (Matrone)	%		27.5	31.2	
Cattle	dig coef %		66.	66.	
Energy	GE Mcal/kg		3.90	4.43	
Cattle	GE dig coef %		57.	57.	
Cattle	DE Mcal/kg		2.22	2.52	
Sheep	DE Mcal/kg	*	2.25	2.55	
Cattle	ME Mcal/kg	*	1.82	2.07	
Sheep	ME Mcal/kg	*	1.84	2.09	
Cattle	TDN %	*	52.0	59.0	
Sheep	TDN %	*	51.0	57.9	

Bermudagrass, coastal, hay, s-c, 32 to 38 days growth, cut 4, (1)

Ref No 1 09 224 United States

Feed Name or Analyses			Mean As Fed	Mean Dry	C.V. ± %
Dry matter	%		90.4	100.0	
Cattle	dig coef %		57.	57.	
Ash	%		5.8	6.4	
Crude fiber	%		27.3	30.2	
Cattle	dig coef %		61.	61.	
Ether extract	%		2.5	2.7	
Cattle	dig coef %		52.	52.	
N-free extract	%		41.9	46.3	
Cattle	dig coef %		52.	52.	
Protein (N x 6.25)	%		13.0	14.4	
Cattle	dig coef %		67.	67.	
Cattle	dig prot %		8.7	9.6	
Goats	dig prot %	*	9.0	10.0	
Horses	dig prot %	*	8.8	9.8	
Rabbits	dig prot %	*	8.8	9.8	
Sheep	dig prot %	*	8.6	9.5	
Lignin (Ellis)	%		8.7	9.6	
Energy	GE Mcal/kg		4.15	4.59	
Cattle	GE dig coef %		49.	49.	
Cattle	DE Mcal/kg		2.03	2.25	
Sheep	DE Mcal/kg	*	2.36	2.61	
Cattle	ME Mcal/kg	*	1.67	1.84	
Sheep	ME Mcal/kg	*	1.93	2.14	
Cattle	TDN %		49.9	55.2	
Sheep	TDN %	*	53.4	59.1	
Calcium	%		0.39	0.43	
Phosphorus	%		0.25	0.27	

Bermudagrass, coastal, hay, s-c, immature, (1)

Ref No 1 00 713 United States

Feed Name or Analyses			Mean As Fed	Mean Dry	C.V. ± %
Dry matter	%		90.4	100.0	
Ash	%		4.1	4.5	
Crude fiber	%		24.7	27.3	
Ether extract	%		1.4	1.6	
N-free extract	%		45.7	50.6	
Protein (N x 6.25)	%		14.5	16.0	
Cattle	dig prot %	*	9.8	10.8	
Goats	dig prot %	*	10.4	11.5	
Horses	dig prot %	*	10.0	11.1	
Rabbits	dig prot %	*	10.0	11.0	
Sheep	dig prot %	*	9.9	10.9	
Energy	GE Mcal/kg				
Cattle	DE Mcal/kg	*	2.21	2.45	
Sheep	DE Mcal/kg	*	2.14	2.37	
Cattle	ME Mcal/kg	*	1.81	2.01	
Sheep	ME Mcal/kg	*	1.75	1.94	
Cattle	TDN %	*	50.2	55.5	
Sheep	TDN %	*	48.6	53.8	

Bermudagrass, coastal, hay, s-c, prebloom, cut 2, (1)

Ref No 1 09 101 United States

Feed Name or Analyses			Mean As Fed	Mean Dry	C.V. ± %
Dry matter	%		88.2	100.0	0
Cattle	dig coef %		59.	59.	6
Ash	%		6.1	7.0	6
Crude fiber	%		27.9	31.7	5
Cattle	dig coef %		60.	60.	
Ether extract	%		2.2	2.5	23
Cattle	dig coef %		38.	38.	
N-free extract	%		40.8	46.2	10
Cattle	dig coef %		56.	56.	
Protein (N x 6.25)	%		11.2	12.7	29
Cattle	dig coef %		63.	63.	16
Cattle	dig prot %		7.0	8.0	
Goats	dig prot %	*	7.4	8.4	
Horses	dig prot %	*	7.3	8.3	
Rabbits	dig prot %	*	7.4	8.4	
Sheep	dig prot %	*	7.0	7.9	
Cell walls (Van Soest)	%		71.6	81.2	
Cellulose (Matrone)	%		28.9	32.8	7
Cattle	dig coef %		64.	64.	6
Fiber, acid detergent (VS)	%		33.4	37.9	
Energy	GE Mcal/kg		3.92	4.45	0
Cattle	GE dig coef %		58.	58.	6
Cattle	DE Mcal/kg		2.27	2.57	
Sheep	DE Mcal/kg	*	2.26	2.56	
Cattle	ME Mcal/kg	*	1.86	2.11	
Sheep	ME Mcal/kg	*	1.85	2.10	
Cattle	TDN %		48.5	55.0	
Sheep	TDN %	*	51.2	58.1	

Bermudagrass, coastal, hay, s-c, prebloom, cut 3, (1)

Ref No 1 09 102 United States

Feed Name or Analyses			Mean As Fed	Mean Dry	C.V. ± %
Dry matter	%		88.4	100.0	
Cattle	dig coef %		61.	61.	
Ash	%		6.3	7.1	
Crude fiber	%		29.4	33.3	
Ether extract	%		2.1	2.4	
N-free extract	%		36.4	41.2	
Protein (N x 6.25)	%		14.1	16.0	
Cattle	dig coef %		68.	68.	

Feed Name or Analyses				Mean As Fed	Mean Dry	C.V. ± %
Cattle	dig prot	%		9.6	10.9	
Goats	dig prot	%	*	10.2	11.5	
Horses	dig prot	%	*	9.8	11.1	
Rabbits	dig prot	%	*	9.7	11.0	
Sheep	dig prot	%	*	9.7	10.9	
Cell walls (Van Soest)		%		72.8	82.3	
Cellulose (Matrone)		%		28.6	32.4	
Cattle	dig coef	%		67.	67.	
Fiber, acid detergent (VS)		%		36.4	41.2	
Energy	GE Mcal/kg			3.89	4.40	
Cattle	GE dig coef	%		57.	57.	
Cattle	DE Mcal/kg			2.22	2.51	
Sheep	DE Mcal/kg		*	2.22	2.51	
Cattle	ME Mcal/kg		*	1.82	2.06	
Sheep	ME Mcal/kg		*	1.82	2.06	
Cattle	TDN	%	*	50.1	56.7	
Sheep	TDN	%	*	50.3	56.9	

Bermudagrass, coastal, hay, s-c, mature, (1)
Ref No 1 00 714 United States

Feed Name or Analyses				Mean As Fed	Mean Dry	C.V. ± %
Dry matter		%		95.5	100.0	
Ash		%		3.8	4.0	
Crude fiber		%		30.4	31.8	
Ether extract		%		1.6	1.7	
N-free extract		%		54.1	56.7	
Protein (N x 6.25)		%		5.5	5.8	
Cattle	dig prot	%	*	1.9	2.0	
Goats	dig prot	%	*	1.9	2.0	
Horses	dig prot	%	*	2.3	2.5	
Rabbits	dig prot	%	*	3.0	3.1	
Sheep	dig prot	%	*	1.7	1.8	
Energy	GE Mcal/kg					
Cattle	DE Mcal/kg		*	2.40	2.51	
Sheep	DE Mcal/kg		*	2.36	2.47	
Cattle	ME Mcal/kg		*	1.97	2.06	
Sheep	ME Mcal/kg		*	1.94	2.03	
Cattle	TDN	%	*	54.4	56.9	
Sheep	TDN	%	*	53.5	56.1	

Bermudagrass, coastal, hay, s-c, cut 1, (1)
Ref No 1 09 096 United States

Feed Name or Analyses				Mean As Fed	Mean Dry	C.V. ± %
Dry matter		%		88.4	100.0	
Cattle	dig coef	%		58.	58.	
Ash		%		5.6	6.3	
Crude fiber		%		26.9	30.4	
Cattle	dig coef	%		69.	69.	
Ether extract		%		1.9	2.1	
Cattle	dig coef	%		43.	43.	
N-free extract		%		37.0	41.9	
Cattle	dig coef	%		51.	51.	
Protein (N x 6.25)		%		17.1	19.3	
Cattle	dig coef	%		69.	69.	
Cattle	dig prot	%		11.8	13.3	
Goats	dig prot	%	*	12.9	14.6	
Horses	dig prot	%	*	12.3	13.9	
Rabbits	dig prot	%	*	12.0	13.6	
Sheep	dig prot	%	*	12.3	13.9	
Cellulose (Matrone)		%		25.1	28.4	
Cattle	dig coef	%		71.	71.	
Energy	GE Mcal/kg			3.95	4.46	
Cattle	GE dig coef	%		57.	57.	
Cattle	DE Mcal/kg			2.25	2.54	
Sheep	DE Mcal/kg		*	2.35	2.66	
Cattle	ME Mcal/kg		*	1.84	2.09	
Sheep	ME Mcal/kg		*	1.93	2.18	

Feed Name or Analyses				Mean As Fed	Mean Dry	C.V. ± %
Cattle	TDN	%		51.0	57.7	
Sheep	TDN	%	*	53.3	60.2	

Bermudagrass, coastal, hay, s-c, cut 4, (1)
Ref No 1 08 622 United States

Feed Name or Analyses				Mean As Fed	Mean Dry	C.V. ± %
Dry matter		%		88.1	100.0	
Cattle	dig coef	%		59.	59.	
Ash		%		6.2	7.0	
Crude fiber		%		27.9	31.7	
Ether extract		%		2.0	2.3	
N-free extract		%		41.2	46.8	
Protein (N x 6.25)		%		10.7	12.2	
Cattle	dig coef	%		63.	63.	
Cattle	dig prot	%		6.8	7.7	
Goats	dig prot	%	*	7.0	7.9	
Horses	dig prot	%	*	6.9	7.9	
Rabbits	dig prot	%	*	7.1	8.1	
Sheep	dig prot	%	*	6.6	7.5	
Cell walls (Van Soest)		%		73.2	83.1	
Cellulose (Matrone)		%		27.5	31.2	
Cattle	dig coef	%		66.	66.	
Fiber, acid detergent (VS)		%		32.4	36.8	
Energy	GE Mcal/kg			3.90	4.43	
Cattle	GE dig coef	%		57.	57.	
Cattle	DE Mcal/kg			2.22	2.52	
Sheep	DE Mcal/kg		*	2.25	2.55	
Cattle	ME Mcal/kg		*	1.82	2.07	
Sheep	ME Mcal/kg		*	1.84	2.09	
Cattle	TDN	%	*	52.0	59.0	
Sheep	TDN	%	*	51.0	57.9	

Bermudagrass, coastal, hay, s-c, gr 1 US, (1)
Ref No 1 00 715 United States

Feed Name or Analyses				Mean As Fed	Mean Dry	C.V. ± %
Dry matter		%			100.0	
Ash		%			4.3	
Crude fiber		%			28.4	
Ether extract		%			2.4	
N-free extract		%			50.2	
Protein (N x 6.25)		%			14.7	
Cattle	dig prot	%	*		9.7	
Goats	dig prot	%	*		10.3	
Horses	dig prot	%	*		10.0	
Rabbits	dig prot	%	*		10.0	
Sheep	dig prot	%	*		9.8	
Energy	GE Mcal/kg				4.32	
Cattle	DE Mcal/kg		*		2.45	
Sheep	DE Mcal/kg		*		2.31	
Cattle	ME Mcal/kg		*		2.01	
Sheep	ME Mcal/kg		*		1.89	
Cattle	TDN	%	*		55.6	
Sheep	TDN	%	*		52.4	

Bermudagrass, coastal, hay, s-c grnd, 32 to 38 days growth, cut 4, (1)
Ref No 1 09 225 United States

Feed Name or Analyses				Mean As Fed	Mean Dry	C.V. ± %
Dry matter		%			100.0	
Cattle	dig coef	%			51.	
Ash		%			6.1	
Crude fiber		%			32.2	
Cattle	dig coef	%			55.	
Ether extract		%			2.4	
N-free extract		%			47.8	
Cattle	dig coef	%			47.	
Protein (N x 6.25)		%			11.6	
Cattle	dig coef	%			65.	
Cattle	dig prot	%			7.4	
Goats	dig prot	%	*		7.3	
Horses	dig prot	%	*		7.3	
Rabbits	dig prot	%	*		7.6	
Sheep	dig prot	%	*		6.9	
Energy	GE Mcal/kg				4.62	
Cattle	GE dig coef	%			49.	
Cattle	DE Mcal/kg				2.26	
Sheep	DE Mcal/kg		*		2.15	
Cattle	ME Mcal/kg		*		1.86	
Sheep	ME Mcal/kg		*		1.76	
Cattle	TDN	%			49.3	
Sheep	TDN	%	*		48.7	
Calcium		%			0.48	
Phosphorus		%			0.27	

Bermudagrass, coastal, hay, s-c grnd, cut 4, (1)
Ref No 1 05 680 United States

Feed Name or Analyses				Mean As Fed	Mean Dry	C.V. ± %
Dry matter		%			100.0	
Ash		%			6.3	
Crude fiber		%			33.1	
Cattle	dig coef	%			56.	
Ether extract		%			2.4	
Cattle	dig coef	%			41.	
N-free extract		%			46.7	
Cattle	dig coef	%			45.	
Protein (N x 6.25)		%			11.5	
Cattle	dig coef	%			66.	
Cattle	dig prot	%			7.6	
Goats	dig prot	%	*		7.3	
Horses	dig prot	%	*		7.3	
Rabbits	dig prot	%	*		7.5	
Sheep	dig prot	%	*		6.9	
Energy	GE Mcal/kg					
Cattle	DE Mcal/kg		*		2.18	
Sheep	DE Mcal/kg		*		2.12	
Cattle	ME Mcal/kg		*		1.78	
Sheep	ME Mcal/kg		*		1.74	
Cattle	TDN	%			49.4	
Sheep	TDN	%	*		48.1	
Calcium		%			0.50	
Phosphorus		%			0.24	

Bermudagrass, coastal, hay, s-c grnd pelleted, 32 to 38 days growth, cut 4, (1)
Ref No 1 09 226 United States

Feed Name or Analyses				Mean As Fed	Mean Dry	C.V. ± %
Dry matter		%			100.0	
Cattle	dig coef	%			46.	
Ash		%			5.7	
Crude fiber		%			28.0	
Cattle	dig coef	%			42.	
Ether extract		%			2.1	
N-free extract		%			52.5	
Cattle	dig coef	%			44.	
Protein (N x 6.25)		%			11.8	
Cattle	dig coef	%			60.	
Cattle	dig prot	%			7.1	
Goats	dig prot	%	*		7.6	
Horses	dig prot	%	*		7.5	
Rabbits	dig prot	%	*		7.8	
Sheep	dig prot	%	*		7.2	
Energy	GE Mcal/kg				4.51	
Cattle	GE dig coef	%			42.	

Continued

Feed Name or Analyses			Mean As Fed	Mean Dry	C.V. ± %
Cattle	DE Mcal/kg			1.89	
Sheep	DE Mcal/kg	*		1.99	
Cattle	ME Mcal/kg	*		1.55	
Sheep	ME Mcal/kg	*		1.63	
Cattle	TDN %			43.7	
Sheep	TDN %	*		45.1	
Calcium	%			0.37	
Phosphorus	%			0.28	

Bermudagrass, coastal, hay, s-c rained on, 25 to 31 days growth, cut 2, (1)

Ref No 1 09 215 United States

Feed Name or Analyses			Mean As Fed	Mean Dry	C.V. ± %
Dry matter	%		91.1	100.0	
Cattle	dig coef %		56.	56.	
Ash	%		6.5	7.1	
Crude fiber	%		28.7	31.5	
Cattle	dig coef %		63.	63.	
Ether extract	%		2.6	2.9	
Cattle	dig coef %		37.	37.	
N-free extract	%		38.0	41.7	
Cattle	dig coef %		52.	52.	
Protein (N x 6.25)	%		15.3	16.8	
Cattle	dig coef %		56.	56.	
Cattle	dig prot %		8.6	9.4	
Goats	dig prot %	*	11.1	12.2	
Horses	dig prot %	*	10.7	11.8	
Rabbits	dig prot %	*	10.6	11.6	
Sheep	dig prot %	*	10.6	11.6	
Lignin (Ellis)	%		8.3	9.1	
Energy	GE Mcal/kg				
Cattle	DE Mcal/kg	*	2.14	2.35	
Sheep	DE Mcal/kg	*	2.07	2.27	
Cattle	ME Mcal/kg	*	1.76	1.93	
Sheep	ME Mcal/kg	*	1.69	1.86	
Cattle	TDN %		48.6	53.4	
Sheep	TDN %	*	46.8	51.4	
Calcium	%		0.48	0.53	
Phosphorus	%		0.25	0.27	

Bermudagrass, coastal, hay, s-c rained on, cut 1, (1)

Ref No 1 09 214 United States

Feed Name or Analyses			Mean As Fed	Mean Dry	C.V. ± %
Dry matter	%		90.8	100.0	
Cattle	dig coef %		54.	54.	
Ash	%		6.7	7.4	
Crude fiber	%		27.1	29.9	
Cattle	dig coef %		61.	61.	
Ether extract	%		2.4	2.6	
Cattle	dig coef %		33.	33.	
N-free extract	%		41.4	45.6	
Cattle	dig coef %		51.	51.	
Protein (N x 6.25)	%		13.2	14.5	
Cattle	dig coef %		52.	52.	
Cattle	dig prot %		6.8	7.5	
Goats	dig prot %	*	9.2	10.1	
Horses	dig prot %	*	8.9	9.8	
Rabbits	dig prot %	*	9.0	9.9	
Sheep	dig prot %	*	8.7	9.6	
Lignin (Ellis)	%		7.5	8.3	
Energy	GE Mcal/kg				
Cattle	DE Mcal/kg	*	2.04	2.25	

(1) dry forages and roughages (2) pasture, range plants, and forages fed green (3) silages (4) energy feeds (5) protein supplements (6) minerals (7) vitamins (8) additives

Feed Name or Analyses			Mean As Fed	Mean Dry	C.V. ± %
Sheep	DE Mcal/kg	*	1.99	2.19	
Cattle	ME Mcal/kg	*	1.67	1.84	
Sheep	ME Mcal/kg	*	1.63	1.80	
Cattle	TDN %		46.3	51.0	
Sheep	TDN %	*	45.0	49.6	
Calcium	%		0.44	0.48	
Phosphorus	%		0.19	0.21	

Bermudagrass, coastal, hay, 40 to 60 lb nitrogen per acre spring s-c, 25 to 31 days growth, cut 2, (1)

Ref No 1 09 231 United States

Feed Name or Analyses			Mean As Fed	Mean Dry	C.V. ± %
Dry matter	%		88.3	100.0	
Cattle	dig coef %		59.	59.	
Ash	%		6.4	7.2	
Crude fiber	%		27.8	31.5	
Ether extract	%		2.3	2.6	
N-free extract	%		42.7	48.4	
Protein (N x 6.25)	%		9.1	10.3	
Cattle	dig coef %		61.	61.	
Cattle	dig prot %		5.5	6.3	
Goats	dig prot %	*	5.4	6.2	
Horses	dig prot %	*	5.5	6.3	
Rabbits	dig prot %	*	5.8	6.6	
Sheep	dig prot %	*	5.1	5.8	
Cellulose (Matrone)	%		29.8	33.7	
Cattle	dig coef %		63.	63.	
Energy	GE Mcal/kg		3.92	4.44	
Cattle	GE dig coef %		59.	59.	
Cattle	DE Mcal/kg		2.31	2.62	
Sheep	DE Mcal/kg	*	2.23	2.52	
Cattle	ME Mcal/kg	*	1.89	2.15	
Sheep	ME Mcal/kg	*	1.83	2.07	
Cattle	TDN %	*	53.2	60.3	
Sheep	TDN %	*	50.5	57.2	

Bermudagrass, coastal, hay, 580 to 600 lb nitrogen per acre s-c grnd, 39 to 45 days growth, (1)

Ref No 1 09 234 United States

Feed Name or Analyses			Mean As Fed	Mean Dry	C.V. ± %
Dry matter	%			100.0	
Sheep	dig coef %			66.	
Ash	%			3.1	
Crude fiber	%			25.8	
Sheep	dig coef %			66.	
Ether extract	%			2.3	
N-free extract	%			57.0	
Sheep	dig coef %			68.	
Protein (N x 6.25)	%			11.8	
Sheep	dig coef %			71.	
Cattle	dig prot %	*		7.2	
Goats	dig prot %	*		7.6	
Horses	dig prot %	*		7.5	
Rabbits	dig prot %	*		7.8	
Sheep	dig prot %			8.4	
Lignin (Ellis)	%			11.2	
Energy	GE Mcal/kg				
Cattle	DE Mcal/kg	*		2.88	
Sheep	DE Mcal/kg	*		2.96	
Cattle	ME Mcal/kg	*		2.36	
Sheep	ME Mcal/kg	*		2.43	
Cattle	TDN %	*		65.3	
Sheep	TDN %			67.2	

Bermudagrass, coastal, hay, 580 to 600 lb nitrogen per acre s-c grnd, 18 to 24 days growth, (1)

Ref No 1 09 236 United States

Feed Name or Analyses			Mean As Fed	Mean Dry	C.V. ± %
Dry matter	%			100.0	
Sheep	dig coef %			70.	
Ash	%			5.1	
Crude fiber	%			25.0	
Sheep	dig coef %			71.	
Ether extract	%			2.9	
N-free extract	%			49.4	
Sheep	dig coef %			65.	
Protein (N x 6.25)	%			17.6	
Sheep	dig coef %			78.	
Cattle	dig prot %	*		12.2	
Goats	dig prot %	*		13.0	
Horses	dig prot %	*		12.5	
Rabbits	dig prot %	*		12.3	
Sheep	dig prot %			13.7	
Lignin (Ellis)	%			9.7	
Energy	GE Mcal/kg				
Cattle	DE Mcal/kg	*		2.65	
Sheep	DE Mcal/kg	*		2.99	
Cattle	ME Mcal/kg	*		2.17	
Sheep	ME Mcal/kg	*		2.45	
Cattle	TDN %	*		60.1	
Sheep	TDN %			67.8	

Bermudagrass, coastal, hay, 580 to 600 lb nitrogen per acre s-c grnd, 53 to 59 days growth, (1)

Ref No 1 09 238 United States

Feed Name or Analyses			Mean As Fed	Mean Dry	C.V. ± %
Dry matter	%			100.0	
Sheep	dig coef %			59.	
Ash	%			4.0	
Crude fiber	%			27.0	
Sheep	dig coef %			59.	
Ether extract	%			1.9	
N-free extract	%			57.2	
Sheep	dig coef %			61.	
Protein (N x 6.25)	%			9.9	
Sheep	dig coef %			63.	
Cattle	dig prot %	*		5.5	
Goats	dig prot %	*		5.8	
Horses	dig prot %	*		5.9	
Rabbits	dig prot %	*		6.3	
Sheep	dig prot %			6.2	
Lignin (Ellis)	%			12.1	
Energy	GE Mcal/kg				
Cattle	DE Mcal/kg	*		2.83	
Sheep	DE Mcal/kg	*		2.61	
Cattle	ME Mcal/kg	*		2.32	
Sheep	ME Mcal/kg	*		2.14	
Cattle	TDN %	*		64.2	
Sheep	TDN %			59.1	

Bermudagrass, coastal, hay, 580 to 600 lb nitrogen per acre s-c grnd, 11 to 17 days growth, (1)

Ref No 1 09 235 United States

Feed Name or Analyses			Mean As Fed	Mean Dry	C.V. ± %
Dry matter	%			100.0	
Sheep	dig coef %			71.	
Ash	%			5.8	

Feed Name or Analyses			Mean As Fed	Mean Dry	C.V. ± %
Crude fiber	%			22.2	
Sheep	dig coef %			72.	
Ether extract	%			3.1	
N-free extract	%			48.3	
Sheep	dig coef %			67.	
Protein (N x 6.25)	%			20.6	
Sheep	dig coef %			79.	
Cattle	dig prot %	*		14.8	
Goats	dig prot %	*		15.8	
Horses	dig prot %	*		15.0	
Rabbits	dig prot %	*		14.6	
Sheep	dig prot %			16.3	
Lignin (Ellis)	%			9.5	
Energy	GE Mcal/kg				
Cattle	DE Mcal/kg	*		2.75	
Sheep	DE Mcal/kg	*		3.07	
Cattle	ME Mcal/kg	*		2.25	
Sheep	ME Mcal/kg	*		2.52	
Cattle	TDN %	*		62.3	
Sheep	TDN %			69.6	

Bermudagrass, coastal, hay, 580 to 600 lb nitrogen per acre s-c grnd, 25 to 31 days growth, (1)

Ref No 1 09 237 United States

Feed Name or Analyses			Mean As Fed	Mean Dry	C.V. ± %
Dry matter	%			100.0	
Sheep	dig coef %			66.	
Ash	%			4.2	
Crude fiber	%			25.2	
Sheep	dig coef %			68.	
Ether extract	%			2.7	
N-free extract	%			51.6	
Sheep	dig coef %			64.	
Protein (N x 6.25)	%			16.3	
Sheep	dig coef %			76.	
Cattle	dig prot %	*		11.1	
Goats	dig prot %	*		11.8	
Horses	dig prot %	*		11.4	
Rabbits	dig prot %	*		11.3	
Sheep	dig prot %			12.4	
Lignin (Ellis)	%			10.3	
Energy	GE Mcal/kg				
Cattle	DE Mcal/kg	*		2.86	
Sheep	DE Mcal/kg	*		2.96	
Cattle	ME Mcal/kg	*		2.34	
Sheep	ME Mcal/kg	*		2.43	
Cattle	TDN %	*		64.9	
Sheep	TDN %			67.2	

Bermudagrass, coastal, hay, 60 to 80 lb nitrogen per acre spring s-c, 32 to 38 days growth, cut 2, (1)

Ref No 1 09 239 United States

Feed Name or Analyses			Mean As Fed	Mean Dry	C.V. ± %
Dry matter	%			100.0	
Cattle	dig coef %			55.	
Ash	%			7.0	
Crude fiber	%			28.9	
Cattle	dig coef %			60.	
Ether extract	%			2.3	
N-free extract	%			53.3	
Cattle	dig coef %			58.	
Protein (N x 6.25)	%			8.5	
Cattle	dig coef %			49.	
Cattle	dig prot %			4.2	
Goats	dig prot %	*		4.5	
Horses	dig prot %	*		4.7	
Rabbits	dig prot %	*		5.2	
Sheep	dig prot %	*		4.2	
Cellulose (Matrone)	%			32.1	
Cattle	dig coef %			57.	
Energy	GE Mcal/kg				
Cattle	DE Mcal/kg	*		2.35	
Sheep	DE Mcal/kg	*		2.55	
Cattle	ME Mcal/kg	*		1.93	
Sheep	ME Mcal/kg	*		2.09	
Cattle	TDN %			53.4	
Sheep	TDN %	*		57.9	

Bermudagrass, coastal, hay, 80 to 100 lb ammonium nitrate per acre after cut 1 s-c, 25 to 31 days growth, cut 2, (1)

Ref No 1 09 232 United States

Feed Name or Analyses			Mean As Fed	Mean Dry	C.V. ± %
Dry matter	%		88.3	100.0	
Cattle	dig coef %		63.	63.	
Ash	%		5.8	6.6	
Crude fiber	%		29.0	32.8	
Ether extract	%		2.6	3.0	
N-free extract	%		36.7	41.6	
Protein (N x 6.25)	%		14.1	16.0	
Cattle	dig coef %		71.	71.	
Cattle	dig prot %		10.0	11.4	
Goats	dig prot %	*	10.1	11.5	
Horses	dig prot %	*	9.8	11.1	
Rabbits	dig prot %	*	9.7	11.0	
Sheep	dig prot %	*	9.6	10.9	
Cellulose (Matrone)	%		28.6	32.4	
Cattle	dig coef %		68.	68.	
Energy	GE Mcal/kg		3.93	4.45	
Cattle	GE dig coef %		61.	61.	
Cattle	DE Mcal/kg		2.40	2.71	
Sheep	DE Mcal/kg	*	2.23	2.52	
Cattle	ME Mcal/kg	*	1.96	2.22	
Sheep	ME Mcal/kg	*	1.82	2.07	
Cattle	TDN %	*	50.4	57.1	
Sheep	TDN %	*	50.5	57.2	

Bermudagrass, coastal, hay, 80 to 100 lb nitrogen per acre after cut 1 s-c, 25 to 31 days growth, cut 2, (1)

Ref No 1 09 233 United States

Feed Name or Analyses			Mean As Fed	Mean Dry	C.V. ± %
Dry matter	%		88.6	100.0	
Cattle	dig coef %		61.	61.	
Ash	%		6.1	6.9	
Crude fiber	%		27.8	31.4	
Ether extract	%		2.6	3.0	
N-free extract	%		38.2	43.1	
Protein (N x 6.25)	%		13.9	15.7	
Cattle	dig coef %		71.	71.	
Cattle	dig prot %		9.8	11.1	
Goats	dig prot %	*	9.9	11.2	
Horses	dig prot %	*	9.5	10.9	
Rabbits	dig prot %	*	9.6	10.8	
Sheep	dig prot %	*	9.4	10.7	
Cellulose (Matrone)	%		27.6	31.1	
Cattle	dig coef %		67.	67.	
Energy	GE Mcal/kg		3.93	4.43	1
Cattle	GE dig coef %		59.	59.	
Cattle	DE Mcal/kg		2.31	2.61	
Sheep	DE Mcal/kg	*	2.26	2.55	
Cattle	ME Mcal/kg	*	1.89	2.14	
Sheep	ME Mcal/kg	*	1.86	2.09	
Cattle	TDN %	*	51.6	58.2	
Sheep	TDN %	*	51.3	57.9	

Bermudagrass, coastal, hay, 80 to 100 lb nitrogen per acre after cut 2 s-c, 25 to 31 days growth, cut 3, (1)

Ref No 1 09 228 United States

Feed Name or Analyses			Mean As Fed	Mean Dry	C.V. ± %
Dry matter	%		88.4	100.0	
Cattle	dig coef %		61.	61.	
Ash	%		6.4	7.2	
Crude fiber	%		29.4	33.3	
Ether extract	%		2.1	2.4	
N-free extract	%		36.3	41.1	
Protein (N x 6.25)	%		14.1	16.0	
Cattle	dig coef %		68.	68.	
Cattle	dig prot %		9.6	10.9	
Goats	dig prot %	*	10.2	11.5	
Horses	dig prot %	*	9.8	11.1	
Rabbits	dig prot %	*	9.7	11.0	
Sheep	dig prot %	*	9.7	10.9	
Cellulose (Matrone)	%		28.6	32.4	
Cattle	dig coef %		67.	67.	
Energy	GE Mcal/kg		3.89	4.40	
Cattle	GE dig coef %		57.	57.	
Cattle	DE Mcal/kg		2.22	2.51	
Sheep	DE Mcal/kg	*	2.21	2.51	
Cattle	ME Mcal/kg	*	1.82	2.06	
Sheep	ME Mcal/kg	*	1.82	2.05	
Cattle	TDN %	*	50.2	56.8	
Sheep	TDN %	*	50.2	56.8	

Bermudagrass, coastal, aerial part, fresh, (2)

Ref No 2 00 719 United States

Feed Name or Analyses			Mean As Fed	Mean Dry	C.V. ± %
Dry matter	%		28.8	100.0	24
Ash	%		1.8	6.3	15
Crude fiber	%		8.2	28.4	7
Ether extract	%		1.1	3.8	25
N-free extract	%		13.4	46.6	6
Protein (N x 6.25)	%		4.3	15.0	16
Cattle	dig prot %	*	3.1	10.7	
Goats	dig prot %	*	3.0	10.6	
Horses	dig prot %	*	3.0	10.3	
Rabbits	dig prot %	*	3.0	10.3	
Sheep	dig prot %	*	3.2	11.0	
Lignin (Ellis)	%		2.4	8.5	
Energy	GE Mcal/kg				
Cattle	DE Mcal/kg	*	0.85	2.96	
Sheep	DE Mcal/kg	*	0.80	2.78	
Cattle	ME Mcal/kg	*	0.70	2.43	
Sheep	ME Mcal/kg	*	0.66	2.28	
Cattle	TDN %	*	19.4	67.2	
Sheep	TDN %	*	18.2	63.0	
Calcium	%		0.14	0.49	15
Phosphorus	%		0.08	0.27	12
Carotene	mg/kg		95.3	330.5	
Vitamin A equivalent	IU/g		158.8	550.9	

Bermudagrass, coastal, aerial part, fresh, immature, (2)

Ref No 2 00 717 United States

Feed Name or Analyses			Mean As Fed	Mean Dry	C.V. ± %
Dry matter	%		46.7	100.0	6
Ash	%		3.5	7.5	12
Crude fiber	%		11.3	24.2	14
Ether extract	%		1.0	2.2	18

Continued

Feed Name or Analyses			Mean As Fed	Mean Dry	C.V. ± %
N-free extract	%		22.3	47.8	
Protein (N x 6.25)	%		8.5	18.3	20
Cattle	dig prot %	*	6.3	13.4	
Goats	dig prot %	*	6.4	13.6	
Horses	dig prot %	*	6.1	13.1	
Rabbits	dig prot %	*	6.0	12.8	
Sheep	dig prot %	*	6.6	14.0	
Energy	GE Mcal/kg				
Cattle	DE Mcal/kg	*	1.36	2.91	
Sheep	DE Mcal/kg	*	1.27	2.71	
Cattle	ME Mcal/kg	*	1.12	2.39	
Sheep	ME Mcal/kg	*	1.04	2.22	
Cattle	TDN %	*	30.8	66.0	
Sheep	TDN %	*	28.7	61.5	

Bermudagrass, coastal, aerial part, fresh, milk stage, (2)
Ref No 2 00 718 United States

Feed Name or Analyses			Mean As Fed	Mean Dry	C.V. ± %
Dry matter	%		49.3	100.0	3
Ash	%		5.2	10.6	10
Crude fiber	%		12.5	25.4	5
Ether extract	%		0.7	1.5	8
N-free extract	%		27.5	55.8	
Protein (N x 6.25)	%		3.3	6.7	6
Cattle	dig prot %	*	1.8	3.6	
Goats	dig prot %	*	1.4	2.8	
Horses	dig prot %	*	1.6	3.2	
Rabbits	dig prot %	*	1.9	3.8	
Sheep	dig prot %	*	1.6	3.2	
Energy	GE Mcal/kg				
Cattle	DE Mcal/kg	*	1.28	2.59	
Sheep	DE Mcal/kg	*	1.37	2.78	
Cattle	ME Mcal/kg	*	1.05	2.12	
Sheep	ME Mcal/kg	*	1.12	2.28	
Cattle	TDN %	*	28.9	58.7	
Sheep	TDN %	*	31.1	63.1	

Bermudagrass, coastal, aerial part, ensiled, 25 to 31 days growth, cut 2, (3)
Ref No 3 09 217 United States

Feed Name or Analyses			Mean As Fed	Mean Dry	C.V. ± %
Dry matter	%		30.0	100.0	
Cattle	dig coef %		58.	58.	
Ash	%		2.0	6.6	
Crude fiber	%		9.6	32.2	
Cattle	dig coef %		68.	68.	
Ether extract	%		1.3	4.4	
N-free extract	%		12.9	43.0	
Cattle	dig coef %		53.	53.	
Protein (N x 6.25)	%		4.2	14.0	
Cattle	dig coef %		62.	62.	
Cattle	dig prot %		2.6	8.6	
Goats	dig prot %	*	2.7	8.9	
Horses	dig prot %	*	2.7	8.9	
Sheep	dig prot %	*	2.7	8.9	
Energy	GE Mcal/kg		1.42	4.74	
Cattle	GE dig coef %		56.	56.	
Cattle	DE Mcal/kg		0.79	2.63	
Sheep	DE Mcal/kg	*	0.83	2.78	
Cattle	ME Mcal/kg	*	0.65	2.16	
Sheep	ME Mcal/kg	*	0.68	2.28	
Cattle	TDN %		17.5	58.3	
Sheep	TDN %	*	18.9	63.1	

(1) dry forages and roughages (2) pasture, range plants, and forages fed green (3) silages (4) energy feeds (5) protein supplements (6) minerals (7) vitamins (8) additives

Feed Name or Analyses			Mean As Fed	Mean Dry	C.V. ± %
Calcium	%		0.15	0.49	
Phosphorus	%		0.08	0.27	

Bermudagrass, coastal, aerial part, ensiled, 32 to 38 day growth, cut 2, (3)
Ref No 3 09 218 United States

Feed Name or Analyses			Mean As Fed	Mean Dry	C.V. ± %
Dry matter	%		24.9	100.0	
Cattle	dig coef %		57.	57.	
Ash	%		1.7	6.7	
Crude fiber	%		8.1	32.7	
Cattle	dig coef %		65.	65.	
Ether extract	%		2.0	8.0	
N-free extract	%		9.9	39.9	
Cattle	dig coef %		47.	47.	
Protein (N x 6.25)	%		3.2	12.7	
Cattle	dig coef %		57.	57.	
Cattle	dig prot %		1.8	7.2	
Goats	dig prot %	*	1.9	7.8	
Horses	dig prot %	*	1.9	7.8	
Sheep	dig prot %	*	1.9	7.8	
Energy	GE Mcal/kg		1.18	4.74	
Cattle	GE dig coef %		54.	54.	
Cattle	DE Mcal/kg		0.64	2.56	
Sheep	DE Mcal/kg	*	0.67	2.69	
Cattle	ME Mcal/kg	*	0.52	2.10	
Sheep	ME Mcal/kg	*	0.55	2.21	
Cattle	TDN %		15.0	60.4	
Sheep	TDN %	*	15.2	61.0	
Calcium	%		0.11	0.46	
Phosphorus	%		0.08	0.32	

Bermudagrass, coastal, aerial part, ensiled, 25 to 31 days growth, cut 3, (3)
Ref No 3 09 219 United States

Feed Name or Analyses			Mean As Fed	Mean Dry	C.V. ± %
Dry matter	%		23.8	100.0	
Cattle	dig coef %		59.	59.	
Ash	%		1.7	7.3	
Crude fiber	%		7.4	31.0	
Cattle	dig coef %		68.	68.	
Ether extract	%		1.2	4.9	
N-free extract	%		10.3	43.3	
Cattle	dig coef %		54.	54.	
Protein (N x 6.25)	%		3.2	13.7	
Cattle	dig coef %		62.	62.	
Cattle	dig prot %		2.0	8.4	
Goats	dig prot %	*	2.1	8.6	
Horses	dig prot %	*	2.1	8.6	
Sheep	dig prot %	*	2.1	8.6	
Energy	GE Mcal/kg		1.12	4.73	
Cattle	GE dig coef %		56.	56.	
Cattle	DE Mcal/kg		0.63	2.65	
Sheep	DE Mcal/kg	*	0.66	2.76	
Cattle	ME Mcal/kg	*	0.52	2.17	
Sheep	ME Mcal/kg	*	0.54	2.27	
Cattle	TDN %		14.1	59.4	
Sheep	TDN %	*	14.9	62.7	
Calcium	%		0.13	0.55	
Phosphorus	%		0.08	0.34	

Bermudagrass, coastal, aerial part, ensiled, 32 to 38 days growth, cut 3, (3)
Ref No 3 09 220 United States

Feed Name or Analyses			Mean As Fed	Mean Dry	C.V. ± %
Dry matter	%		31.8	100.0	
Cattle	dig coef %		61.	61.	
Ash	%		1.8	5.8	
Crude fiber	%		9.6	30.3	
Cattle	dig coef %		68.	68.	
Ether extract	%		1.2	3.8	
N-free extract	%		14.7	46.2	
Cattle	dig coef %		57.	57.	
Protein (N x 6.25)	%		4.4	13.9	
Cattle	dig coef %		64.	64.	
Cattle	dig prot %		2.8	8.9	
Goats	dig prot %	*	2.8	8.9	
Horses	dig prot %	*	2.8	8.9	
Sheep	dig prot %	*	2.8	8.9	
Energy	GE Mcal/kg		1.50	4.72	
Cattle	GE dig coef %		58.	58.	
Cattle	DE Mcal/kg		0.87	2.74	
Sheep	DE Mcal/kg	*	0.90	2.83	
Cattle	ME Mcal/kg	*	0.71	2.24	
Sheep	ME Mcal/kg	*	0.74	2.32	
Cattle	TDN %		19.2	60.5	
Sheep	TDN %	*	20.4	64.3	
Calcium	%		0.13	0.40	
Phosphorus	%		0.07	0.21	

Bermudagrass, coastal, aerial part, wilted ensiled, (3)
Ref No 3 09 097 United States

Feed Name or Analyses			Mean As Fed	Mean Dry	C.V. ± %
Dry matter	%		39.3	100.0	
Cattle	dig coef %		57.	57.	
Ash	%		3.2	8.1	
Crude fiber	%		12.5	31.9	
Ether extract	%		1.3	3.4	
N-free extract	%		16.6	42.2	
Protein (N x 6.25)	%		5.7	14.4	
Cattle	dig coef %		66.	66.	
Cattle	dig prot %		3.7	9.5	
Goats	dig prot %	*	3.7	9.3	
Horses	dig prot %	*	3.7	9.3	
Sheep	dig prot %	*	3.7	9.3	
Cellulose (Matrone)	%		11.2	28.4	
Cattle	dig coef %		67.	67.	
Energy	GE Mcal/kg		1.76	4.47	
Cattle	GE dig coef %		57.	57.	
Cattle	DE Mcal/kg		1.00	2.55	
Sheep	DE Mcal/kg	*	1.09	2.77	
Cattle	ME Mcal/kg	*	0.82	2.09	
Sheep	ME Mcal/kg	*	0.89	2.27	
Cattle	TDN %	*	22.1	56.3	
Sheep	TDN %	*	24.7	62.7	

Bermudagrass, coastal, aerial part, 40 to 60 lb nitrogen per acre after cut 3 wilted ensiled, 25 to 31 days growth, cut 4, (3)
Ref No 3 09 230 United States

Feed Name or Analyses			Mean As Fed	Mean Dry	C.V. ± %
Dry matter	%		39.3	100.0	
Cattle	dig coef %		57.	57.	
Ash	%		3.2	8.1	
Crude fiber	%		12.5	31.7	
Ether extract	%		1.3	3.4	
N-free extract	%		16.7	42.4	
Protein (N x 6.25)	%		5.7	14.4	
Cattle	dig coef %		66.	66.	
Cattle	dig prot %		3.7	9.5	
Goats	dig prot %	*	3.7	9.3	
Horses	dig prot %	*	3.7	9.3	
Sheep	dig prot %	*	3.7	9.3	

Feed Name or Analyses		Mean As Fed	Mean Dry	C.V. ± %
Cellulose (Matrone) %		11.2	28.4	
Cattle dig coef %		67.	67.	
Energy GE Mcal/kg		1.76	4.47	
Cattle GE dig coef %		57.	57.	
Cattle DE Mcal/kg		1.00	2.55	
Sheep DE Mcal/kg	*	1.09	2.77	
Cattle ME Mcal/kg	*	0.82	2.09	
Sheep ME Mcal/kg	*	0.89	2.27	
Cattle TDN %	*	22.2	56.4	
Sheep TDN %	*	24.7	62.8	

Bermudagrass, coastal, aerial part w corn added, ensiled, cut 4, (3)

Ref No 3 09 088 United States

Feed Name or Analyses		Mean As Fed	Mean Dry	C.V. ± %
Dry matter %		34.2	100.0	
Cattle dig coef %		66.	66.	
Ash %		2.6	7.6	
Crude fiber %		10.2	29.7	
Ether extract %		1.1	3.3	
N-free extract %		15.7	45.8	
Protein (N x 6.25) %		4.7	13.6	
Cattle dig coef %		64.	64.	
Cattle dig prot %		3.0	8.7	
Goats dig prot %	*	2.9	8.6	
Horses dig prot %	*	2.9	8.6	
Sheep dig prot %	*	2.9	8.6	
Cellulose (Matrone) %		9.1	26.6	
Cattle dig coef %		64.	64.	
Energy GE Mcal/kg		1.47	4.29	
Cattle GE dig coef %		55.	55.	
Cattle DE Mcal/kg		0.81	2.36	
Sheep DE Mcal/kg	*	0.96	2.81	
Cattle ME Mcal/kg	*	0.66	1.93	
Sheep ME Mcal/kg	*	0.79	2.30	
Cattle TDN %	*	20.0	58.5	
Sheep TDN %	*	21.8	63.7	

Bermudagrass, coastal, aerial part w grnd corn grain added, ensiled, 32 to 38 days growth, (3)

Ref No 3 09 221 United States

Feed Name or Analyses		Mean As Fed	Mean Dry	C.V. ± %
Dry matter %			100.0	
Cattle dig coef %			58.	
Ash %			5.4	
Crude fiber %			30.4	
Cattle dig coef %			62.	
Ether extract %			4.9	
N-free extract %			48.6	
Cattle dig coef %			56.	
Protein (N x 6.25) %			10.7	
Cattle dig coef %			63.	
Cattle dig prot %			6.7	
Goats dig prot %	*		5.9	
Horses dig prot %	*		5.9	
Sheep dig prot %	*		5.9	
Energy GE Mcal/kg			4.77	
Cattle GE dig coef %			56.	
Cattle DE Mcal/kg			2.67	
Sheep DE Mcal/kg	*		2.84	
Cattle ME Mcal/kg	*		2.19	
Sheep ME Mcal/kg	*		2.33	
Cattle TDN %			58.9	
Sheep TDN %	*		64.5	
Calcium %			0.43	
Phosphorus %			0.29	

Bermudagrass, coastal, aerial part w grnd corn grain added, ensiled, cut 1, (3)

Ref No 3 09 216 United States

Feed Name or Analyses		Mean As Fed	Mean Dry	C.V. ± %
Dry matter %			100.0	
Cattle dig coef %			59.	
Ash %			6.7	
Crude fiber %			28.7	
Cattle dig coef %			64.	
Ether extract %			3.2	
N-free extract %			50.1	
Cattle dig coef %			59.	
Protein (N x 6.25) %			11.3	
Cattle dig coef %			57.	
Cattle dig prot %			6.4	
Goats dig prot %	*		6.5	
Horses dig prot %	*		6.5	
Sheep dig prot %	*		6.5	
Energy GE Mcal/kg				
Cattle DE Mcal/kg	*		2.57	
Sheep DE Mcal/kg	*		2.86	
Cattle ME Mcal/kg	*		2.11	
Sheep ME Mcal/kg	*		2.35	
Cattle TDN %			58.3	
Sheep TDN %	*		64.9	
Calcium %			0.48	
Phosphorus %			0.31	

Bermudagrass, coastal, aerial pt w 100 lb corn added per ton, 40 to 60 lb nitrogen per acre after cut 3 ensiled, 25 to 31 days growth, cut 4, (3)

Ref No 3 09 229 United States

Feed Name or Analyses		Mean As Fed	Mean Dry	C.V. ± %
Dry matter %		34.2	100.0	
Cattle dig coef %		56.	56.	
Ash %		2.6	7.6	
Crude fiber %		10.2	29.7	
Ether extract %		1.1	3.3	
N-free extract %		15.7	45.8	
Protein (N x 6.25) %		4.7	13.6	
Cattle dig coef %		64.	64.	
Cattle dig prot %		3.0	8.7	
Goats dig prot %	*	2.9	8.6	
Horses dig prot %	*	2.9	8.6	
Sheep dig prot %	*	2.9	8.6	
Cellulose (Matrone) %		9.1	26.6	
Cattle dig coef %		64.	64.	
Energy GE Mcal/kg		1.47	4.29	
Cattle GE dig coef %		55.	55.	
Cattle DE Mcal/kg		0.81	2.36	
Sheep DE Mcal/kg	*	0.96	2.81	
Cattle ME Mcal/kg	*	0.66	1.93	
Sheep ME Mcal/kg	*	0.79	2.30	
Cattle TDN %	*	20.0	58.5	
Sheep TDN %	*	21.8	63.7	

BERMUDAGRASS, MIDLAND. Cynodon dactylon

Bermudagrass, midland, hay, s-c, (1)

Ref No 1 09 078 United States

Feed Name or Analyses		Mean As Fed	Mean Dry	C.V. ± %
Dry matter %		87.3	100.0	1
Horses dig coef %		46.	46.	7
Organic matter %		81.1	92.9	0
Ash %		6.2	7.1	5
Crude fiber %		26.7	30.6	5
Horses dig coef %		41.	41.	15
Ether extract %		2.4	2.7	7
N-free extract %		44.8	51.3	2
Horses dig coef %		49.	49.	9
Protein (N x 6.25) %		7.3	8.4	6
Cattle dig prot %	*	3.6	4.2	
Goats dig prot %	*	3.8	4.4	
Horses dig prot %	*	4.0	4.6	
Rabbits dig prot %	*	4.5	5.1	
Sheep dig prot %	*	3.5	4.1	
Cell contents (Van Soest) %		23.1	26.5	5
Horses dig coef %		58.	58.	10
Cell walls (Van Soest) %		64.2	73.5	2
Horses dig coef %		42.	42.	7
Cellulose (Matrone) %		27.8	31.8	2
Horses dig coef %		41.	41.	10
Fiber, acid detergent (VS) %		31.6	36.3	2
Horses dig coef %		35.	35.	9
Hemicellulose %		27.9	32.0	4
Horses dig coef %		51.	51.	21
Holocellulose %		55.7	63.8	2
Soluble carbohydrates %		11.8	13.6	11
Lignin (Ellis) %		4.0	4.6	6
Horses dig coef %		-7.	-7.	98
Energy GE Mcal/kg		3.95	4.52	0
Horses GE dig coef %		43.	43.	6
Cattle DE Mcal/kg	*	2.40	2.75	
Horses DE Mcal/kg		1.70	1.95	6
Sheep DE Mcal/kg	*	2.19	2.51	
Cattle ME Mcal/kg	*	1.96	2.25	
Horses ME Mcal/kg	*	1.39	1.59	
Sheep ME Mcal/kg	*	1.80	2.06	
Cattle TDN %	*	54.3	62.3	
Sheep TDN %	*	49.7	56.9	
Calcium %		0.42	0.48	6
Magnesium %		0.30	0.34	7
Phosphorus %		0.16	0.19	10
Potassium %		1.77	2.03	7
Carotene mg/kg		31.7	36.3	17
Vitamin A equivalent IU/g		52.9	60.6	

BERMUDAGRASS-DALLISGRASS. Cynodon dactylon, Paspalum dilatatum

Bermudagrass-Dallisgrass, aerial part, fresh, (2)

Ref No 2 00 720 United States

Feed Name or Analyses		Mean As Fed	Mean Dry	C.V. ± %
Dry matter %			100.0	
Cobalt mg/kg			0.090	

BERMUDAGRASS-LESPEDEZA. Cynodon dactylon, Lespedeza spp

Bermudagrass-Lespedeza, aerial part, fresh, (2)

Ref No 2 00 721 United States

Feed Name or Analyses		Mean As Fed	Mean Dry	C.V. ± %
Dry matter %			100.0	
Ash %			7.9	16
Crude fiber %			31.5	9
Ether extract %			1.9	31
N-free extract %			49.6	
Protein (N x 6.25) %			9.1	13
Cattle dig prot %	*		5.6	
Goats dig prot %	*		5.1	
Horses dig prot %	*		5.3	
Rabbits dig prot %	*		5.7	
Sheep dig prot %	*		5.5	

Continued

Feed Name or Analyses			Mean As Fed	Mean Dry	C.V. ± %
Energy	GE Mcal/kg				
Cattle	DE Mcal/kg	*		2.91	
Sheep	DE Mcal/kg	*		2.92	
Cattle	ME Mcal/kg	*		2.39	
Sheep	ME Mcal/kg	*		2.40	
Cattle	TDN %	*		66.0	
Sheep	TDN %	*		66.3	

BIRCH, SWEET. Betula lenta

Birch, sweet, browse, immature, (2)

Ref No 2 00 724 United States

Feed Name or Analyses			Mean As Fed	Mean Dry	C.V. ± %
Dry matter	%		92.4	100.0	
Ash	%		7.1	7.7	
Crude fiber	%		15.6	16.9	
Ether extract	%		7.9	8.6	
N-free extract	%		35.7	38.7	
Protein (N x 6.25)	%		26.0	28.1	
Cattle	dig prot %	*	20.1	21.8	
Goats	dig prot %	*	21.0	22.8	
Horses	dig prot %	*	19.8	21.4	
Rabbits	dig prot %	*	18.8	20.4	
Sheep	dig prot %	*	21.4	23.2	
Calcium	%		1.12	1.21	
Manganese	mg/kg		1355.3	1467.6	
Phosphorus	%		0.30	0.32	

Birch, sweet, browse, mid-bloom, (2)

Ref No 2 00 725 United States

Feed Name or Analyses			Mean As Fed	Mean Dry	C.V. ± %
Dry matter	%		93.7	100.0	
Calcium	%		1.08	1.15	
Manganese	mg/kg		136.1	145.3	
Phosphorus	%		0.17	0.18	

BITTERBRUSH. Purshia tridentata

Bitterbrush, browse, (2)

Ref No 2 00 727 United States

Feed Name or Analyses			Mean As Fed	Mean Dry	C.V. ± %
Dry matter	%		92.6	100.0	
Ash	%		4.4	4.7	
Crude fiber	%		14.9	16.1	
Sheep	dig coef %		70.	70.	
Ether extract	%		3.1	3.4	
Sheep	dig coef %		71.	71.	
N-free extract	%		57.8	62.4	
Sheep	dig coef %		86.	86.	
Protein (N x 6.25)	%		12.4	13.4	
Sheep	dig coef %		82.	82.	
Cattle	dig prot %	*	8.6	9.3	
Goats	dig prot %	*	8.4	9.1	
Horses	dig prot %	*	8.2	8.9	
Rabbits	dig prot %	*	8.3	9.0	
Sheep	dig prot %		10.2	11.0	
Energy	GE Mcal/kg				
Cattle	DE Mcal/kg	*	2.64	2.85	
Sheep	DE Mcal/kg	*	3.32	3.59	
Cattle	ME Mcal/kg	*	2.16	2.33	
Sheep	ME Mcal/kg	*	2.72	2.94	
Cattle	TDN %	*	59.8	64.5	
Sheep	TDN %		75.3	81.4	

(1) dry forages and roughages (2) pasture, range plants, and forages fed green (3) silages (4) energy feeds (5) protein supplements (6) minerals (7) vitamins (8) additives

BITTERBRUSH, ANTELOPE. Purshia tridentata

Bitterbrush, antelope, aerial part, fresh, full bloom, (2)

Ref No 2 00 728 United States

Feed Name or Analyses			Mean As Fed	Mean Dry	C.V. ± %
Dry matter	%			100.0	
Calcium	%			1.58	62
Magnesium	%			0.24	20
Phosphorus	%			0.28	24
Sulphur	%			0.14	19

Bitterbrush, antelope, browse, (2)

Ref No 2 00 732 United States

Feed Name or Analyses			Mean As Fed	Mean Dry	C.V. ± %
Dry matter	%			100.0	
Ash	%			4.9	39
Crude fiber	%			21.4	26
Ether extract	%			5.0	23
N-free extract	%			56.9	
Protein (N x 6.25)	%			11.8	30
Cattle	dig prot %	*		7.9	
Goats	dig prot %	*		7.6	
Horses	dig prot %	*		7.5	
Rabbits	dig prot %	*		7.8	
Sheep	dig prot %	*		8.0	
Energy	GE Mcal/kg				
Cattle	DE Mcal/kg	*		2.95	
Sheep	DE Mcal/kg	*		3.05	
Cattle	ME Mcal/kg	*		2.42	
Sheep	ME Mcal/kg	*		2.50	
Cattle	TDN %	*		66.9	
Sheep	TDN %	*		69.1	
Sulphur	%			0.13	33

Bitterbrush, antelope, browse, early bloom, (2)

Ref No 2 00 729 United States

Feed Name or Analyses			Mean As Fed	Mean Dry	C.V. ± %
Dry matter	%			100.0	
Ash	%			4.5	
Crude fiber	%			21.9	
Ether extract	%			4.0	
N-free extract	%			54.2	
Protein (N x 6.25)	%			15.4	
Cattle	dig prot %	*		11.0	
Goats	dig prot %	*		10.9	
Horses	dig prot %	*		10.6	
Rabbits	dig prot %	*		10.6	
Sheep	dig prot %	*		11.3	
Energy	GE Mcal/kg				
Cattle	DE Mcal/kg	*		3.22	
Sheep	DE Mcal/kg	*		2.92	
Cattle	ME Mcal/kg	*		2.64	
Sheep	ME Mcal/kg	*		2.40	
Cattle	TDN %	*		73.1	
Sheep	TDN %	*		66.3	
Calcium	%			3.88	
Phosphorus	%			0.64	

Bitterbrush, antelope, browse, mid-bloom, (2)

Ref No 2 00 730 United States

Feed Name or Analyses			Mean As Fed	Mean Dry	C.V. ± %
Dry matter	%			100.0	
Ash	%			6.0	41
Crude fiber	%			23.8	21
Ether extract	%			5.2	25
N-free extract	%			53.5	
Protein (N x 6.25)	%			11.5	20
Cattle	dig prot %	*		7.7	
Goats	dig prot %	*		7.3	
Horses	dig prot %	*		7.3	
Rabbits	dig prot %	*		7.5	
Sheep	dig prot %	*		7.7	
Energy	GE Mcal/kg				
Cattle	DE Mcal/kg	*		2.87	
Sheep	DE Mcal/kg	*		2.99	
Cattle	ME Mcal/kg	*		2.35	
Sheep	ME Mcal/kg	*		2.45	
Cattle	TDN %	*		65.1	
Sheep	TDN %	*		67.8	

Bitterbrush, antelope, browse, mature, (2)

Ref No 2 00 731 United States

Feed Name or Analyses			Mean As Fed	Mean Dry	C.V. ± %
Dry matter	%			100.0	
Ash	%			3.3	25
Crude fiber	%			26.1	20
Ether extract	%			5.7	11
N-free extract	%			56.7	
Protein (N x 6.25)	%			8.2	31
Cattle	dig prot %	*		4.9	
Goats	dig prot %	*		4.2	
Horses	dig prot %	*		4.5	
Rabbits	dig prot %	*		5.0	
Sheep	dig prot %	*		4.6	
Energy	GE Mcal/kg				
Cattle	DE Mcal/kg	*		2.87	
Sheep	DE Mcal/kg	*		3.08	
Cattle	ME Mcal/kg	*		2.35	
Sheep	ME Mcal/kg	*		2.53	
Cattle	TDN %	*		65.1	
Sheep	TDN %	*		69.9	
Calcium	%			0.81	
Phosphorus	%			0.15	

BLACKFOOT, PLAINS. Melampodium leucanthum

Blackfoot, plains, aerial part, fresh, (2)

Ref No 2 00 734 United States

Feed Name or Analyses			Mean As Fed	Mean Dry	C.V. ± %
Dry matter	%			100.0	
Ash	%			9.9	
Crude fiber	%			15.9	
Ether extract	%			10.0	
N-free extract	%			50.7	
Protein (N x 6.25)	%			13.5	
Cattle	dig prot %	*		9.4	
Goats	dig prot %	*		9.2	
Horses	dig prot %	*		9.0	
Rabbits	dig prot %	*		9.1	
Sheep	dig prot %	*		9.6	
Calcium	%			1.73	
Magnesium	%			0.39	
Phosphorus	%			0.18	
Potassium	%			2.90	

Black grass hay —
see Rush, saltmeadow, hay, s-c, (1)

Feed Name or Analyses			Mean As Fed	Mean Dry	C.V. ± %

BLACKMANGROVE. Avicennia nitida

Blackmangrove, leaves, s-c grnd, (1)

Ref No 1 00 735 — United States

Feed Name or Analyses			As Fed	Dry	C.V. ± %
Dry matter	%			100.0	
Ash	%			15.7	
Crude fiber	%			23.9	
Ether extract	%			4.4	
N-free extract	%			45.3	
Protein (N x 6.25)	%			10.7	
Cattle	dig prot %	*		6.2	
Goats	dig prot %	*		6.5	
Horses	dig prot %	*		6.6	
Rabbits	dig prot %	*		6.9	
Sheep	dig prot %	*		6.2	
Carotene	mg/kg			48.9	
Vitamin A equivalent	IU/g			81.6	

BLADDERNUT, EUROPEAN. Staphylea pinnata

Bladdernut, European, seeds w hulls, (5)

Ref No 5 09 121 — United States

Feed Name or Analyses			As Fed	Dry	C.V. ± %
Dry matter	%			100.0	
Protein (N x 6.25)	%			35.0	
Alanine	%			1.26	
Arginine	%			3.33	
Aspartic acid	%			2.77	
Glutamic acid	%			5.99	
Glycine	%			1.44	
Histidine	%			0.49	
Hydroxyproline	%			0.00	
Isoleucine	%			2.03	
Leucine	%			1.58	
Lysine	%			1.02	
Methionine	%			0.21	
Phenylalanine	%			0.91	
Proline	%			1.19	
Serine	%			1.09	
Tyrosine	%			0.77	
Valine	%			0.81	

BLADDERPOD. Lesquerella spp

Bladderpod, aerial part, fresh, immature, (2)

Ref No 2 00 736 — United States

Feed Name or Analyses			As Fed	Dry	C.V. ± %
Dry matter	%			100.0	
Ash	%			14.9	
Crude fiber	%			15.3	
Ether extract	%			1.6	
N-free extract	%			41.2	
Protein (N x 6.25)	%			27.0	
Cattle	dig prot %	*		20.8	
Goats	dig prot %	*		21.8	
Horses	dig prot %	*		20.5	
Rabbits	dig prot %	*		19.5	
Sheep	dig prot %	*		22.2	
Calcium	%			5.02	
Magnesium	%			0.33	
Phosphorus	%			0.35	
Potassium	%			2.06	

Bladderpod, aerial part, fresh, early bloom, (2)

Ref No 2 00 737 — United States

Feed Name or Analyses			As Fed	Dry	C.V. ± %
Dry matter	%			100.0	
Ash	%			10.6	
Crude fiber	%			17.4	
Ether extract	%			2.2	
N-free extract	%			48.4	
Protein (N x 6.25)	%			21.4	
Cattle	dig prot %	*		16.1	
Goats	dig prot %	*		16.5	
Horses	dig prot %	*		15.7	
Rabbits	dig prot %	*		15.2	
Sheep	dig prot %	*		16.9	
Calcium	%			4.64	
Magnesium	%			0.28	
Phosphorus	%			0.25	
Potassium	%			1.68	

BLADDERSENNA, COMMON. Colutea arborescens

Bladdersenna, common, seeds w hulls, (5)

Ref No 5 09 064 — United States

Feed Name or Analyses			As Fed	Dry	C.V. ± %
Dry matter	%			100.0	
Protein (N x 6.25)	%			44.0	
Alanine	%			1.18	
Arginine	%			3.64	
Aspartic acid	%			2.89	
Glutamic acid	%			6.06	
Glycine	%			1.34	
Histidine	%			0.83	
Hydroxyproline	%			0.10	
Isoleucine	%			1.16	
Leucine	%			1.96	
Lysine	%			1.43	
Methionine	%			0.36	
Phenylalanine	%			1.35	
Proline	%			1.31	
Serine	%			1.49	
Threonine	%			1.13	
Tyrosine	%			0.99	
Valine	%			1.39	

Bladekelp –
see Seaweed, bladekelp, entire plant, dehy, (1)

Blood albumin, dried –
see Animal, blood albumin, dehy, (5)

Blood flour –
see Animal, blood, spray dehy, (5)

Blood meal (AAFCO) (CFA) –
see Animal, blood, dehy grnd, (5)

BLUEBELL. Mertensia spp

Bluebell, aerial part, fresh, (2)

Ref No 2 00 739 — United States

Feed Name or Analyses			As Fed	Dry	C.V. ± %
Dry matter	%			100.0	
Ash	%			11.6	
Crude fiber	%			17.1	
Ether extract	%			3.3	
N-free extract	%			42.7	
Protein (N x 6.25)	%			25.3	
Cattle	dig prot %	*		19.4	
Goats	dig prot %	*		20.2	
Horses	dig prot %	*		19.0	
Rabbits	dig prot %	*		18.2	
Sheep	dig prot %	*		20.6	

BLUEGRASS. Poa spp

Bluegrass, hay, fan air dried, mature, (1)

Ref No 1 00 741 — United States

Feed Name or Analyses			As Fed	Dry	C.V. ± %
Dry matter	%		92.2	100.0	
Ash	%		5.5	6.0	
Crude fiber	%		29.0	31.5	
Ether extract	%		2.9	3.1	
N-free extract	%		45.3	49.1	
Protein (N x 6.25)	%		9.5	10.3	
Cattle	dig prot %	*	5.4	5.9	
Goats	dig prot %	*	5.7	6.2	
Horses	dig prot %	*	5.8	6.3	
Rabbits	dig prot %	*	6.1	6.6	
Sheep	dig prot %	*	5.4	5.8	
Energy	GE Mcal/kg				
Cattle	DE Mcal/kg	*	2.42	2.63	
Sheep	DE Mcal/kg	*	2.33	2.53	
Cattle	ME Mcal/kg	*	1.99	2.16	
Sheep	ME Mcal/kg	*	1.91	2.07	
Cattle	TDN %	*	55.0	59.6	
Sheep	TDN %	*	52.9	57.3	
Carotene	mg/kg		8.1	8.8	
Riboflavin	mg/kg		9.1	9.9	
Vitamin A equivalent	IU/g		13.6	14.7	

Bluegrass, hay, s-c, (1)

Ref No 1 00 744 — United States

Feed Name or Analyses			As Fed	Dry	C.V. ± %
Dry matter	%		89.7	100.0	4
Ash	%		6.6	7.3	
Crude fiber	%		25.8	28.7	
Sheep	dig coef %	*	66.	66.	
Ether extract	%		2.8	3.2	
Sheep	dig coef %	*	49.	49.	
N-free extract	%		44.7	49.8	
Sheep	dig coef %	*	68.	68.	
Protein (N x 6.25)	%		9.8	11.0	
Sheep	dig coef %	*	63.	63.	
Cattle	dig prot %	*	5.8	6.4	
Goats	dig prot %	*	6.1	6.8	
Horses	dig prot %	*	6.1	6.8	
Rabbits	dig prot %	*	6.4	7.1	
Sheep	dig prot %		6.2	6.9	
Energy	GE Mcal/kg				
Cattle	DE Mcal/kg	*	2.34	2.61	
Sheep	DE Mcal/kg	*	2.50	2.79	
Cattle	ME Mcal/kg	*	1.92	2.14	
Sheep	ME Mcal/kg	*	2.05	2.29	
Cattle	TDN %	*	53.0	59.1	
Sheep	TDN %		56.7	63.2	
Calcium	%		0.35	0.39	20
Copper	mg/kg		8.9	9.9	17
Iron	%		0.023	0.026	45
Magnesium	%		0.19	0.21	31
Manganese	mg/kg		83.1	92.6	57
Phosphorus	%		0.24	0.27	24
Potassium	%		1.54	1.72	17
Carotene	mg/kg		222.6	248.0	39

Continued

Feed Name or Analyses			Mean As Fed	Mean Dry	C.V. ± %
Riboflavin	mg/kg		9.9	11.0	14
Vitamin A equivalent	IU/g		371.0	413.4	

Bluegrass, hay, s-c, immature, (1)
Ref No 1 00 742 United States

Feed Name or Analyses			Mean As Fed	Mean Dry	C.V. ± %
Dry matter	%		93.9	100.0	2
Ash	%		8.2	8.7	12
Crude fiber	%		25.3	26.9	11
Ether extract	%		3.0	3.2	29
N-free extract	%		41.8	44.5	
Protein (N x 6.25)	%		15.7	16.7	16
Cattle	dig prot %	*	10.7	11.4	
Goats	dig prot %	*	11.4	12.1	
Horses	dig prot %	*	11.0	11.7	
Rabbits	dig prot %	*	10.9	11.6	
Sheep	dig prot %	*	10.8	11.5	
Energy	GE Mcal/kg				
Cattle	DE Mcal/kg	*	2.54	2.70	
Sheep	DE Mcal/kg	*	2.45	2.61	
Cattle	ME Mcal/kg	*	2.08	2.21	
Sheep	ME Mcal/kg	*	2.01	2.14	
Cattle	TDN %	*	57.5	61.3	
Sheep	TDN %	*	55.6	59.3	

Bluegrass, hay, s-c, full bloom, (1)
Ref No 1 00 743 United States

Feed Name or Analyses			Mean As Fed	Mean Dry	C.V. ± %
Dry matter	%		88.9	100.0	
Ash	%		6.9	7.8	25
Crude fiber	%		30.1	33.9	11
Sheep	dig coef %		67.	67.	
Ether extract	%		2.3	2.6	34
Sheep	dig coef %		54.	54.	
N-free extract	%		40.2	45.3	
Sheep	dig coef %		66.	66.	
Protein (N x 6.25)	%		9.3	10.5	6
Sheep	dig coef %		62.	62.	
Cattle	dig prot %	*	5.4	6.0	
Goats	dig prot %	*	5.6	6.4	
Horses	dig prot %	*	5.7	6.4	
Rabbits	dig prot %	*	6.0	6.8	
Sheep	dig prot %		5.8	6.5	
Energy	GE Mcal/kg				
Cattle	DE Mcal/kg	*	2.22	2.50	
Sheep	DE Mcal/kg	*	2.44	2.74	
Cattle	ME Mcal/kg	*	1.82	2.05	
Sheep	ME Mcal/kg	*	2.00	2.25	
Cattle	TDN %	*	50.4	56.8	
Sheep	TDN %		55.3	62.2	

Bluegrass, aerial part, fresh, (2)
Ref No 2 00 756 United States

Feed Name or Analyses			Mean As Fed	Mean Dry	C.V. ± %
Dry matter	%		32.1	100.0	39
Carotene	mg/kg		65.7	204.8	45
Niacin	mg/kg		21.2	65.9	
Riboflavin	mg/kg		3.5	11.0	
Thiamine	mg/kg		2.8	8.8	
Vitamin A equivalent	IU/g		109.6	341.4	

(1) dry forages and roughages (2) pasture, range plants, and forages fed green (3) silages (4) energy feeds (5) protein supplements (6) minerals (7) vitamins (8) additives

Bluegrass, aerial part, fresh, immature, (2)
Ref No 2 00 747 United States

Feed Name or Analyses			Mean As Fed	Mean Dry	C.V. ± %
Dry matter	%		30.9	100.0	29
Calcium	%		0.17	0.55	36
Copper	mg/kg		4.4	14.1	
Iron	%		0.010	0.032	
Magnesium	%		0.06	0.18	9
Manganese	mg/kg		24.8	80.2	
Phosphorus	%		0.13	0.42	46
Potassium	%		0.77	2.48	17
Carotene	mg/kg		123.9	401.0	20
Vitamin A equivalent	IU/g		206.6	668.5	

Bluegrass, aerial part, fresh, immature, pure stand, (2)
Ref No 2 00 746 United States

Feed Name or Analyses			Mean As Fed	Mean Dry	C.V. ± %
Dry matter	%			100.0	
Carotene	mg/kg			441.1	5
Vitamin A equivalent	IU/g			735.4	

Bluegrass, aerial part, fresh, pre-bloom, pure stand, (2)
Ref No 2 00 748 United States

Feed Name or Analyses			Mean As Fed	Mean Dry	C.V. ± %
Dry matter	%			100.0	
Carotene	mg/kg			309.5	11
Vitamin A equivalent	IU/g			516.0	

Bluegrass, aerial part, fresh, early bloom, (2)
Ref No 2 00 749 United States

Feed Name or Analyses			Mean As Fed	Mean Dry	C.V. ± %
Dry matter	%		35.7	100.0	7
Ash	%		2.8	7.9	12
Crude fiber	%		10.1	28.2	10
Ether extract	%		1.4	3.9	24
N-free extract	%		16.2	45.3	
Protein (N x 6.25)	%		5.2	14.7	18
Cattle	dig prot %	*	3.7	10.4	
Goats	dig prot %	*	3.7	10.3	
Horses	dig prot %	*	3.6	10.0	
Rabbits	dig prot %	*	3.6	10.0	
Sheep	dig prot %	*	3.8	10.7	
Energy	GE Mcal/kg				
Cattle	DE Mcal/kg	*	1.08	3.03	
Sheep	DE Mcal/kg	*	0.99	2.77	
Cattle	ME Mcal/kg	*	0.89	2.48	
Sheep	ME Mcal/kg	*	0.81	2.27	
Cattle	TDN %	*	24.5	68.6	
Sheep	TDN %	*	22.4	62.8	
Calcium	%		0.15	0.42	22
Magnesium	%		0.04	0.11	29
Phosphorus	%		0.10	0.29	22
Potassium	%		0.76	2.14	19
Carotene	mg/kg		99.6	278.9	28
Vitamin A equivalent	IU/g		166.0	464.9	

Bluegrass, aerial part, fresh, mid-bloom, (2)
Ref No 2 00 750 United States

Feed Name or Analyses			Mean As Fed	Mean Dry	C.V. ± %
Dry matter	%		29.3	100.0	
Ash	%		2.2	7.5	7
Crude fiber	%		8.6	29.3	8
Ether extract	%		1.2	4.0	22
N-free extract	%		13.7	46.7	
Protein (N x 6.25)	%		3.7	12.5	14
Cattle	dig prot %	*	2.5	8.5	
Goats	dig prot %	*	2.4	8.2	
Horses	dig prot %	*	2.4	8.1	
Rabbits	dig prot %	*	2.4	8.3	
Sheep	dig prot %	*	2.5	8.6	
Energy	GE Mcal/kg				
Cattle	DE Mcal/kg	*	0.86	2.95	
Sheep	DE Mcal/kg	*	0.83	2.83	
Cattle	ME Mcal/kg	*	0.71	2.42	
Sheep	ME Mcal/kg	*	0.68	2.32	
Cattle	TDN %	*	19.6	66.8	
Sheep	TDN %	*	18.8	64.2	
Calcium	%		0.12	0.40	
Phosphorus	%		0.09	0.31	
Potassium	%		0.58	1.97	

Bluegrass, aerial part, fresh, full bloom, (2)
Ref No 2 00 752 United States

Feed Name or Analyses			Mean As Fed	Mean Dry	C.V. ± %
Dry matter	%		32.9	100.0	
Ash	%		1.5	4.6	
Crude fiber	%		11.4	34.6	
Protein (N x 6.25)	%		2.4	7.3	13
Cattle	dig prot %	*	1.3	4.1	
Goats	dig prot %	*	1.1	3.4	
Horses	dig prot %	*	1.2	3.7	
Rabbits	dig prot %	*	1.4	4.3	
Sheep	dig prot %	*	1.2	3.8	
Calcium	%		0.10	0.31	25
Phosphorus	%		0.09	0.27	27
Potassium	%		0.45	1.37	
Carotene	mg/kg		47.6	144.6	44
Vitamin A equivalent	IU/g		79.3	241.1	

Bluegrass, aerial part, fresh, full bloom, pure stand, (2)
Ref No 2 00 751 United States

Feed Name or Analyses			Mean As Fed	Mean Dry	C.V. ± %
Dry matter	%			100.0	
Carotene	mg/kg			140.7	19
Vitamin A equivalent	IU/g			234.5	

Bluegrass, aerial part, fresh, mature, (2)
Ref No 2 00 754 United States

Feed Name or Analyses			Mean As Fed	Mean Dry	C.V. ± %
Dry matter	%		41.6	100.0	13
Carotene	mg/kg		30.4	73.2	26
Vitamin A equivalent	IU/g		50.8	122.0	

Bluegrass, aerial part, fresh, mature, pure stand, (2)
Ref No 2 00 753 United States

Feed Name or Analyses			Mean As Fed	Mean Dry	C.V. ± %
Dry matter	%			100.0	
Carotene	mg/kg			72.3	28
Vitamin A equivalent	IU/g			120.5	

Bluegrass, aerial part, fresh, over ripe, (2)
Ref No 2 00 755 United States

Feed Name or Analyses			Mean As Fed	Mean Dry	C.V. ± %
Dry matter	%		85.6	100.0	
Ash	%		5.0	5.8	39
Crude fiber	%		35.3	41.2	6
Ether extract	%		2.5	2.9	71
N-free extract	%		39.6	46.3	

Feed Name or Analyses				Mean As Fed	Mean Dry	C.V. ± %
Protein (N x 6.25)		%		3.3	3.8	31
Cattle	dig prot	%	*	1.0	1.1	
Goats	dig prot	%	*	0.1	0.1	
Horses	dig prot	%	*	0.6	0.8	
Rabbits	dig prot	%	*	1.4	1.6	
Sheep	dig prot	%	*	0.5	0.5	
Energy	GE	Mcal/kg				
Cattle	DE	Mcal/kg	*	1.68	1.97	
Sheep	DE	Mcal/kg	*	1.90	2.22	
Cattle	ME	Mcal/kg	*	1.38	1.61	
Sheep	ME	Mcal/kg	*	1.56	1.82	
Cattle	TDN	%	*	38.2	44.6	
Sheep	TDN	%	*	43.1	50.3	
Calcium		%		0.25	0.29	26
Phosphorus		%		0.19	0.22	75
Potassium		%		0.66	0.77	
Carotene		mg/kg		17.6	20.5	74
Vitamin A equivalent		IU/g		29.3	34.2	

Bluegrass, aerial part, fresh, pure stand, (2)

Ref No 2 00 757 United States

Feed Name or Analyses				Mean As Fed	Mean Dry	C.V. ± %
Dry matter		%			100.0	
Carotene		mg/kg			247.1	40
Vitamin A equivalent		IU/g			412.0	

Bluegrass, aerial part, fresh fertilized, (2)

Ref No 2 00 758 United States

Feed Name or Analyses				Mean As Fed	Mean Dry	C.V. ± %
Dry matter		%		21.7	100.0	
Ash		%		2.2	10.3	
Crude fiber		%		5.8	26.5	
Cattle	dig coef	%		71.	71.	
Ether extract		%		1.0	4.8	
Cattle	dig coef	%		51.	51.	
N-free extract		%		8.5	39.3	
Cattle	dig coef	%		62.	62.	
Protein (N x 6.25)		%		4.1	19.1	
Cattle	dig coef	%		74.	74.	
Cattle	dig prot	%		3.1	14.1	
Goats	dig prot	%	*	3.1	14.4	
Horses	dig prot	%	*	3.0	13.7	
Rabbits	dig prot	%	*	2.9	13.4	
Sheep	dig prot	%	*	3.2	14.8	
Energy	GE	Mcal/kg				
Cattle	DE	Mcal/kg	*	0.60	2.77	
Sheep	DE	Mcal/kg	*	0.61	2.80	
Cattle	ME	Mcal/kg	*	0.49	2.27	
Sheep	ME	Mcal/kg	*	0.50	2.29	
Cattle	TDN	%		13.6.	62.8	
Sheep	TDN	%	*	13.8	63.4	

BLUEGRASS, CANADA, *Poa compressa*

Bluegrass, Canada, hay, fan air dried, (1)

Ref No 1 00 759 United States

Feed Name or Analyses				Mean As Fed	Mean Dry	C.V. ± %
Dry matter		%		91.3	100.0	
Ash		%		6.5	7.1	
Crude fiber		%		30.8	33.7	
Ether extract		%		1.9	2.1	
N-free extract		%		36.5	40.0	
Protein (N x 6.25)		%		15.6	17.1	
Cattle	dig prot	%	*	10.7	11.7	
Goats	dig prot	%	*	11.4	12.5	
Horses	dig prot	%	*	11.0	12.0	
Rabbits	dig prot	%	*	10.8	11.9	
Sheep	dig prot	%	*	10.9	11.9	
Energy	GE	Mcal/kg				
Cattle	DE	Mcal/kg	*	2.19	2.40	
Sheep	DE	Mcal/kg	*	2.30	2.52	
Cattle	ME	Mcal/kg	*	1.80	1.97	
Sheep	ME	Mcal/kg	*	1.88	2.06	
Cattle	TDN	%	*	49.7	54.5	
Sheep	TDN	%	*	52.1	57.0	

Bluegrass, Canada, hay, s-c, (1)

Ref No 1 00 762 United States

Feed Name or Analyses				Mean As Fed	Mean Dry	C.V. ± %
Dry matter		%		91.8	100.0	2
Ash		%		6.5	7.1	12
Crude fiber		%		27.2	29.7	7
Sheep	dig coef	%	*	66.	66.	
Ether extract		%		2.5	2.7	10
Sheep	dig coef	%	*	49.	49.	
N-free extract		%		46.9	51.2	
Sheep	dig coef	%	*	68.	68.	
Protein (N x 6.25)		%		8.6	9.4	37
Sheep	dig coef	%	*	63.	63.	
Cattle	dig prot	%	*	4.7	5.1	
Goats	dig prot	%	*	4.9	5.3	
Horses	dig prot	%	*	5.1	5.5	
Rabbits	dig prot	%	*	5.4	5.9	
Sheep	dig prot	%		5.4	5.9	
Fatty acids		%		2.4	2.6	
Energy	GE	Mcal/kg				
Cattle	DE	Mcal/kg	*	2.57	2.80	
Sheep	DE	Mcal/kg	*	2.56	2.79	
Cattle	ME	Mcal/kg	*	2.11	2.29	
Sheep	ME	Mcal/kg	*	2.10	2.29	
Cattle	NE_m	Mcal/kg	*	1.29	1.41	
Cattle	NE_{gain}	Mcal/kg	*	0.75	0.82	
Cattle	$NE_{lactating\ cows}$	Mcal/kg	*	1.51	1.64	
Cattle	TDN	%	*	58.2	63.5	
Sheep	TDN	%		58.0	63.3	
Calcium		%		0.28	0.30	19
Magnesium		%		0.30	0.33	
Manganese		mg/kg		85.0	92.6	
Phosphorus		%		0.24	0.26	18
Potassium		%		1.73	1.88	

Bluegrass, Canada, hay, s-c, immature, (1)

Ref No 1 00 760 United States

Feed Name or Analyses				Mean As Fed	Mean Dry	C.V. ± %
Dry matter		%		96.7	100.0	0
Ash		%		9.0	9.3	4
Crude fiber		%		24.9	25.8	1
Sheep	dig coef	%	*	66.	66.	
Ether extract		%		3.0	3.1	2
Sheep	dig coef	%	*	49.	49.	
N-free extract		%		43.0	44.5	
Sheep	dig coef	%	*	68.	68.	
Protein (N x 6.25)		%		16.7	17.3	15
Sheep	dig coef	%	*	63.	63.	
Cattle	dig prot	%	*	11.5	11.9	
Goats	dig prot	%	*	12.3	12.7	
Horses	dig prot	%	*	11.8	12.2	
Rabbits	dig prot	%	*	11.6	12.0	
Sheep	dig prot	%		10.5	10.9	
Energy	GE	Mcal/kg				
Cattle	DE	Mcal/kg	*	2.66	2.75	
Sheep	DE	Mcal/kg	*	2.63	2.72	
Cattle	ME	Mcal/kg	*	2.18	2.25	
Sheep	ME	Mcal/kg	*	2.15	2.23	
Cattle	NE_m	Mcal/kg	*	1.53	1.58	
Cattle	NE_{gain}	Mcal/kg	*	0.97	1.00	
Cattle	$NE_{lactating\ cows}$	Mcal/kg	*	1.80	1.86	
Cattle	TDN	%	*	60.3	62.3	
Sheep	TDN	%	*	59.6	61.6	

Bluegrass, Canada, hay, s-c, cut 1, (1)

Ref No 1 00 761 United States

Feed Name or Analyses				Mean As Fed	Mean Dry	C.V. ± %
Dry matter		%		91.1	100.0	
Ash		%		6.4	7.0	
Crude fiber		%		28.4	31.2	
Ether extract		%		2.3	2.5	
N-free extract		%		42.0	46.1	
Protein (N x 6.25)		%		12.0	13.2	
Cattle	dig prot	%	*	7.6	8.4	
Goats	dig prot	%	*	8.1	8.9	
Horses	dig prot	%	*	8.0	8.7	
Rabbits	dig prot	%	*	8.1	8.9	
Sheep	dig prot	%	*	7.7	8.4	
Energy	GE	Mcal/kg				
Cattle	DE	Mcal/kg	*	2.28	2.50	
Sheep	DE	Mcal/kg	*	2.33	2.55	
Cattle	ME	Mcal/kg	*	1.87	2.05	
Sheep	ME	Mcal/kg	*	1.91	2.09	
Cattle	TDN	%	*	51.7	56.7	
Sheep	TDN	%	*	52.8	57.9	

Bluegrass, Canada, aerial part, fresh, (2)

Ref No 2 00 764 United States

Feed Name or Analyses				Mean As Fed	Mean Dry	C.V. ± %
Dry matter		%		31.5	100.0	11
Ash		%		2.6	8.4	7
Crude fiber		%		9.0	28.7	14
Ether extract		%		1.2	3.7	10
N-free extract		%		14.6	46.2	
Protein (N x 6.25)		%		4.1	13.0	22
Cattle	dig prot	%	*	2.8	9.0	
Goats	dig prot	%	*	2.7	8.7	
Horses	dig prot	%	*	2.7	8.6	
Rabbits	dig prot	%	*	2.7	8.7	
Sheep	dig prot	%	*	2.9	9.1	
Energy	GE	Mcal/kg				
Cattle	DE	Mcal/kg	*	0.92	2.93	
Sheep	DE	Mcal/kg	*	0.89	2.82	
Cattle	ME	Mcal/kg	*	0.76	2.41	
Sheep	ME	Mcal/kg	*	0.73	2.31	
Cattle	NE_m	Mcal/kg	*	0.49	1.54	
Cattle	NE_{gain}	Mcal/kg	*	0.31	0.97	
Cattle	$NE_{lactating\ cows}$	Mcal/kg	*	0.58	1.83	
Cattle	TDN	%	*	20.9	66.5	
Sheep	TDN	%	*	20.1	63.9	
Calcium		%		0.13	0.41	9
Magnesium		%		0.05	0.16	
Manganese		mg/kg		24.9	79.1	6
Phosphorus		%		0.11	0.36	19
Potassium		%		0.57	1.82	

Bluegrass, Canada, aerial part, fresh, immature, (2)

Ref No 2 00 763 United States

Feed Name or Analyses				Mean As Fed	Mean Dry	C.V. ± %
Dry matter		%		25.9	100.0	7
Ash		%		2.4	9.1	7
Crude fiber		%		6.6	25.5	16
Ether extract		%		1.0	3.7	10
N-free extract		%		11.1	43.0	

Continued

Feed Name or Analyses			Mean As Fed	Mean Dry	C.V. ± %
Protein (N x 6.25)	%		4.8	18.7	7
Cattle	dig prot %	*	3.6	13.8	
Goats	dig prot %	*	3.6	14.0	
Horses	dig prot %	*	3.5	13.4	
Rabbits	dig prot %	*	3.4	13.1	
Sheep	dig prot %	*	3.7	14.4	
Energy	GE Mcal/kg				
Cattle	DE Mcal/kg	*	0.70	2.69	
Sheep	DE Mcal/kg	*	0.72	2.76	
Cattle	ME Mcal/kg	*	0.57	2.21	
Sheep	ME Mcal/kg	*	0.59	2.27	
Cattle	NE_m Mcal/kg	*	0.41	1.58	
Cattle	NE_{gain} Mcal/kg	*	0.26	1.00	
Cattle	$NE_{lactating\ cows}$ Mcal/kg	*	0.48	1.86	
Cattle	TDN %	*	15.8	61.1	
Sheep	TDN %	*	16.2	62.7	

BLUEGRASS, FOWL. Poa palustris

Bluegrass, fowl, hay, s-c, (1)
Ref No 1 00 767 United States

Feed Name or Analyses			Mean As Fed	Mean Dry	C.V. ± %
Dry matter	%		85.0	100.0	
Ash	%		7.0	8.2	18
Crude fiber	%		32.3	38.0	6
Ether extract	%		1.6	1.9	36
N-free extract	%		37.2	43.8	
Protein (N x 6.25)	%		6.9	8.1	17
Cattle	dig prot %	*	3.4	4.0	
Goats	dig prot %	*	3.5	4.1	
Horses	dig prot %	*	3.7	4.4	
Rabbits	dig prot %	*	4.2	4.9	
Sheep	dig prot %	*	3.3	3.8	
Energy	GE Mcal/kg				
Cattle	DE Mcal/kg	*	1.88	2.22	
Sheep	DE Mcal/kg	*	1.99	2.35	
Cattle	ME Mcal/kg	*	1.54	1.82	
Sheep	ME Mcal/kg	*	1.64	1.92	
Cattle	TDN %	*	42.7	50.3	
Sheep	TDN %	*	45.2	53.2	
Calcium	%		0.22	0.26	
Phosphorus	%		0.22	0.26	
Carotene	mg/kg		106.3	125.0	
Vitamin A equivalent	IU/g		177.1	208.4	

Bluegrass, fowl, hay, s-c, immature, (1)
Ref No 1 00 765 United States

Feed Name or Analyses			Mean As Fed	Mean Dry	C.V. ± %
Dry matter	%			100.0	
Ash	%			6.4	
Crude fiber	%			34.8	
Ether extract	%			3.2	
N-free extract	%			45.9	
Protein (N x 6.25)	%			9.7	
Cattle	dig prot %	*		5.3	
Goats	dig prot %	*		5.6	
Horses	dig prot %	*		5.8	
Rabbits	dig prot %	*		6.2	
Sheep	dig prot %	*		5.3	
Energy	GE Mcal/kg				
Cattle	DE Mcal/kg	*		2.41	
Sheep	DE Mcal/kg	*		2.44	
Cattle	ME Mcal/kg	*		1.98	
Sheep	ME Mcal/kg	*		2.00	
Cattle	TDN %	*		54.7	
Sheep	TDN %	*		55.4	

Bluegrass, fowl, hay, s-c, full bloom, (1)
Ref No 1 00 766 United States

Feed Name or Analyses			Mean As Fed	Mean Dry	C.V. ± %
Dry matter	%			100.0	
Ash	%			9.8	
Crude fiber	%			40.0	
Ether extract	%			1.6	
N-free extract	%			39.6	
Protein (N x 6.25)	%			9.0	
Cattle	dig prot %	*		4.7	
Goats	dig prot %	*		5.0	
Horses	dig prot %	*		5.2	
Rabbits	dig prot %	*		5.6	
Sheep	dig prot %	*		4.6	
Energy	GE Mcal/kg				
Cattle	DE Mcal/kg	*		2.11	
Sheep	DE Mcal/kg	*		2.31	
Cattle	ME Mcal/kg	*		1.73	
Sheep	ME Mcal/kg	*		1.89	
Cattle	TDN %	*		47.9	
Sheep	TDN %	*		52.4	

BLUEGRASS, KENTUCKY. Poa pratensis

Bluegrass, Kentucky, hay, fan air dried, (1)
Ref No 1 00 769 United States

Feed Name or Analyses			Mean As Fed	Mean Dry	C.V. ± %
Dry matter	%		92.7	100.0	
Ash	%		5.1	5.5	
Crude fiber	%		28.2	30.4	
Ether extract	%		3.3	3.6	
N-free extract	%		49.7	53.6	
Protein (N x 6.25)	%		6.4	6.9	
Cattle	dig prot %	*	2.7	2.9	
Goats	dig prot %	*	2.8	3.0	
Horses	dig prot %	*	3.1	3.4	
Rabbits	dig prot %	*	3.7	4.0	
Sheep	dig prot %	*	2.6	2.8	
Energy	GE Mcal/kg				
Cattle	DE Mcal/kg	*	2.47	2.66	
Sheep	DE Mcal/kg	*	2.31	2.50	
Cattle	ME Mcal/kg	*	2.02	2.18	
Sheep	ME Mcal/kg	*	1.90	2.05	
Cattle	TDN %	*	55.9	60.3	
Sheep	TDN %	*	52.5	56.6	
a-tocopherol	mg/kg		341.8	384.5	

Bluegrass, Kentucky, hay, fan air dried, mature, (1)
Ref No 1 00 768 United States

Feed Name or Analyses			Mean As Fed	Mean Dry	C.V. ± %
Dry matter	%		93.0	100.0	
Ash	%		5.1	5.5	
Crude fiber	%		30.0	32.3	
Ether extract	%		2.8	3.0	
N-free extract	%		49.0	52.7	
Protein (N x 6.25)	%		6.0	6.5	
Cattle	dig prot %	*	2.4	2.6	
Goats	dig prot %	*	2.4	2.6	
Horses	dig prot %	*	2.8	3.0	
Rabbits	dig prot %	*	3.4	3.7	
Sheep	dig prot %	*	2.2	2.4	
Energy	GE Mcal/kg				
Cattle	DE Mcal/kg	*	2.40	2.58	
Sheep	DE Mcal/kg	*	2.28	2.45	
Cattle	ME Mcal/kg	*	1.97	2.12	
Sheep	ME Mcal/kg	*	1.87	2.01	
Cattle	TDN %	*	54.5	58.6	
Sheep	TDN %	*	51.7	55.6	

Bluegrass, Kentucky, hay, s-c, (1)
Ref No 1 00 776 United States

Feed Name or Analyses			Mean As Fed	Mean Dry	C.V. ± %
Dry matter	%		88.9	100.0	4
Ash	%		5.9	6.6	
Crude fiber	%		26.7	30.0	
Sheep	dig coef %		67.	67.	
Ether extract	%		3.0	3.4	
Sheep	dig coef %		50.	50.	
N-free extract	%		44.2	49.7	
Sheep	dig coef %		64.	64.	
Protein (N x 6.25)	%		9.1	10.2	
Sheep	dig coef %		58.	58.	
Cattle	dig prot %	*	5.1	5.8	
Goats	dig prot %	*	5.4	6.1	
Horses	dig prot %	*	5.5	6.2	
Rabbits	dig prot %	*	5.8	6.6	
Sheep	dig prot %		5.3	5.9	
Cellulose (Matrone)	%		24.3	27.3	
Lignin (Ellis)	%		4.9	5.5	
Energy	GE Mcal/kg				
Cattle	DE Mcal/kg	*	2.47	2.78	
Sheep	DE Mcal/kg	*	2.42	2.72	
Cattle	ME Mcal/kg	*	2.02	2.28	
Sheep	ME Mcal/kg	*	1.98	2.23	
Cattle	TDN %	*	56.0	63.0	
Sheep	TDN %		54.9	61.7	
Calcium	%		0.40	0.45	
Chlorine	%		0.55	0.62	
Copper	mg/kg		8.8	9.9	
Iron	%		0.025	0.028	
Magnesium	%		0.19	0.21	
Manganese	mg/kg		76.1	85.6	
Phosphorus	%		0.27	0.30	
Potassium	%		1.66	1.87	
Sodium	%		0.10	0.11	
Sulphur	%		0.12	0.13	
Carotene	mg/kg		300.3	337.7	13
Riboflavin	mg/kg		9.8	11.0	14
Vitamin A equivalent	IU/g		500.5	563.0	

Bluegrass, Kentucky, hay, s-c, immature, (1)
Ref No 1 00 770 United States

Feed Name or Analyses			Mean As Fed	Mean Dry	C.V. ± %
Dry matter	%		91.6	100.0	1
Crude fiber	%				
Sheep	dig coef %	*	67.	67.	
Ether extract	%				
Sheep	dig coef %	*	50.	50.	
N-free extract	%				
Sheep	dig coef %	*	64.	64.	
Protein (N x 6.25)	%				
Sheep	dig coef %	*	58.	58.	
Calcium	%		0.45	0.49	
Phosphorus	%		0.35	0.38	

Bluegrass, Kentucky, hay, s-c, early bloom, (1)
Ref No 1 05 681 United States

Feed Name or Analyses			Mean As Fed	Mean Dry	C.V. ± %
Dry matter	%			100.0	
Ash	%			5.0	

(1) dry forages and roughages (2) pasture, range plants, and forages fed green (3) silages (4) energy feeds (5) protein supplements (6) minerals (7) vitamins (8) additives

Feed Name or Analyses			Mean As Fed	Mean Dry	C.V. ± %
Crude fiber	%			28.3	
Ether extract	%			4.0	
N-free extract	%			50.5	
Protein (N x 6.25)	%			12.2	
Cattle	dig prot %	*		7.5	
Goats	dig prot %	*		7.9	
Horses	dig prot %	*		7.9	
Rabbits	dig prot %	*		8.1	
Sheep	dig prot %	*		7.5	
Cellulose (Matrone)	%			27.5	
Lignin (Ellis)	%			5.8	
Energy	GE Mcal/kg				
Cattle	DE Mcal/kg	*		2.84	
Sheep	DE Mcal/kg	*		2.63	
Cattle	ME Mcal/kg	*		2.33	
Sheep	ME Mcal/kg	*		2.15	
Cattle	TDN %	*		64.4	
Sheep	TDN %	*		59.6	

Bluegrass, Kentucky, hay, s-c, mid-bloom, (1)

Ref No 1 00 771 United States

Feed Name or Analyses			Mean As Fed	Mean Dry	C.V. ± %
Dry matter	%		88.1	100.0	1
Ash	%		7.0	7.9	12
Crude fiber	%		27.9	31.7	5
Sheep	dig coef %	*	67.	67.	
Ether extract	%		3.4	3.9	6
Sheep	dig coef %	*	50.	50.	
N-free extract	%		40.4	45.9	
Sheep	dig coef %	*	64.	64.	
Protein (N x 6.25)	%		9.3	10.6	11
Sheep	dig coef %	*	58.	58.	
Cattle	dig prot %	*	5.4	6.1	
Goats	dig prot %	*	5.7	6.4	
Horses	dig prot %	*	5.8	6.5	
Rabbits	dig prot %	*	6.0	6.9	
Sheep	dig prot %		5.4	6.1	
Energy	GE Mcal/kg				
Cattle	DE Mcal/kg	*	2.40	2.72	
Sheep	DE Mcal/kg	*	2.38	2.70	
Cattle	ME Mcal/kg	*	1.97	2.23	
Sheep	ME Mcal/kg	*	1.95	2.21	
Cattle	TDN %	*	54.4	61.8	
Sheep	TDN %		53.9	61.2	
Calcium	%		0.30	0.34	
Phosphorus	%		0.21	0.24	

Bluegrass, Kentucky, hay, s-c, full bloom, (1)

Ref No 1 00 772 United States

Feed Name or Analyses			Mean As Fed	Mean Dry	C.V. ± %
Dry matter	%		92.1	100.0	
Ash	%		5.4	5.9	
Crude fiber	%		29.9	32.5	
Sheep	dig coef %	*	67.	67.	
Ether extract	%		3.0	3.3	
Sheep	dig coef %	*	50.	50.	
N-free extract	%		45.5	49.4	
Sheep	dig coef %	*	64.	64.	
Protein (N x 6.25)	%		8.2	8.9	
Sheep	dig coef %	*	58.	58.	
Cattle	dig prot %	*	4.3	4.6	
Goats	dig prot %	*	4.5	4.9	
Horses	dig prot %	*	4.7	5.1	
Rabbits	dig prot %	*	5.1	5.5	
Sheep	dig prot %		4.8	5.2	
Energy	GE Mcal/kg				
Cattle	DE Mcal/kg	*	2.37	2.57	
Sheep	DE Mcal/kg	*	2.53	2.75	
Cattle	ME Mcal/kg	*	1.94	2.11	
Sheep	ME Mcal/kg	*	2.07	2.25	
Cattle	TDN %	*	53.8	58.4	
Sheep	TDN %		57.3	62.3	
Calcium	%		0.24	0.26	
Phosphorus	%		0.25	0.27	
Potassium	%		1.40	1.52	

Bluegrass, Kentucky, hay, s-c, dough stage, (1)

Ref No 1 00 773 United States

Feed Name or Analyses			Mean As Fed	Mean Dry	C.V. ± %
Dry matter	%		87.3	100.0	
Ash	%		6.4	7.3	
Crude fiber	%		31.0	35.5	
Ether extract	%		2.5	2.9	
N-free extract	%		42.8	49.0	
Protein (N x 6.25)	%		4.6	5.3	
Cattle	dig prot %	*	1.3	1.5	
Goats	dig prot %	*	1.3	1.5	
Horses	dig prot %	*	1.8	2.0	
Rabbits	dig prot %	*	2.4	2.8	
Sheep	dig prot %	*	1.2	1.3	
Energy	GE Mcal/kg				
Cattle	DE Mcal/kg	*	2.12	2.43	
Sheep	DE Mcal/kg	*	2.05	2.35	
Cattle	ME Mcal/kg	*	1.74	1.99	
Sheep	ME Mcal/kg	*	1.68	1.93	
Cattle	TDN %	*	48.2	55.2	
Sheep	TDN %	*	46.6	53.4	
Calcium	%		0.23	0.26	13
Chlorine	%		0.38	0.44	
Iron	%		0.015	0.017	18
Magnesium	%		0.10	0.11	27
Manganese	mg/kg		55.8	63.9	13
Phosphorus	%		0.20	0.23	21
Potassium	%		1.48	1.69	11
Sodium	%		0.13	0.15	
Sulphur	%		0.16	0.18	

Ref No 1 00 773 Canada

Feed Name or Analyses			Mean As Fed	Mean Dry	C.V. ± %
Dry matter	%			100.0	
Protein (N x 6.25)	%			7.1	
Cattle	dig prot %	*		3.1	
Goats	dig prot %	*		3.2	
Horses	dig prot %	*		3.6	
Rabbits	dig prot %	*		4.2	
Sheep	dig prot %	*		2.9	
Phosphorus	%			0.20	
Sulphur	%			0.17	

Bluegrass, Kentucky, hay, s-c, mature, (1)

Ref No 1 00 774 United States

Feed Name or Analyses			Mean As Fed	Mean Dry	C.V. ± %
Dry matter	%		90.2	100.0	
Ash	%		6.5	7.3	31
Crude fiber	%		30.0	33.3	6
Sheep	dig coef %	*	67.	67.	
Ether extract	%		2.6	2.9	15
Sheep	dig coef %	*	50.	50.	
N-free extract	%		45.3	50.3	
Sheep	dig coef %	*	64.	64.	
Protein (N x 6.25)	%		5.6	6.3	16
Sheep	dig coef %	*	58.	58.	
Cattle	dig prot %	*	2.1	2.4	
Goats	dig prot %	*	2.2	2.4	
Horses	dig prot %	*	2.6	2.8	
Rabbits	dig prot %	*	3.2	3.5	
Sheep	dig prot %		3.3	3.6	
Fatty acids	%		2.6	2.9	
Energy	GE Mcal/kg				
Cattle	DE Mcal/kg	*	2.33	2.58	
Sheep	DE Mcal/kg	*	2.44	2.71	
Cattle	ME Mcal/kg	*	1.91	2.12	
Sheep	ME Mcal/kg	*	2.00	2.22	
Cattle	TDN %	*	52.8	58.6	
Sheep	TDN %		55.3	61.4	
Calcium	%		0.24	0.26	
Chlorine	%		0.39	0.44	
Iron	%		0.016	0.017	
Magnesium	%		0.10	0.11	
Manganese	mg/kg		57.6	63.9	
Phosphorus	%		0.21	0.23	
Potassium	%		1.53	1.70	
Sodium	%		0.13	0.15	
Sulphur	%		0.17	0.18	

Bluegrass, Kentucky, hay, s-c, cut 2, (1)

Ref No 1 00 775 United States

Feed Name or Analyses			Mean As Fed	Mean Dry	C.V. ± %
Dry matter	%			100.0	
Iron	%			0.043	
Manganese	mg/kg			59.1	

Bluegrass, Kentucky, aerial part, fresh, (2)

Ref No 2 00 786 United States

Feed Name or Analyses			Mean As Fed	Mean Dry	C.V. ± %
Dry matter	%		29.0	100.0	10
Organic matter	%		27.3	94.2	1
Ash	%		3.1	10.8	
Crude fiber	%		7.5	25.9	17
Cattle	dig coef %		62.	62.	18
Sheep	dig coef %		62.	62.	
Ether extract	%		1.4	4.8	19
Cattle	dig coef %		67.	67.	15
Sheep	dig coef %		54.	54.	
N-free extract	%		12.4	42.9	6
Cattle	dig coef %		71.	71.	12
Sheep	dig coef %		67.	67.	
Protein (N x 6.25)	%		4.5	15.6	23
Cattle	dig coef %		67.	67.	16
Sheep	dig coef %		69.	69.	
Cattle	dig prot %		3.0	10.4	
Goats	dig prot %	*	3.2	11.1	
Horses	dig prot %	*	3.1	10.8	
Rabbits	dig prot %	*	3.1	10.7	
Sheep	dig prot %		3.1	10.8	
Lignin (Ellis)	%		1.7	6.0	24
Energy	GE Mcal/kg				
Cattle	DE Mcal/kg	*	0.82	2.83	
Sheep	DE Mcal/kg	*	0.78	2.69	
Cattle	ME Mcal/kg	*	0.67	2.32	
Sheep	ME Mcal/kg	*	0.64	2.21	
Cattle	TDN %		18.6	64.2	
Sheep	TDN %		17.7	61.1	
Calcium	%		0.15	0.53	
Copper	mg/kg		4.0	13.9	
Iron	%		0.005	0.017	
Magnesium	%		0.07	0.23	
Manganese	mg/kg		23.3	80.3	
Phosphorus	%		0.12	0.43	
Potassium	%		0.57	1.95	
Sodium	%		0.07	0.23	
Sulphur	%		0.19	0.66	
Carotene	mg/kg		58.0	200.2	47

Continued

Feed Name or Analyses			Mean As Fed	Mean Dry	C.V. ± %
Niacin	mg/kg		19.1	65.9	
Riboflavin	mg/kg		3.2	11.0	
Thiamine	mg/kg		2.6	8.8	
a-tocopherol	mg/kg		180.8	624.3	
Vitamin A equivalent	IU/g		96.6	333.7	

Ref No 2 00 786 Canada

Feed Name or Analyses			Mean As Fed	Mean Dry	C.V. ± %
Dry matter	%			100.0	
Crude fiber	%				
Cattle	dig coef %			62.	
Sheep	dig coef %			62.	
Ether extract	%				
Cattle	dig coef %			67.	
Sheep	dig coef %			54.	
N-free extract	%				
Cattle	dig coef %			71.	
Sheep	dig coef %			67.	
Protein (N x 6.25)	%				
Cattle	dig coef %			67.	
Sheep	dig coef %			69.	
Energy	GE Mcal/kg				
Cattle	DE Mcal/kg	*		2.83	
Sheep	DE Mcal/kg	*		2.69	
Cattle	ME Mcal/kg	*		2.32	
Sheep	ME Mcal/kg	*		2.21	
Cattle	TDN %			64.2	
Sheep	TDN %			61.1	
Copper	mg/kg			10.9	
Molybdenum	mg/kg			0.00	
Sulphur	%			0.95	

Bluegrass, Kentucky, aerial part, fresh, immature, (2)

Ref No 2 00 778 United States

Feed Name or Analyses			Mean As Fed	Mean Dry	C.V. ± %
Dry matter	%		30.5	100.0	26
Energy	GE Mcal/kg				
Cattle	NE_m Mcal/kg	*	0.48	1.59	
Cattle	NE_{gain} Mcal/kg	*	0.31	1.02	
Cattle	$NE_{lactating\ cows}$ Mcal/kg	*	0.58	1.90	
Sulphur	%		0.20	0.66	
Zinc	mg/kg		52.0	170.4	
Carotene	mg/kg		116.8	382.9	24
Vitamin A equivalent	IU/g		194.7	638.4	

Bluegrass, Kentucky, aerial part, fresh, immature, pure stand, (2)

Ref No 2 00 777 United States

Feed Name or Analyses			Mean As Fed	Mean Dry	C.V. ± %
Dry matter	%		31.4	100.0	8
Ash	%		3.0	9.4	13
Crude fiber	%		8.0	25.4	11
Ether extract	%		1.1	3.5	12
N-free extract	%		13.9	44.2	
Protein (N x 6.25)	%		5.5	17.5	19
Cattle	dig prot %	*	4.0	12.8	
Goats	dig prot %	*	4.0	12.9	
Horses	dig prot %	*	3.9	12.4	
Rabbits	dig prot %	*	3.8	12.2	
Sheep	dig prot %	*	4.2	13.3	
Energy	GE Mcal/kg				
Cattle	DE Mcal/kg	*	0.96	3.07	
Sheep	DE Mcal/kg	*	0.87	2.79	

(1) dry forages and roughages (2) pasture, range plants, and forages fed green (3) silages (4) energy feeds (5) protein supplements (6) minerals (7) vitamins (8) additives

Feed Name or Analyses			Mean As Fed	Mean Dry	C.V. ± %
Cattle	ME Mcal/kg	*	0.79	2.52	
Sheep	ME Mcal/kg	*	0.72	2.28	
Cattle	TDN %	*	21.9	69.6	
Sheep	TDN %	*	19.8	63.2	
Calcium	%		0.14	0.43	10
Magnesium	%		0.05	0.17	10
Phosphorus	%		0.13	0.41	17
Potassium	%		0.71	2.26	
Carotene	mg/kg		151.3	481.9	
Vitamin A equivalent	IU/g		252.3	803.4	

Bluegrass, Kentucky, aerial part, fresh, early bloom, (2)

Ref No 2 00 779 United States

Feed Name or Analyses			Mean As Fed	Mean Dry	C.V. ± %
Dry matter	%		34.8	100.0	7
Ash	%		2.5	7.1	13
Crude fiber	%		9.5	27.4	10
Ether extract	%		1.4	3.9	23
N-free extract	%		15.6	44.9	
Protein (N x 6.25)	%		5.8	16.6	12
Cattle	dig prot %	*	4.2	12.0	
Goats	dig prot %	*	4.2	12.1	
Horses	dig prot %	*	4.1	11.7	
Rabbits	dig prot %	*	4.0	11.5	
Sheep	dig prot %	*	4.4	12.5	
Hemicellulose	%		5.7	16.4	
Energy	GE Mcal/kg				
Cattle	DE Mcal/kg	*	1.10	3.16	
Sheep	DE Mcal/kg	*	1.04	2.99	
Cattle	ME Mcal/kg	*	0.90	2.60	
Sheep	ME Mcal/kg	*	0.85	2.45	
Cattle	NE_m Mcal/kg	*	0.53	1.51	
Cattle	NE_{gain} Mcal/kg	*	0.33	0.94	
Cattle	$NE_{lactating\ cows}$ Mcal/kg	*	0.62	1.78	
Cattle	TDN %	*	25.0	71.8	
Sheep	TDN %	*	23.6	67.8	
Calcium	%		0.16	0.46	14
Magnesium	%		0.04	0.11	29
Phosphorus	%		0.14	0.39	15
Potassium	%		0.70	2.01	10

Bluegrass, Kentucky, aerial part, fresh, mid-bloom, (2)

Ref No 2 00 780 United States

Feed Name or Analyses			Mean As Fed	Mean Dry	C.V. ± %
Dry matter	%		31.7	100.0	
Ash	%		2.4	7.6	7
Crude fiber	%		9.2	29.2	8
Ether extract	%		1.2	3.9	22
N-free extract	%		14.6	46.1	
Protein (N x 6.25)	%		4.2	13.2	5
Cattle	dig prot %	*	2.9	9.1	
Goats	dig prot %	*	2.8	8.9	
Horses	dig prot %	*	2.8	8.7	
Rabbits	dig prot %	*	2.8	8.9	
Sheep	dig prot %	*	2.9	9.3	
Hemicellulose	%		5.5	17.3	
Energy	GE Mcal/kg				
Cattle	DE Mcal/kg	*	0.84	2.65	
Sheep	DE Mcal/kg	*	0.89	2.81	
Cattle	ME Mcal/kg	*	0.69	2.17	
Sheep	ME Mcal/kg	*	0.73	2.30	
Cattle	TDN %	*	19.1	60.2	
Sheep	TDN %	*	20.2	63.6	
Calcium	%		0.10	0.31	
Magnesium	%		0.03	0.11	
Phosphorus	%		0.10	0.33	
Potassium	%		0.63	1.99	

Bluegrass, Kentucky, aerial part, fresh, full bloom, (2)

Ref No 2 00 781 United States

Feed Name or Analyses			Mean As Fed	Mean Dry	C.V. ± %
Dry matter	%		32.9	100.0	
Calcium	%		0.11	0.32	22
Phosphorus	%		0.10	0.29	20

Bluegrass, Kentucky, aerial part, fresh, milk stage, (2)

Ref No 2 00 782 United States

Feed Name or Analyses			Mean As Fed	Mean Dry	C.V. ± %
Dry matter	%			100.0	
Ash	%			7.3	4
Crude fiber	%			30.3	3
Ether extract	%			3.6	1
N-free extract	%			47.2	
Protein (N x 6.25)	%			11.6	7
Cattle	dig prot %	*		7.8	
Goats	dig prot %	*		7.4	
Horses	dig prot %	*	6.8	7.9	
Horses	dig prot %	*		7.4	
Rabbits	dig prot %	*		7.6	
Sheep	dig prot %	*		7.8	
Hemicellulose	%			18.4	
Energy	GE Mcal/kg				
Cattle	DE Mcal/kg	*		2.64	
Sheep	DE Mcal/kg	*		2.52	
Cattle	ME Mcal/kg	*		2.16	
Sheep	ME Mcal/kg	*		2.07	
Cattle	TDN %	*		54.8	
Sheep	TDN %	*		57.2	

Bluegrass, Kentucky, aerial part, fresh, dough stage, (2)

Ref No 2 00 783 United States

Feed Name or Analyses			Mean As Fed	Mean Dry	C.V. ± %
Dry matter	%		42.2	100.0	
Ash	%		2.8	6.6	
Crude fiber	%		14.7	34.8	
Ether extract	%		1.3	3.1	
N-free extract	%		19.4	46.0	
Protein (N x 6.25)	%		4.0	9.5	
Cattle	dig prot %	*	2.5	5.9	
Goats	dig prot %	*	2.3	5.4	
Horses	dig prot %	*	2.4	5.6	
Rabbits	dig prot %	*	2.5	6.0	
Sheep	dig prot %	*	2.5	5.8	
Energy	GE Mcal/kg				
Cattle	DE Mcal/kg	*	1.06	2.52	
Sheep	DE Mcal/kg	*	1.01	2.40	
Cattle	ME Mcal/kg	*	0.87	2.07	
Sheep	ME Mcal/kg	*	0.83	1.97	
Cattle	TDN %	*	24.1	57.2	
Sheep	TDN %	*	23.0	54.4	
Calcium	%		0.08	0.19	28
Chlorine	%		0.17	0.40	19
Iron	%		0.012	0.028	25
Magnesium	%		0.03	0.07	46
Manganese	mg/kg		12.3	29.1	
Phosphorus	%		0.12	0.29	23
Potassium	%		0.83	1.96	8
Sodium	%		0.06	0.14	21
Sulphur	%		0.07	0.17	15

Feed Name or Analyses			Mean As Fed	Mean Dry	C.V. ± %

Bluegrass, Kentucky, aerial part, fresh, mature, (2)
Ref No 2 00 784 United States

Analyses	Unit		As Fed	Dry	C.V. ± %
Dry matter	%		41.6	100.0	13
Ash	%		2.6	6.2	15
Crude fiber	%		13.4	32.2	12
Ether extract	%		1.3	3.1	30
N-free extract	%		20.4	49.0	
Protein (N x 6.25)	%		4.0	9.5	15
Cattle	dig prot %	*	2.5	6.0	
Goats	dig prot %	*	2.3	5.4	
Horses	dig prot %	*	2.3	5.6	
Rabbits	dig prot %	*	2.5	6.0	
Sheep	dig prot %	*	2.4	5.8	
Hemicellulose	%		8.6	20.7	
Energy	GE Mcal/kg				
Cattle	DE Mcal/kg	*	1.24	2.97	
Sheep	DE Mcal/kg	*	1.02	2.46	
Cattle	ME Mcal/kg	*	1.02	2.44	
Sheep	ME Mcal/kg	*	0.84	2.02	
Cattle	TDN %	*	28.0	67.3	
Sheep	TDN %	*	23.2	55.8	
Carotene	mg/kg		38.2	91.7	
Vitamin A equivalent	IU/g		63.6	152.9	

Bluegrass, Kentucky, aerial part, fresh, over ripe, (2)
Ref No 2 00 785 United States

Analyses	Unit		As Fed	Dry	C.V. ± %
Dry matter	%			100.0	
Ash	%			6.3	
Crude fiber	%			42.1	
Ether extract	%			1.3	
N-free extract	%			47.0	
Protein (N x 6.25)	%			3.3	
Cattle	dig prot %	*		0.7	
Goats	dig prot %	*		-0.3	
Horses	dig prot %	*		0.3	
Rabbits	dig prot %	*		1.2	
Sheep	dig prot %	*		0.1	
Energy	GE Mcal/kg				
Cattle	DE Mcal/kg	*		2.42	
Sheep	DE Mcal/kg	*		2.49	
Cattle	ME Mcal/kg	*		1.98	
Sheep	ME Mcal/kg	*		2.04	
Cattle	TDN %	*		54.8	
Sheep	TDN %	*		56.4	
Carotene	mg/kg			40.8	
Vitamin A equivalent	IU/g			68.0	

Bluegrass, Kentucky, aerial part, ensiled, (3)
Ref No 3 00 789 United States

Analyses	Unit		As Fed	Dry	C.V. ± %
Dry matter	%		35.7	100.0	17
Sheep	dig coef %		64.	64.	
Ash	%		3.1	8.7	37
Crude fiber	%		11.0	31.0	12
Sheep	dig coef %		71.	71.	
Ether extract	%		1.5	4.1	22
N-free extract	%		14.8	41.5	
Sheep	dig coef %		64.	64.	
Protein (N x 6.25)	%		5.3	14.8	20
Sheep	dig coef %		65.	65.	
Cattle	dig prot %	*	3.5	9.7	
Goats	dig prot %	*	3.5	9.7	
Horses	dig prot %	*	3.5	9.7	
Sheep	dig prot %		3.4	9.6	
Energy	GE Mcal/kg				
Cattle	DE Mcal/kg	*	0.92	2.57	
Sheep	DE Mcal/kg	*	1.01	2.84	
Cattle	ME Mcal/kg	*	0.75	2.11	
Sheep	ME Mcal/kg	*	0.83	2.33	
Cattle	TDN %	*	20.8	58.3	
Sheep	TDN %		23.0	64.4	
Calcium	%		0.25	0.70	40
Phosphorus	%		0.15	0.41	30

Bluegrass, Kentucky, aerial part, ensiled, early bloom, (3)
Ref No 3 00 788 United States

Analyses	Unit		As Fed	Dry	C.V. ± %
Dry matter	%		41.4	100.0	
Ash	%		3.6	8.8	
Crude fiber	%		13.5	32.6	
Ether extract	%		1.7	4.2	
N-free extract	%		16.8	40.5	
Protein (N x 6.25)	%		5.8	13.9	
Cattle	dig prot %	*	3.7	8.9	
Goats	dig prot %	*	3.7	8.9	
Horses	dig prot %	*	3.7	8.9	
Sheep	dig prot %	*	3.7	8.9	
Energy	GE Mcal/kg				
Cattle	DE Mcal/kg	*	1.06	2.56	
Sheep	DE Mcal/kg	*	1.13	2.72	
Cattle	ME Mcal/kg	*	0.87	2.10	
Sheep	ME Mcal/kg	*	0.93	2.23	
Cattle	TDN %	*	24.0	58.0	
Sheep	TDN %	*	25.6	61.8	

Bluegrass, Kentucky, aerial part w molasses added ensiled, (3)
Ref No 3 00 791 United States

Analyses	Unit		As Fed	Dry	C.V. ± %
Dry matter	%		38.3	100.0	10
Ash	%		3.3	8.5	13
Crude fiber	%		11.4	29.8	14
Ether extract	%		1.6	4.2	24
N-free extract	%		16.1	42.2	
Protein (N x 6.25)	%		5.9	15.3	22
Cattle	dig prot %	*	3.9	10.1	
Goats	dig prot %	*	3.9	10.1	
Horses	dig prot %	*	3.9	10.1	
Sheep	dig prot %	*	3.9	10.1	
Energy	GE Mcal/kg				
Cattle	DE Mcal/kg	*	0.99	2.59	
Sheep	DE Mcal/kg	*	0.94	2.46	
Cattle	ME Mcal/kg	*	0.81	2.12	
Sheep	ME Mcal/kg	*	0.77	2.02	
Cattle	TDN %	*	22.5	58.7	
Sheep	TDN %	*	21.3	55.8	
Calcium	%		0.25	0.65	45
Phosphorus	%		0.15	0.38	26

Bluegrass, Kentucky, aerial part w molasses added, ensiled, full bloom, (3)
Ref No 3 00 790 United States

Analyses	Unit		As Fed	Dry	C.V. ± %
Dry matter	%		39.3	100.0	
Ash	%		3.3	8.5	
Crude fiber	%		12.5	31.9	
Ether extract	%		1.7	4.4	
N-free extract	%		16.3	41.5	
Protein (N x 6.25)	%		5.4	13.7	
Cattle	dig prot %	*	3.4	8.7	
Goats	dig prot %	*	3.4	8.7	
Horses	dig prot %	*	3.4	8.7	
Sheep	dig prot %	*	3.4	8.7	
Energy	GE Mcal/kg				
Cattle	DE Mcal/kg	*	1.02	2.59	
Sheep	DE Mcal/kg	*	1.08	2.74	
Cattle	ME Mcal/kg	*	0.84	2.12	
Sheep	ME Mcal/kg	*	0.88	2.24	
Cattle	TDN %	*	23.1	58.8	
Sheep	TDN %	*	24.4	62.0	

BLUEGRASS, MUTTON. Poa fendleriana

Bluegrass, mutton, hay, s-c, (1)
Ref No 1 00 793 United States

Analyses	Unit		As Fed	Dry	C.V. ± %
Dry matter	%		94.2	100.0	0
Calcium	%		0.39	0.41	23
Phosphorus	%		0.32	0.34	28
Carotene	mg/kg		48.6	51.6	82
Vitamin A equivalent	IU/g		81.0	86.0	

Bluegrass, mutton, hay, s-c, mature, (1)
Ref No 1 00 792 United States

Analyses	Unit		As Fed	Dry	C.V. ± %
Dry matter	%		94.0	100.0	
Carotene	mg/kg		7.0	7.5	
Vitamin A equivalent	IU/g		11.7	12.5	

Bluegrass, mutton, aerial part, fresh, (2)
Ref No 2 00 794 United States

Analyses	Unit		As Fed	Dry	C.V. ± %
Dry matter	%			100.0	
Ash	%			11.1	23
Crude fiber	%			33.5	4
Ether extract	%			2.4	20
N-free extract	%			46.0	
Protein (N x 6.25)	%			7.0	21
Cattle	dig prot %	*		3.8	
Goats	dig prot %	*		3.1	
Horses	dig prot %	*		3.5	
Rabbits	dig prot %	*		4.1	
Sheep	dig prot %	*		3.5	
Energy	GE Mcal/kg				
Cattle	DE Mcal/kg	*		2.58	
Sheep	DE Mcal/kg	*		2.69	
Cattle	ME Mcal/kg	*		2.12	
Sheep	ME Mcal/kg	*		2.21	
Cattle	TDN %	*		58.5	
Sheep	TDN %	*		61.0	

BLUEGRASS, NODDING. Poa reflexa

Bluegrass, nodding, hay, s-c, immature, (1)
Ref No 1 00 795 United States

Analyses	Unit		As Fed	Dry	C.V. ± %
Dry matter	%		94.4	100.0	
Ash	%		5.9	6.2	
Crude fiber	%		27.6	29.2	
Ether extract	%		2.2	2.3	
N-free extract	%		49.8	52.8	
Protein (N x 6.25)	%		9.0	9.5	
Cattle	dig prot %	*	4.9	5.2	
Goats	dig prot %	*	5.1	5.4	
Horses	dig prot %	*	5.3	5.6	

Continued

Feed Name or Analyses				Mean As Fed	Mean Dry	C.V. ± %
Rabbits	dig prot	%	*	5.7	6.0	
Sheep	dig prot	%	*	4.8	5.1	
Energy	GE	Mcal/kg				
Cattle	DE	Mcal/kg	*	2.63	2.78	
Sheep	DE	Mcal/kg	*	2.42	2.57	
Cattle	ME	Mcal/kg	*	2.15	2.28	
Sheep	ME	Mcal/kg	*	1.99	2.11	
Cattle	TDN	%	*	59.5	63.1	
Sheep	TDN	%	*	55.0	58.2	

BLUEGRASS, PINE. Poa scabrella

Bluegrass, pine, aerial part, fresh, (2)

Ref No 2 00 801 United States

Feed Name or Analyses				As Fed	Dry	C.V. ± %
Dry matter		%			100.0	
Ash		%			5.3	35
Crude fiber		%			35.7	14
Protein (N x 6.25)		%			9.9	64
Cattle	dig prot	%	*		6.3	
Goats	dig prot	%	*		5.8	
Horses	dig prot	%	*		5.9	
Rabbits	dig prot	%	*		6.3	
Sheep	dig prot	%	*		6.2	
Calcium		%			0.36	56
Phosphorus		%			0.25	48
Potassium		%			1.63	44

Bluegrass, pine, aerial part, fresh, immature, (2)

Ref No 2 00 796 United States

Feed Name or Analyses				As Fed	Dry	C.V. ± %
Dry matter		%			100.0	
Ash		%			8.5	7
Crude fiber		%			28.6	
Protein (N x 6.25)		%			20.8	18
Cattle	dig prot	%	*		15.6	
Goats	dig prot	%	*		16.0	
Horses	dig prot	%	*		15.2	
Rabbits	dig prot	%	*		14.7	
Sheep	dig prot	%	*		16.4	
Calcium		%			0.78	
Phosphorus		%			0.44	13
Potassium		%			2.84	3

Bluegrass, pine, aerial part, fresh, early bloom, (2)

Ref No 2 00 797 United States

Feed Name or Analyses				As Fed	Dry	C.V. ± %
Dry matter		%			100.0	
Ash		%			7.0	
Crude fiber		%			29.9	
Protein (N x 6.25)		%			17.0	
Cattle	dig prot	%	*		12.3	
Goats	dig prot	%	*		12.4	
Horses	dig prot	%	*		12.0	
Rabbits	dig prot	%	*		11.8	
Sheep	dig prot	%	*		12.8	
Calcium		%			0.57	
Phosphorus		%			0.41	
Potassium		%			2.80	

Bluegrass, pine, aerial part, fresh, full bloom, (2)

Ref No 2 00 798 United States

Feed Name or Analyses				As Fed	Dry	C.V. ± %
Dry matter		%			100.0	
Ash		%			4.6	
Crude fiber		%			34.6	
Protein (N x 6.25)		%			7.7	
Cattle	dig prot	%	*		4.4	
Goats	dig prot	%	*		3.7	
Horses	dig prot	%	*		4.1	
Rabbits	dig prot	%	*		4.6	
Sheep	dig prot	%	*		4.2	
Calcium		%			0.33	
Phosphorus		%			0.24	
Potassium		%			1.37	

Bluegrass, pine, aerial part, fresh, mature, (2)

Ref No 2 00 799 United States

Feed Name or Analyses				As Fed	Dry	C.V. ± %
Dry matter		%			100.0	
Ash		%			4.0	
Crude fiber		%			37.8	
Protein (N x 6.25)		%			4.5	
Cattle	dig prot	%	*		1.7	
Goats	dig prot	%	*		0.8	
Horses	dig prot	%	*		1.4	
Rabbits	dig prot	%	*		2.1	
Sheep	dig prot	%	*		1.2	
Calcium		%			0.23	
Phosphorus		%			0.18	
Potassium		%			1.03	

Bluegrass, pine, aerial part, fresh, over ripe, (2)

Ref No 2 00 800 United States

Feed Name or Analyses				As Fed	Dry	C.V. ± %
Dry matter		%			100.0	
Ash		%			3.7	22
Crude fiber		%			41.1	7
Protein (N x 6.25)		%			3.3	32
Cattle	dig prot	%	*		0.7	
Goats	dig prot	%	*		-0.3	
Horses	dig prot	%	*		0.3	
Rabbits	dig prot	%	*		1.2	
Sheep	dig prot	%	*		0.1	
Calcium		%			0.19	17
Phosphorus		%			0.11	32
Potassium		%			0.77	16

BLUEGRASS, SANDBERG. Poa secunda

Bluegrass, Sandberg, hay, s-c, (1)

Ref No 1 00 802 United States

Feed Name or Analyses				As Fed	Dry	C.V. ± %
Dry matter		%		93.6	100.0	
Ash		%		6.0	6.4	
Crude fiber		%		30.5	32.6	
Sheep	dig coef	%		45.	45.	
Ether extract		%		2.9	3.1	
Sheep	dig coef	%		50.	50.	
N-free extract		%		46.1	49.3	
Sheep	dig coef	%		60.	60.	
Protein (N x 6.25)		%		8.0	8.6	
Sheep	dig coef	%		64.	64.	
Cattle	dig prot	%	*	4.1	4.4	
Goats	dig prot	%	*	4.3	4.6	
Horses	dig prot	%	*	4.5	4.8	

Feed Name or Analyses				As Fed	Dry	C.V. ± %
Rabbits	dig prot	%	*	5.0	5.3	
Sheep	dig prot	%		5.2	5.5	
Energy	GE	Mcal/kg				
Cattle	DE	Mcal/kg	*	2.42	2.58	
Sheep	DE	Mcal/kg	*	2.20	2.35	
Cattle	ME	Mcal/kg	*	1.98	2.12	
Sheep	ME	Mcal/kg	*	1.80	1.92	
Cattle	TDN	%	*	54.8	58.6	
Sheep	TDN	%		49.8	53.2	

Bluegrass, Sandberg, aerial part, fresh, immature, (2)

Ref No 2 00 803 United States

Feed Name or Analyses				As Fed	Dry	C.V. ± %
Dry matter		%			100.0	
Ash		%			3.7	
Crude fiber		%			41.1	
Protein (N x 6.25)		%			15.5	
Cattle	dig prot	%	*		11.1	
Goats	dig prot	%	*		11.0	
Horses	dig prot	%	*		10.7	
Rabbits	dig prot	%	*		10.6	
Sheep	dig prot	%	*		11.4	
Calcium		%			0.50	
Phosphorus		%			0.40	

Bluegrass, Sandberg, aerial part, fresh, early bloom, (2)

Ref No 2 00 804 United States

Feed Name or Analyses				As Fed	Dry	C.V. ± %
Dry matter		%			100.0	
Ash		%			8.2	
Crude fiber		%			33.2	
Ether extract		%			2.9	
N-free extract		%			44.7	
Protein (N x 6.25)		%			11.0	
Cattle	dig prot	%	*		7.2	
Goats	dig prot	%	*		6.8	
Horses	dig prot	%	*		6.9	
Rabbits	dig prot	%	*		7.2	
Sheep	dig prot	%	*		7.2	
Energy	GE	Mcal/kg				
Cattle	DE	Mcal/kg	*		2.93	
Sheep	DE	Mcal/kg	*		2.80	
Cattle	ME	Mcal/kg	*		2.40	
Sheep	ME	Mcal/kg	*		2.30	
Cattle	TDN	%	*		66.5	
Sheep	TDN	%	*		63.6	
Calcium		%			0.45	
Phosphorus		%			0.30	
Carotene		mg/kg			136.9	
Vitamin A equivalent		IU/g			228.2	

Bluegrass, Sandberg, aerial part, fresh, mid-bloom, (2)

Ref No 2 00 805 United States

Feed Name or Analyses				As Fed	Dry	C.V. ± %
Dry matter		%			100.0	
Calcium		%			0.44	
Phosphorus		%			0.19	

Bluegrass, Sandberg, aerial part, fresh, full bloom, (2)

Ref No 2 00 806 United States

Feed Name or Analyses				As Fed	Dry	C.V. ± %
Dry matter		%			100.0	
Calcium		%			0.45	
Phosphorus		%			0.17	

(1) dry forages and roughages (3) silages (6) minerals
(2) pasture, range plants, and forages fed green (4) energy feeds (7) vitamins
(5) protein supplements (8) additives

Bluegrass, Sandberg, aerial part, fresh, mature, (2)
Ref No 2 00 807 United States

Feed Name or Analyses			Mean As Fed	Mean Dry	C.V. ± %
Dry matter	%			100.0	
Calcium	%			0.39	
Phosphorus	%			0.25	
Carotene	mg/kg			42.5	
Vitamin A equivalent	IU/g			70.9	

Bluegrass, Sandberg, aerial part, fresh, over ripe, (2)
Ref No 2 00 808 United States

Feed Name or Analyses			Mean As Fed	Mean Dry	C.V. ± %
Dry matter	%			100.0	
Ash	%			9.6	
Crude fiber	%			41.3	
Ether extract	%			2.3	
N-free extract	%			42.6	
Protein (N x 6.25)	%			4.2	
Cattle	dig prot %	*		1.5	
Goats	dig prot %	*		0.5	
Horses	dig prot %	*		1.1	
Rabbits	dig prot %	*		1.9	
Sheep	dig prot %	*		0.9	
Energy	GE Mcal/kg				
Cattle	DE Mcal/kg	*		2.61	
Sheep	DE Mcal/kg	*		2.86	
Cattle	ME Mcal/kg	*		2.14	
Sheep	ME Mcal/kg	*		2.35	
Cattle	TDN %	*		59.3	
Sheep	TDN %	*		64.9	
Calcium	%			0.37	
Phosphorus	%			0.23	
Carotene	mg/kg			13.7	
Vitamin A equivalent	IU/g			22.8	

BLUEGRASS, SKYLINE. Poa epilus

Bluegrass, skyline, hay, s-c, over ripe, (1)
Ref No 1 00 810 United States

Feed Name or Analyses			Mean As Fed	Mean Dry	C.V. ± %
Dry matter	%		94.6	100.0	
Ash	%		6.4	6.8	
Crude fiber	%		36.6	38.7	
Ether extract	%		1.5	1.6	
N-free extract	%		46.8	49.5	
Protein (N x 6.25)	%		3.2	3.4	
Cattle	dig prot %	*	-0.0	-0.0	
Goats	dig prot %	*	-0.2	-0.2	
Horses	dig prot %	*	0.4	0.4	
Rabbits	dig prot %	*	1.2	1.3	
Sheep	dig prot %	*	-0.3	-0.3	
Energy	GE Mcal/kg				
Cattle	DE Mcal/kg	*	2.03	2.14	
Sheep	DE Mcal/kg	*	2.14	2.27	
Cattle	ME Mcal/kg	*	1.66	1.76	
Sheep	ME Mcal/kg	*	1.76	1.86	
Cattle	TDN %	*	45.9	48.6	
Sheep	TDN %	*	48.6	51.4	

BLUEGRASS, TEXAS. Poa arachnifera

Bluegrass, Texas, aerial part, fresh, (2)
Ref No 2 00 813 United States

Feed Name or Analyses			Mean As Fed	Mean Dry	C.V. ± %
Dry matter	%		54.6	100.0	36
Ash	%		4.4	8.1	16
Crude fiber	%		17.7	32.4	17
Ether extract	%		2.6	4.7	21
N-free extract	%		23.2	42.4	
Protein (N x 6.25)	%		6.8	12.4	36
Cattle	dig prot %	*	4.6	8.4	
Goats	dig prot %	*	4.4	8.1	
Horses	dig prot %	*	4.4	8.1	
Rabbits	dig prot %	*	4.5	8.2	
Sheep	dig prot %	*	4.7	8.5	
Energy	GE Mcal/kg				
Cattle	DE Mcal/kg	*	1.57	2.88	
Sheep	DE Mcal/kg	*	1.50	2.75	
Cattle	ME Mcal/kg	*	1.29	2.36	
Sheep	ME Mcal/kg	*	1.23	2.25	
Cattle	TDN %	*	35.7	65.4	
Sheep	TDN %	*	34.0	62.4	

Bluegrass, Texas, aerial part, fresh, immature, (2)
Ref No 2 00 811 United States

Feed Name or Analyses			Mean As Fed	Mean Dry	C.V. ± %
Dry matter	%		49.2	100.0	
Ash	%		4.3	8.7	
Crude fiber	%		14.1	28.7	
Ether extract	%		2.2	4.5	
N-free extract	%		21.8	44.4	
Protein (N x 6.25)	%		6.7	13.7	
Cattle	dig prot %	*	4.7	9.5	
Goats	dig prot %	*	4.6	9.3	
Horses	dig prot %	*	4.5	9.2	
Rabbits	dig prot %	*	4.5	9.2	
Sheep	dig prot %	*	4.8	9.8	
Energy	GE Mcal/kg				
Cattle	DE Mcal/kg	*	1.42	2.89	
Sheep	DE Mcal/kg	*	1.37	2.78	
Cattle	ME Mcal/kg	*	1.17	2.37	
Sheep	ME Mcal/kg	*	1.12	2.28	
Cattle	TDN %	*	32.2	65.5	
Sheep	TDN %	*	31.0	63.0	
Calcium	%		0.24	0.49	22
Phosphorus	%		0.11	0.22	46

Bluegrass, Texas, aerial part, fresh, over ripe, (2)
Ref No 2 00 812 United States

Feed Name or Analyses			Mean As Fed	Mean Dry	C.V. ± %
Dry matter	%		85.6	100.0	
Ash	%		5.9	6.9	
Crude fiber	%		34.4	40.2	
Ether extract	%		5.2	6.1	
N-free extract	%		35.1	41.0	
Protein (N x 6.25)	%		5.0	5.8	
Cattle	dig prot %	*	2.4	2.8	
Goats	dig prot %	*	1.7	2.0	
Horses	dig prot %	*	2.1	2.5	
Rabbits	dig prot %	*	2.7	3.1	
Sheep	dig prot %	*	2.1	2.4	
Energy	GE Mcal/kg				
Cattle	DE Mcal/kg	*	1.88	2.20	
Sheep	DE Mcal/kg	*	1.92	2.24	
Cattle	ME Mcal/kg	*	1.54	1.80	
Sheep	ME Mcal/kg	*	1.57	1.84	
Cattle	TDN %	*	42.7	49.9	
Sheep	TDN %	*	43.5	50.9	
Calcium	%		0.33	0.39	
Phosphorus	%		0.13	0.15	

BLUEGRASS-CLOVER, WHITE. Poa spp, Trifolium repens

Bluegrass-Clover, white, aerial part w molasses added, ensiled, (3)
Ref No 3 08 357 United States

Feed Name or Analyses			Mean As Fed	Mean Dry	C.V. ± %
Dry matter	%		35.0	100.0	
Ash	%		3.6	10.3	
Crude fiber	%		8.9	25.4	
Ether extract	%		1.8	5.1	
N-free extract	%		13.9	39.7	
Protein (N x 6.25)	%		6.8	19.4	
Cattle	dig prot %	*	4.9	13.9	
Goats	dig prot %	*	4.9	13.9	
Horses	dig prot %	*	4.9	13.9	
Sheep	dig prot %	*	4.9	13.9	
Energy	GE Mcal/kg				
Cattle	DE Mcal/kg	*	0.95	2.72	
Sheep	DE Mcal/kg	*	1.01	2.88	
Cattle	ME Mcal/kg	*	0.78	2.23	
Sheep	ME Mcal/kg	*	0.83	2.36	
Cattle	TDN %	*	21.6	61.7	
Sheep	TDN %	*	22.9	65.3	

BLUEGRASS, KENTUCKY-CLOVER, RED. Poa pratensis, Trifolium pratense

Bluegrass, Kentucky-Clover, red, hay, s-c, full bloom, (1)
Ref No 1 00 814 United States

Feed Name or Analyses			Mean As Fed	Mean Dry	C.V. ± %
Dry matter	%		86.6	100.0	
Ash	%		5.9	6.8	
Crude fiber	%		28.9	33.4	
Ether extract	%		2.5	2.9	
N-free extract	%		38.7	44.7	
Protein (N x 6.25)	%		10.6	12.2	
Cattle	dig prot %	*	6.5	7.5	
Goats	dig prot %	*	6.9	7.9	
Rabbits	dig prot %	*	7.0	8.1	
Sheep	dig prot %	*	6.5	7.5	
Energy	GE Mcal/kg				
Cattle	DE Mcal/kg	*	2.16	2.49	
Sheep	DE Mcal/kg	*	2.18	2.51	
Cattle	ME Mcal/kg	*	1.77	2.05	
Sheep	ME Mcal/kg	*	1.79	2.06	
Cattle	TDN %	*	49.0	56.6	
Sheep	TDN %	*	49.4	57.0	

BLUEGRASS, KENTUCKY-CLOVER, WHITE. Poa pratensis, Trifolium repens

Bluegrass, Kentucky-Clover, white, hay, s-c, (1)
Ref No 1 00 815 United States

Feed Name or Analyses			Mean As Fed	Mean Dry	C.V. ± %
Dry matter	%		88.1	100.0	
Ash	%		6.5	7.4	
Crude fiber	%		30.0	34.1	
Ether extract	%		1.9	2.1	
N-free extract	%		37.4	42.5	
Protein (N x 6.25)	%		12.2	13.9	
Cattle	dig prot %	*	7.9	9.0	
Goats	dig prot %	*	8.4	9.5	
Horses	dig prot %	*	8.2	9.3	
Rabbits	dig prot %	*	8.3	9.4	
Sheep	dig prot %	*	8.0	9.0	

Continued

Feed Name or Analyses			Mean As Fed	Mean Dry	C.V. ± %
Energy	GE Mcal/kg				
Cattle	DE Mcal/kg	*	2.13	2.42	
Sheep	DE Mcal/kg	*	2.18	2.47	
Cattle	ME Mcal/kg	*	1.75	1.99	
Sheep	ME Mcal/kg	*	1.79	2.03	
Cattle	TDN %	*	48.4	54.9	
Sheep	TDN %	*	49.4	56.1	

Bluegrass, Kentucky-Clover, white, aerial part, fresh, (2)
Ref No 2 08 356 United States

Feed Name or Analyses			Mean As Fed	Mean Dry	C.V. ± %
Dry matter	%		24.4	100.0	9
Ash	%		2.7	11.0	13
Crude fiber	%		4.5	18.3	7
Ether extract	%		0.9	3.5	24
N-free extract	%		11.4	46.5	
Protein (N x 6.25)	%		5.0	20.6	7
Cattle	dig prot %	*	3.8	15.4	
Goats	dig prot %	*	3.9	15.8	
Horses	dig prot %	*	3.7	15.0	
Rabbits	dig prot %	*	3.6	14.6	
Sheep	dig prot %	*	4.0	16.2	
Energy	GE Mcal/kg				
Cattle	DE Mcal/kg	*	0.75	3.05	
Sheep	DE Mcal/kg	*	0.75	3.08	
Cattle	ME Mcal/kg	*	0.61	2.50	
Sheep	ME Mcal/kg	*	0.62	2.53	
Cattle	TDN %	*	16.9	69.3	
Sheep	TDN %	*	17.0	69.9	
Calcium	%		0.31	1.29	
Phosphorus	%		0.11	0.46	

Bluegrass, Kentucky-Clover, white, aerial part, ensiled, immature, (3)
Ref No 3 00 816 United States

Feed Name or Analyses			Mean As Fed	Mean Dry	C.V. ± %
Dry matter	%		35.0	100.0	
Ash	%		3.6	10.3	
Crude fiber	%		8.9	25.4	
Ether extract	%		1.8	5.1	
N-free extract	%		13.9	39.8	
Protein (N x 6.25)	%		6.8	19.4	
Cattle	dig prot %	*	4.8	13.8	
Goats	dig prot %	*	4.8	13.8	
Horses	dig prot %	*	4.8	13.8	
Sheep	dig prot %	*	4.8	13.8	
Energy	GE Mcal/kg				
Cattle	DE Mcal/kg	*	0.95	2.72	
Sheep	DE Mcal/kg	*	0.87	2.48	
Cattle	ME Mcal/kg	*	0.78	2.23	
Sheep	ME Mcal/kg	*	0.71	2.03	
Cattle	TDN %	*	21.6	61.8	
Sheep	TDN %	*	19.7	56.3	

BLUEGRASS, KENTUCKY-TIMOTHY. Poa pratensis, Phelum pratense

Bluegrass, Kentucky-Timothy, hay, s-c, (1)
Ref No 1 00 817 United States

Feed Name or Analyses			Mean As Fed	Mean Dry	C.V. ± %
Dry matter	%		90.0	100.0	
Ash	%		4.9	5.4	
Crude fiber	%		27.9	31.0	
Ether extract	%		2.9	3.2	
N-free extract	%		46.4	51.5	
Protein (N x 6.25)	%		8.0	8.9	
Cattle	dig prot %	*	4.2	4.6	
Goats	dig prot %	*	4.4	4.9	
Horses	dig prot %	*	4.6	5.1	
Rabbits	dig prot %	*	5.0	5.5	
Sheep	dig prot %	*	4.1	4.6	
Energy	GE Mcal/kg				
Cattle	DE Mcal/kg	*	2.39	2.66	
Sheep	DE Mcal/kg	*	2.27	2.52	
Cattle	ME Mcal/kg	*	1.96	2.18	
Sheep	ME Mcal/kg	*	1.86	2.07	
Cattle	TDN %	*	54.3	60.3	
Sheep	TDN %	*	51.4	57.1	

(1) dry forages and roughages	(3) silages	(6) minerals
(2) pasture, range plants, and forages fed green	(4) energy feeds	(7) vitamins
	(5) protein supplements	(8) additives

BLUESTEM. Andropogon spp

Bluestem, hay, fan air dried, (1)
Ref No 1 00 818 United States

Feed Name or Analyses			Mean As Fed	Mean Dry	C.V. ± %
Dry matter	%		92.3	100.0	1
Cattle	dig coef %		53.	53.	9
Ash	%		6.4	7.0	16
Crude fiber	%		31.5	34.1	4
Cattle	dig coef %		63.	63.	7
Ether extract	%		2.1	2.3	13
Cattle	dig coef %		35.	35.	40
N-free extract	%		48.3	52.4	2
Cattle	dig coef %		55.	55.	9
Protein (N x 6.25)	%		3.9	4.3	21
Cattle	dig coef %		18.	18.	91
Cattle	dig prot %		0.7	0.8	
Goats	dig prot %	*	0.5	0.6	
Horses	dig prot %	*	1.1	1.2	
Rabbits	dig prot %	*	1.8	2.0	
Sheep	dig prot %	*	0.4	0.4	
Energy	GE Mcal/kg				
Cattle	DE Mcal/kg	*	2.14	2.32	
Sheep	DE Mcal/kg	*	2.19	2.37	
Cattle	ME Mcal/kg	*	1.75	1.90	
Sheep	ME Mcal/kg	*	1.79	1.95	
Cattle	TDN %		48.5	52.6	10
Sheep	TDN %	*	49.6	53.8	
Carotene	mg/kg		61.2	66.4	
Vitamin A equivalent	IU/g		102.0	110.6	

Bluestem, hay, s-c, (1)
Ref No 1 00 819 United States

Feed Name or Analyses			Mean As Fed	Mean Dry	C.V. ± %
Dry matter	%		90.9	100.0	
Ash	%		6.4	7.0	
Crude fiber	%		31.2	34.3	
Ether extract	%		2.2	2.4	
N-free extract	%		46.2	50.8	
Protein (N x 6.25)	%		4.9	5.4	
Cattle	dig prot %	*	1.5	1.6	
Goats	dig prot %	*	1.5	1.6	
Horses	dig prot %	*	1.9	2.1	
Rabbits	dig prot %	*	2.6	2.8	
Sheep	dig prot %	*	1.3	1.4	
Fatty acids	%		2.3	2.5	
Energy	GE Mcal/kg				
Cattle	DE Mcal/kg	*	1.87	2.06	
Sheep	DE Mcal/kg	*	1.76	1.94	
Cattle	ME Mcal/kg	*	1.54	1.69	
Sheep	ME Mcal/kg	*	1.45	1.59	
Cattle	TDN %	*	42.2	46.4	
Sheep	TDN %	*	40.1	44.1	
Chlorine	%		0.04	0.04	
Sodium	%		0.01	0.01	
Carotene	mg/kg		37.9	41.7	46
Vitamin A equivalent	IU/g		63.1	69.5	

Bluestem, aerial part, fresh, (2)
Ref No 2 00 827 United States

Feed Name or Analyses			Mean As Fed	Mean Dry	C.V. ± %
Dry matter	%		45.2	100.0	29
Ash	%		3.8	8.5	
Crude fiber	%		16.0	35.4	
Ether extract	%		1.1	2.5	
N-free extract	%		20.4	45.2	
Protein (N x 6.25)	%		3.8	8.4	
Cattle	dig prot %	*	2.3	5.0	
Goats	dig prot %	*	2.0	4.4	
Horses	dig prot %	*	2.1	4.7	
Rabbits	dig prot %	*	2.3	5.2	
Sheep	dig prot %	*	2.2	4.8	
Energy	GE Mcal/kg				
Cattle	DE Mcal/kg	*	1.12	2.48	
Sheep	DE Mcal/kg	*	1.16	2.57	
Cattle	ME Mcal/kg	*	0.92	2.03	
Sheep	ME Mcal/kg	*	0.95	2.11	
Cattle	TDN %	*	25.4	56.2	
Sheep	TDN %	*	26.4	58.4	
Carotene	mg/kg		53.5	118.4	68
Vitamin A equivalent	IU/g		89.2	197.4	

Bluestem, aerial part, fresh, immature, (2)
Ref No 2 00 821 United States

Feed Name or Analyses			Mean As Fed	Mean Dry	C.V. ± %
Dry matter	%		29.8	100.0	11
Ash	%		2.5	8.5	
Crude fiber	%		7.8	26.1	
Ether extract	%		0.8	2.8	
N-free extract	%		15.2	51.1	
Protein (N x 6.25)	%		3.4	11.5	
Cattle	dig prot %	*	2.3	7.7	
Goats	dig prot %	*	2.2	7.3	
Horses	dig prot %	*	2.2	7.3	
Rabbits	dig prot %	*	2.3	7.6	
Sheep	dig prot %	*	2.3	7.7	
Energy	GE Mcal/kg				
Cattle	DE Mcal/kg	*	0.73	2.46	
Sheep	DE Mcal/kg	*	0.76	2.54	
Cattle	ME Mcal/kg	*	0.60	2.02	
Sheep	ME Mcal/kg	*	0.62	2.08	
Cattle	NE_m Mcal/kg	*	0.41	1.38	
Cattle	NE_{gain} Mcal/kg	*	0.24	0.79	
Cattle	TDN %	*	16.6	55.7	
Sheep	TDN %	*	17.1	57.5	
Calcium	%		0.16	0.52	23
Copper	mg/kg		11.0	36.8	
Iron	%		0.021	0.070	8
Manganese	mg/kg		24.9	83.4	17
Phosphorus	%		0.05	0.16	60
Potassium	%		0.40	1.35	10
Carotene	mg/kg		65.4	219.1	35
Vitamin A equivalent	IU/g		109.0	365.3	

Bluestem, aerial part, fresh, early bloom, (2)
Ref No 2 00 822 United States

Feed Name or Analyses			Mean As Fed	Mean Dry	C.V. ± %
Dry matter	%		34.2	100.0	
Ash	%		3.3	9.6	19

Feed Name or Analyses			Mean As Fed	Mean Dry	C.V. ± %
Crude fiber	%		10.3	30.2	7
Ether extract	%		1.0	2.8	30
N-free extract	%		16.0	46.9	
Protein (N x 6.25)	%		3.6	10.5	21
Cattle	dig prot %	*	2.3	6.8	
Goats	dig prot %	*	2.2	6.4	
Horses	dig prot %	*	2.2	6.4	
Rabbits	dig prot %	*	2.3	6.8	
Sheep	dig prot %	*	2.3	6.8	
Energy	GE Mcal/kg				
Cattle	DE Mcal/kg	*	0.96	2.80	
Sheep	DE Mcal/kg	*	0.98	2.87	
Cattle	ME Mcal/kg	*	0.79	2.30	
Sheep	ME Mcal/kg	*	0.80	2.35	
Cattle	TDN %	*	21.7	63.4	
Sheep	TDN %	*	22.3	65.1	
Carotene	mg/kg		64.5	188.9	
Vitamin A equivalent	IU/g		107.6	315.0	

Bluestem, aerial part, fresh, mid-bloom, (2)

Ref No 2 00 823 United States

Feed Name or Analyses			Mean As Fed	Mean Dry	C.V. ± %
Dry matter	%			100.0	
Ash	%			10.1	12
Crude fiber	%			31.0	12
Ether extract	%			2.2	20
N-free extract	%			49.3	
Protein (N x 6.25)	%			7.4	16
Cattle	dig prot %	*		4.2	
Goats	dig prot %	*		3.5	
Horses	dig prot %	*		3.8	
Rabbits	dig prot %	*		4.4	
Sheep	dig prot %	*		3.9	
Energy	GE Mcal/kg				
Cattle	DE Mcal/kg	*		2.44	
Sheep	DE Mcal/kg	*		2.52	
Cattle	ME Mcal/kg	*		2.00	
Sheep	ME Mcal/kg	*		2.07	
Cattle	TDN %	*		55.3	
Sheep	TDN %	*		57.2	
Calcium	%			0.59	17
Phosphorus	%			0.16	45

Bluestem, aerial part, fresh, full bloom, (2)

Ref No 2 00 824 United States

Feed Name or Analyses			Mean As Fed	Mean Dry	C.V. ± %
Dry matter	%		70.4	100.0	
Calcium	%		0.31	0.44	28
Phosphorus	%		0.09	0.13	77
Carotene	mg/kg		55.7	79.1	15
Vitamin A equivalent	IU/g		92.9	131.9	

Bluestem, aerial part, fresh, mature, (2)

Ref No 2 00 825 United States

Feed Name or Analyses			Mean As Fed	Mean Dry	C.V. ± %
Dry matter	%		61.1	100.0	20
Ash	%		2.4	3.9	
Crude fiber	%		21.0	34.4	
Ether extract	%		1.7	2.8	
N-free extract	%		31.6	51.8	
Protein (N x 6.25)	%		4.3	7.1	
Cattle	dig prot %	*	2.4	3.9	
Goats	dig prot %	*	1.9	3.2	
Horses	dig prot %	*	2.2	3.5	
Rabbits	dig prot %	*	2.5	4.1	
Sheep	dig prot %	*	2.2	3.6	
Energy	GE Mcal/kg				
Cattle	DE Mcal/kg	*	1.45	2.37	
Sheep	DE Mcal/kg	*	1.48	2.43	
Cattle	ME Mcal/kg	*	1.18	1.94	
Sheep	ME Mcal/kg	*	1.22	1.99	
Cattle	NE_m Mcal/kg	*	0.81	1.33	
Cattle	NE_{gain} Mcal/kg	*	0.45	0.73	
Cattle	TDN %	*	32.8	53.7	
Sheep	TDN %	*	33.7	55.1	
Calcium	%		0.24	0.40	15
Copper	mg/kg		9.8	16.1	
Iron	%		0.040	0.065	
Magnesium	%		0.04	0.06	
Manganese	mg/kg		22.5	36.9	
Phosphorus	%		0.08	0.12	44
Potassium	%		0.31	0.51	

Bluestem, aerial part, fresh, over ripe, (2)

Ref No 2 00 826 United States

Feed Name or Analyses			Mean As Fed	Mean Dry	C.V. ± %
Dry matter	%		79.7	100.0	6
Calcium	%		0.24	0.30	21
Copper	mg/kg		17.6	22.0	
Iron	%		0.031	0.039	
Manganese	mg/kg		37.2	46.7	
Phosphorus	%		0.06	0.08	74
Carotene	mg/kg		2.1	2.6	40
Vitamin A equivalent	IU/g		3.5	4.4	

Bluestem, aerial part, fresh weathered, mature, (2)

Ref No 2 08 358 United States

Feed Name or Analyses			Mean As Fed	Mean Dry	C.V. ± %
Dry matter	%		84.7	100.0	
Ash	%		7.1	8.4	
Crude fiber	%		29.7	35.1	
Ether extract	%		1.6	1.9	
N-free extract	%		43.6	51.5	
Protein (N x 6.25)	%		2.7	3.2	
Cattle	dig prot %	*	0.5	0.6	
Goats	dig prot %	*	-0.3	-0.4	
Horses	dig prot %	*	0.2	0.2	
Rabbits	dig prot %	*	1.0	1.1	
Sheep	dig prot %	*	0.0	0.0	
Energy	GE Mcal/kg				
Cattle	DE Mcal/kg	*	1.74	2.05	
Sheep	DE Mcal/kg	*	1.60	1.89	
Cattle	ME Mcal/kg	*	1.42	1.68	
Sheep	ME Mcal/kg	*	1.31	1.55	
Cattle	TDN %	*	39.4	46.5	
Sheep	TDN %	*	36.2	42.8	
Calcium	%		0.36	0.43	
Copper	mg/kg		18.7	22.1	
Iron	%		0.033	0.039	
Manganese	mg/kg		39.7	46.9	
Phosphorus	%		0.05	0.06	

Bluestem, shoots, fresh, (2)

Ref No 2 00 828 United States

Feed Name or Analyses			Mean As Fed	Mean Dry	C.V. ± %
Dry matter	%			100.0	
Alanine	%			1.80	
Arginine	%			3.00	
Aspartic acid	%			2.50	
Glutamic acid	%			1.60	
Histidine	%			4.00	
Leucine	%			2.70	
Lysine	%			3.00	
Proline	%			2.00	
Serine	%			4.00	
Threonine	%			1.00	
Valine	%			1.50	

Bluestem, aerial part, ensiled, (3)

Ref No 3 00 829 United States

Feed Name or Analyses			Mean As Fed	Mean Dry	C.V. ± %
Dry matter	%		35.3	100.0	
Carotene	mg/kg		8.5	24.0	
Vitamin A equivalent	IU/g		14.1	40.1	

BLUESTEM, BIG. Andropogon gerardii

Bluestem, big, hay, s-c, (1)

Ref No 1 00 830 United States

Feed Name or Analyses			Mean As Fed	Mean Dry	C.V. ± %
Dry matter	%			100.0	
Carotene	mg/kg			37.0	71
Vitamin A equivalent	IU/g			61.7	

Bluestem, big, aerial part, fresh, (2)

Ref No 2 00 837 United States

Feed Name or Analyses			Mean As Fed	Mean Dry	C.V. ± %
Dry matter	%		54.6	100.0	24
Ash	%		3.5	6.4	12
Crude fiber	%		18.4	33.7	8
Ether extract	%		1.3	2.3	15
N-free extract	%		28.2	51.7	3
Protein (N x 6.25)	%		3.2	5.9	18
Cattle	dig prot %	*	1.6	2.9	
Goats	dig prot %	*	1.1	2.0	
Horses	dig prot %	*	1.4	2.5	
Rabbits	dig prot %	*	1.7	3.2	
Sheep	dig prot %	*	1.3	2.4	
Energy	GE Mcal/kg				
Cattle	DE Mcal/kg	*	1.31	2.40	
Sheep	DE Mcal/kg	*	1.40	2.57	
Cattle	ME Mcal/kg	*	1.08	1.97	
Sheep	ME Mcal/kg	*	1.15	2.11	
Cattle	TDN %	*	29.7	54.4	
Sheep	TDN %	*	31.8	58.2	
Calcium	%		0.17	0.32	11
Phosphorus	%		0.05	0.08	23
Carotene	mg/kg		61.6	113.0	54
Vitamin A equivalent	IU/g		102.7	188.3	

Bluestem, big, aerial part, fresh, immature, (2)

Ref No 2 00 831 United States

Feed Name or Analyses			Mean As Fed	Mean Dry	C.V. ± %
Dry matter	%		21.6	100.0	
Ash	%		1.9	8.7	
Crude fiber	%		5.9	27.5	
Ether extract	%		0.6	2.9	
N-free extract	%		9.8	45.3	
Protein (N x 6.25)	%		3.4	15.6	
Cattle	dig prot %	*	2.4	11.2	
Goats	dig prot %	*	2.4	11.1	
Horses	dig prot %	*	2.3	10.8	
Rabbits	dig prot %	*	2.3	10.7	
Sheep	dig prot %	*	2.5	11.5	
Energy	GE Mcal/kg				
Cattle	DE Mcal/kg	*	0.57	2.64	
Sheep	DE Mcal/kg	*	0.54	2.49	
Cattle	ME Mcal/kg	*	0.47	2.16	

Continued

Feed Name or Analyses		Mean As Fed	Mean Dry	C.V. ± %
Sheep	ME Mcal/kg *	0.44	2.04	
Cattle	TDN % *	12.9	59.8	
Sheep	TDN % *	12.2	56.4	
Calcium	%	0.09	0.43	
Phosphorus	%	0.05	0.21	51
Carotene	mg/kg	49.0	226.6	25
Vitamin A equivalent	IU/g	81.6	377.8	

Bluestem, big, aerial part, fresh, early bloom, (2)
Ref No 2 00 832 United States

Feed Name or Analyses		Mean As Fed	Mean Dry	C.V. ± %
Dry matter	%		100.0	
Ash	%		9.5	18
Crude fiber	%		30.8	5
Ether extract	%		2.6	16
N-free extract	%		47.0	
Protein (N x 6.25)	%		10.1	10
Cattle	dig prot % *		6.5	
Goats	dig prot % *		6.0	
Horses	dig prot % *		6.1	
Rabbits	dig prot % *		6.5	
Sheep	dig prot % *		6.4	
Calcium	%		0.41	12
Phosphorus	%		0.17	20

Bluestem, big, aerial part, fresh, full bloom, (2)
Ref No 2 00 833 United States

Feed Name or Analyses		Mean As Fed	Mean Dry	C.V. ± %
Dry matter	%	67.2	100.0	
Ash	%	4.4	6.6	1
Crude fiber	%	23.0	34.3	7
Ether extract	%	1.5	2.3	19
N-free extract	%	33.7	50.2	
Protein (N x 6.25)	%	4.4	6.6	15
Cattle	dig prot % *	2.4	3.5	
Goats	dig prot % *	1.8	2.7	
Horses	dig prot % *	2.1	3.1	
Rabbits	dig prot % *	2.5	3.8	
Sheep	dig prot % *	2.1	3.1	
Energy	GE Mcal/kg			
Cattle	DE Mcal/kg *	1.63	2.42	
Sheep	DE Mcal/kg *	1.69	2.52	
Cattle	ME Mcal/kg *	1.33	1.98	
Sheep	ME Mcal/kg *	1.39	2.07	
Cattle	TDN % *	37.0	55.0	
Sheep	TDN % *	38.4	57.2	
Calcium	%	0.22	0.32	
Phosphorus	%	0.07	0.10	
Carotene	mg/kg	55.3	82.2	1
Vitamin A equivalent	IU/g	92.1	137.1	

Bluestem, big, aerial part, fresh, milk stage, (2)
Ref No 2 00 834 United States

Feed Name or Analyses		Mean As Fed	Mean Dry	C.V. ± %
Dry matter	%		100.0	
Calcium	%		0.25	
Phosphorus	%		0.12	

(1) dry forages and roughages (2) pasture, range plants, and forages fed green (3) silages (4) energy feeds (5) protein supplements (6) minerals (7) vitamins (8) additives

Bluestem, big, aerial part, fresh, mature, (2)
Ref No 2 00 835 United States

Feed Name or Analyses		Mean As Fed	Mean Dry	C.V. ± %
Dry matter	%	46.8	100.0	
Ash	%	4.1	8.8	23
Crude fiber	%	15.3	32.6	13
Ether extract	%	1.0	2.2	15
N-free extract	%	24.1	51.5	
Protein (N x 6.25)	%	2.3	4.9	17
Cattle	dig prot % *	1.0	2.1	
Goats	dig prot % *	0.5	1.1	
Horses	dig prot % *	0.8	1.7	
Rabbits	dig prot % *	1.1	2.5	
Sheep	dig prot % *	0.7	1.6	
Energy	GE Mcal/kg			
Cattle	DE Mcal/kg *	1.11	2.37	
Sheep	DE Mcal/kg *	1.16	2.47	
Cattle	ME Mcal/kg *	0.91	1.94	
Sheep	ME Mcal/kg *	0.95	2.03	
Cattle	TDN % *	25.1	53.7	
Sheep	TDN % *	26.3	56.1	

Bluestem, big, aerial part, fresh, over ripe, (2)
Ref No 2 00 836 United States

Feed Name or Analyses		Mean As Fed	Mean Dry	C.V. ± %
Dry matter	%		100.0	
Ash	%		6.9	
Crude fiber	%		36.5	
Ether extract	%		1.8	
N-free extract	%		51.9	
Protein (N x 6.25)	%		2.9	
Cattle	dig prot % *		0.4	
Goats	dig prot % *		-0.6	
Horses	dig prot % *		0.0	
Rabbits	dig prot % *		0.9	
Sheep	dig prot % *		-0.2	
Energy	GE Mcal/kg			
Cattle	DE Mcal/kg *		2.35	
Sheep	DE Mcal/kg *		2.47	
Cattle	ME Mcal/kg *		1.93	
Sheep	ME Mcal/kg *		2.03	
Cattle	TDN % *		53.3	
Sheep	TDN % *		56.0	
Calcium	%		0.29	
Phosphorus	%		0.11	
Carotene	mg/kg		3.7	26
Vitamin A equivalent	IU/g		6.2	

BLUESTEM, BUSHY. Andropogon glomeratus

Bluestem, bushy, aerial part, fresh, immature, (2)
Ref No 2 00 838 United States

Feed Name or Analyses		Mean As Fed	Mean Dry	C.V. ± %
Dry matter	%		100.0	
Ash	%		10.6	
Crude fiber	%		31.3	
Ether extract	%		2.1	
N-free extract	%		45.4	
Protein (N x 6.25)	%		10.6	
Cattle	dig prot % *		6.9	
Goats	dig prot % *		6.4	
Horses	dig prot % *		6.5	
Rabbits	dig prot % *		6.9	
Sheep	dig prot % *		6.9	
Energy	GE Mcal/kg			
Cattle	DE Mcal/kg *		2.42	
Sheep	DE Mcal/kg *		2.51	
Cattle	ME Mcal/kg *		1.98	
Sheep	ME Mcal/kg *		2.06	
Cattle	TDN % *		54.9	
Sheep	TDN % *		57.0	
Calcium	%		0.46	12
Phosphorus	%		0.20	18

Bluestem, bushy, aerial part, fresh, mature, (2)
Ref No 2 00 839 United States

Feed Name or Analyses		Mean As Fed	Mean Dry	C.V. ± %
Dry matter	%		100.0	
Ash	%		7.4	
Crude fiber	%		34.1	
Ether extract	%		1.5	
N-free extract	%		52.4	
Protein (N x 6.25)	%		4.6	
Cattle	dig prot % *		1.8	
Goats	dig prot % *		0.9	
Horses	dig prot % *		1.4	
Rabbits	dig prot % *		2.2	
Sheep	dig prot % *		1.3	
Energy	GE Mcal/kg			
Cattle	DE Mcal/kg *		2.37	
Sheep	DE Mcal/kg *		2.47	
Cattle	ME Mcal/kg *		1.94	
Sheep	ME Mcal/kg *		2.03	
Cattle	TDN % *		53.8	
Sheep	TDN % *		56.1	
Calcium	%		0.30	
Phosphorus	%		0.11	

BLUESTEM, CANE. Andropogon barbinodis

Bluestem, cane, aerial part, fresh, full bloom, (2)
Ref No 2 00 840 United States

Feed Name or Analyses		Mean As Fed	Mean Dry	C.V. ± %
Dry matter	%		100.0	
Ash	%		8.5	
Crude fiber	%		33.0	
Ether extract	%		2.1	
N-free extract	%		49.2	
Protein (N x 6.25)	%		7.2	
Cattle	dig prot % *		4.0	
Goats	dig prot % *		3.3	
Horses	dig prot % *		3.6	
Rabbits	dig prot % *		4.2	
Sheep	dig prot % *		3.7	
Energy	GE Mcal/kg			
Cattle	DE Mcal/kg *		2.60	
Cattle	ME Mcal/kg *		2.13	
Sheep	ME Mcal/kg *		2.06	
Cattle	TDN % *		58.9	
Sheep	TDN % *		56.9	
Calcium	%		0.69	
Phosphorus	%		0.39	

BLUESTEM, INDIA. Andropogon carricosus

Bluestem, India, aerial part, fresh, late bloom, (2)
Ref No 2 00 841 United States

Feed Name or Analyses		Mean As Fed	Mean Dry	C.V. ± %
Dry matter	%	20.0	100.0	
Ash	%	0.9	4.3	
Crude fiber	%	7.9	39.7	
Sheep	dig coef %	72.	72.	
Ether extract	%	0.3	1.7	
Sheep	dig coef %	60.	60.	

Feed Name or Analyses			Mean As Fed	Mean Dry	C.V. ± %
N-free extract	%		9.3	46.7	
Sheep	dig coef %		62.	62.	
Protein (N x 6.25)	%		1.5	7.6	
Sheep	dig coef %		50.	50.	
Cattle	dig prot %	*	0.9	4.4	
Goats	dig prot %	*	0.7	3.7	
Horses	dig prot %	*	0.8	4.0	
Rabbits	dig prot %	*	0.9	4.5	
Sheep	dig prot %		0.8	3.8	
Energy	GE Mcal/kg				
Cattle	DE Mcal/kg	*	0.54	2.71	
Sheep	DE Mcal/kg	*	0.56	2.81	
Cattle	ME Mcal/kg	*	0.44	2.22	
Sheep	ME Mcal/kg	*	0.46	2.30	
Cattle	TDN %	*	12.3	61.5	
Sheep	TDN %		12.7	63.6	

BLUESTEM, LITTLE. Andropogon scoparius

Bluestem, little, hay, s-c, (1)

Ref No 1 00 842 United States

Feed Name or Analyses			Mean As Fed	Mean Dry	C.V. ± %
Dry matter	%			100.0	
Calcium	%			0.48	15
Phosphorus	%			0.07	28
Carotene	mg/kg			36.2	60
Vitamin A equivalent	IU/g			60.3	

Bluestem, little, aerial part, fresh, (2)

Ref No 2 00 849 United States

Feed Name or Analyses			Mean As Fed	Mean Dry	C.V. ± %
Dry matter	%		58.4	100.0	22
Ash	%		3.8	6.5	13
Crude fiber	%		19.5	33.3	7
Ether extract	%		1.3	2.2	17
N-free extract	%		30.7	52.6	2
Protein (N x 6.25)	%		3.2	5.5	20
Cattle	dig prot %	*	1.5	2.5	
Goats	dig prot %	*	1.0	1.6	
Horses	dig prot %	*	1.3	2.2	
Rabbits	dig prot %	*	1.7	2.9	
Sheep	dig prot %	*	1.2	2.1	
Energy	GE Mcal/kg				
Cattle	DE Mcal/kg	*	1.39	2.38	
Sheep	DE Mcal/kg	*	1.51	2.58	
Cattle	ME Mcal/kg	*	1.74	1.95	
Sheep	ME Mcal/kg	*	1.24	2.12	
Cattle	TDN %	*	31.5	53.9	
Sheep	TDN %	*	34.3	58.7	
Calcium	%		0.19	0.32	16
Phosphorus	%		0.04	0.07	23
Carotene	mg/kg		55.7	95.3	55
Vitamin A equivalent	IU/g		92.8	158.9	

Bluestem, little, aerial part, fresh, immature, (2)

Ref No 2 00 843 United States

Feed Name or Analyses			Mean As Fed	Mean Dry	C.V. ± %
Dry matter	%		29.4	100.0	
Calcium	%		0.13	0.44	14
Phosphorus	%		0.04	0.13	31

Bluestem, little, aerial part, fresh, early bloom, (2)

Ref No 2 00 844 United States

Feed Name or Analyses			Mean As Fed	Mean Dry	C.V. ± %
Dry matter	%		27.7	100.0	
Ash	%		2.4	8.5	
Crude fiber	%		7.2	26.0	
Ether extract	%		0.7	2.5	
N-free extract	%		13.0	47.0	
Protein (N x 6.25)	%		4.4	16.0	
Cattle	dig prot %	*	3.2	11.5	
Goats	dig prot %	*	3.2	11.5	
Horses	dig prot %	*	3.1	11.1	
Rabbits	dig prot %	*	3.1	11.0	
Sheep	dig prot %	*	3.3	11.9	
Energy	GE Mcal/kg				
Cattle	DE Mcal/kg	*	0.71	2.57	
Sheep	DE Mcal/kg	*	0.65	2.34	
Cattle	ME Mcal/kg	*	0.58	2.11	
Sheep	ME Mcal/kg	*	0.53	1.92	
Cattle	TDN %	*	16.1	58.2	
Sheep	TDN %	*	14.7	53.1	
Calcium	%		0.11	0.38	
Phosphorus	%		0.03	0.11	
Carotene	mg/kg		43.2	156.1	
Vitamin A equivalent	IU/g		72.1	260.2	

Bluestem, little, aerial part, fresh, mid-bloom, (2)

Ref No 2 00 845 United States

Feed Name or Analyses			Mean As Fed	Mean Dry	C.V. ± %
Dry matter	%			100.0	
Ash	%			9.9	
Crude fiber	%			31.6	
Ether extract	%			2.0	
N-free extract	%			49.3	
Protein (N x 6.25)	%			7.2	
Cattle	dig prot %	*		4.0	
Goats	dig prot %	*		3.3	
Horses	dig prot %	*		3.6	
Rabbits	dig prot %	*		4.2	
Sheep	dig prot %	*		3.7	
Energy	GE Mcal/kg				
Cattle	DE Mcal/kg	*		2.34	
Sheep	DE Mcal/kg	*		2.52	
Cattle	ME Mcal/kg	*		1.92	
Sheep	ME Mcal/kg	*		2.07	
Cattle	TDN %	*		53.1	
Sheep	TDN %	*		57.2	

Bluestem, little, aerial part, fresh, full bloom, (2)

Ref No 2 00 846 United States

Feed Name or Analyses			Mean As Fed	Mean Dry	C.V. ± %
Dry matter	%		62.2	100.0	
Calcium	%		0.21	0.34	22
Phosphorus	%		0.06	0.09	70
Carotene	mg/kg		27.0	43.4	
Vitamin A equivalent	IU/g		45.0	72.4	

Bluestem, little, aerial part, fresh, mature, (2)

Ref No 2 00 847 United States

Feed Name or Analyses			Mean As Fed	Mean Dry	C.V. ± %
Dry matter	%		51.8	100.0	
Calcium	%		0.20	0.39	19
Phosphorus	%		0.04	0.08	63

Bluestem, little, aerial part, fresh, over ripe, (2)

Ref No 2 00 848 United States

Feed Name or Analyses			Mean As Fed	Mean Dry	C.V. ± %
Dry matter	%			100.0	
Calcium	%			0.33	16
Phosphorus	%			0.07	67
Carotene	mg/kg			1.1	
Vitamin A equivalent	IU/g			1.8	

BLUESTEM, PITTED. Andropogon pertusus

Bluestem, pitted, hay, s-c, late bloom, (1)

Ref No 1 00 850 United States

Feed Name or Analyses			Mean As Fed	Mean Dry	C.V. ± %
Dry matter	%		94.3	100.0	
Ash	%		10.1	10.7	
Crude fiber	%		37.0	39.2	
Cattle	dig coef %		60.	60.	
Ether extract	%		1.1	1.2	
Cattle	dig coef %		21.	21.	
N-free extract	%		43.3	45.9	
Cattle	dig coef %		48.	48.	
Protein (N x 6.25)	%		2.8	3.0	
Cattle	dig coef %		2.	2.	
Cattle	dig prot %		0.1	0.1	
Goats	dig prot %	*	-0.5	-0.5	
Horses	dig prot %	*	0.1	0.1	
Rabbits	dig prot %	*	0.9	1.0	
Sheep	dig prot %	*	-0.6	-0.6	
Energy	GE Mcal/kg				
Cattle	DE Mcal/kg	*	1.92	2.04	
Sheep	DE Mcal/kg	*	2.10	2.23	
Cattle	ME Mcal/kg	*	1.57	1.67	
Sheep	ME Mcal/kg	*	1.72	1.82	
Cattle	TDN %		43.5	46.2	
Sheep	TDN %	*	47.6	50.5	

BLUESTEM, SAND. Andropogon hallii

Bluestem, sand, aerial part, fresh, (2)

Ref No 2 00 856 United States

Feed Name or Analyses			Mean As Fed	Mean Dry	C.V. ± %
Dry matter	%		35.8	100.0	
Ash	%		2.7	7.6	39
Crude fiber	%		12.0	33.5	20
Ether extract	%		0.8	2.3	19
N-free extract	%		18.2	50.8	
Protein (N x 6.25)	%		2.1	5.8	56
Cattle	dig prot %	*	1.0	2.8	
Goats	dig prot %	*	0.7	2.0	
Horses	dig prot %	*	0.9	2.5	
Rabbits	dig prot %	*	1.1	3.1	
Sheep	dig prot %	*	0.9	2.4	
Energy	GE Mcal/kg				
Cattle	DE Mcal/kg	*	0.91	2.53	
Sheep	DE Mcal/kg	*	0.93	2.61	
Cattle	ME Mcal/kg	*	0.74	2.07	
Sheep	ME Mcal/kg	*	0.77	2.14	
Cattle	TDN %	*	20.5	57.3	
Sheep	TDN %	*	21.2	59.1	
Calcium	%		0.15	0.43	40
Phosphorus	%		0.05	0.13	42

Bluestem, sand, aerial part, fresh, immature, (2)

Ref No 2 00 851 United States

Feed Name or Analyses			Mean As Fed	Mean Dry	C.V. ± %
Dry matter	%		24.1	100.0	
Ash	%		2.7	11.2	
Crude fiber	%		6.9	28.7	
Ether extract	%		0.7	2.7	
N-free extract	%		11.1	45.9	

Continued

Feed Name or Analyses			Mean As Fed	Mean Dry	C.V. ± %
Protein (N x 6.25)	%		2.8	11.5	
Cattle	dig prot %	*	1.8	7.7	
Goats	dig prot %	*	1.8	7.3	
Horses	dig prot %	*	1.8	7.3	
Rabbits	dig prot %	*	1.8	7.5	
Sheep	dig prot %	*	1.9	7.7	
Energy	GE Mcal/kg				
Cattle	DE Mcal/kg	*	0.66	2.73	
Sheep	DE Mcal/kg	*	0.69	2.85	
Cattle	ME Mcal/kg	*	0.54	2.24	
Sheep	ME Mcal/kg	*	0.56	2.34	
Cattle	TDN %	*	14.9	62.0	
Sheep	TDN %	*	15.6	64.6	

Bluestem, sand, aerial part, fresh, mid-bloom, (2)
Ref No 2 00 852 United States

Feed Name or Analyses			Mean As Fed	Mean Dry	C.V. ± %
Dry matter	%			100.0	
Calcium	%			0.74	
Phosphorus	%			0.20	

Bluestem, sand, aerial part, fresh, full bloom, (2)
Ref No 2 00 853 United States

Feed Name or Analyses			Mean As Fed	Mean Dry	C.V. ± %
Dry matter	%			100.0	
Ash	%			8.5	
Crude fiber	%			34.6	
Ether extract	%			2.7	
N-free extract	%			49.0	
Protein (N x 6.25)	%			5.2	
Cattle	dig prot %	*		2.3	
Goats	dig prot %	*		1.4	
Horses	dig prot %	*		1.9	
Rabbits	dig prot %	*		2.7	
Sheep	dig prot %	*		1.8	
Energy	GE Mcal/kg				
Cattle	DE Mcal/kg	*		2.44	
Sheep	DE Mcal/kg	*		2.54	
Cattle	ME Mcal/kg	*		2.00	
Sheep	ME Mcal/kg	*		2.08	
Cattle	TDN %	*		55.4	
Sheep	TDN %	*		57.6	
Calcium	%			0.62	
Phosphorus	%			0.11	

Bluestem, sand, aerial part, fresh, mature, (2)
Ref No 2 00 854 United States

Feed Name or Analyses			Mean As Fed	Mean Dry	C.V. ± %
Dry matter	%		41.7	100.0	
Ash	%		2.6	6.2	56
Crude fiber	%		14.6	35.1	5
Ether extract	%		1.0	2.3	13
N-free extract	%		22.0	52.7	
Protein (N x 6.25)	%		1.5	3.7	8
Cattle	dig prot %	*	0.4	1.0	
Goats	dig prot %	*	0.0	0.0	
Horses	dig prot %	*	0.3	0.7	
Rabbits	dig prot %	*	0.6	1.5	
Sheep	dig prot %	*	0.2	0.4	
Energy	GE Mcal/kg				
Cattle	DE Mcal/kg	*	1.00	2.41	
Sheep	DE Mcal/kg	*	1.09	2.61	
Cattle	ME Mcal/kg	*	0.83	1.98	
Sheep	ME Mcal/kg	*	0.89	2.14	
Cattle	TDN %	*	22.8	54.7	
Sheep	TDN %	*	24.7	59.3	
Calcium	%		0.12	0.29	13
Phosphorus	%		0.05	0.11	9

Bluestem, sand, aerial part, fresh, over ripe, (2)
Ref No 2 00 855 United States

Feed Name or Analyses			Mean As Fed	Mean Dry	C.V. ± %
Dry matter	%			100.0	
Ash	%			3.5	
Crude fiber	%			42.7	
Ether extract	%			1.7	
N-free extract	%			50.1	
Protein (N x 6.25)	%			2.0	
Cattle	dig prot %	*		-0.3	
Goats	dig prot %	*		-1.5	
Horses	dig prot %	*		-0.7	
Rabbits	dig prot %	*		0.2	
Sheep	dig prot %	*		-1.0	
Energy	GE Mcal/kg				
Cattle	DE Mcal/kg	*		2.51	
Cattle	ME Mcal/kg	*		2.55	
Cattle	ME Mcal/kg	*		2.06	
Sheep	ME Mcal/kg	*		2.09	
Cattle	TDN %	*		57.0	
Sheep	TDN %	*		57.9	
Calcium	%			0.24	
Phosphorus	%			0.04	

BLUESTEM, SILVER. Andropogon saccharoides

Bluestem, silver, aerial part, fresh, (2)
Ref No 2 00 861 United States

Feed Name or Analyses			Mean As Fed	Mean Dry	C.V. ± %
Dry matter	%			100.0	
Ash	%			10.8	12
Crude fiber	%			33.8	6
Ether extract	%			1.8	12
N-free extract	%			46.5	
Protein (N x 6.25)	%			7.1	33
Cattle	dig prot %	*		3.9	
Goats	dig prot %	*		3.2	
Horses	dig prot %	*		3.6	
Rabbits	dig prot %	*		4.2	
Sheep	dig prot %	*		3.6	
Energy	GE Mcal/kg				
Cattle	DE Mcal/kg	*		2.39	
Sheep	DE Mcal/kg	*		2.47	
Cattle	ME Mcal/kg	*		1.95	
Sheep	ME Mcal/kg	*		2.03	
Cattle	TDN %	*		54.1	
Sheep	TDN %	*		56.0	
Calcium	%			0.49	14
Magnesium	%			0.11	
Phosphorus	%			0.15	28
Potassium	%			1.09	

Bluestem, silver, aerial part, fresh, immature, (2)
Ref No 2 00 857 United States

Feed Name or Analyses			Mean As Fed	Mean Dry	C.V. ± %
Dry matter	%			100.0	
Ash	%			12.6	13
Crude fiber	%			30.3	4
Ether extract	%			2.1	8
N-free extract	%			43.2	
Protein (N x 6.25)	%			11.8	10
Cattle	dig prot %	*		7.9	
Goats	dig prot %	*		7.6	
Horses	dig prot %	*		7.5	
Rabbits	dig prot %	*		7.8	
Sheep	dig prot %	*		8.0	
Energy	GE Mcal/kg				
Cattle	DE Mcal/kg	*		2.27	
Sheep	DE Mcal/kg	*		2.36	
Cattle	ME Mcal/kg	*		1.86	
Sheep	ME Mcal/kg	*		1.86	
Sheep	ME Mcal/kg	*		1.94	
Cattle	TDN %	*		51.4	
Sheep	TDN %	*		53.6	
Calcium	%			0.57	42
Phosphorus	%			0.22	5

Bluestem, silver, aerial part, fresh, mid–bloom, (2)
Ref No 2 00 858 United States

Feed Name or Analyses			Mean As Fed	Mean Dry	C.V. ± %
Dry matter	%			100.0	
Ash	%			10.7	1
Crude fiber	%			34.2	1
Ether extract	%			1.9	9
N-free extract	%			46.6	
Protein (N x 6.25)	%			6.6	7
Cattle	dig prot %	*		3.5	
Goats	dig prot %	*		2.7	
Horses	dig prot %	*		3.1	
Rabbits	dig prot %	*		3.8	
Sheep	dig prot %	*		3.1	
Energy	GE Mcal/kg				
Cattle	DE Mcal/kg	*		2.34	
Sheep	DE Mcal/kg	*		2.42	
Cattle	ME Mcal/kg	*		1.92	
Sheep	ME Mcal/kg	*		1.98	
Cattle	TDN %	*		53.0	
Sheep	TDN %	*		54.8	
Calcium	%			0.53	13
Phosphorus	%			0.17	10

Bluestem, silver, aerial part, fresh, full bloom, (2)
Ref No 2 00 859 United States

Feed Name or Analyses			Mean As Fed	Mean Dry	C.V. ± %
Dry matter	%			100.0	
Ash	%			9.8	3
Crude fiber	%			34.7	0
Ether extract	%			1.9	6
N-free extract	%			47.9	
Protein (N x 6.25)	%			5.7	5
Cattle	dig prot %	*		2.7	
Goats	dig prot %	*		1.9	
Horses	dig prot %	*		2.4	
Rabbits	dig prot %	*		3.1	
Sheep	dig prot %	*		2.3	
Energy	GE Mcal/kg				
Cattle	DE Mcal/kg	*		2.39	
Sheep	DE Mcal/kg	*		2.55	
Cattle	ME Mcal/kg	*		1.96	
Sheep	ME Mcal/kg	*		2.09	
Cattle	TDN %	*		54.2	
Sheep	TDN %	*		57.9	
Calcium	%			0.46	14
Phosphorus	%			0.12	7

(1) dry forages and roughages (3) silages (6) minerals
(2) pasture, range plants, and forages fed green (4) energy feeds (7) vitamins
(5) protein supplements (8) additives

Bluestem, silver, aerial part, fresh, mature, (2)

Ref No 2 00 860 United States

Feed Name or Analyses			As Fed	Dry	C.V. ± %
Dry matter	%			100.0	
Ash	%			9.7	4
Crude fiber	%			36.5	5
Ether extract	%			1.5	10
N-free extract	%			48.8	
Protein (N x 6.25)	%			3.5	7
Cattle	dig prot %	*		0.9	
Goats	dig prot %	*		-0.1	
Horses	dig prot %	*		0.5	
Rabbits	dig prot %	*		1.4	
Sheep	dig prot %	*		0.3	
Energy	GE Mcal/kg				
Cattle	DE Mcal/kg	*		2.37	
Sheep	DE Mcal/kg	*		2.56	
Cattle	ME Mcal/kg	*		1.94	
Sheep	ME Mcal/kg	*		2.10	
Cattle	TDN %	*		53.7	
Sheep	TDN %	*		58.0	
Calcium	%			0.44	8
Magnesium	%			0.00	8
Phosphorus	%			0.08	19
Potassium	%			0.49	

BLUESTEM, SLENDER. Andropogon tener

Bluestem, slender, aerial part, fresh, (2)

Ref No 2 00 866 United States

Feed Name or Analyses			As Fed	Dry	C.V. ± %
Dry matter	%			100.0	
Ash	%			9.0	5
Crude fiber	%			33.8	5
Ether extract	%			2.4	15
N-free extract	%			48.5	
Protein (N x 6.25)	%			6.3	30
Cattle	dig prot %	*		3.2	
Goats	dig prot %	*		2.4	
Horses	dig prot %	*		2.9	
Rabbits	dig prot %	*		3.5	
Sheep	dig prot %	*		2.9	
Energy	GE Mcal/kg				
Cattle	DE Mcal/kg	*		2.68	
Sheep	DE Mcal/kg	*		2.57	
Cattle	ME Mcal/kg	*		2.03	
Sheep	ME Mcal/kg	*		2.11	
Cattle	TDN %	*		56.3	
Sheep	TDN %	*		58.4	
Calcium	%			0.49	12
Phosphorus	%			0.10	38

Bluestem, slender, aerial part, fresh, immature, (2)

Ref No 2 00 862 United States

Feed Name or Analyses			As Fed	Dry	C.V. ± %
Dry matter	%			100.0	
Calcium	%			0.53	6
Phosphorus	%			0.15	14

Bluestem, slender, aerial part, fresh, mid–bloom, (2)

Ref No 2 00 863 United States

Feed Name or Analyses			As Fed	Dry	C.V. ± %
Dry matter	%			100.0	
Ash	%			9.3	
Crude fiber	%			32.3	
Ether extract	%			3.0	
N-free extract	%			48.8	
Protein (N x 6.25)	%			6.6	
Cattle	dig prot %	*		3.5	
Goats	dig prot %	*		2.7	
Horses	dig prot %	*		3.1	
Rabbits	dig prot %	*		3.8	
Sheep	dig prot %	*		3.1	
Energy	GE Mcal/kg				
Cattle	DE Mcal/kg	*		2.35	
Sheep	DE Mcal/kg	*		2.54	
Cattle	ME Mcal/kg	*		1.93	
Sheep	ME Mcal/kg	*		2.08	
Cattle	TDN %	*		53.4	
Sheep	TDN %	*		57.6	

Bluestem, slender, aerial part, fresh, full bloom, (2)

Ref No 2 00 864 United States

Feed Name or Analyses			As Fed	Dry	C.V. ± %
Dry matter	%			100.0	
Ash	%			8.9	2
Crude fiber	%			34.6	5
Ether extract	%			2.1	12
N-free extract	%			48.2	
Protein (N x 6.25)	%			6.2	3
Cattle	dig prot %	*		3.2	
Goats	dig prot %	*		2.3	
Horses	dig prot %	*		2.8	
Rabbits	dig prot %	*		3.5	
Sheep	dig prot %	*		2.8	
Energy	GE Mcal/kg				
Cattle	DE Mcal/kg	*		2.31	
Sheep	DE Mcal/kg	*		2.50	
Cattle	ME Mcal/kg	*		1.89	
Sheep	ME Mcal/kg	*		2.05	
Cattle	TDN %	*		52.3	
Sheep	TDN %	*		56.8	
Calcium	%			0.42	4
Phosphorus	%			0.12	25

Bluestem, slender, aerial part, fresh, mature, (2)

Ref No 2 00 865 United States

Feed Name or Analyses			As Fed	Dry	C.V. ± %
Dry matter	%			100.0	
Ash	%			8.6	0
Crude fiber	%			35.4	2
Ether extract	%			2.1	8
N-free extract	%			50.3	
Protein (N x 6.25)	%			3.6	23
Cattle	dig prot %	*		1.0	
Goats	dig prot %	*		0.0	
Horses	dig prot %	*		0.6	
Rabbits	dig prot %	*		1.4	
Sheep	dig prot %	*		0.3	
Energy	GE Mcal/kg				
Cattle	DE Mcal/kg	*		2.61	
Sheep	DE Mcal/kg	*		2.58	
Cattle	ME Mcal/kg	*		2.14	
Sheep	ME Mcal/kg	*		2.12	
Cattle	TDN %	*		59.2	
Sheep	TDN %	*		58.6	
Phosphorus	%			0.05	16

BLUESTEM, TEXAS. Andropogon cirratus

Bluestem, Texas, aerial part, fresh, mature, (2)

Ref No 2 00 867 United States

Feed Name or Analyses			As Fed	Dry	C.V. ± %
Dry matter	%			100.0	
Ash	%			4.8	
Crude fiber	%			36.5	
Ether extract	%			1.8	
N-free extract	%			53.2	
Protein (N x 6.25)	%			3.7	
Cattle	dig prot %	*		1.0	
Goats	dig prot %	*		0.0	
Horses	dig prot %	*		0.7	
Rabbits	dig prot %	*		1.5	
Sheep	dig prot %	*		0.4	
Energy	GE Mcal/kg				
Cattle	DE Mcal/kg	*		2.45	
Sheep	DE Mcal/kg	*		2.61	
Cattle	ME Mcal/kg	*		2.01	
Sheep	ME Mcal/kg	*		2.14	
Cattle	TDN %	*		55.6	
Sheep	TDN %	*		59.2	
Calcium	%			0.29	
Magnesium	%			0.12	
Phosphorus	%			0.07	
Potassium	%			0.52	

BLUESTEM, YELLOW. Andropogon ischaemum

Bluestem, yellow, aerial part, fresh, (2)

Ref No 2 00 871 United States

Feed Name or Analyses			As Fed	Dry	C.V. ± %
Dry matter	%			100.0	
Ash	%			8.5	53
Crude fiber	%			33.5	17
Ether extract	%			2.0	28
N-free extract	%			48.0	
Protein (N x 6.25)	%			8.0	77
Cattle	dig prot %	*		4.7	
Goats	dig prot %	*		4.0	
Horses	dig prot %	*		4.3	
Rabbits	dig prot %	*		4.8	
Sheep	dig prot %	*		4.4	
Energy	GE Mcal/kg				
Cattle	DE Mcal/kg	*		2.39	
Sheep	DE Mcal/kg	*		2.47	
Cattle	ME Mcal/kg	*		1.96	
Sheep	ME Mcal/kg	*		2.03	
Cattle	TDN %	*		54.1	
Sheep	TDN %	*		56.0	
Calcium	%			0.44	62
Phosphorus	%			0.25	77

Bluestem, yellow, aerial part, fresh, immature, (2)

Ref No 2 00 868 United States

Feed Name or Analyses			As Fed	Dry	C.V. ± %
Dry matter	%			100.0	
Ash	%			14.7	
Crude fiber	%			23.9	
Ether extract	%			2.3	
N-free extract	%			40.1	
Protein (N x 6.25)	%			19.0	
Cattle	dig prot %	*		14.0	
Goats	dig prot %	*		14.3	

Continued

Feed Name or Analyses			Mean As Fed	Mean Dry	C.V. ± %
Horses	dig prot %	*		13.7	
Rabbits	dig prot %	*		13.3	
Sheep	dig prot %	*		14.7	
Energy	GE Mcal/kg				
Cattle	DE Mcal/kg	*		2.86	
Sheep	DE Mcal/kg	*		2.53	
Cattle	ME Mcal/kg	*		2.35	
Sheep	ME Mcal/kg	*		2.07	
Cattle	TDN %	*		64.9	
Sheep	TDN %	*		57.3	
Calcium	%			0.88	
Phosphorus	%			0.55	

Bluestem, yellow, aerial part, fresh, mid-bloom, (2)
Ref No 2 00 869 United States

Feed Name or Analyses			Mean As Fed	Mean Dry	C.V. ± %
Dry matter	%			100.0	
Ash	%			12.6	
Crude fiber	%			29.7	
Ether extract	%			1.7	
N-free extract	%			46.5	
Protein (N x 6.25)	%			9.5	
Cattle	dig prot %	*		6.0	
Goats	dig prot %	*		5.4	
Horses	dig prot %	*		5.6	
Rabbits	dig prot %	*		6.0	
Sheep	dig prot %	*		5.8	
Cattle	DE Mcal/kg	*		2.27	
Sheep	DE Mcal/kg	*		2.46	
Cattle	ME Mcal/kg	*		1.86	
Sheep	ME Mcal/kg	*		2.01	
Cattle	TDN %	*		51.4	
Sheep	TDN %	*		55.6	
Energy	GE Mcal/kg				
Calcium	%			0.54	
Phosphorus	%			0.34	

Bluestem, yellow, aerial part, fresh, mature, (2)
Ref No 2 00 870 United States

Feed Name or Analyses			Mean As Fed	Mean Dry	C.V. ± %
Dry matter	%			100.0	
Ash	%			6.8	
Crude fiber	%			38.0	
Ether extract	%			1.7	
N-free extract	%			49.8	
Protein (N x 6.25)	%			3.7	
Cattle	dig prot %	*		1.0	
Goats	dig prot %	*		0.0	
Horses	dig prot %	*		0.7	
Rabbits	dig prot %	*		1.5	
Sheep	dig prot %	*		0.4	
Energy	GE Mcal/kg				
Cattle	DE Mcal/kg	*		2.34	
Sheep	DE Mcal/kg	*		2.55	
Cattle	ME Mcal/kg	*		1.92	
Sheep	ME Mcal/kg	*		2.09	
Cattle	TDN %	*		53.1	
Sheep	TDN %	*		57.9	
Calcium	%			0.27	
Phosphorus	%			0.11	

(1) dry forages and roughages (2) pasture, range plants, and forages fed green (3) silages (4) energy feeds (5) protein supplements (6) minerals (7) vitamins (8) additives

BLUESTEM, YELLOWSEDGE. Andropogon virginicus

Bluestem, yellowsedge, aerial part, fresh, immature, (2)
Ref No 2 00 872 United States

Feed Name or Analyses			Mean As Fed	Mean Dry	C.V. ± %
Dry matter	%			100.0	
Ash	%			10.9	
Crude fiber	%			30.9	
Ether extract	%			1.8	
N-free extract	%			42.9	
Protein (N x 6.25)	%			13.5	
Cattle	dig prot %	*		9.4	
Goats	dig prot %	*		9.2	
Horses	dig prot %	*		9.0	
Rabbits	dig prot %	*		9.1	
Sheep	dig prot %	*		9.6	
Energy	GE Mcal/kg				
Cattle	DE Mcal/kg	*		2.49	
Sheep	DE Mcal/kg	*		2.30	
Cattle	ME Mcal/kg	*		2.04	
Sheep	ME Mcal/kg	*		1.89	
Cattle	TDN %	*		56.5	
Sheep	TDN %	*		52.2	
Calcium	%			0.50	
Phosphorus	%			0.21	

BLUESTEM–GRASS. Andropogon spp, Scientific name not used

Bluestem-Grass, aerial part, fresh, (2)
Ref No 2 00 874 United States

Feed Name or Analyses			Mean As Fed	Mean Dry	C.V. ± %
Dry matter	%		45.7	100.0	22
Ash	%		3.7	8.2	11
Crude fiber	%		14.5	31.8	7
Ether extract	%		1.0	2.1	24
N-free extract	%		23.8	52.0	
Protein (N x 6.25)	%		2.7	5.9	25
Cattle	dig prot %	*	1.3	2.9	
Goats	dig prot %	*	0.9	2.1	
Horses	dig prot %	*	1.2	2.5	
Rabbits	dig prot %	*	1.5	3.2	
Sheep	dig prot %	*	1.1	2.5	
Energy	GE Mcal/kg				
Cattle	DE Mcal/kg	*	1.08	2.36	
Sheep	DE Mcal/kg	*	1.18	2.59	
Cattle	ME Mcal/kg	*	0.89	1.94	
Sheep	ME Mcal/kg	*	0.97	2.12	
Cattle	TDN %	*	24.4	53.5	
Sheep	TDN %	*	26.9	58.8	

Bluestem-Grass, aerial part, fresh, over ripe, (2)
Ref No 2 00 873 United States

Feed Name or Analyses			Mean As Fed	Mean Dry	C.V. ± %
Dry matter	%		73.5	100.0	5
Ash	%		6.6	9.0	9
Crude fiber	%		24.5	33.3	5
Ether extract	%		1.2	1.7	14
N-free extract	%		38.6	52.5	
Protein (N x 6.25)	%		2.6	3.5	15
Cattle	dig prot %	*	0.6	0.9	
Goats	dig prot %	*	-0.1	-0.1	
Horses	dig prot %	*	0.4	0.5	
Rabbits	dig prot %	*	1.0	1.4	
Sheep	dig prot %	*	0.2	0.3	
Energy	GE Mcal/kg				
Cattle	DE Mcal/kg	*	1.90	2.59	
Sheep	DE Mcal/kg	*	1.93	2.63	
Cattle	ME Mcal/kg	*	1.56	2.12	
Sheep	ME Mcal/kg	*	1.59	2.16	
Cattle	TDN %	*	43.1	58.7	
Sheep	TDN %	*	43.9	59.7	

Bone charcoal (AAFCO) –
see Animal, bone charcoal, retort-charred grnd, mn 14% phosphorus, (6)

Bone meal, steamed (AAFCO) –
see Animal, bone, steamed dehy grnd, (6)

Bone meal, steamed –
see Animal, bone, steamed dehy grnd, mn 12% phosphorus, (6)

Bone phosphate (AAFCO) –
see Animal, bone phosphate, precipitated dehy, mn 17% phosphorus, (6)

Bone raw –
see Animal, bone, raw dehy grnd, (6)

BORAGE, COMMON. Borago officinalis

Borage, common, seeds w hulls, (5)
Ref No 5 09 132 United States

Feed Name or Analyses			Mean As Fed	Mean Dry	C.V. ± %
Dry matter	%			100.0	
Protein (N x 6.25)	%			21.0	
Alanine	%			0.74	
Arginine	%			1.72	
Aspartic acid	%			1.72	
Glutamic acid	%			3.07	
Glycine	%			0.97	
Histidine	%			0.42	
Hydroxyproline	%			0.59	
Isoleucine	%			0.82	
Leucine	%			1.16	
Lysine	%			0.88	
Methionine	%			0.36	
Phenylalanine	%			0.76	
Proline	%			0.84	
Serine	%			0.88	
Threonine	%			0.69	
Tyrosine	%			0.74	
Valine	%			0.90	

BOX, COMMON. Buxus sempervirens

Box, common, seeds w hulls, (5)
Ref No 5 09 036 United States

Feed Name or Analyses			Mean As Fed	Mean Dry	C.V. ± %
Dry matter	%			100.0	
Protein (N x 6.25)	%			21.0	
Alanine	%			0.82	
Arginine	%			1.83	
Aspartic acid	%			1.85	
Glutamic acid	%			3.11	
Glycine	%			0.80	
Histidine	%			0.34	
Hydroxyproline	%			0.13	
Isoleucine	%			0.74	
Leucine	%			1.30	
Lysine	%			0.78	

Feed Name or Analyses		Mean As Fed	Mean Dry	C.V. ± %
Methionine	%		0.25	
Phenylalanine	%		0.78	
Proline	%		0.82	
Serine	%		0.78	
Threonine	%		0.65	
Tyrosine	%		0.59	
Valine	%		0.90	

Bran (CFA) –
see Wheat, bran, dry milled, (4)

BREAD. Scientific name not used

Bread, dehy, (4)
Ref No 4 07 944 United States

Feed Name or Analyses		Mean As Fed	Mean Dry	C.V. ± %
Dry matter	%	86.0	100.0	
Ash	%	1.5	1.7	
Crude fiber	%	0.5	0.6	
Ether extract	%	1.0	1.2	
N-free extract	%	72.0	83.7	
Protein (N x 6.25)	%	11.0	12.8	
Cattle	dig prot %	* 6.7	7.8	
Goats	dig prot %	* 7.7	9.0	
Horses	dig prot %	* 7.7	9.0	
Sheep	dig prot %	* 7.7	9.0	
Energy	GE Mcal/kg			
Cattle	DE Mcal/kg	* 3.41	3.96	
Sheep	DE Mcal/kg	* 3.25	3.78	
Swine	DE kcal/kg	*3630.	4221.	
Cattle	ME Mcal/kg	* 2.80	3.25	
Chickens	ME_n kcal/kg	3527.	4101.	
Sheep	ME Mcal/kg	* 2.66	3.10	
Swine	ME kcal/kg	*3391.	3943.	
Cattle	TDN %	* 77.2	89.8	
Sheep	TDN %	* 73.7	85.6	
Swine	TDN %	* 82.3	95.7	
Calcium	%	0.05	0.06	
Phosphorus	%	0.10	0.12	
Choline	mg/kg	882.	1025.	
Niacin	mg/kg	28.7	33.3	
Pantothenic acid	mg/kg	11.0	12.8	
Riboflavin	mg/kg	2.0	2.3	
Arginine	%	0.30	0.35	
Cystine	%	0.12	0.14	
Lysine	%	0.20	0.23	
Methionine	%	0.09	0.10	
Tryptophan	%	0.08	0.09	

BREAD, WHITE. Scientific name not used

Bread, white, enriched, (4)
Ref No 4 08 359 United States

Feed Name or Analyses		Mean As Fed	Mean Dry	C.V. ± %
Dry matter	%	64.1	100.0	
Ash	%	1.3	2.0	
Crude fiber	%	0.3	0.5	
Ether extract	%	2.0	3.1	
N-free extract	%	52.0	81.1	
Protein (N x 6.25)	%	8.5	13.3	
Cattle	dig prot %	* 5.3	8.2	
Goats	dig prot %	* 6.0	9.4	
Horses	dig prot %	* 6.0	9.4	
Sheep	dig prot %	* 6.0	9.4	
Energy	GE Mcal/kg			
Cattle	DE Mcal/kg	* 2.53	3.94	
Sheep	DE Mcal/kg	* 2.43	3.79	
Swine	DE kcal/kg	*2715.	4236.	
Cattle	ME Mcal/kg	* 2.07	3.23	
Sheep	ME Mcal/kg	* 1.99	3.11	
Swine	ME kcal/kg	*2534.	3953.	
Cattle	TDN %	* 57.3	89.4	
Sheep	TDN %	* 55.1	86.0	
Swine	TDN %	* 61.6	96.1	
Calcium	%	0.06	0.09	
Phosphorus	%	0.10	0.16	
Potassium	%	0.10	0.19	

Brewers dried grains (AAFCO) (CFA) –
see Grains, brewers grains, dehy, mx 3% dried spent hops, (5)

Brewers dried yeast (AAFCO) –
see Yeast, brewers saccharomyces, dehy grnd, mn 40% protein, (7)

Brewers dried yeast (CFA) –
see Yeast, brewers saccharomyces, dehy grnd, (7)

Brewers rice (AAFCO) –
see Rice, groats, polished and broken, (4)

BRISTLEGRASS. Setaria spp

Bristlegrass, aerial part, fresh, (2)
Ref No 2 00 876 United States

Feed Name or Analyses		Mean As Fed	Mean Dry	C.V. ± %
Dry matter	%		100.0	
Ash	%		12.3	17
Crude fiber	%		31.9	10
Ether extract	%		2.1	15
N-free extract	%		41.9	
Protein (N x 6.25)	%		11.8	46
Cattle	dig prot %	*	7.9	
Goats	dig prot %	*	7.6	
Horses	dig prot %	*	7.5	
Rabbits	dig prot %	*	7.8	
Sheep	dig prot %	*	8.0	
Calcium	%		0.37	34
Copper	mg/kg		6.6	
Magnesium	%		0.28	
Manganese	mg/kg		42.1	
Phosphorus	%		0.19	48
Potassium	%		5.82	

BRISTLEGRASS, GREEN. Setaria viridis

Bristlegrass, green, aerial part, fresh, (2)
Ref No 2 00 877 United States

Feed Name or Analyses		Mean As Fed	Mean Dry	C.V. ± %
Dry matter	%		100.0	
Ash	%		10.8	
Crude fiber	%		31.3	
Ether extract	%		2.5	
N-free extract	%		45.8	
Protein (N x 6.25)	%		9.6	
Cattle	dig prot %	*	6.1	
Goats	dig prot %	*	5.5	
Horses	dig prot %	*	5.7	
Rabbits	dig prot %	*	6.1	
Sheep	dig prot %	*	5.9	
Calcium	%		0.30	
Phosphorus	%		0.21	

BRISTLEGRASS, PLAINS. Setaria macrostachya

Bristlegrass, plains, aerial part, fresh, early bloom, (2)
Ref No 2 00 878 United States

Feed Name or Analyses		Mean As Fed	Mean Dry	C.V. ± %
Dry matter	%		100.0	
Ash	%		11.8	
Crude fiber	%		27.6	
Ether extract	%		2.3	
N-free extract	%		40.8	
Protein (N x 6.25)	%		17.5	
Cattle	dig prot %	*	12.8	
Goats	dig prot %	*	12.9	
Horses	dig prot %	*	12.4	
Rabbits	dig prot %	*	12.2	
Sheep	dig prot %	*	13.3	
Calcium	%		0.50	
Cobalt	mg/kg		0.489	
Copper	mg/kg		6.6	
Magnesium	%		0.28	
Manganese	mg/kg		42.1	
Phosphorus	%		0.23	
Potassium	%		5.82	
Carotene	mg/kg		43.2	
Vitamin A equivalent	IU/g		72.0	

Bristlegrass, plains, aerial part, fresh, mid–bloom, (2)
Ref No 2 00 879 United States

Feed Name or Analyses		Mean As Fed	Mean Dry	C.V. ± %
Dry matter	%		100.0	
Ash	%		14.9	
Crude fiber	%		35.2	
Ether extract	%		2.0	
N-free extract	%		37.6	
Protein (N x 6.25)	%		10.3	
Cattle	dig prot %	*	6.6	
Goats	dig prot %	*	6.2	
Horses	dig prot %	*	6.3	
Rabbits	dig prot %	*	6.6	
Sheep	dig prot %	*	6.6	
Calcium	%		0.33	
Phosphorus	%		0.20	

Bristlegrass, plains, aerial part wo lower stems, fresh, early leaf, (2)
Ref No 2 08 722 United States

Feed Name or Analyses		Mean As Fed	Mean Dry	C.V. ± %
Dry matter	%		100.0	
Organic matter	%		89.9	2
Ash	%		10.1	14
Crude fiber	%		30.0	7
Ether extract	%		1.9	17
N-free extract	%		47.7	3
Protein (N x 6.25)	%		10.2	10
Cattle	dig prot %	*	6.6	
Goats	dig prot %	*	6.1	
Horses	dig prot %	*	6.2	
Rabbits	dig prot %	*	6.6	
Sheep	dig prot %	*	6.5	
Calcium	%		0.43	31
Phosphorus	%		0.09	23
Carotene	mg/kg		68.3	
Vitamin A equivalent	IU/g		113.9	

Feed Name or Analyses		Mean As Fed	Mean Dry	C.V. ± %

Bristlegrass, plains, aerial part wo lower stems, fresh, mid–bloom, (2)

Ref No 2 08 720 United States

Analyses		As Fed	Dry	C.V. ± %
Dry matter	%		100.0	
Organic matter	%		88.8	
Ash	%		11.3	
Crude fiber	%		32.6	
Ether extract	%		1.7	
N-free extract	%		46.8	
Protein (N x 6.25)	%		7.7	
Cattle	dig prot % *		4.4	
Goats	dig prot % *		3.7	
Horses	dig prot % *		4.0	
Rabbits	dig prot % *		4.6	
Sheep	dig prot % *		4.1	
Calcium	%		0.37	
Phosphorus	%		0.16	

Bristlegrass, plains, aerial part wo lower stems, fresh, mature, (2)

Ref No 2 08 721 United States

Analyses		As Fed	Dry	C.V. ± %
Dry matter	%		100.0	
Organic matter	%		85.1	
Ash	%		14.9	
Crude fiber	%		28.9	
Ether extract	%		1.7	
N-free extract	%		43.8	
Protein (N x 6.25)	%		10.7	
Cattle	dig prot % *		7.0	
Goats	dig prot % *		6.5	
Horses	dig prot % *		6.6	
Rabbits	dig prot % *		6.9	
Sheep	dig prot % *		7.0	
Calcium	%		0.63	
Phosphorus	%		0.12	

Bristlegrass, plains, aerial part wo lower stems, fresh, dormant, (2)

Ref No 2 08 717 United States

Analyses		As Fed	Dry	C.V. ± %
Dry matter	%		100.0	
Organic matter	%		88.3	4
Ash	%		11.7	27
Crude fiber	%		31.5	7
Ether extract	%		1.6	40
N-free extract	%		46.8	5
Protein (N x 6.25)	%		8.4	14
Cattle	dig prot % *		5.0	
Goats	dig prot % *		4.4	
Horses	dig prot % *		4.7	
Rabbits	dig prot % *		5.1	
Sheep	dig prot % *		4.8	
Calcium	%		0.57	52
Phosphorus	%		0.08	30
Carotene	mg/kg		7.1	72
Vitamin A equivalent	IU/g		11.8	

(1) dry forages and roughages (2) pasture, range plants, and forages fed green (3) silages (4) energy feeds (5) protein supplements (6) minerals (7) vitamins (8) additives

BRISTLEGRASS, YELLOW. Setaria lutescens

Bristlegrass, yellow, aerial part, fresh, mature, (2)

Ref No 2 00 880 United States

Analyses		As Fed	Dry	C.V. ± %
Dry matter	%		100.0	
Ash	%		9.5	
Crude fiber	%		34.6	
Ether extract	%		1.7	
N-free extract	%		48.4	
Protein (N x 6.25)	%		5.8	
Cattle	dig prot % *		2.8	
Goats	dig prot % *		2.0	
Horses	dig prot % *		2.5	
Rabbits	dig prot % *		3.1	
Sheep	dig prot % *		2.4	
Calcium	%		0.26	
Phosphorus	%		0.08	

BRISTLETHISTLE. Carduus spp

Bristlethistle, aerial part, fresh, (2)

Ref No 2 00 881 United States

Analyses		As Fed	Dry	C.V. ± %
Dry matter	%		100.0	
Ash	%		10.8	
Crude fiber	%		30.4	
Ether extract	%		4.2	
N-free extract	%		45.3	
Protein (N x 6.25)	%		9.3	
Cattle	dig prot % *		5.8	
Goats	dig prot % *		5.2	
Horses	dig prot % *		5.4	
Rabbits	dig prot % *		5.8	
Sheep	dig prot % *		5.7	

Bristly carrot –
see Carrot, southwestern, aerial part, fresh, (2)

BROADBEAN. Vicia faba

Broadbean, seeds w hulls, (5)

Ref No 5 09 262 United States

Analyses		As Fed	Dry	C.V. ± %
Dry matter	%		100.0	
Protein (N x 6.25)	%		25.0	
Alanine	%		0.93	
Arginine	%		1.98	
Aspartic acid	%		2.53	
Glutamic acid	%		3.73	
Glycine	%		0.95	
Histidine	%		0.60	
Hydroxyproline	%		0.03	
Isoleucine	%		0.95	
Leucine	%		1.68	
Lysine	%		1.53	
Methionine	%		0.15	
Phenylalanine	%		0.98	
Proline	%		0.98	
Serine	%		1.08	
Threonine	%		0.88	
Tyrosine	%		0.80	
Valine	%		1.08	

BROCCOLI. Brassica oleracea botrytis

Broccoli, leaves, fresh, (4)

Ref No 2 00 882 United States

Analyses		As Fed	Dry	C.V. ± %
Dry matter	%	13.5	100.0	23
Crude fiber	%	1.8	13.5	8
Ether extract	%	0.7	5.5	20
Protein (N x 6.25)	%	3.9	28.8	16
Cattle	dig prot % *	3.0	22.5	
Goats	dig prot % *	3.2	23.6	
Horses	dig prot % *	3.2	23.6	
Sheep	dig prot % *	3.2	23.6	
Carotene	mg/kg	45.2	335.9	35
Riboflavin	mg/kg	2.2	16.3	25
Thiamine	mg/kg	3.4	25.4	5
a-tocopherol	mg/kg	56.1	416.7	
Vitamin A equivalent	IU/g	75.4	559.9	
Arginine	%	0.28	2.10	
Histidine	%	0.09	0.70	
Isoleucine	%	0.19	1.40	
Leucine	%	0.38	2.80	
Lysine	%	0.27	2.00	
Methionine	%	0.11	0.80	
Phenylalanine	%	0.35	2.60	
Threonine	%	0.20	1.50	
Tryptophan	%	0.08	0.60	
Valine	%	0.27	2.00	

Broccoli, stems, fresh, (4)

Ref No 2 00 884 United States

Analyses		As Fed	Dry	C.V. ± %
Dry matter	%	45.1	100.0	
Crude fiber	%	7.5	16.6	
Ether extract	%	1.5	3.3	
Protein (N x 6.25)	%	9.5	21.1	
Cattle	dig prot % *	6.9	15.4	
Goats	dig prot % *	7.5	16.6	
Horses	dig prot % *	7.5	16.6	
Sheep	dig prot % *	7.5	16.6	
Carotene	mg/kg	22.6	50.0	
Riboflavin	mg/kg	3.9	8.6	
Vitamin A equivalent	IU/g	37.6	83.4	

Broccoli, stems w heads, fresh, immature, (4)

Broccoli spears, raw

Ref No 2 08 186 United States

Analyses		As Fed	Dry	C.V. ± %
Dry matter	%	10.9	100.0	
Ash	%	1.1	10.1	
Crude fiber	%	1.5	13.8	
Ether extract	%	0.3	2.8	
N-free extract	%	4.4	40.4	
Protein (N x 6.25)	%	3.6	33.0	
Cattle	dig prot % *	2.9	26.3	
Goats	dig prot % *	3.0	27.5	
Horses	dig prot % *	3.0	27.5	
Sheep	dig prot % *	3.0	27.5	
Energy	GE Mcal/kg	0.32	2.94	
Calcium	%	0.10	0.92	
Iron	%	0.001	0.009	
Phosphorus	%	0.08	0.73	
Potassium	%	0.38	3.49	
Sodium	%	0.02	0.18	
Niacin	mg/kg	9.0	82.6	
Riboflavin	mg/kg	2.3	21.1	

Feed Name or Analyses			Mean As Fed	Mean Dry	C.V. ± %
Thiamine	mg/kg		1.0	9.2	
Vitamin A	IU/g		25.0	229.4	

Broccoli spears, raw –
see Broccoli, stems w heads, fresh, immature, (4)

Broken rice (AAFCO) –
see Rice, groats, polished and broken, (4)

BROME, Bromus spp

Brome, aerial part, fresh, pre–bloom, (2)
Ref No 2 00 886 United States

Feed Name or Analyses			As Fed	Dry	C.V. ± %
Dry matter	%		37.5	100.0	
Ash	%		2.8	7.5	
Crude fiber	%		10.6	28.3	
Sheep	dig coef %		83.	83.	
Ether extract	%		1.2	3.2	
Sheep	dig coef %		69.	69.	
N-free extract	%		18.3	48.9	
Sheep	dig coef %		85.	85.	
Protein (N x 6.25)	%		4.5	12.1	
Sheep	dig coef %		79.	79.	
Cattle	dig prot %	*	2.8	7.4	
Goats	dig prot %	*	2.9	7.8	
Horses	dig prot %	*	2.9	7.8	
Rabbits	dig prot %	*	3.0	8.0	
Sheep	dig prot %		3.6	9.6	
Energy	GE Mcal/kg				
Cattle	DE Mcal/kg	*	1.15	3.07	
Sheep	DE Mcal/kg	*	1.32	3.51	
Cattle	ME Mcal/kg	*	0.94	2.52	
Sheep	ME Mcal/kg	*	1.08	2.88	
Cattle	TDN %	*	26.1	68.7	
Sheep	TDN %		29.8	79.6	

Brome, hay, s-c, (1)
Ref No 1 00 890 United States

Feed Name or Analyses			As Fed	Dry	C.V. ± %
Dry matter	%		91.4	100.0	2
Ash	%		7.2	7.9	17
Crude fiber	%		29.1	31.9	16
Cattle	dig coef %	*	47.	47.	
Ether extract	%		2.4	2.7	38
Cattle	dig coef %	*	32.	32.	
N-free extract	%		43.7	47.8	
Cattle	dig coef %	*	49.	49.	
Protein (N x 6.25)	%		8.9	9.7	44
Cattle	dig coef %	*	42.	42.	
Cattle	dig prot %		3.7	4.1	
Goats	dig prot %	*	5.2	5.6	
Horses	dig prot %	*	5.3	5.8	
Rabbits	dig prot %	*	5.6	6.2	
Sheep	dig prot %	*	4.8	5.3	
Cellulose (Matrone)	%		25.3	27.7	
Lignin (Ellis)	%		5.3	5.8	
Energy	GE Mcal/kg				
Cattle	DE Mcal/kg	*	1.79	1.96	
Sheep	DE Mcal/kg	*	2.08	2.28	
Cattle	ME Mcal/kg	*	1.47	1.61	
Sheep	ME Mcal/kg	*	1.71	1.87	
Cattle	NE_m Mcal/kg	*	0.99	1.08	
Cattle	NE_{gain} Mcal/kg	*	0.24	0.26	
Cattle	$NE_{lactating\ cows}$ Mcal/kg	*	0.79	0.86	
Cattle	TDN %		40.6	44.4	
Sheep	TDN %	*	47.3	51.7	

Brome, hay, s-c, pre-bloom, (1)
Ref No 1 00 887 United States

Feed Name or Analyses			As Fed	Dry	C.V. ± %
Dry matter	%		88.1	100.0	
Ash	%		8.3	9.4	
Crude fiber	%		29.4	33.4	
Sheep	dig coef %		62.	62.	
Ether extract	%		2.2	2.6	
Sheep	dig coef %		39.	39.	
N-free extract	%		38.9	44.2	
Sheep	dig coef %		62.	62.	
Protein (N x 6.25)	%		9.3	10.5	
Sheep	dig coef %		55.	55.	
Cattle	dig prot %	*	5.3	6.0	
Goats	dig prot %	*	5.6	6.4	
Horses	dig prot %	*	5.7	6.4	
Rabbits	dig prot %	*	6.0	6.8	
Sheep	dig prot %		5.1	5.8	
Energy	GE Mcal/kg				
Cattle	DE Mcal/kg	*	2.30	2.61	
Sheep	DE Mcal/kg	*	2.16	2.45	
Cattle	ME Mcal/kg	*	1.89	2.14	
Sheep	ME Mcal/kg	*	1.77	2.01	
Cattle	TDN %	*	52.2	59.2	
Sheep	TDN %		49.1	55.7	

Ref No 1 00 887 Canada

Feed Name or Analyses			As Fed	Dry	C.V. ± %
Dry matter	%			100.0	
Crude fiber	%				
Sheep	dig coef %			62.	
Ether extract	%				
Sheep	dig coef %			39.	
N-free extract	%				
Sheep	dig coef %			62.	
Protein (N x 6.25)	%				
Sheep	dig coef %			55.	
Energy	GE Mcal/kg				
Sheep	DE Mcal/kg	*		2.45	
Sheep	ME Mcal/kg	*		2.01	
Sheep	TDN %			55.7	
Calcium	%			0.32	
Magnesium	%			0.09	
Potassium	%			2.32	
Sodium	%			0.02	

Brome, hay, s-c, early bloom, (1)
Ref No 1 05 679 United States

Feed Name or Analyses			As Fed	Dry	C.V. ± %
Dry matter	%			100.0	
Ash	%			5.6	
Crude fiber	%			34.8	
Ether extract	%			2.2	
N-free extract	%			47.2	
Protein (N x 6.25)	%			10.3	
Cattle	dig prot %	*		5.9	
Goats	dig prot %	*		6.2	
Horses	dig prot %	*		6.3	
Rabbits	dig prot %	*		6.6	
Sheep	dig prot %	*		5.8	
Cellulose (Matrone)	%			33.8	
Lignin (Ellis)	%			6.1	
Energy	GE Mcal/kg				
Cattle	DE Mcal/kg	*		2.33	
Sheep	DE Mcal/kg	*		2.47	
Cattle	ME Mcal/kg	*		1.91	
Sheep	ME Mcal/kg	*		2.02	
Cattle	TDN %	*		52.9	
Sheep	TDN %	*		56.0	

Ref No 1 05 679 Canada

Feed Name or Analyses			As Fed	Dry	C.V. ± %
Dry matter	%		96.4	100.0	
Crude fiber	%		33.8	35.1	
Protein (N x 6.25)	%		15.9	16.5	
Cattle	dig prot %	*	10.8	11.2	
Goats	dig prot %	*	11.5	11.9	
Horses	dig prot %	*	11.1	11.5	
Rabbits	dig prot %	*	11.0	11.4	
Sheep	dig prot %	*	11.0	11.4	
Calcium	%		0.29	0.30	
Phosphorus	%		0.34	0.35	

Brome, hay, s-c, late bloom, (1)
Ref No 1 00 888 United States

Feed Name or Analyses			As Fed	Dry	C.V. ± %
Dry matter	%		85.4	100.0	
Ash	%		7.2	8.4	
Crude fiber	%		30.4	35.6	
Sheep	dig coef %		56.	56.	
Ether extract	%		2.0	2.3	
Sheep	dig coef %		28.	28.	
N-free extract	%		40.6	47.5	
Sheep	dig coef %		64.	64.	
Protein (N x 6.25)	%		5.3	6.2	
Sheep	dig coef %		53.	53.	
Cattle	dig prot %	*	2.0	2.3	
Goats	dig prot %	*	2.0	2.3	
Horses	dig prot %	*	2.4	2.8	
Rabbits	dig prot %	*	3.0	3.5	
Sheep	dig prot %		2.8	3.3	
Energy	GE Mcal/kg				
Cattle	DE Mcal/kg	*	2.08	2.44	
Sheep	DE Mcal/kg	*	2.07	2.43	
Cattle	ME Mcal/kg	*	1.71	2.00	
Sheep	ME Mcal/kg	*	1.70	1.99	
Cattle	TDN %	*	47.2	55.3	
Sheep	TDN %		47.0	55.1	

Brome, hay, s-c, mature, (1)
Ref No 1 00 889 United States

Feed Name or Analyses			As Fed	Dry	C.V. ± %
Dry matter	%		90.1	100.0	
Ash	%		5.6	6.2	
Crude fiber	%		32.6	36.2	
Sheep	dig coef %		46.	46.	
Ether extract	%		1.7	1.9	
Sheep	dig coef %		53.	53.	
N-free extract	%		44.1	48.9	
Sheep	dig coef %		57.	57.	
Protein (N x 6.25)	%		6.1	6.7	
Sheep	dig coef %		21.	21.	
Cattle	dig prot %	*	2.5	2.8	
Goats	dig prot %	*	2.6	2.8	
Horses	dig prot %	*	2.9	3.2	
Rabbits	dig prot %	*	3.5	3.9	
Sheep	dig prot %		1.3	1.4	
Cellulose (Matrone)	%		31.9	35.4	
Energy	GE Mcal/kg				
Cattle	DE Mcal/kg	*	2.05	2.28	
Sheep	DE Mcal/kg	*	1.91	2.12	
Cattle	ME Mcal/kg	*	1.68	1.87	
Sheep	ME Mcal/kg	*	1.57	1.74	

Continued

Feed Name or Analyses			Mean As Fed	Mean Dry	C.V. ± %
Cattle	TDN %	*	46.5	51.6	
Sheep	TDN %		43.4	48.2	

Ref No 1 00 889 Canada

Feed Name or Analyses			Mean As Fed	Mean Dry	C.V. ± %
Dry matter	%		93.4	100.0	2
Crude fiber	%		34.5	36.9	12
Sheep	dig coef %		46.	46.	
Ether extract	%				
Sheep	dig coef %		53.	53.	
N-free extract	%				
Sheep	dig coef %		57.	57.	
Protein (N x 6.25)	%		5.0	5.3	50
Sheep	dig coef %		21.	21.	
Cattle	dig prot %	*	1.4	1.5	
Goats	dig prot %	*	1.4	1.5	
Horses	dig prot %	*	1.9	2.0	
Rabbits	dig prot %	*	2.6	2.8	
Sheep	dig prot %		1.0	1.1	
Energy	GE Mcal/kg				
Sheep	DE Mcal/kg	*	1.99	2.14	
Sheep	ME Mcal/kg	*	1.64	1.75	
Sheep	TDN %		45.2	48.5	
Calcium	%		0.38	0.41	56
Phosphorus	%		0.08	0.08	56
Carotene	mg/kg		13.3	14.2	
Vitamin A equivalent	IU/g		22.1	23.7	

Brome, aerial part, fresh, (2)

Ref No 2 00 900 United States

Feed Name or Analyses			Mean As Fed	Mean Dry	C.V. ± %
Dry matter	%		36.0	100.0	34
Ash	%		3.5	9.7	
Crude fiber	%		11.6	32.1	
Ether extract	%		1.0	2.8	
N-free extract	%		15.6	43.2	
Protein (N x 6.25)	%		4.4	12.2	
Cattle	dig prot %	*	3.0	8.3	
Goats	dig prot %	*	2.9	7.9	
Horses	dig prot %	*	2.8	7.9	
Rabbits	dig prot %	*	2.9	8.1	
Sheep	dig prot %	*	3.0	8.3	
Energy	GE Mcal/kg				
Cattle	DE Mcal/kg	*	1.04	2.89	
Sheep	DE Mcal/kg	*	1.00	2.77	
Cattle	ME Mcal/kg	*	0.85	2.37	
Sheep	ME Mcal/kg	*	0.82	2.27	
Cattle	TDN %	*	23.6	65.6	
Sheep	TDN %	*	22.6	62.8	
Phosphorus	%		0.10	0.28	
Potassium	%		0.79	2.19	
Carotene	mg/kg		111.9	310.8	39
Riboflavin	mg/kg		2.8	7.7	26
Thiamine	mg/kg		1.1	3.1	28
Vitamin A equivalent	IU/g		186.5	518.2	

Brome, aerial part, fresh, immature, (2)

Ref No 2 00 892 United States

Feed Name or Analyses			Mean As Fed	Mean Dry	C.V. ± %
Dry matter	%		34.9	100.0	18
Ash	%		3.7	10.7	
Crude fiber	%		7.2	20.8	
Sheep	dig coef %		85.	85.	
Ether extract	%		1.6	4.5	
Sheep	dig coef %		66.	66.	
N-free extract	%		17.2	49.4	
Sheep	dig coef %		89.	89.	
Protein (N x 6.25)	%		5.1	14.6	
Sheep	dig coef %		82.	82.	
Cattle	dig prot %	*	3.6	10.3	
Goats	dig prot %	*	3.5	10.2	
Horses	dig prot %	*	3.5	9.9	
Rabbits	dig prot %	*	3.5	9.9	
Sheep	dig prot %		4.2	12.0	
Energy	GE Mcal/kg				
Cattle	DE Mcal/kg	*	1.18	3.37	
Sheep	DE Mcal/kg	*	1.23	3.54	
Cattle	ME Mcal/kg	*	0.96	2.76	
Sheep	ME Mcal/kg	*	1.01	2.90	
Cattle	NE_m Mcal/kg	*	0.53	1.52	
Cattle	NE_{gain} Mcal/kg	*	0.33	0.95	
Cattle	$NE_{lactating\ cows}$ Mcal/kg	*	0.61	1.75	
Cattle	TDN %	*	26.7	76.4	
Sheep	TDN %		28.0	80.3	
Carotene	mg/kg		160.1	459.4	19
Vitamin A equivalent	IU/g		266.9	765.9	

Brome, aerial part, fresh, early bloom, (2)

Ref No 2 00 893 United States

Feed Name or Analyses			Mean As Fed	Mean Dry	C.V. ± %
Dry matter	%		30.0	100.0	11
Carotene	mg/kg		55.2	183.9	
Vitamin A equivalent	IU/g		92.0	306.5	

Brome, aerial part, fresh, mid–bloom, (2)

Ref No 2 00 894 United States

Feed Name or Analyses			Mean As Fed	Mean Dry	C.V. ± %
Dry matter	%			100.0	
Ash	%			6.7	13
Crude fiber	%			30.4	6
Ether extract	%			3.0	6
N-free extract	%			49.2	
Protein (N x 6.25)	%			10.7	15
Cattle	dig prot %	*		7.0	
Goats	dig prot %	*		6.5	
Horses	dig prot %	*		6.6	
Rabbits	dig prot %	*		6.9	
Sheep	dig prot %	*		7.0	
Energy	GE Mcal/kg				
Cattle	DE Mcal/kg	*		2.98	
Sheep	DE Mcal/kg	*		2.89	
Cattle	ME Mcal/kg	*		2.45	
Sheep	ME Mcal/kg	*		2.37	
Cattle	TDN %	*		67.7	
Sheep	TDN %	*		65.6	

Brome, aerial part, fresh, full bloom, (2)

Ref No 2 00 895 United States

Feed Name or Analyses			Mean As Fed	Mean Dry	C.V. ± %
Dry matter	%		27.1	100.0	
Ash	%		2.0	7.5	33
Crude fiber	%		8.2	30.3	19
Ether extract	%		0.9	3.3	79
N-free extract	%		13.1	48.3	
Protein (N x 6.25)	%		2.9	10.6	23
Cattle	dig prot %	*	1.9	6.9	
Goats	dig prot %	*	1.7	6.4	
Horses	dig prot %	*	1.8	6.5	
Rabbits	dig prot %	*	1.9	6.9	
Sheep	dig prot %	*	1.9	6.9	
Energy	GE Mcal/kg				
Cattle	DE Mcal/kg	*	0.79	2.91	
Sheep	DE Mcal/kg	*	0.78	2.88	
Cattle	ME Mcal/kg	*	0.65	2.38	
Sheep	ME Mcal/kg	*	0.64	2.37	
Cattle	TDN %	*	17.9	65.9	
Sheep	TDN %	*	17.7	65.4	
Carotene	mg/kg		22.4	82.7	
Vitamin A equivalent	IU/g		37.3	137.8	

Ref No 2 00 895 Canada

Feed Name or Analyses			Mean As Fed	Mean Dry	C.V. ± %
Dry matter	%		29.0	100.0	
Ash	%		2.0	6.8	
Crude fiber	%		8.7	30.0	
Ether extract	%		0.7	2.4	
N-free extract	%		13.9	48.0	
Protein (N x 6.25)	%		3.7	12.8	
Cattle	dig prot %	*	2.5	8.8	
Goats	dig prot %	*	2.5	8.5	
Horses	dig prot %	*	2.4	8.4	
Rabbits	dig prot %	*	2.5	8.6	
Sheep	dig prot %	*	2.6	8.9	
Energy	GE Mcal/kg				
Cattle	DE Mcal/kg	*	0.90	3.12	
Sheep	DE Mcal/kg	*	0.82	2.83	
Cattle	ME Mcal/kg	*	0.74	2.56	
Sheep	ME Mcal/kg	*	0.67	2.32	
Cattle	TDN %	*	20.5	70.7	
Sheep	TDN %	*	18.6	64.2	

Brome, aerial part, fresh, milk stage, (2)

Ref No 2 00 896 United States

Feed Name or Analyses			Mean As Fed	Mean Dry	C.V. ± %
Dry matter	%			100.0	
Ash	%			6.0	24
Crude fiber	%			29.7	7
Ether extract	%			2.9	7
N-free extract	%			52.1	
Protein (N x 6.25)	%			9.3	12
Cattle	dig prot %	*		5.8	
Goats	dig prot %	*		5.2	
Horses	dig prot %	*		5.4	
Rabbits	dig prot %	*		5.8	
Sheep	dig prot %	*		5.7	
Energy	GE Mcal/kg				
Cattle	DE Mcal/kg	*		2.96	
Sheep	DE Mcal/kg	*		2.97	
Cattle	ME Mcal/kg	*		2.43	
Sheep	ME Mcal/kg	*		2.43	
Cattle	TDN %	*		67.2	
Sheep	TDN %	*		67.3	

Brome, aerial part, fresh, dough stage, (2)

Ref No 2 00 897 United States

Feed Name or Analyses			Mean As Fed	Mean Dry	C.V. ± %
Dry matter	%		38.1	100.0	
Ash	%		2.7	7.1	22
Crude fiber	%		10.8	28.3	28
Ether extract	%		0.9	2.3	23
N-free extract	%		20.9	54.8	
Protein (N x 6.25)	%		2.9	7.5	9
Cattle	dig prot %	*	1.6	4.3	
Goats	dig prot %	*	1.4	3.6	
Horses	dig prot %	*	1.5	3.9	
Rabbits	dig prot %	*	1.7	4.5	
Sheep	dig prot %	*	1.5	4.0	

(1) dry forages and roughages (2) pasture, range plants, and forages fed green (3) silages (4) energy feeds (5) protein supplements (6) minerals (7) vitamins (8) additives

Feed Name or Analyses			Mean As Fed	Mean Dry	C.V. ± %
Energy	GE Mcal/kg				
Cattle	DE Mcal/kg	*	1.08	2.83	
Sheep	DE Mcal/kg	*	1.16	3.06	
Cattle	ME Mcal/kg	*	0.88	2.32	
Sheep	ME Mcal/kg	*	0.95	2.51	
Cattle	TDN %	*	24.4	64.1	
Sheep	TDN %	*	26.4	69.3	
Calcium	%		0.13	0.35	
Phosphorus	%		0.10	0.27	
Potassium	%		0.50	1.30	

Brome, aerial part, fresh, over ripe, (2)

Ref No 2 00 899 United States

Feed Name or Analyses			Mean As Fed	Mean Dry	C.V. ± %
Dry matter	%		75.0	100.0	11
Calcium	%		0.21	0.28	44
Magnesium	%		0.11	0.14	
Phosphorus	%		0.14	0.18	38
Potassium	%		0.92	1.23	16

Brome, heads, fresh, (2)

Ref No 2 00 901 United States

Feed Name or Analyses			Mean As Fed	Mean Dry	C.V. ± %
Dry matter	%		44.2	100.0	12
Ash	%		2.0	4.5	12
Crude fiber	%		12.6	28.4	14
Ether extract	%		0.8	1.9	10
N-free extract	%		24.5	55.5	
Protein (N x 6.25)	%		4.3	9.7	10
Cattle	dig prot %	*	2.7	6.1	
Goats	dig prot %	*	2.5	5.6	
Horses	dig prot %	*	2.5	5.8	
Rabbits	dig prot %	*	2.7	6.2	
Sheep	dig prot %	*	2.7	6.0	
Energy	GE Mcal/kg				
Cattle	DE Mcal/kg	*	1.39	3.13	
Sheep	DE Mcal/kg	*	1.33	3.01	
Cattle	ME Mcal/kg	*	1.14	2.57	
Sheep	ME Mcal/kg	*	1.09	2.47	
Cattle	TDN %	*	31.4	71.1	
Sheep	TDN %	*	30.2	68.3	
Calcium	%		0.15	0.34	20
Magnesium	%		0.08	0.17	20
Phosphorus	%		0.14	0.32	12
Potassium	%		0.41	0.93	

Brome, stems, fresh, (2)

Ref No 2 00 902 United States

Feed Name or Analyses			Mean As Fed	Mean Dry	C.V. ± %
Dry matter	%		56.5	100.0	20
Calcium	%		0.13	0.23	61
Phosphorus	%		0.10	0.18	78
Potassium	%		0.97	1.72	

Brome, aerial pt, ensiled, (3)

Ref No 3 00 903 United States

Feed Name or Analyses			Mean As Fed	Mean Dry	C.V. ± %
Dry matter	%			100.0	
Carotene	mg/kg			71.2	37
Vitamin A equivalent	IU/g			118.7	

BROME, AUSTRALIAN. Bromus arenarius

Brome, Australian, aerial part, fresh, full bloom, (2)

Ref No 2 00 904 United States

Feed Name or Analyses			Mean As Fed	Mean Dry	C.V. ± %
Dry matter	%			100.0	
Ash	%			5.5	
Crude fiber	%			31.4	
Protein (N x 6.25)	%			8.0	
Cattle	dig prot %	*		4.7	
Goats	dig prot %	*		4.0	
Horses	dig prot %	*		4.3	
Rabbits	dig prot %	*		4.8	
Sheep	dig prot %	*		4.4	
Calcium	%			0.36	
Phosphorus	%			0.27	
Potassium	%			1.41	

Brome, Australian, aerial part, fresh, milk stage, (2)

Ref No 2 00 905 United States

Feed Name or Analyses			Mean As Fed	Mean Dry	C.V. ± %
Dry matter	%			100.0	
Ash	%			6.0	
Crude fiber	%			28.2	
Protein (N x 6.25)	%			6.3	
Cattle	dig prot %	*		3.2	
Goats	dig prot %	*		2.4	
Horses	dig prot %	*		2.9	
Rabbits	dig prot %	*		3.5	
Sheep	dig prot %	*		2.9	
Calcium	%			0.29	
Phosphorus	%			0.22	
Potassium	%			1.08	

BROME,CHEATGRASS. Bromus tectorum

Brome, cheatgrass, hay, s-c, (1)

Ref No 1 00 907 United States

Feed Name or Analyses			Mean As Fed	Mean Dry	C.V. ± %
Dry matter	%		90.3	100.0	2
Ash	%		7.0	7.7	7
Crude fiber	%		30.0	33.3	5
Ether extract	%		2.0	2.2	13
N-free extract	%		44.1	48.9	
Protein (N x 6.25)	%		7.1	7.9	8
Cattle	dig prot %	*	3.4	3.7	
Goats	dig prot %	*	3.5	3.9	
Horses	dig prot %	*	3.8	4.2	
Rabbits	dig prot %	*	4.3	4.7	
Sheep	dig prot %	*	3.3	3.6	
Fatty acids	%		2.1	2.3	
Energy	GE Mcal/kg				
Cattle	DE Mcal/kg	*	2.31	2.56	
Sheep	DE Mcal/kg	*	2.21	2.44	
Cattle	ME Mcal/kg	*	1.89	2.10	
Sheep	ME Mcal/kg	*	1.81	2.00	
Cattle	TDN %	*	52.4	58.1	
Sheep	TDN %	*	50.0	55.4	
Calcium	%		0.29	0.32	
Phosphorus	%		0.25	0.27	
Potassium	%		1.45	1.60	

Brome, cheatgrass, hay, s-c, early bloom, (1)

Ref No 1 00 906 United States

Feed Name or Analyses			Mean As Fed	Mean Dry	C.V. ± %
Dry matter	%		86.5	100.0	
Ash	%		5.9	6.8	
Crude fiber	%		32.1	37.1	
Sheep	dig coef %		55.	55.	
Ether extract	%		1.9	2.2	
Sheep	dig coef %		61.	61.	
N-free extract	%		39.1	45.2	
Sheep	dig coef %		60.	60.	
Protein (N x 6.25)	%		7.5	8.7	
Sheep	dig coef %		54.	54.	
Cattle	dig prot %	*	3.9	4.5	
Goats	dig prot %	*	4.0	4.7	
Horses	dig prot %	*	4.3	4.9	
Rabbits	dig prot %	*	4.7	5.4	
Sheep	dig prot %		4.1	4.7	
Energy	GE Mcal/kg				
Cattle	DE Mcal/kg	*	1.93	2.23	
Sheep	DE Mcal/kg	*	2.11	2.44	
Cattle	ME Mcal/kg	*	1.58	1.83	
Sheep	ME Mcal/kg	*	1.73	2.00	
Cattle	TDN %	*	43.7	50.6	
Sheep	TDN %		47.8	55.2	

Brome, cheatgrass, aerial part, fresh, (2)

Ref No 2 00 913 United States

Feed Name or Analyses			Mean As Fed	Mean Dry	C.V. ± %
Dry matter	%			100.0	
Ash	%			9.4	17
Crude fiber	%			24.4	53
Ether extract	%			2.0	32
N-free extract	%			54.5	
Protein (N x 6.25)	%			9.7	50
Cattle	dig prot %	*		6.1	
Goats	dig prot %	*		5.6	
Horses	dig prot %	*		5.8	
Rabbits	dig prot %	*		6.2	
Sheep	dig prot %	*		6.0	
Energy	GE Mcal/kg				
Cattle	DE Mcal/kg	*		2.77	
Sheep	DE Mcal/kg	*		3.04	
Cattle	ME Mcal/kg	*		2.27	
Sheep	ME Mcal/kg	*		2.49	
Cattle	TDN %	*		62.9	
Sheep	TDN %	*		68.9	

Brome, cheatgrass, aerial part, fresh, immature, (2)

Ref No 2 00 908 United States

Feed Name or Analyses			Mean As Fed	Mean Dry	C.V. ± %
Dry matter	%			100.0	
Ash	%			9.6	
Crude fiber	%			22.9	
Ether extract	%			2.7	
N-free extract	%			49.0	
Protein (N x 6.25)	%			15.8	
Cattle	dig prot %	*		11.3	
Goats	dig prot %	*		11.3	
Horses	dig prot %	*		10.9	
Rabbits	dig prot %	*		10.9	
Sheep	dig prot %	*		11.7	
Energy	GE Mcal/kg				
Cattle	DE Mcal/kg	*		3.00	
Sheep	DE Mcal/kg	*		2.84	

Continued

Feed Name or Analyses			Mean As Fed	Mean Dry	C.V. ± %
Cattle	ME Mcal/kg	*		2.46	
Sheep	ME Mcal/kg	*		2.33	
Cattle	TDN %	*		68.1	
Sheep	TDN %	*		64.4	
Calcium	%			0.64	
Phosphorus	%			0.28	

Brome, cheatgrass, aerial part, fresh, full bloom, (2)

Ref No 2 00 909 United States

Analyses			As Fed	Dry	C.V. ± %
Dry matter	%			100.0	
Ash	%			8.4	
Crude fiber	%			31.7	
Ether extract	%			2.4	
N-free extract	%			47.5	
Protein (N x 6.25)	%			10.0	
Cattle	dig prot %	*		6.4	
Goats	dig prot %	*		5.9	
Horses	dig prot %	*		6.0	
Rabbits	dig prot %	*		6.4	
Sheep	dig prot %	*		6.3	
Energy	GE Mcal/kg				
Cattle	DE Mcal/kg	*		2.89	
Sheep	DE Mcal/kg	*		2.88	
Cattle	ME Mcal/kg	*		2.37	
Sheep	ME Mcal/kg	*		2.36	
Cattle	TDN %	*		65.5	
Sheep	TDN %	*		65.2	
Calcium	%			0.41	
Phosphorus	%			0.25	

Brome, cheatgrass, aerial part, fresh, mature, (2)

Ref No 2 00 911 United States

Analyses			As Fed	Dry	C.V. ± %
Dry matter	%			100.0	
Ash	%			8.7	
Crude fiber	%			34.8	
Ether extract	%			1.3	
N-free extract	%			49.9	
Protein (N x 6.25)	%			5.3	
Cattle	dig prot %	*		2.4	
Goats	dig prot %	*		1.5	
Horses	dig prot %	*		2.0	
Rabbits	dig prot %	*		2.8	
Sheep	dig prot %	*		1.9	
Energy	GE Mcal/kg				
Cattle	DE Mcal/kg	*		2.74	
Sheep	DE Mcal/kg	*		2.99	
Cattle	ME Mcal/kg	*		2.24	
Sheep	ME Mcal/kg	*		2.45	
Cattle	TDN %	*		62.0	
Sheep	TDN %	*		67.7	
Calcium	%			0.38	
Phosphorus	%			0.27	

(1) dry forages and roughages (2) pasture, range plants, and forages fed green (3) silages (4) energy feeds (5) protein supplements (6) minerals (7) vitamins (8) additives

BROME, CHESS. Bromus secalinus

Brome, chess, seeds, (4)

Ref No 4 08 360 United States

Analyses			As Fed	Dry	C.V. ± %
Dry matter	%		89.6	100.0	
Ash	%		3.6	4.0	
Crude fiber	%		8.2	9.2	
Ether extract	%		1.7	1.9	
N-free extract	%		66.4	74.1	
Protein (N x 6.25)	%		9.7	10.8	
Cattle	dig prot %	*	5.3	6.0	
Goats	dig prot %	*	6.4	7.2	
Horses	dig prot %	*	6.4	7.2	
Sheep	dig prot %	*	6.4	7.2	
Energy	GE Mcal/kg				
Cattle	DE Mcal/kg	*	3.31	3.69	
Sheep	DE Mcal/kg	*	3.14	3.51	
Swine	DE kcal/kg	*	2737.	3055.	
Cattle	ME Mcal/kg	*	2.71	3.03	
Sheep	ME Mcal/kg	*	2.58	2.88	
Swine	ME kcal/kg	*	2568.	2866.	
Cattle	TDN %	*	75.1	83.8	
Sheep	TDN %	*	71.3	79.6	
Swine	TDN %	*	62.1	69.3	

BROME, FOXTAIL. Bromus rubens

Brome, foxtail, aerial part, dehy grnd, milk stage, (1)

Ref No 1 08 242 United States

Analyses			As Fed	Dry	C.V. ± %
Dry matter	%		91.4	100.0	
Organic matter	%		87.0	95.2	
Ash	%		4.4	4.8	
Crude fiber	%		25.7	28.1	
Ether extract	%		1.5	1.6	
N-free extract	%		54.3	59.4	
Protein (N x 6.25)	%		5.6	6.1	
Cattle	dig prot %	*	2.0	2.2	
Goats	dig prot %	*	2.1	2.3	
Horses	dig prot %	*	2.5	2.7	
Rabbits	dig prot %	*	3.1	3.4	
Sheep	dig prot %	*	1.9	2.0	
Silicon	%		2.58	2.82	

Brome, foxtail, aerial part, dehy grnd, over ripe, (1)

Ref No 1 08 243 United States

Analyses			As Fed	Dry	C.V. ± %
Dry matter	%		92.4	100.0	
Organic matter	%		87.6	94.8	
Ash	%		4.8	5.2	
Crude fiber	%		28.7	31.1	
Ether extract	%		1.5	1.6	
N-free extract	%		51.7	56.0	
Protein (N x 6.25)	%		5.6	6.1	
Cattle	dig prot %	*	2.1	2.2	
Goats	dig prot %	*	2.1	2.3	
Horses	dig prot %	*	2.5	2.7	
Rabbits	dig prot %	*	3.1	3.4	
Sheep	dig prot %	*	1.9	2.0	
Silicon	%		2.59	2.80	

Brome, foxtail, aerial part, dehy grnd, dormant, (1)

Ref No 1 08 244 United States

Analyses			As Fed	Dry	C.V. ± %
Dry matter	%		92.6	100.0	0
Organic matter	%		87.8	94.8	1
Ash	%		4.9	5.3	14
Crude fiber	%		34.4	37.2	12
Ether extract	%		1.1	1.2	20
N-free extract	%		49.4	53.4	5
Protein (N x 6.25)	%		2.8	3.0	37
Cattle	dig prot %	*	-0.3	-0.3	
Goats	dig prot %	*	-0.5	-0.5	
Horses	dig prot %	*	0.1	0.1	
Rabbits	dig prot %	*	0.9	1.0	
Sheep	dig prot %	*	-0.6	-0.6	
Energy	GE Mcal/kg				
Cattle	DE Mcal/kg	*	2.01	2.17	
Sheep	DE Mcal/kg	*	2.14	2.31	
Cattle	ME Mcal/kg	*	1.65	1.78	
Sheep	ME Mcal/kg	*	1.75	1.89	
Cattle	TDN %	*	45.7	49.3	
Sheep	TDN %	*	48.5	52.3	
Silicon	%		3.22	3.48	20

Brome, foxtail, hay, s-c, (1)

Ref No 1 05 646 United States

Analyses			As Fed	Dry	C.V. ± %
Dry matter	%		93.5	100.0	
Ash	%		10.0	10.7	
Crude fiber	%		16.1	17.2	
Ether extract	%		4.2	4.5	
Silicon	%		1.79	1.92	

Brome, foxtail, leaves, dehy grnd, early leaf, (1)

Ref No 1 08 245 United States

Analyses			As Fed	Dry	C.V. ± %
Dry matter	%		93.1	100.0	
Organic matter	%		82.0	88.2	
Ash	%		11.1	11.9	
Crude fiber	%		14.0	15.0	
Ether extract	%		4.9	5.3	
N-free extract	%		37.3	40.1	
Protein (N x 6.25)	%		25.9	27.8	
Cattle	dig prot %	*	19.6	21.0	
Goats	dig prot %	*	20.9	22.5	
Horses	dig prot %	*	19.7	21.1	
Rabbits	dig prot %	*	18.7	20.1	
Sheep	dig prot %	*	20.0	21.5	
Silicon	%		1.27	1.37	

Brome, foxtail, leaves, dehy grnd, pre-bloom, (1)

Ref No 1 08 246 United States

Analyses			As Fed	Dry	C.V. ± %
Dry matter	%		93.2	100.0	
Organic matter	%		85.3	91.5	
Ash	%		7.9	8.5	
Crude fiber	%		25.5	27.4	
Ether extract	%		3.3	3.5	
N-free extract	%		44.5	47.7	
Protein (N x 6.25)	%		12.0	12.9	
Cattle	dig prot %	*	7.6	8.1	
Goats	dig prot %	*	8.0	8.6	
Horses	dig prot %	*	7.9	8.5	
Rabbits	dig prot %	*	8.0	8.6	

Feed Name or Analyses			Mean As Fed	Mean Dry	C.V. ± %
Sheep	dig prot %	*	7.6	8.1	
Silicon	%		1.85	1.99	

Brome, foxtail, leaves, dehy grnd, early bloom, (1)

Ref No 1 08 247 United States

Feed Name or Analyses			As Fed	Dry	C.V. ± %
Dry matter	%		93.5	100.0	
Organic matter	%		84.4	90.3	
Ash	%		8.9	9.5	
Crude fiber	%		21.2	22.7	
Ether extract	%		3.8	4.1	
N-free extract	%		45.5	48.7	
Protein (N x 6.25)	%		14.1	15.0	
Cattle	dig prot %	*	9.3	10.0	
Goats	dig prot %	*	9.9	10.6	
Horses	dig prot %	*	9.6	10.3	
Rabbits	dig prot %	*	9.6	10.3	
Sheep	dig prot %	*	9.4	10.1	
Silicon	%		2.31	2.47	

Brome, foxtail, leaves, dehy grnd, dormant, (1)

Ref No 1 08 248 United States

Feed Name or Analyses			As Fed	Dry	C.V. ± %
Dry matter	%		92.6	100.0	
Organic matter	%		81.1	87.6	
Ash	%		11.5	12.4	
Protein (N x 6.25)	%		3.3	3.6	
Cattle	dig prot %	*	0.1	0.1	
Goats	dig prot %	*	0.0	0.0	
Horses	dig prot %	*	0.5	0.6	
Rabbits	dig prot %	*	1.3	1.4	
Sheep	dig prot %	*	-0.1	-0.1	
Silicon	%		6.70	7.24	

Brome, foxtail, leaves w some stems, dehy grnd, milk stage, (1)

Ref No 1 08 249 United States

Feed Name or Analyses			As Fed	Dry	C.V. ± %
Dry matter	%		93.2	100.0	
Organic matter	%		82.7	88.7	
Ash	%		10.5	11.3	
Silicon	%		4.76	5.11	

Brome, foxtail, aerial part, fresh, (2)

Ref No 2 00 918 United States

Feed Name or Analyses			As Fed	Dry	C.V. ± %
Dry matter	%			100.0	
Ash	%			6.5	22
Crude fiber	%			31.6	13
Ether extract	%			1.1	
N-free extract	%			54.9	
Protein (N x 6.25)	%			6.0	98
Cattle	dig prot %	*		2.9	
Goats	dig prot %	*		2.1	
Horses	dig prot %	*		2.6	
Rabbits	dig prot %	*		3.3	
Sheep	dig prot %	*		2.5	
Cellulose (Matrone)	%			40.3	
Other carbohydrates	%			40.3	
Lignin (Ellis)	%			9.2	
Calcium	%			0.28	40
Phosphorus	%			0.23	37
Potassium	%			1.45	39
Silicon	%			3.30	

Brome, foxtail, aerial part, fresh, immature, (2)

Ref No 2 00 914 United States

Feed Name or Analyses			As Fed	Dry	C.V. ± %
Dry matter	%			100.0	
Calcium	%			0.57	
Phosphorus	%			0.35	
Potassium	%			2.90	

Brome, foxtail, aerial part, fresh, early bloom, (2)

Ref No 2 00 915 United States

Feed Name or Analyses			As Fed	Dry	C.V. ± %
Dry matter	%			100.0	
Ash	%			9.5	
Crude fiber	%			28.0	
Protein (N x 6.25)	%			14.3	
Cattle	dig prot %	*		10.0	
Goats	dig prot %	*		9.9	
Horses	dig prot %	*		9.7	
Rabbits	dig prot %	*		9.7	
Sheep	dig prot %	*		10.3	
Calcium	%			0.39	
Phosphorus	%			0.34	
Potassium	%			2.66	

Brome, foxtail, aerial part, fresh, mature, (2)

Ref No 2 00 916 United States

Feed Name or Analyses			As Fed	Dry	C.V. ± %
Dry matter	%			100.0	
Ash	%			6.8	11
Crude fiber	%			30.8	6
Protein (N x 6.25)	%			7.1	8
Cattle	dig prot %	*		3.9	
Goats	dig prot %	*		3.2	
Horses	dig prot %	*		3.6	
Rabbits	dig prot %	*		4.2	
Sheep	dig prot %	*		3.6	
Calcium	%			0.26	18
Phosphorus	%			0.23	21
Potassium	%			1.22	20

Brome, foxtail, aerial part, fresh, over ripe, (2)

Ref No 2 00 917 United States

Feed Name or Analyses			As Fed	Dry	C.V. ± %
Dry matter	%			100.0	
Ash	%			5.9	
Crude fiber	%			37.6	
Protein (N x 6.25)	%			4.4	
Cattle	dig prot %	*		1.6	
Goats	dig prot %	*		0.7	
Horses	dig prot %	*		1.3	
Rabbits	dig prot %	*		2.1	
Sheep	dig prot %	*		1.1	
Calcium	%			0.21	
Phosphorus	%			0.12	
Potassium	%			1.11	

BROME, JAPANESE, Bromus japonicus

Brome, Japanese, aerial part, fresh, immature, (2)

Ref No 2 00 919 United States

Feed Name or Analyses			As Fed	Dry	C.V. ± %
Dry matter	%			100.0	
Ash	%			10.1	
Crude fiber	%			28.7	
Ether extract	%			3.1	
N-free extract	%			42.0	
Protein (N x 6.25)	%			16.1	
Cattle	dig prot %	*		11.6	
Goats	dig prot %	*		11.6	
Horses	dig prot %	*		11.2	
Rabbits	dig prot %	*		11.1	
Sheep	dig prot %	*		12.0	
Calcium	%			0.40	
Phosphorus	%			0.26	

BROME, LINCOLN. Bromus spp

Brome, Lincoln, hay, s-c, (1)

Ref No 1 09 079 United States

Feed Name or Analyses			As Fed	Dry	C.V. ± %
Dry matter	%		88.8	100.0	1
Horses	dig coef %		49.	49.	11
Organic matter	%		81.9	92.2	0
Ash	%		6.9	7.8	4
Crude fiber	%		33.7	37.9	7
Horses	dig coef %		49.	49.	15
Ether extract	%		2.4	2.7	11
N-free extract	%		35.4	39.8	8
Horses	dig coef %		46.	46.	15
Protein (N x 6.25)	%		10.4	11.7	6
Cattle	dig prot %	*	6.3	7.1	
Goats	dig prot %	*	6.7	7.5	
Horses	dig prot %	*	6.6	7.5	
Rabbits	dig prot %	*	6.9	7.7	
Sheep	dig prot %	*	6.3	7.1	
Cell contents (Van Soest)	%		21.3	24.0	6
Horses	dig coef %		42.	42.	18
Cell walls (Van Soest)	%		67.4	76.0	2
Horses	dig coef %		51.	51.	10
Cellulose (Matrone)	%		32.3	36.3	2
Horses	dig coef %		49.	49.	10
Fiber, acid detergent (VS)	%		36.2	40.7	2
Horses	dig coef %		43.	43.	14
Hemicellulose	%		24.9	28.1	7
Horses	dig coef %		57.	57.	8
Holocellulose	%		57.2	64.4	2
Soluble carbohydrates	%		7.9	8.9	27
Lignin (Ellis)	%		3.9	4.4	12
Horses	dig coef %		-11.	-11.	98
Energy	GE Mcal/kg		3.99	4.50	0
Horses	GE dig coef %		43.	43.	15
Cattle	DE Mcal/kg	*	1.95	2.20	
Horses	DE Mcal/kg		1.70	1.92	14
Sheep	DE Mcal/kg	*	2.14	2.40	
Cattle	ME Mcal/kg	*	1.60	1.81	
Horses	ME Mcal/kg	*	1.40	1.57	
Sheep	ME Mcal/kg	*	1.75	1.97	
Cattle	TDN %	*	44.3	49.9	
Sheep	TDN %	*	48.4	54.5	
Calcium	%		0.33	0.38	14
Magnesium	%		0.12	0.13	19
Phosphorus	%		0.22	0.25	11
Potassium	%		2.68	3.02	7
Carotene	mg/kg		7.8	8.8	39
Vitamin A equivalent	IU/g		13.0	14.6	

Brome, Lincoln, aerial part, fresh, (2)

Ref No 2 05 562 United States

Feed Name or Analyses			As Fed	Dry	C.V. ± %
Dry matter	%			100.0	
Organic matter	%			92.1	1

Continued

Feed Name or Analyses			Mean As Fed	Mean Dry	C.V. ± %
Crude fiber	%			23.0	14
Sheep	dig coef %			74.	12
Ether extract	%			6.3	10
Sheep	dig coef %			61.	17
N-free extract	%			43.7	8
Sheep	dig coef %			76.	10
Protein (N x 6.25)	%			19.0	17
Sheep	dig coef %			74.	12
Cattle	dig prot %	*		14.1	
Goats	dig prot %	*		14.3	
Horses	dig prot %	*		13.7	
Rabbits	dig prot %	*		13.4	
Sheep	dig prot %			14.1	
Hemicellulose	%			22.2	
Lignin (Ellis)	%			5.0	35
Energy	GE Mcal/kg				
Cattle	DE Mcal/kg	*		3.04	
Sheep	DE Mcal/kg	*		3.23	
Cattle	ME Mcal/kg	*		2.49	
Sheep	ME Mcal/kg	*		2.64	
Cattle	TDN %	*		68.8	
Sheep	TDN %			73.2	9

BROME, MANCHAR. Bromus inermis

Brome, manchar, hay, s-c, full bloom, (1)

Ref No 1 08 668 United States

Feed Name or Analyses			Mean As Fed	Mean Dry	C.V. ± %
Dry matter	%			100.0	
Sheep	dig coef %			57.	
Ash	%			6.5	
Crude fiber	%			36.8	
Ether extract	%			1.0	
N-free extract	%			47.8	
Protein (N x 6.25)	%			7.9	
Cattle	dig prot %	*		3.8	
Goats	dig prot %	*		3.9	
Horses	dig prot %	*		4.2	
Rabbits	dig prot %	*		4.8	
Sheep	dig prot %	*		3.7	
Lignin (Ellis)	%			8.7	
Energy	GE Mcal/kg			4.37	
Sheep	GE dig coef %			55.	
Cattle	DE Mcal/kg	*		2.20	
Sheep	DE Mcal/kg			2.40	
Cattle	ME Mcal/kg	*		1.80	
Sheep	ME Mcal/kg	*		1.97	
Cattle	TDN %	*		49.8	
Sheep	TDN %	*		54.1	

Brome, manchar, hay, s-c, chopped, full bloom, (1)

Ref No 1 08 775 United States

Feed Name or Analyses			Mean As Fed	Mean Dry	C.V. ± %
Dry matter	%			100.0	
Sheep	dig coef %			54.	
Ash	%			8.5	
Crude fiber	%			36.7	
Ether extract	%			1.6	
N-free extract	%			44.7	
Protein (N x 6.25)	%			8.5	
Cattle	dig prot %	*		4.3	
Goats	dig prot %	*		4.5	
Horses	dig prot %	*		4.7	

(1) dry forages and roughages (2) pasture, range plants, and forages fed green (3) silages (4) energy feeds (5) protein supplements (6) minerals (7) vitamins (8) additives

Feed Name or Analyses			Mean As Fed	Mean Dry	C.V. ± %
Rabbits	dig prot %	*		5.2	
Sheep	dig prot %	*		4.2	
Lignin (Ellis)	%			8.8	
Energy	GE Mcal/kg			4.22	
Cattle	DE Mcal/kg	*		2.31	
Sheep	DE Mcal/kg	*		2.38	
Cattle	ME Mcal/kg	*		1.89	
Sheep	ME Mcal/kg	*		1.95	
Cattle	TDN %	*		52.3	
Sheep	TDN %	*		54.0	

BROME, MOUNTAIN. Bromus carinatus

Brome, mountain, hay, s-c, (1)

Ref No 1 00 920 United States

Feed Name or Analyses			Mean As Fed	Mean Dry	C.V. ± %
Dry matter	%		93.3	100.0	
Ash	%		7.7	8.2	
Crude fiber	%		31.5	33.8	
Sheep	dig coef %		53.	53.	
Ether extract	%		1.0	1.1	
Sheep	dig coef %		16.	16.	
N-free extract	%		44.4	47.6	
Sheep	dig coef %		67.	67.	
Protein (N x 6.25)	%		8.7	9.3	
Sheep	dig coef %		68.	68.	
Cattle	dig prot %	*	4.7	5.0	
Goats	dig prot %	*	4.9	5.2	
Horses	dig prot %	*	5.1	5.4	
Rabbits	dig prot %	*	5.5	5.8	
Sheep	dig prot %		5.9	6.3	
Energy	GE Mcal/kg				
Cattle	DE Mcal/kg	*	2.31	2.48	
Sheep	DE Mcal/kg	*	2.33	2.49	
Cattle	ME Mcal/kg	*	1.90	2.03	
Sheep	ME Mcal/kg	*	1.91	2.04	
Cattle	TDN %	*	52.4	56.2	
Sheep	TDN %		52.7	56.5	

Brome, mountain, aerial part, fresh, (2)

Ref No 2 00 924 United States

Feed Name or Analyses			Mean As Fed	Mean Dry	C.V. ± %
Dry matter	%		56.2	100.0	13
Calcium	%		0.22	0.39	20
Magnesium	%		0.10	0.17	
Phosphorus	%		0.13	0.24	40

Brome, mountain, aerial part, fresh, immature, (2)

Ref No 2 00 921 United States

Feed Name or Analyses			Mean As Fed	Mean Dry	C.V. ± %
Dry matter	%			100.0	
Ash	%			9.7	
Crude fiber	%			29.0	
Ether extract	%			2.0	
N-free extract	%			46.3	
Protein (N x 6.25)	%			13.0	
Cattle	dig prot %	*		8.9	
Goats	dig prot %	*		8.7	
Horses	dig prot %	*		8.6	
Rabbits	dig prot %	*		8.7	
Sheep	dig prot %	*		9.1	
Energy	GE Mcal/kg				
Cattle	DE Mcal/kg	*		2.96	
Sheep	DE Mcal/kg	*		2.82	
Cattle	ME Mcal/kg	*		2.42	
Sheep	ME Mcal/kg	*		2.31	
Cattle	TDN %	*		67.0	

Feed Name or Analyses			Mean As Fed	Mean Dry	C.V. ± %
Sheep	TDN %	*		63.9	
Calcium	%			0.54	
Magnesium	%			0.19	
Phosphorus	%			0.43	
Sulphur	%			0.24	

Brome, mountain, aerial part, fresh, mature, (2)

Ref No 2 00 922 United States

Feed Name or Analyses			Mean As Fed	Mean Dry	C.V. ± %
Dry matter	%		56.2	100.0	13
Ash	%		2.8	5.0	11
Crude fiber	%		24.2	43.0	4
Ether extract	%		1.1	2.0	6
N-free extract	%		25.3	45.1	
Protein (N x 6.25)	%		2.8	4.9	10
Cattle	dig prot %	*	1.2	2.1	
Goats	dig prot %	*	0.6	1.1	
Horses	dig prot %	*	0.9	1.7	
Rabbits	dig prot %	*	1.4	2.5	
Sheep	dig prot %	*	0.9	1.6	
Energy	GE Mcal/kg				
Cattle	DE Mcal/kg	*	1.68	2.98	
Sheep	DE Mcal/kg	*	1.61	2.86	
Cattle	ME Mcal/kg	*	1.37	2.45	
Sheep	ME Mcal/kg	*	1.32	2.35	
Cattle	TDN %	*	38.0	67.6	
Sheep	TDN %	*	36.5	64.9	
Calcium	%		0.20	0.36	11
Magnesium	%		0.08	0.14	
Phosphorus	%		0.11	0.20	11

Brome, mountain, aerial part, fresh, over ripe, (2)

Ref No 2 00 923 United States

Feed Name or Analyses			Mean As Fed	Mean Dry	C.V. ± %
Dry matter	%			100.0	
Ash	%			5.8	
Crude fiber	%			44.2	
Ether extract	%			1.8	
N-free extract	%			45.2	
Protein (N x 6.25)	%			3.0	
Cattle	dig prot %	*		0.4	
Goats	dig prot %	*		-0.5	
Horses	dig prot %	*		0.1	
Rabbits	dig prot %	*		1.0	
Sheep	dig prot %	*		-0.1	
Energy	GE Mcal/kg				
Cattle	DE Mcal/kg	*		2.86	
Sheep	DE Mcal/kg	*		2.89	
Cattle	ME Mcal/kg	*		2.34	
Sheep	ME Mcal/kg	*		2.37	
Cattle	TDN %	*		64.9	
Sheep	TDN %	*		65.6	
Sulphur	%			0.10	

Brome, mountain, heads, fresh, (2)

Ref No 2 00 925 United States

Feed Name or Analyses			Mean As Fed	Mean Dry	C.V. ± %
Dry matter	%		44.2	100.0	12
Ash	%		1.9	4.4	5
Crude fiber	%		12.8	28.9	7
Ether extract	%		0.8	1.9	10
N-free extract	%		24.4	55.2	
Protein (N x 6.25)	%		4.2	9.6	7
Cattle	dig prot %	*	2.7	6.1	
Goats	dig prot %	*	2.4	5.5	
Horses	dig prot %	*	2.5	5.7	

Feed Name or Analyses			Mean As Fed	Mean Dry	C.V. ± %
Rabbits	dig prot %	*	2.7	6.1	
Sheep	dig prot %	*	2.6	5.9	
Energy	GE Mcal/kg				
Cattle	DE Mcal/kg	*	1.39	3.14	
Sheep	DE Mcal/kg	*	1.33	3.01	
Cattle	ME Mcal/kg	*	1.14	2.58	
Sheep	ME Mcal/kg	*	1.09	2.47	
Cattle	TDN %	*	31.5	71.2	
Sheep	TDN %	*	30.1	68.2	
Calcium	%		0.15	0.34	19
Magnesium	%		0.08	0.17	20
Phosphorus	%		0.14	0.32	12
Sulphur	%		0.08	0.17	43

Brome, mountain, leaves, fresh, (2)
Ref No 2 00 927 United States

Feed Name or Analyses			As Fed	Dry	C.V. ± %
Dry matter	%		55.0	100.0	14
Ash	%		5.0	9.0	5
Crude fiber	%		17.2	31.2	6
Ether extract	%		3.1	5.6	7
N-free extract	%		24.4	44.4	
Protein (N x 6.25)	%		5.4	9.8	16
Cattle	dig prot %	*	3.4	6.2	
Goats	dig prot %	*	3.1	5.7	
Horses	dig prot %	*	3.2	5.9	
Rabbits	dig prot %	*	3.4	6.2	
Sheep	dig prot %	*	3.4	6.1	
Energy	GE Mcal/kg				
Cattle	DE Mcal/kg	*	1.44	2.63	
Sheep	DE Mcal/kg	*	1.56	2.85	
Cattle	ME Mcal/kg	*	1.18	2.15	
Sheep	ME Mcal/kg	*	1.28	2.33	
Cattle	TDN %	*	32.8	59.6	
Sheep	TDN %	*	35.5	64.5	
Calcium	%		0.53	0.97	5
Phosphorus	%		0.12	0.22	19

Brome, mountain, leaves, fresh, immature, (2)
Ref No 2 00 926 United States

Feed Name or Analyses			As Fed	Dry	C.V. ± %
Dry matter	%		35.1	100.0	9
Phosphorus	%		0.11	0.30	10

Brome, mountain, stems, fresh, mature, (2)
Ref No 2 00 928 United States

Feed Name or Analyses			As Fed	Dry	C.V. ± %
Dry matter	%		50.0	100.0	13
Ash	%		2.1	4.2	6
Crude fiber	%		24.2	48.4	3
Ether extract	%		0.5	1.0	13
N-free extract	%		21.7	43.4	
Protein (N x 6.25)	%		1.5	3.0	21
Cattle	dig prot %	*	0.2	0.4	
Goats	dig prot %	*	-0.2	-0.5	
Horses	dig prot %	*	0.0	0.1	
Rabbits	dig prot %	*	0.5	1.0	
Sheep	dig prot %	*	0.0	-0.1	
Calcium	%		0.12	0.23	61
Phosphorus	%		0.09	0.18	82

Brome, mountain, stems, fresh, over ripe, (2)
Ref No 2 00 929 United States

Feed Name or Analyses			As Fed	Dry	C.V. ± %
Dry matter	%		80.3	100.0	7
Calcium	%		0.16	0.20	25
Phosphorus	%		0.10	0.12	38

BROME, NODDING. Bromus anomalus

Brome, nodding, aerial part, fresh, immature, (2)
Ref No 2 00 930 United States

Feed Name or Analyses			As Fed	Dry	C.V. ± %
Dry matter	%			100.0	
Ash	%			11.7	
Crude fiber	%			20.2	
Ether extract	%			4.4	
N-free extract	%			37.7	
Protein (N x 6.25)	%			26.0	
Cattle	dig prot %	*		20.0	
Goats	dig prot %	*		20.8	
Horses	dig prot %	*		19.6	
Rabbits	dig prot %	*		18.7	
Sheep	dig prot %	*		21.2	
Energy	GE Mcal/kg				
Cattle	DE Mcal/kg	*		2.80	
Sheep	DE Mcal/kg	*		2.99	
Cattle	ME Mcal/kg	*		2.30	
Sheep	ME Mcal/kg	*		2.45	
Cattle	TDN %	*		63.4	
Sheep	TDN %	*		67.8	

BROME, RESCUE. Bromus catharticus

Brome, rescue, hay, s-c, (1)
Ref No 1 00 931 United States

Feed Name or Analyses			As Fed	Dry	C.V. ± %
Dry matter	%		90.2	100.0	
Ash	%		8.1	9.0	
Crude fiber	%		24.6	27.3	
Ether extract	%		3.2	3.5	
N-free extract	%		44.5	49.3	
Protein (N x 6.25)	%		9.8	10.9	
Cattle	dig prot %	*	5.7	6.4	
Goats	dig prot %	*	6.1	6.7	
Horses	dig prot %	*	6.1	6.8	
Rabbits	dig prot %	*	6.4	7.1	
Sheep	dig prot %	*	5.7	6.3	
Energy	GE Mcal/kg				
Cattle	DE Mcal/kg	*	2.24	2.48	
Sheep	DE Mcal/kg	*	2.35	2.60	
Cattle	ME Mcal/kg	*	1.83	2.03	
Sheep	ME Mcal/kg	*	1.93	2.13	
Cattle	TDN %	*	50.8	56.3	
Sheep	TDN %	*	53.2	59.0	

Brome, rescue, aerial part, fresh, (2)
Ref No 2 08 361 United States

Feed Name or Analyses			As Fed	Dry	C.V. ± %
Dry matter	%		28.9	100.0	
Ash	%		4.0	13.8	
Crude fiber	%		6.7	23.2	
Ether extract	%		1.0	3.5	
N-free extract	%		12.2	42.2	
Protein (N x 6.25)	%		5.0	17.3	
Cattle	dig prot %	*	3.6	12.6	
Goats	dig prot %	*	3.7	12.7	
Horses	dig prot %	*	3.5	12.2	
Rabbits	dig prot %	*	3.5	12.0	
Sheep	dig prot %	*	3.8	13.1	
Energy	GE Mcal/kg				
Cattle	DE Mcal/kg	*	0.80	2.75	
Sheep	DE Mcal/kg	*	0.87	3.03	
Cattle	ME Mcal/kg	*	0.65	2.26	
Sheep	ME Mcal/kg	*	0.72	2.48	
Cattle	TDN %	*	18.0	62.4	
Sheep	TDN %	*	19.8	68.6	
Calcium	%		0.15	0.52	
Phosphorus	%		0.08	0.28	

Brome, rescue, aerial part, fresh, immature, (2)
Ref No 2 00 932 United States

Feed Name or Analyses			As Fed	Dry	C.V. ± %
Dry matter	%		29.4	100.0	11
Ash	%		4.1	14.1	35
Crude fiber	%		7.0	23.9	13
Ether extract	%		0.9	3.2	21
N-free extract	%		12.2	41.6	
Protein (N x 6.25)	%		5.1	17.2	27
Cattle	dig prot %	*	3.7	12.5	
Goats	dig prot %	*	3.7	12.6	
Horses	dig prot %	*	3.6	12.1	
Rabbits	dig prot %	*	3.5	11.9	
Sheep	dig prot %	*	3.8	13.0	
Energy	GE Mcal/kg				
Cattle	DE Mcal/kg	*	0.81	2.75	
Sheep	DE Mcal/kg	*	0.88	3.00	
Cattle	ME Mcal/kg	*	0.66	2.26	
Sheep	ME Mcal/kg	*	0.72	2.46	
Cattle	TDN %	*	18.4	62.4	
Sheep	TDN %	*	20.0	68.0	
Calcium	%		0.16	0.56	5
Magnesium	%		0.05	0.17	
Phosphorus	%		0.09	0.29	31
Potassium	%		1.13	3.85	
Carotene	mg/kg		119.4	406.1	
Vitamin A equivalent	IU/g		199.0	676.9	

Brome, rescue, aerial part, fresh, pre–bloom, (2)
Ref No 2 08 362 United States

Feed Name or Analyses			As Fed	Dry	C.V. ± %
Dry matter	%		30.0	100.0	
Ash	%		2.2	7.3	
Crude fiber	%		9.2	30.7	
Ether extract	%		1.0	3.3	
N-free extract	%		13.4	44.7	
Protein (N x 6.25)	%		4.2	14.0	
Cattle	dig prot %	*	2.9	9.8	
Goats	dig prot %	*	2.9	9.6	
Horses	dig prot %	*	2.8	9.4	
Rabbits	dig prot %	*	2.8	9.5	
Sheep	dig prot %	*	3.0	10.0	
Energy	GE Mcal/kg				
Cattle	DE Mcal/kg	*	0.93	3.09	
Sheep	DE Mcal/kg	*	0.86	2.87	
Cattle	ME Mcal/kg	*	0.76	2.53	
Sheep	ME Mcal/kg	*	0.70	2.35	
Cattle	TDN %	*	21.0	70.0	
Sheep	TDN %	*	19.5	65.0	

Brome, rescue, aerial part, fresh, early bloom, (2)

Ref No 2 00 933 United States

Feed Name or Analyses			Mean As Fed	Mean Dry	C.V. ± %
Dry matter	%			100.0	
Ash	%			7.2	
Crude fiber	%			26.8	
Ether extract	%			2.0	
N-free extract	%			51.7	
Protein (N x 6.25)	%			12.3	
Cattle	dig prot %	*		8.3	
Goats	dig prot %	*		8.0	
Horses	dig prot %	*		8.0	
Rabbits	dig prot %	*		8.2	
Sheep	dig prot %	*		8.5	
Energy	GE Mcal/kg				
Cattle	DE Mcal/kg	*		3.07	
Sheep	DE Mcal/kg	*		2.92	
Cattle	ME Mcal/kg	*		2.51	
Sheep	ME Mcal/kg	*		2.39	
Cattle	TDN %	*		69.5	
Sheep	TDN %	*		66.2	
Calcium	%			0.35	
Magnesium	%			0.11	
Phosphorus	%			0.30	
Potassium	%			1.97	

BROME, RIPGUT. Bromus rigidus

Brome, ripgut, aerial part, dehy grnd, (1)

Ref No 1 08 252 United States

Feed Name or Analyses			Mean As Fed	Mean Dry	C.V. ± %
Dry matter	%		91.7	100.0	
Organic matter	%		86.9	94.8	
Ash	%		4.8	5.2	
Crude fiber	%		28.5	31.1	
Ether extract	%		1.6	1.7	
N-free extract	%		53.1	57.9	
Protein (N x 6.25)	%		3.8	4.1	
Cattle	dig prot %	*	0.4	0.5	
Goats	dig prot %	*	0.4	0.4	
Horses	dig prot %	*	0.9	1.0	
Rabbits	dig prot %	*	1.7	1.8	
Sheep	dig prot %	*	0.2	0.2	
Energy	GE Mcal/kg				
Cattle	DE Mcal/kg	*	2.11	2.30	
Sheep	DE Mcal/kg	*	2.25	2.45	
Cattle	ME Mcal/kg	*	1.73	1.89	
Sheep	ME Mcal/kg	*	1.84	2.01	
Cattle	TDN %	*	47.8	52.1	
Sheep	TDN %	*	51.0	55.6	
Silicon	%		2.38	2.59	

Brome, ripgut, aerial part, dehy grnd, over ripe, (1)

Ref No 1 08 250 United States

Feed Name or Analyses			Mean As Fed	Mean Dry	C.V. ± %
Dry matter	%		93.5	100.0	
Organic matter	%		87.7	93.8	
Ash	%		5.8	6.2	
Crude fiber	%		31.8	34.0	
Ether extract	%		1.6	1.7	
N-free extract	%		51.2	54.8	
Protein (N x 6.25)	%		3.1	3.3	
Cattle	dig prot %	*	-0.1	-0.1	
Goats	dig prot %	*	-0.2	-0.3	
Horses	dig prot %	*	0.3	0.3	
Rabbits	dig prot %	*	1.1	1.2	
Sheep	dig prot %	*	-0.3	-0.4	
Energy	GE Mcal/kg				
Cattle	DE Mcal/kg	*	2.31	2.47	
Sheep	DE Mcal/kg	*	2.21	2.37	
Cattle	ME Mcal/kg	*	1.89	2.02	
Sheep	ME Mcal/kg	*	1.82	1.94	
Cattle	TDN %	*	52.3	56.0	
Sheep	TDN %	*	50.2	53.7	
Silicon	%		2.71	2.90	

Brome, ripgut, aerial part, dehy grnd, dormant, (1)

Ref No 1 08 251 United States

Feed Name or Analyses			Mean As Fed	Mean Dry	C.V. ± %
Dry matter	%		92.6	100.0	0
Organic matter	%		87.6	94.6	2
Ash	%		5.0	5.4	27
Crude fiber	%		34.7	37.5	11
Ether extract	%		1.1	1.2	26
N-free extract	%		49.3	53.2	5
Protein (N x 6.25)	%		2.5	2.7	36
Cattle	dig prot %	*	-0.6	-0.7	
Goats	dig prot %	*	-0.8	-0.9	
Horses	dig prot %	*	-0.1	-0.1	
Rabbits	dig prot %	*	0.7	0.7	
Sheep	dig prot %	*	-0.9	-0.9	
Energy	GE Mcal/kg				
Cattle	DE Mcal/kg	*	2.00	2.16	
Sheep	DE Mcal/kg	*	2.12	2.29	
Cattle	ME Mcal/kg	*	1.64	1.77	
Sheep	ME Mcal/kg	*	1.74	1.88	
Cattle	TDN %	*	45.3	48.9	
Sheep	TDN %	*	48.1	52.0	
Silicon	%		2.95	3.18	33

Brome, ripgut, hay, s-c, (1)

Ref No 1 05 647 United States

Feed Name or Analyses			Mean As Fed	Mean Dry	C.V. ± %
Dry matter	%		93.8	100.0	
Ash	%		8.5	9.1	
Crude fiber	%		22.3	23.8	
Ether extract	%		2.7	2.9	
Silicon	%		1.77	1.89	

Brome, ripgut, leaves, dehy grnd, early leaf, (1)

Ref No 1 08 253 United States

Feed Name or Analyses			Mean As Fed	Mean Dry	C.V. ± %
Dry matter	%		93.0	100.0	
Organic matter	%		82.2	88.4	
Ash	%		10.8	11.6	
Crude fiber	%		15.3	16.4	
Ether extract	%		4.5	4.8	
N-free extract	%		40.0	43.0	
Protein (N x 6.25)	%		22.5	24.2	
Cattle	dig prot %	*	16.6	17.9	
Goats	dig prot %	*	17.8	19.1	
Horses	dig prot %	*	16.8	18.1	
Rabbits	dig prot %	*	16.1	17.4	
Sheep	dig prot %	*	17.0	18.3	
Energy	GE Mcal/kg				
Cattle	DE Mcal/kg	*	2.53	2.72	
Sheep	DE Mcal/kg	*	2.60	2.80	
Cattle	ME Mcal/kg	*	2.07	2.23	
Sheep	ME Mcal/kg	*	2.14	2.30	
Cattle	TDN %	*	57.4	61.7	
Sheep	TDN %	*	59.1	63.5	
Silicon	%		1.01	1.09	

Brome, ripgut, leaves, dehy grnd, pre-bloom, (1)

Ref No 1 08 254 United States

Feed Name or Analyses			Mean As Fed	Mean Dry	C.V. ± %
Dry matter	%		93.4	100.0	
Organic matter	%		85.4	91.4	
Ash	%		8.1	8.7	
Crude fiber	%		21.8	23.3	
Ether extract	%		2.8	3.1	
N-free extract	%		47.9	51.3	
Protein (N x 6.25)	%		12.8	13.8	
Cattle	dig prot %	*	8.3	8.8	
Goats	dig prot %	*	8.8	9.4	
Horses	dig prot %	*	8.6	9.2	
Rabbits	dig prot %	*	8.7	9.3	
Sheep	dig prot %	*	8.3	8.9	
Energy	GE Mcal/kg				
Cattle	DE Mcal/kg	*	2.68	2.87	
Sheep	DE Mcal/kg	*	2.53	2.71	
Cattle	ME Mcal/kg	*	2.20	2.35	
Sheep	ME Mcal/kg	*	2.07	2.22	
Cattle	TDN %	*	60.8	65.1	
Sheep	TDN %	*	57.4	61.4	
Silicon	%		1.50	1.61	

Brome, ripgut, leaves, dehy grnd, early bloom, (1)

Ref No 1 08 255 United States

Feed Name or Analyses			Mean As Fed	Mean Dry	C.V. ± %
Dry matter	%		93.0	100.0	
Organic matter	%		82.4	88.6	
Ash	%		10.6	11.4	
Ether extract	%		4.5	4.8	
Protein (N x 6.25)	%		8.6	9.2	
Cattle	dig prot %	*	4.6	4.9	
Goats	dig prot %	*	4.8	5.1	
Horses	dig prot %	*	5.0	5.3	
Rabbits	dig prot %	*	5.4	5.8	
Sheep	dig prot %	*	4.5	4.8	
Silicon	%		3.10	3.33	

Brome, ripgut, leaves, dehy grnd, mid–bloom, (1)

Ref No 1 08 256 United States

Feed Name or Analyses			Mean As Fed	Mean Dry	C.V. ± %
Dry matter	%		92.9	100.0	
Organic matter	%		82.5	88.8	
Ash	%		10.4	11.2	
Crude fiber	%		18.4	19.8	
Ether extract	%		2.9	3.1	
N-free extract	%		46.7	50.3	
Protein (N x 6.25)	%		14.5	15.6	
Cattle	dig prot %	*	9.7	10.4	
Goats	dig prot %	*	10.3	11.1	
Horses	dig prot %	*	10.0	10.8	
Rabbits	dig prot %	*	10.0	10.7	
Sheep	dig prot %	*	9.8	10.6	
Energy	GE Mcal/kg				
Cattle	DE Mcal/kg	*	2.55	2.75	
Sheep	DE Mcal/kg	*	2.53	2.73	
Cattle	ME Mcal/kg	*	2.09	2.25	
Sheep	ME Mcal/kg	*	2.08	2.24	
Cattle	TDN %	*	58.0	62.4	

(1) dry forages and roughages (3) sitages (6) minerals
(2) pasture, range plants, and forages fed green (4) energy feeds (7) vitamins
(5) protein supplements (8) additives

Feed Name or Analyses			Mean As Fed	Mean Dry	C.V. ± %
Sheep	TDN %	*	57.4	61.8	
Silicon	%		2.35	2.53	

Brome, ripgut, leaves, dehy grnd, dormant, (1)
Ref No 1 08 257 United States

Feed Name or Analyses			Mean As Fed	Mean Dry	C.V. ± %
Dry matter	%		92.8	100.0	
Organic matter	%		80.4	86.6	
Ash	%		12.4	13.4	
Protein (N x 6.25)	%		3.7	4.0	
Cattle	dig prot %	*	0.4	0.4	
Goats	dig prot %	*	0.3	0.3	
Horses	dig prot %	*	0.9	0.9	
Rabbits	dig prot %	*	1.6	1.8	
Sheep	dig prot %	*	0.1	0.2	
Silicon	%		6.30	6.79	

Brome, ripgut, leaves w some stems, dehy grnd, mid–bloom, (1)
Ref No 1 08 258 United States

Feed Name or Analyses			Mean As Fed	Mean Dry	C.V. ± %
Dry matter	%		93.3	100.0	
Organic matter	%		85.1	91.2	
Ash	%		8.2	8.8	
Crude fiber	%		18.0	19.3	
Ether extract	%		3.4	3.6	
N-free extract	%		54.8	58.7	
Protein (N x 6.25)	%		9.0	9.6	
Cattle	dig prot %	*	4.9	5.3	
Goats	dig prot %	*	5.1	5.5	
Horses	dig prot %	*	5.3	5.7	
Rabbits	dig prot %	*	5.7	6.1	
Sheep	dig prot %	*	4.8	5.2	
Energy	GE Mcal/kg				
Cattle	DE Mcal/kg	*	2.51	2.69	
Sheep	DE Mcal/kg	*	2.57	2.75	
Cattle	ME Mcal/kg	*	2.06	2.21	
Sheep	ME Mcal/kg	*	2.11	2.26	
Cattle	TDN %	*	56.9	61.0	
Sheep	TDN %	*	58.2	62.4	
Silicon	%		3.50	3.75	

Brome, ripgut, aerial part, fresh, (2)
Ref No 2 00 938 United States

Feed Name or Analyses			Mean As Fed	Mean Dry	C.V. ± %
Dry matter	%			100.0	
Ash	%			8.5	35
Crude fiber	%			26.6	22
Protein (N x 6.25)	%			14.0	66
Cattle	dig prot %	*		9.8	
Goats	dig prot %	*		9.6	
Horses	dig prot %	*		9.4	
Rabbits	dig prot %	*		9.5	
Sheep	dig prot %	*		10.0	
Calcium	%			0.48	40
Phosphorus	%			0.39	37
Potassium	%			2.92	50

Brome, ripgut, aerial part, fresh, immature, (2)
Ref No 2 00 934 United States

Feed Name or Analyses			Mean As Fed	Mean Dry	C.V. ± %
Dry matter	%			100.0	
Ash	%			13.9	
Crude fiber	%			19.5	22
Protein (N x 6.25)	%			27.3	20
Cattle	dig prot %	*		21.1	
Goats	dig prot %	*		22.0	
Horses	dig prot %	*		20.7	
Rabbits	dig prot %	*		19.7	
Sheep	dig prot %	*		22.4	
Calcium	%			0.77	32
Phosphorus	%			0.56	18
Potassium	%			5.21	14

Brome, ripgut, aerial part, fresh, pre-bloom, (2)
Ref No 2 00 935 United States

Feed Name or Analyses			Mean As Fed	Mean Dry	C.V. ± %
Dry matter	%			100.0	
Ash	%			9.3	
Crude fiber	%			28.3	
Protein (N x 6.25)	%			14.4	
Cattle	dig prot %	*		10.1	
Goats	dig prot %	*		10.0	
Horses	dig prot %	*		9.8	
Rabbits	dig prot %	*		9.8	
Sheep	dig prot %	*		10.4	
Calcium	%			0.48	
Phosphorus	%			0.46	
Potassium	%			3.14	

Brome, ripgut, aerial part, fresh, mature, (2)
Ref No 2 00 936 United States

Feed Name or Analyses			Mean As Fed	Mean Dry	C.V. ± %
Dry matter	%			100.0	
Ash	%			5.5	
Crude fiber	%			26.9	
Protein (N x 6.25)	%			6.1	
Cattle	dig prot %	*		3.1	
Goats	dig prot %	*		2.3	
Horses	dig prot %	*		2.7	
Rabbits	dig prot %	*		3.4	
Sheep	dig prot %	*		2.7	
Calcium	%			0.31	
Phosphorus	%			0.26	
Potassium	%			1.30	

Brome, ripgut, aerial part, fresh, over ripe, (2)
Ref No 2 00 937 United States

Feed Name or Analyses			Mean As Fed	Mean Dry	C.V. ± %
Dry matter	%			100.0	
Ash	%			6.2	3
Crude fiber	%			32.2	9
Ether extract	%			1.3	
N-free extract	%			57.0	
Protein (N x 6.25)	%			3.4	27
Cattle	dig prot %	*		0.8	
Goats	dig prot %	*		-0.1	
Horses	dig prot %	*		0.4	
Rabbits	dig prot %	*		1.3	
Sheep	dig prot %	*		0.2	
Cellulose (Matrone)	%			38.4	
Other carbohydrates	%			37.3	
Lignin (Ellis)	%			8.7	
Energy	GE Mcal/kg				
Cattle	DE Mcal/kg	*		2.79	
Sheep	DE Mcal/kg	*		2.61	
Cattle	ME Mcal/kg	*		2.29	
Sheep	ME Mcal/kg	*		2.14	
Cattle	TDN %	*		63.3	
Sheep	TDN %	*		59.2	
Calcium	%			0.30	6
Phosphorus	%			0.23	19
Potassium	%			1.29	3
Silicon	%			2.30	

Brome, ripgut, heads, fresh, (2)
Ref No 2 00 939 United States

Feed Name or Analyses			Mean As Fed	Mean Dry	C.V. ± %
Dry matter	%			100.0	
Ash	%			4.0	
Crude fiber	%			17.6	
Protein (N x 6.25)	%			11.1	
Cattle	dig prot %	*		7.3	
Goats	dig prot %	*		6.9	
Horses	dig prot %	*		7.0	
Rabbits	dig prot %	*		7.2	
Sheep	dig prot %	*		7.3	
Calcium	%			0.22	
Phosphorus	%			0.30	
Potassium	%			0.86	

BROME, RUSSIAN. Bromus tomentellus

Brome, Russian, hay, s-c, full bloom, (1)
Ref No 1 08 777 United States

Feed Name or Analyses			Mean As Fed	Mean Dry	C.V. ± %
Dry matter	%			100.0	
Sheep	dig coef %			68.	
Ash	%			7.3	
Crude fiber	%			33.6	
Ether extract	%			1.4	
N-free extract	%			47.7	
Protein (N x 6.25)	%			10.0	
Cattle	dig prot %	*		5.6	
Goats	dig prot %	*		5.9	
Horses	dig prot %	*		6.0	
Rabbits	dig prot %	*		6.4	
Sheep	dig prot %	*		5.5	
Lignin (Ellis)	%			6.5	
Energy	GE Mcal/kg			4.34	
Sheep	GE dig coef %			66.	
Cattle	DE Mcal/kg	*		2.46	
Sheep	DE Mcal/kg			2.86	
Cattle	ME Mcal/kg	*		2.02	
Sheep	ME Mcal/kg	*		2.35	
Cattle	TDN %	*		55.8	
Sheep	TDN %	*		56.3	

Brome, Russian, hay, s-c, chopped, full bloom, (1)
Ref No 1 08 776 United States

Feed Name or Analyses			Mean As Fed	Mean Dry	C.V. ± %
Dry matter	%			100.0	
Sheep	dig coef %			65.	
Ash	%			7.9	
Crude fiber	%			34.7	
Ether extract	%			2.1	
N-free extract	%			45.2	
Protein (N x 6.25)	%			10.1	
Cattle	dig prot %	*		5.7	
Goats	dig prot %	*		6.0	
Horses	dig prot %	*		6.1	
Rabbits	dig prot %	*		6.5	
Sheep	dig prot %	*		5.6	
Lignin (Ellis)	%			6.5	
Energy	GE Mcal/kg			4.38	
Cattle	DE Mcal/kg	*		2.43	
Sheep	DE Mcal/kg	*		2.45	
Cattle	ME Mcal/kg	*		2.00	

Continued

Feed Name or Analyses			Mean As Fed	Mean Dry	C.V. ± %
Sheep	ME Mcal/kg	*		2.01	
Cattle	TDN %	*		55.2	
Sheep	TDN %	*		55.6	

BROME, SMOOTH. Bromus inermis

Brome, smooth, hay, fan air dried, early leaf, (1)

Ref No 1 08 749 United States

Feed Name or Analyses			Mean As Fed	Mean Dry	C.V. ± %
Dry matter	%			100.0	
Lignin (Ellis)	%			3.5	

Brome, smooth, hay, fan air dried, immature, (1)

Ref No 1 08 747 United States

Feed Name or Analyses			Mean As Fed	Mean Dry	C.V. ± %
Dry matter	%			100.0	
Lignin (Ellis)	%			4.3	

Brome, smooth, hay, fan air dried, pre-bloom, (1)

Ref No 1 08 746 United States

Feed Name or Analyses			Mean As Fed	Mean Dry	C.V. ± %
Dry matter	%			100.0	
Lignin (Ellis)	%			4.1	

Brome, smooth, hay, fan air dried, early bloom, (1)

Ref No 1 08 748 United States

Feed Name or Analyses			Mean As Fed	Mean Dry	C.V. ± %
Dry matter	%			100.0	
Lignin (Ellis)	%			6.6	

Brome, smooth, hay, fan air dried, mid-bloom, (1)

Ref No 1 07 742 United States

Feed Name or Analyses			Mean As Fed	Mean Dry	C.V. ± %
Dry matter	%			100.0	
Sheep	dig coef %			52.	
Ash	%			6.1	
Protein (N x 6.25)	%			10.8	
Sheep	dig coef %			61.	
Cattle	dig prot %	*		6.3	
Goats	dig prot %	*		6.6	
Horses	dig prot %	*		6.7	
Rabbits	dig prot %	*		7.0	
Sheep	dig prot %			6.6	
Cellulose (Matrone)	%			35.1	
Sheep	dig coef %			53.	
Lignin (Ellis)	%			8.0	
Energy	GE Mcal/kg			4.40	
Sheep	GE dig coef %			49.	
Sheep	DE Mcal/kg			2.16	
Sheep	ME Mcal/kg	*		1.77	
Sheep	TDN %	*		49.0	

Brome, smooth, hay, fan air dried chopped, pre-bloom, (1)

Ref No 1 07 741 United States

Feed Name or Analyses			Mean As Fed	Mean Dry	C.V. ± %
Dry matter	%			100.0	
Sheep	dig coef %			68.	
Ash	%			8.1	
Protein (N x 6.25)	%			19.9	
Sheep	dig coef %			76.	
Cattle	dig prot %	*		14.2	
Goats	dig prot %	*		15.1	
Horses	dig prot %	*		14.4	
Rabbits	dig prot %	*		14.0	
Sheep	dig prot %			15.1	
Cellulose (Matrone)	%			30.8	
Sheep	dig coef %			72.	
Energy	GE Mcal/kg			4.37	
Sheep	GE dig coef %			65.	
Sheep	DE Mcal/kg			2.84	
Sheep	ME Mcal/kg	*		2.33	
Sheep	TDN %	*		64.4	

(1) dry forages and roughages (2) pasture, range plants, and forages fed green (3) silages (4) energy feeds (5) protein supplements (6) minerals (7) vitamins (8) additives

Brome, smooth, hay, s-c, (1)

Ref No 1 00 947 United States

Feed Name or Analyses			Mean As Fed	Mean Dry	C.V. ± %
Dry matter	%		89.7	100.0	3
Rabbits	dig coef %		36.	36.	
Ash	%		8.3	9.2	
Crude fiber	%		28.5	31.8	
Sheep	dig coef %	*	55.	55.	
Ether extract	%		2.1	2.4	
Sheep	dig coef %	*	36.	36.	
N-free extract	%		40.3	44.9	
Sheep	dig coef %	*	62.	62.	
Protein (N x 6.25)	%		10.5	11.7	
Rabbits	dig coef %		53.	53.	
Sheep	dig coef %	*	46.	46.	
Cattle	dig prot %	*	6.4	7.1	
Goats	dig prot %	*	6.7	7.5	
Horses	dig prot %	*	6.7	7.5	
Rabbits	dig prot %		5.6	6.2	
Sheep	dig prot %		4.8	5.4	
Cellulose (Matrone)	%		27.8	31.0	
Hemicellulose	%		14.4	16.0	
Pentosans	%		10.2	11.4	
Sugars, total	%		3.3	3.7	63
Lignin (Ellis)	%		8.1	9.0	34
Energy	GE Mcal/kg		4.01	4.47	2
Cattle	DE Mcal/kg	*	2.20	2.45	
Sheep	DE Mcal/kg	*	2.08	2.32	
Cattle	ME Mcal/kg	*	1.80	2.01	
Sheep	ME Mcal/kg	*	1.71	1.90	
Cattle	TDN %	*	49.8	55.5	
Sheep	TDN %		47.2	52.6	
Calcium	%		0.32	0.36	
Chlorine	%		0.48	0.54	
Copper	mg/kg		7.7	8.6	
Iron	%		0.011	0.012	
Magnesium	%		0.13	0.15	
Manganese	mg/kg		52.1	58.0	
Phosphorus	%		0.17	0.19	
Potassium	%		1.86	2.07	
Sodium	%		0.57	0.63	
Sulphur	%		0.17	0.19	

Brome, smooth, hay, s-c, immature, (1)

Ref No 1 00 940 United States

Feed Name or Analyses			Mean As Fed	Mean Dry	C.V. ± %
Dry matter	%		89.3	100.0	2
Ash	%		9.0	10.1	
Crude fiber	%		22.9	25.7	
Ether extract	%		2.6	2.9	
N-free extract	%		35.0	39.2	
Protein (N x 6.25)	%		19.7	22.1	
Cattle	dig prot %	*	14.4	16.1	
Goats	dig prot %	*	15.3	17.2	
Horses	dig prot %	*	14.6	16.3	
Rabbits	dig prot %	*	14.1	15.7	
Sheep	dig prot %	*	14.6	16.4	
Energy	GE Mcal/kg		4.01	4.49	
Sheep	GE dig coef %		70.	70.	
Cattle	DE Mcal/kg	*	2.46	2.75	
Sheep	DE Mcal/kg		2.81	3.14	
Cattle	ME Mcal/kg	*	2.01	2.26	
Sheep	ME Mcal/kg	*	2.30	2.58	
Cattle	TDN %	*	55.7	62.4	
Sheep	TDN %	*	53.5	59.9	
Calcium	%		0.58	0.65	24
Cobalt	mg/kg		0.081	0.090	58
Copper	mg/kg		13.4	15.0	33
Iron	%		0.016	0.018	41
Magnesium	%		0.28	0.31	8
Manganese	mg/kg		31.7	35.5	10
Phosphorus	%		0.33	0.37	43
Potassium	%		2.13	2.39	6
Carotene	mg/kg		57.9	64.8	76
Vitamin A equivalent	IU/g		96.4	108.0	

Ref No 1 00 940 Canada

Feed Name or Analyses			Mean As Fed	Mean Dry	C.V. ± %
Dry matter	%			100.0	
Sheep	dig coef %			72.	
Ash	%			7.4	
Crude fiber	%			28.5	
Energy	GE Mcal/kg			4.49	
Sheep	GE dig coef %			70.	
Sheep	DE Mcal/kg			3.14	
Sheep	ME Mcal/kg	*		2.58	
Sheep	TDN %	*		71.2	
Calcium	%			0.33	
Magnesium	%			0.10	
Potassium	%			2.74	
Sodium	%			0.02	

Brome, smooth, hay, s-c, pre-bloom, (1)

Ref No 1 08 363 United States

Feed Name or Analyses			Mean As Fed	Mean Dry	C.V. ± %
Dry matter	%		90.3	100.0	
Ash	%		7.9	8.7	
Crude fiber	%		28.2	31.2	
Ether extract	%		2.2	2.4	
N-free extract	%		41.1	45.5	
Protein (N x 6.25)	%		10.9	12.1	
Cattle	dig prot %	*	6.7	7.4	
Goats	dig prot %	*	7.1	7.8	
Horses	dig prot %	*	7.0	7.8	
Rabbits	dig prot %	*	7.2	8.0	
Sheep	dig prot %	*	6.7	7.4	
Energy	GE Mcal/kg				
Cattle	DE Mcal/kg	*	2.45	2.72	
Sheep	DE Mcal/kg	*	2.30	2.55	
Cattle	ME Mcal/kg	*	2.01	2.23	
Sheep	ME Mcal/kg	*	1.89	2.09	
Cattle	TDN %	*	55.7	61.6	
Sheep	TDN %	*	52.2	57.8	

Brome, smooth, hay, s-c, early bloom, (1)

Ref No 1 00 941 United States

Feed Name or Analyses			Mean As Fed	Mean Dry	C.V. ± %
Dry matter	%		90.3	100.0	
Ash	%		7.9	8.7	

Feed Name or Analyses				As Fed	Dry	C.V. ± %
Crude fiber		%		28.2	31.2	
Sheep	dig coef	%	*	55.	55.	
Ether extract		%		2.2	2.4	
Sheep	dig coef	%	*	36.	36.	
N-free extract		%		41.2	45.6	
Sheep	dig coef	%	*	62.	62.	
Protein (N x 6.25)		%		10.9	12.1	
Sheep	dig coef	%	*	46.	46.	
Cattle	dig prot	%	*	6.7	7.4	
Goats	dig prot	%	*	7.1	7.8	
Horses	dig prot	%	*	7.0	7.8	
Rabbits	dig prot	%	*	7.2	8.0	
Sheep	dig prot	%		5.0	5.6	
Energy	GE Mcal/kg		*	4.03	4.46	
Sheep	GE dig coef	%	*	62.	62.	
Cattle	DE Mcal/kg		*	2.45	2.72	
Sheep	DE Mcal/kg		*	2.50	2.77	
Cattle	ME Mcal/kg		*	2.01	2.23	
Sheep	ME Mcal/kg		*	2.05	2.27	
Cattle	TDN	%	*	55.6	61.6	
Sheep	TDN	%	*	47.8	52.9	

Ref No 1 00 941 Canada

Feed Name or Analyses				As Fed	Dry	C.V. ± %
Dry matter		%			100.0	
Sheep	dig coef	%			65.	
Ash		%			6.3	
Crude fiber		%			31.1	
Energy	GE Mcal/kg				4.46	
Sheep	GE dig coef	%			62.	
Sheep	DE Mcal/kg				2.77	
Sheep	ME Mcal/kg		*		2.27	
Sheep	TDN	%			53.7	
Calcium		%			0.24	
Magnesium		%			0.08	
Potassium		%			1.81	
Sodium		%			0.02	

Brome, smooth, hay, s-c, early bloom, cut 1, (1)
Ref No 1 08 769 United States

Feed Name or Analyses				As Fed	Dry	C.V. ± %
Dry matter		%		91.7	100.0	1
Horses	dig coef	%		48.	48.	5
Organic matter		%		85.8	93.5	0
Ash		%		6.0	6.5	3
Crude fiber		%		36.0	39.3	2
Horses	dig coef	%		45.	45.	14
Ether extract		%				
Horses	dig coef	%		27.	27.	46
N-free extract		%		35.2	38.4	1
Horses	dig coef	%		57.	57.	13
Protein (N x 6.25)		%		12.7	13.8	2
Horses	dig coef	%		58.	58.	6
Cattle	dig prot	%	*	8.2	8.9	
Goats	dig prot	%	*	8.7	9.5	
Horses	dig prot	%		7.3	8.0	
Rabbits	dig prot	%	*	8.6	9.3	
Sheep	dig prot	%	*	8.2	9.0	
Cell contents (Van Soest)		%		22.7	24.8	5
Horses	dig coef	%		43.	43.	8
Cell walls (Van Soest)		%		68.7	75.0	2
Horses	dig coef	%		50.	50.	4
Cellulose (Matrone)		%		34.8	37.9	2
Horses	dig coef	%		49.	49.	5
Fiber, acid detergent (VS)		%		39.0	42.6	2
Horses	dig coef	%		43.	43.	5
Hemicellulose		%		22.6	24.6	4
Horses	dig coef	%		53.	53.	5
Holocellulose		%		57.5	62.7	2

Feed Name or Analyses				As Fed	Dry	C.V. ± %
Soluble carbohydrates		%		9.3	10.2	5
Lignin (Ellis)		%		4.3	4.7	3
Horses	dig coef	%		24.	24.	41
Energy	GE Mcal/kg			4.10	4.47	0
Horses	GE dig coef	%		40.	40.	7
Cattle	DE Mcal/kg		*	2.12	2.31	
Horses	DE Mcal/kg			1.62	1.77	
Sheep	DE Mcal/kg		*	2.16	2.35	
Cattle	ME Mcal/kg		*	1.74	1.89	
Horses	ME Mcal/kg		*	1.33	1.45	
Sheep	ME Mcal/kg		*	1.77	1.93	
Cattle	TDN	%	*	48.0	52.4	
Sheep	TDN	%	*	49.0	53.4	
Calcium		%		0.34	0.37	24
Magnesium		%		0.20	0.22	11
Phosphorus		%		0.18	0.20	6
Potassium		%		0.22	0.24	4
Carotene	mg/kg			10.1	11.0	9
Vitamin A equivalent	IU/g			16.8	18.4	

Brome, smooth, hay, s-c, early bloom, cut 1, gr 1 US, (1)
Ref No 1 08 654 Canada

Feed Name or Analyses				As Fed	Dry	C.V. ± %
Dry matter		%			100.0	
Ash		%			8.9	
Crude fiber		%			33.1	
Ether extract		%			2.4	
N-free extract		%			40.7	
Protein (N x 6.25)		%			14.9	
Cattle	dig prot	%	*		9.8	
Goats	dig prot	%	*		10.5	
Horses	dig prot	%	*		10.2	
Rabbits	dig prot	%	*		10.2	
Sheep	dig prot	%	*		9.9	
Cellulose (Matrone)		%			32.4	
Hexosans		%			11.6	
Pentosans		%			6.8	
Lignin (Ellis)		%			9.5	
Energy	GE Mcal/kg				4.45	
Cattle	DE Mcal/kg		*		2.59	
Sheep	DE Mcal/kg		*		2.44	
Cattle	ME Mcal/kg		*		2.12	
Sheep	ME Mcal/kg		*		2.00	
Cattle	TDN	%	*		58.7	
Sheep	TDN	%	*		55.3	

Brome, smooth, hay, s-c, mid-bloom, (1)
Ref No 1 05 633 United States

Feed Name or Analyses				As Fed	Dry	C.V. ± %
Dry matter		%		87.6	100.0	
Ash		%		7.8	8.9	
Crude fiber		%		31.2	35.6	
Ether extract		%		1.9	2.2	
N-free extract		%		37.9	43.3	
Protein (N x 6.25)		%		8.8	10.0	
Cattle	dig prot	%	*	4.9	5.6	
Goats	dig prot	%	*	5.2	5.9	
Horses	dig prot	%	*	5.3	6.1	
Rabbits	dig prot	%	*	5.6	6.4	
Sheep	dig prot	%	*	4.9	5.6	
Lignin (Ellis)		%		6.7	7.6	
Energy	GE Mcal/kg					
Cattle	DE Mcal/kg		*	2.11	2.41	
Sheep	DE Mcal/kg		*	2.12	2.42	
Cattle	ME Mcal/kg		*	1.73	1.98	
Sheep	ME Mcal/kg		*	1.74	1.99	

Feed Name or Analyses				As Fed	Dry	C.V. ± %
Cattle	TDN	%	*	47.9	54.7	
Sheep	TDN	%	*	48.1	54.9	

Brome, smooth, hay, s-c, full bloom, (1)
Ref No 1 00 942 United States

Feed Name or Analyses				As Fed	Dry	C.V. ± %
Dry matter		%		90.1	100.0	2
Ash		%		5.8	6.4	23
Crude fiber		%		32.5	36.1	3
Sheep	dig coef	%	*	55.	55.	
Ether extract		%		1.7	1.9	14
Sheep	dig coef	%	*	36.	36.	
N-free extract		%		40.7	45.2	
Sheep	dig coef	%	*	62.	62.	
Protein (N x 6.25)		%		9.5	10.6	10
Sheep	dig coef	%	*	46.	46.	
Cattle	dig prot	%	*	5.5	6.1	
Goats	dig prot	%	*	5.8	6.4	
Horses	dig prot	%	*	5.8	6.5	
Rabbits	dig prot	%	*	6.1	6.8	
Sheep	dig prot	%		4.4	4.9	
Energy	GE Mcal/kg			4.07	4.52	
Sheep	GE dig coef	%		55.	55.	
Cattle	DE Mcal/kg		*	2.03	2.26	
Sheep	DE Mcal/kg			2.24	2.49	
Cattle	ME Mcal/kg		*	1.67	1.85	
Sheep	ME Mcal/kg		*	1.84	2.04	
Cattle	TDN	%	*	46.1	51.2	
Sheep	TDN	%		48.8	54.2	
Cobalt	mg/kg			0.087	0.097	

Ref No 1 00 942 Canada

Feed Name or Analyses				As Fed	Dry	C.V. ± %
Dry matter		%			100.0	
Sheep	dig coef	%			60.	
Ash		%			4.9	
Crude fiber		%			32.9	
Energy	GE Mcal/kg				4.52	
Sheep	GE dig coef	%			55.	
Sheep	DE Mcal/kg				2.49	
Sheep	ME Mcal/kg		*		2.04	
Sheep	TDN	%			54.6	
Calcium		%			0.28	
Magnesium		%			0.08	
Potassium		%			1.43	
Sodium		%			0.02	

Brome, smooth, hay, s-c, milk stage, (1)
Ref No 1 00 943 United States

Feed Name or Analyses				As Fed	Dry	C.V. ± %
Dry matter		%		91.6	100.0	3
Ash		%		6.6	7.2	10
Crude fiber		%		31.6	34.5	7
Ether extract		%		2.0	2.2	12
N-free extract		%		45.5	49.7	
Protein (N x 6.25)		%		5.8	6.3	4
Cattle	dig prot	%	*	2.2	2.4	
Goats	dig prot	%	*	2.3	2.5	
Horses	dig prot	%	*	2.7	2.9	
Rabbits	dig prot	%	*	3.3	3.6	
Sheep	dig prot	%	*	2.1	2.3	
Energy	GE Mcal/kg					
Cattle	DE Mcal/kg		*	2.25	2.46	
Sheep	DE Mcal/kg		*	2.19	2.40	
Cattle	ME Mcal/kg		*	1.85	2.02	
Sheep	ME Mcal/kg		*	1.80	1.97	
Cattle	TDN	%	*	51.0	55.7	

Continued

Feed Name or Analyses			Mean As Fed	Mean Dry	C.V. ± %
Sheep	TDN %	*	49.8	54.4	
Calcium	%		0.28	0.31	
Phosphorus	%		0.13	0.14	

Ref No 1 00 943 Canada

Feed Name or Analyses			Mean As Fed	Mean Dry	C.V. ± %
Dry matter	%			100.0	
Calcium	%			0.26	
Magnesium	%			0.07	
Potassium	%			1.45	
Sodium	%			0.01	

Brome, smooth, hay, s-c, mature, (1)

Ref No 1 00 944 United States

Feed Name or Analyses			Mean As Fed	Mean Dry	C.V. ± %
Dry matter	%		92.9	100.0	2
Crude fiber	%				
Sheep	dig coef %	*	55.	55.	
Ether extract	%				
Sheep	dig coef %	*	36.	36.	
N-free extract	%				
Sheep	dig coef %	*	62.	62.	
Protein (N x 6.25)	%				
Sheep	dig coef %	*	46.	46.	
Pentosans	%		10.6	11.4	
Lignin (Ellis)	%		13.2	14.2	
Energy	GE Mcal/kg		4.13	4.45	
Sheep	GE dig coef %		60.	60.	
Sheep	DE Mcal/kg		2.48	2.67	
Sheep	ME Mcal/kg	*	2.03	2.19	
Sheep	TDN %	*	56.3	60.6	
Calcium	%		0.40	0.43	20
Chlorine	%		0.12	0.13	
Cobalt	mg/kg		0.133	0.143	
Copper	mg/kg		6.3	6.8	
Iron	%		0.009	0.010	
Magnesium	%		0.18	0.19	51
Manganese	mg/kg		98.3	105.8	
Phosphorus	%		0.20	0.22	64
Potassium	%		2.56	2.76	20
Carotene	mg/kg		4.5	4.9	49
Vitamin A equivalent	IU/g		7.5	8.1	

Ref No 1 00 944 Canada

Feed Name or Analyses			Mean As Fed	Mean Dry	C.V. ± %
Dry matter	%			100.0	
Sheep	dig coef %			63.	
Ash	%			6.2	
Crude fiber	%			30.2	
Energy	GE Mcal/kg			4.45	
Sheep	GE dig coef %			60.	
Sheep	DE Mcal/kg			2.67	
Sheep	ME Mcal/kg	*		2.19	
Sheep	TDN %	*	56.3	60.6	
Calcium	%			0.20	
Magnesium	%			0.06	
Potassium	%			1.24	
Sodium	%			0.01	

Brome, smooth, hay, s-c, cut 1, (1)

Ref No 1 00 945 United States

Feed Name or Analyses			Mean As Fed	Mean Dry	C.V. ± %
Dry matter	%		89.5	100.0	0
Ash	%		6.2	6.9	16
Crude fiber	%		28.6	32.0	12
Ether extract	%		2.3	2.6	21
N-free extract	%		39.3	44.0	
Protein (N x 6.25)	%		13.1	14.6	35
Cattle	dig prot %	*	8.6	9.6	
Goats	dig prot %	*	9.1	10.2	
Horses	dig prot %	*	8.9	9.9	
Rabbits	dig prot %	*	8.9	9.9	
Sheep	dig prot %	*	8.6	9.7	
Energy	GE Mcal/kg				
Cattle	DE Mcal/kg	*	2.21	2.47	
Sheep	DE Mcal/kg	*	2.27	2.54	
Cattle	ME Mcal/kg	*	1.82	2.03	
Sheep	ME Mcal/kg	*	1.86	2.08	
Cattle	TDN %	*	50.2	56.1	
Sheep	TDN %	*	51.6	57.6	
Calcium	%		0.34	0.38	16
Chlorine	%		0.23	0.26	
Cobalt	mg/kg		0.077	0.086	59
Copper	mg/kg		11.0	12.3	33
Iron	%		0.013	0.015	43
Magnesium	%		0.24	0.27	26
Manganese	mg/kg		36.1	40.3	23
Phosphorus	%		0.22	0.25	20
Potassium	%		1.81	2.02	15
Carotene	mg/kg		28.4	31.7	36
Vitamin A equivalent	IU/g		47.4	52.9	

Brome, smooth, hay, s-c, cut 2, (1)

Ref No 1 00 946 United States

Feed Name or Analyses			Mean As Fed	Mean Dry	C.V. ± %
Dry matter	%		90.6	100.0	
Calcium	%		0.29	0.32	
Chlorine	%		0.24	0.26	
Cobalt	mg/kg		0.050	0.055	
Copper	mg/kg		6.0	6.6	
Iron	%		0.006	0.007	
Magnesium	%		0.11	0.12	
Manganese	mg/kg		65.9	72.8	
Phosphorus	%		0.28	0.31	
Potassium	%		2.22	2.45	

Ref No 1 00 946 Canada

Feed Name or Analyses			Mean As Fed	Mean Dry	C.V. ± %
Dry matter	%			100.0	
Protein (N x 6.25)	%			10.7	
Cattle	dig prot %	*		6.2	
Goats	dig prot %	*		6.5	
Horses	dig prot %	*		6.6	
Rabbits	dig prot %	*		6.9	
Sheep	dig prot %	*		6.2	

Brome, smooth, hay, s-c, good quality, (1)

Ref No 1 00 948 United States

Feed Name or Analyses			Mean As Fed	Mean Dry	C.V. ± %
Dry matter	%			100.0	
Choline	mg/kg			677.	

Brome, smooth, hay, s-c, gr 1 US, (1)

Ref No 1 00 950 United States

Feed Name or Analyses			Mean As Fed	Mean Dry	C.V. ± %
Dry matter	%		88.3	100.0	2
Ash	%		6.6	7.5	20
Crude fiber	%		27.5	31.2	17
Ether extract	%		2.5	2.8	20
N-free extract	%		36.2	41.0	
Protein (N x 6.25)	%		15.5	17.5	46
Cattle	dig prot %	*	10.7	12.1	
Goats	dig prot %	*	11.4	12.9	
Horses	dig prot %	*	10.9	12.4	
Rabbits	dig prot %	*	10.8	12.2	
Sheep	dig prot %	*	10.8	12.3	
Energy	GE Mcal/kg				
Cattle	DE Mcal/kg	*	2.22	2.51	
Sheep	DE Mcal/kg	*	2.25	2.55	
Cattle	ME Mcal/kg	*	1.82	2.06	
Sheep	ME Mcal/kg	*	1.85	2.09	
Cattle	TDN %	*	50.3	57.0	
Sheep	TDN %	*	51.1	57.8	
Calcium	%		0.38	0.43	
Cobalt	mg/kg		0.158	0.179	
Copper	mg/kg		13.2	15.0	
Iron	%		0.027	0.031	
Magnesium	%		0.31	0.35	
Manganese	mg/kg		27.4	31.1	
Phosphorus	%		0.31	0.35	
Potassium	%		2.19	2.48	
Carotene	mg/kg		32.5	36.8	
Vitamin A equivalent	IU/g		54.2	61.4	

Brome, smooth, hay, s-c, gr 2 US, (1)

Ref No 1 00 951 United States

Feed Name or Analyses			Mean As Fed	Mean Dry	C.V. ± %
Dry matter	%		90.8	100.0	0
Ash	%		5.9	6.6	7
Crude fiber	%		30.7	33.8	9
Ether extract	%		2.4	2.6	21
N-free extract	%		41.8	46.1	
Protein (N x 6.25)	%		10.0	11.0	17
Cattle	dig prot %	*	5.9	6.5	
Goats	dig prot %	*	6.2	6.8	
Horses	dig prot %	*	6.2	6.9	
Rabbits	dig prot %	*	6.5	7.2	
Sheep	dig prot %	*	5.8	6.4	
Energy	GE Mcal/kg				
Cattle	DE Mcal/kg	*	2.23	2.45	
Sheep	DE Mcal/kg	*	2.26	2.49	
Cattle	ME Mcal/kg	*	1.83	2.01	
Sheep	ME Mcal/kg	*	1.85	2.04	
Cattle	TDN %	*	50.5	55.7	
Sheep	TDN %	*	51.3	56.5	
Calcium	%		0.36	0.40	9
Cobalt	mg/kg		0.044	0.049	41
Copper	mg/kg		13.6	15.0	38
Iron	%		0.011	0.012	16
Magnesium	%		0.26	0.29	11
Manganese	mg/kg		34.2	37.7	9
Phosphorus	%		0.23	0.25	16
Potassium	%		2.13	2.35	7
Carotene	mg/kg		37.2	41.0	13
Vitamin A equivalent	IU/g		62.0	68.4	

Brome, smooth, hay, s-c, gr 3 US, (1)

Ref No 1 00 952 United States

Feed Name or Analyses			Mean As Fed	Mean Dry	C.V. ± %
Dry matter	%		90.0	100.0	1
Ash	%		5.4	6.0	7
Crude fiber	%		31.9	35.4	2
Ether extract	%		1.6	1.8	4
N-free extract	%		40.8	45.3	
Protein (N x 6.25)	%		10.4	11.5	5
Cattle	dig prot %	*	6.2	6.9	
Goats	dig prot %	*	6.6	7.3	
Horses	dig prot %	*	6.6	7.3	

(1) dry forages and roughages (3) silages (6) minerals
(2) pasture, range plants, and forages fed green (4) energy feeds (7) vitamins
(5) protein supplements (8) additives

Feed Name or Analyses			Mean As Fed	Mean Dry	C.V. ± %
Rabbits	dig prot %	*	6.8	7.5	
Sheep	dig prot %	*	6.2	6.9	
Energy	GE Mcal/kg				
Cattle	DE Mcal/kg	*	2.05	2.28	
Sheep	DE Mcal/kg	*	2.23	2.47	
Cattle	ME Mcal/kg	*	1.68	1.87	
Sheep	ME Mcal/kg	*	1.83	2.03	
Cattle	TDN %	*	46.4	51.6	
Sheep	TDN %	*	50.5	56.1	
Calcium	%		0.32	0.35	24
Copper	mg/kg		8.5	9.5	14
Iron	%		0.011	0.012	13
Magnesium	%		0.20	0.22	43
Manganese	mg/kg		43.5	48.3	26
Phosphorus	%		0.21	0.23	19
Potassium	%		1.39	1.54	3
Carotene	mg/kg		11.1	12.3	
Vitamin A equivalent	IU/g		18.5	20.6	

Brome, smooth, hay, s-c, poor quality, (1)
Ref No 1 00 953 United States

Feed Name or Analyses			Mean As Fed	Mean Dry	C.V. ± %
Dry matter	%			100.0	
Choline	mg/kg			401.	

Brome, smooth, hay, s-c, gr sample US, (1)
Ref No 1 00 949 United States

Feed Name or Analyses			Mean As Fed	Mean Dry	C.V. ± %
Dry matter	%		90.0	100.0	
Ash	%		5.4	6.0	
Crude fiber	%		32.6	36.2	
Ether extract	%		2.1	2.3	
N-free extract	%		41.8	46.4	
Protein (N x 6.25)	%		8.2	9.1	
Cattle	dig prot %	*	4.3	4.8	
Goats	dig prot %	*	4.5	5.1	
Horses	dig prot %	*	4.7	5.3	
Rabbits	dig prot %	*	5.1	5.7	
Sheep	dig prot %	*	4.3	4.7	
Energy	GE Mcal/kg				
Cattle	DE Mcal/kg	*	2.04	2.26	
Sheep	DE Mcal/kg	*	2.17	2.41	
Cattle	ME Mcal/kg	*	1.67	1.86	
Sheep	ME Mcal/kg	*	1.78	1.98	
Cattle	TDN %	*	46.2	51.3	
Sheep	TDN %	*	49.3	54.8	
Calcium	%		0.33	0.37	
Cobalt	mg/kg		0.087	0.097	
Copper	mg/kg		7.1	7.9	
Iron	%		0.007	0.008	
Magnesium	%		0.14	0.16	
Manganese	mg/kg		33.9	37.7	
Phosphorus	%		0.15	0.17	
Potassium	%		1.43	1.59	
Carotene	mg/kg		18.5	20.5	
Vitamin A equivalent	IU/g		30.8	34.2	

BROME, SMOOTH. Bromus inermus

Brome, smooth, hay, s-c baled fan-air dried w heat, immature, cut 1, (1)
Ref No 1 08 860 United States

Feed Name or Analyses			Mean As Fed	Mean Dry	C.V. ± %
Dry matter	%		86.5	100.0	
Sheep	dig coef %		64.	64.	
Ash	%		6.4	7.4	
Crude fiber	%		27.2	31.5	
Ether extract	%		3.1	3.6	
N-free extract	%		32.5	37.6	
Protein (N x 6.25)	%		17.2	19.9	
Cattle	dig prot %	*	12.3	14.2	
Goats	dig prot %	*	13.1	15.1	
Horses	dig prot %	*	12.5	14.4	
Rabbits	dig prot %	*	12.1	14.0	
Sheep	dig prot %	*	12.5	14.4	
Cell walls (Van Soest)	%		54.2	62.7	
Sheep	dig coef %		64.	64.	
Fiber, acid detergent (VS)	%		30.4	35.2	
Sheep	dig coef %		57.	57.	
Hemicellulose	%		23.8	27.5	
Lignin (Ellis)	%		2.5	2.9	
Sheep	dig coef %		-53.	-53.	
Energy	GE Mcal/kg		3.86	4.47	
Sheep	GE dig coef %		61.	61.	
Cattle	DE Mcal/kg	*	2.24	2.59	
Sheep	DE Mcal/kg		2.36	2.72	
Cattle	ME Mcal/kg	*	1.84	2.12	
Sheep	ME Mcal/kg	*	1.93	2.23	
Cattle	TDN %	*	50.8	58.8	
Sheep	TDN %	*	49.6	57.4	
Dry matter intake	$g/w^{0.75}$		91.604	105.900	
Gain, sheep	kg/day		0.106	0.123	

Brome, smooth, hay, s-c baled fan-air dried w heat immature, cut 2, (1)
Ref No 1 08 859 United States

Feed Name or Analyses			Mean As Fed	Mean Dry	C.V. ± %
Dry matter	%		86.5	100.0	
Sheep	dig coef %		61.	61.	
Ash	%		7.4	8.6	
Sheep	dig coef %		48.	55.	
Crude fiber	%		22.4	25.9	
Sheep	dig coef %		62.	62.	
Ether extract	%		3.6	4.2	
Sheep	dig coef %		48.	48.	
N-free extract	%		36.5	42.3	
Sheep	dig coef %		54.	54.	
Protein (N x 6.25)	%		16.6	19.2	
Sheep	dig coef %		71.	71.	
Cattle	dig prot %	*	11.7	13.6	
Goats	dig prot %	*	12.5	14.5	
Horses	dig prot %	*	12.0	13.8	
Rabbits	dig prot %	*	11.7	13.5	
Sheep	dig prot %		11.8	13.6	
Cell walls (Van Soest)	%		51.8	59.9	
Sheep	dig coef %		61.	61.	
Cellulose (Matrone)	%		27.8	32.2	
Fiber, acid detergent (VS)	%		29.1	33.7	
Sheep	dig coef %		50.	50.	
Hemicellulose	%		22.7	26.3	
Lignin (Ellis)	%		3.2	3.8	
Sheep	dig coef %		-20.	-20.	
Energy	GE Mcal/kg		3.98	4.60	
Sheep	GE dig coef %		59.	59.	
Cattle	DE Mcal/kg	*	2.23	2.58	
Sheep	DE Mcal/kg		2.33	2.69	
Cattle	ME Mcal/kg	*	1.83	2.12	
Sheep	ME Mcal/kg	*	1.91	2.21	
Cattle	TDN %	*	50.6	58.5	
Sheep	TDN %		49.2	57.0	
Dry matter intake	$g/w^{0.75}$		81.106	93.800	
Gain, sheep	kg/day		0.137	0.159	

BROME, SMOOTH. Bromus inermis

Brome, smooth, aerial part, fresh, (2)
Ref No 2 00 963 United States

Feed Name or Analyses			Mean As Fed	Mean Dry	C.V. ± %
Dry matter	%		36.2	100.0	35
Cobalt	mg/kg		0.029	0.079	
Carotene	mg/kg		114.2	315.5	41
Riboflavin	mg/kg		2.8	7.7	26
Thiamine	mg/kg		1.1	3.1	28
Vitamin A equivalent	IU/g		190.4	525.9	
Vitamin D_2	IU/g		0.0	0.1	

Brome, smooth, aerial part, fresh, immature, (2)
Ref No 2 00 956 United States

Feed Name or Analyses			Mean As Fed	Mean Dry	C.V. ± %
Dry matter	%		30.0	100.0	21
Ash	%		2.9	9.6	
Crude fiber	%		7.0	23.2	
Ether extract	%		1.2	4.0	
N-free extract	%		12.8	42.8	
Protein (N x 6.25)	%		6.1	20.4	
Cattle	dig prot %	*	4.6	15.2	
Goats	dig prot %	*	4.7	15.6	
Horses	dig prot %	*	4.5	14.8	
Rabbits	dig prot %	*	4.3	14.4	
Sheep	dig prot %	*	4.8	16.0	
Energy	GE Mcal/kg				
Cattle	DE Mcal/kg	*	0.94	3.15	
Sheep	DE Mcal/kg	*	0.97	3.23	
Cattle	ME Mcal/kg	*	0.77	2.58	
Sheep	ME Mcal/kg	*	0.80	2.65	
Cattle	TDN %	*	21.4	71.4	
Sheep	TDN %	*	22.0	73.2	
Calcium	%		0.17	0.55	
Magnesium	%		0.10	0.32	
Phosphorus	%		0.13	0.45	
Potassium	%		0.95	3.16	
Carotene	mg/kg		174.9	582.9	
Vitamin A equivalent	IU/g		291.5	971.7	

Brome, smooth, aerial part, fresh, early bloom, (2)
Ref No 2 00 957 United States

Feed Name or Analyses			Mean As Fed	Mean Dry	C.V. ± %
Dry matter	%		30.0	100.0	11
Ash	%		2.2	7.4	16
Crude fiber	%		8.8	29.2	8
Ether extract	%		1.2	4.1	11
N-free extract	%		13.6	45.3	
Protein (N x 6.25)	%		4.2	14.0	21
Cattle	dig prot %	*	2.9	9.8	
Goats	dig prot %	*	2.9	9.6	
Horses	dig prot %	*	2.8	9.4	
Rabbits	dig prot %	*	2.8	9.5	
Sheep	dig prot %	*	3.0	10.0	
Hemicellulose	%		5.4	18.0	
Energy	GE Mcal/kg				
Cattle	DE Mcal/kg	*	0.91	3.02	
Sheep	DE Mcal/kg	*	0.83	2.78	
Cattle	ME Mcal/kg	*	0.74	2.48	
Sheep	ME Mcal/kg	*	0.68	2.28	
Cattle	TDN %	*	20.6	68.5	
Sheep	TDN %	*	18.9	63.0	

Brome, smooth, aerial part, fresh, full bloom, (2)
Ref No 2 00 958 United States

Feed Name or Analyses			Mean As Fed	Mean Dry	C.V. ± %
Dry matter	%		27.1	100.0	
Ash	%		1.9	7.2	4
Crude fiber	%		8.3	30.8	5
Ether extract	%		0.9	3.3	6
N-free extract	%		13.2	48.6	
Protein (N x 6.25)	%		2.8	10.2	6
Cattle	dig prot %	*	1.8	6.6	
Goats	dig prot %	*	1.6	6.1	
Horses	dig prot %	*	1.7	6.2	
Rabbits	dig prot %	*	1.8	6.5	
Sheep	dig prot %	*	1.8	6.5	
Hemicellulose	%		5.6	20.7	
Energy	GE Mcal/kg				
Cattle	DE Mcal/kg	*	0.79	2.91	
Sheep	DE Mcal/kg	*	0.78	2.89	
Cattle	ME Mcal/kg	*	0.65	2.39	
Sheep	ME Mcal/kg	*	0.64	2.37	
Cattle	TDN %	*	17.9	66.1	
Sheep	TDN %	*	17.8	65.6	
Calcium	%		0.09	0.33	
Phosphorus	%		0.08	0.28	10
Potassium	%		0.63	2.33	17

Brome, smooth, aerial part, fresh, milk stage, (2)
Ref No 2 00 959 United States

Feed Name or Analyses			Mean As Fed	Mean Dry	C.V. ± %
Dry matter	%			100.0	
Ash	%			6.7	4
Crude fiber	%			30.6	4
Ether extract	%			2.9	7
N-free extract	%			51.0	
Protein (N x 6.25)	%			8.8	6
Cattle	dig prot %	*		5.4	
Goats	dig prot %	*		4.8	
Horses	dig prot %	*		5.0	
Rabbits	dig prot %	*		5.5	
Sheep	dig prot %	*		5.2	
Energy	GE Mcal/kg				
Cattle	DE Mcal/kg	*		2.90	
Sheep	DE Mcal/kg	*		2.96	
Cattle	ME Mcal/kg	*		2.38	
Sheep	ME Mcal/kg	*		2.42	
Cattle	TDN %	*		65.8	
Sheep	TDN %	*		67.1	

Brome, smooth, aerial part, fresh, dough stage, (2)
Ref No 2 00 960 United States

Feed Name or Analyses			Mean As Fed	Mean Dry	C.V. ± %
Dry matter	%		38.1	100.0	
Ash	%		2.6	6.9	13
Crude fiber	%		11.9	31.3	4
Ether extract	%		0.9	2.3	24
N-free extract	%		19.9	52.2	
Protein (N x 6.25)	%		2.8	7.3	7
Cattle	dig prot %	*	1.6	4.1	
Goats	dig prot %	*	1.3	3.4	
Horses	dig prot %	*	1.4	3.7	
Rabbits	dig prot %	*	1.6	4.3	
Sheep	dig prot %	*	1.4	3.8	
Energy	GE Mcal/kg				
Cattle	DE Mcal/kg	*	1.09	2.86	
Sheep	DE Mcal/kg	*	1.14	3.00	
Cattle	ME Mcal/kg	*	0.89	2.34	
Sheep	ME Mcal/kg	*	0.94	2.46	
Cattle	TDN %	*	24.7	64.8	
Sheep	TDN %	*	25.9	68.1	

(1) dry forages and roughages (3) silages (6) minerals
(2) pasture, range plants, and forages fed green (4) energy feeds (7) vitamins
(5) protein supplements (8) additives

Brome, smooth, aerial part, fresh, mature, (2)
Ref No 2 08 364 United States

Feed Name or Analyses			Mean As Fed	Mean Dry	C.V. ± %
Dry matter	%		55.1	100.0	4
Ash	%		3.8	6.9	14
Crude fiber	%		19.4	35.2	6
Ether extract	%		1.3	2.3	19
N-free extract	%		27.4	49.7	
Protein (N x 6.25)	%		3.2	5.8	41
Cattle	dig prot %	*	1.6	2.9	
Goats	dig prot %	*	1.1	2.0	
Horses	dig prot %	*	1.4	2.5	
Rabbits	dig prot %	*	1.8	3.2	
Sheep	dig prot %	*	1.3	2.4	
Energy	GE Mcal/kg				
Cattle	DE Mcal/kg	*	1.31	2.38	
Sheep	DE Mcal/kg	*	1.34	2.44	
Cattle	ME Mcal/kg	*	1.07	1.95	
Sheep	ME Mcal/kg	*	1.10	2.00	
Cattle	TDN %	*	29.7	53.9	
Sheep	TDN %	*	30.5	55.4	
Calcium	%		0.14	0.26	13
Copper	mg/kg		1.2	2.2	
Phosphorus	%		0.10	0.18	27
Potassium	%		0.19	0.34	

Brome, smooth, aerial part, fresh, over ripe, (2)
Ref No 2 00 961 United States

Feed Name or Analyses			Mean As Fed	Mean Dry	C.V. ± %
Dry matter	%			100.0	
Phosphorus	%			0.12	

Brome, smooth, aerial part, fresh, cut 1, (2)
Ref No 2 00 962 United States

Feed Name or Analyses			Mean As Fed	Mean Dry	C.V. ± %
Dry matter	%			100.0	
Phosphorus	%			0.27	9
Potassium	%			2.33	17

Brome, smooth, aerial part, fresh weathered, mature, (2)
Ref No 2 08 365 United States

Feed Name or Analyses			Mean As Fed	Mean Dry	C.V. ± %
Dry matter	%		85.0	100.0	
Ash	%		5.0	5.9	
Crude fiber	%		32.4	38.1	
Ether extract	%		1.1	1.3	
N-free extract	%		43.1	50.7	
Protein (N x 6.25)	%		3.4	4.0	
Cattle	dig prot %	*	1.1	1.3	
Goats	dig prot %	*	0.2	0.3	
Horses	dig prot %	*	0.8	0.9	
Rabbits	dig prot %	*	1.5	1.8	
Sheep	dig prot %	*	0.6	0.7	
Energy	GE Mcal/kg				
Cattle	DE Mcal/kg	*	1.81	2.13	
Sheep	DE Mcal/kg	*	1.75	2.06	
Cattle	ME Mcal/kg	*	1.48	1.74	
Sheep	ME Mcal/kg	*	1.44	1.69	
Cattle	TDN %	*	41.0	48.2	
Sheep	TDN %	*	39.8	46.8	
Phosphorus	%		0.03	0.04	

Brome, smooth, aerial part, ensiled, (3)
Ref No 3 00 964 United States

Feed Name or Analyses			Mean As Fed	Mean Dry	C.V. ± %
Dry matter	%			100.0	
Ash	%			6.5	12
Crude fiber	%			32.7	8
Ether extract	%			3.3	35
N-free extract	%			47.8	
Protein (N x 6.25)	%			9.7	15
Cattle	dig prot %	*		5.0	
Goats	dig prot %	*		5.0	
Horses	dig prot %	*		5.0	
Sheep	dig prot %	*		5.0	
Energy	GE Mcal/kg				
Cattle	DE Mcal/kg	*		2.58	
Sheep	DE Mcal/kg	*		2.40	
Cattle	ME Mcal/kg	*		2.12	
Sheep	ME Mcal/kg	*		1.97	
Cattle	TDN %	*		58.6	
Sheep	TDN %	*		54.5	
Sulphur	%			0.54	48
Carotene	mg/kg			95.0	13
Vitamin A equivalent	IU/g			158.4	

Brome, smooth, aerial part, ensiled, early bloom, (3)
Ref No 3 07 894 United States

Feed Name or Analyses			Mean As Fed	Mean Dry	C.V. ± %
Dry matter	%		25.1	100.0	
Ash	%		1.9	7.6	
Crude fiber	%		9.4	37.3	
Ether extract	%		0.9	3.5	
N-free extract	%		10.6	42.4	
Protein (N x 6.25)	%		2.3	9.2	
Cattle	dig prot %	*	1.2	4.6	
Goats	dig prot %	*	1.2	4.6	
Horses	dig prot %	*	1.2	4.6	
Sheep	dig prot %	*	1.2	4.6	
Energy	GE Mcal/kg		1.12	4.45	
Cattle	DE Mcal/kg	*	0.62	2.49	
Sheep	DE Mcal/kg	*	0.58	2.33	
Cattle	ME Mcal/kg	*	0.51	2.04	
Sheep	ME Mcal/kg	*	0.48	1.91	
Cattle	TDN %	*	14.1	56.4	
Sheep	TDN %	*	13.3	52.8	

Brome, smooth, aerial part, wilted ensiled, (3)
Ref No 3 00 965 United States

Feed Name or Analyses			Mean As Fed	Mean Dry	C.V. ± %
Dry matter	%			100.0	
Ash	%			6.9	5
Crude fiber	%			31.5	3
Ether extract	%			3.8	27
N-free extract	%			46.4	
Protein (N x 6.25)	%			11.4	5
Cattle	dig prot %	*		6.6	
Goats	dig prot %	*		6.6	
Horses	dig prot %	*		6.6	
Sheep	dig prot %	*		6.6	
Energy	GE Mcal/kg				
Cattle	DE Mcal/kg	*		2.60	
Sheep	DE Mcal/kg	*		2.39	

Feed Name or Analyses			Mean As Fed	Mean Dry	C.V. ± %
Cattle	ME Mcal/kg	*		2.13	
Sheep	ME Mcal/kg	*		1.96	
Cattle	TDN %	*		59.0	
Sheep	TDN %	*		54.1	

Brome, smooth, aerial part w sulfur dioxide preservative added, wilted ensiled, (3)

Ref No 3 00 966 United States

Feed Name or Analyses			Mean As Fed	Mean Dry	C.V. ± %
Dry matter	%			100.0	
Ash	%			6.8	8
Crude fiber	%			32.3	5
Ether extract	%			3.3	39
N-free extract	%			47.7	
Protein (N x 6.25)	%			9.9	16
Cattle	dig prot %	*		5.2	
Goats	dig prot %	*		5.2	
Horses	dig prot %	*		5.2	
Sheep	dig prot %	*		5.2	
Sulphur	%			0.61	29

BROME, SMOOTH CANADA COMMON. Bromus inermus

Brome, smooth Canada common, hay, fan-air dried w heat, midbloom, cut 1, (1)

Ref No 1 08 888 United States

Feed Name or Analyses			Mean As Fed	Mean Dry	C.V. ± %
Dry matter	%		94.2	100.0	
Goats	dig coef %		71.	71.	
Protein (N x 6.25)	%		22.0	23.4	
Goats	dig coef %		80.	80.	
Cattle	dig prot %	*	16.2	17.2	
Goats	dig prot %		17.6	18.7	
Horses	dig prot %	*	16.4	17.4	
Rabbits	dig prot %	*	15.8	16.7	
Sheep	dig prot %	*	16.5	17.6	
Fiber, acid detergent (VS)	%		29.6	31.4	
Goats	dig coef %		64.	64.	
Lignin (Van Soest)	%		3.1	3.3	
Dry matter intake	g/W$^{0.75}$		61.607	65.400	

BROME, SMOOTH CANADIAN. Bromus inermus

Brome, smooth Canadian, hay, s-c, (1)

Ref No 1 08 858 United States

Feed Name or Analyses			Mean As Fed	Mean Dry	C.V. ± %
Dry matter	%			100.0	
Sheep	dig coef %			67.	
Ash	%			8.7	
Ether extract	%			3.7	
Protein (N x 6.25)	%			24.2	
Cattle	dig prot %	*		17.9	
Goats	dig prot %	*		19.1	
Horses	dig prot %	*		18.1	
Rabbits	dig prot %	*		17.4	
Sheep	dig prot %	*		18.3	
Fiber, acid detergent (VS)	%			30.3	
Lignin (Ellis)	%			3.8	
Energy	GE Mcal/kg			4.63	
Sheep	GE dig coef %			64.	
Sheep	DE Mcal/kg			2.96	
Sheep	ME Mcal/kg	*		2.43	
Sheep	TDN %			63.7	
Nutritive value index (NVI)	%			53.	
Phosphorus	%			0.36	
Sulphur	%			0.32	

Feed Name or Analyses			Mean As Fed	Mean Dry	C.V. ± %
Dry matter intake	g/W$^{0.75}$			64.100	
Gain, sheep	kg/day			0.086	

Brome, smooth Canadian, hay, s-c, cut 1, (1)

Ref No 1 08 855 United States

Feed Name or Analyses			Mean As Fed	Mean Dry	C.V. ± %
Dry matter	%			100.0	
Sheep	dig coef %			62.	
Ash	%			8.4	
Ether extract	%			3.1	
Protein (N x 6.25)	%			22.1	
Cattle	dig prot %	*		16.1	
Goats	dig prot %	*		17.2	
Horses	dig prot %	*		16.3	
Rabbits	dig prot %	*		15.7	
Sheep	dig prot %	*		16.4	
Fiber, acid detergent (VS)	%			31.7	
Lignin (Ellis)	%			4.1	
Energy	GE Mcal/kg			4.41	
Sheep	DE Mcal/kg	*		2.57	
Sheep	ME Mcal/kg	*		2.10	
Sheep	TDN %			58.2	
Phosphorus	%			0.31	
Sulphur	%			0.30	
Dry matter intake	g/W$^{0.75}$			82.500	
Gain, sheep	kg/day			0.054	

Brome, smooth Canadian, hay, s-c, cut 2, (1)

Ref No 1 08 856 United States

Feed Name or Analyses			Mean As Fed	Mean Dry	C.V. ± %
Dry matter	%			100.0	
Sheep	dig coef %			61.	
Ash	%			7.8	
Ether extract	%			4.3	
Protein (N x 6.25)	%			23.5	
Cattle	dig prot %	*		17.2	
Goats	dig prot %	*		18.4	
Horses	dig prot %	*		17.4	
Rabbits	dig prot %	*		16.8	
Sheep	dig prot %	*		17.6	
Fiber, acid detergent (VS)	%			31.6	
Lignin (Ellis)	%			5.5	
Energy	GE Mcal/kg			4.89	
Sheep	GE dig coef %			56.	
Sheep	DE Mcal/kg			2.74	
Sheep	ME Mcal/kg	*		2.25	
Sheep	TDN %			59.1	
Nutritive value index (NVI)	%			44.	
Phosphorus	%			0.29	
Sulphur	%			0.32	
Dry matter intake	g/W$^{0.75}$			68.625	
Gain, sheep	kg/day			0.070	

Brome, smooth Canadian, hay, s-c, cut 3, (1)

Ref No 1 08 857 United States

Feed Name or Analyses			Mean As Fed	Mean Dry	C.V. ± %
Dry matter	%			100.0	
Sheep	dig coef %			65.	
Ash	%			9.2	
Ether extract	%			4.8	
Protein (N x 6.25)	%			32.1	
Cattle	dig prot %	*		24.7	
Goats	dig prot %	*		26.5	
Horses	dig prot %	*		24.8	
Rabbits	dig prot %	*		23.5	
Sheep	dig prot %	*		25.4	
Fiber, acid detergent (VS)	%			27.0	

Feed Name or Analyses			Mean As Fed	Mean Dry	C.V. ± %
Lignin (Ellis)	%			5.2	
Energy	GE Mcal/kg			4.68	
Sheep	DE Mcal/kg	*		2.79	
Sheep	ME Mcal/kg	*		2.29	
Sheep	TDN %			63.2	
Phosphorus	%			0.37	
Sulphur	%			0.43	
Dry matter intake	g/W$^{0.75}$			80.300	
Gain, sheep	kg/day			0.086	

BROME, SMOOTH LINCOLN. Bromus inermus

Brome, smooth Lincoln, hay, s-c, immature, cut 2, (1)

Ref No 1 08 872 United States

Feed Name or Analyses			Mean As Fed	Mean Dry	C.V. ± %
Dry matter	%		89.7	100.0	
Sheep	dig coef %		52.	52.	
Organic matter	%		81.3	90.6	
Ash	%		8.4	9.4	
Protein (N x 6.25)	%		15.4	17.2	
Cattle	dig prot %	*	10.6	11.8	
Goats	dig prot %	*	11.3	12.6	
Horses	dig prot %	*	10.9	12.1	
Rabbits	dig prot %	*	10.7	11.9	
Sheep	dig prot %	*	10.8	12.0	
Cell walls (Van Soest)	%		63.2	70.5	
Fiber, acid detergent (VS)	%		39.0	43.5	
Lignin (Van Soest)	%		5.1	5.7	
Energy	GE Mcal/kg		3.99	4.45	
Nutritive value index (NVI)	%		37.	41.	
Aluminum	mg/kg		226.04	252.00	
Barium	mg/kg		10.764	12.000	
Boron	mg/kg		2.691	3.000	
Calcium	%		0.29	0.32	
Cobalt	mg/kg		0.242	0.270	
Copper	mg/kg		6.3	7.0	
Iron	%		0.024	0.027	
Magnesium	%		0.11	0.12	
Manganese	mg/kg		60.1	67.0	
Molybdenum	mg/kg		0.54	0.60	
Phosphorus	%		0.19	0.21	
Potassium	%		1.36	1.52	
Silicon	%		0.44	0.49	
Sodium	%		0.01	0.01	
Strontium	mg/kg		6.279	7.000	
Zinc	mg/kg		23.3	26.0	
Dry matter intake	g/W$^{0.75}$		56.421	62.900	

BROME, SMOOTH SAC. Bromus inermus

Brome, smooth sac, hay, fan-air dried w heat, midbloom, cut 1, (1)

Ref No 1 08 889 United States

Feed Name or Analyses			Mean As Fed	Mean Dry	C.V. ± %
Dry matter	%		95.8	100.0	
Goats	dig coef %		70.	70.	
Protein (N x 6.25)	%		19.4	20.2	
Goats	dig coef %		79.	79.	
Cattle	dig prot %	*	13.8	14.4	
Goats	dig prot %		15.3	16.0	
Horses	dig prot %	*	14.1	14.7	
Rabbits	dig prot %	*	13.7	14.3	
Sheep	dig prot %	*	14.1	14.7	
Fiber, acid detergent (VS)	%		33.0	34.4	
Goats	dig coef %		64.	64.	

Continued

Feed Name or Analyses			Mean As Fed	Mean Dry	C.V. ± %
Lignin (Van Soest)	%		3.7	3.8	
Dry matter intake	$g/W^{.75}$		58.630	61.200	

BROME, SOFT. Bromus mollis

Brome, soft, aerial part, dehy grnd, over ripe, (1)

Ref No 1 08 294 United States

Feed Name or Analyses			Mean As Fed	Mean Dry	C.V. ± %
Dry matter	%		92.9	100.0	
Organic matter	%		88.0	94.7	
Ash	%		4.9	5.3	
Crude fiber	%		28.8	31.0	
Ether extract	%		1.3	1.4	
N-free extract	%		53.6	57.7	
Protein (N x 6.25)	%		4.3	4.6	
Cattle	dig prot %	*	0.9	0.9	
Goats	dig prot %	*	0.8	0.9	
Horses	dig prot %	*	1.3	1.4	
Rabbits	dig prot %	*	2.1	2.2	
Sheep	dig prot %	*	0.6	0.7	
Energy	GE Mcal/kg				
Cattle	DE Mcal/kg	*	2.44	2.62	
Sheep	DE Mcal/kg	*	2.29	2.46	
Cattle	ME Mcal/kg	*	2.00	2.15	
Sheep	ME Mcal/kg	*	1.88	2.02	
Cattle	TDN %	*	55.3	59.5	
Sheep	TDN %	*	51.9	55.8	
Silicon	%		1.30	1.40	

Brome, soft, aerial part, dehy grnd, dormant, (1)

Ref No 1 08 293 United States

Feed Name or Analyses			Mean As Fed	Mean Dry	C.V. ± %
Dry matter	%		92.7	100.0	1
Organic matter	%		88.1	95.0	1
Ash	%		4.6	5.0	20
Crude fiber	%		33.1	35.7	11
Ether extract	%		1.0	1.0	39
N-free extract	%		51.1	55.2	4
Protein (N x 6.25)	%		2.9	3.1	49
Cattle	dig prot %	*	-0.2	-0.2	
Goats	dig prot %	*	-0.4	-0.4	
Horses	dig prot %	*	0.2	0.2	
Rabbits	dig prot %	*	1.0	1.1	
Sheep	dig prot %	*	-0.5	-0.5	
Energy	GE Mcal/kg				
Cattle	DE Mcal/kg	*	2.10	2.26	
Sheep	DE Mcal/kg	*	2.17	2.34	
Cattle	ME Mcal/kg	*	1.72	1.86	
Sheep	ME Mcal/kg	*	1.78	1.92	
Cattle	TDN %	*	47.6	51.4	
Sheep	TDN %	*	49.2	53.2	
Silicon	%		3.05	3.29	35

Brome, soft, leaves, dehy grnd, early leaf, (1)

Ref No 1 08 295 United States

Feed Name or Analyses			Mean As Fed	Mean Dry	C.V. ± %
Dry matter	%		93.1	100.0	
Organic matter	%		82.5	88.6	
Ash	%		10.6	11.4	
Crude fiber	%		14.3	15.4	
Ether extract	%		4.6	4.9	
N-free extract	%		38.5	41.4	
Protein (N x 6.25)	%		25.0	26.9	
Cattle	dig prot %	*	18.8	20.2	
Goats	dig prot %	*	20.2	21.7	
Horses	dig prot %	*	19.0	20.4	
Rabbits	dig prot %	*	18.1	19.4	
Sheep	dig prot %	*	19.3	20.7	
Energy	GE Mcal/kg				
Cattle	DE Mcal/kg	*	2.68	2.88	
Sheep	DE Mcal/kg	*	2.65	2.85	
Cattle	ME Mcal/kg	*	2.20	2.36	
Sheep	ME Mcal/kg	*	2.18	2.34	
Cattle	TDN %	*	60.8	65.3	
Sheep	TDN %	*	60.2	64.7	
Silicon	%		1.30	1.40	

(1) dry forages and roughages (2) pasture, range plants, and forages fed green (3) silages (4) energy feeds (5) protein supplements (6) minerals (7) vitamins (8) additives

Brome, soft, leaves, dehy grnd, immature, (1)

Ref No 1 08 297 United States

Feed Name or Analyses			Mean As Fed	Mean Dry	C.V. ± %
Dry matter	%		93.8	100.0	
Organic matter	%		85.8	91.6	
Ash	%		7.9	8.5	
Crude fiber	%		16.5	17.6	
Ether extract	%		3.9	4.2	
N-free extract	%		47.5	50.7	
Protein (N x 6.25)	%		17.9	19.1	
Cattle	dig prot %	*	12.6	13.5	
Goats	dig prot %	*	13.5	14.4	
Horses	dig prot %	*	12.9	13.7	
Rabbits	dig prot %	*	12.6	13.4	
Sheep	dig prot %	*	12.8	13.7	
Silicon	%		1.85	1.98	

Brome, soft, leaves, dehy grnd, pre-bloom, (1)

Ref No 1 08 298 United States

Feed Name or Analyses			Mean As Fed	Mean Dry	C.V. ± %
Dry matter	%		93.1	100.0	
Organic matter	%		85.8	92.2	
Ash	%		7.3	7.8	
Crude fiber	%		24.3	26.1	
Ether extract	%		3.5	3.8	
N-free extract	%		46.5	49.9	
Protein (N x 6.25)	%		11.5	12.4	
Cattle	dig prot %	*	7.1	7.7	
Goats	dig prot %	*	7.6	8.1	
Horses	dig prot %	*	7.5	8.1	
Rabbits	dig prot %	*	7.7	8.2	
Sheep	dig prot %	*	7.2	7.7	
Silicon	%		1.68	1.80	

Brome, soft, leaves, dehy grnd, early bloom, (1)

Ref No 1 08 299 United States

Feed Name or Analyses			Mean As Fed	Mean Dry	C.V. ± %
Dry matter	%		93.2	100.0	
Organic matter	%		82.6	88.6	
Ash	%		10.6	11.4	
Protein (N x 6.25)	%		11.7	12.6	
Cattle	dig prot %	*	7.3	7.9	
Goats	dig prot %	*	7.8	8.3	
Horses	dig prot %	*	7.7	8.2	
Rabbits	dig prot %	*	7.8	8.4	
Sheep	dig prot %	*	7.3	7.9	
Silicon	%		3.30	3.54	

Brome, soft, leaves, dehy grnd, milk stage, (1)

Ref No 1 08 296 United States

Feed Name or Analyses			Mean As Fed	Mean Dry	C.V. ± %
Dry matter	%		92.0	100.0	
Organic matter	%		74.3	80.8	
Ash	%		17.7	19.2	
Protein (N x 6.25)	%		4.5	4.9	
Cattle	dig prot %	*	1.1	1.2	
Goats	dig prot %	*	1.0	1.1	
Horses	dig prot %	*	1.6	1.7	
Rabbits	dig prot %	*	2.3	2.5	
Sheep	dig prot %	*	0.9	1.0	
Silicon	%		13.70	14.89	

Brome, soft, leaves w some stems, dehy grnd, milk stage, (1)

Ref No 1 08 300 United States

Feed Name or Analyses			Mean As Fed	Mean Dry	C.V. ± %
Dry matter	%		93.5	100.0	
Organic matter	%		83.9	89.7	
Ash	%		9.6	10.3	
Protein (N x 6.25)	%		13.2	14.1	
Cattle	dig prot %	*	8.6	9.2	
Goats	dig prot %	*	9.1	9.7	
Horses	dig prot %	*	8.9	9.5	
Rabbits	dig prot %	*	8.9	9.6	
Sheep	dig prot %	*	8.6	9.2	
Silicon	%		4.10	4.39	

Brome, soft, aerial part, fresh, immature, (2)

Ref No 2 00 967 United States

Feed Name or Analyses			Mean As Fed	Mean Dry	C.V. ± %
Dry matter	%			100.0	
Ash	%			12.2	13
Crude fiber	%			24.2	13
Protein (N x 6.25)	%			19.2	19
Cattle	dig prot %	*		14.2	
Goats	dig prot %	*		14.5	
Horses	dig prot %	*		13.8	
Rabbits	dig prot %	*		13.5	
Sheep	dig prot %	*		14.9	
Calcium	%			0.59	30
Phosphorus	%			0.39	27
Potassium	%			4.00	14

Brome, soft, aerial part, fresh, early bloom, (2)

Ref No 2 00 968 United States

Feed Name or Analyses			Mean As Fed	Mean Dry	C.V. ± %
Dry matter	%			100.0	
Calcium	%			0.36	
Phosphorus	%			0.48	
Potassium	%			2.70	

Brome, soft, aerial part, fresh, full bloom, (2)

Ref No 2 00 969 United States

Feed Name or Analyses			Mean As Fed	Mean Dry	C.V. ± %
Dry matter	%			100.0	
Calcium	%			0.31	
Phosphorus	%			0.38	
Potassium	%			1.46	

Brome, soft, aerial part, fresh, dough stage, (2)

Ref No 2 00 970 United States

Feed Name or Analyses			Mean As Fed	Mean Dry	C.V. ± %
Dry matter	%			100.0	
Calcium	%			0.26	
Phosphorus	%			0.27	
Potassium	%			1.30	

Brome, soft, aerial part, fresh, mature, (2)

Ref No 2 00 971 United States

Feed Name or Analyses			Mean As Fed	Mean Dry	C.V. ± %
Dry matter	%			100.0	
Ash	%			6.3	44
Crude fiber	%			31.2	7
Ether extract	%			1.0	
N-free extract	%			56.7	
Protein (N x 6.25)	%			4.9	12
Cattle	dig prot %	*		2.0	
Goats	dig prot %	*		1.1	
Horses	dig prot %	*		1.6	
Rabbits	dig prot %	*		2.4	
Sheep	dig prot %	*		1.5	
Cellulose (Matrone)	%			40.9	
Other carbohydrates	%			40.3	
Lignin (Ellis)	%			8.9	
Energy	GE Mcal/kg				
Cattle	DE Mcal/kg	*		2.86	
Sheep	DE Mcal/kg	*		3.11	
Cattle	ME Mcal/kg	*		2.34	
Sheep	ME Mcal/kg	*		2.55	
Cattle	TDN %	*		64.9	
Sheep	TDN %	*		70.6	
Silicon	%			1.80	

Brome, soft, heads, fresh, (2)

Ref No 2 00 973 United States

Feed Name or Analyses			Mean As Fed	Mean Dry	C.V. ± %
Dry matter	%			100.0	
Ash	%			6.3	
Crude fiber	%			17.2	
Protein (N x 6.25)	%			12.5	
Cattle	dig prot %	*		8.5	
Goats	dig prot %	*		8.2	
Horses	dig prot %	*		8.1	
Rabbits	dig prot %	*		8.3	
Sheep	dig prot %	*		8.6	

Brome, soft, stems, fresh, (2)

Ref No 2 00 974 United States

Feed Name or Analyses			Mean As Fed	Mean Dry	C.V. ± %
Dry matter	%			100.0	
Ash	%			4.9	
Crude fiber	%			42.6	
Protein (N x 6.25)	%			1.3	
Cattle	dig prot %	*		-0.9	
Goats	dig prot %	*		-2.1	
Horses	dig prot %	*		-1.3	
Rabbits	dig prot %	*		-0.2	
Sheep	dig prot %	*		-1.7	
Calcium	%			0.18	
Phosphorus	%			0.19	
Potassium	%			1.72	

BROME, TEXAS. Bromus texensis

Brome, Texas, aerial part, fresh, (2)

Ref No 2 00 975 United States

Feed Name or Analyses			Mean As Fed	Mean Dry	C.V. ± %
Dry matter	%			100.0	
Ash	%			14.1	
Crude fiber	%			29.7	
Ether extract	%			2.1	
N-free extract	%			43.2	
Protein (N x 6.25)	%			10.9	
Cattle	dig prot %	*		7.2	
Goats	dig prot %	*		6.7	
Horses	dig prot %	*		6.8	
Rabbits	dig prot %	*		7.1	
Sheep	dig prot %	*		7.1	
Energy	GE Mcal/kg				
Cattle	DE Mcal/kg	*		2.56	
Sheep	DE Mcal/kg	*		2.82	
Cattle	ME Mcal/kg	*		2.10	
Sheep	ME Mcal/kg	*		2.31	
Cattle	TDN %	*		58.1	
Sheep	TDN %	*		63.9	
Calcium	%			0.60	
Phosphorus	%			0.20	

BROME–ALFALFA. Bromus spp, Medicago sativa

Brome-Alfalfa, hay, s-c, immature, (1)

Ref No 1 00 976 United States

Feed Name or Analyses			Mean As Fed	Mean Dry	C.V. ± %
Dry matter	%		90.9	100.0	
Calcium	%		0.67	0.74	
Copper	mg/kg		14.0	15.4	
Iron	%		0.008	0.009	
Magnesium	%		0.26	0.29	
Manganese	mg/kg		22.8	25.1	
Phosphorus	%		0.18	0.20	
Potassium	%		1.10	1.21	

Brome-Alfalfa, hay, s-c, full bloom, (1)

Ref No 1 00 977 United States

Feed Name or Analyses			Mean As Fed	Mean Dry	C.V. ± %
Dry matter	%		90.2	100.0	0
Calcium	%		0.57	0.64	17
Cobalt	mg/kg		0.082	0.090	57
Copper	mg/kg		9.7	10.8	38
Iron	%		0.010	0.012	18
Magnesium	%		0.23	0.26	14
Manganese	mg/kg		22.5	24.9	23
Phosphorus	%		0.18	0.21	19
Potassium	%		1.52	1.69	25
Carotene	mg/kg		22.7	25.1	33
Vitamin A equivalent	IU/g		37.8	41.9	

Brome-Alfalfa, hay, s-c, cut 1, (1)

Ref No 1 00 978 United States

Feed Name or Analyses			Mean As Fed	Mean Dry	C.V. ± %
Dry matter	%		90.7	100.0	0
Calcium	%		0.52	0.58	13
Cobalt	mg/kg		0.053	0.059	92
Copper	mg/kg		7.9	8.7	50
Iron	%		0.009	0.010	17
Magnesium	%		0.22	0.24	15
Manganese	mg/kg		17.1	18.9	40
Phosphorus	%		0.21	0.24	16
Potassium	%		1.80	1.99	17
Carotene	mg/kg		16.2	17.9	32
Vitamin A equivalent	IU/g		27.0	29.8	
Vitamin D_2	IU/g		1.5	1.6	

Brome-Alfalfa, hay, s-c, gr 1 US, (1)

Ref No 1 00 979 United States

Feed Name or Analyses			Mean As Fed	Mean Dry	C.V. ± %
Dry matter	%		90.9	100.0	
Calcium	%		0.53	0.58	
Cobalt	mg/kg		0.066	0.073	
Copper	mg/kg		11.8	13.0	
Iron	%		0.012	0.013	
Magnesium	%		0.19	0.21	
Manganese	mg/kg		26.1	28.7	
Phosphorus	%		0.18	0.20	
Potassium	%		1.67	1.84	

Brome-Alfalfa, hay, s-c, gr 2 US, (1)

Ref No 1 00 980 United States

Feed Name or Analyses			Mean As Fed	Mean Dry	C.V. ± %
Dry matter	%		90.5	100.0	1
Ash	%		5.3	5.9	
Crude fiber	%		31.2	34.5	
Ether extract	%		1.9	2.1	
N-free extract	%		40.8	45.1	
Protein (N x 6.25)	%		11.2	12.4	
Cattle	dig prot %	*	6.9	7.7	
Goats	dig prot %	*	7.4	8.1	
Horses	dig prot %	*	7.3	8.1	
Rabbits	dig prot %	*	7.5	8.2	
Sheep	dig prot %	*	7.0	7.7	
Energy	GE Mcal/kg				
Cattle	DE Mcal/kg	*	2.12	2.34	
Sheep	DE Mcal/kg	*	2.27	2.51	
Cattle	ME Mcal/kg	*	1.74	1.92	
Sheep	ME Mcal/kg	*	1.86	2.06	
Cattle	TDN %	*	48.0	53.1	
Sheep	TDN %	*	51.4	56.8	
Calcium	%		0.66	0.73	5
Cobalt	mg/kg		0.056	0.062	98
Copper	mg/kg		12.6	13.9	22
Iron	%		0.009	0.010	14
Magnesium	%		0.26	0.29	7
Manganese	mg/kg		23.3	25.8	5
Phosphorus	%		0.17	0.19	8
Potassium	%		1.16	1.28	10
Carotene	mg/kg		22.1	24.5	33
Vitamin A equivalent	IU/g		36.9	40.8	

Brome-Alfalfa, hay, s-c, gr 3 US, (1)

Ref No 1 00 981 United States

Feed Name or Analyses			Mean As Fed	Mean Dry	C.V. ± %
Dry matter	%		90.9	100.0	
Ash	%		6.0	6.6	
Crude fiber	%		33.5	36.8	
Ether extract	%		10.4	11.4	
N-free extract	%		31.2	34.3	
Protein (N x 6.25)	%		9.9	10.9	
Cattle	dig prot %	*	5.8	6.4	
Goats	dig prot %	*	6.1	6.7	
Horses	dig prot %	*	6.2	6.8	
Rabbits	dig prot %	*	6.4	7.1	
Sheep	dig prot %	*	5.8	6.3	
Energy	GE Mcal/kg				
Cattle	DE Mcal/kg	*	2.37	2.61	

Continued

Feed Name or Analyses			Mean As Fed	Mean Dry	C.V. ± %
Sheep	DE Mcal/kg	*	2.15	2.36	
Cattle	ME Mcal/kg	*	1.94	2.14	
Sheep	ME Mcal/kg	*	1.76	1.94	
Cattle	TDN %	*	53.7	59.1	
Sheep	TDN %	*	48.7	53.6	
Calcium	%		0.47	0.52	
Cobalt	mg/kg		0.044	0.049	
Copper	mg/kg		5.0	5.5	
Iron	%		0.008	0.009	
Magnesium	%		0.22	0.24	
Manganese	mg/kg		11.8	13.0	
Phosphorus	%		0.24	0.26	
Potassium	%		2.07	2.28	

Brome–Alfalfa, aerial part, fresh, (2)

Ref No 2 00 983 United States

Analyses			As Fed	Dry	C.V. ± %
Dry matter	%			100.0	
Calcium	%			1.42	
Magnesium	%			0.28	
Phosphorus	%			0.38	

BROME–CLOVER, LADINO. Bromus spp, Trifolium repens

Brome–Clover, ladino, aerial part, fresh, (2)

Ref No 2 00 984 United States

Analyses			As Fed	Dry	C.V. ± %
Dry matter	%			100.0	
Calcium	%			0.81	15
Magnesium	%			0.28	20
Potassium	%			3.46	14
Sodium	%			0.04	54

BROME–TIMOTHY. Bromus spp, Phleum pratense

Brome-Timothy, hay, s-c, immature, (1)

Ref No 1 00 985 United States

Analyses			As Fed	Dry	C.V. ± %
Dry matter	%		90.0	100.0	
Ash	%		7.3	8.1	
Crude fiber	%		27.1	30.1	
Ether extract	%		4.2	4.7	
N-free extract	%		35.4	39.3	
Protein (N x 6.25)	%		16.0	17.8	
Cattle	dig prot %	*	11.1	12.4	
Goats	dig prot %	*	11.9	13.2	
Horses	dig prot %	*	11.4	12.6	
Rabbits	dig prot %	*	11.2	12.4	
Sheep	dig prot %	*	11.3	12.5	
Energy	GE Mcal/kg				
Cattle	DE Mcal/kg	*	2.38	2.65	
Sheep	DE Mcal/kg	*	2.24	2.49	
Cattle	ME Mcal/kg	*	1.95	2.17	
Sheep	ME Mcal/kg	*	1.84	2.04	
Cattle	TDN %	*	54.0	60.0	
Sheep	TDN %	*	50.8	56.5	

(1) dry forages and roughages (2) pasture, range plants, and forages fed green (3) silages (4) energy feeds (5) protein supplements (6) minerals (7) vitamins (8) additives

BROME–VETCH. Bromus spp, Vicia spp

Brome–Vetch, aerial part, ensiled, (3)

Ref No 3 00 986 United States

Analyses			As Fed	Dry	C.V. ± %
Dry matter	%		19.8	100.0	
Ash	%		1.9	9.5	
Crude fiber	%		6.9	34.8	
Ether extract	%		0.9	4.7	
N-free extract	%		7.9	39.9	
Protein (N x 6.25)	%		2.2	11.1	
Cattle	dig prot %	*	1.2	6.3	
Goats	dig prot %	*	1.2	6.3	
Horses	dig prot %	*	1.2	6.3	
Sheep	dig prot %	*	1.2	6.3	
Energy	GE Mcal/kg				
Cattle	DE Mcal/kg	*	0.53	2.67	
Sheep	DE Mcal/kg	*	0.53	2.70	
Cattle	ME Mcal/kg	*	0.43	2.19	
Sheep	ME Mcal/kg	*	0.44	2.21	
Cattle	TDN %	*	12.0	60.5	
Sheep	TDN %	*	12.1	61.2	

Broomcorn millet grain – see Millet, proso, grain, (4)

BRUSSELS SPROUTS. Brassica oleracea gemmifera

Brussels sprouts, heads, fresh, (2)

Ref No 2 08 187 United States

Analyses			As Fed	Dry	C.V. ± %
Dry matter	%		14.8	100.0	
Ash	%		1.2	8.1	
Crude fiber	%		1.6	10.8	
Ether extract	%		0.4	2.7	
N-free extract	%		6.7	45.3	
Protein (N x 6.25)	%		4.9	33.1	
Cattle	dig prot %	*	3.9	26.0	
Goats	dig prot %	*	4.1	27.4	
Horses	dig prot %	*	3.8	25.6	
Rabbits	dig prot %	*	3.6	24.2	
Sheep	dig prot %	*	4.1	27.8	
Calcium	%		0.04	0.27	
Iron	%		0.002	0.014	
Phosphorus	%		0.08	0.54	
Potassium	%		0.39	2.64	
Sodium	%		0.01	0.07	
Ascorbic acid	mg/kg		1020.0	6891.9	
Niacin	mg/kg		9.0	60.8	
Riboflavin	mg/kg		1.6	10.8	
Thiamine	mg/kg		1.0	6.8	
Vitamin A	IU/g		5.5	37.2	

BUCKWHEAT. Fagopyrum spp

Buckwheat, grain w added hulls, high gr, (1)
Buckwheat feed, high grade

Ref No 5 08 003 United States

Analyses			As Fed	Dry	C.V. ± %
Dry matter	%		89.3	100.0	
Ash	%		4.2	4.7	
Crude fiber	%		18.2	20.4	
Ether extract	%		4.9	5.5	
N-free extract	%		43.5	48.7	
Protein (N x 6.25)	%		18.5	20.7	
Cattle	dig prot %	*	13.3	14.9	
Goats	dig prot %	*	14.2	15.9	
Horses	dig prot %	*	13.5	15.1	
Rabbits	dig prot %	*	13.1	14.7	
Sheep	dig prot %	*	13.5	15.2	
Energy	GE Mcal/kg				
Cattle	DE Mcal/kg	*	2.54	2.85	
Sheep	DE Mcal/kg	*	2.59	2.90	
Cattle	ME Mcal/kg	*	2.09	2.34	
Sheep	ME Mcal/kg	*	2.12	2.37	
Cattle	TDN %	*	57.7	64.6	
Sheep	TDN %	*	58.6	65.7	
Phosphorus	%		0.48	0.54	
Potassium	%		0.66	0.74	

Buckwheat, grain w added hulls, low gr, (1)
Buckwheat feed, low grade

Ref No 1 08 002 United States

Analyses			As Fed	Dry	C.V. ± %
Dry matter	%		88.3	100.0	
Ash	%		3.2	3.6	
Crude fiber	%		28.6	32.4	
Ether extract	%		3.4	3.9	
N-free extract	%		39.8	45.1	
Protein (N x 6.25)	%		13.3	15.1	
Cattle	dig prot %	*	8.8	10.0	
Goats	dig prot %	*	9.4	10.6	
Horses	dig prot %	*	9.1	10.3	
Rabbits	dig prot %	*	9.1	10.3	
Sheep	dig prot %	*	8.9	10.1	
Energy	GE Mcal/kg				
Cattle	DE Mcal/kg	*	1.38	1.56	
Sheep	DE Mcal/kg	*	1.52	1.72	
Cattle	ME Mcal/kg	*	1.13	1.28	
Sheep	ME Mcal/kg	*	1.24	1.41	
Cattle	TDN %	*	31.3	35.4	
Sheep	TDN %	*	34.5	39.1	
Phosphorus	%		0.37	0.42	
Potassium	%		0.68	0.77	

Buckwheat, hulls, (1)
Buckwheat hulls (AAFCO)

Ref No 1 00 987 United States

Analyses			As Fed	Dry	C.V. ± %
Dry matter	%		87.8	100.0	2
Ash	%		1.4	1.6	22
Crude fiber	%		42.6	48.5	8
Sheep	dig coef %		8.	8.	
Ether extract	%		0.8	0.9	16
Sheep	dig coef %		9.	9.	
N-free extract	%		39.9	45.5	
Sheep	dig coef %		25.	25.	
Protein (N x 6.25)	%		3.0	3.5	13
Sheep	dig coef %		7.	7.	
Cattle	dig prot %	*	0.0	0.0	
Goats	dig prot %	*	-0.1	-0.1	
Horses	dig prot %	*	0.4	0.5	
Rabbits	dig prot %	*	1.2	1.3	
Sheep	dig prot %		0.2	0.2	
Cellulose (Matrone)	%		39.0	44.4	
Pentosans	%		21.1	24.0	
Lignin (Ellis)	%		27.7	31.5	
Fatty acids	%		1.0	1.1	
Energy	GE Mcal/kg				
Cattle	DE Mcal/kg	*	0.64	0.73	
Sheep	DE Mcal/kg	*	0.61	0.69	
Cattle	ME Mcal/kg	*	0.53	0.60	
Sheep	ME Mcal/kg	*	0.50	0.57	
Cattle	TDN %	*	14.6	16.6	
Sheep	TDN %		13.8	15.7	

Feed Name or Analyses		Mean As Fed	Mean Dry	C.V. ± %
Calcium	%	0.26	0.29	
Phosphorus	%	0.02	0.02	
Potassium	%	0.27	0.31	

Buckwheat, straw, (1)

Ref No 1 00 988 United States

Feed Name or Analyses		Mean As Fed	Mean Dry	C.V. ± %
Dry matter	%	88.3	100.0	
Ash	%	8.3	9.4	
Crude fiber	%	36.1	40.9	
Ether extract	%	1.0	1.1	
N-free extract	%	38.7	43.8	
Protein (N x 6.25)	%	4.3	4.9	
Cattle	dig prot %	* 1.0	1.1	
Goats	dig prot %	* 1.0	1.1	
Horses	dig prot %	* 1.5	1.7	
Rabbits	dig prot %	* 2.1	2.4	
Sheep	dig prot %	* 0.8	0.9	
Fatty acids	%	1.0	1.1	
Energy	GE Mcal/kg			
Cattle	DE Mcal/kg	* 1.82	2.06	
Sheep	DE Mcal/kg	* 1.64	1.86	
Cattle	ME Mcal/kg	* 1.49	1.69	
Sheep	ME Mcal/kg	* 1.35	1.53	
Cattle	TDN %	* 41.2	46.6	
Sheep	TDN %	* 37.3	42.3	
Calcium	%	1.24	1.41	22
Phosphorus	%	0.11	0.12	98
Potassium	%	2.42	2.74	32

Buckwheat, aerial part, fresh, (2)

Ref No 2 00 989 United States

Feed Name or Analyses		Mean As Fed	Mean Dry	C.V. ± %
Dry matter	%	36.6	100.0	
Ash	%	3.6	9.8	
Crude fiber	%	8.0	21.9	
Ether extract	%	0.9	2.5	
N-free extract	%	19.5	53.2	
Protein (N x 6.25)	%	4.6	12.6	
Cattle	dig prot %	* 3.1	8.6	
Goats	dig prot %	* 3.0	8.3	
Horses	dig prot %	* 3.0	8.2	
Rabbits	dig prot %	* 3.1	8.4	
Sheep	dig prot %	* 3.2	8.7	
Energy	GE Mcal/kg			
Cattle	DE Mcal/kg	* 1.04	2.84	
Sheep	DE Mcal/kg	* 0.95	2.59	
Cattle	ME Mcal/kg	* 0.85	2.32	
Sheep	ME Mcal/kg	* 0.78	2.12	
Cattle	TDN %	* 23.5	64.3	
Sheep	TDN %	* 21.5	58.8	

Buckwheat, flour, coarse bolt, (4)

Ref No 4 00 990 United States

Feed Name or Analyses		Mean As Fed	Mean Dry	C.V. ± %
Dry matter	%	88.8	100.0	2
Ash	%	2.2	2.5	98
Crude fiber	%	1.3	1.5	72
Ether extract	%	2.4	2.7	22
N-free extract	%	71.7	80.8	
Protein (N x 6.25)	%	11.1	12.5	19
Cattle	dig prot %	* 6.7	7.5	
Goats	dig prot %	* 7.7	8.7	
Horses	dig prot %	* 7.7	8.7	
Sheep	dig prot %	* 7.7	8.7	
Energy	GE Mcal/kg			
Cattle	DE Mcal/kg	* 3.21	3.62	
Sheep	DE Mcal/kg	* 3.33	3.76	
Swine	DE kcal/kg	*3604.	4061.	
Cattle	ME Mcal/kg	* 2.64	2.97	
Sheep	ME Mcal/kg	* 2.73	3.08	
Swine	ME kcal/kg	*3369.	3796.	
Cattle	TDN %	* 73.0	82.2	
Sheep	TDN %	* 75.6	85.2	
Swine	TDN %	* 81.7	92.1	
Calcium	%	0.02	0.03	46
Copper	mg/kg	5.3	5.9	
Iron	%	0.008	0.009	98
Manganese	mg/kg	14.2	16.0	
Phosphorus	%	0.21	0.24	25
Potassium	%	0.16	0.18	
Niacin	mg/kg	29.5	33.3	66
Riboflavin	mg/kg	1.5	1.7	23
Thiamine	mg/kg	6.1	6.9	20
Arginine	%	0.89	1.01	
Histidine	%	0.20	0.22	
Isoleucine	%	0.40	0.45	
Leucine	%	0.60	0.67	
Lysine	%	0.60	0.67	
Methionine	%	0.20	0.22	
Phenylalanine	%	0.40	0.45	
Threonine	%	0.40	0.45	
Tryptophan	%	0.20	0.22	
Valine	%	0.60	0.67	

Buckwheat, grain, (4)

Ref No 4 00 994 United States

Feed Name or Analyses		Mean As Fed	Mean Dry	C.V. ± %
Dry matter	%	87.7	100.0	4
Ash	%	2.1	2.4	13
Crude fiber	%	10.4	11.9	19
Horses	dig coef %	6.	6.	
Sheep	dig coef %	45.	45.	
Swine	dig coef %	60.	60.	
Ether extract	%	2.5	2.8	15
Horses	dig coef %	50.	50.	
Sheep	dig coef %	80.	80.	
Swine	dig coef %	24.	24.	
N-free extract	%	62.0	70.7	3
Horses	dig coef %	80.	80.	
Sheep	dig coef %	73.	73.	
Swine	dig coef %	85.	85.	
Protein (N x 6.25)	%	10.6	12.1	5
Horses	dig coef %	65.	65.	
Sheep	dig coef %	72.	72.	
Swine	dig coef %	72.	72.	
Cattle	dig prot %	* 6.3	7.2	
Goats	dig prot %	* 7.3	8.4	
Horses	dig prot %	6.9	7.9	
Sheep	dig prot %	7.7	8.7	
Swine	dig prot %	7.7	8.7	
Energy	GE Mcal/kg	3.96	4.52	
Cattle	DE Mcal/kg	* 3.12	3.56	
Horses	DE Mcal/kg	* 2.64	3.02	
Sheep	DE Mcal/kg	* 2.74	3.12	
Swine	DE kcal/kg	*2999.	3420.	
Cattle	ME Mcal/kg	* 2.56	2.92	
Horses	ME Mcal/kg	* 2.17	2.47	
Sheep	ME Mcal/kg	* 2.25	2.56	
Swine	ME kcal/kg	*2805.	3199.	
Cattle	TDN %	* 70.8	80.7	
Horses	TDN %	60.0	68.4	
Sheep	TDN %	62.1	70.9	
Swine	TDN %	68.0	77.6	
Calcium	%	0.09	0.10	
Cobalt	mg/kg	0.048	0.055	24
Copper	mg/kg	4.3	4.9	
Iron	%	0.004	0.005	
Manganese	mg/kg	33.8	38.6	
Phosphorus	%	0.31	0.35	
Potassium	%	0.45	0.51	
Zinc	mg/kg	8.7	9.9	
Choline	mg/kg	439.	501.	
Niacin	mg/kg	20.0	22.8	
Pantothenic acid	mg/kg	12.5	14.3	
Riboflavin	mg/kg	1.5	1.8	
Arginine	%	1.03	1.18	
Cystine	%	0.20	0.23	
Histidine	%	0.26	0.30	
Isoleucine	%	0.35	0.40	
Leucine	%	0.53	0.60	
Lysine	%	0.61	0.69	
Methionine	%	0.20	0.23	
Phenylalanine	%	0.44	0.50	
Threonine	%	0.44	0.50	
Tryptophan	%	0.19	0.21	
Valine	%	0.53	0.60	

Ref No 4 00 994 Canada

Feed Name or Analyses		Mean As Fed	Mean Dry	C.V. ± %
Dry matter	%	84.6	100.0	
Crude fiber	%	12.1	14.3	
Horses	dig coef %	6.	6.	
Sheep	dig coef %	45.	45.	
Swine	dig coef %	60.	60.	
Ether extract	%			
Horses	dig coef %	50.	50.	
Sheep	dig coef %	80.	80.	
Swine	dig coef %	24.	24.	
N-free extract	%			
Horses	dig coef %	80.	80.	
Sheep	dig coef %	73.	73.	
Swine	dig coef %	85.	85.	
Protein (N x 6.25)	%	12.0	14.2	
Horses	dig coef %	65.	65.	
Sheep	dig coef %	72.	72.	
Swine	dig coef %	72.	72.	
Cattle	dig prot %	* 7.6	9.0	
Goats	dig prot %	* 8.7	10.2	
Horses	dig prot %	7.8	9.2	
Sheep	dig prot %	8.6	10.2	
Swine	dig prot %	8.6	10.2	
Energy	GE Mcal/kg	3.82	4.52	
Horses	DE Mcal/kg	* 2.54	3.00	
Sheep	DE Mcal/kg	* 2.64	3.12	
Swine	DE kcal/kg	*2888.	3413.	
Chickens	ME_n kcal/kg	2607.	3082.	
Horses	ME Mcal/kg	* 2.08	2.46	
Sheep	ME Mcal/kg	* 2.16	2.56	
Swine	ME kcal/kg	*2689.	3179.	
Horses	TDN %	57.6	68.1	
Sheep	TDN %	59.9	70.7	
Swine	TDN %	65.5	77.4	
Calcium	%	0.06	0.07	
Phosphorus	%	0.41	0.48	

Buckwheat, grain, low protein low fiber, (4)

Ref No 4 00 993 United States

Feed Name or Analyses		Mean As Fed	Mean Dry	C.V. ± %
Dry matter	%	87.5	100.0	
Ash	%	2.2	2.5	
Crude fiber	%	11.8	13.5	
Swine	dig coef %	28.	28.	

Continued

Feed Name or Analyses		Mean As Fed	Mean Dry	C.V. ± %
Ether extract	%	2.0	2.3	
Swine	dig coef %	89.	89.	
N-free extract	%	59.9	68.5	
Swine	dig coef %	90.	90.	
Protein (N x 6.25)	%	11.6	13.2	
Swine	dig coef %	81.	81.	
Cattle	dig prot %	* 7.1	8.1	
Goats	dig prot %	* 8.2	9.3	
Horses	dig prot %	* 8.2	9.3	
Sheep	dig prot %	* 8.2	9.3	
Swine	dig prot %	9.4	10.7	
Energy	GE Mcal/kg			
Cattle	DE Mcal/kg	* 2.98	3.40	
Sheep	DE Mcal/kg	* 2.97	3.39	
Swine	DE kcal/kg	*3115.	3559.	
Cattle	ME Mcal/kg	* 2.44	2.79	
Sheep	ME Mcal/kg	* 2.43	2.78	
Swine	ME kcal/kg	*2907.	3322.	
Cattle	TDN %	* 67.5	77.2	
Sheep	TDN %	* 67.3	76.9	
Swine	TDN %	70.6	80.7	

Buckwheat, groats, (4)

Ref No 4 00 995 United States

Feed Name or Analyses		Mean As Fed	Mean Dry	C.V. ± %
Dry matter	%	88.8	100.0	
Ash	%	2.0	2.2	
Crude fiber	%	1.3	1.5	
Ether extract	%	3.1	3.5	
N-free extract	%	68.7	77.4	
Protein (N x 6.25)	%	13.7	15.4	
Cattle	dig prot %	* 9.0	10.2	
Goats	dig prot %	* 10.1	11.4	
Horses	dig prot %	* 10.1	11.4	
Sheep	dig prot %	* 10.1	11.4	
Energy	GE Mcal/kg			
Cattle	DE Mcal/kg	* 3.40	3.83	
Sheep	DE Mcal/kg	* 3.30	3.72	
Swine	DE kcal/kg	*3685.	4152.	
Cattle	ME Mcal/kg	* 2.79	3.14	
Sheep	ME Mcal/kg	* 2.71	3.05	
Swine	ME kcal/kg	*3422.	3856.	
Cattle	TDN %	* 77.1	86.8	
Sheep	TDN %	* 74.9	84.4	
Swine	TDN %	* 83.6	94.2	
Calcium	%	0.04	0.04	
Copper	mg/kg	6.3	7.1	
Iron	%	0.002	0.002	
Magnesium	%	0.39	0.44	
Manganese	mg/kg	17.7	20.0	
Phosphorus	%	0.40	0.45	
Potassium	%	0.35	0.39	
Sulphur	%	0.21	0.23	

Buckwheat, flour by-product wo hulls, coarse sift, (5)

Buckwheat middlings (CFA)

Buckwheat shorts (CFA)

Ref No 5 00 992 United States

Feed Name or Analyses		Mean As Fed	Mean Dry	C.V. ± %
Dry matter	%	88.0	100.0	
Ash	%	4.4	5.1	
Crude fiber	%	7.3	8.4	
Cattle	dig coef %	-4.	-4.	
Sheep	dig coef %	5.	5.	
Ether extract	%	7.1	8.1	
Cattle	dig coef %	74.	74.	
Sheep	dig coef %	89.	89.	
N-free extract	%	41.9	47.6	
Cattle	dig coef %	90.	90.	
Sheep	dig coef %	83.	83.	
Protein (N x 6.25)	%	27.2	30.9	
Cattle	dig coef %	91.	91.	
Sheep	dig coef %	85.	85.	
Cattle	dig prot %	24.7	28.1	
Sheep	dig prot %	23.1	26.3	
Energy	GE Mcal/kg			
Cattle	DE Mcal/kg	* 3.26	3.70	
Sheep	DE Mcal/kg	* 3.20	3.63	
Swine	DE kcal/kg	*3399.	3862.	
Cattle	ME Mcal/kg	* 2.67	3.04	
Sheep	ME Mcal/kg	* 2.62	2.98	
Swine	ME kcal/kg	*3052.	3468.	
Cattle	TDN %	73.9	84.0	
Sheep	TDN %	72.5	82.4	
Swine	TDN %	* 77.1	87.6	

Buckwheat, flour by-product wo hulls, coarse sift, mx 10% fiber, (5)

Buckwheat middlings (AAFCO)

Ref No 5 00 991 United States

Feed Name or Analyses		Mean As Fed	Mean Dry	C.V. ± %
Dry matter	%	88.7	100.0	
Ash	%	4.9	5.5	
Crude fiber	%	7.4	8.3	
Cattle	dig coef %	* -4.	-4.	
Sheep	dig coef %	* 5.	5.	
Ether extract	%	7.3	8.2	
Cattle	dig coef %	* 74.	74.	
Sheep	dig coef %	* 89.	89.	
N-free extract	%	39.4	44.4	
Cattle	dig coef %	* 90.	90.	
Sheep	dig coef %	* 83.	83.	
Protein (N x 6.25)	%	29.7	33.5	
Cattle	dig coef %	* 91.	91.	
Sheep	dig coef %	* 85.	85.	
Cattle	dig prot %	27.0	30.5	
Sheep	dig prot %	25.2	28.5	
Energy	GE Mcal/kg			
Cattle	DE Mcal/kg	* 3.27	3.69	
Sheep	DE Mcal/kg	* 3.22	3.63	
Swine	DE kcal/kg	*3367.	3796.	
Cattle	ME Mcal/kg	* 2.69	3.03	
Sheep	ME Mcal/kg	* 2.64	2.97	
Swine	ME kcal/kg	*3003.	3386.	
Cattle	TDN %	74.3	83.7	
Sheep	TDN %	72.9	82.2	
Swine	TDN %	* 76.4	86.1	
Phosphorus	%	1.02	1.15	
Potassium	%	0.98	1.10	

BUCKWHEAT, COMMON, Fagopyrum esculentum

Buckwheat, common, seeds w hulls, (4)

Ref No 4 09 265 United States

Feed Name or Analyses		Mean As Fed	Mean Dry	C.V. ± %
Dry matter	%		100.0	
Protein (N x 6.25)	%		13.0	
Alanine	%		0.53	
Arginine	%		1.14	
Aspartic acid	%		1.14	
Glutamic acid	%		2.04	
Glycine	%		0.70	
Histidine	%		0.29	
Hydroxyproline	%		0.04	
Isoleucine	%		0.44	
Leucine	%		0.77	
Lysine	%		0.73	
Methionine	%		0.25	
Phenylalanine	%		0.52	
Proline	%		0.46	
Serine	%		0.57	
Threonine	%		0.48	
Tyrosine	%		0.31	
Valine	%		0.60	

BUCKWHEAT, TARTARY, Fagopyrum tataricum

Buckwheat, tartary, grain, (4)

Ref No 4 00 996 United States

Feed Name or Analyses		Mean As Fed	Mean Dry	C.V. ± %
Dry matter	%	87.8	100.0	2
Ash	%	2.1	2.3	12
Crude fiber	%	10.1	11.5	25
Swine	dig coef %	* 60.	60.	
Ether extract	%	2.6	3.0	9
Swine	dig coef %	* 24.	24.	
N-free extract	%	62.0	70.7	
Swine	dig coef %	* 85.	85.	
Protein (N x 6.25)	%	11.0	12.5	14
Swine	dig coef %	* 72.	72.	
Cattle	dig prot %	* 6.6	7.5	
Goats	dig prot %	* 7.7	8.7	
Horses	dig prot %	* 7.7	8.7	
Sheep	dig prot %	* 7.7	8.7	
Swine	dig prot %	7.9	9.0	
Energy	GE Mcal/kg			
Cattle	DE Mcal/kg	* 2.68	3.05	
Sheep	DE Mcal/kg	* 2.66	3.03	
Swine	DE kcal/kg	*3004.	3421.	
Cattle	ME Mcal/kg	* 2.20	2.50	
Sheep	ME Mcal/kg	* 2.18	2.48	
Swine	ME kcal/kg	*2807.	3198.	
Cattle	TDN %	* 60.7	69.1	
Sheep	TDN %	* 60.4	68.8	
Swine	TDN %	68.1	77.6	
Calcium	%	0.13	0.15	
Phosphorus	%	0.31	0.35	
Potassium	%	0.44	0.50	

Buckwheat, tartary, groats, (4)

Ref No 4 00 997 United States

Feed Name or Analyses		Mean As Fed	Mean Dry	C.V. ± %
Dry matter	%	88.0	100.0	2
Ash	%	2.2	2.5	9
Crude fiber	%	1.8	2.0	13
Ether extract	%	3.4	3.9	8
N-free extract	%	66.5	75.6	
Protein (N x 6.25)	%	14.1	16.0	9
Cattle	dig prot %	* 9.4	10.7	
Goats	dig prot %	* 10.5	11.9	
Horses	dig prot %	* 10.5	11.9	
Sheep	dig prot %	* 10.5	11.9	
Energy	GE Mcal/kg			
Cattle	DE Mcal/kg	* 2.92	3.32	
Sheep	DE Mcal/kg	* 3.00	3.41	
Swine	DE kcal/kg	*3069.	3487.	
Cattle	ME Mcal/kg	* 2.39	2.72	

(1) dry forages and roughages (2) pasture, range plants, and forages fed green (3) silages (4) energy feeds (5) protein supplements (6) minerals (7) vitamins (8) additives

Feed Name or Analyses			Mean As Fed	Mean Dry	C.V. ± %
Sheep	ME Mcal/kg	*	2.46	2.80	
Swine	ME kcal/kg	*	2848.	3236.	
Cattle	TDN %	*	66.2	75.2	
Sheep	TDN %	*	68.0	77.3	
Swine	TDN %	*	69.6	79.1	

BUCKWHEAT, WILD. Polygonum convolulus

Buckwheat, wild, grain, (4)
Ref No 4 08 813 Canada

Feed Name or Analyses			Mean As Fed	Mean Dry	C.V. ± %
Dry matter	%		90.6	100.0	
Ash	%		2.1	2.3	
Crude fiber	%		8.6	9.5	
Ether extract	%		2.8	3.1	
N-free extract	%		67.0	73.9	
Protein (N x 6.25)	%		10.1	11.2	
Cattle	dig prot %	*	5.7	6.3	
Goats	dig prot %	*	6.8	7.5	
Horses	dig prot %	*	6.8	7.5	
Sheep	dig prot %	*	6.8	7.5	
Cellulose (Matrone)	%		8.5	9.4	
Energy	GE Mcal/kg				
Cattle	DE Mcal/kg	*	2.91	3.21	
Sheep	DE Mcal/kg	*	2.83	3.12	
Swine	DE kcal/kg	*	2918.	3221.	
Cattle	ME Mcal/kg	*	2.38	2.63	
Sheep	ME Mcal/kg	*	2.32	2.56	
Swine	ME kcal/kg	*	2736.	3019.	
Cattle	TDN %	*	66.0	72.9	
Sheep	TDN %	*	64.1	70.7	
Swine	TDN %	*	66.2	73.1	
Copper	mg/kg		3.7	4.1	
Magnesium	%		0.22	0.25	
Manganese	mg/kg		18.9	20.9	
Selenium	mg/kg		0.280	0.309	
Zinc	mg/kg		20.4	22.5	
Niacin	mg/kg		35.0	38.6	
Pantothenic acid	mg/kg		12.6	13.9	
Riboflavin	mg/kg		1.2	1.3	
Thiamine	mg/kg		0.3	0.3	
Vitamin B6	mg/kg		2.76	3.05	
Alanine	%		0.45	0.50	
Arginine	%		0.63	0.70	
Aspartic acid	%		0.63	0.70	
Glutamic acid	%		1.00	1.10	
Glycine	%		0.45	0.50	
Histidine	%		0.18	0.20	
Isoleucine	%		0.36	0.40	
Leucine	%		0.54	0.60	
Lysine	%		0.36	0.40	
Methionine	%		0.09	0.10	
Phenylalanine	%		0.36	0.40	
Proline	%		0.36	0.40	
Serine	%		0.27	0.30	
Threonine	%		0.18	0.20	
Tyrosine	%		0.27	0.30	
Valine	%		0.36	0.40	

Buckwheat feed, high grade –
see Buckwheat, grain w added hulls, high gr, (1)

Buckwheat feed, low grade –
see Buckwheat, grain w added hulls, low gr, (1)

Buckwheat middlings (AAFCO) –
see Buckwheat, flour by-product wo hulls, coarse sift, mx 10% fiber, (5)

Buckwheat middlings (CFA) –
see Buckwheat, flour by-product wo hulls, coarse sift, (5)

Buckwheat shorts (CFA) –
see Buckwheat, flour by-product wo hulls, coarse sift, (5)

BUFFALOGOURD. Cucurbita foetidissima

Buffalogourd, browse, mature, (2)
Ref No 2 00 998 United States

Feed Name or Analyses			Mean As Fed	Mean Dry	C.V. ± %
Dry matter	%			100.0	
Ash	%			21.5	
Crude fiber	%			10.2	
Ether extract	%			4.0	
N-free extract	%			39.0	
Protein (N x 6.25)	%			25.3	
Cattle	dig prot %	*		19.4	
Goats	dig prot %	*		20.2	
Horses	dig prot %	*		19.0	
Rabbits	dig prot %	*		18.2	
Sheep	dig prot %	*		20.6	
Calcium	%			7.76	
Magnesium	%			1.10	
Phosphorus	%			0.26	
Potassium	%			2.53	

BUFFALOGRASS. Buchloe dactyloides

Buffalograss, aerial part, dormant, (1)
Ref No 1 00 999 United States

Feed Name or Analyses			Mean As Fed	Mean Dry	C.V. ± %
Dry matter	%			100.0	
Carotene	mg/kg			27.8	98
Vitamin A equivalent	IU/g			46.3	

Buffalograss, hay, s-c, (1)
Ref No 1 01 003 United States

Feed Name or Analyses			Mean As Fed	Mean Dry	C.V. ± %
Dry matter	%		90.8	100.0	3
Ash	%		10.8	11.9	
Crude fiber	%		25.7	28.3	
Cattle	dig coef %	*	65.	65.	
Sheep	dig coef %		60.	60.	
Ether extract	%		1.5	1.7	
Cattle	dig coef %	*	62.	62.	
Sheep	dig coef %		37.	37.	
N-free extract	%		45.8	50.4	
Cattle	dig coef %	*	62.	62.	
Sheep	dig coef %		59.	59.	
Protein (N x 6.25)	%		6.9	7.6	
Cattle	dig coef %	*	54.	54.	
Sheep	dig coef %		54.	54.	
Cattle	dig prot %		3.7	4.1	
Goats	dig prot %	*	3.3	3.7	
Horses	dig prot %	*	3.6	4.0	
Rabbits	dig prot %	*	4.1	4.6	
Sheep	dig prot %		3.7	4.1	
Fatty acids	%		1.8	2.0	
Energy	GE Mcal/kg				
Cattle	DE Mcal/kg	*	2.25	2.47	
Sheep	DE Mcal/kg	*	2.09	2.30	
Cattle	ME Mcal/kg	*	1.84	2.03	
Sheep	ME Mcal/kg	*	1.72	1.89	
Cattle	TDN %		51.0	56.1	
Sheep	TDN %		47.5	52.3	
Calcium	%		0.56	0.62	20
Phosphorus	%		0.12	0.13	18
Potassium	%		1.39	1.53	

Buffalograss, hay, s-c, immature, (1)
Ref No 1 01 000 United States

Feed Name or Analyses			Mean As Fed	Mean Dry	C.V. ± %
Dry matter	%		91.9	100.0	
Ash	%		11.7	12.8	
Crude fiber	%		24.1	26.2	
Cattle	dig coef %		65.	65.	
Sheep	dig coef %	*	60.	60.	
Ether extract	%		2.0	2.2	
Cattle	dig coef %		62.	62.	
Sheep	dig coef %	*	37.	37.	
N-free extract	%		43.1	46.9	
Cattle	dig coef %		62.	62.	
Sheep	dig coef %	*	59.	59.	
Protein (N x 6.25)	%		11.0	12.0	
Cattle	dig coef %		54.	54.	
Sheep	dig coef %	*	54.	54.	
Cattle	dig prot %		5.9	6.5	
Goats	dig prot %	*	7.1	7.7	
Horses	dig prot %	*	7.1	7.7	
Rabbits	dig prot %	*	7.3	7.9	
Sheep	dig prot %		5.9	6.5	
Energy	GE Mcal/kg				
Cattle	DE Mcal/kg	*	2.25	2.45	
Sheep	DE Mcal/kg	*	2.09	2.28	
Cattle	ME Mcal/kg	*	1.85	2.01	
Sheep	ME Mcal/kg	*	1.72	1.87	
Cattle	TDN %		51.1	55.6	
Sheep	TDN %		47.5	51.7	

Buffalograss, hay, s-c, full bloom, (1)
Ref No 1 01 001 United States

Feed Name or Analyses			Mean As Fed	Mean Dry	C.V. ± %
Dry matter	%		88.9	100.0	
Ash	%		13.8	15.5	
Crude fiber	%		23.3	26.2	
Cattle	dig coef %	*	65.	65.	
Sheep	dig coef %	*	60.	60.	
Ether extract	%		1.9	2.1	
Cattle	dig coef %	*	62.	62.	
Sheep	dig coef %	*	37.	37.	
N-free extract	%		41.2	46.4	
Cattle	dig coef %	*	62.	62.	
Sheep	dig coef %	*	59.	59.	
Protein (N x 6.25)	%		8.7	9.8	
Cattle	dig coef %	*	54.	54.	
Sheep	dig coef %	*	54.	54.	
Cattle	dig prot %		4.7	5.3	
Goats	dig prot %	*	5.1	5.7	
Horses	dig prot %	*	5.2	5.9	
Rabbits	dig prot %	*	5.5	6.2	
Sheep	dig prot %		4.7	5.3	
Energy	GE Mcal/kg				
Cattle	DE Mcal/kg	*	2.12	2.38	
Sheep	DE Mcal/kg	*	1.97	2.21	
Cattle	ME Mcal/kg	*	1.74	1.95	
Sheep	ME Mcal/kg	*	1.61	1.81	
Cattle	TDN %	*	48.0	54.0	
Sheep	TDN %	*	44.6	50.1	

Feed Name or Analyses			Mean As Fed	Mean Dry	C.V. ± %
Buffalograss, hay, s-c, mature, (1)					
Ref No 1 01 002				United States	
Dry matter	%		92.9	100.0	
Ash	%		13.4	14.4	
Crude fiber	%		25.5	27.5	
Cattle	dig coef %	*	65.	65.	
Sheep	dig coef %	*	60.	60.	
Ether extract	%		1.6	1.7	
Cattle	dig coef %	*	62.	62.	
Sheep	dig coef %	*	37.	37.	
N-free extract	%		47.1	50.7	
Cattle	dig coef %	*	62.	62.	
Sheep	dig coef %	*	59.	59.	
Protein (N x 6.25)	%		5.3	5.7	
Cattle	dig coef %	*	54.	54.	
Sheep	dig coef %	*	54.	54.	
Cattle	dig prot %		2.9	3.1	
Goats	dig prot %	*	1.7	1.9	
Horses	dig prot %	*	2.2	2.4	
Rabbits	dig prot %	*	2.9	3.1	
Sheep	dig prot %		2.9	3.1	
Energy	GE Mcal/kg				
Cattle	DE Mcal/kg	*	2.24	2.41	
Sheep	DE Mcal/kg	*	2.09	2.24	
Cattle	ME Mcal/kg	*	1.84	1.98	
Sheep	ME Mcal/kg	*	1.71	1.84	
Cattle	TDN %	*	50.9	54.8	
Sheep	TDN %	*	47.3	50.9	
Buffalograss, aerial part, fresh, (2)					
Ref No 2 01 010				United States	
Dry matter	%		48.9	100.0	23
Ash	%		6.1	12.5	
Crude fiber	%		13.0	26.7	
Ether extract	%		0.9	1.9	
N-free extract	%		24.1	49.3	
Protein (N x 6.25)	%		4.7	9.5	
Cattle	dig prot %	*	2.9	6.0	
Goats	dig prot %	*	2.7	5.5	
Horses	dig prot %	*	2.7	5.6	
Rabbits	dig prot %	*	2.9	6.0	
Sheep	dig prot %	*	2.9	5.9	
Energy	GE Mcal/kg				
Cattle	DE Mcal/kg	*	1.26	2.59	
Sheep	DE Mcal/kg	*	1.20	2.46	
Cattle	ME Mcal/kg	*	1.04	2.12	
Sheep	ME Mcal/kg	*	0.99	2.02	
Cattle	NE_m Mcal/kg	*	0.62	1.26	
Cattle	NE_{gain} Mcal/kg	*	0.31	0.63	
Cattle	TDN %	*	28.7	58.6	
Sheep	TDN %	*	27.3	55.9	
Calcium	%		0.28	0.57	16
Magnesium	%		0.07	0.14	23
Phosphorus	%		0.08	0.16	32
Potassium	%		0.35	0.71	34
Carotene	mg/kg		45.8	93.7	55
Vitamin A equivalent	IU/g		76.3	156.2	

(1) dry forages and roughages (2) pasture, range plants, and forages fed green (3) silages (4) energy feeds (5) protein supplements (6) minerals (7) vitamins (8) additives

Feed Name or Analyses			Mean As Fed	Mean Dry	C.V. ± %
Buffalograss, aerial part, fresh, immature, (2)					
Ref No 2 01 004				United States	
Dry matter	%		43.0	100.0	
Calcium	%		0.24	0.56	12
Phosphorus	%		0.10	0.23	13
Carotene	mg/kg		53.8	125.2	43
Vitamin A equivalent	IU/g		89.8	208.7	
Buffalograss, aerial part, fresh, early bloom, (2)					
Ref No 2 01 005				United States	
Dry matter	%		44.8	100.0	
Ash	%		4.7	10.6	22
Crude fiber	%		12.5	27.8	4
Ether extract	%		0.9	2.0	30
N-free extract	%		22.0	49.0	
Protein (N x 6.25)	%		4.7	10.6	19
Cattle	dig prot %	*	3.1	6.9	
Goats	dig prot %	*	2.9	6.4	
Horses	dig prot %	*	2.9	6.5	
Rabbits	dig prot %	*	3.1	6.9	
Sheep	dig prot %	*	3.1	6.9	
Energy	GE Mcal/kg				
Cattle	DE Mcal/kg	*	1.04	2.33	
Sheep	DE Mcal/kg	*	1.11	2.47	
Cattle	ME Mcal/kg	*	0.86	1.91	
Sheep	ME Mcal/kg	*	0.91	2.03	
Carotene	mg/kg		51.1	114.0	20
Vitamin A equivalent	IU/g		85.1	190.0	
Buffalograss, aerial part, fresh, full bloom, (2)					
Ref No 2 01 006				United States	
Dry matter	%		54.5	100.0	
Ash	%		6.3	11.5	13
Crude fiber	%		15.8	29.0	7
Ether extract	%		0.6	1.1	48
N-free extract	%		26.5	48.6	
Protein (N x 6.25)	%		5.3	9.8	19
Cattle	dig prot %	*	3.4	6.2	
Goats	dig prot %	*	3.1	5.7	
Horses	dig prot %	*	3.2	5.9	
Rabbits	dig prot %	*	3.4	6.2	
Sheep	dig prot %	*	3.3	6.1	
Energy	GE Mcal/kg				
Cattle	DE Mcal/kg	*	1.25	2.30	
Sheep	DE Mcal/kg	*	1.35	2.48	
Cattle	ME Mcal/kg	*	1.03	1.89	
Sheep	ME Mcal/kg	*	1.11	2.03	
Cattle	TDN %	*	28.4	52.1	
Sheep	TDN %	*	30.6	56.2	
Carotene	mg/kg		40.4	74.1	36
Vitamin A equivalent	IU/g		67.3	123.5	
Buffalograss, aerial part, fresh, dough stage, (2)					
Ref No 2 01 007				United States	
Dry matter	%			100.0	
Carotene	mg/kg			62.4	24
Vitamin A equivalent	IU/g			104.0	

Feed Name or Analyses			Mean As Fed	Mean Dry	C.V. ± %
Buffalograss, aerial part, fresh, mature, (2)					
Ref No 2 01 008				United States	
Dry matter	%		72.3	100.0	
Ash	%		11.0	15.2	21
Crude fiber	%		22.2	30.7	4
Ether extract	%		1.2	1.7	36
N-free extract	%		33.6	46.5	
Protein (N x 6.25)	%		4.3	5.9	14
Cattle	dig prot %	*	2.1	2.9	
Goats	dig prot %	*	1.5	2.1	
Horses	dig prot %	*	1.8	2.5	
Rabbits	dig prot %	*	2.3	3.2	
Sheep	dig prot %	*	1.8	2.5	
Energy	GE Mcal/kg				
Cattle	DE Mcal/kg	*	1.65	2.28	
Sheep	DE Mcal/kg	*	1.60	2.21	
Cattle	ME Mcal/kg	*	1.35	1.87	
Sheep	ME Mcal/kg	*	1.31	1.81	
Cattle	TDN %	*	37.4	51.7	
Sheep	TDN %	*	36.2	50.1	
Carotene	mg/kg		46.4	64.2	9
Vitamin A equivalent	IU/g		77.3	106.9	
Buffalograss, aerial part, fresh, over ripe, (2)					
Ref No 2 01 009				United States	
Dry matter	%		69.8	100.0	
Calcium	%		0.28	0.40	27
Magnesium	%		0.08	0.12	
Phosphorus	%		0.11	0.16	44
Potassium	%		0.25	0.36	
Carotene	mg/kg		31.5	45.2	34
Vitamin A equivalent	IU/g		52.6	75.3	

BUFFALOGRASS–GAMAGRASS. Buchloe dactyloides, Tripsacum spp

Feed Name or Analyses			Mean As Fed	Mean Dry	C.V. ± %
Buffalograss–Gamagrass, aerial part, fresh, milk stage, (2)					
Ref No 2 01 011				United States	
Dry matter	%			100.0	
Carotene	mg/kg			36.8	
Vitamin A equivalent	IU/g			61.4	

BUFFALOGRASS–GRAMA. Buchloe dactyloides, Bouteloua spp

Feed Name or Analyses			Mean As Fed	Mean Dry	C.V. ± %
Buffalograss–Grama, aerial part, fresh, (2)					
Ref No 2 01 013				United States	
Dry matter	%			100.0	
Ash	%			11.9	16
Crude fiber	%			31.8	8
Ether extract	%			0.9	40
N-free extract	%			48.1	
Protein (N x 6.25)	%			7.3	14
Cattle	dig prot %	*		4.1	
Goats	dig prot %	*		3.4	
Horses	dig prot %	*		3.7	
Rabbits	dig prot %	*		4.3	
Sheep	dig prot %	*		3.8	
Carotene	mg/kg			54.5	34
Vitamin A equivalent	IU/g			90.8	

Feed Name or Analyses			Mean As Fed	Mean Dry	C.V. ± %

Buffalograss–Grama, aerial part, fresh, full bloom, (2)

Ref No 2 01 012 United States

			As Fed	Dry	C.V.
Dry matter	%			100.0	
Ash	%			10.6	
Crude fiber	%			33.6	
Ether extract	%			1.1	
N-free extract	%			47.7	
Protein (N x 6.25)	%			7.0	
Cattle	dig prot %	*		3.8	
Goats	dig prot %	*		3.1	
Horses	dig prot %	*		3.5	
Rabbits	dig prot %	*		4.1	
Sheep	dig prot %	*		3.5	
Carotene	mg/kg			68.6	
Vitamin A equivalent	IU/g			114.3	

BUFFALOGRASS–WHEATGRASS, BLUESTEM. Buchloe dactyloides, Agropyron smithii

Buffalograss–Wheatgrass, bluestem, aerial part, fresh, (2)

Ref No 2 01 016 United States

			As Fed	Dry	C.V.
Dry matter	%			100.0	
Carotene	mg/kg			74.5	31
Vitamin A equivalent	IU/g			124.2	

Buffalograss–Wheatgrass, bluestem, aerial part, fresh, mid-bloom, (2)

Ref No 2 01 014 United States

			As Fed	Dry	C.V.
Dry matter	%			100.0	
Ash	%			8.9	15
Crude fiber	%			37.6	6
Ether extract	%			1.6	27
N-free extract	%			44.3	
Protein (N x 6.25)	%			7.6	11
Cattle	dig prot %	*		4.4	
Goats	dig prot %	*		3.7	
Horses	dig prot %	*		4.0	
Rabbits	dig prot %	*		4.5	
Sheep	dig prot %	*		4.1	

Buffalograss–Wheatgrass, bluestem, aerial part, fresh, over ripe, (2)

Ref No 2 01 015 United States

			As Fed	Dry	C.V.
Dry matter	%			100.0	
Ash	%			10.1	
Crude fiber	%			38.6	
Ether extract	%			1.5	
N-free extract	%			43.2	
Protein (N x 6.25)	%			6.6	
Cattle	dig prot %	*		3.5	
Goats	dig prot %	*		2.7	
Horses	dig prot %	*		3.1	
Rabbits	dig prot %	*		3.8	
Sheep	dig prot %	*		3.1	

BUGLOSS, CAPE. Anchusa capensis

Bugloss, cape, seeds w hulls, (5)

Ref No 5 09 135 United States

			As Fed	Dry	C.V.
Dry matter	%			100.0	
Protein (N x 6.25)	%			19.0	
Alanine	%			0.86	
Arginine	%			1.50	
Aspartic acid	%			1.77	
Glutamic acid	%			2.91	
Glycine	%			1.16	
Histidine	%			0.42	
Hydroxyproline	%			0.00	
Isoleucine	%			0.80	
Leucine	%			1.10	
Lysine	%			0.93	
Methionine	%			0.30	
Phenylalanine	%			0.74	
Proline	%			0.68	
Serine	%			0.91	
Threonine	%			0.72	
Tyrosine	%			0.67	
Valine	%			0.87	

BULLGRASS. Muhlenbergia emersleyi

Bullgrass, hay, s-c, (1)

Ref No 1 01 017 United States

			As Fed	Dry	C.V.
Dry matter	%		91.4	100.0	
Ash	%		8.1	8.9	
Crude fiber	%		31.2	34.1	
Ether extract	%		1.4	1.5	
N-free extract	%		43.8	48.0	
Protein (N x 6.25)	%		6.9	7.5	
Cattle	dig prot %	*	3.1	3.4	
Goats	dig prot %	*	3.2	3.6	
Horses	dig prot %	*	3.6	3.9	
Rabbits	dig prot %	*	4.1	4.5	
Sheep	dig prot %	*	3.0	3.3	
Phosphorus	%		0.11	0.12	

BULRUSH. Scirpus spp

Bulrush, flowers, s-c, (1)

Ref No 1 01 018 United States

			As Fed	Dry	C.V.
Dry matter	%		91.4	100.0	
Ash	%		4.9	5.4	
Crude fiber	%		21.4	23.4	
Ether extract	%		2.3	2.5	
N-free extract	%		56.0	61.3	
Protein (N x 6.25)	%		6.8	7.4	
Cattle	dig prot %	*	3.1	3.3	
Goats	dig prot %	*	3.2	3.5	
Horses	dig prot %	*	3.5	3.8	
Rabbits	dig prot %	*	4.0	4.4	
Sheep	dig prot %	*	2.9	3.2	

Bulrush, hay, s-c, (1)

Ref No 1 01 019 United States

			As Fed	Dry	C.V.
Dry matter	%		90.2	100.0	
Ash	%		12.4	13.7	
Crude fiber	%		25.6	28.4	
Ether extract	%		2.2	2.4	
N-free extract	%		42.1	46.7	
Protein (N x 6.25)	%		7.9	8.8	
Cattle	dig prot %	*	4.1	4.6	
Goats	dig prot %	*	4.3	4.8	
Horses	dig prot %	*	4.5	5.0	
Rabbits	dig prot %	*	4.9	5.5	
Sheep	dig prot %	*	4.0	4.5	
Calcium	%		0.69	0.76	
Magnesium	%		0.51	0.57	
Phosphorus	%		0.14	0.16	

Bulrush, leaves, s-c, (1)

Ref No 1 01 020 United States

			As Fed	Dry	C.V.
Dry matter	%		90.6	100.0	
Ash	%		15.1	16.7	
Crude fiber	%		20.8	23.0	
Ether extract	%		1.4	1.5	
N-free extract	%		42.7	47.1	
Protein (N x 6.25)	%		10.6	11.7	
Cattle	dig prot %	*	6.4	7.1	
Goats	dig prot %	*	6.8	7.5	
Horses	dig prot %	*	6.8	7.5	
Rabbits	dig prot %	*	7.0	7.7	
Sheep	dig prot %	*	6.4	7.1	

BULRUSH, SEA. Scirpus maritimus

Bulrush, sea, hay, s-c, (1)

Ref No 1 01 021 United States

			As Fed	Dry	C.V.
Dry matter	%		77.0	100.0	
Ash	%		8.1	10.5	
Crude fiber	%		23.9	31.0	
Sheep	dig coef %		52.	52.	
Ether extract	%		1.7	2.2	
Sheep	dig coef %		52.	52.	
N-free extract	%		35.4	46.0	
Sheep	dig coef %		38.	38.	
Protein (N x 6.25)	%		7.9	10.3	
Sheep	dig coef %		43.	43.	
Cattle	dig prot %	*	4.5	5.9	
Goats	dig prot %	*	4.8	6.2	
Horses	dig prot %	*	4.8	6.3	
Rabbits	dig prot %	*	5.1	6.6	
Sheep	dig prot %		3.4	4.4	
Energy	GE Mcal/kg				
Cattle	DE Mcal/kg	*	1.49	1.94	
Sheep	DE Mcal/kg	*	1.38	1.79	
Cattle	ME Mcal/kg	*	1.22	1.59	
Sheep	ME Mcal/kg	*	1.13	1.47	
Cattle	TDN %	*	33.8	43.9	
Sheep	TDN %		31.3	40.6	

BUMELIA. Bumelia spp

Bumelia, browse, (2)

Ref No 2 01 022 United States

			As Fed	Dry	C.V.
Dry matter	%		20.3	100.0	13
Ash	%		1.2	5.9	3
Crude fiber	%		4.4	21.9	4
Ether extract	%		0.7	3.3	57
N-free extract	%		12.0	59.3	

Continued

Feed Name or Analyses			Mean As Fed	Mean Dry	C.V. ± %
Protein (N x 6.25)	%		1.9	9.6	10
Cattle	dig prot %	*	1.2	6.1	
Goats	dig prot %	*	1.1	5.5	
Horses	dig prot %	*	1.2	5.7	
Rabbits	dig prot %	*	1.2	6.1	
Sheep	dig prot %	*	1.2	5.9	

BUMELIA, TOUGH. Bumelia spp

Bumelia, tough, browse, (2)

Ref No 2 01 023 United States

Analyses			As Fed	Dry	C.V. ± %
Dry matter	%			100.0	
Ash	%			6.6	
Crude fiber	%			25.3	
Ether extract	%			10.3	
N-free extract	%			44.8	
Protein (N x 6.25)	%			13.0	
Cattle	dig prot %	*		8.9	
Goats	dig prot %	*		8.7	
Horses	dig prot %	*		8.6	
Rabbits	dig prot %	*		8.7	
Sheep	dig prot %	*		9.1	
Calcium	%			3.04	
Magnesium	%			0.24	
Phosphorus	%			0.12	
Potassium	%			0.57	

BUNDLEFLOWER, RAYADO. Desmanthus virgatus

Bundleflower, rayado, aerial part, fresh, (2)

Desmanthus

Dwarf koa

Ref No 2 01 024 United States

Analyses			As Fed	Dry	C.V. ± %
Dry matter	%		42.3	100.0	6
Ash	%		2.4	5.7	3
Crude fiber	%		17.6	41.6	21
Ether extract	%		0.9	2.1	9
N-free extract	%		16.3	38.4	
Protein (N x 6.25)	%		5.1	12.2	21
Cattle	dig prot %	*	3.5	8.2	
Goats	dig prot %	*	3.4	7.9	
Horses	dig prot %	*	3.3	7.9	
Rabbits	dig prot %	*	3.4	8.1	
Sheep	dig prot %	*	3.5	8.3	
Fatty acids	%		0.9	2.0	
Energy	GE Mcal/kg				
Cattle	DE Mcal/kg	*	1.03	2.44	
Sheep	DE Mcal/kg	*	0.92	2.18	
Cattle	ME Mcal/kg	*	0.85	2.00	
Sheep	ME Mcal/kg	*	0.76	1.79	
Cattle	TDN %	*	23.4	55.4	
Sheep	TDN %	*	20.9	49.5	
Calcium	%		0.74	1.75	
Magnesium	%		0.14	0.33	
Phosphorus	%		0.15	0.35	
Potassium	%		0.77	1.81	

(1) dry forages and roughages (3) silages (6) minerals
(2) pasture, range plants, and forages fed green (4) energy feeds (7) vitamins
(5) protein supplements (8) additives

Bundleflower, rayado, stem leaves pods, fresh, (2)

Ref No 2 01 025 United States

Feed Name or Analyses			Mean As Fed	Mean Dry	C.V. ± %
Dry matter	%		38.3	100.0	
Ash	%		2.2	5.8	
Crude fiber	%		15.1	39.3	
Cattle	dig coef %		44.	44.	
Ether extract	%		0.9	2.4	
Cattle	dig coef %		47.	47.	
N-free extract	%		15.7	41.0	
Cattle	dig coef %		59.	59.	
Protein (N x 6.25)	%		4.4	11.5	
Cattle	dig coef %		44.	44.	
Cattle	dig prot %		1.9	5.1	
Goats	dig prot %	*	2.8	7.3	
Horses	dig prot %	*	2.8	7.3	
Rabbits	dig prot %	*	2.9	7.5	
Sheep	dig prot %	*	3.0	7.7	
Energy	GE Mcal/kg				
Cattle	DE Mcal/kg	*	0.83	2.16	
Sheep	DE Mcal/kg	*	0.81	2.12	
Cattle	ME Mcal/kg	*	0.68	1.77	
Sheep	ME Mcal/kg	*	0.67	1.74	
Cattle	TDN %		18.8	49.1	
Sheep	TDN %	*	18.4	48.1	

BURCLOVER. Medicago denticulata

Burclover, hay, s-c, (1)

Ref No 1 08 588 United States

Analyses			As Fed	Dry	C.V. ± %
Dry matter	%		92.1	100.0	
Ash	%		10.1	11.0	
Crude fiber	%		22.9	24.9	
Ether extract	%		2.9	3.1	
N-free extract	%		37.8	41.0	
Protein (N x 6.25)	%		18.4	20.0	
Cattle	dig prot %	*	13.1	14.2	
Goats	dig prot %	*	14.0	15.2	
Horses	dig prot %	*	13.3	14.5	
Rabbits	dig prot %	*	13.0	14.1	
Sheep	dig prot %	*	13.3	14.5	
Fatty acids	%		2.9	3.1	
Energy	GE Mcal/kg				
Cattle	DE Mcal/kg	*	2.53	2.75	
Sheep	DE Mcal/kg	*	2.40	2.61	
Cattle	ME Mcal/kg	*	2.07	2.25	
Sheep	ME Mcal/kg	*	1.97	2.14	
Cattle	TDN %	*	57.4	62.3	
Sheep	TDN %	*	54.5	59.2	
Calcium	%		1.32	1.43	
Phosphorus	%		0.45	0.49	
Potassium	%		2.96	3.21	

Burclover, hay, s-c, full bloom, (1)

Ref No 1 01 026 United States

Analyses			As Fed	Dry	C.V. ± %
Dry matter	%		88.4	100.0	
Ash	%		7.6	8.6	
Crude fiber	%		22.3	25.2	
Sheep	dig coef %		52.	52.	
Ether extract	%		2.4	2.7	
Sheep	dig coef %		40.	40.	
N-free extract	%		40.5	45.8	
Sheep	dig coef %		65.	65.	
Protein (N x 6.25)	%		15.6	17.7	
Sheep	dig coef %		69.	69.	

Feed Name or Analyses			Mean As Fed	Mean Dry	C.V. ± %
Cattle	dig prot %	*	10.8	12.3	
Goats	dig prot %	*	11.6	13.1	
Horses	dig prot %	*	11.1	12.6	
Rabbits	dig prot %	*	10.9	12.3	
Sheep	dig prot %		10.8	12.2	
Energy	GE Mcal/kg				
Cattle	DE Mcal/kg	*	2.40	2.72	
Sheep	DE Mcal/kg	*	2.24	2.54	
Cattle	ME Mcal/kg	*	1.97	2.23	
Sheep	ME Mcal/kg	*	1.84	2.08	
Cattle	TDN %	*	54.4	61.6	
Sheep	TDN %		50.8	57.5	

Burclover, hay, s-c, late bloom, (1)

Ref No 1 01 027 United States

Analyses			As Fed	Dry	C.V. ± %
Dry matter	%		90.3	100.0	
Ash	%		12.1	13.4	
Crude fiber	%		20.9	23.1	
Sheep	dig coef %		64.	64.	
Ether extract	%		2.1	2.3	
Sheep	dig coef %		5.	5.	
N-free extract	%		31.8	35.2	
Sheep	dig coef %		76.	76.	
Protein (N x 6.25)	%		23.5	26.0	
Sheep	dig coef %		81.	81.	
Cattle	dig prot %	*	17.6	19.5	
Goats	dig prot %	*	18.8	20.8	
Horses	dig prot %	*	17.7	19.6	
Rabbits	dig prot %	*	16.9	18.7	
Sheep	dig prot %		19.0	21.1	
Energy	GE Mcal/kg				
Cattle	DE Mcal/kg	*	2.63	2.92	
Sheep	DE Mcal/kg	*	2.50	2.77	
Cattle	ME Mcal/kg	*	2.16	2.39	
Sheep	ME Mcal/kg	*	2.05	2.27	
Cattle	TDN %	*	59.7	66.1	
Sheep	TDN %		56.8	62.9	

Burclover, aerial part, fresh, (2)

Ref No 2 08 366 United States

Analyses			As Fed	Dry	C.V. ± %
Dry matter	%		20.8	100.0	
Ash	%		2.3	11.1	
Crude fiber	%		3.9	18.8	
Ether extract	%		1.7	8.2	
N-free extract	%		7.8	37.5	
Protein (N x 6.25)	%		5.1	24.5	
Cattle	dig prot %	*	3.9	18.7	
Goats	dig prot %	*	4.0	19.4	
Horses	dig prot %	*	3.8	18.3	
Rabbits	dig prot %	*	3.7	17.6	
Sheep	dig prot %	*	4.1	19.8	
Energy	GE Mcal/kg				
Cattle	DE Mcal/kg	*	0.61	2.94	
Sheep	DE Mcal/kg	*	0.66	3.17	
Cattle	ME Mcal/kg	*	0.50	2.41	
Sheep	ME Mcal/kg	*	0.54	2.60	
Cattle	TDN %	*	13.9	66.8	
Sheep	TDN %	*	15.0	72.0	

BURCLOVER, CALIFORNIA. Medicago hispida

Burclover, California, aerial part, dehy grnd, dormant, (1)

Ref No 1 08 210 United States

Feed Name or Analyses			Mean As Fed	Mean Dry	C.V. ± %
Dry matter	%		91.9	100.0	1
Organic matter	%		80.4	87.5	9
Ash	%		11.5	12.5	60
Crude fiber	%		32.5	35.3	10
Ether extract	%		1.3	1.4	30
N-free extract	%		38.6	42.0	15
Protein (N x 6.25)	%		8.0	8.7	30
Cattle	dig prot %	*	4.1	4.5	
Goats	dig prot %	*	4.3	4.7	
Horses	dig prot %	*	4.5	5.0	
Rabbits	dig prot %	*	5.0	5.4	
Sheep	dig prot %	*	4.1	4.4	
Energy	GE Mcal/kg				
Cattle	DE Mcal/kg	*	2.36	2.57	
Sheep	DE Mcal/kg	*	2.19	2.39	
Cattle	ME Mcal/kg	*	1.93	2.10	
Sheep	ME Mcal/kg	*	1.80	1.96	
Cattle	TDN %	*	53.5	58.2	
Sheep	TDN %	*	49.7	54.1	
Silicon	%		4.31	4.69	98

Burclover, California, hay, s-c, (1)

Ref No 1 01 030 United States

Feed Name or Analyses			Mean As Fed	Mean Dry	C.V. ± %
Dry matter	%		89.7	100.0	1
Ash	%		8.7	9.7	17
Crude fiber	%		22.5	25.1	9
Sheep	dig coef %	*	53.	53.	
Ether extract	%		2.5	2.8	10
Sheep	dig coef %	*	36.	36.	
N-free extract	%		38.9	43.4	
Sheep	dig coef %	*	66.	66.	
Protein (N x 6.25)	%		17.0	19.0	16
Sheep	dig coef %	*	70.	70.	
Cattle	dig prot %	*	12.0	13.4	
Goats	dig prot %	*	12.8	14.3	
Horses	dig prot %	*	12.3	13.7	
Rabbits	dig prot %	*	12.0	13.3	
Sheep	dig prot %		11.9	13.3	
Energy	GE Mcal/kg				
Cattle	DE Mcal/kg	*	2.48	2.76	
Sheep	DE Mcal/kg	*	2.27	2.54	
Cattle	ME Mcal/kg	*	2.03	2.27	
Sheep	ME Mcal/kg	*	1.87	2.08	
Cattle	TDN %	*	56.2	62.7	
Sheep	TDN %	*	51.6	57.5	

Burclover, California, hay, s-c, immature, (1)

Ref No 1 01 028 United States

Feed Name or Analyses			Mean As Fed	Mean Dry	C.V. ± %
Dry matter	%		90.3	100.0	
Ash	%		12.1	13.4	
Crude fiber	%		20.9	23.1	
Sheep	dig coef %	*	53.	53.	
Ether extract	%		2.1	2.3	
Sheep	dig coef %	*	36.	36.	
N-free extract	%		31.8	35.2	
Sheep	dig coef %	*	66.	66.	
Protein (N x 6.25)	%		23.5	26.0	
Sheep	dig coef %	*	70.	70.	
Cattle	dig prot %	*	17.6	19.5	
Goats	dig prot %	*	18.8	20.8	
Horses	dig prot %	*	17.7	19.6	
Rabbits	dig prot %	*	16.9	18.7	
Sheep	dig prot %		16.4	18.2	
Energy	GE Mcal/kg				
Cattle	DE Mcal/kg	*	2.23	2.47	
Sheep	DE Mcal/kg	*	2.21	2.45	
Cattle	ME Mcal/kg	*	1.83	2.03	
Sheep	ME Mcal/kg	*	1.81	2.01	
Cattle	TDN %	*	50.7	56.1	
Sheep	TDN %	*	50.2	55.5	

Burclover, California, hay, s-c, full bloom, (1)

Ref No 1 01 029 United States

Feed Name or Analyses			Mean As Fed	Mean Dry	C.V. ± %
Dry matter	%		88.4	100.0	1
Ash	%		7.6	8.6	13
Crude fiber	%		22.3	25.2	9
Sheep	dig coef %	*	53.	53.	
Ether extract	%		2.4	2.7 7	16
Sheep	dig coef %	*	36.	36.	
N-free extract	%		40.5	45.8	
Sheep	dig coef %	*	66.	66.	
Protein (N x 6.25)	%		15.6	17.7	10
Sheep	dig coef %	*	70.	70.	
Cattle	dig prot %	*	10.8	12.3	
Goats	dig prot %	*	11.6	13.1	
Horses	dig prot %	*	11.1	12.6	
Rabbits	dig prot %	*	10.9	12.3	
Sheep	dig prot %		11.0	12.4	
Energy	GE Mcal/kg				
Cattle	DE Mcal/kg	*	2.40	2.72	
Sheep	DE Mcal/kg	*	2.27	2.56	
Cattle	ME Mcal/kg	*	1.97	2.23	
Sheep	ME Mcal/kg	*	1.86	2.10	
Cattle	TDN %	*	54.4	61.6	
Sheep	TDN %	*	51.4	58.2	

Burclover, California, leaves, dehy grnd, (1)

Ref No 1 08 226 United States

Feed Name or Analyses			Mean As Fed	Mean Dry	C.V. ± %
Dry matter	%		93.0	100.0	
Organic matter	%		84.5	90.9	
Ash	%		8.5	9.1	
Crude fiber	%		6.0	6.4	
Ether extract	%		4.9	5.3	
N-free extract	%		44.7	48.1	
Protein (N x 6.25)	%		28.9	31.1	
Cattle	dig prot %	*	22.2	23.9	
Goats	dig prot %	*	23.8	25.6	
Horses	dig prot %	*	22.3	23.9	
Rabbits	dig prot %	*	21.1	22.7	
Sheep	dig prot %	*	22.8	24.5	
Silicon	%		0.03	0.03	

Burclover, California, leaves, dehy grnd, early leaf, (1)

Ref No 1 08 224 United States

Feed Name or Analyses			Mean As Fed	Mean Dry	C.V. ± %
Dry matter	%		92.5	100.0	
Organic matter	%		83.2	89.9	
Ash	%		9.3	10.0	
Crude fiber	%		6.7	7.2	
Ether extract	%		4.7	5.1	
N-free extract	%		38.9	42.1	
Protein (N x 6.25)	%		32.9	35.6	
Cattle	dig prot %	*	25.7	27.8	
Goats	dig prot %	*	27.5	29.8	
Horses	dig prot %	*	25.7	27.8	
Rabbits	dig prot %	*	24.2	26.2	
Sheep	dig prot %	*	26.4	28.5	
Silicon	%		0.05	0.05	

Burclover, California, leaves, dehy grnd, pre-bloom, (1)

Ref No 1 08 229 United States

Feed Name or Analyses			Mean As Fed	Mean Dry	C.V. ± %
Dry matter	%		92.9	100.0	
Organic matter	%		84.6	91.1	
Ash	%		8.3	8.9	
Protein (N x 6.25)	%		27.2	29.3	
Cattle	dig prot %	*	20.7	22.3	
Goats	dig prot %	*	22.2	23.9	
Horses	dig prot %	*	20.8	22.4	
Rabbits	dig prot %	*	19.8	21.3	
Sheep	dig prot %	*	21.2	22.9	

Burclover, California, leaves, dehy grnd, mid-bloom, (1)

Ref No 1 08 207 United States

Feed Name or Analyses			Mean As Fed	Mean Dry	C.V. ± %
Dry matter	%		93.4	100.0	
Organic matter	%		84.1	90.0	
Ash	%		9.3	10.0	
Crude fiber	%		7.6	8.1	
Ether extract	%		4.6	4.9	
N-free extract	%		38.8	41.5	
Protein (N x 6.25)	%		33.2	35.5	
Cattle	dig prot %	*	25.9	27.7	
Goats	dig prot %	*	27.7	29.7	
Horses	dig prot %	*	25.8	27.7	
Rabbits	dig prot %	*	24.4	26.1	
Sheep	dig prot %	*	26.5	28.4	
Silicon	%		0.07	0.08	

Burclover, California, leaves, dehy grnd, dormant, (1)

Ref No 1 08 231 United States

Feed Name or Analyses			Mean As Fed	Mean Dry	C.V. ± %
Dry matter	%		92.0	100.0	
Organic matter	%		81.6	88.7	
Ash	%		10.9	11.9	
Crude fiber	%		19.0	20.7	
Ether extract	%		2.5	2.7	
N-free extract	%		45.9	49.9	
Protein (N x 6.25)	%		13.6	14.8	
Cattle	dig prot %	*	9.0	9.8	
Goats	dig prot %	*	9.5	10.4	
Horses	dig prot %	*	9.3	10.1	
Rabbits	dig prot %	*	9.3	10.1	
Sheep	dig prot %	*	9.1	9.8	
Silicon	%		0.05	0.05	

Burclover, California, leaves w flowers, dehy grnd, mid-bloom, (1)

Ref No 1 08 221 United States

Feed Name or Analyses			Mean As Fed	Mean Dry	C.V. ± %
Dry matter	%		92.3	100.0	
Organic matter	%		84.4	91.4	
Ash	%		7.9	8.6	
Protein (N x 6.25)	%		30.1	32.6	
Cattle	dig prot %	*	23.2	25.2	
Goats	dig prot %	*	24.9	27.0	

Continued

Feed Name or Analyses	Mean As Fed	Mean Dry	C.V. ± %
Horses dig prot %	* 23.3	25.2	
Rabbits dig prot %	* 22.0	23.8	
Sheep dig prot %	* 23.8	25.8	
Silicon %	0.00	0.00	

Burclover, California, aerial part, fresh, immature, (2)

Ref No 2 01 032 United States

Feed Name or Analyses	Mean As Fed	Mean Dry	C.V. ± %
Dry matter %		100.0	
Ash %		9.6	
Crude fiber %		17.9	
Ether extract %		2.9	27
N-free extract %		41.6	
Protein (N x 6.25) %		28.0	
Cattle dig prot %	*	21.7	
Goats dig prot %	*	22.7	
Horses dig prot %	*	21.3	
Rabbits dig prot %	*	20.3	
Sheep dig prot %	*	23.1	

Burclover, California, aerial part, fresh, full bloom, (2)

Ref No 2 01 033 United States

Feed Name or Analyses	Mean As Fed	Mean Dry	C.V. ± %
Dry matter %		100.0	
Ash %		7.9	
Crude fiber %		20.2	
Protein (N x 6.25) %		23.5	
Cattle dig prot %	*	17.9	
Goats dig prot %	*	18.5	
Horses dig prot %	*	17.5	
Rabbits dig prot %	*	16.8	
Sheep dig prot %	*	18.9	

Burclover, California, aerial part, fresh, mature, (2)

Ref No 2 01 034 United States

Feed Name or Analyses	Mean As Fed	Mean Dry	C.V. ± %
Dry matter %		100.0	
Ash %		6.1	
Crude fiber %		30.7	
Protein (N x 6.25) %		16.7	
Cattle dig prot %	*	12.1	
Goats dig prot %	*	12.1	
Horses dig prot %	*	11.7	
Rabbits dig prot %	*	11.6	
Sheep dig prot %	*	12.6	

BURCUMBER, WALL. Sicyos angulata

Burcumber, wall, seeds w hulls, (5)

Ref No 5 09 122 United States

Feed Name or Analyses	Mean As Fed	Mean Dry	C.V. ± %
Dry matter %		100.0	
Protein (N x 6.25) %		31.0	
Alanine %		1.30	
Arginine %		3.41	
Aspartic acid %		3.10	
Glutamic acid %		4.62	
Glycine %		1.86	
Histidine %		1.09	
Hydroxyproline %		0.06	
Isoleucine %		1.12	
Leucine %		1.83	
Lysine %		1.43	
Methionine %		0.56	
Phenylalanine %		1.33	
Proline %		1.02	
Serine %		1.36	
Threonine %		0.78	
Tyrosine %		0.93	
Valine %		1.33	

BURNET, SALAD. Sanguisorba minor

Burnet, salad, aerial part, fresh, (2)

Ref No 2 01 039 United States

Feed Name or Analyses	Mean As Fed	Mean Dry	C.V. ± %
Dry matter %		100.0	
Ash %		6.5	13
Crude fiber %		18.0	40
Ether extract %		2.0	30
N-free extract %		62.4	
Protein (N x 6.25) %		11.1	44
Cattle dig prot %	*	7.3	
Goats dig prot %	*	6.9	
Horses dig prot %	*	7.0	
Rabbits dig prot %	*	7.2	
Sheep dig prot %	*	7.3	

Burnet, salad, aerial part, fresh, pre-bloom, (2)

Ref No 2 01 036 United States

Feed Name or Analyses	Mean As Fed	Mean Dry	C.V. ± %
Dry matter %		100.0	
Ash %		7.3	
Crude fiber %		7.6	
Ether extract %		2.4	
N-free extract %		65.3	
Protein (N x 6.25) %		17.4	
Cattle dig prot %	*	12.7	
Goats dig prot %	*	12.8	
Horses dig prot %	*	12.3	
Rabbits dig prot %	*	12.1	
Sheep dig prot %	*	13.2	

Burnet, salad, aerial part, fresh, mid-bloom, (2)

Ref No 2 01 037 United States

Feed Name or Analyses	Mean As Fed	Mean Dry	C.V. ± %
Dry matter %		100.0	
Ash %		6.7	
Crude fiber %		17.0	
Ether extract %		2.3	
N-free extract %		61.6	
Protein (N x 6.25) %		12.4	
Cattle dig prot %	*	8.4	
Goats dig prot %	*	8.1	
Horses dig prot %	*	8.1	
Rabbits dig prot %	*	8.2	
Sheep dig prot %	*	8.5	

Burnet, salad, aerial part, fresh, mature, (2)

Ref No 2 01 038 United States

Feed Name or Analyses	Mean As Fed	Mean Dry	C.V. ± %
Dry matter %		100.0	
Ash %		5.8	
Crude fiber %		24.4	
Ether extract %		1.6	
N-free extract %		61.6	
Protein (N x 6.25) %		6.6	
Cattle dig prot %	*	3.5	
Goats dig prot %	*	2.7	
Horses dig prot %	*	3.1	
Rabbits dig prot %	*	3.8	
Sheep dig prot %	*	3.1	

BURROFAT, TREE. Isomeris arborea

Burrofat, tree, seeds w hulls, (5)

Ref No 5 09 029 United States

Feed Name or Analyses	Mean As Fed	Mean Dry	C.V. ± %
Dry matter %		100.0	
Protein (N x 6.25) %		37.0	
Alanine %		1.52	
Arginine %		4.29	
Aspartic acid %		2.78	
Glutamic acid %		6.07	
Glycine %		1.33	
Histidine %		0.96	
Hydroxyproline %		0.04	
Isoleucine %		1.33	
Leucine %		2.11	
Lysine %		1.07	
Methionine %		0.48	
Phenylalanine %		1.33	
Proline %		1.63	
Serine %		1.00	
Threonine %		1.11	
Tyrosine %		0.70	
Valine %		1.81	

BURROGRASS. Scleropogon spp

Burrograss, aerial part, fresh, (2)

Ref No 2 01 040 United States

Feed Name or Analyses	Mean As Fed	Mean Dry	C.V. ± %
Dry matter %		100.0	
Ash %		9.4	
Crude fiber %		31.4	
Protein (N x 6.25) %		7.1	
Cattle dig prot %	*	3.9	
Goats dig prot %	*	3.2	
Horses dig prot %	*	3.6	
Rabbits dig prot %	*	4.2	
Sheep dig prot %	*	3.6	
Calcium %		0.51	
Cobalt mg/kg		0.295	
Copper mg/kg		8.8	
Manganese mg/kg		71.4	
Phosphorus %		0.11	
Carotene mg/kg		18.7	
Vitamin A equivalent IU/g		31.2	

BURROGRASS. Scleropogon brevifolius

Burrograss, aerial part wo lower stems, fresh early leaf, (2)

Ref No 2 08 729 United States

Feed Name or Analyses	Mean As Fed	Mean Dry	C.V. ± %
Dry matter %		100.0	
Organic matter %		80.6	9
Ash %		20.9	35
Crude fiber %		27.4	12
Ether extract %		1.5	43
N-free extract %		42.9	10

(1) dry forages and roughages (2) pasture, range plants, and forages fed green (3) silages (4) energy feeds (5) protein supplements (6) minerals (7) vitamins (8) additives

Feed Name or Analyses			Mean As Fed	Mean Dry	C.V. ± %
Protein (N x 6.25)	%			7.3	29
Cattle	dig prot %	*		4.1	
Goats	dig prot %	*		3.3	
Horses	dig prot %	*		3.7	
Rabbits	dig prot %	*		4.3	
Sheep	dig prot %	*		3.8	
Phosphorus	%			0.12	41
Carotene	mg/kg			11.5	
Vitamin A equivalent	IU/g			19.1	

Burrograss, aerial part wo lower stems, fresh, mid-bloom, (2)

Ref No 2 08 730 United States

Feed Name or Analyses			Mean As Fed	Mean Dry	C.V. ± %
Dry matter	%			100.0	
Organic matter	%			84.5	
Ash	%			15.6	
Crude fiber	%			30.5	
Ether extract	%			1.2	
N-free extract	%			45.4	
Protein (N x 6.25)	%			7.4	
Cattle	dig prot %	*		4.1	
Goats	dig prot %	*		3.4	
Horses	dig prot %	*		3.8	
Rabbits	dig prot %	*		4.3	
Sheep	dig prot %	*		3.8	
Calcium	%			0.21	
Phosphorus	%			0.11	

Burrograss, aerial part wo lower stems, fresh mature, (2)

Ref No 2 08 731 United States

Feed Name or Analyses			Mean As Fed	Mean Dry	C.V. ± %
Dry matter	%			100.0	
Organic matter	%			81.2	
Ash	%			18.8	
Crude fiber	%			30.7	
Ether extract	%			1.5	
N-free extract	%			42.3	
Protein (N x 6.25)	%			6.7	
Cattle	dig prot %	*		3.6	
Goats	dig prot %	*		2.8	
Horses	dig prot %	*		3.2	
Rabbits	dig prot %	*		3.8	
Sheep	dig prot %	*		3.2	
Calcium	%			0.22	
Phosphorus	%			0.13	

Burrograss, aerial part wo lower stems, fresh dormant, (2)

Ref No 2 08 728 United States

Feed Name or Analyses			Mean As Fed	Mean Dry	C.V. ± %
Dry matter	%			100.0	
Organic matter	%			75.1	36
Ash	%			13.9	30
Crude fiber	%			33.0	11
Ether extract	%			1.3	21
N-free extract	%			45.5	6
Protein (N x 6.25)	%			6.3	17
Cattle	dig prot %	*		3.2	
Goats	dig prot %	*		2.4	
Horses	dig prot %	*		2.8	
Rabbits	dig prot %	*		3.5	
Sheep	dig prot %	*		2.8	
Calcium	%			0.20	23
Phosphorus	%			0.08	29

Feed Name or Analyses			Mean As Fed	Mean Dry	C.V. ± %
Carotene	mg/kg			14.0	68
Vitamin A equivalent	IU/g			23.3	

BUSHSUNFLOWER. Simsia exauriculata

Bushsunflower, aerial part, fresh, pre-bloom, (2)

Ref No 2 01 041 United States

Feed Name or Analyses			Mean As Fed	Mean Dry	C.V. ± %
Dry matter	%			100.0	
Ash	%			15.7	
Crude fiber	%			18.9	
Ether extract	%			1.3	
N-free extract	%			43.7	
Protein (N x 6.25)	%			20.4	
Cattle	dig prot %	*		15.2	
Goats	dig prot %	*		15.6	
Horses	dig prot %	*		14.8	
Rabbits	dig prot %	*		14.4	
Sheep	dig prot %	*		16.0	
Calcium	%			4.42	
Magnesium	%			0.62	
Phosphorus	%			0.23	
Potassium	%			3.53	

Bushsunflower, aerial part, fresh, mature, (2)

Ref No 2 01 042 United States

Feed Name or Analyses			Mean As Fed	Mean Dry	C.V. ± %
Dry matter	%			100.0	
Ash	%			13.4	
Crude fiber	%			21.6	
Ether extract	%			1.4	
N-free extract	%			47.2	
Protein (N x 6.25)	%			16.4	
Cattle	dig prot %	*		11.8	
Goats	dig prot %	*		11.9	
Horses	dig prot %	*		11.5	
Rabbits	dig prot %	*		11.3	
Sheep	dig prot %	*		12.3	
Calcium	%			3.74	
Magnesium	%			0.92	
Phosphorus	%			0.21	
Potassium	%			3.07	

BUTTERCUP. Ranunculus spp

Buttercup, aerial part, s-c, full bloom, (1)

Ref No 1 01 043 United States

Feed Name or Analyses			Mean As Fed	Mean Dry	C.V. ± %
Dry matter	%		89.6	100.0	
Ash	%		6.0	6.7	
Crude fiber	%		30.5	34.0	
Sheep	dig coef %		41.	41.	
Ether extract	%		3.3	3.7	
Sheep	dig coef %		70.	70.	
N-free extract	%		40.8	45.5	
Sheep	dig coef %		67.	67.	
Protein (N x 6.25)	%		9.0	10.1	
Sheep	dig coef %		56.	56.	
Cattle	dig prot %	*	5.1	5.7	
Goats	dig prot %	*	5.4	6.0	
Horses	dig prot %	*	5.5	6.1	
Rabbits	dig prot %	*	5.8	6.5	
Sheep	dig prot %		5.1	5.7	
Energy	GE Mcal/kg				
Cattle	DE Mcal/kg	*	2.24	2.50	
Sheep	DE Mcal/kg	*	2.21	2.47	
Cattle	ME Mcal/kg	*	1.84	2.05	

Feed Name or Analyses			Mean As Fed	Mean Dry	C.V. ± %
Sheep	ME Mcal/kg	*	1.81	2.02	
Cattle	TDN %	*	50.8	56.7	
Sheep	TDN %		50.1	55.9	

BUTTERFLYBUSH, ORANGEEYE. Buddleia davidii

Butterflybush, orangeeye, seeds w hulls, (5)

Ref No 5 09 113 United States

Feed Name or Analyses			Mean As Fed	Mean Dry	C.V. ± %
Dry matter	%			100.0	
Protein (N x 6.25)	%			19.0	
Alanine	%			0.80	
Arginine	%			1.92	
Aspartic acid	%			1.52	
Glutamic acid	%			3.27	
Glycine	%			0.93	
Histidine	%			0.44	
Hydroxyproline	%			0.02	
Isoleucine	%			0.80	
Leucine	%			1.18	
Lysine	%			0.68	
Methionine	%			0.44	
Phenylalanine	%			0.87	
Proline	%			0.74	
Serine	%			0.86	
Threonine	%			0.70	
Tyrosine	%			0.61	
Valine	%			0.87	

Buttermilk, concentrated –
see Cattle, buttermilk, condensed, mn 27% total solids 0.055% fat mx 0.14% ash per 1% solids, (5)

Buttermilk, condensed –
see Cattle, buttermilk, condensed, mn 27% total solids 0.055% fat mx 0.14% ash per 1% solids, (5)

Buttermilk, dried –
see Cattle, buttermilk, dehy, feed gr mx 8% moisture mx 13% ash mn 5% fat, (5)

Buttermilk, evaporated –
see Cattle, buttermilk, condensed, mn 27% total solids 0.055% fat mx 0.14% ash per 1% solids, (5)

CABBAGE. Brassica oleracea capitata

Cabbage, aerial part, fresh, (2)

Ref No 2 01 046 United States

Feed Name or Analyses			Mean As Fed	Mean Dry	C.V. ± %
Dry matter	%		9.6	100.0	
Ash	%		1.0	10.3	
Crude fiber	%		1.0	10.5	
Sheep	dig coef %		91.	91.	
Ether extract	%		0.2	2.6	
Sheep	dig coef %		70.	70.	
N-free extract	%		5.3	55.8	
Sheep	dig coef %		96.	96.	
Protein (N x 6.25)	%		2.0	20.8	
Sheep	dig coef %		86.	86.	
Cattle	dig prot %	*	1.5	15.5	
Goats	dig prot %	*	1.5	15.9	
Horses	dig prot %	*	1.5	15.2	
Rabbits	dig prot %	*	1.4	14.7	
Sheep	dig prot %		1.7	17.9	

Continued

Feed Name or Analyses			Mean As Fed	Mean Dry	C.V. ± %
Energy	GE Mcal/kg		0.30	3.16	
Cattle	DE Mcal/kg	*	0.36	3.80	
Sheep	DE Mcal/kg	*	0.36	3.75	
Cattle	ME Mcal/kg	*	0.30	3.12	
Sheep	ME Mcal/kg	*	0.29	3.08	
Cattle	TDN %	*	8.3	86.2	
Sheep	TDN %		8.1	85.1	
Calcium	%		0.06	0.64	
Chlorine	%		0.05	0.53	
Copper	mg/kg		1.3	14.1	
Iron	%		0.001	0.008	
Magnesium	%		0.02	0.21	
Manganese	mg/kg		2.9	30.5	
Phosphorus	%		0.03	0.35	
Potassium	%		0.27	2.81	
Sodium	%		0.02	0.18	
Sulphur	%		0.11	1.17	
Ascorbic acid	mg/kg		591.6	6184.2	
Niacin	mg/kg		3.8	39.5	
Riboflavin	mg/kg		0.6	6.6	
Thiamine	mg/kg		0.6	6.6	
Vitamin A	IU/g		1.6	17.1	

Cabbage, leaves, fresh, (2)

Ref No 2 01 047 United States

Analyses			As Fed	Dry	C.V. ± %
Dry matter	%		14.8	100.0	
Ash	%		2.3	15.6	
Crude fiber	%		2.1	14.3	
Sheep	dig coef %		81.	81.	
Ether extract	%		0.4	2.6	
Sheep	dig coef %		44.	44.	
N-free extract	%		7.8	53.0	
Sheep	dig coef %		87.	87.	
Protein (N x 6.25)	%		2.1	14.4	
Sheep	dig coef %		69.	69.	
Cattle	dig prot %	*	1.5	10.2	
Goats	dig prot %	*	1.5	10.0	
Horses	dig prot %	*	1.4	9.8	
Rabbits	dig prot %	*	1.4	9.8	
Sheep	dig prot %		1.5	10.0	
Energy	GE Mcal/kg				
Cattle	DE Mcal/kg	*	0.44	2.94	
Sheep	DE Mcal/kg	*	0.46	3.10	
Cattle	ME Mcal/kg	*	0.36	2.41	
Sheep	ME Mcal/kg	*	0.38	2.54	
Cattle	TDN %	*	9.9	66.7	
Sheep	TDN %		10.4	70.3	
Calcium	%		0.09	0.63	
Phosphorus	%		0.03	0.21	

Cabbage, whole wo outside leaves, fresh, (2)

Ref No 2 01 048 United States

Analyses			As Fed	Dry	C.V. ± %
Dry matter	%		8.7	100.0	
Ash	%		0.8	8.7	
Crude fiber	%		0.9	10.8	
Sheep	dig coef %		112.	112.	
Ether extract	%		0.2	1.9	
Sheep	dig coef %		43.	43.	
N-free extract	%		5.2	60.3	
Sheep	dig coef %		102.	102.	

(1) dry forages and roughages (2) pasture, range plants, and forages fed green (3) silages (4) energy feeds (5) protein supplements (6) minerals (7) vitamins (8) additives

Feed Name or Analyses			Mean As Fed	Mean Dry	C.V. ± %
Protein (N x 6.25)	%		1.6	18.2	
Sheep	dig coef %		86.	86.	
Cattle	dig prot %	*	1.2	13.4	
Goats	dig prot %	*	1.2	13.6	
Horses	dig prot %	*	1.1	13.0	
Rabbits	dig prot %	*	1.1	12.7	
Sheep	dig prot %		1.4	15.7	
Energy	GE Mcal/kg				
Cattle	DE Mcal/kg	*	0.33	3.83	
Sheep	DE Mcal/kg	*	0.35	4.02	
Cattle	ME Mcal/kg	*	0.27	3.14	
Sheep	ME Mcal/kg	*	0.29	3.30	
Cattle	TDN %	*	7.6	86.8	
Sheep	TDN %		7.9	91.2	
Calcium	%		0.06	0.66	
Phosphorus	%		0.03	0.39	
Potassium	%		0.28	3.29	

CABBAGE, WHITE. Brassica oleracea capitata

Cabbage, white, leaves, fresh, (2)

Ref No 2 01 050 United States

Analyses			As Fed	Dry	C.V. ± %
Dry matter	%		8.5	100.0	
Ash	%		0.8	9.5	
Crude fiber	%		1.1	12.6	41
Ether extract	%		0.3	4.0	68
N-free extract	%		4.2	49.9	
Protein (N x 6.25)	%		2.0	24.0	17
Cattle	dig prot %	*	1.6	18.3	
Goats	dig prot %	*	1.6	19.0	
Horses	dig prot %	*	1.5	17.9	
Rabbits	dig prot %	*	1.5	17.2	
Sheep	dig prot %	*	1.6	19.4	
Energy	GE Mcal/kg				
Cattle	DE Mcal/kg	*	0.28	3.24	
Sheep	DE Mcal/kg	*	0.27	3.23	
Cattle	ME Mcal/kg	*	0.23	2.66	
Sheep	ME Mcal/kg	*	0.22	2.65	
Cattle	TDN %	*	6.2	73.5	
Sheep	TDN %	*	6.2	73.2	

CACAO. Theobroma cacao

Cacao, hulls, (1)

Ref No 1 01 051 United States

Analyses			As Fed	Dry	C.V. ± %
Dry matter	%		90.0	100.0	
Ash	%		5.9	6.5	
Crude fiber	%		12.2	13.5	
Ether extract	%		10.9	12.1	
N-free extract	%		46.9	52.1	
Protein (N x 6.25)	%		14.2	15.8	
Cattle	dig prot %	*	9.6	10.6	
Goats	dig prot %	*	10.2	11.3	
Horses	dig prot %	*	9.8	10.9	
Rabbits	dig prot %	*	9.8	10.9	
Sheep	dig prot %	*	9.7	10.7	

Cacao, pods, (1)

Ref No 1 01 052 United States

Analyses			As Fed	Dry	C.V. ± %
Dry matter	%		91.7	100.0	
Ash	%		7.0	7.6	
Crude fiber	%		31.7	34.6	
Ether extract	%		1.3	1.4	
N-free extract	%		44.2	48.2	

Feed Name or Analyses			Mean As Fed	Mean Dry	C.V. ± %
Protein (N x 6.25)	%		7.5	8.2	
Cattle	dig prot %	*	3.7	4.0	
Goats	dig prot %	*	3.9	4.2	
Horses	dig prot %	*	4.1	4.5	
Rabbits	dig prot %	*	4.6	5.0	
Sheep	dig prot %	*	3.6	3.9	
Energy	GE Mcal/kg				
Cattle	DE Mcal/kg	*	2.22	2.42	
Sheep	DE Mcal/kg	*	2.23	2.43	
Cattle	ME Mcal/kg	*	1.82	1.98	
Sheep	ME Mcal/kg	*	1.83	1.99	
Cattle	TDN %	*	50.3	54.9	
Sheep	TDN %	*	50.5	55.1	

Cacao, pods, s-c, (1)

Cocoa pods, dried

Ref No 1 08 387 United States

Analyses			As Fed	Dry	C.V. ± %
Dry matter	%		91.7	100.0	
Ash	%		7.0	7.6	
Crude fiber	%		31.7	34.6	
Ether extract	%		1.3	1.4	
N-free extract	%		44.2	48.2	
Protein (N x 6.25)	%		7.5	8.2	
Cattle	dig prot %	*	3.7	4.0	
Goats	dig prot %	*	3.8	4.2	
Horses	dig prot %	*	4.1	4.5	
Rabbits	dig prot %	*	4.6	5.0	
Sheep	dig prot %	*	3.6	3.9	
Energy	GE Mcal/kg				
Cattle	DE Mcal/kg	*	2.22	2.42	
Sheep	DE Mcal/kg	*	2.23	2.43	
Cattle	ME Mcal/kg	*	1.82	1.99	
Sheep	ME Mcal/kg	*	1.83	1.99	
Cattle	TDN %	*	50.4	55.0	
Sheep	TDN %	*	50.5	55.1	

Cacao, shells, (1)

Ref No 1 01 053 United States

Analyses			As Fed	Dry	C.V. ± %
Dry matter	%		95.0	100.0	
Ash	%		4.5	4.7	
Crude fiber	%		15.6	16.4	
Ether extract	%		6.7	7.1	
N-free extract	%		53.4	56.2	
Protein (N x 6.25)	%		14.8	15.6	
Cattle	dig prot %	*	9.9	10.4	
Goats	dig prot %	*	10.6	11.1	
Horses	dig prot %	*	10.2	10.8	
Rabbits	dig prot %	*	10.2	10.7	
Sheep	dig prot %	*	10.0	10.6	
Vitamin D_2	IU/g		2.7	2.8	

Cacao, shells, grnd, (1)

Cocoa shells

Ref No 1 01 568 United States

Analyses			As Fed	Dry	C.V. ± %
Dry matter	%		93.6	100.0	
Ash	%		8.8	9.4	
Crude fiber	%		17.2	18.4	
Sheep	dig coef %		26.	26.	
Ether extract	%		3.9	4.2	
Sheep	dig coef %		90.	90.	
N-free extract	%		48.3	51.6	
Sheep	dig coef %		65.	65.	
Protein (N x 6.25)	%		15.3	16.4	
Sheep	dig coef %		27.	27.	

Feed Name or Analyses			Mean As Fed	Mean Dry	C.V. ± %
Cattle	dig prot %	*	10.4	11.1	
Goats	dig prot %	*	11.1	11.9	
Horses	dig prot %	*	10.7	11.5	
Rabbits	dig prot %	*	10.6	11.3	
Sheep	dig prot %		4.1	4.4	
Energy	GE Mcal/kg				
Cattle	DE Mcal/kg	*	2.04	2.18	
Sheep	DE Mcal/kg	*	2.11	2.26	
Cattle	ME Mcal/kg	*	1.68	1.79	
Sheep	ME Mcal/kg	*	1.73	1.85	
Cattle	TDN %	*	46.2	49.4	
Sheep	TDN %		47.9	51.2	
Phosphorus	%		0.58	0.62	
Potassium	%		2.13	2.27	

Cacao, seeds, grnd, (5)
Cocoa meal
Ref No 5 01 569 United States

Feed Name or Analyses			Mean As Fed	Mean Dry	C.V. ± %
Dry matter	%		94.9	100.0	
Ash	%		5.8	6.1	
Crude fiber	%		5.3	5.6	
Ether extract	%		16.4	17.3	
Cattle	dig coef %		89.	89.	
N-free extract	%		43.0	45.3	
Cattle	dig coef %		40.	40.	
Protein (N x 6.25)	%		24.5	25.8	
Cattle	dig coef %		37.	37.	
Cattle	dig prot %		9.1	9.5	
Energy	GE Mcal/kg				
Cattle	DE Mcal/kg	*	2.56	2.70	
Sheep	DE Mcal/kg	*	2.51	2.65	
Swine	DE kcal/kg	*	2749.	2897.	
Cattle	ME Mcal/kg	*	2.10	2.21	
Sheep	ME Mcal/kg	*	2.06	2.17	
Swine	ME kcal/kg	*	2496.	2630.	
Cattle	TDN %	*	58.0	61.2	
Sheep	TDN %	*	57.0	60.1	
Swine	TDN %	*	62.3	65.7	

CACTUS. Cactaceae (family)

Cactus, aerial part, fresh, (2)
Ref No 2 01 054 United States

Feed Name or Analyses			Mean As Fed	Mean Dry	C.V. ± %
Dry matter	%		31.7	100.0	
Ash	%		5.1	16.1	51
Crude fiber	%		8.6	27.2	
Ether extract	%		0.7	2.1	
N-free extract	%		15.6	49.3	
Protein (N x 6.25)	%		1.7	5.3	33
Cattle	dig prot %	*	0.8	2.4	
Goats	dig prot %	*	0.5	1.5	
Horses	dig prot %	*	0.6	2.0	
Rabbits	dig prot %	*	0.9	2.8	
Sheep	dig prot %	*	0.6	1.9	
Calcium	%		0.93	2.94	
Cobalt	mg/kg		0.075	0.236	
Copper	mg/kg		6.0	19.0	
Manganese	mg/kg		79.3	250.0	
Phosphorus	%		0.03	0.08	
Carotene	mg/kg		5.8	18.3	
Vitamin A equivalent	IU/g		9.7	30.5	

Cactus, aerial part wo spines, fresh, (2)
Ref No 2 01 055 United States

Feed Name or Analyses			Mean As Fed	Mean Dry	C.V. ± %
Dry matter	%			100.0	
Carotene	mg/kg			33.1	
Vitamin A equivalent	IU/g			55.1	

Cactus, aerial part, ensiled, (3)
Ref No 3 01 056 United States

Feed Name or Analyses			Mean As Fed	Mean Dry	C.V. ± %
Dry matter	%		9.6	100.0	
Ash	%		2.0	20.8	
Crude fiber	%		1.7	17.7	
Ether extract	%		0.4	4.2	
N-free extract	%		5.0	52.1	
Protein (N x 6.25)	%		0.5	5.2	
Cattle	dig prot %	*	0.1	1.0	
Goats	dig prot %	*	0.1	1.0	
Horses	dig prot %	*	0.1	1.0	
Sheep	dig prot %	*	0.1	1.0	

CACTUS, MEXICALI. Opuntia cholla

Cactus, Mexicali, aerial part, fresh, (2)
Cactus, cane
Ref No 2 08 367 United States

Feed Name or Analyses			Mean As Fed	Mean Dry	C.V. ± %
Dry matter	%		21.0	100.0	
Ash	%		3.4	16.2	
Crude fiber	%		3.3	15.7	
Ether extract	%		0.4	1.9	
N-free extract	%		12.4	59.0	
Protein (N x 6.25)	%		1.5	7.1	
Cattle	dig prot %	*	0.8	4.0	
Goats	dig prot %	*	0.7	3.2	
Horses	dig prot %	*	0.8	3.6	
Rabbits	dig prot %	*	0.9	4.2	
Sheep	dig prot %	*	0.8	3.6	
Energy	GE Mcal/kg				
Cattle	DE Mcal/kg	*	0.55	2.64	
Sheep	DE Mcal/kg	*	0.52	2.49	
Cattle	ME Mcal/kg	*	0.45	2.16	
Sheep	ME Mcal/kg	*	0.43	2.04	
Cattle	TDN %	*	12.6	59.9	
Sheep	TDN %	*	11.8	56.4	
Calcium	%		0.01	0.05	
Potassium	%		0.17	0.81	

Cactus, Mexicali, fruit, fresh, (2)
Cactus, cane, fruit
Ref No 2 08 368 United States

Feed Name or Analyses			Mean As Fed	Mean Dry	C.V. ± %
Dry matter	%		18.6	100.0	
Ash	%		2.7	14.5	
Crude fiber	%		3.2	17.2	
Ether extract	%		0.8	4.3	
N-free extract	%		10.4	55.9	
Protein (N x 6.25)	%		1.5	8.1	
Cattle	dig prot %	*	0.9	4.7	
Goats	dig prot %	*	0.8	4.1	
Horses	dig prot %	*	0.8	4.4	
Rabbits	dig prot %	*	0.9	4.9	
Sheep	dig prot %	*	0.8	4.5	
Energy	GE Mcal/kg				
Cattle	DE Mcal/kg	*	0.51	2.75	
Sheep	DE Mcal/kg	*	0.48	2.59	
Cattle	ME Mcal/kg	*	0.42	2.25	
Sheep	ME Mcal/kg	*	0.39	2.12	
Cattle	TDN %	*	11.6	62.3	
Sheep	TDN %	*	10.9	58.7	

Cactus, Mexicali, stems, fresh, (2)
Cactus, cane, stems
Ref No 2 08 369 United States

Feed Name or Analyses			Mean As Fed	Mean Dry	C.V. ± %
Dry matter	%		21.7	100.0	
Ash	%		3.8	17.5	
Crude fiber	%		3.4	15.7	
Ether extract	%		0.4	1.8	
N-free extract	%		12.6	58.1	
Protein (N x 6.25)	%		1.5	6.9	
Cattle	dig prot %	*	0.8	3.8	
Goats	dig prot %	*	0.7	3.0	
Horses	dig prot %	*	0.7	3.4	
Rabbits	dig prot %	*	0.9	4.0	
Sheep	dig prot %	*	0.7	3.4	
Energy	GE Mcal/kg				
Cattle	DE Mcal/kg	*	0.56	2.58	
Sheep	DE Mcal/kg	*	0.53	2.43	
Cattle	ME Mcal/kg	*	0.46	2.12	
Sheep	ME Mcal/kg	*	0.43	1.99	
Cattle	TDN %	*	12.7	58.6	
Sheep	TDN %	*	12.0	55.1	
Phosphorus	%		0.04	0.18	
Potassium	%		0.40	1.84	

CACTUS, PRICKLYPEAR. Opuntia spp

Cactus, pricklypear, aerial part, s-c, (1)
Ref No 1 01 057 United States

Feed Name or Analyses			Mean As Fed	Mean Dry	C.V. ± %
Dry matter	%		91.9	100.0	
Ash	%		16.5	17.9	
Crude fiber	%		12.5	13.6	
Cattle	dig coef %		48.	48.	
Ether extract	%		1.5	1.6	
Cattle	dig coef %		55.	55.	
N-free extract	%		57.2	62.2	
Cattle	dig coef %		76.	76.	
Protein (N x 6.25)	%		4.3	4.7	
Cattle	dig coef %		68.	68.	
Cattle	dig prot %		2.9	3.2	
Goats	dig prot %	*	0.9	0.9	
Horses	dig prot %	*	1.4	1.5	
Rabbits	dig prot %	*	2.1	2.3	
Sheep	dig prot %	*	0.7	0.8	
Energy	GE Mcal/kg				
Cattle	DE Mcal/kg	*	2.39	2.60	
Sheep	DE Mcal/kg	*	2.32	2.52	
Cattle	ME Mcal/kg	*	1.96	2.13	
Sheep	ME Mcal/kg	*	1.90	2.07	
Cattle	TDN %		54.2	59.0	
Sheep	TDN %	*	52.5	57.1	

Cactus, pricklypear, aerial part, fresh, (2)
Ref No 2 01 061 United States

Feed Name or Analyses			Mean As Fed	Mean Dry	C.V. ± %
Dry matter	%		20.6	100.0	43
Ash	%		3.9	18.9	15
Crude fiber	%		2.8	13.4	6
Cattle	dig coef %		42.	42.	
Sheep	dig coef %		13.	13.	

Continued

507

Feed Name or Analyses			Mean As Fed	Mean Dry	C.V. ± %
Ether extract	%		0.5	2.3	51
Cattle	dig coef %		74.	74.	
Sheep	dig coef %		74.	74.	
N-free extract	%		12.5	60.6	5
Cattle	dig coef %		79.	79.	
Sheep	dig coef %		76.	76.	
Protein (N x 6.25)	%		1.0	4.8	22
Cattle	dig coef %		34.	34.	
Sheep	dig coef %		50.	50.	
Cattle	dig prot %		0.3	1.6	
Goats	dig prot %	*	0.2	1.0	
Horses	dig prot %	*	0.3	1.6	
Rabbits	dig prot %	*	0.5	2.3	
Sheep	dig prot %		0.5	2.4	
Energy	GE Mcal/kg				
Cattle	DE Mcal/kg	*	0.53	2.59	
Sheep	DE Mcal/kg	*	0.49	2.38	
Cattle	ME Mcal/kg	*	0.44	2.13	
Sheep	ME Mcal/kg	*	0.40	1.95	
Cattle	NE_m Mcal/kg	*	0.26	1.28	
Cattle	NE_{gain} Mcal/kg	*	0.13	0.65	
Cattle	TDN %		12.1	58.8	
Sheep	TDN %		11.1	54.0	
Calcium	%		1.89	9.16	
Phosphorus	%		0.02	0.12	
Potassium	%		0.52	2.53	

Cactus, pricklypear, aerial part, fresh, immature, (2)

Ref No 2 01 058 United States

Feed Name or Analyses			As Fed	Dry	C.V. ± %
Dry matter	%		12.9	100.0	
Ash	%		2.6	20.2	
Crude fiber	%		1.2	9.3	
Ether extract	%		0.4	3.1	
N-free extract	%		7.8	60.4	
Protein (N x 6.25)	%		0.9	7.0	
Cattle	dig prot %	*	0.5	3.8	
Goats	dig prot %	*	0.4	3.1	
Horses	dig prot %	*	0.4	3.5	
Rabbits	dig prot %	*	0.5	4.1	
Sheep	dig prot %	*	0.5	3.5	
Energy	GE Mcal/kg				
Cattle	DE Mcal/kg	*	0.34	2.63	
Sheep	DE Mcal/kg	*	0.32	2.49	
Cattle	ME Mcal/kg	*	0.28	2.16	
Sheep	ME Mcal/kg	*	0.26	2.04	
Cattle	TDN %	*	7.7	59.7	
Sheep	TDN %	*	7.3	56.4	

Cactus, pricklypear, aerial part, fresh, mature, (2)

Ref No 2 01 059 United States

Feed Name or Analyses			As Fed	Dry	C.V. ± %
Dry matter	%		19.2	100.0	
Ash	%		3.1	16.4	38
Crude fiber	%		2.6	13.7	
Ether extract	%		0.3	1.8	
N-free extract	%		12.3	63.9	
Protein (N x 6.25)	%		0.8	4.2	16
Cattle	dig prot %	*	0.3	1.5	
Goats	dig prot %	*	0.1	0.5	
Horses	dig prot %	*	0.2	1.1	
Rabbits	dig prot %	*	0.4	1.9	
Sheep	dig prot %	*	0.2	0.9	
Energy	GE Mcal/kg				
Cattle	DE Mcal/kg	*	0.47	2.44	
Sheep	DE Mcal/kg	*	0.44	2.27	
Cattle	ME Mcal/kg	*	0.38	2.00	
Sheep	ME Mcal/kg	*	0.36	1.86	
Cattle	TDN %	*	10.6	55.3	
Sheep	TDN %	*	9.9	51.4	
Chlorine	%		0.04	0.21	
Sodium	%		0.06	0.30	
Sulphur	%		0.04	0.23	
Carotene	mg/kg		1.1	6.0	
Vitamin A equivalent	IU/g		1.9	9.9	

Cactus, pricklypear, aerial part, fresh, over ripe, (2)

Ref No 2 01 060 United States

Feed Name or Analyses			As Fed	Dry	C.V. ± %
Dry matter	%		27.3	100.0	
Ash	%		6.6	24.2	
Protein (N x 6.25)	%		0.6	2.3	
Cattle	dig prot %	*	0.0	-0.1	
Goats	dig prot %	*	-0.3	-1.2	
Horses	dig prot %	*	0.0	-0.4	
Rabbits	dig prot %	*	0.1	0.4	
Sheep	dig prot %	*	-0.1	-0.8	

Cactus, pricklypear, joints, fresh, (2)

Ref No 2 01 063 United States

Feed Name or Analyses			As Fed	Dry	C.V. ± %
Dry matter	%		17.4	100.0	
Ash	%		3.8	21.6	
Crude fiber	%		2.0	11.4	
Ether extract	%		0.4	2.2	
N-free extract	%		10.3	58.8	
Protein (N x 6.25)	%		1.0	5.9	
Cattle	dig prot %	*	0.5	2.9	
Goats	dig prot %	*	0.4	2.1	
Horses	dig prot %	*	0.4	2.6	
Rabbits	dig prot %	*	0.6	3.3	
Sheep	dig prot %	*	0.4	2.5	
Energy	GE Mcal/kg				
Cattle	DE Mcal/kg	*	0.47	2.70	
Sheep	DE Mcal/kg	*	0.44	2.54	
Cattle	ME Mcal/kg	*	0.38	2.21	
Sheep	ME Mcal/kg	*	0.36	2.08	
Cattle	TDN %	*	10.6	61.2	
Sheep	TDN %	*	10.0	57.6	
Calcium	%		1.70	9.76	24
Magnesium	%		0.21	1.20	26
Phosphorus	%		0.02	0.13	49
Potassium	%		0.36	2.09	35

Cactus, pricklypear, joints wo spines, fresh, (2)

Ref No 2 01 064 United States

Feed Name or Analyses			As Fed	Dry	C.V. ± %
Dry matter	%		19.7	100.0	
Ash	%		5.0	25.2	
Crude fiber	%		3.2	16.0	
Ether extract	%		0.2	1.2	
N-free extract	%		10.8	54.8	
Protein (N x 6.25)	%		0.6	2.8	
Cattle	dig prot %	*	0.1	0.3	
Goats	dig prot %	*	-0.1	-0.7	
Horses	dig prot %	*	0.0	0.0	
Rabbits	dig prot %	*	0.2	0.8	
Sheep	dig prot %	*	0.0	-0.3	
Energy	GE Mcal/kg				
Cattle	DE Mcal/kg	*	0.55	2.81	
Sheep	DE Mcal/kg	*	0.52	2.65	
Cattle	ME Mcal/kg	*	0.45	2.30	
Sheep	ME Mcal/kg	*	0.43	2.17	
Cattle	TDN %	*	12.5	63.7	
Sheep	TDN %	*	11.8	60.1	
Calcium	%		1.95	9.90	
Magnesium	%		0.31	1.56	
Phosphorus	%		0.02	0.10	
Potassium	%		0.42	2.15	

Cactus, cane –
see Cactus, Mexicali, aerial part, fresh, (2)

Cactus, cane, fruit –
see Cactus, Mexicali, fruit, fresh, (2)

Cactus, cane, stems –
see Cactus, Mexicali, stems, fresh, (2)

CADILLO. Urena lobata

Cadillo, seeds w hulls, (5)

Ref No 5 09 031 United States

Feed Name or Analyses			As Fed	Dry	C.V. ± %
Dry matter	%			100.0	
Protein (N x 6.25)	%			29.0	
Alanine	%			0.99	
Arginine	%			2.93	
Aspartic acid	%			2.49	
Glutamic acid	%			4.09	
Glycine	%			1.62	
Histidine	%			0.75	
Hydroxyproline	%			0.00	
Isoleucine	%			0.75	
Leucine	%			1.45	
Lysine	%			0.87	
Methionine	%			0.41	
Phenylalanine	%			1.02	
Proline	%			0.87	
Serine	%			1.22	
Threonine	%			0.75	
Tyrosine	%			0.81	
Valine	%			0.99	

CALCITE. Scientific name not used

Calcite, $CaCO_3$ grnd, mn 33% calcium, (6)
Calcite (AAFCO)

Ref No 6 01 067 United States

Feed Name or Analyses			As Fed	Dry	C.V. ± %
Dry matter	%		99.9	100.0	0
Calcium	%		34.97	35.01	7
Phosphorus	%		0.01	0.01	

CALCIUM CARBONATE

Calcium carbonate, $CaCO_3$, commercial mn 38% calcium, (6)
Calcium carbonate (AAFCO)

Ref No 6 01 069 United States

Feed Name or Analyses			As Fed	Dry	C.V. ± %
Dry matter	%		99.6	100.0	
Calcium	%		36.59	36.74	7
Chlorine	%		0.04	0.04	
Magnesium	%		0.50	0.50	89
Phosphorus	%		0.04	0.04	

(1) dry forages and roughages (3) silages (6) minerals
(2) pasture, range plants, and forages fed green (4) energy feeds (7) vitamins
(5) protein supplements (8) additives

Feed Name or Analyses		Mean As Fed	Mean Dry	C.V. ± %
Sodium	%	0.02	0.02	
Sulphur	%	0.09	0.09	

Calcium carbonate, $CaCO_3$, cp, (6)
Ref No 6 01 070 United States

Feed Name or Analyses		Mean As Fed	Mean Dry	C.V. ± %
Dry matter	%		100.0	
Calcium	%		40.04	
Chlorine	%		0.00	
Iron	%		0.002	
Magnesium	%		0.02	
Potassium	%		0.01	
Sodium	%		0.10	

CALCIUM PHOSPHATE, DIBASIC

Calcium phosphate, dibasic, commercial, (6)
Dicalcium phosphate (AAFCO)
Ref No 6 01 080 United States

Feed Name or Analyses		Mean As Fed	Mean Dry	C.V. ± %
Dry matter	%	96.0	100.0	3
Ash	%	90.0	93.8	6
Calcium	%	23.35	24.32	10
Fluorine	mg/kg	784.00	816.67	64
Phosphorus	%	18.21	18.97	5

Calcium phosphate, dibasic, cp, (6)
Ref No 6 01 081 United States

Feed Name or Analyses		Mean As Fed	Mean Dry	C.V. ± %
Dry matter	%		100.0	
Calcium	%		29.46	
Chlorine	%		0.00	
Iron	%		0.005	
Phosphorus	%		22.77	

CALCIUM PHOSPHATE, MONOBASIC

Calcium phosphate, monobasic, $CaH_4(PO_4)_2.H_2O$, cp, (6)
Ref No 6 01 083 United States

Feed Name or Analyses		Mean As Fed	Mean Dry	C.V. ± %
Dry matter	%	92.9	100.0	
Calcium	%	15.90	17.12	
Chlorine	%	0.00	0.00	
Iron	%	0.002	0.002	
Phosphorus	%	24.57	26.45	

CALCIUM SULFATE

Calcium sulfate, anhydrous, $CaSO_4$, commercial, (6)
Calcium sulfate, anhydrous (AAFCO)
Ref No 6 01 087 United States

Feed Name or Analyses		Mean As Fed	Mean Dry	C.V. ± %
Dry matter	%	84.8	100.0	1
Calcium	%	21.96	25.90	2
Iron	%	0.170	0.201	
Magnesium	%	2.21	2.61	
Phosphorus	%	0.01	0.01	

Calcium sulfate, anhydrous, $CaSO_4$, cp, (6)
Ref No 6 01 088 United States

Feed Name or Analyses		Mean As Fed	Mean Dry	C.V. ± %
Dry matter	%		100.0	
Calcium	%		29.44	
Chlorine	%		0.00	
Iron	%		0.003	
Sulphur	%		23.55	

Calcium sulfate, $CaSO_4.2H_2O$, cp, (6)
Ref No 6 01 089 United States

Feed Name or Analyses		Mean As Fed	Mean Dry	C.V. ± %
Dry matter	%	79.1	100.0	
Calcium	%	23.28	29.43	
Chlorine	%	0.00	0.00	
Iron	%	0.002	0.003	
Sulphur	%	18.62	23.54	

CALLIANDRA, FALSEMESQUITE. Calliandra eriophylla

Calliandra, falsemesquite, seeds w hulls, (5)
Ref No 5 09 025 United States

Feed Name or Analyses		Mean As Fed	Mean Dry	C.V. ± %
Dry matter	%		100.0	
Protein (N x 6.25)	%		39.0	
Alanine	%		1.52	
Arginine	%		2.77	
Aspartic acid	%		3.90	
Glutamic acid	%		6.86	
Glycine	%		1.52	
Histidine	%		0.94	
Hydroxyproline	%		0.20	
Isoleucine	%		1.40	
Leucine	%		2.77	
Lysine	%		2.34	
Methionine	%		0.35	
Phenylalanine	%		1.48	
Proline	%		1.72	
Serine	%		1.64	
Threonine	%		1.25	
Tyrosine	%		1.40	
Valine	%		1.68	

CAMELINA. Camelina sativa

Camelina, seeds, solv-extd grnd, (5)
Camelina meal
Ref No 5 08 168 Canada

Feed Name or Analyses		Mean As Fed	Mean Dry	C.V. ± %
Dry matter	%		100.0	
Ash	%		5.7	
Crude fiber	%		10.6	
Ether extract	%		5.1	
N-free extract	%		32.7	
Protein (N x 6.25)	%		46.0	
Energy	GE Mcal/kg		4.80	1
Cattle	DE Mcal/kg *		3.38	
Sheep	DE Mcal/kg *		3.50	
Swine	DE kcal/kg *		3632.	
Cattle	ME Mcal/kg *		2.77	
Sheep	ME Mcal/kg *		2.87	
Swine	ME kcal/kg *		3149.	
Cattle	TDN % *		76.6	
Sheep	TDN % *		79.4	
Swine	TDN % *		82.4	
Arginine	%		7.31	
Cystine	%		0.55	
Histidine	%		1.57	
Isoleucine	%		2.53	
Leucine	%		4.41	
Lysine	%		3.00	
Methionine	%		1.12	
Phenylalanine	%		2.83	
Threonine	%		2.84	
Tryptophan	%		0.95	
Valine	%		3.02	

Camelina meal –
see Camelina, seeds, solv-extd grnd, (5)

Canada rapeseed meal (CFA) –
see Rape, Canada, seeds, cooked mech-extd grnd, Can 1 mx 6% fat, (5)

Canada rapeseed pre-press solvent extracted meal (CFA) –
see Rape, Canada, seeds, cooked pre-press solv-extd grnd, Can 1 mx 1% fat, (5)

CANARYGRASS. Phalaris arundinacea

Canarygrass, hay, fan air dried w heat, immature, cut 2, (1)
Ref No 1 07 720 United States

Feed Name or Analyses		Mean As Fed	Mean Dry	C.V. ± %
Dry matter	%	91.8	100.0	
Cattle	dig coef %	63.	63.	
Sheep	dig coef %	63.	63.	
Ash	%	8.6	9.4	
Crude fiber	%	23.9	26.1	
Cattle	dig coef %	64.	64.	
Sheep	dig coef %	62.	62.	
Ether extract	%	3.2	3.5	
Cattle	dig coef %	36.	36.	
Sheep	dig coef %	36.	36.	
N-free extract	%	42.8	46.6	
Cattle	dig coef %	68.	68.	
Sheep	dig coef %	67.	67.	
Protein (N x 6.25)	%	13.3	14.6	
Cattle	dig coef %	65.	65.	
Sheep	dig coef %	71.	71.	
Cattle	dig prot %	8.7	9.5	
Goats	dig prot % *	9.3	10.1	
Horses	dig prot % *	9.1	9.9	
Rabbits	dig prot % *	9.1	9.9	
Sheep	dig prot %	9.5	10.3	
Cellulose (Crampton)	%			
Cattle	dig coef %	68.	68.	
Sheep	dig coef %	67.	67.	
Energy	GE Mcal/kg			
Cattle	DE Mcal/kg *	2.45	2.67	
Sheep	DE Mcal/kg *	2.45	2.67	
Cattle	ME Mcal/kg *	2.01	2.19	
Sheep	ME Mcal/kg *	2.01	2.19	
Cattle	TDN %	55.6	60.6	
Sheep	TDN %	55.5	60.5	

CANARYGRASS. Phalaris canariensis

Canarygrass, hay, s-c, (1)
Ref No 1 01 091 United States

Feed Name or Analyses		Mean As Fed	Mean Dry	C.V. ± %
Dry matter	%	92.1	100.0	3
Crude fiber	%			
Sheep	dig coef %	54.	54.	
Ether extract	%			
Sheep	dig coef %	35.	35.	
N-free extract	%			
Sheep	dig coef %	64.	64.	

Continued

Feed Name or Analyses		Mean As Fed	Mean Dry	C.V. ± %
Protein (N x 6.25)	%			
Sheep	dig coef %	60.	60.	
Energy	GE Mcal/kg	4.02	4.37	
Sheep	GE dig coef %	56.	56.	
Sheep	DE Mcal/kg	2.25	2.45	
Sheep	ME Mcal/kg *	1.85	2.01	
Sheep	TDN %	49.3	53.5	
Calcium	%	0.33	0.36	24
Copper	mg/kg	11.0	11.9	
Iron	%	0.014	0.015	
Magnesium	%	0.24	0.26	
Manganese	mg/kg	85.1	92.4	39
Phosphorus	%	0.24	0.26	35
Potassium	%	2.16	2.35	

Ref No 1 01 091 Canada

Feed Name or Analyses		As Fed	Dry	C.V. ± %
Dry matter	%	91.9	100.0	
Sheep	dig coef %	57.	57.	
Organic matter	%	83.6	91.0	
Crude fiber	%	32.5	35.3	
Sheep	dig coef %	54.	54.	
Ether extract	%	1.0	1.1	
Sheep	dig coef %	35.	35.	
N-free extract	%	41.0	44.6	
Sheep	dig coef %	64.	64.	
Protein (N x 6.25)	%	7.7	8.4	
Sheep	dig coef %	60.	60.	
Cattle	dig prot % *	3.8	4.2	
Goats	dig prot % *	4.0	4.4	
Horses	dig prot % *	4.3	4.6	
Rabbits	dig prot % *	4.7	5.1	
Sheep	dig prot %	4.6	5.0	
Lignin (Ellis)	%	9.9	10.8	
Sheep	dig coef %	3.	3.	
Energy	GE Mcal/kg	4.02	4.37	
Sheep	GE dig coef %	56.	56.	
Cattle	DE Mcal/kg *	2.28	2.48	
Sheep	DE Mcal/kg	2.25	2.45	
Cattle	ME Mcal/kg *	1.87	2.03	
Sheep	ME Mcal/kg *	1.84	2.01	
Cattle	TDN % *	51.7	56.2	
Sheep	TDN %	49.2	53.5	
Potassium	%	0.50	0.54	

Canarygrass, aerial part, fresh, (2)
Ref No 2 01 093 United States

Feed Name or Analyses		As Fed	Dry	C.V. ± %
Dry matter	%	25.8	100.0	21
Ash	%	2.4	9.4	38
Crude fiber	%	6.9	26.8	15
Ether extract	%	1.0	3.7	30
N-free extract	%	12.1	47.0	
Protein (N x 6.25)	%	3.4	13.1	31
Cattle	dig prot % *	2.3	9.0	
Goats	dig prot % *	2.3	8.8	
Horses	dig prot % *	2.2	8.7	
Rabbits	dig prot % *	2.3	8.8	
Sheep	dig prot % *	2.4	9.2	
Energy	GE Mcal/kg			
Cattle	DE Mcal/kg *	0.74	2.85	
Sheep	DE Mcal/kg *	0.73	2.84	
Cattle	ME Mcal/kg *	0.60	2.34	
Sheep	ME Mcal/kg *	0.60	2.33	
Cattle	TDN % *	16.7	64.6	
Sheep	TDN % *	16.6	64.5	
Calcium	%	0.11	0.44	31
Magnesium	%	0.07	0.27	
Phosphorus	%	0.07	0.29	40
Potassium	%	0.82	3.17	

Canarygrass, aerial part, fresh, immature, (2)
Ref No 2 01 092 United States

Feed Name or Analyses		As Fed	Dry	C.V. ± %
Dry matter	%	24.5	100.0	8
Ash	%	3.2	13.2	31
Crude fiber	%	5.5	22.5	9
Ether extract	%	1.1	4.3	26
N-free extract	%	10.4	42.6	
Protein (N x 6.25)	%	4.3	17.4	24
Cattle	dig prot % *	3.1	12.7	
Goats	dig prot % *	3.1	12.8	
Horses	dig prot % *	3.0	12.3	
Rabbits	dig prot % *	3.0	12.1	
Sheep	dig prot % *	3.2	13.2	
Energy	GE Mcal/kg			
Cattle	DE Mcal/kg *	0.67	2.74	
Sheep	DE Mcal/kg *	0.64	2.62	
Cattle	ME Mcal/kg *	0.55	2.25	
Sheep	ME Mcal/kg *	0.53	2.15	
Cattle	TDN % *	15.2	62.1	
Sheep	TDN % *	14.6	59.4	
Calcium	%	0.12	0.51	15
Magnesium	%	0.07	0.27	
Phosphorus	%	0.08	0.33	30
Potassium	%	0.78	3.17	

CANARYGRASS, CAROLINA. Phalaris caroliniana

Canarygrass, Carolina, aerial part, fresh, immature, (2)
Ref No 2 01 094 United States

Feed Name or Analyses		As Fed	Dry	C.V. ± %
Dry matter	%		100.0	
Ash	%		15.6	21
Crude fiber	%		28.8	2
Ether extract	%		2.6	22
N-free extract	%		42.7	
Protein (N x 6.25)	%		10.3	6
Cattle	dig prot % *		6.6	
Goats	dig prot % *		6.2	
Horses	dig prot % *		6.3	
Rabbits	dig prot % *		6.6	
Sheep	dig prot % *		6.6	
Energy	GE Mcal/kg			
Cattle	DE Mcal/kg *		2.39	
Sheep	DE Mcal/kg *		2.25	
Cattle	ME Mcal/kg *		1.96	
Sheep	ME Mcal/kg *		1.84	
Cattle	TDN % *		54.3	
Sheep	TDN % *		51.0	
Calcium	%		0.49	7
Phosphorus	%		0.27	16

CANARYGRASS, LITTLESEED. Phalaris minor

Canarygrass, littleseed, hay, s-c, immature, (1)
Ref No 1 01 095 United States

Feed Name or Analyses		As Fed	Dry	C.V. ± %
Dry matter	%	92.7	100.0	
Calcium	%	0.42	0.45	
Phosphorus	%	0.26	0.28	

Canarygrass, littleseed, aerial part, fresh, immature, (2)
Ref No 2 01 096 United States

Feed Name or Analyses		As Fed	Dry	C.V. ± %
Dry matter	%		100.0	
Ash	%		19.2	
Crude fiber	%		26.3	
Ether extract	%		2.6	
N-free extract	%		41.6	
Protein (N x 6.25)	%		10.3	
Cattle	dig prot % *		6.6	
Goats	dig prot % *		6.2	
Horses	dig prot % *		6.3	
Rabbits	dig prot % *		6.6	
Sheep	dig prot % *		6.6	

CANARYGRASS, REED. Phalaris arundinacea

Canarygrass, reed, aerial part, dehy, pre-bloom, (1)
Ref No 1 05 671 United States

Feed Name or Analyses		As Fed	Dry	C.V. ± %
Dry matter	%		100.0	
Sheep	dig coef %		64.	
Ash	%		9.0	
Protein (N x 6.25)	%		18.4	
Sheep	dig coef %		75.	
Cattle	dig prot % *		12.9	
Goats	dig prot % *		13.7	
Horses	dig prot % *		13.2	
Rabbits	dig prot % *		12.9	
Sheep	dig prot %		13.8	
Cellulose (Matrone)	%		31.2	
Sheep	dig coef %		68.	
Energy	GE Mcal/kg		4.30	
Sheep	GE dig coef %		62.	
Sheep	DE Mcal/kg		2.67	
Sheep	ME Mcal/kg *		2.19	
Sheep	TDN % *		60.6	

Canarygrass, reed, aerial part, dehy chopped, mid-bloom, (1)
Ref No 1 05 672 United States

Feed Name or Analyses		As Fed	Dry	C.V. ± %
Dry matter	%		100.0	
Sheep	dig coef %		58.	
Ash	%		8.2	
Protein (N x 6.25)	%		13.0	
Sheep	dig coef %		69.	
Cattle	dig prot % *		8.2	
Goats	dig prot % *		8.7	
Horses	dig prot % *		8.6	
Rabbits	dig prot % *		8.7	
Sheep	dig prot %		9.0	
Cellulose (Matrone)	%		34.3	
Sheep	dig coef %		61.	
Energy	GE Mcal/kg		4.36	
Sheep	GE dig coef %		56.	
Sheep	DE Mcal/kg		2.44	

(1) dry forages and roughages (2) pasture, range plants, and forages fed green (3) silages (4) energy feeds (5) protein supplements (6) minerals (7) vitamins (8) additives

Feed Name or Analyses			Mean As Fed	Mean Dry	C.V. ± %
Sheep	ME Mcal/kg	*		2.00	
Sheep	TDN %	*		55.3	

Canarygrass, reed, hay, s-c, (1)

Ref No 1 01 104 United States

Feed Name or Analyses			Mean As Fed	Mean Dry	C.V. ± %
Dry matter	%		88.8	100.0	2
Horses	dig coef %		47.	47.	10
Organic matter	%		81.4	91.7	0
Ash	%		7.3	8.2	10
Crude fiber	%		30.0	33.8	12
Horses	dig coef %		47.	47.	8
Sheep	dig coef %		55.	55.	
Ether extract	%		2.8	3.1	20
Sheep	dig coef %		13.	13.	
N-free extract	%		38.5	43.4	10
Horses	dig coef %		43.	43.	26
Sheep	dig coef %		53.	53.	
Protein (N x 6.25)	%		10.2	11.5	12
Sheep	dig coef %		63.	63.	
Cattle	dig prot %	*	6.1	6.9	
Goats	dig prot %	*	6.4	7.3	
Horses	dig prot %	*	6.5	7.3	
Rabbits	dig prot %	*	6.7	7.5	
Sheep	dig prot %		6.4	7.2	
Cell contents (Van Soest)	%		26.2	29.5	4
Horses	dig coef %		53.	53.	13
Cell walls (Van Soest)	%		62.6	70.5	2
Horses	dig coef %		45.	45.	12
Cellulose (Matrone)	%		29.1	32.8	8
Horses	dig coef %		42.	42.	10
Fiber, acid detergent (VS)	%		32.5	36.6	2
Horses	dig coef %		38.	38.	13
Hemicellulose	%		25.0	28.2	3
Horses	dig coef %		47.	47.	13
Holocellulose	%		55.0	61.9	2
Soluble carbohydrates	%		10.6	11.9	8
Lignin (Ellis)	%		2.7	3.1	23
Horses	dig coef %		-20.	-20.	83
Fatty acids	%		2.2	2.5	
Energy	GE Mcal/kg		3.97	4.47	0
Horses	GE dig coef %		45.	45.	7
Cattle	DE Mcal/kg	*	2.07	2.33	
Horses	DE Mcal/kg		1.79	2.01	6
Sheep	DE Mcal/kg	*	1.95	2.19	
Cattle	ME Mcal/kg	*	1.70	1.91	
Horses	ME Mcal/kg	*	1.47	1.65	
Sheep	ME Mcal/kg	*	1.60	1.80	
Cattle	NE_m Mcal/kg	*	1.06	1.19	
Cattle	NE_{gain} Mcal/kg	*	0.46	0.52	
Cattle	TDN %	*	46.9	52.8	
Sheep	TDN %		44.2	49.7	
Calcium	%		0.37	0.41	18
Magnesium	%		0.21	0.24	9
Phosphorus	%		0.23	0.25	14
Potassium	%		2.80	3.15	6
Carotene	mg/kg		6.2	7.0	34
Vitamin A equivalent	IU/g		10.4	11.7	

Canarygrass, reed, hay, s-c, immature, (1)

Ref No 1 01 097 United States

Feed Name or Analyses			Mean As Fed	Mean Dry	C.V. ± %
Dry matter	%		91.7	100.0	
Ash	%		6.7	7.3	12
Crude fiber	%		27.8	30.3	3
Sheep	dig coef %	*	55.	55.	
Ether extract	%		2.7	2.9	22
Sheep	dig coef %	*	13.	13.	
N-free extract	%		40.9	44.6	
Sheep	dig coef %	*	53.	53.	
Protein (N x 6.25)	%		13.7	14.9	8
Sheep	dig coef %	*	63.	63.	
Cattle	dig prot %	*	9.0	9.8	
Goats	dig prot %	*	9.6	10.5	
Horses	dig prot %	*	9.3	10.2	
Rabbits	dig prot %	*	9.3	10.2	
Sheep	dig prot %		8.6	9.4	
Energy	GE Mcal/kg				
Cattle	DE Mcal/kg	*	2.17	2.37	
Sheep	DE Mcal/kg	*	2.04	2.23	
Cattle	ME Mcal/kg	*	1.79	1.94	
Sheep	ME Mcal/kg	*	1.68	1.83	
Cattle	TDN %	*	49.3	53.7	
Sheep	TDN %		46.3	50.5	
Calcium	%		0.32	0.35	42
Phosphorus	%		0.26	0.28	16
Carotene	mg/kg		188.8	205.9	
Riboflavin	mg/kg		10.5	11.5	
Thiamine	mg/kg		4.0	4.4	
Vitamin A equivalent	IU/g		314.8	343.3	

Canarygrass, reed, hay, s-c, early bloom, (1)

Ref No 1 01 098 United States

Feed Name or Analyses			Mean As Fed	Mean Dry	C.V. ± %
Dry matter	%		91.5	100.0	
Ash	%		5.2	5.7	
Crude fiber	%		26.7	29.2	
Ether extract	%		3.1	3.4	
N-free extract	%		47.7	52.1	
Protein (N x 6.25)	%		8.8	9.6	
Cattle	dig prot %	*	4.8	5.3	
Goats	dig prot %	*	5.0	5.5	
Horses	dig prot %	*	5.2	5.7	
Rabbits	dig prot %	*	5.6	6.1	
Sheep	dig prot %	*	4.7	5.2	
Cellulose (Matrone)	%		26.1	28.5	
Lignin (Ellis)	%		4.9	5.4	
Energy	GE Mcal/kg				
Cattle	DE Mcal/kg	*	2.19	2.39	
Sheep	DE Mcal/kg	*	2.06	2.25	
Cattle	ME Mcal/kg	*	1.39	1.96	
Sheep	ME Mcal/kg	*	1.68	1.84	
Cattle	TDN %	*	49.7	54.3	
Sheep	TDN %	*	46.7	51.0	
Calcium	%		0.31	0.34	
Cobalt	mg/kg		0.020	0.022	
Copper	mg/kg		10.9	11.9	
Iron	%		0.014	0.015	
Magnesium	%		0.24	0.26	
Manganese	mg/kg		131.9	144.2	
Phosphorus	%		0.43	0.47	
Potassium	%		2.40	2.62	
Sodium	%		0.36	0.39	

Canarygrass, reed, hay, s-c, early bloom, cut 1, (1)

Ref No 1 08 767 United States

Feed Name or Analyses			Mean As Fed	Mean Dry	C.V. ± %
Dry matter	%		91.4	100.0	1
Horses	dig coef %		50.	50.	3
Organic matter	%		84.1	92.0	0
Ash	%		7.3	8.0	3
Crude fiber	%		32.3	35.4	4
Horses	dig coef %		39.	39.	5
Ether extract	%				
Horses	dig coef %		44.	44.	23
N-free extract	%		38.5	42.1	1
Horses	dig coef %		66.	66.	4
Protein (N x 6.25)	%		10.2	11.2	5
Horses	dig coef %		62.	62.	7
Cattle	dig prot %	*	6.1	6.6	
Goats	dig prot %	*	6.4	7.0	
Horses	dig prot %		6.3	6.9	
Rabbits	dig prot %	*	6.7	7.3	
Sheep	dig prot %	*	6.0	6.6	
Cell contents (Van Soest)	%		35.5	38.9	6
Horses	dig coef %		69.	69.	2
Cell walls (Van Soest)	%		55.9	61.1	4
Horses	dig coef %		38.	38.	7
Cellulose (Matrone)	%		31.4	34.4	3
Horses	dig coef %		42.	42.	6
Fiber, acid detergent (VS)	%		35.6	39.0	3
Horses	dig coef %		34.	34.	7
Hemicellulose	%		16.0	17.6	9
Horses	dig coef %		42.	42.	12
Holocellulose	%		47.4	51.9	4
Soluble carbohydrates	%		19.2	21.0	6
Lignin (Ellis)	%		4.2	4.7	5
Horses	dig coef %		15.	15.	35
Energy	GE Mcal/kg		4.10	4.49	1
Horses	GE dig coef %		43.	43.	4
Cattle	DE Mcal/kg	*	2.22	2.43	
Horses	DE Mcal/kg		1.76	1.92	
Sheep	DE Mcal/kg	*	2.23	2.44	
Cattle	ME Mcal/kg	*	1.82	1.99	
Horses	ME Mcal/kg	*	1.44	1.58	
Sheep	ME Mcal/kg	*	1.83	2.00	
Cattle	TDN %	*	50.3	55.0	
Sheep	TDN %	*	50.7	55.5	
Calcium	%		0.43	0.47	7
Magnesium	%		0.37	0.40	5
Phosphorus	%		0.20	0.22	6
Potassium	%		0.26	0.28	4
Carotene	mg/kg		20.4	22.3	16
Vitamin A equivalent	IU/g		34.0	37.2	

Canarygrass, reed, hay, s-c, mid-bloom, (1)

Ref No 1 01 099 United States

Feed Name or Analyses			Mean As Fed	Mean Dry	C.V. ± %
Dry matter	%		91.9	100.0	
Calcium	%		0.33	0.36	
Phosphorus	%		0.28	0.31	
Potassium	%		1.90	2.07	

Canarygrass, reed, hay, s-c, milk stage, (1)

Ref No 1 01 100 United States

Feed Name or Analyses			Mean As Fed	Mean Dry	C.V. ± %
Dry matter	%		90.3	100.0	2
Calcium	%		0.26	0.29	
Phosphorus	%		0.20	0.22	
Carotene	mg/kg		84.8	93.9	
Riboflavin	mg/kg		7.6	8.4	
Thiamine	mg/kg		3.4	3.7	
Vitamin A equivalent	IU/g		141.4	156.6	

Canarygrass, reed, hay, s-c, mature, (1)

Ref No 1 01 101 United States

Feed Name or Analyses			Mean As Fed	Mean Dry	C.V. ± %
Dry matter	%		91.5	100.0	2
Crude fiber	%				
Cattle	dig coef %	*	51.	51.	
Sheep	dig coef %	*	55.	55.	

Continued

Feed Name or Analyses				Mean As Fed	Mean Dry	C.V. ± %
Ether extract		%				
Cattle	dig coef	%	*	34.	34.	
Sheep	dig coef	%	*	13.	13.	
N-free extract		%				
Sheep	dig coef	%	*	53.	53.	
Protein (N x 6.25)		%				
Cattle	dig coef	%	*	59.	59.	
Sheep	dig coef	%	*	63.	63.	
Sugars, total		%		1.2	1.3	
Energy		GE Mcal/kg		4.05	4.42	
Cattle	GE dig coef	%		48.	48.	
Cattle		DE Mcal/kg		1.94	2.12	
Sheep		DE Mcal/kg	*	2.03	2.22	
Cattle		ME Mcal/kg	*	1.59	1.74	
Sheep		ME Mcal/kg	*	1.66	1.82	
Cattle		TDN %	*	44.0	48.1	
Sheep		TDN %		46.0	50.3	
Calcium		%		0.28	0.31	
Phosphorus		%		0.16	0.17	
Carotene		mg/kg		82.5	90.2	76
Riboflavin		mg/kg		7.4	8.0	29
Thiamine		mg/kg		3.1	3.4	22
Vitamin A equivalent		IU/g		137.6	150.3	

Ref No 1 01 101 Canada

Feed Name or Analyses				Mean As Fed	Mean Dry	C.V. ± %
Dry matter		%		93.2	100.0	2
Ash		%		7.3	7.8	
Crude fiber		%		32.5	34.9	10
Cattle	dig coef	%		51.	51.	
Ether extract		%		1.3	1.3	
Cattle	dig coef	%		34.	34.	
N-free extract		%		42.5	45.6	
Protein (N x 6.25)		%		9.6	10.3	47
Cattle	dig coef	%		59.	59.	
Cattle	dig prot	%		5.7	6.1	
Goats	dig prot	%	*	5.8	6.2	
Horses	dig prot	%	*	5.9	6.3	
Rabbits	dig prot	%	*	6.2	6.6	
Sheep	dig prot	%	*	5.4	5.8	
Cellulose (Matrone)		%		33.6	36.0	
Cattle	dig coef	%		56.	56.	
Energy		GE Mcal/kg		4.12	4.42	
Cattle	GE dig coef	%		48.	48.	
Cattle		DE Mcal/kg		1.98	2.12	
Sheep		DE Mcal/kg	*	2.07	2.22	
Cattle		ME Mcal/kg	*	1.62	1.74	
Sheep		ME Mcal/kg	*	1.69	1.82	
Cattle		TDN %	*	50.4	54.0	
Sheep		TDN %		46.9	50.3	
Calcium		%		0.26	0.28	39
Copper		mg/kg		4.8	5.1	
Iron		%	*	.034	.036	
Manganese		mg/kg		58.9	63.1	
Phosphorus		%		0.15	0.16	92
Carotene		mg/kg		32.8	35.2	
Vitamin A equivalent		IU/g		54.6	58.6	

Canarygrass, reed, hay, s-c, cut 1, (1)

Ref No 1 01 102 United States

Feed Name or Analyses				Mean As Fed	Mean Dry	C.V. ± %
Dry matter		%		91.1	100.0	2
Ash		%		5.0	5.5	12
Crude fiber		%		33.6	36.9	10
Ether extract		%		1.5	1.6	10
N-free extract		%		45.1	49.5	
Protein (N x 6.25)		%		5.9	6.5	13
Cattle	dig prot	%	*	2.3	2.6	
Goats	dig prot	%	*	2.4	2.6	
Horses	dig prot	%	*	2.8	3.0	
Rabbits	dig prot	%	*	3.4	3.7	
Sheep	dig prot	%	*	2.2	2.4	
Energy		GE Mcal/kg				
Cattle		DE Mcal/kg	*	1.99	2.19	
Sheep		DE Mcal/kg	*	2.15	2.37	
Cattle		ME Mcal/kg	*	1.63	1.79	
Sheep		ME Mcal/kg	*	1.77	1.94	
Cattle		TDN %	*	45.2	49.6	
Sheep		TDN %	*	48.9	53.6	
Calcium		%		0.33	0.36	
Phosphorus		%		0.28	0.31	
Potassium		%		1.89	2.07	

Canarygrass, reed, hay, s-c, cut 2, (1)

Ref No 1 01 103 United States

Feed Name or Analyses				Mean As Fed	Mean Dry	C.V. ± %
Dry matter		%		91.1	100.0	
Crude fiber		%				
Cattle	dig coef	%		74.	74.	
Ether extract		%				
Cattle	dig coef	%		40.	40.	
N-free extract		%				
Cattle	dig coef	%		65.	65.	
Protein (N x 6.25)		%				
Cattle	dig coef	%		77.	77.	
Energy		GE Mcal/kg		4.47	4.91	
Cattle	GE dig coef	%		65.	65.	
Cattle		DE Mcal/kg		2.91	3.19	
Cattle		ME Mcal/kg	*	2.38	2.62	
Cattle		TDN %		58.4	64.1	
Calcium		%		0.40	0.44	
Manganese		mg/kg		115.1	126.3	
Phosphorus		%		0.16	0.18	

Ref No 1 01 103 Great Britain

Feed Name or Analyses				Mean As Fed	Mean Dry	C.V. ± %
Dry matter		%		89.2	100.0	
Cattle	dig coef	%		68.	68.	
Ash		%		10.1	11.3	
Crude fiber		%		29.1	32.6	
Cattle	dig coef	%		74.	74.	
Ether extract		%		2.3	2.6	
Cattle	dig coef	%		40.	40.	
N-free extract		%		26.4	29.6	
Cattle	dig coef	%		65.	65.	
Protein (N x 6.25)		%		21.3	23.9	
Cattle	dig coef	%		77.	77.	
Cattle	dig prot	%		16.4	18.4	
Goats	dig prot	%	*	16.8	18.8	
Horses	dig prot	%	*	15.9	17.8	
Rabbits	dig prot	%	*	15.3	17.1	
Sheep	dig prot	%	*	16.0	18.0	
Cellulose (Matrone)		%		20.3	22.8	
Cattle	dig coef	%		74.	74.	
Pentosans		%		19.0	21.3	
Lignin (Ellis)		%		5.6	6.3	
Cattle	dig coef	%		6.	6.	
Energy		GE Mcal/kg		4.38	4.91	
Cattle	GE dig coef	%		65.	65.	
Cattle		DE Mcal/kg		2.85	3.19	
Sheep		DE Mcal/kg	*	2.15	2.41	
Cattle		ME Mcal/kg	*	2.33	2.62	
Sheep		ME Mcal/kg	*	1.76	1.98	
Cattle		TDN %		57.2	64.1	
Sheep		TDN %	*	48.8	54.7	

Canarygrass, reed, hay, s-c baled fan-air dried w heat, immature, cut 1, (1)

Ref No 1 08 865 United States

Feed Name or Analyses				Mean As Fed	Mean Dry	C.V. ± %
Dry matter		%		86.1	100.0	
Sheep	dig coef	%		64.	64.	
Ash		%		7.7	9.0	
Sheep	dig coef	%		48.	56.	
Crude fiber		%		23.7	27.5	
Sheep	dig coef	%		65.	65.	
Ether extract		%		3.2	3.7	
Sheep	dig coef	%		39.	39.	
N-free extract		%		34.9	40.5	
Sheep	dig coef	%		60.	60.	
Protein (N x 6.25)		%		16.6	19.3	
Sheep	dig coef	%		74.	74.	
Cattle	dig prot	%	*	11.8	13.7	
Goats	dig prot	%	*	12.5	14.6	
Horses	dig prot	%	*	12.0	13.9	
Rabbits	dig prot	%	*	11.7	13.6	
Sheep	dig prot	%		12.3	14.3	
Cell walls (Van Soest)		%		49.9	58.0	
Sheep	dig coef	%		59.	59.	
Cellulose (Matrone)		%		25.2	29.3	
Sheep	dig coef	%		65.	65.	
Fiber, acid detergent (VS)		%		27.6	32.0	
Sheep	dig coef	%		54.	54.	
Hemicellulose		%		22.3	26.0	
Sheep	dig coef	%		69.	69.	
Lignin (Ellis)		%		2.7	3.2	
Sheep	dig coef	%		21.	21.	
Energy		GE Mcal/kg		3.90	4.52	
Sheep	GE dig coef	%		62.	62.	
Cattle		DE Mcal/kg	*	2.39	2.77	
Sheep		DE Mcal/kg		2.40	2.78	
Cattle		ME Mcal/kg	*	1.96	2.27	
Sheep		ME Mcal/kg	*	1.96	2.28	
Cattle		TDN %	*	54.1	62.9	
Sheep		TDN %		51.4	59.7	
Dry matter intake		g/w⋅75		70.817	82.250	
Gain, sheep		kg/day		0.086	0.100	

Canarygrass, reed, hay, s-c baled fan-air dried w heat, immature, cut 2, (1)

Ref No 1 08 863 United States

Feed Name or Analyses				Mean As Fed	Mean Dry	C.V. ± %
Dry matter		%		87.2	100.0	
Sheep	dig coef	%		58.	58.	
Ash		%		8.2	9.4	
Sheep	dig coef	%		37.	42.	
Crude fiber		%		22.3	25.6	
Sheep	dig coef	%		53.	53.	
Ether extract		%		3.5	4.0	
Sheep	dig coef	%		51.	51.	
N-free extract		%		37.9	43.5	
Sheep	dig coef	%		49.	49.	
Protein (N x 6.25)		%		15.3	17.5	
Sheep	dig coef	%		73.	73.	
Cattle	dig prot	%	*	10.5	12.1	
Goats	dig prot	%	*	11.2	12.9	
Horses	dig prot	%	*	10.8	12.4	
Rabbits	dig prot	%	*	10.6	12.2	
Sheep	dig prot	%		11.1	12.8	
Cell walls (Van Soest)		%		51.8	59.4	
Sheep	dig coef	%		56.	56.	

(1) dry forages and roughages (2) pasture, range plants, and forages fed green (3) silages (4) energy feeds (5) protein supplements (6) minerals (7) vitamins (8) additives

Feed Name or Analyses			Mean As Fed	Mean Dry	C.V. ± %
Cellulose (Matrone)	%		27.5	31.5	
Fiber, acid detergent (VS)	%		29.4	33.8	
Sheep	dig coef %		45.	45.	
Hemicellulose	%		22.3	25.6	
Lignin (Ellis)	%		3.1	3.6	
Sheep	dig coef %		-40.	-40.	
Energy	GE Mcal/kg		3.93	4.51	
Sheep	GE dig coef %		57.	57.	
Cattle	DE Mcal/kg	*	2.11	2.42	
Sheep	DE Mcal/kg		2.22	2.55	
Cattle	ME Mcal/kg	*	1.73	1.98	
Sheep	ME Mcal/kg	*	1.82	2.09	
Cattle	TDN %	*	47.9	54.9	
Sheep	TDN %		45.6	52.2	
Dry matter intake	$g/W^{0.75}$		63.947	73.333	
Gain, sheep	kg/day		0.106	0.121	

Canarygrass, reed, aerial part, fresh, (2)

Ref No 2 01 113 United States

Feed Name or Analyses			Mean As Fed	Mean Dry	C.V. ± %
Dry matter	%		25.4	100.0	22
Ash	%		2.5	10.0	
Crude fiber	%		5.6	22.0	
Ether extract	%		1.1	4.4	
N-free extract	%		11.8	46.4	
Protein (N x 6.25)	%		4.4	17.2	
Cattle	dig prot %	*	3.2	12.5	
Goats	dig prot %	*	3.2	12.6	
Horses	dig prot %	*	3.1	12.1	
Rabbits	dig prot %	*	3.0	11.9	
Sheep	dig prot %	*	3.3	13.0	
Energy	GE Mcal/kg				
Cattle	DE Mcal/kg	*	0.70	2.75	
Sheep	DE Mcal/kg	*	0.68	2.66	
Cattle	ME Mcal/kg	*	0.57	2.26	
Sheep	ME Mcal/kg	*	0.55	2.18	
Cattle	NE_m Mcal/kg	*	0.37	1.45	
Cattle	NE_{gain} Mcal/kg	*	0.22	0.87	
Cattle	TDN %	*	15.8	62.3	
Sheep	TDN %	*	15.3	60.4	
Calcium	%		0.09	0.36	46
Phosphorus	%		0.08	0.33	43
Potassium	%		0.92	3.64	

Canarygrass, reed, aerial part, fresh, immature, (2)

Ref No 2 01 105 United States

Feed Name or Analyses			Mean As Fed	Mean Dry	C.V. ± %
Dry matter	%		24.5	100.0	8
Ash	%		2.5	10.2	20
Crude fiber	%		5.2	21.3	10
Ether extract	%		1.2	4.7	22
N-free extract	%		11.0	44.7	
Protein (N x 6.25)	%		4.7	19.1	15
Cattle	dig prot %	*	3.5	14.1	
Goats	dig prot %	*	3.5	14.4	
Horses	dig prot %	*	3.4	13.7	
Rabbits	dig prot %	*	3.3	13.4	
Sheep	dig prot %	*	3.6	14.8	
Energy	GE Mcal/kg				
Cattle	DE Mcal/kg	*	0.54	2.22	
Sheep	DE Mcal/kg	*	0.54	2.19	
Cattle	ME Mcal/kg	*	0.45	1.82	
Sheep	ME Mcal/kg	*	0.44	1.80	
Cattle	TDN %	*	12.3	50.3	
Sheep	TDN %	*	12.2	49.7	
Calcium	%		0.14	0.57	
Phosphorus	%		0.12	0.49	
Potassium	%		0.89	3.64	

Canarygrass, reed, aerial part, fresh, early bloom, (2)

Ref No 2 01 106 United States

Feed Name or Analyses			Mean As Fed	Mean Dry	C.V. ± %
Dry matter	%		25.5	100.0	
Ash	%		1.9	7.3	24
Crude fiber	%		8.1	32.0	8
Ether extract	%		0.8	3.2	29
N-free extract	%		12.0	47.2	
Protein (N x 6.25)	%		2.6	10.3	14
Cattle	dig prot %	*	1.7	6.6	
Goats	dig prot %	*	1.6	6.2	
Horses	dig prot %	*	1.6	6.3	
Rabbits	dig prot %	*	1.7	6.6	
Sheep	dig prot %	*	1.7	6.6	
Energy	GE Mcal/kg				
Cattle	DE Mcal/kg	*	0.58	2.27	
Sheep	DE Mcal/kg	*	0.55	2.15	
Cattle	ME Mcal/kg	*	0.47	1.86	
Sheep	ME Mcal/kg	*	0.45	1.76	
Cattle	TDN %	*	13.1	51.4	
Sheep	TDN %	*	12.4	48.7	
Calcium	%		0.07	0.27	
Phosphorus	%		0.05	0.20	

Canarygrass, reed, aerial part, fresh, mid-bloom, (2)

Ref No 2 01 107 United States

Feed Name or Analyses			Mean As Fed	Mean Dry	C.V. ± %
Dry matter	%		27.7	100.0	
Ash	%		2.3	8.2	27
Crude fiber	%		8.1	29.4	7
Ether extract	%		0.9	3.3	22
N-free extract	%		12.9	46.7	
Protein (N x 6.25)	%		3.4	12.4	12
Cattle	dig prot %	*	2.3	8.4	
Goats	dig prot %	*	2.3	8.1	
Horses	dig prot %	*	2.2	8.1	
Rabbits	dig prot %	*	2.3	8.2	
Sheep	dig prot %	*	2.4	8.5	
Energy	GE Mcal/kg				
Cattle	DE Mcal/kg	*	0.65	2.35	
Sheep	DE Mcal/kg	*	0.61	2.22	
Cattle	ME Mcal/kg	*	0.53	1.93	
Sheep	ME Mcal/kg	*	0.50	1.82	
Cattle	TDN %	*	14.7	53.2	
Sheep	TDN %	*	13.9	50.3	

Canarygrass, reed, aerial part, fresh, full bloom, (2)

Ref No 2 01 108 United States

Feed Name or Analyses			Mean As Fed	Mean Dry	C.V. ± %
Dry matter	%			100.0	
Ash	%			8.2	5
Crude fiber	%			31.0	3
Ether extract	%			3.4	8
N-free extract	%			47.7	
Protein (N x 6.25)	%			9.7	15
Cattle	dig prot %	*		6.1	
Goats	dig prot %	*		5.6	
Horses	dig prot %	*		5.8	
Rabbits	dig prot %	*		6.2	
Sheep	dig prot %	*		6.0	
Energy	GE Mcal/kg				
Cattle	DE Mcal/kg	*		2.21	
Sheep	DE Mcal/kg	*		2.13	
Cattle	ME Mcal/kg	*		1.81	
Sheep	ME Mcal/kg	*		1.75	
Cattle	TDN %	*		50.1	
Sheep	TDN %	*		48.3	

Canarygrass, reed, aerial part, fresh, milk stage, (2)

Ref No 2 01 109 United States

Feed Name or Analyses			Mean As Fed	Mean Dry	C.V. ± %
Dry matter	%		35.2	100.0	
Ash	%		2.9	8.1	21
Crude fiber	%		11.2	31.7	8
Ether extract	%		1.1	3.1	18
N-free extract	%		16.7	47.5	
Protein (N x 6.25)	%		3.4	9.6	13
Cattle	dig prot %	*	2.1	6.1	
Goats	dig prot %	*	1.9	5.5	
Horses	dig prot %	*	2.0	5.7	
Rabbits	dig prot %	*	2.1	6.1	
Sheep	dig prot %	*	2.1	5.9	
Energy	GE Mcal/kg				
Cattle	DE Mcal/kg	*	0.80	2.26	
Sheep	DE Mcal/kg	*	0.78	2.22	
Cattle	ME Mcal/kg	*	0.65	1.85	
Sheep	ME Mcal/kg	*	0.64	1.82	
Cattle	TDN %	*	18.0	51.2	
Sheep	TDN %	*	17.7	56.3	

Canarygrass, reed, aerial part, fresh, mature, (2)

Ref No 2 01 111 United States

Feed Name or Analyses			Mean As Fed	Mean Dry	C.V. ± %
Dry matter	%		30.0	100.0	
Ash	%		2.1	7.0	12
Crude fiber	%		9.9	33.1	8
Ether extract	%		0.9	2.9	5
N-free extract	%		14.8	49.2	
Protein (N x 6.25)	%		2.3	7.8	9
Cattle	dig prot %	*	1.4	4.5	
Goats	dig prot %	*	1.2	3.8	
Horses	dig prot %	*	1.2	4.2	
Rabbits	dig prot %	*	1.4	4.7	
Sheep	dig prot %	*	1.3	4.3	
Energy	GE Mcal/kg				
Cattle	DE Mcal/kg	*	0.68	2.27	
Sheep	DE Mcal/kg	*	0.70	2.34	
Cattle	ME Mcal/kg	*	0.56	1.86	
Sheep	ME Mcal/kg	*	0.58	1.92	
Cattle	TDN %	*	15.4	51.4	
Sheep	TDN %	*	15.8	53.1	
Calcium	%		0.08	0.27	
Phosphorus	%		0.06	0.20	

Canarygrass, reed, aerial part, fresh, cut 1, (2)

Ref No 2 01 112 United States

Feed Name or Analyses			Mean As Fed	Mean Dry	C.V. ± %
Dry matter	%		19.3	100.0	
Ash	%		1.7	8.8	9
Crude fiber	%		5.5	28.3	11
Ether extract	%		0.7	3.6	15
N-free extract	%		9.1	47.4	
Protein (N x 6.25)	%		2.3	11.9	23
Cattle	dig prot %	*	1.5	8.0	
Goats	dig prot %	*	1.5	7.7	
Horses	dig prot %	*	1.5	7.6	
Rabbits	dig prot %	*	1.5	7.9	
Sheep	dig prot %	*	1.6	8.1	
Hemicellulose	%		3.9	20.0	6

Continued

Feed Name or Analyses			Mean As Fed	Mean Dry	C.V. ± %
Energy	GE Mcal/kg				
Cattle	DE Mcal/kg	*	0.45	2.31	
Sheep	DE Mcal/kg	*	0.44	2.28	
Cattle	ME Mcal/kg	*	0.36	1.89	
Sheep	ME Mcal/kg	*	0.36	1.89	
Cattle	TDN %	*	10.1	52.5	
Sheep	TDN %	*	10.00	51.6	
Calcium	%		0.13	0.67	
Phosphorus	%		0.10	0.52	

Canarygrass, reed, aerial part, ensiled, (3)
Ref No 3 01 114 United States

Feed Name or Analyses			Mean As Fed	Mean Dry	C.V. ± %
Dry matter	%		27.5	100.0	20
Ash	%		2.0	7.3	9
Crude fiber	%		9.1	33.1	17
Ether extract	%		0.7	2.5	31
N-free extract	%		13.2	48.2	
Protein (N x 6.25)	%		2.4	8.8	22
Cattle	dig prot %	*	1.2	4.2	
Goats	dig prot %	*	1.2	4.2	
Horses	dig prot %	*	1.2	4.2	
Sheep	dig prot %	*	1.2	4.2	
Energy	GE Mcal/kg				
Cattle	DE Mcal/kg	*	0.61	2.22	
Sheep	DE Mcal/kg	*	0.59	2.13	
Cattle	ME Mcal/kg	*	0.50	1.82	
Sheep	ME Mcal/kg	*	0.48	1.75	
Cattle	TDN %	*	13.8	50.3	
Sheep	TDN %	*	13.3	48.2	

CANARYGRASS, REED COMMON. Phalaris arundinacea

Canarygrass, reed common, hay, baled fan-air dried w heat, mature, cut 3, (1)
Ref No 1 08 902 United States

Feed Name or Analyses			Mean As Fed	Mean Dry	C.V. ± %
Dry matter	%			100.0	
Sheep	dig coef %			39.	
Crude fiber	%			24.6	
Protein (N x 6.25)	%			10.9	
Cattle	dig prot %	*		6.4	
Goats	dig prot %	*		6.7	
Horses	dig prot %	*		6.8	
Rabbits	dig prot %	*		7.1	
Sheep	dig prot %	*		6.3	
Dry matter intake	g/W$^{0.75}$			56.000	

Canarygrass, reed common, hay, s-c, cut 1, (1)
Ref No 1 08 881 United States

Feed Name or Analyses			Mean As Fed	Mean Dry	C.V. ± %
Dry matter	%			100.0	
Sheep	dig coef %			65.	
Ash	%			8.0	
Ether extract	%			3.6	
Protein (N x 6.25)	%			21.7	
Cattle	dig prot %	*		15.7	
Goats	dig prot %	*		16.8	
Horses	dig prot %	*		16.0	
Rabbits	dig prot %	*		15.4	
Sheep	dig prot %	*		16.0	
Fiber, acid detergent (VS)	%			28.5	
Lignin (Ellis)	%			3.1	
Energy	GE Mcal/kg			4.62	
Sheep	DE Mcal/kg	*		2.77	
Sheep	ME Mcal/kg	*		2.27	
Sheep	TDN %			62.8	
Phosphorus	%			0.26	
Sulphur	%			0.37	
Dry matter intake	g/W$^{0.75}$			75.100	
Gain, sheep	kg/day			0.052	

Canarygrass, reed common, hay, s-c, cut 2, (1)
Ref No 1 08 882 United States

Feed Name or Analyses			Mean As Fed	Mean Dry	C.V. ± %
Dry matter	%			100.0	
Sheep	dig coef %			56.	
Ash	%			6.5	
Ether extract	%			3.2	
Protein (N x 6.25)	%			20.1	
Cattle	dig prot %	*		14.3	
Goats	dig prot %	*		15.3	
Horses	dig prot %	*		14.6	
Rabbits	dig prot %	*		14.2	
Sheep	dig prot %	*		14.6	
Fiber, acid detergent (VS)	%			32.2	
Lignin (Ellis)	%			4.2	
Energy	GE Mcal/kg			4.61	
Sheep	DE Mcal/kg	*		2.37	
Sheep	ME Mcal/kg	*		1.95	
Sheep	TDN %			53.8	
Phosphorus	%			0.24	
Sulphur	%			0.34	
Dry matter intake	g/W$^{0.75}$			69.600	
Gain, sheep	kg/day			0.090	

Canarygrass, reed common, hay, s-c, cut 3, (1)
Ref No 1 08 861 United States

Feed Name or Analyses			Mean As Fed	Mean Dry	C.V. ± %
Dry matter	%			100.0	
Sheep	dig coef %			69.	
Ash	%			10.1	
Ether extract	%			3.5	
Protein (N x 6.25)	%			29.6	
Cattle	dig prot %	*		22.6	
Goats	dig prot %	*		24.2	
Horses	dig prot %	*		22.7	
Rabbits	dig prot %	*		21.5	
Sheep	dig prot %	*		23.1	
Fiber, acid detergent (VS)	%			28.8	
Lignin (Ellis)	%			2.8	
Energy	GE Mcal/kg			4.60	
Sheep	DE Mcal/kg	*		2.79	
Sheep	ME Mcal/kg	*		2.29	
Sheep	TDN %			63.2	
Phosphorus	%			0.41	
Sulphur	%			0.55	
Dry matter intake	g/W$^{0.75}$			75.900	
Gain, sheep	kg/day			0.090	

CANARYGRASS, REED SIBERIAN. Phalaris arundinacea

Canarygrass, reed Siberian, hay, s-c baled fan-air dried w heat, immature, cut 1, (1)
Ref No 1 08 864 United States

Feed Name or Analyses			Mean As Fed	Mean Dry	C.V. ± %
Dry matter	%		83.5	100.0	
Sheep	dig coef %		62.	62.	
Ash	%		8.0	9.6	
Sheep	dig coef %		49.	59.	
Crude fiber	%		25.0	29.9	
Sheep	dig coef %		64.	64.	
Ether extract	%		2.3	2.7	
Sheep	dig coef %		37.	37.	
N-free extract	%		34.0	40.7	
Sheep	dig coef %		62.	62.	
Protein (N x 6.25)	%		14.2	17.0	
Sheep	dig coef %		70.	70.	
Cattle	dig prot %	*	9.8	11.7	
Goats	dig prot %	*	10.4	12.5	
Horses	dig prot %	*	10.0	12.0	
Rabbits	dig prot %	*	9.9	11.8	
Sheep	dig prot %		10.0	11.9	
Cell walls (Van Soest)	%		51.6	61.8	
Sheep	dig coef %		62.	62.	
Cellulose (Matrone)	%		26.7	32.0	
Sheep	dig coef %		67.	67.	
Fiber, acid detergent (VS)	%		28.9	34.6	
Sheep	dig coef %		55.	55.	
Hemicellulose	%		22.7	27.2	
Sheep	dig coef %		71.	71.	
Lignin (Ellis)	%		2.8	3.4	
Sheep	dig coef %		24.	24.	
Energy	GE Mcal/kg		3.72	4.45	
Sheep	GE dig coef %		60.	60.	
Cattle	DE Mcal/kg	*	2.26	2.70	
Sheep	DE Mcal/kg		2.23	2.67	
Cattle	ME Mcal/kg	*	1.85	2.22	
Sheep	ME Mcal/kg	*	1.83	2.19	
Cattle	TDN %	*	51.2	61.3	
Sheep	TDN %		48.8	58.4	
Dry matter intake	g/W$^{0.75}$		50.510	60.467	
Gain, sheep	kg/day		0.063	0.076	

Canarygrass, reed Siberian, hay, s-c baled fan-air dried w heat, immature, cut 2, (1)
Ref No 1 08 862 United States

Feed Name or Analyses			Mean As Fed	Mean Dry	C.V. ± %
Dry matter	%		61.3	100.0	
Sheep	dig coef %		55.	55.	
Ash	%		5.6	9.1	
Sheep	dig coef %		23.	37.	
Crude fiber	%		15.8	25.8	
Sheep	dig coef %		51.	51.	
Ether extract	%		2.2	3.7	
Sheep	dig coef %		53.	53.	
N-free extract	%		26.8	43.8	
Sheep	dig coef %		52.	52.	
Protein (N x 6.25)	%		10.9	17.7	
Sheep	dig coef %		76.	76.	
Cattle	dig prot %	*	7.5	12.3	
Goats	dig prot %	*	8.0	13.1	
Horses	dig prot %	*	7.7	12.6	
Rabbits	dig prot %	*	7.6	12.3	
Sheep	dig prot %		8.2	13.5	
Cell walls (Van Soest)	%		35.7	58.3	
Sheep	dig coef %		48.	48.	

(1) dry forages and roughages (3) silages (6) minerals
(2) pasture, range plants, and forages fed green (4) energy feeds (7) vitamins
(5) protein supplements (8) additives

Feed Name or Analyses			Mean As Fed	Mean Dry	C.V. ± %
Cellulose (Matrone)	%		18.3	29.8	
Fiber, acid detergent (VS)	%		21.4	34.9	
Sheep	dig coef %		40.	40.	
Hemicellulose	%		14.3	23.4	
Lignin (Ellis)	%		2.5	4.0	
Sheep	dig coef %		-22.	-22.	
Energy	GE Mcal/kg		2.74	4.47	
Sheep	GE dig coef %		54.	54.	
Cattle	DE Mcal/kg	*	1.49	2.43	
Sheep	DE Mcal/kg		1.47	2.39	
Cattle	ME Mcal/kg	*	1.22	1.99	
Sheep	ME Mcal/kg	*	1.20	1.96	
Cattle	TDN %	*	33.8	55.1	
Sheep	TDN %		32.9	53.7	
Dry matter intake	g/w.75		34.328	56.000	
Gain, sheep	kg/day		0.033	0.055	

CANARYGRASS, TIMOTHY. Phalaris angusta

Canarygrass, timothy, aerial part, fresh, immature, (2)
Ref No 2 01 115 United States

			As Fed	Dry	C.V.
Dry matter	%			100.0	
Ash	%			9.3	
Crude fiber	%			25.6	
Ether extract	%			3.0	
N-free extract	%			42.0	
Protein (N x 6.25)	%			20.1	
Cattle	dig prot %	*		15.0	
Goats	dig prot %	*		15.3	
Horses	dig prot %	*		14.6	
Rabbits	dig prot %	*		14.2	
Sheep	dig prot %	*		15.7	
Calcium	%			0.51	
Magnesium	%			0.27	
Phosphorus	%			0.25	
Potassium	%			2.70	

CANARYGRASS, REED-ALFALFA. Phalaris arundinacea, Medicago sativa

Canarygrass, reed-Alfalfa, aerial part, fresh, (2)
Ref No 2 01 116 United States

		As Fed	Dry	C.V.
Dry matter	%		100.0	
Calcium	%		0.45	24
Magnesium	%		0.31	10
Potassium	%		3.06	10
Sodium	%		0.02	35

CANARYGRASS, REED-CLOVER, LADINO. Phalaris arundinacea, Trifolium repens

Canarygrass, reed-Clover, ladino, aerial part, fresh, (2)
Ref No 2 01 117 United States

		As Fed	Dry	C.V.
Dry matter	%		100.0	
Calcium	%		0.67	38
Magnesium	%		0.37	28
Potassium	%		3.29	10
Sodium	%		0.04	34

Candied copra –
see Coconut, meats w molasses added, extn unspecified grnd, (4)

CANE. Arundinaria spp

Cane, leaves, fresh, (2)
Ref No 2 01 118 United States

			As Fed	Dry	C.V.
Dry matter	%			100.0	
Ash	%			9.0	
Crude fiber	%			36.7	
Ether extract	%			2.2	
N-free extract	%			40.3	
Protein (N x 6.25)	%			11.8	33
Cattle	dig prot %	*		7.9	
Goats	dig prot %	*		7.6	
Horses	dig prot %	*		7.5	
Rabbits	dig prot %	*		7.8	
Sheep	dig prot %	*		8.0	

CANE, JAPANESE. Arundinaria japonica

Cane, Japanese, aerial part, s-c, mature, (1)
Ref No 1 01 119 United States

		As Fed	Dry	C.V.
Dry matter	%	94.1	100.0	2
Calcium	%	0.35	0.37	6
Magnesium	%	0.13	0.14	23
Phosphorus	%	0.14	0.15	16
Potassium	%	0.64	0.68	24

Cane, Japanese, stalks, (1)
Ref No 1 01 120 United States

			As Fed	Dry	C.V.
Dry matter	%		95.2	100.0	
Ash	%		3.4	3.6	
Crude fiber	%		26.7	28.0	
Ether extract	%		1.0	1.1	
N-free extract	%		61.5	64.6	
Protein (N x 6.25)	%		2.6	2.7	
Cattle	dig prot %	*	-0.6	-0.6	
Goats	dig prot %	*	-0.8	-0.8	
Horses	dig prot %	*	-0.1	-0.1	
Rabbits	dig prot %	*	0.7	0.8	
Sheep	dig prot %	*	-0.9	-0.9	

CANE, SWITCH. Arundinaria tecta

Cane, switch, aerial part, fresh, (2)
Ref No 2 01 123 United States

			As Fed	Dry	C.V.
Dry matter	%			100.0	
Ash	%			10.6	43
Crude fiber	%			29.4	12
Ether extract	%			3.5	30
N-free extract	%			41.4	
Protein (N x 6.25)	%			15.1	34
Cattle	dig prot %	*		10.7	
Goats	dig prot %	*		10.6	
Horses	dig prot %	*		10.3	
Rabbits	dig prot %	*		10.3	
Sheep	dig prot %	*		11.1	
Calcium	%			0.21	37
Copper	mg/kg			7.9	9
Iron	%			0.020	98
Manganese	mg/kg			73.0	29
Phosphorus	%			0.19	43

Cane, switch, aerial part, fresh, immature, (2)
Ref No 2 01 121 United States

		As Fed	Dry	C.V.
Dry matter	%		100.0	
Calcium	%		0.29	27
Cobalt	mg/kg		0.042	36
Copper	mg/kg		7.9	9
Iron	%		0.020	98
Manganese	mg/kg		73.0	29
Phosphorus	%		0.28	3

Cane, switch, aerial part, fresh, mature, (2)
Ref No 2 01 122 United States

		As Fed	Dry	C.V.
Dry matter	%		100.0	
Calcium	%		0.15	
Phosphorus	%		0.05	

Cane, switch, leaves, fresh, (2)
Ref No 2 01 124 United States

			As Fed	Dry	C.V.
Dry matter	%			100.0	
Protein (N x 6.25)	%			12.1	26
Cattle	dig prot %	*		8.2	
Goats	dig prot %	*		7.8	
Horses	dig prot %	*		7.8	
Rabbits	dig prot %	*		8.0	
Sheep	dig prot %	*		8.3	
Calcium	%			0.24	30
Cobalt	mg/kg			0.046	
Manganese	mg/kg			174.6	67
Phosphorus	%			0.06	63

Cane bagasse, dried –
see Sugarcane, pulp, dehy, (1)

Cane molasses (AAFCO) –
see Sugarcane, molasses, mn 48% invert sugar mn 79.5 degrees brix, (4)

Cane, molasses, ammoniated –
see Sugarcane, molasses, ammoniated, (5)

Cane molasses, dried –
see Sugarcane, molasses, dehy, (4)

CANE-SORGHUM, JOHNSONGRASS. Arundinaria spp, Sorghum halepense

Cane-Sorghum, johnsongrass, aerial part, ensiled, (3)
Ref No 3 01 125 United States

		As Fed	Dry	C.V.
Dry matter	%	37.8	100.0	
Vitamin D_2	IU/g	0.2	0.5	

CANNA. Canna spp

Canna, aerial part, fresh, early leaf, (2)
Ref No 2 01 126 United States

		As Fed	Dry	C.V.
Dry matter	%	16.5	100.0	
Ash	%	2.7	16.5	
Crude fiber	%	3.2	19.6	
Sheep	dig coef %	35.	35.	

Continued

Feed Name or Analyses				Mean As Fed	Mean Dry	C.V. ± %
Ether extract		%		0.8	5.1	
Sheep	dig coef	%		27.	27.	
N-free extract		%		8.0	48.6	
Sheep	dig coef	%		66.	66.	
Protein (N x 6.25)		%		1.7	10.2	
Sheep	dig coef	%		44.	44.	
Cattle	dig prot	%	*	1.1	6.6	
Goats	dig prot	%	*	1.0	6.1	
Horses	dig prot	%	*	1.0	6.2	
Rabbits	dig prot	%	*	1.1	6.5	
Sheep	dig prot	%		0.7	4.5	
Energy	GE Mcal/kg					
Cattle	DE Mcal/kg		*	0.34	2.09	
Sheep	DE Mcal/kg		*	0.34	2.05	
Cattle	ME Mcal/kg		*	0.28	1.71	
Sheep	ME Mcal/kg		*	0.28	1.68	
Cattle	TDN	%	*	7.8	47.4	
Sheep	TDN	%		7.7	46.5	

Canna, tubers, (4)
Ref No 4 01 127 United States

Feed Name or Analyses				Mean As Fed	Mean Dry	C.V. ± %
Dry matter		%		30.7	100.0	
Ash		%		2.3	7.4	
Crude fiber		%		1.0	3.4	
Sheep	dig coef	%		89.	89.	
Ether extract		%		0.2	0.8	
Sheep	dig coef	%		63.	63.	
N-free extract		%		26.0	84.8	
Sheep	dig coef	%		90.	90.	
Protein (N x 6.25)		%		1.1	3.6	
Sheep	dig coef	%		44.	44.	
Cattle	dig prot	%	*	-0.1	-0.6	
Goats	dig prot	%	*	0.2	0.5	
Horses	dig prot	%	*	0.2	0.5	
Sheep	dig prot	%		0.5	1.6	
Energy	GE Mcal/kg					
Cattle	DE Mcal/kg		*	1.07	3.50	
Sheep	DE Mcal/kg		*	1.11	3.62	
Cattle	ME Mcal/kg		*	0.88	2.87	
Sheep	ME Mcal/kg		*	0.91	2.97	
Cattle	TDN	%	*	24.3	79.3	
Sheep	TDN	%		25.2	82.1	

CAPEJASMINE. Gardenia jasminoides

Capejasmine, seeds w hulls, (4)
Ref No 4 09 005 United States

Feed Name or Analyses				Mean As Fed	Mean Dry	C.V. ± %
Dry matter		%			100.0	
Protein (N x 6.25)		%			14.0	
Cattle	dig prot	%	*		8.9	
Goats	dig prot	%	*		10.1	
Horses	dig prot	%	*		10.1	
Sheep	dig prot	%	*		10.1	
Alanine		%			0.63	
Arginine		%			1.26	
Aspartic acid		%			1.29	
Glutamic acid		%			2.70	
Glycine		%			0.74	
Histidine		%			0.34	
Isoleucine		%			0.56	
Leucine		%			1.08	
Lysine		%			0.50	
Methionine		%			0.18	
Phenylalanine		%			0.60	
Proline		%			0.57	
Serine		%			0.62	
Threonine		%			0.45	
Tyrosine		%			0.41	
Valine		%			0.76	

(1) dry forages and roughages (2) pasture, range plants, and forages fed green (3) silages (4) energy feeds (5) protein supplements (6) minerals (7) vitamins (8) additives

CAPEWILLOW. Chilopsis linearis

Capewillow, seeds w hulls, (5)
Ref No 5 09 006 United States

Feed Name or Analyses				Mean As Fed	Mean Dry	C.V. ± %
Dry matter		%			100.0	
Protein (N x 6.25)		%			22.0	
Alanine		%			0.90	
Arginine		%			2.20	
Aspartic acid		%			1.80	
Glutamic acid		%			3.67	
Glycine		%			1.01	
Histidine		%			0.51	
Hydroxyproline		%			0.09	
Isoleucine		%			0.84	
Leucine		%			1.39	
Lysine		%			0.97	
Methionine		%			0.29	
Phenylalanine		%			0.99	
Proline		%			0.77	
Serine		%			0.84	
Threonine		%			0.79	
Tyrosine		%			0.75	
Valine		%			1.01	

CARAWAY. Carum carvi

Caraway, seeds, extn unspecified grnd, (5)
Ref No 5 01 130 United States

Feed Name or Analyses				Mean As Fed	Mean Dry	C.V. ± %
Dry matter		%		86.2	100.0	
Ash		%		6.7	7.8	
Crude fiber		%		13.7	15.9	
Cattle	dig coef	%		84.	84.	
Ether extract		%		13.9	16.1	
Cattle	dig coef	%		96.	96.	
N-free extract		%		30.4	35.3	
Cattle	dig coef	%		75.	75.	
Protein (N x 6.25)		%		21.5	24.9	
Cattle	dig coef	%		60.	60.	
Cattle	dig prot	%		12.9	14.9	
Energy	GE Mcal/kg					
Cattle	DE Mcal/kg		*	3.40	3.95	
Sheep	DE Mcal/kg		*	3.25	3.77	
Swine	DE kcal/kg		*	3531.	4096.	
Cattle	ME Mcal/kg		*	2.79	3.24	
Sheep	ME Mcal/kg		*	2.66	3.09	
Swine	ME kcal/kg		*	3213.	3727.	
Cattle	TDN	%		77.2	89.5	
Sheep	TDN	%	*	73.6	85.4	
Swine	TDN	%	*	80.1	92.9	

CARDOON. Cynara cardunculus

Cardoon, seeds w hulls, (5)
Ref No 5 09 114 United States

Feed Name or Analyses				Mean As Fed	Mean Dry	C.V. ± %
Dry matter		%			100.0	
Protein (N x 6.25)		%			27.0	
Alanine		%			1.32	
Arginine		%			1.30	
Aspartic acid		%			2.97	
Glutamic acid		%			5.64	
Glycine		%			1.30	
Histidine		%			0.68	
Isoleucine		%			1.22	
Leucine		%			1.97	
Lysine		%			1.13	
Methionine		%			0.49	
Phenylalanine		%			1.49	
Proline		%			1.19	
Serine		%			1.49	
Threonine		%			1.03	
Tyrosine		%			0.97	
Valine		%			1.73	

CAROB BEAN. Ceratonia siliqua

Carob bean, pods, (4)
Ref No 4 01 132 United States

Feed Name or Analyses				Mean As Fed	Mean Dry	C.V. ± %
Dry matter		%		83.9	100.0	
Ash		%		3.5	4.2	
Crude fiber		%		7.9	9.5	
Sheep	dig coef	%		52.	52.	
Ether extract		%		1.5	1.7	
Sheep	dig coef	%		66.	66.	
N-free extract		%		66.2	79.0	
Sheep	dig coef	%		79.	79.	
Protein (N x 6.25)		%		4.8	5.7	
Sheep	dig coef	%		17.	17.	
Cattle	dig prot	%	*	1.0	1.2	
Goats	dig prot	%	*	2.0	2.4	
Horses	dig prot	%	*	2.0	2.4	
Sheep	dig prot	%		0.8	1.0	
Energy	GE Mcal/kg					
Cattle	DE Mcal/kg		*	2.72	3.24	
Sheep	DE Mcal/kg		*	2.62	3.12	
Cattle	ME Mcal/kg		*	2.23	2.66	
Sheep	ME Mcal/kg		*	2.15	2.56	
Cattle	TDN	%	*	61.8	73.6	
Sheep	TDN	%		59.4	70.9	

Carob bean, seeds, grnd, (4)
Ref No 4 01 133 United States

Feed Name or Analyses				Mean As Fed	Mean Dry	C.V. ± %
Dry matter		%		87.5	100.0	
Ash		%		2.9	3.3	
Crude fiber		%		8.1	9.3	
Sheep	dig coef	%		77.	77.	
Ether extract		%		1.9	2.1	
Sheep	dig coef	%		68.	68.	
N-free extract		%		63.2	72.2	
Sheep	dig coef	%		90.	90.	
Protein (N x 6.25)		%		11.4	13.0	
Sheep	dig coef	%		52.	52.	
Cattle	dig prot	%	*	7.0	8.0	
Goats	dig prot	%	*	8.0	9.2	
Horses	dig prot	%	*	8.0	9.2	

Feed Name or Analyses			Mean As Fed	Mean Dry	C.V. ± %
Sheep	dig prot %		5.9	6.8	
Energy	GE Mcal/kg		1.77	2.03	
Cattle	DE Mcal/kg	*	2.99	3.42	
Sheep	DE Mcal/kg	*	3.17	3.62	
Swine	DE kcal/kg	*	3256.	3721.	
Cattle	ME Mcal/kg	*	2.45	2.80	
Sheep	ME Mcal/kg	*	2.60	2.97	
Swine	ME kcal/kg	*	3041.	3475.	
Cattle	TDN %	*	67.8	77.5	
Sheep	TDN %		71.9	82.2	
Swine	TDN %	*	73.8	84.4	
Calcium	%		0.34	0.39	
Phosphorus	%		0.08	0.09	

Carob bean, seeds w pods, (4)
Ref No 4 08 370 United States

Feed Name or Analyses			As Fed	Dry	C.V. ± %
Dry matter	%		87.8	100.0	
Ash	%		2.5	2.8	
Crude fiber	%		8.7	9.9	
Ether extract	%		2.6	3.0	
N-free extract	%		68.5	78.0	
Protein (N x 6.25)	%		5.5	6.3	
Cattle	dig prot %	*	1.6	1.8	
Goats	dig prot %	*	2.6	3.0	
Horses	dig prot %	*	2.6	3.0	
Sheep	dig prot %	*	2.6	3.0	
Energy	GE Mcal/kg				
Cattle	DE Mcal/kg	*	3.06	3.48	
Sheep	DE Mcal/kg	*	3.18	3.62	
Swine	DE kcal/kg	*	2670.	3041.	
Cattle	ME Mcal/kg	*	2.50	2.85	
Sheep	ME Mcal/kg	*	2.60	2.97	
Swine	ME kcal/kg	*	2529.	2881.	
Cattle	TDN %	*	69.4	79.0	
Sheep	TDN %	*	72.0	82.0	
Swine	TDN %	*	60.5	69.0	

Carob bean, gluten, fresh etiolated, (5)
Ref No 5 01 129 United States

Feed Name or Analyses			As Fed	Dry	C.V. ± %
Dry matter	%		90.0	100.0	
Alanine	%		1.90	2.11	
Arginine	%		11.50	12.78	
Aspartic acid	%		5.60	6.22	
Cystine	%		1.10	1.22	
Glutamic acid	%		18.20	20.22	
Glycine	%		3.70	4.11	
Histidine	%		2.30	2.56	
Isoleucine	%		2.90	3.22	
Leucine	%		4.50	5.00	
Lysine	%		3.90	4.33	
Methionine	%		0.50	0.56	
Phenylalanine	%		2.40	2.67	
Proline	%		2.40	2.67	
Threonine	%		2.30	2.56	
Tryptophan	%		0.50	0.56	
Tyrosine	%		2.00	2.22	
Valine	%		3.10	3.44	

Carolina allspice –
see Sweetshrub, common, seeds w hulls, (5)

CARPETGRASS. Axonopus spp

Carpetgrass, hay, s-c, (1)
Ref No 1 01 138 United States

Feed Name or Analyses			As Fed	Dry	C.V. ± %
Dry matter	%		91.9	100.0	1
Ash	%		10.2	11.1	
Crude fiber	%		31.7	34.5	
Ether extract	%		2.2	2.4	
N-free extract	%		40.8	44.4	
Protein (N x 6.25)	%		7.0	7.6	
Cattle	dig prot %	*	3.2	3.5	
Goats	dig prot %	*	3.4	3.7	
Horses	dig prot %	*	3.7	4.0	
Rabbits	dig prot %	*	4.2	4.5	
Sheep	dig prot %	*	3.1	3.4	
Fatty acids	%		2.2	2.4	
Energy	GE Mcal/kg				
Cattle	DE Mcal/kg	*	1.93	2.10	
Sheep	DE Mcal/kg	*	1.96	2.13	
Cattle	ME Mcal/kg	*	1.58	1.72	
Sheep	ME Mcal/kg	*	1.61	1.75	
Cattle	TDN %	*	43.7	47.6	
Sheep	TDN %	*	44.3	48.2	
Calcium	%		0.39	0.43	16
Phosphorus	%		0.12	0.14	22

Carpetgrass, hay, s-c, mid-bloom, (1)
Ref No 1 01 135 United States

Feed Name or Analyses			As Fed	Dry	C.V. ± %
Dry matter	%		93.0	100.0	
Ash	%		10.7	11.5	
Crude fiber	%		26.3	28.3	
Ether extract	%		1.8	1.9	
N-free extract	%		46.5	50.0	
Protein (N x 6.25)	%		7.7	8.3	
Cattle	dig prot %	*	3.8	4.1	
Goats	dig prot %	*	4.0	4.3	
Horses	dig prot %	*	4.3	4.6	
Rabbits	dig prot %	*	4.7	5.1	
Sheep	dig prot %	*	3.7	4.0	
Energy	GE Mcal/kg				
Cattle	DE Mcal/kg	*	2.03	2.18	
Sheep	DE Mcal/kg	*	1.94	2.09	
Cattle	ME Mcal/kg	*	1.66	1.79	
Sheep	ME Mcal/kg	*	1.59	1.71	
Cattle	TDN %	*	46.0	49.5	
Sheep	TDN %	*	44.1	47.4	
Calcium	%		0.34	0.37	
Phosphorus	%		0.15	0.16	

Carpetgrass, hay, s-c, full bloom, (1)
Ref No 1 01 136 United States

Feed Name or Analyses			As Fed	Dry	C.V. ± %
Dry matter	%		92.4	100.0	
Ash	%		8.4	9.1	
Crude fiber	%		30.7	33.2	
Ether extract	%		1.3	1.4	
N-free extract	%		45.1	48.8	
Protein (N x 6.25)	%		6.9	7.5	
Cattle	dig prot %	*	3.2	3.4	
Goats	dig prot %	*	3.3	3.6	
Horses	dig prot %	*	3.6	3.9	
Rabbits	dig prot %	*	4.1	4.5	
Sheep	dig prot %	*	3.0	3.3	
Energy	GE Mcal/kg				
Cattle	DE Mcal/kg	*	1.99	2.15	
Sheep	DE Mcal/kg	*	1.85	2.00	
Cattle	ME Mcal/kg	*	1.63	1.76	
Sheep	ME Mcal/kg	*	1.52	1.64	
Cattle	TDN %	*	45.1	48.8	
Sheep	TDN %	*	41.9	45.3	
Calcium	%		0.37	0.40	
Phosphorus	%		0.08	0.09	

Carpetgrass, hay, s-c, mature, (1)
Ref No 1 01 137 United States

Feed Name or Analyses			As Fed	Dry	C.V. ± %
Dry matter	%		91.0	100.0	
Ash	%		7.3	8.0	
Crude fiber	%		31.0	34.1	
Ether extract	%		1.3	1.4	
N-free extract	%		46.0	50.6	
Protein (N x 6.25)	%		5.4	5.9	
Cattle	dig prot %	*	1.9	2.0	
Goats	dig prot %	*	1.9	2.1	
Horses	dig prot %	*	2.3	2.5	
Rabbits	dig prot %	*	2.9	3.2	
Sheep	dig prot %	*	1.7	1.9	
Energy	GE Mcal/kg				
Cattle	DE Mcal/kg	*	1.87	2.05	
Sheep	DE Mcal/kg	*	1.78	1.96	
Cattle	ME Mcal/kg	*	1.53	1.68	
Sheep	ME Mcal/kg	*	1.49	1.61	
Cattle	TDN %	*	42.4	46.6	
Sheep	TDN %	*	40.4	44.4	
Calcium	%		0.31	0.34	
Phosphorus	%		0.10	0.11	

Carpetgrass, aerial part, fresh, (2)
Ref No 2 01 140 United States

Feed Name or Analyses			As Fed	Dry	C.V. ± %
Dry matter	%		31.6	100.0	33
Ash	%		4.0	12.8	
Crude fiber	%		8.4	26.4	
Ether extract	%		0.5	1.6	
N-free extract	%		15.8	50.0	
Protein (N x 6.25)	%		2.9	9.2	
Cattle	dig prot %	*	1.8	5.7	
Goats	dig prot %	*	1.6	5.1	
Horses	dig prot %	*	1.7	5.3	
Rabbits	dig prot %	*	1.8	5.8	
Sheep	dig prot %	*	1.8	5.6	
Energy	GE Mcal/kg				
Cattle	DE Mcal/kg	*	0.87	2.74	
Sheep	DE Mcal/kg	*	0.94	2.97	
Cattle	ME Mcal/kg	*	0.71	2.25	
Sheep	ME Mcal/kg	*	0.77	2.44	
Cattle	TDN %	*	19.6	62.1	
Sheep	TDN %	*	21.3	67.4	
Calcium	%		0.13	0.40	
Chlorine	%		0.13	0.42	15
Cobalt	mg/kg		0.029	0.090	74
Iron	%		0.009	0.028	
Magnesium	%		0.08	0.24	
Phosphorus	%		0.05	0.16	
Potassium	%		0.29	0.92	
Sodium	%		0.16	0.52	
Sulphur	%		0.03	0.09	42
Carotene	mg/kg		45.8	144.8	42
Vitamin A equivalent	IU/g		76.4	241.5	

CARPETGRASS, TROPICAL. Axonopus compressus

Carpetgrass, tropical, aerial part, fresh, (2)

Ref No 2 01 141 United States

Feed Name or Analyses			Mean As Fed	Mean Dry	C.V. ± %
Dry matter	%		39.9	100.0	
Ash	%		3.4	8.4	
Crude fiber	%		10.4	26.1	
Cattle	dig coef %		55.	55.	
Ether extract	%		0.6	1.6	
Cattle	dig coef %		84.	84.	
N-free extract	%		22.0	55.1	
Cattle	dig coef %		84.	84.	
Protein (N x 6.25)	%		3.5	8.8	
Cattle	dig coef %		50.	50.	
Cattle	dig prot %		1.8	4.4	
Goats	dig prot %	*	1.9	4.8	
Horses	dig prot %	*	2.0	5.0	
Rabbits	dig prot %	*	2.2	5.5	
Sheep	dig prot %	*	2.1	5.2	
Energy	GE Mcal/kg				
Cattle	DE Mcal/kg	*	1.20	3.00	
Sheep	DE Mcal/kg	*	1.22	3.05	
Cattle	ME Mcal/kg	*	0.98	2.46	
Sheep	ME Mcal/kg	*	1.00	2.50	
Cattle	TDN %		27.2	68.1	
Sheep	TDN %	*	27.6	69.2	

CARPETGRASS-DALLISGRASS. Axonopus spp, Paspalum dilatatum

Carpetgrass-Dallisgrass, aerial part, fresh, (2)

Ref No 2 01 142 United States

Feed Name or Analyses			Mean As Fed	Mean Dry	C.V. ± %
Dry matter	%			100.0	
Cobalt	mg/kg			0.097	66

CARROT. Daucus spp

Carrot, aerial part, fresh, (2)

Ref No 2 08 371 United States

Feed Name or Analyses			Mean As Fed	Mean Dry	C.V. ± %
Dry matter	%		16.0	100.0	
Ash	%		2.4	15.0	
Crude fiber	%		2.9	18.1	
Ether extract	%		0.6	3.8	
N-free extract	%		8.0	50.0	
Protein (N x 6.25)	%		2.1	13.1	
Cattle	dig prot %	*	1.4	9.0	
Goats	dig prot %	*	1.4	8.8	
Horses	dig prot %	*	1.4	8.7	
Rabbits	dig prot %	*	1.4	8.8	
Sheep	dig prot %	*	1.5	9.2	
Energy	GE Mcal/kg				
Cattle	DE Mcal/kg	*	0.52	3.28	
Sheep	DE Mcal/kg	*	0.54	3.37	
Cattle	ME Mcal/kg	*	0.43	2.69	
Sheep	ME Mcal/kg	*	0.44	2.76	
Cattle	TDN %	*	11.9	74.6	
Sheep	TDN %	*	12.2	76.5	
Calcium	%		0.31	1.94	
Phosphorus	%		0.03	0.19	
Potassium	%		0.30	1.88	

Carrot, leaves, fresh, (2)

Ref No 2 01 143 United States

Feed Name or Analyses			Mean As Fed	Mean Dry	C.V. ± %
Dry matter	%		16.5	100.0	34
Ash	%		2.5	15.0	
Crude fiber	%		1.8	11.1	23
Ether extract	%		0.8	4.6	29
N-free extract	%		8.7	52.9	
Protein (N x 6.25)	%		2.7	16.4	29
Cattle	dig prot %	*	2.0	11.8	
Goats	dig prot %	*	2.0	11.9	
Horses	dig prot %	*	1.9	11.5	
Rabbits	dig prot %	*	1.9	11.3	
Sheep	dig prot %	*	2.0	12.3	
Energy	GE Mcal/kg				
Cattle	DE Mcal/kg	*	0.52	3.18	
Sheep	DE Mcal/kg	*	0.49	2.98	
Cattle	ME Mcal/kg	*	0.43	2.61	
Sheep	ME Mcal/kg	*	0.40	2.45	
Cattle	TDN %	*	11.9	72.1	
Sheep	TDN %	*	11.2	67.6	
Zinc	mg/kg		0.7	4.0	
Carotene	mg/kg		28.3	171.5	31
Riboflavin	mg/kg		1.5	9.3	30
a-tocopherol	mg/kg		130.3	789.9	
Vitamin A equivalent	IU/g		47.2	285.9	
Arginine	%		0.15	0.90	
Histidine	%		0.03	0.20	
Isoleucine	%		0.17	1.00	
Leucine	%		0.25	1.50	
Lysine	%		0.15	0.90	
Methionine	%		0.07	0.40	
Phenylalanine	%		0.23	1.40	
Tryptophan	%		0.05	0.30	
Valine	%		0.20	1.20	

Carrot, stems, fresh, (2)

Ref No 2 01 144 United States

Feed Name or Analyses			Mean As Fed	Mean Dry	C.V. ± %
Dry matter	%		12.0	100.0	
Crude fiber	%		2.2	18.1	6
Ether extract	%		0.7	5.5	
Protein (N x 6.25)	%		1.2	10.4	5
Cattle	dig prot %	*	0.8	6.7	
Goats	dig prot %	*	0.8	6.3	
Horses	dig prot %	*	0.8	6.4	
Rabbits	dig prot %	*	0.8	6.7	
Sheep	dig prot %	*	0.8	6.7	
Carotene	mg/kg		9.2	76.7	48
Riboflavin	mg/kg		0.8	6.6	14
Vitamin A equivalent	IU/g		15.3	127.9	

Carrot, roots, fresh, (4)

Ref No 2 01 146 United States

Feed Name or Analyses			Mean As Fed	Mean Dry	C.V. ± %
Dry matter	%		12.9	100.0	17
Ash	%		1.2	9.7	24
Crude fiber	%		1.2	9.1	10
Cattle	dig coef %		84.	84.	
Horses	dig coef %		79.	79.	
Sheep	dig coef %		108.	108.	
Swine	dig coef %		85.	85.	
Ether extract	%		0.2	1.4	19
Cattle	dig coef %		105.	105.	
Horses	dig coef %		52.	52.	
Sheep	dig coef %		73.	73.	
Swine	dig coef %		68.	68.	
N-free extract	%		9.0	69.6	6
Cattle	dig coef %		95.	95.	
Horses	dig coef %		98.	98.	
Sheep	dig coef %		97.	97.	
Swine	dig coef %		94.	94.	
Protein (N x 6.25)	%		1.3	10.3	25
Cattle	dig coef %		50.	50.	
Horses	dig coef %		89.	89.	
Sheep	dig coef %		76.	76.	
Swine	dig coef %		71.	71.	
Cattle	dig prot %		0.7	5.1	
Goats	dig prot %	*	0.9	6.7	
Horses	dig prot %		1.2	9.1	
Sheep	dig prot %		1.0	7.8	
Swine	dig prot %		0.9	7.3	
Energy	GE Mcal/kg				
Cattle	DE Mcal/kg	*	0.47	3.62	
Horses	DE Mcal/kg	*	0.49	3.78	
Sheep	DE Mcal/kg	*	0.50	3.85	
Swine	DE kcal/kg	*	469.	3638.	
Cattle	ME Mcal/kg	*	0.38	2.97	
Horses	ME Mcal/kg	*	0.40	3.10	
Sheep	ME Mcal/kg	*	0.41	3.16	
Swine	ME kcal/kg	*	440.	3417.	
Cattle	TDN %		10.6	82.1	
Horses	TDN %		11.0	85.6	
Sheep	TDN %		11.3	87.4	
Swine	TDN %		10.6	82.5	
Calcium	%		0.04	0.37	
Chlorine	%		0.06	0.50	
Copper	mg/kg		1.3	11.1	
Iron	%		0.001	0.011	
Magnesium	%		0.02	0.17	
Manganese	mg/kg		3.7	31.5	
Phosphorus	%		0.04	0.32	
Potassium	%		0.30	2.50	
Sodium	%		0.12	1.00	
Sulphur	%		0.02	0.17	
Ascorbic acid	mg/kg		80.3	678.0	
Niacin	mg/kg		6.0	50.8	
Riboflavin	mg/kg		0.5	4.2	
Thiamine	mg/kg		0.6	5.1	
Vitamin A	IU/g		110.5	932.2	

CARROT, SOUTHWESTERN. Daucus pusillus

Carrot, southwestern, aerial part, fresh, (2)

Rattlesnake weed

Bristly carrot

Ref No 2 01 149 United States

Feed Name or Analyses			Mean As Fed	Mean Dry	C.V. ± %
Dry matter	%			100.0	
Ash	%			9.8	
Crude fiber	%			18.9	
Ether extract	%			2.5	
N-free extract	%			63.9	
Protein (N x 6.25)	%			4.9	
Cattle	dig prot %	*		2.0	
Goats	dig prot %	*		1.1	
Horses	dig prot %	*		1.7	
Rabbits	dig prot %	*		2.4	
Sheep	dig prot %	*		1.5	
Other carbohydrates	%			31.0	
Lignin (Ellis)	%			12.9	
Energy	GE Mcal/kg				
Cattle	DE Mcal/kg	*		3.27	
Sheep	DE Mcal/kg	*		3.32	

(1) dry forages and roughages (2) pasture, range plants, and forages fed green (3) silages (4) energy feeds (5) protein supplements (6) minerals (7) vitamins (8) additives

Feed Name or Analyses			Mean As Fed	Mean Dry	C.V. ± %
Cattle	ME Mcal/kg	*		2.68	
Sheep	ME Mcal/kg	*		2.72	
Cattle	TDN %	*		74.2	
Sheep	TDN %	*		75.2	
Silicon	%			0.27	

Carrot, southwestern, leaves, fresh, (2)

Ref No 2 01 148 United States

Feed Name or Analyses			As Fed	Dry	C.V. ± %
Dry matter	%			100.0	
Calcium	%			2.35	
Magnesium	%			0.30	
Phosphorus	%			0.35	
Potassium	%			2.59	

Carrot, southwestern, leaves, fresh, mature, (2)

Ref No 2 01 147 United States

Feed Name or Analyses			As Fed	Dry	C.V. ± %
Dry matter	%			100.0	
Calcium	%			1.88	
Magnesium	%			0.28	
Phosphorus	%			0.28	
Potassium	%			3.31	

Casein (AAFCO) –

see Cattle, casein, milk acid-precipitated dehy, mn 80% protein, (5)

Casein, dried –

see Cattle, casein, milk acid-precipitated dehy, mn 80% protein, (5)

CASSAVA. Manihot spp

Cassava, roots, (4)

Ref No 2 01 150 United States

Feed Name or Analyses			As Fed	Dry	C.V. ± %
Dry matter	%		32.4	100.0	
Ash	%		1.3	3.9	
Crude fiber	%		1.5	4.6	
Sheep	dig coef %		53.	53.	
Ether extract	%		0.3	1.0	
Sheep	dig coef %		51.	51.	
N-free extract	%		28.1	86.9	
Sheep	dig coef %		90.	90.	
Protein (N x 6.25)	%		1.2	3.6	
Sheep	dig coef %		-1.	-1.	
Cattle	dig prot %	*	-0.1	-0.5	
Goats	dig prot %	*	0.2	0.6	
Horses	dig prot %	*	0.2	0.6	
Sheep	dig prot %		0.0	-0.0	
Energy	GE Mcal/kg				
Cattle	DE Mcal/kg	*	1.12	3.47	
Sheep	DE Mcal/kg	*	1.16	3.60	
Swine	DE kcal/kg	*	1106.	3419.	
Cattle	ME Mcal/kg	*	0.92	2.85	
Sheep	ME Mcal/kg	*	0.95	2.95	
Swine	ME kcal/kg	*	1054.	3257.	
Cattle	TDN %	*	25.5	78.8	
Sheep	TDN %		26.4	81.7	
Swine	TDN %	*	25.1	77.6	
Phosphorus	%		0.04	0.12	
Potassium	%		0.33	1.01	

Cassava, roots, dehy grnd, (4)

Cassava meal

Ref No 4 01 152 United States

Feed Name or Analyses			As Fed	Dry	C.V. ± %
Dry matter	%		90.9	100.0	3
Ash	%		2.1	2.3	8
Crude fiber	%		4.5	4.9	30
Ether extract	%		0.7	0.7	37
N-free extract	%		81.1	89.2	
Protein (N x 6.25)	%		2.6	2.9	93
Cattle	dig prot %	*	-1.1	-1.3	
Goats	dig prot %	*	0.0	0.0	
Horses	dig prot %	*	0.0	0.0	
Sheep	dig prot %	*	0.0	0.0	
Energy	GE Mcal/kg				
Cattle	DE Mcal/kg	*	3.18	3.50	
Sheep	DE Mcal/kg	*	3.46	3.81	
Swine	DE kcal/kg	*	3220.	3544.	
Cattle	ME Mcal/kg	*	2.61	2.87	
Sheep	ME Mcal/kg	*	2.84	3.12	
Swine	ME kcal/kg	*	3073.	3382.	
Cattle	TDN %	*	72.2	79.4	
Sheep	TDN %	*	78.5	86.4	
Swine	TDN %	*	73.0	80.4	
Phosphorus	%		0.03	0.03	
Potassium	%		0.24	0.26	

Cassava, seeds, extn unspecified grnd, (4)

Ref No 4 01 151 United States

Feed Name or Analyses			As Fed	Dry	C.V. ± %
Dry matter	%		87.2	100.0	
Ash	%		1.9	2.2	
Crude fiber	%		2.2	2.5	
Sheep	dig coef %		32.	32.	
Ether extract	%		0.5	0.6	
Sheep	dig coef %		110.	110.	
N-free extract	%		81.2	93.1	
Sheep	dig coef %		92.	92.	
Protein (N x 6.25)	%		1.4	1.6	
Sheep	dig coef %		-78.	-78.	
Cattle	dig prot %	*	-2.1	-2.4	
Goats	dig prot %	*	-1.0	-1.2	
Horses	dig prot %	*	-1.0	-1.2	
Sheep	dig prot %		-1.0	-1.2	
Energy	GE Mcal/kg				
Cattle	DE Mcal/kg	*	3.24	3.72	
Sheep	DE Mcal/kg	*	3.33	3.82	
Swine	DE kcal/kg	*	3302.	3787.	
Cattle	ME Mcal/kg	*	2.66	3.05	
Sheep	ME Mcal/kg	*	2.73	3.13	
Swine	ME kcal/kg	*	3160.	3623.	
Cattle	TDN %	*	73.5	84.3	
Sheep	TDN %		75.6	86.7	
Swine	TDN %	*	74.9	85.9	

Cassava, starch by-product, dehy, (4)

Tapioca meal

Manihot meal

Manioc meal

Ref No 4 08 572 United States

Feed Name or Analyses			As Fed	Dry	C.V. ± %
Dry matter	%		86.8	100.0	
Ash	%		1.8	2.1	
Crude fiber	%		4.6	5.3	
Ether extract	%		0.7	0.8	
N-free extract	%		78.8	90.8	
Protein (N x 6.25)	%		0.9	1.0	
Cattle	dig prot %	*	-2.5	-2.9	
Goats	dig prot %	*	-1.5	-1.7	
Horses	dig prot %	*	-1.5	-1.7	
Sheep	dig prot %	*	-1.5	-1.7	
Energy	GE Mcal/kg				
Cattle	DE Mcal/kg	*	3.11	3.58	
Sheep	DE Mcal/kg	*	3.33	3.84	
Swine	DE kcal/kg	*	3037.	3499.	
Cattle	ME Mcal/kg	*	2.55	2.94	
Sheep	ME Mcal/kg	*	2.73	3.15	
Swine	ME kcal/kg	*	2909.	3351.	
Cattle	TDN %	*	70.4	81.1	
Sheep	TDN %	*	75.5	87.0	
Swine	TDN %	*	68.9	79.4	
Phosphorus	%		0.03	0.03	
Potassium	%		0.23	0.27	

CASSAVA, COMMON. Manihot esculenta

Cassava, common, leaves, fresh, (2)

Ref No 2 01 153 United States

Feed Name or Analyses			As Fed	Dry	C.V. ± %
Dry matter	%		20.9	100.0	
Ash	%		2.3	10.8	
Crude fiber	%		2.0	9.6	
Ether extract	%		0.6	3.0	
N-free extract	%		11.1	52.9	
Protein (N x 6.25)	%		5.0	23.7	
Cattle	dig prot %	*	3.8	18.0	
Goats	dig prot %	*	3.9	18.7	
Horses	dig prot %	*	3.7	17.7	
Rabbits	dig prot %	*	3.5	17.0	
Sheep	dig prot %	*	4.0	19.1	
Energy	GE Mcal/kg				
Cattle	DE Mcal/kg	*	0.66	3.18	
Sheep	DE Mcal/kg	*	0.70	3.36	
Cattle	ME Mcal/kg	*	0.55	2.61	
Sheep	ME Mcal/kg	*	0.58	2.75	
Cattle	TDN %	*	15.1	72.2	
Sheep	TDN %	*	15.9	76.2	
Calcium	%		0.08	0.38	
Iron	%		0.003	0.012	
Phosphorus	%		0.08	0.38	
Carotene	mg/kg		80.4	384.9	
Vitamin A equivalent	IU/g		134.1	641.7	

Cassava, common, root flour, fine grnd, (4)

Tapioca flour

Ref No 4 01 154 United States

Feed Name or Analyses			As Fed	Dry	C.V. ± %
Dry matter	%		88.3	100.0	
Ash	%		2.1	2.4	
Crude fiber	%		2.5	2.8	
Swine	dig coef %		76.	76.	
Ether extract	%		0.4	0.5	
Swine	dig coef %		23.	23.	
N-free extract	%		81.0	91.7	
Swine	dig coef %		99.	99.	
Protein (N x 6.25)	%		2.3	2.6	
Swine	dig coef %		68.	68.	
Cattle	dig prot %	*	-1.3	-1.5	
Goats	dig prot %	*	-0.2	-0.3	
Horses	dig prot %	*	-0.2	-0.3	
Sheep	dig prot %	*	-0.2	-0.3	
Swine	dig prot %		1.6	1.8	

Continued

Feed Name or Analyses		Mean As Fed	Dry	C.V. ± %
Energy	GE Mcal/kg			
Cattle	DE Mcal/kg	* 3.21	3.63	
Sheep	DE Mcal/kg	* 3.41	3.87	
Swine	DE kcal/kg	*3696.	4186.	
Cattle	ME Mcal/kg	* 2.63	2.98	
Chickens	ME_n kcal/kg	2970.	3364.	
Sheep	ME Mcal/kg	* 2.80	3.17	
Swine	ME kcal/kg	*3529.	3997.	
Cattle	TDN %	* 72.8	82.4	
Sheep	TDN %	* 77.4	87.7	
Swine	TDN %	83.8	94.9	

Cassava meal –
see Cassava, roots, dehy grnd, (4)

CASTORBEAN. Ricinus communis

Castorbean, seeds with hulls, toxicity extd grnd, (5)

Ref No 5 01 155 United States

Feed Name or Analyses		As Fed	Dry	C.V. ± %
Dry matter	%	86.7	100.0	
Ash	%	7.5	8.6	
Crude fiber	%	35.5	41.0	
Cattle	dig coef %	-1.	-1.	
Ether extract	%	1.0	1.2	
Cattle	dig coef %	82.	82.	
N-free extract	%	16.6	19.2	
Cattle	dig coef %	10.	10.	
Protein (N x 6.25)	%	26.0	30.0	
Cattle	dig coef %	77.	77.	
Cattle	dig prot %	20.0	23.1	
Energy	GE Mcal/kg			
Cattle	DE Mcal/kg	* 1.01	1.16	
Cattle	ME Mcal/kg	* 0.83	0.96	
Cattle	TDN %	22.9	26.4	
Alanine	%	0.88	1.01	
Arginine	%	2.77	3.20	
Aspartic acid	%	2.14	2.47	
Glutamic acid	%	4.06	4.68	
Glycine	%	0.86	0.99	
Histidine	%	0.47	0.55	
Hydroxyproline	%	0.02	0.03	
Isoleucine	%	1.01	1.17	
Leucine	%	1.40	1.61	
Lysine	%	0.77	0.88	
Methionine	%	0.36	0.42	
Phenylalanine	%	0.81	0.94	
Proline	%	0.81	0.94	
Serine	%	1.13	1.30	
Threonine	%	0.72	0.83	
Tyrosine	%	0.68	0.78	
Valine	%	1.26	1.46	

CATTAIL. Typha spp

Cattail, hay, s-c, (1)

Ref No 1 01 156 United States

Feed Name or Analyses		As Fed	Dry	C.V. ± %
Dry matter	%	90.8	100.0	
Ash	%	8.2	9.0	
Crude fiber	%	30.8	33.9	
Ether extract	%	1.7	1.9	
N-free extract	%	44.3	48.8	
Protein (N x 6.25)	%	5.8	6.4	
Cattle	dig prot %	* 2.3	2.5	
Goats	dig prot %	* 2.3	2.5	
Horses	dig prot %	* 2.7	3.0	
Rabbits	dig prot %	* 3.3	3.6	
Sheep	dig prot %	* 2.1	2.3	

CATTAIL, NARROWLEAF. Typha angustifolia

Cattail, narrowleaf, hay, s-c, (1)

Ref No 1 08 372 United States

Feed Name or Analyses		As Fed	Dry	C.V. ± %
Dry matter	%	90.8	100.0	
Ash	%	8.2	9.0	
Crude fiber	%	30.8	33.9	
Ether extract	%	1.7	1.9	
N-free extract	%	44.3	48.8	
Protein (N x 6.25)	%	5.8	6.4	
Cattle	dig prot %	* 2.2	2.5	
Goats	dig prot %	* 2.3	2.5	
Horses	dig prot %	* 2.7	3.0	
Rabbits	dig prot %	* 3.3	3.6	
Sheep	dig prot %	* 2.1	2.3	
Energy	GE Mcal/kg			
Cattle	DE Mcal/kg	* 1.93	2.13	
Sheep	DE Mcal/kg	* 1.78	1.96	
Cattle	ME Mcal/kg	* 1.59	1.75	
Sheep	ME Mcal/kg	* 1.46	1.61	
Cattle	TDN %	* 43.8	48.2	
Sheep	TDN %	* 40.3	44.4	

Cattail millet grain –
see Millet, pearl, grain, (4)

CATTLE. Bos spp

Cattle, lactose, crude, (4)
Lactose

Ref No 4 07 881 United States

Feed Name or Analyses		As Fed	Dry	C.V. ± %
Dry matter	%	92.2	100.0	
Ash	%	0.1	0.1	
Energy	GE Mcal/kg	3.67	3.98	
Swine	DE kcal/kg	3384.	3670.	
Swine	ME kcal/kg	3301.	3580.	

Cattle, tallow, raw, (4)

Ref No 4 08 127 United States

Feed Name or Analyses		As Fed	Dry	C.V. ± %
Dry matter	%	96.0	100.0	
Ash	%	0.1	0.1	
Ether extract	%	94.0	97.9	
Protein (N x 6.25)	%	1.5	1.6	
Cattle	dig prot %	* -2.3	-2.4	
Goats	dig prot %	* -1.2	-1.2	
Horses	dig prot %	* -1.2	-1.2	
Sheep	dig prot %	* -1.2	-1.2	
Energy	GE Mcal/kg	8.54	8.90	

Cattle, whey, (4)

Ref No 4 08 134 United States

Feed Name or Analyses		As Fed	Dry	C.V. ± %
Dry matter	%	6.9	100.0	
Ash	%	0.7	9.4	
Crude fiber	%	0.0	0.0	
Ether extract	%	0.3	4.3	
N-free extract	%	5.0	73.2	
Protein (N x 6.25)	%	0.9	13.0	
Cattle	dig prot %	* 0.6	8.0	
Goats	dig prot %	* 0.6	9.2	
Horses	dig prot %	* 0.6	9.2	
Sheep	dig prot %	* 0.6	9.2	
Energy	GE Mcal/kg	0.26	3.77	
Cattle	DE Mcal/kg	* 0.27	3.85	
Sheep	DE Mcal/kg	* 0.25	3.61	
Swine	DE kcal/kg	* 243.	3518.	
Cattle	ME Mcal/kg	* 0.22	3.16	
Sheep	ME Mcal/kg	* 0.20	2.96	
Swine	ME kcal/kg	* 227.	3284.	
Cattle	TDN %	* 6.0	87.3	
Sheep	TDN %	* 5.7	81.9	
Swine	TDN %	* 5.5	79.8	
Calcium	%	0.05	0.72	
Iron	%	0.001	0.015	
Phosphorus	%	0.04	0.65	
Potassium	%	0.19	2.75	
Niacin	mg/kg	1.0	14.5	
Riboflavin	mg/kg	1.4	20.3	
Thiamine	mg/kg	0.3	4.3	
Vitamin A	IU/g	0.1	1.4	

Cattle, whey, condensed, mn solids declared, (4)
Condensed whey (AAFCO)
Whey, condensed
Whey, evaporated
Whey, semisolid

Ref No 4 01 180 United States

Feed Name or Analyses		As Fed	Dry	C.V. ± %
Dry matter	%	63.6	100.0	28
Ash	%	6.4	10.0	11
Crude fiber	%	0.2	0.4	98
Ether extract	%	0.6	0.9	71
N-free extract	%	47.7	75.0	
Protein (N x 6.25)	%	8.7	13.7	15
Cattle	dig prot %	* 5.5	8.6	
Goats	dig prot %	* 6.2	9.8	
Horses	dig prot %	* 6.2	9.8	
Sheep	dig prot %	* 6.2	9.8	
Energy	GE Mcal/kg			
Cattle	DE Mcal/kg	* 2.31	3.63	
Sheep	DE Mcal/kg	* 2.24	3.52	
Swine	DE kcal/kg	*2152.	3382.	
Cattle	ME Mcal/kg	* 1.90	2.98	
Sheep	ME Mcal/kg	* 1.84	2.89	
Swine	ME kcal/kg	*2006.	3153.	
Cattle	TDN %	* 52.3	82.3	
Sheep	TDN %	* 50.8	79.9	
Swine	TDN %	* 48.8	76.7	
Calcium	%	0.38	0.60	10
Phosphorus	%	0.58	0.91	35
Pantothenic acid	mg/kg	14.6	23.0	
Riboflavin	mg/kg	16.0	25.2	
Thiamine	mg/kg	3.3	5.2	

Cattle, whey, dehy, mn 65 lactose, (4)
Dried whey (AAFCO)
Whey, dried

Ref No 4 01 182 United States

Feed Name or Analyses		As Fed	Dry	C.V. ± %
Dry matter	%	93.2	100.0	2
Ash	%	8.4	9.0	13
Crude fiber	%	0.2	0.2	48
Ether extract	%	1.1	1.2	98
N-free extract	%	68.6	73.6	8

(1) dry forages and roughages
(2) pasture, range plants, and forages fed green
(3) silages
(4) energy feeds
(5) protein supplements
(6) minerals
(7) vitamins
(8) additives

Feed Name or Analyses		Mean As Fed	Mean Dry	C.V. ± %
Protein (N x 6.25)	%	14.9	16.0	34
Cattle	dig prot %	* 10.0	10.7	
Goats	dig prot %	* 11.1	11.9	
Horses	dig prot %	* 11.1	11.9	
Sheep	dig prot %	* 11.1	11.9	
Linoleic	%	.010	.011	
Energy	GE Mcal/kg	3.46	3.71	
Cattle	DE Mcal/kg	* 34.3	36.8	
Sheep	DE Mcal/kg	* 3.28	3.52	
Swine	DE kcal/kg	3085.	3310.	
Cattle	ME Mcal/kg	* 2.81	3.02	
Chickens	ME_n kcal/kg	1882.	2020.	
Sheep	ME Mcal/kg	* 2.69	2.89	
Swine	ME kcal/kg	2973.	3190.	
Cattle	$NE_{lactating\ cows}$ Mcal/kg	* 2.14	2.30	
Cattle	TDN %	* 77.8	83.5	
Sheep	TDN %	* 74.4	79.8	
Swine	TDN %	* 75.0	80.4	
Calcium	%	0.91	0.98	28
Cobalt	mg/kg	0.108	0.116	48
Copper	mg/kg	47.9	51.4	36
Iron	%	0.013	0.014	69
Magnesium	%	0.13	0.14	81
Manganese	mg/kg	4.3	4.6	97
Phosphorus	%	0.76	0.81	20
Sulphur	%	1.04	1.12	
Biotin	mg/kg	0.37	0.40	38
Choline	mg/kg	1684.	1806.	28
Folic acid	mg/kg	0.88	0.94	81
Niacin	mg/kg	9.6	10.3	54
Pantothenic acid	mg/kg	44.0	47.2	29
Riboflavin	mg/kg	26.8	28.7	48
Thiamine	mg/kg	4.3	4.6	21
Vitamin B_6	mg/kg	3.96	4.24	25
Vitamin B_{12}	μg/kg	16.7	17.9	64
Vitamin A	IU/g	0.5	0.5	
Arginine	%	0.33	0.36	20
Cysteine	%	0.40	0.43	
Cystine	%	0.30	0.32	
Glutamic acid	%	1.73	1.86	6
Histidine	%	0.20	0.21	
Isoleucine	%	0.89	0.96	12
Leucine	%	1.24	1.33	30
Lysine	%	1.00	1.07	26
Methionine	%	0.20	0.21	31
Phenylalanine	%	0.35	0.37	11
Serine	%	0.30	0.32	
Threonine	%	1.28	1.38	9
Tryptophan	%	0.17	0.18	39
Tyrosine	%	0.20	0.21	56
Valine	%	0.69	0.74	20

Ref No 4 01 182 Canada

Feed Name or Analyses		Mean As Fed	Mean Dry	C.V. ± %
Dry matter	%	92.8	100.0	1
Crude fiber	%	0.1	0.1	
Ether extract	%	1.8	1.9	
Protein (N x 6.25)	%	13.7	14.8	18
Cattle	dig prot %	* 8.9	9.6	
Goats	dig prot %	* 10.0	10.8	
Horses	dig prot %	* 10.0	10.8	
Sheep	dig prot %	* 10.0	10.8	
Energy	GE Mcal/kg	3.44	3.71	
Swine	DE kcal/kg	3071.	3310.	
Chickens	ME_n kcal/kg	1848.	1991.	
Swine	ME kcal/kg	2960.	3190.	
Calcium	%	1.18	1.27	
Phosphorus	%	0.79	0.85	
Potassium	%	0.90	0.97	
Salt (NaCl)	%	7.51	8.09	

Feed Name or Analyses		Mean As Fed	Mean Dry	C.V. ± %
Sodium	%	2.79	3.01	
Pantothenic acid	mg/kg	45.6	49.2	19
Riboflavin	mg/kg	26.4	28.5	10

Cattle, whey, skimmed, (4)

Ref No 4 08 373 United States

Feed Name or Analyses		Mean As Fed	Mean Dry	C.V. ± %
Dry matter	%	6.6	100.0	
Ash	%	0.7	10.6	
Crude fiber	%	0.0	0.0	
Ether extract	%	0.0	0.5	
N-free extract	%	5.0	75.3	
Protein (N x 6.25)	%	0.9	13.6	
Cattle	dig prot %	* 0.6	8.5	
Goats	dig prot %	* 0.6	9.7	
Horses	dig prot %	* 0.6	9.7	
Sheep	dig prot %	* 0.6	9.7	
Energy	GE Mcal/kg			
Cattle	DE Mcal/kg	* 0.26	3.95	
Sheep	DE Mcal/kg	* 0.26	3.95	
Swine	DE kcal/kg	* 250.	2792.	
Cattle	ME Mcal/kg	* 0.21	3.24	
Sheep	ME Mcal/kg	* 0.21	3.24	
Swine	ME kcal/kg	* 234.	3538.	
Cattle	TDN %	* 5.9	89.5	
Sheep	TDN %	* 5.9	89.6	
Swine	TDN %	* 5.7	86.0	

Cattle, whey low lactose, condensed, mn solids declared, (4)

Condensed whey-product (AAFCO)

Whey-product, condensed

Whey-product, evaporated

Ref No 4 01 185 United States

Feed Name or Analyses		Mean As Fed	Mean Dry	C.V. ± %
Dry matter	%	54.8	100.0	10
Ash	%	5.6	10.2	20
Crude fiber	%	0.0	0.0	20
Ether extract	%	0.7	1.3	31
N-free extract	%	40.2	73.4	
Protein (N x 6.25)	%	8.3	15.1	20
Cattle	dig prot %	* 5.4	9.9	
Goats	dig prot %	* 6.1	11.1	
Horses	dig prot %	* 6.1	11.1	
Sheep	dig prot %	* 6.1	11.1	
Energy	GE Mcal/kg			
Cattle	DE Mcal/kg	* 2.04	3.72	
Sheep	DE Mcal/kg	* 2.16	3.95	
Swine	DE kcal/kg	*2119.	3867.	
Cattle	ME Mcal/kg	* 1.67	3.05	
Sheep	ME Mcal/kg	* 1.78	3.24	
Swine	ME kcal/kg	*1971.	3596.	
Cattle	TDN %	* 46.3	84.5	
Sheep	TDN %	* 49.1	89.6	
Swine	TDN %	* 48.1	87.7	
Calcium	%	0.52	0.95	25
Phosphorus	%	0.79	1.44	36
Biotin	mg/kg	0.29	0.52	7
Carotene	mg/kg	1.1	2.0	
Niacin	mg/kg	3.5	6.4	12
Pantothenic acid	mg/kg	14.6	26.6	
Riboflavin	mg/kg	13.9	25.3	29
Thiamine	mg/kg	3.1	5.6	3
Vitamin B_6	mg/kg	1.76	3.22	6
Vitamin A equivalent	IU/g	1.8	3.4	

Cattle, whey low lactose, dehy, mn lactose declared, (4)

Dried whey-product (AAFCO)

Whey product, dried

Ref No 4 01 186 United States

Feed Name or Analyses		Mean As Fed	Mean Dry	C.V. ± %
Dry matter	%	91.0	100.0	2
Ash	%	14.3	15.7	10
Crude fiber	%	0.2	0.3	56
Ether extract	%	1.2	1.4	68
N-free extract	%	59.5	65.4	
Protein (N x 6.25)	%	15.7	17.3	10
Cattle	dig prot %	* 10.8	11.9	
Goats	dig prot %	* 11.9	13.0	
Horses	dig prot %	* 11.9	13.0	
Sheep	dig prot %	* 11.9	13.0	
Linoleic	%	.010	.011	
Energy	GE Mcal/kg			
Cattle	DE Mcal/kg	* 2.82	3.10	
Sheep	DE Mcal/kg	* 3.01	3.31	
Swine	DE kcal/kg	*3001.	3298.	
Cattle	ME Mcal/kg	* 2.31	2.54	
Chickens	ME_n kcal/kg	2020.	2219.	
Sheep	ME Mcal/kg	* 2.47	2.71	
Swine	ME kcal/kg	*2776.	3051.	
Cattle	TDN %	*	70.4	
Sheep	TDN %	* 68.2	75.0	
Swine	TDN %	*	74.8	
Calcium	%	1.55	1.70	28
Phosphorus	%	0.99	1.08	22
Choline	mg/kg	1769.	1945.	
Niacin	mg/kg	59.9	65.9	
Pantothenic acid	mg/kg	69.0	75.8	
Riboflavin	mg/kg	55.5	61.0	
Arginine	%	0.50	0.55	
Cystine	%	0.35	0.39	
Lysine	%	1.30	1.43	
Methionine	%	0.22	0.24	
Tryptophan	%	0.22	0.24	

Cattle, whey wo albumin low lactose, condensed, (4)

Condensed whey solubles (AAFCO)

Whey solubles, condensed

Whey solubles, evaporated

Ref No 4 01 188 United States

Feed Name or Analyses		Mean As Fed	Mean Dry	C.V. ± %
Dry matter	%	51.3	100.0	7
Ash	%	12.4	24.2	9
Protein (N x 6.25)	%	7.2	14.0	9
Cattle	dig prot %	* 4.6	8.9	
Goats	dig prot %	* 5.2	10.1	
Horses	dig prot %	* 5.2	10.1	
Sheep	dig prot %	* 5.2	10.1	

Cattle, whey wo albumin low lactose, dehy, (4)

Dried whey solubles (AAFCO)

Whey solubles, dried

Ref No 4 01 189 United States

Feed Name or Analyses		Mean As Fed	Mean Dry	C.V. ± %
Dry matter	%	96.3	100.0	
Ash	%	13.7	14.2	
Crude fiber	%	0.0	0.0	
Ether extract	%	1.9	2.0	
N-free extract	%	63.6	66.0	

Continued

Feed Name or Analyses			Mean As Fed	Mean Dry	C.V. ± %
Protein (N x 6.25)	%		17.1	17.8	
Cattle	dig prot %	*	11.9	12.3	
Goats	dig prot %	*	13.0	13.5	
Horses	dig prot %	*	13.0	13.5	
Sheep	dig prot %	*	13.0	13.5	
Energy	GE Mcal/kg				
Cattle	DE Mcal/kg	*	3.31	3.44	
Sheep	DE Mcal/kg	*	3.23	3.35	
Swine	DE kcal/kg	*	3371.	3501.	
Cattle	ME Mcal/kg	*	2.72	2.82	
Sheep	ME Mcal/kg	*	2.65	2.75	
Swine	ME kcal/kg	*	3115.	3235.	
Cattle	TDN %	*	75.2	78.1	
Sheep	TDN %	*	73.3	76.1	
Swine	TDN %	*	76.5	79.4	
Arginine	%		1.00	1.04	
Histidine	%		0.10	0.10	
Isoleucine	%		0.30	0.31	
Leucine	%		0.20	0.21	
Methionine	%		0.10	0.10	
Phenylalanine	%		0.10	0.10	
Threonine	%		0.50	0.52	
Tryptophan	%		0.10	0.10	
Tyrosine	%		0.20	0.21	
Valine	%		0.30	0.31	

Cattle, buttermilk, condensed, mn 27% total solids w mn 0.055% fat mx 0.14% ash per 1% solids, (5)

Condensed buttermilk (AAFCO)
Buttermilk, concentrated
Buttermilk, condensed
Buttermilk, evaporated

Ref No 5 01 159 United States

Feed Name or Analyses			Mean As Fed	Mean Dry	C.V. ± %
Dry matter	%		29.2	100.0	12
Ash	%		3.6	12.5	19
Crude fiber	%		0.1	0.2	98
Sheep	dig coef %	*	1.	1.	
Ether extract	%		2.4	8.1	39
Sheep	dig coef %	*	98.	98.	
N-free extract	%		12.3	42.4	
Sheep	dig coef %	*	94.	94.	
Protein (N x 6.25)	%		10.8	36.9	8
Sheep	dig coef %	*	90.	90.	
Sheep	dig prot %		9.7	33.2	
Energy	GE Mcal/kg				
Cattle	DE Mcal/kg	*	1.09	3.74	
Sheep	DE Mcal/kg	*	1.17	4.00	
Swine	DE kcal/kg	*	1225.	4202.	
Cattle	ME Mcal/kg	*	0.89	3.07	
Sheep	ME Mcal/kg	*	0.96	3.28	
Swine	ME kcal/kg	*	1085.	3721.	
Cattle	TDN %	*	24.7	84.9	
Sheep	TDN %		26.5	90.8	
Swine	TDN %	*	27.8	95.3	
Calcium	%		0.44	1.51	
Chlorine	%		0.12	0.41	
Magnesium	%		0.19	0.65	
Phosphorus	%		0.26	0.89	
Potassium	%		0.23	0.79	
Sodium	%		0.31	1.06	
Sulphur	%		0.03	0.10	
Riboflavin	mg/kg		14.5	49.8	17

Cattle, buttermilk, cultured, (5)

Ref No 5 08 188 United States

Feed Name or Analyses			Mean As Fed	Mean Dry	C.V. ± %
Dry matter	%		9.5	100.0	
Ash	%		0.8	7.9	
Crude fiber	%		0.0	0.0	
Ether extract	%		0.4	3.7	
N-free extract	%		4.8	50.8	
Protein (N x 6.25)	%		3.5	37.6	
Energy	GE Mcal/kg		0.36	3.79	
Cattle	DE Mcal/kg	*	0.41	4.27	
Sheep	DE Mcal/kg	*	0.37	3.95	
Swine	DE kcal/kg	*	387.	4093.	
Cattle	ME Mcal/kg	*	0.33	3.50	
Sheep	ME Mcal/kg	*	0.31	3.24	
Swine	ME kcal/kg	*	342.	3618.	
Cattle	TDN %	*	9.2	96.8	
Sheep	TDN %	*	8.5	89.6	
Swine	TDN %	*	8.8	92.8	
Calcium	%		0.13	1.38	
Chlorine	%		0.04	0.43	
Magnesium	%		0.06	0.64	
Phosphorus	%		0.09	0.95	
Potassium	%		0.10	1.11	
Sodium	%		0.11	1.22	
Sulphur	%		0.01	0.11	
Ascorbic acid	mg/kg		9.9	105.3	
Niacin	mg/kg		1.0	10.5	
Riboflavin	mg/kg		1.8	18.9	
Thiamine	mg/kg		0.4	4.2	

Cattle, buttermilk, dehy, feed gr mx 8% moisture mx 13% ash mn 5% fat, (5)

Dried buttermilk, feed grade (AAFCO)
Buttermilk, dried

Ref No 5 01 160 United States

Feed Name or Analyses			Mean As Fed	Mean Dry	C.V. ± %
Dry matter	%		93.0	100.0	2
Ash	%		9.0	9.6	14
Crude fiber	%		0.4	0.4	52
Sheep	dig coef %		1.	1.	
Ether extract	%		5.2	5.6	32
Sheep	dig coef %		98.	98.	
Swine	dig coef %		-29.	-29.	
N-free extract	%		46.7	50.3	5
Sheep	dig coef %		94.	94.	
Swine	dig coef %		98.	98.	
Protein (N x 6.25)	%		31.8	34.2	4
Sheep	dig coef %		90.	90.	
Swine	dig coef %		93.	93.	
Sheep	dig prot %		28.6	30.7	
Swine	dig prot %		29.5	31.8	
Energy	GE Mcal/kg		3.70	3.98	
Cattle	DE Mcal/kg	*	3.51	3.77	
Sheep	DE Mcal/kg	*	3.70	3.98	
Swine	DE kcal/kg	*	3410.	3668.	
Cattle	ME Mcal/kg	*	2.88	3.09	
Chickens	ME_n kcal/kg		2754.	2962.	
Sheep	ME Mcal/kg	*	3.03	3.26	
Swine	ME kcal/kg	*	3039.	3269.	
Cattle	$NE_{lactating\ cows}$ Mcal/kg	*	2.23	2.40	
Cattle	TDN %	*	79.6	85.6	
Sheep	TDN %		83.9	90.3	
Swine	TDN %		77.3	83.2	
Calcium	%		1.33	1.43	25
Chlorine	%		0.36	0.38	
Iron	%		0.001	0.001	
Magnesium	%		0.48	0.52	19
Manganese	mg/kg		3.6	3.8	71
Phosphorus	%		0.95	1.02	8
Potassium	%		0.99	1.06	
Sodium	%		0.80	0.86	
Sulphur	%		0.08	0.09	
Biotin	mg/kg		0.29	0.31	24
Choline	mg/kg		1822.	1960.	32
Folic acid	mg/kg		0.40	0.43	
Niacin	mg/kg		8.6	9.3	51
Pantothenic acid	mg/kg		30.4	32.7	32
Riboflavin	mg/kg		26.3	28.3	34
Thiamine	mg/kg		3.0	3.2	19
Vitamin B_6	mg/kg		2.44	2.62	21
Vitamin B_{12}	μg/kg		18.4	19.8	
Vitamin A	IU/g		2.1	2.3	
Arginine	%		1.07	1.15	
Cystine	%		0.40	0.43	
Glutamic acid	%		7.34	7.89	14
Glycine	%		0.20	0.22	
Histidine	%		0.80	0.86	
Isoleucine	%		2.16	2.32	
Leucine	%		3.11	3.34	
Lysine	%		2.20	2.36	
Methionine	%		0.72	0.77	
Phenylalanine	%		1.42	1.53	
Serine	%		1.41	1.51	
Threonine	%		1.44	1.55	
Tryptophan	%		0.46	0.50	
Tyrosine	%		1.01	1.08	
Valine	%		2.39	2.57	

Ref No 5 01 160 Canada

Feed Name or Analyses			Mean As Fed	Mean Dry	C.V. ± %
Dry matter	%		92.1	100.0	1
Crude fiber	%				
Sheep	dig coef %		1.	1.	
Ether extract	%		4.1	4.5	80
Sheep	dig coef %		98.	98.	
Swine	dig coef %		-29.	-29.	
N-free extract	%				
Sheep	dig coef %		94.	94.	
Swine	dig coef %		98.	98.	
Protein (N x 6.25)	%		31.5	34.2	5
Sheep	dig coef %		90.	90.	
Swine	dig coef %		93.	93.	
Sheep	dig prot %		28.3	30.8	
Swine	dig prot %		29.3	31.8	
Energy	GE Mcal/kg		3.67	3.98	
Sheep	DE Mcal/kg	*	3.63	3.95	
Swine	DE kcal/kg	*	3379.	3668.	
Chickens	ME_n kcal/kg		2753.	2989.	
Sheep	ME Mcal/kg	*	2.98	3.24	
Swine	ME kcal/kg	*	3010.	3268.	
Sheep	TDN %		82.4	89.5	
Swine	TDN %		76.6	83.2	
Pantothenic acid	mg/kg		40.0	43.5	10
Riboflavin	mg/kg		33.0	35.8	17

Cattle, carcass residue, dry rendered dehy grnd, (5)

Meat, cattle, meal
Beef scrap

Ref No 5 01 161 United States

Feed Name or Analyses			Mean As Fed	Mean Dry	C.V. ± %
Dry matter	%		92.7	100.0	2
Ash	%		22.4	24.2	23
Crude fiber	%		1.9	2.0	89
Ether extract	%		10.7	11.5	30
Sheep	dig coef %		97.	97.	
N-free extract	%		0.4	0.5	

(1) dry forages and roughages (2) pasture, range plants, and forages fed green (3) silages (4) energy feeds (5) protein supplements (6) minerals (7) vitamins (8) additives

Feed Name or Analyses		Mean As Fed	Mean Dry	C.V. ± %
Protein (N x 6.25)	%	57.3	61.8	5
Sheep	dig coef %	82.	82.	
Sheep	dig prot %	47.0	50.7	
Energy	GE Mcal/kg			
Cattle	DE Mcal/kg	* 2.87	3.09	
Sheep	DE Mcal/kg	* 3.20	3.46	
Swine	DE kcal/kg	*3056.	3296.	
Cattle	ME Mcal/kg	* 2.35	2.54	
Sheep	ME Mcal/kg	* 2.63	2.83	
Swine	ME kcal/kg	*2552.	2753.	
Cattle	TDN %	* 65.1	70.2	
Sheep	TDN %	72.7	78.4	
Swine	TDN %	* 69.3	74.8	
Calcium	%	5.41	5.84	32
Cobalt	mg/kg	0.119	0.129	
Copper	mg/kg	21.7	23.4	
Iron	%	0.120	0.130	
Manganese	mg/kg	9.5	10.3	91
Phosphorus	%	2.83	3.06	16
Niacin	mg/kg	32.8	35.3	
Pantothenic acid	mg/kg	8.9	9.6	
Riboflavin	mg/kg	7.8	8.4	45

Cattle, casein, milk acid-precipitated dehy, mn 80% protein, (5)
Casein (AAFCO)
Casein, dried
Ref No 5 01 162 United States

Feed Name or Analyses		Mean As Fed	Mean Dry	C.V. ± %
Dry matter	%	90.7	100.0	2
Ash	%	2.9	3.2	34
Crude fiber	%	0.2	0.2	21
Ether extract	%	0.8	0.9	48
Sheep	dig coef %	55.	55.	
Swine	dig coef %	* -29.	-29.	
N-free extract	%	5.2	5.7	
Sheep	dig coef %	111.	111.	
Swine	dig coef %	* 98.	98.	
Protein (N x 6.25)	%	81.6	90.0	4
Sheep	dig coef %	97.	97.	
Swine	dig coef %	* 93.	93.	
Sheep	dig prot %	79.1	87.3	
Swine	dig prot %	75.9	83.7	
Energy	GE Mcal/kg			
Cattle	DE Mcal/kg	* 3.24	3.57	
Sheep	DE Mcal/kg	* 3.83	4.22	
Swine	DE kcal/kg	*3479.	3836.	
Cattle	ME Mcal/kg	* 2.66	2.93	
Chickens	ME_n kcal/kg	4051.	4466.	
Sheep	ME Mcal/kg	* 3.14	3.46	
Swine	ME kcal/kg	*2706.	2984.	
Cattle	TDN %	* 73.5	81.0	
Sheep	TDN %	86.8	95.7	
Swine	TDN %	* 78.9	87.0	
Calcium	%	0.61	0.67	25
Manganese	mg/kg	4.4	4.9	11
Phosphorus	%	0.99	1.10	29
Biotin	mg/kg	0.04	0.05	80
Choline	mg/kg	210.	232.	
Folic acid	mg/kg	0.51	0.56	98
Niacin	mg/kg	1.3	1.5	98
Pantothenic acid	mg/kg	2.7	2.9	88
Riboflavin	mg/kg	1.5	1.7	98
Thiamine	mg/kg	0.4	0.5	98
Vitamin B_6	mg/kg	0.44	0.49	53
Alanine	%	2.31	2.54	33
Arginine	%	3.41	3.76	6
Aspartic acid	%	6.02	6.64	12
Cysteine	%	0.50	0.55	
Cystine	%	0.30	0.33	
Glutamic acid	%	19.45	21.46	6
Glycine	%	1.50	1.66	67
Histidine	%	2.51	2.77	10
Isoleucine	%	5.72	6.31	14
Leucine	%	8.62	9.51	13
Lysine	%	7.02	7.74	9
Methionine	%	2.71	2.99	8
Phenylalanine	%	4.61	5.09	7
Proline	%	9.33	10.29	17
Serine	%	5.21	5.75	19
Threonine	%	3.81	4.20	10
Tryptophan	%	1.00	1.11	11
Tyrosine	%	4.71	5.20	14
Valine	%	6.82	7.52	8

Ref No 5 01 162 Canada

Feed Name or Analyses		Mean As Fed	Mean Dry	C.V. ± %
Dry matter	%		100.0	
Ether extract	%			
Sheep	dig coef %		55.	
N-free extract	%			
Sheep	dig coef %		111.	
Protein (N x 6.25)	%		88.4	
Sheep	dig coef %		97.	
Sheep	dig prot %		85.8	
Energy	GE Mcal/kg			
Sheep	DE Mcal/kg	*	4.22	
Sheep	ME Mcal/kg	*	3.46	
Sheep	TDN %		95.7	
Calcium	%		0.02	
Copper	mg/kg		4.4	
Phosphorus	%		0.57	

Cattle, cheese, cottage wet, (5)
Cottage cheese
Ref No 5 08 001 United States

Feed Name or Analyses		Mean As Fed	Mean Dry	C.V. ± %
Dry matter	%	21.0	100.0	
Ash	%	1.0	4.8	
Ether extract	%	0.3	1.4	
N-free extract	%	2.7	12.9	
Protein (N x 6.25)	%	17.0	81.0	
Energy	GE Mcal/kg	0.86	4.10	
Cattle	DE Mcal/kg	* 0.75	3.58	
Sheep	DE Mcal/kg	* 0.72	3.45	
Swine	DE kcal/kg	* 656.	3126.	
Cattle	ME Mcal/kg	* 0.62	2.94	
Sheep	ME Mcal/kg	* 0.59	2.83	
Swine	ME kcal/kg	* 523.	2490.	
Cattle	TDN %	* 17.1	81.2	
Sheep	TDN %	* 16.4	78.1	
Swine	TDN %	* 14.9	70.9	
Calcium	%	0.09	0.43	
Iron	%	0.000	0.000	
Phosphorus	%	0.18	0.86	
Potassium	%	0.07	0.33	
Sodium	%	0.29	1.38	
Niacin	mg/kg	1.0	4.8	
Riboflavin	mg/kg	2.8	13.3	
Thiamine	mg/kg	0.3	1.4	
Vitamin A	IU/g	0.1	0.5	

Cattle, cheese rind, (5)
Cheese rind (AAFCO)
Ref No 5 01 163 United States

Feed Name or Analyses		Mean As Fed	Mean Dry	C.V. ± %
Dry matter	%	82.8	100.0	8
Ash	%	7.5	9.0	23
Crude fiber	%	0.2	0.3	41
Ether extract	%	19.3	23.3	29
N-free extract	%	11.4	13.7	
Protein (N x 6.25)	%	44.4	53.7	20
Energy	GE Mcal/kg			
Cattle	DE Mcal/kg	* 3.38	4.08	
Sheep	DE Mcal/kg	* 3.29	3.98	
Swine	DE kcal/kg	*3457.	4175.	
Cattle	ME Mcal/kg	* 2.77	3.35	
Sheep	ME Mcal/kg	* 2.70	3.26	
Swine	ME kcal/kg	*2945.	3557.	
Cattle	TDN %	* 76.7	92.6	
Sheep	TDN %	* 74.6	90.2	
Swine	TDN %	* 78.4	94.7	
Calcium	%	0.96	1.15	23
Chlorine	%	0.59	0.71	14
Magnesium	%	0.02	0.03	16
Phosphorus	%	0.54	0.66	16
Potassium	%	0.27	0.32	98
Sodium	%	0.79	0.95	21

Cattle, hearts, raw, (5)
Ref No 5 01 164 United States

Feed Name or Analyses		Mean As Fed	Mean Dry	C.V. ± %
Dry matter	%	23.7	100.0	32
Ash	%	1.1	4.6	30
Crude fiber	%	0.0	0.0	30
Ether extract	%	8.5	36.0	
N-free extract	%	-3.4	-14.7	
Protein (N x 6.25)	%	17.6	74.3	25
Energy	GE Mcal/kg	1.38	5.82	
Cattle	DE Mcal/kg	* 0.93	3.94	
Sheep	DE Mcal/kg	* 0.97	4.09	
Swine	DE kcal/kg	* 989.	4175.	
Cattle	ME Mcal/kg	* 0.77	3.23	
Sheep	ME Mcal/kg	* 0.79	3.35	
Swine	ME kcal/kg	* 802.	3382.	
Cattle	TDN %	* 21.2	89.3	
Sheep	TDN %	* 22.0	92.8	
Swine	TDN %	* 22.4	94.7	
Calcium	%	0.01	0.04	
Iron	%	0.004	0.018	
Phosphorus	%	0.19	0.81	23
Potassium	%	0.20	0.86	
Sodium	%	0.09	0.38	
Ascorbic acid	mg/kg	21.0	88.9	
Niacin	mg/kg	78.9	333.3	
Riboflavin	mg/kg	9.3	39.1	
Thiamine	mg/kg	5.6	23.6	
Vitamin A	IU/g	0.2	0.9	

Cattle, kidneys, raw, (5)
Ref No 5 01 165 United States

Feed Name or Analyses		Mean As Fed	Mean Dry	C.V. ± %
Dry matter	%	24.1	100.0	
Ash	%	1.1	4.6	
Crude fiber	%	0.0	0.0	
Ether extract	%	6.7	27.8	
N-free extract	%	0.9	3.7	
Protein (N x 6.25)	%	15.4	63.9	

Continued

Feed Name or Analyses		Mean As Fed	Mean Dry	C.V. ± %
Energy	GE Mcal/kg	1.30	5.39	
Cattle	DE Mcal/kg	* 0.95	3.94	
Sheep	DE Mcal/kg	* 0.99	4.11	
Swine	DE kcal/kg	*1948.	4096.	
Cattle	ME Mcal/kg	* 0.78	3.23	
Sheep	ME Mcal/kg	* 0.81	3.37	
Swine	ME kcal/kg	* 820.	3404.	
Cattle	TDN %	* 21.5	89.3	
Sheep	TDN %	* 22.5	93.2	
Swine	TDN %	* 22.4	92.9	
Calcium	%	0.01	0.05	
Iron	%	0.007	0.031	
Phosphorus	%	0.22	0.91	
Potassium	%	0.22	0.93	
Sodium	%	0.18	0.73	
Ascorbic acid	mg/kg	150.0	622.4	
Niacin	mg/kg	64.0	265.6	
Riboflavin	mg/kg	25.5	105.8	
Thiamine	mg/kg	3.6	14.9	
Vitamin A	IU/g	6.9	28.6	

Cattle, livers, raw, (5)
Beef liver
Ref No 5 01 166 — United States

Feed Name or Analyses		Mean As Fed	Mean Dry	C.V. ± %
Dry matter	%	27.2	100.0	
Ash	%	1.3	4.8	
Crude fiber	%	0.0	0.0	
Ether extract	%	3.4	12.5	
N-free extract	%	2.5	9.1	
Protein (N x 6.25)	%	20.0	73.6	
Energy	GE Mcal/kg	1.26	4.62	
Cattle	DE Mcal/kg	* 0.96	3.52	
Sheep	DE Mcal/kg	* 1.01	3.72	
Swine	DE kcal/kg	*1021.	3752.	
Cattle	ME Mcal/kg	* 0.79	2.89	
Sheep	ME Mcal/kg	* 0.83	3.05	
Swine	ME kcal/kg	* 828.	3043.	
Cattle	TDN %	* 21.7	79.8	
Sheep	TDN %	* 22.9	84.4	
Swine	TDN %	* 23.1	85.1	
Calcium	%	0.01	0.04	
Iron	%	0.006	0.022	
Phosphorus	%	0.23	0.85	
Potassium	%	0.25	0.93	
Sodium	%	0.12	0.45	
Ascorbic acid	mg/kg	278.0	1023.1	
Niacin	mg/kg	122.0	448.8	
Riboflavin	mg/kg	29.2	107.6	
Thiamine	mg/kg	2.2	8.3	
Vitamin A	IU/g	393.7	1448.8	

Cattle, lungs, raw, (5)
Ref No 5 07 941 — United States

Feed Name or Analyses		Mean As Fed	Mean Dry	C.V. ± %
Dry matter	%	21.2	100.0	
Ash	%	1.0	4.7	
Ether extract	%	2.3	10.8	
Protein (N x 6.25)	%	17.6	83.0	
Energy	GE Mcal/kg	0.96	4.53	
Phosphorus	%	0.22	1.02	
Niacin	mg/kg	62.0	292.5	

(1) dry forages and roughages (2) pasture, range plants, and forages fed green (3) silages (4) energy feeds (5) protein supplements (6) minerals (7) vitamins (8) additives

Cattle, meat wo fat, raw, (5)
Ref No 5 07 921 — United States

Feed Name or Analyses		Mean As Fed	Mean Dry	C.V. ± %
Dry matter	%	23.3	100.0	
Ash	%	1.1	4.7	
Protein (N x 6.25)	%	21.5	92.4	
Calcium	%	0.01	0.05	
Phosphorus	%	0.23	0.99	

Cattle, milk, dehy, feed gr mx 8% moisture mn 26% fat, (5)
Dried whole milk, feed grade (AAFCO)
Milk, whole, dried
Ref No 5 01 167 — United States

Feed Name or Analyses		Mean As Fed	Mean Dry	C.V. ± %
Dry matter	%	96.2	100.0	1
Ash	%	5.6	5.8	17
Crude fiber	%	0.1	0.1	98
Ether extract	%	26.7	27.8	4
N-free extract	%	38.3	39.8	
Protein (N x 6.25)	%	25.5	26.5	11
Energy	GE Mcal/kg	4.92	5.12	
Cattle	DE Mcal/kg	* 4.85	5.05	
Sheep	DE Mcal/kg	* 4.68	4.86	
Swine	DE kcal/kg	*5884.	6119.	
Cattle	ME Mcal/kg	* 3.98	4.14	
Sheep	ME Mcal/kg	* 3.84	3.99	
Swine	ME kcal/kg	*5334.	5547.	
Cattle	TDN %	* 110.1	114.5	
Sheep	TDN %	* 106.1	110.3	
Swine	TDN %	* 133.5	138.8	
Calcium	%	0.90	0.94	4
Chlorine	%	1.49	1.55	
Copper	mg/kg	0.9	0.9	72
Iron	%	0.009	0.009	98
Manganese	mg/kg	0.5	0.5	
Phosphorus	%	0.72	0.74	7
Potassium	%	1.13	1.18	
Sodium	%	0.38	0.40	
Ascorbic acid	mg/kg	58.9	61.2	
Biotin	mg/kg	0.38	0.40	[illegible]
Carotene	mg/kg	7.2	7.5	
Niacin	mg/kg	7.7	8.0	87
Pantothenic acid	mg/kg	23.3	24.2	22
Riboflavin	mg/kg	17.2	17.9	50
Thiamine	mg/kg	3.3	3.5	46
Vitamin B_6	mg/kg	4.75	4.94	33
Vitamin A	IU/g	11.1	11.5	
Vitamin A equivalent	IU/g	12.1	12.6	
Vitamin D_2	IU/g	0.3	0.4	
Arginine	%	0.92	0.96	
Histidine	%	0.72	0.75	
Isoleucine	%	1.33	1.39	
Leucine	%	2.57	2.67	
Lysine	%	2.26	2.35	
Methionine	%	0.62	0.64	
Phenylalanine	%	1.33	1.39	
Threonine	%	1.03	1.07	
Tryptophan	%	0.41	0.43	
Tyrosine	%	1.33	1.39	
Valine	%	1.74	1.81	

Cattle, milk, fresh, (5)
Milk, cattle, fresh
Ref No 5 01 168 — United States

Feed Name or Analyses		Mean As Fed	Mean Dry	C.V. ± %
Dry matter	%	12.6	100.0	2
Ash	%	0.7	5.8	8
Crude fiber	%	0.0	0.0	8
Ether extract	%	3.6	28.5	2
Cattle	dig coef %	100.	100.	
Swine	dig coef %	96.	96.	
N-free extract	%	4.8	37.9	2
Cattle	dig coef %	99.	99.	
Swine	dig coef %	97.	97.	
Protein (N x 6.25)	%	3.5	27.8	3
Cattle	dig coef %	96.	96.	
Swine	dig coef %	97.	97.	
Cattle	dig prot %	3.3	26.7	
Swine	dig prot %	3.4	26.9	
Energy	GE Mcal/kg	0.65	5.16	
Cattle	DE Mcal/kg	* 0.71	5.66	
Sheep	DE Mcal/kg	* 0.68	5.42	
Swine	DE kcal/kg	* 688.	5476.	
Cattle	ME Mcal/kg	* 0.58	4.64	
Sheep	ME Mcal/kg	* 0.56	4.44	
Swine	ME kcal/kg	* 622.	4950.	
Cattle	NE_m Mcal/kg	* 0.58	4.59	
Cattle	NE_{gain} Mcal/kg	* 0.25	2.01	
Cattle	$NE_{lactating\ cows}$ Mcal/kg	* 0.51	4.01	
Cattle	TDN %	16.1	128.3	
Sheep	TDN %	* 15.5	123.00	
Swine	TDN %	15.6	124.2	
Calcium	%	0.12	0.93	
Chlorine	%	0.20	1.56	
Copper	mg/kg	0.0	0.3	
Phosphorus	%	0.09	0.75	
Potassium	%	0.14	1.11	
Sodium	%	0.05	0.39	
Ascorbic acid	mg/kg	9.9	78.7	
Niacin	mg/kg	1.0	7.9	
Riboflavin	mg/kg	1.7	13.4	
Thiamine	mg/kg	0.3	2.4	
Vitamin A	IU/g	1.4	11.4	

Cattle, milk, skimmed, (5)
Ref No 5 01 169 — United States

Feed Name or Analyses		Mean As Fed	Mean Dry	C.V. ± %
Dry matter	%	9.8	100.0	
Ash	%	0.7	7.4	
Crude fiber	%	0.0	0.0	
Ether extract	%	0.5	4.6	
Sheep	dig coef %	110.	110.	
N-free extract	%	5.0	51.5	
Sheep	dig coef %	101.	101.	
Protein (N x 6.25)	%	3.6	36.6	
Sheep	dig coef %	94.	94.	
Sheep	dig prot %	3.4	34.4	
Energy	GE Mcal/kg			
Cattle	DE Mcal/kg	* 0.38	3.85	
Sheep	DE Mcal/kg	* 0.40	4.07	
Swine	DE kcal/kg	* 412.	4203.	
Cattle	ME Mcal/kg	* 0.31	3.15	
Sheep	ME Mcal/kg	* 0.33	3.34	
Swine	ME kcal/kg	* 365.	3724.	
Cattle	TDN %	* 8.6	87.3	
Sheep	TDN %	9.1	92.4	
Swine	TDN %	* 9.3	95.3	
Calcium	%	0.13	1.29	

Feed Name or Analyses		Mean As Fed	Mean Dry	C.V. ± %
Phosphorus	%	0.10	0.99	
Potassium	%	0.15	1.49	

Cattle, milk, skimmed centrifugal, (5)

Ref No 5 01 170 United States

Feed Name or Analyses		Mean As Fed	Mean Dry	C.V. ± %
Dry matter	%	9.3	100.0	
Ash	%	0.7	7.5	
Crude fiber	%	0.0	0.0	
Cattle	dig coef %	96.	96.	
Ether extract	%	0.1	0.8	
Cattle	dig coef %	38.	38.	
Swine	dig coef %	151.	151.	
N-free extract	%	5.3	56.5	
Cattle	dig coef %	100.	100.	
Swine	dig coef %	102.	102.	
Protein (N x 6.25)	%	3.3	35.1	
Cattle	dig coef %	96.	96.	
Swine	dig coef %	98.	98.	
Cattle	dig prot %	3.1	33.7	
Swine	dig prot %	3.2	34.4	
Energy	GE Mcal/kg			
Cattle	DE Mcal/kg	* 0.37	4.01	
Sheep	DE Mcal/kg	* 0.37	3.96	
Swine	DE kcal/kg	* 404.	4330.	
Cattle	ME Mcal/kg	* 0.31	3.29	
Sheep	ME Mcal/kg	* 0.30	3.25	
Swine	ME kcal/kg	* 359.	3849.	
Cattle	NE_m Mcal/kg	* 0.22	2.32	
Cattle	NE_{gain} Mcal/kg	* 0.14	1.50	
Cattle	$NE_{lactating\ cows}$ Mcal/kg	* 0.25	2.67	
Cattle	TDN %	8.5	91.0	
Sheep	TDN %	* 8.4	89.8	
Swine	TDN %	9.2	98.2	
Calcium	%	0.13	1.37	
Copper	mg/kg	1.1	11.6	
Iron	%	0.000	0.005	
Magnesium	%	0.01	0.11	
Manganese	mg/kg	0.2	2.3	
Phosphorus	%	0.10	1.05	
Potassium	%	0.15	1.58	
Sulphur	%	0.03	0.32	

Cattle, milk, skimmed dehy, mx 8% moisture, (5)

Dried skimmed milk, feed grade (AAFCO)

Milk, skimmed, dried

Ref No 5 01 175 United States

Feed Name or Analyses		Mean As Fed	Mean Dry	C.V. ± %
Dry matter	%	94.3	100.0	2
Ash	%	8.0	8.5	3
Crude fiber	%	0.3	0.3	69
Ether extract	%	1.0	1.1	23
Sheep	dig coef %	102.	102.	
Swine	dig coef %	122.	122.	
N-free extract	%	51.1	54.2	1
Sheep	dig coef %	93.	93.	
Swine	dig coef %	97.	97.	
Protein (N x 6.25)	%	34.0	36.0	2
Sheep	dig coef %	90.	90.	
Swine	dig coef %	98.	98.	
Sheep	dig prot %	30.6	32.4	
Swine	dig prot %	33.3	35.3	
Energy	GE Mcal/kg	3.50	3.72	
Cattle	DE Mcal/kg	* 3.41	3.62	
Sheep	DE Mcal/kg	* 3.57	3.78	
Swine	DE kcal/kg	*3750.	3977.	
Cattle	ME Mcal/kg	* 2.80	2.97	
Chickens	ME_n kcal/kg	2529.	2682.	
Sheep	ME Mcal/kg	* 2.92	3.10	
Swine	ME kcal/kg	*3327.	3528.	
Cattle	TDN %	* 77.4	82.1	
Sheep	TDN %	80.9	85.8	
Swine	TDN %	85.0	90.2	
Calcium	%	1.27	1.35	10
Cobalt	mg/kg	0.111	0.117	74
Copper	mg/kg	11.5	12.2	51
Iron	%	0.003	0.004	78
Magnesium	%	0.12	0.12	
Manganese	mg/kg	2.2	2.3	54
Phosphorus	%	1.03	1.09	9
Potassium	%	1.61	1.71	
Sodium	%	0.52	0.55	
Sulphur	%	0.32	0.34	
Ascorbic acid	mg/kg	68.0	72.2	
Biotin	mg/kg	0.33	0.35	25
Choline	mg/kg	1442.	1530.	35
Folic acid	mg/kg	0.62	0.66	16
Niacin	mg/kg	10.6	11.3	26
Pantothenic acid	mg/kg	34.0	36.1	21
Riboflavin	mg/kg	19.3	20.5	27
Thiamine	mg/kg	3.5	3.7	12
α-tocopherol	mg/kg	9.2	9.7	
Vitamin B_6	mg/kg	3.98	4.23	13
Vitamin B_{12}	μg/kg	42.1	44.6	40
Vitamin A	IU/g	0.3	0.3	
Vitamin D_2	IU/g	0.4	0.4	
Arginine	%	1.16	1.23	16
Cystine	%	0.45	0.48	18
Glutamic acid	%	6.83	7.24	
Glycine	%	0.20	0.21	
Histidine	%	0.90	0.96	19
Isoleucine	%	2.31	2.45	18
Leucine	%	3.31	3.51	4
Lysine	%	2.57	2.73	5
Methionine	%	0.91	0.96	17
Phenylalanine	%	1.51	1.60	15
Serine	%	1.61	1.70	
Threonine	%	1.41	1.49	7
Tryptophan	%	0.43	0.45	15
Tyrosine	%	1.31	1.38	29
Valine	%	2.21	2.34	3

Ref No 5 01 175 Canada

Feed Name or Analyses		Mean As Fed	Mean Dry	C.V. ± %
Dry matter	%	93.9	100.0	0
Ether extract	%			
Sheep	dig coef %	102.	102.	
Swine	dig coef %	122.	122.	
N-free extract	%			
Sheep	dig coef %	93.	93.	
Swine	dig coef %	97.	97.	
Protein (N x 6.25)	%			
Sheep	dig coef %	90.	90.	
Swine	dig coef %	98.	98.	
Energy	GE Mcal/kg	3.49	3.72	
Sheep	DE Mcal/kg	* 3.55	3.78	
Swine	DE kcal/kg	*3735.	3977.	
Chickens	ME_n kcal/kg	2538.	2702.	
Sheep	ME Mcal/kg	* 2.91	3.10	
Sheep	TDN %	80.6	85.8	
Swine	TDN %	84.7	90.2	
Iron	%	.002	.002	
Pantothenic acid	mg/kg	39.0	41.6	13
Riboflavin	mg/kg	18.2	19.4	23

Cattle, spleen, raw, (5)

Cattle, melts, raw

Ref No 5 07 942 United States

Feed Name or Analyses		Mean As Fed	Mean Dry	C.V. ± %
Dry matter	%	23.1	100.0	
Ash	%	1.4	6.1	
Ether extract	%	3.0	13.0	
Protein (N x 6.25)	%	18.1	78.4	
Energy	GE Mcal/kg	1.04	4.50	
Iron	%	0.011	0.048	
Phosphorus	%	0.27	1.17	
Niacin	mg/kg	82.0	355.0	
Riboflavin	mg/kg	3.7	16.0	

Cattle, tongues, raw, (5)

Ref No 5 08 129 United States

Feed Name or Analyses		Mean As Fed	Mean Dry	C.V. ± %
Dry matter	%	32.0	100.0	
Ash	%	0.9	2.8	
Ether extract	%	15.0	46.9	
N-free extract	%	0.4	1.3	
Protein (N x 6.25)	%	16.4	51.3	
Energy	GE Mcal/kg	2.07	6.47	
Cattle	DE Mcal/kg	* 1.94	6.06	
Sheep	DE Mcal/kg	* 1.80	5.63	
Swine	DE kcal/kg	*2291.	7161.	
Cattle	ME Mcal/kg	* 1.59	4.97	
Swine	ME kcal/kg	*1963.	6133.	
Sheep	ME Mcal/kg	* 1.48	4.62	
Cattle	TDN %	* 44.0	137.4	
Sheep	TDN %	* 40.9	127.8	
Swine	TDN %	* 52.0	162.4	
Calcium	%	0.01	0.03	
Iron	%	0.002	0.006	
Phosphorus	%	0.18	0.56	
Potassium	%	0.20	0.63	
Sodium	%	0.07	0.22	
Niacin	mg/kg	50.0	156.3	
Riboflavin	mg/kg	2.9	9.1	
Thiamine	mg/kg	1.2	3.8	

Cattle, whey albumin, heat-and acid-precipitated dehy, mn 75% protein, (5)

Dried milk albumin (AAFCO)

Milk, albumin, dried

Lactalbumin, dried

Ref No 5 01 177 United States

Feed Name or Analyses		Mean As Fed	Mean Dry	C.V. ± %
Dry matter	%	92.2	100.0	2
Ash	%	29.1	31.6	20
Crude fiber	%	0.9	0.9	32
Ether extract	%	1.0	1.1	69
N-free extract	%	12.8	13.9	
Protein (N x 6.25)	%	48.4	52.5	9
Energy	GE Mcal/kg			
Cattle	DE Mcal/kg	* 2.58	2.80	
Sheep	DE Mcal/kg	* 2.45	2.65	
Swine	DE kcal/kg	*2700.	2928.	
Cattle	ME Mcal/kg	* 2.12	2.30	
Sheep	ME Mcal/kg	* 2.01	2.18	
Swine	ME kcal/kg	*2306.	2501.	
Cattle	TDN %	* 58.5	63.4	
Sheep	TDN %	* 55.5	60.2	
Swine	TDN %	* 61.2	66.4	
Calcium	%	10.87	11.79	8
Phosphorus	%	4.03	4.37	14

Continued

Feed Name or Analyses			Mean As Fed	Mean Dry	C.V. ± %
Niacin	mg/kg		2.0	2.1	
Pantothenic acid	mg/kg		7.3	7.9	
Riboflavin	mg/kg		8.8	9.6	
Thiamine	mg/kg		0.7	0.7	

Cattle, whey w protein, (5)

Ref No 5 01 187 — United States

Feed Name or Analyses			Mean As Fed	Mean Dry	C.V. ± %
Dry matter	%		86.2	100.0	
Ash	%		27.0	31.3	
Ether extract	%		4.4	5.1	
Swine	dig coef %		57.	57.	
N-free extract	%		15.4	17.9	
Swine	dig coef %		12.	12.	
Protein (N x 6.25)	%		39.4	45.7	
Swine	dig coef %		91.	91.	
Swine	dig prot %		35.8	41.6	
Energy	GE Mcal/kg				
Cattle	DE Mcal/kg	*	1.71	1.98	
Sheep	DE Mcal/kg	*	1.87	2.17	
Swine	DE kcal/kg	*	1912.	2218.	
Cattle	ME Mcal/kg	*	1.40	1.62	
Sheep	ME Mcal/kg	*	1.55	1.80	
Swine	ME kcal/kg	*	1659.	1924.	
Cattle	TDN %	*	38.8	45.0	
Sheep	TDN %	*	42.5	47.3	
Swine	TDN %		43.4	50.3	

Cattle, manure, dehy grnd, (7)

Ref No 7 01 190 — United States

Feed Name or Analyses			Mean As Fed	Mean Dry	C.V. ± %
Dry matter	%		93.5	100.0	3
Ash	%		17.9	19.1	22
Crude fiber	%		26.6	28.4	18
Ether extract	%		2.7	2.9	64
N-free extract	%		34.1	36.5	
Protein (N x 6.25)	%		12.2	13.0	19
Calcium	%		1.89	2.02	24
Phosphorus	%		0.66	0.71	33

CATTLE, HEREFORD. Bos taurus

Cattle, Hereford, meat w fat, raw, (5)

Ref No 5 07 920 — United States

Feed Name or Analyses			Mean As Fed	Mean Dry	C.V. ± %
Dry matter	%		52.1	100.0	
Ash	%		0.7	1.3	
Ether extract	%		36.6	70.2	
Protein (N x 6.25)	%		14.4	27.6	

CATTLE, HOLSTEIN. Bos taurus

Cattle, holstein, meat w fat, raw, (5)

Ref No 5 07 919 — United States

Feed Name or Analyses			Mean As Fed	Mean Dry	C.V. ± %
Dry matter	%		44.3	100.0	
Ash	%		0.8	1.8	
Ether extract	%		25.4	57.3	
Protein (N x 6.25)	%		16.9	38.1	

(1) dry forages and roughages (2) pasture, range plants, and forages fed green (3) silages (4) energy feeds (5) protein supplements (6) minerals (7) vitamins (8) additives

Cattle, melts, raw —
see Cattle, spleen, raw, (5)

CAULIFLOWER. Brassica oleracea botrytis

Cauliflower, heads, fresh, (4)

Ref No 2 08 189 — United States

Feed Name or Analyses			Mean As Fed	Mean Dry	C.V. ± %
Dry matter	%		9.0	100.0	
Ash	%		0.9	10.0	
Crude fiber	%		1.0	11.1	
Ether extract	%		0.2	2.2	
N-free extract	%		4.2	46.7	
Protein (N x 6.25)	%		2.7	30.0	
Cattle	dig prot %	*	2.1	23.6	
Goats	dig prot %	*	2.2	24.7	
Horses	dig prot %	*	2.2	24.7	
Sheep	dig prot %	*	2.2	24.7	
Energy	GE Mcal/kg		0.27	3.00	
Calcium	%		0.02	0.22	
Iron	%		0.001	0.011	
Phosphorus	%		0.06	0.67	
Potassium	%		0.30	3.33	
Sodium	%		0.01	0.11	
Ascorbic acid	mg/kg		780.0	8666.7	
Niacin	mg/kg		7.0	77.8	
Riboflavin	mg/kg		1.0	11.1	
Thiamine	mg/kg		1.1	12.2	
Vitamin A	IU/g		0.6	6.7	

Cauliflower, leaves, fresh, (4)

Ref No 2 01 192 — United States

Feed Name or Analyses			Mean As Fed	Mean Dry	C.V. ± %
Dry matter	%			100.0	
Crude fiber	%			9.5	
Ether extract	%			4.1	
Protein (N x 6.25)	%			26.6	
Cattle	dig prot %	*		20.4	
Goats	dig prot %	*		21.6	
Horses	dig prot %	*		21.6	
Sheep	dig prot %	*		21.6	
Carotene	mg/kg			185.0	
Riboflavin	mg/kg			23.1	
Vitamin A equivalent	IU/g			308.3	

Cauliflower, stems, fresh, (4)

Ref No 2 01 193 — United States

Feed Name or Analyses			Mean As Fed	Mean Dry	C.V. ± %
Dry matter	%			100.0	
Crude fiber	%			17.3	
Protein (N x 6.25)	%			17.1	
Cattle	dig prot %	*		11.7	
Goats	dig prot %	*		12.9	
Horses	dig prot %	*		12.9	
Sheep	dig prot %	*		12.9	
Carotene	mg/kg			28.0	
Riboflavin	mg/kg			9.3	
Vitamin A equivalent	IU/g			46.7	

CEIBA. Ceiba acuminata

Ceiba, seeds w hulls, (5)

Ref No 5 09 030 — United States

Feed Name or Analyses			Mean As Fed	Mean Dry	C.V. ± %
Dry matter	%			100.0	
Protein (N x 6.25)	%			39.0	

Feed Name or Analyses			Mean As Fed	Mean Dry	C.V. ± %
Alanine	%			1.68	
Arginine	%			4.41	
Aspartic acid	%			3.16	
Glutamic acid	%			8.39	
Glycine	%			1.33	
Histidine	%			0.74	
Hydroxyproline	%			0.00	
Isoleucine	%			1.21	
Leucine	%			2.26	
Lysine	%			1.76	
Methionine	%			0.55	
Phenylalanine	%			1.99	
Proline	%			1.05	
Serine	%			1.68	
Threonine	%			0.94	
Tyrosine	%			0.90	
Valine	%			1.83	

CELERY. Apium graveolens

Celery, aerial part, fresh, (4)

Ref No 2 01 195 — United States

Feed Name or Analyses			Mean As Fed	Mean Dry	C.V. ± %
Dry matter	%		5.9	100.0	
Ash	%		1.0	16.9	
Crude fiber	%		0.6	10.2	
Ether extract	%		0.1	1.7	
N-free extract	%		3.3	55.9	
Protein (N x 6.25)	%		0.9	15.3	
Cattle	dig prot %	*	0.6	10.0	
Goats	dig prot %	*	0.7	11.2	
Horses	dig prot %	*	0.7	11.2	
Sheep	dig prot %	*	0.7	11.2	
Energy	GE Mcal/kg		0.17	2.88	
Calcium	%		0.04	0.66	
Iron	%		0.000	0.005	
Phosphorus	%		0.03	0.47	
Potassium	%		0.34	5.78	
Sodium	%		0.13	2.14	
Ascorbic acid	mg/kg		90.0	1525.4	
Niacin	mg/kg		3.0	50.8	
Riboflavin	mg/kg		0.3	5.1	
Thiamine	mg/kg		0.3	5.1	
Vitamin A	IU/g		2.4	40.7	

Celery, leaves, fresh, (4)

Ref No 2 01 196 — United States

Feed Name or Analyses			Mean As Fed	Mean Dry	C.V. ± %
Dry matter	%		13.9	100.0	32
Crude fiber	%		1.1	7.9	33
Ether extract	%		0.9	6.2	22
Protein (N x 6.25)	%		3.9	28.2	6
Cattle	dig prot %	*	3.0	21.9	
Goats	dig prot %	*	3.2	23.1	
Horses	dig prot %	*	3.2	23.1	
Sheep	dig prot %	*	3.2	23.1	
Carotene	mg/kg		37.4	269.0	40
Riboflavin	mg/kg		2.4	17.0	13
Vitamin A equivalent	IU/g		62.3	448.4	
Arginine	%		0.14	1.00	
Histidine	%		0.06	0.40	
Isoleucine	%		0.14	1.00	
Leucine	%		0.25	1.80	
Lysine	%		0.08	0.60	
Methionine	%		0.08	0.60	
Phenylalanine	%		0.17	1.20	
Threonine	%		0.13	0.90	

Feed Name or Analyses		Mean As Fed	Mean Dry	C.V. ± %
Tryptophan	%	0.04	0.30	
Valine	%	0.18	1.30	

Celery, stalks, fresh, (4)
Ref No 2 01 197 United States

Feed Name or Analyses		Mean As Fed	Mean Dry	C.V. ± %
Dry matter	%	5.9	100.0	
Crude fiber	%	0.9	14.5	
Ether extract	%	0.2	3.3	
Protein (N x 6.25)	%	0.7	12.3	
Cattle	dig prot %	* 0.4	7.3	
Goats	dig prot %	* 0.5	8.5	
Horses	dig prot %	* 0.5	8.5	
Sheep	dig prot %	* 0.5	8.5	
Carotene	mg/kg	0.6	10.4	
Riboflavin	mg/kg	0.4	6.2	
Vitamin A equivalent	IU/g	1.0	17.3	

CENTIPEDEGRASS. Eremochloa ophiuroides

Centipedegrass, aerial part, fresh, (2)
Ref No 2 01 198 United States

Feed Name or Analyses		Mean As Fed	Mean Dry	C.V. ± %
Dry matter	%		100.0	
Ash	%		5.6	
Crude fiber	%		31.6	
Ether extract	%		2.8	
N-free extract	%		48.7	
Protein (N x 6.25)	%		11.3	
Cattle	dig prot %	*	7.5	
Goats	dig prot %	*	7.1	
Horses	dig prot %	*	7.1	
Rabbits	dig prot %	*	7.4	
Sheep	dig prot %	*	7.5	
Calcium	%		0.58	
Cobalt	mg/kg		0.035	
Iron	%		0.013	
Magnesium	%		0.50	
Phosphorus	%		0.26	

Cereals, young, dehy –
see Grass, cereal, aerial part, dehy, immature, (1)

CERITONIA, SILIQUA. Ceritonia siliqua

Ceritonia, siliqua, seeds w hulls, (5)
Ref No 5 09 306 United States

Feed Name or Analyses		Mean As Fed	Mean Dry	C.V. ± %
Dry matter	%		100.0	
Protein (N x 6.25)	%		47.0	
Alanine	%		1.93	
Arginine	%		5.55	
Aspartic acid	%		4.23	
Glutamic acid	%		13.16	
Glycine	%		2.49	
Histidine	%		1.18	
Hydroxyproline	%		0.05	
Isoleucine	%		1.65	
Leucine	%		3.06	
Lysine	%		2.63	
Methionine	%		0.47	
Phenylalanine	%		1.50	
Proline	%		1.88	
Serine	%		2.35	
Threonine	%		1.69	

Feed Name or Analyses		Mean As Fed	Mean Dry	C.V. ± %
Tyrosine	%		1.65	
Valine	%		2.07	

CHAENOSTOMA. Chaenostoma spp

Chaenostoma, aerial part, s-c, late bloom, (1)
Ref No 1 01 200 United States

Feed Name or Analyses		Mean As Fed	Mean Dry	C.V. ± %
Dry matter	%	87.0	100.0	
Ash	%	4.6	5.3	
Crude fiber	%	35.6	40.9	
Sheep	dig coef %	28.	28.	
Ether extract	%	1.0	1.2	
Sheep	dig coef %	29.	29.	
N-free extract	%	41.2	47.3	
Sheep	dig coef %	60.	60.	
Protein (N x 6.25)	%	4.6	5.3	
Sheep	dig coef %	36.	36.	
Cattle	dig prot %	* 1.3	1.5	
Goats	dig prot %	* 1.3	1.5	
Horses	dig prot %	* 1.8	2.0	
Rabbits	dig prot %	* 2.4	2.8	
Sheep	dig prot %	1.7	1.9	
Energy	GE Mcal/kg			
Cattle	DE Mcal/kg	* 1.64	1.88	
Sheep	DE Mcal/kg	* 1.63	1.87	
Cattle	ME Mcal/kg	* 1.34	1.54	
Sheep	ME Mcal/kg	* 1.34	1.54	
Cattle	TDN %	* 37.1	42.7	
Sheep	TDN %	37.0	42.5	

CHAMISE. Adenostoma fasiculatum

Chamise, aerial part, fan air dried grnd pelleted, early leaf, (1)
Ref No 1 07 860 United States

Feed Name or Analyses		Mean As Fed	Mean Dry	C.V. ± %
Dry matter	%	90.0	100.0	
Ash	%	3.0	3.3	
Crude fiber	%	18.6	20.7	
Sheep	dig coef %	13.	13.	
Ether extract	%	4.9	5.4	
Sheep	dig coef %	26.	26.	
N-free extract	%	57.2	63.5	
Sheep	dig coef %	63.	63.	
Protein (N x 6.25)	%	6.4	7.1	
Sheep	dig coef %	30.	30.	
Cattle	dig prot %	* 2.8	3.1	
Goats	dig prot %	* 2.9	3.2	
Horses	dig prot %	* 3.2	3.6	
Rabbits	dig prot %	* 3.7	4.2	
Sheep	dig prot %	1.9	2.1	
Energy	GE Mcal/kg			
Cattle	DE Mcal/kg	* 2.01	2.23	
Sheep	DE Mcal/kg	* 1.90	2.12	
Cattle	ME Mcal/kg	* 1.65	1.83	
Sheep	ME Mcal/kg	* 1.56	1.73	
Cattle	TDN %	* 45.5	50.6	
Sheep	TDN %	43.2	48.0	

Charcoal, vegetable –
see Plant, charcoal, (6)

CHARLOCK. Brassica kaber

Charlock, seeds, extn unspecified grnd, (5)
Ref No 5 01 204 United States

Feed Name or Analyses		Mean As Fed	Mean Dry	C.V. ± %
Dry matter	%	91.6	100.0	
Ash	%	10.0	10.9	
Crude fiber	%	13.5	14.7	
Sheep	dig coef %	-2.	-2.	
Ether extract	%	5.2	5.7	
Sheep	dig coef %	97.	97.	
N-free extract	%	32.5	35.5	
Sheep	dig coef %	71.	71.	
Protein (N x 6.25)	%	30.4	33.2	
Sheep	dig coef %	81.	81.	
Sheep	dig prot %	24.6	26.9	
Energy	GE Mcal/kg			
Cattle	DE Mcal/kg	* 2.76	3.02	
Sheep	DE Mcal/kg	* 2.59	2.83	
Swine	DE kcal/kg	*2840.	3100.	
Cattle	ME Mcal/kg	* 2.27	2.47	
Sheep	ME Mcal/kg	* 2.12	2.32	
Swine	ME kcal/kg	*2535.	2768.	
Cattle	TDN %	* 62.7	68.4	
Sheep	TDN %	58.7	64.1	
Swine	TDN %	* 64.4	70.3	

Cheese rind (AAFCO) –
see Cattle, cheese rind, (5)

CHERRY, BLACK. Prunus serotina

Cherry, black, leaves, s-c, (1)
Ref No 1 01 206 United States

Feed Name or Analyses		Mean As Fed	Mean Dry	C.V. ± %
Dry matter	%	76.1	100.0	
Ash	%	3.1	4.1	
Crude fiber	%	17.2	22.6	
Sheep	dig coef %	40.	40.	
Ether extract	%	2.7	3.5	
Sheep	dig coef %	33.	33.	
N-free extract	%	41.3	54.3	
Sheep	dig coef %	72.	72.	
Protein (N x 6.25)	%	11.8	15.5	
Sheep	dig coef %	58.	58.	
Cattle	dig prot %	* 7.9	10.4	
Goats	dig prot %	* 8.4	11.0	
Horses	dig prot %	* 8.1	10.7	
Rabbits	dig prot %	* 8.1	10.6	
Sheep	dig prot %	6.8	9.0	
Energy	GE Mcal/kg			
Cattle	DE Mcal/kg	* 2.01	2.64	
Sheep	DE Mcal/kg	* 2.00	2.63	
Cattle	ME Mcal/kg	* 1.65	2.16	
Sheep	ME Mcal/kg	* 1.64	2.16	
Cattle	TDN %	* 45.6	59.9	
Sheep	TDN %	45.5	59.7	

CHESTNUT. Castanea spp

Chestnut, meats, fresh, (4)
Ref No 2 01 207 United States

Feed Name or Analyses		Mean As Fed	Mean Dry	C.V. ± %
Dry matter	%	47.5	100.0	
Ash	%	1.0	2.1	
Crude fiber	%	1.1	2.3	

Continued

Feed Name or Analyses		Mean As Fed	Mean Dry	C.V. ± %
Ether extract	%	1.5	3.2	
N-free extract	%	41.0	86.3	
Protein (N x 6.25)	%	2.9	6.1	
Cattle	dig prot %	* 0.8	1.6	
Goats	dig prot %	* 1.3	2.8	
Horses	dig prot %	* 1.3	2.8	
Sheep	dig prot %	* 1.3	2.8	
Energy	GE Mcal/kg	1.94	4.08	
Cattle	DE Mcal/kg	* 1.83	3.86	
Sheep	DE Mcal/kg	* 1.84	3.86	
Swine	DE kcal/kg	*1862.	3921.	
Cattle	ME Mcal/kg	* 1.51	3.17	
Sheep	ME Mcal/kg	* 1.51	3.17	
Swine	ME kcal/kg	*1765.	3716.	
Cattle	TDN %	* 41.6	87.5	
Sheep	TDN %	* 41.6	87.6	
Swine	TDN %	* 42.2	88.9	
Calcium	%	0.03	0.06	
Iron	%	0.002	0.004	
Phosphorus	%	0.09	0.19	
Potassium	%	0.45	0.96	
Sodium	%	0.01	0.01	
Niacin	mg/kg	6.0	12.6	
Riboflavin	mg/kg	2.2	4.6	
Thiamine	mg/kg	2.2	4.6	

Chestnut, seed residue wo starch w shells, (4)

Ref No 4 01 208 United States

Feed Name or Analyses		Mean As Fed	Mean Dry	C.V. ± %
Dry matter	%	89.0	100.0	
Ash	%	1.4	1.6	
Crude fiber	%	15.4	17.3	
Sheep	dig coef %	14.	14.	
Ether extract	%	3.7	4.2	
Sheep	dig coef %	63.	63.	
N-free extract	%	62.7	70.4	
Sheep	dig coef %	62.	62.	
Protein (N x 6.25)	%	5.8	6.5	
Sheep	dig coef %	-19.	-19.	
Cattle	dig prot %	* 1.8	2.0	
Goats	dig prot %	* 2.8	3.2	
Horses	dig prot %	* 2.8	3.2	
Sheep	dig prot %	-1.1	-1.2	
Energy	GE Mcal/kg			
Cattle	DE Mcal/kg	* 2.05	2.30	
Sheep	DE Mcal/kg	* 1.99	2.24	
Swine	DE kcal/kg	*2146.	2412.	
Cattle	ME Mcal/kg	* 1.68	1.89	
Sheep	ME Mcal/kg	* 1.63	1.83	
Swine	ME kcal/kg	*2032.	2284.	
Cattle	TDN %	* 46.4	52.1	
Sheep	TDN %	45.1	50.7	
Swine	TDN %	* 48.7	54.7	

CHICKEN. Gallus domesticus

Chicken, carcass, raw, (5)

Whole eviscerated chicken (AAFCO)

Ref No 5 08 095 United States

Feed Name or Analyses		Mean As Fed	Mean Dry	C.V. ± %
Dry matter	%	32.5	100.0	
Ash	%	1.0	3.1	
Ether extract	%	11.9	36.6	
Protein (N x 6.25)	%	19.6	60.3	
Energy	GE Mcal/kg	1.91	5.88	
Calcium	%	0.01	0.04	
Iron	%	0.002	0.005	
Phosphorus	%	0.19	0.60	
Niacin	mg/kg	73.0	224.6	
Riboflavin	mg/kg	2.1	6.5	
Thiamine	mg/kg	0.8	2.5	
Vitamin A	IU/g	7.6	23.4	

Chicken, eggs wo shells, dehy, (5)

Ref No 5 01 214 United States

Feed Name or Analyses		Mean As Fed	Mean Dry	C.V. ± %
Dry matter	%	95.9	100.0	
Ash	%	3.6	3.8	
Crude fiber	%	0.0	0.0	
Ether extract	%	41.2	43.0	
N-free extract	%	4.1	4.3	
Protein (N x 6.25)	%	47.0	49.0	
Energy	GE Mcal/kg	5.92	6.17	
Cattle	DE Mcal/kg	* 5.50	5.73	
Sheep	DE Mcal/kg	* 4.40	4.59	
Swine	DE kcal/kg	*6516.	6794.	
Cattle	ME Mcal/kg	* 4.51	4.70	
Sheep	ME Mcal/kg	* 3.61	3.76	
Swine	ME kcal/kg	*5610.	5850.	
Cattle	TDN %	* 124.7	130.0	
Sheep	TDN %	* 99.8	104.1	
Swine	TDN %	* 147.8	154.1	
Calcium	%	0.19	0.20	
Iron	%	0.009	0.009	
Phosphorus	%	0.80	0.83	
Potassium	%	0.46	0.48	
Sodium	%	0.43	0.45	
Ascorbic acid	mg/kg	0.0	0.0	
Niacin	mg/kg	2.0	2.1	
Riboflavin	mg/kg	12.0	12.5	
Thiamine	mg/kg	3.3	3.4	
Vitamin A	IU/g	42.9	44.7	

Ref No 5 01 214 Canada

Feed Name or Analyses		Mean As Fed	Mean Dry	C.V. ± %
Dry matter	%	94.7	100.0	
Ether extract	%	35.6	37.6	
Protein (N x 6.25)	%	45.2	47.7	
Energy	GE Mcal/kg	5.85	6.17	
Calcium	%	0.17	0.18	
Phosphorus	%	0.68	0.72	

Chicken, eggs wo shells, raw, (5)

Ref No 5 08 114 United States

Feed Name or Analyses		Mean As Fed	Mean Dry	C.V. ± %
Dry matter	%	26.3	100.0	
Ash	%	1.0	3.8	
Ether extract	%	11.5	43.7	
N-free extract	%	0.9	3.4	
Protein (N x 6.25)	%	12.9	49.0	
Energy	GE Mcal/kg	1.63	6.20	
Cattle	DE Mcal/kg	* 1.52	5.77	
Sheep	DE Mcal/kg	* 1.21	4.60	
Swine	DE kcal/kg	*1800.	6845.	
Cattle	ME Mcal/kg	* 1.24	4.73	
Sheep	ME Mcal/kg	* 0.99	3.77	
Swine	ME kcal/kg	*1550.	5893.	
Cattle	TDN %	* 34.4	130.8	
Sheep	TDN %	* 27.4	104.3	
Swine	TDN %	* 40.8	155.3	
Calcium	%	0.05	0.19	
Iron	%	0.002	0.008	
Phosphorus	%	0.20	0.76	
Potassium	%	0.13	0.49	
Sodium	%	0.12	0.46	
Niacin	mg/kg	1.0	3.8	
Riboflavin	mg/kg	3.0	11.4	
Thiamine	mg/kg	1.1	4.2	
Vitamin A	IU/g	11.8	44.9	

Chicken, gizzards, raw, (5)

Ref No 5 07 948 United States

Feed Name or Analyses		Mean As Fed	Mean Dry	C.V. ± %
Dry matter	%	25.0	100.0	
Ash	%	1.5	6.0	
Crude fiber	%	0.0	0.0	
Ether extract	%	2.7	10.8	
N-free extract	%	0.7	2.8	
Protein (N x 6.25)	%	20.1	80.4	
Energy	GE Mcal/kg	1.13	4.52	
Cattle	DE Mcal/kg	* 0.99	3.96	
Sheep	DE Mcal/kg	* 0.89	3.55	
Swine	DE kcal/kg	* 936.	3743.	
Cattle	ME Mcal/kg	* 0.81	3.24	
Sheep	ME Mcal/kg	* 0.73	2.91	
Swine	ME kcal/kg	* 746.	2986.	
Cattle	TDN %	* 22.4	89.7	
Sheep	TDN %	* 20.1	80.5	
Swine	TDN %	* 21.2	84.9	
Calcium	%	0.01	0.04	
Iron	%	0.003	0.012	
Phosphorus	%	0.11	0.42	
Potassium	%	0.24	0.96	
Sodium	%	0.07	0.26	
Niacin	mg/kg	45.0	180.0	
Riboflavin	mg/kg	2.0	8.0	
Thiamine	mg/kg	0.3	1.2	

CHICKEN, BROILERS. Gallus domesticus

Chicken, broilers, whole, raw, (5)

Ref No 5 07 945 United States

Feed Name or Analyses		Mean As Fed	Mean Dry	C.V. ± %
Dry matter	%	24.3	100.0	
Ash	%	0.8	3.3	
Ether extract	%	4.9	20.2	
Protein (N x 6.25)	%	18.6	76.5	
Energy	GE Mcal/kg	1.24	5.10	
Calcium	%	0.01	0.04	
Iron	%	0.002	0.008	
Phosphorus	%	0.20	0.82	
Niacin	mg/kg	56.0	230.5	
Riboflavin	mg/kg	3.8	15.6	
Thiamine	mg/kg	0.7	2.9	
Vitamin A	IU/g	7.3	30.0	

CHICKEN, CULL HENS. Gallus Domesticus

Chicken, cull hens, whole wo feathers, raw, grnd, (5)

Ref No 5 07 760 United States

Feed Name or Analyses		Mean As Fed	Mean Dry	C.V. ± %
Dry matter	%	57.9	100.0	
Ash	%	3.3	5.7	
Crude fiber	%	0.5	0.9	
Ether extract	%	20.4	35.2	
N-free extract	%	17.7	30.6	
Protein (N x 6.25)	%	16.0	27.6	

(1) dry forages and roughages (2) pasture, range plants, and forages fed green (3) sitages (4) energy feeds (5) protein supplements (6) minerals (7) vitamins (8) additives

Feed Name or Analyses			Mean As Fed	Mean Dry	C.V. ± %
Calcium	%		0.98	1.69	
Phosphorus	%		0.61	1.05	

CHICKPEA, GRAM. Cicer arientinum

Chickpea, gram, straw, (1)
Ref No 1 01 217 — United States

Analyses			As Fed	Dry	C.V.
Dry matter	%		90.6	100.0	
Ash	%		12.0	13.3	
Crude fiber	%		40.2	44.4	
Cattle	dig coef %		40.	40.	
Ether extract	%		0.5	0.5	
Cattle	dig coef %		0.	0.	
N-free extract	%		32.4	35.8	
Cattle	dig coef %		47.	47.	
Protein (N x 6.25)	%		5.4	6.0	
Cattle	dig coef %		40.	40.	
Cattle	dig prot %		2.2	2.4	
Goats	dig prot %	*	2.0	2.2	
Horses	dig prot %	*	2.4	2.6	
Rabbits	dig prot %	*	3.0	3.3	
Sheep	dig prot %	*	1.8	2.0	
Energy	GE Mcal/kg				
Cattle	DE Mcal/kg	*	1.48	1.63	
Sheep	DE Mcal/kg	*	1.54	1.70	
Cattle	ME Mcal/kg	*	1.21	1.34	
Sheep	ME Mcal/kg	*	1.26	1.39	
Cattle	TDN %		33.5	37.0	
Sheep	TDN %	*	35.0	38.6	

Chickpea, gram, seeds, (5)
Gram
Garbanzos
Ref No 5 01 218 — United States

Analyses			As Fed	Dry	C.V.
Dry matter	%		89.1	100.0	
Ash	%		2.9	3.3	
Crude fiber	%		7.0	7.9	
Cattle	dig coef %		166.	166.	
Sheep	dig coef %		59.	59.	
Ether extract	%		3.9	4.4	
Cattle	dig coef %		52.	52.	
Sheep	dig coef %		88.	88.	
N-free extract	%		55.7	62.5	
Cattle	dig coef %		89.	89.	
Sheep	dig coef %		88.	88.	
Protein (N x 6.25)	%		19.5	21.9	
Cattle	dig coef %		72.	72.	
Sheep	dig coef %		78.	78.	
Cattle	dig prot %		14.0	15.8	
Sheep	dig prot %		15.2	17.1	
Energy	GE Mcal/kg		3.59	4.03	
Cattle	DE Mcal/kg	*	3.52	3.95	
Sheep	DE Mcal/kg	*	3.36	3.77	
Swine	DE kcal/kg	*	3865.	4340.	
Cattle	ME Mcal/kg	*	2.89	3.24	
Sheep	ME Mcal/kg	*	2.75	3.09	
Swine	ME kcal/kg	*	3539.	3974.	
Cattle	TDN %		79.9	89.7	
Sheep	TDN %		76.1	85.5	
Swine	TDN %	*	87.7	98.4	
Calcium	%		0.15	0.17	
Iron	%		0.007	0.008	
Phosphorus	%		0.33	0.37	
Potassium	%		0.79	0.89	
Sodium	%		0.03	0.03	
Niacin	mg/kg		19.9	22.4	
Riboflavin	mg/kg		1.5	1.7	
Thiamine	mg/kg		3.1	3.5	
Vitamin A	IU/g		0.5	0.6	

CHICKWEED. Stellaria media

Chickweed, aerial part, fresh, (2)
Ref No 2 01 219 — United States

Analyses			As Fed	Dry	C.V.
Dry matter	%		8.4	100.0	
Carotene	mg/kg		3.6	42.8	
Niacin	mg/kg		5.1	60.6	
Riboflavin	mg/kg		1.4	16.1	
Thiamine	mg/kg		0.1	1.5	
Vitamin A equivalent	IU/g		6.0	71.3	

CHICORY. Cichorium spp

Chicory, aerial part, fresh, (2)
Ref No 2 01 220 — United States

Analyses			As Fed	Dry	C.V.
Dry matter	%		17.4	100.0	
Ash	%		3.1	18.0	
Crude fiber	%		2.7	15.3	
Ether extract	%		0.4	2.2	
N-free extract	%		8.7	50.1	
Protein (N x 6.25)	%		2.5	14.4	
Cattle	dig prot %	*	1.8	10.1	
Goats	dig prot %	*	1.7	10.0	
Horses	dig prot %	*	1.7	9.8	
Rabbits	dig prot %	*	1.7	9.8	
Sheep	dig prot %	*	1.8	10.4	
Calcium	%		0.33	1.89	
Magnesium	%		0.13	0.73	
Phosphorus	%		0.04	0.22	

Chicory, roots, s-c, (4)
Ref No 4 01 221 — United States

Analyses			As Fed	Dry	C.V.
Dry matter	%		88.2	100.0	
Ash	%		3.8	4.3	
Crude fiber	%		4.1	4.6	
Sheep	dig coef %		85.	85.	
Ether extract	%		0.5	0.6	
Sheep	dig coef %		51.	51.	
N-free extract	%		75.2	85.3	
Sheep	dig coef %		95.	95.	
Protein (N x 6.25)	%		4.6	5.2	
Sheep	dig coef %		73.	73.	
Cattle	dig prot %	*	0.7	0.8	
Goats	dig prot %	*	1.8	2.0	
Horses	dig prot %	*	1.8	2.0	
Sheep	dig prot %		3.3	3.8	
Energy	GE Mcal/kg				
Cattle	DE Mcal/kg	*	3.36	3.81	
Sheep	DE Mcal/kg	*	3.48	3.94	
Swine	DE kcal/kg	*	2999.	3400.	
Cattle	ME Mcal/kg	*	2.75	3.12	
Sheep	ME Mcal/kg	*	2.85	3.23	
Swine	ME kcal/kg	*	2848.	3229.	
Cattle	TDN %	*	76.2	86.4	
Sheep	TDN %		78.9	89.4	
Swine	TDN %	*	68.0	77.1	

CHINESELANTERN. Quincula spp

Chineselantern, aerial part, fresh, mid-bloom, (2)
Ref No 2 01 222 — United States

Analyses			As Fed	Dry	C.V.
Dry matter	%			100.0	
Ash	%			17.7	
Crude fiber	%			14.5	
Ether extract	%			4.1	
N-free extract	%			32.2	
Protein (N x 6.25)	%			31.5	
Cattle	dig prot %	*		24.7	
Goats	dig prot %	*		25.9	
Horses	dig prot %	*		24.3	
Rabbits	dig prot %	*		23.0	
Sheep	dig prot %	*		26.3	
Calcium	%			2.00	
Magnesium	%			0.52	
Phosphorus	%			0.34	
Potassium	%			5.59	

Chipped rice (AAFCO) –
see Rice, groats, polished and broken, (4)

Chipped rice (AAFCO) –
see Rice, groats, polished and broken, (4)

CHOKECHERRY, BLACK. Prunus virginiana melanocarpa

Chokecherry, black, browse, immature, (2)
Ref No 2 01 226 — United States

Analyses			As Fed	Dry	C.V.
Dry matter	%			100.0	
Ash	%			7.1	
Crude fiber	%			10.8	
Ether extract	%			3.2	
N-free extract	%			63.0	
Protein (N x 6.25)	%			15.9	
Cattle	dig prot %	*		11.4	
Goats	dig prot %	*		11.4	
Horses	dig prot %	*		11.0	
Rabbits	dig prot %	*		10.9	
Sheep	dig prot %	*		11.8	
Calcium	%			1.80	
Magnesium	%			0.42	
Phosphorus	%			0.41	
Sulphur	%			0.16	

Chokecherry, black, browse, mature, (2)
Ref No 2 01 227 — United States

Analyses			As Fed	Dry	C.V.
Dry matter	%			100.0	
Ash	%			10.2	
Crude fiber	%			11.5	
Ether extract	%			6.5	
N-free extract	%			63.7	
Protein (N x 6.25)	%			8.1	
Cattle	dig prot %	*		4.8	
Goats	dig prot %	*		4.1	
Horses	dig prot %	*		4.4	
Rabbits	dig prot %	*		4.9	
Sheep	dig prot %	*		4.5	
Calcium	%			3.03	
Magnesium	%			0.54	

Continued

Feed Name or Analyses			Mean As Fed	Mean Dry	C.V. ± %
Phosphorus	%			0.67	
Sulphur	%			0.09	

Chopped alfalfa hay (AAFCO) –
see Alfalfa, hay, s-c chopped, (1)

CHUFA. Cyperus esculentus

Chufa, hay, s-c, (1)
Ref No 1 01 231 United States

Feed Name or Analyses			As Fed	Dry	C.V. ± %
Dry matter	%			100.0	
Calcium	%			0.67	
Magnesium	%			0.74	
Phosphorus	%			0.15	
Potassium	%			0.74	

Chufa, tubers, fresh, (4)
Ref No 2 08 374 United States

Feed Name or Analyses			As Fed	Dry	C.V. ± %
Dry matter	%		26.5	100.0	
Ash	%		1.8	6.8	
Crude fiber	%		2.0	7.5	
Ether extract	%		1.8	6.8	
N-free extract	%		18.8	70.9	
Protein (N x 6.25)	%		2.1	7.9	
Cattle	dig prot %	*	0.9	3.3	
Goats	dig prot %	*	1.2	4.5	
Horses	dig prot %	*	1.2	4.5	
Sheep	dig prot %	*	1.2	4.5	
Energy	GE Mcal/kg				
Cattle	DE Mcal/kg	*	0.80	3.03	
Sheep	DE Mcal/kg	*	0.26	2.86	
Swine	DE kcal/kg	*	782.	2950.	
Cattle	ME Mcal/kg	*	0.66	2.48	
Sheep	ME Mcal/kg	*	0.62	2.35	
Swine	ME kcal/kg	*	738.	2784.	
Cattle	TDN %	*	18.2	68.7	
Sheep	TDN %	*	17.2	64.9	
Swine	TDN %	*	17.7	66.9	
Calcium	%		0.01	0.04	
Phosphorus	%		0.07	0.26	
Potassium	%		0.14	0.53	

CINQUEFOIL. Potentilla spp

Cinquefoil, aerial part, fresh, immature, (2)
Ref No 2 01 232 United States

Feed Name or Analyses			As Fed	Dry	C.V. ± %
Dry matter	%			100.0	
Ash	%			8.1	
Crude fiber	%			11.3	
Ether extract	%			3.9	
N-free extract	%			54.6	
Protein (N x 6.25)	%			22.1	
Cattle	dig prot %	*		16.7	
Goats	dig prot %	*		17.2	
Horses	dig prot %	*		16.3	
Rabbits	dig prot %	*		15.7	
Sheep	dig prot %	*		17.6	

(1) dry forages and roughages (2) pasture, range plants, and forages fed green (3) silages (4) energy feeds (5) protein supplements (6) minerals (7) vitamins (8) additives

CITRUS. Citrus spp

Citrus, pulp, ensiled, (3)
Ref No 3 01 234 United States

Feed Name or Analyses			As Fed	Dry	C.V. ± %
Dry matter	%		19.5	100.0	
Ash	%		1.0	5.3	
Crude fiber	%		3.1	15.9	
Ether extract	%		1.7	8.8	
N-free extract	%		12.3	62.9	
Protein (N x 6.25)	%		1.4	7.1	
Cattle	dig prot %	*	0.5	2.7	
Goats	dig prot %	*	0.5	2.7	
Horses	dig prot %	*	0.5	2.7	
Sheep	dig prot %	*	0.5	2.7	
Energy	GE Mcal/kg				
Cattle	DE Mcal/kg	*	0.71	3.65	
Sheep	DE Mcal/kg	*	0.73	3.72	
Cattle	ME Mcal/kg	*	0.58	2.99	
Sheep	ME Mcal/kg	*	0.59	3.05	
Cattle	NE_m Mcal/kg	*	0.42	2.15	
Cattle	NE_{gain} Mcal/kg	*	0.28	1.42	
Cattle	$NE_{lactating\ cows}$ Mcal/kg	*	0.49	2.49	
Cattle	TDN %	*	16.1	82.8	
Sheep	TDN %	*	16.4	84.3	

Citrus, pulp, pressed ensiled, (3)
Ref No 3 08 375 United States

Feed Name or Analyses			As Fed	Dry	C.V. ± %
Dry matter	%		22.8	100.0	
Ash	%		1.3	5.7	
Crude fiber	%		3.5	15.4	
Ether extract	%		2.4	10.5	
N-free extract	%		13.9	61.0	
Protein (N x 6.25)	%		1.7	7.5	
Cattle	dig prot %	*	0.7	3.0	
Goats	dig prot %	*	0.7	3.0	
Horses	dig prot %	*	0.7	3.0	
Sheep	dig prot %	*	0.7	3.0	
Energy	GE Mcal/kg				
Cattle	DE Mcal/kg	*	0.86	3.77	
Sheep	DE Mcal/kg	*	0.85	3.74	
Cattle	ME Mcal/kg	*	0.71	3.09	
Sheep	ME Mcal/kg	*	0.70	3.07	
Cattle	TDN %	*	19.5	85.6	
Sheep	TDN %	*	19.4	84.9	

Citrus, pulp, fresh, (4)
Ref No 2 08 376 United States

Feed Name or Analyses			As Fed	Dry	C.V. ± %
Dry matter	%		18.3	100.0	
Ash	%		1.4	7.7	
Crude fiber	%		2.3	12.6	
Ether extract	%		0.6	3.3	
N-free extract	%		12.8	69.9	
Protein (N x 6.25)	%		1.2	6.6	
Cattle	dig prot %	*	0.4	2.0	
Goats	dig prot %	*	0.6	3.2	
Horses	dig prot %	*	0.6	3.2	
Sheep	dig prot %	*	0.6	3.2	
Energy	GE Mcal/kg				
Cattle	DE Mcal/kg	*	0.67	3.67	
Sheep	DE Mcal/kg	*	0.62	3.40	
Swine	DE kcal/kg	*	414.	2261.	
Cattle	ME Mcal/kg	*	0.55	3.01	
Sheep	ME Mcal/kg	*	0.51	2.79	
Cattle	TDN %	*	15.2	83.2	
Sheep	TDN %	*	14.1	77.1	

Citrus, pulp fines, shredded dehy, (4)
Dried citrus meal (AAFCO)
Citrus meal, dried
Ref No 4 01 235 United States

Feed Name or Analyses			As Fed	Dry	C.V. ± %
Dry matter	%		91.5	100.0	
Ash	%		6.3	6.9	
Crude fiber	%		13.1	14.3	
Ether extract	%		3.2	3.5	
N-free extract	%		62.6	68.4	
Protein (N x 6.25)	%		6.3	6.9	
Cattle	dig prot %	*	2.1	2.3	
Goats	dig prot %	*	3.2	3.5	
Horses	dig prot %	*	3.2	3.5	
Sheep	dig prot %	*	3.2	3.5	
Energy	GE Mcal/kg				
Cattle	DE Mcal/kg	*	3.48	3.80	
Sheep	DE Mcal/kg	*	3.28	3.59	
Cattle	ME Mcal/kg	*	2.85	3.12	
Sheep	ME Mcal/kg	*	2.69	2.94	
Cattle	TDN %	*	79.0	86.3	
Sheep	TDN %	*	74.5	81.4	
Calcium	%		1.98	2.16	
Phosphorus	%		0.10	0.11	

Citrus, pulp w molasses, dehy grnd, (4)
Citrus pulp with molasses, dried
Ref No 4 08 079 United States

Feed Name or Analyses			As Fed	Dry	C.V. ± %
Dry matter	%		90.3	100.0	
Ash	%		6.5	7.2	
Crude fiber	%		10.3	11.5	
Ether extract	%		3.6	4.0	
N-free extract	%		63.6	70.5	
Protein (N x 6.25)	%		6.2	6.9	
Cattle	dig prot %	*	2.1	2.3	
Goats	dig prot %	*	3.2	3.5	
Horses	dig prot %	*	3.2	3.5	
Sheep	dig prot %	*	3.2	3.5	
Energy	GE Mcal/kg				
Cattle	DE Mcal/kg	*	3.40	3.76	
Sheep	DE Mcal/kg	*	3.11	3.45	
Cattle	ME Mcal/kg	*	2.78	3.08	
Sheep	ME Mcal/kg	*	2.55	2.83	
Cattle	TDN %	*	77.0	85.3	
Sheep	TDN %	*	70.6	78.3	
Calcium	%		1.66	1.84	
Phosphorus	%		0.11	0.12	

Citrus, pulp wo fines, ammoniated shredded dehy, (4)
Ref No 4 01 238 United States

Feed Name or Analyses			As Fed	Dry	C.V. ± %
Dry matter	%		87.2	100.0	4
Ash	%		4.6	5.3	12
Crude fiber	%		13.2	15.1	9
Ether extract	%		5.5	6.4	26
N-free extract	%		51.8	59.4	
Protein (N x 6.25)	%		12.1	13.8	13
Cattle	dig prot %	*	7.6	8.7	
Goats	dig prot %	*	8.6	9.9	
Horses	dig prot %	*	8.6	9.9	
Sheep	dig prot %	*	8.6	9.9	

Feed Name or Analyses			Mean As Fed	Mean Dry	C.V. ± %
Energy	GE Mcal/kg				
Cattle	DE Mcal/kg	*	3.18	3.65	
Sheep	DE Mcal/kg	*	2.88	3.31	
Cattle	ME Mcal/kg	*	2.61	2.99	
Sheep	ME Mcal/kg	*	2.37	2.71	
Cattle	TDN %	*	72.2	82.8	
Sheep	TDN %	*	65.4	75.1	
Calcium	%		1.66	1.90	6
Magnesium	%		0.07	0.08	66
Phosphorus	%		0.12	0.14	24

Citrus, pulp wo fines, shredded dehy, (4)

Dried citrus pulp (AAFCO)

Citrus pulp, dried

Ref No 4 01 237 United States

Feed Name or Analyses			Mean As Fed	Mean Dry	C.V. ± %
Dry matter	%		90.2	100.0	
Ash	%		6.3	7.0	
Crude fiber	%		11.6	12.9	
Ether extract	%		3.4	3.8	
N-free extract	%		62.5	69.3	
Protein (N x 6.25)	%		6.4	7.1	
Cattle	dig prot %	*	2.3	2.5	
Goats	dig prot %	*	3.4	3.7	
Horses	dig prot %	*	3.4	3.7	
Sheep	dig prot %	*	3.4	3.7	
Energy	GE Mcal/kg				
Cattle	DE Mcal/kg	*	3.37	3.74	
Sheep	DE Mcal/kg	*	3.08	3.41	
Cattle	ME Mcal/kg	*	2.77	3.07	
Sheep	ME Mcal/kg	*	2.52	2.80	
Cattle	NE_m Mcal/kg	*	1.78	1.97	
Cattle	NE_{gain} Mcal/kg	*	1.19	1.32	
Cattle	$NE_{lactating\ cows}$ Mcal/kg	*	1.89	2.09	
Cattle	TDN %	*	76.6	84.9	
Sheep	TDN %	*	69.7	77.3	
Swine	TDN %	*	47.4	52.5	
Calcium	%		2.00	2.22	
Copper	mg/kg		5.7	6.4	
Iron	%		0.016	0.018	
Magnesium	%		0.16	0.18	
Manganese	mg/kg		6.8	7.6	
Phosphorus	%		0.14	0.15	
Potassium	%		0.62	0.69	
Choline	mg/kg		770.	854.	
Niacin	mg/kg		21.6	23.9	
Pantothenic acid	mg/kg		13.0	14.4	
Riboflavin	mg/kg		2.4	2.7	

Citrus, syrup, mn 45% invert sugar mn 71 degrees brix, (4)

Citrus molasses (AAFCO)

Ref No 4 01 241 United States

Feed Name or Analyses			Mean As Fed	Mean Dry	C.V. ± %
Dry matter	%		67.7	100.0	6
Ash	%		5.4	8.0	30
Crude fiber	%		0.0	0.0	30
Ether extract	%		0.2	0.3	98
N-free extract	%		56.4	83.4	
Sheep	dig coef %	*	90.	90.	
Protein (N x 6.25)	%		5.7	8.4	44
Sheep	dig coef %	*	51.	51.	
Cattle	dig prot %	*	2.5	3.7	
Goats	dig prot %	*	3.3	4.9	
Horses	dig prot %	*	3.3	4.9	
Sheep	dig prot %		2.9	4.3	
Pectins	%		1.0	1.5	11
Pentosans	%		1.6	2.3	

Feed Name or Analyses			Mean As Fed	Mean Dry	C.V. ± %
Energy	GE Mcal/kg				
Cattle	DE Mcal/kg	*	2.32	3.42	
Sheep	DE Mcal/kg	*	2.19	3.24	
Swine	DE kcal/kg	*	2400.	3546.	
Cattle	ME Mcal/kg	*	1.90	2.80	
Sheep	ME Mcal/kg	*	1.80	2.66	
Swine	ME kcal/kg	*	2264.	3344.	
Cattle	NE_m Mcal/kg	*	1.33	1.97	
Cattle	NE_{gain} Mcal/kg	*	0.89	1.32	
Cattle	$NE_{lactating\ cows}$ Mcal/kg	*	1.41	2.09	
Cattle	TDN %	*	52.5	77.6	
Sheep	TDN %	*	49.8	73.5	
Swine	TDN %	*	54.4	80.4	
Calcium	%		1.20	1.77	31
Chlorine	%		0.07	0.10	
Cobalt	mg/kg		0.108	0.159	25
Copper	mg/kg		73.1	108.0	68
Iron	%		0.034	0.050	68
Magnesium	%		0.14	0.21	70
Manganese	mg/kg		26.1	38.5	43
Phosphorus	%		0.12	0.18	98
Potassium	%		0.09	0.13	
Sodium	%		0.27	0.40	
Zinc	mg/kg		92.8	137.0	
Niacin	mg/kg		27.8	41.0	16
Pantothenic acid	mg/kg		13.1	19.3	56
Riboflavin	mg/kg		6.4	9.5	60

Citrus, seeds, mech-extd grnd, (5)

Citrus seed meal, mechanical extracted (AAFCO)

Ref No 5 01 239 United States

Feed Name or Analyses			Mean As Fed	Mean Dry	C.V. ± %
Dry matter	%		88.2	100.0	4
Ash	%		6.0	6.8	15
Crude fiber	%		10.1	11.5	15
Ether extract	%		10.2	11.6	42
N-free extract	%		30.6	34.8	
Protein (N x 6.25)	%		31.2	35.4	25
Energy	GE Mcal/kg				
Cattle	DE Mcal/kg	*	3.21	3.64	
Sheep	DE Mcal/kg	*	3.23	3.67	
Swine	DE kcal/kg	*	3730.	4231.	
Cattle	ME Mcal/kg	*	2.63	2.98	
Sheep	ME Mcal/kg	*	2.65	3.01	
Swine	ME kcal/kg	*	3314.	3760.	
Cattle	TDN %	*	72.7	82.5	
Sheep	TDN %	*	73.3	83.1	
Swine	TDN %	*	84.6	96.0	
Calcium	%		1.10	1.25	23
Copper	mg/kg		6.6	7.5	
Iron	%		0.029	0.033	
Magnesium	%		0.60	0.68	30
Manganese	mg/kg		7.5	8.5	
Phosphorus	%		0.67	0.75	61
Potassium	%		1.31	1.49	
Zinc	mg/kg		7.5	8.5	

Citrus, syrup, ammoniated, (5)

Citrus molasses, ammoniated

Ref No 5 01 240 United States

Feed Name or Analyses			Mean As Fed	Mean Dry	C.V. ± %
Dry matter	%		60.7	100.0	1
Ash	%		4.7	7.7	33
Crude fiber	%		0.0	0.0	33
Ether extract	%		2.1	3.5	38
N-free extract	%		32.5	53.5	
Protein (N x 6.25)	%		21.4	35.2	32

Feed Name or Analyses			Mean As Fed	Mean Dry	C.V. ± %
Energy	GE Mcal/kg				
Cattle	DE Mcal/kg	*	2.30	3.79	
Sheep	DE Mcal/kg	*	2.42	3.99	
Swine	DE kcal/kg	*	2513.	4141.	
Cattle	ME Mcal/kg	*	1.88	3.11	
Sheep	ME Mcal/kg	*	1.99	3.27	
Swine	ME kcal/kg	*	2234.	3681.	
Cattle	TDN %	*	52.1	85.9	
Sheep	TDN %	*	54.9	90.5	
Swine	TDN %	*	57.0	93.9	
Calcium	%		1.02	1.68	14
Magnesium	%		0.08	0.13	47
Phosphorus	%		0.16	0.26	40

CITRUS, GRAPEFRUIT. Citrus paradisi

Citrus, grapefruit, fruit, fresh, (4)

Ref No 2 01 242 United States

Feed Name or Analyses			Mean As Fed	Mean Dry	C.V. ± %
Dry matter	%		13.6	100.0	
Ash	%		0.5	3.7	
Crude fiber	%		1.4	10.3	
Ether extract	%		0.6	4.4	
N-free extract	%		10.0	73.5	
Protein (N x 6.25)	%		1.1	8.1	
Cattle	dig prot %	*	0.5	3.4	
Goats	dig prot %	*	0.6	4.6	
Horses	dig prot %	*	0.6	4.6	
Sheep	dig prot %	*	0.6	4.6	
Energy	GE Mcal/kg				
Cattle	DE Mcal/kg	*	0.52	3.85	
Sheep	DE Mcal/kg	*	0.49	3.57	
Cattle	ME Mcal/kg	*	0.43	3.16	
Sheep	ME Mcal/kg	*	0.40	2.93	
Cattle	TDN %	*	11.9	87.3	
Sheep	TDN %	*	11.0	81.0	
Calcium	%		0.07	0.51	
Phosphorus	%		0.02	0.15	

Citrus, grapefruit, pulp, shredded wet, (4)

Grapefruit pulp, wet

Ref No 2 01 243 United States

Feed Name or Analyses			Mean As Fed	Mean Dry	C.V. ± %
Dry matter	%		16.2	100.0	
Ash	%		0.6	3.7	
Crude fiber	%		2.7	16.7	
Ether extract	%		1.4	8.7	
N-free extract	%		9.8	61.0	
Protein (N x 6.25)	%		1.6	9.9	
Cattle	dig prot %	*	0.8	5.1	
Goats	dig prot %	*	1.0	6.3	
Horses	dig prot %	*	1.0	6.3	
Sheep	dig prot %	*	1.0	6.3	
Energy	GE Mcal/kg				
Cattle	DE Mcal/kg	*	0.61	3.75	
Sheep	DE Mcal/kg	*	0.60	3.69	
Cattle	ME Mcal/kg	*	0.50	3.03	
Sheep	ME Mcal/kg	*	0.49	3.02	
Cattle	TDN %	*	13.8	85.1	
Sheep	TDN %	*	13.5	83.6	
Calcium	%		0.19	1.18	
Magnesium	%		0.09	0.56	
Phosphorus	%		0.17	1.02	

Feed Name or Analyses			Mean As Fed	Mean Dry	C.V. ± %

Citrus, grapefruit, pulp wo fines, shredded dehy, (4)
Grapefruit pulp, dried
Ref No 4 01 244 United States

Analyses			As Fed	Dry	C.V. ± %
Dry matter	%		89.5	100.0	3
Ash	%		5.3	6.0	26
Crude fiber	%		13.1	14.6	13
Ether extract	%		3.1	3.5	62
N-free extract	%		61.1	68.3	
Protein (N x 6.25)	%		6.9	7.7	21
Cattle	dig prot %	*	2.7	3.1	
Goats	dig prot %	*	3.8	4.3	
Horses	dig prot %	*	3.8	4.3	
Sheep	dig prot %	*	3.8	4.3	
Energy	GE Mcal/kg				
Cattle	DE Mcal/kg	*	3.24	3.62	
Sheep	DE Mcal/kg	*	3.02	3.37	
Cattle	ME Mcal/kg	*	2.66	2.97	
Sheep	ME Mcal/kg	*	2.48	2.77	
Cattle	TDN %	*	73.5	82.1	
Sheep	TDN %	*	68.5	76.5	
Calcium	%		1.33	1.48	55
Magnesium	%		0.39	0.43	43
Phosphorus	%		0.16	0.18	23

Citrus, grapefruit, seeds, extn unspecified grnd, (5)
Ref No 5 01 245 United States

Analyses			As Fed	Dry	C.V. ± %
Dry matter	%		96.6	100.0	
Ash	%		4.0	4.1	
Crude fiber	%		26.5	27.4	
Ether extract	%		14.0	14.5	
N-free extract	%		30.5	31.6	
Protein (N x 6.25)	%		21.6	22.4	
Calcium	%		0.35	0.36	
Iron	%		0.001	0.001	
Magnesium	%		0.39	0.40	
Phosphorus	%		0.55	0.57	
Sulphur	%		0.09	0.09	

CITRUS, LEMON. Citrus limon

Citrus, lemon, pulp wo fines, shredded dehy, (4)
Lemon pulp, dried
Ref No 4 01 247 United States

Analyses			As Fed	Dry	C.V. ± %
Dry matter	%		92.9	100.0	
Ash	%		5.3	5.7	25
Crude fiber	%		14.8	15.9	6
Sheep	dig coef %		60.	60.	
Ether extract	%		1.4	1.5	72
Sheep	dig coef %		28.	28.	
N-free extract	%		65.0	70.0	
Sheep	dig coef %		92.	92.	
Protein (N x 6.25)	%		6.5	7.0	8
Sheep	dig coef %		46.	46.	
Cattle	dig prot %	*	2.2	2.4	
Goats	dig prot %	*	3.4	3.6	
Horses	dig prot %	*	3.4	3.6	
Sheep	dig prot %		3.0	3.2	
Energy	GE Mcal/kg				
Cattle	DE Mcal/kg	*	3.17	3.41	
Sheep	DE Mcal/kg	*	3.20	3.44	
Cattle	ME Mcal/kg	*	2.60	2.80	
Sheep	ME Mcal/kg	*	2.62	2.82	
Cattle	TDN %	*	71.9	77.4	
Sheep	TDN %		72.5	78.0	

CITRUS, LIME. Citrus aurantifolia

Citrus, lime, pulp wo fines, shredded dehy, (4)
Lime pulp, dried
Ref No 4 01 249 United States

Analyses			As Fed	Dry	C.V. ± %
Dry matter	%		84.9	100.0	
Ash	%		5.1	6.0	
Crude fiber	%		15.2	17.9	
Ether extract	%		2.9	3.4	
N-free extract	%		54.0	63.6	
Protein (N x 6.25)	%		7.7	9.1	
Cattle	dig prot %	*	3.7	4.3	
Goats	dig prot %	*	4.7	5.5	
Horses	dig prot %	*	4.7	5.5	
Sheep	dig prot %	*	4.7	5.5	
Energy	GE Mcal/kg				
Cattle	DE Mcal/kg	*	2.61	3.07	
Sheep	DE Mcal/kg	*	2.76	3.25	
Cattle	ME Mcal/kg	*	2.14	2.52	
Sheep	ME Mcal/kg	*	2.26	2.67	
Cattle	TDN %	*	59.2	69.7	
Sheep	TDN %	*	62.6	73.7	

CITRUS, ORANGE. Citrus sinensis

Citrus, orange, pulp, ensiled, (3)
Ref No 3 01 250 United States

Analyses			As Fed	Dry	C.V. ± %
Dry matter	%		16.0	100.0	
Ash	%		0.8	5.0	
Crude fiber	%		2.7	16.7	
Sheep	dig coef %		77.	77.	
Ether extract	%		0.3	1.9	
Sheep	dig coef %		65.	65.	
N-free extract	%		11.0	69.0	
Sheep	dig coef %		94.	94.	
Protein (N x 6.25)	%		1.2	7.5	
Sheep	dig coef %		53.	53.	
Cattle	dig prot %	*	0.5	3.0	
Goats	dig prot %	*	0.5	3.0	
Horses	dig prot %	*	0.5	3.0	
Sheep	dig prot %		0.6	4.0	
Energy	GE Mcal/kg				
Cattle	DE Mcal/kg	*	0.58	3.65	
Sheep	DE Mcal/kg	*	0.59	3.70	
Cattle	ME Mcal/kg	*	0.48	2.99	
Sheep	ME Mcal/kg	*	0.48	3.04	
Cattle	TDN %	*	13.2	82.7	
Sheep	TDN %		13.4	84.0	

Citrus, orange, fruit, fresh, (4)
Ref No 2 01 251 United States

Analyses			As Fed	Dry	C.V. ± %
Dry matter	%		15.9	100.0	
Ash	%		0.7	4.4	
Crude fiber	%		1.8	11.3	
Ether extract	%		0.3	1.9	
N-free extract	%		11.9	74.8	
Protein (N x 6.25)	%		1.2	7.5	
Cattle	dig prot %	*	0.5	2.9	
Goats	dig prot %	*	0.7	4.2	
Horses	dig prot %	*	0.7	4.2	
Sheep	dig prot %	*	0.7	4.2	
Energy	GE Mcal/kg				
Cattle	DE Mcal/kg	*	0.64	4.03	
Sheep	DE Mcal/kg	*	0.61	3.81	
Cattle	ME Mcal/kg	*	0.52	3.30	
Sheep	ME Mcal/kg	*	0.50	3.12	
Cattle	TDN %	*	14.5	91.3	
Sheep	TDN %	*	13.7	86.4	
Calcium	%		0.09	0.57	
Phosphorus	%		0.02	0.13	

Citrus, orange, fruit, fresh, cull, (4)
Ref No 2 01 252 United States

Analyses			As Fed	Dry	C.V. ± %
Dry matter	%		12.8	100.0	
Ash	%		0.6	4.7	
Crude fiber	%		1.2	9.4	
Sheep	dig coef %		82.	82.	
Ether extract	%		0.1	0.8	
Sheep	dig coef %		44.	44.	
N-free extract	%		9.9	77.3	
Sheep	dig coef %		99.	99.	
Protein (N x 6.25)	%		1.0	7.8	
Sheep	dig coef %		64.	64.	
Cattle	dig prot %	*	0.4	3.2	
Goats	dig prot %	*	0.6	4.4	
Horses	dig prot %	*	0.6	4.4	
Sheep	dig prot %		0.6	5.0	
Energy	GE Mcal/kg				
Cattle	DE Mcal/kg	*	0.53	4.17	
Sheep	DE Mcal/kg	*	0.51	3.97	
Cattle	ME Mcal/kg	*	0.44	3.42	
Sheep	ME Mcal/kg	*	0.42	3.25	
Cattle	TDN %	*	12.1	94.6	
Sheep	TDN %		11.5	90.0	

Citrus, orange, pulp, shredded wet, (4)
Orange pulp, wet
Ref No 2 01 253 United States

Analyses			As Fed	Dry	C.V. ± %
Dry matter	%		25.5	100.0	41
Ash	%		1.2	4.7	
Crude fiber	%		3.3	13.0	
Ether extract	%		0.5	1.8	
N-free extract	%		18.2	71.5	
Protein (N x 6.25)	%		2.3	8.9	22
Cattle	dig prot %	*	1.1	4.2	
Goats	dig prot %	*	1.4	5.4	
Horses	dig prot %	*	1.4	5.4	
Sheep	dig prot %	*	1.4	5.4	
Calcium	%		0.05	0.21	
Chlorine	%		0.01	0.05	
Cobalt	mg/kg		0.053	0.209	
Copper	mg/kg		7.9	31.1	
Iron	%		0.006	0.023	
Magnesium	%		0.06	0.25	
Manganese	mg/kg		10.4	41.0	
Phosphorus	%		0.07	0.28	
Potassium	%		0.27	1.05	
Sodium	%		0.01	0.02	

(1) dry forages and roughages (3) silages (6) minerals
(2) pasture, range plants, and forages fed green (4) energy feeds (7) vitamins
(5) protein supplements (8) additives

Citrus, orange, pulp wo fines, ammoniated shredded dehy, (4)

Orange pulp, ammoniated

Ref No 4 01 255 United States

Feed Name or Analyses			Mean As Fed	Dry	C.V. ± %
Dry matter	%		89.0	100.0	2
Ash	%		5.6	6.3	55
Crude fiber	%		12.4	13.9	12
Ether extract	%		3.3	3.7	17
N-free extract	%		53.0	59.6	
Protein (N x 6.25)	%		14.7	16.5	20
Cattle	dig prot %	*	10.0	11.2	
Goats	dig prot %	*	11.0	12.4	
Horses	dig prot %	*	11.0	12.4	
Sheep	dig prot %	*	11.0	12.4	

Citrus, orange, pulp wo fines, shredded dehy, (4)

Orange pulp, dried

Ref No 4 01 254 United States

Feed Name or Analyses			Mean As Fed	Dry	C.V. ± %
Dry matter	%		88.2	100.0	2
Ash	%		3.7	4.2	41
Crude fiber	%		8.5	9.6	24
Sheep	dig coef %		84.	84.	
Ether extract	%		1.7	1.9	40
Sheep	dig coef %		49.	49.	
N-free extract	%		66.9	75.9	
Sheep	dig coef %		95.	95.	
Protein (N x 6.25)	%		7.5	8.5	12
Sheep	dig coef %		79.	79.	
Cattle	dig prot %	*	3.3	3.8	
Goats	dig prot %	*	4.4	5.0	
Horses	dig prot %	*	4.4	5.0	
Sheep	dig prot %		5.9	6.7	
Energy	GE Mcal/kg				
Cattle	DE Mcal/kg	*	3.63	4.11	
Sheep	DE Mcal/kg	*	3.46	3.92	
Cattle	ME Mcal/kg	*	2.97	3.37	
Sheep	ME Mcal/kg	*	2.84	3.21	
Cattle	TDN %	*	82.2	93.2	
Sheep	TDN %		78.4	88.9	
Calcium	%		0.62	0.71	
Phosphorus	%		0.10	0.11	

CITRUS, TANGERINE. Citrus reticulata

Citrus, tangerine, pulp wo fines, shredded dehy, (4)

Tangerine pulp, dried

Ref No 4 01 257 United States

Feed Name or Analyses			Mean As Fed	Dry	C.V. ± %
Dry matter	%		87.5	100.0	
Ash	%		4.4	5.0	
Crude fiber	%		9.6	11.0	
Ether extract	%		5.0	5.7	
N-free extract	%		61.4	70.2	
Protein (N x 6.25)	%		7.1	8.1	
Cattle	dig prot %	*	3.0	3.5	
Goats	dig prot %	*	4.1	4.7	
Horses	dig prot %	*	4.1	4.7	
Sheep	dig prot %	*	4.1	4.7	
Calcium	%		1.37	1.57	
Magnesium	%		0.12	0.14	
Phosphorus	%		0.12	0.14	

Citrus meal, dried –
see Citrus, pulp fines, shredded dehy, (4)

Citrus molasses (AAFCO) –
see Citrus, syrup, mn 45% invert sugar mn 71 degrees brix, (4)

Citrus molasses, ammoniated –
see Citrus, syrup, ammoniated, (5)

Citrus pulp, dried –
see Citrus, pulp wo fines, shredded dehy, (4)

Citrus pulp with molasses, dried –
see Citrus, pulp w molasses, dehy grnd, (4)

Citrus seed meal, mechanical extracted (AAFCO) –
see Citrus, seeds, mech-extd grnd, (5)

CLAMS. Mercenaria mercenaria, Mya arenaria, Siliqua patula

Clams, meat, raw, (5)

Ref No 5 01 258 United States

Feed Name or Analyses			Mean As Fed	Dry	C.V. ± %
Dry matter	%		14.1	100.0	
Ash	%		2.6	18.4	
Ether extract	%		0.9	6.4	
N-free extract	%		2.5	17.7	
Protein (N x 6.25)	%		8.1	57.4	
Energy	GE Mcal/kg		0.53	3.76	
Cattle	DE Mcal/kg	*	0.46	3.27	
Sheep	DE Mcal/kg	*	0.46	3.27	
Swine	DE kcal/kg	*	476.	3378.	
Cattle	ME Mcal/kg	*	0.38	2.68	
Sheep	ME Mcal/kg	*	0.38	2.68	
Swine	ME kcal/kg	*	402.	2851.	
Cattle	TDN %	*	10.4	74.1	
Sheep	TDN %	*	10.5	74.1	
Swine	TDN %	*	10.8	76.6	
Phosphorus	%		0.20	1.40	

Clams, shells, grnd, (6)

Ref No 6 01 259 United States

Feed Name or Analyses			Mean As Fed	Dry	C.V. ± %
Dry matter	%		99.4	100.0	
Ash	%		66.9	67.3	
Protein (N x 6.25)	%		1.4	1.4	
Calcium	%		36.39	36.61	2
Chlorine	%		0.02	0.02	
Iron	%		0.464	0.467	
Magnesium	%		0.62	0.62	
Manganese	mg/kg		336.4	338.5	
Phosphorus	%		0.03	0.03	58
Sodium	%		0.51	0.51	

CLARKIA, ROSE. Clarkia elegans

Clarkia, rose, seeds w hulls, (5)

Ref No 5 09 011 United States

Feed Name or Analyses			Mean As Fed	Dry	C.V. ± %
Dry matter	%			100.0	
Protein (N x 6.25)	%			29.0	
Alanine	%			1.33	
Arginine	%			3.51	
Aspartic acid	%			2.76	
Glutamic acid	%			5.37	
Glycine	%			1.77	
Histidine	%			0.70	
Isoleucine	%			0.96	
Leucine	%			2.03	
Lysine	%			0.81	
Methionine	%			0.81	
Phenylalanine	%			1.39	
Proline	%			0.87	
Serine	%			1.48	
Threonine	%			0.78	
Tyrosine	%			0.78	
Valine	%			1.51	

CLEMATIS, DRUMMOND. Clematis drummondi

Clematis, drummond, aerial part, fresh, mid-bloom, (2)

Ref No 2 01 260 United States

Feed Name or Analyses			Mean As Fed	Dry	C.V. ± %
Dry matter	%			100.0	
Ash	%			5.4	
Crude fiber	%			27.6	
Ether extract	%			2.0	
N-free extract	%			49.2	
Protein (N x 6.25)	%			15.8	
Cattle	dig prot %	*		11.3	
Goats	dig prot %	*		11.3	
Horses	dig prot %	*		10.9	
Rabbits	dig prot %	*		10.9	
Sheep	dig prot %	*		11.7	
Calcium	%			1.47	
Magnesium	%			0.19	
Phosphorus	%			0.21	
Potassium	%			1.55	

Clipped oat by-product (AAFCO) –
see Oats, grain clippings, (1)

CLOVER. Trifolium spp

Clover, aerial part, dehy, full bloom, (1)

Ref No 1 01 261 United States

Feed Name or Analyses			Mean As Fed	Dry	C.V. ± %
Dry matter	%		85.0	100.0	
Ash	%		6.8	8.0	
Crude fiber	%		26.0	30.6	
Sheep	dig coef %		72.	72.	
Ether extract	%		1.8	2.1	
Sheep	dig coef %		13.	13.	
N-free extract	%		38.7	45.5	
Sheep	dig coef %		77.	77.	
Protein (N x 6.25)	%		11.7	13.8	
Sheep	dig coef %		63.	63.	
Cattle	dig prot %	*	7.6	8.9	
Goats	dig prot %	*	8.0	9.4	
Horses	dig prot %	*	7.9	9.2	
Rabbits	dig prot %	*	7.9	9.3	
Sheep	dig prot %		7.4	8.7	
Energy	GE Mcal/kg				
Cattle	DE Mcal/kg	*	2.38	2.80	
Sheep	DE Mcal/kg	*	2.49	2.93	
Cattle	ME Mcal/kg	*	1.96	2.30	
Sheep	ME Mcal/kg	*	2.04	2.40	
Cattle	TDN %	*	54.0	63.5	
Sheep	TDN %		56.4	66.4	

Clover, hay, dehy, (1)

Ref No 1 01 264 United States

Feed Name or Analyses				Mean As Fed	Mean Dry	C.V. ± %
Dry matter		%		87.1	100.0	4
Ash		%		6.3	7.3	6
Crude fiber		%		23.0	26.5	24
Sheep	dig coef	%		73.	73.	
Ether extract		%		3.1	3.6	31
Sheep	dig coef	%		33.	33.	
N-free extract		%		40.7	46.8	
Sheep	dig coef	%		73.	73.	
Protein (N x 6.25)		%		13.9	16.0	26
Sheep	dig coef	%		65.	65.	
Cattle	dig prot	%	*	9.4	10.8	
Goats	dig prot	%	*	10.0	11.4	
Horses	dig prot	%	*	9.6	11.1	
Rabbits	dig prot	%	*	9.6	11.0	
Sheep	dig prot	%		9.0	10.4	
Energy	GE Mcal/kg					
Cattle	DE Mcal/kg		*	2.32	2.67	
Sheep	DE Mcal/kg		*	2.55	2.93	
Cattle	ME Mcal/kg		*	1.91	2.19	
Sheep	ME Mcal/kg		*	2.09	2.40	
Cattle	TDN	%	*	52.7	60.6	
Sheep	TDN	%		57.9	66.5	

Clover, hay, dehy, early bloom, (1)

Ref No 1 01 263 United States

Feed Name or Analyses				Mean As Fed	Mean Dry	C.V. ± %
Dry matter		%		86.9	100.0	5
Ash		%		6.8	7.8	
Crude fiber		%		19.9	22.9	25
Cattle	dig coef	%		52.	52.	
Ether extract		%		2.8	3.2	38
Cattle	dig coef	%		44.	44.	
N-free extract		%		42.6	49.1	
Cattle	dig coef	%		76.	76.	
Protein (N x 6.25)		%		14.8	17.1	23
Cattle	dig coef	%		60.	60.	
Cattle	dig prot	%		8.9	10.2	
Goats	dig prot	%	*	10.8	12.5	
Horses	dig prot	%	*	10.4	12.0	
Rabbits	dig prot	%	*	10.3	11.8	
Sheep	dig prot	%	*	10.3	11.9	
Energy	GE Mcal/kg					
Cattle	DE Mcal/kg		*	2.40	2.76	
Sheep	DE Mcal/kg		*	2.40	2.77	
Cattle	ME Mcal/kg		*	1.97	2.26	
Sheep	ME Mcal/kg		*	1.97	2.27	
Cattle	TDN	%		54.4	62.6	
Sheep	TDN	%	*	54.5	62.8	

Clover, hay, fan air dried, (1)

Ref No 1 01 267 United States

Feed Name or Analyses				Mean As Fed	Mean Dry	C.V. ± %
Dry matter		%		96.0	100.0	
Ash		%		7.8	8.1	
Crude fiber		%		32.1	33.4	
Sheep	dig coef	%		62.	62.	
Ether extract		%		1.8	1.9	
Sheep	dig coef	%		41.	41.	
N-free extract		%		42.2	44.0	
Sheep	dig coef	%		66.	66.	
Protein (N x 6.25)		%		12.1	12.6	
Sheep	dig coef	%		64.	64.	
Cattle	dig prot	%	*	7.5	7.9	
Goats	dig prot	%	*	8.0	8.3	
Horses	dig prot	%	*	7.9	8.2	
Rabbits	dig prot	%	*	8.1	8.4	
Sheep	dig prot	%		7.7	8.1	
Energy	GE Mcal/kg					
Cattle	DE Mcal/kg		*	2.40	2.50	
Sheep	DE Mcal/kg		*	2.52	2.63	
Cattle	ME Mcal/kg		*	1.97	2.05	
Sheep	ME Mcal/kg		*	2.07	2.15	
Cattle	TDN	%	*	54.5	56.8	
Sheep	TDN	%		57.2	59.6	
Vitamin D2	IU/g			0.5	0.5	

(1) dry forages and roughages (2) pasture, range plants, and forages fed green (3) silages (4) energy feeds (5) protein supplements (6) minerals (7) vitamins (8) additives

Clover, hay, fan air dried, late bloom, cut 2, (1)

Ref No 1 01 265 United States

Feed Name or Analyses				Mean As Fed	Mean Dry	C.V. ± %
Dry matter		%		85.4	100.0	
Ash		%		7.3	8.5	
Crude fiber		%		19.7	23.1	
Sheep	dig coef	%		46.	46.	
Ether extract		%		3.8	4.4	
Sheep	dig coef	%		60.	60.	
N-free extract		%		37.2	43.6	
Sheep	dig coef	%		62.	62.	
Protein (N x 6.25)		%		17.4	20.4	
Sheep	dig coef	%		67.	67.	
Cattle	dig prot	%	*	12.5	14.6	
Goats	dig prot	%	*	13.3	15.6	
Horses	dig prot	%	*	12.7	14.8	
Rabbits	dig prot	%	*	12.3	14.4	
Sheep	dig prot	%		11.7	13.7	
Energy	GE Mcal/kg					
Cattle	DE Mcal/kg		*	2.27	2.66	
Sheep	DE Mcal/kg		*	2.16	2.52	
Cattle	ME Mcal/kg		*	1.86	2.18	
Sheep	ME Mcal/kg		*	1.77	2.07	
Cattle	TDN	%	*	51.5	60.3	
Sheep	TDN	%		48.9	57.3	

Clover, hay, fan air dried, cut 2, (1)

Ref No 1 01 266 United States

Feed Name or Analyses				Mean As Fed	Mean Dry	C.V. ± %
Dry matter		%		88.0	100.0	
Ash		%		7.5	8.5	
Crude fiber		%		23.2	26.4	
Sheep	dig coef	%		48.	48.	
Ether extract		%		3.3	3.7	
Sheep	dig coef	%		57.	57.	
N-free extract		%		37.8	43.0	
Sheep	dig coef	%		62.	62.	
Protein (N x 6.25)		%		16.2	18.4	
Sheep	dig coef	%		65.	65.	
Cattle	dig prot	%	*	11.3	12.9	
Goats	dig prot	%	*	12.1	13.7	
Horses	dig prot	%	*	11.6	13.2	
Rabbits	dig prot	%	*	11.3	12.9	
Sheep	dig prot	%		10.5	12.0	
Energy	GE Mcal/kg					
Cattle	DE Mcal/kg		*	2.39	2.72	
Sheep	DE Mcal/kg		*	2.17	2.47	
Cattle	ME Mcal/kg		*	1.96	2.23	
Sheep	ME Mcal/kg		*	1.78	2.03	
Cattle	TDN	%	*	54.2	61.6	
Sheep	TDN	%		49.3	56.0	

Clover, hay, s-c, (1)

Ref No 1 01 278 United States

Feed Name or Analyses				Mean As Fed	Mean Dry	C.V. ± %
Dry matter		%		85.0	100.0	4
Ash		%		6.2	7.4	15
Crude fiber		%		28.8	33.8	8
Cattle	dig coef	%		51.	51.	
Horses	dig coef	%		38.	38.	
Sheep	dig coef	%		52.	52.	
Ether extract		%		1.9	2.2	14
Cattle	dig coef	%		53.	53.	
Horses	dig coef	%		29.	29.	
Sheep	dig coef	%		38.	38.	
N-free extract		%		36.3	42.8	4
Cattle	dig coef	%		66.	66.	
Horses	dig coef	%		64.	64.	
Sheep	dig coef	%		66.	66.	
Protein (N x 6.25)		%		11.8	13.8	18
Cattle	dig coef	%		56.	56.	
Horses	dig coef	%		56.	56.	
Sheep	dig coef	%		59.	59.	
Cattle	dig prot	%		6.6	7.7	
Goats	dig prot	%	*	8.0	9.5	
Horses	dig prot	%		6.6	7.7	
Rabbits	dig prot	%	*	7.9	9.3	
Sheep	dig prot	%		6.9	8.2	
Lignin (Ellis)		%		10.6	12.5	
Energy	GE Mcal/kg					
Cattle	DE Mcal/kg		*	2.09	2.45	
Horses	DE Mcal/kg		*	1.85	2.18	
Sheep	DE Mcal/kg		*	2.09	2.46	
Cattle	ME Mcal/kg		*	1.71	2.01	
Horses	ME Mcal/kg		*	1.52	1.79	
Sheep	ME Mcal/kg		*	1.72	2.02	
Cattle	TDN	%		47.3	55.6	
Horses	TDN	%		42.0	49.4	
Sheep	TDN	%		47.5	55.9	
Calcium		%		1.08	1.27	15
Copper		mg/kg		8.2	9.7	
Iron		%		0.026	0.030	
Magnesium		%		0.26	0.31	
Manganese		mg/kg		37.3	43.9	20
Phosphorus		%		0.20	0.23	16
Potassium		%		2.03	2.39	
Biotin		mg/kg		0.07	0.09	26
α-tocopherol		mg/kg		108.9	128.1	
Vitamin D2		IU/g		1.5	1.8	29

Ref No 1 01 278 Canada

Feed Name or Analyses				Mean As Fed	Mean Dry	C.V. ± %
Dry matter		%		89.0	100.0	
Crude fiber		%		24.7	27.7	
Cattle	dig coef	%		51.	51.	
Horses	dig coef	%		38.	38.	
Sheep	dig coef	%		52.	52.	
Ether extract		%				
Cattle	dig coef	%		53.	53.	
Horses	dig coef	%		29.	29.	
Sheep	dig coef	%		38.	38.	
N-free extract		%				
Cattle	dig coef	%		66.	66.	
Horses	dig coef	%		64.	64.	
Sheep	dig coef	%		66.	66.	
Protein (N x 6.25)		%		14.7	16.5	
Cattle	dig coef	%		56.	56.	
Horses	dig coef	%		56.	56.	

Feed Name or Analyses			Mean As Fed	Mean Dry	C.V. ± %
Sheep dig coef	%		59.	59.	
Cattle dig prot	%		8.2	9.2	
Goats dig prot	%	*	10.6	12.0	
Horses dig prot	%		8.2	9.2	
Rabbits dig prot	%	*	10.1	11.4	
Sheep dig prot	%		8.7	9.7	
Energy	GE Mcal/kg				
Cattle	DE Mcal/kg	*	2.19	2.46	
Horses	DE Mcal/kg	*	1.95	2.19	
Sheep	DE Mcal/kg	*	2.20	2.47	
Cattle	ME Mcal/kg	*	1.79	2.02	
Horses	ME Mcal/kg	*	1.60	1.80	
Sheep	ME Mcal/kg	*	1.80	2.03	
Cattle	TDN %		49.6	55.8	
Horses	TDN %		44.3	49.8	
Sheep	TDN %		49.8	56.0	
Calcium	%		1.21	1.36	
Phosphorus	%		0.20	0.22	
Carotene	mg/kg		34.0	38.3	
Vitamin A equivalent	IU/g		56.7	63.8	

Clover, hay, s-c, early bloom, (1)

Ref No 1 01 270 United States

Analyses			As Fed	Dry	C.V. ± %
Dry matter	%		86.9	100.0	4
Ash	%		9.0	10.3	24
Crude fiber	%		21.3	24.5	23
Ether extract	%		2.3	2.7	35
N-free extract	%		38.2	44.0	
Protein (N x 6.25)	%		16.1	18.5	15
Cattle dig prot	%	*	11.3	13.0	
Goats dig prot	%	*	12.0	13.8	
Horses dig prot	%	*	11.5	13.2	
Rabbits dig prot	%	*	11.3	13.0	
Sheep dig prot	%	*	11.4	13.2	
Energy	GE Mcal/kg				
Cattle	DE Mcal/kg	*	2.44	2.80	
Sheep	DE Mcal/kg	*	2.31	2.65	
Cattle	ME Mcal/kg	*	2.00	2.30	
Sheep	ME Mcal/kg	*	1.89	2.18	
Cattle	TDN %	*	55.2	63.6	
Sheep	TDN %	*	52.3	60.2	

Clover, hay, s-c, early bloom, cut 2, (1)

Ref No 1 01 269 United States

Analyses			As Fed	Dry	C.V. ± %
Dry matter	%		84.9	100.0	
Ash	%		5.9	7.0	
Crude fiber	%		28.4	33.4	
Sheep dig coef	%		68.	68.	
Ether extract	%		1.6	1.9	
Sheep dig coef	%		2.	2.	
N-free extract	%		36.3	42.7	
Sheep dig coef	%		45.	45.	
Protein (N x 6.25)	%		12.7	15.0	
Sheep dig coef	%		60.	60.	
Cattle dig prot	%	*	8.4	9.9	
Goats dig prot	%	*	9.0	10.6	
Horses dig prot	%	*	8.7	10.3	
Rabbits dig prot	%	*	8.7	10.3	
Sheep dig prot	%		7.6	9.0	
Energy	GE Mcal/kg				
Cattle	DE Mcal/kg	*	2.04	2.41	
Sheep	DE Mcal/kg	*	1.91	2.25	
Cattle	ME Mcal/kg	*	1.68	1.98	
Sheep	ME Mcal/kg	*	1.57	1.84	

Feed Name or Analyses			Mean As Fed	Mean Dry	C.V. ± %
Cattle	TDN %	*	46.4	54.6	
Sheep	TDN %		43.3	51.0	

Clover, hay, s-c, full bloom, (1)

Ref No 1 01 271 United States

Analyses			As Fed	Dry	C.V. ± %
Dry matter	%		86.1	100.0	2
Ash	%		6.5	7.6	30
Crude fiber	%		25.8	30.0	16
Cattle dig coef	%		50.	50.	
Horses dig coef	%		39.	39.	
Sheep dig coef	%		56.	56.	
Ether extract	%		2.4	2.8	45
Cattle dig coef	%		61.	61.	
Horses dig coef	%		31.	31.	
Sheep dig coef	%		29.	29.	
N-free extract	%		38.8	45.1	
Cattle dig coef	%		77.	77.	
Horses dig coef	%		67.	67.	
Sheep dig coef	%		58.	58.	
Protein (N x 6.25)	%		12.5	14.6	14
Cattle dig coef	%		67.	67.	
Horses dig coef	%		60.	60.	
Sheep dig coef	%		56.	56.	
Cattle dig prot	%		8.4	9.7	
Goats dig prot	%	*	8.7	10.1	
Horses dig prot	%		7.5	8.7	
Rabbits dig prot	%	*	8.5	9.9	
Sheep dig prot	%		7.0	8.1	
Energy	GE Mcal/kg				
Cattle	DE Mcal/kg	*	2.40	2.79	
Horses	DE Mcal/kg	*	2.00	2.32	
Sheep	DE Mcal/kg	*	2.01	2.33	
Cattle	ME Mcal/kg	*	1.97	2.29	
Horses	ME Mcal/kg	*	1.64	1.90	
Sheep	ME Mcal/kg	*	1.65	1.91	
Cattle	TDN %		54.5	63.3	
Horses	TDN %		45.3	52.6	
Sheep	TDN %		45.6	52.9	
Calcium	%		1.38	1.60	17
Iron	%		0.015	0.018	
Magnesium	%		0.40	0.46	25
Manganese	mg/kg		53.3	61.9	
Phosphorus	%		0.22	0.26	12
Potassium	%		1.58	1.84	7

Clover, hay, s-c, late bloom, (1)

Ref No 1 01 273 United States

Analyses			As Fed	Dry	C.V. ± %
Dry matter	%		88.1	100.0	
Ash	%		6.3	7.1	
Crude fiber	%		30.0	34.1	
Sheep dig coef	%		45.	45.	
Ether extract	%		2.9	3.3	
Sheep dig coef	%		52.	52.	
N-free extract	%		38.9	44.2	
Sheep dig coef	%		64.	64.	
Protein (N x 6.25)	%		10.0	11.3	
Sheep dig coef	%		59.	59.	
Cattle dig prot	%	*	5.9	6.7	
Goats dig prot	%	*	6.3	7.1	
Horses dig prot	%	*	6.3	7.1	
Rabbits dig prot	%	*	6.5	7.4	
Sheep dig prot	%		5.9	6.7	
Energy	GE Mcal/kg				
Cattle	DE Mcal/kg	*	2.19	2.48	
Sheep	DE Mcal/kg	*	2.10	2.39	
Cattle	ME Mcal/kg	*	1.79	2.04	

Feed Name or Analyses			Mean As Fed	Mean Dry	C.V. ± %
Sheep	ME Mcal/kg	*	1.73	1.96	
Cattle	TDN %	*	49.6	56.3	
Sheep	TDN %		47.7	54.2	

Clover, hay, s-c, late bloom, cut 2, (1)

Ref No 1 01 272 United States

Analyses			As Fed	Dry	C.V. ± %
Dry matter	%		84.9	100.0	
Ash	%		7.2	8.5	
Crude fiber	%		21.0	24.7	
Sheep dig coef	%		49.	49.	
Ether extract	%		3.6	4.2	
Sheep dig coef	%		59.	59.	
N-free extract	%		37.4	44.1	
Sheep dig coef	%		64.	64.	
Protein (N x 6.25)	%		15.7	18.5	
Sheep dig coef	%		63.	63.	
Cattle dig prot	%	*	11.0	13.0	
Goats dig prot	%	*	11.7	13.8	
Horses dig prot	%	*	11.2	13.2	
Rabbits dig prot	%	*	11.0	13.0	
Sheep dig prot	%		9.9	11.7	
Energy	GE Mcal/kg				
Cattle	DE Mcal/kg	*	2.36	2.78	
Sheep	DE Mcal/kg	*	2.15	2.54	
Cattle	ME Mcal/kg	*	1.94	2.28	
Sheep	ME Mcal/kg	*	1.77	2.08	
Cattle	TDN %	*	53.6	63.1	
Sheep	TDN %		48.9	57.6	

Clover, hay, s-c, milk stage, (1)

Ref No 1 01 274 United States

Analyses			As Fed	Dry	C.V. ± %
Dry matter	%		89.6	100.0	3
Ash	%		7.3	8.1	6
Crude fiber	%		26.1	29.1	12
Ether extract	%		2.6	2.9	21
N-free extract	%		39.6	44.2	
Protein (N x 6.25)	%		14.1	15.7	15
Cattle dig prot	%	*	9.4	10.5	
Goats dig prot	%	*	10.0	11.2	
Horses dig prot	%	*	9.7	10.9	
Rabbits dig prot	%	*	9.7	10.8	
Sheep dig prot	%	*	9.5	10.7	
Energy	GE Mcal/kg				
Cattle	DE Mcal/kg	*	2.34	2.61	
Sheep	DE Mcal/kg	*	2.31	2.58	
Cattle	ME Mcal/kg	*	1.92	2.14	
Sheep	ME Mcal/kg	*	1.89	2.11	
Cattle	TDN %	*	53.0	59.2	
Sheep	TDN %	*	52.3	58.4	
Calcium	%		0.96	1.07	
Iron	%		0.012	0.013	
Magnesium	%		0.24	0.27	
Manganese	mg/kg		56.9	63.5	43
Phosphorus	%		0.24	0.27	
Potassium	%		1.75	1.95	

Clover, hay, s-c, mature, (1)

Ref No 1 01 275 United States

Analyses			As Fed	Dry	C.V. ± %
Dry matter	%		91.4	100.0	2
Calcium	%		1.15	1.26	14
Iron	%		0.022	0.024	74
Magnesium	%		0.33	0.36	15
Phosphorus	%		0.21	0.23	29

Continued

Feed Name or Analyses			Mean As Fed	Mean Dry	C.V. ± %
Potassium	%		1.79	1.96	31
Carotene	mg/kg		3.0	3.3	49
Vitamin A equivalent	IU/g		5.0	5.5	

Ref No 1 01 275 Canada

Feed Name or Analyses			Mean As Fed	Mean Dry	C.V. ± %
Dry matter	%		87.9	100.0	
Crude fiber	%		24.3	27.6	
Protein (N x 6.25)	%		11.1	12.6	
Cattle	dig prot %	*	6.9	7.9	
Goats	dig prot %	*	7.3	8.3	
Horses	dig prot %	*	7.3	8.3	
Rabbits	dig prot %	*	7.4	8.4	
Sheep	dig prot %	*	6.9	7.9	
Calcium	%		1.13	1.29	
Phosphorus	%		0.17	0.19	
Carotene	mg/kg		88.2	100.3	
Vitamin A equivalent	IU/g		147.0	167.2	

Clover, hay, s-c, cut 1, (1)
Ref No 1 01 276 United States

Feed Name or Analyses			Mean As Fed	Mean Dry	C.V. ± %
Dry matter	%		88.9	100.0	2
Sugars, total	%		5.9	6.6	
Lignin (Ellis)	%		11.2	12.6	29
Calcium	%		1.14	1.28	16
Cobalt	mg/kg		0.147	0.165	51
Copper	mg/kg		8.6	9.7	
Iron	%		0.027	0.030	
Magnesium	%		0.28	0.31	
Manganese	mg/kg		39.0	43.9	20
Phosphorus	%		0.21	0.24	15
Potassium	%		2.13	2.39	
Carotene	mg/kg		20.2	22.7	73
Niacin	mg/kg		9.8	11.0	
Vitamin A equivalent	IU/g		33.7	37.9	
Vitamin D2	IU/g		1.7	2.0	12

Clover, hay, s-c, cut 2, (1)
Ref No 1 01 277 United States

Feed Name or Analyses			Mean As Fed	Mean Dry	C.V. ± %
Dry matter	%		87.6	100.0	2
Ash	%		7.2	8.2	20
Crude fiber	%		25.5	29.2	12
Sheep	dig coef %		55.	55.	
Ether extract	%		2.5	2.9	19
Sheep	dig coef %		45.	45.	
N-free extract	%		37.9	43.3	
Sheep	dig coef %		60.	60.	
Protein (N x 6.25)	%		14.5	16.5	9
Sheep	dig coef %		62.	62.	
Cattle	dig prot %	*	9.8	11.2	
Goats	dig prot %	*	10.5	12.0	
Horses	dig prot %	*	10.1	11.5	
Rabbits	dig prot %	*	10.0	11.4	
Sheep	dig prot %		9.0	10.2	
Energy	GE Mcal/kg				
Cattle	DE Mcal/kg	*	2.28	2.61	
Sheep	DE Mcal/kg	*	2.13	2.43	
Cattle	ME Mcal/kg	*	1.87	2.14	
Sheep	ME Mcal/kg	*	1.75	1.99	
Cattle	TDN %	*	51.8	59.1	
Sheep	TDN %		48.3	55.1	

(1) dry forages and roughages (3) silages (6) minerals
(2) pasture, range plants, and forages fed green (4) energy feeds (7) vitamins
(5) protein supplements (8) additives

Feed Name or Analyses			Mean As Fed	Mean Dry	C.V. ± %
Carotene	mg/kg		31.6	36.0	98
Vitamin A equivalent	IU/g		52.6	60.1	

Clover, hay, s-c, gr 1 US, (1)
Ref No 1 01 280 United States

Feed Name or Analyses			Mean As Fed	Mean Dry	C.V. ± %
Dry matter	%		91.7	100.0	2
Lignin (Ellis)	%		13.2	14.4	20
Calcium	%		1.25	1.36	15
Cobalt	mg/kg		0.157	0.171	54
Manganese	mg/kg		37.4	40.8	16
Phosphorus	%		0.20	0.22	21
Carotene	mg/kg		26.3	28.7	98
Vitamin A equivalent	IU/g		43.8	47.8	
Vitamin D2	IU/g		1.4	1.5	11

Clover, hay, s-c, gr 2 US, (1)
Ref No 1 01 281 United States

Feed Name or Analyses			Mean As Fed	Mean Dry	C.V. ± %
Dry matter	%		86.8	100.0	3
Sugars, total	%		4.3	5.0	
Lignin (Ellis)	%		10.7	12.3	13
Calcium	%		0.95	1.10	17
Cobalt	mg/kg		0.157	0.181	43
Copper	mg/kg		9.4	10.8	
Iron	%		0.031	0.036	
Magnesium	%		0.40	0.46	
Manganese	mg/kg		36.7	42.3	23
Phosphorus	%		0.20	0.23	15
Potassium	%		1.97	2.27	
Carotene	mg/kg		14.0	16.1	79
Niacin	mg/kg		9.6	11.0	
Vitamin A equivalent	IU/g		23.3	26.8	
Vitamin D2	IU/g		1.9	2.2	4

Clover, hay, s-c, gr 3 US, (1)
Ref No 1 01 282 United States

Feed Name or Analyses			Mean As Fed	Mean Dry	C.V. ± %
Dry matter	%		91.1	100.0	
Lignin (Ellis)	%		16.9	18.5	13
Calcium	%		0.92	1.01	33
Iron	%		0.026	0.029	
Magnesium	%		0.36	0.40	
Phosphorus	%		0.20	0.22	13
Potassium	%		1.65	1.81	
Carotene	mg/kg		18.1	19.8	98
Vitamin A equivalent	IU/g		30.1	33.1	
Vitamin D2	IU/g		1.8	2.0	

Clover, hay, s-c, gr sample US, (1)
Ref No 1 01 279 United States

Feed Name or Analyses			Mean As Fed	Mean Dry	C.V. ± %
Dry matter	%			100.0	
Lignin (Ellis)	%			19.8	
Calcium	%			1.30	11
Cobalt	mg/kg			0.119	
Manganese	mg/kg			102.3	
Phosphorus	%			0.21	7
Carotene	mg/kg			9.3	45
Vitamin A equivalent	IU/g			15.4	

Clover, hay, s-c, grnd, (1)
Ref No 1 01 284 United States

Feed Name or Analyses			Mean As Fed	Mean Dry	C.V. ± %
Dry matter	%		83.8	100.0	
Ash	%		7.4	8.8	
Crude fiber	%		21.5	25.7	
Sheep	dig coef %		34.	34.	
Ether extract	%		1.2	1.4	
Sheep	dig coef %		36.	36.	
N-free extract	%		42.2	50.3	
Sheep	dig coef %		69.	69.	
Protein (N x 6.25)	%		11.6	13.8	
Sheep	dig coef %		53.	53.	
Cattle	dig prot %	*	7.5	8.9	
Goats	dig prot %	*	7.9	9.4	
Horses	dig prot %	*	7.7	9.2	
Rabbits	dig prot %	*	7.8	9.3	
Sheep	dig prot %		6.1	7.3	
Energy	GE Mcal/kg				
Cattle	DE Mcal/kg	*	2.06	2.46	
Sheep	DE Mcal/kg	*	1.92	2.29	
Cattle	ME Mcal/kg	*	1.69	2.02	
Sheep	ME Mcal/kg	*	1.57	1.88	
Cattle	TDN %	*	46.7	55.7	
Sheep	TDN %		43.5	51.9	

Clover, hay, s-c on riders, (1)
Ref No 1 01 287 United States

Feed Name or Analyses			Mean As Fed	Mean Dry	C.V. ± %
Dry matter	%		90.2	100.0	
Ash	%		6.6	7.3	
Crude fiber	%		28.7	31.8	
Sheep	dig coef %		66.	66.	
Ether extract	%		2.0	2.2	
Sheep	dig coef %		45.	45.	
N-free extract	%		40.7	45.1	
Sheep	dig coef %		66.	66.	
Protein (N x 6.25)	%		12.3	13.6	
Sheep	dig coef %		65.	65.	
Cattle	dig prot %	*	7.9	8.7	
Goats	dig prot %	*	8.3	9.2	
Horses	dig prot %	*	8.2	9.1	
Rabbits	dig prot %	*	8.3	9.2	
Sheep	dig prot %		8.0	8.8	
Energy	GE Mcal/kg				
Cattle	DE Mcal/kg	*	2.24	2.48	
Sheep	DE Mcal/kg	*	2.46	2.73	
Cattle	ME Mcal/kg	*	1.84	2.04	
Sheep	ME Mcal/kg	*	2.02	2.24	
Cattle	TDN %	*	50.8	56.3	
Sheep	TDN %		55.8	61.8	

Clover, hay, s-c on riders, early bloom, (1)
Ref No 1 01 285 United States

Feed Name or Analyses			Mean As Fed	Mean Dry	C.V. ± %
Dry matter	%		83.0	100.0	
Ash	%		6.6	8.0	
Crude fiber	%		24.8	29.9	
Sheep	dig coef %		74.	74.	
Ether extract	%		1.8	2.2	
Sheep	dig coef %		12.	12.	
N-free extract	%		35.9	43.3	
Sheep	dig coef %		55.	55.	
Protein (N x 6.25)	%		13.8	16.6	
Sheep	dig coef %		67.	67.	
Cattle	dig prot %	*	9.4	11.3	
Goats	dig prot %	*	10.0	12.0	

Feed Name or Analyses			Mean As Fed	Mean Dry	C.V. ± %
Horses	dig prot %	*	9.6	11.6	
Rabbits	dig prot %	*	9.5	11.5	
Sheep	dig prot %		9.2	11.1	
Energy	GE Mcal/kg				
Cattle	DE Mcal/kg	*	2.12	2.55	
Sheep	DE Mcal/kg	*	2.11	2.54	
Cattle	ME Mcal/kg	*	1.73	2.09	
Sheep	ME Mcal/kg	*	1.73	2.08	
Cattle	TDN %	*	48.0	57.8	
Sheep	TDN %		47.9	57.7	

Clover, hay, s-c on riders, full bloom, (1)

Ref No 1 01 286 United States

Feed Name or Analyses			Mean As Fed	Mean Dry	C.V. ± %
Dry matter	%		83.3	100.0	
Ash	%		4.9	5.9	
Crude fiber	%		27.2	32.6	
Sheep	dig coef %		82.	82.	
Ether extract	%		1.9	2.3	
Sheep	dig coef %		40.	40.	
N-free extract	%		39.1	46.9	
Sheep	dig coef %		63.	63.	
Protein (N x 6.25)	%		10.2	12.3	
Sheep	dig coef %		59.	59.	
Cattle	dig prot %	*	6.3	7.6	
Goats	dig prot %	*	6.7	8.0	
Horses	dig prot %	*	6.6	8.0	
Rabbits	dig prot %	*	6.8	8.2	
Sheep	dig prot %		6.0	7.3	
Energy	GE Mcal/kg				
Cattle	DE Mcal/kg	*	2.36	2.83	
Sheep	DE Mcal/kg	*	2.41	2.89	
Cattle	ME Mcal/kg	*	1.93	2.32	
Sheep	ME Mcal/kg	*	1.98	2.37	
Cattle	TDN %	*	53.5	64.2	
Sheep	TDN %		54.6	65.6	

Clover, hay, s-c weathered, (1)

Ref No 1 01 288 United States

Feed Name or Analyses			Mean As Fed	Mean Dry	C.V. ± %
Dry matter	%		93.4	100.0	
Ash	%		9.4	10.1	
Crude fiber	%		31.6	33.8	
Sheep	dig coef %		57.	57.	
Ether extract	%		1.3	1.4	
Sheep	dig coef %		10.	10.	
N-free extract	%		36.8	39.4	
Sheep	dig coef %		58.	58.	
Protein (N x 6.25)	%		14.3	15.3	
Sheep	dig coef %		61.	61.	
Cattle	dig prot %	*	9.5	10.2	
Goats	dig prot %	*	10.1	10.8	
Horses	dig prot %	*	9.8	10.5	
Rabbits	dig prot %	*	9.8	10.5	
Sheep	dig prot %		8.7	9.3	
Energy	GE Mcal/kg				
Cattle	DE Mcal/kg	*	2.34	2.51	
Sheep	DE Mcal/kg	*	2.13	2.28	
Cattle	ME Mcal/kg	*	1.92	2.06	
Sheep	ME Mcal/kg	*	1.75	1.87	
Cattle	TDN %	*	53.1	56.8	
Sheep	TDN %		48.3	51.8	

Clover, stubble, dehy, (1)

Ref No 1 01 290 United States

Feed Name or Analyses			Mean As Fed	Mean Dry	C.V. ± %
Dry matter	%		89.5	100.0	
Ash	%		8.2	9.2	
Crude fiber	%		21.3	23.8	
Sheep	dig coef %		53.	53.	
Ether extract	%		2.8	3.1	
Sheep	dig coef %		58.	58.	
N-free extract	%		40.4	45.1	
Sheep	dig coef %		72.	72.	
Protein (N x 6.25)	%		16.8	18.8	
Sheep	dig coef %		65.	65.	
Cattle	dig prot %	*	11.8	13.2	
Goats	dig prot %	*	12.6	14.1	
Horses	dig prot %	*	12.1	13.5	
Rabbits	dig prot %	*	11.8	13.2	
Sheep	dig prot %		10.9	12.2	
Energy	GE Mcal/kg				
Cattle	DE Mcal/kg	*	2.50	2.79	
Sheep	DE Mcal/kg	*	2.42	2.71	
Cattle	ME Mcal/kg	*	2.05	2.29	
Sheep	ME Mcal/kg	*	1.99	2.22	
Cattle	TDN %	*	56.6	63.3	
Sheep	TDN %		54.9	61.4	

Clover, aerial part, fresh, (2)

Ref No 2 01 299 United States

Feed Name or Analyses			Mean As Fed	Mean Dry	C.V. ± %
Dry matter	%		19.0	100.0	3
Ash	%		1.9	10.2	39
Crude fiber	%		4.6	24.4	19
Cattle	dig coef %		59.	59.	
Sheep	dig coef %		52.	52.	
Ether extract	%		0.8	4.4	27
Cattle	dig coef %		66.	66.	
Sheep	dig coef %		67.	67.	
N-free extract	%		8.1	42.7	
Cattle	dig coef %		80.	80.	
Sheep	dig coef %		71.	71.	
Protein (N x 6.25)	%		3.5	18.4	27
Cattle	dig coef %		70.	70.	
Sheep	dig coef %		67.	67.	
Cattle	dig prot %		2.4	12.9	
Goats	dig prot %	*	2.6	13.7	
Horses	dig prot %	*	2.5	13.2	
Rabbits	dig prot %	*	2.4	12.9	
Sheep	dig prot %		2.3	12.3	
Energy	GE Mcal/kg				
Cattle	DE Mcal/kg	*	0.57	2.99	
Sheep	DE Mcal/kg	*	0.52	2.73	
Cattle	ME Mcal/kg	*	0.47	2.45	
Sheep	ME Mcal/kg	*	0.42	2.24	
Cattle	TDN %		12.9	67.9	
Sheep	TDN %		11.7	61.9	
Chlorine	%		0.13	0.67	20
Cobalt	mg/kg		0.026	0.135	82
Sodium	%		0.05	0.27	41
Sulphur	%		0.04	0.23	33
Zinc	mg/kg		7.3	38.6	53
Carotene	mg/kg		53.8	283.5	35
Niacin	mg/kg		16.5	87.1	37
Riboflavin	mg/kg		4.5	23.6	84
Thiamine	mg/kg		1.6	8.4	39
Vitamin A equivalent	IU/g		89.6	472.6	

Clover, aerial part, fresh, early bloom, (2)

Ref No 2 01 292 United States

Feed Name or Analyses			Mean As Fed	Mean Dry	C.V. ± %
Dry matter	%		20.4	100.0	28
Ash	%		2.0	9.9	10
Crude fiber	%		4.9	23.9	17
Ether extract	%		0.8	4.0	16
N-free extract	%		8.8	43.1	
Protein (N x 6.25)	%		3.9	19.1	19
Cattle	dig prot %	*	2.9	14.1	
Goats	dig prot %	*	2.9	14.4	
Horses	dig prot %	*	2.8	13.7	
Rabbits	dig prot %	*	2.7	13.4	
Sheep	dig prot %	*	3.0	14.8	
Energy	GE Mcal/kg				
Cattle	DE Mcal/kg	*	0.63	3.07	
Sheep	DE Mcal/kg	*	0.58	2.86	
Cattle	ME Mcal/kg	*	0.51	2.52	
Sheep	ME Mcal/kg	*	0.48	2.34	
Cattle	TDN %	*	14.2	69.6	
Sheep	TDN %	*	13.2	64.8	
Calcium	%		0.29	1.43	40
Magnesium	%		0.06	0.30	
Phosphorus	%		0.06	0.31	24
Potassium	%		0.58	2.83	22

Clover, aerial part, fresh, full bloom, (2)

Ref No 2 01 293 United States

Feed Name or Analyses			Mean As Fed	Mean Dry	C.V. ± %
Dry matter	%		26.1	100.0	12
Calcium	%		0.33	1.26	29
Chlorine	%		0.15	0.57	
Iron	%		0.007	0.028	
Magnesium	%		0.10	0.37	28
Manganese	mg/kg		49.3	188.9	
Phosphorus	%		0.08	0.31	21
Potassium	%		0.56	2.13	16
Sodium	%		0.12	0.47	
Sulphur	%		0.05	0.18	21

Clover, aerial part, fresh, late bloom, (2)

Ref No 2 01 294 United States

Feed Name or Analyses			Mean As Fed	Mean Dry	C.V. ± %
Dry matter	%		19.0	100.0	
Ash	%		2.3	11.9	
Crude fiber	%		4.4	23.4	
Cattle	dig coef %		53.	53.	
Sheep	dig coef %		44.	44.	
Ether extract	%		1.1	5.6	
Cattle	dig coef %		65.	65.	
Sheep	dig coef %		72.	72.	
N-free extract	%		7.6	40.0	
Cattle	dig coef %		78.	78.	
Sheep	dig coef %		71.	71.	
Protein (N x 6.25)	%		3.6	19.1	
Cattle	dig coef %		67.	67.	
Sheep	dig coef %		62.	62.	
Cattle	dig prot %		2.4	12.8	
Goats	dig prot %	*	2.7	14.4	
Horses	dig prot %	*	2.6	13.7	
Rabbits	dig prot %	*	2.5	13.4	
Sheep	dig prot %		2.2	11.8	
Energy	GE Mcal/kg				
Cattle	DE Mcal/kg	*	0.54	2.85	
Sheep	DE Mcal/kg	*	0.50	2.63	
Cattle	ME Mcal/kg	*	0.44	2.34	

Continued

Feed Name or Analyses				Mean As Fed	Mean Dry	C.V. ± %
Sheep	ME Mcal/kg		*	0.41	2.16	
Cattle	TDN	%		12.3	64.6	
Sheep	TDN	%		11.3	59.6	

Clover, aerial part, fresh, milk stage, (2)
Ref No 2 01 295 United States

Feed Name or Analyses				Mean As Fed	Mean Dry	C.V. ± %
Dry matter		%			100.0	
Calcium		%			1.38	30
Iron		%			0.015	
Magnesium		%			0.29	
Manganese		mg/kg			275.6	
Phosphorus		%			0.29	17
Potassium		%			2.10	43

Clover, aerial part, fresh, mature, (2)
Ref No 2 01 296 United States

Feed Name or Analyses				Mean As Fed	Mean Dry	C.V. ± %
Dry matter		%		22.8	100.0	
Ash		%		1.6	7.0	
Crude fiber		%		7.4	32.4	
Sheep	dig coef	%		44.	44.	
Ether extract		%		1.4	6.3	
Sheep	dig coef	%		74.	74.	
N-free extract		%		9.7	42.4	
Sheep	dig coef	%		69.	69.	
Protein (N x 6.25)		%		2.7	11.9	
Sheep	dig coef	%		54.	54.	
Cattle	dig prot	%	*	1.8	8.0	
Goats	dig prot	%	*	1.7	7.7	
Horses	dig prot	%	*	1.7	7.6	
Rabbits	dig prot	%	*	1.8	7.9	
Sheep	dig prot	%		1.5	6.4	
Energy	GE Mcal/kg					
Cattle	DE Mcal/kg		*	0.64	2.83	
Sheep	DE Mcal/kg		*	0.61	2.66	
Cattle	ME Mcal/kg		*	0.53	2.32	
Sheep	ME Mcal/kg		*	0.50	2.18	
Cattle	TDN	%	*	14.6	64.1	
Sheep	TDN	%		13.8	60.4	

Clover, aerial part, fresh, cut 1, (2)
Ref No 2 01 297 United States

Feed Name or Analyses				Mean As Fed	Mean Dry	C.V. ± %
Dry matter		%		13.6	100.0	8
Ash		%		2.2	16.0	17
Crude fiber		%		2.5	18.7	14
Ether extract		%		0.6	4.1	13
N-free extract		%		5.4	40.0	
Protein (N x 6.25)		%		2.9	21.2	24
Cattle	dig prot	%	*	2.2	15.9	
Goats	dig prot	%	*	2.2	16.3	
Horses	dig prot	%	*	2.1	15.5	
Rabbits	dig prot	%	*	2.0	15.0	
Sheep	dig prot	%	*	2.3	16.7	
Energy	GE Mcal/kg					
Cattle	DE Mcal/kg		*	0.37	2.72	
Sheep	DE Mcal/kg		*	0.44	3.25	
Cattle	ME Mcal/kg		*	0.30	2.23	
Sheep	ME Mcal/kg		*	0.36	2.66	
Cattle	TDN	%	*	8.4	61.8	
Sheep	TDN	%	*	10.0	73.7	

(1) dry forages and roughages (2) pasture, range plants, and forages fed green (3) silages (4) energy feeds (5) protein supplements (6) minerals (7) vitamins (8) additives

Feed Name or Analyses				Mean As Fed	Mean Dry	C.V. ± %
Calcium		%		0.18	1.32	19
Magnesium		%		0.04	0.32	33
Phosphorus		%		0.05	0.36	24
Potassium		%		0.34	2.49	23

Clover, aerial part, fresh, cut 2, (2)
Ref No 2 01 298 United States

Feed Name or Analyses				Mean As Fed	Mean Dry	C.V. ± %
Dry matter		%		22.3	100.0	28
Ash		%		2.4	10.6	25
Crude fiber		%		5.5	24.8	14
Sheep	dig coef	%		53.	53.	
Ether extract		%		1.0	4.7	9
Sheep	dig coef	%		61.	61.	
N-free extract		%		9.4	42.1	
Sheep	dig coef	%		65.	65.	
Protein (N x 6.25)		%		4.0	17.9	19
Sheep	dig coef	%		62.	62.	
Cattle	dig prot	%	*	2.9	13.1	
Goats	dig prot	%	*	3.0	13.3	
Horses	dig prot	%	*	2.8	12.7	
Rabbits	dig prot	%	*	2.8	12.5	
Sheep	dig prot	%		2.5	11.1	
Energy	GE Mcal/kg					
Cattle	DE Mcal/kg		*	0.62	2.79	
Sheep	DE Mcal/kg		*	0.57	2.56	
Cattle	ME Mcal/kg		*	0.51	2.29	
Sheep	ME Mcal/kg		*	0.47	2.10	
Cattle	TDN	%	*	14.1	63.3	
Sheep	TDN	%		12.9	58.0	

Clover, aerial part, fresh, pure stand, (2)
Ref No 2 01 300 United States

Feed Name or Analyses				Mean As Fed	Mean Dry	C.V. ± %
Dry matter		%		18.0	100.0	16
Calcium		%		0.33	1.81	20
Magnesium		%		0.07	0.37	19
Phosphorus		%		0.07	0.39	13

Clover, aerial part, ensiled, (3)
Ref No 3 01 305 United States

Feed Name or Analyses				Mean As Fed	Mean Dry	C.V. ± %
Dry matter		%		21.6	100.0	9
Ash		%		2.3	10.5	18
Crude fiber		%		6.1	28.3	12
Cattle	dig coef	%		57.	57.	
Horses	dig coef	%		35.	35.	
Sheep	dig coef	%		59.	59.	
Ether extract		%		0.9	4.4	17
Cattle	dig coef	%		73.	73.	
Horses	dig coef	%		43.	43.	
Sheep	dig coef	%		60.	60.	
N-free extract		%		8.8	41.0	8
Cattle	dig coef	%		70.	70.	
Horses	dig coef	%		43.	43.	
Sheep	dig coef	%		64.	64.	
Protein (N x 6.25)		%		3.4	15.8	15
Cattle	dig coef	%		64.	64.	
Horses	dig coef	%		7.	7.	
Sheep	dig coef	%		65.	65.	
Cattle	dig prot	%		2.2	10.1	
Goats	dig prot	%	*	2.3	10.6	
Horses	dig prot	%		0.2	1.1	
Sheep	dig prot	%		2.2	10.2	
Energy	GE Mcal/kg					
Cattle	DE Mcal/kg		*	0.59	2.73	
Horses	DE Mcal/kg		*	0.31	1.45	
Sheep	DE Mcal/kg		*	0.56	2.59	
Cattle	ME Mcal/kg		*	0.48	2.24	
Horses	ME Mcal/kg		*	0.26	1.19	
Sheep	ME Mcal/kg		*	0.46	2.12	
Cattle	TDN	%		13.4	61.9	
Horses	TDN	%		7.1	32.9	
Sheep	TDN	%		12.7	58.7	
Carotene		mg/kg		23.7	109.8	98
Vitamin A equivalent		IU/g		39.5	183.0	

Ref No 3 01 3 5 Canada

Feed Name or Analyses				Mean As Fed	Mean Dry	C.V. ± %
Dry matter		%		25.8	100.0	9
Crude fiber		%		13.5	52.5	97
Cattle	dig coef	%		57.	57.	
Horses	dig coef	%		35.	35.	
Sheep	dig coef	%		59.	59.	
Ether extract		%		1.7	6.5	
Cattle	dig coef	%		73.	73.	
Horses	dig coef	%		43.	43.	
Sheep	dig coef	%		60.	60.	
N-free extract		%				
Cattle	dig coef	%		70.	70.	
Horses	dig coef	%		43.	43.	
Sheep	dig coef	%		64.	64.	
Protein (N x 6.25)		%		5.7	22.1	73
Cattle	dig coef	%		64.	64.	
Horses	dig coef	%		7.	7.	
Sheep	dig coef	%		65.	65.	
Cattle	dig prot	%		3.6	14.1	
Goats	dig prot	%	*	4.2	16.3	
Horses	dig prot	%		0.4	1.5	
Sheep	dig prot	%		3.7	14.3	
Energy	GE Mcal/kg					
Cattle	DE Mcal/kg		*	0.69	2.67	
Horses	DE Mcal/kg		*	0.35	1.37	
Sheep	DE Mcal/kg		*	0.66	2.57	
Cattle	ME Mcal/kg		*	0.57	2.19	
Horses	ME Mcal/kg		*	0.29	1.13	
Sheep	ME Mcal/kg		*	0.54	2.11	
Cattle	TDN	%		15.6	60.7	
Horses	TDN	%		8.0	31.2	
Sheep	TDN	%		15.0	58.4	
Calcium		%		0.53	2.06	58
Phosphorus		%		0.11	0.41	94
Carotene		mg/kg		19.0	73.7	
Vitamin A equivalent		IU/g		31.7	122.8	

Clover, aerial part, ensiled, early bloom, (3)
Ref No 3 01 301 United States

Feed Name or Analyses				Mean As Fed	Mean Dry	C.V. ± %
Dry matter		%		20.5	100.0	
Ash		%		2.0	9.6	
Crude fiber		%		5.9	29.0	
Sheep	dig coef	%		52.	52.	
Ether extract		%		0.8	4.1	
Sheep	dig coef	%		61.	61.	
N-free extract		%		8.4	40.9	
Sheep	dig coef	%		66.	66.	
Protein (N x 6.25)		%		3.4	16.4	
Sheep	dig coef	%		60.	60.	
Cattle	dig prot	%	*	2.3	11.1	
Goats	dig prot	%	*	2.3	11.1	
Horses	dig prot	%	*	2.3	11.1	
Sheep	dig prot	%		2.0	9.8	
Energy	GE Mcal/kg					
Cattle	DE Mcal/kg		*	0.54	2.63	
Sheep	DE Mcal/kg		*	0.52	2.54	
Cattle	ME Mcal/kg		*	0.44	2.15	

Feed Name or Analyses			Mean As Fed	Mean Dry	C.V. ± %
Sheep	ME Mcal/kg	*	0.43	2.08	
Cattle	TDN %	*	12.2	59.5	
Sheep	TDN %		11.8	57.5	

Clover, aerial part, ensiled, full bloom, (3)

Ref No 3 01 303 United States

Feed Name or Analyses			Mean As Fed	Mean Dry	C.V. ± %
Dry matter	%		17.1	100.0	
Ash	%		1.6	9.5	
Crude fiber	%		4.9	28.4	
Cattle	dig coef %		52.	52.	
Sheep	dig coef %		59.	59.	
Ether extract	%		0.9	5.1	
Cattle	dig coef %		78.	78.	
Sheep	dig coef %		45.	45.	
N-free extract	%		6.9	40.6	
Cattle	dig coef %		79.	79.	
Sheep	dig coef %		70.	70.	
Protein (N x 6.25)	%		2.8	16.5	
Cattle	dig coef %		70.	70.	
Sheep	dig coef %		67.	67.	
Cattle	dig prot %		2.0	11.5	
Goats	dig prot %	*	1.9	11.2	
Horses	dig prot %	*	1.9	11.2	
Sheep	dig prot %		1.9	11.0	
Energy	GE Mcal/kg				
Cattle	DE Mcal/kg	*	0.51	2.97	
Sheep	DE Mcal/kg	*	0.46	2.71	
Cattle	ME Mcal/kg	*	0.42	2.43	
Sheep	ME Mcal/kg	*	0.38	2.22	
Cattle	TDN %		11.5	67.3	
Sheep	TDN %		10.5	61.4	

Clover, aerial part, ensiled, full bloom, cut 2, (3)

Ref No 3 01 302 United States

Feed Name or Analyses			Mean As Fed	Mean Dry	C.V. ± %
Dry matter	%		23.4	100.0	
Ash	%		2.3	10.0	
Crude fiber	%		6.1	26.0	
Cattle	dig coef %		50.	50.	
Ether extract	%		1.2	5.2	
Cattle	dig coef %		78.	78.	
N-free extract	%		9.8	41.8	
Cattle	dig coef %		74.	74.	
Protein (N x 6.25)	%		4.0	17.0	
Cattle	dig coef %		65.	65.	
Cattle	dig prot %		2.6	11.1	
Goats	dig prot %	*	2.7	11.7	
Horses	dig prot %	*	2.7	11.7	
Sheep	dig prot %	*	2.7	11.7	
Energy	GE Mcal/kg				
Cattle	DE Mcal/kg	*	0.66	2.83	
Sheep	DE Mcal/kg	*	0.61	2.61	
Cattle	ME Mcal/kg	*	0.54	2.32	
Sheep	ME Mcal/kg	*	0.50	2.14	
Cattle	TDN %		15.0	64.1	
Sheep	TDN %	*	13.9	59.2	

Clover, aerial part, ensiled, cut 2, (3)

Ref No 3 01 304 United States

Feed Name or Analyses			Mean As Fed	Mean Dry	C.V. ± %
Dry matter	%		21.6	100.0	12
Ash	%		2.5	11.7	24
Crude fiber	%		6.4	29.7	13
Ether extract	%		1.0	4.6	16
N-free extract	%		8.0	37.0	
Protein (N x 6.25)	%		3.7	17.0	14
Cattle	dig prot %	*	2.5	11.7	
Goats	dig prot %	*	2.5	11.7	
Horses	dig prot %	*	2.5	11.7	
Sheep	dig prot %	*	2.5	11.7	
Energy	GE Mcal/kg				
Cattle	DE Mcal/kg	*	0.59	2.71	
Sheep	DE Mcal/kg	*	0.55	2.55	
Cattle	ME Mcal/kg	*	0.48	2.22	
Sheep	ME Mcal/kg	*	0.45	2.09	
Cattle	TDN %	*	13.3	61.4	
Sheep	TDN %	*	12.5	57.8	

Clover, seed screenings, (5)

Ref No 5 01 289 United States

Feed Name or Analyses			Mean As Fed	Mean Dry	C.V. ± %
Dry matter	%		88.1	100.0	1
Ash	%		11.4	13.0	
Crude fiber	%		11.6	13.1	
Cattle	dig coef %		21.	21.	
Sheep	dig coef %		54.	54.	
Swine	dig coef %		66.	66.	
Ether extract	%		6.8	7.7	
Cattle	dig coef %		85.	85.	
Sheep	dig coef %		75.	75.	
Swine	dig coef %		45.	45.	
N-free extract	%		29.2	33.1	
Cattle	dig coef %		77.	77.	
Sheep	dig coef %		80.	80.	
Swine	dig coef %		83.	83.	
Protein (N x 6.25)	%		29.2	33.1	
Cattle	dig coef %		78.	78.	
Sheep	dig coef %		81.	81.	
Swine	dig coef %		70.	70.	
Cattle	dig prot %		22.8	25.8	
Sheep	dig prot %		23.6	26.8	
Swine	dig prot %		20.4	23.2	
Energy	GE Mcal/kg		4.07	4.61	
Cattle	DE Mcal/kg	*	2.67	3.03	
Sheep	DE Mcal/kg	*	2.85	3.24	
Swine	DE kcal/kg	*	2608.	2959.	
Cattle	ME Mcal/kg	*	2.19	2.49	
Sheep	ME Mcal/kg	*	2.34	2.65	
Swine	ME kcal/kg	*	2329.	2643.	
Cattle	TDN %		60.6	68.8	
Sheep	TDN %		64.7	73.4	
Swine	TDN %		59.2	67.1	

CLOVER, ALSIKE. Trifolium hybridum

Clover, alsike, aerial part wo seeds, threshed, mature, (1)

Ref No 1 08 178 Canada

Feed Name or Analyses			Mean As Fed	Mean Dry	C.V. ± %
Dry matter	%		90.2	100.0	
Crude fiber	%		39.3	43.6	
Protein (N x 6.25)	%		8.0	8.9	
Cattle	dig prot %	*	4.2	4.6	
Goats	dig prot %	*	4.4	4.8	
Horses	dig prot %	*	4.6	5.1	
Rabbits	dig prot %	*	5.0	5.5	
Sheep	dig prot %	*	4.1	4.5	
Calcium	%		1.03	1.14	
Phosphorus	%		0.13	0.14	
Carotene	mg/kg		19.8	22.0	
Vitamin A equivalent	IU/g		33.1	36.7	

Clover, alsike, hay, dehy, (1)

Ref No 1 01 306 United States

Feed Name or Analyses			Mean As Fed	Mean Dry	C.V. ± %
Dry matter	%		93.9	100.0	
Carotene	mg/kg		134.4	143.1	
Vitamin A equivalent	IU/g		224.0	238.5	

Clover, alsike, hay, s-c, (1)

Ref No 1 01 313 United States

Feed Name or Analyses			Mean As Fed	Mean Dry	C.V. ± %
Dry matter	%		87.4	100.0	2
Ash	%		7.6	8.7	
Crude fiber	%		26.3	30.1	
Cattle	dig coef %	*	56.	56.	
Sheep	dig coef %	*	50.	50.	
Ether extract	%		2.4	2.7	
Cattle	dig coef %	*	58.	58.	
Sheep	dig coef %	*	40.	40.	
N-free extract	%		38.8	44.3	
Cattle	dig coef %	*	69.	69.	
Sheep	dig coef %	*	66.	66.	
Protein (N x 6.25)	%		12.4	14.2	
Cattle	dig coef %	*	63.	63.	
Sheep	dig coef %	*	67.	67.	
Cattle	dig prot %		7.8	8.9	
Goats	dig prot %	*	8.5	9.8	
Horses	dig prot %	*	8.3	9.5	
Rabbits	dig prot %	*	8.4	9.6	
Sheep	dig prot %		8.3	9.5	
Hemicellulose	%		11.0	12.6	
Fatty acids	%		2.1	2.4	
Energy	GE Mcal/kg		3.87	4.42	
Cattle	DE Mcal/kg	*	2.31	2.64	
Sheep	DE Mcal/kg	*	2.17	2.48	
Cattle	ME Mcal/kg	*	1.89	2.17	
Sheep	ME Mcal/kg	*	1.78	2.03	
Cattle	NE_m Mcal/kg	*	1.13	1.29	
Cattle	NE_{gain} Mcal/kg	*	0.58	0.66	
Cattle	$NE_{lactating\ cows}$ Mcal/kg	*	1.28	1.46	
Cattle	TDN %		52.4	59.9	
Sheep	TDN %		49.2	56.2	
Calcium	%		1.13	1.29	
Chlorine	%		0.68	0.78	7
Copper	mg/kg		5.2	6.0	
Iodine	mg/kg		0.160	0.183	
Iron	%		0.039	0.045	
Magnesium	%		0.28	0.32	
Manganese	mg/kg		102.3	117.0	
Phosphorus	%		0.23	0.26	
Potassium	%		2.40	2.74	
Sodium	%		0.40	0.46	7
Sulphur	%		0.19	0.21	12

Clover, alsike, hay, s-c, immature, (1)

Ref No 1 01 307 United States

Feed Name or Analyses			Mean As Fed	Mean Dry	C.V. ± %
Dry matter	%		92.2	100.0	
Ash	%		7.5	8.1	
Crude fiber	%		17.8	19.3	
Cattle	dig coef %	*	56.	56.	
Sheep	dig coef %	*	50.	50.	
Ether extract	%		2.3	2.5	
Cattle	dig coef %	*	58.	58.	
Sheep	dig coef %	*	40.	40.	
N-free extract	%		39.4	42.7	
Cattle	dig coef %	*	69.	69.	

Continued

Feed Name or Analyses				Mean As Fed	Mean Dry	C.V. ± %
Sheep	dig coef	%	*	66.	66.	
Protein (N x 6.25)		%		25.3	27.4	
Cattle	dig coef	%	*	63.	63.	
Sheep	dig coef	%	*	67.	67.	
Cattle	dig prot	%		15.9	17.3	
Goats	dig prot	%	*	20.4	22.1	
Horses	dig prot	%	*	19.2	20.8	
Rabbits	dig prot	%	*	18.3	19.8	
Sheep	dig prot	%		16.9	18.4	
Energy	GE Mcal/kg					
Cattle	DE Mcal/kg		*	2.47	2.68	
Sheep	DE Mcal/kg		*	2.38	2.58	
Cattle	ME Mcal/kg		*	2.03	2.20	
Sheep	ME Mcal/kg		*	1.95	2.11	
Cattle	TDN	%		56.1	60.8	
Sheep	TDN	%		53.9	58.4	
Calcium		%		1.35	1.46	11
Phosphorus		%		0.38	0.41	73
Carotene	mg/kg			237.4	257.5	15
Riboflavin	mg/kg			19.1	20.7	
Thiamine	mg/kg			5.1	5.5	
Vitamin A equivalent	IU/g			395.8	429.2	

Clover, alsike, hay, s-c, early bloom, (1)

Ref No 1 01 308 United States

Feed Name or Analyses				Mean As Fed	Mean Dry	C.V. ± %
Dry matter		%		82.2	100.0	6
Ash		%		7.8	9.6	6
Crude fiber		%		23.5	28.7	3
Cattle	dig coef	%	*	56.	56.	
Sheep	dig coef	%		52.	52.	
Ether extract		%		2.1	2.5	43
Cattle	dig coef	%	*	58.	58.	
Sheep	dig coef	%		28.	28.	
N-free extract		%		35.2	42.9	
Cattle	dig coef	%	*	69.	69.	
Sheep	dig coef	%		63.	63.	
Protein (N x 6.25)		%		13.5	16.5	20
Cattle	dig coef	%	*	63.	63.	
Sheep	dig coef	%		66.	66.	
Cattle	dig prot	%		8.5	10.4	
Goats	dig prot	%	*	9.8	11.9	
Horses	dig prot	%	*	9.4	11.5	
Rabbits	dig prot	%	*	9.3	11.4	
Sheep	dig prot	%		8.9	10.9	
Energy	GE Mcal/kg					
Cattle	DE Mcal/kg		*	2.15	2.61	
Sheep	DE Mcal/kg		*	1.97	2.40	
Cattle	ME Mcal/kg		*	1.76	2.14	
Sheep	ME Mcal/kg		*	1.61	1.96	
Cattle	TDN	%		48.7	59.2	
Sheep	TDN	%		44.6	54.3	

Clover, alsike, hay, s-c, mid-bloom, (1)

Ref No 1 08 377 United States

Feed Name or Analyses				Mean As Fed	Mean Dry	C.V. ± %
Dry matter		%		89.0	100.0	
Ash		%		7.8	8.8	
Crude fiber		%		26.9	30.2	
Ether extract		%		3.2	3.6	
N-free extract		%		37.7	42.4	
Protein (N x 6.25)		%		13.4	15.1	
Cattle	dig prot	%	*	8.9	10.0	

(1) dry forages and roughages (3) silages (6) minerals
(2) pasture, range plants, and forages fed green (4) energy feeds (7) vitamins
(5) protein supplements (8) additives

Feed Name or Analyses				Mean As Fed	Mean Dry	C.V. ± %
Goats	dig prot	%	*	9.4	10.6	
Horses	dig prot	%	*	9.2	10.3	
Rabbits	dig prot	%	*	9.2	10.3	
Sheep	dig prot	%	*	9.0	10.1	
Energy	GE Mcal/kg					
Cattle	DE Mcal/kg		*	2.41	2.71	
Sheep	DE Mcal/kg		*	2.21	2.49	
Cattle	ME Mcal/kg		*	1.98	2.22	
Sheep	ME Mcal/kg		*	1.81	2.04	
Cattle	TDN	%	*	54.7	61.4	
Sheep	TDN	%	*	50.2	56.4	
Calcium		%		1.32	1.48	
Phosphorus		%		0.25	0.28	
Potassium		%		2.27	2.55	

Clover, alsike, hay, s-c, full bloom, (1)

Ref No 1 01 309 United States

Feed Name or Analyses				Mean As Fed	Mean Dry	C.V. ± %
Dry matter		%		90.3	100.0	2
Ash		%		7.6	8.4	28
Crude fiber		%		27.7	30.7	9
Cattle	dig coef	%	*	56.	56.	
Sheep	dig coef	%		54.	54.	
Ether extract		%		3.4	3.8	24
Cattle	dig coef	%	*	58.	58.	
Sheep	dig coef	%		55.	55.	
N-free extract		%		38.5	42.7	
Cattle	dig coef	%	*	69.	69.	
Sheep	dig coef	%		70.	70.	
Protein (N x 6.25)		%		13.1	14.5	16
Cattle	dig coef	%	*	63.	63.	
Sheep	dig coef	%		67.	67.	
Cattle	dig prot	%		8.2	9.1	
Goats	dig prot	%	*	9.1	10.1	
Horses	dig prot	%	*	8.9	9.8	
Rabbits	dig prot	%	*	8.9	9.9	
Sheep	dig prot	%		8.8	9.7	
Energy	GE Mcal/kg					
Cattle	DE Mcal/kg		*	2.42	2.68	
Sheep	DE Mcal/kg		*	2.42	2.68	
Cattle	ME Mcal/kg		*	1.98	2.19	
Sheep	ME Mcal/kg		*	1.99	2.20	
Cattle	TDN	%		54.8	60.7	
Sheep	TDN	%		54.9	60.8	

Clover, alsike, hay, s-c, late bloom, (1)

Ref No 1 01 310 United States

Feed Name or Analyses				Mean As Fed	Mean Dry	C.V. ± %
Dry matter		%		82.3	100.0	
Ash		%		6.9	8.4	
Crude fiber		%		24.3	29.5	
Sheep	dig coef	%		44.	44.	
Ether extract		%		2.4	2.9	
Sheep	dig coef	%		35.	35.	
N-free extract		%		36.7	44.6	
Sheep	dig coef	%		64.	64.	
Protein (N x 6.25)		%		12.0	14.6	
Sheep	dig coef	%		67.	67.	
Cattle	dig prot	%	*	7.9	9.6	
Goats	dig prot	%	*	8.4	10.2	
Horses	dig prot	%	*	8.2	9.9	
Rabbits	dig prot	%	*	8.2	9.9	
Sheep	dig prot	%		8.1	9.8	
Energy	GE Mcal/kg					
Cattle	DE Mcal/kg		*	2.15	2.62	
Sheep	DE Mcal/kg		*	1.94	2.36	
Cattle	ME Mcal/kg		*	1.77	2.15	
Sheep	ME Mcal/kg		*	1.59	1.94	

Feed Name or Analyses				Mean As Fed	Mean Dry	C.V. ± %
Cattle	TDN	%	*	48.8	59.3	
Sheep	TDN	%		44.1	53.6	

Clover, alsike, hay, s-c, milk stage, (1)

Ref No 1 01 311 United States

Feed Name or Analyses				Mean As Fed	Mean Dry	C.V. ± %
Dry matter		%		82.3	100.0	
Ash		%		6.9	8.4	
Crude fiber		%		25.9	31.5	
Cattle	dig coef	%	*	56.	56.	
Ether extract		%		2.4	2.9	
Cattle	dig coef	%	*	58.	58.	
N-free extract		%		35.1	42.6	
Cattle	dig coef	%	*	69.	69.	
Protein (N x 6.25)		%		12.0	14.6	
Cattle	dig coef	%	*	63.	63.	
Cattle	dig prot	%		7.6	9.2	
Goats	dig prot	%	*	8.4	10.2	
Horses	dig prot	%	*	8.2	9.9	
Rabbits	dig prot	%	*	8.2	9.9	
Sheep	dig prot	%	*	8.0	9.7	
Energy	GE Mcal/kg					
Cattle	DE Mcal/kg		*	2.18	2.65	
Sheep	DE Mcal/kg		*	2.05	2.49	
Cattle	ME Mcal/kg		*	1.79	2.17	
Sheep	ME Mcal/kg		*	1.68	2.04	
Cattle	TDN	%	*	49.4	60.0	
Sheep	TDN	%	*	46.4	56.4	

Clover, alsike, hay, s-c, cut 1, (1)

Ref No 1 01 312 United States

Feed Name or Analyses				Mean As Fed	Mean Dry	C.V. ± %
Dry matter		%		93.4	100.0	
Ash		%		7.6	8.1	
Crude fiber		%		30.9	33.1	
Ether extract		%		2.8	3.0	
N-free extract		%		39.0	41.8	
Protein (N x 6.25)		%		13.1	14.0	
Cattle	dig prot	%	*	8.5	9.1	
Goats	dig prot	%	*	9.0	9.6	
Horses	dig prot	%	*	8.8	9.4	
Rabbits	dig prot	%	*	8.9	9.5	
Sheep	dig prot	%	*	8.5	9.1	
Energy	GE Mcal/kg					
Cattle	DE Mcal/kg		*	2.35	2.51	
Sheep	DE Mcal/kg		*	2.28	2.44	
Cattle	ME Mcal/kg		*	1.93	2.06	
Sheep	ME Mcal/kg		*	1.87	2.00	
Cattle	TDN	%	*	53.3	57.0	
Sheep	TDN	%	*	51.7	55.4	
Calcium		%		0.99	1.06	
Magnesium		%		0.35	0.37	
Phosphorus		%		0.23	0.25	
Potassium		%		2.67	2.86	

Clover, alsike, aerial part, fresh, (2)

Ref No 2 01 316 United States

Feed Name or Analyses				Mean As Fed	Mean Dry	C.V. ± %
Dry matter		%		22.5	100.0	1
Ash		%		2.1	9.3	14
Crude fiber		%		5.2	23.3	12
Cattle	dig coef	%	*	59.	59.	
Ether extract		%		0.8	3.6	9
Cattle	dig coef	%	*	66.	66.	
N-free extract		%		10.3	45.7	
Cattle	dig coef	%	*	80.	80.	

606

Feed Name or Analyses		Mean As Fed	Mean Dry	C.V. ± %
Protein (N x 6.25)	%	4.1	18.1	14
Cattle	dig coef %	* 70.	70.	
Cattle	dig prot %	2.8	12.6	
Goats	dig prot %	* 3.0	13.4	
Horses	dig prot %	* 2.9	12.9	
Rabbits	dig prot %	* 2.8	12.6	
Sheep	dig prot %	* 3.1	13.8	
Energy	GE Mcal/kg			
Cattle	DE Mcal/kg	* 0.68	3.01	
Sheep	DE Mcal/kg	* 0.64	2.86	
Cattle	ME Mcal/kg	* 0.56	2.47	
Sheep	ME Mcal/kg	* 0.53	2.35	
Cattle	TDN %	15.4	68.3	
Sheep	TDN %	* 14.6	64.9	
Calcium	%	0.31	1.36	12
Chlorine	%	0.17	0.77	6
Copper	mg/kg	1.3	6.0	9
Iron	%	0.010	0.043	24
Magnesium	%	0.07	0.32	18
Manganese	mg/kg	26.3	117.1	12
Phosphorus	%	0.06	0.29	20
Potassium	%	0.61	2.70	5
Sodium	%	0.10	0.45	7
Sulphur	%	0.05	0.22	21
Zinc	mg/kg	13.5	60.2	
Riboflavin	mg/kg	4.4	19.6	22
Thiamine	mg/kg	2.0	8.8	12

Clover, alsike, aerial part, fresh, immature, (2)

Ref No 2 01 314 United States

Feed Name or Analyses		Mean As Fed	Mean Dry	C.V. ± %
Dry matter	%	18.8	100.0	10
Ash	%	2.4	12.8	10
Crude fiber	%	3.3	17.5	11
Cattle	dig coef %	* 59.	59.	
Sheep	dig coef %	* 52.	52.	
Ether extract	%	0.6	3.2	5
Cattle	dig coef %	* 66.	66.	
Sheep	dig coef %	* 67.	67.	
N-free extract	%	8.0	42.4	
Cattle	dig coef %	* 80.	80.	
Sheep	dig coef %	* 71.	71.	
Protein (N x 6.25)	%	4.5	24.1	8
Cattle	dig coef %	* 70.	70.	
Sheep	dig coef %	* 67.	67.	
Cattle	dig prot %	3.2	16.9	
Goats	dig prot %	* 3.6	19.0	
Horses	dig prot %	* 3.4	18.0	
Rabbits	dig prot %	* 3.2	17.3	
Sheep	dig prot %	3.0	16.1	
Energy	GE Mcal/kg			
Cattle	DE Mcal/kg	* 0.55	2.90	
Sheep	DE Mcal/kg	* 0.50	2.65	
Cattle	ME Mcal/kg	* 0.45	2.38	
Sheep	ME Mcal/kg	* 0.41	2.18	
Cattle	TDN %	12.4	65.9	
Sheep	TDN %	11.3	60.2	
Calcium	%	0.22	1.19	13
Magnesium	%	0.06	0.34	
Phosphorus	%	0.08	0.42	17
Potassium	%	0.43	2.31	

Clover, alsike, aerial part, fresh, mid-bloom, (2)

Ref No 2 08 378 United States

Feed Name or Analyses		Mean As Fed	Mean Dry	C.V. ± %
Dry matter	%	23.6	100.0	
Ash	%	2.3	9.7	
Crude fiber	%	6.4	27.1	
Ether extract	%	0.7	3.0	
N-free extract	%	10.5	44.5	
Protein (N x 6.25)	%	3.7	15.7	
Cattle	dig prot %	* 2.6	11.2	
Goats	dig prot %	* 2.6	11.2	
Horses	dig prot %	* 2.6	10.8	
Rabbits	dig prot %	* 2.5	10.8	
Sheep	dig prot %	* 2.7	11.6	
Energy	GE Mcal/kg			
Cattle	DE Mcal/kg	* 0.71	3.01	
Sheep	DE Mcal/kg	* 0.65	2.75	
Cattle	ME Mcal/kg	* 0.58	2.46	
Sheep	ME Mcal/kg	* 0.53	2.25	
Cattle	TDN %	* 16.1	68.2	
Sheep	TDN %	* 14.7	62.3	
Calcium	%	0.30	1.27	
Chlorine	%	0.18	0.76	
Copper	mg/kg	1.3	5.6	
Iron	%	0.011	0.047	
Magnesium	%	0.07	0.30	
Manganese	mg/kg	27.6	116.8	
Phosphorus	%	0.06	0.25	
Potassium	%	0.61	2.58	
Sodium	%	0.11	0.47	
Sulphur	%	0.03	0.13	

Clover, alsike, aerial part, fresh, full bloom, (2)

Ref No 2 01 315 United States

Feed Name or Analyses		Mean As Fed	Mean Dry	C.V. ± %
Dry matter	%	23.6	100.0	6
Ash	%	2.3	9.7	10
Crude fiber	%	6.4	27.1	6
Cattle	dig coef %	* 59.	59.	
Sheep	dig coef %	* 52.	52.	
Ether extract	%	0.7	3.0	7
Cattle	dig coef %	* 66.	66.	
Sheep	dig coef %	* 67.	67.	
N-free extract	%	10.5	44.5	
Cattle	dig coef %	* 80.	80.	
Sheep	dig coef %	* 71.	71.	
Protein (N x 6.25)	%	3.7	15.7	8
Cattle	dig coef %	* 70.	70.	
Sheep	dig coef %	* 67.	67.	
Cattle	dig prot %	2.6	11.0	
Goats	dig prot %	* 2.6	11.2	
Horses	dig prot %	* 2.6	10.9	
Rabbits	dig prot %	* 2.5	10.8	
Sheep	dig prot %	2.5	10.5	
Energy	GE Mcal/kg			
Cattle	DE Mcal/kg	* 0.70	2.96	
Sheep	DE Mcal/kg	* 0.63	2.68	
Cattle	ME Mcal/kg	* 0.57	2.42	
Sheep	ME Mcal/kg	* 0.52	2.20	
Cattle	TDN %	15.8	67.0	
Sheep	TDN %	14.3	60.7	
Calcium	%	0.31	1.30	10
Iron	%	0.011	0.047	
Manganese	mg/kg	27.6	116.8	
Phosphorus	%	0.07	0.29	22
Potassium	%	0.61	2.58	

Clover, alsike, aerial part, ensiled, late bloom, (3)

Ref No 3 01 317 United States

Feed Name or Analyses		Mean As Fed	Mean Dry	C.V. ± %
Dry matter	%	21.2	100.0	
Ash	%	2.2	10.4	
Crude fiber	%	7.9	37.2	
Sheep	dig coef %	55.	55.	
Ether extract	%	0.8	3.9	
Sheep	dig coef %	54.	54.	
N-free extract	%	7.7	36.4	
Sheep	dig coef %	56.	56.	
Protein (N x 6.25)	%	2.6	12.1	
Sheep	dig coef %	40.	40.	
Cattle	dig prot %	* 1.5	7.2	
Goats	dig prot %	* 1.5	7.2	
Horses	dig prot %	* 1.5	7.2	
Sheep	dig prot %	1.0	4.8	
Energy	GE Mcal/kg			
Cattle	DE Mcal/kg	* 0.54	2.52	
Sheep	DE Mcal/kg	* 0.47	2.22	
Cattle	ME Mcal/kg	* 0.44	2.07	
Sheep	ME Mcal/kg	* 0.39	1.82	
Cattle	TDN %	* 12.1	57.2	
Sheep	TDN %	10.7	50.4	

Clover, alsike, seeds, (5)

Ref No 5 01 318 United States

Feed Name or Analyses		Mean As Fed	Mean Dry	C.V. ± %
Dry matter	%	91.5	100.0	
Ash	%	5.4	5.9	
Crude fiber	%	12.0	13.1	
Sheep	dig coef %	86.	86.	
Ether extract	%	4.8	5.3	
Sheep	dig coef %	85.	85.	
N-free extract	%	37.9	41.4	
Sheep	dig coef %	86.	86.	
Protein (N x 6.25)	%	31.4	34.3	
Sheep	dig coef %	75.	75.	
Sheep	dig prot %	23.5	25.7	
Energy	GE Mcal/kg			
Cattle	DE Mcal/kg	* 3.04	3.32	
Sheep	DE Mcal/kg	* 3.34	3.65	
Swine	DE kcal/kg	*3461.	3782.	
Cattle	ME Mcal/kg	* 2.49	2.72	
Sheep	ME Mcal/kg	* 2.74	2.99	
Swine	ME kcal/kg	*3083.	3369.	
Cattle	TDN %	* 68.9	75.3	
Sheep	TDN %	75.7	82.7	
Swine	TDN %	* 78.5	85.8	

CLOVER, BURDOCK. Trifolium lappaceum

Clover, burdock, hay, s-c, (1)

Ref No 1 01 322 United States

Feed Name or Analyses		Mean As Fed	Mean Dry	C.V. ± %
Dry matter	%	94.4	100.0	2
Ash	%	9.4	10.0	12
Crude fiber	%	24.2	25.6	39
Ether extract	%	3.0	3.2	28
N-free extract	%	38.4	40.7	
Protein (N x 6.25)	%	19.4	20.5	29
Cattle	dig prot %	* 13.9	14.7	
Goats	dig prot %	* 14.8	15.7	
Horses	dig prot %	* 14.1	14.9	
Rabbits	dig prot %	* 13.7	14.5	
Sheep	dig prot %	* 14.1	15.0	
Calcium	%	1.26	1.34	13
Iron	%	0.059	0.063	
Magnesium	%	0.29	0.31	
Phosphorus	%	0.22	0.23	27

Clover, burdock, hay, s-c, immature, (1)

Ref No 1 01 319 United States

Feed Name or Analyses			Mean As Fed	Mean Dry	C.V. ± %
Dry matter	%		94.8	100.0	
Ash	%		10.6	11.2	
Crude fiber	%		15.6	16.5	
Ether extract	%		3.7	3.9	
N-free extract	%		39.3	41.5	
Protein (N x 6.25)	%		25.5	26.9	
Cattle	dig prot %	*	19.2	20.2	
Goats	dig prot %	*	20.5	21.7	
Horses	dig prot %	*	19.3	20.4	
Rabbits	dig prot %	*	18.4	19.4	
Sheep	dig prot %	*	19.6	20.7	

Clover, burdock, hay, s-c, full blm, (1)

Ref No 1 01 320 United States

Feed Name or Analyses			Mean As Fed	Mean Dry	C.V. ± %
Dry matter	%		97.3	100.0	
Ash	%		10.6	10.9	
Crude fiber	%		21.9	22.5	
Ether extract	%		3.0	3.1	
N-free extract	%		41.0	42.1	
Protein (N x 6.25)	%		20.8	21.4	
Cattle	dig prot %	*	15.1	15.5	
Goats	dig prot %	*	16.1	16.5	
Horses	dig prot %	*	15.3	15.7	
Rabbits	dig prot %	*	14.8	15.2	
Sheep	dig prot %	*	15.3	15.8	

Clover, burdock, hay, s-c, mature, (1)

Ref No 1 01 321 United States

Feed Name or Analyses			Mean As Fed	Mean Dry	C.V. ± %
Dry matter	%		92.9	100.0	
Ash	%		8.4	9.0	
Crude fiber	%		38.3	41.2	
Ether extract	%		1.8	1.9	
N-free extract	%		33.5	36.1	
Protein (N x 6.25)	%		11.0	11.8	
Cattle	dig prot %	*	6.7	7.2	
Goats	dig prot %	*	7.0	7.6	
Horses	dig prot %	*	7.0	7.5	
Rabbits	dig prot %	*	7.2	7.8	
Sheep	dig prot %	*	6.6	7.2	

Clover, burdock, aerial part, fresh, (2)

Ref No 2 01 323 United States

Feed Name or Analyses			Mean As Fed	Mean Dry	C.V. ± %
Dry matter	%			100.0	
Carotene	mg/kg			225.5	
Vitamin A equivalent	IU/g			376.0	

CLOVER, CRIMSON. Trifolium incarnatum

Clover, crimson, aerial part, dehy grnd, over ripe, (1)

Ref No 1 08 211 United States

Feed Name or Analyses			Mean As Fed	Mean Dry	C.V. ± %
Dry matter	%		92.5	100.0	
Organic matter	%		84.5	91.3	
Ash	%		8.0	8.6	
Crude fiber	%		33.5	36.2	
Ether extract	%		1.2	1.3	
N-free extract	%		43.1	46.6	
Protein (N x 6.25)	%		6.8	7.3	
Cattle	dig prot %	*	3.0	3.3	
Goats	dig prot %	*	3.1	3.4	
Horses	dig prot %	*	3.4	3.7	
Rabbits	dig prot %	*	4.0	4.3	
Sheep	dig prot %	*	2.9	3.1	
Energy	GE Mcal/kg				
Cattle	DE Mcal/kg	*	2.17	2.35	
Sheep	DE Mcal/kg	*	2.20	2.37	
Cattle	ME Mcal/kg	*	1.78	1.93	
Sheep	ME Mcal/kg	*	1.80	1.95	
Cattle	TDN %	*	49.3	53.2	
Sheep	TDN %	*	49.8	53.8	
Silicon	%		1.57	1.70	

Clover, crimson, aerial part, dehy grnd, dormant, (1)

Ref No 1 08 205 United States

Feed Name or Analyses			Mean As Fed	Mean Dry	C.V. ± %
Dry matter	%		92.4	100.0	1
Organic matter	%		81.7	88.4	8
Ash	%		10.7	11.6	61
Crude fiber	%		33.9	36.7	17
Ether extract	%		1.2	1.3	21
N-free extract	%		38.6	41.7	11
Protein (N x 6.25)	%		8.1	8.7	12
Cattle	dig prot %	*	4.2	4.5	
Goats	dig prot %	*	4.3	4.7	
Horses	dig prot %	*	4.6	4.9	
Rabbits	dig prot %	*	5.0	5.4	
Sheep	dig prot %	*	4.1	4.4	
Energy	GE Mcal/kg				
Cattle	DE Mcal/kg	*	2.24	2.42	
Sheep	DE Mcal/kg	*	2.19	2.37	
Cattle	ME Mcal/kg	*	1.83	1.99	
Sheep	ME Mcal/kg	*	1.79	1.94	
Cattle	TDN %	*	50.7	54.9	
Sheep	TDN %	*	49.6	53.7	
Silicon	%		4.82	5.22	98

Clover, crimson, hay, fan air dried, (1)

Ref No 1 01 324 United States

Feed Name or Analyses			Mean As Fed	Mean Dry	C.V. ± %
Dry matter	%		80.4	100.0	
Ash	%		7.3	9.1	
Crude fiber	%		25.9	32.2	
Sheep	dig coef %		46.	46.	
Ether extract	%		1.9	2.4	
Sheep	dig coef %		35.	35.	
N-free extract	%		29.7	37.0	
Sheep	dig coef %		62.	62.	
Protein (N x 6.25)	%		15.5	19.3	
Sheep	dig coef %		69.	69.	
Cattle	dig prot %	*	11.0	13.7	
Goats	dig prot %	*	11.7	14.6	
Horses	dig prot %	*	11.2	13.9	
Rabbits	dig prot %	*	10.9	13.6	
Sheep	dig prot %		10.7	13.3	
Energy	GE Mcal/kg				
Cattle	DE Mcal/kg	*	2.03	2.53	
Sheep	DE Mcal/kg	*	1.88	2.34	
Cattle	ME Mcal/kg	*	1.67	2.07	
Sheep	ME Mcal/kg	*	1.54	1.91	
Cattle	TDN %	*	46.1	57.3	
Sheep	TDN %		42.6	53.0	

Clover, crimson, hay, s-c, (1)

Ref No 1 01 328 United States

Feed Name or Analyses			Mean As Fed	Mean Dry	C.V. ± %
Dry matter	%		88.8	100.0	3
Ash	%		8.2	9.2	14
Crude fiber	%		24.6	27.7	35
Cattle	dig coef %		58.	58.	
Sheep	dig coef %		45.	45.	
Ether extract	%		2.4	2.7	33
Cattle	dig coef %		51.	51.	
Sheep	dig coef %		47.	47.	
N-free extract	%		38.8	43.7	
Cattle	dig coef %		70.	70.	
Sheep	dig coef %		61.	61.	
Protein (N x 6.25)	%		14.8	16.7	10
Cattle	dig coef %		70.	70.	
Sheep	dig coef %		67.	67.	
Cattle	dig prot %		10.4	11.7	
Goats	dig prot %	*	10.8	12.2	
Horses	dig prot %	*	10.4	11.7	
Rabbits	dig prot %	*	10.3	11.6	
Sheep	dig prot %		9.9	11.2	
Fatty acids	%		2.2	2.5	
Energy	GE Mcal/kg				
Cattle	DE Mcal/kg	*	2.41	2.71	
Sheep	DE Mcal/kg	*	2.08	2.34	
Cattle	ME Mcal/kg	*	1.97	2.22	
Sheep	ME Mcal/kg	*	1.71	1.92	
Cattle	NE_m Mcal/kg	*	1.15	1.29	
Cattle	NE_{gain} Mcal/kg	*	0.59	0.66	
Cattle	$NE_{lactating\ cows}$ Mcal/kg	*	1.30	1.46	
Cattle	TDN %		54.6	61.4	
Sheep	TDN %		47.2	53.2	
Calcium	%		1.22	1.37	
Chlorine	%		0.56	0.63	13
Iodine	mg/kg		0.059	0.066	
Magnesium	%		0.26	0.29	
Manganese	mg/kg		218.6	246.1	
Phosphorus	%		0.24	0.27	
Potassium	%		2.77	3.12	
Silicon	%		0.04	0.05	
Sodium	%		0.35	0.39	16
Sulphur	%		0.25	0.28	9

Clover, crimson, hay, s-c, early bloom, (1)

Ref No 1 01 325 United States

Feed Name or Analyses			Mean As Fed	Mean Dry	C.V. ± %
Dry matter	%		84.9	100.0	
Ash	%		14.2	16.7	
Crude fiber	%		19.9	23.5	
Cattle	dig coef %	*	58.	58.	
Sheep	dig coef %		69.	69.	
Ether extract	%		2.4	2.8	
Cattle	dig coef %	*	51.	51.	
Sheep	dig coef %		49.	49.	
N-free extract	%		28.4	33.5	
Cattle	dig coef %	*	70.	70.	
Sheep	dig coef %		78.	78.	
Protein (N x 6.25)	%		20.0	23.6	
Cattle	dig coef %	*	70.	70.	
Sheep	dig coef %		73.	73.	
Cattle	dig prot %		14.0	16.5	
Goats	dig prot %	*	15.8	18.6	
Horses	dig prot %	*	14.9	17.6	
Rabbits	dig prot %	*	14.3	16.9	
Sheep	dig prot %		14.6	17.2	
Energy	GE Mcal/kg				
Cattle	DE Mcal/kg	*	2.12	2.50	

(1) dry forages and roughages (2) pasture, range plants, and forages fed green (3) silages (4) energy feeds (5) protein supplements (6) minerals (7) vitamins (8) additives

Feed Name or Analyses				Mean As Fed	Mean Dry	C.V. ± %
Sheep	DE Mcal/kg		*	2.34	2.76	
Cattle	ME Mcal/kg		*	1.74	2.05	
Sheep	ME Mcal/kg		*	1.92	2.26	
Cattle	TDN	%		48.2	56.7	
Sheep	TDN	%		53.1	62.6	

Clover, crimson, hay, s-c, late bloom, (1)
Ref No 1 01 326 United States

Feed Name or Analyses				Mean As Fed	Mean Dry	C.V. ± %
Dry matter		%		86.1	100.0	
Ash		%		7.2	8.4	
Crude fiber		%		29.7	34.5	
Sheep	dig coef	%		45.	45.	
Ether extract		%		1.9	2.2	
Sheep	dig coef	%		47.	47.	
N-free extract		%		33.1	38.5	
Sheep	dig coef	%		61.	61.	
Protein (N x 6.25)		%		14.1	16.4	
Sheep	dig coef	%		66.	66.	
Cattle	dig prot	%	*	9.6	11.1	
Goats	dig prot	%	*	10.2	11.9	
Horses	dig prot	%	*	9.9	11.5	
Rabbits	dig prot	%	*	9.8	11.3	
Sheep	dig prot	%		9.3	10.8	
Energy	GE Mcal/kg					
Cattle	DE Mcal/kg		*	2.10	2.44	
Sheep	DE Mcal/kg		*	1.98	2.30	
Cattle	ME Mcal/kg		*	1.73	2.00	
Sheep	ME Mcal/kg		*	1.62	1.89	
Cattle	TDN	%	*	47.7	55.4	
Sheep	TDN	%		44.9	52.2	

Clover, crimson, hay, s-c, milk stage, (1)
Ref No 1 01 327 United States

Feed Name or Analyses				Mean As Fed	Mean Dry	C.V. ± %
Dry matter		%		86.1	100.0	2
Ash		%		7.2	8.4	18
Crude fiber		%		31.3	36.4	5
Cattle	dig coef	%	*	58.	58.	
Sheep	dig coef	%	*	45.	45.	
Ether extract		%		1.8	2.1	13
Cattle	dig coef	%	*	51.	51.	
Sheep	dig coef	%	*	47.	47.	
N-free extract		%		31.6	36.7	
Cattle	dig coef	%	*	70.	70.	
Sheep	dig coef	%	*	61.	61.	
Protein (N x 6.25)		%		14.1	16.4	4
Cattle	dig coef	%	*	70.	70.	
Sheep	dig coef	%	*	67.	67.	
Cattle	dig prot	%		9.9	11.5	
Goats	dig prot	%	*	10.2	11.9	
Horses	dig prot	%	*	9.9	11.5	
Rabbits	dig prot	%	*	9.8	11.3	
Sheep	dig prot	%		9.5	11.0	
Energy	GE Mcal/kg					
Cattle	DE Mcal/kg		*	2.30	2.68	
Sheep	DE Mcal/kg		*	1.97	2.29	
Cattle	ME Mcal/kg		*	1.89	2.19	
Sheep	ME Mcal/kg		*	1.62	1.88	
Cattle	TDN	%	*	52.3	60.7	
Sheep	TDN	%	*	44.8	52.0	

Clover, crimson, leaves, (1)
Ref No 1 01 329 United States

Feed Name or Analyses				Mean As Fed	Mean Dry	C.V. ± %
Dry matter		%			100.0	
Crude fiber		%				
Cattle	dig coef	%	*		58.	
Sheep	dig coef	%	*		52.	
Ether extract		%				
Cattle	dig coef	%	*		51.	
Sheep	dig coef	%	*		67.	
N-free extract		%				
Cattle	dig coef	%	*		70.	
Sheep	dig coef	%	*		71.	
Protein (N x 6.25)		%				
Cattle	dig coef	%	*		70.	
Sheep	dig coef	%	*		67.	

Clover, crimson, leaves, dehy grnd, early leaf, (1)
Ref No 1 08 208 United States

Feed Name or Analyses				Mean As Fed	Mean Dry	C.V. ± %
Dry matter		%		93.1	100.0	
Organic matter		%		84.0	90.2	
Ash		%		9.1	9.8	
Crude fiber		%		9.1	9.8	
Ether extract		%		4.2	4.5	
N-free extract		%		41.7	44.8	
Protein (N x 6.25)		%		29.0	31.1	
Cattle	dig prot	%	*	22.2	23.9	
Goats	dig prot	%	*	23.8	25.6	
Horses	dig prot	%	*	22.3	23.9	
Rabbits	dig prot	%	*	21.1	22.7	
Sheep	dig prot	%	*	22.8	24.5	
Silicon		%		0.22	0.24	

Clover, crimson, leaves, dehy grnd, immature, (1)
Ref No 1 08 206 United States

Feed Name or Analyses				Mean As Fed	Mean Dry	C.V. ± %
Dry matter		%		92.9	100.0	
Organic matter		%		82.7	89.1	
Ash		%		10.2	11.0	
Crude fiber		%		10.1	10.9	
Ether extract		%		3.5	3.8	
N-free extract		%		44.3	47.7	
Protein (N x 6.25)		%		24.8	26.7	
Cattle	dig prot	%	*	18.6	20.1	
Goats	dig prot	%	*	19.9	21.5	
Horses	dig prot	%	*	18.8	20.2	
Rabbits	dig prot	%	*	17.9	19.3	
Sheep	dig prot	%	*	19.1	20.5	
Silicon		%		0.07	0.08	

Clover, crimson, leaves, dehy grnd, full bloom, (1)
Ref No 1 08 219 United States

Feed Name or Analyses				Mean As Fed	Mean Dry	C.V. ± %
Dry matter		%		92.8	100.0	
Organic matter		%		82.1	88.5	
Ash		%		10.7	11.5	
Crude fiber		%		10.1	10.9	
Ether extract		%		3.9	4.3	
N-free extract		%		45.5	49.1	
Protein (N x 6.25)		%		22.6	24.3	
Cattle	dig prot	%	*	16.7	18.0	
Goats	dig prot	%	*	17.8	19.2	
Horses	dig prot	%	*	16.9	18.2	
Rabbits	dig prot	%	*	16.2	17.4	
Sheep	dig prot	%	*	17.0	18.4	
Silicon		%		0.00	0.00	

Clover, crimson, leaves, dehy grnd, dormant, (1)
Ref No 1 08 213 United States

Feed Name or Analyses				Mean As Fed	Mean Dry	C.V. ± %
Dry matter		%		91.7	100.0	
Organic matter		%		80.1	87.4	
Ash		%		11.6	12.7	
Crude fiber		%		18.1	19.8	
Ether extract		%		2.3	2.5	
N-free extract		%		52.8	57.7	
Protein (N x 6.25)		%		6.8	7.4	
Cattle	dig prot	%	*	3.1	3.3	
Goats	dig prot	%	*	3.2	3.5	
Horses	dig prot	%	*	3.5	3.8	
Rabbits	dig prot	%	*	4.0	4.4	
Sheep	dig prot	%	*	2.9	3.2	
Silicon		%		0.01	0.01	

Clover, crimson, stems, (1)
Ref No 1 01 331 United States

Feed Name or Analyses				Mean As Fed	Mean Dry	C.V. ± %
Dry matter		%		17.3	100.0	
Calcium		%		0.14	0.82	
Iron		%		0.002	0.014	
Magnesium		%		0.03	0.19	
Manganese		mg/kg		5.1	29.8	
Phosphorus		%		0.03	0.19	
Potassium		%		0.36	2.08	

Clover, crimson, stems, cut 2, (1)
Ref No 1 01 330 United States

Feed Name or Analyses				Mean As Fed	Mean Dry	C.V. ± %
Dry matter		%		87.5	100.0	3
Ash		%		10.9	12.5	8
Crude fiber		%		25.6	29.3	0
Ether extract		%		3.9	4.4	7
N-free extract		%		36.1	41.3	
Protein (N x 6.25)		%		10.9	12.5	11
Cattle	dig prot	%	*	6.8	7.8	
Goats	dig prot	%	*	7.2	8.2	
Horses	dig prot	%	*	7.1	8.1	
Rabbits	dig prot	%	*	7.3	8.3	
Sheep	dig prot	%	*	6.8	7.8	

Clover, crimson, straw, (1)
Ref No 1 08 379 United States

Feed Name or Analyses				Mean As Fed	Mean Dry	C.V. ± %
Dry matter		%		87.7	100.0	
Ash		%		7.0	8.0	
Crude fiber		%		38.8	44.2	
Ether extract		%		1.5	1.7	
N-free extract		%		32.9	37.5	
Protein (N x 6.25)		%		7.5	8.6	
Cattle	dig prot	%	*	3.8	4.3	
Goats	dig prot	%	*	4.0	4.5	
Horses	dig prot	%	*	4.2	4.8	
Rabbits	dig prot	%	*	4.6	5.3	
Sheep	dig prot	%	*	3.7	4.2	
Energy	GE Mcal/kg					
Cattle	DE Mcal/kg		*	1.71	1.95	
Sheep	DE Mcal/kg		*	1.84	2.10	
Cattle	ME Mcal/kg		*	1.40	1.60	
Sheep	ME Mcal/kg		*	1.51	1.72	
Cattle	TDN	%	*	38.9	44.3	
Sheep	TDN	%	*	41.7	47.6	

Feed Name or Analyses			As Fed	Dry	C.V. ± %

Clover, crimson, aerial part, fresh, (2)
Ref No 2 01 336 — United States

Feed Name or Analyses			As Fed	Dry	C.V. ± %
Dry matter	%		17.6	100.0	4
Ash	%		1.7	9.5	3
Crude fiber	%		4.9	27.7	7
Cattle	dig coef %	*	59.	59.	
Sheep	dig coef %	*	52.	52.	
Ether extract	%		0.6	3.3	21
Cattle	dig coef %	*	66.	66.	
Sheep	dig coef %	*	67.	67.	
N-free extract	%		7.5	42.6	
Cattle	dig coef %	*	80.	80.	
Sheep	dig coef %	*	71.	71.	
Protein (N x 6.25)	%		3.0	17.0	2
Cattle	dig coef %	*	70.	70.	
Sheep	dig coef %	*	67.	67.	
Cattle	dig prot %		2.1	11.9	
Goats	dig prot %	*	2.2	12.4	
Horses	dig prot %	*	2.1	11.9	
Rabbits	dig prot %	*	2.1	11.8	
Sheep	dig prot %		2.0	11.4	
Energy	GE Mcal/kg				
Cattle	DE Mcal/kg	*	0.52	2.96	
Sheep	DE Mcal/kg	*	0.47	2.69	
Cattle	ME Mcal/kg	*	0.43	2.43	
Sheep	ME Mcal/kg	*	0.39	2.20	
Cattle	TDN %		11.8	67.1	
Sheep	TDN %		10.7	60.9	
Calcium	%		0.24	1.38	
Chlorine	%		0.11	0.61	18
Magnesium	%		0.05	0.29	
Manganese	mg/kg		43.3	245.8	
Phosphorus	%		0.05	0.29	
Potassium	%		0.55	3.10	
Sodium	%		0.07	0.40	18
Sulphur	%		0.05	0.28	14

Clover, crimson, aerial part, fresh, full bloom, (2)
Ref No 2 01 332 — United States

Feed Name or Analyses			As Fed	Dry	C.V. ± %
Dry matter	%			100.0	
Calcium	%			1.41	5
Iron	%			0.018	
Magnesium	%			0.30	8
Manganese	mg/kg			387.8	
Phosphorus	%			0.31	17
Potassium	%			2.36	16

Clover, crimson, aerial part, fresh, late bloom, (2)
Ref No 2 01 333 — United States

Feed Name or Analyses			As Fed	Dry	C.V. ± %
Dry matter	%		15.9	100.0	
Ash	%		1.5	9.4	
Crude fiber	%		4.3	27.0	
Sheep	dig coef %		56.	56.	
Ether extract	%		0.7	4.4	
Sheep	dig coef %		66.	66.	
N-free extract	%		6.6	41.6	
Sheep	dig coef %		74.	74.	
Protein (N x 6.25)	%		2.8	17.6	
Sheep	dig coef %		77.	77.	
Cattle	dig prot %	*	2.0	12.9	
Goats	dig prot %	*	2.1	13.0	
Horses	dig prot %	*	2.0	12.5	
Rabbits	dig prot %	*	1.9	12.3	
Sheep	dig prot %		2.2	13.6	
Energy	GE Mcal/kg				
Cattle	DE Mcal/kg	*	0.48	3.03	
Sheep	DE Mcal/kg	*	0.46	2.91	
Cattle	ME Mcal/kg	*	0.39	2.48	
Sheep	ME Mcal/kg	*	0.38	2.39	
Cattle	TDN %	*	10.9	68.7	
Sheep	TDN %		10.5	66.0	

(1) dry forages and roughages (2) pasture, range plants, and forages fed green (3) silages (4) energy feeds (5) protein supplements (6) minerals (7) vitamins (8) additives

Clover, crimson, aerial part, fresh, cut 1, (2)
Ref No 2 01 334 — United States

Feed Name or Analyses			As Fed	Dry	C.V. ± %
Dry matter	%			100.0	
Calcium	%			1.12	
Magnesium	%			0.28	
Phosphorus	%			0.36	
Potassium	%			2.10	

Clover, crimson, aerial part, fresh, cut 3, (2)
Ref No 2 01 335 — United States

Feed Name or Analyses			As Fed	Dry	C.V. ± %
Dry matter	%			100.0	
Calcium	%			1.42	
Magnesium	%			0.31	
Phosphorus	%			0.31	
Potassium	%			2.49	

Clover, crimson, aerial part, ensiled, (3)
Ref No 3 01 339 — United States

Feed Name or Analyses			As Fed	Dry	C.V. ± %
Dry matter	%		22.2	100.0	
Ash	%		5.2	23.4	
Crude fiber	%		4.1	18.6	
Horses	dig coef %		43.	43.	
Ether extract	%		1.8	8.2	
Horses	dig coef %		64.	64.	
N-free extract	%		6.3	28.3	
Horses	dig coef %		83.	83.	
Protein (N x 6.25)	%		4.8	21.5	
Horses	dig coef %		75.	75.	
Cattle	dig prot %	*	3.5	15.8	
Goats	dig prot %	*	3.5	15.8	
Horses	dig prot %		3.6	16.1	
Sheep	dig prot %	*	3.5	15.8	
Energy	GE Mcal/kg				
Cattle	DE Mcal/kg	*	0.65	2.91	
Horses	DE Mcal/kg	*	0.58	2.62	
Sheep	DE Mcal/kg	*	0.53	2.39	
Cattle	ME Mcal/kg	*	0.53	2.39	
Horses	ME Mcal/kg	*	14.7	66.1	
Sheep	ME Mcal/kg	*	0.52	2.34	
Cattle	TDN %	*	14.7	66.1	
Horses	TDN %		13.2	59.4	
Sheep	TDN %	*	14.3	64.6	

CLOVER, EGYPTIAN. Trifolium alexandrinum

Clover, Egyptian, hay, s-c, (1)
Ref No 1 01 340 — United States

Feed Name or Analyses			As Fed	Dry	C.V. ± %
Dry matter	%		89.5	100.0	1
Ash	%		11.6	13.0	8
Crude fiber	%		21.8	24.4	6
Cattle	dig coef %		49.	49.	
Ether extract	%		2.2	2.5	24
Cattle	dig coef %		29.	29.	
N-free extract	%		40.2	44.9	
Cattle	dig coef %		77.	77.	
Protein (N x 6.25)	%		13.6	15.2	9
Cattle	dig coef %		70.	70.	
Cattle	dig prot %		9.5	10.7	
Goats	dig prot %	*	9.6	10.8	
Horses	dig prot %	*	9.4	10.5	
Rabbits	dig prot %	*	9.3	10.4	
Sheep	dig prot %	*	9.2	10.2	
Fatty acids	%		3.1	3.4	
Energy	GE Mcal/kg				
Cattle	DE Mcal/kg	*	2.32	2.59	
Sheep	DE Mcal/kg	*	2.28	2.54	
Cattle	ME Mcal/kg	*	1.90	2.13	
Sheep	ME Mcal/kg	*	1.87	2.09	
Cattle	TDN %		52.6	58.8	
Sheep	TDN %	*	51.6	57.7	
Calcium	%		2.27	2.54	
Magnesium	%		0.36	0.40	
Phosphorus	%		0.26	0.29	
Potassium	%		2.03	2.26	

Clover, Egyptian, aerial part, fresh, (2)
Ref No 2 01 349 — United States

Feed Name or Analyses			As Fed	Dry	C.V. ± %
Dry matter	%		15.7	100.0	19
Ash	%		2.5	16.1	17
Crude fiber	%		3.5	22.6	13
Cattle	dig coef %	*	60.	60.	
Sheep	dig coef %		60.	60.	
Ether extract	%		0.6	3.7	13
Cattle	dig coef %	*	50.	50.	
Sheep	dig coef %		58.	58.	
N-free extract	%		6.2	39.9	
Cattle	dig coef %	*	80.	80.	
Sheep	dig coef %		79.	79.	
Protein (N x 6.25)	%		2.8	17.7	10
Cattle	dig coef %	*	81.	81.	
Sheep	dig coef %		77.	77.	
Cattle	dig prot %		2.2	14.3	
Goats	dig prot %	*	2.0	13.0	
Horses	dig prot %	*	2.0	12.5	
Rabbits	dig prot %	*	1.9	12.3	
Sheep	dig prot %		2.1	13.6	
Energy	GE Mcal/kg				
Cattle	DE Mcal/kg	*	0.44	2.82	
Sheep	DE Mcal/kg	*	0.44	2.80	
Cattle	ME Mcal/kg	*	0.36	2.31	
Sheep	ME Mcal/kg	*	0.36	2.30	
Cattle	TDN %		10.0	64.0	
Sheep	TDN %		9.9	63.5	
Calcium	%		0.56	3.56	
Phosphorus	%		0.05	0.32	
Potassium	%		0.36	2.31	

Ref No 2 01 349 — Egypt

Feed Name or Analyses			As Fed	Dry	C.V. ± %
Dry matter	%		14.6	100.0	
Sheep	dig coef %		73.	73.	
Ash	%		1.5	10.3	
Crude fiber	%		3.8	26.0	
Sheep	dig coef %		60.	60.	
Ether extract	%		0.5	3.4	
Sheep	dig coef %		62.	62.	

Feed Name or Analyses				Mean As Fed	Mean Dry	C.V. ± %
N-free extract		%		6.7	45.9	
Sheep	dig coef	%		81.	81.	
Protein (N x 6.25)		%		2.1	14.4	
Sheep	dig coef	%		78.	78.	
Cattle	dig prot	%	*	1.5	10.1	
Goats	dig prot	%	*	1.5	10.0	
Horses	dig prot	%	*	1.4	9.7	
Rabbits	dig prot	%	*	1.4	9.8	
Sheep	dig prot	%		1.6	11.2	
Energy	GE Mcal/kg					
Cattle	DE Mcal/kg		*	0.42	2.86	
Sheep	DE Mcal/kg		*	0.44	3.03	
Cattle	ME Mcal/kg		*	0.34	2.35	
Sheep	ME Mcal/kg		*	0.36	2.49	
Cattle	TDN	%		9.5	64.9	
Sheep	TDN	%		10.0	68.8	

Clover, Egyptian, aerial part, fresh, immature, (2)

Ref No 2 01 341 United States

Feed Name or Analyses				Mean As Fed	Mean Dry	C.V. ± %
Dry matter		%		11.7	100.0	10
Ash		%		2.3	19.8	3
Crude fiber		%		2.5	21.6	1
Cattle	dig coef	%	*	60.	60.	
Sheep	dig coef	%	*	60.	60.	
Ether extract		%		0.4	3.8	7
Cattle	dig coef	%	*	50.	50.	
Sheep	dig coef	%	*	58.	58.	
N-free extract		%		4.0	34.5	
Cattle	dig coef	%	*	80.	80.	
Sheep	dig coef	%	*	79.	79.	
Protein (N x 6.25)		%		2.4	20.3	5
Cattle	dig coef	%	*	81.	81.	
Sheep	dig coef	%	*	77.	77.	
Cattle	dig prot	%		1.9	16.4	
Goats	dig prot	%	*	1.8	15.5	
Horses	dig prot	%	*	1.7	14.8	
Rabbits	dig prot	%	*	1.7	14.3	
Sheep	dig prot	%		1.8	15.6	
Energy	GE Mcal/kg					
Cattle	DE Mcal/kg		*	0.32	2.70	
Sheep	DE Mcal/kg		*	0.31	2.68	
Cattle	ME Mcal/kg		*	0.26	2.22	
Sheep	ME Mcal/kg		*	0.26	2.20	
Cattle	TDN	%	*	7.2	61.3	
Sheep	TDN	%	*	7.1	60.8	

Clover, Egyptian, aerial part, fresh, pre-bloom, cut 1, (2)

Ref No 2 01 342 United States

Feed Name or Analyses				Mean As Fed	Mean Dry	C.V. ± %
Dry matter		%		11.8	100.0	
Ash		%		2.4	20.6	
Crude fiber		%		2.6	22.0	
Sheep	dig coef	%		72.	72.	
Ether extract		%		0.4	3.6	
Sheep	dig coef	%		61.	61.	
N-free extract		%		4.1	35.0	
Sheep	dig coef	%		82.	82.	
Protein (N x 6.25)		%		2.2	18.8	
Sheep	dig coef	%		79.	79.	
Cattle	dig prot	%	*	1.6	13.9	
Goats	dig prot	%	*	1.7	14.1	
Horses	dig prot	%	*	1.6	13.5	
Rabbits	dig prot	%	*	1.6	13.2	
Sheep	dig prot	%		1.8	14.9	
Energy	GE Mcal/kg					
Cattle	DE Mcal/kg		*	0.33	2.80	
Sheep	DE Mcal/kg		*	0.33	2.84	
Cattle	ME Mcal/kg		*	0.27	2.30	
Sheep	ME Mcal/kg		*	0.27	2.33	
Cattle	TDN	%	*	7.5	63.6	
Sheep	TDN	%		7.6	64.3	

Clover, Egyptian, aerial part, fresh, pre-bloom, cut 2, (2)

Ref No 2 01 343 United States

Feed Name or Analyses				Mean As Fed	Mean Dry	C.V. ± %
Dry matter		%		8.6	100.0	
Ash		%		1.6	18.4	
Crude fiber		%		1.8	21.2	
Sheep	dig coef	%		60.	60.	
Ether extract		%		0.4	4.6	
Sheep	dig coef	%		63.	63.	
N-free extract		%		2.9	33.4	
Sheep	dig coef	%		82.	82.	
Protein (N x 6.25)		%		1.9	22.4	
Sheep	dig coef	%		80.	80.	
Cattle	dig prot	%	*	1.5	16.9	
Goats	dig prot	%	*	1.5	17.5	
Horses	dig prot	%	*	1.4	16.5	
Rabbits	dig prot	%	*	1.4	16.0	
Sheep	dig prot	%		1.5	17.9	
Energy	GE Mcal/kg					
Cattle	DE Mcal/kg		*	0.22	2.61	
Sheep	DE Mcal/kg		*	0.24	2.85	
Cattle	ME Mcal/kg		*	0.18	2.14	
Sheep	ME Mcal/kg		*	0.20	2.33	
Cattle	TDN	%	*	5.1	59.3	
Sheep	TDN	%		5.6	64.5	

Clover, Egyptian, aerial part, fresh, pre-bloom, cut 3, (2)

Ref No 2 01 344 United States

Feed Name or Analyses				Mean As Fed	Mean Dry	C.V. ± %
Dry matter		%		13.2	100.0	
Ash		%		2.7	20.1	
Crude fiber		%		2.8	21.5	
Sheep	dig coef	%		57.	57.	
Ether extract		%		0.5	3.5	
Sheep	dig coef	%		56.	56.	
N-free extract		%		4.6	34.9	
Sheep	dig coef	%		75.	75.	
Protein (N x 6.25)		%		2.6	20.0	
Sheep	dig coef	%		76.	76.	
Cattle	dig prot	%	*	2.0	14.9	
Goats	dig prot	%	*	2.0	15.2	
Horses	dig prot	%	*	1.9	14.5	
Rabbits	dig prot	%	*	1.9	14.1	
Sheep	dig prot	%		2.0	15.2	
Energy	GE Mcal/kg					
Cattle	DE Mcal/kg		*	0.32	2.46	
Sheep	DE Mcal/kg		*	0.34	2.56	
Cattle	ME Mcal/kg		*	0.27	2.02	
Sheep	ME Mcal/kg		*	0.28	2.10	
Cattle	TDN	%	*	7.4	55.8	
Sheep	TDN	%		7.7	58.0	

Clover, Egyptian, aerial part, fresh, early bloom, cut 3, (2)

Ref No 2 01 345 United States

Feed Name or Analyses				Mean As Fed	Mean Dry	C.V. ± %
Dry matter		%		17.2	100.0	10
Ash		%		1.9	10.8	7
Crude fiber		%		4.2	24.2	10
Cattle	dig coef	%	*	60.	60.	
Sheep	dig coef	%		55.	55.	
Ether extract		%		0.6	3.4	7
Cattle	dig coef	%	*	50.	50.	
Sheep	dig coef	%		66.	66.	
N-free extract		%		7.2	41.8	
Cattle	dig coef	%	*	80.	80.	
Sheep	dig coef	%		86.	86.	
Protein (N x 6.25)		%		3.4	19.8	8
Cattle	dig coef	%	*	81.	81.	
Sheep	dig coef	%		84.	84.	
Cattle	dig prot	%		2.8	16.0	
Goats	dig prot	%	*	2.6	15.0	
Horses	dig prot	%	*	2.5	14.3	
Rabbits	dig prot	%	*	2.4	14.0	
Sheep	dig prot	%		2.9	16.6	
Energy	GE Mcal/kg					
Cattle	DE Mcal/kg		*	0.51	2.99	
Sheep	DE Mcal/kg		*	0.54	3.13	
Cattle	ME Mcal/kg		*	0.42	2.45	
Sheep	ME Mcal/kg		*	0.44	2.56	
Cattle	TDN	%		11.7	67.8	
Sheep	TDN	%		12.2	70.9	

Clover, Egyptian, aerial part, fresh, cut 1, (2)

Ref No 2 01 346 United States

Feed Name or Analyses				Mean As Fed	Mean Dry	C.V. ± %
Dry matter		%		12.3	100.0	9
Ash		%		2.4	19.8	6
Crude fiber		%		2.6	21.4	6
Sheep	dig coef	%		71.	71.	
Ether extract		%		0.5	3.7	5
Sheep	dig coef	%		59.	59.	
N-free extract		%		4.5	36.4	
Sheep	dig coef	%		82.	82.	
Protein (N x 6.25)		%		2.3	18.7	8
Sheep	dig coef	%		79.	79.	
Cattle	dig prot	%	*	1.7	13.8	
Goats	dig prot	%	*	1.7	14.0	
Horses	dig prot	%	*	1.6	13.4	
Rabbits	dig prot	%	*	1.6	13.1	
Sheep	dig prot	%		1.8	14.8	
Energy	GE Mcal/kg					
Cattle	DE Mcal/kg		*	0.33	2.66	
Sheep	DE Mcal/kg		*	0.35	2.85	
Cattle	ME Mcal/kg		*	0.27	2.18	
Sheep	ME Mcal/kg		*	0.29	2.34	
Cattle	TDN	%	*	7.4	60.4	
Sheep	TDN	%		8.0	64.7	

Clover, Egyptian, aerial part, fresh, cut 2, (2)

Ref No 2 01 347 United States

Feed Name or Analyses				Mean As Fed	Mean Dry	C.V. ± %
Dry matter		%		10.2	100.0	27
Ash		%		1.7	17.1	8
Crude fiber		%		2.1	21.2	7
Sheep	dig coef	%		60.	60.	
Ether extract		%		0.5	4.7	7
Sheep	dig coef	%		62.	62.	
N-free extract		%		3.7	36.2	
Sheep	dig coef	%		81.	81.	
Protein (N x 6.25)		%		2.1	21.0	9
Sheep	dig coef	%		78.	78.	
Cattle	dig prot	%	*	1.6	15.7	
Goats	dig prot	%	*	1.6	16.1	
Horses	dig prot	%	*	1.6	15.3	

Continued

Feed Name or Analyses				Mean As Fed	Mean Dry	C.V. ± %
Rabbits	dig prot	%	*	1.5	14.8	
Sheep	dig prot	%		1.7	16.3	
Energy		GE Mcal/kg				
Cattle		DE Mcal/kg	*	0.27	2.62	
Sheep		DE Mcal/kg	*	0.29	2.86	
Cattle		ME Mcal/kg	*	0.22	2.15	
Sheep		ME Mcal/kg	*	0.24	2.34	
Cattle		TDN %	*	6.0	59.5	
Sheep		TDN %		6.6	64.8	

Clover, Egyptian, aerial part, fresh, cut 3, (2)
Ref No 2 01 348 United States

Feed Name or Analyses				Mean As Fed	Mean Dry	C.V. ± %
Dry matter		%		16.2	100.0	15
Ash		%		2.6	16.4	13
Crude fiber		%		3.7	22.8	10
Sheep	dig coef	%		57.	57.	
Ether extract		%		0.6	3.4	4
Sheep	dig coef	%		57.	57.	
N-free extract		%		6.2	38.1	
Sheep	dig coef	%		77.	77.	
Protein (N x 6.25)		%		3.1	19.4	5
Sheep	dig coef	%		77.	77.	
Cattle	dig prot	%	*	2.3	14.4	
Goats	dig prot	%	*	2.4	14.7	
Horses	dig prot	%	*	2.3	14.0	
Rabbits	dig prot	%	*	2.2	13.6	
Sheep	dig prot	%		2.4	14.9	
Energy		GE Mcal/kg				
Cattle		DE Mcal/kg	*	0.44	2.69	
Sheep		DE Mcal/kg	*	0.44	2.72	
Cattle		ME Mcal/kg	*	0.36	2.21	
Sheep		ME Mcal/kg	*	0.36	2.23	
Cattle		TDN %	*	9.9	61.1	
Sheep		TDN %		10.0	61.6	

Clover, Egyptian, aerial part, ensiled, (3)
Ref No 3 01 353 United States

Feed Name or Analyses				Mean As Fed	Mean Dry	C.V. ± %
Dry matter		%		24.1	100.0	10
Ash		%		3.6	15.0	12
Crude fiber		%		7.8	32.4	11
Ether extract		%		0.9	3.7	21
N-free extract		%		7.7	32.0	
Protein (N x 6.25)		%		4.1	17.0	9
Cattle	dig prot	%	*	2.8	11.7	
Goats	dig prot	%	*	2.8	11.7	
Horses	dig prot	%	*	2.8	11.7	
Sheep	dig prot	%	*	2.8	11.7	
Energy		GE Mcal/kg				
Cattle		DE Mcal/kg	*	0.42	1.76	
Sheep		DE Mcal/kg	*	0.40	1.66	
Cattle		ME Mcal/kg	*	0.35	1.44	
Sheep		ME Mcal/kg	*	0.33	1.36	
Cattle		TDN %	*	9.6	40.0	
Sheep		TDN %	*	9.1	37.7	

Clover, Egyptian, aerial part, ensiled, cut 2, (3)
Ref No 3 01 351 United States

Feed Name or Analyses				Mean As Fed	Mean Dry	C.V. ± %
Dry matter		%		20.9	100.0	
Ash		%		3.5	16.9	
Crude fiber		%		7.4	35.4	
Ether extract		%		0.8	3.7	
N-free extract		%		5.7	27.3	
Protein (N x 6.25)		%		3.5	16.7	
Cattle	dig prot	%	*	2.4	11.4	
Goats	dig prot	%	*	2.4	11.4	
Horses	dig prot	%	*	2.4	11.4	
Sheep	dig prot	%	*	2.4	11.4	

Clover, Egyptian, aerial part, ensiled, cut 3, (3)
Ref No 3 01 352 United States

Feed Name or Analyses				Mean As Fed	Mean Dry	C.V. ± %
Dry matter		%		27.4	100.0	
Ash		%		3.7	13.4	
Crude fiber		%		8.1	29.5	
Ether extract		%		0.9	3.4	
N-free extract		%		9.9	36.3	
Protein (N x 6.25)		%		4.8	17.4	
Cattle	dig prot	%	*	3.3	12.0	
Goats	dig prot	%	*	3.3	12.0	
Horses	dig prot	%	*	3.3	12.0	
Sheep	dig prot	%	*	3.3	12.0	

CLOVER, HOP. Trifolium agrarium

Clover, hop, hay, s-c, (1)
Ref No 1 01 357 United States

Feed Name or Analyses				Mean As Fed	Mean Dry	C.V. ± %
Dry matter		%		92.8	100.0	
Calcium		%		0.97	1.05	14
Iron		%		0.026	0.028	55
Magnesium		%		0.25	0.27	28
Manganese		mg/kg		66.5	71.6	11
Phosphorus		%		0.32	0.34	15
Potassium		%		2.01	2.17	12

Clover, hop, hay, s-c, immature, (1)
Ref No 1 01 354 United States

Feed Name or Analyses				Mean As Fed	Mean Dry	C.V. ± %
Dry matter		%			100.0	
Ash		%			9.4	
Protein (N x 6.25)		%			21.2	5
Cattle	dig prot	%	*		15.3	
Goats	dig prot	%	*		16.3	
Horses	dig prot	%	*		15.5	
Rabbits	dig prot	%	*		15.0	
Sheep	dig prot	%	*		15.6	
Calcium		%			1.18	7
Iron		%			0.052	
Magnesium		%			0.32	20
Manganese		mg/kg			77.6	
Phosphorus		%			0.36	9
Potassium		%			2.30	7

Clover, hop, hay, s-c, full bloom, (1)
Ref No 1 01 355 United States

Feed Name or Analyses				Mean As Fed	Mean Dry	C.V. ± %
Dry matter		%			100.0	
Ash		%			7.3	
Protein (N x 6.25)		%			14.4	
Cattle	dig prot	%	*		9.4	
Goats	dig prot	%	*		10.0	
Horses	dig prot	%	*		9.8	
Rabbits	dig prot	%	*		9.8	
Sheep	dig prot	%	*		9.5	
Calcium		%			1.10	
Iron		%			0.018	
Magnesium		%			0.22	
Manganese		mg/kg			61.9	
Phosphorus		%			0.35	
Potassium		%			1.72	

Clover, hop, hay, s-c, milk stage, (1)
Ref No 1 01 356 United States

Feed Name or Analyses				Mean As Fed	Mean Dry	C.V. ± %
Dry matter		%			100.0	
Calcium		%			1.03	
Iron		%			0.013	
Magnesium		%			0.13	
Manganese		mg/kg			59.5	
Phosphorus		%			0.24	
Potassium		%			1.66	

Clover, hop, aerial part, fresh, (2)
Ref No 2 01 361 United States

Feed Name or Analyses				Mean As Fed	Mean Dry	C.V. ± %
Dry matter		%		25.1	100.0	12
Ash		%		1.6	6.5	10
Crude fiber		%		5.0	20.0	12
Ether extract		%		1.0	3.8	15
N-free extract		%		13.2	52.5	
Protein (N x 6.25)		%		4.3	17.1	16
Cattle	dig prot	%	*	3.1	12.5	
Goats	dig prot	%	*	3.1	12.6	
Horses	dig prot	%	*	3.0	12.1	
Rabbits	dig prot	%	*	3.0	11.9	
Sheep	dig prot	%	*	3.2	13.0	
Energy		GE Mcal/kg				
Cattle		DE Mcal/kg	*	0.80	3.17	
Sheep		DE Mcal/kg	*	0.73	2.91	
Cattle		ME Mcal/kg	*	0.65	2.60	
Sheep		ME Mcal/kg	*	0.60	2.39	
Cattle		TDN %	*	18.0	72.0	
Sheep		TDN %	*	16.6	66.1	
Calcium		%		0.27	1.09	11
Cobalt		mg/kg		0.011	0.044	
Iron		%		0.006	0.025	76
Magnesium		%		0.05	0.19	13
Manganese		mg/kg		24.9	99.2	61
Phosphorus		%		0.08	0.33	27
Potassium		%		0.49	1.94	23

Clover, hop, aerial part, fresh, immature, (2)
Ref No 2 01 358 United States

Feed Name or Analyses				Mean As Fed	Mean Dry	C.V. ± %
Dry matter		%			100.0	
Calcium		%			1.26	
Iron		%			0.034	
Magnesium		%			0.21	
Manganese		mg/kg			124.8	
Phosphorus		%			0.39	
Potassium		%			2.09	

Clover, hop, aerial part, fresh, full bloom, (2)
Ref No 2 01 359 United States

Feed Name or Analyses				Mean As Fed	Mean Dry	C.V. ± %
Dry matter		%			100.0	
Calcium		%			1.13	
Iron		%			0.018	
Magnesium		%			0.17	
Manganese		mg/kg			62.2	

(1) dry forages and roughages (2) pasture, range plants, and forages fed green (3) silages (4) energy feeds (5) protein supplements (6) minerals (7) vitamins (8) additives

Feed Name or Analyses			Mean As Fed	Mean Dry	C.V. ± %
Phosphorus	%			0.25	
Potassium	%			1.67	

Clover, hop, aerial part, fresh, milk stage, (2)

Ref No 2 01 360 United States

Feed Name or Analyses			Mean As Fed	Mean Dry	C.V. ± %
Dry matter	%			100.0	
Calcium	%			1.11	
Iron	%			0.013	
Magnesium	%			0.18	
Manganese	mg/kg			85.1	
Phosphorus	%			0.30	
Potassium	%			1.56	

CLOVER, KENLAND. Trifolium pratense

Clover, kenland, aerial part, fresh, (2)

Ref No 2 07 908 United States

Feed Name or Analyses			Mean As Fed	Mean Dry	C.V. ± %
Dry matter	%			100.0	
Organic matter	%			92.8	
Crude fiber	%			15.3	
Sheep	dig coef %			54.	
Ether extract	%			6.0	
Sheep	dig coef %			69.	
N-free extract	%			50.0	
Sheep	dig coef %			84.	
Protein (N x 6.25)	%			21.6	
Sheep	dig coef %			78.	
Cattle	dig prot %	*		16.3	
Goats	dig prot %	*		16.7	
Horses	dig prot %	*		15.9	
Rabbits	dig prot %	*		15.3	
Sheep	dig prot %			16.8	
Lignin (Ellis)	%			5.0	
Energy	GE Mcal/kg				
Cattle	DE Mcal/kg	*		3.17	
Sheep	DE Mcal/kg	*		3.35	
Cattle	ME Mcal/kg	*		2.60	
Sheep	ME Mcal/kg	*		2.74	
Cattle	TDN %	*		71.9	
Sheep	TDN %			75.9	

CLOVER, KENTUCKY 215. Trifolium spp

Clover, Kentucky 215, aerial part, fresh, (2)

Ref No 2 07 909 United States

Feed Name or Analyses			Mean As Fed	Mean Dry	C.V. ± %
Dry matter	%			100.0	
Organic matter	%			90.9	
Crude fiber	%			12.5	
Sheep	dig coef %			50.	
Ether extract	%			5.8	
Sheep	dig coef %			72.	
N-free extract	%			51.6	
Sheep	dig coef %			82.	
Protein (N x 6.25)	%			21.0	
Sheep	dig coef %			77.	
Cattle	dig prot %	*		15.7	
Goats	dig prot %	*		16.2	
Horses	dig prot %	*		15.4	
Rabbits	dig prot %	*		14.9	
Sheep	dig prot %			16.2	
Lignin (Ellis)	%			5.1	
Energy	GE Mcal/kg				
Cattle	DE Mcal/kg	*		3.00	
Sheep	DE Mcal/kg	*		3.27	
Cattle	ME Mcal/kg	*		2.46	
Sheep	ME Mcal/kg	*		2.68	
Cattle	TDN %	*		68.1	
Sheep	TDN %			74.1	

CLOVER, LADINO. Trifolium repens

Clover, ladino, aerial part, dehy, (1)

Ref No 1 01 367 United States

Feed Name or Analyses			Mean As Fed	Mean Dry	C.V. ± %
Dry matter	%		92.1	100.0	2
Crude fiber	%		14.9	16.2	18
Ether extract	%		4.8	5.2	6
Protein (N x 6.25)	%		21.8	23.7	7
Cattle	dig prot %	*	16.1	17.5	
Goats	dig prot %	*	17.2	18.7	
Horses	dig prot %	*	16.3	17.7	
Rabbits	dig prot %	*	15.6	17.0	
Sheep	dig prot %	*	16.4	17.8	

Clover, ladino, aerial part, dehy, immature, (1)

Ref No 1 01 363 United States

Feed Name or Analyses			Mean As Fed	Mean Dry	C.V. ± %
Dry matter	%		90.4	100.0	
Crude fiber	%		13.3	14.7	
Ether extract	%		5.1	5.6	
Protein (N x 6.25)	%		23.6	26.1	
Cattle	dig prot %	*	17.7	19.5	
Goats	dig prot %	*	18.9	20.9	
Horses	dig prot %	*	17.8	19.7	
Rabbits	dig prot %	*	17.0	18.8	
Sheep	dig prot %	*	18.1	20.0	

Clover, ladino, aerial part, dehy, early bloom, (1)

Ref No 1 01 364 United States

Feed Name or Analyses			Mean As Fed	Mean Dry	C.V. ± %
Dry matter	%		92.5	100.0	3
Crude fiber	%		15.2	16.4	19
Ether extract	%		4.7	5.1	5
Protein (N x 6.25)	%		21.6	23.4	4
Cattle	dig prot %	*	15.9	17.2	
Goats	dig prot %	*	17.0	18.4	
Horses	dig prot %	*	16.1	17.4	
Rabbits	dig prot %	*	15.5	16.7	
Sheep	dig prot %	*	16.2	17.6	

Clover, ladino, aerial part, dehy, cut 1, (1)

Ref No 1 01 365 United States

Feed Name or Analyses			Mean As Fed	Mean Dry	C.V. ± %
Dry matter	%		92.5	100.0	3
Crude fiber	%		16.7	18.0	2
Ether extract	%		4.8	5.2	6
Protein (N x 6.25)	%		21.3	23.0	4
Cattle	dig prot %	*	15.6	16.9	
Goats	dig prot %	*	16.7	18.0	
Horses	dig prot %	*	15.8	17.1	
Rabbits	dig prot %	*	15.2	16.4	
Sheep	dig prot %	*	15.9	17.2	

Clover, ladino, aerial part, dehy, cut 2, (1)

Ref No 1 01 366 United States

Feed Name or Analyses			Mean As Fed	Mean Dry	C.V. ± %
Dry matter	%		91.4	100.0	
Crude fiber	%		14.0	15.3	
Ether extract	%		4.8	5.2	
Protein (N x 6.25)	%		22.9	25.1	
Cattle	dig prot %	*	17.1	18.7	
Goats	dig prot %	*	18.3	20.0	
Horses	dig prot %	*	17.2	18.8	
Rabbits	dig prot %	*	16.5	18.0	
Sheep	dig prot %	*	17.4	19.1	

Clover, ladino, hay, fan air dried, early bloom, (1)

Ref No 1 01 362 United States

Feed Name or Analyses			Mean As Fed	Mean Dry	C.V. ± %
Dry matter	%		87.9	100.0	
Ash	%		9.1	10.4	
Crude fiber	%		17.1	19.5	
Ether extract	%		1.6	1.8	
N-free extract	%		40.3	45.8	
Protein (N x 6.25)	%		19.8	22.5	
Cattle	dig prot %	*	14.4	16.4	
Goats	dig prot %	*	15.4	17.6	
Horses	dig prot %	*	14.6	16.6	
Rabbits	dig prot %	*	14.1	16.0	
Sheep	dig prot %	*	14.7	16.8	
Energy	GE Mcal/kg				
Cattle	DE Mcal/kg	*	2.53	2.88	
Sheep	DE Mcal/kg	*	2.51	2.86	
Cattle	ME Mcal/kg	*	2.08	2.36	
Sheep	ME Mcal/kg	*	2.06	2.34	
Cattle	TDN %	*	57.4	65.4	
Sheep	TDN %	*	56.9	64.8	

Clover, ladino, hay, s-c, (1)

Ref No 1 01 378 United States

Feed Name or Analyses			Mean As Fed	Mean Dry	C.V. ± %
Dry matter	%		89.5	100.0	
Ash	%		8.3	9.3	
Crude fiber	%		21.6	24.1	
Cattle	dig coef %	*	56.	56.	
Sheep	dig coef %	*	56.	56.	
Ether extract	%		1.7	1.9	
Cattle	dig coef %	*	58.	58.	
Sheep	dig coef %	*	43.	43.	
N-free extract	%		39.4	44.0	
Cattle	dig coef %	*	69.	69.	
Sheep	dig coef %	*	64.	64.	
Protein (N x 6.25)	%		18.5	20.7	
Cattle	dig coef %	*	63.	63.	
Sheep	dig coef %	*	62.	62.	
Cattle	dig prot %		11.7	13.0	
Goats	dig prot %	*	14.2	15.8	
Horses	dig prot %	*	13.5	15.1	
Rabbits	dig prot %	*	13.1	14.6	
Sheep	dig prot %		11.5	12.8	
Fatty acids	%		1.7	1.9	
Energy	GE Mcal/kg				
Cattle	DE Mcal/kg	*	2.34	2.62	
Sheep	DE Mcal/kg	*	2.22	2.48	
Cattle	ME Mcal/kg	*	1.92	2.15	
Sheep	ME Mcal/kg	*	1.82	2.04	
Cattle	NE_m Mcal/kg	*	1.17	1.31	
Cattle	NE_{gain} Mcal/kg	*	0.62	0.69	
Cattle	$NE_{lactating\ cows}$ Mcal/kg	*	1.33	1.49	
Cattle	TDN %		53.2	59.4	
Sheep	TDN %		50.4	56.3	
Calcium	%		1.53	1.71	
Phosphorus	%		0.29	0.32	
Potassium	%		2.17	2.42	

Clover, ladino, hay, s-c, immature, (1)

Ref No 1 01 368 United States

Feed Name or Analyses			Mean As Fed	Mean Dry	C.V. ± %
Dry matter	%		88.4	100.0	3
Ash	%		9.6	10.9	
Crude fiber	%		16.1	18.2	
Cattle	dig coef %	*	56.	56.	
Sheep	dig coef %	*	72.	72.	
Ether extract	%		2.3	2.6	
Cattle	dig coef %	*	58.	58.	
Sheep	dig coef %	*	61.	61.	
N-free extract	%		39.8	45.0	
Cattle	dig coef %	*	69.	69.	
Sheep	dig coef %	*	82.	82.	
Protein (N x 6.25)	%		20.6	23.3	
Cattle	dig coef %	*	63.	63.	
Sheep	dig coef %	*	79.	79.	
Cattle	dig prot %		13.0	14.7	
Goats	dig prot %	*	16.2	18.3	
Horses	dig prot %	*	15.3	17.3	
Rabbits	dig prot %	*	14.7	16.7	
Sheep	dig prot %		16.3	18.4	
Energy	GE Mcal/kg				
Cattle	DE Mcal/kg	*	2.31	2.62	
Sheep	DE Mcal/kg	*	2.80	3.17	
Cattle	ME Mcal/kg	*	1.89	2.14	
Sheep	ME Mcal/kg	*	2.30	2.60	
Cattle	TDN %		52.4	59.3	
Sheep	TDN %		63.6	72.0	
Calcium	%		1.24	1.40	4
Phosphorus	%		0.31	0.35	52

Clover, ladino, hay, s-c, pre-bloom, (1)

Ref No 1 08 381 United States

Feed Name or Analyses			Mean As Fed	Mean Dry	C.V. ± %
Dry matter	%		88.4	100.0	
Ash	%		10.4	11.8	
Crude fiber	%		18.6	21.0	
Ether extract	%		1.4	1.6	
N-free extract	%		36.6	41.4	
Protein (N x 6.25)	%		21.4	24.2	
Cattle	dig prot %	*	15.8	17.9	
Goats	dig prot %	*	16.9	19.1	
Horses	dig prot %	*	16.0	18.1	
Rabbits	dig prot %	*	15.3	17.4	
Sheep	dig prot %	*	16.2	18.3	
Energy	GE Mcal/kg				
Cattle	DE Mcal/kg	*	2.61	2.95	
Sheep	DE Mcal/kg	*	2.46	2.78	
Cattle	ME Mcal/kg	*	2.14	2.42	
Sheep	ME Mcal/kg	*	2.01	2.28	
Cattle	TDN %	*	59.1	66.9	
Sheep	TDN %	*	55.7	63.0	
Calcium	%		1.53	1.73	
Chlorine	%		0.29	0.33	
Copper	mg/kg		7.7	8.7	
Iron	%		0.016	0.018	
Magnesium	%		0.46	0.52	
Manganese	mg/kg		160.1	181.1	
Phosphorus	%		0.29	0.33	
Potassium	%		2.17	2.45	
Sulphur	%		0.19	0.21	

(1) dry forages and roughages (2) pasture, range plants, and forages fed green (3) silages (4) energy feeds (5) protein supplements (6) minerals (7) vitamins (8) additives

Clover, ladino, hay, s-c, early bloom, (1)

Ref No 1 01 369 United States

Feed Name or Analyses			Mean As Fed	Mean Dry	C.V. ± %
Dry matter	%		94.5	100.0	
Ash	%		9.0	9.5	
Crude fiber	%		19.4	20.6	
Cattle	dig coef %	*	56.	56.	
Sheep	dig coef %	*	61.	61.	
Ether extract	%		1.9	2.0	
Cattle	dig coef %	*	58.	58.	
Sheep	dig coef %	*	33.	33.	
N-free extract	%		44.9	47.5	
Cattle	dig coef %	*	69.	69.	
Sheep	dig coef %	*	69.	69.	
Protein (N x 6.25)	%		19.2	20.4	
Cattle	dig coef %	*	63.	63.	
Sheep	dig coef %	*	69.	69.	
Cattle	dig prot %		12.1	12.8	
Goats	dig prot %	*	14.7	15.6	
Horses	dig prot %	*	14.0	14.8	
Rabbits	dig prot %	*	13.6	14.4	
Sheep	dig prot %		13.2	13.9	
Lignin (Ellis)	%		5.9	6.3	
Energy	GE Mcal/kg				
Cattle	DE Mcal/kg	*	2.49	2.63	
Sheep	DE Mcal/kg	*	2.52	2.66	
Cattle	ME Mcal/kg	*	2.04	2.16	
Sheep	ME Mcal/kg	*	2.06	2.18	
Cattle	TDN %		56.4	59.7	
Sheep	TDN %		57.1	60.4	
Calcium	%		1.23	1.31	4
Phosphorus	%		0.29	0.31	14
Carotene	mg/kg		19.0	20.1	
Vitamin A equivalent	IU/g		31.7	33.5	

Clover, ladino, hay, s-c, mid-bloom, (1)

Ref No 1 01 370 United States

Feed Name or Analyses			Mean As Fed	Mean Dry	C.V. ± %
Dry matter	%		88.4	100.0	
Carotene	mg/kg		35.7	40.3	
Vitamin A equivalent	IU/g		59.5	67.3	

Clover, ladino, hay, s-c, milk stage, (1)

Ref No 1 01 371 United States

Feed Name or Analyses			Mean As Fed	Mean Dry	C.V. ± %
Dry matter	%			100.0	
Calcium	%			0.97	
Iron	%			0.012	
Magnesium	%			0.40	
Manganese	mg/kg			20.9	
Phosphorus	%			0.31	
Potassium	%			2.76	

Clover, ladino, hay, s-c, mature, (1)

Ref No 1 01 372 United States

Feed Name or Analyses			Mean As Fed	Mean Dry	C.V. ± %
Dry matter	%		91.1	100.0	1
Ash	%		7.7	8.4	
Crude fiber	%		25.2	27.7	
Ether extract	%		1.8	2.0	
N-free extract	%		42.0	46.1	
Protein (N x 6.25)	%		14.4	15.8	
Cattle	dig prot %	*	9.7	10.6	
Goats	dig prot %	*	10.3	11.3	
Horses	dig prot %	*	10.0	10.9	
Rabbits	dig prot %	*	9.9	10.9	
Sheep	dig prot %	*	9.8	10.7	
Lignin (Ellis)	%		11.7	12.9	13
Energy	GE Mcal/kg				
Cattle	DE Mcal/kg	*	2.39	2.62	
Sheep	DE Mcal/kg	*	2.40	2.64	
Cattle	ME Mcal/kg	*	1.96	2.15	
Sheep	ME Mcal/kg	*	1.97	2.16	
Cattle	TDN %	*	54.1	59.4	
Sheep	TDN %	*	54.5	59.9	
Calcium	%		1.28	1.40	11
Iron	%		0.018	0.020	36
Magnesium	%		0.34	0.37	13
Phosphorus	%		0.23	0.25	31
Potassium	%		1.97	2.16	24
Carotene	mg/kg		3.4	3.7	48
Vitamin A equivalent	IU/g		5.7	6.2	

Clover, ladino, hay, s-c, cut 1, (1)

Ref No 1 01 373 United States

Feed Name or Analyses			Mean As Fed	Mean Dry	C.V. ± %
Dry matter	%		89.1	100.0	2
Ash	%		8.9	10.0	13
Crude fiber	%		19.5	21.9	16
Ether extract	%		2.8	3.1	38
N-free extract	%		39.7	44.6	
Protein (N x 6.25)	%		18.2	20.4	14
Cattle	dig prot %	*	13.0	14.6	
Goats	dig prot %	*	13.9	15.6	
Horses	dig prot %	*	13.2	14.8	
Rabbits	dig prot %	*	12.8	14.4	
Sheep	dig prot %	*	13.2	14.9	
Sugars, total	%		5.9	6.6	
Energy	GE Mcal/kg				
Cattle	DE Mcal/kg	*	2.55	2.87	
Sheep	DE Mcal/kg	*	2.44	2.74	
Cattle	ME Mcal/kg	*	2.09	2.35	
Sheep	ME Mcal/kg	*	2.00	2.25	
Cattle	TDN %	*	57.9	65.0	
Sheep	TDN %	*	55.3	62.1	
Calcium	%		1.11	1.25	
Cobalt	mg/kg		0.147	0.165	
Copper	mg/kg		8.6	9.7	
Iron	%		0.027	0.030	
Magnesium	%		0.30	0.34	
Manganese	mg/kg		47.7	53.6	
Phosphorus	%		0.24	0.28	
Potassium	%		2.23	2.50	
Carotene	mg/kg		14.7	16.5	
Niacin	mg/kg		9.8	11.0	
Vitamin A equivalent	IU/g		24.5	27.6	

Clover, ladino, hay, s-c, cut 2, (1)

Ref No 1 01 374 United States

Feed Name or Analyses			Mean As Fed	Mean Dry	C.V. ± %
Dry matter	%		91.5	100.0	2
Ash	%		9.6	10.5	13
Crude fiber	%		18.3	20.0	21
Ether extract	%		3.4	3.7	38
N-free extract	%		38.7	42.3	
Protein (N x 6.25)	%		21.5	23.5	10
Cattle	dig prot %	*	15.8	17.3	
Goats	dig prot %	*	16.9	18.5	
Horses	dig prot %	*	16.0	17.5	
Rabbits	dig prot %	*	15.4	16.8	
Sheep	dig prot %	*	16.1	17.6	
Energy	GE Mcal/kg				
Cattle	DE Mcal/kg	*	2.70	2.95	
Sheep	DE Mcal/kg	*	2.54	2.77	

Feed Name or Analyses			Mean As Fed	Mean Dry	C.V. ± %
Cattle	ME Mcal/kg	*	2.21	2.42	
Sheep	ME Mcal/kg	*	2.08	2.27	
Cattle	TDN %	*	61.2	66.9	
Sheep	TDN %	*	57.5	62.9	
Calcium	%		1.18	1.29	6
Magnesium	%		0.71	0.78	
Phosphorus	%		0.30	0.33	4
Potassium	%		3.14	3.43	
Carotene	mg/kg		133.1	145.5	
Vitamin A equivalent	IU/g		221.9	242.6	

Clover, ladino, hay, s-c, gr 1 US, (1)

Ref No 1 01 375 United States

Feed Name or Analyses			Mean As Fed	Mean Dry	C.V. ± %
Dry matter	%		90.4	100.0	
Carotene	mg/kg		106.0	117.3	
Vitamin A equivalent	IU/g		176.7	195.5	

Clover, ladino, hay, s-c, gr 2 US, (1)

Ref No 1 01 376 United States

Feed Name or Analyses			Mean As Fed	Mean Dry	C.V. ± %
Dry matter	%		84.5	100.0	
Calcium	%		1.07	1.27	
Copper	mg/kg		9.1	10.8	
Iron	%		0.038	0.045	
Magnesium	%		0.33	0.39	
Manganese	mg/kg		49.0	58.0	
Phosphorus	%		0.24	0.28	
Potassium	%		2.30	2.72	
Niacin	mg/kg		9.3	11.0	

Clover, ladino, hay, s-c, pure stand, (1)

Ref No 1 01 377 United States

Feed Name or Analyses			Mean As Fed	Mean Dry	C.V. ± %
Dry matter	%			100.0	
Alanine	%			1.60	6
Arginine	%			1.10	6
Aspartic acid	%			3.70	2
Cystine	%			0.40	
Glutamic acid	%			1.70	6
Glycine	%			1.00	6
Histidine	%			0.50	13
Isoleucine	%			1.20	5
Leucine	%			2.10	6
Lysine	%			1.20	5
Methionine	%			0.30	
Phenylalanine	%			1.20	5
Proline	%			1.50	
Serine	%			1.00	6
Threonine	%			1.30	5
Tryptophan	%			0.50	13
Tyrosine	%			0.70	9
Valine	%			1.30	5

Clover, ladino, aerial part, fresh, (2)

Ref No 2 01 383 United States

Feed Name or Analyses			Mean As Fed	Mean Dry	C.V. ± %
Dry matter	%		18.0	100.0	30
Organic matter	%		16.4	90.8	1
Ash	%		1.9	10.8	20
Crude fiber	%		2.5	14.1	12
Cattle	dig coef %	*	68.	68.	
Sheep	dig coef %	*	67.	67.	20
Ether extract	%		0.9	4.8	18
Cattle	dig coef %	*	74.	74.	
Sheep	dig coef %	*	54.	54.	45
N-free extract	%		8.2	45.5	8
Cattle	dig coef %	*	83.	83.	
Sheep	dig coef %	*	82.	82.	9
Protein (N x 6.25)	%		4.5	24.7	11
Cattle	dig coef %	*	79.	79.	
Sheep	dig coef %	*	78.	78.	12
Cattle	dig prot %		3.5	19.5	
Goats	dig prot %	*	3.5	19.6	
Horses	dig prot %	*	3.3	18.5	
Rabbits	dig prot %	*	3.2	17.8	
Sheep	dig prot %		3.5	19.4	
Lignin (Ellis)	%		0.7	4.1	36
Energy	GE Mcal/kg				
Cattle	DE Mcal/kg	*	0.60	3.31	
Sheep	DE Mcal/kg	*	0.57	3.17	
Cattle	ME Mcal/kg	*	0.49	2.71	
Sheep	ME Mcal/kg	*	0.47	2.60	
Cattle	TDN %		13.5	75.0	
Sheep	TDN %		12.9	71.8	
Calcium	%		0.23	1.27	
Iron	%		0.006	0.036	
Magnesium	%		0.09	0.48	
Manganese	mg/kg		12.9	71.7	
Phosphorus	%		0.08	0.42	
Potassium	%		0.34	1.87	
Sodium	%		0.02	0.12	
Sulphur	%		0.02	0.12	
Carotene	mg/kg		57.5	319.4	29
Riboflavin	mg/kg		4.4	24.3	20
Vitamin A equivalent	IU/g		95.9	532.5	

Clover, ladino, aerial part, fresh, immature, (2)

Ref No 2 01 380 United States

Feed Name or Analyses			Mean As Fed	Mean Dry	C.V. ± %
Dry matter	%		20.6	100.0	30
Crude fiber	%				
Cattle	dig coef %	*	59.	59.	
Sheep	dig coef %	*	52.	52.	
Ether extract	%				
Cattle	dig coef %	*	66.	66.	
Sheep	dig coef %	*	67.	67.	
N-free extract	%				
Cattle	dig coef %	*	80.	80.	
Sheep	dig coef %	*	71.	71.	
Protein (N x 6.25)	%				
Cattle	dig coef %	*	70.	70.	
Sheep	dig coef %	*	67.	67.	
Sodium	%		0.02	0.12	45
Sulphur	%		0.03	0.16	
Carotene	mg/kg		72.4	352.5	25
Vitamin A equivalent	IU/g		120.8	587.6	

Clover, ladino, aerial part, fresh, immature, pure stand, (2)

Ref No 2 01 379 United States

Feed Name or Analyses			Mean As Fed	Mean Dry	C.V. ± %
Dry matter	%		18.1	100.0	19
Ash	%		2.4	13.5	19
Crude fiber	%		2.5	14.0	4
Ether extract	%		0.5	2.5	22
N-free extract	%		7.7	42.8	
Protein (N x 6.25)	%		4.9	27.2	10
Cattle	dig prot %	*	3.8	21.0	
Goats	dig prot %	*	4.0	21.9	
Horses	dig prot %	*	3.7	20.6	
Rabbits	dig prot %	*	3.6	19.7	
Sheep	dig prot %	*	4.0	22.3	
Energy	GE Mcal/kg				
Cattle	DE Mcal/kg	*	0.59	3.25	
Sheep	DE Mcal/kg	*	0.58	3.21	
Cattle	ME Mcal/kg	*	0.48	2.66	
Sheep	ME Mcal/kg	*	0.48	2.64	
Cattle	TDN %	*	13.3	73.7	
Sheep	TDN %	*	13.2	72.9	
Calcium	%		0.35	1.93	24
Magnesium	%		0.08	0.42	12
Phosphorus	%		0.06	0.35	7

Clover, ladino, aerial part, fresh, early bloom, (2)

Ref No 2 01 381 United States

Feed Name or Analyses			Mean As Fed	Mean Dry	C.V. ± %
Dry matter	%		35.4	100.0	
Calcium	%		0.46	1.29	
Magnesium	%		0.11	0.30	
Phosphorus	%		0.11	0.31	
Potassium	%		1.12	3.17	

Clover, ladino, aerial part, fresh, cut 1, (2)

Ref No 2 01 382 United States

Feed Name or Analyses			Mean As Fed	Mean Dry	C.V. ± %
Dry matter	%			100.0	
Crude fiber	%			15.5	10
Protein (N x 6.25)	%			24.1	8
Cattle	dig prot %	*		18.4	
Goats	dig prot %	*		19.0	
Horses	dig prot %	*		18.0	
Rabbits	dig prot %	*		17.3	
Sheep	dig prot %	*		19.5	
Calcium	%			1.26	15
Magnesium	%			0.34	34
Phosphorus	%			0.33	21
Potassium	%			2.64	26

Clover, ladino, aerial part, ensiled, (3)

Ref No 3 01 384 United States

Feed Name or Analyses			Mean As Fed	Mean Dry	C.V. ± %
Dry matter	%		24.9	100.0	6
Sheep	dig coef %		71.	71.	
Ash	%		2.4	9.8	12
Crude fiber	%		5.3	21.2	15
Sheep	dig coef %		65.	65.	
Ether extract	%		1.0	4.1	24
N-free extract	%		10.3	41.5	
Sheep	dig coef %		83.	83.	
Protein (N x 6.25)	%		5.9	23.5	18
Sheep	dig coef %		72.	72.	
Cattle	dig prot %	*	4.4	17.6	
Goats	dig prot %	*	4.4	17.6	
Horses	dig prot %	*	4.4	17.6	
Sheep	dig prot %		4.2	16.9	
Energy	GE Mcal/kg				
Cattle	DE Mcal/kg	*	0.75	3.01	
Sheep	DE Mcal/kg	*	0.76	3.06	
Cattle	ME Mcal/kg	*	0.62	2.47	
Sheep	ME Mcal/kg	*	0.63	2.51	
Cattle	TDN %	*	17.0	68.3	
Sheep	TDN %		17.3	69.5	

Feed Name or Analyses			Mean As Fed	Mean Dry	C.V. ± %

Clover, ladino, aerial part w molasses added, ensiled, early bloom, (3)
Ref No 3 01 385 United States

Analyses			As Fed	Dry	C.V. ± %
Dry matter	%		26.8	100.0	6
Ash	%		2.8	10.4	12
Crude fiber	%		5.4	20.1	7
Ether extract	%		1.0	3.7	12
N-free extract	%		10.7	40.1	
Protein (N x 6.25)	%		6.9	25.7	4
Cattle	dig prot %	*	5.2	19.6	
Goats	dig prot %	*	5.2	19.6	
Horses	dig prot %	*	5.2	19.6	
Sheep	dig prot %	*	5.2	19.6	
Energy	GE Mcal/kg				
Cattle	DE Mcal/kg	*	0.69	2.57	
Sheep	DE Mcal/kg	*	0.64	2.38	
Cattle	ME Mcal/kg	*	0.56	2.10	
Sheep	ME Mcal/kg	*	0.52	1.95	
Cattle	TDN %	*	15.6	58.2	
Sheep	TDN %	*	14.5	54.0	

CLOVER, MAMMOTH RED. Trifolium pratense

Clover, mammoth red, hay, s-c, (1)
Ref No 1 01 386 United States

Analyses			As Fed	Dry	C.V. ± %
Dry matter	%		88.6	100.0	3
Ash	%		6.9	7.8	13
Crude fiber	%		28.9	32.6	11
Cattle	dig coef %	*	56.	56.	
Sheep	dig coef %	*	56.	56.	
Ether extract	%		3.3	3.7	17
Cattle	dig coef %	*	58.	58.	
Sheep	dig coef %	*	43.	43.	
N-free extract	%		37.5	42.3	
Cattle	dig coef %	*	69.	69.	
Sheep	dig coef %	*	64.	64.	
Protein (N x 6.25)	%		12.0	13.5	9
Cattle	dig coef %	*	63.	63.	
Sheep	dig coef %	*	62.	62.	
Cattle	dig prot %		7.6	8.5	
Goats	dig prot %	*	8.2	9.2	
Horses	dig prot %	*	8.0	9.0	
Rabbits	dig prot %	*	8.1	9.1	
Sheep	dig prot %		7.4	8.4	
Fatty acids	%		3.4	3.9	
Energy	GE Mcal/kg				
Cattle	DE Mcal/kg	*	2.38	2.68	
Sheep	DE Mcal/kg	*	2.24	2.53	
Cattle	ME Mcal/kg	*	1.95	2.20	
Sheep	ME Mcal/kg	*	1.84	2.07	
Cattle	TDN %		53.9	60.9	
Sheep	TDN %		50.8	57.4	
Calcium	%		1.68	1.90	4
Phosphorus	%		0.25	0.28	15

Clover, mammoth red, aerial part, fresh, (2)
Ref No 2 01 387 United States

Analyses			As Fed	Dry	C.V. ± %
Dry matter	%		24.9	100.0	7
Ash	%		2.3	9.3	15
Crude fiber	%		7.3	29.4	12
Cattle	dig coef %	*	59.	59.	
Sheep	dig coef %	*	52.	52.	
Ether extract	%		0.5	2.1	9
Cattle	dig coef %	*	66.	66.	
Sheep	dig coef %	*	67.	67.	
N-free extract	%		10.9	43.6	
Cattle	dig coef %	*	80.	80.	
Sheep	dig coef %	*	71.	71.	
Protein (N x 6.25)	%		3.9	15.6	17
Cattle	dig coef %	*	70.	70.	
Sheep	dig coef %	*	67.	67.	
Cattle	dig prot %		2.7	10.9	
Goats	dig prot %	*	2.8	11.1	
Horses	dig prot %	*	2.7	10.8	
Rabbits	dig prot %	*	2.7	10.7	
Sheep	dig prot %		2.6	10.5	
Energy	GE Mcal/kg				
Cattle	DE Mcal/kg	*	0.73	2.92	
Sheep	DE Mcal/kg	*	0.66	2.64	
Cattle	ME Mcal/kg	*	0.60	2.40	
Sheep	ME Mcal/kg	*	0.54	2.16	
Cattle	TDN %		16.5	66.3	
Sheep	TDN %		14.9	59.9	

(1) dry forages and roughages (2) pasture, range plants, and forages fed green (3) silages (4) energy feeds (5) protein supplements (6) minerals (7) vitamins (8) additives

CLOVER, PERSIAN. Trifolium resupinatum

Clover, Persian, hay, s-c, (1)
Ref No 1 01 389 United States

Analyses			As Fed	Dry	C.V. ± %
Dry matter	%		90.0	100.0	
Ash	%		9.3	10.3	
Crude fiber	%		27.7	30.8	
Ether extract	%		1.6	1.8	
N-free extract	%		36.7	40.8	
Protein (N x 6.25)	%		14.7	16.3	
Cattle	dig prot %	*	10.0	11.1	
Goats	dig prot %	*	10.6	11.8	
Horses	dig prot %	*	10.2	11.4	
Rabbits	dig prot %	*	10.1	11.3	
Sheep	dig prot %	*	10.1	11.2	
Fatty acids	%		1.6	1.8	
Energy	GE Mcal/kg				
Cattle	DE Mcal/kg	*	2.35	2.61	
Sheep	DE Mcal/kg	*	2.24	2.49	
Cattle	ME Mcal/kg	*	1.92	2.14	
Sheep	ME Mcal/kg	*	1.84	2.04	
Cattle	TDN %	*	53.2	59.1	
Sheep	TDN %	*	50.8	56.4	
Calcium	%		1.49	1.66	
Phosphorus	%		0.21	0.23	

Clover, Persian, leaves, (1)
Ref No 1 01 390 United States

Analyses			As Fed	Dry	C.V. ± %
Dry matter	%		90.0	100.0	
Ash	%		9.5	10.6	
Crude fiber	%		17.5	19.4	
Ether extract	%		2.0	2.2	
N-free extract	%		40.2	44.7	
Protein (N x 6.25)	%		20.8	23.1	
Cattle	dig prot %	*	15.3	16.9	
Goats	dig prot %	*	16.3	18.1	
Horses	dig prot %	*	15.4	17.1	
Rabbits	dig prot %	*	14.9	16.5	
Sheep	dig prot %	*	15.6	17.3	
Energy	GE Mcal/kg				
Cattle	DE Mcal/kg	*	2.62	2.91	
Sheep	DE Mcal/kg	*	2.56	2.84	
Cattle	ME Mcal/kg	*	2.15	2.38	
Sheep	ME Mcal/kg	*	2.10	2.33	
Cattle	TDN %	*	59.3	65.9	
Sheep	TDN %	*	58.0	64.4	

Clover, Persian, stems, (1)
Ref No 1 01 391 United States

Analyses			As Fed	Dry	C.V. ± %
Dry matter	%		90.0	100.0	
Ash	%		9.6	10.7	
Crude fiber	%		35.8	39.8	
Ether extract	%		1.2	1.3	
N-free extract	%		34.1	37.9	
Protein (N x 6.25)	%		9.3	10.3	
Cattle	dig prot %	*	5.3	5.9	
Goats	dig prot %	*	5.6	6.2	
Horses	dig prot %	*	5.6	6.3	
Rabbits	dig prot %	*	6.0	6.6	
Sheep	dig prot %	*	5.2	5.8	
Energy	GE Mcal/kg				
Cattle	DE Mcal/kg	*	1.99	2.21	
Sheep	DE Mcal/kg	*	2.10	2.33	
Cattle	ME Mcal/kg	*	1.63	1.81	
Sheep	ME Mcal/kg	*	1.72	1.91	
Cattle	TDN %	*	45.1	50.1	
Sheep	TDN %	*	47.6	52.9	

CLOVER, RED. Trifolium pratense

Clover, red, hay, fan air dried, (1)
Ref No 1 01 393 United States

Analyses			As Fed	Dry	C.V. ± %
Dry matter	%		84.9	100.0	
Ash	%		9.2	10.8	
Crude fiber	%		26.2	30.9	
Sheep	dig coef %		61.	61.	
Ether extract	%		2.3	2.7	
Sheep	dig coef %		21.	21.	
N-free extract	%		28.4	33.4	
Sheep	dig coef %		59.	59.	
Protein (N x 6.25)	%		18.8	22.2	
Sheep	dig coef %		65.	65.	
Cattle	dig prot %	*	13.7	16.2	
Goats	dig prot %	*	14.7	17.3	
Horses	dig prot %	*	13.9	16.4	
Rabbits	dig prot %	*	13.4	15.8	
Sheep	dig prot %		12.3	14.4	
Energy	GE Mcal/kg				
Cattle	DE Mcal/kg	*	2.24	2.63	
Sheep	DE Mcal/kg	*	2.03	2.39	
Cattle	ME Mcal/kg	*	1.83	2.16	
Sheep	ME Mcal/kg	*	1.67	1.96	
Cattle	TDN %	*	50.7	59.7	
Sheep	TDN %		46.1	54.3	

Clover, red, hay, fan air dried, cut 1, (1)
Ref No 1 01 392 United States

Analyses			As Fed	Dry	C.V. ± %
Dry matter	%		87.6	100.0	
Ash	%		7.0	8.0	
Crude fiber	%		30.1	34.4	
Sheep	dig coef %		54.	54.	
Ether extract	%		2.5	2.9	
Sheep	dig coef %		12.	12.	
N-free extract	%		36.1	41.2	
Sheep	dig coef %		72.	72.	

Feed Name or Analyses			Mean As Fed	Mean Dry	C.V. ± %
Protein (N x 6.25)	%		11.8	13.5.	
Sheep	dig coef %		66.	66.	
Cattle	dig prot %	*	7.6	8.6	
Goats	dig prot %	*	8.0	9.2	
Horses	dig prot %	*	7.9	9.0	
Rabbits	dig prot %	*	8.0	9.1	
Sheep	dig prot %		7.8	8.9	
Energy	GE Mcal/kg				
Cattle	DE Mcal/kg	*	2.17	2.47	
Sheep	DE Mcal/kg	*	2.24	2.55	
Cattle	ME Mcal/kg	*	1.78	2.03	
Sheep	ME Mcal/kg	*	1.83	2.09	
Cattle	TDN %	*	49.2	56.1	
Sheep	TDN %		50.7	57.9	

Clover, red, hay, s-c, (1)

Ref No 1 01 415 United States

Feed Name or Analyses			Mean As Fed	Mean Dry	C.V. ± %
Dry matter	%		78.0	100.0	27
Ash	%		6.2	7.9	
Crude fiber	%		25.2	32.3	
Cattle	dig coef %		54.	54.	
Horses	dig coef %		35.	35.	
Sheep	dig coef %		52.	52.	
Ether extract	%		2.0	2.6	
Cattle	dig coef %		58.	58.	
Horses	dig coef %		28.	28.	
Sheep	dig coef %		59.	59.	
N-free extract	%		33.2	42.6	
Cattle	dig coef %		68.	68.	
Horses	dig coef %		61.	61.	
Sheep	dig coef %		67.	67.	
Protein (N x 6.25)	%		11.4	14.7	
Cattle	dig coef %		60.	60.	
Horses	dig coef %		54.	54.	
Sheep	dig coef %		61.	61.	
Cattle	dig prot %		6.9	8.8	
Goats	dig prot %	*	8.0	10.2	
Horses	dig prot %		6.2	7.9	
Rabbits	dig prot %	*	7.8	10.0	
Sheep	dig prot %		7.0	9.0	
Hemicellulose	%		6.9	8.8	
Energy	GE Mcal/kg				
Cattle	DE Mcal/kg	*	2.01	2.58	
Horses	DE Mcal/kg	*	1.61	2.06	
Sheep	DE Mcal/kg	*	1.98	2.54	
Cattle	ME Mcal/kg	*	1.65	2.12	
Horses	ME Mcal/kg	*	1.32	1.69	
Sheep	ME Mcal/kg	*	1.63	2.09	
Cattle	NE_m Mcal/kg	*	0.98	1.26	
Cattle	NE_{gain} Mcal/kg	*	0.48	0.62	
Cattle $NE_{lactating\ cows}$	Mcal/kg	*	1.10	1.41	
Cattle	TDN %		45.7	58.5	
Horses	TDN %		36.5	46.8	
Sheep	TDN %		45.0	57.7	
Calcium	%		1.13	1.45	
Chlorine	%		0.29	0.37	
Copper	mg/kg		8.4	10.7	
Iron	%		0.009	0.011	
Magnesium	%		0.33	0.42	
Manganese	mg/kg		60.0	76.9	
Phosphorus	%		0.18	0.23	
Potassium	%		1.46	1.87	
Sodium	%		0.16	0.20	
Sulphur	%		0.12	0.16	
Biotin	mg/kg		0.09	0.11	

Clover, red, hay, s-c, immature, (1)

Ref No 1 01 394 United States

Feed Name or Analyses			Mean As Fed	Mean Dry	C.V. ± %
Dry matter	%		87.3	100.0	4
Ash	%		8.5	9.7	23
Crude fiber	%		17.8	20.4	23
Cattle	dig coef %	*	54.	54.	
Sheep	dig coef %	*	57.	57.	
Ether extract	%		3.4	3.9	30
Cattle	dig coef %	*	58.	58.	
Sheep	dig coef %	*	59.	59.	
N-free extract	%		38.9	44.6	
Cattle	dig coef %	*	68.	68.	
Sheep	dig coef %	*	82.	82.	
Protein (N x 6.25)	%		18.7	21.4	9
Cattle	dig coef %	*	60.	60.	
Sheep	dig coef %	*	62.	62.	
Cattle	dig prot %		11.2	12.8	
Goats	dig prot %	*	14.4	16.5	
Horses	dig prot %	*	13.7	15.7	
Rabbits	dig prot %	*	13.3	15.2	
Sheep	dig prot %		11.6	13.3	
Energy	GE Mcal/kg				
Cattle	DE Mcal/kg	*	2.28	2.61	
Sheep	DE Mcal/kg	*	2.57	2.94	
Cattle	ME Mcal/kg	*	1.87	2.14	
Sheep	ME Mcal/kg	*	2.10	2.41	
Cattle	TDN %		51.7	59.3	
Sheep	TDN %		58.2	66.6	
Calcium	%		1.55	1.77	13
Magnesium	%		0.45	0.51	23
Phosphorus	%		0.27	0.31	19
Potassium	%		2.24	2.57	5
Carotene	mg/kg		216.9	248.5	20
Riboflavin	mg/kg		14.2	16.3	
Thiamine	mg/kg		3.7	4.2	
Vitamin A equivalent	IU/g		361.6	414.2	

Clover, red, hay, s-c, pre-bloom, cut 2, (1)

Ref No 1 01 395 United States

Feed Name or Analyses			Mean As Fed	Mean Dry	C.V. ± %
Dry matter	%		84.6	100.0	
Ash	%		7.8	9.2	
Crude fiber	%		23.3	27.5	
Sheep	dig coef %		61.	61.	
Ether extract	%		1.9	2.2	
Sheep	dig coef %		7.	7.	
N-free extract	%		34.8	41.1	
Sheep	dig coef %		65.	65.	
Protein (N x 6.25)	%		16.9	20.0	
Sheep	dig coef %		72.	72.	
Cattle	dig prot %	*	12.1	14.3	
Goats	dig prot %	*	12.9	15.2	
Horses	dig prot %	*	12.3	14.5	
Rabbits	dig prot %	*	11.9	14.1	
Sheep	dig prot %		12.2	14.4	
Energy	GE Mcal/kg				
Cattle	DE Mcal/kg	*	2.24	2.65	
Sheep	DE Mcal/kg	*	2.17	2.57	
Cattle	ME Mcal/kg	*	1.84	2.17	
Sheep	ME Mcal/kg	*	1.78	2.11	
Cattle	TDN %	*	50.8	60.0	
Sheep	TDN %		49.3	58.2	

Clover, red, hay, s-c, pre-bloom, cut 3, (1)

Ref No 1 01 396 United States

Feed Name or Analyses			Mean As Fed	Mean Dry	C.V. ± %
Dry matter	%		82.6	100.0	
Ash	%		10.2	12.4	
Crude fiber	%		20.8	25.2	
Sheep	dig coef %		69.	69.	
Ether extract	%		1.8	2.2	
Sheep	dig coef %		7.	7.	
N-free extract	%		31.7	38.4	
Sheep	dig coef %		54.	54.	
Protein (N x 6.25)	%		18.0	21.8	
Sheep	dig coef %		69.	69.	
Cattle	dig prot %	*	13.1	15.8	
Goats	dig prot %	*	14.0	16.9	
Horses	dig prot %	*	13.2	16.0	
Rabbits	dig prot %	*	12.8	15.5	
Sheep	dig prot %		12.4	15.0	
Energy	GE Mcal/kg				
Cattle	DE Mcal/kg	*	1.97	2.39	
Sheep	DE Mcal/kg	*	1.95	2.36	
Cattle	ME Mcal/kg	*	1.62	1.96	
Sheep	ME Mcal/kg	*	1.60	1.93	
Cattle	TDN %	*	44.9	54.3	
Sheep	TDN %		44.2	53.5	

Clover, red, hay, s-c, early bloom, (1)

Ref No 1 01 400 United States

Feed Name or Analyses			Mean As Fed	Mean Dry	C.V. ± %
Dry matter	%		86.4	100.0	3
Ash	%		8.1	9.4	10
Crude fiber	%		23.5	27.2	5
Cattle	dig coef %		57.	57.	
Sheep	dig coef %		49.	49.	
Ether extract	%		1.9	2.2	9
Cattle	dig coef %		53.	53.	
Sheep	dig coef %		42.	42.	
N-free extract	%		40.3	46.7	
Cattle	dig coef %		75.	75.	
Sheep	dig coef %		67.	67.	
Protein (N x 6.25)	%		12.5	14.5	15
Cattle	dig coef %		63.	63.	
Sheep	dig coef %		62.	62.	
Cattle	dig prot %		7.9	9.1	
Goats	dig prot %	*	8.7	10.1	
Horses	dig prot %	*	8.5	9.8	
Rabbits	dig prot %	*	8.5	9.8	
Sheep	dig prot %		7.7	9.0	
Energy	GE Mcal/kg				
Cattle	DE Mcal/kg	*	2.37	2.75	
Sheep	DE Mcal/kg	*	2.12	2.45	
Cattle	ME Mcal/kg	*	1.94	2.25	
Sheep	ME Mcal/kg	*	1.74	2.01	
Cattle	TDN %		53.8	62.3	
Sheep	TDN %		48.1	55.7	
Calcium	%		1.44	1.67	
Phosphorus	%		0.32	0.37	

Clover, red, hay, s-c, early bloom, cut 1, (1)

Ref No 1 01 398 United States

Feed Name or Analyses			Mean As Fed	Mean Dry	C.V. ± %
Dry matter	%		88.6	100.0	
Ash	%		9.9	11.2	
Crude fiber	%		25.1	28.3	
Sheep	dig coef %		60.	60.	

Continued

Feed Name or Analyses			Mean As Fed	Mean Dry	C.V. ± %
Ether extract	%		1.8	2.0	
Sheep	dig coef %		52.	52.	
N-free extract	%		36.1	40.7	
Sheep	dig coef %		68.	68.	
Protein (N x 6.25)	%		15.8	17.8	
Sheep	dig coef %		64.	64.	
Cattle	dig prot %	*	10.9	12.4	
Goats	dig prot %	*	11.7	13.2	
Horses	dig prot %	*	11.2	12.6	
Rabbits	dig prot %	*	11.0	12.4	
Sheep	dig prot %		10.1	11.4	
Energy	GE Mcal/kg				
Cattle	DE Mcal/kg	*	2.40	2.71	
Sheep	DE Mcal/kg	*	2.28	2.57	
Cattle	ME Mcal/kg	*	1.97	2.22	
Sheep	ME Mcal/kg	*	1.87	2.11	
Cattle	TDN %	*	54.5	61.5	
Sheep	TDN %		51.7	58.4	

Clover, red, hay, s-c, early bloom, cut 2, (1)

Ref No 1 01 399 United States

Feed Name or Analyses			Mean As Fed	Mean Dry	C.V. ± %
Dry matter	%		88.1	100.0	
Ash	%		7.9	9.0	
Crude fiber	%		26.3	29.8	
Sheep	dig coef %		46.	46.	
Ether extract	%		1.7	1.9	
Sheep	dig coef %		55.	55.	
N-free extract	%		38.8	44.0	
Sheep	dig coef %		69.	69.	
Protein (N x 6.25)	%		13.5	15.3	
Sheep	dig coef %		60.	60.	
Cattle	dig prot %	*	9.0	10.2	
Goats	dig prot %	*	9.5	10.8	
Horses	dig prot %	*	9.3	10.5	
Rabbits	dig prot %	*	9.2	10.5	
Sheep	dig prot %		8.1	9.2	
Energy	GE Mcal/kg				
Cattle	DE Mcal/kg	*	2.28	2.59	
Sheep	DE Mcal/kg	*	2.16	2.45	
Cattle	ME Mcal/kg	*	1.87	2.12	
Sheep	ME Mcal/kg	*	1.77	2.01	
Cattle	TDN %	*	51.7	58.7	
Sheep	TDN %		49.0	55.6	

Clover, red, hay, s-c, mid-bloom, (1)

Ref No 1 01 401 United States

Feed Name or Analyses			Mean As Fed	Mean Dry	C.V. ± %
Dry matter	%		88.4	100.0	2
Ash	%		7.3	8.2	21
Crude fiber	%		25.3	28.6	11
Cattle	dig coef %		52.	52.	
Sheep	dig coef %	*	52.	52.	
Ether extract	%		3.8	4.3	19
Cattle	dig coef %		73.	73.	
Sheep	dig coef %	*	59.	59.	
N-free extract	%		38.2	43.3	
Cattle	dig coef %		75.	75.	
Sheep	dig coef %	*	67.	67.	
Protein (N x 6.25)	%		13.8	15.6	14
Cattle	dig coef %		70.	70.	
Sheep	dig coef %	*	61.	61.	
Cattle	dig prot %		9.6	10.9	
Goats	dig prot %	*	9.8	11.1	
Horses	dig prot %	*	9.5	10.8	
Rabbits	dig prot %	*	9.5	10.7	
Sheep	dig prot %		8.4	9.5	
Fatty acids	%		3.5	4.0	
Energy	GE Mcal/kg				
Cattle	DE Mcal/kg	*	2.54	2.88	
Sheep	DE Mcal/kg	*	2.30	2.61	
Cattle	ME Mcal/kg	*	2.09	2.36	
Sheep	ME Mcal/kg	*	1.89	2.14	
Cattle	TDN %		57.7	65.3	
Sheep	TDN %		52.2	59.1	
Calcium	%		1.58	1.79	
Phosphorus	%		0.26	0.29	
Potassium	%		1.47	1.66	

Ref No 1 01 401 Canada

Feed Name or Analyses			Mean As Fed	Mean Dry	C.V. ± %
Dry matter	%			100.0	
Crude fiber	%				
Cattle	dig coef %			52.	
Ether extract	%				
Cattle	dig coef %			73.	
N-free extract	%				
Cattle	dig coef %			75.	
Protein (N x 6.25)	%			21.9	
Cattle	dig coef %			70.	
Cattle	dig prot %			15.3	
Goats	dig prot %	*		16.9	
Horses	dig prot %	*		16.1	
Rabbits	dig prot %	*		15.5	
Sheep	dig prot %	*		16.2	
Energy	GE Mcal/kg				
Cattle	DE Mcal/kg	*		2.87	
Sheep	DE Mcal/kg	*		2.60	
Cattle	ME Mcal/kg	*		2.36	
Sheep	ME Mcal/kg	*		2.13	
Cattle	TDN %			65.2	
Sheep	TDN %			58.9	
Calcium	%			1.56	
Phosphorus	%			0.29	
Potassium	%			1.24	

Clover, red, hay, s-c, full bloom, (1)

Ref No 1 01 403 United States

Feed Name or Analyses			Mean As Fed	Mean Dry	C.V. ± %
Dry matter	%		86.1	100.0	6
Ash	%		7.4	8.6	
Crude fiber	%		19.7	22.9	
Cattle	dig coef %	*	54.	54.	
Sheep	dig coef %		52.	52.	
Ether extract	%		2.7	3.1	
Cattle	dig coef %	*	58.	58.	
Sheep	dig coef %		70.	70.	
N-free extract	%		41.8	48.5	
Cattle	dig coef %	*	68.	68.	
Sheep	dig coef %		72.	72.	
Protein (N x 6.25)	%		14.6	16.9	
Cattle	dig coef %	*	60.	60.	
Sheep	dig coef %		65.	65.	
Cattle	dig prot %		8.7	10.1	
Goats	dig prot %	*	10.6	12.3	
Horses	dig prot %	*	10.2	11.9	
Rabbits	dig prot %	*	10.1	11.7	
Sheep	dig prot %		9.5	11.0	
Energy	GE Mcal/kg		3.77	4.37	
Cattle	GE dig coef %		53.	53.	
Cattle	DE Mcal/kg		2.00	2.32	
Sheep	DE Mcal/kg	*	2.38	2.76	
Cattle	ME Mcal/kg	*	1.64	1.90	
Sheep	ME Mcal/kg	*	1.95	2.27	
Cattle	TDN %		51.3	59.5	
Sheep	TDN %		54.0	62.7	
Calcium	%		1.43	1.66	10
Magnesium	%		0.44	0.51	16
Phosphorus	%		0.22	0.25	13
Potassium	%		1.59	1.85	7
Sulphur	%		0.16	0.19	

Ref No 1 01 403 Canada

Feed Name or Analyses			Mean As Fed	Mean Dry	C.V. ± %
Dry matter	%			100.0	
Ash	%			8.0	
Crude fiber	%			27.4	
Cattle	dig coef %			37.	
Sheep	dig coef %			52.	
Ether extract	%			1.2	
Cattle	dig coef %			18.	
Sheep	dig coef %			70.	
N-free extract	%			48.8	
Sheep	dig coef %			72.	
Protein (N x 6.25)	%			14.7	
Cattle	dig coef %			56.	
Sheep	dig coef %			65.	
Cattle	dig prot %			8.2	
Goats	dig prot %	*		10.3	
Horses	dig prot %	*		10.0	
Rabbits	dig prot %	*		10.0	
Sheep	dig prot %			9.6	
Cellulose (Matrone)	%			31.5	
Cattle	dig coef %			50.	
Energy	GE Mcal/kg			4.37	
Cattle	GE dig coef %			53.	
Cattle	DE Mcal/kg			2.32	
Sheep	DE Mcal/kg	*		2.68	
Cattle	ME Mcal/kg	*		1.90	
Sheep	ME Mcal/kg	*		2.20	
Cattle	TDN %			54.2	
Sheep	TDN %			60.8	

Clover, red, hay, s-c, full bloom, cut 1, (1)

Ref No 1 01 402 United States

Feed Name or Analyses			Mean As Fed	Mean Dry	C.V. ± %
Dry matter	%		89.3	100.0	2
Horses	dig coef %		60.	60.	2
Organic matter	%		80.5	90.1	0
Ash	%		8.7	9.7	4
Crude fiber	%		26.4	29.6	3
Horses	dig coef %		44.	44.	8
Sheep	dig coef %		61.	61.	
Ether extract	%		2.6	2.9	
Horses	dig coef %		48.	48.	10
Sheep	dig coef %		20.	20.	
N-free extract	%		38.7	43.3	2
Horses	dig coef %		77.	77.	5
Sheep	dig coef %		68.	68.	
Protein (N x 6.25)	%		12.9	14.4	6
Horses	dig coef %		60.	60.	3
Sheep	dig coef %		75.	75.	
Cattle	dig prot %	*	8.4	9.4	
Goats	dig prot %	*	9.0	10.0	
Horses	dig prot %		7.7	8.6	
Rabbits	dig prot %	*	8.8	9.8	
Sheep	dig prot %		9.7	10.8	
Cell contents (Van Soest)	%		47.6	53.3	1
Horses	dig coef %		76.	76.	2
Cell walls (Van Soest)	%		41.7	46.7	1
Horses	dig coef %		42.	42.	4

(1) dry forages and roughages (2) pasture, range plants, and forages fed green (3) silages (4) energy feeds (5) protein supplements (6) minerals (7) vitamins (8) additives

Feed Name or Analyses			Mean As Fed	Mean Dry	C.V. ± %
Cellulose (Matrone)	%		25.6	28.6	3
Horses	dig coef %		50.	50.	6
Fiber, acid detergent (VS)	%		31.2	34.9	3
Horses	dig coef %		40.	40.	7
Hemicellulose	%		6.5	7.3	17
Horses	dig coef %		52.	52.	14
Holocellulose	%		32.1	36.0	2
Soluble carbohydrates	%		27.7	31.0	2
Lignin (Ellis)	%		5.6	6.3	3
Horses	dig coef %		23.	23.	37
Energy	GE Mcal/kg		3.77	4.22	1
Horses	GE dig coef %		51.	51.	4
Cattle	DE Mcal/kg	*	2.39	2.67	
Horses	DE Mcal/kg		1.92	2.15	
Sheep	DE Mcal/kg	*	2.35	2.63	
Cattle	ME Mcal/kg	*	1.96	2.19	
Horses	ME Mcal/kg	*	1.58	1.77	
Sheep	ME Mcal/kg	*	1.93	2.16	
Cattle	TDN %	*	54.1	60.6	
Horses	TDN %		52.0	58.3	
Sheep	TDN %		53.3	59.7	
Calcium	%		0.70	0.79	66
Magnesium	%		0.67	0.75	4
Phosphorus	%		0.17	0.19	7
Potassium	%		0.24	0.27	6
Carotene	mg/kg		27.9	31.3	24
Vitamin A equivalent	IU/g		46.5	52.1	

Clover, red, hay, s-c, milk stage, (1)
Ref No 1 01 404 United States

Feed Name or Analyses			Mean As Fed	Mean Dry	C.V. ± %
Dry matter	%		87.5	100.0	
Ash	%		6.0	6.8	
Sheep	dig coef %		35.	40.	
Crude fiber	%		25.6	29.2	
Sheep	dig coef %		40.	40.	
Ether extract	%		1.9	2.2	
Sheep	dig coef %		23.	23.	
N-free extract	%		43.9	50.1	
Sheep	dig coef %		64.	64.	
Protein (N x 6.25)	%		10.2	11.7	
Sheep	dig coef %		44.	44.	
Cattle	dig prot %	*	6.2	7.1	
Goats	dig prot %	*	6.5	7.5	
Horses	dig prot %	*	6.5	7.5	
Rabbits	dig prot %	*	6.7	7.7	
Sheep	dig prot %		4.5	5.1	
Energy	GE Mcal/kg				
Cattle	DE Mcal/kg	*	2.00	2.29	
Sheep	DE Mcal/kg	*	1.93	2.21	
Cattle	ME Mcal/kg	*	1.64	1.88	
Sheep	ME Mcal/kg	*	1.58	1.81	
Cattle	TDN %	*	45.5	52.0	
Sheep	TDN %		43.8	50.0	
Calcium	%		1.26	1.44	
Phosphorus	%		0.27	0.31	
Potassium	%		1.24	1.42	

Clover, red, hay, s-c, mature, (1)
Ref No 1 01 405 United States

Feed Name or Analyses			Mean As Fed	Mean Dry	C.V. ± %
Dry matter	%		91.2	100.0	3
Ash	%		6.1	6.7	11
Crude fiber	%		31.4	34.4	13
Cattle	dig coef %	*	54.	54.	
Sheep	dig coef %	*	52.	52.	
Ether extract	%		2.2	2.4	36
Cattle	dig coef %	*	58.	58.	
Sheep	dig coef %	*	59.	59.	
N-free extract	%		42.0	46.0	
Cattle	dig coef %	*	68.	68.	
Sheep	dig coef %	*	67.	67.	
Protein (N x 6.25)	%		9.6	10.5	17
Cattle	dig coef %	*	60.	60.	
Sheep	dig coef %	*	61.	61.	
Cattle	dig prot %		5.7	6.3	
Goats	dig prot %	*	5.8	6.4	
Horses	dig prot %	*	5.9	6.4	
Rabbits	dig prot %	*	6.2	6.8	
Sheep	dig prot %		5.8	6.4	
Sugars, total	%		2.2	2.4	
Lignin (Ellis)	%		16.4	18.0	
Energy	GE Mcal/kg				
Cattle	DE Mcal/kg	*	2.38	2.61	
Sheep	DE Mcal/kg	*	2.34	2.57	
Cattle	ME Mcal/kg	*	1.96	2.14	
Sheep	ME Mcal/kg	*	1.92	2.11	
Cattle	TDN %		54.1	59.3	
Sheep	TDN %		53.2	58.3	
Calcium	%		0.99	1.09	
Iron	%		0.026	0.029	
Magnesium	%		0.32	0.35	
Phosphorus	%		0.19	0.21	
Potassium	%		1.55	1.70	
Carotene	mg/kg		2.2	2.4	
Vitamin A equivalent	IU/g		3.7	4.0	

Ref No 1 01 405 Canada

Feed Name or Analyses			Mean As Fed	Mean Dry	C.V. ± %
Dry matter	%		92.2	100.0	
Crude fiber	%		30.4	33.0	
Protein (N x 6.25)	%		11.7	12.7	
Cattle	dig prot %	*	7.3	7.9	
Goats	dig prot %	*	7.8	8.4	
Horses	dig prot %	*	7.7	8.3	
Rabbits	dig prot %	*	7.8	8.5	
Sheep	dig prot %	*	7.3	8.0	
Energy	GE Mcal/kg				
Cattle	DE Mcal/kg	*	2.41	2.61	
Sheep	DE Mcal/kg	*	2.37	2.57	
Cattle	ME Mcal/kg	*	1.98	2.14	
Sheep	ME Mcal/kg	*	1.95	2.11	
Cattle	TDN %		54.7	59.3	
Sheep	TDN %		53.8	58.4	
Calcium	%		1.15	1.25	
Phosphorus	%		0.20	0.22	
Carotene	mg/kg		6.0	6.5	
Vitamin A equivalent	IU/g		9.9	10.8	

Clover, red, hay, s-c, cut 1, (1)
Ref No 1 01 406 United States

Feed Name or Analyses			Mean As Fed	Mean Dry	C.V. ± %
Dry matter	%		91.4	100.0	3
Ash	%		7.9	8.7	
Crude fiber	%		27.9	30.5	
Cattle	dig coef %		58.	58.	
Sheep	dig coef %		56.	56.	
Ether extract	%		2.7	2.9	
Cattle	dig coef %		52.	52.	
Sheep	dig coef %		36.	36.	
N-free extract	%		39.1	42.8	
Cattle	dig coef %		71.	71.	
Sheep	dig coef %		70.	70.	
Protein (N x 6.25)	%		13.8	15.2	
Cattle	dig coef %		63.	63.	
Sheep	dig coef %		64.	64.	
Cattle	dig prot %		8.7	9.5	
Goats	dig prot %	*	9.8	10.7	
Horses	dig prot %	*	9.5	10.4	
Rabbits	dig prot %	*	9.5	10.4	
Sheep	dig prot %		8.9	9.7	
Lignin (Ellis)	%		14.4	15.8	21
Energy	GE Mcal/kg				
Cattle	DE Mcal/kg	*	2.46	2.69	
Sheep	DE Mcal/kg	*	2.38	2.61	
Cattle	ME Mcal/kg	*	2.02	2.21	
Sheep	ME Mcal/kg	*	1.95	2.14	
Cattle	TDN %		55.8	61.0	
Sheep	TDN %		54.0	59.1	
Carotene	mg/kg		10.5	11.5	98
Vitamin A equivalent	IU/g		17.5	19.1	

Clover, red, hay, s-c, cut 2, (1)
Ref No 1 01 407 United States

Feed Name or Analyses			Mean As Fed	Mean Dry	C.V. ± %
Dry matter	%		89.7	100.0	2
Ash	%		7.4	8.2	21
Crude fiber	%		25.0	27.9	13
Cattle	dig coef %	*	54.	54.	
Sheep	dig coef %	*	52.	52.	
Ether extract	%		2.7	3.0	15
Cattle	dig coef %	*	58.	58.	
Sheep	dig coef %	*	59.	59.	
N-free extract	%		40.1	44.7	
Cattle	dig coef %	*	68.	68.	
Sheep	dig coef %	*	67.	67.	
Protein (N x 6.25)	%		14.5	16.2	9
Cattle	dig coef %	*	60.	60.	
Sheep	dig coef %	*	61.	61.	
Cattle	dig prot %		8.7	9.7	
Goats	dig prot %	*	10.5	11.7	
Horses	dig prot %	*	10.1	11.3	
Rabbits	dig prot %	*	10.0	11.2	
Sheep	dig prot %		8.9	9.9	
Lignin (Ellis)	%		12.6	14.1	24
Fatty acids	%		3.0	3.3	
Energy	GE Mcal/kg				
Cattle	DE Mcal/kg	*	2.34	2.61	
Sheep	DE Mcal/kg	*	2.31	2.57	
Cattle	ME Mcal/kg	*	1.92	2.14	
Sheep	ME Mcal/kg	*	1.89	2.11	
Cattle	TDN %		53.0	59.1	
Sheep	TDN %		52.3	58.3	
Calcium	%		1.30	1.45	
Phosphorus	%		0.21	0.24	
Carotene	mg/kg		22.9	25.6	66
Vitamin A equivalent	IU/g		38.2	42.6	

Clover, red, hay, s-c, cut 3, (1)
Ref No 1 01 408 United States

Feed Name or Analyses			Mean As Fed	Mean Dry	C.V. ± %
Dry matter	%		90.7	100.0	4
Ash	%		8.3	9.2	29
Crude fiber	%		25.7	28.3	13
Cattle	dig coef %	*	54.	54.	
Sheep	dig coef %	*	52.	52.	
Ether extract	%		2.1	2.3	21
Cattle	dig coef %	*	58.	58.	
Sheep	dig coef %	*	59.	59.	
N-free extract	%		38.6	42.6	
Cattle	dig coef %	*	68.	68.	
Sheep	dig coef %	*	67.	67.	
Protein (N x 6.25)	%		16.0	17.6	30
Cattle	dig coef %	*	60.	60.	

Continued

Feed Name or Analyses		Mean As Fed	Mean Dry	C.V. ± %
Sheep dig coef %	*	61.	61.	
Cattle dig prot %		9.6	10.6	
Goats dig prot %	*	11.8	13.0	
Horses dig prot %	*	11.3	12.5	
Rabbits dig prot %	*	11.1	12.3	
Sheep dig prot %		9.7	10.7	
Lignin (Ellis) %		13.8	15.2	
Energy GE Mcal/kg				
Cattle DE Mcal/kg	*	2.31	2.55	
Sheep DE Mcal/kg	*	2.28	2.52	
Cattle ME Mcal/kg	*	1.90	2.09	
Sheep ME Mcal/kg	*	1.87	2.06	
Cattle TDN %		52.4	57.8	
Sheep TDN %		51.7	57.0	
Calcium %		1.27	1.40	
Iron %		0.021	0.023	
Magnesium %		0.22	0.24	
Phosphorus %		0.11	0.12	
Potassium %		1.26	1.39	
Carotene mg/kg		10.8	11.9	
Vitamin A equivalent IU/g		18.0	19.8	

Clover, red, hay, s-c, gr 1 US, (1)
Ref No 1 01 410 United States

Feed Name or Analyses		Mean As Fed	Mean Dry	C.V. ± %
Dry matter %		91.7	100.0	2
Lignin (Ellis) %		13.2	14.4	22
Calcium %		1.34	1.46	15
Iron %		0.029	0.032	98
Magnesium %		0.32	0.35	19
Phosphorus %		0.18	0.20	21
Potassium %		1.55	1.69	25
Carotene mg/kg		21.8	23.8	28
Vitamin A equivalent IU/g		36.4	39.7	

Clover, red, hay, s-c, gr 2 US, (1)
Ref No 1 01 411 United States

Feed Name or Analyses		Mean As Fed	Mean Dry	C.V. ± %
Dry matter %		91.8	100.0	2
Lignin (Ellis) %		11.1	12.1	
Calcium %		1.31	1.43	
Iron %		0.017	0.018	
Magnesium %		0.55	0.60	
Phosphorus %		0.19	0.21	
Potassium %		1.27	1.38	
Carotene mg/kg		16.4	17.9	66
Vitamin A equivalent IU/g		27.3	29.8	

Clover, red, hay, s-c, gr 3 US, (1)
Ref No 1 01 412 United States

Feed Name or Analyses		Mean As Fed	Mean Dry	C.V. ± %
Dry matter %		90.7	100.0	
Ash %		5.6	6.2	
Crude fiber %		24.0	26.5	
Ether extract %		2.4	2.6	
N-free extract %		43.7	48.2	
Protein (N x 6.25) %		15.0	16.5	
Cattle dig prot %	*	10.2	11.2	
Goats dig prot %	*	10.8	12.0	
Horses dig prot %	*	10.5	11.5	
Rabbits dig prot %	*	10.3	11.4	
Sheep dig prot %	*	10.3	11.4	

(1) dry forages and roughages (2) pasture, range plants, and forages fed green (3) silages (4) energy feeds (5) protein supplements (6) minerals (7) vitamins (8) additives

Feed Name or Analyses		Mean As Fed	Mean Dry	C.V. ± %
Lignin (Ellis) %		17.0	18.7	
Energy GE Mcal/kg				
Cattle DE Mcal/kg	*	2.34	2.58	
Sheep DE Mcal/kg	*	2.49	2.74	
Cattle ME Mcal/kg	*	1.92	2.12	
Sheep ME Mcal/kg	*	2.04	2.25	
Cattle TDN %	*	53.1	58.5	
Sheep TDN %	*	56.4	62.1	
Calcium %		1.15	1.27	
Iron %		0.026	0.029	
Magnesium %		0.36	0.40	
Phosphorus %		0.18	0.20	
Potassium %		1.64	1.81	
Carotene mg/kg		11.8	13.0	
Vitamin A equivalent IU/g		19.7	21.7	

Clover, red, hay, s-c, leafy, (1)
Ref No 1 01 413 United States

Feed Name or Analyses		Mean As Fed	Mean Dry	C.V. ± %
Dry matter %		88.3	100.0	2
Ash %		7.1	8.0	11
Crude fiber %		23.4	26.5	11
Ether extract %		2.9	3.3	19
N-free extract %		41.4	46.9	
Protein (N x 6.25) %		13.5	15.2	5
Cattle dig prot %	*	9.0	10.1	
Goats dig prot %	*	9.5	10.8	
Horses dig prot %	*	9.2	10.5	
Rabbits dig prot %	*	9.2	10.4	
Sheep dig prot %	*	9.0	10.2	
Fatty acids %		2.9	3.3	
Energy GE Mcal/kg				
Cattle DE Mcal/kg	*	2.38	2.70	
Sheep DE Mcal/kg	*	2.33	2.64	
Cattle ME Mcal/kg	*	1.95	2.21	
Sheep ME Mcal/kg	*	1.91	2.16	
Cattle TDN %	*	54.0	61.1	
Sheep TDN %	*	52.8	59.8	
Calcium %		1.47	1.66	11
Phosphorus %		0.21	0.24	14
Potassium %		1.36	1.54	

Clover, red, hay, s-c, stemmy, (1)
Ref No 1 01 414 United States

Feed Name or Analyses		Mean As Fed	Mean Dry	C.V. ± %
Dry matter %		88.6	100.0	2
Ash %		6.0	6.7	17
Crude fiber %		33.5	37.8	3
Ether extract %		2.2	2.5	20
N-free extract %		36.4	41.1	
Protein (N x 6.25) %		10.5	11.8	11
Cattle dig prot %	*	6.4	7.2	
Goats dig prot %	*	6.7	7.6	
Horses dig prot %	*	6.7	7.6	
Rabbits dig prot %	*	6.9	7.8	
Sheep dig prot %	*	6.4	7.2	
Energy GE Mcal/kg				
Cattle DE Mcal/kg	*	2.07	2.34	
Sheep DE Mcal/kg	*	2.14	2.42	
Cattle ME Mcal/kg	*	1.70	1.92	
Sheep ME Mcal/kg	*	1.76	1.98	
Cattle TDN %	*	47.0	53.0	
Sheep TDN %	*	48.6	54.8	
Calcium %		1.12	1.27	7
Phosphorus %		0.20	0.23	13
Potassium %		1.12	1.26	

Clover, red, hay, s-c, 25-31% fiber, (1)
Ref No 1 08 385 United States

Feed Name or Analyses		Mean As Fed	Mean Dry	C.V. ± %
Dry matter %		88.3	100.0	
Ash %		6.2	7.0	
Crude fiber %		27.3	30.9	
Ether extract %		2.5	2.8	
N-free extract %		40.7	46.1	
Protein (N x 6.25) %		11.6	13.1	
Cattle dig prot %	*	7.3	8.3	
Goats dig prot %	*	7.8	8.8	
Horses dig prot %	*	7.7	8.7	
Rabbits dig prot %	*	7.8	8.8	
Sheep dig prot %	*	7.4	8.4	
Energy GE Mcal/kg				
Cattle DE Mcal/kg	*	2.29	2.59	
Sheep DE Mcal/kg	*	2.25	2.55	
Cattle ME Mcal/kg	*	1.88	2.13	
Sheep ME Mcal/kg	*	1.85	2.09	
Cattle TDN %	*	51.9	58.8	
Sheep TDN %	*	51.0	57.8	
Calcium %		1.21	1.37	
Phosphorus %		0.21	0.24	
Potassium %		1.45	1.64	

Clover, red, hay, s-c, gr sample US, (1)
Ref No 1 01 409 United States

Feed Name or Analyses		Mean As Fed	Mean Dry	C.V. ± %
Dry matter %			100.0	
Ash %			6.7	
Crude fiber %			36.9	
Ether extract %			2.4	
N-free extract %			40.6	
Protein (N x 6.25) %			13.4	
Cattle dig prot %	*		8.5	
Goats dig prot %	*		9.1	
Horses dig prot %	*		8.9	
Rabbits dig prot %	*		9.0	
Sheep dig prot %	*		8.6	
Energy GE Mcal/kg				
Cattle DE Mcal/kg	*		2.33	
Sheep DE Mcal/kg	*		2.40	
Cattle ME Mcal/kg	*		1.91	
Sheep ME Mcal/kg	*		1.97	
Cattle TDN %	*		52.9	
Sheep TDN %	*		54.5	

Clover, red, hay, s-c, on riders, early bloom, (1)
Ref No 1 01 416 United States

Feed Name or Analyses		Mean As Fed	Mean Dry	C.V. ± %
Dry matter %		81.4	100.0	
Ash %		8.2	10.1	
Crude fiber %		20.6	25.3	
Cattle dig coef %		51.	51.	
Ether extract %		2.0	2.5	
Cattle dig coef %		58.	58.	
N-free extract %		34.6	42.5	
Cattle dig coef %		70.	70.	
Protein (N x 6.25) %		16.0	19.6	
Cattle dig coef %		71.	71.	
Cattle dig prot %		11.3	13.9	
Goats dig prot %	*	12.1	14.8	
Horses dig prot %	*	11.5	14.2	
Rabbits dig prot %	*	11.2	13.8	
Sheep dig prot %	*	11.5	14.2	
Energy GE Mcal/kg				
Cattle DE Mcal/kg	*	2.15	2.64	

Feed Name or Analyses			Mean As Fed	Mean Dry	C.V. ± %
Sheep	DE Mcal/kg	*	2.16	2.65	
Cattle	ME Mcal/kg	*	1.76	2.16	
Sheep	ME Mcal/kg	*	1.77	2.17	
Cattle	TDN %		48.7	59.8	
Sheep	TDN %	*	49.0	60.1	

Clover, red, hay, s-c on riders, full bloom, cut 1, (1)
Ref No 1 01 417 United States

Feed Name or Analyses			Mean As Fed	Mean Dry	C.V. ± %
Dry matter	%		83.8	100.0	
Ash	%		6.8	8.2	
Crude fiber	%		25.0	29.9	
Cattle	dig coef %		47.	47.	
Sheep	dig coef %		60.	60.	
Ether extract	%		2.3	2.8	
Cattle	dig coef %		65.	65.	
Sheep	dig coef %		16.	16.	
N-free extract	%		36.2	43.3	
Cattle	dig coef %		68.	68.	
Sheep	dig coef %		68.	68.	
Protein (N x 6.25)	%		13.4	16.0	
Cattle	dig coef %		65.	65.	
Sheep	dig coef %		66.	66.	
Cattle	dig prot %		8.7	10.4	
Goats	dig prot %	*	9.6	11.5	
Horses	dig prot %	*	9.3	11.1	
Rabbits	dig prot %	*	9.2	11.0	
Sheep	dig prot %		8.8	10.6	
Energy	GE Mcal/kg				
Cattle	DE Mcal/kg	*	2.14	2.55	
Sheep	DE Mcal/kg	*	2.17	2.60	
Cattle	ME Mcal/kg	*	1.75	2.09	
Sheep	ME Mcal/kg	*	1.78	2.13	
Cattle	TDN %		48.5	57.9	
Sheep	TDN %		49.3	58.9	

Clover, red, hay, s-c on riders, late bloom, cut 1, (1)
Ref No 1 01 418 United States

Feed Name or Analyses			Mean As Fed	Mean Dry	C.V. ± %
Dry matter	%		82.4	100.0	
Ash	%		5.6	6.8	
Crude fiber	%		23.7	28.8	
Cattle	dig coef %		40.	40.	
Ether extract	%		2.4	2.9	
Cattle	dig coef %		60.	60.	
N-free extract	%		39.8	48.3	
Cattle	dig coef %		66.	66.	
Protein (N x 6.25)	%		10.9	13.2	
Cattle	dig coef %		59.	59.	
Cattle	dig prot %		6.4	7.8	
Goats	dig prot %	*	7.3	8.9	
Horses	dig prot %	*	7.2	8.7	
Rabbits	dig prot %	*	7.3	8.9	
Sheep	dig prot %	*	6.9	8.4	
Energy	GE Mcal/kg				
Cattle	DE Mcal/kg	*	2.00	2.43	
Sheep	DE Mcal/kg	*	2.16	2.62	
Cattle	ME Mcal/kg	*	1.64	1.99	
Sheep	ME Mcal/kg	*	1.77	2.15	
Cattle	TDN %		45.4	55.1	
Sheep	TDN %	*	48.9	59.4	

Clover, red, hay, s-c on riders, cut 1, (1)
Ref No 1 01 419 United States

Feed Name or Analyses			Mean As Fed	Mean Dry	C.V. ± %
Dry matter	%		84.3	100.0	
Ash	%		6.5	7.7	
Crude fiber	%		24.6	29.2	
Cattle	dig coef %		46.	46.	
Sheep	dig coef %		63.	63.	
Ether extract	%		2.5	3.0	
Cattle	dig coef %		61.	61.	
Sheep	dig coef %		26.	26.	
N-free extract	%		38.5	45.7	
Cattle	dig coef %		68.	68.	
Sheep	dig coef %		72.	72.	
Protein (N x 6.25)	%		12.3	14.6	
Cattle	dig coef %		65.	65.	
Sheep	dig coef %		66.	66.	
Cattle	dig prot %		8.0	9.5	
Goats	dig prot %	*	8.5	10.1	
Horses	dig prot %	*	8.3	9.9	
Rabbits	dig prot %	*	8.3	9.9	
Sheep	dig prot %		8.1	9.6	
Energy	GE Mcal/kg				
Cattle	DE Mcal/kg	*	2.15	2.56	
Sheep	DE Mcal/kg	*	2.32	2.76	
Cattle	ME Mcal/kg	*	1.77	2.10	
Sheep	ME Mcal/kg	*	1.91	2.26	
Cattle	TDN %		48.8	58.0	
Sheep	TDN %		52.7	62.6	

Clover, red, leaves, s-c, mid-bloom, (1)
Ref No 1 01 421 United States

Feed Name or Analyses			Mean As Fed	Mean Dry	C.V. ± %
Dry matter	%			100.0	
Carotene	mg/kg			291.0	
Vitamin A equivalent	IU/g			485.1	

Clover, red, leaves, s-c, milk stage, (1)
Ref No 1 01 422 United States

Feed Name or Analyses			Mean As Fed	Mean Dry	C.V. ± %
Dry matter	%			100.0	
Carotene	mg/kg			165.3	
Vitamin A equivalent	IU/g			275.6	

Clover, red, leaves, s-c, cut 1, (1)
Ref No 1 01 423 United States

Feed Name or Analyses			Mean As Fed	Mean Dry	C.V. ± %
Dry matter	%			100.0	
Carotene	mg/kg			228.2	34
Vitamin A equivalent	IU/g			380.4	

Clover, red, stems, s-c, mid-bloom, (1)
Ref No 1 01 424 United States

Feed Name or Analyses			Mean As Fed	Mean Dry	C.V. ± %
Dry matter	%			100.0	
Carotene	mg/kg			34.2	
Vitamin A equivalent	IU/g			57.0	

Clover, red, stems, s-c, full bloom, (1)
Ref No 1 01 425 United States

Feed Name or Analyses			Mean As Fed	Mean Dry	C.V. ± %
Dry matter	%			100.0	
Carotene	mg/kg			26.5	
Vitamin A equivalent	IU/g			44.1	

Clover, red, stems, s-c, cut 1, (1)
Ref No 1 01 426 United States

Feed Name or Analyses			Mean As Fed	Mean Dry	C.V. ± %
Dry matter	%			100.0	
Carotene	mg/kg			30.4	18
Vitamin A equivalent	IU/g			50.7	

Clover, red, aerial part, fresh, (2)
Ref No 2 01 434 United States

Feed Name or Analyses			Mean As Fed	Mean Dry	C.V. ± %
Dry matter	%		22.7	100.0	11
Ash	%		2.0	8.7	15
Crude fiber	%		5.0	22.3	16
Cattle	dig coef %	*	53.	53.	
Sheep	dig coef %	*	52.	52.	
Ether extract	%		1.0	4.2	13
Cattle	dig coef %	*	58.	58.	
Sheep	dig coef %	*	67.	67.	
N-free extract	%		10.5	46.5	
Cattle	dig coef %	*	76.	76.	
Sheep	dig coef %	*	71.	71.	
Protein (N x 6.25)	%		4.2	18.3	24
Cattle	dig coef %	*	65.	65.	
Sheep	dig coef %	*	67.	67.	
Cattle	dig prot %		2.7	11.9	
Goats	dig prot %	*	3.1	13.7	
Horses	dig prot %	*	3.0	13.1	
Rabbits	dig prot %	*	2.9	12.8	
Sheep	dig prot %		2.8	12.3	
Energy	GE Mcal/kg				
Cattle	DE Mcal/kg	*	0.65	2.84	
Sheep	DE Mcal/kg	*	0.63	2.79	
Cattle	ME Mcal/kg	*	0.53	2.33	
Sheep	ME Mcal/kg	*	0.52	2.28	
Cattle	TDN %		14.6	64.5	
Sheep	TDN %		14.3	63.2	
Calcium	%		0.41	1.80	
Chlorine	%		0.17	0.77	26
Cobalt	mg/kg		0.032	0.141	23
Copper	mg/kg		2.0	9.0	
Iron	%		0.007	0.033	
Magnesium	%		0.10	0.43	
Manganese	mg/kg		28.0	123.3	
Phosphorus	%		0.06	0.26	
Potassium	%		0.56	2.46	
Sodium	%		0.05	0.20	21
Sulphur	%		0.04	0.17	31
Carotene	mg/kg		41.8	184.3	30
Niacin	mg/kg		18.2	80.2	26
Riboflavin	mg/kg		4.4	19.2	33
Thiamine	mg/kg		1.5	6.6	33
Vitamin A equivalent	IU/g		69.7	307.2	

Clover, red, aerial part, fresh, immature, (2)
Ref No 2 01 427 United States

Feed Name or Analyses			Mean As Fed	Mean Dry	C.V. ± %
Dry matter	%			100.0	
Crude fiber	%				
Cattle	dig coef %	*		58.	
Sheep	dig coef %	*		52.	
Ether extract	%				
Cattle	dig coef %	*		75.	
Sheep	dig coef %	*		67.	
N-free extract	%				
Cattle	dig coef %	*		79.	
Sheep	dig coef %	*		71.	

Continued

Feed Name or Analyses			Mean As Fed	Mean Dry	C.V. ± %
Protein (N x 6.25)	%				
Cattle	dig coef %	*		72.	
Sheep	dig coef %	*		67.	

Clover, red, aerial part, fresh, early bloom, (2)
Ref No 2 01 428 United States

Feed Name or Analyses			Mean As Fed	Mean Dry	C.V. ± %
Dry matter	%		19.7	100.0	13
Ash	%		2.0	10.2	6
Crude fiber	%		4.6	23.3	22
Cattle	dig coef %		58.	58.	
Sheep	dig coef %	*	52.	52.	
Ether extract	%		1.0	5.0	10
Cattle	dig coef %		75.	75.	
Sheep	dig coef %	*	67.	67.	
N-free extract	%		8.3	42.3	
Cattle	dig coef %		79.	79.	
Sheep	dig coef %	*	71.	71.	
Protein (N x 6.25)	%		3.8	19.4	25
Cattle	dig coef %		72.	72.	
Sheep	dig coef %	*	67.	67.	
Cattle	dig prot %		2.7	13.9	
Goats	dig prot %	*	2.9	14.6	
Horses	dig prot %	*	2.7	14.0	
Rabbits	dig prot %	*	2.7	13.6	
Sheep	dig prot %		2.5	13.0	
Energy	GE Mcal/kg				
Cattle	DE Mcal/kg	*	0.60	3.05	
Sheep	DE Mcal/kg	*	0.54	2.76	
Cattle	ME Mcal/kg	*	0.49	2.50	
Sheep	ME Mcal/kg	*	0.44	2.26	
Cattle	NE_m Mcal/kg	*	0.31	1.56	
Cattle	NE_{gain} Mcal/kg	*	0.20	0.99	
Cattle	$NE_{lactating\ cows}$ Mcal/kg	*	0.36	1.83	
Cattle	TDN %		13.6	69.1	
Sheep	TDN %		12.3	62.5	
Calcium	%		0.44	2.26	
Phosphorus	%		0.07	0.38	
Potassium	%		0.49	2.49	

Clover, red, aerial part, fresh, mid-bloom, (2)
Ref No 2 07 725 Canada

Feed Name or Analyses			Mean As Fed	Mean Dry	C.V. ± %
Dry matter	%		23.5	100.0	
Ash	%		2.0	8.3	
Crude fiber	%		5.3	22.6	
Ether extract	%		0.3	1.4	
N-free extract	%		12.2	52.1	
Protein (N x 6.25)	%		3.7	15.6	
Cattle	dig prot %	*	2.6	11.2	
Goats	dig prot %	*	2.6	11.1	
Horses	dig prot %	*	2.5	10.8	
Rabbits	dig prot %	*	2.5	10.7	
Sheep	dig prot %	*	2.7	11.5	
Energy	GE Mcal/kg				
Cattle	DE Mcal/kg	*	0.74	3.16	
Sheep	DE Mcal/kg	*	0.68	2.89	
Cattle	ME Mcal/kg	*	0.61	2.59	
Sheep	ME Mcal/kg	*	0.56	2.37	
Cattle	TDN %	*	16.8	71.7	
Sheep	TDN %	*	15.4	65.5	

(1) dry forages and roughages (2) pasture, range plants, and forages fed green (3) silages (4) energy feeds (5) protein supplements (6) minerals (7) vitamins (8) additives

Clover, red, aerial part, fresh, full bloom, (2)
Ref No 2 01 429 United States

Feed Name or Analyses			Mean As Fed	Mean Dry	C.V. ± %
Dry matter	%		27.6	100.0	2
Ash	%		2.0	7.3	5
Crude fiber	%		8.2	29.6	4
Cattle	dig coef %	*	53.	53.	
Sheep	dig coef %	*	52.	52.	
Ether extract	%		1.1	4.0	9
Cattle	dig coef %	*	58.	58.	
Sheep	dig coef %	*	67.	67.	
N-free extract	%		12.2	44.2	
Cattle	dig coef %	*	76.	76.	
Sheep	dig coef %	*	71.	71.	
Protein (N x 6.25)	%		4.1	14.9	7
Cattle	dig coef %	*	65.	65.	
Sheep	dig coef %	*	67.	67.	
Cattle	dig prot %		2.7	9.7	
Goats	dig prot %	*	2.9	10.5	
Horses	dig prot %	*	2.8	10.2	
Rabbits	dig prot %	*	2.8	10.2	
Sheep	dig prot %		2.8	10.0	
Energy	GE Mcal/kg				
Cattle	DE Mcal/kg	*	0.78	2.83	
Sheep	DE Mcal/kg	*	0.76	2.77	
Cattle	ME Mcal/kg	*	0.64	2.32	
Sheep	ME Mcal/kg	*	0.63	2.27	
Cattle	NE_m Mcal/kg	*	0.38	1.39	
Cattle	NE_{gain} Mcal/kg	*	0.22	0.80	
Cattle	$NE_{lactating\ cows}$ Mcal/kg	*	0.44	1.60	
Cattle	TDN %		17.7	64.2	
Sheep	TDN %		17.3	62.8	
Calcium	%		0.28	1.01	33
Magnesium	%		0.14	0.51	4
Phosphorus	%		0.07	0.27	9
Potassium	%		0.54	1.96	4

Ref No 2 01 429 Canada

Feed Name or Analyses			Mean As Fed	Mean Dry	C.V. ± %
Dry matter	%		21.7	100.0	
Ash	%		1.8	8.3	
Crude fiber	%		4.9	22.5	
Ether extract	%		0.4	1.8	
N-free extract	%		11.6	53.3	
Protein (N x 6.25)	%		3.1	14.1	
Cattle	dig prot %	*	2.1	9.9	
Goats	dig prot %	*	2.1	9.7	
Horses	dig prot %	*	2.1	9.5	
Rabbits	dig prot %	*	2.1	9.6	
Sheep	dig prot %	*	2.2	10.1	
Energy	GE Mcal/kg				
Cattle	DE Mcal/kg	*	0.61	2.82	
Sheep	DE Mcal/kg	*	0.60	2.74	
Cattle	ME Mcal/kg	*	0.50	2.32	
Sheep	ME Mcal/kg	*	0.49	2.25	
Cattle	TDN %		13.9	64.1	
Sheep	TDN %		13.5	62.2	

Clover, red, aerial part, fresh, late bloom, (2)
Ref No 2 07 724 Canada

Feed Name or Analyses			Mean As Fed	Mean Dry	C.V. ± %
Dry matter	%		23.4	100.0	
Ash	%		2.0	8.6	
Crude fiber	%		5.5	23.4	
Ether extract	%		0.3	1.4	
N-free extract	%		12.0	51.4	
Protein (N x 6.25)	%		3.6	15.2	
Cattle	dig prot %	*	2.5	10.8	
Goats	dig prot %	*	2.5	10.7	
Horses	dig prot %	*	2.4	10.4	
Rabbits	dig prot %	*	2.4	10.4	
Sheep	dig prot %	*	2.6	11.2	
Energy	GE Mcal/kg				
Cattle	DE Mcal/kg	*	0.64	2.74	
Sheep	DE Mcal/kg	*	0.67	2.88	
Cattle	ME Mcal/kg	*	0.53	2.25	
Sheep	ME Mcal/kg	*	0.55	2.36	
Cattle	TDN %	*	14.5	62.1	
Sheep	TDN %	*	15.3	65.3	

Clover, red, aerial part, fresh, milk stage, (2)
Ref No 2 01 430 United States

Feed Name or Analyses			Mean As Fed	Mean Dry	C.V. ± %
Dry matter	%			100.0	
Calcium	%			1.45	
Iron	%			0.016	
Magnesium	%			0.36	
Manganese	mg/kg			466.1	
Phosphorus	%			0.27	
Potassium	%			1.53	

Clover, red, aerial part, fresh, cut 1, (2)
Ref No 2 01 431 United States

Feed Name or Analyses			Mean As Fed	Mean Dry	C.V. ± %
Dry matter	%		15.2	100.0	6
Ash	%		1.6	10.3	6
Crude fiber	%		3.3	21.9	15
Cattle	dig coef %		66.	66.	
Ether extract	%		0.7	4.5	10
Cattle	dig coef %		71.	71.	
N-free extract	%		6.8	44.7	
Cattle	dig coef %		85.	85.	
Protein (N x 6.25)	%		2.9	18.8	37
Cattle	dig coef %		76.	76.	
Cattle	dig prot %		2.2	14.3	
Goats	dig prot %	*	2.1	14.1	
Horses	dig prot %	*	2.1	13.5	
Rabbits	dig prot %	*	2.0	13.2	
Sheep	dig prot %	*	2.2	14.5	
Energy	GE Mcal/kg				
Cattle	DE Mcal/kg	*	0.50	3.25	
Sheep	DE Mcal/kg	*	0.45	2.97	
Cattle	ME Mcal/kg	*	0.41	2.67	
Sheep	ME Mcal/kg	*	0.37	2.44	
Cattle	TDN %		11.2	73.8	
Sheep	TDN %	*	10.3	67.4	
Calcium	%		0.28	1.87	
Phosphorus	%		0.07	0.46	
Potassium	%		0.37	2.45	

Clover, red, aerial part, fresh, cut 2, (2)
Ref No 2 01 432 United States

Feed Name or Analyses			Mean As Fed	Mean Dry	C.V. ± %
Dry matter	%		25.6	100.0	17
Ash	%		2.1	8.2	10
Crude fiber	%		6.7	26.4	16
Cattle	dig coef %		53.	53.	
Ether extract	%		0.9	3.6	3
Cattle	dig coef %		58.	58.	
N-free extract	%		11.6	45.5	
Cattle	dig coef %		76.	76.	
Protein (N x 6.25)	%		4.2	16.4	25
Cattle	dig coef %		65.	65.	
Cattle	dig prot %		2.7	10.6	
Goats	dig prot %	*	3.0	11.8	

Feed Name or Analyses			Mean As Fed	Mean Dry	C.V. ± %
Horses	dig prot %	*	2.9	11.4	
Rabbits	dig prot %	*	2.9	11.3	
Sheep	dig prot %	*	3.1	12.2	
Energy	GE Mcal/kg				
Cattle	DE Mcal/kg	*	0.72	2.82	
Sheep	DE Mcal/kg	*	0.70	2.75	
Cattle	ME Mcal/kg	*	0.59	2.31	
Sheep	ME Mcal/kg	*	0.58	2.26	
Cattle	NE_m Mcal/kg	*	0.36	1.39	
Cattle	NE_{gain} Mcal/kg	*	0.20	0.80	
Cattle	$NE_{lactating\ cows}$ Mcal/kg	*	0.41	1.60	
Cattle	TDN %		16.3	63.9	
Sheep	TDN %	*	15.9	62.4	
Calcium	%		0.42	1.64	
Phosphorus	%		0.09	0.36	
Potassium	%		0.62	2.44	

Clover, red, aerial part, fresh, cut 3, (2)
Ref No 2 01 433 United States

Feed Name or Analyses			Mean As Fed	Mean Dry	C.V. ± %
Dry matter	%		17.0	100.0	
Ash	%		1.8	10.6	
Crude fiber	%		2.8	16.5	
Ether extract	%		1.0	5.9	
N-free extract	%		6.5	38.2	
Protein (N x 6.25)	%		4.9	28.8	
Cattle	dig prot %	*	3.8	22.4	
Goats	dig prot %	*	4.0	23.4	
Horses	dig prot %	*	3.7	22.0	
Rabbits	dig prot %	*	3.6	20.9	
Sheep	dig prot %	*	4.1	23.8	
Energy	GE Mcal/kg				
Cattle	DE Mcal/kg	*	0.51	3.02	
Sheep	DE Mcal/kg	*	0.53	3.11	
Cattle	ME Mcal/kg	*	0.42	2.48	
Sheep	ME Mcal/kg	*	0.43	2.55	
Cattle	TDN %	*	11.7	68.6	
Sheep	TDN %	*	12.0	70.6	
Calcium	%		0.31	1.83	
Phosphorus	%		0.07	0.41	
Potassium	%		0.36	2.12	

Clover, red, aerial part, ensiled, (3)
Ref No 3 01 441 United States

Feed Name or Analyses			Mean As Fed	Mean Dry	C.V. ± %
Dry matter	%		26.2	100.0	18
Ash	%		2.5	9.6	15
Crude fiber	%		8.5	32.5	6
Cattle	dig coef %		68.	68.	
Sheep	dig coef %		55.	55.	
Ether extract	%		1.0	3.6	26
Cattle	dig coef %		72.	72.	
Sheep	dig coef %		57.	57.	
N-free extract	%		10.4	39.7	12
Cattle	dig coef %		55.	55.	
Sheep	dig coef %		58.	58.	
Protein (N x 6.25)	%		3.8	14.5	14
Cattle	dig coef %		58.	58.	
Sheep	dig coef %		67.	67.	
Cattle	dig prot %		2.2	8.4	
Goats	dig prot %	*	2.5	9.4	
Horses	dig prot %	*	2.5	9.4	
Sheep	dig prot %		2.6	9.7	
Energy	GE Mcal/kg				
Cattle	DE Mcal/kg	*	0.67	2.57	
Sheep	DE Mcal/kg	*	0.64	2.44	
Cattle	ME Mcal/kg	*	0.55	2.11	
Sheep	ME Mcal/kg	*	0.52	2.00	
Cattle	TDN %		15.3	58.3	
Sheep	TDN %		14.5	55.3	
Calcium	%		0.42	1.61	20
Chlorine	%		0.23	0.86	
Copper	mg/kg		2.9	11.0	12
Iron	%		0.009	0.033	17
Magnesium	%		0.10	0.40	15
Manganese	mg/kg		34.1	130.0	11
Phosphorus	%		0.06	0.23	26
Potassium	%		0.46	1.76	5
Sodium	%		0.06	0.23	
Sulphur	%		0.04	0.16	23
Carotene	mg/kg		54.1	206.4	81
Vitamin A equivalent	IU/g		90.1	344.0	

Ref No 3 01 441 Canada

Feed Name or Analyses			Mean As Fed	Mean Dry	C.V. ± %
Dry matter	%		24.3	100.0	
Crude fiber	%		8.3	34.4	
Cattle	dig coef %		68.	68.	
Sheep	dig coef %		55.	55.	
Ether extract	%				
Cattle	dig coef %		72.	72.	
Sheep	dig coef %		57.	57.	
N-free extract	%				
Cattle	dig coef %		55.	55.	
Sheep	dig coef %		58.	58.	
Protein (N x 6.25)	%		3.0	12.3	
Cattle	dig coef %		58.	58.	
Sheep	dig coef %		67.	67.	
Cattle	dig prot %		1.7	7.1	
Goats	dig prot %	*	1.8	7.4	
Horses	dig prot %	*	1.8	7.4	
Sheep	dig prot %		2.0	8.3	
Energy	GE Mcal/kg				
Cattle	DE Mcal/kg	*	0.62	2.57	
Sheep	DE Mcal/kg	*	0.59	2.43	
Cattle	ME Mcal/kg	*	0.51	2.11	
Sheep	ME Mcal/kg	*	0.48	2.00	
Cattle	TDN %		14.1	58.3	
Sheep	TDN %		13.4	55.2	
Calcium	%		0.38	1.58	
Phosphorus	%		0.05	0.20	
Carotene	mg/kg		11.4	47.1	
Vitamin A equivalent	IU/g		19.0	78.5	

Clover, red, aerial part, ensiled, immature, (3)
Ref No 3 01 435 United States

Feed Name or Analyses			Mean As Fed	Mean Dry	C.V. ± %
Dry matter	%		31.4	100.0	15
Ash	%		2.7	8.6	12
Crude fiber	%		9.3	29.6	10
Ether extract	%		1.0	3.1	21
N-free extract	%		14.0	44.6	
Protein (N x 6.25)	%		4.4	14.1	14
Cattle	dig prot %	*	2.8	9.0	
Goats	dig prot %	*	2.8	9.0	
Horses	dig prot %	*	2.8	9.0	
Sheep	dig prot %	*	2.8	9.0	
Energy	GE Mcal/kg				
Cattle	DE Mcal/kg	*	0.82	2.60	
Sheep	DE Mcal/kg	*	0.87	2.79	
Cattle	ME Mcal/kg	*	0.67	2.13	
Sheep	ME Mcal/kg	*	0.72	2.28	
Cattle	TDN %	*	18.5	59.0	
Sheep	TDN %	*	19.8	63.2	
Carotene	mg/kg		126.5	403.0	
Vitamin A equivalent	IU/g		210.9	671.8	

Clover, red, aerial part, ensiled, early bloom, (3)
Ref No 3 01 436 United States

Feed Name or Analyses			Mean As Fed	Mean Dry	C.V. ± %
Dry matter	%		24.8	100.0	25
Ash	%		2.6	10.5	13
Crude fiber	%		7.6	30.6	6
Cattle	dig coef %		53.	53.	
Sheep	dig coef %		46.	46.	
Ether extract	%		0.9	3.4	7
Cattle	dig coef %		64.	64.	
Sheep	dig coef %		62.	62.	
N-free extract	%		9.7	39.0	
Cattle	dig coef %		64.	64.	
Sheep	dig coef %		55.	55.	
Protein (N x 6.25)	%		4.1	16.5	11
Cattle	dig coef %		56.	56.	
Sheep	dig coef %		68.	68.	
Cattle	dig prot %		2.3	9.2	
Goats	dig prot %	*	2.8	11.2	
Horses	dig prot %	*	2.8	11.2	
Sheep	dig prot %		2.8	11.2	
Energy	GE Mcal/kg				
Cattle	DE Mcal/kg	*	0.61	2.44	
Sheep	DE Mcal/kg	*	0.56	2.27	
Cattle	ME Mcal/kg	*	0.50	2.00	
Sheep	ME Mcal/kg	*	0.46	1.86	
Cattle	TDN %		13.7	55.3	
Sheep	TDN %		12.8	51.5	

Clover, red, aerial part, ensiled, mid-bloom, (3)
Ref No 3 01 437 United States

Feed Name or Analyses			Mean As Fed	Mean Dry	C.V. ± %
Dry matter	%		31.4	100.0	6
Ash	%		2.4	7.8	9
Crude fiber	%		10.5	33.3	5
Cattle	dig coef %	*	68.	68.	
Sheep	dig coef %	*	55.	55.	
Ether extract	%		0.9	2.8	13
Cattle	dig coef %	*	72.	72.	
Sheep	dig coef %	*	57.	57.	
N-free extract	%		13.7	43.6	
Cattle	dig coef %	*	55.	55.	
Sheep	dig coef %	*	58.	58.	
Protein (N x 6.25)	%		3.9	12.5	4
Cattle	dig coef %	*	58.	58.	
Sheep	dig coef %	*	67.	67.	
Cattle	dig prot %		2.3	7.3	
Goats	dig prot %	*	2.4	7.6	
Horses	dig prot %	*	2.4	7.6	
Sheep	dig prot %		2.6	8.4	
Energy	GE Mcal/kg				
Cattle	DE Mcal/kg	*	0.81	2.58	
Sheep	DE Mcal/kg	*	0.77	2.45	
Cattle	ME Mcal/kg	*	0.66	2.11	
Sheep	ME Mcal/kg	*	0.63	2.01	
Cattle	TDN %	*	18.3	58.4	
Sheep	TDN %	*	17.4	55.6	

Clover, red, aerial part, ensiled, mid-bloom, cut 2, (3)
Ref No 3 07 902 United States

Feed Name or Analyses			Mean As Fed	Mean Dry	C.V. ± %
Dry matter	%			100.0	
Sheep	dig coef %			62.	
Ash	%			6.6	
Crude fiber	%			26.8	
Sheep	dig coef %			49.	

Continued

Feed Name or Analyses			Mean As Fed	Mean Dry	C.V. ± %
Ether extract	%			4.2	
Sheep	dig coef %			68.	
N-free extract	%			44.9	
Sheep	dig coef %			71.	
Protein (N x 6.25)	%			17.6	
Sheep	dig coef %			63.	
Cattle	dig prot %	*		12.2	
Goats	dig prot %	*		12.2	
Horses	dig prot %	*		12.2	
Sheep	dig prot %			11.0	
Energy	GE Mcal/kg			4.48	
Sheep	GE dig coef %			60.	
Cattle	DE Mcal/kg	*		2.50	
Sheep	DE Mcal/kg			2.67	
Cattle	ME Mcal/kg	*		2.05	
Sheep	ME Mcal/kg			2.22	
Cattle	TDN %	*		56.7	
Sheep	TDN %			62.2	

Clover, red, aerial part, ensiled, full bloom, (3)

Ref No 3 01 438 United States

Feed Name or Analyses			Mean As Fed	Mean Dry	C.V. ± %
Dry matter	%		14.9	100.0	
Ash	%		1.4	9.6	
Crude fiber	%		4.2	28.0	
Sheep	dig coef %		72.	72.	
Ether extract	%		0.5	3.4	
Sheep	dig coef %		51.	51.	
N-free extract	%		6.3	42.2	
Sheep	dig coef %		62.	62.	
Protein (N x 6.25)	%		2.5	16.8	
Sheep	dig coef %		72.	72.	
Cattle	dig prot %	*	1.7	11.5	
Goats	dig prot %	*	1.7	11.5	
Horses	dig prot %	*	1.7	11.5	
Sheep	dig prot %		1.8	12.1	
Energy	GE Mcal/kg				
Cattle	DE Mcal/kg	*	0.39	2.61	
Sheep	DE Mcal/kg	*	0.41	2.75	
Cattle	ME Mcal/kg	*	0.32	2.14	
Sheep	ME Mcal/kg	*	0.34	2.25	
Cattle	TDN %	*	8.8	59.2	
Sheep	TDN %		9.3	62.3	

Ref No 3 01 438 Canada

Feed Name or Analyses			Mean As Fed	Mean Dry	C.V. ± %
Dry matter	%		24.3	100.0	
Crude fiber	%		8.3	34.2	
Sheep	dig coef %		72.	72.	
Ether extract	%				
Sheep	dig coef %		51.	51.	
N-free extract	%				
Sheep	dig coef %		62.	62.	
Protein (N x 6.25)	%		3.2	13.2	
Sheep	dig coef %		72.	72.	
Cattle	dig prot %	*	2.0	8.2	
Goats	dig prot %	*	2.0	8.2	
Horses	dig prot %	*	2.0	8.2	
Sheep	dig prot %		2.3	9.5	
Energy	GE Mcal/kg				
Sheep	DE Mcal/kg	*	0.67	2.75	
Sheep	ME Mcal/kg	*	0.55	2.26	
Sheep	TDN %		15.2	62.4	
Calcium	%		0.28	1.15	
Phosphorus	%		0.09	0.37	

(1) dry forages and roughages (2) pasture, range plants, and forages fed green (3) silages (4) energy feeds (5) protein supplements (6) minerals (7) vitamins (8) additives

Clover, red, aerial part, ensiled, cut 1, (3)

Ref No 3 01 439 United States

Feed Name or Analyses			Mean As Fed	Mean Dry	C.V. ± %
Dry matter	%		24.6	100.0	5
Ash	%		2.1	8.6	10
Crude fiber	%		7.1	29.0	5
Sheep	dig coef %		49.	49.	
Ether extract	%		0.9	3.9	20
Sheep	dig coef %		34.	34.	
N-free extract	%		10.6	43.0	
Sheep	dig coef %		70.	70.	
Protein (N x 6.25)	%		3.8	15.6	14
Sheep	dig coef %		62.	62.	
Cattle	dig prot %	*	2.6	10.4	
Goats	dig prot %	*	2.6	10.4	
Horses	dig prot %	*	2.6	10.4	
Sheep	dig prot %		2.4	9.7	
Energy	GE Mcal/kg				
Cattle	DE Mcal/kg	*	0.64	2.59	
Sheep	DE Mcal/kg	*	0.62	2.51	
Cattle	ME Mcal/kg	*	0.52	2.13	
Sheep	ME Mcal/kg	*	0.51	2.06	
Cattle	TDN %	*	14.4	58.8	
Sheep	TDN %		14.0	56.9	

Clover, red, aerial part, ensiled, cut 2, (3)

Ref No 3 01 440 United States

Feed Name or Analyses			Mean As Fed	Mean Dry	C.V. ± %
Dry matter	%		18.3	100.0	
Ash	%		1.9	10.6	
Crude fiber	%		5.0	27.5	
Sheep	dig coef %		52.	52.	
Ether extract	%		0.9	4.9	
Sheep	dig coef %		49.	49.	
N-free extract	%		6.9	37.8	
Sheep	dig coef %		67.	67.	
Protein (N x 6.25)	%		3.5	19.3	
Sheep	dig coef %		72.	72.	
Cattle	dig prot %	*	2.5	13.8	
Goats	dig prot %	*	2.5	13.8	
Horses	dig prot %	*	2.5	13.8	
Sheep	dig prot %		2.5	13.9	
Energy	GE Mcal/kg				
Cattle	DE Mcal/kg	*	0.48	2.64	
Sheep	DE Mcal/kg	*	0.47	2.59	
Cattle	ME Mcal/kg	*	0.40	2.17	
Sheep	ME Mcal/kg	*	0.39	2.13	
Cattle	TDN %	*	10.9	59.9	
Sheep	TDN %		10.7	58.8	

Clover, red, aerial part, wilted ensiled, (3)

Ref No 3 01 444 United States

Feed Name or Analyses			Mean As Fed	Mean Dry	C.V. ± %
Dry matter	%		33.1	100.0	7
Ash	%		2.7	8.3	8
Crude fiber	%		10.3	31.2	7
Ether extract	%		1.0	2.9	7
N-free extract	%		14.7	44.4	
Protein (N x 6.25)	%		4.4	13.2	6
Cattle	dig prot %	*	2.7	8.2	
Goats	dig prot %	*	2.7	8.2	
Horses	dig prot %	*	2.7	8.2	
Sheep	dig prot %	*	2.7	8.2	
Energy	GE Mcal/kg				
Cattle	DE Mcal/kg	*	0.84	2.55	
Sheep	DE Mcal/kg	*	0.92	2.79	
Cattle	ME Mcal/kg	*	0.69	2.09	
Sheep	ME Mcal/kg	*	0.76	2.29	
Cattle	TDN %	*	19.2	57.9	
Sheep	TDN %	*	20.9	63.3	

Clover, red, aerial part, wilted ensiled, early bloom, (3)

Ref No 3 01 442 United States

Feed Name or Analyses			Mean As Fed	Mean Dry	C.V. ± %
Dry matter	%		35.7	100.0	5
Ash	%		3.3	9.2	3
Crude fiber	%		10.1	28.4	4
Ether extract	%		1.1	3.0	7
N-free extract	%		16.2	45.4	
Protein (N x 6.25)	%		5.0	14.0	2
Cattle	dig prot %	*	3.2	8.9	
Goats	dig prot %	*	3.2	8.9	
Horses	dig prot %	*	3.2	8.9	
Sheep	dig prot %	*	3.2	8.9	
Energy	GE Mcal/kg				
Cattle	DE Mcal/kg	*	0.96	2.69	
Sheep	DE Mcal/kg	*	0.99	2.78	
Cattle	ME Mcal/kg	*	0.79	2.20	
Sheep	ME Mcal/kg	*	0.82	2.28	
Cattle	TDN %	*	21.7	60.9	
Sheep	TDN %	*	22.5	63.2	

Clover, red, aerial part, wilted ensiled, mid-bloom, (3)

Ref No 3 01 443 United States

Feed Name or Analyses			Mean As Fed	Mean Dry	C.V. ± %
Dry matter	%		31.3	100.0	2
Ash	%		2.4	7.8	3
Crude fiber	%		10.4	33.2	2
Cattle	dig coef %		54.	54.	
Ether extract	%		0.9	2.9	5
Cattle	dig coef %		64.	64.	
N-free extract	%		13.6	43.4	
Cattle	dig coef %		64.	64.	
Protein (N x 6.25)	%		4.0	12.9	4
Cattle	dig coef %		56.	56.	
Cattle	dig prot %		2.3	7.2	
Goats	dig prot %	*	2.5	7.9	
Horses	dig prot %	*	2.5	7.9	
Sheep	dig prot %	*	2.5	7.9	
Energy	GE Mcal/kg				
Cattle	DE Mcal/kg	*	0.79	2.51	
Sheep	DE Mcal/kg	*	0.87	2.79	
Cattle	ME Mcal/kg	*	0.64	2.06	
Sheep	ME Mcal/kg	*	0.72	2.29	
Cattle	TDN %		17.8	57.0	
Sheep	TDN %	*	19.8	63.3	

Clover, red, aerial part w AIV preservative added, ensiled, (3)

Ref No 3 01 446 United States

Feed Name or Analyses			Mean As Fed	Mean Dry	C.V. ± %
Dry matter	%		21.1	100.0	
Ash	%		1.9	9.1	
Crude fiber	%		6.3	29.7	
Cattle	dig coef %		63.	63.	
Sheep	dig coef %	*	55.	55.	
Ether extract	%		1.3	6.2	
Cattle	dig coef %		78.	78.	
Sheep	dig coef %	*	57.	57.	
N-free extract	%		8.1	38.3	
Cattle	dig coef %		78.	78.	
Sheep	dig coef %	*	58.	58.	
Protein (N x 6.25)	%		3.5	16.8	
Cattle	dig coef %		75.	75.	

Feed Name or Analyses			Mean As Fed	Mean Dry	C.V. ± %
Sheep	dig coef %	*	67.	67.	
Cattle	dig prot %		2.7	12.6	
Goats	dig prot %	*	2.4	11.5	
Horses	dig prot %	*	2.4	11.5	
Sheep	dig prot %		2.4	11.3	
Energy	GE Mcal/kg				
Cattle	DE Mcal/kg	*	0.67	3.17	
Sheep	DE Mcal/kg	*	0.54	2.54	
Cattle	ME Mcal/kg	*	0.55	2.60	
Sheep	ME Mcal/kg	*	0.44	2.09	
Cattle	TDN %		15.2	72.0	
Sheep	TDN %		12.2	57.7	

Clover, red, aerial part w phosphoric acid preservative added, ensiled, (3)

Ref No 3 01 448 United States

Feed Name or Analyses			Mean As Fed	Mean Dry	C.V. ± %
Dry matter	%		34.1	100.0	4
Ash	%		3.2	9.4	10
Crude fiber	%		10.2	30.0	9
Cattle	dig coef %	*	54.	54.	
Sheep	dig coef %	*	55.	55.	
Ether extract	%		1.0	3.0	9
Cattle	dig coef %	*	65.	65.	
Sheep	dig coef %	*	57.	57.	
N-free extract	%		15.0	44.0	
Cattle	dig coef %	*	65.	65.	
Sheep	dig coef %	*	58.	58.	
Protein (N x 6.25)	%		4.6	13.6	10
Cattle	dig coef %	*	53.	53.	
Sheep	dig coef %	*	67.	67.	
Cattle	dig prot %		2.5	7.2	
Goats	dig prot %	*	2.9	8.6	
Horses	dig prot %	*	2.9	8.6	
Sheep	dig prot %		3.1	9.1	
Energy	GE Mcal/kg				
Cattle	DE Mcal/kg	*	0.85	2.49	
Sheep	DE Mcal/kg	*	0.83	2.42	
Cattle	ME Mcal/kg	*	0.70	2.04	
Sheep	ME Mcal/kg	*	0.68	1.99	
Cattle	TDN %	*	19.2	56.4	
Sheep	TDN %	*	18.7	55.0	

Clover, red, aerial part, w phosphoric acid preservative added, ensiled, early bloom, (3)

Ref No 3 01 447 United States

Feed Name or Analyses			Mean As Fed	Mean Dry	C.V. ± %
Dry matter	%		36.9	100.0	
Ash	%		3.4	9.3	
Crude fiber	%		11.1	30.0	
Cattle	dig coef %		54.	54.	
Ether extract	%		1.1	2.9	
Cattle	dig coef %		65.	65.	
N-free extract	%		16.3	44.3	
Cattle	dig coef %		65.	65.	
Protein (N x 6.25)	%		5.0	13.5	
Cattle	dig coef %		53.	53.	
Cattle	dig prot %		2.6	7.2	
Goats	dig prot %	*	3.1	8.5	
Horses	dig prot %	*	3.1	8.5	
Sheep	dig prot %	*	3.1	8.5	
Energy	GE Mcal/kg				
Cattle	DE Mcal/kg	*	0.92	2.49	
Sheep	DE Mcal/kg	*	0.96	2.60	
Cattle	ME Mcal/kg	*	0.75	2.04	
Sheep	ME Mcal/kg	*	0.79	2.13	
Cattle	TDN %		20.8	56.4	
Sheep	TDN %	*	21.7	58.9	

Clover, red, aerial part w molasses added, ensiled, (3)

Ref No 3 01 451 United States

Feed Name or Analyses			Mean As Fed	Mean Dry	C.V. ± %
Dry matter	%		33.9	100.0	
Ash	%		2.8	8.4	
Crude fiber	%		10.3	30.3	
Cattle	dig coef %		54.	54.	
Sheep	dig coef %	*	68.	68.	
Ether extract	%		1.0	2.9	
Cattle	dig coef %		61.	61.	
Sheep	dig coef %	*	74.	74.	
N-free extract	%		15.3	45.1	
Cattle	dig coef %		66.	66.	
Sheep	dig coef %	*	61.	61.	
Protein (N x 6.25)	%		4.5	13.4	
Cattle	dig coef %		57.	57.	
Sheep	dig coef %	*	72.	72.	
Cattle	dig prot %		2.6	7.6	
Goats	dig prot %	*	2.8	8.4	
Horses	dig prot %	*	2.8	8.4	
Sheep	dig prot %		3.3	9.6	
Energy	GE Mcal/kg				
Cattle	DE Mcal/kg	*	0.86	2.54	
Sheep	DE Mcal/kg	*	0.93	2.76	
Cattle	ME Mcal/kg	*	0.71	2.09	
Sheep	ME Mcal/kg	*	0.77	2.26	
Cattle	TDN %		19.6	57.7	
Sheep	TDN %		21.2	62.5	

Clover, red, aerial part w molasses added, ensiled, early bloom, (3)

Ref No 3 01 449 United States

Feed Name or Analyses			Mean As Fed	Mean Dry	C.V. ± %
Dry matter	%		34.1	100.0	10
Ash	%		3.1	9.0	12
Crude fiber	%		9.3	27.4	6
Cattle	dig coef %		50.	50.	
Ether extract	%		1.0	2.9	11
Cattle	dig coef %		60.	60.	
N-free extract	%		15.9	46.6	
Cattle	dig coef %		72.	72.	
Protein (N x 6.25)	%		4.8	14.2	6
Cattle	dig coef %		56.	56.	
Cattle	dig prot %		2.7	8.0	
Goats	dig prot %	*	3.1	9.1	
Horses	dig prot %	*	3.1	9.1	
Sheep	dig prot %	*	3.1	9.1	
Energy	GE Mcal/kg				
Cattle	DE Mcal/kg	*	0.89	2.60	
Sheep	DE Mcal/kg	*	0.96	2.80	
Cattle	ME Mcal/kg	*	0.73	2.14	
Sheep	ME Mcal/kg	*	0.78	2.30	
Cattle	TDN %		20.1	59.1	
Sheep	TDN %	*	21.7	63.6	

Clover, red, aerial part w molasses added, ensiled, mid-bloom, (3)

Ref No 3 01 450 United States

Feed Name or Analyses			Mean As Fed	Mean Dry	C.V. ± %
Dry matter	%		32.1	100.0	2
Ash	%		2.6	8.0	3
Crude fiber	%		10.9	34.0	2
Cattle	dig coef %		57.	57.	
Ether extract	%		0.9	2.7	4
Cattle	dig coef %		62.	62.	
N-free extract	%		13.9	43.5	
Cattle	dig coef %		61.	61.	
Protein (N x 6.25)	%		3.8	11.9	2
Cattle	dig coef %		58.	58.	
Cattle	dig prot %		2.2	6.9	
Goats	dig prot %	*	2.3	7.0	
Horses	dig prot %	*	2.3	7.0	
Sheep	dig prot %	*	2.3	7.0	
Energy	GE Mcal/kg				
Cattle	DE Mcal/kg	*	0.80	2.49	
Sheep	DE Mcal/kg	*	0.89	2.79	
Cattle	ME Mcal/kg	*	0.65	2.04	
Sheep	ME Mcal/kg	*	0.73	2.29	
Cattle	TDN %		18.1	56.5	
Sheep	TDN %	*	20.3	63.2	

Clover, red, aerial part w molasses added, ensiled, mid-bloom, cut 2, (3)

Ref No 3 07 900 United States

Feed Name or Analyses			Mean As Fed	Mean Dry	C.V. ± %
Dry matter	%			100.0	
Sheep	dig coef %			66.	
Ash	%			6.5	
Crude fiber	%			22.8	
Sheep	dig coef %			50.	
Ether extract	%			3.1	
Sheep	dig coef %			64.	
N-free extract	%			52.2	
Sheep	dig coef %			76.	
Protein (N x 6.25)	%			15.4	
Sheep	dig coef %			59.	
Cattle	dig prot %	*		10.2	
Goats	dig prot %	*		10.2	
Horses	dig prot %	*		10.2	
Sheep	dig prot %			9.1	
Energy	GE Mcal/kg			4.29	
Sheep	GE dig coef %			62.	
Cattle	DE Mcal/kg	*		2.71	
Sheep	DE Mcal/kg			2.66	
Cattle	ME Mcal/kg	*		2.22	
Sheep	ME Mcal/kg			2.24	
Cattle	TDN %	*		61.4	
Sheep	TDN %			64.6	

Clover, red, aerial part w phosphorus pentachloride preservative added, ensiled, (3)

Ref No 3 01 452 United States

Feed Name or Analyses			Mean As Fed	Mean Dry	C.V. ± %
Dry matter	%		22.0	100.0	
Ash	%		2.5	11.4	
Crude fiber	%		5.1	23.4	
Sheep	dig coef %		68.	68.	
Ether extract	%		0.8	3.8	
Sheep	dig coef %		72.	72.	
N-free extract	%		9.5	43.1	
Sheep	dig coef %		64.	64.	
Protein (N x 6.25)	%		4.0	18.3	
Sheep	dig coef %		70.	70.	
Cattle	dig prot %	*	2.8	12.8	
Goats	dig prot %	*	2.8	12.8	
Horses	dig prot %	*	2.8	12.8	
Sheep	dig prot %		2.8	12.8	
Energy	GE Mcal/kg				
Cattle	DE Mcal/kg	*	0.63	2.85	
Sheep	DE Mcal/kg	*	0.61	2.75	

Continued

Feed Name or Analyses			Mean As Fed	Mean Dry	C.V. ± %
Cattle	ME Mcal/kg	*	0.51	2.34	
Sheep	ME Mcal/kg	*	0.50	2.26	
Cattle	TDN %	*	14.2	64.6	
Sheep	TDN %		13.7	62.5	

Clover, red, aerial part w sodium bisulfite preservative added, ensiled, mid-bloom, cut 2, (3)

Ref No 3 07 901 United States

Feed Name or Analyses			As Fed	Dry	C.V. ± %
Dry matter	%			100.0	
Sheep	dig coef %			64.	
Ash	%			6.9	
Crude fiber	%			26.5	
Sheep	dig coef %			53.	
Ether extract	%			3.8	
Sheep	dig coef %			67.	
N-free extract	%			46.3	
Sheep	dig coef %			72.	
Protein (N x 6.25)	%			16.5	
Sheep	dig coef %			63.	
Cattle	dig prot %	*		11.2	
Goats	dig prot %	*		11.2	
Horses	dig prot %	*		11.2	
Sheep	dig prot %			10.4	
Energy	GE Mcal/kg			4.47	
Sheep	GE dig coef %			61.	
Cattle	DE Mcal/kg	*		2.56	
Sheep	DE Mcal/kg			2.73	
Cattle	ME Mcal/kg	*		2.10	
Sheep	ME Mcal/kg			2.29	
Cattle	TDN %	*		58.1	
Sheep	TDN %			63.5	

Clover, red, aerial part w sulfur dioxide preservative added, ensiled, (3)

Ref No 3 01 445 United States

Feed Name or Analyses			As Fed	Dry	C.V. ± %
Dry matter	%		22.0	100.0	
Ash	%		2.5	11.3	
Crude fiber	%		5.2	23.8	
Sheep	dig coef %		70.	70.	
Ether extract	%		0.9	4.0	
Sheep	dig coef %		71.	71.	
N-free extract	%		9.4	42.5	
Sheep	dig coef %		64.	64.	
Protein (N x 6.25)	%		4.0	18.4	
Sheep	dig coef %		73.	73.	
Cattle	dig prot %	*	2.8	12.9	
Goats	dig prot %	*	2.8	12.9	
Horses	dig prot %	*	2.8	12.9	
Sheep	dig prot %		3.0	13.4	
Energy	GE Mcal/kg				
Cattle	DE Mcal/kg	*	0.62	2.83	
Sheep	DE Mcal/kg	*	0.62	2.81	
Cattle	ME Mcal/kg	*	0.51	2.32	
Sheep	ME Mcal/kg	*	0.51	2.30	
Cattle	TDN %	*	14.1	64.3	
Sheep	TDN %		14.0	63.7	

Clover, red, seeds, (5)

Ref No 5 08 004 United States

Feed Name or Analyses			As Fed	Dry	C.V. ± %
Dry matter	%		87.5	100.0	
Ash	%		6.7	7.7	
Crude fiber	%		9.2	10.5	
Ether extract	%		7.8	8.9	
N-free extract	%		31.2	35.7	
Protein (N x 6.25)	%		32.6	37.3	
Energy	GE Mcal/kg				
Cattle	DE Mcal/kg	*	3.07	3.51	
Sheep	DE Mcal/kg	*	3.15	3.60	
Swine	DE kcal/kg	*	3508.	4009.	
Cattle	ME Mcal/kg	*	2.52	2.88	
Sheep	ME Mcal/kg	*	2.58	2.95	
Swine	ME kcal/kg	*	3104.	3547.	
Cattle	TDN %	*	69.7	79.6	
Sheep	TDN %	*	71.5	81.7	
Swine	TDN %	*	79.6	90.9	

Ref No 5 08 004 Canada

Feed Name or Analyses			As Fed	Dry	C.V. ± %
Dry matter	%		88.2	100.0	
Protein (N x 6.25)	%		31.9	36.2	
Calcium	%		0.29	0.33	
Phosphorus	%		0.54	0.61	

Clover, red, seed screenings, (5)

Ref No 5 08 005 United States

Feed Name or Analyses			As Fed	Dry	C.V. ± %
Dry matter	%		90.5	100.0	
Ash	%		5.9	6.5	
Crude fiber	%		10.2	11.3	
Ether extract	%		5.9	6.5	
N-free extract	%		40.3	44.5	
Protein (N x 6.25)	%		28.2	31.2	
Energy	GE Mcal/kg				
Cattle	DE Mcal/kg	*	3.12	3.45	
Sheep	DE Mcal/kg	*	3.32	3.67	
Swine	DE kcal/kg	*	3612.	3991.	
Cattle	ME Mcal/kg	*	2.56	2.83	
Sheep	ME Mcal/kg	*	2.72	3.01	
Swine	ME kcal/kg	*	3240.	3580.	
Cattle	TDN %	*	70.9	78.3	
Sheep	TDN %	*	75.2	83.1	
Swine	TDN %	*	81.9	90.5	

Ref No 5 08 005 Canada

Feed Name or Analyses			As Fed	Dry	C.V. ± %
Dry matter	%		90.0	100.0	
Crude fiber	%		15.2	16.9	
Protein (N x 6.25)	%		28.9	32.1	
Calcium	%		0.41	0.46	
Phosphorus	%		0.67	0.74	
Carotene	mg/kg		2.6	2.9	
Vitamin A equivalent	IU/g		4.4	4.9	

CLOVER, ROSE. Trifolium hirtum

Clover, rose, aerial part, dehy grnd, mid-bloom, (1)

Ref No 1 08 215 United States

Feed Name or Analyses			As Fed	Dry	C.V. ± %
Dry matter	%		92.2	100.0	
Organic matter	%		82.0	88.9	
Ash	%		10.2	11.1	
Crude fiber	%		10.0	10.9	
Ether extract	%		3.4	3.7	
N-free extract	%		39.8	43.2	
Protein (N x 6.25)	%		28.7	31.1	
Cattle	dig prot %	*	22.0	23.9	
Goats	dig prot %	*	23.6	25.6	
Horses	dig prot %	*	22.1	23.9	
Rabbits	dig prot %	*	20.9	22.7	
Sheep	dig prot %	*	22.6	24.5	
Silicon	%		0.15	0.16	

Clover, rose, aerial part, dehy grnd, dormant, (1)

Ref No 1 08 216 United States

Feed Name or Analyses			As Fed	Dry	C.V. ± %
Dry matter	%		91.6	100.0	1
Organic matter	%		83.8	91.5	2
Ash	%		7.8	8.5	18
Crude fiber	%		34.2	37.4	10
Ether extract	%		1.2	1.3	24
N-free extract	%		40.7	44.5	8
Protein (N x 6.25)	%		7.6	8.3	33
Cattle	dig prot %	*	3.8	4.2	
Goats	dig prot %	*	4.0	4.3	
Horses	dig prot %	*	4.2	4.6	
Rabbits	dig prot %	*	4.7	5.1	
Sheep	dig prot %	*	3.7	4.0	
Energy	GE Mcal/kg				
Cattle	DE Mcal/kg	*	2.06	2.25	
Sheep	DE Mcal/kg	*	2.17	2.37	
Cattle	ME Mcal/kg	*	1.69	1.84	
Sheep	ME Mcal/kg	*	1.78	1.94	
Cattle	TDN %	*	46.7	51.0	
Sheep	TDN %	*	49.2	53.7	
Silicon	%		1.17	1.28	98

Clover, rose, leaves, dehy grnd, (1)

Ref No 1 08 225 United States

Feed Name or Analyses			As Fed	Dry	C.V. ± %
Dry matter	%		93.4	100.0	
Organic matter	%		82.4	88.2	
Ash	%		10.5	11.2	
Crude fiber	%		9.3	10.0	
Ether extract	%		3.6	3.9	
N-free extract	%		43.8	46.9	
Protein (N x 6.25)	%		26.2	28.0	
Cattle	dig prot %	*	19.8	21.2	
Goats	dig prot %	*	21.2	22.7	
Horses	dig prot %	*	19.9	21.3	
Rabbits	dig prot %	*	18.9	20.3	
Sheep	dig prot %	*	20.3	21.7	
Silicon	%		0.12	0.13	

Clover, rose, leaves, dehy grnd, immature, (1)

Ref No 1 08 220 United States

Feed Name or Analyses			As Fed	Dry	C.V. ± %
Dry matter	%		92.3	100.0	
Organic matter	%		82.2	89.1	
Ash	%		10.1	10.9	
Protein (N x 6.25)	%		24.4	26.5	
Cattle	dig prot %	*	18.3	19.8	
Goats	dig prot %	*	19.6	21.2	
Horses	dig prot %	*	18.4	20.0	
Rabbits	dig prot %	*	17.6	19.1	
Sheep	dig prot %	*	18.7	20.3	
Silicon	%		0.00	0.00	

(1) dry forages and roughages (2) pasture, range plants, and forages fed green (3) silages (4) energy feeds (5) protein supplements (6) minerals (7) vitamins (8) additives

Feed Name or Analyses			Mean As Fed	Mean Dry	C.V. ± %
Clover, rose, leaves, dehy grnd, early bloom, (1)					
Ref No 1 08 228				United States	
Dry matter	%		91.2	100.0	
Organic matter	%		79.9	87.6	
Ash	%		11.3	12.4	
Crude fiber	%		17.2	18.9	
Ether extract	%		2.7	3.0	
N-free extract	%		47.0	51.5	
Protein (N x 6.25)	%		13.0	14.2	
Cattle dig prot	%	*	8.4	9.2	
Goats dig prot	%	*	8.9	9.8	
Horses dig prot	%	*	8.7	9.6	
Rabbits dig prot	%	*	8.8	9.6	
Sheep dig prot	%	*	8.5	9.3	
Silicon	%		0.18	0.20	
Clover, rose, leaves, dehy grnd, full bloom, (1)					
Ref No 1 08 227				United States	
Dry matter	%		92.6	100.0	
Organic matter	%		82.0	88.5	
Ash	%		10.6	11.5	
Crude fiber	%		10.4	11.2	
Ether extract	%		3.8	4.1	
N-free extract	%		40.2	43.4	
Protein (N x 6.25)	%		27.6	29.8	
Cattle dig prot	%	*	21.1	22.7	
Goats dig prot	%	*	22.6	24.4	
Horses dig prot	%	*	21.1	22.8	
Rabbits dig prot	%	*	20.1	21.7	
Sheep dig prot	%	*	21.6	23.3	
Silicon	%		0.03	0.03	
Clover, rose, leaves, dehy grnd, dormant, (1)					
Ref No 1 08 230				United States	
Dry matter	%		91.6	100.0	
Organic matter	%		78.9	86.1	
Ash	%		12.7	13.9	
Crude fiber	%		17.8	19.4	
Ether extract	%		1.7	1.9	
N-free extract	%		44.2	48.2	
Protein (N x 6.25)	%		15.2	16.6	
Cattle dig prot	%	*	10.4	11.3	
Goats dig prot	%	*	11.0	12.0	
Horses dig prot	%	*	10.6	11.6	
Rabbits dig prot	%	*	10.5	11.5	
Sheep dig prot	%	*	10.5	11.5	
Silicon	%		0.05	0.05	

CLOVER, SUBTERRANEAN. Trifolium subterraneum

Feed Name or Analyses			Mean As Fed	Mean Dry	C.V. ± %
Clover, subterranean, aerial part, dehy grnd, mid-bloom, (1)					
Ref No 1 08 615				United States	
Dry matter	%		92.5	100.0	
Organic matter	%		82.2	88.9	
Ash	%		10.3	11.1	
Crude fiber	%		9.3	10.1	
Ether extract	%		3.4	3.7	
N-free extract	%		41.3	44.6	
Protein (N x 6.25)	%		28.2	30.5	
Cattle dig prot	%	*	21.6	23.4	
Goats dig prot	%	*	23.1	25.0	
Horses dig prot	%	*	21.7	23.4	
Rabbits dig prot	%	*	20.5	22.2	
Sheep dig prot	%	*	22.1	23.9	
Silicon	%		0.28	0.30	
Clover, subterranean, aerial part, dehy grnd, dormant, (1)					
Ref No 1 08 217				United States	
Dry matter	%		91.8	100.0	1
Organic matter	%		79.6	86.6	7
Ash	%		12.3	13.4	45
Crude fiber	%		26.7	29.1	12
Ether extract	%		2.0	2.2	6
N-free extract	%		41.8	45.5	14
Protein (N x 6.25)	%		9.1	9.9	18
Cattle dig prot	%	*	5.1	5.5	
Goats dig prot	%	*	5.3	5.8	
Horses dig prot	%	*	5.5	5.9	
Rabbits dig prot	%	*	5.8	6.3	
Sheep dig prot	%	*	5.0	5.5	
Silicon	%		4.36	4.74	98
Clover, subterranean, leaves, dehy grnd, early leaf, (1)					
Ref No 1 08 209				United States	
Dry matter	%		93.1	100.0	
Organic matter	%		82.7	88.8	
Ash	%		10.4	11.2	
Crude fiber	%		8.4	9.0	
Ether extract	%		3.8	4.1	
N-free extract	%		45.1	48.4	
Protein (N x 6.25)	%		25.4	27.3	
Cattle dig prot	%	*	19.2	20.6	
Goats dig prot	%	*	20.5	22.0	
Horses dig prot	%	*	19.3	20.7	
Rabbits dig prot	%	*	18.4	19.7	
Sheep dig prot	%	*	19.6	21.1	
Silicon	%		0.50	0.54	
Clover, subterranean, leaves, dehy grnd, immature, (1)					
Ref No 1 08 218				United States	
Dry matter	%		92.6	100.0	
Organic matter	%		81.6	88.1	
Ash	%		10.1	10.9	
Protein (N x 6.25)	%		20.5	22.1	
Cattle dig prot	%	*	14.9	16.1	
Goats dig prot	%	*	15.9	17.2	
Horses dig prot	%	*	15.1	16.3	
Rabbits dig prot	%	*	14.6	15.7	
Sheep dig prot	%	*	15.2	16.4	
Clover, subterranean, leaves, dehy grnd, pre-bloom, (1)					
Ref No 1 08 212				United States	
Dry matter	%		91.9	100.0	
Organic matter	%		82.1	89.3	
Ash	%		9.8	10.7	
Protein (N x 6.25)	%		11.5	12.5	
Cattle dig prot	%	*	7.1	7.8	
Goats dig prot	%	*	7.6	8.2	
Horses dig prot	%	*	7.5	8.1	
Rabbits dig prot	%	*	7.6	8.3	
Sheep dig prot	%	*	7.2	7.8	
Silicon	%		0.09	0.10	
Clover, subterranean, leaves, dehy grnd, mid-bloom, (1)					
Ref No 1 08 222				United States	
Dry matter	%		92.2	100.0	
Organic matter	%		82.4	89.4	
Ash	%		9.8	10.6	
Crude fiber	%		11.8	12.8	
Ether extract	%		3.6	3.9	
N-free extract	%		46.5	50.4	
Protein (N x 6.25)	%		20.6	22.3	
Cattle dig prot	%	*	15.0	16.3	
Goats dig prot	%	*	16.0	17.4	
Horses dig prot	%	*	15.2	16.5	
Rabbits dig prot	%	*	14.6	15.9	
Sheep dig prot	%	*	15.3	16.6	
Silicon	%		0.25	0.27	
Clover, subterranean, leaves, dehy grnd, dormant, (1)					
Ref No 1 08 223				United States	
Dry matter	%		91.6	100.0	
Organic matter	%		77.4	84.5	
Ash	%		14.2	15.5	
Crude fiber	%		15.9	17.4	
Ether extract	%		1.7	1.9	
N-free extract	%		49.1	53.6	
Protein (N x 6.25)	%		10.6	11.6	
Cattle dig prot	%	*	6.4	7.0	
Goats dig prot	%	*	6.8	7.4	
Horses dig prot	%	*	6.8	7.4	
Rabbits dig prot	%	*	7.0	7.6	
Sheep dig prot	%	*	6.4	7.0	
Silicon	%		0.75	0.82	

CLOVER, TOMCAT. Trifolium tridentatum

Feed Name or Analyses			Mean As Fed	Mean Dry	C.V. ± %
Clover, tomcat, aerial part, fresh, (2)					
Ref No 2 01 457				United States	
Dry matter	%			100.0	
Ash	%			10.2	35
Crude fiber	%			24.8	30
Protein (N x 6.25)	%			14.3	62
Cattle dig prot	%	*		10.0	
Goats dig prot	%	*		9.9	
Horses dig prot	%	*		9.7	
Rabbits dig prot	%	*		9.7	
Sheep dig prot	%	*		10.3	
Calcium	%			2.11	7
Phosphorus	%			0.18	68
Potassium	%			2.24	64
Clover, tomcat, aerial part, fresh, immature, (2)					
Ref No 2 01 453				United States	
Dry matter	%			100.0	
Ash	%			16.9	
Crude fiber	%			16.3	
Protein (N x 6.25)	%			28.1	
Cattle dig prot	%	*		21.8	
Goats dig prot	%	*		22.8	
Horses dig prot	%	*		21.4	

Continued

Feed Name or Analyses				Mean As Fed	Mean Dry	C.V. ± %
Rabbits	dig prot	%	*		20.4	
Sheep	dig prot	%	*		23.2	
Calcium		%			2.30	
Phosphorus		%			0.35	
Potassium		%			4.90	

Clover, tomcat, aerial part, fresh, full bloom, (2)
Ref No 2 01 454 — United States

Feed Name or Analyses				Mean As Fed	Mean Dry	C.V. ± %
Dry matter		%			100.0	
Ash		%			9.7	
Crude fiber		%			20.4	
Protein (N x 6.25)		%			18.8	
Cattle	dig prot	%	*		13.9	
Goats	dig prot	%	*		14.1	
Horses	dig prot	%	*		13.5	
Rabbits	dig prot	%	*		13.2	
Sheep	dig prot	%	*		14.5	
Calcium		%			2.12	
Phosphorus		%			0.28	
Potassium		%			2.08	

Clover, tomcat, aerial part, fresh, milk stage, (2)
Ref No 2 01 455 — United States

Feed Name or Analyses				Mean As Fed	Mean Dry	C.V. ± %
Dry matter		%			100.0	
Ash		%			8.4	
Crude fiber		%			22.3	
Protein (N x 6.25)		%			17.9	
Cattle	dig prot	%	*		13.1	
Goats	dig prot	%	*		13.3	
Horses	dig prot	%	*		12.7	
Rabbits	dig prot	%	*		12.5	
Sheep	dig prot	%	*		13.7	
Calcium		%			1.90	
Phosphorus		%			0.24	
Potassium		%			1.91	

Clover, tomcat, aerial part, fresh, mature, (2)
Ref No 2 01 456 — United States

Feed Name or Analyses				Mean As Fed	Mean Dry	C.V. ± %
Dry matter		%			100.0	
Ash		%			8.7	
Crude fiber		%			32.4	
Protein (N x 6.25)		%			6.9	
Cattle	dig prot	%	*		3.8	
Goats	dig prot	%	*		3.0	
Horses	dig prot	%	*		3.4	
Rabbits	dig prot	%	*		4.0	
Sheep	dig prot	%	*		3.4	
Phosphorus		%			0.06	
Potassium		%			1.51	

CLOVER, WHITE. Trifolium repens

Clover, white, hay, fan air dried, (1)
Ref No 1 01 458 — United States

Feed Name or Analyses				Mean As Fed	Mean Dry	C.V. ± %
Dry matter		%			100.0	
Carotene		mg/kg			88.4	
Vitamin A equivalent		IU/g			147.4	

(1) dry forages and roughages (2) pasture, range plants, and forages fed green (3) silages (4) energy feeds (5) protein supplements (6) minerals (7) vitamins (8) additives

Clover, white, hay, s-c, (1)
Ref No 1 01 464 — United States

Feed Name or Analyses				Mean As Fed	Mean Dry	C.V. ± %
Dry matter		%		90.7	100.0	1
Ash		%		9.0	9.9	17
Crude fiber		%		22.0	24.2	4
Cattle	dig coef	%	*	56.	56.	
Sheep	dig coef	%	*	56.	56.	
Ether extract		%		2.4	2.6	29
Cattle	dig coef	%	*	58.	58.	
Sheep	dig coef	%	*	43.	43.	
N-free extract		%		40.3	44.4	
Cattle	dig coef	%	*	69.	69.	
Sheep	dig coef	%	*	64.	64.	
Protein (N x 6.25)		%		17.0	18.8	15
Cattle	dig coef	%	*	63.	63.	
Sheep	dig coef	%	*	62.	62.	
Cattle	dig prot	%		10.7	11.8	
Goats	dig prot	%	*	12.8	14.1	
Horses	dig prot	%	*	12.2	13.5	
Rabbits	dig prot	%	*	11.9	13.2	
Sheep	dig prot	%		10.6	11.6	
Fatty acids		%		2.5	2.7	
Energy		GE Mcal/kg				
Cattle		DE Mcal/kg	*	2.38	2.62	
Sheep		DE Mcal/kg	*	2.25	2.48	
Cattle		ME Mcal/kg	*	1.95	2.15	
Sheep		ME Mcal/kg	*	1.84	2.03	
Cattle		TDN %		54.0	59.5	
Sheep		TDN %		51.0	56.2	
Calcium		%		1.72	1.90	
Phosphorus		%		0.29	0.32	
Carotene		mg/kg		55.6	61.3	
α-tocopherol		mg/kg		116.2	128.1	
Vitamin A equivalent		IU/g		92.7	102.2	

Clover, white, hay, s-c, immature, (1)
Ref No 1 01 459 — United States

Feed Name or Analyses				Mean As Fed	Mean Dry	C.V. ± %
Dry matter		%		88.1	100.0	2
Ash		%		7.6	8.6	16
Crude fiber		%		17.4	19.8	15
Cattle	dig coef	%	*	56.	56.	
Sheep	dig coef	%	*	56.	56.	
Ether extract		%		3.3	3.8	16
Cattle	dig coef	%	*	58.	58.	
Sheep	dig coef	%	*	43.	43.	
N-free extract		%		35.4	40.2	
Cattle	dig coef	%	*	69.	69.	
Sheep	dig coef	%	*	64.	64.	
Protein (N x 6.25)		%		24.4	27.7	4
Cattle	dig coef	%	*	63.	63.	
Sheep	dig coef	%	*	62.	62.	
Cattle	dig prot	%		15.3	17.4	
Goats	dig prot	%	*	19.7	22.4	
Horses	dig prot	%	*	18.5	21.0	
Rabbits	dig prot	%	*	17.6	20.0	
Sheep	dig prot	%		15.1	17.1	
Energy		GE Mcal/kg				
Cattle		DE Mcal/kg	*	2.37	2.70	
Sheep		DE Mcal/kg	*	2.24	2.54	
Cattle		ME Mcal/kg	*	1.95	2.21	
Sheep		ME Mcal/kg	*	1.83	2.08	
Cattle		TDN %	*	53.9	61.1	
Sheep		TDN %	*	50.7	57.6	

Clover, white, hay, s-c, early bloom, (1)
Ref No 1 01 460 — United States

Feed Name or Analyses				Mean As Fed	Mean Dry	C.V. ± %
Dry matter		%			100.0	
Ash		%			13.2	
Protein (N x 6.25)		%			23.0	5
Cattle	dig prot	%	*		16.8	
Goats	dig prot	%	*		18.0	
Horses	dig prot	%	*		17.0	
Rabbits	dig prot	%	*		16.4	
Sheep	dig prot	%	*		17.2	

Clover, white, hay, s-c, full bloom, (1)
Ref No 1 01 461 — United States

Feed Name or Analyses				Mean As Fed	Mean Dry	C.V. ± %
Dry matter		%		92.4	100.0	1
Ash		%		13.2	14.3	10
Crude fiber		%		19.4	21.0	7
Ether extract		%		1.6	1.7	32
N-free extract		%		39.6	42.9	
Protein (N x 6.25)		%		18.6	20.1	5
Cattle	dig prot	%	*	13.3	14.3	
Goats	dig prot	%	*	14.1	15.3	
Horses	dig prot	%	*	13.5	14.6	
Rabbits	dig prot	%	*	13.1	14.2	
Sheep	dig prot	%	*	13.5	14.6	
Energy		GE Mcal/kg				
Cattle		DE Mcal/kg	*	2.37	2.57	
Sheep		DE Mcal/kg	*	2.45	2.65	
Cattle		ME Mcal/kg	*	1.95	2.11	
Sheep		ME Mcal/kg	*	2.01	2.18	
Cattle		TDN %	*	54.0	58.4	
Sheep		TDN %	*	55.6	60.2	

Clover, white, hay, s-c, late bloom, (1)
Ref No 1 01 462 — United States

Feed Name or Analyses				Mean As Fed	Mean Dry	C.V. ± %
Dry matter		%		87.6	100.0	
Ash		%		7.4	8.5	
Crude fiber		%		24.9	28.4	
Sheep	dig coef	%		61.	61.	
Ether extract		%		3.7	4.2	
Sheep	dig coef	%		51.	51.	
N-free extract		%		36.4	41.6	
Sheep	dig coef	%		70.	70.	
Protein (N x 6.25)		%		15.2	17.3	
Sheep	dig coef	%		73.	73.	
Cattle	dig prot	%	*	10.4	11.9	
Goats	dig prot	%	*	11.1	12.7	
Horses	dig prot	%	*	10.7	12.2	
Rabbits	dig prot	%	*	10.5	12.0	
Sheep	dig prot	%		11.1	12.6	
Energy		GE Mcal/kg				
Cattle		DE Mcal/kg	*	2.36	2.69	
Sheep		DE Mcal/kg	*	2.47	2.82	
Cattle		ME Mcal/kg	*	1.93	2.21	
Sheep		ME Mcal/kg	*	2.02	2.31	
Cattle		TDN %	*	53.5	61.0	
Sheep		TDN %		56.0	63.9	

Clover, white, hay, s-c, mature, (1)
Ref No 1 01 463 — United States

Feed Name or Analyses				Mean As Fed	Mean Dry	C.V. ± %
Dry matter		%		92.2	100.0	
Ash		%		4.5	4.9	

Feed Name or Analyses			Mean As Fed	Mean Dry	C.V. ± %
Crude fiber	%		33.3	36.1	
Cattle	dig coef %	*	56.	56.	
Sheep	dig coef %	*	56.	56.	
Ether extract	%		1.7	1.8	
Cattle	dig coef %	*	58.	58.	
Sheep	dig coef %	*	43.	43.	
N-free extract	%		39.7	43.1	
Cattle	dig coef %	*	69.	69.	
Sheep	dig coef %	*	64.	64.	
Protein (N x 6.25)	%		13.0	14.1	
Cattle	dig coef %	*	63.	63.	
Sheep	dig coef %	*	62.	62.	
Cattle	dig prot %		8.2	8.9	
Goats	dig prot %	*	9.0	9.7	
Horses	dig prot %	*	8.8	9.5	
Rabbits	dig prot %	*	8.8	9.6	
Sheep	dig prot %		8.1	8.7	
Energy	GE Mcal/kg				
Cattle	DE Mcal/kg	*	2.49	2.70	
Sheep	DE Mcal/kg	*	2.37	2.57	
Cattle	ME Mcal/kg	*	2.04	2.21	
Sheep	ME Mcal/kg	*	1.94	2.11	
Cattle	TDN %	*	56.4	61.2	
Sheep	TDN %	*	53.7	58.3	

Clover, white, aerial part, fresh, (2)

Ref No 2 01 468 United States

Feed Name or Analyses			Mean As Fed	Mean Dry	C.V. ± %
Dry matter	%		17.7	100.0	8
Ash	%		2.1	11.9	36
Crude fiber	%		2.8	15.7	19
Cattle	dig coef %	*	59.	59.	
Sheep	dig coef %	*	52.	52.	
Ether extract	%		0.6	3.3	24
Cattle	dig coef %	*	66.	66.	
Sheep	dig coef %	*	67.	67.	
N-free extract	%		7.2	40.9	
Cattle	dig coef %	*	80.	80.	
Sheep	dig coef %	*	71.	71.	
Protein (N x 6.25)	%		5.0	28.2	13
Cattle	dig coef %	*	70.	70.	
Sheep	dig coef %	*	57.	57.	
Cattle	dig prot %		3.5	19.8	
Goats	dig prot %	*	4.0	22.9	
Horses	dig prot %	*	3.8	21.5	
Rabbits	dig prot %	*	3.6	20.5	
Sheep	dig prot %		2.8	16.1	
Energy	GE Mcal/kg				
Cattle	DE Mcal/kg	*	0.52	2.94	
Sheep	DE Mcal/kg	*	0.45	2.57	
Cattle	ME Mcal/kg	*	0.43	2.41	
Sheep	ME Mcal/kg	*	0.37	2.10	
Cattle	TDN %		11.8	66.6	
Sheep	TDN %		10.3	58.2	
Calcium	%		0.25	1.40	
Chlorine	%		0.11	0.61	16
Iron	%		0.006	0.034	
Magnesium	%		0.08	0.45	
Manganese	mg/kg		54.3	307.2	
Phosphorus	%		0.09	0.51	
Potassium	%		0.38	2.13	
Sodium	%		0.07	0.39	15
Sulphur	%		0.06	0.33	15
Carotene	mg/kg		26.3	149.0	
Niacin	mg/kg		11.1	62.8	
Riboflavin	mg/kg		15.9	90.2	
Thiamine	mg/kg		2.5	14.1	

Feed Name or Analyses			Mean As Fed	Mean Dry	C.V. ± %
α-tocopherol	mg/kg		54.6	308.6	
Vitamin A equivalent	IU/g		43.9	248.4	

Clover, white, aerial part, fresh, immature, (2)

Ref No 2 01 465 United States

Feed Name or Analyses			Mean As Fed	Mean Dry	C.V. ± %
Dry matter	%			100.0	
Crude fiber	%				
Cattle	dig coef %	*		59.	
Sheep	dig coef %	*		52.	
Ether extract	%				
Cattle	dig coef %	*		66.	
Sheep	dig coef %	*		67.	
N-free extract	%				
Cattle	dig coef %	*		80.	
Sheep	dig coef %	*		71.	
Protein (N x 6.25)	%				
Cattle	dig coef %	*		70.	
Sheep	dig coef %	*		67.	

Clover, white, aerial part, fresh, full bloom, pure stand, (2)

Ref No 2 01 466 United States

Feed Name or Analyses			Mean As Fed	Mean Dry	C.V. ± %
Dry matter	%		17.9	100.0	9
Ash	%		2.2	12.5	11
Crude fiber	%		2.5	14.1	13
Ether extract	%		0.5	3.0	10
N-free extract	%		7.5	42.1	
Protein (N x 6.25)	%		5.1	28.3	11
Cattle	dig prot %	*	3.9	21.9	
Goats	dig prot %	*	4.1	23.0	
Horses	dig prot %	*	3.9	21.6	
Rabbits	dig prot %	*	3.7	20.5	
Sheep	dig prot %	*	4.2	23.4	
Energy	GE Mcal/kg				
Cattle	DE Mcal/kg	*	0.60	3.34	
Sheep	DE Mcal/kg	*	0.57	3.18	
Cattle	ME Mcal/kg	*	0.49	2.74	
Sheep	ME Mcal/kg	*	0.47	2.61	
Cattle	TDN %	*	13.5	75.7	
Sheep	TDN %	*	12.9	72.1	
Calcium	%		0.30	1.70	10
Magnesium	%		0.06	0.32	28
Phosphorus	%		0.08	0.44	8

Clover, white, aerial part, fresh, milk stage, (2)

Ref No 2 01 467 United States

Feed Name or Analyses			Mean As Fed	Mean Dry	C.V. ± %
Dry matter	%			100.0	
Ash	%			10.8	
Crude fiber	%			27.1	
Ether extract	%			2.2	
N-free extract	%			43.7	
Protein (N x 6.25)	%			16.2	
Cattle	dig prot %	*		11.7	
Goats	dig prot %	*		11.7	
Horses	dig prot %	*		11.3	
Rabbits	dig prot %	*		11.2	
Sheep	dig prot %	*		12.1	
Energy	GE Mcal/kg				
Cattle	DE Mcal/kg	*		3.01	
Sheep	DE Mcal/kg	*		2.73	
Cattle	ME Mcal/kg	*		2.47	
Sheep	ME Mcal/kg	*		2.24	
Cattle	TDN %	*		68.3	
Sheep	TDN %	*		61.8	

Feed Name or Analyses			Mean As Fed	Mean Dry	C.V. ± %
Calcium	%			1.05	
Magnesium	%			0.34	
Phosphorus	%			0.34	

Clover, white, aerial part, ensiled, immature, (3)

Ref No 3 01 469 United States

Feed Name or Analyses			Mean As Fed	Mean Dry	C.V. ± %
Dry matter	%		32.4	100.0	19
Ash	%		4.0	12.4	16
Crude fiber	%		6.6	20.3	18
Ether extract	%		1.6	5.0	32
N-free extract	%		13.2	40.8	
Protein (N x 6.25)	%		7.0	21.5	11
Cattle	dig prot %	*	5.1	15.8	
Goats	dig prot %	*	5.1	15.8	
Horses	dig prot %	*	5.1	15.8	
Sheep	dig prot %	*	5.1	15.8	
Energy	GE Mcal/kg				
Cattle	DE Mcal/kg	*	0.81	2.50	
Sheep	DE Mcal/kg	*	0.81	2.49	
Cattle	ME Mcal/kg	*	0.66	2.05	
Sheep	ME Mcal/kg	*	0.66	2.04	
Cattle	TDN %	*	18.4	56.8	
Sheep	TDN %	*	18.3	56.4	
Calcium	%		0.44	1.37	23
Magnesium	%		0.01	0.03	43
Phosphorus	%		0.12	0.36	24

Clover, alyce, fresh –

see Alyceclover, aerial part, fresh, (2)

CLOVER-BLUEGRASS. Trifolium spp, Poa spp

Clover-Bluegrass, aerial part, fresh, (2)

Ref No 2 01 470 United States

Feed Name or Analyses			Mean As Fed	Mean Dry	C.V. ± %
Dry matter	%			100.0	
Cobalt	mg/kg			0.060	

CLOVER, CRIMSON-RYEGRASS. Trifolium incarnatum, Lolium spp

Clover, crimson-Ryegrass, aerial part, fresh, (2)

Ref No 2 08 380 United States

Feed Name or Analyses			Mean As Fed	Mean Dry	C.V. ± %
Dry matter	%		18.3	100.0	
Ash	%		1.9	10.4	
Crude fiber	%		3.4	18.6	
Ether extract	%		1.1	6.0	
N-free extract	%		8.0	43.7	
Protein (N x 6.25)	%		3.9	21.3	
Cattle	dig prot %	*	2.9	16.0	
Goats	dig prot %	*	3.0	16.4	
Horses	dig prot %	*	2.9	15.6	
Rabbits	dig prot %	*	2.8	15.1	
Sheep	dig prot %	*	3.1	16.9	
Energy	GE Mcal/kg				
Cattle	DE Mcal/kg	*	0.54	2.97	
Sheep	DE Mcal/kg	*	0.57	3.12	
Cattle	ME Mcal/kg	*	0.45	2.44	
Sheep	ME Mcal/kg	*	0.47	2.56	
Cattle	TDN %	*	12.3	67.4	
Sheep	TDN %	*	13.0	70.9	
Calcium	%		0.12	0.66	
Phosphorus	%		0.12	0.66	

Feed Name or Analyses				Mean As Fed	Mean Dry	C.V. ± %

Clover, crimson-Ryegrass, aerial part, fresh, immature, (2)

Ref No 2 01 496 United States

Feed Name or Analyses				As Fed	Dry	C.V. ± %
Dry matter		%		18.3	100.0	0
Ash		%		1.9	10.4	
Crude fiber		%		3.4	18.6	
Ether extract		%		1.1	6.0	
N-free extract		%		8.0	43.7	
Protein (N x 6.25)		%		3.9	21.3	
Cattle	dig prot	%	*	2.9	16.0	
Goats	dig prot	%	*	3.0	16.4	
Horses	dig prot	%	*	2.9	15.6	
Rabbits	dig prot	%	*	2.8	15.1	
Sheep	dig prot	%	*	3.1	16.8	
Energy	GE	Mcal/kg				
Cattle	DE	Mcal/kg	*	0.54	2.97	
Sheep	DE	Mcal/kg	*	0.57	3.12	
Cattle	ME	Mcal/kg	*	0.45	2.44	
Sheep	ME	Mcal/kg	*	0.47	2.56	
Cattle	TDN	%	*	12.3	67.4	
Sheep	TDN	%	*	13.0	70.9	
Cobalt		mg/kg		0.015	0.084	25
Carotene		mg/kg		75.0	409.6	
Vitamin A equivalent		IU/g		125.0	682.8	

CLOVER, CRIMSON-RYEGRASS, ITALIAN. Trifolium incarnatum, Lolium multiflorum

Clover, crimson-Ryegrass, Italian, aerial part, fresh, (2)

Ref No 2 01 500 United States

Feed Name or Analyses				As Fed	Dry	C.V. ± %
Dry matter		%		18.2	100.0	22
Ash		%		1.9	10.4	
Crude fiber		%		3.4	18.7	
Ether extract		%		1.1	6.0	
N-free extract		%		8.1	44.4	
Protein (N x 6.25)		%		3.7	20.5	
Cattle	dig prot	%	*	2.8	15.3	
Goats	dig prot	%	*	2.9	15.7	
Horses	dig prot	%	*	2.7	14.9	
Rabbits	dig prot	%	*	2.6	14.5	
Sheep	dig prot	%	*	2.9	16.1	
Energy	GE	Mcal/kg				
Cattle	DE	Mcal/kg	*	0.53	2.93	
Sheep	DE	Mcal/kg	*	0.57	3.13	
Cattle	ME	Mcal/kg	*	0.44	2.40	
Sheep	ME	Mcal/kg	*	0.47	2.57	
Cattle	TDN	%	*	12.1	66.5	
Sheep	TDN	%	*	12.9	71.0	
Calcium		%		0.12	0.65	24
Magnesium		%		0.04	0.23	19
Phosphorus		%		0.07	0.37	25
Potassium		%		0.34	1.88	38

Clover, crimson-Ryegrass, Italian, aerial part, fresh, immature, (2)

Ref No 2 01 497 United States

Feed Name or Analyses				As Fed	Dry	C.V. ± %
Dry matter		%			100.0	
Calcium		%			0.88	14
Magnesium		%			0.23	23
Phosphorus		%			0.52	25
Potassium		%			1.91	46

Clover, crimson-Ryegrass, Italian, aerial part, fresh, cut 2, (2)

Ref No 2 01 498 United States

Feed Name or Analyses				As Fed	Dry	C.V. ± %
Dry matter		%			100.0	
Calcium		%			0.91	
Magnesium		%			0.26	
Phosphorus		%			0.29	
Potassium		%			1.16	

Clover, crimson-Ryegrass, Italian, aerial part, fresh, cut 4, (2)

Ref No 2 01 499 United States

Feed Name or Analyses				As Fed	Dry	C.V. ± %
Dry matter		%			100.0	
Calcium		%			0.80	
Magnesium		%			0.23	
Phosphorus		%			0.29	
Potassium		%			1.83	

CLOVER-GRASS. Trifolium spp, Scientific name not used

Clover-Grass, hay, fan air dried, (1)

Ref No 1 01 472 United States

Feed Name or Analyses				As Fed	Dry	C.V. ± %
Dry matter		%		90.0	100.0	
Carotene		mg/kg		23.0	25.6	24
Vitamin A equivalent		IU/g		38.4	42.6	

Clover-Grass, hay, fan air dried, cut 2, (1)

Ref No 1 01 471 United States

Feed Name or Analyses				As Fed	Dry	C.V. ± %
Dry matter		%		90.0	100.0	
Carotene		mg/kg		29.8	33.1	13
Vitamin A equivalent		IU/g		49.6	55.1	

Clover-Grass, hay, s-c, (1)

Ref No 1 01 473 United States

Feed Name or Analyses				As Fed	Dry	C.V. ± %
Dry matter		%		88.7	100.0	1
Ash		%		6.3	7.1	3
Crude fiber		%		29.3	33.0	2
Ether extract		%		2.4	2.7	13
N-free extract		%		40.2	45.3	
Protein (N x 6.25)		%		10.5	11.8	5
Cattle	dig prot	%	*	6.4	7.2	
Goats	dig prot	%	*	6.7	7.6	
Horses	dig prot	%	*	6.7	7.6	
Rabbits	dig prot	%	*	6.9	7.8	
Sheep	dig prot	%	*	6.3	7.2	
Energy	GE	Mcal/kg				
Cattle	DE	Mcal/kg	*	2.24	2.53	
Sheep	DE	Mcal/kg	*	2.23	2.52	
Cattle	ME	Mcal/kg	*	1.84	2.07	
Sheep	ME	Mcal/kg	*	1.83	2.06	
Cattle	TDN	%	*	50.9	57.4	
Sheep	TDN	%	*	50.6	57.0	
Calcium		%		0.87	0.99	6
Chlorine		%		0.63	0.71	
Copper		mg/kg		7.0	7.9	14
Iron		%		0.022	0.025	17
Magnesium		%		0.25	0.28	21
Manganese		mg/kg		91.8	103.5	10
Phosphorus		%		0.20	0.23	14
Potassium		%		1.43	1.62	2
Sodium		%		0.17	0.19	
Sulphur		%		0.13	0.14	
Carotene		mg/kg		14.5	16.3	46
Vitamin A equivalent		IU/g		24.1	27.2	

Clover-Grass, hay, s-c, gr 3 US, (1)

Ref No 1 01 474 United States

Feed Name or Analyses				As Fed	Dry	C.V. ± %
Dry matter		%			100.0	
Carotene		mg/kg			9.5	56
Vitamin A equivalent		IU/g			15.8	

Clover-Grass, aerial part, fresh, (2)

Ref No 2 01 476 United States

Feed Name or Analyses				As Fed	Dry	C.V. ± %
Dry matter		%		23.8	100.0	
Ash		%		1.7	7.3	
Crude fiber		%		6.4	26.9	
Cattle	dig coef	%		70.	70.	
Ether extract		%		0.8	3.5	
Cattle	dig coef	%		47.	47.	
N-free extract		%		11.2	47.1	
Cattle	dig coef	%		69.	69.	
Protein (N x 6.25)		%		3.6	15.2	
Cattle	dig coef	%		63.	63.	
Cattle	dig prot	%		2.3	9.6	
Goats	dig prot	%	*	2.5	10.7	
Horses	dig prot	%	*	2.5	10.4	
Rabbits	dig prot	%	*	2.5	10.4	
Sheep	dig prot	%	*	2.6	11.1	
Energy	GE	Mcal/kg				
Cattle	DE	Mcal/kg	*	0.68	2.85	
Sheep	DE	Mcal/kg	*	0.67	2.80	
Cattle	ME	Mcal/kg	*	0.56	2.34	
Sheep	ME	Mcal/kg	*	0.55	2.29	
Cattle	TDN	%		15.4	64.7	
Sheep	TDN	%	*	15.1	63.4	
Calcium		%		0.21	0.87	
Magnesium		%		0.20	0.85	
Phosphorus		%		0.08	0.32	
Potassium		%		0.84	3.55	
Carotene		mg/kg		36.9	155.0	
Vitamin A equivalent		IU/g		61.5	258.4	

Clover-Grass, aerial part, fresh, immature, (2)

Ref No 2 01 475 United States

Feed Name or Analyses				As Fed	Dry	C.V. ± %
Dry matter		%		23.8	100.0	7
Ash		%		1.6	6.7	11
Crude fiber		%		5.9	24.8	13
Ether extract		%		0.9	3.6	9
N-free extract		%		11.5	48.3	
Protein (N x 6.25)		%		4.0	16.6	17
Cattle	dig prot	%	*	2.9	12.0	
Goats	dig prot	%	*	2.9	12.0	
Horses	dig prot	%	*	2.8	11.6	
Rabbits	dig prot	%	*	2.7	11.5	
Sheep	dig prot	%	*	3.0	12.5	
Energy	GE	Mcal/kg				
Cattle	DE	Mcal/kg	*	0.76	3.19	
Sheep	DE	Mcal/kg	*	0.73	3.06	
Cattle	ME	Mcal/kg	*	0.62	2.62	
Sheep	ME	Mcal/kg	*	0.60	2.51	

(1) dry forages and roughages (2) pasture, range plants, and forages fed green (3) silages (4) energy feeds (5) protein supplements (6) minerals (7) vitamins (8) additives

Feed Name or Analyses		Mean As Fed	Mean Dry	C.V. ± %
Cattle	TDN % *	17.2	72.3	
Sheep	TDN % *	16.5	69.5	
Calcium	%	0.27	1.15	
Phosphorus	%	0.08	0.35	
Potassium	%	0.84	3.55	

Clover-Grass, aerial part, ensiled, (3)
Ref No 3 01 477 United States

Feed Name or Analyses		Mean As Fed	Mean Dry	C.V. ± %
Dry matter	%	23.4	100.0	12
Ash	%	2.1	8.9	10
Crude fiber	%	7.7	32.8	15
Ether extract	%	1.1	4.6	17
N-free extract	%	9.7	41.5	
Protein (N x 6.25)	%	2.9	12.2	21
Cattle	dig prot % *	1.7	7.3	
Goats	dig prot % *	1.7	7.3	
Horses	dig prot % *	1.7	7.3	
Sheep	dig prot % *	1.7	7.3	
Energy	GE Mcal/kg			
Cattle	DE Mcal/kg *	0.62	2.66	
Sheep	DE Mcal/kg *	0.64	2.72	
Cattle	ME Mcal/kg *	0.51	2.18	
Sheep	ME Mcal/kg *	0.52	2.23	
Cattle	TDN % *	14.1	60.3	
Sheep	TDN % *	14.5	61.8	
Calcium	%	0.27	1.15	
Cobalt	mg/kg	0.053	0.225	
Copper	mg/kg	2.6	11.2	
Iron	%	0.025	0.106	
Magnesium	%	0.10	0.42	
Manganese	mg/kg	11.4	48.7	
Phosphorus	%	0.06	0.25	
Potassium	%	0.37	1.60	
Carotene	mg/kg	45.2	193.1	18
Vitamin A equivalent	IU/g	75.3	321.9	

CLOVER, LADINO-BROME. Trifolium repens, Bromus spp

Clover, ladino-Brome, aerial part, fresh, (2)
Ref No 2 01 501 United States

Feed Name or Analyses		Mean As Fed	Mean Dry	C.V. ± %
Dry matter	%		100.0	
Calcium	%		1.04	15
Magnesium	%		0.37	14
Potassium	%		3.51	9
Sodium	%		0.10	32

CLOVER, LADINO-CANARYGRASS, REED. Trifolium repens, Phalaris arundinacea

Clover, ladino-Canarygrass, reed, aerial part, fresh, (2)
Ref No 2 01 502 United States

Feed Name or Analyses		Mean As Fed	Mean Dry	C.V. ± %
Dry matter	%		100.0	
Calcium	%		0.78	
Magnesium	%		0.39	
Potassium	%		3.28	
Sodium	%		0.08	

CLOVER, LADINO-FESCUE, MEADOW. Trifolium repens, Festuca elatior

Clover, ladino-Fescue, meadow, hay, s-c, immature, (1)
Ref No 1 01 503 United States

Feed Name or Analyses		Mean As Fed	Mean Dry	C.V. ± %
Dry matter	%	94.7	100.0	2
Ash	%	8.4	8.9	7
Crude fiber	%	20.3	21.4	13
Ether extract	%	2.4	2.5	8
N-free extract	%	41.2	43.5	
Protein (N x 6.25)	%	22.4	23.7	6
Cattle	dig prot % *	16.5	17.5	
Goats	dig prot % *	17.7	18.7	
Horses	dig prot % *	16.7	17.7	
Rabbits	dig prot % *	16.1	17.0	
Sheep	dig prot % *	16.9	17.8	
Energy	GE Mcal/kg			
Cattle	DE Mcal/kg *	2.64	2.79	
Sheep	DE Mcal/kg *	2.69	2.84	
Cattle	ME Mcal/kg *	2.17	2.29	
Sheep	ME Mcal/kg *	2.20	2.33	
Cattle	TDN % *	59.9	63.3	
Sheep	TDN % *	60.9	64.4	

Clover, ladino-Fescue, meadow, aerial part, fresh, (2)
Ref No 2 08 382 United States

Feed Name or Analyses		Mean As Fed	Mean Dry	C.V. ± %
Dry matter	%	25.4	100.0	
Ash	%	2.6	10.2	
Crude fiber	%	5.3	20.9	
Ether extract	%	1.4	5.5	
N-free extract	%	11.1	43.7	
Protein (N x 6.25)	%	5.0	19.7	
Cattle	dig prot % *	3.7	14.6	
Goats	dig prot % *	3.8	14.9	
Horses	dig prot % *	3.6	14.2	
Rabbits	dig prot % *	3.5	13.9	
Sheep	dig prot % *	3.9	15.3	
Energy	GE Mcal/kg			
Cattle	DE Mcal/kg *	0.75	2.95	
Sheep	DE Mcal/kg *	0.77	3.03	
Cattle	ME Mcal/kg *	0.61	2.42	
Sheep	ME Mcal/kg *	0.63	2.49	
Cattle	TDN % *	17.0	66.9	
Sheep	TDN % *	17.5	68.8	
Calcium	%	0.19	0.75	
Phosphorus	%	0.09	0.35	

Clover, ladino-Fescue, meadow, aerial part, fresh, immature, (2)
Ref No 2 01 504 United States

Feed Name or Analyses		Mean As Fed	Mean Dry	C.V. ± %
Dry matter	%	25.0	100.0	11
Ash	%	2.6	10.4	10
Crude fiber	%	5.0	20.1	18
Ether extract	%	1.4	5.6	11
N-free extract	%	11.0	43.8	
Protein (N x 6.25)	%	5.0	20.1	14
Cattle	dig prot % *	3.7	15.0	
Goats	dig prot % *	3.8	15.3	
Horses	dig prot % *	3.6	14.6	
Rabbits	dig prot % *	3.5	14.2	
Sheep	dig prot % *	3.9	15.7	
Energy	GE Mcal/kg			
Cattle	DE Mcal/kg *	0.74	2.95	
Sheep	DE Mcal/kg *	0.77	3.07	
Cattle	ME Mcal/kg *	0.60	2.42	
Sheep	ME Mcal/kg *	0.63	2.52	
Cattle	TDN % *	16.7	66.9	
Sheep	TDN % *	17.4	69.6	
Carotene	mg/kg	64.4	257.7	
Vitamin A equivalent	IU/g	107.4	429.6	

CLOVER, LADINO-GRASS. Trifolium repens, Scientific name not used

Clover, ladino-Grass, hay, dehy, (1)
Ref No 1 01 507 United States

Feed Name or Analyses		Mean As Fed	Mean Dry	C.V. ± %
Dry matter	%	87.8	100.0	
Ash	%	7.2	8.2	
Crude fiber	%	20.1	22.9	
Ether extract	%	2.8	3.2	
N-free extract	%	44.2	50.3	
Protein (N x 6.25)	%	13.5	15.4	
Cattle	dig prot % *	9.0	10.3	
Goats	dig prot % *	9.6	10.9	
Horses	dig prot % *	9.3	10.6	
Rabbits	dig prot % *	9.3	10.6	
Sheep	dig prot % *	9.1	10.4	
Energy	GE Mcal/kg			
Cattle	DE Mcal/kg *	2.45	2.79	
Sheep	DE Mcal/kg *	2.41	2.74	
Cattle	ME Mcal/kg *	2.01	2.29	
Sheep	ME Mcal/kg *	1.97	2.25	
Cattle	TDN % *	55.6	63.4	
Sheep	TDN % *	54.6	62.2	

Clover, ladino-Grass, hay, fan air dried, (1)
Ref No 1 01 506 United States

Feed Name or Analyses		Mean As Fed	Mean Dry	C.V. ± %
Dry matter	%	92.4	100.0	
Ash	%	7.8	8.4	
Crude fiber	%	24.8	26.8	
Ether extract	%	2.3	2.5	
N-free extract	%	39.8	43.1	
Protein (N x 6.25)	%	17.7	19.2	
Cattle	dig prot % *	12.5	13.6	
Goats	dig prot % *	13.4	14.5	
Horses	dig prot % *	12.8	13.8	
Rabbits	dig prot % *	12.5	13.5	
Sheep	dig prot % *	12.7	13.8	
Energy	GE Mcal/kg			
Cattle	DE Mcal/kg *	2.45	2.65	
Sheep	DE Mcal/kg *	2.47	2.67	
Cattle	ME Mcal/kg *	2.01	2.17	
Sheep	ME Mcal/kg *	2.02	2.19	
Cattle	TDN % *	55.5	60.1	
Sheep	TDN % *	55.9	60.5	

Clover, ladino-Grass, hay, fan air dried, immature, (1)
Ref No 1 01 505 United States

Feed Name or Analyses		Mean As Fed	Mean Dry	C.V. ± %
Dry matter	%	88.3	100.0	
Carotene	mg/kg	179.1	202.8	
Vitamin A equivalent	IU/g	298.5	338.1	

Feed Name or Analyses			Mean As Fed	Mean Dry	C.V. ± %

Clover, ladino-Grass, hay, s-c, (1)

Ref No 1 01 511 United States

Feed Name or Analyses			As Fed	Dry	C.V. ± %
Dry matter	%		89.2	100.0	1
Ash	%		7.3	8.2	14
Crude fiber	%		19.9	22.3	24
Ether extract	%		2.2	2.5	27
N-free extract	%		43.2	48.5	
Protein (N x 6.25)	%		16.5	18.5	15
Cattle	dig prot %	*	11.6	13.0	
Goats	dig prot %	*	12.4	13.9	
Horses	dig prot %	*	11.8	13.3	
Rabbits	dig prot %	*	11.6	13.0	
Sheep	dig prot %	*	11.8	13.2	
Energy	GE Mcal/kg				
Cattle	DE Mcal/kg	*	2.46	2.76	
Sheep	DE Mcal/kg	*	2.50	2.81	
Cattle	ME Mcal/kg	*	2.02	2.26	
Sheep	ME Mcal/kg	*	2.05	2.30	
Cattle	TDN %	*	55.9	62.6	
Sheep	TDN %	*	56.8	63.6	
Calcium	%		0.95	1.07	31
Copper	mg/kg		14.4	16.1	
Iron	%		0.015	0.017	37
Magnesium	%		0.30	0.34	22
Manganese	mg/kg		39.7	44.5	
Phosphorus	%		0.22	0.24	18
Potassium	%		1.79	2.00	16
Carotene	mg/kg		56.8	63.7	98
Riboflavin	mg/kg		8.3	9.3	
Thiamine	mg/kg		13.2	14.8	
Vitamin A equivalent	IU/g		94.7	106.2	

Clover, ladino-Grass, hay, s-c, immature, (1)

Ref No 1 01 508 United States

Feed Name or Analyses			As Fed	Dry	C.V. ± %
Dry matter	%		89.9	100.0	
Ash	%		7.6	8.5	
Crude fiber	%		13.9	15.5	
Ether extract	%		2.2	2.4	
N-free extract	%		45.9	51.1	
Protein (N x 6.25)	%		20.2	22.5	
Cattle	dig prot %	*	14.8	16.4	
Goats	dig prot %	*	15.8	17.6	
Horses	dig prot %	*	15.0	16.6	
Rabbits	dig prot %	*	14.4	16.0	
Sheep	dig prot %	*	15.1	16.8	
Energy	GE Mcal/kg				
Cattle	DE Mcal/kg	*	2.64	2.93	
Sheep	DE Mcal/kg	*	2.72	3.02	
Cattle	ME Mcal/kg	*	2.16	2.41	
Sheep	ME Mcal/kg	*	2.23	2.48	
Cattle	TDN %	*	59.8	66.5	
Sheep	TDN %	*	61.7	68.6	

Clover, ladino-Grass, hay, s-c, mature, (1)

Ref No 1 01 509 United States

Feed Name or Analyses			As Fed	Dry	C.V. ± %
Dry matter	%		91.5	100.0	
Ash	%		5.7	6.2	
Crude fiber	%		31.5	34.4	
Ether extract	%		1.9	2.1	
N-free extract	%		42.1	46.0	
Protein (N x 6.25)	%		10.3	11.3	
Cattle	dig prot %	*	6.2	6.7	
Goats	dig prot %	*	6.5	7.1	
Horses	dig prot %	*	6.5	7.1	
Rabbits	dig prot %	*	6.8	7.4	
Sheep	dig prot %	*	6.1	6.7	
Energy	GE Mcal/kg				
Cattle	DE Mcal/kg	*	2.17	2.37	
Sheep	DE Mcal/kg	*	2.28	2.49	
Cattle	ME Mcal/kg	*	1.78	1.95	
Sheep	ME Mcal/kg	*	1.87	2.04	
Cattle	TDN %	*	49.2	53.8	
Sheep	TDN %	*	51.6	56.4	
Calcium	%		0.69	0.75	
Iron	%		0.013	0.014	
Magnesium	%		0.22	0.24	
Phosphorus	%		0.20	0.22	
Potassium	%		1.60	1.75	
Carotene	mg/kg		4.4	4.9	
Vitamin A equivalent	IU/g		7.4	8.1	

Clover, ladino-Grass, hay, s-c, cut 1, (1)

Ref No 1 01 510 United States

Feed Name or Analyses			As Fed	Dry	C.V. ± %
Dry matter	%		90.8	100.0	3
Calcium	%		0.99	1.09	53
Cobalt	mg/kg		0.124	0.137	
Copper	mg/kg		11.2	12.3	
Iron	%		0.015	0.017	45
Magnesium	%		0.31	0.35	37
Manganese	mg/kg		40.6	44.8	
Phosphorus	%		0.20	0.22	35
Potassium	%		1.66	1.83	29
Carotene	mg/kg		16.2	17.9	98
Riboflavin	mg/kg		8.4	9.3	
Thiamine	mg/kg		13.4	14.8	
Vitamin A equivalent	IU/g		27.0	29.8	

Clover, ladino-Grass, aerial part, fresh, (2)

Ref No 2 08 383 United States

Feed Name or Analyses			As Fed	Dry	C.V. ± %
Dry matter	%		20.0	100.0	
Ash	%		2.4	12.0	
Crude fiber	%		4.5	22.5	
Ether extract	%		0.5	2.5	
N-free extract	%		9.1	45.5	
Protein (N x 6.25)	%		3.5	17.5	
Cattle	dig prot %	*	2.6	12.8	
Goats	dig prot %	*	2.6	12.9	
Horses	dig prot %	*	2.5	12.4	
Rabbits	dig prot %	*	2.4	12.2	
Sheep	dig prot %	*	2.7	13.3	
Energy	GE Mcal/kg				
Cattle	DE Mcal/kg	*	0.59	2.94	
Sheep	DE Mcal/kg	*	0.59	2.96	
Cattle	ME Mcal/kg	*	0.48	2.41	
Sheep	ME Mcal/kg	*	0.48	2.42	
Cattle	TDN %	*	13.3	66.7	
Sheep	TDN %	*	13.4	67.0	
Calcium	%		0.30	1.50	
Phosphorus	%		0.08	0.40	

Clover, ladino-Grass, aerial part, fresh, early bloom, (2)

Ref No 2 01 513 United States

Feed Name or Analyses			As Fed	Dry	C.V. ± %
Dry matter	%		20.0	100.0	10
Ash	%		2.2	11.2	7
Crude fiber	%		4.8	23.9	7
Ether extract	%		0.6	2.8	15
N-free extract	%		9.4	46.8	
Protein (N x 6.25)	%		3.1	15.3	12
Cattle	dig prot %	*	2.2	10.9	
Goats	dig prot %	*	2.2	10.8	
Horses	dig prot %	*	2.1	10.5	
Rabbits	dig prot %	*	2.1	10.5	
Sheep	dig prot %	*	2.2	11.2	
Energy	GE Mcal/kg				
Cattle	DE Mcal/kg	*	0.58	2.88	
Sheep	DE Mcal/kg	*	0.56	2.82	
Cattle	ME Mcal/kg	*	0.47	2.36	
Sheep	ME Mcal/kg	*	0.46	2.31	
Cattle	TDN %	*	13.0	65.2	
Sheep	TDN %	*	12.8	63.8	

Clover, ladino-Grass, aerial part, ensiled, (3)

Ref No 3 08 384 United States

Feed Name or Analyses			As Fed	Dry	C.V. ± %
Dry matter	%		29.9	100.0	
Ash	%		2.6	8.7	
Crude fiber	%		7.5	25.1	
Ether extract	%		1.5	5.0	
N-free extract	%		12.9	43.1	
Protein (N x 6.25)	%		5.4	18.1	
Cattle	dig prot %	*	3.8	12.6	
Goats	dig prot %	*	3.8	12.6	
Horses	dig prot %	*	3.8	12.6	
Sheep	dig prot %	*	3.8	12.6	
Energy	GE Mcal/kg				
Cattle	DE Mcal/kg	*	0.81	2.71	
Sheep	DE Mcal/kg	*	0.86	2.88	
Cattle	ME Mcal/kg	*	0.66	2.22	
Sheep	ME Mcal/kg	*	0.71	2.36	
Cattle	TDN %	*	18.4	61.4	
Sheep	TDN %	*	19.5	65.3	
Calcium	%		0.31	1.04	
Magnesium	%		0.09	0.30	
Phosphorus	%		0.07	0.23	

Clover, ladino-Grass, aerial part, ensiled, immature, (3)

Ref No 3 01 514 United States

Feed Name or Analyses			As Fed	Dry	C.V. ± %
Dry matter	%		27.6	100.0	9
Ash	%		2.4	8.7	13
Crude fiber	%		6.9	25.1	9
Ether extract	%		1.4	5.0	13
N-free extract	%		11.9	43.1	
Protein (N x 6.25)	%		5.0	18.1	8
Cattle	dig prot %	*	3.5	12.7	
Goats	dig prot %	*	3.5	12.7	
Horses	dig prot %	*	3.5	12.7	
Sheep	dig prot %	*	3.5	12.7	
Energy	GE Mcal/kg				
Cattle	DE Mcal/kg	*	0.75	2.71	
Sheep	DE Mcal/kg	*	0.67	2.44	
Cattle	ME Mcal/kg	*	0.61	2.22	
Sheep	ME Mcal/kg	*	0.55	2.00	
Cattle	TDN %	*	16.9	61.4	
Sheep	TDN %	*	15.3	55.3	

(1) dry forages and roughages
(2) pasture, range plants, and forages fed green
(3) silages
(4) energy feeds
(5) protein supplements
(6) minerals
(7) vitamins
(8) additives

Feed Name or Analyses			Mean As Fed	Mean Dry	C.V. ± %
Carotene	mg/kg		31.8	115.1	
Vitamin A equivalent	IU/g		52.9	191.8	

CLOVER, LADINO-RYEGRASS, PERENNIAL. Trifolium repens, Lolium perenne

Clover, ladino-Ryegrass, perennial, aerial part, fresh, (2)

Ref No 2 01 515 United States

Feed Name or Analyses			Mean As Fed	Mean Dry	C.V. ± %
Dry matter	%			100.0	
Calcium	%			1.01	
Magnesium	%			0.45	
Potassium	%			3.47	
Sodium	%			0.13	

CLOVER, LADINO-TIMOTHY. Trifolium repens, Phleum pratense

Clover, ladino-Timothy, hay, dehy, immature, (1)

Ref No 1 01 517 United States

Feed Name or Analyses			Mean As Fed	Mean Dry	C.V. ± %
Dry matter	%		90.0	100.0	
Carotene	mg/kg		145.0	161.2	
Vitamin A equivalent	IU/g		241.8	268.6	

Clover, ladino-Timothy, hay, fan air dried, (1)

Ref No 1 01 516 United States

Feed Name or Analyses			Mean As Fed	Mean Dry	C.V. ± %
Dry matter	%		90.0	100.0	
Carotene	mg/kg		16.7	18.5	
Vitamin A equivalent	IU/g		27.8	30.9	

Clover, ladino-Timothy, hay, s-c, (1)

Ref No 1 01 519 United States

Feed Name or Analyses			Mean As Fed	Mean Dry	C.V. ± %
Dry matter	%		90.3	100.0	2
Carotene	mg/kg		25.7	28.4	98
Vitamin A equivalent	IU/g		42.8	47.4	

Clover, ladino-Timothy, hay, s-c, cut 2, (1)

Ref No 1 01 518 United States

Feed Name or Analyses			Mean As Fed	Mean Dry	C.V. ± %
Dry matter	%		90.6	100.0	
Carotene	mg/kg		19.8	21.8	
Vitamin A equivalent	IU/g		33.0	36.4	

Clover, ladino-Timothy, aerial part, fresh, cut 1, (2)

Ref No 2 01 520 United States

Feed Name or Analyses			Mean As Fed	Mean Dry	C.V. ± %
Dry matter	%		21.9	100.0	
Ash	%		2.3	10.5	
Crude fiber	%		5.7	26.0	
Ether extract	%		0.4	1.8	
N-free extract	%		10.1	46.2	
Protein (N x 6.25)	%		3.4	15.5	
Cattle	dig prot %	*	2.4	11.1	
Goats	dig prot %	*	2.4	11.0	
Horses	dig prot %	*	2.3	10.7	
Rabbits	dig prot %	*	2.3	10.6	
Sheep	dig prot %	*	2.5	11.4	
Energy	GE Mcal/kg				
Cattle	DE Mcal/kg	*	0.66	3.01	
Sheep	DE Mcal/kg	*	0.61	2.78	
Cattle	ME Mcal/kg	*	0.54	2.47	
Sheep	ME Mcal/kg	*	0.50	2.28	
Cattle	TDN %	*	15.0	68.4	
Sheep	TDN %	*	13.8	63.1	
Calcium	%		0.26	1.19	
Magnesium	%		0.07	0.32	
Phosphorus	%		0.06	0.27	
Potassium	%		0.61	2.79	

Clover, ladino-Timothy, aerial part, fresh, cut 2, (2)

Ref No 2 01 521 United States

Feed Name or Analyses			Mean As Fed	Mean Dry	C.V. ± %
Dry matter	%		37.6	100.0	
Ash	%		3.2	8.5	
Crude fiber	%		7.0	18.6	
Ether extract	%		1.5	4.0	
N-free extract	%		18.1	48.2	
Protein (N x 6.25)	%		7.8	20.7	
Cattle	dig prot %	*	5.8	15.5	
Goats	dig prot %	*	6.0	15.9	
Horses	dig prot %	*	5.7	15.1	
Rabbits	dig prot %	*	5.5	14.7	
Sheep	dig prot %	*	6.1	16.3	
Energy	GE Mcal/kg				
Cattle	DE Mcal/kg	*	1.20	3.20	
Sheep	DE Mcal/kg	*	1.13	3.00	
Cattle	ME Mcal/kg	*	0.99	2.62	
Sheep	ME Mcal/kg	*	0.92	2.46	
Cattle	TDN %	*	27.2	72.5	
Sheep	TDN %	*	25.6	68.0	
Carotene	mg/kg		86.0	228.6	
Vitamin A equivalent	IU/g		143.3	381.1	

Clover, ladino-Timothy, aerial part, ensiled, (3)

Ref No 3 01 526 United States

Feed Name or Analyses			Mean As Fed	Mean Dry	C.V. ± %
Dry matter	%		29.4	100.0	16
Ash	%		2.8	9.5	17
Crude fiber	%		7.0	23.8	18
Ether extract	%		1.4	4.9	26
N-free extract	%		12.3	42.0	
Protein (N x 6.25)	%		5.8	19.8	27
Cattle	dig prot %	*	4.2	14.2	
Goats	dig prot %	*	4.2	14.2	
Horses	dig prot %	*	4.2	14.2	
Sheep	dig prot %	*	4.2	14.2	
Energy	GE Mcal/kg				
Cattle	DE Mcal/kg	*	0.80	2.72	
Sheep	DE Mcal/kg	*	0.72	2.44	
Cattle	ME Mcal/kg	*	0.65	2.23	
Sheep	ME Mcal/kg	*	0.59	2.00	
Cattle	TDN %	*	18.1	61.6	
Sheep	TDN %	*	16.3	55.3	
Calcium	%		0.36	1.21	11
Magnesium	%		0.09	0.30	21
Phosphorus	%		0.06	0.20	32
Potassium	%		0.65	2.22	9
Carotene	mg/kg		42.3	144.0	31
Vitamin A equivalent	IU/g		70.6	240.0	

Clover, ladino-Timothy, aerial part, ensiled, cut 1, (3)

Ref No 3 01 522 United States

Feed Name or Analyses			Mean As Fed	Mean Dry	C.V. ± %
Dry matter	%		26.2	100.0	
Ash	%		2.0	7.7	
Crude fiber	%		7.9	30.0	
Ether extract	%		1.2	4.5	
N-free extract	%		12.1	46.0	
Protein (N x 6.25)	%		3.1	11.8	
Cattle	dig prot %	*	1.8	6.9	
Goats	dig prot %	*	1.8	6.9	
Horses	dig prot %	*	1.8	6.9	
Sheep	dig prot %	*	1.8	6.9	
Energy	GE Mcal/kg				
Cattle	DE Mcal/kg	*	0.71	2.73	
Sheep	DE Mcal/kg	*	0.73	2.79	
Cattle	ME Mcal/kg	*	0.59	2.23	
Sheep	ME Mcal/kg	*	0.60	2.29	
Cattle	TDN %	*	16.2	61.8	
Sheep	TDN %	*	16.6	63.2	

Clover, ladino-Timothy, aerial part, ensiled, cut 2, (3)

Ref No 3 01 523 United States

Feed Name or Analyses			Mean As Fed	Mean Dry	C.V. ± %
Dry matter	%		30.6	100.0	
Ash	%		3.1	10.2	
Crude fiber	%		6.8	22.3	
Ether extract	%		1.6	5.1	
N-free extract	%		12.3	40.1	
Protein (N x 6.25)	%		6.8	22.3	
Cattle	dig prot %	*	5.0	16.5	
Goats	dig prot %	*	5.0	16.5	
Horses	dig prot %	*	5.0	16.5	
Sheep	dig prot %	*	5.0	16.5	
Energy	GE Mcal/kg				
Cattle	DE Mcal/kg	*	0.83	2.71	
Sheep	DE Mcal/kg	*	0.75	2.44	
Cattle	ME Mcal/kg	*	0.68	2.22	
Sheep	ME Mcal/kg	*	0.61	2.00	
Cattle	TDN %	*	18.8	61.4	
Sheep	TDN %	*	17.0	55.4	

Clover, ladino-Timothy, aerial part w phosphoric acid preservative added, ensiled, (3)

Ref No 3 01 524 United States

Feed Name or Analyses			Mean As Fed	Mean Dry	C.V. ± %
Dry matter	%		32.5	100.0	
Ash	%		3.3	10.2	
Crude fiber	%		7.8	24.0	
Ether extract	%		2.2	6.8	
N-free extract	%		12.8	39.3	
Protein (N x 6.25)	%		6.4	19.7	
Cattle	dig prot %	*	4.6	14.1	
Goats	dig prot %	*	4.6	14.1	
Horses	dig prot %	*	4.6	14.1	
Sheep	dig prot %	*	4.6	14.1	
Energy	GE Mcal/kg				
Cattle	DE Mcal/kg	*	0.93	2.86	
Sheep	DE Mcal/kg	*	0.85	2.63	
Cattle	ME Mcal/kg	*	0.76	2.34	
Sheep	ME Mcal/kg	*	0.70	2.16	
Cattle	TDN %	*	21.0	64.8	
Sheep	TDN %	*	19.4	59.6	

Clover, ladino-Timothy, aerial part w molasses added, ensiled, (3)

Ref No 3 01 525 United States

Feed Name or Analyses			Mean As Fed	Mean Dry	C.V. ± %
Dry matter	%		26.1	100.0	
Ash	%		2.4	9.2	
Crude fiber	%		7.0	26.8	
Ether extract	%		1.6	6.1	
N-free extract	%		11.2	43.0	

Continued

Feed Name or Analyses			Mean As Fed	Mean Dry	C.V. ± %
Protein (N x 6.25)	%		3.9	14.9	
Cattle	dig prot %	*	2.5	9.8	
Goats	dig prot %	*	2.5	9.8	
Horses	dig prot %	*	2.5	9.8	
Sheep	dig prot %	*	2.5	9.8	
Energy	GE Mcal/kg				
Cattle	DE Mcal/kg	*	0.75	2.88	
Sheep	DE Mcal/kg	*	0.71	2.71	
Cattle	ME Mcal/kg	*	0.62	2.36	
Sheep	ME Mcal/kg	*	0.58	2.22	
Cattle	TDN %	*	17.0	65.3	
Sheep	TDN %	*	16.1	61.5	

CLOVER, RED-ALFALFA. Trifolium pratense, Medicago sativa

Clover, Red-Alfalfa, aerial part, ensiled, (3)

Ref No 3 01 527 United States

Feed Name or Analyses			Mean As Fed	Mean Dry	C.V. ± %
Dry matter	%		25.1	100.0	
Ash	%		2.1	8.4	
Crude fiber	%		8.3	33.1	
Ether extract	%		1.0	4.0	
N-free extract	%		10.1	40.3	
Protein (N x 6.25)	%		3.6	14.2	
Cattle	dig prot %	*	2.3	9.1	
Goats	dig prot %	*	2.3	9.1	
Horses	dig prot %	*	2.3	9.1	
Sheep	dig prot %	*	2.3	9.1	
Energy	GE Mcal/kg				
Cattle	DE Mcal/kg	*	0.62	2.49	
Sheep	DE Mcal/kg	*	0.69	2.73	
Cattle	ME Mcal/kg	*	0.51	2.04	
Sheep	ME Mcal/kg	*	0.56	2.24	
Cattle	TDN %	*	14.2	56.5	
Sheep	TDN %	*	15.6	62.0	
Carotene	mg/kg		25.3	100.8	
Vitamin A equivalent	IU/g		42.2	168.0	

Clover, red-Alfalfa, aerial part w sulfur dioxide preservative added, ensiled, (3)

Ref No 3 01 528 United States

Feed Name or Analyses			Mean As Fed	Mean Dry	C.V. ± %
Dry matter	%		23.8	100.0	
Ash	%		2.0	8.2	
Crude fiber	%		7.5	31.4	
Ether extract	%		1.0	4.0	
N-free extract	%		10.1	42.3	
Protein (N x 6.25)	%		3.4	14.1	
Cattle	dig prot %	*	2.1	9.0	
Goats	dig prot %	*	2.1	9.0	
Horses	dig prot %	*	2.1	9.0	
Sheep	dig prot %	*	2.1	9.0	
Energy	GE Mcal/kg				
Cattle	DE Mcal/kg	*	0.61	2.55	
Sheep	DE Mcal/kg	*	0.66	2.76	
Cattle	ME Mcal/kg	*	0.50	2.09	
Sheep	ME Mcal/kg	*	0.54	2.26	
Cattle	TDN %	*	13.8	57.9	
Sheep	TDN %	*	14.9	62.5	

(1) dry forages and roughages (2) pasture, range plants, and forages fed green (3) silages (4) energy feeds (5) protein supplements (6) minerals (7) vitamins (8) additives

CLOVER, RED-GRASS. Trifolium pratense, Scientific name not used

Clover, red-Grass, hay, s-c, (1)

Ref No 1 01 536 United States

Feed Name or Analyses			Mean As Fed	Mean Dry	C.V. ± %
Dry matter	%		88.1	100.0	1
Ash	%		6.5	7.4	14
Crude fiber	%		29.3	33.3	9
Ether extract	%		1.6	1.9	11
N-free extract	%		38.3	43.5	
Protein (N x 6.25)	%		12.3	14.0	17
Cattle	dig prot %	*	8.0	9.1	
Goats	dig prot %	*	8.5	9.6	
Horses	dig prot %	*	8.3	9.4	
Rabbits	dig prot %	*	8.4	9.5	
Sheep	dig prot %	*	8.0	9.1	
Energy	GE Mcal/kg				
Cattle	DE Mcal/kg	*	2.14	2.43	
Sheep	DE Mcal/kg	*	2.21	2.51	
Cattle	ME Mcal/kg	*	1.76	1.99	
Sheep	ME Mcal/kg	*	1.81	2.06	
Cattle	TDN %	*	48.6	55.2	
Sheep	TDN %	*	50.1	56.9	
Calcium	%		1.06	1.20	14
Cobalt	mg/kg		0.132	0.150	14
Copper	mg/kg		14.2	16.1	18
Iron	%		0.015	0.017	16
Magnesium	%		0.36	0.41	27
Manganese	mg/kg		53.4	60.6	34
Phosphorus	%		0.20	0.23	8
Potassium	%		1.26	1.43	15
Carotene	mg/kg		11.2	12.7	38
Vitamin A equivalent	IU/g		18.6	21.1	

Clover, red-Grass, hay, s-c, immature, (1)

Ref No 1 01 529 United States

Feed Name or Analyses			Mean As Fed	Mean Dry	C.V. ± %
Dry matter	%		91.0	100.0	
Carotene	mg/kg		39.1	43.0	
Vitamin A equivalent	IU/g		65.2	71.7	

Clover, red-Grass, hay, s-c, early bloom, (1)

Ref No 1 01 530 United States

Feed Name or Analyses			Mean As Fed	Mean Dry	C.V. ± %
Dry matter	%		91.5	100.0	
Carotene	mg/kg		30.3	33.1	
Vitamin A equivalent	IU/g		50.4	55.1	

Clover, red-Grass, hay, s-c, mid-bloom, (1)

Ref No 1 01 531 United States

Feed Name or Analyses			Mean As Fed	Mean Dry	C.V. ± %
Dry matter	%		89.2	100.0	2
Ash	%		7.6	8.5	
Crude fiber	%		26.0	29.1	
Ether extract	%		1.6	1.8	
N-free extract	%		39.2	44.0	
Protein (N x 6.25)	%		14.8	16.6	
Cattle	dig prot %	*	10.1	11.3	
Goats	dig prot %	*	10.7	12.0	
Horses	dig prot %	*	10.4	11.6	
Rabbits	dig prot %	*	10.2	11.5	
Sheep	dig prot %	*	10.2	11.5	
Energy	GE Mcal/kg				
Cattle	DE Mcal/kg	*	2.30	2.57	
Sheep	DE Mcal/kg	*	2.33	2.61	
Cattle	ME Mcal/kg	*	1.88	2.11	
Sheep	ME Mcal/kg	*	1.91	2.14	
Cattle	TDN %	*	52.1	58.4	
Sheep	TDN %	*	52.7	59.1	
Calcium	%		1.20	1.34	7
Cobalt	mg/kg		0.108	0.121	
Copper	mg/kg		15.6	17.5	10
Iron	%		0.016	0.018	6
Magnesium	%		0.37	0.41	31
Manganese	mg/kg		68.1	76.4	21
Phosphorus	%		0.20	0.23	3
Potassium	%		1.41	1.59	7

Clover, red-Grass, hay, s-c, full bloom, (1)

Ref No 1 01 532 United States

Feed Name or Analyses			Mean As Fed	Mean Dry	C.V. ± %
Dry matter	%		88.7	100.0	2
Ash	%		5.9	6.6	22
Crude fiber	%		29.9	33.7	8
Ether extract	%		1.6	1.8	14
N-free extract	%		37.4	42.2	
Protein (N x 6.25)	%		14.0	15.8	8
Cattle	dig prot %	*	9.4	10.6	
Goats	dig prot %	*	10.0	11.3	
Horses	dig prot %	*	9.7	10.9	
Rabbits	dig prot %	*	9.6	10.9	
Sheep	dig prot %	*	9.5	10.7	
Energy	GE Mcal/kg				
Cattle	DE Mcal/kg	*	2.10	2.37	
Sheep	DE Mcal/kg	*	2.25	2.54	
Cattle	ME Mcal/kg	*	1.73	1.95	
Sheep	ME Mcal/kg	*	1.85	2.08	
Cattle	TDN %	*	47.7	53.8	
Sheep	TDN %	*	51.1	57.6	
Cobalt	mg/kg		0.137	0.154	15
Carotene	mg/kg		10.0	11.2	32
Vitamin A equivalent	IU/g		16.6	18.7	

Clover, red-Grass, hay, s-c, mature, (1)

Ref No 1 01 533 United States

Feed Name or Analyses			Mean As Fed	Mean Dry	C.V. ± %
Dry matter	%		91.4	100.0	
Ash	%		6.8	7.4	
Crude fiber	%		32.4	35.4	
Ether extract	%		1.6	1.8	
N-free extract	%		40.5	44.3	
Protein (N x 6.25)	%		10.1	11.1	
Cattle	dig prot %	*	6.0	6.6	
Goats	dig prot %	*	6.3	6.9	
Horses	dig prot %	*	6.4	7.0	
Rabbits	dig prot %	*	6.6	7.2	
Sheep	dig prot %	*	6.0	6.5	
Energy	GE Mcal/kg				
Cattle	DE Mcal/kg	*	2.14	2.34	
Sheep	DE Mcal/kg	*	2.25	2.46	
Cattle	ME Mcal/kg	*	1.75	1.92	
Sheep	ME Mcal/kg	*	1.84	2.02	
Cattle	TDN %	*	48.5	53.0	
Sheep	TDN %	*	50.9	55.7	
Calcium	%		0.55	0.60	
Phosphorus	%		0.23	0.25	

Clover, red-Grass, hay, s-c, cut 1, (1)

Ref No 1 01 534 United States

Feed Name or Analyses			Mean As Fed	Mean Dry	C.V. ± %
Dry matter	%		88.6	100.0	2
Ash	%		8.2	9.3	33

Feed Name or Analyses			Mean As Fed	Mean Dry	C.V. ± %
Crude fiber	%		28.1	31.7	7
Ether extract	%		1.6	1.8	11
N-free extract	%		36.7	41.4	
Protein (N x 6.25)	%		14.0	15.8	6
Cattle	dig prot %	*	9.4	10.6	
Goats	dig prot %	*	10.0	11.3	
Horses	dig prot %	*	9.7	10.9	
Rabbits	dig prot %	*	9.6	10.9	
Sheep	dig prot %	*	9.5	10.7	
Energy	GE Mcal/kg				
Cattle	DE Mcal/kg	*	2.25	2.54	
Sheep	DE Mcal/kg	*	2.21	2.49	
Cattle	ME Mcal/kg	*	1.85	2.09	
Sheep	ME Mcal/kg	*	1.81	2.05	
Cattle	TDN %	*	51.1	57.7	
Sheep	TDN %	*	50.1	56.6	
Calcium	%		1.17	1.32	15
Cobalt	mg/kg		0.125	0.141	14
Copper	mg/kg		14.6	16.5	8
Iron	%		0.016	0.018	11
Magnesium	%		0.47	0.53	22
Manganese	mg/kg		62.3	70.3	44
Phosphorus	%		0.20	0.23	2
Potassium	%		1.33	1.50	21
Carotene	mg/kg		10.0	11.2	47
Vitamin A equivalent	IU/g		16.6	18.7	

Clover, red-Grass, hay, s-c, cut 2, (1)

Ref No 1 01 535 United States

Feed Name or Analyses			Mean As Fed	Mean Dry	C.V. ± %
Dry matter	%		89.4	100.0	
Ash	%		5.6	6.3	
Crude fiber	%		30.8	34.5	
Ether extract	%		1.5	1.7	
N-free extract	%		37.0	41.4	
Protein (N x 6.25)	%		14.4	16.1	
Cattle	dig prot %	*	9.7	10.9	
Goats	dig prot %	*	10.4	11.6	
Horses	dig prot %	*	10.0	11.2	
Rabbits	dig prot %	*	9.9	11.1	
Sheep	dig prot %	*	9.8	11.0	
Energy	GE Mcal/kg				
Cattle	DE Mcal/kg	*	2.09	2.34	
Sheep	DE Mcal/kg	*	2.26	2.53	
Cattle	ME Mcal/kg	*	1.71	1.92	
Sheep	ME Mcal/kg	*	1.85	2.07	
Cattle	TDN %	*	47.4	53.0	
Sheep	TDN %	*	51.3	57.4	
Calcium	%		1.05	1.18	9
Cobalt	mg/kg		0.136	0.152	17
Copper	mg/kg		16.0	17.9	9
Iron	%		0.013	0.015	7
Magnesium	%		0.30	0.33	10
Manganese	mg/kg		48.9	54.7	34
Phosphorus	%		0.20	0.22	4
Potassium	%		1.25	1.40	5
Carotene	mg/kg		8.3	9.3	8
Vitamin A equivalent	IU/g		13.8	15.4	

Clover, red-Grass, hay, s-c, gr 1 US, (1)

Ref No 1 01 538 United States

Feed Name or Analyses			Mean As Fed	Mean Dry	C.V. ± %
Dry matter	%		89.4	100.0	
Ash	%		7.6	8.5	
Crude fiber	%		26.0	29.1	
Ether extract	%		1.6	1.8	
N-free extract	%		39.3	44.0	
Protein (N x 6.25)	%		14.8	16.6	
Cattle	dig prot %	*	10.1	11.3	
Goats	dig prot %	*	10.8	12.0	
Horses	dig prot %	*	10.4	11.6	
Rabbits	dig prot %	*	10.3	11.5	
Sheep	dig prot %	*	10.2	11.5	
Energy	GE Mcal/kg				
Cattle	DE Mcal/kg	*	2.30	2.57	
Sheep	DE Mcal/kg	*	2.33	2.61	
Cattle	ME Mcal/kg	*	1.89	2.11	
Sheep	ME Mcal/kg	*	1.91	2.14	
Cattle	TDN %	*	52.2	58.4	
Sheep	TDN %	*	52.9	59.1	
Calcium	%		1.22	1.36	12
Cobalt	mg/kg		0.120	0.135	10
Copper	mg/kg		15.2	17.0	6
Iron	%		0.015	0.017	17
Magnesium	%		0.31	0.35	17
Manganese	mg/kg		62.5	69.9	45
Phosphorus	%		0.21	0.23	11
Potassium	%		1.48	1.66	10

Clover, red-Grass, hay, s-c, gr 2 US, (1)

Ref No 1 01 539 United States

Feed Name or Analyses			Mean As Fed	Mean Dry	C.V. ± %
Dry matter	%		88.8	100.0	2
Ash	%		5.9	6.7	
Crude fiber	%		27.6	31.1	
Ether extract	%		1.6	1.9	
N-free extract	%		41.3	46.6	
Protein (N x 6.25)	%		12.3	13.9	
Cattle	dig prot %	*	7.9	8.9	
Goats	dig prot %	*	8.4	9.5	
Horses	dig prot %	*	8.2	9.3	
Rabbits	dig prot %	*	8.3	9.4	
Sheep	dig prot %	*	8.0	9.0	
Energy	GE Mcal/kg				
Cattle	DE Mcal/kg	*	2.18	2.46	
Sheep	DE Mcal/kg	*	2.31	2.60	
Cattle	ME Mcal/kg	*	1.79	2.02	
Sheep	ME Mcal/kg	*	1.89	2.13	
Cattle	TDN %	*	49.5	55.8	
Sheep	TDN %	*	52.3	58.9	
Calcium	%		0.91	1.03	30
Cobalt	mg/kg		0.143	0.161	
Copper	mg/kg		13.3	15.0	
Iron	%		0.014	0.016	
Magnesium	%		0.58	0.65	
Manganese	mg/kg		17.8	20.1	66
Phosphorus	%		0.20	0.22	13
Potassium	%		1.05	1.18	
Carotene	mg/kg		15.5	17.4	12
Vitamin A equivalent	IU/g		25.8	29.0	

Clover, red-Grass, hay, s-c, gr 3 US, (1)

Ref No 1 01 540 United States

Feed Name or Analyses			Mean As Fed	Mean Dry	C.V. ± %
Dry matter	%		89.2	100.0	
Ash	%		6.0	6.7	
Crude fiber	%		30.7	34.4	
Ether extract	%		1.4	1.6	
N-free extract	%		38.4	43.1	
Protein (N x 6.25)	%		12.7	14.2	
Cattle	dig prot %	*	8.2	9.2	
Goats	dig prot %	*	8.7	9.8	
Horses	dig prot %	*	8.6	9.6	
Rabbits	dig prot %	*	8.6	9.6	
Sheep	dig prot %	*	8.3	9.3	
Energy	GE Mcal/kg				
Cattle	DE Mcal/kg	*	2.11	2.36	
Sheep	DE Mcal/kg	*	2.24	2.51	
Cattle	ME Mcal/kg	*	1.73	1.94	
Sheep	ME Mcal/kg	*	1.84	2.06	
Cattle	TDN %	*	47.8	53.5	
Sheep	TDN %	*	50.8	56.9	
Calcium	%		1.00	1.12	10
Cobalt	mg/kg		0.140	0.157	17
Copper	mg/kg		14.4	16.1	23
Iron	%		0.014	0.016	4
Magnesium	%		0.33	0.37	7
Manganese	mg/kg		55.3	61.9	27
Phosphorus	%		0.21	0.23	10
Potassium	%		1.21	1.36	3
Carotene	mg/kg		6.5	7.3	31
Vitamin A equivalent	IU/g		10.8	12.1	

Clover, red-Grass, hay, s-c, gr sample US, (1)

Ref No 1 01 537 United States

Feed Name or Analyses			Mean As Fed	Mean Dry	C.V. ± %
Dry matter	%		87.1	100.0	
Ash	%		9.0	10.3	
Crude fiber	%		31.1	35.7	
Cattle	dig coef %	*	47.	47.	
Ether extract	%		1.6	1.8	
Cattle	dig coef %	*	36.	36.	
N-free extract	%		35.4	40.7	
Cattle	dig coef %	*	61.	61.	
Protein (N x 6.25)	%		10.0	11.5	
Cattle	dig coef %	*	62.	62.	
Cattle	dig prot %		6.2	7.1	
Goats	dig prot %	*	6.3	7.3	
Horses	dig prot %	*	6.4	7.3	
Rabbits	dig prot %	*	6.6	7.5	
Sheep	dig prot %	*	6.0	6.9	
Energy	GE Mcal/kg				
Cattle	DE Mcal/kg	*	1.93	2.21	
Sheep	DE Mcal/kg	*	2.12	2.44	
Cattle	ME Mcal/kg	*	1.58	1.81	
Sheep	ME Mcal/kg	*	1.74	2.00	
Cattle	TDN %		43.7	50.2	
Sheep	TDN %	*	48.2	55.3	
Calcium	%		0.61	0.70	
Phosphorus	%		0.26	0.30	
Carotene	mg/kg		8.6	9.9	
Vitamin A equivalent	IU/g		14.4	16.5	

Clover, red-Grass, aerial part, wilted ensiled, (3)

Ref No 3 01 542 United States

Feed Name or Analyses			Mean As Fed	Mean Dry	C.V. ± %
Dry matter	%		35.5	100.0	
Ash	%		3.0	8.5	
Crude fiber	%		11.1	31.3	
Ether extract	%		1.1	3.1	
N-free extract	%		15.9	44.8	
Protein (N x 6.25)	%		4.4	12.3	
Cattle	dig prot %	*	2.6	7.4	
Goats	dig prot %	*	2.6	7.4	
Horses	dig prot %	*	2.6	7.4	
Sheep	dig prot %	*	2.6	7.4	
Energy	GE Mcal/kg				
Cattle	DE Mcal/kg	*	0.93	2.62	
Sheep	DE Mcal/kg	*	0.99	2.79	
Cattle	ME Mcal/kg	*	0.76	2.15	
Sheep	ME Mcal/kg	*	0.81	2.28	
Cattle	TDN %	*	21.1	59.3	

Continued

Feed Name or Analyses			Mean As Fed	Mean Dry	C.V. ± %
Sheep	TDN %	*	22.4	63.2	
Calcium	%		0.39	1.11	
Phosphorus	%		0.07	0.20	

CLOVER, RED-POTATO. Trifolium pratense, Solanum tuberosum

Clover, red-Potato, aerial part, ensiled, (3)

Ref No 3 01 543 United States

Feed Name or Analyses			Mean As Fed	Mean Dry	C.V. ± %
Dry matter	%		19.2	100.0	
Ash	%		1.7	8.9	
Crude fiber	%		1.8	9.5	
Ether extract	%		0.3	1.4	
N-free extract	%		12.8	66.8	
Protein (N x 6.25)	%		2.6	13.4	
Cattle	dig prot %	*	1.6	8.4	
Goats	dig prot %	*	1.6	8.4	
Horses	dig prot %	*	1.6	8.4	
Sheep	dig prot %	*	1.6	8.4	

CLOVER, RED-RYEGRASS. Trifolium pratense, Lolium spp

Clover, red-Ryegrass, hay, s-c, (1)

Ref No 1 01 544 United States

Feed Name or Analyses			Mean As Fed	Mean Dry	C.V. ± %
Dry matter	%		84.7	100.0	2
Ash	%		5.7	6.7	10
Crude fiber	%		26.4	31.2	7
Ether extract	%		1.5	1.8	15
N-free extract	%		39.9	47.1	
Protein (N x 6.25)	%		11.2	13.2	11
Cattle	dig prot %	*	7.1	8.4	
Goats	dig prot %	*	7.5	8.9	
Horses	dig prot %	*	7.4	8.7	
Rabbits	dig prot %	*	7.5	8.9	
Sheep	dig prot %	*	7.1	8.4	
Energy	GE Mcal/kg				
Cattle	DE Mcal/kg	*	2.08	2.46	
Sheep	DE Mcal/kg	*	2.20	2.59	
Cattle	ME Mcal/kg	*	1.71	2.02	
Sheep	ME Mcal/kg	*	1.80	2.13	
Cattle	TDN %	*	47.3	55.8	
Sheep	TDN %	*	49.8	58.8	

CLOVER, RED-TIMOTHY. Trifolium pratense, Phleum pratense

Clover, red-Timothy, hay, s-c, (1)

Ref No 1 01 547 United States

Feed Name or Analyses			Mean As Fed	Mean Dry	C.V. ± %
Dry matter	%		89.5	100.0	2
Ash	%		7.2	8.1	13
Crude fiber	%		26.2	29.3	10
Ether extract	%		1.9	2.1	27
N-free extract	%		42.5	47.5	
Protein (N x 6.25)	%		11.6	13.0	18
Cattle	dig prot %	*	7.3	8.2	
Goats	dig prot %	*	7.8	8.7	
Horses	dig prot %	*	7.7	8.6	
Rabbits	dig prot %	*	7.8	8.7	

(1) dry forages and roughages (2) pasture, range plants, and forages fed green (3) silages (4) energy feeds (5) protein supplements (6) minerals (7) vitamins (8) additives

Feed Name or Analyses			Mean As Fed	Mean Dry	C.V. ± %
Sheep	dig prot %	*	7.4	8.2	
Energy	GE Mcal/kg				
Cattle	DE Mcal/kg	*	2.51	2.81	
Sheep	DE Mcal/kg	*	2.34	2.61	
Cattle	ME Mcal/kg	*	2.06	2.30	
Sheep	ME Mcal/kg	*	1.92	2.14	
Cattle	TDN %	*	57.0	63.7	
Sheep	TDN %	*	53.0	59.2	

Clover, red-Timothy, hay, s-c, early bloom, (1)

Ref No 1 01 545 United States

Feed Name or Analyses			Mean As Fed	Mean Dry	C.V. ± %
Dry matter	%		91.4	100.0	2
Ash	%		8.0	8.7	6
Crude fiber	%		23.7	25.9	9
Ether extract	%		2.1	2.3	5
N-free extract	%		44.1	48.3	
Protein (N x 6.25)	%		13.5	14.8	8
Cattle	dig prot %	*	8.9	9.8	
Goats	dig prot %	*	9.5	10.4	
Horses	dig prot %	*	9.2	10.1	
Rabbits	dig prot %	*	9.2	10.1	
Sheep	dig prot %	*	9.0	9.8	
Energy	GE Mcal/kg				
Cattle	DE Mcal/kg	*	2.47	2.70	
Sheep	DE Mcal/kg	*	2.44	2.66	
Cattle	ME Mcal/kg	*	2.02	2.21	
Sheep	ME Mcal/kg	*	2.00	2.18	
Cattle	TDN %	*	56.0	61.2	
Sheep	TDN %	*	55.2	60.4	

Clover, red-Timothy, hay, s-c, cut 2, (1)

Ref No 1 01 546 United States

Feed Name or Analyses			Mean As Fed	Mean Dry	C.V. ± %
Dry matter	%		92.1	100.0	0
Ash	%		8.2	8.9	5
Crude fiber	%		29.1	31.6	3
Ether extract	%		1.9	2.1	8
N-free extract	%		38.9	42.2	
Protein (N x 6.25)	%		14.0	15.2	8
Cattle	dig prot %	*	9.3	10.1	
Goats	dig prot %	*	9.9	10.7	
Horses	dig prot %	*	9.6	10.4	
Rabbits	dig prot %	*	9.6	10.4	
Sheep	dig prot %	*	9.4	10.2	
Energy	GE Mcal/kg				
Cattle	DE Mcal/kg	*	2.34	2.54	
Sheep	DE Mcal/kg	*	2.30	2.50	
Cattle	ME Mcal/kg	*	1.92	2.09	
Sheep	ME Mcal/kg	*	1.89	2.05	
Cattle	TDN %	*	53.2	57.7	
Sheep	TDN %	*	52.2	56.7	

Clover, red-Timothy, aerial part, ensiled, mature (3)

Ref No 3 01 548 United States

Feed Name or Analyses			Mean As Fed	Mean Dry	C.V. ± %
Dry matter	%		30.4	100.0	25
Ash	%		2.9	9.5	17
Crude fiber	%		9.1	29.9	18
Ether extract	%		1.3	4.3	41
N-free extract	%		12.8	42.1	
Protein (N x 6.25)	%		4.3	14.2	8
Cattle	dig prot %	*	2.8	9.1	
Goats	dig prot %	*	2.8	9.1	
Horses	dig prot %	*	2.8	9.1	
Sheep	dig prot %	*	2.8	9.1	

Feed Name or Analyses			Mean As Fed	Mean Dry	C.V. ± %
Energy	GE Mcal/kg				
Cattle	DE Mcal/kg	*	0.82	2.70	
Sheep	DE Mcal/kg	*	0.83	2.73	
Cattle	ME Mcal/kg	*	0.67	2.21	
Sheep	ME Mcal/kg	*	0.68	2.24	
Cattle	TDN %	*	18.6	61.2	
Sheep	TDN %	*	18.8	61.9	
Calcium	%		0.34	1.12	
Phosphorus	%		0.03	0.10	
Carotene	mg/kg		6.5	21.4	
Vitamin A equivalent	IU/g		10.8	35.6	

Clover, red-Timothy, aerial part, ensiled, cut 1, (3)

Ref No 3 01 549 United States

Feed Name or Analyses			Mean As Fed	Mean Dry	C.V. ± %
Dry matter	%		22.7	100.0	
Ash	%		2.3	10.3	
Crude fiber	%		7.0	30.8	
Ether extract	%		0.9	4.0	
N-free extract	%		9.4	41.2	
Protein (N x 6.25)	%		3.1	13.7	
Cattle	dig prot %	*	2.0	8.7	
Goats	dig prot %	*	2.0	8.7	
Horses	dig prot %	*	2.0	8.7	
Sheep	dig prot %	*	2.0	8.7	
Energy	GE Mcal/kg				
Cattle	DE Mcal/kg	*	0.62	2.71	
Sheep	DE Mcal/kg	*	0.62	2.71	
Cattle	ME Mcal/kg	*	0.51	2.23	
Sheep	ME Mcal/kg	*	0.50	2.22	
Cattle	TDN %	*	14.0	61.6	
Sheep	TDN %	*	14.0	61.5	

Clover, red-Timothy, aerial part, wilted ensiled, (3)

Ref No 3 01 550 United States

Feed Name or Analyses			Mean As Fed	Mean Dry	C.V. ± %
Dry matter	%		37.9	100.0	
Ash	%		3.5	9.2	
Crude fiber	%		10.3	27.2	
Ether extract	%		1.9	4.9	
N-free extract	%		16.6	43.8	
Protein (N x 6.25)	%		5.6	14.9	
Cattle	dig prot %	*	3.7	9.8	
Goats	dig prot %	*	3.7	9.8	
Horses	dig prot %	*	3.7	9.8	
Sheep	dig prot %	*	3.7	9.8	
Energy	GE Mcal/kg				
Cattle	DE Mcal/kg	*	1.06	2.80	
Sheep	DE Mcal/kg	*	1.04	2.74	
Cattle	ME Mcal/kg	*	0.87	2.29	
Sheep	ME Mcal/kg	*	0.85	2.25	
Cattle	TDN %	*	24.0	63.4	
Sheep	TDN %	*	23.5	62.1	

Clover, red-Timothy, aerial part w phosphoric acid preservative added, ensiled, (3)

Ref No 3 01 551 United States

Feed Name or Analyses			Mean As Fed	Mean Dry	C.V. ± %
Dry matter	%		25.3	100.0	
Ash	%		2.3	9.0	
Crude fiber	%		8.3	33.0	
Ether extract	%		1.4	5.6	
N-free extract	%		9.5	37.6	
Protein (N x 6.25)	%		3.7	14.8	
Cattle	dig prot %	*	2.4	9.7	
Goats	dig prot %	*	2.4	9.7	
Horses	dig prot %	*	2.4	9.7	

Feed Name or Analyses			Mean As Fed	Dry	C.V. ± %
Sheep	dig prot %	*	2.4	9.7	
Energy	GE Mcal/kg				
Cattle	DE Mcal/kg	*	0.65	2.58	
Sheep	DE Mcal/kg	*	0.68	2.67	
Cattle	ME Mcal/kg	*	0.54	2.12	
Sheep	ME Mcal/kg	*	0.55	2.19	
Cattle	TDN %	*	14.8	58.6	
Sheep	TDN %	*	15.3	60.6	

Clover, red-Timothy, aerial part w molasses added, ensiled, (3)

Ref No 3 01 552 United States

Feed Name or Analyses			Mean As Fed	Dry	C.V. ± %
Dry matter	%		40.9	100.0	
Ash	%		3.9	9.6	
Crude fiber	%		10.9	26.7	
Ether extract	%		1.6	3.8	
N-free extract	%		18.6	45.4	
Protein (N x 6.25)	%		5.9	14.5	
Cattle	dig prot %	*	3.8	9.4	
Goats	dig prot %	*	3.8	9.4	
Horses	dig prot %	*	3.8	9.4	
Sheep	dig prot %	*	3.8	9.4	
Energy	GE Mcal/kg				
Cattle	DE Mcal/kg	*	1.14	2.80	
Sheep	DE Mcal/kg	*	1.13	2.77	
Cattle	ME Mcal/kg	*	0.94	2.29	
Sheep	ME Mcal/kg	*	0.93	2.27	
Cattle	TDN %	*	26.0	63.5	
Sheep	TDN %	*	25.7	62.7	

CLOVER-RYEGRASS. Trifolium spp, Lolium spp

Clover-Ryegrass, hay, s-c, (1)

Ref No 1 01 481 United States

Feed Name or Analyses			Mean As Fed	Dry	C.V. ± %
Dry matter	%		85.6	100.0	1
Ash	%		9.4	11.0	5
Crude fiber	%		26.1	30.5	9
Ether extract	%		1.8	2.1	18
N-free extract	%		36.1	42.2	
Protein (N x 6.25)	%		12.2	14.2	9
Cattle	dig prot %	*	7.9	9.2	
Goats	dig prot %	*	8.4	9.8	
Horses	dig prot %	*	8.2	9.6	
Rabbits	dig prot %	*	8.2	9.6	
Sheep	dig prot %	*	8.0	9.3	
Energy	GE Mcal/kg				
Cattle	DE Mcal/kg	*	2.28	2.67	
Sheep	DE Mcal/kg	*	2.09	2.45	
Cattle	ME Mcal/kg	*	1.87	2.19	
Sheep	ME Mcal/kg	*	1.72	2.01	
Cattle	TDN %	*	51.8	60.5	
Sheep	TDN %	*	47.5	55.5	

Clover-Ryegrass, hay, s-c, milk stage, (1)

Ref No 1 01 480 United States

Feed Name or Analyses			Mean As Fed	Dry	C.V. ± %
Dry matter	%		83.8	100.0	
Ash	%		8.4	10.0	
Crude fiber	%		29.8	35.6	
Ether extract	%		1.2	1.4	
N-free extract	%		34.6	41.3	
Protein (N x 6.25)	%		9.8	11.7	
Cattle	dig prot %	*	5.9	7.1	
Goats	dig prot %	*	6.3	7.5	
Horses	dig prot %	*	6.3	7.5	
Rabbits	dig prot %	*	6.5	7.7	
Sheep	dig prot %	*	5.9	7.1	
Energy	GE Mcal/kg				
Cattle	DE Mcal/kg	*	2.02	2.41	
Sheep	DE Mcal/kg	*	2.05	2.45	
Cattle	ME Mcal/kg	*	1.66	1.98	
Sheep	ME Mcal/kg	*	1.68	2.01	
Cattle	TDN %	*	45.8	54.6	
Sheep	TDN %	*	46.5	55.5	

Clover-Ryegrass, aerial part, ensiled, early bloom, (3)

Ref No 3 01 482 United States

Feed Name or Analyses			Mean As Fed	Dry	C.V. ± %
Dry matter	%		31.8	100.0	
Ash	%		2.9	9.1	
Crude fiber	%		10.2	32.1	
Ether extract	%		1.0	3.3	
N-free extract	%		13.4	42.2	
Protein (N x 6.25)	%		4.2	13.3	
Cattle	dig prot %	*	2.6	8.3	
Goats	dig prot %	*	2.6	8.3	
Horses	dig prot %	*	2.6	8.3	
Sheep	dig prot %	*	2.6	8.3	
Energy	GE Mcal/kg				
Cattle	DE Mcal/kg	*	0.82	2.58	
Sheep	DE Mcal/kg	*	0.87	2.75	
Cattle	ME Mcal/kg	*	0.67	2.11	
Sheep	ME Mcal/kg	*	0.72	2.26	
Cattle	TDN %	*	18.6	58.4	
Sheep	TDN %	*	19.8	62.4	

CLOVER-TIMOTHY. Trifolium spp, Phleum pratense

Clover-Timothy, hay, s-c, (1)

Ref No 1 01 487 United States

Feed Name or Analyses			Mean As Fed	Dry	C.V. ± %
Dry matter	%		88.5	100.0	2
Chlorine	%		0.58	0.65	
Cobalt	mg/kg		0.119	0.135	78
Sodium	%		0.17	0.19	
Sulphur	%		0.12	0.13	
Carotene	mg/kg		14.0	15.9	90
Vitamin A equivalent	IU/g		23.4	26.5	

Clover-Timothy, hay, s-c, immature, (1)

Ref No 1 01 483 United States

Feed Name or Analyses			Mean As Fed	Dry	C.V. ± %
Dry matter	%		83.2	100.0	2
Ash	%		5.6	6.7	9
Crude fiber	%		23.7	28.5	11
Ether extract	%		2.0	2.4	12
N-free extract	%		39.6	47.6	
Protein (N x 6.25)	%		12.3	14.8	14
Cattle	dig prot %	*	8.1	9.8	
Goats	dig prot %	*	8.6	10.4	
Horses	dig prot %	*	8.4	10.1	
Rabbits	dig prot %	*	8.4	10.1	
Sheep	dig prot %	*	8.2	9.8	
Energy	GE Mcal/kg				
Cattle	DE Mcal/kg	*	2.12	2.55	
Sheep	DE Mcal/kg	*	2.21	2.66	
Cattle	ME Mcal/kg	*	1.74	2.09	
Sheep	ME Mcal/kg	*	1.81	2.18	
Cattle	TDN %	*	48.1	57.8	
Sheep	TDN %	*	50.2	60.3	

Clover-Timothy, hay, s-c, full bloom, (1)

Ref No 1 01 484 United States

Feed Name or Analyses			Mean As Fed	Dry	C.V. ± %
Dry matter	%			100.0	
Ash	%			6.3	
Crude fiber	%			35.8	
Ether extract	%			3.4	
N-free extract	%			44.0	
Protein (N x 6.25)	%			10.5	
Cattle	dig prot %	*		6.0	
Goats	dig prot %	*		6.4	
Horses	dig prot %	*		6.4	
Rabbits	dig prot %	*		6.8	
Sheep	dig prot %	*		6.0	
Energy	GE Mcal/kg				
Cattle	DE Mcal/kg	*		2.34	
Sheep	DE Mcal/kg	*		2.44	
Cattle	ME Mcal/kg	*		1.92	
Sheep	ME Mcal/kg	*		2.00	
Cattle	TDN %	*		53.0	
Sheep	TDN %	*		55.2	

Clover-Timothy, hay, s-c, mature, (1)

Ref No 1 01 485 United States

Feed Name or Analyses			Mean As Fed	Dry	C.V. ± %
Dry matter	%			100.0	
Calcium	%			0.85	
Cobalt	mg/kg			0.079	
Manganese	mg/kg			28.0	
Phosphorus	%			0.19	33
Carotene	mg/kg			3.1	
Vitamin A equivalent	IU/g			5.1	

Clover-Timothy, hay, s-c, cut 1, (1)

Ref No 1 01 486 United States

Feed Name or Analyses			Mean As Fed	Dry	C.V. ± %
Dry matter	%			100.0	
Calcium	%			1.10	29
Cobalt	mg/kg			0.130	49
Manganese	mg/kg			48.3	23
Phosphorus	%			0.22	29
Carotene	mg/kg			19.0	66
Vitamin A equivalent	IU/g			31.6	

Clover-Timothy, hay, s-c, gr 1 US, (1)

Ref No 1 01 489 United States

Feed Name or Analyses			Mean As Fed	Dry	C.V. ± %
Dry matter	%			100.0	
Ash	%			6.4	23
Crude fiber	%			35.3	5
Ether extract	%			3.4	18
N-free extract	%			43.1	
Protein (N x 6.25)	%			11.8	20
Cattle	dig prot %	*		7.2	
Goats	dig prot %	*		7.6	
Horses	dig prot %	*		7.5	
Rabbits	dig prot %	*		7.8	
Sheep	dig prot %	*		7.2	
Energy	GE Mcal/kg				
Cattle	DE Mcal/kg	*		2.36	
Sheep	DE Mcal/kg	*		2.47	
Cattle	ME Mcal/kg	*		1.94	
Sheep	ME Mcal/kg	*		2.02	
Cattle	TDN %	*		53.6	
Sheep	TDN %	*		56.0	

Continued

Feed Name or Analyses			Mean As Fed	Mean Dry	C.V. ± %
Calcium	%			1.18	18
Cobalt	mg/kg			0.097	45
Manganese	mg/kg			47.1	36
Phosphorus	%			0.28	10
Carotene	mg/kg			23.8	34
Vitamin A equivalent	IU/g			39.7	

Clover-Timothy, hay, s-c, gr 2 US, (1)

Ref No 1 01 490 United States

Feed Name or Analyses			Mean As Fed	Mean Dry	C.V. ± %
Dry matter	%		88.1	100.0	
Ash	%		5.3	6.0	
Crude fiber	%		31.2	35.4	
Ether extract	%		2.2	2.5	
N-free extract	%		38.9	44.1	
Protein (N x 6.25)	%		10.6	12.0	21
Cattle	dig prot %	*	6.5	7.3	
Goats	dig prot %	*	6.8	7.8	
Horses	dig prot %	*	6.8	7.7	
Rabbits	dig prot %	*	7.0	7.9	
Sheep	dig prot %	*	6.5	7.3	
Energy	GE Mcal/kg				
Cattle	DE Mcal/kg	*	2.03	2.30	
Sheep	DE Mcal/kg	*	2.18	2.48	
Cattle	ME Mcal/kg	*	1.66	1.89	
Sheep	ME Mcal/kg	*	1.79	2.03	
Cattle	TDN %	*	46.0	52.2	
Sheep	TDN %	*	49.5	56.2	
Calcium	%		0.77	0.87	17
Cobalt	mg/kg		0.149	0.169	84
Manganese	mg/kg		37.4	42.4	16
Phosphorus	%		0.17	0.19	20
Carotene	mg/kg		10.2	11.6	77
Vitamin A equivalent	IU/g		17.0	19.3	

Clover-Timothy, hay, s-c, gr 3 US, (1)

Ref No 1 01 491 United States

Feed Name or Analyses			Mean As Fed	Mean Dry	C.V. ± %
Dry matter	%			100.0	
N-free extract	%			44.1	
Protein (N x 6.25)	%			10.6	
Cattle	dig prot %	*		6.1	
Goats	dig prot %	*		6.4	
Horses	dig prot %	*		6.5	
Rabbits	dig prot %	*		6.9	
Sheep	dig prot %	*		6.1	
Calcium	%			0.86	9
Cobalt	mg/kg			0.073	
Manganese	mg/kg			39.9	
Phosphorus	%			0.19	15
Carotene	mg/kg			4.9	66
Vitamin A equivalent	IU/g			8.1	

Clover-Timothy, hay, s-c, 35% timothy, (1)

Ref No 1 08 738 United States

Feed Name or Analyses			Mean As Fed	Mean Dry	C.V. ± %
Dry matter	%		90.0	100.0	
Ash	%		5.7	6.3	
Crude fiber	%		32.1	35.7	
Ether extract	%		2.3	2.6	
N-free extract	%		39.2	43.6	

(1) dry forages and roughages (2) pasture, range plants, and forages fed green (3) silages (4) energy feeds (5) protein supplements (6) minerals (7) vitamins (8) additives

Feed Name or Analyses			Mean As Fed	Mean Dry	C.V. ± %
Protein (N x 6.25)	%		10.7	11.9	
Cattle	dig prot %	*	6.5	7.2	
Goats	dig prot %	*	6.9	7.7	
Horses	dig prot %	*	6.9	7.6	
Rabbits	dig prot %	*	7.1	7.8	
Sheep	dig prot %	*	6.5	7.2	
Energy	GE Mcal/kg				
Cattle	DE Mcal/kg	*	2.07	2.30	
Sheep	DE Mcal/kg	*	2.22	2.47	
Cattle	ME Mcal/kg	*	1.70	1.88	
Sheep	ME Mcal/kg	*	1.82	2.02	
Cattle	TDN %	*	46.9	52.1	
Sheep	TDN %	*	50.4	56.0	

Clover-Timothy, hay, s-c, gr sample US, (1)

Ref No 1 01 488 United States

Feed Name or Analyses			Mean As Fed	Mean Dry	C.V. ± %
Dry matter	%			100.0	
N-free extract	%			44.1	
Protein (N x 6.25)	%			14.2	
Cattle	dig prot %	*		9.2	
Goats	dig prot %	*		9.8	
Horses	dig prot %	*		9.6	
Rabbits	dig prot %	*		9.6	
Sheep	dig prot %	*		9.3	
Calcium	%			1.37	7
Cobalt	mg/kg			0.229	
Manganese	mg/kg			35.7	
Phosphorus	%			0.24	20

Clover-Timothy, aerial part, ensiled, (3)

Ref No 3 01 492 United States

Feed Name or Analyses			Mean As Fed	Mean Dry	C.V. ± %
Dry matter	%		33.9	100.0	7
Ash	%		2.7	8.1	8
Crude fiber	%		8.8	26.0	5
Ether extract	%		2.2	6.5	17
N-free extract	%		15.1	44.6	
Protein (N x 6.25)	%		5.0	14.8	7
Cattle	dig prot %	*	3.3	9.7	
Goats	dig prot %	*	3.3	9.7	
Horses	dig prot %	*	3.3	9.7	
Sheep	dig prot %	*	3.3	9.7	
Energy	GE Mcal/kg				
Cattle	DE Mcal/kg	*	0.98	2.88	
Sheep	DE Mcal/kg	*	0.93	2.74	
Cattle	ME Mcal/kg	*	0.80	2.36	
Sheep	ME Mcal/kg	*	0.76	2.25	
Cattle	TDN %	*	22.1	65.3	
Sheep	TDN %	*	21.1	62.1	

Cover-Timothy, aerial part w phosphoric acid preservative added, ensiled, (3)

Ref No 3 01 493 United States

Feed Name or Analyses			Mean As Fed	Mean Dry	C.V. ± %
Dry matter	%		31.9	100.0	
Ash	%		2.7	8.6	
Crude fiber	%		9.0	28.2	
Ether extract	%		2.3	7.3	
N-free extract	%		12.9	40.3	
Protein (N x 6.25)	%		5.0	15.6	
Cattle	dig prot %	*	3.3	10.4	
Goats	dig prot %	*	3.3	10.4	
Horses	dig prot %	*	3.3	10.4	
Sheep	dig prot %	*	3.3	10.4	
Energy	GE Mcal/kg				
Cattle	DE Mcal/kg	*	0.90	2.82	

Feed Name or Analyses			Mean As Fed	Mean Dry	C.V. ± %
Sheep	DE Mcal/kg	*	0.81	2.53	
Cattle	ME Mcal/kg	*	0.74	2.31	
Sheep	ME Mcal/kg	*	0.66	2.08	
Cattle	TDN %	*	20.4	63.9	
Sheep	TDN %	*	18.3	57.4	

Clover-Timothy, aerial part w molasses added, ensiled, (3)

Ref No 3 01 494 United States

Feed Name or Analyses			Mean As Fed	Mean Dry	C.V. ± %
Dry matter	%		34.3	100.0	6
Ash	%		2.7	8.0	8
Crude fiber	%		8.7	25.4	3
Ether extract	%		2.2	6.3	19
N-free extract	%		15.7	45.7	
Protein (N x 6.25)	%		5.0	14.6	6
Cattle	dig prot %	*	3.3	9.5	
Goats	dig prot %	*	3.3	9.5	
Horses	dig prot %	*	3.3	9.5	
Sheep	dig prot %	*	3.3	9.5	
Energy	GE Mcal/kg				
Cattle	DE Mcal/kg	*	0.99	2.90	
Sheep	DE Mcal/kg	*	0.94	2.75	
Cattle	ME Mcal/kg	*	0.81	2.38	
Sheep	ME Mcal/kg	*	0.77	2.26	
Cattle	TDN %	*	22.5	65.7	
Sheep	TDN %	*	21.4	62.5	

CLOVER-VETCH. Trifolium spp, Vicia spp

Clover-Vetch, aerial part, fresh, (2)

Ref No 2 01 495 United States

Feed Name or Analyses			Mean As Fed	Mean Dry	C.V. ± %
Dry matter	%			100.0	
Cobalt	mg/kg			0.099	

COBALT ACETATE

Cobalt acetate, $Co(CH_3CO_2)_2.4H_2O$, cp, (6)

Ref No 6 01 553 United States

Feed Name or Analyses			Mean As Fed	Mean Dry	C.V. ± %
Dry matter	%		71.1	100.0	
Chlorine	%		0.00	0.00	
Cobalt	%		23.6	33.3	
Copper	%		0.0	0.0	
Iron	%		0.001	0.001	
Sulphur	%		0.07	0.10	
Zinc	%		0.0	0.0	

COBALT CHLORIDE

Cobalt chloride, hydrated, $CoCl_2 \cdot 6H_2O$, cp, (6)

Ref No 6 01 557 United States

Feed Name or Analyses			Mean As Fed	Mean Dry	C.V. ± %
Dry matter	%		54.6	100.0	
Chlorine	%		29.80	54.58	
Cobalt	%		24.8	45.3	
Copper	%		0.0	0.0	
Iron	%		0.002	0.004	
Sulphur	%		0.07	0.13	
Zinc	%		0.0	0.0	

COBALT OXIDE

Cobalt oxide, Co_2O_3, cp, (6)

Ref No 6 01 559 United States

Feed Name or Analyses			Mean As Fed	Mean Dry	C.V. ± %
Dry matter	%			100.0	
Chlorine	%			0.01	
Cobalt	%			72.0	
Iron	%			0.050	
Sulphur	%			0.20	

COBALT SULFATE

Cobalt sulfate, heptahydrate, $CoSO_4 \cdot 7H_2O$, cp, (6)

Ref No 6 01 563 United States

Feed Name or Analyses			Mean As Fed	Mean Dry	C.V. ± %
Dry matter	%		55.2	100.0	
Chlorine	%		0.00	0.00	
Cobalt	%		21.0	38.0	
Copper	%		0.0	0.0	
Iron	%		0.001	0.002	
Sulphur	%		11.40	20.65	
Zinc	%		0.0	0.0	

COBALTOUS CARBONATE

Cobaltous carbonate, $CoCO_3$, cp, (6)

Ref No 6 01 565 United States

Feed Name or Analyses			Mean As Fed	Mean Dry	C.V. ± %
Dry matter	%			100.0	
Chlorine	%			0.01	
Cobalt	%			72.0	
Iron	%			0.050	
Sulphur	%			0.20	

COCKCHAFER. Melolontha vulgaris

Cockchafer, s-c, (5)

Ref No 5 01 567 United States

Feed Name or Analyses			Mean As Fed	Mean Dry	C.V. ± %
Dry matter	%		85.3	100.0	
Ash	%		7.0	8.2	
Crude fiber	%		13.1	15.4	
Swine	dig coef %		-1.	-1.	
Ether extract	%		5.6	6.6	
Swine	dig coef %		90.	90.	
N-free extract	%		4.4	5.2	
Swine	dig coef %		44.	44.	
Protein (N x 6.25)	%		55.1	64.6	
Swine	dig coef %		69.	69.	
Swine	dig prot %		38.0	44.6	
Energy	GE Mcal/kg				
Cattle	DE Mcal/kg	*	2.54	2.98	
Sheep	DE Mcal/kg	*	2.52	2.95	
Swine	DE kcal/kg	*	2254.	2642.	
Cattle	ME Mcal/kg	*	2.09	2.44	
Sheep	ME Mcal/kg	*	2.06	2.42	
Swine	ME kcal/kg	*	1869.	2192.	
Cattle	TDN %	*	57.7	67.6	
Sheep	TDN %	*	57.1	67.0	
Swine	TDN %		51.1	59.9	

COCKSCOMB, COMMON FEATHER. Celosia cristata

Cockscomb, common feather, seeds w hulls, (4)

Ref No 4 09 043 United States

Feed Name or Analyses			Mean As Fed	Mean Dry	C.V. ± %
Dry matter	%			100.0	
Protein (N x 6.25)	%			18.0	
Cattle	dig prot %	*		12.5	
Goats	dig prot %	*		13.7	
Horses	dig prot %	*		13.7	
Sheep	dig prot %	*		13.7	
Alanine	%			0.72	
Arginine	%			1.94	
Aspartic acid	%			1.48	
Glutamic acid	%			2.97	
Glycine	%			1.17	
Histidine	%			0.47	
Hydroxyproline	%			0.09	
Isoleucine	%			0.65	
Leucine	%			1.03	
Lysine	%			0.81	
Methionine	%			0.34	
Phenylalanine	%			0.68	
Proline	%			0.40	
Serine	%			0.68	
Threonine	%			0.61	
Tyrosine	%			0.68	
Valine	%			0.79	

Cocoa meal —
see Cacao, seeds, grnd, (5)

Cocoa pods, dried —
see Cacao, pods, s-c, (1)

Cocoa shells —
see Cacao, shells, grnd, (1)

COCONUT. Cocos nucifera

Coconut, meats, dehy, (4)

Ref No 4 08 190 United States

Feed Name or Analyses			Mean As Fed	Mean Dry	C.V. ± %
Dry matter	%		96.5	100.0	
Ash	%		1.4	1.5	
Crude fiber	%		3.9	4.0	
Ether extract	%		64.9	67.3	
N-free extract	%		19.1	19.8	
Protein (N x 6.25)	%		7.2	7.5	
Cattle	dig prot %	*	2.8	2.9	
Goats	dig prot %	*	3.9	4.1	
Horses	dig prot %	*	3.9	4.1	
Sheep	dig prot %	*	3.9	4.1	
Energy	GE Mcal/kg		6.62	6.86	
Cattle	DE Mcal/kg	*	4.65	4.82	
Sheep	DE Mcal/kg	*	4.55	4.72	
Swine	DE kcal/kg	*	4502.	4666.	
Cattle	ME Mcal/kg	*	3.81	3.95	
Sheep	ME Mcal/kg	*	3.73	3.87	
Swine	ME kcal/kg	*	4254.	4409.	
Cattle	TDN %	*	105.5	109.3	
Sheep	TDN %	*	103.3	107.0	
Swine	TDN %	*	102.1	105.8	
Calcium	%		0.03	0.03	
Iron	%		0.003	0.003	
Phosphorus	%		0.19	0.20	
Potassium	%		0.59	0.61	
Niacin	mg/kg		6.0	6.2	
Riboflavin	mg/kg		0.4	0.4	
Thiamine	mg/kg		0.6	0.6	

Coconut, meats, fresh, (4)

Ref No 4 01 574 United States

Feed Name or Analyses			Mean As Fed	Mean Dry	C.V. ± %
Dry matter	%		49.1	100.0	
Ash	%		0.9	1.8	
Crude fiber	%		4.0	8.1	
Ether extract	%		35.3	71.9	
N-free extract	%		5.4	11.0	
Protein (N x 6.25)	%		3.5	7.1	
Cattle	dig prot %	*	1.3	2.6	
Goats	dig prot %	*	1.9	3.8	
Horses	dig prot %	*	1.9	3.8	
Sheep	dig prot %	*	1.9	3.8	
Energy	GE Mcal/kg		3.46	7.05	
Cattle	DE Mcal/kg	*	2.34	4.77	
Sheep	DE Mcal/kg	*	2.29	4.66	
Swine	DE kcal/kg	*	2085.	4247.	
Cattle	ME Mcal/kg	*	1.92	3.91	
Sheep	ME Mcal/kg	*	1.88	3.82	
Swine	ME kcal/kg	*	1972.	4016.	
Cattle	TDN %	*	53.1	108.1	
Sheep	TDN %	*	51.9	105.7	
Swine	TDN %	*	47.3	96.3	
Calcium	%		0.01	0.03	
Iron	%		0.002	0.004	
Phosphorus	%		0.10	0.19	
Potassium	%		0.26	0.52	
Sodium	%		0.02	0.05	
Ascorbic acid	mg/kg		30.0	61.1	
Niacin	mg/kg		5.0	10.2	
Riboflavin	mg/kg		0.2	0.4	
Thiamine	mg/kg		0.5	1.0	

Coconut, meats, mech-extd grnd, high fat, (4)

Ref No 4 08 388 United States

Feed Name or Analyses			Mean As Fed	Mean Dry	C.V. ± %
Dry matter	%		93.1	100.0	
Ash	%		6.5	7.0	
Crude fiber	%		11.3	12.1	
Ether extract	%		12.0	12.9	
N-free extract	%		42.9	46.1	
Protein (N x 6.25)	%		20.4	21.9	
Cattle	dig prot %	*	15.0	16.1	
Goats	dig prot %	*	16.1	17.3	
Horses	dig prot %	*	16.1	17.3	
Sheep	dig prot %	*	16.1	17.3	
Energy	GE Mcal/kg				
Cattle	DE Mcal/kg	*	3.68	3.95	
Sheep	DE Mcal/kg	*	3.62	3.89	
Swine	DE kcal/kg	*	4266.	4582.	
Cattle	ME Mcal/kg	*	3.02	3.24	
Sheep	ME Mcal/kg	*	2.97	3.19	
Swine	ME kcal/kg	*	3906.	4196.	
Cattle	TDN %	*	83.3	89.5	
Sheep	TDN %	*	82.2	88.3	
Swine	TDN %	*	96.8	103.9	

Coconut, meats w molasses added, extn unspecified grnd, (4)

Candied copra

Ref No 4 08 389 United States

Feed Name or Analyses		Mean As Fed	Mean Dry	C.V. ± %
Dry matter	%	85.7	100.0	
Ash	%	6.9	8.1	
Crude fiber	%	6.1	7.1	
Ether extract	%	2.5	2.9	
N-free extract	%	57.4	67.0	
Protein (N x 6.25)	%	12.8	14.9	
Cattle	dig prot %	* 8.3	9.7	
Goats	dig prot %	* 9.4	10.9	
Horses	dig prot %	* 9.4	10.9	
Sheep	dig prot %	* 9.4	10.9	
Energy	GE Mcal/kg			
Cattle	DE Mcal/kg	* 2.67	3.12	
Sheep	DE Mcal/kg	* 2.91	3.39	
Swine	DE kcal/kg	*2504.	2922.	
Cattle	ME Mcal/kg	* 2.19	2.56	
Sheep	ME Mcal/kg	* 2.38	2.78	
Swine	ME kcal/kg	*2328.	2717.	
Cattle	TDN %	* 60.6	70.7	
Sheep	TDN %	* 65.9	76.9	
Swine	TDN %	* 56.8	66.3	

Coconut, meats, extn unspecified grnd, (5)

Coconut meal (CFA)

Copra meal (CFA)

Coconut oil meal, extraction unspecified

Copra oil meal, extraction unspecified

Ref No 5 01 570 United States

Feed Name or Analyses		Mean As Fed	Mean Dry	C.V. ± %
Dry matter	%	88.2	100.0	2
Ash	%	5.9	6.7	14
Crude fiber	%	12.3	14.0	27
Cattle	dig coef %	63.	63.	
Swine	dig coef %	60.	60.	
Ether extract	%	7.4	8.4	24
Cattle	dig coef %	98.	98.	
Swine	dig coef %	83.	83.	
N-free extract	%	39.9	45.2	
Cattle	dig coef %	78.	78.	
Swine	dig coef %	89.	89.	
Protein (N x 6.25)	%	22.7	25.7	7
Cattle	dig coef %	81.	81.	
Swine	dig coef %	73.	73.	
Cattle	dig prot %	18.4	20.8	
Swine	dig prot %	16.6	18.8	
Energy	GE Mcal/kg			
Cattle	DE Mcal/kg	* 3.25	3.68	
Sheep	DE Mcal/kg	* 3.14	3.56	
Swine	DE kcal/kg	*3235.	3666.	
Cattle	ME Mcal/kg	* 2.66	3.02	
Chickens	ME_n kcal/kg	1489.	1689.	
Sheep	ME Mcal/kg	* 2.58	2.92	
Swine	ME kcal/kg	*2937.	3329.	
Cattle	TDN %	73.7	83.5	
Sheep	TDN %	* 71.2	80.7	
Swine	TDN %	73.4	83.1	
Calcium	%	0.17	0.20	34
Copper	mg/kg	9.2	10.4	
Iron	%	0.066	0.075	
Magnesium	%	0.35	0.39	
Manganese	mg/kg	73.2	82.9	24
Phosphorus	%	0.59	0.67	12
Potassium	%	1.85	2.09	9
Sulphur	%	0.33	0.37	
Choline	mg/kg	1352.	1533.	8
Folic acid	mg/kg	0.77	0.87	
Niacin	mg/kg	37.3	42.3	
Pantothenic acid	mg/kg	5.1	5.8	72
Riboflavin	mg/kg	4.1	4.6	
Thiamine	mg/kg	1.1	1.2	
Arginine	%	2.13	2.41	6
Histidine	%	0.29	0.33	11
Isoleucine	%	0.97	1.10	16
Leucine	%	1.35	1.54	16
Lysine	%	0.48	0.55	12
Methionine	%	0.29	0.33	20
Phenylalanine	%	0.77	0.88	12
Threonine	%	0.58	0.66	5
Tryptophan	%	0.19	0.22	
Tyrosine	%	0.58	0.66	27
Valine	%	0.97	1.10	6

(1) dry forages and roughages (2) pasture, range plants, and forages fed green (3) silages (4) energy feeds (5) protein supplements (6) minerals (7) vitamins (8) additives

Coconut, meats, extn unspecified steamed grnd, (5)

Ref No 5 01 571 United States

Feed Name or Analyses		Mean As Fed	Mean Dry	C.V. ± %
Dry matter	%	90.8	100.0	
Ash	%	6.4	7.0	
Crude fiber	%	12.4	13.7	
Sheep	dig coef %	42.	42.	
Ether extract	%	12.7	14.0	
Sheep	dig coef %	100.	100.	
N-free extract	%	39.5	43.5	
Sheep	dig coef %	85.	85.	
Protein (N x 6.25)	%	19.8	21.8	
Sheep	dig coef %	84.	84.	
Sheep	dig prot %	16.6	18.3	
Energy	GE Mcal/kg			
Cattle	DE Mcal/kg	* 3.60	3.97	
Sheep	DE Mcal/kg	* 3.70	4.08	
Swine	DE kcal/kg	*4170.	4592.	
Cattle	ME Mcal/kg	* 2.96	3.26	
Sheep	ME Mcal/kg	* 3.04	3.35	
Swine	ME kcal/kg	*3819.	4206.	
Cattle	TDN %	* 81.8	90.1	
Sheep	TDN %	84.0	92.5	
Swine	TDN %	* 94.6	104.2	

Coconut, meats, mech-extd grnd, (5)

Coconut meal, mechanical extracted (AAFCO)

Copra meal, mechanical extracted (AAFCO)

Coconut oil meal, expeller extracted

Coconut oil meal, hydraulic extracted

Copra oil meal, expeller extracted

Copra oil meal, hydraulic extracted

Ref No 5 01 572 United States

Feed Name or Analyses		Mean As Fed	Mean Dry	C.V. ± %
Dry matter	%	92.7	100.0	2
Ash	%	6.6	7.2	17
Crude fiber	%	11.3	12.2	13
Cattle	dig coef %	* 63.	63.	
Sheep	dig coef %	* 42.	42.	
Ether extract	%	6.7	7.2	29
Cattle	dig coef %	* 98.	98.	
Sheep	dig coef %	* 100.	100.	
N-free extract	%	47.4	51.1	
Cattle	dig coef %	* 78.	78.	
Sheep	dig coef %	* 85.	85.	
Protein (N x 6.25)	%	20.7	22.3	4
Cattle	dig coef %	* 81.	81.	
Sheep	dig coef %	* 84.	84.	
Cattle	dig prot %	16.7	18.1	
Sheep	dig prot %	17.4	18.7	
Energy	GE Mcal/kg			
Cattle	DE Mcal/kg	* 3.33	3.59	
Sheep	DE Mcal/kg	* 3.41	3.68	
Swine	DE kcal/kg	*3855.	4160.	
Cattle	ME Mcal/kg	* 2.73	2.95	
Chickens	ME_n kcal/kg	1773.	1913.	
Sheep	ME Mcal/kg	* 2.80	3.02	
Swine	ME kcal/kg	*3527.	3807.	
Cattle	NE_m Mcal/kg	* 1.74	1.88	
Cattle	NE_{gain} Mcal/kg	* 1.16	1.25	
Cattle	$NE_{lactating\ cows}$ Mcal/kg	* 2.07	2.23	
Cattle	TDN %	75.5	81.5	
Sheep	TDN %	77.4	83.5	
Swine	TDN %	* 87.4	94.4	
Calcium	%	0.21	0.23	21
Cobalt	mg/kg	0.128	0.138	
Copper	mg/kg	14.1	15.2	
Iron	%	0.132	0.142	
Magnesium	%	0.31	0.33	
Manganese	mg/kg	65.4	70.6	24
Phosphorus	%	0.62	0.67	9
Potassium	%	1.53	1.65	
Sodium	%	0.04	0.04	
Sulphur	%	0.34	0.37	
Choline	mg/kg	1092.	1178.	
Folic acid	mg/kg	1.39	1.50	
Niacin	mg/kg	27.9	30.2	27
Pantothenic acid	mg/kg	6.2	6.7	15
Riboflavin	mg/kg	3.3	3.6	22
Thiamine	mg/kg	0.7	0.7	
Arginine	%	2.41	2.60	
Lysine	%	0.70	0.76	
Methionine	%	0.33	0.36	
Tryptophan	%	0.20	0.22	

Coconut, meats, solv-extd grnd, (5)

Copra meal, solvent extracted (AAFCO)

Coconut oil meal, solvent extracted

Copra oil meal, solvent extracted

Coconut meal, solvent extracted (AAFCO)

Ref No 5 01 573 United States

Feed Name or Analyses		Mean As Fed	Mean Dry	C.V. ± %
Dry matter	%	91.2	100.0	1
Ash	%	6.3	6.9	40
Crude fiber	%	14.0	15.4	19
Cattle	dig coef %	* 63.	63.	
Sheep	dig coef %	* 42.	42.	
Ether extract	%	2.1	2.3	56
Cattle	dig coef %	* 98.	98.	
Sheep	dig coef %	* 100.	100.	
N-free extract	%	47.5	52.1	
Cattle	dig coef %	* 78.	78.	
Sheep	dig coef %	* 85.	85.	
Protein (N x 6.25)	%	21.3	23.4	2
Cattle	dig coef %	* 71.	71.	
Sheep	dig coef %	* 84.	84.	
Cattle	dig prot %	15.1	16.6	
Sheep	dig prot %	17.9	19.7	
Lignin (Ellis)	%	0.7	0.8	
Energy	GE Mcal/kg			
Cattle	DE Mcal/kg	* 2.89	3.17	
Sheep	DE Mcal/kg	* 3.04	3.33	
Swine	DE kcal/kg	*3341.	3665.	

Feed Name or Analyses		Mean As Fed	Mean Dry	C.V. ± %
Cattle	ME Mcal/kg	* 2.37	2.60	
Sheep	ME Mcal/kg	* 2.49	2.73	
Swine	ME kcal/kg	*3050.	3345.	
Cattle	NE_m Mcal/kg	* 1.51	1.66	
Cattle	NE_{gain} Mcal/kg	* 0.98	1.08	
Cattle	$NE_{lactating\ cows}$ Mcal/kg	* 1.80	1.97	
Cattle	TDN %	65.6	72.0	
Sheep	TDN %	68.9	75.5	
Swine	TDN %	* 75.8	83.1	
Calcium	%	0.17	0.19	98
Chlorine	%	0.03	0.03	
Phosphorus	%	0.61	0.67	8
Choline	mg/kg	1263.	1385.	
Niacin	mg/kg	31.0	34.0	
Pantothenic acid	mg/kg	5.8	6.3	
Riboflavin	mg/kg	3.5	3.9	
Arginine	%	2.41	2.65	
Lysine	%	0.70	0.77	
Methionine	%	0.33	0.36	
Tryptophan	%	0.20	0.22	

Coconut meal (CFA) –
see Coconut, meats, extn unspecified grnd, (5)

Coconut meal, mechanical extracted (AAFCO) –
see Coconut, meats, mech-extd grnd, (5)

Coconut meal, mechanical extracted (AAFCO) –
see Coconut, meats, mech-extd grnd, (5)

Coconut meal, mechanical extracted (AAFCO) –
see Coconut, meats, mech-extd grnd, (5)

Coconut meal, solvent extracted (AAFCO) –
see Coconut, meats, solv-extd grnd, (5)

Coconut oil meal, expeller extracted –
see Coconut, meats, mech-extd grnd, (5)

Coconut oil meal, extraction unspecified –
see Coconut, meats, extn unspecified grnd, (5)

Coconut oil meal, hydraulic extracted –
see Coconut, meats, mech-extd grnd, (5)

Coconut oil meal, solvent extracted –
see Coconut, meats, solv-extd grnd, (5)

Cod liver oil (CFA) –
see Fish, cod, liver oil, (7)

Cod liver oil meal –
see Fish, cod, livers, dehy grnd, (5)

COFFEE. Coffea spp

Coffee, grounds, (1)
Ref No 1 01 576 United States

Feed Name or Analyses		Mean As Fed	Mean Dry	C.V. ± %
Dry matter	%	73.7	100.0	32
Ash	%	1.2	1.6	50
Crude fiber	%	21.5	29.2	9
Ether extract	%	9.3	12.6	14
N-free extract	%	31.5	42.7	
Protein (N x 6.25)	%	10.2	13.8	7
Cattle	dig prot %	* 6.6	8.9	
Goats	dig prot %	* 7.0	9.5	
Horses	dig prot %	* 6.8	9.3	
Rabbits	dig prot %	* 6.9	9.4	
Sheep	dig prot %	* 6.6	9.0	
Calcium	%	0.09	0.12	25
Phosphorus	%	0.06	0.08	42

Coffee, grounds w chicory residue, dehy, (1)
Ref No 1 01 575 United States

Feed Name or Analyses		Mean As Fed	Mean Dry	C.V. ± %
Dry matter	%	92.8	100.0	
Ash	%	5.5	5.9	
Crude fiber	%	36.2	39.0	
Sheep	dig coef %	38.	38.	
Ether extract	%	6.6	7.1	
Sheep	dig coef %	82.	82.	
N-free extract	%	31.4	33.8	
Sheep	dig coef %	14.	14.	
Protein (N x 6.25)	%	13.2	14.2	
Sheep	dig coef %	1.	1.	
Cattle	dig prot %	* 8.6	9.2	
Goats	dig prot %	* 9.1	9.8	
Horses	dig prot %	* 8.9	9.6	
Rabbits	dig prot %	* 8.9	9.6	
Sheep	dig prot %	0.1	0.1	
Energy	GE Mcal/kg			
Sheep	DE Mcal/kg	* 1.34	1.45	
Sheep	ME Mcal/kg	* 1.10	1.19	
Sheep	TDN %	30.4	32.8	

Coffee, hulls, grnd, (1)
Ref No 1 01 577 United States

Feed Name or Analyses		Mean As Fed	Mean Dry	C.V. ± %
Dry matter	%	90.0	100.0	
Ash	%	4.9	5.4	
Crude fiber	%	32.6	36.2	
Ether extract	%	7.4	8.2	
N-free extract	%	29.5	32.8	
Protein (N x 6.25)	%	15.6	17.3	
Cattle	dig prot %	* 10.8	12.0	
Goats	dig prot %	* 11.5	12.7	
Horses	dig prot %	* 11.0	12.2	
Rabbits	dig prot %	* 10.8	12.1	
Sheep	dig prot %	* 10.9	12.1	
Lignin (Ellis)	%	22.0	24.4	

COLUBRINA, TEXAS. Colubrina texensis

Colubrina, Texas, browse, immature, (2)
Ref No 2 01 578 United States

Feed Name or Analyses		Mean As Fed	Mean Dry	C.V. ± %
Dry matter	%		100.0	
Ash	%		8.7	
Crude fiber	%		11.8	
Ether extract	%		4.5	
N-free extract	%		55.5	
Protein (N x 6.25)	%		19.5	
Cattle	dig prot %	*	14.5	
Goats	dig prot %	*	14.8	
Horses	dig prot %	*	14.1	
Rabbits	dig prot %	*	13.7	
Sheep	dig prot %	*	15.2	
Calcium	%		3.29	
Magnesium	%		0.11	
Phosphorus	%		0.17	
Potassium	%		1.28	

COLUMBINE, ALPINE. Aquilegia alpina

Columbine, alpine, seeds w hulls, (5)
Ref No 5 09 047 United States

Feed Name or Analyses		Mean As Fed	Mean Dry	C.V. ± %
Dry matter	%		100.0	
Protein (N x 6.25)	%		21.0	
Alanine	%		0.78	
Arginine	%		2.06	
Aspartic acid	%		1.91	
Glutamic acid	%		4.31	
Glycine	%		0.99	
Histidine	%		0.48	
Hydroxyproline	%		0.19	
Isoleucine	%		0.74	
Leucine	%		1.26	
Lysine	%		0.92	
Methionine	%		0.27	
Phenylalanine	%		0.71	
Proline	%		1.39	
Serine	%		0.92	
Threonine	%		0.67	
Tyrosine	%		0.76	
Valine	%		0.92	

Colza oil meal, extraction unspecified –
see Rape, seeds, extn unspecified grnd, (5)

COMANDRA, COMMON. Comandra pallida

Comandra, common, seeds w hulls, (4)
Ref No 4 09 293 United States

Feed Name or Analyses		Mean As Fed	Mean Dry	C.V. ± %
Dry matter	%		100.0	
Protein (N x 6.25)	%		8.0	
Cattle	dig prot %	*	3.4	
Goats	dig prot %	*	4.6	
Horses	dig prot %	*	4.6	
Sheep	dig prot %	*	4.6	
Alanine	%		0.30	
Arginine	%		0.64	
Aspartic acid	%		0.66	
Glutamic acid	%		0.88	
Glycine	%		0.30	
Histidine	%		0.14	
Hydroxyproline	%		0.11	
Isoleucine	%		0.24	
Leucine	%		0.43	
Lysine	%		0.28	
Methionine	%		0.09	
Phenylalanine	%		0.22	
Proline	%		0.31	
Serine	%		0.32	
Threonine	%		0.24	
Tyrosine	%		0.18	
Valine	%		0.27	

COMFREY, PRICKLY. Symphytum asperrimum

Comfrey, prickly, aerial part, fresh, (2)
Ref No 2 01 579 United States

Feed Name or Analyses		Mean As Fed	Mean Dry	C.V. ± %
Dry matter	%	13.2	100.0	14
Ash	%	2.4	18.2	15
Crude fiber	%	1.8	13.8	13
Ether extract	%	0.3	2.4	9

Continued

Feed Name or Analyses			Mean As Fed	Mean Dry	C.V. ± %
N-free extract	%		6.0	45.8	
Protein (N x 6.25)	%		2.6	19.7	10
Cattle	dig prot %	*	1.9	14.6	
Goats	dig prot %	*	2.0	14.9	
Horses	dig prot %	*	1.9	14.2	
Rabbits	dig prot %	*	1.8	13.9	
Sheep	dig prot %	*	2.0	15.3	
Energy	GE Mcal/kg				
Cattle	DE Mcal/kg	*	0.34	2.57	
Sheep	DE Mcal/kg	*	0.38	2.88	
Cattle	ME Mcal/kg	*	0.28	2.11	
Sheep	ME Mcal/kg	*	0.31	2.36	
Cattle	TDN %	*	7.7	58.3	
Sheep	TDN %	*	8.6	65.3	
Phosphorus	%		0.07	0.55	
Potassium	%		0.60	4.53	

CONDALIA, LOTEWOOD. Condalia obtusifolia

Condalia, lotewood, browse, immature, (2)
Ref No 2 01 580 United States

Feed Name or Analyses			Mean As Fed	Mean Dry	C.V. ± %
Dry matter	%			100.0	
Ash	%			7.7	
Crude fiber	%			10.1	
Ether extract	%			2.4	
N-free extract	%			57.3	
Protein (N x 6.25)	%			22.5	
Cattle	dig prot %	*		17.0	
Goats	dig prot %	*		17.6	
Horses	dig prot %	*		16.6	
Rabbits	dig prot %	*		16.0	
Sheep	dig prot %	*		18.0	
Calcium	%			1.50	
Magnesium	%			0.31	
Phosphorus	%			0.22	
Potassium	%			2.39	

Condalia, lotewood, browse, mature, (2)
Ref No 2 01 581 United States

Feed Name or Analyses			Mean As Fed	Mean Dry	C.V. ± %
Dry matter	%			100.0	
Ash	%			5.4	
Crude fiber	%			22.6	
Ether extract	%			2.6	
N-free extract	%			54.3	
Protein (N x 6.25)	%			15.1	
Cattle	dig prot %	*		10.7	
Goats	dig prot %	*		10.6	
Horses	dig prot %	*		10.3	
Rabbits	dig prot %	*		10.3	
Sheep	dig prot %	*		11.1	

Condensed beet solubles product (AAFCO) –
see Beet, sugar, solubles w low potassium salts and glutamic acid, condensed, (5)

Condensed buttermilk (AAFCO) –
see Cattle, buttermilk, condensed, mn 27% total solids 0.055% fat mx 0.14% ash per 1% solids, (5)

Condensed fermented corn extractives (AAFCO) –
see Corn, steepwater solubles, fermented condensed, (4)

Condensed fish solubles (AAFCO) –
see Fish, stickwater solubles, condensed, mn 30% protein, (5)

Condensed fish solubles (CFA) –
see Fish, stickwater solubles, cooked condensed, (5)

Condensed whey (AAFCO) –
see Cattle, whey, condensed, mn solids declared, (4)

Condensed whey-product (AAFCO) –
see Cattle, whey low lactose, condensed, mn solids declared, (4)

Condensed whey solubles (AAFCO) –
see Cattle, whey wo albumin low lactose, condensed, (4)

CONEFLOWER, PINEWOODS. Rudbeckia bicolor

Coneflower, pinewoods, seeds w hulls, (5)
Ref No 5 09 116 United States

Feed Name or Analyses			Mean As Fed	Mean Dry	C.V. ± %
Dry matter	%			100.0	
Protein (N x 6.25)	%			29.0	
Alanine	%			1.04	
Arginine	%			2.78	
Aspartic acid	%			2.58	
Glutamic acid	%			5.77	
Glycine	%			1.33	
Histidine	%			0.61	
Hydroxyproline	%			0.32	
Isoleucine	%			1.04	
Leucine	%			1.68	
Lysine	%			0.87	
Methionine	%			0.70	
Phenylalanine	%			1.31	
Proline	%			1.10	
Serine	%			1.02	
Threonine	%			0.84	
Tyrosine	%			0.70	
Valine	%			1.31	

CONIFER. Coniferae (family)

Conifer, cones, grnd, (1)
Ref No 1 01 582 United States

Feed Name or Analyses			Mean As Fed	Mean Dry	C.V. ± %
Dry matter	%		79.8	100.0	
Ash	%		2.7	3.4	
Crude fiber	%		34.1	42.7	
Sheep	dig coef %		51.	51.	
Ether extract	%		7.7	9.6	
Sheep	dig coef %		45.	45.	
N-free extract	%		27.0	33.8	
Sheep	dig coef %		50.	50.	
Protein (N x 6.25)	%		8.4	10.5	
Sheep	dig coef %		10.	10.	
Cattle	dig prot %	*	4.8	6.0	
Goats	dig prot %	*	5.1	6.4	
Horses	dig prot %	*	5.1	6.4	
Rabbits	dig prot %	*	5.4	6.8	
Sheep	dig prot %		0.8	1.1	
Energy	GE Mcal/kg				
Cattle	DE Mcal/kg	*	1.57	1.97	
Sheep	DE Mcal/kg	*	1.74	2.18	
Cattle	ME Mcal/kg	*	1.29	1.61	
Sheep	ME Mcal/kg	*	1.43	1.79	
Cattle	TDN %	*	35.6	44.6	
Sheep	TDN %		39.5	49.4	

COPPERLEAF, SLENDER. Acalypha spp

Copperleaf, slender, browse, early bloom, (2)
Ref No 2 01 584 United States

Feed Name or Analyses			Mean As Fed	Mean Dry	C.V. ± %
Dry matter	%			100.0	
Calcium	%			2.29	
Magnesium	%			0.37	
Phosphorus	%			0.19	
Potassium	%			1.89	

Copra meal (CFA) –
see Coconut, meats, extn unspecified grnd, (5)

Copra meal, mechanical extracted (AAFCO) –
see Coconut, meats, mech-extd grnd, (5)

Copra oil meal, expeller extracted –
see Coconut, meats, mech-extd grnd, (5)

Copra oil meal, extraction unspecified –
see Coconut, meats, extn unspecified grnd, (5)

Copra oil meal, hydraulic extracted –
see Coconut, meats, mech-extd grnd, (5)

Copra oil meal, solvent extracted –
see Coconut, meats, solv-extd grnd, (5)

CORDGRASS. Spartina spp

Cordgrass, hay, s-c, (1)
Ref No 1 01 585 United States

Feed Name or Analyses			Mean As Fed	Mean Dry	C.V. ± %
Dry matter	%		81.6	100.0	
Ash	%		5.9	7.2	
Crude fiber	%		29.4	36.0	
Sheep	dig coef %		57.	57.	
Ether extract	%		1.7	2.1	
Sheep	dig coef %		47.	47.	
N-free extract	%		39.2	48.0	
Sheep	dig coef %		50.	50.	
Protein (N x 6.25)	%		5.5	6.7	
Sheep	dig coef %		43.	43.	
Cattle	dig prot %	*	2.2	2.7	
Goats	dig prot %	*	2.3	2.8	
Horses	dig prot %	*	2.6	3.2	
Rabbits	dig prot %	*	3.1	3.8	
Sheep	dig prot %		2.4	2.9	
Energy	GE Mcal/kg				
Cattle	DE Mcal/kg	*	1.91	2.34	
Sheep	DE Mcal/kg	*	1.79	2.19	
Cattle	ME Mcal/kg	*	1.57	1.92	
Sheep	ME Mcal/kg	*	1.46	1.79	
Cattle	TDN %	*	43.4	53.2	
Sheep	TDN %		40.5	49.6	

(1) dry forages and roughages (2) pasture, range plants, and forages fed green (3) silages (4) energy feeds (5) protein supplements (6) minerals (7) vitamins (8) additives

Feed Name or Analyses		As Fed	Dry	C.V. ± %

Cordgrass, aerial part, fresh, (2)
Ref No 2 01 588 United States

Feed Name or Analyses		As Fed	Dry	C.V. ± %
Dry matter	%		100.0	
Ash	%		8.6	23
Crude fiber	%		33.8	5
Ether extract	%		2.3	15
N-free extract	%		48.4	
Protein (N x 6.25)	%		6.9	32
Cattle	dig prot % *		3.8	
Goats	dig prot % *		3.0	
Horses	dig prot % *		3.4	
Rabbits	dig prot % *		4.0	
Sheep	dig prot % *		3.4	
Calcium	%		0.38	24
Phosphorus	%		0.11	26

Cordgrass, aerial part, fresh, immature, (2)
Ref No 2 01 586 United States

Feed Name or Analyses		As Fed	Dry	C.V. ± %
Dry matter	%		100.0	
Ash	%		10.5	17
Crude fiber	%		32.0	4
Ether extract	%		2.5	9
N-free extract	%		43.9	
Protein (N x 6.25)	%		11.1	12
Cattle	dig prot % *		7.3	
Goats	dig prot % *		6.9	
Horses	dig prot % *		7.0	
Rabbits	dig prot % *		7.2	
Sheep	dig prot % *		7.3	

Cordgrass, aerial part, fresh, mature, (2)
Ref No 2 01 587 United States

Feed Name or Analyses		As Fed	Dry	C.V. ± %
Dry matter	%		100.0	
Ash	%		8.0	19
Crude fiber	%		34.4	4
Ether extract	%		2.3	11
N-free extract	%		49.9	
Protein (N x 6.25)	%		5.4	16
Cattle	dig prot % *		2.5	
Goats	dig prot % *		1.6	
Horses	dig prot % *		2.1	
Rabbits	dig prot % *		2.8	
Sheep	dig prot % *		2.0	
Calcium	%		0.34	18
Phosphorus	%		0.10	22

CORDGRASS, GULF. Spartina spartinae

Cordgrass, gulf, aerial part, fresh, immature, (2)
Ref No 2 01 589 United States

Feed Name or Analyses		As Fed	Dry	C.V. ± %
Dry matter	%		100.0	
Ash	%		10.0	14
Crude fiber	%		32.1	3
Ether extract	%		2.5	8
N-free extract	%		44.3	
Protein (N x 6.25)	%		11.1	11
Cattle	dig prot % *		7.3	
Goats	dig prot % *		6.9	
Horses	dig prot % *		7.0	
Rabbits	dig prot % *		7.2	
Sheep	dig prot % *		7.3	
Calcium	%		0.56	13
Phosphorus	%		0.16	15

Cordgrass, gulf, aerial part, fresh, mature, (2)
Ref No 2 01 590 United States

Feed Name or Analyses		As Fed	Dry	C.V. ± %
Dry matter	%		100.0	
Ash	%		8.3	6
Crude fiber	%		35.1	3
Ether extract	%		2.3	8
N-free extract	%		48.9	
Protein (N x 6.25)	%		5.4	12
Cattle	dig prot % *		2.5	
Goats	dig prot % *		1.6	
Horses	dig prot % *		2.1	
Rabbits	dig prot % *		2.8	
Sheep	dig prot % *		2.0	
Calcium	%		0.32	13
Phosphorus	%		0.10	14

CORDGRASS, MARSHHAY. Spartina patens

Cordgrass, marshhay, aerial part, fresh, immature, (2)
Ref No 2 01 591 United States

Feed Name or Analyses		As Fed	Dry	C.V. ± %
Dry matter	%		100.0	
Ash	%		11.4	
Crude fiber	%		31.9	
Ether extract	%		2.4	
N-free extract	%		43.2	
Protein (N x 6.25)	%		11.1	
Cattle	dig prot % *		7.3	
Goats	dig prot % *		6.9	
Horses	dig prot % *		7.0	
Rabbits	dig prot % *		7.2	
Sheep	dig prot % *		7.3	

Cordgrass, marshhay, aerial part, fresh, mature, (2)
Ref No 2 01 592 United States

Feed Name or Analyses		As Fed	Dry	C.V. ± %
Dry matter	%		100.0	
Ash	%		8.5	6
Crude fiber	%		33.3	0
Ether extract	%		2.4	2
N-free extract	%		50.8	
Protein (N x 6.25)	%		5.0	3
Cattle	dig prot % *		2.1	
Goats	dig prot % *		1.2	
Horses	dig prot % *		1.8	
Rabbits	dig prot % *		2.5	
Sheep	dig prot % *		1.7	
Calcium	%		0.37	16
Phosphorus	%		0.09	30

CORDGRASS, PRAIRIE. Spartina pectinata

Cordgrass, prairie, aerial part, fresh, mature, (2)
Ref No 2 01 593 United States

Feed Name or Analyses		As Fed	Dry	C.V. ± %
Dry matter	%		100.0	
Ash	%		5.3	
Crude fiber	%		35.6	
Ether extract	%		2.1	
N-free extract	%		49.9	
Protein (N x 6.25)	%		7.1	
Cattle	dig prot % *		3.9	
Goats	dig prot % *		3.2	
Horses	dig prot % *		3.6	
Rabbits	dig prot % *		4.2	
Sheep	dig prot % *		3.6	
Calcium	%		0.25	
Phosphorus	%		0.11	

CORN. Zea mays

Corn, aerial part, dehy, (1)
Dehydrated corn plant (AAFCO)
Ref No 1 02 768 United States

Feed Name or Analyses		As Fed	Dry	C.V. ± %
Dry matter	%	89.6	100.0	2
Ash	%	4.8	5.4	33
Crude fiber	%	21.3	23.8	23
Sheep	dig coef %	50.	50.	
Ether extract	%	2.6	2.9	12
Sheep	dig coef %	67.	67.	
N-free extract	%	53.6	59.8	
Sheep	dig coef %	67.	67.	
Protein (N x 6.25)	%	7.3	8.2	9
Sheep	dig coef %	45.	45.	
Cattle	dig prot % *	3.6	4.0	
Goats	dig prot % *	3.8	4.2	
Horses	dig prot % *	4.0	4.5	
Rabbits	dig prot % *	4.5	5.0	
Sheep	dig prot %	3.3	3.7	
Energy	GE Mcal/kg			
Cattle	DE Mcal/kg *	2.46	2.74	
Sheep	DE Mcal/kg *	2.37	2.64	
Cattle	ME Mcal/kg *	2.02	2.25	
Sheep	ME Mcal/kg *	1.94	2.17	
Cattle	TDN % *	55.7	62.2	
Sheep	TDN %	53.7	59.9	

Corn, aerial part, dehy, pre-bloom, (1)
Ref No 1 02 767 United States

Feed Name or Analyses		As Fed	Dry	C.V. ± %
Dry matter	%	91.6	100.0	
Ash	%	6.0	6.6	
Crude fiber	%	24.4	26.6	
Cattle	dig coef %	71.	71.	
Sheep	dig coef %	42.	42.	
Ether extract	%	2.5	2.8	
Cattle	dig coef %	60.	60.	
Sheep	dig coef %	73.	73.	
N-free extract	%	51.4	56.1	
Cattle	dig coef %	72.	72.	
Sheep	dig coef %	70.	70.	
Protein (N x 6.25)	%	7.3	8.0	
Cattle	dig coef %	49.	49.	
Sheep	dig coef %	43.	43.	
Cattle	dig prot %	3.6	3.9	
Goats	dig prot % *	3.6	4.0	
Horses	dig prot % *	3.9	4.3	
Rabbits	dig prot % *	4.4	4.8	
Sheep	dig prot %	3.1	3.4	
Energy	GE Mcal/kg			
Cattle	DE Mcal/kg *	2.70	2.95	
Sheep	DE Mcal/kg *	2.36	2.57	
Cattle	ME Mcal/kg *	2.22	2.42	
Sheep	ME Mcal/kg *	1.93	2.11	
Cattle	TDN %	61.3	66.9	
Sheep	TDN %	53.5	58.4	

Feed Name or Analyses			Mean As Fed	Mean Dry	C.V. ± %

Corn, aerial part, s-c, (1)
Corn fodder, sun-cured
Ref No 1 02 775 United States

Feed Name or Analyses			As Fed	Dry	C.V. ± %
Dry matter	%		78.9	100.0	12
Ash	%		5.2	6.6	18
Crude fiber	%		21.0	26.7	9
Cattle	dig coef %	*	69.	69.	
Sheep	dig coef %	*	68.	68.	
Ether extract	%		1.9	2.4	3
Cattle	dig coef %	*	73.	73.	
Sheep	dig coef %	*	65.	65.	
N-free extract	%		43.9	55.6	7
Cattle	dig coef %	*	70.	70.	
Sheep	dig coef %	*	67.	67.	
Protein (N x 6.25)	%		6.8	8.6	12
Cattle	dig coef %	*	46.	46.	
Sheep	dig coef %	*	53.	53.	
Cattle	dig prot %		3.1	4.0	
Goats	dig prot %	*	3.6	4.6	
Horses	dig prot %	*	3.8	4.8	
Rabbits	dig prot %	*	4.2	5.3	
Sheep	dig prot %		3.6	4.6	
Energy	GE Mcal/kg				
Cattle	DE Mcal/kg	*	2.27	2.88	
Sheep	DE Mcal/kg	*	2.21	2.80	
Cattle	ME Mcal/kg	*	1.86	2.36	
Sheep	ME Mcal/kg	*	1.81	2.30	
Cattle	NE_m Mcal/kg	*	1.14	1.44	
Cattle	NE_{gain} Mcal/kg	*	0.48	0.61	
Cattle	$NE_{lactating\ cows}$ Mcal/kg	*	1.29	1.64	
Cattle	TDN %		51.5	65.3	
Sheep	TDN %		50.1	63.5	
Calcium	%		0.24	0.30	
Chlorine	%		0.15	0.19	
Copper	mg/kg		3.8	4.8	
Iron	%		0.008	0.010	
Magnesium	%		0.14	0.18	
Manganese	mg/kg		53.8	68.2	
Phosphorus	%		0.14	0.18	
Potassium	%		0.78	0.99	
Sodium	%		0.03	0.03	
Sulphur	%		0.11	0.14	

Corn, aerial part, s-c, late bloom, (1)
Ref No 1 02 769 United States

Feed Name or Analyses			As Fed	Dry	C.V. ± %
Dry matter	%		87.5	100.0	
Ash	%		7.8	8.9	
Crude fiber	%		24.7	28.2	
Sheep	dig coef %		69.	69.	
Ether extract	%		1.8	2.1	
Sheep	dig coef %		60.	60.	
N-free extract	%		44.5	50.8	
Sheep	dig coef %		67.	67.	
Protein (N x 6.25)	%		8.8	10.0	
Sheep	dig coef %		55.	55.	
Cattle	dig prot %	*	4.9	5.6	
Goats	dig prot %	*	5.2	5.9	
Horses	dig prot %	*	5.3	6.0	
Rabbits	dig prot %	*	5.6	6.4	
Sheep	dig prot %		4.8	5.5	
Energy	GE Mcal/kg				
Cattle	DE Mcal/kg	*	2.40	2.24	
Sheep	DE Mcal/kg	*	2.39	2.73	
Cattle	ME Mcal/kg	*	1.97	2.25	
Sheep	ME Mcal/kg	*	1.96	2.24	
Cattle	TDN %	*	54.3	62.1	
Sheep	TDN %		54.1	61.8	

Corn, aerial part, s-c, milk stage, (1)
Ref No 1 02 770 United States

Feed Name or Analyses			As Fed	Dry	C.V. ± %
Dry matter	%		77.7	100.0	
Ash	%		6.3	8.1	
Crude fiber	%		25.9	33.4	
Cattle	dig coef %		70.	70.	
Sheep	dig coef %		57.	57.	
Ether extract	%		2.6	3.4	
Cattle	dig coef %		75.	75.	
Sheep	dig coef %		66.	66.	
N-free extract	%		38.8	50.0	
Cattle	dig coef %		63.	63.	
Sheep	dig coef %		61.	61.	
Protein (N x 6.25)	%		4.1	5.3	
Cattle	dig coef %		29.	29.	
Sheep	dig coef %		36.	36.	
Cattle	dig prot %		1.2	1.5	
Goats	dig prot %	*	1.1	1.5	
Horses	dig prot %	*	1.5	2.0	
Rabbits	dig prot %	*	2.1	2.7	
Sheep	dig prot %		1.5	1.9	
Energy	GE Mcal/kg				
Cattle	DE Mcal/kg	*	2.12	2.73	
Sheep	DE Mcal/kg	*	1.93	2.49	
Cattle	ME Mcal/kg	*	1.74	2.24	
Sheep	ME Mcal/kg	*	1.58	2.04	
Cattle	TDN %		48.2	62.0	
Sheep	TDN %		43.8	56.4	

Corn, aerial part, s-c, dough stage, (1)
Ref No 1 02 771 United States

Feed Name or Analyses			As Fed	Dry	C.V. ± %
Dry matter	%		62.5	100.0	
Ash	%		4.2	6.8	
Crude fiber	%		18.4	29.4	
Cattle	dig coef %		80.	80.	
Sheep	dig coef %		67.	67.	
Ether extract	%		1.5	2.4	
Cattle	dig coef %		72.	72.	
Sheep	dig coef %		67.	67.	
N-free extract	%		32.5	52.0	
Cattle	dig coef %		74.	74.	
Sheep	dig coef %		69.	69.	
Protein (N x 6.25)	%		6.0	9.6	
Cattle	dig coef %		57.	57.	
Sheep	dig coef %		54.	54.	
Cattle	dig prot %		3.4	5.4	
Goats	dig prot %	*	3.4	5.5	
Horses	dig prot %	*	3.5	5.6	
Rabbits	dig prot %	*	3.8	6.0	
Sheep	dig prot %		3.2	5.2	
Energy	GE Mcal/kg				
Cattle	DE Mcal/kg	*	1.96	3.14	
Sheep	DE Mcal/kg	*	1.77	2.83	
Cattle	ME Mcal/kg	*	1.61	2.57	
Sheep	ME Mcal/kg	*	1.45	2.32	
Cattle	TDN %		44.5	71.2	
Sheep	TDN %		40.2	64.2	

Corn, aerial part, s-c, mature, (1)
Corn fodder, sun-cured, mature
Ref No 1 02 772 United States

Feed Name or Analyses			As Fed	Dry	C.V. ± %
Dry matter	%		68.1	100.0	
Ash	%		4.5	6.6	
Crude fiber	%		17.5	25.8	
Cattle	dig coef %		70.	70.	
Sheep	dig coef %		80.	80.	
Ether extract	%		1.5	2.3	
Cattle	dig coef %		71.	71.	
Sheep	dig coef %		72.	72.	
N-free extract	%		37.3	54.8	
Cattle	dig coef %		73.	73.	
Sheep	dig coef %		70.	70.	
Protein (N x 6.25)	%		7.2	10.6	
Cattle	dig coef %		50.	50.	
Sheep	dig coef %		64.	64.	
Cattle	dig prot %		3.6	5.3	
Goats	dig prot %	*	4.4	6.4	
Horses	dig prot %	*	4.4	6.5	
Rabbits	dig prot %	*	4.7	6.9	
Sheep	dig prot %		4.6	6.8	
Energy	GE Mcal/kg				
Cattle	DE Mcal/kg	*	2.01	2.95	
Sheep	DE Mcal/kg	*	2.08	3.06	
Cattle	ME Mcal/kg	*	1.65	2.42	
Sheep	ME Mcal/kg	*	1.71	2.51	
Cattle	TDN %		45.5	66.9	
Sheep	TDN %		47.2	69.4	

Corn, aerial part, s-c, drouth stricken wo ears, (1)
Ref No 1 08 390 United States

Feed Name or Analyses			As Fed	Dry	C.V. ± %
Dry matter	%		85.0	100.0	
Ash	%		6.5	7.6	
Crude fiber	%		25.7	30.2	
Ether extract	%		0.8	0.9	
N-free extract	%		41.2	48.5	
Protein (N x 6.25)	%		10.8	12.7	
Cattle	dig prot %	*	6.8	7.9	
Goats	dig prot %	*	7.2	8.4	
Horses	dig prot %	*	7.1	8.3	
Rabbits	dig prot %	*	7.2	8.5	
Sheep	dig prot %	*	6.8	8.0	
Energy	GE Mcal/kg				
Cattle	DE Mcal/kg	*	2.03	2.39	
Sheep	DE Mcal/kg	*	2.21	2.60	
Cattle	ME Mcal/kg	*	1.67	1.96	
Sheep	ME Mcal/kg	*	1.81	2.13	
Cattle	TDN %	*	46.1	54.2	
Sheep	TDN %	*	50.1	58.9	

Corn, aerial part wo ears wo husks, s-c, (1)
Corn stover, sun-cured
Ref No 1 02 778 United States

Feed Name or Analyses			As Fed	Dry	C.V. ± %
Dry matter	%		79.4	100.0	15
Ash	%		5.5	6.9	9
Crude fiber	%		27.2	34.3	3
Cattle	dig coef %	*	68.	68.	
Sheep	dig coef %	*	62.	62.	
Ether extract	%		1.3	1.6	22
Cattle	dig coef %	*	54.	54.	
Sheep	dig coef %	*	60.	60.	
N-free extract	%		40.4	50.9	1
Cattle	dig coef %	*	59.	59.	

(1) dry forages and roughages (2) pasture, range plants, and forages fed green (3) silages (4) energy feeds (5) protein supplements (6) minerals (7) vitamins (8) additives

Feed Name or Analyses			Mean As Fed	Mean Dry	C.V. ± %
Sheep	dig coef %	*	58.	58.	
Protein (N x 6.25)	%		5.0	6.3	13
Cattle	dig coef %	*	36.	36.	
Sheep	dig coef %	*	31.	31.	
Cattle	dig prot %		1.8	2.3	
Goats	dig prot %	*	1.9	2.4	
Horses	dig prot %	*	2.3	2.9	
Rabbits	dig prot %	*	2.8	3.5	
Sheep	dig prot %		1.5	1.9	
Energy	GE Mcal/kg				
Cattle	DE Mcal/kg	*	2.02	2.54	
Sheep	DE Mcal/kg	*	1.92	2.42	
Cattle	ME Mcal/kg	*	1.66	2.09	
Sheep	ME Mcal/kg	*	1.58	1.99	
Cattle	TDN %		45.8	57.7	
Sheep	TDN %		43.6	54.9	
Calcium	%		0.47	0.60	
Phosphorus	%		0.08	0.10	
Potassium	%		1.30	1.64	

Corn, aerial part wo ears wo husks, s-c, mature, (1)
Corn stover, sun-cured, mature
Ref No 1 02 776 United States

Feed Name or Analyses			Mean As Fed	Mean Dry	C.V. ± %
Dry matter	%		84.4	100.0	
Ash	%		6.2	7.3	
Crude fiber	%		28.2	33.4	
Cattle	dig coef %		73.	73.	
Sheep	dig coef %		72.	72.	
Ether extract	%		1.1	1.3	
Cattle	dig coef %		61.	61.	
Sheep	dig coef %		36.	36.	
N-free extract	%		43.2	51.2	
Cattle	dig coef %		63.	63.	
Sheep	dig coef %		64.	64.	
Protein (N x 6.25)	%		5.7	6.8	
Cattle	dig coef %		40.	40.	
Sheep	dig coef %		53.	53.	
Cattle	dig prot %		2.3	2.7	
Goats	dig prot %	*	2.4	2.9	
Horses	dig prot %	*	2.8	3.3	
Rabbits	dig prot %	*	3.3	3.9	
Sheep	dig prot %		3.0	3.6	
Energy	GE Mcal/kg				
Cattle	DE Mcal/kg	*	2.26	2.68	
Sheep	DE Mcal/kg	*	2.29	2.71	
Cattle	ME Mcal/kg	*	1.86	2.20	
Sheep	ME Mcal/kg	*	1.88	2.22	
Cattle	NE_m Mcal/kg	*	1.02	1.21	
Cattle	NE_{gain} Mcal/kg	*	0.46	0.55	
Cattle	$NE_{lactating\ cows}$ Mcal/kg	*	1.19	1.41	
Cattle	TDN %		51.4	60.9	
Sheep	TDN %		51.9	61.5	
Calcium	%		0.50	0.60	
Chlorine	%		0.26	0.31	
Copper	mg/kg		4.3	5.1	
Iron	%		0.019	0.022	
Magnesium	%		0.38	0.45	
Manganese	mg/kg		114.6	135.8	
Phosphorus	%		0.08	0.10	
Potassium	%		1.39	1.64	
Sodium	%		0.06	0.07	
Sulphur	%		0.14	0.17	

Corn, cobs, (1)
Ref No 1 02 783 United States

Feed Name or Analyses			Mean As Fed	Mean Dry	C.V. ± %
Dry matter	%		90.4	100.0	0
Ash	%		1.6	1.8	
Crude fiber	%		32.1	35.5	
Sheep	dig coef %	*	51.	51.	
Ether extract	%		0.4	0.4	
Sheep	dig coef %	*	-169.	-169.	
N-free extract	%		54.0	59.7	
Sheep	dig coef %	*	57.	57.	
Protein (N x 6.25)	%		2.3	2.5	
Sheep	dig coef %	*	-25.	-25.	
Cattle	dig prot %	*	-0.7	-0.8	
Goats	dig prot %	*	-0.9	-1.0	
Horses	dig prot %	*	-0.2	-0.2	
Rabbits	dig prot %	*	0.6	0.6	
Sheep	dig prot %		-0.5	-0.6	
Sugars, total	%		0.2	0.2	
Energy	GE Mcal/kg		4.00	4.42	
Cattle	DE Mcal/kg	*	1.92	2.13	
Sheep	DE Mcal/kg	*	1.99	2.20	
Cattle	ME Mcal/kg	*	1.58	1.74	
Chickens	ME_n kcal/kg		1094.	1210.	
Sheep	ME Mcal/kg	*	1.63	1.80	
Cattle	TDN %	*	43.6	48.2	
Sheep	TDN %		45.0	49.8	
Calcium	%		0.11	0.12	
Cobalt	mg/kg		0.122	0.135	
Phosphorus	%		0.04	0.04	
Sulphur	%		0.42	0.47	
Biotin	mg/kg		0.04	0.04	25
Carotene	mg/kg		0.6	0.7	
Niacin	mg/kg		7.4	8.2	34
Pantothenic acid	mg/kg		3.8	4.2	27
Riboflavin	mg/kg		1.2	1.3	59
Vitamin B6	mg/kg		1.79	1.98	25
Vitamin A equivalent	IU/g		1.0	1.1	

Corn, cobs, dehy fine grnd, 10 sieve mx 10% water, (1)
Dehydrated fine ground corn cob (AAFCO)
Ref No 1 02 781 Canada

Feed Name or Analyses			Mean As Fed	Mean Dry	C.V. ± %
Dry matter	%		95.5	100.0	
Ash	%		1.2	1.3	
Crude fiber	%		37.9	39.7	
Ether extract	%		0.9	0.9	
N-free extract	%		52.0	54.4	
Protein (N x 6.25)	%		3.5	3.7	
Cattle	dig prot %	*	0.1	0.1	
Goats	dig prot %	*	0.0	0.0	
Horses	dig prot %	*	0.6	0.7	
Rabbits	dig prot %	*	1.5	1.5	
Sheep	dig prot %	*	0.0	0.0	
Energy	GE Mcal/kg		4.02	4.21	
Cattle	DE Mcal/kg	*	1.72	1.81	
Sheep	DE Mcal/kg	*	2.19	2.29	
Cattle	ME Mcal/kg	*	1.41	1.48	
Chickens	ME_n kcal/kg		1748.	1830.	
Sheep	ME Mcal/kg	*	1.80	1.88	
Cattle	TDN %	*	39.1	40.9	
Sheep	TDN %	*	49.7	52.0	

Corn, cobs, grnd, (1)
Ground corn cob (AAFCO)
Ref No 1 02 782 United States

Feed Name or Analyses			Mean As Fed	Mean Dry	C.V. ± %
Dry matter	%		89.8	100.0	6
Ash	%		1.6	1.8	8
Crude fiber	%		31.1	34.6	8
Horses	dig coef %		5.	5.	
Sheep	dig coef %		55.	55.	
Ether extract	%		0.7	0.7	59
Horses	dig coef %		0.	0.	59
Sheep	dig coef %		-67.	-67.	
N-free extract	%		53.7	59.8	5
Horses	dig coef %		45.	45.	
Sheep	dig coef %		55.	55.	
Protein (N x 6.25)	%		2.8	3.1	12
Horses	dig coef %		16.	16.	
Sheep	dig coef %		-14.	-14.	
Cattle	dig prot %	*	-0.2	-0.3	
Goats	dig prot %	*	-0.4	-0.4	
Horses	dig prot %		0.4	0.5	
Rabbits	dig prot %	*	1.0	1.1	
Sheep	dig prot %		-0.3	-0.4	
Fatty acids	%		0.4	0.4	
Energy	GE Mcal/kg				
Cattle	DE Mcal/kg	*	1.98	2.20	
Horses	DE Mcal/kg	*	1.15	1.28	
Sheep	DE Mcal/kg	*	1.98	2.20	
Cattle	ME Mcal/kg	*	1.62	1.80	
Horses	ME Mcal/kg	*	0.95	1.05	
Sheep	ME Mcal/kg	*	1.62	1.81	
Cattle	NE_m Mcal/kg	*	0.95	1.06	
Cattle	NE_{gain} Mcal/kg	*	0.22	0.25	
Cattle	$NE_{lactating\ cows}$ Mcal/kg	*	0.87	0.97	
Cattle	TDN %	*	44.8	49.9	
Horses	TDN %		26.2	29.1	
Sheep	TDN %		44.9	50.0	
Calcium	%		0.11	0.12	
Magnesium	%		0.06	0.07	
Phosphorus	%		0.04	0.04	
Potassium	%		0.81	0.91	

Ref No 1 02 782 Canada

Feed Name or Analyses			Mean As Fed	Mean Dry	C.V. ± %
Dry matter	%		88.3	100.0	
Crude fiber	%		37.9	42.9	
Horses	dig coef %		5.	5.	
Sheep	dig coef %		55.	55.	
Ether extract	%		0.9	1.0	
Horses	dig coef %		0.	0.	
Sheep	dig coef %		-67.	-67.	
N-free extract	%				
Horses	dig coef %		45.	45.	
Sheep	dig coef %		55.	55.	
Protein (N x 6.25)	%		3.5	4.0	
Horses	dig coef %		16.	16.	
Sheep	dig coef %		-14.	-14.	
Cattle	dig prot %	*	0.3	0.4	
Goats	dig prot %	*	0.2	0.3	
Horses	dig prot %		0.6	0.6	
Rabbits	dig prot %	*	1.5	1.7	
Sheep	dig prot %		-0.4	-0.5	
Energy	GE Mcal/kg				
Horses	DE Mcal/kg	*	1.11	1.26	
Sheep	DE Mcal/kg	*	1.94	2.20	
Horses	ME Mcal/kg	*	0.91	1.03	
Sheep	ME Mcal/kg	*	1.59	1.80	

Continued

Feed Name or Analyses				Mean As Fed	Mean Dry	C.V. ± %
Horses		TDN %		25.2	28.5	
Sheep		TDN %		44.0	49.8	

Corn, husks, (1)

Ref No 1 02 786 United States

Feed Name or Analyses				Mean As Fed	Mean Dry	C.V. ± %
Dry matter		%			100.0	
Crude fiber		%				
Cattle	dig coef	%	*		80.	
Sheep	dig coef	%	*		69.	
Ether extract		%				
Cattle	dig coef	%	*		32.	
Sheep	dig coef	%	*		39.	
N-free extract		%				
Cattle	dig coef	%	*		75.	
Sheep	dig coef	%	*		60.	
Protein (N x 6.25)		%				
Cattle	dig coef	%	*		30.	
Sheep	dig coef	%	*		12.	
Calcium		%			0.18	
Phosphorus		%			0.14	

Corn, husks, s-c, (1)

Ref No 1 02 785 United States

Feed Name or Analyses				Mean As Fed	Mean Dry	C.V. ± %
Dry matter		%		88.6	100.0	
Ash		%		3.2	3.6	
Crude fiber		%		29.3	33.0	
Sheep	dig coef	%		69.	69.	
Ether extract		%		0.8	0.9	
Sheep	dig coef	%		39.	39.	
N-free extract		%		52.0	58.7	
Sheep	dig coef	%		60.	60.	
Protein (N x 6.25)		%		3.3	3.8	
Sheep	dig coef	%		12.	12.	
Cattle	dig prot	%	*	0.2	0.2	
Goats	dig prot	%	*	0.1	0.1	
Horses	dig prot	%	*	0.6	0.7	
Rabbits	dig prot	%	*	1.4	1.6	
Sheep	dig prot	%		0.4	0.5	
Energy		GE Mcal/kg				
Cattle		DE Mcal/kg	*	2.12	2.39	
Sheep		DE Mcal/kg	*	2.31	2.61	
Cattle		ME Mcal/kg	*	1.74	1.96	
Sheep		ME Mcal/kg	*	1.90	2.14	
Cattle		TDN %	*	48.0	54.2	
Sheep		TDN %		52.5	59.3	
Calcium		%		0.16	0.18	
Phosphorus		%		0.13	0.14	
Potassium		%		0.57	0.65	

Corn, husks, s-c, mature, (1)

Ref No 1 02 784 United States

Feed Name or Analyses				Mean As Fed	Mean Dry	C.V. ± %
Dry matter		%		91.9	100.0	
Ash		%		3.4	3.7	
Crude fiber		%		32.8	35.7	
Cattle	dig coef	%		80.	80.	
Ether extract		%		0.8	0.9	
Cattle	dig coef	%		32.	32.	
N-free extract		%		51.6	56.1	
Cattle	dig coef	%		75.	75.	

Feed Name or Analyses				Mean As Fed	Mean Dry	C.V. ± %
Protein (N x 6.25)		%		3.3	3.6	
Cattle	dig coef	%		30.	30.	
Cattle	dig prot	%		1.0	1.1	
Goats	dig prot	%	*	0.0	0.0	
Horses	dig prot	%	*	0.5	0.6	
Rabbits	dig prot	%	*	1.3	1.4	
Sheep	dig prot	%	*	-0.1	-0.1	
Energy		GE Mcal/kg				
Cattle		DE Mcal/kg	*	2.93	3.19	
Sheep		DE Mcal/kg	*	2.17	2.36	
Cattle		ME Mcal/kg	*	2.40	2.62	
Sheep		ME Mcal/kg	*	1.78	1.94	
Cattle		TDN %		66.5	72.4	
Sheep		TDN %	*	49.2	53.5	

Corn, husks w leaves, s-c, (1)

Ref No 1 02 790 United States

Feed Name or Analyses				Mean As Fed	Mean Dry	C.V. ± %
Dry matter		%		91.1	100.0	
Ash		%		6.7	7.4	
Crude fiber		%		30.0	32.9	
Cattle	dig coef	%		73.	73.	
Ether extract		%		2.3	2.5	
Cattle	dig coef	%		58.	58.	
N-free extract		%		45.6	50.0	
Cattle	dig coef	%		66.	66.	
Protein (N x 6.25)		%		6.6	7.2	
Cattle	dig coef	%		48.	48.	
Cattle	dig prot	%		3.1	3.5	
Goats	dig prot	%	*	3.0	3.3	
Horses	dig prot	%	*	3.3	3.6	
Rabbits	dig prot	%	*	3.9	4.2	
Sheep	dig prot	%	*	2.8	3.0	
Energy		GE Mcal/kg				
Cattle		DE Mcal/kg	*	2.56	2.81	
Sheep		DE Mcal/kg	*	2.22	2.44	
Cattle		ME Mcal/kg	*	2.10	2.30	
Sheep		ME Mcal/kg	*	1.82	2.00	
Cattle		TDN %		58.1	63.7	
Sheep		TDN %	*	50.5	55.4	

Corn, leaves, s-c, (1)

Ref No 1 02 788 United States

Feed Name or Analyses				Mean As Fed	Mean Dry	C.V. ± %
Dry matter		%		86.1	100.0	
Ash		%		7.1	8.2	
Crude fiber		%		23.0	26.7	
Sheep	dig coef	%		67.	67.	
Ether extract		%		3.1	3.6	
Sheep	dig coef	%		61.	61.	
N-free extract		%		44.4	51.6	
Sheep	dig coef	%		61.	61.	
Protein (N x 6.25)		%		8.5	9.9	
Sheep	dig coef	%		43.	43.	
Cattle	dig prot	%	*	4.7	5.5	
Goats	dig prot	%	*	5.0	5.8	
Horses	dig prot	%	*	5.1	5.9	
Rabbits	dig prot	%	*	5.4	6.3	
Sheep	dig prot	%		3.7	4.3	
Energy		GE Mcal/kg				
Cattle		DE Mcal/kg	*	2.29	2.66	
Sheep		DE Mcal/kg	*	2.22	2.58	
Cattle		ME Mcal/kg	*	1.88	2.18	
Sheep		ME Mcal/kg	*	1.82	2.12	
Cattle		TDN %	*	51.9	60.3	
Sheep		TDN %		50.4	58.6	
Calcium		%		0.59	0.69	

Feed Name or Analyses				Mean As Fed	Mean Dry	C.V. ± %
Phosphorus		%		0.22	0.25	
Potassium		%		1.54	1.79	

Corn, leaves, s-c, mature, (1)

Ref No 1 02 787 United States

Feed Name or Analyses				Mean As Fed	Mean Dry	C.V. ± %
Dry matter		%		92.6	100.0	
Ash		%		10.5	11.3	
Crude fiber		%		27.0	29.2	
Cattle	dig coef	%		78.	78.	
Ether extract		%		2.2	2.4	
Cattle	dig coef	%		56.	56.	
N-free extract		%		47.6	51.4	
Cattle	dig coef	%		68.	68.	
Protein (N x 6.25)		%		5.3	5.7	
Cattle	dig coef	%		34.	34.	
Cattle	dig prot	%		1.8	1.9	
Goats	dig prot	%	*	1.7	1.9	
Horses	dig prot	%	*	2.2	2.4	
Rabbits	dig prot	%	*	2.8	3.1	
Sheep	dig prot	%	*	1.6	1.7	
Energy		GE Mcal/kg				
Cattle		DE Mcal/kg	*	2.56	2.76	
Sheep		DE Mcal/kg	*	2.29	2.47	
Cattle		ME Mcal/kg	*	2.10	2.27	
Sheep		ME Mcal/kg	*	1.87	2.02	
Cattle		TDN %		58.1	62.7	
Sheep		TDN %	*	51.9	56.0	

Corn, leaves, s-c, grnd, (1)

Ref No 1 02 789 United States

Feed Name or Analyses				Mean As Fed	Mean Dry	C.V. ± %
Dry matter		%		91.2	100.0	2
Ash		%		10.1	11.1	
Crude fiber		%		25.3	27.7	
Ether extract		%		2.3	2.5	
N-free extract		%		45.0	49.3	
Protein (N x 6.25)		%		8.6	9.4	
Cattle	dig prot	%	*	4.6	5.1	
Goats	dig prot	%	*	4.9	5.3	
Horses	dig prot	%	*	5.0	5.5	
Rabbits	dig prot	%	*	5.4	5.9	
Sheep	dig prot	%	*	4.6	5.0	
Energy		GE Mcal/kg				
Cattle		DE Mcal/kg	*	2.43	2.66	
Sheep		DE Mcal/kg	*	2.34	2.56	
Cattle		ME Mcal/kg	*	1.99	2.18	
Sheep		ME Mcal/kg	*	1.92	2.10	
Cattle		TDN %	*	55.1	60.4	
Sheep		TDN %	*	53.0	58.1	

Corn, silk, s-c, (1)

Ref No 1 02 791 United States

Feed Name or Analyses				Mean As Fed	Mean Dry	C.V. ± %
Dry matter		%			100.0	
Vitamin D_2		IU/g			5.4	

Corn, stalks w tassels, s-c, (1)

Corn, stalks, shanks, tassels, dried

Ref No 1 08 606 United States

Feed Name or Analyses				Mean As Fed	Mean Dry	C.V. ± %
Dry matter		%		82.8	100.0	
Ash		%		5.3	6.4	
Crude fiber		%		28.0	33.8	
Ether extract		%		1.5	1.8	
N-free extract		%		43.3	52.3	

(1) dry forages and roughages (2) pasture, range plants, and forages fed green (3) silages (4) energy feeds (5) protein supplements (6) minerals (7) vitamins (8) additives

Feed Name or Analyses			Mean As Fed	Mean Dry	C.V. ± %
Protein (N x 6.25)	%		4.7	5.7	
Cattle	dig prot %	*	1.5	1.9	
Goats	dig prot %	*	1.5	1.9	
Horses	dig prot %	*	1.9	2.3	
Rabbits	dig prot %	*	2.5	3.1	
Sheep	dig prot %	*	1.4	1.7	
Energy	GE Mcal/kg				
Cattle	DE Mcal/kg	*	1.85	2.24	
Sheep	DE Mcal/kg	*	1.76	2.13	
Cattle	ME Mcal/kg	*	1.52	1.84	
Sheep	ME Mcal/kg	*	1.45	1.75	
Cattle	TDN %	*	42.1	50.9	
Sheep	TDN %	*	40.0	48.3	
Calcium	%		0.32	0.39	
Phosphorus	%		0.23	0.28	
Potassium	%		1.79	2.16	

Corn, stems, immature, (1)

Ref No 1 02 792 United States

Feed Name or Analyses			Mean As Fed	Mean Dry	C.V. ± %
Dry matter	%		92.7	100.0	
Ash	%		9.2	9.9	
Crude fiber	%		26.9	29.0	
Ether extract	%		1.9	2.0	
N-free extract	%		44.4	47.9	
Protein (N x 6.25)	%		10.4	11.2	
Cattle	dig prot %	*	6.2	6.6	
Goats	dig prot %	*	6.5	7.0	
Horses	dig prot %	*	6.5	7.0	
Rabbits	dig prot %	*	6.8	7.3	
Sheep	dig prot %	*	6.1	6.6	

Corn, stems, dough stage, (1)

Ref No 1 02 793 United States

Feed Name or Analyses			Mean As Fed	Mean Dry	C.V. ± %
Dry matter	%		92.4	100.0	
Ash	%		8.7	9.4	
Crude fiber	%		27.4	29.7	
Ether extract	%		1.5	1.6	
N-free extract	%		48.6	52.6	
Protein (N x 6.25)	%		6.2	6.7	
Cattle	dig prot %	*	2.5	2.7	
Goats	dig prot %	*	2.6	2.8	
Horses	dig prot %	*	3.0	3.2	
Rabbits	dig prot %	*	3.6	3.8	
Sheep	dig prot %	*	2.4	2.6	

Corn, stems, over ripe, (1)

Ref No 1 02 794 United States

Feed Name or Analyses			Mean As Fed	Mean Dry	C.V. ± %
Dry matter	%		91.3	100.0	1
Ash	%		2.9	3.2	11
Crude fiber	%		53.0	58.0	3
Ether extract	%		1.4	1.5	14
N-free extract	%		29.7	32.5	
Protein (N x 6.25)	%		4.4	4.8	4
Cattle	dig prot %	*	1.0	1.1	
Goats	dig prot %	*	0.9	1.0	
Horses	dig prot %	*	1.5	1.6	
Rabbits	dig prot %	*	2.2	2.4	
Sheep	dig prot %	*	0.8	0.9	

Corn, top of aerial part, s-c, (1)

Ref No 1 02 766 United States

Feed Name or Analyses			Mean As Fed	Mean Dry	C.V. ± %
Dry matter	%		84.1	100.0	
Ash	%		6.5	7.7	
Crude fiber	%		27.7	32.9	
Cattle	dig coef %		71.	71.	
Ether extract	%		1.9	2.3	
Cattle	dig coef %		67.	67.	
N-free extract	%		42.4	50.4	
Cattle	dig coef %		58.	58.	
Protein (N x 6.25)	%		5.6	6.7	
Cattle	dig coef %		39.	39.	
Cattle	dig prot %		2.2	2.6	
Goats	dig prot %	*	2.4	2.8	
Horses	dig prot %	*	2.7	3.2	
Rabbits	dig prot %	*	3.2	3.9	
Sheep	dig prot %	*	2.2	2.6	
Energy	GE Mcal/kg				
Cattle	DE Mcal/kg	*	2.17	2.59	
Sheep	DE Mcal/kg	*	2.05	2.43	
Cattle	ME Mcal/kg	*	1.78	2.12	
Sheep	ME Mcal/kg	*	1.68	2.00	
Cattle	TDN %		49.3	58.6	
Sheep	TDN %	*	46.4	55.2	

Corn, top of aerial part, s-c, mature, (1)

Ref No 1 02 765 United States

Feed Name or Analyses			Mean As Fed	Mean Dry	C.V. ± %
Dry matter	%		84.0	100.0	
Ash	%		6.3	7.5	
Crude fiber	%		28.6	34.0	
Cattle	dig coef %		20.	20.	
Ether extract	%		2.2	2.6	
Cattle	dig coef %		64.	64.	
N-free extract	%		42.4	50.5	
Cattle	dig coef %		54.	54.	
Protein (N x 6.25)	%		4.5	5.4	
Cattle	dig coef %		22.	22.	
Cattle	dig prot %		1.0	1.2	
Goats	dig prot %	*	1.3	1.6	
Horses	dig prot %	*	1.8	2.1	
Rabbits	dig prot %	*	2.4	2.8	
Sheep	dig prot %	*	1.2	1.4	
Energy	GE Mcal/kg				
Cattle	DE Mcal/kg	*	1.44	1.72	
Sheep	DE Mcal/kg	*	2.01	2.39	
Cattle	ME Mcal/kg	*	1.18	1.41	
Sheep	ME Mcal/kg	*	1.64	1.96	
Cattle	TDN %		32.8	39.0	
Sheep	TDN %	*	45.5	54.1	

Corn, aerial part, close-planted fresh, (2)

Ref No 2 02 799 United States

Feed Name or Analyses			Mean As Fed	Mean Dry	C.V. ± %
Dry matter	%		18.2	100.0	
Ash	%		1.1	6.0	
Crude fiber	%		4.7	25.6	
Sheep	dig coef %		69.	69.	
Ether extract	%		0.8	4.4	
Sheep	dig coef %		81.	81.	
N-free extract	%		9.6	52.6	
Sheep	dig coef %		76.	76.	
Protein (N x 6.25)	%		2.1	11.4	
Sheep	dig coef %		67.	67.	
Cattle	dig prot %	*	1.4	7.6	
Goats	dig prot %	*	1.3	7.2	
Horses	dig prot %	*	1.3	7.2	
Rabbits	dig prot %	*	1.4	7.5	
Sheep	dig prot %		1.4	7.6	
Energy	GE Mcal/kg				
Cattle	DE Mcal/kg	*	0.53	2.93	
Sheep	DE Mcal/kg	*	0.59	3.23	
Cattle	ME Mcal/kg	*	0.44	2.40	
Sheep	ME Mcal/kg	*	0.48	2.65	
Cattle	TDN %	*	12.1	66.5	
Sheep	TDN %		13.3	73.3	

Corn, aerial part, close-planted fresh, late bloom, (2)

Ref No 2 02 797 United States

Feed Name or Analyses			Mean As Fed	Mean Dry	C.V. ± %
Dry matter	%		13.8	100.0	
Ash	%		1.0	7.4	
Crude fiber	%		3.6	25.8	
Sheep	dig coef %		71.	71.	
Ether extract	%		0.7	5.1	
Sheep	dig coef %		79.	79.	
N-free extract	%		6.5	46.9	
Sheep	dig coef %		75.	75.	
Protein (N x 6.25)	%		2.0	14.8	
Sheep	dig coef %		74.	74.	
Cattle	dig prot %	*	1.4	10.5	
Goats	dig prot %	*	1.4	10.4	
Horses	dig prot %	*	1.4	10.1	
Rabbits	dig prot %	*	1.4	10.1	
Sheep	dig prot %		1.5	11.0	
Energy	GE Mcal/kg				
Cattle	DE Mcal/kg	*	0.41	2.97	
Sheep	DE Mcal/kg	*	0.45	3.24	
Cattle	ME Mcal/kg	*	0.34	2.43	
Sheep	ME Mcal/kg	*	0.37	2.66	
Cattle	TDN %	*	9.3	67.2	
Sheep	TDN %		10.1	73.5	

Corn, aerial part, close-planted fresh, milk stage, (2)

Ref No 2 02 798 United States

Feed Name or Analyses			Mean As Fed	Mean Dry	C.V. ± %
Dry matter	%		22.7	100.0	
Ash	%		1.0	4.6	
Crude fiber	%		5.7	25.3	
Sheep	dig coef %		66.	66.	
Ether extract	%		0.8	3.7	
Sheep	dig coef %		82.	82.	
N-free extract	%		13.3	58.4	
Sheep	dig coef %		78.	78.	
Protein (N x 6.25)	%		1.8	8.0	
Sheep	dig coef %		60.	60.	
Cattle	dig prot %	*	1.1	4.7	
Goats	dig prot %	*	0.9	4.0	
Horses	dig prot %	*	1.0	4.3	
Rabbits	dig prot %	*	1.1	4.8	
Sheep	dig prot %		1.1	4.8	
Energy	GE Mcal/kg				
Cattle	DE Mcal/kg	*	0.66	2.90	
Sheep	DE Mcal/kg	*	0.74	3.26	
Cattle	ME Mcal/kg	*	0.54	2.38	
Sheep	ME Mcal/kg	*	0.61	2.67	
Cattle	TDN %	*	14.9	65.8	
Sheep	TDN %		16.8	73.9	

Corn, aerial part, fresh, (2)
Corn fodder, fresh

Ref No 2 02 806 — United States

Feed Name or Analyses			Mean As Fed	Mean Dry	C.V. ± %
Dry matter	%		22.6	100.0	
Ash	%		1.4	6.4	
Crude fiber	%		5.5	24.5	
Cattle	dig coef %	*	64.	64.	
Sheep	dig coef %		63.	63.	
Ether extract	%		0.6	2.6	
Cattle	dig coef %	*	74.	74.	
Sheep	dig coef %		76.	76.	
N-free extract	%		13.0	57.5	
Cattle	dig coef %	*	73.	73.	
Sheep	dig coef %		74.	74.	
Protein (N x 6.25)	%		2.0	9.1	
Cattle	dig coef %	*	59.	59.	
Sheep	dig coef %		59.	59.	
Cattle	dig prot %		1.2	5.3	
Goats	dig prot %	*	1.1	5.0	
Horses	dig prot %	*	1.2	5.2	
Rabbits	dig prot %	*	1.3	5.7	
Sheep	dig prot %		1.2	5.3	
Energy	GE Mcal/kg				
Cattle	DE Mcal/kg	*	0.67	2.96	
Sheep	DE Mcal/kg	*	0.67	2.99	
Cattle	ME Mcal/kg	*	0.55	2.42	
Sheep	ME Mcal/kg	*	0.55	2.45	
Cattle	TDN %		15.1	67.0	
Sheep	TDN %		15.3	67.8	

Corn, aerial part, fresh, full bloom, (2)

Ref No 2 02 800 — United States

Feed Name or Analyses			Mean As Fed	Mean Dry	C.V. ± %
Dry matter	%		24.1	100.0	
Ash	%		1.3	5.5	
Crude fiber	%		6.8	28.3	
Sheep	dig coef %		67.	67.	
Ether extract	%		1.0	4.3	
Sheep	dig coef %		81.	81.	
N-free extract	%		12.6	52.3	
Sheep	dig coef %		58.	58.	
Protein (N x 6.25)	%		2.3	9.6	
Sheep	dig coef %		65.	65.	
Cattle	dig prot %	*	1.5	6.1	
Goats	dig prot %	*	1.3	5.5	
Horses	dig prot %	*	1.4	5.7	
Rabbits	dig prot %	*	1.5	6.1	
Sheep	dig prot %		1.5	6.2	
Energy	GE Mcal/kg				
Cattle	DE Mcal/kg	*	0.70	2.91	
Sheep	DE Mcal/kg	*	0.67	2.79	
Cattle	ME Mcal/kg	*	0.57	2.38	
Sheep	ME Mcal/kg	*	0.55	2.29	
Cattle	TDN %	*	15.9	65.9	
Sheep	TDN %		15.3	63.4	

Corn, aerial part, fresh, late bloom, (2)

Ref No 2 02 801 — United States

Feed Name or Analyses			Mean As Fed	Mean Dry	C.V. ± %
Dry matter	%		19.0	100.0	
Ash	%		1.2	6.1	
Crude fiber	%		5.1	26.9	
Sheep	dig coef %		67.	67.	
Ether extract	%		0.6	3.1	
Sheep	dig coef %		68.	68.	
N-free extract	%		9.9	52.1	
Sheep	dig coef %		70.	70.	
Protein (N x 6.25)	%		2.2	11.8	
Sheep	dig coef %		64.	64.	
Cattle	dig prot %	*	1.5	7.9	
Goats	dig prot %	*	1.4	7.6	
Horses	dig prot %	*	1.4	7.5	
Rabbits	dig prot %	*	1.5	7.8	
Sheep	dig prot %		1.4	7.6	
Energy	GE Mcal/kg				
Cattle	DE Mcal/kg	*	0.58	3.04	
Sheep	DE Mcal/kg	*	0.56	2.94	
Cattle	ME Mcal/kg	*	0.47	2.49	
Sheep	ME Mcal/kg	*	0.46	2.41	
Cattle	TDN %	*	13.1	69.0	
Sheep	TDN %		12.7	66.8	

(1) dry forages and roughages (2) pasture, range plants, and forages fed green (3) silages (4) energy feeds (5) protein supplements (6) minerals (7) vitamins (8) additives

Corn, aerial part, fresh, milk stage, (2)

Ref No 2 02 802 — United States

Feed Name or Analyses			Mean As Fed	Mean Dry	C.V. ± %
Dry matter	%		25.4	100.0	39
Ash	%		1.2	4.7	11
Crude fiber	%		5.8	22.9	14
Sheep	dig coef %		58.	58.	
Ether extract	%		0.7	2.8	24
Sheep	dig coef %		75.	75.	
N-free extract	%		15.6	61.3	
Sheep	dig coef %		75.	75.	
Protein (N x 6.25)	%		2.1	8.4	12
Sheep	dig coef %		57.	57.	
Cattle	dig prot %	*	1.3	5.0	
Goats	dig prot %	*	1.1	4.4	
Horses	dig prot %	*	1.2	4.7	
Rabbits	dig prot %	*	1.3	5.2	
Sheep	dig prot %		1.2	4.8	
Energy	GE Mcal/kg				
Cattle	DE Mcal/kg	*	0.75	2.95	
Sheep	DE Mcal/kg	*	0.77	3.03	
Cattle	ME Mcal/kg	*	0.62	2.42	
Sheep	ME Mcal/kg	*	0.63	2.49	
Cattle	TDN %	*	17.0	67.0	
Sheep	TDN %		17.5	68.7	

Corn, aerial part, fresh, dough stage, (2)

Ref No 2 02 803 — United States

Feed Name or Analyses			Mean As Fed	Mean Dry	C.V. ± %
Dry matter	%		28.1	100.0	
Ash	%		1.3	4.6	
Crude fiber	%		5.9	21.1	
Sheep	dig coef %		59.	59.	
Ether extract	%		0.7	2.6	
Sheep	dig coef %		77.	77.	
N-free extract	%		18.0	64.0	
Sheep	dig coef %		77.	77.	
Protein (N x 6.25)	%		2.2	7.7	
Sheep	dig coef %		53.	53.	
Cattle	dig prot %	*	1.2	4.4	
Goats	dig prot %	*	1.1	3.7	
Horses	dig prot %	*	1.1	4.1	
Rabbits	dig prot %	*	1.3	4.6	
Sheep	dig prot %		1.1	4.1	
Energy	GE Mcal/kg				
Cattle	DE Mcal/kg	*	0.82	2.92	
Sheep	DE Mcal/kg	*	0.87	3.10	
Cattle	ME Mcal/kg	*	0.67	2.39	
Sheep	ME Mcal/kg	*	0.71	2.54	
Cattle	TDN %	*	18.6	66.2	
Sheep	TDN %		19.8	70.3	

Corn, aerial part, fresh, mature, (2)

Ref No 2 02 804 — United States

Feed Name or Analyses			Mean As Fed	Mean Dry	C.V. ± %
Dry matter	%		33.6	100.0	13
Ash	%		1.4	4.3	
Crude fiber	%		8.1	24.1	
Cattle	dig coef %		75.	75.	
Sheep	dig coef %		52.	52.	
Ether extract	%		0.9	2.6	
Cattle	dig coef %		83.	83.	
Sheep	dig coef %		78.	78.	
N-free extract	%		20.6	61.5	
Cattle	dig coef %		77.	77.	
Sheep	dig coef %		77.	77.	
Protein (N x 6.25)	%		2.6	7.6	
Cattle	dig coef %		73.	73.	
Sheep	dig coef %		53.	53.	
Cattle	dig prot %		1.9	5.5	
Goats	dig prot %	*	1.2	3.7	
Horses	dig prot %	*	1.3	4.0	
Rabbits	dig prot %	*	1.5	4.5	
Sheep	dig prot %		1.4	4.0	
Energy	GE Mcal/kg				
Cattle	DE Mcal/kg	*	1.12	3.34	
Sheep	DE Mcal/kg	*	1.01	3.01	
Cattle	ME Mcal/kg	*	0.92	2.74	
Sheep	ME Mcal/kg	*	0.83	2.47	
Cattle	TDN %		25.4	75.7	
Sheep	TDN %		23.0	68.4	
Calcium	%		0.09	0.28	18
Copper	mg/kg		3.1	9.3	
Iron	%		0.005	0.014	
Magnesium	%		0.04	0.13	43
Manganese	mg/kg		10.4	31.1	
Phosphorus	%		0.08	0.24	18
Potassium	%		0.33	0.99	20
Carotene	mg/kg		12.7	37.7	36
Vitamin A equivalent	IU/g		21.1	62.8	

Corn, aerial part, fresh, over ripe, (2)

Ref No 2 02 805 — United States

Feed Name or Analyses			Mean As Fed	Mean Dry	C.V. ± %
Dry matter	%			100.0	
Ash	%			4.8	23
Protein (N x 6.25)	%			3.1	10
Cattle	dig prot %	*		0.5	
Goats	dig prot %	*		-0.4	
Horses	dig prot %	*		0.2	
Rabbits	dig prot %	*		1.1	
Sheep	dig prot %	*		0.0	

Corn, aerial part, fresh wide-planted, late bloom, (2)

Ref No 2 02 808 — United States

Feed Name or Analyses			Mean As Fed	Mean Dry	C.V. ± %
Dry matter	%		20.1	100.0	
Ash	%		0.8	3.9	
Crude fiber	%		5.7	28.5	
Sheep	dig coef %		61.	61.	
Ether extract	%		0.5	2.3	
Sheep	dig coef %		72.	72.	
N-free extract	%		11.9	59.2	
Sheep	dig coef %		74.	74.	

Feed Name or Analyses			Mean As Fed	Mean Dry	C.V. ± %
Protein (N x 6.25)	%		1.2	6.1	
Sheep	dig coef %		56.	56.	
Cattle	dig prot %	*	0.6	3.1	
Goats	dig prot %	*	0.5	2.3	
Horses	dig prot %	*	0.5	2.7	
Rabbits	dig prot %	*	0.7	3.4	
Sheep	dig prot %		0.7	3.4	
Energy	GE Mcal/kg				
Cattle	DE Mcal/kg	*	0.60	2.97	
Sheep	DE Mcal/kg	*	0.61	3.01	
Cattle	ME Mcal/kg	*	0.49	2.44	
Sheep	ME Mcal/kg	*	0.50	2.47	
Cattle	TDN %	*	13.5	67.4	
Sheep	TDN %		13.7	68.3	

Corn, aerial part wo ears wo husks, fresh, (2)

Ref No 2 02 809 United States

Feed Name or Analyses			Mean As Fed	Mean Dry	C.V. ± %
Dry matter	%		22.7	100.0	
Ash	%		1.4	6.2	
Crude fiber	%		6.0	26.4	
Ether extract	%		0.4	1.8	
N-free extract	%		13.6	59.9	
Protein (N x 6.25)	%		1.3	5.7	
Cattle	dig prot %	*	0.6	2.8	
Goats	dig prot %	*	0.4	1.9	
Horses	dig prot %	*	0.5	2.4	
Rabbits	dig prot %	*	0.7	3.1	
Sheep	dig prot %	*	0.5	2.3	
Energy	GE Mcal/kg				
Cattle	DE Mcal/kg	*	0.59	2.58	
Sheep	DE Mcal/kg	*	0.56	2.47	
Cattle	ME Mcal/kg	*	0.48	2.12	
Sheep	ME Mcal/kg	*	0.46	2.03	
Cattle	TDN %	*	13.3	58.5	
Sheep	TDN %	*	12.7	56.1	
Calcium	%		0.14	0.62	
Chlorine	%		0.07	0.31	
Copper	mg/kg		1.1	4.9	
Iron	%		0.005	0.022	
Magnesium	%		0.05	0.22	
Manganese	mg/kg		30.9	136.0	
Phosphorus	%		0.02	0.09	
Potassium	%		0.37	1.63	
Sodium	%		0.01	0.04	
Sulphur	%		0.04	0.18	

Ref No 2 02 8 9 Canada

Feed Name or Analyses			Mean As Fed	Mean Dry	C.V. ± %
Dry matter	%		32.9	100.0	
Protein (N x 6.25)	%		6.3	19.1	
Cattle	dig prot %	*	4.7	14.2	
Goats	dig prot %	*	4.7	14.4	
Horses	dig prot %	*	4.5	13.8	
Rabbits	dig prot %	*	4.4	13.5	
Sheep	dig prot %	*	4.9	14.8	

Corn, cannery residue, fresh, (2)

Ref No 2 02 810 United States

Feed Name or Analyses			Mean As Fed	Mean Dry	C.V. ± %
Dry matter	%		28.2	100.0	
Ash	%		1.5	5.2	
Crude fiber	%		8.5	30.0	
Cattle	dig coef %		25.	25.	
Ether extract	%		0.6	2.0	
Cattle	dig coef %		44.	44.	
N-free extract	%		15.5	54.9	
Cattle	dig coef %		40.	40.	
Protein (N x 6.25)	%		2.2	7.9	
Cattle	dig coef %		7.	7.	
Cattle	dig prot %		0.2	0.6	
Goats	dig prot %	*	1.1	3.9	
Horses	dig prot %	*	1.2	4.2	
Rabbits	dig prot %	*	1.3	4.8	
Sheep	dig prot %	*	1.2	4.4	
Energy	GE Mcal/kg				
Cattle	DE Mcal/kg	*	0.40	1.41	
Sheep	DE Mcal/kg	*	0.46	1.62	
Cattle	ME Mcal/kg	*	0.33	1.16	
Sheep	ME Mcal/kg	*	0.38	1.33	
Cattle	TDN %		9.0	32.0	
Sheep	TDN %	*	10.4	36.8	
Carotene	mg/kg		3.8	13.4	
Vitamin A equivalent	IU/g		6.3	22.4	

Corn, husks w leaves, fresh, (2)

Ref No 2 02 813 United States

Feed Name or Analyses			Mean As Fed	Mean Dry	C.V. ± %
Dry matter	%			100.0	
Calcium	%			0.47	
Potassium	%			2.18	

Corn, leaves, fresh, (2)

Ref No 2 02 812 United States

Feed Name or Analyses			Mean As Fed	Mean Dry	C.V. ± %
Dry matter	%		52.3	100.0	5
Carotene	mg/kg		302.3	578.0	
Pantothenic acid	mg/kg		4.7	9.0	
Riboflavin	mg/kg		2.9	5.5	
Vitamin A equivalent	IU/g		504.0	963.6	
Vitamin D2	IU/g		2.8	5.4	

Corn, leaves w top of aerial part, fresh, (2)

Ref No 2 08 391 United States

Feed Name or Analyses			Mean As Fed	Mean Dry	C.V. ± %
Dry matter	%		49.7	100.0	2
Ash	%		3.7	7.5	7
Crude fiber	%		15.1	30.4	5
Ether extract	%		1.5	3.0	9
N-free extract	%		24.8	49.9	
Protein (N x 6.25)	%		4.6	9.2	4
Cattle	dig prot %	*	2.8	5.7	
Goats	dig prot %	*	2.6	5.2	
Horses	dig prot %	*	2.7	5.4	
Rabbits	dig prot %	*	2.9	5.8	
Sheep	dig prot %	*	2.8	5.6	
Energy	GE Mcal/kg				
Cattle	DE Mcal/kg	*	1.42	2.86	
Sheep	DE Mcal/kg	*	1.46	2.93	
Cattle	ME Mcal/kg	*	1.17	2.35	
Sheep	ME Mcal/kg	*	1.20	2.41	
Cattle	TDN %	*	32.3	64.9	
Sheep	TDN %	*	33.1	66.6	

Corn, stalks, fresh, (2)

Ref No 2 02 814 United States

Feed Name or Analyses			Mean As Fed	Mean Dry	C.V. ± %
Dry matter	%		56.2	100.0	4
Ash	%		3.3	5.8	30
Crude fiber	%		21.8	38.8	19
Ether extract	%		1.0	1.7	18
N-free extract	%		26.8	47.8	
Protein (N x 6.25)	%		3.3	5.9	28
Cattle	dig prot %	*	1.6	2.9	
Goats	dig prot %	*	1.2	2.1	
Horses	dig prot %	*	1.4	2.5	
Rabbits	dig prot %	*	1.8	3.2	
Sheep	dig prot %	*	1.4	2.5	
Hemicellulose	%		14.8	26.4	3
Cobalt	mg/kg		0.051	0.090	
Carotene	mg/kg		11.8	20.9	
Vitamin A equivalent	IU/g		19.6	34.9	

Corn, stalks w tassels, fresh, (2)

Ref No 2 02 815 United States

Feed Name or Analyses			Mean As Fed	Mean Dry	C.V. ± %
Dry matter	%			100.0	
Calcium	%			0.39	11
Magnesium	%			0.23	16
Phosphorus	%			0.30	18
Potassium	%			2.31	17

Corn, tassels, fresh, immature, (2)

Ref No 2 02 816 United States

Feed Name or Analyses			Mean As Fed	Mean Dry	C.V. ± %
Dry matter	%		39.5	100.0	
Ash	%		1.8	4.5	
Crude fiber	%		7.8	19.8	
Ether extract	%		2.3	5.7	
N-free extract	%		20.5	52.0	
Protein (N x 6.25)	%		7.1	18.0	
Cattle	dig prot %	*	5.2	13.2	
Goats	dig prot %	*	5.3	13.4	
Horses	dig prot %	*	5.1	12.8	
Rabbits	dig prot %	*	5.0	12.6	
Sheep	dig prot %	*	5.4	13.8	
Carotene	mg/kg		5.1	13.0	
Niacin	mg/kg		16.5	41.7	23
Pantothenic acid	mg/kg		9.1	23.1	26
Riboflavin	mg/kg		3.2	8.2	
Thiamine	mg/kg		3.3	8.4	
Vitamin B6	mg/kg		1.39	3.53	
Vitamin A equivalent	IU/g		8.6	21.7	
Vitamin D2	IU/g		1.1	2.7	

Corn, aerial part, close-planted ensiled, (3)

Ref No 3 02 827 United States

Feed Name or Analyses			Mean As Fed	Mean Dry	C.V. ± %
Dry matter	%		21.0	100.0	
Ash	%		1.7	8.0	
Crude fiber	%		5.3	25.4	
Sheep	dig coef %		48.	48.	
Ether extract	%		0.6	2.8	
Sheep	dig coef %		63.	63.	
N-free extract	%		12.0	57.1	
Sheep	dig coef %		64.	64.	
Protein (N x 6.25)	%		1.4	6.7	
Sheep	dig coef %		26.	26.	
Cattle	dig prot %	*	0.5	2.3	
Goats	dig prot %	*	0.5	2.3	
Horses	dig prot %	*	0.5	2.3	
Sheep	dig prot %		0.4	1.7	
Energy	GE Mcal/kg				
Cattle	DE Mcal/kg	*	0.52	2.48	
Sheep	DE Mcal/kg	*	0.50	2.40	
Cattle	ME Mcal/kg	*	0.43	2.03	
Sheep	ME Mcal/kg	*	0.41	1.97	
Cattle	TDN %	*	11.8	56.3	
Sheep	TDN %		11.4	54.4	

Feed Name or Analyses			Mean As Fed	Mean Dry	C.V. ± %

Corn, aerial part, ensiled, (3)
Corn fodder silage

Ref No 3 02 822 United States

Feed Name or Analyses			As Fed	Dry	C.V. ± %
Dry matter	%		23.2	100.0	27
Dry matter	%		24.1	100.0	14
Dry matter	%		31.7	100.0	
Sheep	dig coef %		67.	67.	
Ash	%		1.4	5.7	25
Crude fiber	%		5.9	24.5	13
Crude fiber	%		6.8	21.6	
Cattle	dig coef %		68.	68.	
Cattle	dig coef %		68.	68.	
Sheep	dig coef %		61.	61.	8
Sheep	dig coef %		61.	61.	
Ether extract	%		0.8	3.1	12
Ether extract	%				
Cattle	dig coef %		75.	75.	
Cattle	dig coef %		75.	75.	
Sheep	dig coef %		77.	77.	
Sheep	dig coef %		77.	77.	
N-free extract	%		14.1	58.5	6
N-free extract	%				
Cattle	dig coef %		72.	72.	
Cattle	dig coef %		72.	72.	
Sheep	dig coef %		74.	74.	5
Sheep	dig coef %		74.	74.	
Protein (N x 6.25)	%		2.0	8.2	8
Protein (N x 6.25)	%		3.0	9.5	
Cattle	dig coef %		47.	47.	
Cattle	dig coef %		47.	47.	
Sheep	dig coef %		54.	54.	10
Sheep	dig coef %		54.	54.	
Cattle	dig prot %		0.9	3.8	
Cattle	dig prot %		1.4	4.5	
Goats	dig prot %	*	0.9	3.6	
Goats	dig prot %	*	1.5	4.9	
Horses	dig prot %	*	0.9	3.6	
Horses	dig prot %	*	1.5	4.9	
Sheep	dig prot %		1.1	4.4	
Sheep	dig prot %		1.6	5.1	
Lignin (Ellis)	%		2.0	8.1	
Energy	GE Mcal/kg				
Energy	GE Mcal/kg				
Cattle	DE Mcal/kg	*	0.72	2.98	
Cattle	DE Mcal/kg	*	0.89	2.81	
Sheep	DE Mcal/kg	*	0.72	2.99	
Sheep	DE Mcal/kg	*	0.89	2.80	
Cattle	ME Mcal/kg	*	0.59	2.44	
Cattle	ME Mcal/kg	*	0.73	2.30	
Sheep	ME Mcal/kg	*	0.59	2.45	
Sheep	ME Mcal/kg	*	0.73	2.30	
Cattle	TDN %		16.3	67.6	
Cattle	TDN %		20.2	63.7	
Sheep	TDN %		16.4	67.9	5
Sheep	TDN %		20.2	63.6	
Carotene	mg/kg		11.0	45.6	70
Niacin	mg/kg		10.4	43.0	36
Vitamin A equivalent	IU/g		18.4	76.1	
Vitamin D_2	IU/g		0.1	0.4	15
pH	pH units		4.61	4.61	10
pH	pH units		4.50	4.50	

Ref No 3 02 822 Canada

Feed Name or Analyses			As Fed	Dry	C.V. ± %
Ash	%		10.6	45.6	
Crude fiber	%		7.7	33.2	98
Cattle	dig coef %		68.	68.	
Sheep	dig coef %		61.	61.	
Ether extract	%		1.5	6.4	
Cattle	dig coef %		75.	75.	
Sheep	dig coef %		77.	77.	
N-free extract	%		0.3	1.4	
Cattle	dig coef %		72.	72.	
Sheep	dig coef %		74.	74.	
Protein (N x 6.25)	%		3.1	13.5	98
Cattle	dig coef %		47.	47.	
Sheep	dig coef %		54.	54.	
Cattle	dig prot %		1.5	6.3	
Goats	dig prot %	*	2.0	8.5	
Horses	dig prot %	*	2.0	8.5	
Sheep	dig prot %		1.7	7.3	
Cellulose (Matrone)	%		17.1	73.7	
Energy	GE Mcal/kg				
Cattle	DE Mcal/kg	*	0.42	1.79	
Sheep	DE Mcal/kg	*	0.41	1.75	
Cattle	ME Mcal/kg	*	0.34	1.47	
Sheep	ME Mcal/kg	*	0.33	1.43	
Cattle	TDN %		9.4	40.6	
Sheep	TDN %		9.2	39.6	
Calcium	%		0.12	0.50	98
Chlorine	%		0.27	1.16	
Phosphorus	%		0.05	0.20	29
Potassium	%		0.20	0.88	
Carotene	mg/kg		8.3	35.6	75
Vitamin A equivalent	IU/g		13.8	59.4	

Corn, aerial part, ensiled, immature, (3)

Ref No 3 02 817 United States

Feed Name or Analyses			As Fed	Dry	C.V. ± %
Dry matter	%		23.6	100.0	15
Ash	%		1.7	7.3	6
Crude fiber	%		5.8	24.6	11
Ether extract	%		0.7	3.1	19
N-free extract	%		13.1	55.3	
Protein (N x 6.25)	%		2.3	9.7	13
Cattle	dig prot %	*	1.2	5.0	
Goats	dig prot %	*	1.2	5.0	
Horses	dig prot %	*	1.2	5.0	
Sheep	dig prot %	*	1.2	5.0	
Energy	GE Mcal/kg				
Cattle	DE Mcal/kg	*	0.70	2.96	
Sheep	DE Mcal/kg	*	0.69	2.90	
Cattle	ME Mcal/kg	*	0.57	2.42	
Sheep	ME Mcal/kg	*	0.56	2.38	
Cattle	TDN %	*	15.8	67.0	
Sheep	TDN %	*	15.6	65.9	
Calcium	%		0.12	0.52	16
Iron	%		0.012	0.049	
Magnesium	%		0.07	0.31	9
Phosphorus	%		0.07	0.31	30
Potassium	%		0.39	1.64	
Carotene	mg/kg		26.2	110.7	26
Vitamin A equivalent	IU/g		43.6	184.5	

Corn, aerial part, ensiled, milk stage, (3)
Corn fodder silage, milk stage

Ref No 3 02 818 United States

Feed Name or Analyses			As Fed	Dry	C.V. ± %
Dry matter	%		23.5	100.0	11
Cattle	dig coef %		62.	62.	6
Ash	%		1.3	5.7	19
Crude fiber	%		7.4	31.4	10
Cattle	dig coef %		71.	71.	
Sheep	dig coef %		69.	69.	
Ether extract	%		0.6	2.7	19
Cattle	dig coef %		74.	74.	
Sheep	dig coef %		85.	85.	
N-free extract	%		12.3	52.3	6
Cattle	dig coef %		68.	68.	
Sheep	dig coef %		76.	76.	
Protein (N x 6.25)	%		1.9	7.9	13
Cattle	dig coef %		48.	48.	12
Sheep	dig coef %		59.	59.	
Cattle	dig prot %		0.9	3.8	
Goats	dig prot %	*	0.8	3.4	
Horses	dig prot %	*	0.8	3.4	
Sheep	dig prot %		1.1	4.7	
Cellulose (Matrone)	%		8.3	35.5	
Cattle	dig coef %		65.	65.	
Energy	GE Mcal/kg		1.05	4.46	
Cattle	GE dig coef %		62.	62.	
Cattle	DE Mcal/kg		0.65	2.75	
Sheep	DE Mcal/kg	*	0.74	3.14	
Cattle	ME Mcal/kg	*	0.53	2.25	
Sheep	ME Mcal/kg	*	0.60	2.57	
Cattle	TDN %		15.5	66.2	
Sheep	TDN %		16.7	71.2	
Calcium	%		0.07	0.28	17
Magnesium	%		0.10	0.41	8
Phosphorus	%		0.06	0.24	25
Potassium	%		0.37	1.57	5
Sodium	%		0.00	0.01	

Ref No 3 02 818 Canada

Feed Name or Analyses			As Fed	Dry	C.V. ± %
Dry matter	%		22.5	100.0	15
Crude fiber	%				
Cattle	dig coef %		71.	71.	
Sheep	dig coef %		69.	69.	
Ether extract	%				
Cattle	dig coef %		74.	74.	
Sheep	dig coef %		85.	85.	
N-free extract	%				
Cattle	dig coef %		68.	68.	
Sheep	dig coef %		76.	76.	
Protein (N x 6.25)	%		1.1	5.1	87
Cattle	dig coef %		48.	48.	
Sheep	dig coef %		59.	59.	
Cattle	dig prot %		0.5	2.4	
Goats	dig prot %	*	0.2	0.8	
Horses	dig prot %	*	0.2	0.8	
Sheep	dig prot %		0.7	3.0	
Cellulose (Matrone)	%		14.0	62.2	75
Energy	GE Mcal/kg		1.00	4.46	
Cattle	GE dig coef %		62.	62.	
Cattle	DE Mcal/kg		0.62	2.75	
Sheep	DE Mcal/kg	*	0.71	3.15	
Cattle	ME Mcal/kg	*	0.51	2.25	
Sheep	ME Mcal/kg	*	0.58	2.58	
Cattle	TDN %		15.0	66.5	
Sheep	TDN %		16.1	71.5	

Corn, aerial part, ensiled, dough stage, (3)
Corn fodder silage, dough stage

Ref No 3 02 819 United States

Feed Name or Analyses			As Fed	Dry	C.V. ± %
Dry matter	%		29.4	100.0	20
Cattle	dig coef %		66.	66.	6
Ash	%		1.4	4.7	23

(1) dry forages and roughages (3) silages (6) minerals
(2) pasture, range plants, and forages fed green (4) energy feeds (7) vitamins
(5) protein supplements (8) additives

Feed Name or Analyses			Mean As Fed	Mean Dry	C.V. ± %
Crude fiber	%		7.4	25.4	18
Cattle	dig coef %		62.	62.	7
Sheep	dig coef %		71.	71.	
Ether extract	%		0.8	2.8	17
Cattle	dig coef %		79.	79.	6
Sheep	dig coef %		78.	78.	
N-free extract	%		17.3	59.0	7
Cattle	dig coef %		73.	73.	5
Sheep	dig coef %		73.	73.	
Protein (N x 6.25)	%		2.4	8.1	8
Cattle	dig coef %		50.	50.	10
Sheep	dig coef %		56.	56.	
Cattle	dig prot %		1.2	4.1	
Goats	dig prot %	*	1.1	3.6	
Horses	dig prot %	*	1.1	3.6	
Sheep	dig prot %		1.3	4.6	
Cellulose (Matrone)	%		10.3	35.2	5
Cattle	dig coef %		64.	64.	5
Energy	GE Mcal/kg		1.29	4.40	1
Cattle	GE dig coef %		61.	61.	3
Cattle	DE Mcal/kg		0.79	2.68	
Sheep	DE Mcal/kg	*	0.91	3.11	
Cattle	ME Mcal/kg	*	0.65	2.20	
Sheep	ME Mcal/kg	*	0.75	2.55	
Cattle	TDN %		19.9	67.8	4
Sheep	TDN %		20.7	70.6	
Boron	mg/kg		2.838	9.667	
Calcium	%		0.06	0.20	
Copper	mg/kg		3.9	13.3	
Iron	%		0.019	0.064	
Magnesium	%		0.08	0.28	
Manganese	mg/kg		10.1	34.3	
Phosphorus	%		0.06	0.20	
Potassium	%		0.38	1.31	
Silicon	%		1.47	5.00	
Sodium	%		0.00	0.01	
Zinc	mg/kg		6.2	21.0	
Carotene	mg/kg		3.8	13.0	
Vitamin A equivalent	IU/g		6.4	21.7	

Ref No 3 02 819 Canada

Feed Name or Analyses			Mean As Fed	Mean Dry	C.V. ± %
Dry matter	%		24.8	100.0	17
Ash	%		5.2	21.0	
Crude fiber	%		11.8	47.6	69
Cattle	dig coef %		62.	62.	
Sheep	dig coef %		71.	71.	
Ether extract	%		2.5	10.0	
Cattle	dig coef %		79.	79.	
Sheep	dig coef %		78.	78.	
N-free extract	%		2.6	10.6	
Cattle	dig coef %		73.	73.	
Sheep	dig coef %		73.	73.	
Protein (N x 6.25)	%		2.7	10.8	98
Cattle	dig coef %		50.	50.	
Sheep	dig coef %		56.	56.	
Cattle	dig prot %		1.3	5.4	
Goats	dig prot %	*	1.5	6.0	
Horses	dig prot %	*	1.5	6.0	
Sheep	dig prot %		1.5	6.1	
Cellulose (Matrone)	%		15.7	63.2	51
Energy	GE Mcal/kg		1.09	4.40	
Cattle	GE dig coef %		61.	61.	
Cattle	DE Mcal/kg		0.67	2.68	
Sheep	DE Mcal/kg	*	0.71	2.87	
Cattle	ME Mcal/kg	*	0.55	2.20	
Sheep	ME Mcal/kg	*	0.58	2.36	
Cattle	TDN %		15.0	60.6	
Sheep	TDN %		16.2	65.2	
Calcium	%		0.39	1.55	29

Feed Name or Analyses			Mean As Fed	Mean Dry	C.V. ± %
Phosphorus	%		0.47	1.89	98
Potassium	%		1.28	5.17	20
Carotene	mg/kg		40.2	161.7	
Vitamin A equivalent	IU/g		66.9	269.6	

Corn, aerial part, ensiled, mature, (3)
Corn fodder silage, mature
Ref No 3 02 820 United States

Feed Name or Analyses			Mean As Fed	Mean Dry	C.V. ± %
Dry matter	%		27.8	100.0	11
Cattle	dig coef %		69.	69.	
Ash	%		1.9	7.0	
Crude fiber	%		7.5	27.2	
Cattle	dig coef %		70.	70.	
Sheep	dig coef %		46.	46.	
Ether extract	%		1.2	4.3	
Cattle	dig coef %		84.	84.	
Sheep	dig coef %		74.	74.	
N-free extract	%		14.9	53.5	
Cattle	dig coef %		73.	73.	
Sheep	dig coef %		76.	76.	
Protein (N x 6.25)	%		2.2	8.0	
Cattle	dig coef %		61.	61.	
Sheep	dig coef %		58.	58.	
Cattle	dig prot %		1.4	4.9	
Goats	dig prot %	*	1.0	3.5	
Horses	dig prot %	*	1.0	3.5	
Sheep	dig prot %		1.3	4.7	
Cellulose (Matrone)	%				
Sheep	dig coef %		42.	42.	
Fatty acids	%				
Acetic	%		0.527	1.900	
Lactic acid	%		2.014	7.255	
Energy	GE Mcal/kg				
Cattle	DE Mcal/kg	*	0.87	3.12	
Sheep	DE Mcal/kg	*	0.80	2.87	
Cattle	ME Mcal/kg	*	0.71	2.56	
Sheep	ME Mcal/kg	*	0.65	2.35	
Cattle	TDN %		19.6	70.7	
Sheep	TDN %		18.0	65.0	
Cobalt	mg/kg		0.017	0.060	41
Carotene	mg/kg		4.3	15.7	42
Vitamin A equivalent	IU/g		7.2	26.1	
pH	pH units		4.05	4.05	

Ref No 3 02 820 Canada

Feed Name or Analyses			Mean As Fed	Mean Dry	C.V. ± %
Dry matter	%		36.1	100.0	
Crude fiber	%				
Cattle	dig coef %		70.	70.	
Sheep	dig coef %		46.	46.	
Ether extract	%				
Cattle	dig coef %		84.	84.	
Sheep	dig coef %		74.	74.	
N-free extract	%				
Cattle	dig coef %		73.	73.	
Sheep	dig coef %		76.	76.	
Protein (N x 6.25)	%		2.7	7.4	
Cattle	dig coef %		61.	61.	
Sheep	dig coef %		58.	58.	
Cattle	dig prot %		1.6	4.5	
Goats	dig prot %	*	1.1	3.0	
Horses	dig prot %	*	1.1	3.0	
Sheep	dig prot %		1.6	4.3	
Cellulose (Matrone)	%		25.9	71.8	
Energy	GE Mcal/kg				
Cattle	DE Mcal/kg	*	1.13	3.12	
Sheep	DE Mcal/kg	*	1.04	2.87	
Cattle	ME Mcal/kg	*	0.92	2.56	

Feed Name or Analyses			Mean As Fed	Mean Dry	C.V. ± %
Sheep	ME Mcal/kg	*	0.85	2.35	
Cattle	TDN %		25.5	70.8	
Sheep	TDN %		23.5	65.0	

Corn, aerial part, ensiled, over ripe, (3)
Ref No 3 02 821 United States

Feed Name or Analyses			Mean As Fed	Mean Dry	C.V. ± %
Dry matter	%		32.9	100.0	
Ash	%		2.2	6.6	
Crude fiber	%		8.4	25.4	
Sheep	dig coef %		65.	65.	
Ether extract	%		1.7	5.2	
Sheep	dig coef %		87.	87.	
N-free extract	%		18.4	55.7	
Sheep	dig coef %		73.	73.	
Protein (N x 6.25)	%		2.3	7.1	
Sheep	dig coef %		42.	42.	
Cattle	dig prot %	*	0.9	2.6	
Goats	dig prot %	*	0.9	2.6	
Horses	dig prot %	*	0.9	2.6	
Sheep	dig prot %		1.0	3.0	
Energy	GE Mcal/kg		1.42	4.32	
Cattle	DE Mcal/kg	*	1.03	3.13	
Sheep	DE Mcal/kg	*	1.02	3.10	
Cattle	ME Mcal/kg	*	0.85	2.57	
Sheep	ME Mcal/kg	*	0.84	2.54	
Cattle	TDN %	*	23.4	71.0	
Sheep	TDN %		23.2	70.3	

Ref No 3 02 821 Canada

Feed Name or Analyses			Mean As Fed	Mean Dry	C.V. ± %
Dry matter	%		28.5	100.0	14
Crude fiber	%				
Sheep	dig coef %		65.	65.	
Ether extract	%				
Sheep	dig coef %		87.	87.	
N-free extract	%				
Sheep	dig coef %		73.	73.	
Protein (N x 6.25)	%		1.9	6.7	36
Sheep	dig coef %		42.	42.	
Cattle	dig prot %	*	0.7	2.4	
Goats	dig prot %	*	0.7	2.4	
Horses	dig prot %	*	0.7	2.4	
Sheep	dig prot %		0.8	2.8	
Cellulose (Matrone)	%		20.1	70.5	32
Energy	GE Mcal/kg		1.23	4.32	
Sheep	DE Mcal/kg	*	0.89	3.10	
Sheep	ME Mcal/kg	*	0.73	2.55	
Sheep	TDN %		20.1	70.4	

Corn, aerial part, ensiled, well eared, (3)
Ref No 3 02 823 United States

Feed Name or Analyses			Mean As Fed	Mean Dry	C.V. ± %
Dry matter	%		25.6	100.0	12
Ash	%		1.4	5.4	19
Crude fiber	%		5.9	23.0	9
Ether extract	%		0.7	2.9	17
N-free extract	%		15.6	60.9	
Protein (N x 6.25)	%		2.0	7.8	14
Cattle	dig prot %	*	0.8	3.3	
Goats	dig prot %	*	0.8	3.3	
Horses	dig prot %	*	0.8	3.3	
Sheep	dig prot %	*	0.8	3.3	
Energy	GE Mcal/kg				
Cattle	DE Mcal/kg	*	0.77	3.00	
Sheep	DE Mcal/kg	*	0.77	2.99	
Cattle	ME Mcal/kg	*	0.63	2.46	

Continued

Feed Name or Analyses			Mean As Fed	Mean Dry	C.V. ± %
Sheep	ME Mcal/kg	*	0.63	2.45	
Cattle	TDN %	*	17.4	68.0	
Sheep	TDN %	*	17.4	67.8	
Sulphur	%		0.02	0.08	14

Corn, aerial part, ensiled trench silo, (3)
Ref No 3 02 824 United States

Feed Name or Analyses			As Fed	Dry	C.V. ± %
Dry matter	%		26.0	100.0	6
Ash	%		1.3	5.0	15
Crude fiber	%		6.2	23.9	6
Ether extract	%		0.8	3.0	24
N-free extract	%		15.6	60.2	
Protein (N x 6.25)	%		2.1	7.9	10
Cattle	dig prot %	*	0.9	3.4	
Goats	dig prot %	*	0.9	3.4	
Horses	dig prot %	*	0.9	3.4	
Sheep	dig prot %	*	0.9	3.4	
Energy	GE Mcal/kg				
Cattle	DE Mcal/kg	*	0.76	2.94	
Sheep	DE Mcal/kg	*	0.78	2.99	
Cattle	ME Mcal/kg	*	0.63	2.41	
Sheep	ME Mcal/kg	*	0.64	2.45	
Cattle	TDN %	*	17.3	66.7	
Sheep	TDN %	*	17.6	67.7	
Calcium	%		0.11	0.41	
Phosphorus	%		0.05	0.18	
Carotene	mg/kg		18.8	72.5	
Vitamin A equivalent	IU/g		31.4	120.9	

Corn, aerial part, steamed ensiled, (3)
Ref No 3 02 825 United States

Feed Name or Analyses			As Fed	Dry	C.V. ± %
Dry matter	%		21.4	100.0	
Ash	%		1.4	6.4	
Crude fiber	%		4.6	21.4	
Cattle	dig coef %		63.	63.	
Ether extract	%		0.8	3.7	
Cattle	dig coef %		79.	79.	
N-free extract	%		13.0	60.7	
Cattle	dig coef %		71.	71.	
Protein (N x 6.25)	%		1.7	7.8	
Cattle	dig coef %		5.	5.	
Cattle	dig prot %		0.1	0.4	
Goats	dig prot %	*	0.7	3.3	
Horses	dig prot %	*	0.7	3.3	
Sheep	dig prot %	*	0.7	3.3	
Energy	GE Mcal/kg				
Cattle	DE Mcal/kg	*	0.60	2.80	
Sheep	DE Mcal/kg	*	0.63	2.96	
Cattle	ME Mcal/kg	*	0.49	2.30	
Sheep	ME Mcal/kg	*	0.52	2.43	
Cattle	TDN %		13.6	63.5	
Sheep	TDN %	*	14.4	67.1	

Corn, aerial part w AIV preservative added, ensiled, (3)
Ref No 3 02 826 United States

Feed Name or Analyses			As Fed	Dry	C.V. ± %
Dry matter	%			100.0	
Manganese	mg/kg			35.1	

(1) dry forages and roughages	(3) silages	(6) minerals
(2) pasture, range plants, and forages fed green	(4) energy feeds	(7) vitamins
	(5) protein supplements	(8) additives

Corn, aerial part w biosil added, ensiled, (3)
Ref No 3 02 832 United States

Feed Name or Analyses			As Fed	Dry	C.V. ± %
Dry matter	%		16.2	100.0	
Ash	%		1.2	7.6	
Crude fiber	%		4.0	24.7	
Sheep	dig coef %		65.	65.	
Ether extract	%		0.5	3.0	
Sheep	dig coef %		82.	82.	
N-free extract	%		9.1	56.3	
Sheep	dig coef %		76.	76.	
Protein (N x 6.25)	%		1.4	8.4	
Sheep	dig coef %		50.	50.	
Cattle	dig prot %	*	0.6	3.9	
Goats	dig prot %	*	0.6	3.9	
Horses	dig prot %	*	0.6	3.9	
Sheep	dig prot %		0.7	4.2	
Energy	GE Mcal/kg				
Cattle	DE Mcal/kg	*	0.49	3.03	
Sheep	DE Mcal/kg	*	0.49	3.02	
Cattle	ME Mcal/kg	*	0.40	2.48	
Sheep	ME Mcal/kg	*	0.40	2.48	
Cattle	TDN %	*	11.1	68.6	
Sheep	TDN %		11.1	68.6	

Corn, aerial part w molasses added, ensiled, (3)
Ref No 3 02 834 United States

Feed Name or Analyses			As Fed	Dry	C.V. ± %
Dry matter	%			100.0	
Ash	%			6.4	8
Crude fiber	%			23.4	9
Ether extract	%			2.9	19
N-free extract	%			59.3	
Protein (N x 6.25)	%			8.0	14
Cattle	dig prot %	*		3.5	
Goats	dig prot %	*		3.5	
Horses	dig prot %	*		3.5	
Sheep	dig prot %	*		3.5	
Energy	GE Mcal/kg				
Cattle	DE Mcal/kg	*		3.03	
Sheep	DE Mcal/kg	*		2.96	
Cattle	ME Mcal/kg	*		2.48	
Sheep	ME Mcal/kg	*		2.43	
Cattle	TDN %	*		68.7	
Sheep	TDN %	*		67.1	
Calcium	%			0.25	15
Phosphorus	%			0.30	17

Corn, aerial part wo ears wo husks, ensiled, (3)
Corn stover silage
Ref No 3 02 836 United States

Feed Name or Analyses			As Fed	Dry	C.V. ± %
Dry matter	%		35.1	100.0	
Ash	%		3.0	8.6	
Crude fiber	%		11.8	33.6	
Cattle	dig coef %	*	67.	67.	
Sheep	dig coef %		60.	60.	
Ether extract	%		0.8	2.4	
Cattle	dig coef %	*	61.	61.	
Sheep	dig coef %		67.	67.	
N-free extract	%		16.8	47.9	
Cattle	dig coef %	*	60.	60.	
Sheep	dig coef %		55.	55.	
Protein (N x 6.25)	%		2.6	7.5	
Cattle	dig coef %	*	40.	40.	
Sheep	dig coef %		38.	38.	
Cattle	dig prot %		1.1	3.0	
Goats	dig prot %	*	1.1	3.0	
Horses	dig prot %	*	1.1	3.0	
Sheep	dig prot %		1.0	2.9	
Energy	GE Mcal/kg				
Cattle	DE Mcal/kg	*	0.89	2.54	
Sheep	DE Mcal/kg	*	0.82	2.34	
Cattle	ME Mcal/kg	*	0.73	2.08	
Sheep	ME Mcal/kg	*	0.67	1.92	
Cattle	NE_m Mcal/kg	*	0.44	1.24	
Cattle	NE_{gain} Mcal/kg	*	0.21	0.59	
Cattle	$NE_{lactating\ cows}$ Mcal/kg	*	0.48	1.38	
Cattle	TDN %		20.2	57.5	
Sheep	TDN %		18.6	53.0	

Corn, aerial part wo ears wo husks, ensiled, mature, (3)
Ref No 3 02 835 United States

Feed Name or Analyses			As Fed	Dry	C.V. ± %
Dry matter	%		26.7	100.0	
Ash	%		2.5	9.2	
Crude fiber	%		8.7	32.6	
Cattle	dig coef %		67.	67.	
Ether extract	%		0.5	2.0	
Cattle	dig coef %		60.	60.	
N-free extract	%		13.3	49.7	
Cattle	dig coef %		56.	56.	
Protein (N x 6.25)	%		1.7	6.5	
Cattle	dig coef %		38.	38.	
Cattle	dig prot %		0.7	2.5	
Goats	dig prot %	*	0.6	2.1	
Horses	dig prot %	*	0.6	2.1	
Sheep	dig prot %	*	0.6	2.1	
Energy	GE Mcal/kg				
Cattle	DE Mcal/kg	*	0.65	2.42	
Sheep	DE Mcal/kg	*	0.63	2.36	
Cattle	ME Mcal/kg	*	0.53	1.98	
Sheep	ME Mcal/kg	*	0.52	1.94	
Cattle	TDN %		14.6	54.8	
Sheep	TDN %	*	14.3	53.5	

Corn, aerial part wo ears wo husks, stack ensiled, (3)
Ref No 3 02 904 United States

Feed Name or Analyses			As Fed	Dry	C.V. ± %
Dry matter	%		35.0	100.0	
Ash	%		2.5	7.0	
Crude fiber	%		9.9	28.3	
Sheep	dig coef %		50.	50.	
Ether extract	%		1.1	3.0	
Sheep	dig coef %		74.	74.	
N-free extract	%		19.2	54.8	
Sheep	dig coef %		61.	61.	
Protein (N x 6.25)	%		2.4	6.9	
Sheep	dig coef %		15.	15.	
Cattle	dig prot %	*	0.9	2.5	
Goats	dig prot %	*	0.9	2.5	
Horses	dig prot %	*	0.9	2.5	
Sheep	dig prot %		0.4	1.0	
Energy	GE Mcal/kg				
Cattle	DE Mcal/kg	*	0.86	2.47	
Sheep	DE Mcal/kg	*	0.83	2.36	
Cattle	ME Mcal/kg	*	0.71	2.03	
Sheep	ME Mcal/kg	*	0.68	1.94	
Cattle	TDN %	*	19.6	56.1	
Sheep	TDN %		18.8	53.6	

Feed Name or Analyses			Mean As Fed	Mean Dry	C.V. ± %
Sheep	ME Mcal/kg	*	1.81	2.59	
Cattle	TDN %	*	52.0	74.3	
Sheep	TDN %	*	50.1	71.5	

Corn, aerial part w sugar added, ensiled, (3)
Ref No 3 02 833 United States

Feed Name or Analyses			Mean As Fed	Mean Dry	C.V. ± %
Dry matter	%		21.0	100.0	
Ash	%		1.2	5.8	
Crude fiber	%		4.2	19.9	
Sheep	dig coef %		58.	58.	
Ether extract	%		0.7	3.2	
Sheep	dig coef %		91.	91.	
N-free extract	%		13.3	63.3	
Sheep	dig coef %		77.	77.	
Protein (N x 6.25)	%		1.6	7.8	
Sheep	dig coef %		54.	54.	
Cattle	dig prot %	*	0.7	3.3	
Goats	dig prot %	*	0.7	3.3	
Horses	dig prot %	*	0.7	3.3	
Sheep	dig prot %		0.9	4.2	
Energy	GE Mcal/kg				
Cattle	DE Mcal/kg	*	0.67	3.17	
Sheep	DE Mcal/kg	*	0.66	3.13	
Cattle	ME Mcal/kg	*	0.55	2.60	
Sheep	ME Mcal/kg	*	0.54	2.57	
Cattle	TDN %	*	15.1	71.9	
Sheep	TDN %		14.9	71.0	

Corn, cannery residue, ensiled, (3)
Ref No 3 02 837 United States

Feed Name or Analyses			Mean As Fed	Mean Dry	C.V. ± %
Dry matter	%		22.5	100.0	
Ash	%		1.2	5.2	
Crude fiber	%		5.4	23.9	
Cattle	dig coef %		70.	70.	
Ether extract	%		1.4	6.0	
Cattle	dig coef %		87.	87.	
N-free extract	%		12.5	55.6	
Cattle	dig coef %		71.	71.	
Protein (N x 6.25)	%		2.1	9.3	
Cattle	dig coef %		56.	56.	
Cattle	dig prot %		1.2	5.2	
Goats	dig prot %	*	1.1	4.7	
Horses	dig prot %	*	1.1	4.7	
Sheep	dig prot %	*	1.1	4.7	
Energy	GE Mcal/kg				
Cattle	DE Mcal/kg	*	0.72	3.23	
Sheep	DE Mcal/kg	*	0.65	2.89	
Cattle	ME Mcal/kg	*	0.59	2.65	
Sheep	ME Mcal/kg	*	0.53	2.37	
Cattle	TDN %		16.4	73.2	
Sheep	TDN %	*	14.7	65.6	

Corn, ears, ensiled, (3)
Ref No 3 07 740 United States

Feed Name or Analyses			Mean As Fed	Mean Dry	C.V. ± %
Dry matter	%		70.0	100.0	
Ash	%		1.6	2.3	
Crude fiber	%		7.3	10.4	
Ether extract	%		2.5	3.6	
N-free extract	%		51.6	73.7	
Protein (N x 6.25)	%		7.1	10.2	
Cattle	dig prot %	*	3.8	5.4	
Goats	dig prot %	*	3.8	5.4	
Horses	dig prot %	*	3.8	5.4	
Sheep	dig prot %	*	3.8	5.4	
Energy	GE Mcal/kg				
Cattle	DE Mcal/kg	*	2.29	3.28	
Sheep	DE Mcal/kg	*	2.21	3.15	
Cattle	ME Mcal/kg	*	1.88	2.69	

Corn, ears, grnd ensiled, mature, (3)
Ref No 3 07 801 United States

Feed Name or Analyses			Mean As Fed	Mean Dry	C.V. ± %
Dry matter	%		54.0	100.0	
Crude fiber	%				
Sheep	dig coef %		35.	35.	
Ether extract	%				
Sheep	dig coef %		88.	88.	
Protein (N x 6.25)	%				
Sheep	dig coef %		65.	65.	
Cellulose (Matrone)	%				
Sheep	dig coef %		36.	36.	
Fatty acids	%				
Acetic	%		0.254	0.470	
Lactic acid	%		1.473	2.730	

Corn, ears w husks, ensiled, (3)
Ref No 3 02 839 United States

Feed Name or Analyses			Mean As Fed	Mean Dry	C.V. ± %
Dry matter	%			100.0	
Ash	%			1.8	13
Crude fiber	%			9.7	5
Ether extract	%			3.5	18
N-free extract	%			75.2	
Protein (N x 6.25)	%			9.8	10
Cattle	dig prot %	*		5.1	
Goats	dig prot %	*		5.1	
Horses	dig prot %	*		5.1	
Sheep	dig prot %	*		5.1	
Energy	GE Mcal/kg				
Cattle	DE Mcal/kg	*		3.30	
Sheep	DE Mcal/kg	*		3.18	
Cattle	ME Mcal/kg	*		2.70	
Sheep	ME Mcal/kg	*		2.60	
Cattle	NE_m Mcal/kg	*		1.60	
Cattle	NE_{gain} Mcal/kg	*		1.03	
Cattle	$NE_{lactating\ cows}$ Mcal/kg	*		1.90	
Cattle	TDN %	*		74.7	
Sheep	TDN %	*		72.0	

Corn, ears w husks, ensiled, milk stage, (3)
Ref No 3 08 392 United States

Feed Name or Analyses			Mean As Fed	Mean Dry	C.V. ± %
Dry matter	%		44.8	100.0	
Ash	%		1.0	2.2	
Crude fiber	%		5.8	12.9	
Ether extract	%		1.7	3.8	
N-free extract	%		32.3	72.1	
Protein (N x 6.25)	%		4.0	8.9	
Cattle	dig prot %	*	1.9	4.3	
Goats	dig prot %	*	1.9	4.3	
Horses	dig prot %	*	1.9	4.3	
Sheep	dig prot %	*	1.9	4.3	
Energy	GE Mcal/kg				
Cattle	DE Mcal/kg	*	1.45	3.24	
Sheep	DE Mcal/kg	*	1.40	3.13	
Cattle	ME Mcal/kg	*	1.19	2.66	
Sheep	ME Mcal/kg	*	1.15	2.57	
Cattle	TDN %	*	32.9	73.5	
Sheep	TDN %	*	31.8	71.1	

Corn, ears w husks, ensiled, dough stage, (3)
Ref No 3 02 838 United States

Feed Name or Analyses			Mean As Fed	Mean Dry	C.V. ± %
Dry matter	%		42.8	100.0	
Ash	%		0.9	2.2	
Crude fiber	%		5.0	11.6	
Cattle	dig coef %		34.	34.	
Ether extract	%		1.7	4.0	
Cattle	dig coef %		80.	80.	
N-free extract	%		31.8	74.4	
Cattle	dig coef %		80.	80.	
Protein (N x 6.25)	%		3.3	7.8	
Cattle	dig coef %		54.	54.	
Cattle	dig prot %		1.8	4.2	
Goats	dig prot %	*	1.4	3.3	
Horses	dig prot %	*	1.4	3.3	
Sheep	dig prot %	*	1.4	3.3	
Energy	GE Mcal/kg				
Cattle	DE Mcal/kg	*	1.41	3.30	
Sheep	DE Mcal/kg	*	1.35	3.15	
Cattle	ME Mcal/kg	*	1.16	2.71	
Sheep	ME Mcal/kg	*	1.11	2.58	
Cattle	TDN %		32.0	74.9	
Sheep	TDN %	*	30.6	71.5	

Corn, ears w husks, ensiled pit silo, (3)
Ref No 3 02 840 United States

Feed Name or Analyses			Mean As Fed	Mean Dry	C.V. ± %
Dry matter	%		43.5	100.0	3
Ash	%		2.1	4.9	48
Crude fiber	%		5.1	11.8	7
Ether extract	%		1.7	3.8	4
N-free extract	%		30.8	70.7	
Protein (N x 6.25)	%		3.8	8.8	9
Cattle	dig prot %	*	1.8	4.2	
Goats	dig prot %	*	1.8	4.2	
Horses	dig prot %	*	1.8	4.2	
Sheep	dig prot %	*	1.8	4.2	
Energy	GE Mcal/kg				
Cattle	DE Mcal/kg	*	1.50	3.44	
Sheep	DE Mcal/kg	*	1.34	3.08	
Cattle	ME Mcal/kg	*	1.23	2.82	
Sheep	ME Mcal/kg	*	1.10	2.52	
Cattle	TDN %	*	34.0	78.1	
Sheep	TDN %	*	30.4	69.8	
Carotene	mg/kg		3.4	7.7	
Vitamin A equivalent	IU/g		5.6	12.9	

Corn, grain, ensiled, (3)
Ref No 3 07 739 United States

Feed Name or Analyses			Mean As Fed	Mean Dry	C.V. ± %
Dry matter	%		70.0	100.0	0
Ash	%		1.3	1.8	8
Crude fiber	%		2.2	3.1	17
Ether extract	%		2.5	3.6	46
N-free extract	%		55.4	79.2	3
Protein (N x 6.25)	%		8.6	12.3	4
Cattle	dig prot %	*	5.2	7.4	
Goats	dig prot %	*	5.2	7.4	
Horses	dig prot %	*	5.2	7.4	
Sheep	dig prot %	*	5.2	7.4	
Energy	GE Mcal/kg				
Cattle	DE Mcal/kg	*	2.42	3.46	
Sheep	DE Mcal/kg	*	2.25	3.21	
Cattle	ME Mcal/kg	*	1.98	2.83	
Sheep	ME Mcal/kg	*	1.84	2.64	

Continued

Feed Name or Analyses		Mean As Fed	Mean Dry	C.V. ± %
Cattle	TDN %	* 54.9	78.4	
Sheep	TDN %	* 51.0	72.9	

Corn, bran, wet or dry milled dehy, (4)
Corn bran (AAFCO)
Corn bran (CFA)
Ref No 4 02 841 United States

Feed Name or Analyses		Mean As Fed	Mean Dry	C.V. ± %
Dry matter	%	88.7	100.0	2
Ash	%	1.9	2.2	32
Crude fiber	%	9.6	10.9	15
Horses	dig coef %	140.	140.	
Sheep	dig coef %	63.	63.	
Ether extract	%	4.5	5.1	53
Horses	dig coef %	49.	49.	
Sheep	dig coef %	78.	78.	
N-free extract	%	64.6	72.8	
Horses	dig coef %	67.	67.	
Sheep	dig coef %	73.	73.	
Protein (N x 6.25)	%	8.0	9.1	29
Horses	dig coef %	52.	52.	
Sheep	dig coef %	57.	57.	
Cattle	dig prot %	* 3.8	4.3	
Goats	dig prot %	* 4.9	5.5	
Horses	dig prot %	4.2	4.7	
Sheep	dig prot %	4.6	5.2	
Energy	GE Mcal/kg	3.99	4.50	
Cattle	DE Mcal/kg	* 3.01	3.39	
Horses	DE Mcal/kg	* 2.91	3.28	
Sheep	DE Mcal/kg	* 2.90	3.27	
Swine	DE kcal/kg	*2739.	3087.	
Cattle	ME Mcal/kg	* 2.47	2.78	
Horses	ME Mcal/kg	* 2.38	2.69	
Sheep	ME Mcal/kg	* 2.38	2.68	
Swine	ME kcal/kg	*2579.	2907.	
Cattle	TDN %	* 68.1	76.8	
Horses	TDN %	66.0	74.3	
Sheep	TDN %	65.8	74.2	
Swine	TDN %	* 62.1	70.0	
Calcium	%	0.03	0.04	44
Chlorine	%	0.05	0.06	
Magnesium	%	0.26	0.29	6
Manganese	mg/kg	15.9	17.9	
Phosphorus	%	0.16	0.18	55
Potassium	%	0.64	0.72	
Sulphur	%	0.07	0.08	
Riboflavin	mg/kg	1.5	1.7	
Thiamine	mg/kg	4.4	4.9	2

Corn, ears, fresh grnd, immature, (4)
Ref No 2 02 847 United States

Feed Name or Analyses		Mean As Fed	Mean Dry	C.V. ± %
Dry matter	%	27.5	100.0	
Ash	%	0.8	2.8	
Crude fiber	%	4.2	15.3	
Sheep	dig coef %	88.	88.	
Ether extract	%	1.0	3.6	
Sheep	dig coef %	78.	78.	
N-free extract	%	18.1	65.9	
Sheep	dig coef %	87.	87.	
Protein (N x 6.25)	%	3.4	12.4	
Sheep	dig coef %	76.	76.	
Cattle	dig prot %	* 2.0	7.4	
Goats	dig prot %	* 2.4	8.6	
Horses	dig prot %	* 2.4	8.6	
Sheep	dig prot %	2.6	9.4	
Energy	GE Mcal/kg			
Cattle	DE Mcal/kg	* 1.01	3.67	
Sheep	DE Mcal/kg	* 1.05	3.82	
Swine	DE kcal/kg	* 710.	2581.	
Cattle	ME Mcal/kg	* 0.83	3.01	
Sheep	ME Mcal/kg	* 0.86	3.13	
Swine	ME kcal/kg	* 664.	2413.	
Cattle	TDN %	* 22.9	83.3	
Sheep	TDN %	23.8	86.5	
Swine	TDN %	* 16.1	58.5	

Corn, ears, fresh grnd, milk stage, (4)
Ref No 2 02 848 United States

Feed Name or Analyses		Mean As Fed	Mean Dry	C.V. ± %
Dry matter	%	23.5	100.0	
Ash	%	0.6	2.6	
Crude fiber	%	2.9	12.2	
Sheep	dig coef %	76.	76.	
Swine	dig coef %	46.	46.	
Ether extract	%	1.0	4.1	
Sheep	dig coef %	77.	77.	
Swine	dig coef %	55.	55.	
N-free extract	%	16.3	69.7	
Sheep	dig coef %	81.	81.	
Swine	dig coef %	82.	82.	
Protein (N x 6.25)	%	2.7	11.4	
Sheep	dig coef %	60.	60.	
Swine	dig coef %	67.	67.	
Cattle	dig prot %	* 1.5	6.5	
Goats	dig prot %	* 1.8	7.7	
Horses	dig prot %	* 1.8	7.7	
Sheep	dig prot %	1.6	6.8	
Swine	dig prot %	1.8	7.6	
Energy	GE Mcal/kg			
Cattle	DE Mcal/kg	* 0.88	3.77	
Sheep	DE Mcal/kg	* 0.82	3.51	
Swine	DE kcal/kg	* 780.	3328.	
Cattle	ME Mcal/kg	* 0.72	3.09	
Sheep	ME Mcal/kg	* 0.68	2.88	
Swine	ME kcal/kg	* 731.	3118.	
Cattle	TDN %	* 20.0	85.5	
Sheep	TDN %	18.7	79.7	
Swine	TDN %	17.7	75.5	

Corn, ears, grnd, (4)
Corn and cob meal (AAFCO)
Ear corn chop (AAFCO)
Ground ear corn (AAFCO)
Ref No 4 02 849 United States

Feed Name or Analyses		Mean As Fed	Mean Dry	C.V. ± %
Dry matter	%	85.4	100.0	2
Ash	%	1.5	1.7	13
Crude fiber	%	7.1	8.3	26
Sheep	dig coef %	52.	52.	
Swine	dig coef %	28.	28.	
Ether extract	%	3.4	4.0	13
Sheep	dig coef %	81.	81.	
Swine	dig coef %	65.	65.	
N-free extract	%	65.4	76.6	2
Sheep	dig coef %	81.	81.	
Swine	dig coef %	86.	86.	
Protein (N x 6.25)	%	8.0	9.3	7
Sheep	dig coef %	55.	55.	
Swine	dig coef %	72.	72.	
Cattle	dig prot %	* 3.9	4.6	
Goats	dig prot %	* 4.9	5.8	
Horses	dig prot %	* 4.9	5.8	
Sheep	dig prot %	4.4	5.1	
Swine	dig prot %	5.7	6.7	
Sugars, total	%	1.0	1.2	
Linoleic	%	1.500	1.756	
Energy	GE Mcal/kg	3.97	4.65	
Cattle	DE Mcal/kg	* 3.20	3.75	
Sheep	DE Mcal/kg	* 2.97	3.48	
Swine	DE kcal/kg	*3044.	3563.	
Cattle	ME Mcal/kg	* 2.63	3.08	
Chickens	ME_n kcal/kg	2800.	3278.	
Sheep	ME Mcal/kg	* 2.43	2.85	
Swine	ME kcal/kg	*2865.	3354.	
Cattle	NE_m Mcal/kg	* 1.90	2.23	
Cattle	NE_{gain} Mcal/kg	* 1.19	1.39	
Cattle	$NE_{lactating\ cows}$ Mcal/kg	* 2.19	2.56	
Cattle	TDN %	* 72.6	85.0	
Sheep	TDN %	67.3	78.8	
Swine	TDN %	69.0	80.8	
Calcium	%	0.04	0.04	33
Cobalt	mg/kg	0.290	0.340	89
Copper	mg/kg	6.6	7.7	
Iron	%	0.007	0.008	
Magnesium	%	0.13	0.15	
Manganese	mg/kg	11.1	13.0	
Phosphorus	%	0.23	0.26	17
Potassium	%	0.42	0.50	
Choline	mg/kg	350.	410.	
Niacin	mg/kg	17.5	20.5	
Pantothenic acid	mg/kg	4.2	4.9	
Riboflavin	mg/kg	0.9	1.0	
α-tocopherol	mg/kg	20.0	85.1	
Vitamin A	IU/g	3.5	4.1	
Arginine	%	0.40	0.46	
Cystine	%	0.14	0.16	
Glycine	%	0.20	0.23	
Lysine	%	0.20	0.23	
Methionine	%	0.14	0.16	
Tryptophan	%	0.07	0.08	
Xanthophylls	mg/kg	8.75	10.24	

Ref No 4 02 849 Canada

Feed Name or Analyses		Mean As Fed	Mean Dry	C.V. ± %
Dry matter	%	66.6	100.0	
Crude fiber	%	8.2	12.3	
Sheep	dig coef %	52.	52.	
Swine	dig coef %	28.	28.	
Ether extract	%	2.0	3.0	
Sheep	dig coef %	81.	81.	
Swine	dig coef %	65.	65.	
N-free extract	%			
Sheep	dig coef %	81.	81.	
Swine	dig coef %	86.	86.	
Protein (N x 6.25)	%	6.7	10.1	
Sheep	dig coef %	55.	55.	
Swine	dig coef %	72.	72.	
Cattle	dig prot %	* 3.5	5.3	
Goats	dig prot %	* 4.3	6.5	
Horses	dig prot %	* 4.3	6.5	
Sheep	dig prot %	3.7	5.5	
Swine	dig prot %	4.8	7.2	
Energy	GE Mcal/kg	3.09	4.65	
Sheep	DE Mcal/kg	* 2.31	3.46	
Swine	DE kcal/kg	*2363.	3548.	
Chickens	ME_n kcal/kg	2183.	3278.	
Sheep	ME Mcal/kg	* 1.89	2.84	
Swine	ME kcal/kg	*2221.	3334.	

(1) dry forages and roughages (3) silages (6) minerals
(2) pasture, range plants, and forages fed green (4) energy feeds (7) vitamins
(5) protein supplements (8) additives

Feed Name or Analyses			Mean As Fed	Mean Dry	C.V. ±%
Sheep	TDN	%	52.3	78.6	
Swine	TDN	%	53.6	80.5	

Corn, ears, milk stage, (4)

Ref No 4 08 393 United States

Feed Name or Analyses			Mean As Fed	Mean Dry	C.V. ±%
Dry matter		%	61.1	100.0	
Ash		%	1.1	1.8	
Crude fiber		%	7.5	12.3	
Ether extract		%	1.8	2.9	
N-free extract		%	45.1	73.8	
Protein (N x 6.25)		%	5.6	9.2	
Cattle	dig prot	%	* 2.7	4.4	
Goats	dig prot	%	* 3.4	5.6	
Horses	dig prot	%	* 3.4	5.6	
Sheep	dig prot	%	* 3.4	5.6	
Energy	GE	Mcal/kg			
Cattle	DE	Mcal/kg	* 2.17	3.55	
Sheep	DE	Mcal/kg	* 2.16	3.53	
Swine	DE	kcal/kg	*1801.	2947.	
Cattle	ME	Mcal/kg	* 1.78	2.91	
Sheep	ME	Mcal/kg	* 1.77	2.89	
Swine	ME	kcal/kg	*1695.	2774.	
Cattle	TDN	%	* 49.2	80.5	
Sheep	TDN	%	* 48.9	80.0	
Swine	TDN	%	* 40.8	66.8	

Corn, ears w husks, grnd, (4)

Corn and cob meal with husks (AAFCO)
Ear corn chop with husks (AAFCO)
Ground ear corn with husks (AAFCO)
Ground snapped corn

Ref No 4 02 850 United States

Feed Name or Analyses			Mean As Fed	Mean Dry	C.V. ±%
Dry matter		%	88.5	100.0	1
Ash		%	2.7	3.0	
Crude fiber		%	10.4	11.8	
Ether extract		%	3.0	3.4	
N-free extract		%	64.7	73.1	
Protein (N x 6.25)		%	7.7	8.7	
Cattle	dig prot	%	* 3.6	4.0	
Goats	dig prot	%	* 4.6	5.2	
Horses	dig prot	%	* 4.6	5.2	
Sheep	dig prot	%	* 4.6	5.2	
Energy	GE	Mcal/kg			
Cattle	DE	Mcal/kg	* 2.94	3.32	
Sheep	DE	Mcal/kg	* 3.12	3.52	
Swine	DE	kcal/kg	*2541.	2871.	
Cattle	ME	Mcal/kg	* 2.41	2.72	
Sheep	ME	Mcal/kg	* 2.55	2.89	
Swine	ME	kcal/kg	*2394.	2705.	
Cattle	TDN	%	* 66.7	75.4	
Sheep	TDN	%	* 70.7	79.8	
Swine	TDN	%	* 57.6	65.1	
Calcium		%	0.06	0.07	49
Magnesium		%	0.13	0.15	21
Manganese		mg/kg	10.7	12.1	
Phosphorus		%	0.34	0.38	14
Potassium		%	0.86	0.97	9
Carotene		mg/kg	4.1	4.6	
Riboflavin		mg/kg	1.4	1.5	
Vitamin A equivalent		IU/g	6.8	7.7	

Corn, ears w husks, grnd, milk stage, (4)

Ref No 4 08 394 United States

Feed Name or Analyses			Mean As Fed	Mean Dry	C.V. ±%
Dry matter		%	55.3	100.0	
Ash		%	1.8	3.3	
Crude fiber		%	7.6	13.7	
Ether extract		%	1.7	3.0	
N-free extract		%	39.3	71.2	
Protein (N x 6.25)		%	4.9	8.8	
Cattle	dig prot	%	* 2.3	4.1	
Goats	dig prot	%	* 2.9	5.3	
Horses	dig prot	%	* 2.9	5.3	
Sheep	dig prot	%	* 2.9	5.3	
Sugars, total		%	1.6	2.9	
Energy	GE	Mcal/kg			
Cattle	DE	Mcal/kg	* 1.84	3.32	
Sheep	DE	Mcal/kg	* 1.91	3.45	
Swine	DE	kcal/kg	*1455.	2633.	
Cattle	ME	Mcal/kg	* 1.50	2.72	
Sheep	ME	Mcal/kg	* 1.56	2.83	
Swine	ME	kcal/kg	*1371.	2481.	
Cattle	TDN	%	* 41.7	75.4	
Sheep	TDN	%	* 43.2	78.2	
Swine	TDN	%	* 33.0	59.7	

Corn, endosperm, (4)

Ref No 4 02 851 United States

Feed Name or Analyses			Mean As Fed	Mean Dry	C.V. ±%
Dry matter		%	90.0	100.0	
Ash		%	1.1	1.2	98
Ether extract		%	0.8	0.9	27
Protein (N x 6.25)		%	8.7	9.7	37
Cattle	dig prot	%	* 4.4	4.9	
Goats	dig prot	%	* 5.5	6.1	
Horses	dig prot	%	* 5.5	6.1	
Sheep	dig prot	%	* 5.5	6.1	

Corn, endosperm oil, mn 85% fatty acids mx 14% unsaponifiable matter 1% insoluble matter, (4)

Corn endosperm oil (AAFCO)

Ref No 4 02 852 United States

Feed Name or Analyses			Mean As Fed	Mean Dry	C.V. ±%
Dry matter		%		100.0	
Ether extract		%		100.0	

Corn, grain, (4)

Ref No 4 02 879 United States

Feed Name or Analyses			Mean As Fed	Mean Dry	C.V. ±%
Dry matter		%	87.0	100.0	2
Ash		%	1.4	1.6	13
Crude fiber		%	2.1	2.4	19
Cattle	dig coef	%	16.	16.	
Horses	dig coef	%	17.	17.	
Sheep	dig coef	%	62.	62.	
Swine	dig coef	%	12.	12.	
Ether extract		%	4.1	4.7	22
Cattle	dig coef	%	85.	85.	
Horses	dig coef	%	70.	70.	
Sheep	dig coef	%	86.	86.	
Swine	dig coef	%	62.	62.	
N-free extract		%	69.9	80.3	2
Cattle	dig coef	%	90.	90.	
Horses	dig coef	%	93.	93.	
Sheep	dig coef	%	96.	96.	
Swine	dig coef	%	80.	80.	
Protein (N x 6.25)		%	9.5	10.9	8
Cattle	dig coef	%	69.	69.	
Horses	dig coef	%	78.	78.	
Sheep	dig coef	%	74.	74.	
Swine	dig coef	%	66.	66.	
Cattle	dig prot	%	6.5	7.5	
Goats	dig prot	%	* 6.3	7.2	
Horses	dig prot	%	7.4	8.5	
Sheep	dig prot	%	7.0	8.1	
Swine	dig prot	%	6.3	7.2	
Starch		%	66.4	76.3	
Energy	GE	Mcal/kg	3.67	4.22	
Cattle	DE	Mcal/kg	* 3.41	3.92	
Horses	DE	Mcal/kg	* 3.49	4.01	
Sheep	DE	Mcal/kg	* 3.67	4.22	
Swine	DE	kcal/kg	*2989.	3436.	
Cattle	ME	Mcal/kg	* 2.79	3.21	
Horses	ME	Mcal/kg	* 2.86	3.29	
Sheep	ME	Mcal/kg	* 3.01	3.46	
Swine	ME	kcal/kg	*2803.	3223.	
Cattle	TDN	%	77.3	88.8	
Horses	TDN	%	79.2	91.1	
Sheep	TDN	%	83.3	95.8	
Swine	TDN	%	67.8	77.9	
Arginine		%	0.35	0.40	32
Cystine		%	0.09	0.10	27
Glutamic acid		%	2.35	2.70	19
Histidine		%	0.17	0.20	38
Isoleucine		%	0.43	0.50	25
Leucine		%	1.04	1.20	38
Lysine		%	0.26	0.30	24
Methionine		%	0.17	0.20	36
Phenylalanine		%	0.43	0.50	24
Serine		%	0.70	0.80	
Threonine		%	0.26	0.30	26
Tryptophan		%	0.09	0.10	
Tyrosine		%	0.43	0.50	44
Valine		%	0.43	0.50	24

Ref No 4 02 879 Canada

Feed Name or Analyses			Mean As Fed	Mean Dry	C.V. ±%
Dry matter		%	88.0	100.0	
Crude fiber		%	1.9	2.2	
Cattle	dig coef	%	16.	16.	
Horses	dig coef	%	17.	17.	
Sheep	dig coef	%	62.	62.	
Swine	dig coef	%	12.	12.	
Ether extract		%	3.9	4.5	
Cattle	dig coef	%	85.	85.	
Horses	dig coef	%	70.	70.	
Sheep	dig coef	%	86.	86.	
Swine	dig coef	%	62.	62.	
N-free extract		%			
Cattle	dig coef	%	90.	90.	
Horses	dig coef	%	93.	93.	
Sheep	dig coef	%	96.	96.	
Swine	dig coef	%	80.	80.	
Protein (N x 6.25)		%	8.7	9.8	
Cattle	dig coef	%	69.	69.	
Horses	dig coef	%	78.	78.	
Sheep	dig coef	%	74.	74.	
Swine	dig coef	%	66.	66.	
Cattle	dig prot	%	6.0	6.8	
Goats	dig prot	%	* 5.5	6.3	
Horses	dig prot	%	6.8	7.7	
Sheep	dig prot	%	6.4	7.3	
Swine	dig prot	%	5.7	6.5	
Energy	GE	Mcal/kg	3.71	4.22	
Cattle	DE	Mcal/kg	* 3.45	3.92	

Continued

Feed Name or Analyses	Mean As Fed	Mean Dry	C.V. ± %
Horses DE Mcal/kg	* 3.54	4.02	
Sheep DE Mcal/kg	* 3.72	4.23	
Swine DE kcal/kg	*3026.	3437.	
Cattle ME Mcal/kg	* 2.83	3.21	
Horses ME Mcal/kg	* 2.90	3.29	
Sheep ME Mcal/kg	* 3.05	3.47	
Swine ME kcal/kg	*2845.	3231.	
Cattle TDN %	78.2	88.9	
Horses TDN %	80.2	91.1	
Sheep TDN %	84.4	95.8	
Swine TDN %	68.6	78.0	
Calcium %	0.04	0.05	
Phosphorus %	0.30	0.35	

Corn, grain, cooked, (4)
Ref No 4 02 853 United States

Feed Name or Analyses	Mean As Fed	Mean Dry	C.V. ± %
Dry matter %	87.0	100.0	
Ash %	1.8	2.1	
Crude fiber %	1.6	1.8	
Swine dig coef %	23.	23.	
Ether extract %	4.5	5.2	
Swine dig coef %	64.	64.	
N-free extract %	69.7	80.1	
Swine dig coef %	92.	92.	
Protein (N x 6.25) %	9.4	10.8	
Swine dig coef %	86.	86.	
Cattle dig prot %	* 5.2	5.9	
Goats dig prot %	* 6.2	7.1	
Horses dig prot %	* 6.2	7.1	
Sheep dig prot %	* 6.2	7.1	
Swine dig prot %	8.1	9.3	
Energy GE Mcal/kg			
Cattle DE Mcal/kg	* 3.14	3.60	
Sheep DE Mcal/kg	* 3.33	3.82	
Swine DE kcal/kg	*3486.	4007.	
Cattle ME Mcal/kg	* 2.57	2.96	
Sheep ME Mcal/kg	* 2.73	3.14	
Swine ME kcal/kg	*3271.	3759.	
Cattle TDN %	* 71.1	81.7	
Sheep TDN %	* 75.5	86.7	
Swine TDN %	79.1	90.9	

Corn, grain, extn unspecified grnd, (4)
Ref No 4 02 857 United States

Feed Name or Analyses	Mean As Fed	Mean Dry	C.V. ± %
Dry matter %	89.2	100.0	
Ash %	3.1	3.5	
Crude fiber %	9.5	10.7	
Sheep dig coef %	82.	82.	
Ether extract %	7.9	8.9	
Sheep dig coef %	82.	82.	
N-free extract %	47.3	53.0	
Sheep dig coef %	78.	78.	
Protein (N x 6.25) %	21.3	23.9	
Sheep dig coef %	72.	72.	
Cattle dig prot %	* 16.0	18.0	
Goats dig prot %	* 17.1	19.1	
Horses dig prot %	* 17.1	19.1	
Sheep dig prot %	15.3	17.2	
Energy GE Mcal/kg			
Cattle DE Mcal/kg	* 3.16	3.54	
Sheep DE Mcal/kg	* 3.29	3.69	

(1) dry forages and roughages (3) silages (6) minerals
(2) pasture, range plants, and forages fed green (4) energy feeds (7) vitamins
(5) protein supplements (8) additives

Feed Name or Analyses	Mean As Fed	Mean Dry	C.V. ± %
Swine DE kcal/kg	*3993.	4477.	
Cattle ME Mcal/kg	* 2.59	2.90	
Sheep ME Mcal/kg	* 2.70	3.03	
Swine ME kcal/kg	*3641.	4082.	
Cattle TDN %	* 71.6	80.3	
Sheep TDN %	74.7	83.7	
Swine TDN %	* 90.6	101.5	
a-tocopherol mg/kg	0.8	0.9	

Corn, grain, extn unspecified grnd, low protein, (4)
Ref No 4 02 856 United States

Feed Name or Analyses	Mean As Fed	Mean Dry	C.V. ± %
Dry matter %	94.4	100.0	
Ash %	4.6	4.9	
Crude fiber %	6.4	6.8	
Cattle dig coef %	21.	21.	
Ether extract %	7.9	8.4	
Cattle dig coef %	98.	98.	
N-free extract %	59.8	63.3	
Cattle dig coef %	80.	80.	
Protein (N x 6.25) %	15.7	16.6	
Cattle dig coef %	63.	63.	
Cattle dig prot %	9.9	10.5	
Goats dig prot %	* 11.7	12.4	
Horses dig prot %	* 11.7	12.4	
Sheep dig prot %	* 11.7	12.4	
Energy GE Mcal/kg			
Cattle DE Mcal/kg	* 3.37	3.57	
Sheep DE Mcal/kg	* 3.34	3.54	
Swine DE kcal/kg	*3196.	3386.	
Cattle ME Mcal/kg	* 2.77	2.93	
Sheep ME Mcal/kg	* 2.74	2.90	
Swine ME kcal/kg	*2961.	3137.	
Cattle TDN %	76.5	81.0	
Sheep TDN %	* 75.8	80.3	
Swine TDN %	* 72.5	76.8	

Corn, grain, flaked, (4)
Flaked corn (AAFCO)
Ref No 4 02 859 United States

Feed Name or Analyses	Mean As Fed	Mean Dry	C.V. ± %
Dry matter %	91.9	100.0	
Ash %	1.9	2.0	
Crude fiber %	0.5	0.5	
Swine dig coef %	30.	30.	
Ether extract %	1.1	1.2	
Swine dig coef %	45.	45.	
N-free extract %	79.5	86.5	
Swine dig coef %	97.	97.	
Protein (N x 6.25) %	9.0	9.8	
Swine dig coef %	95.	95.	
Cattle dig prot %	* 4.6	5.0	
Goats dig prot %	* 5.7	6.2	
Horses dig prot %	* 5.7	6.2	
Sheep dig prot %	* 5.7	6.2	
Swine dig prot %	8.6	9.3	
Energy GE Mcal/kg			
Cattle DE Mcal/kg	* 3.23	3.51	
Sheep DE Mcal/kg	* 3.51	3.82	
Swine DE kcal/kg	*3831.	4169.	
Cattle ME Mcal/kg	* 2.65	2.88	
Chickens ME_n kcal/kg	3412.	3713.	
Sheep ME Mcal/kg	* 2.88	3.14	
Swine ME kcal/kg	*3602.	3920.	
Cattle TDN %	* 73.2	79.6	
Sheep TDN %	* 79.7	86.7	
Swine TDN %	86.9	94.5	
Phosphorus %	0.10	0.11	

Feed Name or Analyses	Mean As Fed	Mean Dry	C.V. ± %
Choline mg/kg	427.	464.	
Niacin mg/kg	8.7	9.5	
Pantothenic acid mg/kg	0.4	0.5	
Riboflavin mg/kg	15.6	16.9	
Arginine %	0.29	0.32	
Cystine %	0.10	0.11	
Lysine %	0.29	0.32	
Methionine %	0.10	0.11	
Tryptophan %	0.07	0.07	

Corn, grain, grnd, mx 4% foreign material, (4)
Corn chop (AAFCO)
Corn meal (AAFCO)
Ground corn (AAFCO)
Ref No 4 02 861 United States

Feed Name or Analyses	Mean As Fed	Mean Dry	C.V. ± %
Dry matter %	87.2	100.0	2
Ash %	1.9	2.2	28
Crude fiber %	2.0	2.3	13
Ether extract %	3.2	3.6	17
N-free extract %	70.8	81.3	1
Protein (N x 6.25) %	9.3	10.7	7
Cattle dig prot %	* 5.1	5.8	
Goats dig prot %	* 6.1	7.0	
Horses dig prot %	* 6.1	7.0	
Sheep dig prot %	* 6.1	7.0	
Energy GE Mcal/kg			
Cattle DE Mcal/kg	* 3.10	3.55	
Sheep DE Mcal/kg	* 3.30	3.79	
Swine DE kcal/kg	*3485.	3998.	
Cattle ME Mcal/kg	* 2.54	2.91	
Sheep ME Mcal/kg	* 2.71	3.11	
Swine ME kcal/kg	*3271.	3752.	
Cattle TDN %	* 70.3	80.6	
Sheep TDN %	* 74.9	85.9	
Swine TDN %	* 79.0	90.7	
Calcium %	0.03	0.03	58
Chlorine %	0.02	0.02	
Copper mg/kg	3.1	3.5	98
Iron %	0.004	0.005	63
Magnesium %	0.11	0.13	19
Manganese mg/kg	6.1	7.0	82
Phosphorus %	0.27	0.31	53
Potassium %	0.31	0.35	23
Sodium %	0.07	0.08	
Sulphur %	0.08	0.09	98
Choline mg/kg	591.	678.	
Alanine %	0.79	0.91	
Arginine %	0.40	0.46	26
Aspartic acid %	0.20	0.23	
Cystine %	0.10	0.11	
Glutamic acid %	2.78	3.19	
Glycine %	0.00	0.00	
Histidine %	0.20	0.23	17
Hydroxyproline %	0.10	0.11	
Isoleucine %	0.40	0.46	19
Leucine %	0.89	1.03	15
Lysine %	0.20	0.23	32
Methionine %	0.10	0.11	65
Phenylalanine %	0.40	0.46	24
Proline %	0.89	1.03	
Serine %	0.10	0.11	
Threonine %	0.30	0.34	25
Tryptophan %	0.10	0.11	
Tyrosine %	0.40	0.46	64
Valine %	0.30	0.34	22

Feed Name or Analyses			Mean As Fed	Mean Dry	C.V. ± %
Ref No 4 02 861			Canada		
Dry matter	%		87.6	100.0	1
Ash	%		1.2	1.4	
Crude fiber	%		1.7	2.0	11
Ether extract	%		3.9	4.4	15
N-free extract	%		71.5	81.7	
Protein (N x 6.25)	%		9.2	10.6	8
Cattle	dig prot %	*	5.0	5.7	
Goats	dig prot %	*	6.0	6.9	
Horses	dig prot %	*	6.0	6.9	
Sheep	dig prot %	*	6.0	6.9	
Energy	GE Mcal/kg				
Cattle	DE Mcal/kg	*	3.14	3.59	
Sheep	DE Mcal/kg	*	3.36	3.83	
Swine	DE kcal/kg	*	3607.	4118.	
Cattle	ME Mcal/kg	*	2.58	2.94	
Sheep	ME Mcal/kg	*	2.75	3.14	
Swine	ME kcal/kg	*	3386.	3865.	
Cattle	TDN %	*	71.3	81.3	
Sheep	TDN %	*	76.1	86.9	
Swine	TDN %	*	81.8	93.4	
Calcium	%		0.09	0.11	98
Copper	mg/kg		3.9	4.4	
Phosphorus	%		0.29	0.33	9
Salt (NaCl)	%		0.12	0.14	

Corn, grain, germinated, (4)

Ref No 4 02 869			United States		
Dry matter	%			100.0	
Biotin	mg/kg			0.53	

Corn, grain, Can 1 CE mn wt 56 lb per bushel mx 2% foreign material, (4)

Ref No 4 02 870			United States		
Dry matter	%			100.0	
Crude fiber	%				
Cattle	dig coef %	*		19.	
Sheep	dig coef %	*		30.	
Swine	dig coef %	*		47.	
Ether extract	%				
Cattle	dig coef %	*		87.	
Sheep	dig coef %	*		87.	
Swine	dig coef %	*		70.	
N-free extract	%				
Cattle	dig coef %	*		91.	
Sheep	dig coef %	*		99.	
Swine	dig coef %	*		93.	
Protein (N x 6.25)	%				
Cattle	dig coef %	*		75.	
Sheep	dig coef %	*		78.	
Swine	dig coef %	*		80.	
Energy	GE Mcal/kg				
Cattle	DE Mcal/kg	*		4.03	
Sheep	DE Mcal/kg	*		4.34	
Swine	DE kcal/kg	*		4074.	
Cattle	ME Mcal/kg	*		3.30	
Sheep	ME Mcal/kg	*		3.56	
Cattle	TDN %	*		91.4	
Sheep	TDN %	*		98.5	
Swine	TDN %	*		92.4	

Ref No 4 02 870			Canada		
Dry matter	%		86.5	100.0	0
Ash	%		1.2	1.4	6
Crude fiber	%		1.7	2.0	4
Ether extract	%		3.9	4.5	5
N-free extract	%		71.0	82.0	
Protein (N x 6.25)	%		8.8	10.1	4
Cattle	dig prot %	*	4.6	5.3	
Goats	dig prot %	*	5.6	6.5	
Horses	dig prot %	*	5.6	6.5	
Sheep	dig prot %	*	5.6	6.5	
Energy	GE Mcal/kg				
Cattle	DE Mcal/kg	*	3.49	4.03	
Sheep	DE Mcal/kg	*	3.76	4.34	
Swine	DE kcal/kg	*	3524.	4074.	
Cattle	ME Mcal/kg	*	2.86	3.30	
Sheep	ME Mcal/kg	*	3.08	3.56	
Swine	ME kcal/kg	*	3311.	3827.	
Cattle	TDN %		79.1	91.4	
Sheep	TDN %		85.2	98.5	
Swine	TDN %		79.9	92.4	
Calcium	%		0.05	0.06	
Phosphorus	%		0.25	0.29	

Corn, grain, Can 1 CW mn wt 56 lb per bushel mx 2% foreign material, (4)

Ref No 4 02 871			United States		
Dry matter	%			100.0	
Crude fiber	%				
Cattle	dig coef %	*		19.	
Sheep	dig coef %	*		30.	
Swine	dig coef %	*		47.	
Ether extract	%				
Cattle	dig coef %	*		87.	
Sheep	dig coef %	*		87.	
Swine	dig coef %	*		70.	
N-free extract	%				
Cattle	dig coef %	*		91.	
Sheep	dig coef %	*		99.	
Swine	dig coef %	*		93.	
Protein (N x 6.25)	%				
Cattle	dig coef %	*		75.	
Sheep	dig coef %	*		78.	
Swine	dig coef %	*		80.	
Energy	GE Mcal/kg				
Cattle	DE Mcal/kg	*		4.03	
Sheep	DE Mcal/kg	*		4.34	
Swine	DE kcal/kg	*		4071.	
Cattle	ME Mcal/kg	*		3.30	
Sheep	ME Mcal/kg	*		3.56	
Cattle	TDN %	*		91.3	
Sheep	TDN %	*		98.4	
Swine	TDN %	*		92.3	

Ref No 4 02 871			Canada		
Dry matter	%		86.5	100.0	0
Ash	%		1.2	1.4	7
Crude fiber	%		1.8	2.1	11
Ether extract	%		3.9	4.5	7
N-free extract	%		70.6	81.6	
Protein (N x 6.25)	%		9.0	10.4	4
Cattle	dig prot %	*	4.8	5.6	
Goats	dig prot %	*	5.9	6.8	
Horses	dig prot %	*	5.9	6.8	
Sheep	dig prot %	*	5.9	6.8	
Energy	GE Mcal/kg				
Cattle	DE Mcal/kg	*	3.48	4.03	
Sheep	DE Mcal/kg	*	3.75	4.34	
Swine	DE kcal/kg	*	3521.	4071.	
Cattle	ME Mcal/kg	*	2.86	3.30	
Sheep	ME Mcal/kg	*	3.08	3.56	
Swine	ME kcal/kg	*	3306.	3822.	
Cattle	TDN %		79.0	91.3	
Sheep	TDN %		85.1	98.4	
Swine	TDN %		79.9	92.3	

Corn, grain, Can 2 CE mn wt 54 lb per bushel mx 3% foreign material, (4)

Ref No 4 02 872			United States		
Dry matter	%			100.0	
Crude fiber	%				
Cattle	dig coef %	*		19.	
Sheep	dig coef %	*		30.	
Swine	dig coef %	*		47.	
Ether extract	%				
Cattle	dig coef %	*		87.	
Sheep	dig coef %	*		87.	
Swine	dig coef %	*		70.	
N-free extract	%				
Cattle	dig coef %	*		91.	
Sheep	dig coef %	*		99.	
Swine	dig coef %	*		93.	
Protein (N x 6.25)	%				
Cattle	dig coef %	*		75.	
Sheep	dig coef %	*		78.	
Swine	dig coef %	*		80.	
Energy	GE Mcal/kg				
Cattle	DE Mcal/kg	*		4.03	
Sheep	DE Mcal/kg	*		4.34	
Swine	DE kcal/kg	*		4075.	
Cattle	ME Mcal/kg	*		3.31	
Sheep	ME Mcal/kg	*		3.56	
Cattle	TDN %	*		91.4	
Sheep	TDN %	*		98.5	
Swine	TDN %	*		92.4	

Ref No 4 02 872			Canada		
Dry matter	%		86.5	100.0	0
Ash	%		1.2	1.4	7
Crude fiber	%		1.8	2.1	7
Ether extract	%		3.9	4.6	9
N-free extract	%		70.9	82.0	
Protein (N x 6.25)	%		8.6	10.0	4
Cattle	dig prot %	*	4.5	5.2	
Goats	dig prot %	*	5.5	6.4	
Horses	dig prot %	*	5.5	6.4	
Sheep	dig prot %	*	5.5	6.4	
Energy	GE Mcal/kg				
Cattle	DE Mcal/kg	*	3.49	4.03	
Sheep	DE Mcal/kg	*	3.76	4.34	
Swine	DE kcal/kg	*	3525.	4075.	
Cattle	ME Mcal/kg	*	2.86	3.31	
Sheep	ME Mcal/kg	*	3.08	3.56	
Swine	ME kcal/kg	*	3313.	3830.	
Cattle	TDN %		79.1	91.4	
Sheep	TDN %		85.2	98.5	
Swine	TDN %		79.9	92.4	
Calcium	%		0.05	0.06	
Phosphorus	%		0.23	0.27	

Corn, grain, Can 2 CW mn wt 54 lb per bushel mx 3% foreign material, (4)

Ref No 4 02 873 United States

Feed Name or Analyses			Mean As Fed	Mean Dry	C.V. ± %
Dry matter	%			100.0	
Crude fiber	%				
Cattle	dig coef %	*		19.	
Sheep	dig coef %	*		30.	
Swine	dig coef %	*		47.	
Ether extract	%				
Cattle	dig coef %	*		87.	
Sheep	dig coef %	*		87.	
Swine	dig coef %	*		70.	
N-free extract	%				
Cattle	dig coef %	*		91.	
Sheep	dig coef %	*		99.	
Swine	dig coef %	*		93.	
Protein (N x 6.25)	%				
Cattle	dig coef %	*		75.	
Swine	dig coef %	*		80.	
Energy	GE Mcal/kg				
Cattle	DE Mcal/kg	*		4.03	
Swine	DE kcal/kg	*		4071.	
Cattle	ME Mcal/kg	*		3.30	
Cattle	TDN %	*		91.4	
Swine	TDN %	*		92.3	

Ref No 4 02 873 Canada

Feed Name or Analyses			Mean As Fed	Mean Dry	C.V. ± %
Dry matter	%		86.5	100.0	0
Ash	%		1.2	1.4	10
Crude fiber	%		1.8	2.1	11
Ether extract	%		4.0	4.7	7
N-free extract	%		70.3	81.2	
Protein (N x 6.25)	%		9.1	10.6	2
Cattle	dig prot %	*	4.9	5.7	
Goats	dig prot %	*	6.0	6.9	
Horses	dig prot %	*	6.0	6.9	
Sheep	dig prot %	*	6.0	6.9	
Energy	GE Mcal/kg				
Cattle	DE Mcal/kg	*	3.48	4.03	
Sheep	DE Mcal/kg	*	3.31	3.83	
Swine	DE kcal/kg	*	3521.	4071.	
Cattle	ME Mcal/kg	*	2.86	3.30	
Sheep	ME Mcal/kg	*	2.72	3.14	
Swine	ME kcal/kg	*	3305.	3821.	
Cattle	TDN %		79.0	91.4	
Sheep	TDN %	*	75.2	86.9	
Swine	TDN %		79.9	92.3	

Corn, grain, Can 3 CE mn wt 52 lb per bushel mx 5% foreign material, (4)

Ref No 4 02 874 United States

Feed Name or Analyses			Mean As Fed	Mean Dry	C.V. ± %
Dry matter	%			100.0	
Crude fiber	%				
Cattle	dig coef %	*		19.	
Sheep	dig coef %	*		30.	
Swine	dig coef %	*		47.	
Ether extract	%				
Cattle	dig coef %	*		87.	
Sheep	dig coef %	*		87.	
Swine	dig coef %	*		70.	
N-free extract	%				
Cattle	dig coef %	*		91.	
Sheep	dig coef %	*		99.	
Swine	dig coef %	*		93.	
Protein (N x 6.25)	%				
Cattle	dig coef %	*		75.	
Sheep	dig coef %	*		78.	
Swine	dig coef %	*		80.	
Energy	GE Mcal/kg				
Cattle	DE Mcal/kg	*		4.04	
Sheep	DE Mcal/kg	*		4.35	
Swine	DE kcal/kg	*		4078.	
Cattle	ME Mcal/kg	*		3.31	
Sheep	ME Mcal/kg	*		3.57	
Cattle	TDN %	*		91.6	
Sheep	TDN %	*		98.6	
Swine	TDN %	*		92.5	

Ref No 4 02 874 Canada

Feed Name or Analyses			Mean As Fed	Mean Dry	C.V. ± %
Dry matter	%		86.5	100.0	0
Ash	%		1.2	1.4	8
Crude fiber	%		1.8	2.1	8
Ether extract	%		4.1	4.7	13
N-free extract	%		70.8	81.9	
Protein (N x 6.25)	%		8.6	10.0	4
Cattle	dig prot %	*	4.5	5.2	
Goats	dig prot %	*	5.5	6.4	
Horses	dig prot %	*	5.5	6.4	
Sheep	dig prot %	*	5.5	6.4	
Energy	GE Mcal/kg				
Cattle	DE Mcal/kg	*	3.49	4.04	
Sheep	DE Mcal/kg	*	3.76	4.35	
Swine	DE kcal/kg	*	3527.	4078.	
Cattle	ME Mcal/kg	*	2.86	3.31	
Sheep	ME Mcal/kg	*	3.09	3.57	
Swine	ME kcal/kg	*	3315.	3833.	
Cattle	TDN %		79.2	91.6	
Sheep	TDN %		85.3	98.6	
Swine	TDN %		80.0	92.5	
Calcium	%		0.05	0.06	
Phosphorus	%		0.21	0.24	

Corn, grain, Can 3 CW mn wt 52 lb per bushel mx 5% foreign material, (4)

Ref No 4 02 875 United States

Feed Name or Analyses			Mean As Fed	Mean Dry	C.V. ± %
Dry matter	%			100.0	
Crude fiber	%				
Cattle	dig coef %	*		19.	
Sheep	dig coef %	*		30.	
Swine	dig coef %	*		47.	
Ether extract	%				
Cattle	dig coef %	*		87.	
Sheep	dig coef %	*		87.	
Swine	dig coef %	*		70.	
N-free extract	%				
Cattle	dig coef %	*		91.	
Sheep	dig coef %	*		99.	
Swine	dig coef %	*		93.	
Protein (N x 6.25)	%				
Cattle	dig coef %	*		75.	
Sheep	dig coef %	*		78.	
Swine	dig coef %	*		80.	
Energy	GE Mcal/kg				
Cattle	DE Mcal/kg	*		4.02	
Sheep	DE Mcal/kg	*		4.33	
Swine	DE kcal/kg	*		4064.	
Cattle	ME Mcal/kg	*		3.29	
Sheep	ME Mcal/kg	*		3.55	
Cattle	TDN %	*		91.1	
Sheep	TDN %	*		98.2	
Swine	TDN %	*		92.2	

Ref No 4 02 875 Canada

Feed Name or Analyses			Mean As Fed	Mean Dry	C.V. ± %
Dry matter	%		86.5	100.0	0
Ash	%		1.3	1.4	7
Crude fiber	%		1.8	2.0	13
Ether extract	%		3.8	4.4	5
N-free extract	%		70.5	81.5	
Protein (N x 6.25)	%		9.2	10.6	5
Cattle	dig prot %	*	5.0	5.8	
Goats	dig prot %	*	6.0	7.0	
Horses	dig prot %	*	6.0	7.0	
Sheep	dig prot %	*	6.0	7.0	
Energy	GE Mcal/kg				
Cattle	DE Mcal/kg	*	3.48	4.02	
Sheep	DE Mcal/kg	*	3.74	4.33	
Swine	DE kcal/kg	*	3515.	4064.	
Cattle	ME Mcal/kg	*	2.85	3.29	
Sheep	ME Mcal/kg	*	3.07	3.55	
Swine	ME kcal/kg	*	3300.	3814.	
Cattle	TDN %		78.8	91.1	
Sheep	TDN %		84.9	98.2	
Swine	TDN %		79.7	92.2	

Corn, grain, Can 4 CE mn wt 50 lb per bushel mx 7% foreign material, (4)

Ref No 4 02 876 United States

Feed Name or Analyses			Mean As Fed	Mean Dry	C.V. ± %
Dry matter	%			100.0	
Crude fiber	%				
Cattle	dig coef %	*		19.	
Sheep	dig coef %	*		30.	
Swine	dig coef %	*		47.	
Ether extract	%				
Cattle	dig coef %	*		87.	
Sheep	dig coef %	*		87.	
Swine	dig coef %	*		70.	
N-free extract	%				
Cattle	dig coef %	*		91.	
Sheep	dig coef %	*		99.	
Swine	dig coef %	*		93.	
Protein (N x 6.25)	%				
Cattle	dig coef %	*		75.	
Sheep	dig coef %	*		78.	
Swine	dig coef %	*		80.	
Energy	GE Mcal/kg				
Cattle	DE Mcal/kg	*		3.99	
Sheep	DE Mcal/kg	*		4.30	
Swine	DE kcal/kg	*		4047.	
Cattle	ME Mcal/kg	*		3.27	
Sheep	ME Mcal/kg	*		3.53	
Cattle	TDN %	*		90.4	
Sheep	TDN %	*		97.6	
Swine	TDN %	*		91.8	

Ref No 4 02 876 Canada

Feed Name or Analyses			Mean As Fed	Mean Dry	C.V. ± %
Dry matter	%		86.5	100.0	
Ash	%		1.2	1.4	
Crude fiber	%		1.8	2.1	
Ether extract	%		3.2	3.7	
N-free extract	%		71.5	82.6	
Protein (N x 6.25)	%		8.8	10.2	
Cattle	dig prot %	*	4.6	5.4	
Goats	dig prot %	*	5.7	6.6	

(1) dry forages and roughages (2) pasture, range plants, and forages fed green (3) silages (4) energy feeds (5) protein supplements (6) minerals (7) vitamins (8) additives

Feed Name or Analyses			Mean As Fed	Mean Dry	C.V. ± %
Horses	dig prot %	*	5.7	6.6	
Sheep	dig prot %	*	5.7	6.6	
Energy	GE Mcal/kg				
Cattle	DE Mcal/kg	*	3.45	3.99	
Sheep	DE Mcal/kg	*	3.72	4.30	
Swine	DE kcal/kg	*	3501.	4047.	
Cattle	ME Mcal/kg	*	2.83	3.27	
Sheep	ME Mcal/kg	*	3.05	3.53	
Swine	ME kcal/kg	*	3289.	3802.	
Cattle	TDN %		78.2	90.4	
Sheep	TDN %		84.4	97.6	
Swine	TDN %		79.4	91.8	
Calcium	%		0.05	0.06	
Phosphorus	%		0.21	0.24	

Corn, grain, Can 4 CW mn wt 50 lb per bushel mx 7% foreign material, (4)

Ref No 4 02 877 United States

Feed Name or Analyses			Mean As Fed	Mean Dry	C.V. ± %
Dry matter	%			100.0	
Crude fiber	%				
Cattle	dig coef %	*		19.	
Sheep	dig coef %	*		30.	
Swine	dig coef %	*		47.	
Ether extract	%				
Cattle	dig coef %	*		87.	
Sheep	dig coef %	*		87.	
Swine	dig coef %	*		70.	
N-free extract	%				
Cattle	dig coef %	*		91.	
Sheep	dig coef %	*		99.	
Swine	dig coef %	*		93.	
Protein (N x 6.25)	%				
Cattle	dig coef %	*		75.	
Sheep	dig coef %	*		78.	
Swine	dig coef %	*		80.	
Energy	GE Mcal/kg				
Cattle	DE Mcal/kg	*		4.02	
Sheep	DE Mcal/kg	*		4.33	
Swine	DE kcal/kg	*		4063.	
Cattle	ME Mcal/kg	*		3.29	
Sheep	ME Mcal/kg	*		3.55	
Cattle	TDN %	*		91.1	
Sheep	TDN %	*		98.2	
Swine	TDN %	*		92.2	

Ref No 4 02 877 Canada

Feed Name or Analyses			Mean As Fed	Mean Dry	C.V. ± %
Dry matter	%		86.5	100.0	0
Ash	%		1.3	1.4	12
Crude fiber	%		1.8	2.1	8
Ether extract	%		3.8	4.4	8
N-free extract	%		70.5	81.5	
Protein (N x 6.25)	%		9.2	10.6	4
Cattle	dig prot %	*	5.0	5.7	
Goats	dig prot %	*	6.0	6.9	
Horses	dig prot %	*	6.0	6.9	
Sheep	dig prot %	*	6.0	6.9	
Energy	GE Mcal/kg				
Cattle	DE Mcal/kg	*	3.47	4.02	
Sheep	DE Mcal/kg	*	3.74	4.33	
Swine	DE kcal/kg	*	3515.	4063.	
Cattle	ME Mcal/kg	*	2.85	3.29	
Sheep	ME Mcal/kg	*	3.07	3.55	
Swine	ME kcal/kg	*	3299.	3814.	
Cattle	TDN %		78.8	91.1	
Sheep	TDN %		84.9	98.2	
Swine	TDN %		79.7	92.2	

Corn, grain, Can 5 CW mn wt 47 lb per bushel mx 12% foreign material, (4)

Ref No 4 02 878 United States

Feed Name or Analyses			Mean As Fed	Mean Dry	C.V. ± %
Dry matter	%			100.0	
Crude fiber	%				
Cattle	dig coef %	*		19.	
Sheep	dig coef %	*		30.	
Swine	dig coef %	*		47.	
Ether extract	%				
Cattle	dig coef %	*		87.	
Sheep	dig coef %	*		87.	
Swine	dig coef %	*		70.	
N-free extract	%				
Cattle	dig coef %	*		91.	
Sheep	dig coef %	*		99.	
Swine	dig coef %	*		93.	
Protein (N x 6.25)	%				
Cattle	dig coef %	*		75.	
Sheep	dig coef %	*		78.	
Swine	dig coef %	*		80.	
Energy	GE Mcal/kg				
Cattle	DE Mcal/kg	*		4.03	
Sheep	DE Mcal/kg	*		4.34	
Swine	DE kcal/kg	*		4069.	
Cattle	ME Mcal/kg	*		3.30	
Sheep	ME Mcal/kg	*		3.56	
Cattle	TDN %	*		91.3	
Sheep	TDN %	*		98.4	
Swine	TDN %	*		92.3	

Ref No 4 02 878 Canada

Feed Name or Analyses			Mean As Fed	Mean Dry	C.V. ± %
Dry matter	%		86.5	100.0	0
Ash	%		1.3	1.5	12
Crude fiber	%		1.8	2.1	12
Ether extract	%		4.1	4.7	14
N-free extract	%		70.0	80.9	
Protein (N x 6.25)	%		9.4	10.8	4
Cattle	dig prot %	*	5.2	6.0	
Goats	dig prot %	*	6.2	7.2	
Horses	dig prot %	*	6.2	7.2	
Sheep	dig prot %	*	6.2	7.2	
Energy	GE Mcal/kg				
Cattle	DE Mcal/kg	*	3.48	4.03	
Sheep	DE Mcal/kg	*	3.75	4.34	
Swine	DE kcal/kg	*	3520.	4069.	
Cattle	ME Mcal/kg	*	2.86	3.30	
Sheep	ME Mcal/kg	*	3.08	3.56	
Swine	ME kcal/kg	*	3302.	3817.	
Cattle	TDN %		79.0	91.3	
Sheep	TDN %		85.1	98.4	
Swine	TDN %		79.8	92.3	

Corn, grain fines, cracked coarse screened passed, (4)

Corn feed meal (AAFCO)

Corn feed meal (CFA)

Ref No 4 02 880 United States

Feed Name or Analyses			Mean As Fed	Mean Dry	C.V. ± %
Dry matter	%		88.0	100.0	2
Ash	%		1.7	2.0	31
Crude fiber	%		2.4	2.7	40
Ether extract	%		4.5	5.1	25
N-free extract	%		70.1	79.7	
Protein (N x 6.25)	%		9.2	10.5	11
Cattle	dig prot %	*	5.0	5.7	
Goats	dig prot %	*	6.0	6.9	
Horses	dig prot %	*	6.0	6.9	
Sheep	dig prot %	*	6.0	6.9	
Energy	GE Mcal/kg				
Cattle	DE Mcal/kg	*	3.49	3.97	
Sheep	DE Mcal/kg	*	3.66	4.16	
Swine	DE kcal/kg	*	3693.	4197.	
Cattle	ME Mcal/kg	*	2.87	3.26	
Chickens	ME_n kcal/kg		3092.	3515.	
Sheep	ME Mcal/kg	*	3.00	3.41	
Swine	ME kcal/kg	*	3468.	3941.	
Cattle	TDN %	*	79.3	90.1	
Sheep	TDN %	*	83.0	94.3	
Swine	TDN %	*	83.8	95.2	
Calcium	%		0.03	0.04	63
Magnesium	%		0.10	0.11	
Manganese	mg/kg		4.8	5.5	47
Phosphorus	%		0.36	0.41	22
Potassium	%		0.29	0.33	
Sodium	%		0.10	0.11	
Sulphur	%		0.11	0.12	
Carotene	mg/kg		2.4	2.7	21
Choline	mg/kg		492.	560.	
Niacin	mg/kg		22.7	25.8	13
Pantothenic acid	mg/kg		5.3	6.0	
Riboflavin	mg/kg		1.1	1.3	78
Thiamine	mg/kg		3.3	3.7	
α-tocopherol	mg/kg		3.5	4.0	50
Vitamin A equivalent	IU/g		4.0	4.6	
Arginine	%		0.50	0.57	
Cystine	%		0.17	0.19	
Glycine	%		0.30	0.34	
Lysine	%		0.20	0.23	
Methionine	%		0.18	0.21	
Tryptophan	%		0.09	0.10	

Corn, grain wo germ, (4)

Ref No 4 02 884 United States

Feed Name or Analyses			Mean As Fed	Mean Dry	C.V. ± %
Dry matter	%		85.4	100.0	
Ash	%		0.8	0.9	
Crude fiber	%		1.1	1.3	
Swine	dig coef %		34.	34.	
Ether extract	%		2.7	3.2	
Swine	dig coef %		68.	68.	
N-free extract	%		71.1	83.2	
Swine	dig coef %		95.	95.	
Protein (N x 6.25)	%		9.7	11.4	
Swine	dig coef %		76.	76.	
Cattle	dig prot %	*	5.5	6.5	
Goats	dig prot %	*	6.6	7.7	
Horses	dig prot %	*	6.6	7.7	
Sheep	dig prot %	*	6.6	7.7	
Swine	dig prot %		7.4	8.7	
Energy	GE Mcal/kg				
Cattle	DE Mcal/kg	*	3.23	3.78	
Sheep	DE Mcal/kg	*	3.27	3.83	
Swine	DE kcal/kg	*	3503.	4102.	
Cattle	ME Mcal/kg	*	2.65	3.10	
Sheep	ME Mcal/kg	*	2.68	3.14	
Swine	ME kcal/kg	*	3283.	3844.	
Cattle	TDN %	*	73.3	85.8	
Sheep	TDN %	*	74.3	86.9	
Swine	TDN %		79.5	93.0	

Corn, grain wo germ, cooked, (4)

Ref No 4 02 883 — United States

Feed Name or Analyses		Mean As Fed	Mean Dry	C.V. ± %
Dry matter	%	93.6	100.0	
Ash	%	0.9	1.0	
Crude fiber	%	1.3	1.4	
Swine	dig coef %	94.	94.	
Ether extract	%	1.6	1.7	
Swine	dig coef %	85.	85.	
N-free extract	%	79.9	85.4	
Swine	dig coef %	99.	99.	
Protein (N x 6.25)	%	9.8	10.5	
Swine	dig coef %	94.	94.	
Cattle	dig prot %	* 5.3	5.7	
Goats	dig prot %	* 6.4	6.9	
Horses	dig prot %	* 6.4	6.9	
Sheep	dig prot %	* 6.4	6.9	
Swine	dig prot %	9.2	9.9	
Energy	GE Mcal/kg			
Cattle	DE Mcal/kg	* 3.61	3.86	
Sheep	DE Mcal/kg	* 3.58	3.82	
Swine	DE kcal/kg	*4085.	4364.	
Cattle	ME Mcal/kg	* 2.97	3.17	
Sheep	ME Mcal/kg	* 2.93	3.13	
Swine	ME kcal/kg	*3835.	4097.	
Cattle	TDN %	* 81.9	87.5	
Sheep	TDN %	* 81.2	86.7	
Swine	TDN %	92.6	99.0	

Corn, grain wo hulls, cracked, (4)

Ref No 4 08 395 — United States

Feed Name or Analyses		Mean As Fed	Mean Dry	C.V. ± %
Dry matter	%	87.6	100.0	
Ash	%	1.1	1.3	
Crude fiber	%	1.2	1.4	
Ether extract	%	2.1	2.4	
N-free extract	%	74.9	85.5	
Protein (N x 6.25)	%	8.3	9.5	
Cattle	dig prot %	* 4.1	4.7	
Goats	dig prot %	* 5.2	5.9	
Horses	dig prot %	* 5.2	5.9	
Sheep	dig prot %	* 5.2	5.9	
Energy	GE Mcal/kg			
Cattle	DE Mcal/kg	* 3.18	3.63	
Sheep	DE Mcal/kg	* 3.37	3.84	
Swine	DE kcal/kg	*3639.	4154.	
Cattle	ME Mcal/kg	* 2.61	2.98	
Sheep	ME Mcal/kg	* 2.76	3.15	
Swine	ME kcal/kg	*3423.	3908.	
Cattle	TDN %	* 72.2	82.4	
Sheep	TDN %	* 76.4	87.2	
Swine	TDN %	* 82.5	94.2	

Corn, grits, cracked fine screened, (4)

Corn grits (AAFCO)
Hominy grits (AAFCO)
Corn grits (CFA)
Hominy grits (CFA)

Ref No 4 02 886 — United States

Feed Name or Analyses		Mean As Fed	Mean Dry	C.V. ± %
Dry matter	%	88.2	100.0	1
Ash	%	0.4	0.5	24
Crude fiber	%	0.5	0.6	28
Ether extract	%	0.7	0.8	48
N-free extract	%	78.1	88.6	
Protein (N x 6.25)	%	8.5	9.6	10
Cattle	dig prot %	* 4.3	4.9	
Goats	dig prot %	* 5.4	6.1	
Horses	dig prot %	* 5.4	6.1	
Sheep	dig prot %	* 5.4	6.1	
Energy	GE Mcal/kg	3.63	4.11	
Cattle	DE Mcal/kg	* 3.45	3.91	
Sheep	DE Mcal/kg	* 3.41	3.86	
Swine	DE kcal/kg	*3590.	4070.	
Cattle	ME Mcal/kg	* 2.33	3.21	
Sheep	ME Mcal/kg	* 2.80	3.17	
Swine	ME kcal/kg	*3378.	3830.	
Cattle	TDN %	* 78.1	88.6	
Sheep	TDN %	* 77.3	87.7	
Swine	TDN %	* 81.4	92.3	
Calcium	%	0.01	0.01	98
Chlorine	%	0.05	0.06	
Iron	%	0.001	0.002	0
Magnesium	%	0.02	0.02	26
Phosphorus	%	0.09	0.10	12
Potassium	%	0.08	0.09	72
Sodium	%	0.01	0.01	
Sulphur	%	0.17	0.19	21
Biotin	mg/kg	0.02	0.02	
Niacin	mg/kg	10.6	12.1	10
Pantothenic acid	mg/kg	2.2	2.5	
Riboflavin	mg/kg	0.6	0.7	98
Thiamine	mg/kg	1.4	1.6	98
Vitamin B_6	mg/kg	2.65	3.00	
Vitamin A	IU/g	4.4	5.0	

Corn, grits by-product, mn 5% fat, (4)

Hominy feed (AAFCO)
Hominy feed (CFA)

Ref No 4 02 887 — United States

Feed Name or Analyses		Mean As Fed	Mean Dry	C.V. ± %
Dry matter	%	89.2	100.0	1
Ash	%	2.7	3.0	6
Crude fiber	%	4.4	5.0	14
Cattle	dig coef %	79.	79.	
Sheep	dig coef %	40.	40.	
Ether extract	%	6.8	7.7	20
Cattle	dig coef %	96.	96.	
Sheep	dig coef %	91.	91.	
N-free extract	%	64.6	72.5	
Cattle	dig coef %	92.	92.	
Sheep	dig coef %	89.	89.	
Protein (N x 6.25)	%	10.6	11.9	4
Cattle	dig coef %	67.	67.	
Sheep	dig coef %	64.	64.	
Cattle	dig prot %	7.1	8.0	
Goats	dig prot %	* 7.3	8.1	
Horses	dig prot %	* 7.3	8.1	
Sheep	dig prot %	6.8	7.6	
Energy	GE Mcal/kg	4.21	4.72	0
Cattle	DE Mcal/kg	* 3.74	4.20	
Sheep	DE Mcal/kg	* 3.53	3.96	
Swine	DE kcal/kg	*3296.	3697.	
Cattle	ME Mcal/kg	* 3.07	3.44	
Chickens	ME_n kcal/kg	2860.	3206.	
Sheep	ME Mcal/kg	* 2.90	3.25	
Swine	ME kcal/kg	*3085.	3460.	
Cattle	NE_m Mcal/kg	* 2.19	2.45	
Cattle	NE_{gain} Mcal/kg	* 1.38	1.55	
Cattle	$NE_{lactating\ cows}$ Mcal/kg	* 2.45	2.75	
Cattle	TDN %	84.9	95.2	
Sheep	TDN %	80.1	89.8	
Swine	TDN %	* 74.8	83.9	
Calcium	%	0.05	0.06	35
Cobalt	mg/kg	0.059	0.066	
Copper	mg/kg	13.2	14.8	47
Iron	%	0.007	0.008	38
Magnesium	%	0.23	0.26	4
Manganese	mg/kg	14.7	16.5	14
Phosphorus	%	0.54	0.61	15
Potassium	%	0.63	0.71	8
Sodium	%	0.13	0.14	
Sulphur	%	0.03	0.03	98
Carotene	mg/kg	9.1	10.2	98
Niacin	mg/kg	50.3	56.5	34
Pantothenic acid	mg/kg	7.4	8.3	11
Riboflavin	mg/kg	2.0	2.2	10
Thiamine	mg/kg	7.8	8.8	58
Vitamin A equivalent	IU/g	15.2	17.0	

Ref No 4 02 887 — Canada

Feed Name or Analyses		Mean As Fed	Mean Dry	C.V. ± %
Dry matter	%		100.0	
Crude fiber	%		5.9	12
Cattle	dig coef %		79.	
Sheep	dig coef %		40.	
Ether extract	%		9.0	19
Cattle	dig coef %		96.	
Sheep	dig coef %		91.	
N-free extract	%			
Cattle	dig coef %		92.	
Sheep	dig coef %		89.	
Protein (N x 6.25)	%		12.4	4
Cattle	dig coef %		67.	
Sheep	dig coef %		64.	
Cattle	dig prot %		8.3	
Goats	dig prot %	*	8.6	
Horses	dig prot %	*	8.6	
Sheep	dig prot %		7.9	
Energy	GE Mcal/kg		4.72	
Cattle	DE Mcal/kg	*	4.23	
Sheep	DE Mcal/kg	*	3.99	
Cattle	ME Mcal/kg	*	3.47	
Sheep	ME Mcal/kg	*	3.27	
Cattle	TDN %		96.0	
Sheep	TDN %		90.4	

Corn, grits by-product, mx 5% fat, (4)

Hominy feed, low fat

Ref No 4 08 396 — United States

Feed Name or Analyses		Mean As Fed	Mean Dry	C.V. ± %
Dry matter	%	89.4	100.0	
Ash	%	2.3	2.6	
Crude fiber	%	3.9	4.4	
Ether extract	%	4.6	5.1	
N-free extract	%	68.5	76.6	
Protein (N x 6.25)	%	10.1	11.3	
Cattle	dig prot %	* 5.7	6.4	
Goats	dig prot %	* 6.8	7.6	
Horses	dig prot %	* 6.8	7.6	
Sheep	dig prot %	* 6.8	7.6	
Energy	GE Mcal/kg			
Cattle	DE Mcal/kg	* 3.23	3.61	
Sheep	DE Mcal/kg	* 3.33	3.73	
Swine	DE kcal/kg	*3364.	3763.	
Cattle	ME Mcal/kg	* 2.65	2.96	
Sheep	ME Mcal/kg	* 2.73	3.06	
Swine	ME kcal/kg	*3153.	3527.	
Cattle	TDN %	* 73.2	81.8	
Sheep	TDN %	* 75.6	84.5	

(1) dry forages and roughages (2) pasture, range plants, and forages fed green (3) silages (4) energy feeds (5) protein supplements (6) minerals (7) vitamins (8) additives

Feed Name or Analyses			As Fed (Mean)	Dry (Mean)	C.V. ± %
Swine	TDN %	*	76.3	85.3	
Phosphorus	%		0.48	0.54	

Corn, molasses, mn 43% dextrose-equivalent mn 50% total dextrose mn 78 degrees brix, (4)
Corn sugar molasses (AAFCO)
Ref No 4 02 888 United States

Feed Name or Analyses			As Fed	Dry	C.V. ± %
Dry matter	%		72.5	100.0	6
Ash	%		8.0	11.0	15
Crude fiber	%		0.0	0.0	15
Ether extract	%		0.0	0.0	15
N-free extract	%		64.2	88.6	
Protein (N x 6.25)	%		0.3	0.4	19
Cattle	dig prot %	*	-2.5	-3.5	
Goats	dig prot %	*	-1.6	-2.3	
Horses	dig prot %	*	-1.6	-2.3	
Sheep	dig prot %	*	-1.6	-2.3	
Energy	GE Mcal/kg				
Cattle	DE Mcal/kg	*	2.67	3.68	
Sheep	DE Mcal/kg	*	2.70	3.73	
Swine	DE kcal/kg	*	2242.	3092.	
Cattle	ME Mcal/kg	*	2.19	3.02	
Sheep	ME Mcal/kg	*	2.22	3.06	
Swine	ME kcal/kg	*	2150.	2965.	
Cattle	TDN %	*	60.5	83.4	
Sheep	TDN %	*	61.3	84.6	
Swine	TDN %	*	50.8	70.1	

Corn, oil, (4)
Ref No 4 07 882 United States

Feed Name or Analyses			As Fed	Dry	C.V. ± %
Dry matter	%			100.0	
Energy	GE Mcal/kg			9.41	
Swine	DE kcal/kg			7620.	
Swine	ME kcal/kg			7340.	

Corn, starch, dehy grnd, (4)
Ref No 4 02 889 United States

Feed Name or Analyses			As Fed	Dry	C.V. ± %
Dry matter	%		90.4	100.0	2
Ash	%		0.2	0.2	43
Crude fiber	%		0.1	0.2	16
Cattle	dig coef %		0.	0.	16
Ether extract	%		0.2	0.2	44
Cattle	dig coef %		0.	0.	44
N-free extract	%		89.3	98.8	
Cattle	dig coef %		96.	96.	
Protein (N x 6.25)	%		0.6	0.7	25
Cattle	dig coef %		0.	0.	25
Goats	dig prot %	*	-1.9	-2.1	
Horses	dig prot %	*	-1.9	-2.1	
Sheep	dig prot %	*	-1.9	-2.1	
Energy	GE Mcal/kg		3.62	4.00	
Cattle	DE Mcal/kg	*	3.78	4.18	
Sheep	DE Mcal/kg	*	3.65	4.04	
Swine	DE kcal/kg		3671.	4060.	
Cattle	ME Mcal/kg	*	3.10	3.43	
Chickens	ME_n kcal/kg		3650.	4038.	
Sheep	ME Mcal/kg	*	2.99	3.31	
Swine	ME kcal/kg		3680.	4070.	
Cattle	TDN %		85.8	94.9	
Sheep	TDN %	*	82.8	91.6	
Swine	TDN %	*	86.7	95.8	

Ref No 4 02 889 Canada

Feed Name or Analyses			As Fed	Dry	C.V. ± %
Dry matter	%		95.0	100.0	
Crude fiber	%				
Cattle	dig coef %		0.	0.	
Ether extract	%				
Cattle	dig coef %		0.	0.	
N-free extract	%				
Cattle	dig coef %		96.	96.	
Protein (N x 6.25)	%				
Cattle	dig coef %		0.	0.	
Energy	GE Mcal/kg		3.80	4.00	
Cattle	DE Mcal/kg	*	3.97	4.18	
Swine	DE kcal/kg		3857.	4060.	
Cattle	ME Mcal/kg	*	3.26	3.43	
Swine	ME kcal/kg		3867.	4070.	
Cattle	TDN %		90.1	94.9	

Corn, steepwater solubles, fermented condensed, (5)
Condensed fermented corn extractives (AAFCO)
Ref No 5 02 890 United States

Feed Name or Analyses			As Fed	Dry	C.V. ± %
Dry matter	%		94.3	100.0	
Rats	dig coef %		97.	97.	
Ash	%		7.7	8.2	
Crude fiber	%		0.0	0.0	
Ether extract	%		1.5	1.6	
N-free extract	%		63.6	67.4	
Protein (N x 6.25)	%		21.5	22.8	
Cattle	dig prot %	*	16.0	17.0	
Goats	dig prot %	*	17.1	18.1	
Horses	dig prot %	*	17.1	18.1	
Sheep	dig prot %	*	17.1	18.1	
Energy	GE Mcal/kg		4.41	4.68	
Cattle	DE Mcal/kg	*	3.56	3.77	
Sheep	DE Mcal/kg	*	3.45	3.66	
Swine	DE kcal/kg	*	4012.	4255.	
Cattle	ME Mcal/kg	*	2.91	3.09	
Sheep	ME Mcal/kg	*	2.83	3.00	
Swine	ME kcal/kg	*	3667.	3889.	
Cattle	TDN %	*	80.7	85.6	
Sheep	TDN %	*	78.4	83.1	
Swine	TDN %	*	91.0	96.5	
Calcium	%		0.90	0.95	
Chlorine	%		0.60	0.64	
Magnesium	%		1.90	2.01	
Phosphorus	%		2.50	2.65	
Potassium	%		3.50	3.71	
Sulphur	%		0.60	0.64	
Arginine	%		0.95	1.01	
Histidine	%		0.80	0.85	
Isoleucine	%		0.78	0.83	
Leucine	%		2.00	2.12	
Lysine	%		0.90	0.95	
Methionine	%		0.48	0.51	
Phenylalanine	%		0.77	0.82	
Threonine	%		0.95	1.00	
Tryptophan	%		0.14	0.15	
Valine	%		1.40	1.48	

Ref No 5 02 890 Canada

Feed Name or Analyses			As Fed	Dry	C.V. ± %
Dry matter	%		95.0	100.0	0
Energy	GE Mcal/kg		4.44	4.68	
Pantothenic acid	mg/kg		17.0	17.9	36
Riboflavin	mg/kg		11.6	12.2	

Corn, cobs furfural residue, ammoniated, (5)
Furfural residue, ammoniated
Ref No 5 08 397 United States

Feed Name or Analyses			As Fed	Dry	C.V. ± %
Dry matter	%		94.3	100.0	
Ash	%		4.9	5.2	
Crude fiber	%		51.7	54.8	
Ether extract	%		0.4	0.4	
N-free extract	%		2.4	2.5	
Protein (N x 6.25)	%		34.9	37.0	
Energy	GE Mcal/kg				
Cattle	DE Mcal/kg	*	1.92	2.04	
Sheep	DE Mcal/kg	*	1.59	1.69	
Swine	DE kcal/kg	*	1463.	1551.	
Cattle	ME Mcal/kg	*	1.57	1.67	
Sheep	ME Mcal/kg	*	1.31	1.38	
Swine	ME kcal/kg	*	1295.	1373.	
Cattle	TDN %	*	43.6	46.2	
Sheep	TDN %	*	36.1	38.3	
Swine	TDN %	*	33.2	35.2	

Corn, distillers grains, dehy, (5)
Corn distillers dried grains (AAFCO)
Corn distillers dried grains (CFA)
Ref No 5 02 842 United States

Feed Name or Analyses			As Fed	Dry	C.V. ± %
Dry matter	%		93.8	100.0	3
Ash	%		2.2	2.4	46
Crude fiber	%		12.6	13.4	15
Sheep	dig coef %		64.	64.	
Ether extract	%		9.3	9.9	16
Sheep	dig coef %		88.	88.	
N-free extract	%		41.9	44.7	
Sheep	dig coef %		72.	72.	
Protein (N x 6.25)	%		27.8	29.7	11
Sheep	dig coef %		72.	72.	
Sheep	dig prot %		20.0	21.4	
Linoleic	%		4.467	4.762	
Energy	GE Mcal/kg		5.08	5.42	0
Cattle	DE Mcal/kg	*	3.48	3.72	
Sheep	DE Mcal/kg	*	3.38	3.60	
Swine	DE kcal/kg	*	4080.	4350.	
Cattle	ME Mcal/kg	*	2.86	3.05	
Chickens	ME_n kcal/kg		2000.	2132.	
Sheep	ME Mcal/kg	*	2.77	2.95	
Swine	ME kcal/kg	*	3673.	3915.	
Cattle	NE_m Mcal/kg	*	1.85	1.99	
Cattle	NE_{gain} Mcal/kg	*	1.25	1.33	
Cattle	$NE_{lactating\ cows}$ Mcal/kg	*	2.19	2.34	
Cattle	TDN %	*	79.0	84.3	
Sheep	TDN %		76.6	81.7	
Swine	TDN %	*	92.5	98.7	
Calcium	%		0.10	0.10	60
Chlorine	%		0.07	0.08	98
Cobalt	mg/kg		0.083	0.088	
Copper	mg/kg		45.2	48.2	1
Iron	%		0.022	0.024	47
Magnesium	%		0.07	0.07	16
Manganese	mg/kg		22.6	24.0	61
Phosphorus	%		0.41	0.43	41
Potassium	%		0.16	0.18	
Sodium	%		0.10	0.10	47
Sulphur	%		0.43	0.46	19
Biotin	mg/kg		0.45	0.48	98
Carotene	mg/kg		3.1	3.3	98
Choline	mg/kg		1342.	1431.	9
Niacin	mg/kg		34.6	36.9	49

Continued

Feed Name or Analyses		Mean As Fed	Mean Dry	C.V. ± %
Pantothenic acid	mg/kg	5.7	6.1	47
Riboflavin	mg/kg	3.0	3.2	70
Thiamine	mg/kg	1.8	1.9	63
Vitamin A	IU/g	5.0	5.3	
Vitamin A equivalent	IU/g	5.2	5.6	
Arginine	%	1.00	1.07	6
Cystine	%	0.39	0.42	
Histidine	%	0.61	0.65	9
Isoleucine	%	1.02	1.08	11
Leucine	%	3.66	3.90	6
Lysine	%	0.85	0.91	7
Methionine	%	0.45	0.48	
Phenylalanine	%	0.61	0.65	
Threonine	%	0.30	0.33	
Tryptophan	%	0.21	0.22	
Tyrosine	%	0.91	0.98	5
Valine	%	1.22	1.30	11
Xanthophylls	mg/kg	10.84	11.55	

Ref No 5 02 842 Canada

Feed Name or Analyses		Mean As Fed	Mean Dry	C.V. ± %
Dry matter	%	94.0	100.0	0
Crude fiber	%	10.4	11.1	14
Sheep	dig coef %	64.	64.	
Ether extract	%			
Sheep	dig coef %	88.	88.	
N-free extract	%			
Sheep	dig coef %	72.	72.	
Protein (N x 6.25)	%	28.1	29.9	4
Sheep	dig coef %	72.	72.	
Sheep	dig prot %	20.2	21.5	
Energy	GE Mcal/kg	5.09	5.42	
Sheep	DE Mcal/kg	* 3.39	3.61	
Chickens	ME_n kcal/kg	1607.	1710.	
Sheep	ME Mcal/kg	* 2.78	2.96	
Sheep	TDN %	76.9	81.8	
Pantothenic acid	mg/kg	15.3	16.3	36
Riboflavin	mg/kg	10.4	11.0	

Corn, distillers grains w solubles, dehy, mn 75% original solids, (5)

Corn distillers dried grains with solubles (AAFCO)

Ref No 5 02 843 United States

Feed Name or Analyses		Mean As Fed	Mean Dry	C.V. ± %
Dry matter	%	92.5	100.0	1
Ash	%	4.6	5.0	17
Crude fiber	%	9.1	9.8	23
Sheep	dig coef %	67.	67.	
Ether extract	%	10.3	11.2	30
Sheep	dig coef %	94.	94.	
N-free extract	%	41.4	44.8	10
Sheep	dig coef %	68.	68.	
Protein (N x 6.25)	%	27.0	29.2	2
Sheep	dig coef %	49.	49.	
Sheep	dig prot %	13.2	14.3	
Pentosans	%	15.3	16.5	
Lignin (Ellis)	%	5.6	6.1	
Linoleic	%	4.564	4.934	
Energy	GE Mcal/kg	4.57	4.94	2
Cattle	DE Mcal/kg	* 3.54	3.83	
Sheep	DE Mcal/kg	* 3.06	3.31	
Swine	DE kcal/kg	*4157.	4495.	
Cattle	ME Mcal/kg	* 2.90	3.14	
Chickens	ME_n kcal/kg	2453.	2652.	
Sheep	ME Mcal/kg	* 2.51	2.71	
Swine	ME kcal/kg	*3746.	4050.	
Cattle	NE_m Mcal/kg	* 2.00	2.16	
Cattle	NE_{gain} Mcal/kg	* 1.31	1.42	
Cattle	$NE_{lactating\ cows}$ Mcal/kg	* 2.30	2.49	
Cattle	TDN %	* 80.2	86.8	
Sheep	TDN %	69.4	75.0	
Swine	TDN %	* 94.3	101.9	
Calcium	%	0.19	0.21	21
Chlorine	%	0.17	0.18	6
Cobalt	mg/kg	0.111	0.120	
Copper	mg/kg	61.8	66.8	30
Iron	%	0.031	0.034	42
Magnesium	%	0.25	0.27	32
Manganese	mg/kg	28.5	30.8	55
Phosphorus	%	0.79	0.85	18
Potassium	%	0.65	0.70	45
Sodium	%	0.36	0.39	93
Sulphur	%	0.30	0.32	21
Biotin	mg/kg	0.67	0.72	53
Carotene	mg/kg	3.8	4.1	74
Choline	mg/kg	2814.	3043.	2
Folic acid	mg/kg	0.82	0.89	42
Niacin	mg/kg	73.8	79.8	17
Pantothenic acid	mg/kg	12.8	13.9	43
Riboflavin	mg/kg	10.1	10.9	30
Thiamine	mg/kg	2.9	3.1	47
α-tocopherol	mg/kg	40.0	43.2	
Vitamin B_6	mg/kg	2.22	2.40	89
Vitamin A	IU/g	3.9	4.2	
Vitamin A equivalent	IU/g	6.3	6.8	
Vitamin D_2	IU/g	0.6	0.6	
Arginine	%	1.03	1.11	15
Cystine	%	0.36	0.39	
Glutamic acid	%	5.53	5.98	9
Glycine	%	0.66	0.72	
Histidine	%	0.70	0.76	13
Isoleucine	%	1.71	1.85	13
Leucine	%	2.21	2.39	14
Lysine	%	0.76	0.83	12
Methionine	%	0.50	0.54	20
Phenylalanine	%	1.71	1.85	15
Serine	%	1.61	1.74	5
Threonine	%	1.01	1.09	3
Tryptophan	%	0.19	0.20	17
Tyrosine	%	0.60	0.65	10
Valine	%	1.61	1.74	6
Xanthophylls	mg/kg	9.82	10.62	

Ref No 5 02 843 Canada

Feed Name or Analyses		Mean As Fed	Mean Dry	C.V. ± %
Dry matter	%	93.0	100.0	0
Crude fib	%			
Sheep	dig coef %	67.	67.	
Ether extract	%			
Sheep	dig coef %	94.	94.	
N-free extract	%			
Sheep	dig coef %	68.	68.	
Protein (N x 6.25)	%			
Sheep	dig coef %	49.	49.	
Energy	GE Mcal/kg	4.59	4.94	
Sheep	DE Mcal/kg	* 3.07	3.31	
Sheep	ME Mcal/kg	* 2.52	2.71	
Sheep	TDN %	69.7	75.0	
Pantothenic acid	mg/kg	18.1	19.4	19
Riboflavin	mg/kg	11.1	12.0	

Corn, distillers solubles, dehy, (5)

Corn distillers dried solubles (AAFCO)

Ref No 5 02 844 United States

Feed Name or Analyses		Mean As Fed	Mean Dry	C.V. ± %
Dry matter	%	93.3	100.0	3
Ash	%	7.5	8.0	22
Crude fiber	%	3.6	3.8	41
Ether extract	%	9.3	10.0	28
N-free extract	%	43.6	46.7	
Protein (N x 6.25)	%	29.4	31.5	7
Pentosans	%	5.8	6.2	
Lignin (Ellis)	%	2.4	2.6	
Linoleic	%	4.659	4.994	
Energy	GE Mcal/kg			
Cattle	DE Mcal/kg	* 3.54	3.80	
Sheep	DE Mcal/kg	* 3.71	3.97	
Swine	DE kcal/kg	*3513.	3765.	
Cattle	ME Mcal/kg	* 2.90	3.11	
Chickens	ME_n kcal/kg	2942.	3153.	
Sheep	ME Mcal/kg	* 3.04	3.26	
Swine	ME kcal/kg	*3147.	3373.	
Cattle	NE_m Mcal/kg	* 2.01	2.15	
Cattle	NE_{gain} Mcal/kg	* 1.32	1.42	
Cattle	$NE_{lactating\ cows}$ Mcal/kg	* 2.32	2.49	
Cattle	TDN %	80.3	86.1	
Sheep	TDN %	* 84.0	90.1	
Swine	TDN %	* 79.7	85.4	
Calcium	%	0.35	0.37	33
Chlorine	%	0.26	0.28	23
Cobalt	mg/kg	0.197	0.211	1
Copper	mg/kg	83.4	89.4	3
Iron	%	0.055	0.059	30
Magnesium	%	0.64	0.69	17
Manganese	mg/kg	74.1	79.4	21
Phosphorus	%	1.38	1.48	26
Potassium	%	1.75	1.88	5
Sodium	%	0.24	0.26	36
Sulphur	%	0.37	0.40	2
Biotin	mg/kg	1.50	1.61	51
Carotene	mg/kg	0.7	0.7	27
Choline	mg/kg	4842.	5189.	3
Folic acid	mg/kg	1.13	1.21	24
Niacin	mg/kg	115.9	124.1	18
Pantothenic acid	mg/kg	21.0	22.5	30
Riboflavin	mg/kg	17.0	18.2	38
Thiamine	mg/kg	6.9	7.3	47
α-tocopherol	mg/kg	55.0	58.9	
Vitamin B_6	mg/kg	8.40	9.00	25
Vitamin A	IU/g	1.1	1.2	
Vitamin A equivalent	IU/g	1.1	1.2	
Arginine	%	1.05	1.13	39
Cystine	%	0.46	0.49	
Glutamic acid	%	4.31	4.62	11
Glycine	%	1.10	1.18	37
Histidine	%	0.70	0.75	5
Isoleucine	%	1.50	1.61	7
Leucine	%	2.11	2.26	42
Lysine	%	0.90	0.97	46
Methionine	%	0.55	0.59	33
Phenylalanine	%	1.50	1.61	10
Serine	%	1.30	1.40	19
Threonine	%	1.00	1.07	18
Tryptophan	%	0.23	0.24	33
Tyrosine	%	0.70	0.75	32
Valine	%	1.50	1.61	10
Xanthophylls	mg/kg	2.21	2.37	

(1) dry forages and roughages (2) pasture, range plants, and forages fed green (3) silages (4) energy feeds (5) protein supplements (6) minerals (7) vitamins (8) additives

Feed Name or Analyses		Mean As Fed	Mean Dry	C.V. ± %
Ref No 5 02 844		**Canada**		
Dry matter	%	91.8	100.0	1
Crude fiber	%	7.7	8.4	
Ether extract	%	7.8	8.4	
Protein (N x 6.25)	%	26.0	28.3	
Energy	GE Mcal/kg			
Cattle	DE Mcal/kg	* 3.49	3.80	
Cattle	ME Mcal/kg	* 2.86	3.11	
Chickens	MEn kcal/kg	2895.	3153.	
Cattle	TDN %	79.0	86.1	
Calcium	%	0.09	0.10	
Phosphorus	%	0.72	0.79	
Salt (NaCl)	%	0.27	0.29	
Pantothenic acid	mg/kg	26.8	29.1	19
Riboflavin	mg/kg	29.8	32.5	

Corn, distillers, stillage, wet, (5)

Feed Name or Analyses		Mean As Fed	Mean Dry	C.V. ± %
Ref No 2 02 846		**United States**		
Dry matter	%	6.5	100.0	17
Ash	%	0.3	4.9	23
Crude fiber	%	0.5	7.4	13
Ether extract	%	0.6	8.8	16
N-free extract	%	3.2	49.3	
Protein (N x 6.25)	%	1.9	29.6	3
Energy	GE Mcal/kg			
Cattle	DE Mcal/kg	* 0.25	3.84	
Sheep	DE Mcal/kg	* 0.26	3.96	
Swine	DE kcal/kg	* 250.	3849.	
Cattle	ME Mcal/kg	* 0.20	3.15	
Sheep	ME Mcal/kg	* 0.21	3.24	
Swine	ME kcal/kg	* 225.	3464.	
Cattle	TDN %	* 5.6	87.0	
Sheep	TDN %	* 5.8	89.7	
Swine	TDN %	* 5.7	87.3	
Calcium	%	0.01	0.14	
Phosphorus	%	0.05	0.82	

Corn, fungal amylase process distillers grains w solubles, dehy, (5)
Distillers dried corn grains w solubles, fungal amylase process

Feed Name or Analyses		Mean As Fed	Mean Dry	C.V. ± %
Ref No 5 08 398		**United States**		
Dry matter	%	94.5	100.0	
Ash	%	4.3	4.6	
Crude fiber	%	6.0	6.3	
Ether extract	%	13.3	14.1	
N-free extract	%	45.3	47.9	
Protein (N x 6.25)	%	25.6	27.1	
Energy	GE Mcal/kg			
Cattle	DE Mcal/kg	* 3.94	4.16	
Sheep	DE Mcal/kg	* 3.84	4.06	
Swine	DE kcal/kg	*4016.	4250.	
Cattle	ME Mcal/kg	* 3.23	3.42	
Sheep	ME Mcal/kg	* 3.15	3.33	
Swine	ME kcal/kg	*3634.	3846.	
Cattle	TDN %	* 89.3	94.5	
Sheep	TDN %	* 87.0	92.1	
Swine	TDN %	* 91.1	96.4	

Corn, germ, (5)

Feed Name or Analyses		Mean As Fed	Mean Dry	C.V. ± %
Ref No 5 02 896		**United States**		
Dry matter	%	90.0	100.0	
Crude fiber	%			
Sheep	dig coef %	* 100.	100.	
Ether extract	%			
Sheep	dig coef %	* 50.	50.	
N-free extract	%			
Sheep	dig coef %	* 90.	90.	
Protein (N x 6.25)	%			
Sheep	dig coef %	* 75.	75.	
Choline	mg/kg	1779.	1977.	
Niacin	mg/kg	30.6	34.0	2
Riboflavin	mg/kg	3.5	3.9	28
Thiamine	mg/kg	24.3	26.9	29
Ref No 5 02 896		**Canada**		
Dry matter	%	93.0	100.0	
Pantothenic acid	mg/kg	5.2	5.6	

Corn, germ, dry milled extn unspecified grnd, (5)
Corn germ meal, (dry milled) (AAFCO)
Corn oil meal

Feed Name or Analyses		Mean As Fed	Mean Dry	C.V. ± %
Ref No 5 02 894		**United States**		
Dry matter	%	91.7	100.0	
Ash	%	3.5	3.8	
Crude fiber	%	15.3	16.6	
Sheep	dig coef %	68.	68.	
Ether extract	%	9.0	9.8	
Sheep	dig coef %	96.	96.	
N-free extract	%	42.6	46.4	
Sheep	dig coef %	72.	72.	
Protein (N x 6.25)	%	21.4	23.3	
Sheep	dig coef %	75.	75.	
Sheep	dig prot %	16.1	17.5	
Linoleic	%	.700	.763	
Energy	GE Mcal/kg			
Cattle	DE Mcal/kg	* 3.23	3.52	
Sheep	DE Mcal/kg	* 3.38	3.68	
Swine	DE kcal/kg	*3308.	3607.	
Cattle	ME Mcal/kg	* 2.65	2.89	
Chickens	MEn kcal/kg	1700.	1854.	
Sheep	ME Mcal/kg	* 2.77	3.02	
Swine	ME kcal/kg	*3020.	3293.	
Cattle	TDN %	* 73.2	79.8	
Sheep	TDN %	76.6	83.5	
Swine	TDN %	* 75.0	81.8	
Phosphorus	%	0.57	0.62	
Potassium	%	0.21	0.23	
Ref No 5 02 894		**Canada**		
Dry matter	%	93.0	100.0	
Crude fiber	%			
Sheep	dig coef %	68.	68.	
Ether extract	%			
Sheep	dig coef %	96.	96.	
N-free extract	%			
Sheep	dig coef %	72.	72.	
Protein (N x 6.25)	%			
Sheep	dig coef %	75.	75.	
Energy	GE Mcal/kg			
Sheep	DE Mcal/kg	* 3.42	3.68	
Sheep	ME Mcal/kg	* 2.81	3.02	
Sheep	TDN %	77.6	83.5	

Feed Name or Analyses		Mean As Fed	Mean Dry	C.V. ± %
Pantothenic acid	mg/kg	17.3	18.6	
Riboflavin	mg/kg	4.2	4.5	

Corn, germ, mech-extd grnd, (5)
Corn oil meal expeller extracted
Corn germ meal (dry milled) mechanical extracted

Feed Name or Analyses		Mean As Fed	Mean Dry	C.V. ± %
Ref No 5 02 867		**United States**		
Dry matter	%	91.6	100.0	
Ash	%	2.4	2.6	
Crude fiber	%	10.2	11.1	
Ether extract	%	7.8	8.5	
N-free extract	%	48.8	53.3	
Protein (N x 6.25)	%	22.4	24.5	
Energy	GE Mcal/kg			
Cattle	DE Mcal/kg	* 3.46	3.78	
Sheep	DE Mcal/kg	* 3.30	3.60	
Swine	DE kcal/kg	*3360.	3668.	
Cattle	ME Mcal/kg	* 2.84	3.10	
Sheep	ME Mcal/kg	* 2.70	2.95	
Swine	ME kcal/kg	*3061.	3342.	
Cattle	TDN %	* 78.5	85.8	
Sheep	TDN %	* 74.7	81.6	
Swine	TDN %	* 76.2	83.2	
Calcium	%	0.06	0.07	
Chlorine	%	0.11	0.12	
Copper	mg/kg	11.7	12.8	
Iron	%	0.021	0.023	
Magnesium	%	0.28	0.31	
Manganese	mg/kg	11.0	12.0	
Phosphorus	%	0.56	0.61	
Potassium	%	0.13	0.14	

Corn, germ, solv-extd grnd, (5)
Corn oil meal, solvent extracted
Corn germ meal (dry milled), solvent extracted

Feed Name or Analyses		Mean As Fed	Mean Dry	C.V. ± %
Ref No 5 02 868		**United States**		
Dry matter	%	90.8	100.0	
Ash	%	3.5	3.8	
Crude fiber	%	10.9	12.0	
Sheep	dig coef %	100.	100.	
Ether extract	%	0.9	1.0	
Sheep	dig coef %	50.	50.	
N-free extract	%	57.9	63.7	
Sheep	dig coef %	90.	90.	
Protein (N x 6.25)	%	17.7	19.5	
Sheep	dig coef %	75.	75.	
Sheep	dig prot %	13.3	14.6	
Energy	GE Mcal/kg			
Cattle	DE Mcal/kg	* 3.05	3.36	
Sheep	DE Mcal/kg	* 3.41	3.75	
Swine	DE kcal/kg	*2838.	3126.	
Cattle	ME Mcal/kg	* 2.50	2.76	
Sheep	ME Mcal/kg	* 2.79	3.08	
Swine	ME kcal/kg	*2614.	2879.	
Cattle	TDN %	* 69.2	76.2	
Sheep	TDN %	77.2	85.1	
Swine	TDN %	* 64.4	70.9	
Calcium	%	0.03	0.03	
Copper	mg/kg	13.0	14.3	
Iron	%	0.032	0.035	
Manganese	mg/kg	16.5	18.2	
Phosphorus	%	0.50	0.55	
Choline	mg/kg	1936.	2132.	
Niacin	mg/kg	42.0	46.3	
Pantothenic acid	mg/kg	3.3	3.6	

Continued

Feed Name or Analyses			Mean As Fed	Mean Dry	C.V. ± %
Riboflavin	mg/kg		3.7	4.1	
a-tocopherol	mg/kg		87.0	95.8	
Arginine	%		1.40	1.54	
Cystine	%		0.40	0.44	
Glycine	%		1.10	1.21	
Lysine	%		1.10	1.21	
Methionine	%		0.43	0.47	
Tryptophan	%		0.25	0.27	

Corn, germ wo solubles, wet milled mech-extd dehy grnd, (5)
Corn germ meal, mechanical extracted, (wet milled) (AAFCO)

Ref No 5 02 897 — United States

Feed Name or Analyses			Mean As Fed	Mean Dry	C.V. ± %
Dry matter	%			100.0	
Crude fiber	%				
Sheep	dig coef %	*		68.	
Ether extract	%				
Sheep	dig coef %	*		96.	
N-free extract	%				
Sheep	dig coef %	*		72.	
Protein (N x 6.25)	%				
Sheep	dig coef %	*		75.	

Corn, glutens, maltose process, (5)

Ref No 5 02 899 — United States

Feed Name or Analyses			Mean As Fed	Mean Dry	C.V. ± %
Dry matter	%		89.7	100.0	
Ash	%		0.7	0.8	
Crude fiber	%		8.2	9.1	
Sheep	dig coef %		82.	82.	
Ether extract	%		8.1	9.0	
Sheep	dig coef %		92.	92.	
N-free extract	%		49.3	55.0	
Sheep	dig coef %		88.	88.	
Protein (N x 6.25)	%		23.4	26.1	
Sheep	dig coef %		86.	86.	
Sheep	dig prot %		20.1	22.4	
Energy	GE Mcal/kg				
Cattle	DE Mcal/kg	*	3.57	3.98	
Sheep	DE Mcal/kg	*	3.83	4.27	
Swine	DE kcal/kg	*	4151.	4628.	
Cattle	ME Mcal/kg	*	2.93	3.27	
Sheep	ME Mcal/kg	*	3.14	3.50	
Swine	ME kcal/kg	*	3766.	4199.	
Cattle	TDN %	*	81.1	90.4	
Sheep	TDN %		87.0	96.9	
Swine	TDN %	*	94.1	105.0	

Corn, gluten, wet milled dehy, (5)
Corn gluten meal (AAFCO)
Corn gluten meal (CFA)

Ref No 5 02 900 — United States

Feed Name or Analyses			Mean As Fed	Mean Dry	C.V. ± %
Dry matter	%		91.0	100.0	1
Ash	%		3.3	3.6	30
Crude fiber	%		4.6	5.1	33
Sheep	dig coef %	*	50.	50.	
Ether extract	%		2.3	2.6	35
Sheep	dig coef %	*	45.	45.	
N-free extract	%		41.7	45.9	11
Sheep	dig coef %	*	93.	93.	
Protein (N x 6.25)	%		39.0	42.9	17
Sheep	dig coef %	*	86.	86.	
Sheep	dig prot %		33.5	36.9	
Linoleic	%		1.200	1.319	
Energy	GE Mcal/kg		4.57	5.02	
Cattle	DE Mcal/kg	*	3.32	3.65	
Sheep	DE Mcal/kg	*	3.40	3.73	
Swine	DE kcal/kg	*	3514.	3862.	
Cattle	ME Mcal/kg	*	2.72	3.00	
Chickens	ME_n kcal/kg		3291.	3618.	
Sheep	ME Mcal/kg	*	2.78	3.06	
Swine	ME kcal/kg	*	3069.	3373.	
Cattle	NE_m Mcal/kg	*	1.81	1.99	
Cattle	NE_{gain} Mcal/kg	*	1.21	1.33	
Cattle	$NE_{lactating\ cows}$ Mcal/kg	*	2.13	2.34	
Cattle	TDN %	*	75.4	82.8	
Sheep	TDN %		77.0	84.7	
Swine	TDN %	*	79.7	87.6	
Calcium	%		0.14	0.16	18
Chlorine	%		0.07	0.08	
Cobalt	mg/kg		0.071	0.078	42
Copper	mg/kg		28.3	31.1	35
Iron	%		0.040	0.044	35
Magnesium	%		0.05	0.05	87
Manganese	mg/kg		10.6	11.6	31
Phosphorus	%		0.46	0.51	35
Potassium	%		0.02	0.03	91
Sodium	%		0.10	0.11	32
Biotin	mg/kg		0.20	0.22	
Carotene	mg/kg		16.4	18.0	61
Choline	mg/kg		330.	363.	22
Folic acid	mg/kg		0.22	0.24	5
Niacin	mg/kg		50.0	55.0	19
Pantothenic acid	mg/kg		10.4	11.4	35
Riboflavin	mg/kg		1.5	1.7	46
Thiamine	mg/kg		0.2	0.2	88
α-tocopherol	mg/kg		24.0	26.4	
Vitamin B_6	mg/kg		7.96	8.75	
Vitamin A	IU/g		26.3	28.9	
Vitamin A equivalent	IU/g		27.3	30.0	
Arginine	%		1.35	1.48	19
Cysteine	%		0.70	0.77	
Cystine	%		0.65	0.71	
Glutamic acid	%		8.32	9.15	13
Glycine	%		1.50	1.65	11
Histidine	%		1.00	1.10	4
Isoleucine	%		2.31	2.54	10
Leucine	%		7.62	8.38	22
Lysine	%		0.75	0.82	10
Methionine	%		1.00	1.10	29
Phenylalanine	%		2.91	3.20	42
Serine	%		1.50	1.65	
Threonine	%		1.40	1.54	19
Tryptophan	%		0.20	0.23	43
Tyrosine	%		1.00	1.10	
Valine	%		2.21	2.43	26
Xanthopyll	mg/kg		65.82	72.36	

(1) dry forages and roughages (2) pasture, range plants, and forages fed green (3) silages (4) energy feeds (5) protein supplements (6) minerals (7) vitamins (8) additives

Ref No 5 02 900 — Canada

Feed Name or Analyses			Mean As Fed	Mean Dry	C.V. ± %
Dry matter	%		91.7	100.0	1
Ash	%		3.7	4.0	
Crude fiber	%		5.6	6.1	23
Ether extract	%		2.3	2.6	12
N-free extract	%		38.3	41.8	
Protein (N x 6.25)	%		41.8	45.6	15
Energy	GE Mcal/kg		4.60	5.02	
Cattle	DE Mcal/kg	*	3.27	3.57	
Sheep	DE Mcal/kg	*	3.41	3.72	
Swine	DE kcal/kg	*	3431.	3741.	
Cattle	ME Mcal/kg	*	2.68	2.93	
Chickens	ME_n kcal/kg		2997.	3268.	
Sheep	ME Mcal/kg	*	2.79	3.05	
Swine	ME kcal/kg	*	2978.	3247.	
Cattle	TDN %	*	74.2	80.9	
Sheep	TDN %		77.3	84.3	
Swine	TDN %	*	77.8	84.8	
Calcium	%		0.14	0.15	69
Phosphorus	%		0.62	0.68	41
Salt (NaCl)	%		0.27	0.29	36
Pantothenic acid	mg/kg		9.6	10.5	14

Corn, gluten w bran, wet milled dehy, (5)
Corn gluten feed (AAFCO)
Corn gluten feed (CFA)

Ref No 5 02 903 — United States

Feed Name or Analyses			Mean As Fed	Mean Dry	C.V. ± %
Dry matter	%		90.4	100.0	1
Ash	%		6.6	7.3	14
Crude fiber	%		7.3	8.1	6
Cattle	dig coef %		70.	70.	
Sheep	dig coef %	*	82.	82.	
Ether extract	%		2.7	2.9	7
Cattle	dig coef %		74.	74.	
Sheep	dig coef %	*	65.	65.	
N-free extract	%		48.1	53.2	1
Cattle	dig coef %		88.	88.	
Sheep	dig coef %	*	90.	90.	
Protein (N x 6.25)	%		25.8	28.6	7
Cattle	dig coef %		86.	86.	
Sheep	dig coef %	*	86.	86.	
Cattle	dig prot %		22.2	24.6	
Sheep	dig prot %		22.2	24.6	
Pentosans	%		18.1	20.0	6
Linoleic	%		1.000	1.106	
Energy	GE Mcal/kg		3.95	4.37	
Cattle	DE Mcal/kg	*	3.27	3.61	
Sheep	DE Mcal/kg	*	3.32	3.67	
Swine	DE kcal/kg	*	3531.	3904.	
Cattle	ME Mcal/kg	*	2.68	2.96	
Chickens	ME_n kcal/kg		1661.	1837.	
Sheep	ME Mcal/kg	*	2.72	3.01	
Swine	ME kcal/kg	*	3186.	3523.	
Cattle	NE_m Mcal/kg	*	1.74	1.93	
Cattle	NE_{gain} Mcal/kg	*	1.17	1.29	
Cattle	$NE_{lactating\ cows}$ Mcal/kg	*	2.05	2.27	
Cattle	TDN %		74.1	81.9	
Sheep	TDN %		75.4	83.3	
Swine	TDN %	*	80.1	88.5	
Calcium	%		0.44	0.49	46
Chlorine	%		0.22	0.24	32
Cobalt	mg/kg		0.088	0.098	68
Copper	mg/kg		47.9	52.9	35
Iron	%		0.046	0.051	7
Magnesium	%		0.29	0.32	44
Manganese	mg/kg		23.8	26.4	29
Phosphorus	%		0.78	0.86	20
Potassium	%		0.57	0.63	20
Sodium	%		0.95	1.05	4
Sulphur	%		0.22	0.24	98
Biotin	mg/kg		0.33	0.37	
Carotene	mg/kg		8.4	9.3	49
Choline	mg/kg		1514.	1674.	53
Folic acid	mg/kg		0.31	0.34	34
Niacin	mg/kg		71.8	79.4	19
Pantothenic acid	mg/kg		17.1	18.9	28
Riboflavin	mg/kg		2.4	2.7	59

Feed Name or Analyses			Mean As Fed	Dry	C.V. ± %
Thiamine	mg/kg		2.0	2.2	75
Vitamin B6	mg/kg		15.00	16.58	
Vitamin A	IU/g		13.1	14.5	
Vitamin A equivalent	IU/g		14.0	15.4	
Arginine	%		0.80	0.88	18
Cystine	%		0.20	0.22	
Glutamic acid	%		4.30	4.76	9
Glycine	%		1.49	1.64	
Histidine	%		0.60	0.66	12
Isoleucine	%		1.20	1.33	24
Leucine	%		2.60	2.88	18
Lysine	%		0.80	0.88	9
Methionine	%		0.30	0.33	12
Phenylalanine	%		0.90	1.00	20
Serine	%		0.80	0.89	
Threonine	%		0.80	0.89	9
Tryptophan	%		0.20	0.22	18
Tyrosine	%		0.90	1.00	28
Valine	%		1.30	1.44	22
Xanthophylls	mg/kg		32.79	36.26	

Ref No 5 02 903 Canada

Feed Name or Analyses			Mean As Fed	Dry	C.V. ± %
Dry matter	%		90.2	100.0	1
Ash	%		6.3	7.0	
Crude fiber	%		8.2	9.1	
Cattle	dig coef %		70.	70.	
Ether extract	%		2.9	3.2	
Cattle	dig coef %		74.	74.	
N-free extract	%		49.0	54.3	
Cattle	dig coef %		88.	88.	
Protein (N x 6.25)	%		23.8	26.4	9
Cattle	dig coef %		86.	86.	
Cattle	dig prot %		20.5	22.7	
Energy	GE Mcal/kg		3.75	4.15	
Cattle	DE Mcal/kg	*	3.27	3.62	
Sheep	DE Mcal/kg	*	3.32	3.68	
Swine	DE kcal/kg	*	3551.	3936.	
Cattle	ME Mcal/kg	*	2.68	2.97	
Chickens	ME_n kcal/kg		1657.	1837.	
Sheep	ME Mcal/kg	*	2.72	3.02	
Swine	ME kcal/kg	*	3220.	3569.	
Cattle	TDN %		74.1	82.2	
Sheep	TDN %		75.3	83.4	
Swine	TDN %	*	80.5	89.3	
Calcium	%		0.17	0.19	
Cobalt	mg/kg		0.081	0.090	
Copper	mg/kg		43.3	48.0	
Iron	%		0.042	0.046	
Magnesium	%		0.58	0.64	
Manganese	mg/kg		21.6	24.0	
Phosphorus	%		0.65	0.72	
Potassium	%		0.54	0.60	
Salt (NaCl)	%		0.17	0.19	
Sodium	%		0.86	0.95	
Pantothenic acid	mg/kg		11.0	12.2	42
Riboflavin	mg/kg		10.0	11.1	
Thiamine	mg/kg		1.8	2.0	
Histidine	%		0.54	0.60	
Isoleucine	%		1.08	1.20	
Lysine	%		0.72	0.80	
Methionine	%		0.27	0.30	

CORN, BLUE. Zea mays

Corn, blue, grain, (4)

Ref No 4 02 905 United States

Feed Name or Analyses			Mean As Fed	Dry	C.V. ± %
Dry matter	%		91.8	100.0	1
Ash	%		1.5	1.6	12
Ether extract	%		4.2	4.6	6
Protein (N x 6.25)	%		9.3	10.1	6
Cattle	dig prot %	*	4.9	5.3	
Goats	dig prot %	*	6.0	6.5	
Horses	dig prot %	*	6.0	6.5	
Sheep	dig prot %	*	6.0	6.5	
Calcium	%		0.01	0.01	
Iron	%		0.006	0.007	
Phosphorus	%		0.30	0.33	
Niacin	mg/kg		5.9	6.4	9
Riboflavin	mg/kg		0.2	0.2	
Thiamine	mg/kg		1.2	1.3	22

CORN, DENT. Zea mays indentata

Corn, dent, aerial part, fresh, (2)

Ref No 2 02 907 United States

Feed Name or Analyses			Mean As Fed	Dry	C.V. ± %
Dry matter	%		24.3	100.0	39
Ash	%		1.3	5.5	9
Crude fiber	%		5.8	23.7	6
Ether extract	%		0.6	2.5	20
N-free extract	%		14.6	60.0	
Protein (N x 6.25)	%		2.0	8.4	6
Cattle	dig prot %	*	1.2	5.0	
Goats	dig prot %	*	1.1	4.4	
Horses	dig prot %	*	1.1	4.7	
Rabbits	dig prot %	*	1.3	5.2	
Sheep	dig prot %	*	1.2	4.8	
Energy	GE Mcal/kg				
Cattle	DE Mcal/kg	*	0.71	2.93	
Sheep	DE Mcal/kg	*	0.76	3.14	
Cattle	ME Mcal/kg	*	0.58	2.40	
Sheep	ME Mcal/kg	*	0.63	2.58	
Cattle	TDN %	*	16.2	66.5	
Sheep	TDN %	*	17.3	71.3	
Calcium	%		0.07	0.28	16
Chlorine	%		0.05	0.21	
Copper	mg/kg		1.1	4.6	46
Iron	%		0.002	0.008	31
Magnesium	%		0.04	0.17	25
Manganese	mg/kg		16.5	67.9	21
Phosphorus	%		0.05	0.21	16
Potassium	%		0.32	1.33	6
Sodium	%		0.01	0.04	
Sulphur	%		0.04	0.17	

Corn, dent, aerial part, fresh, milk stage, (2)

Ref No 2 02 906 United States

Feed Name or Analyses			Mean As Fed	Dry	C.V. ± %
Dry matter	%		17.5	100.0	7
Ash	%		1.1	6.4	7
Crude fiber	%		4.7	26.8	2
Ether extract	%		0.4	2.3	5
N-free extract	%		9.7	55.2	
Protein (N x 6.25)	%		1.6	9.3	7
Cattle	dig prot %	*	1.0	5.8	
Goats	dig prot %	*	0.9	5.3	
Horses	dig prot %	*	1.0	5.5	
Rabbits	dig prot %	*	1.0	5.9	
Sheep	dig prot %	*	1.0	5.7	
Energy	GE Mcal/kg				
Cattle	DE Mcal/kg	*	0.52	2.95	
Sheep	DE Mcal/kg	*	0.53	3.03	
Cattle	ME Mcal/kg	*	0.42	2.42	
Sheep	ME Mcal/kg	*	0.44	2.49	
Cattle	TDN %	*	11.7	67.0	
Sheep	TDN %	*	12.0	68.8	

Corn, dent, aerial part, fresh, mature, (2)

Ref No 2 08 399 United States

Feed Name or Analyses			Mean As Fed	Dry	C.V. ± %
Dry matter	%		32.3	100.0	
Ash	%		1.5	4.7	
Crude fiber	%		7.1	21.9	
Ether extract	%		0.8	2.6	
N-free extract	%		20.3	63.0	
Protein (N x 6.25)	%		2.5	7.9	
Cattle	dig prot %	*	1.5	4.6	
Goats	dig prot %	*	1.3	3.9	
Horses	dig prot %	*	1.4	4.2	
Rabbits	dig prot %	*	1.5	4.8	
Sheep	dig prot %	*	1.4	4.3	
Energy	GE Mcal/kg				
Cattle	DE Mcal/kg	*	0.95	2.93	
Sheep	DE Mcal/kg	*	1.04	3.21	
Cattle	ME Mcal/kg	*	0.78	2.40	
Sheep	ME Mcal/kg	*	0.85	2.63	
Cattle	TDN %	*	21.5	66.4	
Sheep	TDN %	*	23.5	72.8	
Calcium	%		0.10	0.29	
Phosphorus	%		0.07	0.22	
Potassium	%		0.39	1.20	

Corn, dent, aerial part, fresh, drouth stricken, (2)

Ref No 2 08 400 United States

Feed Name or Analyses			Mean As Fed	Dry	C.V. ± %
Dry matter	%		26.4	100.0	
Ash	%		1.6	6.1	
Crude fiber	%		6.8	25.8	
Ether extract	%		0.4	1.5	
N-free extract	%		15.0	56.8	
Protein (N x 6.25)	%		2.6	9.8	
Cattle	dig prot %	*	1.7	6.3	
Goats	dig prot %	*	1.5	5.7	
Horses	dig prot %	*	1.6	5.9	
Rabbits	dig prot %	*	1.7	6.3	
Sheep	dig prot %	*	1.6	6.2	
Energy	GE Mcal/kg				
Cattle	DE Mcal/kg	*	0.80	3.04	
Sheep	DE Mcal/kg	*	0.81	3.05	
Cattle	ME Mcal/kg	*	0.66	2.49	
Sheep	ME Mcal/kg	*	0.66	2.50	
Cattle	TDN %	*	18.2	69.0	
Sheep	TDN %	*	18.3	69.2	

Corn, dent, aerial part, ensiled, (3)

Ref No 3 02 912 United States

Feed Name or Analyses			Mean As Fed	Dry	C.V. ± %
Dry matter	%		26.4	100.0	16
Calcium	%		0.09	0.35	23
Copper	mg/kg		1.3	5.1	32
Iron	%		0.005	0.018	70
Magnesium	%		0.06	0.21	33
Manganese	mg/kg		17.8	67.2	30
Phosphorus	%		0.06	0.23	35

Continued

Feed Name or Analyses			Mean As Fed	Mean Dry	C.V. ± %
Potassium	%		0.31	1.17	36
Carotene	mg/kg		13.7	52.0	24
Niacin	mg/kg		13.3	50.3	22
Vitamin A equivalent	IU/g		22.9	86.7	

Corn, dent, aerial part, ensiled, immature, (3)

Ref No 3 02 908 United States

Feed Name or Analyses			Mean As Fed	Mean Dry	C.V. ± %
Dry matter	%		19.9	100.0	18
Ash	%		1.1	5.7	
Crude fiber	%		6.1	30.9	
Ether extract	%		0.5	2.6	
N-free extract	%		10.4	52.6	
Protein (N x 6.25)	%		1.6	8.2	
Cattle	dig prot %	*	0.7	3.7	
Goats	dig prot %	*	0.7	3.7	
Horses	dig prot %	*	0.7	3.7	
Sheep	dig prot %	*	0.7	3.7	
Energy	GE Mcal/kg				
Cattle	DE Mcal/kg	*	0.52	2.64	
Sheep	DE Mcal/kg	*	0.58	2.91	
Cattle	ME Mcal/kg	*	0.43	2.17	
Sheep	ME Mcal/kg	*	0.47	2.39	
Cattle	TDN %	*	11.9	59.9	
Sheep	TDN %	*	13.1	66.0	
Calcium	%		0.11	0.54	17
Iron	%		0.010	0.049	
Magnesium	%		0.06	0.30	
Phosphorus	%		0.07	0.34	19

Corn, dent, aerial part, ensiled, milk stage, (3)

Ref No 3 08 402 United States

Feed Name or Analyses			Mean As Fed	Mean Dry	C.V. ± %
Dry matter	%		20.3	100.0	
Ash	%		1.3	6.4	
Crude fiber	%		5.8	28.6	
Ether extract	%		0.6	3.0	
N-free extract	%		10.8	53.2	
Protein (N x 6.25)	%		1.8	8.9	
Cattle	dig prot %	*	0.9	4.3	
Goats	dig prot %	*	0.9	4.3	
Horses	dig prot %	*	0.9	4.3	
Sheep	dig prot %	*	0.9	4.3	
Energy	GE Mcal/kg				
Cattle	DE Mcal/kg	*	0.56	2.77	
Sheep	DE Mcal/kg	*	0.59	2.90	
Cattle	ME Mcal/kg	*	0.46	2.27	
Sheep	ME Mcal/kg	*	0.48	2.38	
Cattle	TDN %	*	12.8	62.9	
Sheep	TDN %	*	13.3	65.8	
Calcium	%		0.11	0.54	
Iron	%		0.010	0.049	
Magnesium	%		0.06	0.30	
Phosphorus	%		0.07	0.34	

Corn, dent, aerial part, ensiled, mature, (3)

Ref No 3 02 910 United States

Feed Name or Analyses			Mean As Fed	Mean Dry	C.V. ± %
Dry matter	%		27.3	100.0	4
Ash	%		1.7	6.2	8
Crude fiber	%		7.0	25.8	11
Ether extract	%		0.8	2.8	19
N-free extract	%		15.6	57.1	
Protein (N x 6.25)	%		2.2	8.2	13
Cattle	dig prot %	*	1.0	3.7	
Goats	dig prot %	*	1.0	3.7	
Horses	dig prot %	*	1.0	3.7	
Sheep	dig prot %	*	1.0	3.7	
Energy	GE Mcal/kg				
Cattle	DE Mcal/kg	*	0.79	2.90	
Sheep	DE Mcal/kg	*	0.80	2.94	
Cattle	ME Mcal/kg	*	0.65	2.38	
Sheep	ME Mcal/kg	*	0.66	2.41	
Cattle	TDN %	*	17.9	65.7	
Sheep	TDN %	*	18.2	66.7	
Calcium	%		0.09	0.35	19
Chlorine	%		0.05	0.18	
Copper	mg/kg		1.3	4.8	40
Iron	%		0.003	0.011	60
Magnesium	%		0.05	0.19	31
Manganese	mg/kg		19.1	70.2	27
Phosphorus	%		0.06	0.23	15
Potassium	%		0.32	1.17	
Sodium	%		0.01	0.04	
Sulphur	%		0.04	0.13	

(1) dry forages and roughages (2) pasture, range plants, and forages fed green (3) silages (4) energy feeds (5) protein supplements (6) minerals (7) vitamins (8) additives

Corn, dent, aerial part, ensiled, mature, few ears, (3)

Ref No 3 08 600 United States

Feed Name or Analyses			Mean As Fed	Mean Dry	C.V. ± %
Dry matter	%		26.3	100.0	
Ash	%		1.9	7.2	
Crude fiber	%		8.5	32.3	
Ether extract	%		0.8	3.0	
N-free extract	%		12.9	49.0	
Protein (N x 6.25)	%		2.2	8.4	
Cattle	dig prot %	*	1.0	3.8	
Goats	dig prot %	*	1.0	3.8	
Horses	dig prot %	*	1.0	3.8	
Sheep	dig prot %	*	1.0	3.8	
Energy	GE Mcal/kg				
Cattle	DE Mcal/kg	*	0.71	2.69	
Sheep	DE Mcal/kg	*	0.75	2.84	
Cattle	ME Mcal/kg	*	0.58	2.21	
Sheep	ME Mcal/kg	*	0.61	2.33	
Cattle	TDN %	*	16.0	61.0	
Sheep	TDN %	*	17.0	64.5	
Calcium	%		0.09	0.34	
Magnesium	%		0.06	0.23	
Phosphorus	%		0.05	0.19	
Potassium	%		0.37	1.41	
Sulphur	%		0.02	0.08	

Corn, dent, aerial part, ensiled, mature, well eared, (3)

Ref No 3 02 909 United States

Feed Name or Analyses			Mean As Fed	Mean Dry	C.V. ± %
Dry matter	%		28.3	100.0	4
Ash	%		1.7	5.9	3
Crude fiber	%		6.8	24.2	7
Ether extract	%		0.8	2.9	23
N-free extract	%		16.6	58.7	
Protein (N x 6.25)	%		2.3	8.2	12
Cattle	dig prot %	*	1.0	3.7	
Goats	dig prot %	*	1.0	3.7	
Horses	dig prot %	*	1.0	3.7	
Sheep	dig prot %	*	1.0	3.7	
Energy	GE Mcal/kg				
Cattle	DE Mcal/kg	*	0.84	2.96	
Sheep	DE Mcal/kg	*	0.84	2.96	
Cattle	ME Mcal/kg	*	0.69	2.43	
Sheep	ME Mcal/kg	*	0.69	2.43	
Cattle	TDN %	*	19.0	67.1	
Sheep	TDN %	*	19.0	67.1	
Calcium	%		0.08	0.30	9
Chlorine	%		0.05	0.18	
Copper	mg/kg		1.3	4.6	16
Iron	%		0.003	0.011	48
Magnesium	%		0.05	0.18	40
Manganese	mg/kg		19.3	68.2	17
Phosphorus	%		0.06	0.23	9
Potassium	%		0.27	0.95	12
Sodium	%		0.01	0.04	
Sulphur	%		0.04	0.14	

Corn, dent, aerial part, ensiled, over ripe, (3)

Ref No 3 02 911 United States

Feed Name or Analyses			Mean As Fed	Mean Dry	C.V. ± %
Dry matter	%		27.0	100.0	4
Chlorine	%		0.05	0.18	42
Sodium	%		0.01	0.04	47
Sulphur	%		0.04	0.13	13

Corn, dent, aerial part, ensiled, drouth stricken, (3)

Ref No 3 08 403 United States

Feed Name or Analyses			Mean As Fed	Mean Dry	C.V. ± %
Dry matter	%		24.6	100.0	
Ash	%		2.2	8.9	
Crude fiber	%		6.4	26.0	
Ether extract	%		0.5	2.0	
N-free extract	%		12.9	52.4	
Protein (N x 6.25)	%		2.6	10.6	
Cattle	dig prot %	*	1.4	5.8	
Goats	dig prot %	*	1.4	5.8	
Horses	dig prot %	*	1.4	5.8	
Sheep	dig prot %	*	1.4	5.8	
Energy	GE Mcal/kg				
Cattle	DE Mcal/kg	*	0.71	2.89	
Sheep	DE Mcal/kg	*	0.71	2.87	
Cattle	ME Mcal/kg	*	0.58	2.37	
Sheep	ME Mcal/kg	*	0.58	2.35	
Cattle	TDN %	*	16.1	65.4	
Sheep	TDN %	*	16.0	65.0	

Corn, dent, aerial part wo ears wo husks, ensiled, mature, (3)

Corn, dent stover silage, mature

Ref No 3 08 401 United States

Feed Name or Analyses			Mean As Fed	Mean Dry	C.V. ± %
Dry matter	%		23.7	100.0	
Ash	%		1.6	6.8	
Crude fiber	%		7.8	32.9	
Ether extract	%		0.7	3.0	
N-free extract	%		12.0	50.6	
Protein (N x 6.25)	%		1.6	6.8	
Cattle	dig prot %	*	0.6	2.4	
Goats	dig prot %	*	0.6	2.4	
Horses	dig prot %	*	0.6	2.4	
Sheep	dig prot %	*	0.6	2.4	
Energy	GE Mcal/kg				
Cattle	DE Mcal/kg	*	0.64	2.71	
Sheep	DE Mcal/kg	*	0.62	2.61	
Cattle	ME Mcal/kg	*	0.53	2.22	
Sheep	ME Mcal/kg	*	0.51	2.14	
Cattle	TDN %	*	14.6	61.5	
Sheep	TDN %	*	14.0	59.1	
Calcium	%		0.08	0.34	

Feed Name or Analyses			Mean As Fed	Mean Dry	C.V. ± %
Phosphorus		%	0.10	0.42	
Potassium		%	0.39	1.65	

Corn, dent, grain, (4)

Ref No 4 02 920 United States

Feed Name or Analyses			Mean As Fed	Mean Dry	C.V. ± %
Dry matter		%	87.2	100.0	8
Ash		%	1.5	1.7	
Crude fiber		%	2.4	2.7	
Cattle	dig coef	%	* 19.	19.	
Sheep	dig coef	%	* 30.	30.	
Swine	dig coef	%	* 47.	47.	
Ether extract		%	3.8	4.3	
Cattle	dig coef	%	* 87.	87.	
Sheep	dig coef	%	* 87.	87.	
Swine	dig coef	%	* 70.	70.	
N-free extract		%	70.7	81.1	
Cattle	dig coef	%	* 91.	91.	
Sheep	dig coef	%	* 99.	99.	
Swine	dig coef	%	* 93.	93.	
Protein (N x 6.25)		%	8.9	10.2	
Cattle	dig coef	%	* 75.	75.	
Sheep	dig coef	%	* 78.	78.	
Swine	dig coef	%	* 80.	80.	
Cattle	dig prot	%	6.7	7.7	
Goats	dig prot	%	* 5.7	6.6	
Horses	dig prot	%	* 5.7	6.6	
Sheep	dig prot	%	6.9	8.0	
Swine	dig prot	%	7.1	8.2	
Energy	GE	Mcal/kg			
Cattle	DE	Mcal/kg	* 3.47	3.98	
Sheep	DE	Mcal/kg	* 3.75	4.30	
Swine	DE	kcal/kg	*3523.	4040.	
Cattle	ME	Mcal/kg	* 2.85	3.27	
Sheep	ME	Mcal/kg	* 3.07	3.52	
Swine	ME	kcal/kg	*3309.	3795.	
Cattle	TDN	%	78.8	90.4	
Sheep	TDN	%	85.0	97.5	
Swine	TDN	%	79.9	91.6	
Calcium		%	0.03	0.03	
Phosphorus		%	0.27	0.31	
Potassium		%	0.43	0.49	
Biotin		mg/kg	0.06	0.07	18
Folic acid		mg/kg	0.21	0.24	48

Corn, dent, grain, dehy, soft, (4)

Ref No 4 08 404 United States

Feed Name or Analyses			Mean As Fed	Mean Dry	C.V. ± %
Dry matter		%	86.4	100.0	
Ash		%	1.7	2.0	
Crude fiber		%	2.4	2.8	
Ether extract		%	2.8	3.2	
N-free extract		%	70.5	81.6	
Protein (N x 6.25)		%	9.0	10.4	
Cattle	dig prot	%	* 4.8	5.6	
Goats	dig prot	%	* 5.9	6.8	
Horses	dig prot	%	* 5.9	6.8	
Sheep	dig prot	%	* 5.9	6.8	
Energy	GE	Mcal/kg			
Cattle	DE	Mcal/kg	* 3.10	3.59	
Sheep	DE	Mcal/kg	* 3.26	3.78	
Swine	DE	kcal/kg	*3418.	3956.	
Cattle	ME	Mcal/kg	* 2.54	2.94	
Sheep	ME	Mcal/kg	* 2.68	3.10	
Swine	ME	kcal/kg	*3209.	3714.	
Cattle	TDN	%	* 70.3	81.3	

Feed Name or Analyses			Mean As Fed	Mean Dry	C.V. ± %
Sheep	TDN	%	* 74.0	85.7	
Swine	TDN	%	* 77.5	89.7	

Corn, dent, grain gr 1 US mn wt 56 lb per bushel, (4)

Ref No 4 02 914 United States

Feed Name or Analyses			Mean As Fed	Mean Dry	C.V. ± %
Dry matter		%	86.3	100.0	2
Ash		%	1.3	1.5	
Crude fiber		%	2.0	2.3	
Cattle	dig coef	%	* 19.	19.	
Sheep	dig coef	%	* 30.	30.	
Swine	dig coef	%	* 47.	47.	
Ether extract		%	4.0	4.6	
Cattle	dig coef	%	* 87.	87.	
Sheep	dig coef	%	* 87.	87.	
Swine	dig coef	%	* 70.	70.	
N-free extract		%	70.3	81.4	
Cattle	dig coef	%	* 91.	91.	
Sheep	dig coef	%	* 99.	99.	
Swine	dig coef	%	* 93.	93.	
Protein (N x 6.25)		%	8.8	10.2	
Cattle	dig coef	%	* 75.	75.	
Sheep	dig coef	%	* 78.	78.	
Swine	dig coef	%	* 80.	80.	
Cattle	dig prot	%	6.6	7.7	
Goats	dig prot	%	* 5.7	6.6	
Horses	dig prot	%	* 5.7	6.6	
Sheep	dig prot	%	6.9	8.0	
Swine	dig prot	%	7.1	8.2	
Starch		%	62.5	72.4	4
Sugars, total		%	1.7	2.0	20
Energy	GE	Mcal/kg			
Cattle	DE	Mcal/kg	* 3.47	4.02	
Sheep	DE	Mcal/kg	* 3.74	4.33	
Swine	DE	kcal/kg	*3509.	4065.	
Cattle	ME	Mcal/kg	* 2.85	3.30	
Sheep	ME	Mcal/kg	* 3.07	3.55	
Swine	ME	kcal/kg	*3296.	3818.	
Cattle	TDN	%	78.7	91.2	
Sheep	TDN	%	84.8	98.2	
Swine	TDN	%	79.6	92.2	
Calcium		%	0.02	0.02	67
Chlorine		%	0.05	0.06	27
Copper		mg/kg	4.1	4.7	12
Iron		%	0.002	0.003	23
Magnesium		%	0.10	0.11	14
Manganese		mg/kg	5.6	6.5	23
Phosphorus		%	0.28	0.32	13
Potassium		%	0.28	0.33	12
Sodium		%	0.01	0.01	
Sulphur		%	0.12	0.14	21

Corn, dent, grain, gr 2 US mn wt 54 lb per bushel, (4)

Ref No 4 02 915 United States

Feed Name or Analyses			Mean As Fed	Mean Dry	C.V. ± %
Dry matter		%	85.4	100.0	0
Ash		%	1.3	1.5	
Crude fiber		%	2.1	2.4	
Cattle	dig coef	%	* 19.	19.	
Sheep	dig coef	%	* 30.	30.	
Ether extract		%	3.9	4.6	
Cattle	dig coef	%	* 87.	87.	
Sheep	dig coef	%	* 87.	87.	
N-free extract		%	69.2	81.1	
Cattle	dig coef	%	* 91.	91.	
Sheep	dig coef	%	* 99.	99.	
Protein (N x 6.25)		%	8.9	10.5	
Cattle	dig coef	%	* 75.	75.	
Sheep	dig coef	%	* 78.	78.	
Cattle	dig prot	%	6.7	7.9	
Goats	dig prot	%	* 5.8	6.8	
Horses	dig prot	%	* 5.8	6.8	
Sheep	dig prot	%	7.0	8.2	
Starch		%	61.7	72.2	4
Sugars, total		%	1.6	1.9	17
Energy	GE	Mcal/kg	3.77	4.41	
Cattle	DE	Mcal/kg	* 3.43	4.01	
Sheep	DE	Mcal/kg	* 3.70	4.33	
Swine	DE	kcal/kg	*3472.	4065.	
Cattle	ME	Mcal/kg	* 2.81	3.29	
Sheep	ME	Mcal/kg	* 3.03	3.55	
Swine	ME	kcal/kg	*3260.	3817.	
Cattle	TDN	%	77.8	91.1	
Sheep	TDN	%	83.8	98.1	
Swine	TDN	%	* 78.8	92.2	
Calcium		%	0.02	0.02	24
Chlorine		%	0.05	0.06	22
Cobalt		mg/kg	0.017	0.020	
Copper		mg/kg	3.2	3.7	10
Iron		%	0.002	0.003	34
Magnesium		%	0.10	0.12	14
Manganese		mg/kg	5.2	6.1	18
Phosphorus		%	0.27	0.32	10
Potassium		%	0.28	0.33	15
Sodium		%	0.01	0.01	
Sulphur		%	0.12	0.14	17

Corn, dent, grain, gr 3 US mn wt 52 lb per bushel, (4)

Ref No 4 02 916 United States

Feed Name or Analyses			Mean As Fed	Mean Dry	C.V. ± %
Dry matter		%	83.8	100.0	0
Ash		%	1.2	1.5	19
Crude fiber		%	2.0	2.4	
Cattle	dig coef	%	* 19.	19.	
Sheep	dig coef	%	* 30.	30.	
Swine	dig coef	%	* 41.	41.	
Ether extract		%	3.7	4.5	10
Cattle	dig coef	%	* 87.	87.	
Sheep	dig coef	%	* 87.	87.	
Swine	dig coef	%	* 72.	72.	
N-free extract		%	68.2	81.4	
Cattle	dig coef	%	* 91.	91.	
Sheep	dig coef	%	* 99.	99.	
Swine	dig coef	%	* 94.	94.	
Protein (N x 6.25)		%	8.6	10.3	6
Cattle	dig coef	%	* 75.	75.	
Sheep	dig coef	%	* 78.	78.	
Swine	dig coef	%	* 81.	81.	
Cattle	dig prot	%	6.5	7.7	
Goats	dig prot	%	* 5.6	6.7	
Horses	dig prot	%	* 5.6	6.7	
Sheep	dig prot	%	6.7	8.0	
Swine	dig prot	%	7.0	8.4	
Energy	GE	Mcal/kg			
Cattle	DE	Mcal/kg	* 3.36	4.01	
Sheep	DE	Mcal/kg	* 3.62	4.32	
Swine	DE	kcal/kg	*3438.	4103.	
Cattle	ME	Mcal/kg	* 2.76	3.29	
Sheep	ME	Mcal/kg	* 2.97	3.55	
Swine	ME	kcal/kg	*3229.	3853.	
Cattle	TDN	%	76.2	91.0	
Sheep	TDN	%	82.2	98.1	
Swine	TDN	%	78.0	93.1	
Calcium		%	0.02	0.02	53
Chlorine		%	0.04	0.05	23
Copper		mg/kg	2.6	3.1	

Continued

Feed Name or Analyses		Mean As Fed	Mean Dry	C.V. ± %
Iron	%	0.002	0.003	48
Magnesium	%	0.11	0.13	17
Manganese	mg/kg	5.1	6.1	17
Phosphorus	%	0.27	0.32	12
Potassium	%	0.28	0.33	18
Sodium	%	0.01	0.01	
Sulphur	%	0.12	0.14	10

Corn, dent, grain, gr 4 US mn wt 49 lb per bushel, (4)

Ref No 4 02 917 United States

Feed Name or Analyses		Mean As Fed	Mean Dry	C.V. ± %
Dry matter	%	82.2	100.0	1
Ash	%	1.2	1.4	15
Crude fiber	%	2.0	2.4	11
Cattle	dig coef %	* 19.	19.	
Sheep	dig coef %	* 30.	30.	
Swine	dig coef %	* 35.	35.	
Ether extract	%	3.6	4.4	7
Cattle	dig coef %	* 87.	87.	
Sheep	dig coef %	* 87.	87.	
Swine	dig coef %	* 67.	67.	
N-free extract	%	66.8	81.2	
Cattle	dig coef %	* 91.	91.	
Sheep	dig coef %	* 99.	99.	
Swine	dig coef %	* 92.	92.	
Protein (N x 6.25)	%	8.6	10.5	5
Cattle	dig coef %	* 75.	75.	
Sheep	dig coef %	* 78.	78.	
Swine	dig coef %	* 76.	76.	
Cattle	dig prot %	6.5	7.8	
Goats	dig prot %	* 5.6	6.8	
Horses	dig prot %	* 5.6	6.8	
Sheep	dig prot %	6.7	8.2	
Swine	dig prot %	6.5	8.0	
Energy	GE Mcal/kg			
Cattle	DE Mcal/kg	* 3.30	4.01	
Sheep	DE Mcal/kg	* 3.55	4.32	
Swine	DE kcal/kg	*3271.	3978.	
Cattle	ME Mcal/kg	* 2.70	3.29	
Sheep	ME Mcal/kg	* 2.91	3.54	
Swine	ME kcal/kg	*3071.	3735.	
Cattle	TDN %	74.8	90.9	
Sheep	TDN %	80.6	98.0	
Swine	TDN %	74.2	90.2	
Calcium	%	0.02	0.03	45
Chlorine	%	0.05	0.06	24
Copper	mg/kg	2.1	2.5	
Iron	%	0.002	0.003	32
Magnesium	%	0.11	0.14	26
Manganese	mg/kg	4.8	5.8	28
Phosphorus	%	0.25	0.31	17
Potassium	%	0.26	0.32	9
Sodium	%	0.01	0.01	
Sulphur	%	0.10	0.12	8

Corn, dent, grain, gr 5 US mn wt 46 lb per bushel, (4)

Ref No 4 02 918 United States

Feed Name or Analyses		Mean As Fed	Mean Dry	C.V. ± %
Dry matter	%	78.5	100.0	
Ash	%	1.1	1.5	25
Crude fiber	%	1.9	2.4	
Ether extract	%	3.4	4.4	10
N-free extract	%	63.8	81.3	
Protein (N x 6.25)	%	8.2	10.4	7
Cattle	dig prot %	* 4.4	5.6	
Goats	dig prot %	* 5.3	6.8	
Horses	dig prot %	* 5.3	6.8	
Sheep	dig prot %	* 5.3	6.8	
Energy	GE Mcal/kg			
Cattle	DE Mcal/kg	* 2.85	3.63	
Sheep	DE Mcal/kg	* 3.00	3.82	
Swine	DE kcal/kg	*3189.	4062.	
Cattle	ME Mcal/kg	* 2.33	2.97	
Sheep	ME Mcal/kg	* 2.46	3.13	
Swine	ME kcal/kg	*2994.	3815.	
Cattle	TDN %	* 64.6	82.2	
Sheep	TDN %	* 68.0	86.6	
Swine	TDN %	* 72.3	92.1	
Calcium	%	0.02	0.03	
Chlorine	%	0.04	0.05	
Copper	mg/kg	1.6	2.0	
Iron	%	0.002	0.003	
Magnesium	%	0.10	0.13	
Manganese	mg/kg	4.4	5.6	
Phosphorus	%	0.24	0.31	
Potassium	%	0.26	0.33	
Sodium	%	0.01	0.01	

(1) dry forages and roughages (2) pasture, range plants, and forages fed green (3) silages (4) energy feeds (5) protein supplements (6) minerals (7) vitamins (8) additives

Corn, dent, grain, soft, (4)

Ref No 4 02 919 United States

Feed Name or Analyses		Mean As Fed	Mean Dry	C.V. ± %
Dry matter	%	69.8	100.0	17
Ash	%	1.3	1.8	21
Crude fiber	%	2.1	3.0	34
Cattle	dig coef %	* 19.	19.	
Sheep	dig coef %	* 30.	30.	
Swine	dig coef %	* 41.	41.	
Ether extract	%	2.8	4.0	14
Cattle	dig coef %	* 87.	87.	
Sheep	dig coef %	* 87.	87.	
Swine	dig coef %	* 71.	71.	
N-free extract	%	56.2	80.5	
Cattle	dig coef %	* 91.	91.	
Sheep	dig coef %	* 99.	99.	
Swine	dig coef %	* 93.	93.	
Protein (N x 6.25)	%	7.5	10.7	18
Cattle	dig coef %	* 75.	75.	
Sheep	dig coef %	* 78.	78.	
Swine	dig coef %	* 80.	80.	
Cattle	dig prot %	5.6	8.0	
Goats	dig prot %	* 4.9	7.0	
Horses	dig prot %	* 4.9	7.0	
Sheep	dig prot %	5.8	8.3	
Swine	dig prot %	6.0	8.6	
Energy	GE Mcal/kg			
Cattle	DE Mcal/kg	* 2.76	3.95	
Sheep	DE Mcal/kg	* 2.98	4.27	
Swine	DE kcal/kg	*2801.	4013.	
Cattle	ME Mcal/kg	* 2.26	3.24	
Sheep	ME Mcal/kg	* 2.44	3.50	
Swine	ME kcal/kg	*2629.	3766.	
Cattle	TDN %	62.6	89.6	
Sheep	TDN %	67.5	96.7	
Swine	TDN %	63.5	91.0	
Phosphorus	%	0.25	0.36	
Potassium	%	0.27	0.39	

CORN, DENT WHITE. Zea mays indentata

Corn, dent white, aerial part, ensiled, (3)

Ref No 3 02 921 United States

Feed Name or Analyses		Mean As Fed	Mean Dry	C.V. ± %
Dry matter	%		100.0	
Copper	mg/kg		9.0	
Iron	%		0.004	
Manganese	mg/kg		5.1	
Phosphorus	%		0.31	
Zinc	mg/kg		20.9	

Corn, dent white, grain, (4)

Ref No 4 02 928 United States

Feed Name or Analyses		Mean As Fed	Mean Dry	C.V. ± %
Dry matter	%	87.5	100.0	0
Crude fiber	%			
Cattle	dig coef %	* 19.	19.	
Sheep	dig coef %	* 30.	30.	
Swine	dig coef %	* 47.	47.	
Ether extract	%			
Cattle	dig coef %	* 87.	87.	
Sheep	dig coef %	* 87.	87.	
Swine	dig coef %	* 70.	70.	
N-free extract	%			
Cattle	dig coef %	* 91.	91.	
Sheep	dig coef %	* 99.	99.	
Swine	dig coef %	* 93.	93.	
Protein (N x 6.25)	%			
Cattle	dig coef %	* 75.	75.	
Sheep	dig coef %	* 78.	78.	
Swine	dig coef %	* 80.	80.	
Calcium	%	0.04	0.04	
Cobalt	mg/kg	0.060	0.068	
Copper	mg/kg	5.8	6.6	26
Iron	%	0.003	0.003	40
Manganese	mg/kg	8.5	9.7	49
Phosphorus	%	0.26	0.30	13
Zinc	mg/kg	23.5	26.9	
Biotin	mg/kg	0.06	0.07	13
Carotene	mg/kg	0.4	0.4	
Niacin	mg/kg	15.0	17.2	29
Pantothenic acid	mg/kg	3.9	4.4	28
Riboflavin	mg/kg	1.4	1.5	20
Thiamine	mg/kg	4.4	5.1	23
Vitamin A equivalent	IU/g	0.6	0.7	
Arginine	%	0.26	0.30	
Cystine	%	0.09	0.10	
Histidine	%	0.18	0.20	
Isoleucine	%	0.44	0.50	
Leucine	%	0.88	1.00	
Lysine	%	0.26	0.30	
Methionine	%	0.09	0.10	
Phenylalanine	%	0.35	0.40	
Threonine	%	0.35	0.40	
Tryptophan	%	0.09	0.10	
Tyrosine	%	0.44	0.50	58
Valine	%	0.35	0.40	

Corn, dent white, grain, gr 1 US mn wt 56 lb per bushel, (4)

Ref No 4 02 923 United States

Feed Name or Analyses		Mean As Fed	Mean Dry	C.V. ± %
Dry matter	%		100.0	
Ash	%		1.4	13
Ether extract	%		4.1	10

Feed Name or Analyses		Mean As Fed	Mean Dry	C.V. ± %
Protein (N x 6.25)	%		9.9	7
Cattle	dig prot %	*	5.1	
Goats	dig prot %	*	6.3	
Horses	dig prot %	*	6.3	
Sheep	dig prot %	*	6.3	
Starch	%		73.4	1
Sugars, total	%		2.2	9

Corn, dent white, grain, gr 2 US mn wt 54 lb per bushel, (4)

Ref No 4 02 924 United States

Feed Name or Analyses		Mean As Fed	Mean Dry	C.V. ± %
Dry matter	%		100.0	
Ash	%		1.3	13
Ether extract	%		4.1	6
Protein (N x 6.25)	%		9.6	8
Cattle	dig prot %	*	4.8	
Goats	dig prot %	*	6.0	
Horses	dig prot %	*	6.0	
Sheep	dig prot %	*	6.0	
Starch	%		73.7	0
Sugars, total	%		2.0	8

CORN, DENT YELLOW. Zea mays indentata

Corn, dent yellow, grain, (4)

Ref No 4 02 935 United States

Feed Name or Analyses		Mean As Fed	Mean Dry	C.V. ± %
Dry matter	%	86.4	100.0	3
Crude fiber	%			
Cattle	dig coef %	* 19.	19.	
Sheep	dig coef %	* 30.	30.	
Ether extract	%			
Cattle	dig coef %	* 87.	87.	
Sheep	dig coef %	* 87.	87.	
N-free extract	%			
Cattle	dig coef %	* 91.	91.	
Sheep	dig coef %	* 99.	99.	
Protein (N x 6.25)	%			
Cattle	dig coef %	* 75.	75.	
Sheep	dig coef %	* 78.	78.	
a-tocopherol	mg/kg	22.0	25.5	
Arginine	%	0.43	0.50	37
Cystine	%	0.09	0.10	43
Glutamic acid	%	2.85	3.30	
Glycine	%	0.43	0.50	
Histidine	%	0.17	0.20	40
Isoleucine	%	0.43	0.50	19
Leucine	%	0.95	1.10	4
Lysine	%	0.17	0.20	37
Methionine	%	0.17	0.20	51
Phenylalanine	%	0.43	0.50	26
Threonine	%	0.35	0.40	21
Tryptophan	%	0.09	0.10	
Tyrosine	%	0.52	0.60	16
Valine	%	0.43	0.50	26

Corn, dent yellow, grain, gr 1 US mn wt 56 lb per bushel, (4)

Ref No 4 02 930 United States

Feed Name or Analyses		Mean As Fed	Mean Dry	C.V. ± %
Dry matter	%	85.8	100.0	1
Crude fiber	%			
Cattle	dig coef %	* 19.	19.	
Sheep	dig coef %	* 30.	30.	
Ether extract	%			
Cattle	dig coef %	* 87.	87.	
Sheep	dig coef %	* 87.	87.	
N-free extract	%			
Cattle	dig coef %	* 91.	91.	
Sheep	dig coef %	* 99.	99.	
Protein (N x 6.25)	%			
Cattle	dig coef %	* 75.	75.	
Sheep	dig coef %	* 78.	78.	
Starch	%	61.9	72.2	4
Sugars, total	%	1.7	2.0	20
Copper	mg/kg	4.2	4.9	
Iron	%	0.002	0.002	
Manganese	mg/kg	5.5	6.4	

Corn, dent yellow, grain, gr 2 US mn wt 54 lb per bushel, (4)

Ref No 4 02 931 United States

Feed Name or Analyses		Mean As Fed	Mean Dry	C.V. ± %
Dry matter	%	88.7	100.0	2
Crude fiber	%			
Cattle	dig coef %	* 19.	19.	
Sheep	dig coef %	* 30.	30.	
Ether extract	%			
Cattle	dig coef %	* 87.	87.	
Sheep	dig coef %	* 87.	87.	
N-free extract	%			
Cattle	dig coef %	* 91.	91.	
Sheep	dig coef %	* 99.	99.	
Protein (N x 6.25)	%			
Cattle	dig coef %	* 75.	75.	
Sheep	dig coef %	* 78.	78.	
Starch	%	63.8	72.0	4
Sugars, total	%	1.7	1.9	17
Linoleic	%	1.872	2.111	
Energy	GE Mcal/kg	3.91	4.41	
Chickens	ME_n kcal/kg	3430.	3811.	
Cattle	NE_m Mcal/kg	* 2.02	2.28	
Cattle	NE_{gain} Mcal/kg	* 1.31	1.48	
Cattle	$NE_{lactating\ cows}$ Mcal/kg	* 2.31	2.60	
Cobalt	mg/kg	0.018	0.020	
Carotene	mg/kg	1.8	2.0	
Niacin	mg/kg	26.2	29.5	
Pantothenic acid	mg/kg	3.9	4.4	
Riboflavin	mg/kg	1.4	1.5	
Thiamine	mg/kg	3.5	4.0	
Vitamin A equivalent	IU/g	2.9	3.3	
Cystine	%	0.09	0.10	
Glutamic acid	%	2.93	3.30	
Methionine	%	0.09	0.10	
Tryptophan	%	0.09	0.10	
Tyrosine	%	0.44	0.50	
Valine	%	0.35	0.40	

Corn, dent yellow, grain, gr 3 US mn wt 52 lb per bushel, (4)

Ref No 4 02 932 United States

Feed Name or Analyses		Mean As Fed	Mean Dry	C.V. ± %
Dry matter	%	86.0	100.0	
Ash	%	1.2	1.4	14
Crude fiber	%	2.1	2.4	
Cattle	dig coef %	* 19.	19.	
Sheep	dig coef %	* 30.	30.	
Ether extract	%	3.7	4.3	8
Cattle	dig coef %	* 87.	87.	
Sheep	dig coef %	* 87.	87.	
N-free extract	%	70.3	81.8	
Cattle	dig coef %	* 91.	91.	
Sheep	dig coef %	* 99.	99.	
Protein (N x 6.25)	%	8.7	10.1	4
Cattle	dig coef %	* 75.	75.	
Sheep	dig coef %	* 78.	78.	
Cattle	dig prot %	6.5	7.6	
Goats	dig prot %	* 5.6	6.5	
Horses	dig prot %	* 5.6	6.5	
Sheep	dig prot %	6.8	7.9	
Starch	%	62.1	72.2	2
Sugars, total	%	1.5	1.8	17
Energy	GE Mcal/kg			
Cattle	DE Mcal/kg	* 3.45	4.01	
Sheep	DE Mcal/kg	* 3.72	4.32	
Swine	DE kcal/kg	*3495.	4064.	
Cattle	ME Mcal/kg	* 2.83	3.29	
Sheep	ME Mcal/kg	* 3.05	3.54	
Swine	ME kcal/kg	*3284.	3819.	
Cattle	NE_m Mcal/kg	* 1.96	2.28	
Cattle	NE_{gain} Mcal/kg	* 1.27	1.48	
Cattle	$NE_{lactating\ cows}$ Mcal/kg	* 2.24	2.60	
Cattle	TDN %	78.2	90.9	
Sheep	TDN %	84.3	98.0	
Swine	TDN %	* 79.3	92.2	
Calcium	%	0.02	0.02	
Iron	%	0.002	0.002	
Manganese	mg/kg	5.5	6.4	
Phosphorus	%	0.25	0.29	

Corn, dent yellow, grain, gr 4 US mn wt 49 lb per bushel, (4)

Ref No 4 02 933 United States

Feed Name or Analyses		Mean As Fed	Mean Dry	C.V. ± %
Dry matter	%	84.3	100.0	
Ash	%	1.2	1.4	16
Crude fiber	%	2.1	2.5	
Cattle	dig coef %	* 19.	19.	
Sheep	dig coef %	* 30.	30.	
Ether extract	%	3.6	4.3	8
Cattle	dig coef %	* 87.	87.	
Sheep	dig coef %	* 87.	87.	
N-free extract	%	68.8	81.6	
Cattle	dig coef %	* 91.	91.	
Sheep	dig coef %	* 99.	99.	
Protein (N x 6.25)	%	8.6	10.2	5
Cattle	dig coef %	* 75.	75.	
Sheep	dig coef %	* 78.	78.	
Cattle	dig prot %	6.4	7.7	
Goats	dig prot %	* 5.5	6.6	
Horses	dig prot %	* 5.5	6.6	
Sheep	dig prot %	6.7	8.0	
Starch	%	60.7	72.0	2
Sugars, total	%	1.5	1.8	16
Energy	GE Mcal/kg			
Cattle	DE Mcal/kg	* 3.37	4.00	
Sheep	DE Mcal/kg	* 3.64	4.32	
Swine	DE kcal/kg	*3419.	4055.	
Cattle	ME Mcal/kg	* 2.77	3.28	
Sheep	ME Mcal/kg	* 2.98	3.54	
Swine	ME kcal/kg	*3211.	3810.	
Cattle	TDN %	76.5	90.8	
Sheep	TDN %	82.5	97.9	
Swine	TDN %	* 77.5	92.0	
Calcium	%	0.04	0.05	
Magnesium	%	0.17	0.20	
Phosphorus	%	0.29	0.34	

Feed Name or Analyses				Mean As Fed	Mean Dry	C.V. ± %

Corn, dent yellow, grain, gr 5 US mn wt 46 lb per bushel, (4)
Ref No 4 02 934 United States

Analyses				As Fed	Dry	C.V. ± %
Dry matter		%		78.5	100.0	
Ash		%		1.1	1.4	26
Crude fiber		%		1.9	2.4	
Cattle	dig coef	%	*	19.	19.	
Sheep	dig coef	%	*	30.	30.	
Ether extract		%		3.3	4.2	11
Cattle	dig coef	%	*	87.	87.	
Sheep	dig coef	%	*	89.	89.	
N-free extract		%		64.3	81.9	
Cattle	dig coef	%	*	91.	91.	
Sheep	dig coef	%	*	99.	99.	
Protein (N x 6.25)		%		7.9	10.1	8
Cattle	dig coef	%	*	75.	75.	
Sheep	dig coef	%	*	78.	78.	
Cattle	dig prot	%		5.9	7.6	
Goats	dig prot	%	*	5.1	6.5	
Horses	dig prot	%	*	5.1	6.5	
Sheep	dig prot	%		6.2	7.9	
Starch		%		56.7	72.2	2
Sugars, total		%		1.3	1.7	16
Energy	GE Mcal/kg					
Cattle	DE Mcal/kg		*	3.14	4.00	
Sheep	DE Mcal/kg		*	3.40	4.32	
Swine	DE kcal/kg		*	3189.	4063.	
Cattle	ME Mcal/kg		*	2.58	3.28	
Sheep	ME Mcal/kg		*	2.78	3.55	
Swine	ME kcal/kg		*	2997.	3818.	
Cattle	TDN	%		71.3	90.8	
Sheep	TDN	%		77.0	98.1	
Swine	TDN	%	*	72.3	92.1	

Corn, dent yellow, grain, gr sample US,(4)
Ref No 4 02 929 United States

Analyses				As Fed	Dry	C.V. ± %
Dry matter		%			100.0	
Ash		%			1.4	17
Ether extract		%			4.2	8
Protein (N x 6.25)		%			10.2	6
Cattle	dig prot	%	*		5.4	
Goats	dig prot	%	*		6.6	
Horses	dig prot	%	*		6.6	
Sheep	dig prot	%	*		6.6	
Starch		%			72.3	2
Sugars, total		%			1.5	16

CORN, FLINT. Zea mays indurata

Corn, flint, aerial part, s-c, dough stage, (1)
Ref No 1 02 936 United States

Analyses				As Fed	Dry	C.V. ± %
Dry matter		%		49.5	100.0	
Ash		%		2.2	4.4	
Crude fiber		%		14.3	28.8	
Cattle	dig coef	%		72.	72.	
Ether extract		%		1.5	3.1	
Cattle	dig coef	%		72.	72.	
N-free extract		%		27.9	56.4	
Cattle	dig coef	%		63.	63.	
Protein (N x 6.25)		%		3.6	7.3	
Cattle	dig coef	%		38.	38.	
Cattle	dig prot	%		1.4	2.8	
Goats	dig prot	%	*	1.7	3.4	
Horses	dig prot	%	*	1.8	3.7	
Rabbits	dig prot	%	*	2.1	4.3	
Sheep	dig prot	%	*	1.5	3.1	
Energy	GE Mcal/kg					
Cattle	DE Mcal/kg		*	1.40	2.82	
Sheep	DE Mcal/kg		*	1.26	2.55	
Cattle	ME Mcal/kg		*	1.15	2.32	
Sheep	ME Mcal/kg		*	1.03	2.09	
Cattle	TDN	%		31.7	64.1	
Sheep	TDN	%	*	28.6	57.8	

Corn, flint, aerial part, fresh, (2)
Ref No 2 02 938 United States

Analyses				As Fed	Dry	C.V. ± %
Dry matter		%		22.7	100.0	33
Ash		%		1.3	5.6	15
Crude fiber		%		5.1	22.4	9
Ether extract		%		0.6	2.8	15
N-free extract		%		13.6	59.9	
Protein (N x 6.25)		%		2.1	9.2	12
Cattle	dig prot	%	*	1.3	5.7	
Goats	dig prot	%	*	1.2	5.1	
Horses	dig prot	%	*	1.2	5.3	
Rabbits	dig prot	%	*	1.3	5.8	
Sheep	dig prot	%	*	1.3	5.5	
Energy	GE Mcal/kg					
Cattle	DE Mcal/kg		*	0.66	2.92	
Sheep	DE Mcal/kg		*	0.71	3.14	
Cattle	ME Mcal/kg		*	0.54	2.40	
Sheep	ME Mcal/kg		*	0.58	2.57	
Cattle	TDN	%	*	15.0	66.3	
Sheep	TDN	%	*	16.1	71.1	
Phosphorus		%		0.04	0.18	
Potassium		%		0.34	1.48	

Corn, flint, aerial part, fresh, milk stage, (2)
Ref No 2 02 937 United States

Analyses				As Fed	Dry	C.V. ± %
Dry matter		%		14.3	100.0	10
Ash		%		1.1	7.6	7
Crude fiber		%		3.8	26.6	4
Ether extract		%		0.3	2.2	9
N-free extract		%		7.5	52.5	
Protein (N x 6.25)		%		1.6	11.2	8
Cattle	dig prot	%	*	1.1	7.4	
Goats	dig prot	%	*	1.0	7.0	
Horses	dig prot	%	*	1.0	7.0	
Rabbits	dig prot	%	*	1.0	7.3	
Sheep	dig prot	%	*	1.1	7.4	
Energy	GE Mcal/kg					
Cattle	DE Mcal/kg		*	0.42	2.97	
Sheep	DE Mcal/kg		*	0.42	2.96	
Cattle	ME Mcal/kg		*	0.35	2.44	
Sheep	ME Mcal/kg		*	0.35	2.43	
Cattle	TDN	%	*	9.6	67.4	
Sheep	TDN	%	*	9.6	67.1	

Corn, flint, aerial part, fresh, mature, (2)
Ref No 2 08 405 United States

Analyses				As Fed	Dry	C.V. ± %
Dry matter		%		33.0	100.0	
Ash		%		1.7	5.2	
Crude fiber		%		6.6	20.2	
Ether extract		%		1.1	3.2	
N-free extract		%		20.7	62.8	
Protein (N x 6.25)		%		2.9	8.7	
Cattle	dig prot	%	*	1.7	5.3	
Goats	dig prot	%	*	1.5	4.7	
Horses	dig prot	%	*	1.6	4.9	
Rabbits	dig prot	%	*	1.8	5.4	
Sheep	dig prot	%	*	1.7	5.1	
Energy	GE Mcal/kg					
Cattle	DE Mcal/kg		*	0.95	2.89	
Sheep	DE Mcal/kg		*	1.06	3.20	
Cattle	ME Mcal/kg		*	0.78	2.37	
Sheep	ME Mcal/kg		*	0.87	2.63	
Cattle	TDN	%	*	21.6	65.4	
Sheep	TDN	%	*	23.9	72.6	

Corn, flint, aerial part, ensiled, (3)
Ref No 3 02 941 United States

Analyses				As Fed	Dry	C.V. ± %
Dry matter		%		23.5	100.0	13
Ash		%		1.5	6.4	13
Crude fiber		%		5.5	23.2	5
Ether extract		%		0.6	2.7	26
N-free extract		%		14.2	60.6	
Protein (N x 6.25)		%		1.7	7.1	16
Cattle	dig prot	%	*	0.6	2.7	
Goats	dig prot	%	*	0.6	2.7	
Horses	dig prot	%	*	0.6	2.7	
Sheep	dig prot	%	*	0.6	2.7	
Energy	GE Mcal/kg					
Cattle	DE Mcal/kg		*	0.72	3.07	
Sheep	DE Mcal/kg		*	0.70	2.97	
Cattle	ME Mcal/kg		*	0.59	2.51	
Sheep	ME Mcal/kg		*	0.57	2.44	
Cattle	TDN	%	*	16.3	69.5	
Sheep	TDN	%	*	15.8	67.4	

Corn, flint, aerial part, ensiled, dough stage, (3)
Ref No 3 02 939 United States

Analyses				As Fed	Dry	C.V. ± %
Dry matter		%		21.4	100.0	10
Ash		%		1.3	6.0	20
Crude fiber		%		5.3	24.9	5
Cattle	dig coef	%		69.	69.	
Sheep	dig coef	%		72.	72.	
Ether extract		%		0.8	3.8	19
Cattle	dig coef	%		86.	86.	
Sheep	dig coef	%		73.	73.	
N-free extract		%		12.0	56.0	
Cattle	dig coef	%		72.	72.	
Sheep	dig coef	%		76.	76.	
Protein (N x 6.25)		%		2.0	9.3	6
Cattle	dig coef	%		59.	59.	
Sheep	dig coef	%		56.	56.	
Cattle	dig prot	%		1.2	5.5	
Goats	dig prot	%	*	1.0	4.7	
Horses	dig prot	%	*	1.0	4.7	
Sheep	dig prot	%		1.1	5.2	
Energy	GE Mcal/kg					
Cattle	DE Mcal/kg		*	0.66	3.10	
Sheep	DE Mcal/kg		*	0.68	3.17	
Cattle	ME Mcal/kg		*	0.54	2.54	
Sheep	ME Mcal/kg		*	0.56	2.60	
Cattle	TDN	%		15.0	70.3	
Sheep	TDN	%		15.4	71.9	

(1) dry forages and roughages (2) pasture, range plants, and forages fed green (3) silages (4) energy feeds (5) protein supplements (6) minerals (7) vitamins (8) additives

Corn, flint, aerial part, ensiled, mature, (3)

Ref No 3 02 940 United States

Feed Name or Analyses			Mean As Fed	Mean Dry	C.V. ± %
Dry matter	%		24.6	100.0	12
Ash	%		1.5	6.2	8
Crude fiber	%		5.7	23.1	5
Ether extract	%		0.6	2.5	10
N-free extract	%		15.2	61.6	
Protein (N x 6.25)	%		1.6	6.6	12
Cattle	dig prot %	*	0.5	2.2	
Goats	dig prot %	*	0.5	2.2	
Horses	dig prot %	*	0.5	2.2	
Sheep	dig prot %	*	0.5	2.2	
Energy	GE Mcal/kg				
Cattle	DE Mcal/kg	*	0.76	3.08	
Sheep	DE Mcal/kg	*	0.73	2.99	
Cattle	ME Mcal/kg	*	0.62	2.52	
Sheep	ME Mcal/kg	*	0.60	2.45	
Cattle	TDN %	*	17.2	69.8	
Sheep	TDN %	*	16.7	67.8	

Corn, flint, grain, (4)

Ref No 4 02 946 United States

Feed Name or Analyses			Mean As Fed	Mean Dry	C.V. ± %
Dry matter	%		88.5	100.0	0
Ash	%		1.5	1.7	14
Crude fiber	%		1.9	2.1	25
Cattle	dig coef %	*	19.	19.	
Sheep	dig coef %	*	94.	94.	
Ether extract	%		4.3	4.9	20
Cattle	dig coef %	*	87.	87.	
Sheep	dig coef %	*	91.	91.	
N-free extract	%		70.9	80.2	
Cattle	dig coef %	*	91.	91.	
Sheep	dig coef %	*	98.	98.	
Protein (N x 6.25)	%		9.9	11.1	15
Cattle	dig coef %	*	75.	75.	
Sheep	dig coef %	*	86.	86.	
Cattle	dig prot %		7.4	8.4	
Goats	dig prot %	*	6.6	7.4	
Horses	dig prot %	*	6.6	7.4	
Sheep	dig prot %		8.5	9.6	
Energy	GE Mcal/kg				
Cattle	DE Mcal/kg	*	3.56	4.02	
Sheep	DE Mcal/kg	*	3.91	4.41	
Swine	DE kcal/kg	*	3616.	4086.	
Cattle	ME Mcal/kg	*	2.92	3.30	
Sheep	ME Mcal/kg	*	3.20	3.62	
Swine	ME kcal/kg	*	3390.	3831.	
Cattle	TDN %		80.8	91.3	
Sheep	TDN %		88.6	100.1	
Swine	TDN %	*	82.0	92.7	
Copper	mg/kg		11.5	13.0	
Iron	%		0.003	0.003	
Manganese	mg/kg		7.0	7.9	
Phosphorus	%		0.27	0.31	
Potassium	%		0.32	0.36	
Niacin	mg/kg		15.8	17.9	
Lysine	%		0.27	0.30	
Methionine	%		0.18	0.20	
Tryptophan	%		0.09	0.10	

CORN, MIXED. Zea mays

Corn, mixed, grain, Can 2 CW mn wt 54 lb per bushel mx 3% foreign material, (4)

Ref No 4 02 955 United States

Feed Name or Analyses			Mean As Fed	Mean Dry	C.V. ± %
Dry matter	%			100.0	
Crude fiber	%				
Cattle	dig coef %	*		19.	
Sheep	dig coef %	*		30.	
Swine	dig coef %	*		47.	
Ether extract	%				
Cattle	dig coef %	*		87.	
Sheep	dig coef %	*		87.	
Swine	dig coef %	*		70.	
N-free extract	%				
Cattle	dig coef %	*		91.	
Sheep	dig coef %	*		99.	
Swine	dig coef %	*		93.	
Protein (N x 6.25)	%				
Cattle	dig coef %	*		75.	
Sheep	dig coef %	*		78.	
Swine	dig coef %	*		80.	
Energy	GE Mcal/kg				
Cattle	DE Mcal/kg	*		3.99	
Sheep	DE Mcal/kg	*		4.30	
Swine	DE kcal/kg	*		4041.	
Cattle	ME Mcal/kg	*		3.27	
Sheep	ME Mcal/kg	*		3.52	
Cattle	TDN %	*		90.4	
Sheep	TDN %	*		97.4	
Swine	TDN %	*		91.6	

Ref No 4 02 955 Canada

Feed Name or Analyses			Mean As Fed	Mean Dry	C.V. ± %
Dry matter	%		86.5	100.0	
Ash	%		1.4	1.6	
Crude fiber	%		2.2	2.5	
Ether extract	%		3.8	4.4	
N-free extract	%		69.0	79.8	
Protein (N x 6.25)	%		10.1	11.7	
Cattle	dig prot %	*	5.8	6.7	
Goats	dig prot %	*	6.9	7.9	
Horses	dig prot %	*	6.9	7.9	
Sheep	dig prot %	*	6.9	7.9	
Energy	GE Mcal/kg				
Cattle	DE Mcal/kg	*	3.45	3.99	
Sheep	DE Mcal/kg	*	3.72	4.30	
Swine	DE kcal/kg	*	3495.	4041.	
Cattle	ME Mcal/kg	*	2.83	3.27	
Sheep	ME Mcal/kg	*	3.05	3.52	
Swine	ME kcal/kg	*	3273.	3784.	
Cattle	TDN %		78.2	90.4	
Sheep	TDN %		84.3	97.4	
Swine	TDN %		79.3	91.6	

Corn, mixed, grain, Can 3 CW mn wt 52 lb per bushel mx 5% foreign material, (4)

Ref No 4 02 956 United States

Feed Name or Analyses			Mean As Fed	Mean Dry	C.V. ± %
Dry matter	%			100.0	
Crude fiber	%				
Cattle	dig coef %	*		19.	
Sheep	dig coef %	*		30.	
Swine	dig coef %	*		47.	
Ether extract	%				
Cattle	dig coef %	*		87.	
Sheep	dig coef %	*		87.	
Swine	dig coef %	*		70.	
N-free extract	%				
Cattle	dig coef %	*		91.	
Sheep	dig coef %	*		99.	
Swine	dig coef %	*		93.	
Protein (N x 6.25)	%				
Cattle	dig coef %	*		75.	
Sheep	dig coef %	*		78.	
Swine	dig coef %	*		80.	
Energy	GE Mcal/kg				
Cattle	DE Mcal/kg	*		3.98	
Sheep	DE Mcal/kg	*		4.28	
Swine	DE kcal/kg	*		4029.	
Cattle	ME Mcal/kg	*		3.26	
Sheep	ME Mcal/kg	*		3.51	
Cattle	TDN %	*		90.2	
Sheep	TDN %	*		97.1	
Swine	TDN %	*		91.4	

Ref No 4 02 956 Canada

Feed Name or Analyses			Mean As Fed	Mean Dry	C.V. ± %
Dry matter	%		86.5	100.0	
Ash	%		1.6	1.8	
Crude fiber	%		2.3	2.7	
Ether extract	%		3.9	4.5	
N-free extract	%		68.1	78.7	
Protein (N x 6.25)	%		10.6	12.3	
Cattle	dig prot %	*	6.3	7.3	
Goats	dig prot %	*	7.3	8.5	
Horses	dig prot %	*	7.3	8.5	
Sheep	dig prot %	*	7.3	8.5	
Energy	GE Mcal/kg				
Cattle	DE Mcal/kg	*	3.44	3.98	
Sheep	DE Mcal/kg	*	3.70	4.28	
Swine	DE kcal/kg	*	3485.	4029.	
Cattle	ME Mcal/kg	*	2.82	3.26	
Sheep	ME Mcal/kg	*	3.04	3.51	
Swine	ME kcal/kg	*	3259.	3768.	
Cattle	TDN %		78.0	90.2	
Sheep	TDN %		84.0	97.1	
Swine	TDN %		79.0	91.4	

CORN, POP. Zea mays everta

Corn, pop, aerial part, fresh, immature, (2)

Ref No 2 08 406 United States

Feed Name or Analyses			Mean As Fed	Mean Dry	C.V. ± %
Dry matter	%		16.9	100.0	
Ash	%		1.0	5.9	
Crude fiber	%		6.0	35.5	
Ether extract	%		0.4	2.4	
N-free extract	%		8.2	48.5	
Protein (N x 6.25)	%		1.3	7.7	
Cattle	dig prot %	*	0.7	4.4	
Goats	dig prot %	*	0.6	3.7	
Horses	dig prot %	*	0.7	4.1	
Rabbits	dig prot %	*	0.8	4.6	
Sheep	dig prot %	*	0.7	4.2	
Energy	GE Mcal/kg				
Cattle	DE Mcal/kg	*	0.50	2.97	
Sheep	DE Mcal/kg	*	0.49	2.91	
Cattle	ME Mcal/kg	*	0.41	2.44	
Sheep	ME Mcal/kg	*	0.40	2.38	
Cattle	TDN %	*	11.4	67.4	
Sheep	TDN %	*	11.1	65.9	

Corn, pop, aerial part, fresh, mature, (2)
Ref No 2 02 963 United States

Feed Name or Analyses		Mean As Fed	Mean Dry	C.V. ± %
Dry matter	%	16.9	100.0	
Ash	%	1.0	5.9	
Crude fiber	%	6.0	35.5	
Ether extract	%	0.4	2.4	
N-free extract	%	8.2	48.5	
Protein (N x 6.25)	%	1.3	7.7	
Cattle	dig prot %	* 0.7	4.4	
Goats	dig prot %	* 0.6	3.7	
Horses	dig prot %	* 0.7	4.1	
Rabbits	dig prot %	* 0.8	4.6	
Sheep	dig prot %	* 0.7	4.2	
Energy	GE Mcal/kg			
Cattle	DE Mcal/kg	* 0.50	2.97	
Sheep	DE Mcal/kg	* 0.49	2.91	
Cattle	ME Mcal/kg	* 0.41	2.44	
Sheep	ME Mcal/kg	* 0.40	2.38	
Cattle	TDN %	* 11.4	67.4	
Sheep	TDN %	* 11.1	65.9	

Corn, pop, grain, (4)
Ref No 4 02 964 United States

Feed Name or Analyses		Mean As Fed	Mean Dry	C.V. ± %
Dry matter	%	89.8	100.0	1
Ash	%	1.5	1.7	
Crude fiber	%	1.9	2.1	
Cattle	dig coef %	* 19.	19.	
Sheep	dig coef %	* 30.	30.	
Swine	dig coef %	* 47.	47.	
Ether extract	%	5.0	5.6	
Cattle	dig coef %	* 87.	87.	
Sheep	dig coef %	* 87.	87.	
Swine	dig coef %	* 70.	70.	
N-free extract	%	69.9	77.9	
Cattle	dig coef %	* 91.	91.	
Sheep	dig coef %	* 99.	99.	
Swine	dig coef %	* 93.	93.	
Protein (N x 6.25)	%	11.5	12.8	
Cattle	dig coef %	* 75.	75.	
Sheep	dig coef %	* 78.	78.	
Swine	dig coef %	* 80.	80.	
Cattle	dig prot %	8.6	9.6	
Goats	dig prot %	* 8.0	8.9	
Horses	dig prot %	* 8.0	8.9	
Sheep	dig prot %	8.9	10.0	
Swine	dig prot %	9.2	10.2	
Energy	GE Mcal/kg			
Cattle	DE Mcal/kg	* 3.63	4.04	
Sheep	DE Mcal/kg	* 3.90	4.35	
Swine	DE kcal/kg	*3658.	4074.	
Cattle	ME Mcal/kg	* 2.98	3.32	
Sheep	ME Mcal/kg	* 3.20	3.56	
Swine	ME kcal/kg	*3417.	3806.	
Cattle	TDN %	82.4	91.7	
Sheep	TDN %	88.5	98.6	
Swine	TDN %	83.0	92.4	
Copper	mg/kg	2.4	2.6	
Iron	%	0.003	0.003	23
Phosphorus	%	0.30	0.33	11
Biotin	mg/kg	0.08	0.09	
Niacin	mg/kg	17.2	19.2	9
Pantothenic acid	mg/kg	3.4	3.7	19
Riboflavin	mg/kg	1.2	1.3	36
Vitamin B6	mg/kg	4.35	4.85	

CORN, SWEET. *Zea mays saccharata*

Corn, sweet, aerial part, s-c, (1)
Ref No 1 08 407 United States

Feed Name or Analyses		Mean As Fed	Mean Dry	C.V. ± %
Dry matter	%	87.7	100.0	
Ash	%	9.0	10.3	
Crude fiber	%	26.4	30.1	
Ether extract	%	1.8	2.1	
N-free extract	%	41.3	47.1	
Protein (N x 6.25)	%	9.2	10.5	
Cattle	dig prot %	* 5.3	6.0	
Goats	dig prot %	* 5.6	6.3	
Horses	dig prot %	* 5.6	6.4	
Rabbits	dig prot %	* 5.9	6.8	
Sheep	dig prot %	* 5.2	6.0	
Energy	GE Mcal/kg			
Cattle	DE Mcal/kg	* 2.51	2.86	
Sheep	DE Mcal/kg	* 2.42	2.76	
Cattle	ME Mcal/kg	* 2.06	2.35	
Sheep	ME Mcal/kg	* 1.98	2.26	
Cattle	TDN %	* 57.0	65.0	
Sheep	TDN %	* 54.9	62.6	
Phosphorus	%	0.17	0.19	
Potassium	%	0.98	1.12	

Corn, sweet, pollen, (1)
Ref No 1 02 965 United States

Feed Name or Analyses		Mean As Fed	Mean Dry	C.V. ± %
Dry matter	%		100.0	
Niacin	mg/kg		68.1	18
Pantothenic acid	mg/kg		11.2	

Corn, sweet, aerial part, fresh, (2)
Ref No 2 02 968 United States

Feed Name or Analyses		Mean As Fed	Mean Dry	C.V. ± %
Dry matter	%	20.2	100.0	51
Ash	%	1.3	6.5	16
Crude fiber	%	4.4	22.0	11
Sheep	dig coef %	64.	64.	
Ether extract	%	0.6	3.2	10
Sheep	dig coef %	75.	75.	
N-free extract	%	11.8	58.4	
Sheep	dig coef %	77.	77.	
Protein (N x 6.25)	%	2.0	10.1	14
Sheep	dig coef %	64.	64.	
Cattle	dig prot %	* 1.3	6.4	
Goats	dig prot %	* 1.2	5.9	
Horses	dig prot %	* 1.2	6.1	
Rabbits	dig prot %	* 1.3	6.4	
Sheep	dig prot %	1.3	6.4	
Energy	GE Mcal/kg			
Cattle	DE Mcal/kg	* 0.58	2.88	
Sheep	DE Mcal/kg	* 0.63	3.12	
Cattle	ME Mcal/kg	* 0.48	2.36	
Sheep	ME Mcal/kg	* 0.52	2.56	
Cattle	TDN %	* 13.2	65.4	
Sheep	TDN %	14.3	70.7	

Corn, sweet, aerial part, fresh, late bloom, (2)
Ref No 2 08 409 United States

Feed Name or Analyses		Mean As Fed	Mean Dry	C.V. ± %
Dry matter	%	10.0	100.0	
Ash	%	1.0	10.0	
Crude fiber	%	2.5	25.0	
Ether extract	%	0.3	3.0	
N-free extract	%	5.2	52.0	
Protein (N x 6.25)	%	1.0	10.0	
Cattle	dig prot %	* 0.6	6.4	
Goats	dig prot %	* 0.6	5.9	
Horses	dig prot %	* 0.6	6.0	
Rabbits	dig prot %	* 0.6	6.4	
Sheep	dig prot %	* 0.6	6.3	
Energy	GE Mcal/kg			
Cattle	DE Mcal/kg	* 0.27	2.69	
Sheep	DE Mcal/kg	* 0.30	2.99	
Cattle	ME Mcal/kg	* 0.22	2.20	
Sheep	ME Mcal/kg	* 0.25	2.45	
Cattle	TDN %	* 6.1	61.0	
Sheep	TDN %	* 6.8	67.9	
Phosphorus	%	0.02	0.20	
Potassium	%	0.16	1.60	

Corn, sweet, aerial part, fresh, milk stage, (2)
Ref No 2 02 966 United States

Feed Name or Analyses		Mean As Fed	Mean Dry	C.V. ± %
Dry matter	%	20.2	100.0	16
Ash	%	1.2	5.9	7
Crude fiber	%	4.1	20.3	8
Sheep	dig coef %	60.	60.	
Ether extract	%	0.6	3.2	8
Sheep	dig coef %	75.	75.	
N-free extract	%	12.5	61.8	
Sheep	dig coef %	77.	77.	
Protein (N x 6.25)	%	1.8	8.8	7
Sheep	dig coef %	62.	62.	
Cattle	dig prot %	* 1.1	5.4	
Goats	dig prot %	* 1.0	4.8	
Horses	dig prot %	* 1.0	5.0	
Rabbits	dig prot %	* 1.1	5.5	
Sheep	dig prot %	1.1	5.5	
Energy	GE Mcal/kg			
Cattle	DE Mcal/kg	* 0.57	2.84	
Sheep	DE Mcal/kg	* 0.63	3.11	
Cattle	ME Mcal/kg	* 0.47	2.33	
Sheep	ME Mcal/kg	* 0.52	2.55	
Cattle	TDN %	* 13.0	64.5	
Sheep	TDN %	14.3	70.6	

Corn, sweet, aerial part, fresh, mature, (2)
Ref No 2 02 967 United States

Feed Name or Analyses		Mean As Fed	Mean Dry	C.V. ± %
Dry matter	%	16.9	100.0	
Ash	%	1.2	7.2	
Crude fiber	%	3.9	23.2	
Sheep	dig coef %	75.	75.	
Ether extract	%	0.5	3.0	
Sheep	dig coef %	74.	74.	
N-free extract	%	9.3	54.9	
Sheep	dig coef %	80.	80.	
Protein (N x 6.25)	%	2.0	11.7	
Sheep	dig coef %	78.	78.	
Cattle	dig prot %	* 1.3	7.8	
Goats	dig prot %	* 1.3	7.5	
Horses	dig prot %	* 1.3	7.4	
Rabbits	dig prot %	* 1.3	7.7	

(1) dry forages and roughages (2) pasture, range plants, and forages fed green (3) silages (4) energy feeds (5) protein supplements (6) minerals (7) vitamins (8) additives

Feed Name or Analyses		Mean As Fed	Mean Dry	C.V. ± %
Sheep	dig prot %	1.5	9.1	
Energy	GE Mcal/kg			
Cattle	DE Mcal/kg *	0.50	2.94	
Sheep	DE Mcal/kg *	0.56	3.33	
Cattle	ME Mcal/kg *	0.41	2.41	
Sheep	ME Mcal/kg *	0.46	2.73	
Cattle	TDN % *	11.3	66.7	
Sheep	TDN %	12.8	75.5	
Phosphorus	%	0.03	0.20	
Potassium	%	0.27	1.58	

Corn, sweet, aerial part wo ears wo husks, fresh, (2)

Ref No 2 02 969 United States

Feed Name or Analyses		Mean As Fed	Mean Dry	C.V. ± %
Dry matter	%	22.0	100.0	
Ash	%	1.4	6.2	
Crude fiber	%	5.7	26.0	
Ether extract	%	0.4	1.8	
N-free extract	%	12.9	58.6	
Protein (N x 6.25)	%	1.6	7.3	
Cattle	dig prot % *	0.9	4.1	
Goats	dig prot % *	0.7	3.4	
Horses	dig prot % *	0.8	3.7	
Rabbits	dig prot % *	1.0	4.3	
Sheep	dig prot % *	0.8	3.8	
Energy	GE Mcal/kg			
Cattle	DE Mcal/kg *	0.64	2.89	
Sheep	DE Mcal/kg *	0.59	2.69	
Cattle	ME Mcal/kg *	0.52	2.37	
Sheep	ME Mcal/kg *	0.49	2.21	
Cattle	TDN % *	14.4	65.5	
Sheep	TDN % *	13.4	61.0	

Corn, sweet, cannery residue, fresh, (2)

Corn, sweet, cannery refuse

Ref No 2 02 975 United States

Feed Name or Analyses		Mean As Fed	Mean Dry	C.V. ± %
Dry matter	%	76.6	100.0	38
Ash	%	2.5	3.3	22
Crude fiber	%	17.2	22.5	27
Cattle	dig coef % *	70.	70.	
Ether extract	%	1.8	2.3	19
Cattle	dig coef % *	87.	87.	
N-free extract	%	48.3	63.1	
Cattle	dig coef % *	71.	71.	
Protein (N x 6.25)	%	6.8	8.9	69
Cattle	dig coef % *	56.	56.	
Cattle	dig prot %	3.8	5.0	
Goats	dig prot % *	3.7	4.8	
Horses	dig prot % *	3.9	5.1	
Rabbits	dig prot % *	4.2	5.5	
Sheep	dig prot % *	4.0	5.3	
Energy	GE Mcal/kg			
Cattle	DE Mcal/kg *	2.37	3.09	
Sheep	DE Mcal/kg *	2.08	2.72	
Cattle	ME Mcal/kg *	1.94	2.53	
Sheep	ME Mcal/kg *	1.71	2.23	
Cattle	NE_m Mcal/kg *	1.19	1.56	
Cattle	NE_{gain} Mcal/kg *	0.76	0.99	
Cattle	$NE_{lactating\ cows}$ Mcal/kg *	1.06	1.39	
Cattle	TDN %	53.7	70.1	
Sheep	TDN % *	47.3	61.7	
Phosphorus	%	0.69	0.90	
Carotene	mg/kg	10.4	13.5	
Vitamin A equivalent	IU/g	17.3	22.5	

Corn, sweet, leaves, fresh, (2)

Ref No 2 02 970 United States

Feed Name or Analyses		Mean As Fed	Mean Dry	C.V. ± %
Dry matter	%	25.0	100.0	
Crude fiber	%	6.7	26.6	
Ether extract	%	1.4	5.5	
Protein (N x 6.25)	%	4.3	17.1	
Cattle	dig prot % *	3.1	12.4	
Goats	dig prot % *	3.1	12.5	
Horses	dig prot % *	3.0	12.0	
Rabbits	dig prot % *	3.0	11.9	
Sheep	dig prot % *	3.2	12.9	
Carotene	mg/kg	144.5	578.0	
Riboflavin	mg/kg	1.4	5.5	
Vitamin A equivalent	IU/g	240.9	963.6	

Corn, sweet, leaves w stalks, fresh, immature, (2)

Ref No 2 02 971 United States

Feed Name or Analyses		Mean As Fed	Mean Dry	C.V. ± %
Dry matter	%	20.9	100.0	15
Niacin	mg/kg	9.0	43.0	11
Pantothenic acid	mg/kg	2.4	11.7	36

Corn, sweet, leaves w stalks, fresh, mature, (2)

Ref No 2 02 972 United States

Feed Name or Analyses		Mean As Fed	Mean Dry	C.V. ± %
Dry matter	%		100.0	
Niacin	mg/kg		13.2	23

Corn, sweet, tassels, fresh, immature, (2)

Ref No 2 02 973 United States

Feed Name or Analyses		Mean As Fed	Mean Dry	C.V. ± %
Dry matter	%	40.0	100.0	26
Niacin	mg/kg	15.8	39.5	23
Pantothenic acid	mg/kg	9.3	23.1	26

Corn, sweet, aerial part, ensiled, (3)

Ref No 3 02 974 United States

Feed Name or Analyses		Mean As Fed	Mean Dry	C.V. ± %
Dry matter	%	21.2	100.0	12
Ash	%	1.5	7.1	35
Crude fiber	%	5.8	27.2	5
Sheep	dig coef %	71.	71.	
Ether extract	%	0.9	4.3	20
Sheep	dig coef %	83.	83.	
N-free extract	%	10.8	51.0	
Sheep	dig coef %	72.	72.	
Protein (N x 6.25)	%	2.2	10.4	9
Sheep	dig coef %	54.	54.	
Cattle	dig prot % *	1.2	5.7	
Goats	dig prot % *	1.2	5.7	
Horses	dig prot % *	1.2	5.7	
Sheep	dig prot %	1.2	5.6	
Energy	GE Mcal/kg			
Cattle	DE Mcal/kg *	0.61	2.87	
Sheep	DE Mcal/kg *	0.65	3.07	
Cattle	ME Mcal/kg *	0.50	2.35	
Sheep	ME Mcal/kg *	0.54	2.52	
Cattle	TDN % *	13.8	65.1	
Sheep	TDN %	14.8	69.7	

Corn, sweet, ears w husks, fresh, (4)

Ref No 4 08 408 United States

Feed Name or Analyses		Mean As Fed	Mean Dry	C.V. ± %
Dry matter	%	37.8	100.0	
Ash	%	0.9	2.4	
Crude fiber	%	4.3	11.4	
Ether extract	%	2.6	6.9	
N-free extract	%	26.2	69.3	
Protein (N x 6.25)	%	3.8	10.1	
Cattle	dig prot % *	2.0	5.2	
Goats	dig prot % *	2.4	6.4	
Horses	dig prot % *	2.4	6.4	
Sheep	dig prot % *	2.4	6.4	
Energy	GE Mcal/kg			
Cattle	DE Mcal/kg *	1.32	3.49	
Sheep	DE Mcal/kg *	1.35	3.58	
Swine	DE kcal/kg	*1151.	3046.	
Cattle	ME Mcal/kg *	1.08	2.86	
Sheep	ME Mcal/kg *	1.11	2.93	
Swine	ME kcal/kg	*1082.	2862.	
Cattle	TDN % *	29.9	79.2	
Sheep	TDN % *	30.7	81.1	
Swine	TDN % *	26.1	69.1	

Corn, sweet, grain, (4)

Ref No 4 02 977 United States

Feed Name or Analyses		Mean As Fed	Mean Dry	C.V. ± %
Dry matter	%	90.9	100.0	1
Ash	%	1.9	2.1	18
Crude fiber	%	2.5	2.8	36
Cattle	dig coef % *	19.	19.	
Sheep	dig coef % *	30.	30.	
Swine	dig coef % *	47.	47.	
Ether extract	%	7.9	8.7	25
Cattle	dig coef % *	87.	87.	
Sheep	dig coef % *	87.	87.	
Swine	dig coef % *	70.	70.	
N-free extract	%	66.9	73.6	
Cattle	dig coef % *	91.	91.	
Sheep	dig coef % *	99.	99.	
Swine	dig coef % *	93.	93.	
Protein (N x 6.25)	%	11.6	12.8	13
Cattle	dig coef % *	75.	75.	
Sheep	dig coef % *	78.	78.	
Swine	dig coef % *	80.	80.	
Cattle	dig prot %	8.7	9.6	
Goats	dig prot % *	8.1	9.0	
Horses	dig prot % *	8.1	9.0	
Sheep	dig prot %	9.1	10.0	
Swine	dig prot %	9.3	10.2	
Energy	GE Mcal/kg			
Cattle	DE Mcal/kg *	3.77	4.15	
Sheep	DE Mcal/kg *	4.04	4.44	
Swine	DE kcal/kg	*3756.	4132.	
Cattle	ME Mcal/kg *	3.09	3.40	
Sheep	ME Mcal/kg *	3.31	3.64	
Swine	ME kcal/kg	*3508.	3860.	
Cattle	TDN %	85.6	94.1	
Sheep	TDN %	91.6	100.7	
Swine	TDN %	85.2	93.7	
Calcium	%	0.01	0.01	
Copper	mg/kg	5.4	6.0	
Iron	%	0.005	0.005	
Phosphorus	%	0.41	0.45	

Corn, sweet, grain, milk stage, (4)

Ref No 4 02 976 United States

Feed Name or Analyses		Mean As Fed	Mean Dry	C.V. ± %
Dry matter	%	29.7	100.0	29
Niacin	mg/kg	19.0	63.9	27
Pantothenic acid	mg/kg	6.7	22.7	35

Corn, sweet, grain, mature, (4)

Ref No 4 08 410 United States

Feed Name or Analyses		Mean As Fed	Mean Dry	C.V. ± %
Dry matter	%	90.7	100.0	
Ash	%	1.8	2.0	
Crude fiber	%	2.4	2.6	
Ether extract	%	7.9	8.7	
N-free extract	%	67.1	74.0	
Protein (N x 6.25)	%	11.5	12.7	
Cattle	dig prot %	* 6.9	7.7	
Goats	dig prot %	* 8.0	8.9	
Horses	dig prot %	* 8.0	8.9	
Sheep	dig prot %	* 8.0	8.9	
Energy	GE Mcal/kg			
Cattle	DE Mcal/kg	* 3.58	3.95	
Sheep	DE Mcal/kg	* 3.55	3.91	
Swine	DE kcal/kg	*3696.	4075.	
Cattle	ME Mcal/kg	* 2.94	3.24	
Sheep	ME Mcal/kg	* 2.91	3.21	
Swine	ME kcal/kg	*3453.	3807.	
Cattle	TDN %	* 81.2	89.5	
Sheep	TDN %	* 80.4	88.6	
Swine	TDN %	* 83.8	92.4	

CORN, WARWICK 210. Zea mays

Corn, warwick 210, aerial part w limestone preservative added, ensiled, milk stage, (3)

Ref No 3 07 914 Canada

Feed Name or Analyses		Mean As Fed	Mean Dry	C.V. ± %
Dry matter	%	19.8	100.0	
Ash	%	2.1	10.8	
Crude fiber	%	6.7	33.8	
pH	pH units	3.70	3.70	

CORN, WHITE. Zea mays

Corn, white, bran, wet or dry milled dehy, (4)

Ref No 4 02 978 United States

Feed Name or Analyses		Mean As Fed	Mean Dry	C.V. ± %
Dry matter	%	88.3	100.0	
Ash	%	1.7	1.9	18
Crude fiber	%	9.1	10.3	
Ether extract	%	4.0	4.5	46
N-free extract	%	66.4	75.2	
Protein (N x 6.25)	%	7.1	8.0	21
Cattle	dig prot %	* 3.0	3.4	
Goats	dig prot %	* 4.1	4.6	
Horses	dig prot %	* 4.1	4.6	
Sheep	dig prot %	* 4.1	4.6	
Energy	GE Mcal/kg			
Cattle	DE Mcal/kg	* 3.00	3.40	
Sheep	DE Mcal/kg	* 2.81	3.18	
Swine	DE kcal/kg	*2777.	3145.	
Cattle	ME Mcal/kg	* 2.46	2.79	
Sheep	ME Mcal/kg	* 2.30	2.61	
Swine	ME kcal/kg	*2620.	2968.	
Cattle	TDN %	* 68.1	77.1	
Sheep	TDN %	* 63.8	73.2	
Swine	TDN %	* 63.0	71.3	
Biotin	mg/kg	0.07	0.07	26
Niacin	mg/kg	37.0	41.9	43
Pantothenic acid	mg/kg	5.7	6.5	
Riboflavin	mg/kg	1.5	1.7	45
Thiamine	mg/kg	5.1	5.7	27
Vitamin B_6	mg/kg	7.50	8.49	7

Corn, white, grain, grnd, (4)

Corn, white, meal

Ref No 4 02 979 United States

Feed Name or Analyses		Mean As Fed	Mean Dry	C.V. ± %
Dry matter	%	87.9	100.0	2
Ash	%	0.6	0.7	78
Crude fiber	%	0.6	0.7	17
Cattle	dig coef %	* 19.	19.	
Sheep	dig coef %	* 30.	30.	
Swine	dig coef %	* 47.	47.	
Ether extract	%	1.2	1.4	22
Cattle	dig coef %	* 87.	87.	
Sheep	dig coef %	* 87.	87.	
Swine	dig coef %	* 70.	70.	
N-free extract	%	76.9	87.5	
Cattle	dig coef %	* 91.	91.	
Sheep	dig coef %	* 99.	99.	
Swine	dig coef %	* 93.	93.	
Protein (N x 6.25)	%	8.6	9.8	7
Cattle	dig coef %	* 75.	75.	
Sheep	dig coef %	* 78.	78.	
Swine	dig coef %	* 80.	80.	
Cattle	dig prot %	6.4	7.3	
Goats	dig prot %	* 5.5	6.2	
Horses	dig prot %	* 5.5	6.2	
Sheep	dig prot %	6.7	7.6	
Swine	dig prot %	6.9	7.8	
Energy	GE Mcal/kg			
Cattle	DE Mcal/kg	* 3.48	3.96	
Sheep	DE Mcal/kg	* 3.76	4.28	
Swine	DE kcal/kg	*3552.	4041.	
Cattle	ME Mcal/kg	* 2.85	3.25	
Chickens	ME_n kcal/kg	3500.	3982.	
Sheep	ME Mcal/kg	* 3.09	3.51	
Swine	ME kcal/kg	*3340.	3800.	
Cattle	TDN %	78.9	89.8	
Sheep	TDN %	85.4	97.1	
Swine	TDN %	80.6	91.7	
Calcium	%	0.02	0.02	
Phosphorus	%	0.16	0.18	92
Niacin	mg/kg	17.6	20.1	
Riboflavin	mg/kg	0.9	1.0	
Thiamine	mg/kg	1.5	1.8	30
Tyrosine	%	0.40	0.46	

Corn, white, grain, Can 1 CE mn wt 56 lb per bushel mx 2% foreign material, (4)

Ref No 4 02 980 United States

Feed Name or Analyses		Mean As Fed	Mean Dry	C.V. ± %
Dry matter	%		100.0	
Crude fiber	%			
Cattle	dig coef %	*	19.	
Sheep	dig coef %	*	30.	
Swine	dig coef %	*	47.	
Ether extract	%			
Cattle	dig coef %	*	87.	
Sheep	dig coef %	*	87.	
Swine	dig coef %	*	70.	
N-free extract	%			
Cattle	dig coef %	*	91.	
Sheep	dig coef %	*	99.	
Swine	dig coef %	*	93.	
Protein (N x 6.25)	%			
Cattle	dig coef %	*	75.	
Sheep	dig coef %	*	78.	
Swine	dig coef %	*	80.	
Energy	GE Mcal/kg			
Cattle	DE Mcal/kg	*	4.01	
Sheep	DE Mcal/kg	*	4.33	
Swine	DE kcal/kg	*	4065.	
Cattle	ME Mcal/kg	*	3.29	
Sheep	ME Mcal/kg	*	3.55	
Cattle	TDN %	*	91.0	
Sheep	TDN %	*	98.2	
Swine	TDN %	*	92.2	

Ref No 4 02 980 Canada

Feed Name or Analyses		Mean As Fed	Mean Dry	C.V. ± %
Dry matter	%	86.5	100.0	
Ash	%	1.1	1.3	
Crude fiber	%	2.4	2.8	
Ether extract	%	3.8	4.4	
N-free extract	%	71.4	82.5	
Protein (N x 6.25)	%	7.8	9.0	
Cattle	dig prot %	* 3.7	4.3	
Goats	dig prot %	* 4.8	5.5	
Horses	dig prot %	* 4.8	5.5	
Sheep	dig prot %	* 4.8	5.5	
Energy	GE Mcal/kg			
Cattle	DE Mcal/kg	* 3.47	4.01	
Sheep	DE Mcal/kg	* 3.74	4.33	
Swine	DE kcal/kg	*3517.	4065.	
Cattle	ME Mcal/kg	* 2.85	3.29	
Sheep	ME Mcal/kg	* 3.07	3.55	
Swine	ME kcal/kg	*3312.	3829.	
Cattle	TDN %	78.7	91.0	
Sheep	TDN %	84.9	98.2	
Swine	TDN %	79.8	92.2	

Corn, white, grain, Can 2 CE mn wt 54 lb per bushel mx 3% foreign material, (4)

Ref No 4 02 981 United States

Feed Name or Analyses		Mean As Fed	Mean Dry	C.V. ± %
Dry matter	%		100.0	
Crude fiber	%			
Cattle	dig coef %	*	19.	
Sheep	dig coef %	*	30.	
Swine	dig coef %	*	47.	
Ether extract	%			
Cattle	dig coef %	*	87.	
Sheep	dig coef %	*	87.	
Swine	dig coef %	*	70.	
N-free extract	%			
Cattle	dig coef %	*	91.	
Sheep	dig coef %	*	99.	
Swine	dig coef %	*	93.	
Protein (N x 6.25)	%			
Cattle	dig coef %	*	75.	
Sheep	dig coef %	*	78.	
Swine	dig coef %	*	80.	
Energy	GE Mcal/kg			
Cattle	DE Mcal/kg	*	3.99	
Sheep	DE Mcal/kg	*	4.30	

(1) dry forages and roughages (2) pasture, range plants, and forages fed green (3) silages (4) energy feeds (5) protein supplements (6) minerals (7) vitamins (8) additives

Feed Name or Analyses		Mean As Fed	Mean Dry	C.V. ± %
Swine	DE kcal/kg	*	4046.	
Cattle	ME Mcal/kg	*	3.27	
Sheep	ME Mcal/kg	*	3.53	
Cattle	TDN %	*	90.4	
Sheep	TDN %	*	97.6	
Swine	TDN %	*	91.8	

Ref No 4 02 981 Canada

Feed Name or Analyses		Mean As Fed	Mean Dry	C.V. ± %
Dry matter	%	86.5	100.0	
Ash	%	1.2	1.4	
Crude fiber	%	2.8	3.2	
Ether extract	%	3.7	4.3	
N-free extract	%	70.6	81.7	
Protein (N x 6.25)	%	8.2	9.4	
Cattle	dig prot %	* 4.0	4.7	
Goats	dig prot %	* 5.1	5.9	
Horses	dig prot %	* 5.1	5.9	
Sheep	dig prot %	* 5.1	5.9	
Energy	GE Mcal/kg			
Cattle	DE Mcal/kg	* 3.45	3.99	
Sheep	DE Mcal/kg	* 3.72	4.30	
Swine	DE kcal/kg	*3499.	4046.	
Cattle	ME Mcal/kg	* 2.83	3.27	
Sheep	ME Mcal/kg	* 3.05	3.53	
Swine	ME kcal/kg	*3293.	3807.	
Cattle	TDN %	78.2	90.4	
Sheep	TDN %	84.4	97.6	
Swine	TDN %	79.4	91.8	

Corn, white, grain wo germ, grnd, (4)

Degermed white corn meal

Ref No 4 02 988 United States

Feed Name or Analyses		Mean As Fed	Mean Dry	C.V. ± %
Dry matter	%	88.7	100.0	1
Ash	%	1.1	1.2	
Crude fiber	%	0.8	0.9	
Cattle	dig coef %	* 19.	19.	
Sheep	dig coef %	* 30.	30.	
Ether extract	%	1.2	1.3	
Cattle	dig coef %	* 87.	87.	
Sheep	dig coef %	* 87.	87.	
N-free extract	%	76.7	86.5	
Cattle	dig coef %	* 91.	91.	
Sheep	dig coef %	* 99.	99.	
Protein (N x 6.25)	%	8.9	10.0	
Cattle	dig coef %	* 75.	75.	
Sheep	dig coef %	* 78.	78.	
Cattle	dig prot %	6.7	7.5	
Goats	dig prot %	* 5.7	6.4	
Horses	dig prot %	* 5.7	6.4	
Sheep	dig prot %	6.9	7.8	
Energy	GE Mcal/kg			
Cattle	DE Mcal/kg	* 3.48	3.92	
Sheep	DE Mcal/kg	* 3.77	4.25	
Swine	DE kcal/kg	*3724.	4199.	
Cattle	ME Mcal/kg	* 2.85	3.22	
Sheep	ME Mcal/kg	* 3.09	3.48	
Swine	ME kcal/kg	*3500.	3946.	
Cattle	TDN %	78.9	89.0	
Sheep	TDN %	85.4	96.3	
Swine	TDN %	* 84.5	95.2	
Calcium	%	0.01	0.01	
Iron	%	0.002	0.002	
Phosphorus	%	0.15	0.17	
Niacin	mg/kg	10.3	11.6	
Pantothenic acid	mg/kg	3.1	3.5	

Feed Name or Analyses		Mean As Fed	Mean Dry	C.V. ± %
Riboflavin	mg/kg	0.7	0.7	53
Thiamine	mg/kg	1.1	1.2	42

Corn, white, grits, cracked fine screened, (4)

Corn, white, flour

Corn, white, hominy grits

Ref No 4 02 989 United States

Feed Name or Analyses		Mean As Fed	Mean Dry	C.V. ± %
Dry matter	%	88.1	100.0	2
Ash	%	0.6	0.7	28
Crude fiber	%	0.6	0.7	14
Ether extract	%	1.8	2.0	17
N-free extract	%	77.0	87.4	
Protein (N x 6.25)	%	8.1	9.2	16
Cattle	dig prot %	* 3.9	4.5	
Goats	dig prot %	* 5.0	5.7	
Horses	dig prot %	* 5.0	5.7	
Sheep	dig prot %	* 5.0	5.7	
Energy	GE Mcal/kg			
Cattle	DE Mcal/kg	* 3.19	3.62	
Sheep	DE Mcal/kg	* 3.42	3.88	
Swine	DE kcal/kg	*3767.	4278.	
Cattle	ME Mcal/kg	* 2.62	2.97	
Sheep	ME Mcal/kg	* 2.80	3.18	
Swine	ME kcal/kg	*3546.	4027.	
Cattle	TDN %	* 72.3	82.2	
Sheep	TDN %	* 77.5	88.0	
Swine	TDN %	* 85.4	97.0	
Calcium	%	0.01	0.01	
Copper	mg/kg	2.0	2.3	
Iron	%	0.001	0.001	
Phosphorus	%	0.12	0.14	
Pantothenic acid	mg/kg	8.8	10.0	
Riboflavin	mg/kg	0.4	0.5	
Thiamine	mg/kg	1.5	1.8	
Arginine	%	0.40	0.45	9
Histidine	%	0.20	0.23	19
Isoleucine	%	0.40	0.45	19
Leucine	%	0.90	1.02	4
Lysine	%	0.40	0.45	9
Methionine	%	0.10	0.11	
Phenylalanine	%	0.40	0.45	
Threonine	%	0.30	0.34	13
Tryptophan	%	0.10	0.11	
Tyrosine	%	0.40	0.45	9
Valine	%	0.40	0.45	9

Corn, white, grits by-product, mn 5% fat, (4)

White hominy feed (AAFCO)

White hominy feed (CFA)

Hominy, white corn, feed

Corn, white, hominy feed

Ref No 4 02 990 United States

Feed Name or Analyses		Mean As Fed	Mean Dry	C.V. ± %
Dry matter	%	90.0	100.0	0
Ash	%	3.0	3.3	22
Crude fiber	%	5.1	5.7	7
Ether extract	%	8.2	9.1	15
N-free extract	%	63.1	70.1	
Protein (N x 6.25)	%	10.5	11.7	5
Cattle	dig prot %	* 6.1	6.8	
Goats	dig prot %	* 7.2	8.0	
Horses	dig prot %	* 7.2	8.0	
Sheep	dig prot %	* 7.2	8.0	
Energy	GE Mcal/kg			
Cattle	DE Mcal/kg	* 3.44	3.82	
Sheep	DE Mcal/kg	* 3.34	3.71	
Swine	DE kcal/kg	*3236.	3597.	

Feed Name or Analyses		Mean As Fed	Mean Dry	C.V. ± %
Cattle	ME Mcal/kg	* 2.82	3.14	
Sheep	ME Mcal/kg	* 2.74	3.05	
Swine	ME kcal/kg	*3030.	3368.	
Cattle	TDN %	* 78.0	86.7	
Sheep	TDN %	* 75.8	84.2	
Swine	TDN %	* 73.4	81.6	
Calcium	%	0.02	0.02	69
Cobalt	mg/kg	0.020	0.022	
Copper	mg/kg	13.3	14.7	53
Iron	%	0.007	0.008	36
Manganese	mg/kg	13.9	15.5	14
Phosphorus	%	0.58	0.65	37
Biotin	mg/kg	0.13	0.15	26
Niacin	mg/kg	46.4	51.6	20
Pantothenic acid	mg/kg	7.3	8.1	7
Riboflavin	mg/kg	2.2	2.5	27
Thiamine	mg/kg	9.7	10.8	40
Vitamin B_6	mg/kg	13.25	14.73	16
Arginine	%	0.40	0.45	
Histidine	%	0.20	0.22	
Isoleucine	%	0.30	0.33	
Leucine	%	0.90	1.00	
Lysine	%	0.40	0.45	
Methionine	%	0.10	0.11	
Phenylalanine	%	0.40	0.45	
Threonine	%	0.30	0.33	
Tryptophan	%	0.10	0.11	
Tyrosine	%	0.40	0.45	
Valine	%	0.40	0.45	

CORN, YELLOW. Zea mays

Corn, yellow, bran, wet or dry milled dehy, (4)

Ref No 4 02 991 United States

Feed Name or Analyses		Mean As Fed	Mean Dry	C.V. ± %
Dry matter	%	90.0	100.0	
Ash	%	1.4	1.6	23
Ether extract	%	3.3	3.7	24
Protein (N x 6.25)	%	7.3	8.1	2
Cattle	dig prot %	* 3.1	3.5	
Goats	dig prot %	* 4.2	4.7	
Horses	dig prot %	* 4.2	4.7	
Sheep	dig prot %	* 4.2	4.7	
Biotin	mg/kg	0.11	0.12	48
Niacin	mg/kg	47.2	52.4	7
Pantothenic acid	mg/kg	4.9	5.4	14
Riboflavin	mg/kg	1.3	1.5	
Thiamine	mg/kg	6.0	6.6	56
Vitamin B_6	mg/kg	8.82	9.80	24

Corn, yellow, grain, (4)

Ref No 4 07 911 United States

Feed Name or Analyses		Mean As Fed	Mean Dry	C.V. ± %
Dry matter	%	87.0	100.0	
Ash	%	1.4	1.6	
Crude fiber	%	2.5	2.9	
Ether extract	%	3.4	3.9	
N-free extract	%	71.3	82.0	
Protein (N x 6.25)	%	8.3	9.6	
Cattle	dig prot %	* 4.2	4.8	
Goats	dig prot %	* 5.2	6.0	
Horses	dig prot %	* 5.2	6.0	
Sheep	dig prot %	* 5.2	6.0	
Energy	GE Mcal/kg			
Cattle	DE Mcal/kg	* 3.27	3.76	
Sheep	DE Mcal/kg	* 3.31	3.81	
Swine	DE kcal/kg	*3461.	3980.	

Continued

Feed Name or Analyses			Mean As Fed	Mean Dry	C.V. ± %
Cattle	ME Mcal/kg	*	2.68	3.08	
Chickens	ME_n kcal/kg		3367.	3870.	
Sheep	ME Mcal/kg	*	2.72	3.13	
Swine	ME kcal/kg	*	3256.	3743.	
Cattle	TDN %	*	74.2	85.3	
Sheep	TDN %	*	75.2	86.4	
Swine	TDN %	*	78.5	90.3	
Calcium	%		0.02	0.02	
Phosphorus	%		0.27	0.31	
Carotene	mg/kg		1.4	1.7	
Choline	mg/kg		436.	502.	
Niacin	mg/kg		21.4	24.6	
Pantothenic acid	mg/kg		5.2	6.0	
Riboflavin	mg/kg		1.1	1.3	
Vitamin A	IU/g		4.4	5.0	
Vitamin A equivalent	IU/g		2.4	2.8	
Arginine	%		0.49	0.57	
Cystine	%		0.17	0.19	
Glycine	%		0.30	0.34	
Lysine	%		0.20	0.23	
Methionine	%		0.18	0.20	
Tryptophan	%		0.09	0.10	
Xanthophylls	mg/kg		10.91	12.54	

Corn, yellow, grain, fan air dried w heat grnd, (4)

Ref No 4 07 800 United States

Feed Name or Analyses			Mean As Fed	Mean Dry	C.V. ± %
Dry matter	%		87.9	100.0	
Crude fiber	%		2.4	2.7	
Ether extract	%		4.0	4.5	
Protein (N x 6.25)	%		8.0	9.1	
Cattle	dig prot %	*	3.8	4.4	
Goats	dig prot %	*	4.9	5.6	
Horses	dig prot %	*	4.9	5.6	
Sheep	dig prot %	*	4.9	5.6	
Carotene	mg/kg		2.7	3.1	
Niacin	mg/kg		24.9	28.3	
Pantothenic acid	mg/kg		5.4	6.1	
Riboflavin	mg/kg		1.3	1.5	
Vitamin A equivalent	IU/g		4.5	5.1	

Corn, yellow, grain, grnd, (4)

Ref No 4 02 992 United States

Feed Name or Analyses			Mean As Fed	Mean Dry	C.V. ± %
Dry matter	%		89.0	100.0	2
Ash	%		1.4	1.6	21
Crude fiber	%		2.2	2.5	24
Cattle	dig coef %	*	19.	19.	
Sheep	dig coef %	*	30.	30.	
Swine	dig coef %	*	47.	47.	
Ether extract	%		3.1	3.5	28
Cattle	dig coef %	*	87.	87.	
Sheep	dig coef %	*	87.	87.	
Swine	dig coef %	*	70.	70.	
N-free extract	%		73.6	82.7	
Cattle	dig coef %	*	91.	91.	
Sheep	dig coef %	*	99.	99.	
Swine	dig coef %	*	93.	93.	
Protein (N x 6.25)	%		8.6	9.7	8
Cattle	dig coef %	*	75.	75.	
Sheep	dig coef %	*	78.	78.	
Swine	dig coef %	*	80.	80.	
Cattle	dig prot %		6.5	7.3	
Goats	dig prot %	*	5.4	6.1	
Horses	dig prot %	*	5.4	6.1	
Sheep	dig prot %		6.7	7.5	
Swine	dig prot %		6.9	7.7	
Energy	GE Mcal/kg		3.98	4.48	
Cattle	DE Mcal/kg	*	3.52	3.96	
Sheep	DE Mcal/kg	*	3.81	4.28	
Swine	DE kcal/kg		3411.	3835.	
Cattle	ME Mcal/kg	*	2.89	3.25	
Sheep	ME Mcal/kg	*	3.12	3.51	
Swine	ME kcal/kg		3349.	3765.	
Cattle	TDN %		79.9	89.9	
Sheep	TDN %		86.3	97.0	
Swine	TDN %		81.3	91.4	
Chlorine	%		0.04	0.05	
Sodium	%		0.00	0.00	
Choline	mg/kg		599.	674.	
Niacin	mg/kg		10.5	11.8	3
Pantothenic acid	mg/kg		6.9	7.8	
Riboflavin	mg/kg		0.7	0.8	46
Thiamine	mg/kg		1.6	1.8	31

Ref No 4 02 992 Canada

Feed Name or Analyses			Mean As Fed	Mean Dry	C.V. ± %
Dry matter	%			100.0	
Crude fiber	%			1.4	
Protein (N x 6.25)	%			8.4	
Swine	dig coef %			84.	
Cattle	dig prot %	*		3.7	
Goats	dig prot %	*		4.9	
Horses	dig prot %	*		4.9	
Sheep	dig prot %	*		4.9	
Swine	dig prot %			7.1	
Energy	GE Mcal/kg			4.48	
Cattle	DE Mcal/kg	*		3.98	
Sheep	DE Mcal/kg	*		4.29	
Swine	DE kcal/kg			3835.	
Cattle	ME Mcal/kg	*		3.26	
Sheep	ME Mcal/kg	*		3.52	
Swine	ME kcal/kg			3765.	
Cattle	TDN %			90.2	
Sheep	TDN %			97.4	
Swine	TDN %			91.8	

Corn, yellow, grain, Can 1 CE mn wt 56 lb per bushel mx 2% foreign material, (4)

Ref No 4 02 993 United States

Feed Name or Analyses			Mean As Fed	Mean Dry	C.V. ± %
Dry matter	%		86.5	100.0	
Crude fiber	%				
Cattle	dig coef %	*	19.	19.	
Sheep	dig coef %	*	30.	30.	
Swine	dig coef %	*	47.	47.	
Ether extract	%				
Cattle	dig coef %	*	87.	87.	
Sheep	dig coef %	*	87.	87.	
Swine	dig coef %	*	70.	70.	
N-free extract	%				
Cattle	dig coef %	*	91.	91.	
Sheep	dig coef %	*	99.	99.	
Swine	dig coef %	*	93.	93.	
Protein (N x 6.25)	%		9.9	11.4	
Cattle	dig coef %	*	75.	75.	
Sheep	dig coef %	*	78.	78.	
Swine	dig coef %	*	80.	80.	
Cattle	dig prot %		7.4	8.6	
Goats	dig prot %	*	6.6	7.7	
Horses	dig prot %	*	6.6	7.7	
Sheep	dig prot %		7.7	8.9	
Swine	dig prot %		7.9	9.1	
Energy	GE Mcal/kg		3.93	4.54	
Cattle	DE Mcal/kg	*	3.46	4.00	
Sheep	DE Mcal/kg	*	3.73	4.32	
Swine	DE kcal/kg	*	3509.	4056.	
Cattle	ME Mcal/kg	*	2.84	3.28	
Chickens	ME_n kcal/kg		3480.	4023.	
Sheep	ME Mcal/kg	*	3.06	3.54	
Swine	ME kcal/kg	*	3288.	3801.	
Cattle	TDN %		78.5	90.8	
Sheep	TDN %		84.7	97.9	
Swine	TDN %		79.6	92.0	

Ref No 4 02 993 Canada

Feed Name or Analyses			Mean As Fed	Mean Dry	C.V. ± %
Dry matter	%		87.8	100.0	3
Ash	%		1.2	1.4	6
Crude fiber	%		2.6	2.9	28
Ether extract	%		4.1	4.6	8
N-free extract	%		70.9	80.7	
Protein (N x 6.25)	%		9.1	10.3	10
Cattle	dig prot %	*	4.8	5.5	
Goats	dig prot %	*	5.9	6.7	
Horses	dig prot %	*	5.9	6.7	
Sheep	dig prot %	*	5.9	6.7	
Cellulose (Matrone)	%		2.1	2.4	
Energy	GE Mcal/kg		3.99	4.54	
Cattle	DE Mcal/kg	*	3.52	4.00	
Sheep	DE Mcal/kg	*	3.79	4.32	
Swine	DE kcal/kg	*	3562.	4056.	
Cattle	ME Mcal/kg	*	2.88	3.28	
Chickens	ME_n kcal/kg		3293.	3750.	
Sheep	ME Mcal/kg	*	3.11	3.54	
Swine	ME kcal/kg	*	3346.	3809.	
Cattle	TDN %		79.8	90.8	
Sheep	TDN %		86.0	97.9	
Swine	TDN %		80.8	92.0	
Copper	mg/kg		2.9	3.3	
Manganese	mg/kg		3.9	4.5	
Selenium	mg/kg		0.067	0.076	
Niacin	mg/kg		33.6	38.2	
Pantothenic acid	mg/kg		7.5	8.6	
Riboflavin	mg/kg		1.2	1.3	
Thiamine	mg/kg		0.8	0.9	
Vitamin B_6	mg/kg		4.26	4.85	
Alanine	%		0.69	0.78	
Arginine	%		0.37	0.42	
Aspartic acid	%		0.62	0.71	
Glutamic acid	%		1.67	1.90	
Glycine	%		0.35	0.40	
Histidine	%		0.21	0.24	
Isoleucine	%		0.21	0.24	
Leucine	%		1.04	1.18	
Lysine	%		0.25	0.28	
Methionine	%		0.15	0.17	
Phenylalanine	%		0.40	0.46	
Proline	%		0.78	0.89	
Serine	%		0.46	0.52	
Threonine	%		0.32	0.37	
Tyrosine	%		0.32	0.36	
Valine	%		0.32	0.36	

Corn, yellow, grain, Can 1 CW mn wt 56 lb per bushel mx 2% foreign material, (4)

Ref No 4 02 994 United States

Feed Name or Analyses			Mean As Fed	Mean Dry	C.V. ± %
Dry matter	%			100.0	
Crude fiber	%				
Cattle	dig coef %	*		19.	

(1) dry forages and roughages (2) pasture, range plants, and forages fed green (3) silages (4) energy feeds (5) protein supplements (6) minerals (7) vitamins (8) additives

Feed Name or Analyses			Mean As Fed	Mean Dry	C.V. ± %
Sheep	dig coef %	*		30.	
Swine	dig coef %	*		47.	
Ether extract	%				
Cattle	dig coef %	*		87.	
Sheep	dig coef %	*		87.	
Swine	dig coef %	*		70.	
N-free extract	%				
Cattle	dig coef %	*		91.	
Sheep	dig coef %	*		99.	
Swine	dig coef %	*		93.	
Protein (N x 6.25)	%				
Cattle	dig coef %	*		75.	
Sheep	dig coef %	*		78.	
Swine	dig coef %	*		80.	
Energy	GE Mcal/kg				
Cattle	DE Mcal/kg	*		3.97	
Sheep	DE Mcal/kg	*		4.28	
Swine	DE kcal/kg	*		4034.	
Cattle	ME Mcal/kg	*		3.26	
Sheep	ME Mcal/kg	*		3.51	
Cattle	TDN %	*		90.0	
Sheep	TDN %	*		97.1	
Swine	TDN %	*		91.5	

Ref No 4 02 994 Canada

Feed Name or Analyses			Mean As Fed	Mean Dry	C.V. ± %
Dry matter	%		86.4	100.0	
Ash	%		1.2	1.4	
Crude fiber	%		3.1	3.6	
Ether extract	%		3.8	4.4	
N-free extract	%		68.9	79.8	
Protein (N x 6.25)	%		9.3	10.8	
Cattle	dig prot %	*	5.1	5.9	
Goats	dig prot %	*	6.1	7.1	
Horses	dig prot %	*	6.1	7.1	
Sheep	dig prot %	*	6.1	7.1	
Energy	GE Mcal/kg				
Cattle	DE Mcal/kg	*	3.43	3.97	
Sheep	DE Mcal/kg	*	3.70	4.28	
Swine	DE kcal/kg	*	3484.	4034.	
Cattle	ME Mcal/kg	*	2.81	3.26	
Sheep	ME Mcal/kg	*	3.03	3.51	
Swine	ME kcal/kg	*	3269.	3785.	
Cattle	TDN %		77.8	90.0	
Sheep	TDN %		83.9	97.1	
Swine	TDN %		79.0	91.5	
Pantothenic acid	mg/kg		5.8	6.7	

Corn, yellow, grain, Can 2 CE mn wt 54 lb per bushel mx 3% foreign material, (4)

Ref No 4 02 995 United States

Feed Name or Analyses			Mean As Fed	Mean Dry	C.V. ± %
Dry matter	%			100.0	
Crude fiber	%				
Cattle	dig coef %	*		19.	
Sheep	dig coef %	*		30.	
Swine	dig coef %	*		47.	
Ether extract	%				
Cattle	dig coef %	*		87.	
Sheep	dig coef %	*		87.	
Swine	dig coef %	*		70.	
N-free extract	%				
Cattle	dig coef %	*		91.	
Sheep	dig coef %	*		99.	
Swine	dig coef %	*		93.	
Protein (N x 6.25)	%				
Cattle	dig coef %	*		75.	
Sheep	dig coef %	*		78.	
Swine	dig coef %	*		80.	
Energy	GE Mcal/kg				
Cattle	DE Mcal/kg	*		4.00	
Sheep	DE Mcal/kg	*		4.32	
Swine	DE kcal/kg	*		4055.	
Cattle	ME Mcal/kg	*		3.28	
Sheep	ME Mcal/kg	*		3.54	
Cattle	TDN %	*		90.8	
Sheep	TDN %	*		97.9	
Swine	TDN %	*		92.0	

Ref No 4 02 995 Canada

Feed Name or Analyses			Mean As Fed	Mean Dry	C.V. ± %
Dry matter	%		87.7	100.0	2
Ash	%		1.3	1.5	5
Crude fiber	%		2.6	3.0	30
Ether extract	%		4.1	4.7	8
N-free extract	%		70.6	80.5	
Protein (N x 6.25)	%		9.1	10.4	11
Cattle	dig prot %	*	4.9	5.5	
Goats	dig prot %	*	5.9	6.7	
Horses	dig prot %	*	5.9	6.7	
Sheep	dig prot %	*	5.9	6.7	
Cellulose (Matrone)	%		2.0	2.3	
Energy	GE Mcal/kg				
Cattle	DE Mcal/kg	*	3.51	4.00	
Sheep	DE Mcal/kg	*	3.79	4.32	
Swine	DE kcal/kg	*	3557.	4055.	
Cattle	ME Mcal/kg	*	2.88	3.28	
Sheep	ME Mcal/kg	*	3.10	3.54	
Swine	ME kcal/kg	*	3340.	3808.	
Cattle	TDN %		79.7	90.8	
Sheep	TDN %		85.9	97.9	
Swine	TDN %		80.7	92.0	
Copper	mg/kg		3.1	3.5	
Iron	%		0.003	0.003	
Magnesium	%		0.14	0.16	
Manganese	mg/kg		4.2	4.8	
Selenium	mg/kg		0.097	0.111	
Zinc	mg/kg		29.6	33.7	
Niacin	mg/kg		36.8	42.0	
Pantothenic acid	mg/kg		8.8	10.0	
Riboflavin	mg/kg		1.5	1.7	
Thiamine	mg/kg		0.8	0.9	
Vitamin B_6	mg/kg		3.83	4.36	
Alanine	%		0.68	0.78	
Arginine	%		0.36	0.41	
Aspartic acid	%		0.60	0.68	
Glutamic acid	%		1.72	1.96	
Glycine	%		0.33	0.37	
Histidine	%		0.21	0.24	
Isoleucine	%		0.20	0.23	
Leucine	%		1.05	1.20	
Lysine	%		0.23	0.26	
Methionine	%		0.14	0.16	
Phenylalanine	%		0.40	0.46	
Proline	%		0.77	0.88	
Serine	%		0.45	0.52	
Threonine	%		0.31	0.35	
Tyrosine	%		0.31	0.35	
Valine	%		0.30	0.34	

Corn, yellow, grain, Can 2 CW mn wt 54 lb per bushel mx 3% foreign material, (4)

Ref No 4 02 996 United States

Feed Name or Analyses			Mean As Fed	Mean Dry	C.V. ± %
Dry matter	%			100.0	
Crude fiber	%				
Cattle	dig coef %	*		19.	
Sheep	dig coef %	*		30.	
Swine	dig coef %	*		47.	
Ether extract	%				
Cattle	dig coef %	*		87.	
Sheep	dig coef %	*		87.	
Swine	dig coef %	*		70.	
N-free extract	%				
Cattle	dig coef %	*		91.	
Sheep	dig coef %	*		99.	
Swine	dig coef %	*		93.	
Protein (N x 6.25)	%				
Cattle	dig coef %	*		75.	
Sheep	dig coef %	*		78.	
Swine	dig coef %	*		80.	
Energy	GE Mcal/kg				
Cattle	DE Mcal/kg	*		3.99	
Sheep	DE Mcal/kg	*		4.30	
Swine	DE kcal/kg	*		4044.	
Cattle	ME Mcal/kg	*		3.27	
Sheep	ME Mcal/kg	*		3.53	
Cattle	TDN %	*		90.4	
Sheep	TDN %	*		97.5	
Swine	TDN %	*		91.7	

Ref No 4 02 996 Canada

Feed Name or Analyses			Mean As Fed	Mean Dry	C.V. ± %
Dry matter	%		86.5	100.0	
Ash	%		1.2	1.4	
Crude fiber	%		2.9	3.3	
Ether extract	%		4.0	4.6	
N-free extract	%		69.2	79.9	
Protein (N x 6.25)	%		9.3	10.7	
Cattle	dig prot %	*	5.1	5.9	
Goats	dig prot %	*	6.1	7.1	
Horses	dig prot %	*	6.1	7.1	
Sheep	dig prot %	*	6.1	7.1	
Energy	GE Mcal/kg				
Cattle	DE Mcal/kg	*	3.45	3.99	
Sheep	DE Mcal/kg	*	3.72	4.30	
Swine	DE kcal/kg	*	3498.	4044.	
Cattle	ME Mcal/kg	*	2.83	3.27	
Sheep	ME Mcal/kg	*	3.05	3.53	
Swine	ME kcal/kg	*	3283.	3795.	
Cattle	TDN %		78.2	90.4	
Sheep	TDN %		84.3	97.5	
Swine	TDN %		79.3	91.7	

Corn, yellow, grain, Can 3 CE mn wt 52 lb per bushel mx 5% foreign material, (4)

Ref No 4 02 997 United States

Feed Name or Analyses			Mean As Fed	Mean Dry	C.V. ± %
Dry matter	%			100.0	
Crude fiber	%				
Cattle	dig coef %	*		19.	
Sheep	dig coef %	*		30.	
Swine	dig coef %	*		47.	
Ether extract	%				
Cattle	dig coef %	*		87.	
Sheep	dig coef %	*		87.	
Swine	dig coef %	*		70.	
N-free extract	%				
Cattle	dig coef %	*		91.	
Sheep	dig coef %	*		99.	
Swine	dig coef %	*		93.	
Protein (N x 6.25)	%				
Cattle	dig coef %	*		75.	
Sheep	dig coef %	*		78.	
Swine	dig coef %	*		80.	

Continued

Feed Name or Analyses			Mean As Fed	Mean Dry	C.V. ± %
Energy	GE Mcal/kg				
Cattle	DE Mcal/kg	*		4.00	
Sheep	DE Mcal/kg	*		4.31	
Swine	DE kcal/kg	*		4052.	
Cattle	ME Mcal/kg	*		3.28	
Sheep	ME Mcal/kg	*		3.54	
Cattle	TDN %	*		90.7	
Sheep	TDN %	*		97.8	
Swine	TDN %	*		91.9	

Ref No 4 02 997 Canada

Feed Name or Analyses			As Fed	Dry	C.V. ± %
Dry matter	%		87.8	100.0	3
Ash	%		1.3	1.5	11
Crude fiber	%		2.6	3.0	27
Ether extract	%		4.0	4.6	10
N-free extract	%		70.9	80.8	
Protein (N x 6.25)	%		8.9	10.2	9
Cattle	dig prot %	*	4.7	5.3	
Goats	dig prot %	*	5.7	6.5	
Horses	dig prot %	*	5.7	6.5	
Sheep	dig prot %	*	5.7	6.5	
Cellulose (Matrone)	%		2.1	2.4	
Energy	GE Mcal/kg				
Cattle	DE Mcal/kg	*	3.51	4.00	
Sheep	DE Mcal/kg	*	3.79	4.31	
Swine	DE kcal/kg	*	3558.	4052.	
Cattle	ME Mcal/kg	*	2.88	3.28	
Sheep	ME Mcal/kg	*	3.10	3.54	
Swine	ME kcal/kg	*	3342.	3807.	
Cattle	TDN %		79.6	90.7	
Sheep	TDN %		85.9	97.8	
Swine	TDN %		80.7	91.9	
Copper	mg/kg		3.1	3.5	
Iron	%		0.003	0.003	
Magnesium	%		0.12	0.14	
Manganese	mg/kg		4.1	4.7	
Selenium	mg/kg		0.039	0.044	
Zinc	mg/kg		23.1	26.3	
Niacin	mg/kg		38.6	43.9	
Pantothenic acid	mg/kg		8.7	9.9	
Riboflavin	mg/kg		1.5	1.8	
Thiamine	mg/kg		0.8	0.9	
Vitamin B6	mg/kg		4.53	5.16	
Alanine	%		0.68	0.77	
Arginine	%		0.36	0.41	
Aspartic acid	%		0.60	0.68	
Glutamic acid	%		1.70	1.93	
Glycine	%		0.34	0.38	
Histidine	%		0.22	0.25	
Isoleucine	%		0.20	0.23	
Leucine	%		1.04	1.19	
Lysine	%		0.24	0.27	
Methionine	%		0.14	0.16	
Phenylalanine	%		0.40	0.45	
Proline	%		0.77	0.88	
Serine	%		0.46	0.53	
Threonine	%		0.30	0.34	
Tyrosine	%		0.30	0.34	
Valine	%		0.31	0.35	

(1) dry forages and roughages (2) pasture, range plants, and forages fed green (3) silages (4) energy feeds (5) protein supplements (6) minerals (7) vitamins (8) additives

Corn, yellow, grain, Can 3 CW mn wt 52 lb per bushel mx 5% foreign material, (4)

Ref No 4 02 998 United States

Feed Name or Analyses			As Fed	Dry	C.V. ± %
Dry matter	%			100.0	
Crude fiber	%				
Cattle	dig coef %	*		19.	
Sheep	dig coef %	*		30.	
Swine	dig coef %	*		47.	
Ether extract	%				
Cattle	dig coef %	*		87.	
Sheep	dig coef %	*		87.	
Swine	dig coef %	*		70.	
N-free extract	%				
Cattle	dig coef %	*		91.	
Sheep	dig coef %	*		99.	
Swine	dig coef %	*		93.	
Protein (N x 6.25)	%				
Cattle	dig coef %	*		75.	
Sheep	dig coef %	*		78.	
Swine	dig coef %	*		80.	
Energy	GE Mcal/kg				
Cattle	DE Mcal/kg	*		3.97	
Sheep	DE Mcal/kg	*		4.29	
Swine	DE kcal/kg	*		4035.	
Cattle	ME Mcal/kg	*		3.26	
Sheep	ME Mcal/kg	*		3.51	
Cattle	TDN %	*		90.1	
Sheep	TDN %	*		97.2	
Swine	TDN %	*		91.5	

Ref No 4 02 998 Canada

Feed Name or Analyses			As Fed	Dry	C.V. ± %
Dry matter	%		86.5	100.0	
Ash	%		1.3	1.5	
Crude fiber	%		2.9	3.3	
Ether extract	%		3.8	4.5	
N-free extract	%		68.9	79.7	
Protein (N x 6.25)	%		9.5	11.0	
Cattle	dig prot %	*	5.3	6.2	
Goats	dig prot %	*	6.4	7.4	
Horses	dig prot %	*	6.4	7.4	
Sheep	dig prot %	*	6.4	7.4	
Energy	GE Mcal/kg				
Cattle	DE Mcal/kg	*	3.44	3.97	
Sheep	DE Mcal/kg	*	3.71	4.29	
Swine	DE kcal/kg	*	3490.	4035.	
Cattle	ME Mcal/kg	*	2.82	3.26	
Sheep	ME Mcal/kg	*	3.04	3.51	
Swine	ME kcal/kg	*	3273.	3783.	
Cattle	TDN %		78.0	90.1	
Sheep	TDN %		84.1	97.2	
Swine	TDN %		79.2	91.5	

Corn, yellow, grain, Can 4 CE mn wt 50 lb per bushel mx 7% foreign material, (4)

Ref No 4 02 999 United States

Feed Name or Analyses			As Fed	Dry	C.V. ± %
Dry matter	%			100.0	
Crude fiber	%				
Cattle	dig coef %	*		19.	
Sheep	dig coef %	*		30.	
Swine	dig coef %	*		47.	
Ether extract	%				
Cattle	dig coef %	*		87.	
Sheep	dig coef %	*		87.	
Swine	dig coef %	*		70.	
N-free extract	%				
Cattle	dig coef %	*		91.	
Sheep	dig coef %	*		99.	
Swine	dig coef %	*		93.	
Protein (N x 6.25)	%				
Cattle	dig coef %	*		75.	
Sheep	dig coef %	*		78.	
Swine	dig coef %	*		80.	
Energy	GE Mcal/kg				
Cattle	DE Mcal/kg	*		3.99	
Sheep	DE Mcal/kg	*		4.30	
Swine	DE kcal/kg	*		4046.	
Cattle	ME Mcal/kg	*		3.27	
Sheep	ME Mcal/kg	*		3.53	
Cattle	TDN %	*		90.5	
Sheep	TDN %	*		97.6	
Swine	TDN %	*		91.8	

Ref No 4 02 999 Canada

Feed Name or Analyses			As Fed	Dry	C.V. ± %
Dry matter	%		88.6	100.0	3
Ash	%		1.2	1.4	
Crude fiber	%		2.5	2.8	
Ether extract	%		3.7	4.2	
N-free extract	%		72.0	81.2	
Protein (N x 6.25)	%		9.2	10.4	9
Cattle	dig prot %	*	4.9	5.5	
Goats	dig prot %	*	6.0	6.7	
Horses	dig prot %	*	6.0	6.7	
Sheep	dig prot %	*	6.0	6.7	
Cellulose (Matrone)	%		2.1	2.4	
Energy	GE Mcal/kg				
Cattle	DE Mcal/kg	*	3.54	3.99	
Sheep	DE Mcal/kg	*	3.81	4.30	
Swine	DE kcal/kg	*	3587.	4046.	
Cattle	ME Mcal/kg	*	2.90	3.27	
Sheep	ME Mcal/kg	*	3.13	3.53	
Swine	ME kcal/kg	*	3368.	3800.	
Cattle	TDN %		80.2	90.5	
Sheep	TDN %		86.5	97.6	
Swine	TDN %		81.3	91.8	
Copper	mg/kg		3.1	3.5	
Iron	%		0.003	0.003	
Magnesium	%		0.13	0.15	
Manganese	mg/kg		4.6	5.2	
Selenium	mg/kg		0.068	0.076	
Zinc	mg/kg		25.1	28.4	
Niacin	mg/kg		46.4	52.3	
Pantothenic acid	mg/kg		8.8	9.9	
Riboflavin	mg/kg		1.9	2.2	
Thiamine	mg/kg		0.8	0.9	
Vitamin B6	mg/kg		4.15	4.68	
Alanine	%		0.62	0.70	
Arginine	%		0.34	0.38	
Aspartic acid	%		0.57	0.64	
Glutamic acid	%		1.57	1.77	
Glycine	%		0.32	0.36	
Histidine	%		0.21	0.24	
Isoleucine	%		0.16	0.18	
Leucine	%		0.93	1.05	
Lysine	%		0.23	0.26	
Methionine	%		0.13	0.15	
Phenylalanine	%		0.38	0.42	
Proline	%		0.71	0.80	
Serine	%		0.44	0.50	
Threonine	%		0.30	0.34	
Tyrosine	%		0.29	0.33	
Valine	%		0.28	0.32	

Feed Name or Analyses			Mean As Fed	Mean Dry	C.V. ± %

Corn, yellow, grain, Can 4 CW mn wt 50 lb per bushel mx 7% foreign material, (4)

Ref No 4 03 000 United States

Analyses	Unit		As Fed	Dry	C.V. ± %
Dry matter	%			100.0	
Crude fiber	%				
Cattle	dig coef %	*		19.	
Sheep	dig coef %	*		30.	
Swine	dig coef %	*		47.	
Ether extract	%				
Cattle	dig coef %	*		87.	
Sheep	dig coef %	*		87.	
Swine	dig coef %	*		70.	
N-free extract	%				
Cattle	dig coef %	*		91.	
Sheep	dig coef %	*		99.	
Swine	dig coef %	*		93.	
Protein (N x 6.25)	%				
Cattle	dig coef %	*		75.	
Sheep	dig coef %	*		78.	
Swine	dig coef %	*		80.	
Energy	GE Mcal/kg				
Cattle	DE Mcal/kg	*		3.98	
Sheep	DE Mcal/kg	*		4.29	
Swine	DE kcal/kg	*		4039.	
Cattle	ME Mcal/kg	*		3.27	
Sheep	ME Mcal/kg	*		3.52	
Cattle	TDN %	*		90.3	
Sheep	TDN %	*		97.4	
Swine	TDN %	*		91.6	

Ref No 4 03 000 Canada

Analyses	Unit		As Fed	Dry	C.V. ± %
Dry matter	%		86.5	100.0	0
Ash	%		1.4	1.6	4
Crude fiber	%		2.6	3.0	31
Ether extract	%		3.9	4.5	8
N-free extract	%		68.9	79.7	
Protein (N x 6.25)	%		9.8	11.3	4
Cattle	dig prot %	*	5.5	6.4	
Goats	dig prot %	*	6.6	7.6	
Horses	dig prot %	*	6.6	7.6	
Sheep	dig prot %	*	6.6	7.6	
Energy	GE Mcal/kg				
Cattle	DE Mcal/kg	*	3.45	3.98	
Sheep	DE Mcal/kg	*	3.71	4.29	
Swine	DE kcal/kg	*	3494.	4039.	
Cattle	ME Mcal/kg	*	2.83	3.27	
Sheep	ME Mcal/kg	*	3.05	3.52	
Swine	ME kcal/kg	*	3274.	3786.	
Cattle	TDN %		78.1	90.3	
Sheep	TDN %		84.2	97.4	
Swine	TDN %		79.2	91.6	

Corn, yellow, grain, Can 5 CE mn wt 47 lb per bushel mx 12% foreign material, (4)

Ref No 4 03 001 United States

Analyses	Unit		As Fed	Dry	C.V. ± %
Dry matter	%			100.0	
Crude fiber	%				
Cattle	dig coef %	*		19.	
Sheep	dig coef %	*		30.	
Swine	dig coef %	*		47.	
Ether extract	%				
Cattle	dig coef %	*		87.	
Sheep	dig coef %	*		87.	
Swine	dig coef %	*		70.	
N-free extract	%				
Cattle	dig coef %	*		91.	
Sheep	dig coef %	*		99.	
Swine	dig coef %	*		93.	
Protein (N x 6.25)	%				
Cattle	dig coef %	*		75.	
Sheep	dig coef %	*		78.	
Swine	dig coef %	*		80.	
Energy	GE Mcal/kg				
Cattle	DE Mcal/kg	*		3.96	
Sheep	DE Mcal/kg	*		4.28	
Swine	DE kcal/kg	*		4029.	
Cattle	ME Mcal/kg	*		3.25	
Sheep	ME Mcal/kg	*		3.51	
Cattle	TDN %	*		89.9	
Sheep	TDN %	*		97.0	
Swine	TDN %	*		91.4	

Ref No 4 03 001 Canada

Analyses	Unit		As Fed	Dry	C.V. ± %
Dry matter	%		86.5	100.0	
Ash	%		1.3	1.4	
Crude fiber	%		3.2	3.8	
Ether extract	%		3.7	4.3	
N-free extract	%		69.6	80.5	
Protein (N x 6.25)	%		8.7	10.1	
Cattle	dig prot %	*	4.5	5.3	
Goats	dig prot %	*	5.6	6.5	
Horses	dig prot %	*	5.6	6.5	
Sheep	dig prot %	*	5.6	6.5	
Energy	GE Mcal/kg				
Cattle	DE Mcal/kg	*	3.43	3.96	
Sheep	DE Mcal/kg	*	3.70	4.28	
Swine	DE kcal/kg	*	3485.	4029.	
Cattle	ME Mcal/kg	*	2.81	3.25	
Sheep	ME Mcal/kg	*	3.03	3.51	
Swine	ME kcal/kg	*	3275.	3786.	
Cattle	TDN %		77.7	89.9	
Sheep	TDN %		83.9	97.0	
Swine	TDN %		79.0	91.4	

Corn, yellow, grain, Can 5 CW mn wt 47 lb per bushel mx 12% foreign material, (4)

Ref No 4 03 002 United States

Analyses	Unit		As Fed	Dry	C.V. ± %
Dry matter	%			100.0	
Crude fiber	%				
Cattle	dig coef %	*		19.	
Sheep	dig coef %	*		30.	
Swine	dig coef %	*		47.	
Ether extract	%				
Cattle	dig coef %	*		87.	
Sheep	dig coef %	*		87.	
Swine	dig coef %	*		70.	
N-free extract	%				
Cattle	dig coef %	*		91.	
Sheep	dig coef %	*		99.	
Swine	dig coef %	*		93.	
Protein (N x 6.25)	%				
Cattle	dig coef %	*		75.	
Sheep	dig coef %	*		78.	
Swine	dig coef %	*		80.	
Energy	GE Mcal/kg				
Cattle	DE Mcal/kg	*		3.93	
Sheep	DE Mcal/kg	*		4.25	
Swine	DE kcal/kg	*		4006.	
Cattle	ME Mcal/kg	*		3.22	
Sheep	ME Mcal/kg	*		3.48	
Cattle	TDN %	*		89.2	
Sheep	TDN %	*		96.3	
Swine	TDN %	*		90.9	

Ref No 4 03 002 Canada

Analyses	Unit		As Fed	Dry	C.V. ± %
Dry matter	%		86.5	100.0	
Ash	%		1.4	1.7	
Crude fiber	%		3.2	3.7	
Ether extract	%		3.4	4.0	
N-free extract	%		68.6	79.3	
Protein (N x 6.25)	%		9.9	11.4	
Cattle	dig prot %	*	5.6	6.5	
Goats	dig prot %	*	6.7	7.7	
Horses	dig prot %	*	6.7	7.7	
Sheep	dig prot %	*	6.7	7.7	
Energy	GE Mcal/kg				
Cattle	DE Mcal/kg	*	3.40	3.93	
Sheep	DE Mcal/kg	*	3.67	4.25	
Swine	DE kcal/kg	*	3465.	4006.	
Cattle	ME Mcal/kg	*	2.79	3.22	
Sheep	ME Mcal/kg	*	3.01	3.48	
Swine	ME kcal/kg	*	3247.	3753.	
Cattle	TDN %		77.1	89.2	
Sheep	TDN %		83.3	96.3	
Swine	TDN %		78.6	90.9	

Corn, yellow, grain, gr 2 US mn wt 54 lb per bushel, (4)

Ref No 4 03 005 United States

Analyses	Unit		As Fed	Dry	C.V. ± %
Dry matter	%		90.0	100.0	
Ash	%		1.4	1.6	
Crude fiber	%		2.3	2.6	
Ether extract	%		4.1	4.6	
N-free extract	%		72.3	80.3	
Protein (N x 6.25)	%		9.9	11.0	
Cattle	dig prot %	*	5.5	6.1	
Goats	dig prot %	*	6.6	7.3	
Horses	dig prot %	*	6.6	7.3	
Sheep	dig prot %	*	6.6	7.3	
Energy	GE Mcal/kg				
Cattle	DE Mcal/kg	*	3.58	3.98	
Sheep	DE Mcal/kg	*	3.42	3.80	
Swine	DE kcal/kg	*	3644.	4049.	
Cattle	ME Mcal/kg	*	2.93	3.26	
Sheep	ME Mcal/kg	*	2.81	3.12	
Swine	ME kcal/kg	*	3417.	3797.	
Cattle	TDN %	*	81.3	90.3	
Sheep	TDN %	*	77.7	86.3	
Swine	TDN %	*	82.6	91.8	

Corn, yellow, grain wo germ, grnd, (4)

Degermed yellow corn meal

Ref No 4 03 009 United States

Analyses	Unit		As Fed	Dry	C.V. ± %
Dry matter	%		88.7	100.0	
Ash	%		1.1	1.2	
Crude fiber	%		0.6	0.7	
Cattle	dig coef %	*	19.	19.	
Sheep	dig coef %	*	30.	30.	
Ether extract	%		1.2	1.4	
Cattle	dig coef %	*	87.	87.	
Sheep	dig coef %	*	87.	87.	
N-free extract	%		77.1	86.9	
Cattle	dig coef %	*	91.	91.	
Sheep	dig coef %	*	99.	99.	

Continued

Feed Name or Analyses			Mean As Fed	Mean Dry	C.V. ± %
Protein (N x 6.25)	%		8.7	9.8	
Cattle	dig coef %	*	75.	75.	
Sheep	dig coef %	*	78.	78.	
Cattle	dig prot %		6.5	7.4	
Goats	dig prot %	*	5.5	6.2	
Horses	dig prot %	*	5.5	6.2	
Sheep	dig prot %		6.8	7.7	
Energy	GE Mcal/kg				
Cattle	DE Mcal/kg	*	3.49	3.93	
Sheep	DE Mcal/kg	*	3.78	4.26	
Swine	DE kcal/kg	*	3743.	4220.	
Cattle	ME Mcal/kg	*	2.86	3.23	
Sheep	ME Mcal/kg	*	3.10	3.49	
Swine	ME kcal/kg	*	3519.	3968.	
Cattle	TDN %		79.1	89.2	
Sheep	TDN %		85.6	96.6	
Swine	TDN %	*	84.9	95.7	
Calcium	%		0.01	0.01	
Copper	mg/kg		0.9	1.0	
Iron	%		0.002	0.002	
Phosphorus	%		0.14	0.16	

Corn, yellow, grits, cracked fine screened, (4)
Corn, yellow, flour
Corn, yellow, grits
Corn, yellow, hominy grits
Ref No 4 03 010 United States

Feed Name or Analyses			Mean As Fed	Mean Dry	C.V. ± %
Dry matter	%		91.6	100.0	
Ash	%		0.9	1.0	
Crude fiber	%		7.6	8.3	
Ether extract	%		2.3	2.5	
N-free extract	%		70.9	77.4	
Protein (N x 6.25)	%		9.9	10.8	17
Cattle	dig prot %	*	5.4	5.9	
Goats	dig prot %	*	6.5	7.1	
Horses	dig prot %	*	6.5	7.1	
Sheep	dig prot %	*	6.5	7.1	
Energy	GE Mcal/kg				
Cattle	DE Mcal/kg	*	3.31	3.61	
Sheep	DE Mcal/kg	*	3.33	3.63	
Swine	DE kcal/kg	*	3182.	3473.	
Cattle	ME Mcal/kg	*	2.72	2.96	
Sheep	ME Mcal/kg	*	2.73	2.98	
Swine	ME kcal/kg	*	2985.	3259.	
Cattle	TDN %	*	75.1	82.0	
Sheep	TDN %	*	75.4	82.3	
Swine	TDN %	*	72.2	78.8	
Calcium	%		0.03	0.03	
Iron	%		0.004	0.004	
Phosphorus	%		0.16	0.17	
Niacin	mg/kg		52.7	57.5	
Riboflavin	mg/kg		0.2	0.2	
Thiamine	mg/kg		2.0	2.2	
Arginine	%		0.60	0.66	15
Histidine	%		0.30	0.33	
Isoleucine	%		0.50	0.55	9
Leucine	%		0.80	0.87	11
Lysine	%		0.40	0.44	11
Methionine	%		0.10	0.11	
Phenylalanine	%		0.40	0.44	22
Threonine	%		0.40	0.44	11
Tryptophan	%		0.10	0.11	

(1) dry forages and roughages (2) pasture, range plants, and forages fed green (3) silages (4) energy feeds (5) protein supplements (6) minerals (7) vitamins (8) additives

Feed Name or Analyses			Mean As Fed	Mean Dry	C.V. ± %
Tyrosine	%		0.40	0.44	11
Valine	%		0.50	0.55	18

Corn, yellow, grits by-product, mn 5% fat, (4)
Yellow hominy feed (AAFCO)
Yellow hominy feed (CFA)
Hominy, yellow corn, feed
Ref No 4 03 011 United States

Feed Name or Analyses			Mean As Fed	Mean Dry	C.V. ± %
Dry matter	%		90.5	100.0	2
Ash	%		2.7	3.0	22
Crude fiber	%		4.8	5.3	22
Ether extract	%		5.8	6.4	18
N-free extract	%		66.0	73.0	
Protein (N x 6.25)	%		11.1	12.3	8
Cattle	dig prot %	*	6.6	7.3	
Goats	dig prot %	*	7.7	8.5	
Horses	dig prot %	*	7.7	8.5	
Sheep	dig prot %	*	7.7	8.5	
Pentosans	%		15.8	17.4	30
Linoleic	%		3.300	3.646	
Energy	GE Mcal/kg				
Cattle	DE Mcal/kg	*	3.22	3.56	
Sheep	DE Mcal/kg	*	3.33	3.69	
Swine	DE kcal/kg	*	3298.	3646.	
Cattle	ME Mcal/kg	*	2.64	2.92	
Chickens	ME_n kcal/kg		2850.	3150.	
Sheep	ME Mcal/kg	*	2.73	3.02	
Swine	ME kcal/kg	*	3084.	3410.	
Cattle	TDN %	*	73.0	80.7	
Sheep	TDN %	*	75.6	83.6	
Swine	TDN %	*	74.8	82.7	
Calcium	%		0.05	0.06	47
Cobalt	mg/kg		0.055	0.061	17
Copper	mg/kg		9.7	10.7	24
Iron	%		0.010	0.011	69
Magnesium	%		0.04	0.04	80
Manganese	mg/kg		16.0	17.7	30
Phosphorus	%		0.52	0.57	26
Potassium	%		0.52	0.57	
Sodium	%		0.13	0.14	
Biotin	mg/kg		0.13	0.15	13
Carotene	mg/kg		6.8	7.5	66
Choline	mg/kg		959.	1060.	16
Folic acid	mg/kg		0.33	0.36	54
Niacin	mg/kg		43.2	47.8	20
Pantothenic acid	mg/kg		8.6	9.5	14
Riboflavin	mg/kg		2.4	2.7	52
Thiamine	mg/kg		7.3	8.0	34
Vitamin B_6	mg/kg		11.87	13.13	18
Vitamin A	IU/g		1.8	2.0	
Vitamin A equivalent	IU/g		11.4	12.6	
Arginine	%		0.50	0.55	
Cystine	%		0.10	0.11	
Glycine	%		0.30	0.33	
Histidine	%		0.20	0.22	
Isoleucine	%		0.40	0.44	
Leucine	%		0.80	0.88	
Lysine	%		0.40	0.44	
Methionine	%		0.10	0.11	
Phenylalanine	%		0.30	0.33	
Threonine	%		0.40	0.44	
Tryptophan	%		0.10	0.11	
Tyrosine	%		0.50	0.55	
Valine	%		0.50	0.55	
Xanthophylls	mg/kg		4.42	4.89	

Ref No 4 03 011 Canada

Feed Name or Analyses			Mean As Fed	Mean Dry	C.V. ± %
Dry matter	%		90.0	100.0	
Crude fiber	%		5.5	6.1	
Ether extract	%		9.1	10.1	
Protein (N x 6.25)	%		11.3	12.6	
Cattle	dig prot %	*	6.8	7.5	
Goats	dig prot %	*	7.9	8.7	
Horses	dig prot %	*	7.9	8.7	
Sheep	dig prot %	*	7.9	8.7	
Energy	GE Mcal/kg				
Chickens	ME_n kcal/kg		2860.	3177.	

Corn and cob meal (AAFCO) –
see Corn, ears, grnd, (4)

Corn and cob meal (AAFCO) –
see Corn, ears, grnd, (4)

Corn and cob meal (AAFCO) –
see Corn, ears, grnd, (4)

Corn and cob meal with husks (AAFCO) –
see Corn, ears w husks, grnd, (4)

Corn bran (AAFCO) (CFA) –
see Corn, bran, wet or dry milled dehy, (4)

Corn chop (AAFCO) –
see Corn, grain, grnd, mx 4% foreign material, (4)

Corn chop (AAFCO) –
see Corn, grain, grnd, mx 4% foreign material, (4)

CORNCOCKLE, COMMON. Agrostemma gethogo

Corncockle, common, seeds w hulls, (4)
Ref No 4 09 046 United States

Feed Name or Analyses			Mean As Fed	Mean Dry	C.V. ± %
Dry matter	%			100.0	
Protein (N x 6.25)	%			15.0	
Cattle	dig prot %	*		9.8	
Goats	dig prot %	*		11.0	
Horses	dig prot %	*		11.0	
Sheep	dig prot %	*		11.0	
Alanine	%			0.54	
Arginine	%			1.38	
Aspartic acid	%			1.20	
Glutamic acid	%			2.22	
Glycine	%			0.92	
Histidine	%			0.35	
Hydroxyproline	%			0.02	
Isoleucine	%			0.48	
Leucine	%			0.83	
Lysine	%			0.72	
Methionine	%			0.24	
Phenylalanine	%			0.57	
Proline	%			0.54	
Serine	%			0.63	
Threonine	%			0.62	
Tyrosine	%			0.51	
Valine	%			0.63	

Corn, dent stover silage, mature –
see Corn, dent, aerial part wo ears wo husks, ensiled, mature, (3)

Feed Name or Analyses			Mean As Fed	Mean Dry	C.V. ± %

Corn distillers dried grains (AAFCO) (CFA) –
see Corn, distillers grains, dehy, (5)

Corn distillers dried grains with solubles (AAFCO) –
see Corn, distillers grains w solubles, dehy, mn 75% original solids, (5)

Corn distillers dried solubles (AAFCO) –
see Corn, distillers solubles, dehy, (5)

Corn feed meal (AAFCO) (CFA) –
see Corn, grain fines, cracked coarse screened passed, (4)

Corn fodder, fresh –
see Corn, aerial part, fresh, (2)

Corn fodder silage –
see Corn, aerial part, ensiled, (3)

Corn fodder silage, dough stage –
see Corn, aerial part, ensiled, dough stage, (3)

Corn fodder silage, mature –
see Corn, aerial part, ensiled, mature, (3)

Corn fodder silage, milk stage –
see Corn, aerial part, ensiled, milk stage, (3)

Corn fodder, sun-cured –
see Corn, aerial part, s-c, (1)

Corn fodder, sun-cured, dough stage –
see Corn,aerial part, s-c, dough stage, (1)

Corn fodder, sun-cured, mature –
see Corn, aerial part, s-c, mature, (1)

Corn germ meal, (dry milled) (AAFCO) –
see Corn, germ, dry milled extn unspec grnd, (5)

Corn germ meal (dry milled) mechanical extracted –
see Corn, germ, mech-extd grnd, (5)

Corn germ meal (dry milled) solvent extracted –
see Corn, germ, solv-extd grnd, (5)

Corn germ meal, mechanical extracted, (wet milled) (AAFCO) –
see Corn, germ wo solubles, wet milled mech-extd dehy grnd, (5)

Corn gluten feed (AAFCO) (CFA) –
see Corn, gluten w bran, wet milled dehy, (5)

Corn gluten meal (AAFCO) (CFA) –
see Corn, gluten, wet milled dehy, (5)

Corn meal (AAFCO) –
see Corn, grain, grnd, mx 4% foreign material, (4)

Corn oil meal –
see Corn, germ, dry milled extn unspecified grnd, (5)

Corn oil meal expeller extracted –
see Corn, germ, mech-extd grnd, (5)

Corn oil meal, solvent extracted –
see Corn, germ, solv-extd grnd, (5)

CORN-PEA. Zea mays, Pisum spp

Corn-Pea, aerial part, ensiled, (3)

Ref No 3 03 012 United States

Analyses	Unit		As Fed	Dry	C.V. ± %
Dry matter	%		22.2	100.0	
Ash	%		2.1	9.3	
Crude fiber	%		6.7	30.4	
Ether extract	%		0.6	2.8	
N-free extract	%		9.6	43.2	
Protein (N x 6.25)	%		3.2	14.3	
Cattle	dig prot %	*	2.0	9.2	
Goats	dig prot %	*	2.0	9.2	
Horses	dig prot %	*	2.0	9.2	
Sheep	dig prot %	*	2.0	9.2	
Energy	GE Mcal/kg				
Cattle	DE Mcal/kg	*	0.57	2.58	
Sheep	DE Mcal/kg	*	0.61	2.77	
Cattle	ME Mcal/kg	*	0.47	2.12	
Sheep	ME Mcal/kg	*	0.50	2.27	
Cattle	TDN %	*	13.0	58.5	
Sheep	TDN %	*	13.9	62.7	

CORN-SORGHUM. Zea mays, Sorghum vulgare

Corn-Sorghum, aerial part, ensiled, (3)

Ref No 3 08 411 United States

Analyses	Unit		As Fed	Dry	C.V. ± %
Dry matter	%		26.0	100.0	
Ash	%		1.4	5.4	
Crude fiber	%		7.1	27.3	
Ether extract	%		0.8	3.1	
N-free extract	%		14.8	56.9	
Protein (N x 6.25)	%		1.9	7.3	
Cattle	dig prot %	*	0.7	2.9	
Goats	dig prot %	*	0.7	2.9	
Horses	dig prot %	*	0.7	2.9	
Sheep	dig prot %	*	0.7	2.9	
Energy	GE Mcal/kg				
Cattle	DE Mcal/kg	*	0.74	2.85	
Sheep	DE Mcal/kg	*	0.77	2.95	
Cattle	ME Mcal/kg	*	0.61	2.34	
Sheep	ME Mcal/kg	*	0.63	2.42	
Cattle	TDN %	*	16.8	64.7	
Sheep	TDN %	*	17.4	66.9	
Calcium	%		0.08	0.31	
Phosphorus	%		0.05	0.19	

Corn-Sorghum, aerial part, ensiled, mature, (3)

Ref No 3 03 013 United States

Analyses	Unit		As Fed	Dry	C.V. ± %
Dry matter	%		34.6	100.0	30
Ash	%		2.1	6.0	7
Crude fiber	%		9.1	26.2	8
Ether extract	%		1.0	2.9	9
N-free extract	%		19.7	57.0	
Protein (N x 6.25)	%		2.7	7.9	16
Cattle	dig prot %	*	1.2	3.4	
Goats	dig prot %	*	1.2	3.4	
Horses	dig prot %	*	1.2	3.4	
Sheep	dig prot %	*	1.2	3.4	
Energy	GE Mcal/kg				
Cattle	DE Mcal/kg	*	1.00	2.89	
Sheep	DE Mcal/kg	*	1.02	2.94	
Cattle	ME Mcal/kg	*	0.82	2.37	
Sheep	ME Mcal/kg	*	0.83	2.41	
Cattle	TDN %	*	22.7	65.6	
Sheep	TDN %	*	23.1	66.7	
Calcium	%		0.13	0.37	
Phosphorus	%		0.07	0.19	
Carotene	mg/kg		1.9	5.5	
Vitamin A equivalent	IU/g		3.2	9.2	

CORN-SOYBEAN. Zea mays, Glycine max

Corn-Soybean, aerial part, fresh, (2)

Ref No 2 03 014 United States

Analyses	Unit		As Fed	Dry	C.V. ± %
Dry matter	%		23.9	100.0	11
Ash	%		1.7	7.0	17
Crude fiber	%		5.5	23.1	8
Ether extract	%		0.8	3.4	11
N-free extract	%		13.2	55.2	
Protein (N x 6.25)	%		2.7	11.3	16
Cattle	dig prot %	*	1.8	7.5	
Goats	dig prot %	*	1.7	7.1	
Horses	dig prot %	*	1.7	7.1	
Rabbits	dig prot %	*	1.8	7.4	
Sheep	dig prot %	*	1.8	7.5	
Energy	GE Mcal/kg				
Cattle	DE Mcal/kg	*	0.69	2.91	
Sheep	DE Mcal/kg	*	0.72	3.02	
Cattle	ME Mcal/kg	*	0.57	2.38	
Sheep	ME Mcal/kg	*	0.59	2.48	
Cattle	TDN %	*	15.7	65.9	
Sheep	TDN %	*	16.3	68.5	
Calcium	%		0.13	0.56	
Phosphorus	%		0.06	0.26	
Potassium	%		0.33	1.39	

Corn-Soybean, aerial part, ensiled, (3)

Ref No 3 03 015 United States

Analyses	Unit		As Fed	Dry	C.V. ± %
Dry matter	%		25.1	100.0	9
Ash	%		1.7	6.6	14
Crude fiber	%		6.6	26.3	7
Cattle	dig coef %		60.	60.	
Ether extract	%		0.9	3.5	62
Cattle	dig coef %		88.	88.	
N-free extract	%		13.5	53.9	
Cattle	dig coef %		80.	80.	
Protein (N x 6.25)	%		2.4	9.7	18
Cattle	dig coef %		62.	62.	
Cattle	dig prot %		1.5	6.0	
Goats	dig prot %	*	1.3	5.0	
Horses	dig prot %	*	1.3	5.0	
Sheep	dig prot %	*	1.3	5.0	
Energy	GE Mcal/kg				
Cattle	DE Mcal/kg	*	0.79	3.17	
Sheep	DE Mcal/kg	*	0.73	2.90	
Cattle	ME Mcal/kg	*	0.65	2.60	
Sheep	ME Mcal/kg	*	0.59	2.37	
Cattle	NE_m Mcal/kg	*	0.40	1.60	
Cattle	NE_{gain} Mcal/kg	*	0.26	1.03	
Cattle	$NE_{lactating\ cows}$ Mcal/kg	*	0.73	2.90	
Cattle	TDN %		18.0	71.8	
Sheep	TDN %	*	16.5	65.7	
Calcium	%		0.21	0.83	63
Phosphorus	%		0.09	0.36	31
Potassium	%		0.28	1.11	
Carotene	mg/kg		49.5	197.5	38
Vitamin A equivalent	IU/g		82.5	329.3	

Feed Name or Analyses		Mean As Fed	Mean Dry	C.V. ± %

Corn-Soybean, aerial part, ensiled, mature, (3)
Ref No 3 08 412 United States

Feed Name or Analyses		As Fed	Dry	C.V. ± %
Dry matter	%	28.3	100.0	
Ash	%	2.4	8.5	
Crude fiber	%	7.3	25.8	
Ether extract	%	1.2	4.2	
N-free extract	%	14.2	50.2	
Protein (N x 6.25)	%	3.2	11.3	
Cattle	dig prot %	* 1.8	6.5	
Goats	dig prot %	* 1.8	6.5	
Horses	dig prot %	* 1.8	6.5	
Sheep	dig prot %	* 1.8	6.5	
Energy	GE Mcal/kg			
Cattle	DE Mcal/kg	* 0.84	2.95	
Sheep	DE Mcal/kg	* 0.87	3.07	
Cattle	ME Mcal/kg	* 0.69	2.42	
Sheep	ME Mcal/kg	* 0.71	2.52	
Cattle	TDN %	* 19.0	67.0	
Sheep	TDN %	* 19.7	69.6	
Calcium	%	0.20	0.71	
Phosphorus	%	0.08	0.28	
Potassium	%	0.29	1.02	

Corn, stalks, shanks, tassels, dried –
see Corn, stalks w tassels, s-c, (1)

Corn stover silage –
see Corn, aerial part wo ears wo husks, ensiled, (3)

Corn stover, sun-cured –
see Corn, aerial part wo ears wo husks, s-c, (1)

Corn stover, sun-cured, mature –
see Corn, aerial part wo ears wo husks, s-c, mature, (1)

Corn sugar molasses (AAFCO) –
see Corn, molasses, mn 43% dextrose-equivalent 50% total dextrose mn 78 degrees brix, (4)

CORN-SUNFLOWER. Zea mays, Helianthus spp

Corn-Sunflower, aerial part, fresh, (2)
Ref No 2 03 016 United States

Feed Name or Analyses		As Fed	Dry	C.V. ± %
Dry matter	%	18.2	100.0	
Ash	%	1.6	8.8	
Crude fiber	%	6.1	33.5	
Ether extract	%	0.6	3.3	
N-free extract	%	8.5	46.7	
Protein (N x 6.25)	%	1.4	7.7	
Cattle	dig prot %	* 0.8	4.4	
Goats	dig prot %	* 0.7	3.7	
Horses	dig prot %	* 0.7	4.1	
Rabbits	dig prot %	* 0.8	4.6	
Sheep	dig prot %	* 0.8	4.2	
Energy	GE Mcal/kg			
Cattle	DE Mcal/kg	* 0.49	2.71	
Sheep	DE Mcal/kg	* 0.53	2.90	
Cattle	ME Mcal/kg	* 0.40	2.22	
Sheep	ME Mcal/kg	* 0.43	2.38	
Cattle	TDN %	* 11.2	61.4	
Sheep	TDN %	* 12.0	65.8	

(1) dry forages and roughages (2) pasture, range plants, and forages fed green (3) silages (4) energy feeds (5) protein supplements (6) minerals (7) vitamins (8) additives

Corn-Sunflower, aerial part, ensiled, (3)
Ref No 3 03 017 United States

Feed Name or Analyses		As Fed	Dry	C.V. ± %
Dry matter	%	23.3	100.0	
Ash	%	2.2	9.4	12
Crude fiber	%	7.2	31.0	2
Ether extract	%	0.6	2.5	4
N-free extract	%	10.3	44.4	
Protein (N x 6.25)	%	3.0	12.7	8
Cattle	dig prot %	* 1.8	7.8	
Goats	dig prot %	* 1.8	7.8	
Horses	dig prot %	* 1.8	7.8	
Sheep	dig prot %	* 1.8	7.8	
Energy	GE Mcal/kg			
Cattle	DE Mcal/kg	* 0.61	2.62	
Sheep	DE Mcal/kg	* 0.65	2.78	
Cattle	ME Mcal/kg	* 0.50	2.15	
Sheep	ME Mcal/kg	* 0.53	2.28	
Cattle	TDN %	* 13.9	59.5	
Sheep	TDN %	* 14.7	63.0	

Corn, sweet, cannery refuse –
see Corn, sweet, cannery residue, fresh, (2)

CORN-VETCH. Zea mays, Vicia spp

Corn-Vetch, aerial part, ensiled, (3)
Ref No 3 03 018 United States

Feed Name or Analyses		As Fed	Dry	C.V. ± %
Dry matter	%	21.1	100.0	
Ash	%	1.9	9.2	
Crude fiber	%	7.0	33.2	
Ether extract	%	0.6	2.8	
N-free extract	%	9.2	43.4	
Protein (N x 6.25)	%	2.4	11.4	
Cattle	dig prot %	* 1.4	6.6	
Goats	dig prot %	* 1.4	6.6	
Horses	dig prot %	* 1.4	6.6	
Sheep	dig prot %	* 1.4	6.6	
Energy	GE Mcal/kg			
Cattle	DE Mcal/kg	* 0.55	2.60	
Sheep	DE Mcal/kg	* 0.58	2.77	
Cattle	ME Mcal/kg	* 0.45	2.13	
Sheep	ME Mcal/kg	* 0.48	2.27	
Cattle	TDN %	* 12.4	59.0	
Sheep	TDN %	* 13.2	62.7	

Corn, white, flour –
see Corn, white, grits, cracked fine screened, (4)

Corn, white, hominy feed –
see Corn, white, grits by-product, mn 5% fat, (4)

Corn, white, hominy grits –
see Corn, white, grits, cracked fine screened, (4)

Corn, white, meal –
see Corn, white, grain, grnd, (4)

Corn, yellow, flour –
see Corn, yellow, grits, cracked fine screened, (4)

Corn, yellow, hominy grits –
see Corn, yellow, grits, cracked fine screened, (4)

CORYDALIS, GOLDEN. Corydalis aurea

Corydalis, golden, leaves, fresh, immature, (2)
Ref No 2 01 594 United States

Feed Name or Analyses		As Fed	Dry	C.V. ± %
Dry matter	%		100.0	
Ash	%		14.5	
Crude fiber	%		9.3	
Ether extract	%		3.4	
N-free extract	%		33.9	
Protein (N x 6.25)	%		38.9	
Cattle	dig prot %	*	31.0	
Goats	dig prot %	*	32.9	
Horses	dig prot %	*	30.6	
Rabbits	dig prot %	*	28.7	
Sheep	dig prot %	*	33.2	
Calcium	%		2.50	
Magnesium	%		0.37	
Phosphorus	%		0.72	
Potassium	%		4.25	

Cottage cheese –
see Cattle, cheese, cottage wet, (5)

COTTON. Gossypium spp

Cotton, bolls, s-c, (1)
Ref No 1 01 596 United States

Feed Name or Analyses		As Fed	Dry	C.V. ± %
Dry matter	%	92.0	100.0	1
Ash	%	7.1	7.7	14
Crude fiber	%	29.3	31.8	28
Sheep	dig coef %	39.	39.	
Ether extract	%	2.4	2.7	20
Sheep	dig coef %	64.	64.	
N-free extract	%	42.8	46.6	
Sheep	dig coef %	59.	59.	
Protein (N x 6.25)	%	10.4	11.3	19
Sheep	dig coef %	26.	26.	
Cattle	dig prot %	* 6.2	6.7	
Goats	dig prot %	* 6.5	7.1	
Horses	dig prot %	* 6.5	7.1	
Rabbits	dig prot %	* 6.8	7.4	
Sheep	dig prot %	2.7	2.9	
Fatty acids	%	2.4	2.6	
Energy	GE Mcal/kg			
Cattle	DE Mcal/kg	* 1.92	2.09	
Sheep	DE Mcal/kg	* 1.89	2.06	
Cattle	ME Mcal/kg	* 1.57	1.71	
Sheep	ME Mcal/kg	* 1.55	1.69	
Cattle	NE_m Mcal/kg	* 0.96	1.06	
Cattle	NE_{gain} Mcal/kg	* 0.18	0.20	
Cattle	$NE_{lactating\ cows}$ Mcal/kg	* 0.72	0.78	
Cattle	TDN %	* 43.5	47.3	
Sheep	TDN %	42.9	46.6	
Calcium	%	0.62	0.67	
Phosphorus	%	0.11	0.11	
Potassium	%	2.35	2.55	

Cotton, bolls process residue, s-c, (1)
Ref No 1 01 605 United States

Feed Name or Analyses		As Fed	Dry	C.V. ± %
Dry matter	%	88.5	100.0	
Ash	%	5.2	5.9	
Crude fiber	%	37.6	42.5	
Cattle	dig coef %	50.	50.	

Feed Name or Analyses		*	Mean As Fed	Mean Dry	C.V. ± %
Ether extract	%		2.0	2.3	
Cattle	dig coef %		55.	55.	
N-free extract	%		35.5	40.2	
Cattle	dig coef %		46.	46.	
Protein (N x 6.25)	%		8.2	9.3	
Cattle	dig coef %		6.	6.	
Cattle	dig prot %		0.5	0.6	
Goats	dig prot %	*	4.6	5.2	
Horses	dig prot %	*	4.8	5.4	
Rabbits	dig prot %	*	5.1	5.8	
Sheep	dig prot %	*	4.3	4.9	
Energy	GE Mcal/kg				
Cattle	DE Mcal/kg	*	1.68	1.90	
Sheep	DE Mcal/kg	*	1.75	1.98	
Cattle	ME Mcal/kg	*	1.38	1.56	
Sheep	ME Mcal/kg	*	1.43	1.62	
Cattle	TDN %		38.2	43.1	
Sheep	TDN %	*	39.8	45.0	
Calcium	%		0.58	0.65	
Phosphorus	%		0.11	0.12	

Cotton, fiber, (1)

Ref No 1 01 597 United States

Feed Name or Analyses		*	Mean As Fed	Mean Dry	C.V. ± %
Dry matter	%		92.0	100.0	
Ash	%		1.2	1.3	
Crude fiber	%		82.8	90.0	
Sheep	dig coef %		91.	91.	
Ether extract	%		0.5	0.5	
Sheep	dig coef %		-28.	-28.	
N-free extract	%		6.4	7.0	
Sheep	dig coef %		-152.	-152.	
Protein (N x 6.25)	%		1.1	1.2	
Sheep	dig coef %		-131.	-131.	
Cattle	dig prot %	*	-1.8	-1.9	
Goats	dig prot %	*	-2.0	-2.2	
Horses	dig prot %	*	-1.2	-1.4	
Rabbits	dig prot %	*	-0.3	-0.3	
Sheep	dig prot %		-1.4	-1.5	
Energy	GE Mcal/kg				
Sheep	DE Mcal/kg	*	2.81	3.05	
Sheep	ME Mcal/kg	*	2.30	2.50	
Sheep	TDN %		63.7	69.3	

Cotton, gin by-product, (1)

Cotton gin trash

Ref No 1 08 413 United States

Feed Name or Analyses		*	Mean As Fed	Mean Dry	C.V. ± %
Dry matter	%		90.3	100.0	
Ash	%		7.7	8.5	
Crude fiber	%		33.1	36.7	
Ether extract	%		1.5	1.7	
N-free extract	%		41.3	45.7	
Protein (N x 6.25)	%		6.7	7.4	
Cattle	dig prot %	*	3.0	3.4	
Goats	dig prot %	*	3.1	3.5	
Horses	dig prot %	*	3.5	3.8	
Rabbits	dig prot %	*	4.0	4.4	
Sheep	dig prot %	*	2.9	3.2	
Energy	GE Mcal/kg				
Cattle	DE Mcal/kg	*	1.79	1.98	
Sheep	DE Mcal/kg	*	1.95	2.16	
Cattle	ME Mcal/kg	*	1.46	1.62	
Sheep	ME Mcal/kg	*	1.60	1.77	
Cattle	TDN %	*	40.5	44.8	
Sheep	TDN %	*	44.3	47.1	

Cotton, leaves, (1)

Ref No 1 01 598 United States

Feed Name or Analyses		*	Mean As Fed	Mean Dry	C.V. ± %
Dry matter	%		92.4	100.0	1
Ash	%		16.1	17.4	5
Crude fiber	%		10.4	11.2	8
Ether extract	%		7.0	7.6	30
N-free extract	%		43.4	47.0	
Protein (N x 6.25)	%		15.5	16.7	8
Cattle	dig prot %	*	10.6	11.4	
Goats	dig prot %	*	11.3	12.2	
Horses	dig prot %	*	10.9	11.7	
Rabbits	dig prot %	*	10.7	11.6	
Sheep	dig prot %	*	10.7	11.6	
Energy	GE Mcal/kg				
Cattle	DE Mcal/kg	*	1.56	1.69	
Sheep	DE Mcal/kg	*	1.48	1.60	
Cattle	ME Mcal/kg	*	1.28	1.39	
Sheep	ME Mcal/kg	*	1.21	1.31	
Cattle	TDN %	*	35.5	38.4	
Sheep	TDN %	*	33.5	36.3	
Calcium	%		3.93	4.25	
Magnesium	%		0.82	0.89	
Manganese	mg/kg		25.7	27.8	
Phosphorus	%		0.22	0.24	82
Potassium	%		1.23	1.33	

Cotton, seed hulls, (1)

Cottonseed hulls (AAFCO)

Ref No 1 01 599 United States

Feed Name or Analyses		*	Mean As Fed	Mean Dry	C.V. ± %
Dry matter	%		90.9	100.0	3
Organic matter	%		88.2	97.1	
Ash	%		2.6	2.9	14
Crude fiber	%		40.9	45.0	
Cattle	dig coef %		48.	48.	
Sheep	dig coef %		53.	53.	
Ether extract	%		1.4	1.6	77
Cattle	dig coef %		78.	78.	
Sheep	dig coef %		76.	76.	
N-free extract	%		34.9	38.4	
Cattle	dig coef %		42.	42.	
Sheep	dig coef %		60.	60.	
Protein (N x 6.25)	%		4.0	4.4	12
Cattle	dig coef %		-17.	-17.	
Sheep	dig coef %		-5.	-5.	
Cattle	dig prot %		-0.6	-0.7	
Goats	dig prot %	*	0.6	0.6	
Horses	dig prot %	*	1.1	1.2	
Rabbits	dig prot %	*	1.8	2.0	
Sheep	dig prot %		-0.1	-0.2	
Lignin (Ellis)	%		20.8	22.9	
Fatty acids	%		0.9	1.0	
Energy	GE Mcal/kg		3.91	4.30	
Cattle	DE Mcal/kg	*	2.07	2.28	
Sheep	DE Mcal/kg	*	1.73	1.91	
Cattle	ME Mcal/kg	*	1.70	1.87	
Sheep	ME Mcal/kg	*	1.42	1.57	
Cattle	NE_m Mcal/kg	*	0.94	1.03	
Cattle	NE_{gain} Mcal/kg	*	0.17	0.19	
Cattle	$NE_{lactating\ cows}$ Mcal/kg	*	0.68	0.75	
Cattle	TDN %		46.9	51.6	
Sheep	TDN %		39.3	43.3	
Calcium	%		0.13	0.14	
Magnesium	%		0.13	0.14	
Phosphorus	%		0.06	0.07	
Potassium	%		0.87	0.96	

Cotton, seed hulls wo lint, (1)

Cottonseed hull bran

Ref No 1 01 600 United States

Feed Name or Analyses		*	Mean As Fed	Mean Dry	C.V. ± %
Dry matter	%		90.9	100.0	1
Ash	%		2.6	2.8	27
Crude fiber	%		38.6	42.4	3
Sheep	dig coef %		46.	46.	
Ether extract	%		0.7	0.8	28
Sheep	dig coef %		76.	76.	
N-free extract	%		45.5	50.0	
Sheep	dig coef %		63.	63.	
Protein (N x 6.25)	%		3.6	4.0	26
Sheep	dig coef %		0.	0.	26
Cattle	dig prot %	*	0.3	0.4	
Goats	dig prot %	*	0.2	0.3	
Horses	dig prot %	*	0.8	0.9	
Rabbits	dig prot %	*	1.6	1.7	
Energy	GE Mcal/kg				
Cattle	DE Mcal/kg	*	1.91	2.10	
Sheep	DE Mcal/kg	*	2.10	2.31	
Cattle	ME Mcal/kg	*	1.56	1.32	
Sheep	ME Mcal/kg	*	1.72	1.89	
Cattle	TDN %	*	43.4	47.7	
Sheep	TDN %		47.6	52.3	
Calcium	%		0.10	0.11	39
Cobalt	mg/kg		0.058	0.064	23
Copper	mg/kg		4.2	4.6	
Iron	%		0.035	0.038	45
Manganese	mg/kg		17.0	18.7	13
Phosphorus	%		0.10	0.11	41
Niacin	mg/kg		7.6	8.4	24
Riboflavin	mg/kg		2.6	2.9	
Thiamine	mg/kg		0.2	0.2	

Cotton, stems, (1)

Ref No 1 01 601 United States

Feed Name or Analyses		*	Mean As Fed	Mean Dry	C.V. ± %
Dry matter	%		92.3	100.0	
Ash	%		4.2	4.6	
Crude fiber	%		43.8	47.5	
Ether extract	%		0.8	0.9	
N-free extract	%		37.8	41.0	
Protein (N x 6.25)	%		5.7	6.2	
Cattle	dig prot %	*	2.1	2.3	
Goats	dig prot %	*	2.1	2.3	
Horses	dig prot %	*	2.6	2.8	
Rabbits	dig prot %	*	3.2	3.4	
Sheep	dig prot %	*	1.9	2.1	
Energy	GE Mcal/kg				
Cattle	DE Mcal/kg	*	1.65	1.79	
Sheep	DE Mcal/kg	*	1.98	2.15	
Cattle	ME Mcal/kg	*	1.36	1.47	
Sheep	ME Mcal/kg	*	1.63	1.76	
Cattle	TDN %	*	37.4	40.5	
Sheep	TDN %	*	45.0	48.7	

Cotton, browse, immature, (2)

Ref No 2 01 602 United States

Feed Name or Analyses		*	Mean As Fed	Mean Dry	C.V. ± %
Dry matter	%			100.0	
Ash	%			6.3	
Crude fiber	%			33.4	
Ether extract	%			4.2	
N-free extract	%			46.3	

Continued

Feed Name or Analyses		Mean As Fed	Mean Dry	C.V. ± %
Protein (N x 6.25)	%		9.8	
Cattle	dig prot %	*	6.2	
Goats	dig prot %	*	5.7	
Horses	dig prot %	*	5.9	
Rabbits	dig prot %	*	6.2	
Sheep	dig prot %	*	6.1	

Cotton, browse, mature, (2)

Ref No 2 01 603 United States

Feed Name or Analyses		Mean As Fed	Mean Dry	C.V. ± %
Dry matter	%		100.0	
Ash	%		5.7	
Crude fiber	%		37.8	
Ether extract	%		2.8	
N-free extract	%		47.2	
Protein (N x 6.25)	%		6.5	
Cattle	dig prot %	*	3.4	
Goats	dig prot %	*	2.6	
Horses	dig prot %	*	3.0	
Rabbits	dig prot %	*	3.7	
Sheep	dig prot %	*	3.0	

Cotton, seed hulls, fresh, (2)

Ref No 1 01 604 United States

Feed Name or Analyses		Mean As Fed	Mean Dry	C.V. ± %
Dry matter	%	90.8	100.0	2
Cellulose (Matrone)	%	54.7	60.3	
Pentosans	%	26.4	29.1	
Lignin (Ellis)	%	24.1	26.6	
Chlorine	%	0.02	0.02	
Cobalt	mg/kg	0.028	0.031	
Sodium	%	0.02	0.02	
Zinc	mg/kg	20.0	22.0	
Riboflavin	mg/kg	3.8	4.2	

Cotton, immature seeds, dehy, (5)

Ref No 5 08 414 United States

Feed Name or Analyses		Mean As Fed	Mean Dry	C.V. ± %
Dry matter	%	93.2	100.0	
Ash	%	3.7	4.0	
Crude fiber	%	24.1	25.9	
Ether extract	%	15.9	17.1	
N-free extract	%	29.0	31.1	
Protein (N x 6.25)	%	20.5	22.0	
Energy	GE Mcal/kg			
Cattle	DE Mcal/kg	* 3.31	3.40	
Sheep	DE Mcal/kg	* 3.21	3.04	
Swine	DE kcal/kg	*3505.	3761.	
Cattle	ME Mcal/kg	* 2.71	2.91	
Sheep	ME Mcal/kg	* 2.63	2.82	
Swine	ME kcal/kg	*3211.	3445.	
Cattle	TDN %	* 75.1	80.6	
Sheep	TDN %	* 72.8	78.1	
Swine	TDN %	* 79.5	85.3	

Cotton, seed flour, coarse bolt, (5)

Ref No 5 01 612 United States

Feed Name or Analyses		Mean As Fed	Mean Dry	C.V. ± %
Dry matter	%	94.3	100.0	1
Organic matter	%	88.0	93.3	
Ash	%	6.4	6.8	3

(1) dry forages and roughages (2) pasture, range plants, and forages fed green (3) silages (4) energy feeds (5) protein supplements (6) minerals (7) vitamins (8) additives

Feed Name or Analyses		Mean As Fed	Mean Dry	C.V. ± %
Crude fiber	%	2.3	2.5	17
Ether extract	%	6.0	6.4	4
N-free extract	%	23.0	24.4	
Protein (N x 6.25)	%	56.6	60.1	5
Sugars, total	%	6.5	6.8	
Energy	GE Mcal/kg			
Cattle	DE Mcal/kg	* 3.49	3.70	
Sheep	DE Mcal/kg	* 3.42	3.62	
Swine	DE kcal/kg	*3517.	3729.	
Cattle	ME Mcal/kg	* 2.86	3.04	
Sheep	ME Mcal/kg	* 2.80	2.97	
Swine	ME kcal/kg	*2949.	3128.	
Cattle	TDN %	* 79.2	84.0	
Sheep	TDN %	* 77.5	82.2	
Swine	TDN %	* 79.8	84.6	
Calcium	%	0.22	0.23	
Niacin	mg/kg	77.2	81.9	
Riboflavin	mg/kg	9.3	9.8	
Thiamine	mg/kg	9.5	10.1	
Arginine	%	6.61	7.01	11
Histidine	%	0.80	0.85	98
Isoleucine	%	0.20	0.21	
Leucine	%	3.50	3.72	
Lysine	%	2.33	2.47	18
Methionine	%	0.80	0.85	21
Phenylalanine	%	3.00	3.18	8
Threonine	%	2.00	2.12	17
Tryptophan	%	0.80	0.85	
Valine	%	2.60	2.76	
Gossypol, total	%	1.25	1.33	
Gossypol, free	%	0.04	0.04	

Cotton, seeds, grnd, (5)

Cottonseed, whole, ground

Ref No 5 01 608 United States

Feed Name or Analyses		Mean As Fed	Mean Dry	C.V. ± %
Dry matter	%	92.7	100.0	
Ash	%	3.5	3.8	
Crude fiber	%	16.9	18.2	
Ether extract	%	22.9	24.7	
N-free extract	%	26.3	28.4	
Protein (N x 6.25)	%	23.1	24.9	
Energy	GE Mcal/kg			
Cattle	DE Mcal/kg	* 3.82	4.12	
Sheep	DE Mcal/kg	* 4.00	4.32	
Swine	DE kcal/kg	*4839.	5220.	
Cattle	ME Mcal/kg	* 3.13	3.38	
Sheep	ME Mcal/kg	* 3.28	3.54	
Swine	ME kcal/kg	*4402.	4749.	
Cattle	NE_m Mcal/kg	* 1.86	2.01	
Cattle	NE_{gain} Mcal/kg	* 1.11	1.20	
Cattle	$NE_{lactating\ cows}$ Mcal/kg	* 2.41	2.60	
Cattle	TDN %	* 86.6	93.5	
Sheep	TDN %	* 90.8	97.9	
Swine	TDN %	* 109.7	118.4	
Calcium	%	0.14	0.15	
Copper	mg/kg	50.0	54.0	
Iron	%	0.014	0.015	
Magnesium	%	0.32	0.35	
Manganese	mg/kg	12.1	13.1	
Phosphorus	%	0.68	0.73	
Potassium	%	1.11	1.20	
Sodium	%	0.29	0.31	
Sulphur	%	0.24	0.26	

Cotton, seeds, mech-extd grnd, (5)

Whole-pressed cottonseed, mechanical extracted (AAFCO)

Ref No 5 01 609 United States

Feed Name or Analyses		Mean As Fed	Mean Dry	C.V. ± %
Dry matter	%	92.4	100.0	
Ash	%	4.6	5.0	
Crude fiber	%	21.4	23.2	
Ether extract	%	5.2	5.6	
N-free extract	%	33.2	35.9	
Protein (N x 6.25)	%	28.0	30.3	
Energy	GE Mcal/kg			
Cattle	DE Mcal/kg	* 2.68	2.90	
Sheep	DE Mcal/kg	* 2.59	2.80	
Swine	DE kcal/kg	*2559.	2769.	
Cattle	ME Mcal/kg	* 2.20	2.38	
Sheep	ME Mcal/kg	* 2.13	2.36	
Swine	ME kcal/kg	*2300.	2489.	
Cattle	NE_m Mcal/kg	* 1.43	1.55	
Cattle	NE_{gain} Mcal/kg	* 0.91	0.98	
Cattle	$NE_{lactating\ cows}$ Mcal/kg	* 1.20	1.30	
Cattle	TDN %	* 60.9	65.9	
Sheep	TDN %	* 58.6	63.4	
Swine	TDN %	* 58.0	62.8	
Calcium	%	0.17	0.18	
Magnesium	%	0.36	0.39	
Phosphorus	%	0.64	0.69	
Potassium	%	1.25	1.35	

Cotton, seeds, roasted, (5)

Ref No 5 01 613 United States

Feed Name or Analyses		Mean As Fed	Mean Dry	C.V. ± %
Dry matter	%	90.7	100.0	
Ash	%	2.3	2.5	
Crude fiber	%	24.0	26.5	
Cattle	dig coef %	66.	66.	
Ether extract	%	22.5	24.8	
Cattle	dig coef %	72.	72.	
N-free extract	%	25.8	28.4	
Cattle	dig coef %	51.	51.	
Protein (N x 6.25)	%	16.1	17.8	
Cattle	dig coef %	47.	47.	
Cattle	dig prot %	7.6	8.4	
Energy	GE Mcal/kg			
Cattle	DE Mcal/kg	* 3.22	3.55	
Sheep	DE Mcal/kg	* 2.95	3.25	
Swine	DE kcal/kg	*2923.	3223.	
Cattle	ME Mcal/kg	* 2.64	2.91	
Sheep	ME Mcal/kg	* 2.42	2.67	
Swine	ME kcal/kg	*2701.	2978.	
Cattle	TDN %	73.0	80.5	
Sheep	TDN %	* 66.9	73.7	
Swine	TDN %	* 66.3	73.1	

Cotton, seeds mech-extd w added hulls, grnd, high fiber, (5)

Ref No 5 01 610 United States

Feed Name or Analyses		Mean As Fed	Mean Dry	C.V. ± %
Dry matter	%	92.5	100.0	
Ash	%	6.0	6.5	
Crude fiber	%	22.3	24.1	
Cattle	dig coef %	74.	74.	
Ether extract	%	8.5	9.2	
Cattle	dig coef %	98.	98.	
N-free extract	%	34.6	37.4	
Cattle	dig coef %	59.	59.	
Protein (N x 6.25)	%	21.1	22.8	
Cattle	dig coef %	85.	85.	

Feed Name or Analyses		Mean As Fed	Mean Dry	C.V. ± %
Cattle	dig prot %	17.9	19.4	
Energy	GE Mcal/kg			
Cattle	DE Mcal/kg	* 3.25	3.51	
Sheep	DE Mcal/kg	* 3.15	3.41	
Swine	DE kcal/kg	*3507.	3791.	
Cattle	ME Mcal/kg	* 2.66	2.88	
Sheep	ME Mcal/kg	* 2.59	2.80	
Swine	ME kcal/kg	*3205.	3465.	
Cattle	TDN %	73.6	79.6	
Sheep	TDN %	* 71.5	77.3	
Swine	TDN %	* 79.5	86.0	

Cotton, seeds w lint, low protein high fiber, (5)

Ref No 5 01 614 United States

Feed Name or Analyses		Mean As Fed	Mean Dry	C.V. ± %
Dry matter	%	92.5	100.0	
Ash	%	4.4	4.8	
Crude fiber	%	24.5	26.5	
Cattle	dig coef %	58.	58.	
Ether extract	%	17.9	19.4	
Cattle	dig coef %	89.	89.	
N-free extract	%	28.6	30.9	
Cattle	dig coef %	52.	52.	
Protein (N x 6.25)	%	17.0	18.4	
Cattle	dig coef %	63.	63.	
Cattle	dig prot %	10.7	11.6	
Energy	GE Mcal/kg			
Cattle	DE Mcal/kg	* 3.34	3.61	
Sheep	DE Mcal/kg	* 3.19	3.45	
Cattle	ME Mcal/kg	* 2.74	2.96	
Sheep	ME Mcal/kg	* 2.62	2.83	
Cattle	TDN %	75.7	81.9	
Sheep	TDN %	* 72.4	78.3	

Cotton, seeds wo hulls, (5)

Ref No 5 08 416 United States

Feed Name or Analyses		Mean As Fed	Mean Dry	C.V. ± %
Dry matter	%	93.6	100.0	
Ash	%	4.5	4.8	
Crude fiber	%	2.3	2.5	
Ether extract	%	33.3	35.6	
N-free extract	%	15.1	16.1	
Protein (N x 6.25)	%	38.4	41.0	
Energy	GE Mcal/kg			
Cattle	DE Mcal/kg	* 4.93	5.26	
Sheep	DE Mcal/kg	* 4.95	5.29	
Swine	DE kcal/kg	*5477.	5851.	
Cattle	ME Mcal/kg	* 4.04	4.32	
Sheep	ME Mcal/kg	* 4.06	4.34	
Swine	ME kcal/kg	*4803.	5131.	
Cattle	TDN %	* 111.7	119.4	
Sheep	TDN %	* 112.3	120.0	
Swine	TDN %	* 124.2	132.7	

Cotton, seeds wo hulls, mech-extd grnd, mn 43% protein mx 13% fiber mn 2% fat, (5)

Ref No 5 01 627 United States

Feed Name or Analyses		Mean As Fed	Mean Dry	C.V. ± %
Dry matter	%	93.6	100.0	1
Ash	%	6.2	6.6	7
Crude fiber	%	11.0	11.7	5
Ether extract	%	5.6	6.0	7
N-free extract	%	27.1	29.0	8
Protein (N x 6.25)	%	43.6	46.6	4
Energy	GE Mcal/kg			
Cattle	DE Mcal/kg	* 3.10	3.32	
Sheep	DE Mcal/kg	* 3.20	3.42	

Feed Name or Analyses		Mean As Fed	Mean Dry	C.V. ± %
Swine	DE kcal/kg	*3364.	3595.	
Cattle	ME Mcal/kg	* 2.55	2.72	
Sheep	ME Mcal/kg	* 2.62	2.80	
Swine	ME kcal/kg	*2913.	3113.	
Cattle	TDN %	* 70.4	75.2	
Sheep	TDN %	* 72.6	77.6	
Swine	TDN %	* 76.3	81.5	
Calcium	%	0.22	0.23	
Chlorine	%	0.01	0.01	
Copper	mg/kg	11.0	11.8	
Iron	%	0.058	0.062	
Magnesium	%	0.59	0.63	
Manganese	mg/kg	9.2	9.9	
Phosphorus	%	1.06	1.13	
Potassium	%	1.41	1.50	
Sodium	%	0.01	0.01	
Sulphur	%	0.48	0.52	

Cotton, seeds wo hulls, pre-press solv-extd grnd, mn 50% protein, (5)

Cottonseed meal, pre-press solvent extracted, 50% protein

Ref No 5 07 874 United States

Feed Name or Analyses		Mean As Fed	Mean Dry	C.V. ± %
Dry matter	%	94.3	100.0	
Ash	%	7.4	7.8	
Crude fiber	%	8.1	8.6	
Cattle	dig coef %	* 57.	57.	
Sheep	dig coef %	* 69.	69.	
Ether extract	%	1.1	1.2	
Cattle	dig coef %	* 92.	92.	
Sheep	dig coef %	* 97.	97.	
N-free extract	%	27.3	28.9	
Cattle	dig coef %	* 80.	80.	
Sheep	dig coef %	* 79.	79.	
Protein (N x 6.25)	%	52.0	55.1	
Cattle	dig coef %	* 81.	81.	
Sheep	dig coef %	* 84.	84.	
Energy	GE Mcal/kg			
Chickens	ME_n kcal/kg	2299.	2438.	
Cattle	NE_m Mcal/kg	*	1.70	
Cattle	NE_{gain} Mcal/kg	*	1.11	
Cattle	$NE_{lactating\ cows}$ Mcal/kg	*	2.01	
Gossypol, total	%	1.02	1.08	
Gossypol, free	%	0.02	0.02	

Cotton, seeds w some hulls, mech-extd caked, mn 36% protein, (5)

Cottonseed cake, mechanical extracted (AAFCO)

Ref No 5 01 623 United States

Feed Name or Analyses		Mean As Fed	Mean Dry	C.V. ± %
Dry matter	%	92.1	100.0	
Ash	%	7.2	7.8	
Crude fiber	%	9.9	10.7	
Ether extract	%	2.3	2.5	
N-free extract	%	26.5	28.8	
Protein (N x 6.25)	%	46.2	50.2	
Cellulose (Matrone)	%	10.1	11.0	
Lignin (Ellis)	%	5.2	5.6	
Energy	GE Mcal/kg			
Cattle	DE Mcal/kg	* 2.88	3.12	
Sheep	DE Mcal/kg	* 3.05	3.31	
Swine	DE kcal/kg	*3003.	3260.	
Cattle	ME Mcal/kg	* 2.36	2.56	
Sheep	ME Mcal/kg	* 2.50	2.71	
Swine	ME kcal/kg	*2578.	2799.	
Cattle	TDN %	* 65.3	70.9	

Feed Name or Analyses		Mean As Fed	Mean Dry	C.V. ± %
Sheep	TDN %	* 69.1	75.0	
Swine	TDN %	* 68.1	73.9	

Cotton, seeds w some hulls, mech-extd grnd, mn 36% protein mx 17% fiber mn 2% fat, (5)

Cottonseed meal, 36% protein

Ref No 5 01 615 United States

Feed Name or Analyses		Mean As Fed	Mean Dry	C.V. ± %
Dry matter	%	91.8	100.0	
Ash	%	6.0	6.5	
Crude fiber	%	14.0	15.3	
Ether extract	%	6.4	7.0	
N-free extract	%	29.5	32.1	
Protein (N x 6.25)	%	35.9	39.1	
Energy	GE Mcal/kg			
Cattle	DE Mcal/kg	* 2.98	3.22	
Sheep	DE Mcal/kg	* 2.91	3.14	
Swine	DE kcal/kg	*3185.	3439.	
Cattle	ME Mcal/kg	* 2.44	2.64	
Sheep	ME Mcal/kg	* 2.38	2.57	
Swine	ME kcal/kg	*2793.	3016.	
Cattle	NE_m Mcal/kg	* 1.54	1.66	
Cattle	NE_{gain} Mcal/kg	* 1.00	1.08	
Cattle	$NE_{lactating\ cows}$ Mcal/kg	* 2.44	2.64	
Cattle	TDN %	* 67.7	73.1	
Sheep	TDN %	* 66.0	71.3	
Swine	TDN %	* 72.2	78.0	
Calcium	%	0.23	0.25	
Chlorine	%	0.02	0.02	
Magnesium	%	0.49	0.53	
Phosphorus	%	0.92	1.00	
Potassium	%	1.40	1.53	
Sodium	%	0.06	0.07	
Sulphur	%	0.26	0.28	

Cotton, seeds w some hulls, mech-extd grnd, mn 41% protein mx 14% fiber mn 2% fat, (5)

Cottonseed meal, 41% protein

Ref No 5 01 617 United States

Feed Name or Analyses		Mean As Fed	Mean Dry	C.V. ± %
Dry matter	%	92.7	100.0	
Ash	%	6.1	6.6	
Crude fiber	%	10.9	11.8	
Cattle	dig coef %	* 57.	57.	
Sheep	dig coef %	* 60.	60.	
Ether extract	%	5.6	6.0	
Cattle	dig coef %	* 92.	92.	
Sheep	dig coef %	* 104.	104.	
N-free extract	%	28.6	30.9	
Cattle	dig coef %	* 80.	80.	
Sheep	dig coef %	* 58.	58.	
Protein (N x 6.25)	%	* 41.4	44.7	
Cattle	dig coef %	* 81.	81.	
Sheep	dig coef %	* 81.	81.	
Cattle	dig prot %	33.5	36.2	
Sheep	dig prot %	33.5	36.2	
Sugars, total	%	6.5	7.0	
Lignin (Ellis)	%	6.4	6.9	
Energy	GE Mcal/kg	4.48	4.84	
Cattle	DE Mcal/kg	3.21	3.47	
Sheep	DE Mcal/kg	3.35	3.63	
Swine	DE kcal/kg	*3361.	3623.	
Cattle	ME Mcal/kg	2.36	2.56	
Chickens	ME_n kcal/kg	2312.	2499.	
Sheep	ME Mcal/kg	2.50	2.70	
Swine	ME kcal/kg	*2929.	3158.	
Cattle	NE_m Mcal/kg	* 1.68	1.81	

Continued

843

Feed Name or Analyses		Mean As Fed	Mean Dry	C.V. ± %
Cattle	NE_{gain} Mcal/kg	* 1.11	1.20	
Cattle	$NE_{lactating\ cows}$ Mcal/kg	* 1.97	2.12	
Cattle	TDN %	73.6	79.4	
Sheep	TDN %	68.6	74.0	
Calcium	%	0.19	0.20	
Chlorine	%	0.06	0.06	
Cobalt	mg/kg	1.9	2.0	
Copper	mg/kg	18.2	19.6	
Iron	%	0.01	0.01	
Magnesium	%	0.49	0.53	
Manganese	mg/kg	23.0	24.8	
Phosphorus	%	1.09	1.18	
Potassium	%	1.25	1.35	
Sodium	%	0.05	0.05	
Sulphur	%	0.40	0.43	
Zinc	mg/kg	78.0	84.1	
Biotin	mg/kg	0.54	0.58	
Choline	mg/kg	2797.	3017.	
Folic acid	mg/kg	3.41	3.68	
Niacin	mg/kg	34.5	37.2	
Pantothenic acid	mg/kg	10.2	11.0	
Riboflavin	mg/kg	5.3	5.7	
Thiamine	mg/kg	5.2	5.6	
α-tocopherol	mg/kg	40.0	43.1	
Vitamin B_6	mg/kg	4.0	4.3	
Alanine	%	1.48	1.60	
Arginine	%	4.15	4.49	
Aspartic acid	%	3.42	3.70	
Cystine	%	0.70	0.76	
Glutamic acid	%	7.92	8.56	
Glycine	%	1.65	1.78	
Histidine	%	1.01	1.09	
Isoleucine	%	1.21	1.31	
Leucine	%	2.12	2.29	
Lysine	%	1.47	1.59	
Methionine	%	0.52	0.56	
Phenylalanine	%	2.04	2.21	
Proline	%	1.33	1.44	
Serine	%	1.56	1.69	
Threonine	%	1.21	1.31	
Tryptophan	%	0.48	0.52	
Tyrosine	%	1.05	1.14	
Valine	%	1.76	1.90	
Gossypol, total	%	0.98	1.06	
Gossypol, free	%	0.03	0.03	

Cotton, seeds w some hulls, pre-press solv-extd grnd, 41% protein, (5)

Cottonseed meal, pre-press solvent extracted 41% protein

Ref No 5 07 872 United States

Feed Name or Analyses		Mean As Fed	Mean Dry	C.V. ± %
Dry matter	%	91.0	100.0	
Ash	%	6.4	7.0	
Crude fiber	%	12.9	14.2	
Cattle	dig coef %	* 57.	57.	
Sheep	dig coef %	* 60.	60.	
Ether extract	%	0.9	1.0	
Ether extract	%	* 92.	92.	
Sheep	dig coef %	* 104.	104.	
N-free extract	%	29.4	32.3	
Cattle	dig coef %	* 80.	80.	
Sheep	dig coef %	* 58.	58.	
Protein (N x 6.25)	%	41.4	45.5	
Cattle	dig coef %	* 81.	81.	
Sheep	dig coef %	81.	81.	
Energy	GE Mcal/kg	4.25	4.67	
Cattle	DE Mcal/kg	3.22	3.54	
Sheep	DE Mcal/kg	3.46	3.80	
Cattle	ME Mcal/kg	2.38	2.62	
Chickens	ME_n kcal/kg	1952.	2145.	
Sheep	ME Mcal/kg	2.66	2.92	
Calcium	%	0.19	0.21	
Cobalt	mg/kg	1.900	2.088	
Copper	mg/kg	18.9	20.8	
Iron	%	0.01	0.01	
Magnesium	%	0.44	0.48	
Manganese	mg/kg	22.0	24.2	
Phosphorus	%	1.02	1.12	
Potassium	%	1.26	1.38	
Sodium	%	0.06	0.06	
Zinc	mg/kg	92.0	101.0	
Biotin	mg/kg	0.52	0.57	
Choline	mg/kg	2810.	3088.	
Folic acid	mg/kg	2.74	3.01	
Niacin	mg/kg	37.8	41.5	
Pantothenic acid	mg/kg	7.8	8.6	
Riboflavin	mg/kg	4.2	4.6	
Thiamine	mg/kg	9.7	10.7	
Vitamin B_6	mg/kg	4.80	5.30	
Alanine	%	1.49	1.64	
Arginine	%	4.20	4.61	
Aspartic acid	%	3.44	3.78	
Cystine	%	0.66	0.72	
Glutamic acid	%	7.67	8.43	
Glycine	%	1.57	1.72	
Histidine	%	1.03	1.13	
Isoleucine	%	1.22	1.34	
Leucine	%	2.21	2.43	
Lysine	%	1.59	1.75	
Methionine	%	0.48	0.53	
Phenylalanine	%	2.05	2.25	
Proline	%	1.45	1.59	
Serine	%	1.59	1.75	
Threonine	%	1.20	1.32	
Tryptophan	%	0.44	0.48	
Tyrosine	%	1.06	1.16	
Valine	%	1.72	1.89	
Gossypol, total	%	1.01	1.11	
Gossypol, free	%	0.04	0.04	

Cotton, seeds w some hulls, solv-extd grnd, mn 36% protein, (5)

Cottonseed meal, solvent extracted (AAFCO)

Ref No 5 01 632 United States

Feed Name or Analyses		Mean As Fed	Mean Dry	C.V. ± %
Dry matter	%	91.4	100.0	2
Ash	%	6.5	7.1	10
Crude fiber	%	11.0	12.0	20
Ether extract	%	1.6	1.8	60
N-free extract	%	30.7	33.6	
Protein (N x 6.25)	%	41.6	45.5	7
Energy	GE Mcal/kg	4.23	4.63	1
Cattle	DE Mcal/kg	* 2.82	3.08	
Sheep	DE Mcal/kg	* 3.05	3.33	
Swine	DE kcal/kg	*2999.	3281.	
Cattle	ME Mcal/kg	* 2.31	2.53	
Chickens	ME_n kcal/kg	1820.	1991.	
Sheep	ME Mcal/kg	* 2.50	2.73	
Swine	ME kcal/kg	*2603.	2848.	
Cattle	TDN %	* 63.9	69.9	
Sheep	TDN %	* 69.1	75.6	
Swine	TDN %	* 68.0	74.4	
Calcium	%	0.15	0.16	34
Chlorine	%	0.04	0.04	20
Cobalt	mg/kg	0.079	0.087	97
Copper	mg/kg	21.4	23.4	41
Iron	%	0.015	0.016	50
Magnesium	%	0.58	0.63	34
Manganese	mg/kg	20.1	21.9	27
Phosphorus	%	1.10	1.20	25
Potassium	%	1.47	1.61	43
Sodium	%	0.05	0.05	98
Sulphur	%	0.21	0.23	
Choline	mg/kg	2868.	3138.	25
Folic acid	mg/kg	0.77	0.84	
Niacin	mg/kg	45.6	49.9	9
Pantothenic acid	mg/kg	17.9	19.5	
Riboflavin	mg/kg	4.6	5.1	18
Thiamine	mg/kg	8.2	8.9	55
α-tocopherol	mg/kg	10.9	12.0	32
Lysine	%	1.60	1.75	
Methionine	%	0.60	0.66	
Tryptophan	%	0.50	0.55	

Cotton, seeds w some hulls, solv-extd grnd, mn 41% protein mx 14% fiber mn 0.5% fat, (5)

Cottonseed meal, solvent extracted, 41% protein

Ref No 5 01 621 United States

Feed Name or Analyses		Mean As Fed	Mean Dry	C.V. ± %
Dry matter	%	91.1	100.0	
Ash	%	6.2	6.8	
Crude fiber	%	11.4	12.5	
Cattle	dig coef %	* 57.	57.	
Sheep	dig coef %	* 60.	60.	
Ether extract	%	2.1	2.3	
Cattle	dig coef %	* 92.	92.	
Sheep	dig coef %	* 104.	104.	
N-free extract	%	31.8	34.9	
Cattle	dig coef %	* 80.	80.	
Sheep	dig coef %	* 58.	58.	
Protein (N x 6.25)	%	41.9	46.0	
Cattle	dig coef %	* 81.	81.	
Sheep	dig coef %	* 81.	81.	
Cattle	dig prot %	33.9	37.3	
Sheep	dig prot %	33.9	37.3	
Energy	GE Mcal/kg	4.35	4.77	
Cattle	DE Mcal/kg	3.28	3.60	
Sheep	DE Mcal/kg	3.45	3.78	
Cattle	ME Mcal/kg	2.50	2.74	
Chickens	ME_n kcal/kg	2098.	2303.	
Sheep	ME Mcal/kg	2.67	2.93	
Cattle	NE_m Mcal/kg	* 1.55	1.69	
Cattle	NE_{gain} Mcal/kg	* 1.02	1.11	
Cattle	$NE_{lactating\ cows}$ Mcal/kg	* 1.85	2.01	
Cattle	TDN %	68.5	75.2	
Sheep	TDN %	62.6	68.7	
Calcium	%	0.16	0.18	
Chlorine	%	0.04	0.04	
Cobalt	mg/kg	1.90	2.10	
Copper	mg/kg	18.1	19.9	
Iron	%	0.01	0.01	
Magnesium	%	0.47	0.52	
Manganese	mg/kg	21.8	23.9	
Phosphorus	%	1.06	1.16	
Potassium	%	1.26	1.38	
Sodium	%	0.06	0.07	
Sulphur	%	0.21	0.23	
Zinc	mg/kg	60.0	66.0	
Biotin	mg/kg	0.54	0.59	
Choline	mg/kg	2784.	3056.	

(1) dry forages and roughages (2) pasture, range plants, and forages fed green (3) silages (4) energy feeds (5) protein supplements (6) minerals (7) vitamins (8) additives

Feed Name or Analyses		Mean As Fed	Mean Dry	C.V. ± %
Folic acid	mg/kg	2.79	3.06	
Niacin	mg/kg	42.4	46.5	
Pantothenic acid	mg/kg	13.8	15.1	
Riboflavin	mg/kg	4.4	4.8	
Thiamine	mg/kg	7.6	8.3	
α-tocopherol	mg/kg	15.0	16.3	
Vitamin B6	mg/kg	4.8	5.3	
Alanine	%	1.47	1.61	
Arginine	%	4.36	4.79	
Aspartic acid	%	3.35	3.68	
Cystine	%	0.67	0.73	
Glutamic acid	%	7.48	8.21	
Glycine	%	1.61	1.77	
Histidine	%	1.02	1.12	
Isoleucine	%	1.20	1.32	
Leucine	%	2.17	2.38	
Lysine	%	1.59	1.74	
Methionine	%	0.49	0.54	
Phenylalanine	%	2.00	2.20	
Proline	%	1.39	1.53	
Serine	%	1.57	1.72	
Threonine	%	1.21	1.33	
Tryptophan	%	0.48	0.53	
Tyrosine	%	1.06	1.16	
Valine	%	1.64	1.80	
Gossypol, total	%	0.99	1.09	
Gossypol, free	%	0.23	0.25	

Cotton, seeds w some hulls, solv-extd grnd, mn 43% protein mx 13% fiber mn 0.5% fat, (5)

Ref No 5 01 630 United States

Analyses		As Fed	Dry	C.V. ± %
Dry matter	%	92.3	100.0	
Ash	%	5.8	6.3	
Crude fiber	%	10.3	11.2	13
Ether extract	%	1.3	1.5	24
N-free extract	%	30.1	32.6	
Protein (N x 6.25)	%	44.8	48.5	5
Energy	GE Mcal/kg			
Cattle	DE Mcal/kg	*2.89	3.14	
Sheep	DE Mcal/kg	*2.86	3.10	
Swine	DE kcal/kg	*3017.	3270.	
Cattle	ME Mcal/kg	*2.37	2.57	
Chickens	ME_n kcal/kg	1856.	2012.	18
Sheep	ME Mcal/kg	*2.34	2.54	
Swine	ME kcal/kg	*2600.	2819.	
Cattle	TDN %	*65.6	71.1	
Sheep	TDN %	*64.8	70.2	
Swine	TDN %	*68.4	74.2	
Calcium	%	0.18	0.20	
Phosphorus	%	1.09	1.18	
Choline	mg/kg	2828.	3065.	
Niacin	mg/kg	45.0	48.8	
Pantothenic acid	mg/kg	17.6	19.1	
Riboflavin	mg/kg	4.6	5.0	
Arginine	%	4.44	4.81	
Cystine	%	0.89	0.96	
Glycine	%	2.37	2.57	
Lysine	%	1.58	1.71	
Methionine	%	0.67	0.73	
Tryptophan	%	0.57	0.62	
Gossypol, free	%	0.04	0.04	21

Cotton, seeds w some hulls wo gossypol, solv-extd grnd, mx 0.04% free gossypol, (5)

Low gossypol cottonseed meal, solvent extracted (AAFCO)

Cottonseed meal, degossypolized, solvent extracted

Ref No 5 01 633 United States

Analyses		As Fed	Dry	C.V. ± %
Dry matter	%	92.7	100.0	1
Organic matter	%	87.1	94.0	
Ash	%	5.8	6.3	11
Crude fiber	%	12.7	13.7	21
Ether extract	%	1.5	1.6	98
N-free extract	%	31.3	33.7	
Protein (N x 6.25)	%	41.5	44.8	7
Sugars, total	%	6.7	7.2	
Energy	GE Mcal/kg			
Cattle	DE Mcal/kg	*2.82	3.04	
Sheep	DE Mcal/kg	*3.07	3.31	
Swine	DE kcal/kg	*3014.	3252.	
Cattle	ME Mcal/kg	*2.31	2.49	
Sheep	ME Mcal/kg	*2.51	2.71	
Swine	ME kcal/kg	*2621.	2828.	
Cattle	TDN %	*64.0	69.0	
Sheep	TDN %	*69.5	75.0	
Swine	TDN %	*68.4	73.7	
Thiamine	mg/kg	20.3	21.9	56
Lysine	%	1.76	1.90	
Gossypol, total	%	0.76	0.82	
Gossypol, free	%	0.04	0.04	

COTTON, GLANDLESS. Gossypium spp

Cotton, glandless, seeds w some hulls, solv-extd grnd, (5)

Ref No 5 08 979 United States

Analyses		As Fed	Dry	C.V. ± %
Dry matter	%	91.2	100.0	
Ash	%	6.4	7.0	
Crude fiber	%	12.5	13.7	
Ether extract	%	2.6	2.8	
N-free extract	%	25.3	27.7	
Protein (N x 6.25)	%	46.2	50.7	
Energy	GE Mcal/kg			
Chickens	ME_n kcal/kg	1997.	2190.	
Gossypol, total	%	0.05	0.05	
Gossypol, free	%	0.02	0.02	

Cotton gin trash –
see Cotton, gin by-product, (1)

Cottonseed cake, mechanical extracted (AAFCO) –
see Cotton, seeds w some hulls, mech-extd caked, mn 36% protein, (5)

Cottonseed hull bran –
see Cotton, seed hulls wo lint, (1)

Cottonseed meal –
see Cotton, seeds extn unspecified grnd, (5)

Cottonseed meal, degossypolized, solvent extracted –
see Cotton, seeds w some hulls wo gossypol, solv-extd grnd, mx 0.04% free gossypol, (5)

Cottonseed meal, expeller extracted –
see Cotton, seeds w some hulls, mech-extd grnd, mn 36% protein, (5)

Cottonseed meal, hydraulic extracted –
see Cotton, seeds w some hulls, mech-extd grnd, mn 36% protein, (5)

Cottonseed meal, mechanical extracted (AAFCO) –
see Cotton, seeds w some hulls, mech-extd grnd, 36% protein, (5)

Cottonseed meal, pre-press solvent extracted 41% protein –
see Cotton, seeds w some hulls, pre-press solv-extd grnd, 41% protein, (5)

Cottonseed meal, pre-press solvent extracted –
see Cotton, seeds wo hulls, pre-press solv-extd grnd, mn 50% protein, (5)

Cottonseed meal, solvent extracted (AAFCO) –
see Cotton, seeds w some hulls, solv-extd grnd, mn 36% protein, (5)

Cottonseed meal, solvent extracted, 41% protein –
see Cotton, seeds w some hulls, solv-extd grnd, mn 41% protein mx 14% fiber mn 0.5% fat, (5)

Cottonseed meal, 36% protein –
see Cotton, seeds w some hulls, mech-extd grnd, mn 36% protein mx 17% fiber mn 2% fat, (5)

Cottonseed meal, 41% protein –
see Cotton, seeds w some hulls, mech-extd grnd, mn 41% protein mx 14% fiber mn 2% fat, (5)

Cottonseed, whole, ground –
see Cotton, seeds, grnd, (5)

COTTONTOP, ARIZONA. Trichachne californica

Cottontop, Arizona, aerial part, fresh, (2)

Ref No 2 01 634 United States

Analyses		As Fed	Dry	C.V. ± %
Dry matter	%		100.0	
Ash	%		11.5	
Crude fiber	%		36.1	
Ether extract	%		1.9	
N-free extract	%		41.6	
Protein (N x 6.25)	%		8.9	
Cattle	dig prot %	*	5.5	
Goats	dig prot %	*	4.9	
Horses	dig prot %	*	5.1	
Rabbits	dig prot %	*	5.5	
Sheep	dig prot %	*	5.3	
Calcium	%		0.45	16
Phosphorus	%		0.17	33

COTTONWOOD, NARROWLEAF. Populus angustifolia

Cottonwood, narrowleaf, leaves, immature, (1)

Ref No 1 01 635 United States

Analyses		As Fed	Dry	C.V. ± %
Dry matter	%	94.1	100.0	
Ash	%	6.6	7.0	
Crude fiber	%	13.1	13.9	

Continued

Feed Name or Analyses				Mean As Fed	Mean Dry	C.V. ± %
Ether extract		%		6.5	6.9	
N-free extract		%		44.6	47.4	
Protein (N x 6.25)		%		23.3	24.8	
Cattle	dig prot	%	*	17.9	19.0	
Goats	dig prot	%	*	18.5	19.7	
Horses	dig prot	%	*	17.5	18.6	
Rabbits	dig prot	%	*	16.8	17.8	
Sheep	dig prot	%	*	18.9	20.1	

Couch grass –
see Quackgrass, aerial part, fresh, (2)

COWPARSNIP, COMMON. Heracleum lanatum

Cowparsnip, common, flowers, immature, (1)
Ref No 1 01 636 United States

Feed Name or Analyses				Mean As Fed	Mean Dry	C.V. ± %
Dry matter		%		91.7	100.0	
Ash		%		10.6	11.6	
Crude fiber		%		15.0	16.4	
Ether extract		%		4.8	5.2	
N-free extract		%		38.1	41.5	
Protein (N x 6.25)		%		23.2	25.3	
Cattle	dig prot	%	*	17.8	19.4	
Goats	dig prot	%	*	18.5	20.2	
Horses	dig prot	%	*	17.4	19.0	
Rabbits	dig prot	%	*	16.7	18.2	
Sheep	dig prot	%	*	18.9	20.6	

Cowparsnip, common, leaves, immature, (1)
Ref No 1 01 637 United States

Feed Name or Analyses				Mean As Fed	Mean Dry	C.V. ± %
Dry matter		%		92.1	100.0	
Ash		%		14.0	15.2	
Crude fiber		%		11.6	12.6	
Ether extract		%		3.1	3.4	
N-free extract		%		42.9	46.6	
Protein (N x 6.25)		%		20.6	22.4	
Cattle	dig prot	%	*	15.6	16.9	
Goats	dig prot	%	*	16.0	17.4	
Horses	dig prot	%	*	15.2	16.5	
Rabbits	dig prot	%	*	14.7	15.9	
Sheep	dig prot	%	*	16.4	17.8	

Cowparsnip, common, seeds w hulls, (5)
Ref No 5 09 140 United States

Feed Name or Analyses				Mean As Fed	Mean Dry	C.V. ± %
Dry matter		%			100.0	
Protein (N x 6.25)		%			21.0	
Alanine		%			0.92	
Arginine		%			1.05	
Aspartic acid		%			2.37	
Glutamic acid		%			2.35	
Glycine		%			0.80	
Histidine		%			0.32	
Hydroxyproline		%			0.17	
Isoleucine		%			0.74	
Leucine		%			1.13	
Lysine		%			0.95	
Methionine		%			0.27	
Phenylalanine		%			0.76	
Proline		%			0.78	
Serine		%			0.76	
Threonine		%			0.74	
Tyrosine		%			0.50	
Valine		%			1.01	

(1) dry forages and roughages (3) silages (6) minerals
(2) pasture, range plants, and forages fed green (4) energy feeds (7) vitamins
(5) protein supplements (8) additives

COWPEA. Vigna spp

Cowpea, hay, s-c, (1)
Ref No 1 01 645 United States

Feed Name or Analyses				Mean As Fed	Mean Dry	C.V. ± %
Dry matter		%		90.5	100.0	1
Ash		%		10.6	11.7	20
Crude fiber		%		24.3	26.9	4
Sheep	dig coef	%		47.	47.	
Ether extract		%		2.6	2.9	12
Sheep	dig coef	%		46.	46.	
N-free extract		%		37.0	40.9	
Sheep	dig coef	%		71.	71.	
Protein (N x 6.25)		%		16.0	17.7	8
Sheep	dig coef	%		69.	69.	
Cattle	dig prot	%	*	11.1	12.2	
Goats	dig prot	%	*	11.8	13.0	
Horses	dig prot	%	*	11.3	12.5	
Rabbits	dig prot	%	*	11.1	12.3	
Sheep	dig prot	%		11.0	12.2	
Fatty acids		%		2.6	2.9	
Energy	GE Mcal/kg					
Cattle	DE Mcal/kg		*	2.35	2.60	
Sheep	DE Mcal/kg		*	2.27	2.51	
Cattle	ME Mcal/kg		*	1.93	2.13	
Sheep	ME Mcal/kg		*	1.86	2.06	
Cattle	NE_m Mcal/kg		*	1.23	1.36	
Cattle	NE_{gain} Mcal/kg		*	0.69	0.76	
Cattle	$NE_{lactating\ cows}$ Mcal/kg		*	1.42	1.57	
Cattle	TDN	%	*	53.3	58.9	
Sheep	TDN	%		51.4	56.8	
Calcium		%		1.37	1.52	
Chlorine		%		0.15	0.17	
Iron		%		0.082	0.091	
Magnesium		%		0.37	0.41	
Phosphorus		%		0.34	0.37	17
Potassium		%		2.17	2.39	6
Sodium		%		0.20	0.22	
Sulphur		%		0.32	0.35	

Cowpea, hay, s-c, immature, (1)
Ref No 1 01 639 United States

Feed Name or Analyses				Mean As Fed	Mean Dry	C.V. ± %
Dry matter		%		89.4	100.0	2
Ash		%		10.7	12.0	8
Crude fiber		%		19.7	22.0	11
Ether extract		%		3.0	3.4	10
N-free extract		%		33.4	37.4	
Protein (N x 6.25)		%		22.5	25.2	7
Cattle	dig prot	%	*	16.8	18.8	
Goats	dig prot	%	*	17.9	20.1	
Horses	dig prot	%	*	16.9	18.9	
Rabbits	dig prot	%	*	16.2	18.1	
Sheep	dig prot	%	*	17.1	19.2	
Energy	GE Mcal/kg					
Cattle	DE Mcal/kg		*	2.24	2.50	
Sheep	DE Mcal/kg		*	2.09	2.34	
Cattle	ME Mcal/kg		*	1.83	2.05	
Sheep	ME Mcal/kg		*	1.72	1.92	
Cattle	TDN	%	*	50.6	56.6	
Sheep	TDN	%	*	47.5	53.1	
Calcium		%		1.07	1.20	19
Iron		%		0.027	0.030	
Magnesium		%		0.38	0.43	21
Phosphorus		%		0.27	0.30	25
Potassium		%		1.39	1.55	35

Cowpea, hay, s-c, full bloom, (1)
Ref No 1 01 640 United States

Feed Name or Analyses				Mean As Fed	Mean Dry	C.V. ± %
Dry matter		%		88.8	100.0	2
Ash		%		9.8	11.0	7
Crude fiber		%		21.5	24.2	6
Ether extract		%		3.0	3.4	8
N-free extract		%		36.4	41.0	
Protein (N x 6.25)		%		18.1	20.4	7
Cattle	dig prot	%	*	13.0	14.6	
Goats	dig prot	%	*	13.8	15.6	
Horses	dig prot	%	*	13.2	14.8	
Rabbits	dig prot	%	*	12.8	14.4	
Sheep	dig prot	%	*	13.2	14.9	
Energy	GE Mcal/kg					
Cattle	DE Mcal/kg		*	2.17	2.44	
Sheep	DE Mcal/kg		*	2.33	2.62	
Cattle	ME Mcal/kg		*	1.78	2.00	
Sheep	ME Mcal/kg		*	1.91	2.15	
Cattle	TDN	%	*	49.2	55.4	
Sheep	TDN	%	*	52.8	59.5	

Cowpea, hay, s-c, late bloom, (1)
Ref No 1 01 641 United States

Feed Name or Analyses				Mean As Fed	Mean Dry	C.V. ± %
Dry matter		%		88.6	100.0	
Ash		%		9.6	10.8	
Crude fiber		%		24.0	27.1	
Sheep	dig coef	%		52.	52.	
Ether extract		%		3.1	3.5	
Sheep	dig coef	%		29.	29.	
N-free extract		%		35.5	40.1	
Sheep	dig coef	%		65.	65.	
Protein (N x 6.25)		%		16.4	18.5	
Sheep	dig coef	%		72.	72.	
Cattle	dig prot	%	*	11.5	12.9	
Goats	dig prot	%	*	12.2	13.8	
Horses	dig prot	%	*	11.7	13.2	
Rabbits	dig prot	%	*	11.5	12.9	
Sheep	dig prot	%		11.8	13.3	
Fatty acids		%		3.2	3.6	
Energy	GE Mcal/kg					
Cattle	DE Mcal/kg		*	2.33	2.63	
Sheep	DE Mcal/kg		*	2.18	2.46	
Cattle	ME Mcal/kg		*	1.91	2.16	
Sheep	ME Mcal/kg		*	1.79	2.02	
Cattle	TDN	%	*	52.9	59.7	
Sheep	TDN	%		49.4	55.7	

Cowpea, hay, s-c, milk stage, (1)
Ref No 1 01 642 United States

Feed Name or Analyses				Mean As Fed	Mean Dry	C.V. ± %
Dry matter		%		87.3	100.0	
Ash		%		9.1	10.4	
Crude fiber		%		26.2	30.0	
Ether extract		%		3.1	3.5	
N-free extract		%		34.3	39.3	
Protein (N x 6.25)		%		14.7	16.8	
Cattle	dig prot	%	*	10.0	11.5	
Goats	dig prot	%	*	10.7	12.2	
Horses	dig prot	%	*	10.3	11.8	
Rabbits	dig prot	%	*	10.2	11.6	
Sheep	dig prot	%	*	10.2	11.6	

Feed Name or Analyses			Mean As Fed	Mean Dry	C.V. ± %
Energy	GE Mcal/kg				
Cattle	DE Mcal/kg	*	2.36	2.70	
Sheep	DE Mcal/kg	*	2.14	2.45	
Cattle	ME Mcal/kg	*	1.93	2.22	
Sheep	ME Mcal/kg	*	1.75	2.01	
Cattle	TDN %	*	53.5	61.3	
Sheep	TDN %	*	48.4	55.5	

Cowpea, hay, s-c, dough stage, (1)

Ref No 1 01 643 United States

Feed Name or Analyses			Mean As Fed	Mean Dry	C.V. ± %
Dry matter	%		91.1	100.0	
Ash	%		9.3	10.2	
Crude fiber	%		30.7	33.7	
Cattle	dig coef %		67.	67.	
Sheep	dig coef %		44.	44.	
Ether extract	%		1.8	2.0	
Cattle	dig coef %		43.	43.	
Sheep	dig coef %		64.	64.	
N-free extract	%		37.3	41.0	
Cattle	dig coef %		70.	70.	
Sheep	dig coef %		76.	76.	
Protein (N x 6.25)	%		12.0	13.1	
Cattle	dig coef %		61.	61.	
Sheep	dig coef %		68.	68.	
Cattle	dig prot %		7.3	8.0	
Goats	dig prot %	*	8.0	8.8	
Horses	dig prot %	*	7.9	8.7	
Rabbits	dig prot %	*	8.0	8.8	
Sheep	dig prot %		8.1	8.9	
Energy	GE Mcal/kg				
Cattle	DE Mcal/kg	*	2.46	2.70	
Sheep	DE Mcal/kg	*	2.32	2.55	
Cattle	ME Mcal/kg	*	2.02	2.21	
Sheep	ME Mcal/kg	*	1.90	2.09	
Cattle	TDN %		55.7	61.2	
Sheep	TDN %		52.6	57.8	

Cowpea, hay, s-c, mature, (1)

Ref No 1 01 644 United States

Feed Name or Analyses			Mean As Fed	Mean Dry	C.V. ± %
Dry matter	%		90.3	100.0	1
Ash	%		6.4	7.1	3
Crude fiber	%		30.7	34.0	3
Ether extract	%		2.5	2.7	4
N-free extract	%		40.5	44.9	
Protein (N x 6.25)	%		10.2	11.3	4
Cattle	dig prot %	*	6.1	6.7	
Goats	dig prot %	*	6.4	7.1	
Horses	dig prot %	*	6.4	7.1	
Rabbits	dig prot %	*	6.7	7.4	
Sheep	dig prot %	*	6.1	6.7	
Fatty acids	%		2.5	2.8	
Energy	GE Mcal/kg				
Cattle	DE Mcal/kg	*	2.22	2.46	
Sheep	DE Mcal/kg	*	2.24	2.49	
Cattle	ME Mcal/kg	*	1.82	2.02	
Sheep	ME Mcal/kg	*	1.84	2.04	
Cattle	TDN %	*	50.4	55.8	
Sheep	TDN %	*	50.9	56.4	

Cowpea, hay wo seeds, s-c, (1)

Ref No 1 01 646 United States

Feed Name or Analyses			Mean As Fed	Mean Dry	C.V. ± %
Dry matter	%			100.0	
Crude fiber	%				
Sheep	dig coef %	*		47.	
Ether extract	%				
Sheep	dig coef %	*		46.	
N-free extract	%				
Sheep	dig coef %	*		71.	
Protein (N x 6.25)	%				
Sheep	dig coef %	*		69.	

Cowpea, hulls, (1)

Ref No 1 01 647 United States

Feed Name or Analyses			Mean As Fed	Mean Dry	C.V. ± %
Dry matter	%		87.1	100.0	
Ash	%		6.2	7.1	
Crude fiber	%		30.8	35.4	
Ether extract	%		0.6	0.7	
N-free extract	%		44.0	50.5	
Protein (N x 6.25)	%		5.5	6.3	
Cattle	dig prot %	*	2.1	2.4	
Goats	dig prot %	*	2.1	2.4	
Horses	dig prot %	*	2.5	2.9	
Rabbits	dig prot %	*	3.1	3.5	
Sheep	dig prot %	*	1.9	2.2	
Energy	GE Mcal/kg				
Cattle	DE Mcal/kg	*	2.03	2.33	
Sheep	DE Mcal/kg	*	2.08	2.39	
Cattle	ME Mcal/kg	*	1.66	1.91	
Sheep	ME Mcal/kg	*	1.71	1.96	
Cattle	TDN %	*	46.0	52.8	
Sheep	TDN %	*	47.2	54.2	

Cowpea, pods w seeds, (1)

Ref No 1 01 648 United States

Feed Name or Analyses			Mean As Fed	Mean Dry	C.V. ± %
Dry matter	%		92.3	100.0	
Ash	%		4.7	5.1	
Crude fiber	%		10.8	11.7	
Ether extract	%		1.3	1.4	
N-free extract	%		53.8	58.3	
Protein (N x 6.25)	%		21.7	23.5	
Cattle	dig prot %	*	16.0	17.3	
Goats	dig prot %	*	17.1	18.5	
Horses	dig prot %	*	16.1	17.5	
Rabbits	dig prot %	*	15.5	16.8	
Sheep	dig prot %	*	16.3	17.6	
Energy	GE Mcal/kg				
Cattle	DE Mcal/kg	*	2.62	2.84	
Sheep	DE Mcal/kg	*	2.83	3.07	
Cattle	ME Mcal/kg	*	2.15	2.33	
Sheep	ME Mcal/kg	*	2.33	2.52	
Cattle	TDN %	*	59.5	64.5	
Sheep	TDN %	*	64.3	69.7	

Cowpea, straw, (1)

Ref No 1 01 649 United States

Feed Name or Analyses			Mean As Fed	Mean Dry	C.V. ± %
Dry matter	%		91.2	100.0	
Ash	%		5.4	5.9	
Crude fiber	%		43.7	48.0	
Ether extract	%		1.2	1.4	
N-free extract	%		33.9	37.2	
Protein (N x 6.25)	%		6.9	7.6	
Cattle	dig prot %	*	3.2	3.5	
Goats	dig prot %	*	3.3	3.7	
Horses	dig prot %	*	3.6	4.0	
Rabbits	dig prot %	*	4.1	4.5	
Sheep	dig prot %	*	3.1	3.4	
Fatty acids	%		1.2	1.3	
Energy	GE Mcal/kg				
Cattle	DE Mcal/kg	*	1.56	1.71	
Sheep	DE Mcal/kg	*	1.73	1.90	
Cattle	ME Mcal/kg	*	1.28	1.40	
Sheep	ME Mcal/kg	*	1.42	1.56	
Cattle	TDN %	*	35.3	38.7	
Sheep	TDN %	*	39.4	43.2	

Cowpea, aerial part, fresh, (2)

Ref No 2 01 655 United States

Feed Name or Analyses			Mean As Fed	Mean Dry	C.V. ± %
Dry matter	%		15.0	100.0	9
Ash	%		1.9	12.6	8
Crude fiber	%		3.7	24.4	13
Sheep	dig coef %		58.	58.	
Ether extract	%		0.6	4.2	13
Sheep	dig coef %		60.	60.	
N-free extract	%		6.4	42.7	
Sheep	dig coef %		82.	82.	
Protein (N x 6.25)	%		2.4	16.1	13
Sheep	dig coef %		74.	74.	
Cattle	dig prot %	*	1.7	11.6	
Goats	dig prot %	*	1.7	11.6	
Horses	dig prot %	*	1.7	11.2	
Rabbits	dig prot %	*	1.7	11.1	
Sheep	dig prot %		1.8	11.9	
Energy	GE Mcal/kg				
Cattle	DE Mcal/kg	*	0.41	2.74	
Sheep	DE Mcal/kg	*	0.44	2.95	
Cattle	ME Mcal/kg	*	0.34	2.25	
Sheep	ME Mcal/kg	*	0.36	2.42	
Cattle	TDN %	*	9.3	62.1	
Sheep	TDN %		10.0	66.8	
Calcium	%		0.23	1.53	
Chlorine	%		0.03	0.18	16
Iron	%		0.012	0.080	
Magnesium	%		0.06	0.43	
Phosphorus	%		0.05	0.31	
Potassium	%		0.25	1.66	
Sodium	%		0.04	0.25	12
Sulphur	%		0.05	0.31	16

Cowpea, aerial part, fresh, immature, (2)

Ref No 2 01 650 United States

Feed Name or Analyses			Mean As Fed	Mean Dry	C.V. ± %
Dry matter	%		17.1	100.0	8
Ash	%		2.0	11.7	9
Crude fiber	%		3.4	19.9	8
Sheep	dig coef %	*	58.	58.	
Ether extract	%		0.7	3.9	4
Sheep	dig coef %	*	60.	60.	
N-free extract	%		7.9	46.3	
Sheep	dig coef %	*	82.	82.	
Protein (N x 6.25)	%		3.1	18.2	7
Sheep	dig coef %	*	74.	74.	
Cattle	dig prot %	*	2.3	13.4	
Goats	dig prot %	*	2.3	13.5	
Horses	dig prot %	*	2.2	13.0	
Rabbits	dig prot %	*	2.2	12.7	
Sheep	dig prot %		2.3	13.5	
Energy	GE Mcal/kg				
Cattle	DE Mcal/kg	*	0.49	2.88	
Sheep	DE Mcal/kg	*	0.51	3.01	
Cattle	ME Mcal/kg	*	0.40	2.36	
Sheep	ME Mcal/kg	*	0.42	2.47	
Cattle	TDN %	*	11.2	65.3	
Sheep	TDN %	*	11.7	68.2	

Feed Name or Analyses				Mean As Fed	Mean Dry	C.V. ± %
Cowpea, aerial part, fresh, pre-bloom, (2)						
Ref No 2 01 651				United States		
Dry matter		%		17.8	100.0	
Ash		%		2.0	11.2	
Crude fiber		%		3.7	20.6	
Sheep	dig coef	%		60.	60.	
Ether extract		%		0.7	3.9	
Sheep	dig coef	%		59.	59.	
N-free extract		%		8.3	46.7	
Sheep	dig coef	%		81.	81.	
Protein (N x 6.25)		%		3.1	17.6	
Sheep	dig coef	%		76.	76.	
Cattle	dig prot	%	*	2.3	12.9	
Goats	dig prot	%	*	2.3	13.0	
Horses	dig prot	%	*	2.2	12.5	
Rabbits	dig prot	%	*	2.2	12.3	
Sheep	dig prot	%		2.4	13.4	
Energy	GE Mcal/kg					
Cattle	DE Mcal/kg		*	0.51	2.89	
Sheep	DE Mcal/kg		*	0.54	3.03	
Cattle	ME Mcal/kg		*	0.42	2.37	
Sheep	ME Mcal/kg		*	0.44	2.49	
Cattle	TDN	%	*	11.7	65.5	
Sheep	TDN	%		12.2	68.7	

Feed Name or Analyses				Mean As Fed	Mean Dry	C.V. ± %
Cowpea, aerial part, fresh, early bloom, (2)						
Ref No 2 0-1 652				United States		
Dry matter		%		13.4	100.0	
Ash		%		1.7	13.0	
Crude fiber		%		3.5	26.3	
Sheep	dig coef	%		58.	58.	
Ether extract		%		0.7	5.2	
Sheep	dig coef	%		59.	59.	
N-free extract		%		5.6	42.0	
Sheep	dig coef	%		84.	84.	
Protein (N x 6.25)		%		1.8	13.5	
Sheep	dig coef	%		72.	72.	
Cattle	dig prot	%	*	1.3	9.4	
Goats	dig prot	%	*	1.2	9.2	
Horses	dig prot	%	*	1.2	9.0	
Rabbits	dig prot	%	*	1.2	9.1	
Sheep	dig prot	%		1.3	9.7	
Energy	GE Mcal/kg					
Cattle	DE Mcal/kg		*	0.39	2.90	
Sheep	DE Mcal/kg		*	0.40	2.96	
Cattle	ME Mcal/kg		*	0.32	2.38	
Sheep	ME Mcal/kg		*	0.33	2.43	
Cattle	TDN	%	*	8.8	65.8	
Sheep	TDN	%		9.0	67.2	

Feed Name or Analyses				Mean As Fed	Mean Dry	C.V. ± %
Cowpea, aerial part, fresh, full bloom, (2)						
Ref No 2 01 653				United States		
Dry matter		%		12.8	100.0	12
Ash		%		1.8	14.2	5
Crude fiber		%		3.2	25.4	8
Sheep	dig coef	%		54.	54.	
Ether extract		%		0.7	5.2	7
Sheep	dig coef	%		66.	66.	
N-free extract		%		4.9	38.6	
Sheep	dig coef	%		74.	74.	
Protein (N x 6.25)		%		2.2	16.8	6
Sheep	dig coef	%		79.	79.	
Cattle	dig prot	%	*	1.6	12.2	
Goats	dig prot	%	*	1.6	12.2	
Horses	dig prot	%	*	1.5	11.8	
Rabbits	dig prot	%	*	1.5	11.6	
Sheep	dig prot	%		1.7	13.3	
Energy	GE Mcal/kg					
Cattle	DE Mcal/kg		*	0.33	2.61	
Sheep	DE Mcal/kg		*	0.36	2.78	
Cattle	ME Mcal/kg		*	0.27	2.14	
Sheep	ME Mcal/kg		*	0.29	2.28	
Cattle	TDN	%	*	7.6	59.3	
Sheep	TDN	%		8.1	63.1	

(1) dry forages and roughages (2) pasture, range plants, and forages fed green (3) silages (4) energy feeds (5) protein supplements (6) minerals (7) vitamins (8) additives

Feed Name or Analyses				Mean As Fed	Mean Dry	C.V. ± %
Cowpea, aerial part, fresh, dough stage, (2)						
Ref No 2 01 654				United States		
Dry matter		%		16.2	100.0	
Ash		%		2.0	12.4	
Crude fiber		%		5.0	30.8	
Sheep	dig coef	%		45.	45.	
Ether extract		%		0.6	3.4	
Sheep	dig coef	%		60.	60.	
N-free extract		%		6.4	39.4	
Sheep	dig coef	%		81.	81.	
Protein (N x 6.25)		%		2.3	14.0	
Sheep	dig coef	%		76.	76.	
Cattle	dig prot	%	*	1.6	9.8	
Goats	dig prot	%	*	1.6	9.6	
Horses	dig prot	%	*	1.5	9.4	
Rabbits	dig prot	%	*	1.5	9.5	
Sheep	dig prot	%		1.7	10.6	
Energy	GE Mcal/kg					
Cattle	DE Mcal/kg		*	0.45	2.75	
Sheep	DE Mcal/kg		*	0.44	2.69	
Cattle	ME Mcal/kg		*	0.37	2.26	
Sheep	ME Mcal/kg		*	0.36	2.21	
Cattle	TDN	%	*	10.1	62.4	
Sheep	TDN	%		9.9	61.0	

Feed Name or Analyses				Mean As Fed	Mean Dry	C.V. ± %
Cowpea, browse, (2)						
Ref No 2 01 638				United States		
Dry matter		%		90.5	100.0	2
Ash		%		10.5	11.6	24
Crude fiber		%		24.7	27.3	17
Ether extract		%		2.6	2.9	19
N-free extract		%		36.0	39.8	
Protein (N x 6.25)		%		16.7	18.4	20
Cattle	dig prot	%	*	12.2	13.5	
Goats	dig prot	%	*	12.4	13.7	
Horses	dig prot	%	*	11.9	13.2	
Rabbits	dig prot	%	*	11.7	12.9	
Sheep	dig prot	%	*	12.8	14.1	
Energy	GE Mcal/kg					
Cattle	DE Mcal/kg		*	2.53	2.79	
Sheep	DE Mcal/kg		*	2.27	2.51	
Cattle	ME Mcal/kg		*	2.07	2.29	
Sheep	ME Mcal/kg		*	1.87	2.06	
Cattle	TDN	%	*	57.3	63.3	
Sheep	TDN	%	*	51.6	57.0	
Arginine		%		1.00	1.10	
Histidine		%		0.45	0.50	
Isoleucine		%		1.18	1.30	
Leucine		%		1.81	2.00	
Lysine		%		1.00	1.10	
Methionine		%		0.45	0.50	
Phenylalanine		%		1.18	1.30	
Threonine		%		1.00	1.10	
Tryptophan		%		0.45	0.50	
Valine		%		1.27	1.40	

Feed Name or Analyses				Mean As Fed	Mean Dry	C.V. ± %
Cowpea, leaves, fresh, (2)						
Ref No 2 01 656				United States		
Dry matter		%		10.2	100.0	
Ash		%		1.4	13.9	
Crude fiber		%		1.4	13.8	
Ether extract		%		0.8	7.5	
N-free extract		%		3.7	36.7	
Protein (N x 6.25)		%		2.9	28.1	
Cattle	dig prot	%	*	2.2	21.8	
Goats	dig prot	%	*	2.3	22.8	
Horses	dig prot	%	*	2.2	21.4	
Rabbits	dig prot	%	*	2.1	20.4	
Sheep	dig prot	%	*	2.4	23.2	
Energy	GE Mcal/kg					
Cattle	DE Mcal/kg		*	0.30	2.94	
Sheep	DE Mcal/kg		*	0.30	2.95	
Cattle	ME Mcal/kg		*	0.25	2.41	
Sheep	ME Mcal/kg		*	0.25	2.42	
Cattle	TDN	%	*	6.8	66.6	
Sheep	TDN	%	*	6.8	67.0	

Feed Name or Analyses				Mean As Fed	Mean Dry	C.V. ± %
Cowpea, pods, fresh, (2)						
Ref No 2 01 657				United States		
Dry matter		%		13.5	100.0	20
Ash		%		0.8	5.8	17
Crude fiber		%		1.5	11.2	12
Ether extract		%		0.4	2.6	22
N-free extract		%		7.3	53.9	
Protein (N x 6.25)		%		3.6	26.5	10
Cattle	dig prot	%	*	2.8	20.4	
Goats	dig prot	%	*	2.9	21.3	
Horses	dig prot	%	*	2.7	20.0	
Rabbits	dig prot	%	*	2.6	19.1	
Sheep	dig prot	%	*	2.9	21.7	
Energy	GE Mcal/kg					
Cattle	DE Mcal/kg		*	0.38	2.80	
Sheep	DE Mcal/kg		*	0.41	3.07	
Cattle	ME Mcal/kg		*	0.31	2.30	
Sheep	ME Mcal/kg		*	0.34	2.52	
Cattle	TDN	%	*	8.6	63.6	
Sheep	TDN	%	*	9.4	69.7	

Feed Name or Analyses				Mean As Fed	Mean Dry	C.V. ± %
Cowpea, aerial part, ensiled, (3)						
Ref No 3 01 658				United States		
Dry matter		%		24.5	100.0	14
Ash		%		3.3	13.4	51
Crude fiber		%		6.6	27.1	14
Cattle	dig coef	%		52.	52.	
Ether extract		%		0.9	3.8	32
Cattle	dig coef	%		63.	63.	
N-free extract		%		10.1	41.2	
Cattle	dig coef	%		72.	72.	
Protein (N x 6.25)		%		3.6	14.6	15
Cattle	dig coef	%		57.	57.	
Cattle	dig prot	%		2.0	8.3	
Goats	dig prot	%	*	2.3	9.5	
Horses	dig prot	%	*	2.3	9.5	

Feed Name or Analyses		Mean As Fed	Mean Dry	C.V. ± %
Sheep	dig prot % *	2.3	9.5	
Energy	GE Mcal/kg			
Cattle	DE Mcal/kg *	0.62	2.53	
Sheep	DE Mcal/kg *	0.65	2.66	
Cattle	ME Mcal/kg *	0.51	2.08	
Sheep	ME Mcal/kg *	0.54	2.19	
Cattle	TDN %	14.1	57.4	
Sheep	TDN % *	14.8	60.4	
Calcium	%	0.37	1.49	13
Iron	%	0.024	0.097	16
Magnesium	%	0.07	0.28	50
Phosphorus	%	0.08	0.33	21
Potassium	%	0.72	2.93	7

Cowpea, aerial part, wilted ensiled, (3)

Ref No 3 01 659 United States

Feed Name or Analyses		Mean As Fed	Mean Dry	C.V. ± %
Dry matter	%	30.7	100.0	8
Ash	%	3.4	11.0	34
Crude fiber	%	8.7	28.3	6
Cattle	dig coef % *	52.	52.	
Ether extract	%	1.2	4.0	21
Cattle	dig coef % *	63.	63.	
N-free extract	%	12.8	41.7	
Cattle	dig coef % *	72.	72.	
Protein (N x 6.25)	%	4.6	15.0	14
Cattle	dig coef % *	57.	57.	
Cattle	dig prot %	2.6	8.6	
Goats	dig prot % *	3.0	9.9	
Horses	dig prot % *	3.0	9.9	
Sheep	dig prot % *	3.0	9.9	
Energy	GE Mcal/kg			
Cattle	DE Mcal/kg *	0.80	2.60	
Sheep	DE Mcal/kg *	0.83	2.70	
Cattle	ME Mcal/kg *	0.65	2.13	
Sheep	ME Mcal/kg *	0.68	2.22	
Cattle	TDN %	18.1	59.0	
Sheep	TDN % *	18.8	61.3	
Calcium	%	0.49	1.60	
Chlorine	%	0.05	0.17	
Iron	%	0.030	0.097	
Magnesium	%	0.13	0.43	
Phosphorus	%	0.10	0.33	
Potassium	%	0.90	2.93	
Sodium	%	0.07	0.23	
Sulphur	%	0.01	0.03	54

Cowpea, aerial part w molasses added, ensiled, (3)

Ref No 3 01 660 United States

Feed Name or Analyses		Mean As Fed	Mean Dry	C.V. ± %
Dry matter	%	22.7	100.0	
Ash	%	5.4	23.8	
Crude fiber	%	5.0	22.0	
Ether extract	%	0.7	3.1	
N-free extract	%	8.5	37.4	
Protein (N x 6.25)	%	3.1	13.7	
Cattle	dig prot % *	2.0	8.7	
Goats	dig prot % *	2.0	8.7	
Horses	dig prot % *	2.0	8.7	
Sheep	dig prot % *	2.0	8.7	
Energy	GE Mcal/kg			
Cattle	DE Mcal/kg *	0.58	2.57	
Sheep	DE Mcal/kg *	0.56	2.47	
Cattle	ME Mcal/kg *	0.48	2.11	
Sheep	ME Mcal/kg *	0.46	2.02	
Cattle	TDN % *	13.3	58.4	
Sheep	TDN % *	12.7	56.0	

Cowpea, seeds, (5)

Ref No 5 01 661 United States

Feed Name or Analyses		Mean As Fed	Mean Dry	C.V. ± %
Dry matter	%	89.0	100.0	
Ash	%	3.5	4.0	
Crude fiber	%	5.0	5.6	
Sheep	dig coef %	63.	63.	
Swine	dig coef %	49.	49.	
Ether extract	%	1.5	1.7	
Sheep	dig coef %	69.	69.	
Swine	dig coef %	38.	38.	
N-free extract	%	55.2	62.0	
Sheep	dig coef %	92.	92.	
Swine	dig coef %	90.	90.	
Protein (N x 6.25)	%	23.8	26.8	
Sheep	dig coef %	82.	82.	
Swine	dig coef %	94.	94.	
Sheep	dig prot %	19.5	21.9	
Swine	dig prot %	22.4	25.2	
Energy	GE Mcal/kg	3.41	3.83	
Cattle	DE Mcal/kg *	3.25	3.65	
Sheep	DE Mcal/kg *	3.34	3.75	
Swine	DE kcal/kg *	3341.	3753.	
Cattle	ME Mcal/kg *	2.67	3.00	
Sheep	ME Mcal/kg *	2.74	3.08	
Swine	ME kcal/kg *	3027.	3400.	
Cattle	TDN % *	73.8	82.9	
Sheep	TDN %	75.7	85.1	
Swine	TDN %	75.8	85.1	
Calcium	%	0.09	0.10	
Chlorine	%	0.04	0.04	
Copper	mg/kg	4.4	5.0	
Iron	%	0.021	0.023	
Magnesium	%	0.26	0.29	
Manganese	mg/kg	40.1	45.1	
Phosphorus	%	0.44	0.50	
Potassium	%	1.16	1.30	
Sodium	%	0.15	0.17	
Sulphur	%	0.25	0.28	
Niacin	mg/kg	21.9	24.6	
Riboflavin	mg/kg	2.1	2.3	
Thiamine	mg/kg	10.4	11.7	
Vitamin A	IU/g	0.3	0.3	

COWPEA, COMMON. Vigna sinensis

Cowpea, common, seeds w hulls, (5)

Ref No 5 09 263 United States

Feed Name or Analyses		Mean As Fed	Mean Dry	C.V. ± %
Dry matter	%		100.0	
Protein (N x 6.25)	%		26.0	
Alanine	%		1.14	
Arginine	%		1.90	
Aspartic acid	%		2.76	
Glutamic acid	%		4.39	
Glycine	%		1.07	
Histidine	%		0.83	
Hydroxyproline	%		0.03	
Isoleucine	%		1.04	
Leucine	%		1.87	
Lysine	%		1.66	
Methionine	%		0.31	
Phenylalanine	%		1.56	
Proline	%		0.91	
Serine	%		1.17	
Threonine	%		0.94	
Tyrosine	%		0.88	
Valine	%		1.22	

COWPEA-CORN. Vigna spp, Zea mays

Cowpea-Corn, aerial part, fresh, (2)

Ref No 2 08 417 United States

Feed Name or Analyses		Mean As Fed	Mean Dry	C.V. ± %
Dry matter	%	20.0	100.0	
Ash	%	1.8	9.0	
Crude fiber	%	5.3	26.5	
Ether extract	%	0.4	2.0	
N-free extract	%	10.4	52.0	
Protein (N x 6.25)	%	2.1	10.5	
Cattle	dig prot % *	1.4	6.8	
Goats	dig prot % *	1.3	6.4	
Horses	dig prot % *	1.3	6.4	
Rabbits	dig prot % *	1.4	6.8	
Sheep	dig prot % *	1.4	6.8	
Energy	GE Mcal/kg			
Cattle	DE Mcal/kg *	0.57	2.86	
Sheep	DE Mcal/kg *	0.59	2.97	
Cattle	ME Mcal/kg *	0.47	2.34	
Sheep	ME Mcal/kg *	0.49	2.43	
Cattle	TDN % *	13.0	64.8	
Sheep	TDN % *	13.5	67.3	

COWPEA-SORGHUM. Vigna spp, Sorghum vulgare

Cowpea-Sorghum, aerial part, ensiled, (3)

Ref No 3 01 662 United States

Feed Name or Analyses		Mean As Fed	Mean Dry	C.V. ± %
Dry matter	%	31.6	100.0	
Ash	%	1.9	5.9	
Crude fiber	%	8.5	27.0	
Ether extract	%	0.7	2.2	
N-free extract	%	18.3	57.8	
Protein (N x 6.25)	%	2.2	7.1	
Cattle	dig prot % *	0.8	2.7	
Goats	dig prot % *	0.8	2.7	
Horses	dig prot % *	0.8	2.7	
Sheep	dig prot % *	0.8	2.7	
Energy	GE Mcal/kg			
Cattle	DE Mcal/kg *	0.90	2.86	
Sheep	DE Mcal/kg *	0.94	2.96	
Cattle	ME Mcal/kg *	0.74	2.34	
Sheep	ME Mcal/kg *	0.77	2.43	
Cattle	TDN % *	20.5	64.8	
Sheep	TDN % *	21.2	67.2	

CRAB. Callinectes sapidus, Cancer spp, Paralithodes camschatica

Crab, process residue, dehy grnd, mn 25% protein salt declared above 3% mx 7% (5)

Crab meal (AAFCO)

Ref No 5 01 663 United States

Feed Name or Analyses		Mean As Fed	Mean Dry	C.V. ± %
Dry matter	%	92.3	100.0	3
Ash	%	40.8	44.2	13
Crude fiber	%	10.5	11.3	11
Ether extract	%	2.0	2.2	46
N-free extract	%	7.6	8.2	
Protein (N x 6.25)	%	31.4	34.1	9
Linoleic	%	.330	.358	

Continued

Feed Name or Analyses			Mean As Fed	Mean Dry	C.V. ± %
Energy	GE Mcal/kg				
Cattle	DE Mcal/kg	*	1.26	1.37	
Sheep	DE Mcal/kg	*	1.20	1.30	
Swine	DE kcal/kg	*	1380.	1495.	
Cattle	ME Mcal/kg	*	1.04	1.12	
Chickens	ME_n kcal/kg		1819.	1970.	
Sheep	ME Mcal/kg	*	0.99	1.07	
Swine	ME kcal/kg	*	1229.	1332.	
Cattle	TDN %	*	28.7	31.1	
Sheep	TDN %	*	27.2	29.5	
Swine	TDN %	*	31.3	33.9	
Calcium	%		14.99	16.24	17
Chlorine	%		1.52	1.65	39
Copper	mg/kg		32.8	35.6	3
Iodine	mg/kg		0.558	0.604	
Iron	%		0.435	0.472	81
Magnesium	%		0.88	0.95	48
Manganese	mg/kg		133.9	145.1	38
Phosphorus	%		1.57	1.70	8
Potassium	%		0.45	0.49	21
Sodium	%		0.94	1.02	3
Sulphur	%		0.32	0.35	
Choline	mg/kg		2011.	2178.	
Niacin	mg/kg		44.2	47.9	
Pantothenic acid	mg/kg		6.6	7.2	
Riboflavin	mg/kg		5.9	6.4	43
Vitamin B_{12}	µg/kg		445.8	483.0	29
Arginine	%		1.65	1.79	8
Cystine	%		0.40	0.43	
Histidine	%		0.50	0.54	16
Isoleucine	%		1.19	1.29	17
Leucine	%		1.59	1.73	12
Lysine	%		1.40	1.52	3
Methionine	%		0.52	0.56	15
Phenylalanine	%		1.19	1.29	20
Threonine	%		1.00	1.08	4
Tryptophan	%		0.32	0.35	14
Tyrosine	%		1.19	1.29	23
Valine	%		1.49	1.62	8

Crab, shells, grnd, (6)
Ref No 6 01 664 — United States

Feed Name or Analyses			Mean As Fed	Mean Dry	C.V. ± %
Dry matter	%		94.8	100.0	
Ash	%		50.4	53.2	
Crude fiber	%		13.1	13.8	
Ether extract	%		0.6	0.6	
N-free extract	%		18.4	19.4	
Protein (N x 6.25)	%		12.3	13.0	
Calcium	%		22.51	23.74	
Phosphorus	%		2.13	2.25	

CRABGRASS. Digitaria spp

Crabgrass, hay, s-c, (1)
Ref No 1 01 667 — United States

Feed Name or Analyses			Mean As Fed	Mean Dry	C.V. ± %
Dry matter	%		89.6	100.0	1
Ash	%		7.0	7.8	17
Crude fiber	%		30.7	34.2	4
Sheep	dig coef %		63.	63.	
Ether extract	%		2.1	2.4	9
Sheep	dig coef %		38.	38.	
N-free extract	%		42.8	47.8	
Sheep	dig coef %		55.	55.	
Protein (N x 6.25)	%		7.0	7.8	7
Sheep	dig coef %		21.	21.	
Cattle	dig prot %	*	3.3	3.7	
Goats	dig prot %	*	3.5	3.9	
Horses	dig prot %	*	3.8	4.2	
Rabbits	dig prot %	*	4.2	4.7	
Sheep	dig prot %		1.5	1.6	
Fatty acids	%		2.4	2.7	
Energy	GE Mcal/kg				
Cattle	DE Mcal/kg	*	2.24	2.50	
Sheep	DE Mcal/kg	*	2.04	2.27	
Cattle	ME Mcal/kg	*	1.84	2.05	
Sheep	ME Mcal/kg	*	1.67	1.86	
Cattle	TDN %	*	50.8	56.7	
Sheep	TDN %		46.2	51.5	

(1) dry forages and roughages
(2) pasture, range plants, and forages fed green
(3) silages
(4) energy feeds
(5) protein supplements
(6) minerals
(7) vitamins
(8) additives

Crabgrass, hay, s-c, over ripe, (1)
Ref No 1 01 665 — United States

Feed Name or Analyses			Mean As Fed	Mean Dry	C.V. ± %
Dry matter	%		89.2	100.0	
Ash	%		7.3	8.2	
Crude fiber	%		33.2	37.2	
Sheep	dig coef %		64.	64.	
Ether extract	%		1.6	1.8	
Sheep	dig coef %		36.	36.	
N-free extract	%		40.2	45.1	
Sheep	dig coef %		53.	53.	
Protein (N x 6.25)	%		6.9	7.7	
Sheep	dig coef %		32.	32.	
Cattle	dig prot %	*	3.2	3.6	
Goats	dig prot %	*	3.3	3.7	
Horses	dig prot %	*	3.6	4.1	
Rabbits	dig prot %	*	4.1	4.6	
Sheep	dig prot %		2.2	2.5	
Energy	GE Mcal/kg				
Cattle	DE Mcal/kg	*	2.03	2.27	
Sheep	DE Mcal/kg	*	2.03	2.28	
Cattle	ME Mcal/kg	*	1.66	1.87	
Sheep	ME Mcal/kg	*	1.67	1.87	
Cattle	TDN %	*	46.0	51.6	
Sheep	TDN %		46.1	51.6	

Crabgrass, hay, s-c, gr sample US, (1)
Ref No 1 01 666 — United States

Feed Name or Analyses			Mean As Fed	Mean Dry	C.V. ± %
Dry matter	%			100.0	
Energy	GE Mcal/kg			4.41	

Crabgrass, aerial part, fresh, (2)
Ref No 2 01 668 — United States

Feed Name or Analyses			Mean As Fed	Mean Dry	C.V. ± %
Dry matter	%		29.6	100.0	12
Ash	%		4.0	13.6	10
Crude fiber	%		8.8	29.7	5
Ether extract	%		0.9	3.2	10
N-free extract	%		13.1	44.1	
Protein (N x 6.25)	%		2.8	9.5	19
Cattle	dig prot %	*	_1.8	6.0	
Goats	dig prot %	*	_1.6	5.4	
Horses	dig prot %	*	_1.7	5.6	
Rabbits	dig prot %	*	_1.8	6.0	
Sheep	dig prot %	*	1.7	5.9	
Energy	GE Mcal/kg				
Cattle	DE Mcal/kg	*	_0.73	2.46	
Sheep	DE Mcal/kg	*	_0.77	2.59	
Cattle	ME Mcal/kg	*	0.60	2.02	
Sheep	ME Mcal/kg	*	0.63	2.12	
Cattle	TDN %	*	16.5	55.8	
Sheep	TDN %	*	17.4	58.8	
Calcium	%		0.17	0.57	
Magnesium	%		0.12	0.40	
Phosphorus	%		0.07	0.23	
Potassium	%		1.25	4.22	

CRABGRASS, HAIRY. Digitaria sanguinalis

Crabgrass, hairy, aerial part, fresh, early bloom, (2)
Ref No 2 01 669 — United States

Feed Name or Analyses			Mean As Fed	Mean Dry	C.V. ± %
Dry matter	%			100.0	
Ash	%			11.8	
Crude fiber	%			30.9	
Ether extract	%			2.9	
N-free extract	%			40.8	
Protein (N x 6.25)	%			13.6	
Cattle	dig prot %	*		9.5	
Goats	dig prot %	*		9.2	
Horses	dig prot %	*		9.1	
Rabbits	dig prot %	*		9.2	
Sheep	dig prot %	*		9.7	
Calcium	%			0.68	
Magnesium	%			0.40	
Phosphorus	%			0.25	
Potassium	%			4.22	

Crab meal (AAFCO) –
see Crab, process residue, dehy grnd, mn 25% protein salt declared above 3% mx 7% (5)

Cracklings –
see Swine, cracklings, dry rendered dehy, (5)

Cracklings, animal –
see Animal, cracklings, dehy grnd, (5)

CRAMBE. Crambe abyssinica

Crambe, seeds wo hulls, solv-extd grnd, (5)
Crambe meal
Ref No 5 08 167 — Canada

Feed Name or Analyses			Mean As Fed	Mean Dry	C.V. ± %
Dry matter	%			100.0	
Ash	%			6.3	
Crude fiber	%			4.6	
Ether extract	%			5.3	
N-free extract	%			34.1	
Protein (N x 6.25)	%			49.8	
Energy	GE Mcal/kg			4.85	4
Cattle	DE Mcal/kg	*		3.63	
Sheep	DE Mcal/kg	*		3.68	
Swine	DE kcal/kg	*		3804.	
Cattle	ME Mcal/kg	*		2.98	
Sheep	ME Mcal/kg	*		3.02	
Swine	ME kcal/kg	*		3269.	
Cattle	TDN %	*		82.4	
Sheep	TDN %	*		83.5	
Swine	TDN %	*		86.3	
Arginine	%			4.99	
Cystine	%			0.53	
Histidine	%			1.80	
Isoleucine	%			2.54	
Leucine	%			4.88	
Lysine	%			3.48	

Feed Name or Analyses			Mean As Fed	Mean Dry	C.V. ± %
Methionine	%			0.90	
Phenylalanine	%			2.90	
Threonine	%			3.57	
Tryptophan	%			0.94	
Valine	%			3.13	
Crambe meal –					
see Crambe, seeds wo hulls, solv-extd grnd, (5)					
CRANBERRIES. Vaccinium macrocarpon					
Cranberries, pulp, dehy, (4)					
Ref No 4 01 671				United States	
Dry matter	%		90.0	100.0	
Ash	%		1.6	1.8	70
Crude fiber	%		23.2	25.8	5
Ether extract	%		14.1	15.7	53
N-free extract	%		42.9	47.7	
Protein (N x 6.25)	%		8.2	9.1	36
Cattle	dig prot %	*	3.9	4.4	
Goats	dig prot %	*	5.0	5.6	
Horses	dig prot %	*	5.0	5.6	
Sheep	dig prot %	*	5.0	5.6	
CRAZYWEED, HARESFOOT. Oxytropis lagopus					
Crazyweed, haresfoot, aerial part, fresh, (2)					
Ref No 2 01 672				United States	
Dry matter	%			100.0	
Ash	%			6.6	
Crude fiber	%			29.3	
Ether extract	%			2.0	
N-free extract	%			45.0	
Protein (N x 6.25)	%			17.1	
Cattle	dig prot %	*		12.4	
Goats	dig prot %	*		12.5	
Horses	dig prot %	*		12.0	
Rabbits	dig prot %	*		11.9	
Sheep	dig prot %	*		12.9	
CRAZYWEED, URAL. Oxytropis uralensis					
Crazyweed, ural, seeds w hulls, (5)					
Ref No 5 09 037				United States	
Dry matter	%			100.0	
Protein (N x 6.25)	%			27.0	
Alanine	%			1.08	
Arginine	%			2.51	
Aspartic acid	%			2.57	
Glutamic acid	%			4.21	
Glycine	%			1.13	
Histidine	%			0.59	
Isoleucine	%			1.00	
Leucine	%			1.67	
Lysine	%			1.24	
Methionine	%			0.27	
Phenylalanine	%			0.97	
Proline	%			0.92	
Serine	%			1.16	
Threonine	%			0.76	
Tyrosine	%			0.81	
Valine	%			1.19	
CREOSOTEBUSH. Larrea spp					
Creosotebush, leaves, fresh alcohol extd, immature, (2)					
Ref No 2 01 673				United States	
Dry matter	%		52.1	100.0	
Ash	%		5.2	9.9	
Crude fiber	%		9.5	18.2	
Ether extract	%		1.5	2.8	
N-free extract	%		29.2	56.0	
Protein (N x 6.25)	%		6.8	13.1	
Cattle	dig prot %	*	4.7	9.0	
Goats	dig prot %	*	4.6	8.8	
Horses	dig prot %	*	4.5	8.7	
Rabbits	dig prot %	*	4.6	8.8	
Sheep	dig prot %	*	4.8	9.2	
Calcium	%		1.31	2.52	
Iron	%		0.021	0.040	
Magnesium	%		0.05	0.10	
Phosphorus	%		0.14	0.26	
Potassium	%		0.72	1.39	
Carotene	mg/kg		17.9	34.4	
Vitamin A equivalent	IU/g		29.9	57.3	
Arginine	%		0.31	0.60	
Aspartic acid	%		0.57	1.10	
Cystine	%		0.10	0.20	
Glutamic acid	%		1.36	2.60	
Glycine	%		0.42	0.80	
Isoleucine	%		0.47	0.90	
Leucine	%		0.78	1.50	
Phenylalanine	%		0.42	0.80	
Tryptophan	%		0.10	0.20	
Tyrosine	%		0.31	0.60	
Valine	%		0.42	0.80	
CREOSOTEBUSH, COVILLE. Larrea tridentata					
Creosotebush, coville, aerial part, fresh, early bloom, (2)					
Ref No 2 01 674				United States	
Dry matter	%		58.7	100.0	
Calcium	%		0.79	1.35	
Chlorine	%		0.09	0.16	
Iron	%		0.058	0.098	
Magnesium	%		0.06	0.10	
Phosphorus	%		0.08	0.14	
Potassium	%		0.51	0.87	
Sodium	%		0.04	0.07	
Sulphur	%		0.16	0.28	
Creosotebush, coville, leaves, fresh, immature, (2)					
Ref No 2 01 675				United States	
Dry matter	%			100.0	
Ash	%			10.3	
Crude fiber	%			14.5	
Ether extract	%			11.4	
N-free extract	%			48.0	
Protein (N x 6.25)	%			15.8	
Cattle	dig prot %	*		11.3	
Goats	dig prot %	*		11.3	
Horses	dig prot %	*		10.9	
Rabbits	dig prot %	*		10.9	
Sheep	dig prot %	*		11.7	
CRINKLEAWN. Trachypogon montufari					
Crinkleawn, aerial part, fresh, mature, (2)					
Ref No 2 01 676				United States	
Dry matter	%			100.0	
Ash	%			6.0	
Crude fiber	%			38.0	
Ether extract	%			2.3	
N-free extract	%			50.2	
Protein (N x 6.25)	%			3.5	
Cattle	dig prot %	*		0.9	
Goats	dig prot %	*		-0.1	
Horses	dig prot %	*		0.5	
Rabbits	dig prot %	*		1.4	
Sheep	dig prot %	*		0.3	
Calcium	%			0.24	
Phosphorus	%			0.11	
CROTALARIA. Crotalaria spp					
Crotalaria, aerial part, fresh, (2)					
Ref No 2 01 680				United States	
Dry matter	%		19.0	100.0	
Niacin	mg/kg		25.0	131.6	
Riboflavin	mg/kg		9.6	50.5	
Thiamine	mg/kg		3.5	18.3	
Crotalaria, aerial part, fresh, immature, (2)					
Ref No 2 01 677				United States	
Dry matter	%		22.9	100.0	7
Ash	%		1.8	7.8	45
Crude fiber	%		8.0	35.1	15
Ether extract	%		0.5	2.0	26
N-free extract	%		8.5	37.2	
Protein (N x 6.25)	%		4.1	17.9	5
Cattle	dig prot %	*	3.0	13.1	
Goats	dig prot %	*	3.0	13.3	
Horses	dig prot %	*	2.9	12.7	
Rabbits	dig prot %	*	2.9	12.5	
Sheep	dig prot %	*	3.1	13.7	
Calcium	%		0.45	1.95	33 1
Magnesium	%		0.15	0.67	46 1
Phosphorus	%		0.12	0.51	41 1
Crotalaria, aerial part, fresh, mid-bloom, (2)					
Ref No 2 01 678				United States	
Dry matter	%		27.0	100.0	13 1
Ash	%		1.5	5.6	63 1
Crude fiber	%		10.9	40.3	11 1
Ether extract	%		0.6	2.1	15 1
N-free extract	%		10.4	38.7	
Protein (N x 6.25)	%		3.6	13.3	7 1
Cattle	dig prot %	*	2.5	9.2	
Goats	dig prot %	*	2.4	9.0	
Horses	dig prot %	*	2.4	8.8	
Rabbits	dig prot %	*	2.4	8.9	
Sheep	dig prot %	*	2.5	9.4	
Calcium	%		0.51	1.88	31 1
Magnesium	%		0.20	0.75	38 1
Phosphorus	%		0.18	0.67	31 1

Feed Name or Analyses		Mean As Fed	Mean Dry	C.V. ± %

Crotalaria, aerial part, fresh, milk stage, (2)
Ref No 2 01 679 United States

Analyses		As Fed	Dry	C.V. ± %
Dry matter	%	28.4	100.0	
Ash	%	1.1	3.9	
Crude fiber	%	13.6	47.9	
Ether extract	%	0.4	1.5	
N-free extract	%	9.9	34.8	
Protein (N x 6.25)	%	3.4	11.9	
Cattle	dig prot % *	2.3	8.0	
Goats	dig prot % *	2.2	7.7	
Horses	dig prot % *	2.2	7.6	
Rabbits	dig prot % *	2.2	7.9	
Sheep	dig prot % *	2.3	8.1	
Calcium	%	0.18	0.62	
Magnesium	%	0.10	0.35	
Phosphorus	%	0.05	0.19	

Crotalaria, aerial part, ensiled, (3)
Ref No 3 01 683 United States

Analyses		As Fed	Dry	C.V. ± %
Dry matter	%	23.4	100.0	13
Calcium	%	0.18	0.77	24
Magnesium	%	0.09	0.37	19
Phosphorus	%	0.07	0.28	16

Crotalaria, aerial part, ensiled, immature, (3)
Ref No 3 01 681 United States

Analyses		As Fed	Dry	C.V. ± %
Dry matter	%	22.5	100.0	6
Ash	%	3.0	13.2	10
Crude fiber	%	7.3	32.6	4
Ether extract	%	0.8	3.6	5
N-free extract	%	7.6	33.7	
Protein (N x 6.25)	%	3.8	16.9	2
Cattle	dig prot % *	2.6	11.6	
Goats	dig prot % *	2.6	11.6	
Horses	dig prot % *	2.6	11.6	
Sheep	dig prot % *	2.6	11.6	
Calcium	%	0.20	0.90	
Magnesium	%	0.07	0.30	
Phosphorus	%	0.07	0.33	

Crotalaria, aerial part, ensiled, mature, (3)
Ref No 3 01 682 United States

Analyses		As Fed	Dry	C.V. ± %
Dry matter	%	22.9	100.0	
Ash	%	3.7	16.1	
Crude fiber	%	9.9	43.2	
Ether extract	%	0.5	2.4	
N-free extract	%	6.9	30.0	
Protein (N x 6.25)	%	1.9	8.3	
Cattle	dig prot % *	0.9	3.8	
Goats	dig prot % *	0.9	3.8	
Horses	dig prot % *	0.9	3.8	
Sheep	dig prot % *	0.9	3.8	
Calcium	%	0.14	0.61	
Magnesium	%	0.05	0.21	
Phosphorus	%	0.07	0.30	

(1) dry forages and roughages (2) pasture, range plants, and forages fed green (3) silages (4) energy feeds (5) protein supplements (6) minerals (7) vitamins (8) additives

CROTALARIA. Crotalaria intermedia

Crotalaria, seeds w hulls, (5)
Ref No 5 09 063 United States

Analyses		As Fed	Dry	C.V. ± %
Dry matter	%		100.0	
Protein (N x 6.25)	%		31.0	
Alanine	%		1.12	
Arginine	%		2.64	
Aspartic acid	%		2.88	
Glutamic acid	%		5.83	
Glycine	%		1.61	
Histidine	%		0.71	
Hydroxyproline	%		0.00	
Isoleucine	%		1.30	
Leucine	%		1.83	
Lysine	%		1.46	
Methionine	%		0.28	
Phenylalanine	%		0.93	
Proline	%		1.21	
Serine	%		1.33	
Threonine	%		1.05	
Tyrosine	%		0.87	
Valine	%		1.21	

CROTALARIA, SHOWY. Crotalaria spectabilis

Crotalaria, showy, seeds w hulls, (5)
Ref No 5 09 062 United States

Analyses		As Fed	Dry	C.V. ± %
Dry matter	%		100.0	
Protein (N x 6.25)	%		31.0	
Alanine	%		1.05	
Arginine	%		3.01	
Aspartic acid	%		2.85	
Glutamic acid	%		5.95	
Glycine	%		1.27	
Histidine	%		0.71	
Isoleucine	%		1.21	
Leucine	%		1.89	
Lysine	%		1.80	
Methionine	%		0.40	
Phenylalanine	%		0.87	
Proline	%		0.99	
Serine	%		1.61	
Threonine	%		0.90	
Tyrosine	%		0.78	
Valine	%		1.40	

CROTALARIA, SLENDERLEAF. Crotalaria intermedia

Crotalaria, slenderleaf, aerial part, fresh, (2)
Ref No 2 01 687 United States

Analyses		As Fed	Dry	C.V. ± %
Dry matter	%	25.6	100.0	8
Ash	%	1.3	5.1	49
Crude fiber	%	10.3	40.3	10
Ether extract	%	0.6	2.2	12
N-free extract	%	9.7	37.9	
Protein (N x 6.25)	%	3.7	14.5	13
Cattle	dig prot % *	2.6	10.2	
Goats	dig prot % *	2.6	10.1	
Horses	dig prot % *	2.5	9.8	
Rabbits	dig prot % *	2.5	9.8	
Sheep	dig prot % *	2.7	10.5	
Energy	GE Mcal/kg			
Cattle	DE Mcal/kg *	0.49	1.90	
Sheep	DE Mcal/kg *	0.46	1.78	
Cattle	ME Mcal/kg *	0.40	1.56	
Sheep	ME Mcal/kg *	0.37	1.46	
Cattle	TDN % *	11.0	43.1	
Sheep	TDN % *	10.3	40.3	
Calcium	%	0.18	0.70	
Phosphorus	%	0.06	0.23	
Potassium	%	0.76	2.97	

Crotalaria, slenderleaf, aerial part, fresh, immature, (2)
Ref No 2 01 684 United States

Analyses		As Fed	Dry	C.V. ± %
Dry matter	%	22.5	100.0	2
Ash	%	1.8	8.1	54
Crude fiber	%	8.3	37.1	12
Ether extract	%	0.4	1.8	29
N-free extract	%	7.9	35.2	
Protein (N x 6.25)	%	4.0	17.8	6
Cattle	dig prot % *	2.9	13.0	
Goats	dig prot % *	3.0	13.2	
Horses	dig prot % *	2.8	12.6	
Rabbits	dig prot % *	2.8	12.4	
Sheep	dig prot % *	3.1	13.6	
Energy	GE Mcal/kg			
Cattle	DE Mcal/kg *	0.43	1.93	
Sheep	DE Mcal/kg *	0.41	1.83	
Cattle	ME Mcal/kg *	0.36	1.58	
Sheep	ME Mcal/kg *	0.34	1.50	
Cattle	TDN % *	9.8	43.74	
Sheep	TDN % *	9.3	41.4	

Crotalaria, slenderleaf, aerial part, fresh, pre-bloom, (2)
Ref No 2 01 685 United States

Analyses		As Fed	Dry	C.V. ± %
Dry matter	%	27.1	100.0	7
Ash	%	1.3	4.9	15
Crude fiber	%	10.9	40.1	4
Ether extract	%	0.6	2.2	4
N-free extract	%	10.6	39.1	
Protein (N x 6.25)	%	3.7	13.7	3
Cattle	dig prot % *	2.6	9.5	
Goats	dig prot % *	2.5	9.3	
Horses	dig prot % *	2.5	9.2	
Rabbits	dig prot % *	2.5	9.2	
Sheep	dig prot % *	2.6	9.8	
Energy	GE Mcal/kg			
Cattle	DE Mcal/kg *	0.53	1.95	
Sheep	DE Mcal/kg *	0.51	1.87	
Cattle	ME Mcal/kg *	0.43	1.60	
Sheep	ME Mcal/kg *	0.41	1.53	
Cattle	TDN % *	12.0	44.3	
Sheep	TDN % *	11.5	42.4	
Calcium	%	0.63	2.33	23
Magnesium	%	0.25	0.93	26
Phosphorus	%	0.22	0.81	27

Crotalaria, slenderleaf, aerial part, fresh, mid-bloom, (2)
Ref No 2 01 686 United States

Analyses		As Fed	Dry	C.V. ± %
Dry matter	%	26.2	100.0	
Calcium	%	0.17	0.66	
Magnesium	%	0.11	0.43	
Phosphorus	%	0.06	0.24	

Feed Name or Analyses			Mean As Fed	Mean Dry	C.V. ± %

Crotalaria, slenderleaf, aerial part, ensiled, (3)
Ref No 3 01 691 United States

			As Fed	Dry	C.V.
Dry matter	%		23.5	100.0	
Ash	%		1.4	6.0	
Crude fiber	%		10.6	45.1	
Ether extract	%		0.7	3.0	
N-free extract	%		7.8	33.2	
Protein (N x 6.25)	%		3.0	12.8	
Cattle	dig prot %	*	1.8	7.8	
Goats	dig prot %	*	1.8	7.8	
Horses	dig prot %	*	1.8	7.8	
Sheep	dig prot %	*	1.8	7.8	
Energy	GE Mcal/kg				
Cattle	DE Mcal/kg	*	0.44	1.87	
Sheep	DE Mcal/kg	*	0.42	1.80	
Cattle	ME Mcal/kg	*	0.36	1.53	
Sheep	ME Mcal/kg	*	0.35	1.48	
Cattle	TDN %	*	9.9	42.3	
Sheep	TDN %	*	9.6	40.8	
Calcium	%		0.18	0.77	
Phosphorus	%		0.06	0.26	

Crotalaria, slenderleaf, aerial part, ensiled, immature, (3)
Ref No 3 01 688 United States

			As Fed	Dry	C.V.
Dry matter	%		23.3	100.0	
Ash	%		3.3	14.1	
Crude fiber	%		7.6	32.6	
Ether extract	%		0.8	3.5	
N-free extract	%		7.6	32.6	
Protein (N x 6.25)	%		4.0	17.2	
Cattle	dig prot %	*	2.8	11.8	
Goats	dig prot %	*	2.8	11.8	
Horses	dig prot %	*	2.8	11.8	
Sheep	dig prot %	*	2.8	11.8	
Energy	GE Mcal/kg				
Cattle	DE Mcal/kg	*	0.43	1.83	
Sheep	DE Mcal/kg	*	0.41	1.78	
Cattle	ME Mcal/kg	*	0.35	1.50	
Sheep	ME Mcal/kg	*	0.34	1.46	
Cattle	TDN %	*	9.7	41.5	
Sheep	TDN %	*	9.4	40.3	

Crotalaria, slenderleaf, aerial part, ensiled, full bloom, (3)
Ref No 3 01 689 United States

			As Fed	Dry	C.V.
Dry matter	%		24.6	100.0	
Ash	%		1.0	4.1	
Crude fiber	%		11.7	47.6	
Ether extract	%		0.7	2.8	
N-free extract	%		7.8	31.7	
Protein (N x 6.25)	%		3.4	13.8	
Cattle	dig prot %	*	2.2	8.8	
Goats	dig prot %	*	2.2	8.8	
Horses	dig prot %	*	2.2	8.8	
Sheep	dig prot %	*	2.2	8.8	
Energy	GE Mcal/kg				
Cattle	DE Mcal/kg	*	0.39	1.60	
Sheep	DE Mcal/kg	*	0.46	1.86	
Cattle	ME Mcal/kg	*	0.32	1.31	
Sheep	ME Mcal/kg	*	0.38	1.53	
Cattle	TDN %	*	8.9	36.3	
Sheep	TDN %	*	10.4	42.1	

Crotalaria, slenderleaf, aerial part, ensiled, milk stage, (3)
Ref No 3 01 690 United States

			As Fed	Dry	C.V.
Dry matter	%		26.9	100.0	
Ash	%		0.9	3.4	
Crude fiber	%		14.3	53.2	
Ether extract	%		0.5	1.9	
N-free extract	%		8.3	30.7	
Protein (N x 6.25)	%		2.9	10.8	
Cattle	dig prot %	*	1.6	6.0	
Goats	dig prot %	*	1.6	6.0	
Horses	dig prot %	*	1.6	6.0	
Sheep	dig prot %	*	1.6	6.0	
Energy	GE Mcal/kg				
Cattle	DE Mcal/kg	*	0.50	1.86	
Sheep	DE Mcal/kg	*	0.53	1.96	
Cattle	ME Mcal/kg	*	0.41	1.53	
Sheep	ME Mcal/kg	*	0.43	1.61	
Cattle	TDN %	*	11.4	42.3	
Sheep	TDN %	*	12.0	44.5	

CROTON. Croton spp

Croton, aerial part, fresh, (2)
Ref No 2 01 694 United States

			As Fed	Dry	C.V.
Dry matter	%			100.0	
Ash	%			7.4	17
Crude fiber	%			23.0	16
Ether extract	%			4.4	28
N-free extract	%			49.2	
Protein (N x 6.25)	%			16.0	29
Cattle	dig prot %	*		11.5	
Goats	dig prot %	*		11.5	
Horses	dig prot %	*		11.1	
Rabbits	dig prot %	*		11.0	
Sheep	dig prot %	*		11.9	
Calcium	%			2.12	18
Magnesium	%			0.38	23
Phosphorus	%			0.16	30
Potassium	%			1.51	38

Croton, aerial part, fresh, immature, (2)
Ref No 2 01 692 United States

			As Fed	Dry	C.V.
Dry matter	%			100.0	
Ash	%			7.9	
Crude fiber	%			20.5	
Ether extract	%			3.5	
N-free extract	%			49.1	
Protein (N x 6.25)	%			19.0	
Cattle	dig prot %	*		14.0	
Goats	dig prot %	*		14.3	
Horses	dig prot %	*		13.7	
Rabbits	dig prot %	*		13.3	
Sheep	dig prot %	*		14.7	
Calcium	%			1.89	
Magnesium	%			0.34	
Phosphorus	%			0.16	
Potassium	%			1.75	

Croton, aerial part, fresh, mature, (2)
Ref No 2 01 693 United States

			As Fed	Dry	C.V.
Dry matter	%			100.0	
Ash	%			6.7	
Crude fiber	%			23.8	
Ether extract	%			4.2	
N-free extract	%			49.0	
Protein (N x 6.25)	%			16.3	
Cattle	dig prot %	*		11.7	
Goats	dig prot %	*		11.8	
Horses	dig prot %	*		11.4	
Rabbits	dig prot %	*		11.3	
Sheep	dig prot %	*		12.2	
Calcium	%			2.29	
Magnesium	%			0.40	
Phosphorus	%			0.18	
Potassium	%			1.37	

CRYPTANTHA. Cryptantha spp

Cryptantha, aerial part, fresh, (2)
Ref No 2 01 700 United States

			As Fed	Dry	C.V.
Dry matter	%			100.0	
Ash	%			13.3	18
Crude fiber	%			27.7	23
Protein (N x 6.25)	%			7.5	72
Cattle	dig prot %	*		4.3	
Goats	dig prot %	*		3.6	
Horses	dig prot %	*		3.9	
Rabbits	dig prot %	*		4.5	
Sheep	dig prot %	*		4.0	
Calcium	%			2.06	24
Phosphorus	%			0.30	31
Potassium	%			2.52	29

Cryptantha, aerial part, fresh, pre-bloom, (2)
Ref No 2 01 695 United States

			As Fed	Dry	C.V.
Dry matter	%			100.0	
Ash	%			16.8	
Crude fiber	%			16.1	
Protein (N x 6.25)	%			17.8	
Cattle	dig prot %	*		13.0	
Goats	dig prot %	*		13.2	
Horses	dig prot %	*		12.6	
Rabbits	dig prot %	*		12.4	
Sheep	dig prot %	*		13.6	
Calcium	%			2.72	
Phosphorus	%			0.39	
Potassium	%			3.81	

Cryptantha, aerial part, fresh, full bloom, (2)
Ref No 2 01 696 United States

			As Fed	Dry	C.V.
Dry matter	%			100.0	
Ash	%			15.0	
Crude fiber	%			25.8	
Protein (N x 6.25)	%			7.8	
Cattle	dig prot %	*		4.5	
Goats	dig prot %	*		3.8	
Horses	dig prot %	*		4.2	
Rabbits	dig prot %	*		4.7	
Sheep	dig prot %	*		4.3	

Continued

Feed Name or Analyses		Mean As Fed	Mean Dry	C.V. ± %
Calcium	%		2.26	
Phosphorus	%		0.35	
Potassium	%		2.70	

Cryptantha, aerial part, fresh, milk stage, (2)
Ref No 2 01 697 United States

Feed Name or Analyses		Mean As Fed	Mean Dry	C.V. ± %
Dry matter	%		100.0	
Calcium	%		2.27	
Phosphorus	%		0.25	
Potassium	%		2.47	

Cryptantha, aerial part, fresh, over ripe, (2)
Ref No 2 01 698 United States

Feed Name or Analyses		Mean As Fed	Mean Dry	C.V. ± %
Dry matter	%		100.0	
Ash	%		12.6	
Crude fiber	%		32.4	
Protein (N x 6.25)	%		6.1	
Cattle	dig prot % *		3.1	
Goats	dig prot % *		2.3	
Horses	dig prot % *		2.7	
Rabbits	dig prot % *		3.4	
Sheep	dig prot % *		2.7	
Calcium	%		2.03	
Phosphorus	%		0.25	
Potassium	%		1.98	

Cryptantha, aerial part, fresh, dormant, (2)
Ref No 2 01 699 United States

Feed Name or Analyses		Mean As Fed	Mean Dry	C.V. ± %
Dry matter	%		100.0	
Ash	%		10.3	
Crude fiber	%		33.2	
Protein (N x 6.25)	%		3.6	
Cattle	dig prot % *		1.0	
Goats	dig prot % *		0.0	
Horses	dig prot % *		0.6	
Rabbits	dig prot % *		1.4	
Sheep	dig prot % *		0.3	
Calcium	%		1.46	
Phosphorus	%		0.25	
Potassium	%		2.00	

CUPGRASS, SOUTHWESTERN. Eriochloa gracilis

Cupgrass, southwestern, aerial part, fresh, (2)
Ref No 2 01 702 United States

Feed Name or Analyses		Mean As Fed	Mean Dry	C.V. ± %
Dry matter	%		100.0	
Ash	%		10.8	
Crude fiber	%		29.5	
Ether extract	%		2.4	
N-free extract	%		44.0	
Protein (N x 6.25)	%		13.3	
Cattle	dig prot % *		9.2	
Goats	dig prot % *		9.0	
Horses	dig prot % *		8.8	
Rabbits	dig prot % *		8.9	
Sheep	dig prot % *		9.4	
Calcium	%		0.68	

(1) dry forages and roughages (2) pasture, range plants, and forages fed green (3) silages (4) energy feeds (5) protein supplements (6) minerals (7) vitamins (8) additives

Feed Name or Analyses		Mean As Fed	Mean Dry	C.V. ± %
Magnesium	%		0.36	
Phosphorus	%		0.22	
Potassium	%		3.04	

CUPHEA, CINNABAR. Cuphea miniata

Cuphea, cinnabar, seeds w hulls, (5)
Ref No 5 09 125 United States

Feed Name or Analyses		Mean As Fed	Mean Dry	C.V. ± %
Dry matter	%		100.0	
Protein (N x 6.25)	%		17.0	
Alanine	%		0.73	
Arginine	%		1.50	
Aspartic acid	%		1.45	
Glutamic acid	%		2.21	
Glycine	%		0.82	
Histidine	%		0.36	
Isoleucine	%		0.65	
Leucine	%		1.07	
Lysine	%		0.71	
Methionine	%		0.32	
Phenylalanine	%		0.66	
Proline	%		0.60	
Serine	%		0.90	
Threonine	%		0.60	
Tyrosine	%		0.39	
Valine	%		0.95	

CUPRIC CARBONATE

Cupric carbonate, $CuCO_3.Cu(OH)_2.H_2O$, cp, (6)
Ref No 6 01 704 United States

Feed Name or Analyses		Mean As Fed	Mean Dry	C.V. ± %
Dry matter	%		100.0	
Chlorine	%		0.00	
Copper	%		55.0	
Iron	%		0.005	
Sulphur	%		0.17	

CUPRIC CHLORIDE

Cupric chloride, $CuCl_2{\cdot}2H_2O$, cp, (6)
Ref No 6 01 706 United States

Feed Name or Analyses		Mean As Fed	Mean Dry	C.V. ± %
Dry matter	%	78.9	100.0	
Chlorine	%	41.59	52.71	
Copper	%	37.3	47.2	
Iron	%	0.005	0.006	
Sulphur	%	0.03	0.04	

CUPRIC OXIDE

Cupric oxide, CuO, cp, (6)
Ref No 6 01 712 United States

Feed Name or Analyses		Mean As Fed	Mean Dry	C.V. ± %
Dry matter	%		100.0	
Chlorine	%		0.01	
Copper	%		79.9	
Sulphur	%		0.13	

CUPRIC SULFATE

Cupric sulfate, $CuSO_4{\cdot}5H_2O$, cp, (6)
Ref No 6 01 720 United States

Feed Name or Analyses		Mean As Fed	Mean Dry	C.V. ± %
Dry matter	%	64.0	100.0	
Chlorine	%	0.00	0.00	
Copper	%	25.4	39.8	
Iron	%	0.003	0.005	
Sulphur	%	12.84	20.06	

CURLYMESQUITE. Hilaria belangeri

Curlymesquite, aerial part, fresh, (2)
Ref No 2 01 728 United States

Feed Name or Analyses		Mean As Fed	Mean Dry	C.V. ± %
Dry matter	%	35.0	100.0	
Ash	%	5.1	14.6	17
Crude fiber	%	9.8	27.9	6
Ether extract	%	0.7	2.1	10
N-free extract	%	16.4	46.8	
Protein (N x 6.25)	%	3.0	8.6	37
Cattle	dig prot % *	1.8	5.2	
Goats	dig prot % *	1.6	4.6	
Horses	dig prot % *	1.7	4.8	
Rabbits	dig prot % *	1.9	5.3	
Sheep	dig prot % *	1.8	5.0	
Energy	GE Mcal/kg			
Cattle	DE Mcal/kg *	0.90	2.58	
Sheep	DE Mcal/kg *	0.97	2.76	
Cattle	ME Mcal/kg *	0.74	2.12	
Sheep	ME Mcal/kg *	0.79	2.26	
Cattle	TDN % *	20.5	58.5	
Sheep	TDN % *	21.9	62.5	
Calcium	%	0.18	0.52	37
Copper	mg/kg	3.5	10.1	
Magnesium	%	0.06	0.16	58
Manganese	mg/kg	16.4	47.0	
Phosphorus	%	0.05	0.15	41
Potassium	%	0.23	0.67	46

Curlymesquite, aerial part, fresh, immature, (2)
Ref No 2 01 723 United States

Feed Name or Analyses		Mean As Fed	Mean Dry	C.V. ± %
Dry matter	%		100.0	
Ash	%		15.7	
Crude fiber	%		25.6	
Ether extract	%		2.5	
N-free extract	%		39.0	
Protein (N x 6.25)	%		17.2	
Cattle	dig prot % *		12.5	
Goats	dig prot % *		12.6	
Horses	dig prot % *		12.1	
Rabbits	dig prot % *		11.9	
Sheep	dig prot % *		13.0	
Energy	GE Mcal/kg			
Cattle	DE Mcal/kg *		2.71	
Sheep	DE Mcal/kg *		2.61	
Cattle	ME Mcal/kg *		2.22	
Sheep	ME Mcal/kg *		2.14	
Cattle	TDN % *		61.4	
Sheep	TDN % *		59.2	
Calcium	%		1.04	
Magnesium	%		0.31	
Phosphorus	%		0.26	
Potassium	%		0.79	

Feed Name or Analyses				Mean As Fed	Mean Dry	C.V. ± %

Curlymesquite, aerial part, fresh, early bloom, (2)

Ref No 2 01 724 United States

Analyses		Unit		As Fed	Dry	C.V. ± %
Dry matter		%			100.0	
Calcium		%			0.72	
Magnesium		%			0.17	
Phosphorus		%			0.19	
Potassium		%			1.18	

Curlymesquite, aerial part, fresh, mid-bloom, (2)

Ref No 2 01 725 United States

Analyses		Unit		As Fed	Dry	C.V. ± %
Dry matter		%			100.0	
Ash		%			13.3	15
Crude fiber		%			27.6	6
Ether extract		%			2.3	13
N-free extract		%			47.0	
Protein (N x 6.25)		%			9.8	4
Cattle	dig prot	%	*		6.2	
Goats	dig prot	%	*		5.7	
Horses	dig prot	%	*		5.9	
Rabbits	dig prot	%	*		6.2	
Sheep	dig prot	%	*		6.1	
Energy	GE Mcal/kg					
Cattle	DE Mcal/kg		*		2.53	
Sheep	DE Mcal/kg		*		2.53	
Cattle	ME Mcal/kg		*		2.07	
Sheep	ME Mcal/kg		*		2.07	
Cattle	TDN	%	*		57.4	
Sheep	TDN	%	*		57.3	
Calcium		%			0.52	7
Magnesium		%			0.13	
Phosphorus		%			0.18	10
Potassium		%			0.88	

Curlymesquite, aerial part, fresh, dough stage, (2)

Ref No 2 01 726 United States

Analyses		Unit		As Fed	Dry	C.V. ± %
Dry matter		%			100.0	
Calcium		%			0.51	
Magnesium		%			0.14	
Phosphorus		%			0.16	
Potassium		%			0.42	

Curlymesquite, aerial part, fresh, over ripe, (2)

Ref No 2 01 727 United States

Analyses		Unit		As Fed	Dry	C.V. ± %
Dry matter		%			100.0	
Ash		%			13.5	
Crude fiber		%			29.3	
Ether extract		%			2.0	
N-free extract		%			50.3	
Protein (N x 6.25)		%			4.9	
Cattle	dig prot	%	*		2.1	
Goats	dig prot	%	*		1.1	
Horses	dig prot	%	*		1.7	
Rabbits	dig prot	%	*		2.5	
Sheep	dig prot	%	*		1.6	
Energy	GE Mcal/kg					
Cattle	DE Mcal/kg		*		2.31	
Sheep	DE Mcal/kg		*		2.44	
Cattle	ME Mcal/kg		*		1.89	
Sheep	ME Mcal/kg		*		2.00	
Cattle	TDN	%	*		52.4	
Sheep	TDN	%	*		55.3	
Calcium		%			0.33	
Cobalt		mg/kg			0.176	
Copper		mg/kg			10.1	
Magnesium		%			0.10	
Manganese		mg/kg			47.0	
Phosphorus		%			0.06	
Potassium		%			0.42	
Carotene		mg/kg			2.0	
Vitamin A equivalent		IU/g			3.3	

Curlymesquite, browse, mature, (2)

Ref No 2 01 729 United States

Analyses		Unit		As Fed	Dry	C.V. ± %
Dry matter		%			100.0	
Ash		%			17.1	16
Crude fiber		%			28.8	4
Ether extract		%			2.2	11
N-free extract		%			45.9	
Protein (N x 6.25)		%			6.0	15
Cattle	dig prot	%	*		3.0	
Goats	dig prot	%	*		2.2	
Horses	dig prot	%	*		2.6	
Rabbits	dig prot	%	*		3.3	
Sheep	dig prot	%	*		2.6	
Energy	GE Mcal/kg					
Cattle	DE Mcal/kg		*		2.11	
Sheep	DE Mcal/kg		*		2.19	
Cattle	ME Mcal/kg		*		1.73	
Sheep	ME Mcal/kg		*		1.80	
Cattle	TDN	%	*		47.9	
Sheep	TDN	%	*		49.7	
Calcium		%			0.55	16
Magnesium		%			0.15	82
Phosphorus		%			0.09	16
Potassium		%			0.39	46

CUTGRASS, CLUBHEAD. Lurisia spp

Cutgrass, clubhead, aerial part, fresh, pre-bloom, (2)

Ref No 2 01 731 United States

Analyses		Unit		As Fed	Dry	C.V. ± %
Dry matter		%		18.0	100.0	
Ash		%		0.9	4.8	
Crude fiber		%		4.9	27.2	
Sheep	dig coef	%		57.	57.	
Ether extract		%		0.3	1.9	
Sheep	dig coef	%		32.	32.	
N-free extract		%		10.0	55.3	
Sheep	dig coef	%		70.	70.	
Protein (N x 6.25)		%		1.9	10.8	
Sheep	dig coef	%		42.	42.	
Cattle	dig prot	%	*	1.3	7.1	
Goats	dig prot	%	*	1.2	6.6	
Horses	dig prot	%	*	1.2	6.7	
Rabbits	dig prot	%	*	1.3	7.0	
Sheep	dig prot	%		0.8	4.5	
Energy	GE Mcal/kg					
Cattle	DE Mcal/kg		*	0.50	2.75	
Sheep	DE Mcal/kg		*	0.48	2.65	
Cattle	ME Mcal/kg		*	0.41	2.26	
Sheep	ME Mcal/kg		*	0.39	2.17	
Cattle	TDN	%	*	11.2	62.3	
Sheep	TDN	%		10.8	60.1	

DALEA. Dalea spp

Dalea, browse, mid-bloom, (2)

Ref No 2 01 732 United States

Analyses		Unit		As Fed	Dry	C.V. ± %
Dry matter		%			100.0	
Ash		%			6.7	
Crude fiber		%			14.5	
Ether extract		%			6.5	
N-free extract		%			53.3	
Protein (N x 6.25)		%			19.0	
Cattle	dig prot	%	*		14.0	
Goats	dig prot	%	*		14.3	
Horses	dig prot	%	*		13.7	
Rabbits	dig prot	%	*		13.3	
Sheep	dig prot	%	*		14.7	
Calcium		%			2.27	
Magnesium		%			0.24	
Phosphorus		%			0.14	
Potassium		%			1.21	

DALEA. Dalea oaxacana

Dalea, seeds w hulls, (5)

Ref No 5 09 302 United States

Analyses		Unit		As Fed	Dry	C.V. ± %
Dry matter		%			100.0	
Protein (N x 6.25)		%			37.0	
Alanine		%			1.33	
Arginine		%			4.07	
Aspartic acid		%			3.11	
Glutamic acid		%			7.18	
Glycine		%			1.81	
Histidine		%			0.78	
Hydroxyproline		%			0.04	
Isoleucine		%			1.30	
Leucine		%			2.11	
Lysine		%			1.85	
Methionine		%			0.41	
Phenylalanine		%			1.44	
Proline		%			1.52	
Serine		%			1.55	
Threonine		%			1.04	
Tyrosine		%			1.07	
Valine		%			1.52	

DALLISGRASS. Paspalum dilatatum

Dallisgrass, aerial part, dormant, (1)

Ref No 1 01 742 United States

Analyses		Unit		As Fed	Dry	C.V. ± %
Dry matter		%			100.0	
Carotene		mg/kg			15.0	
Vitamin A equivalent		IU/g			25.0	

Dallisgrass, hay, s-c, (1)

Ref No 1 01 737 United States

Analyses		Unit		As Fed	Dry	C.V. ± %
Dry matter		%		90.9	100.0	1
Calcium		%		0.46	0.51	42
Iron		%		0.011	0.012	
Magnesium		%		0.67	0.74	
Phosphorus		%		0.18	0.20	58

Dallisgrass, hay, s-c, immature, (1)
Ref No 1 01 733 United States

Feed Name or Analyses			Mean As Fed	Mean Dry	C.V. ± %
Dry matter	%		90.6	100.0	
Calcium	%		0.72	0.80	
Phosphorus	%		0.19	0.21	

Dallisgrass, hay, s-c, mid-bloom, (1)
Ref No 1 01 734 United States

Feed Name or Analyses			Mean As Fed	Mean Dry	C.V. ± %
Dry matter	%		90.7	100.0	
Ash	%		8.8	9.7	
Crude fiber	%		29.1	32.1	
Ether extract	%		2.2	2.4	
N-free extract	%		44.1	48.6	
Protein (N x 6.25)	%		6.5	7.2	
Cattle	dig prot %	*	2.9	3.2	
Goats	dig prot %	*	3.0	3.3	
Horses	dig prot %	*	3.3	3.6	
Rabbits	dig prot %	*	3.8	4.2	
Sheep	dig prot %	*	2.7	3.0	
Energy	GE Mcal/kg				
Cattle	DE Mcal/kg	*	2.33	2.57	
Sheep	DE Mcal/kg	*	2.22	2.44	
Cattle	ME Mcal/kg	*	1.91	2.11	
Sheep	ME Mcal/kg	*	1.82	2.00	
Cattle	TDN %	*	52.9	58.3	
Sheep	TDN %	*	50.3	55.4	
Calcium	%		0.39	0.43	
Phosphorus	%		0.15	0.17	

Dallisgrass, hay, s-c, mature, (1)
Ref No 1 01 735 United States

Feed Name or Analyses			Mean As Fed	Mean Dry	C.V. ± %
Dry matter	%		91.4	100.0	
Calcium	%		0.33	0.36	
Phosphorus	%		0.07	0.08	

Dallisgrass, hay, s-c, over ripe, (1)
Ref No 1 01 736 United States

Feed Name or Analyses			Mean As Fed	Mean Dry	C.V. ± %
Dry matter	%		91.4	100.0	
Ash	%		9.0	9.8	
Crude fiber	%		33.8	37.0	
Ether extract	%		1.6	1.8	
N-free extract	%		43.0	47.0	
Protein (N x 6.25)	%		4.0	4.4	
Cattle	dig prot %	*	0.7	0.8	
Goats	dig prot %	*	0.6	0.7	
Horses	dig prot %	*	1.2	1.3	
Rabbits	dig prot %	*	1.9	2.1	
Sheep	dig prot %	*	0.5	0.5	
Energy	GE Mcal/kg				
Cattle	DE Mcal/kg	*	2.18	2.39	
Sheep	DE Mcal/kg	*	2.10	2.30	
Cattle	ME Mcal/kg	*	1.79	1.96	
Sheep	ME Mcal/kg	*	1.72	1.88	
Cattle	TDN %	*	49.5	54.2	
Sheep	TDN %	*	47.6	52.1	

(1) dry forages and roughages (2) pasture, range plants, and forages fed green (3) silages (4) energy feeds (5) protein supplements (6) minerals (7) vitamins (8) additives

Dallisgrass, aerial part, fresh, (2)
Ref No 2 01 741 United States

Feed Name or Analyses			Mean As Fed	Mean Dry	C.V. ± %
Dry matter	%		25.0	100.0	14
Ash	%		3.2	12.8	
Crude fiber	%		7.2	28.8	
Ether extract	%		0.6	2.4	
N-free extract	%		11.0	44.0	
Protein (N x 6.25)	%		3.0	12.0	
Cattle	dig prot %	*	2.0	8.1	
Goats	dig prot %	*	1.9	7.8	
Horses	dig prot %	*	1.9	7.7	
Rabbits	dig prot %	*	2.0	7.9	
Sheep	dig prot %	*	2.0	8.2	
Energy	GE Mcal/kg				
Cattle	DE Mcal/kg	*	0.67	2.68	
Sheep	DE Mcal/kg	*	0.70	2.81	
Cattle	ME Mcal/kg	*	0.55	2.19	
Sheep	ME Mcal/kg	*	0.58	2.31	
Cattle	TDN %	*	15.2	60.7	
Sheep	TDN %	*	15.9	63.8	
Calcium	%		0.14	0.56	
Cobalt	mg/kg		0.018	0.073	90
Iron	%		0.004	0.016	
Magnesium	%		0.10	0.40	
Phosphorus	%		0.05	0.20	
Potassium	%		0.43	1.72	
Sodium	%		0.09	0.34	
Carotene	mg/kg		75.6	302.3	40
Vitamin A equivalent	IU/g		126.0	503.9	

Dallisgrass, aerial part, fresh, immature, (2)
Ref No 2 01 738 United States

Feed Name or Analyses			Mean As Fed	Mean Dry	C.V. ± %
Dry matter	%		25.5	100.0	
Ash	%		2.7	10.5	
Crude fiber	%		7.7	30.1	
Ether extract	%		0.7	2.7	
N-free extract	%		8.5	33.5	
Protein (N x 6.25)	%		5.9	23.2	
Cattle	dig prot %	*	4.5	17.6	
Goats	dig prot %	*	4.6	18.2	
Horses	dig prot %	*	4.4	17.2	
Rabbits	dig prot %	*	4.2	16.6	
Sheep	dig prot %	*	4.7	18.6	
Energy	GE Mcal/kg				
Cattle	DE Mcal/kg	*	0.71	2.80	
Sheep	DE Mcal/kg	*	0.70	2.74	
Cattle	ME Mcal/kg	*	0.59	2.30	
Sheep	ME Mcal/kg	*	0.57	2.25	
Cattle	TDN %	*	16.2	63.5	
Sheep	TDN %	*	15.8	62.1	
Calcium	%		0.17	0.65	19
Phosphorus	%		0.11	0.42	14
Carotene	mg/kg		108.8	426.6	
Vitamin A equivalent	IU/g		181.3	711.1	

Dallisgrass, aerial part, fresh, early bloom, (2)
Ref No 2 01 743 United States

Feed Name or Analyses			Mean As Fed	Mean Dry	C.V. ± %
Dry matter	%		18.0	100.0	
Ash	%		0.8	4.4	
Crude fiber	%		6.2	34.2	
Sheep	dig coef %		72.	72.	
Ether extract	%		0.3	1.6	
Sheep	dig coef %		51.	51.	
N-free extract	%		9.5	52.9	
Sheep	dig coef %		71.	71.	
Protein (N x 6.25)	%		1.2	6.9	
Sheep	dig coef %		48.	48.	
Cattle	dig prot %	*	0.7	3.8	
Goats	dig prot %	*	0.5	3.0	
Horses	dig prot %	*	0.6	3.4	
Rabbits	dig prot %	*	0.7	4.0	
Sheep	dig prot %		0.6	3.3	
Energy	GE Mcal/kg				
Cattle	DE Mcal/kg	*	0.55	3.07	
Sheep	DE Mcal/kg	*	0.53	2.97	
Cattle	ME Mcal/kg	*	0.45	2.52	
Sheep	ME Mcal/kg	*	0.44	2.43	
Cattle	TDN %	*	12.5	69.7	
Sheep	TDN %		12.1	67.3	

Dallisgrass, aerial part, fresh, mid-bloom, (2)
Ref No 2 08 418 United States

Feed Name or Analyses			Mean As Fed	Mean Dry	C.V. ± %
Dry matter	%		30.0	100.0	
Ash	%		2.9	9.7	
Crude fiber	%		9.7	32.3	
Ether extract	%		0.7	2.3	
N-free extract	%		14.6	48.7	
Protein (N x 6.25)	%		2.1	7.0	
Cattle	dig prot %	*	1.2	3.8	
Goats	dig prot %	*	0.9	3.1	
Horses	dig prot %	*	1.0	3.5	
Rabbits	dig prot %	*	1.2	4.1	
Sheep	dig prot %	*	1.1	3.5	
Energy	GE Mcal/kg				
Cattle	DE Mcal/kg	*	0.80	2.67	
Sheep	DE Mcal/kg	*	0.76	2.53	
Cattle	ME Mcal/kg	*	0.66	2.19	
Sheep	ME Mcal/kg	*	0.62	2.07	
Cattle	TDN %	*	18.2	60.5	
Sheep	TDN %	*	17.2	57.3	
Calcium	%		0.14	0.47	
Phosphorus	%		0.04	0.13	

Dallisgrass, aerial part, fresh, full bloom, (2)
Ref No 2 01 739 United States

Feed Name or Analyses			Mean As Fed	Mean Dry	C.V. ± %
Dry matter	%		30.0	100.0	
Ash	%		2.9	9.7	
Crude fiber	%		9.7	32.2	
Ether extract	%		0.7	2.4	
N-free extract	%		14.6	48.6	
Protein (N x 6.25)	%		2.1	7.1	
Cattle	dig prot %	*	1.2	3.9	
Goats	dig prot %	*	1.0	3.2	
Horses	dig prot %	*	1.1	3.6	
Rabbits	dig prot %	*	1.2	4.2	
Sheep	dig prot %	*	1.1	3.6	
Energy	GE Mcal/kg				
Cattle	DE Mcal/kg	*	0.80	2.67	
Sheep	DE Mcal/kg	*	0.75	2.51	
Cattle	ME Mcal/kg	*	0.66	2.19	
Sheep	ME Mcal/kg	*	0.62	2.06	
Cattle	TDN %	*	18.1	60.4	
Sheep	TDN %	*	17.1	56.9	

Dallisgrass, aerial part, fresh, mature, (2)

Ref No 2 01 740 United States

Feed Name or Analyses			Mean As Fed	Mean Dry	C.V. ± %
Dry matter	%		28.0	100.0	
Ash	%		1.2	4.4	
Crude fiber	%		9.6	34.2	
Ether extract	%		0.4	1.6	
N-free extract	%		14.8	52.9	
Protein (N x 6.25)	%		1.9	6.9	
Cattle	dig prot %	*	1.1	3.8	
Goats	dig prot %	*	0.8	3.0	
Horses	dig prot %	*	0.9	3.4	
Rabbits	dig prot %	*	1.1	4.0	
Sheep	dig prot %	*	1.0	3.4	
Energy	GE Mcal/kg				
Cattle	DE Mcal/kg	*	0.74	2.63	
Sheep	DE Mcal/kg	*	0.71	2.55	
Cattle	ME Mcal/kg	*	0.60	2.16	
Sheep	ME Mcal/kg	*	0.59	2.09	
Cattle	TDN %	*	16.7	59.7	
Sheep	TDN %	*	16.2	57.9	

DANDELION. Taraxacum spp

Dandelion, aerial part, fresh, (2)

Ref No 2 01 744 United States

Feed Name or Analyses			Mean As Fed	Mean Dry	C.V. ± %
Dry matter	%			100.0	
Zinc	mg/kg			9.7	

DANDELION, COMMON. Taraxacum officinale

Dandelion, common, hay, s-c, early bloom, (1)

Ref No 1 01 745 United States

Feed Name or Analyses			Mean As Fed	Mean Dry	C.V. ± %
Dry matter	%		88.6	100.0	
Ash	%		12.0	13.5	
Crude fiber	%		15.0	16.9	
Sheep	dig coef %		78.	78.	
Ether extract	%		4.2	4.7	
Sheep	dig coef %		45.	45.	
N-free extract	%		42.8	48.3	
Sheep	dig coef %		76.	76.	
Protein (N x 6.25)	%		14.7	16.6	
Sheep	dig coef %		72.	72.	
Cattle	dig prot %	*	10.0	11.3	
Goats	dig prot %	*	10.7	12.0	
Horses	dig prot %	*	10.3	11.6	
Rabbits	dig prot %	*	10.2	11.5	
Sheep	dig prot %		10.6	12.0	
Energy	GE Mcal/kg				
Cattle	DE Mcal/kg	*	2.67	3.01	
Sheep	DE Mcal/kg	*	2.60	2.94	
Cattle	ME Mcal/kg	*	2.19	2.47	
Sheep	ME Mcal/kg	*	2.13	2.41	
Cattle	TDN %	*	60.5	68.3	
Sheep	TDN %		59.0	66.6	

Dandelion, common, aerial part, fresh, immature, (2)

Ref No 2 01 746 United States

Feed Name or Analyses			Mean As Fed	Mean Dry	C.V. ± %
Dry matter	%		13.7	100.0	
Ash	%		1.8	12.9	
Crude fiber	%		1.5	11.1	
Ether extract	%		0.7	4.8	
N-free extract	%		6.6	48.2	
Protein (N x 6.25)	%		3.2	23.0	
Cattle	dig prot %	*	2.4	17.4	
Goats	dig prot %	*	2.5	18.0	
Horses	dig prot %	*	2.3	17.1	
Rabbits	dig prot %	*	2.3	16.4	
Sheep	dig prot %	*	2.5	18.4	
Energy	GE Mcal/kg				
Cattle	DE Mcal/kg	*	0.40	2.90	
Sheep	DE Mcal/kg	*	0.38	2.80	
Cattle	ME Mcal/kg	*	0.33	2.38	
Sheep	ME Mcal/kg	*	0.32	2.30	
Cattle	TDN %	*	9.0	65.9	
Sheep	TDN %	*	8.7	63.5	
Copper	mg/kg		2.3	17.0	
Iron	%		0.008	0.062	
Manganese	mg/kg		7.2	52.9	
Phosphorus	%		0.06	0.44	

Dandelion, common, leaves, fresh, (2)

Ref No 2 01 748 United States

Feed Name or Analyses			Mean As Fed	Mean Dry	C.V. ± %
Dry matter	%		14.5	100.0	
Ash	%		1.8	12.5	
Crude fiber	%		1.6	11.1	
Ether extract	%		0.7	4.9	
N-free extract	%		7.6	52.8	
Protein (N x 6.25)	%		2.7	18.8	
Cattle	dig prot %	*	2.0	13.8	
Goats	dig prot %	*	2.0	14.1	
Horses	dig prot %	*	1.9	13.4	
Rabbits	dig prot %	*	1.9	13.1	
Sheep	dig prot %	*	2.1	14.5	
Energy	GE Mcal/kg		0.45	3.13	
Cattle	DE Mcal/kg	*	0.39	2.72	
Sheep	DE Mcal/kg	*	0.38	2.61	
Cattle	ME Mcal/kg	*	0.32	2.23	
Sheep	ME Mcal/kg	*	0.31	2.14	
Cattle	TDN %	*	8.9	61.6	
Sheep	TDN %	*	8.6	59.1	
Calcium	%		0.19	1.30	
Iron	%		0.003	0.022	
Phosphorus	%		0.07	0.46	
Potassium	%		0.40	2.76	
Sodium	%		0.08	0.53	
Ascorbic acid	mg/kg		351.2	2430.6	
Carotene	mg/kg		28.8	199.5	98
Riboflavin	mg/kg		2.6	18.1	
Thiamine	mg/kg		1.9	13.2	
Vitamin A	IU/g		140.5	972.2	
Vitamin A equivalent	IU/g		48.1	332.6	

Dandelion, common, leaves, fresh, immature, (2)

Ref No 2 01 747 United States

Feed Name or Analyses			Mean As Fed	Mean Dry	C.V. ± %
Dry matter	%		14.5	100.0	
Carotene	mg/kg		97.0	669.1	
Vitamin A equivalent	IU/g		161.7	1115.4	

DANTHONIA, POVERTY. Danthonia spp

Danthonia, poverty, hay, s-c, (1)

Ref No 1 01 751 United States

Feed Name or Analyses			Mean As Fed	Mean Dry	C.V. ± %
Dry matter	%		91.7	100.0	
Ash	%		3.9	4.2	
Crude fiber	%		29.5	32.2	
Sheep	dig coef %		68.	68.	
Ether extract	%		2.9	3.2	
Sheep	dig coef %		50.	50.	
N-free extract	%		47.6	51.9	
Sheep	dig coef %		65.	65.	
Protein (N x 6.25)	%		7.8	8.5	
Sheep	dig coef %		58.	58.	
Cattle	dig prot %	*	3.9	4.3	
Goats	dig prot %	*	4.1	4.5	
Horses	dig prot %	*	4.4	4.7	
Rabbits	dig prot %	*	4.8	5.2	
Sheep	dig prot %		4.5	4.9	
Energy	GE Mcal/kg				
Cattle	DE Mcal/kg	*	2.32	2.52	
Sheep	DE Mcal/kg	*	2.59	2.83	
Cattle	ME Mcal/kg	*	1.90	2.07	
Sheep	ME Mcal/kg	*	2.13	2.32	
Cattle	TDN %	*	52.5	57.3	
Sheep	TDN %		58.8	64.2	

Danthonia, poverty, hay, s-c, mid-bloom, (1)

Ref No 1 01 749 United States

Feed Name or Analyses			Mean As Fed	Mean Dry	C.V. ± %
Dry matter	%		71.7	100.0	
Ash	%		2.7	3.8	
Crude fiber	%		24.4	34.1	
Sheep	dig coef %		65.	65.	
Ether extract	%		2.1	2.9	
Sheep	dig coef %		38.	38.	
N-free extract	%		37.1	51.7	
Sheep	dig coef %		62.	62.	
Protein (N x 6.25)	%		5.4	7.5	
Sheep	dig coef %		49.	49.	
Cattle	dig prot %	*	2.5	3.4	
Goats	dig prot %	*	2.6	3.6	
Horses	dig prot %	*	2.8	3.9	
Rabbits	dig prot %	*	3.2	4.5	
Sheep	dig prot %		2.6	3.7	
Energy	GE Mcal/kg				
Cattle	DE Mcal/kg	*	1.86	2.59	
Sheep	DE Mcal/kg	*	1.91	2.66	
Cattle	ME Mcal/kg	*	1.52	2.12	
Sheep	ME Mcal/kg	*	1.57	2.18	
Cattle	TDN %	*	42.1	58.7	
Sheep	TDN %		43.3	60.4	

Danthonia, poverty, hay, s-c, full bloom, (1)

Ref No 1 01 750 United States

Feed Name or Analyses			Mean As Fed	Mean Dry	C.V. ± %
Dry matter	%		91.7	100.0	
Ash	%		4.2	4.6	
Crude fiber	%		27.7	30.2	
Sheep	dig coef %		71.	71.	
Ether extract	%		3.3	3.6	
Sheep	dig coef %		63.	63.	
N-free extract	%		47.8	52.1	
Sheep	dig coef %		69.	69.	
Protein (N x 6.25)	%		8.7	9.5	
Sheep	dig coef %		68.	68.	
Cattle	dig prot %	*	4.7	5.2	
Goats	dig prot %	*	5.0	5.4	
Horses	dig prot %	*	5.1	5.6	
Rabbits	dig prot %	*	5.5	6.0	
Sheep	dig prot %		5.9	6.5	
Energy	GE Mcal/kg				
Cattle	DE Mcal/kg	*	2.47	2.70	
Sheep	DE Mcal/kg	*	2.79	3.04	
Cattle	ME Mcal/kg	*	2.03	2.21	

Continued

Feed Name or Analyses				Mean As Fed	Mean Dry	C.V. ± %
Sheep	ME	Mcal/kg	*	2.29	2.49	
Cattle	TDN	%	*	56.1	61.1	
Sheep	TDN	%		63.2	69.0	

DATES, EGYPT. Phoenix dactylifera

Dates, Egypt, seeds, grnd, mature, (4)

Ref No 4 05 701 Egypt

Feed Name or Analyses				Mean As Fed	Mean Dry	C.V. ± %
Dry matter		%		90.3	100.0	
Sheep	dig coef	%		58.	58.	
Ash		%		2.9	3.2	
Crude fiber		%		14.2	15.7	
Sheep	dig coef	%		53.	53.	
Ether extract		%		8.1	9.0	
Sheep	dig coef	%		91.	91.	
N-free extract		%		59.1	65.4	
Sheep	dig coef	%		73.	73.	
Protein (N x 6.25)		%		6.0	6.6	
Sheep	dig coef	%		0.	0.	
Cattle	dig prot	%	*	1.9	2.1	
Goats	dig prot	%	*	3.0	3.3	
Horses	dig prot	%	*	3.0	3.3	
Energy	GE	Mcal/kg				
Cattle	DE	Mcal/kg	*	3.12	3.45	
Sheep	DE	Mcal/kg	*	2.97	3.28	
Swine	DE	kcal/kg	*	3041.	3368.	
Cattle	ME	Mcal/kg	*	2.56	2.83	
Sheep	ME	Mcal/kg	*	2.43	2.69	
Swine	ME	kcal/kg	*	2880.	3184.	
Cattle	TDN	%	*	70.7	28.3	
Sheep	TDN	%		67.3	74.5	
Swine	TDN	%	*	69.0	76.4	

DAYFLOWER. Commelina spp

Dayflower, aerial part, fresh, (2)

Ref No 2 01 753 United States

Feed Name or Analyses				Mean As Fed	Mean Dry	C.V. ± %
Dry matter		%		12.1	100.0	
Ash		%		1.7	13.9	
Crude fiber		%		3.3	27.0	
Cattle	dig coef	%		64.	64.	
Ether extract		%		0.3	2.1	
Cattle	dig coef	%		54.	54.	
N-free extract		%		5.6	46.1	
Cattle	dig coef	%		79.	79.	
Protein (N x 6.25)		%		1.3	10.9	
Cattle	dig coef	%		65.	65.	
Cattle	dig prot	%		0.9	7.1	
Goats	dig prot	%	*	0.8	6.7	
Horses	dig prot	%	*	0.8	6.8	
Rabbits	dig prot	%	*	0.9	7.1	
Sheep	dig prot	%	*	0.9	7.1	
Energy	GE	Mcal/kg				
Cattle	DE	Mcal/kg	*	0.34	2.79	
Sheep	DE	Mcal/kg	*	0.35	2.88	
Cattle	ME	Mcal/kg	*	0.28	2.29	
Sheep	ME	Mcal/kg	*	0.29	2.36	
Cattle	TDN	%		7.7	63.3	
Sheep	TDN	%	*	7.9	65.3	

DAYLILY, TAWNY. Hemeracallis fulva

Daylily, tawny, seeds w hulls, (5)

Ref No 5 09 270 United States

Feed Name or Analyses				Mean As Fed	Mean Dry	C.V. ± %
Dry matter		%			100.0	
Protein (N x 6.25)		%			39.0	
Alanine		%			1.37	
Arginine		%			2.81	
Aspartic acid		%			3.94	
Glutamic acid		%			5.42	
Glycine		%			1.29	
Histidine		%			0.62	
Hydroxyproline		%			0.00	
Isoleucine		%			1.44	
Leucine		%			1.99	
Lysine		%			1.29	
Methionine		%			0.47	
Phenylalanine		%			1.40	
Proline		%			1.09	
Serine		%			1.44	
Threonine		%			0.94	
Tyrosine		%			0.78	
Valine		%			1.79	

DEERVETCH. Lotus spp

Deervetch, aerial part, fresh, (2)

Ref No 2 01 757 United States

Feed Name or Analyses				Mean As Fed	Mean Dry	C.V. ± %
Dry matter		%		15.4	100.0	
Calcium		%		0.30	1.92	32
Phosphorus		%		0.04	0.28	74
Potassium		%		0.32	2.08	39

Deervetch, aerial part, fresh, full bloom, (2)

Ref No 2 01 755 United States

Feed Name or Analyses				Mean As Fed	Mean Dry	C.V. ± %
Dry matter		%			100.0	
Calcium		%			2.22	31
Phosphorus		%			0.27	27
Potassium		%			2.09	27

Deervetch, aerial part, fresh, milk stage, (2)

Ref No 2 01 756 United States

Feed Name or Analyses				Mean As Fed	Mean Dry	C.V. ± %
Dry matter		%			100.0	
Ash		%			7.8	18
Crude fiber		%			25.9	13
Ether extract		%			1.7	23
N-free extract		%			46.7	
Protein (N x 6.25)		%			17.9	13
Cattle	dig prot	%	*		13.1	
Goats	dig prot	%	*		13.3	
Horses	dig prot	%	*		12.7	
Rabbits	dig prot	%	*		12.5	
Sheep	dig prot	%	*		13.7	

DEERVETCH, BIG. Lotus crassifolius

Deervetch, big, hay, s-c, (1)

Ref No 1 01 758 United States

Feed Name or Analyses				Mean As Fed	Mean Dry	C.V. ± %
Dry matter		%		89.3	100.0	
Ash		%		7.3	8.2	
Crude fiber		%		21.9	24.5	
Ether extract		%		2.7	3.0	
N-free extract		%		45.5	50.9	
Protein (N x 6.25)		%		12.0	13.4	
Cattle	dig prot	%	*	7.6	8.5	
Goats	dig prot	%	*	8.1	9.1	
Horses	dig prot	%	*	8.0	8.9	
Rabbits	dig prot	%	*	8.1	9.0	
Sheep	dig prot	%	*	7.7	8.6	

DEERVETCH, BIRDSFOOT. Lotus corniculatus

Deervetch, birdsfoot, hay, fan air dried, (1)

Ref No 1 01 759 United States

Feed Name or Analyses				Mean As Fed	Mean Dry	C.V. ± %
Dry matter		%		91.2	100.0	
Carotene		mg/kg		43.4	47.6	
Vitamin A equivalent		IU/g		72.4	79.4	

Deervetch, birdsfoot, hay, s-c, (1)

Ref No 1 01 765 United States

Feed Name or Analyses				Mean As Fed	Mean Dry	C.V. ± %
Dry matter		%		91.0	100.0	1
Sodium		%		1.17	1.29	

Deervetch, birdsfoot, hay, s-c, immature, (1)

Ref No 1 01 760 United States

Feed Name or Analyses				Mean As Fed	Mean Dry	C.V. ± %
Dry matter		%			100.0	
Ash		%			7.5	
Crude fiber		%			12.9	
Ether extract		%			5.1	
N-free extract		%			46.7	
Protein (N x 6.25)		%			27.8	
Cattle	dig prot	%	*		21.0	
Goats	dig prot	%	*		22.5	
Horses	dig prot	%	*		21.1	
Rabbits	dig prot	%	*		20.1	
Sheep	dig prot	%	*		21.5	
Calcium		%			1.53	17
Phosphorus		%			0.28	24
Carotene		mg/kg			298.5	
Riboflavin		mg/kg			16.1	
Thiamine		mg/kg			6.8	
Vitamin A equivalent		IU/g			497.6	

Deervetch, birdsfoot, hay, s-c, early bloom, (1)

Ref No 1 01 761 United States

Feed Name or Analyses				Mean As Fed	Mean Dry	C.V. ± %
Dry matter		%		86.8	100.0	
Ash		%		6.4	7.4	
Crude fiber		%		24.7	28.5	
Ether extract		%		2.9	3.3	
N-free extract		%		33.8	38.9	
Protein (N x 6.25)		%		19.0	21.9	
Cattle	dig prot	%	*	13.8	15.9	
Goats	dig prot	%	*	14.7	17.0	
Horses	dig prot	%	*	14.0	16.1	
Rabbits	dig prot	%	*	13.5	15.6	
Sheep	dig prot	%	*	14.1	16.2	

(1) dry forages and roughages (2) pasture, range plants, and forages fed green (3) silages (4) energy feeds (5) protein supplements (6) minerals (7) vitamins (8) additives

Feed Name or Analyses			Mean As Fed	Mean Dry	C.V. ± %

Deervetch, birdsfoot, hay, s-c, mid-bloom, (1)
Ref No 1 01 762 United States

Analyses			As Fed	Dry	C.V. ± %
Dry matter	%		90.5	100.0	
Ash	%		6.0	6.6	
Crude fiber	%		28.4	31.4	
Ether extract	%		2.1	2.3	
N-free extract	%		40.9	45.2	
Protein (N x 6.25)	%		13.1	14.5	
Cattle	dig prot %	*	8.6	9.5	
Goats	dig prot %	*	9.1	10.1	
Horses	dig prot %	*	8.9	9.8	
Rabbits	dig prot %	*	8.9	9.9	
Sheep	dig prot %	*	8.7	9.6	
Energy	GE Mcal/kg				
Cattle	DE Mcal/kg	*	2.23	2.46	
Sheep	DE Mcal/kg	*	2.33	2.58	
Cattle	ME Mcal/kg	*	1.83	2.02	
Sheep	ME Mcal/kg	*	1.91	2.11	
Cattle	TDN %	*	50.6	55.9	
Sheep	TDN %	*	52.9	58.5	
Carotene	mg/kg		72.6	80.2	
Vitamin A equivalent	IU/g		121.1	133.8	

Deervetch, birdsfoot, hay, s-c, cut 1, (1)
Ref No 1 01 763 United States

Analyses			As Fed	Dry	C.V. ± %
Dry matter	%		92.1	100.0	
Ash	%		6.4	6.9	
Crude fiber	%		25.7	27.9	
Ether extract	%		2.2	2.4	
N-free extract	%		41.1	44.6	
Protein (N x 6.25)	%		16.8	18.2	
Cattle	dig prot %	*	11.7	12.7	
Goats	dig prot %	*	12.5	13.5	
Horses	dig prot %	*	12.0	13.0	
Rabbits	dig prot %	*	11.7	12.7	
Sheep	dig prot %	*	11.9	12.9	
Carotene	mg/kg		119.0	129.2	
Vitamin A equivalent	IU/g		198.3	215.4	

Deervetch, birdsfoot, hay, s-c, cut 2, (1)
Ref No 1 01 764 United States

Analyses			As Fed	Dry	C.V. ± %
Dry matter	%		90.5	100.0	
Ash	%		6.4	7.1	6
Crude fiber	%		20.1	22.2	40
Ether extract	%		3.3	3.7	37
N-free extract	%		41.4	45.8	
Protein (N x 6.25)	%		19.2	21.2	30
Cattle	dig prot %	*	13.8	15.3	
Goats	dig prot %	*	14.8	16.3	
Horses	dig prot %	*	14.1	15.5	
Rabbits	dig prot %	*	13.6	15.0	
Sheep	dig prot %	*	14.1	15.6	
Calcium	%		1.55	1.71	
Phosphorus	%		0.35	0.39	
Carotene	mg/kg		263.2	290.8	
Vitamin A equivalent	IU/g		438.7	484.7	

Deervetch, birdsfoot, aerial part, fresh, immature, (2)
Ref No 2 01 766 United States

Analyses			As Fed	Dry	C.V. ± %
Dry matter	%			100.0	
Ash	%			10.0	4
Crude fiber	%			17.2	14
Ether extract	%			2.4	5
N-free extract	%			43.7	
Protein (N x 6.25)	%			26.7	7
Cattle	dig prot %	*		20.6	
Goats	dig prot %	*		21.5	
Horses	dig prot %	*		20.2	
Rabbits	dig prot %	*		19.3	
Sheep	dig prot %	*		21.9	

Deervetch, birdsfoot, aerial part, fresh, full bloom, (2)
Ref No 2 01 767 United States

Analyses			As Fed	Dry	C.V. ± %
Dry matter	%			100.0	
Ash	%			7.6	
Crude fiber	%			21.6	
Ether extract	%			2.0	
N-free extract	%			48.9	
Protein (N x 6.25)	%			19.9	
Cattle	dig prot %	*		14.8	
Goats	dig prot %	*		15.1	
Horses	dig prot %	*		14.4	
Rabbits	dig prot %	*		14.0	
Sheep	dig prot %	*		15.5	

Deervetch, birdsfoot, aerial part, fresh, milk stage, (2)
Ref No 2 01 768 United States

Analyses			As Fed	Dry	C.V. ± %
Dry matter	%			100.0	
Ash	%			7.4	9
Crude fiber	%			27.1	3
Ether extract	%			1.7	23
N-free extract	%			45.1	
Protein (N x 6.25)	%			18.7	3
Cattle	dig prot %	*		13.8	
Goats	dig prot %	*		14.0	
Horses	dig prot %	*		13.4	
Rabbits	dig prot %	*		13.1	
Sheep	dig prot %	*		14.4	

DEERVETCH, BROOM. Lotus scoparius

Deervetch, broom, seeds w hulls, (5)
Ref No 5 09 058 United States

Analyses			As Fed	Dry	C.V. ± %
Dry matter	%			100.0	
Protein (N x 6.25)	%			35.0	
Alanine	%			1.12	
Arginine	%			3.19	
Aspartic acid	%			2.91	
Glutamic acid	%			4.87	
Glycine	%			1.82	
Histidine	%			0.77	
Hydroxyproline	%			0.04	
Isoleucine	%			1.05	
Leucine	%			1.79	
Lysine	%			1.37	
Methionine	%			0.35	
Phenylalanine	%			1.19	
Proline	%			1.23	
Serine	%			1.47	
Threonine	%			0.98	
Tyrosine	%			0.98	
Valine	%			1.16	

DEERVETCH, CHILEAN. Lotus subpinnatus

Deervetch, Chilean, aerial part, fresh, (2)
Ref No 2 01 773 United States

Analyses			As Fed	Dry	C.V. ± %
Dry matter	%			100.0	
Calcium	%			2.12	35
Phosphorus	%			0.34	71
Potassium	%			2.35	33

Deervetch, Chilean, aerial part, fresh, immature, (2)
Ref No 2 01 769 United States

Analyses			As Fed	Dry	C.V. ± %
Dry matter	%			100.0	
Ash	%			13.1	
Crude fiber	%			16.7	
Protein (N x 6.25)	%			23.4	
Cattle	dig prot %	*		17.8	
Goats	dig prot %	*		18.4	
Horses	dig prot %	*		17.4	
Rabbits	dig prot %	*		16.7	
Sheep	dig prot %	*		18.8	
Calcium	%			2.15	
Phosphorus	%			0.48	46
Potassium	%			2.54	

Deervetch, Chilean, aerial part, fresh, full bloom, (2)
Ref No 2 01 770 United States

Analyses			As Fed	Dry	C.V. ± %
Dry matter	%			100.0	
Ash	%			11.8	25
Crude fiber	%			19.4	9
Protein (N x 6.25)	%			18.0	7
Cattle	dig prot %	*		13.2	
Goats	dig prot %	*		13.4	
Horses	dig prot %	*		12.8	
Rabbits	dig prot %	*		12.6	
Sheep	dig prot %	*		13.8	
Calcium	%			2.54	31
Phosphorus	%			0.33	13
Potassium	%			2.56	12

Deervetch, Chilean, aerial part, fresh, milk stage, (2)
Ref No 2 01 771 United States

Analyses			As Fed	Dry	C.V. ± %
Dry matter	%			100.0	
Calcium	%			2.00	
Phosphorus	%			0.40	
Potassium	%			2.66	

Deervetch, Chilean, aerial part, fresh, dormant, (2)
Ref No 2 01 772 United States

Analyses			As Fed	Dry	C.V. ± %
Dry matter	%			100.0	
Ash	%			6.7	
Crude fiber	%			31.1	
Protein (N x 6.25)	%			4.4	
Cattle	dig prot %	*		1.6	
Goats	dig prot %	*		0.7	
Horses	dig prot %	*		1.3	
Rabbits	dig prot %	*		2.1	
Sheep	dig prot %	*		1.1	
Calcium	%			1.20	

Continued

Feed Name or Analyses			Mean As Fed	Mean Dry	C.V. ± %
Phosphorus	%			0.07	
Potassium	%			1.80	

DEERVETCH, SHRUBBY. Lotus rigidus

Deervetch, shrubby, seeds w hulls, (5)

Ref No 5 09 057 United States

Feed Name or Analyses			Mean As Fed	Mean Dry	C.V. ± %
Dry matter	%			100.0	
Protein (N x 6.25)	%			38.0	
Alanine	%			1.22	
Arginine	%			4.29	
Aspartic acid	%			3.27	
Glutamic acid	%			5.74	
Glycine	%			1.94	
Histidine	%			0.84	
Hydroxyproline	%			0.08	
Isoleucine	%			1.10	
Leucine	%			1.86	
Lysine	%			1.37	
Methionine	%			0.38	
Phenylalanine	%			1.29	
Proline	%			1.25	
Serine	%			1.63	
Threonine	%			1.03	
Tyrosine	%			0.99	
Valine	%			1.25	

DEERVETCH, SPANISHCLOVER. Lotus americanus

Deervetch, Spanishclover, aerial part, fresh, (2)

Ref No 2 01 777 United States

Feed Name or Analyses			Mean As Fed	Mean Dry	C.V. ± %
Dry matter	%		7.4	100.0	
Ash	%		0.5	7.0	26
Crude fiber	%		1.7	23.4	16
Sheep	dig coef %		34.	34.	
Ether extract	%		0.1	1.9	
Sheep	dig coef %		37.	37.	
N-free extract	%		4.0	54.4	
Sheep	dig coef %		56.	56.	
Protein (N x 6.25)	%		1.0	13.4	38
Sheep	dig coef %		19.	19.	
Cattle	dig prot %	*	0.7	9.3	
Goats	dig prot %	*	0.7	9.1	
Horses	dig prot %	*	0.7	8.9	
Rabbits	dig prot %	*	0.7	9.0	
Sheep	dig prot %		0.2	2.5	
Energy	GE Mcal/kg				
Cattle	DE Mcal/kg	*	0.13	1.76	
Sheep	DE Mcal/kg	*	0.14	1.88	
Cattle	ME Mcal/kg	*	0.11	1.44	
Sheep	ME Mcal/kg	*	0.11	1.54	
Cattle	TDN %	*	3.0	40.6	
Sheep	TDN %		3.1	42.5	
Calcium	%		0.12	1.65	20
Phosphorus	%		0.02	0.21	46
Potassium	%		0.12	1.68	42

(1) dry forages and roughages (2) pasture, range plants, and forages fed green (3) silages (4) energy feeds (5) protein supplements (6) minerals (7) vitamins (8) additives

Deervetch, Spanishclover, aerial part, fresh, immature, (2)

Ref No 2 01 774 United States

Feed Name or Analyses			Mean As Fed	Mean Dry	C.V. ± %
Dry matter	%			100.0	
Ash	%			10.3	
Crude fiber	%			17.2	
Protein (N x 6.25)	%			22.2	
Cattle	dig prot %	*		16.8	
Goats	dig prot %	*		17.3	
Horses	dig prot %	*		16.4	
Rabbits	dig prot %	*		15.8	
Sheep	dig prot %	*		17.7	
Calcium	%			1.80	
Phosphorus	%			0.34	
Potassium	%			2.55	

Deervetch, Spanishclover, aerial part, fresh, dough stage, (2)

Ref No 2 01 775 United States

Feed Name or Analyses			Mean As Fed	Mean Dry	C.V. ± %
Dry matter	%			100.0	
Ash	%			8.1	
Crude fiber	%			19.9	
Protein (N x 6.25)	%			17.2	
Cattle	dig prot %	*		12.5	
Goats	dig prot %	*		12.6	
Horses	dig prot %	*		12.1	
Rabbits	dig prot %	*		11.9	
Sheep	dig prot %	*		13.0	
Calcium	%			1.80	
Phosphorus	%			0.20	
Potassium	%			1.47	

Deervetch, Spanishclover, aerial part, fresh, mature, (2)

Ref No 2 01 776 United States

Feed Name or Analyses			Mean As Fed	Mean Dry	C.V. ± %
Dry matter	%			100.0	
Ash	%			6.1	
Crude fiber	%			28.5	
Protein (N x 6.25)	%			7.9	
Cattle	dig prot %	*		4.6	
Goats	dig prot %	*		3.9	
Horses	dig prot %	*		4.2	
Rabbits	dig prot %	*		4.8	
Sheep	dig prot %	*		4.4	
Calcium	%			1.32	
Phosphorus	%			0.08	
Potassium	%			0.70	

Deervetch, Spanishclover, stems, fresh, mature, (2)

Ref No 2 01 778 United States

Feed Name or Analyses			Mean As Fed	Mean Dry	C.V. ± %
Dry matter	%			100.0	
Ash	%			2.4	
Crude fiber	%			43.1	
Protein (N x 6.25)	%			4.0	
Cattle	dig prot %	*		1.3	
Goats	dig prot %	*		0.3	
Horses	dig prot %	*		0.9	
Rabbits	dig prot %	*		1.8	
Sheep	dig prot %	*		0.7	
Calcium	%			0.74	
Phosphorus	%			0.02	
Potassium	%			0.13	

DEERVETCH, WETLAND. Lotus spp

Deervetch, wetland, aerial part, fresh, immature, (2)

Ref No 2 01 779 United States

Feed Name or Analyses			Mean As Fed	Mean Dry	C.V. ± %
Dry matter	%		23.6	100.0	
Ash	%		1.9	8.1	17
Crude fiber	%		5.2	22.2	19
Ether extract	%		0.6	2.5	42
N-free extract	%		10.8	45.6	
Protein (N x 6.25)	%		5.1	21.6	18
Cattle	dig prot %	*	3.8	16.3	
Goats	dig prot %	*	3.9	16.7	
Horses	dig prot %	*	3.7	15.9	
Rabbits	dig prot %	*	3.6	15.3	
Sheep	dig prot %	*	4.0	17.1	
Calcium	%		0.41	1.76	
Cobalt	mg/kg		0.049	0.209	
Phosphorus	%		0.05	0.20	
Potassium	%		0.43	1.84	
Carotene	mg/kg		98.3	417.1	
Niacin	mg/kg		44.4	188.5	
Riboflavin	mg/kg		9.3	39.2	
Thiamine	mg/kg		1.8	7.7	
Vitamin A equivalent	IU/g		163.9	695.3	

Defatted wheat germ meal (AAFCO) –
see Wheat, germ, extn unspecified grnd, mn 30% protein, (5)

Defluorinated phosphate (CFA) –
see Phosphate, defluorinated grnd, mn 1 part fluorine per 100 parts phosphorus, (6)

Degermed white corn meal –
see Corn, white, grain wo germ, grnd, (4)

Degermed yellow corn meal –
see Corn, yellow, grain wo germ, grnd, (4)

Dehydrated alfalfa meal (AAFCO) –
see Alfalfa, aerial part, dehy grnd, (1)

Dehydrated alfalfa stem meal (AAFCO) –
see Alfalfa, stems, dehy grnd, (1)

Dehydrated corn plant (AAFCO) –
see Corn, aerial part, dehy, (1)

Dehydrated fine ground corn cob (AAFCO) –
see Corn, cobs, dehy fine grnd, 10 sieve mx 10% water, (1)

Desmanthus –
see Bundleflower, rayado, aerial part, fresh, (2)

DEVILSCLAWS, SWEET. Proboscidea fragrans

Devilsclaws, sweet, browse, mature, (2)

Ref No 2 01 781 United States

Feed Name or Analyses			Mean As Fed	Mean Dry	C.V. ± %
Dry matter	%			100.0	
Ash	%			25.7	
Crude fiber	%			6.2	
Ether extract	%			9.5	
N-free extract	%			40.5	
Protein (N x 6.25)	%			18.1	
Cattle	dig prot %	*		13.3	

Feed Name or Analyses			Mean As Fed	Mean Dry	C.V. ± %
Goats	dig prot %	*		13.4	
Horses	dig prot %	*		12.9	
Rabbits	dig prot %	*		12.6	
Sheep	dig prot %	*		13.9	
Calcium	%			2.19	
Magnesium	%			0.52	
Phosphorus	%			0.25	
Potassium	%			3.01	

DEVILS-WALKING STICK. Aralia spinosa

Devils-walking stick, seeds w hulls, (4)

Ref No 4 09 141 United States

Feed Name or Analyses			Mean As Fed	Mean Dry	C.V. ± %
Dry matter	%			100.0	
Protein (N x 6.25)	%			17.0	
Alanine	%			0.75	
Arginine	%			1.29	
Aspartic acid	%			1.63	
Glutamic acid	%			3.50	
Glycine	%			0.83	
Histidine	%			0.46	
Hydroxyproline	%			0.14	
Isoleucine	%			0.75	
Leucine	%			1.24	
Lysine	%			0.78	
Methionine	%			0.36	
Phenylalanine	%			0.71	
Proline	%			0.92	
Serine	%			0.90	
Threonine	%			0.61	
Tyrosine	%			0.56	
Valine	%			0.87	

Dicalcium phosphate (AAFCO) –
see Calcium phosphate, dibasic, commercial, (6)

Digester tankage –
see Animal, carcass residue w blood, dry or wet rendered dehy grnd, mn 9% indigestible material mx 4.4% phosphorus (5)

Digester tankage w $CaCO_3$ added –
see Animal, carcass residue w blood w calcium carbonate added, dry or wet rendered dehy grnd, (5)

Digester tankage with bone –
see Animal, carcass residue w blood w bone, dry or wet rendered dehy grnd, mn 9% indigestible material mn 4.4% phosphorus (5)

Digester tankage w rumen contents –
see Animal, carcass residue w blood w rumen contents, dry or wet rendered dehy grnd, (5)

Distillers dried corn grains w solubles, fungal amylase process –
see Corn, fungal amylase process distillers grains w solubles, dehy, (5)

Distillers dried grains w solubles –
see Grains, distillers grains w solubles, dehy, (5)

Distillers stillage, wet –
see Grains, distillers stillage, wet, (5)

Distillery stillage, strained –
see Grains, distillers stillage, wet strained, (5)

DOCK. Rumex spp

Dock, leaves, fresh, (2)

Ref No 2 01 793 United States

Feed Name or Analyses			Mean As Fed	Mean Dry	C.V. ± %
Dry matter	%			100.0	
Carotene	mg/kg			132.5	
Vitamin A equivalent	IU/g			220.9	

DOCK, CURLY. Rumex crispus

Dock, curly, aerial part, fresh, (2)

Ref No 2 01 794 United States

Feed Name or Analyses			Mean As Fed	Mean Dry	C.V. ± %
Dry matter	%			100.0	
Ash	%			3.6	
Crude fiber	%			23.7	
Ether extract	%			3.2	
N-free extract	%			56.5	
Protein (N x 6.25)	%			13.0	
Cattle	dig prot %	*		8.9	
Goats	dig prot %	*		8.7	
Horses	dig prot %	*		8.6	
Rabbits	dig prot %	*		8.7	
Sheep	dig prot %	*		9.1	

DODDER. Cuscuta spp

Dodder, aerial part, fresh, (2)

Ref No 2 01 795 United States

Feed Name or Analyses			Mean As Fed	Mean Dry	C.V. ± %
Dry matter	%			100.0	
Ash	%			2.5	
Crude fiber	%			19.6	
Ether extract	%			3.4	
N-free extract	%			62.5	
Protein (N x 6.25)	%			12.0	
Cattle	dig prot %	*		8.1	
Goats	dig prot %	*		7.8	
Horses	dig prot %	*		7.7	
Rabbits	dig prot %	*		7.9	
Sheep	dig prot %	*		8.2	

Dodder, stems, fresh, (2)

Ref No 2 01 796 United States

Feed Name or Analyses			Mean As Fed	Mean Dry	C.V. ± %
Dry matter	%			100.0	
Ash	%			2.9	
Crude fiber	%			24.4	
Ether extract	%			1.7	
N-free extract	%			63.4	
Protein (N x 6.25)	%			7.6	
Cattle	dig prot %	*		4.4	
Goats	dig prot %	*		3.7	
Horses	dig prot %	*		4.0	
Rabbits	dig prot %	*		4.5	
Sheep	dig prot %	*		4.1	
Calcium	%			0.11	
Magnesium	%			0.06	
Phosphorus	%			0.13	
Potassium	%			1.24	

DOGTAIL, CRESTED. Cynosurus cristatus

Dogtail, crested, aerial part, fresh, (2)

Ref No 2 08 567 United States

Feed Name or Analyses			Mean As Fed	Mean Dry	C.V. ± %
Dry matter	%			100.0	
Ash	%			6.2	
Ether extract	%			1.2	
Protein (N x 6.25)	%			3.1	
Cattle	dig prot %	*		0.5	
Goats	dig prot %	*		-0.4	
Horses	dig prot %	*		0.2	
Rabbits	dig prot %	*		1.1	
Sheep	dig prot %	*		-0.0	
Other carbohydrates	%			39.3	
Lignin (Ellis)	%			9.4	
Silicon	%			1.80	

DOGTOOTHGRASS. Cynodon spp

Dogtoothgrass, hay, s-c, (1)

Ref No 1 01 799 United States

Feed Name or Analyses			Mean As Fed	Mean Dry	C.V. ± %
Dry matter	%		85.6	100.0	
Ash	%		8.1	9.5	
Crude fiber	%		29.4	34.4	
Cattle	dig coef %		56.	56.	
Sheep	dig coef %		57.	57.	
Ether extract	%		1.4	1.6	
Cattle	dig coef %		28.	28.	
Sheep	dig coef %		41.	41.	
N-free extract	%		39.4	46.1	
Cattle	dig coef %		56.	56.	
Sheep	dig coef %		50.	50.	
Protein (N x 6.25)	%		7.2	8.5	
Cattle	dig coef %		49.	49.	
Sheep	dig coef %		52.	52.	
Cattle	dig prot %		3.5	4.1	
Goats	dig prot %	*	3.8	4.4	
Horses	dig prot %	*	4.0	4.7	
Rabbits	dig prot %	*	4.4	5.2	
Sheep	dig prot %		3.8	4.4	
Energy	GE Mcal/kg				
Cattle	DE Mcal/kg	*	1.89	2.21	
Sheep	DE Mcal/kg	*	1.83	2.14	
Cattle	ME Mcal/kg	*	1.55	1.82	
Sheep	ME Mcal/kg	*	1.50	1.75	
Cattle	TDN %		43.0	50.2	
Sheep	TDN %		41.5	48.5	

Dogtoothgrass, hay, s-c, milk stage, (1)

Ref No 1 01 797 United States

Feed Name or Analyses			Mean As Fed	Mean Dry	C.V. ± %
Dry matter	%		85.5	100.0	
Ash	%		8.3	9.7	
Crude fiber	%		29.2	34.2	
Sheep	dig coef %		54.	54.	
Ether extract	%		2.1	2.5	
Sheep	dig coef %		57.	57.	
N-free extract	%		38.5	45.0	
Sheep	dig coef %		48.	48.	
Protein (N x 6.25)	%		7.4	8.6	
Sheep	dig coef %		45.	45.	
Cattle	dig prot %	*	3.8	4.4	
Goats	dig prot %	*	3.9	4.6	
Horses	dig prot %	*	4.1	4.8	

Continued

Feed Name or Analyses			Mean As Fed	Mean Dry	C.V. ± %
Rabbits	dig prot %	*	4.5	5.3	
Sheep	dig prot %		3.3	3.9	
Energy	GE Mcal/kg				
Cattle	DE Mcal/kg	*	1.83	2.14	
Sheep	DE Mcal/kg	*	1.78	2.08	
Cattle	ME Mcal/kg	*	1.50	1.75	
Sheep	ME Mcal/kg	*	1.46	1.70	
Cattle	TDN %	*	41.5	48.5	
Sheep	TDN %		40.3	47.1	

Dogtoothgrass, hay, s-c, cut 2, (1)
Ref No 1 01 798 United States

Feed Name or Analyses			Mean As Fed	Mean Dry	C.V. ± %
Dry matter	%		82.4	100.0	
Ash	%		8.2	10.0	
Crude fiber	%		27.5	33.4	
Cattle	dig coef %		55.	55.	
Ether extract	%		3.0	3.6	
Cattle	dig coef %		67.	67.	
N-free extract	%		37.1	45.0	
Cattle	dig coef %		48.	48.	
Protein (N x 6.25)	%		6.6	8.0	
Cattle	dig coef %		47.	47.	
Cattle	dig prot %		3.1	3.8	
Goats	dig prot %	*	3.3	4.0	
Horses	dig prot %	*	3.6	4.3	
Rabbits	dig prot %	*	4.0	4.8	
Sheep	dig prot %	*	3.1	3.7	
Energy	GE Mcal/kg				
Cattle	DE Mcal/kg	*	1.79	2.17	
Sheep	DE Mcal/kg	*	1.99	2.42	
Cattle	ME Mcal/kg	*	1.46	1.78	
Sheep	ME Mcal/kg	*	1.63	1.98	
Cattle	TDN %		40.5	49.2	
Sheep	TDN %	*	45.2	54.8	

Dogtoothgrass, aerial part, fresh, pre-bloom, (2)
Ref No 2 01 800 United States

Feed Name or Analyses			Mean As Fed	Mean Dry	C.V. ± %
Dry matter	%		20.0	100.0	
Ash	%		1.6	7.9	
Crude fiber	%		4.8	24.0	
Sheep	dig coef %		66.	66.	
Ether extract	%		0.6	3.0	
Sheep	dig coef %		45.	45.	
N-free extract	%		9.0	44.9	
Sheep	dig coef %		66.	66.	
Protein (N x 6.25)	%		4.0	20.2	
Sheep	dig coef %		76.	76.	
Cattle	dig prot %	*	3.0	15.1	
Goats	dig prot %	*	3.1	15.4	
Horses	dig prot %	*	2.9	14.7	
Rabbits	dig prot %	*	2.9	14.3	
Sheep	dig prot %		3.1	15.4	
Energy	GE Mcal/kg				
Cattle	DE Mcal/kg	*	0.58	2.88	
Sheep	DE Mcal/kg	*	0.56	2.82	
Cattle	ME Mcal/kg	*	0.47	2.36	
Sheep	ME Mcal/kg	*	0.46	2.31	
Cattle	TDN %	*	13.1	65.3	
Sheep	TDN %		12.8	63.9	

(1) dry forages and roughages (2) pasture, range plants, and forages fed green (3) silages (4) energy feeds (5) protein supplements (6) minerals (7) vitamins (8) additives

Dogtoothgrass, aerial part, fresh, early bloom, (2)
Ref No 2 01 801 United States

Feed Name or Analyses			Mean As Fed	Mean Dry	C.V. ± %
Dry matter	%		20.0	100.0	
Ash	%		1.2	5.9	
Crude fiber	%		6.4	32.2	
Sheep	dig coef %		57.	57.	
Ether extract	%		0.3	1.4	
Sheep	dig coef %		21.	21.	
N-free extract	%		9.8	48.9	
Sheep	dig coef %		60.	60.	
Protein (N x 6.25)	%		2.3	11.6	
Sheep	dig coef %		70.	70.	
Cattle	dig prot %	*	1.6	7.8	
Goats	dig prot %	*	1.5	7.4	
Horses	dig prot %	*	1.5	7.4	
Rabbits	dig prot %	*	1.5	7.6	
Sheep	dig prot %		1.6	8.1	
Energy	GE Mcal/kg				
Cattle	DE Mcal/kg	*	0.48	2.41	
Sheep	DE Mcal/kg	*	0.50	2.49	
Cattle	ME Mcal/kg	*	0.40	1.98	
Sheep	ME Mcal/kg	*	0.41	2.04	
Cattle	TDN %	*	10.9	54.7	
Sheep	TDN %		11.3	56.5	

DOGTOOTHGRASS, GIANT. Cynodon spp

Dogtoothgrass, giant, aerial part, fresh, (2)
Ref No 2 01 802 United States

Feed Name or Analyses			Mean As Fed	Mean Dry	C.V. ± %
Dry matter	%		20.0	100.0	
Ash	%		1.9	9.5	
Crude fiber	%		6.4	32.1	
Cattle	dig coef %		53.	53.	
Sheep	dig coef %		61.	61.	
Ether extract	%		0.4	2.0	
Cattle	dig coef %		61.	61.	
Sheep	dig coef %		33.	33.	
N-free extract	%		9.1	45.3	
Cattle	dig coef %		45.	45.	
Sheep	dig coef %		59.	59.	
Protein (N x 6.25)	%		2.2	11.2	
Cattle	dig coef %		50.	50.	
Sheep	dig coef %		72.	72.	
Cattle	dig prot %		1.1	5.6	
Goats	dig prot %	*	1.4	7.0	
Horses	dig prot %	*	1.4	7.0	
Rabbits	dig prot %	*	1.5	7.3	
Sheep	dig prot %		1.6	8.1	
Energy	GE Mcal/kg				
Cattle	DE Mcal/kg	*	0.40	2.01	
Sheep	DE Mcal/kg	*	0.49	2.46	
Cattle	ME Mcal/kg	*	0.33	1.65	
Sheep	ME Mcal/kg	*	0.40	2.02	
Cattle	TDN %		9.1	45.7	
Sheep	TDN %		11.2	55.8	

DOGWOOD. Cornus florida

Dogwood, browse, mid-bloom, (2)
Ref No 2 01 803 United States

Feed Name or Analyses			Mean As Fed	Mean Dry	C.V. ± %
Dry matter	%			100.0	
Ash	%			9.2	8
Crude fiber	%			16.0	11
Ether extract	%			5.4	22
N-free extract	%			57.3	
Protein (N x 6.25)	%			12.1	4
Cattle	dig prot %	*		8.2	
Goats	dig prot %	*		7.8	
Horses	dig prot %	*		7.8	
Rabbits	dig prot %	*		8.0	
Sheep	dig prot %	*		8.3	
Calcium	%			2.59	5
Cobalt	mg/kg			0.099	85
Copper	mg/kg			5.1	
Iron	%			0.006	40
Manganese	mg/kg			43.7	53
Phosphorus	%			0.15	19

DOLICHOS. Dolichos spp

Dolichos, hay, s-c, milk stage, (1)
Ref No 1 01 804 United States

Feed Name or Analyses			Mean As Fed	Mean Dry	C.V. ± %
Dry matter	%		90.2	100.0	
Ash	%		8.0	8.9	
Crude fiber	%		34.0	37.7	
Sheep	dig coef %		51.	51.	
Ether extract	%		2.0	2.2	
Sheep	dig coef %		61.	61.	
N-free extract	%		35.6	39.5	
Sheep	dig coef %		64.	64.	
Protein (N x 6.25)	%		10.6	11.7	
Sheep	dig coef %		58.	58.	
Cattle	dig prot %	*	6.4	7.1	
Goats	dig prot %	*	6.7	7.5	
Horses	dig prot %	*	6.7	7.5	
Rabbits	dig prot %	*	6.9	7.7	
Sheep	dig prot %		6.1	6.8	
Energy	GE Mcal/kg				
Cattle	DE Mcal/kg	*	2.02	2.24	
Sheep	DE Mcal/kg	*	2.16	2.39	
Cattle	ME Mcal/kg	*	1.66	1.84	
Sheep	ME Mcal/kg	*	1.77	1.96	
Cattle	TDN %	*	45.9	50.9	
Sheep	TDN %		49.0	54.3	

DOLICHOS, HYACINTH. Dolichos lablab

Dolichos, hyacinth, hay, s-c, (1)
Ref No 1 01 805 United States

Feed Name or Analyses			Mean As Fed	Mean Dry	C.V. ± %
Dry matter	%		90.2	100.0	
Ash	%		7.1	7.9	
Crude fiber	%		33.6	37.2	
Sheep	dig coef %		53.	53.	
Ether extract	%		1.6	1.7	
Sheep	dig coef %		56.	56.	
N-free extract	%		34.2	37.9	
Sheep	dig coef %		64.	64.	
Protein (N x 6.25)	%		13.8	15.3	
Sheep	dig coef %		65.	65.	
Cattle	dig prot %	*	9.2	10.2	
Goats	dig prot %	*	9.8	10.8	
Horses	dig prot %	*	9.5	10.5	
Rabbits	dig prot %	*	9.5	10.5	
Sheep	dig prot %		9.0	9.9	
Energy	GE Mcal/kg				
Cattle	DE Mcal/kg	*	2.11	2.33	
Sheep	DE Mcal/kg	*	2.23	2.47	
Cattle	ME Mcal/kg	*	1.73	1.91	
Sheep	ME Mcal/kg	*	1.83	2.03	

Feed Name or Analyses		Mean As Fed	Mean Dry	C.V. ± %
Cattle	TDN %	* 47.7	52.9	
Sheep	TDN %	50.6	56.1	

DOLICHOS, TWINFLOWER. Dolichos biflorus

Dolichos, twinflower, aerial part, fresh, (2)

Ref No 2 01 806 United States

Dry matter	%		100.0	
Ash	%		11.1	
Crude fiber	%		30.2	
Ether extract	%		0.2	
N-free extract	%		42.1	
Protein (N x 6.25)	%		16.4	
Cattle	dig prot %	*	11.8	
Goats	dig prot %	*	11.9	
Horses	dig prot %	*	11.5	
Rabbits	dig prot %	*	11.3	
Sheep	dig prot %	*	12.3	
Energy	GE Mcal/kg			
Cattle	DE Mcal/kg	*	2.40	
Sheep	DE Mcal/kg	*	2.67	
Cattle	ME Mcal/kg	*	1.97	
Sheep	ME Mcal/kg	*	2.19	
Cattle	TDN %	*	54.5	
Sheep	TDN %	*	60.6	

DOMORPHOTHECA, SINUATA. Domorphotheca sinuata

Domorphotheca, sinuata, seeds w hulls, (5)

Ref No 5 09 304 United States

Dry matter	%		100.0	
Protein (N x 6.25)	%		42.0	
Alanine	%		1.55	
Aspartic acid	%		3.91	
Glutamic acid	%		7.01	
Glycine	%		2.27	
Histidine	%		0.84	
Hydroxyproline	%		0.04	
Isoleucine	%		1.64	
Leucine	%		2.44	
Lysine	%		1.60	
Methionine	%		0.59	
Phenylalanine	%		1.39	
Proline	%		1.76	
Serine	%		1.55	
Threonine	%		1.39	
Tyrosine	%		1.18	
Valine	%		1.89	

DOUMPALM. Hyphaene spp

Doumpalm, kernels, extn unspecified grnd, high fiber low protein, (1)

Ref No 1 01 807 United States

Dry matter	%	88.4	100.0	
Ash	%	1.6	1.8	
Crude fiber	%	36.4	41.2	
Swine	dig coef %	86.	86.	
Ether extract	%	7.3	8.3	
Swine	dig coef %	88.	88.	
N-free extract	%	38.4	43.4	
Swine	dig coef %	73.	73.	

Feed Name or Analyses		Mean As Fed	Mean Dry	C.V. ± %
Protein (N x 6.25)	%	4.7	5.3	
Swine	dig coef %	-10.	-10.	
Cattle	dig prot %	* 0.8	0.9	
Goats	dig prot %	* 1.9	2.1	
Horses	dig prot %	* 1.9	2.1	
Sheep	dig prot %	* 1.9	2.1	
Swine	dig prot %	-0.4	-0.5	
Energy	GE Mcal/kg			
Cattle	DE Mcal/kg	* 3.32	3.76	
Sheep	DE Mcal/kg	* 3.29	3.72	
Swine	DE kcal/kg	*3234.	3658.	
Cattle	ME Mcal/kg	* 2.72	3.08	
Sheep	ME Mcal/kg	* 2.70	3.05	
Swine	ME kcal/kg	*3070.	3473.	
Cattle	TDN %	* 72.6	82.1	
Sheep	TDN %	* 74.6	84.4	
Swine	TDN %	73.3	83.0	

DRABA. Draba spp

Draba, aerial part, fresh, mid-bloom, (2)

Ref No 2 01 808 United States

Dry matter	%		100.0	
Ash	%		14.4	
Crude fiber	%		17.6	
Ether extract	%		2.1	
N-free extract	%		50.1	
Protein (N x 6.25)	%		15.8	
Cattle	dig prot %	*	11.3	
Goats	dig prot %	*	11.3	
Horses	dig prot %	*	10.9	
Rabbits	dig prot %	*	10.9	
Sheep	dig prot %	*	11.7	
Calcium	%		3.31	
Magnesium	%		0.23	
Phosphorus	%		0.33	
Potassium	%		1.78	

DRACENA, GIANT. Cordyline australis

Dracena, giant, seeds w hulls, (5)

Ref No 5 09 016 United States

Dry matter	%		100.0	
Protein (N x 6.25)	%		20.0	
Alanine	%		0.74	
Arginine	%		2.42	
Aspartic acid	%		1.68	
Glutamic acid	%		3.90	
Glycine	%		0.88	
Histidine	%		0.48	
Hydroxyproline	%		0.06	
Isoleucine	%		0.76	
Leucine	%		1.22	
Lysine	%		0.74	
Methionine	%		0.46	
Phenylalanine	%		0.84	
Proline	%		0.82	
Serine	%		0.92	
Threonine	%		0.60	
Tyrosine	%		0.64	
Valine	%		0.98	

Feed Name or Analyses		Mean As Fed	Mean Dry	C.V. ± %

DREGS. Scientific name not used

Dregs, dehy grnd, (5)

Ref No 5 01 809 United States

Dry matter	%	94.5	100.0	
Ash	%	1.6	1.7	
Crude fiber	%	4.9	5.2	
Sheep	dig coef %	23.	23.	
Ether extract	%	14.0	14.8	
Sheep	dig coef %	97.	97.	
N-free extract	%	26.9	28.5	
Sheep	dig coef %	70.	70.	
Protein (N x 6.25)	%	47.1	49.8	
Sheep	dig coef %	65.	65.	
Sheep	dig prot %	30.6	32.4	
Energy	GE Mcal/kg			
Cattle	DE Mcal/kg	* 3.62	3.83	
Sheep	DE Mcal/kg	* 3.58	3.78	
Cattle	ME Mcal/kg	* 2.97	3.04	
Sheep	ME Mcal/kg	* 2.93	3.10	
Cattle	TDN %	* 82.0	86.8	
Sheep	TDN %	81.1	85.8	

Dried apple pectin pulp (AAFCO) –
see Apples, pulp wo pectin, dehy, (4)

Dried apple pomace (AAFCO) –
see Apples, pulp, dehy grnd, (4)

Dried bakery product (AAFCO) –
see Bakery, refuse, dehy grnd, mx salt declared above 3% (4)

Dried beet product (AAFCO) –
see Beet, sugar, pulp w Steffens filtrate, dehy, (4)

Dried beet pulp (AAFCO) (CFA) –
see Beet, sugar, pulp, dehy, (4)

Dried buttermilk, feed grade (AAFCO) –
see Cattle, buttermilk, dehy, feed gr mx 8% moisture mx 13% ash mn 5% fat, (5)

Dried citrus meal (AAFCO) –
see Citrus, pulp fines, shredded dehy, (4)

Dried citrus pulp (AAFCO) –
see Citrus, pulp wo fines, shredded dehy, (4)

Dried distillers solubles (CFA) –
see Grains, distillers solubles, dehy, (5)

Dried fish solubles (AAFCO) –
see Fish, stickwater solubles, cooked dehy, mn 60% protein, (5)

Dried grain fermentation solubles (CFA) –
see Grains, fermentation solubles, dehy, (5)

Dried kelp (AAFCO) –
see Seaweed, kelp, entire plant, dehy, (1)

Dried meat solubles (AAFCO) –
see Animal, stickwater solubles, vacuum dehy, (5)

Feed Name or Analyses			Mean As Fed	Mean Dry	C.V. ± %
Dried milk albumin (AAFCO) – see Cattle, whey albumin, heat-and acid-precipitated dehy, mn 75% protein, (5)					
Dried molasses fermentation solubles (CFA) – see Sugarcane, molasses fermentation solubles, dehy, (5)					
Dried skimmed milk, feed grade (AAFCO) – see Cattle, milk, skimmed dehy, mx 8% moisture, (5)					
Dried spent hops (AAFCO) – see Hops, spent, dehy, (1)					
Dried tomato pomace (AAFCO) – see Tomato, pulp, dehy, (5)					
Dried whey (AAFCO) – see Cattle, whey, dehy, mn 65% lactose, (4)					
Dried whey-product (AAFCO) – see Cattle, whey low lactose, dehy mn lactose declared, (4)					
Dried whey solubles (AAFCO) – see Cattle, whey wo albumin low lactose, dehy, (4)					
Dried whole milk, feed grade (AAFCO) – see Cattle, milk, dehy, feed gr mx 8% moisture mn 26% fat, (5)					
Dried yeast (AAFCO) – see Yeast, primary Saccharomyces, dehy, mn 40% protein, (7)					
Dried yeast (AAFCO) – see Yeast, primary saccharomyces, dehy, mn 40% protein, (7)					

DROPSEED. Sporobolus spp

Dropseed, aerial part, fresh, (2)
Ref No 2 01 813 United States

Analyses	Unit		As Fed	Dry	C.V. ± %
Dry matter	%			100.0	
Carotene	mg/kg			24.9	
Vitamin A equivalent	IU/g			41.5	

Dropseed, aerial part, fresh, early bloom, (2)
Ref No 2 01 810 United States

Analyses	Unit		As Fed	Dry	C.V. ± %
Dry matter	%		36.0	100.0	
Ash	%		3.7	10.3	11
Crude fiber	%		11.1	30.9	9
Ether extract	%		0.8	2.1	29
N-free extract	%		15.7	43.5	
Protein (N x 6.25)	%		4.8	13.2	15
Cattle	dig prot %	*	3.3	9.1	
Goats	dig prot %	*	3.2	8.9	
Horses	dig prot %	*	3.1	8.7	
Rabbits	dig prot %	*	3.2	8.9	
Sheep	dig prot %	*	3.3	9.3	
Energy	GE Mcal/kg				
Cattle	DE Mcal/kg	*	1.06	2.94	
Sheep	DE Mcal/kg	*	0.99	2.76	
Cattle	ME Mcal/kg	*	0.87	2.41	
Sheep	ME Mcal/kg	*	0.81	2.26	
Cattle	TDN %	*	24.0	66.6	
Sheep	TDN %	*	22.5	62.6	

Dropseed, aerial part, fresh, mature, (2)
Ref No 2 01 811 United States

Analyses	Unit		As Fed	Dry	C.V. ± %
Dry matter	%		56.8	100.0	
Ash	%		3.5	6.1	12
Crude fiber	%		20.2	35.6	7
Ether extract	%		0.9	1.5	24
N-free extract	%		28.7	50.5	
Protein (N x 6.25)	%		3.6	6.3	18
Cattle	dig prot %	*	1.8	3.2	
Goats	dig prot %	*	1.4	2.4	
Horses	dig prot %	*	1.6	2.9	
Rabbits	dig prot %	*	2.0	3.5	
Sheep	dig prot %	*	1.6	2.9	
Energy	GE Mcal/kg				
Cattle	DE Mcal/kg	*	1.68	2.95	
Sheep	DE Mcal/kg	*	1.68	2.96	
Cattle	ME Mcal/kg	*	1.37	2.42	
Sheep	ME Mcal/kg	*	1.38	2.43	
Cattle	TDN %	*	38.0	66.9	
Sheep	TDN %	*	38.2	67.2	
Carotene	mg/kg		20.2	35.5	64
Vitamin A equivalent	IU/g		33.6	59.2	

Dropseed, aerial part, fresh, over ripe, (2)
Ref No 2 01 812 United States

Analyses	Unit		As Fed	Dry	C.V. ± %
Dry matter	%			100.0	
Calcium	%			0.20	52
Cobalt	mg/kg			0.340	
Copper	mg/kg			39.9	
Manganese	mg/kg			85.5	
Phosphorus	%			0.06	68

DROPSEED, MESA. Sporobolus flexuosus

Dropseed, mesa, aerial part, fresh, (2)
Ref No 2 01 819 United States

Analyses	Unit		As Fed	Dry	C.V. ± %
Dry matter	%		42.5	100.0	17
Ash	%		2.4	5.7	27
Crude fiber	%		15.8	37.1	11
Ether extract	%		0.6	1.5	44
N-free extract	%		21.0	49.3	
Protein (N x 6.25)	%		2.7	6.4	55
Cattle	dig prot %	*	1.4	3.3	
Goats	dig prot %	*	1.1	2.5	
Horses	dig prot %	*	1.3	3.0	
Rabbits	dig prot %	*	1.5	3.6	
Sheep	dig prot %	*	1.3	3.0	
Energy	GE Mcal/kg				
Cattle	DE Mcal/kg	*	1.27	2.99	
Sheep	DE Mcal/kg	*	1.25	2.93	
Cattle	ME Mcal/kg	*	1.04	2.45	
Sheep	ME Mcal/kg	*	1.02	2.40	
Cattle	TDN %	*	28.9	67.9	
Sheep	TDN %	*	28.3	66.5	
Calcium	%		0.08	0.19	33
Phosphorus	%		0.04	0.09	45
Carotene	mg/kg		13.0	30.6	
Vitamin A equivalent	IU/g		21.7	51.1	

Dropseed, mesa, aerial part, fresh, early bloom, (2)
Ref No 2 01 814 United States

Analyses	Unit		As Fed	Dry	C.V. ± %
Dry matter	%			100.0	
Ash	%			8.2	
Crude fiber	%			27.0	
Ether extract	%			3.6	
N-free extract	%			43.7	
Protein (N x 6.25)	%			17.5	
Cattle	dig prot %	*		12.8	
Goats	dig prot %	*		12.9	
Horses	dig prot %	*		12.4	
Rabbits	dig prot %	*		12.2	
Sheep	dig prot %	*		13.3	
Energy	GE Mcal/kg				
Cattle	DE Mcal/kg	*		2.72	
Sheep	DE Mcal/kg	*		2.68	
Cattle	ME Mcal/kg	*		2.23	
Sheep	ME Mcal/kg	*		2.20	
Cattle	TDN %	*		61.7	
Sheep	TDN %	*		60.9	

Dropseed, mesa, aerial part, fresh, mid-bloom, (2)
Ref No 2 08 419 United States

Analyses	Unit		As Fed	Dry	C.V. ± %
Dry matter	%		35.0	100.0	
Ash	%		2.2	6.3	
Crude fiber	%		13.5	38.6	
Ether extract	%		0.6	1.7	
N-free extract	%		15.6	44.6	
Protein (N x 6.25)	%		3.1	8.9	
Cattle	dig prot %	*	1.9	5.4	
Goats	dig prot %	*	1.7	4.8	
Horses	dig prot %	*	1.8	5.0	
Rabbits	dig prot %	*	1.9	5.5	
Sheep	dig prot %	*	1.8	5.2	
Energy	GE Mcal/kg				
Cattle	DE Mcal/kg	*	0.99	2.83	
Sheep	DE Mcal/kg	*	0.98	2.80	
Cattle	ME Mcal/kg	*	0.81	2.32	
Sheep	ME Mcal/kg	*	0.80	2.30	
Cattle	TDN %	*	22.4	64.1	
Sheep	TDN %	*	22.2	63.6	
Calcium	%		0.12	0.34	
Phosphorus	%		0.06	0.17	

Dropseed, mesa, aerial part, fresh, full bloom, (2)
Ref No 2 01 815 United States

Analyses	Unit		As Fed	Dry	C.V. ± %
Dry matter	%		35.0	100.0	
Ash	%		2.5	7.1	25
Crude fiber	%		11.5	32.8	17
Ether extract	%		0.6	1.7	13
N-free extract	%		16.7	47.6	
Protein (N x 6.25)	%		3.8	10.8	21
Cattle	dig prot %	*	2.5	7.1	
Goats	dig prot %	*	2.3	6.6	
Horses	dig prot %	*	2.3	6.7	
Rabbits	dig prot %	*	2.5	7.0	
Sheep	dig prot %	*	2.5	7.1	
Energy	GE Mcal/kg				
Cattle	DE Mcal/kg	*	0.92	2.63	
Sheep	DE Mcal/kg	*	1.00	2.85	
Cattle	ME Mcal/kg	*	0.76	2.16	

(1) dry forages and roughages (2) pasture, range plants, and forages fed green (3) silages (4) energy feeds (5) protein supplements (6) minerals (7) vitamins (8) additives

404

Feed Name or Analyses			Mean As Fed	Mean Dry	C.V. ± %
Sheep	ME Mcal/kg	*	0.82	2.34	
Cattle	TDN %	*	20.9	59.6	
Sheep	TDN %	*	22.6	64.6	
Calcium	%		0.11	0.31	
Phosphorus	%		0.05	0.15	
Carotene	mg/kg		24.2	69.2	
Vitamin A equivalent	IU/g		40.4	115.4	

Dropseed, mesa, aerial part, fresh, dough stage, (2)

Ref No 2 01 816 United States

Feed Name or Analyses			Mean As Fed	Mean Dry	C.V. ± %
Dry matter	%		50.0	100.0	
Ash	%		2.5	5.0	
Crude fiber	%		18.7	37.4	
Ether extract	%		0.6	1.2	
N-free extract	%		24.6	49.2	
Protein (N x 6.25)	%		3.6	7.2	
Cattle	dig prot %	*	2.0	4.0	
Goats	dig prot %	*	1.6	3.3	
Horses	dig prot %	*	1.8	3.6	
Rabbits	dig prot %	*	2.1	4.2	
Sheep	dig prot %	*	1.9	3.7	
Energy	GE Mcal/kg				
Cattle	DE Mcal/kg	*	1.20	2.39	
Sheep	DE Mcal/kg	*	1.24	2.47	
Cattle	ME Mcal/kg	*	0.98	1.96	
Sheep	ME Mcal/kg	*	1.02	2.03	
Cattle	TDN %	*	27.0	54.1	
Sheep	TDN %	*	28.0	56.0	
Calcium	%		0.11	0.23	
Phosphorus	%		0.08	0.16	
Potassium	%		0.09	0.18	

Dropseed, mesa, aerial part, fresh, mature, (2)

Ref No 2 01 817 United States

Feed Name or Analyses			Mean As Fed	Mean Dry	C.V. ± %
Dry matter	%		50.0	100.0	
Ash	%		2.9	5.7	11
Crude fiber	%		17.5	34.9	5
Ether extract	%		0.8	1.5	32
N-free extract	%		25.2	50.3	
Protein (N x 6.25)	%		3.8	7.6	20
Cattle	dig prot %	*	2.2	4.4	
Goats	dig prot %	*	1.8	3.7	
Horses	dig prot %	*	2.0	4.0	
Rabbits	dig prot %	*	2.3	4.5	
Sheep	dig prot %	*	2.0	4.1	
Energy	GE Mcal/kg				
Cattle	DE Mcal/kg	*	1.28	2.56	
Sheep	DE Mcal/kg	*	1.25	2.50	
Cattle	ME Mcal/kg	*	1.05	2.10	
Sheep	ME Mcal/kg	*	1.02	2.05	
Cattle	TDN %	*	29.0	58.0	
Sheep	TDN %	*	28.3	56.6	
Calcium	%		0.11	0.22	23
Phosphorus	%		0.06	0.12	43
Carotene	mg/kg		20.7	41.4	47
Vitamin A equivalent	IU/g		34.5	69.1	

Dropseed, mesa, aerial part, fresh, over ripe, (2)

Ref No 2 01 818 United States

Feed Name or Analyses			Mean As Fed	Mean Dry	C.V. ± %
Dry matter	%			100.0	
Ash	%			5.3	18
Crude fiber	%			39.4	5
Ether extract	%			1.3	48
N-free extract	%			49.5	
Protein (N x 6.25)	%			4.5	28
Cattle	dig prot %	*		1.7	
Goats	dig prot %	*		0.8	
Horses	dig prot %	*		1.4	
Rabbits	dig prot %	*		2.1	
Sheep	dig prot %	*		1.2	
Energy	GE Mcal/kg				
Cattle	DE Mcal/kg	*		2.52	
Sheep	DE Mcal/kg	*		2.52	
Cattle	ME Mcal/kg	*		2.07	
Sheep	ME Mcal/kg	*		2.07	
Cattle	TDN %	*		57.1	
Sheep	TDN %	*		57.1	
Calcium	%			0.15	33
Phosphorus	%			0.06	74
Carotene	mg/kg			7.9	
Vitamin A equivalent	IU/g			13.2	

DROPSEED, PRAIRIE. Sporobolus heterolepis

Dropseed, prairie, aerial part, fresh, mature, (2)

Ref No 2 01 820 United States

Feed Name or Analyses			Mean As Fed	Mean Dry	C.V. ± %
Dry matter	%			100.0	
Ash	%			5.1	
Crude fiber	%			33.5	
Ether extract	%			1.4	
N-free extract	%			54.3	
Protein (N x 6.25)	%			5.7	
Cattle	dig prot %	*		2.7	
Goats	dig prot %	*		1.9	
Horses	dig prot %	*		2.4	
Rabbits	dig prot %	*		3.1	
Sheep	dig prot %	*		2.3	
Energy	GE Mcal/kg				
Cattle	DE Mcal/kg	*		2.53	
Sheep	DE Mcal/kg	*		2.60	
Cattle	ME Mcal/kg	*		2.07	
Sheep	ME Mcal/kg	*		2.13	
Cattle	TDN %	*		57.4	
Sheep	TDN %	*		59.0	

DROPSEED, SAND. Sporobolus cryptandrus

Dropseed, sand, aerial part, fresh, early bloom, (2)

Sand dropseed

Ref No 2 01 821 United States

Feed Name or Analyses			Mean As Fed	Mean Dry	C.V. ± %
Dry matter	%		36.0	100.0	
Ash	%		3.9	10.8	12
Crude fiber	%		11.4	31.8	5
Ether extract	%		0.6	1.8	15
N-free extract	%		15.6	43.4	
Protein (N x 6.25)	%		4.4	12.2	10
Cattle	dig prot %	*	3.0	8.3	
Goats	dig prot %	*	2.9	7.9	
Horses	dig prot %	*	2.8	7.9	
Rabbits	dig prot %	*	2.9	8.1	
Sheep	dig prot %	*	3.0	8.4	
Energy	GE Mcal/kg				
Cattle	DE Mcal/kg	*	0.88	2.44	
Sheep	DE Mcal/kg	*	0.84	2.33	
Cattle	ME Mcal/kg	*	0.72	2.00	
Sheep	ME Mcal/kg	*	0.69	1.91	
Cattle	TDN %	*	19.9	55.3	
Sheep	TDN %	*	19.0	52.9	
Calcium	%		0.16	0.45	33
Magnesium	%		0.09	0.24	
Phosphorus	%		0.05	0.14	34
Potassium	%		0.55	1.53	

Dropseed, sand, aerial part, fresh, mature, (2)

Sand dropseed

Ref No 2 01 822 United States

Feed Name or Analyses			Mean As Fed	Mean Dry	C.V. ± %
Dry matter	%		56.8	100.0	
Ash	%		3.6	6.3	15
Crude fiber	%		20.8	36.7	6
Ether extract	%		0.8	1.4	31
N-free extract	%		28.3	49.9	
Protein (N x 6.25)	%		3.2	5.7	10
Cattle	dig prot %	*	1.6	2.7	
Goats	dig prot %	*	1.1	1.9	
Horses	dig prot %	*	1.3	2.4	
Rabbits	dig prot %	*	1.7	3.1	
Sheep	dig prot %	*	1.3	2.3	
Energy	GE Mcal/kg				
Cattle	DE Mcal/kg	*	1.41	2.48	
Sheep	DE Mcal/kg	*	1.43	2.52	
Cattle	ME Mcal/kg	*	1.15	2.03	
Sheep	ME Mcal/kg	*	1.18	2.07	
Cattle	TDN %	*	32.0	56.3	
Sheep	TDN %	*	32.4	57.1	
Calcium	%		0.22	0.39	40
Phosphorus	%		0.07	0.13	30

Dropseed, sand, aerial part, fresh, over ripe, (2)

Sand dropseed

Ref No 2 01 823 United States

Feed Name or Analyses			Mean As Fed	Mean Dry	C.V. ± %
Dry matter	%		49.9	100.0	
Ash	%		4.1	8.3	51
Crude fiber	%		19.4	39.0	10
Ether extract	%		0.7	1.5	38
N-free extract	%		22.7	45.6	
Protein (N x 6.25)	%		2.9	5.8	11
Cattle	dig prot %	*	1.4	2.8	
Goats	dig prot %	*	1.0	1.9	
Horses	dig prot %	*	1.2	2.4	
Rabbits	dig prot %	*	1.6	3.1	
Sheep	dig prot %	*	1.2	2.3	
Energy	GE Mcal/kg				
Cattle	DE Mcal/kg	*	1.18	2.37	
Sheep	DE Mcal/kg	*	1.22	2.44	
Cattle	ME Mcal/kg	*	0.97	1.94	
Sheep	ME Mcal/kg	*	1.00	2.00	
Cattle	TDN %	*	26.8	53.7	
Sheep	TDN %	*	27.6	55.4	
Calcium	%		0.15	0.31	23
Phosphorus	%		0.03	0.07	23
Carotene	mg/kg		0.3	0.7	
Vitamin A equivalent	IU/g		0.6	1.1	

Dropseed, sand, aerial part wo lower stems, fresh, early leaf, (2)

Ref No 2 08 694 United States

Feed Name or Analyses			Mean As Fed	Mean Dry	C.V. ± %
Dry matter	%			100.0	
Organic matter	%			91.0	2
Ash	%			9.0	25
Crude fiber	%			33.7	6
Ether extract	%			1.6	18
N-free extract	%			48.0	5

Continued

Feed Name or Analyses			Mean As Fed	Mean Dry	C.V. ± %
Protein (N x 6.25)	%			7.7	24
Cattle	dig prot %	*		4.4	
Goats	dig prot %	*		3.7	
Horses	dig prot %	*		4.0	
Rabbits	dig prot %	*		4.6	
Sheep	dig prot %	*		4.1	
Calcium	%			0.37	24
Phosphorus	%			0.11	48
Potassium	%			0.16	
Carotene	mg/kg			66.2	52
Vitamin A equivalent	IU/g			110.4	

Dropseed, sand, aerial part wo lower stems, fresh, mid-bloom, (2)

Ref No 2 08 695 United States

Analyses			As Fed	Dry	C.V. ± %
Dry matter	%			100.0	
Organic matter	%			92.3	1
Ash	%			7.7	15
Crude fiber	%			33.8	9
Ether extract	%			1.5	28
N-free extract	%			49.2	5
Protein (N x 6.25)	%			7.8	18
Cattle	dig prot %	*		4.5	
Goats	dig prot %	*		3.9	
Horses	dig prot %	*		4.2	
Rabbits	dig prot %	*		4.7	
Sheep	dig prot %	*		4.3	
Calcium	%			0.40	25
Phosphorus	%			0.16	33

Dropseed, sand, aerial part wo lower stems, fresh, mature, (2)

Ref No 2 08 696 United States

Analyses			As Fed	Dry	C.V. ± %
Dry matter	%			100.0	
Organic matter	%			90.6	5
Ash	%			9.4	52
Crude fiber	%			33.3	7
Ether extract	%			1.5	25
N-free extract	%			49.3	7
Protein (N x 6.25)	%			6.4	25
Cattle	dig prot %	*		3.3	
Goats	dig prot %	*		2.5	
Horses	dig prot %	*		3.0	
Rabbits	dig prot %	*		3.6	
Sheep	dig prot %	*		2.9	
Calcium	%			0.32	29
Phosphorus	%			0.13	33

Dropseed, sand, aerial part wo lower stems, fresh, dormant, (2)

Ref No 2 08 693 United States

Analyses			As Fed	Dry	C.V. ± %
Dry matter	%			100.0	
Organic matter	%			90.7	3
Ash	%			9.3	25
Crude fiber	%			36.3	6
Ether extract	%			1.1	21
N-free extract	%			47.9	4
Protein (N x 6.25)	%			5.5	30
Cattle	dig prot %	*		2.5	

Analyses			As Fed	Dry	C.V. ± %
Goats	dig prot %	*		1.7	
Horses	dig prot %	*		2.2	
Rabbits	dig prot %	*		2.9	
Sheep	dig prot %	*		2.1	
Calcium	%			0.40	34
Phosphorus	%			0.11	98
Carotene	mg/kg			12.2	98
Vitamin A equivalent	IU/g			20.3	

DROPSEED, TALL. Sporobolus asper

Dropseed, tall, aerial part, fresh, over ripe, (2)

Ref No 2 01 824 United States

Analyses			As Fed	Dry	C.V. ± %
Dry matter	%			100.0	
Ash	%			3.2	
Crude fiber	%			37.9	
Ether extract	%			1.9	
N-free extract	%			52.8	
Protein (N x 6.25)	%			4.2	
Cattle	dig prot %	*		1.5	
Goats	dig prot %	*		0.5	
Horses	dig prot %	*		1.1	
Rabbits	dig prot %	*		1.9	
Sheep	dig prot %	*		0.9	
Calcium	%			0.29	
Phosphorus	%			0.11	

Drumfish, whole, raw –
see Fish, redfish, whole, raw, (5)

Dwarf koa –
see Bundleflower, rayado, aerial part, fresh, (2)

Ear corn chop (AAFCO) –
see Corn, ears, grnd, (4)

Ear corn chop (AAFCO) –
see Corn, ears, grnd, (4)

EMMER. Triticum dicoccum

Emmer, hay, s-c, (1)

Ref No 1 01 827 United States

Analyses			As Fed	Dry	C.V. ± %
Dry matter	%		90.0	100.0	
Ash	%		9.1	10.1	
Crude fiber	%		32.8	36.4	
Ether extract	%		2.0	2.2	
N-free extract	%		36.4	40.5	
Protein (N x 6.25)	%		9.7	10.8	
Cattle	dig prot %	*	5.7	6.3	
Goats	dig prot %	*	6.0	6.6	
Horses	dig prot %	*	6.0	6.7	
Rabbits	dig prot %	*	6.3	7.0	
Sheep	dig prot %	*	5.6	6.2	
Fatty acids	%		2.0	2.2	
Energy	GE Mcal/kg				
Cattle	DE Mcal/kg	*	1.92	2.13	
Sheep	DE Mcal/kg	*	1.89	2.10	
Cattle	ME Mcal/kg	*	1.58	1.75	
Sheep	ME Mcal/kg	*	1.55	1.72	
Cattle	TDN %	*	43.4	48.2	
Sheep	TDN %	*	42.8	47.6	

Emmer, hulls, (1)

Ref No 1 01 828 United States

Analyses			As Fed	Dry	C.V. ± %
Dry matter	%		95.2	100.0	
Ash	%		8.5	8.9	
Crude fiber	%		33.4	35.1	
Ether extract	%		1.3	1.4	
N-free extract	%		47.6	50.0	
Protein (N x 6.25)	%		4.4	4.6	
Cattle	dig prot %	*	0.9	0.9	
Goats	dig prot %	*	0.8	0.9	
Horses	dig prot %	*	1.4	1.4	
Rabbits	dig prot %	*	2.1	2.2	
Sheep	dig prot %	*	0.7	0.7	

Emmer, grain, (4)

Ref No 4 01 830 United States

Analyses			As Fed	Dry	C.V. ± %
Dry matter	%		90.7	100.0	1
Ash	%		3.5	3.9	10
Crude fiber	%		9.6	10.6	3
Sheep	dig coef %		29.	29.	
Ether extract	%		2.0	2.2	7
Sheep	dig coef %		87.	87.	
N-free extract	%		63.4	69.9	
Sheep	dig coef %		88.	88.	
Protein (N x 6.25)	%		12.2	13.5	12
Sheep	dig coef %		80.	80.	
Cattle	dig prot %	*	7.6	8.4	
Goats	dig prot %	*	8.7	9.6	
Horses	dig prot %	*	8.7	9.6	
Sheep	dig prot %		9.8	10.8	
Energy	GE Mcal/kg				
Cattle	DE Mcal/kg	*	3.00	3.31	
Sheep	DE Mcal/kg	*	3.18	3.51	
Swine	DE kcal/kg	*	2690.	2966.	
Cattle	ME Mcal/kg	*	2.46	2.71	
Sheep	ME Mcal/kg	*	2.61	2.88	
Swine	ME kcal/kg	*	2509.	2767.	
Cattle	TDN %	*	68.0	75.0	
Sheep	TDN %		72.2	79.6	
Swine	TDN %	*	61.0	67.3	
Copper	mg/kg		31.4	34.6	47
Iron	%		0.005	0.006	7
Manganese	mg/kg		78.0	86.0	38
Phosphorus	%		0.36	0.40	
Potassium	%		0.47	0.52	

ENGELMANNDAISY. Engelmannia pinnatifida

Engelmanndaisy, browse, immature, (2)

Ref No 2 01 832 United States

Analyses			As Fed	Dry	C.V. ± %
Dry matter	%			100.0	
Ash	%			17.4	
Crude fiber	%			9.9	
Ether extract	%			1.2	
N-free extract	%			48.3	
Protein (N x 6.25)	%			23.2	
Cattle	dig prot %	*		17.6	
Goats	dig prot %	*		18.2	
Horses	dig prot %	*		17.2	
Rabbits	dig prot %	*		16.6	
Sheep	dig prot %	*		18.6	
Calcium	%			2.63	
Magnesium	%			0.33	

(1) dry forages and roughages (2) pasture, range plants, and forages fed green (3) silages (4) energy feeds (5) protein supplements (6) minerals (7) vitamins (8) additives

Feed Name or Analyses			Mean As Fed	Mean Dry	C.V. ± %
Phosphorus	%			0.37	
Potassium	%			4.73	

EPHEDRA. Ephedra spp

Ephedra, browse, mature, (2)
Ref No 2 01 834 United States

			As Fed	Dry	C.V.
Dry matter	%		63.4	100.0	
Ash	%		3.4	5.4	18
Crude fiber	%		24.7	38.9	8
Ether extract	%		2.5	4.0	95
N-free extract	%		28.0	44.2	
Protein (N x 6.25)	%		4.8	7.5	9
Cattle	dig prot %	*	2.7	4.3	
Goats	dig prot %	*	2.3	3.6	
Horses	dig prot %	*	2.5	3.9	
Rabbits	dig prot %	*	2.8	4.5	
Sheep	dig prot %	*	2.5	4.0	

EPHEDRA, LONGLEAF. Ephedra trifurca

Ephedra, longleaf, browse, mature, (2)
Ref No 2 01 835 United States

			As Fed	Dry	C.V.
Dry matter	%		63.4	100.0	
Ash	%		4.9	7.7	
Protein (N x 6.25)	%		3.8	6.0	
Cattle	dig prot %	*	1.9	3.0	
Goats	dig prot %	*	1.4	2.2	
Horses	dig prot %	*	1.7	2.6	
Rabbits	dig prot %	*	2.1	3.3	
Sheep	dig prot %	*	1.6	2.6	
Calcium	%		2.10	3.31	
Chlorine	%		0.03	0.05	
Iron	%		0.008	0.013	
Magnesium	%		0.15	0.24	
Manganese	mg/kg		196.4	309.7	
Phosphorus	%		0.05	0.08	
Potassium	%		0.27	0.42	
Sodium	%		0.04	0.06	
Sulphur	%		0.23	0.36	

EPHEDRA, NEVADA. Ephedra nevadensis

Ephedra, Nevada, aerial part, fresh, mature, (2)
Ref No 2 01 836 United States

			As Fed	Dry	C.V.
Dry matter	%			100.0	
Calcium	%			5.81	35
Magnesium	%			1.28	11
Phosphorus	%			0.50	65

EPHEDRA, VINE. Ephedra antisyphilitica

Ephedra, vine, stems, fresh, (2)
Ref No 2 01 837 United States

			As Fed	Dry	C.V.
Dry matter	%			100.0	
Ash	%			6.3	
Crude fiber	%			33.7	
Ether extract	%			1.5	
N-free extract	%			51.5	
Protein (N x 6.25)	%			7.0	
Cattle	dig prot %	*		3.8	
Goats	dig prot %	*		3.1	
Horses	dig prot %	*		3.5	
Rabbits	dig prot %	*		4.1	
Sheep	dig prot %	*		3.5	
Calcium	%			3.06	
Magnesium	%			0.17	
Phosphorus	%			0.08	
Potassium	%			0.62	

ERIOGONUM. Eriogonum spp

Eriogonum, aerial part, fresh, mature, (2)
Ref No 2 01 838 United States

			As Fed	Dry	C.V.
Dry matter	%			100.0	
Ash	%			13.7	
Protein (N x 6.25)	%			6.6	
Cattle	dig prot %	*		3.5	
Goats	dig prot %	*		2.7	
Horses	dig prot %	*		3.1	
Rabbits	dig prot %	*		3.8	
Sheep	dig prot %	*		3.1	
Calcium	%			1.03	
Cobalt	mg/kg			0.163	
Copper	mg/kg			79.4	
Manganese	mg/kg			352.1	
Phosphorus	%			0.09	
Carotene	mg/kg			9.9	
Vitamin A equivalent	IU/g			16.5	

Eriogonum, flowers, fresh, (2)
Ref No 2 01 839 United States

			As Fed	Dry	C.V.
Dry matter	%			100.0	
Ash	%			6.5	
Crude fiber	%			13.5	
Ether extract	%			2.2	
N-free extract	%			60.8	
Protein (N x 6.25)	%			17.0	
Cattle	dig prot %	*		12.3	
Goats	dig prot %	*		12.4	
Horses	dig prot %	*		12.0	
Rabbits	dig prot %	*		11.8	
Sheep	dig prot %	*		12.8	

EUONYMUS, WINGED. Euonymus alatus

Eyonymus, winged, seeds w hulls, (5)
Ref No 5 09 144 United States

			As Fed	Dry	C.V.
Dry matter	%			100.0	
Protein (N x 6.25)	%			20.0	
Alanine	%			0.70	
Arginine	%			1.98	
Aspartic acid	%			1.56	
Glutamic acid	%			2.76	
Glycine	%			0.66	
Histidine	%			0.40	
Hydroxyproline	%			0.00	
Isoleucine	%			0.60	
Leucine	%			1.04	
Lysine	%			0.92	
Methionine	%			0.26	
Phenylalanine	%			0.66	
Proline	%			0.66	
Serine	%			0.76	
Threonine	%			0.50	
Tyrosine	%			0.48	
Valine	%			0.80	

EUPHORBIA. Euphorbia spp

Euphorbia, aerial part, fresh, immature, (2)
Ref No 2 01 844 United States

			As Fed	Dry	C.V.
Dry matter	%			100.0	
Ash	%			11.0	
Crude fiber	%			13.0	
Ether extract	%			4.1	
N-free extract	%			57.5	
Protein (N x 6.25)	%			14.4	
Cattle	dig prot %	*		10.1	
Goats	dig prot %	*		10.0	
Horses	dig prot %	*		9.8	
Rabbits	dig prot %	*		9.8	
Sheep	dig prot %	*		10.4	
Calcium	%			1.19	
Magnesium	%			0.27	
Phosphorus	%			0.26	
Potassium	%			1.94	

EUPHORBIA, MARGINATA. Euphorbia marginata

Euphorbia, marginata, seeds w hulls, (4)
Ref No 4 09 143 United States

			As Fed	Dry	C.V.
Dry matter	%			100.0	
Protein (N x 6.25)	%			16.0	
Alanine	%			0.61	
Arginine	%			1.63	
Aspartic acid	%			1.68	
Glutamic acid	%			2.14	
Glycine	%			0.72	
Histidine	%			0.34	
Hydroxyproline	%			0.06	
Isoleucine	%			0.58	
Leucine	%			0.83	
Lysine	%			0.62	
Methionine	%			0.24	
Phenylalanine	%			0.59	
Proline	%			0.50	
Serine	%			0.70	
Threonine	%			0.45	
Tyrosine	%			0.35	
Valine	%			0.70	

EVENING PRIMROSE. Oenothera spp

Evening primrose, aerial part, fresh, (2)
Ref No 2 01 846 United States

			As Fed	Dry	C.V.
Dry matter	%			100.0	
Ash	%			7.9	31
Crude fiber	%			20.0	19
Ether extract	%			3.0	32
N-free extract	%			54.7	
Protein (N x 6.25)	%			14.4	39
Cattle	dig prot %	*		10.1	
Goats	dig prot %	*		10.0	
Horses	dig prot %	*		9.8	
Rabbits	dig prot %	*		9.8	
Sheep	dig prot %	*		10.4	
Calcium	%			1.85	30

Continued

Feed Name or Analyses		Mean As Fed	Mean Dry	C.V. ± %
Magnesium	%		0.34	17
Phosphorus	%		0.26	37
Potassium	%		2.14	51

Evening primrose, aerial part, fresh, immature, (2)
Ref No 2 01 845 United States

Dry matter	%		100.0	
Calcium	%		1.93	
Magnesium	%		0.38	
Phosphorus	%		0.34	
Potassium	%		2.94	

EVENINGPRIMROSE, COMMON. Oenothera biennis

Eveningprimrose, common, seeds w hulls, (4)
Ref No 4 09 142 United States

Dry matter	%		100.0	
Protein (N x 6.25)	%		16.0	
Alanine	%		0.53	
Arginine	%		1.58	
Aspartic acid	%		1.14	
Glutamic acid	%		2.72	
Glycine	%		0.93	
Histidine	%		0.34	
Hydroxyproline	%		0.34	
Isoleucine	%		0.51	
Leucine	%		0.93	
Lysine	%		0.32	
Methionine	%		0.27	
Phenylalanine	%		0.66	
Proline	%		0.54	
Serine	%		0.72	
Threonine	%		0.38	
Tyrosine	%		0.38	
Valine	%		0.69	

FALLOWWOOD. Ximenia americana

Fallowwood, seeds w hulls, (5)
Ref No 5 09 018 United States

Dry matter	%		100.0	
Protein (N x 6.25)	%		20.0	
Alanine	%		0.84	
Arginine	%		2.00	
Aspartic acid	%		1.78	
Glutamic acid	%		2.88	
Glycine	%		0.74	
Histidine	%		0.52	
Hydroxyproline	%		0.44	
Isoleucine	%		0.72	
Leucine	%		1.58	
Lysine	%		1.04	
Methionine	%		0.24	
Phenylalanine	%		0.80	
Proline	%		0.92	
Serine	%		0.76	
Threonine	%		0.78	
Tyrosine	%		0.54	
Valine	%		1.00	

(1) dry forages and roughages (2) pasture, range plants, and forages fed green (3) silages (4) energy feeds (5) protein supplements (6) minerals (7) vitamins (8) additives

Feed Name or Analyses		Mean As Fed	Mean Dry	C.V. ± %

FALSE BUFFALOGRASS. Munroa squarrosa

False buffalograss, aerial part, fresh, immature, (2)
Ref No 2 01 847 United States

Dry matter	%		100.0	
Ash	%		13.9	
Crude fiber	%		17.3	
Ether extract	%		1.7	
N-free extract	%		54.6	
Protein (N x 6.25)	%		12.5	
Cattle	dig prot %	*	8.5	
Goats	dig prot %	*	8.2	
Horses	dig prot %	*	8.1	
Rabbits	dig prot %	*	8.3	
Sheep	dig prot %	*	8.6	
Calcium	%		0.48	
Phosphorus	%		0.16	

FALSEFLAX. Camelina spp

Falseflax, seeds, extn unspecified grnd, (5)
Ref No 5 01 849 United States

Dry matter	%	89.8	100.0	
Ash	%	11.3	12.6	
Crude fiber	%	11.0	12.3	
Sheep	dig coef %	35.	35.	
Ether extract	%	6.1	6.8	
Sheep	dig coef %	96.	96.	
N-free extract	%	30.1	33.5	
Sheep	dig coef %	91.	91.	
Protein (N x 6.25)	%	31.3	34.8	
Sheep	dig coef %	83.	83.	
Sheep	dig prot %	25.9	28.9	
Energy	GE Mcal/kg			
Cattle	DE Mcal/kg	* 2.78	3.09	
Sheep	DE Mcal/kg	* 3.10	3.46	
Swine	DE kcal/kg	*3245.	3614.	
Cattle	ME Mcal/kg	* 2.28	2.54	
Sheep	ME Mcal/kg	* 2.54	2.83	
Swine	ME kcal/kg	*2887.	3215.	
Cattle	TDN %	* 63.0	70.2	
Sheep	TDN %	70.4	78.4	
Swine	TDN %	* 73.6	82.0	

FALSE PENNYROYAL, DRUMMOND. Hedeoma drummondii

False pennyroyal, drummond, aerial part, fresh, (2)
Ref No 2 01 848 United States

Dry matter	%		100.0	
Ash	%		8.1	
Crude fiber	%		30.1	
Ether extract	%		3.8	
N-free extract	%		47.2	
Protein (N x 6.25)	%		10.8	
Cattle	dig prot %	*	7.1	
Goats	dig prot %	*	6.6	
Horses	dig prot %	*	6.7	
Rabbits	dig prot %	*	7.0	
Sheep	dig prot %	*	7.1	
Calcium	%		1.99	
Magnesium	%		0.31	

Feed Name or Analyses		Mean As Fed	Mean Dry	C.V. ± %
Phosphorus	%		0.19	
Potassium	%		1.73	

FAREWELL-TO-SPRING. Godetia amoena

Farewell-to-spring, aerial part, fresh, (2)
Ref No 2 01 852 United States

Dry matter	%		100.0	
Ash	%		8.0	12
Crude fiber	%		19.1	26
Protein (N x 6.25)	%		6.1	10
Cattle	dig prot %	*	3.1	
Goats	dig prot %	*	2.3	
Horses	dig prot %	*	2.7	
Rabbits	dig prot %	*	3.4	
Sheep	dig prot %	*	2.7	

Farewell-to-spring, aerial part, fresh, full bloom, (2)
Ref No 2 01 850 United States

Dry matter	%		100.0	
Calcium	%		2.15	
Phosphorus	%		0.27	
Potassium	%		1.74	

Farewell-to-spring, aerial part, fresh, mature, (2)
Ref No 2 01 851 United States

Dry matter	%		100.0	
Calcium	%		1.58	
Phosphorus	%		0.26	
Potassium	%		1.15	

Farina –
see Wheat, grits, cracked fine screened, (4)

Feather meal –
see Poultry, feathers, hydrolyzed dehy grnd, mn 75% of protein digestible, (5)

Feed flour, mx 2.0 fbr (CFA) –
see Wheat, flour, coarse bolt, feed gr mx 2% fiber, (4)

Feeding bone meal (CFA) –
see Animal, bone, cooked dehy grnd, mn 10% phosphorus, (6)

Feeding meat and bone tankage (CFA) –
see Animal-Poultry, carcass residue w bone mx 35% blood, dry or wet rendered dehy grnd, mn 40% protein, (5)

Feeding oat meal (AAFCO) –
see Oats, cereal by-product, mx 4% fiber, (4)

Feeding steamed bone meal (CFA) –
see Animal, bone, steamed dehy grnd, mn 12% phosphorus, (6)

Feeding tankage (CFA) –
see Animal-Poultry, carcass residue mx 35% blood, dry or wet rendered dehy grnd, mn 50% protein, (5)

Feed Name or Analyses			Mean As Fed	Mean Dry	C.V. ± %

Feed screenings no 1 —
see Grains, screenings, gr 1 mn 35% grain mx 7% fiber mx 6% foreign material mx 8% wild oats, (4)

Feed screenings no 2 —
see Grains, screenings, gr 2 mx 11% fiber mx 10% foreign material mx 49% wild oats, (4)

FELICIA, ROUGHLEAF. Felicia spp

Felicia, roughleaf, aerial part, s-c, (1)
Ref No 1 01 853 United States

				As Fed	Dry	C.V.
Dry matter		%		87.0	100.0	
Ash		%		8.4	9.6	
Crude fiber		%		23.5	27.0	
Sheep	dig coef	%		23.	23.	
Ether extract		%		3.2	3.7	
Sheep	dig coef	%		43.	43.	
N-free extract		%		41.1	47.2	
Sheep	dig coef	%		60.	60.	
Protein (N x 6.25)		%		10.9	12.5	
Sheep	dig coef	%		66.	66.	
Cattle	dig prot	%	*	6.8	7.8	
Goats	dig prot	%	*	7.2	8.2	
Horses	dig prot	%	*	7.1	8.1	
Rabbits	dig prot	%	*	7.2	8.3	
Sheep	dig prot	%		7.2	8.3	
Energy	GE Mcal/kg					
Cattle	DE Mcal/kg		*	1.94	2.23	
Sheep	DE Mcal/kg		*	1.78	2.04	
Cattle	ME Mcal/kg		*	1.59	1.83	
Sheep	ME Mcal/kg		*	1.46	1.68	
Cattle	TDN	%	*	44.0	50.6	
Sheep	TDN	%		40.3	46.4	

FENNEL. Foeniculum vulgare

Fennel, seeds, extn unspecified grnd, (4)
Ref No 4 01 854 United States

				As Fed	Dry	C.V.
Dry matter		%		89.2	100.0	
Ash		%		9.9	11.1	
Crude fiber		%		13.9	15.6	
Cattle	dig coef	%		45.	45.	
Ether extract		%		14.9	16.7	
Cattle	dig coef	%		94.	94.	
N-free extract		%		34.5	38.7	
Cattle	dig coef	%		72.	72.	
Protein (N x 6.25)		%		16.0	17.9	
Cattle	dig coef	%		38.	38.	
Cattle	dig prot	%		6.1	6.8	
Goats	dig prot	%	*	12.2	13.6	
Horses	dig prot	%	*	12.2	13.6	
Sheep	dig prot	%	*	12.2	13.6	
Energy	GE Mcal/kg					
Cattle	DE Mcal/kg		*	3.03	3.40	
Sheep	DE Mcal/kg		*	2.86	3.20	
Cattle	ME Mcal/kg		*	2.48	2.78	
Sheep	ME Mcal/kg		*	2.34	2.63	
Cattle	TDN	%		68.7	77.0	
Sheep	TDN	%	*	64.8	72.6	

FERRIC OXIDE

Ferric oxide, red, Fe_2O_3, cp, (6)
Iron oxide
Ref No 6 02 432 United States

			As Fed	Dry	C.V.
Dry matter	%			100.0	
Copper	%			0.0	
Iron	%			69.940	
Manganese	%			0.1	
Sulphur	%			0.13	
Zinc	%			0.0	

FERROUS CHLORIDE

Ferrous chloride, $FeCl_2.4H_2O$, cp, (6)
Ref No 6 01 866 United States

		As Fed	Dry	C.V.
Dry matter	%	63.8	100.0	
Chlorine	%	35.67	55.91	
Copper	%	0.0	0.0	
Iron	%	28.090	44.028	
Sulphur	%	0.03	0.05	
Zinc	%	0.0	0.0	

FERROUS SULFATE

Ferrous sulfate, $FeSO_4.7H_2O$, cp, (6)
Ref No 6 01 870 United States

		As Fed	Dry	C.V.
Dry matter	%	54.7	100.0	
Chlorine	%	0.00	0.00	
Copper	%	0.0	0.0	
Iron	%	20.080	36.709	
Manganese	%	0.0	0.1	
Sulphur	%	11.54	21.10	
Zinc	%	0.0	0.0	

FESCUE. Festuca spp

Fescue, hay, s-c, early bloom, (1)
Ref No 1 01 871 United States

				As Fed	Dry	C.V.
Dry matter		%		85.4	100.0	
Ash		%		8.8	10.3	
Crude fiber		%		27.8	32.5	
Sheep	dig coef	%		64.	64.	
Ether extract		%		1.7	2.0	
Sheep	dig coef	%		48.	48.	
N-free extract		%		39.9	46.7	
Sheep	dig coef	%		58.	58.	
Protein (N x 6.25)		%		7.3	8.5	
Sheep	dig coef	%		55.	55.	
Cattle	dig prot	%	*	3.7	4.3	
Goats	dig prot	%	*	3.8	4.5	
Horses	dig prot	%	*	4.1	4.7	
Rabbits	dig prot	%	*	4.5	5.2	
Sheep	dig prot	%		4.0	4.7	
Energy	GE Mcal/kg					
Cattle	DE Mcal/kg		*	2.14	2.50	
Sheep	DE Mcal/kg		*	2.06	2.41	
Cattle	ME Mcal/kg		*	1.75	2.05	
Sheep	ME Mcal/kg		*	1.69	1.98	
Cattle	TDN	%	*	48.4	56.7	
Sheep	TDN	%		46.7	54.7	

Fescue, aerial part, fresh, (2)
Ref No 2 01 879 United States

				As Fed	Dry	C.V.
Dry matter		%		36.0	100.0	
Ash		%		1.9	5.2	
Crude fiber		%		12.5	34.7	
Ether extract		%		0.5	1.5	
N-free extract		%		19.3	53.7	
Protein (N x 6.25)		%		1.7	4.8	
Cattle	dig prot	%	*	0.7	2.0	
Goats	dig prot	%	*	0.4	1.1	
Horses	dig prot	%	*	0.6	1.6	
Rabbits	dig prot	%	*	0.9	2.4	
Sheep	dig prot	%	*	0.5	1.5	
Other carbohydrates		%		15.7	43.5	
Lignin (Ellis)		%		2.9	8.1	
Energy	GE Mcal/kg					
Cattle	DE Mcal/kg		*	1.05	2.93	
Sheep	DE Mcal/kg		*	0.94	2.61	
Cattle	ME Mcal/kg		*	0.86	2.40	
Sheep	ME Mcal/kg		*	0.77	2.14	
Cattle	TDN	%	*	23.9	66.4	
Sheep	TDN	%	*	21.3	59.1	
Calcium		%		0.13	0.36	
Phosphorus		%		0.09	0.25	
Silicon		%		0.67	1.85	

Fescue, aerial part, fresh, immature, (2)
Ref No 2 01 872 United States

				As Fed	Dry	C.V.
Dry matter		%		26.1	100.0	14
Ash		%		2.3	8.7	11
Crude fiber		%		6.6	25.3	12
Ether extract		%		1.0	3.7	26
N-free extract		%		11.8	45.1	
Protein (N x 6.25)		%		4.5	17.2	20
Cattle	dig prot	%	*	3.3	12.5	
Goats	dig prot	%	*	3.3	12.6	
Horses	dig prot	%	*	3.2	12.1	
Rabbits	dig prot	%	*	3.1	11.9	
Sheep	dig prot	%	*	3.4	13.0	
Energy	GE Mcal/kg					
Cattle	DE Mcal/kg		*	0.72	2.76	
Sheep	DE Mcal/kg		*	0.69	2.66	
Cattle	ME Mcal/kg		*	0.59	2.26	
Sheep	ME Mcal/kg		*	0.57	2.18	
Cattle	TDN	%	*	16.3	62.6	
Sheep	TDN	%	*	15.7	60.3	
Calcium		%		0.14	0.55	20
Copper		mg/kg		7.6	29.1	
Iron		%		0.011	0.043	
Magnesium		%		0.05	0.21	68
Manganese		mg/kg		21.4	82.0	
Phosphorus		%		0.10	0.40	26
Potassium		%		0.56	2.13	16

Fescue, aerial part, fresh, early bloom, (2)
Ref No 2 01 873 United States

				As Fed	Dry	C.V.
Dry matter		%		29.0	100.0	32
Ash		%		2.8	9.6	13
Crude fiber		%		7.7	26.6	13
Ether extract		%		1.4	4.9	24
N-free extract		%		12.6	43.6	
Protein (N x 6.25)		%		4.4	15.3	11
Cattle	dig prot	%	*	3.2	10.9	

Continued

Feed Name or Analyses			Mean As Fed	Mean Dry	C.V. ± %
Goats	dig prot %	*	3.1	10.8	
Horses	dig prot %	*	3.1	10.5	
Rabbits	dig prot %	*	3.0	10.5	
Sheep	dig prot %	*	3.3	11.2	
Energy	GE Mcal/kg				
Cattle	DE Mcal/kg	*	0.78	2.68	
Sheep	DE Mcal/kg	*	0.80	2.75	
Cattle	ME Mcal/kg	*	0.64	2.20	
Sheep	ME Mcal/kg	*	0.65	2.26	
Cattle	TDN %	*	17.7	60.9	
Sheep	TDN %	*	18.1	62.4	
Calcium	%		0.15	0.52	16
Magnesium	%		0.04	0.15	
Phosphorus	%		0.11	0.37	18
Potassium	%		0.56	1.92	

Fescue, aerial part, fresh, full bloom, (2)
Ref No 2 01 874 United States

Feed Name or Analyses			Mean As Fed	Mean Dry	C.V. ± %
Dry matter	%			100.0	
Ash	%			7.5	10
Crude fiber	%			31.9	7
Ether extract	%			2.9	9
N-free extract	%			48.1	
Protein (N x 6.25)	%			9.6	13
Cattle	dig prot %	*		6.1	
Goats	dig prot %	*		5.5	
Horses	dig prot %	*		5.7	
Rabbits	dig prot %	*		6.1	
Sheep	dig prot %	*		5.9	
Energy	GE Mcal/kg				
Cattle	DE Mcal/kg	*		2.90	
Sheep	DE Mcal/kg	*		2.89	
Cattle	ME Mcal/kg	*		2.38	
Sheep	ME Mcal/kg	*		2.37	
Cattle	TDN %	*		65.7	
Sheep	TDN %	*		65.6	

Fescue, aerial part, fresh, mature, (2)
Ref No 2 01 876 United States

Feed Name or Analyses			Mean As Fed	Mean Dry	C.V. ± %
Dry matter	%			100.0	
Ash	%			6.7	10
Crude fiber	%			33.1	2
Ether extract	%			2.4	16
N-free extract	%			51.1	
Protein (N x 6.25)	%			6.7	14
Cattle	dig prot %	*		3.6	
Goats	dig prot %	*		2.8	
Horses	dig prot %	*		3.2	
Rabbits	dig prot %	*		3.8	
Sheep	dig prot %	*		3.2	
Energy	GE Mcal/kg				
Cattle	DE Mcal/kg	*		2.85	
Sheep	DE Mcal/kg	*		2.80	
Cattle	ME Mcal/kg	*		2.34	
Sheep	ME Mcal/kg	*		2.30	
Cattle	TDN %	*		64.6	
Sheep	TDN %	*		63.5	

(1) dry forages and roughages (2) pasture, range plants, and forages fed green (3) silages (4) energy feeds (5) protein supplements (6) minerals (7) vitamins (8) additives

Fescue, aerial part, fresh, mature, pure stand, (2)
Ref No 2 01 875 United States

Feed Name or Analyses			Mean As Fed	Mean Dry	C.V. ± %
Dry matter	%			100.0	
Carotene	mg/kg			89.7	14
Vitamin A equivalent	IU/g			149.6	

Fescue, aerial part, fresh, over ripe, (2)
Ref No 2 01 877 United States

Feed Name or Analyses			Mean As Fed	Mean Dry	C.V. ± %
Dry matter	%			100.0	
Ash	%			7.9	26
Crude fiber	%			39.3	5
Ether extract	%			3.1	18
N-free extract	%			45.9	
Protein (N x 6.25)	%			3.8	18
Cattle	dig prot %	*		1.1	
Goats	dig prot %	*		0.1	
Horses	dig prot %	*		0.8	
Rabbits	dig prot %	*		1.6	
Sheep	dig prot %	*		0.5	
Energy	GE Mcal/kg				
Cattle	DE Mcal/kg	*		2.63	
Sheep	DE Mcal/kg	*		2.49	
Cattle	ME Mcal/kg	*		2.16	
Sheep	ME Mcal/kg	*		2.04	
Cattle	TDN %	*		59.8	
Sheep	TDN %	*		56.4	
Calcium	%			0.31	24
Phosphorus	%			0.11	61
Potassium	%			0.90	37

Fescue, aerial part, fresh, cut 1, (2)
Ref No 2 01 878 United States

Feed Name or Analyses			Mean As Fed	Mean Dry	C.V. ± %
Dry matter	%			100.0	
Ash	%			8.6	8
Crude fiber	%			32.7	10
Ether extract	%			3.3	17
N-free extract	%			44.8	
Protein (N x 6.25)	%			10.6	29
Cattle	dig prot %	*		6.9	
Goats	dig prot %	*		6.4	
Horses	dig prot %	*		6.5	
Rabbits	dig prot %	*		6.9	
Sheep	dig prot %	*		6.9	
Energy	GE Mcal/kg				
Cattle	DE Mcal/kg	*		2.86	
Sheep	DE Mcal/kg	*		2.82	
Cattle	ME Mcal/kg	*		2.34	
Sheep	ME Mcal/kg	*		2.31	
Cattle	TDN %	*		64.8	
Sheep	TDN %	*		63.9	

Fescue, leaves, fresh, (2)
Ref No 2 01 880 United States

Feed Name or Analyses			Mean As Fed	Mean Dry	C.V. ± %
Dry matter	%			100.0	
Ash	%			7.2	11
Crude fiber	%			26.9	15
Ether extract	%			3.0	8
N-free extract	%			51.0	
Protein (N x 6.25)	%			11.9	19
Cattle	dig prot %	*		8.0	
Goats	dig prot %	*		7.7	
Horses	dig prot %	*		7.6	
Rabbits	dig prot %	*		7.9	
Sheep	dig prot %	*		8.1	
Energy	GE Mcal/kg				
Cattle	DE Mcal/kg	*		2.98	
Sheep	DE Mcal/kg	*		2.92	
Cattle	ME Mcal/kg	*		2.44	
Sheep	ME Mcal/kg	*		2.39	
Cattle	TDN %	*		67.6	
Sheep	TDN %	*		66.2	
Calcium	%			0.62	47
Magnesium	%			0.07	21
Phosphorus	%			0.28	35

Fescue, aerial part, ensiled, (3)
Ref No 3 09 183 United States

Feed Name or Analyses			Mean As Fed	Mean Dry	C.V. ± %
Dry matter	%			100.0	
Sheep	dig coef %			61.	
Ash	%			11.6	
Crude fiber	%			28.1	
Sheep	dig coef %			67.	
Ether extract	%			3.7	
N-free extract	%			46.4	
Sheep	dig coef %			64.	
Protein (N x 6.25)	%			10.2	
Sheep	dig coef %			65.	
Cattle	dig prot %	*		5.5	
Goats	dig prot %	*		5.5	
Horses	dig prot %	*		5.5	
Sheep	dig prot %			6.6	
Energy	GE Mcal/kg				
Cattle	DE Mcal/kg	*		2.65	
Sheep	DE Mcal/kg	*		2.57	
Cattle	ME Mcal/kg	*		2.17	
Sheep	ME Mcal/kg	*		2.10	
Cattle	TDN %	*		60.1	
Sheep	TDN %			58.2	

FESCUE, ALTA. Festuca arundinacea

Fescue, alta, hay, s-c, (1)
Ref No 1 05 684 United States

Feed Name or Analyses			Mean As Fed	Mean Dry	C.V. ± %
Dry matter	%		89.0	100.0	2
Horses	dig coef %		46.	46.	11
Organic matter	%		83.1	93.4	0
Ash	%		5.9	6.6	5
Crude fiber	%		32.6	36.6	13
Horses	dig coef %		43.	43.	11
Ether extract	%		2.0	2.2	16
N-free extract	%		40.9	45.9	10
Horses	dig coef %		48.	48.	22
Protein (N x 6.25)	%		7.7	8.7	9
Cattle	dig prot %	*	4.0	4.5	
Goats	dig prot %	*	4.2	4.7	
Horses	dig prot %	*	4.4	4.9	
Rabbits	dig prot %	*	4.8	5.4	
Sheep	dig prot %	*	3.9	4.4	
Cell contents (Van Soest)	%		27.3	30.7	5
Horses	dig coef %		64.	64.	2
Cell walls (Van Soest)	%		61.7	69.3	2
Horses	dig coef %		40.	40.	19
Cellulose (Matrone)	%		30.4	34.2	8
Horses	dig coef %		35.	35.	44
Fiber, acid detergent (VS)	%		35.6	40.0	1
Horses	dig coef %		35.	35.	20
Hemicellulose	%		23.7	26.6	5
Horses	dig coef %		45.	45.	22

Feed Name or Analyses			Mean As Fed	Mean Dry	C.V. ± %
Holocellulose	%		55.4	62.3	3
Soluble carbohydrates	%		14.4	16.2	7
Lignin (Ellis)	%		4.7	5.3	28
Horses	dig coef %		-5.	-5.	98
Energy	GE Mcal/kg		3.93	4.42	1
Horses	GE dig coef %		44.	44.	7
Cattle	DE Mcal/kg	*	2.01	2.26	
Horses	DE Mcal/kg		1.71	1.92	5
Sheep	DE Mcal/kg	*	2.13	2.40	
Cattle	ME Mcal/kg	*	1.65	1.85	
Horses	ME Mcal/kg	*	1.40	1.58	
Sheep	ME Mcal/kg	*	1.75	1.96	
Cattle	TDN %	*	45.6	51.2	
Sheep	TDN %	*	48.4	54.3	
Calcium	%		0.34	0.39	19
Magnesium	%		0.22	0.24	15
Phosphorus	%		0.21	0.24	4
Potassium	%		2.12	2.38	3
Carotene	mg/kg		18.5	20.7	24
Vitamin A equivalent	IU/g		30.8	34.6	

Fescue, alta, hay, s-c, early bloom, (1)

Ref No 1 05 682 United States

Feed Name or Analyses			Mean As Fed	Mean Dry	C.V. ± %
Dry matter	%		95.6	100.0	
Ash	%		8.8	9.2	
Crude fiber	%		26.0	27.2	
Sheep	dig coef %		66.	66.	
Ether extract	%		2.8	3.0	
Sheep	dig coef %		34.	34.	
N-free extract	%		45.2	47.3	
Sheep	dig coef %		71.	71.	
Protein (N x 6.25)	%		12.8	13.4	
Sheep	dig coef %		78.	78.	
Cattle	dig prot %	*	8.2	8.5	
Goats	dig prot %	*	8.7	9.0	
Horses	dig prot %	*	8.5	8.9	
Rabbits	dig prot %	*	8.6	9.0	
Sheep	dig prot %		10.0	10.4	
Cellulose (Matrone)	%		28.1	29.4	
Lignin (Ellis)	%		6.0	6.2	
Energy	GE Mcal/kg				
Cattle	DE Mcal/kg	*	2.66	2.79	
Sheep	DE Mcal/kg	*	2.71	2.83	
Cattle	ME Mcal/kg	*	2.18	2.29	
Sheep	ME Mcal/kg	*	2.22	2.32	
Cattle	TDN %	*	60.4	63.2	
Sheep	TDN %		61.4	64.2	
Calcium	%		0.40	0.42	
Phosphorus	%		0.23	0.24	
Carotene	mg/kg		25.6	26.8	
Vitamin A equivalent	IU/g		42.6	44.6	

Fescue, alta, hay, s-c, early bloom, cut 1, (1)

Ref No 1 08 768 United States

Feed Name or Analyses			Mean As Fed	Mean Dry	C.V. ± %
Dry matter	%		92.2	100.0	1
Horses	dig coef %		47.	47.	2
Organic matter	%		85.7	92.9	0
Ash	%		6.6	7.1	2
Crude fiber	%		35.9	38.9	1
Horses	dig coef %		42.	42.	5
Ether extract	%		1.8	1.9	
Horses	dig coef %		30.	30.	37
N-free extract	%		39.2	42.5	2
Horses	dig coef %		55.	55.	6
Protein (N x 6.25)	%		8.8	9.5	3
Horses	dig coef %		57.	57.	13
Cattle	dig prot %	*	4.8	5.2	
Goats	dig prot %	*	5.0	5.4	
Horses	dig prot %		5.0	5.4	
Rabbits	dig prot %	*	5.5	6.0	
Sheep	dig prot %	*	4.7	5.1	
Cell contents (Van Soest)	%		26.1	28.3	
Horses	dig coef %		59.	59.	
Cell walls (Van Soest)	%		65.8	71.4	2
Horses	dig coef %		42.	42.	1
Cellulose (Matrone)	%		34.3	37.2	2
Horses	dig coef %		43.	43.	6
Fiber, acid detergent (VS)	%		38.3	41.5	2
Horses	dig coef %		39.	39.	9
Hemicellulose	%		23.1	25.1	6
Horses	dig coef %		44.	44.	8
Holocellulose	%		57.5	62.4	2
Soluble carbohydrates	%		13.7	14.9	7
Lignin (Ellis)	%		4.1	4.4	5
Horses	dig coef %		21.	21.	21
Fatty acids	%				
Caprylic	%		27.038	29.320	
Energy	GE Mcal/kg		4.01	4.34	0
Horses	GE dig coef %		38.	38.	12
Cattle	DE Mcal/kg	*	1.93	2.09	
Horses	DE Mcal/kg		1.51	1.64	
Sheep	DE Mcal/kg	*	2.17	2.36	
Cattle	ME Mcal/kg	*	1.58	1.71	
Horses	ME Mcal/kg	*	1.24	1.34	
Sheep	ME Mcal/kg	*	1.78	1.93	
Cattle	TDN %	*	43.7	47.4	
Horses	TDN %		42.9	46.5	
Sheep	TDN %	*	49.3	53.5	
Calcium	%		0.36	0.39	15
Magnesium	%		0.40	0.44	4
Phosphorus	%		0.18	0.20	6
Potassium	%		0.23	0.25	3
Carotene	mg/kg		17.6	19.1	9
Vitamin A equivalent	IU/g		29.4	31.9	

Fescue, alta, hay, s-c, full bloom, (1)

Ref No 1 08 657 United States

Feed Name or Analyses			Mean As Fed	Mean Dry	C.V. ± %
Dry matter	%			100.0	
Sheep	dig coef %			53.	
Ash	%			7.5	
Crude fiber	%			34.7	
Ether extract	%			0.6	
N-free extract	%			52.5	
Protein (N x 6.25)	%			4.7	
Cattle	dig prot %	*		1.0	
Goats	dig prot %	*		0.9	
Horses	dig prot %	*		1.5	
Rabbits	dig prot %	*		2.3	
Sheep	dig prot %	*		0.8	
Lignin (Ellis)	%			7.9	
Energy	GE Mcal/kg			4.21	
Cattle	DE Mcal/kg	*		2.41	
Sheep	DE Mcal/kg			2.69	
Cattle	ME Mcal/kg	*		1.98	
Sheep	ME Mcal/kg	*		2.20	
Cattle	TDN %	*		54.6	
Sheep	TDN %	*		53.9	

Ref No 1 08 657 Canada

Feed Name or Analyses			Mean As Fed	Mean Dry	C.V. ± %
Dry matter	%			100.0	
Sheep	dig coef %			66.	
Ash	%			12.9	
Crude fiber	%			25.6	
Ether extract	%			2.6	
N-free extract	%			38.4	
Protein (N x 6.25)	%			20.5	
Cattle	dig prot %	*		14.7	
Goats	dig prot %	*		15.7	
Horses	dig prot %	*		14.9	
Rabbits	dig prot %	*		14.5	
Sheep	dig prot %	*		15.0	
Cellulose (Matrone)	%			27.3	
Hexosans	%			7.4	
Pentosans	%			7.0	
Lignin (Ellis)	%			7.8	
Energy	GE Mcal/kg			4.18	
Cattle	DE Mcal/kg	*		2.50	
Sheep	DE Mcal/kg			2.69	
Cattle	ME Mcal/kg	*		2.05	
Sheep	ME Mcal/kg	*		2.20	
Cattle	TDN %	*		56.6	
Sheep	TDN %	*		57.6	

Fescue, alta, hay, s-c, chopped, early bloom, (1)

Ref No 1 08 658 United States

Feed Name or Analyses			Mean As Fed	Mean Dry	C.V. ± %
Dry matter	%			100.0	
Sheep	dig coef %			59.	
Ash	%			10.3	
Crude fiber	%			35.4	
Ether extract	%			1.2	
N-free extract	%			44.4	
Protein (N x 6.25)	%			8.7	
Cattle	dig prot %	*		4.5	
Goats	dig prot %	*		4.7	
Horses	dig prot %	*		4.9	
Rabbits	dig prot %	*		5.4	
Sheep	dig prot %	*		4.4	
Lignin (Ellis)	%			7.2	
Energy	GE Mcal/kg			4.20	
Cattle	DE Mcal/kg	*		2.46	
Sheep	DE Mcal/kg			2.66	
Cattle	ME Mcal/kg	*		2.02	
Sheep	ME Mcal/kg	*		2.18	
Cattle	TDN %	*		55.8	
Sheep	TDN %	*		54.5	

Ref No 1 08 658 Canada

Feed Name or Analyses			Mean As Fed	Mean Dry	C.V. ± %
Dry matter	%			100.0	
Sheep	dig coef %			61.	
Ash	%			8.7	
Crude fiber	%			37.6	
Ether extract	%			2.2	
N-free extract	%			36.6	
Protein (N x 6.25)	%			14.9	
Cattle	dig prot %	*		9.8	
Goats	dig prot %	*		10.5	
Horses	dig prot %	*		10.2	
Rabbits	dig prot %	*		10.2	
Sheep	dig prot %	*		9.9	
Cellulose (Matrone)	%			38.0	
Hexosans	%			5.8	
Pentosans	%			4.1	
Lignin (Ellis)	%			10.6	
Energy	GE Mcal/kg			4.26	
Cattle	DE Mcal/kg	*		2.45	
Sheep	DE Mcal/kg			2.66	
Cattle	ME Mcal/kg	*		2.01	
Sheep	ME Mcal/kg	*		2.18	

Continued

Feed Name or Analyses	Mean As Fed	Mean Dry	C.V. ± %
Cattle TDN % *		55.5	
Sheep TDN % *		52.7	

Fescue, alta, aerial part, fresh, (2)
Ref No 2 01 889 United States

Feed Name or Analyses	Mean As Fed	Mean Dry	C.V. ± %
Dry matter %	23.9	100.0	10
Ash %	2.0	8.3	8
Crude fiber %	7.1	29.6	10
Ether extract %	0.8	3.3	17
N-free extract %	11.3	47.2	
Protein (N x 6.25) %	2.8	11.6	22
Cattle dig prot % *	1.9	7.8	
Goats dig prot % *	1.8	7.4	
Horses dig prot % *	1.8	7.4	
Rabbits dig prot % *	1.8	7.6	
Sheep dig prot % *	1.9	7.8	
Energy GE Mcal/kg			
Cattle DE Mcal/kg *	0.69	2.90	
Sheep DE Mcal/kg *	0.68	2.85	
Cattle ME Mcal/kg *	0.57	2.38	
Sheep ME Mcal/kg *	0.56	2.34	
Cattle TDN % *	15.7	65.7	
Sheep TDN % *	15.5	64.7	

Fescue, alta, aerial part, fresh, immature, (2)
Ref No 2 01 882 United States

Feed Name or Analyses	Mean As Fed	Mean Dry	C.V. ± %
Dry matter %	23.9	100.0	10
Ash %	2.4	10.2	6
Crude fiber %	5.3	22.1	4
Ether extract %	1.0	4.2	4
N-free extract %	10.9	45.4	
Protein (N x 6.25) %	4.3	18.1	6
Cattle dig prot % *	3.2	13.3	
Goats dig prot % *	3.2	13.4	
Horses dig prot % *	3.1	12.9	
Rabbits dig prot % *	3.0	12.6	
Sheep dig prot % *	3.3	13.9	
Energy GE Mcal/kg			
Cattle DE Mcal/kg *	0.71	2.97	
Sheep DE Mcal/kg *	0.71	2.96	
Cattle ME Mcal/kg *	0.58	2.44	
Sheep ME Mcal/kg *	0.58	2.43	
Cattle TDN % *	16.1	67.4	
Sheep TDN % *	16.0	67.1	

Fescue, alta, aerial part, fresh, early bloom, (2)
Ref No 2 01 883 United States

Feed Name or Analyses	Mean As Fed	Mean Dry	C.V. ± %
Dry matter %		100.0	
Ash %		8.9	6
Crude fiber %		28.7	7
Ether extract %		3.7	15
N-free extract %		45.1	
Protein (N x 6.25) %		13.6	9
Cattle dig prot % *		9.5	
Goats dig prot % *		9.2	
Horses dig prot % *		9.1	
Rabbits dig prot % *		9.2	
Sheep dig prot % *		9.7	
Energy GE Mcal/kg			
Cattle DE Mcal/kg *		2.92	
Sheep DE Mcal/kg *		2.79	
Cattle ME Mcal/kg *		2.40	
Sheep ME Mcal/kg *		2.29	
Cattle TDN % *		66.3	
Sheep TDN % *		63.3	

Fescue, alta, aerial part, fresh, mid-bloom, (2)
Ref No 2 01 884 United States

Feed Name or Analyses	Mean As Fed	Mean Dry	C.V. ± %
Dry matter %		100.0	
Ash %		8.2	5
Crude fiber %		30.3	4
Ether extract %		3.0	9
N-free extract %		48.0	
Protein (N x 6.25) %		10.5	8
Cattle dig prot % *		6.8	
Goats dig prot % *		6.4	
Horses dig prot % *		6.4	
Rabbits dig prot % *		6.8	
Sheep dig prot % *		6.8	
Hemicellulose %		19.9	
Energy GE Mcal/kg			
Cattle DE Mcal/kg *		2.88	
Sheep DE Mcal/kg *		2.88	
Cattle ME Mcal/kg *		2.36	
Sheep ME Mcal/kg *		2.36	
Cattle TDN % *		65.2	
Sheep TDN % *		65.4	

Fescue, alta, aerial part, fresh, full bloom, (2)
Ref No 2 01 885 United States

Feed Name or Analyses	Mean As Fed	Mean Dry	C.V. ± %
Dry matter %		100.0	
Ash %		8.1	
Crude fiber %		31.4	
Ether extract %		3.2	
N-free extract %		48.3	
Protein (N x 6.25) %		9.0	
Cattle dig prot % *		5.5	
Goats dig prot % *		5.0	
Horses dig prot % *		5.2	
Rabbits dig prot % *		5.6	
Sheep dig prot % *		5.4	
Energy GE Mcal/kg			
Cattle DE Mcal/kg *		2.80	
Sheep DE Mcal/kg *		2.91	
Cattle ME Mcal/kg *		2.30	
Sheep ME Mcal/kg *		2.39	
Cattle TDN % *		63.6	
Sheep TDN % *		66.0	

Fescue, alta, aerial part, fresh, dough stage, (2)
Ref No 2 01 886 United States

Feed Name or Analyses	Mean As Fed	Mean Dry	C.V. ± %
Dry matter %		100.0	
Ash %		7.2	3
Crude fiber %		32.7	1
Ether extract %		2.5	6
N-free extract %		50.1	
Protein (N x 6.25) %		7.5	17
Cattle dig prot % *		4.3	
Goats dig prot % *		3.6	
Horses dig prot % *		3.9	
Rabbits dig prot % *		4.5	
Sheep dig prot % *		4.0	
Energy GE Mcal/kg			
Cattle DE Mcal/kg *		2.85	
Sheep DE Mcal/kg *		2.96	
Cattle ME Mcal/kg *		2.33	
Sheep ME Mcal/kg *		2.42	
Cattle TDN % *		64.6	
Sheep TDN % *		67.1	

Fescue, alta, aerial part, fresh, mature, (2)
Ref No 2 01 887 United States

Feed Name or Analyses	Mean As Fed	Mean Dry	C.V. ± %
Dry matter %		100.0	
Ash %		7.2	
Crude fiber %		33.2	
Ether extract %		2.4	
N-free extract %		50.5	
Protein (N x 6.25) %		6.7	
Cattle dig prot % *		3.6	
Goats dig prot % *		2.8	
Horses dig prot % *		3.2	
Rabbits dig prot % *		3.8	
Sheep dig prot % *		3.2	
Energy GE Mcal/kg			
Cattle DE Mcal/kg *		2.82	
Sheep DE Mcal/kg *		2.98	
Cattle ME Mcal/kg *		2.31	
Sheep ME Mcal/kg *		2.44	
Cattle TDN % *		63.9	
Sheep TDN % *		67.5	

Fescue, alta, aerial part, fresh, cut 1, (2)
Ref No 2 01 888 United States

Feed Name or Analyses	Mean As Fed	Mean Dry	C.V. ± %
Dry matter %		100.0	
Ash %		8.5	7
Crude fiber %		31.8	5
Ether extract %		3.3	17
N-free extract %		44.8	
Protein (N x 6.25) %		11.6	16
Cattle dig prot % *		7.8	
Goats dig prot % *		7.4	
Horses dig prot % *		7.4	
Rabbits dig prot % *		7.6	
Sheep dig prot % *		7.8	
Hemicellulose %		19.9	
Energy GE Mcal/kg			
Cattle DE Mcal/kg *		2.90	
Sheep DE Mcal/kg *		2.80	
Cattle ME Mcal/kg *		2.38	
Sheep ME Mcal/kg *		2.30	
Cattle TDN % *		65.9	
Sheep TDN % *		63.6	

FESCUE, CHEWINGS. Festuca rubra commutata

Fescue, chewings, hay, s-c, (1)
Ref No 1 01 890 United States

Feed Name or Analyses	Mean As Fed	Mean Dry	C.V. ± %
Dry matter %		100.0	
Protein (N x 6.25) %		13.9	10
Cattle dig prot % *		9.0	
Goats dig prot % *		9.5	
Horses dig prot % *		9.3	
Rabbits dig prot % *		9.4	
Sheep dig prot % *		9.0	
Calcium %		0.50	6
Magnesium %		0.29	15

(1) dry forages and roughages (2) pasture, range plants, and forages fed green (3) silages (4) energy feeds (5) protein supplements (6) minerals (7) vitamins (8) additives

Feed Name or Analyses				Mean As Fed	Mean Dry	C.V. ± %
Phosphorus		%			0.38	12
Potassium		%			1.58	12

FESCUE, FOXTAIL. Festuca megalura

Fescue, foxtail, aerial part, fresh, (2)
Ref No 2 01 895 United States

Feed Name or Analyses				Mean As Fed	Mean Dry	C.V. ± %
Dry matter		%			100.0	
Calcium		%			0.36	27
Phosphorus		%			0.27	38
Potassium		%			1.41	39

Fescue, foxtail, aerial part,fresh, immature, (2)
Ref No 2 01 891 United States

Feed Name or Analyses				Mean As Fed	Mean Dry	C.V. ± %
Dry matter		%			100.0	
Ash		%			9.5	6
Crude fiber		%			24.0	23
Protein (N x 6.25)		%			15.5	21
Cattle	dig prot	%	*		11.1	
Goats	dig prot	%	*		11.0	
Horses	dig prot	%	*		10.7	
Rabbits	dig prot	%	*		10.6	
Sheep	dig prot	%	*		11.4	
Calcium		%			0.48	11
Phosphorus		%			0.36	17
Potassium		%			2.22	12

Fescue, foxtail, aerial part, fresh, full bloom, (2)
Ref No 2 01 892 United States

Feed Name or Analyses				Mean As Fed	Mean Dry	C.V. ± %
Dry matter		%			100.0	
Ash		%			5.6	
Crude fiber		%			33.1	
Protein (N x 6.25)		%			7.6	
Cattle	dig prot	%	*		4.4	
Goats	dig prot	%	*		3.7	
Horses	dig prot	%	*		4.0	
Rabbits	dig prot	%	*		4.5	
Sheep	dig prot	%	*		4.1	
Calcium		%			0.29	
Phosphorus		%			0.30	
Potassium		%			1.09	

Fescue, foxtail, aerial part, fresh, mature, (2)
Ref No 2 01 893 United States

Feed Name or Analyses				Mean As Fed	Mean Dry	C.V. ± %
Dry matter		%			100.0	
Ash		%			6.3	11
Crude fiber		%			32.9	3
Protein (N x 6.25)		%			6.8	18
Cattle	dig prot	%	*		3.7	
Goats	dig prot	%	*		2.9	
Horses	dig prot	%	*		3.3	
Rabbits	dig prot	%	*		3.9	
Sheep	dig prot	%	*		3.3	
Calcium		%			0.31	9
Phosphorus		%			0.23	24
Potassium		%			0.89	11

Fescue, foxtail, aerial part, fresh, over ripe, (2)
Ref No 2 01 894 United States

Feed Name or Analyses				Mean As Fed	Mean Dry	C.V. ± %
Dry matter		%			100.0	
Ash		%			5.7	24
Crude fiber		%			40.1	8
Protein (N x 6.25)		%			3.0	31
Cattle	dig prot	%	*		0.4	
Goats	dig prot	%	*		-0.5	
Horses	dig prot	%	*		0.1	
Rabbits	dig prot	%	*		1.0	
Sheep	dig prot	%	*		-0.1	
Calcium		%			0.25	8
Phosphorus		%			0.15	65
Potassium		%			0.90	37

FESCUE, IDAHO. Festuca idahoensis

Fescue, Idaho, hay, s-c, over ripe, (1)
Ref No 1 01 896 United States

Feed Name or Analyses				Mean As Fed	Mean Dry	C.V. ± %
Dry matter		%		87.4	100.0	
Ash		%		10.1	11.6	
Crude fiber		%		30.9	35.4	
Sheep	dig coef	%		61.	61.	
Ether extract		%		2.3	2.6	
Sheep	dig coef	%		41.	41.	
N-free extract		%		40.7	46.6	
Sheep	dig coef	%		51.	51.	
Protein (N x 6.25)		%		3.3	3.8	
Sheep	dig coef	%		11.	11.	
Cattle	dig prot	%	*	0.2	0.2	
Goats	dig prot	%	*	0.1	0.1	
Horses	dig prot	%	*	0.7	0.8	
Rabbits	dig prot	%	*	1.4	1.6	
Sheep	dig prot	%		0.4	0.4	
Energy	GE Mcal/kg					
Cattle	DE Mcal/kg		*	1.91	2.18	
Sheep	DE Mcal/kg		*	1.86	2.12	
Cattle	ME Mcal/kg		*	1.56	1.79	
Sheep	ME Mcal/kg		*	1.52	1.74	
Cattle	TDN	%	*	43.3	49.5	
Sheep	TDN	%		42.1	48.2	

Fescue, Idaho, aerial part, (2)
Ref No 2 01 899 United States

Feed Name or Analyses				Mean As Fed	Mean Dry	C.V. ± %
Dry matter		%			100.0	
Carotene		mg/kg			92.6	65
Vitamin A equivalent		IU/g			154.4	

Fescue, Idaho, aerial part, fresh, over ripe, (2)
Ref No 2 01 897 United States

Feed Name or Analyses				Mean As Fed	Mean Dry	C.V. ± %
Dry matter		%			100.0	
Ash		%			7.9	26
Crude fiber		%			39.3	5
Ether extract		%			3.1	18
N-free extract		%			45.9	
Protein (N x 6.25)		%			3.8	18
Cattle	dig prot	%	*		1.1	
Goats	dig prot	%	*		0.1	
Horses	dig prot	%	*		0.8	
Rabbits	dig prot	%	*		1.6	
Sheep	dig prot	%	*		0.5	

Feed Name or Analyses				Mean As Fed	Mean Dry	C.V. ± %
Energy	GE Mcal/kg					
Cattle	DE Mcal/kg		*		2.63	
Sheep	DE Mcal/kg		*		2.49	
Cattle	ME Mcal/kg		*		2.16	
Sheep	ME Mcal/kg		*		2.04	
Cattle	TDN	%	*		59.8	
Sheep	TDN	%	*		56.4	
Calcium		%			0.36	21
Phosphorus		%			0.08	21

Fescue, Idaho, aerial part, fresh, cut 1, (2)
Ref No 2 01 898 United States

Feed Name or Analyses				Mean As Fed	Mean Dry	C.V. ± %
Dry matter		%			100.0	
Ash		%			8.7	15
Crude fiber		%			37.5	11
Ether extract		%			3.1	18
N-free extract		%			45.1	
Protein (N x 6.25)		%			5.6	64
Cattle	dig prot	%	*		2.7	
Goats	dig prot	%	*		1.8	
Horses	dig prot	%	*		2.3	
Rabbits	dig prot	%	*		3.0	
Sheep	dig prot	%	*		2.2	
Energy	GE Mcal/kg					
Cattle	DE Mcal/kg		*		2.66	
Sheep	DE Mcal/kg		*		2.58	
Cattle	ME Mcal/kg		*		2.18	
Sheep	ME Mcal/kg		*		2.12	
Cattle	TDN	%	*		60.2	
Sheep	TDN	%	*		58.6	
Calcium		%			0.38	19
Phosphorus		%			0.09	31

FESCUE, KENTUCKY 31. Festuca arundinacea

Fescue, Kentucky 31, hay, fan air dried w heat, early bloom, cut 1, (1)
Ref No 1 07 706 United States

Feed Name or Analyses				Mean As Fed	Mean Dry	C.V. ± %
Dry matter		%			100.0	
Sheep	dig coef	%			63.	
Ash		%			8.3	
Crude fiber		%			24.8	
Sheep	dig coef	%			66.	
Ether extract		%			1.8	
Sheep	dig coef	%			33.	
N-free extract		%			48.7	
Sheep	dig coef	%			64.	
Protein (N x 6.25)		%			16.5	
Sheep	dig coef	%			65.	
Cattle	dig prot	%	*		11.2	
Goats	dig prot	%	*		11.9	
Horses	dig prot	%	*		11.5	
Rabbits	dig prot	%	*		11.4	
Sheep	dig prot	%			10.7	
Cellulose (Crampton)		%				
Sheep	dig coef	%			69.	
Soluble carbohydrates		%			13.5	
Lignin (Ellis)		%			7.7	
Energy	GE Mcal/kg					
Cattle	DE Mcal/kg		*		2.75	
Sheep	DE Mcal/kg		*		2.62	
Cattle	ME Mcal/kg		*		2.25	
Sheep	ME Mcal/kg		*		2.15	
Cattle	TDN	%	*		62.4	
Sheep	TDN	%			59.5	

Feed Name or Analyses		As Fed	Dry	C.V. ± %

Fescue, Kentucky 31, hay, fan air dried w heat, cut 1, (1)
Ref No 1 07 704 United States

Feed Name or Analyses		As Fed	Dry	C.V. ± %
Dry matter	%		100.0	
Sheep	dig coef %		66.	4
Organic matter	%		91.3	
Ash	%		7.8	8
Crude fiber	%		27.1	5
Sheep	dig coef %		70.	5
Ether extract	%		4.2	13
Sheep	dig coef %		50.	16
N-free extract	%		46.0	5
Sheep	dig coef %		66.	6
Protein (N x 6.25)	%		15.0	11
Sheep	dig coef %		68.	6
Cattle	dig prot % *		9.9	
Goats	dig prot % *		10.5	
Horses	dig prot % *		10.2	
Rabbits	dig prot % *		10.2	
Sheep	dig prot %		10.1	
Cellulose (Crampton)	%			
Sheep	dig coef %		73.	
Soluble carbohydrates	%		14.1	13
Lignin (Ellis)	%		8.1	
Energy	GE Mcal/kg			
Cattle	DE Mcal/kg *		2.77	
Sheep	DE Mcal/kg *		2.82	
Cattle	ME Mcal/kg *		2.27	
Sheep	ME Mcal/kg *		2.32	
Cattle	TDN % *		62.9	
Sheep	TDN %		64.0	

Fescue, Kentucky 31, hay, s-c, immature, (1)
Ref No 1 09 184 United States

Feed Name or Analyses		As Fed	Dry	C.V. ± %
Dry matter	%		100.0	
Sheep	dig coef %		76.	7
Ash	%		8.8	2
Crude fiber	%		25.0	25
Sheep	dig coef %		81.	9
Ether extract	%		6.7	9
N-free extract	%		39.0	25
Sheep	dig coef %		71.	16
Protein (N x 6.25)	%		20.6	25
Sheep	dig coef %		74.	11
Cattle	dig prot % *		14.7	
Goats	dig prot % *		15.7	
Horses	dig prot % *		15.0	
Rabbits	dig prot % *		14.5	
Sheep	dig prot %		15.3	
Cellulose (Matrone)	%		27.4	
Sheep	dig coef %		66.	
Lignin (Ellis)	%		3.8	
Sheep	dig coef %		8.	
Energy	GE Mcal/kg		4.45	98
Sheep	GE dig coef %		76.	
Cattle	DE Mcal/kg *		2.95	
Sheep	DE Mcal/kg		3.36	
Cattle	ME Mcal/kg *		2.42	
Sheep	ME Mcal/kg *		2.75	
Cattle	TDN % *		66.9	
Sheep	TDN % *		76.2	

(1) dry forages and roughages (3) silages (6) minerals
(2) pasture, range plants, and forages fed green (4) energy feeds (7) vitamins
(5) protein supplements (8) additives

Fescue, Kentucky 31, hay, s-c, prebloom, (1)
Ref No 1 09 185 United States

Feed Name or Analyses		As Fed	Dry	C.V. ± %
Dry matter	%		100.0	
Sheep	dig coef %		74.	6
Ash	%		9.0	5
Crude fiber	%		24.6	18
Sheep	dig coef %		79.	10
Ether extract	%		7.2	12
N-free extract	%		37.8	14
Sheep	dig coef %		74.	6
Protein (N x 6.25)	%		21.4	10
Sheep	dig coef %		76.	5
Cattle	dig prot % *		15.5	
Goats	dig prot % *		16.5	
Horses	dig prot % *		15.7	
Rabbits	dig prot % *		15.2	
Sheep	dig prot %		16.2	
Cellulose (Matrone)	%		29.6	
Sheep	dig coef %		82.	
Lignin (Ellis)	%		4.1	
Sheep	dig coef %		13.	
Energy	GE Mcal/kg		4.48	1
Sheep	GE dig coef %		74.	
Cattle	DE Mcal/kg *		2.99	
Sheep	DE Mcal/kg		3.30	
Cattle	ME Mcal/kg *		2.45	
Sheep	ME Mcal/kg *		2.71	
Cattle	TDN % *		67.8	
Sheep	TDN % *		74.8	

Fescue, Kentucky 31, hay, s-c, early bloom, (1)
Ref No 1 09 186 United States

Feed Name or Analyses		As Fed	Dry	C.V. ± %
Dry matter	%		100.0	
Sheep	dig coef %		70.	9
Ash	%		9.0	4
Crude fiber	%		25.1	21
Sheep	dig coef %		74.	5
Ether extract	%		6.6	14
N-free extract	%		39.9	20
Sheep	dig coef %		72.	11
Protein (N x 6.25)	%		19.4	12
Sheep	dig coef %		70.	12
Cattle	dig prot % *		13.8	
Goats	dig prot % *		14.7	
Horses	dig prot % *		14.0	
Rabbits	dig prot % *		13.7	
Sheep	dig prot %		13.6	
Cellulose (Matrone)	%		29.0	
Sheep	dig coef %		78.	
Lignin (Ellis)	%		4.2	
Sheep	dig coef %		19.	
Energy	GE Mcal/kg		4.30	3
Sheep	GE dig coef %		63.	
Cattle	DE Mcal/kg *		2.96	
Sheep	DE Mcal/kg		2.69	
Cattle	ME Mcal/kg *		2.43	
Sheep	ME Mcal/kg *		2.21	
Cattle	TDN % *		67.1	
Sheep	TDN % *		61.0	

Fescue, Kentucky 31, hay, s-c, midbloom, (1)
Ref No 1 09 187 United States

Feed Name or Analyses		As Fed	Dry	C.V. ± %
Dry matter	%		100.0	
Sheep	dig coef %		67.	11
Ash	%		8.9	28
Crude fiber	%		27.9	15
Sheep	dig coef %		69.	16
Ether extract	%		5.7	24
N-free extract	%		42.8	11
Sheep	dig coef %		69.	12
Protein (N x 6.25)	%		14.7	31
Sheep	dig coef %		67.	11
Cattle	dig prot % *		9.6	
Goats	dig prot % *		10.2	
Horses	dig prot % *		10.0	
Rabbits	dig prot % *		10.0	
Sheep	dig prot %		9.8	
Cellulose (Matrone)	%		31.8	3
Sheep	dig coef %		78.	5
Lignin (Ellis)	%		4.6	8
Sheep	dig coef %		11.	56
Energy	GE Mcal/kg		4.29	2
Sheep	GE dig coef %		70.	4
Cattle	DE Mcal/kg *		2.87	
Sheep	DE Mcal/kg		2.99	
Cattle	ME Mcal/kg *		2.35	
Sheep	ME Mcal/kg *		2.45	
Cattle	TDN % *		65.0	
Sheep	TDN % *		67.8	

Fescue, Kentucky 31, hay, s-c, full bloom, (1)
Ref No 1 09 188 United States

Feed Name or Analyses		As Fed	Dry	C.V. ± %
Dry matter	%		100.0	
Sheep	dig coef %		72.	
Ash	%		9.0	
Crude fiber	%		24.7	
Sheep	dig coef %		77.	
Ether extract	%		5.9	
N-free extract	%		45.1	
Sheep	dig coef %		74.	
Protein (N x 6.25)	%		15.4	
Sheep	dig coef %		70.	
Cattle	dig prot % *		10.2	
Goats	dig prot % *		10.9	
Horses	dig prot % *		10.6	
Rabbits	dig prot % *		10.5	
Sheep	dig prot %		10.7	
Cellulose (Matrone)	%		30.3	
Sheep	dig coef %		78.	
Lignin (Ellis)	%		4.4	
Sheep	dig coef %		6.	
Energy	GE Mcal/kg		4.32	2
Sheep	GE dig coef %		72.	
Cattle	DE Mcal/kg *		2.96	
Sheep	DE Mcal/kg		3.09	
Cattle	ME Mcal/kg *		2.43	
Sheep	ME Mcal/kg *		2.53	
Cattle	TDN % *		67.1	
Sheep	TDN % *		70.1	

Fescue, Kentucky 31, hay, s-c, mature, (1)
Ref No 1 09 189 United States

Feed Name or Analyses		As Fed	Dry	C.V. ± %
Dry matter	%		100.0	
Sheep	dig coef %		72.	7
Ash	%		7.4	4
Crude fiber	%		25.6	22
Sheep	dig coef %		77.	7
Ether extract	%		5.2	15
N-free extract	%		47.3	7
Sheep	dig coef %		74.	12

Feed Name or Analyses			Mean As Fed	Mean Dry	C.V. ± %
Protein (N x 6.25)	%			14.6	23
Sheep	dig coef %			67.	15
Cattle	dig prot %	*		9.6	
Goats	dig prot %	*		10.2	
Horses	dig prot %	*		9.9	
Rabbits	dig prot %	*		9.9	
Sheep	dig prot %			9.8	
Cellulose (Matrone)	%			29.2	
Sheep	dig coef %			79.	
Lignin (Ellis)	%			4.6	
Sheep	dig coef %			11.	
Energy	GE Mcal/kg			4.38	0
Sheep	GE dig coef %			72.	
Cattle	DE Mcal/kg	*		2.84	
Sheep	DE Mcal/kg			3.14	
Cattle	ME Mcal/kg	*		2.33	
Sheep	ME Mcal/kg	*		2.58	
Cattle	TDN %	*		64.5	
Sheep	TDN %	*		71.2	

Fescue, Kentucky 31, hay, s-c, over ripe, (1)

Ref No 1 09 190 United States

Feed Name or Analyses			Mean As Fed	Mean Dry	C.V. ± %
Dry matter	%			100.0	
Sheep	dig coef %			62.	9
Ash	%			6.8	4
Crude fiber	%			32.0	6
Sheep	dig coef %			66.	10
Ether extract	%			4.4	6
N-free extract	%			47.3	1
Sheep	dig coef %			62.	9
Protein (N x 6.25)	%			9.5	22
Sheep	dig coef %			54.	21
Cattle	dig prot %	*		5.1	
Goats	dig prot %	*		5.4	
Horses	dig prot %	*		5.6	
Rabbits	dig prot %	*		6.0	
Sheep	dig prot %			5.1	
Cellulose (Matrone)	%			36.3	2
Sheep	dig coef %			65.	11
Lignin (Ellis)	%			7.5	6
Sheep	dig coef %			14.	61
Energy	GE Mcal/kg			4.29	4
Sheep	GE dig coef %			59.	8
Cattle	DE Mcal/kg	*		2.69	
Sheep	DE Mcal/kg			2.53	
Cattle	ME Mcal/kg	*		2.20	
Sheep	ME Mcal/kg	*		2.07	
Cattle	TDN %	*		60.9	
Sheep	TDN %	*		57.4	

Fescue, Kentucky 31, aerial part, fresh, (2)

Ref No 2 01 902 United States

Feed Name or Analyses			Mean As Fed	Mean Dry	C.V. ± %
Dry matter	%		29.5	100.0	28
Organic matter	%		26.9	91.2	1
Ash	%		2.9	9.7	16
Crude fiber	%		7.2	24.6	10
Cattle	dig coef %		62.	62.	20
Sheep	dig coef %		71.	71.	10
Ether extract	%		1.6	5.5	14
Cattle	dig coef %		73.	73.	12
Sheep	dig coef %		60.	60.	24
N-free extract	%		13.4	45.5	5
Cattle	dig coef %		72.	72.	13
Sheep	dig coef %		70.	70.	13
Protein (N x 6.25)	%		4.4	14.8	22
Cattle	dig coef %		68.	68.	10
Sheep	dig coef %		68.	68.	16
Cattle	dig prot %		3.0	10.1	
Goats	dig prot %	*	3.1	10.4	
Horses	dig prot %	*	3.0	10.1	
Rabbits	dig prot %	*	3.0	10.1	
Sheep	dig prot %		2.9	10.0	
Lignin (Ellis)	%		1.7	5.7	28
Energy	GE Mcal/kg				
Cattle	DE Mcal/kg	*	0.87	2.95	
Sheep	DE Mcal/kg	*	0.86	2.93	
Cattle	ME Mcal/kg	*	0.71	2.42	
Sheep	ME Mcal/kg	*	0.71	2.40	
Cattle	TDN %		19.7	66.9	
Sheep	TDN %		19.6	66.5	
Calcium	%		0.15	0.50	16
Phosphorus	%		0.11	0.37	18
Carotene	mg/kg		64.3	218.0	43
Vitamin A equivalent	IU/g		107.2	363.5	

Fescue, Kentucky 31, aerial part, fresh, immature, (2)

Ref No 2 01 900 United States

Feed Name or Analyses			Mean As Fed	Mean Dry	C.V. ± %
Dry matter	%		27.6	100.0	15
Ash	%		2.6	9.4	19
Crude fiber	%		5.9	21.4	11
Ether extract	%		1.7	6.1	14
N-free extract	%		11.3	41.0	
Protein (N x 6.25)	%		6.1	22.1	14
Cattle	dig prot %	*	4.6	16.7	
Goats	dig prot %	*	4.7	17.2	
Horses	dig prot %	*	4.5	16.3	
Rabbits	dig prot %	*	4.3	15.7	
Sheep	dig prot %	*	4.9	17.6	
Energy	GE Mcal/kg				
Cattle	DE Mcal/kg	*	0.85	3.09	
Sheep	DE Mcal/kg	*	0.82	2.97	
Cattle	ME Mcal/kg	*	0.70	2.54	
Sheep	ME Mcal/kg	*	0.67	2.44	
Cattle	TDN %	*	19.4	70.1	
Sheep	TDN %	*	18.6	67.4	
Calcium	%		0.15	0.53	8
Phosphorus	%		0.11	0.39	7
Carotene	mg/kg		84.3	305.6	29
Vitamin A equivalent	IU/g		140.6	509.4	

Fescue, Kentucky 31, aerial part, fresh, early bloom, (2)

Ref No 2 01 901 United States

Feed Name or Analyses			Mean As Fed	Mean Dry	C.V. ± %
Dry matter	%		28.6	100.0	33
Ash	%		2.9	10.1	15
Crude fiber	%		7.1	24.7	12
Ether extract	%		1.7	5.9	19
N-free extract	%		12.2	42.6	
Protein (N x 6.25)	%		4.8	16.7	7
Cattle	dig prot %	*	3.5	12.1	
Goats	dig prot %	*	3.5	12.1	
Horses	dig prot %	*	3.3	11.7	
Rabbits	dig prot %	*	3.3	11.6	
Sheep	dig prot %	*	3.6	12.6	
Energy	GE Mcal/kg				
Cattle	DE Mcal/kg	*	0.81	2.82	
Sheep	DE Mcal/kg	*	0.78	2.72	
Cattle	ME Mcal/kg	*	0.66	2.31	
Sheep	ME Mcal/kg	*	0.64	2.23	
Cattle	TDN %	*	18.3	64.0	
Sheep	TDN %	*	17.7	61.7	
Calcium	%		0.15	0.52	17
Phosphorus	%		0.11	0.37	19
Carotene	mg/kg		63.1	220.7	40
Vitamin A equivalent	IU/g		105.2	367.9	

FESCUE, MEADOW. Festuca elatior

Fescue, meadow, hay, s-c, (1)

Fescue hay, tall

Ref No 1 01 912 United States

Feed Name or Analyses			Mean As Fed	Mean Dry	C.V. ± %
Dry matter	%		86.5	100.0	4
Ash	%		8.0	9.2	
Crude fiber	%		28.1	32.5	
Sheep	dig coef %		72.	72.	
Ether extract	%		2.2	2.6	
Sheep	dig coef %		36.	36.	
N-free extract	%		40.0	46.2	
Sheep	dig coef %		64.	64.	
Protein (N x 6.25)	%		8.2	9.5	
Sheep	dig coef %		58.	58.	
Cattle	dig prot %	*	4.5	5.2	
Goats	dig prot %	*	4.7	5.4	
Horses	dig prot %	*	4.8	5.6	
Rabbits	dig prot %	*	5.2	6.0	
Sheep	dig prot %		4.7	5.5	
Energy	GE Mcal/kg				
Cattle	DE Mcal/kg	*	2.32	2.68	
Sheep	DE Mcal/kg	*	2.31	2.67	
Cattle	ME Mcal/kg	*	1.90	2.20	
Sheep	ME Mcal/kg	*	1.89	2.19	
Cattle	NE_m Mcal/kg	*	1.15	1.33	
Cattle	NE_{gain} Mcal/kg	*	0.62	0.72	
Cattle	$NE_{lactating\ cows}$ Mcal/kg	*	1.31	1.52	
Cattle	TDN %	*	52.6	60.8	
Sheep	TDN %		52.4	60.5	
Phosphorus	%		0.19	0.22	
Potassium	%		1.39	1.60	
Carotene	mg/kg		61.8	71.4	
α-tocopherol	mg/kg		117.3	135.6	
Vitamin A equivalent	IU/g		103.0	119.1	

Fescue, meadow, hay, s-c, immature, (1)

Ref No 1 01 903 United States

Feed Name or Analyses			Mean As Fed	Mean Dry	C.V. ± %
Dry matter	%		91.0	100.0	5
Ash	%		8.0	8.8	7
Crude fiber	%		23.4	25.7	10
Sheep	dig coef %	*	66.	66.	
Ether extract	%		4.0	4.4	27
Sheep	dig coef %	*	50.	50.	
N-free extract	%		39.1	43.0	
Sheep	dig coef %	*	70.	70.	
Protein (N x 6.25)	%		16.5	18.1	23
Sheep	dig coef %	*	61.	61.	
Cattle	dig prot %	*	11.5	12.6	
Goats	dig prot %	*	12.2	13.4	
Horses	dig prot %	*	11.7	12.9	
Rabbits	dig prot %	*	11.5	12.6	
Sheep	dig prot %		10.0	11.0	
Lignin (Ellis)	%		3.0	3.3	
Energy	GE Mcal/kg				
Cattle	DE Mcal/kg	*	2.53	2.78	
Sheep	DE Mcal/kg	*	2.53	2.78	
Cattle	ME Mcal/kg	*	2.07	2.28	
Sheep	ME Mcal/kg	*	2.07	2.28	
Cattle	TDN %	*	57.4	63.1	
Sheep	TDN %		57.3	63.1	
Calcium	%		0.66	0.73	

Continued

Feed Name or Analyses				Mean As Fed	Mean Dry	C.V. ± %
Magnesium		%		0.62	0.68	
Phosphorus		%		0.45	0.49	

Fescue, meadow, hay, s-c, pre-bloom, (1)

Ref No 1 01 904 United States

Feed Name or Analyses				Mean As Fed	Mean Dry	C.V. ± %
Dry matter		%		83.0	100.0	
Ash		%		7.1	8.6	
Crude fiber		%		24.7	29.8	
Sheep	dig coef	%		66.	66.	
Ether extract		%		2.2	2.6	
Sheep	dig coef	%		50.	50.	
N-free extract		%		39.6	47.7	
Sheep	dig coef	%		70.	70.	
Protein (N x 6.25)		%		9.4	11.3	
Sheep	dig coef	%		61.	61.	
Cattle	dig prot	%	*	5.6	6.7	
Goats	dig prot	%	*	5.9	7.1	
Horses	dig prot	%	*	5.9	7.1	
Rabbits	dig prot	%	*	6.1	7.4	
Sheep	dig prot	%		5.7	6.9	
Energy	GE Mcal/kg					
Cattle	DE Mcal/kg		*	2.35	2.83	
Sheep	DE Mcal/kg		*	2.30	2.77	
Cattle	ME Mcal/kg		*	1.93	2.32	
Sheep	ME Mcal/kg		*	1.89	2.27	
Cattle	TDN	%	*	53.3	64.2	
Sheep	TDN	%		52.2	62.9	

Fescue, meadow, hay, s-c, early bloom, (1)

Ref No 1 01 905 United States

Feed Name or Analyses				Mean As Fed	Mean Dry	C.V. ± %
Dry matter		%		84.7	100.0	
Ash		%		7.5	8.9	
Crude fiber		%		30.7	36.2	
Sheep	dig coef	%		63.	63.	
Ether extract		%		1.2	1.4	
Sheep	dig coef	%		55.	55.	
N-free extract		%		40.1	47.4	
Sheep	dig coef	%		56.	56.	
Protein (N x 6.25)		%		5.2	6.1	
Sheep	dig coef	%		49.	49.	
Cattle	dig prot	%	*	1.9	2.2	
Goats	dig prot	%	*	1.9	2.3	
Horses	dig prot	%	*	2.3	2.7	
Rabbits	dig prot	%	*	2.9	3.4	
Sheep	dig prot	%		2.5	3.0	
Energy	GE Mcal/kg					
Cattle	DE Mcal/kg		*	2.01	2.38	
Sheep	DE Mcal/kg		*	2.02	2.38	
Cattle	ME Mcal/kg		*	1.65	1.95	
Sheep	ME Mcal/kg		*	1.66	1.95	
Cattle	TDN	%	*	45.7	53.9	
Sheep	TDN	%		45.8	54.1	

Fescue, meadow, hay, s-c, full bloom, (1)

Ref No 1 01 906 United States

Feed Name or Analyses				Mean As Fed	Mean Dry	C.V. ± %
Dry matter		%		84.4	100.0	
Ash		%		7.8	9.2	
Crude fiber		%		24.5	29.0	
Sheep	dig coef	%		54.	54.	
Ether extract		%		2.2	2.6	
Sheep	dig coef	%		73.	73.	
N-free extract		%		43.1	51.1	
Sheep	dig coef	%		64.	64.	
Protein (N x 6.25)		%		6.8	8.1	
Sheep	dig coef	%		54.	54.	
Cattle	dig prot	%	*	3.3	4.0	
Goats	dig prot	%	*	3.5	4.1	
Horses	dig prot	%	*	3.7	4.4	
Rabbits	dig prot	%	*	4.2	4.9	
Sheep	dig prot	%		3.7	4.4	
Energy	GE Mcal/kg					
Cattle	DE Mcal/kg		*	2.04	2.42	
Sheep	DE Mcal/kg		*	2.12	2.51	
Cattle	ME Mcal/kg		*	1.67	1.98	
Sheep	ME Mcal/kg		*	1.74	2.06	
Cattle	TDN	%	*	46.3	54.9	
Sheep	TDN	%		48.1	57.0	

Ref No 1 01 906 Canada

Feed Name or Analyses				Mean As Fed	Mean Dry	C.V. ± %
Dry matter		%		95.1	100.0	
Crude fiber		%		32.2	33.9	
Sheep	dig coef	%		54.	54.	
Ether extract		%				
Sheep	dig coef	%		73.	73.	
N-free extract		%				
Sheep	dig coef	%		64.	64.	
Protein (N x 6.25)		%		14.6	15.4	
Sheep	dig coef	%		54.	54.	
Cattle	dig prot	%	*	9.7	10.2	
Goats	dig prot	%	*	10.4	10.9	
Horses	dig prot	%	*	10.0	10.6	
Rabbits	dig prot	%	*	10.0	10.5	
Sheep	dig prot	%		7.9	8.3	
Energy	GE Mcal/kg					
Sheep	DE Mcal/kg		*	2.37	2.50	
Sheep	ME Mcal/kg		*	1.95	2.05	
Sheep	TDN	%		53.8	56.6	
Calcium		%		0.47	0.49	
Phosphorus		%		0.37	0.39	

Fescue, meadow, hay, s-c, milk stage, (1)

Ref No 1 01 907 United States

Feed Name or Analyses				Mean As Fed	Mean Dry	C.V. ± %
Dry matter		%		92.3	100.0	2
Ash		%		6.0	6.5	7
Crude fiber		%		31.7	34.3	7
Ether extract		%		2.5	2.7	29
N-free extract		%		45.0	48.7	
Protein (N x 6.25)		%		7.2	7.8	3
Cattle	dig prot	%	*	3.4	3.7	
Goats	dig prot	%	*	3.5	3.8	
Horses	dig prot	%	*	3.8	4.2	
Rabbits	dig prot	%	*	4.3	4.7	
Sheep	dig prot	%	*	3.3	3.6	
Energy	GE Mcal/kg					
Cattle	DE Mcal/kg		*	2.26	2.45	
Sheep	DE Mcal/kg		*	2.24	2.43	
Cattle	ME Mcal/kg		*	1.86	2.01	
Sheep	ME Mcal/kg		*	1.84	1.99	
Cattle	TDN	%	*	51.4	55.6	
Sheep	TDN	%	*	50.8	55.0	

Fescue, meadow, hay, s-c, mature, (1)

Ref No 1 01 908 United States

Feed Name or Analyses				Mean As Fed	Mean Dry	C.V. ± %
Dry matter		%		86.3	100.0	
Ash		%		9.5	11.0	
Crude fiber		%		29.9	34.6	
Sheep	dig coef	%		52.	52.	
Ether extract		%		2.0	2.3	
Sheep	dig coef	%		57.	57.	
N-free extract		%		39.7	46.0	
Sheep	dig coef	%		57.	57.	
Protein (N x 6.25)		%		5.3	6.1	
Sheep	dig coef	%		43.	43.	
Cattle	dig prot	%	*	1.9	2.2	
Goats	dig prot	%	*	1.9	2.3	
Horses	dig prot	%	*	2.3	2.7	
Rabbits	dig prot	%	*	2.9	3.4	
Sheep	dig prot	%		2.3	2.6	
Energy	GE Mcal/kg					
Cattle	DE Mcal/kg		*	1.81	2.10	
Sheep	DE Mcal/kg		*	1.89	2.20	
Cattle	ME Mcal/kg		*	1.48	1.72	
Sheep	ME Mcal/kg		*	1.55	1.80	
Cattle	TDN	%	*	41.1	47.6	
Sheep	TDN	%		43.0	49.8	

Fescue, meadow, hay, s-c, over ripe, (1)

Ref No 1 01 909 United States

Feed Name or Analyses				Mean As Fed	Mean Dry	C.V. ± %
Dry matter		%		92.4	100.0	2
Ash		%		6.2	6.7	12
Crude fiber		%		37.8	40.9	9
Sheep	dig coef	%	*	52.	52.	
Ether extract		%		1.6	1.7	25
Sheep	dig coef	%	*	57.	57.	
N-free extract		%		41.3	44.7	
Sheep	dig coef	%	*	57.	57.	
Protein (N x 6.25)		%		5.5	6.0	16
Sheep	dig coef	%	*	43.	43.	
Cattle	dig prot	%	*	2.0	2.1	
Goats	dig prot	%	*	2.0	2.2	
Horses	dig prot	%	*	2.4	2.6	
Rabbits	dig prot	%	*	3.1	3.3	
Sheep	dig prot	%		2.4	2.6	
Energy	GE Mcal/kg					
Cattle	DE Mcal/kg		*	2.17	2.35	
Sheep	DE Mcal/kg		*	2.10	2.27	
Cattle	ME Mcal/kg		*	1.78	1.93	
Sheep	ME Mcal/kg		*	1.72	1.86	
Cattle	TDN	%	*	49.2	53.2	
Sheep	TDN	%	*	47.6	51.5	

Fescue, meadow, hay, s-c, cut 1, (1)

Ref No 1 01 910 United States

Feed Name or Analyses				Mean As Fed	Mean Dry	C.V. ± %
Dry matter		%		92.7	100.0	2
Ash		%		5.6	6.0	5
Crude fiber		%		38.4	41.4	11
Ether extract		%		1.6	1.7	34
N-free extract		%		41.2	44.4	
Protein (N x 6.25)		%		6.0	6.5	25
Cattle	dig prot	%	*	2.4	2.6	
Goats	dig prot	%	*	2.4	2.6	
Horses	dig prot	%	*	2.8	3.0	
Rabbits	dig prot	%	*	3.4	3.7	
Sheep	dig prot	%	*	2.2	2.4	

(1) dry forages and roughages
(2) pasture, range plants, and forages fed green
(3) silages
(4) energy feeds
(5) protein supplements
(6) minerals
(7) vitamins
(8) additives

Feed Name or Analyses			Mean As Fed	Mean Dry	C.V. ± %
Energy	GE Mcal/kg				
Cattle	DE Mcal/kg	*	2.03	2.19	
Sheep	DE Mcal/kg	*	2.10	2.27	
Cattle	ME Mcal/kg	*	1.67	1.80	
Sheep	ME Mcal/kg	*	1.72	1.86	
Cattle	TDN %	*	46.1	49.7	
Sheep	TDN %	*	47.6	51.4	

Fescue, meadow, hay, s-c, cut 3, (1)

Ref No 1 01 911 United States

Feed Name or Analyses			As Fed	Dry	C.V. ± %
Dry matter	%		90.6	100.0	
Ash	%		5.4	6.0	
Crude fiber	%		33.1	36.5	
Ether extract	%		1.7	1.9	
N-free extract	%		43.5	48.0	
Protein (N x 6.25)	%		6.9	7.6	
Cattle	dig prot %	*	3.2	3.5	
Goats	dig prot %	*	3.3	3.7	
Horses	dig prot %	*	3.6	4.0	
Rabbits	dig prot %	*	4.1	4.5	
Sheep	dig prot %	*	3.1	3.4	
Energy	GE Mcal/kg				
Cattle	DE Mcal/kg	*	2.03	2.24	
Sheep	DE Mcal/kg	*	2.16	2.39	
Cattle	ME Mcal/kg	*	1.66	1.84	
Sheep	ME Mcal/kg	*	1.77	1.96	
Cattle	TDN %	*	46.0	50.8	
Sheep	TDN %	*	49.0	54.1	

Fescue, meadow, aerial part, fresh, (2)

Fescue forage, tall

Ref No 2 01 920 United States

Feed Name or Analyses			As Fed	Dry	C.V. ± %
Dry matter	%		28.5	100.0	5
Ash	%		2.5	8.6	18
Crude fiber	%		8.2	28.6	15
Ether extract	%		1.2	4.2	35
N-free extract	%		13.1	46.1	
Protein (N x 6.25)	%		3.5	12.4	43
Cattle	dig prot %	*	2.4	8.4	
Goats	dig prot %	*	2.3	8.2	
Horses	dig prot %	*	2.3	8.1	
Rabbits	dig prot %	*	2.4	8.3	
Sheep	dig prot %	*	2.4	8.6	
Energy	GE Mcal/kg				
Cattle	DE Mcal/kg	*	0.81	2.85	
Sheep	DE Mcal/kg	*	0.81	2.83	
Cattle	ME Mcal/kg	*	0.67	2.34	
Sheep	ME Mcal/kg	*	0.66	2.32	
Cattle	TDN %	*	18.4	64.6	
Sheep	TDN %	*	18.3	64.2	
Calcium	%		0.12	0.43	
Cobalt	mg/kg		0.038	0.135	42
Magnesium	%		0.07	0.23	
Phosphorus	%		0.08	0.30	
Potassium	%		0.67	2.34	
Carotene	mg/kg		96.1	337.5	36
Riboflavin	mg/kg		2.4	8.6	49
Thiamine	mg/kg		3.4	11.9	55
α-tocopherol	mg/kg		47.0	165.1	51
Vitamin A equivalent	IU/g		160.3	562.7	

Feed Name or Analyses | Mean As Fed | Dry | C.V. ± %

Fescue, meadow, aerial part, fresh, immature, (2)

Fescue, tall, fresh, immature

Ref No 2 08 420 United States

Feed Name or Analyses			As Fed	Dry	C.V. ± %
Dry matter	%		25.0	100.0	
Ash	%		2.1	8.4	
Crude fiber	%		5.9	23.6	
Ether extract	%		1.3	5.2	
N-free extract	%		11.3	45.2	
Protein (N x 6.25)	%		4.4	17.6	
Cattle	dig prot %	*	3.2	12.9	
Goats	dig prot %	*	3.2	13.0	
Horses	dig prot %	*	3.1	12.5	
Rabbits	dig prot %	*	3.1	12.3	
Sheep	dig prot %	*	3.3	13.4	
Energy	GE Mcal/kg				
Cattle	DE Mcal/kg	*	0.75	3.01	
Sheep	DE Mcal/kg	*	0.72	2.88	
Cattle	ME Mcal/kg	*	0.62	2.47	
Sheep	ME Mcal/kg	*	0.59	2.36	
Cattle	TDN %	*	17.1	68.3	
Sheep	TDN %	*	16.3	65.3	
Calcium	%		0.19	0.76	
Phosphorus	%		0.12	0.48	

Fescue, meadow, aerial part, fresh, immature, pure stand, (2)

Ref No 2 01 913 United States

Feed Name or Analyses			As Fed	Dry	C.V. ± %
Dry matter	%		24.4	100.0	
Carotene	mg/kg		123.4	505.7	25
a-tocopherol	mg/kg		67.7	277.6	10
Vitamin A equivalent	IU/g		205.7	843.1	

Fescue, meadow, aerial part, fresh, pre-bloom, pure stand, (2)

Ref No 2 01 914 United States

Feed Name or Analyses			As Fed	Dry	C.V. ± %
Dry matter	%			100.0	
Carotene	mg/kg			449.7	24
Vitamin A equivalent	IU/g			749.7	

Fescue, meadow, aerial part, fresh, early bloom, (2)

Ref No 2 01 915 United States

Feed Name or Analyses			As Fed	Dry	C.V. ± %
Dry matter	%		26.8	100.0	
Ash	%		2.1	7.7	
Crude fiber	%		7.7	28.9	
Ether extract	%		1.4	5.3	
N-free extract	%		11.4	42.6	
Protein (N x 6.25)	%		4.2	15.5	
Cattle	dig prot %	*	3.0	11.1	
Goats	dig prot %	*	3.0	11.0	
Horses	dig prot %	*	2.9	10.7	
Rabbits	dig prot %	*	2.9	10.6	
Sheep	dig prot %	*	3.1	11.4	
Energy	GE Mcal/kg				
Cattle	DE Mcal/kg	*	0.80	2.99	
Sheep	DE Mcal/kg	*	0.73	2.71	
Cattle	ME Mcal/kg	*	0.66	2.45	
Sheep	ME Mcal/kg	*	0.60	2.22	
Cattle	TDN %	*	18.2	67.9	
Sheep	TDN %	*	16.5	61.5	
Carotene	mg/kg		113.9	424.8	25

Feed Name or Analyses | Mean As Fed | Dry | C.V. ± %

Feed Name or Analyses			As Fed	Dry	C.V. ± %
a-tocopherol	mg/kg		44.5	166.0	
Vitamin A equivalent	IU/g		189.8	708.2	

Fescue, meadow, aerial part, fresh, full bloom, (2)

Ref No 2 01 917 United States

Feed Name or Analyses			As Fed	Dry	C.V. ± %
Dry matter	%		24.4	100.0	40
Carotene	mg/kg		107.1	438.9	29
Riboflavin	mg/kg		3.8	15.4	
Thiamine	mg/kg		5.3	21.6	
α-tocopherol	mg/kg		24.1	98.8	8
Vitamin A equivalent	IU/g		178.5	731.7	

Fescue, meadow, aerial part, fresh, full bloom, pure stand, (2)

Ref No 2 01 916 United States

Feed Name or Analyses			As Fed	Dry	C.V. ± %
Dry matter	%			100.0	
Carotene	mg/kg			254.0	18
Vitamin A equivalent	IU/g			423.4	

Fescue, meadow, aerial part, fresh, mature, (2)

Ref No 2 01 918 United States

Feed Name or Analyses			As Fed	Dry	C.V. ± %
Dry matter	%			100.0	
Ash	%			6.6	
Crude fiber	%			31.5	
Ether extract	%			1.9	
N-free extract	%			52.8	
Protein (N x 6.25)	%			7.2	
Cattle	dig prot %	*		4.0	
Goats	dig prot %	*		3.3	
Horses	dig prot %	*		3.6	
Rabbits	dig prot %	*		4.2	
Sheep	dig prot %	*		3.7	
Energy	GE Mcal/kg				
Cattle	DE Mcal/kg	*		2.90	
Sheep	DE Mcal/kg	*		3.01	
Cattle	ME Mcal/kg	*		2.38	
Sheep	ME Mcal/kg	*		2.47	
Cattle	TDN %	*		65.8	
Sheep	TDN %	*		68.2	
Carotene	mg/kg			89.7	14
Riboflavin	mg/kg			5.5	
Thiamine	mg/kg			15.2	
Vitamin A equivalent	IU/g			149.6	

Fescue, meadow, aerial part, fresh, over ripe, (2)

Ref No 2 01 919 United States

Feed Name or Analyses			As Fed	Dry	C.V. ± %
Dry matter	%			100.0	
Ash	%			6.2	
Crude fiber	%			32.1	
Ether extract	%			0.8	
N-free extract	%			57.1	
Protein (N x 6.25)	%			3.8	
Cattle	dig prot %	*		1.1	
Goats	dig prot %	*		0.1	
Horses	dig prot %	*		0.8	
Rabbits	dig prot %	*		1.6	
Sheep	dig prot %	*		0.5	
α-tocopherol	mg/kg			24.5	

Fescue, meadow, aerial part, fresh, pure stand, (2)

Ref No 2 01 921 United States

Feed Name or Analyses			Mean As Fed	Mean Dry	C.V. ± %
Dry matter	%			100.0	
Carotene	mg/kg			351.6	35
Vitamin A equivalent	IU/g			586.2	

Fescue, meadow, aerial part, ensiled, (3)

Fescue silage, tall

Ref No 3 01 925 United States

Feed Name or Analyses			Mean As Fed	Mean Dry	C.V. ± %
Dry matter	%		32.8	100.0	12
Ash	%		3.4	10.3	12
Crude fiber	%		9.4	28.8	7
Ether extract	%		1.1	3.5	11
N-free extract	%		14.6	44.5	
Protein (N x 6.25)	%		4.2	12.9	12
Cattle	dig prot %	*	2.6	7.9	
Goats	dig prot %	*	2.6	7.9	
Horses	dig prot %	*	2.6	7.9	
Sheep	dig prot %	*	2.6	7.9	
Energy	GE Mcal/kg				
Cattle	DE Mcal/kg	*	0.92	2.81	
Sheep	DE Mcal/kg	*	0.90	2.75	
Cattle	ME Mcal/kg	*	0.76	2.31	
Sheep	ME Mcal/kg	*	0.74	2.25	
Cattle	TDN %	*	20.9	63.8	
Sheep	TDN %	*	20.4	62.3	

Fescue, meadow, aerial part, ensiled, immature, (3)

Ref No 3 01 922 United States

Feed Name or Analyses			Mean As Fed	Mean Dry	C.V. ± %
Dry matter	%			100.0	
Ash	%			10.0	
Crude fiber	%			28.9	
Ether extract	%			3.5	
N-free extract	%			42.2	
Protein (N x 6.25)	%			15.4	
Cattle	dig prot %	*		10.2	
Goats	dig prot %	*		10.2	
Horses	dig prot %	*		10.2	
Sheep	dig prot %	*		10.2	
Energy	GE Mcal/kg				
Cattle	DE Mcal/kg	*		2.67	
Sheep	DE Mcal/kg	*		2.47	
Cattle	ME Mcal/kg	*		2.19	
Sheep	ME Mcal/kg	*		2.03	
Cattle	TDN %	*		60.5	
Sheep	TDN %	*		56.1	

Fescue, meadow, aerial part, ensiled, early bloom, (3)

Ref No 3 01 923 United States

Feed Name or Analyses			Mean As Fed	Mean Dry	C.V. ± %
Dry matter	%		34.4	100.0	6
Ash	%		3.2	9.2	12
Crude fiber	%		10.1	29.5	5
Ether extract	%		1.1	3.3	7
N-free extract	%		14.9	43.4	
Protein (N x 6.25)	%		5.0	14.6	7
Cattle	dig prot %	*	3.3	9.5	
Goats	dig prot %	*	3.3	9.5	
Horses	dig prot %	*	3.3	9.5	
Sheep	dig prot %	*	3.3	9.5	
Energy	GE Mcal/kg				
Cattle	DE Mcal/kg	*	0.90	2.63	
Sheep	DE Mcal/kg	*	0.95	2.76	
Cattle	ME Mcal/kg	*	0.74	2.15	
Sheep	ME Mcal/kg	*	0.78	2.26	
Cattle	TDN %	*	20.5	59.6	
Sheep	TDN %	*	21.5	62.6	

(1) dry forages and roughages (2) pasture, range plants, and forages fed green (3) silages (4) energy feeds (5) protein supplements (6) minerals (7) vitamins (8) additives

Fescue, meadow, aerial part, ensiled, full bloom, (3)

Ref No 3 01 924 United States

Feed Name or Analyses			Mean As Fed	Mean Dry	C.V. ± %
Dry matter	%		24.8	100.0	
Ash	%		2.0	7.9	
Crude fiber	%		7.1	28.5	
Ether extract	%		0.7	2.8	
N-free extract	%		12.4	49.8	
Protein (N x 6.25)	%		2.7	11.0	
Cattle	dig prot %	*	1.5	6.2	
Goats	dig prot %	*	1.5	6.2	
Horses	dig prot %	*	1.5	6.2	
Sheep	dig prot %	*	1.5	6.2	
Energy	GE Mcal/kg				
Cattle	DE Mcal/kg	*	0.68	2.74	
Sheep	DE Mcal/kg	*	0.71	2.85	
Cattle	ME Mcal/kg	*	0.56	2.25	
Sheep	ME Mcal/kg	*	0.58	2.34	
Cattle	TDN %	*	15.4	62.3	
Sheep	TDN %	*	16.0	64.6	

Fescue, meadow, aerial part w molasses added, ensiled, (3)

Ref No 3 01 926 United States

Feed Name or Analyses			Mean As Fed	Mean Dry	C.V. ± %
Dry matter	%		33.6	100.0	2
Ash	%		3.7	10.9	5
Crude fiber	%		9.6	28.5	3
Ether extract	%		1.2	3.6	6
N-free extract	%		14.9	44.2	
Protein (N x 6.25)	%		4.3	12.8	15
Cattle	dig prot %	*	2.6	7.9	
Goats	dig prot %	*	2.6	7.9	
Horses	dig prot %	*	2.6	7.9	
Sheep	dig prot %	*	2.6	7.9	
Energy	GE Mcal/kg				
Cattle	DE Mcal/kg	*	0.96	2.87	
Sheep	DE Mcal/kg	*	0.92	2.73	
Cattle	ME Mcal/kg	*	0.79	2.35	
Sheep	ME Mcal/kg	*	0.75	2.24	
Cattle	TDN %	*	21.8	65.0	
Sheep	TDN %	*	20.8	62.0	

FESCUE, RED. Festuca rubra

Fescue, red, hay, s-c, (1)

Ref No 1 01 927 United States

Feed Name or Analyses			Mean As Fed	Mean Dry	C.V. ± %
Dry matter	%		85.4	100.0	
Ash	%		7.3	8.6	
Crude fiber	%		26.2	30.7	
Sheep	dig coef %		68.	68.	
Ether extract	%		1.9	2.2	
Sheep	dig coef %		49.	49.	
N-free extract	%		42.0	49.2	
Sheep	dig coef %		61.	61.	
Protein (N x 6.25)	%		7.9	9.3	
Sheep	dig coef %		52.	52.	
Cattle	dig prot %	*	4.3	5.0	
Goats	dig prot %	*	4.5	5.2	
Horses	dig prot %	*	4.6	5.4	
Rabbits	dig prot %	*	5.0	5.8	
Sheep	dig prot %		4.1	4.8	
Energy	GE Mcal/kg				
Cattle	DE Mcal/kg	*	2.36	2.77	
Sheep	DE Mcal/kg	*	2.19	2.56	
Cattle	ME Mcal/kg	*	1.94	2.27	
Sheep	ME Mcal/kg	*	1.80	2.10	
Cattle	TDN %	*	53.6	62.8	
Sheep	TDN %		49.7	58.1	

Ref No 1 01 927 Canada

Feed Name or Analyses			Mean As Fed	Mean Dry	C.V. ± %
Dry matter	%		93.5	100.0	1
Crude fiber	%		27.9	29.8	4
Sheep	dig coef %		68.	68.	
Ether extract	%				
Sheep	dig coef %		49.	49.	
N-free extract	%				
Sheep	dig coef %		61.	61.	
Protein (N x 6.25)	%		7.4	7.9	17
Sheep	dig coef %		52.	52.	
Cattle	dig prot %	*	3.6	3.8	
Goats	dig prot %	*	3.7	3.9	
Horses	dig prot %	*	4.0	4.3	
Rabbits	dig prot %	*	4.5	4.8	
Sheep	dig prot %		3.9	4.1	
Energy	GE Mcal/kg				
Sheep	DE Mcal/kg	*	2.40	2.57	
Sheep	ME Mcal/kg	*	1.97	2.10	
Sheep	TDN %		54.4	58.2	
Calcium	%		0.58	0.62	12
Phosphorus	%		0.07	0.08	20
Carotene	mg/kg		31.6	33.8	
Vitamin A equivalent	IU/g		52.6	56.3	

Fescue, red, aerial part, fresh, immature, (2)

Ref No 2 01 928 United States

Feed Name or Analyses			Mean As Fed	Mean Dry	C.V. ± %
Dry matter	%		29.5	100.0	
Calcium	%		0.16	0.54	
Phosphorus	%		0.13	0.44	

Fescue, red, aerial part, fresh, early bloom, (2)

Ref No 2 01 929 United States

Feed Name or Analyses			Mean As Fed	Mean Dry	C.V. ± %
Dry matter	%		29.5	100.0	
Ash	%		2.8	9.5	
Crude fiber	%		8.2	27.8	
Ether extract	%		0.9	3.1	
N-free extract	%		13.5	45.7	
Protein (N x 6.25)	%		4.1	13.9	
Cattle	dig prot %	*	2.9	9.7	
Goats	dig prot %	*	2.8	9.5	
Horses	dig prot %	*	2.8	9.3	
Rabbits	dig prot %	*	2.8	9.4	
Sheep	dig prot %	*	2.9	9.9	
Energy	GE Mcal/kg				
Cattle	DE Mcal/kg	*	0.86	2.93	
Sheep	DE Mcal/kg	*	0.83	2.80	
Cattle	ME Mcal/kg	*	0.71	2.40	
Sheep	ME Mcal/kg	*	0.68	2.29	

Feed Name or Analyses			Mean As Fed	Mean Dry	C.V. ± %
Cattle	TDN %	*	19.6	66.5	
Sheep	TDN %	*	18.7	63.5	

FESCUE, SHEEP. Festuca ovina

Fescue, sheep, hay, s-c, early bloom, (1)

Ref No 1 01 930 United States

Analyses			As Fed	Dry	C.V. ± %
Dry matter	%			100.0	
N-free extract	%			46.2	
Protein (N x 6.25)	%			13.4	8
Cattle	dig prot %	*		8.5	
Goats	dig prot %	*		9.1	
Horses	dig prot %	*		8.9	
Rabbits	dig prot %	*		9.0	
Sheep	dig prot %	*		8.6	
Calcium	%			0.49	16
Magnesium	%			0.27	22
Phosphorus	%			0.35	19
Potassium	%			1.52	6

Fescue, sheep, aerial part, fresh, (2)

Ref No 2 08 421 United States

Analyses			As Fed	Dry	C.V. ± %
Dry matter	%		33.0	100.0	5
Ash	%		2.7	8.3	8
Crude fiber	%		8.9	26.9	6
Ether extract	%		1.1	3.3	6
N-free extract	%		14.8	44.9	
Protein (N x 6.25)	%		5.5	16.7	11
Cattle	dig prot %	*	4.0	12.1	
Goats	dig prot %	*	4.0	12.1	
Horses	dig prot %	*	3.9	11.7	
Rabbits	dig prot %	*	3.8	11.5	
Sheep	dig prot %	*	4.1	12.5	
Energy	GE Mcal/kg				
Cattle	DE Mcal/kg	*	0.92	2.80	
Sheep	DE Mcal/kg	*	0.90	2.73	
Cattle	ME Mcal/kg	*	0.76	2.30	
Sheep	ME Mcal/kg	*	0.74	2.24	
Cattle	TDN %	*	20.9	63.4	
Sheep	TDN %	*	20.4	62.0	
Calcium	%		0.10	0.30	
Phosphorus	%		0.08	0.24	
Potassium	%		0.37	1.12	

Fescue, sheep, aerial part, fresh, immature, (2)

Ref No 2 01 932 United States

Analyses			As Fed	Dry	C.V. ± %
Dry matter	%		33.2	100.0	5
Ash	%		2.8	8.4	4
Crude fiber	%		9.0	27.1	5
Ether extract	%		1.1	3.3	6
N-free extract	%		14.5	43.8	
Protein (N x 6.25)	%		5.8	17.4	7
Cattle	dig prot %	*	4.2	12.7	
Goats	dig prot %	*	4.2	12.8	
Horses	dig prot %	*	4.1	12.3	
Rabbits	dig prot %	*	4.0	12.1	
Sheep	dig prot %	*	4.4	13.2	
Energy	GE Mcal/kg				
Cattle	DE Mcal/kg	*	0.90	2.70	
Sheep	DE Mcal/kg	*	0.89	2.68	
Cattle	ME Mcal/kg	*	0.73	3.21	
Sheep	ME Mcal/kg	*	0.73	2.20	
Cattle	TDN %	*	20.3	61.0	
Sheep	TDN %	*	20.2	60.7	
Calcium	%		0.16	0.49	15
Copper	mg/kg		9.7	29.1	
Iron	%		0.014	0.043	
Magnesium	%		0.05	0.16	27
Manganese	mg/kg		27.2	82.0	
Phosphorus	%		0.12	0.36	9
Potassium	%		0.61	1.85	14

Fescue, sheep, aerial part, fresh, immature, pure stand, (2)

Ref No 2 01 931 United States

Analyses			As Fed	Dry	C.V. ± %
Dry matter	%		35.7	100.0	8
Ash	%		3.1	8.7	5
Crude fiber	%		10.4	29.2	7
Ether extract	%		1.1	3.0	13
N-free extract	%		16.5	46.2	
Protein (N x 6.25)	%		4.6	12.9	20
Cattle	dig prot %	*	3.2	8.9	
Goats	dig prot %	*	3.1	8.6	
Horses	dig prot %	*	3.0	8.5	
Rabbits	dig prot %	*	3.1	8.6	
Sheep	dig prot %	*	3.2	9.0	
Energy	GE Mcal/kg				
Cattle	DE Mcal/kg	*	1.05	2.95	
Sheep	DE Mcal/kg	*	1.00	2.81	
Cattle	ME Mcal/kg	*	0.86	2.42	
Sheep	ME Mcal/kg	*	0.82	2.31	
Cattle	TDN %	*	23.9	67.0	
Sheep	TDN %	*	22.8	63.8	
Calcium	%		0.20	0.57	17
Magnesium	%		0.05	0.15	9
Phosphorus	%		0.12	0.33	18
Potassium	%		0.65	1.82	

Fescue, sheep, aerial part, fresh, early bloom, (2)

Ref No 2 01 933 United States

Analyses			As Fed	Dry	C.V. ± %
Dry matter	%		34.1	100.0	
Ash	%		2.9	8.4	
Crude fiber	%		9.9	29.0	
Ether extract	%		1.1	3.2	
N-free extract	%		15.8	46.3	
Protein (N x 6.25)	%		4.5	13.1	
Cattle	dig prot %	*	3.1	9.0	
Goats	dig prot %	*	3.0	8.8	
Horses	dig prot %	*	3.0	8.7	
Rabbits	dig prot %	*	3.0	8.8	
Sheep	dig prot %	*	3.1	9.2	
Energy	GE Mcal/kg				
Cattle	DE Mcal/kg	*	1.01	2.97	
Sheep	DE Mcal/kg	*	0.96	2.81	
Cattle	ME Mcal/kg	*	0.83	2.43	
Sheep	ME Mcal/kg	*	0.79	2.31	
Cattle	TDN %	*	22.9	67.3	
Sheep	TDN %	*	21.8	63.8	

Fescue, sheep, aerial part, fresh, full bloom, (2)

Ref No 2 01 934 United States

Analyses			As Fed	Dry	C.V. ± %
Dry matter	%			100.0	
Ash	%			6.7	
Crude fiber	%			29.8	
Ether extract	%			2.5	
N-free extract	%			50.6	
Protein (N x 6.25)	%			10.4	
Cattle	dig prot %	*		6.7	
Goats	dig prot %	*		6.3	
Horses	dig prot %	*		6.4	
Rabbits	dig prot %	*		6.7	
Sheep	dig prot %	*		6.7	
Energy	GE Mcal/kg				
Cattle	DE Mcal/kg	*		3.00	
Sheep	DE Mcal/kg	*		2.92	
Cattle	ME Mcal/kg	*		2.46	
Sheep	ME Mcal/kg	*		2.40	
Cattle	TDN %	*		68.0	
Sheep	TDN %	*		66.3	

FESCUE, SIXWEEKS. Festuca octoflora

Fescue, sixweeks, aerial part, fresh, (2)

Ref No 2 01 936 United States

Analyses			As Fed	Dry	C.V. ± %
Dry matter	%			100.0	
Ash	%			10.0	
Crude fiber	%			27.4	
Ether extract	%			2.5	
N-free extract	%			47.1	
Protein (N x 6.25)	%			13.0	
Cattle	dig prot %	*		8.9	
Goats	dig prot %	*		8.7	
Horses	dig prot %	*		8.6	
Rabbits	dig prot %	*		8.7	
Sheep	dig prot %	*		9.1	
Energy	GE Mcal/kg				
Cattle	DE Mcal/kg	*		2.89	
Sheep	DE Mcal/kg	*		2.84	
Cattle	ME Mcal/kg	*		2.37	
Sheep	ME Mcal/kg	*		2.33	
Cattle	TDN %	*		65.5	
Sheep	TDN %	*		64.4	
Calcium	%			0.44	7
Magnesium	%			0.18	
Phosphorus	%			0.22	49
Potassium	%			1.72	

Fescue, sixweeks, aerial part, fresh, immature, (2)

Ref No 2 01 935 United States

Analyses			As Fed	Dry	C.V. ± %
Dry matter	%			100.0	
Ash	%			9.1	
Crude fiber	%			23.5	
Ether extract	%			3.2	
N-free extract	%			44.5	
Protein (N x 6.25)	%			19.7	
Cattle	dig prot %	*		14.6	
Goats	dig prot %	*		14.9	
Horses	dig prot %	*		14.3	
Rabbits	dig prot %	*		13.9	
Sheep	dig prot %	*		15.4	
Energy	GE Mcal/kg				
Cattle	DE Mcal/kg	*		2.76	
Sheep	DE Mcal/kg	*		2.81	
Cattle	ME Mcal/kg	*		2.26	
Sheep	ME Mcal/kg	*		2.31	
Cattle	TDN %	*		62.6	
Sheep	TDN %	*		63.8	
Calcium	%			0.45	
Magnesium	%			0.21	
Phosphorus	%			0.37	
Potassium	%			2.35	

FESCUE, TALL ALTA. Festuca arundinacea

Fescue, tall alta, hay, s-c baled, mature, cut 4, (1)

Ref No 1 08 899 — United States

Feed Name or Analyses			Mean As Fed	Mean Dry	C.V. ± %
Dry matter	%			100.0	
Sheep	dig coef %			63.	
Crude fiber	%			25.4	
Protein (N x 6.25)	%			14.0	
Cattle	dig prot %	*		9.1	
Goats	dig prot %	*		9.6	
Horses	dig prot %	*		9.4	
Rabbits	dig prot %	*		9.5	
Sheep	dig prot %	*		9.1	
Dry matter intake	g/W^0.75			76.000	

Fescue, tall alta, hay, s-c fan-air dried w heat, mature after frost, cut 4, (1)

Ref No 1 08 900 — United States

Feed Name or Analyses			Mean As Fed	Mean Dry	C.V. ± %
Dry matter	%			100.0	
Sheep	dig coef %			63.	
Crude fiber	%			25.5	
Protein (N x 6.25)	%			14.4	
Cattle	dig prot %	*		9.4	
Goats	dig prot %	*		10.0	
Horses	dig prot %	*		9.8	
Rabbits	dig prot %	*		9.8	
Sheep	dig prot %	*		9.5	
Dry matter intake	g/W^0.75			77.140	

FESCUE, TALL KY31. Festuca arundinacea

Fescue, tall Ky31, hay, s-c, immature, cut 2, (1)

Ref No 1 08 873 — United States

Feed Name or Analyses			Mean As Fed	Mean Dry	C.V. ± %
Dry matter	%		89.6	100.0	
Sheep	dig coef %		63.	63.	
Organic matter	%		80.9	90.3	
Ash	%		8.7	9.7	
Protein (N x 6.25)	%		14.5	16.2	
Cattle	dig prot %	*	9.8	11.0	
Goats	dig prot %	*	10.5	11.7	
Horses	dig prot %	*	10.1	11.3	
Rabbits	dig prot %	*	10.0	11.2	
Sheep	dig prot %	*	9.9	11.1	
Cell walls (Van Soest)	%		52.9	59.0	
Fiber, acid detergent (VS)	%		29.0	32.4	
Lignin (Van Soest)	%		2.3	2.6	
Energy	GE Mcal/kg		3.84	4.29	
Aluminum	mg/kg		29.57	33.00	
Barium	mg/kg		13.440	15.000	
Boron	mg/kg		2.688	3.000	
Calcium	%		0.37	0.41	
Cobalt	mg/kg		2.419	2.700	
Copper	mg/kg		7.2	8.0	
Iron	%		0.006	0.006	
Magnesium	%		0.27	0.30	
Manganese	mg/kg		92.3	103.0	
Molybdenum	mg/kg		0.54	0.60	
Phosphorus	%		0.24	0.27	
Potassium	%		2.24	2.50	
Silicon	%		0.29	0.32	
Sodium	%		0.04	0.04	
Strontium	mg/kg		6.272	7.000	
Zinc	mg/kg		18.8	21.0	
Dry matter intake	g/W^0.75		63.258	70.600	

FESCUE, WESTERN. Festuca occidentalis

Fescue, western, hay, s-c, (1)

Ref No 1 08 422 — United States

Feed Name or Analyses			Mean As Fed	Mean Dry	C.V. ± %
Dry matter	%		90.0	100.0	
Ash	%		5.7	6.3	
Crude fiber	%		31.0	34.4	
Ether extract	%		2.0	2.2	
N-free extract	%		42.8	47.6	
Protein (N x 6.25)	%		8.5	9.4	
Cattle	dig prot %	*	4.6	5.1	
Goats	dig prot %	*	4.8	5.4	
Horses	dig prot %	*	5.0	5.5	
Rabbits	dig prot %	*	5.4	6.0	
Sheep	dig prot %	*	4.5	5.0	
Energy	GE Mcal/kg				
Cattle	DE Mcal/kg	*	2.16	2.40	
Sheep	DE Mcal/kg	*	2.21	2.46	
Cattle	ME Mcal/kg	*	1.77	1.97	
Sheep	ME Mcal/kg	*	1.81	2.01	
Cattle	TDN %	*	49.0	54.4	
Sheep	TDN %	*	50.1	55.7	

Fescue, alta-Clover, ladino, aerial part, fresh, (2)

Ref No 2 01 943 — United States

Feed Name or Analyses			Mean As Fed	Mean Dry	C.V. ± %
Dry matter	%		23.9	100.0	33
Ash	%		2.5	10.6	14
Crude fiber	%		4.8	20.1	19
Ether extract	%		1.3	5.6	17
N-free extract	%		10.0	41.9	
Protein (N x 6.25)	%		5.2	21.8	20
Cattle	dig prot %	*	3.9	16.4	
Goats	dig prot %	*	4.0	16.9	
Horses	dig prot %	*	3.8	16.0	
Rabbits	dig prot %	*	3.7	15.5	
Sheep	dig prot %	*	4.1	17.3	
Energy	GE Mcal/kg				
Cattle	DE Mcal/kg	*	0.72	3.02	
Sheep	DE Mcal/kg	*	0.73	3.05	
Cattle	ME Mcal/kg	*	0.59	2.48	
Sheep	ME Mcal/kg	*	0.60	2.50	
Cattle	TDN %	*	16.4	68.5	
Sheep	TDN %	*	16.5	69.2	
Calcium	%		0.18	0.77	17
Magnesium	%		0.09	0.38	
Phosphorus	%		0.08	0.33	21

Fescue, alta-Clover, ladino, aerial part, fresh, immature, (2)

Ref No 2 01 941 — United States

Feed Name or Analyses			Mean As Fed	Mean Dry	C.V. ± %
Dry matter	%		22.7	100.0	30
Ash	%		2.5	10.8	13
Crude fiber	%		4.4	19.3	17
Ether extract	%		1.3	5.7	18
N-free extract	%		9.4	41.5	
Protein (N x 6.25)	%		5.2	22.7	17
Cattle	dig prot %	*	3.9	17.2	
Goats	dig prot %	*	4.0	17.7	
Horses	dig prot %	*	3.8	16.8	
Rabbits	dig prot %	*	3.7	16.2	
Sheep	dig prot %	*	4.1	18.1	
Energy	GE Mcal/kg				
Cattle	DE Mcal/kg	*	0.69	3.04	
Sheep	DE Mcal/kg	*	0.70	3.08	
Cattle	ME Mcal/kg	*	0.57	2.49	
Sheep	ME Mcal/kg	*	0.57	2.53	
Cattle	TDN %	*	15.6	68.9	
Sheep	TDN %	*	15.9	69.9	
Calcium	%		0.18	0.81	10
Magnesium	%		0.10	0.43	
Phosphorus	%		0.08	0.34	18
Carotene	mg/kg		62.8	276.7	36
Vitamin A equivalent	IU/g		104.7	461.2	

Fescue, alta-Clover, ladino, aerial part, fresh, early bloom, (2)

Ref No 2 01 942 — United States

Feed Name or Analyses			Mean As Fed	Mean Dry	C.V. ± %
Dry matter	%		25.7	100.0	31
Ash	%		2.5	9.8	25
Crude fiber	%		5.6	21.9	14
Ether extract	%		1.5	5.7	23
N-free extract	%		11.4	44.2	
Protein (N x 6.25)	%		4.7	18.4	8
Cattle	dig prot %	*	3.5	13.5	
Goats	dig prot %	*	3.5	13.7	
Horses	dig prot %	*	3.4	13.2	
Rabbits	dig prot %	*	3.3	12.9	
Sheep	dig prot %	*	3.6	14.1	
Energy	GE Mcal/kg				
Cattle	DE Mcal/kg	*	0.75	2.91	
Sheep	DE Mcal/kg	*	0.77	3.00	
Cattle	ME Mcal/kg	*	0.61	2.39	
Sheep	ME Mcal/kg	*	0.63	2.46	
Cattle	TDN %	*	17.0	66.1	
Sheep	TDN %	*	17.5	68.1	
Calcium	%		0.19	0.73	25
Magnesium	%		0.08	0.33	
Phosphorus	%		0.07	0.29	13
Carotene	mg/kg		43.1	167.5	
Vitamin A equivalent	IU/g		71.8	279.3	

FESCUE-CLOVER, LADINO. Festuca spp, Trifolium repens

Fescue-Clover, ladino, aerial part w molasses added, ensiled, (3)

Ref No 3 01 938 — United States

Feed Name or Analyses			Mean As Fed	Mean Dry	C.V. ± %
Dry matter	%		25.7	100.0	3
Ash	%		2.3	8.8	
Crude fiber	%		7.8	30.2	
Ether extract	%		0.7	2.9	
N-free extract	%		11.9	46.4	
Protein (N x 6.25)	%		3.0	11.7	
Cattle	dig prot %	*	1.8	6.9	
Goats	dig prot %	*	1.8	6.9	
Horses	dig prot %	*	1.8	6.9	
Sheep	dig prot %	*	1.8	6.9	
Energy	GE Mcal/kg				
Cattle	DE Mcal/kg	*	0.69	2.69	
Sheep	DE Mcal/kg	*	0.72	2.80	
Cattle	ME Mcal/kg	*	0.57	2.21	
Sheep	ME Mcal/kg	*	0.59	2.30	
Cattle	TDN %	*	15.7	61.1	
Sheep	TDN %	*	16.3	63.5	

(1) dry forages and roughages (2) pasture, range plants, and forages fed green (3) silages (4) energy feeds (5) protein supplements (6) minerals (7) vitamins (8) additives

Feed Name or Analyses			Mean As Fed	Mean Dry	C.V. ± %
Calcium	%		0.26	1.03	12
Phosphorus	%		0.06	0.23	8
Carotene	mg/kg		62.1	241.6	
Vitamin A equivalent	IU/g		103.5	402.8	

Fescue-Clover, ladino, aerial part w molasses added, ensiled, full bloom, (3)

Ref No 3 01 937 United States

Analyses			As Fed	Dry	C.V. ± %
Dry matter	%		25.8	100.0	
Carotene	mg/kg		52.9	205.0	
Vitamin A equivalent	IU/g		88.2	341.8	

Fescue, forage, tall –
see Fescue, meadow, aerial part, fresh, (2)

FESCUE-GRASS. Festuca spp, Scientific name not used

Fescue-Grass, hay, s-c, (1)

Ref No 1 01 939 United States

Analyses			As Fed	Dry	C.V. ± %
Dry matter	%		85.5	100.0	
Ash	%		7.0	8.2	
Crude fiber	%		25.7	30.1	
Ether extract	%		1.7	2.0	
N-free extract	%		42.7	49.9	
Protein (N x 6.25)	%		8.4	9.8	
Cattle	dig prot %	*	4.6	5.4	
Goats	dig prot %	*	4.9	5.7	
Horses	dig prot %	*	5.0	5.9	
Rabbits	dig prot %	*	5.3	6.2	
Sheep	dig prot %	*	4.6	5.4	
Energy	GE Mcal/kg				
Cattle	DE Mcal/kg	*	2.00	2.34	
Sheep	DE Mcal/kg	*	2.17	2.54	
Cattle	ME Mcal/kg	*	1.64	1.92	
Sheep	ME Mcal/kg	*	1.78	2.08	
Cattle	TDN %	*	45.4	53.1	
Sheep	TDN %	*	49.3	57.6	

Fescue hay, tall –
see Fescue, meadow, hay, s-c, (1)

FESCUE, MEADOW-CLOVER, LADINO. Festuca elatior, Trifolium repens

Fescue, meadow-Clover, ladino, hay, s-c, (1)

Ref No 1 01 944 United States

Analyses			As Fed	Dry	C.V. ± %
Dry matter	%		90.0	100.0	
Ash	%		10.4	11.6	
Crude fiber	%		19.4	21.6	
Ether extract	%		2.0	2.2	
N-free extract	%		42.3	47.1	
Protein (N x 6.25)	%		15.8	17.6	
Cattle	dig prot %	*	10.9	12.2	
Goats	dig prot %	*	11.7	13.0	
Horses	dig prot %	*	11.2	12.5	
Rabbits	dig prot %	*	11.0	12.2	
Sheep	dig prot %	*	11.1	12.3	
Fatty acids	%		2.0	2.2	
Energy	GE Mcal/kg				
Cattle	DE Mcal/kg	*	2.63	2.92	
Sheep	DE Mcal/kg	*	2.43	2.70	
Cattle	ME Mcal/kg	*	2.15	2.39	
Sheep	ME Mcal/kg	*	2.00	2.22	

Analyses			As Fed	Dry	C.V. ± %
Cattle	TDN %	*	59.6	66.2	
Sheep	TDN %	*	55.2	61.3	

Fescue, meadow-Clover, ladino, aerial part, fresh, (2)

Ref No 2 01 947 United States

Analyses			As Fed	Dry	C.V. ± %
Dry matter	%		26.0	100.0	27
Ash	%		2.3	9.0	17
Crude fiber	%		6.1	23.3	24
Ether extract	%		1.3	5.1	23
N-free extract	%		11.9	45.7	
Protein (N x 6.25)	%		4.4	16.9	21
Cattle	dig prot %	*	3.2	12.3	
Goats	dig prot %	*	3.2	12.3	
Horses	dig prot %	*	3.1	11.9	
Rabbits	dig prot %	*	3.0	11.7	
Sheep	dig prot %	*	3.3	12.7	
Energy	GE Mcal/kg				
Cattle	DE Mcal/kg	*	0.77	2.94	
Sheep	DE Mcal/kg	*	0.72	2.77	
Cattle	ME Mcal/kg	*	0.63	2.41	
Sheep	ME Mcal/kg	*	0.59	2.27	
Cattle	TDN %	*	17.4	66.7	
Sheep	TDN %	*	16.3	62.8	
Calcium	%		0.17	0.64	13
Phosphorus	%		0.10	0.38	11

Fescue, meadow-Clover, ladino, aerial part, fresh, immature, (2)

Ref No 2 01 945 United States

Analyses			As Fed	Dry	C.V. ± %
Dry matter	%		20.8	100.0	
Ash	%		2.0	9.8	
Crude fiber	%		4.3	20.9	
Ether extract	%		1.3	6.1	
N-free extract	%		9.0	43.4	
Protein (N x 6.25)	%		4.1	19.8	
Cattle	dig prot %	*	3.1	14.7	
Goats	dig prot %	*	3.1	15.0	
Horses	dig prot %	*	3.0	14.3	
Rabbits	dig prot %	*	2.9	14.0	
Sheep	dig prot %	*	3.2	15.4	
Energy	GE Mcal/kg				
Cattle	DE Mcal/kg	*	0.61	2.95	
Sheep	DE Mcal/kg	*	0.63	3.04	
Cattle	ME Mcal/kg	*	0.50	2.42	
Sheep	ME Mcal/kg	*	0.52	2.49	
Cattle	TDN %	*	13.9	66.9	
Sheep	TDN %	*	14.3	68.9	
Calcium	%		0.14	0.65	16
Phosphorus	%		0.08	0.40	10
Carotene	mg/kg		53.4	256.8	57
Vitamin A equivalent	IU/g		89.1	428.1	

Fescue, meadow-Clover, ladino, aerial part, fresh, early bloom, (2)

Ref No 2 01 946 United States

Analyses			As Fed	Dry	C.V. ± %
Dry matter	%		37.4	100.0	
Calcium	%		0.24	0.63	
Phosphorus	%		0.12	0.33	
Carotene	mg/kg		57.6	154.1	
Vitamin A equivalent	IU/g		96.1	256.9	

FESCUE-ORCHARDGRASS. Festuca spp, Dactylis glomerata

Fescue-Orchardgrass, hay, s-c, (1)

Ref No 1 01 940 United States

Analyses			As Fed	Dry	C.V. ± %
Dry matter	%		84.6	100.0	
Ash	%		5.5	6.5	
Crude fiber	%		29.0	34.3	
Ether extract	%		1.4	1.6	
N-free extract	%		39.6	46.8	
Protein (N x 6.25)	%		9.1	10.8	
Cattle	dig prot %	*	5.3	6.3	
Goats	dig prot %	*	5.6	6.6	
Horses	dig prot %	*	5.7	6.7	
Rabbits	dig prot %	*	5.9	7.0	
Sheep	dig prot %	*	5.3	6.3	
Energy	GE Mcal/kg				
Cattle	DE Mcal/kg	*	2.01	2.38	
Sheep	DE Mcal/kg	*	2.10	2.48	
Cattle	ME Mcal/kg	*	1.65	1.95	
Sheep	ME Mcal/kg	*	1.72	2.04	
Cattle	TDN %	*	45.6	53.9	
Sheep	TDN %	*	47.7	56.3	

Fescue silage, tall –
see Fescue, meadow, aerial part, ensiled, (3)

Fescue, tall, fresh, immature –
see Fescue, meadow, aerial part, fresh, immature, (2)

FIDDLENECK, DOUGLAS. Amsinckia douglasiana

Fiddleneck, douglas, aerial part, fresh, (2)

Ref No 2 01 954 United States

Analyses			As Fed	Dry	C.V. ± %
Dry matter	%			100.0	
Ash	%			16.2	16
Crude fiber	%			23.0	29
Protein (N x 6.25)	%			13.9	40
Cattle	dig prot %	*		9.7	
Goats	dig prot %	*		9.5	
Horses	dig prot %	*		9.3	
Rabbits	dig prot %	*		9.4	
Sheep	dig prot %	*		9.9	
Calcium	%			2.28	16
Phosphorus	%			0.52	52
Potassium	%			4.42	31

Fiddleneck, douglas, aerial part, fresh, early leaf, (2)

Ref No 2 01 948 United States

Analyses			As Fed	Dry	C.V. ± %
Dry matter	%			100.0	
Ash	%			19.2	12
Crude fiber	%			13.8	23
Protein (N x 6.25)	%			22.2	14
Cattle	dig prot %	*		16.8	
Goats	dig prot %	*		17.3	
Horses	dig prot %	*		16.4	
Rabbits	dig prot %	*		15.8	
Sheep	dig prot %	*		17.7	
Calcium	%			2.34	18
Phosphorus	%			0.72	36
Potassium	%			6.25	13

Feed Name or Analyses	Mean As Fed	Mean Dry	C.V. ± %

Fiddleneck, douglas, aerial part, fresh, early bloom, (2)

Ref No 2 01 949 United States

Feed Name or Analyses	As Fed	Dry	C.V. ± %
Dry matter %		100.0	
Ash %		15.3	
Protein (N x 6.25) %		14.1	
Cattle dig prot % *		9.9	
Goats dig prot % *		9.7	
Horses dig prot % *		9.5	
Rabbits dig prot % *		9.6	
Sheep dig prot % *		10.1	
Calcium %		2.28	
Phosphorus %		0.67	
Potassium %		4.78	

Fiddleneck, douglas, aerial part, fresh, full bloom, (2)

Ref No 2 01 950 United States

Feed Name or Analyses	As Fed	Dry	C.V. ± %
Dry matter %		100.0	
Ash %		15.4	10
Crude fiber %		22.2	8
Protein (N x 6.25) %		12.6	12
Cattle dig prot % *		8.6	
Goats dig prot % *		8.3	
Horses dig prot % *		8.2	
Rabbits dig prot % *		8.4	
Sheep dig prot % *		8.7	
Calcium %		2.26	12
Phosphorus %		0.50	31
Potassium %		3.88	22

Fiddleneck, douglas, aerial part, fresh, milk stage, (2)

Ref No 2 01 951 United States

Feed Name or Analyses	As Fed	Dry	C.V. ± %
Dry matter %		100.0	
Calcium %		2.46	
Phosphorus %		0.50	
Potassium %		3.70	

Fiddleneck, douglas, aerial part, fresh, dough stage, (2)

Ref No 2 01 952 United States

Feed Name or Analyses	As Fed	Dry	C.V. ± %
Dry matter %		100.0	
Ash %		16.1	
Crude fiber %		24.4	
Protein (N x 6.25) %		11.5	
Cattle dig prot % *		7.7	
Goats dig prot % *		7.3	
Horses dig prot % *		7.3	
Rabbits dig prot % *		7.5	
Sheep dig prot % *		7.7	

Fiddleneck, douglas, aerial part, fresh, dormant, (2)

Ref No 2 01 953 United States

Feed Name or Analyses	As Fed	Dry	C.V. ± %
Dry matter %		100.0	
Ash %		13.3	7
Crude fiber %		30.9	9
Protein (N x 6.25) %		4.5	19
Cattle dig prot % *		1.7	
Goats dig prot % *		0.8	
Horses dig prot % *		1.4	
Rabbits dig prot % *		2.1	
Sheep dig prot % *		1.2	
Calcium %		2.21	19
Phosphorus %		0.30	53
Potassium %		3.09	16

FIGS, SYCOMORE. Ficus sycomorus

Figs, sycomore, leaves, fresh, (2)

Ref No 2 01 956 United States

Feed Name or Analyses	As Fed	Dry	C.V. ± %
Dry matter %	29.5	100.0	
Ash %	4.5	15.4	
Crude fiber %	4.2	14.3	
Sheep dig coef %	32.	32.	
Ether extract %	0.6	2.1	
Sheep dig coef %	59.	59.	
N-free extract %	16.2	54.8	
Sheep dig coef %	57.	57.	
Protein (N x 6.25) %	4.0	13.4	
Sheep dig coef %	18.	18.	
Cattle dig prot % *	2.7	9.3	
Goats dig prot % *	2.7	9.1	
Horses dig prot % *	2.6	8.9	
Rabbits dig prot % *	2.7	9.0	
Sheep dig prot %	0.7	2.4	
Energy GE Mcal/kg			
Cattle DE Mcal/kg *	0.55	1.85	
Sheep DE Mcal/kg *	0.53	1.81	
Cattle ME Mcal/kg *	0.45	1.52	
Sheep ME Mcal/kg *	0.44	1.48	
Cattle TDN % *	12.4	42.0	
Sheep TDN %	12.1	41.0	

FINGERGRASS. Digitaria spp

Fingergrass, hay, s-c, (1)

Ref No 1 01 957 United States

Feed Name or Analyses	As Fed	Dry	C.V. ± %
Dry matter %	93.4	100.0	
Ash %	5.9	6.3	
Crude fiber %	34.5	36.9	
Sheep dig coef %	68.	68.	
Ether extract %	1.9	2.0	
Sheep dig coef %	41.	41.	
N-free extract %	42.3	45.3	
Sheep dig coef %	60.	60.	
Protein (N x 6.25) %	8.9	9.5	
Sheep dig coef %	56.	56.	
Cattle dig prot % *	4.8	5.2	
Goats dig prot % *	5.1	5.4	
Horses dig prot % *	5.2	5.6	
Rabbits dig prot % *	5.6	6.0	
Sheep dig prot %	5.0	5.3	
Energy GE Mcal/kg			
Cattle DE Mcal/kg *	2.30	2.46	
Sheep DE Mcal/kg *	2.45	2.62	
Cattle ME Mcal/kg *	1.89	2.02	
Sheep ME Mcal/kg *	2.01	2.15	
Cattle TDN % *	52.0	55.7	
Sheep TDN %	55.5	59.4	
Calcium %	0.35	0.37	
Iron %	0.007	0.007	
Magnesium %	0.58	0.62	
Phosphorus %	0.25	0.27	

Fingergrass, aerial part, fresh, pre-bloom, (2)

Ref No 2 01 958 United States

Feed Name or Analyses	As Fed	Dry	C.V. ± %
Dry matter %	18.0	100.0	
Ash %	1.6	8.9	
Crude fiber %	5.4	30.2	
Sheep dig coef %	78.	78.	
Ether extract %	0.6	3.2	
Sheep dig coef %	71.	71.	
N-free extract %	7.7	42.8	
Sheep dig coef %	76.	76.	
Protein (N x 6.25) %	2.7	14.9	
Sheep dig coef %	78.	78.	
Cattle dig prot % *	1.9	10.6	
Goats dig prot % *	1.9	10.5	
Horses dig prot % *	1.8	10.2	
Rabbits dig prot % *	1.8	10.2	
Sheep dig prot %	2.1	11.6	
Energy GE Mcal/kg			
Cattle DE Mcal/kg *	0.55	3.03	
Sheep DE Mcal/kg *	0.58	3.21	
Cattle ME Mcal/kg *	0.45	2.49	
Sheep ME Mcal/kg *	0.47	2.63	
Cattle TDN % *	12.4	68.8	
Sheep TDN %	13.1	72.8	

Fingergrass, aerial part, fresh, early bloom, (2)

Ref No 2 01 959 United States

Feed Name or Analyses	As Fed	Dry	C.V. ± %
Dry matter %	18.0	100.0	
Ash %	1.2	6.6	
Crude fiber %	6.0	33.6	
Sheep dig coef %	73.	73.	
Ether extract %	0.4	2.0	
Sheep dig coef %	34.	34.	
N-free extract %	8.7	48.6	
Sheep dig coef %	69.	69.	
Protein (N x 6.25) %	1.7	9.2	
Sheep dig coef %	57.	57.	
Cattle dig prot % *	1.0	5.7	
Goats dig prot % *	0.9	5.1	
Horses dig prot % *	1.0	5.3	
Rabbits dig prot % *	1.0	5.8	
Sheep dig prot %	0.9	5.2	
Energy GE Mcal/kg			
Cattle DE Mcal/kg *	0.54	3.01	
Sheep DE Mcal/kg *	0.51	2.86	
Cattle ME Mcal/kg *	0.44	2.47	
Sheep ME Mcal/kg *	0.42	2.34	
Cattle TDN % *	12.3	68.3	
Sheep TDN %	11.7	64.8	

Fingergrass, aerial part, fresh, late bloom, (2)

Ref No 2 01 960 United States

Feed Name or Analyses	As Fed	Dry	C.V. ± %
Dry matter %	20.0	100.0	
Ash %	0.9	4.4	
Crude fiber %	7.6	38.0	
Sheep dig coef %	66.	66.	
Ether extract %	0.5	2.4	
Sheep dig coef %	55.	55.	
N-free extract %	9.6	47.9	
Sheep dig coef %	56.	56.	

(1) dry forages and roughages (2) pasture, range plants, and forages fed green (3) silages (4) energy feeds (5) protein supplements (6) minerals (7) vitamins (8) additives

Feed Name or Analyses			Mean As Fed	Dry	C.V. ± %
Protein (N x 6.25)	%		1.5	7.3	
Sheep	dig coef %		51.	51.	
Cattle	dig prot %	*	0.8	4.1	
Goats	dig prot %	*	0.7	3.4	
Horses	dig prot %	*	0.7	3.7	
Rabbits	dig prot %	*	0.9	4.3	
Sheep	dig prot %		0.7	3.7	
Energy	GE Mcal/kg				
Cattle	DE Mcal/kg	*	0.53	2.63	
Sheep	DE Mcal/kg	*	0.52	2.58	
Cattle	ME Mcal/kg	*	0.43	2.16	
Sheep	ME Mcal/kg	*	0.42	2.12	
Cattle	TDN %	*	11.9	59.7	
Sheep	TDN %		11.7	58.6	

FINGERGRASS, PENTZ. Digitaria spp

Fingergrass, pentz, aerial part, fresh, (2)

Ref No 2 01 961 United States

Feed Name or Analyses			Mean As Fed	Dry	C.V. ± %
Dry matter	%		18.0	100.0	
Ash	%		1.2	6.6	
Crude fiber	%		6.1	33.9	
Sheep	dig coef %		72.	72.	
Ether extract	%		0.5	2.5	
Sheep	dig coef %		53.	53.	
N-free extract	%		8.4	46.5	
Sheep	dig coef %		67.	67.	
Protein (N x 6.25)	%		1.9	10.5	
Sheep	dig coef %		62.	62.	
Cattle	dig prot %	*	1.2	6.8	
Goats	dig prot %	*	1.1	6.4	
Horses	dig prot %	*	1.2	6.4	
Rabbits	dig prot %	*	1.2	6.8	
Sheep	dig prot %		1.2	6.5	
Energy	GE Mcal/kg				
Cattle	DE Mcal/kg	*	0.55	3.04	
Sheep	DE Mcal/kg	*	0.52	2.87	
Cattle	ME Mcal/kg	*	0.45	2.50	
Sheep	ME Mcal/kg	*	0.42	2.35	
Cattle	TDN %	*	12.4	69.0	
Sheep	TDN %		11.7	65.1	

FINGERHUTHIA, SESLERIAEFORMIS. Fingerhuthia sesleriaeformis

Fingerhuthia, sesleriaeformis, seeds w hulls, (5)

Ref No 5 09 299 United States

Feed Name or Analyses		Mean As Fed	Dry	C.V. ± %
Dry matter	%		100.0	
Protein (N x 6.25)	%		33.0	
Alanine	%		1.65	
Arginine	%		1.02	
Aspartic acid	%		1.78	
Glutamic acid	%		10.92	
Glycine	%		0.86	
Histidine	%		0.56	
Hydroxyproline	%		0.00	
Isoleucine	%		1.32	
Leucine	%		2.94	
Lysine	%		0.59	
Methionine	%		0.66	
Phenylalanine	%		1.58	
Proline	%		1.75	
Serine	%		1.09	
Threonine	%		1.12	
Tyrosine	%		1.02	
Valine	%		1.78	

FIR. Abies spp

Fir, needles, extn unspecified grnd, (1)

Ref No 1 01 962 United States

Feed Name or Analyses			Mean As Fed	Dry	C.V. ± %
Dry matter	%		79.9	100.0	
Ash	%		2.1	2.6	
Crude fiber	%		37.3	46.7	
Sheep	dig coef %		43.	43.	
Ether extract	%		4.9	6.1	
Sheep	dig coef %		24.	24.	
N-free extract	%		27.0	33.8	
Sheep	dig coef %		25.	25.	
Protein (N x 6.25)	%		8.6	10.8	
Sheep	dig coef %		-5.	-5.	
Cattle	dig prot %	*	5.0	6.3	
Goats	dig prot %	*	5.3	6.6	
Horses	dig prot %	*	5.4	6.7	
Rabbits	dig prot %	*	5.6	7.0	
Sheep	dig prot %		-0.4	-0.5	
Energy	GE Mcal/kg				
Cattle	DE Mcal/kg	*	1.19	1.49	
Sheep	DE Mcal/kg	*	1.10	1.37	
Cattle	ME Mcal/kg	*	0.98	1.22	
Sheep	ME Mcal/kg	*	0.90	1.13	
Cattle	TDN %	*	27.0	33.8	
Sheep	TDN %		24.9	31.2	

Fir, needles, grnd, (1)

Ref No 1 01 963 United States

Feed Name or Analyses			Mean As Fed	Dry	C.V. ± %
Dry matter	%		86.8	100.0	
Ash	%		3.0	3.5	
Crude fiber	%		35.6	41.0	
Sheep	dig coef %		36.	36.	
Ether extract	%		6.4	7.4	
Sheep	dig coef %		48.	48.	
N-free extract	%		33.0	38.0	
Sheep	dig coef %		29.	29.	
Protein (N x 6.25)	%		8.8	10.1	
Sheep	dig coef %		52.	52.	
Cattle	dig prot %	*	4.9	5.7	
Goats	dig prot %	*	5.2	6.0	
Horses	dig prot %	*	5.3	6.1	
Rabbits	dig prot %	*	5.6	6.5	
Sheep	dig prot %		4.6	5.3	
Energy	GE Mcal/kg				
Cattle	DE Mcal/kg	*	1.74	2.01	
Sheep	DE Mcal/kg	*	1.49	1.72	
Cattle	ME Mcal/kg	*	1.43	1.65	
Sheep	ME Mcal/kg	*	1.22	1.41	
Cattle	TDN %	*	39.5	45.5	
Sheep	TDN %		33.9	39.0	

FIREWEED. Epilobium angustifolium

Fireweed, leaves, fresh, over ripe, (2)

Ref No 2 01 964 United States

Feed Name or Analyses			Mean As Fed	Dry	C.V. ± %
Dry matter	%			100.0	
Ash	%			7.6	
Crude fiber	%			12.7	
Ether extract	%			7.1	
N-free extract	%			69.8	
Protein (N x 6.25)	%			2.8	
Cattle	dig prot %	*		0.3	
Goats	dig prot %	*		-0.7	
Horses	dig prot %	*		0.0	
Rabbits	dig prot %	*		0.8	
Sheep	dig prot %	*		-0.3	

FISH. Scientific name not used

Fish, glue by-product, dehy grnd, salt declared above 3% mx 7%, (5)

Fish residue meal (AAFCO)

Ref No 5 01 966 United States

Feed Name or Analyses			Mean As Fed	Dry	C.V. ± %
Dry matter	%		94.3	100.0	
Ash	%		22.6	24.0	
Ether extract	%		2.6	2.8	
Sheep	dig coef %		101.	101.	
Protein (N x 6.25)	%		69.0	73.2	
Sheep	dig coef %		75.	75.	
Sheep	dig prot %		51.8	54.9	
Energy	GE Mcal/kg				
Sheep	DE Mcal/kg	*	2.55	2.70	
Sheep	ME Mcal/kg	*	2.09	2.22	
Sheep	TDN %		57.8	61.3	

Fish, livers, extn unspecified dehy grnd, salt declared above 4%, (5)

Fish liver meal (CFA)

Ref No 5 01 968 United States

Feed Name or Analyses			Mean As Fed	Dry	C.V. ± %
Dry matter	%		92.8	100.0	
Ash	%		6.1	6.6	
Crude fiber	%		1.2	1.3	
Ether extract	%		17.3	18.6	
N-free extract	%		5.4	5.8	
Protein (N x 6.25)	%		62.8	67.7	
Energy	GE Mcal/kg				
Cattle	DE Mcal/kg	*	3.58	3.86	
Sheep	DE Mcal/kg	*	3.50	3.77	
Swine	DE kcal/kg	*	4167.	4490.	
Cattle	ME Mcal/kg	*	2.94	3.17	
Sheep	ME Mcal/kg	*	2.87	3.09	
Swine	ME kcal/kg	*	3431.	3697.	
Cattle	TDN %	*	81.2	87.5	
Sheep	TDN %	*	79.3	85.5	
Swine	TDN %	*	94.5	101.8	

Fish, stickwater solubles, condensed, mn 30% protein, (5)

Condensed fish solubles (AAFCO)

Ref No 5 01 969 United States

Feed Name or Analyses			Mean As Fed	Dry	C.V. ± %
Dry matter	%		50.1	100.0	
Ash	%		9.4	18.8	
Crude fiber	%		0.1	0.2	
Ether extract	%		7.7	15.4	
Swine	dig coef %	*	99.	99.	
N-free extract	%		2.6	5.2	
Protein (N x 6.25)	%		30.3	60.5	
Swine	dig coef %	*	96.	96.	
Swine	dig prot %		29.1	58.1	
Energy	GE Mcal/kg				
Cattle	DE Mcal/kg	*	1.82	3.64	
Sheep	DE Mcal/kg	*	1.68	3.35	
Swine	DE kcal/kg	*	1965.	3922.	

Continued

Feed Name or Analyses		Mean As Fed	Mean Dry	C.V. ± %
Cattle	ME Mcal/kg	* 1.50	2.98	
Chickens	ME_n kcal/kg	1477.	2948.	
Sheep	ME Mcal/kg	* 1.37	2.74	
Swine	ME kcal/kg	*1646.	3286.	
Cattle	TDN %	* 41.4	82.6	
Sheep	TDN %	* 38.0	75.9	
Swine	TDN %	* 44.6	89.0	
Calcium	%	0.17	0.34	
Chlorine	%	2.62	5.23	
Copper	mg/kg	41.7	83.2	
Iron	%	0.003	0.006	
Magnesium	%	0.02	0.04	
Manganese	mg/kg	16.8	33.4	
Phosphorus	%	0.82	1.64	
Sodium	%	2.13	4.25	
Sulphur	%	0.12	0.24	
Choline	mg/kg	2998.	5985.	
Niacin	mg/kg	169.1	337.5	
Pantothenic acid	mg/kg	35.5	70.8	
Riboflavin	mg/kg	14.6	29.0	
Arginine	%	1.50	2.99	
Cystine	%	0.40	0.80	
Glycine	%	3.00	5.99	
Lysine	%	1.60	3.19	
Methionine	%	0.80	1.60	
Tryptophan	%	0.30	0.60	

Fish, stickwater solubles, cooked condensed, (5)
Condensed fish solubles (CFA)
Ref No 5 01 970 United States

Feed Name or Analyses		Mean As Fed	Mean Dry	C.V. ± %
Dry matter	%	54.4	100.0	12
Ash	%	13.1	24.0	23
Crude fiber	%	0.6	1.2	47
Ether extract	%	5.1	9.3	74
Sheep	dig coef %	80.	80.	
Swine	dig coef %	99.	99.	
N-free extract	%	-0.0	-0.2	
Protein (N x 6.25)	%	35.7	65.7	10
Sheep	dig coef %	89.	89.	
Swine	dig coef %	96.	96.	
Sheep	dig prot %	31.8	58.5	
Swine	dig prot %	34.3	63.1	
Energy	GE Mcal/kg			
Cattle	DE Mcal/kg	* 1.64	3.02	
Sheep	DE Mcal/kg	* 1.81	3.32	
Swine	DE kcal/kg	*1911.	3514.	
Cattle	ME Mcal/kg	* 1.35	2.48	
Sheep	ME Mcal/kg	* 1.48	2.73	
Swine	ME kcal/kg	*1580.	2907.	
Cattle	TDN %	* 37.3	68.5	
Sheep	TDN %	41.0	75.4	
Swine	TDN %	43.3	79.7	
Calcium	%	0.66	1.21	98
Chlorine	%	4.08	7.51	37
Cobalt	mg/kg	0.076	0.140	77
Copper	mg/kg	52.0	95.6	59
Iodine	mg/kg	1.194	2.196	
Iron	%	0.037	0.067	87
Magnesium	%	0.02	0.04	
Manganese	mg/kg	12.8	23.6	84
Phosphorus	%	0.75	1.39	37
Potassium	%	1.88	3.47	19
Sodium	%	3.29	6.06	61
Sulphur	%	0.13	0.24	
Zinc	mg/kg	41.3	76.0	9
Biotin	mg/kg	0.17	0.31	
Carotene	mg/kg	1.4	2.6	
Choline	mg/kg	4346.	7993.	
Niacin	mg/kg	182.0	334.8	60
Pantothenic acid	mg/kg	38.2	70.3	26
Riboflavin	mg/kg	15.7	28.8	51
Thiamine	mg/kg	5.9	10.9	50
Vitamin B_6	mg/kg	13.05	24.01	8
Vitamin B_{12}	µg/kg	711.1	1307.9	
Vitamin A equivalent	IU/g	2.4	4.4	
Arginine	%	2.58	4.75	30
Cystine	%	1.83	3.37	84
Glutamic acid	%	6.46	11.88	4
Glycine	%	5.28	9.70	25
Histidine	%	2.69	4.95	39
Isoleucine	%	1.72	3.17	32
Leucine	%	2.69	4.95	30
Lysine	%	2.91	5.35	33
Methionine	%	1.08	1.98	26
Phenylalanine	%	1.51	2.77	23
Serine	%	1.18	2.18	41
Threonine	%	1.29	2.38	27
Tryptophan	%	0.86	1.58	61
Tyrosine	%	0.54	0.99	48
Valine	%	1.72	3.17	22

Fish, stickwater solubles, cooked dehy, mn 60% protein, (5)
Dried fish solubles (AAFCO)
Fish solubles, dried
Ref No 5 01 971 United States

Feed Name or Analyses		Mean As Fed	Mean Dry	C.V. ± %
Dry matter	%	93.7	100.0	1
Ash	%	14.8	15.8	72
Crude fiber	%	0.5	0.5	98
Ether extract	%	9.3	9.9	26
N-free extract	%	3.5	3.8	
Protein (N x 6.25)	%	65.5	69.9	14
Linoleic	%	.120	.128	
Energy	GE Mcal/kg			
Cattle	DE Mcal/kg	* 3.25	3.46	
Sheep	DE Mcal/kg	* 3.04	3.24	
Swine	DE kcal/kg	*3232.	3449.	
Cattle	ME Mcal/kg	* 2.66	2.84	
Chickens	ME_n kcal/kg	2850.	3041.	
Sheep	ME Mcal/kg	* 2.49	2.66	
Swine	ME kcal/kg	*2646.	2824.	
Cattle	TDN %	* 73.6	78.6	
Sheep	TDN %	* 68.8	73.5	
Swine	TDN %	* 73.3	78.2	
Calcium	%	1.27	1.36	59
Phosphorus	%	1.69	1.80	34
Choline	mg/kg	5535.	5907.	
Niacin	mg/kg	280.1	298.9	21
Pantothenic acid	mg/kg	56.8	60.6	
Riboflavin	mg/kg	17.8	19.0	
Arginine	%	2.66	2.84	3
Cystine	%	0.76	0.81	
Glycine	%	5.71	6.10	
Histidine	%	2.66	2.84	72
Isoleucine	%	1.74	1.86	24
Leucine	%	2.76	2.95	72
Lysine	%	3.06	3.27	57
Methionine	%	1.22	1.30	27
Phenylalanine	%	1.33	1.42	31
Serine	%	2.05	2.19	
Threonine	%	1.23	1.31	13
Tryptophan	%	0.64	0.69	98
Tyrosine	%	0.72	0.77	58
Valine	%	1.95	2.08	51

Fish, viscera, dehy grnd, mn 50% liver mn 40% mg riboflavin per kg, (5)
Fish liver and glandular meal (AAFCO)
Ref No 5 01 973 United States

Feed Name or Analyses		Mean As Fed	Mean Dry	C.V. ± %
Dry matter	%	87.7	100.0	4
Ash	%	6.5	7.4	
Crude fiber	%	18.1	20.6	
Ether extract	%	15.4	17.6	49
N-free extract	%	0.6	0.7	
Protein (N x 6.25)	%	47.1	53.7	
Energy	GE Mcal/kg			
Cattle	DE Mcal/kg	* 2.92	3.33	
Sheep	DE Mcal/kg	* 2.71	3.09	
Swine	DE kcal/kg	*3349.	3819.	
Cattle	ME Mcal/kg	* 2.40	2.73	
Sheep	ME Mcal/kg	* 2.22	2.53	
Swine	ME kcal/kg	*2852.	3252.	
Cattle	TDN %	* 66.3	75.5	
Sheep	TDN %	* 61.4	70.0	
Swine	TDN %	* 76.0	86.6	
Riboflavin	mg/kg	34.4	39.2	13

Fish, whole or cutting, cooked dehy grnd, mn 9% oil salt declared above 4%, (5)
Oily fish meal (CFA)
Ref No 5 01 974 Canada

Feed Name or Analyses		Mean As Fed	Mean Dry	C.V. ± %
Dry matter	%	93.4	100.0	
Ether extract	%	14.1	15.1	
Protein (N x 6.25)	%	55.8	59.7	
Salt (NaCl)	%	0.50	0.53	

Fish, whole or cuttings, cooked extn unspecified dehy grnd, mx 5% fat, (5)
Ref No 5 01 983 United States

Feed Name or Analyses		Mean As Fed	Mean Dry	C.V. ± %
Dry matter	%	85.8	100.0	
Ash	%	26.3	30.7	
Ether extract	%	2.3	2.7	
Sheep	dig coef %	92.	92.	
Protein (N x 6.25)	%	56.9	66.3	
Sheep	dig coef %	90.	90.	
Sheep	dig prot %	51.2	59.7	
Energy	GE Mcal/kg			
Sheep	DE Mcal/kg	* 2.47	2.87	
Sheep	ME Mcal/kg	* 2.02	2.36	
Sheep	TDN %	55.9	65.2	

Fish, whole or cuttings, cooked extn unspecified dehy grnd, 5-9% fat, (5)
Ref No 5 01 984 United States

Feed Name or Analyses		Mean As Fed	Mean Dry	C.V. ± %
Dry matter	%	80.2	100.0	
Ash	%	19.3	24.1	
Ether extract	%	6.0	7.5	
Sheep	dig coef %	96.	96.	
Protein (N x 6.25)	%	52.5	65.5	
Sheep	dig coef %	87.	87.	
Sheep	dig prot %	45.7	57.0	
Energy	GE Mcal/kg			
Sheep	DE Mcal/kg	* 2.59	3.23	

(1) dry forages and roughages (2) pasture, range plants, and forages fed green (3) silages (4) energy feeds (5) protein supplements (6) minerals (7) vitamins (8) additives

Feed Name or Analyses		Mean As Fed	Mean Dry	C.V. ± %
Sheep	ME Mcal/kg	* 2.12	2.65	
Sheep	TDN %	58.7	73.2	

Fish, whole or cuttings, cooked homogenized condensed, mn 50% solids mx 7% salt, (5)

Ref No 5 01 975 United States

Analyses		As Fed	Dry	C.V. ± %
Dry matter	%	52.2	100.0	5
Ash	%	14.5	27.8	15
Crude fiber	%	0.1	0.2	41
Ether extract	%	8.4	16.1	28
N-free extract	%	0.1	0.2	
Protein (N x 6.25)	%	29.1	55.7	7
Energy	GE Mcal/kg			
Cattle	DE Mcal/kg	* 1.70	3.26	
Sheep	DE Mcal/kg	* 1.59	3.04	
Swine	DE kcal/kg	*1918.	3675.	
Cattle	ME Mcal/kg	* 1.39	2.67	
Sheep	ME Mcal/kg	* 1.30	2.49	
Swine	ME kcal/kg	*1625.	3114.	
Cattle	TDN %	* 38.5	73.8	
Sheep	TDN %	* 36.0	68.9	
Swine	TDN %	* 43.5	83.3	
Calcium	%	1.60	3.07	33
Chlorine	%	5.21	9.98	
Phosphorus	%	0.98	1.88	18
Choline	mg/kg	1755.	3362.	14
Niacin	mg/kg	38.1	73.1	21
Pantothenic acid	mg/kg	9.5	18.2	33
Riboflavin	mg/kg	2.9	5.5	48

Fish, whole or cuttings, cooked mech-extd dehy grnd, (5)

Fish meal (CFA)

Ref No 5 01 976 United States

Analyses		As Fed	Dry	C.V. ± %
Dry matter	%	92.3	100.0	
Protein (N x 6.25)	%	67.3	72.9	
Energy	GE Mcal/kg	4.02	4.36	
Chickens	ME_n kcal/kg	2607.	2825.	
Pantothenic acid	mg/kg	15.7	17.0	

Ref No 5 01 976 Canada

Analyses		As Fed	Dry	C.V. ± %
Dry matter	%	91.7	100.0	2
Ash	%	23.1	25.2	
Crude fiber	%	1.2	1.3	98
Ether extract	%	5.9	6.4	40
N-free extract	%	-4.1	-4.5	
Protein (N x 6.25)	%	65.7	71.6	7
Energy	GE Mcal/kg	4.00	4.36	
Cattle	DE Mcal/kg	* 2.56	2.79	
Sheep	DE Mcal/kg	* 2.50	2.73	
Swine	DE kcal/kg	*2486.	2713.	
Cattle	ME Mcal/kg	* 2.10	2.29	
Chickens	ME_n kcal/kg	2589.	2825.	
Sheep	ME Mcal/kg	* 2.05	2.24	
Swine	ME kcal/kg	*2027.	2212.	
Cattle	TDN %	* 58.0	63.3	
Sheep	TDN %	* 56.7	61.8	
Swine	TDN %	* 56.4	61.5	
Calcium	%	6.89	7.52	31
Phosphorus	%	3.69	4.02	21
Salt (NaCl)	%	1.88	2.05	41
Niacin	mg/kg	83.8	91.4	
Pantothenic acid	mg/kg	17.8	19.4	
Riboflavin	mg/kg	4.8	5.3	

Feed Name or Analyses		Mean As Fed	Mean Dry	C.V. ± %
Thiamine	mg/kg	0.5	0.5	
Tryptophan	%	0.88	0.96	

Fish, whole or cuttings, cooked mech-extd dehy grnd, mn oil specified, (5)

Ref No 5 01 978 United States

Analyses		As Fed	Dry	C.V. ± %
Dry matter	%	90.2	100.0	
Ash	%	15.1	16.7	
Ether extract	%	10.6	11.8	
Sheep	dig coef %	99.	99.	
N-free extract	%	2.3	2.6	
Sheep	dig coef %	15.	15.	
Protein (N x 6.25)	%	62.1	68.9	
Sheep	dig coef %	89.	89.	
Sheep	dig prot %	55.3	61.3	
Energy	GE Mcal/kg			
Cattle	DE Mcal/kg	* 3.19	3.54	
Sheep	DE Mcal/kg	* 3.50	3.88	
Swine	DE kcal/kg	*3236.	3588.	
Cattle	ME Mcal/kg	* 2.62	2.90	
Sheep	ME Mcal/kg	* 2.87	3.18	
Swine	ME kcal/kg	*2656.	2945.	
Cattle	TDN %	* 72.4	80.3	
Sheep	TDN %	79.4	88.0	
Swine	TDN %	* 73.4	81.4	

Fish, whole or cuttings, cooked mech-extd dehy grnd, salt declared above 3% mx 7%, (5)

Fish meal (AAFCO)

Ref No 5 01 977 United States

Analyses		As Fed	Dry	C.V. ± %
Dry matter	%	88.4	100.0	2
Ash	%	20.2	22.8	19
Crude fiber	%	1.0	1.1	72
Ether extract	%	5.6	6.3	39
Sheep	dig coef %	95.	95.	
Swine	dig coef %	81.	81.	
N-free extract	%	2.2	2.5	
Protein (N x 6.25)	%	59.4	67.2	9
Sheep	dig coef %	89.	89.	
Swine	dig coef %	92.	92.	
Sheep	dig prot %	52.9	59.8	
Swine	dig prot %	54.7	61.8	
Energy	GE Mcal/kg			
Cattle	DE Mcal/kg	* 2.60	2.94	
Sheep	DE Mcal/kg	* 2.87	3.25	
Swine	DE kcal/kg	*2802.	3170.	
Cattle	ME Mcal/kg	* 2.13	2.41	
Chickens	ME_n kcal/kg	2359.	2669.	
Sheep	ME Mcal/kg	* 2.35	2.66	
Swine	ME kcal/kg	*2309.	2613.	
Cattle	TDN %	* 58.9	66.7	
Sheep	TDN %	65.0	73.6	
Swine	TDN %	63.5	71.9	
Calcium	%	5.48	6.20	41
Chlorine	%	1.21	1.37	98
Cobalt	mg/kg	0.106	0.119	59
Copper	mg/kg	14.6	16.5	22
Iron	%	0.036	0.041	48
Magnesium	%	0.21	0.24	18
Manganese	mg/kg	22.8	25.8	96
Phosphorus	%	3.33	3.77	27
Potassium	%	0.39	0.44	24
Sodium	%	1.07	1.21	32
Sulphur	%	0.24	0.27	
Choline	mg/kg	3510.	3972.	25
Niacin	mg/kg	60.8	68.8	19
Pantothenic acid	mg/kg	8.7	9.8	33
Riboflavin	mg/kg	6.5	7.4	44
Thiamine	mg/kg	1.3	1.4	92
α-tocopherol	mg/kg	18.5	20.9	
Vitamin B_6	mg/kg	14.14	16.00	43
Vitamin B_{12}	μg/kg	249.5	282.3	21
Arginine	%	3.73	4.23	17
Cystine	%	0.57	0.65	13
Glutamic acid	%	8.04	9.10	
Glycine	%	3.93	4.44	12
Histidine	%	1.53	1.73	25
Isoleucine	%	3.64	4.12	17
Leucine	%	4.69	5.31	23
Lysine	%	5.17	5.85	34
Methionine	%	1.72	1.95	22
Phenylalanine	%	2.68	3.03	23
Threonine	%	2.49	2.82	17
Tryptophan	%	0.67	0.76	27
Tyrosine	%	1.91	2.17	36
Valine	%	3.26	3.68	19

Fish, whole or cuttings, cooked mech-extd dehy grnd, 50% protein, (5)

Ref No 5 01 979 United States

Analyses		As Fed	Dry	C.V. ± %
Dry matter	%	88.1	100.0	
Ash	%	27.0	30.6	
Crude fiber	%	0.7	0.8	
Ether extract	%	5.8	6.6	
Sheep	dig coef %	95.	95.	
Swine	dig coef %	66.	66.	
N-free extract	%	2.8	3.1	
Protein (N x 6.25)	%	51.9	58.9	
Sheep	dig coef %	89.	89.	
Swine	dig coef %	92.	92.	
Sheep	dig prot %	46.2	52.4	
Swine	dig prot %	47.8	54.2	
Energy	GE Mcal/kg			
Cattle	DE Mcal/kg	* 2.32	2.63	
Sheep	DE Mcal/kg	* 2.55	2.89	
Swine	DE kcal/kg	*2561.	2906.	
Cattle	ME Mcal/kg	* 1.90	2.16	
Chickens	ME_n kcal/kg	1899.	2154.	
Sheep	ME Mcal/kg	* 2.09	2.37	
Swine	ME kcal/kg	*2154.	2444.	
Cattle	TDN %	* 52.6	59.7	
Sheep	TDN %	57.8	65.6	
Swine	TDN %	58.1	65.9	
Calcium	%	7.27	8.25	
Phosphorus	%	3.25	3.69	
Choline	mg/kg	1920.	2178.	
Niacin	mg/kg	25.3	28.7	
Pantothenic acid	mg/kg	3.6	4.1	
Riboflavin	mg/kg	4.2	4.8	
Arginine	%	3.35	3.80	
Cystine	%	0.61	0.69	
Glycine	%	3.83	4.34	
Lysine	%	3.25	3.69	
Methionine	%	0.62	0.71	
Tryptophan	%	0.33	0.38	

Fish, whole or cuttings, cooked mech-extd dehy grnd, 55% protein, (5)

Ref No 5 01 980 United States

Analyses		As Fed	Dry	C.V. ± %
Dry matter	%	88.5	100.0	
Ash	%	23.7	26.8	

Continued

Feed Name or Analyses		Mean As Fed	Mean Dry	C.V. ± %
Ether extract	%	5.7	6.4	
Sheep	dig coef %	101.	101.	
Protein (N x 6.25)	%	56.6	64.0	
Sheep	dig coef %	88.	88.	
Sheep	dig prot %	49.8	56.3	
Energy	GE Mcal/kg			
Sheep	DE Mcal/kg	* 2.77	3.13	
Sheep	ME Mcal/kg	* 2.27	2.56	
Sheep	TDN %	62.7	70.9	

Fish, whole or cuttings, cooked mech-extd dehy grnd, 60% protein, (5)

Ref No 5 01 981 United States

Feed Name or Analyses		Mean As Fed	Mean Dry	C.V. ± %
Dry matter	%	90.4	100.0	
Ash	%	21.5	23.8	
Ether extract	%	6.4	7.1	
Sheep	dig coef %	96.	96.	
Protein (N x 6.25)	%	62.4	69.0	
Sheep	dig coef %	90.	90.	
Sheep	dig prot %	56.1	62.1	
Energy	GE Mcal/kg			
Sheep	DE Mcal/kg	* 3.09	3.41	
Sheep	ME Mcal/kg	* 2.53	2.80	
Sheep	TDN %	70.0	77.4	

Fish, whole or cuttings, cooked mech-extd dehy grnd, 65% protein, (5)

Ref No 5 01 982 United States

Feed Name or Analyses		Mean As Fed	Mean Dry	C.V. ± %
Dry matter	%	90.8	100.0	
Ash	%	17.3	19.0	
Ether extract	%	7.9	8.7	
Sheep	dig coef %	89.	89.	
Protein (N x 6.25)	%	67.2	74.0	
Sheep	dig coef %	88.	88.	
Sheep	dig prot %	59.1	65.1	
Energy	GE Mcal/kg			
Sheep	DE Mcal/kg	* 3.30	3.64	
Sheep	ME Mcal/kg	* 2.71	2.98	
Sheep	TDN %	74.9	82.5	

FISH, ALEWIFE. Pomolobus pseudoharengus

Fish, alewife, whole, raw, (5)

Ref No 5 07 964 United States

Feed Name or Analyses		Mean As Fed	Mean Dry	C.V. ± %
Dry matter	%	25.6	100.0	
Ash	%	1.5	5.9	
Crude fiber	%	0.0	0.0	
Ether extract	%	4.9	19.1	
N-free extract	%	-0.1	-0.7	
Protein (N x 6.25)	%	19.4	75.8	
Energy	GE Mcal/kg	1.27	4.96	
Cattle	DE Mcal/kg	* 1.12	4.38	
Sheep	DE Mcal/kg	* 0.96	3.75	
Swine	DE kcal/kg	*1136.	4438.	
Cattle	ME Mcal/kg	* 0.92	3.59	
Sheep	ME Mcal/kg	* 0.79	3.07	
Swine	ME kcal/kg	* 917.	3581.	
Cattle	TDN %	* 25.4	99.2	
Sheep	TDN %	* 21.8	85.0	
Swine	TDN %	* 25.8	100.7	
Phosphorus	%	0.22	0.85	

FISH, ANCHOVY. Engraulis spp

Fish, anchovy, whole or cuttings, cooked mech-extd dehy grnd, (5)

Fish meal, anchovy

Ref No 5 01 985 United States

Feed Name or Analyses		Mean As Fed	Mean Dry	C.V. ± %
Dry matter	%	92.0	100.0	1
Ash	%	15.0	16.3	13
Ether extract	%	5.3	5.8	49
Total fat, official	%	10.1	11.0	22
Protein (N x 6.25)	%	65.6	71.3	4
Linoleic	%	.060	.065	
Energy	GE Mcal/kg			
Chickens	ME_n kcal/kg	2640.	2870.	
Aluminum	mg/kg	75.08	81.60	44
Barium	mg/kg	5.915	6.429	50
Boron	mg/kg	12.755	13.862	27
Calcium	%	3.74	4.07	17
Chromium	mg/kg	6.983	7.589	40
Copper	mg/kg	9.7	10.5	33
Iron	%	0.023	0.025	58
Magnesium	%	0.25	0.27	21
Manganese	mg/kg	9.3	10.1	47
Phosphorus	%	2.46	2.67	19
Potassium	%	0.74	0.81	35
Selenium	mg/kg	1.383	1.504	24
Sodium	%	0.95	1.04	45
Strontium	mg/kg	73.527	79.909	45
Zinc	mg/kg	105.5	114.7	16
α-tocopherol	mg/kg	3.4	3.7	
Alanine	%	4.09	4.44	7
Arginine	%	3.82	4.15	7
Aspartic acid	%	6.28	6.83	8
Cystine	%	0.60	0.65	15
Glutamic acid	%	8.54	9.28	6
Glycine	%	3.66	3.98	6
Histidine	%	1.60	1.74	13
Isoleucine	%	3.12	3.39	6
Leucine	%	5.01	5.44	7
Lysine	%	5.09	5.53	8
Methionine	%	2.00	2.17	6
Phenylalanine	%	2.82	3.06	8
Proline	%	2.65	2.88	6
Serine	%	2.41	2.62	8
Threonine	%	2.80	3.04	7
Tryptophan	%	0.76	0.83	13
Tyrosine	%	2.26	2.46	8
Valine	%	3.51	3.81	6

FISH, CARP. Cyprinus carpio

Fish, carp, whole, raw, (5)

Ref No 5 01 986 United States

Feed Name or Analyses		Mean As Fed	Mean Dry	C.V. ± %
Dry matter	%	22.2	100.0	
Ash	%	1.1	5.0	
Ether extract	%	4.2	18.9	
Protein (N x 6.25)	%	18.0	81.1	
Energy	GE Mcal/kg	1.15	5.18	
Calcium	%	0.05	0.23	
Iron	%	0.001	0.004	
Phosphorus	%	0.25	1.14	
Potassium	%	0.29	1.29	
Sodium	%	0.05	0.23	
Ascorbic acid	mg/kg	10.0	45.0	
Niacin	mg/kg	15.0	67.6	
Riboflavin	mg/kg	0.4	1.8	
Thiamine	mg/kg	0.1	0.5	
Vitamin A	IU/g	1.7	7.7	

Ref No 5 01 986 Canada

Feed Name or Analyses		Mean As Fed	Mean Dry	C.V. ± %
Dry matter	%	31.7	100.0	
Ash	%	3.0	9.4	
Ether extract	%	9.7	30.7	
Protein (N x 6.25)	%	18.8	59.5	
Energy	GE Mcal/kg	1.64	5.18	

Fish, carp, whole or cuttings, cooked mech-extd dehy grnd, (5)

Ref No 5 01 987 United States

Feed Name or Analyses		Mean As Fed	Mean Dry	C.V. ± %
Dry matter	%	90.0	100.0	
Protein (N x 6.25)	%	52.7	58.6	1
Methionine	%	1.40	1.56	

FISH, CATFISH. Ictalurus spp

Fish, catfish, whole, raw, (5)

Ref No 5 07 965 United States

Feed Name or Analyses		Mean As Fed	Mean Dry	C.V. ± %
Dry matter	%	22.0	100.0	
Ash	%	1.3	5.9	
Ether extract	%	3.1	14.1	
Protein (N x 6.25)	%	17.6	80.0	
Energy	GE Mcal/kg	1.03	4.68	
Iron	%	0.000	0.002	
Potassium	%	0.33	1.50	
Sodium	%	0.06	0.27	
Niacin	mg/kg	17.0	77.3	
Riboflavin	mg/kg	0.3	1.4	
Thiamine	mg/kg	0.4	1.8	

FISH, COD. Gadus morrhua

Fish, cod, fillets, (5)

Ref No 5 09 071 Canada

Feed Name or Analyses		Mean As Fed	Mean Dry	C.V. ± %
Dry matter	%	20.0	100.0	
Niacin	mg/kg	25.7	128.5	
Pantothenic acid	mg/kg	1.1	5.5	
Riboflavin	mg/kg	0.2	1.0	
Thiamine	mg/kg	0.5	2.5	
Vitamin B_{12}	μg/kg	49.0	245.0	

Fish, cod, fillets, frozen, (5)

Ref No 5 09 069 United States

Feed Name or Analyses		Mean As Fed	Mean Dry	C.V. ± %
Dry matter	%	20.0	100.0	
Niacin	mg/kg	10.1	50.5	
Pantothenic acid	mg/kg	1.7	8.3	
Riboflavin	mg/kg	0.8	4.2	

Ref No 5 09 069 Canada

Feed Name or Analyses		Mean As Fed	Mean Dry	C.V. ± %
Dry matter	%	20.0	100.0	
Niacin	mg/kg	23.2	116.0	
Pantothenic acid	mg/kg	1.0	4.8	
Riboflavin	mg/kg	0.2	0.9	

(1) dry forages and roughages (2) pasture, range plants, and forages fed green (3) silages (4) energy feeds (5) protein supplements (6) minerals (7) vitamins (8) additives

Feed Name or Analyses		Mean As Fed	Mean Dry	C.V. ± %
Thiamine	mg/kg	0.6	2.9	
Vitamin B12	μg/kg	63.0	315.0	

Fish, cod, fillets, smoked, (5)
Ref No 5 09 068 Canada

Feed Name or Analyses		Mean As Fed	Mean Dry	C.V. ± %
Dry matter	%	20.0	100.0	
Niacin	mg/kg	29.8	149.0	
Pantothenic acid	mg/kg	1.6	8.0	
Riboflavin	mg/kg	0.3	1.5	
Thiamine	mg/kg	0.6	3.0	
Vitamin B12	μg/kg	118.0	590.0	

FISH, COD. Melanogrammus aeglefinus

Fish, cod, flesh, raw, (5)
Ref No 5 09 281 Canada

Feed Name or Analyses		Mean As Fed	Mean Dry	C.V. ± %
Dry matter	%	19.6	100.0	
Ash	%	1.4	7.1	
Ether extract	%	0.2	1.0	91
Protein (N x 6.25)	%	17.4	88.8	
Fatty acids	%			
Arachidonic	%	0.007	0.035	54
Docosahexaenoic	%	0.048	0.247	51
Docosapentaenoic	%	0.002	0.012	93
Eicosapentaenoic	%	0.026	0.135	49
Eicosenoic	%	0.003	0.013	88
Linoleic	%	0.001	0.004	98
Myristic	%	0.002	0.009	86
Oleic	%	0.021	0.106	67
Palmitic	%	0.032	0.166	53
Palmitoleic	%	0.004	0.022	91
Stearic	%	0.005	0.027	58
Niacin	mg/kg	31.6	161.5	
Pantothenic acid	mg/kg	1.1	5.5	
Riboflavin	mg/kg	0.4	2.0	
Thiamine	mg/kg	0.7	3.5	
Vitamin B12	μg/kg	50.9	260.0	

FISH, COD. Gadus morrhua, Gadus macrocephalus

Fish, cod, livers, dehy grnd, (5)
Cod liver oil meal
Ref No 5 08 423 United States

Feed Name or Analyses		Mean As Fed	Mean Dry	C.V. ± %
Dry matter	%	92.5	100.0	
Ash	%	2.9	3.1	
Crude fiber	%	0.7	0.8	
Ether extract	%	28.9	31.2	
N-free extract	%	9.6	10.4	
Protein (N x 6.25)	%	50.4	54.5	
Energy	GE Mcal/kg			
Cattle	DE Mcal/kg	* 4.77	5.15	
Sheep	DE Mcal/kg	* 4.00	4.32	
Swine	DE kcal/kg	*5405.	5843.	
Cattle	ME Mcal/kg	* 3.91	4.23	
Sheep	ME Mcal/kg	* 3.28	3.54	
Swine	ME kcal/kg	*4594.	4966.	
Cattle	TDN %	* 108.1	116.9	
Sheep	TDN %	* 90.7	98.0	
Swine	TDN %	* 122.6	132.5	
Calcium	%	0.16	0.17	
Phosphorus	%	0.69	0.75	

Fish, cod, whole, raw, (5)
Ref No 5 01 988 United States

Feed Name or Analyses		Mean As Fed	Mean Dry	C.V. ± %
Dry matter	%	18.8	100.0	
Ash	%	1.2	6.4	
Ether extract	%	0.3	1.6	
Protein (N x 6.25)	%	17.6	93.6	
Energy	GE Mcal/kg	0.78	4.15	
Calcium	%	0.01	0.05	
Iron	%	0.000	0.002	
Phosphorus	%	0.19	1.03	
Potassium	%	0.38	2.03	
Sodium	%	0.07	0.37	
Ascorbic acid	mg/kg	20.0	106.4	
Niacin	mg/kg	22.0	117.0	
Riboflavin	mg/kg	0.7	3.7	
Thiamine	mg/kg	0.6	3.2	
Vitamin A	IU/g	0.0	0.0	

Fish, cod, whole or cuttings, air dried grnd, (5)
Ref No 5 01 990 United States

Feed Name or Analyses		Mean As Fed	Mean Dry	C.V. ± %
Dry matter	%	83.0	100.0	
Ash	%	29.1	35.1	
Ether extract	%	1.8	2.2	
Sheep	dig coef %	96.	96.	
Protein (N x 6.25)	%	51.6	62.2	
Sheep	dig coef %	92.	92.	
Sheep	dig prot %	47.5	57.2	
Energy	GE Mcal/kg			
Sheep	DE Mcal/kg	* 2.27	2.73	
Sheep	ME Mcal/kg	* 1.86	2.24	
Sheep	TDN %	51.4	61.9	

Fish, cod, whole or cuttings, air dried grnd, 50% protein, (5)
Ref No 5 01 989 United States

Feed Name or Analyses		Mean As Fed	Mean Dry	C.V. ± %
Dry matter	%	86.1	100.0	
Ash	%	29.8	34.6	
Ether extract	%	2.0	2.3	
Swine	dig coef %	85.	85.	
Protein (N x 6.25)	%	54.0	62.7	
Swine	dig coef %	91.	91.	
Swine	dig prot %	49.1	57.1	
Energy	GE Mcal/kg			
Swine	DE kcal/kg	*2335.	2712.	
Swine	ME kcal/kg	*1946.	2260.	
Swine	TDN %	53.0	61.5	

Fish, cod, whole or cuttings, steamed dehy grnd, (5)
Ref No 5 01 992 United States

Feed Name or Analyses		Mean As Fed	Mean Dry	C.V. ± %
Dry matter	%	86.8	100.0	
Ash	%	21.2	24.4	
Ether extract	%	3.6	4.2	
Sheep	dig coef %	82.	82.	
Protein (N x 6.25)	%	62.9	72.5	
Sheep	dig coef %	95.	95.	
Sheep	dig prot %	59.8	68.9	
Energy	GE Mcal/kg			
Sheep	DE Mcal/kg	* 2.93	3.38	
Sheep	ME Mcal/kg	* 2.40	2.77	
Sheep	TDN %	66.5	76.6	

Fish, cod, whole or cuttings, steamed dehy grnd, 60% protein, (5)
Ref No 5 01 991 United States

Feed Name or Analyses		Mean As Fed	Mean Dry	C.V. ± %
Dry matter	%	85.5	100.0	
Ash	%	20.7	24.2	
Ether extract	%	1.2	1.4	
Swine	dig coef %	176.	176.	
Protein (N x 6.25)	%	62.9	73.6	
Swine	dig coef %	97.	97.	
Swine	dig prot %	61.0	71.4	
Energy	GE Mcal/kg			
Swine	DE kcal/kg	*2899.	3391.	
Swine	ME kcal/kg	*2352.	2751.	
Swine	TDN %	65.7	76.9	

Fish, cod, liver oil, (7)
Cod liver oil (CFA)
Ref No 7 01 994 Canada

Feed Name or Analyses		Mean As Fed	Mean Dry	C.V. ± %
Dry matter	%	100.0	100.0	
Ether extract	%	63.3	63.3	
Fatty acids	%			
Arachidonic	%	0.810	0.810	
Docosahexaenoic	%	7.340	7.340	
Docosapentaenoic	%	1.000	1.000	
Eicosapentaenoic	%	6.160	6.160	
Eicosenoic	%	5.635	5.635	
Linoleic	%	1.300	1.300	
Linolenic	%	0.600	0.600	
Myristic	%	2.015	2.015	
Octadecatetraenoic	%	1.000	1.000	
Oleic	%	13.695	13.695	
Palmitic	%	7.730	7.730	
Palmitoleic	%	6.165	6.165	
Stearic	%	1.460	1.460	

FISH, DOGFISH. Squalus acanthias

Fish, dogfish, whole or cuttings, cooked mech-extd dehy grnd, (5)
Ref No 5 01 995 United States

Feed Name or Analyses		Mean As Fed	Mean Dry	C.V. ± %
Dry matter	%	98.0	100.0	
Ash	%	10.4	10.6	
Crude fiber	%	0.0	0.0	
Ether extract	%	7.9	8.1	
N-free extract	%	6.3	6.4	
Protein (N x 6.25)	%	73.4	74.9	
Energy	GE Mcal/kg			
Cattle	DE Mcal/kg	* 3.56	3.63	
Sheep	DE Mcal/kg	* 3.32	3.39	
Swine	DE kcal/kg	*3395.	3465.	
Cattle	ME Mcal/kg	* 2.92	2.98	
Sheep	ME Mcal/kg	* 2.72	2.78	
Swine	ME kcal/kg	*2746.	2802.	
Cattle	TDN %	* 80.7	82.4	
Sheep	TDN %	* 75.2	76.8	
Swine	TDN %	* 77.0	78.6	
Calcium	%	1.65	1.68	
Phosphorus	%	1.42	1.45	
Choline	mg/kg	3461.	3532.	
Cysteine	%	1.40	1.43	10
Methionine	%	1.30	1.33	33

Feed Name or Analyses		Mean As Fed	Mean Dry	C.V. ± %

FISH, FLOUNDER. Bothidae (family), Pleuronectidae (family)

Fish, flounder, whole or cuttings, cooked mech-extd dehy grnd, (5)

Ref No 5 01 997 United States

Analyses		As Fed	Dry	C.V.
Dry matter	%	90.0	100.0	
Ash	%	18.9	21.0	
Ether extract	%	11.0	12.2	
Protein (N x 6.25)	%	58.9	65.4	
Calcium	%	1.35	1.50	
Phosphorus	%	0.81	0.90	

FISH, HADDOCK. Melanogrammus aeglefinus

Fish, haddock, flesh, frozen, (5)

Ref No 5 09 070 Canada

Analyses		As Fed	Dry	C.V.
Dry matter	%	20.0	100.0	
Niacin	mg/kg	35.9	179.5	
Pantothenic acid	mg/kg	1.0	4.8	
Riboflavin	mg/kg	0.2	0.8	
Thiamine	mg/kg	0.4	2.1	
Vitamin B_{12}	μg/kg	112.0	560.0	

Fish, haddock, flesh, smoked, (5)

Ref No 5 09 073 Canada

Analyses		As Fed	Dry	C.V.
Dry matter	%	20.0	100.0	
Niacin	mg/kg	45.6	228.0	
Pantothenic acid	mg/kg	1.1	5.7	
Riboflavin	mg/kg	0.2	1.2	
Thiamine	mg/kg	0.6	3.2	
Vitamin B_{12}	μg/kg	129.0	645.0	

Fish, haddock, whole, raw, (5)

Ref No 5 07 966 United States

Analyses		As Fed	Dry	C.V.
Dry matter	%	19.5	100.0	
Ash	%	1.4	7.2	
Ether extract	%	0.1	0.5	
Protein (N x 6.25)	%	18.3	93.8	
Energy	GE Mcal/kg	0.79	4.05	
Calcium	%	0.02	0.12	
Iron	%	0.001	0.004	
Phosphorus	%	0.20	1.01	
Potassium	%	0.30	1.56	
Sodium	%	0.06	0.31	
Niacin	mg/kg	3.0	15.4	
Riboflavin	mg/kg	0.1	0.4	
Thiamine	mg/kg	0.0	0.2	

Ref No 5 07 966 Canada

Analyses		As Fed	Dry	C.V.
Dry matter	%	19.9	100.0	
Ash	%	1.3	6.6	
Ether extract	%	0.7	3.6	
Protein (N x 6.25)	%	18.5	93.4	
Energy	GE Mcal/kg	0.80	4.05	
Niacin	mg/kg	33.6	169.5	
Pantothenic acid	mg/kg	1.4	6.9	
Riboflavin	mg/kg	0.2	1.1	
Thiamine	mg/kg	0.3	1.3	
Vitamin B_{12}	μg/kg	125.1	630.0	

FISH, HALIBUT. Hippoglossus spp

Fish, halibut, whole, raw, (5)

Ref No 5 01 998 United States

Analyses		As Fed	Dry	C.V.
Dry matter	%	23.7	100.0	
Ash	%	1.2	5.2	
Ether extract	%	3.5	14.8	
Protein (N x 6.25)	%	19.2	80.8	
Energy	GE Mcal/kg	1.14	4.78	
Calcium	%	0.01	0.06	
Iron	%	0.001	0.003	
Phosphorus	%	0.20	0.86	
Potassium	%	0.45	1.91	
Sodium	%	0.05	0.23	
Niacin	mg/kg	83.8	353.2	
Riboflavin	mg/kg	0.7	3.0	
Thiamine	mg/kg	0.4	1.7	
Vitamin A	IU/g	4.4	18.7	

Ref No 5 01 998 Canada

Analyses		As Fed	Dry	C.V.
Dry matter	%	23.7	100.0	
Ash	%	1.3	5.5	
Ether extract	%	1.2	5.1	
Protein (N x 6.25)	%	21.9	92.4	
Energy	GE Mcal/kg	1.13	4.78	

FISH, HERRING. Clupea harengus harengus, Clupea harengus pallasi

Fish, herring, whole, raw, (5)

Ref No 5 01 999 United States

Analyses		As Fed	Dry	C.V.
Dry matter	%	25.8	100.0	
Ash	%	1.6	6.3	
Ether extract	%	6.3	24.5	
Protein (N x 6.25)	%	18.2	70.4	
Energy	GE Mcal/kg	1.35	5.22	
Iron	%	0.001	0.005	
Phosphorus	%	0.25	0.96	
Potassium	%	0.53	2.04	
Sodium	%	0.09	0.36	
Niacin	mg/kg	37.5	145.4	
Riboflavin	mg/kg	1.6	6.1	
Thiamine	mg/kg	0.2	0.8	
Vitamin A	IU/g	1.1	4.2	

Fish, herring, whole or cuttings, cooked extn unspecified dehy grnd, mn oil specified, (5)

Ref No 5 02 001 United States

Analyses		As Fed	Dry	C.V.
Dry matter	%	90.0	100.0	
Ash	%	15.3	17.0	
Ether extract	%	9.5	10.5	
Sheep	dig coef %	99.	99.	
Protein (N x 6.25)	%	64.9	72.1	
Sheep	dig coef %	92.	92.	
Sheep	dig prot %	59.7	66.3	
Linoleic	%	.150	.167	
Energy	GE Mcal/kg			
Sheep	DE Mcal/kg	* 3.56	3.96	
Chickens	ME_n kcal/kg	3190.	3544.	
Sheep	ME Mcal/kg	* 2.92	3.24	
Sheep	TDN %	80.7	89.7	

Fish, herring, whole or cuttings, cooked extn unspecified dehy grnd, mn 3% salt, (5)

Ref No 5 02 002 United States

Analyses		As Fed	Dry	C.V.
Dry matter	%	90.4	100.0	
Ash	%	25.0	27.6	
Ether extract	%	5.7	6.3	
Sheep	dig coef %	99.	99.	
N-free extract	%			
Sheep	dig coef %	-29.	-29.	
Protein (N x 6.25)	%	56.0	62.0	
Sheep	dig coef %	88.	88.	
Sheep	dig prot %	49.3	54.6	
Energy	GE Mcal/kg			
Sheep	DE Mcal/kg	* 2.73	3.02	
Sheep	ME Mcal/kg	* 2.24	2.48	
Sheep	TDN %	62.0	68.6	

Fish, herring, whole or cuttings, cooked extn unspecified dehy grnd, mx 3% salt, (5)

Ref No 5 02 003 United States

Analyses		As Fed	Dry	C.V.
Dry matter	%	91.6	100.0	
Ash	%	16.1	17.6	
Ether extract	%	8.4	9.2	
Sheep	dig coef %	104.	104.	
N-free extract	%			
Sheep	dig coef %	58.	58.	
Protein (N x 6.25)	%	65.4	71.4	
Sheep	dig coef %	88.	88.	
Sheep	dig prot %	57.6	62.8	
Energy	GE Mcal/kg			
Sheep	DE Mcal/kg	* 3.41	3.72	
Sheep	ME Mcal/kg	* 2.80	3.05	
Sheep	TDN %	77.3	84.4	

Fish, herring, whole or cuttings, cooked extn unspecified dehy grnd, 65% protein mn oil specified, (5)

Ref No 5 02 004 United States

Analyses		As Fed	Dry	C.V.
Dry matter	%	90.8	100.0	
Ash	%	12.9	14.2	
Ether extract	%	10.7	11.8	
Swine	dig coef %	94.	94.	
Protein (N x 6.25)	%	68.6	75.6	
Swine	dig coef %	94.	94.	
Swine	dig prot %	64.5	71.1	
Energy	GE Mcal/kg			
Swine	DE kcal/kg	*3843.	4233.	
Swine	ME kcal/kg	*3103.	3417.	
Swine	TDN %	87.2	96.0	

Fish, herring, whole or cuttings, cooked extn unspecified dehy grnd, 70% protein low fat, (5)

Ref No 5 02 005 United States

Analyses		As Fed	Dry	C.V.
Dry matter	%	87.6	100.0	
Ash	%	15.2	17.4	
Ether extract	%	1.1	1.3	
Swine	dig coef %	89.	89.	

(1) dry forages and roughages (2) pasture, range plants, and forages fed green (3) silages (4) energy feeds (5) protein supplements (6) minerals (7) vitamins (8) additives

Feed Name or Analyses		Mean As Fed	Mean Dry	C.V. ± %
Protein (N x 6.25)	%	73.7	84.1	
Swine	dig coef %	93.	93.	
Swine	dig prot %	68.5	78.2	
Energy	GE Mcal/kg			
Swine	DE kcal/kg	*3125.	3567.	
Swine	ME kcal/kg	*2469.	2818.	
Swine	TDN %	70.9	80.9	

Fish, herring, whole or cuttings, cooked mech-extd dehy grnd, (5)
Fish meal, herring
Ref No 5 02 000 United States

Feed Name or Analyses		Mean As Fed	Mean Dry	C.V. ± %
Dry matter	%	93.0	100.0	2
Ash	%	10.7	11.5	14
Crude fiber	%	0.6	0.6	60
Ether extract	%	9.5	10.2	17
Sheep	dig coef %	97.	97.	
Total fat, official	%	12.6	13.6	13
N-free extract	%	0.2	0.2	
Sheep	dig coef %	-20.	-20.	
Protein (N x 6.25)	%	72.0	77.4	4
Sheep	dig coef %	88.	88.	
Sheep	dig prot %	63.3	68.1	
Energy	GE Mcal/kg			
Cattle	DE Mcal/kg	* 3.39	3.64	
Sheep	DE Mcal/kg	* 3.44	3.70	
Swine	DE kcal/kg	*3583.	3853.	
Cattle	ME Mcal/kg	* 2.78	2.99	
Chickens	ME_n kcal/kg	2959.	3183.	
Sheep	ME Mcal/kg	* 2.82	3.04	
Swine	ME kcal/kg	*2881.	3098.	
Cattle	TDN %	* 76.8	82.6	
Sheep	TDN %	78.1	84.0	
Swine	TDN %	* 81.3	87.4	
Aluminum	mg/kg	34.74	37.37	24
Barium	mg/kg	1.599	1.720	32
Boron	mg/kg	6.375	6.858	24
Calcium	%	2.08	2.24	25
Chlorine	%	1.06	1.14	25
Chromium	mg/kg	3.297	3.547	48
Copper	mg/kg	5.5	5.9	32
Iron	%	0.015	0.016	51
Magnesium	%	0.14	0.15	14
Manganese	mg/kg	5.4	5.8	45
Phosphorus	%	1.55	1.67	18
Potassium	%	1.12	1.20	22
Selenium	mg/kg	1.942	2.089	19
Sodium	%	0.62	0.67	35
Strontium	mg/kg	36.870	39.664	48
Zinc	mg/kg	120.4	129.5	20
Choline	mg/kg	4015.	4319.	24
Folic acid	mg/kg	2.40	2.58	93
Niacin	mg/kg	89.1	95.9	42
Pantothenic acid	mg/kg	11.5	12.3	48
Riboflavin	mg/kg	9.0	9.7	17
Vitamin B_{12}	µg/kg	220.7	237.4	62
Alanine	%	4.38	4.71	4
Arginine	%	4.22	4.54	9
Aspartic acid	%	6.62	7.12	5
Cysteine	%	1.41	1.52	4
Cystine	%	0.84	0.91	33
Glutamic acid	%	9.20	9.90	4
Glycine	%	4.13	4.44	12
Histidine	%	1.53	1.64	11
Isoleucine	%	3.24	3.49	6
Leucine	%	5.35	5.75	4
Lysine	%	5.80	6.24	11
Methionine	%	2.14	2.30	6
Phenylalanine	%	2.87	3.08	5
Proline	%	2.79	3.00	5
Serine	%	2.67	2.88	5
Threonine	%	2.96	3.18	7
Tryptophan	%	0.83	0.90	11
Tyrosine	%	2.36	2.54	7
Valine	%	3.86	4.15	7

Ref No 5 02 000 Canada

Feed Name or Analyses		Mean As Fed	Mean Dry	C.V. ± %
Dry matter	%	93.2	100.0	1
Ash	%	11.1	11.9	8
Crude fiber	%	0.6	0.6	18
Ether extract	%	8.7	9.3	17
Sheep	dig coef %	97.	97.	
N-free extract	%	0.0	0.0	
Sheep	dig coef %	-20.	-20.	
Protein (N x 6.25)	%	72.9	78.2	2
Sheep	dig coef %	88.	88.	
Sheep	dig prot %	64.2	68.9	
Energy	GE Mcal/kg			
Cattle	DE Mcal/kg	* 3.34	3.58	
Sheep	DE Mcal/kg	* 3.45	3.70	
Swine	DE kcal/kg	*3591.	3853.	
Cattle	ME Mcal/kg	* 2.74	2.93	
Chickens	ME_n kcal/kg	3302.	3406.	8
Sheep	ME Mcal/kg	* 2.83	3.04	
Swine	ME kcal/kg	*2880.	3090.	
Cattle	TDN %	* 75.6	81.1	
Sheep	TDN %	78.3	84.0	
Swine	TDN %	* 81.5	87.4	
Calcium	%	2.49	2.67	14
Cobalt	mg/kg	0.097	0.104	
Copper	mg/kg	7.5	8.1	
Iodine	mg/kg	6.794	7.288	
Iron	%	0.013	0.014	
Magnesium	%	0.13	0.14	
Manganese	mg/kg	1.4	1.5	
Molybdenum	mg/kg	3.53	3.78	
Phosphorus	%	2.04	2.18	10
Potassium	%	1.17	1.25	
Salt (NaCl)	%	1.56	1.68	
Sodium	%	0.56	0.60	
Sulphur	%	0.69	0.74	
Zinc	mg/kg	174.7	187.5	
Biotin	mg/kg	0.55	0.59	21
Choline	mg/kg	6118.	6563.	14
Folic acid	mg/kg	0.34	0.36	
Niacin	mg/kg	89.5	96.0	15
Pantothenic acid	mg/kg	19.7	21.1	22
Riboflavin	mg/kg	10.5	11.2	25
Thiamine	mg/kg	0.3	0.3	
α-tocopherol	mg/kg	27.0	29.0	
Vitamin K	mg/kg	2.19	2.35	83
Vitamin B_{12}	µg/kg	422.3	453.0	23
Alanine	%	5.05	5.42	
Arginine	%	5.63	6.04	5
Aspartic acid	%	8.36	8.97	
Cystine	%	0.76	0.81	7
Glutamic acid	%	9.95	10.68	
Glycine	%	5.14	5.52	6
Histidine	%	2.06	2.21	15
Isoleucine	%	3.13	3.36	9
Leucine	%	5.50	5.90	11
Lysine	%	6.06	6.50	7
Chicks available	%	0.37	0.39	
Methionine	%	2.07	2.22	10
Phenylalanine	%	2.77	2.97	14
Proline	%	3.75	4.02	
Serine	%	3.30	3.54	
Threonine	%	3.22	3.46	22
Tryptophan	%	0.77	0.82	28
Tyrosine	%	2.20	2.36	20
Valine	%	5.42	5.81	11

Fish, herring, oil from whole or parts of fish, (7)
Herring oil (AAFCO)
Ref No 7 08 048 Canada

Feed Name or Analyses		Mean As Fed	Mean Dry	C.V. ± %
Dry matter	%	100.0	100.0	
Fatty acids	%			
Docosahexaenoic	%	3.960	3.960	35
Eicosapentaenoic	%	6.240	6.240	32
Eicosenoic	%	15.820	15.820	17
Linoleic	%	1.667	1.667	
Myristic	%	5.900	5.900	23
Octadecatetraenoic	%	1.575	1.575	
Oleic	%	12.360	12.360	41
Palmitic	%	11.240	11.240	12
Palmitoleic	%	9.040	9.040	20
Stearic	%	1.550	1.550	

FISH, HERRING NORWEGIAN. *Clupea harengus, Clupea sprattus*

Fish, herring Norwegian, whole or cuttings, cooked mech-extd grnd, (5)
Ref No 5 08 987 United States

Feed Name or Analyses		Mean As Fed	Mean Dry	C.V. ± %
Dry matter	%	93.4	100.0	1
Ash	%	10.1	10.8	11
Ether extract	%	6.1	6.5	30
Total fat, official	%	10.6	11.3	13
Protein (N x 6.25)	%	75.3	80.7	2
Aluminum	mg/kg	33.54	35.90	61
Barium	mg/kg	3.031	3.245	98
Boron	mg/kg	6.347	6.796	26
Calcium	%	1.95	2.09	28
Chromium	mg/kg	4.219	4.517	27
Copper	mg/kg	5.5	5.8	24
Iron	%	0.015	0.016	44
Magnesium	%	0.11	0.12	13
Manganese	mg/kg	2.4	2.6	40
Phosphorus	%	1.51	1.61	23
Potassium	%	1.21	1.30	14
Selenium	mg/kg	2.789	2.986	24
Sodium	%	0.42	0.45	44
Strontium	mg/kg	71.411	76.454	64
Zinc	mg/kg	120.8	129.3	7
Alanine	%	6.52	6.98	7
Arginine	%	4.39	4.70	7
Aspartic acid	%	4.45	4.76	5
Cystine	%	0.72	0.77	11
Glutamic acid	%	9.21	9.86	7
Glycine	%	4.62	4.95	7
Histidine	%	1.88	2.02	15
Isoleucine	%	3.41	3.65	7
Leucine	%	5.19	5.55	12
Lysine	%	5.70	6.10	7
Methionine	%	2.13	2.28	5
Phenylalanine	%	2.80	2.99	5
Proline	%	3.14	3.36	5
Serine	%	2.63	2.81	7
Threonine	%	2.96	3.17	7
Tryptophan	%	0.78	0.83	6
Tyrosine	%	2.36	2.53	16
Valine	%	4.05	4.34	8

Feed Name or Analyses		Mean As Fed	Mean Dry	C.V. ± %

FISH, HERRING PACIFIC. Clupea pallasi

Fish, herring Pacific, whole or cuttings, cooked extn unspecified grnd, (5)

Ref No 5 08 985 United States

Analyses		As Fed	Dry	C.V. ± %
Dry matter	%	92.3	100.0	1
Ash	%	10.8	11.7	15
Crude fiber	%	23.3	25.2	
Ether extract	%	12.2	13.3	26
N-free extract	%	-23.2	-25.2	
Protein (N x 6.25)	%	69.4	75.2	4
Lignin (Ellis)	%	5.5	6.0	
Energy	GE Mcal/kg			
Cattle	DE Mcal/kg	* 2.38	2.58	
Sheep	DE Mcal/kg	* 2.16	2.34	
Swine	DE kcal/kg	*2451.	2656.	
Cattle	ME Mcal/kg	* 1.95	2.11	
Sheep	ME Mcal/kg	* 1.77	1.92	
Swine	ME kcal/kg	*1981.	2147.	
Cattle	TDN %	* 53.9	58.5	
Sheep	TDN %	* 49.1	53.2	
Swine	TDN %	* 55.6	60.2	
Silicon	%	9.97	10.80	

FISH, MACKEREL. Scomber scombrus, Scomber japonicus

Fish, mackerel, whole or cuttings, cooked mech-extd dehy grnd, (5)

Ref No 5 02 006 United States

Analyses		As Fed	Dry	C.V. ± %
Dry matter	%	91.7	100.0	2
Ash	%	20.3	22.2	10
Ether extract	%	9.5	10.4	20
Protein (N x 6.25)	%	60.3	65.8	7
Niacin	mg/kg	63.2	68.9	
Riboflavin	mg/kg	5.6	6.1	
Vitamin B_{12}	μg/kg	406.6	443.5	

FISH, MACKEREL ATLANTIC. Scomber scombrus

Fish, mackerel Atlantic, whole, raw, (5)

Ref No 5 07 971 United States

Analyses		As Fed	Dry	C.V. ± %
Dry matter	%	32.8	100.0	
Ash	%	1.6	4.9	
Ether extract	%	12.2	37.2	
Protein (N x 6.25)	%	19.0	57.9	
Energy	GE Mcal/kg	1.91	5.82	
Calcium	%	0.00	0.02	
Iron	%	0.001	0.003	
Phosphorus	%	0.24	0.73	
Niacin	mg/kg	82.0	250.0	
Riboflavin	mg/kg	3.3	10.1	
Thiamine	mg/kg	1.5	4.6	
Vitamin A	IU/g	4.5	13.7	

Ref No 5 07 971 Canada

Analyses		As Fed	Dry	C.V. ± %
Dry matter	%	31.2	100.0	
Ash	%	1.6	5.0	
Ether extract	%	9.9	31.8	
Protein (N x 6.25)	%	23.0	73.8	
Energy	GE Mcal/kg	1.82	5.82	

FISH, MACKEREL PACIFIC. Scomber japonicus

Fish, mackerel Pacific, whole, raw, (5)

Ref No 5 07 972 United States

Analyses		As Fed	Dry	C.V. ± %
Dry matter	%	30.2	100.0	
Ash	%	1.4	4.6	
Ether extract	%	7.3	24.2	
Protein (N x 6.25)	%	21.9	72.5	
Energy	GE Mcal/kg	1.59	5.26	
Calcium	%	0.01	0.03	
Iron	%	0.002	0.007	
Phosphorus	%	0.27	0.91	
Vitamin A	IU/g	1.2	4.0	

FISH, MENHADEN. Brevoortia tyrannus

Fish, menhaden, stickwater solubles, cooked condensed, (5)

Ref No 5 02 007 United States

Analyses		As Fed	Dry	C.V. ± %
Dry matter	%	51.0	100.0	2
Ether extract	%	3.9	7.6	
Protein (N x 6.25)	%	30.4	59.6	13
Niacin	mg/kg	239.0	468.6	
Riboflavin	mg/kg	9.5	18.6	4

Fish, menhaden, stickwater solubles, cooked dehy, (5)

Ref No 5 02 008 United States

Analyses		As Fed	Dry	C.V. ± %
Dry matter	%	90.0	100.0	
Protein (N x 6.25)	%	54.7	60.8	
Methionine	%	0.50	0.56	

Fish, menhaden, whole or cuttings, cooked mech-extd dehy grnd, (5)

Fish meal, menhaden

Ref No 5 02 009 United States

Analyses		As Fed	Dry	C.V. ± %
Dry matter	%	91.4	100.0	1
Ash	%	19.0	20.8	12
Crude fiber	%	0.6	0.6	96
Ether extract	%	9.8	10.7	18
Sheep	dig coef %	108.	108.	
Total fat, official	%	12.4	13.5	13
N-free extract	%	1.6	1.8	
Protein (N x 6.25)	%	60.4	66.1	5
Sheep	dig coef %	81.	81.	
Sheep	dig prot %	48.9	53.5	
Linoleic	%	.120	.131	
Energy	GE Mcal/kg	4.39	4.80	
Cattle	DE Mcal/kg	* 2.99	3.27	
Sheep	DE Mcal/kg	* 2.57	2.81	
Swine	DE kcal/kg	3282.	3590.	
Cattle	ME Mcal/kg	* 2.45	2.68	
Chickens	ME_n kcal/kg	2898.	3171.	
Sheep	ME Mcal/kg	* 2.11	2.30	
Swine	ME kcal/kg	2843.	3110.	
Cattle	TDN %	* 67.8	74.2	
Sheep	TDN %	58.2	63.7	
Swine	TDN %	* 69.7	76.2	
Aluminum	mg/kg	351.78	384.83	39
Barium	mg/kg	20.302	22.210	73
Boron	mg/kg	14.152	15.482	17
Calcium	%	5.14	5.62	10
Chlorine	%	0.32	0.34	98
Chromium	mg/kg	10.965	11.995	12
Cobalt	mg/kg	0.197	0.215	22
Copper	mg/kg	11.0	12.0	31
Iodine	mg/kg	2.004	2.193	
Iron	%	0.045	0.050	32
Magnesium	%	0.14	0.16	9
Manganese	mg/kg	34.1	37.3	44
Phosphorus	%	2.91	3.18	8
Potassium	%	0.73	0.80	18
Selenium	mg/kg	2.224	2.433	32
Sodium	%	0.34	0.37	16
Strontium	mg/kg	63.312	69.260	32
Zinc	mg/kg	150.6	164.7	7
Choline	mg/kg	2826.	3092.	
Niacin	mg/kg	55.4	60.6	12
Pantothenic acid	mg/kg	8.7	9.5	
Riboflavin	mg/kg	4.8	5.2	28
Thiamine	mg/kg	0.7	0.7	66
a-tocopherol	mg/kg	9.0	9.8	
Vitamin B_{12}	μg/kg	76.1	83.2	98
Alanine	%	4.00	4.38	5
Arginine	%	4.09	4.48	8
Aspartic acid	%	6.12	6.70	6
Cystine	%	0.63	0.68	16
Glutamic acid	%	8.57	9.38	5
Glycine	%	4.56	4.99	5
Histidine	%	1.56	1.71	8
Isoleucine	%	3.01	3.29	9
Leucine	%	4.83	5.28	6
Lysine	%	5.14	5.62	7
Methionine	%	1.89	2.07	7
Phenylalanine	%	2.67	2.92	6
Proline	%	3.09	3.38	6
Serine	%	2.44	2.67	7
Threonine	%	2.69	2.94	6
Tryptophan	%	0.72	0.78	15
Tyrosine	%	2.15	2.35	10
Valine	%	3.45	3.77	6

FISH, MULLET. Mugilidae (family), Mullidae (family)

Fish, mullet, whole, cooked grnd, (5)

Ref No 5 08 107 Canada

Analyses		As Fed	Dry	C.V. ± %
Dry matter	%	26.1	100.0	
Ether extract	%	3.7	14.3	
Protein (N x 6.25)	%	18.3	69.9	
Calcium	%	1.13	4.33	
Phosphorus	%	0.73	2.81	

Fish, mullet, whole, raw, (5)

Goatfish, whole, raw

Ref No 5 08 118 United States

Analyses		As Fed	Dry	C.V. ± %
Dry matter	%	27.4	100.0	
Ash	%	1.3	4.7	
Ether extract	%	6.9	25.2	
Protein (N x 6.25)	%	19.6	71.5	
Energy	GE Mcal/kg	1.46	5.33	
Calcium	%	0.03	0.11	
Iron	%	0.002	0.007	
Phosphorus	%	0.22	0.80	
Potassium	%	0.29	1.06	

(1) dry forages and roughages (2) pasture, range plants, and forages fed green (3) silages (4) energy feeds (5) protein supplements (6) minerals (7) vitamins (8) additives

Feed Name or Analyses		Mean As Fed	Mean Dry	C.V. ± %
Sodium	%	0.08	0.29	
Niacin	mg/kg	52.0	189.8	
Riboflavin	mg/kg	0.8	2.9	
Thiamine	mg/kg	0.7	2.6	

FISH, PILCHARD. Sardinops spp

Fish, pilchard, whole or cuttings, cooked mech-extd dehy grnd, (5)
Fish meal, pilchard
Ref No 5 02 010 United States

Feed Name or Analyses		Mean As Fed	Mean Dry	C.V. ± %
Dry matter	%	91.8	100.0	
Ash	%	23.3	25.4	
Ether extract	%	5.1	5.6	
Sheep	dig coef %	99.	99.	
Swine	dig coef %	* 176.	176.	
Protein (N x 6.25)	%	60.3	65.7	
Sheep	dig coef %	82.	82.	
Swine	dig coef %	* 97.	97.	
Sheep	dig prot %	49.5	53.9	
Swine	dig prot %	58.5	63.7	
Linoleic	%	.070	.076	
Energy	GE Mcal/kg			
Sheep	DE Mcal/kg	* 2.68	2.92	
Chickens	ME_n kcal/kg	2640.	2876.	
Sheep	ME Mcal/kg	* 2.20	2.39	
Sheep	TDN %	60.8	66.2	
α-tocopherol	mg/kg	9.0	9.8	

FISH, REDFISH. Sciaenops ocellata

Fish, redfish, whole, raw, (5)
Drumfish, whole, raw
Ocean perch, whole, raw
Ref No 5 08 191 United States

Feed Name or Analyses		Mean As Fed	Mean Dry	C.V. ± %
Dry matter	%	21.0	100.0	
Ash	%	1.1	5.4	
Ether extract	%	2.1	10.2	
Protein (N x 6.25)	%	17.9	84.9	
Energy	GE Mcal/kg	0.96	4.55	
Calcium	%	0.02	0.10	
Iron	%	0.001	0.005	
Phosphorus	%	0.22	1.03	
Potassium	%	0.28	1.32	
Sodium	%	0.07	0.33	
Niacin	mg/kg	28.4	135.2	
Riboflavin	mg/kg	0.7	3.2	
Thiamine	mg/kg	1.3	6.3	

Fish, redfish, whole or cuttings, cooked mech-extd dehy grnd, (5)
Fish meal, drum
Fish meal, redfish
Ref No 5 07 973 United States

Feed Name or Analyses		Mean As Fed	Mean Dry	C.V. ± %
Dry matter	%	94.1	100.0	
Ash	%	24.4	25.9	
Crude fiber	%	0.9	1.0	
Ether extract	%	11.1	11.8	
N-free extract	%	0.9	1.0	
Protein (N x 6.25)	%	56.8	60.4	
Linoleic	%	.140	.149	
Energy	GE Mcal/kg	4.94	5.25	
Cattle	DE Mcal/kg	* 3.17	3.37	
Sheep	DE Mcal/kg	* 3.27	3.47	
Swine	DE kcal/kg	*3124.	3320.	
Cattle	ME Mcal/kg	* 2.60	2.76	
Chickens	ME_n kcal/kg	2970.	3170.	
Sheep	ME Mcal/kg	* 2.68	2.85	
Swine	ME kcal/kg	*2618.	2782.	
Cattle	TDN %	* 71.9	76.4	
Sheep	TDN %	* 74.2	78.8	
Swine	TDN %	* 70.8	75.3	
Calcium	%	4.01	4.26	
Phosphorus	%	2.44	2.59	

Ref No 5 07 973 Canada

Feed Name or Analyses		Mean As Fed	Mean Dry	C.V. ± %
Dry matter	%	24.7	100.0	
Ash	%	3.0	12.3	
Crude fiber	%	0.0	0.0	
Ether extract	%	5.5	22.3	
N-free extract	%	1.1	4.4	
Protein (N x 6.25)	%	15.1	61.0	
Energy	GE Mcal/kg	1.30	5.25	
Cattle	DE Mcal/kg	* 0.83	3.37	
Sheep	DE Mcal/kg	* 0.92	3.73	
Swine	DE kcal/kg	* 873.	3536.	
Cattle	ME Mcal/kg	* 0.68	2.76	
Sheep	ME Mcal/kg	* 0.72	3.06	
Swine	ME kcal/kg	* 731.	2960.	
Cattle	TDN %	* 18.9	76.5	
Sheep	TDN %	* 20.9	84.6	
Swine	TDN %	* 19.8	80.2	

FISH, ROCKFISH. Sebastodes spp

Fish, rockfish, whole, raw, (5)
Ref No 5 07 974 United States

Feed Name or Analyses		Mean As Fed	Mean Dry	C.V. ± %
Dry matter	%	21.1	100.0	
Ash	%	1.2	5.7	
Ether extract	%	1.8	8.5	
Protein (N x 6.25)	%	18.9	89.6	
Energy	GE Mcal/kg	0.97	4.60	
Potassium	%	0.39	1.85	
Sodium	%	0.06	0.28	
Riboflavin	mg/kg	1.2	5.7	
Thiamine	mg/kg	0.6	2.8	

FISH, SALMON. Oncorhynchus spp, Salmo spp

Fish, salmon, whole, raw, (5)
Ref No 5 02 011 United States

Feed Name or Analyses		Mean As Fed	Mean Dry	C.V. ± %
Dry matter	%	32.1	100.0	
Ash	%	1.3	4.0	
Ether extract	%	10.2	31.9	
Protein (N x 6.25)	%	21.2	66.2	
Energy	GE Mcal/kg	1.83	5.71	
Calcium	%	0.13	0.41	
Iron	%	0.001	0.003	
Phosphorus	%	0.23	0.71	
Potassium	%	0.42	1.31	9
Sodium	%	0.06	0.17	29
Ascorbic acid	mg/kg	45.0	140.3	
Niacin	mg/kg	63.4	197.8	
Riboflavin	mg/kg	1.0	3.1	56
Thiamine	mg/kg	1.3	3.9	33
Vitamin A	IU/g	2.2	6.8	

Fish, salmon, whole or cuttings, cooked mech-extd dehy grnd, (5)
Fish meal, salmon
Ref No 5 02 012 United States

Feed Name or Analyses		Mean As Fed	Mean Dry	C.V. ± %
Dry matter	%	92.9	100.0	3
Ash	%	17.9	19.2	16
Crude fiber	%	0.3	0.3	93
Ether extract	%	9.7	10.4	45
Swine	dig coef %	* 176.	176.	
N-free extract	%	6.3	6.8	
Protein (N x 6.25)	%	58.7	63.2	8
Swine	dig coef %	* 97.	97.	
Swine	dig prot %	56.9	61.3	
Energy	GE Mcal/kg			
Cattle	DE Mcal/kg	* 3.13	3.37	
Sheep	DE Mcal/kg	* 2.98	3.21	
Swine	DE kcal/kg	*3242.	3490.	
Cattle	ME Mcal/kg	* 2.57	2.76	
Sheep	ME Mcal/kg	* 2.44	2.63	
Swine	ME kcal/kg	*2699.	2905.	
Cattle	TDN %	* 71.0	76.4	
Sheep	TDN %	* 67.5	72.7	
Swine	TDN %	* 73.5	79.2	
Calcium	%	5.47	5.88	25
Cobalt	mg/kg	0.064	0.069	6
Copper	mg/kg	11.9	12.8	0
Iron	%	0.018	0.019	0
Magnesium	%	0.36	0.39	
Manganese	mg/kg	7.9	8.5	
Phosphorus	%	3.46	3.72	26
Choline	mg/kg	2775.	2987.	
Niacin	mg/kg	24.9	26.8	1
Pantothenic acid	mg/kg	6.8	7.3	
Riboflavin	mg/kg	5.7	6.2	55
Thiamine	mg/kg	0.9	0.9	21
Arginine	%	5.19	5.59	
Cystine	%	0.70	0.75	
Glycine	%	5.19	5.59	
Lysine	%	7.59	8.17	
Methionine	%	1.60	1.72	
Tryptophan	%	0.50	0.54	

Ref No 5 02 012 Canada

Feed Name or Analyses		Mean As Fed	Mean Dry	C.V. ± %
Dry matter	%	93.3	100.0	
Ether extract	%	11.5	12.4	
Protein (N x 6.25)	%	63.5	68.1	

FISH, SARDINE. Clupea spp, Sardinops spp

Fish, sardine, stickwater solubles, cooked condensed, (5)
Ref No 5 02 014 United States

Feed Name or Analyses		Mean As Fed	Mean Dry	C.V. ± %
Dry matter	%	49.7	100.0	3
Ash	%	10.2	20.5	9
Crude fiber	%	0.0	0.0	9
Ether extract	%	9.4	18.9	
Swine	dig coef %	* 99.	99.	
N-free extract	%	0.6	1.2	
Protein (N x 6.25)	%	29.5	59.4	9
Swine	dig coef %	* 96.	96.	
Swine	dig prot %	28.3	57.0	
Energy	GE Mcal/kg			
Cattle	DE Mcal/kg	* 1.86	3.73	
Sheep	DE Mcal/kg	* 1.67	3.35	

Continued

Feed Name or Analyses		Mean As Fed	Mean Dry	C.V. ± %
Swine	DE kcal/kg	*2051.	4127.	
Cattle	ME Mcal/kg	* 1.52	3.06	
Sheep	ME Mcal/kg	* 1.37	2.75	
Swine	ME kcal/kg	*1723.	3467.	
Cattle	TDN %	* 42.1	84.7	
Sheep	TDN %	* 37.8	76.0	
Swine	TDN %	* 46.5	93.6	
Calcium	%	0.14	0.28	
Chlorine	%	0.28	0.56	
Copper	mg/kg	25.8	51.9	
Iodine	mg/kg	4.934	9.927	98
Iron	%	0.002	0.004	
Manganese	mg/kg	24.9	50.1	
Phosphorus	%	0.83	1.67	
Potassium	%	0.18	0.36	
Sodium	%	0.18	0.36	
Sulphur	%	0.11	0.22	
Biotin	mg/kg	0.13	0.27	
Choline	mg/kg	3009.	6055.	31
Niacin	mg/kg	355.8	715.9	15
Pantothenic acid	mg/kg	41.2	82.9	4
Riboflavin	mg/kg	16.8	33.7	16
Thiamine	mg/kg	4.0	8.0	
Vitamin B12	µg/kg	1041.0	2094.6	
Arginine	%	1.50	3.02	
Cystine	%	0.20	0.40	
Histidine	%	2.00	4.02	
Isoleucine	%	0.90	1.81	
Leucine	%	1.60	3.22	
Lysine	%	1.60	3.22	
Methionine	%	0.90	1.81	73
Phenylalanine	%	0.80	1.61	
Threonine	%	0.80	1.61	
Tryptophan	%	0.10	0.20	
Valine	%	1.00	2.01	

Fish, sardine, whole, raw, (5)

Ref No 5 08 124 United States

Feed Name or Analyses		Mean As Fed	Mean Dry	C.V. ± %
Dry matter	%	29.3	100.0	
Ash	%	2.4	8.2	
Ether extract	%	8.6	29.4	
Protein (N x 6.25)	%	19.2	65.5	
Energy	GE Mcal/kg	1.60	5.46	
Calcium	%	0.03	0.11	
Iron	%	0.002	0.006	
Phosphorus	%	0.22	0.73	

Fish, sardines, whole or cuttings, cooked mech-extd dehy grnd, (5)

Fish meal, sardine

Ref No 5 02 015 United States

Feed Name or Analyses		Mean As Fed	Mean Dry	C.V. ± %
Dry matter	%	92.9	100.0	1
Ash	%	15.7	16.9	16
Crude fiber	%	0.9	0.9	59
Ether extract	%	5.4	5.8	27
Swine	dig coef %	* 176.	176.	
N-free extract	%	5.8	6.3	
Protein (N x 6.25)	%	65.2	70.1	6
Swine	dig coef %	* 97.	97.	
Swine	dig prot %	63.2	68.0	

(1) dry forages and roughages (2) pasture, range plants, and forages fed green (3) silages (4) energy feeds (5) protein supplements (6) minerals (7) vitamins (8) additives

Feed Name or Analyses		Mean As Fed	Mean Dry	C.V. ± %
Energy	GE Mcal/kg			
Cattle	DE Mcal/kg	* 2.97	3.20	
Sheep	DE Mcal/kg	* 2.89	3.11	
Swine	DE kcal/kg	*2868.	3086.	
Cattle	ME Mcal/kg	* 2.44	2.62	
Chickens	ME_n kcal/kg	2908.	3129.	
Sheep	ME Mcal/kg	* 2.37	2.55	
Swine	ME kcal/kg	*2347.	2525.	
Cattle	TDN %	* 67.4	72.5	
Sheep	TDN %	* 65.5	70.5	
Swine	TDN %	* 65.0	70.0	
Calcium	%	4.50	4.84	13
Chlorine	%	0.41	0.44	
Cobalt	mg/kg	0.183	0.196	2
Copper	mg/kg	20.1	21.7	0
Iron	%	0.030	0.032	0
Magnesium	%	0.10	0.11	66
Manganese	mg/kg	22.1	23.8	1
Phosphorus	%	2.61	2.81	14
Potassium	%	0.33	0.35	
Sodium	%	0.18	0.19	
Choline	mg/kg	2983.	3209.	10
Niacin	mg/kg	62.5	67.3	11
Pantothenic acid	mg/kg	9.3	10.0	69
Riboflavin	mg/kg	6.0	6.4	21
Thiamine	mg/kg	0.4	0.5	
Vitamin B12	µg/kg	171.9	185.0	35
Arginine	%	3.27	3.52	31
Cystine	%	1.01	1.08	8
Glutamic acid	%	2.99	3.22	
Glycine	%	4.58	4.93	4
Histidine	%	1.79	1.93	26
Lysine	%	5.99	6.44	34
Methionine	%	1.99	2.14	2
Threonine	%	2.59	2.79	
Tryptophan	%	0.68	0.73	85
Valine	%	4.09	4.40	

FISH, SARDINE PACIFIC. Sardinops caerulea

Fish, sardine Pacific, whole or cuttings, cooked extn unspecified grnd, (5)

Ref No 5 08 986 United States

Feed Name or Analyses		Mean As Fed	Mean Dry	C.V. ± %
Dry matter	%	91.8	100.0	2
Ash	%	20.3	22.1	14
Ether extract	%	7.9	8.6	12
Protein (N x 6.25)	%	60.0	65.4	6

FISH, SHAD. Alosa sapidissima

Fish, shad, flesh, raw, (5)

Ref No 5 09 066 Canada

Feed Name or Analyses		Mean As Fed	Mean Dry	C.V. ± %
Dry matter	%	27.1	100.0	
Ether extract	%	17.2	63.5	
Protein (N x 6.25)	%	18.3	67.5	

FISH, SHAD. Alosa sapidissima, Dorosoma cepedianum

Fish, shad, whole, raw, (5)

Ref No 5 02 017 United States

Feed Name or Analyses		Mean As Fed	Mean Dry	C.V. ± %
Dry matter	%	29.6	100.0	
Ash	%	1.3	4.4	
Ether extract	%	10.0	33.8	

Feed Name or Analyses		Mean As Fed	Mean Dry	C.V. ± %
Protein (N x 6.25)	%	18.6	62.8	
Energy	GE Mcal/kg	1.70	5.74	
Calcium	%	0.02	0.07	
Iron	%	0.001	0.003	
Phosphorus	%	0.26	0.88	
Potassium	%	0.33	1.11	
Sodium	%	0.05	0.17	
Niacin	mg/kg	84.0	283.8	
Riboflavin	mg/kg	2.4	8.1	
Thiamine	mg/kg	1.5	5.1	

FISH, SHARK. Selachii (order)

Fish, shark, whole or cuttings, cooked mech-extd dehy grnd, (5)

Fish meal, shark

Ref No 5 02 018 United States

Feed Name or Analyses		Mean As Fed	Mean Dry	C.V. ± %
Dry matter	%	91.2	100.0	4
Ash	%	13.4	14.8	13
Crude fiber	%	0.4	0.5	80
Ether extract	%	2.6	2.9	28
Swine	dig coef %	* 176.	176.	
N-free extract	%	1.4	1.5	
Protein (N x 6.25)	%	73.2	80.4	8
Swine	dig coef %	* 97.	97.	
Swine	dig prot %	71.1	78.0	
Energy	GE Mcal/kg			
Cattle	DE Mcal/kg	* 2.45	2.69	
Sheep	DE Mcal/kg	* 2.36	2.54	
Swine	DE kcal/kg	*2526.	2772.	
Cattle	ME Mcal/kg	* 2.02	02.21	
Sheep	ME Mcal/kg	* 1.93	2.12	
Swine	ME kcal/kg	*2015.	2211.	
Cattle	TDN %	* 55.6	61.0	
Sheep	TDN %	* 53.5	58.7	
Swine	TDN %	* 57.3	62.9	
Calcium	%	3.48	3.82	23
Copper	mg/kg	51.1	56.1	
Iron	%	0.016	0.018	
Magnesium	%	0.17	0.19	36
Manganese	mg/kg	90.2	99.0	
Phosphorus	%	1.87	2.06	16
Sodium	%	0.33	0.36	
Zinc	mg/kg	112.7	123.7	
Methionine	%	0.80	0.88	

FISH, SMELT. Asmerus spp

Fish, smelt, whole, raw, (5)

Ref No 5 07 975 United States

Feed Name or Analyses		Mean As Fed	Mean Dry	C.V. ± %
Dry matter	%	21.0	100.0	
Ash	%	1.1	5.2	
Ether extract	%	2.1	10.0	
Protein (N x 6.25)	%	18.6	88.6	
Energy	GE Mcal/kg	0.98	4.67	
Iron	%	0.000	0.000	
Phosphorus	%	0.27	1.29	
Niacin	mg/kg	14.0	66.7	
Riboflavin	mg/kg	1.2	5.7	
Thiamine	mg/kg	0.1	0.5	

Feed Name or Analyses		Mean As Fed	Dry	C.V. ± %

FISH, STICKLEBACK. Gasterosteidae (family)

Fish, stickleback, whole or cuttings, cooked extn unspecified dehy grnd, (5)

Ref No 5 02 020 — United States

Analyses	Unit	As Fed	Dry	C.V. ± %
Dry matter	%	93.9	100.0	
Ash	%	28.4	30.2	
Ether extract	%	2.9	3.1	
Sheep	dig coef %	111.	111.	
Protein (N x 6.25)	%	62.1	66.1	
Sheep	dig coef %	88.	88.	
Sheep	dig prot %	54.6	58.2	
Energy	GE Mcal/kg			
Sheep	DE Mcal/kg	* 2.73	2.91	
Sheep	ME Mcal/kg	* 2.24	2.38	
Sheep	TDN %	61.9	65.9	

FISH, STURGEON. Acipenser oxyrhynchus

Fish, sturgeon, whole, raw, (5)

Ref No 5 08 126 — United States

Analyses	Unit	As Fed	Dry	C.V. ± %
Dry matter	%	21.3	100.0	
Ash	%	1.4	6.6	
Ether extract	%	1.9	8.9	
Protein (N x 6.25)	%	18.1	85.0	
Energy	GE Mcal/kg	0.94	4.41	

FISH, SUCKER. Catostomidae (family)

Fish, sucker, whole, cooked grnd, (5)

Ref No 5 08 108 — Canada

Analyses	Unit	As Fed	Dry	C.V. ± %
Dry matter	%	21.1	100.0	
Ether extract	%	2.9	13.7	
Protein (N x 6.25)	%	16.4	77.7	

FISH, SUCKER. Carpiodes spp, Castostomus spp

Fish, sucker, whole, raw, (5)

Ref No 5 02 021 — United States

Analyses	Unit	As Fed	Dry	C.V. ± %
Dry matter	%	23.6	100.0	
Ash	%	1.2	5.1	
Ether extract	%	1.8	7.6	
Protein (N x 6.25)	%	20.6	87.3	
Energy	GE Mcal/kg	1.04	4.41	
Phosphorus	%	0.22	0.93	
Potassium	%	0.34	1.44	
Sodium	%	0.06	0.25	
Niacin	mg/kg	12.0	50.8	

Ref No 5 02 021 — Canada

Analyses	Unit	As Fed	Dry	C.V. ± %
Dry matter	%	19.4	100.0	
Ash	%	1.3	6.7	
Ether extract	%	1.3	6.7	
Protein (N x 6.25)	%	18.0	92.8	
Energy	GE Mcal/kg	0.85	4.41	

FISH, SWORDFISH. Xiphias fladius

Fish, swordfish, flesh, raw, (5)

Ref No 5 09 072 — Canada

Analyses	Unit	As Fed	Dry	C.V. ± %
Dry matter	%	24.0	100.0	
Ash	%	1.9	7.9	
Ether extract	%	2.1	8.8	
Protein (N x 6.25)	%	20.8	86.7	

FISH, TUNA. Thunnus thynnus, Thunnus albacares

Fish, tuna, stickwater solubles, cooked condensed, (5)

Ref No 5 02 022 — United States

Analyses	Unit	As Fed	Dry	C.V. ± %
Dry matter	%	51.8	100.0	2
Ash	%	9.5	18.3	11
Crude fiber	%	0.0	0.0	11
Ether extract	%	4.5	8.7	27
Swine	dig coef %	* 99.	99.	
N-free extract	%	2.9	5.6	
Protein (N x 6.25)	%	34.9	67.4	3
Swine	dig coef %	* 96.	96.	
Swine	dig prot %	33.5	64.7	
Energy	GE Mcal/kg			
Cattle	DE Mcal/kg	* 1.72	3.32	
Sheep	DE Mcal/kg	* 1.64	3.17	
Swine	DE kcal/kg	*1724.	3329.	
Cattle	ME Mcal/kg	* 1.41	2.72	
Sheep	ME Mcal/kg	* 1.35	2.60	
Swine	ME kcal/kg	*1421.	2742.	
Cattle	TDN %	* 39.0	75.4	
Sheep	TDN %	* 37.2	71.9	
Swine	TDN %	* 39.1	75.5	
Potassium	%	1.76	3.40	9
Sodium	%	1.68	3.24	27
Folic acid	mg/kg	0.18	0.34	44
Niacin	mg/kg	314.2	606.5	8
Riboflavin	mg/kg	21.4	41.3	10
Thiamine	mg/kg	23.6	45.5	22
Vitamin B_{12}	μg/kg	483.2	932.9	4
Arginine	%	1.20	2.32	
Glutamic acid	%	1.60	3.09	
Glycine	%	0.70	1.35	
Leucine	%	1.20	2.32	
Lysine	%	1.40	2.70	
Valine	%	0.90	1.74	

Fish, tuna, whole, raw, (5)

Ref No 5 08 130 — United States

Analyses	Unit	As Fed	Dry	C.V. ± %
Dry matter	%	29.0	100.0	
Ash	%	1.4	4.7	
Ether extract	%	3.5	12.2	
Protein (N x 6.25)	%	25.0	86.0	
Energy	GE Mcal/kg	1.39	4.79	
Iron	%	0.001	0.003	
Sodium	%	0.04	0.14	

Ref No 5 08 130 — Canada

Analyses	Unit	As Fed	Dry	C.V. ± %
Dry matter	%	29.9	100.0	
Ash	%	1.2	4.0	
Ether extract	%	6.9	23.1	
Protein (N x 6.25)	%	23.7	79.3	
Energy	GE Mcal/kg	1.43	4.79	

Fish, tuna, whole or cuttings, cooked mech-extd dehy grnd, (5)

Fish meal, tuna

Ref No 5 02 023 — United States

Analyses	Unit	As Fed	Dry	C.V. ± %
Dry matter	%	93.2	100.0	2
Ash	%	22.8	24.5	20
Crude fiber	%	0.7	0.8	46
Ether extract	%	7.0	7.5	32
Cattle	dig coef %	97.	97.	
Swine	dig coef %	* 176.	176.	
Total fat, official	%	10.8	11.6	18
N-free extract	%	3.4	3.6	
Protein (N x 6.25)	%	59.4	63.7	9
Cattle	dig coef %	76.	76.	
Swine	dig coef %	* 97.	97.	
Cattle	dig prot %	45.1	48.4	
Swine	dig prot %	57.6	61.8	
Energy	GE Mcal/kg			
Cattle	DE Mcal/kg	* 2.76	2.96	
Sheep	DE Mcal/kg	* 2.71	2.91	
Swine	DE kcal/kg	*2813.	3017.	
Cattle	ME Mcal/kg	* 2.26	2.43	
Chickens	ME_n kcal/kg	1944.	2086.	
Sheep	ME Mcal/kg	* 2.22	2.39	
Swine	ME kcal/kg	*2339.	2508.	
Cattle	TDN %	62.6	67.1	
Sheep	TDN %	* 61.5	66.0	
Swine	TDN %	* 63.8	68.4	
Aluminum	mg/kg	144.50	154.97	15
Barium	mg/kg	4.490	4.816	35
Boron	mg/kg	16.279	17.458	17
Calcium	%	8.03	8.61	28
Chromium	mg/kg	17.425	18.688	15
Copper	mg/kg	10.7	11.5	26
Iron	%	0.037	0.040	27
Magnesium	%	0.23	0.25	17
Manganese	mg/kg	8.6	9.2	29
Phosphorus	%	4.33	4.65	23
Potassium	%	0.73	0.78	12
Selenium	mg/kg	4.595	4.928	21
Sodium	%	0.73	0.78	26
Strontium	mg/kg	198.822	213.232	2
Zinc	mg/kg	211.3	226.7	17
Choline	mg/kg	2738.	2936.	
Niacin	mg/kg	141.4	151.7	22
Pantothenic acid	mg/kg	7.1	7.6	
Riboflavin	mg/kg	6.8	7.3	14
Vitamin B_{12}	μg/kg	300.7	322.5	22
Alanine	%	3.58	3.84	7
Arginine	%	3.74	4.01	31
Aspartic acid	%	4.93	5.28	10
Cystine	%	0.45	0.48	25
Glutamic acid	%	6.32	6.78	8
Glycine	%	4.28	4.59	7
Histidine	%	1.81	1.94	14
Isoleucine	%	2.36	2.53	10
Leucine	%	3.81	4.09	9
Lysine	%	4.18	4.48	22
Methionine	%	1.45	1.55	14
Phenylalanine	%	2.17	2.33	10
Proline	%	2.87	3.08	6
Serine	%	2.21	2.37	20
Threonine	%	2.30	2.47	10
Tryptophan	%	0.59	0.63	27
Tyrosine	%	1.74	1.86	10
Valine	%	2.77	2.97	12

Feed Name or Analyses		Mean As Fed	Mean Dry	C.V. ± %

FISH, WHITE. Gadidae (family), Lophiidae (family), Rajidae (family)

Fish, white, whole or cuttings, cooked mech-extd dehy grnd, mx 4% oil, (5)
White fish meal (CFA)
Fish, cod, meal
Fish, cusk, meal
Fish, haddock, meal
Fish, hake, meal
Fish, pollock, meal
Fish, monkfish, meal
Fish, skate, meal

Ref No 5 02 025 United States

Analyses		As Fed	Dry	C.V.
Dry matter	%	90.6	100.0	1
Ash	%	24.0	26.5	10
Crude fiber	%	0.5	0.6	46
Ether extract	%	4.2	4.6	10
Sheep	dig coef %	90.	90.	
Swine	dig coef %	* 176.	176.	
N-free extract	%	1.0	1.1	
Protein (N x 6.25)	%	60.9	67.2	6
Sheep	dig coef %	93.	93.	
Swine	dig coef %	* 97.	97.	
Sheep	dig prot %	56.6	62.5	
Swine	dig prot %	59.0	65.2	
Linoleic	%	.070	.077	
Energy	GE Mcal/kg			
Cattle	DE Mcal/kg	* 2.84	3.14	
Sheep	DE Mcal/kg	* 2.87	3.17	
Swine	DE kcal/kg	*2804.	3095.	
Cattle	ME Mcal/kg	* 2.33	2.57	
Chickens	ME_n kcal/kg	2640.	2914.	
Sheep	ME Mcal/kg	* 2.36	2.60	
Swine	ME kcal/kg	*2310.	2550.	
Cattle	TDN %	* 64.6	70.3	
Sheep	TDN %	65.1	71.9	
Swine	TDN %	* 63.6	70.2	
Calcium	%	7.27	8.02	15
Copper	mg/kg	2.8	3.0	
Iron	%	0.016	0.018	
Manganese	mg/kg	14.2	15.7	44
Phosphorus	%	3.63	4.00	18
Choline	mg/kg	8838.	9755.	
Niacin	mg/kg	69.1	76.3	24
Pantothenic acid	mg/kg	8.7	9.6	63
Riboflavin	mg/kg	8.9	9.9	36
Thiamine	mg/kg	1.7	1.9	79
α-tocopherol	mg/kg	9.0	9.9	
Vitamin B_6	mg/kg	8.50	9.39	54
Vitamin B_{12}	μg/kg	98.3	108.5	

Ref No 5 02 025 Canada

Analyses		As Fed	Dry	C.V.
Dry matter	%	91.0	100.0	0
Ether extract	%			
Sheep	dig coef %	90.	90.	
Protein (N x 6.25)	%			
Sheep	dig coef %	93.	93.	
Energy	GE Mcal/kg			
Sheep	DE Mcal/kg	* 2.88	3.17	
Sheep	ME Mcal/kg	* 2.37	2.60	
Sheep	TDN %	65.4	71.9	

(1) dry forages and roughages (2) pasture, range plants, and forages fed green (3) silages (4) energy feeds (5) protein supplements (6) minerals (7) vitamins (8) additives

Analyses		As Fed	Dry	C.V.
Pantothenic acid	mg/kg	10.1	11.1	16
Riboflavin	mg/kg	9.9	10.8	13
Thiamine	mg/kg	10.4	11.4	

FISH, WHITING. Gadus merlangus

Fish, whiting, whole or cuttings, cooked mech-extd dehy grnd, (5)

Ref No 5 02 026 United States

Analyses		As Fed	Dry	C.V.
Dry matter	%	29.9	100.0	
Ash	%	5.4	18.1	
Crude fiber	%	1.7	5.7	
Ether extract	%	4.0	13.4	
Protein (N x 6.25)	%	18.8	62.9	
Energy	GE Mcal/kg			
Cattle	DE Mcal/kg	* 0.99	3.30	
Sheep	DE Mcal/kg	* 0.92	3.08	
Swine	DE kcal/kg	*1051.	3516.	
Cattle	ME Mcal/kg	* 0.81	2.70	
Sheep	ME Mcal/kg	* 0.75	2.52	
Swine	ME kcal/kg	* 876.	2929.	
Cattle	TDN %	* 22.4	74.8	
Sheep	TDN %	* 20.9	69.8	
Swine	TDN %	* 23.8	79.7	

FISH, WOLFFISH. Anarhichas lupus

Fish, wolffish, flesh, raw, (5)

Ref No 5 09 067 Canada

Analyses		As Fed	Dry	C.V.
Dry matter	%	20.3	100.0	
Ash	%	1.2	5.9	
Ether extract	%	2.1	10.3	
Protein (N x 6.25)	%	18.5	91.1	

Fish, cod, meal – see Fish, white, whole or cuttings, cooked mech-extd dehy grnd, mx 4% oil, (5)

Fish, cusk, meal – see Fish, white, whole or cuttings, cooked mech-extd dehy grnd, mx 4% oil, (5)

Fish, haddock, meal – see Fish, white, whole or cuttings, cooked mech-extd dehy grnd, mx 4% oil, (5)

Fish, hake, meal – see Fish, white, whole or cuttings, cooked mech-extd dehy grnd, mx 4% oil, (5)

Fish liver and glandular meal (AAFCO) – see Fish, viscera, dehy grnd, mn 50% liver mn 40% mg riboflavin per kg (5)

Fish liver meal (CFA) – see Fish, livers, extn unspecified dehy grnd, salt declared above 4%, (5)

Fish meal (AAFCO) – see Fish, whole or cuttings, cooked mech-extd dehy grnd, salt declared above 3% mx 7%, (5)

Fish meal (CFA) – see Fish, whole or cuttings, cooked mech-extd dehy grnd, (5)

Fish meal, anchovy – see Fish, anchovy, whole or cuttings, cooked mech-extd dehy grnd, (5)

Fish meal, drum – see Fish, redfish, whole or cuttings, cooked mech-extd dehy grnd, (5)

Fish meal, herring – see Fish, herring, whole or cuttings, cooked mech-extd dehy grnd, (5)

Fish meal, menhaden – see Fish, menhaden, whole or cuttings, cooked mech-extd dehy grnd, (5)

Fish meal, pilchard – see Fish, pilchard, whole or cuttings, cooked mech-extd dehy grnd, (5)

Fish meal, redfish – see Fish, redfish, whole or cuttings, cooked mech-extd dehy grnd, (5)

Fish meal, salmon – see Fish, salmon, whole or cuttings, cooked mech-extd dehy grnd, (5)

Fish meal, sardine – see Fish, sardine, whole or cuttings, cooked mech-extd dehy grnd, (5)

Fish, meal, shark – see Fish, shark, whole or cuttings, cooked mech-extd dehy grnd, (5)

Fish meal, tuna – see Fish, tuna, whole or cuttings, cooked mech-extd dehy grnd, (5)

Fish, monkfish, meal – see Fish, white, whole or cuttings, cooked mech-extd dehy grnd, mx 4% oil, (5)

Fish, pollock, meal – see Fish, white, whole or cuttings, cooked mech-extd dehy grnd, mx 4% oil, (5)

Fish residue meal (AAFCO) – see Fish, glue by-product, dehy grnd, salt declared above 3% mx 7%, (5)

Fish, skate, meal – see Fish, white, whole or cuttings, cooked mech-extd dehy grnd, mx 4% oil, (5)

Fish solubles, dried – see Fish, stickwater solubles, cooked dehy, mn 60% protein, (5)

FLAMBOYANTTREE. Delonix regia

Flamboyanttree, seeds w hulls, (5)

Ref No 5 09 022 United States

Analyses		As Fed	Dry	C.V.
Dry matter	%		100.0	
Protein (N x 6.25)	%		58.0	
Alanine	%		2.15	
Arginine	%		6.96	

Feed Name or Analyses			Mean As Fed	Mean Dry	C.V. ± %
Aspartic acid	%			4.81	
Glutamic acid	%			12.24	
Glycine	%			2.20	
Histidine	%			1.28	
Hydroxyproline	%			0.00	
Isoleucine	%			1.86	
Leucine	%			3.60	
Lysine	%			2.61	
Methionine	%			0.64	
Phenylalanine	%			2.26	
Proline	%			2.09	
Serine	%			2.44	
Threonine	%			1.51	
Tyrosine	%			1.62	
Valine	%			2.44	

FLATPEA. Lathyrus sylvestris

Flatpea, hay, s-c, (1)

Ref No 1 02 030 United States

Feed Name or Analyses			Mean As Fed	Mean Dry	C.V. ± %
Dry matter	%		90.0	100.0	2
Ash	%		6.5	7.2	3
Crude fiber	%		26.4	29.3	9
Ether extract	%		3.3	3.7	10
N-free extract	%		30.1	33.5	
Protein (N x 6.25)	%		23.7	26.3	13
Cattle	dig prot %	*	17.8	19.8	
Goats	dig prot %	*	19.0	21.1	
Horses	dig prot %	*	17.9	19.9	
Rabbits	dig prot %	*	17.1	19.0	
Sheep	dig prot %	*	18.2	20.2	
Fatty acids	%		3.1	3.5	
Energy	GE Mcal/kg				
Cattle	DE Mcal/kg	*	2.29	2.54	
Sheep	DE Mcal/kg	*	2.38	2.64	
Cattle	ME Mcal/kg	*	1.88	2.09	
Sheep	ME Mcal/kg	*	1.95	2.16	
Cattle	TDN %	*	51.9	57.7	
Sheep	TDN %	*	53.9	59.9	
Phosphorus	%		0.29	0.33	
Potassium	%		1.97	2.19	

Flatpea, hay, s-c, immature, (1)

Ref No 1 02 027 United States

Feed Name or Analyses			Mean As Fed	Mean Dry	C.V. ± %
Dry matter	%		86.2	100.0	2
Ash	%		6.0	7.0	3
Crude fiber	%		22.9	26.6	10
Ether extract	%		3.2	3.7	5
N-free extract	%		26.1	30.3	
Protein (N x 6.25)	%		27.9	32.4	4
Cattle	dig prot %	*	21.5	25.0	
Goats	dig prot %	*	23.1	26.8	
Horses	dig prot %	*	21.6	25.0	
Rabbits	dig prot %	*	20.4	23.7	
Sheep	dig prot %	*	22.1	25.6	
Energy	GE Mcal/kg				
Cattle	DE Mcal/kg	*	2.22	2.58	
Sheep	DE Mcal/kg	*	2.38	2.76	
Cattle	ME Mcal/kg	*	1.82	2.11	
Sheep	ME Mcal/kg	*	1.95	2.27	
Cattle	TDN %	*	50.4	58.4	
Sheep	TDN %	*	54.0	62.7	
Calcium	%		0.45	0.52	
Phosphorus	%		0.42	0.49	

Flatpea, hay, s-c, milk stage, (1)

Ref No 1 02 028 United States

Feed Name or Analyses			Mean As Fed	Mean Dry	C.V. ± %
Dry matter	%		86.2	100.0	2
Ash	%		5.9	6.9	3
Crude fiber	%		26.2	30.4	4
Ether extract	%		4.0	4.6	5
N-free extract	%		29.0	33.7	
Protein (N x 6.25)	%		21.0	24.4	6
Cattle	dig prot %	*	15.6	18.1	
Goats	dig prot %	*	16.7	19.3	
Horses	dig prot %	*	15.7	18.2	
Rabbits	dig prot %	*	15.1	17.5	
Sheep	dig prot %	*	15.9	18.5	
Energy	GE Mcal/kg				
Cattle	DE Mcal/kg	*	2.20	2.55	
Sheep	DE Mcal/kg	*	2.22	2.57	
Cattle	ME Mcal/kg	*	1.80	2.09	
Sheep	ME Mcal/kg	*	1.82	2.11	
Cattle	TDN %	*	49.9	57.8	
Sheep	TDN %	*	50.3	58.3	
Phosphorus	%		0.26	0.30	

Flatpea, hay, s-c, cut 1, (1)

Ref No 1 02 029 United States

Feed Name or Analyses			Mean As Fed	Mean Dry	C.V. ± %
Dry matter	%		86.2	100.0	2
Calcium	%		0.50	0.58	13
Phosphorus	%		0.35	0.41	28

Flatpea, aerial part, fresh, (2)

Ref No 2 02 033 United States

Feed Name or Analyses			Mean As Fed	Mean Dry	C.V. ± %
Dry matter	%		22.5	100.0	7
Ash	%		1.6	7.1	6
Crude fiber	%		6.2	27.7	5
Ether extract	%		0.8	3.7	8
N-free extract	%		7.4	32.7	
Protein (N x 6.25)	%		6.5	28.9	10
Cattle	dig prot %	*	5.1	22.5	
Goats	dig prot %	*	5.3	23.5	
Horses	dig prot %	*	5.0	22.1	
Rabbits	dig prot %	*	4.7	21.0	
Sheep	dig prot %	*	5.4	24.0	
Energy	GE Mcal/kg				
Cattle	DE Mcal/kg	*	0.85	3.80	
Sheep	DE Mcal/kg	*	0.56	2.47	
Cattle	ME Mcal/kg	*	0.70	3.11	
Sheep	ME Mcal/kg	*	0.46	2.03	
Cattle	TDN %	*	19.4	86.1	
Sheep	TDN %	*	12.6	56.1	
Calcium	%		0.11	0.49	
Phosphorus	%		0.11	0.49	

Flatpea, aerial part, fresh, immature, (2)

Ref No 2 02 031 United States

Feed Name or Analyses			Mean As Fed	Mean Dry	C.V. ± %
Dry matter	%			100.0	
Ash	%			7.2	3
Crude fiber	%			25.6	8
Ether extract	%			4.3	10
N-free extract	%			26.9	
Protein (N x 6.25)	%			36.0	5
Cattle	dig prot %	*		28.5	
Goats	dig prot %	*		30.1	
Horses	dig prot %	*		28.1	
Rabbits	dig prot %	*		26.5	
Sheep	dig prot %	*		30.5	
Energy	GE Mcal/kg				
Cattle	DE Mcal/kg	*		2.76	
Sheep	DE Mcal/kg	*		2.49	
Cattle	ME Mcal/kg	*		2.26	
Sheep	ME Mcal/kg	*		2.04	
Cattle	TDN %	*		62.5	
Sheep	TDN %	*		56.4	
Calcium	%			0.48	11
Phosphorus	%			0.49	19

Flatpea, aerial part, fresh, mid-bloom, (2)

Ref No 2 02 032 United States

Feed Name or Analyses			Mean As Fed	Mean Dry	C.V. ± %
Dry matter	%			100.0	
Ash	%			5.9	
Crude fiber	%			29.0	
Ether extract	%			3.6	
N-free extract	%			33.4	
Protein (N x 6.25)	%			28.1	
Cattle	dig prot %	*		21.8	
Goats	dig prot %	*		22.8	
Horses	dig prot %	*		21.4	
Rabbits	dig prot %	*		20.4	
Sheep	dig prot %	*		23.2	
Energy	GE Mcal/kg				
Cattle	DE Mcal/kg	*		2.52	
Sheep	DE Mcal/kg	*		2.39	
Cattle	ME Mcal/kg	*		2.07	
Sheep	ME Mcal/kg	*		1.96	
Cattle	TDN %	*		57.2	
Sheep	TDN %	*		54.1	

FLAX. Linum usitatissimum

Flax, capsule chaff, (1)

Ref No 1 02 039 United States

Feed Name or Analyses			Mean As Fed	Mean Dry	C.V. ± %
Dry matter	%		88.9	100.0	
Ash	%		10.9	12.3	
Crude fiber	%		31.1	35.1	
Sheep	dig coef %		32.	32.	
Swine	dig coef %		19.	19.	
Ether extract	%		6.8	7.7	
Sheep	dig coef %		84.	84.	
Swine	dig coef %		79.	79.	
N-free extract	%		30.4	34.3	
Sheep	dig coef %		39.	39.	
Swine	dig coef %		36.	36.	
Protein (N x 6.25)	%		9.6	10.8	
Sheep	dig coef %		61.	61.	
Swine	dig coef %		43.	43.	
Cattle	dig prot %	*	5.6	6.3	
Goats	dig prot %	*	5.9	6.6	
Horses	dig prot %	*	6.0	6.7	
Rabbits	dig prot %	*	6.2	7.0	
Sheep	dig prot %		5.9	6.6	
Swine	dig prot %		4.1	4.6	
Energy	GE Mcal/kg				
Cattle	DE Mcal/kg	*	2.50	2.81	
Sheep	DE Mcal/kg	*	1.79	2.01	
Swine	DE kcal/kg	*	1459.	1642.	
Cattle	ME Mcal/kg	*	2.05	2.31	
Sheep	ME Mcal/kg	*	1.47	1.65	
Swine	ME kcal/kg	*	1368.	1540.	
Cattle	TDN %	*	56.7	63.8	

Continued

Feed Name or Analyses				Mean As Fed	Mean Dry	C.V. ± %
Sheep	TDN	%		40.5	45.6	
Swine	TDN	%		33.1	37.2	

Ref No 1 02 039 Canada

Feed Name or Analyses				As Fed	Dry	C.V. ± %
Dry matter		%		89.3	100.0	
Crude fiber		%		30.5	34.1	
Sheep	dig coef	%		32.	32.	
Swine	dig coef	%		19.	19.	
Ether extract		%		3.6	4.1	
Sheep	dig coef	%		84.	84.	
Swine	dig coef	%		79.	79.	
N-free extract		%				
Sheep	dig coef	%		39.	39.	
Swine	dig coef	%		36.	36.	
Protein (N x 6.25)		%		8.6	9.7	
Sheep	dig coef	%		61.	61.	
Swine	dig coef	%		43.	43.	
Cattle	dig prot	%	*	4.7	5.3	
Goats	dig prot	%	*	5.0	5.6	
Horses	dig prot	%	*	5.1	5.7	
Rabbits	dig prot	%	*	5.5	6.1	
Sheep	dig prot	%		5.3	5.9	
Swine	dig prot	%		3.7	4.2	
Energy	GE Mcal/kg					
Sheep	DE Mcal/kg		*	1.72	1.93	
Swine	DE kcal/kg		*	1401.	1569.	
Sheep	ME Mcal/kg		*	1.41	1.58	
Swine	ME kcal/kg		*	1317.	1476.	
Sheep	TDN	%		39.0	43.7	
Swine	TDN	%		31.8	35.6	
Calcium		%		0.85	0.96	
Phosphorus		%		0.10	0.11	

Flax, capsule chaff w molasses added, (1)

Ref No 1 02 040 United States

Feed Name or Analyses				As Fed	Dry	C.V. ± %
Dry matter		%		80.6	100.0	
Ash		%		9.5	11.8	
Crude fiber		%		14.9	18.5	
Sheep	dig coef	%		28.	28.	
Swine	dig coef	%		20.	20.	
Ether extract		%		2.6	3.3	
Sheep	dig coef	%		72.	72.	
Swine	dig coef	%		46.	46.	
N-free extract		%		44.1	54.7	
Sheep	dig coef	%		78.	78.	
Swine	dig coef	%		74.	74.	
Protein (N x 6.25)		%		9.6	11.9	
Sheep	dig coef	%		64.	64.	
Swine	dig coef	%		44.	44.	
Cattle	dig prot	%	*	5.8	7.2	
Goats	dig prot	%	*	6.1	7.6	
Horses	dig prot	%	*	6.1	7.6	
Rabbits	dig prot	%	*	6.3	7.8	
Sheep	dig prot	%		6.1	7.6	
Swine	dig prot	%		4.2	5.2	
Energy	GE Mcal/kg					
Cattle	DE Mcal/kg		*	3.08	3.82	
Sheep	DE Mcal/kg		*	2.16	2.68	
Swine	DE kcal/kg		*	1874.	2326.	
Cattle	ME Mcal/kg		*	2.52	3.13	
Sheep	ME Mcal/kg		*	1.77	2.19	
Swine	ME kcal/kg		*	1755.	2177.	

(1) dry forages and roughages (2) pasture, range plants, and forages fed green (3) silages (4) energy feeds (5) protein supplements (6) minerals (7) vitamins (8) additives

Feed Name or Analyses				Mean As Fed	Mean Dry	C.V. ± %
Cattle	TDN	%	*	69.8	86.6	
Sheep	TDN	%		48.9	60.7	
Swine	TDN	%		42.5	52.7	

Flax, fiber by-product, s-c, (1)

Ref No 1 02 035 United States

Feed Name or Analyses				As Fed	Dry	C.V. ± %
Dry matter		%		90.9	100.0	
Ash		%		6.3	6.9	
Crude fiber		%		37.8	41.6	
Cattle	dig coef	%		48.	48.	
Sheep	dig coef	%		19.	19.	
Ether extract		%		2.7	3.0	
Cattle	dig coef	%		75.	75.	
Sheep	dig coef	%		65.	65.	
N-free extract		%		34.2	37.6	
Cattle	dig coef	%		43.	43.	
Sheep	dig coef	%		38.	38.	
Protein (N x 6.25)		%		9.8	10.8	
Cattle	dig coef	%		64.	64.	
Sheep	dig coef	%		48.	48.	
Cattle	dig prot	%		6.3	6.9	
Goats	dig prot	%	*	6.1	6.7	
Horses	dig prot	%	*	6.1	6.7	
Rabbits	dig prot	%	*	6.4	7.0	
Sheep	dig prot	%		4.7	5.2	
Energy	GE Mcal/kg					
Cattle	DE Mcal/kg		*	1.93	2.12	
Sheep	DE Mcal/kg		*	1.26	1.38	
Cattle	ME Mcal/kg		*	1.58	1.74	
Sheep	ME Mcal/kg		*	1.03	1.13	
Cattle	TDN	%		43.7	48.1	
Sheep	TDN	%		28.5	31.4	

Flax, fiber by-product, s-c grnd pelleted, (1)

Ref No 1 07 718 United States

Feed Name or Analyses				As Fed	Dry	C.V. ± %
Dry matter		%		89.3	100.0	
Sheep	dig coef	%		34.	34.	
Organic matter		%		75.6	84.7	
Ash		%		13.7	15.3	
Crude fiber		%		24.1	27.0	
Sheep	dig coef	%		14.	14.	
Ether extract		%		2.9	3.2	
Sheep	dig coef	%		48.	48.	
N-free extract		%		36.9	41.3	
Sheep	dig coef	%		49.	49.	
Protein (N x 6.25)		%		11.8	13.2	
Sheep	dig coef	%		54.	54.	
Cattle	dig prot	%	*	7.5	8.4	
Goats	dig prot	%	*	7.9	8.9	
Horses	dig prot	%	*	7.8	8.7	
Rabbits	dig prot	%	*	7.9	8.9	
Sheep	dig prot	%		6.4	7.1	
Energy	GE Mcal/kg					
Cattle	DE Mcal/kg		*	1.55	1.74	
Sheep	DE Mcal/kg		*	1.36	1.53	
Cattle	ME Mcal/kg		*	1.28	1.43	
Sheep	ME Mcal/kg		*	1.12	1.25	
Cattle	TDN	%	*	35.3	39.5	
Sheep	TDN	%		30.9	34.6	

Flax, fiber by-product, mn 9% protein mx 35% fiber, (1)

Flax plant product (AAFCO)

Ref No 1 02 036 United States

Feed Name or Analyses				Mean As Fed	Mean Dry	C.V. ± %
Dry matter		%		91.6	100.0	
Ash		%		5.5	6.1	
Crude fiber		%		32.6	35.7	5
Ether extract		%		2.7	3.0	41
N-free extract		%		40.3	44.0	
Protein (N x 6.25)		%		10.3	11.3	64
Cattle	dig prot	%	*	6.1	6.7	
Goats	dig prot	%	*	6.5	7.1	
Horses	dig prot	%	*	6.5	7.1	
Rabbits	dig prot	%	*	6.7	7.4	
Sheep	dig prot	%	*	6.1	6.7	
Fatty acids		%		2.9	3.1	
Energy	GE Mcal/kg					
Cattle	DE Mcal/kg		*	1.31	1.43	
Sheep	DE Mcal/kg		*	1.44	1.57	
Cattle	ME Mcal/kg		*	1.07	1.17	
Sheep	ME Mcal/kg		*	1.18	1.29	
Cattle	TDN	%	*	29.7	32.4	
Sheep	TDN	%	*	32.7	35.7	
Calcium		%		1.21	1.32	
Phosphorus		%		0.22	0.24	

Flax, hulls, (1)

Ref No 1 02 037 United States

Feed Name or Analyses				As Fed	Dry	C.V. ± %
Dry matter		%		91.0	100.0	0
Ash		%		9.2	10.1	8
Crude fiber		%		28.6	31.5	3
Sheep	dig coef	%		31.	31.	
Ether extract		%		1.3	1.5	22
Sheep	dig coef	%		38.	38.	
N-free extract		%		44.1	48.5	
Sheep	dig coef	%		52.	52.	
Protein (N x 6.25)		%		7.7	8.5	32
Sheep	dig coef	%		20.	20.	
Cattle	dig prot	%	*	3.9	4.3	
Goats	dig prot	%	*	4.1	4.5	
Horses	dig prot	%	*	4.3	4.7	
Rabbits	dig prot	%	*	4.8	5.2	
Sheep	dig prot	%		1.5	1.7	
Energy	GE Mcal/kg					
Cattle	DE Mcal/kg		*	1.70	1.87	
Sheep	DE Mcal/kg		*	1.52	1.67	
Cattle	ME Mcal/kg		*	1.39	1.53	
Sheep	ME Mcal/kg		*	1.25	1.37	
Cattle	TDN	%	*	38.6	42.4	
Sheep	TDN	%		34.5	37.9	

Flax, straw, (1)

Ref No 1 02 038 United States

Feed Name or Analyses				As Fed	Dry	C.V. ± %
Dry matter		%		92.9	100.0	2
Ash		%		6.8	7.4	14
Crude fiber		%		42.7	46.0	3
Sheep	dig coef	%		48.	48.	
Ether extract		%		3.1	3.4	11
Sheep	dig coef	%		36.	36.	
N-free extract		%		32.9	35.5	
Sheep	dig coef	%		38.	38.	
Protein (N x 6.25)		%		7.2	7.8	10
Sheep	dig coef	%		6.	6.	
Cattle	dig prot	%	*	3.4	3.7	

Feed Name or Analyses		Mean As Fed	Mean Dry	C.V. ± %
Goats	dig prot %	* 3.5	3.8	
Horses	dig prot %	* 3.8	4.1	
Rabbits	dig prot %	* 4.3	4.7	
Sheep	dig prot %	0.4	0.5	
Fatty acids	%	3.2	3.4	
Energy	GE Mcal/kg	4.22	4.54	
Sheep	GE dig coef %	42.	42.	
Cattle	DE Mcal/kg	* 1.54	1.66	
Sheep	DE Mcal/kg	1.77	1.91	
Cattle	ME Mcal/kg	* 1.26	1.36	
Sheep	ME Mcal/kg	* 1.45	1.56	
Cattle	TDN %	* 34.9	37.6	
Sheep	TDN %	36.0	38.8	
Calcium	%	0.67	0.72	
Chlorine	%	0.25	0.27	
Magnesium	%	0.29	0.31	
Manganese	mg/kg	7.6	8.2	
Phosphorus	%	0.10	0.11	
Potassium	%	1.62	1.74	
Sulphur	%	0.25	0.27	

Ref No 1 02 038 — Canada

Feed Name or Analyses		Mean As Fed	Mean Dry	C.V. ± %
Dry matter	%	93.3	100.0	2
Sheep	dig coef %	44.	44.	
Ash	%	2.8	3.0	
Crude fiber	%	48.9	52.5	15
Sheep	dig coef %	48.	48.	
Ether extract	%	1.5	1.7	
Sheep	dig coef %	36.	36.	
N-free extract	%	35.4	37.9	
Sheep	dig coef %	38.	38.	
Protein (N x 6.25)	%	4.6	5.0	45
Sheep	dig coef %	6.	6.	
Cattle	dig prot %	* 1.2	1.2	
Goats	dig prot %	* 1.1	1.2	
Horses	dig prot %	* 1.6	1.7	
Rabbits	dig prot %	* 2.3	2.5	
Sheep	dig prot %	0.3	0.3	
Energy	GE Mcal/kg	4.23	4.54	
Sheep	GE dig coef %	42.	42.	
Cattle	DE Mcal/kg	* 1.64	1.76	
Sheep	DE Mcal/kg	1.78	1.91	
Cattle	ME Mcal/kg	* 1.34	1.44	
Sheep	ME Mcal/kg	* 1.46	1.56	
Cattle	TDN %	* 37.3	40.0	
Sheep	TDN %	38.5	41.2	
Calcium	%	0.48	0.52	22
Phosphorus	%	0.06	0.06	98

Flax, seed screenings, (4)

Ref No 4 02 056 — United States

Feed Name or Analyses		Mean As Fed	Mean Dry	C.V. ± %
Dry matter	%	91.5	100.0	1
Ash	%	7.1	7.8	16
Crude fiber	%	13.0	14.2	17
Ether extract	%	9.9	10.9	16
N-free extract	%	45.6	49.9	
Protein (N x 6.25)	%	15.8	17.3	10
Cattle	dig prot %	* 10.9	11.9	
Goats	dig prot %	* 12.0	13.1	
Horses	dig prot %	* 12.0	13.1	
Sheep	dig prot %	* 12.0	13.1	
Energy	GE Mcal/kg	4.33	4.73	
Cattle	DE Mcal/kg	* 2.67	2.92	
Sheep	DE Mcal/kg	* 2.59	2.83	
Swine	DE kcal/kg	*2146.	2345.	
Cattle	ME Mcal/kg	* 2.19	2.39	
Sheep	ME Mcal/kg	* 2.12	2.32	
Swine	ME kcal/kg	*1985.	2169.	
Cattle	NE_m Mcal/kg	* 1.26	1.38	
Cattle	NE_{gain} Mcal/kg	* 0.72	0.79	
Cattle	$NE_{lactating\ cows}$ Mcal/kg	* 1.46	1.60	
Cattle	TDN %	* 60.7	66.3	
Sheep	TDN %	* 58.7	64.1	
Swine	TDN %	* 48.7	53.2	
Calcium	%	0.37	0.40	
Phosphorus	%	0.43	0.47	

Ref No 4 02 056 — Canada

Feed Name or Analyses		Mean As Fed	Mean Dry	C.V. ± %
Dry matter	%	90.8	100.0	
Crude fiber	%	8.2	9.0	
Protein (N x 6.25)	%	19.1	21.0	
Cattle	dig prot %	* 13.9	15.3	
Goats	dig prot %	* 15.0	16.5	
Horses	dig prot %	* 15.0	16.5	
Sheep	dig prot %	* 15.0	16.5	
Energy	GE Mcal/kg	4.30	4.73	
Calcium	%	0.25	0.27	
Phosphorus	%	0.42	0.46	

Flax, seeds, (5)

Ref No 5 02 052 — United States

Feed Name or Analyses		Mean As Fed	Mean Dry	C.V. ± %
Dry matter	%	93.2	100.0	
Ash	%	4.9	5.2	
Crude fiber	%	6.1	6.6	
Sheep	dig coef %	30.	30.	
Ether extract	%	35.0	37.5	
Sheep	dig coef %	87.	87.	
N-free extract	%	24.1	25.8	
Sheep	dig coef %	42.	42.	
Protein (N x 6.25)	%	23.2	24.8	
Sheep	dig coef %	84.	84.	
Sheep	dig prot %	19.4	20.9	
Energy	GE Mcal/kg			
Cattle	DE Mcal/kg	* 4.87	5.22	
Sheep	DE Mcal/kg	* 4.40	4.73	
Swine	DE kcal/kg	*4902.	5260.	
Cattle	ME Mcal/kg	* 3.99	4.28	
Sheep	ME Mcal/kg	* 3.61	3.87	
Swine	ME kcal/kg	*4461.	4787.	
Cattle	TDN %	* 110.4	118.4	
Sheep	TDN %	99.9	107.2	
Swine	TDN %	* 111.2	119.3	
Calcium	%	0.22	0.23	
Iron	%	0.009	0.010	
Magnesium	%	0.40	0.43	
Manganese	mg/kg	60.7	65.1	
Phosphorus	%	0.52	0.55	
Potassium	%	0.78	0.84	
Sulphur	%	0.23	0.25	

Flax, seeds, grnd, commercial, (5)

Flaxseed meal (CFA)

Ground flaxseed (CFA)

Ref No 5 02 042 — Canada

Feed Name or Analyses		Mean As Fed	Mean Dry	C.V. ± %
Dry matter	%	95.7	100.0	
Protein (N x 6.25)	%	17.8	18.6	
Calcium	%	0.28	0.29	
Phosphorus	%	0.55	0.57	

Flax, seeds, mech-extd grnd, (5)

Ref No 5 02 044 — United States

Feed Name or Analyses		Mean As Fed	Mean Dry	C.V. ± %
Dry matter	%	89.9	100.0	
Ash	%	6.2	6.9	
Crude fiber	%	8.1	9.1	
Cattle	dig coef %	-10.	-10.	
Sheep	dig coef %	50.	50.	
Ether extract	%	8.4	9.4	
Cattle	dig coef %	93.	93.	
Sheep	dig coef %	92.	92.	
N-free extract	%	33.5	37.3	
Cattle	dig coef %	78.	78.	
Sheep	dig coef %	79.	79.	
Protein (N x 6.25)	%	33.6	37.4	
Cattle	dig coef %	85.	85.	
Sheep	dig coef %	84.	84.	
Cattle	dig prot %	28.6	31.8	
Sheep	dig prot %	28.2	31.4	
Energy	GE Mcal/kg			
Cattle	DE Mcal/kg	* 3.15	3.51	
Sheep	DE Mcal/kg	* 3.36	3.74	
Swine	DE kcal/kg	*3719.	4139.	
Cattle	ME Mcal/kg	* 2.58	2.88	
Sheep	ME Mcal/kg	* 2.76	3.07	
Swine	ME kcal/kg	*3290.	3661.	
Cattle	TDN %	71.4	79.5	
Sheep	TDN %	76.2	84.8	
Swine	TDN %	* 84.4	93.9	

Flax, seeds, mech-extd grnd, mx 10% fiber, (5)

Linseed meal, mechanical extracted (AAFCO)

Linseed meal (CFA)

Linseed oil meal, expeller extracted

Linseed oil meal, hydraulic extracted

Linseed oil meal, old process

Ref No 5 02 045 — United States

Feed Name or Analyses		Mean As Fed	Mean Dry	C.V. ± %
Dry matter	%	91.1	100.0	1
Ash	%	5.6	6.2	4
Crude fiber	%	8.9	9.7	13
Cattle	dig coef %	28.	28.	
Sheep	dig coef %	* 50.	50.	
Swine	dig coef %	20.	20.	
Ether extract	%	5.1	5.6	17
Cattle	dig coef %	89.	89.	
Sheep	dig coef %	* 92.	92.	
Swine	dig coef %	62.	62.	
N-free extract	%	35.6	39.1	5
Cattle	dig coef %	82.	82.	
Sheep	dig coef %	* 79.	79.	
Swine	dig coef %	80.	80.	
Protein (N x 6.25)	%	35.9	39.4	3
Cattle	dig coef %	88.	88.	
Sheep	dig coef %	* 84.	84.	
Swine	dig coef %	90.	90.	
Cattle	dig prot %	31.6	34.6	
Sheep	dig prot %	30.1	33.1	
Swine	dig prot %	32.3	35.4	
Energy	GE Mcal/kg	4.26	4.68	1
Cattle	DE Mcal/kg	* 3.24	3.56	
Sheep	DE Mcal/kg	* 3.23	3.55	
Swine	DE kcal/kg	*3075.	3374.	
Cattle	ME Mcal/kg	* 2.66	2.92	
Chickens	ME_n kcal/kg	1520.	1668.	
Sheep	ME Mcal/kg	* 2.65	2.91	
Swine	ME kcal/kg	*2707.	2970.	

Continued

Feed Name or Analyses		Mean As Fed	Mean Dry	C.V. ± %
Cattle	NE_m Mcal/kg	* 1.73	1.90	
Cattle	NE_{gain} Mcal/kg	* 1.16	1.27	
Cattle	$NE_{lactating\ cows}$ Mcal/kg	* 2.03	2.23	
Cattle	TDN %	73.6	80.7	
Sheep	TDN %	73.3	80.5	
Swine	TDN %	69.7	76.5	
Calcium	%	0.39	0.43	12
Chlorine	%	0.04	0.04	20
Cobalt	mg/kg	0.429	0.471	36
Copper	mg/kg	42.1	46.2	16
Iron	%	0.017	0.019	31
Magnesium	%	0.56	0.62	6
Manganese	mg/kg	32.5	35.7	39
Phosphorus	%	0.87	0.95	3
Potassium	%	1.22	1.34	10
Sodium	%	0.10	0.11	36
Sulphur	%	0.39	0.43	6
Carotene	mg/kg	0.1	0.1	
Choline	mg/kg	1757.	1928.	8
Niacin	mg/kg	38.2	41.9	14
Pantothenic acid	mg/kg	15.0	16.5	29
Riboflavin	mg/kg	3.4	3.8	19
Thiamine	mg/kg	5.1	5.6	41
Vitamin A equivalent	IU/g	0.2	0.2	
Arginine	%	3.20	3.51	
Cystine	%	0.66	0.72	
Lysine	%	1.20	1.32	
Methionine	%	0.65	0.71	
Tryptophan	%	0.56	0.61	

Ref No 5 02 045 — Canada

Feed Name or Analyses		Mean As Fed	Mean Dry	C.V. ± %
Dry matter	%	91.0	100.0	1
Crude fiber	%	8.0	8.8	
Cattle	dig coef %	28.	28.	
Swine	dig coef %	20.	20.	
Ether extract	%	5.2	5.7	
Cattle	dig coef %	89.	89.	
Swine	dig coef %	62.	62.	
N-free extract	%			
Cattle	dig coef %	82.	82.	
Swine	dig coef %	80.	80.	
Protein (N x 6.25)	%	34.0	37.3	
Cattle	dig coef %	88.	88.	
Swine	dig coef %	90.	90.	
Cattle	dig prot %	29.9	32.8	
Swine	dig prot %	30.6	33.6	
Energy	GE Mcal/kg	4.26	4.68	
Cattle	DE Mcal/kg	* 3.24	3.56	
Sheep	DE Mcal/kg	* 3.23	3.55	
Swine	DE kcal/kg	*3071.	3374.	
Cattle	ME Mcal/kg	* 2.66	2.92	
Chickens	ME_n kcal/kg	1518.	1668.	
Sheep	ME Mcal/kg	* 2.65	2.91	
Swine	ME kcal/kg	*2716.	2985.	
Cattle	TDN %	73.5	80.7	
Sheep	TDN %	73.3	80.5	
Swine	TDN %	69.6	76.5	
Sulphur	%	0.28	0.30	
Niacin	mg/kg	30.5	33.6	
Pantothenic acid	mg/kg	12.7	13.9	9
Riboflavin	mg/kg	2.2	2.4	
Thiamine	mg/kg	2.9	3.1	
Tryptophan	%	0.63	0.69	

(1) dry forages and roughages (2) pasture, range plants, and forages fed green (3) silages (4) energy feeds (5) protein supplements (6) minerals (7) vitamins (8) additives

Flax, seeds, mech-extd grnd, mx 10% fat, (5)

Ref No 5 02 046 — United States

Feed Name or Analyses		Mean As Fed	Mean Dry	C.V. ± %
Dry matter	%	89.1	100.0	
Ash	%	5.8	6.5	
Crude fiber	%	8.2	9.2	
Cattle	dig coef %	-9.	-9.	
Ether extract	%	7.3	8.2	
Cattle	dig coef %	94.	94.	
N-free extract	%	34.2	38.4	
Cattle	dig coef %	78.	78.	
Protein (N x 6.25)	%	33.6	37.7	
Cattle	dig coef %	84.	84.	
Cattle	dig prot %	28.2	31.7	
Energy	GE Mcal/kg			
Cattle	DE Mcal/kg	* 3.07	3.44	
Sheep	DE Mcal/kg	* 3.29	3.69	
Swine	DE kcal/kg	*3617.	4060.	
Cattle	ME Mcal/kg	* 2.51	2.82	
Sheep	ME Mcal/kg	* 2.70	3.03	
Swine	ME kcal/kg	*3197.	3588.	
Cattle	TDN %	69.5	78.0	
Sheep	TDN %	* 74.6	83.7	
Swine	TDN %	* 82.0	92.1	

Flax, seeds, mech-extd grnd, 10% fat, (5)

Ref No 5 02 047 — United States

Feed Name or Analyses		Mean As Fed	Mean Dry	C.V. ± %
Dry matter	%	92.0	100.0	
Ash	%	5.3	5.8	
Crude fiber	%	7.6	8.3	
Cattle	dig coef %	-10.	-10.	
Ether extract	%	11.7	12.7	
Cattle	dig coef %	92.	92.	
N-free extract	%	35.0	38.0	
Cattle	dig coef %	75.	75.	
Protein (N x 6.25)	%	32.4	35.2	
Cattle	dig coef %	85.	85.	
Cattle	dig prot %	27.5	29.9	
Energy	GE Mcal/kg			
Cattle	DE Mcal/kg	* 3.40	3.69	
Sheep	DE Mcal/kg	* 3.56	3.86	
Swine	DE kcal/kg	*3545.	3853.	
Cattle	ME Mcal/kg	* 2.79	3.03	
Sheep	ME Mcal/kg	* 2.92	3.17	
Swine	ME kcal/kg	*3151.	3425.	
Cattle	TDN %	77.1	83.8	
Sheep	TDN %	* 80.6	87.6	
Swine	TDN %	* 80.4	87.4	

Flax, seeds, solv-extd grnd, mx 10% fiber, (5)

Linseed meal, solvent extracted (AAFCO)

Solvent extracted linseed meal (CFA)

Linseed oil meal, solvent extracted

Ref No 5 02 048 — United States

Feed Name or Analyses		Mean As Fed	Mean Dry	C.V. ± %
Dry matter	%	90.1	100.0	2
Ash	%	5.6	6.3	11
Crude fiber	%	8.7	9.7	9
Cattle	dig coef %	* 28.	28.	
Sheep	dig coef %	43.	43.	
Ether extract	%	1.7	1.9	44
Cattle	dig coef %	* 89.	89.	
Sheep	dig coef %	92.	92.	
N-free extract	%	38.0	42.2	
Cattle	dig coef %	* 82.	82.	
Sheep	dig coef %	85.	85.	
Protein (N x 6.25)	%	36.0	40.0	6
Cattle	dig coef %	* 88.	88.	
Sheep	dig coef %	87.	87.	
Cattle	dig prot %	31.7	35.2	
Sheep	dig prot %	31.3	34.8	
Energy	GE Mcal/kg	4.19	4.65	
Cattle	DE Mcal/kg	* 3.03	3.36	
Sheep	DE Mcal/kg	* 3.12	3.47	
Swine	DE kcal/kg	*3199.	3552.	
Cattle	ME Mcal/kg	* 2.48	2.76	
Chickens	ME_n kcal/kg	1406.	1561.	
Sheep	ME Mcal/kg	* 2.56	2.85	
Swine	ME kcal/kg	*2812.	3123.	
Cattle	NE_m Mcal/kg	* 1.56	1.73	
Cattle	NE_{gain} Mcal/kg	* 1.03	1.14	
Cattle	$NE_{lactating\ cows}$ Mcal/kg	* 1.85	2.05	
Cattle	TDN %	68.7	76.2	
Sheep	TDN %	70.9	78.7	
Swine	TDN %	* 72.5	80.6	
Calcium	%	0.40	0.44	17
Chlorine	%	0.04	0.04	
Cobalt	mg/kg	0.168	0.187	52
Copper	mg/kg	25.5	28.4	10
Iron	%	0.033	0.036	57
Magnesium	%	0.59	0.66	1
Manganese	mg/kg	37.3	41.4	32
Phosphorus	%	0.82	0.91	19
Potassium	%	1.37	1.52	16
Sodium	%	0.14	0.15	85
Choline	mg/kg	1223.	1358.	
Niacin	mg/kg	30.0	33.3	
Pantothenic acid	mg/kg	12.1	13.4	
Riboflavin	mg/kg	2.8	3.2	20
Thiamine	mg/kg	9.4	10.4	25
Arginine	%	3.19	3.54	
Cystine	%	0.66	0.73	
Lysine	%	1.20	1.33	
Methionine	%	0.60	0.66	
Tryptophan	%	0.56	0.62	

Ref No 5 02 048 — Canada

Feed Name or Analyses		Mean As Fed	Mean Dry	C.V. ± %
Dry matter	%		100.0	
Crude fiber	%			
Sheep	dig coef %		43.	
Ether extract	%			
Sheep	dig coef %		92.	
N-free extract	%			
Sheep	dig coef %		85.	
Protein (N x 6.25)	%			
Sheep	dig coef %		87.	
Energy	GE Mcal/kg		4.65	
Cattle	DE Mcal/kg	*	3.36	
Sheep	DE Mcal/kg	*	3.47	
Cattle	ME Mcal/kg	*	2.76	
Chickens	ME_n kcal/kg		1561.	
Sheep	ME Mcal/kg	*	2.85	
Cattle	TDN %		76.2	
Sheep	TDN %		78.7	
Niacin	mg/kg		27.8	
Pantothenic acid	mg/kg		15.7	
Riboflavin	mg/kg		1.2	
Thiamine	mg/kg		3.3	

Feed Name or Analyses		Mean As Fed	Mean Dry	C.V. ± %

Flax, seed screenings, extn unspecified grnd, (5)
Flax seed screenings oil feed (CFA)
Ref No 5 02 053 United States

		As Fed	Dry	C.V. ± %
Dry matter	%	91.3	100.0	1
Ash	%	8.1	8.9	21
Crude fiber	%	11.3	12.3	17
Ether extract	%	6.7	7.4	11
N-free extract	%	39.1	42.8	
Protein (N x 6.25)	%	26.1	28.6	6
Energy	GE Mcal/kg			
Cattle	DE Mcal/kg	* 2.54	2.78	
Sheep	DE Mcal/kg	* 2.61	2.86	
Swine	DE kcal/kg	*2802.	3069.	
Cattle	ME Mcal/kg	* 2.08	2.28	
Sheep	ME Mcal/kg	* 2.15	2.35	
Swine	ME kcal/kg	*2527.	2768.	
Cattle	NE_m Mcal/kg	* 1.15	1.26	
Cattle	NE_{gain} Mcal/kg	* 0.58	0.63	
Cattle	$NE_{lactating\ cows}$ Mcal/kg	* 1.29	1.41	
Cattle	TDN %	* 57.6	63.1	
Sheep	TDN %	* 59.3	64.9	
Swine	TDN %	* 63.5	69.6	
Calcium	%	0.44	0.48	
Phosphorus	%	0.64	0.70	

Flax, seed screenings, mech-extd grnd, (5)
Flaxseed screenings meal, mechanical extracted (AAFCO)
Ref No 5 02 054 United States

		As Fed	Dry	C.V. ± %
Dry matter	%	92.7	100.0	
Ash	%	6.6	7.1	
Crude fiber	%	10.1	10.9	
Ether extract	%	5.3	5.7	
N-free extract	%	36.9	39.8	
Protein (N x 6.25)	%	33.8	36.5	
Energy	GE Mcal/kg			
Cattle	DE Mcal/kg	* 2.44	2.63	
Sheep	DE Mcal/kg	* 2.49	2.69	
Swine	DE kcal/kg	*2714.	2928.	
Cattle	ME Mcal/kg	* 2.00	2.16	
Sheep	ME Mcal/kg	* 2.05	2.21	
Swine	ME kcal/kg	*2405.	2594.	
Cattle	NE_m Mcal/kg	* 1.13	1.22	
Cattle	NE_{gain} Mcal/kg	* 0.52	0.56	
Cattle	$NE_{lactating\ cows}$ Mcal/kg	* 1.83	1.97	
Cattle	TDN %	* 55.2	59.6	
Sheep	TDN %	* 56.5	61.0	
Swine	TDN %	* 61.6	66.4	
Calcium	%	0.43	0.46	
Phosphorus	%	0.65	0.70	

Flax plant product (AAFCO) –
see Flax, fiber by-product, mn 9% protein mx 35% fiber, (1)

Flaxseed meal (CFA) –
see Flax, seeds, grnd, commercial, (5)

Flaxseed screenings meal, mechanical extracted (AAFCO) –
see Flax, seed screenings, mech-extd grnd, (5)

see Flax, seed screenings, extn unspecified grnd, (5)

FLEABANE, HORSEWEED. Erigeron canadensis

Fleabane, horseweed, aerial part, fresh, (2)
Ref No 2 02 058 United States

		As Fed	Dry	C.V. ± %
Dry matter	%		100.0	
Ash	%		8.2	
Crude fiber	%		26.1	
Ether extract	%		1.8	
N-free extract	%		49.0	
Protein (N x 6.25)	%		14.9	
Cattle	dig prot %	*	10.6	
Goats	dig prot %	*	10.5	
Horses	dig prot %	*	10.2	
Rabbits	dig prot %	*	10.2	
Sheep	dig prot %	*	10.9	
Calcium	%		1.01	
Magnesium	%		0.21	
Phosphorus	%		0.28	
Potassium	%		2.61	

FLUFFGRASS. Triodia pulchella

Fluffgrass, aerial part, fresh, (2)
Ref No 2 02 061 United States

		As Fed	Dry	C.V. ± %
Dry matter	%		100.0	
Ash	%		22.7	
Crude fiber	%		29.1	
Ether extract	%		1.4	
N-free extract	%		39.2	
Protein (N x 6.25)	%		7.6	
Cattle	dig prot %	*	4.4	
Goats	dig prot %	*	3.7	
Horses	dig prot %	*	4.0	
Rabbits	dig prot %	*	4.5	
Sheep	dig prot %	*	4.1	

Fluffgrass, aerial part, fresh, full bloom, (2)
Ref No 2 02 059 United States

		As Fed	Dry	C.V. ± %
Dry matter	%		100.0	
Calcium	%		1.39	
Phosphorus	%		0.09	

Fluffgrass, aerial part, fresh, mature, (2)
Ref No 2 02 060 United States

		As Fed	Dry	C.V. ± %
Dry matter	%		100.0	
Calcium	%		0.99	
Phosphorus	%		0.06	

FLUFFWEED, FRENCH. Filago gallica

Fluffweed, French, aerial part, fresh, (2)
Ref No 2 02 065 United States

		As Fed	Dry	C.V. ± %
Dry matter	%		100.0	
Ash	%		9.8	20
Crude fiber	%		28.1	17
Protein (N x 6.25)	%		10.5	49
Cattle	dig prot %	*	6.8	
Goats	dig prot %	*	6.4	
Horses	dig prot %	*	6.4	
Rabbits	dig prot %	*	6.8	
Sheep	dig prot %	*	6.8	
Calcium	%		1.05	18
Phosphorus	%		0.33	35
Potassium	%		2.31	39

Fluffweed, French, aerial part, fresh, immature, (2)
Ref No 2 02 062 United States

		As Fed	Dry	C.V. ± %
Dry matter	%		100.0	
Ash	%		12.2	
Crude fiber	%		21.1	
Protein (N x 6.25)	%		17.3	
Cattle	dig prot %	*	12.6	
Goats	dig prot %	*	12.7	
Horses	dig prot %	*	12.2	
Rabbits	dig prot %	*	12.0	
Sheep	dig prot %	*	13.1	
Calcium	%		1.30	
Phosphorus	%		0.43	
Potassium	%		3.42	

Fluffweed, French, aerial part, fresh, full bloom, (2)
Ref No 2 02 063 United States

		As Fed	Dry	C.V. ± %
Dry matter	%		100.0	
Ash	%		7.4	
Crude fiber	%		31.2	
Protein (N x 6.25)	%		7.6	
Cattle	dig prot %	*	4.4	
Goats	dig prot %	*	3.7	
Horses	dig prot %	*	4.0	
Rabbits	dig prot %	*	4.5	
Sheep	dig prot %	*	4.1	
Calcium	%		0.93	
Phosphorus	%		0.33	
Potassium	%		1.94	

Fluffweed, French, aerial part, fresh, over ripe, (2)
Ref No 2 02 064 United States

		As Fed	Dry	C.V. ± %
Dry matter	%		100.0	
Ash	%		9.7	
Crude fiber	%		34.0	
Protein (N x 6.25)	%		4.6	
Cattle	dig prot %	*	1.8	
Goats	dig prot %	*	0.9	
Horses	dig prot %	*	1.4	
Rabbits	dig prot %	*	2.2	
Sheep	dig prot %	*	1.3	
Calcium	%		0.84	
Phosphorus	%		0.17	
Potassium	%		1.21	

FORESTIERA, NEW MEXICO. Forestiera neomexicana

Forestiera, New Mexico, browse, (2)
Ref No 2 02 070 United States

		As Fed	Dry	C.V. ± %
Dry matter	%		100.0	
Ash	%		7.3	
Crude fiber	%		12.8	
Ether extract	%		2.2	
N-free extract	%		63.8	
Protein (N x 6.25)	%		13.9	
Cattle	dig prot %	*	9.7	

Continued

Feed Name or Analyses		Mean As Fed	Mean Dry	C.V. ± %
Goats	dig prot % *		9.5	
Horses	dig prot % *		9.3	
Rabbits	dig prot % *		9.4	
Sheep	dig prot % *		9.9	

Forestiera, New Mexico, browse, immature, (2)
Ref No 2 02 067 United States

Feed Name or Analyses		As Fed	Dry	C.V. ± %
Dry matter	%		100.0	
Ash	%		6.4	
Crude fiber	%		9.4	
Ether extract	%		2.9	
N-free extract	%		61.6	
Protein (N x 6.25)	%		19.7	
Cattle	dig prot % *		14.6	
Goats	dig prot % *		14.9	
Horses	dig prot % *		14.3	
Rabbits	dig prot % *		13.9	
Sheep	dig prot % *		15.4	

Forestiera, New Mexico, browse, mid-bloom, (2)
Ref No 2 02 068 United States

Feed Name or Analyses		As Fed	Dry	C.V. ± %
Dry matter	%		100.0	
Ash	%		7.9	
Crude fiber	%		16.2	
Ether extract	%		1.9	
N-free extract	%		61.3	
Protein (N x 6.25)	%		12.7	
Cattle	dig prot % *		8.7	
Goats	dig prot % *		8.4	
Horses	dig prot % *		8.3	
Rabbits	dig prot % *		8.5	
Sheep	dig prot % *		8.8	
Calcium	%		1.00	
Magnesium	%		0.28	
Phosphorus	%		0.25	
Potassium	%		2.55	

Forestiera, New Mexico, browse, mature, (2)
Ref No 2 02 069 United States

Feed Name or Analyses		As Fed	Dry	C.V. ± %
Dry matter	%		100.0	
Ash	%		7.6	
Crude fiber	%		12.8	
Ether extract	%		1.9	
N-free extract	%		68.3	
Protein (N x 6.25)	%		9.4	
Cattle	dig prot % *		5.9	
Goats	dig prot % *		5.3	
Horses	dig prot % *		5.5	
Rabbits	dig prot % *		5.9	
Sheep	dig prot % *		5.8	
Calcium	%		1.32	
Magnesium	%		0.28	
Phosphorus	%		0.12	
Potassium	%		2.47	

(1) dry forages and roughages (2) pasture, range plants, and forages fed green (3) silages (4) energy feeds (5) protein supplements (6) minerals (7) vitamins (8) additives

FORGETMENOT, ALPINE. Myosotis alpestris

Forgetmenot, alpine, seeds w hulls, (4)
Ref No 4 09 013 United States

Feed Name or Analyses		As Fed	Dry	C.V. ± %
Dry matter	%		100.0	
Protein (N x 6.25)	%		17.0	
Cattle	dig prot % *		11.6	
Goats	dig prot % *		12.8	
Horses	dig prot % *		12.8	
Sheep	dig prot % *		12.8	
Alanine	%		0.63	
Arginine	%		0.78	
Aspartic acid	%		1.46	
Glutamic acid	%		2.98	
Glycine	%		0.73	
Histidine	%		0.37	
Isoleucine	%		0.78	
Leucine	%		1.19	
Lysine	%		0.75	
Methionine	%		0.48	
Phenylalanine	%		0.66	
Serine	%		1.02	
Threonine	%		0.65	
Tyrosine	%		0.61	
Valine	%		0.82	

FOUROCLOCK, COMMON. Mirabilis jalapa

Fouroclock, common, seeds w hulls, (4)
Ref No 4 09 044 United States

Feed Name or Analyses		As Fed	Dry	C.V. ± %
Dry matter	%		100.0	
Protein (N x 6.25)	%		17.0	
Cattle	dig prot % *		11.6	
Goats	dig prot % *		12.8	
Horses	dig prot % *		12.8	
Sheep	dig prot % *		12.8	
Alanine	%		0.60	
Arginine	%		1.09	
Aspartic acid	%		1.50	
Glutamic acid	%		2.50	
Glycine	%		1.00	
Histidine	%		0.46	
Hydroxyproline	%		0.12	
Isoleucine	%		0.61	
Leucine	%		0.97	
Lysine	%		0.82	
Methionine	%		0.24	
Phenylalanine	%		0.71	
Proline	%		0.66	
Serine	%		0.68	
Threonine	%		0.60	
Tyrosine	%		0.51	
Valine	%		0.78	

FOXTAIL. Alopecurus pratensis

Foxtail, hay, s-c, immature, (1)
Ref No 1 02 071 United States

Feed Name or Analyses		As Fed	Dry	C.V. ± %
Dry matter	%	85.7	100.0	
Ash	%	6.0	7.0	
Crude fiber	%	28.4	33.1	
Sheep	dig coef %	63.	63.	
Ether extract	%	2.8	3.3	
Sheep	dig coef %	44.	44.	
N-free extract	%	38.1	44.5	
Sheep	dig coef %	63.	63.	
Protein (N x 6.25)	%	10.4	12.1	
Sheep	dig coef %	63.	63.	
Cattle	dig prot % *	6.4	7.4	
Goats	dig prot % *	6.7	7.8	
Horses	dig prot % *	6.7	7.8	
Rabbits	dig prot % *	6.9	8.0	
Sheep	dig prot %	6.5	7.6	
Energy	GE Mcal/kg			
Cattle	DE Mcal/kg *	2.18	2.54	
Sheep	DE Mcal/kg *	2.26	2.64	
Cattle	ME Mcal/kg *	1.79	2.09	
Sheep	ME Mcal/kg *	1.85	2.16	
Cattle	TDN % *	49.4	57.7	
Sheep	TDN %	51.2	59.8	

FOXTAIL, MEADOW. Alopecurus pratensis

Foxtail, meadow, hay, s-c, (1)
Ref No 1 02 072 United States

Feed Name or Analyses		As Fed	Dry	C.V. ± %
Dry matter	%	88.1	100.0	
Ash	%	8.2	9.4	
Crude fiber	%	25.1	28.5	
Sheep	dig coef %	71.	71.	
Ether extract	%	1.9	2.2	
Sheep	dig coef %	39.	39.	
N-free extract	%	39.9	45.4	
Sheep	dig coef %	71.	71.	
Protein (N x 6.25)	%	12.9	14.7	
Sheep	dig coef %	67.	67.	
Cattle	dig prot % *	8.5	9.6	
Goats	dig prot % *	9.0	10.2	
Horses	dig prot % *	8.8	10.0	
Rabbits	dig prot % *	8.8	10.0	
Sheep	dig prot %	8.6	9.8	
Energy	GE Mcal/kg			
Cattle	DE Mcal/kg *	2.34	2.66	
Sheep	DE Mcal/kg *	2.49	2.83	
Cattle	ME Mcal/kg *	1.92	2.18	
Sheep	ME Mcal/kg *	2.04	2.32	
Cattle	TDN % *	53.0	60.2	
Sheep	TDN %	56.5	64.1	

Foxtail, meadow, aerial part, fresh, immature, (2)
Ref No 2 02 073 United States

Feed Name or Analyses		As Fed	Dry	C.V. ± %
Dry matter	%	26.1	100.0	
Ash	%	2.8	10.7	
Crude fiber	%	5.6	21.5	
Ether extract	%	1.2	4.6	
N-free extract	%	12.0	46.0	
Protein (N x 6.25)	%	4.5	17.2	
Cattle	dig prot % *	3.3	12.5	
Goats	dig prot % *	3.3	12.6	
Horses	dig prot % *	3.2	12.1	
Rabbits	dig prot % *	3.1	11.9	
Sheep	dig prot % *	3.4	13.0	
Energy	GE Mcal/kg			
Cattle	DE Mcal/kg *	0.75	2.86	
Sheep	DE Mcal/kg *	0.79	3.03	
Cattle	ME Mcal/kg *	0.61	2.35	
Sheep	ME Mcal/kg *	0.65	2.48	
Cattle	TDN % *	17.0	64.9	
Sheep	TDN % *	17.9	68.6	

Feed Name or Analyses		Mean As Fed	Mean Dry	C.V. ± %
Calcium	%	0.15	0.57	
Phosphorus	%	0.12	0.46	

Furfural residue, ammoniated –
see Corn, cobs furfural residue, ammoniated, (5)

Furze –
see Gorse, common, browse, s-c, (1)

GAILLARDIA, COMMON PERENNIAL. Gaillardia aristosa

Gaillardia, common perennial, seeds w hulls, (5)
Ref No 5 09 127 United States

			As Fed	Dry	C.V.
Dry matter		%		100.0	
Protein (N x 6.25)		%		46.0	
Alanine		%		1.79	
Arginine		%		4.23	
Aspartic acid		%		4.42	
Glutamic acid		%		9.48	
Glycine		%		2.25	
Histidine		%		1.20	
Isoleucine		%		1.89	
Leucine		%		2.58	
Lysine		%		1.66	
Methionine		%		0.97	
Phenylalanine		%		2.16	
Proline		%		1.56	
Serine		%		1.93	
Threonine		%		1.56	
Tyrosine		%		1.06	
Valine		%		2.58	

GAILLARDIA, ROSERING. Gaillardia pulchella

Gaillardia, rosering, aerial part, fresh, immature, (2)
Ref No 2 02 076 United States

			As Fed	Dry	C.V.
Dry matter		%		100.0	
Calcium		%		2.68	
Magnesium		%		0.20	
Phosphorus		%		0.28	
Potassium		%		3.27	

Gaillardia, rosering, aerial part, fresh, mature, (2)
Ref No 2 02 077 United States

			As Fed	Dry	C.V.
Dry matter		%		100.0	
Ash		%		11.3	
Crude fiber		%		28.6	
Ether extract		%		4.7	
N-free extract		%		42.4	
Protein (N x 6.25)		%		13.0	
Cattle	dig prot	% *		8.9	
Goats	dig prot	% *		8.7	
Horses	dig prot	% *		8.6	
Rabbits	dig prot	% *		8.7	
Sheep	dig prot	% *		9.1	
Energy	GE Mcal/kg				
Cattle	DE Mcal/kg	*		2.67	
Sheep	DE Mcal/kg	*		2.77	
Cattle	ME Mcal/kg	*		2.19	
Sheep	ME Mcal/kg	*		2.27	
Cattle	TDN	% *		60.6	
Sheep	TDN	% *		62.8	

GALLETA. Hilaria Jamesii

Galleta, aerial part, fresh, (2)
Ref No 2 05 593 United States

			As Fed	Dry	C.V.
Dry matter		%		100.0	
Calcium		%		0.67	42
Magnesium		%		0.15	32
Phosphorus		%		0.11	37

Galleta, aerial part, fresh, mid-bloom, (2)
Ref No 2 02 078 United States

			As Fed	Dry	C.V.
Dry matter		%		100.0	
Ash		%		11.3	
Crude fiber		%		32.2	
Ether extract		%		1.9	
N-free extract		%		44.7	
Protein (N x 6.25)		%		9.9	
Cattle	dig prot	% *		6.3	
Goats	dig prot	% *		5.8	
Horses	dig prot	% *		5.9	
Rabbits	dig prot	% *		6.3	
Sheep	dig prot	% *		6.2	
Energy	GE Mcal/kg				
Cattle	DE Mcal/kg	*		2.73	
Sheep	DE Mcal/kg	*		2.84	
Cattle	ME Mcal/kg	*		2.24	
Sheep	ME Mcal/kg	*		2.33	
Cattle	TDN	% *		62.0	
Sheep	TDN	% *		64.4	

Galleta, aerial part, fresh, mature, (2)
Ref No 2 02 079 United States

			As Fed	Dry	C.V.
Dry matter		%		100.0	
Ash		%		14.3	16
Crude fiber		%		32.9	3
Ether extract		%		1.5	37
N-free extract		%		46.4	
Protein (N x 6.25)		%		4.9	38
Cattle	dig prot	% *		2.1	
Goats	dig prot	% *		1.1	
Horses	dig prot	% *		1.7	
Rabbits	dig prot	% *		2.5	
Sheep	dig prot	% *		1.6	
Calcium		%		1.16	
Phosphorus		%		0.08	
Carotene		mg/kg		0.4	
Vitamin A equivalent		IU/g		0.7	

Galleta, aerial part, fresh, over ripe, (2)
Ref No 2 02 080 United States

			As Fed	Dry	C.V.
Dry matter		%		100.0	
Ash		%		14.0	9
Crude fiber		%		33.0	3
Ether extract		%		1.2	24
N-free extract		%		48.0	
Protein (N x 6.25)		%		3.8	22
Cattle	dig prot	% *		1.1	
Goats	dig prot	% *		0.1	
Horses	dig prot	% *		0.8	
Rabbits	dig prot	% *		1.6	
Sheep	dig prot	% *		0.5	

Feed Name or Analyses		Mean As Fed	Mean Dry	C.V. ± %
Calcium	%		0.59	43
Magnesium	%		0.15	32
Phosphorus	%		0.08	16

Galleta, aerial part wo lower stems, fresh, (2)
Ref No 2 08 684 United States

			As Fed	Dry	C.V.
Dry matter		%		100.0	
Organic matter		%		90.0	3
Ash		%		10.0	23
Crude fiber		%		31.8	8
Ether extract		%		1.5	26
N-free extract		%		50.2	15
Protein (N x 6.25)		%		6.5	66
Cattle	dig prot	% *		3.4	
Goats	dig prot	% *		2.6	
Horses	dig prot	% *		3.0	
Rabbits	dig prot	% *		3.7	
Sheep	dig prot	% *		3.0	
Calcium		%		0.24	61
Phosphorus		%		0.09	70
Carotene		mg/kg		12.7	98
Vitamin A equivalent		IU/g		21.2	

Galleta, aerial part wo lower stems, fresh, early leaf, (2)
Ref No 2 08 688 United States

			As Fed	Dry	C.V.
Dry matter		%		100.0	
Organic matter		%		88.0	3
Ash		%		12.0	21
Crude fiber		%		30.9	7
Ether extract		%		1.5	25
N-free extract		%		48.0	6
Protein (N x 6.25)		%		7.6	35
Cattle	dig prot	% *		4.3	
Goats	dig prot	% *		3.6	
Horses	dig prot	% *		4.0	
Rabbits	dig prot	% *		4.5	
Sheep	dig prot	% *		4.1	
Calcium		%		0.36	27
Phosphorus		%		0.13	44
Carotene		mg/kg		69.4	
Vitamin A equivalent		IU/g		115.8	

Galleta, aerial part wo lower stems, fresh, mid-bloom, (2)
Ref No 2 08 685 United States

			As Fed	Dry	C.V.
Dry matter		%		100.0	
Organic matter		%		90.2	
Ash		%		9.8	
Crude fiber		%		30.7	
Ether extract		%		1.5	
N-free extract		%		51.5	
Protein (N x 6.25)		%		6.5	
Cattle	dig prot	% *		3.4	
Goats	dig prot	% *		2.6	
Horses	dig prot	% *		3.0	
Rabbits	dig prot	% *		3.7	
Sheep	dig prot	% *		3.0	
Calcium		%		0.38	
Phosphorus		%		0.15	

Galleta, aerial part wo lower stems, fresh, mature, (2)
Ref No 2 08 686 United States

Feed Name or Analyses	Unit		As Fed	Dry	C.V. ± %
Dry matter	%			100.0	
Organic matter	%			90.6	
Ash	%			9.4	
Crude fiber	%			32.0	
Ether extract	%			1.7	
N-free extract	%			51.7	
Protein (N x 6.25)	%			5.2	
Cattle	dig prot %	*		2.3	
Goats	dig prot %	*		1.4	
Horses	dig prot %	*		1.9	
Rabbits	dig prot %	*		2.7	
Sheep	dig prot %	*		1.8	
Calcium	%			0.58	
Phosphorus	%			0.11	

Galleta, aerial part wo lower stems, fresh, dormant, (2)
Ref No 2 08 687 United States

Feed Name or Analyses	Unit		As Fed	Dry	C.V. ± %
Dry matter	%			100.0	
Organic matter	%			88.8	5
Ash	%			12.6	19
Crude fiber	%			32.0	6
Ether extract	%			1.4	22
N-free extract	%			50.0	5
Protein (N x 6.25)	%			4.0	15
Cattle	dig prot %	*		1.3	
Goats	dig prot %	*		0.3	
Horses	dig prot %	*		0.9	
Rabbits	dig prot %	*		1.8	
Sheep	dig prot %	*		0.7	
Calcium	%			0.34	21
Iodine	mg/kg			23.810	
Phosphorus	%			0.08	28
Carotene	mg/kg			3.4	
Vitamin A equivalent	IU/g			5.6	

GAMAGRASS. Tripsacum spp

Gamagrass, hay, s-c, (1)
Ref No 1 02 081 United States

Feed Name or Analyses	Unit		As Fed	Dry	C.V. ± %
Dry matter	%		89.3	100.0	2
Ash	%		6.3	7.0	9
Crude fiber	%		29.3	32.8	7
Ether extract	%		1.8	2.0	7
N-free extract	%		44.3	49.6	
Protein (N x 6.25)	%		7.6	8.5	35
Cattle	dig prot %	*	3.8	4.3	
Goats	dig prot %	*	4.0	4.5	
Horses	dig prot %	*	4.2	4.7	
Rabbits	dig prot %	*	4.7	5.2	
Sheep	dig prot %	*	3.7	4.2	
Fatty acids	%		1.8	2.0	
Energy	GE Mcal/kg				
Cattle	DE Mcal/kg	*	1.88	2.11	
Sheep	DE Mcal/kg	*	1.81	2.03	
Cattle	ME Mcal/kg	*	1.54	1.73	
Sheep	ME Mcal/kg	*	1.48	1.66	
Cattle	TDN %	*	42.7	47.8	
Sheep	TDN %	*	41.1	46.0	

(1) dry forages and roughages (2) pasture, range plants, and forages fed green (3) silages (4) energy feeds (5) protein supplements (6) minerals (7) vitamins (8) additives

Gamagrass, aerial part, fresh, pre-bloom, cut 2, (2)
Ref No 2 02 082 United States

Feed Name or Analyses	Unit		As Fed	Dry	C.V. ± %
Dry matter	%		30.5	100.0	
Ash	%		2.4	8.0	
Crude fiber	%		10.2	33.4	
Sheep	dig coef %		60.	60.	
Ether extract	%		0.8	2.5	
Sheep	dig coef %		77.	77.	
N-free extract	%		15.8	51.8	
Sheep	dig coef %		63.	63.	
Protein (N x 6.25)	%		1.3	4.3	
Sheep	dig coef %		54.	54.	
Cattle	dig prot %	*	0.5	1.5	
Goats	dig prot %	*	0.2	0.6	
Horses	dig prot %	*	0.4	1.2	
Rabbits	dig prot %	*	0.6	2.0	
Sheep	dig prot %		0.7	2.3	
Energy	GE Mcal/kg				
Cattle	DE Mcal/kg	*	0.81	2.64	
Sheep	DE Mcal/kg	*	0.80	2.62	
Cattle	ME Mcal/kg	*	0.66	2.17	
Sheep	ME Mcal/kg	*	0.65	2.15	
Cattle	TDN %	*	18.3	59.9	
Sheep	TDN %		18.1	59.3	

GAMAGRASS, EASTERN. Tripsacum dactyloides

Gamagrass, eastern, aerial part, fresh, (2)
Ref No 2 02 086 United States

Feed Name or Analyses	Unit		As Fed	Dry	C.V. ± %
Dry matter	%			100.0	
Ash	%			10.2	21
Crude fiber	%			30.2	4
Ether extract	%			2.0	22
N-free extract	%			49.1	
Protein (N x 6.25)	%			8.5	29
Cattle	dig prot %	*		5.1	
Goats	dig prot %	*		4.5	
Horses	dig prot %	*		4.7	
Rabbits	dig prot %	*		5.2	
Sheep	dig prot %	*		4.9	
Energy	GE Mcal/kg				
Cattle	DE Mcal/kg	*		2.71	
Sheep	DE Mcal/kg	*		2.94	
Cattle	ME Mcal/kg	*		2.22	
Sheep	ME Mcal/kg	*		2.41	
Cattle	TDN %	*		61.5	
Sheep	TDN %	*		66.8	
Calcium	%			0.62	28
Phosphorus	%			0.31	46

Gamagrass, eastern, aerial part, fresh, immature, (2)
Ref No 2 02 083 United States

Feed Name or Analyses	Unit		As Fed	Dry	C.V. ± %
Dry matter	%			100.0	
Ash	%			11.8	10
Crude fiber	%			29.2	3
Ether extract	%			2.2	10
N-free extract	%			44.7	
Protein (N x 6.25)	%			12.1	10
Cattle	dig prot %	*		8.2	
Goats	dig prot %	*		7.8	
Horses	dig prot %	*		7.8	
Rabbits	dig prot %	*		8.0	
Sheep	dig prot %	*		8.3	
Energy	GE Mcal/kg				
Cattle	DE Mcal/kg	*		2.76	
Sheep	DE Mcal/kg	*		2.81	
Cattle	ME Mcal/kg	*		2.27	
Sheep	ME Mcal/kg	*		2.31	
Cattle	TDN %	*		62.7	
Sheep	TDN %	*		63.8	
Calcium	%			0.65	22
Phosphorus	%			0.50	12

Gamagrass, eastern, aerial part, fresh, full bloom, (2)
Ref No 2 02 084 United States

Feed Name or Analyses	Unit		As Fed	Dry	C.V. ± %
Dry matter	%			100.0	
Ash	%			10.6	
Crude fiber	%			29.5	
Ether extract	%			2.0	
N-free extract	%			50.2	
Protein (N x 6.25)	%			7.7	
Cattle	dig prot %	*		4.4	
Goats	dig prot %	*		3.7	
Horses	dig prot %	*		4.1	
Rabbits	dig prot %	*		4.6	
Sheep	dig prot %	*		4.2	
Energy	GE Mcal/kg				
Cattle	DE Mcal/kg	*		2.64	
Sheep	DE Mcal/kg	*		2.98	
Cattle	ME Mcal/kg	*		2.16	
Sheep	ME Mcal/kg	*		2.45	
Cattle	TDN %	*		59.9	
Sheep	TDN %	*		67.6	

Gamagrass, eastern, aerial part, fresh, mature, (2)
Ref No 2 02 085 United States

Feed Name or Analyses	Unit		As Fed	Dry	C.V. ± %
Dry matter	%			100.0	
Ash	%			10.5	
Crude fiber	%			30.4	
Ether extract	%			2.0	
N-free extract	%			52.0	
Protein (N x 6.25)	%			5.1	
Cattle	dig prot %	*		2.2	
Goats	dig prot %	*		1.3	
Horses	dig prot %	*		1.9	
Rabbits	dig prot %	*		2.6	
Sheep	dig prot %	*		1.7	
Energy	GE Mcal/kg				
Cattle	DE Mcal/kg	*		2.52	
Sheep	DE Mcal/kg	*		2.61	
Cattle	ME Mcal/kg	*		2.07	
Sheep	ME Mcal/kg	*		2.14	
Cattle	TDN %	*		57.3	
Sheep	TDN %	*		57.3	
Calcium	%			0.50	
Phosphorus	%			0.08	

GAMAGRASS, FLORIDA. Tripsacum floridanum

Gamagrass, Florida, hay, s-c, (1)
Ref No 1 02 087 United States

Feed Name or Analyses	Unit		As Fed	Dry	C.V. ± %
Dry matter	%		92.3	100.0	
Ash	%		6.8	7.4	
Crude fiber	%		26.0	28.2	
Sheep	dig coef %		57.	57.	
Ether extract	%		1.8	1.9	
Sheep	dig coef %		57.	57.	

Feed Name or Analyses				Mean As Fed	Mean Dry	C.V. ± %
N-free extract		%		49.3	53.4	
Sheep	dig coef	%		55.	55.	
Protein (N x 6.25)		%		8.4	9.1	
Sheep	dig coef	%		51.	51.	
Cattle	dig prot	%	*	4.4	4.8	
Goats	dig prot	%	*	4.7	5.1	
Horses	dig prot	%	*	4.9	5.3	
Rabbits	dig prot	%	*	5.3	5.7	
Sheep	dig prot	%		4.3	4.6	
Energy	GE	Mcal/kg				
Cattle	DE	Mcal/kg	*	2.26	2.45	
Sheep	DE	Mcal/kg	*	2.14	2.32	
Cattle	ME	Mcal/kg	*	1.86	2.01	
Sheep	ME	Mcal/kg	*	1.75	1.90	
Cattle	TDN	%	*	51.3	55.6	
Sheep	TDN	%		48.5	52.5	

GAMAGRASS, GUATEMALA. Tripsacum laxum

Gamagrass, Guatemala, hay, s-c, (1)

Ref No 1 02 088 United States

Feed Name or Analyses				Mean As Fed	Mean Dry	C.V. ± %
Dry matter		%			100.0	
Ash		%			5.7	
Crude fiber		%			30.0	
Ether extract		%			2.2	
N-free extract		%			46.4	
Protein (N x 6.25)		%			15.7	
Cattle	dig prot	%	*		10.5	
Goats	dig prot	%	*		11.2	
Horses	dig prot	%	*		10.9	
Rabbits	dig prot	%	*		10.8	
Sheep	dig prot	%	*		10.7	
Energy	GE	Mcal/kg				
Cattle	DE	Mcal/kg	*		2.45	
Sheep	DE	Mcal/kg	*		2.67	
Cattle	ME	Mcal/kg	*		2.01	
Sheep	ME	Mcal/kg	*		2.19	
Cattle	TDN	%	*		55.6	
Sheep	TDN	%	*		60.4	
Calcium		%			0.89	
Iron		%			0.012	
Magnesium		%			0.81	
Phosphorus		%			0.40	

Gamagrass, Guatemala, aerial part, fresh, (2)

Ref No 2 02 091 United States

Feed Name or Analyses				Mean As Fed	Mean Dry	C.V. ± %
Dry matter		%		27.0	100.0	
Ash		%		3.3	12.4	
Crude fiber		%		8.9	32.8	
Ether extract		%		0.3	1.1	
N-free extract		%		12.8	47.3	
Protein (N x 6.25)		%		1.7	6.4	
Cattle	dig prot	%	*	0.9	3.3	
Goats	dig prot	%	*	0.7	2.5	
Horses	dig prot	%	*	0.8	3.0	
Rabbits	dig prot	%	*	1.0	3.6	
Sheep	dig prot	%	*	0.8	3.0	
Energy	GE	Mcal/kg				
Cattle	DE	Mcal/kg	*	0.69	2.54	
Sheep	DE	Mcal/kg	*	0.68	2.50	
Cattle	ME	Mcal/kg	*	0.56	2.09	
Sheep	ME	Mcal/kg	*	0.55	2.05	
Cattle	TDN	%	*	15.6	57.7	
Sheep	TDN	%	*	15.3	56.8	

Gamagrass, Guatemala, aerial part, fresh, pre-bloom, cut 1, (2)

Ref No 2 02 089 United States

Feed Name or Analyses				Mean As Fed	Mean Dry	C.V. ± %
Dry matter		%		27.0	100.0	
Ash		%		2.2	8.1	
Crude fiber		%		9.5	35.0	
Sheep	dig coef	%		66.	66.	
Ether extract		%		0.8	2.9	
Sheep	dig coef	%		76.	76.	
N-free extract		%		13.2	49.0	
Sheep	dig coef	%		68.	68.	
Protein (N x 6.25)		%		1.4	5.0	
Sheep	dig coef	%		58.	58.	
Cattle	dig prot	%	*	0.6	2.1	
Goats	dig prot	%	*	0.3	1.2	
Horses	dig prot	%	*	0.5	1.8	
Rabbits	dig prot	%	*	0.7	2.5	
Sheep	dig prot	%		0.8	2.9	
Energy	GE	Mcal/kg				
Cattle	DE	Mcal/kg	*	0.72	2.66	
Sheep	DE	Mcal/kg	*	0.77	2.83	
Cattle	ME	Mcal/kg	*	0.59	2.18	
Sheep	ME	Mcal/kg	*	0.63	2.32	
Cattle	TDN	%	*	16.3	60.3	
Sheep	TDN	%		17.4	64.3	

Gamagrass, Guatemala, aerial part, fresh, cut 1, (2)

Ref No 2 02 090 United States

Feed Name or Analyses				Mean As Fed	Mean Dry	C.V. ± %
Dry matter		%		21.9	100.0	
Ash		%		2.2	10.1	
Crude fiber		%		7.7	35.2	
Cattle	dig coef	%		61.	61.	
Ether extract		%		0.5	2.2	
Cattle	dig coef	%		53.	53.	
N-free extract		%		9.4	43.1	
Cattle	dig coef	%		54.	54.	
Protein (N x 6.25)		%		2.1	9.4	
Cattle	dig coef	%		56.	56.	
Cattle	dig prot	%		1.2	5.3	
Goats	dig prot	%	*	1.2	5.3	
Horses	dig prot	%	*	1.2	5.5	
Rabbits	dig prot	%	*	1.3	5.9	
Sheep	dig prot	%	*	1.3	5.8	
Energy	GE	Mcal/kg				
Cattle	DE	Mcal/kg	*	0.51	2.32	
Sheep	DE	Mcal/kg	*	0.52	2.36	
Cattle	ME	Mcal/kg	*	0.42	1.90	
Sheep	ME	Mcal/kg	*	0.42	1.94	
Cattle	TDN	%		11.5	52.6	
Sheep	TDN	%	*	11.7	53.5	

GARBAGE. Scientific name not used

Garbage, cooked dehy, high fat, (4)

Ref No 4 07 863 United States

Feed Name or Analyses				Mean As Fed	Mean Dry	C.V. ± %
Dry matter		%		95.9	100.0	
Ash		%		12.9	13.5	
Crude fiber		%		20.0	20.9	
Ether extract		%		23.7	24.7	
N-free extract		%		21.8	22.7	
Protein (N x 6.25)		%		17.5	18.2	
Cattle	dig prot	%	*	12.2	12.8	
Goats	dig prot	%	*	13.4	14.0	
Horses	dig prot	%	*	13.4	14.0	
Sheep	dig prot	%	*	13.4	14.0	
Energy	GE	Mcal/kg				
Cattle	DE	Mcal/kg	*	3.75	3.91	
Sheep	DE	Mcal/kg	*	3.39	3.53	
Swine	DE	kcal/kg	*	3772.	3993.	
Cattle	ME	Mcal/kg	*	3.08	3.21	
Sheep	ME	Mcal/kg	*	2.77	2.89	
Swine	ME	kcal/kg	*	3481.	3630.	
Cattle	TDN	%	*	85.0	88.6	
Sheep	TDN	%	*	76.8	80.1	
Swine	TDN	%	*	85.5	89.2	
Phosphorus		%		0.33	0.34	
Potassium		%		0.62	0.65	

Garbage, cooked dehy, low fat, (4)

Ref No 4 07 862 United States

Feed Name or Analyses				Mean As Fed	Mean Dry	C.V. ± %
Dry matter		%		92.3	100.0	
Ash		%		14.1	15.3	
Crude fiber		%		13.5	14.6	
Ether extract		%		3.5	3.8	
N-free extract		%		38.1	41.3	
Protein (N x 6.25)		%		23.1	25.0	
Cattle	dig prot	%	*	17.5	19.0	
Goats	dig prot	%	*	18.6	20.2	
Horses	dig prot	%	*	18.6	20.2	
Sheep	dig prot	%	*	18.6	20.2	
Energy	GE	Mcal/kg				
Cattle	DE	Mcal/kg	*	2.57	2.78	
Sheep	DE	Mcal/kg	*	2.58	2.80	
Swine	DE	kcal/kg	*	3131.	3392.	
Cattle	ME	Mcal/kg	*	2.10	2.28	
Sheep	ME	Mcal/kg	*	2.12	2.29	
Swine	ME	kcal/kg	*	2848.	3085.	
Cattle	TDN	%	*	58.2	63.1	
Sheep	TDN	%	*	58.6	63.4	
Swine	TDN	%	*	71.0	76.9	

GARBAGE, HOTEL AND RESTAURANT. Scientific name not used

Garbage, hotel and restaurant, cooked dehy grnd, (4)

Ref No 4 07 879 United States

Feed Name or Analyses				Mean As Fed	Mean Dry	C.V. ± %
Dry matter		%		53.6	100.0	
Ash		%		3.5	6.6	
Crude fiber		%		1.6	2.9	
Ether extract		%		14.5	27.1	
N-free extract		%		24.5	45.8	
Protein (N x 6.25)		%		9.5	17.7	
Cattle	dig prot	%	*	6.5	12.2	
Goats	dig prot	%	*	7.2	13.4	
Horses	dig prot	%	*	7.2	13.4	
Sheep	dig prot	%	*	7.2	13.4	
Energy	GE	Mcal/kg		2.85	5.33	
Cattle	DE	Mcal/kg	*	2.05	3.83	
Sheep	DE	Mcal/kg	*	2.06	3.85	
Swine	DE	kcal/kg	*	2077.	3878.	
Cattle	ME	Mcal/kg	*	1.68	3.14	
Sheep	ME	Mcal/kg	*	1.69	3.16	
Swine	ME	kcal/kg	*	1920.	3585.	
Cattle	TDN	%	*	46.5	86.9	
Sheep	TDN	%	*	46.8	87.3	
Swine	TDN	%	*	47.1	88.0	
Calcium		%		0.32	0.60	
Copper		mg/kg		9.6	18.0	

Continued

Feed Name or Analyses		Mean As Fed	Mean Dry	C.V. ± %
Iron	%	0.013	0.024	
Magnesium	%	0.17	0.32	
Manganese	mg/kg	4.8	9.0	
Phosphorus	%	0.21	0.40	

Garbage, hotel and restaurant, cooked wet grnd, (4)

Ref No 4 07 865 United States

Feed Name or Analyses		Mean As Fed	Mean Dry	C.V. ± %
Dry matter	%	26.3	100.0	
Ash	%	1.4	5.3	
Crude fiber	%	0.7	2.7	
Ether extract	%	5.9	22.4	
N-free extract	%	14.0	53.2	
Protein (N x 6.25)	%	4.3	16.3	
Cattle	dig prot %	* 2.9	11.0	
Goats	dig prot %	* 3.2	12.2	
Horses	dig prot %	* 3.2	12.2	
Sheep	dig prot %	* 3.2	12.2	
Energy	GE Mcal/kg			
Cattle	DE Mcal/kg	* 0.99	3.75	
Sheep	DE Mcal/kg	* 1.01	3.85	
Swine	DE kcal/kg	*1041.	3957.	
Cattle	ME Mcal/kg	* 0.81	3.07	
Sheep	ME Mcal/kg	* 0.83	3.16	
Swine	ME kcal/kg	* 965.	3668.	
Cattle	TDN %	* 22.4	85.0	
Sheep	TDN %	* 23.0	87.3	
Swine	TDN %	* 23.6	89.8	
Calcium	%	0.11	0.42	
Phosphorus	%	0.07	0.27	

GARBAGE, INSTITUTIONAL. Scientific name not used

Garbage, institutional, cooked dehy grnd, (4)

Ref No 4 07 878 United States

Feed Name or Analyses		Mean As Fed	Mean Dry	C.V. ± %
Dry matter	%	42.8	100.0	
Ash	%	2.4	5.6	
Crude fiber	%	1.1	2.6	
Ether extract	%	6.8	16.0	
N-free extract	%	26.0	60.9	
Protein (N x 6.25)	%	6.4	14.9	
Cattle	dig prot %	* 4.2	9.7	
Goats	dig prot %	* 4.7	10.9	
Horses	dig prot %	* 4.7	10.9	
Sheep	dig prot %	* 4.7	10.9	
Energy	GE Mcal/kg	2.06	4.82	
Cattle	DE Mcal/kg	* 1.56	3.64	
Sheep	DE Mcal/kg	* 1.62	3.78	
Swine	DE kcal/kg	*1637.	3828.	
Cattle	ME Mcal/kg	* 1.28	2.98	
Sheep	ME Mcal/kg	* 1.33	3.10	
Swine	ME kcal/kg	*1522.	3560.	
Cattle	TDN %	* 35.3	82.5	
Sheep	TDN %	* 36.7	85.7	
Swine	TDN %	* 37.1	86.8	
Calcium	%	0.15	0.35	
Copper	mg/kg	10.1	23.7	
Iron	%	0.016	0.038	
Magnesium	%	0.04	0.09	
Manganese	mg/kg	7.8	18.3	
Phosphorus	%	0.13	0.30	

(1) dry forages and roughages (2) pasture, range plants, and forages fed green (3) silages (4) energy feeds (5) protein supplements (6) minerals (7) vitamins (8) additives

Garbage, institutional, cooked wet grnd, (4)

Ref No 4 07 867 United States

Feed Name or Analyses		Mean As Fed	Mean Dry	C.V. ± %
Dry matter	%	18.9	100.0	
Swine	dig coef %	94.	94.	
Ash	%	1.2	6.3	
Crude fiber	%	0.5	2.6	
Swine	dig coef %	77.	77.	
Ether extract	%	3.6	19.0	
Swine	dig coef %	91.	91.	
N-free extract	%	10.5	55.6	
Swine	dig coef %	98.	98.	
Protein (N x 6.25)	%	3.1	16.4	
Swine	dig coef %	89.	89.	
Cattle	dig prot %	* 2.1	11.1	
Goats	dig prot %	* 2.3	12.3	
Horses	dig prot %	* 2.3	12.3	
Sheep	dig prot %	* 2.3	12.3	
Swine	dig prot %	2.8	14.6	
Energy	GE Mcal/kg	0.93	4.92	
Swine	GE dig coef %	93.	93.	
Cattle	DE Mcal/kg	* 0.68	3.58	
Sheep	DE Mcal/kg	* 0.71	3.77	
Swine	DE kcal/kg	865.	4576.	
Cattle	ME Mcal/kg	* 0.56	2.94	
Sheep	ME Mcal/kg	* 0.58	3.09	
Swine	ME kcal/kg	* 802.	4242.	
Cattle	TDN %	* 15.4	81.3	
Sheep	TDN %	* 16.2	85.6	
Swine	TDN %	20.8	110.1	

GARBAGE, MILITARY. Scientific name not used

Garbage, military, cooked dehy grnd, (4)

Ref No 4 07 877 United States

Feed Name or Analyses		Mean As Fed	Mean Dry	C.V. ± %
Dry matter	%	48.3	100.0	
Ash	%	2.9	6.1	
Crude fiber	%	1.2	2.5	
Ether extract	%	16.0	33.1	
N-free extract	%	20.0	41.4	
Protein (N x 6.25)	%	8.2	17.0	
Cattle	dig prot %	* 5.6	11.6	
Goats	dig prot %	* 6.2	12.8	
Horses	dig prot %	* 6.2	12.8	
Sheep	dig prot %	* 6.2	12.8	
Energy	GE Mcal/kg	2.73	5.67	
Cattle	DE Mcal/kg	* 2.05	4.25	
Sheep	DE Mcal/kg	* 1.92	3.97	
Swine	DE kcal/kg	*1950.	4041.	
Cattle	ME Mcal/kg	* 1.68	3.49	
Sheep	ME Mcal/kg	* 1.57	3.26	
Swine	ME kcal/kg	*1806.	3741.	
Cattle	TDN %	* 46.6	96.5	
Sheep	TDN %	* 43.5	90.1	
Swine	TDN %	* 44.2	91.6	
Calcium	%	0.30	0.61	
Copper	mg/kg	18.5	38.3	
Iron	%	0.018	0.038	
Magnesium	%	0.05	0.10	
Manganese	mg/kg	12.9	26.7	
Phosphorus	%	0.28	0.58	

Garbage, military, cooked wet grnd, (4)

Ref No 4 07 866 United States

Feed Name or Analyses		Mean As Fed	Mean Dry	C.V. ± %
Dry matter	%	31.6	100.0	
Swine	dig coef %	96.	96.	
Ash	%	2.2	7.0	
Crude fiber	%	0.8	2.5	
Swine	dig coef %	87.	87.	
Ether extract	%	12.6	39.9	
Swine	dig coef %	96.	96.	
N-free extract	%	10.7	33.9	
Swine	dig coef %	99.	99.	
Protein (N x 6.25)	%	5.3	16.8	
Swine	dig coef %	90.	90.	
Cattle	dig prot %	* 3.6	11.4	
Goats	dig prot %	* 4.0	12.6	
Horses	dig prot %	* 4.0	12.6	
Sheep	dig prot %	* 4.0	12.6	
Swine	dig prot %	4.8	15.1	
Energy	GE Mcal/kg	1.92	6.08	
Swine	GE dig coef %	96.	96.	
Cattle	DE Mcal/kg	* 1.48	4.69	
Sheep	DE Mcal/kg	* 1.42	4.49	
Swine	DE kcal/kg	1843.	5833.	
Cattle	ME Mcal/kg	* 1.22	3.85	
Sheep	ME Mcal/kg	* 1.16	3.68	
Swine	ME kcal/kg	*1707.	5402.	
Cattle	TDN %	* 33.6	106.4	
Sheep	TDN %	* 32.2	101.8	
Swine	TDN %	43.3	136.9	

GARBAGE, MUNICIPAL. Scientific name not used

Garbage, municipal, cooked dehy grnd, (4)

Ref No 4 07 876 United States

Feed Name or Analyses		Mean As Fed	Mean Dry	C.V. ± %
Dry matter	%	52.7	100.0	
Ash	%	4.8	9.2	
Crude fiber	%	4.0	7.7	
Ether extract	%	12.0	22.7	
N-free extract	%	21.7	41.2	
Protein (N x 6.25)	%	10.2	19.3	
Cattle	dig prot %	* 7.2	13.7	
Goats	dig prot %	* 7.9	14.9	
Horses	dig prot %	* 7.9	14.9	
Sheep	dig prot %	* 7.9	14.9	
Energy	GE Mcal/kg	2.69	5.10	
Cattle	DE Mcal/kg	* 1.86	3.53	
Sheep	DE Mcal/kg	* 1.87	3.55	
Swine	DE kcal/kg	*2061.	3911.	
Cattle	ME Mcal/kg	* 1.53	2.90	
Sheep	ME Mcal/kg	* 1.53	2.91	
Swine	ME kcal/kg	*1898.	3602.	
Cattle	TDN %	* 42.2	80.1	
Sheep	TDN %	* 42.3	80.4	
Swine	TDN %	* 46.7	88.7	
Calcium	%	0.84	1.60	
Copper	mg/kg	16.8	32.0	
Iron	%	0.017	0.032	
Magnesium	%	0.05	0.09	
Manganese	mg/kg	13.2	25.0	
Phosphorus	%	0.23	0.45	

Garbage, municipal, cooked wet grnd, (4)

Ref No 4 07 864 United States

Feed Name or Analyses				Mean As Fed	Mean Dry	C.V. ± %
Dry matter		%		22.5	100.0	
Swine	dig coef	%		76.	76.	
Ash		%		2.4	10.8	
Crude fiber		%		1.8	8.2	
Swine	dig coef	%		40.	40.	
Ether extract		%		5.2	23.3	
Swine	dig coef	%		89.	89.	
N-free extract		%		8.9	39.7	
Swine	dig coef	%		77.	77.	
Protein (N x 6.25)		%		4.0	18.0	
Swine	dig coef	%		82.	82.	
Cattle	dig prot	%	*	2.8	12.6	
Goats	dig prot	%	*	3.1	13.7	
Horses	dig prot	%	*	3.1	13.7	
Sheep	dig prot	%	*	3.1	13.7	
Swine	dig prot	%		3.3	14.8	
Energy	GE	Mcal/kg		1.21	5.37	
Swine	GE dig coef	%		78.	78.	
Cattle	DE	Mcal/kg	*	0.76	3.40	
Sheep	DE	Mcal/kg	*	0.79	3.52	
Swine	DE	kcal/kg		941.	4191.	
Cattle	ME	Mcal/kg	*	0.63	2.79	
Sheep	ME	Mcal/kg	*	0.65	2.88	
Swine	ME	kcal/kg	*	869.	3871.	
Cattle	TDN	%	*	17.4	77.2	
Sheep	TDN	%	*	17.9	79.7	
Swine	TDN	%		21.4	95.3	

Garbanzos —
see Chickpea, gram, seeds, (5)

GAURA. Gaura spp

Gaura, aerial part, fresh, mature, (2)

Ref No 2 02 094 United States

Feed Name or Analyses				Mean As Fed	Mean Dry	C.V. ± %
Dry matter		%			100.0	
Ash		%			12.8	
Crude fiber		%			15.7	
Ether extract		%			2.0	
N-free extract		%			44.1	
Protein (N x 6.25)		%			25.4	
Cattle	dig prot	%	*		19.5	
Goats	dig prot	%	*		20.3	
Horses	dig prot	%	*		19.1	
Rabbits	dig prot	%	*		18.3	
Sheep	dig prot	%	*		20.7	

GAYFEATHER, DOTTED. Liatris punctata

Gayfeather, dotted, aerial part, fresh, mature, (2)

Ref No 2 02 095 United States

Feed Name or Analyses				Mean As Fed	Mean Dry	C.V. ± %
Dry matter		%			100.0	
Ash		%			10.0	
Crude fiber		%			29.4	
Ether extract		%			8.8	
N-free extract		%			44.3	
Protein (N x 6.25)		%			7.5	
Cattle	dig prot	%	*		4.3	
Goats	dig prot	%	*		3.6	
Horses	dig prot	%	*		3.9	
Rabbits	dig prot	%	*		4.5	
Sheep	dig prot	%	*		4.0	

GERANIUM. Geranium spp

Geranium, aerial part, fresh, (2)

Ref No 2 02 096 United States

Feed Name or Analyses				Mean As Fed	Mean Dry	C.V. ± %
Dry matter		%			100.0	
Ash		%			8.4	20
Crude fiber		%			11.4	23
Ether extract		%			4.2	32
N-free extract		%			62.1	
Protein (N x 6.25)		%			13.9	42
Cattle	dig prot	%	*		9.7	
Goats	dig prot	%	*		9.5	
Horses	dig prot	%	*		9.3	
Rabbits	dig prot	%	*		9.4	
Sheep	dig prot	%	*		9.9	

GERANIUM, CAROLINA. Geranium carolinianum

Geranium, Carolina, aerial part, fresh, (2)

Ref No 2 02 097 United States

Feed Name or Analyses				Mean As Fed	Mean Dry	C.V. ± %
Dry matter		%			100.0	
Ash		%			13.0	
Crude fiber		%			16.6	
Ether extract		%			2.0	
N-free extract		%			55.1	
Protein (N x 6.25)		%			13.3	
Cattle	dig prot	%	*		9.2	
Goats	dig prot	%	*		9.0	
Horses	dig prot	%	*		8.8	
Rabbits	dig prot	%	*		8.9	
Sheep	dig prot	%	*		9.4	
Calcium		%			1.34	
Magnesium		%			0.25	
Phosphorus		%			0.57	
Potassium		%			2.16	

GERANIUM, FREMONT. Geranium fremontii

Geranium, fremont, aerial part, fresh, immature, (2)

Ref No 2 02 098 United States

Feed Name or Analyses				Mean As Fed	Mean Dry	C.V. ± %
Dry matter		%			100.0	
Ash		%			7.2	
Crude fiber		%			8.4	
Ether extract		%			3.3	
N-free extract		%			62.3	
Protein (N x 6.25)		%			18.8	
Cattle	dig prot	%	*		13.9	
Goats	dig prot	%	*		14.1	
Horses	dig prot	%	*		13.5	
Rabbits	dig prot	%	*		13.2	
Sheep	dig prot	%	*		14.5	
Calcium		%			1.25	
Magnesium		%			0.24	
Phosphorus		%			0.48	
Sulphur		%			0.24	

Geranium, fremont, aerial part, fresh, mature, (2)

Ref No 2 02 099 United States

Feed Name or Analyses				Mean As Fed	Mean Dry	C.V. ± %
Dry matter		%			100.0	
Ash		%			9.2	
Crude fiber		%			14.1	
Ether extract		%			5.5	
N-free extract		%			63.6	
Protein (N x 6.25)		%			7.6	
Cattle	dig prot	%	*		4.4	
Goats	dig prot	%	*		3.7	
Horses	dig prot	%	*		4.0	
Rabbits	dig prot	%	*		4.5	
Sheep	dig prot	%	*		4.1	
Calcium		%			2.25	
Magnesium		%			0.36	
Phosphorus		%			0.49	
Sulphur		%			0.17	

GERANIUM, RICHARDSON. Geranium richardsoni

Geranium, Richardson, aerial part, fresh, early leaf, (2)

Ref No 2 08 837 United States

Feed Name or Analyses				Mean As Fed	Mean Dry	C.V. ± %
Dry matter		%			100.0	
Crude fiber		%			13.0	
Ether extract		%			3.2	
N-free extract		%			53.0	
Protein (N x 6.25)		%			19.0	
Cattle	dig prot	%	*		14.0	
Goats	dig prot	%	*		14.3	
Horses	dig prot	%	*		13.7	
Rabbits	dig prot	%	*		13.3	
Sheep	dig prot	%	*		14.7	

Geranium, Richardson, aerial part, fresh, dough stage, (2)

Ref No 2 08 836 United States

Feed Name or Analyses				Mean As Fed	Mean Dry	C.V. ± %
Dry matter		%			100.0	
Crude fiber		%			14.0	
Ether extract		%			3.0	
N-free extract		%			50.0	
Protein (N x 6.25)		%			13.0	
Cattle	dig prot	%	*		8.9	
Goats	dig prot	%	*		8.7	
Horses	dig prot	%	*		8.6	
Rabbits	dig prot	%	*		8.7	
Sheep	dig prot	%	*		9.1	

GERANIUM, STICKY. Geranium viscosissimum

Geranium, sticky, aerial part, fresh, immature, (2)

Ref No 2 02 100 United States

Feed Name or Analyses				Mean As Fed	Mean Dry	C.V. ± %
Dry matter		%			100.0	
Ash		%			8.3	
Crude fiber		%			8.2	
Ether extract		%			2.4	
N-free extract		%			52.2	
Protein (N x 6.25)		%			28.9	
Cattle	dig prot	%	*		22.5	
Goats	dig prot	%	*		23.5	
Horses	dig prot	%	*		22.1	
Rabbits	dig prot	%	*		21.0	
Sheep	dig prot	%	*		23.9	

Feed Name or Analyses		Mean As Fed	Mean Dry	C.V. ± %

GIANTHYSSOP, NETTLELEAF. Agastache urticifolia

Gianthyssop, nettleleaf, aerial part, fresh, early leaf, (2)

Ref No 2 08 825 United States

Dry matter	%		100.0	
Crude fiber	%		13.0	
Ether extract	%		3.0	
N-free extract	%		46.0	
Protein (N x 6.25)	%		19.0	
Cattle	dig prot % *		14.0	
Goats	dig prot % *		14.3	
Horses	dig prot % *		13.7	
Rabbits	dig prot % *		13.3	
Sheep	dig prot % *		14.7	

Gianthyssop, nettleleaf, aerial part, fresh, immature, (2)

Ref No 2 02 101 United States

Dry matter	%		100.0	
Ash	%		11.9	
Crude fiber	%		15.8	
Ether extract	%		3.0	
N-free extract	%		46.5	
Protein (N x 6.25)	%		22.8	
Cattle	dig prot % *		17.3	
Goats	dig prot % *		17.8	
Horses	dig prot % *		16.9	
Rabbits	dig prot % *		16.3	
Sheep	dig prot % *		18.2	
Calcium	%		1.49	
Magnesium	%		0.43	
Phosphorus	%		0.42	
Sulphur	%		0.30	

Gianthyssop, nettleleaf, aerial part, fresh, mid-bloom, (2)

Ref No 2 08 826 United States

Dry matter	%		100.0	
Crude fiber	%		25.0	
Ether extract	%		2.8	
N-free extract	%		44.0	
Protein (N x 6.25)	%		11.0	
Cattle	dig prot % *		7.2	
Goats	dig prot % *		6.8	
Horses	dig prot % *		6.9	
Rabbits	dig prot % *		7.2	
Sheep	dig prot % *		7.2	

Gianthyssop, nettleleaf, aerial part, fresh, mature, (2)

Ref No 2 02 102 United States

Dry matter	%		100.0	
Ash	%		9.3	
Crude fiber	%		25.0	
Ether extract	%		5.2	
N-free extract	%		52.2	
Protein (N x 6.25)	%		8.3	
Cattle	dig prot % *		4.9	
Goats	dig prot % *		4.3	
Horses	dig prot % *		4.6	
Rabbits	dig prot % *		5.1	
Sheep	dig prot % *		4.7	
Calcium	%		1.98	
Magnesium	%		0.47	
Phosphorus	%		0.31	
Sulphur	%		0.17	

GIANTREED. Arundo donax

Giantreed, aerial part, s-c, (1)

Ref No 1 02 103 United States

Dry matter	%	81.9	100.0	
Ash	%	7.5	9.2	
Crude fiber	%	30.7	37.5	
Sheep	dig coef %	41.	41.	
Ether extract	%	1.1	1.3	
Sheep	dig coef %	35.	35.	
N-free extract	%	36.4	44.4	
Sheep	dig coef %	27.	27.	
Protein (N x 6.25)	%	6.2	7.6	
Sheep	dig coef %	36.	36.	
Cattle	dig prot % *	2.9	3.5	
Goats	dig prot % *	3.0	3.7	
Horses	dig prot % *	3.3	4.0	
Rabbits	dig prot % *	3.7	4.5	
Sheep	dig prot %	2.2	2.7	
Energy	GE Mcal/kg			
Cattle	DE Mcal/kg *	1.36	1.66	
Sheep	DE Mcal/kg *	1.12	1.37	
Cattle	ME Mcal/kg *	1.11	1.36	
Sheep	ME Mcal/kg *	0.92	1.13	
Cattle	TDN % *	30.8	37.6	
Sheep	TDN %	25.5	31.1	

Giantreed, aerial part, fresh, (2)

Ref No 2 02 107 United States

Dry matter	%		100.0	
Ash	%		9.2	
Crude fiber	%		37.5	
Ether extract	%		1.3	
N-free extract	%		45.1	
Protein (N x 6.25)	%		6.9	52
Cattle	dig prot % *		3.8	
Goats	dig prot % *		3.0	
Horses	dig prot % *		3.4	
Rabbits	dig prot % *		4.0	
Sheep	dig prot % *		3.4	
Calcium	%		0.49	37
Magnesium	%		0.23	21
Phosphorus	%		0.11	31
Potassium	%		2.55	

Giantreed, aerial part, fresh, immature, (2)

Ref No 2 02 104 United States

Dry matter	%		100.0	
Protein (N x 6.25)	%		12.2	
Cattle	dig prot % *		8.3	
Goats	dig prot % *		7.9	
Horses	dig prot % *		7.9	
Rabbits	dig prot % *		8.1	
Sheep	dig prot % *		8.4	
Calcium	%		0.67	
Magnesium	%		0.20	
Phosphorus	%		0.15	
Potassium	%		3.19	

Giantreed, aerial part, fresh, full bloom, (2)

Ref No 2 02 105 United States

Dry matter	%		100.0	
Protein (N x 6.25)	%		6.9	
Cattle	dig prot % *		3.8	
Goats	dig prot % *		3.0	
Horses	dig prot % *		3.4	
Rabbits	dig prot % *		4.0	
Sheep	dig prot % *		3.4	
Calcium	%		0.43	
Magnesium	%		0.30	
Phosphorus	%		0.11	
Potassium	%		2.42	

Giantreed, aerial part, fresh, over ripe, (2)

Ref No 2 02 106 United States

Dry matter	%		100.0	
Protein (N x 6.25)	%		3.5	
Cattle	dig prot % *		0.9	
Goats	dig prot % *		-0.1	
Horses	dig prot % *		0.5	
Rabbits	dig prot % *		1.4	
Sheep	dig prot % *		0.3	
Calcium	%		0.43	
Magnesium	%		0.20	
Phosphorus	%		0.09	
Potassium	%		2.04	

GILIA. Gilia spp

Gilia, aerial part, fresh, (2)

Ref No 2 02 111 United States

Dry matter	%		100.0	
Ash	%		5.1	27
Crude fiber	%		24.4	22
Ether extract	%		11.0	
N-free extract	%		52.4	
Protein (N x 6.25)	%		7.1	98
Cattle	dig prot % *		3.9	
Goats	dig prot % *		3.2	
Horses	dig prot % *		3.6	
Rabbits	dig prot % *		4.2	
Sheep	dig prot % *		3.6	
Other carbohydrates	%		34.9	
Lignin (Ellis)	%		10.1	
Calcium	%		1.44	20
Phosphorus	%		0.45	59
Potassium	%		2.60	69
Silicon	%		0.40	

Gilia, aerial part, fresh, immature, (2)

Ref No 2 02 108 United States

Dry matter	%		100.0	
Ash	%		9.1	
Crude fiber	%		21.0	39

(1) dry forages and roughages (2) pasture, range plants, and forages fed green (3) silages (4) energy feeds (5) protein supplements (6) minerals (7) vitamins (8) additives

Feed Name or Analyses		Mean As Fed	Mean Dry	C.V. ± %
Protein (N x 6.25)	%		19.1	41
Cattle	dig prot % *		14.1	
Goats	dig prot % *		14.4	
Horses	dig prot % *		13.7	
Rabbits	dig prot % *		13.4	
Sheep	dig prot % *		14.8	
Calcium	%		1.69	15
Phosphorus	%		0.69	52
Potassium	%		3.96	57

Gilia, aerial part, fresh, full bloom, (2)

Ref No 2 02 109 United States

Feed Name or Analyses		Mean As Fed	Mean Dry	C.V. ± %
Dry matter	%		100.0	
Ash	%		8.6	8
Crude fiber	%		26.1	6
Protein (N x 6.25)	%		6.8	21
Cattle	dig prot % *		3.7	
Goats	dig prot % *		2.9	
Horses	dig prot % *		3.3	
Rabbits	dig prot % *		3.9	
Sheep	dig prot % *		3.3	
Calcium	%		1.50	14
Phosphorus	%		0.28	49
Potassium	%		1.75	42

Gilia, aerial part, fresh, dormant, (2)

Ref No 2 02 110 United States

Feed Name or Analyses		Mean As Fed	Mean Dry	C.V. ± %
Dry matter	%		100.0	
Ash	%		7.3	20
Crude fiber	%		27.8	10
Protein (N x 6.25)	%		3.7	31
Cattle	dig prot % *		1.0	
Goats	dig prot % *		0.0	
Horses	dig prot % *		0.7	
Rabbits	dig prot % *		1.5	
Sheep	dig prot % *		0.4	
Calcium	%		1.31	24
Phosphorus	%		0.27	36
Potassium	%		1.58	

GILIA, BIRDSEYE. Gilia tricolor

Gilia, birdseye, aerial part, fresh, (2)

Ref No 2 02 117 United States

Feed Name or Analyses		Mean As Fed	Mean Dry	C.V. ± %
Dry matter	%		100.0	
Ash	%		8.8	13
Crude fiber	%		23.1	26
Protein (N x 6.25)	%		12.0	68
Cattle	dig prot % *		8.1	
Goats	dig prot % *		7.8	
Horses	dig prot % *		7.7	
Rabbits	dig prot % *		7.9	
Sheep	dig prot % *		8.2	
Calcium	%		1.42	22
Phosphorus	%		0.54	51
Potassium	%		3.08	59

Gilia, birdseye, aerial part, fresh, immature, (2)

Ref No 2 02 112 United States

Feed Name or Analyses		Mean As Fed	Mean Dry	C.V. ± %
Dry matter	%		100.0	
Ash	%		8.8	
Crude fiber	%		14.4	

Feed Name or Analyses		Mean As Fed	Mean Dry	C.V. ± %
Protein (N x 6.25)	%		24.4	
Cattle	dig prot % *		18.6	
Goats	dig prot % *		19.3	
Horses	dig prot % *		18.2	
Rabbits	dig prot % *		17.5	
Sheep	dig prot % *		19.7	
Calcium	%		1.79	
Phosphorus	%		0.95	
Potassium	%		5.16	

Gilia, birdseye, aerial part, fresh, mid-bloom, (2)

Ref No 2 02 113 United States

Feed Name or Analyses		Mean As Fed	Mean Dry	C.V. ± %
Dry matter	%		100.0	
Ash	%		9.8	
Crude fiber	%		20.7	
Protein (N x 6.25)	%		12.5	
Cattle	dig prot % *		8.5	
Goats	dig prot % *		8.2	
Horses	dig prot % *		8.1	
Rabbits	dig prot % *		8.3	
Sheep	dig prot % *		8.6	
Phosphorus	%		0.59	
Potassium	%		2.73	

Gilia, birdseye, aerial part, fresh, full bloom, (2)

Ref No 2 02 114 United States

Feed Name or Analyses		Mean As Fed	Mean Dry	C.V. ± %
Dry matter	%		100.0	
Ash	%		8.6	
Crude fiber	%		24.9	
Protein (N x 6.25)	%		8.0	
Cattle	dig prot % *		4.7	
Goats	dig prot % *		4.0	
Horses	dig prot % *		4.3	
Rabbits	dig prot % *		4.8	
Sheep	dig prot % *		4.4	
Calcium	%		1.45	
Phosphorus	%		0.38	
Potassium	%		2.32	

Gilia, birdseye, aerial part, fresh, mature, (2)

Ref No 2 02 115 United States

Feed Name or Analyses		Mean As Fed	Mean Dry	C.V. ± %
Dry matter	%		100.0	
Ash	%		9.5	
Crude fiber	%		28.2	
Protein (N x 6.25)	%		6.7	
Cattle	dig prot % *		3.6	
Goats	dig prot % *		2.8	
Horses	dig prot % *		3.2	
Rabbits	dig prot % *		3.8	
Sheep	dig prot % *		3.2	

Gilia, birdseye, aerial part, fresh, dormant, (2)

Ref No 2 02 116 United States

Feed Name or Analyses		Mean As Fed	Mean Dry	C.V. ± %
Dry matter	%		100.0	
Ash	%		7.8	
Crude fiber	%		28.5	
Protein (N x 6.25)	%		3.4	
Cattle	dig prot % *		0.8	
Goats	dig prot % *		-0.2	
Horses	dig prot % *		0.4	
Rabbits	dig prot % *		1.3	
Sheep	dig prot % *		0.2	

Feed Name or Analyses		Mean As Fed	Mean Dry	C.V. ± %
Calcium	%		1.36	
Phosphorus	%		0.29	
Potassium	%		1.86	

GILIA, VISCID. Gilia viscidula

Gilia, viscid, aerial part, fresh, (2)

Ref No 2 02 121 United States

Feed Name or Analyses		Mean As Fed	Mean Dry	C.V. ± %
Dry matter	%		100.0	
Ash	%		8.2	23
Crude fiber	%		27.1	5
Protein (N x 6.25)	%		7.7	37
Cattle	dig prot % *		4.4	
Goats	dig prot % *		3.7	
Horses	dig prot % *		4.1	
Rabbits	dig prot % *		4.6	
Sheep	dig prot % *		4.2	
Calcium	%		1.48	18
Phosphorus	%		0.23	24
Potassium	%		1.54	33

Gilia, viscid, aerial part, fresh, immature, (2)

Ref No 2 02 118 United States

Feed Name or Analyses		Mean As Fed	Mean Dry	C.V. ± %
Dry matter	%		100.0	
Ash	%		9.1	
Crude fiber	%		27.7	
Protein (N x 6.25)	%		11.1	
Cattle	dig prot % *		7.3	
Goats	dig prot % *		6.9	
Horses	dig prot % *		7.0	
Rabbits	dig prot % *		7.2	
Sheep	dig prot % *		7.3	
Calcium	%		1.58	
Phosphorus	%		0.30	
Potassium	%		2.15	

Gilia, viscid, aerial part, fresh, full bloom, (2)

Ref No 2 02 119 United States

Feed Name or Analyses		Mean As Fed	Mean Dry	C.V. ± %
Dry matter	%		100.0	
Ash	%		8.6	
Crude fiber	%		27.2	
Protein (N x 6.25)	%		5.7	
Cattle	dig prot % *		2.7	
Goats	dig prot % *		1.9	
Horses	dig prot % *		2.4	
Rabbits	dig prot % *		3.1	
Sheep	dig prot % *		2.3	
Calcium	%		1.55	
Phosphorus	%		0.18	
Potassium	%		1.19	

Gilia, viscid, aerial part, fresh, dormant, (2)

Ref No 2 02 120 United States

Feed Name or Analyses		Mean As Fed	Mean Dry	C.V. ± %
Dry matter	%		100.0	
Ash	%		5.6	
Crude fiber	%		25.8	
Protein (N x 6.25)	%		4.8	
Cattle	dig prot % *		2.0	
Goats	dig prot % *		1.0	
Horses	dig prot % *		1.6	
Rabbits	dig prot % *		2.4	

Continued

Feed Name or Analyses			Mean As Fed	Mean Dry	C.V. ± %
Sheep dig prot	%	*		1.5	
Calcium	%			1.14	
Phosphorus	%			0.19	
Potassium	%			1.03	

GLOBEMALLOW, NARROWLEAF. Sphaeralcea angustifolis

Globemallow, narrowleaf, aerial part, fresh, (2)

Ref No 2 02 124 United States

Analyses			As Fed	Dry	C.V.
Dry matter	%			100.0	
Ash	%			11.7	
Crude fiber	%			23.2	
Ether extract	%			2.4	
N-free extract	%			42.3	
Protein (N x 6.25)	%			20.4	
Cattle dig prot	%	*		15.2	
Goats dig prot	%	*		15.6	
Horses dig prot	%	*		14.8	
Rabbits dig prot	%	*		14.4	
Sheep dig prot	%	*		16.0	

Globemallow, narrowleaf, aerial part, fresh, full bloom, (2)

Ref No 2 02 123 United States

Analyses			As Fed	Dry	C.V.
Dry matter	%			100.0	
Calcium	%			3.34	
Magnesium	%			0.36	
Phosphorus	%			0.31	
Potassium	%			2.69	

GLUCOSE. Scientific name not used

Glucose, (4)

Ref No 4 02 125 United States

Analyses			As Fed	Dry	C.V.
Dry matter	%		90.2	100.0	
Energy	GE Mcal/kg		3.35	3.72	1
Swine	DE kcal/kg		3385.	3753.	
Swine	ME kcal/kg		3333.	3695.	

GOAT. Copra hircus

Goat, milk, fresh, (5)

Milk, goat, fresh

Ref No 5 02 128 United States

Analyses			As Fed	Dry	C.V.
Dry matter	%		13.2	100.0	
Ash	%		0.8	6.1	
Crude fiber	%		0.0	0.0	
Ether extract	%		4.1	31.1	
N-free extract	%		4.7	35.6	
Protein (N x 6.25)	%		3.6	27.3	
Energy	GE Mcal/kg				
Cattle	DE Mcal/kg	*	0.69	5.19	
Sheep	DE Mcal/kg	*	0.68	5.15	
Swine	DE kcal/kg	*	835.	6328.	
Cattle	ME Mcal/kg	*	0.56	4.26	
Sheep	ME Mcal/kg	*	0.56	4.22	
Swine	ME kcal/kg	*	756.	5726.	
Cattle	TDN %	*	15.5	117.7	
Sheep	TDN %	*	15.4	116.9	
Swine	TDN %	*	18.9	143.5	
Calcium	%		0.13	0.98	
Phosphorus	%		0.11	0.83	
Potassium	%		0.18	1.36	

(1) dry forages and roughages (2) pasture, range plants, and forages fed green (3) silages (4) energy feeds (5) protein supplements (6) minerals (7) vitamins (8) additives

Goatfish, whole, raw –

see Fish, mullet, whole, raw, (5)

Goat nut –

see Jojoba, California, seeds, (4)

GODETIA, SPOTTED. Godetia quadrivulnera

Godetia, spotted, aerial part, fresh, (2)

Ref No 2 02 130 United States

Analyses			As Fed	Dry	C.V.
Dry matter	%			100.0	
Calcium	%			1.79	9
Phosphorus	%			0.32	44
Potassium	%			1.32	18

Godetia, spotted, aerial part, fresh, full bloom, (2)

Ref No 2 02 129 United States

Analyses			As Fed	Dry	C.V.
Dry matter	%			100.0	
Ash	%			7.5	12
Crude fiber	%			23.8	16
Protein (N x 6.25)	%			6.8	12
Cattle dig prot	%	*		3.7	
Goats dig prot	%	*		2.9	
Horses dig prot	%	*		3.3	
Rabbits dig prot	%	*		3.9	
Sheep dig prot	%	*		3.3	
Calcium	%			1.84	8
Phosphorus	%			0.38	36
Potassium	%			1.46	12

Godetia, spotted, aerial part, over ripe, (2)

Ref No 2 02 131 United States

Analyses			As Fed	Dry	C.V.
Dry matter	%			100.0	
Ash	%			6.9	
Crude fiber	%			32.6	
Protein (N x 6.25)	%			4.0	
Cattle dig prot	%	*		1.3	
Goats dig prot	%	*		0.3	
Horses dig prot	%	*		0.9	
Rabbits dig prot	%	*		1.8	
Sheep dig prot	%	*		0.7	
Calcium	%			1.76	
Phosphorus	%			0.24	
Potassium	%			1.14	

GOLDENROD. Solidago spp

Goldenrod, aerial part, fresh, (2)

Ref No 2 02 132 United States

Analyses			As Fed	Dry	C.V.
Dry matter	%		39.2	100.0	
Ash	%		2.8	7.1	
Crude fiber	%		4.9	12.6	
Ether extract	%		5.6	14.2	
N-free extract	%		22.5	57.4	
Protein (N x 6.25)	%		3.4	8.7	
Cattle dig prot	%	*	2.1	5.3	
Goats dig prot	%	*	1.8	4.7	
Horses dig prot	%	*	1.9	4.9	
Rabbits dig prot	%	*	2.1	5.4	
Sheep dig prot	%	*	2.0	5.1	
Calcium	%		0.44	1.13	
Phosphorus	%		0.20	0.52	

GOOBER, CONGO. Voandzeia subterranea

Goober, congo, nuts, (4)

Ref No 4 02 133 United States

Analyses			As Fed	Dry	C.V.
Dry matter	%		85.0	100.0	
Ash	%		3.4	4.0	
Crude fiber	%		10.7	12.6	
Sheep dig coef	%		26.	26.	
Ether extract	%		3.9	4.6	
Sheep dig coef	%		100.	100.	
N-free extract	%		49.0	57.7	
Sheep dig coef	%		84.	84.	
Protein (N x 6.25)	%		17.9	21.1	
Sheep dig coef	%		84.	84.	
Cattle dig prot	%	*	13.1	15.4	
Goats dig prot	%	*	14.1	16.6	
Horses dig prot	%	*	14.1	16.6	
Sheep dig prot	%		15.1	17.7	
Energy	GE Mcal/kg				
Cattle	DE Mcal/kg	*	2.64	3.10	
Sheep	DE Mcal/kg	*	2.99	3.52	
Swine	DE kcal/kg	*	3509.	4128.	
Cattle	ME Mcal/kg	*	2.16	2.54	
Sheep	ME Mcal/kg	*	2.45	2.89	
Swine	ME kcal/kg	*	3219.	3787.	
Cattle	TDN %	*	59.8	70.3	
Sheep	TDN %		67.8	79.8	
Swine	TDN %	*	79.6	93.6	

GOOSEFOOT, LAMBSQUARTERS. Chenopodium album

Goosefoot, lambsquarters, aerial part, fresh, (2)

Ref No 2 02 137 United States

Analyses			As Fed	Dry	C.V.
Dry matter	%		14.5	100.0	
Niacin	mg/kg		11.8	81.3	
Riboflavin	mg/kg		4.7	32.2	
Thiamine	mg/kg		1.1	7.5	

Goosefoot, lambsquarters, aerial part, fresh, immature, (2)

Ref No 2 02 135 United States

Analyses			As Fed	Dry	C.V.
Dry matter	%			100.0	
Ash	%			27.6	
Crude fiber	%			11.2	
Ether extract	%			1.7	
N-free extract	%			32.6	
Protein (N x 6.25)	%			26.9	
Cattle dig prot	%	*		20.8	
Goats dig prot	%	*		21.7	
Horses dig prot	%	*		20.4	
Rabbits dig prot	%	*		19.4	
Sheep dig prot	%	*		22.1	
Calcium	%			3.33	
Magnesium	%			0.78	

Feed Name or Analyses			Mean As Fed	Mean Dry	C.V. ± %
Phosphorus	%			0.51	
Potassium	%			8.71	

Goosefoot, lambsquarters, aerial part, fresh, mid-bloom, (2)

Ref No 2 02 136 United States

Feed Name or Analyses			Mean As Fed	Mean Dry	C.V. ± %
Dry matter	%			100.0	
Ash	%			6.4	
Crude fiber	%			19.3	
Ether extract	%			7.2	
N-free extract	%			52.0	
Protein (N x 6.25)	%			15.1	
Cattle	dig prot %	*		10.7	
Goats	dig prot %	*		10.6	
Horses	dig prot %	*		10.3	
Rabbits	dig prot %	*		10.3	
Sheep	dig prot %	*		11.1	

Goosefoot, lambsquarters, leaves, fresh, (2)

Ref No 2 02 138 United States

Feed Name or Analyses			Mean As Fed	Mean Dry	C.V. ± %
Dry matter	%			100.0	
Carotene	mg/kg			59.7	
Vitamin A equivalent	IU/g			99.6	

Goosefoot, lambsquarters, seeds w hulls, (4)

Ref No 4 09 040 United States

Feed Name or Analyses			Mean As Fed	Mean Dry	C.V. ± %
Dry matter	%			100.0	
Protein (N x 6.25)	%			19.0	
Cattle	dig prot %	*		13.5	
Goats	dig prot %	*		14.6	
Horses	dig prot %	*		14.6	
Sheep	dig prot %	*		14.6	
Alanine	%			0.63	
Arginine	%			1.69	
Aspartic acid	%			1.35	
Glutamic acid	%			2.77	
Glycine	%			1.01	
Histidine	%			0.48	
Hydroxyproline	%			0.10	
Isoleucine	%			0.61	
Leucine	%			0.99	
Lysine	%			0.78	
Methionine	%			0.32	
Phenylalanine	%			0.68	
Proline	%			0.51	
Serine	%			0.68	
Threonine	%			0.53	
Tyrosine	%			0.49	
Valine	%			0.72	

Goosefoot, lambsquarters, seeds, (5)

Lambs quarters seed

Ref No 5 08 424 United States

Feed Name or Analyses			Mean As Fed	Mean Dry	C.V. ± %
Dry matter	%		90.0	100.0	
Ash	%		9.6	10.7	
Crude fiber	%		15.1	16.8	
Ether extract	%		4.5	5.0	
N-free extract	%		40.2	44.7	
Protein (N x 6.25)	%		20.6	22.9	
Energy	GE Mcal/kg				
Cattle	DE Mcal/kg	*	2.66	2.95	
Sheep	DE Mcal/kg	*	2.63	2.92	
Swine	DE kcal/kg	*	3270.	3634.	
Cattle	ME Mcal/kg	*	2.18	2.42	
Sheep	ME Mcal/kg	*	2.16	2.40	
Swine	ME kcal/kg	*	2988.	3320.	
Cattle	TDN %	*	60.2	66.9	
Sheep	TDN %	*	59.7	66.3	
Swine	TDN %	*	74.2	82.4	

GOOSEFOOT, QUINOA. Chenopodium quinoa

Goosefoot, quinoa, seeds w hulls, (4)

Ref No 4 09 041 United States

Feed Name or Analyses			Mean As Fed	Mean Dry	C.V. ± %
Dry matter	%			100.0	
Protein (N x 6.25)	%			13.0	
Cattle	dig prot %	*		8.0	
Goats	dig prot %	*		9.1	
Horses	dig prot %	*		9.1	
Sheep	dig prot %	*		9.1	
Alanine	%			0.61	
Arginine	%			0.91	
Aspartic acid	%			0.95	
Glutamic acid	%			1.55	
Glycine	%			0.68	
Histidine	%			0.31	
Hydroxyproline	%			0.07	
Isoleucine	%			0.47	
Leucine	%			0.78	
Lysine	%			0.73	
Methionine	%			0.26	
Phenylalanine	%			0.53	
Proline	%			0.40	
Serine	%			0.48	
Threonine	%			0.46	
Tyrosine	%			0.36	
Valine	%			0.59	

GOOSEGRASS. Eleusine indica

Goosegrass, hay, s-c, (1)

Ref No 1 02 140 United States

Feed Name or Analyses			Mean As Fed	Mean Dry	C.V. ± %
Dry matter	%		92.1	100.0	
Ash	%		9.1	9.9	
Crude fiber	%		33.0	35.8	
Sheep	dig coef %		44.	44.	
Ether extract	%		1.0	1.1	
Sheep	dig coef %		100.	100.	
N-free extract	%		45.8	49.7	
Sheep	dig coef %		34.	34.	
Protein (N x 6.25)	%		3.2	3.5	
Cattle	dig prot %	*	0.0	0.0	
Goats	dig prot %	*	-0.1	-0.1	
Horses	dig prot %	*	0.5	0.5	
Rabbits	dig prot %	*	1.3	1.4	
Sheep	dig prot %	*	-0.2	-0.2	
Energy	GE Mcal/kg				
Cattle	DE Mcal/kg	*	1.57	1.70	
Sheep	DE Mcal/kg	*	1.43	1.55	
Cattle	ME Mcal/kg	*	1.28	1.39	
Sheep	ME Mcal/kg	*	1.17	1.27	
Cattle	TDN %	*	35.6	38.6	
Sheep	TDN %		32.3	35.1	

Goosegrass, hay, s-c, mature, (1)

Ref No 1 02 139 United States

Feed Name or Analyses			Mean As Fed	Mean Dry	C.V. ± %
Dry matter	%		92.1	100.0	
Ash	%		9.1	9.9	
Crude fiber	%		33.0	35.8	
Ether extract	%		1.0	1.1	
N-free extract	%		45.8	49.7	
Protein (N x 6.25)	%		3.2	3.5	
Cattle	dig prot %	*	0.0	0.0	
Goats	dig prot %	*	-0.1	-0.1	
Horses	dig prot %	*	0.5	0.5	
Rabbits	dig prot %	*	1.3	1.4	
Sheep	dig prot %	*	-0.2	-0.2	
Energy	GE Mcal/kg				
Cattle	DE Mcal/kg	*	1.46	1.58	
Sheep	DE Mcal/kg	*	1.32	1.43	
Cattle	ME Mcal/kg	*	1.20	1.30	
Sheep	ME Mcal/kg	*	1.08	1.17	
Cattle	TDN %	*	33.0	35.8	
Sheep	TDN %	*	29.8	32.4	

GORSE, COMMON. Ulex europaeus

Gorse, common, browse, s-c, (1)

Furze

Ref No 1 08 425 United States

Feed Name or Analyses			Mean As Fed	Mean Dry	C.V. ± %
Dry matter	%		94.5	100.0	
Ash	%		7.0	7.4	
Crude fiber	%		38.5	40.7	
Ether extract	%		2.0	2.1	
N-free extract	%		35.4	37.5	
Protein (N x 6.25)	%		11.6	12.3	
Cattle	dig prot %	*	7.2	7.6	
Goats	dig prot %	*	7.6	8.0	
Horses	dig prot %	*	7.5	8.0	
Rabbits	dig prot %	*	7.7	8.1	
Sheep	dig prot %	*	7.2	7.6	
Energy	GE Mcal/kg				
Cattle	DE Mcal/kg	*	1.55	1.64	
Sheep	DE Mcal/kg	*	1.43	1.51	
Cattle	ME Mcal/kg	*	1.27	1.34	
Sheep	ME Mcal/kg	*	1.17	1.24	
Cattle	TDN %	*	35.2	37.2	
Sheep	TDN %	*	32.4	34.3	

GRAINS. Scientific name not used

Grains, distillers grains w potato flakes added, dehy, 1 part distil dried grain 2 part potato flakes, (4)

Ref No 4 02 145 United States

Feed Name or Analyses			Mean As Fed	Mean Dry	C.V. ± %
Dry matter	%		88.7	100.0	
Ash	%		8.6	9.7	
Crude fiber	%		6.2	7.0	
Swine	dig coef %		68.	68.	
Ether extract	%		0.2	0.2	
Swine	dig coef %		-689.	-689.	
N-free extract	%		57.7	65.0	
Swine	dig coef %		93.	93.	
Protein (N x 6.25)	%		16.1	18.1	
Swine	dig coef %		60.	60.	
Cattle	dig prot %	*	11.2	12.6	
Goats	dig prot %	*	12.3	13.8	

Continued

Feed Name or Analyses		Mean As Fed	Mean Dry	C.V. ± %
Horses	dig prot %	* 12.3	13.8	
Sheep	dig prot %	* 12.3	13.8	
Swine	dig prot %	9.6	10.9	
Energy	GE Mcal/kg			
Cattle	DE Mcal/kg	* 2.68	3.02	
Sheep	DE Mcal/kg	* 2.88	3.25	
Swine	DE kcal/kg	*2854.	3217.	
Cattle	ME Mcal/kg	* 2.20	2.48	
Sheep	ME Mcal/kg	* 2.37	2.67	
Swine	ME kcal/kg	*2635.	2971.	
Cattle	TDN %	* 60.8	68.5	
Sheep	TDN %	* 65.4	73.8	
Swine	TDN %	64.7	73.0	

Grains, mill dust, (4)

Ref No 4 02 143 United States

Feed Name or Analyses		Mean As Fed	Mean Dry	C.V. ± %
Dry matter	%	92.0	100.0	
Ash	%	12.0	13.0	
Crude fiber	%	16.8	18.3	
Ether extract	%	1.7	1.8	
N-free extract	%	51.0	55.4	
Protein (N x 6.25)	%	10.5	11.4	
Cattle	dig prot %	* 6.0	6.5	
Goats	dig prot %	* 7.1	7.7	
Horses	dig prot %	* 7.1	7.7	
Sheep	dig prot %	* 7.1	7.7	

Grains, screenings, uncleaned, mn 12% grain mx 3% wild oats mx 17% buckwheat and large seeds mx 68% small weed seeds chaff hulls dust scourings noxious seeds, (4)

Uncleaned screenings (CFA)

Ref No 4 02 153 Canada

Feed Name or Analyses		Mean As Fed	Mean Dry	C.V. ± %
Dry matter	%	92.1	100.0	0
Ash	%	8.6	9.3	
Crude fiber	%	14.2	15.5	
Ether extract	%	5.8	6.3	
N-free extract	%	49.7	54.0	
Protein (N x 6.25)	%	13.8	15.0	10
Cattle	dig prot %	* 9.0	9.8	
Goats	dig prot %	* 10.1	11.0	
Horses	dig prot %	* 10.1	11.0	
Sheep	dig prot %	* 10.1	11.0	
Cellulose (Matrone)	%	4.7	5.1	
Energy	GE Mcal/kg			
Cattle	NE_m Mcal/kg	* 1.30	1.41	
Cattle	NE_{gain} Mcal/kg	* 0.76	0.82	
Cattle	$NE_{lactating\ cows}$ Mcal/kg	* 1.51	1.64	
Copper	mg/kg	5.0	5.5	
Magnesium	%	0.20	0.22	
Selenium	mg/kg	0.805	0.874	
Zinc	mg/kg	34.0	36.9	
Niacin	mg/kg	72.5	78.7	
Pantothenic acid	mg/kg	21.2	23.1	
Riboflavin	mg/kg	0.6	0.7	
Thiamine	mg/kg	0.6	0.6	
Vitamin B_6	mg/kg	2.20	2.39	
Alanine	%	0.54	0.59	
Arginine	%	0.67	0.73	
Aspartic acid	%	0.81	0.88	
Glutamic acid	%	3.69	4.01	

(1) dry forages and roughages (2) pasture, range plants, and forages fed green (3) silages (4) energy feeds (5) protein supplements (6) minerals (7) vitamins (8) additives

Feed Name or Analyses		Mean As Fed	Mean Dry	C.V. ± %
Glycine	%	0.61	0.66	
Histidine	%	0.30	0.33	
Isoleucine	%	0.45	0.49	
Leucine	%	0.90	0.98	
Lysine	%	0.42	0.46	
Methionine	%	0.19	0.21	
Phenylalanine	%	0.58	0.63	
Proline	%	1.15	1.25	
Serine	%	0.67	0.73	
Threonine	%	0.44	0.48	
Tyrosine	%	0.58	0.63	
Valine	%	0.58	0.63	

Grains, screenings, gr 1 mn 35% grain mx 7% fiber mx 6% foreign material mx 8% wild oats, (4)

No 1 feed screenings (CFA)

Feed screenings no 1

Ref No 4 02 154 Canada

Feed Name or Analyses		Mean As Fed	Mean Dry	C.V. ± %
Dry matter	%	89.3	100.0	
Ash	%	2.1	2.4	
Crude fiber	%	6.2	6.9	
Ether extract	%	2.5	2.8	
N-free extract	%	63.7	71.3	
Protein (N x 6.25)	%	14.8	16.6	
Cattle	dig prot %	* 10.0	11.3	
Goats	dig prot %	* 11.1	12.4	
Horses	dig prot %	* 11.1	12.4	
Sheep	dig prot %	* 11.1	12.4	
Cellulose (Matrone)	%	3.3	3.6	
Energy	GE Mcal/kg			
Cattle	DE Mcal/kg	* 2.19	2.45	
Sheep	DE Mcal/kg	* 2.37	2.65	
Swine	DE kcal/kg	*2398.	2685.	
Cattle	ME Mcal/kg	* 1.79	2.01	
Sheep	ME Mcal/kg	* 1.94	2.17	
Swine	ME kcal/kg	*2220.	2486.	
Cattle	TDN %	* 49.6	55.6	
Sheep	TDN %	* 53.7	60.1	
Swine	TDN %	* 54.4	60.9	
Calcium	%	0.14	0.16	
Copper	mg/kg	7.3	8.2	
Magnesium	%	0.18	0.20	
Manganese	mg/kg	28.2	31.6	
Phosphorus	%	0.43	0.48	
Selenium	mg/kg	0.887	0.993	
Zinc	mg/kg	30.7	34.4	
Niacin	mg/kg	69.0	77.3	
Pantothenic acid	mg/kg	12.8	14.3	
Riboflavin	mg/kg	1.4	1.6	
Thiamine	mg/kg	0.6	0.7	
Vitamin B_6	mg/kg	2.12	2.37	
Alanine	%	0.55	0.62	
Arginine	%	0.74	0.82	
Aspartic acid	%	0.78	0.88	
Glutamic acid	%	4.30	4.81	
Glycine	%	0.64	0.71	
Histidine	%	0.32	0.36	
Isoleucine	%	0.56	0.63	
Leucine	%	0.99	1.11	
Lysine	%	0.40	0.45	
Methionine	%	0.17	0.19	
Phenylalanine	%	0.84	0.95	
Proline	%	1.34	1.51	
Serine	%	0.64	0.71	
Threonine	%	0.36	0.41	
Tyrosine	%	0.32	0.36	
Valine	%	0.66	0.74	

Grains, screenings, gr 2 mx 11% fiber mx 10% foreign material mx 49% wild oats, (4)

No 2 feed screenings (CFA)

Feed screenings no 2

Ref No 4 02 155 Canada

Feed Name or Analyses		Mean As Fed	Mean Dry	C.V. ± %
Dry matter	%	91.1	100.0	
Ash	%	2.5	2.7	
Crude fiber	%	7.6	8.3	
Ether extract	%	3.9	4.3	
N-free extract	%	64.7	71.0	
Protein (N x 6.25)	%	12.5	13.7	
Cattle	dig prot %	* 7.8	8.6	
Goats	dig prot %	* 8.9	9.8	
Horses	dig prot %	* 8.9	9.8	
Sheep	dig prot %	* 8.9	9.8	
Cellulose (Matrone)	%	7.2	7.9	
Energy	GE Mcal/kg			
Cattle	DE Mcal/kg	* 2.31	2.54	
Sheep	DE Mcal/kg	* 2.43	2.67	
Swine	DE kcal/kg	*2651.	2910.	
Cattle	ME Mcal/kg	* 1.89	2.08	
Sheep	ME Mcal/kg	* 2.00	2.19	
Swine	ME kcal/kg	*2471.	2712.	
Cattle	TDN %	* 52.4	57.5	
Sheep	TDN %	* 55.2	60.6	
Swine	TDN %	* 60.1	66.0	
Copper	mg/kg	3.9	4.3	
Magnesium	%	0.22	0.24	
Manganese	mg/kg	132.6	145.5	
Selenium	mg/kg	0.380	0.417	
Zinc	mg/kg	22.7	24.9	
Niacin	mg/kg	40.0	43.9	
Pantothenic acid	mg/kg	10.9	12.0	
Riboflavin	mg/kg	1.2	1.3	
Thiamine	mg/kg	0.3	0.3	
Vitamin B_6	mg/kg	2.69	2.95	
Alanine	%	0.47	0.52	
Arginine	%	0.82	0.90	
Aspartic acid	%	0.90	0.99	
Glutamic acid	%	1.82	2.00	
Glycine	%	0.53	0.58	
Histidine	%	0.26	0.29	
Isoleucine	%	0.49	0.54	
Leucine	%	0.74	0.81	
Lysine	%	0.48	0.53	
Methionine	%	0.12	0.13	
Phenylalanine	%	0.50	0.55	
Proline	%	0.62	0.68	
Serine	%	0.50	0.55	
Threonine	%	0.38	0.42	
Tyrosine	%	0.24	0.26	
Valine	%	0.60	0.66	

Grains, screenings, mn 70% grain mx 6.5% ash, (4)

Grain screenings (AAFCO)

Ref No 4 02 156 United States

Feed Name or Analyses		Mean As Fed	Mean Dry	C.V. ± %
Dry matter	%	90.2	100.0	0
Ash	%	8.5	9.5	19
Crude fiber	%	13.0	14.5	22
Ether extract	%	5.0	5.5	7
N-free extract	%	50.2	55.7	8
Protein (N x 6.25)	%	13.4	14.8	7
Cattle	dig prot %	* 8.7	9.6	
Goats	dig prot %	* 9.8	10.8	
Horses	dig prot %	* 9.8	10.8	
Sheep	dig prot %	* 9.8	10.8	

Feed Name or Analyses	Mean As Fed	Mean Dry	C.V. ± %
Energy GE Mcal/kg			
Cattle DE Mcal/kg	* 2.40	2.66	
Sheep DE Mcal/kg	* 2.47	2.74	
Swine DE kcal/kg	*1830.	2029.	
Cattle ME Mcal/kg	* 1.97	2.18	
Chickens ME_n kcal/kg	1267.	1405.	
Sheep ME Mcal/kg	* 2.03	2.25	
Swine ME kcal/kg	*1702.	1887.	
Cattle TDN %	* 54.4	60.3	
Sheep TDN %	* 56.0	62.1	
Swine TDN %	* 41.5	46.0	
Calcium %	0.35	0.39	
Phosphorus %	0.33	0.36	

Ref No 4 02 156 Canada

Feed Name or Analyses	Mean As Fed	Mean Dry	C.V. ± %
Dry matter %	89.3	100.0	
Protein (N x 6.25) %	14.3	16.0	
Cattle dig prot %	* 9.6	10.7	
Goats dig prot %	* 10.6	11.9	
Horses dig prot %	* 10.6	11.9	
Sheep dig prot %	* 10.6	11.9	
Energy GE Mcal/kg			
Chickens ME_n kcal/kg	1255.	1405.	
Calcium %	0.18	0.20	
Phosphorus %	0.29	0.32	

Grains, screenings, refuse mx 100% small weed seeds chaff hulls dust scourings noxious seeds, (4)

Refuse screenings (CFA)

Ref No 4 02 151 United States

Feed Name or Analyses	Mean As Fed	Mean Dry	C.V. ± %
Dry matter %	90.4	100.0	
Sheep dig coef %	57.	57.	
Ash %	9.1	10.1	
Crude fiber %	21.9	24.3	
Sheep dig coef %	47.	47.	
Ether extract %	5.2	5.8	
Sheep dig coef %	78.	78.	
N-free extract %	41.6	46.0	
Sheep dig coef %	71.	71.	
Protein (N x 6.25) %	12.4	13.8	
Sheep dig coef %	73.	73.	
Cattle dig prot %	* 7.8	8.7	
Goats dig prot %	* 8.9	9.8	
Horses dig prot %	* 8.9	9.8	
Sheep dig prot %	9.1	10.0	
Energy GE Mcal/kg			
Cattle DE Mcal/kg	* 2.50	2.76	
Sheep DE Mcal/kg	* 2.56	2.84	
Swine DE kcal/kg	* 824.	912.	
Cattle ME Mcal/kg	* 2.04	2.26	
Sheep ME Mcal/kg	* 2.10	2.33	
Swine ME kcal/kg	* 768.	850.	
Cattle NE_m Mcal/kg	* 0.99	1.09	
Cattle NE_{gain} Mcal/kg	* 0.29	0.32	
Cattle $NE_{lactating\ cows}$ Mcal/kg	* 1.01	1.12	
Cattle TDN %	* 56.5	62.5	
Sheep TDN %	58.1	64.3	
Swine TDN %	* 18.7	20.7	
Calcium %	0.20	0.22	
Phosphorus %	0.20	0.22	

Ref No 4 02 151 Canada

Feed Name or Analyses	Mean As Fed	Mean Dry	C.V. ± %
Dry matter %	91.7	100.0	2
Ash %	10.3	11.3	
Crude fiber %	13.4	14.6	55
Sheep dig coef %	47.	47.	
Ether extract %	4.5	4.9	67
Sheep dig coef %	78.	78.	
N-free extract %	51.1	55.7	
Sheep dig coef %	71.	71.	
Protein (N x 6.25) %	12.4	13.5	24
Sheep dig coef %	73.	73.	
Cattle dig prot %	* 7.7	8.4	
Goats dig prot %	* 8.8	9.6	
Horses dig prot %	* 8.8	9.6	
Sheep dig prot %	9.0	9.9	
Cellulose (Matrone) %	9.8	10.7	
Energy GE Mcal/kg			
Cattle DE Mcal/kg	* 2.67	2.91	
Sheep DE Mcal/kg	* 2.62	2.86	
Swine DE kcal/kg	*1641.	1790.	
Cattle ME Mcal/kg	* 2.19	2.39	
Sheep ME Mcal/kg	* 2.15	2.34	
Swine ME kcal/kg	*1530.	1670.	
Cattle TDN %	* 60.4	65.9	
Sheep TDN %	59.4	64.8	
Swine TDN %	* 37.2	40.6	
Calcium %	0.23	0.25	
Phosphorus %	0.29	0.32	
Selenium mg/kg	0.767	0.837	
Niacin mg/kg	47.8	52.2	
Pantothenic acid mg/kg	23.2	25.3	
Riboflavin mg/kg	0.7	0.7	
Thiamine mg/kg	0.5	0.5	
Vitamin B_6 mg/kg	2.41	2.63	
Alanine %	0.65	0.71	
Arginine %	0.69	0.75	
Aspartic acid %	0.95	1.03	
Glutamic acid %	3.11	3.39	
Glycine %	0.60	0.65	
Histidine %	0.30	0.33	
Isoleucine %	0.53	0.58	
Leucine %	0.99	1.08	
Lysine %	0.49	0.53	
Methionine %	0.15	0.16	
Phenylalanine %	0.65	0.71	
Proline %	1.10	1.20	
Serine %	0.58	0.63	
Threonine %	0.47	0.51	
Tyrosine %	0.32	0.35	
Valine %	0.64	0.70	

Grains, brewers grains, dehy, mx 3% dried spent hops, (5)

Brewers dried grains (AAFCO)

Brewers dried grains (CFA)

Ref No 5 02 141 United States

Feed Name or Analyses	Mean As Fed	Mean Dry	C.V. ± %
Dry matter %	91.0	100.0	3
Ash %	3.8	4.2	11
Crude fiber %	14.7	16.1	7
Cattle dig coef %	41.	41.	
Horses dig coef %	18.	18.	
Sheep dig coef %	47.	47.	
Swine dig coef %	2.	2.	
Ether extract %	6.6	7.2	9
Cattle dig coef %	89.	89.	
Horses dig coef %	46.	46.	
Sheep dig coef %	86.	86.	
Swine dig coef %	60.	60.	
N-free extract %	40.2	44.2	8
Cattle dig coef %	55.	55.	
Horses dig coef %	44.	44.	
Sheep dig coef %	61.	61.	
Swine dig coef %	27.	27.	
Protein (N x 6.25) %	25.8	28.3	15
Cattle dig coef %	74.	74.	
Horses dig coef %	77.	77.	
Sheep dig coef %	74.	74.	
Swine dig coef %	79.	79.	
Cattle dig prot %	19.1	21.0	
Horses dig prot %	19.8	21.8	
Sheep dig prot %	19.1	21.0	
Swine dig prot %	20.3	22.4	
Pentosans %	22.9	25.2	7
Linoleic %	.300	.330	
Energy GE Mcal/kg	4.57	5.03	
Cattle DE Mcal/kg	* 2.66	2.92	
Horses DE Mcal/kg	* 2.07	2.28	
Sheep DE Mcal/kg	* 2.79	3.06	
Swine DE kcal/kg	*1779.	1956.	
Cattle ME Mcal/kg	* 2.18	2.40	
Chickens ME_n kcal/kg	2112.	2321.	
Horses ME Mcal/kg	* 1.70	1.87	
Sheep ME Mcal/kg	* 2.28	2.51	
Swine ME kcal/kg	*1606.	1766.	
Cattle NE_m Mcal/kg	* 1.29	1.42	
Cattle NE_{gain} Mcal/kg	* 0.76	0.83	
Cattle $NE_{lactating\ cows}$ Mcal/kg	* 1.52	1.67	
Cattle TDN %	60.3	66.3	
Horses TDN %	47.0	51.6	
Sheep TDN %	63.2	69.4	
Swine TDN %	40.4	44.4	
Calcium %	0.27	0.30	40
Chlorine %	0.18	0.19	50
Cobalt mg/kg	0.061	0.067	28
Copper mg/kg	21.0	23.1	49
Iron %	0.025	0.027	30
Magnesium %	0.14	0.15	69
Manganese mg/kg	37.0	40.7	21
Phosphorus %	0.48	0.53	14
Potassium %	0.09	0.10	62
Sodium %	0.26	0.28	
Sulphur %	0.30	0.33	67
Biotin mg/kg	0.00	0.00	21
Choline mg/kg	1559.	1714.	19
Folic acid mg/kg	0.22	0.24	22
Niacin mg/kg	42.7	46.9	23
Pantothenic acid mg/kg	8.4	9.3	46
Riboflavin mg/kg	1.5	1.7	81
Thiamine mg/kg	0.7	0.7	51
Vitamin B_6 mg/kg	0.65	0.72	27
Arginine %	1.33	1.47	
Cystine %	0.29	0.32	
Histidine %	0.61	0.68	
Isoleucine %	1.49	1.64	
Leucine %	2.55	2.80	
Lysine %	0.99	1.09	
Methionine %	0.46	0.50	
Phenylalanine %	1.44	1.58	
Threonine %	1.03	1.13	
Tryptophan %	0.34	0.38	
Tyrosine %	1.18	1.30	
Valine %	1.67	1.84	

Ref No 5 02 141 Canada

Feed Name or Analyses	Mean As Fed	Mean Dry	C.V. ± %
Dry matter %	92.4	100.0	
Ash %	3.9	4.2	
Crude fiber %	12.7	13.8	
Cattle dig coef %	41.	41.	
Horses dig coef %	18.	18.	

Continued

Feed Name or Analyses		Mean		C.V.
		As Fed	Dry	± %
Sheep	dig coef %	47.	47.	
Swine	dig coef %	2.	2.	
Ether extract	%	8.1	8.8	
Cattle	dig coef %	89.	89.	
Horses	dig coef %	46.	46.	
Sheep	dig coef %	86.	86.	
Swine	dig coef %	60.	60.	
N-free extract	%	42.8	46.4	
Cattle	dig coef %	55.	55.	
Horses	dig coef %	44.	44.	
Sheep	dig coef %	61.	61.	
Swine	dig coef %	27.	27.	
Protein (N x 6.25)	%	24.8	26.8	
Cattle	dig coef %	74.	74.	
Horses	dig coef %	77.	77.	
Sheep	dig coef %	74.	74.	
Swine	dig coef %	79.	79.	
Cattle	dig prot %	18.3	19.8	
Horses	dig prot %	19.1	20.6	
Sheep	dig prot %	18.3	19.8	
Swine	dig prot %	19.6	21.2	
Energy	GE Mcal/kg	4.62	5.00	
Cattle	DE Mcal/kg	* 2.80	3.03	
Horses	DE Mcal/kg	* 2.14	2.32	
Sheep	DE Mcal/kg	* 2.92	3.16	
Swine	DE kcal/kg	*1869.	2023.	
Cattle	ME Mcal/kg	* 2.29	2.48	
Chickens	ME_n kcal/kg	1727.	1870.	
Horses	ME Mcal/kg	* 1.76	1.90	
Sheep	ME Mcal/kg	* 2.39	2.59	
Swine	ME kcal/kg	*1693.	1833.	
Cattle	TDN %	63.4	68.7	
Horses	TDN %	48.6	52.7	
Sheep	TDN %	66.2	71.7	
Swine	TDN %	42.4	45.9	
Calcium	%	0.39	0.42	
Phosphorus	%	0.64	0.69	
Salt (NaCl)	%	0.07	0.08	

Grains, brewers grains, wet, (5)

Ref No 5 02 142 United States

Feed Name or Analyses		As Fed	Dry	C.V. ± %
Dry matter	%	23.8	100.0	
Ash	%	1.1	4.8	
Crude fiber	%	3.8	16.1	
Cattle	dig coef %	39.	39.	
Ether extract	%	1.5	6.5	
Cattle	dig coef %	84.	84.	
N-free extract	%	11.8	49.6	
Cattle	dig coef %	64.	64.	
Protein (N x 6.25)	%	5.5	23.0	
Cattle	dig coef %	73.	73.	
Cattle	dig prot %	4.0	16.8	
Energy	GE Mcal/kg			
Cattle	DE Mcal/kg	* 0.70	2.96	
Sheep	DE Mcal/kg	* 0.74	3.13	
Swine	DE kcal/kg	* 956.	4026.	
Cattle	ME Mcal/kg	* 0.58	2.42	
Sheep	ME Mcal/kg	* 0.61	2.57	
Swine	ME kcal/kg	* 874.	3678.	
Cattle	TDN %	15.9	67.1	
Sheep	TDN %	* 16.9	71.0	
Swine	TDN %	* 21.7	91.3	
Calcium	%	0.07	0.30	

(1) dry forages and roughages (3) silages (6) minerals
(2) pasture, range plants, and forages fed green (4) energy feeds (7) vitamins
(5) protein supplements (8) additives

Feed Name or Analyses		Mean		C.V.
		As Fed	Dry	± %
Phosphorus	%	0.12	0.51	
Potassium	%	0.02	0.08	

Ref No 5 02 142 Canada

Feed Name or Analyses		As Fed	Dry	C.V. ± %
Dry matter	%	16.2	100.0	
Crude fiber	%	2.2	13.6	
Cattle	dig coef %	39.	39.	
Ether extract	%	3.0	18.5	
Cattle	dig coef %	84.	84.	
N-free extract	%			
Cattle	dig coef %	64.	64.	
Protein (N x 6.25)	%	4.2	25.9	
Cattle	dig coef %	73.	73.	
Cattle	dig prot %	3.1	18.9	
Energy	GE Mcal/kg			
Cattle	DE Mcal/kg	* 0.52	3.19	
Cattle	ME Mcal/kg	* 0.42	2.62	
Cattle	TDN %	11.7	72.4	

Grains, distillers grains, dehy, (5)

Ref No 5 02 144 United States

Feed Name or Analyses		As Fed	Dry	C.V. ± %
Dry matter	%	92.5	100.0	1
Ash	%	1.6	1.7	32
Crude fiber	%	12.8	13.8	9
Sheep	dig coef %	70.	70.	
Ether extract	%	7.4	8.0	9
Sheep	dig coef %	94.	94.	
N-free extract	%	43.4	46.9	
Sheep	dig coef %	75.	75.	
Protein (N x 6.25)	%	27.4	29.6	5
Sheep	dig coef %	69.	69.	
Sheep	dig prot %	18.9	20.4	
Pentosans	%	14.1	15.3	1
Fatty acids	%			
Linoleic	%	2.9	3.1	
Energy	GE Mcal/kg	5.00	5.41	0
Cattle	DE Mcal/kg	* 3.37	3.64	
Sheep	DE Mcal/kg	* 3.35	3.62	
Swine	DE kcal/kg	*3915.	4231.	
Cattle	ME Mcal/kg	* 2.76	2.99	
Sheep	ME Mcal/kg	* 2.75	2.97	
Swine	ME kcal/kg	*3525.	3809.	
Cattle	TDN %	* 76.4	82.6	
Sheep	TDN %	76.1	82.2	
Swine	TDN %	* 88.8	95.9	
Boron	mg/kg	3.332	3.600	
Calcium	%	0.16	0.17	42
Chlorine	%	0.05	0.05	
Cobalt	mg/kg	0.091	0.099	31
Copper	mg/kg	45.2	48.9	52
Iron	%	0.032	0.035	26
Magnesium	%	0.14	0.15	98
Manganese	mg/kg	53.6	57.9	23
Phosphorus	%	1.06	1.15	14
Potassium	%	1.19	1.28	
Sodium	%	0.04	0.04	
Sulphur	%	0.45	0.49	
Strontium	mg/kg	1.111	1.200	
Zinc	mg/kg	194.3	210.0	
Carotene	mg/kg	7.8	8.4	14
Niacin	mg/kg	46.8	50.5	33
Pantothenic acid	mg/kg	11.6	12.5	19
Riboflavin	mg/kg	3.8	4.1	48
Thiamine	mg/kg	2.5	2.6	42
Vitamin A equivalent	IU/g	13.0	14.0	
Arginine	%	1.01	1.09	11
Histidine	%	0.51	0.55	

Feed Name or Analyses		Mean		C.V.
		As Fed	Dry	± %
Isoleucine	%	1.62	1.75	6
Leucine	%	2.73	2.95	12
Lysine	%	0.81	0.87	17
Methionine	%	0.40	0.44	
Phenylalanine	%	1.11	1.20	12
Threonine	%	0.81	0.87	22
Tryptophan	%	0.20	0.22	22
Tyrosine	%	0.91	0.98	15
Valine	%	1.21	1.31	22

Ref No 5 02 144 Canada

Feed Name or Analyses		As Fed	Dry	C.V. ± %
Dry matter	%	95.6	100.0	
Crude fiber	%	13.0	13.6	
Sheep	dig coef %	70.	70.	
Ether extract	%			
Sheep	dig coef %	94.	94.	
N-free extract	%			
Sheep	dig coef %	75.	75.	
Protein (N x 6.25)	%	21.7	22.7	
Sheep	dig coef %	69.	69.	
Sheep	dig prot %	15.0	15.7	
Energy	GE Mcal/kg	5.17	5.41	
Sheep	DE Mcal/kg	* 3.46	3.62	
Sheep	ME Mcal/kg	* 2.84	2.97	
Sheep	TDN %	78.6	82.2	
Calcium	%	0.08	0.08	
Phosphorus	%	0.25	0.26	

Grains, distillers grains w solubles, dehy, (5)

Distillers dried grains w solubles

Ref No 5 07 987 United States

Feed Name or Analyses		As Fed	Dry	C.V. ± %
Dry matter	%	91.2	100.0	2
Ash	%	4.3	4.8	15
Crude fiber	%	7.8	8.6	13
Ether extract	%	8.4	9.2	20
N-free extract	%	41.9	45.9	
Protein (N x 6.25)	%	28.8	31.6	37
Fatty acids	%			
Linoleic	%	4.6	5.0	
Energy	GE Mcal/kg	5.05	5.53	
Cattle	DE Mcal/kg	* 3.46	3.79	
Sheep	DE Mcal/kg	* 3.55	3.89	
Swine	DE kcal/kg	*3983.	4366.	
Cattle	ME Mcal/kg	* 2.84	3.11	
Sheep	ME Mcal/kg	* 2.91	3.19	
Swine	ME kcal/kg	*3570.	3913.	
Cattle	TDN %	* 78.5	86.0	
Sheep	TDN %	* 80.5	88.3	
Swine	TDN %	* 90.3	99.0	
Aluminum	mg/kg	10.49	11.50	
Boron	mg/kg	1.368	1.500	
Calcium	%	0.27	0.29	3
Copper	mg/kg	73.0	80.0	
Iron	%	0.032	0.035	
Magnesium	%	0.34	0.37	12
Manganese	mg/kg	111.6	122.3	
Phosphorus	%	0.78	0.86	19
Potassium	%	0.86	0.95	
Sodium	%	0.05	0.05	
Strontium	mg/kg	13.685	15.000	
Zinc	mg/kg	442.5	485.0	
Biotin	mg/kg	0.40	0.44	75
Choline	mg/kg	4005.	4390.	12
Niacin	mg/kg	81.3	89.1	
Pantothenic acid	mg/kg	12.3	13.5	
Riboflavin	mg/kg	9.6	10.6	26
Thiamine	mg/kg	4.0	4.4	

Feed Name or Analyses		Mean As Fed	Mean Dry	C.V. ± %
Vitamin B6	mg/kg	4.93	5.40	
Arginine	%	1.12	1.22	8
Cysteine	%	0.54	0.60	
Cystine	%	0.52	0.57	
Glutamic acid	%	0.61	0.67	
Histidine	%	0.81	0.89	
Isoleucine	%	1.93	2.12	
Leucine	%	2.34	2.56	
Lysine	%	0.81	0.89	
Chicks available	%	0.58	0.63	14
Methionine	%	0.46	0.50	6
Phenylalanine	%	1.93	2.12	
Serine	%	1.73	1.89	
Threonine	%	1.12	1.22	
Tryptophan	%	0.20	0.22	
Tyrosine	%	5.69	6.24	
Valine	%	1.83	2.00	

Grains, distillers solubles, dehy, (5)

Dried distillers solubles (CFA)

Ref No 5 02 147 United States

Feed Name or Analyses		Mean As Fed	Mean Dry	C.V. ± %
Dry matter	%	92.1	100.0	1
Ash	%	6.2	6.8	6
Crude fiber	%	3.4	3.7	18
Ether extract	%	8.9	9.6	9
N-free extract	%	44.8	48.6	
Protein (N x 6.25)	%	28.8	31.2	4
Cellulose (Matrone)	%	5.1	5.6	
Fatty acids	%			
Linoleic	%	2.7	2.9	
Energy	GE Mcal/kg	4.53	4.92	
Cattle	DE Mcal/kg	* 3.64	3.96	
Sheep	DE Mcal/kg	* 3.43	3.72	
Swine	DE kcal/kg	*3679.	3995.	
Cattle	ME Mcal/kg	* 2.99	3.24	
Chickens	MEn kcal/kg	3049.	3311.	
Sheep	ME Mcal/kg	* 2.81	3.05	
Swine	ME kcal/kg	*3301.	3584.	
Cattle	TDN %	* 82.6	89.7	
Sheep	TDN %	* 77.7	84.4	
Swine	TDN %	* 83.4	90.6	
Boron	mg/kg	1.658	1.800	
Calcium	%	0.21	0.22	19
Cobalt	mg/kg	0.196	0.213	
Copper	mg/kg	71.7	77.9	
Iron	%	0.030	0.033	
Magnesium	%	0.46	0.50	
Manganese	mg/kg	64.2	69.7	39
Phosphorus	%	1.23	1.33	26
Potassium	%	1.84	2.00	
Sodium	%	0.15	0.16	
Strontium	mg/kg	3.592	3.900	
Zinc	mg/kg	138.1	150.0	
Biotin	mg/kg	2.85	3.09	
Carotene	mg/kg	1.1	1.2	
Choline	mg/kg	4254.	4619.	26
Niacin	mg/kg	143.2	155.5	19
Pantothenic acid	mg/kg	25.4	27.5	18
Riboflavin	mg/kg	11.3	12.3	55
Thiamine	mg/kg	6.9	7.5	23
Vitamin B6	mg/kg	8.67	9.42	
Vitamin B12	µg/kg	2.9	3.1	
Vitamin A equivalent	IU/g	1.9	2.0	
Arginine	%	8.68	9.42	
Cystine	%	0.40	0.44	
Glutamic acid	%	1.82	1.97	
Histidine	%	0.71	0.77	11
Isoleucine	%	1.41	1.53	24

Feed Name or Analyses		Mean As Fed	Mean Dry	C.V. ± %
Leucine	%	2.82	3.07	26
Lysine	%	1.21	1.31	27
Methionine	%	0.50	0.55	19
Phenylalanine	%	1.41	1.53	11
Serine	%	0.61	0.66	
Threonine	%	1.11	1.20	7
Tryptophan	%	0.20	0.22	30
Tyrosine	%	0.91	0.99	26
Valine	%	1.61	1.75	19

Grains, distillers stillage, wet, (5)

Distillers stillage, wet

Ref No 5 02 148 United States

Feed Name or Analyses		Mean As Fed	Mean Dry	C.V. ± %
Dry matter	%	6.2	100.0	
Ash	%	0.3	4.8	
Crude fiber	%	0.5	8.1	
Ether extract	%	0.6	9.7	
N-free extract	%	2.9	46.8	
Protein (N x 6.25)	%	1.9	30.6	
Energy	GE Mcal/kg			
Cattle	DE Mcal/kg	* 0.18	2.96	
Sheep	DE Mcal/kg	* 0.19	3.05	
Swine	DE kcal/kg	* 206.	3324.	
Cattle	ME Mcal/kg	* 0.15	2.43	
Sheep	ME Mcal/kg	* 0.16	2.50	
Swine	ME kcal/kg	* 185.	2985.	
Cattle	TDN %	* 4.2	67.1	
Sheep	TDN %	* 4.3	69.1	
Swine	TDN %	* 4.7	75.4	

Grains, distillers stillage, wet strained, (5)

Distillery stillage, strained

Ref No 5 08 429 United States

Feed Name or Analyses		Mean As Fed	Mean Dry	C.V. ± %
Dry matter	%	3.8	100.0	
Ash	%	0.3	7.9	
Crude fiber	%	0.2	5.3	
Ether extract	%	0.4	10.5	
N-free extract	%	1.8	47.4	
Protein (N x 6.25)	%	1.1	28.9	
Energy	GE Mcal/kg			
Cattle	DE Mcal/kg	* 0.11	2.82	
Sheep	DE Mcal/kg	* 0.11	2.91	
Swine	DE kcal/kg	* 122.	3205.	
Cattle	ME Mcal/kg	* 0.09	2.31	
Sheep	ME Mcal/kg	* 0.09	2.39	
Swine	ME kcal/kg	* 110.	2891.	
Cattle	TDN %	* 2.4	64.0	
Sheep	TDN %	* 2.5	65.9	
Swine	TDN %	* 2.8	72.7	
Calcium	%	0.00	0.11	
Phosphorus	%	0.05	1.32	

Grains, fermentation solubles, dehy, (5)

Dried grain fermentation solubles (CFA)

Ref No 5 02 150 United States

Feed Name or Analyses		Mean As Fed	Mean Dry	C.V. ± %
Dry matter	%	94.7	100.0	
Ash	%	6.5	6.9	
Crude fiber	%	7.2	7.6	
Ether extract	%	5.1	5.4	
N-free extract	%	43.8	46.3	
Protein (N x 6.25)	%	32.1	33.9	
Energy	GE Mcal/kg			
Cattle	DE Mcal/kg	* 3.36	3.55	
Sheep	DE Mcal/kg	* 3.55	3.75	

Feed Name or Analyses		Mean As Fed	Mean Dry	C.V. ± %
Swine	DE kcal/kg	*3787.	3999.	
Cattle	ME Mcal/kg	* 2.76	2.91	
Sheep	ME Mcal/kg	* 2.91	3.08	
Swine	ME kcal/kg	*3376.	3565.	
Cattle	TDN %	* 76.2	80.5	
Sheep	TDN %	* 80.6	85.1	
Swine	TDN %	* 85.9	90.7	

Grains, yeast fermentation grains, dehy, (7)

Yeast dried grains (AAFCO)

Ref No 7 02 159 United States

Feed Name or Analyses		Mean As Fed	Mean Dry	C.V. ± %
Dry matter	%	92.4	100.0	
Ash	%	2.4	2.6	
Crude fiber	%	18.6	20.1	
Sheep	dig coef %	58.	58.	
Ether extract	%	6.1	6.6	
Sheep	dig coef %	87.	87.	
N-free extract	%	46.5	50.3	
Sheep	dig coef %	56.	56.	
Protein (N x 6.25)	%	18.8	20.4	
Sheep	dig coef %	64.	64.	
Sheep	dig prot %	12.1	13.1	
Energy	GE Mcal/kg			
Cattle	DE Mcal/kg	* 2.43	2.63	
Sheep	DE Mcal/kg	* 2.68	2.90	
Swine	DE kcal/kg	*2498.	2703.	
Cattle	ME Mcal/kg	* 1.99	2.16	
Sheep	ME Mcal/kg	* 2.20	2.38	
Swine	ME kcal/kg	*2295.	2484.	
Cattle	TDN %	* 55.1	59.6	
Sheep	TDN %	60.8	65.8	
Swine	TDN %	* 56.6	61.3	

Grain screenings (AAFCO) –
see Grains, screenings, mn 70% grain mx 6.5% ash, (4)

Grain sorghum distillers dried grains (AAFCO) –
see Sorghum, grain variety, distillers grains, dehy, (5)

Grain sorghum distillers dried grains with solubles (AAFCO) –
see Sorghum, grain variety, distillers grains w solubles, dehy, mn 75% original solids, (5)

Grain sorghum distillers dried solubles (AAFCO) –
see Sorghum, grain variety, distillers solubles, dehy, (5)

Grain sorghum fodder, sun-cured –
see Sorghum, grain variety, aerial part, s-c, (1)

Grain sorghum gluten feed (AAFCO) –
see Sorghum, grain variety, gluten w bran, wet milled dehy, (5)

Grain sorghum gluten meal (AAFCO) –
see Sorghum, grain variety, gluten, wet milled dehy, (5)

Gram –
see Chickpea, gram, seeds, (5)

GRAMA. Bouteloua spp

Grama, aerial part, dormant, (1)
Ref No 1 02 160 — United States

Feed Name or Analyses		As Fed	Dry	C.V. ± %
Dry matter	%		100.0	
Carotene	mg/kg		11.2	
Vitamin A equivalent	IU/g		18.7	

Grama, hay, s-c, (1)
Ref No 1 02 162 — United States

Feed Name or Analyses		As Fed	Dry	C.V. ± %
Dry matter	%	89.2	100.0	5
Ash	%	8.4	9.5	21
Crude fiber	%	29.1	32.6	7
Ether extract	%	1.5	1.7	16
N-free extract	%	44.6	50.0	
Protein (N x 6.25)	%	5.6	6.2	27
Cattle	dig prot % *	2.1	2.3	
Goats	dig prot % *	2.1	2.4	
Horses	dig prot % *	2.5	2.8	
Rabbits	dig prot % *	3.1	3.5	
Sheep	dig prot % *	1.9	2.2	
Fatty acids	%	1.5	1.7	
Energy	GE Mcal/kg			
Cattle	DE Mcal/kg *	1.78	1.99	
Sheep	DE Mcal/kg *	1.68	1.88	
Cattle	ME Mcal/kg *	1.45	1.63	
Sheep	ME Mcal/kg *	1.37	1.54	
Cattle	TDN % *	40.2	45.1	
Sheep	TDN % *	38.0	42.6	
Calcium	%	0.34	0.38	
Phosphorus	%	0.18	0.20	

Grama, hay, s-c, over ripe, (1)
Ref No 1 02 161 — United States

Feed Name or Analyses		As Fed	Dry	C.V. ± %
Dry matter	%	89.6	100.0	8
Ash	%	9.5	10.6	16
Crude fiber	%	30.1	33.6	7
Ether extract	%	1.4	1.6	14
N-free extract	%	44.4	49.6	
Protein (N x 6.25)	%	4.1	4.6	12
Cattle	dig prot % *	0.8	0.9	
Goats	dig prot % *	0.8	0.9	
Horses	dig prot % *	1.3	1.4	
Rabbits	dig prot % *	2.0	2.2	
Sheep	dig prot % *	0.6	0.7	
Energy	GE Mcal/kg			
Cattle	DE Mcal/kg *	1.70	1.90	
Sheep	DE Mcal/kg *	1.57	1.75	
Cattle	ME Mcal/kg *	1.40	1.56	
Sheep	ME Mcal/kg *	1.29	1.44	
Cattle	TDN % *	38.7	43.2	
Sheep	TDN % *	35.6	39.7	

Grama, aerial part, fresh, (2)
Ref No 2 02 168 — United States

Feed Name or Analyses		As Fed	Dry	C.V. ± %
Dry matter	%	56.9	100.0	17
Carotene	mg/kg	27.6	48.5	
Vitamin A equivalent	IU/g	46.0	80.9	

(1) dry forages and roughages (2) pasture, range plants, and forages fed green (3) silages (4) energy feeds (5) protein supplements (6) minerals (7) vitamins (8) additives

Grama, aerial part, fresh, mid-bloom, (2)
Ref No 2 02 164 — United States

Feed Name or Analyses		As Fed	Dry	C.V. ± %
Dry matter	%		100.0	
Ash	%		15.1	4
Crude fiber	%		28.9	4
Ether extract	%		1.9	8
N-free extract	%		45.8	
Protein (N x 6.25)	%		8.3	3
Cattle	dig prot % *		4.9	
Goats	dig prot % *		4.3	
Horses	dig prot % *		4.6	
Rabbits	dig prot % *		5.1	
Sheep	dig prot % *		4.7	
Energy	GE Mcal/kg			
Cattle	DE Mcal/kg *		2.38	
Sheep	DE Mcal/kg *		2.15	
Cattle	ME Mcal/kg *		1.95	
Sheep	ME Mcal/kg *		1.76	
Cattle	TDN % *		53.9	
Sheep	TDN % *		48.7	

Grama, aerial part, fresh, full bloom, (2)
Ref No 2 02 165 — United States

Feed Name or Analyses		As Fed	Dry	C.V. ± %
Dry matter	%		100.0	
Ash	%		13.8	19
Crude fiber	%		30.6	5
Ether extract	%		1.7	11
N-free extract	%		46.8	
Protein (N x 6.25)	%		7.1	8
Cattle	dig prot % *		3.9	
Goats	dig prot % *		3.2	
Horses	dig prot % *		3.6	
Rabbits	dig prot % *		4.2	
Sheep	dig prot % *		3.6	
Energy	GE Mcal/kg			
Cattle	DE Mcal/kg *		2.37	
Sheep	DE Mcal/kg *		2.29	
Cattle	ME Mcal/kg *		1.94	
Sheep	ME Mcal/kg *		1.88	
Cattle	TDN % *		53.7	
Sheep	TDN % *		51.9	
Calcium	%		0.50	19
Cobalt	mg/kg		0.240	
Copper	mg/kg		8.2	
Manganese	mg/kg		11.9	
Phosphorus	%		0.13	31
Carotene	mg/kg		115.3	
Vitamin A equivalent	IU/g		192.2	

Grama, aerial part, fresh, mature, (2)
Ref No 2 02 166 — United States

Feed Name or Analyses		As Fed	Dry	C.V. ± %
Dry matter	%	63.4	100.0	11
Energy	GE Mcal/kg			
Cattle	NE_m Mcal/kg *	0.79	1.24	
Cattle	NE_{gain} Mcal/kg *	0.37	0.59	
Cobalt	mg/kg	0.115	0.181	
Carotene	mg/kg	19.3	30.4	
Vitamin A equivalent	IU/g	32.2	50.7	

Grama, aerial part, fresh, over ripe, (2)
Ref No 2 02 167 — United States

Feed Name or Analyses		As Fed	Dry	C.V. ± %
Dry matter	%		100.0	
Cobalt	mg/kg		0.090	
Carotene	mg/kg		22.7	
Vitamin A equivalent	IU/g		37.9	

GRAMA, BLACK. Bouteloua eriopoda

Grama, black, aerial part, fresh, (2)
Ref No 2 02 174 — United States

Feed Name or Analyses		As Fed	Dry	C.V. ± %
Dry matter	%	56.1	100.0	6
Cobalt	mg/kg	0.093	0.165	
Carotene	mg/kg	22.3	39.7	40
Vitamin A equivalent	IU/g	37.1	66.2	

Grama, black, aerial part, fresh, immature, (2)
Ref No 2 02 169 — United States

Feed Name or Analyses		As Fed	Dry	C.V. ± %
Dry matter	%	49.6	100.0	
Ash	%	4.2	8.5	8
Crude fiber	%	15.0	30.3	3
Ether extract	%	0.8	1.7	17
N-free extract	%	23.8	48.0	
Protein (N x 6.25)	%	5.7	11.5	16
Cattle	dig prot % *	3.8	7.7	
Goats	dig prot % *	3.6	7.3	
Horses	dig prot % *	3.6	7.3	
Rabbits	dig prot % *	3.8	7.6	
Sheep	dig prot % *	3.8	7.7	
Energy	GE Mcal/kg			
Cattle	DE Mcal/kg *	1.26	2.55	
Sheep	DE Mcal/kg *	1.20	2.42	
Cattle	ME Mcal/kg *	1.04	2.09	
Sheep	ME Mcal/kg *	0.98	1.98	
Cattle	TDN % *	28.7	57.9	
Sheep	TDN % *	27.2	54.8	
Calcium	%	0.19	0.38	21
Phosphorus	%	0.08	0.16	24

Grama, black, aerial part, fresh, early bloom, (2)
Ref No 2 02 170 — United States

Feed Name or Analyses		As Fed	Dry	C.V. ± %
Dry matter	%		100.0	
Ash	%		9.5	11
Crude fiber	%		30.7	1
Ether extract	%		1.4	7
N-free extract	%		49.1	
Protein (N x 6.25)	%		9.3	0
Cattle	dig prot % *		5.8	
Goats	dig prot % *		5.2	
Horses	dig prot % *		5.4	
Rabbits	dig prot % *		5.8	
Sheep	dig prot % *		5.7	
Energy	GE Mcal/kg			
Cattle	DE Mcal/kg *		2.52	
Sheep	DE Mcal/kg *		2.60	
Cattle	ME Mcal/kg *		2.07	
Sheep	ME Mcal/kg *		2.13	
Cattle	TDN % *		57.1	
Sheep	TDN % *		58.9	
Calcium	%		0.42	25
Phosphorus	%		0.16	8

Feed Name or Analyses			Mean As Fed	Mean Dry	C.V. ± %

Grama, black, aerial part, fresh, dough stage, (2)
Ref No 2 02 171 United States

Dry matter	%		50.0	100.0	
Ash	%		4.6	9.2	
Crude fiber	%		15.9	31.8	
Ether extract	%		0.9	1.8	
N-free extract	%		24.9	49.8	
Protein (N x 6.25)	%		3.7	7.4	
Cattle	dig prot %	*	2.1	4.2	
Goats	dig prot %	*	1.7	3.5	
Horses	dig prot %	*	1.9	3.8	
Rabbits	dig prot %	*	2.2	4.4	
Sheep	dig prot %	*	1.9	3.9	
Energy	GE Mcal/kg				
Cattle	DE Mcal/kg	*	1.08	2.16	
Sheep	DE Mcal/kg	*	1.14	2.28	
Cattle	ME Mcal/kg	*	0.88	1.77	
Sheep	ME Mcal/kg	*	0.94	1.87	
Cattle	TDN %	*	24.6	49.1	
Sheep	TDN %	*	25.8	51.6	
Calcium	%		0.15	0.30	
Phosphorus	%		0.06	0.12	

Grama, black, aerial part, fresh, mature, (2)
Ref No 2 02 172 United States

Dry matter	%		64.5	100.0	8
Ash	%		4.3	6.7	9
Crude fiber	%		21.4	33.1	3
Ether extract	%		1.0	1.6	10
N-free extract	%		33.7	52.2	
Protein (N x 6.25)	%		4.1	6.4	13
Cattle	dig prot %	*	2.2	3.3	
Goats	dig prot %	*	1.6	2.6	
Horses	dig prot %	*	1.9	3.0	
Rabbits	dig prot %	*	2.3	3.6	
Sheep	dig prot %	*	1.9	3.0	
Energy	GE Mcal/kg				
Cattle	DE Mcal/kg	*	1.30	2.01	
Sheep	DE Mcal/kg	*	1.37	2.12	
Cattle	ME Mcal/kg	*	1.06	1.65	
Sheep	ME Mcal/kg	*	1.12	1.74	
Cattle	TDN %	*	29.4	45.6	
Sheep	TDN %	*	31.0	48.1	
Calcium	%		0.14	0.22	22
Phosphorus	%		0.07	0.11	21
Carotene	mg/kg		16.8	26.0	56
Vitamin A equivalent	IU/g		28.0	43.4	

Grama, black, aerial part, fresh, over ripe, (2)
Ref No 2 02 173 United States

Dry matter	%			100.0	
Ash	%			6.3	10
Crude fiber	%			33.9	5
Ether extract	%			1.7	10
N-free extract	%			52.9	
Protein (N x 6.25)	%			5.2	15
Cattle	dig prot %	*		2.3	
Goats	dig prot %	*		1.4	
Horses	dig prot %	*		1.9	
Rabbits	dig prot %	*		2.7	
Sheep	dig prot %	*		1.8	
Energy	GE Mcal/kg				
Cattle	DE Mcal/kg	*		1.97	
Sheep	DE Mcal/kg	*		2.15	
Cattle	ME Mcal/kg	*		1.62	
Sheep	ME Mcal/kg	*		1.76	
Cattle	TDN %	*		44.7	
Sheep	TDN %	*		48.8	
Calcium	%			0.15	30
Phosphorus	%			0.06	37
Carotene	mg/kg			23.8	32
Vitamin A equivalent	IU/g			39.7	

Grama, black, aerial part wo lower stems, fresh, (2)
Ref No 2 08 727 United States

Dry matter	%			100.0	
Organic matter	%			87.4	
Ash	%			12.6	
Crude fiber	%			30.7	
Ether extract	%			1.3	
N-free extract	%			50.9	
Protein (N x 6.25)	%			4.5	
Cattle	dig prot %	*		1.7	
Goats	dig prot %	*		0.8	
Horses	dig prot %	*		1.4	
Rabbits	dig prot %	*		2.1	
Sheep	dig prot %	*		1.2	
Energy	GE Mcal/kg				
Cattle	DE Mcal/kg	*		2.41	
Sheep	DE Mcal/kg	*		2.19	
Cattle	ME Mcal/kg	*		1.97	
Sheep	ME Mcal/kg	*		1.80	
Cattle	TDN %	*		54.6	
Sheep	TDN %	*		49.7	
Calcium	%			0.21	
Phosphorus	%			0.10	
Carotene	mg/kg			0.0	

Grama, black, aerial part wo lower stems, fresh, early leaf, (2)
Ref No 2 08 706 United States

Dry matter	%			100.0	
Organic matter	%			90.9	2
Ash	%			9.1	19
Crude fiber	%			31.7	4
Ether extract	%			1.6	20
N-free extract	%			51.0	4
Protein (N x 6.25)	%			6.7	12
Cattle	dig prot %	*		3.5	
Goats	dig prot %	*		2.8	
Horses	dig prot %	*		3.2	
Rabbits	dig prot %	*		3.8	
Sheep	dig prot %	*		3.2	
Energy	GE Mcal/kg				
Cattle	DE Mcal/kg	*		2.73	
Sheep	DE Mcal/kg	*		2.49	
Cattle	ME Mcal/kg	*		2.24	
Sheep	ME Mcal/kg	*		2.04	
Cattle	TDN %	*		61.9	
Sheep	TDN %	*		56.4	
Calcium	%			0.38	43
Phosphorus	%			0.10	41
Carotene	mg/kg			42.0	52
Vitamin A equivalent	IU/g			70.0	

Grama, black, aerial part wo lower stems, fresh, mid-bloom, (2)
Ref No 2 08 707 United States

Dry matter	%			100.0	
Organic matter	%			92.1	1
Ash	%			7.9	13
Crude fiber	%			33.7	5
Ether extract	%			1.6	22
N-free extract	%			50.4	4
Protein (N x 6.25)	%			6.5	17
Cattle	dig prot %	*		3.4	
Goats	dig prot %	*		2.6	
Horses	dig prot %	*		3.0	
Rabbits	dig prot %	*		3.6	
Sheep	dig prot %	*		3.0	
Energy	GE Mcal/kg				
Cattle	DE Mcal/kg	*		2.82	
Sheep	DE Mcal/kg	*		2.59	
Cattle	ME Mcal/kg	*		2.31	
Sheep	ME Mcal/kg	*		2.12	
Cattle	TDN %	*		64.0	
Sheep	TDN %	*		58.7	
Calcium	%			0.35	30
Phosphorus	%			0.10	23
Carotene	mg/kg			19.5	79
Vitamin A equivalent	IU/g			32.5	

Grama, black, aerial part wo lower stems, fresh, mature, (2)
Ref No 2 08 708 United States

Dry matter	%			100.0	
Organic matter	%			90.7	3
Ash	%			9.3	28
Crude fiber	%			32.8	6
Ether extract	%			1.6	29
N-free extract	%			50.1	5
Protein (N x 6.25)	%			6.2	26
Cattle	dig prot %	*		3.2	
Goats	dig prot %	*		2.4	
Horses	dig prot %	*		2.8	
Rabbits	dig prot %	*		3.5	
Sheep	dig prot %	*		2.8	
Energy	GE Mcal/kg				
Cattle	DE Mcal/kg	*		2.71	
Sheep	DE Mcal/kg	*		2.55	
Cattle	ME Mcal/kg	*		2.22	
Sheep	ME Mcal/kg	*		2.09	
Cattle	TDN %	*		61.4	
Sheep	TDN %	*		57.8	
Calcium	%			0.32	34
Phosphorus	%			0.11	29
Carotene	mg/kg			17.4	
Vitamin A equivalent	IU/g			28.9	

Grama, black, aerial part wo lower stems, fresh, dormant, (2)
Ref No 2 08 705 United States

Dry matter	%			100.0	
Organic matter	%			91.0	2
Ash	%			8.8	20
Crude fiber	%			32.1	6
Ether extract	%			1.6	25
N-free extract	%			51.9	4

Continued

Feed Name or Analyses		Mean As Fed	Mean Dry	C.V. ± %
Protein (N x 6.25)	%		5.6	15
Cattle	dig prot % *		2.6	
Goats	dig prot % *		1.7	
Horses	dig prot % *		2.2	
Rabbits	dig prot % *		3.0	
Sheep	dig prot % *		2.2	
Energy	GE Mcal/kg			
Cattle	DE Mcal/kg *		2.70	
Sheep	DE Mcal/kg *		2.59	
Cattle	ME Mcal/kg *		2.21	
Sheep	ME Mcal/kg *		2.12	
Cattle	TDN % *		61.2	
Sheep	TDN % *		58.7	
Calcium	%		0.27	47
Phosphorus	%		0.08	31
Biotin	mg/kg		12.13	
Carotene	mg/kg		14.8	70
Vitamin A equivalent	IU/g		24.6	

GRAMA, BLUE. Bouteloua gracilis

Grama, blue, aerial part, fresh weathered, mature, (1)

Ref No 1 08 597 United States

Feed Name or Analyses		Mean As Fed	Mean Dry	C.V. ± %
Dry matter	%	85.0	100.0	
Ash	%	6.7	7.9	
Crude fiber	%	32.8	38.6	
Ether extract	%	0.9	1.1	
N-free extract	%	41.6	48.9	
Protein (N x 6.25)	%	3.0	3.5	
Cattle	dig prot % *	0.0	0.0	
Goats	dig prot % *	0.0	0.0	
Horses	dig prot % *	0.4	0.5	
Rabbits	dig prot % *	1.2	1.4	
Sheep	dig prot % *	-0.1	-0.2	
Energy	GE Mcal/kg			
Cattle	DE Mcal/kg *	1.47	1.73	
Sheep	DE Mcal/kg *	1.56	1.83	
Cattle	ME Mcal/kg *	1.21	1.42	
Sheep	ME Mcal/kg *	1.28	1.50	
Cattle	TDN % *	33.3	39.2	
Sheep	TDN % *	35.2	41.4	
Calcium	%	0.24	0.28	
Phosphorus	%	0.06	0.07	

Grama, blue, hay, s-c, early bloom, (1)

Ref No 1 02 175 United States

Feed Name or Analyses		Mean As Fed	Mean Dry	C.V. ± %
Dry matter	%		100.0	
Ash	%		12.0	
Crude fiber	%		30.1	
Ether extract	%		2.2	
N-free extract	%		44.4	
Protein (N x 6.25)	%		11.3	
Cattle	dig prot % *		6.7	
Goats	dig prot % *		7.1	
Horses	dig prot % *		7.1	
Rabbits	dig prot % *		7.4	
Sheep	dig prot % *		6.7	
Energy	GE Mcal/kg			
Cattle	DE Mcal/kg *		1.78	
Sheep	DE Mcal/kg *		1.55	
Cattle	ME Mcal/kg *		1.46	
Sheep	ME Mcal/kg *		1.27	
Cattle	TDN % *		40.3	
Sheep	TDN % *		35.1	

Grama, blue, hay, s-c, milk stage, (1)

Ref No 1 02 176 United States

Feed Name or Analyses		Mean As Fed	Mean Dry	C.V. ± %
Dry matter	%		100.0	
Ash	%		10.6	
Crude fiber	%		32.5	
Ether extract	%		2.0	
N-free extract	%		45.7	
Protein (N x 6.25)	%		9.2	
Cattle	dig prot % *		4.9	
Goats	dig prot % *		5.1	
Horses	dig prot % *		5.3	
Rabbits	dig prot % *		5.8	
Sheep	dig prot % *		4.8	
Energy	GE Mcal/kg			
Cattle	DE Mcal/kg *		1.83	
Sheep	DE Mcal/kg *		1.58	
Cattle	ME Mcal/kg *		1.50	
Sheep	ME Mcal/kg *		1.30	
Cattle	TDN % *		41.5	
Sheep	TDN % *		35.9	

Grama, blue, hay, s-c, mature, (1)

Ref No 1 02 177 United States

Feed Name or Analyses		Mean As Fed	Mean Dry	C.V. ± %
Dry matter	%	90.6	100.0	
Calcium	%	0.39	0.43	26
Iron	%	0.033	0.036	62
Magnesium	%	0.13	0.14	
Manganese	mg/kg	15.0	16.5	30
Phosphorus	%	0.14	0.16	28

Grama, blue, hay, s-c, over ripe, (1)

Ref No 1 02 178 United States

Feed Name or Analyses		Mean As Fed	Mean Dry	C.V. ± %
Dry matter	%		100.0	
Carotene	mg/kg		13.2	
Vitamin A equivalent	IU/g		22.1	

Grama, blue, aerial part, fresh, (2)

Ref No 2 02 184 United States

Feed Name or Analyses		Mean As Fed	Mean Dry	C.V. ± %
Dry matter	%	57.2	100.0	18
Cobalt	mg/kg	0.124	0.216	79
Carotene	mg/kg	57.0	99.6	
Vitamin A equivalent	IU/g	95.0	166.1	

Grama, blue, aerial part, fresh, immature, (2)

Ref No 2 02 180 United States

Feed Name or Analyses		Mean As Fed	Mean Dry	C.V. ± %
Dry matter	%	47.6	100.0	
Ash	%	5.3	11.1	
Crude fiber	%	13.7	28.7	
Ether extract	%	0.9	1.8	
N-free extract	%	22.3	46.9	
Protein (N x 6.25)	%	5.5	11.5	
Cattle	dig prot % *	3.7	7.7	
Goats	dig prot % *	3.5	7.3	
Horses	dig prot % *	3.5	7.3	
Rabbits	dig prot % *	3.6	7.6	
Sheep	dig prot % *	3.7	7.7	
Energy	GE Mcal/kg			
Cattle	DE Mcal/kg *	1.12	2.36	
Sheep	DE Mcal/kg *	1.15	2.42	
Cattle	ME Mcal/kg *	0.92	1.94	
Sheep	ME Mcal/kg *	0.94	1.98	
Cattle	TDN % *	25.5	53.5	
Sheep	TDN % *	26.1	54.9	
Calcium	%	0.19	0.40	
Phosphorus	%	0.08	0.16	
Carotene	mg/kg	165.8	348.3	
Vitamin A equivalent	IU/g	276.4	580.7	

Grama, blue, aerial part, fresh, full bloom, (2)

Ref No 2 02 181 United States

Feed Name or Analyses		Mean As Fed	Mean Dry	C.V. ± %
Dry matter	%	48.6	100.0	
Carotene	mg/kg	79.2	162.9	
Vitamin A equivalent	IU/g	132.0	271.6	

Grama, blue, aerial part, fresh, dough stage, (2)

Ref No 2 08 426 United States

Feed Name or Analyses		Mean As Fed	Mean Dry	C.V. ± %
Dry matter	%	64.6	100.0	
Ash	%	8.3	12.8	
Crude fiber	%	18.9	29.3	
Ether extract	%	1.0	1.5	
N-free extract	%	31.2	48.3	
Protein (N x 6.25)	%	5.2	8.0	
Cattle	dig prot % *	3.1	4.7	
Goats	dig prot % *	2.6	4.1	
Horses	dig prot % *	2.8	4.4	
Rabbits	dig prot % *	3.2	4.9	
Sheep	dig prot % *	2.9	4.5	
Energy	GE Mcal/kg			
Cattle	DE Mcal/kg *	1.10	1.70	
Sheep	DE Mcal/kg *	1.30	2.01	
Cattle	ME Mcal/kg *	0.90	1.39	
Sheep	ME Mcal/kg *	1.05	1.65	
Cattle	TDN % *	24.9	38.5	
Sheep	TDN % *	29.5	45.7	
Calcium	%	0.14	0.22	
Phosphorus	%	0.08	0.12	

Grama, blue, aerial part, fresh, mature, (2)

Ref No 2 02 182 United States

Feed Name or Analyses		Mean As Fed	Mean Dry	C.V. ± %
Dry matter	%	62.3	100.0	13
Ash	%	8.8	14.2	33
Crude fiber	%	18.7	30.0	8
Ether extract	%	1.1	1.7	55
N-free extract	%	28.8	46.2	
Protein (N x 6.25)	%	4.9	7.9	15
Cattle	dig prot % *	2.9	4.6	
Goats	dig prot % *	2.4	3.9	
Horses	dig prot % *	2.6	4.2	
Rabbits	dig prot % *	3.0	4.8	
Sheep	dig prot % *	2.7	4.4	
Energy	GE Mcal/kg			
Cattle	DE Mcal/kg *	1.13	1.81	
Sheep	DE Mcal/kg *	1.27	2.04	
Cattle	ME Mcal/kg *	0.92	1.48	
Sheep	ME Mcal/kg *	1.04	1.69	
Cattle	TDN % *	25.5	41.0	
Sheep	TDN % *	28.8	46.3	

(1) dry forages and roughages (2) pasture, range plants, and forages fed green (3) silages (4) energy feeds (5) protein supplements (6) minerals (7) vitamins (8) additives

Feed Name or Analyses		Mean As Fed	Mean Dry	C.V. ± %
Carotene	mg/kg	24.9	39.9	
Vitamin A equivalent	IU/g	41.4	66.5	

Grama, blue, aerial part, fresh, dormant, (2)
Ref No 2 05 588 United States

Feed Name or Analyses		Mean As Fed	Mean Dry	C.V. ± %
Dry matter	%	20.6	100.0	
Ash	%	4.0	19.6	
Crude fiber	%	8.2	39.7	
Ether extract	%	0.6	2.7	
N-free extract	%	6.5	31.7	
Protein (N x 6.25)	%	1.3	6.3	
Cattle	dig prot % *	0.7	3.2	
Goats	dig prot % *	0.5	2.4	
Horses	dig prot % *	0.6	2.9	
Rabbits	dig prot % *	0.7	3.5	
Sheep	dig prot % *	0.6	2.9	
Energy	GE Mcal/kg			
Cattle	DE Mcal/kg *	0.33	1.58	
Sheep	DE Mcal/kg *	0.39	1.81	
Cattle	ME Mcal/kg *	0.27	1.30	
Sheep	ME Mcal/kg *	0.30	1.48	
Cattle	TDN % *	7.4	35.9	
Sheep	TDN % *	8.4	41.0	

Grama, blue, aerial part, fresh, cut 2, (2)
Ref No 2 02 183 United States

Feed Name or Analyses		Mean As Fed	Mean Dry	C.V. ± %
Dry matter	%	41.7	100.0	5
Ash	%	5.9	14.1	38
Crude fiber	%	12.4	29.8	7
Ether extract	%	1.2	2.9	40
N-free extract	%	17.7	42.5	
Protein (N x 6.25)	%	4.5	10.7	44
Cattle	dig prot % *	2.9	7.0	
Goats	dig prot % *	2.7	6.5	
Horses	dig prot % *	2.8	6.6	
Rabbits	dig prot % *	2.9	6.9	
Sheep	dig prot % *	2.9	7.0	
Calcium	%	0.18	0.42	38
Phosphorus	%	0.19	0.45	43

Grama, blue, aerial part wo lower stems, fresh, early leaf, (2)
Ref No 2 08 672 United States

Feed Name or Analyses		Mean As Fed	Mean Dry	C.V. ± %
Dry matter	%		100.0	
Organic matter	%		89.1	7
Ash	%		10.0	25
Crude fiber	%		30.3	5
Ether extract	%		1.5	24
N-free extract	%		50.5	8
Protein (N x 6.25)	%		7.6	28
Cattle	dig prot % *		4.4	
Goats	dig prot % *		3.7	
Horses	dig prot % *		4.0	
Rabbits	dig prot % *		4.6	
Sheep	dig prot % *		4.1	
Calcium	%		0.38	37
Phosphorus	%		0.12	35
Carotene	mg/kg		74.6	53
Vitamin A equivalent	IU/g		124.3	

Grama, blue, aerial part wo lower stems, fresh, mid-bloom, (2)
Ref No 2 08 675 United States

Feed Name or Analyses		Mean As Fed	Mean Dry	C.V. ± %
Dry matter	%		100.0	
Organic matter	%		90.9	2
Ash	%		9.1	18
Crude fiber	%		32.0	5
Ether extract	%		1.6	20
N-free extract	%		50.3	3
Protein (N x 6.25)	%		6.9	19
Cattle	dig prot % *		3.8	
Goats	dig prot % *		3.0	
Horses	dig prot % *		3.4	
Rabbits	dig prot % *		4.0	
Sheep	dig prot % *		3.4	
Calcium	%		0.33	26
Phosphorus	%		0.14	30
Carotene	mg/kg		65.3	39
Vitamin A equivalent	IU/g		108.8	

Grama, blue, aerial part wo lower stems, fresh, mature, (2)
Ref No 2 08 674 United States

Feed Name or Analyses		Mean As Fed	Mean Dry	C.V. ± %
Dry matter	%		100.0	
Organic matter	%		89.4	3
Ash	%		10.6	26
Crude fiber	%		31.9	7
Ether extract	%		1.8	75
N-free extract	%		49.4	4
Protein (N x 6.25)	%		6.2	25
Cattle	dig prot % *		3.2	
Goats	dig prot % *		2.4	
Horses	dig prot % *		2.8	
Rabbits	dig prot % *		3.5	
Sheep	dig prot % *		2.8	
Calcium	%		0.33	29
Phosphorus	%		0.12	20
Potassium	%		0.06	
Carotene	mg/kg		31.7	68
Vitamin A equivalent	IU/g		52.8	

Grama, blue, aerial part wo lower stems, fresh, dormant, (2)
Ref No 2 08 673 United States

Feed Name or Analyses		Mean As Fed	Mean Dry	C.V. ± %
Dry matter	%		100.0	
Organic matter	%		90.9	4
Ash	%		9.7	30
Crude fiber	%		32.6	12
Ether extract	%		1.3	26
N-free extract	%		51.9	5
Protein (N x 6.25)	%		4.5	25
Cattle	dig prot % *		1.7	
Goats	dig prot % *		0.7	
Horses	dig prot % *		1.3	
Rabbits	dig prot % *		2.1	
Sheep	dig prot % *		1.2	
Calcium	%		0.28	51
Phosphorus	%		0.08	44
Carotene	mg/kg		20.1	98
Vitamin A equivalent	IU/g		33.6	

Grama, blue, leaves, fresh, mature, (2)
Ref No 2 02 185 United States

Feed Name or Analyses		Mean As Fed	Mean Dry	C.V. ± %
Dry matter	%		100.0	
Magnesium	%		0.09	
Phosphorus	%		0.56	

Grama, blue, stems, fresh, mature, (2)
Ref No 2 02 186 United States

Feed Name or Analyses		Mean As Fed	Mean Dry	C.V. ± %
Dry matter	%		100.0	
Ash	%		7.6	
Crude fiber	%		37.1	
Ether extract	%		2.0	
N-free extract	%		48.9	
Protein (N x 6.25)	%		4.4	
Cattle	dig prot % *		1.6	
Goats	dig prot % *		0.7	
Horses	dig prot % *		1.3	
Rabbits	dig prot % *		2.1	
Sheep	dig prot % *		1.1	
Calcium	%		0.10	
Magnesium	%		0.07	

GRAMA, HAIRY. Bouteloua hirsuta

Grama, hairy, aerial part, fresh, (2)
Ref No 2 02 189 United States

Feed Name or Analyses		Mean As Fed	Mean Dry	C.V. ± %
Dry matter	%		100.0	
Ash	%		17.6	23
Crude fiber	%		29.0	7
Ether extract	%		1.6	15
N-free extract	%		44.6	
Protein (N x 6.25)	%		7.2	24
Cattle	dig prot % *		4.0	
Goats	dig prot % *		3.3	
Horses	dig prot % *		3.6	
Rabbits	dig prot % *		4.2	
Sheep	dig prot % *		3.7	
Calcium	%		0.58	17
Copper	mg/kg		5.1	
Magnesium	%		0.14	
Manganese	mg/kg		43.0	
Phosphorus	%		0.09	49
Potassium	%		0.87	
Carotene	mg/kg		76.3	
Vitamin A equivalent	IU/g		127.2	

Grama, hairy, aerial part, fresh, immature, (2)
Ref No 2 02 187 United States

Feed Name or Analyses		Mean As Fed	Mean Dry	C.V. ± %
Dry matter	%		100.0	
Calcium	%		0.66	4
Phosphorus	%		0.14	

Grama, hairy, aerial part, fresh, mature, (2)
Ref No 2 02 188 United States

Feed Name or Analyses		Mean As Fed	Mean Dry	C.V. ± %
Dry matter	%		100.0	
Ash	%		20.5	
Crude fiber	%		28.3	
Ether extract	%		1.4	
N-free extract	%		44.2	

Continued

Feed Name or Analyses			Mean As Fed	Mean Dry	C.V. ± %
Protein (N x 6.25)	%			5.6	
Cattle	dig prot %	*		2.7	
Goats	dig prot %	*		1.8	
Horses	dig prot %	*		2.3	
Rabbits	dig prot %	*		3.0	
Sheep	dig prot %	*		2.2	
Calcium	%			0.52	
Phosphorus	%			0.08	31

GRAMA, RED. Bouteloua trifida

Grama, red, aerial part, fresh, milk stage, (2)

Ref No 2 02 190 United States

Dry matter	%			100.0	
Ash	%			8.7	
Crude fiber	%			33.0	
Ether extract	%			2.3	
N-free extract	%			48.4	
Protein (N x 6.25)	%			7.6	
Cattle	dig prot %	*		4.4	
Goats	dig prot %	*		3.7	
Horses	dig prot %	*		4.0	
Rabbits	dig prot %	*		4.5	
Sheep	dig prot %	*		4.1	
Calcium	%			0.61	
Magnesium	%			0.14	
Phosphorus	%			0.10	
Potassium	%			0.60	

GRAMA, SIDEOATS. Bouteloua curtipendula

Grama, sideoats, aerial part, fresh, (2)

Ref No 2 02 196 United States

Dry matter	%		60.2	100.0	
Cobalt	mg/kg		0.066	0.110	
Carotene	mg/kg		15.5	25.8	
Vitamin A equivalent	IU/g		25.9	43.0	

Grama, sideoats, aerial part, fresh, immature, (2)

Ref No 2 02 191 United States

Dry matter	%		41.8	100.0	
Ash	%		5.3	12.7	31
Crude fiber	%		11.9	28.4	8
Ether extract	%		0.8	2.0	22
N-free extract	%		18.9	45.3	
Protein (N x 6.25)	%		4.8	11.6	13
Cattle	dig prot %	*	3.2	7.8	
Goats	dig prot %	*	3.1	7.4	
Horses	dig prot %	*	3.1	7.4	
Rabbits	dig prot %	*	3.2	7.6	
Sheep	dig prot %	*	3.3	7.8	
Calcium	%		0.28	0.66	18
Phosphorus	%		0.08	0.18	22

(1) dry forages and roughages (2) pasture, range plants, and forages fed green (3) silages (4) energy feeds (5) protein supplements (6) minerals (7) vitamins (8) additives

Grama, sideoats, aerial part, fresh, mid-bloom, (2)

Ref No 2 02 192 United States

Dry matter	%			100.0	
Ash	%			14.6	0
Crude fiber	%			28.9	5
Ether extract	%			1.9	13
N-free extract	%			46.2	
Protein (N x 6.25)	%			8.4	9
Cattle	dig prot %	*		5.0	
Goats	dig prot %	*		4.4	
Horses	dig prot %	*		4.7	
Rabbits	dig prot %	*		5.2	
Sheep	dig prot %	*		4.8	
Calcium	%			0.70	
Phosphorus	%			0.12	

Grama, sideoats, aerial part, fresh, full bloom, (2)

Ref No 2 02 193 United States

Dry matter	%			100.0	
Ash	%			13.6	15
Crude fiber	%			30.8	3
Ether extract	%			1.7	8
N-free extract	%			46.8	
Protein (N x 6.25)	%			7.1	10
Cattle	dig prot %	*		3.9	
Goats	dig prot %	*		3.2	
Horses	dig prot %	*		3.6	
Rabbits	dig prot %	*		4.2	
Sheep	dig prot %	*		3.6	
Calcium	%			0.51	5
Phosphorus	%			0.10	28

Grama, sideoats, aerial part, fresh, mature, (2)

Ref No 2 02 194 United States

Dry matter	%		62.1	100.0	
Ash	%		8.6	13.8	16
Crude fiber	%		19.5	31.4	10
Ether extract	%		1.1	1.7	29
N-free extract	%		30.1	48.4	
Protein (N x 6.25)	%		2.9	4.7	20
Cattle	dig prot %	*	1.2	1.9	
Goats	dig prot %	*	0.6	0.9	
Horses	dig prot %	*	0.9	1.5	
Rabbits	dig prot %	*	1.4	2.3	
Sheep	dig prot %	*	0.9	1.4	
Calcium	%		0.22	0.36	17
Copper	mg/kg		11.9	19.2	
Magnesium	%		0.07	0.12	
Manganese	mg/kg		40.8	65.7	
Phosphorus	%		0.05	0.08	30
Potassium	%		0.22	0.35	

Grama, sideoats, aerial part, fresh, over ripe, (2)

Ref No 2 02 195 United States

Dry matter	%		76.7	100.0	
Ash	%		9.1	11.9	18
Crude fiber	%		26.4	34.4	5
Ether extract	%		1.2	1.6	39
N-free extract	%		37.7	49.1	
Protein (N x 6.25)	%		2.3	3.0	32
Cattle	dig prot %	*	0.3	0.4	
Goats	dig prot %	*	-0.4	-0.5	
Horses	dig prot %	*	0.1	0.1	
Rabbits	dig prot %	*	0.8	1.0	
Sheep	dig prot %	*	-0.1	-0.1	
Calcium	%		0.17	0.22	38
Copper	mg/kg		5.6	7.3	
Manganese	mg/kg		30.3	39.5	
Phosphorus	%		0.05	0.07	34
Carotene	mg/kg		0.0	0.0	

Grama, sideoats, aerial part wo lower stems, fresh, early leaf, (2)

Ref No 2 08 677 United States

Dry matter	%			100.0	
Organic matter	%			89.4	4
Ash	%			11.1	21
Crude fiber	%			30.3	8
Ether extract	%			1.8	26
N-free extract	%			51.0	8
Protein (N x 6.25)	%			5.7	28
Cattle	dig prot %	*		2.8	
Goats	dig prot %	*		1.9	
Horses	dig prot %	*		2.4	
Rabbits	dig prot %	*		3.1	
Sheep	dig prot %	*		2.3	
Calcium	%			0.38	37
Phosphorus	%			0.12	60
Carotene	mg/kg			29.9	98
Vitamin A equivalent	IU/g			49.9	

Grama, sideoats, aerial part wo lower stems, fresh, mid-bloom, (2)

Ref No 2 08 678 United States

Dry matter	%			100.0	
Organic matter	%			89.6	3
Ash	%			9.6	12
Crude fiber	%			32.7	5
Ether extract	%			1.7	17
N-free extract	%			50.4	4
Protein (N x 6.25)	%			5.6	25
Cattle	dig prot %	*		2.7	
Goats	dig prot %	*		1.8	
Horses	dig prot %	*		2.3	
Rabbits	dig prot %	*		3.0	
Sheep	dig prot %	*		2.2	
Calcium	%			0.28	20
Phosphorus	%			0.12	45
Carotene	mg/kg			0.0	45

Grama, sideoats, aerial part wo lower stems, fresh, mature, (2)

Ref No 2 08 679 United States

Dry matter	%			100.0	
Organic matter	%			89.2	2
Ash	%			10.8	19
Crude fiber	%			33.4	8
Ether extract	%			1.5	23
N-free extract	%			50.0	3
Protein (N x 6.25)	%			4.3	24
Cattle	dig prot %	*		1.5	
Goats	dig prot %	*		0.6	
Horses	dig prot %	*		1.2	
Rabbits	dig prot %	*		2.0	
Sheep	dig prot %	*		1.0	

Feed Name or Analyses			Mean As Fed	Mean Dry	C.V. ± %
Calcium	%			0.32	33
Phosphorus	%			0.11	28

Grama, sideoats, aerial part wo lower stems, fresh, dormant, (2)

Ref No 2 08 676 United States

Feed Name or Analyses			Mean As Fed	Mean Dry	C.V. ± %
Dry matter	%			100.0	
Organic matter	%			89.8	3
Ash	%			10.3	25
Crude fiber	%			32.8	8
Ether extract	%			1.4	29
N-free extract	%			51.8	5
Protein (N x 6.25)	%			3.8	35
Cattle	dig prot %	*		1.1	
Goats	dig prot %	*		0.1	
Horses	dig prot %	*		0.7	
Rabbits	dig prot %	*		1.6	
Sheep	dig prot %	*		0.5	
Calcium	%			0.24	57
Phosphorus	%			0.07	40
Carotene	mg/kg			8.6	98
Vitamin A equivalent	IU/g			14.3	

GRAMA, SLENDER. Bouteloua filiformis

Grama, slender, aerial part, fresh, (2)

Ref No 2 02 197 United States

Feed Name or Analyses			Mean As Fed	Mean Dry	C.V. ± %
Dry matter	%			100.0	
Ash	%			14.9	
Crude fiber	%			30.4	
Ether extract	%			2.0	
N-free extract	%			43.7	
Protein (N x 6.25)	%			9.0	
Cattle	dig prot %	*		5.5	
Goats	dig prot %	*		5.0	
Horses	dig prot %	*		5.2	
Rabbits	dig prot %	*		5.6	
Sheep	dig prot %	*		5.4	

GRAMA, TEXAS. Bouteloua rigidiseta

Grama, Texas, aerial part, fresh, (2)

Ref No 2 02 202 United States

Feed Name or Analyses			Mean As Fed	Mean Dry	C.V. ± %
Dry matter	%			100.0	
Ash	%			14.4	17
Crude fiber	%			30.9	6
Ether extract	%			1.9	9
N-free extract	%			44.8	
Protein (N x 6.25)	%			8.0	15
Cattle	dig prot %	*		4.7	
Goats	dig prot %	*		4.0	
Horses	dig prot %	*		4.3	
Rabbits	dig prot %	*		4.8	
Sheep	dig prot %	*		4.4	
Calcium	%			0.62	17
Phosphorus	%			0.10	33

Grama, Texas, aerial part, fresh, immature, (2)

Ref No 2 02 198 United States

Feed Name or Analyses			Mean As Fed	Mean Dry	C.V. ± %
Dry matter	%			100.0	
Ash	%			12.6	9
Crude fiber	%			29.6	5
Ether extract	%			2.0	11
N-free extract	%			46.6	
Protein (N x 6.25)	%			9.2	3
Cattle	dig prot %	*		5.7	
Goats	dig prot %	*		5.1	
Horses	dig prot %	*		5.3	
Rabbits	dig prot %	*		5.8	
Sheep	dig prot %	*		5.6	
Calcium	%			0.70	15
Phosphorus	%			0.16	13

Grama, Texas, aerial part, fresh, mid-bloom, (2)

Ref No 2 02 199 United States

Feed Name or Analyses			Mean As Fed	Mean Dry	C.V. ± %
Dry matter	%			100.0	
Ash	%			16.2	
Crude fiber	%			31.0	
Ether extract	%			1.8	
N-free extract	%			42.8	
Protein (N x 6.25)	%			8.2	
Cattle	dig prot %	*		4.9	
Goats	dig prot %	*		4.2	
Horses	dig prot %	*		4.5	
Rabbits	dig prot %	*		5.0	
Sheep	dig prot %	*		4.6	
Calcium	%			0.69	
Phosphorus	%			0.10	

Grama, Texas, aerial part, fresh, full bloom, (2)

Ref No 2 02 200 United States

Feed Name or Analyses			Mean As Fed	Mean Dry	C.V. ± %
Dry matter	%			100.0	
Calcium	%			0.65	
Phosphorus	%			0.09	

Grama, Texas, aerial part, fresh, mature, (2)

Ref No 2 02 201 United States

Feed Name or Analyses			Mean As Fed	Mean Dry	C.V. ± %
Dry matter	%			100.0	
Calcium	%			0.52	
Phosphorus	%			0.09	

Grape concentrate –
see Grapes, juice, evaporated, (4)

Grapefruit pulp, dried –
see Citrus, grapefruit, pulp wo fines, shredded dehy, (4)

Grapefruit pulp, wet –
see Citrus, grapefruit, pulp, shredded wet, (4)

Grape marc, dried –
see Grapes, pulp, dehy grnd, (1)

Grape marc, fresh –
see Grapes, pulp, fresh, (2)

Grape marc meal, molasses added –
see Grapes, pulp w molasses added, dehy grnd, (1)

Grape pomace, dried –
see Grapes, pulp, dehy grnd, (1)

Grape pomace, fresh –
see Grapes, pulp, fresh, (2)

GRAPES. Vitis spp

Grapes, pulp, dehy grnd, (1)
Grape marc, dried
Grape pomace, dried

Ref No 1 02 208 United States

Feed Name or Analyses			Mean As Fed	Mean Dry	C.V. ± %
Dry matter	%		90.7	100.0	1
Ash	%		10.7	11.7	39
Crude fiber	%		27.5	30.3	5
Sheep	dig coef %		20.	20.	
Ether extract	%		6.6	7.3	15
Sheep	dig coef %		62.	62.	
N-free extract	%		34.4	38.0	7
Sheep	dig coef %		25.	25.	
Protein (N x 6.25)	%		11.5	12.7	7
Sheep	dig coef %		14.	14.	
Cattle	dig prot %	*	7.2	7.9	
Goats	dig prot %	*	7.6	8.4	
Horses	dig prot %	*	7.5	8.3	
Rabbits	dig prot %	*	7.7	8.5	
Sheep	dig prot %		1.6	1.7	
Energy	GE Mcal/kg				
Cattle	DE Mcal/kg	*	1.22	1.34	
Sheep	DE Mcal/kg	*	1.09	1.20	
Cattle	ME Mcal/kg	*	1.00	1.10	
Sheep	ME Mcal/kg	*	0.89	0.98	
Cattle	TDN %	*	27.6	30.4	
Sheep	TDN %		24.7	27.2	

Grapes, pulp w molasses added, dehy grnd, (1)
Grape marc meal, molasses added

Ref No 1 02 207 United States

Feed Name or Analyses			Mean As Fed	Mean Dry	C.V. ± %
Dry matter	%		79.6	100.0	
Ash	%		8.0	10.1	
Crude fiber	%		21.8	27.4	
Cattle	dig coef %		30.	30.	
Horses	dig coef %		20.	20.	
Ether extract	%		6.0	7.5	
Cattle	dig coef %		44.	44.	
Horses	dig coef %		20.	20.	
N-free extract	%		34.2	43.0	
Cattle	dig coef %		52.	52.	
Horses	dig coef %		54.	54.	
Protein (N x 6.25)	%		9.7	12.2	
Cattle	dig coef %		21.	21.	
Horses	dig coef %		12.	12.	
Cattle	dig prot %		2.0	2.6	
Goats	dig prot %	*	6.3	7.9	
Horses	dig prot %		1.2	1.5	
Rabbits	dig prot %	*	6.4	8.0	
Sheep	dig prot %	*	5.9	7.5	
Energy	GE Mcal/kg				
Cattle	DE Mcal/kg	*	1.42	1.79	
Sheep	DE Mcal/kg	*	1.36	1.71	
Horses	DE Mcal/kg	*	1.18	1.48	
Cattle	ME Mcal/kg	*	1.17	1.46	
Horses	ME Mcal/kg	*	0.96	1.21	
Sheep	ME Mcal/kg	*	1.11	1.40	
Cattle	TDN %		32.3	40.5	
Horses	TDN %		26.7	33.5	
Sheep	TDN %	*	30.8	38.7	

Grapes, pulp, fresh, (2)
Grape marc, fresh
Grape pomace, fresh
Ref No 2 02 206 — United States

Feed Name or Analyses			Mean As Fed	Mean Dry	C.V. ± %
Dry matter	%		37.5	100.0	
Ash	%		2.7	7.3	
Crude fiber	%		9.7	26.0	
Horses	dig coef %		20.	20.	
Sheep	dig coef %		26.	26.	
Ether extract	%		1.8	4.9	
Horses	dig coef %		50.	50.	
Sheep	dig coef %		49.	49.	
N-free extract	%		18.0	48.1	
Horses	dig coef %		34.	34.	
Sheep	dig coef %		37.	37.	
Protein (N x 6.25)	%		5.2	13.9	
Horses	dig coef %		22.	22.	
Sheep	dig coef %		16.	16.	
Cattle	dig prot %	*	3.6	9.7	
Goats	dig prot %	*	3.6	9.5	
Horses	dig prot %		1.1	3.0	
Rabbits	dig prot %	*	3.5	9.4	
Sheep	dig prot %		0.8	2.2	
Energy	GE Mcal/kg				
Cattle	DE Mcal/kg	*	0.61	1.62	
Horses	DE Mcal/kg	*	0.50	1.32	
Sheep	DE Mcal/kg	*	0.53	1.42	
Cattle	ME Mcal/kg	*	0.50	1.33	
Horses	ME Mcal/kg	*	0.41	1.09	
Sheep	ME Mcal/kg	*	0.43	1.16	
Cattle	TDN %	*	13.8	36.8	
Horses	TDN %		11.3	30.0	
Sheep	TDN %		12.0	32.1	

Grapes, fruit, dehy, cull, (4)
Raisins, cull
Ref No 4 08 427 — United States

Feed Name or Analyses			Mean As Fed	Mean Dry	C.V. ± %
Dry matter	%		84.8	100.0	
Ash	%		3.0	3.5	
Crude fiber	%		4.4	5.2	
Ether extract	%		0.9	1.1	
N-free extract	%		73.1	86.2	
Protein (N x 6.25)	%		3.4	4.0	
Cattle	dig prot %	*	-0.2	-0.2	
Goats	dig prot %	*	0.8	0.9	
Horses	dig prot %	*	0.8	0.9	
Sheep	dig prot %	*	0.8	0.9	
Energy	GE Mcal/kg				
Cattle	DE Mcal/kg	*	1.90	2.24	
Sheep	DE Mcal/kg	*	1.68	1.98	
Swine	DE kcal/kg	*	1765.	2081.	
Cattle	ME Mcal/kg	*	1.56	1.84	
Sheep	ME Mcal/kg	*	1.37	1.62	
Swine	ME kcal/kg	*	1680.	1981.	
Cattle	TDN %	*	43.1	50.8	
Sheep	TDN %	*	38.2	45.0	
Swine	TDN %	*	40.0	47.2	

(1) dry forages and roughages (2) pasture, range plants, and forages fed green (3) silages (4) energy feeds (5) protein supplements (6) minerals (7) vitamins (8) additives

Grapes, juice, evaporated, (4)
Grape concentrate
Ref No 4 08 569 — United States

Feed Name or Analyses			Mean As Fed	Mean Dry	C.V. ± %
Dry matter	%		67.4	100.0	
Ash	%		1.0	1.5	
Ether extract	%		0.1	0.2	
Protein (N x 6.25)	%		1.1	1.6	
Cattle	dig prot %	*	-1.6	-2.4	
Goats	dig prot %	*	-0.8	-1.2	
Horses	dig prot %	*	-0.8	-1.2	
Sheep	dig prot %	*	-0.8	-1.2	

Grapes, raisin syrup by-product, (4)
Raisin pulp dried
Ref No 4 08 428 — United States

Feed Name or Analyses			Mean As Fed	Mean Dry	C.V. ± %
Dry matter	%		89.4	100.0	
Ash	%		5.5	6.2	
Crude fiber	%		16.1	18.0	
Ether extract	%		7.8	8.7	
N-free extract	%		50.4	56.4	
Protein (N x 6.25)	%		9.6	10.7	
Cattle	dig prot %	*	5.3	5.9	
Goats	dig prot %	*	6.3	7.1	
Horses	dig prot %	*	6.3	7.1	
Sheep	dig prot %	*	6.3	7.1	
Energy	GE Mcal/kg				
Cattle	DE Mcal/kg	*	2.26	2.53	
Sheep	DE Mcal/kg	*	2.15	2.41	
Swine	DE kcal/kg	*	1929.	2156.	
Cattle	ME Mcal/kg	*	1.85	2.07	
Sheep	ME Mcal/kg	*	1.77	1.98	
Swine	ME kcal/kg	*	1808.	2022.	
Cattle	TDN %	*	51.2	57.3	
Sheep	TDN %	*	48.8	54.6	
Swine	TDN %	*	43.7	48.9	

GRASS. Scientific name not used

Grass, aerial part, dehy, (1)
Ref No 1 02 241 — United States

Feed Name or Analyses			Mean As Fed	Mean Dry	C.V. ± %
Dry matter	%		86.9	100.0	1
Ash	%		8.3	9.6	8
Crude fiber	%		19.1	22.0	6
Ether extract	%		3.0	3.5	11
N-free extract	%		39.9	45.9	
Protein (N x 6.25)	%		16.5	19.0	9
Cattle	dig prot %	*	11.6	13.4	
Goats	dig prot %	*	12.4	14.3	
Horses	dig prot %	*	11.9	13.7	
Rabbits	dig prot %	*	11.6	13.3	
Sheep	dig prot %	*	11.8	13.6	
Energy	GE Mcal/kg				
Cattle	DE Mcal/kg	*	2.49	2.87	
Sheep	DE Mcal/kg	*	2.37	2.73	
Cattle	ME Mcal/kg	*	2.05	2.35	
Sheep	ME Mcal/kg	*	1.94	2.24	
Cattle	TDN %	*	56.6	65.1	
Sheep	TDN %	*	53.7	61.8	
Calcium	%		0.56	0.65	
Phosphorus	%		0.33	0.38	
Potassium	%		2.53	2.91	

Grass, aerial part, dehy, immature, (1)
Ref No 1 02 239 — United States

Feed Name or Analyses			Mean As Fed	Mean Dry	C.V. ± %
Dry matter	%		89.0	100.0	
Ash	%		8.5	9.5	
Crude fiber	%		16.6	18.7	
Cattle	dig coef %		73.	73.	
Ether extract	%		3.5	3.9	
Cattle	dig coef %		40.	40.	
N-free extract	%		39.9	44.8	
Cattle	dig coef %		80.	80.	
Protein (N x 6.25)	%		20.6	23.1	
Cattle	dig coef %		72.	72.	
Cattle	dig prot %		14.8	16.6	
Goats	dig prot %	*	16.1	18.1	
Horses	dig prot %	*	15.3	17.1	
Rabbits	dig prot %	*	14.7	16.5	
Sheep	dig prot %	*	15.4	17.3	
Energy	GE Mcal/kg				
Cattle	DE Mcal/kg	*	2.73	3.07	
Sheep	DE Mcal/kg	*	2.53	2.84	
Cattle	ME Mcal/kg	*	2.24	2.52	
Sheep	ME Mcal/kg	*	2.07	2.33	
Cattle	TDN %		62.0	69.6	
Sheep	TDN %	*	57.3	64.4	

Grass, aerial part, dehy grnd, (1)
Ref No 1 02 211 — United States

Feed Name or Analyses			Mean As Fed	Mean Dry	C.V. ± %
Dry matter	%		86.8	100.0	
Ash	%		8.5	9.8	
Crude fiber	%		18.8	21.7	
Sheep	dig coef %		79.	79.	
Ether extract	%		3.1	3.6	
Sheep	dig coef %		72.	72.	
N-free extract	%		39.7	45.7	
Sheep	dig coef %		80.	80.	
Protein (N x 6.25)	%		16.7	19.2	
Sheep	dig coef %		77.	77.	
Sheep	dig prot %	*	12.8	14.8	
Energy	GE Mcal/kg				
Sheep	DE Mcal/kg	*	2.85	3.28	
Sheep	ME Mcal/kg	*	2.33	2.69	
Sheep	TDN %	*	64.5	74.3	
α-tocopherol	mg/kg		241.1	459.7	40
Arginine	%		1.12	1.30	
Glutamic acid	%		3.27	3.80	
Histidine	%		0.52	0.60	
Isoleucine	%		1.55	1.80	
Leucine	%		2.24	2.60	
Lysine	%		1.21	1.40	
Methionine	%		0.34	0.40	
Phenylalanine	%		1.46	1.70	
Serine	%		1.03	1.20	
Threonine	%		1.12	1.30	
Tryptophan	%		0.34	0.40	
Tyrosine	%		0.52	0.60	
Valine	%		1.72	2.00	

Grass, aerial part, dehy grnd, early leaf, (1)
Ref No 1 02 209 — United States

Feed Name or Analyses			Mean As Fed	Mean Dry	C.V. ± %
Dry matter	%		87.7	100.0	
Ash	%		11.1	12.6	
Crude fiber	%		13.5	15.4	
Sheep	dig coef %		78.	78.	

Feed Name or Analyses			Mean As Fed	Mean Dry	C.V. ± %
Ether extract	%		5.6	6.4	
Sheep	dig coef %		71.	71.	
N-free extract	%		34.6	39.5	
Sheep	dig coef %		79.	79.	
Protein (N x 6.25)	%		22.9	26.1	
Sheep	dig coef %		78.	78.	
Cattle	dig prot %	*	17.1	19.5	
Goats	dig prot %	*	18.3	20.9	
Horses	dig prot %	*	17.3	19.7	
Rabbits	dig prot %	*	16.5	18.8	
Sheep	dig prot %		17.9	20.4	
Energy	GE Mcal/kg				
Cattle	DE Mcal/kg	*	2.86	3.26	
Sheep	DE Mcal/kg	*	2.85	3.25	
Cattle	ME Mcal/kg	*	2.34	2.67	
Sheep	ME Mcal/kg	*	2.34	2.67	
Cattle	TDN %	*	64.9	74.0	
Sheep	TDN %		64.7	73.8	

Grass, hay, s-c, (1)

Ref No 1 02 250 United States

Feed Name or Analyses			Mean As Fed	Mean Dry	C.V. ± %
Dry matter	%		89.3	100.0	2
Cattle	dig coef %		62.	62.	
Sheep	dig coef %		62.	62.	
Ash	%		6.9	7.7	
Crude fiber	%		29.6	33.1	
Cattle	dig coef %		62.	62.	
Horses	dig coef %		44.	44.	
Sheep	dig coef %		66.	66.	
Ether extract	%		2.3	2.6	
Cattle	dig coef %		28.	28.	
Horses	dig coef %		14.	14.	
Sheep	dig coef %		46.	46.	
N-free extract	%		42.0	47.0	
Cattle	dig coef %		50.	50.	
Horses	dig coef %		45.	45.	
Sheep	dig coef %		64.	64.	
Protein (N x 6.25)	%		8.6	9.6	
Cattle	dig coef %		57.	57.	35
Horses	dig coef %		64.	64.	
Sheep	dig coef %		65.	65.	12
Cattle	dig prot %		4.9	5.5	
Goats	dig prot %	*	5.0	5.6	
Horses	dig prot %		5.5	6.2	
Rabbits	dig prot %	*	5.5	6.1	
Sheep	dig prot %		5.6	6.3	
Hemicellulose	%		20.5	23.0	3
Energy	GE Mcal/kg				
Cattle	DE Mcal/kg		2.36	2.65	
Horses	DE Mcal/kg	*	1.68	1.88	
Sheep	DE Mcal/kg		2.39	2.67	
Cattle	ME Mcal/kg	*	1.94	2.17	
Horses	ME Mcal/kg	*	1.38	1.54	
Sheep	ME Mcal/kg	*	1.96	2.19	
Cattle	TDN %		45.7	51.1	
Horses	TDN %		38.1	42.7	
Sheep	TDN %		54.4	60.9	
Calcium	%		0.48	0.54	
Iron	%		0.056	0.063	
Magnesium	%		0.16	0.18	
Manganese	mg/kg		66.2	74.1	
Phosphorus	%		0.21	0.24	
Potassium	%		1.20	1.35	
Sulphur	%		0.15	0.17	
Carotene	mg/kg		20.3	22.7	98
Vitamin A equivalent	IU/g		33.8	37.9	

Grass, hay, s-c, immature, (1)

Ref No 1 02 212 United States

Feed Name or Analyses			Mean As Fed	Mean Dry	C.V. ± %
Dry matter	%		88.1	100.0	2
Ash	%		9.6	10.9	10
Crude fiber	%		21.1	23.9	14
Cattle	dig coef %		77.	77.	
Sheep	dig coef %		72.	72.	
Ether extract	%		2.8	3.2	40
Cattle	dig coef %		60.	60.	
Sheep	dig coef %		27.	27.	
N-free extract	%		39.9	45.3	6
Cattle	dig coef %		73.	73.	
Sheep	dig coef %		74.	74.	
Protein (N x 6.25)	%		14.7	16.7	17
Cattle	dig coef %		72.	72.	
Sheep	dig coef %		67.	67.	
Cattle	dig prot %		10.6	12.0	
Goats	dig prot %	*	10.7	12.1	
Horses	dig prot %	*	10.3	11.7	
Rabbits	dig prot %	*	10.2	11.6	
Sheep	dig prot %		9.8	11.1	
Energy	GE Mcal/kg				
Cattle	DE Mcal/kg	*	2.64	2.99	
Sheep	DE Mcal/kg	*	2.48	2.81	
Cattle	ME Mcal/kg	*	2.16	2.45	
Sheep	ME Mcal/kg	*	2.03	2.30	
Cattle	TDN %		59.8	67.8	
Sheep	TDN %		56.2	63.7	
Calcium	%		0.40	0.46	
Phosphorus	%		0.12	0.13	
Potassium	%		0.72	0.82	

Grass, hay, s-c, early bloom, (1)

Ref No 1 02 243 United States

Feed Name or Analyses			Mean As Fed	Mean Dry	C.V. ± %
Dry matter	%		89.4	100.0	
Calcium	%		0.47	0.53	
Phosphorus	%		0.25	0.28	

Grass, hay, s-c, full bloom, (1)

Ref No 1 02 244 United States

Feed Name or Analyses			Mean As Fed	Mean Dry	C.V. ± %
Dry matter	%		89.3	100.0	1
Ash	%		6.6	7.4	14
Crude fiber	%		28.7	32.1	5
Ether extract	%		2.1	2.3	10
N-free extract	%		43.5	48.7	
Protein (N x 6.25)	%		8.5	9.5	13
Cattle	dig prot %	*	4.6	5.2	
Goats	dig prot %	*	4.8	5.4	
Horses	dig prot %	*	5.0	5.6	
Rabbits	dig prot %	*	5.4	6.0	
Sheep	dig prot %	*	4.5	5.1	
Energy	GE Mcal/kg				
Cattle	DE Mcal/kg	*	2.34	2.62	
Sheep	DE Mcal/kg	*	2.23	2.50	
Cattle	ME Mcal/kg	*	1.92	2.15	
Sheep	ME Mcal/kg	*	1.83	2.05	
Cattle	TDN %	*	53.0	59.4	
Sheep	TDN %	*	50.6	56.7	
Calcium	%		0.51	0.57	48
Magnesium	%		0.16	0.18	43
Phosphorus	%		0.21	0.24	29
Potassium	%		1.20	1.34	
Sulphur	%		0.11	0.12	
Carotene	mg/kg		12.2	13.7	
Vitamin A equivalent	IU/g		20.3	22.8	

Grass, hay, s-c, late bloom, poor quality, (1)

Ref No 1 02 245 United States

Feed Name or Analyses			Mean As Fed	Mean Dry	C.V. ± %
Dry matter	%		93.4	100.0	
Ash	%		10.5	11.2	
Crude fiber	%		34.0	36.4	
Cattle	dig coef %		62.	62.	
Ether extract	%		0.9	1.0	
Cattle	dig coef %		17.	17.	
N-free extract	%		45.1	48.3	
Cattle	dig coef %		50.	50.	
Protein (N x 6.25)	%		2.9	3.1	
Cattle	dig coef %		12.	12.	
Cattle	dig prot %		0.3	0.4	
Goats	dig prot %	*	-0.4	-0.4	
Horses	dig prot %	*	0.2	0.2	
Rabbits	dig prot %	*	1.0	1.1	
Sheep	dig prot %	*	-0.5	-0.5	
Energy	GE Mcal/kg				
Cattle	DE Mcal/kg	*	1.95	2.09	
Sheep	DE Mcal/kg	*	2.13	2.28	
Cattle	ME Mcal/kg	*	1.60	1.72	
Sheep	ME Mcal/kg	*	1.75	1.87	
Cattle	TDN %		44.3	47.5	
Sheep	TDN %	*	48.4	51.8	

Grass, hay, s-c, mature, (1)

Ref No 1 02 246 United States

Feed Name or Analyses			Mean As Fed	Mean Dry	C.V. ± %
Dry matter	%		90.9	100.0	1
Ash	%		4.3	4.7	7
Crude fiber	%		31.3	34.4	5
Ether extract	%		1.8	2.0	17
N-free extract	%		50.4	55.4	
Protein (N x 6.25)	%		3.2	3.5	12
Cattle	dig prot %	*	0.0	0.0	
Goats	dig prot %	*	-0.1	-0.1	
Horses	dig prot %	*	0.5	0.5	
Rabbits	dig prot %	*	1.2	1.4	
Sheep	dig prot %	*	-0.2	-0.2	
Energy	GE Mcal/kg				
Cattle	DE Mcal/kg	*	2.17	2.39	
Sheep	DE Mcal/kg	*	2.16	2.37	
Cattle	ME Mcal/kg	*	1.78	1.96	
Sheep	ME Mcal/kg	*	1.77	1.94	
Cattle	TDN %	*	49.2	54.1	
Sheep	TDN %	*	48.9	53.8	
Carotene	mg/kg		1.4	1.5	
Vitamin A equivalent	IU/g		2.3	2.6	

Grass, hay, s-c, over ripe, poor quality, (1)

Ref No 1 02 247 United States

Feed Name or Analyses			Mean As Fed	Mean Dry	C.V. ± %
Dry matter	%		52.5	100.0	11
Ash	%		4.6	8.8	
Crude fiber	%		19.4	36.9	
Cattle	dig coef %		60.	60.	
Ether extract	%		0.9	1.7	
Cattle	dig coef %		40.	40.	
N-free extract	%		25.4	48.4	
Cattle	dig coef %		46.	46.	
Protein (N x 6.25)	%		2.2	4.2	
Cattle	dig coef %		19.	19.	

Continued

Feed Name or Analyses			Mean As Fed	Mean Dry	C.V. ± %
Cattle	dig prot %		0.4	0.8	
Goats	dig prot %	*	0.3	0.5	
Horses	dig prot %	*	0.6	1.1	
Rabbits	dig prot %	*	1.0	1.9	
Sheep	dig prot %	*	0.2	0.3	
Energy	GE Mcal/kg				
Cattle	DE Mcal/kg	*	1.08	2.06	
Sheep	DE Mcal/kg	*	1.21	2.30	
Cattle	ME Mcal/kg	*	0.89	1.69	
Sheep	ME Mcal/kg	*	0.99	1.89	
Cattle	TDN %		24.5	46.7	
Sheep	TDN %	*	27.4	52.2	

Grass, hay, s-c, cut 1, (1)
Ref No 1 02 248 United States

Analyses			As Fed	Dry	C.V.
Dry matter	%		89.6	100.0	2
Calcium	%		0.47	0.53	27
Magnesium	%		0.22	0.25	34
Phosphorus	%		0.15	0.17	14
Potassium	%		1.35	1.51	17
Sulphur	%		0.12	0.13	41

Grass, hay, s-c, cut 2, (1)
Ref No 1 02 249 United States

Analyses			As Fed	Dry	C.V.
Dry matter	%		89.2	100.0	2
Ash	%		7.1	8.0	9
Crude fiber	%		24.7	27.7	3
Cattle	dig coef %		68.	68.	
Ether extract	%		3.3	3.7	12
Cattle	dig coef %		32.	32.	
N-free extract	%		42.2	47.4	
Cattle	dig coef %		63.	63.	
Protein (N x 6.25)	%		11.8	13.2	9
Cattle	dig coef %		60.	60.	
Cattle	dig prot %		7.1	7.9	
Goats	dig prot %	*	8.0	8.9	
Horses	dig prot %	*	7.8	8.8	
Rabbits	dig prot %	*	7.9	8.9	
Sheep	dig prot %	*	7.5	8.4	
Energy	GE Mcal/kg				
Cattle	DE Mcal/kg	*	2.33	2.62	
Sheep	DE Mcal/kg	*	2.30	2.58	
Cattle	ME Mcal/kg	*	1.91	2.14	
Sheep	ME Mcal/kg	*	1.88	2.11	
Cattle	TDN %		52.9	59.3	
Sheep	TDN %	*	52.1	58.4	
Calcium	%		0.79	0.89	
Phosphorus	%		0.31	0.35	
Potassium	%		1.15	1.29	

Grass, hay, s-c, gr 2 US, (1)
Ref No 1 02 252 United States

Analyses			As Fed	Dry	C.V.
Dry matter	%		89.6	100.0	0
Ash	%		7.7	8.6	9
Crude fiber	%		31.2	34.8	3
Ether extract	%		2.0	2.2	9
N-free extract	%		38.9	43.4	
Protein (N x 6.25)	%		9.9	11.0	10
Cattle	dig prot %	*	5.8	6.5	
Goats	dig prot %	*	6.1	6.8	
Horses	dig prot %	*	6.2	6.9	
Rabbits	dig prot %	*	6.4	7.2	
Sheep	dig prot %	*	5.8	6.4	
Energy	GE Mcal/kg				
Cattle	DE Mcal/kg	*	2.20	2.45	
Sheep	DE Mcal/kg	*	2.20	2.46	
Cattle	ME Mcal/kg	*	1.80	2.01	
Sheep	ME Mcal/kg	*	1.81	2.02	
Cattle	TDN %	*	49.8	55.6	
Sheep	TDN %	*	50.0	55.7	
Calcium	%		0.58	0.65	38
Cobalt	mg/kg		0.081	0.090	44
Magnesium	%		0.26	0.29	
Manganese	mg/kg		62.3	69.4	59
Phosphorus	%		0.17	0.19	28
Potassium	%		1.49	1.66	
Sulphur	%		0.15	0.17	
Carotene	mg/kg		27.3	30.4	85
Vitamin A equivalent	IU/g		45.5	50.7	

(1) dry forages and roughages (2) pasture, range plants, and forages fed green (3) silages (4) energy feeds (5) protein supplements (6) minerals (7) vitamins (8) additives

Grass, hay, s-c, gr 3 US, (1)
Ref No 1 02 253 United States

Analyses			As Fed	Dry	C.V.
Dry matter	%		89.2	100.0	1
Ash	%		6.1	6.8	6
Crude fiber	%		30.2	33.9	9
Ether extract	%		2.4	2.7	4
N-free extract	%		42.1	47.2	
Protein (N x 6.25)	%		8.4	9.4	11
Cattle	dig prot %	*	4.5	5.1	
Goats	dig prot %	*	4.8	5.3	
Horses	dig prot %	*	4.9	5.5	
Rabbits	dig prot %	*	5.3	5.9	
Sheep	dig prot %	*	4.5	5.0	
Energy	GE Mcal/kg				
Cattle	DE Mcal/kg	*	2.21	2.48	
Sheep	DE Mcal/kg	*	2.19	2.46	
Cattle	ME Mcal/kg	*	1.81	2.03	
Sheep	ME Mcal/kg	*	1.80	2.02	
Cattle	TDN %	*	50.1	56.2	
Sheep	TDN %	*	49.8	55.8	
Calcium	%		0.38	0.43	18
Cobalt	mg/kg		0.049	0.055	
Manganese	mg/kg		49.6	55.6	63
Phosphorus	%		0.17	0.19	24
Carotene	mg/kg		40.3	45.2	76
Vitamin A equivalent	IU/g		67.2	75.3	

Grass, hay, s-c, mn 30% fiber, (1)
Ref No 1 02 254 United States

Analyses			As Fed	Dry	C.V.
Dry matter	%		83.9	100.0	
Ash	%		6.5	7.8	
Crude fiber	%		31.2	37.2	
Horses	dig coef %		52.	52.	
Ether extract	%		2.0	2.4	
Horses	dig coef %		11.	11.	
N-free extract	%		34.6	41.2	
Horses	dig coef %		46.	46.	
Protein (N x 6.25)	%		9.6	11.4	
Horses	dig coef %		61.	61.	
Cattle	dig prot %	*	5.7	6.8	
Goats	dig prot %	*	6.0	7.2	
Horses	dig prot %		5.8	7.0	
Rabbits	dig prot %	*	6.3	7.5	
Sheep	dig prot %	*	5.7	6.8	
Energy	GE Mcal/kg				
Cattle	DE Mcal/kg	*	1.88	2.24	
Horses	DE Mcal/kg	*	1.70	2.02	
Sheep	DE Mcal/kg	*	2.03	2.42	
Cattle	ME Mcal/kg	*	1.54	1.84	
Horses	ME Mcal/kg	*	1.39	1.66	
Sheep	ME Mcal/kg	*	1.66	1.98	
Cattle	TDN %	*	42.7	50.9	
Horses	TDN %		38.5	45.8	
Sheep	TDN %	*	46.0	54.8	

Grass, hay, s-c, mn 33% fiber, (1)
Ref No 1 02 255 United States

Analyses			As Fed	Dry	C.V.
Dry matter	%		92.2	100.0	
Ash	%		9.0	9.8	
Crude fiber	%		34.4	37.3	
Cattle	dig coef %		62.	62.	
Ether extract	%		1.1	1.2	
Cattle	dig coef %		29.	29.	
N-free extract	%		44.3	48.1	
Cattle	dig coef %		47.	47.	
Protein (N x 6.25)	%		3.3	3.6	
Cattle	dig coef %		14.	14.	
Cattle	dig prot %		0.5	0.5	
Goats	dig prot %	*	0.0	0.0	
Horses	dig prot %	*	0.5	0.6	
Rabbits	dig prot %	*	1.3	1.4	
Sheep	dig prot %	*	-0.1	-0.1	
Energy	GE Mcal/kg				
Cattle	DE Mcal/kg	*	1.91	2.07	
Sheep	DE Mcal/kg	*	2.10	2.28	
Cattle	ME Mcal/kg	*	1.57	1.70	
Sheep	ME Mcal/kg	*	1.73	1.87	
Cattle	TDN %		43.4	47.0	
Sheep	TDN %	*	47.7	51.7	

Grass, hay, s-c, mx 25% fiber, (1)
Ref No 1 02 256 United States

Analyses			As Fed	Dry	C.V.
Dry matter	%		88.8	100.0	
Ash	%		8.3	9.4	
Crude fiber	%		19.4	21.8	
Cattle	dig coef %		71.	71.	
Ether extract	%		3.4	3.8	
Cattle	dig coef %		38.	38.	
N-free extract	%		39.9	44.9	
Cattle	dig coef %		75.	75.	
Protein (N x 6.25)	%		17.8	20.1	
Cattle	dig coef %		68.	68.	
Cattle	dig prot %		12.1	13.7	
Goats	dig prot %	*	13.6	15.3	
Horses	dig prot %	*	13.0	14.6	
Rabbits	dig prot %	*	12.6	14.2	
Sheep	dig prot %	*	13.0	14.6	
Energy	GE Mcal/kg				
Cattle	DE Mcal/kg	*	2.59	2.91	
Sheep	DE Mcal/kg	*	2.43	2.74	
Cattle	ME Mcal/kg	*	2.12	2.39	
Sheep	ME Mcal/kg	*	1.99	2.24	
Cattle	TDN %		58.7	66.1	
Sheep	TDN %	*	55.1	62.0	

Grass, hay, s-c, mx 30% fiber, (1)

Ref No 1 02 258 United States

Feed Name or Analyses				Mean As Fed	Mean Dry	C.V. ± %
Dry matter		%		86.3	100.0	
Ash		%		6.0	7.0	
Crude fiber		%		26.8	31.1	
Horses	dig coef	%		41.	41.	
Ether extract		%		2.6	3.0	
Horses	dig coef	%		15.	15.	
N-free extract		%		39.6	45.9	
Horses	dig coef	%		45.	45.	
Protein (N x 6.25)		%		11.2	13.0	
Horses	dig coef	%		66.	66.	
Cattle	dig prot	%	*	7.1	8.2	
Goats	dig prot	%	*	7.5	8.7	
Horses	dig prot	%		7.4	8.6	
Rabbits	dig prot	%	*	7.5	8.7	
Sheep	dig prot	%	*	7.1	8.2	
Energy	GE	Mcal/kg				
Cattle	DE	Mcal/kg	*	2.30	2.67	
Horses	DE	Mcal/kg	*	1.64	1.90	
Sheep	DE	Mcal/kg	*	2.22	2.57	
Cattle	ME	Mcal/kg	*	1.89	2.19	
Horses	ME	Mcal/kg	*	1.34	1.55	
Sheep	ME	Mcal/kg	*	1.82	2.11	
Cattle	TDN	%	*	52.2	60.5	
Horses	TDN	%		37.1	43.0	
Sheep	TDN	%	*	50.4	58.4	

Grass, hay, s-c, poor quality, (1)

Ref No 1 02 257 United States

Feed Name or Analyses				Mean As Fed	Mean Dry	C.V. ± %
Dry matter		%		92.2	100.0	
Ash		%		9.0	9.8	
Crude fiber		%		34.4	37.3	
Cattle	dig coef	%		62.	62.	
Ether extract		%		1.1	1.2	
Cattle	dig coef	%		29.	29.	
N-free extract		%		44.3	48.1	
Cattle	dig coef	%		47.	47.	
Protein (N x 6.25)		%		3.3	3.6	
Cattle	dig coef	%		14.	14.	
Cattle	dig prot	%		0.5	0.5	
Goats	dig prot	%	*	0.0	0.0	
Horses	dig prot	%	*	0.5	0.6	
Rabbits	dig prot	%	*	1.3	1.4	
Sheep	dig prot	%	*	-0.1	-0.1	
Energy	GE	Mcal/kg				
Cattle	DE	Mcal/kg	*	1.91	2.07	
Sheep	DE	Mcal/kg	*	2.10	2.28	
Cattle	ME	Mcal/kg	*	1.57	1.70	
Sheep	ME	Mcal/kg	*	1.73	1.87	
Cattle	TDN	%		43.4	47.0	
Sheep	TDN	%	*	47.7	51.7	

Grass, hay, s-c, gr sample US, (1)

Ref No 1 02 251 United States

Feed Name or Analyses				Mean As Fed	Mean Dry	C.V. ± %
Dry matter		%		89.1	100.0	
Ash		%		8.0	9.0	
Crude fiber		%		30.2	33.9	
Ether extract		%		2.0	2.2	
N-free extract		%		38.3	43.0	
Protein (N x 6.25)		%		10.6	11.9	
Cattle	dig prot	%	*	6.5	7.2	
Goats	dig prot	%	*	6.8	7.7	
Horses	dig prot	%	*	6.8	7.6	
Rabbits	dig prot	%	*	7.0	7.9	
Sheep	dig prot	%	*	6.5	7.2	
Energy	GE	Mcal/kg				
Cattle	DE	Mcal/kg	*	2.25	2.52	
Sheep	DE	Mcal/kg	*	2.22	2.49	
Cattle	ME	Mcal/kg	*	1.84	2.07	
Sheep	ME	Mcal/kg	*	1.82	2.04	
Cattle	TDN	%	*	51.0	57.2	
Sheep	TDN	%	*	50.3	56.4	
Calcium		%		0.45	0.50	26
Cobalt		mg/kg		0.071	0.079	
Manganese		mg/kg		40.9	45.9	
Phosphorus		%		0.20	0.22	23
Carotene		mg/kg		3.3	3.7	
Vitamin A equivalent		IU/g		5.6	6.2	

Grass, hay, s-c weathered, mature, (1)

Ref No 1 08 595 United States

Feed Name or Analyses				Mean As Fed	Mean Dry	C.V. ± %
Dry matter		%		90.0	100.0	
Ash		%		6.3	7.0	
Crude fiber		%		34.1	37.9	
Ether extract		%		1.8	2.0	
N-free extract		%		44.5	49.4	
Protein (N x 6.25)		%		3.3	3.7	
Cattle	dig prot	%	*	0.1	0.1	
Goats	dig prot	%	*	0.0	0.0	
Horses	dig prot	%	*	0.6	0.6	
Rabbits	dig prot	%	*	1.4	1.5	
Sheep	dig prot	%	*	0.0	0.0	
Energy	GE	Mcal/kg				
Cattle	DE	Mcal/kg	*	2.00	2.22	
Sheep	DE	Mcal/kg	*	2.06	2.28	
Cattle	ME	Mcal/kg	*	1.64	1.82	
Sheep	ME	Mcal/kg	*	1.69	1.87	
Cattle	TDN	%	*	45.4	50.4	
Sheep	TDN	%	*	46.6	51.8	
Calcium		%		0.33	0.37	
Phosphorus		%		0.09	0.10	

Grass, straw, s-c, (1)

Ref No 1 08 430 United States

Feed Name or Analyses				Mean As Fed	Mean Dry	C.V. ± %
Dry matter		%		85.0	100.0	
Ash		%		5.7	6.7	
Crude fiber		%		35.0	41.2	
Ether extract		%		2.0	2.4	
N-free extract		%		37.8	44.5	
Protein (N x 6.25)		%		4.5	5.3	
Cattle	dig prot	%	*	1.3	1.5	
Goats	dig prot	%	*	1.3	1.5	
Horses	dig prot	%	*	1.7	2.0	
Rabbits	dig prot	%	*	2.3	2.8	
Sheep	dig prot	%	*	1.1	1.3	
Energy	GE	Mcal/kg				
Cattle	DE	Mcal/kg	*	1.67	1.97	
Sheep	DE	Mcal/kg	*	1.90	2.24	
Cattle	ME	Mcal/kg	*	1.37	1.61	
Sheep	ME	Mcal/kg	*	1.56	1.84	
Cattle	TDN	%	*	37.9	44.6	
Sheep	TDN	%	*	43.2	50.8	

Grass, aerial part, fresh, (2)

Ref No 2 02 260 United States

Feed Name or Analyses				Mean As Fed	Mean Dry	C.V. ± %
Dry matter		%		32.1	100.0	37
Ash		%		2.8	8.7	27
Crude fiber		%		7.9	24.6	33
Ether extract		%		1.1	3.6	40
N-free extract		%		15.2	47.5	6
Protein (N x 6.25)		%		5.0	15.6	60
Cattle	dig prot	%	*	3.6	11.2	
Goats	dig prot	%	*	3.6	11.2	
Horses	dig prot	%	*	3.5	10.8	
Rabbits	dig prot	%	*	3.4	10.7	
Sheep	dig prot	%	*	3.7	11.6	
Energy	GE	Mcal/kg				
Cattle	DE	Mcal/kg	*	0.87	2.72	
Sheep	DE	Mcal/kg	*	0.90	2.81	
Cattle	ME	Mcal/kg	*	0.72	2.23	
Sheep	ME	Mcal/kg	*	0.74	2.30	
Cattle	TDN	%	*	19.8	61.7	
Sheep	TDN	%	*	20.4	63.7	
Calcium		%		0.12	0.36	
Chlorine		%		0.10	0.31	
Magnesium		%		0.03	0.10	
Phosphorus		%		0.06	0.17	
Potassium		%		0.39	1.21	
Sodium		%		0.01	0.03	
Niacin		mg/kg		52.6	164.0	
Riboflavin		mg/kg		5.1	15.9	

Grass, aerial part, fresh, early bloom (2)

Ref No 1 08 431 United States

Feed Name or Analyses				Mean As Fed	Mean Dry	C.V. ± %
Dry matter		%		30.8	100.0	
Ash		%		1.8	5.8	
Crude fiber		%		10.6	34.4	
Ether extract		%		1.3	4.2	
N-free extract		%		14.1	45.8	
Protein (N x 6.25)		%		3.0	9.7	
Cattle	dig prot	%	*	1.7	5.4	
Goats	dig prot	%	*	1.7	5.6	
Horses	dig prot	%	*	1.8	5.8	
Rabbits	dig prot	%	*	1.9	6.2	
Sheep	dig prot	%	*	1.6	5.3	
Energy	GE	Mcal/kg				
Cattle	DE	Mcal/kg	*	0.94	3.06	
Sheep	DE	Mcal/kg	*	0.92	2.98	
Cattle	ME	Mcal/kg	*	0.77	2.51	
Sheep	ME	Mcal/kg	*	0.75	2.44	
Cattle	TDN	%	*	21.4	69.4	
Sheep	TDN	%	*	20.8	67.5	
Calcium		%		0.17	0.55	
Phosphorus		%		0.07	0.23	
Potassium		%		0.41	1.33	

Grass, aerial part, fresh, dough stage, (2)

Ref No 2 07 799 United States

Feed Name or Analyses				Mean As Fed	Mean Dry	C.V. ± %
Dry matter		%		28.5	100.0	
Cattle	dig coef	%		60.	60.	
Ash		%		2.1	7.3	
Crude fiber		%		9.9	34.7	
Cattle	dig coef	%		61.	61.	
Ether extract		%		1.4	4.9	
Cattle	dig coef	%		72.	72.	
N-free extract		%		12.4	43.6	
Cattle	dig coef	%		61.	61.	
Protein (N x 6.25)		%		2.7	9.6	
Cattle	dig coef	%		54.	54.	
Cattle	dig prot	%		1.5	5.2	
Goats	dig prot	%	*	1.6	5.5	
Horses	dig prot	%	*	1.6	5.6	

Continued

Feed Name or Analyses				As Fed (Mean)	Dry (Mean)	C.V. ± %
Rabbits	dig prot	%	*	1.7	6.0	
Sheep	dig prot	%	*	1.7	5.9	
Lignin (Ellis)		%		2.5	8.8	
Energy	GE	Mcal/kg		1.25	4.41	98
Cattle	GE dig coef	%		58.	58.	
Cattle	DE	Mcal/kg		0.72	2.53	
Sheep	DE	Mcal/kg	*	0.80	2.81	
Cattle	ME	Mcal/kg	*	0.59	2.08	
Sheep	ME	Mcal/kg	*	0.66	2.30	
Cattle	TDN	%		17.2	60.6	
Sheep	TDN	%	*	18.1	63.7	

Grass, aerial part, fresh, mature, (2)
Ref No 2 08 432 United States

Feed Name or Analyses				As Fed (Mean)	Dry (Mean)	C.V. ± %
Dry matter		%		53.6	100.0	
Ash		%		3.1	5.8	
Crude fiber		%		19.4	36.2	
Ether extract		%		1.0	1.8	
N-free extract		%		27.2	50.7	
Protein (N x 6.25)		%		2.9	5.4	
Cattle	dig prot	%	*	1.3	2.5	
Goats	dig prot	%	*	0.9	1.6	
Horses	dig prot	%	*	1.1	2.1	
Rabbits	dig prot	%	*	1.5	2.9	
Sheep	dig prot	%	*	1.1	2.1	
Energy	GE	Mcal/kg				
Cattle	DE	Mcal/kg	*	1.18	2.20	
Sheep	DE	Mcal/kg	*	1.13	2.10	
Cattle	ME	Mcal/kg	*	0.96	1.80	
Sheep	ME	Mcal/kg	*	0.92	1.72	
Cattle	TDN	%	*	26.7	49.8	
Sheep	TDN	%	*	25.5	47.6	
Calcium		%		0.16	0.31	
Phosphorus		%		0.02	0.05	

Grass, aerial part, fresh, cut 2, (2)
Ref No 2 08 433 United States

Feed Name or Analyses				As Fed (Mean)	Dry (Mean)	C.V. ± %
Dry matter		%		28.2	100.0	
Ash		%		2.4	8.5	
Crude fiber		%		7.3	25.9	
Ether extract		%		1.5	5.3	
N-free extract		%		12.3	43.6	
Protein (N x 6.25)		%		4.7	16.7	
Cattle	dig prot	%	*	3.4	12.1	
Goats	dig prot	%	*	3.4	12.1	
Horses	dig prot	%	*	3.3	11.7	
Rabbits	dig prot	%	*	3.3	11.5	
Sheep	dig prot	%	*	3.5	12.5	
Energy	GE	Mcal/kg				
Cattle	DE	Mcal/kg	*	0.84	2.97	
Sheep	DE	Mcal/kg	*	0.77	2.73	
Cattle	ME	Mcal/kg	*	0.69	2.44	
Sheep	ME	Mcal/kg	*	0.63	2.23	
Cattle	TDN	%	*	19.0	67.4	
Sheep	TDN	%	*	17.4	61.8	
Calcium		%		0.25	0.89	
Phosphorus		%		0.10	0.35	
Potassium		%		0.36	1.28	

Grass, aerial part, fresh weathered, mature, (2)
Ref No 2 08 434 United States

Feed Name or Analyses				As Fed (Mean)	Dry (Mean)	C.V. ± %
Dry matter		%		90.8	100.0	
Ash		%		7.2	7.9	
Crude fiber		%		33.7	37.1	
Ether extract		%		1.6	1.8	
N-free extract		%		45.6	50.2	
Protein (N x 6.25)		%		2.7	3.0	
Cattle	dig prot	%	*	0.4	0.4	
Goats	dig prot	%	*	-0.5	-0.6	
Horses	dig prot	%	*	0.0	0.1	
Rabbits	dig prot	%	*	0.9	1.0	
Sheep	dig prot	%	*	-0.1	-0.1	
Energy	GE	Mcal/kg				
Cattle	DE	Mcal/kg	*	2.10	2.32	
Sheep	DE	Mcal/kg	*	2.07	2.28	
Cattle	ME	Mcal/kg	*	1.72	1.90	
Sheep	ME	Mcal/kg	*	1.70	1.87	
Cattle	TDN	%	*	47.7	52.5	
Sheep	TDN	%	*	47.0	51.8	
Calcium		%		0.28	0.31	
Chlorine		%		0.05	0.06	
Copper		mg/kg		88.0	96.9	
Iron		%		0.036	0.040	
Magnesium		%		0.23	0.25	
Manganese		mg/kg		64.4	70.9	
Phosphorus		%		0.04	0.04	
Potassium		%		0.43	0.47	
Sodium		%		0.01	0.01	

Grass, juice, fresh condensed, (2)
Ref No 2 02 216 United States

Feed Name or Analyses				As Fed (Mean)	Dry (Mean)	C.V. ± %
Dry matter		%			100.0	
Vitamin B_{12}		μg/kg			0.7	40

Grass, aerial part, ensiled, (3)
Ref No 3 02 221 United States

Feed Name or Analyses				As Fed (Mean)	Dry (Mean)	C.V. ± %
Dry matter		%		27.9	100.0	11
Ash		%		3.1	11.2	
Crude fiber		%		8.0	28.6	
Sheep	dig coef	%		76.	76.	
Ether extract		%		1.3	4.6	
Sheep	dig coef	%		59.	59.	
N-free extract		%		11.3	40.7	
Sheep	dig coef	%		66.	66.	
Protein (N x 6.25)		%		4.2	14.9	
Sheep	dig coef	%		59.	59.	
Cattle	dig prot	%	*	2.7	9.8	
Goats	dig prot	%	*	2.7	9.8	
Horses	dig prot	%	*	2.7	9.8	
Sheep	dig prot	%		2.4	8.8	
Energy	GE	Mcal/kg				
Cattle	DE	Mcal/kg	*	0.79	2.83	
Sheep	DE	Mcal/kg	*	0.78	2.80	
Cattle	ME	Mcal/kg	*	0.65	2.32	
Sheep	ME	Mcal/kg	*	0.64	2.30	
Cattle	TDN	%	*	17.9	64.2	
Sheep	TDN	%		17.7	63.5	
Chlorine		%		0.20	0.73	
Cobalt		mg/kg		0.020	0.071	
Carotene		mg/kg		12.6	45.2	82
Vitamin A equivalent		IU/g		21.0	75.3	

Grass, aerial part, ensiled, immature, (3)
Ref No 3 02 217 United States

Feed Name or Analyses				As Fed (Mean)	Dry (Mean)	C.V. ± %
Dry matter		%		28.3	100.0	21
Ash		%		2.5	8.7	7
Crude fiber		%		9.0	31.9	8
Ether extract		%		1.1	4.0	10
N-free extract		%		11.9	42.2	
Protein (N x 6.25)		%		3.7	13.2	15
Cattle	dig prot	%	*	2.3	8.2	
Goats	dig prot	%	*	2.3	8.2	
Horses	dig prot	%	*	2.3	8.2	
Sheep	dig prot	%	*	2.3	8.2	
Energy	GE	Mcal/kg				
Cattle	DE	Mcal/kg	*	0.74	2.60	
Sheep	DE	Mcal/kg	*	0.78	2.75	
Cattle	ME	Mcal/kg	*	0.60	2.14	
Sheep	ME	Mcal/kg	*	0.64	2.25	
Cattle	TDN	%	*	16.7	59.1	
Sheep	TDN	%	*	17.6	62.3	

Grass, aerial part, ensiled, early bloom, (3)
Ref No 3 02 218 United States

Feed Name or Analyses				As Fed (Mean)	Dry (Mean)	C.V. ± %
Dry matter		%		23.4	100.0	14
Ash		%		2.1	9.1	16
Crude fiber		%		7.3	31.0	6
Ether extract		%		0.6	2.7	29
N-free extract		%		10.6	45.1	
Protein (N x 6.25)		%		2.8	12.1	6
Cattle	dig prot	%	*	1.7	7.2	
Goats	dig prot	%	*	1.7	7.2	
Horses	dig prot	%	*	1.7	7.2	
Sheep	dig prot	%	*	1.7	7.2	
Energy	GE	Mcal/kg				
Cattle	DE	Mcal/kg	*	0.62	2.65	
Sheep	DE	Mcal/kg	*	0.65	2.79	
Cattle	ME	Mcal/kg	*	0.51	2.17	
Sheep	ME	Mcal/kg	*	0.53	2.28	
Cattle	TDN	%	*	14.0	60.0	
Sheep	TDN	%	*	14.8	63.2	

Grass, aerial part, ensiled, dormant, (3)
Ref No 3 02 219 United States

Feed Name or Analyses				As Fed (Mean)	Dry (Mean)	C.V. ± %
Dry matter		%		21.0	100.0	
Ash		%		2.3	11.0	
Crude fiber		%		5.4	25.5	
Sheep	dig coef	%		66.	66.	
Ether extract		%		1.1	5.0	
Sheep	dig coef	%		49.	49.	
N-free extract		%		9.6	45.6	
Sheep	dig coef	%		53.	53.	
Protein (N x 6.25)		%		2.7	12.9	
Sheep	dig coef	%		62.	62.	
Cattle	dig prot	%	*	1.7	7.9	
Goats	dig prot	%	*	1.7	7.9	
Horses	dig prot	%	*	1.7	7.9	
Sheep	dig prot	%		1.7	8.0	
Energy	GE	Mcal/kg				
Cattle	DE	Mcal/kg	*	0.54	2.58	
Sheep	DE	Mcal/kg	*	0.50	2.40	
Cattle	ME	Mcal/kg	*	0.45	2.12	
Sheep	ME	Mcal/kg	*	0.41	1.97	
Cattle	TDN	%	*	12.3	58.6	
Sheep	TDN	%		11.4	54.5	

(1) dry forages and roughages (2) pasture, range plants, and forages fed green (3) silages (4) energy feeds (5) protein supplements (6) minerals (7) vitamins (8) additives

Grass, aerial part, ensiled, cut 1, (3)

Ref No 3 02 220 United States

Feed Name or Analyses			Mean As Fed	Mean Dry	C.V. ± %
Dry matter	%		28.2	100.0	13
Ash	%		2.5	9.0	10
Crude fiber	%		10.3	36.7	13
Ether extract	%		1.1	3.8	11
N-free extract	%		11.4	40.3	
Protein (N x 6.25)	%		2.9	10.2	17
Cattle	dig prot %	*	1.5	5.5	
Goats	dig prot %	*	1.5	5.5	
Horses	dig prot %	*	1.5	5.5	
Sheep	dig prot %	*	1.5	5.5	
Lignin (Ellis)	%		2.6	9.3	21
Energy	GE Mcal/kg				
Cattle	DE Mcal/kg	*	0.72	2.55	
Sheep	DE Mcal/kg	*	0.77	2.72	
Cattle	ME Mcal/kg	*	0.59	2.09	
Sheep	ME Mcal/kg	*	0.63	2.23	
Cattle	TDN %	*	16.3	57.9	
Sheep	TDN %	*	17.4	61.7	
pH	pH units		4.83	4.83	16

Grass, aerial part, ensiled, cut 2, (3)

Ref No 3 07 794 United States

Feed Name or Analyses			Mean As Fed	Mean Dry	C.V. ± %
Dry matter	%		29.2	100.0	
Crude fiber	%		6.6	22.6	
Protein (N x 6.25)	%		3.7	12.6	
Cattle	dig prot %	*	2.2	7.7	
Goats	dig prot %	*	2.2	7.7	
Horses	dig prot %	*	2.2	7.7	
Sheep	dig prot %	*	2.2	7.7	
pH	pH units		4.70	4.70	

Grass, aerial part, ensiled trench silo, (3)

Ref No 3 02 222 United States

Feed Name or Analyses			Mean As Fed	Mean Dry	C.V. ± %
Dry matter	%		23.8	100.0	7
Protein (N x 6.25)	%		3.0	12.6	12
Cattle	dig prot %	*	1.8	7.7	
Goats	dig prot %	*	1.8	7.7	
Horses	dig prot %	*	1.8	7.7	
Sheep	dig prot %	*	1.8	7.7	
Carotene	mg/kg		8.6	35.9	32
Vitamin A equivalent	IU/g		14.3	59.9	

Grass, aerial part w AIV preservative added, ensiled, (3)

Ref No 3 02 224 United States

Feed Name or Analyses			Mean As Fed	Mean Dry	C.V. ± %
Dry matter	%		23.1	100.0	4
Ash	%		2.9	12.6	8
Crude fiber	%		5.9	25.6	6
Sheep	dig coef %		79.	79.	
Ether extract	%		0.9	4.0	13
Sheep	dig coef %		55.	55.	
N-free extract	%		9.7	42.1	
Sheep	dig coef %		71.	71.	
Protein (N x 6.25)	%		3.6	15.8	10
Sheep	dig coef %		67.	67.	
Cattle	dig prot %	*	2.4	10.6	
Goats	dig prot %	*	2.4	10.6	
Horses	dig prot %	*	2.4	10.6	
Sheep	dig prot %		2.4	10.6	
Energy	GE Mcal/kg				
Cattle	DE Mcal/kg	*	0.68	2.95	
Sheep	DE Mcal/kg	*	0.67	2.89	
Cattle	ME Mcal/kg	*	0.56	2.42	
Sheep	ME Mcal/kg	*	0.55	2.37	
Cattle	TDN %	*	15.5	67.0	
Sheep	TDN %		15.2	65.6	

Grass, aerial part w AIV preservative added, ensiled, early bloom, (3)

Ref No 3 02 223 United States

Feed Name or Analyses			Mean As Fed	Mean Dry	C.V. ± %
Dry matter	%		23.0	100.0	
Ash	%		2.2	9.5	
Crude fiber	%		6.9	30.1	
Sheep	dig coef %		76.	76.	
Ether extract	%		0.5	2.3	
Sheep	dig coef %		47.	47.	
N-free extract	%		10.9	47.4	
Sheep	dig coef %		72.	72.	
Protein (N x 6.25)	%		2.5	10.7	
Sheep	dig coef %		56.	56.	
Cattle	dig prot %	*	1.4	5.9	
Goats	dig prot %	*	1.4	5.9	
Horses	dig prot %	*	1.4	5.9	
Sheep	dig prot %		1.4	6.0	
Energy	GE Mcal/kg				
Cattle	DE Mcal/kg	*	0.63	2.75	
Sheep	DE Mcal/kg	*	0.66	2.88	
Cattle	ME Mcal/kg	*	0.52	2.26	
Sheep	ME Mcal/kg	*	0.54	2.37	
Cattle	TDN %	*	14.4	62.4	
Sheep	TDN %		15.0	65.4	

Grass, aerial part w AIV preservative and sugar added, ensiled, (3)

Ref No 3 02 225 United States

Feed Name or Analyses			Mean As Fed	Mean Dry	C.V. ± %
Dry matter	%		19.9	100.0	
Ash	%		2.1	10.5	
Crude fiber	%		5.1	25.6	
Sheep	dig coef %		82.	82.	
Ether extract	%		0.9	4.5	
Sheep	dig coef %		65.	65.	
N-free extract	%		8.7	43.9	
Sheep	dig coef %		72.	72.	
Protein (N x 6.25)	%		3.1	15.5	
Sheep	dig coef %		72.	72.	
Cattle	dig prot %	*	2.1	10.3	
Goats	dig prot %	*	2.1	10.3	
Horses	dig prot %	*	2.1	10.3	
Sheep	dig prot %		2.2	11.2	
Energy	GE Mcal/kg				
Cattle	DE Mcal/kg	*	0.57	2.88	
Sheep	DE Mcal/kg	*	0.62	3.10	
Cattle	ME Mcal/kg	*	0.47	2.36	
Sheep	ME Mcal/kg	*	0.51	2.54	
Cattle	TDN %	*	13.0	65.4	
Sheep	TDN %		14.0	70.3	

Grass, aerial part w barley added, ensiled, (3)

Ref No 3 02 226 United States

Feed Name or Analyses			Mean As Fed	Mean Dry	C.V. ± %
Dry matter	%		33.4	100.0	
Ash	%		2.2	6.5	
Crude fiber	%		8.5	25.4	
Ether extract	%		1.3	4.0	
N-free extract	%		16.3	48.9	
Protein (N x 6.25)	%		5.1	15.2	
Cattle	dig prot %	*	3.4	10.0	
Goats	dig prot %	*	3.4	10.0	
Horses	dig prot %	*	3.4	10.0	
Sheep	dig prot %	*	3.4	10.0	
Energy	GE Mcal/kg				
Cattle	DE Mcal/kg	*	0.89	2.66	
Sheep	DE Mcal/kg	*	0.94	2.81	
Cattle	ME Mcal/kg	*	0.73	2.18	
Sheep	ME Mcal/kg	*	0.77	2.30	
Cattle	TDN %	*	20.1	60.3	
Sheep	TDN %	*	21.3	63.7	
Calcium	%		0.23	0.70	
Phosphorus	%		0.12	0.35	

Grass, aerial part w citrus pulp fines added, ensiled, (3)

Ref No 3 02 227 United States

Feed Name or Analyses			Mean As Fed	Mean Dry	C.V. ± %
Dry matter	%		29.8	100.0	
Ash	%		2.7	9.2	
Crude fiber	%		9.8	32.9	
Ether extract	%		1.3	4.2	
N-free extract	%		12.2	41.0	
Protein (N x 6.25)	%		3.8	12.7	
Cattle	dig prot %	*	2.3	7.8	
Goats	dig prot %	*	2.3	7.8	
Horses	dig prot %	*	2.3	7.8	
Sheep	dig prot %	*	2.3	7.8	
Energy	GE Mcal/kg				
Cattle	DE Mcal/kg	*	0.78	2.63	
Sheep	DE Mcal/kg	*	0.81	2.72	
Cattle	ME Mcal/kg	*	0.64	2.15	
Sheep	ME Mcal/kg	*	0.67	2.23	
Cattle	TDN %	*	17.7	59.5	
Sheep	TDN %	*	18.4	61.7	

Grass, aerial part w hydrochloric acid preservative added, ensiled, (3)

Ref No 3 02 228 United States

Feed Name or Analyses			Mean As Fed	Mean Dry	C.V. ± %
Dry matter	%		23.7	100.0	
Ash	%		2.7	11.6	
Crude fiber	%		7.5	31.6	
Sheep	dig coef %		71.	71.	
Ether extract	%		1.3	5.4	
Sheep	dig coef %		47.	47.	
N-free extract	%		8.8	37.3	
Sheep	dig coef %		63.	63.	
Protein (N x 6.25)	%		3.3	14.1	
Sheep	dig coef %		61.	61.	
Cattle	dig prot %	*	2.1	9.0	
Goats	dig prot %	*	2.1	9.0	
Horses	dig prot %	*	2.1	9.0	
Sheep	dig prot %		2.0	8.6	
Energy	GE Mcal/kg				
Cattle	DE Mcal/kg	*	0.67	2.81	
Sheep	DE Mcal/kg	*	0.63	2.66	
Cattle	ME Mcal/kg	*	0.55	2.30	
Sheep	ME Mcal/kg	*	0.52	2.18	
Cattle	TDN %	*	15.1	63.7	
Sheep	TDN %		14.3	60.2	

Grass, aerial part w phosphoric acid preservative added, ensiled, (3)

Ref No 3 02 229 United States

Feed Name or Analyses			Mean As Fed	Mean Dry	C.V. ± %
Dry matter	%		26.8	100.0	9
Ash	%		2.0	7.5	5
Crude fiber	%		8.6	32.1	9
Ether extract	%		0.8	3.0	20
N-free extract	%		12.4	46.2	
Protein (N x 6.25)	%		3.0	11.2	16
Cattle	dig prot %	*	1.7	6.4	
Goats	dig prot %	*	1.7	6.4	
Horses	dig prot %	*	1.7	6.4	
Sheep	dig prot %	*	1.7	6.4	
Energy	GE Mcal/kg				
Cattle	DE Mcal/kg	*	0.69	2.57	
Sheep	DE Mcal/kg	*	0.75	2.82	
Cattle	ME Mcal/kg	*	0.57	2.11	
Sheep	ME Mcal/kg	*	0.62	2.31	
Cattle	TDN %	*	15.6	58.4	
Sheep	TDN %	*	17.1	63.9	

Grass, aerial part w lactic acid bacteria and dried whey added, ensiled, (3)

Ref No 3 02 231 United States

Feed Name or Analyses			Mean As Fed	Mean Dry	C.V. ± %
Dry matter	%		24.2	100.0	
Ash	%		2.3	9.5	
Crude fiber	%		6.1	25.0	
Sheep	dig coef %		82.	82.	
Ether extract	%		1.0	4.3	
Sheep	dig coef %		69.	69.	
N-free extract	%		11.6	48.0	
Sheep	dig coef %		80.	80.	
Protein (N x 6.25)	%		3.2	13.2	
Sheep	dig coef %		71.	71.	
Cattle	dig prot %	*	2.0	8.2	
Goats	dig prot %	*	2.0	8.2	
Horses	dig prot %	*	2.0	8.2	
Sheep	dig prot %		2.3	9.4	
Energy	GE Mcal/kg				
Cattle	DE Mcal/kg	*	0.71	2.95	
Sheep	DE Mcal/kg	*	0.80	3.30	
Cattle	ME Mcal/kg	*	0.59	2.42	
Sheep	ME Mcal/kg	*	0.66	2.71	
Cattle	TDN %	*	16.2	67.0	
Sheep	TDN %		18.1	74.9	

Grass, aerial part w lactic acid bacteria and dried whey added, ensiled, pre-bloom, (3)

Ref No 3 02 230 United States

Feed Name or Analyses			Mean As Fed	Mean Dry	C.V. ± %
Dry matter	%		23.9	100.0	
Ash	%		2.3	9.5	
Crude fiber	%		6.0	25.1	
Sheep	dig coef %		82.	82.	
Ether extract	%		1.0	4.0	
Sheep	dig coef %		69.	69.	
N-free extract	%		11.6	48.4	
Sheep	dig coef %		80.	80.	
Protein (N x 6.25)	%		3.1	13.0	
Sheep	dig coef %		70.	70.	

(1) dry forages and roughages (2) pasture, range plants, and forages fed green (3) silages (4) energy feeds (5) protein supplements (6) minerals (7) vitamins (8) additives

Feed Name or Analyses			Mean As Fed	Mean Dry	C.V. ± %
Cattle	dig prot %	*	1.9	8.0	
Goats	dig prot %	*	1.9	8.0	
Horses	dig prot %	*	1.9	8.0	
Sheep	dig prot %		2.2	9.1	
Energy	GE Mcal/kg				
Cattle	DE Mcal/kg	*	0.70	2.94	
Sheep	DE Mcal/kg	*	0.79	3.29	
Cattle	ME Mcal/kg	*	0.58	2.41	
Sheep	ME Mcal/kg	*	0.64	2.70	
Cattle	TDN %	*	16.0	66.8	
Sheep	TDN %		17.8	74.6	

Grass, aerial part w molasses added, ensiled, (3)

Ref No 3 02 261 United States

Feed Name or Analyses			Mean As Fed	Mean Dry	C.V. ± %
Dry matter	%		25.8	100.0	15
Ash	%		3.8	14.6	14
Crude fiber	%		6.1	23.8	15
Sheep	dig coef %		79.	79.	
Ether extract	%		1.4	5.3	11
Sheep	dig coef %		67.	67.	
N-free extract	%		10.3	39.8	
Sheep	dig coef %		74.	74.	
Protein (N x 6.25)	%		4.3	16.5	7
Sheep	dig coef %		71.	71.	
Cattle	dig prot %	*	2.9	11.2	
Goats	dig prot %	*	2.9	11.2	
Horses	dig prot %	*	2.9	11.2	
Sheep	dig prot %		3.0	11.7	
Energy	GE Mcal/kg				
Cattle	DE Mcal/kg	*	0.82	3.18	
Sheep	DE Mcal/kg	*	0.77	3.00	
Cattle	ME Mcal/kg	*	0.67	2.61	
Sheep	ME Mcal/kg	*	0.63	2.46	
Cattle	TDN %	*	18.6	72.1	
Sheep	TDN %		17.5	67.9	
Calcium	%		0.27	1.04	
Phosphorus	%		0.07	0.28	

Grass, aerial part w molasses added, ensiled, early leaf, (3)

Ref No 3 02 232 United States

Feed Name or Analyses			Mean As Fed	Mean Dry	C.V. ± %
Dry matter	%		21.3	100.0	
Ash	%		3.6	17.1	
Crude fiber	%		4.5	21.0	
Sheep	dig coef %		86.	86.	
Ether extract	%		1.7	8.0	
Sheep	dig coef %		71.	71.	
N-free extract	%		6.7	31.3	
Sheep	dig coef %		71.	71.	
Protein (N x 6.25)	%		4.8	22.6	
Sheep	dig coef %		72.	72.	
Cattle	dig prot %	*	3.6	16.8	
Goats	dig prot %	*	3.6	16.8	
Horses	dig prot %	*	3.6	16.8	
Sheep	dig prot %		3.5	16.3	
Energy	GE Mcal/kg				
Cattle	DE Mcal/kg	*	0.70	3.28	
Sheep	DE Mcal/kg	*	0.65	3.06	
Cattle	ME Mcal/kg	*	0.57	2.69	
Sheep	ME Mcal/kg	*	0.53	2.51	
Cattle	TDN %	*	15.9	74.5	
Sheep	TDN %		14.8	69.3	

Grass, aerial part w molasses added, ensiled, pre-bloom, (3)

Ref No 3 02 233 United States

Feed Name or Analyses			Mean As Fed	Mean Dry	C.V. ± %
Dry matter	%		24.3	100.0	
Ash	%		2.2	9.0	
Crude fiber	%		6.4	26.5	
Sheep	dig coef %		79.	79.	
Ether extract	%		0.8	3.1	
Sheep	dig coef %		65.	65.	
N-free extract	%		11.9	48.9	
Sheep	dig coef %		80.	80.	
Protein (N x 6.25)	%		3.0	12.5	
Sheep	dig coef %		69.	69.	
Cattle	dig prot %	*	1.8	7.6	
Goats	dig prot %	*	1.8	7.6	
Horses	dig prot %	*	1.8	7.6	
Sheep	dig prot %		2.1	8.6	
Energy	GE Mcal/kg				
Cattle	DE Mcal/kg	*	0.74	3.05	
Sheep	DE Mcal/kg	*	0.78	3.23	
Cattle	ME Mcal/kg	*	0.61	2.50	
Sheep	ME Mcal/kg	*	0.64	2.65	
Cattle	TDN %	*	16.8	69.2	
Sheep	TDN %		17.8	73.2	

Grass, aerial part w molasses and acid added, ensiled, (3)

Ref No 3 02 236 United States

Feed Name or Analyses			Mean As Fed	Mean Dry	C.V. ± %
Dry matter	%		21.1	100.0	
Ash	%		2.3	11.0	
Crude fiber	%		6.1	28.7	
Sheep	dig coef %		83.	83.	
Ether extract	%		1.0	4.6	
Sheep	dig coef %		65.	65.	
N-free extract	%		8.4	39.8	
Sheep	dig coef %		72.	72.	
Protein (N x 6.25)	%		3.4	15.9	
Sheep	dig coef %		71.	71.	
Cattle	dig prot %	*	2.3	10.7	
Goats	dig prot %	*	2.3	10.7	
Horses	dig prot %	*	2.3	10.7	
Sheep	dig prot %		2.4	11.3	
Energy	GE Mcal/kg				
Cattle	DE Mcal/kg	*	0.64	3.03	
Sheep	DE Mcal/kg	*	0.66	3.11	
Cattle	ME Mcal/kg	*	0.52	2.48	
Sheep	ME Mcal/kg	*	0.54	2.55	
Cattle	TDN %	*	14.5	68.7	
Sheep	TDN %		14.9	70.5	

Grass, aerial part w molasses and acid added, ensiled, pre-bloom, (3)

Ref No 3 02 235 United States

Feed Name or Analyses			Mean As Fed	Mean Dry	C.V. ± %
Dry matter	%		25.8	100.0	
Ash	%		2.2	8.7	
Crude fiber	%		7.7	29.8	
Sheep	dig coef %		80.	80.	
Ether extract	%		0.9	3.3	
Sheep	dig coef %		59.	59.	
N-free extract	%		11.7	45.3	
Sheep	dig coef %		75.	75.	
Protein (N x 6.25)	%		3.3	12.9	
Sheep	dig coef %		71.	71.	
Cattle	dig prot %	*	2.0	7.9	

Feed Name or Analyses			Mean As Fed	Mean Dry	C.V. ± %
Goats	dig prot %	*	2.0	7.9	
Horses	dig prot %	*	2.0	7.9	
Sheep	dig prot %		2.4	9.2	
Energy	GE Mcal/kg				
Cattle	DE Mcal/kg	*	0.78	3.01	
Sheep	DE Mcal/kg	*	0.81	3.15	
Cattle	ME Mcal/kg	*	0.64	2.47	
Sheep	ME Mcal/kg	*	0.67	2.58	
Cattle	TDN %	*	17.6	68.3	
Sheep	TDN %		18.4	71.4	

Grass, aerial part w salt added, ensiled, mature, (3)

Ref No 3 02 238 United States

Feed Name or Analyses			Mean As Fed	Mean Dry	C.V. ± %
Dry matter	%		24.1	100.0	
Ash	%		2.9	12.0	
Crude fiber	%		7.1	29.5	
Sheep	dig coef %		70.	70.	
Ether extract	%		0.6	2.5	
Sheep	dig coef %		54.	54.	
N-free extract	%		11.2	46.6	
Sheep	dig coef %		68.	68.	
Protein (N x 6.25)	%		2.3	9.4	
Sheep	dig coef %		37.	37.	
Cattle	dig prot %	*	1.1	4.8	
Goats	dig prot %	*	1.1	4.8	
Horses	dig prot %	*	1.1	4.8	
Sheep	dig prot %		0.8	3.5	
Energy	GE Mcal/kg				
Cattle	DE Mcal/kg	*	0.67	2.77	
Sheep	DE Mcal/kg	*	0.63	2.59	
Cattle	ME Mcal/kg	*	0.55	2.27	
Sheep	ME Mcal/kg	*	0.51	2.13	
Cattle	TDN %	*	15.1	62.8	
Sheep	TDN %		14.2	58.9	

Grass, aerial part w sodium metabisulfite preservative added, ensiled, (3)

Ref No 3 02 237 United States

Feed Name or Analyses			Mean As Fed	Mean Dry	C.V. ± %
Dry matter	%		26.5	100.0	9
Ash	%		2.2	8.3	
Crude fiber	%		9.1	34.4	5
Ether extract	%		1.1	4.0	
N-free extract	%		11.0	41.5	
Protein (N x 6.25)	%		3.1	11.8	9
Cattle	dig prot %	*	1.8	6.9	
Goats	dig prot %	*	1.8	6.9	
Horses	dig prot %	*	1.8	6.9	
Sheep	dig prot %	*	1.8	6.9	
Energy	GE Mcal/kg				
Cattle	DE Mcal/kg	*	0.67	2.55	
Sheep	DE Mcal/kg	*	0.73	2.74	
Cattle	ME Mcal/kg	*	0.55	2.09	
Sheep	ME Mcal/kg	*	0.60	2.25	
Cattle	TDN %	*	15.3	57.7	
Sheep	TDN %	*	16.5	62.2	

GRASS, CAN EASTERN 6-7 YR POST SEEDING. Scientific name not used

Grass, Can Eastern 6-7 yr post seeding, hay, s-c, (1)

Ref No 1 08 232 Canada

Feed Name or Analyses			Mean As Fed	Mean Dry	C.V. ± %
Dry matter	%			100.0	
Ash	%			7.4	
Crude fiber	%			32.4	
Ether extract	%			2.7	
N-free extract	%			48.5	
Protein (N x 6.25)	%			9.0	
Sheep	dig coef %			49.	
Cattle	dig prot %	*		4.7	
Goats	dig prot %	*		5.0	
Horses	dig prot %	*		5.2	
Rabbits	dig prot %	*		5.6	
Sheep	dig prot %			4.4	
Energy	GE Mcal/kg				
Cattle	DE Mcal/kg	*		2.45	
Sheep	DE Mcal/kg	*		2.29	
Cattle	ME Mcal/kg	*		2.01	
Sheep	ME Mcal/kg	*		1.88	
Cattle	TDN %	*		55.6	
Sheep	TDN %			52.0	

GRASS, CEREAL. Scientific name not used

Grass, cereal, aerial part, dehy, immature, (1)

Cereals, young, dehy

Ref No 1 08 435 United States

Feed Name or Analyses			Mean As Fed	Mean Dry	C.V. ± %
Dry matter	%		92.8	100.0	
Ash	%		14.4	15.5	
Crude fiber	%		16.1	17.3	
Ether extract	%		4.7	5.1	
N-free extract	%		33.1	35.7	
Protein (N x 6.25)	%		24.5	26.4	
Cattle	dig prot %	*	18.4	19.8	
Goats	dig prot %	*	19.7	21.2	
Horses	dig prot %	*	18.5	19.9	
Rabbits	dig prot %	*	17.7	19.1	
Sheep	dig prot %	*	18.8	20.3	
Energy	GE Mcal/kg				
Cattle	DE Mcal/kg	*	2.51	2.71	
Sheep	DE Mcal/kg	*	2.44	2.63	
Cattle	ME Mcal/kg	*	2.06	2.22	
Sheep	ME Mcal/kg	*	2.00	2.16	
Cattle	TDN %	*	57.0	61.4	
Sheep	TDN %	*	55.3	59.6	
Calcium	%		0.66	0.71	
Manganese	mg/kg		88.0	94.8	
Phosphorus	%		0.46	0.50	

GRASS-ALFALFA. Scientific name not used, Medicago sativa

Grass-Alfalfa, hay, s-c, 15% alfalfa, (1)

Ref No 1 08 741 United States

Feed Name or Analyses			Mean As Fed	Mean Dry	C.V. ± %
Dry matter	%		90.0	100.0	
Ash	%		6.3	7.0	
Crude fiber	%		31.7	35.2	
Ether extract	%		2.1	2.3	
N-free extract	%		40.3	44.8	
Protein (N x 6.25)	%		9.6	10.7	
Cattle	dig prot %	*	5.6	6.2	
Goats	dig prot %	*	5.9	6.5	
Horses	dig prot %	*	5.9	6.6	
Rabbits	dig prot %	*	6.2	6.9	
Sheep	dig prot %	*	5.5	6.1	
Energy	GE Mcal/kg				
Cattle	DE Mcal/kg	*	2.13	2.36	
Sheep	DE Mcal/kg	*	2.21	2.45	
Cattle	ME Mcal/kg	*	1.74	1.94	
Sheep	ME Mcal/kg	*	1.81	2.01	
Cattle	TDN %	*	48.2	53.6	
Sheep	TDN %	*	50.1	55.6	

GRASS-CLOVER. Scientific name not used, Trifolium spp

Grass-Clover, hay, s-c, (1)

Ref No 1 08 436 United States

Feed Name or Analyses			Mean As Fed	Mean Dry	C.V. ± %
Dry matter	%		89.0	100.0	
Ash	%		6.8	7.6	
Crude fiber	%		25.4	28.6	
Ether extract	%		2.7	3.0	
N-free extract	%		39.8	44.7	
Protein (N x 6.25)	%		14.3	16.1	
Cattle	dig prot %	*	9.6	10.8	
Goats	dig prot %	*	10.3	11.5	
Horses	dig prot %	*	9.9	11.2	
Rabbits	dig prot %	*	9.8	11.1	
Sheep	dig prot %	*	9.8	11.0	
Energy	GE Mcal/kg				
Cattle	DE Mcal/kg	*	2.38	2.67	
Sheep	DE Mcal/kg	*	2.32	2.61	
Cattle	ME Mcal/kg	*	1.95	2.19	
Sheep	ME Mcal/kg	*	1.90	2.14	
Cattle	TDN %	*	54.0	60.7	
Sheep	TDN %	*	52.6	59.1	
Calcium	%		0.70	0.78	
Phosphorus	%		0.28	0.31	
Potassium	%		1.83	2.06	

Grass-Clover, aerial part, fresh, (2)

Ref No 2 08 437 United States

Feed Name or Analyses			Mean As Fed	Mean Dry	C.V. ± %
Dry matter	%		33.0	100.0	
Ash	%		1.2	3.6	
Crude fiber	%		2.4	7.3	
Ether extract	%		0.9	2.7	
N-free extract	%		25.1	76.1	
Protein (N x 6.25)	%		3.4	10.3	
Cattle	dig prot %	*	2.2	6.6	
Goats	dig prot %	*	2.0	6.2	
Horses	dig prot %	*	2.1	6.3	
Rabbits	dig prot %	*	2.2	6.6	
Sheep	dig prot %	*	2.2	6.6	
Energy	GE Mcal/kg				
Cattle	DE Mcal/kg	*	0.86	2.61	
Sheep	DE Mcal/kg	*	0.91	2.75	
Cattle	ME Mcal/kg	*	0.71	2.14	
Sheep	ME Mcal/kg	*	0.75	2.26	
Cattle	TDN %	*	19.5	59.1	
Sheep	TDN %	*	20.6	62.3	

GRASS-CLOVER, LADINO. Scientific name not used, Trifolium repens

Grass-Clover, ladino, aerial part, fresh, (2)

Ref No 2 08 438 United States

Feed Name or Analyses			Mean As Fed	Mean Dry	C.V. ± %
Dry matter	%		20.0	100.0	
Ash	%		2.1	10.5	
Crude fiber	%		5.0	25.0	
Ether extract	%		0.6	3.0	
N-free extract	%		9.6	48.0	
Protein (N x 6.25)	%		2.7	13.5	
Cattle	dig prot %	*	1.9	9.4	

Continued

Feed Name or Analyses			Mean As Fed	Mean Dry	C.V. ± %
Goats	dig prot %	*	1.8	9.2	
Horses	dig prot %	*	1.8	9.0	
Rabbits	dig prot %	*	1.8	9.1	
Sheep	dig prot %	*	1.9	9.6	
Energy	GE Mcal/kg				
Cattle	DE Mcal/kg	*	0.57	2.83	
Sheep	DE Mcal/kg	*	0.57	2.86	
Cattle	ME Mcal/kg	*	0.46	2.32	
Sheep	ME Mcal/kg	*	0.47	2.35	
Cattle	TDN %	*	12.8	64.1	
Sheep	TDN %	*	13.0	64.9	
Calcium	%		0.16	0.80	
Phosphorus	%		0.07	0.35	

GRASS-LEGUME. Scientific name not used

Grass-Legume, hay, fan air dried, (1)

Ref No 1 02 272 United States

Feed Name or Analyses			Mean As Fed	Mean Dry	C.V. ± %
Dry matter	%		89.2	100.0	1
Ash	%		5.2	5.8	6
Crude fiber	%		31.4	35.2	7
Ether extract	%		1.8	2.0	8
N-free extract	%		42.0	47.1	
Protein (N x 6.25)	%		8.8	9.9	5
Cattle	dig prot %	*	4.9	5.5	
Goats	dig prot %	*	5.2	5.8	
Horses	dig prot %	*	5.3	5.9	
Rabbits	dig prot %	*	5.6	6.3	
Sheep	dig prot %	*	4.8	5.4	
Energy	GE Mcal/kg				
Cattle	DE Mcal/kg	*	2.06	2.31	
Sheep	DE Mcal/kg	*	2.19	2.45	
Cattle	ME Mcal/kg	*	1.69	1.89	
Sheep	ME Mcal/kg	*	1.79	2.01	
Cattle	TDN %	*	46.7	52.3	
Sheep	TDN %	*	49.6	55.6	
Carotene	mg/kg		34.6	38.8	
Vitamin A equivalent	IU/g		57.7	64.7	
Vitamin D_2	IU/g		1.0	1.1	

Grass-Legume, hay, s-c, (1)

Ref No 1 02 301 United States

Feed Name or Analyses			Mean As Fed	Mean Dry	C.V. ± %
Dry matter	%		89.5	100.0	2
Ash	%		5.7	6.4	14
Crude fiber	%		30.9	34.5	7
Ether extract	%		1.8	2.0	17
N-free extract	%		41.3	46.1	
Protein (N x 6.25)	%		9.8	11.0	16
Cattle	dig prot %	*	5.8	6.4	
Goats	dig prot %	*	6.1	6.8	
Horses	dig prot %	*	6.1	6.8	
Rabbits	dig prot %	*	6.4	7.1	
Sheep	dig prot %	*	5.7	6.4	
Energy	GE Mcal/kg				
Cattle	DE Mcal/kg	*	2.12	2.37	
Sheep	DE Mcal/kg	*	2.22	2.48	
Cattle	ME Mcal/kg	*	1.74	1.94	
Sheep	ME Mcal/kg	*	1.82	2.03	
Cattle	TDN %	*	48.1	53.7	
Sheep	TDN %	*	50.3	56.2	
Calcium	%		0.82	0.92	31

(1) dry forages and roughages (2) pasture, range plants, and forages fed green (3) silages (4) energy feeds (5) protein supplements (6) minerals (7) vitamins (8) additives

Feed Name or Analyses			Mean As Fed	Mean Dry	C.V. ± %
Cobalt	mg/kg		0.120	0.135	77
Manganese	mg/kg		39.5	44.1	67
Phosphorus	%		0.18	0.21	26
Potassium	%		1.45	1.62	
Sodium	%		0.12	0.13	36
Sulphur	%		0.13	0.14	21
Carotene	mg/kg		10.9	12.1	98
Vitamin A equivalent	IU/g		18.1	20.2	
Vitamin D_2	IU/g		1.9	2.1	38

Grass-Legume, hay, s-c, immature, (1)

Ref No 1 02 275 United States

Feed Name or Analyses			Mean As Fed	Mean Dry	C.V. ± %
Dry matter	%		90.0	100.0	2
Ash	%		8.0	8.9	15
Crude fiber	%		27.1	30.1	11
Ether extract	%		2.9	3.2	43
N-free extract	%		38.4	42.7	
Protein (N x 6.25)	%		13.6	15.1	13
Cattle	dig prot %	*	9.0	10.1	
Goats	dig prot %	*	9.6	10.7	
Horses	dig prot %	*	9.3	10.4	
Rabbits	dig prot %	*	9.3	10.4	
Sheep	dig prot %	*	9.1	10.1	
Energy	GE Mcal/kg				
Cattle	DE Mcal/kg	*	2.37	2.63	
Sheep	DE Mcal/kg	*	2.25	2.50	
Cattle	ME Mcal/kg	*	1.94	2.16	
Sheep	ME Mcal/kg	*	1.85	2.05	
Cattle	TDN %	*	53.7	59.7	
Sheep	TDN %	*	51.1	56.7	
Calcium	%		0.95	1.05	38
Cobalt	mg/kg		0.387	0.430	
Copper	mg/kg		6.0	6.6	
Iron	%		0.007	0.008	
Magnesium	%		0.14	0.15	
Manganese	mg/kg		79.8	88.6	
Phosphorus	%		0.28	0.32	38
Potassium	%		1.95	2.17	
Carotene	mg/kg		122.0	135.6	69
Vitamin A equivalent	IU/g		203.4	226.0	

Grass-Legume, hay, s-c, immature, mn 30% legume, (1)

Ref No 1 02 273 United States

Feed Name or Analyses			Mean As Fed	Mean Dry	C.V. ± %
Dry matter	%		92.0	100.0	
Ash	%		6.7	7.3	5
Crude fiber	%		29.6	32.2	7
Ether extract	%		2.6	2.8	39
N-free extract	%		38.0	41.3	
Protein (N x 6.25)	%		15.1	16.4	6
Cattle	dig prot %	*	10.3	11.1	
Goats	dig prot %	*	10.9	11.9	
Horses	dig prot %	*	10.5	11.5	
Rabbits	dig prot %	*	10.4	11.3	
Sheep	dig prot %	*	10.4	11.3	
Energy	GE Mcal/kg				
Cattle	DE Mcal/kg	*	2.29	2.48	
Sheep	DE Mcal/kg	*	2.32	2.52	
Cattle	ME Mcal/kg	*	1.87	2.04	
Sheep	ME Mcal/kg	*	1.90	2.07	
Cattle	TDN %	*	51.8	56.3	
Sheep	TDN %	*	52.6	57.2	

Grass-Legume, hay, s-c, immature, mx 30% legume, (1)

Ref No 1 02 274 United States

Feed Name or Analyses			Mean As Fed	Mean Dry	C.V. ± %
Dry matter	%		88.9	100.0	1
Ash	%		6.8	7.6	21
Crude fiber	%		28.4	32.0	12
Ether extract	%		3.5	3.9	38
N-free extract	%		37.3	42.0	
Protein (N x 6.25)	%		12.9	14.5	14
Cattle	dig prot %	*	8.4	9.5	
Goats	dig prot %	*	9.0	10.1	
Horses	dig prot %	*	8.7	9.8	
Rabbits	dig prot %	*	8.8	9.9	
Sheep	dig prot %	*	8.5	9.6	
Energy	GE Mcal/kg				
Cattle	DE Mcal/kg	*	2.27	2.56	
Sheep	DE Mcal/kg	*	2.19	2.46	
Cattle	ME Mcal/kg	*	1.87	2.10	
Sheep	ME Mcal/kg	*	1.80	2.02	
Cattle	TDN %	*	51.6	58.0	
Sheep	TDN %	*	49.7	55.9	
Calcium	%		1.02	1.15	
Manganese	mg/kg		128.2	144.2	
Phosphorus	%		0.34	0.38	
Carotene	mg/kg		82.9	93.3	
Vitamin A equivalent	IU/g		138.2	155.5	

Grass-Legume, hay, s-c, pre-bloom, mn 30% legume, (1)

Ref No 1 02 276 United States

Feed Name or Analyses			Mean As Fed	Mean Dry	C.V. ± %
Dry matter	%		90.0	100.0	
Carotene	mg/kg		29.6	32.8	
Vitamin A equivalent	IU/g		49.3	54.8	

Grass-Legume, hay, s-c, mid-bloom, mx 30% legume, (1)

Ref No 1 02 277 United States

Feed Name or Analyses			Mean As Fed	Mean Dry	C.V. ± %
Dry matter	%		92.8	100.0	2
Ash	%		6.5	7.0	13
Crude fiber	%		30.3	32.7	3
Ether extract	%		2.3	2.4	12
N-free extract	%		44.0	47.4	
Protein (N x 6.25)	%		9.7	10.5	21
Cattle	dig prot %	*	5.6	6.0	
Goats	dig prot %	*	5.9	6.3	
Horses	dig prot %	*	6.0	6.4	
Rabbits	dig prot %	*	6.3	6.8	
Sheep	dig prot %	*	5.5	6.0	
Energy	GE Mcal/kg				
Cattle	DE Mcal/kg	*	2.37	2.55	
Sheep	DE Mcal/kg	*	2.32	2.50	
Cattle	ME Mcal/kg	*	1.94	2.09	
Sheep	ME Mcal/kg	*	1.90	2.05	
Cattle	TDN %	*	53.7	57.9	
Sheep	TDN %	*	52.7	56.8	
Calcium	%		0.62	0.67	28
Manganese	mg/kg		66.7	71.9	
Phosphorus	%		0.19	0.21	21

Grass-Legume, hay, s-c, full bloom, (1)

Ref No 1 02 280 United States

Feed Name or Analyses			Mean As Fed	Mean Dry	C.V. ± %
Dry matter	%		90.0	100.0	1
Ash	%		5.2	5.8	7
Crude fiber	%		33.4	37.1	8
Ether extract	%		1.8	2.0	13
N-free extract	%		40.4	44.9	
Protein (N x 6.25)	%		9.2	10.2	17
Cattle	dig prot %	*	5.2	5.8	
Goats	dig prot %	*	5.5	6.1	
Horses	dig prot %	*	5.6	6.2	
Rabbits	dig prot %	*	5.9	6.5	
Sheep	dig prot %	*	5.1	5.7	
Energy	GE Mcal/kg				
Cattle	DE Mcal/kg	*	1.95	2.16	
Sheep	DE Mcal/kg	*	2.17	2.42	
Cattle	ME Mcal/kg	*	1.60	1.77	
Sheep	ME Mcal/kg	*	1.78	1.98	
Cattle	TDN %	*	44.2	49.1	
Sheep	TDN %	*	49.3	54.8	

Grass-Legume, hay, s-c, full bloom, mn 30% legume, (1)

Ref No 1 02 278 United States

Feed Name or Analyses			Mean As Fed	Mean Dry	C.V. ± %
Dry matter	%		90.6	100.0	0
Crude fiber	%		29.4	32.5	4
Ether extract	%		1.6	1.8	16
Protein (N x 6.25)	%		11.4	12.6	7
Cattle	dig prot %	*	7.1	7.9	
Goats	dig prot %	*	7.5	8.3	
Horses	dig prot %	*	7.5	8.2	
Rabbits	dig prot %	*	7.6	8.4	
Sheep	dig prot %	*	7.1	7.9	

Grass-Legume, hay, s-c, full bloom, mx 30% legume, (1)

Ref No 1 02 279 United States

Feed Name or Analyses			Mean As Fed	Mean Dry	C.V. ± %
Dry matter	%		89.8	100.0	2
Ash	%		5.2	5.8	7
Crude fiber	%		34.9	38.9	6
Ether extract	%		1.9	2.1	14
N-free extract	%		39.4	43.9	
Protein (N x 6.25)	%		8.4	9.3	15
Cattle	dig prot %	*	4.5	5.0	
Goats	dig prot %	*	4.7	5.2	
Horses	dig prot %	*	4.9	5.4	
Rabbits	dig prot %	*	5.3	5.8	
Sheep	dig prot %	*	4.4	4.9	
Energy	GE Mcal/kg				
Cattle	DE Mcal/kg	*	1.98	2.20	
Sheep	DE Mcal/kg	*	2.12	2.36	
Cattle	ME Mcal/kg	*	1.62	1.80	
Sheep	ME Mcal/kg	*	1.74	1.94	
Cattle	TDN %	*	44.7	49.8	
Sheep	TDN %	*	48.1	53.6	

Grass-Legume, hay, s-c, milk stage, mx 30% legume, (1)

Ref No 1 02 281 United States

Feed Name or Analyses			Mean As Fed	Mean Dry	C.V. ± %
Dry matter	%		89.8	100.0	2
Ash	%		4.5	5.0	2
Crude fiber	%		34.8	38.7	4
Ether extract	%		2.2	2.5	27
N-free extract	%		40.4	45.0	
Protein (N x 6.25)	%		7.9	8.8	1
Cattle	dig prot %	*	4.1	4.6	
Goats	dig prot %	*	4.3	4.8	
Horses	dig prot %	*	4.5	5.0	
Rabbits	dig prot %	*	4.9	5.5	
Sheep	dig prot %	*	4.0	4.5	
Energy	GE Mcal/kg				
Cattle	DE Mcal/kg	*	2.02	2.25	
Sheep	DE Mcal/kg	*	2.12	2.36	
Cattle	ME Mcal/kg	*	1.65	1.84	
Sheep	ME Mcal/kg	*	1.74	1.94	
Cattle	TDN %	*	45.9	51.1	
Sheep	TDN %	*	48.1	53.6	
Carotene	mg/kg		12.5	13.9	23
Vitamin A equivalent	IU/g		20.8	23.2	

Grass-Legume, hay, s-c, cut 1, mn 30% legume, (1)

Ref No 1 02 282 United States

Feed Name or Analyses			Mean As Fed	Mean Dry	C.V. ± %
Dry matter	%		89.6	100.0	1
Ash	%		5.6	6.2	16
Crude fiber	%		30.1	33.6	9
Ether extract	%		2.0	2.3	8
N-free extract	%		42.2	47.2	
Protein (N x 6.25)	%		9.7	10.8	12
Cattle	dig prot %	*	5.6	6.3	
Goats	dig prot %	*	5.9	6.6	
Horses	dig prot %	*	6.0	6.7	
Rabbits	dig prot %	*	6.3	7.0	
Sheep	dig prot %	*	5.6	6.3	
Energy	GE Mcal/kg				
Cattle	DE Mcal/kg	*	2.19	2.44	
Sheep	DE Mcal/kg	*	2.24	2.50	
Cattle	ME Mcal/kg	*	1.79	2.00	
Sheep	ME Mcal/kg	*	1.83	2.05	
Cattle	TDN %	*	49.6	55.4	
Sheep	TDN %	*	50.7	56.6	
Calcium	%		0.44	0.49	
Magnesium	%		0.17	0.19	
Phosphorus	%		0.18	0.20	
Potassium	%		1.16	1.29	

Grass-Legume, hay, s-c, cut 1, mx 30% legume, (1)

Ref No 1 02 283 United States

Feed Name or Analyses			Mean As Fed	Mean Dry	C.V. ± %
Dry matter	%		90.4	100.0	3
Ash	%		5.3	5.9	7
Crude fiber	%		34.6	38.3	6
Ether extract	%		1.9	2.1	10
N-free extract	%		40.7	45.0	
Protein (N x 6.25)	%		7.9	8.7	9
Cattle	dig prot %	*	4.0	4.5	
Goats	dig prot %	*	4.2	4.7	
Horses	dig prot %	*	4.4	4.9	
Rabbits	dig prot %	*	4.9	5.4	
Sheep	dig prot %	*	4.0	4.4	
Energy	GE Mcal/kg				
Cattle	DE Mcal/kg	*	2.15	2.38	
Sheep	DE Mcal/kg	*	2.14	2.37	
Cattle	ME Mcal/kg	*	1.76	1.95	
Sheep	ME Mcal/kg	*	1.75	1.94	
Cattle	TDN %	*	48.7	53.9	
Sheep	TDN %	*	48.5	53.6	
Carotene	mg/kg		8.0	8.8	54
Vitamin A equivalent	IU/g		13.3	14.7	

Grass-Legume, hay, s-c, cut 2, (1)

Ref No 1 02 286 United States

Feed Name or Analyses			Mean As Fed	Mean Dry	C.V. ± %
Dry matter	%		89.6	100.0	1
Ash	%		6.6	7.4	14
Crude fiber	%		28.1	31.4	8
Ether extract	%		2.8	3.1	22
N-free extract	%		37.5	41.9	
Protein (N x 6.25)	%		14.5	16.2	10
Cattle	dig prot %	*	9.8	11.0	
Goats	dig prot %	*	10.5	11.7	
Horses	dig prot %	*	10.1	11.3	
Rabbits	dig prot %	*	10.0	11.2	
Sheep	dig prot %	*	9.9	11.1	
Energy	GE Mcal/kg				
Cattle	DE Mcal/kg	*	2.26	2.52	
Sheep	DE Mcal/kg	*	2.26	2.53	
Cattle	ME Mcal/kg	*	1.85	2.07	
Sheep	ME Mcal/kg	*	1.86	2.07	
Cattle	TDN %	*	51.3	57.2	
Sheep	TDN %	*	51.4	57.4	
Carotene	mg/kg		42.8	47.8	55
Vitamin A equivalent	IU/g		71.4	79.7	

Grass-Legume, hay, s-c, cut 2, mx 30% legume, (1)

Ref No 1 02 285 United States

Feed Name or Analyses			Mean As Fed	Mean Dry	C.V. ± %
Dry matter	%		89.4	100.0	
Calcium	%		0.96	1.07	15
Cobalt	mg/kg		0.116	0.130	18
Manganese	mg/kg		36.5	40.8	38
Phosphorus	%		0.21	0.24	23

Grass-Legume, hay, s-c, gr sample US mn 30% legume, (1)

Ref No 1 02 288 United States

Feed Name or Analyses			Mean As Fed	Mean Dry	C.V. ± %
Dry matter	%			100.0	
N-free extract	%			43.7	
Protein (N x 6.25)	%			9.9	31
Cattle	dig prot %	*		5.5	
Goats	dig prot %	*		5.8	
Horses	dig prot %	*		5.9	
Rabbits	dig prot %	*		6.3	
Sheep	dig prot %	*		5.5	
Calcium	%			0.82	29
Cobalt	mg/kg			0.110	45
Manganese	mg/kg			42.5	31
Phosphorus	%			0.20	32
Vitamin D2	IU/g			1.4	

Grass-Legume, hay, s-c, gr sample US mx 30% legume, (1)

Ref No 1 02 289 United States

Feed Name or Analyses			Mean As Fed	Mean Dry	C.V. ± %
Dry matter	%			100.0	
Protein (N x 6.25)	%			8.3	34
Cattle	dig prot %	*		4.1	
Goats	dig prot %	*		4.3	
Horses	dig prot %	*		4.6	
Rabbits	dig prot %	*		5.1	
Sheep	dig prot %	*		4.0	

Continued

Feed Name or Analyses			Mean As Fed	Mean Dry	C.V. ± %
Calcium	%			0.58	39
Cobalt	mg/kg			0.132	62
Manganese	mg/kg			85.1	60
Phosphorus	%			0.19	43
Carotene	mg/kg			8.8	91
Vitamin A equivalent	IU/g			14.7	

Grass-Legume, hay, s-c, gr 1 US, (1)
Ref No 1 02 290 United States

Feed Name or Analyses			Mean As Fed	Mean Dry	C.V. ± %
Dry matter	%		90.0	100.0	
Ash	%		6.0	6.7	12
Crude fiber	%		32.0	35.6	9
Ether extract	%		2.6	2.9	33
N-free extract	%		37.4	41.5	
Protein (N x 6.25)	%		12.0	13.3	14
Cattle	dig prot %	*	7.6	8.5	
Goats	dig prot %	*	8.1	9.0	
Horses	dig prot %	*	7.9	8.8	
Rabbits	dig prot %	*	8.0	8.9	
Sheep	dig prot %	*	7.7	8.5	
Energy	GE Mcal/kg				
Cattle	DE Mcal/kg	*	2.15	2.39	
Sheep	DE Mcal/kg	*	2.18	2.42	
Cattle	ME Mcal/kg	*	1.76	1.96	
Sheep	ME Mcal/kg	*	1.79	1.98	
Cattle	TDN %	*	48.7	54.2	
Sheep	TDN %	*	49.4	54.9	
Calcium	%		0.76	0.84	27
Cobalt	mg/kg		0.081	0.090	23
Manganese	mg/kg		35.9	39.9	18
Phosphorus	%		0.22	0.24	14
Carotene	mg/kg		18.3	20.3	51
Vitamin A equivalent	IU/g		30.4	33.8	
Vitamin D2	IU/g		1.8	2.0	

Grass-Legume, hay, s-c, gr 1 US mn 30% legume, (1)
Ref No 1 02 291 United States

Feed Name or Analyses			Mean As Fed	Mean Dry	C.V. ± %
Dry matter	%		90.0	100.0	
Ash	%		6.0	6.7	10
Crude fiber	%		31.3	34.8	11
Ether extract	%		2.8	3.1	35
N-free extract	%		37.2	41.3	
Protein (N x 6.25)	%		12.7	14.1	11
Cattle	dig prot %	*	8.2	9.2	
Goats	dig prot %	*	8.7	9.7	
Horses	dig prot %	*	8.6	9.5	
Rabbits	dig prot %	*	8.6	9.6	
Sheep	dig prot %	*	8.3	9.2	
Energy	GE Mcal/kg				
Cattle	DE Mcal/kg	*	2.17	2.41	
Sheep	DE Mcal/kg	*	2.20	2.44	
Cattle	ME Mcal/kg	*	1.78	1.98	
Sheep	ME Mcal/kg	*	1.80	2.00	
Cattle	TDN %	*	49.3	54.7	
Sheep	TDN %	*	49.8	55.4	
Calcium	%		0.81	0.90	28
Cobalt	mg/kg		0.089	0.099	
Manganese	mg/kg		39.9	44.3	
Phosphorus	%		0.23	0.25	15
Carotene	mg/kg		22.6	25.1	44
Vitamin A equivalent	IU/g		37.7	41.9	

Grass-Legume, hay, s-c, gr 1 US mx 30% legume, (1)
Ref No 1 02 292 United States

Feed Name or Analyses			Mean As Fed	Mean Dry	C.V. ± %
Dry matter	%			100.0	
Ash	%			6.7	18
Crude fiber	%			37.2	1
Ether extract	%			2.7	13
N-free extract	%			41.6	
Protein (N x 6.25)	%			11.8	21
Cattle	dig prot %	*		7.2	
Goats	dig prot %	*		7.6	
Horses	dig prot %	*		7.5	
Rabbits	dig prot %	*		7.8	
Sheep	dig prot %	*		7.2	
Energy	GE Mcal/kg				
Cattle	DE Mcal/kg	*		2.21	
Sheep	DE Mcal/kg	*		2.43	
Cattle	ME Mcal/kg	*		1.81	
Sheep	ME Mcal/kg	*		1.99	
Cattle	TDN %	*		50.1	
Sheep	TDN %	*		55.1	
Calcium	%			0.69	
Cobalt	mg/kg			0.084	
Manganese	mg/kg			37.0	
Phosphorus	%			0.22	
Carotene	mg/kg			14.1	
Vitamin A equivalent	IU/g			23.5	

Grass-Legume, hay, s-c, gr 2 US, (1)
Ref No 1 02 293 United States

Feed Name or Analyses			Mean As Fed	Mean Dry	C.V. ± %
Dry matter	%		87.2	100.0	
Ash	%		5.1	5.8	10
Crude fiber	%		33.5	38.4	4
Ether extract	%		2.0	2.3	21
N-free extract	%		37.8	43.4	
Protein (N x 6.25)	%		8.8	10.1	28
Cattle	dig prot %	*	5.0	5.7	
Goats	dig prot %	*	5.2	6.0	
Horses	dig prot %	*	5.3	6.1	
Rabbits	dig prot %	*	5.6	6.5	
Sheep	dig prot %	*	4.9	5.6	
Energy	GE Mcal/kg				
Cattle	DE Mcal/kg	*	2.01	2.31	
Sheep	DE Mcal/kg	*	2.08	2.39	
Cattle	ME Mcal/kg	*	1.65	1.89	
Sheep	ME Mcal/kg	*	1.71	1.96	
Cattle	TDN %	*	45.7	52.4	
Sheep	TDN %	*	47.2	54.1	
Calcium	%		0.63	0.72	29
Cobalt	mg/kg		0.112	0.128	70
Manganese	mg/kg		45.2	51.8	67
Phosphorus	%		0.17	0.20	25
Carotene	mg/kg		9.6	11.0	60
Vitamin A equivalent	IU/g		16.0	18.4	
Vitamin D2	IU/g		2.5	2.8	

Grass-Legume, hay, s-c, gr 2 US mn 30% legume, (1)
Ref No 1 02 294 United States

Feed Name or Analyses			Mean As Fed	Mean Dry	C.V. ± %
Dry matter	%		87.2	100.0	
Ash	%		5.1	5.9	5
Crude fiber	%		33.2	38.1	2
Ether extract	%		1.9	2.2	27
N-free extract	%		38.2	43.8	
Protein (N x 6.25)	%		8.8	10.1	20
Cattle	dig prot %	*	5.0	5.7	
Goats	dig prot %	*	5.2	6.0	
Horses	dig prot %	*	5.3	6.1	
Rabbits	dig prot %	*	5.6	6.5	
Sheep	dig prot %	*	4.9	5.6	
Energy	GE Mcal/kg				
Cattle	DE Mcal/kg	*	2.06	2.36	
Sheep	DE Mcal/kg	*	2.09	2.39	
Cattle	ME Mcal/kg	*	1.69	1.94	
Sheep	ME Mcal/kg	*	1.71	1.96	
Cattle	TDN %	*	46.7	53.6	
Sheep	TDN %	*	47.3	54.3	

Grass-Legume, hay, s-c, gr 2 US mx 30% legume, (1)
Ref No 1 02 295 United States

Feed Name or Analyses			Mean As Fed	Mean Dry	C.V. ± %
Dry matter	%			100.0	
Calcium	%			0.60	37
Cobalt	mg/kg			0.126	81
Manganese	mg/kg			68.8	60
Phosphorus	%			0.19	31
Carotene	mg/kg			11.2	75
Vitamin A equivalent	IU/g			18.7	

Grass-Legume, hay, s-c, gr 3 US, (1)
Ref No 1 02 296 United States

Feed Name or Analyses			Mean As Fed	Mean Dry	C.V. ± %
Dry matter	%		88.1	100.0	
Ash	%		4.3	4.9	
Crude fiber	%		35.4	40.2	
Ether extract	%		1.6	1.8	
N-free extract	%		39.4	44.7	
Protein (N x 6.25)	%		7.4	8.4	22
Cattle	dig prot %	*	3.7	4.2	
Goats	dig prot %	*	3.9	4.4	
Horses	dig prot %	*	4.1	4.7	
Rabbits	dig prot %	*	4.5	5.2	
Sheep	dig prot %	*	3.6	4.1	
Energy	GE Mcal/kg				
Cattle	DE Mcal/kg	*	1.89	2.15	
Sheep	DE Mcal/kg	*	2.05	2.33	
Cattle	ME Mcal/kg	*	1.55	1.26	
Sheep	ME Mcal/kg	*	1.68	1.91	
Cattle	TDN %	*	42.9	48.7	
Sheep	TDN %	*	46.5	52.8	
Calcium	%		0.53	0.60	37
Cobalt	mg/kg		0.111	0.126	
Manganese	mg/kg		47.2	53.6	44
Phosphorus	%		0.16	0.18	17
Carotene	mg/kg		15.1	17.2	33
Vitamin A equivalent	IU/g		25.3	28.7	
Vitamin D2	IU/g		2.9	3.3	

Grass-Legume, hay, s-c, gr 3 US mn 30% legume, (1)
Ref No 1 02 297 United States

Feed Name or Analyses			Mean As Fed	Mean Dry	C.V. ± %
Dry matter	%			100.0	
N-free extract	%			43.7	
Protein (N x 6.25)	%			10.5	16
Cattle	dig prot %	*		6.0	
Goats	dig prot %	*		6.4	
Horses	dig prot %	*		6.4	
Rabbits	dig prot %	*		6.8	
Sheep	dig prot %	*		6.0	
Calcium	%			0.88	34

(1) dry forages and roughages (2) pasture, range plants, and forages fed green (3) silages (4) energy feeds (5) protein supplements (6) minerals (7) vitamins (8) additives

Feed Name or Analyses			Mean As Fed	Mean Dry	C.V. ± %
Cobalt	mg/kg			0.141	
Manganese	mg/kg			26.2	
Phosphorus	%			0.20	20

Grass-Legume, hay, s-c, gr 3 US mx 30% legume, (1)

Ref No 1 02 298 United States

Analyses			As Fed	Dry	C.V. ± %
Dry matter	%			100.0	
Ash	%			4.9	
Crude fiber	%			40.2	
Ether extract	%			1.8	
N-free extract	%			45.3	
Protein (N x 6.25)	%			7.8	20
Cattle	dig prot %	*		3.7	
Goats	dig prot %	*		3.8	
Horses	dig prot %	*		4.2	
Rabbits	dig prot %	*		4.7	
Sheep	dig prot %	*		3.6	
Energy	GE Mcal/kg				
Cattle	DE Mcal/kg	*		2.13	
Sheep	DE Mcal/kg	*		2.32	
Cattle	ME Mcal/kg	*		1.75	
Sheep	ME Mcal/kg	*		1.90	
Cattle	TDN %	*		48.4	
Sheep	TDN %	*		52.6	
Calcium	%			0.53	9
Manganese	mg/kg			59.1	26
Phosphorus	%			0.17	16

Grass-Legume, hay, s-c, mn 30% legume, (1)

Ref No 1 02 299 United States

Analyses			As Fed	Dry	C.V. ± %
Dry matter	%		89.8	100.0	1
Cobalt	mg/kg		0.103	0.115	43
Sulphur	%		0.13	0.14	28
Carotene	mg/kg		19.8	22.0	72
Vitamin A equivalent	IU/g		33.0	36.8	

Grass-Legume, hay, s-c, mx 30% legume, (1)

Ref No 1 02 300 United States

Analyses			As Fed	Dry	C.V. ± %
Dry matter	%		88.2	100.0	
Ash	%		5.3	6.0	
Crude fiber	%		31.3	35.5	
Ether extract	%		1.8	2.0	
N-free extract	%		41.4	46.9	
Protein (N x 6.25)	%		8.4	9.5	
Cattle	dig prot %	*	4.6	5.2	
Goats	dig prot %	*	4.8	5.4	
Horses	dig prot %	*	5.0	5.6	
Rabbits	dig prot %	*	5.3	6.0	
Sheep	dig prot %	*	4.5	5.1	
Energy	GE Mcal/kg				
Cattle	DE Mcal/kg	*	2.03	2.30	
Sheep	DE Mcal/kg	*	2.15	2.44	
Cattle	ME Mcal/kg	*	1.66	1.89	
Sheep	ME Mcal/kg	*	1.76	2.00	
Cattle	TDN %	*	46.0	52.2	
Sheep	TDN %	*	48.8	55.3	
Calcium	%		0.59	0.67	
Cobalt	mg/kg		0.109	0.124	75
Phosphorus	%		0.18	0.20	
Potassium	%		1.47	1.67	

Grass-Legume, aerial part, fresh, (2)

Ref No 2 08 439 United States

Analyses			As Fed	Dry	C.V. ± %
Dry matter	%		23.6	100.0	
Ash	%		2.2	9.3	
Crude fiber	%		5.7	24.3	
Ether extract	%		0.9	3.6	
N-free extract	%		10.3	43.9	
Protein (N x 6.25)	%		4.5	18.9	
Cattle	dig prot %	*	3.3	14.0	
Goats	dig prot %	*	3.3	14.2	
Horses	dig prot %	*	3.2	13.6	
Rabbits	dig prot %	*	3.1	13.3	
Sheep	dig prot %	*	3.4	14.6	
Energy	GE Mcal/kg				
Cattle	DE Mcal/kg	*	0.74	3.13	
Sheep	DE Mcal/kg	*	0.66	2.81	
Cattle	ME Mcal/kg	*	0.60	2.56	
Sheep	ME Mcal/kg	*	0.54	2.31	
Cattle	TDN %	*	16.7	70.9	
Sheep	TDN %	*	15.0	63.8	
Calcium	%		0.16	0.70	
Chlorine	%		0.11	0.48	
Iron	%		0.012	0.052	
Magnesium	%		0.07	0.32	
Manganese	mg/kg		131.3	557.7	
Phosphorus	%		0.09	0.36	
Potassium	%		0.45	1.92	
Sulphur	%		0.06	0.24	

Grass-Legume, aerial part, fresh, few legumes, (2)

Ref No 2 08 440 United States

Analyses			As Fed	Dry	C.V. ± %
Dry matter	%		23.5	100.0	
Ash	%		2.5	10.7	
Crude fiber	%		5.9	25.0	
Ether extract	%		0.9	3.7	
N-free extract	%		10.7	45.5	
Protein (N x 6.25)	%		3.5	15.0	
Cattle	dig prot %	*	2.5	10.7	
Goats	dig prot %	*	2.5	10.6	
Horses	dig prot %	*	2.4	10.3	
Rabbits	dig prot %	*	2.4	10.3	
Sheep	dig prot %	*	2.6	11.0	
Energy	GE Mcal/kg				
Cattle	DE Mcal/kg	*	0.67	2.84	
Sheep	DE Mcal/kg	*	0.66	2.79	
Cattle	ME Mcal/kg	*	0.55	2.33	
Sheep	ME Mcal/kg	*	0.54	2.29	
Cattle	TDN %	*	15.2	64.5	
Sheep	TDN %	*	14.9	63.4	
Calcium	%		0.13	0.57	
Phosphorus	%		0.07	0.32	
Potassium	%		0.32	1.36	

Grass-Legume, aerial part, ensiled, (3)

Ref No 3 02 303 United States

Analyses			As Fed	Dry	C.V. ± %
Dry matter	%		28.9	100.0	5
Ash	%		2.2	7.6	11
Crude fiber	%		9.3	32.0	9
Ether extract	%		0.9	3.2	26
N-free extract	%		13.4	46.2	
Protein (N x 6.25)	%		3.2	11.0	7
Cattle	dig prot %	*	1.8	6.2	
Goats	dig prot %	*	1.8	6.2	
Horses	dig prot %	*	1.8	6.2	

Analyses			As Fed	Dry	C.V. ± %
Sheep	dig prot %	*	1.8	6.2	
Energy	GE Mcal/kg				
Cattle	DE Mcal/kg	*	0.75	2.60	
Sheep	DE Mcal/kg	*	0.81	2.81	
Cattle	ME Mcal/kg	*	0.62	2.13	
Sheep	ME Mcal/kg	*	0.67	2.31	
Cattle	NE_m Mcal/kg	*	0.34	1.19	
Cattle	NE_{gain} Mcal/kg	*	0.14	0.50	
Cattle	$NE_{lactating\ cows}$ Mcal/kg	*	0.38	1.30	
Cattle	TDN %	*	17.1	59.0	
Sheep	TDN %	*	18.4	63.8	
Calcium	%		0.27	0.92	17
Chlorine	%		0.31	1.06	
Phosphorus	%		0.08	0.26	16
Sulphur	%		0.20	0.69	42
Carotene	mg/kg		66.5	229.9	26
Niacin	mg/kg		13.2	45.6	
Vitamin A equivalent	IU/g		110.9	383.3	

Grass-Legume, aerial part, ensiled, immature, (3)

Ref No 3 02 302 United States

Analyses			As Fed	Dry	C.V. ± %
Dry matter	%		27.3	100.0	6
Chlorine	%		0.29	1.06	
Sulphur	%		0.21	0.78	
Carotene	mg/kg		75.6	277.6	17
Vitamin A equivalent	IU/g		126.1	462.7	

Grass-Legume, aerial part, ensiled, considerable legumes, (3)

Ref No 3 08 598 United States

Analyses			As Fed	Dry	C.V. ± %
Dry matter	%		25.6	100.0	
Ash	%		1.7	6.6	
Crude fiber	%		8.6	33.6	
Ether extract	%		0.9	3.5	
N-free extract	%		10.8	42.2	
Protein (N x 6.25)	%		3.6	14.1	
Cattle	dig prot %	*	2.3	9.0	
Goats	dig prot %	*	2.3	9.0	
Horses	dig prot %	*	2.3	9.0	
Sheep	dig prot %	*	2.3	9.0	
Energy	GE Mcal/kg				
Cattle	DE Mcal/kg	*	0.66	2.57	
Sheep	DE Mcal/kg	*	0.71	2.79	
Cattle	ME Mcal/kg	*	0.54	2.11	
Sheep	ME Mcal/kg	*	0.59	2.29	
Cattle	TDN %	*	14.9	58.3	
Sheep	TDN %	*	16.2	63.2	

Grass-Legume, aerial part, ensiled, few legumes, (3)

Ref No 3 08 599 United States

Analyses			As Fed	Dry	C.V. ± %
Dry matter	%		27.6	100.0	
Ash	%		2.5	9.1	
Crude fiber	%		9.7	35.1	
Ether extract	%		1.1	4.0	
N-free extract	%		11.1	40.2	
Protein (N x 6.25)	%		3.2	11.6	
Cattle	dig prot %	*	1.9	6.8	
Goats	dig prot %	*	1.9	6.8	
Horses	dig prot %	*	1.9	6.8	
Sheep	dig prot %	*	1.9	6.8	
Energy	GE Mcal/kg				
Cattle	DE Mcal/kg	*	0.71	2.57	

Continued

Feed Name or Analyses			Mean As Fed	Mean Dry	C.V. ± %
Sheep	DE Mcal/kg	*	0.75	2.72	
Cattle	ME Mcal/kg	*	0.58	2.10	
Sheep	ME Mcal/kg	*	0.62	2.23	
Cattle	TDN %	*	16.1	58.2	
Sheep	TDN %	*	17.0	61.7	

Grass-Legume, aerial part, wilted ensiled, (3)
Ref No 3 02 304 United States

Analyses			As Fed	Dry	C.V. ± %
Dry matter	%		33.0	100.0	9
Ash	%		3.0	9.2	15
Crude fiber	%		9.6	29.2	12
Ether extract	%		1.2	3.7	16
N-free extract	%		14.4	43.6	
Protein (N x 6.25)	%		4.7	14.3	11
Cattle	dig prot %	*	3.0	9.2	
Goats	dig prot %	*	3.0	9.2	
Horses	dig prot %	*	3.0	9.2	
Sheep	dig prot %	*	3.0	9.2	
Energy	GE Mcal/kg				
Cattle	DE Mcal/kg	*	0.88	2.68	
Sheep	DE Mcal/kg	*	0.91	2.76	
Cattle	ME Mcal/kg	*	0.72	2.19	
Sheep	ME Mcal/kg	*	0.75	2.26	
Cattle	TDN %	*	20.0	60.7	
Sheep	TDN %	*	20.6	62.5	
Calcium	%		0.24	0.73	
Phosphorus	%		0.12	0.35	

Grass, Legume, aerial part, wilted ensiled, mid-bloom, (3)
Ref No 3 02 265 United States

Analyses			As Fed	Dry	C.V. ± %
Dry matter	%		31.2	100.0	9
Ash	%		2.6	8.2	4
Crude fiber	%		10.4	33.2	2
Ether extract	%		1.0	3.3	6
N-free extract	%		13.4	42.9	
Protein (N x 6.25)	%		3.9	12.4	6
Cattle	dig prot %	*	2.3	7.5	
Goats	dig prot %	*	2.3	7.5	
Horses	dig prot %	*	2.3	7.5	
Sheep	dig prot %	*	2.3	7.5	
Energy	GE Mcal/kg				
Cattle	DE Mcal/kg	*	0.79	2.52	
Sheep	DE Mcal/kg	*	0.86	2.77	
Cattle	ME Mcal/kg	*	0.65	2.07	
Sheep	ME Mcal/kg	*	0.71	2.27	
Cattle	TDN %	*	17.9	57.2	
Sheep	TDN %	*	19.6	62.8	

Grass-Legume, aerial part, wilted ensiled, cut 2, (3)
Ref No 3 02 266 United States

Analyses			As Fed	Dry	C.V. ± %
Dry matter	%		30.8	100.0	7
Ash	%		2.8	9.1	7
Crude fiber	%		9.1	29.7	11
Ether extract	%		1.1	3.7	9
N-free extract	%		13.6	44.2	
Protein (N x 6.25)	%		4.1	13.3	10
Cattle	dig prot %	*	2.6	8.3	
Goats	dig prot %	*	2.6	8.3	
Horses	dig prot %	*	2.6	8.3	
Sheep	dig prot %	*	2.6	8.3	
Energy	GE Mcal/kg				
Cattle	DE Mcal/kg	*	0.83	2.70	
Sheep	DE Mcal/kg	*	0.85	2.76	
Cattle	ME Mcal/kg	*	0.68	2.21	
Sheep	ME Mcal/kg	*	0.70	2.27	
Cattle	TDN %	*	18.8	61.2	
Sheep	TDN %	*	19.3	62.7	

Grass-Legume, aerial part, wilted ensiled, considerable legumes, (3)
Ref No 3 08 441 United States

Analyses			As Fed	Dry	C.V. ± %
Dry matter	%		33.3	100.0	
Ash	%		3.8	11.4	
Crude fiber	%		8.8	26.4	
Ether extract	%		1.3	3.9	
N-free extract	%		14.2	42.6	
Protein (N x 6.25)	%		5.2	15.6	
Cattle	dig prot %	*	3.5	10.4	
Goats	dig prot %	*	3.5	10.4	
Horses	dig prot %	*	3.5	10.4	
Sheep	dig prot %	*	3.5	10.4	
Energy	GE Mcal/kg				
Cattle	DE Mcal/kg	*	0.88	2.63	
Sheep	DE Mcal/kg	*	0.83	2.50	
Cattle	ME Mcal/kg	*	0.72	2.16	
Sheep	ME Mcal/kg	*	0.68	2.05	
Cattle	TDN %	*	19.8	59.6	
Sheep	TDN %	*	18.9	56.7	

Grass-Legume, aerial part, wilted ensiled, few legumes, (3)
Ref No 3 08 442 United States

Analyses			As Fed	Dry	C.V. ± %
Dry matter	%		37.3	100.0	
Ash	%		2.7	7.2	
Crude fiber	%		14.1	37.8	
Ether extract	%		1.1	2.9	
N-free extract	%		15.4	41.3	
Protein (N x 6.25)	%		4.0	10.7	
Cattle	dig prot %	*	2.2	6.0	
Goats	dig prot %	*	2.2	6.0	
Horses	dig prot %	*	2.2	6.0	
Sheep	dig prot %	*	2.2	6.0	
Energy	GE Mcal/kg				
Cattle	DE Mcal/kg	*	0.91	2.43	
Sheep	DE Mcal/kg	*	1.00	2.68	
Cattle	ME Mcal/kg	*	0.74	1.99	
Sheep	ME Mcal/kg	*	0.82	2.20	
Cattle	TDN %	*	20.6	55.1	
Sheep	TDN %	*	22.7	60.9	
Calcium	%		0.32	0.86	
Phosphorus	%		0.08	0.21	

Grass-Legume, aerial part w barley added, ensiled, (3)
Ref No 3 02 305 United States

Analyses			As Fed	Dry	C.V. ± %
Dry matter	%		34.0	100.0	4
Ash	%		2.2	6.4	
Crude fiber	%		8.6	25.4	
Ether extract	%		1.4	4.0	
N-free extract	%		16.7	49.0	
Protein (N x 6.25)	%		5.2	15.2	
Cattle	dig prot %	*	3.4	10.0	
Goats	dig prot %	*	3.4	10.0	
Horses	dig prot %	*	3.4	10.0	
Sheep	dig prot %	*	3.4	10.0	
Energy	GE Mcal/kg				
Cattle	DE Mcal/kg	*	0.90	2.65	
Sheep	DE Mcal/kg	*	0.88	2.59	
Cattle	ME Mcal/kg	*	0.74	2.18	
Sheep	ME Mcal/kg	*	0.72	2.12	
Cattle	TDN %	*	20.5	60.2	
Sheep	TDN %	*	2.00	58.7	
Calcium	%		0.26	0.75	16
Phosphorus	%		0.12	0.35	3

Grass-Legume, aerial part w citrus pulp fines added, ensiled, (3)
Ref No 3 02 267 United States

Analyses			As Fed	Dry	C.V. ± %
Dry matter	%		26.2	100.0	
Ash	%		1.6	6.0	
Crude fiber	%		8.9	34.1	
Ether extract	%		1.3	4.8	
N-free extract	%		10.7	40.9	
Protein (N x 6.25)	%		3.7	14.2	
Cattle	dig prot %	*	2.4	9.1	
Goats	dig prot %	*	2.4	9.1	
Horses	dig prot %	*	2.4	9.1	
Sheep	dig prot %	*	2.4	9.1	
Energy	GE Mcal/kg				
Cattle	DE Mcal/kg	*	0.62	2.36	
Sheep	DE Mcal/kg	*	0.67	2.54	
Cattle	ME Mcal/kg	*	0.51	1.93	
Sheep	ME Mcal/kg	*	0.54	2.08	
Cattle	TDN %	*	14.0	53.5	
Sheep	TDN %	*	15.1	57.6	

Grass-Legume, aerial part w corn grits by-product added, ensiled, (3)
Ref No 3 02 270 United States

Analyses			As Fed	Dry	C.V. ± %
Dry matter	%		30.7	100.0	
Ash	%		1.6	5.1	
Crude fiber	%		9.2	30.1	
Ether extract	%		2.0	6.5	
N-free extract	%		14.4	46.8	
Protein (N x 6.25)	%		3.5	11.5	
Cattle	dig prot %	*	2.0	6.7	
Goats	dig prot %	*	2.0	6.7	
Horses	dig prot %	*	2.0	6.7	
Sheep	dig prot %	*	2.0	6.7	
Energy	GE Mcal/kg				
Cattle	DE Mcal/kg	*	0.83	2.70	
Sheep	DE Mcal/kg	*	0.86	2.80	
Cattle	ME Mcal/kg	*	0.68	2.22	
Sheep	ME Mcal/kg	*	0.71	2.30	
Cattle	TDN %	*	18.8	61.3	
Sheep	TDN %	*	19.5	63.6	

Grass-Legume, aerial part w grain added, ensiled, (3)
Ref No 3 02 268 United States

Analyses			As Fed	Dry	C.V. ± %
Dry matter	%		33.8	100.0	
Ash	%		2.2	6.5	
Crude fiber	%		8.6	25.4	
Ether extract	%		1.3	3.8	
N-free extract	%		16.6	49.2	

(1) dry forages and roughages (2) pasture, range plants, and forages fed green (3) silages (4) energy feeds (5) protein supplements (6) minerals (7) vitamins (8) additives

Feed Name or Analyses			Mean As Fed	Mean Dry	C.V. ± %
Protein (N x 6.25)	%		5.1	15.1	
Cattle	dig prot %	*	3.4	9.9	
Goats	dig prot %	*	3.4	9.9	
Horses	dig prot %	*	3.4	9.9	
Sheep	dig prot %	*	3.4	9.9	
Energy	GE Mcal/kg				
Cattle	DE Mcal/kg	*	0.90	2.65	
Sheep	DE Mcal/kg	*	0.87	2.58	
Cattle	ME Mcal/kg	*	0.74	2.18	
Sheep	ME Mcal/kg	*	0.72	2.12	
Cattle	TDN %	*	20.3	60.2	
Sheep	TDN %	*	19.8	58.5	

Grass-Legume, aerial part w grain added, wilted ensiled, (3)

Ref No 3 08 443 United States

			As Fed	Dry	C.V.
Dry matter	%		33.8	100.0	
Ash	%		2.2	6.5	
Crude fiber	%		8.6	25.4	
Ether extract	%		1.3	3.8	
N-free extract	%		16.6	49.1	
Protein (N x 6.25)	%		5.1	15.1	
Cattle	dig prot %	*	3.4	9.9	
Goats	dig prot %	*	3.4	9.9	
Horses	dig prot %	*	3.4	9.9	
Sheep	dig prot %	*	3.4	9.9	
Energy	GE Mcal/kg				
Cattle	DE Mcal/kg	*	0.90	2.65	
Sheep	DE Mcal/kg	*	0.93	2.74	
Cattle	ME Mcal/kg	*	0.74	2.18	
Sheep	ME Mcal/kg	*	0.26	2.25	
Cattle	TDN %	*	20.3	60.2	
Sheep	TDN %	*	21.0	61.1	
Calcium	%		0.25	0.74	
Phosphorus	%		0.12	0.36	

Grass-Legume, aerial part w grnd corn grain added, ensiled, (3)

Ref No 3 02 306 United States

		As Fed	Dry	C.V.
Dry matter	%	30.0	100.0	
Sulphur	%	0.08	0.27	

Grass-Legume, aerial part w phosphoric acid preservative added, ensiled, (3)

Ref No 3 02 269 United States

			As Fed	Dry	C.V.
Dry matter	%		27.0	100.0	6
Ash	%		2.0	7.4	6
Crude fiber	%		8.8	32.5	5
Ether extract	%		0.8	3.0	10
N-free extract	%		12.4	46.1	
Protein (N x 6.25)	%		3.0	11.0	11
Cattle	dig prot %	*	1.7	6.2	
Goats	dig prot %	*	1.7	6.2	
Horses	dig prot %	*	1.7	6.2	
Sheep	dig prot %	*	1.7	6.2	
Energy	GE Mcal/kg				
Cattle	DE Mcal/kg	*	0.69	2.56	
Sheep	DE Mcal/kg	*	0.76	2.82	
Cattle	ME Mcal/kg	*	0.57	2.10	
Sheep	ME Mcal/kg	*	0.62	2.31	
Cattle	TDN %	*	15.7	58.1	
Sheep	TDN %	*	17.3	63.9	

Grass-Legume, aerial part w molasses added, ensiled, (3)

Ref No 3 02 309 United States

			As Fed	Dry	C.V.
Dry matter	%		30.3	100.0	11
Ash	%		2.2	7.4	12
Crude fiber	%		9.4	31.1	9
Ether extract	%		1.0	3.3	20
N-free extract	%		14.2	47.0	
Protein (N x 6.25)	%		3.4	11.2	13
Cattle	dig prot %	*	1.9	6.4	
Goats	dig prot %	*	1.9	6.4	
Horses	dig prot %	*	1.9	6.4	
Sheep	dig prot %	*	1.9	6.4	
Energy	GE Mcal/kg				
Cattle	DE Mcal/kg	*	0.79	2.63	
Sheep	DE Mcal/kg	*	0.78	2.56	
Cattle	ME Mcal/kg	*	0.65	2.15	
Sheep	ME Mcal/kg	*	0.64	2.10	
Cattle	NE_m Mcal/kg	*	0.37	1.22	
Cattle	NE_{gain} Mcal/kg	*	0.17	0.56	
Cattle	$NE_{lactating\ cows}$ Mcal/kg	*	0.41	1.34	
Cattle	TDN %	*	18.0	59.6	
Sheep	TDN %	*	17.6	58.1	
Calcium	%		0.28	0.92	16
Phosphorus	%		0.08	0.26	17

Grass-Legume, aerial part w molasses added, ensiled, full bloom, (3)

Ref No 3 02 307 United States

		As Fed	Dry	C.V.
Dry matter	%	30.7	100.0	3
Calcium	%	0.32	1.04	8
Phosphorus	%	0.10	0.34	11

Grass-Legume, aerial part w molasses added, ensiled, cut 2, (3)

Ref No 3 02 308 United States

		As Fed	Dry	C.V.
Dry matter	%	31.6	100.0	
Calcium	%	0.35	1.11	
Phosphorus	%	0.12	0.38	

Grass-Legume, aerial part w molasses added, wilted ensiled, (3)

Ref No 3 08 445 United States

			As Fed	Dry	C.V.
Dry matter	%		31.3	100.0	
Ash	%		2.4	7.6	
Crude fiber	%		9.7	31.0	
Ether extract	%		1.0	3.3	
N-free extract	%		14.5	46.2	
Protein (N x 6.25)	%		3.7	11.9	
Cattle	dig prot %	*	2.2	7.0	
Goats	dig prot %	*	2.2	7.0	
Horses	dig prot %	*	2.2	7.0	
Sheep	dig prot %	*	2.2	7.0	
Energy	GE Mcal/kg				
Cattle	DE Mcal/kg	*	0.82	2.61	
Sheep	DE Mcal/kg	*	0.78	2.49	
Cattle	ME Mcal/kg	*	0.67	2.14	
Sheep	ME Mcal/kg	*	0.64	2.04	
Cattle	TDN %	*	18.5	59.3	
Sheep	TDN %	*	17.7	56.5	
Phosphorus	%		0.08	0.24	

Grass-Legume, aerial part w sodium bisulfite preservative added, ensiled, (3)

Ref No 3 02 310 United States

		As Fed	Dry	C.V.
Dry matter	%	25.5	100.0	
Sulphur	%	0.19	0.76	

Grass-Legume, aerial part w sulfur dioxide preservative added, ensiled, (3)

Ref No 3 02 271 United States

			As Fed	Dry	C.V.
Dry matter	%		24.8	100.0	
Ash	%		1.5	6.2	
Crude fiber	%		8.6	34.6	
Ether extract	%		1.0	4.0	
N-free extract	%		10.7	43.2	
Protein (N x 6.25)	%		3.0	12.0	
Cattle	dig prot %	*	1.8	7.1	
Goats	dig prot %	*	1.8	7.1	
Horses	dig prot %	*	1.8	7.1	
Sheep	dig prot %	*	1.8	7.1	
Energy	GE Mcal/kg				
Cattle	DE Mcal/kg	*	0.60	2.41	
Sheep	DE Mcal/kg	*	0.58	2.35	
Cattle	ME Mcal/kg	*	0.49	1.98	
Sheep	ME Mcal/kg	*	0.48	1.93	
Cattle	TDN %	*	13.6	54.7	
Sheep	TDN %	*	13.2	53.4	

Grass-Legume, aerial part w sulfur dioxide preservative added, ensiled, full bloom, (3)

Ref No 3 02 311 United States

		As Fed	Dry	C.V.
Dry matter	%	27.7	100.0	
Sulphur	%	0.22	0.78	

GRASS-LEGUME, CAN EASTERN 4-5 YR POST SEEDING. Scientific name not used

Grass-Legume, Can Eastern 4-5 yr post seeding, hay, s-c, mid-bloom, (1)

Ref No 1 08 195 Canada

			As Fed	Dry	C.V.
Dry matter	%		92.5	100.0	
In vitro	dig coef %		49.	53.	
Protein (N x 6.25)	%		9.5	10.3	
Cattle	dig prot %	*	5.4	5.8	
Goats	dig prot %	*	5.7	6.1	
Horses	dig prot %	*	5.8	6.2	
Rabbits	dig prot %	*	6.1	6.6	
Sheep	dig prot %	*	5.3	5.8	
Cellulose (Matrone)	%		33.1	35.8	
Nutritive value index (NVI)	%		32.	34.	

Grass-Legume, Can Eastern 4-5 yr post seeding, hay, s-c, full bloom, (1)

Ref No 1 08 196 Canada

			As Fed	Dry	C.V.
Dry matter	%		92.5	100.0	0
In vitro	dig coef %		50.	54.	3
Protein (N x 6.25)	%		8.6	9.2	21
Cattle	dig prot %	*	4.6	4.9	
Goats	dig prot %	*	4.8	5.2	
Horses	dig prot %	*	5.0	5.4	

Continued

Feed Name or Analyses			Mean As Fed	Mean Dry	C.V. ± %
Rabbits	dig prot %	*	5.4	5.8	
Sheep	dig prot %	*	4.5	4.9	
Cellulose (Matrone)	%		32.9	35.5	5
Nutritive value index (NVI)	%		35.	38.	13

Grass-Legume, Can Eastern 4-5 yr post seeding, hay, s-c, late bloom, (1)
Ref No 1 08 192 Canada

Feed Name or Analyses			Mean As Fed	Mean Dry	C.V. ± %
Dry matter	%		92.5	100.0	0
In vitro	dig coef %		49.	53.	4
Protein (N x 6.25)	%		7.5	8.1	18
Cattle	dig prot %	*	3.7	4.0	
Goats	dig prot %	*	3.8	4.1	
Horses	dig prot %	*	4.1	4.4	
Rabbits	dig prot %	*	4.6	4.9	
Sheep	dig prot %	*	3.6	3.8	
Cellulose (Matrone)	%		32.0	34.6	7
Nutritive value index (NVI)	%		33.	36.	16

Grass-Legume, CE 4-5 years post seeding, hay, s-c, mx 75% grass, (1)
Ref No 1 09 082 Canada

Feed Name or Analyses			Mean As Fed	Mean Dry	C.V. ± %
Dry matter	%			100.0	
Ash	%			7.7	
Crude fiber	%			31.7	
Ether extract	%			2.5	
N-free extract	%			46.7	
Protein (N x 6.25)	%			11.4	
Sheep	dig coef %			57.	
Cattle	dig prot %	*		6.8	
Goats	dig prot %	*		7.2	
Horses	dig prot %	*		7.2	
Rabbits	dig prot %	*		7.5	
Sheep	dig prot %			6.5	
Energy	GE Mcal/kg				
Cattle	DE Mcal/kg	*		2.44	
Sheep	DE Mcal/kg	*		2.34	
Cattle	ME Mcal/kg	*		2.00	
Sheep	ME Mcal/kg	*		1.92	
Cattle	TDN %	*		55.4	
Sheep	TDN %			53.0	

GRASS-LEGUME. Scientific name not used

Grass-Legume, hay, s-c, gr sample US, (1)
Ref No 1 02 287 United States

Feed Name or Analyses			Mean As Fed	Mean Dry	C.V. ± %
Dry matter	%			100.0	
Calcium	%			0.66	35
Cobalt	mg/kg			0.124	56
Manganese	mg/kg			69.7	63
Phosphorus	%			0.20	37
Carotene	mg/kg			8.4	89
Vitamin A equivalent	IU/g			14.0	

Gray molly –
see Summercypress, gray aerial part, fresh, mature, (2)

Gray molly –
see Summercypress, gray, aerial part, fresh, dormant, (2)

GREASEWOOD. Sarcobatus spp

Greasewood, browse, (2)
Ref No 2 02 312 United States

Feed Name or Analyses			Mean As Fed	Mean Dry	C.V. ± %
Dry matter	%		50.0	100.0	
Ash	%		7.3	14.6	
Crude fiber	%		11.7	23.5	
Ether extract	%		1.7	3.4	
N-free extract	%		18.6	37.3	
Protein (N x 6.25)	%		10.7	21.4	
Cattle	dig prot %	*	8.0	16.0	
Goats	dig prot %	*	8.2	16.5	
Horses	dig prot %	*	7.8	15.7	
Rabbits	dig prot %	*	7.6	15.2	
Sheep	dig prot %	*	8.4	16.9	
Energy	GE Mcal/kg				
Cattle	DE Mcal/kg	*	1.01	2.02	
Sheep	DE Mcal/kg	*	1.04	2.09	
Cattle	ME Mcal/kg	*	0.83	1.66	
Sheep	ME Mcal/kg	*	0.86	1.71	
Cattle	TDN %	*	23.0	45.9	
Sheep	TDN %	*	23.6	47.3	
Calcium	%		0.46	0.91	
Cobalt	mg/kg		0.030	0.060	
Copper	mg/kg		7.8	15.7	
Manganese	mg/kg		12.9	25.8	
Phosphorus	%		0.09	0.18	
Carotene	mg/kg		21.7	43.4	
Vitamin A equivalent	IU/g		36.2	72.4	

Greasewood, buds, fresh, (2)
Ref No 2 02 313 United States

Feed Name or Analyses			Mean As Fed	Mean Dry	C.V. ± %
Dry matter	%			100.0	
Ash	%			16.3	
Crude fiber	%			9.3	
Ether extract	%			3.3	
N-free extract	%			36.8	
Protein (N x 6.25)	%			34.3	
Cattle	dig prot %	*		27.0	
Goats	dig prot %	*		28.6	
Horses	dig prot %	*		26.7	
Rabbits	dig prot %	*		25.1	
Sheep	dig prot %	*		29.0	
Energy	GE Mcal/kg				
Cattle	DE Mcal/kg	*		2.44	
Sheep	DE Mcal/kg	*		2.55	
Cattle	ME Mcal/kg	*		2.00	
Sheep	ME Mcal/kg	*		2.09	
Cattle	TDN %	*		55.4	
Sheep	TDN %	*		57.8	

GREENTHREAD. Thelesperma spp

Greenthread, browse, mid-bloom, (2)
Ref No 2 02 314 United States

Feed Name or Analyses			Mean As Fed	Mean Dry	C.V. ± %
Dry matter	%			100.0	
Ash	%			5.2	
Crude fiber	%			33.6	
Ether extract	%			1.6	
N-free extract	%			49.6	
Protein (N x 6.25)	%			10.0	
Cattle	dig prot %	*		6.4	
Goats	dig prot %	*		5.9	
Horses	dig prot %	*		6.0	
Rabbits	dig prot %	*		6.4	
Sheep	dig prot %	*		6.3	
Calcium	%			1.30	
Magnesium	%			0.19	
Phosphorus	%			0.15	
Potassium	%			1.60	

GROMWELL, WAYSIDE. Lithospermum ruderale

Gromwell, wayside, aerial part, fresh, (2)
Ref No 2 02 319 United States

Feed Name or Analyses			Mean As Fed	Mean Dry	C.V. ± %
Dry matter	%			100.0	
Ash	%			14.3	
Crude fiber	%			12.6	
Ether extract	%			3.9	
N-free extract	%			41.7	
Protein (N x 6.25)	%			27.5	
Cattle	dig prot %	*		21.3	
Goats	dig prot %	*		22.2	
Horses	dig prot %	*		20.9	
Rabbits	dig prot %	*		19.9	
Sheep	dig prot %	*		22.6	

Gromwell, wayside, aerial part, fresh, mature, (2)
Ref No 2 02 315 United States

Feed Name or Analyses			Mean As Fed	Mean Dry	C.V. ± %
Dry matter	%			100.0	
Ash	%			21.1	
Ether extract	%			4.0	
Protein (N x 6.25)	%			8.0	
Cattle	dig prot %	*		4.7	
Goats	dig prot %	*		4.0	
Horses	dig prot %	*		4.3	
Rabbits	dig prot %	*		4.8	
Sheep	dig prot %	*		4.4	

Gromwell, wayside, browse, immature, (2)
Ref No 2 02 316 United States

Feed Name or Analyses			Mean As Fed	Mean Dry	C.V. ± %
Dry matter	%			100.0	
Ash	%			15.3	
Crude fiber	%			15.8	
Ether extract	%			3.0	
N-free extract	%			48.7	
Protein (N x 6.25)	%			17.2	
Cattle	dig prot %	*		12.5	
Goats	dig prot %	*		12.6	
Horses	dig prot %	*		12.1	
Rabbits	dig prot %	*		11.9	
Sheep	dig prot %	*		13.0	
Calcium	%			2.52	
Magnesium	%			0.28	
Phosphorus	%			0.36	
Sulphur	%			0.29	

Gromwell, wayside, leaves, fresh, (2)
Ref No 2 02 317 United States

Feed Name or Analyses			Mean As Fed	Mean Dry	C.V. ± %
Dry matter	%			100.0	
Ash	%			26.9	
Ether extract	%			5.6	

(1) dry forages and roughages
(2) pasture, range plants, and forages fed green
(3) silages
(4) energy feeds
(5) protein supplements
(6) minerals
(7) vitamins
(8) additives

Feed Name or Analyses			Mean As Fed	Mean Dry	C.V. ± %
Protein (N x 6.25)		%		10.4	
Cattle	dig prot	% *		6.7	
Goats	dig prot	% *		6.3	
Horses	dig prot	% *		6.4	
Rabbits	dig prot	% *		6.7	
Sheep	dig prot	% *		6.7	
Calcium		%		5.93	
Phosphorus		%		0.41	

Gromwell, wayside, stems, fresh, (2)

Ref No 2 02 318 United States

Feed Name or Analyses			Mean As Fed	Mean Dry	C.V. ± %
Dry matter		%		100.0	
Ash		%		13.2	
Ether extract		%		1.8	
Protein (N x 6.25)		%		4.2	
Cattle	dig prot	% *		1.5	
Goats	dig prot	% *		0.5	
Horses	dig prot	% *		1.1	
Rabbits	dig prot	% *		1.9	
Sheep	dig prot	% *		0.9	
Calcium		%		2.14	
Phosphorus		%		0.25	

Ground corn (AAFCO) –
see Corn, grain, grnd, mx 4% foreign material, (4)

Ground corn cob (AAFCO) –
see Corn, cobs, grnd, (1)

Ground ear corn (AAFCO) –
see Corn, ears, grnd, (4)

Ground flaxseed (CFA) –
see Flax, seeds, grnd, commercial, (5)

Ground grain sorghum (AAFCO) –
see Sorghum, grain variety, grain, grnd, (4)

Groundnut oil meal, solvent extracted –
see Peanut, kernels, solv-extd grnd, mx 7% fiber, (5)

Ground paddy rice (AAFCO) –
see Rice, grain w hulls, grnd, (4)

Ground rough rice (AAFCO) –
see Rice, grain w hulls, grnd, (4)

GROUNDSEL, ARROWLEAF. Senecio triangularis

Groundsel, arrowleaf, aerial part, fresh, immature, (2)

Ref No 2 02 320 United States

Feed Name or Analyses			Mean As Fed	Mean Dry	C.V. ± %
Dry matter		%		100.0	
Ash		%		13.1	
Crude fiber		%		16.5	
Ether extract		%		3.0	
N-free extract		%		38.3	
Protein (N x 6.25)		%		29.1	
Cattle	dig prot	% *		22.6	
Goats	dig prot	% *		23.7	
Horses	dig prot	% *		22.2	
Rabbits	dig prot	% *		21.1	
Sheep	dig prot	% *		24.1	

Groundsel, butterweed, aerial part, fresh, (2)

Ref No 2 02 325 United States

Feed Name or Analyses			Mean As Fed	Mean Dry	C.V. ± %
Dry matter		%		100.0	
Ash		%		13.6	22
Crude fiber		%		13.3	30
Ether extract		%		3.6	32
N-free extract		%		48.6	
Protein (N x 6.25)		%		20.9	35
Cattle	dig prot	% *		15.7	
Goats	dig prot	% *		16.1	
Horses	dig prot	% *		15.3	
Rabbits	dig prot	% *		14.8	
Sheep	dig prot	% *		16.5	

Groundsel, butterweed, aerial part, fresh, immature, (2)

Ref No 2 02 321 United States

Feed Name or Analyses			Mean As Fed	Mean Dry	C.V. ± %
Dry matter		%		100.0	
Ash		%		14.7	
Crude fiber		%		12.5	
Ether extract		%		3.8	
N-free extract		%		43.8	
Protein (N x 6.25)		%		25.2	
Cattle	dig prot	% *		19.3	
Goats	dig prot	% *		20.1	
Horses	dig prot	% *		18.9	
Rabbits	dig prot	% *		18.1	
Sheep	dig prot	% *		20.5	
Calcium		%		1.56	
Magnesium		%		0.37	
Phosphorus		%		0.60	
Sulphur		%		0.53	

Groundsel, butterweed, aerial part, fresh, full bloom, (2)

Ref No 2 02 323 United States

Feed Name or Analyses			Mean As Fed	Mean Dry	C.V. ± %
Dry matter		%		100.0	
Ash		%		13.9	
Crude fiber		%		21.2	
Ether extract		%		4.6	
N-free extract		%		51.6	
Protein (N x 6.25)		%		8.7	
Cattle	dig prot	% *		5.3	
Goats	dig prot	% *		4.7	
Horses	dig prot	% *		4.9	
Rabbits	dig prot	% *		5.4	
Sheep	dig prot	% *		5.1	
Calcium		%		3.32	
Magnesium		%		0.46	
Phosphorus		%		0.47	
Sulphur		%		0.59	

Groundsel, butterweed, aerial part, fresh, over ripe, (2)

Ref No 2 02 324 United States

Feed Name or Analyses			Mean As Fed	Mean Dry	C.V. ± %
Dry matter		%		100.0	
Ash		%		6.5	
Ether extract		%		0.7	
Protein (N x 6.25)		%		3.3	
Cattle	dig prot	% *		0.7	
Goats	dig prot	% *		0.3	
Horses	dig prot	% *		0.3	
Rabbits	dig prot	% *		1.2	
Sheep	dig prot	% *		0.1	
Calcium		%		0.88	
Phosphorus		%		0.19	

Groundsel, butterweed, leaves, fresh, (2)

Ref No 2 02 326 United States

Feed Name or Analyses			Mean As Fed	Mean Dry	C.V. ± %
Dry matter		%		100.0	
Ash		%		12.0	
Ether extract		%		6.0	
Protein (N x 6.25)		%		13.2	
Cattle	dig prot	% *		9.1	
Goats	dig prot	% *		8.9	
Horses	dig prot	% *		8.7	
Rabbits	dig prot	% *		8.9	
Sheep	dig prot	% *		9.3	
Calcium		%		2.26	
Phosphorus		%		0.46	

Ground snapped corn –
see Corn, ears w husks, grnd, (4)

Ground soybean hay (AAFCO) –
see Soybean, hay, s-c, grnd, mx 33% fiber, (1)

GUAR. Cyamopsis tetragonoloba

Guar, hay, s-c, (1)

Ref No 1 02 327 United States

Feed Name or Analyses			Mean As Fed	Mean Dry	C.V. ± %
Dry matter		%	91.9	100.0	1
Ash		%	12.2	13.3	2
Crude fiber		%	20.4	22.2	3
Sheep	dig coef	%	45.	45.	
Ether extract		%	1.5	1.6	10
Sheep	dig coef	%	16.	16.	
N-free extract		%	41.9	45.6	
Sheep	dig coef	%	73.	73.	
Protein (N x 6.25)		%	15.9	17.3	4
Sheep	dig coef	%	75.	75.	
Cattle	dig prot	% *	11.0	11.9	
Goats	dig prot	% *	11.7	12.7	
Horses	dig prot	% *	11.2	12.2	
Rabbits	dig prot	% *	11.1	12.0	
Sheep	dig prot	%	11.9	13.0	
Fatty acids		%	1.7	1.8	
Energy	GE Mcal/kg				
Cattle	DE Mcal/kg	*	2.31	2.51	
Sheep	DE Mcal/kg	*	2.30	2.51	
Cattle	ME Mcal/kg	*	1.89	2.06	
Sheep	ME Mcal/kg	*	1.89	2.05	
Cattle	TDN	% *	52.4	57.0	
Sheep	TDN	%	52.2	56.8	

Guar, leaves, grnd, immature, (1)

Ref No 1 02 328 United States

Feed Name or Analyses			Mean As Fed	Mean Dry	C.V. ± %
Dry matter		%	92.7	100.0	
Ash		%	14.6	15.7	
Crude fiber		%	10.0	10.8	
Ether extract		%	2.7	2.9	
N-free extract		%	36.7	39.6	

Continued

Feed Name or Analyses			Mean As Fed	Mean Dry	C.V. ± %
rotein (N x 6.25)	%		28.7	31.0	
Cattle dig prot	%	*	22.0	23.8	
Goats dig prot	%	*	23.6	25.5	
Horses dig prot	%	*	22.1	23.8	
Rabbits dig prot	%	*	21.0	22.6	
Sheep dig prot	%	*	22.6	24.4	
Energy	GE Mcal/kg				
Cattle	DE Mcal/kg	*	2.38	2.57	
Sheep	DE Mcal/kg	*	2.35	2.54	
Cattle	ME Mcal/kg	*	1.96	2.11	
Sheep	ME Mcal/kg	*	1.93	2.08	
Cattle	TDN %	*	54.1	58.4	
Sheep	TDN %	*	53.3	57.5	

Guar, stems, (1)
Ref No 1 02 329 United States

Feed Name or Analyses			Mean As Fed	Mean Dry	C.V. ± %
Dry matter	%		93.9	100.0	
Ash	%		8.3	8.8	
Crude fiber	%		39.2	41.7	
Ether extract	%		0.9	1.0	
N-free extract	%		32.9	35.0	
Protein (N x 6.25)	%		12.7	13.5	
Cattle dig prot	%	*	8.1	8.6	
Goats dig prot	%	*	8.6	9.2	
Horses dig prot	%	*	8.4	9.0	
Rabbits dig prot	%	*	8.5	9.1	
Sheep dig prot	%	*	8.2	8.7	
Energy	GE Mcal/kg				
Cattle	DE Mcal/kg	*	2.10	2.23	
Sheep	DE Mcal/kg	*	2.10	2.23	
Cattle	ME Mcal/kg	*	1.72	1.83	
Sheep	ME Mcal/kg	*	1.72	1.83	
Cattle	TDN %	*	47.5	50.6	
Sheep	TDN %	*	47.6	50.6	

Guar, aerial part, fresh, (2)
Ref No 2 02 333 United States

Feed Name or Analyses			Mean As Fed	Mean Dry	C.V. ± %
Dry matter	%		19.2	100.0	
Ash	%		2.3	12.2	35
Crude fiber	%		4.4	22.8	14
Ether extract	%		0.5	2.5	39
N-free extract	%		9.2	47.9	
Protein (N x 6.25)	%		2.8	14.6	12
Cattle dig prot	%	*	2.0	10.3	
Goats dig prot	%	*	2.0	10.2	
Horses dig prot	%	*	1.9	9.9	
Rabbits dig prot	%	*	1.9	9.9	
Sheep dig prot	%	*	2.0	10.6	
Energy	GE Mcal/kg				
Cattle	DE Mcal/kg	*	0.45	2.35	
Sheep	DE Mcal/kg	*	0.46	2.42	
Cattle	ME Mcal/kg	*	0.37	1.93	
Sheep	ME Mcal/kg	*	0.38	1.98	
Cattle	TDN %	*	10.2	53.2	
Sheep	TDN %	*	10.5	54.8	

(1) dry forages and roughages (3) silages (6) minerals
(2) pasture, range plants, and forages fed green (4) energy feeds (7) vitamins
(5) protein supplements (8) additives

Guar, aerial part, fresh, immature, (2)
Ref No 2 02 330 United States

Feed Name or Analyses			Mean As Fed	Mean Dry	C.V. ± %
Dry matter	%			100.0	
Ash	%			9.8	
Crude fiber	%			21.5	
Ether extract	%			3.2	
N-free extract	%			50.9	
Protein (N x 6.25)	%			14.6	
Cattle dig prot	%	*		10.3	
Goats dig prot	%	*		10.2	
Horses dig prot	%	*		9.9	
Rabbits dig prot	%	*		9.9	
Sheep dig prot	%	*		10.6	
Energy	GE Mcal/kg				
Cattle	DE Mcal/kg	*		2.44	
Sheep	DE Mcal/kg	*		2.46	
Cattle	ME Mcal/kg	*		2.00	
Sheep	ME Mcal/kg	*		2.02	
Cattle	TDN %	*		55.4	
Sheep	TDN %	*		55.9	
Calcium	%			1.90	
Magnesium	%			0.42	
Phosphorus	%			0.17	
Potassium	%			1.92	

Guar, aerial part, fresh, pre-bloom, (2)
Ref No 2 02 331 United States

Feed Name or Analyses			Mean As Fed	Mean Dry	C.V. ± %
Dry matter	%		19.2	100.0	
Ash	%		3.3	17.0	
Crude fiber	%		4.4	22.7	
Cattle dig coef	%		26.	26.	
Ether extract	%		0.4	1.9	
Cattle dig coef	%		39.	39.	
N-free extract	%		8.1	42.4	
Cattle dig coef	%		70.	70.	
Protein (N x 6.25)	%		3.1	16.0	
Cattle dig coef	%		77.	77.	
Cattle dig prot	%		2.4	12.3	
Goats dig prot	%	*	2.2	11.5	
Horses dig prot	%	*	2.1	11.1	
Rabbits dig prot	%	*	2.1	11.0	
Sheep dig prot	%	*	2.3	11.9	
Energy	GE Mcal/kg				
Cattle	DE Mcal/kg	*	0.42	2.19	
Sheep	DE Mcal/kg	*	0.44	2.31	
Cattle	ME Mcal/kg	*	0.34	1.79	
Sheep	ME Mcal/kg	*	0.36	1.89	
Cattle	TDN %		9.5	49.6	
Sheep	TDN %	*	10.1	52.5	

Guar, aerial part, fresh, mature, (2)
Ref No 2 02 332 United States

Feed Name or Analyses			Mean As Fed	Mean Dry	C.V. ± %
Dry matter	%			100.0	
Ash	%			7.5	
Crude fiber	%			25.8	
Ether extract	%			2.4	
N-free extract	%			52.4	
Protein (N x 6.25)	%			11.9	
Cattle dig prot	%	*		8.0	
Goats dig prot	%	*		7.7	
Horses dig prot	%	*		7.6	
Rabbits dig prot	%	*		7.9	
Sheep dig prot	%	*		8.1	
Calcium	%			1.98	

Feed Name or Analyses			Mean As Fed	Mean Dry	C.V. ± %
Magnesium	%			0.29	
Phosphorus	%			0.16	
Potassium	%			1.91	

GUINEAGRASS. Panicum maximum

Guineagrass, hay, s-c, (1)
Ref No 1 02 336 United States

Feed Name or Analyses			Mean As Fed	Mean Dry	C.V. ± %
Dry matter	%		89.3	100.0	2
Ash	%		9.6	10.7	16
Crude fiber	%		28.1	31.5	8
Ether extract	%		1.1	1.2	15
N-free extract	%		44.3	49.6	
Protein (N x 6.25)	%		6.3	7.0	29
Cattle dig prot	%	*	2.7	3.0	
Goats dig prot	%	*	2.8	3.1	
Horses dig prot	%	*	3.1	3.5	
Rabbits dig prot	%	*	3.6	4.1	
Sheep dig prot	%	*	2.5	2.8	
Energy	GE Mcal/kg				
Cattle	DE Mcal/kg	*	2.09	2.34	
Sheep	DE Mcal/kg	*	2.19	2.45	
Cattle	ME Mcal/kg	*	1.71	1.92	
Sheep	ME Mcal/kg	*	1.80	2.01	
Cattle	TDN %	*	47.4	52.1	
Sheep	TDN %	*	49.7	55.7	

Guineagrass, hay, s-c, milk stage, (1)
Ref No 1 02 335 United States

Feed Name or Analyses			Mean As Fed	Mean Dry	C.V. ± %
Dry matter	%		88.9	100.0	2
Ash	%		9.7	10.9	16
Crude fiber	%		27.2	30.6	2
Ether extract	%		1.1	1.2	16
N-free extract	%		45.3	51.0	
Protein (N x 6.25)	%		5.6	6.3	12
Cattle dig prot	%	*	2.1	2.4	
Goats dig prot	%	*	2.2	2.4	
Horses dig prot	%	*	2.6	2.9	
Rabbits dig prot	%	*	3.1	3.5	
Sheep dig prot	%	*	2.0	2.2	
Energy	GE Mcal/kg				
Cattle	DE Mcal/kg	*	2.08	2.34	
Sheep	DE Mcal/kg	*	2.19	2.46	
Cattle	ME Mcal/kg	*	1.71	1.92	
Sheep	ME Mcal/kg	*	1.79	2.02	
Cattle	TDN %	*	47.2	53.1	
Sheep	TDN %	*	49.6	55.8	

Guineagrass, aerial part, fresh, (2)
Ref No 2 02 345 United States

Feed Name or Analyses			Mean As Fed	Mean Dry	C.V. ± %
Dry matter	%		27.0	100.0	7
Ash	%		3.0	11.1	17
Crude fiber	%		10.1	37.5	8
Cattle dig coef	%		57.	57.	
Sheep dig coef	%		59.	59.	
Ether extract	%		0.4	1.4	32
Cattle dig coef	%		46.	46.	
Sheep dig coef	%		52.	52.	
N-free extract	%		11.9	44.1	
Cattle dig coef	%		57.	57.	
Sheep dig coef	%		56.	56.	
Protein (N x 6.25)	%		1.6	5.9	24
Cattle dig coef	%		59.	59.	
Sheep dig coef	%		56.	56.	

Feed Name or Analyses			Mean As Fed	Mean Dry	C.V. ± %
Cattle	dig prot %		0.9	3.5	
Goats	dig prot %	*	0.6	2.1	
Horses	dig prot %	*	0.7	2.5	
Rabbits	dig prot %	*	0.9	3.2	
Sheep	dig prot %		0.9	3.3	
Energy	GE Mcal/kg				
Cattle	DE Mcal/kg	*	0.61	2.27	
Sheep	DE Mcal/kg	*	0.62	2.28	
Cattle	ME Mcal/kg	*	0.50	1.86	
Sheep	ME Mcal/kg	*	0.50	1.87	
Cattle	TDN %		13.9	51.4	
Sheep	TDN %		14.0	51.7	
Carotene	mg/kg		63.0	233.7	23
Vitamin A equivalent	IU/g		105.1	389.6	

Guineagrass, aerial part, fresh, immature, (2)

Ref No 2 02 337 United States

Feed Name or Analyses			Mean As Fed	Mean Dry	C.V. ± %
Dry matter	%		20.2	100.0	24
Ash	%		1.6	7.9	12
Crude fiber	%		6.6	32.8	7
Ether extract	%		0.3	1.6	12
N-free extract	%		9.8	48.5	
Protein (N x 6.25)	%		1.9	9.2	16
Cattle	dig prot %	*	1.2	5.7	
Goats	dig prot %	*	1.0	5.1	
Horses	dig prot %	*	1.1	5.3	
Rabbits	dig prot %	*	1.2	5.8	
Sheep	dig prot %	*	1.1	5.6	
Energy	GE Mcal/kg				
Cattle	DE Mcal/kg	*	0.50	2.50	
Sheep	DE Mcal/kg	*	0.50	2.46	
Cattle	ME Mcal/kg	*	0.41	2.05	
Sheep	ME Mcal/kg	*	0.41	2.03	
Cattle	TDN %	*	11.5	56.8	
Sheep	TDN %	*	11.3	55.7	
Carotene	mg/kg		56.1	277.8	
Vitamin A equivalent	IU/g		93.5	463.1	

Guineagrass, aerial part, fresh, pre-bloom, (2)

Ref No 2 02 338 United States

Feed Name or Analyses			Mean As Fed	Mean Dry	C.V. ± %
Dry matter	%		24.2	100.0	
Ash	%		1.9	7.9	
Crude fiber	%		7.9	32.8	
Sheep	dig coef %		69.	69.	
Ether extract	%		0.4	1.6	
Sheep	dig coef %		21.	21.	
N-free extract	%		11.7	48.5	
Sheep	dig coef %		61.	61.	
Protein (N x 6.25)	%		2.2	9.2	
Sheep	dig coef %		52.	52.	
Cattle	dig prot %	*	1.4	5.7	
Goats	dig prot %	*	1.2	5.1	
Horses	dig prot %	*	1.3	5.3	
Rabbits	dig prot %	*	1.4	5.8	
Sheep	dig prot %		1.2	4.8	
Energy	GE Mcal/kg				
Cattle	DE Mcal/kg	*	0.60	2.50	
Sheep	DE Mcal/kg	*	0.62	2.55	
Cattle	ME Mcal/kg	*	0.50	2.05	
Sheep	ME Mcal/kg	*	0.51	2.09	
Cattle	TDN %	*	13.7	56.8	
Sheep	TDN %		14.0	57.8	

Guineagrass, aerial part, fresh, early bloom, (2)

Ref No 2 02 339 United States

Feed Name or Analyses			Mean As Fed	Mean Dry	C.V. ± %
Dry matter	%		24.2	100.0	
Ash	%		1.4	5.6	
Crude fiber	%		10.1	41.8	
Sheep	dig coef %		61.	61.	
Ether extract	%		0.4	1.5	
Sheep	dig coef %		45.	45.	
N-free extract	%		11.0	45.5	
Sheep	dig coef %		51.	51.	
Protein (N x 6.25)	%		1.4	5.6	
Sheep	dig coef %		49.	49.	
Cattle	dig prot %	*	0.6	2.7	
Goats	dig prot %	*	0.4	1.8	
Horses	dig prot %	*	0.6	2.3	
Rabbits	dig prot %	*	0.7	3.0	
Sheep	dig prot %		0.7	2.7	
Energy	GE Mcal/kg				
Cattle	DE Mcal/kg	*	0.62	2.56	
Sheep	DE Mcal/kg	*	0.57	2.34	
Cattle	ME Mcal/kg	*	0.51	2.10	
Sheep	ME Mcal/kg	*	0.46	1.91	
Cattle	TDN %	*	14.0	58.0	
Sheep	TDN %		12.8	53.0	

Guineagrass, aerial part, fresh, full bloom, (2)

Ref No 2 02 341 United States

Feed Name or Analyses			Mean As Fed	Mean Dry	C.V. ± %
Dry matter	%		29.7	100.0	6
Ash	%		3.2	10.8	8
Crude fiber	%		10.2	34.4	8
Ether extract	%		0.7	2.5	11
N-free extract	%		13.6	45.7	
Protein (N x 6.25)	%		2.0	6.6	9
Cattle	dig prot %	*	1.0	3.5	
Goats	dig prot %	*	0.8	2.7	
Horses	dig prot %	*	0.9	3.1	
Rabbits	dig prot %	*	1.1	3.8	
Sheep	dig prot %	*	0.9	3.1	
Energy	GE Mcal/kg				
Cattle	DE Mcal/kg	*	0.77	2.58	
Sheep	DE Mcal/kg	*	0.73	2.46	
Cattle	ME Mcal/kg	*	0.63	2.12	
Sheep	ME Mcal/kg	*	0.60	2.02	
Cattle	TDN %	*	17.4	58.5	
Sheep	TDN %	*	16.6	55.9	

Guineagrass, aerial part, fresh, full bloom, cut 1, (2)

Ref No 2 02 340 United States

Feed Name or Analyses			Mean As Fed	Mean Dry	C.V. ± %
Dry matter	%		29.7	100.0	
Ash	%		3.2	10.8	
Crude fiber	%		10.5	35.4	
Sheep	dig coef %		57.	57.	
Ether extract	%		0.7	2.5	
Sheep	dig coef %		57.	57.	
N-free extract	%		13.6	45.7	
Sheep	dig coef %		56.	56.	
Protein (N x 6.25)	%		1.7	5.6	
Sheep	dig coef %		58.	58.	
Cattle	dig prot %	*	0.8	2.7	
Goats	dig prot %	*	0.5	1.8	
Horses	dig prot %	*	0.7	2.3	
Rabbits	dig prot %	*	0.9	3.0	
Sheep	dig prot %		1.0	3.2	
Energy	GE Mcal/kg				
Cattle	DE Mcal/kg	*	0.75	2.54	
Sheep	DE Mcal/kg	*	0.68	2.30	
Cattle	ME Mcal/kg	*	0.62	2.08	
Sheep	ME Mcal/kg	*	0.56	1.89	
Cattle	TDN %	*	17.1	57.6	
Sheep	TDN %		15.5	52.2	

Guineagrass, aerial part, fresh, mature, (2)

Ref No 2 02 342 United States

Feed Name or Analyses			Mean As Fed	Mean Dry	C.V. ± %
Dry matter	%		26.2	100.0	7
Ash	%		3.2	12.3	7
Crude fiber	%		11.0	41.9	12
Ether extract	%		0.3	1.2	16
N-free extract	%		10.4	39.7	
Protein (N x 6.25)	%		1.3	4.9	15
Cattle	dig prot %	*	0.5	2.1	
Goats	dig prot %	*	0.3	1.1	
Horses	dig prot %	*	0.4	1.7	
Rabbits	dig prot %	*	0.6	2.5	
Sheep	dig prot %	*	0.4	1.6	
Energy	GE Mcal/kg				
Cattle	DE Mcal/kg	*	0.67	2.55	
Sheep	DE Mcal/kg	*	0.63	2.42	
Cattle	ME Mcal/kg	*	0.55	2.09	
Sheep	ME Mcal/kg	*	0.52	1.98	
Cattle	TDN %	*	15.1	57.8	
Sheep	TDN %	*	14.4	54.8	

Guineagrass, aerial part, fresh, over ripe, (2)

Ref No 2 02 343 United States

Feed Name or Analyses			Mean As Fed	Mean Dry	C.V. ± %
Dry matter	%			100.0	
Calcium	%			2.09	
Phosphorus	%			0.59	

Guineagrass, aerial part, fresh, cut 1, (2)

Ref No 2 02 344 United States

Feed Name or Analyses			Mean As Fed	Mean Dry	C.V. ± %
Dry matter	%		29.7	100.0	6
Ash	%		3.2	10.8	8
Crude fiber	%		10.2	34.4	8
Ether extract	%		0.7	2.5	11
N-free extract	%		13.6	45.7	
Protein (N x 6.25)	%		2.0	6.6	9
Cattle	dig prot %	*	1.0	3.5	
Goats	dig prot %	*	0.8	2.7	
Horses	dig prot %	*	0.9	3.1	
Rabbits	dig prot %	*	1.1	3.8	
Sheep	dig prot %	*	0.9	3.1	
Energy	GE Mcal/kg				
Cattle	DE Mcal/kg	*	0.77	2.58	
Sheep	DE Mcal/kg	*	0.73	2.46	
Cattle	ME Mcal/kg	*	0.63	2.12	
Sheep	ME Mcal/kg	*	0.60	2.02	
Cattle	TDN %	*	17.4	58.5	
Sheep	TDN %	*	16.6	55.9	

Guineagrass, aerial part, ensiled, early bloom, (3)

Ref No 3 02 347 United States

Feed Name or Analyses			Mean As Fed	Mean Dry	C.V. ± %
Dry matter	%		20.6	100.0	
Ash	%		4.0	19.6	

Continued

Feed Name or Analyses				Mean As Fed	Mean Dry	C.V. ± %
Crude fiber		%		8.2	39.7	
Sheep	dig coef	%		72.	72.	
Ether extract		%		0.6	2.7	
Sheep	dig coef	%		35.	35.	
N-free extract		%		6.5	31.7	
Sheep	dig coef	%		45.	45.	
Protein (N x 6.25)		%		1.3	6.3	
Sheep	dig coef	%		31.	31.	
Cattle	dig prot	%	*	0.4	2.0	
Goats	dig prot	%	*	0.4	2.0	
Horses	dig prot	%	*	0.4	2.0	
Sheep	dig prot	%		0.4	2.0	
Energy	GE Mcal/kg					
Cattle	DE Mcal/kg		*	0.46	2.25	
Sheep	DE Mcal/kg		*	0.43	2.07	
Cattle	ME Mcal/kg		*	0.38	1.84	
Sheep	ME Mcal/kg		*	0.35	1.70	
Cattle	TDN	%	*	10.5	51.0	
Sheep	TDN	%		9.7	46.9	

Guineagrass, aerial part, ensiled, milk stage, (3)
Ref No 3 02 348 United States

Feed Name or Analyses			Mean As Fed	Mean Dry	C.V. ± %
Dry matter	%			100.0	
Carotene	mg/kg			93.7	30
Vitamin A equivalent	IU/g			156.2	

GUINEAGRASS-KUDZU. Panicum maximum, Pueraria spp

Guineagrass-kudzu, aerial part, fresh, (2)
Ref No 2 02 350 United States

Feed Name or Analyses				Mean As Fed	Mean Dry	C.V. ± %
Dry matter		%		22.9	100.0	14
Ash		%		2.3	10.0	14
Crude fiber		%		7.8	34.2	8
Ether extract		%		0.3	1.4	34
N-free extract		%		10.3	44.8	
Protein (N x 6.25)		%		2.2	9.6	17
Cattle	dig prot	%	*	1.4	6.1	
Goats	dig prot	%	*	1.3	5.5	
Horses	dig prot	%	*	1.3	5.7	
Rabbits	dig prot	%	*	1.4	6.1	
Sheep	dig prot	%	*	1.4	5.9	

HACKBERRY, NETLEAF. Celtis reticulata

Hackberry, netleaf, browse, (2)
Ref No 2 02 351 United States

Feed Name or Analyses				Mean As Fed	Mean Dry	C.V. ± %
Dry matter		%			100.0	
Ash		%			15.8	
Crude fiber		%			18.5	
Ether extract		%			1.8	
N-free extract		%			48.1	
Protein (N x 6.25)		%			15.8	
Cattle	dig prot	%	*		11.3	
Goats	dig prot	%	*		11.3	
Horses	dig prot	%	*		10.9	
Rabbits	dig prot	%	*		10.9	
Sheep	dig prot	%	*		11.7	
Calcium		%			4.91	

(1) dry forages and roughages (2) pasture, range plants, and forages fed green (3) silages (4) energy feeds (5) protein supplements (6) minerals (7) vitamins (8) additives

Feed Name or Analyses		Mean As Fed	Mean Dry	C.V. ± %
Magnesium	%		0.46	
Phosphorus	%		0.18	
Potassium	%		1.38	

HAIRGRASS, SILVER. Aira carophyllea

Hairgrass, silver, aerial part, fresh, (2)
Ref No 2 08 566 United States

Feed Name or Analyses				Mean As Fed	Mean Dry	C.V. ± %
Dry matter		%			100.0	
Ash		%			6.7	
Ether extract		%			1.9	
Protein (N x 6.25)		%			2.7	
Cattle	dig prot	%	*		0.2	
Goats	dig prot	%	*		0.9	
Horses	dig prot	%	*		0.1	
Rabbits	dig prot	%	*		0.7	
Sheep	dig prot	%	*		0.4	
Other carbohydrates		%			39.8	
Lignin (Ellis)		%			8.6	
Silicon		%			2.90	

HARDINGGRASS. Phalaris tuberosa stenoptera

Hardinggrass, hay, s-c, (1)
Ref No 1 02 353 United States

Feed Name or Analyses				Mean As Fed	Mean Dry	C.V. ± %
Dry matter		%		93.2	100.0	
Ash		%		6.4	6.8	
Crude fiber		%		32.6	35.0	
Ether extract		%		1.7	1.9	
N-free extract		%		46.7	50.1	
Protein (N x 6.25)		%		5.8	6.2	
Cattle	dig prot	%	*	2.1	2.3	
Goats	dig prot	%	*	2.2	2.3	
Horses	dig prot	%	*	2.6	2.8	
Rabbits	dig prot	%	*	3.2	3.4	
Sheep	dig prot	%	*	2.0	2.1	
Fatty acids		%		1.6	1.7	
Energy	GE Mcal/kg					
Cattle	DE Mcal/kg		*	2.23	2.40	
Sheep	DE Mcal/kg		*	2.23	2.39	
Cattle	ME Mcal/kg		*	1.83	1.97	
Sheep	ME Mcal/kg		*	1.83	1.96	
Cattle	TDN	%	*	50.7	54.4	
Sheep	TDN	%	*	50.5	54.2	

Hardinggrass, hay, s-c, mature, (1)
Ref No 1 02 352 United States

Feed Name or Analyses				Mean As Fed	Mean Dry	C.V. ± %
Dry matter		%			100.0	
Ash		%			5.4	
Crude fiber		%			34.3	
Ether extract		%			2.2	
N-free extract		%			54.3	
Protein (N x 6.25)		%			3.8	
Cattle	dig prot	%	*		0.2	
Goats	dig prot	%	*		0.1	
Horses	dig prot	%	*		0.8	
Rabbits	dig prot	%	*		1.6	
Sheep	dig prot	%	*		0.0	
Energy	GE Mcal/kg					
Cattle	DE Mcal/kg		*		2.43	
Sheep	DE Mcal/kg		*		2.37	
Cattle	ME Mcal/kg		*		1.99	
Sheep	ME Mcal/kg		*		1.95	

Feed Name or Analyses				Mean As Fed	Mean Dry	C.V. ± %
Cattle	TDN	%	*		55.0	
Sheep	TDN	%	*		53.8	

Hardinggrass, aerial part, fresh, (2)
Ref No 2 02 354 United States

Feed Name or Analyses		Mean As Fed	Mean Dry	C.V. ± %
Dry matter	%		100.0	
Calcium	%		0.29	
Phosphorus	%		0.12	

HAWKSBEARD, GRAY. Crepis acuminata

Hawksbeard, gray, aerial part w heads, s-c, dormant, (1)
Ref No 1 02 355 United States

Feed Name or Analyses				Mean As Fed	Mean Dry	C.V. ± %
Dry matter		%		93.3	100.0	
Ash		%		8.8	9.4	
Crude fiber		%		23.5	25.2	
Sheep	dig coef	%		36.	36.	
Ether extract		%		3.4	3.6	
Sheep	dig coef	%		33.	33.	
N-free extract		%		49.3	52.8	
Sheep	dig coef	%		78.	78.	
Protein (N x 6.25)		%		8.4	9.0	
Sheep	dig coef	%		63.	63.	
Cattle	dig prot	%	*	4.4	4.7	
Goats	dig prot	%	*	4.6	5.0	
Horses	dig prot	%	*	4.8	5.2	
Rabbits	dig prot	%	*	5.2	5.6	
Sheep	dig prot	%		5.3	5.7	
Energy	GE Mcal/kg					
Cattle	DE Mcal/kg		*	2.48	2.66	
Sheep	DE Mcal/kg		*	2.41	2.58	
Cattle	ME Mcal/kg		*	2.03	2.18	
Sheep	ME Mcal/kg		*	1.98	2.12	
Cattle	TDN	%	*	56.3	60.3	
Sheep	TDN	%		54.7	58.6	

HAWKSBEARD, TAPERTIP. Crepis acuminata

Hawksbeard, tapertip, aerial part, fresh, (2)
Ref No 2 02 358 United States

Feed Name or Analyses				Mean As Fed	Mean Dry	C.V. ± %
Dry matter		%			100.0	
Protein (N x 6.25)		%			14.1	34
Cattle	dig prot	%	*		9.9	
Goats	dig prot	%	*		9.7	
Horses	dig prot	%	*		9.5	
Rabbits	dig prot	%	*		9.6	
Sheep	dig prot	%	*		10.1	
Calcium		%			1.47	37
Phosphorus		%			0.40	33

Hawksbeard, tapertip, aerial part, fresh, immature, (2)
Ref No 2 02 356 United States

Feed Name or Analyses				Mean As Fed	Mean Dry	C.V. ± %
Dry matter		%			100.0	
Protein (N x 6.25)		%			16.4	18
Cattle	dig prot	%	*		11.8	
Goats	dig prot	%	*		11.9	
Horses	dig prot	%	*		11.5	
Rabbits	dig prot	%	*		11.3	
Sheep	dig prot	%	*		12.3	

Hawksbeard, tapertip, aerial part, fresh, early bloom, (2)

Ref No 2 02 357 — United States

Feed Name or Analyses			Mean As Fed	Mean Dry	C.V. ± %
Dry matter	%			100.0	
Calcium	%			1.81	
Phosphorus	%			0.47	

HAY-POTATO. Scientific name not used, Solanum tuberosum

Hay-Potato, aerial part and tubers, ensiled, (3)

Ref No 3 03 792 — United States

Feed Name or Analyses			Mean As Fed	Mean Dry	C.V. ± %
Dry matter	%		33.7	100.0	
Ash	%		2.0	5.9	
Crude fiber	%		6.0	17.8	
Ether extract	%		0.8	2.4	
N-free extract	%		21.1	62.6	
Protein (N x 6.25)	%		3.8	11.3	
Cattle	dig prot %	*	2.2	6.5	
Goats	dig prot %	*	2.2	6.5	
Horses	dig prot %	*	2.2	6.5	
Sheep	dig prot %	*	2.2	6.5	
Energy	GE Mcal/kg				
Cattle	DE Mcal/kg	*	0.95	2.83	
Sheep	DE Mcal/kg	*	0.94	2.79	
Cattle	ME Mcal/kg	*	0.78	2.32	
Sheep	ME Mcal/kg	*	0.77	2.29	
Cattle	TDN %	*	21.6	64.1	
Sheep	TDN %	*	21.3	63.2	

HEATH, CROSSLEAF. Erica tetralix

Heath, crossleaf, aerial part, s-c, (1)

Ref No 1 02 359 — United States

Feed Name or Analyses			Mean As Fed	Mean Dry	C.V. ± %
Dry matter	%		88.8	100.0	
Ash	%		5.8	6.5	
Crude fiber	%		25.9	29.2	
Sheep	dig coef %		12.	12.	
Ether extract	%		10.1	11.4	
Sheep	dig coef %		12.	12.	
N-free extract	%		41.6	46.9	
Sheep	dig coef %		50.	50.	
Protein (N x 6.25)	%		5.3	6.0	
Sheep	dig coef %		15.	15.	
Cattle	dig prot %	*	1.9	2.1	
Goats	dig prot %	*	1.9	2.2	
Horses	dig prot %	*	2.3	2.6	
Rabbits	dig prot %	*	2.9	3.3	
Sheep	dig prot %		0.8	0.9	
Energy	GE Mcal/kg				
Cattle	DE Mcal/kg	*	1.31	1.48	
Sheep	DE Mcal/kg	*	1.14	1.28	
Cattle	ME Mcal/kg	*	1.07	1.21	
Sheep	ME Mcal/kg	*	0.93	1.05	
Cattle	TDN %	*	29.7	33.5	
Sheep	TDN %		25.8	29.1	

HEATHER, SCOTCH. Calluna vulgaris

Heather, Scotch, aerial part, s-c, (1)

Ref No 1 02 360 — United States

Feed Name or Analyses			Mean As Fed	Mean Dry	C.V. ± %
Dry matter	%		89.0	100.0	
Ash	%		7.6	8.5	
Crude fiber	%		19.9	22.4	
Sheep	dig coef %		21.	21.	
Ether extract	%		7.9	8.9	
Sheep	dig coef %		16.	16.	
N-free extract	%		48.1	54.0	
Sheep	dig coef %		68.	68.	
Protein (N x 6.25)	%		5.5	6.2	
Sheep	dig coef %		6.	6.	
Cattle	dig prot %	*	2.1	2.3	
Goats	dig prot %	*	2.1	2.3	
Horses	dig prot %	*	2.5	2.8	
Rabbits	dig prot %	*	3.1	3.5	
Sheep	dig prot %		0.3	0.4	
Energy	GE Mcal/kg				
Cattle	DE Mcal/kg	*	1.99	2.24	
Sheep	DE Mcal/kg	*	1.77	1.98	
Cattle	ME Mcal/kg	*	1.64	1.84	
Sheep	ME Mcal/kg	*	1.45	1.63	
Cattle	TDN %	*	45.1	50.7	
Sheep	TDN %		40.1	45.0	

HELIANTHELLA, ONEFLOWER. Helianthella uniflora

Helianthella, oneflower, aerial part, fresh, (2)

Ref No 2 02 362 — United States

Feed Name or Analyses			Mean As Fed	Mean Dry	C.V. ± %
Dry matter	%			100.0	
Ash	%			9.4	
Ether extract	%			4.3	
Protein (N x 6.25)	%			7.4	
Cattle	dig prot %	*		4.2	
Goats	dig prot %	*		3.5	
Horses	dig prot %	*		3.8	
Rabbits	dig prot %	*		4.4	
Sheep	dig prot %	*		3.9	
Calcium	%			2.03	
Phosphorus	%			0.31	

Helianthella, oneflower, aerial part, fresh, early leaf, (2)

One-flowered sunflower

Ref No 2 08 838 — United States

Feed Name or Analyses			Mean As Fed	Mean Dry	C.V. ± %
Dry matter	%			100.0	
Crude fiber	%			13.0	
Ether extract	%			3.0	
N-free extract	%			44.0	
Protein (N x 6.25)	%			20.0	
Cattle	dig prot %	*		14.9	
Goats	dig prot %	*		15.2	
Horses	dig prot %	*		14.5	
Rabbits	dig prot %	*		14.1	
Sheep	dig prot %	*		15.6	

Helianthella, oneflower, aerial part, fresh, immature, (2)

Ref No 2 02 361 — United States

Feed Name or Analyses			Mean As Fed	Mean Dry	C.V. ± %
Dry matter	%			100.0	
Ash	%			15.2	
Crude fiber	%			15.0	
Ether extract	%			1.3	
N-free extract	%			48.5	
Protein (N x 6.25)	%			20.0	
Cattle	dig prot %	*		14.9	
Goats	dig prot %	*		15.2	
Horses	dig prot %	*		14.5	
Rabbits	dig prot %	*		14.1	
Sheep	dig prot %	*		15.6	

Helianthella, oneflower, aerial part, fresh, mature, (2)

One-flowered sunflower

Ref No 2 08 839 — United States

Feed Name or Analyses			Mean As Fed	Mean Dry	C.V. ± %
Dry matter	%			100.0	
Crude fiber	%			21.0	
Ether extract	%			4.0	
N-free extract	%			50.0	
Protein (N x 6.25)	%			13.0	
Cattle	dig prot %	*		8.9	
Goats	dig prot %	*		8.7	
Horses	dig prot %	*		8.6	
Rabbits	dig prot %	*		8.7	
Sheep	dig prot %	*		9.1	

Helianthella, oneflower, aerial part, fresh, dormant, (2)

One-flowered sunflower

Ref No 2 08 840 — United States

Feed Name or Analyses			Mean As Fed	Mean Dry	C.V. ± %
Dry matter	%			100.0	
Crude fiber	%			25.0	
Ether extract	%			3.0	
N-free extract	%			50.0	
Protein (N x 6.25)	%			5.0	
Cattle	dig prot %	*		2.1	
Goats	dig prot %	*		1.2	
Horses	dig prot %	*		1.8	
Rabbits	dig prot %	*		2.5	
Sheep	dig prot %	*		1.7	

Helianthella, oneflower, heads, fresh, (2)

Ref No 2 02 363 — United States

Feed Name or Analyses			Mean As Fed	Mean Dry	C.V. ± %
Dry matter	%			100.0	
Ash	%			9.6	
Ether extract	%			6.1	
Protein (N x 6.25)	%			10.7	
Cattle	dig prot %	*		7.0	
Goats	dig prot %	*		6.5	
Horses	dig prot %	*		6.6	
Rabbits	dig prot %	*		6.9	
Sheep	dig prot %	*		7.0	
Calcium	%			1.97	
Phosphorus	%			0.34	

Feed Name or Analyses		Mean As Fed	Mean Dry	C.V. ± %

Helianthella, oneflower, leaves, fresh, (2)

Ref No 2 02 364 United States

			As Fed	Dry	C.V.
Dry matter		%		100.0	
Ash		%		12.5	
Ether extract		%		6.5	
Protein (N x 6.25)		%		9.8	
Cattle	dig prot	% *		6.2	
Goats	dig prot	% *		5.7	
Horses	dig prot	% *		5.9	
Rabbits	dig prot	% *		6.2	
Sheep	dig prot	% *		6.1	
Calcium		%		2.86	
Phosphorus		%		0.40	

Helianthella, oneflower, stems, fresh, (2)

Ref No 2 02 365 United States

			As Fed	Dry	C.V.
Dry matter		%		100.0	
Ash		%		5.0	
Ether extract		%		1.0	
Protein (N x 6.25)		%		2.8	
Cattle	dig prot	% *		0.3	
Goats	dig prot	% *		0.7	
Horses	dig prot	% *		0.0	
Rabbits	dig prot	% *		0.8	
Sheep	dig prot	% *		0.3	
Calcium		%		1.03	
Phosphorus		%		0.17	

HEMP. Cannabis sativa

Hemp, seed screenings, (4)

Ref No 4 02 366 United States

			As Fed	Dry	C.V.
Dry matter		%	88.3	100.0	
Ash		%	16.2	18.4	
Crude fiber		%	22.0	24.9	
Sheep	dig coef	%	27.	27.	
Swine	dig coef	%	7.	7.	
Ether extract		%	4.3	4.9	
Sheep	dig coef	%	25.	25.	
Swine	dig coef	%	6.	6.	
N-free extract		%	26.7	30.2	
Sheep	dig coef	%	43.	43.	
Swine	dig coef	%	25.	25.	
Protein (N x 6.25)		%	19.1	21.7	
Sheep	dig coef	%	58.	58.	
Swine	dig coef	%	10.	10.	
Cattle	dig prot	% *	14.0	15.9	
Goats	dig prot	% *	15.1	17.1	
Horses	dig prot	% *	15.1	17.1	
Sheep	dig prot	%	11.1	12.6	
Swine	dig prot	%	2.0	2.3	
Energy	GE Mcal/kg				
Cattle	DE Mcal/kg	*	1.43	1.62	
Sheep	DE Mcal/kg	*	1.36	1.54	
Swine	DE kcal/kg	*	295.	334.	
Cattle	ME Mcal/kg	*	1.17	1.33	
Sheep	ME Mcal/kg	*	1.12	1.27	
Swine	ME kcal/kg	*	270.	306.	
Cattle	TDN	% *	32.4	36.7	

(1) dry forages and roughages (2) pasture, range plants, and forages fed green (3) silages (4) energy feeds (5) protein supplements (6) minerals (7) vitamins (8) additives

Feed Name or Analyses		Mean As Fed	Mean Dry	C.V. ± %

			As Fed	Dry	C.V.
Sheep	TDN	%	30.9	35.0	
Swine	TDN	%	6.7	7.6	

Hemp, seeds, extn unspecified grnd, (5)

Hempseed oil meal, extraction unspecified

Ref No 5 02 367 United States

			As Fed	Dry	C.V.
Dry matter		%	91.6	100.0	2
Ash		%	8.7	9.5	9
Crude fiber		%	24.1	26.3	6
Sheep	dig coef	% *	5.	5.	
Ether extract		%	5.9	6.4	33
Sheep	dig coef	% *	80.	80.	
N-free extract		%	22.2	24.3	
Sheep	dig coef	% *	26.	26.	
Protein (N x 6.25)		%	30.7	33.5	3
Sheep	dig coef	% *	81.	81.	
Sheep	dig prot	%	24.9	27.2	
Energy	GE Mcal/kg				
Cattle	DE Mcal/kg	*	1.92	2.10	
Sheep	DE Mcal/kg	*	1.87	2.04	
Swine	DE kcal/kg	*	2060.	2249.	
Cattle	ME Mcal/kg	*	1.58	1.72	
Sheep	ME Mcal/kg	*	1.53	1.68	
Swine	ME kcal/kg	*	1837.	2006.	
Cattle	TDN	% *	43.7	47.7	
Sheep	TDN	%	42.4	46.3	
Swine	TDN	% *	46.7	51.0	
Calcium		%	0.24	0.26	7
Magnesium		%	0.76	0.84	
Phosphorus		%	0.61	0.67	38
Pantothenic acid		mg/kg	3.5	3.8	98
Riboflavin		mg/kg	2.8	3.1	

HEMP, DARK GREEN. Cannabis spp

Hemp, dark green, seeds, extn unspecified coarse grnd, (5)

Ref No 5 02 369 United States

			As Fed	Dry	C.V.
Dry matter		%	91.1	100.0	
Ash		%	9.3	10.2	
Crude fiber		%	24.3	26.7	
Sheep	dig coef	%	13.	13.	
Ether extract		%	7.3	8.0	
Sheep	dig coef	%	88.	88.	
N-free extract		%	19.7	21.6	
Sheep	dig coef	%	14.	14.	
Protein (N x 6.25)		%	30.5	33.5	
Sheep	dig coef	%	84.	84.	
Sheep	dig prot	%	25.6	28.1	
Energy	GE Mcal/kg				
Cattle	DE Mcal/kg	*	2.14	2.35	
Sheep	DE Mcal/kg	*	2.03	2.23	
Swine	DE kcal/kg	*	2354.	2584.	
Cattle	ME Mcal/kg	*	1.76	1.93	
Sheep	ME Mcal/kg	*	1.66	1.82	
Swine	ME kcal/kg	*	2100.	2305.	
Cattle	TDN	% *	48.5	53.2	
Sheep	TDN	%	46.0	50.5	
Swine	TDN	% *	53.4	58.6	

Feed Name or Analyses		Mean As Fed	Mean Dry	C.V. ± %

HEMP, LIGHT BROWN. Cannabis spp

Hemp, light brown, seeds, extn unspecified fine grnd, (5)

Ref No 5 02 370 United States

			As Fed	Dry	C.V.
Dry matter		%	92.8	100.0	
Ash		%	8.7	9.4	
Crude fiber		%	23.3	25.1	
Sheep	dig coef	%	6.	6.	
Ether extract		%	5.1	5.5	
Sheep	dig coef	%	82.	82.	
N-free extract		%	24.3	26.2	
Sheep	dig coef	%	15.	15.	
Protein (N x 6.25)		%	31.4	33.8	
Sheep	dig coef	%	84.	84.	
Sheep	dig prot	%	26.3	28.4	
Energy	GE Mcal/kg				
Cattle	DE Mcal/kg	*	1.98	2.13	
Sheep	DE Mcal/kg	*	1.80	1.94	
Swine	DE kcal/kg	*	2337.	2518.	
Cattle	ME Mcal/kg	*	1.62	1.75	
Sheep	ME Mcal/kg	*	1.48	1.59	
Swine	ME kcal/kg	*	2084.	2246.	
Cattle	TDN	% *	44.7	48.2	
Sheep	TDN	%	40.8	44.0	
Swine	TDN	% *	53.0	57.1	

Hempseed oil meal, extraction unspecified — see Hemp, seeds, extn unspecified grnd, (5)

HERONBILL. Erodium spp

Heronbill, hay, s-c, (1)

Ref No 1 02 372 United States

			As Fed	Dry	C.V.
Dry matter		%	85.6	100.0	
Ash		%	7.9	9.2	
Crude fiber		%	29.3	34.2	
Sheep	dig coef	%	45.	45.	
Ether extract		%	1.4	1.6	
Sheep	dig coef	%	51.	51.	
N-free extract		%	43.5	50.8	
Sheep	dig coef	%	58.	58.	
Protein (N x 6.25)		%	3.6	4.2	
Sheep	dig coef	%	66.	66.	
Cattle	dig prot	% *	0.5	0.6	
Goats	dig prot	% *	0.4	0.5	
Horses	dig prot	% *	0.9	1.1	
Rabbits	dig prot	% *	1.6	1.9	
Sheep	dig prot	%	2.4	2.8	
Energy	GE Mcal/kg				
Cattle	DE Mcal/kg	*	2.01	2.35	
Sheep	DE Mcal/kg	*	1.87	2.18	
Cattle	ME Mcal/kg	*	1.65	1.93	
Sheep	ME Mcal/kg	*	1.53	1.79	
Cattle	TDN	% *	45.6	53.3	
Sheep	TDN	%	42.3	49.5	

Heronbill, hay, s-c, mature, (1)

Ref No 1 02 371 United States

			As Fed	Dry	C.V.
Dry matter		%	85.6	100.0	
Ash		%	7.9	9.2	
Crude fiber		%	29.3	34.2	
Ether extract		%	1.4	1.6	
N-free extract		%	43.5	50.8	

Feed Name or Analyses			Mean As Fed	Mean Dry	C.V. ± %
Protein (N x 6.25)	%		3.6	4.2	
Cattle	dig prot %	*	0.5	0.6	
Goats	dig prot %	*	0.4	0.5	
Horses	dig prot %	*	0.9	1.1	
Rabbits	dig prot %	*	1.6	1.9	
Sheep	dig prot %	*	0.3	0.3	
Energy	GE Mcal/kg				
Cattle	DE Mcal/kg	*	1.93	2.26	
Sheep	DE Mcal/kg	*	2.02	2.36	
Cattle	ME Mcal/kg	*	1.58	1.85	
Sheep	ME Mcal/kg	*	1.66	1.93	
Cattle	TDN %	*	43.9	51.3	
Sheep	TDN %	*	45.8	53.5	

Heronbill, aerial part, fresh, (2)
Ref No 2 02 376 United States

			As Fed	Dry	C.V.
Dry matter	%			100.0	
Carotene	mg/kg			105.6	
Vitamin A equivalent	IU/g			176.0	

Heronbill, aerial part, fresh, mature, (2)
Ref No 2 02 374 United States

			As Fed	Dry	C.V.
Dry matter	%			100.0	
Ash	%			9.5	
Crude fiber	%			35.3	
Ether extract	%			1.7	
N-free extract	%			49.6	
Protein (N x 6.25)	%			3.9	
Cattle	dig prot %	*		1.2	
Goats	dig prot %	*		0.2	
Horses	dig prot %	*		0.8	
Rabbits	dig prot %	*		1.7	
Sheep	dig prot %	*		0.6	
Energy	GE Mcal/kg				
Cattle	DE Mcal/kg	*		2.59	
Sheep	DE Mcal/kg	*		2.37	
Cattle	ME Mcal/kg	*		2.13	
Sheep	ME Mcal/kg	*		1.94	
Cattle	TDN %	*		58.8	
Sheep	TDN %	*		53.7	

Heronbill, aerial part, fresh, over ripe, (2)
Ref No 2 02 375 United States

			As Fed	Dry	C.V.
Dry matter	%			100.0	
Calcium	%			1.73	13
Phosphorus	%			0.18	53
Potassium	%			2.42	17

HERONBILL, BIG. Erodium botrys

Heronbill, big, aerial part, dehy grnd, dormant, (1)
Ref No 1 08 259 United States

			As Fed	Dry	C.V.
Dry matter	%		91.1	100.0	
Organic matter	%		80.0	87.8	
Ash	%		11.1	12.2	
Crude fiber	%		31.1	34.2	
Ether extract	%		1.2	1.4	
N-free extract	%		43.5	47.7	
Protein (N x 6.25)	%		4.1	4.6	
Cattle	dig prot %	*	0.8	0.9	
Goats	dig prot %	*	0.7	0.8	
Horses	dig prot %	*	1.3	1.4	

Feed Name or Analyses			Mean As Fed	Mean Dry	C.V. ± %
Rabbits	dig prot %	*	2.0	2.2	
Sheep	dig prot %	*	0.6	0.7	
Energy	GE Mcal/kg				
Cattle	DE Mcal/kg	*	2.18	2.39	
Sheep	DE Mcal/kg	*	2.14	2.35	
Cattle	ME Mcal/kg	*	1.79	1.96	
Sheep	ME Mcal/kg	*	1.75	1.92	
Cattle	TDN %	*	49.3	54.1	
Sheep	TDN %	*	48.5	53.2	
Silicon	%		2.99	3.29	

Heronbill, big, leaves, dehy grnd, (1)
Ref No 1 08 262 United States

			As Fed	Dry	C.V.
Dry matter	%		93.0	100.0	
Organic matter	%		80.9	87.0	
Ash	%		12.1	13.0	
Protein (N x 6.25)	%		13.7	14.7	
Cattle	dig prot %	*	9.0	9.7	
Goats	dig prot %	*	9.6	10.3	
Horses	dig prot %	*	9.3	10.0	
Rabbits	dig prot %	*	9.3	10.0	
Sheep	dig prot %	*	9.1	9.8	
Silicon	%		1.40	1.51	

Heronbill, big, leaves, dehy grnd, mid-bloom, (1)
Ref No 1 08 260 United States

			As Fed	Dry	C.V.
Dry matter	%		92.5	100.0	
Organic matter	%		80.3	86.9	
Ash	%		12.2	13.2	
Crude fiber	%		6.7	7.2	
Ether extract	%		2.4	2.6	
N-free extract	%		57.3	62.0	
Protein (N x 6.25)	%		13.9	15.1	
Cattle	dig prot %	*	9.2	10.0	
Goats	dig prot %	*	9.8	10.6	
Horses	dig prot %	*	9.5	10.3	
Rabbits	dig prot %	*	9.5	10.3	
Sheep	dig prot %	*	9.3	10.1	
Energy	GE Mcal/kg				
Cattle	DE Mcal/kg	*	2.42	2.62	
Sheep	DE Mcal/kg	*	2.33	2.52	
Cattle	ME Mcal/kg	*	1.99	2.15	
Sheep	ME Mcal/kg	*	1.91	2.07	
Cattle	TDN %	*	54.9	59.4	
Sheep	TDN %	*	52.8	57.1	
Silicon	%		3.20	3.47	

Heronbill, big, leaves, dehy grnd, dormant, (1)
Ref No 1 08 261 United States

			As Fed	Dry	C.V.
Dry matter	%		90.2	100.0	
Organic matter	%		78.2	86.8	
Ash	%		11.9	13.3	
Crude fiber	%		13.2	14.6	
Ether extract	%		3.2	3.6	
N-free extract	%		58.2	64.6	
Protein (N x 6.25)	%		3.6	4.0	
Cattle	dig prot %	*	0.3	0.4	
Goats	dig prot %	*	0.2	0.2	
Horses	dig prot %	*	0.8	0.9	
Rabbits	dig prot %	*	1.6	1.7	
Sheep	dig prot %	*	0.1	0.1	
Silicon	%		1.84	2.04	

Feed Name or Analyses			Mean As Fed	Mean Dry	C.V. ± %

Heronbill, big, leaves w some stems, dehy grnd, (1)
Ref No 1 08 263 United States

			As Fed	Dry	C.V.
Dry matter	%		92.4	100.0	
Organic matter	%		83.7	90.6	
Ash	%		8.7	9.4	
Protein (N x 6.25)	%		9.8	10.6	
Cattle	dig prot %	*	5.7	6.1	
Goats	dig prot %	*	6.0	6.4	
Horses	dig prot %	*	6.0	6.5	
Rabbits	dig prot %	*	6.3	6.9	
Sheep	dig prot %	*	5.6	6.1	
Silicon	%		0.40	0.43	

Heronbill, big, aerial part, fresh, (2)
Ref No 2 02 381 United States

			As Fed	Dry	C.V.
Dry matter	%			100.0	
Ash	%			10.7	30
Crude fiber	%			21.4	27
Protein (N x 6.25)	%			12.0	57
Cattle	dig prot %	*		8.1	
Goats	dig prot %	*		7.8	
Horses	dig prot %	*		7.7	
Rabbits	dig prot %	*		7.9	
Sheep	dig prot %	*		8.2	
Calcium	%			1.83	22
Phosphorus	%			0.31	46
Potassium	%			2.80	31

Heronbill, big, aerial part, fresh, immature, (2)
Ref No 2 02 377 United States

			As Fed	Dry	C.V.
Dry matter	%			100.0	
Ash	%			12.9	23
Crude fiber	%			11.9	21
Protein (N x 6.25)	%			23.5	23
Cattle	dig prot %	*		17.9	
Goats	dig prot %	*		18.5	
Horses	dig prot %	*		17.5	
Rabbits	dig prot %	*		16.8	
Sheep	dig prot %	*		18.9	
Calcium	%			2.29	11
Phosphorus	%			0.47	24
Potassium	%			4.10	19

Heronbill, big, aerial part, fresh, full bloom, (2)
Ref No 2 02 378 United States

			As Fed	Dry	C.V.
Dry matter	%			100.0	
Ash	%			9.3	19
Crude fiber	%			18.0	20
Protein (N x 6.25)	%			12.6	46
Cattle	dig prot %	*		8.6	
Goats	dig prot %	*		8.3	
Horses	dig prot %	*		8.2	
Rabbits	dig prot %	*		8.4	
Sheep	dig prot %	*		8.7	
Calcium	%			1.55	13
Phosphorus	%			0.39	16
Potassium	%			2.87	25

Feed Name or Analyses			Mean As Fed	Mean Dry	C.V. ± %

Heronbill, big, aerial part, fresh, mature, (2)
Ref No 2 02 379 United States

			As Fed	Dry	C.V.
Dry matter		%		100.0	
Ash		%		8.3	30
Crude fiber		%		27.0	11
Protein (N x 6.25)		%		7.5	21
Cattle	dig prot	% *		4.3	
Goats	dig prot	% *		3.6	
Horses	dig prot	% *		3.9	
Rabbits	dig prot	% *		4.5	
Sheep	dig prot	% *		4.0	
Calcium		%		1.57	14
Phosphorus		%		0.24	51
Potassium		%		2.12	22

Heronbill, big, aerial part, fresh, over ripe, (2)
Ref No 2 02 380 United States

			As Fed	Dry	C.V.
Dry matter		%		100.0	
Ash		%		11.4	4
Crude fiber		%		29.1	7
Protein (N x 6.25)		%		4.2	26
Cattle	dig prot	% *		1.5	
Goats	dig prot	% *		0.5	
Horses	dig prot	% *		1.i	
Rabbits	dig prot	% *		1.9	
Sheep	dig prot	% *		0.9	
Calcium		%		1.75	11
Phosphorus		%		0.18	56
Potassium		%		2.47	18

Heronbill, big, aerial part, fresh leached, (2)
Ref No 2 02 385 United States

			As Fed	Dry	C.V.
Dry matter		%		100.0	
Ash		%		8.1	26
Crude fiber		%		23.9	32
Protein (N x 6.25)		%		12.6	65
Cattle	dig prot	% *		8.6	
Goats	dig prot	% *		8.3	
Horses	dig prot	% *		8.2	
Rabbits	dig prot	% *		8.4	
Sheep	dig prot	% *		8.7	
Calcium		%		1.62	13
Phosphorus		%		0.25	53

Heronbill, big, aerial part, fresh leached, immature, (2)
Ref No 2 02 382 United States

		As Fed	Dry	C.V.
Dry matter	%		100.0	
Calcium	%		1.90	
Phosphorus	%		0.39	

Heronbill, big, aerial part, fresh leached, mid-bloom, (2)
Ref No 2 02 383 United States

		As Fed	Dry	C.V.
Dry matter	%		100.0	
Calcium	%		1.40	
Phosphorus	%		0.34	

Heronbill, big, aerial part, fresh leached, mature, (2)
Ref No 2 02 384 United States

		As Fed	Dry	C.V.
Dry matter	%		100.0	
Calcium	%		1.80	
Phosphorus	%		0.11	

HERONBILL, MUSK. Erodium moschatum

Heronbill, musk, leaves, dehy grnd, (1)
Ref No 1 08 265 United States

			As Fed	Dry	C.V.
Dry matter		%	91.6	100.0	
Organic matter		%	77.5	84.6	
Ash		%	14.1	15.4	
Protein (N x 6.25)		%	13.6	14.8	
Cattle	dig prot	% *	8.9	9.8	
Goats	dig prot	% *	9.5	10.4	
Horses	dig prot	% *	9.2	10.1	
Rabbits	dig prot	% *	9.2	10.1	
Sheep	dig prot	% *	9.0	9.8	
Silicon		%	0.50	0.55	

Heronbill, musk, leaves, dehy grnd, early leaf, (1)
Ref No 1 08 264 United States

			As Fed	Dry	C.V.
Dry matter		%	93.6	100.0	
Organic matter		%	76.3	81.6	
Ash		%	17.3	18.5	
Crude fiber		%	7.4	8.0	
Ether extract		%	3.4	3.6	
N-free extract		%	41.5	44.3	
Protein (N x 6.25)		%	24.1	25.7	
Cattle	dig prot	% *	18.0	19.2	
Goats	dig prot	% *	19.2	20.5	
Horses	dig prot	% *	18.1	19.3	
Rabbits	dig prot	% *	17.3	18.5	
Sheep	dig prot	% *	18.4	19.6	
Silicon		%	6.82	7.29	

Heronbill, musk, aerial part, fresh, (2)
Ref No 2 02 389 United States

			As Fed	Dry	C.V.
Dry matter		%		100.0	
Ash		%		15.0	9
Crude fiber		%		18.1	46
Protein (N x 6.25)		%		16.2	53
Cattle	dig prot	% *		11.7	
Goats	dig prot	% *		11.7	
Horses	dig prot	% *		11.3	
Rabbits	dig prot	% *		11.2	
Sheep	dig prot	% *		12.1	
Calcium		%		2.21	26
Phosphorus		%		0.44	43
Potassium		%		3.45	35

Heronbill, musk, aerial part, fresh, immature, (2)
Ref No 2 02 386 United States

			As Fed	Dry	C.V.
Dry matter		%		100.0	
Ash		%		14.8	
Crude fiber		%		10.3	
Protein (N x 6.25)		%		27.6	
Cattle	dig prot	% *		21.4	
Goats	dig prot	% *		22.3	
Horses	dig prot	% *		21.0	
Rabbits	dig prot	% *		20.0	
Sheep	dig prot	% *		22.7	
Calcium		%		2.94	
Phosphorus		%		0.61	
Potassium		%		4.49	

Heronbill, musk, aerial part, fresh, pre-bloom, (2)
Ref No 2 02 387 United States

			As Fed	Dry	C.V.
Dry matter		%		100.0	
Ash		%		15.5	
Crude fiber		%		14.6	
Protein (N x 6.25)		%		19.5	
Cattle	dig prot	% *		14.5	
Goats	dig prot	% *		14.8	
Horses	dig prot	% *		14.1	
Rabbits	dig prot	% *		13.7	
Sheep	dig prot	% *		15.2	
Calcium		%		2.60	
Phosphorus		%		0.54	
Potassium		%		3.33	

Heronbill, musk, aerial part, fresh, over ripe, (2)
Ref No 2 02 388 United States

			As Fed	Dry	C.V.
Dry matter		%		100.0	
Ash		%		14.7	
Crude fiber		%		32.0	
Protein (N x 6.25)		%		4.4	
Cattle	dig prot	% *		1.6	
Goats	dig prot	% *		0.7	
Horses	dig prot	% *		1.3	
Rabbits	dig prot	% *		2.1	
Sheep	dig prot	% *		1.1	
Calcium		%		1.67	
Phosphorus		%		0.17	
Potassium		%		2.07	

HERONBILL, TEXAS.. Erodium texanum

Heronbill, Texas, aerial part, fresh, immature, (2)
Ref No 2 02 390 United States

			As Fed	Dry	C.V.
Dry matter		%		100.0	
Ash		%		11.2	
Crude fiber		%		8.1	
Ether extract		%		2.4	
N-free extract		%		53.9	
Protein (N x 6.25)		%		24.4	
Cattle	dig prot	% *		18.6	
Goats	dig prot	% *		19.3	
Horses	dig prot	% *		18.2	
Rabbits	dig prot	% *		17.5	
Sheep	dig prot	% *		19.7	
Calcium		%		1.96	
Magnesium		%		0.23	

(1) dry forages and roughages (2) pasture, range plants, and forages fed green (3) silages (4) energy feeds (5) protein supplements (6) minerals (7) vitamins (8) additives

Feed Name or Analyses		Mean As Fed	Mean Dry	C.V. ± %
Phosphorus	%		0.26	
Potassium	%		2.28	

Herring oil (AAFCO) –
see Fish, herring, oil from whole or parts of fish, (7)

Hershey millet grain –
see Millet, proso, grain, (4)

HOARYPEA, SMOOTHPOD. Tephrosia leicarpa

Hoarypea, smoothpod, seeds w hulls, (5)
Ref No 5 09 085 United States

Feed Name or Analyses		Mean As Fed	Mean Dry	C.V. ± %
Dry matter	%		100.0	
Protein (N x 6.25)	%		44.0	
Alanine	%		1.32	
Arginine	%		3.17	
Aspartic acid	%		3.92	
Glutamic acid	%		5.81	
Glycine	%		1.41	
Histidine	%		0.97	
Hydroxyproline	%		0.04	
Isoleucine	%		1.32	
Leucine	%		2.73	
Lysine	%		2.20	
Methionine	%		0.40	
Phenylalanine	%		1.63	
Proline	%		1.67	
Serine	%		1.67	
Threonine	%		1.14	
Tyrosine	%		1.14	
Valine	%		1.45	

Hog millet grain –
see Millet, proso, grain, (4)

HOLLYHOCK. Althaea rosea

Hollyhock, seeds w hulls, (5)
Ref No 5 09 124 United States

Feed Name or Analyses		Mean As Fed	Mean Dry	C.V. ± %
Dry matter	%		100.0	
Protein (N x 6.25)	%		26.0	
Alanine	%		0.75	
Arginine	%		1.79	
Aspartic acid	%		1.74	
Glutamic acid	%		3.09	
Glycine	%		1.14	
Histidine	%		0.62	
Hydroxyproline	%		0.03	
Isoleucine	%		0.60	
Leucine	%		1.07	
Lysine	%		1.01	
Methionine	%		0.31	
Phenylalanine	%		0.83	
Proline	%		0.68	
Serine	%		0.81	
Threonine	%		0.60	
Tyrosine	%		0.57	
Valine	%		0.86	

Hominy feed (AAFCO) (CFA) –
see Corn, grits by-product, mn 5% fat, (4)

Hominy feed, low fat –
see Corn, grits by-product, mx 5% fat, (4)

Hominy grits (AAFCO) (CFA) –
see Corn, grits, cracked fine screened, (4)

Hominy, white corn, feed –
see Corn, white, grits by-product, mn 5% fat, (4)

Hominy, yellow corn, feed –
see Corn, yellow, grits by-product, mn 5% fat, (4)

HONEYLOCUST, COMMON. Gleditsia triacanthos

Honeylocust, common, seeds w pods, grnd, (4)
Ref No 4 08 446 United States

Feed Name or Analyses		Mean As Fed	Mean Dry	C.V. ± %
Dry matter	%	88.4	100.0	
Ash	%	3.5	4.0	
Crude fiber	%	16.1	18.2	
Ether extract	%	2.4	2.7	
N-free extract	%	57.1	64.6	
Protein (N x 6.25)	%	9.3	10.5	
Cattle	dig prot % *	5.0	5.7	
Goats	dig prot % *	6.1	6.9	
Horses	dig prot % *	6.1	6.9	
Sheep	dig prot % *	6.1	6.9	
Energy	GE Mcal/kg			
Cattle	DE Mcal/kg *	2.86	3.24	
Sheep	DE Mcal/kg *	2.89	3.27	
Swine	DE kcal/kg *	1865.	2110.	
Cattle	ME Mcal/kg *	2.35	2.66	
Sheep	ME Mcal/kg *	2.37	2.68	
Swine	ME kcal/kg *	1751.	1981.	
Cattle	TDN % *	65.0	73.5	
Sheep	TDN % *	65.5	74.1	
Swine	TDN % *	42.3	47.9	

HONEYSUCKLE, BEARBERRY. Lonicera involucrata

Honeysuckle, bearberry, browse, immature, (2)
Ref No 2 02 394 United States

Feed Name or Analyses		Mean As Fed	Mean Dry	C.V. ± %
Dry matter	%		100.0	
Ash	%		6.8	
Crude fiber	%		10.5	
Ether extract	%		2.4	
N-free extract	%		62.1	
Protein (N x 6.25)	%		18.2	
Cattle	dig prot % *		13.4	
Goats	dig prot % *		13.5	
Horses	dig prot % *		13.0	
Rabbits	dig prot % *		12.7	
Sheep	dig prot % *		14.0	
Calcium	%		1.05	
Magnesium	%		0.33	
Phosphorus	%		0.42	
Sulphur	%		0.26	

Honeysuckle, bearberry, browse, full bloom, (2)
Ref No 2 02 395 United States

Feed Name or Analyses		Mean As Fed	Mean Dry	C.V. ± %
Dry matter	%		100.0	
Ash	%		11.0	
Crude fiber	%		11.6	
Ether extract	%		5.3	
N-free extract	%		62.5	
Protein (N x 6.25)	%		9.6	
Cattle	dig prot % *		6.1	
Goats	dig prot % *		5.5	
Horses	dig prot % *		5.7	
Rabbits	dig prot % *		6.1	
Sheep	dig prot % *		5.9	
Calcium	%		2.73	
Magnesium	%		0.62	
Phosphorus	%		0.62	
Sulphur	%		0.42	

HONEYSUCKLE, JAPANESE. Lonicera japonica

Honeysuckle, Japanese, browse, fresh, (2)
Ref No 2 08 447 United States

Feed Name or Analyses		Mean As Fed	Mean Dry	C.V. ± %
Dry matter	%	40.6	100.0	
Ash	%	3.3	8.1	
Crude fiber	%	9.4	23.2	
Ether extract	%	2.0	4.9	
N-free extract	%	21.8	53.7	
Protein (N x 6.25)	%	4.1	10.1	
Cattle	dig prot % *	2.6	6.5	
Goats	dig prot % *	2.4	6.0	
Horses	dig prot % *	2.5	6.1	
Rabbits	dig prot % *	2.6	6.5	
Sheep	dig prot % *	2.6	6.4	
Energy	GE Mcal/kg			
Cattle	DE Mcal/kg *	0.99	2.44	
Sheep	DE Mcal/kg *	1.06	2.62	
Cattle	ME Mcal/kg *	0.81	2.00	
Sheep	ME Mcal/kg *	0.87	2.15	
Cattle	TDN % *	22.5	55.3	
Sheep	TDN % *	24.1	59.4	

HOPS. Humulus spp

Hops, spent dehy, (1)
Dried spent hops (AAFCO)
Ref No 1 02 396 United States

Feed Name or Analyses		Mean As Fed	Mean Dry	C.V. ± %
Dry matter	%	93.1	100.0	1
Ash	%	5.6	6.0	30
Crude fiber	%	22.6	24.3	27
Sheep	dig coef %	18.	18.	
Ether extract	%	4.7	5.1	24
Sheep	dig coef %	47.	47.	
N-free extract	%	37.2	39.9	
Sheep	dig coef %	41.	41.	
Protein (N x 6.25)	%	23.0	24.8	29
Sheep	dig coef %	30.	30.	
Cattle	dig prot % *	17.1	18.4	
Goats	dig prot % *	18.3	19.7	
Horses	dig prot % *	17.3	18.5	
Rabbits	dig prot % *	16.6	17.8	
Sheep	dig prot %	6.9	7.4	
Fatty acids	%	3.6	3.8	
Energy	GE Mcal/kg			
Cattle	DE Mcal/kg *	1.31	1.41	
Sheep	DE Mcal/kg *	1.38	1.48	
Cattle	ME Mcal/kg *	1.08	1.16	
Sheep	ME Mcal/kg *	1.13	1.21	
Cattle	TDN % *	29.7	31.9	
Sheep	TDN %	31.2	33.5	

HOPSAGE, SPINY. Grayia spinosa

Hopsage, spiny, aerial part, fresh, (2)

Ref No 2 02 398 United States

Feed Name or Analyses			Mean As Fed	Mean Dry	C.V. ± %
Dry matter	%			100.0	
Ash	%			5.1	9
Crude fiber	%			38.3	7
Ether extract	%			2.0	19
N-free extract	%			45.7	
Protein (N x 6.25)	%			8.9	9
Cattle	dig prot %	*		5.5	
Goats	dig prot %	*		4.9	
Horses	dig prot %	*		5.1	
Rabbits	dig prot %	*		5.5	
Sheep	dig prot %	*		5.3	
Calcium	%			1.10	21
Phosphorus	%			0.19	31

Hopsage, spiny, aerial part, fresh, early leaf, (2)

Ref No 2 08 829 United States

Feed Name or Analyses			Mean As Fed	Mean Dry	C.V. ± %
Dry matter	%			100.0	
Crude fiber	%			36.0	
Ether extract	%			2.0	
N-free extract	%			44.0	
Protein (N x 6.25)	%			8.0	
Cattle	dig prot %	*		4.7	
Goats	dig prot %	*		4.0	
Horses	dig prot %	*		4.3	
Rabbits	dig prot %	*		4.8	
Sheep	dig prot %	*		4.4	

Hopsage, spiny, aerial part, fresh, over ripe, (2)

Ref No 2 02 397 United States

Feed Name or Analyses			Mean As Fed	Mean Dry	C.V. ± %
Dry matter	%			100.0	
Calcium	%			1.08	27
Phosphorus	%			0.17	37

Hopsage, spiny, aerial part, fresh, dormant, (2)

Ref No 2 08 830 United States

Feed Name or Analyses			Mean As Fed	Mean Dry	C.V. ± %
Dry matter	%			100.0	
Crude fiber	%			35.0	
Ether extract	%			2.5	
N-free extract	%			44.0	
Protein (N x 6.25)	%			9.0	
Cattle	dig prot %	*		5.5	
Goats	dig prot %	*		5.0	
Horses	dig prot %	*		5.2	
Rabbits	dig prot %	*		5.6	
Sheep	dig prot %	*		5.4	

Hopsage, spiny, leaves, fresh, (2)

Ref No 2 02 399 United States

Feed Name or Analyses			Mean As Fed	Mean Dry	C.V. ± %
Dry matter	%			100.0	
Ash	%			30.6	
Crude fiber	%			9.2	
Ether extract	%			1.7	
N-free extract	%			48.3	
Protein (N x 6.25)	%			10.2	
Cattle	dig prot %	*		6.6	
Goats	dig prot %	*		6.1	
Horses	dig prot %	*		6.2	
Rabbits	dig prot %	*		6.5	
Sheep	dig prot %	*		6.5	

Hopsage, spiny, pods, fresh, (2)

Ref No 2 02 400 United States

Feed Name or Analyses			Mean As Fed	Mean Dry	C.V. ± %
Dry matter	%			100.0	
Ash	%			23.0	
Crude fiber	%			22.2	
Ether extract	%			1.2	
N-free extract	%			44.8	
Protein (N x 6.25)	%			8.8	
Cattle	dig prot %	*		5.4	
Goats	dig prot %	*		4.8	
Horses	dig prot %	*		5.0	
Rabbits	dig prot %	*		5.5	
Sheep	dig prot %	*		5.2	

Ref No 5 09 004 United States

Feed Name or Analyses			Mean As Fed	Mean Dry	C.V. ± %
Dry matter	%		90.0	100.0	
Ash	%		3.2	3.6	
Crude fiber	%		0.0	0.0	
Ether extract	%		12.8	14.2	
N-free extract	%		56.3	62.6	
Protein (N x 6.25)	%		17.6	19.6	
Energy	GE Mcal/kg				
Cattle	DE Mcal/kg	*	4.12	4.57	
Sheep	DE Mcal/kg	*	3.42	3.80	
Swine	DE kcal/kg	*	3928.	4364.	
Cattle	ME Mcal/kg	*	3.37	3.75	
Sheep	ME Mcal/kg	*	2.81	3.12	
Swine	ME kcal/kg	*	3615.	4017.	
Cattle	TDN %	*	93.3	103.7	
Sheep	TDN %	*	77.6	86.2	
Swine	TDN %	*	89.1	99.0	
Calcium	%		0.81	0.90	
Magnesium	%		0.08	0.09	
Phosphorus	%		0.45	0.50	
Sodium	%		0.18	0.20	
Ascorbic acid	mg/kg		801.6	890.7	
Folic acid	mg/kg		0.01	0.01	
Niacin	mg/kg		4.0	4.4	
Pantothenic acid	mg/kg		24.0	26.7	
Riboflavin	mg/kg		1.6	1.8	
Thiamine	mg/kg		2.4	2.6	
Vitamin B_6	mg/kg		2.38	2.65	
Vitamin B_{12}	µg/kg		10.9	12.1	
Vitamin A	IU/g		3.6	4.0	

Horse, milk, fresh, (5)

Ref No 5 02 401 United States

Feed Name or Analyses			Mean As Fed	Mean Dry	C.V. ± %
Dry matter	%		9.4	100.0	
Ash	%		0.4	4.3	
Crude fiber	%		0.0	0.0	
Ether extract	%		1.1	11.7	
N-free extract	%		5.9	62.8	
Protein (N x 6.25)	%		2.0	21.3	
Energy	GE Mcal/kg				
Cattle	DE Mcal/kg	*	0.42	4.42	
Sheep	DE Mcal/kg	*	0.43	4.60	
Swine	DE kcal/kg	*	486.	5175.	
Cattle	ME Mcal/kg	*	0.34	3.62	
Sheep	ME Mcal/kg	*	0.35	3.77	
Swine	ME kcal/kg	*	446.	4745.	
Cattle	TDN %	*	9.4	100.1	
Sheep	TDN %	*	9.8	104.3	
Swine	TDN %	*	11.0	117.4	
Calcium	%		0.08	0.85	
Phosphorus	%		0.05	0.53	
Potassium	%		0.08	0.85	

HORSEBEAN. Vicia faba equina

Horsebean, hay, s-c, (1)

Ref No 1 02 402 United States

Feed Name or Analyses			Mean As Fed	Mean Dry	C.V. ± %
Dry matter	%		91.5	100.0	
Ash	%		5.5	6.0	
Crude fiber	%		22.0	24.0	
Ether extract	%		0.8	0.9	
N-free extract	%		49.8	54.5	
Protein (N x 6.25)	%		13.4	14.6	
Cattle	dig prot %	*	8.8	9.6	
Goats	dig prot %	*	9.3	10.2	
Horses	dig prot %	*	9.1	9.9	
Rabbits	dig prot %	*	9.1	10.0	
Sheep	dig prot %	*	8.9	9.7	
Fatty acids	%		0.8	0.9	
Energy	GE Mcal/kg				
Cattle	DE Mcal/kg	*	2.36	2.58	
Sheep	DE Mcal/kg	*	2.22	2.43	
Cattle	ME Mcal/kg	*	1.93	2.11	
Sheep	ME Mcal/kg	*	1.82	1.99	
Cattle	TDN %	*	53.5	58.4	
Sheep	TDN %	*	50.5	55.2	

Horsebean, straw, (1)

Ref No 1 02 404 United States

Feed Name or Analyses			Mean As Fed	Mean Dry	C.V. ± %
Dry matter	%		86.8	100.0	
Ash	%		7.6	8.7	
Crude fiber	%		36.0	41.5	
Sheep	dig coef %		41.	41.	
Ether extract	%		1.3	1.5	
Sheep	dig coef %		57.	57.	
N-free extract	%		33.6	38.7	
Sheep	dig coef %		64.	64.	
Protein (N x 6.25)	%		8.4	9.6	
Sheep	dig coef %		47.	47.	
Cattle	dig prot %	*	4.6	5.3	
Goats	dig prot %	*	4.8	5.5	
Horses	dig prot %	*	4.9	5.7	
Rabbits	dig prot %	*	5.3	6.1	
Sheep	dig prot %		3.9	4.5	
Fatty acids	%		1.4	1.6	
Energy	GE Mcal/kg				
Cattle	DE Mcal/kg	*	1.69	1.95	
Sheep	DE Mcal/kg	*	1.84	2.12	
Cattle	ME Mcal/kg	*	1.38	1.60	
Sheep	ME Mcal/kg	*	1.51	1.74	
Cattle	TDN %	*	38.3	44.1	
Sheep	TDN %		41.8	48.2	

(1) dry forages and roughages (2) pasture, range plants, and forages fed green (3) silages (4) energy feeds (5) protein supplements (6) minerals (7) vitamins (8) additives

Feed Name or Analyses			Mean As Fed	Mean Dry	C.V. ± %

Horsebean, straw, full bloom, (1)
Ref No 1 02 403 — United States

Analyses	Unit		As Fed	Dry	C.V. ± %
Dry matter	%		12.2	100.0	
Ash	%		2.5	20.8	
Crude fiber	%		3.8	31.1	
Sheep	dig coef %		54.	54.	
Ether extract	%		0.3	2.3	
Sheep	dig coef %		62.	62.	
N-free extract	%		3.4	28.1	
Sheep	dig coef %		78.	78.	
Protein (N x 6.25)	%		2.2	17.7	
Sheep	dig coef %		78.	78.	
Cattle	dig prot %	*	1.5	12.3	
Goats	dig prot %	*	1.6	13.1	
Horses	dig prot %	*	1.5	12.6	
Rabbits	dig prot %	*	1.5	12.3	
Sheep	dig prot %		1.7	13.8	
Energy	GE Mcal/kg				
Cattle	DE Mcal/kg	*	0.32	2.61	
Sheep	DE Mcal/kg	*	0.30	2.46	
Cattle	ME Mcal/kg	*	0.26	2.14	
Sheep	ME Mcal/kg	*	0.25	2.01	
Cattle	TDN %	*	7.2	59.3	
Sheep	TDN %		6.8	55.7	

Horsebean, aerial part, fresh, (2)
Ref No 2 02 405 — United States

Analyses	Unit		As Fed	Dry	C.V. ± %
Dry matter	%		18.0	100.0	
Ash	%		1.7	9.6	
Crude fiber	%		4.2	23.1	
Ether extract	%		0.5	2.8	
N-free extract	%		7.9	44.1	
Protein (N x 6.25)	%		3.7	20.3	
Cattle	dig prot %	*	2.7	15.2	
Goats	dig prot %	*	2.8	15.5	
Horses	dig prot %	*	2.7	14.8	
Rabbits	dig prot %	*	2.6	14.3	
Sheep	dig prot %	*	2.9	15.9	
Energy	GE Mcal/kg				
Cattle	DE Mcal/kg	*	0.53	2.97	
Sheep	DE Mcal/kg	*	0.51	2.82	
Cattle	ME Mcal/kg	*	0.44	2.44	
Sheep	ME Mcal/kg	*	0.42	2.31	
Cattle	TDN %	*	12.1	67.3	
Sheep	TDN %	*	11.5	64.0	
Calcium	%		0.17	0.92	
Phosphorus	%		0.05	0.29	
Potassium	%		0.37	2.07	

Horsebean, aerial part, ensiled, (3)
Ref No 3 02 406 — United States

Analyses	Unit		As Fed	Dry	C.V. ± %
Dry matter	%		21.2	100.0	
Ash	%		1.4	6.6	
Crude fiber	%		5.7	26.9	
Ether extract	%		0.5	2.4	
N-free extract	%		10.3	48.5	
Protein (N x 6.25)	%		3.3	15.6	
Cattle	dig prot %	*	2.2	10.4	
Goats	dig prot %	*	2.2	10.4	
Horses	dig prot %	*	2.2	10.4	
Sheep	dig prot %	*	2.2	10.4	
Energy	GE Mcal/kg				
Cattle	DE Mcal/kg	*	0.53	2.49	
Sheep	DE Mcal/kg	*	0.50	2.34	
Cattle	ME Mcal/kg	*	0.43	2.05	
Sheep	ME Mcal/kg	*	0.41	1.92	
Cattle	TDN %	*	12.0	56.6	
Sheep	TDN %	*	11.2	53.0	
Calcium	%		0.19	0.90	
Phosphorus	%		0.06	0.28	
Potassium	%		0.44	2.08	

Horsebean, seeds, (5)
Ref No 5 02 407 — United States

Analyses	Unit		As Fed	Dry	C.V. ± %
Dry matter	%		87.3	100.0	
Ash	%		3.5	4.0	
Crude fiber	%		7.7	8.8	
Horses	dig coef %		69.	69.	
Ether extract	%		1.3	1.5	
Horses	dig coef %		11.	11.	
N-free extract	%		49.3	56.5	
Horses	dig coef %		91.	91.	
Protein (N x 6.25)	%		25.5	29.2	
Horses	dig coef %		76.	76.	
Horses	dig prot %		19.4	22.2	
Energy	GE Mcal/kg				
Cattle	DE Mcal/kg	*	3.03	3.47	
Horses	DE Mcal/kg	*	3.08	3.53	
Sheep	DE Mcal/kg	*	3.34	3.83	
Swine	DE kcal/kg	*	3402.	3899.	
Cattle	ME Mcal/kg	*	2.49	2.85	
Horses	ME Mcal/kg	*	2.53	2.90	
Sheep	ME Mcal/kg	*	2.74	3.14	
Swine	ME kcal/kg	*	3065.	3513.	
Cattle	TDN %	*	68.8	78.8	
Horses	TDN %		69.9	80.1	
Sheep	TDN %	*	75.7	86.8	
Swine	TDN %	*	77.2	88.4	
Calcium	%		0.13	0.15	
Phosphorus	%		0.54	0.62	
Potassium	%		1.16	1.33	

HORSEBRUSH, COTTONTHORN. Tetradymia spinosa

Horsebrush, cottonhorn, aerial part, fresh, (2)
Ref No 2 02 408 — United States

Analyses	Unit		As Fed	Dry	C.V. ± %
Dry matter	%			100.0	
Ash	%			5.4	
Crude fiber	%			36.8	
Ether extract	%			6.3	
N-free extract	%			42.8	
Protein (N x 6.25)	%			8.7	
Cattle	dig prot %	*		5.3	
Goats	dig prot %	*		4.7	
Horses	dig prot %	*		4.9	
Rabbits	dig prot %	*		5.4	
Sheep	dig prot %	*		5.1	
Calcium	%			0.94	
Phosphorus	%			0.25	

HORSECHESTNUT. Aesculus hippocastanum

Horsechestnut, meats, (4)
Ref No 4 02 409 — United States

Analyses	Unit		As Fed	Dry	C.V. ± %
Dry matter	%		90.4	100.0	
Ash	%		2.2	2.4	
Crude fiber	%		2.5	2.8	
Sheep	dig coef %		304.	304.	
Ether extract	%		6.5	7.2	
Sheep	dig coef %		50.	50.	
N-free extract	%		72.1	79.8	
Sheep	dig coef %		76.	76.	
Protein (N x 6.25)	%		7.1	7.8	
Sheep	dig coef %		25.	25.	
Cattle	dig prot %	*	2.9	3.2	
Goats	dig prot %	*	4.0	4.4	
Horses	dig prot %	*	4.0	4.4	
Sheep	dig prot %		1.8	2.0	
Energy	GE Mcal/kg				
Cattle	DE Mcal/kg	*	2.66	2.94	
Sheep	DE Mcal/kg	*	2.48	2.74	
Swine	DE kcal/kg	*	2746.	3038.	
Cattle	ME Mcal/kg	*	2.18	2.41	
Sheep	ME Mcal/kg	*	2.03	2.25	
Swine	ME kcal/kg	*	2593.	2868.	
Cattle	TDN %	*	60.2	66.6	
Sheep	TDN %		56.2	62.2	
Swine	TDN %	*	62.3	68.9	

HORSETAIL, FIELD. Equisetum arvense

Horsetail, field, aerial part, fresh, (2)
Ref No 2 02 410 — United States

Analyses	Unit		As Fed	Dry	C.V. ± %
Dry matter	%			100.0	
Ash	%			18.5	
Crude fiber	%			23.5	
Ether extract	%			2.4	
N-free extract	%			50.3	
Protein (N x 6.25)	%			5.3	
Cattle	dig prot %	*		2.4	
Goats	dig prot %	*		1.5	
Horses	dig prot %	*		2.0	
Rabbits	dig prot %	*		2.8	
Sheep	dig prot %	*		1.9	

HUAJILLO. Acacia berlandieri

Huajillo, leaves, fresh, (2)
Ref No 2 02 411 — United States

Analyses	Unit		As Fed	Dry	C.V. ± %
Dry matter	%			100.0	
Carotene	mg/kg			104.9	
Vitamin A equivalent	IU/g			174.9	

Hulled oats (CFA) —
see Oats, groats, (4)

HURRAHGRASS. Paspalum pubescens

Hurrahgrass, aerial part, fresh, mature, (2)
Ref No 2 02 412 — United States

Analyses	Unit		As Fed	Dry	C.V. ± %
Dry matter	%			100.0	
Ash	%			9.2	
Crude fiber	%			35.6	
Ether extract	%			1.4	
N-free extract	%			48.9	
Protein (N x 6.25)	%			4.9	
Cattle	dig prot %	*		2.1	
Goats	dig prot %	*		1.1	
Horses	dig prot %	*		1.7	
Rabbits	dig prot %	*		2.5	
Sheep	dig prot %	*		1.6	

Continued

Feed Name or Analyses		Mean As Fed	Mean Dry	C.V. ± %
Calcium	%		0.32	
Phosphorus	%		0.08	

HYACINTH. Hyacinthus spp

Hyacinth, hay, s-c, (1)
Ref No 1 02 413 United States

Dry matter	%	90.2	100.0	
Ash	%	6.8	7.5	
Crude fiber	%	33.6	37.2	
Ether extract	%	1.4	1.6	
N-free extract	%	33.6	37.3	
Protein (N x 6.25)	%	14.8	16.4	
Cattle dig prot	% *	10.1	11.1	
Goats dig prot	% *	10.7	11.9	
Horses dig prot	% *	10.3	11.5	
Rabbits dig prot	% *	10.2	11.3	
Sheep dig prot	% *	10.2	11.3	

Hyacinth, browse, (2)
Ref No 2 02 414 United States

Dry matter	%	12.9	100.0	
Ash	%	1.6	12.5	
Crude fiber	%	2.3	18.0	
Ether extract	%	0.4	3.3	
N-free extract	%	7.5	58.2	
Protein (N x 6.25)	%	1.0	8.0	
Cattle dig prot	% *	0.6	4.7	
Goats dig prot	% *	0.5	4.0	
Horses dig prot	% *	0.6	4.3	
Rabbits dig prot	% *	0.6	4.8	
Sheep dig prot	% *	0.6	4.4	

Hydrolyzed animal fat, feed grade (AAFCO) — see Animal, fat, hydrolyzed, feed gr mn 85% fatty acids mx 6% unsaponifiable matter mx 1% insoluble matter, (4)

Hydrolyzed poultry feathers (AAFCO) — see Poultry, feathers, hydrolyzed dehy grnd, mn 75% of protein digestible, (5)

Hydrolyzed poultry feathers (CFA) — see Poultry, feathers, hydrolyzed dehy grnd, mn 75% of protein digestible, (5)

HYMENOXYS, BITTERWEED. Hymenoxys odorata

Hymenoxys, bitterweed, browse, immature, (2)
Ref No 2 02 417 United States

Dry matter	%		100.0	
Ash	%		12.4	
Crude fiber	%		11.0	
Ether extract	%		3.3	
N-free extract	%		40.0	
Protein (N x 6.25)	%		33.3	
Cattle dig prot	% *		26.2	
Goats dig prot	% *		27.6	
Horses dig prot	% *		25.8	

(1) dry forages and roughages (2) pasture, range plants, and forages fed green (3) silages (4) energy feeds (5) protein supplements (6) minerals (7) vitamins (8) additives

Feed Name or Analyses		Mean As Fed	Mean Dry	C.V. ± %
Rabbits dig prot	% *		24.4	
Sheep dig prot	% *		28.0	
Calcium	%		1.84	
Magnesium	%		0.36	
Phosphorus	%		0.46	
Potassium	%		4.59	

Hymenoxys, bitterweed, browse, mid-bloom, (2)
Ref No 2 02 418 United States

Dry matter	%		100.0	
Calcium	%		2.10	
Magnesium	%		0.21	
Phosphorus	%		0.46	
Potassium	%		2.58	

Hymenoxys, bitterweed, browse, milk stage, (2)
Ref No 2 02 419 United States

Dry matter	%		100.0	
Ash	%		8.2	
Crude fiber	%		23.9	
Ether extract	%		6.1	
N-free extract	%		49.5	
Protein (N x 6.25)	%		12.3	
Cattle dig prot	% *		8.3	
Goats dig prot	% *		8.0	
Horses dig prot	% *		8.0	
Rabbits dig prot	% *		8.2	
Sheep dig prot	% *		8.5	
Calcium	%		1.04	
Magnesium	%		0.25	
Phosphorus	%		0.18	
Potassium	%		2.54	

ICE CREAM CONE. Scientific name not used

Ice cream cone, grnd, (4)
Ref No 4 02 420 United States

Dry matter	%	90.0	100.0	
Ash	%	1.5	1.7	80
Crude fiber	%	1.6	1.8	98
Ether extract	%	1.4	1.6	74
N-free extract	%	76.4	84.9	
Protein (N x 6.25)	%	9.1	10.1	4
Cattle dig prot	% *	4.8	5.3	
Goats dig prot	% *	5.9	6.5	
Horses dig prot	% *	5.9	6.5	
Sheep dig prot	% *	5.9	6.5	

INDIANGRASS. Sorghastrum spp

Indiangrass, aerial part, fresh, (2)
Ref No 2 08 770 United States

Dry matter	%	56.9	100.0	21
Ash	%	4.1	7.2	13
Crude fiber	%	19.7	34.6	7
Ether extract	%	1.3	2.3	14
N-free extract	%	28.9	50.8	4
Protein (N x 6.25)	%	2.9	5.1	20
Cattle dig prot	% *	1.3	2.2	
Goats dig prot	% *	0.8	1.3	
Horses dig prot	% *	1.1	1.9	
Rabbits dig prot	% *	1.5	2.6	

Feed Name or Analyses		Mean As Fed	Mean Dry	C.V. ± %
Sheep dig prot	% *	1.0	1.7	
Calcium	%	0.19	0.33	16
Phosphorus	%	0.04	0.08	21
Carotene	mg/kg	55.5	97.5	59
Vitamin A equivalent	IU/g	92.6	162.6	

INDIANGRASS, YELLOW. Sorghastrum nutans

Indiangrass, yellow, aerial part, fresh, immature, (2)
Ref No 2 02 421 United States

Dry matter	%		100.0	
Ash	%		9.8	7
Crude fiber	%		30.4	3
Ether extract	%		2.6	8
N-free extract	%		45.4	
Protein (N x 6.25)	%		11.8	2
Cattle dig prot	% *		7.9	
Goats dig prot	% *		7.6	
Horses dig prot	% *		7.5	
Rabbits dig prot	% *		7.8	
Sheep dig prot	% *		8.0	
Calcium	%		0.63	9
Phosphorus	%		0.20	22

Indiangrass, yellow, aerial part, fresh, mature, (2)
Ref No 2 02 422 United States

Dry matter	%		100.0	
Ash	%		7.6	30
Crude fiber	%		34.5	10
Ether extract	%		1.9	23
N-free extract	%		51.2	
Protein (N x 6.25)	%		4.8	29
Cattle dig prot	% *		2.0	
Goats dig prot	% *		1.0	
Horses dig prot	% *		1.6	
Rabbits dig prot	% *		2.4	
Sheep dig prot	% *		1.5	
Calcium	%		0.30	29
Phosphorus	%		0.08	61

Indiangrass, yellow, aerial part, fresh, over ripe, (2)
Ref No 2 02 423 United States

Dry matter	%		100.0	
Ash	%		6.2	
Crude fiber	%		40.0	
Ether extract	%		2.0	
N-free extract	%		49.3	
Protein (N x 6.25)	%		2.5	
Cattle dig prot	% *		0.0	
Goats dig prot	% *		1.0	
Horses dig prot	% *		0.2	
Rabbits dig prot	% *		0.6	
Sheep dig prot	% *		0.6	
Carotene	mg/kg		1.1	
Vitamin A equivalent	IU/g		1.8	

INDIGO. Indigofera spp

Indigo, hay, s-c, cut 1, (1)

Ref No 1 02 425 — United States

Feed Name or Analyses			Mean As Fed	Mean Dry	C.V. ± %
Dry matter	%		89.0	100.0	
Ash	%		5.1	5.7	
Crude fiber	%		22.2	24.9	
Ether extract	%		2.2	2.5	
N-free extract	%		48.9	54.9	
Protein (N x 6.25)	%		10.7	12.0	
Cattle	dig prot %	*	6.5	7.3	
Goats	dig prot %	*	6.9	7.8	
Horses	dig prot %	*	6.9	7.7	
Rabbits	dig prot %	*	7.1	7.9	
Sheep	dig prot %	*	6.5	7.3	
Calcium	%		1.34	1.50	
Magnesium	%		0.50	0.56	
Phosphorus	%		0.19	0.21	

INDIGO, HAIRY. Indigofera hirsuta

Indigo, hairy, aerial part, ensiled, (3)

Ref No 3 02 426 — United States

Feed Name or Analyses			Mean As Fed	Mean Dry	C.V. ± %
Dry matter	%		29.4	100.0	20
Ash	%		1.9	6.5	26
Crude fiber	%		9.6	32.5	16
Ether extract	%		0.7	2.3	21
N-free extract	%		13.5	45.8	
Protein (N x 6.25)	%		3.8	12.9	19
Cattle	dig prot %	*	2.3	7.9	
Goats	dig prot %	*	2.3	7.9	
Horses	dig prot %	*	2.3	7.9	
Sheep	dig prot %	*	2.3	7.9	
Calcium	%		0.40	1.35	20
Magnesium	%		0.07	0.25	56
Phosphorus	%		0.06	0.20	20

Indigo, hairy, aerial part w molasses added, ensiled, (3)

Ref No 3 02 427 — United States

Feed Name or Analyses			Mean As Fed	Mean Dry	C.V. ± %
Dry matter	%		31.7	100.0	
Ash	%		2.3	7.2	
Crude fiber	%		7.1	22.5	
Ether extract	%		0.9	2.8	
N-free extract	%		16.0	50.4	
Protein (N x 6.25)	%		5.4	17.1	
Cattle	dig prot %	*	3.7	11.8	
Goats	dig prot %	*	3.7	11.8	
Horses	dig prot %	*	3.7	11.8	
Sheep	dig prot %	*	3.7	11.8	
Calcium	%		0.53	1.67	
Magnesium	%		0.13	0.42	
Phosphorus	%		0.08	0.26	

INDIGO, NINELEAF. Indigofera endecaphylla

Indigo, nineleaf, browse, (2)

Ref No 2 02 428 — United States

Feed Name or Analyses			Mean As Fed	Mean Dry	C.V. ± %
Dry matter	%		38.6	100.0	
Ash	%		2.1	5.4	
Crude fiber	%		12.9	33.5	
Ether extract	%		0.8	2.2	
N-free extract	%		17.7	45.9	
Protein (N x 6.25)	%		5.0	13.0	
Cattle	dig prot %	*	3.5	8.9	
Goats	dig prot %	*	3.4	8.7	
Horses	dig prot %	*	3.3	8.6	
Rabbits	dig prot %	*	3.4	8.7	
Sheep	dig prot %	*	3.5	9.1	

IRIS, GERMAN. Iris germanica

Iris, German, seeds w hulls, (5)

Ref No 4 09 267 — United States

Feed Name or Analyses			Mean As Fed	Mean Dry	C.V. ± %
Dry matter	%			100.0	
Protein (N x 6.25)	%			15.0	
Alanine	%			0.44	
Arginine	%			1.16	
Aspartic acid	%			1.37	
Glutamic acid	%			2.16	
Glycine	%			0.69	
Histidine	%			0.32	
Hydroxyproline	%			0.86	
Isoleucine	%			0.44	
Leucine	%			0.78	
Lysine	%			0.53	
Methionine	%			0.23	
Phenylalanine	%			0.66	
Proline	%			0.54	
Serine	%			0.84	
Threonine	%			0.44	
Tyrosine	%			0.69	
Valine	%			0.62	

Iron oxide —
see Ferric oxide, red, Fe_2O_3, cp, (6)

Irradiated dried yeast (AAFCO) —
see Yeast, irradiated, dehy, (7)

Ivory nut meal —
see Ivorypalm, common, nut meats, extn unspecified grnd, (4)

IVORYPALM, COMMON. Phytelephas macrocarpa

Ivorypalm, common, nut meats, extn unspecified grnd, (4)

Ivory nut meal

Ref No 4 02 433 — United States

Feed Name or Analyses			Mean As Fed	Mean Dry	C.V. ± %
Dry matter	%		89.2	100.0	
Ash	%		1.1	1.3	
Crude fiber	%		7.1	8.0	
Sheep	dig coef %		82.	82.	
Ether extract	%		0.9	1.0	
Sheep	dig coef %		22.	22.	
N-free extract	%		75.3	84.4	
Sheep	dig coef %		94.	94.	
Protein (N x 6.25)	%		4.8	5.3	
Sheep	dig coef %		15.	15.	
Cattle	dig prot %	*	0.8	0.9	
Goats	dig prot %	*	1.9	2.1	
Horses	dig prot %	*	1.9	2.1	
Sheep	dig prot %		0.7	0.8	
Energy	GE Mcal/kg				
Cattle	DE Mcal/kg	*	3.58	4.01	
Sheep	DE Mcal/kg	*	3.43	3.84	
Swine	DE kcal/kg	*	3473.	3893.	
Cattle	ME Mcal/kg	*	2.93	3.29	
Sheep	ME Mcal/kg	*	2.81	3.15	
Swine	ME kcal/kg	*	3295.	3694.	
Cattle	TDN %	*	81.2	91.0	
Sheep	TDN %		77.8	87.2	
Swine	TDN %	*	78.8	88.3	

JACKBEAN. Canavalia spp

Jackbean, aerial part, fresh, (2)

Ref No 2 02 434 — United States

Feed Name or Analyses			Mean As Fed	Mean Dry	C.V. ± %
Dry matter	%		23.2	100.0	
Ash	%		2.7	11.6	
Crude fiber	%		6.4	27.6	
Ether extract	%		0.5	2.2	
N-free extract	%		8.4	36.2	
Protein (N x 6.25)	%		5.2	22.4	
Cattle	dig prot %	*	3.9	16.9	
Goats	dig prot %	*	4.1	17.5	
Horses	dig prot %	*	3.8	16.6	
Rabbits	dig prot %	*	3.7	16.0	
Sheep	dig prot %	*	4.1	17.9	
Energy	GE Mcal/kg				
Cattle	DE Mcal/kg	*	0.65	2.82	
Sheep	DE Mcal/kg	*	0.62	2.67	
Cattle	ME Mcal/kg	*	0.54	2.31	
Sheep	ME Mcal/kg	*	0.51	2.19	
Cattle	TDN %	*	14.9	64.2	
Sheep	TDN %	*	14.0	60.4	

Jackbean, seeds, (5)

Ref No 5 02 435 — United States

Feed Name or Analyses			Mean As Fed	Mean Dry	C.V. ± %
Dry matter	%		89.1	100.0	
Ash	%		2.9	3.2	
Crude fiber	%		8.5	9.5	
Sheep	dig coef %		75.	75.	
Ether extract	%		2.8	3.1	
Sheep	dig coef %		75.	75.	
N-free extract	%		47.6	53.4	
Sheep	dig coef %		98.	98.	
Protein (N x 6.25)	%		27.4	30.7	
Sheep	dig coef %		84.	84.	
Sheep	dig prot %		23.0	25.8	
Energy	GE Mcal/kg				
Cattle	DE Mcal/kg	*	3.56	3.99	
Sheep	DE Mcal/kg	*	3.56	3.99	
Swine	DE kcal/kg	*	3554.	3989.	
Cattle	ME Mcal/kg	*	2.91	3.27	
Sheep	ME Mcal/kg	*	2.92	3.28	
Swine	ME kcal/kg	*	3191.	3581.	
Cattle	TDN %	*	80.6	90.5	
Sheep	TDN %		80.7	90.6	
Swine	TDN %	*	80.6	90.5	

JACKBEAN, COMMON. Canalavia ensiformis

Jackbean, common, seeds w hulls, (5)

Ref No 5 09 024 — United States

Feed Name or Analyses			Mean As Fed	Mean Dry	C.V. ± %
Dry matter	%			100.0	
Protein (N x 6.25)	%			30.0	
Alanine	%			1.11	
Arginine	%			1.35	
Aspartic acid	%			2.70	

Continued

Feed Name or Analyses		Mean As Fed	Mean Dry	C.V. ± %
Glutamic acid	%		2.73	
Glycine	%		0.99	
Histidine	%		0.72	
Hydroxyproline	%		0.09	
Isoleucine	%		1.05	
Leucine	%		1.92	
Lysine	%		1.53	
Methionine	%		0.30	
Phenylalanine	%		1.20	
Proline	%		1.08	
Serine	%		1.29	
Threonine	%		1.17	
Tyrosine	%		0.93	
Valine	%		1.20	

JACKINTHEPULPIT, INDIAN. Arisaema triphyllum

Jackinthepulpit, Indian, seeds w hulls, (4)
Ref No 4 09 272 United States

Feed Name or Analyses		Mean As Fed	Mean Dry	C.V. ± %
Dry matter	%		100.0	
Protein (N x 6.25)	%		15.0	
Alanine	%		0.83	
Arginine	%		1.49	
Aspartic acid	%		0.96	
Glutamic acid	%		2.40	
Glycine	%		0.81	
Histidine	%		0.35	
Hydroxyproline	%		0.02	
Isoleucine	%		0.47	
Leucine	%		0.87	
Lysine	%		0.68	
Methionine	%		0.24	
Phenylalanine	%		0.66	
Proline	%		0.53	
Serine	%		0.60	
Threonine	%		0.45	
Tyrosine	%		0.38	
Valine	%		0.72	

Jajoba nut –
see Jojoba, California, seeds, (4)

Japanesemillet, aerial part, fresh –
see Millet, Japanese, aerial part, fresh, (2)

JETBEAD. Rhodotypos tetraptala

Jetbead, seeds w hulls, (5)
Ref No 5 09 297 United States

Feed Name or Analyses		Mean As Fed	Mean Dry	C.V. ± %
Dry matter	%		100.0	
Protein (N x 6.25)	%		49.0	
Alanine	%		1.52	
Arginine	%		5.49	
Aspartic acid	%		4.41	
Glutamic acid	%		10.14	
Glycine	%		2.30	
Histidine	%		0.98	
Hydroxyproline	%		0.00	
Isoleucine	%		1.13	
Leucine	%		2.84	
Lysine	%		1.42	
Methionine	%		0.29	
Phenylalanine	%		1.27	
Proline	%		1.27	
Serine	%		1.72	
Threonine	%		0.88	
Tyrosine	%		1.47	
Valine	%		1.72	

JOBSTEARS. Coix lacrymajobi

Jobstears, seeds, (4)
Adlay
Ref No 4 08 448 United States

Feed Name or Analyses		Mean As Fed	Mean Dry	C.V. ± %
Dry matter	%	89.2	100.0	
Ash	%	2.6	2.9	
Crude fiber	%	8.4	9.4	
Ether extract	%	6.1	6.8	
N-free extract	%	58.5	65.6	
Protein (N x 6.25)	%	13.6	15.2	
Cattle	dig prot % *	8.9	10.0	
Goats	dig prot % *	10.0	11.2	
Horses	dig prot % *	10.0	11.2	
Sheep	dig prot % *	10.0	11.2	
Energy	GE Mcal/kg			
Cattle	DE Mcal/kg *	2.85	3.20	
Sheep	DE Mcal/kg *	3.14	3.52	
Swine	DE kcal/kg *	2924.	3277.	
Cattle	ME Mcal/kg *	2.34	2.62	
Sheep	ME Mcal/kg *	2.58	2.89	
Swine	ME kcal/kg *	2717.	3045.	
Cattle	TDN % *	64.7	72.6	
Sheep	TDN % *	71.3	79.9	
Swine	TDN % *	66.3	74.3	

Johnsongrass, aerial part, ensiled –
see Sorghum, johnsongrass, aerial part, ensiled, (3)

Johnsongrass hay, fan air dried –
see Sorghum, johnsongrass, hay, fan air dried, (1)

JOJOBA, CALIFORNIA. Simmondsia chinensis

Jojoba, California, seeds, (4)
Jajoba nut
Goat nut
Ref No 4 08 568 United States

Feed Name or Analyses		Mean As Fed	Mean Dry	C.V. ± %
Dry matter	%	95.1	100.0	
Ash	%	1.4	1.5	
Crude fiber	%	6.1	6.4	
Ether extract	%	51.0	53.7	
N-free extract	%	22.5	23.7	
Protein (N x 6.25)	%	14.0	14.8	
Cattle	dig prot % *	9.1	9.6	
Goats	dig prot % *	10.2	10.8	
Horses	dig prot % *	10.2	10.8	
Sheep	dig prot % *	10.2	10.8	

JUNEGRASS. Koeleria cristata

Junegrass, hay, s-c, (1)
Ref No 1 02 436 United States

Feed Name or Analyses		Mean As Fed	Mean Dry	C.V. ± %
Dry matter	%	88.9	100.0	1
Ash	%	6.9	7.8	12
Crude fiber	%	30.4	34.2	12
Ether extract	%	2.5	2.9	17
N-free extract	%	40.9	46.0	
Protein (N x 6.25)	%	8.1	9.1	19
Cattle	dig prot % *	4.3	4.8	
Goats	dig prot % *	4.5	5.0	
Horses	dig prot % *	4.7	5.2	
Rabbits	dig prot % *	5.1	5.7	
Sheep	dig prot % *	4.2	4.7	
Energy	GE Mcal/kg			
Cattle	DE Mcal/kg *	1.88	2.11	
Sheep	DE Mcal/kg *	1.77	1.99	
Cattle	ME Mcal/kg *	1.54	1.73	
Sheep	ME Mcal/kg *	1.45	1.63	
Cattle	TDN % *	42.5	47.8	
Sheep	TDN % *	40.2	45.2	

Junegrass, aerial part, fresh, (2)
Ref No 2 02 441 United States

Feed Name or Analyses		Mean As Fed	Mean Dry	C.V. ± %
Dry matter	%		100.0	
Ash	%		8.1	19
Crude fiber	%		37.8	14
Ether extract	%		2.1	40
N-free extract	%		44.7	
Protein (N x 6.25)	%		7.3	84
Cattle	dig prot % *		4.1	
Goats	dig prot % *		3.4	
Horses	dig prot % *		3.7	
Rabbits	dig prot % *		4.3	
Sheep	dig prot % *		3.8	
Energy	GE Mcal/kg			
Cattle	DE Mcal/kg *		2.58	
Sheep	DE Mcal/kg *		2.54	
Cattle	ME Mcal/kg *		2.12	
Sheep	ME Mcal/kg *		2.08	
Cattle	TDN % *		58.5	
Sheep	TDN % *		57.6	
Calcium	%		0.32	11
Phosphorus	%		0.24	54
Carotene	mg/kg		85.1	54
Vitamin A equivalent	IU/g		141.9	

Junegrass, aerial part, fresh, immature, (2)
Ref No 2 02 437 United States

Feed Name or Analyses		Mean As Fed	Mean Dry	C.V. ± %
Dry matter	%		100.0	
Ash	%		7.8	
Crude fiber	%		25.8	
Ether extract	%		2.3	
N-free extract	%		40.3	
Protein (N x 6.25)	%		23.8	
Cattle	dig prot % *		18.1	
Goats	dig prot % *		18.8	
Horses	dig prot % *		17.7	
Rabbits	dig prot % *		17.0	
Sheep	dig prot % *		19.2	
Energy	GE Mcal/kg			
Cattle	DE Mcal/kg *		2.50	
Sheep	DE Mcal/kg *		2.59	
Cattle	ME Mcal/kg *		2.05	
Sheep	ME Mcal/kg *		2.12	
Cattle	TDN % *		56.7	
Sheep	TDN % *		58.8	

(1) dry forages and roughages (2) pasture, range plants, and forages fed green (3) silages (4) energy feeds (5) protein supplements (6) minerals (7) vitamins (8) additives

Junegrass, aerial part, fresh, full bloom, (2)

Ref No 2 02 438 United States

Feed Name or Analyses			As Fed	Dry	C.V. ± %
Dry matter	%			100.0	
Ash	%			9.6	
Crude fiber	%			37.3	
Ether extract	%			1.9	
N-free extract	%			42.0	
Protein (N x 6.25)	%			9.2	
Cattle	dig prot %	*		5.7	
Goats	dig prot %	*		5.1	
Horses	dig prot %	*		5.3	
Rabbits	dig prot %	*		5.8	
Sheep	dig prot %	*		5.6	
Energy	GE Mcal/kg				
Cattle	DE Mcal/kg	*		2.41	
Sheep	DE Mcal/kg	*		2.33	
Cattle	ME Mcal/kg	*		1.98	
Sheep	ME Mcal/kg	*		1.91	
Cattle	TDN %	*		54.7	
Sheep	TDN %	*		52.9	

Junegrass, aerial part, fresh, mature, (2)

Ref No 2 02 439 United States

Feed Name or Analyses			As Fed	Dry	C.V. ± %
Dry matter	%			100.0	
Ash	%			8.1	
Crude fiber	%			38.6	
Ether extract	%			2.5	
N-free extract	%			45.0	
Protein (N x 6.25)	%			5.8	
Cattle	dig prot %	*		2.8	
Goats	dig prot %	*		2.0	
Horses	dig prot %	*		2.5	
Rabbits	dig prot %	*		3.1	
Sheep	dig prot %	*		2.4	
Energy	GE Mcal/kg				
Cattle	DE Mcal/kg	*		2.31	
Sheep	DE Mcal/kg	*		2.44	
Cattle	ME Mcal/kg	*		1.89	
Sheep	ME Mcal/kg	*		2.00	
Cattle	TDN %	*		52.5	
Sheep	TDN %	*		55.3	
Calcium	%			0.33	
Phosphorus	%			0.31	
Carotene	mg/kg			58.2	
Vitamin A equivalent	IU/g			97.0	

Junegrass, aerial part, fresh, over ripe, (2)

Ref No 2 02 440 United States

Feed Name or Analyses			As Fed	Dry	C.V. ± %
Dry matter	%			100.0	
Ash	%			7.7	26
Crude fiber	%			41.0	6
Ether extract	%			1.8	47
N-free extract	%			45.7	
Protein (N x 6.25)	%			3.8	58
Cattle	dig prot %	*		1.1	
Goats	dig prot %	*		0.1	
Horses	dig prot %	*		0.8	
Rabbits	dig prot %	*		1.6	
Sheep	dig prot %	*		0.5	
Energy	GE Mcal/kg				
Cattle	DE Mcal/kg	*		2.31	
Sheep	DE Mcal/kg	*		2.47	
Cattle	ME Mcal/kg	*		1.89	
Sheep	ME Mcal/kg	*		2.03	
Cattle	TDN %	*		52.3	
Sheep	TDN %	*		56.0	
Calcium	%			0.29	
Phosphorus	%			0.14	
Carotene	mg/kg			33.3	
Vitamin A equivalent	IU/g			55.5	

JUNGLERICE. Echinochloa colonum

Junglerice, aerial part, fresh, (2)

Ref No 2 02 443 United States

Feed Name or Analyses			As Fed	Dry	C.V. ± %
Dry matter	%			100.0	
Calcium	%			1.33	
Magnesium	%			1.42	
Phosphorus	%			0.19	
Potassium	%			2.10	

Junglerice, aerial part, fresh, mature, (2)

Ref No 2 02 442 United States

Feed Name or Analyses			As Fed	Dry	C.V. ± %
Dry matter	%			100.0	
Ash	%			13.6	
Crude fiber	%			29.6	
Ether extract	%			2.3	
N-free extract	%			37.1	
Protein (N x 6.25)	%			17.4	
Cattle	dig prot %	*		12.7	
Goats	dig prot %	*		12.8	
Horses	dig prot %	*		12.3	
Rabbits	dig prot %	*		12.1	
Sheep	dig prot %	*		13.2	
Calcium	%			0.83	
Magnesium	%			2.67	
Phosphorus	%			0.24	
Potassium	%			3.67	

JUNIPER, PINCHOT. Juniperus pinchotii

Juniper, pinchot, browse, (2)

Ref No 2 02 444 United States

Feed Name or Analyses			As Fed	Dry	C.V. ± %
Dry matter	%			100.0	
Ash	%			4.8	
Crude fiber	%			19.9	
Ether extract	%			10.3	
N-free extract	%			56.4	
Protein (N x 6.25)	%			8.6	
Cattle	dig prot %	*		5.2	
Goats	dig prot %	*		4.6	
Horses	dig prot %	*		4.8	
Rabbits	dig prot %	*		5.3	
Sheep	dig prot %	*		5.0	

JUNIPER, UTAH. Juniperus osteosperma

Juniper, Utah, browse, (2)

Ref No 2 02 445 United States

Feed Name or Analyses			As Fed	Dry	C.V. ± %
Dry matter	%			100.0	
Ash	%			4.5	
Crude fiber	%			22.0	
Ether extract	%			16.3	
N-free extract	%			50.8	
Protein (N x 6.25)	%			6.4	
Cattle	dig prot %	*		3.3	
Goats	dig prot %	*		2.5	
Horses	dig prot %	*		3.0	
Rabbits	dig prot %	*		3.6	
Sheep	dig prot %	*		3.0	
Calcium	%			1.59	
Magnesium	%			0.25	
Phosphorus	%			0.17	
Thiamine	mg/kg			2.4	

KALE. Brassica oleracea acephala

Kale, aerial part, fresh, (2)

Ref No 2 02 446 United States

Feed Name or Analyses			As Fed	Dry	C.V. ± %
Dry matter	%		11.6	100.0	
Ash	%		1.8	15.8	
Crude fiber	%		1.6	13.6	
Cattle	dig coef %		59.	59.	
Ether extract	%		0.5	4.5	
Cattle	dig coef %		66.	66.	
N-free extract	%		5.2	45.4	
Cattle	dig coef %		76.	76.	
Protein (N x 6.25)	%		2.4	20.8	
Cattle	dig coef %		81.	81.	
Cattle	dig prot %		1.9	16.8	
Goats	dig prot %	*	1.8	15.9	
Horses	dig prot %	*	1.8	15.2	
Rabbits	dig prot %	*	1.7	14.7	
Sheep	dig prot %	*	1.9	16.3	
Energy	GE Mcal/kg				
Cattle	DE Mcal/kg	*	0.34	2.91	
Sheep	DE Mcal/kg	*	0.35	3.02	
Cattle	ME Mcal/kg	*	0.28	2.38	
Sheep	ME Mcal/kg	*	0.29	2.48	
Cattle	TDN %		7.6	65.9	
Sheep	TDN %	*	8.0	68.6	
Calcium	%		0.19	1.61	
Phosphorus	%		0.06	0.51	

Kale, leaves, fresh, (2)

Ref No 2 02 447 United States

Feed Name or Analyses			As Fed	Dry	C.V. ± %
Dry matter	%		15.4	100.0	38
Carotene	mg/kg		55.1	357.5	55
Riboflavin	mg/kg		2.5	16.1	20
Vitamin A equivalent	IU/g		91.8	595.9	
Arginine	%		0.20	1.30	
Histidine	%		0.06	0.40	
Isoleucine	%		0.12	0.80	
Leucine	%		0.25	1.60	
Lysine	%		0.12	0.80	
Methionine	%		0.03	0.20	
Phenylalanine	%		0.17	1.10	
Threonine	%		0.14	0.90	
Tryptophan	%		0.05	0.30	
Valine	%		0.18	1.20	

Kale, leaves w stems, fresh, (2)

Ref No 2 02 449 United States

Feed Name or Analyses			As Fed	Dry	C.V. ± %
Dry matter	%		12.5	100.0	
Ash	%		1.5	12.0	
Crude fiber	%		1.2	9.5	
Ether extract	%		0.8	6.4	

Continued

Feed Name or Analyses				Mean As Fed	Mean Dry	C.V. ± %
N-free extract		%		5.6	45.1	
Protein (N x 6.25)		%		3.4	27.0	
Cattle	dig prot	%	*	2.6	20.8	
Goats	dig prot	%	*	2.7	21.8	
Horses	dig prot	%	*	2.6	20.5	
Rabbits	dig prot	%	*	2.4	19.5	
Sheep	dig prot	%	*	2.8	22.2	
Energy	GE Mcal/kg			0.38	3.04	
Cattle	DE Mcal/kg		*	0.38	3.04	
Sheep	DE Mcal/kg		*	0.38	3.04	
Cattle	ME Mcal/kg		*	0.31	2.50	
Sheep	ME Mcal/kg		*	0.31	2.49	
Cattle	TDN	%	*	8.6	69.0	
Sheep	TDN	%	*	8.6	68.9	
Calcium		%		0.18	1.43	
Iron		%		0.002	0.018	
Phosphorus		%		0.07	0.58	
Potassium		%		0.38	3.02	
Sodium		%		0.08	0.60	
Carotene		mg/kg		19.5	155.6	
Riboflavin		mg/kg		1.4	10.8	
Vitamin A		IU/g		89.0	712.0	
Vitamin A equivalent		IU/g		32.4	259.5	

Kale, stems, fresh, (2)

Ref No 2 02 450 United States

Feed Name or Analyses				Mean As Fed	Mean Dry	C.V. ± %
Dry matter		%		13.3	100.0	
Ash		%		1.2	8.8	
Crude fiber		%		2.0	15.3	16
Sheep	dig coef	%		51.	51.	
Ether extract		%		0.3	2.5	21
Sheep	dig coef	%		69.	69.	
N-free extract		%		8.0	60.1	
Sheep	dig coef	%		88.	88.	
Protein (N x 6.25)		%		1.8	13.3	28
Sheep	dig coef	%		70.	70.	
Cattle	dig prot	%	*	1.2	9.2	
Goats	dig prot	%	*	1.2	9.0	
Horses	dig prot	%	*	1.2	8.8	
Rabbits	dig prot	%	*	1.2	8.9	
Sheep	dig prot	%		1.2	9.3	
Energy	GE Mcal/kg					
Cattle	DE Mcal/kg		*	0.38	2.88	
Sheep	DE Mcal/kg		*	0.43	3.26	
Cattle	ME Mcal/kg		*	0.31	2.36	
Sheep	ME Mcal/kg		*	0.35	2.67	
Cattle	TDN	%	*	8.7	65.3	
Sheep	TDN	%		9.8	73.9	
Carotene		mg/kg		5.6	42.1	72
Riboflavin		mg/kg		1.3	9.7	25
α-tocopherol		mg/kg		1.7	12.6	
Vitamin A equivalent		IU/g		9.3	70.2	

Kale, aerial part, ensiled, (3)

Ref No 3 02 451 United States

Feed Name or Analyses				Mean As Fed	Mean Dry	C.V. ± %
Dry matter		%		13.3	100.0	12
Ash		%		2.0	15.1	9
Crude fiber		%		2.7	20.1	5
Ether extract		%		0.5	3.4	17
N-free extract		%		6.4	48.0	

(1) dry forages and roughages (2) pasture, range plants, and forages fed green (3) silages (4) energy feeds (5) protein supplements (6) minerals (7) vitamins (8) additives

Feed Name or Analyses				Mean As Fed	Mean Dry	C.V. ± %
Protein (N x 6.25)		%		1.8	13.4	13
Cattle	dig prot	%	*	1.1	8.4	
Goats	dig prot	%	*	1.1	8.4	
Horses	dig prot	%	*	1.1	8.4	
Sheep	dig prot	%	*	1.1	8.4	
Energy	GE Mcal/kg					
Cattle	DE Mcal/kg		*	0.40	2.97	
Sheep	DE Mcal/kg		*	0.36	2.71	
Cattle	ME Mcal/kg		*	0.32	2.44	
Sheep	ME Mcal/kg		*	0.30	2.22	
Cattle	TDN	%	*	9.0	67.3	
Sheep	TDN	%	*	8.2	61.4	

KALE, BLUE THICKSTEM. Brassica oleracea acephala

Kale, blue thickstem, aerial part, fresh, (2)

Ref No 2 02 452 United States

Feed Name or Analyses				Mean As Fed	Mean Dry	C.V. ± %
Dry matter		%		9.3	100.0	
Ash		%		1.3	14.2	
Crude fiber		%		2.0	21.0	
Sheep	dig coef	%		57.	57.	
Ether extract		%		0.2	1.9	
Sheep	dig coef	%		100.	100.	
N-free extract		%		4.1	44.4	
Sheep	dig coef	%		83.	83.	
Protein (N x 6.25)		%		1.7	18.5	
Sheep	dig coef	%		92.	92.	
Cattle	dig prot	%	*	1.3	13.6	
Goats	dig prot	%	*	1.3	13.8	
Horses	dig prot	%	*	1.2	13.2	
Rabbits	dig prot	%	*	1.2	13.0	
Sheep	dig prot	%		1.6	17.0	
Energy	GE Mcal/kg					
Cattle	DE Mcal/kg		*	0.27	2.87	
Sheep	DE Mcal/kg		*	0.29	3.09	
Cattle	ME Mcal/kg		*	0.22	2.35	
Sheep	ME Mcal/kg		*	0.24	2.54	
Cattle	TDN	%	*	6.1	65.1	
Sheep	TDN	%		6.5	70.1	

KALE, MARROW. Brassica oleracea acephala

Kale, marrow, aerial part, s-c, (1)

Ref No 1 02 453 United States

Feed Name or Analyses				Mean As Fed	Mean Dry	C.V. ± %
Dry matter		%		88.8	100.0	
Ash		%		13.1	14.8	
Crude fiber		%		17.0	19.2	
Cattle	dig coef	%		53.	53.	
Sheep	dig coef	%		60.	60.	
Ether extract		%		2.0	2.3	
Cattle	dig coef	%		25.	25.	
Sheep	dig coef	%		57.	57.	
N-free extract		%		44.8	50.5	
Cattle	dig coef	%		77.	77.	
Sheep	dig coef	%		90.	90.	
Protein (N x 6.25)		%		11.8	13.3	
Cattle	dig coef	%		68.	68.	
Sheep	dig coef	%		72.	72.	
Cattle	dig prot	%		8.0	9.0	
Goats	dig prot	%	*	7.9	8.9	
Horses	dig prot	%	*	7.8	8.8	
Rabbits	dig prot	%	*	7.9	8.9	
Sheep	dig prot	%		8.5	9.5	
Energy	GE Mcal/kg					
Cattle	DE Mcal/kg		*	2.32	2.62	
Sheep	DE Mcal/kg		*	2.72	3.06	

Feed Name or Analyses				Mean As Fed	Mean Dry	C.V. ± %
Cattle	ME Mcal/kg		*	1.90	2.15	
Sheep	ME Mcal/kg		*	2.23	2.51	
Cattle	TDN	%		52.7	59.3	
Sheep	TDN	%		61.6	69.4	

Kale, marrow, aerial part, fresh, (2)

Ref No 2 02 456 United States

Feed Name or Analyses				Mean As Fed	Mean Dry	C.V. ± %
Dry matter		%		14.1	100.0	
Ash		%		1.9	13.6	
Crude fiber		%		2.3	16.0	
Cattle	dig coef	%		48.	48.	
Sheep	dig coef	%		70.	70.	
Ether extract		%		0.4	2.9	
Cattle	dig coef	%		66.	66.	
Sheep	dig coef	%		60.	60.	
N-free extract		%		7.5	53.4	
Cattle	dig coef	%		88.	88.	
Sheep	dig coef	%		89.	89.	
Protein (N x 6.25)		%		2.0	14.2	
Cattle	dig coef	%		75.	75.	
Sheep	dig coef	%		78.	78.	
Cattle	dig prot	%		1.5	10.7	
Goats	dig prot	%	*	1.4	9.8	
Horses	dig prot	%	*	1.4	9.6	
Rabbits	dig prot	%	*	1.4	9.6	
Sheep	dig prot	%		1.6	11.1	
Energy	GE Mcal/kg					
Cattle	DE Mcal/kg		*	0.43	3.07	
Sheep	DE Mcal/kg		*	0.46	3.23	
Cattle	ME Mcal/kg		*	0.36	2.51	
Sheep	ME Mcal/kg		*	0.37	2.65	
Cattle	TDN	%		9.8	69.5	
Sheep	TDN	%		10.4	73.3	

Kale, marrow, aerial part, fresh, early leaf, (2)

Ref No 2 02 454 United States

Feed Name or Analyses				Mean As Fed	Mean Dry	C.V. ± %
Dry matter		%		13.8	100.0	
Ash		%		1.8	13.2	
Crude fiber		%		2.5	18.3	
Sheep	dig coef	%		61.	61.	
Ether extract		%		0.4	2.9	
Sheep	dig coef	%		57.	57.	
N-free extract		%		7.0	50.7	
Sheep	dig coef	%		89.	89.	
Protein (N x 6.25)		%		2.1	15.0	
Sheep	dig coef	%		79.	79.	
Cattle	dig prot	%	*	1.5	10.6	
Goats	dig prot	%	*	1.5	10.6	
Horses	dig prot	%	*	1.4	10.3	
Rabbits	dig prot	%	*	1.4	10.3	
Sheep	dig prot	%		1.6	11.9	
Energy	GE Mcal/kg					
Cattle	DE Mcal/kg		*	0.42	3.02	
Sheep	DE Mcal/kg		*	0.44	3.15	
Cattle	ME Mcal/kg		*	0.34	2.48	
Sheep	ME Mcal/kg		*	0.36	2.59	
Cattle	TDN	%	*	9.4	68.4	
Sheep	TDN	%		9.9	71.5	

Kale, marrow, aerial part, fresh, pre-bloom, (2)

Ref No 2 02 455 United States

Feed Name or Analyses				Mean As Fed	Mean Dry	C.V. ± %
Dry matter		%		12.8	100.0	
Ash		%		2.0	15.8	

Feed Name or Analyses				Mean As Fed	Mean Dry	C.V. ± %
Crude fiber		%		1.5	11.5	
Sheep	dig coef	%		74.	74.	
Ether extract		%		0.4	3.4	
Sheep	dig coef	%		68.	68.	
N-free extract		%		6.0	47.2	
Sheep	dig coef	%		87.	87.	
Protein (N x 6.25)		%		2.8	22.1	
Sheep	dig coef	%		88.	88.	
Cattle	dig prot	%	*	2.1	16.7	
Goats	dig prot	%	*	2.2	17.2	
Horses	dig prot	%	*	2.1	16.3	
Rabbits	dig prot	%	*	2.0	15.7	
Sheep	dig prot	%		2.5	19.4	
Energy	GE Mcal/kg					
Cattle	DE Mcal/kg		*	0.40	3.14	
Sheep	DE Mcal/kg		*	0.42	3.27	
Cattle	ME Mcal/kg		*	0.33	2.57	
Sheep	ME Mcal/kg		*	0.34	2.68	
Cattle	TDN	%	*	9.1	71.2	
Sheep	TDN	%		9.5	74.2	

Kale, marrow, leaves, fresh, (2)

Ref No 2 02 457 United States

				As Fed	Dry	C.V.
Dry matter		%		17.9	100.0	
Ash		%		2.1	11.6	
Crude fiber		%		1.6	8.7	
Sheep	dig coef	%		90.	90.	
Ether extract		%		0.6	3.5	
Sheep	dig coef	%		47.	47.	
N-free extract		%		10.4	57.9	
Sheep	dig coef	%		90.	90.	
Protein (N x 6.25)		%		3.3	18.3	
Sheep	dig coef	%		83.	83.	
Cattle	dig prot	%	*	2.4	13.4	
Goats	dig prot	%	*	2.4	13.6	
Horses	dig prot	%	*	2.3	13.1	
Rabbits	dig prot	%	*	2.3	12.8	
Sheep	dig prot	%		2.7	15.2	
Energy	GE Mcal/kg					
Cattle	DE Mcal/kg		*	0.58	3.26	
Sheep	DE Mcal/kg		*	0.62	3.48	
Cattle	ME Mcal/kg		*	0.48	2.67	
Sheep	ME Mcal/kg		*	0.51	2.85	
Cattle	TDN	%	*	13.2	74.0	
Sheep	TDN	%		14.1	78.8	

Kale, marrow, stems, fresh, (2)

Ref No 2 08 449 United States

				As Fed	Dry	C.V.
Dry matter		%		13.8	100.0	
Ash		%		1.9	13.8	
Crude fiber		%		2.2	15.9	
Ether extract		%		0.4	2.9	
N-free extract		%		7.1	51.4	
Protein (N x 6.25)		%		2.2	15.9	
Cattle	dig prot	%	*	1.6	11.4	
Goats	dig prot	%	*	1.6	11.4	
Horses	dig prot	%	*	1.5	11.1	
Rabbits	dig prot	%	*	1.5	11.0	
Sheep	dig prot	%	*	1.6	11.8	
Energy	GE Mcal/kg					
Cattle	DE Mcal/kg		*	0.43	3.10	
Sheep	DE Mcal/kg		*	0.46	3.37	
Cattle	ME Mcal/kg		*	0.35	2.54	
Sheep	ME Mcal/kg		*	0.38	2.76	

Feed Name or Analyses				Mean As Fed	Mean Dry	C.V. ± %
Cattle	TDN	%	*	9.7	70.4	
Sheep	TDN	%	*	10.6	76.5	

Kale, marrow, aerial part, ensiled, (3)

Ref No 3 02 459 United States

				As Fed	Dry	C.V.
Dry matter		%		13.3	100.0	
Ash		%		2.0	15.1	
Crude fiber		%		2.5	19.0	
Sheep	dig coef	%		68.	68.	
Ether extract		%		0.5	4.0	
Sheep	dig coef	%		65.	65.	
N-free extract		%		6.2	46.7	
Sheep	dig coef	%		88.	88.	
Protein (N x 6.25)		%		2.0	15.3	
Sheep	dig coef	%		80.	80.	
Cattle	dig prot	%	*	1.3	10.1	
Goats	dig prot	%	*	1.3	10.1	
Horses	dig prot	%	*	1.3	10.1	
Sheep	dig prot	%		1.6	12.1	
Energy	GE Mcal/kg					
Cattle	DE Mcal/kg		*	0.41	3.08	
Sheep	DE Mcal/kg		*	0.42	3.16	
Cattle	ME Mcal/kg		*	0.34	2.53	
Sheep	ME Mcal/kg		*	0.34	2.59	
Cattle	TDN	%	*	9.3	69.8	
Sheep	TDN	%		9.5	71.6	

Kale, marrow, aerial part, ensiled, pre-bloom, (3)

Ref No 3 02 458 United States

				As Fed	Dry	C.V.
Dry matter		%		15.9	100.0	
Ash		%		2.5	15.8	
Crude fiber		%		3.8	23.6	
Sheep	dig coef	%		75.	75.	
Ether extract		%		0.5	3.1	
Sheep	dig coef	%		71.	71.	
N-free extract		%		7.2	45.2	
Sheep	dig coef	%		84.	84.	
Protein (N x 6.25)		%		2.0	12.3	
Sheep	dig coef	%		78.	78.	
Cattle	dig prot	%	*	1.2	7.4	
Goats	dig prot	%	*	1.2	7.4	
Horses	dig prot	%	*	1.2	7.4	
Sheep	dig prot	%		1.5	9.6	
Energy	GE Mcal/kg					
Cattle	DE Mcal/kg		*	0.48	3.02	
Sheep	DE Mcal/kg		*	0.49	3.10	
Cattle	ME Mcal/kg		*	0.39	2.48	
Sheep	ME Mcal/kg		*	0.40	2.54	
Cattle	TDN	%	*	10.9	68.5	
Sheep	TDN	%		11.2	70.2	

KALE, THOUSAND HEAD. Brassica oleracea acephala

Kale, thousand head, aerial part, fresh, early leaf, (2)

Ref No 2 02 460 United States

				As Fed	Dry	C.V.
Dry matter		%		15.8	100.0	
Ash		%		1.7	10.6	
Crude fiber		%		3.1	19.9	
Sheep	dig coef	%		59.	59.	
Ether extract		%		0.4	2.4	
Sheep	dig coef	%		44.	44.	
N-free extract		%		8.5	53.5	
Sheep	dig coef	%		89.	89.	

Feed Name or Analyses				Mean As Fed	Mean Dry	C.V. ± %
Protein (N x 6.25)		%		2.1	13.6	
Sheep	dig coef	%		77.	77.	
Cattle	dig prot	%	*	1.5	9.5	
Goats	dig prot	%	*	1.5	9.2	
Horses	dig prot	%	*	1.4	9.1	
Rabbits	dig prot	%	*	1.4	9.2	
Sheep	dig prot	%		1.7	10.5	
Energy	GE Mcal/kg					
Cattle	DE Mcal/kg		*	0.49	3.08	
Sheep	DE Mcal/kg		*	0.50	3.18	
Cattle	ME Mcal/kg		*	0.40	2.53	
Sheep	ME Mcal/kg		*	0.41	2.61	
Cattle	TDN	%	*	11.0	69.8	
Sheep	TDN	%		11.4	72.2	

KALIHARIGRASS. Schmidtia kalihariensis

Kaliharigrass, aerial part, fresh, mature, (2)

Ref No 2 02 461 United States

				As Fed	Dry	C.V.
Dry matter		%		26.0	100.0	
Ash		%		2.0	7.5	
Crude fiber		%		10.3	39.7	
Sheep	dig coef	%		72.	72.	
Ether extract		%		0.6	2.3	
Sheep	dig coef	%		56.	56.	
N-free extract		%		12.4	47.8	
Sheep	dig coef	%		66.	66.	
Protein (N x 6.25)		%		0.7	2.7	
Sheep	dig coef	%		11.	11.	
Cattle	dig prot	%	*	0.0	0.2	
Goats	dig prot	%	*	-0.1	-0.8	
Horses	dig prot	%	*	0.0	-0.1	
Rabbits	dig prot	%	*	0.2	0.8	
Sheep	dig prot	%		0.1	0.3	
Energy	GE Mcal/kg					
Cattle	DE Mcal/kg		*	0.69	2.66	
Sheep	DE Mcal/kg		*	0.73	2.79	
Cattle	ME Mcal/kg		*	0.57	2.18	
Sheep	ME Mcal/kg		*	0.60	2.29	
Cattle	TDN	%	*	15.7	60.4	
Sheep	TDN	%		16.5	63.3	

KANGAROOGRASS. Themeda spp

Kangaroograss, aerial part, s-c, mature, (1)

Ref No 1 02 462 United States

				As Fed	Dry	C.V.
Dry matter		%		87.0	100.0	
Ash		%		11.5	13.2	
Crude fiber		%		29.2	33.6	
Sheep	dig coef	%		52.	52.	
Ether extract		%		1.5	1.7	
Sheep	dig coef	%		44.	44.	
N-free extract		%		42.1	48.4	
Sheep	dig coef	%		49.	49.	
Protein (N x 6.25)		%		2.7	3.1	
Sheep	dig coef	%		4.	4.	
Cattle	dig prot	%	*	-0.2	-0.3	
Goats	dig prot	%	*	-0.4	-0.4	
Horses	dig prot	%	*	0.1	0.2	
Rabbits	dig prot	%	*	0.9	1.1	
Sheep	dig prot	%		0.1	0.1	
Energy	GE Mcal/kg					
Cattle	DE Mcal/kg		*	1.75	2.01	
Sheep	DE Mcal/kg		*	1.65	1.90	
Cattle	ME Mcal/kg		*	1.44	1.65	

Continued

Feed Name or Analyses			Mean As Fed	Mean Dry	C.V. ± %
Sheep	ME Mcal/kg	*	1.35	1.55	
Cattle	TDN %	*	39.6	45.5	
Sheep	TDN %		37.4	43.0	

Kangaroograss, aerial part, fresh, (2)

Ref No 2 02 465. United States

Feed Name or Analyses			Mean As Fed	Mean Dry	C.V. ± %
Dry matter	%		17.5	100.0	
Ash	%		2.2	12.3	
Crude fiber	%		5.5	31.3	
Sheep	dig coef %		62.	62.	
Ether extract	%		0.5	3.1	
Sheep	dig coef %		46.	46.	
N-free extract	%		7.9	45.1	
Sheep	dig coef %		60.	60.	
Protein (N x 6.25)	%		1.4	8.2	
Sheep	dig coef %		57.	57.	
Cattle	dig prot %	*	0.9	4.9	
Goats	dig prot %	*	0.7	4.2	
Horses	dig prot %	*	0.8	4.5	
Rabbits	dig prot %	*	0.9	5.0	
Sheep	dig prot %		0.8	4.7	
Energy	GE Mcal/kg				
Cattle	DE Mcal/kg	*	0.44	2.50	
Sheep	DE Mcal/kg	*	0.42	2.40	
Cattle	ME Mcal/kg	*	0.36	2.05	
Sheep	ME Mcal/kg	*	0.34	1.96	
Cattle	TDN %	*	9.9	56.6	
Sheep	TDN %		9.5	54.3	

Kangaroograss, aerial part, fresh, early leaf, (2)

Ref No 2 02 463 United States

Feed Name or Analyses			Mean As Fed	Mean Dry	C.V. ± %
Dry matter	%		15.0	100.0	
Ash	%		1.9	12.6	
Crude fiber	%		4.5	30.3	
Sheep	dig coef %		64.	64.	
Ether extract	%		0.5	3.2	
Sheep	dig coef %		41.	41.	
N-free extract	%		6.6	44.3	
Sheep	dig coef %		62.	62.	
Protein (N x 6.25)	%		1.4	9.6	
Sheep	dig coef %		62.	62.	
Cattle	dig prot %	*	0.9	6.1	
Goats	dig prot %	*	0.8	5.5	
Horses	dig prot %	*	0.9	5.7	
Rabbits	dig prot %	*	0.9	6.1	
Sheep	dig prot %		0.9	6.0	
Energy	GE Mcal/kg				
Cattle	DE Mcal/kg	*	0.38	2.53	
Sheep	DE Mcal/kg	*	0.37	2.46	
Cattle	ME Mcal/kg	*	0.31	2.07	
Sheep	ME Mcal/kg	*	0.30	2.02	
Cattle	TDN %	*	8.6	57.4	
Sheep	TDN %		8.4	55.8	

Kangaroograss, aerial part, fresh, full bloom, (2)

Ref No 2 02 464 United States

Feed Name or Analyses			Mean As Fed	Mean Dry	C.V. ± %
Dry matter	%		20.0	100.0	
Ash	%		2.4	12.0	

(1) dry forages and roughages (2) pasture, range plants, and forages fed green (3) silages (4) energy feeds (5) protein supplements (6) minerals (7) vitamins (8) additives

Feed Name or Analyses			Mean As Fed	Mean Dry	C.V. ± %
Crude fiber	%		6.5	32.3	
Sheep	dig coef %		59.	59.	
Ether extract	%		0.6	3.0	
Sheep	dig coef %		50.	50.	
N-free extract	%		9.2	45.8	
Sheep	dig coef %		57.	57.	
Protein (N x 6.25)	%		1.4	6.9	
Sheep	dig coef %		52.	52.	
Cattle	dig prot %	*	0.8	3.8	
Goats	dig prot %	*	0.6	3.0	
Horses	dig prot %	*	0.7	3.4	
Rabbits	dig prot %	*	0.8	4.0	
Sheep	dig prot %		0.7	3.6	
Energy	GE Mcal/kg				
Cattle	DE Mcal/kg	*	0.49	2.47	
Sheep	DE Mcal/kg	*	0.46	2.30	
Cattle	ME Mcal/kg	*	0.40	2.02	
Sheep	ME Mcal/kg	*	0.38	1.88	
Cattle	TDN %	*	11.2	55.9	
Sheep	TDN %		10.4	52.1	

KAPOK. Ceiba pentandra

Kapok, hulls, (1)

Ref No 1 02 466 United States

Feed Name or Analyses			Mean As Fed	Mean Dry	C.V. ± %
Dry matter	%		90.0	100.0	
Ash	%		4.5	5.0	
Crude fiber	%		36.8	40.9	
Ether extract	%		4.5	5.0	
N-free extract	%		36.2	40.2	
Protein (N x 6.25)	%		8.0	8.9	
Cattle	dig prot %	*	4.2	4.6	
Goats	dig prot %	*	4.4	4.9	
Horses	dig prot %	*	4.6	5.1	
Rabbits	dig prot %	*	5.0	5.5	
Sheep	dig prot %	*	4.1	4.6	
Energy	GE Mcal/kg				
Cattle	DE Mcal/kg	*	1.79	1.99	
Sheep	DE Mcal/kg	*	1.67	1.86	
Cattle	ME Mcal/kg	*	1.47	1.63	
Sheep	ME Mcal/kg	*	1.38	1.53	
Cattle	TDN %	*	40.6	45.1	
Sheep	TDN %	*	37.9	42.1	

Kapok, seeds, extn unspecified grnd, 22% protein, (5)

Ref No 5 02 467 United States

Feed Name or Analyses			Mean As Fed	Mean Dry	C.V. ± %
Dry matter	%		87.0	100.0	
Ash	%		7.7	8.9	
Crude fiber	%		19.0	21.8	
Cattle	dig coef %		11.	11.	
Ether extract	%		4.9	5.6	
Cattle	dig coef %		51.	51.	
N-free extract	%		33.0	37.9	
Cattle	dig coef %		71.	71.	
Protein (N x 6.25)	%		22.4	25.8	
Cattle	dig coef %		41.	41.	
Cattle	dig prot %		9.2	10.6	
Energy	GE Mcal/kg				
Cattle	DE Mcal/kg	*	1.78	2.04	
Sheep	DE Mcal/kg	*	1.86	2.14	
Swine	DE kcal/kg	*	1964.	2257.	
Cattle	ME Mcal/kg	*	1.46	1.67	
Sheep	ME Mcal/kg	*	1.52	1.75	
Swine	ME kcal/kg	*	1783.	2049.	
Cattle	TDN %		40.3	46.3	

Feed Name or Analyses			Mean As Fed	Mean Dry	C.V. ± %
Sheep	TDN %	*	42.2	48.5	
Swine	TDN %	*	44.5	51.2	

Kapok, seeds, extn unspecified grnd, 30% protein (5)

Ref No 5 02 468 United States

Feed Name or Analyses			Mean As Fed	Mean Dry	C.V. ± %
Dry matter	%		87.0	100.0	
Ash	%		6.4	7.4	
Crude fiber	%		20.0	23.0	
Sheep	dig coef %		10.	10.	
Ether extract	%		7.2	8.3	
Sheep	dig coef %		97.	97.	
N-free extract	%		22.7	26.1	
Sheep	dig coef %		51.	51.	
Protein (N x 6.25)	%		30.6	35.2	
Sheep	dig coef %		89.	89.	
Sheep	dig prot %		27.3	31.3	
Energy	GE Mcal/kg				
Cattle	DE Mcal/kg	*	2.51	2.89	
Sheep	DE Mcal/kg	*	2.50	2.87	
Swine	DE kcal/kg	*	3015.	3466.	
Cattle	ME Mcal/kg	*	2.06	2.37	
Sheep	ME Mcal/kg	*	2.05	2.35	
Swine	ME kcal/kg	*	2680.	3081.	
Cattle	TDN %	*	57.0	65.5	
Sheep	TDN %		56.6	65.1	
Swine	TDN %	*	68.4	78.6	

Kapok, seeds, extn unspecified grnd, 40% protein, (5)

Ref No 5 02 469 United States

Feed Name or Analyses			Mean As Fed	Mean Dry	C.V. ± %
Dry matter	%		89.3	100.0	
Ash	%		7.7	8.6	
Crude fiber	%		17.1	19.1	
Sheep	dig coef %		12.	12.	
Ether extract	%		7.9	8.8	
Sheep	dig coef %		94.	94.	
N-free extract	%		14.6	16.3	
Sheep	dig coef %		42.	42.	
Protein (N x 6.25)	%		42.1	47.2	
Sheep	dig coef %		88.	88.	
Sheep	dig prot %		37.1	41.5	
Energy	GE Mcal/kg				
Cattle	DE Mcal/kg	*	2.66	2.98	
Sheep	DE Mcal/kg	*	2.73	3.05	
Swine	DE kcal/kg	*	3000.	3360.	
Cattle	ME Mcal/kg	*	2.18	2.44	
Sheep	ME Mcal/kg	*	2.24	2.51	
Swine	ME kcal/kg	*	2594.	2905.	
Cattle	TDN %	*	60.3	67.6	
Sheep	TDN %		61.9	69.3	
Swine	TDN %	*	68.0	76.2	

KARROOBUSH. Phymaspermum parvifolium

Karroobush, aerial part, s-c, late bloom, (1)

Ref No 1 02 470 United States

Feed Name or Analyses			Mean As Fed	Mean Dry	C.V. ± %
Dry matter	%		87.0	100.0	
Ash	%		9.2	10.6	
Crude fiber	%		29.0	33.3	
Sheep	dig coef %		13.	13.	
Ether extract	%		5.1	5.9	
Sheep	dig coef %		49.	49.	

Feed Name or Analyses			Mean As Fed	Mean Dry	C.V. ± %
N-free extract	%		34.3	39.4	
Sheep	dig coef %		50.	50.	
Protein (N x 6.25)	%		9.4	10.8	
Sheep	dig coef %		60.	60.	
Cattle	dig prot %	*	5.5	6.3	
Goats	dig prot %	*	5.8	6.6	
Horses	dig prot %	*	5.8	6.7	
Rabbits	dig prot %	*	6.1	7.0	
Sheep	dig prot %		5.6	6.5	
Energy	GE Mcal/kg				
Cattle	DE Mcal/kg	*	1.52	1.75	
Sheep	DE Mcal/kg	*	1.42	1.63	
Cattle	ME Mcal/kg	*	1.25	1.44	
Sheep	ME Mcal/kg	*	1.16	1.34	
Cattle	TDN %	*	34.5	39.7	
Sheep	TDN %		32.2	37.0	

Kelp, dried –
see Seaweed, kelp, entire plant, dehy, (1)

KIKUYUGRASS. Pennisetum clandestinum

Kikuyugrass, aerial part, fresh, immature, (2)
Ref No 2 05 666 United States

			As Fed	Dry	C.V.
Dry matter	%			100.0	
Ash	%			10.9	
Crude fiber	%			22.9	
Ether extract	%			1.9	
N-free extract	%			46.6	
Protein (N x 6.25)	%			17.7	
Cattle	dig prot %	*		12.9	
Goats	dig prot %	*		13.1	
Horses	dig prot %	*		12.6	
Rabbits	dig prot %	*		12.3	
Sheep	dig prot %	*		13.5	
Energy	GE Mcal/kg				
Cattle	DE Mcal/kg	*		3.06	
Sheep	DE Mcal/kg	*		2.88	
Cattle	ME Mcal/kg	*		2.51	
Sheep	ME Mcal/kg	*		2.36	
Cattle	TDN %	*		69.5	
Sheep	TDN %	*		65.3	
Calcium	%			0.43	
Iron	%			0.023	
Magnesium	%			0.28	
Phosphorus	%			0.25	
Potassium	%			4.59	
Carotene	mg/kg			256.6	
Niacin	mg/kg			49.8	
Riboflavin	mg/kg			20.5	
Vitamin A equivalent	IU/g			427.8	

Kikuyugrass, aerial part, fresh, pre-bloom, (2)
Ref No 2 05 665 Guatemala

			As Fed	Dry	C.V.
Dry matter	%			100.0	
Ash	%			12.6	
Crude fiber	%			39.3	
Sheep	dig coef %			63.	
Ether extract	%			4.3	
Sheep	dig coef %			57.	
N-free extract	%			28.6	
Sheep	dig coef %			73.	
Protein (N x 6.25)	%			15.2	
Sheep	dig coef %			62.	
Cattle	dig prot %	*		10.8	
Goats	dig prot %	*		10.7	

Feed Name or Analyses			Mean As Fed	Mean Dry	C.V. ± %
Horses	dig prot %	*		10.4	
Rabbits	dig prot %	*		10.4	
Sheep	dig prot %			9.4	
Energy	GE Mcal/kg				
Cattle	DE Mcal/kg	*		2.81	
Sheep	DE Mcal/kg	*		2.67	
Cattle	ME Mcal/kg	*		2.31	
Sheep	ME Mcal/kg	*		2.19	
Cattle	TDN %	*		63.8	
Sheep	TDN %			60.6	

KNOTGRASS. Paspalum distichum

Knotgrass, aerial part, fresh, (2)
Ref No 2 02 475 United States

			As Fed	Dry	C.V.
Dry matter	%			100.0	
Ash	%			12.5	16
Crude fiber	%			28.8	12
Ether extract	%			2.3	19
N-free extract	%			46.8	
Protein (N x 6.25)	%			9.6	18
Cattle	dig prot %	*		6.1	
Goats	dig prot %	*		5.5	
Horses	dig prot %	*		5.7	
Rabbits	dig prot %	*		6.1	
Sheep	dig prot %	*		5.9	
Energy	GE Mcal/kg				
Cattle	DE Mcal/kg	*		2.58	
Sheep	DE Mcal/kg	*		2.90	
Cattle	ME Mcal/kg	*		2.12	
Sheep	ME Mcal/kg	*		2.38	
Cattle	TDN %	*		58.6	
Sheep	TDN %	*		65.8	
Calcium	%			0.67	28
Magnesium	%			0.18	
Phosphorus	%			0.38	37
Potassium	%			2.39	

Knotgrass, aerial part, fresh, immature, (2)
Ref No 2 02 472 United States

			As Fed	Dry	C.V.
Dry matter	%			100.0	
Ash	%			13.1	
Crude fiber	%			29.3	
Ether extract	%			2.2	
N-free extract	%			45.1	
Protein (N x 6.25)	%			10.3	
Cattle	dig prot %	*		6.6	
Goats	dig prot %	*		6.2	
Horses	dig prot %	*		6.3	
Rabbits	dig prot %	*		6.6	
Sheep	dig prot %	*		6.6	
Energy	GE Mcal/kg				
Cattle	DE Mcal/kg	*		2.59	
Sheep	DE Mcal/kg	*		2.86	
Cattle	ME Mcal/kg	*		2.12	
Sheep	ME Mcal/kg	*		2.35	
Cattle	TDN %	*		58.7	
Sheep	TDN %	*		64.9	
Calcium	%			0.83	
Phosphorus	%			0.52	

Feed Name or Analyses			Mean As Fed	Mean Dry	C.V. ± %

Knotgrass, aerial part, fresh, mid-bloom, (2)
Ref No 2 02 473 United States

			As Fed	Dry	C.V.
Dry matter	%			100.0	
Ash	%			13.8	
Crude fiber	%			24.9	
Ether extract	%			2.0	
N-free extract	%			52.1	
Protein (N x 6.25)	%			7.2	
Cattle	dig prot %	*		4.0	
Goats	dig prot %	*		3.3	
Horses	dig prot %	*		3.6	
Rabbits	dig prot %	*		4.2	
Sheep	dig prot %	*		3.7	
Energy	GE Mcal/kg				
Cattle	DE Mcal/kg	*		2.57	
Sheep	DE Mcal/kg	*		2.74	
Cattle	ME Mcal/kg	*		2.11	
Sheep	ME Mcal/kg	*		2.25	
Cattle	TDN %	*		58.4	
Sheep	TDN %	*		62.1	

Knotgrass, aerial part, fresh, full bloom, (2)
Ref No 2 02 474 United States

			As Fed	Dry	C.V.
Dry matter	%			100.0	
Calcium	%			0.57	
Phosphorus	%			0.23	

KNOTWEED. Polygonum spp

Knotweed, aerial part, fresh, (2)
Koa haole –
see Leadtree, whitepopinac, aerial part, fresh, (2)
Ref No 2 02 476 United States

			As Fed	Dry	C.V.
Dry matter	%			100.0	
Ash	%			2.8	
Crude fiber	%			19.0	
Ether extract	%			2.9	
N-free extract	%			66.0	
Protein (N x 6.25)	%			9.3	
Cattle	dig prot %	*		5.8	
Goats	dig prot %	*		5.2	
Horses	dig prot %	*		5.4	
Rabbits	dig prot %	*		5.8	
Sheep	dig prot %	*		5.7	
Energy	GE Mcal/kg				
Cattle	DE Mcal/kg	*		2.62	
Sheep	DE Mcal/kg	*		2.70	
Cattle	ME Mcal/kg	*		2.15	
Sheep	ME Mcal/kg	*		2.21	
Cattle	TDN %	*		59.5	
Sheep	TDN %	*		61.3	

Koa haole –
see Leadtree, whitepopinac, hay, s-c, (1)

Kochia scoparia hay –
see Summercypress, belvedere, hay, s-c, (1)

Feed Name or Analyses			Mean As Fed	Mean Dry	C.V. ± %

KOHLRABI. Brassica oleracea gongylodes

Kohlrabi, aerial part, fresh, (2)
Ref No 2 02 477 United States

Analyses	Unit		As Fed	Dry	C.V. ± %
Dry matter	%		9.0	100.0	
Ash	%		1.3	14.4	
Crude fiber	%		1.3	14.4	
Ether extract	%		0.1	1.1	
N-free extract	%		4.3	47.8	
Protein (N x 6.25)	%		2.0	22.2	
Cattle	dig prot %	*	1.5	16.8	
Goats	dig prot %	*	1.6	17.3	
Horses	dig prot %	*	1.5	16.4	
Rabbits	dig prot %	*	1.4	15.8	
Sheep	dig prot %	*	1.6	17.7	
Calcium	%		0.08	0.89	
Phosphorus	%		0.07	0.78	
Potassium	%		0.37	4.11	

KUDZU. Pueraria spp

Kudzu, hay, s-c, (1)
Ref No 1 02 478 United States

Analyses	Unit		As Fed	Dry	C.V. ± %
Dry matter	%		91.5	100.0	1
Ash	%		6.7	7.3	
Crude fiber	%		29.9	32.7	
Ether extract	%		2.3	2.6	
N-free extract	%		37.0	40.4	
Protein (N x 6.25)	%		15.5	16.9	
Cattle	dig prot %	*	10.6	11.6	
Goats	dig prot %	*	11.3	12.4	
Horses	dig prot %	*	10.9	11.9	
Rabbits	dig prot %	*	10.7	11.7	
Sheep	dig prot %	*	10.8	11.8	
Fatty acids	%		2.3	2.6	
Energy	GE Mcal/kg				
Cattle	DE Mcal/kg	*	2.25	2.46	
Sheep	DE Mcal/kg	*	2.30	2.52	
Cattle	ME Mcal/kg	*	1.84	2.02	
Sheep	ME Mcal/kg	*	1.89	2.06	
Cattle	TDN %	*	51.0	55.8	
Sheep	TDN %	*	52.2	57.1	
Calcium	%		2.15	2.35	
Magnesium	%		0.73	0.80	
Phosphorus	%		0.32	0.35	
Carotene	mg/kg		40.3	44.1	
Riboflavin	mg/kg		7.7	8.4	
Vitamin A equivalent	IU/g		67.2	73.5	

Kudzu, leaves, dehy grnd, (1)
Ref No 1 02 483 United States

Analyses	Unit		As Fed	Dry	C.V. ± %
Dry matter	%			100.0	
Calcium	%			2.61	
Phosphorus	%			0.30	

(1) dry forages and roughages (2) pasture, range plants, and forages fed green (3) silages (4) energy feeds (5) protein supplements (6) minerals (7) vitamins (8) additives

Kudzu, dehy grnd, (1)
Ref No 1 02 479 United States

Analyses	Unit		As Fed	Dry	C.V. ± %
Dry matter	%		91.8	100.0	3
Ash	%		7.1	7.7	15
Crude fiber	%		25.2	27.5	19
Ether extract	%		5.4	5.9	12
N-free extract	%		34.6	37.7	
Protein (N x 6.25)	%		19.5	21.2	14
Cattle	dig prot %	*	14.0	15.3	
Goats	dig prot %	*	15.0	16.3	
Horses	dig prot %	*	14.3	15.5	
Rabbits	dig prot %	*	13.8	15.0	
Sheep	dig prot %	*	14.3	15.6	
Energy	GE Mcal/kg				
Cattle	DE Mcal/kg	*	2.27	2.47	
Sheep	DE Mcal/kg	*	2.35	2.56	
Cattle	ME Mcal/kg	*	1.86	2.03	
Sheep	ME Mcal/kg	*	1.93	2.10	
Cattle	TDN %	*	51.4	56.0	
Sheep	TDN %	*	53.3	58.0	
Carotene	mg/kg		333.1	362.9	26
Riboflavin	mg/kg		11.7	12.8	
Vitamin A equivalent	IU/g		555.3	604.9	

Kudzu, grnd, (1)
Ref No 1 02 480 United States

Analyses	Unit		As Fed	Dry	C.V. ± %
Dry matter	%		92.5	100.0	2
Ash	%		7.4	8.0	13
Crude fiber	%		24.8	26.8	17
Ether extract	%		4.4	4.8	28
N-free extract	%		37.5	40.5	
Protein (N x 6.25)	%		18.4	19.9	15
Cattle	dig prot %	*	13.1	14.2	
Goats	dig prot %	*	14.0	15.1	
Horses	dig prot %	*	13.3	14.4	
Rabbits	dig prot %	*	13.0	14.0	
Sheep	dig prot %	*	13.3	14.4	
Energy	GE Mcal/kg				
Cattle	DE Mcal/kg	*	2.52	2.73	
Sheep	DE Mcal/kg	*	2.41	2.60	
Cattle	ME Mcal/kg	*	2.07	2.23	
Sheep	ME Mcal/kg	*	1.97	2.13	
Cattle	TDN %	*	57.2	61.8	
Sheep	TDN %	*	54.6	59.0	
Carotene	mg/kg		292.8	316.6	47
Riboflavin	mg/kg		9.6	10.4	
Vitamin A equivalent	IU/g		488.2	527.7	

Kudzu, aerial part, fresh, (2)
Ref No 2 02 482 United States

Analyses	Unit		As Fed	Dry	C.V. ± %
Dry matter	%		26.4	100.0	12
Ash	%		2.0	7.6	4
Crude fiber	%		8.1	30.5	7
Ether extract	%		0.7	2.7	14
N-free extract	%		11.0	41.6	
Protein (N x 6.25)	%		4.6	17.6	8
Cattle	dig prot %	*	3.4	12.8	
Goats	dig prot %	*	3.4	13.0	
Horses	dig prot %	*	3.3	12.5	
Rabbits	dig prot %	*	3.2	12.2	
Sheep	dig prot %	*	3.5	13.4	
Energy	GE Mcal/kg				
Cattle	DE Mcal/kg	*	0.78	2.95	
Sheep	DE Mcal/kg	*	0.75	2.83	
Cattle	ME Mcal/kg	*	0.64	2.42	
Sheep	ME Mcal/kg	*	0.61	2.32	
Cattle	TDN %	*	17.7	67.0	
Sheep	TDN %	*	16.9	64.1	
Calcium	%		0.83	3.14	
Phosphorus	%		0.06	0.23	

Kudzu, aerial part, fresh, milk stage, (2)
Ref No 2 02 481 United States

Analyses	Unit		As Fed	Dry	C.V. ± %
Dry matter	%		21.5	100.0	13
Ash	%		1.7	8.0	2
Crude fiber	%		7.4	34.2	4
Ether extract	%		0.5	2.1	12
N-free extract	%		8.3	38.5	
Protein (N x 6.25)	%		3.7	17.2	5
Cattle	dig prot %	*	2.7	12.5	
Goats	dig prot %	*	2.7	12.6	
Horses	dig prot %	*	2.6	12.1	
Rabbits	dig prot %	*	2.6	11.9	
Sheep	dig prot %	*	2.8	13.0	
Energy	GE Mcal/kg				
Cattle	DE Mcal/kg	*	0.62	2.87	
Sheep	DE Mcal/kg	*	0.60	2.79	
Cattle	ME Mcal/kg	*	0.51	2.35	
Sheep	ME Mcal/kg	*	0.49	2.29	
Cattle	TDN %	*	14.0	65.1	
Sheep	TDN %	*	13.6	63.2	

Kudzu, aerial part w molasses added, ensiled, (3)
Ref No 3 02 485 United States

Analyses	Unit		As Fed	Dry	C.V. ± %
Dry matter	%		29.0	100.0	2
Ash	%		2.6	9.0	5
Crude fiber	%		10.9	37.7	1
Ether extract	%		0.6	1.9	6
N-free extract	%		11.5	39.8	
Protein (N x 6.25)	%		3.4	11.6	4
Cattle	dig prot %	*	2.0	6.8	
Goats	dig prot %	*	2.0	6.8	
Horses	dig prot %	*	2.0	6.8	
Sheep	dig prot %	*	2.0	6.8	
Energy	GE Mcal/kg				
Cattle	DE Mcal/kg	*	0.78	2.70	
Sheep	DE Mcal/kg	*	0.80	2.75	
Cattle	ME Mcal/kg	*	0.64	2.21	
Sheep	ME Mcal/kg	*	0.65	2.25	
Cattle	TDN %	*	17.8	61.3	
Sheep	TDN %	*	18.1	62.4	
Calcium	%		0.48	1.67	3
Phosphorus	%		0.06	0.20	18
Carotene	mg/kg		82.0	282.9	61
Vitamin A equivalent	IU/g		136.7	471.5	

KUHNIA, GLUTINOSA. Kuhnia glutinosa

Kuhnia, glutinosa, seeds w hulls (5)
Ref No 5 09 296 United States

Analyses	Unit		As Fed	Dry	C.V. ± %
Dry matter	%			100.0	
Protein (N x 6.25)	%			33.0	
Alanine	%			1.19	
Arginine	%			1.95	
Aspartic acid	%			2.81	
Glutamic acid	%			6.01	
Glycine	%			1.58	
Histidine	%			0.63	

Feed Name or Analyses			Mean As Fed	Mean Dry	C.V. ± %
Hydroxyproline	%			0.20	
Isoleucine	%			1.29	
Leucine	%			1.98	
Lysine	%			1.52	
Methionine	%			0.59	
Phenylalanine	%			1.29	
Proline	%			1.39	
Serine	%			1.45	
Threonine	%			1.16	
Tyrosine	%			0.76	
Valine	%			1.65	

LACEFLOWER, SKY BLUE. Didiscus coeritea

Laceflower, sky blue, seeds w hulls, (5)

Ref No 5 09 012 United States

Analyses			As Fed	Dry	C.V. ± %
Dry matter	%			100.0	
Protein (N x 6.25)	%			22.0	
Alanine	%			0.92	
Arginine	%			1.67	
Aspartic acid	%			2.22	
Glutamic acid	%			4.20	
Glycine	%			1.28	
Histidine	%			0.53	
Hydroxyproline	%			0.11	
Isoleucine	%			0.99	
Leucine	%			1.39	
Lysine	%			0.75	
Methionine	%			0.35	
Phenylalanine	%			0.92	
Proline	%			1.03	
Serine	%			0.99	
Threonine	%			0.79	
Tyrosine	%			0.66	
Valine	%			1.08	

Lactalbumin, dried –
see Cattle, whey albumin, heat- and acid-precipitated dehy, mn 75% protein, (5)

Lactose –
see Cattle, lactose, crude, (4)

Lambs quarters seed –
see Goosefoot, lambsquarters, seeds, (5)

Lard –
see Swine, lard, (4)

LARKSPUR. Delphinium spp

Larkspur, aerial part, fresh, (2)

Ref No 2 02 487 United States

Analyses			As Fed	Dry	C.V. ± %
Dry matter	%			100.0	
Calcium	%			1.40	
Phosphorus	%			0.35	

LEADPLANT, AMORPHA. Amorpha canescens

Leadplant, amorpha, aerial part, fresh, (2)

Ref No 2 02 489 United States

Analyses			As Fed	Dry	C.V. ± %
Dry matter	%		55.7	100.0	
Ash	%		5.8	10.4	
Crude fiber	%		12.0	21.5	
Ether extract	%		4.7	8.5	
N-free extract	%		28.6	51.4	
Protein (N x 6.25)	%		4.6	8.2	
Cattle	dig prot %	*	2.7	4.9	
Goats	dig prot %	*	2.3	4.2	
Horses	dig prot %	*	2.5	4.5	
Rabbits	dig prot %	*	2.8	5.0	
Sheep	dig prot %	*	2.6	4.6	

Leadplant, amorpha, aerial part, fresh, immature, (2)

Ref No 2 02 488 United States

Analyses			As Fed	Dry	C.V. ± %
Dry matter	%			100.0	
Calcium	%			2.80	
Phosphorus	%			0.92	

LEADTREE, LITTLELEAF. Leucaena retusa

Leadtree, littleleaf, aerial part, fresh, immature, (2)

Ref No 2 02 490 United States

Analyses			As Fed	Dry	C.V. ± %
Dry matter	%			100.0	
Calcium	%			2.81	
Magnesium	%			0.22	
Phosphorus	%			0.12	
Potassium	%			1.30	

Leadtree, littleleaf, browse, immature, (2)

Ref No 2 02 491 United States

Analyses			As Fed	Dry	C.V. ± %
Dry matter	%			100.0	
Ash	%			6.4	
Crude fiber	%			24.4	
Ether extract	%			3.9	
N-free extract	%			47.7	
Protein (N x 6.25)	%			17.6	
Cattle	dig prot %	*		12.9	
Goats	dig prot %	*		13.0	
Horses	dig prot %	*		12.5	
Rabbits	dig prot %	*		12.3	
Sheep	dig prot %	*		13.4	

LEADTREE, WHITEPOPINAC. Leucaena glauca

Leadtree, whitepopinac, hay, s-c, (1)

Koa haole

Ref No 1 02 492 United States

Analyses			As Fed	Dry	C.V. ± %
Dry matter	%		88.7	100.0	
Ash	%		5.1	5.7	
Crude fiber	%		29.8	33.6	
Ether extract	%		1.9	2.1	
N-free extract	%		39.2	44.2	
Protein (N x 6.25)	%		12.7	14.3	
Cattle	dig prot %	*	8.3	9.3	
Goats	dig prot %	*	8.8	9.9	
Horses	dig prot %	*	8.6	9.7	
Rabbits	dig prot %	*	8.6	9.7	
Sheep	dig prot %	*	8.3	9.4	
Energy	GE Mcal/kg				
Cattle	DE Mcal/kg	*	2.10	2.36	
Sheep	DE Mcal/kg	*	2.26	2.55	
Cattle	ME Mcal/kg	*	1.72	1.94	
Sheep	ME Mcal/kg	*	1.86	2.09	
Cattle	TDN %	*	47.5	53.6	
Sheep	TDN %	*	51.3	57.9	

Leadtree, whitepopinac, aerial part, fresh, (2)

Koa haole

Ref No 2 08 451 United States

Analyses			As Fed	Dry	C.V. ± %
Dry matter	%		29.4	100.0	
Ash	%		1.8	6.1	
Crude fiber	%		9.7	33.0	
Ether extract	%		0.6	2.0	
N-free extract	%		12.0	40.8	
Protein (N x 6.25)	%		5.3	18.0	
Cattle	dig prot %	*	3.9	13.2	
Goats	dig prot %	*	3.9	13.4	
Horses	dig prot %	*	3.8	12.8	
Rabbits	dig prot %	*	3.7	12.6	
Sheep	dig prot %	*	4.1	13.8	
Energy	GE Mcal/kg				
Cattle	DE Mcal/kg	*	0.80	2.71	
Sheep	DE Mcal/kg	*	0.73	2.47	
Cattle	ME Mcal/kg	*	0.65	2.22	
Sheep	ME Mcal/kg	*	0.60	2.03	
Cattle	TDN %	*	18.1	61.4	
Sheep	TDN %	*	16.5	56.1	

Leadtree, whitepopinac, aerial part, fresh, dough stage, (2)

Ref No 2 02 493 United States

Analyses			As Fed	Dry	C.V. ± %
Dry matter	%		30.7	100.0	
Ash	%		2.7	8.9	
Crude fiber	%		7.4	24.2	
Cattle	dig coef %		35.	35.	
Ether extract	%		0.8	2.7	
Cattle	dig coef %		36.	36.	
N-free extract	%		12.3	40.0	
Cattle	dig coef %		74.	74.	
Protein (N x 6.25)	%		7.4	24.2	
Cattle	dig coef %		65.	65.	
Cattle	dig prot %		4.8	15.7	
Goats	dig prot %	*	5.9	19.1	
Horses	dig prot %	*	5.5	18.1	
Rabbits	dig prot %	*	5.3	17.4	
Sheep	dig prot %	*	6.0	19.5	
Energy	GE Mcal/kg				
Cattle	DE Mcal/kg	*	0.76	2.47	
Sheep	DE Mcal/kg	*	0.83	2.70	
Cattle	ME Mcal/kg	*	0.62	2.02	
Sheep	ME Mcal/kg	*	0.68	2.21	
Cattle	TDN %		17.2	56.0	
Sheep	TDN %	*	18.8	61.2	

Leadtree, whitepopinac, browse, (2)

Ref No 2 02 495 United States

Analyses			As Fed	Dry	C.V. ± %
Dry matter	%		32.3	100.0	7
Ash	%		2.2	6.7	9
Crude fiber	%		12.2	37.7	10
Ether extract	%		0.7	2.1	17
N-free extract	%		11.4	35.2	
Protein (N x 6.25)	%		5.9	18.3	13
Cattle	dig prot %	*	4.3	13.4	
Goats	dig prot %	*	4.4	13.6	
Horses	dig prot %	*	4.2	13.1	

Continued

Feed Name or Analyses		As Fed	Dry	C.V. ± %
Rabbits	dig prot % *	4.1	12.8	
Sheep	dig prot % *	4.5	14.0	

Leadtree, whitepopinac, browse, mature, (2)
Ref No 2 02 494 United States

Feed Name or Analyses		As Fed	Dry	C.V. ± %
Dry matter	%	32.7	100.0	8
Ash	%	2.5	7.6	9
Crude fiber	%	14.2	43.4	7
Ether extract	%	0.4	1.2	19
N-free extract	%	11.4	35.0	
Protein (N x 6.25)	%	4.2	12.8	7
Cattle	dig prot % *	2.9	8.8	
Goats	dig prot % *	2.8	8.5	
Horses	dig prot % *	2.7	8.4	
Rabbits	dig prot % *	2.8	8.6	
Sheep	dig prot % *	2.9	8.9	

Leadtree, whitepopinac, aerial part, ensiled, (3)
Ref No 3 02 496 United States

Feed Name or Analyses		As Fed	Dry	C.V. ± %
Dry matter	%	42.2	100.0	
Ash	%	3.0	7.2	
Crude fiber	%	16.6	39.3	
Ether extract	%	1.3	3.0	
N-free extract	%	16.3	38.7	
Protein (N x 6.25)	%	5.0	11.8	
Cattle	dig prot % *	2.9	6.9	
Goats	dig prot % *	2.9	6.9	
Horses	dig prot % *	2.9	6.9	
Sheep	dig prot % *	2.9	6.9	

LEEK. Allium porrum

Leek, seeds w hulls, (5)
Ref No 5 09 271 United States

Feed Name or Analyses		As Fed	Dry	C.V. ± %
Dry matter	%		100.0	
Protein (N x 6.25)	%		28.0	
Alanine	%		1.06	
Arginine	%		3.02	
Aspartic acid	%		1.79	
Glutamic acid	%		5.77	
Glycine	%		1.12	
Histidine	%		0.56	
Hydroxyproline	%		0.22	
Isoleucine	%		0.78	
Leucine	%		1.34	
Lysine	%		1.26	
Methionine	%		0.90	
Phenylalanine	%		1.01	
Proline	%		0.92	
Serine	%		1.15	
Threonine	%		0.87	
Tyrosine	%		0.73	
Valine	%		1.51	

(1) dry forages and roughages (2) pasture, range plants, and forages fed green (3) silages (4) energy feeds (5) protein supplements (6) minerals (7) vitamins (8) additives

LEGUME. Scientific name not used

Legume, aerial part, ensiled, (3)
Ref No 3 07 796 United States

Feed Name or Analyses		As Fed	Dry	C.V. ± %
Dry matter	%	39.2	100.0	
Protein (N x 6.25)	%	4.7	11.9	
Cattle	dig prot % *	2.8	7.0	
Goats	dig prot % *	2.8	7.0	
Horses	dig prot % *	2.8	7.0	
Sheep	dig prot % *	2.8	7.0	
Lignin (Ellis)	%	3.8	9.7	
pH	pH units	4.50	4.50	

Legume, aerial part, ensiled, cut 1, (3)
Ref No 3 07 795 United States

Feed Name or Analyses		As Fed	Dry	C.V. ± %
Dry matter	%	24.3	100.0	21
Crude fiber	%	7.8	32.3	
Protein (N x 6.25)	%	3.2	13.3	14
Cattle	dig prot % *	2.0	8.3	
Goats	dig prot % *	2.0	8.3	
Horses	dig prot % *	2.0	8.3	
Sheep	dig prot % *	2.0	8.3	
Lignin (Ellis)	%	2.2	9.3	
pH	pH units	4.87	4.87	11

Legume, aerial part, ensiled, cut 2, (3)
Ref No 3 07 797 United States

Feed Name or Analyses		As Fed	Dry	C.V. ± %
Dry matter	%	24.8	100.0	
Crude fiber	%	7.3	29.4	
Protein (N x 6.25)	%	4.3	17.4	
Cattle	dig prot % *	3.0	12.0	
Goats	dig prot % *	3.0	12.0	
Horses	dig prot % *	3.0	12.0	
Sheep	dig prot % *	3.0	12.0	
pH	pH units	4.75	4.75	

Legume, aerial part, ensiled, cut 3, (3)
Ref No 3 07 798 United States

Feed Name or Analyses		As Fed	Dry	C.V. ± %
Dry matter	%	24.1	100.0	
Protein (N x 6.25)	%	4.8	20.0	
Cattle	dig prot % *	3.5	14.4	
Goats	dig prot % *	3.5	14.4	
Horses	dig prot % *	3.5	14.4	
Sheep	dig prot % *	3.5	14.4	

LEGUME, CAN EASTERN 1 YR POST SEEDING. Scientific name not used

Legume, Can Eastern 1 yr post seeding, hay, s-c, Early bloom, (1)
Ref No 1 08 200 Canada

Feed Name or Analyses		As Fed	Dry	C.V. ± %
Dry matter	%	92.5	100.0	
In vitro	dig coef %	55.	59.	
Protein (N x 6.25)	%	18.6	20.2	
Cattle	dig prot % *	13.3	14.4	
Goats	dig prot % *	14.2	15.4	
Horses	dig prot % *	13.5	14.6	
Rabbits	dig prot % *	13.2	14.2	
Sheep	dig prot % *	13.6	14.7	
Cellulose (Matrone)	%	29.4	31.8	
Nutritive value index (NVI)	%	56.	61.	

Legume, Can Eastern 1 yr post seeding, hay, s-c, mid-bloom, (1)
Ref No 1 08 193 Canada

Feed Name or Analyses		As Fed	Dry	C.V. ± %
Dry matter	%	92.5	100.0	0
In vitro	dig coef %	54.	58.	5
Protein (N x 6.25)	%	16.2	17.5	12
Cattle	dig prot % *	11.2	12.1	
Goats	dig prot % *	11.9	12.9	
Horses	dig prot % *	11.5	12.4	
Rabbits	dig prot % *	11.3	12.2	
Sheep	dig prot % *	11.4	12.3	
Cellulose (Matrone)	%	29.0	31.4	7
Nutritive value index (NVI)	%	52.	56.	18

Legume, Can Eastern 1 yr post seeding, hay, s-c, full bloom, (1)
Ref No 1 08 197 Canada

Feed Name or Analyses		As Fed	Dry	C.V. ± %
Dry matter	%	92.5	100.0	0
In vitro	dig coef %	52.	56.	5
Protein (N x 6.25)	%	12.5	13.6	24
Cattle	dig prot % *	8.0	8.7	
Goats	dig prot % *	8.5	9.2	
Horses	dig prot % *	8.4	9.0	
Rabbits	dig prot % *	8.4	9.1	
Sheep	dig prot % *	8.1	8.7	
Cellulose (Matrone)	%	31.2	33.8	10
Nutritive value index (NVI)	%	44.	47.	21

Legume, Can Eastern 1 yr post seeding, hay, s-c, late bloom, (1)
Ref No 1 08 201 Canada

Feed Name or Analyses		As Fed	Dry	C.V. ± %
Dry matter	%	92.5	100.0	0
In vitro	dig coef %	47.	51.	8
Protein (N x 6.25)	%	12.9	14.0	35
Cattle	dig prot % *	8.4	9.0	
Goats	dig prot % *	8.9	9.6	
Horses	dig prot % *	8.7	9.4	
Rabbits	dig prot % *	8.7	9.5	
Sheep	dig prot % *	8.4	9.1	
Cellulose (Matrone)	%	32.3	35.0	4
Nutritive value index (NVI)	%	29.	32.	28

LEGUME-GRASS. Scientific name not used

Legume-Grass, hay, s-c, (1)
Ref No 1 08 450 United States

Feed Name or Analyses		As Fed	Dry	C.V. ± %
Dry matter	%	91.4	100.0	
Ash	%	6.6	7.2	
Crude fiber	%	24.7	27.0	
Ether extract	%	2.3	2.5	
N-free extract	%	43.5	47.6	
Protein (N x 6.25)	%	14.3	15.6	
Cattle	dig prot % *	9.6	10.5	
Goats	dig prot % *	10.2	11.2	
Horses	dig prot % *	9.9	10.8	
Rabbits	dig prot % *	9.8	10.7	
Sheep	dig prot % *	9.7	10.6	
Energy	GE Mcal/kg			
Cattle	DE Mcal/kg *	2.45	2.68	

Feed Name or Analyses			Mean As Fed	Mean Dry	C.V. ± %
Sheep	DE Mcal/kg	*	2.45	2.68	
Cattle	ME Mcal/kg	*	2.01	2.20	
Sheep	ME Mcal/kg	*	2.01	2.20	
Cattle	TDN %	*	55.5	60.8	
Sheep	TDN %	*	55.6	60.9	
Calcium	%		0.98	1.07	
Phosphorus	%		0.22	0.24	

LEGUME-GRASS, CAN EASTERN 2-3 YR POST SEEDING. Scientific name not used

Legume-Grass, Can Eastern 2-3 yr post seeding, hay, s-c, early bloom, (1)

Ref No 1 08 202 Canada

Feed Name or Analyses			Mean As Fed	Mean Dry	C.V. ± %
Dry matter	%		92.5	100.0	
In vitro	dig coef %		50.	54.	
Protein (N x 6.25)	%		17.9	19.4	
Cattle	dig prot %	*	12.7	13.7	
Goats	dig prot %	*	13.5	14.6	
Horses	dig prot %	*	12.9	14.0	
Rabbits	dig prot %	*	12.6	13.6	
Sheep	dig prot %	*	12.9	13.9	
Cellulose (Matrone)	%		31.1	33.6	
Nutritive value index (NVI)	%		37.	40.	

Legume-Grass, Can Eastern 2-3 yr post seeding, hay, s-c, mid-bloom, (1)

Ref No 1 08 198 Canada

Feed Name or Analyses			Mean As Fed	Mean Dry	C.V. ± %
Dry matter	%		92.5	100.0	0
In vitro	dig coef %		52.	56.	3
Protein (N x 6.25)	%		12.5	13.5	9
Cattle	dig prot %	*	8.0	8.7	
Goats	dig prot %	*	8.5	9.2	
Horses	dig prot %	*	8.3	9.0	
Rabbits	dig prot %	*	8.4	9.1	
Sheep	dig prot %	*	8.1	8.7	
Cellulose (Matrone)	%		32.5	35.2	5
Nutritive value index (NVI)	%		42.	46.	12

Legume-Grass, Can Eastern 2-3 yr post seeding, hay, s-c, full bloom, (1)

Ref No 1 08 194 Canada

Feed Name or Analyses			Mean As Fed	Mean Dry	C.V. ± %
Dry matter	%		92.5	100.0	0
In vitro	dig coef %		50.	55.	4
Protein (N x 6.25)	%		10.0	10.9	21
Cattle	dig prot %	*	5.9	6.3	
Goats	dig prot %	*	6.2	6.7	
Horses	dig prot %	*	6.2	6.7	
Rabbits	dig prot %	*	6.5	7.1	
Sheep	dig prot %	*	5.8	6.3	
Cellulose (Matrone)	%		32.9	35.6	5
Nutritive value index (NVI)	%		38.	41.	16

Legume-Grass, Can Eastern 2-3 yr post seeding, hay, s-c, late bloom, (1)

Ref No 1 08 199 Canada

Feed Name or Analyses			Mean As Fed	Mean Dry	C.V. ± %
Dry matter	%		92.5	100.0	0
In vitro	dig coef %		50.	54.	3
Protein (N x 6.25)	%		8.8	9.5	18
Cattle	dig prot %	*	4.8	5.2	
Goats	dig prot %	*	5.0	5.4	
Horses	dig prot %	*	5.2	5.6	
Rabbits	dig prot %	*	5.5	6.0	
Sheep	dig prot %	*	4.7	5.1	
Cellulose (Matrone)	%		31.5	34.1	5
Nutritive value index (NVI)	%		39.	42.	15

Legume-Grass, CE 1 year post seeding, hay, s-c, mx 25% grass, (1)

Ref No 1 09 080 Canada

Feed Name or Analyses			Mean As Fed	Mean Dry	C.V. ± %
Dry matter	%			100.0	
Ash	%			8.4	
Crude fiber	%			30.3	
Ether extract	%			2.1	
N-free extract	%			43.1	
Protein (N x 6.25)	%			16.1	
Sheep	dig coef %			66.	
Cattle	dig prot %	*		10.9	
Goats	dig prot %	*		11.6	
Horses	dig prot %	*		11.2	
Rabbits	dig prot %	*		11.1	
Sheep	dig prot %			10.6	
Energy	GE Mcal/kg				
Cattle	DE Mcal/kg	*		2.62	
Sheep	DE Mcal/kg	*		2.43	
Cattle	ME Mcal/kg	*		2.15	
Sheep	ME Mcal/kg	*		1.99	
Cattle	TDN %	*		59.5	
Sheep	TDN %			55.0	

Legume-Grass, CE 2-3 years post seeding, hay, s-c, mx 50% grass, (1)

Ref No 1 09 081 Canada

Feed Name or Analyses			Mean As Fed	Mean Dry	C.V. ± %
Dry matter	%			100.0	
Ash	%			7.9	
Crude fiber	%			31.9	
Ether extract	%			2.4	
N-free extract	%			45.2	
Protein (N x 6.25)	%			12.6	
Sheep	dig coef %			60.	
Cattle	dig prot %	*		7.9	
Goats	dig prot %	*		8.3	
Horses	dig prot %	*		8.2	
Rabbits	dig prot %	*		8.4	
Sheep	dig prot %			7.6	
Energy	GE Mcal/kg				
Cattle	DE Mcal/kg	*		2.63	
Sheep	DE Mcal/kg	*		2.38	
Cattle	ME Mcal/kg	*		2.15	
Sheep	ME Mcal/kg	*		1.95	
Cattle	TDN %	*		59.6	
Sheep	TDN %			54.0	

Lemon pulp, dried —
see Citrus, lemon, pulp wo fines, shredded dehy, (4)

LENTIL. Lens spp

Lentil, seeds, (5)

Ref No 5 02 506 United States

Feed Name or Analyses			Mean As Fed	Mean Dry	C.V. ± %
Dry matter	%		88.5	100.0	
Ash	%		2.6	2.9	
Crude fiber	%		3.4	3.8	
Sheep	dig coef %		52.	52.	
Ether extract	%		1.1	1.3	
Sheep	dig coef %		62.	62.	
N-free extract	%		56.6	64.0	
Sheep	dig coef %		90.	90.	
Protein (N x 6.25)	%		24.8	28.0	
Sheep	dig coef %		79.	79.	
Sheep	dig prot %		19.6	22.1	
Energy	GE Mcal/kg		3.38	3.82	
Cattle	DE Mcal/kg	*	3.34	3.77	
Sheep	DE Mcal/kg	*	3.25	3.68	
Swine	DE kcal/kg	*	3700.	4183.	
Cattle	ME Mcal/kg	*	2.74	3.09	
Sheep	ME Mcal/kg	*	2.67	3.02	
Swine	ME kcal/kg	*	3343.	3779.	
Cattle	TDN %	*	75.7	85.5	
Sheep	TDN %		73.8	83.5	
Swine	TDN %	*	83.9	94.9	
Calcium	%		0.08	0.09	
Iron	%		0.007	0.008	
Phosphorus	%		0.38	0.42	
Potassium	%		0.79	0.89	
Sodium	%		0.03	0.03	
Niacin	mg/kg		19.9	22.5	
Riboflavin	mg/kg		2.2	2.5	
Thiamine	mg/kg		3.7	4.2	
Vitamin A	IU/g		0.6	0.7	

LENTIL, COMMON. Lens culinaris

Lentil, common, husks, s-c, (1)

Ref No 1 02 507 United States

Feed Name or Analyses			Mean As Fed	Mean Dry	C.V. ± %
Dry matter	%		87.0	100.0	
Ash	%		3.1	3.6	
Crude fiber	%		25.3	29.1	
Sheep	dig coef %		67.	67.	
Ether extract	%		0.7	0.8	
Sheep	dig coef %		96.	96.	
N-free extract	%		46.9	53.9	
Sheep	dig coef %		70.	70.	
Protein (N x 6.25)	%		11.0	12.6	
Sheep	dig coef %		12.	12.	
Cattle	dig prot %	*	6.8	7.9	
Goats	dig prot %	*	7.2	8.3	
Horses	dig prot %	*	7.2	8.2	
Rabbits	dig prot %	*	7.3	8.4	
Sheep	dig prot %		1.3	1.5	
Energy	GE Mcal/kg				
Cattle	DE Mcal/kg	*	2.25	2.59	
Sheep	DE Mcal/kg	*	2.32	2.67	
Cattle	ME Mcal/kg	*	1.84	2.12	
Sheep	ME Mcal/kg	*	1.90	2.19	
Cattle	TDN %	*	51.0	58.6	
Sheep	TDN %		52.6	60.5	

LEPTOTAENIA, CARROTLEAF. Leptotaenia multifida

Leptotaenia, carrotleaf, aerial part, s-c, (1)

Ref No 1 02 508 United States

Feed Name or Analyses			Mean As Fed	Mean Dry	C.V. ± %
Dry matter	%		93.8	100.0	
Ash	%		8.9	9.5	
Crude fiber	%		21.7	23.1	
Sheep	dig coef %		47.	47.	
Ether extract	%		7.0	7.5	
Sheep	dig coef %		82.	82.	

Continued

Feed Name or Analyses			Mean As Fed	Mean Dry	C.V. ± %
N-free extract	%		46.3	49.4	
Sheep	dig coef %		83.	83.	
Protein (N x 6.25)	%		9.8	10.5	
Sheep	dig coef %		71.	71.	
Cattle	dig prot %	*	5.7	6.0	
Goats	dig prot %	*	6.0	6.4	
Horses	dig prot %	*	6.0	6.4	
Rabbits	dig prot %	*	6.4	6.8	
Sheep	dig prot %		7.0	7.5	
Energy	GE Mcal/kg				
Cattle	DE Mcal/kg	*	3.11	3.32	
Sheep	DE Mcal/kg	*	3.03	3.23	
Cattle	ME Mcal/kg	*	2.55	2.72	
Sheep	ME Mcal/kg	*	2.48	2.64	
Cattle	TDN %	*	70.5	75.2	
Sheep	TDN %		68.6	73.2	

LESPEDEZA. Lespedeza spp

Lespedeza, hay, s-c, (1)
Ref No 1 02 522 United States

Feed Name or Analyses			Mean As Fed	Mean Dry	C.V. ± %
Dry matter	%		91.7	100.0	2
Ash	%		6.6	7.2	
Crude fiber	%		28.0	30.5	
Ether extract	%		2.8	3.0	
N-free extract	%		41.7	45.5	
Protein (N x 6.25)	%		12.7	13.8	
Cattle	dig prot %	*	8.2	8.9	
Goats	dig prot %	*	8.6	9.4	
Horses	dig prot %	*	8.5	9.2	
Rabbits	dig prot %	*	8.5	9.3	
Sheep	dig prot %	*	8.2	8.9	
Hemicellulose	%		8.5	9.3	
Lignin (Ellis)	%		11.4	12.4	
Energy	GE Mcal/kg				
Cattle	DE Mcal/kg	*	2.34	2.55	
Sheep	DE Mcal/kg	*	2.34	2.55	
Cattle	ME Mcal/kg	*	1.91	2.09	
Sheep	ME Mcal/kg	*	1.92	2.09	
Cattle	TDN %	*	53.0	57.8	
Sheep	TDN %	*	53.1	57.9	
Chlorine	%		0.05	0.05	43
Cobalt	mg/kg		0.184	0.201	74
Iodine	mg/kg		0.145	0.159	
Sodium	%		0.06	0.07	93
Sulphur	%		0.17	0.19	26
Zinc	mg/kg		26.9	29.3	98
Arginine	%		0.46	0.50	23
Histidine	%		0.37	0.40	14
Isoleucine	%		1.65	1.80	11
Leucine	%		1.10	1.20	12
Lysine	%		0.92	1.00	12
Methionine	%		0.09	0.10	
Threonine	%		0.55	0.60	10
Tryptophan	%		0.18	0.20	14
Valine	%		0.73	0.80	36

Lespedeza, hay, s-c, immature, (1)
Ref No 1 02 509 United States

Feed Name or Analyses			Mean As Fed	Mean Dry	C.V. ± %
Dry matter	%		93.6	100.0	1
Carotene	mg/kg		136.0	145.3	
Vitamin A equivalent	IU/g		226.7	242.2	

(1) dry forages and roughages (2) pasture, range plants, and forages fed green (3) silages (4) energy feeds (5) protein supplements (6) minerals (7) vitamins (8) additives

Lespedeza, hay, s-c, mid-bloom, (1)
Ref No 1 02 511 United States

Feed Name or Analyses			Mean As Fed	Mean Dry	C.V. ± %
Dry matter	%		94.1	100.0	1
Cellulose (Matrone)	%		29.2	31.0	
Energy	GE Mcal/kg				
Cattle	NE_m Mcal/kg	*	1.15	1.22	
Cattle	NE_{gain} Mcal/kg	*	0.52	0.55	
Cattle	$NE_{lactating\ cows}$ Mcal/kg	*	1.26	1.34	
Calcium	%		1.12	1.19	11
Iron	%		0.030	0.032	17
Magnesium	%		0.25	0.27	22
Manganese	mg/kg		252.1	267.9	19
Phosphorus	%		0.24	0.26	38
Potassium	%		0.99	1.05	19

Lespedeza, hay, s-c, mature, (1)
Ref No 1 02 514 United States

Feed Name or Analyses			Mean As Fed	Mean Dry	C.V. ± %
Dry matter	%		90.7	100.0	1
Ash	%		5.6	6.2	6
Crude fiber	%		36.2	39.9	1
Ether extract	%		2.1	2.3	6
N-free extract	%		38.7	42.7	
Protein (N x 6.25)	%		8.1	8.9	8
Cattle	dig prot %	*	4.2	4.6	
Goats	dig prot %	*	4.4	4.9	
Horses	dig prot %	*	4.6	5.1	
Rabbits	dig prot %	*	5.0	5.5	
Sheep	dig prot %	*	4.1	4.6	
Cellulose (Matrone)	%		30.0	33.1	
Lignin (Ellis)	%		20.9	23.1	
Energy	GE Mcal/kg				
Cattle	DE Mcal/kg	*	1.81	2.00	
Sheep	DE Mcal/kg	*	2.11	2.33	
Cattle	ME Mcal/kg	*	1.49	1.64	
Sheep	ME Mcal/kg	*	1.73	1.91	
Cattle	TDN %	*	41.1	45.4	
Sheep	TDN %	*	47.9	52.9	
Calcium	%		0.63	0.69	
Iron	%		0.015	0.016	45
Magnesium	%		0.24	0.26	
Manganese	mg/kg		120.9	133.4	
Phosphorus	%		0.10	0.11	
Potassium	%		0.73	0.81	
Carotene	mg/kg		10.6	11.7	
Vitamin A equivalent	IU/g		17.7	19.5	

Lespedeza, hay, s-c, over ripe, (1)
Ref No 1 02 515 United States

Feed Name or Analyses			Mean As Fed	Mean Dry	C.V. ± %
Dry matter	%		92.5	100.0	
Carotene	mg/kg		0.8	0.9	
Vitamin A equivalent	IU/g		1.4	1.5	

Lespedeza, hay, s-c, cut 2, (1)
Ref No 1 02 517 United States

Feed Name or Analyses			Mean As Fed	Mean Dry	C.V. ± %
Dry matter	%		91.5	100.0	4
Ash	%		4.8	5.2	6
Crude fiber	%		30.5	33.3	6
Ether extract	%		2.7	2.9	41
N-free extract	%		40.4	44.1	
Protein (N x 6.25)	%		13.3	14.5	10
Cattle	dig prot %	*	8.7	9.5	
Goats	dig prot %	*	9.2	10.1	
Horses	dig prot %	*	9.0	9.8	
Rabbits	dig prot %	*	9.0	9.9	
Sheep	dig prot %	*	8.8	9.6	
Energy	GE Mcal/kg				
Cattle	DE Mcal/kg	*	2.18	2.38	
Sheep	DE Mcal/kg	*	2.34	2.55	
Cattle	ME Mcal/kg	*	1.79	1.95	
Sheep	ME Mcal/kg	*	1.92	2.09	
Cattle	TDN %	*	49.4	54.0	
Sheep	TDN %	*	53.0	57.9	
Calcium	%		0.93	1.02	17
Iron	%		0.018	0.020	23
Magnesium	%		0.20	0.22	19
Manganese	mg/kg		93.0	101.6	16
Phosphorus	%		0.22	0.24	17
Potassium	%		0.92	1.01	20

Lespedeza, hay, s-c, cut 3, (1)
Ref No 1 02 518 United States

Feed Name or Analyses			Mean As Fed	Mean Dry	C.V. ± %
Dry matter	%		92.3	100.0	1
Ash	%		5.1	5.5	13
Crude fiber	%		23.9	25.9	15
Ether extract	%		4.4	4.8	10
N-free extract	%		45.5	49.3	
Protein (N x 6.25)	%		13.4	14.5	8
Cattle	dig prot %	*	8.8	9.5	
Goats	dig prot %	*	9.3	10.1	
Horses	dig prot %	*	9.1	9.8	
Rabbits	dig prot %	*	9.1	9.9	
Sheep	dig prot %	*	8.8	9.6	
Energy	GE Mcal/kg				
Cattle	DE Mcal/kg	*	2.47	2.67	
Sheep	DE Mcal/kg	*	2.48	2.69	
Cattle	ME Mcal/kg	*	2.02	2.19	
Sheep	ME Mcal/kg	*	2.04	2.21	
Cattle	TDN %	*	55.9	60.6	
Sheep	TDN %	*	56.3	61.0	
Calcium	%		1.10	1.19	22
Iron	%		0.030	0.032	17
Magnesium	%		0.22	0.24	20
Manganese	mg/kg		114.4	123.9	24
Phosphorus	%		0.23	0.25	19
Potassium	%		0.86	0.93	23

Lespedeza, hay, s-c, gr 2 US, (1)
Ref No 1 02 520 United States

Feed Name or Analyses			Mean As Fed	Mean Dry	C.V. ± %
Dry matter	%			100.0	
Vitamin D_2	IU/g			2.4	

Lespedeza, hay, s-c, gr 3 US, (1)
Ref No 1 02 521 United States

Feed Name or Analyses			Mean As Fed	Mean Dry	C.V. ± %
Dry matter	%		90.6	100.0	1
Ash	%		3.7	4.1	7
Crude fiber	%		32.3	35.7	1
Ether extract	%		2.2	2.4	3
N-free extract	%		41.6	45.9	
Protein (N x 6.25)	%		10.8	11.9	5
Cattle	dig prot %	*	6.6	7.2	
Goats	dig prot %	*	6.9	7.7	
Horses	dig prot %	*	6.9	7.6	
Rabbits	dig prot %	*	7.1	7.9	
Sheep	dig prot %	*	6.6	7.2	
Energy	GE Mcal/kg				
Cattle	DE Mcal/kg	*	1.99	2.20	

Feed Name or Analyses			Mean As Fed	Mean Dry	C.V. ± %
Sheep	DE Mcal/kg	*	2.25	2.48	
Cattle	ME Mcal/kg	*	1.63	1.80	
Sheep	ME Mcal/kg	*	1.84	2.04	
Cattle	TDN %	*	45.1	49.8	
Sheep	TDN %	*	51.0	56.3	

Lespedeza, hay, s-c, gr sample US, (1)

Ref No 1 02 519 United States

			As Fed	Dry	C.V.
Dry matter	%		91.1	100.0	
Ash	%		3.7	4.1	
Crude fiber	%		35.7	39.2	
Ether extract	%		2.0	2.2	
N-free extract	%		40.9	44.9	
Protein (N x 6.25)	%		8.7	9.6	
Cattle	dig prot %	*	4.8	5.3	
Goats	dig prot %	*	5.0	5.5	
Horses	dig prot %	*	5.2	5.7	
Rabbits	dig prot %	*	5.5	6.1	
Sheep	dig prot %	*	4.7	5.2	
Energy	GE Mcal/kg				
Cattle	DE Mcal/kg	*	1.78	1.95	
Sheep	DE Mcal/kg	*	2.16	2.37	
Cattle	ME Mcal/kg	*	1.46	1.60	
Sheep	ME Mcal/kg	*	1.77	1.95	
Cattle	TDN %	*	40.4	44.3	
Sheep	TDN %	*	49.0	53.8	

Lespedeza, leaves, immature, (1)

Ref No 1 02 524 United States

			As Fed	Dry	C.V.
Dry matter	%			100.0	
Ash	%			6.7	8
Protein (N x 6.25)	%			20.7	10
Cattle	dig prot %	*		14.9	
Goats	dig prot %	*		15.9	
Horses	dig prot %	*		15.1	
Rabbits	dig prot %	*		14.7	
Sheep	dig prot %	*		15.1	

Lespedeza, leaves, cut 1, (1)

Ref No 1 02 525 United States

			As Fed	Dry	C.V.
Dry matter	%			100.0	
Ash	%			7.1	11
Protein (N x 6.25)	%			20.3	12
Cattle	dig prot %	*		14.5	
Goats	dig prot %	*		15.5	
Horses	dig prot %	*		14.8	
Rabbits	dig prot %	*		14.3	
Sheep	dig prot %	*		14.8	

Lespedeza, leaves, cut 2, (1)

Ref No 1 02 526 United States

			As Fed	Dry	C.V.
Dry matter	%			100.0	
Ash	%			6.0	4
Protein (N x 6.25)	%			18.1	9
Cattle	dig prot %	*		12.6	
Goats	dig prot %	*		13.4	
Horses	dig prot %	*		12.9	
Rabbits	dig prot %	*		12.6	
Sheep	dig prot %	*		12.8	

Lespedeza, leaves, cut 3, (1)

Ref No 1 02 527 United States

			As Fed	Dry	C.V.
Dry matter	%			100.0	
Ash	%			6.6	
Protein (N x 6.25)	%			18.0	
Cattle	dig prot %	*		12.5	
Goats	dig prot %	*		13.4	
Horses	dig prot %	*		12.8	
Rabbits	dig prot %	*		12.6	
Sheep	dig prot %	*		12.7	

Lespedeza, stems, immature, (1)

Ref No 1 02 530 United States

			As Fed	Dry	C.V.
Dry matter	%			100.0	
Ash	%			5.5	
Protein (N x 6.25)	%			12.0	
Cattle	dig prot %	*		7.3	
Goats	dig prot %	*		7.8	
Horses	dig prot %	*		7.7	
Rabbits	dig prot %	*		7.9	
Sheep	dig prot %	*		7.3	
Calcium	%			0.88	
Iron	%			0.035	
Magnesium	%			0.29	
Manganese	mg/kg			97.2	
Phosphorus	%			0.17	
Potassium	%			1.02	

Lespedeza, stems, early bloom, (1)

Ref No 1 02 531 United States

			As Fed	Dry	C.V.
Dry matter	%			100.0	
Ash	%			4.6	
Protein (N x 6.25)	%			9.7	
Cattle	dig prot %	*		5.3	
Goats	dig prot %	*		5.6	
Horses	dig prot %	*		5.8	
Rabbits	dig prot %	*		6.2	
Sheep	dig prot %	*		5.3	

Lespedeza, stems, full bloom, (1)

Ref No 1 02 532 United States

		As Fed	Dry	C.V.
Dry matter	%	92.9	100.0	2
Calcium	%	0.72	0.77	12
Iron	%	0.029	0.031	58
Magnesium	%	0.18	0.19	36
Manganese	mg/kg	80.1	86.2	26
Phosphorus	%	0.15	0.16	33
Potassium	%	0.92	0.99	10

Lespedeza, stems, milk stage, (1)

Ref No 1 02 533 United States

			As Fed	Dry	C.V.
Dry matter	%			100.0	
Ash	%			4.7	11
Protein (N x 6.25)	%			9.6	11
Cattle	dig prot %	*		5.3	
Goats	dig prot %	*		5.5	
Horses	dig prot %	*		5.7	
Rabbits	dig prot %	*		6.1	
Sheep	dig prot %	*		5.2	
Calcium	%			0.59	7
Iron	%			0.017	7
Magnesium	%			0.14	28
Manganese	mg/kg			84.2	52
Phosphorus	%			0.12	46
Potassium	%			0.78	21

Lespedeza, stems, dough stage, (1)

Ref No 1 02 534 United States

			As Fed	Dry	C.V.
Dry matter	%			100.0	
Ash	%			3.4	
Protein (N x 6.25)	%			8.1	
Cattle	dig prot %	*		4.0	
Goats	dig prot %	*		4.1	
Horses	dig prot %	*		4.4	
Rabbits	dig prot %	*		4.9	
Sheep	dig prot %	*		3.8	

Lespedeza, stems, cut 1, (1)

Ref No 1 02 535 United States

			As Fed	Dry	C.V.
Dry matter	%			100.0	
Ash	%			5.0	22
Protein (N x 6.25)	%			10.5	27
Cattle	dig prot %	*		6.0	
Goats	dig prot %	*		6.4	
Horses	dig prot %	*		6.4	
Rabbits	dig prot %	*		6.8	
Sheep	dig prot %	*		6.0	

Lespedeza, stems, cut 2, (1)

Ref No 1 02 536 United States

			As Fed	Dry	C.V.
Dry matter	%			100.0	
Ash	%			4.3	11
Protein (N x 6.25)	%			9.9	12
Cattle	dig prot %	*		5.5	
Goats	dig prot %	*		5.8	
Horses	dig prot %	*		5.9	
Rabbits	dig prot %	*		6.3	
Sheep	dig prot %	*		5.5	

Lespedeza, stems, cut 3, (1)

Ref No 1 02 537 United States

			As Fed	Dry	C.V.
Dry matter	%			100.0	
Ash	%			4.7	
Protein (N x 6.25)	%			9.5	
Cattle	dig prot %	*		5.2	
Goats	dig prot %	*		5.4	
Horses	dig prot %	*		5.6	
Rabbits	dig prot %	*		6.0	
Sheep	dig prot %	*		5.1	

Lespedeza, straw, (1)

Ref No 1 08 452 United States

			As Fed	Dry	C.V.
Dry matter	%		90.0	100.0	
Ash	%		4.6	5.1	
Crude fiber	%		29.2	32.4	
Ether extract	%		2.3	2.6	
N-free extract	%		47.1	52.3	
Protein (N x 6.25)	%		6.8	7.6	
Cattle	dig prot %	*	3.1	3.5	

Continued

Feed Name or Analyses		Mean As Fed	Mean Dry	C.V. ± %
Goats	dig prot %	* 3.2	3.6	
Horses	dig prot %	* 3.6	3.9	
Rabbits	dig prot %	* 4.1	4.5	
Sheep	dig prot %	* 3.0	3.3	
Energy	GE Mcal/kg			
Cattle	DE Mcal/kg	* 1.92	2.13	
Sheep	DE Mcal/kg	* 1.99	2.21	
Cattle	ME Mcal/kg	* 1.58	1.75	
Sheep	ME Mcal/kg	* 1.63	1.81	
Cattle	TDN %	* 43.5	48.3	
Sheep	TDN %	* 45.2	50.2	

Lespedeza, aerial part, fresh, immature, (2)
Ref No 2 02 539 United States

Analyses		As Fed	Dry	C.V. ± %
Dry matter	%	31.1	100.0	19
Ash	%	3.3	10.6	18
Crude fiber	%	8.5	27.3	15
Ether extract	%	0.8	2.7	24
N-free extract	%	12.7	40.7	
Protein (N x 6.25)	%	5.8	18.7	12
Cattle	dig prot %	* 4.3	13.8	
Goats	dig prot %	* 4.4	14.0	
Horses	dig prot %	* 4.2	13.4	
Rabbits	dig prot %	* 4.1	13.1	
Sheep	dig prot %	* 4.5	14.4	

Lespedeza, aerial part, fresh, early bloom, (2)
Ref No 2 02 540 United States

Analyses		As Fed	Dry	C.V. ± %
Dry matter	%		100.0	
Energy	GE Mcal/kg			
Cattle	NE$_m$ Mcal/kg	*	1.47	
Cattle	NE$_{gain}$ Mcal/kg	*	0.90	
Cattle	NE$_{lactating\ cows}$ Mcal/kg	*	1.71	
Calcium	%		1.35	
Iron	%		0.025	
Magnesium	%		0.27	
Manganese	mg/kg		208.6	
Phosphorus	%		0.21	
Potassium	%		1.12	

Lespedeza, aerial part, fresh, full bloom, (2)
Ref No 2 02 541 United States

Analyses		As Fed	Dry	C.V. ± %
Dry matter	%	34.8	100.0	
Calcium	%	0.42	1.21	17
Iron	%	0.009	0.027	24
Magnesium	%	0.09	0.25	41
Manganese	mg/kg	61.4	176.4	70
Phosphorus	%	0.09	0.27	32
Potassium	%	0.35	1.01	12

Lespedeza, aerial part, fresh, mature, (2)
Ref No 2 02 542 United States

Analyses		As Fed	Dry	C.V. ± %
Dry matter	%	36.3	100.0	7
Ash	%	2.7	7.4	12
Crude fiber	%	16.3	44.9	3
Ether extract	%	0.8	2.1	15
N-free extract	%	11.9	32.8	
Protein (N x 6.25)	%	4.6	12.8	8
Cattle	dig prot %	* 3.2	8.8	
Goats	dig prot %	* 3.1	8.5	
Horses	dig prot %	* 3.0	8.4	
Rabbits	dig prot %	* 3.1	8.6	
Sheep	dig prot %	* 3.2	8.9	
Energy	GE Mcal/kg			
Cattle	DE Mcal/kg	* 0.93	2.57	
Sheep	DE Mcal/kg	* 0.91	2.51	
Cattle	ME Mcal/kg	* 0.77	2.11	
Sheep	ME Mcal/kg	* 0.75	2.05	
Cattle	NE$_m$ Mcal/kg	* 0.59	1.63	
Cattle	NE$_{gain}$ Mcal/kg	* 0.38	1.06	
Cattle	NE$_{lactating\ cows}$ Mcal/kg	* 0.70	1.93	
Cattle	TDN %	* 21.2	58.4	
Sheep	TDN %	* 20.6	56.8	
Calcium	%	0.37	1.02	
Iron	%	0.008	0.022	
Magnesium	%	0.06	0.16	
Manganese	mg/kg	30.9	85.1	
Phosphorus	%	0.11	0.31	
Potassium	%	0.28	0.77	

Lespedeza, aerial part, ensiled, (3)
Ref No 3 02 544 United States

Analyses		As Fed	Dry	C.V. ± %
Dry matter	%	37.6	100.0	18
Ash	%	2.2	5.9	9
Crude fiber	%	11.8	31.3	8
Ether extract	%	1.0	2.7	26
N-free extract	%	17.2	45.8	
Protein (N x 6.25)	%	5.4	14.3	12
Cattle	dig prot %	* 3.5	9.2	
Goats	dig prot %	* 3.5	9.2	
Horses	dig prot %	* 3.5	9.2	
Sheep	dig prot %	* 3.5	9.2	
Energy	GE Mcal/kg			
Cattle	DE Mcal/kg	* 0.88	2.35	
Sheep	DE Mcal/kg	* 0.91	2.43	
Cattle	ME Mcal/kg	* 0.72	1.93	
Sheep	ME Mcal/kg	* 0.75	1.99	
Cattle	TDN %	* 20.1	53.3	
Sheep	TDN %	* 20.7	55.1	

Lespedeza, aerial part w molasses added, ensiled, (3)
Ref No 3 02 545 United States

Analyses		As Fed	Dry	C.V. ± %
Dry matter	%	36.2	100.0	16
Ash	%	2.2	6.0	9
Crude fiber	%	10.7	29.5	3
Ether extract	%	1.2	3.3	18
N-free extract	%	16.9	46.6	
Protein (N x 6.25)	%	5.3	14.6	8
Cattle	dig prot %	* 3.4	9.5	
Goats	dig prot %	* 3.4	9.5	
Horses	dig prot %	* 3.4	9.5	
Sheep	dig prot %	* 3.4	9.5	
Energy	GE Mcal/kg			
Cattle	DE Mcal/kg	* 0.89	2.45	
Sheep	DE Mcal/kg	* 0.90	2.48	
Cattle	ME Mcal/kg	* 0.73	2.01	
Sheep	ME Mcal/kg	* 0.73	2.03	
Cattle	TDN %	* 20.1	55.6	
Sheep	TDN %	* 20.4	56.3	

Lespedeza, seeds, (5)
Ref No 5 02 546 United States

Analyses		As Fed	Dry	C.V. ± %
Dry matter	%	90.0	100.0	
Ash	%	5.4	6.0	
Crude fiber	%	10.7	11.9	
Cattle	dig coef %	28.	28.	
Ether extract	%	7.3	8.1	
Cattle	dig coef %	48.	48.	
N-free extract	%	32.7	36.3	
Cattle	dig coef %	76.	76.	
Protein (N x 6.25)	%	33.9	37.7	
Cattle	dig coef %	77.	77.	
Cattle	dig prot %	26.1	29.0	
Energy	GE Mcal/kg			
Cattle	DE Mcal/kg	* 2.73	3.03	
Sheep	DE Mcal/kg	* 2.84	3.15	
Swine	DE kcal/kg	*3163.	3514.	
Cattle	ME Mcal/kg	* 2.24	2.48	
Sheep	ME Mcal/kg	* 2.32	2.58	
Swine	ME kcal/kg	*2795.	3106.	
Cattle	TDN %	61.8	68.7	
Sheep	TDN %	* 64.4	71.54	
Swine	TDN %	* 71.7	79.7	

LESPEDEZA, ANNUAL. Lespedeza spp

Lespedeza, annual, aerial part, fresh, mature, (2)
Ref No 2 08 596 United States

Analyses		As Fed	Dry	C.V. ± %
Dry matter	%	37.2	100.0	
Ash	%	2.9	7.8	
Crude fiber	%	16.1	43.3	
Ether extract	%	0.9	2.4	
N-free extract	%	12.2	32.8	
Protein (N x 6.25)	%	5.1	13.7	
Cattle	dig prot %	* 3.3	8.8	
Goats	dig prot %	* 3.5	9.4	
Horses	dig prot %	* 3.4	9.2	
Rabbits	dig prot %	* 3.4	9.3	
Sheep	dig prot %	* 3.3	8.9	
Energy	GE Mcal/kg			
Cattle	DE Mcal/kg	* 0.70	1.88	
Sheep	DE Mcal/kg	* 0.64	1.73	
Cattle	ME Mcal/kg	* 0.57	1.54	
Sheep	ME Mcal/kg	* 0.53	1.42	
Cattle	TDN %	* 15.8	42.6	
Sheep	TDN %	* 14.6	39.3	
Calcium	%	0.38	1.02	
Phosphorus	%	0.06	0.16	
Potassium	%	0.34	0.91	

Lespedeza, annual, hay, s-c, (1)
Ref No 1 08 591 United States

Analyses		As Fed	Dry	C.V. ± %
Dry matter	%	90.0	100.0	
Ash	%	5.3	5.9	
Crude fiber	%	26.9	29.9	
Ether extract	%	2.5	2.8	
N-free extract	%	42.2	46.9	
Protein (N x 6.25)	%	13.1	14.6	
Cattle	dig prot %	* 8.6	9.5	
Goats	dig prot %	* 9.1	10.1	
Horses	dig prot %	* 8.9	9.9	
Rabbits	dig prot %	* 8.9	9.9	
Sheep	dig prot %	* 8.7	9.6	
Fatty acids	%	2.5	2.8	

(1) dry forages and roughages (2) pasture, range plants, and forages fed green (3) silages (4) energy feeds (5) protein supplements (6) minerals (7) vitamins (8) additives

Feed Name or Analyses			Mean As Fed	Mean Dry	C.V. ± %
Energy	GE Mcal/kg				
Cattle	DE Mcal/kg	*	2.03	2.26	
Sheep	DE Mcal/kg	*	1.97	2.19	
Cattle	ME Mcal/kg	*	1.66	1.85	
Sheep	ME Mcal/kg	*	1.62	1.80	
Cattle	TDN %	*	46.1	51.2	
Sheep	TDN %	*	44.7	49.7	
Calcium	%		0.96	1.07	
Iron	%		0.026	0.029	
Magnesium	%		0.22	0.24	
Manganese	mg/kg		131.0	145.5	
Phosphorus	%		0.18	0.20	
Potassium	%		0.94	1.04	

Lespedeza, annual, hay, s-c, pre-bloom, (1)

Ref No 1 08 594 United States

Feed Name or Analyses			Mean As Fed	Mean Dry	C.V. ± %
Dry matter	%		89.1	100.0	
Ash	%		6.4	7.2	
Crude fiber	%		22.7	25.5	
Ether extract	%		2.7	3.0	
N-free extract	%		43.0	48.3	
Protein (N x 6.25)	%		14.3	16.0	
Cattle	dig prot %	*	9.7	10.8	
Goats	dig prot %	*	10.3	11.5	
Horses	dig prot %	*	9.9	11.2	
Rabbits	dig prot %	*	9.9	11.1	
Sheep	dig prot %	*	9.8	11.0	
Fatty acids	%		2.7	3.0	
Energy	GE Mcal/kg				
Cattle	DE Mcal/kg	*	2.25	2.52	
Sheep	DE Mcal/kg	*	2.15	2.41	
Cattle	ME Mcal/kg	*	1.84	2.07	
Sheep	ME Mcal/kg	*	1.76	1.98	
Cattle	TDN %	*	51.0	57.2	
Sheep	TDN %	*	48.6	54.6	
Calcium	%		1.03	1.16	
Iron	%		0.030	0.034	
Magnesium	%		0.21	0.24	
Manganese	mg/kg		159.0	178.4	
Phosphorus	%		0.20	0.22	
Potassium	%		1.07	1.20	

Lespedeza, annual, hay, s-c, mid-bloom, (1)

Ref No 1 08 589 United States

Feed Name or Analyses			Mean As Fed	Mean Dry	C.V. ± %
Dry matter	%		89.1	100.0	
Ash	%		5.1	5.7	
Crude fiber	%		26.5	29.7	
Ether extract	%		1.7	1.9	
N-free extract	%		42.8	48.0	
Protein (N x 6.25)	%		13.0	14.6	
Cattle	dig prot %	*	8.5	9.6	
Goats	dig prot %	*	9.1	10.2	
Horses	dig prot %	*	8.8	9.9	
Rabbits	dig prot %	*	8.9	9.9	
Sheep	dig prot %	*	8.6	9.7	
Fatty acids	%		1.8	2.0	
Energy	GE Mcal/kg				
Cattle	DE Mcal/kg	*	2.00	2.24	
Sheep	DE Mcal/kg	*	2.12	2.38	
Cattle	ME Mcal/kg	*	1.64	1.84	
Sheep	ME Mcal/kg	*	1.74	1.95	
Cattle	TDN %	*	45.2	50.7	
Sheep	TDN %	*	48.0	53.9	
Calcium	%		1.00	1.12	
Iron	%		0.025	0.028	
Magnesium	%		0.19	0.21	
Manganese	mg/kg		134.9	151.4	
Phosphorus	%		0.19	0.21	
Potassium	%		0.94	1.06	

Lespedeza, annual, hay, s-c, milk stage, (1)

Ref No 1 08 590 United States

Feed Name or Analyses			Mean As Fed	Mean Dry	C.V. ± %
Dry matter	%		89.1	100.0	
Ash	%		4.5	5.1	
Crude fiber	%		32.6	36.6	
Ether extract	%		1.9	2.1	
N-free extract	%		38.6	43.3	
Protein (N x 6.25)	%		11.5	12.9	
Cattle	dig prot %	*	7.2	8.1	
Goats	dig prot %	*	7.7	8.6	
Horses	dig prot %	*	7.6	8.5	
Rabbits	dig prot %	*	7.7	8.6	
Sheep	dig prot %	*	7.3	8.1	
Fatty acids	%		1.9	2.1	
Energy	GE Mcal/kg				
Cattle	DE Mcal/kg	*	1.69	1.90	
Sheep	DE Mcal/kg	*	1.88	2.11	
Cattle	ME Mcal/kg	*	1.39	1.56	
Sheep	ME Mcal/kg	*	1.54	1.73	
Cattle	TDN %	*	38.4	43.1	
Sheep	TDN %	*	42.6	47.8	
Calcium	%		0.90	1.01	
Iron	%		0.030	0.034	
Magnesium	%		0.19	0.21	
Manganese	mg/kg		168.0	188.5	
Phosphorus	%		0.15	0.17	
Potassium	%		0.82	0.92	

Lespedeza, annual, leaves, s-c, (1)

Ref No 1 08 592 United States

Feed Name or Analyses			Mean As Fed	Mean Dry	C.V. ± %
Dry matter	%		89.2	100.0	
Ash	%		6.4	7.2	
Crude fiber	%		19.7	22.1	
Ether extract	%		2.9	3.3	
N-free extract	%		43.1	48.3	
Protein (N x 6.25)	%		17.1	19.2	
Cattle	dig prot %	*	12.1	13.5	
Goats	dig prot %	*	12.9	14.4	
Horses	dig prot %	*	12.3	13.8	
Rabbits	dig prot %	*	12.0	13.5	
Sheep	dig prot %	*	12.3	13.8	
Fatty acids	%		2.9	3.3	
Energy	GE Mcal/kg				
Cattle	DE Mcal/kg	*	2.52	2.82	
Sheep	DE Mcal/kg	*	2.52	2.83	
Cattle	ME Mcal/kg	*	2.06	2.31	
Sheep	ME Mcal/kg	*	2.07	2.32	
Cattle	TDN %	*	57.1	64.0	
Sheep	TDN %	*	57.2	64.1	
Calcium	%		1.30	1.46	
Phosphorus	%		0.20	0.22	
Potassium	%		0.92	1.03	

Lespedeza, annual, stems, (1)

Ref No 1 08 593 United States

Feed Name or Analyses			Mean As Fed	Mean Dry	C.V. ± %
Dry matter	%		89.2	100.0	
Ash	%		3.7	4.1	
Crude fiber	%		38.5	43.2	
Ether extract	%		1.0	1.1	
N-free extract	%		37.7	42.3	
Protein (N x 6.25)	%		8.3	9.3	
Cattle	dig prot %	*	4.5	5.0	
Goats	dig prot %	*	4.7	5.2	
Horses	dig prot %	*	4.8	5.4	
Rabbits	dig prot %	*	5.2	5.9	
Sheep	dig prot %	*	4.4	4.9	
Fatty acids	%		1.0	1.1	
Energy	GE Mcal/kg				
Cattle	DE Mcal/kg	*	1.77	1.98	
Sheep	DE Mcal/kg	*	1.87	2.10	
Cattle	ME Mcal/kg	*	1.44	1.62	
Sheep	ME Mcal/kg	*	1.53	1.72	
Cattle	TDN %	*	40.0	44.9	
Sheep	TDN %	*	42.5	47.6	
Calcium	%		0.64	0.72	
Phosphorus	%		0.13	0.15	
Potassium	%		0.89	1.00	

Lespedeza, annual, aerial part, fresh, pre-bloom, (2)

Ref No 2 08 453 United States

Feed Name or Analyses			Mean As Fed	Mean Dry	C.V. ± %
Dry matter	%		25.0	100.0	
Ash	%		3.2	12.8	
Crude fiber	%		8.0	32.0	
Ether extract	%		0.5	2.0	
N-free extract	%		9.2	36.8	
Protein (N x 6.25)	%		4.1	16.4	
Cattle	dig prot %	*	3.0	11.8	
Goats	dig prot %	*	3.0	11.9	
Horses	dig prot %	*	2.9	11.5	
Rabbits	dig prot %	*	2.8	11.3	
Sheep	dig prot %	*	3.1	12.3	
Energy	GE Mcal/kg				
Cattle	DE Mcal/kg	*	0.64	2.54	
Sheep	DE Mcal/kg	*	0.65	2.59	
Cattle	ME Mcal/kg	*	0.52	2.08	
Sheep	ME Mcal/kg	*	0.53	2.12	
Cattle	TDN %	*	14.4	57.6	
Sheep	TDN %	*	14.7	58.7	
Calcium	%		0.28	1.12	
Iron	%		0.008	0.032	
Magnesium	%		0.06	0.24	
Manganese	mg/kg		44.5	178.1	
Phosphorus	%		0.07	0.28	
Potassium	%		0.32	1.28	

Lespedeza, annual, aerial part, fresh, mid-bloom, (2)

Ref No 2 08 454 United States

Feed Name or Analyses			Mean As Fed	Mean Dry	C.V. ± %
Dry matter	%		28.5	100.0	
Ash	%		2.4	8.4	
Crude fiber	%		12.9	45.3	
Ether extract	%		0.6	2.1	
N-free extract	%		9.0	31.6	
Protein (N x 6.25)	%		3.6	12.6	
Cattle	dig prot %	*	2.5	8.6	
Goats	dig prot %	*	2.4	8.3	
Horses	dig prot %	*	2.4	8.3	
Rabbits	dig prot %	*	2.4	8.4	
Sheep	dig prot %	*	2.5	8.8	
Energy	GE Mcal/kg				
Cattle	DE Mcal/kg	*	0.66	2.31	
Sheep	DE Mcal/kg	*	0.61	2.14	
Cattle	ME Mcal/kg	*	0.54	1.89	
Sheep	ME Mcal/kg	*	0.50	1.75	
Cattle	TDN %	*	15.0	52.5	
Sheep	TDN %	*	13.9	48.6	

Continued

Feed Name or Analyses		Mean As Fed	Mean Dry	C.V. ± %
Calcium	%	0.33	1.16	
Phosphorus	%	0.06	0.21	
Potassium	%	0.30	1.05	

Lespedeza, annual, aerial part, ensiled, (3)

Ref No 3 08 455 United States

Feed Name or Analyses		Mean As Fed	Mean Dry	C.V. ± %
Dry matter	%	30.2	100.0	
Ash	%	1.8	6.0	
Crude fiber	%	9.5	31.5	
Ether extract	%	0.8	2.6	
N-free extract	%	13.8	45.7	
Protein (N x 6.25)	%	4.3	14.2	
Cattle	dig prot %	* 2.8	9.2	
Goats	dig prot %	* 2.8	9.2	
Horses	dig prot %	* 2.8	9.2	
Sheep	dig prot %	* 2.8	9.2	
Energy	GE Mcal/kg			
Cattle	DE Mcal/kg	* 0.71	2.35	
Sheep	DE Mcal/kg	* 0.65	2.16	
Cattle	ME Mcal/kg	* 0.58	1.93	
Sheep	ME Mcal/kg	* 0.53	1.77	
Cattle	TDN %	* 16.1	53.3	
Sheep	TDN %	* 14.8	49.1	

Lespedeza, annual, seeds, (5)

Ref No 5 08 456 United States

Feed Name or Analyses		Mean As Fed	Mean Dry	C.V. ± %
Dry matter	%	91.7	100.0	
Ash	%	5.1	5.6	
Crude fiber	%	9.6	10.5	
Ether extract	%	7.6	8.3	
N-free extract	%	32.8	35.8	
Protein (N x 6.25)	%	36.6	39.9	
Energy	GE Mcal/kg			
Cattle	DE Mcal/kg	* 2.87	3.13	
Sheep	DE Mcal/kg	* 2.94	3.21	
Swine	DE kcal/kg	*3266.	3562.	
Cattle	ME Mcal/kg	* 2.36	2.57	
Sheep	ME Mcal/kg	* 2.41	2.63	
Swine	ME kcal/kg	*2871.	3131.	
Cattle	TDN %	* 65.1	71.0	
Sheep	TDN %	* 66.7	72.7	
Swine	TDN %	* 74.1	80.8	

LESPEDEZA, BICOLOR. Lespedeza bicolor

Lespedeza, bicolor, leaves, milk stage, (1)

Ref No 1 02 547 United States

Feed Name or Analyses		Mean As Fed	Mean Dry	C.V. ± %
Dry matter	%	93.6	100.0	
Calcium	%	1.24	1.32	
Iron	%	0.032	0.034	
Magnesium	%	0.23	0.25	
Manganese	mg/kg	385.1	411.4	
Phosphorus	%	0.21	0.22	
Potassium	%	0.92	0.98	

Lespedeza, bicolor, aerial part, fresh, (2)

Ref No 2 02 552 United States

Feed Name or Analyses		Mean As Fed	Mean Dry	C.V. ± %
Dry matter	%		100.0	
Calcium	%		1.51	19
Iron	%		0.039	52
Magnesium	%		0.33	28
Manganese	mg/kg		293.7	30
Phosphorus	%		0.29	89
Potassium	%		1.20	30

Lespedeza, bicolor, aerial part, fresh, immature, (2)

Ref No 2 02 548 United States

Feed Name or Analyses		Mean As Fed	Mean Dry	C.V. ± %
Dry matter	%		100.0	
Calcium	%		1.63	
Iron	%		0.034	
Magnesium	%		0.38	
Manganese	mg/kg		343.7	
Phosphorus	%		0.48	
Potassium	%		1.65	

Lespedeza, bicolor, aerial part, fresh, early bloom, (2)

Ref No 2 02 549 United States

Feed Name or Analyses		Mean As Fed	Mean Dry	C.V. ± %
Dry matter	%		100.0	
Calcium	%		1.57	
Iron	%		0.030	
Magnesium	%		0.33	
Manganese	mg/kg		213.2	
Phosphorus	%		0.24	
Potassium	%		1.21	

Lespedeza, bicolor, aerial part, fresh, full bloom, (2)

Ref No 2 02 550 United States

Feed Name or Analyses		Mean As Fed	Mean Dry	C.V. ± %
Dry matter	%		100.0	
Calcium	%		1.49	
Iron	%		0.032	
Magnesium	%		0.34	
Manganese	mg/kg		386.5	
Phosphorus	%		0.25	
Potassium	%		1.01	

Lespedeza, bicolor, aerial part, fresh, milk stage, (2)

Ref No 2 02 551 United States

Feed Name or Analyses		Mean As Fed	Mean Dry	C.V. ± %
Dry matter	%		100.0	
Calcium	%		1.42	9
Iron	%		0.049	48
Magnesium	%		0.30	26
Manganese	mg/kg		258.8	47
Phosphorus	%		0.23	9
Potassium	%		1.01	12

LESPEDEZA, COMMON. Lespedeza striata

Lespedeza, common, hay, s-c, (1)

Ref No 1 02 563 United States

Feed Name or Analyses		Mean As Fed	Mean Dry	C.V. ± %
Dry matter	%	89.6	100.0	2
Carotene	mg/kg	49.6	55.3	55
Riboflavin	mg/kg	8.9	9.9	29
Vitamin A equivalent	IU/g	82.7	92.2	

Lespedeza, common, hay, s-c, immature, (1)

Ref No 1 02 553 United States

Feed Name or Analyses		Mean As Fed	Mean Dry	C.V. ± %
Dry matter	%	89.1	100.0	1
Ash	%	6.4	7.2	
Crude fiber	%	22.7	25.5	
Ether extract	%	2.7	3.0	
N-free extract	%	43.0	48.3	
Protein (N x 6.25)	%	14.3	16.0	
Cattle	dig prot %	* 9.6	10.8	
Goats	dig prot %	* 10.2	11.5	
Horses	dig prot %	* 9.9	11.1	
Rabbits	dig prot %	* 9.8	11.0	
Sheep	dig prot %	* 9.7	10.9	
Energy	GE Mcal/kg			
Cattle	DE Mcal/kg	* 1.99	2.23	
Sheep	DE Mcal/kg	* 2.03	2.28	
Cattle	ME Mcal/kg	* 1.63	1.83	
Sheep	ME Mcal/kg	* 1.67	1.87	
Cattle	TDN %	* 45.1	50.6	
Sheep	TDN %	* 46.0	51.6	
Calcium	%	1.09	1.22	6
Iron	%	0.029	0.032	14
Magnesium	%	0.23	0.26	17
Manganese	mg/kg	145.2	162.9	8
Phosphorus	%	0.26	0.29	18
Potassium	%	0.95	1.07	8

Lespedeza, common, hay, s-c, mid-bloom, (1)

Ref No 1 02 554 United States

Feed Name or Analyses		Mean As Fed	Mean Dry	C.V. ± %
Dry matter	%	94.2	100.0	1
Calcium	%	1.16	1.23	7
Iron	%	0.032	0.034	11
Magnesium	%	0.25	0.27	16
Manganese	mg/kg	294.3	312.4	5
Phosphorus	%	0.16	0.17	13
Potassium	%	0.96	1.02	8

Lespedeza, common, hay, s-c, full bloom, (1)

Ref No 1 02 555 United States

Feed Name or Analyses		Mean As Fed	Mean Dry	C.V. ± %
Dry matter	%	89.4	100.0	1
Ash	%	5.0	5.6	8
Crude fiber	%	27.4	30.7	5
Ether extract	%	1.9	2.1	11
N-free extract	%	42.5	47.6	
Protein (N x 6.25)	%	12.5	14.0	6
Cattle	dig prot %	* 8.1	9.1	
Goats	dig prot %	* 8.6	9.6	
Horses	dig prot %	* 8.4	9.4	
Rabbits	dig prot %	* 8.5	9.5	
Sheep	dig prot %	* 8.2	9.1	
Lignin (Ellis)	%	16.5	18.5	10
Energy	GE Mcal/kg	4.26	4.77	2
Cattle	DE Mcal/kg	* 1.90	2.13	

(1) dry forages and roughages (2) pasture, range plants, and forages fed green (3) silages (4) energy feeds (5) protein supplements (6) minerals (7) vitamins (8) additives

Feed Name or Analyses			Mean As Fed	Mean Dry	C.V. ± %
Sheep	DE Mcal/kg	*	1.97	2.20	
Cattle	ME Mcal/kg	*	1.56	1.75	
Sheep	ME Mcal/kg	*	1.61	1.80	
Cattle	TDN %	*	43.1	48.2	
Sheep	TDN %	*	44.6	49.9	
Calcium	%		1.04	1.16	6
Iron	%		0.029	0.032	15
Magnesium	%		0.21	0.24	22
Manganese	mg/kg		135.4	151.5	7
Phosphorus	%		0.18	0.20	17
Potassium	%		0.92	1.03	5

Lespedeza, common, hay, s-c, milk stage, (1)

Ref No 1 02 556 United States

Feed Name or Analyses			Mean As Fed	Mean Dry	C.V. ± %
Dry matter	%		87.3	100.0	2
Cellulose (Matrone)	%		26.9	30.8	
Lignin (Ellis)	%		18.7	21.4	5
Calcium	%		0.87	1.00	6
Iron	%		0.025	0.029	41
Magnesium	%		0.19	0.22	20
Manganese	mg/kg		176.0	201.7	24
Phosphorus	%		0.16	0.18	18
Potassium	%		0.76	0.87	10

Lespedeza, common, hay, s-c, cut 1, (1)

Ref No 1 02 557 United States

Feed Name or Analyses			Mean As Fed	Mean Dry	C.V. ± %
Dry matter	%		91.4	100.0	2
Ash	%		4.8	5.3	10
Crude fiber	%		25.2	27.6	10
Ether extract	%		3.8	4.2	3
N-free extract	%		43.7	47.8	
Protein (N x 6.25)	%		13.8	15.1	4
Cattle	dig prot %	*	9.2	10.0	
Goats	dig prot %	*	9.7	10.6	
Horses	dig prot %	*	9.5	10.3	
Rabbits	dig prot %	*	9.4	10.3	
Sheep	dig prot %	*	9.2	10.1	
Energy	GE Mcal/kg				
Cattle	DE Mcal/kg	*	2.07	2.26	
Sheep	DE Mcal/kg	*	2.04	2.23	
Cattle	ME Mcal/kg	*	1.69	1.85	
Sheep	ME Mcal/kg	*	1.67	1.83	
Cattle	TDN %	*	46.8	51.2	
Sheep	TDN %	*	46.2	50.6	
Calcium	%		1.01	1.10	7
Iron	%		0.029	0.032	9
Magnesium	%		0.20	0.22	17
Manganese	mg/kg		148.1	162.0	7
Phosphorus	%		0.18	0.20	20
Potassium	%		0.95	1.04	7

Lespedeza, common, hay, s-c, cut 2, (1)

Ref No 1 02 558 United States

Feed Name or Analyses			Mean As Fed	Mean Dry	C.V. ± %
Dry matter	%		91.8	100.0	
Ash	%		4.7	5.1	
Crude fiber	%		28.3	30.8	
Ether extract	%		4.2	4.6	
N-free extract	%		41.1	44.8	
Protein (N x 6.25)	%		13.5	14.7	
Cattle	dig prot %	*	8.9	9.7	
Goats	dig prot %	*	9.4	10.3	
Horses	dig prot %	*	9.2	10.0	
Rabbits	dig prot %	*	9.2	10.0	
Sheep	dig prot %	*	9.0	9.8	

Feed Name or Analyses			Mean As Fed	Mean Dry	C.V. ± %
Energy	GE Mcal/kg				
Cattle	DE Mcal/kg	*	1.90	2.07	
Sheep	DE Mcal/kg	*	1.96	2.13	
Cattle	ME Mcal/kg	*	1.56	1.70	
Sheep	ME Mcal/kg	*	1.61	1.75	
Cattle	TDN %	*	43.1	47.0	
Sheep	TDN %	*	44.2	48.2	
Calcium	%		1.20	1.31	
Phosphorus	%		0.30	0.33	

Lespedeza, common, hay, s-c, cut 3, (1)

Ref No 1 02 559 United States

Feed Name or Analyses			Mean As Fed	Mean Dry	C.V. ± %
Dry matter	%		92.8	100.0	
Ash	%		4.0	4.3	
Crude fiber	%		27.7	29.8	
Ether extract	%		4.0	4.3	
N-free extract	%		44.7	48.2	
Protein (N x 6.25)	%		12.4	13.4	
Cattle	dig prot %	*	7.9	8.5	
Goats	dig prot %	*	8.4	9.1	
Horses	dig prot %	*	8.3	8.9	
Rabbits	dig prot %	*	8.4	9.0	
Sheep	dig prot %	*	8.0	8.6	
Energy	GE Mcal/kg				
Cattle	DE Mcal/kg	*	1.91	2.06	
Sheep	DE Mcal/kg	*	2.03	2.19	
Cattle	ME Mcal/kg	*	1.57	1.69	
Sheep	ME Mcal/kg	*	1.67	1.80	
Cattle	TDN %	*	43.3	46.7	
Sheep	TDN %	*	46.1	49.7	
Calcium	%		1.00	1.08	
Phosphorus	%		0.30	0.32	

Lespedeza, common, hay, s-c, gr 3 US, (1)

Ref No 1 02 562 United States

Feed Name or Analyses			Mean As Fed	Mean Dry	C.V. ± %
Dry matter	%		90.6	100.0	1
Ash	%		3.7	4.1	7
Crude fiber	%		32.3	35.7	2
Ether extract	%		2.2	2.4	4
N-free extract	%		41.6	45.9	
Protein (N x 6.25)	%		10.8	11.9	7
Cattle	dig prot %	*	6.6	7.2	
Goats	dig prot %	*	6.9	7.7	
Horses	dig prot %	*	6.9	7.6	
Rabbits	dig prot %	*	7.1	7.9	
Sheep	dig prot %	*	6.6	7.2	
Energy	GE Mcal/kg				
Cattle	DE Mcal/kg	*	1.99	2.20	
Sheep	DE Mcal/kg	*	2.25	2.48	
Cattle	ME Mcal/kg	*	1.63	1.80	
Sheep	ME Mcal/kg	*	1.84	2.04	
Cattle	TDN %	*	45.1	49.8	
Sheep	TDN %	*	51.0	56.3	

Lespedeza, common, hay, s-c, gr sample US, (1)

Ref No 1 02 560 United States

Feed Name or Analyses			Mean As Fed	Mean Dry	C.V. ± %
Dry matter	%		91.1	100.0	
Ash	%		4.0	4.4	
Crude fiber	%		34.2	37.5	
Ether extract	%		1.9	2.1	
N-free extract	%		43.0	47.2	
Protein (N x 6.25)	%		8.0	8.8	
Cattle	dig prot %	*	4.2	4.6	
Goats	dig prot %	*	4.3	4.8	

Feed Name or Analyses			Mean As Fed	Mean Dry	C.V. ± %
Horses	dig prot %	*	4.6	5.0	
Rabbits	dig prot %	*	5.0	5.5	
Sheep	dig prot %	*	4.1	4.5	
Energy	GE Mcal/kg				
Cattle	DE Mcal/kg	*	1.91	2.09	
Sheep	DE Mcal/kg	*	2.18	2.39	
Cattle	ME Mcal/kg	*	1.56	1.72	
Sheep	ME Mcal/kg	*	1.79	1.96	
Cattle	TDN %	*	43.3	47.5	
Sheep	TDN %	*	49.5	54.3	

Lespedeza, common, leaves, (1)

Ref No 1 02 564 United States

Feed Name or Analyses			Mean As Fed	Mean Dry	C.V. ± %
Dry matter	%		89.9	100.0	3
Ash	%		6.3	7.0	9
Crude fiber	%		20.6	22.9	7
Ether extract	%		2.8	3.1	10
N-free extract	%		42.5	47.3	
Protein (N x 6.25)	%		17.7	19.7	7
Cattle	dig prot %	*	12.6	14.0	
Goats	dig prot %	*	13.4	14.9	
Horses	dig prot %	*	12.8	14.3	
Rabbits	dig prot %	*	12.5	13.9	
Sheep	dig prot %	*	12.8	14.2	
Energy	GE Mcal/kg				
Cattle	DE Mcal/kg	*	2.44	2.72	
Sheep	DE Mcal/kg	*	2.53	2.82	
Cattle	ME Mcal/kg	*	2.00	2.23	
Sheep	ME Mcal/kg	*	2.08	2.31	
Cattle	TDN %	*	55.4	61.6	
Sheep	TDN %	*	57.4	63.9	

Lespedeza, common, stems, (1)

Ref No 1 02 565 United States

Feed Name or Analyses			Mean As Fed	Mean Dry	C.V. ± %
Dry matter	%		89.9	100.0	1
Ash	%		3.6	4.0	11
Crude fiber	%		37.9	42.2	6
Ether extract	%		1.2	1.3	20
N-free extract	%		38.6	42.9	
Protein (N x 6.25)	%		8.6	9.6	13
Cattle	dig prot %	*	4.7	5.3	
Goats	dig prot %	*	5.0	5.5	
Horses	dig prot %	*	5.1	5.7	
Rabbits	dig prot %	*	5.5	6.1	
Sheep	dig prot %	*	4.7	5.2	
Energy	GE Mcal/kg				
Cattle	DE Mcal/kg	*	1.52	1.69	
Sheep	DE Mcal/kg	*	1.68	1.87	
Cattle	ME Mcal/kg	*	1.25	1.39	
Sheep	ME Mcal/kg	*	1.38	1.53	
Cattle	TDN %	*	34.5	38.3	
Sheep	TDN %	*	38.2	42.5	

Lespedeza, common, aerial part, fresh, (2)

Ref No 2 02 568 United States

Feed Name or Analyses			Mean As Fed	Mean Dry	C.V. ± %
Dry matter	%		27.6	100.0	17
Ash	%		3.1	11.2	18
Crude fiber	%		10.0	36.1	14
Ether extract	%		0.6	2.1	12
N-free extract	%		9.7	35.3	
Protein (N x 6.25)	%		4.2	15.3	12
Cattle	dig prot %	*	3.0	10.9	
Goats	dig prot %	*	3.0	10.8	

Continued

Feed Name or Analyses			Mean As Fed	Mean Dry	C.V. ± %
Horses	dig prot %	*	2.9	10.5	
Rabbits	dig prot %	*	2.9	10.5	
Sheep	dig prot %	*	3.1	11.2	
Energy	GE Mcal/kg				
Cattle	DE Mcal/kg	*	0.84	3.03	
Sheep	DE Mcal/kg	*	0.71	2.56	
Cattle	ME Mcal/kg	*	0.68	2.48	
Sheep	ME Mcal/kg	*	0.58	2.10	
Cattle	TDN %	*	18.9	68.6	
Sheep	TDN %	*	16.0	58.0	
Calcium	%		0.31	1.13	5
Iron	%		0.009	0.032	25
Magnesium	%		0.07	0.27	19
Manganese	mg/kg		49.2	178.1	33
Phosphorus	%		0.07	0.27	9
Potassium	%		0.32	1.16	19

Lespedeza, common, aerial part, fresh, immature, (2)
Ref No 2 02 566 United States

Feed Name or Analyses			Mean As Fed	Mean Dry	C.V. ± %
Dry matter	%		25.0	100.0	7
Ash	%		3.2	12.8	11
Crude fiber	%		8.0	32.0	5
Ether extract	%		0.5	2.0	11
N-free extract	%		9.2	36.8	
Protein (N x 6.25)	%		4.1	16.4	7
Cattle	dig prot %	*	3.0	11.8	
Goats	dig prot %	*	3.0	11.9	
Horses	dig prot %	*	2.9	11.5	
Rabbits	dig prot %	*	2.8	11.3	
Sheep	dig prot %	*	3.1	12.3	
Energy	GE Mcal/kg				
Cattle	DE Mcal/kg	*	0.67	2.69	
Sheep	DE Mcal/kg	*	0.65	2.59	
Cattle	ME Mcal/kg	*	0.55	2.21	
Sheep	ME Mcal/kg	*	0.53	2.12	
Cattle	TDN %	*	15.2	61.0	
Sheep	TDN %	*	14.7	58.7	

Lespedeza, common, aerial part, fresh, mature, (2)
Ref No 2 02 567 United States

Feed Name or Analyses			Mean As Fed	Mean Dry	C.V. ± %
Dry matter	%		37.2	100.0	
Ash	%		2.9	7.8	
Crude fiber	%		16.1	43.3	
Ether extract	%		0.9	2.4	
N-free extract	%		12.2	32.8	
Protein (N x 6.25)	%		5.1	13.7	
Cattle	dig prot %	*	3.5	9.5	
Goats	dig prot %	*	3.5	9.3	
Horses	dig prot %	*	3.4	9.2	
Rabbits	dig prot %	*	3.4	9.2	
Sheep	dig prot %	*	3.6	9.8	
Energy	GE Mcal/kg				
Cattle	DE Mcal/kg	*	0.94	2.52	
Sheep	DE Mcal/kg	*	0.93	2.50	
Cattle	ME Mcal/kg	*	0.77	2.07	
Sheep	ME Mcal/kg	*	0.76	2.05	
Cattle	TDN %	*	21.2	57.1	
Sheep	TDN %	*	21.1	56.7	

(1) dry forages and roughages (2) pasture, range plants, and forages fed green (3) silages (4) energy feeds (5) protein supplements (6) minerals (7) vitamins (8) additives

Lespedeza, common, aerial part, ensiled, (3)
Ref No 3 02 569 United States

Feed Name or Analyses			Mean As Fed	Mean Dry	C.V. ± %
Dry matter	%		43.6	100.0	16
Ash	%		2.7	6.2	5
Crude fiber	%		13.6	31.3	8
Ether extract	%		1.0	2.4	15
N-free extract	%		19.8	45.5	
Protein (N x 6.25)	%		6.4	14.6	10
Cattle	dig prot %	*	4.1	9.5	
Goats	dig prot %	*	4.1	9.5	
Horses	dig prot %	*	4.1	9.5	
Sheep	dig prot %	*	4.1	9.5	
Energy	GE Mcal/kg				
Cattle	DE Mcal/kg	*	1.02	2.34	
Sheep	DE Mcal/kg	*	1.05	2.40	
Cattle	ME Mcal/kg	*	0.84	1.92	
Sheep	ME Mcal/kg	*	0.86	1.97	
Cattle	TDN %	*	23.1	53.0	
Sheep	TDN %	*	23.8	54.5	

Lespedeza, common, aerial part w grnd corn grain added, ensiled, (3)
Ref No 3 02 570 United States

Feed Name or Analyses			Mean As Fed	Mean Dry	C.V. ± %
Dry matter	%		52.8	100.0	
Ash	%		3.2	6.1	
Crude fiber	%		18.4	34.8	
Ether extract	%		1.0	1.9	
N-free extract	%		23.4	44.3	
Protein (N x 6.25)	%		6.8	12.9	
Cattle	dig prot %	*	4.2	7.9	
Goats	dig prot %	*	4.2	7.9	
Horses	dig prot %	*	4.2	7.9	
Sheep	dig prot %	*	4.2	7.9	
Energy	GE Mcal/kg				
Cattle	DE Mcal/kg	*	1.18	2.24	
Sheep	DE Mcal/kg	*	1.27	2.40	
Cattle	ME Mcal/kg	*	0.97	1.84	
Sheep	ME Mcal/kg	*	1.04	1.97	
Cattle	TDN %	*	26.8	50.8	
Sheep	TDN %	*	28.7	54.4	

Lespedeza, common, aerial part w molasses added, ensiled, (3)
Ref No 3 02 571 United States

Feed Name or Analyses			Mean As Fed	Mean Dry	C.V. ± %
Dry matter	%		42.4	100.0	
Ash	%		2.8	6.6	
Crude fiber	%		12.3	29.0	
Ether extract	%		1.1	2.6	
N-free extract	%		19.9	46.9	
Protein (N x 6.25)	%		6.3	14.9	
Cattle	dig prot %	*	4.1	9.8	
Goats	dig prot %	*	4.1	9.8	
Horses	dig prot %	*	4.1	9.8	
Sheep	dig prot %	*	4.1	9.8	
Energy	GE Mcal/kg				
Cattle	DE Mcal/kg	*	1.04	2.45	
Sheep	DE Mcal/kg	*	1.07	2.53	
Cattle	ME Mcal/kg	*	0.85	2.01	
Sheep	ME Mcal/kg	*	0.88	2.07	
Cattle	TDN %	*	23.6	55.6	
Sheep	TDN %	*	24.3	57.3	

LESPEDEZA, KOBE. Lespedeza striata kobe

Lespedeza, kobe, hay, s-c, immature, (1)
Ref No 1 02 572 United States

Feed Name or Analyses			Mean As Fed	Mean Dry	C.V. ± %
Dry matter	%			100.0	
Calcium	%			1.32	4
Iron	%			0.035	8
Magnesium	%			0.27	18
Manganese	mg/kg			263.2	7
Phosphorus	%			0.23	15
Potassium	%			1.15	5

Lespedeza, kobe, hay, s-c, early bloom, (1)
Ref No 1 02 573 United States

Feed Name or Analyses			Mean As Fed	Mean Dry	C.V. ± %
Dry matter	%		92.4	100.0	2
Ash	%		5.5	6.0	9
Crude fiber	%		29.6	32.0	3
Ether extract	%		3.7	4.0	9
N-free extract	%		38.5	41.7	
Protein (N x 6.25)	%		15.1	16.3	7
Cattle	dig prot %	*	10.2	11.1	
Goats	dig prot %	*	10.9	11.8	
Horses	dig prot %	*	10.5	11.4	
Rabbits	dig prot %	*	10.4	11.3	
Sheep	dig prot %	*	10.3	11.2	
Energy	GE Mcal/kg				
Cattle	DE Mcal/kg	*	2.30	2.49	
Sheep	DE Mcal/kg	*	2.34	2.53	
Cattle	ME Mcal/kg	*	1.88	2.04	
Sheep	ME Mcal/kg	*	1.92	2.07	
Cattle	TDN %	*	52.1	56.4	
Sheep	TDN %	*	53.0	57.4	
Calcium	%		1.08	1.17	6
Iron	%		0.037	0.040	6
Magnesium	%		0.23	0.25	8
Manganese	mg/kg		236.1	255.5	6
Phosphorus	%		0.22	0.24	12
Potassium	%		1.02	1.10	7

Lespedeza, kobe, hay, s-c, mid-bloom, (1)
Ref No 1 02 574 United States

Feed Name or Analyses			Mean As Fed	Mean Dry	C.V. ± %
Dry matter	%		93.9	100.0	
Sheep	dig coef %		62.	62.	
Ash	%		3.8	4.1	
Crude fiber	%		26.2	27.9	
Sheep	dig coef %		59.	59.	
Ether extract	%		2.8	3.0	
N-free extract	%		51.1	54.4	
Sheep	dig coef %		68.	68.	
Protein (N x 6.25)	%		10.0	10.6	
Sheep	dig coef %		57.	57.	
Cattle	dig prot %	*	5.7	6.1	
Goats	dig prot %	*	6.1	6.4	
Horses	dig prot %	*	6.1	6.5	
Rabbits	dig prot %	*	6.4	6.9	
Sheep	dig prot %		5.7	6.0	
Energy	GE Mcal/kg				
Cattle	DE Mcal/kg	*	2.64	2.81	
Sheep	DE Mcal/kg	*	2.57	2.73	
Cattle	ME Mcal/kg	*	2.16	2.30	
Sheep	ME Mcal/kg	*	2.10	2.24	
Cattle	TDN %	*	59.8	63.7	
Sheep	TDN %		58.2	62.0	
Calcium	%		1.11	1.18	

Feed Name or Analyses			Mean As Fed	Mean Dry	C.V. ± %
Iron	%		0.029	0.031	
Magnesium	%		0.27	0.29	
Manganese	mg/kg		193.1	205.7	
Phosphorus	%		0.32	0.34	
Potassium	%		0.89	0.95	

Lespedeza, kobe, hay, s-c, full bloom, (1)

Ref No 1 02 575 United States

Feed Name or Analyses			Mean As Fed	Mean Dry	C.V. ± %
Dry matter	%		96.4	100.0	2
Ash	%		5.1	5.3	2
Crude fiber	%		32.4	33.6	3
Ether extract	%		3.4	3.5	5
N-free extract	%		41.5	43.0	
Protein (N x 6.25)	%		14.1	14.6	2
Cattle	dig prot %	*	9.2	9.6	
Goats	dig prot %	*	9.8	10.2	
Horses	dig prot %	*	9.6	9.9	
Rabbits	dig prot %	*	9.6	9.9	
Sheep	dig prot %	*	9.3	9.7	
Energy	GE Mcal/kg				
Cattle	DE Mcal/kg	*	2.31	2.40	
Sheep	DE Mcal/kg	*	2.43	2.52	
Cattle	ME Mcal/kg	*	1.90	1.97	
Sheep	ME Mcal/kg	*	1.99	2.06	
Cattle	TDN %	*	52.5	54.5	
Sheep	TDN %	*	55.0	57.1	

Lespedeza, kobe, hay, s-c, milk stage, (1)

Ref No 1 02 576 United States

Feed Name or Analyses			Mean As Fed	Mean Dry	C.V. ± %
Dry matter	%		93.6	100.0	
Ash	%		5.5	5.9	8
Crude fiber	%		32.9	35.2	5
Ether extract	%		3.5	3.7	5
N-free extract	%		39.9	42.6	
Protein (N x 6.25)	%		11.8	12.6	12
Cattle	dig prot %	*	7.3	7.9	
Goats	dig prot %	*	7.8	8.3	
Horses	dig prot %	*	7.7	8.2	
Rabbits	dig prot %	*	7.9	8.4	
Sheep	dig prot %	*	7.4	7.9	
Energy	GE Mcal/kg				
Cattle	DE Mcal/kg	*	2.20	2.35	
Sheep	DE Mcal/kg	*	2.32	2.48	
Cattle	ME Mcal/kg	*	1.81	1.93	
Sheep	ME Mcal/kg	*	1.91	2.04	
Cattle	TDN %	*	50.0	53.4	
Sheep	TDN %	*	52.7	56.3	
Calcium	%		0.93	0.99	7
Iron	%		0.030	0.032	15
Magnesium	%		0.19	0.20	17
Manganese	mg/kg		204.1	218.0	10
Phosphorus	%		0.20	0.21	25
Potassium	%		0.94	1.00	12

Lespedeza, kobe, hay, s-c, cut 1, (1)

Ref No 1 02 577 United States

Feed Name or Analyses			Mean As Fed	Mean Dry	C.V. ± %
Dry matter	%		94.7	100.0	2
Ash	%		5.3	5.6	9
Crude fiber	%		29.9	31.6	4
Ether extract	%		3.5	3.7	7
N-free extract	%		42.4	44.8	
Protein (N x 6.25)	%		13.5	14.3	15
Cattle	dig prot %	*	8.8	9.3	
Goats	dig prot %	*	9.4	9.9	
Horses	dig prot %	*	9.2	9.7	
Rabbits	dig prot %	*	9.2	9.7	
Sheep	dig prot %	*	8.9	9.4	
Energy	GE Mcal/kg				
Cattle	DE Mcal/kg	*	2.35	2.48	
Sheep	DE Mcal/kg	*	2.42	2.56	
Cattle	ME Mcal/kg	*	1.92	2.03	
Sheep	ME Mcal/kg	*	1.99	2.10	
Cattle	TDN %	*	53.2	56.2	
Sheep	TDN %	*	54.9	58.0	

Lespedeza, kobe, hay, s-c, gr 1 US, (1)

Ref No 1 02 578 United States

Feed Name or Analyses			Mean As Fed	Mean Dry	C.V. ± %
Dry matter	%			100.0	
Ash	%			6.1	
Crude fiber	%			30.8	
Ether extract	%			3.6	
N-free extract	%			47.3	
Protein (N x 6.25)	%			12.2	
Cattle	dig prot %	*		7.5	
Goats	dig prot %	*		7.9	
Horses	dig prot %	*		7.9	
Rabbits	dig prot %	*		8.1	
Sheep	dig prot %	*		7.5	
Energy	GE Mcal/kg				
Cattle	DE Mcal/kg	*		2.69	
Sheep	DE Mcal/kg	*		2.57	
Cattle	ME Mcal/kg	*		2.20	
Sheep	ME Mcal/kg	*		2.11	
Cattle	TDN %	*		60.9	
Sheep	TDN %	*		58.3	

Lespedeza, kobe, hay, s-c, gr 2 US, (1)

Ref No 1 02 579 United States

Feed Name or Analyses			Mean As Fed	Mean Dry	C.V. ± %
Dry matter	%			100.0	
Ash	%			3.8	24
Crude fiber	%			35.0	8
Ether extract	%			2.9	27
N-free extract	%			46.6	
Protein (N x 6.25)	%			11.7	9
Cattle	dig prot %	*		7.1	
Goats	dig prot %	*		7.5	
Horses	dig prot %	*		7.5	
Rabbits	dig prot %	*		7.7	
Sheep	dig prot %	*		7.1	
Energy	GE Mcal/kg			4.41	
Cattle	DE Mcal/kg	*		2.26	
Sheep	DE Mcal/kg	*		2.49	
Cattle	ME Mcal/kg	*		1.85	
Sheep	ME Mcal/kg	*		2.04	
Cattle	TDN %	*		51.2	
Sheep	TDN %	*		56.6	

Lespedeza, kobe, aerial part, fresh, (2)

Ref No 2 02 581 United States

Feed Name or Analyses			Mean As Fed	Mean Dry	C.V. ± %
Dry matter	%			100.0	
Calcium	%			0.76	
Cobalt	mg/kg			0.187	
Iron	%			0.029	
Magnesium	%			0.30	
Manganese	mg/kg			130.1	
Phosphorus	%			0.21	
Sodium	%			0.31	

LESPEDEZA, KOREAN. Lespedeza stipulacea

Lespedeza, Korean, hay, s-c, (1)

Ref No 1 02 592 United States

Feed Name or Analyses			Mean As Fed	Mean Dry	C.V. ± %
Dry matter	%			100.0	
Crude fiber	%				
Cattle	dig coef %	*		54.	
Ether extract	%				
Cattle	dig coef %	*		29.	
N-free extract	%				
Cattle	dig coef %	*		69.	
Protein (N x 6.25)	%				
Cattle	dig coef %	*		49.	

Lespedeza, Korean, hay, s-c, immature, (1)

Ref No 1 02 582 United States

Feed Name or Analyses			Mean As Fed	Mean Dry	C.V. ± %
Dry matter	%		90.4	100.0	1
Ash	%		6.1	6.7	25
Crude fiber	%		26.0	28.8	9
Ether extract	%		2.8	3.1	11
N-free extract	%		41.1	45.5	
Protein (N x 6.25)	%		14.4	15.9	16
Cattle	dig prot %	*	9.7	10.7	
Goats	dig prot %	*	10.3	11.4	
Horses	dig prot %	*	10.0	11.0	
Rabbits	dig prot %	*	9.9	10.9	
Sheep	dig prot %	*	9.8	10.8	
Energy	GE Mcal/kg				
Cattle	DE Mcal/kg	*	2.32	2.57	
Sheep	DE Mcal/kg	*	2.38	2.63	
Cattle	ME Mcal/kg	*	1.90	2.10	
Sheep	ME Mcal/kg	*	1.95	2.16	
Cattle	TDN %	*	52.6	58.2	
Sheep	TDN %	*	53.9	59.7	
Calcium	%		1.33	1.47	
Iron	%		0.027	0.030	
Magnesium	%		0.33	0.37	
Manganese	mg/kg		98.1	108.5	
Phosphorus	%		0.19	0.21	
Potassium	%		0.95	1.05	

Lespedeza, Korean, hay, s-c, pre-bloom, (1)

Ref No 1 02 583 United States

Feed Name or Analyses			Mean As Fed	Mean Dry	C.V. ± %
Dry matter	%		87.0	100.0	
Ash	%		4.6	5.3	
Crude fiber	%		28.0	32.2	
Cattle	dig coef %		54.	54.	
Ether extract	%		2.8	3.2	
Cattle	dig coef %		29.	29.	
N-free extract	%		40.8	46.9	
Cattle	dig coef %		69.	69.	
Protein (N x 6.25)	%		10.8	12.4	
Cattle	dig coef %		49.	49.	
Cattle	dig prot %		5.3	6.1	
Goats	dig prot %	*	7.1	8.1	
Horses	dig prot %	*	7.0	8.1	
Rabbits	dig prot %	*	7.2	8.2	
Sheep	dig prot %	*	6.7	7.7	
Energy	GE Mcal/kg				
Cattle	DE Mcal/kg	*	2.22	2.55	
Sheep	DE Mcal/kg	*	2.22	2.55	
Cattle	ME Mcal/kg	*	1.82	2.09	
Sheep	ME Mcal/kg	*	1.82	2.09	

Continued

Feed Name or Analyses			Mean As Fed	Mean Dry	C.V. ± %
Cattle	TDN %		50.4	57.9	
Sheep	TDN %	*	50.3	57.9	

Lespedeza, Korean, hay, s-c, early bloom, (1)
Ref No 1 02 584 United States

Feed Name or Analyses			Mean As Fed	Mean Dry	C.V. ± %
Dry matter	%		93.9	100.0	1
Ash	%		6.8	7.2	21
Crude fiber	%		28.1	29.9	7
Ether extract	%		4.3	4.6	30
N-free extract	%		40.4	43.0	
Protein (N x 6.25)	%		14.4	15.3	4
Cattle	dig prot %	*	9.6	10.2	
Goats	dig prot %	*	10.2	10.8	
Horses	dig prot %	*	9.9	10.5	
Rabbits	dig prot %	*	9.8	10.5	
Sheep	dig prot %	*	9.7	10.3	
Energy	GE Mcal/kg				
Cattle	DE Mcal/kg	*	2.46	2.62	
Sheep	DE Mcal/kg	*	2.37	2.52	
Cattle	ME Mcal/kg	*	2.02	2.15	
Sheep	ME Mcal/kg	*	1.94	2.07	
Cattle	TDN %	*	55.9	59.5	
Sheep	TDN %	*	53.6	57.1	

Lespedeza, Korean, hay, s-c, mid-bloom, (1)
Ref No 1 02 585 United States

Feed Name or Analyses			Mean As Fed	Mean Dry	C.V. ± %
Dry matter	%			100.0	
Calcium	%			1.19	
Phosphorus	%			0.31	

Lespedeza, Korean, hay, s-c, full bloom, (1)
Ref No 1 02 586 United States

Feed Name or Analyses			Mean As Fed	Mean Dry	C.V. ± %
Dry matter	%		91.0	100.0	1
Ash	%		5.6	6.2	16
Crude fiber	%		29.3	32.2	4
Ether extract	%		3.2	3.5	29
N-free extract	%		39.9	43.8	
Protein (N x 6.25)	%		13.0	14.3	2
Cattle	dig prot %	*	8.5	9.3	
Goats	dig prot %	*	9.0	9.9	
Horses	dig prot %	*	8.8	9.7	
Rabbits	dig prot %	*	8.8	9.7	
Sheep	dig prot %	*	8.6	9.4	
Energy	GE Mcal/kg				
Cattle	DE Mcal/kg	*	2.26	2.48	
Sheep	DE Mcal/kg	*	2.30	2.52	
Cattle	ME Mcal/kg	*	1.85	2.03	
Sheep	ME Mcal/kg	*	1.88	2.07	
Cattle	TDN %	*	51.1	56.2	
Sheep	TDN %	*	52.1	57.2	
Calcium	%		0.90	0.99	6
Phosphorus	%		0.22	0.24	20
Potassium	%		0.87	0.96	

(1) dry forages and roughages (2) pasture, range plants, and forages fed green (3) silages (4) energy feeds (5) protein supplements (6) minerals (7) vitamins (8) additives

Lespedeza, Korean, hay, s-c, milk stage, (1)
Ref No 1 02 587 United States

Feed Name or Analyses			Mean As Fed	Mean Dry	C.V. ± %
Dry matter	%		88.0	100.0	2
Ash	%		4.8	5.4	3
Crude fiber	%		32.4	36.8	6
Cattle	dig coef %		50.	50.	
Ether extract	%		2.4	2.8	6
Cattle	dig coef %		10.	10.	
N-free extract	%		37.4	42.5	
Cattle	dig coef %		46.	46.	
Protein (N x 6.25)	%		11.1	12.6	13
Cattle	dig coef %		41.	41.	
Cattle	dig prot %		4.5	5.2	
Goats	dig prot %	*	7.3	8.3	
Horses	dig prot %	*	7.2	8.2	
Rabbits	dig prot %	*	7.4	8.4	
Sheep	dig prot %	*	6.9	7.9	
Energy	GE Mcal/kg				
Cattle	DE Mcal/kg	*	1.70	1.93	
Sheep	DE Mcal/kg	*	1.78	2.02	
Cattle	ME Mcal/kg	*	1.39	1.58	
Sheep	ME Mcal/kg	*	1.46	1.66	
Cattle	TDN %		38.5	43.7	
Sheep	TDN %	*	40.3	45.8	
Calcium	%		1.00	1.14	
Iron	%		0.041	0.047	
Magnesium	%		0.19	0.22	
Manganese	mg/kg		81.5	92.6	
Phosphorus	%		0.18	0.20	
Potassium	%		0.96	1.09	

Lespedeza, Korean, hay, s-c, dough stage, (1)
Ref No 1 02 588 United States

Feed Name or Analyses			Mean As Fed	Mean Dry	C.V. ± %
Dry matter	%			100.0	
Calcium	%			1.08	12
Iron	%			0.026	
Magnesium	%			0.23	
Manganese	mg/kg			154.1	
Phosphorus	%			0.26	24
Potassium	%			0.86	

Lespedeza, Korean, hay, s-c, cut 1, (1)
Ref No 1 02 589 United States

Feed Name or Analyses			Mean As Fed	Mean Dry	C.V. ± %
Dry matter	%		92.2	100.0	3
Calcium	%		1.12	1.22	10
Phosphorus	%		0.30	0.32	12
Potassium	%		1.01	1.10	21

Lespedeza, Korean, hay, s-c, gr 1 US, (1)
Ref No 1 02 590 United States

Feed Name or Analyses			Mean As Fed	Mean Dry	C.V. ± %
Dry matter	%			100.0	
Calcium	%			1.07	
Phosphorus	%			0.31	

Lespedeza, Korean, hay, s-c, gr 2 US, (1)
Ref No 1 02 591 United States

Feed Name or Analyses			Mean As Fed	Mean Dry	C.V. ± %
Dry matter	%			100.0	
Calcium	%			1.04	
Phosphorus	%			0.33	

Lespedeza, Korean, aerial part, fresh, immature, (2)
Ref No 2 02 593 United States

Feed Name or Analyses			Mean As Fed	Mean Dry	C.V. ± %
Dry matter	%		32.5	100.0	19
Calcium	%		0.44	1.36	14
Phosphorus	%		0.18	0.55	33
Potassium	%		0.51	1.56	10

Lespedeza, Korean, aerial part, fresh, mid-bloom, (2)
Ref No 2 02 594 United States

Feed Name or Analyses			Mean As Fed	Mean Dry	C.V. ± %
Dry matter	%		27.7	100.0	13
Ash	%		2.6	9.4	13
Crude fiber	%		7.8	28.3	26
Ether extract	%		1.1	3.8	20
N-free extract	%		11.2	40.3	
Protein (N x 6.25)	%		5.0	18.2	19
Cattle	dig prot %	*	3.7	13.4	
Goats	dig prot %	*	3.8	13.5	
Horses	dig prot %	*	3.6	13.0	
Rabbits	dig prot %	*	3.5	12.7	
Sheep	dig prot %	*	3.9	14.0	
Energy	GE Mcal/kg				
Cattle	DE Mcal/kg	*	0.86	3.11	
Sheep	DE Mcal/kg	*	0.74	2.67	
Cattle	ME Mcal/kg	*	0.71	2.55	
Sheep	ME Mcal/kg	*	0.61	2.19	
Cattle	TDN %	*	19.5	70.5	
Sheep	TDN %	*	16.8	60.6	
Calcium	%		0.37	1.35	
Phosphorus	%		0.11	0.38	
Potassium	%		0.45	1.62	

Lespedeza, Korean, aerial part, fresh, full bloom, (2)
Ref No 2 02 595 United States

Feed Name or Analyses			Mean As Fed	Mean Dry	C.V. ± %
Dry matter	%		34.8	100.0	
Calcium	%		0.38	1.10	
Phosphorus	%		0.14	0.40	

Lespedeza, Korean, aerial part, fresh, mature, (2)
Ref No 2 02 596 United States

Feed Name or Analyses			Mean As Fed	Mean Dry	C.V. ± %
Dry matter	%		35.3	100.0	7
Ash	%		2.6	7.4	13
Crude fiber	%		15.9	45.1	3
Ether extract	%		0.7	2.1	16
N-free extract	%		11.5	32.7	
Protein (N x 6.25)	%		4.5	12.7	6
Cattle	dig prot %	*	3.1	8.7	
Goats	dig prot %	*	3.0	8.4	
Horses	dig prot %	*	2.9	8.3	
Rabbits	dig prot %	*	3.0	8.5	
Sheep	dig prot %	*	3.1	8.8	
Energy	GE Mcal/kg				
Cattle	DE Mcal/kg	*	0.91	2.57	
Sheep	DE Mcal/kg	*	0.88	2.50	
Cattle	ME Mcal/kg	*	0.74	2.11	
Sheep	ME Mcal/kg	*	0.73	2.05	
Cattle	TDN %	*	20.5	58.2	
Sheep	TDN %	*	20.1	56.8	
Calcium	%		0.35	1.00	
Phosphorus	%		0.07	0.20	

Lespedeza, Korean, aerial part, fresh, over ripe, (2)

Ref No 2 02 597 United States

Feed Name or Analyses			Mean As Fed	Mean Dry	C.V. ± %
Dry matter	%		37.2	100.0	
Calcium	%		0.39	1.05	
Phosphorus	%		0.03	0.09	

LESPEDEZA, SERICEA. Lespedeza cuneata

Lespedeza, sericea, aerial part, dehy grnd, (1)

Ref No 1 02 608 United States

Feed Name or Analyses			Mean As Fed	Mean Dry	C.V. ± %
Dry matter	%		95.5	100.0	2
Ash	%		4.4	4.6	10
Crude fiber	%		32.8	34.3	7
Ether extract	%		3.4	3.6	24
N-free extract	%		42.4	44.4	
Protein (N x 6.25)	%		12.5	13.1	16
Cattle	dig prot %	*	7.9	8.3	
Goats	dig prot %	*	8.4	8.8	
Horses	dig prot %	*	8.3	8.7	
Rabbits	dig prot %	*	8.4	8.8	
Sheep	dig prot %	*	7.9	8.3	
Energy	GE Mcal/kg				
Cattle	DE Mcal/kg	*	2.26	2.36	
Sheep	DE Mcal/kg	*	2.40	2.51	
Cattle	ME Mcal/kg	*	1.85	1.94	
Sheep	ME Mcal/kg	*	1.97	2.06	
Cattle	TDN %	*	51.2	53.6	
Sheep	TDN %	*	54.4	56.9	
Carotene	mg/kg		172.0	180.1	16
Vitamin A equivalent	IU/g		286.7	300.3	

Lespedeza, sericea, hay, s-c, (1)

Ref No 1 02 607 United States

Feed Name or Analyses			Mean As Fed	Mean Dry	C.V. ± %
Dry matter	%		90.8	100.0	3
Ash	%		4.9	5.4	
Crude fiber	%		28.4	31.3	
Ether extract	%		1.8	2.0	
N-free extract	%		43.1	47.4	
Protein (N x 6.25)	%		12.6	13.9	
Cattle	dig prot %	*	8.2	9.0	
Goats	dig prot %	*	8.7	9.5	
Horses	dig prot %	*	8.5	9.3	
Rabbits	dig prot %	*	8.5	9.4	
Sheep	dig prot %	*	8.2	9.0	
Fatty acids	%		1.8	2.0	
Energy	GE Mcal/kg		4.13	4.55	
Cattle	DE Mcal/kg	*	1.79	1.97	
Sheep	DE Mcal/kg	*	2.00	2.20	
Cattle	ME Mcal/kg	*	1.47	1.62	
Sheep	ME Mcal/kg	*	1.63	1.80	
Cattle	TDN %	*	40.5	44.6	
Sheep	TDN %	*	45.2	49.8	
Calcium	%		0.94	1.03	
Iron	%		0.026	0.029	
Magnesium	%		0.20	0.22	
Manganese	mg/kg		91.6	100.8	
Phosphorus	%		0.22	0.25	
Potassium	%		1.00	1.10	
Carotene	mg/kg		35.8	39.5	68
Riboflavin	mg/kg		8.8	9.7	32
Vitamin A equivalent	IU/g		59.7	65.8	

Lespedeza, sericea, hay, s-c, immature, (1)

Ref No 1 02 599 United States

Feed Name or Analyses			Mean As Fed	Mean Dry	C.V. ± %
Dry matter	%		94.2	100.0	1
Sheep	dig coef %		57.	57.	
Ash	%		5.5	5.9	15
Crude fiber	%		20.2	21.4	5
Sheep	dig coef %		50.	50.	
Ether extract	%		5.1	5.5	18
Sheep	dig coef %		67.	67.	
N-free extract	%		45.8	48.7	3
Sheep	dig coef %		60.	60.	
Protein (N x 6.25)	%		17.5	18.6	3
Sheep	dig coef %		55.	55.	
Cattle	dig prot %	*	12.3	13.0	
Goats	dig prot %	*	13.1	13.9	
Horses	dig prot %	*	12.5	13.3	
Rabbits	dig prot %	*	12.3	13.0	
Sheep	dig prot %		9.5	10.1	
Energy	GE Mcal/kg				
Cattle	DE Mcal/kg	*	2.65	2.81	
Sheep	DE Mcal/kg	*	2.41	2.56	
Cattle	ME Mcal/kg	*	2.17	2.31	
Sheep	ME Mcal/kg	*	1.98	2.10	
Cattle	TDN %	*	60.1	63.8	
Sheep	TDN %		54.7	58.1	
Calcium	%		0.94	1.00	16
Iron	%		0.057	0.060	
Magnesium	%		0.23	0.24	
Manganese	mg/kg		73.1	77.6	
Phosphorus	%		0.25	0.27	33
Potassium	%		0.93	0.99	27

Lespedeza, sericea, hay, s-c, prebloom, (1)

Ref No 1 09 172 United States

Feed Name or Analyses			Mean As Fed	Mean Dry	C.V. ± %
Dry matter	%			100.0	
Sheep	dig coef %			57.	
Ash	%			5.5	
Crude fiber	%			22.2	
Sheep	dig coef %			52.	
Ether extract	%			5.8	
N-free extract	%			48.0	
Sheep	dig coef %			61.	
Protein (N x 6.25)	%			18.6	
Sheep	dig coef %			52.	
Cattle	dig prot %	*		13.0	
Goats	dig prot %	*		13.9	
Horses	dig prot %	*		13.3	
Rabbits	dig prot %	*		13.0	
Sheep	dig prot %			9.7	
Energy	GE Mcal/kg				
Cattle	DE Mcal/kg	*		2.86	
Sheep	DE Mcal/kg	*		2.79	
Cattle	ME Mcal/kg	*		2.34	
Sheep	ME Mcal/kg	*		2.29	
Cattle	TDN %	*		64.8	
Sheep	TDN %	*		63.3	

Lespedeza, sericea, hay, s-c, early bloom, (1)

Ref No 1 02 600 United States

Feed Name or Analyses			Mean As Fed	Mean Dry	C.V. ± %
Dry matter	%		94.6	100.0	3
Sheep	dig coef %		56.	56.	
Ash	%		5.1	5.4	10
Crude fiber	%		22.3	23.6	11
Sheep	dig coef %		53.	53.	
Ether extract	%		5.1	5.4	16
N-free extract	%		45.7	48.4	2
Sheep	dig coef %		59.	59.	
Protein (N x 6.25)	%		16.3	17.2	13
Sheep	dig coef %		51.	51.	
Cattle	dig prot %	*	11.2	11.9	
Goats	dig prot %	*	12.0	12.6	
Horses	dig prot %	*	11.5	12.2	
Rabbits	dig prot %	*	11.3	12.0	
Sheep	dig prot %		8.3	8.8	
Energy	GE Mcal/kg				
Cattle	DE Mcal/kg	*	2.59	2.74	
Sheep	DE Mcal/kg	*	2.61	2.76	
Cattle	ME Mcal/kg	*	2.13	2.25	
Sheep	ME Mcal/kg	*	2.14	2.26	
Cattle	TDN %	*	58.8	62.2	
Sheep	TDN %	*	59.1	62.5	
Calcium	%		1.46	1.54	
Phosphorus	%		0.25	0.26	
Potassium	%		0.65	0.69	

Lespedeza, sericea, hay, s-c, midbloom, (1)

Ref No 1 09 173 United States

Feed Name or Analyses			Mean As Fed	Mean Dry	C.V. ± %
Dry matter	%			100.0	
Sheep	dig coef %			53.	7
Ash	%			5.2	6
Crude fiber	%			23.2	7
Sheep	dig coef %			45.	9
Ether extract	%			5.1	8
N-free extract	%			49.7	2
Sheep	dig coef %			58.	5
Protein (N x 6.25)	%			16.9	8
Sheep	dig coef %			45.	15
Cattle	dig prot %	*		11.5	
Goats	dig prot %	*		12.3	
Horses	dig prot %	*		11.8	
Rabbits	dig prot %	*		11.7	
Sheep	dig prot %			7.5	
Energy	GE Mcal/kg				
Cattle	DE Mcal/kg	*		2.80	
Sheep	DE Mcal/kg	*		2.78	
Cattle	ME Mcal/kg	*		2.29	
Sheep	ME Mcal/kg	*		2.28	
Cattle	TDN %	*		63.5	
Sheep	TDN %	*		63.2	

Lespedeza, sericea, hay, s-c, full bloom, (1)

Ref No 1 02 601 United States

Feed Name or Analyses			Mean As Fed	Mean Dry	C.V. ± %
Dry matter	%		94.4	100.0	0
Sheep	dig coef %		47.	47.	
Ash	%		4.1	4.3	
Crude fiber	%		23.1	24.5	
Sheep	dig coef %		34.	34.	
Ether extract	%		5.0	5.3	
N-free extract	%		47.9	50.8	
Sheep	dig coef %		55.	55.	
Protein (N x 6.25)	%		14.3	15.2	
Sheep	dig coef %		36.	36.	
Cattle	dig prot %	*	9.5	10.1	
Goats	dig prot %	*	10.1	10.7	
Horses	dig prot %	*	9.9	10.4	
Rabbits	dig prot %	*	9.8	10.4	
Sheep	dig prot %		5.2	5.5	
Energy	GE Mcal/kg				
Cattle	DE Mcal/kg	*	2.53	2.68	

Continued

Feed Name or Analyses			Mean As Fed	Mean Dry	C.V. ± %
Sheep	DE Mcal/kg	*	2.61	2.76	
Cattle	ME Mcal/kg	*	2.07	2.19	
Sheep	ME Mcal/kg	*	2.14	2.27	
Cattle	TDN %	*	57.3	60.7	
Sheep	TDN %	*	59.2	62.7	
Calcium	%		0.95	1.01	
Phosphorus	%		0.21	0.22	

Lespedeza, sericea, hay, s-c, mature, (1)
Ref No 1 09 174 United States

Feed Name or Analyses			Mean As Fed	Mean Dry	C.V. ± %
Dry matter	%			100.0	
Sheep	dig coef %			46.	
Ash	%			4.8	
Crude fiber	%			26.0	
Sheep	dig coef %			41.	
Ether extract	%			5.4	
N-free extract	%			47.9	
Sheep	dig coef %			51.	
Protein (N x 6.25)	%			16.0	
Sheep	dig coef %			40.	
Cattle	dig prot %	*		10.8	
Goats	dig prot %	*		11.5	
Horses	dig prot %	*		11.1	
Rabbits	dig prot %	*		11.0	
Sheep	dig prot %			6.3	
Energy	GE Mcal/kg				
Cattle	DE Mcal/kg	*		2.72	
Sheep	DE Mcal/kg	*		2.70	
Cattle	ME Mcal/kg	*		2.23	
Sheep	ME Mcal/kg	*		2.22	
Cattle	TDN %	*		61.7	
Sheep	TDN %	*		61.3	

Lespedeza, sericea, hay, s-c, over ripe, (1)
Ref No 1 09 175 United States

Feed Name or Analyses			Mean As Fed	Mean Dry	C.V. ± %
Dry matter	%			100.0	
Sheep	dig coef %			49.	4
Ash	%			4.8	
Crude fiber	%			30.7	3
Sheep	dig coef %			45.	12
Ether extract	%			4.4	3
Sheep	dig coef %			50.	
N-free extract	%			45.0	2
Sheep	dig coef %			49.	18
Protein (N x 6.25)	%			15.0	10
Sheep	dig coef %			45.	7
Cattle	dig prot %	*		10.0	
Goats	dig prot %	*		10.6	
Horses	dig prot %	*		10.3	
Rabbits	dig prot %	*		10.3	
Sheep	dig prot %			6.7	
Energy	GE Mcal/kg				
Cattle	DE Mcal/kg	*		2.18	
Sheep	DE Mcal/kg	*		2.09	
Cattle	ME Mcal/kg	*		1.79	
Sheep	ME Mcal/kg	*		1.72	
Cattle	TDN %	*		49.5	
Sheep	TDN %			47.5	

(1) dry forages and roughages (2) pasture, range plants, and forages fed green (3) silages (4) energy feeds (5) protein supplements (6) minerals (7) vitamins (8) additives

Lespedeza, sericea,hay, s-c, cut 1, (1)
Ref No 1 02 602 United States

Feed Name or Analyses			Mean As Fed	Mean Dry	C.V. ± %
Dry matter	%		92.9	100.0	3
Calcium	%		0.96	1.03	9
Iron	%		0.033	0.035	27
Magnesium	%		0.21	0.23	22
Manganese	mg/kg		101.4	109.1	17
Phosphorus	%		0.23	0.25	28
Potassium	%		1.06	1.14	17

Lespedeza, sericea, hay, s-c, cut 2, (1)
Ref No 1 02 603 United States

Feed Name or Analyses			Mean As Fed	Mean Dry	C.V. ± %
Dry matter	%		91.4	100.0	5
Ash	%		4.8	5.2	6
Crude fiber	%		31.6	34.6	4
Ether extract	%		1.9	2.1	23
N-free extract	%		39.9	43.7	
Protein (N x 6.25)	%		13.2	14.4	11
Cattle	dig prot %	*	8.6	9.4	
Goats	dig prot %	*	9.1	10.0	
Horses	dig prot %	*	8.9	9.8	
Rabbits	dig prot %	*	8.9	9.8	
Sheep	dig prot %	*	8.7	9.5	
Energy	GE Mcal/kg				
Cattle	DE Mcal/kg	*	2.11	2.31	
Sheep	DE Mcal/kg	*	2.32	2.54	
Cattle	ME Mcal/kg	*	1.73	1.90	
Sheep	ME Mcal/kg	*	1.91	2.09	
Cattle	TDN %	*	47.9	52.4	
Sheep	TDN %	*	52.7	57.7	
Calcium	%		0.90	0.99	18
Iron	%		0.018	0.020	23
Magnesium	%		0.20	0.22	19
Manganese	mg/kg		92.9	101.6	16
Phosphorus	%		0.21	0.23	12
Potassium	%		0.92	1.01	20

Lespedeza, sericea, hay, s-c, cut 3 (1)
Ref No 1 02 604 United States

Feed Name or Analyses			Mean As Fed	Mean Dry	C.V. ± %
Dry matter	%		91.8	100.0	
Ash	%		5.6	6.1	
Crude fiber	%		20.2	22.0	
Ether extract	%		4.9	5.3	
N-free extract	%		47.3	51.5	
Protein (N x 6.25)	%		13.9	15.1	
Cattle	dig prot %	*	9.2	10.0	
Goats	dig prot %	*	9.8	10.6	
Horses	dig prot %	*	9.5	10.3	
Rabbits	dig prot %	*	9.5	10.3	
Sheep	dig prot %	*	9.3	10.1	
Energy	GE Mcal/kg				
Cattle	DE Mcal/kg	*	2.59	2.82	
Sheep	DE Mcal/kg	*	2.54	2.77	
Cattle	ME Mcal/kg	*	2.12	2.31	
Sheep	ME Mcal/kg	*	2.08	2.27	
Cattle	TDN %	*	58.7	63.9	
Sheep	TDN %	*	57.6	62.8	
Calcium	%		1.11	1.21	24
Iron	%		0.029	0.032	17
Magnesium	%		0.22	0.24	20
Manganese	mg/kg		113.7	123.9	24
Phosphorus	%		0.21	0.23	14
Potassium	%		0.85	0.93	23

Lespedeza, sericea, hay, s-c, gr 2 US, (1)
Ref No 1 02 606 United States

Feed Name or Analyses			Mean As Fed	Mean Dry	C.V. ± %
Dry matter	%		90.2	100.0	
Ash	%		4.5	5.0	
Crude fiber	%		29.2	32.4	
Ether extract	%		2.3	2.6	
N-free extract	%		44.6	49.5	
Protein (N x 6.25)	%		9.5	10.5	
Cattle	dig prot %	*	5.4	6.0	
Goats	dig prot %	*	5.7	6.4	
Horses	dig prot %	*	5.8	6.4	
Rabbits	dig prot %	*	6.1	6.8	
Sheep	dig prot %	*	5.4	6.0	
Energy	GE Mcal/kg				
Cattle	DE Mcal/kg	*	2.25	2.50	
Sheep	DE Mcal/kg	*	2.28	2.52	
Cattle	ME Mcal/kg	*	1.85	2.05	
Sheep	ME Mcal/kg	*	1.87	2.07	
Cattle	TDN %	*	51.1	56.7	
Sheep	TDN %	*	51.6	57.2	

Lespedeza, sericea, hay, s-c, gr sample US, (1)
Ref No 1 02 605 United States

Feed Name or Analyses			Mean As Fed	Mean Dry	C.V. ± %
Dry matter	%			100.0	
Ash	%			3.4	
Crude fiber	%			42.6	
Ether extract	%			2.3	
N-free extract	%			40.4	
Protein (N x 6.25)	%			11.3	
Cattle	dig prot %	*		6.7	
Goats	dig prot %	*		7.1	
Horses	dig prot %	*		7.1	
Rabbits	dig prot %	*		7.4	
Sheep	dig prot %	*		6.7	
Energy	GE Mcal/kg				
Cattle	DE Mcal/kg	*		2.15	
Sheep	DE Mcal/kg	*		2.33	
Cattle	ME Mcal/kg	*		1.76	
Sheep	ME Mcal/kg	*		1.91	
Cattle	TDN %	*		48.7	
Sheep	TDN %	*		52.9	

Lespedeza, sericea, hay, s-c grnd, cut 1, gr 2 US, (1)
Ref No 1 07 820 United States

Feed Name or Analyses			Mean As Fed	Mean Dry	C.V. ± %
Dry matter	%			100.0	
Cattle	dig coef %			46.	14
Ash	%			5.4	23
Crude fiber	%			32.5	11
Ether extract	%			2.6	26
N-free extract	%			45.8	8
Protein (N x 6.25)	%			13.6	15
Cattle	dig coef %			27.	37
Cattle	dig prot %			3.7	
Goats	dig prot %	*		9.3	
Horses	dig prot %	*		9.1	
Rabbits	dig prot %	*		9.2	
Sheep	dig prot %	*		8.8	
Cellulose (Matrone)	%			37.8	10
Lignin (Ellis)	%			20.6	12
Energy	GE Mcal/kg				
Cattle	DE Mcal/kg	*		2.47	
Sheep	DE Mcal/kg	*		2.57	
Cattle	ME Mcal/kg	*		2.03	
Sheep	ME Mcal/kg	*		2.11	

Feed Name or Analyses			Mean As Fed	Mean Dry	C.V. ± %
Cattle	TDN %	*		56.1	
Sheep	TDN %	*		58.3	
Tannin	%			3.48	22

Lespedeza, sericea, stems, dehy grnd, (1)

Ref No 1 05 677 United States

Analyses			As Fed	Dry	C.V. ± %
Dry matter	%			100.0	
Ash	%			6.1	
Crude fiber	%			22.9	
Rabbits	dig coef %			13.	
Ether extract	%			3.3	
Rabbits	dig coef %			47.	
N-free extract	%			51.4	
Rabbits	dig coef %			50.	
Protein (N x 6.25)	%			16.4	
Rabbits	dig coef %			39.	
Cattle	dig prot %	*		11.1	
Goats	dig prot %	*		11.8	
Horses	dig prot %	*		11.4	
Rabbits	dig prot %			6.4	
Sheep	dig prot %	*		11.2	
Cellulose (Matrone)	%			29.6	
Rabbits	dig coef %			25.	
Energy	GE Mcal/kg				
Cattle	DE Mcal/kg	*		2.77	
Rabbits	DE kcal/kg	*		1687.	
Sheep	DE Mcal/kg	*		2.82	
Cattle	ME Mcal/kg	*		2.27	
Sheep	ME Mcal/kg	*		2.32	
Cattle	TDN %	*		62.9	
Rabbits	TDN %			38.3	
Sheep	TDN %	*		64.0	

Lespedeza, sericea, aerial part, fresh, (2)

Ref No 2 02 611 United States

Analyses			As Fed	Dry	C.V. ± %
Dry matter	%		32.8	100.0	12
Ash	%		2.0	6.2	7
Crude fiber	%		7.4	22.7	7
Ether extract	%		1.2	3.8	13
N-free extract	%		16.2	49.3	
Protein (N x 6.25)	%		5.9	18.0	14
Cattle	dig prot %	*	4.3	13.2	
Goats	dig prot %	*	4.4	13.4	
Horses	dig prot %	*	4.2	12.8	
Rabbits	dig prot %	*	4.1	12.6	
Sheep	dig prot %	*	4.5	13.8	
Energy	GE Mcal/kg				
Cattle	DE Mcal/kg	*	0.92	2.79	
Sheep	DE Mcal/kg	*	0.86	2.62	
Cattle	ME Mcal/kg	*	0.75	2.29	
Sheep	ME Mcal/kg	*	0.71	2.15	
Cattle	TDN %	*	20.7	63.2	
Sheep	TDN %	*	19.5	59.4	
Calcium	%		0.42	1.27	16
Cobalt	mg/kg		0.023	0.071	
Iron	%		0.008	0.024	30
Magnesium	%		0.07	0.22	23
Manganese	mg/kg		34.1	103.8	41
Phosphorus	%		0.10	0.29	35
Potassium	%		0.39	1.20	24

Feed Name or Analyses			Mean As Fed	Mean Dry	C.V. ± %

Lespedeza, sericea, aerial part, fresh, immature, (2)

Ref No 2 02 609 United States

Analyses			As Fed	Dry	C.V. ± %
Dry matter	%		32.8	100.0	18
Calcium	%		0.46	1.41	16
Phosphorus	%		0.10	0.32	34
Potassium	%		0.46	1.40	18

Lespedeza, sericea, aerial part, fresh, full bloom, (2)

Ref No 2 02 610 United States

Analyses			As Fed	Dry	C.V. ± %
Dry matter	%			100.0	
Calcium	%			1.16	4
Iron	%			0.025	29
Magnesium	%			0.22	25
Manganese	mg/kg			104.9	44
Phosphorus	%			0.26	30
Potassium	%			0.98	15

Lespedeza, sericea, leaves, fresh, (2)

Ref No 2 02 612 United States

Analyses			As Fed	Dry	C.V. ± %
Dry matter	%			100.0	
Ash	%			5.5	10
Crude fiber	%			23.7	9
Ether extract	%			3.5	27
N-free extract	%			49.5	
Protein (N x 6.25)	%			17.8	11
Cattle	dig prot %	*		13.0	
Goats	dig prot %	*		13.2	
Horses	dig prot %	*		12.6	
Rabbits	dig prot %	*		12.4	
Sheep	dig prot %	*		13.6	
Energy	GE Mcal/kg				
Cattle	DE Mcal/kg	*		2.79	
Sheep	DE Mcal/kg	*		2.71	
Cattle	ME Mcal/kg	*		2.29	
Sheep	ME Mcal/kg	*		2.23	
Cattle	TDN %	*		63.2	
Sheep	TDN %	*		61.6	

Lespedeza, sericea, stems, fresh, (2)

Ref No 2 02 613 United States

Analyses			As Fed	Dry	C.V. ± %
Dry matter	%			100.0	
Ash	%			3.2	22
Crude fiber	%			53.6	4
Ether extract	%			1.4	26
N-free extract	%			33.3	
Protein (N x 6.25)	%			8.5	17
Cattle	dig prot %	*		5.1	
Goats	dig prot %	*		4.5	
Horses	dig prot %	*		4.7	
Rabbits	dig prot %	*		5.2	
Sheep	dig prot %	*		4.9	
Energy	GE Mcal/kg				
Cattle	DE Mcal/kg	*		2.60	
Sheep	DE Mcal/kg	*		2.53	
Cattle	ME Mcal/kg	*		2.13	
Sheep	ME Mcal/kg	*		2.08	
Cattle	TDN %	*		59.0	
Sheep	TDN %	*		57.4	

Feed Name or Analyses			Mean As Fed	Mean Dry	C.V. ± %

Lespedeza, sericea, aerial part, ensiled, (3)

Ref No 3 02 614 United States

Analyses			As Fed	Dry	C.V. ± %
Dry matter	%		30.4	100.0	10
Ash	%		1.7	5.5	9
Crude fiber	%		9.5	31.3	6
Ether extract	%		0.9	3.1	23
N-free extract	%		14.0	46.1	
Protein (N x 6.25)	%		4.3	14.0	13
Cattle	dig prot %	*	2.7	8.9	
Goats	dig prot %	*	2.7	8.9	
Horses	dig prot %	*	2.7	8.9	
Sheep	dig prot %	*	2.7	8.9	
Energy	GE Mcal/kg				
Cattle	DE Mcal/kg	*	0.72	2.37	
Sheep	DE Mcal/kg	*	0.73	2.41	
Cattle	ME Mcal/kg	*	0.59	1.94	
Sheep	ME Mcal/kg	*	0.60	1.98	
Cattle	TDN %	*	16.3	53.7	
Sheep	TDN %	*	16.6	54.6	

Lespedeza, sericea, aerial part w molasses added, ensiled, (3)

Ref No 3 02 615 United States

Analyses			As Fed	Dry	C.V. ± %
Dry matter	%		30.0	100.0	
Ash	%		1.6	5.3	
Crude fiber	%		9.0	30.0	
Ether extract	%		1.2	4.0	
N-free extract	%		13.9	46.4	
Protein (N x 6.25)	%		4.3	14.3	
Cattle	dig prot %	*	2.8	9.2	
Goats	dig prot %	*	2.8	9.2	
Horses	dig prot %	*	2.8	9.2	
Sheep	dig prot %	*	2.8	9.2	
Energy	GE Mcal/kg				
Cattle	DE Mcal/kg	*	0.73	2.44	
Sheep	DE Mcal/kg	*	0.74	2.48	
Cattle	ME Mcal/kg	*	0.60	2.00	
Sheep	ME Mcal/kg	*	0.61	2.03	
Cattle	TDN %	*	16.6	55.4	
Sheep	TDN %	*	16.9	56.3	

Lespedeza, sericea, seeds, (5)

Ref No 5 08 457 United States

Analyses			As Fed	Dry	C.V. ± %
Dry matter	%		92.3	100.0	
Ash	%		3.8	4.1	
Crude fiber	%		13.5	14.6	
Ether extract	%		4.2	4.6	
N-free extract	%		37.3	40.4	
Protein (N x 6.25)	%		33.5	36.3	
Energy	GE Mcal/kg				
Cattle	DE Mcal/kg	*	2.61	2.83	
Sheep	DE Mcal/kg	*	2.69	2.91	
Swine	DE kcal/kg	*	2816.	3051.	
Cattle	ME Mcal/kg	*	2.14	2.32	
Sheep	ME Mcal/kg	*	2.21	2.39	
Swine	ME kcal/kg	*	2498.	2706.	
Cattle	TDN %	*	59.3	64.3	
Sheep	TDN %	*	61.0	66.1	
Swine	TDN %	*	63.9	69.2	

Feed Name or Analyses		Mean As Fed	Mean Dry	C.V. ± %

LESPEDEZA-BERMUDAGRASS. Lespedeza spp, Cynodon dactylon

Lespedeza-Bermudagrass, aerial part, fresh, milk stage, (2)
Ref No 2 02 616 United States

Analyses	Unit	As Fed	Dry	C.V. ± %
Dry matter	%		100.0	
Ash	%		8.0	
Crude fiber	%		32.2	
Ether extract	%		1.6	
N-free extract	%		46.8	
Protein (N x 6.25)	%		11.4	
Cattle	dig prot % *		7.6	
Goats	dig prot % *		7.2	
Horses	dig prot % *		7.2	
Rabbits	dig prot % *		7.5	
Sheep	dig prot % *		7.6	
Energy	GE Mcal/kg			
Cattle	DE Mcal/kg *		2.60	
Sheep	DE Mcal/kg *		2.39	
Cattle	ME Mcal/kg *		2.13	
Sheep	ME Mcal/kg *		1.96	
Cattle	TDN % *		59.0	
Sheep	TDN % *		54.2	
Carotene	mg/kg		163.1	
Vitamin A equivalent	IU/g		272.0	

LESPEDEZA-BLUESTEM. Lespedeza spp, Andropogon spp

Lespedeza-Bluestem, aerial part, fresh, dough stage, (2)
Ref No 2 02 617 United States

Analyses	Unit	As Fed	Dry	C.V. ± %
Dry matter	%		100.0	
Carotene	mg/kg		84.2	
Vitamin A equivalent	IU/g		140.4	

LESSINGIA, GERMAN. Lessingia germanorum

Lessingia, German, aerial part, fresh, (2)
Ref No 2 02 623 United States

Analyses	Unit	As Fed	Dry	C.V. ± %
Dry matter	%		100.0	
Ash	%		7.9	48
Crude fiber	%		25.5	14
Protein (N x 6.25)	%		7.6	71
Cattle	dig prot % *		4.4	
Goats	dig prot % *		3.7	
Horses	dig prot % *		4.0	
Rabbits	dig prot % *		4.5	
Sheep	dig prot % *		4.1	
Calcium	%		0.99	22
Phosphorus	%		0.28	75
Potassium	%		2.18	61

Lessingia, German, aerial part, fresh, immature, (2)
Ref No 2 02 618 United States

Analyses	Unit	As Fed	Dry	C.V. ± %
Dry matter	%		100.0	
Ash	%		13.9	
Crude fiber	%		21.6	
Protein (N x 6.25)	%		16.1	
Cattle	dig prot % *		11.6	
Goats	dig prot % *		11.6	
Horses	dig prot % *		11.2	
Rabbits	dig prot % *		11.1	
Sheep	dig prot % *		12.0	
Calcium	%		1.28	
Phosphorus	%		0.62	
Potassium	%		4.02	

(1) dry forages and roughages (2) pasture, range plants, and forages fed green (3) silages (4) energy feeds (5) protein supplements (6) minerals (7) vitamins (8) additives

Lessingia, German, aerial part, fresh, early bloom, (2)
Ref No 2 02 619 United States

Analyses	Unit	As Fed	Dry	C.V. ± %
Dry matter	%		100.0	
Ash	%		8.7	
Crude fiber	%		23.8	
Protein (N x 6.25)	%		9.2	
Cattle	dig prot % *		5.7	
Goats	dig prot % *		5.1	
Horses	dig prot % *		5.3	
Rabbits	dig prot % *		5.8	
Sheep	dig prot % *		5.6	
Calcium	%		1.06	
Phosphorus	%		0.32	
Potassium	%		2.69	

Lessingia, German, aerial part, fresh, full bloom, (2)
Ref No 2 02 620 United States

Analyses	Unit	As Fed	Dry	C.V. ± %
Dry matter	%		100.0	
Ash	%		6.2	
Crude fiber	%		25.0	
Protein (N x 6.25)	%		4.8	
Cattle	dig prot % *		2.0	
Goats	dig prot % *		1.0	
Horses	dig prot % *		1.6	
Rabbits	dig prot % *		2.4	
Sheep	dig prot % *		1.5	
Calcium	%		0.81	
Phosphorus	%		0.18	
Potassium	%		1.50	

Lessingia, German, aerial part, fresh, mature, (2)
Ref No 2 02 621 United States

Analyses	Unit	As Fed	Dry	C.V. ± %
Dry matter	%		100.0	
Ash	%		5.3	
Crude fiber	%		28.3	
Protein (N x 6.25)	%		4.2	
Cattle	dig prot % *		1.5	
Goats	dig prot % *		0.5	
Horses	dig prot % *		1.1	
Rabbits	dig prot % *		1.9	
Sheep	dig prot % *		0.9	

Lessingia, German, aerial part, fresh, dormant, (2)
Ref No 2 02 622 United States

Analyses	Unit	As Fed	Dry	C.V. ± %
Dry matter	%		100.0	
Ash	%		4.3	
Crude fiber	%		30.7	
Protein (N x 6.25)	%		2.4	
Cattle	dig prot % *		0.0	
Goats	dig prot % *		1.1	
Horses	dig prot % *		0.3	
Rabbits	dig prot % *		0.5	
Sheep	dig prot % *		-0.7	
Calcium	%		0.92	
Phosphorus	%		0.09	
Potassium	%		0.67	

LETTUCE. Lactuca sativa

Lettuce, aerial part, fresh, (2)
Ref No 2 02 624 United States

Analyses	Unit	As Fed	Dry	C.V. ± %
Dry matter	%	5.4	100.0	
Ash	%	0.9	15.9	
Crude fiber	%	0.6	11.2	
Ether extract	%	0.2	4.1	
N-free extract	%	2.5	46.9	
Protein (N x 6.25)	%	1.2	22.0	
Cattle	dig prot % *	0.9	16.6	
Goats	dig prot % *	0.9	17.0	
Horses	dig prot % *	0.9	16.2	
Rabbits	dig prot % *	0.8	15.6	
Sheep	dig prot % *	0.9	17.5	
Energy	GE Mcal/kg	0.16	2.94	2
Cattle	DE Mcal/kg *	0.12	2.26	
Sheep	DE Mcal/kg *	0.13	2.40	
Cattle	ME Mcal/kg *	0.10	1.85	
Sheep	ME Mcal/kg *	0.11	1.97	
Cattle	TDN % *	2.8	51.3	
Sheep	TDN % *	2.9	54.4	
Calcium	%	0.05	0.86	
Iron	%	0.001	0.025	
Phosphorus	%	0.02	0.46	
Potassium	%	0.24	4.52	
Sodium	%	0.01	0.17	
Ascorbic acid	mg/kg	119.9	2241.5	
Niacin	mg/kg	3.5	65.3	
Riboflavin	mg/kg	0.7	13.1	
Thiamine	mg/kg	0.6	10.6	
Vitamin A	IU/g	12.1	226.2	

Lettuce, refuse, fresh, (2)
Ref No 2 02 625 United States

Analyses	Unit	As Fed	Dry	C.V. ± %
Dry matter	%		100.0	
Ash	%		20.8	
Crude fiber	%		19.1	
Ether extract	%		6.1	
N-free extract	%		41.2	
Protein (N x 6.25)	%		12.8	
Cattle	dig prot % *		8.8	
Goats	dig prot % *		8.5	
Horses	dig prot % *		8.4	
Rabbits	dig prot % *		8.6	
Sheep	dig prot % *		8.9	
Energy	GE Mcal/kg			

LICHEN. Lichen planus

Lichen, aerial part, fresh, (2)
Ref No 2 02 626 United States

Analyses	Unit	As Fed	Dry	C.V. ± %
Dry matter	%		100.0	
Ash	%		20.7	
Crude fiber	%		13.2	
Ether extract	%		4.8	
N-free extract	%		54.9	

Feed Name or Analyses				Mean As Fed	Mean Dry	C.V. ± %
Protein (N x 6.25)		%			6.4	
Cattle	dig prot	%	*		3.3	
Goats	dig prot	%	*		2.5	
Horses	dig prot	%	*		3.0	
Rabbits	dig prot	%	*		3.6	
Sheep	dig prot	%	*		3.0	
Calcium		%			3.70	
Phosphorus		%			0.09	

LIGUSTICUM. Ligusticum spp

Ligusticum, aerial part, fresh, immature, (2)

Ref No 2 02 629 United States

Feed Name or Analyses				Mean As Fed	Mean Dry	C.V. ± %
Dry matter		%			100.0	
Ash		%			11.5	
Crude fiber		%			12.8	
Ether extract		%			3.0	
N-free extract		%			44.2	
Protein (N x 6.25)		%			28.5	
Cattle	dig prot	%	*		22.1	
Goats	dig prot	%	*		23.2	
Horses	dig prot	%	*		21.7	
Rabbits	dig prot	%	*		20.7	
Sheep	dig prot	%	*		23.6	

Ligusticum, leaves, fresh, immature, (2)

Ref No 2 02 628 United States

Feed Name or Analyses				Mean As Fed	Mean Dry	C.V. ± %
Dry matter		%			100.0	
Ash		%			12.7	
Crude fiber		%			14.0	
Ether extract		%			3.2	
N-free extract		%			52.1	
Protein (N x 6.25)		%			18.0	
Cattle	dig prot	%	*		13.2	
Goats	dig prot	%	*		13.4	
Horses	dig prot	%	*		12.8	
Rabbits	dig prot	%	*		12.6	
Sheep	dig prot	%	*		13.8	

LIGUSTICUM, PORTER. Ligusticum porteri

Ligusticum, porter, flowers, fresh, immature, (2)

Ref No 2 02 630 United States

Feed Name or Analyses				Mean As Fed	Mean Dry	C.V. ± %
Dry matter		%			100.0	
Ash		%			11.2	
Crude fiber		%			15.9	
Ether extract		%			2.7	
N-free extract		%			51.3	
Protein (N x 6.25)		%			18.9	
Cattle	dig prot	%	*		14.0	
Goats	dig prot	%	*		14.2	
Horses	dig prot	%	*		13.6	
Rabbits	dig prot	%	*		13.3	
Sheep	dig prot	%	*		14.6	

Ligusticum, porter, leaves, fresh, immature, (2)

Ref No 2 02 631 United States

Feed Name or Analyses				Mean As Fed	Mean Dry	C.V. ± %
Dry matter		%			100.0	
Ash		%			12.8	
Crude fiber		%			14.4	
Ether extract		%			3.0	
N-free extract		%			53.7	

Feed Name or Analyses				Mean As Fed	Mean Dry	C.V. ± %
Protein (N x 6.25)		%			16.1	
Cattle	dig prot	%	*		11.6	
Goats	dig prot	%	*		11.6	
Horses	dig prot	%	*		11.2	
Rabbits	dig prot	%	*		11.1	
Sheep	dig prot	%	*		12.0	

Lime pulp, dried —
see Citrus, lime, pulp wo fines, shredded dehy, (4)

LIMESTONE. Scientific name not applicable

Limestone, grnd, mn 33% calcium, (6)
Limestone, ground (AAFCO)

Ref No 6 02 632 United States

Feed Name or Analyses				Mean As Fed	Mean Dry	C.V. ± %
Dry matter		%		99.9	100.0	0
Ash		%		96.8	96.9	1
Calcium		%		35.84	35.89	4
Chlorine		%		0.02	0.03	
Iron		%		0.349	0.350	47
Magnesium		%		2.06	2.06	98
Manganese		mg/kg		269.2	269.6	98
Phosphorus		%		0.01	0.02	98
Potassium		%		0.11	0.12	
Sodium		%		0.06	0.06	
Sulphur		%		0.04	0.04	

Limestone, dolomitic, mn 10% magnesium, (6)
Limestone, magnesium (AAFCO)
Limestone, dolomitic (AAFCO)

Ref No 6 02 633 United States

Feed Name or Analyses				Mean As Fed	Mean Dry	C.V. ± %
Dry matter		%		99.8	100.0	
Calcium		%		22.26	22.30	
Chlorine		%		0.12	0.12	
Iron		%		0.077	0.077	
Magnesium		%		9.97	9.99	
Phosphorus		%		0.04	0.04	
Potassium		%		0.36	0.36	

Limestone, magnesium (AAFCO) —
see Limestone, dolomitic, mn 10% magnesium, (6)

Linseed meal (CFA) —
see Flax, seeds, mech-extd grnd, mx 0.5% acid insoluble ash, (5)

Linseed meal, mechanical extracted (AAFCO) —
see Flax, seeds, mech-extd grnd, mx 0.5% acid insoluble ash, (5)

Linseed meal, solvent extracted (AAFCO) —
see Flax, seeds, solv-extd grnd, mx 0.5% acid insoluble ash, (5)

Linseed oil meal, expeller extracted —
see Flax, seeds, mech-extd grnd, mx 0.5% acid insoluble ash, (5)

Linseed oil meal, hydraulic extracted —
see Flax, seeds, mech-extd grnd, mx 0.5% acid insoluble ash, (5)

Linseed oil meal, old process —
see Flax, seeds, mech-extd grnd, mx 0.5% acid insoluble ash, (5)

Linseed oil meal, solvent extracted —
see Flax, seeds, solv-extd grnd, mx 0.5% acid insoluble ash, (5)

LITCHI. Litchi chinensis

Litchi, meats, dehy, (4)
Lychees

Ref No 4 08 117 United States

Feed Name or Analyses				Mean As Fed	Mean Dry	C.V. ± %
Dry matter		%		77.7	100.0	
Ash		%		2.0	2.6	
Crude fiber		%		1.4	1.8	
Ether extract		%		1.2	1.5	
N-free extract		%		69.3	89.2	
Protein (N x 6.25)		%		3.8	4.9	
Cattle	dig prot	%	*	0.4	0.5	
Goats	dig prot	%	*	1.3	1.7	
Horses	dig prot	%	*	1.3	1.7	
Sheep	dig prot	%	*	1.3	1.7	
Energy	GE	Mcal/kg		2.77	3.56	
Cattle	DE	Mcal/kg	*	2.04	2.62	
Sheep	DE	Mcal/kg	*	1.97	2.54	
Swine	DE	kcal/kg	*	1991.	2562.	
Cattle	ME	Mcal/kg	*	1.67	2.15	
Sheep	ME	Mcal/kg	*	1.62	2.08	
Swine	ME	kcal/kg	*	1893.	2436.	
Cattle	TDN	%	*	46.2	59.4	
Sheep	TDN	%	*	44.8	57.6	
Swine	TDN	%	*	45.1	58.1	
Calcium		%		0.03	0.04	
Iron		%		0.002	0.003	
Phosphorus		%		0.18	0.23	
Potassium		%		1.10	1.42	
Sodium		%		0.00	0.00	

Liver and glandular meal —
see Animal, liver w gland tissue, dehy grnd, mn 50% liver, (5)

Liver meal —
see Animal, livers, dehy grnd, (5)

LOBELIA, EDGING. Lobelia erinus

Lobelia, edging, seeds w hulls, (5)

Ref No 5 09 119 United States

Feed Name or Analyses				Mean As Fed	Mean Dry	C.V. ± %
Dry matter		%			100.0	
Protein (N x 6.25)		%			22.0	
Alanine		%			0.92	
Arginine		%			2.57	
Aspartic acid		%			2.31	
Glutamic acid		%			4.58	
Glycine		%			1.06	
Histidine		%			0.51	
Hydroxyproline		%			0.11	
Isoleucine		%			0.95	
Leucine		%			1.45	
Lysine		%			0.77	
Methionine		%			0.48	
Phenylalanine		%			0.99	
Proline		%			0.95	
Serine		%			0.97	
Threonine		%			0.77	
Tyrosine		%			0.73	
Valine		%			1.12	

Feed Name or Analyses			Mean As Fed	Mean Dry	C.V. ± %

LOBSTER. Homarus americanus

Lobster, meat, raw, (5)
Ref No 5 02 634 United States

Analyses	Unit		As Fed	Dry	C.V. ± %
Dry matter	%		21.5	100.0	
Ash	%		2.2	10.2	
Ether extract	%		1.9	8.8	
N-free extract	%		0.5	2.3	
Protein (N x 6.25)	%		16.9	78.6	
Energy	GE Mcal/kg		0.91	4.23	
Cattle	DE Mcal/kg	*	0.70	3.24	
Sheep	DE Mcal/kg	*	0.67	3.10	
Swine	DE kcal/kg	*	721.	3355.	
Cattle	ME Mcal/kg	*	0.57	2.66	
Sheep	ME Mcal/kg	*	0.55	2.54	
Swine	ME kcal/kg	*	578.	2687.	
Cattle	TDN %	*	15.8	73.4	
Sheep	TDN %	*	15.1	70.3	
Swine	TDN %	*	16.4	76.1	
Calcium	%		0.03	0.13	
Iron	%		0.001	0.003	
Phosphorus	%		0.18	0.85	
Niacin	mg/kg		15.0	69.8	
Riboflavin	mg/kg		0.5	2.3	
Thiamine	mg/kg		4.0	18.6	

Lobster, process residue, dehy grnd, (5)
Ref No 5 02 635 United States

Analyses	Unit		As Fed	Dry	C.V. ± %
Dry matter	%		95.7	100.0	
Ash	%		30.7	32.1	
Crude fiber	%		15.7	16.4	
Ether extract	%		4.7	4.9	
N-free extract	%		12.8	13.4	
Protein (N x 6.25)	%		31.8	33.2	
Energy	GE Mcal/kg				
Cattle	DE Mcal/kg	*	2.02	2.11	
Sheep	DE Mcal/kg	*	2.20	2.30	
Swine	DE kcal/kg	*	2358.	2464.	
Cattle	ME Mcal/kg	*	1.66	1.73	
Sheep	ME Mcal/kg	*	1.81	1.89	
Swine	ME kcal/kg	*	2105.	2200.	
Cattle	TDN %	*	45.7	47.8	
Sheep	TDN %	*	50.0	52.2	
Swine	TDN %	*	53.5	55.9	
Calcium	%		15.10	15.78	
Phosphorus	%		1.18	1.23	

LOCO. Astragalus spp

Loco, aerial part, fresh, (2)
Ref No 2 02 637 United States

Analyses	Unit		As Fed	Dry	C.V. ± %
Dry matter	%			100.0	
Ash	%			13.1	69
Crude fiber	%			20.9	27
Ether extract	%			2.2	51
N-free extract	%			42.6	
Protein (N x 6.25)	%			21.2	19
Cattle	dig prot %	*		15.9	
Goats	dig prot %	*		16.3	
Horses	dig prot %	*		15.5	
Rabbits	dig prot %	*		15.0	
Sheep	dig prot %	*		16.7	
Calcium	%			1.18	44
Cobalt	mg/kg			0.020	
Copper	mg/kg			34.0	
Magnesium	%			0.23	
Manganese	mg/kg			226.0	
Phosphorus	%			0.30	34
Potassium	%			2.80	
Carotene	mg/kg			7.9	
Vitamin A equivalent	IU/g			13.2	

(1) dry forages and roughages (2) pasture, range plants, and forages fed green (3) silages (4) energy feeds (5) protein supplements (6) minerals (7) vitamins (8) additives

Loco, aerial part, fresh, immature, (2)
Ref No 2 02 636 United States

Analyses	Unit		As Fed	Dry	C.V. ± %
Dry matter	%			100.0	
Ash	%			8.7	
Crude fiber	%			19.1	
Ether extract	%			2.7	
N-free extract	%			46.3	
Protein (N x 6.25)	%			23.2	
Cattle	dig prot %	*		17.6	
Goats	dig prot %	*		18.2	
Horses	dig prot %	*		17.2	
Rabbits	dig prot %	*		16.6	
Sheep	dig prot %	*		18.6	
Calcium	%			1.65	
Magnesium	%			0.23	
Phosphorus	%			0.39	
Potassium	%			2.80	

LOCUST, NEW MEXICO. Robinia neomexicana

Locust, New Mexico, seeds w hulls, (5)
Ref No 5 09 055 United States

Analyses	Unit		As Fed	Dry	C.V. ± %
Dry matter	%			100.0	
Protein (N x 6.25)	%			41.0	
Alanine	%			1.15	
Arginine	%			3.28	
Aspartic acid	%			2.67	
Glutamic acid	%			5.08	
Glycine	%			1.48	
Histidine	%			0.82	
Hydroxyproline	%			0.04	
Isoleucine	%			0.98	
Leucine	%			1.97	
Lysine	%			1.48	
Methionine	%			0.29	
Phenylalanine	%			1.15	
Proline	%			1.23	
Serine	%			1.44	
Threonine	%			0.98	
Tyrosine	%			0.94	
Valine	%			1.31	

LOMATIUM. Lomatium spp

Lomatium, aerial part, fresh, immature, (2)
Ref No 2 02 639 United States

Analyses	Unit		As Fed	Dry	C.V. ± %
Dry matter	%			100.0	
Ash	%			10.5	
Crude fiber	%			14.6	
Ether extract	%			5.0	
N-free extract	%			46.9	
Protein (N x 6.25)	%			23.0	
Cattle	dig prot %	*		17.4	
Goats	dig prot %	*		18.0	
Horses	dig prot %	*		17.1	
Rabbits	dig prot %	*		16.4	
Sheep	dig prot %	*		18.4	

LONGTOMGRASS. Paspalum lividum

Longtomgrass, aerial part, fresh, (2)
Ref No 2 02 641 United States

Analyses	Unit		As Fed	Dry	C.V. ± %
Dry matter	%			100.0	
Ash	%			10.0	7
Crude fiber	%			30.2	5
Ether extract	%			1.6	10
N-free extract	%			52.8	
Protein (N x 6.25)	%			5.4	19
Cattle	dig prot %	*		2.5	
Goats	dig prot %	*		1.6	
Horses	dig prot %	*		2.1	
Rabbits	dig prot %	*		2.8	
Sheep	dig prot %	*		2.0	
Calcium	%			0.44	17
Phosphorus	%			0.14	31

Longtomgrass, aerial part, fresh, mature, (2)
Ref No 2 02 640 United States

Analyses	Unit		As Fed	Dry	C.V. ± %
Dry matter	%			100.0	
Ash	%			8.9	
Crude fiber	%			30.9	
Ether extract	%			1.4	
N-free extract	%			54.8	
Protein (N x 6.25)	%			4.0	
Cattle	dig prot %	*		1.3	
Goats	dig prot %	*		0.3	
Horses	dig prot %	*		0.9	
Rabbits	dig prot %	*		1.8	
Sheep	dig prot %	*		0.7	
Calcium	%			0.37	18
Phosphorus	%			0.09	27

LOVEGRASS. Eragrostis spp

Lovegrass, hay, s-c, (1)
Ref No 1 02 646 United States

Analyses	Unit		As Fed	Dry	C.V. ± %
Dry matter	%			100.0	
Crude fiber	%				
Sheep	dig coef %	*		66.	
Ether extract	%				
Sheep	dig coef %	*		48.	
N-free extract	%				
Sheep	dig coef %	*		60.	
Protein (N x 6.25)	%				
Sheep	dig coef %	*		62.	

Lovegrass, hay, s-c, immature, (1)
Ref No 1 02 642 United States

Analyses	Unit		As Fed	Dry	C.V. ± %
Dry matter	%		89.0	100.0	
Ash	%		7.7	8.7	
Crude fiber	%		28.4	31.9	
Sheep	dig coef %	*	66.	66.	

Feed Name or Analyses			Mean As Fed	Mean Dry	C.V. ± %
Ether extract	%		1.5	1.7	
Sheep	dig coef %	*	48.	48.	
N-free extract	%		42.1	47.3	
Sheep	dig coef %	*	60.	60.	
Protein (N x 6.25)	%		9.3	10.4	
Sheep	dig coef %	*	62.	62.	
Cattle	dig prot %	*	5.3	5.9	
Goats	dig prot %	*	5.6	6.3	
Horses	dig prot %	*	5.7	6.4	
Rabbits	dig prot %	*	6.0	6.7	
Sheep	dig prot %		5.7	6.4	
Energy	GE Mcal/kg				
Cattle	DE Mcal/kg	*	2.36	2.65	
Sheep	DE Mcal/kg	*	2.26	2.54	
Cattle	ME Mcal/kg	*	1.94	2.18	
Sheep	ME Mcal/kg	*	1.86	2.09	
Cattle	TDN %	*	53.6	60.2	
Sheep	TDN %		51.4	57.7	
Calcium	%		0.36	0.40	
Magnesium	%		0.09	0.10	
Phosphorus	%		0.16	0.18	

Lovegrass, hay, s-c, pre-bloom, (1)
Ref No 1 02 643 United States

Feed Name or Analyses			Mean As Fed	Mean Dry	C.V. ± %
Dry matter	%		87.0	100.0	
Ash	%		7.6	8.7	
Crude fiber	%		27.8	31.9	
Sheep	dig coef %		70.	70.	
Ether extract	%		1.5	1.7	
Sheep	dig coef %		40.	40.	
N-free extract	%		41.2	47.3	
Sheep	dig coef %		61.	61.	
Protein (N x 6.25)	%		9.0	10.4	
Sheep	dig coef %		68.	68.	
Cattle	dig prot %	*	5.2	5.9	
Goats	dig prot %	*	5.4	6.3	
Horses	dig prot %	*	5.5	6.4	
Rabbits	dig prot %	*	5.8	6.7	
Sheep	dig prot %		6.2	7.1	
Energy	GE Mcal/kg				
Cattle	DE Mcal/kg	*	2.31	2.65	
Sheep	DE Mcal/kg	*	2.29	2.64	
Cattle	ME Mcal/kg	*	1.89	2.18	
Sheep	ME Mcal/kg	*	1.88	2.16	
Cattle	TDN %	*	52.3	60.2	
Sheep	TDN %		52.0	59.8	

Lovegrass, hay, s-c, full bloom, (1)
Ref No 1 02 644 United States

Feed Name or Analyses			Mean As Fed	Mean Dry	C.V. ± %
Dry matter	%		91.8	100.0	
Calcium	%		0.28	0.30	
Magnesium	%		0.06	0.06	
Phosphorus	%		0.11	0.12	

Lovegrass, hay, s-c, mature, (1)
Ref No 1 02 645 United States

Feed Name or Analyses			Mean As Fed	Mean Dry	C.V. ± %
Dry matter	%		87.0	100.0	
Ash	%		8.4	9.6	
Crude fiber	%		28.9	33.2	
Sheep	dig coef %		63.	63.	
Ether extract	%		1.6	1.8	
Sheep	dig coef %		52.	52.	
N-free extract	%		42.4	48.7	
Sheep	dig coef %		59.	59.	
Protein (N x 6.25)	%		5.8	6.7	
Sheep	dig coef %		59.	59.	
Cattle	dig prot %	*	2.4	2.7	
Goats	dig prot %	*	2.4	2.8	
Horses	dig prot %	*	2.8	3.2	
Rabbits	dig prot %	*	3.3	3.8	
Sheep	dig prot %		3.4	4.0	
Energy	GE Mcal/kg				
Cattle	DE Mcal/kg	*	2.29	2.64	
Sheep	DE Mcal/kg	*	2.14	2.46	
Cattle	ME Mcal/kg	*	1.88	2.16	
Sheep	ME Mcal/kg	*	1.75	2.01	
Cattle	TDN %	*	52.0	59.8	
Sheep	TDN %		48.5	55.7	

Lovegrass, aerial part, fresh, immature, (2)
Ref No 2 02 647 United States

Feed Name or Analyses			Mean As Fed	Mean Dry	C.V. ± %
Dry matter	%		42.7	100.0	5
Ash	%		2.8	6.5	26
Crude fiber	%		13.1	30.6	6
Ether extract	%		1.3	3.1	12
N-free extract	%		20.2	47.2	
Protein (N x 6.25)	%		5.4	12.6	13
Cattle	dig prot %	*	3.7	8.6	
Goats	dig prot %	*	3.6	8.3	
Horses	dig prot %	*	3.5	8.2	
Rabbits	dig prot %	*	3.6	8.4	
Sheep	dig prot %	*	3.7	8.7	
Energy	GE Mcal/kg				
Cattle	DE Mcal/kg	*	1.24	2.90	
Sheep	DE Mcal/kg	*	1.20	2.82	
Cattle	ME Mcal/kg	*	1.02	2.38	
Sheep	ME Mcal/kg	*	0.99	2.31	
Cattle	TDN %	*	28.1	65.8	
Sheep	TDN %	*	27.3	64.0	
Calcium	%		0.20	0.47	37
Phosphorus	%		0.10	0.24	16

Lovegrass, aerial part, fresh, full bloom, (2)
Ref No 2 02 648 United States

Feed Name or Analyses			Mean As Fed	Mean Dry	C.V. ± %
Dry matter	%		44.6	100.0	4
Ash	%		2.7	6.0	27
Crude fiber	%		14.5	32.5	2
Ether extract	%		1.3	3.0	12
N-free extract	%		22.1	49.6	
Protein (N x 6.25)	%		4.0	8.9	14
Cattle	dig prot %	*	2.4	5.5	
Goats	dig prot %	*	2.2	4.9	
Horses	dig prot %	*	2.3	5.1	
Rabbits	dig prot %	*	2.5	5.5	
Sheep	dig prot %	*	2.4	5.3	
Energy	GE Mcal/kg				
Cattle	DE Mcal/kg	*	1.32	2.96	
Sheep	DE Mcal/kg	*	1.30	2.92	
Cattle	ME Mcal/kg	*	1.08	2.43	
Sheep	ME Mcal/kg	*	1.07	2.39	
Cattle	TDN %	*	29.9	67.1	
Sheep	TDN %	*	29.5	66.2	
Calcium	%		0.15	0.33	19
Phosphorus	%		0.08	0.18	18

Lovegrass, aerial part, fresh, mature, (2)
Ref No 2 02 649 United States

Feed Name or Analyses			Mean As Fed	Mean Dry	C.V. ± %
Dry matter	%		42.8	100.0	5
Ash	%		2.7	6.2	29
Crude fiber	%		14.7	34.3	7
Ether extract	%		1.2	2.7	22
N-free extract	%		21.5	50.4	
Protein (N x 6.25)	%		2.7	6.4	16
Cattle	dig prot %	*	1.4	3.3	
Goats	dig prot %	*	1.1	2.5	
Horses	dig prot %	*	1.3	3.0	
Rabbits	dig prot %	*	1.5	3.6	
Sheep	dig prot %	*	1.3	3.0	
Energy	GE Mcal/kg				
Cattle	DE Mcal/kg	*	1.22	2.86	
Sheep	DE Mcal/kg	*	1.27	2.97	
Cattle	ME Mcal/kg	*	1.00	2.34	
Sheep	ME Mcal/kg	*	1.04	2.44	
Cattle	TDN %	*	27.7	64.8	
Sheep	TDN %	*	28.8	67.4	
Calcium	%		0.12	0.28	20
Phosphorus	%		0.05	0.11	25

Lovegrass, aerial part, fresh, over ripe, (2)
Ref No 2 02 650 United States

Feed Name or Analyses			Mean As Fed	Mean Dry	C.V. ± %
Dry matter	%			100.0	
Ash	%			5.1	
Crude fiber	%			36.3	
Ether extract	%			2.0	
N-free extract	%			53.3	
Protein (N x 6.25)	%			3.3	
Cattle	dig prot %	*		0.7	
Goats	dig prot %	*		-0.3	
Horses	dig prot %	*		0.3	
Rabbits	dig prot %	*		1.2	
Sheep	dig prot %	*		0.1	
Energy	GE Mcal/kg				
Cattle	DE Mcal/kg	*		2.74	
Sheep	DE Mcal/kg	*		2.80	
Cattle	ME Mcal/kg	*		2.25	
Sheep	ME Mcal/kg	*		2.30	
Cattle	TDN %	*		62.1	
Sheep	TDN %	*		63.4	

LOVEGRASS, BOER. Eragrostis chloromelas

Lovegrass, boer, aerial part, fresh, (2)
Ref No 2 02 652 United States

Feed Name or Analyses			Mean As Fed	Mean Dry	C.V. ± %
Dry matter	%			100.0	
Ash	%			7.8	
Crude fiber	%			34.5	
Ether extract	%			3.0	
N-free extract	%			44.3	
Protein (N x 6.25)	%			10.4	
Cattle	dig prot %	*		6.7	
Goats	dig prot %	*		6.3	
Horses	dig prot %	*		6.4	
Rabbits	dig prot %	*		6.7	
Sheep	dig prot %	*		6.7	

Lovegrass, boer, leaves, fresh, mature, (2)
Ref No 2 02 653 United States

Feed Name or Analyses				Mean As Fed	Mean Dry	C.V. ± %
Dry matter		%			100.0	
Ash		%			7.8	
Crude fiber		%			33.9	
Ether extract		%			2.6	
N-free extract		%			49.2	
Protein (N x 6.25)		%			6.5	
Cattle	dig prot	%	*		3.4	
Goats	dig prot	%	*		2.6	
Horses	dig prot	%	*		3.0	
Rabbits	dig prot	%	*		3.7	
Sheep	dig prot	%	*		3.0	

LOVEGRASS, LEHMANN. Eragrostis lehmanniana

Lovegrass, lehmann, hay, s-c, (1)
Ref No 1 02 654 United States

Feed Name or Analyses				Mean As Fed	Mean Dry	C.V. ± %
Dry matter		%		87.0	100.0	
Ash		%		8.1	9.3	
Crude fiber		%		27.7	31.8	
Sheep	dig coef	%		66.	66.	
Ether extract		%		1.6	1.8	
Sheep	dig coef	%		48.	48.	
N-free extract		%		42.8	49.2	
Sheep	dig coef	%		60.	60.	
Protein (N x 6.25)		%		6.9	7.9	
Sheep	dig coef	%		62.	62.	
Cattle	dig prot	%	*	3.3	3.8	
Goats	dig prot	%	*	3.4	3.9	
Horses	dig prot	%	*	3.7	4.2	
Rabbits	dig prot	%	*	4.1	4.8	
Sheep	dig prot	%		4.3	4.9	
Energy	GE Mcal/kg					
Cattle	DE Mcal/kg		*	2.36	2.72	
Sheep	DE Mcal/kg		*	2.20	2.53	
Cattle	ME Mcal/kg		*	1.94	2.23	
Sheep	ME Mcal/kg		*	1.80	2.07	
Cattle	TDN	%	*	53.6	61.6	
Sheep	TDN	%		49.9	57.4	

Lovegrass, lehmann, aerial part, fresh, mature, (2)
Ref No 2 02 655 United States

Feed Name or Analyses				Mean As Fed	Mean Dry	C.V. ± %
Dry matter		%			100.0	
Ash		%			7.6	
Crude fiber		%			35.0	
Ether extract		%			1.7	
N-free extract		%			49.0	
Protein (N x 6.25)		%			6.7	
Cattle	dig prot	%	*		3.6	
Goats	dig prot	%	*		2.8	
Horses	dig prot	%	*		3.2	
Rabbits	dig prot	%	*		3.8	
Sheep	dig prot	%	*		3.2	
Energy	GE Mcal/kg					
Cattle	DE Mcal/kg		*		2.85	
Sheep	DE Mcal/kg		*		2.94	
Cattle	ME Mcal/kg		*		2.34	
Sheep	ME Mcal/kg		*		2.41	

(1) dry forages and roughages (2) pasture, range plants, and forages fed green (3) silages (4) energy feeds (5) protein supplements (6) minerals (7) vitamins (8) additives

Feed Name or Analyses				Mean As Fed	Mean Dry	C.V. ± %
Cattle	TDN	%	*		64.7	
Sheep	TDN	%	*		66.7	

LOVEGRASS, MOURNING. Eragrostis lugens

Lovegrass, mourning, aerial part, fresh, immature, (2)
Ref No 2 02 656 United States

Feed Name or Analyses				Mean As Fed	Mean Dry	C.V. ± %
Dry matter		%			100.0	
Ash		%			10.8	
Crude fiber		%			29.4	
Ether extract		%			2.1	
N-free extract		%			44.5	
Protein (N x 6.25)		%			13.2	
Cattle	dig prot	%	*		9.1	
Goats	dig prot	%	*		8.9	
Horses	dig prot	%	*		8.7	
Rabbits	dig prot	%	*		8.9	
Sheep	dig prot	%	*		9.3	
Energy	GE Mcal/kg					
Cattle	DE Mcal/kg		*		2.89	
Sheep	DE Mcal/kg		*		2.78	
Cattle	ME Mcal/kg		*		2.37	
Sheep	ME Mcal/kg		*		2.28	
Cattle	TDN	%	*		65.6	
Sheep	TDN	%	*		63.2	
Calcium		%			0.37	
Phosphorus		%			0.16	

LOVEGRASS, PLAINS. Eragrostis intermedia

Lovegrass, plains, aerial part, fresh, (2)
Ref No 2 02 659 United States

Feed Name or Analyses				Mean As Fed	Mean Dry	C.V. ± %
Dry matter		%			100.0	
Ash		%			11.5	2
Crude fiber		%			32.8	1
Ether extract		%			2.0	13
N-free extract		%			43.8	
Protein (N x 6.25)		%			9.9	29
Cattle	dig prot	%	*		6.3	
Goats	dig prot	%	*		5.8	
Horses	dig prot	%	*		5.9	
Rabbits	dig prot	%	*		6.3	
Sheep	dig prot	%	*		6.2	
Energy	GE Mcal/kg					
Cattle	DE Mcal/kg		*		2.72	
Sheep	DE Mcal/kg		*		2.82	
Cattle	ME Mcal/kg		*		2.23	
Sheep	ME Mcal/kg		*		2.31	
Cattle	TDN	%	*		61.6	
Sheep	TDN	%	*		64.0	

Lovegrass, plains, aerial part, fresh, immature, (2)
Ref No 2 02 657 United States

Feed Name or Analyses				Mean As Fed	Mean Dry	C.V. ± %
Dry matter		%			100.0	
Calcium		%			0.41	
Phosphorus		%			0.17	

Lovegrass, plains, aerial part, fresh, full bloom, (2)
Ref No 2 02 658 United States

Feed Name or Analyses				Mean As Fed	Mean Dry	C.V. ± %
Dry matter		%			100.0	
Ash		%			11.9	
Crude fiber		%			33.3	

Feed Name or Analyses				Mean As Fed	Mean Dry	C.V. ± %
Ether extract		%			2.0	
N-free extract		%			46.9	
Protein (N x 6.25)		%			5.9	
Cattle	dig prot	%	*		2.9	
Goats	dig prot	%	*		2.1	
Horses	dig prot	%	*		2.5	
Rabbits	dig prot	%	*		3.2	
Sheep	dig prot	%	*		2.5	
Energy	GE Mcal/kg					
Cattle	DE Mcal/kg		*		2.50	
Sheep	DE Mcal/kg		*		2.53	
Cattle	ME Mcal/kg		*		2.05	
Sheep	ME Mcal/kg		*		2.07	
Cattle	TDN	%	*		56.6	
Sheep	TDN	%	*		57.4	
Calcium		%			0.41	
Phosphorus		%			0.13	

LOVEGRASS, RED. Eragrostis secundiflora

Lovegrass, red, aerial part, fresh, (2)
Ref No 2 02 663 United States

Feed Name or Analyses				Mean As Fed	Mean Dry	C.V. ± %
Dry matter		%			100.0	
Ash		%			10.1	20
Crude fiber		%			37.1	7
Ether extract		%			2.0	11
N-free extract		%			44.3	
Protein (N x 6.25)		%			6.5	35
Cattle	dig prot	%	*		3.4	
Goats	dig prot	%	*		2.6	
Horses	dig prot	%	*		3.0	
Rabbits	dig prot	%	*		3.7	
Sheep	dig prot	%	*		3.0	
Energy	GE Mcal/kg					
Cattle	DE Mcal/kg		*		2.24	
Sheep	DE Mcal/kg		*		2.42	
Cattle	ME Mcal/kg		*		1.84	
Sheep	ME Mcal/kg		*		1.98	
Cattle	TDN	%	*		50.7	
Sheep	TDN	%	*		55.0	

Lovegrass, red, aerial part, fresh, immature, (2)
Ref No 2 02 660 United States

Feed Name or Analyses				Mean As Fed	Mean Dry	C.V. ± %
Dry matter		%			100.0	
Ash		%			10.6	
Crude fiber		%			35.1	
Ether extract		%			2.3	
N-free extract		%			40.9	
Protein (N x 6.25)		%			11.1	
Cattle	dig prot	%	*		7.3	
Goats	dig prot	%	*		6.9	
Horses	dig prot	%	*		7.0	
Rabbits	dig prot	%	*		7.2	
Sheep	dig prot	%	*		7.3	
Energy	GE Mcal/kg					
Cattle	DE Mcal/kg		*		2.39	
Sheep	DE Mcal/kg		*		2.29	
Cattle	ME Mcal/kg		*		1.96	
Sheep	ME Mcal/kg		*		1.88	
Cattle	TDN	%	*		54.3	
Sheep	TDN	%	*		52.0	
Calcium		%			0.59	
Phosphorus		%			0.24	

Feed Name or Analyses			Mean As Fed	Mean Dry	C.V. ± %

Lovegrass, red, aerial part, fresh, full bloom, (2)
Ref No 2 02 661 United States

Feed Name or Analyses			As Fed	Dry	C.V. ± %
Dry matter	%			100.0	
Ash	%			13.0	
Crude fiber	%			34.6	
Ether extract	%			2.1	
N-free extract	%			44.2	
Protein (N x 6.25)	%			6.1	
Cattle	dig prot %	*		3.1	
Goats	dig prot %	*		2.3	
Horses	dig prot %	*		2.7	
Rabbits	dig prot %	*		3.4	
Sheep	dig prot %	*		2.7	
Energy	GE Mcal/kg				
Cattle	DE Mcal/kg	*		2.44	
Sheep	DE Mcal/kg	*		2.54	
Cattle	ME Mcal/kg	*		2.00	
Sheep	ME Mcal/kg	*		2.08	
Cattle	TDN %	*		55.3	
Sheep	TDN %	*		57.5	
Calcium	%			0.34	
Phosphorus	%			0.12	

Lovegrass, red, aerial part, fresh, mature, (2)
Ref No 2 02 662 United States

Feed Name or Analyses			As Fed	Dry	C.V. ± %
Dry matter	%			100.0	
Ash	%			8.1	20
Crude fiber	%			39.5	1
Ether extract	%			2.0	17
N-free extract	%			46.7	
Protein (N x 6.25)	%			3.7	3
Cattle	dig prot %	*		1.0	
Goats	dig prot %	*		0.0	
Horses	dig prot %	*		0.7	
Rabbits	dig prot %	*		1.5	
Sheep	dig prot %	*		0.4	
Energy	GE Mcal/kg				
Cattle	DE Mcal/kg	*		2.25	
Sheep	DE Mcal/kg	*		2.41	
Cattle	ME Mcal/kg	*		1.84	
Sheep	ME Mcal/kg	*		1.98	
Cattle	TDN %	*		51.0	
Sheep	TDN %	*		54.7	
Calcium	%			0.31	
Phosphorus	%			0.09	

LOVEGRASS, SAND. Eragrostis trichodes

Lovegrass, sand, aerial part, fresh, (2)
Ref No 2 02 668 United States

Feed Name or Analyses			As Fed	Dry	C.V. ± %
Dry matter	%		46.3	100.0	
Ash	%		2.5	5.5	42
Crude fiber	%		15.5	33.4	17
Ether extract	%		1.0	2.1	25
N-free extract	%		25.0	54.0	
Protein (N x 6.25)	%		2.3	5.0	62
Cattle	dig prot %	*	1.0	2.1	
Goats	dig prot %	*	0.6	1.2	
Horses	dig prot %	*	0.8	1.8	
Rabbits	dig prot %	*	1.2	2.5	
Sheep	dig prot %	*	0.8	1.7	
Energy	GE Mcal/kg				
Cattle	DE Mcal/kg	*	1.33	2.87	
Sheep	DE Mcal/kg	*	1.42	3.06	
Cattle	ME Mcal/kg	*	1.09	2.35	
Sheep	ME Mcal/kg	*	1.16	2.51	
Cattle	TDN %	*	30.1	65.0	
Sheep	TDN %	*	32.1	69.3	

Lovegrass, sand, aerial part, fresh, immature, (2)
Ref No 2 02 664 United States

Feed Name or Analyses			As Fed	Dry	C.V. ± %
Dry matter	%		36.7	100.0	
Ash	%		3.1	8.4	
Crude fiber	%		10.2	27.8	
Ether extract	%		0.9	2.5	
N-free extract	%		18.9	51.5	
Protein (N x 6.25)	%		3.6	9.8	
Cattle	dig prot %	*	2.3	6.2	
Goats	dig prot %	*	2.1	5.7	
Horses	dig prot %	*	2.1	5.9	
Rabbits	dig prot %	*	2.3	6.2	
Sheep	dig prot %	*	2.2	6.1	
Energy	GE Mcal/kg				
Cattle	DE Mcal/kg	*	0.88	2.40	
Sheep	DE Mcal/kg	*	0.92	2.53	
Cattle	ME Mcal/kg	*	0.72	1.97	
Sheep	ME Mcal/kg	*	0.76	2.07	
Cattle	TDN %	*	20.0	54.4	
Sheep	TDN %	*	21.0	57.3	
Calcium	%		0.18	0.49	
Phosphorus	%		0.07	0.18	

Lovegrass, sand, aerial part, fresh, full bloom, (2)
Ref No 2 02 665 United States

Feed Name or Analyses			As Fed	Dry	C.V. ± %
Dry matter	%		49.9	100.0	
Calcium	%		0.19	0.38	
Phosphorus	%		0.09	0.18	

Lovegrass, sand, aerial part, fresh, mature, (2)
Ref No 2 02 666 United States

Feed Name or Analyses			As Fed	Dry	C.V. ± %
Dry matter	%		52.3	100.0	
Ash	%		1.6	3.0	
Crude fiber	%		17.9	34.3	
Ether extract	%		1.4	2.6	
N-free extract	%		28.9	55.3	
Protein (N x 6.25)	%		2.5	4.8	
Cattle	dig prot %	*	1.0	2.0	
Goats	dig prot %	*	0.5	1.0	
Horses	dig prot %	*	0.8	1.6	
Rabbits	dig prot %	*	1.2	2.4	
Sheep	dig prot %	*	0.8	1.5	
Energy	GE Mcal/kg				
Cattle	DE Mcal/kg	*	1.33	2.55	
Sheep	DE Mcal/kg	*	1.38	2.63	
Cattle	ME Mcal/kg	*	1.09	2.09	
Sheep	ME Mcal/kg	*	1.13	2.16	
Cattle	TDN %	*	30.3	57.9	
Sheep	TDN %	*	31.2	59.6	

Lovegrass, sand, aerial part, fresh, over ripe, (2)
Ref No 2 02 667 United States

Feed Name or Analyses			As Fed	Dry	C.V. ± %
Dry matter	%			100.0	
Ash	%			5.3	
Crude fiber	%			36.3	
Ether extract	%			2.1	
N-free extract	%			53.2	
Protein (N x 6.25)	%			3.1	
Cattle	dig prot %	*		0.5	
Goats	dig prot %	*		-0.4	
Horses	dig prot %	*		0.2	
Rabbits	dig prot %	*		1.1	
Sheep	dig prot %	*		0.0	
Energy	GE Mcal/kg				
Cattle	DE Mcal/kg	*		2.37	
Sheep	DE Mcal/kg	*		2.49	
Cattle	ME Mcal/kg	*		1.94	
Sheep	ME Mcal/kg	*		2.04	
Cattle	TDN %	*		53.7	
Sheep	TDN %	*		56.5	
Calcium	%			0.37	
Phosphorus	%			0.12	

LOVEGRASS, SPREADING. Eragrostis diffusa

Lovegrass, spreading, aerial part, fresh, immature, (2)
Ref No 2 02 669 United States

Feed Name or Analyses			As Fed	Dry	C.V. ± %
Dry matter	%			100.0	
Ash	%			8.4	
Crude fiber	%			30.8	
Ether extract	%			3.7	
N-free extract	%			45.3	
Protein (N x 6.25)	%			11.8	
Cattle	dig prot %	*		7.9	
Goats	dig prot %	*		7.6	
Horses	dig prot %	*		7.5	
Rabbits	dig prot %	*		7.8	
Sheep	dig prot %	*		8.0	
Energy	GE Mcal/kg				
Cattle	DE Mcal/kg	*		2.44	
Sheep	DE Mcal/kg	*		2.37	
Cattle	ME Mcal/kg	*		2.00	
Sheep	ME Mcal/kg	*		1.94	
Cattle	TDN %	*		55.4	
Sheep	TDN %	*		53.8	
Calcium	%			0.38	
Phosphorus	%			0.20	

LOVEGRASS, TUMBLE. Eragrostis sessilispica

Lovegrass, tumble, aerial part, fresh, immature, (2)
Ref No 2 02 670 United States

Feed Name or Analyses			As Fed	Dry	C.V. ± %
Dry matter	%			100.0	
Calcium	%			0.43	
Phosphorus	%			0.14	

Lovegrass, tumble, aerial part, fresh, full bloom, (2)
Ref No 2 02 671 United States

Feed Name or Analyses			As Fed	Dry	C.V. ± %
Dry matter	%			100.0	
Calcium	%			0.34	
Phosphorus	%			0.14	

LOVEGRASS, WEEPING. Eragrostis curvula

Lovegrass, weeping, hay, s-c, (1)
Ref No 1 02 672 United States

Feed Name or Analyses			As Fed	Dry	C.V. ± %
Dry matter	%		91.2	100.0	
Ash	%		4.9	5.4	

Continued

Feed Name or Analyses			Mean As Fed	Mean Dry	C.V. ± %
Crude fiber	%		30.9	33.9	
Ether extract	%		2.8	3.1	
N-free extract	%		43.4	47.5	
Protein (N x 6.25)	%		9.2	10.1	
Cattle	dig prot %	*	5.2	5.7	
Goats	dig prot %	*	5.5	6.0	
Horses	dig prot %	*	5.6	6.1	
Rabbits	dig prot %	*	5.9	6.5	
Sheep	dig prot %	*	5.1	5.6	
Fatty acids	%		2.8	3.1	
Energy	GE Mcal/kg				
Cattle	DE Mcal/kg	*	2.22	2.43	
Sheep	DE Mcal/kg	*	2.26	2.48	
Cattle	ME Mcal/kg	*	1.82	1.99	
Sheep	ME Mcal/kg	*	1.85	2.03	
Cattle	TDN %	*	50.2	55.1	
Sheep	TDN %	*	51.3	56.2	
Calcium	%		0.33	0.36	12
Phosphorus	%		0.06	0.07	34

Lovegrass, weeping, hay, s-c, immature, (1)
Ref No 1 09 177 United States

Feed Name or Analyses			Mean As Fed	Mean Dry	C.V. ± %
Dry matter	%			100.0	
Sheep	dig coef %			51.	
Ash	%			3.6	
Crude fiber	%			29.4	
Sheep	dig coef %			59.	
Ether extract	%			5.0	
N-free extract	%			49.3	
Sheep	dig coef %			48.	
Protein (N x 6.25)	%			12.7	
Sheep	dig coef %			54.	
Cattle	dig prot %	*		7.9	
Goats	dig prot %	*		8.4	
Horses	dig prot %	*		8.3	
Rabbits	dig prot %	*		8.5	
Sheep	dig prot %			6.8	
Lignin (Ellis)	%			5.4	
Sheep	dig coef %			3.	
Energy	GE Mcal/kg				
Cattle	DE Mcal/kg	*		2.37	
Sheep	DE Mcal/kg	*		2.17	
Cattle	ME Mcal/kg	*		1.94	
Sheep	ME Mcal/kg	*		1.78	
Cattle	TDN %	*		53.7	
Sheep	TDN %	*		49.3	

Lovegrass, weeping, hay, s-c, prebloom, (1)
Ref No 1 09 178 United States

Feed Name or Analyses			Mean As Fed	Mean Dry	C.V. ± %
Dry matter	%			100.0	
Sheep	dig coef %			51.	
Ash	%			3.5	
Crude fiber	%			29.8	
Sheep	dig coef %			57.	
Ether extract	%			4.9	
N-free extract	%			50.2	
Sheep	dig coef %			49.	
Protein (N x 6.25)	%			11.6	
Sheep	dig coef %			49.	
Cattle	dig prot %	*		7.0	
Goats	dig prot %	*		7.4	
Horses	dig prot %	*		7.4	
Rabbits	dig prot %	*		7.6	
Sheep	dig prot %			5.6	
Lignin (Ellis)	%			5.6	
Sheep	dig coef %			21.	
Energy	GE Mcal/kg				
Cattle	DE Mcal/kg	*		2.71	
Sheep	DE Mcal/kg	*		2.59	
Cattle	ME Mcal/kg	*		2.22	
Sheep	ME Mcal/kg	*		2.12	
Cattle	TDN %	*		61.5	
Sheep	TDN %	*		58.7	

Lovegrass, weeping, hay, s-c, early bloom, (1)
Ref No 1 09 179 United States

Feed Name or Analyses			Mean As Fed	Mean Dry	C.V. ± %
Dry matter	%			100.0	
Sheep	dig coef %			54.	
Ash	%			3.8	
Crude fiber	%			29.3	
Sheep	dig coef %			57.	
Ether extract	%			4.6	
N-free extract	%			49.5	
Sheep	dig coef %			51.	
Protein (N x 6.25)	%			12.9	
Sheep	dig coef %			58.	
Cattle	dig prot %	*		8.1	
Goats	dig prot %	*		8.6	
Horses	dig prot %	*		8.5	
Rabbits	dig prot %	*		8.6	
Sheep	dig prot %			7.5	
Lignin (Ellis)	%			5.6	
Sheep	dig coef %			20.	
Energy	GE Mcal/kg				
Cattle	DE Mcal/kg	*		2.73	
Sheep	DE Mcal/kg	*		2.62	
Cattle	ME Mcal/kg	*		2.24	
Sheep	ME Mcal/kg	*		2.15	
Cattle	TDN %	*		62.0	
Sheep	TDN %	*		59.5	

Lovegrass, weeping, hay, s-c, midbloom, (1)
Ref No 1 09 180 United States

Feed Name or Analyses			Mean As Fed	Mean Dry	C.V. ± %
Dry matter	%			100.0	
Sheep	dig coef %			60.	
Ash	%			3.4	
Crude fiber	%			30.3	
Sheep	dig coef %			63.	
Ether extract	%			4.5	
N-free extract	%			52.3	
Sheep	dig coef %			60.	
Protein (N x 6.25)	%			9.6	
Sheep	dig coef %			52.	
Cattle	dig prot %	*		5.2	
Goats	dig prot %	*		5.5	
Horses	dig prot %	*		5.6	
Rabbits	dig prot %	*		6.0	
Sheep	dig prot %			5.0	
Lignin (Ellis)	%			4.7	
Sheep	dig coef %			17.	
Energy	GE Mcal/kg				
Cattle	DE Mcal/kg	*		2.67	
Sheep	DE Mcal/kg	*		2.55	
Cattle	ME Mcal/kg	*		2.19	
Sheep	ME Mcal/kg	*		2.09	
Cattle	TDN %	*		60.6	
Sheep	TDN %	*		57.8	

Lovegrass, weeping, hay, s-c, mature, (1)
Ref No 1 09 181 United States

Feed Name or Analyses			Mean As Fed	Mean Dry	C.V. ± %
Dry matter	%			100.0	
Sheep	dig coef %			53.	
Ash	%			3.2	
Crude fiber	%			32.9	
Sheep	dig coef %			60.	
Ether extract	%			4.7	
N-free extract	%			52.9	
Sheep	dig coef %			54.	
Protein (N x 6.25)	%			6.3	
Sheep	dig coef %			32.	
Cattle	dig prot %	*		2.4	
Goats	dig prot %	*		2.4	
Horses	dig prot %	*		2.9	
Rabbits	dig prot %	*		3.5	
Sheep	dig prot %			2.0	
Lignin (Ellis)	%			5.7	
Sheep	dig coef %			11.	
Energy	GE Mcal/kg				
Cattle	DE Mcal/kg	*		2.52	
Sheep	DE Mcal/kg	*		2.44	
Cattle	ME Mcal/kg	*		2.06	
Sheep	ME Mcal/kg	*		2.00	
Cattle	TDN %	*		57.1	
Sheep	TDN %	*		55.4	

Lovegrass, weeping, hay, s-c, over ripe, (1)
Ref No 1 09 182 United States

Feed Name or Analyses			Mean As Fed	Mean Dry	C.V. ± %
Dry matter	%			100.0	
Sheep	dig coef %			57.	
Ash	%			2.9	
Crude fiber	%			33.1	
Sheep	dig coef %			65.	
Ether extract	%			4.3	
N-free extract	%			53.9	
Sheep	dig coef %			57.	
Protein (N x 6.25)	%			5.8	
Sheep	dig coef %			35.	
Cattle	dig prot %	*		2.0	
Goats	dig prot %	*		2.0	
Horses	dig prot %	*		2.5	
Rabbits	dig prot %	*		3.1	
Sheep	dig prot %			2.0	
Lignin (Ellis)	%			5.0	
Sheep	dig coef %			12.	
Energy	GE Mcal/kg				
Cattle	DE Mcal/kg	*		2.48	
Sheep	DE Mcal/kg	*		2.43	
Cattle	ME Mcal/kg	*		2.03	
Sheep	ME Mcal/kg	*		2.00	
Cattle	TDN %	*		56.2	
Sheep	TDN %	*		55.2	

Lovegrass, weeping, aerial part, fresh, (2)
Ref No 2 02 677 United States

Feed Name or Analyses			Mean As Fed	Mean Dry	C.V. ± %
Dry matter	%		44.3	100.0	
Ash	%		2.5	5.6	
Crude fiber	%		14.2	32.1	
Ether extract	%		1.4	3.2	
N-free extract	%		22.0	49.7	

(1) dry forages and roughages (2) pasture, range plants, and forages fed green (3) silages (4) energy feeds (5) protein supplements (6) minerals (7) vitamins (8) additives

Feed Name or Analyses			Mean As Fed	Mean Dry	C.V. ± %
Protein (N x 6.25)	%		4.2	9.5	
Cattle	dig prot %	*	2.6	5.9	
Goats	dig prot %	*	2.4	5.4	
Horses	dig prot %	*	2.5	5.6	
Rabbits	dig prot %	*	2.7	6.0	
Sheep	dig prot %	*	2.6	5.8	
Energy	GE Mcal/kg				
Cattle	DE Mcal/kg	*	1.25	2.83	
Sheep	DE Mcal/kg	*	1.19	2.69	
Cattle	ME Mcal/kg	*	1.03	2.32	
Sheep	ME Mcal/kg	*	0.98	2.21	
Cattle	TDN %	*	28.5	64.3	
Sheep	TDN %	*	27.0	60.9	
Calcium	%		0.14	0.32	
Phosphorus	%		0.08	0.18	

Lovegrass, weeping, aerial part, fresh, immature, (2)
Ref No 2 02 673 United States

Feed Name or Analyses			Mean As Fed	Mean Dry	C.V. ± %
Dry matter	%		43.0	100.0	7
Ash	%		2.5	5.9	19
Crude fiber	%		13.1	30.5	4
Ether extract	%		1.4	3.2	8
N-free extract	%		20.4	47.5	
Protein (N x 6.25)	%		5.5	12.9	9
Cattle	dig prot %	*	3.8	8.9	
Goats	dig prot %	*	3.7	8.6	
Horses	dig prot %	*	3.6	8.5	
Rabbits	dig prot %	*	3.7	8.6	
Sheep	dig prot %	*	3.9	9.0	
Energy	GE Mcal/kg				
Cattle	DE Mcal/kg	*	1.24	2.88	
Sheep	DE Mcal/kg	*	1.21	2.82	
Cattle	ME Mcal/kg	*	1.01	2.36	
Sheep	ME Mcal/kg	*	0.99	2.31	
Cattle	TDN %	*	28.1	65.4	
Sheep	TDN %	*	27.5	63.9	
Calcium	%		0.19	0.44	26
Phosphorus	%		0.09	0.20	11

Lovegrass, weeping, aerial part, fresh, full bloom, (2)
Ref No 2 02 674 United States

Feed Name or Analyses			Mean As Fed	Mean Dry	C.V. ± %
Dry matter	%		44.5	100.0	3
Ash	%		2.5	5.7	4
Crude fiber	%		14.5	32.5	2
Ether extract	%		1.4	3.1	3
N-free extract	%		22.0	49.5	
Protein (N x 6.25)	%		4.1	9.2	6
Cattle	dig prot %	*	2.5	5.7	
Goats	dig prot %	*	2.3	5.1	
Horses	dig prot %	*	2.4	5.3	
Rabbits	dig prot %	*	2.6	5.8	
Sheep	dig prot %	*	2.5	5.6	
Energy	GE Mcal/kg				
Cattle	DE Mcal/kg	*	1.33	2.99	
Sheep	DE Mcal/kg	*	1.30	2.91	
Cattle	ME Mcal/kg	*	1.09	2.45	
Sheep	ME Mcal/kg	*	1.06	2.39	
Cattle	TDN %	*	30.2	67.8	
Sheep	TDN %	*	29.4	66.0	
Calcium	%		0.14	0.32	17
Phosphorus	%		0.08	0.18	15
Carotene	mg/kg		112.7	253.3	4
Vitamin A equivalent	IU/g		187.9	422.3	

Lovegrass, weeping, aerial part, fresh, mature, (2)
Ref No 2 02 675 United States

Feed Name or Analyses			Mean As Fed	Mean Dry	C.V. ± %
Dry matter	%		46.8	100.0	4
Ash	%		2.5	5.2	11
Crude fiber	%		15.3	32.7	8
Ether extract	%		1.4	3.0	8
N-free extract	%		24.4	52.2	
Protein (N x 6.25)	%		3.2	6.9	9
Cattle	dig prot %	*	1.8	3.8	
Goats	dig prot %	*	1.4	3.0	
Horses	dig prot %	*	1.6	3.4	
Rabbits	dig prot %	*	1.9	4.0	
Sheep	dig prot %	*	1.6	3.4	
Energy	GE Mcal/kg				
Cattle	DE Mcal/kg	*	1.16	2.47	
Sheep	DE Mcal/kg	*	1.20	2.56	
Cattle	ME Mcal/kg	*	0.95	2.03	
Sheep	ME Mcal/kg	*	0.98	2.10	
Cattle	TDN %	*	26.3	56.1	
Sheep	TDN %	*	27.1	58.0	
Calcium	%		0.10	0.22	
Phosphorus	%		0.05	0.11	
Carotene	mg/kg		88.1	188.1	
Vitamin A equivalent	IU/g		146.8	313.5	

Lovegrass, weeping, aerial part, fresh, over ripe, (2)
Ref No 2 02 676 United States

Feed Name or Analyses			Mean As Fed	Mean Dry	C.V. ± %
Dry matter	%			100.0	
Ash	%			4.9	
Crude fiber	%			36.3	
Ether extract	%			1.8	
N-free extract	%			53.3	
Protein (N x 6.25)	%			3.7	
Cattle	dig prot %	*		1.0	
Goats	dig prot %	*		0.0	
Horses	dig prot %	*		0.7	
Rabbits	dig prot %	*		1.5	
Sheep	dig prot %	*		0.4	
Energy	GE Mcal/kg				
Cattle	DE Mcal/kg	*		2.35	
Sheep	DE Mcal/kg	*		2.48	
Cattle	ME Mcal/kg	*		1.92	
Sheep	ME Mcal/kg	*		2.03	
Cattle	TDN %	*		53.4	
Sheep	TDN %	*		56.3	

LOVEGRASS, WICHITA. Eragrostis beyrichii

Lovegrass, wichita, aerial part, fresh, mature, (2)
Ref No 2 02 678 United States

Feed Name or Analyses			Mean As Fed	Mean Dry	C.V. ± %
Dry matter	%			100.0	
Ash	%			11.7	
Crude fiber	%			32.7	
Ether extract	%			1.6	
N-free extract	%			46.8	
Protein (N x 6.25)	%			7.2	
Cattle	dig prot %	*		4.0	
Goats	dig prot %	*		3.3	
Horses	dig prot %	*		3.6	
Rabbits	dig prot %	*		4.2	
Sheep	dig prot %	*		3.7	
Energy	GE Mcal/kg				
Cattle	DE Mcal/kg	*		2.60	
Sheep	DE Mcal/kg	*		2.92	
Cattle	ME Mcal/kg	*		2.13	
Sheep	ME Mcal/kg	*		2.40	
Cattle	TDN %	*		58.9	
Sheep	TDN %	*		66.3	

Low gossypol cottonseed meal, solvent extracted (AAFCO) –
see Cotton, seeds w some hulls wo gossypol, solv-extd grnd, mx 0.04% free gossypol, (5)

LUPINE. Lupinus spp

Lupine, hay, dehy, early bloom, (1)
Ref No 1 02 679 United States

Feed Name or Analyses			Mean As Fed	Mean Dry	C.V. ± %
Dry matter	%		90.5	100.0	1
Ash	%		9.7	10.7	10
Crude fiber	%		29.2	32.3	7
Ether extract	%		2.9	3.2	16
N-free extract	%		30.4	33.6	
Protein (N x 6.25)	%		18.3	20.2	6
Cattle	dig prot %	*	13.1	14.4	
Goats	dig prot %	*	13.9	15.4	
Horses	dig prot %	*	13.3	14.7	
Rabbits	dig prot %	*	12.9	14.3	
Sheep	dig prot %	*	13.3	14.7	
Energy	GE Mcal/kg				
Cattle	DE Mcal/kg	*	2.37	2.62	
Sheep	DE Mcal/kg	*	2.17	2.40	
Cattle	ME Mcal/kg	*	1.95	2.15	
Sheep	ME Mcal/kg	*	1.78	1.97	
Cattle	TDN %	*	53.8	59.5	
Sheep	TDN %	*	49.2	54.4	

Lupine, hay, s-c, (1)
Ref No 1 02 682 United States

Feed Name or Analyses			Mean As Fed	Mean Dry	C.V. ± %
Dry matter	%		87.4	100.0	3
Ash	%		8.0	9.2	40
Crude fiber	%		26.2	30.0	18
Sheep	dig coef %		67.	67.	
Ether extract	%		3.0	3.4	24
Sheep	dig coef %		53.	53.	
N-free extract	%		31.8	36.4	
Sheep	dig coef %		70.	70.	
Protein (N x 6.25)	%		18.4	21.1	24
Sheep	dig coef %		69.	69.	
Cattle	dig prot %	*	13.3	15.2	
Goats	dig prot %	*	14.2	16.2	
Horses	dig prot %	*	13.5	15.4	
Rabbits	dig prot %	*	13.0	14.9	
Sheep	dig prot %		12.7	14.5	
Energy	GE Mcal/kg				
Cattle	DE Mcal/kg	*	2.29	2.62	
Sheep	DE Mcal/kg	*	2.47	2.83	
Cattle	ME Mcal/kg	*	1.88	2.15	
Sheep	ME Mcal/kg	*	2.03	2.32	
Cattle	TDN %	*	52.0	59.5	
Sheep	TDN %		56.0	64.2	

Lupine, hay, s-c, full bloom, (1)
Ref No 1 02 680 United States

Feed Name or Analyses			Mean As Fed	Mean Dry	C.V. ± %
Dry matter	%		90.9	100.0	
Ash	%		14.1	15.5	

Continued

Feed Name or Analyses				Mean As Fed	Mean Dry	C.V. ± %
Crude fiber		%		26.5	29.1	
Sheep	dig coef	%	*	67.	67.	
Ether extract		%		3.2	3.5	
Sheep	dig coef	%	*	53.	53.	
N-free extract		%		34.1	37.5	
Sheep	dig coef	%	*	70.	70.	
Protein (N x 6.25)		%		13.1	14.4	
Sheep	dig coef	%	*	69.	69.	
Cattle	dig prot	%	*	8.6	9.4	
Goats	dig prot	%	*	9.1	10.0	
Horses	dig prot	%	*	8.9	9.8	
Rabbits	dig prot	%	*	8.9	9.8	
Sheep	dig prot	%		9.0	9.9	
Energy	GE Mcal/kg					
Cattle	DE Mcal/kg		*	2.46	2.71	
Sheep	DE Mcal/kg		*	2.40	2.64	
Cattle	ME Mcal/kg		*	2.02	2.22	
Sheep	ME Mcal/kg		*	1.97	2.16	
Cattle	TDN	%	*	55.9	61.5	
Sheep	TDN	%	*	54.4	59.9	

Lupine, hay, s-c, late bloom, (1)
Ref No 1 02 681 United States

Feed Name or Analyses				Mean As Fed	Mean Dry	C.V. ± %
Dry matter		%		81.0	100.0	
Ash		%		4.0	4.9	
Crude fiber		%		24.5	30.2	
Sheep	dig coef	%		73.	73.	
Ether extract		%		1.9	2.4	
Sheep	dig coef	%		30.	30.	
N-free extract		%		28.1	34.7	
Sheep	dig coef	%		62.	62.	
Protein (N x 6.25)		%		22.5	27.8	
Sheep	dig coef	%		74.	74.	
Cattle	dig prot	%	*	17.0	21.0	
Goats	dig prot	%	*	18.2	22.5	
Horses	dig prot	%	*	17.1	21.1	
Rabbits	dig prot	%	*	16.3	20.1	
Sheep	dig prot	%		16.7	20.6	
Energy	GE Mcal/kg					
Cattle	DE Mcal/kg		*	2.27	2.80	
Sheep	DE Mcal/kg		*	2.35	2.90	
Cattle	ME Mcal/kg		*	1.86	2.30	
Sheep	ME Mcal/kg		*	1.93	2.38	
Cattle	TDN	%	*	51.4	63.5	
Sheep	TDN	%		53.3	65.8	

Lupine, straw, (1)
Ref No 1 02 683 United States

Feed Name or Analyses				Mean As Fed	Mean Dry	C.V. ± %
Dry matter		%		84.9	100.0	4
Ash		%		2.5	3.0	13
Crude fiber		%		45.1	53.1	8
Sheep	dig coef	%		43.	43.	
Ether extract		%		1.1	1.3	15
Sheep	dig coef	%		51.	51.	
N-free extract		%		30.8	36.3	
Sheep	dig coef	%		53.	53.	
Protein (N x 6.25)		%		5.3	6.3	11
Sheep	dig coef	%		42.	42.	
Cattle	dig prot	%	*	2.0	2.4	
Goats	dig prot	%	*	2.1	2.4	
Horses	dig prot	%	*	2.4	2.9	

(1) dry forages and roughages (2) pasture, range plants, and forages fed green (3) silages (4) energy feeds (5) protein supplements (6) minerals (7) vitamins (8) additives

Feed Name or Analyses				Mean As Fed	Mean Dry	C.V. ± %
Rabbits	dig prot	%	*	3.0	3.5	
Sheep	dig prot	%		2.2	2.6	
Energy	GE Mcal/kg					
Cattle	DE Mcal/kg		*	1.49	1.76	
Sheep	DE Mcal/kg		*	1.72	2.03	
Cattle	ME Mcal/kg		*	1.22	1.44	
Sheep	ME Mcal/kg		*	1.41	1.66	
Cattle	TDN	%	*	34.0	40.0	
Sheep	TDN	%		39.1	46.0	

Lupine, aerial part, fresh, (2)
Ref No 2 02 691 United States

Feed Name or Analyses				Mean As Fed	Mean Dry	C.V. ± %
Dry matter		%			100.0	
Crude fiber		%				
Cattle	dig coef	%	*		56.	
Sheep	dig coef	%	*		64.	
Ether extract		%				
Cattle	dig coef	%	*		61.	
Sheep	dig coef	%	*		47.	
N-free extract		%				
Cattle	dig coef	%	*		74.	
Sheep	dig coef	%	*		76.	
Protein (N x 6.25)		%				
Cattle	dig coef	%	*		74.	
Sheep	dig coef	%	*		76.	

Lupine, aerial part, fresh, immature, (2)
Ref No 2 02 684 United States

Feed Name or Analyses				Mean As Fed	Mean Dry	C.V. ± %
Dry matter		%			100.0	
Calcium		%			1.76	23
Phosphorus		%			0.47	10
Potassium		%			3.51	19

Lupine, aerial part, fresh, mid-bloom, (2)
Ref No 2 02 686 United States

Feed Name or Analyses				Mean As Fed	Mean Dry	C.V. ± %
Dry matter		%			100.0	
Ash		%			8.9	
Crude fiber		%			18.2	
Cattle	dig coef	%	*		56.	
Sheep	dig coef	%	*		64.	
Ether extract		%			2.8	
Cattle	dig coef	%	*		61.	
Sheep	dig coef	%	*		47.	
N-free extract		%			47.3	
Cattle	dig coef	%	*		74.	
Sheep	dig coef	%	*		76.	
Protein (N x 6.25)		%			22.8	
Cattle	dig coef	%	*		74.	
Sheep	dig coef	%	*		76.	
Cattle	dig prot	%			16.9	
Goats	dig prot	%	*		17.8	
Horses	dig prot	%	*		16.9	
Rabbits	dig prot	%	*		16.3	
Sheep	dig prot	%			17.3	
Energy	GE Mcal/kg					
Cattle	DE Mcal/kg		*		2.91	
Sheep	DE Mcal/kg		*		2.99	
Cattle	ME Mcal/kg		*		2.38	
Sheep	ME Mcal/kg		*		2.45	
Cattle	TDN	%	*		65.9	
Sheep	TDN	%	*		67.9	

Lupine, aerial part, fresh, full bloom, (2)
Ref No 2 02 687 United States

Feed Name or Analyses				Mean As Fed	Mean Dry	C.V. ± %
Dry matter		%			100.0	
Crude fiber		%				
Cattle	dig coef	%	*		56.	
Sheep	dig coef	%	*		64.	
Ether extract		%				
Cattle	dig coef	%	*		61.	
Sheep	dig coef	%	*		47.	
N-free extract		%				
Cattle	dig coef	%	*		74.	
Sheep	dig coef	%	*		76.	
Protein (N x 6.25)		%				
Cattle	dig coef	%	*		74.	
Sheep	dig coef	%	*		76.	
Calcium		%			1.58	29
Phosphorus		%			0.35	30
Potassium		%			2.79	16

Lupine, aerial part, fresh, milk stage, (2)
Ref No 2 02 688 United States

Feed Name or Analyses				Mean As Fed	Mean Dry	C.V. ± %
Dry matter		%			100.0	
Calcium		%			1.09	22
Phosphorus		%			0.21	41
Potassium		%			1.89	25

Lupine, aerial part, fresh, mature, (2)
Ref No 2 02 689 United States

Feed Name or Analyses				Mean As Fed	Mean Dry	C.V. ± %
Dry matter		%		17.8	100.0	
Ash		%		1.2	6.5	28
Crude fiber		%		5.7	32.1	20
Cattle	dig coef	%	*	56.	56.	
Sheep	dig coef	%	*	64.	64.	
Ether extract		%		0.5	2.6	
Cattle	dig coef	%	*	61.	61.	
Sheep	dig coef	%	*	47.	47.	
N-free extract		%		8.4	47.3	
Cattle	dig coef	%	*	74.	74.	
Sheep	dig coef	%	*	76.	76.	
Protein (N x 6.25)		%		2.0	11.5	49
Cattle	dig coef	%	*	74.	74.	
Sheep	dig coef	%	*	76.	76.	
Cattle	dig prot	%		1.5	8.5	
Goats	dig prot	%	*	1.3	7.3	
Horses	dig prot	%	*	1.3	7.3	
Rabbits	dig prot	%	*	1.3	7.5	
Sheep	dig prot	%		1.6	8.7	
Energy	GE Mcal/kg					
Cattle	DE Mcal/kg		*	0.51	2.87	
Sheep	DE Mcal/kg		*	0.53	3.00	
Cattle	ME Mcal/kg		*	0.42	2.35	
Sheep	ME Mcal/kg		*	0.44	2.46	
Cattle	TDN	%		11.6	65.1	
Sheep	TDN	%		12.1	68.0	
Calcium		%		0.17	0.97	
Phosphorus		%		0.02	0.12	
Potassium		%		0.29	1.61	

Lupine, aerial part, fresh, over ripe, (2)
Ref No 2 02 690 United States

Feed Name or Analyses				Mean As Fed	Mean Dry	C.V. ± %
Dry matter		%			100.0	
Ash		%			7.1	18

Feed Name or Analyses		Mean As Fed	Mean Dry	C.V. ± %
Crude fiber	%		34.2	7
Protein (N x 6.25)	%		5.9	27
Cattle	dig prot % *		2.9	
Goats	dig prot % *		2.1	
Horses	dig prot % *		2.5	
Rabbits	dig prot % *		3.2	
Sheep	dig prot % *		2.5	
Calcium	%		0.92	33
Phosphorus	%		0.11	50
Potassium	%		2.20	28

Lupine, leaves, fresh, immature, (2)
Ref No 2 02 692 United States

Feed Name or Analyses		Mean As Fed	Mean Dry	C.V. ± %
Dry matter	%	13.2	100.0	
Ash	%	1.3	9.5	50
Crude fiber	%	2.9	22.2	33
Ether extract	%	0.6	4.5	46
N-free extract	%	4.7	35.7	
Protein (N x 6.25)	%	3.7	28.1	20
Cattle	dig prot % *	2.9	21.8	
Goats	dig prot % *	3.0	22.8	
Horses	dig prot % *	2.8	21.4	
Rabbits	dig prot % *	2.7	20.4	
Sheep	dig prot % *	3.1	23.2	
Energy	GE Mcal/kg			
Cattle	DE Mcal/kg *	0.40	3.01	
Sheep	DE Mcal/kg *	0.37	2.81	
Cattle	ME Mcal/kg *	0.32	2.44	
Sheep	ME Mcal/kg *	0.30	2.30	
Cattle	TDN % *	9.0	68.2	
Sheep	TDN % *	8.4	63.7	

Lupine, leaves, fresh, mid-bloom, (2)
Ref No 2 02 693 United States

Feed Name or Analyses		Mean As Fed	Mean Dry	C.V. ± %
Dry matter	%		100.0	
Ash	%		5.8	
Crude fiber	%		25.3	
Ether extract	%		3.2	
N-free extract	%		45.1	
Protein (N x 6.25)	%		20.6	
Cattle	dig prot % *		15.4	
Goats	dig prot % *		15.8	
Horses	dig prot % *		15.0	
Rabbits	dig prot % *		14.6	
Sheep	dig prot % *		16.2	
Energy	GE Mcal/kg			
Cattle	DE Mcal/kg *		2.80	
Sheep	DE Mcal/kg *		2.61	
Cattle	ME Mcal/kg *		2.30	
Sheep	ME Mcal/kg *		2.14	
Cattle	TDN % *		63.5	
Sheep	TDN % *		59.3	

Lupine, leaves, fresh, full bloom, (2)
Ref No 2 02 694 United States

Feed Name or Analyses		Mean As Fed	Mean Dry	C.V. ± %
Dry matter	%		100.0	
Ash	%		6.7	22
Crude fiber	%		29.6	5
Ether extract	%		2.9	13
N-free extract	%		42.1	
Protein (N x 6.25)	%		18.7	8
Cattle	dig prot % *		13.8	
Goats	dig prot % *		14.0	
Horses	dig prot % *		13.4	

Feed Name or Analyses		Mean As Fed	Mean Dry	C.V. ± %
Rabbits	dig prot % *		13.1	
Sheep	dig prot % *		14.4	
Energy	GE Mcal/kg			
Cattle	DE Mcal/kg *		2.66	
Sheep	DE Mcal/kg *		2.49	
Cattle	ME Mcal/kg *		2.18	
Sheep	ME Mcal/kg *		2.04	
Cattle	TDN % *		60.3	
Sheep	TDN % *		56.5	

Lupine, leaves, fresh, mature, (2)
Ref No 2 02 695 United States

Feed Name or Analyses		Mean As Fed	Mean Dry	C.V. ± %
Dry matter	%		100.0	
Ash	%		11.4	
Crude fiber	%		26.4	
Ether extract	%		2.3	
N-free extract	%		46.5	
Protein (N x 6.25)	%		13.4	
Cattle	dig prot % *		9.3	
Goats	dig prot % *		9.1	
Horses	dig prot % *		8.9	
Rabbits	dig prot % *		9.0	
Sheep	dig prot % *		9.5	
Energy	GE Mcal/kg			
Cattle	DE Mcal/kg *		2.82	
Sheep	DE Mcal/kg *		2.83	
Cattle	ME Mcal/kg *		2.31	
Sheep	ME Mcal/kg *		2.32	
Cattle	TDN % *		64.0	
Sheep	TDN % *		64.2	

Lupine, leaves, fresh, over ripe, (2)
Ref No 2 02 696 United States

Feed Name or Analyses		Mean As Fed	Mean Dry	C.V. ± %
Dry matter	%		100.0	
Ash	%		7.0	
Crude fiber	%		35.4	
Ether extract	%		2.2	
N-free extract	%		45.9	
Protein (N x 6.25)	%		9.5	
Cattle	dig prot % *		6.0	
Goats	dig prot % *		5.4	
Horses	dig prot % *		5.6	
Rabbits	dig prot % *		6.0	
Sheep	dig prot % *		5.8	
Energy	GE Mcal/kg			
Cattle	DE Mcal/kg *		2.56	
Sheep	DE Mcal/kg *		2.39	
Cattle	ME Mcal/kg *		2.10	
Sheep	ME Mcal/kg *		1.96	
Cattle	TDN % *		58.1	
Sheep	TDN % *		54.3	
Calcium	%		0.58	
Phosphorus	%		0.09	

Lupine, leaves, fresh, cut 1, (2)
Ref No 2 02 697 United States

Feed Name or Analyses		Mean As Fed	Mean Dry	C.V. ± %
Dry matter	%		100.0	
Ash	%		6.4	12
Crude fiber	%		30.1	16
Ether extract	%		2.6	17
N-free extract	%		45.9	
Protein (N x 6.25)	%		15.0	35
Cattle	dig prot % *		10.6	
Goats	dig prot % *		10.6	

Feed Name or Analyses		Mean As Fed	Mean Dry	C.V. ± %
Horses	dig prot % *		10.3	
Rabbits	dig prot % *		10.3	
Sheep	dig prot % *		11.0	
Energy	GE Mcal/kg			
Cattle	DE Mcal/kg *		2.80	
Sheep	DE Mcal/kg *		2.75	
Cattle	ME Mcal/kg *		2.30	
Sheep	ME Mcal/kg *		2.26	
Cattle	TDN % *		63.5	
Sheep	TDN % *		62.5	
Calcium	%		0.95	44
Phosphorus	%		0.18	57

Lupine, leaves w stems, fresh, (2)
Ref No 2 02 699 United States

Feed Name or Analyses		Mean As Fed	Mean Dry	C.V. ± %
Dry matter	%		100.0	
Niacin	mg/kg		27.3	
Riboflavin	mg/kg		7.5	
Thiamine	mg/kg		4.2	

Lupine, aerial part, ensiled, (3)
Ref No 3 02 704 United States

Feed Name or Analyses		Mean As Fed	Mean Dry	C.V. ± %
Dry matter	%	16.9	100.0	14
Ash	%	1.4	8.1	39
Crude fiber	%	6.2	36.7	7
Sheep	dig coef %	54.	54.	
Ether extract	%	0.6	3.5	15
Sheep	dig coef %	64.	64.	
N-free extract	%	5.7	33.6	
Sheep	dig coef %	72.	72.	
Protein (N x 6.25)	%	3.1	18.2	12
Sheep	dig coef %	73.	73.	
Cattle	dig prot % *	2.1	12.7	
Goats	dig prot % *	2.1	12.7	
Horses	dig prot % *	2.1	12.7	
Sheep	dig prot %	2.2	13.2	
Energy	GE Mcal/kg			
Cattle	DE Mcal/kg *	0.43	2.54	
Sheep	DE Mcal/kg *	0.46	2.72	
Cattle	ME Mcal/kg *	0.35	2.08	
Sheep	ME Mcal/kg *	0.38	2.23	
Cattle	TDN % *	9.8	57.7	
Sheep	TDN %	10.4	61.7	

Lupine, aerial part, ensiled, early bloom, (3)
Ref No 3 02 700 United States

Feed Name or Analyses		Mean As Fed	Mean Dry	C.V. ± %
Dry matter	%	17.5	100.0	
Ash	%	1.8	10.0	
Crude fiber	%	5.7	32.4	
Ether extract	%	0.5	2.9	
N-free extract	%	6.4	36.7	
Protein (N x 6.25)	%	3.2	18.0	
Cattle	dig prot % *	2.2	12.6	
Goats	dig prot % *	2.2	12.6	
Horses	dig prot % *	2.2	12.6	
Sheep	dig prot % *	2.2	12.6	
Energy	GE Mcal/kg			
Cattle	DE Mcal/kg *	0.41	2.36	
Sheep	DE Mcal/kg *	0.43	2.48	
Cattle	ME Mcal/kg *	0.34	1.94	
Sheep	ME Mcal/kg *	0.36	2.03	
Cattle	TDN % *	9.4	53.5	
Sheep	TDN % *	9.8	56.2	

Feed Name or Analyses				Mean As Fed	Mean Dry	C.V. ± %

Lupine, aerial part, ensiled, full bloom, (3)

Ref No 3 02 701 United States

Analyses				As Fed	Dry	C.V. ± %
Dry matter		%		15.6	100.0	10
Ash		%		2.5	16.1	18
Crude fiber		%		4.9	31.5	12
Sheep	dig coef	%		59.	59.	
Ether extract		%		0.5	3.3	10
Sheep	dig coef	%		44.	44.	
N-free extract		%		5.3	34.0	
Sheep	dig coef	%		78.	78.	
Protein (N x 6.25)		%		2.4	15.3	19
Sheep	dig coef	%		69.	69.	
Cattle	dig prot	%	*	1.6	10.1	
Goats	dig prot	%	*	1.6	10.1	
Horses	dig prot	%	*	1.6	10.1	
Sheep	dig prot	%		1.6	10.5	
Energy	GE Mcal/kg					
Cattle	DE Mcal/kg		*	0.42	2.67	
Sheep	DE Mcal/kg		*	0.40	2.59	
Cattle	ME Mcal/kg		*	0.34	2.19	
Sheep	ME Mcal/kg		*	0.33	2.13	
Cattle	TDN	%	*	9.4	60.5	
Sheep	TDN	%		9.1	58.8	

Lupine, aerial part, ensiled, milk stage, (3)

Ref No 3 02 702 United States

Analyses				As Fed	Dry	C.V. ± %
Dry matter		%		16.8	100.0	
Ash		%		1.8	11.0	
Crude fiber		%		6.7	39.9	
Sheep	dig coef	%		55.	55.	
Ether extract		%		0.5	3.0	
Sheep	dig coef	%		49.	49.	
N-free extract		%		5.5	32.7	
Sheep	dig coef	%		61.	61.	
Protein (N x 6.25)		%		2.3	13.4	
Sheep	dig coef	%		55.	55.	
Cattle	dig prot	%	*	1.4	8.4	
Goats	dig prot	%	*	1.4	8.4	
Horses	dig prot	%	*	1.4	8.4	
Sheep	dig prot	%		1.2	7.4	
Energy	GE Mcal/kg					
Cattle	DE Mcal/kg		*	0.39	2.33	
Sheep	DE Mcal/kg		*	0.39	2.32	
Cattle	ME Mcal/kg		*	0.32	1.91	
Sheep	ME Mcal/kg		*	0.32	1.90	
Cattle	TDN	%	*	8.9	52.9	
Sheep	TDN	%		8.8	52.6	

Lupine, aerial part, ensiled, dough stage, (3)

Ref No 3 02 703 United States

Analyses				As Fed	Dry	C.V. ± %
Dry matter		%		17.8	100.0	9
Ash		%		2.1	11.7	10
Crude fiber		%		7.0	39.6	4
Cattle	dig coef	%		50.	50.	
Ether extract		%		0.4	2.4	16
Cattle	dig coef	%		60.	60.	
N-free extract		%		5.8	32.4	
Cattle	dig coef	%		59.	59.	
Protein (N x 6.25)		%		2.5	14.0	10
Cattle	dig coef	%		69.	69.	
Cattle	dig prot	%		1.7	9.7	
Goats	dig prot	%	*	1.6	8.9	
Horses	dig prot	%	*	1.6	8.9	
Sheep	dig prot	%	*	1.6	8.9	
Energy	GE Mcal/kg					
Cattle	DE Mcal/kg		*	0.41	2.28	
Sheep	DE Mcal/kg		*	0.39	2.19	
Cattle	ME Mcal/kg		*	0.33	1.87	
Sheep	ME Mcal/kg		*	0.32	1.80	
Cattle	TDN	%		9.2	51.7	
Sheep	TDN	%	*	8.8	49.7	

Lupine, aerial part w AIV preservative added, ensiled, (3)

Ref No 3 02 706 United States

Analyses				As Fed	Dry	C.V. ± %
Dry matter		%		17.2	100.0	12
Ash		%		1.9	11.2	14
Crude fiber		%		6.3	36.8	6
Cattle	dig coef	%		52.	52.	
Ether extract		%		0.4	2.5	13
Cattle	dig coef	%		60.	60.	
N-free extract		%		5.9	34.1	
Cattle	dig coef	%		65.	65.	
Protein (N x 6.25)		%		2.6	15.4	8
Cattle	dig coef	%		72.	72.	
Cattle	dig prot	%		1.9	11.1	
Goats	dig prot	%	*	1.8	10.2	
Horses	dig prot	%	*	1.8	10.2	
Sheep	dig prot	%	*	1.8	10.2	
Energy	GE Mcal/kg					
Cattle	DE Mcal/kg		*	0.42	2.46	
Sheep	DE Mcal/kg		*	0.44	2.54	
Cattle	ME Mcal/kg		*	0.35	2.02	
Sheep	ME Mcal/kg		*	0.36	2.09	
Cattle	TDN	%		9.6	55.8	
Sheep	TDN	%	*	9.9	57.7	

Lupine, aerial part w AIV preservative added, ensiled, late bloom, (3)

Ref No 3 02 705 United States

Analyses				As Fed	Dry	C.V. ± %
Dry matter		%		17.5	100.0	
Ash		%		1.8	10.0	
Crude fiber		%		5.7	32.4	
Cattle	dig coef	%		54.	54.	
Ether extract		%		0.5	2.9	
Cattle	dig coef	%		59.	59.	
N-free extract		%		6.4	36.7	
Cattle	dig coef	%		71.	71.	
Protein (N x 6.25)		%		3.2	18.0	
Cattle	dig coef	%		75.	75.	
Cattle	dig prot	%		2.4	13.5	
Goats	dig prot	%	*	2.2	12.6	
Horses	dig prot	%	*	2.2	12.6	
Sheep	dig prot	%	*	2.2	12.6	
Energy	GE Mcal/kg					
Cattle	DE Mcal/kg		*	0.47	2.69	
Sheep	DE Mcal/kg		*	0.43	2.48	
Cattle	ME Mcal/kg		*	0.39	2.20	
Sheep	ME Mcal/kg		*	0.36	2.03	
Cattle	TDN	%		10.7	60.9	
Sheep	TDN	%	*	9.8	56.2	

Lupine, seeds, (5)

Ref No 5 02 707 United States

Analyses				As Fed	Dry	C.V. ± %
Dry matter		%		90.0	100.0	
Ash		%		3.9	4.4	
Crude fiber		%		14.8	16.5	
Sheep	dig coef	%		97.	97.	
Swine	dig coef	%		40.	40.	
Ether extract		%		4.5	5.0	
Sheep	dig coef	%		80.	80.	
Swine	dig coef	%		50.	50.	
N-free extract		%		28.2	31.4	
Sheep	dig coef	%		85.	85.	
Swine	dig coef	%		87.	87.	
Protein (N x 6.25)		%		38.5	42.8	
Sheep	dig coef	%		88.	88.	
Swine	dig coef	%		87.	87.	
Sheep	dig prot	%		33.9	37.7	
Swine	dig prot	%		33.5	37.2	
Energy	GE Mcal/kg					
Cattle	DE Mcal/kg		*	2.85	3.17	
Sheep	DE Mcal/kg		*	3.54	3.94	
Swine	DE kcal/kg		*	3044.	3384.	
Cattle	ME Mcal/kg		*	2.34	2.60	
Sheep	ME Mcal/kg		*	2.90	3.23	
Swine	ME kcal/kg		*	2659.	2956.	
Cattle	TDN	%	*	64.6	71.8	
Sheep	TDN	%		80.3	89.3	
Swine	TDN	%		69.0	76.8	

LUPINE, BENTHAM. Lupinus benthami

Lupine, bentham, aerial part, fresh, (2)

Ref No 2 02 713 United States

Analyses				As Fed	Dry	C.V. ± %
Dry matter		%			100.0	
Ash		%			10.4	16
Crude fiber		%			23.9	26
Protein (N x 6.25)		%			18.3	36
Cattle	dig prot	%	*		13.4	
Goats	dig prot	%	*		13.6	
Horses	dig prot	%	*		13.1	
Rabbits	dig prot	%	*		12.8	
Sheep	dig prot	%	*		14.0	
Calcium		%			1.43	29
Phosphorus		%			0.31	36
Potassium		%			3.07	17

Lupine, bentham, aerial part, fresh, immature, (2)

Ref No 2 02 708 United States

Analyses				As Fed	Dry	C.V. ± %
Dry matter		%			100.0	
Ash		%			12.7	
Crude fiber		%			18.0	
Protein (N x 6.25)		%			27.6	
Cattle	dig prot	%	*		21.4	
Goats	dig prot	%	*		22.3	
Horses	dig prot	%	*		21.0	
Rabbits	dig prot	%	*		20.0	
Sheep	dig prot	%	*		22.7	
Calcium		%			1.65	
Phosphorus		%			0.43	
Potassium		%			3.79	

(1) dry forages and roughages (2) pasture, range plants, and forages fed green (3) silages (4) energy feeds (5) protein supplements (6) minerals (7) vitamins (8) additives

Feed Name or Analyses	Mean As Fed	Mean Dry	C.V. ± %

Lupine, bentham, aerial part, fresh, early bloom, (2)
Ref No 2 02 709 United States

Dry matter	%			100.0	
Ash	%			12.5	
Crude fiber	%			18.5	
Protein (N x 6.25)	%			21.9	
Cattle	dig prot %	*		16.5	
Goats	dig prot %	*		17.0	
Horses	dig prot %	*		16.1	
Rabbits	dig prot %	*		15.6	
Sheep	dig prot %	*		17.4	
Calcium	%			1.67	
Phosphorus	%			0.42	
Potassium	%			3.87	

Lupine, bentham, aerial part, fresh, full bloom, (2)
Ref No 2 02 710 United States

Dry matter	%			100.0	
Ash	%			10.7	12
Crude fiber	%			21.3	19
Protein (N x 6.25)	%			19.0	17
Cattle	dig prot %	*		14.0	
Goats	dig prot %	*		14.3	
Horses	dig prot %	*		13.7	
Rabbits	dig prot %	*		13.3	
Sheep	dig prot %	*		14.7	
Calcium	%			1.70	16
Phosphorus	%			0.31	34
Potassium	%			2.77	15

Lupine, bentham, aerial part, fresh, milk stage, (2)
Ref No 2 02 711 United States

Dry matter	%			100.0	
Ash	%			8.5	
Crude fiber	%			26.4	
Protein (N x 6.25)	%			16.4	
Cattle	dig prot %	*		11.8	
Goats	dig prot %	*		11.9	
Horses	dig prot %	*		11.5	
Rabbits	dig prot %	*		11.3	
Sheep	dig prot %	*		12.3	
Calcium	%			1.31	
Phosphorus	%			0.28	
Potassium	%			2.42	

Lupine, bentham, aerial part, fresh, over ripe, (2)
Ref No 2 02 712 United States

Dry matter	%			100.0	
Ash	%			7.6	
Crude fiber	%			36.8	
Protein (N x 6.25)	%			6.7	
Cattle	dig prot %	*		3.6	
Goats	dig prot %	*		2.8	
Horses	dig prot %	*		3.2	
Rabbits	dig prot %	*		3.8	
Sheep	dig prot %	*		3.2	
Calcium	%			0.68	
Phosphorus	%			0.14	
Potassium	%			2.41	

LUPINE, BICOLOR. Lupinus bicolor

Lupine, bicolor, aerial part, fresh, (2)
Ref No 2 02 718 United States

Dry matter	%			100.0	
Ash	%			9.1	27
Crude fiber	%			24.9	28
Protein (N x 6.25)	%			17.1	57
Cattle	dig prot %	*		12.4	
Goats	dig prot %	*		12.5	
Horses	dig prot %	*		12.0	
Rabbits	dig prot %	*		11.9	
Sheep	dig prot %	*		12.9	
Calcium	%			1.41	28
Phosphorus	%			0.29	55
Potassium	%			2.79	36

Lupine, bicolor, aerial part, fresh, immature, (2)
Ref No 2 02 714 United States

Dry matter	%			100.0	
Ash	%			11.4	
Crude fiber	%			15.0	
Protein (N x 6.25)	%			31.7	
Cattle	dig prot %	*		24.8	
Goats	dig prot %	*		26.1	
Horses	dig prot %	*		24.4	
Rabbits	dig prot %	*		23.1	
Sheep	dig prot %	*		26.5	
Calcium	%			1.54	
Phosphorus	%			0.53	
Potassium	%			4.13	

Lupine, bicolor, aerial part, fresh, full bloom, (2)
Ref No 2 02 715 United States

Dry matter	%			100.0	
Ash	%			10.7	
Crude fiber	%			20.6	
Protein (N x 6.25)	%			24.5	
Cattle	dig prot %	*		18.7	
Goats	dig prot %	*		19.4	
Horses	dig prot %	*		18.3	
Rabbits	dig prot %	*		17.6	
Sheep	dig prot %	*		19.8	
Calcium	%			1.69	
Phosphorus	%			0.41	
Potassium	%			3.07	

Lupine, bicolor, aerial part, fresh, mature, (2)
Ref No 2 02 716 United States

Dry matter	%			100.0	
Ash	%			8.8	
Crude fiber	%			25.5	
Protein (N x 6.25)	%			12.1	
Cattle	dig prot %	*		8.2	
Goats	dig prot %	*		7.8	
Horses	dig prot %	*		7.8	
Rabbits	dig prot %	*		8.0	
Sheep	dig prot %	*		8.3	

Lupine, bicolor, aerial part, fresh, over ripe, (2)
Ref No 2 02 717 United States

Dry matter	%			100.0	
Ash	%			6.9	23
Crude fiber	%			32.9	5
Protein (N x 6.25)	%			5.5	16
Cattle	dig prot %	*		2.6	
Goats	dig prot %	*		1.7	
Horses	dig prot %	*		2.2	
Rabbits	dig prot %	*		2.9	
Sheep	dig prot %	*		2.1	
Calcium	%			1.04	26
Phosphorus	%			0.10	34
Potassium	%			2.09	36

LUPINE, BITTER. Lupinus spp

Lupine, bitter, seeds, (5)
Ref No 5 02 721 United States

Dry matter	%		84.2	100.0	
Ash	%		3.4	4.0	
Crude fiber	%		13.8	16.4	
Sheep	dig coef %		99.	99.	
Ether extract	%		4.5	5.4	
Sheep	dig coef %		84.	84.	
N-free extract	%		29.1	34.6	
Sheep	dig coef %		90.	90.	
Protein (N x 6.25)	%		33.3	39.6	
Sheep	dig coef %		91.	91.	
Sheep	dig prot %		30.3	36.0	
Energy	GE Mcal/kg				
Cattle	DE Mcal/kg	*	2.71	3.22	
Sheep	DE Mcal/kg	*	3.48	4.13	
Swine	DE kcal/kg	*	3041.	3611.	
Cattle	ME Mcal/kg	*	2.23	2.64	
Sheep	ME Mcal/kg	*	2.85	3.38	
Swine	ME kcal/kg	*	2676.	3178.	
Cattle	TDN %	*	61.6	73.1	
Sheep	TDN %		78.8	93.6	
Swine	TDN %	*	69.0	81.9	

Lupine, bitter, seeds, bitterness extd, (5)
Ref No 5 02 719 United States

Dry matter	%		86.5	100.0	
Ash	%		2.7	3.1	
Crude fiber	%		17.3	20.0	
Sheep	dig coef %		91.	91.	
Ether extract	%		3.8	4.4	
Sheep	dig coef %		78.	78.	
N-free extract	%		27.1	31.3	
Sheep	dig coef %		86.	86.	
Protein (N x 6.25)	%		35.6	41.2	
Sheep	dig coef %		87.	87.	
Sheep	dig prot %		31.0	35.8	
Energy	GE Mcal/kg				
Cattle	DE Mcal/kg	*	2.63	3.04	
Sheep	DE Mcal/kg	*	3.38	3.91	
Swine	DE kcal/kg	*	2928.	3386.	
Cattle	ME Mcal/kg	*	2.15	2.49	
Sheep	ME Mcal/kg	*	2.77	3.21	
Swine	ME kcal/kg	*	2568.	2968.	
Cattle	TDN %	*	59.5	68.8	

Continued

Feed Name or Analyses			Mean As Fed	Mean Dry	C.V. ± %
Sheep	TDN %		76.7	88.7	
Swine	TDN %	*	66.4	76.8	

LUPINE, BLUE. Lupinus angustifolius

Lupine, blue, hay, s-c, immature, (1)

Ref No 1 02 722 United States

Feed Name or Analyses			Mean As Fed	Mean Dry	C.V. ± %
Dry matter	%		88.7	100.0	
Ash	%		3.3	3.7	
Crude fiber	%		17.8	20.1	
Ether extract	%		4.6	5.2	
N-free extract	%		34.1	38.5	
Protein (N x 6.25)	%		28.8	32.5	
Cattle	dig prot %	*	22.3	25.1	
Goats	dig prot %	*	23.8	26.9	
Horses	dig prot %	*	22.3	25.1	
Rabbits	dig prot %	*	21.1	23.8	
Sheep	dig prot %	*	22.8	25.7	
Calcium	%		0.52	0.59	
Magnesium	%		0.33	0.37	
Phosphorus	%		0.42	0.47	

Lupine, blue, leaves, fresh, immature, (2)

Ref No 2 02 723 United States

Feed Name or Analyses			Mean As Fed	Mean Dry	C.V. ± %
Dry matter	%		8.8	100.0	
Ash	%		1.3	15.3	
Crude fiber	%		1.9	21.9	
Ether extract	%		0.2	2.2	
N-free extract	%		4.0	45.1	
Protein (N x 6.25)	%		1.4	15.5	
Cattle	dig prot %	*	1.0	11.1	
Goats	dig prot %	*	1.0	11.0	
Horses	dig prot %	*	0.9	10.7	
Rabbits	dig prot %	*	0.9	10.6	
Sheep	dig prot %	*	1.0	11.4	
Energy	GE Mcal/kg				
Cattle	DE Mcal/kg	*	0.23	2.64	
Sheep	DE Mcal/kg	*	0.25	2.81	
Cattle	ME Mcal/kg	*	0.19	2.16	
Sheep	ME Mcal/kg	*	0.20	2.30	
Cattle	TDN %	*	5.3	59.9	
Sheep	TDN %	*	5.6	63.7	
Calcium	%		0.10	1.18	
Magnesium	%		0.05	0.54	
Phosphorus	%		0.05	0.60	

Lupine, blue, leaves, fresh, mature, (2)

Ref No 2 02 724 United States

Feed Name or Analyses			Mean As Fed	Mean Dry	C.V. ± %
Dry matter	%		13.2	100.0	
Ash	%		1.7	12.7	
Crude fiber	%		1.8	13.8	
Ether extract	%		0.8	5.8	
N-free extract	%		4.7	35.9	
Protein (N x 6.25)	%		4.2	31.8	
Cattle	dig prot %	*	3.3	24.9	
Goats	dig prot %	*	3.5	26.2	
Horses	dig prot %	*	3.2	24.5	
Rabbits	dig prot %	*	3.1	23.2	
Sheep	dig prot %	*	3.5	26.6	

(1) dry forages and roughages (2) pasture, range plants, and forages fed green (3) silages (4) energy feeds (5) protein supplements (6) minerals (7) vitamins (8) additives

Feed Name or Analyses			Mean As Fed	Mean Dry	C.V. ± %
Calcium	%		0.11	0.86	
Magnesium	%		0.05	0.37	
Phosphorus	%		0.06	0.43	

LUPINE, LODGEPOLE. Lupinus parviflorus

Lupine, lodgepole, aerial part, fresh, (2)

Ref No 2 02 729 United States

Feed Name or Analyses			Mean As Fed	Mean Dry	C.V. ± %
Dry matter	%			100.0	
Ash	%			6.2	16
Crude fiber	%			27.0	15
Protein (N x 6.25)	%			19.4	52
Cattle	dig prot %	*		14.4	
Goats	dig prot %	*		14.7	
Horses	dig prot %	*		14.0	
Rabbits	dig prot %	*		13.6	
Sheep	dig prot %	*		15.1	

Lupine, lodgepole, aerial part, fresh, immature, (2)

Ref No 2 02 725 United States

Feed Name or Analyses			Mean As Fed	Mean Dry	C.V. ± %
Dry matter	%			100.0	
Ash	%			7.9	
Crude fiber	%			26.7	
Protein (N x 6.25)	%			30.5	
Cattle	dig prot %	*		23.8	
Goats	dig prot %	*		25.0	
Horses	dig prot %	*		23.4	
Rabbits	dig prot %	*		22.2	
Sheep	dig prot %	*		25.4	

Lupine, lodgepole, aerial part, fresh, full bloom, (2)

Ref No 2 02 726 United States

Feed Name or Analyses			Mean As Fed	Mean Dry	C.V. ± %
Dry matter	%			100.0	
Ash	%			6.2	
Crude fiber	%			23.4	
Protein (N x 6.25)	%			19.5	
Cattle	dig prot %	*		14.5	
Goats	dig prot %	*		14.8	
Horses	dig prot %	*		14.1	
Rabbits	dig prot %	*		13.7	
Sheep	dig prot %	*		15.2	

Lupine, lodgepole, aerial part, fresh, milk stage, (2)

Ref No 2 02 727 United States

Feed Name or Analyses			Mean As Fed	Mean Dry	C.V. ± %
Dry matter	%			100.0	
Ash	%			5.6	
Crude fiber	%			26.1	
Protein (N x 6.25)	%			19.3	
Cattle	dig prot %	*		14.3	
Goats	dig prot %	*		14.6	
Horses	dig prot %	*		13.9	
Rabbits	dig prot %	*		13.6	
Sheep	dig prot %	*		15.0	

Lupine, lodgepole, aerial part, fresh, mature, (2)

Ref No 2 02 728 United States

Feed Name or Analyses			Mean As Fed	Mean Dry	C.V. ± %
Dry matter	%			100.0	
Ash	%			5.4	
Crude fiber	%			32.1	

Feed Name or Analyses			Mean As Fed	Mean Dry	C.V. ± %
Protein (N x 6.25)	%			8.6	
Cattle	dig prot %	*		5.2	
Goats	dig prot %	*		4.6	
Horses	dig prot %	*		4.8	
Rabbits	dig prot %	*		5.3	
Sheep	dig prot %	*		5.0	

LUPINE, SWEET. Lupinus albus

Lupine, sweet, hay, dehy, (1)

Ref No 1 02 731 United States

Feed Name or Analyses			Mean As Fed	Mean Dry	C.V. ± %
Dry matter	%		91.6	100.0	1
Ash	%		10.4	11.3	5
Crude fiber	%		27.1	29.6	7
Sheep	dig coef %		59.	59.	
Ether extract	%		3.1	3.4	5
Sheep	dig coef %		72.	72.	
N-free extract	%		32.1	35.1	
Sheep	dig coef %		77.	77.	
Protein (N x 6.25)	%		19.0	20.8	5
Sheep	dig coef %		69.	69.	
Cattle	dig prot %	*	13.7	14.9	
Goats	dig prot %	*	14.6	15.9	
Horses	dig prot %	*	13.9	15.1	
Rabbits	dig prot %	*	13.5	14.7	
Sheep	dig prot %		13.1	14.3	
Energy	GE Mcal/kg				
Cattle	DE Mcal/kg	*	2.50	2.73	
Sheep	DE Mcal/kg	*	2.59	2.83	
Cattle	ME Mcal/kg	*	2.05	2.23	
Sheep	ME Mcal/kg	*	2.13	2.32	
Cattle	TDN %	*	56.6	61.8	
Sheep	TDN %		58.8	64.2	

Lupine, sweet, hay, dehy, early bloom, (1)

Ref No 1 02 730 United States

Feed Name or Analyses			Mean As Fed	Mean Dry	C.V. ± %
Dry matter	%		91.2	100.0	
Ash	%		9.7	10.6	
Crude fiber	%		31.2	34.2	
Sheep	dig coef %		61.	61.	
Ether extract	%		2.9	3.2	
Sheep	dig coef %		69.	69.	
N-free extract	%		30.3	33.2	
Sheep	dig coef %		77.	77.	
Protein (N x 6.25)	%		17.1	18.8	
Sheep	dig coef %		68.	68.	
Cattle	dig prot %	*	12.1	13.2	
Goats	dig prot %	*	12.9	14.1	
Horses	dig prot %	*	12.3	13.5	
Rabbits	dig prot %	*	12.0	13.2	
Sheep	dig prot %		11.7	12.8	
Energy	GE Mcal/kg				
Cattle	DE Mcal/kg	*	2.35	2.58	
Sheep	DE Mcal/kg	*	2.58	2.83	
Cattle	ME Mcal/kg	*	1.93	2.11	
Sheep	ME Mcal/kg	*	2.12	2.32	
Cattle	TDN %	*	53.3	58.4	
Sheep	TDN %		58.5	64.2	

Lupine, sweet, hay, s-c, (1)

Ref No 1 02 734 United States

Feed Name or Analyses			Mean As Fed	Mean Dry	C.V. ± %
Dry matter	%		88.4	100.0	2
Ash	%		8.0	9.1	13

Feed Name or Analyses			Mean As Fed	Mean Dry	C.V. ± %
Crude fiber	%		29.5	33.4	14
Sheep	dig coef %		58.	58.	
Ether extract	%		2.9	3.3	17
Sheep	dig coef %		40.	40.	
N-free extract	%		31.5	35.6	
Sheep	dig coef %		69.	69.	
Protein (N x 6.25)	%		16.5	18.7	9
Sheep	dig coef %		71.	71.	
Cattle	dig prot %	*	11.6	13.1	
Goats	dig prot %	*	12.4	14.0	
Horses	dig prot %	*	11.8	13.4	
Rabbits	dig prot %	*	11.6	13.1	
Sheep	dig prot %		11.7	13.3	
Energy	GE Mcal/kg				
Cattle	DE Mcal/kg	*	2.24	2.54	
Sheep	DE Mcal/kg	*	2.34	2.65	
Cattle	ME Mcal/kg	*	1.84	2.08	
Sheep	ME Mcal/kg	*	1.92	2.17	
Cattle	TDN %	*	50.8	57.5	
Sheep	TDN %		53.1	60.2	

Lupine, sweet, hay, s-c, full bloom, (1)

Ref No 1 02 732 United States

Feed Name or Analyses			Mean As Fed	Mean Dry	C.V. ± %
Dry matter	%		85.7	100.0	
Ash	%		6.2	7.2	
Crude fiber	%		33.2	38.7	
Sheep	dig coef %		57.	57.	
Ether extract	%		1.6	1.9	
Sheep	dig coef %		32.	32.	
N-free extract	%		30.6	35.7	
Sheep	dig coef %		68.	68.	
Protein (N x 6.25)	%		14.1	16.5	
Sheep	dig coef %		80.	80.	
Cattle	dig prot %	*	9.6	11.2	
Goats	dig prot %	*	10.2	12.0	
Horses	dig prot %	*	9.9	11.5	
Rabbits	dig prot %	*	9.8	11.4	
Sheep	dig prot %		11.3	13.2	
Energy	GE Mcal/kg				
Cattle	DE Mcal/kg	*	2.18	2.54	
Sheep	DE Mcal/kg	*	2.30	2.69	
Cattle	ME Mcal/kg	*	1.85	2.16	
Sheep	ME Mcal/kg	*	1.89	2.20	
Cattle	TDN %	*	49.3	57.5	
Sheep	TDN %		52.2	60.9	

Lupine, sweet, hay, s-c, milk stage, (1)

Ref No 1 02 733 United States

Feed Name or Analyses			Mean As Fed	Mean Dry	C.V. ± %
Dry matter	%		85.7	100.0	
Ash	%		6.3	7.4	
Crude fiber	%		36.9	43.0	
Sheep	dig coef %		51.	51.	
Ether extract	%		1.5	1.8	
Sheep	dig coef %		12.	12.	
N-free extract	%		27.2	31.7	
Sheep	dig coef %		64.	64.	
Protein (N x 6.25)	%		13.8	16.1	
Sheep	dig coef %		72.	72.	
Cattle	dig prot %	*	9.3	10.9	
Goats	dig prot %	*	9.9	11.6	
Horses	dig prot %	*	9.6	11.2	
Rabbits	dig prot %	*	9.5	11.1	
Sheep	dig prot %		9.9	11.6	
Energy	GE Mcal/kg				
Cattle	DE Mcal/kg	*	1.85	2.16	
Sheep	DE Mcal/kg	*	2.05	2.39	

Feed Name or Analyses			Mean As Fed	Mean Dry	C.V. ± %
Cattle	ME Mcal/kg	*	1.52	1.77	
Sheep	ME Mcal/kg	*	1.68	1.96	
Cattle	TDN %	*	42.0	49.0	
Sheep	TDN %		46.5	54.3	

Lupine, sweet, pods, s-c, (1)

Ref No 1 02 735 United States

Feed Name or Analyses			Mean As Fed	Mean Dry	C.V. ± %
Dry matter	%		89.7	100.0	
Ash	%		2.8	3.1	
Crude fiber	%		40.2	44.8	
Sheep	dig coef %		93.	93.	
Ether extract	%		1.8	2.0	
Sheep	dig coef %		80.	80.	
N-free extract	%		30.9	34.4	
Sheep	dig coef %		85.	85.	
Protein (N x 6.25)	%		14.1	15.7	
Sheep	dig coef %		75.	75.	
Cattle	dig prot %	*	9.5	10.5	
Goats	dig prot %	*	10.1	11.2	
Horses	dig prot %	*	9.7	10.9	
Rabbits	dig prot %	*	9.7	10.8	
Sheep	dig prot %		10.6	11.8	
Energy	GE Mcal/kg				
Cattle	DE Mcal/kg	*	2.71	3.02	
Sheep	DE Mcal/kg	*	3.41	3.80	
Cattle	ME Mcal/kg	*	2.22	2.48	
Sheep	ME Mcal/kg	*	2.80	3.12	
Cattle	TDN %	*	70.3	78.4	
Sheep	TDN %		77.4	86.3	

Lupine, sweet, aerial part, fresh, (2)

Ref No 2 02 740 United States

Feed Name or Analyses			Mean As Fed	Mean Dry	C.V. ± %
Dry matter	%		13.4	100.0	17
Ash	%		1.3	9.8	23
Crude fiber	%		4.2	31.2	14
Cattle	dig coef %		56.	56.	
Sheep	dig coef %		64.	64.	
Swine	dig coef %		58.	58.	
Ether extract	%		0.4	2.8	15
Cattle	dig coef %		61.	61.	
Sheep	dig coef %		47.	47.	
Swine	dig coef %		4.	4.	
N-free extract	%		5.3	39.3	6
Cattle	dig coef %		74.	74.	
Sheep	dig coef %		76.	76.	
Swine	dig coef %		71.	71.	
Protein (N x 6.25)	%		2.3	16.9	10
Cattle	dig coef %		74.	74.	
Sheep	dig coef %		76.	76.	
Swine	dig coef %		68.	68.	
Cattle	dig prot %		1.7	12.5	
Goats	dig prot %	*	1.7	12.4	
Horses	dig prot %	*	1.6	11.9	
Rabbits	dig prot %	*	1.6	11.8	
Sheep	dig prot %		1.7	12.9	
Swine	dig prot %		1.5	11.5	
Energy	GE Mcal/kg				
Cattle	DE Mcal/kg	*	0.37	2.77	
Sheep	DE Mcal/kg	*	0.39	2.89	
Swine	DE kcal/kg	*	342.	2547.	
Cattle	ME Mcal/kg	*	0.31	2.27	
Sheep	ME Mcal/kg	*	0.32	2.37	
Swine	ME kcal/kg	*	316.	2358.	
Cattle	TDN %		8.4	62.9	
Sheep	TDN %		8.8	65.7	
Swine	TDN %		7.8	57.8	

Feed Name or Analyses			Mean As Fed	Mean Dry	C.V. ± %
Phosphorus	%		0.03	0.23	
Potassium	%		0.32	2.41	

Lupine, sweet, aerial part, fresh, full bloom, (2)

Ref No 2 02 736 United States

Feed Name or Analyses			Mean As Fed	Mean Dry	C.V. ± %
Dry matter	%		12.8	100.0	3
Ash	%		1.0	7.9	6
Crude fiber	%		4.4	34.4	8
Swine	dig coef %		54.	54.	
Ether extract	%		0.4	2.8	9
Swine	dig coef %		2.	2.	
N-free extract	%		5.1	40.3	
Swine	dig coef %		75.	75.	
Protein (N x 6.25)	%		1.9	14.7	4
Swine	dig coef %		65.	65.	
Cattle	dig prot %	*	1.3	10.4	
Goats	dig prot %	*	1.3	10.3	
Horses	dig prot %	*	1.3	10.0	
Rabbits	dig prot %	*	1.3	10.0	
Sheep	dig prot %	*	1.4	10.7	
Swine	dig prot %		1.2	9.6	
Energy	GE Mcal/kg				
Cattle	DE Mcal/kg	*	0.35	2.72	
Sheep	DE Mcal/kg	*	0.34	2.65	
Swine	DE kcal/kg	*	329.	2577.	
Cattle	ME Mcal/kg	*	0.29	2.23	
Sheep	ME Mcal/kg	*	0.28	2.17	
Swine	ME kcal/kg	*	306.	2397.	
Cattle	TDN %	*	7.9	61.6	
Sheep	TDN %	*	7.7	60.1	
Swine	TDN %		7.5	58.4	

Lupine, sweet, aerial part, fresh, late bloom, (2)

Ref No 2 02 737 United States

Feed Name or Analyses			Mean As Fed	Mean Dry	C.V. ± %
Dry matter	%		11.4	100.0	
Ash	%		1.4	12.5	
Crude fiber	%		3.5	30.5	
Cattle	dig coef %		56.	56.	
Sheep	dig coef %		65.	65.	
Ether extract	%		0.3	2.7	
Cattle	dig coef %		57.	57.	
Sheep	dig coef %		54.	54.	
N-free extract	%		4.4	38.2	
Cattle	dig coef %		74.	74.	
Sheep	dig coef %		82.	82.	
Protein (N x 6.25)	%		1.8	16.1	
Cattle	dig coef %		74.	74.	
Sheep	dig coef %		74.	74.	
Cattle	dig prot %		1.4	11.9	
Goats	dig prot %	*	1.3	11.6	
Horses	dig prot %	*	1.3	11.2	
Rabbits	dig prot %	*	1.3	11.1	
Sheep	dig prot %		1.4	11.9	
Energy	GE Mcal/kg				
Cattle	DE Mcal/kg	*	0.31	2.68	
Sheep	DE Mcal/kg	*	0.33	2.93	
Cattle	ME Mcal/kg	*	0.25	2.20	
Sheep	ME Mcal/kg	*	0.27	2.40	
Cattle	TDN %		6.9	60.7	
Sheep	TDN %		7.6	66.3	

Feed Name or Analyses		Mean As Fed	Mean Dry	C.V. ± %

Lupine, sweet, aerial part, fresh, milk stage, (2)
Ref No 2 02 738 United States

Feed Name or Analyses			As Fed	Dry	C.V. ± %
Dry matter	%		14.2	100.0	
Ash	%		0.8	5.9	
Crude fiber	%		5.3	37.1	
Sheep	dig coef %		74.	74.	
Swine	dig coef %		61.	61.	
Ether extract	%		0.3	2.2	
Sheep	dig coef %		17.	17.	
Swine	dig coef %		7.	7.	
N-free extract	%		5.5	39.0	
Sheep	dig coef %		67.	67.	
Swine	dig coef %		66.	66.	
Protein (N x 6.25)	%		2.3	15.9	
Sheep	dig coef %		80.	80.	
Swine	dig coef %		70.	70.	
Cattle	dig prot %	*	1.6	11.4	
Goats	dig prot %	*	1.6	11.4	
Horses	dig prot %	*	1.6	11.0	
Rabbits	dig prot %	*	1.6	10.9	
Sheep	dig prot %		1.8	12.7	
Swine	dig prot %		1.6	11.1	
Energy	GE Mcal/kg				
Cattle	DE Mcal/kg	*	0.43	3.03	
Sheep	DE Mcal/kg	*	0.42	2.96	
Swine	DE kcal/kg	*	374.	2637.	
Cattle	ME Mcal/kg	*	0.35	2.48	
Sheep	ME Mcal/kg	*	0.34	2.43	
Swine	ME kcal/kg	*	347.	2447.	
Cattle	TDN %	*	9.6	68.7	
Sheep	TDN %		9.5	67.1	
Swine	TDN %		8.5	59.8	

Lupine, sweet, aerial part, fresh, mature, (2)
Ref No 2 02 739 United States

Feed Name or Analyses			As Fed	Dry	C.V. ± %
Dry matter	%		17.8	100.0	
Ash	%		1.2	6.5	
Crude fiber	%		6.9	38.5	
Sheep	dig coef %		63.	63.	
Ether extract	%		0.5	2.6	
Sheep	dig coef %		41.	41.	
N-free extract	%		6.4	35.8	
Sheep	dig coef %		69.	69.	
Protein (N x 6.25)	%		3.0	16.6	
Sheep	dig coef %		81.	81.	
Cattle	dig prot %	*	2.1	12.0	
Goats	dig prot %	*	2.1	12.0	
Horses	dig prot %	*	2.1	11.6	
Rabbits	dig prot %	*	2.0	11.5	
Sheep	dig prot %		2.4	13.4	
Energy	GE Mcal/kg				
Cattle	DE Mcal/kg	*	0.53	2.95	
Sheep	DE Mcal/kg	*	0.51	2.86	
Cattle	ME Mcal/kg	*	0.43	2.42	
Sheep	ME Mcal/kg	*	0.42	2.34	
Cattle	TDN %	*	11.9	66.8	
Sheep	TDN %		11.5	64.8	

(1) dry forages and roughages
(2) pasture, range plants, and forages fed green
(3) silages
(4) energy feeds
(5) protein supplements
(6) minerals
(7) vitamins
(8) additives

Lupine, sweet, aerial part, ensiled, (3)
Ref No 3 02 744 United States

Feed Name or Analyses			As Fed	Dry	C.V. ± %
Dry matter	%		15.8	100.0	19
Ash	%		1.6	10.1	24
Crude fiber	%		5.7	36.3	5
Sheep	dig coef %		62.	62.	
Ether extract	%		0.5	3.3	8
Sheep	dig coef %		52.	52.	
N-free extract	%		5.2	33.1	
Sheep	dig coef %		70.	70.	
Protein (N x 6.25)	%		2.7	17.3	15
Sheep	dig coef %		73.	73.	
Cattle	dig prot %	*	1.9	11.9	
Goats	dig prot %	*	1.9	11.9	
Horses	dig prot %	*	1.9	11.9	
Sheep	dig prot %		2.0	12.6	
Energy	GE Mcal/kg				
Cattle	DE Mcal/kg	*	0.50	2.66	
Sheep	DE Mcal/kg	*	0.43	2.74	
Cattle	ME Mcal/kg	*	0.34	2.18	
Sheep	ME Mcal/kg	*	0.35	2.24	
Cattle	TDN %	*	9.5	60.4	
Sheep	TDN %		9.8	62.1	

Lupine, sweet, aerial part, ensiled, full bloom, (3)
Ref No 3 02 742 United States

Feed Name or Analyses			As Fed	Dry	C.V. ± %
Dry matter	%		12.1	100.0	6
Ash	%		1.4	11.8	29
Crude fiber	%		4.2	35.2	5
Sheep	dig coef %		67.	67.	
Ether extract	%		0.4	3.4	11
Sheep	dig coef %		51.	51.	
N-free extract	%		3.9	32.2	
Sheep	dig coef %		75.	75.	
Protein (N x 6.25)	%		2.1	17.5	20
Sheep	dig coef %		69.	69.	
Cattle	dig prot %	*	1.5	12.1	
Goats	dig prot %	*	1.5	12.1	
Horses	dig prot %	*	1.5	12.1	
Sheep	dig prot %		1.5	12.1	
Energy	GE Mcal/kg				
Cattle	DE Mcal/kg	*	0.29	2.40	
Sheep	DE Mcal/kg	*	0.34	2.81	
Cattle	ME Mcal/kg	*	0.24	1.96	
Sheep	ME Mcal/kg	*	0.28	2.30	
Cattle	TDN %	*	6.5	54.3	
Sheep	TDN %		7.7	63.6	

Lupine, sweet, aerial part, ensiled, dough stage, (3)
Ref No 3 02 743 United States

Feed Name or Analyses			As Fed	Dry	C.V. ± %
Dry matter	%		19.7	100.0	7
Ash	%		1.7	8.9	5
Crude fiber	%		7.3	37.0	3
Sheep	dig coef %		56.	56.	
Ether extract	%		0.6	3.3	5
Sheep	dig coef %		53.	53.	
N-free extract	%		6.6	33.3	
Sheep	dig coef %		65.	65.	
Protein (N x 6.25)	%		3.5	17.6	6
Sheep	dig coef %		78.	78.	
Cattle	dig prot %	*	2.4	12.2	
Goats	dig prot %	*	2.4	12.2	
Horses	dig prot %	*	2.4	12.2	
Sheep	dig prot %		2.7	13.7	

Feed Name or Analyses			As Fed	Dry	C.V. ± %
Energy	GE Mcal/kg				
Cattle	DE Mcal/kg	*	0.51	2.58	
Sheep	DE Mcal/kg	*	0.52	2.64	
Cattle	ME Mcal/kg	*	0.42	2.12	
Sheep	ME Mcal/kg	*	0.43	2.17	
Cattle	TDN %	*	11.5	58.6	
Sheep	TDN %		11.8	60.0	

LUPINE, SWEET YELLOW. Lupinus spp

Lupine, sweet yellow, seeds, (5)
Ref No 5 08 458 United States

Feed Name or Analyses			As Fed	Dry	C.V. ± %
Dry matter	%		88.9	100.0	
Ash	%		4.5	5.1	
Crude fiber	%		14.0	15.7	
Ether extract	%		4.9	5.5	
N-free extract	%		25.7	28.9	
Protein (N x 6.25)	%		39.8	44.8	
Energy	GE Mcal/kg				
Cattle	DE Mcal/kg	*	3.22	3.62	
Sheep	DE Mcal/kg	*	3.35	3.77	
Swine	DE kcal/kg	*	3492.	3928.	
Cattle	ME Mcal/kg	*	2.64	2.97	
Sheep	ME Mcal/kg	*	2.75	3.09	
Swine	ME kcal/kg	*	3038.	3417.	
Cattle	TDN %	*	73.0	82.1	
Sheep	TDN %	*	26.1	85.6	
Swine	TDN %	*	79.2	89.1	
Calcium	%		0.23	0.26	
Phosphorus	%		0.39	0.44	
Potassium	%		0.81	0.91	

LUPINE, TAILCUP. Lupinus caudatus

Lupine, tailcup, aerial part, fresh, (2)
Ref No 2 02 746 United States

Feed Name or Analyses			As Fed	Dry	C.V. ± %
Dry matter	%			100.0	
Ash	%			8.5	
Ether extract	%			2.2	
Protein (N x 6.25)	%			12.4	
Cattle	dig prot %	*		8.4	
Goats	dig prot %	*		8.1	
Horses	dig prot %	*		8.1	
Rabbits	dig prot %	*		8.2	
Sheep	dig prot %	*		8.5	
Calcium	%			2.54	
Phosphorus	%			0.44	

Lupine, tailcup, aerial part, fresh, early leaf, (2)
Ref No 2 08 831 United States

Feed Name or Analyses			As Fed	Dry	C.V. ± %
Dry matter	%			100.0	
Crude fiber	%			26.0	
Ether extract	%			3.0	
N-free extract	%			48.0	
Protein (N x 6.25)	%			23.0	
Cattle	dig prot %	*		17.4	
Goats	dig prot %	*		18.0	
Horses	dig prot %	*		17.1	
Rabbits	dig prot %	*		16.4	
Sheep	dig prot %	*		18.4	

Feed Name or Analyses			Mean As Fed	Mean Dry	C.V. ± %

Lupine, tailcup, aerial part, fresh, midbloom, (2)
Ref No 2 08 832 United States

Dry matter	%			100.0	
Crude fiber	%			23.0	
Ether extract	%			2.3	
N-free extract	%			41.0	
Protein (N x 6.25)	%			17.0	
Cattle	dig prot %	*		12.3	
Goats	dig prot %	*		12.4	
Horses	dig prot %	*		12.0	
Rabbits	dig prot %	*		11.8	
Sheep	dig prot %	*		12.8	
Energy	GE Mcal/kg				
Cattle	DE Mcal/kg	*		2.62	
Sheep	DE Mcal/kg	*		2.71	
Cattle	ME Mcal/kg	*		2.15	
Sheep	ME Mcal/kg	*		2.22	
Cattle	TDN %	*		59.5	
Sheep	TDN %	*		61.5	

Lupine, tailcup, aerial part, fresh, dough stage, (2)
Ref No 2 08 833 United States

Dry matter	%			100.0	
Crude fiber	%			20.0	
Ether extract	%			2.5	
N-free extract	%			49.0	
Protein (N x 6.25)	%			13.0	
Cattle	dig prot %	*		8.9	
Goats	dig prot %	*		8.7	
Horses	dig prot %	*		8.6	
Rabbits	dig prot %	*		8.7	
Sheep	dig prot %	*		9.1	

Lupine, tailcup, aerial part, fresh, mature (2)
Ref No 2 08 834 United States

Dry matter	%			100.0	
Crude fiber	%			26.0	
Ether extract	%			2.0	
N-free extract	%			42.0	
Protein (N x 6.25)	%			17.0	
Cattle	dig prot %	*		12.3	
Goats	dig prot %	*		12.4	
Horses	dig prot %	*		12.0	
Rabbits	dig prot %	*		11.8	
Sheep	dig prot %	*		12.8	
Energy	GE Mcal/kg				
Cattle	DE Mcal/kg	*		2.91	
Sheep	DE Mcal/kg	*		2.70	
Cattle	ME Mcal/kg	*		2.39	
Sheep	ME Mcal/kg	*		2.21	
Cattle	TDN %	*		66.0	
Sheep	TDN %	*		61.2	

Lupine, tailcup, aerial part, fresh, over ripe, (2)
Ref No 2 08 835 United States

Dry matter	%			100.0	
Crude fiber	%			34.5	
Ether extract	%			2.0	
N-free extract	%			46.0	
Protein (N x 6.25)	%			9.5	
Cattle	dig prot %	*		6.0	
Goats	dig prot %	*		5.4	
Horses	dig prot %	*		5.6	
Rabbits	dig prot %	*		6.0	
Sheep	dig prot %	*		5.8	

Lupine, tailcup, leaves, fresh, (2)
Ref No 2 02 749 United States

Dry matter	%			100.0	
Ash	%			7.3	23
Crude fiber	%			30.2	5
Ether extract	%			3.1	12
N-free extract	%			39.5	
Protein (N x 6.25)	%			19.9	19
Cattle	dig prot %	*		14.8	
Goats	dig prot %	*		15.1	
Horses	dig prot %	*		14.4	
Rabbits	dig prot %	*		14.0	
Sheep	dig prot %	*		15.5	
Calcium	%			1.89	55
Phosphorus	%			0.23	46

Lupine, tailcup, leaves, fresh, immature, (2)
Ref No 2 02 747 United States

Dry matter	%			100.0	
Ash	%			6.4	
Crude fiber	%			29.7	
Ether extract	%			3.3	
N-free extract	%			36.2	
Protein (N x 6.25)	%			24.4	
Cattle	dig prot %	*		18.6	
Goats	dig prot %	*		19.3	
Horses	dig prot %	*		18.2	
Rabbits	dig prot %	*		17.5	
Sheep	dig prot %	*		19.7	
Calcium	%			3.28	
Phosphorus	%			0.37	

Lupine, tailcup, leaves, fresh, full bloom, (2)
Ref No 2 02 748 United States

Dry matter	%			100.0	
Ash	%			7.0	22
Crude fiber	%			30.9	7
Ether extract	%			3.1	5
N-free extract	%			40.2	
Protein (N x 6.25)	%			18.8	9
Cattle	dig prot %	*		13.9	
Goats	dig prot %	*		14.1	
Horses	dig prot %	*		13.5	
Rabbits	dig prot %	*		13.2	
Sheep	dig prot %	*		14.5	
Calcium	%			1.57	
Phosphorus	%			0.20	

Lupine, tailcup, stems, fresh, (2)
Ref No 2 02 750 United States

Dry matter	%			100.0	
Ash	%			5.0	
Ether extract	%			1.3	
Protein (N x 6.25)	%			6.3	
Cattle	dig prot %	*		3.2	
Goats	dig prot %	*		2.4	
Horses	dig prot %	*		2.9	
Rabbits	dig prot %	*		3.5	
Sheep	dig prot %	*		2.9	
Calcium	%			1.00	
Phosphorus	%			0.10	

LUPINE, TEXAS. Lupinus subcarnosus

Lupine, Texas, aerial part, fresh, (2)
Ref No 2 02 751 United States

Dry matter	%			100.0	
Calcium	%			2.57	
Magnesium	%			0.29	
Phosphorus	%			0.24	
Potassium	%			1.71	

Lupine, Texas, leaves, fresh, (2)
Ref No 2 02 752 United States

Dry matter	%			100.0	
Ash	%			9.3	
Crude fiber	%			22.5	
Ether extract	%			2.3	
N-free extract	%			49.7	
Protein (N x 6.25)	%			16.2	
Cattle	dig prot %	*		11.7	
Goats	dig prot %	*		11.7	
Horses	dig prot %	*		11.3	
Rabbits	dig prot %	*		11.2	
Sheep	dig prot %	*		12.1	

LUPINE, YELLOW. Lupinus luteus

Lupine, yellow, hay, s-c, early leaf, (1)
Ref No 1 02 753 United States

Dry matter	%		89.4	100.0	
Ash	%		4.5	5.0	
Crude fiber	%		17.5	19.6	
Ether extract	%		4.6	5.2	
N-free extract	%		30.9	34.6	
Protein (N x 6.25)	%		31.8	35.6	
Cattle	dig prot %	*	24.8	27.8	
Goats	dig prot %	*	26.6	29.8	
Horses	dig prot %	*	24.8	27.8	
Rabbits	dig prot %	*	23.4	26.2	
Sheep	dig prot %	*	25.5	28.5	
Calcium	%		0.57	0.64	
Magnesium	%		0.34	0.38	
Phosphorus	%		0.73	0.82	

Lychees –
see Litchi, meats, dehy, (4)

LYSILOMA. Lysiloma desmostachya

Lysiloma, seeds w hulls, (5)
Ref No 5 09 021 United States

Dry matter	%			100.0	
Protein (N x 6.25)	%			41.0	
Alanine	%			1.07	
Arginine	%			1.56	
Aspartic acid	%			2.54	
Glutamic acid	%			3.53	

Continued

Feed Name or Analyses			Mean As Fed	Mean Dry	C.V. ± %
Glycine	%			1.07	
Hydroxyproline	%			0.08	
Isoleucine	%			0.98	
Leucine	%			2.01	
Lysine	%			1.80	
Methionine	%			0.25	
Phenylalanine	%			1.03	
Proline	%			1.35	
Serine	%			1.23	
Threonine	%			0.86	
Tyrosine	%			1.03	
Valine	%			1.11	

MAGNESIUM CARBONATE

Magnesium carbonate, cp, (6)
Ref No 6 02 755 United States

Analyses			As Fed	Dry	C.V. ± %
Dry matter	%		81.8	100.0	
Calcium	%		0.02	0.02	
Chlorine	%		0.00	0.00	
Iron	%		0.002	0.002	
Magnesium	%		25.20	30.81	

MAGNESIUM OXIDE

Magnesium oxide, MgO, cp, (6)
Ref No 6 02 757 United States

Analyses			As Fed	Dry	C.V. ± %
Dry matter	%			100.0	
Calcium	%			0.05	
Chlorine	%			0.01	
Iron	%			0.010	
Magnesium	%			60.31	
Manganese	%			0.0	
Potassium	%			0.01	
Sodium	%			0.50	

MAGNESIUM SULFATE

Magnesium sulfate, $MgSO_4 \cdot 7H_2O$, cp, (6)
Ref No 6 02 759 United States

Analyses			As Fed	Dry	C.V. ± %
Dry matter	%		48.9	100.0	
Calcium	%		0.02	0.04	
Iron	%		0.000	0.001	
Magnesium	%		9.87	20.18	
Manganese	%		0.0	0.0	
Potassium	%		0.00	0.01	
Sodium	%		0.00	0.01	
Sulphur	%		13.00	26.58	

MAHONIA, LAREDO. Mahonia trifoliata

Mahonia, laredo, browse, (2)
Ref No 2 02 761 United States

Analyses			As Fed	Dry	C.V. ± %
Dry matter	%			100.0	
Ash	%			3.2	
Crude fiber	%			32.6	
Ether extract	%			2.6	
N-free extract	%			50.6	
Protein (N x 6.25)	%			11.0	
Cattle	dig prot %	*		7.2	
Goats	dig prot %	*		6.8	
Horses	dig prot %	*		6.9	
Rabbits	dig prot %	*		7.2	
Sheep	dig prot %	*		7.2	

(1) dry forages and roughages (2) pasture, range plants, and forages fed green (3) silages (4) energy feeds (5) protein supplements (6) minerals (7) vitamins (8) additives

Mahonia, laredo, browse, mature, (2)
Ref No 2 02 760 United States

Analyses			As Fed	Dry	C.V. ± %
Dry matter	%			100.0	
Calcium	%			0.67	
Magnesium	%			0.15	
Phosphorus	%			0.16	
Potassium	%			0.80	

MAIDENCANE. Panicum hemitomon

Maidencane, aerial part, fresh, (2)
Ref No 2 02 764 United States

Analyses			As Fed	Dry	C.V. ± %
Dry matter	%			100.0	
Ash	%			10.3	
Crude fiber	%			29.9	
Ether extract	%			1.9	
N-free extract	%			41.4	
Protein (N x 6.25)	%			16.5	16
Cattle	dig prot %	*		11.9	
Goats	dig prot %	*		12.0	
Horses	dig prot %	*		11.5	
Rabbits	dig prot %	*		11.4	
Sheep	dig prot %	*		12.4	

Maidencane, aerial part, fresh, immature, (2)
Ref No 2 02 762 United States

Analyses			As Fed	Dry	C.V. ± %
Dry matter	%			100.0	
Ash	%			5.3	
Crude fiber	%			33.4	
Ether extract	%			1.4	
N-free extract	%			39.8	
Protein (N x 6.25)	%			20.1	
Cattle	dig prot %	*		15.0	
Goats	dig prot %	*		15.3	
Horses	dig prot %	*		14.6	
Rabbits	dig prot %	*		14.2	
Sheep	dig prot %	*		15.7	

Maidencane, aerial part, fresh, early bloom, (2)
Ref No 2 02 763 United States

Analyses			As Fed	Dry	C.V. ± %
Dry matter	%			100.0	
Protein (N x 6.25)	%			17.1	4
Cattle	dig prot %	*		12.4	
Goats	dig prot %	*		12.5	
Horses	dig prot %	*		12.0	
Rabbits	dig prot %	*		11.9	
Sheep	dig prot %	*		12.9	

Maize—
see Corn

MALLOW. Malva spp

Mallow, hay, s-c, (1)
Ref No 1 03 019 United States

Analyses			As Fed	Dry	C.V. ± %
Dry matter	%		81.7	100.0	
Ash	%		9.8	12.0	
Crude fiber	%		28.0	34.3	
Sheep	dig coef %		48.	48.	
Ether extract	%		1.8	2.2	
Sheep	dig coef %		78.	78.	
N-free extract	%		29.4	36.0	
Sheep	dig coef %		55.	55.	
Protein (N x 6.25)	%		12.7	15.5	
Sheep	dig coef %		69.	69.	
Cattle	dig prot %	*	8.5	10.4	
Goats	dig prot %	*	9.0	11.0	
Horses	dig prot %	*	8.7	10.7	
Rabbits	dig prot %	*	8.7	10.6	
Sheep	dig prot %		8.7	10.7	
Energy	GE Mcal/kg				
Cattle	DE Mcal/kg	*	1.95	2.39	
Sheep	DE Mcal/kg	*	1.83	2.24	
Cattle	ME Mcal/kg	*	1.60	1.96	
Sheep	ME Mcal/kg	*	1.50	1.84	
Cattle	TDN %	*	44.3	54.2	
Sheep	TDN %		41.5	50.8	

Mallow, straw, (1)
Ref No 1 03 020 United States

Analyses			As Fed	Dry	C.V. ± %
Dry matter	%		87.9	100.0	
Ash	%		8.6	9.8	
Crude fiber	%		43.2	49.1	
Sheep	dig coef %		38.	38.	
Ether extract	%		0.6	0.7	
Sheep	dig coef %		68.	68.	
N-free extract	%		29.7	33.8	
Sheep	dig coef %		49.	49.	
Protein (N x 6.25)	%		5.8	6.6	
Sheep	dig coef %		61.	61.	
Cattle	dig prot %	*	2.3	2.7	
Goats	dig prot %	*	2.4	2.7	
Horses	dig prot %	*	2.8	3.1	
Rabbits	dig prot %	*	3.3	3.8	
Sheep	dig prot %		3.5	4.0	
Energy	GE Mcal/kg				
Cattle	DE Mcal/kg	*	1.26	1.43	
Sheep	DE Mcal/kg	*	1.56	1.78	
Cattle	ME Mcal/kg	*	1.03	1.17	
Sheep	ME Mcal/kg	*	1.28	1.46	
Cattle	TDN %	*	28.5	32.5	
Sheep	TDN %		35.4	40.3	

Mallow, aerial part, fresh, (2)
Ref No 2 03 023 United States

Analyses			As Fed	Dry	C.V. ± %
Dry matter	%		13.2	100.0	
Ash	%		2.0	15.0	
Crude fiber	%		3.3	24.9	
Sheep	dig coef %		52.	52.	
Swine	dig coef %		33.	33.	
Ether extract	%		0.4	3.3	
Sheep	dig coef %		71.	71.	
Swine	dig coef %		29.	29.	
N-free extract	%		4.4	33.0	
Sheep	dig coef %		78.	78.	

Feed Name or Analyses		Mean As Fed	Mean Dry	C.V. ± %
Swine	dig coef %	65.	65.	
Protein (N x 6.25)	%	3.1	23.9	
Sheep	dig coef %	82.	82.	
Swine	dig coef %	75.	75.	
Cattle	dig prot %	* 2.4	18.2	
Goats	dig prot %	* 2.5	18.8	
Horses	dig prot %	* 2.3	17.8	
Rabbits	dig prot %	* 2.3	17.1	
Sheep	dig prot %	2.6	19.6	
Swine	dig prot %	2.4	17.9	
Energy	GE Mcal/kg			
Cattle	DE Mcal/kg	* 0.40	3.02	
Sheep	DE Mcal/kg	* 0.37	2.80	
Swine	DE kcal/kg	* 289.	2191.	
Cattle	ME Mcal/kg	* 0.33	2.48	
Sheep	ME Mcal/kg	* 0.30	2.30	
Swine	ME kcal/kg	* 264.	1998.	
Cattle	TDN %	* 9.1	68.6	
Sheep	TDN %	8.4	63.5	
Swine	TDN %	6.6	49.7	

Mallow, aerial part, fresh, cut 1, (2)

Ref No 2 03 021 United States

Feed Name or Analyses		Mean As Fed	Mean Dry	C.V. ± %
Dry matter	%	14.5	100.0	
Ash	%	1.8	12.6	
Crude fiber	%	5.3	36.3	
Sheep	dig coef %	39.	39.	
Ether extract	%	0.3	2.2	
Sheep	dig coef %	58.	58.	
N-free extract	%	4.5	31.3	
Sheep	dig coef %	69.	69.	
Protein (N x 6.25)	%	2.6	17.6	
Sheep	dig coef %	75.	75.	
Cattle	dig prot %	* 1.9	12.9	
Goats	dig prot %	* 1.9	13.0	
Horses	dig prot %	* 1.8	12.5	
Rabbits	dig prot %	* 1.8	12.3	
Sheep	dig prot %	1.9	13.2	
Energy	GE Mcal/kg			
Cattle	DE Mcal/kg	* 0.34	2.36	
Sheep	DE Mcal/kg	* 0.33	2.29	
Cattle	ME Mcal/kg	* 0.28	1.94	
Sheep	ME Mcal/kg	* 0.27	1.87	
Cattle	TDN %	* 7.8	53.5	
Sheep	TDN %	7.5	51.8	

Mallow, aerial part, fresh, cut 2, (2)

Ref No 2 03 022 United States

Feed Name or Analyses		Mean As Fed	Mean Dry	C.V. ± %
Dry matter	%	17.1	100.0	
Ash	%	2.8	16.6	
Crude fiber	%	4.5	26.2	
Sheep	dig coef %	53.	53.	
Ether extract	%	0.5	2.8	
Sheep	dig coef %	64.	64.	
N-free extract	%	5.8	34.1	
Sheep	dig coef %	80.	80.	
Protein (N x 6.25)	%	3.5	20.3	
Sheep	dig coef %	78.	78.	
Cattle	dig prot %	* 2.6	15.1	
Goats	dig prot %	* 2.7	15.5	
Horses	dig prot %	* 2.5	14.8	
Rabbits	dig prot %	* 2.5	14.3	
Sheep	dig prot %	2.7	15.8	
Energy	GE Mcal/kg			
Cattle	DE Mcal/kg	* 0.48	2.79	
Sheep	DE Mcal/kg	* 0.46	2.69	

Feed Name or Analyses		Mean As Fed	Mean Dry	C.V. ± %
Cattle	ME Mcal/kg	* 0.39	2.29	
Sheep	ME Mcal/kg	* 0.38	2.21	
Cattle	TDN %	* 10.8	63.3	
Sheep	TDN %	10.4	61.0	

Mallow, seeds, (5)

Ref No 5 03 027 United States

Feed Name or Analyses		Mean As Fed	Mean Dry	C.V. ± %
Dry matter	%	91.6	100.0	
Ash	%	6.2	6.8	
Crude fiber	%	23.4	25.5	
Sheep	dig coef %	17.	17.	
Ether extract	%	15.2	16.6	
Sheep	dig coef %	98.	98.	
N-free extract	%	25.2	27.5	
Sheep	dig coef %	60.	60.	
Protein (N x 6.25)	%	21.6	23.6	
Sheep	dig coef %	77.	77.	
Sheep	dig prot %	16.6	18.2	
Energy	GE Mcal/kg			
Cattle	DE Mcal/kg	* 2.97	3.25	
Sheep	DE Mcal/kg	* 3.05	3.33	
Swine	DE kcal/kg	* 3865.	4219.	
Cattle	ME Mcal/kg	* 2.44	2.66	
Sheep	ME Mcal/kg	* 2.50	2.73	
Swine	ME kcal/kg	* 3526.	3849.	
Cattle	TDN %	* 67.4	73.6	
Sheep	TDN %	69.3	75.6	
Swine	TDN %	* 87.7	95.7	

Mallow, seeds, roasted, (5)

Ref No 5 03 025 United States

Feed Name or Analyses		Mean As Fed	Mean Dry	C.V. ± %
Dry matter	%	94.9	100.0	
Ash	%	6.2	6.5	
Crude fiber	%	22.3	23.5	
Sheep	dig coef %	19.	19.	
Ether extract	%	16.3	17.2	
Sheep	dig coef %	99.	99.	
N-free extract	%	27.4	28.9	
Sheep	dig coef %	61.	61.	
Protein (N x 6.25)	%	22.7	23.9	
Sheep	dig coef %	76.	76.	
Sheep	dig prot %	17.2	18.2	
Energy	GE Mcal/kg			
Cattle	DE Mcal/kg	* 3.21	3.39	
Sheep	DE Mcal/kg	* 3.29	3.46	
Swine	DE kcal/kg	* 4135.	4357.	
Cattle	ME Mcal/kg	* 2.63	2.78	
Sheep	ME Mcal/kg	* 2.70	2.84	
Swine	ME kcal/kg	* 3770.	3973.	
Cattle	TDN %	* 72.9	76.8	
Sheep	TDN %	74.6	78.6	
Swine	TDN %	* 93.8	98.8	

Mallow, seeds, solv-extd grnd, (5)

Ref No 5 03 026 United States

Feed Name or Analyses		Mean As Fed	Mean Dry	C.V. ± %
Dry matter	%	92.2	100.0	
Ash	%	11.0	11.9	
Crude fiber	%	23.8	25.8	
Sheep	dig coef %	33.	33.	
Ether extract	%	2.6	2.8	
Sheep	dig coef %	77.	77.	
N-free extract	%	28.8	31.2	
Sheep	dig coef %	66.	66.	

Feed Name or Analyses		Mean As Fed	Mean Dry	C.V. ± %
Protein (N x 6.25)	%	26.1	28.3	
Sheep	dig coef %	79.	79.	
Sheep	dig prot %	20.6	22.4	
Energy	GE Mcal/kg			
Cattle	DE Mcal/kg	* 2.13	2.31	
Sheep	DE Mcal/kg	* 2.29	2.48	
Swine	DE kcal/kg	* 2677.	2904.	
Cattle	ME Mcal/kg	* 1.75	1.90	
Sheep	ME Mcal/kg	* 1.88	2.04	
Swine	ME kcal/kg	* 2417.	2621.	
Cattle	TDN %	* 48.4	52.5	
Sheep	TDN %	51.9	56.3	
Swine	TDN %	* 60.7	65.9	

MALLOW, CURLY. *Malva crispa*

Mallow, curly, hay, fresh, milk stage, (1)

Ref No 1 03 028 United States

Feed Name or Analyses		Mean As Fed	Mean Dry	C.V. ± %
Dry matter	%	29.2	100.0	
Ash	%	3.2	11.0	
Crude fiber	%	8.2	28.2	
Sheep	dig coef %	73.	73.	
Ether extract	%	1.7	5.9	
Sheep	dig coef %	93.	93.	
N-free extract	%	12.2	41.9	
Sheep	dig coef %	86.	86.	
Protein (N x 6.25)	%	3.8	13.0	
Sheep	dig coef %	85.	85.	
Cattle	dig prot %	* 2.4	8.2	
Goats	dig prot %	* 2.5	8.7	
Horses	dig prot %	* 2.5	8.6	
Rabbits	dig prot %	* 2.5	8.7	
Sheep	dig prot %	3.2	11.1	
Energy	GE Mcal/kg			
Cattle	DE Mcal/kg	* 0.93	3.17	
Sheep	DE Mcal/kg	* 1.03	3.53	
Cattle	ME Mcal/kg	* 0.76	2.60	
Sheep	ME Mcal/kg	* 0.84	2.89	
Cattle	TDN %	* 21.0	71.9	
Sheep	TDN %	23.4	80.0	

Mallow, curly, hay, s-c, (1)

Ref No 1 03 029 United States

Feed Name or Analyses		Mean As Fed	Mean Dry	C.V. ± %
Dry matter	%	81.3	100.0	
Ash	%	10.7	13.1	
Crude fiber	%	23.3	28.6	
Sheep	dig coef %	54.	54.	
Ether extract	%	1.4	1.7	
Sheep	dig coef %	72.	72.	
N-free extract	%	30.9	38.0	
Sheep	dig coef %	60.	60.	
Protein (N x 6.25)	%	15.1	18.6	
Sheep	dig coef %	78.	78.	
Cattle	dig prot %	* 10.6	13.0	
Goats	dig prot %	* 11.3	13.9	
Horses	dig prot %	* 10.8	13.3	
Rabbits	dig prot %	* 10.6	13.0	
Sheep	dig prot %	11.8	14.5	
Energy	GE Mcal/kg			
Cattle	DE Mcal/kg	* 2.07	2.55	
Sheep	DE Mcal/kg	* 1.99	2.45	
Cattle	ME Mcal/kg	* 1.70	2.09	
Sheep	ME Mcal/kg	* 1.63	2.01	
Cattle	TDN %	* 47.0	57.8	
Sheep	TDN %	45.1	55.5	

Mallow, curly, aerial part, fresh, pre-bloom, (2)

Ref No 2 03 030 United States

Feed Name or Analyses			Mean As Fed	Mean Dry	C.V. ± %
Dry matter	%		15.4	100.0	
Ash	%		2.6	16.8	
Crude fiber	%		2.7	17.8	
Sheep	dig coef %		62.	62.	
Ether extract	%		0.5	3.5	
Sheep	dig coef %		75.	75.	
N-free extract	%		5.9	38.2	
Sheep	dig coef %		84.	84.	
Protein (N x 6.25)	%		3.6	23.7	
Sheep	dig coef %		85.	85.	
Cattle	dig prot %	*	2.8	18.0	
Goats	dig prot %	*	2.9	18.7	
Horses	dig prot %	*	2.7	17.7	
Rabbits	dig prot %	*	2.6	17.0	
Sheep	dig prot %		3.1	20.1	
Energy	GE Mcal/kg				
Cattle	DE Mcal/kg	*	0.44	2.83	
Sheep	DE Mcal/kg	*	0.47	3.05	
Cattle	ME Mcal/kg	*	0.36	2.32	
Sheep	ME Mcal/kg	*	0.39	2.50	
Cattle	TDN %	*	9.9	64.1	
Sheep	TDN %		10.7	69.2	

Mallow, curly, aerial part, fresh, full bloom, (2)

Ref No 2 03 031 United States

Feed Name or Analyses			Mean As Fed	Mean Dry	C.V. ± %
Dry matter	%		14.6	100.0	
Ash	%		2.5	16.8	
Crude fiber	%		3.1	21.2	
Sheep	dig coef %		40.	40.	
Ether extract	%		0.4	2.7	
Sheep	dig coef %		72.	72.	
N-free extract	%		5.4	36.9	
Sheep	dig coef %		73.	73.	
Protein (N x 6.25)	%		3.3	22.4	
Sheep	dig coef %		85.	85.	
Cattle	dig prot %	*	2.5	16.9	
Goats	dig prot %	*	2.5	17.5	
Horses	dig prot %	*	2.4	16.5	
Rabbits	dig prot %	*	2.3	16.0	
Sheep	dig prot %		2.8	19.0	
Energy	GE Mcal/kg				
Cattle	DE Mcal/kg	*	0.42	2.84	
Sheep	DE Mcal/kg	*	0.38	2.59	
Cattle	ME Mcal/kg	*	0.34	2.33	
Sheep	ME Mcal/kg	*	0.31	2.13	
Cattle	TDN %	*	9.4	64.5	
Sheep	TDN %		8.6	58.8	

Mallow, curly, hay, fresh, (2)

Ref No 2 03 032 United States

Feed Name or Analyses			Mean As Fed	Mean Dry	C.V. ± %
Dry matter	%		16.7	100.0	
Ash	%		2.7	16.1	
Crude fiber	%		3.3	19.8	
Sheep	dig coef %		50.	50.	
Ether extract	%		0.6	3.8	
Sheep	dig coef %		78.	78.	
N-free extract	%		6.3	37.8	
Sheep	dig coef %		79.	79.	
Protein (N x 6.25)	%		3.8	22.5	
Sheep	dig coef %		85.	85.	
Cattle	dig prot %	*	2.8	17.0	
Goats	dig prot %	*	2.9	17.6	
Horses	dig prot %	*	2.8	16.6	
Rabbits	dig prot %	*	2.7	16.0	
Sheep	dig prot %		3.2	19.1	
Energy	GE Mcal/kg				
Cattle	DE Mcal/kg	*	0.47	2.81	
Sheep	DE Mcal/kg	*	0.48	2.89	
Cattle	ME Mcal/kg	*	0.38	2.30	
Sheep	ME Mcal/kg	*	0.40	2.37	
Cattle	TDN %	*	10.6	63.7	
Sheep	TDN %		10.9	65.6	

(1) dry forages and roughages (2) pasture, range plants, and forages fed green (3) silages (4) energy feeds (5) protein supplements (6) minerals (7) vitamins (8) additives

MALLOW, LITTLE. Malva parviflora

Mallow, little, seeds w hulls, (5)

Ref No 5 09 126 United States

Feed Name or Analyses			Mean As Fed	Mean Dry	C.V. ± %
Dry matter	%			100.0	
Protein (N x 6.25)	%			21.0	
Alanine	%			0.80	
Arginine	%			1.51	
Aspartic acid	%			2.06	
Glutamic acid	%			2.73	
Glycine	%			1.05	
Histidine	%			0.61	
Hydroxyproline	%			0.15	
Isoleucine	%			0.63	
Leucine	%			1.09	
Lysine	%			1.07	
Methionine	%			0.34	
Phenylalanine	%			0.76	
Proline	%			0.69	
Serine	%			0.97	
Threonine	%			0.69	
Tyrosine	%			0.57	
Valine	%			0.82	

MALLOW, SMOOTH. Malva spp

Mallow, smooth, hay, s-c, (1)

Ref No 1 03 033 United States

Feed Name or Analyses			Mean As Fed	Mean Dry	C.V. ± %
Dry matter	%		78.9	100.0	
Ash	%		9.0	11.4	
Crude fiber	%		29.6	37.5	
Sheep	dig coef %		52.	52.	
Ether extract	%		1.9	2.4	
Sheep	dig coef %		81.	81.	
N-free extract	%		29.4	37.2	
Sheep	dig coef %		56.	56.	
Protein (N x 6.25)	%		9.1	11.5	
Sheep	dig coef %		68.	68.	
Cattle	dig prot %	*	5.4	6.9	
Goats	dig prot %	*	5.8	7.3	
Horses	dig prot %	*	5.8	7.3	
Rabbits	dig prot %	*	6.0	7.5	
Sheep	dig prot %		6.2	7.8	
Energy	GE Mcal/kg				
Cattle	DE Mcal/kg	*	1.87	2.37	
Sheep	DE Mcal/kg	*	1.83	2.32	
Cattle	ME Mcal/kg	*	1.53	1.95	
Sheep	ME Mcal/kg	*	1.50	1.90	
Cattle	TDN %	*	42.5	53.8	
Sheep	TDN %		41.4	52.5	

Malt cleanings (AAFCO) –
see Barley, malt sprout cleanings w hulls, dehy, mx 24% protein, (5)

Malt hulls (AAFCO) –
see Barley, malt hulls, (1)

Malt sprouts (AAFCO) –
see Barley, malt sprouts w hulls, dehy, mn 24% protein, (5)

Malt sprouts (CFA) –
see Barley, malt sprouts w hulls, dehy, (5)

MANGANESE CARBONATE

Manganese carbonate, $MnCO_3$, cp, (6)

Ref No 6 03 037 United States

Feed Name or Analyses			Mean As Fed	Mean Dry	C.V. ± %
Dry matter	%			100.0	
Chlorine	%			0.02	
Iron	%			0.002	
Manganese	%			43.0	
Sulphur	%			0.07	
Zinc	%			0.1	

MANGANESE CHLORIDE

Manganese chloride, $MnCl_2 \cdot 4H_2O$, cp, (6)

Ref No 6 03 039 United States

Feed Name or Analyses			Mean As Fed	Mean Dry	C.V. ± %
Dry matter	%		63.6	100.0	
Chlorine	%		35.82	56.32	
Iron	%		0.001	0.001	
Manganese	%		27.8	43.6	
Sulphur	%		0.07	0.11	
Zinc	%		0.0	0.0	

MANGANESE DIOXIDE

Manganese dioxide, MnO_2, cp, (6)

Ref No 6 03 043 United States

Feed Name or Analyses			Mean As Fed	Mean Dry	C.V. ± %
Dry matter	%			100.0	
Chlorine	%			0.01	
Iron	%			0.050	
Manganese	%			63.2	
Sulphur	%			0.02	

Mangel, roots –
see Beet, mangels, roots, (4)

MANGROVE, AMERICAN. Rhizophora mangle

Mangrove, American, leaves, grnd, (1)

Ref No 1 03 057 United States

Feed Name or Analyses			Mean As Fed	Mean Dry	C.V. ± %
Dry matter	%			100.0	
Ash	%			8.9	
Crude fiber	%			14.5	
Ether extract	%			3.4	
N-free extract	%			62.5	

Feed Name or Analyses			Mean As Fed	Mean Dry	C.V. ± %
Protein (N x 6.25)	%			10.7	
Cattle	dig prot %	*		6.2	
Goats	dig prot %	*		6.5	
Horses	dig prot %	*		6.6	
Rabbits	dig prot %	*		6.9	
Sheep	dig prot %	*		6.2	

Manihot meal —
see Cassava, starch by-product, dehy, (4)

Manioc meal —
see Cassava, starch by-product, dehy, (4)

MANNAGRASS, AMERICAN. Glyceria grandis

Managrass, American, aerial part, fresh, (2)
Ref No 2 03 058 United States

			As Fed	Dry	C.V.
Dry matter	%			100.0	
Ash	%			8.4	
Crude fiber	%			32.8	
Ether extract	%			2.4	
N-free extract	%			43.9	
Protein (N x 6.25)	%			12.5	
Cattle	dig prot %	*		8.5	
Goats	dig prot %	*		8.2	
Horses	dig prot %	*		8.1	
Rabbits	dig prot %	*		8.3	
Sheep	dig prot %	*		8.6	
Calcium	%			0.26	
Phosphorus	%			0.55	
Carotene	mg/kg			138.2	
Vitamin A equivalent	IU/g			230.4	

MAPLE, RED. Acer rubrum

Maple, red, browse, (2)
Ref No 2 03 062 United States

			As Fed	Dry	C.V.
Dry matter	%			100.0	
Ash	%			4.7	8
Crude fiber	%			17.8	14
Ether extract	%			6.0	34
N-free extract	%			56.2	
Protein (N x 6.25)	%			15.3	36
Cattle	dig prot %	*		10.9	
Goats	dig prot %	*		10.8	
Horses	dig prot %	*		10.5	
Rabbits	dig prot %	*		10.5	
Sheep	dig prot %	*		11.2	
Calcium	%			0.85	17
Cobalt	mg/kg			0.110	15
Copper	mg/kg			6.4	22
Iron	%			0.029	80
Manganese	mg/kg			775.1	24
Phosphorus	%			0.22	49

Maple, red, browse, immature, (2)
Ref No 2 03 060 United States

			As Fed	Dry	C.V.
Dry matter	%			100.0	
Ash	%			5.2	
Crude fiber	%			14.7	
Ether extract	%			3.6	
N-free extract	%			52.5	
Protein (N x 6.25)	%			24.0	
Cattle	dig prot %	*		18.3	
Goats	dig prot %	*		19.0	
Horses	dig prot %	*		17.9	
Rabbits	dig prot %	*		17.2	
Sheep	dig prot %	*		19.4	

Maple, red, browse, mature, (2)
Ref No 2 03 061 United States

			As Fed	Dry	C.V.
Dry matter	%			100.0	
Ash	%			4.5	9
Crude fiber	%			19.4	10
Ether extract	%			7.2	13
N-free extract	%			58.0	
Protein (N x 6.25)	%			10.9	4
Cattle	dig prot %	*		7.2	
Goats	dig prot %	*		6.7	
Horses	dig prot %	*		6.8	
Rabbits	dig prot %	*		7.1	
Sheep	dig prot %	*		7.1	

MARABU. Dichrostachys glomerata

Marabu, pods, s-c, (1)
Ref No 1 03 063 United States

			As Fed	Dry	C.V.
Dry matter	%		90.2	100.0	
Ash	%		4.9	5.4	
Crude fiber	%		24.0	26.6	
Sheep	dig coef %		55.	55.	
Ether extract	%		1.3	1.4	
Sheep	dig coef %		69.	69.	
N-free extract	%		50.3	55.8	
Sheep	dig coef %		67.	67.	
Protein (N x 6.25)	%		9.7	10.8	
Sheep	dig coef %		37.	37.	
Cattle	dig prot %	*	5.7	6.3	
Goats	dig prot %	*	6.0	6.6	
Horses	dig prot %	*	6.0	6.7	
Rabbits	dig prot %	*	6.3	7.0	
Sheep	dig prot %		3.6	4.0	
Energy	GE Mcal/kg				
Cattle	DE Mcal/kg	*	2.42	2.68	
Sheep	DE Mcal/kg	*	2.31	2.57	
Cattle	ME Mcal/kg	*	1.98	2.20	
Sheep	ME Mcal/kg	*	1.90	2.10	
Cattle	TDN %	*	54.8	60.7	
Sheep	TDN %		52.5	58.2	

MARSDENIA, EDULIS. Marsdenia edulis

Marsdenia, edulis, seeds w hulls, (5)
Ref No 5 09 305 United States

			As Fed	Dry	C.V.
Dry matter	%			100.0	
Protein (N x 6.25)	%			27.0	
Alanine	%			1.19	
Arginine	%			2.65	
Aspartic acid	%			1.97	
Glutamic acid	%			6.86	
Glycine	%			1.54	
Histidine	%			0.76	
Isoleucine	%			1.19	
Leucine	%			2.08	
Lysine	%			0.89	
Methionine	%			0.35	
Phenylalanine	%			1.46	
Proline	%			0.92	
Serine	%			1.49	
Threonine	%			0.84	
Tyrosine	%			0.89	
Valine	%			1.38	

MARSHALLIA. Marshallia caespitosa

Marshallia, seeds w hulls, (5)
Ref No 5 09 290 United States

			As Fed	Dry	C.V.
Dry matter	%			100.0	
Protein (N x 6.25)	%			31.0	
Alanine	%			1.33	
Arginine	%			2.36	
Aspartic acid	%			2.85	
Glutamic acid	%			5.67	
Glycine	%			1.58	
Histidine	%			0.68	
Hydroxyproline	%			0.09	
Isoleucine	%			1.49	
Leucine	%			1.95	
Lysine	%			1.40	
Methionine	%			0.65	
Phenylalanine	%			1.33	
Proline	%			1.61	
Serine	%			1.30	
Threonine	%			1.05	
Tyrosine	%			0.78	
Valine	%			1.55	

Marsh hay —
see Native plants, marsh, hay, s-c, (1)

MARTYNIA. Martynia parviflora

Martynia, seeds w hulls, (5)
Ref No 5 09 129 United States

			As Fed	Dry	C.V.
Dry matter	%			100.0	
Protein (N x 6.25)	%			30.0	
Alanine	%			1.32	
Arginine	%			3.57	
Aspartic acid	%			1.77	
Glutamic acid	%			4.35	
Glycine	%			0.99	
Histidine	%			0.75	
Hydroxyproline	%			0.03	
Isoleucine	%			0.93	
Leucine	%			1.65	
Lysine	%			0.84	
Methionine	%			0.51	
Phenylalanine	%			1.02	
Proline	%			0.78	
Serine	%			0.87	
Threonine	%			1.02	
Tyrosine	%			0.96	
Valine	%			1.05	

Meadow hay —
see Native plants, intermountain, hay, s-c, (1)

MEADOWRUE. Thalictrum revolutum

Meadowrue, seeds w hulls, (5)
Ref No 5 09 048 United States

Feed Name or Analyses				Mean As Fed	Mean Dry	C.V. ± %
Dry matter		%			100.0	
Protein (N x 6.25)		%			28.0	
Alanine		%			1.01	
Arginine		%			2.38	
Aspartic acid		%			2.44	
Glutamic acid		%			5.15	
Glycine		%			1.01	
Histidine		%			0.50	
Hydroxyproline		%			0.34	
Isoleucine		%			0.92	
Leucine		%			1.48	
Lysine		%			1.20	
Methionine		%			0.36	
Phenylalanine		%			0.78	
Proline		%			1.37	
Serine		%			1.26	
Threonine		%			0.90	
Tyrosine		%			0.98	
Valine		%			1.15	

MEADOWRUE, FENDLER. Thalictrum fendleri

Meadowrue, fendler, aerial part, fresh, (2)
Ref No 2 03 064 United States

Feed Name or Analyses				Mean As Fed	Mean Dry	C.V. ± %
Dry matter		%			100.0	
Ash		%			9.6	
Ether extract		%			3.1	
Protein (N x 6.25)		%			8.9	
Cattle	dig prot	%	*		5.5	
Goats	dig prot	%	*		4.9	
Horses	dig prot	%	*		5.1	
Rabbits	dig prot	%	*		5.5	
Sheep	dig prot	%	*		5.3	
Calcium		%			1.73	
Phosphorus		%			0.50	

Meadowrue, fendler, leaves, fresh, (2)
Ref No 2 03 065 United States

Feed Name or Analyses				Mean As Fed	Mean Dry	C.V. ± %
Dry matter		%			100.0	
Ash		%			12.1	
Ether extract		%			5.0	
Protein (N x 6.25)		%			16.2	
Cattle	dig prot	%	*		11.7	
Goats	dig prot	%	*		11.7	
Horses	dig prot	%	*		11.3	
Rabbits	dig prot	%	*		11.2	
Sheep	dig prot	%	*		12.1	
Calcium		%			1.70	
Phosphorus		%			0.60	

(1) dry forages and roughages (2) pasture, range plants, and forages fed green (3) silages (4) energy feeds (5) protein supplements (6) minerals (7) vitamins (8) additives

Meadowrue, fendler, stems, fresh, (2)
Ref No 2 03 066 United States

Feed Name or Analyses				Mean As Fed	Mean Dry	C.V. ± %
Dry matter		%			100.0	
Ash		%			7.4	
Ether extract		%			1.4	
Protein (N x 6.25)		%			4.4	
Cattle	dig prot	%	*		1.6	
Goats	dig prot	%	*		0.7	
Horses	dig prot	%	*		1.3	
Rabbits	dig prot	%	*		2.1	
Sheep	dig prot	%	*		1.1	
Calcium		%			0.77	
Phosphorus		%			0.41	

MEADOWRUE, SIERRA. Thalictrum polycarpum

Meadowrue, sierra, seeds w hulls, (5)
Ref No 5 09 109 United States

Feed Name or Analyses				Mean As Fed	Mean Dry	C.V. ± %
Dry matter		%			100.0	
Protein (N x 6.25)		%			25.0	
Alanine		%			0.85	
Arginine		%			1.95	
Aspartic acid		%			2.13	
Glutamic acid		%			4.60	
Glycine		%			0.95	
Histidine		%			0.53	
Hydroxyproline		%			0.03	
Isoleucine		%			0.85	
Leucine		%			1.28	
Lysine		%			1.15	
Methionine		%			0.35	
Phenylalanine		%			0.65	
Proline		%			1.15	
Serine		%			0.98	
Threonine		%			0.83	
Tyrosine		%			0.88	
Valine		%			0.75	

Meat and bone meal (AAFCO) –
see Animal, carcass residue w bone, dry rendered dehy grnd, mn 9% indigestible material mn 4.4% phosphorus, (5)

Meat and bone meal (CFA) –
see Animal-Poultry, carcass residue mx 5% blood w bone, dry or wet rendered dehy grnd, mn 40% protein, (5)

Meat and bone meal digester tankage –
see Animal, carcass residue w blood w bone, dry or wet rendered dehy grnd, mn 9% indigestible material mn 4.4% phosphorus, (5)

Meat and bone meal tankage (AAFCO) –
see Animal, carcass residue w blood w bone, dry or wet rendered dehy grnd, mn 9% indigestible material mn 4.4% phosphorus, (5)

Meat and bone scrap (CFA) –
see Animal-Poultry, carcass residue mx 5% blood w bone, dry or wet rendered dehy grnd, mn 40% protein, (5)

Meat and bone scrap –
see Animal, carcass residue w bone, dry rendered dehy grnd, mn 9% indigestible material mn 4.4% phosphorus, (5)

Meat and bone scrap, solvent extracted –
see Animal, carcass residue w bone, dry rendered solv-extd dehy grnd, (5)

Meat, cattle, meal –
see Cattle, carcass residue, dry rendered dehy grnd, (5)

Meat meal (AAFCO) –
see Animal, carcass residue, dry rendered dehy grnd, mn 9% indigestible material mx 4.4% phosphorus, (5)

Meat meal (CFA) –
see Animal-Poultry, carcass residue mx 5% blood, dry or wet rendered dehy grnd, mn 50% protein, (5)

Meat meal tankage (AAFCO) –
see Animal, carcass residue w blood, dry or wet rendered dehy grnd, mn 9% indigestible material mx 4.4% phosphorus, (5)

Meat scrap (CFA) –
see Animal-Vegetable, mixed, fat or oil, feed gr, (4)
see Animal-Poultry, carcass residue mx 5% blood, dry or wet rendered dehy grnd, mn 50% protein, (5)

Meat scrap –
see Animal, carcass residue, dry rendered dehy grnd, mn 9% indigestible material mx 4.4% phosphorus, (5)

Meat solubles, dried –
see Animal, stickwater solubles, vacuum dehy, (5)

MEDIC. Medica lupulina

Medic, seed hulls, grnd, (4)
Ref No 4 03 067 United States

Feed Name or Analyses				Mean As Fed	Mean Dry	C.V. ± %
Dry matter		%		86.4	100.0	
Ash		%		6.7	7.8	
Crude fiber		%		22.7	26.3	
Sheep	dig coef	%		68.	68.	
Ether extract		%		1.9	2.2	
Sheep	dig coef	%		51.	51.	
N-free extract		%		40.6	47.0	
Sheep	dig coef	%		47.	47.	
Protein (N x 6.25)		%		14.4	16.7	
Sheep	dig coef	%		50.	50.	
Cattle	dig prot	%	*	9.8	11.4	
Goats	dig prot	%	*	10.8	12.5	
Horses	dig prot	%	*	10.8	12.5	
Sheep	dig prot	%		7.2	8.4	
Energy	GE Mcal/kg					
Cattle	DE Mcal/kg		*	1.84	2.13	
Sheep	DE Mcal/kg		*	1.94	2.24	
Cattle	ME Mcal/kg		*	1.51	1.75	
Sheep	ME Mcal/kg		*	1.59	1.84	
Cattle	TDN	%	*	41.7	48.3	
Sheep	TDN	%		43.9	50.8	

Feed Name or Analyses			Mean As Fed	Mean Dry	C.V. ± %

MEDIC, BLACK. Medica lupulina

Medic, black, aerial part, fresh, (2)

Ref No 2 03 070 United States

Feed Name or Analyses			As Fed	Dry	C.V. ± %
Dry matter	%		22.7	100.0	
Ash	%		2.3	10.2	
Crude fiber	%		5.6	24.7	
Ether extract	%		0.8	3.4	
N-free extract	%		9.1	40.2	
Protein (N x 6.25)	%		4.9	21.6	15
Cattle	dig prot %	*	3.7	16.2	
Goats	dig prot %	*	3.8	16.7	
Horses	dig prot %	*	3.6	15.8	
Rabbits	dig prot %	*	3.5	15.3	
Sheep	dig prot %	*	3.9	17.1	
Energy	GE Mcal/kg				
Cattle	DE Mcal/kg	*	0.59	2.60	
Sheep	DE Mcal/kg	*	0.63	2.78	
Cattle	ME Mcal/kg	*	0.48	2.13	
Sheep	ME Mcal/kg	*	0.52	2.28	
Cattle	TDN %	*	13.4	58.9	
Sheep	TDN %	*	14.3	63.1	

Medic, black, aerial part, fresh, immature, (2)

Ref No 2 03 068 United States

Feed Name or Analyses			As Fed	Dry	C.V. ± %
Dry matter	%			100.0	
N-free extract	%			38.5	
Protein (N x 6.25)	%			25.6	5
Cattle	dig prot %	*		19.7	
Goats	dig prot %	*		20.4	
Horses	dig prot %	*		19.3	
Rabbits	dig prot %	*		18.4	
Sheep	dig prot %	*		20.8	
Calcium	%			1.33	
Magnesium	%			0.45	
Phosphorus	%			0.44	
Potassium	%			2.28	

Medic, black, aerial part, fresh, full bloom, (2)

Ref No 2 03 069 United States

Feed Name or Analyses			As Fed	Dry	C.V. ± %
Dry matter	%			100.0	
Ash	%			10.5	
Crude fiber	%			24.7	
Ether extract	%			2.9	
N-free extract	%			45.2	
Protein (N x 6.25)	%			16.7	
Cattle	dig prot %	*		12.1	
Goats	dig prot %	*		12.1	
Horses	dig prot %	*		11.7	
Rabbits	dig prot %	*		11.6	
Sheep	dig prot %	*		12.6	
Energy	GE Mcal/kg				
Cattle	DE Mcal/kg	*		2.63	
Sheep	DE Mcal/kg	*		2.76	
Cattle	ME Mcal/kg	*		2.16	
Sheep	ME Mcal/kg	*		2.26	
Cattle	TDN %	*		59.6	
Sheep	TDN %	*		62.5	
Calcium	%			0.60	
Phosphorus	%			0.20	

Ref No 5 09 112 United States

Feed Name or Analyses			As Fed	Dry	C.V. ± %
Dry matter	%			100.0	
Protein (N x 6.25)	%			33.0	
Alanine	%			1.19	
Arginine	%			2.48	
Aspartic acid	%			3.10	
Glutamic acid	%			4.32	
Glycine	%			1.39	
Histidine	%			0.86	
Hydroxyproline	%			0.03	
Isoleucine	%			1.16	
Leucine	%			1.95	
Lysine	%			1.55	
Methionine	%			0.40	
Phenylalanine	%			1.32	
Proline	%			1.19	
Serine	%			1.39	
Threonine	%			1.12	
Tyrosine	%			0.92	
Valine	%			1.32	

MELIC, COAST RANGE. Melica imperfecta

Melic, coast range, aerial part, fresh, (2)

Ref No 2 03 075 United States

Feed Name or Analyses			As Fed	Dry	C.V. ± %
Dry matter	%			100.0	
Ash	%			6.5	33
Crude fiber	%			30.3	8
Ether extract	%			2.6	
N-free extract	%			49.6	
Protein (N x 6.25)	%			11.0	57
Cattle	dig prot %	*		7.2	
Goats	dig prot %	*		6.8	
Horses	dig prot %	*		6.9	
Rabbits	dig prot %	*		7.2	
Sheep	dig prot %	*		7.2	
Calcium	%			0.39	52
Phosphorus	%			0.26	29
Potassium	%			1.94	38

Melic, coast range, aerial part, fresh, immature, (2)

Ref No 2 03 071 United States

Feed Name or Analyses			As Fed	Dry	C.V. ± %
Dry matter	%			100.0	
Ash	%			10.6	
Crude fiber	%			26.6	
Protein (N x 6.25)	%			24.0	
Cattle	dig prot %	*		18.3	
Goats	dig prot %	*		19.0	
Horses	dig prot %	*		17.9	
Rabbits	dig prot %	*		17.2	
Sheep	dig prot %	*		19.4	
Calcium	%			0.75	
Phosphorus	%			0.40	
Potassium	%			3.21	

Melic, coast range, aerial part, fresh, full bloom, (2)

Ref No 2 03 072 United States

Feed Name or Analyses			As Fed	Dry	C.V. ± %
Dry matter	%			100.0	
Ash	%			8.4	
Crude fiber	%			29.5	
Protein (N x 6.25)	%			20.0	
Cattle	dig prot %	*		14.9	
Goats	dig prot %	*		15.2	
Horses	dig prot %	*		14.5	
Rabbits	dig prot %	*		14.1	
Sheep	dig prot %	*		15.6	
Calcium	%			0.43	
Phosphorus	%			0.39	
Potassium	%			2.65	

Melic, coast range, aerial part, fresh, milk stage, (2)

Ref No 2 03 073 United States

Feed Name or Analyses			As Fed	Dry	C.V. ± %
Dry matter	%			100.0	
Ash	%			6.9	15
Crude fiber	%			29.9	6
Ether extract	%			2.6	
N-free extract	%			47.9	
Protein (N x 6.25)	%			12.7	10
Cattle	dig prot %	*		8.7	
Goats	dig prot %	*		8.4	
Horses	dig prot %	*		8.3	
Rabbits	dig prot %	*		8.5	
Sheep	dig prot %	*		8.8	
Calcium	%			0.34	
Phosphorus	%			0.30	
Potassium	%			1.99	

Melic, coast range, aerial part, fresh, over ripe, (2)

Ref No 2 03 074 United States

Feed Name or Analyses			As Fed	Dry	C.V. ± %
Dry matter	%			100.0	
Ash	%			4.5	12
Crude fiber	%			31.7	6
Protein (N x 6.25)	%			3.9	25
Cattle	dig prot %	*		1.2	
Goats	dig prot %	*		0.2	
Horses	dig prot %	*		0.8	
Rabbits	dig prot %	*		1.7	
Sheep	dig prot %	*		0.6	
Calcium	%			0.29	13
Phosphorus	%			0.17	20
Potassium	%			1.35	15

MELONS, PIE. Curcurbita spp

Melons, pie, fruit w seeds, fresh, (2)

Stock melons

Ref No 2 08 459 United States

Feed Name or Analyses			As Fed	Dry	C.V. ± %
Dry matter	%		6.1	100.0	
Ash	%		0.4	6.6	
Crude fiber	%		1.4	23.0	
Ether extract	%		0.2	3.3	
N-free extract	%		3.4	55.7	
Protein (N x 6.25)	%		0.7	11.5	
Cattle	dig prot %	*	0.4	6.6	
Goats	dig prot %	*	0.5	7.8	
Horses	dig prot %	*	0.5	7.8	
Sheep	dig prot %	*	0.5	7.8	
Fatty acids	%		0.2	3.3	
Energy	GE Mcal/kg				
Cattle	DE Mcal/kg	*	0.21	3.51	
Sheep	DE Mcal/kg	*	0.20	3.36	
Swine	DE kcal/kg	*	192.	3144.	
Cattle	ME Mcal/kg	*	0.18	2.88	
Sheep	ME Mcal/kg	*	0.17	2.76	
Swine	ME kcal/kg	*	180.	2946.	

Continued

Feed Name or Analyses		Mean As Fed	Mean Dry	C.V. ± %
Cattle	TDN %	* 4.9	74.6	
Sheep	TDN %	* 4.6	76.2	
Swine	TDN %	* 4.3	71.3	

MESQUITE. Prosopis spp

Mesquite, seeds w pods, (1)

Ref No 1 08 579 United States

Feed Name or Analyses		Mean As Fed	Mean Dry	C.V. ± %
Dry matter	%	94.0	100.0	
Ash	%	4.5	4.8	
Crude fiber	%	26.3	28.0	
Ether extract	%	2.8	3.0	
N-free extract	%	47.4	50.4	
Protein (N x 6.25)	%	13.0	13.8	
Cattle	dig prot %	* 8.4	8.9	
Goats	dig prot %	* 8.9	9.5	
Horses	dig prot %	* 8.7	9.3	
Rabbits	dig prot %	* 8.8	9.3	
Sheep	dig prot %	* 8.4	9.0	
Energy	GE Mcal/kg			
Cattle	DE Mcal/kg	* 2.91	3.10	
Sheep	DE Mcal/kg	* 3.02	3.21	
Cattle	ME Mcal/kg	* 2.39	2.54	
Sheep	ME Mcal/kg	* 2.47	2.63	
Cattle	TDN %	* 66.2	70.4	
Sheep	TDN %	* 68.4	72.8	

Mesquite, browse, (2)

Ref No 2 03 079 United States

Feed Name or Analyses		Mean As Fed	Mean Dry	C.V. ± %
Dry matter	%		100.0	
Carotene	mg/kg		44.1	
Vitamin A equivalent	IU/g		73.5	

MESQUITE, COMMON. Prosopis juliflora

Mesquite, common, browse, (2)

Ref No 2 03 081 United States

Feed Name or Analyses		Mean As Fed	Mean Dry	C.V. ± %
Dry matter	%		100.0	
Ash	%		5.9	28
Crude fiber	%		27.3	7
Ether extract	%		3.4	19
N-free extract	%		42.3	
Protein (N x 6.25)	%		21.1	11
Cattle	dig prot %	*	15.8	
Goats	dig prot %	*	16.2	
Horses	dig prot %	*	15.4	
Rabbits	dig prot %	*	15.0	
Sheep	dig prot %	*	16.7	
Calcium	%		1.94	54
Magnesium	%		0.23	21
Phosphorus	%		0.19	28
Potassium	%		1.41	6

Mesquite, common, browse, immature, (2)

Ref No 2 03 080 United States

Feed Name or Analyses		Mean As Fed	Mean Dry	C.V. ± %
Dry matter	%		100.0	
Ash	%		4.7	
Crude fiber	%		24.7	
Ether extract	%		2.9	
N-free extract	%		44.2	
Protein (N x 6.25)	%		23.5	
Cattle	dig prot %	*	17.9	
Goats	dig prot %	*	18.5	
Horses	dig prot %	*	17.5	
Rabbits	dig prot %	*	16.8	
Sheep	dig prot %	*	18.9	

Mesquite, common, pods, fresh, (2)

Ref No 2 03 083 United States

Feed Name or Analyses		Mean As Fed	Mean Dry	C.V. ± %
Dry matter	%		100.0	
Ash	%		3.6	
Crude fiber	%		28.8	
Ether extract	%		1.9	
N-free extract	%		51.3	
Protein (N x 6.25)	%		14.4	
Cattle	dig prot %	*	10.1	
Goats	dig prot %	*	10.0	
Horses	dig prot %	*	9.8	
Rabbits	dig prot %	*	9.8	
Sheep	dig prot %	*	10.4	

Mesquite, common, pods, fresh, mature, (2)

Ref No 2 03 082 United States

Feed Name or Analyses		Mean As Fed	Mean Dry	C.V. ± %
Dry matter	%		100.0	
Calcium	%		0.66	
Magnesium	%		0.12	
Phosphorus	%		0.21	
Potassium	%		1.42	

Mesquite, common, seeds, (4)

Ref No 4 03 085 United States

Feed Name or Analyses		Mean As Fed	Mean Dry	C.V. ± %
Dry matter	%	94.6	100.0	
Ash	%	4.8	5.1	
Crude fiber	%	27.5	29.1	
Sheep	dig coef %	59.	59.	
Ether extract	%	1.9	2.0	
Sheep	dig coef %	95.	95.	
N-free extract	%	48.4	51.2	
Sheep	dig coef %	81.	81.	
Protein (N x 6.25)	%	11.9	12.6	
Sheep	dig coef %	90.	90.	
Cattle	dig prot %	* 7.2	7.6	
Goats	dig prot %	* 8.3	8.8	
Horses	dig prot %	* 8.3	8.8	
Sheep	dig prot %	10.7	11.3	
Energy	GE Mcal/kg			
Cattle	DE Mcal/kg	* 3.14	3.32	
Sheep	DE Mcal/kg	* 3.10	3.27	
Cattle	ME Mcal/kg	*2636.	2786.	
Cattle	ME Mcal/kg	* 2.57	2.72	
Sheep	ME Mcal/kg	* 2.54	2.68	
Swine	ME kcal/kg	*2023.	2139.	
Cattle	TDN %	* 71.3	75.4	
Sheep	TDN %	70.2	74.3	
Swine	TDN %	* 59.8	63.2	

Mesquite, common, seeds, low protein, (4)

Ref No 4 03 084 United States

Feed Name or Analyses		Mean As Fed	Mean Dry	C.V. ± %
Dry matter	%	89.2	100.0	
Ash	%	3.7	4.2	
Crude fiber	%	20.7	23.2	
Cattle	dig coef %	54.	54.	
Ether extract	%	1.3	1.5	
Cattle	dig coef %	75.	75.	
N-free extract	%	54.9	61.6	
Cattle	dig coef %	70.	70.	
Protein (N x 6.25)	%	8.5	9.5	
Cattle	dig coef %	70.	70.	
Cattle	dig prot %	5.9	6.7	
Goats	dig prot %	* 5.3	5.9	
Horses	dig prot %	* 5.3	5.9	
Sheep	dig prot %	* 5.3	5.9	
Energy	GE Mcal/kg			
Cattle	DE Mcal/kg	* 2.55	2.86	
Sheep	DE Mcal/kg	* 2.60	2.92	
Swine	DE kcal/kg	*2367.	2654.	
Cattle	ME Mcal/kg	* 2.09	2.34	
Sheep	ME Mcal/kg	* 2.13	2.39	
Swine	ME kcal/kg	*2227.	2497.	
Cattle	TDN %	57.8	64.8	
Sheep	TDN %	* 59.1	66.3	
Swine	TDN %	* 53.7	60.2	

METHIONINE. Scientific name not applicable

Methionine, mn 95%
DL-2-amino-4-methylthiobutyric acid, (5)
Methionine (AAFCO)

Ref No 5 03 086 United States

Feed Name or Analyses		Mean As Fed	Mean Dry	C.V. ± %
Dry matter	%	98.0	100.0	
Methionine	%	98.00	100.00	

MEXICANBUCKEYE. Ungnadia speciosa

Mexicanbuckeye, browse, (2)

Ref No 2 03 089 United States

Feed Name or Analyses		Mean As Fed	Mean Dry	C.V. ± %
Dry matter	%		100.0	
Ash	%		6.4	
Crude fiber	%		23.4	
Ether extract	%		2.4	
N-free extract	%		55.0	
Protein (N x 6.25)	%		12.8	
Cattle	dig prot %	*	8.8	
Goats	dig prot %	*	8.5	
Horses	dig prot %	*	8.4	
Rabbits	dig prot %	*	8.6	
Sheep	dig prot %	*	8.9	
Calcium	%		2.42	
Magnesium	%		0.35	
Phosphorus	%		0.17	
Potassium	%		1.23	

Middlings, mx 4.5 fbr (CFA) –
see Wheat, flour by-product, fine sift, mx 4% fiber, (4)

Milk, albumin, dried –
see Cattle, whey albumin, heat-and acid-precipitated dehy, mn 75% protein, (5)

(1) dry forages and roughages
(2) pasture, range plants, and forages fed green
(3) silages
(4) energy feeds
(5) protein supplements
(6) minerals
(7) vitamins
(8) additives

Feed Name or Analyses		Mean As Fed	Mean Dry	C.V. ± %

Milk, cattle, fresh –
see Cattle, milk, fresh, (5)

Milk, goat, fresh –
see Goat, milk, fresh, (5)

Milk, skimmed, dried –
see Cattle, milk, skimmed dehy, mx 8% moisture, (5)

MILKVETCH, CHINESE. Astragalus sinicus

Milkvetch, Chinese, hay, s-c, (1)
Ref No 1 03 090 United States

Analyses		As Fed	Dry	C.V.
Dry matter	%	86.4	100.0	
Ash	%	6.1	7.1	
Crude fiber	%	26.3	30.4	
Sheep	dig coef %	43.	43.	
Ether extract	%	2.9	3.3	
Sheep	dig coef %	68.	68.	
N-free extract	%	32.9	38.1	
Sheep	dig coef %	69.	69.	
Protein (N x 6.25)	%	18.2	21.1	
Sheep	dig coef %	69.	69.	
Cattle	dig prot % *	13.1	15.2	
Goats	dig prot % *	14.0	16.2	
Horses	dig prot % *	13.3	15.4	
Rabbits	dig prot % *	12.9	15.0	
Sheep	dig prot %	12.6	14.6	
Energy	GE Mcal/kg			
Cattle	DE Mcal/kg *	2.18	2.52	
Sheep	DE Mcal/kg *	2.25	2.60	
Cattle	ME Mcal/kg *	1.79	2.07	
Sheep	ME Mcal/kg *	1.84	2.13	
Cattle	TDN % *	49.4	57.2	
Sheep	TDN %	50.9	59.0	

MILKVETCH, GROUNDPLUM. Astragalus crossicarpus

Milkvetch, groundplum, seeds w hulls, (5)
Ref No 5 09 054 United States

Analyses		As Fed	Dry	C.V.
Dry matter	%		100.0	
Protein (N x 6.25)	%		45.0	
Alanine	%		1.17	
Arginine	%		3.83	
Aspartic acid	%		2.88	
Glutamic acid	%		5.94	
Glycine	%		1.80	
Histidine	%		0.81	
Hydroxyproline	%		0.05	
Isoleucine	%		1.13	
Leucine	%		1.89	
Lysine	%		1.40	
Methionine	%		0.32	
Phenylalanine	%		1.08	
Proline	%		1.31	
Serine	%		1.53	
Threonine	%		1.17	
Tyrosine	%		0.95	
Valine	%		1.35	

MILKVETCH, MEXICAN. Astragalus mexicanus

Milkvetch, Mexican, seeds w hulls, (5)
Ref No 5 09 303 United States

Analyses		As Fed	Dry	C.V.
Dry matter	%		100.0	
Protein (N x 6.25)	%		39.0	
Alanine	%		1.29	
Arginine	%		3.98	
Aspartic acid	%		3.00	
Glutamic acid	%		6.32	
Glycine	%		1.79	
Histidine	%		0.90	
Hydroxyproline	%		0.23	
Isoleucine	%		1.17	
Leucine	%		1.95	
Lysine	%		1.48	
Methionine	%		0.31	
Phenylalanine	%		1.17	
Proline	%		1.37	
Serine	%		1.64	
Threonine	%		1.13	
Tyrosine	%		0.98	
Valine	%		1.37	

MILKWEED, COMMON. Asclepias syriaca

Milkweed, common, seeds w hulls, (5)
Ref No 5 09 137 United States

Analyses		As Fed	Dry	C.V.
Dry matter	%		100.0	
Protein (N x 6.25)	%		37.0	
Alanine	%		1.33	
Arginine	%		3.55	
Aspartic acid	%		2.78	
Glutamic acid	%		8.07	
Glycine	%		1.92	
Histidine	%		0.85	
Hydroxyproline	%		0.04	
Isoleucine	%		1.30	
Leucine	%		2.29	
Lysine	%		1.81	
Methionine	%		0.52	
Phenylalanine	%		1.78	
Proline	%		1.52	
Serine	%		1.52	
Threonine	%		1.00	
Tyrosine	%		1.26	
Valine	%		1.59	

Milk, whole, dried –
see Cattle, milk, dehy, feed gr mx 8% moisture mn 26% fat, (5)

MILLET. Setaria spp

Millet, hay, s-c, (1)
Ref No 1 03 093 United States

Analyses		As Fed	Dry	C.V.
Dry matter	%	90.6	100.0	
Ash	%	9.0	9.9	
Crude fiber	%	27.9	30.8	
Sheep	dig coef %	67.	67.	
Ether extract	%	1.6	1.8	
Sheep	dig coef %	55.	55.	
N-free extract	%	47.8	52.8	
Sheep	dig coef %	60.	60.	
Protein (N x 6.25)	%	4.3	4.7	
Sheep	dig coef %	29.	29.	
Cattle	dig prot % *	0.9	1.0	
Goats	dig prot % *	0.9	0.9	
Horses	dig prot % *	1.4	1.5	
Rabbits	dig prot % *	2.1	2.3	
Sheep	dig prot %	1.2	1.4	
Energy	GE Mcal/kg			
Cattle	DE Mcal/kg *	2.14	2.36	
Sheep	DE Mcal/kg *	2.23	2.47	
Cattle	ME Mcal/kg *	1.76	1.94	
Sheep	ME Mcal/kg *	1.83	2.02	
Cattle	TDN % *	48.6	53.6	
Sheep	TDN %	50.7	55.9	

Millet, hay, s-c, mature, (1)
Ref No 1 03 091 United States

Analyses		As Fed	Dry	C.V.
Dry matter	%		100.0	
Crude fiber	%			
Sheep	dig coef % *		67.	
Ether extract	%			
Sheep	dig coef % *		55.	
N-free extract	%			
Sheep	dig coef % *		60.	
Protein (N x 6.25)	%			
Sheep	dig coef % *		29.	

Millet, hay, s-c, cut 1, (1)
Ref No 1 03 092 United States

Analyses		As Fed	Dry	C.V.
Dry matter	%	88.6	100.0	3
Calcium	%	0.29	0.33	13
Chlorine	%	0.12	0.13	
Magnesium	%	0.20	0.23	15
Manganese	mg/kg	118.7	134.0	11
Phosphorus	%	0.18	0.20	32
Potassium	%	1.72	1.94	7
Sodium	%	0.09	0.10	
Sulphur	%	0.14	0.16	

Millet, hulls, (1)
Ref No 1 03 094 United States

Analyses		As Fed	Dry	C.V.
Dry matter	%	88.4	100.0	
Ash	%	9.5	10.8	
Crude fiber	%	45.8	51.8	
Sheep	dig coef %	4.	4.	
Ether extract	%	1.2	1.4	
Sheep	dig coef %	112.	112.	
N-free extract	%	27.9	31.6	
Sheep	dig coef %	9.	9.	
Protein (N x 6.25)	%	3.9	4.4	
Sheep	dig coef %	19.	19.	
Cattle	dig prot % *	0.7	0.8	
Goats	dig prot % *	0.6	0.7	
Horses	dig prot % *	1.1	1.3	
Rabbits	dig prot % *	1.8	2.1	
Sheep	dig prot %	0.7	0.8	
Energy	GE Mcal/kg			
Cattle	DE Mcal/kg *	0.53	0.60	
Sheep	DE Mcal/kg *	0.36	0.41	
Cattle	ME Mcal/kg *	0.43	0.49	
Sheep	ME Mcal/kg *	0.30	0.34	
Cattle	TDN % *	12.0	13.6	
Sheep	TDN %	8.2	9.3	

Feed Name or Analyses		Mean As Fed	Mean Dry	C.V. ± %

Millet, straw, (1)
Ref No 1 03 095 United States

Feed Name or Analyses		As Fed	Dry	C.V. ± %
Dry matter	%	89.5	100.0	2
Ash	%	5.5	6.1	4
Crude fiber	%	37.3	41.7	4
Ether extract	%	1.6	1.8	12
N-free extract	%	41.3	46.2	
Protein (N x 6.25)	%	3.8	4.2	8
Cattle	dig prot %	* 0.5	0.6	
Goats	dig prot %	* 0.4	0.5	
Horses	dig prot %	* 1.0	1.1	
Rabbits	dig prot %	* 1.7	1.9	
Sheep	dig prot %	* 0.3	0.3	
Fatty acids	%	1.6	1.8	
Energy	GE Mcal/kg			
Cattle	DE Mcal/kg	* 1.78	1.99	
Sheep	DE Mcal/kg	* 1.99	2.22	
Cattle	ME Mcal/kg	* 1.46	1.63	
Sheep	ME Mcal/kg	* 1.63	1.82	
Cattle	TDN %	* 40.4	45.1	
Sheep	TDN %	* 45.0	50.4	
Calcium	%	0.08	0.09	
Potassium	%	1.43	1.60	

Millet, aerial part, ensiled, (3)
Ref No 3 03 096 United States

Feed Name or Analyses		As Fed	Dry	C.V. ± %
Dry matter	%	31.5	100.0	6
Ash	%	3.0	9.4	8
Crude fiber	%	9.8	31.2	7
Ether extract	%	1.0	3.3	22
N-free extract	%	15.0	47.7	
Protein (N x 6.25)	%	2.7	8.4	10
Cattle	dig prot %	* 1.2	3.9	
Goats	dig prot %	* 1.2	3.9	
Horses	dig prot %	* 1.2	3.9	
Sheep	dig prot %	* 1.2	3.9	
Energy	GE Mcal/kg			
Cattle	DE Mcal/kg	* 0.75	2.39	
Sheep	DE Mcal/kg	* 0.80	2.54	
Cattle	ME Mcal/kg	* 0.62	1.96	
Sheep	ME Mcal/kg	* 0.66	2.08	
Cattle	TDN %	* 17.1	54.3	
Sheep	TDN %	* 18.1	57.6	
Calcium	%	0.13	0.42	
Phosphorus	%	0.08	0.26	
Carotene	mg/kg	19.2	61.1	
Vitamin A equivalent	IU/g	32.1	101.8	

Millet, aerial part w molasses added, ensiled, (3)
Ref No 3 03 097 United States

Feed Name or Analyses		As Fed	Dry	C.V. ± %
Dry matter	%	32.0	100.0	
Calcium	%	0.10	0.31	
Phosphorus	%	0.09	0.28	

(1) dry forages and roughages
(2) pasture, range plants, and forages fed green
(3) silages
(4) energy feeds
(5) protein supplements
(6) minerals
(7) vitamins
(8) additives

Millet, grain, (4)
Ref No 4 03 098 United States

Feed Name or Analyses		As Fed	Dry	C.V. ± %
Dry matter	%	89.9	100.0	1
Ash	%	2.9	3.3	36
Crude fiber	%	6.4	7.2	24
Sheep	dig coef %	38.	38.	
Ether extract	%	4.0	4.5	18
Sheep	dig coef %	56.	56.	
N-free extract	%	64.3	71.6	
Sheep	dig coef %	65.	65.	
Protein (N x 6.25)	%	12.2	13.5	7
Sheep	dig coef %	50.	50.	
Cattle	dig prot %	* 7.6	8.4	
Goats	dig prot %	* 8.7	9.6	
Horses	dig prot %	* 8.7	9.6	
Sheep	dig prot %	6.1	6.8	
Pentosans	%	6.5	7.2	
Starch	%	46.5	51.8	
Energy	GE Mcal/kg			
Cattle	DE Mcal/kg	* 2.65	2.95	
Sheep	DE Mcal/kg	* 2.44	2.72	
Swine	DE kcal/kg	*3073.	3420.	
Cattle	ME Mcal/kg	* 2.18	2.42	
Sheep	ME Mcal/kg	* 2.00	2.23	
Swine	ME kcal/kg	*2866.	3190.	
Cattle	TDN %	* 60.1	66.8	
Sheep	TDN %	55.4	61.6	
Swine	TDN %	* 69.7	77.6	
Calcium	%	0.05	0.06	
Chlorine	%	0.14	0.16	
Cobalt	mg/kg	0.044	0.049	
Copper	mg/kg	21.6	24.0	
Iron	%	0.004	0.005	
Magnesium	%	0.16	0.18	
Manganese	mg/kg	29.1	32.4	98
Phosphorus	%	0.28	0.31	30
Potassium	%	0.43	0.48	
Sodium	%	0.04	0.04	
Sulphur	%	0.13	0.14	
Zinc	mg/kg	13.9	15.4	
Choline	mg/kg	788.	877.	21
Niacin	mg/kg	52.5	58.4	28
Pantothenic acid	mg/kg	7.3	8.2	55
Riboflavin	mg/kg	1.6	1.8	23
Thiamine	mg/kg	6.5	7.3	11

MILLET, FOXTAIL. Setaria italica

Millet, foxtail, hay, s-c, (1)
Ref No 1 03 099 United States

Feed Name or Analyses		As Fed	Dry	C.V. ± %
Dry matter	%	85.8	100.0	3
Ash	%	7.0	8.1	14
Crude fiber	%	25.0	29.2	4
Sheep	dig coef %	68.	68.	
Ether extract	%	2.8	3.3	12
Sheep	dig coef %	64.	64.	
N-free extract	%	42.5	49.5	
Sheep	dig coef %	67.	67.	
Protein (N x 6.25)	%	8.5	9.9	6
Sheep	dig coef %	60.	60.	
Cattle	dig prot %	* 4.7	5.5	
Goats	dig prot %	* 5.0	5.8	
Horses	dig prot %	* 5.1	5.9	
Rabbits	dig prot %	* 5.4	6.3	
Sheep	dig prot %	5.1	5.9	
Fatty acids	%	2.6	3.1	
Energy	GE Mcal/kg			
Cattle	DE Mcal/kg	* 2.16	2.52	
Sheep	DE Mcal/kg	* 2.41	2.81	
Cattle	ME Mcal/kg	* 1.78	2.07	
Sheep	ME Mcal/kg	* 1.97	2.30	
Cattle	TDN %	* 49.0	57.1	
Sheep	TDN %	54.6	63.7	
Calcium	%	0.28	0.33	
Chlorine	%	0.11	0.13	
Magnesium	%	0.20	0.23	
Manganese	mg/kg	118.5	138.2	
Phosphorus	%	0.16	0.18	
Potassium	%	1.66	1.94	
Sodium	%	0.09	0.10	
Sulphur	%	0.14	0.16	

Millet, foxtail, aerial part, fresh, (2)
Ref No 2 03 101 United States

Feed Name or Analyses		As Fed	Dry	C.V. ± %
Dry matter	%	28.4	100.0	17
Ash	%	2.4	8.6	
Crude fiber	%	9.1	32.0	
Sheep	dig coef %	67.	67.	
Ether extract	%	0.9	3.1	
Sheep	dig coef %	59.	59.	
N-free extract	%	13.2	46.7	
Sheep	dig coef %	67.	67.	
Protein (N x 6.25)	%	2.7	9.6	
Sheep	dig coef %	61.	61.	
Cattle	dig prot %	* 1.7	6.1	
Goats	dig prot %	* 1.6	5.6	
Horses	dig prot %	* 1.6	5.7	
Rabbits	dig prot %	* 1.7	6.1	
Sheep	dig prot %	1.7	5.9	
Energy	GE Mcal/kg			
Cattle	DE Mcal/kg	* 0.80	2.81	
Sheep	DE Mcal/kg	* 0.78	2.76	
Cattle	ME Mcal/kg	* 0.65	2.31	
Sheep	ME Mcal/kg	* 0.64	2.27	
Cattle	TDN %	* 18.1	63.8	
Sheep	TDN %	17.8	62.7	
Calcium	%	0.09	0.32	
Phosphorus	%	0.05	0.19	
Potassium	%	0.55	1.94	

Millet, foxtail, aerial part, fresh, late bloom, (2)
Ref No 2 03 100 United States

Feed Name or Analyses		As Fed	Dry	C.V. ± %
Dry matter	%	26.2	100.0	
Ash	%	2.3	8.8	
Crude fiber	%	8.1	30.9	
Sheep	dig coef %	71.	71.	
Ether extract	%	0.8	3.1	
Sheep	dig coef %	61.	61.	
N-free extract	%	12.6	48.0	
Sheep	dig coef %	68.	68.	
Protein (N x 6.25)	%	2.4	9.2	
Sheep	dig coef %	61.	61.	
Cattle	dig prot %	* 1.5	5.7	
Goats	dig prot %	* 1.3	5.1	
Horses	dig prot %	* 1.4	5.3	
Rabbits	dig prot %	* 1.5	5.8	
Sheep	dig prot %	1.5	5.6	
Energy	GE Mcal/kg			
Cattle	DE Mcal/kg	* 0.73	2.77	
Sheep	DE Mcal/kg	* 0.74	2.84	
Cattle	ME Mcal/kg	* 0.60	2.27	
Sheep	ME Mcal/kg	* 0.61	2.33	

Feed Name or Analyses		Mean As Fed	Mean Dry	C.V. ± %
Cattle	TDN %	* 16.5	62.8	
Sheep	TDN %	16.9	64.4	

Millet, foxtail, grain, (4)

Ref No 4 03 102 United States

Feed Name or Analyses		Mean As Fed	Mean Dry	C.V. ± %
Dry matter	%	89.3	100.0	2
Ash	%	3.6	4.0	27
Crude fiber	%	8.3	9.3	14
Sheep	dig coef %	* 38.	38.	
Swine	dig coef %	* 33.	33.	
Ether extract	%	4.1	4.6	12
Sheep	dig coef %	* 56.	56.	
Swine	dig coef %	* 59.	59.	
N-free extract	%	61.2	68.5	
Sheep	dig coef %	* 65.	65.	
Swine	dig coef %	* 92.	92.	
Protein (N x 6.25)	%	12.1	13.5	6
Sheep	dig coef %	* 50.	50.	
Swine	dig coef %	* 68.	68.	
Cattle	dig prot %	* 7.5	8.4	
Goats	dig prot %	* 8.6	9.6	
Horses	dig prot %	* 8.6	9.6	
Sheep	dig prot %	6.0	6.8	
Swine	dig prot %	8.2	9.2	
Energy	GE Mcal/kg			
Cattle	DE Mcal/kg	* 3.03	3.39	
Sheep	DE Mcal/kg	* 2.39	2.67	
Swine	DE kcal/kg	*3206.	3590.	
Cattle	ME Mcal/kg	* 2.48	2.78	
Sheep	ME Mcal/kg	* 1.96	2.19	
Swine	ME kcal/kg	*2990.	3348.	
Cattle	TDN %	* 68.7	76.9	
Sheep	TDN %	54.2	60.6	
Swine	TDN %	72.7	81.4	
Phosphorus	%	0.20	0.22	
Potassium	%	0.31	0.35	

MILLET, JAPANESE. Echinochloa crusgalli

Millet, Japanese, hay, dehy, (1)

Ref No 1 03 104 United States

Feed Name or Analyses		Mean As Fed	Mean Dry	C.V. ± %
Dry matter	%	89.1	100.0	1
Ash	%	9.0	10.1	10
Crude fiber	%	19.4	21.8	10
Cattle	dig coef %	75.	75.	
Ether extract	%	2.3	2.6	5
Cattle	dig coef %	69.	69.	
N-free extract	%	46.6	52.3	
Cattle	dig coef %	76.	76.	
Protein (N x 6.25)	%	11.8	13.3	7
Cattle	dig coef %	60.	60.	
Cattle	dig prot %	7.1	8.0	
Goats	dig prot %	* 7.9	8.9	
Horses	dig prot %	* 7.8	8.8	
Rabbits	dig prot %	* 7.9	8.9	
Sheep	dig prot %	* 7.5	8.5	
Energy	GE Mcal/kg			
Cattle	DE Mcal/kg	* 2.67	3.00	
Sheep	DE Mcal/kg	* 2.49	2.79	
Cattle	ME Mcal/kg	* 2.19	2.46	
Sheep	ME Mcal/kg	* 2.04	2.29	
Cattle	TDN %	60.6	68.0	
Sheep	TDN %	* 56.4	63.3	

Millet, Japanese, hay, dehy, early bloom, (1)

Ref No 1 03 103 United States

Feed Name or Analyses		Mean As Fed	Mean Dry	C.V. ± %
Dry matter	%	89.2	100.0	2
Ash	%	8.2	9.2	15
Crude fiber	%	17.1	19.2	13
Ether extract	%	2.5	2.8	3
N-free extract	%	50.2	56.3	
Protein (N x 6.25)	%	11.2	12.5	10
Cattle	dig prot %	* 6.9	7.8	
Goats	dig prot %	* 7.3	8.2	
Horses	dig prot %	* 7.3	8.1	
Rabbits	dig prot %	* 7.4	8.3	
Sheep	dig prot %	* 6.9	7.8	
Energy	GE Mcal/kg			
Cattle	DE Mcal/kg	* 2.45	2.75	
Sheep	DE Mcal/kg	* 2.50	2.80	
Cattle	ME Mcal/kg	* 2.02	2.26	
Sheep	ME Mcal/kg	* 2.05	2.30	
Cattle	TDN %	* 55.7	62.4	
Sheep	TDN %	* 56.7	63.6	

Millet, Japanese, hay, s-c, (1)

Ref No 1 03 105 United States

Feed Name or Analyses		Mean As Fed	Mean Dry	C.V. ± %
Dry matter	%	87.3	100.0	1
Ash	%	8.5	9.8	11
Crude fiber	%	26.3	30.2	13
Ether extract	%	1.7	2.0	15
N-free extract	%	41.9	48.0	
Protein (N x 6.25)	%	8.8	10.1	11
Cattle	dig prot %	* 5.0	5.7	
Goats	dig prot %	* 5.2	6.0	
Horses	dig prot %	* 5.4	6.1	
Rabbits	dig prot %	* 5.7	6.5	
Sheep	dig prot %	* 4.9	5.7	
Fatty acids	%	1.6	1.8	
Energy	GE Mcal/kg			
Cattle	DE Mcal/kg	* 2.10	2.40	
Sheep	DE Mcal/kg	* 2.21	2.54	
Cattle	ME Mcal/kg	* 1.72	1.97	
Sheep	ME Mcal/kg	* 1.81	2.08	
Cattle	TDN %	* 47.6	54.5	
Sheep	TDN %	* 50.2	57.5	
Calcium	%	0.20	0.23	
Potassium	%	2.11	2.42	

Millet, Japanese, aerial part, fresh, (2)

Japanesemillet, aerial part, fresh

Ref No 2 03 108 United States

Feed Name or Analyses		Mean As Fed	Mean Dry	C.V. ± %
Dry matter	%	21.5	100.0	
Ash	%	2.0	9.3	
Crude fiber	%	6.2	28.8	
Sheep	dig coef %	* 62.	62.	
Ether extract	%	0.5	2.6	
Sheep	dig coef %	* 68.	68.	
N-free extract	%	10.5	48.8	
Sheep	dig coef %	* 67.	67.	
Protein (N x 6.25)	%	2.3	10.5	
Sheep	dig coef %	* 50.	50.	
Cattle	dig prot %	* 1.5	6.8	
Goats	dig prot %	* 1.4	6.3	
Horses	dig prot %	* 1.4	6.4	
Rabbits	dig prot %	* 1.5	6.8	
Sheep	dig prot %	1.1	5.2	
Energy	GE Mcal/kg			
Cattle	DE Mcal/kg	* 0.61	2.82	
Sheep	DE Mcal/kg	* 0.57	2.63	
Cattle	ME Mcal/kg	* 0.50	2.31	
Sheep	ME Mcal/kg	* 0.46	2.16	
Cattle	TDN %	* 13.7	63.9	
Sheep	TDN %	12.8	59.7	
Calcium	%	0.11	0.51	
Phosphorus	%	0.07	0.32	
Potassium	%	0.52	2.40	

Millet, Japanese, aerial part, fresh, late bloom, (2)

Ref No 2 03 106 United States

Feed Name or Analyses		Mean As Fed	Mean Dry	C.V. ± %
Dry matter	%	24.4	100.0	
Ash	%	2.1	8.6	
Crude fiber	%	7.6	31.1	
Sheep	dig coef %	59.	59.	
Ether extract	%	0.7	2.9	
Sheep	dig coef %	60.	60.	
N-free extract	%	12.2	50.0	
Sheep	dig coef %	64.	64.	
Protein (N x 6.25)	%	1.8	7.4	
Sheep	dig coef %	57.	57.	
Cattle	dig prot %	* 1.0	4.2	
Goats	dig prot %	* 0.8	3.5	
Horses	dig prot %	* 0.9	3.8	
Rabbits	dig prot %	* 1.1	4.4	
Sheep	dig prot %	1.0	4.2	
Energy	GE Mcal/kg			
Cattle	DE Mcal/kg	* 0.66	2.71	
Sheep	DE Mcal/kg	* 0.63	2.58	
Cattle	ME Mcal/kg	* 0.54	2.22	
Sheep	ME Mcal/kg	* 0.52	2.11	
Cattle	TDN %	* 15.0	61.5	
Sheep	TDN %	14.3	58.5	

Millet, Japanese, aerial part, fresh, milk stage, (2)

Ref No 2 03 107 United States

Feed Name or Analyses		Mean As Fed	Mean Dry	C.V. ± %
Dry matter	%	29.7	100.0	
Ash	%	2.1	7.1	
Crude fiber	%	9.2	30.9	
Sheep	dig coef %	63.	63.	
Ether extract	%	0.9	3.1	
Sheep	dig coef %	70.	70.	
N-free extract	%	15.7	53.0	
Sheep	dig coef %	68.	68.	
Protein (N x 6.25)	%	1.8	5.9	
Sheep	dig coef %	49.	49.	
Cattle	dig prot %	* 0.9	2.9	
Goats	dig prot %	* 0.6	2.1	
Horses	dig prot %	* 0.8	2.5	
Rabbits	dig prot %	* 1.0	3.2	
Sheep	dig prot %	0.8	2.9	
Energy	GE Mcal/kg			
Cattle	DE Mcal/kg	* 0.81	2.72	
Sheep	DE Mcal/kg	* 0.82	2.77	
Cattle	ME Mcal/kg	* 0.66	2.23	
Sheep	ME Mcal/kg	* 0.67	2.27	
Cattle	TDN %	* 18.3	61.6	
Sheep	TDN %	18.6	62.9	

Millet, Japanese, aerial part, ensiled, (3)
Ref No 3 03 109 United States

Feed Name or Analyses				Mean As Fed	Mean Dry	C.V. ± %
Dry matter		%		22.9	100.0	
Ash		%		1.6	7.0	
Crude fiber		%		8.5	37.1	
Ether extract		%		0.6	2.6	
N-free extract		%		10.4	45.4	
Protein (N x 6.25)		%		1.8	7.9	
Cattle	dig prot	%	*	0.8	3.4	
Goats	dig prot	%	*	0.8	3.4	
Horses	dig prot	%	*	0.8	3.4	
Sheep	dig prot	%	*	0.8	3.4	
Energy	GE Mcal/kg					
Cattle	DE Mcal/kg		*	0.57	2.47	
Sheep	DE Mcal/kg		*	0.58	2.53	
Cattle	ME Mcal/kg		*	0.46	2.03	
Sheep	ME Mcal/kg		*	0.47	2.07	
Cattle	TDN	%	*	12.9	56.1	
Sheep	TDN	%	*	13.1	57.3	

Millet, Japanese, grain, (4)
Ref No 4 03 110 United States

Feed Name or Analyses				Mean As Fed	Mean Dry	C.V. ± %
Dry matter		%		89.8	100.0	2
Ash		%		4.6	5.1	20
Crude fiber		%		13.9	15.5	12
Ether extract		%		5.1	5.6	6
N-free extract		%		55.7	62.1	
Protein (N x 6.25)		%		10.5	11.7	2
Cattle	dig prot	%	*	6.1	6.8	
Goats	dig prot	%	*	7.1	8.0	
Horses	dig prot	%	*	7.1	8.0	
Sheep	dig prot	%	*	7.1	8.0	
Energy	GE Mcal/kg					
Cattle	DE Mcal/kg		*	3.25	3.62	
Sheep	DE Mcal/kg		*	2.99	3.33	
Swine	DE kcal/kg		*	2871.	3197.	
Cattle	ME Mcal/kg		*	2.67	2.97	
Sheep	ME Mcal/kg		*	2.45	2.73	
Swine	ME kcal/kg		*	2687.	2992.	
Cattle	TDN	%	*	73.6	82.0	
Sheep	TDN	%	*	67.8	75.5	
Swine	TDN	%	*	65.1	72.5	
Copper		mg/kg		17.0	19.0	
Iron		%		0.004	0.005	
Manganese		mg/kg		66.3	73.9	
Phosphorus		%		0.40	0.44	
Potassium		%		0.33	0.37	

MILLET, PEARL. Pennisetum glaucum

Millet, pearl, hay, s-c, (1)
Ref No 1 03 112 United States

Feed Name or Analyses				Mean As Fed	Mean Dry	C.V. ± %
Dry matter		%		87.6	100.0	1
Ash		%		8.9	10.2	6
Crude fiber		%		32.3	36.9	7
Ether extract		%		1.8	2.0	12
N-free extract		%		37.2	42.5	
Protein (N x 6.25)		%		7.3	8.4	21
Cattle	dig prot	%	*	3.7	4.2	
Goats	dig prot	%	*	3.8	4.4	
Horses	dig prot	%	*	4.1	4.7	
Rabbits	dig prot	%	*	4.5	5.1	
Sheep	dig prot	%	*	3.6	4.1	
Fatty acids		%		1.7	1.9	
Energy	GE Mcal/kg					
Cattle	DE Mcal/kg		*	2.09	2.38	
Sheep	DE Mcal/kg		*	2.07	2.36	
Cattle	ME Mcal/kg		*	1.71	1.96	
Sheep	ME Mcal/kg		*	1.69	1.94	
Cattle	TDN	%	*	47.4	54.1	
Sheep	TDN	%	*	46.9	53.5	

(1) dry forages and roughages (2) pasture, range plants, and forages fed green (3) silages (4) energy feeds (5) protein supplements (6) minerals (7) vitamins (8) additives

Millet, pearl, aerial part, fresh, (2)
Ref No 2 03 115 United States

Feed Name or Analyses				Mean As Fed	Mean Dry	C.V. ± %
Dry matter		%		20.7	100.0	9
Ash		%		1.9	9.2	12
Crude fiber		%		6.4	31.1	8
Cattle	dig coef	%	*	60.	60.	
Ether extract		%		0.6	2.9	22
Cattle	dig coef	%	*	67.	67.	
N-free extract		%		9.7	46.8	
Cattle	dig coef	%	*	69.	69.	
Protein (N x 6.25)		%		2.1	10.1	18
Cattle	dig coef	%	*	62.	62.	
Cattle	dig prot	%		1.3	6.3	
Goats	dig prot	%	*	1.2	6.0	
Horses	dig prot	%	*	1.3	6.1	
Rabbits	dig prot	%	*	1.3	6.5	
Sheep	dig prot	%	*	1.3	6.4	
Energy	GE Mcal/kg					
Cattle	DE Mcal/kg		*	0.56	2.71	
Sheep	DE Mcal/kg		*	0.59	2.87	
Cattle	ME Mcal/kg		*	0.46	2.22	
Sheep	ME Mcal/kg		*	0.49	2.35	
Cattle	TDN	%		12.7	61.5	
Sheep	TDN	%	*	13.4	65.1	

Millet, pearl, aerial part, fresh, full bloom, (2)
Ref No 2 03 113 United States

Feed Name or Analyses				Mean As Fed	Mean Dry	C.V. ± %
Dry matter		%		21.6	100.0	
Ash		%		2.4	10.9	
Crude fiber		%		6.9	31.8	
Ether extract		%		0.3	1.5	
N-free extract		%		10.6	48.9	
Protein (N x 6.25)		%		1.5	6.9	
Cattle	dig prot	%	*	0.8	3.8	
Goats	dig prot	%	*	0.6	3.0	
Horses	dig prot	%	*	0.7	3.4	
Rabbits	dig prot	%	*	0.9	4.0	
Sheep	dig prot	%	*	0.7	3.4	
Energy	GE Mcal/kg					
Cattle	DE Mcal/kg		*	0.57	2.63	
Sheep	DE Mcal/kg		*	0.54	2.52	
Cattle	ME Mcal/kg		*	0.47	2.16	
Sheep	ME Mcal/kg		*	0.45	2.07	
Cattle	TDN	%	*	12.9	59.7	
Sheep	TDN	%	*	12.4	57.2	

Millet, pearl, aerial part, fresh, milk stage, (2)
Ref No 2 03 114 United States

Feed Name or Analyses				Mean As Fed	Mean Dry	C.V. ± %
Dry matter		%		21.6	100.0	
Ash		%		2.4	10.9	
Crude fiber		%		6.9	31.8	
Cattle	dig coef	%		60.	60.	
Ether extract		%		0.3	1.5	
Cattle	dig coef	%		67.	67.	
N-free extract		%		10.6	48.9	
Cattle	dig coef	%		69.	69.	
Protein (N x 6.25)		%		1.5	6.9	
Cattle	dig coef	%		62.	62.	
Cattle	dig prot	%		0.9	4.3	
Goats	dig prot	%	*	0.6	3.0	
Horses	dig prot	%	*	0.7	3.4	
Rabbits	dig prot	%	*	0.9	4.0	
Sheep	dig prot	%	*	0.7	3.4	
Energy	GE Mcal/kg					
Cattle	DE Mcal/kg		*	0.57	2.62	
Sheep	DE Mcal/kg		*	0.55	2.55	
Cattle	ME Mcal/kg		*	0.46	2.15	
Sheep	ME Mcal/kg		*	0.45	2.09	
Cattle	TDN	%		12.8	59.4	
Sheep	TDN	%	*	12.5	57.8	

Millet, pearl, aerial part, ensiled, immature, cut 1, (3)
Ref No 3 03 116 United States

Feed Name or Analyses				Mean As Fed	Mean Dry	C.V. ± %
Dry matter		%			100.0	
Calcium		%			0.34	12
Magnesium		%			0.33	25
Phosphorus		%			0.26	8
Potassium		%			1.63	4

Millet, pearl, aerial part, ensiled, cut 1, (3)
Ref No 3 03 117 United States

Feed Name or Analyses				Mean As Fed	Mean Dry	C.V. ± %
Dry matter		%			100.0	
Sodium		%			0.02	20

Millet, pearl, grain, (4)
Cattail millet grain
Ref No 4 03 118 United States

Feed Name or Analyses				Mean As Fed	Mean Dry	C.V. ± %
Dry matter		%		89.1	100.0	1
Ash		%		1.8	2.1	14
Crude fiber		%		2.5	2.8	15
Sheep	dig coef	%	*	14.	14.	
Swine	dig coef	%		35.	35.	
Ether extract		%		4.3	4.8	9
Sheep	dig coef	%	*	84.	84.	
Swine	dig coef	%		64.	64.	
N-free extract		%		68.3	76.6	
Sheep	dig coef	%	*	88.	88.	
Swine	dig coef	%		62.	62.	
Protein (N x 6.25)		%		12.3	13.8	8
Sheep	dig coef	%	*	65.	65.	
Swine	dig coef	%		61.	61.	
Cattle	dig prot	%	*	7.7	8.6	
Goats	dig prot	%	*	8.8	9.8	
Horses	dig prot	%	*	8.8	9.8	
Sheep	dig prot	%		8.0	8.9	
Swine	dig prot	%		7.5	8.4	
Energy	GE Mcal/kg			3.35	3.76	
Cattle	DE Mcal/kg		*	2.78	3.11	
Sheep	DE Mcal/kg		*	3.37	3.78	
Swine	DE kcal/kg		*	2506.	2812.	
Cattle	ME Mcal/kg		*	2.28	2.55	
Sheep	ME Mcal/kg		*	2.77	3.10	
Swine	ME kcal/kg		*	2336.	2621.	
Cattle	TDN	%	*	62.9	70.6	
Sheep	TDN	%		76.5	85.8	

Feed Name or Analyses		Mean As Fed	Mean Dry	C.V. ± %
Swine	TDN %	56.8	63.8	
Calcium	%	0.05	0.06	
Iron	%	0.001	0.001	
Phosphorus	%	0.38	0.43	

MILLET, PROSO. Panicum miliaceum

Millet, proso, hay, s-c, (1)

Proso, millet, hay s-c

Ref No 1 03 119 United States

Feed Name or Analyses		Mean As Fed	Mean Dry	C.V. ± %
Dry matter	%	90.3	100.0	1
Ash	%	7.6	8.4	7
Crude fiber	%	24.4	27.0	4
Ether extract	%	2.1	2.4	8
N-free extract	%	46.8	51.8	
Protein (N x 6.25)	%	9.4	10.4	8
Cattle	dig prot %	* 5.4	5.9	
Goats	dig prot %	* 5.7	6.3	
Horses	dig prot %	* 5.7	6.4	
Rabbits	dig prot %	* 6.0	6.7	
Sheep	dig prot %	* 5.3	5.9	
Fatty acids	%	2.2	2.4	
Energy	GE Mcal/kg			
Cattle	DE Mcal/kg	* 2.23	2.47	
Sheep	DE Mcal/kg	* 2.36	2.61	
Cattle	ME Mcal/kg	* 1.83	2.03	
Sheep	ME Mcal/kg	* 1.93	2.14	
Cattle	TDN %	* 50.7	56.1	
Sheep	TDN %	* 53.5	59.3	

Millet, proso, aerial part, fresh, (2)

Ref No 2 03 811 United States

Feed Name or Analyses		Mean As Fed	Mean Dry	C.V. ± %
Dry matter	%	25.1	100.0	13
Ash	%	1.9	7.4	15
Crude fiber	%	7.4	29.5	15
Ether extract	%	0.6	2.5	24
N-free extract	%	13.1	52.4	
Protein (N x 6.25)	%	2.1	8.2	21
Cattle	dig prot %	* 1.2	4.9	
Goats	dig prot %	* 1.1	4.3	
Horses	dig prot %	* 1.1	4.5	
Rabbits	dig prot %	* 1.3	5.0	
Sheep	dig prot %	* 1.2	4.7	
Energy	GE Mcal/kg			
Cattle	DE Mcal/kg	* 0.71	2.84	
Sheep	DE Mcal/kg	* 0.70	2.79	
Cattle	ME Mcal/kg	* 0.58	2.33	
Sheep	ME Mcal/kg	* 0.57	2.29	
Cattle	TDN %	* 16.1	64.5	
Sheep	TDN %	* 15.9	63.2	

Millet, proso, aerial part wo heads, ensiled, (3)

Ref No 3 08 460 United States

Feed Name or Analyses		Mean As Fed	Mean Dry	C.V. ± %
Dry matter	%	26.7	100.0	
Ash	%	2.2	8.2	
Crude fiber	%	9.5	35.6	
Ether extract	%	0.4	1.5	
N-free extract	%	13.3	49.8	
Protein (N x 6.25)	%	1.3	4.9	
Cattle	dig prot %	* 0.2	0.7	
Goats	dig prot %	* 0.2	0.7	
Horses	dig prot %	* 0.2	0.7	
Sheep	dig prot %	* 0.2	0.7	
Energy	GE Mcal/kg			
Cattle	DE Mcal/kg	* 0.59	2.20	
Sheep	DE Mcal/kg	* 0.59	2.20	
Cattle	ME Mcal/kg	* 0.48	1.80	
Sheep	ME Mcal/kg	* 0.48	1.80	
Cattle	TDN %	* 13.3	49.8	
Sheep	TDN %	* 13.3	49.9	

Millet, proso, grain, (4)

Broomcorn millet grain

Hershey millet grain

Hog millet grain

Proso millet grain

Ref No 4 03 120 United States

Feed Name or Analyses		Mean As Fed	Mean Dry	C.V. ± %
Dry matter	%	90.0	100.0	1
Ash	%	2.7	3.0	18
Crude fiber	%	6.5	7.2	46
Sheep	dig coef %	* 14.	14.	
Ether extract	%	3.6	4.0	18
Sheep	dig coef %	* 84.	84.	
N-free extract	%	65.7	73.0	6
Sheep	dig coef %	* 88.	88.	
Protein (N x 6.25)	%	11.5	12.8	10
Sheep	dig coef %	* 65.	65.	
Cattle	dig prot %	* 7.0	7.7	
Goats	dig prot %	* 8.0	8.9	
Horses	dig prot %	* 8.0	8.9	
Sheep	dig prot %	7.5	8.3	
Energy	GE Mcal/kg	3.47	3.86	
Cattle	DE Mcal/kg	* 3.06	3.40	
Sheep	DE Mcal/kg	* 3.22	3.58	
Swine	DE kcal/kg	*3084.	3427.	
Cattle	ME Mcal/kg	* 2.51	2.79	
Chickens	ME_n kcal/kg	2984.	3316.	
Sheep	ME Mcal/kg	* 2.64	2.93	
Swine	ME kcal/kg	*2881.	3202.	
Cattle	TDN %	* 69.4	77.1	
Sheep	TDN %	73.0	81.2	
Swine	TDN %	* 69.9	77.7	
Calcium	%	0.04	0.04	
Iron	%	0.007	0.008	
Magnesium	%	0.16	0.18	
Phosphorus	%	0.30	0.34	
Potassium	%	0.43	0.48	
Choline	mg/kg	439.	488.	
Niacin	mg/kg	23.3	25.9	
Pantothenic acid	mg/kg	11.0	12.2	
Riboflavin	mg/kg	3.8	4.2	
Thiamine	mg/kg	7.3	8.1	
Arginine	%	0.33	0.37	
Cysteine	%	0.25	0.28	
Glycine	%	0.30	0.33	
Histidine	%	0.18	0.20	
Isoleucine	%	0.45	0.50	
Leucine	%	1.08	1.20	
Lysine	%	0.23	0.26	
Methionine	%	0.28	0.32	
Phenylalanine	%	0.54	0.60	
Threonine	%	0.36	0.40	
Tryptophan	%	0.17	0.19	
Valine	%	0.54	0.60	

Millet, proso, grain, high fiber, (4)

Ref No 4 03 812 United States

Feed Name or Analyses		Mean As Fed	Mean Dry	C.V. ± %
Dry matter	%	88.5	100.0	1
Ash	%	3.4	3.9	8
Crude fiber	%	9.0	10.2	31
Swine	dig coef %	33.	33.	
Ether extract	%	3.8	4.4	7
Swine	dig coef %	59.	59.	
N-free extract	%	61.1	69.1	
Swine	dig coef %	92.	92.	
Protein (N x 6.25)	%	11.1	12.5	3
Swine	dig coef %	68.	68.	
Cattle	dig prot %	* 6.6	7.5	
Goats	dig prot %	* 7.7	8.7	
Horses	dig prot %	* 7.7	8.7	
Sheep	dig prot %	* 7.7	8.7	
Swine	dig prot %	7.5	8.5	
Energy	GE Mcal/kg			
Cattle	DE Mcal/kg	* 3.16	3.58	
Sheep	DE Mcal/kg	* 3.08	3.49	
Swine	DE kcal/kg	*3165.	3579.	
Cattle	ME Mcal/kg	* 2.59	2.93	
Sheep	ME Mcal/kg	* 2.53	2.86	
Swine	ME kcal/kg	*2959.	3345.	
Cattle	TDN %	* 71.7	81.1	
Sheep	TDN %	* 69.9	79.1	
Swine	TDN %	71.8	81.2	

MILLETTIA. Millettia ovalifolia

Millettia, seeds w hulls, (5)

Ref No 5 09 020 United States

Feed Name or Analyses		Mean As Fed	Mean Dry	C.V. ± %
Dry matter	%		100.0	
Protein (N x 6.25)	%		27.0	
Alanine	%		0.84	
Arginine	%		1.13	
Aspartic acid	%		2.35	
Glutamic acid	%		3.08	
Glycine	%		1.08	
Histidine	%		0.43	
Hydroxyproline	%		0.00	
Isoleucine	%		0.76	
Leucine	%		1.67	
Lysine	%		1.38	
Methionine	%		0.22	
Phenylalanine	%		1.08	
Proline	%		1.03	
Serine	%		1.08	
Threonine	%		0.84	
Tyrosine	%		0.73	
Valine	%		1.05	

Milo gluten feed –
see Sorghum, milo, gluten w bran, wet milled dehy grnd, (5)

Milo gluten meal –
see Sorghum, milo, gluten, wet milled dehy grnd, (5)

Milo grits –
see Sorghum, milo, grits, cracked fine screened, (4)

MIMOSA. Mimosa spp

Mimosa, browse, immature, (2)

Ref No 2 03 122 United States

Feed Name or Analyses		Mean As Fed	Mean Dry	C.V. ± %
Dry matter	%		100.0	
Ash	%		7.1	

Continued

Feed Name or Analyses				Mean As Fed	Mean Dry	C.V. ± %
Crude fiber		%			21.1	
Ether extract		%			3.4	
N-free extract		%			47.8	
Protein (N x 6.25)		%			20.6	
Cattle	dig prot	%	*		15.4	
Goats	dig prot	%	*		15.8	
Horses	dig prot	%	*		15.0	
Rabbits	dig prot	%	*		14.6	
Sheep	dig prot	%	*		16.2	
Calcium		%			2.38	
Magnesium		%			0.32	
Phosphorus		%			0.17	
Potassium		%			1.14	

MINT. Mentha spp

Mint, hay, s-c, (1)

Ref No 1 03 124 — United States

Feed Name or Analyses				Mean As Fed	Mean Dry	C.V. ± %
Dry matter		%		87.5	100.0	
Ash		%		7.7	8.8	
Crude fiber		%		20.4	23.3	
Ether extract		%		2.1	2.4	
N-free extract		%		44.5	50.9	
Protein (N x 6.25)		%		12.8	14.6	
Cattle	dig prot	%	*	8.4	9.6	
Goats	dig prot	%	*	8.9	10.2	
Horses	dig prot	%	*	8.7	9.9	
Rabbits	dig prot	%	*	8.7	9.9	
Sheep	dig prot	%	*	8.4	9.7	
Fatty acids		%		2.1	2.4	
Energy	GE Mcal/kg					
Cattle	DE Mcal/kg		*	2.21	2.53	
Sheep	DE Mcal/kg		*	2.11	2.41	
Cattle	ME Mcal/kg		*	1.81	2.07	
Sheep	ME Mcal/kg		*	1.73	1.98	
Cattle	TDN	%	*	50.2	57.4	
Sheep	TDN	%	*	47.9	54.7	
Calcium		%		1.50	1.71	
Phosphorus		%		0.19	0.22	

MISTLETOE, AMERICAN. Phoradendron flavescens

Mistletow, American, browse, (2)

Ref No 2 03 125 — United States

Feed Name or Analyses				Mean As Fed	Mean Dry	C.V. ± %
Dry matter		%		41.0	100.0	
Ash		%		2.4	5.9	
Crude fiber		%		8.1	19.8	
Ether extract		%		2.3	5.6	
N-free extract		%		19.1	46.7	
Protein (N x 6.25)		%		9.0	22.0	
Cattle	dig prot	%	*	6.8	16.6	
Goats	dig prot	%	*	7.0	17.1	
Horses	dig prot	%	*	6.6	16.2	
Rabbits	dig prot	%	*	6.4	15.7	
Sheep	dig prot	%	*	7.2	17.5	

Molasses (CFA) –
see Beet, sugar, molasses, mn 48% invert sugar mn 79.5 degrees brix, (4)

(1) dry forages and roughages
(2) pasture, range plants, and forages fed green
(3) silages
(4) energy feeds
(5) protein supplements
(6) minerals
(7) vitamins
(8) additives

Molasses, cane –
see Sugarcane, molasses, mn 48% invert sugar mn 79.5 degrees brix, (4)

Molasses, cane, ammoniated –
see Sugarcane, molasses, ammoniated, (5)

Molasses, cane, dried –
see Sugarcane, molasses, dehy, (4)

Molasses distillers condensed solubles (AAFCO) –
see Sugarcane, molasses distillation solubles, condensed, (4)

Molasses distillers dried solubles (AAFCO) –
see Sugarcane, molasses distillation solubles, dehy, (4)

Molasses distillers dried yeast (AAFCO) –
see Yeast, molasses distillers Saccharomyces, dehy, mn 40% protein, (7)

Molasses dried fermentation solubles –
see Sugarcane, molasses distillation solubles, dehy, (4)

Molasses fermentation solubles, dried –
see Sugarcane, molasses fermentation solubles, dehy, (5)

MOLASSESGRASS. Melinis minutiflora

Molassesgrass, aerial part, fresh, (2)

Ref No 2 03 130 — United States

Feed Name or Analyses				Mean As Fed	Mean Dry	C.V. ± %
Dry matter		%		32.6	100.0	13
Ash		%		2.5	7.6	24
Crude fiber		%		12.9	39.6	9
Sheep	dig coef	%		60.	60.	
Ether extract		%		0.8	2.6	8
Sheep	dig coef	%		46.	46.	
N-free extract		%		15.0	46.0	
Sheep	dig coef	%		51.	51.	
Protein (N x 6.25)		%		1.4	4.2	98
Sheep	dig coef	%		38.	38.	
Cattle	dig prot	%	*	0.5	1.5	
Goats	dig prot	%	*	0.2	0.5	
Horses	dig prot	%	*	0.4	1.1	
Rabbits	dig prot	%	*	0.6	1.9	
Sheep	dig prot	%		0.5	1.6	
Energy	GE Mcal/kg					
Cattle	DE Mcal/kg		*	0.75	2.31	
Sheep	DE Mcal/kg		*	0.74	2.27	
Cattle	ME Mcal/kg		*	0.62	1.89	
Sheep	ME Mcal/kg		*	0.61	1.86	
Cattle	TDN	%	*	17.0	52.3	
Sheep	TDN	%		16.8	51.5	

Molassesgrass, aerial part, fresh, immature, (2)

Ref No 2 03 126 — United States

Feed Name or Analyses				Mean As Fed	Mean Dry	C.V. ± %
Dry matter		%			100.0	
Calcium		%			0.69	
Phosphorus		%			0.33	

Molassesgrass, aerial part, fresh, pre-bloom, cut 2, (2)

Ref No 2 03 127 — United States

Feed Name or Analyses				Mean As Fed	Mean Dry	C.V. ± %
Dry matter		%		35.0	100.0	
Ash		%		2.5	7.0	
Crude fiber		%		14.1	40.2	
Sheep	dig coef	%		52.	52.	
Ether extract		%		0.8	2.4	
Sheep	dig coef	%		44.	44.	
N-free extract		%		16.6	47.4	
Sheep	dig coef	%		48.	48.	
Protein (N x 6.25)		%		1.1	3.0	
Sheep	dig coef	%		16.	16.	
Cattle	dig prot	%	*	0.2	0.4	
Goats	dig prot	%	*	-0.1	-0.5	
Horses	dig prot	%	*	0.0	0.1	
Rabbits	dig prot	%	*	0.3	1.0	
Sheep	dig prot	%		0.2	0.5	
Energy	GE Mcal/kg					
Cattle	DE Mcal/kg		*	0.77	2.21	
Sheep	DE Mcal/kg		*	0.72	2.05	
Cattle	ME Mcal/kg		*	0.64	1.82	
Sheep	ME Mcal/kg		*	0.59	1.68	
Cattle	TDN	%	*	17.5	50.1	
Sheep	TDN	%		16.3	46.5	

Molassesgrass, aerial part, fresh, full bloom, (2)

Ref No 2 03 128 — United States

Feed Name or Analyses				Mean As Fed	Mean Dry	C.V. ± %
Dry matter		%		31.6	100.0	
Ash		%		2.5	7.8	
Crude fiber		%		12.1	38.4	
Sheep	dig coef	%		56.	56.	
Ether extract		%		0.8	2.5	
Sheep	dig coef	%		37.	37.	
N-free extract		%		15.2	48.2	
Sheep	dig coef	%		55.	55.	
Protein (N x 6.25)		%		1.0	3.1	
Sheep	dig coef	%		18.	18.	
Cattle	dig prot	%	*	0.2	0.5	
Goats	dig prot	%	*	-0.1	-0.4	
Horses	dig prot	%	*	0.1	0.2	
Rabbits	dig prot	%	*	0.3	1.1	
Sheep	dig prot	%		0.2	0.6	
Energy	GE Mcal/kg					
Cattle	DE Mcal/kg		*	0.73	2.31	
Sheep	DE Mcal/kg		*	0.71	2.23	
Cattle	ME Mcal/kg		*	0.60	1.89	
Sheep	ME Mcal/kg		*	0.58	1.83	
Cattle	TDN	%	*	16.6	52.5	
Sheep	TDN	%		16.0	50.7	

Molassesgrass, aerial part, fresh, mature, (2)

Ref No 2 03 129 — United States

Feed Name or Analyses				Mean As Fed	Mean Dry	C.V. ± %
Dry matter		%		36.3	100.0	14
Ash		%		2.7	7.5	5
Crude fiber		%		14.4	39.8	11
Ether extract		%		0.8	2.3	11
N-free extract		%		16.8	46.2	
Protein (N x 6.25)		%		1.5	4.2	8
Cattle	dig prot	%	*	0.5	1.5	
Goats	dig prot	%	*	0.2	0.5	
Horses	dig prot	%	*	0.4	1.1	
Rabbits	dig prot	%	*	0.7	1.9	
Sheep	dig prot	%	*	0.3	0.9	

Feed Name or Analyses			Mean As Fed	Mean Dry	C.V. ± %
Energy	GE Mcal/kg				
Cattle	DE Mcal/kg	*	0.83	2.30	
Sheep	DE Mcal/kg	*	0.90	2.48	
Cattle	ME Mcal/kg	*	0.69	1.89	
Sheep	ME Mcal/kg	*	0.74	2.03	
Cattle	TDN %	*	18.9	52.1	
Sheep	TDN %	*	20.4	56.2	

MONKEYFLOWER, LEWIS. Mimulus lewisii

Monkeyflower, lewis, aerial part, fresh, (2)

Ref No 2 03 132 United States

Feed Name or Analyses			As Fed	Dry	C.V. ± %
Dry matter	%			100.0	
Ash	%			13.9	
Crude fiber	%			17.1	
Ether extract	%			5.9	
N-free extract	%			47.5	
Protein (N x 6.25)	%			15.6	
Cattle	dig prot %	*		11.2	
Goats	dig prot %	*		11.1	
Horses	dig prot %	*		10.8	
Rabbits	dig prot %	*		10.7	
Sheep	dig prot %	*		11.5	

MOORGRASS. Molinia caerulea

Moorgrass, hay, s-c, (1)

Ref No 1 03 133 United States

Feed Name or Analyses			As Fed	Dry	C.V. ± %
Dry matter	%		87.3	100.0	
Ash	%		4.7	5.4	
Crude fiber	%		22.9	26.2	
Sheep	dig coef %		66.	66.	
Ether extract	%		2.0	2.3	
Sheep	dig coef %		28.	28.	
N-free extract	%		44.3	50.8	
Sheep	dig coef %		65.	65.	
Protein (N x 6.25)	%		13.4	15.3	
Sheep	dig coef %		72.	72.	
Cattle	dig prot %	*	8.9	10.2	
Goats	dig prot %	*	9.5	10.8	
Horses	dig prot %	*	9.2	10.5	
Rabbits	dig prot %	*	9.2	10.5	
Sheep	dig prot %		9.6	11.0	
Energy	GE Mcal/kg				
Cattle	DE Mcal/kg	*	2.23	2.55	
Sheep	DE Mcal/kg	*	2.42	2.77	
Cattle	ME Mcal/kg	*	1.82	2.09	
Sheep	ME Mcal/kg	*	1.98	2.27	
Cattle	TDN %	*	50.5	57.8	
Sheep	TDN %		54.8	62.8	

MORNINGGLORY, COMMON. Ipomoea purpurea

Morningglory, common, seeds w hulls, (5)

Ref No 5 09 115 United States

Feed Name or Analyses			As Fed	Dry	C.V. ± %
Dry matter	%			100.0	
Protein (N x 6.25)	%			23.0	
Alanine	%			0.94	
Arginine	%			1.40	
Aspartic acid	%			1.86	
Glutamic acid	%			3.36	
Glycine	%			0.97	
Histidine	%			0.51	
Hydroxyproline	%			0.28	
Isoleucine	%			0.74	
Leucine	%			1.38	
Lysine	%			1.06	
Methionine	%			0.32	
Phenylalanine	%			0.92	
Proline	%			0.74	
Serine	%			0.97	
Threonine	%			0.81	
Tyrosine	%			0.64	
Valine	%			0.99	

MOSS, SPHAGNUM. Sphagnum spp

Moss, sphagnum, entire plant w molasses added, s-c, (1)

Ref No 1 03 135 United States

Feed Name or Analyses			As Fed	Dry	C.V. ± %
Dry matter	%		82.3	100.0	
Ash	%		7.5	9.1	
Crude fiber	%		7.4	9.0	
Sheep	dig coef %		-45.	-45.	
Ether extract	%		0.5	0.6	
Sheep	dig coef %		-6.	-6.	
N-free extract	%		57.6	70.0	
Sheep	dig coef %		72.	72.	
Protein (N x 6.25)	%		9.3	11.3	
Sheep	dig coef %		42.	42.	
Cattle	dig prot %	*	5.5	6.7	
Goats	dig prot %	*	5.8	7.1	
Horses	dig prot %	*	5.9	7.1	
Rabbits	dig prot %	*	6.1	7.4	
Sheep	dig prot %		3.9	4.7	
Energy	GE Mcal/kg				
Cattle	DE Mcal/kg	*	1.98	2.40	
Sheep	DE Mcal/kg	*	1.85	2.24	
Cattle	ME Mcal/kg	*	1.62	1.97	
Sheep	ME Mcal/kg	*	1.51	1.84	
Cattle	TDN %	*	44.9	54.5	
Sheep	TDN %		41.9	50.9	

MOUNTAINLAUREL. Kalmia latifolia

Mountainlaurel, browse, (2)

Ref No 2 03 136 United States

Feed Name or Analyses			As Fed	Dry	C.V. ± %
Dry matter	%			100.0	
Ash	%			3.6	13
Crude fiber	%			14.7	8
Ether extract	%			9.5	8
N-free extract	%			63.5	
Protein (N x 6.25)	%			8.7	7
Cattle	dig prot %	*		5.3	
Goats	dig prot %	*		4.7	
Horses	dig prot %	*		4.9	
Rabbits	dig prot %	*		5.4	
Sheep	dig prot %	*		5.1	
Calcium	%			0.96	12
Cobalt	mg/kg			0.057	46
Copper	mg/kg			4.4	26
Iron	%			0.005	24
Manganese	mg/kg			651.2	37
Phosphorus	%			0.11	14

MUHLY. Muhlenbergia spp

Muhly, aerial part, fresh, immature, (2)

Ref No 2 03 137 United States

Feed Name or Analyses			As Fed	Dry	C.V. ± %
Dry matter	%			100.0	
Ash	%			12.4	11
Crude fiber	%			28.3	5
Ether extract	%			1.8	7
N-free extract	%			41.2	
Protein (N x 6.25)	%			16.3	13
Cattle	dig prot %	*		11.7	
Goats	dig prot %	*		11.8	
Horses	dig prot %	*		11.4	
Rabbits	dig prot %	*		11.3	
Sheep	dig prot %	*		12.2	
Calcium	%			0.75	30
Phosphorus	%			0.36	25

MUHLY, BUSH. Muhlenbergia porteri

Muhly, bush, hay, s-c, (1)

Ref No 1 03 139 United States

Feed Name or Analyses			As Fed	Dry	C.V. ± %
Dry matter	%		93.3	100.0	
Ash	%		16.4	17.6	
Crude fiber	%		27.9	29.9	
Sheep	dig coef %		44.	44.	
Ether extract	%		1.6	1.7	
Sheep	dig coef %		68.	68.	
N-free extract	%		42.3	45.3	
Sheep	dig coef %		36.	36.	
Protein (N x 6.25)	%		5.1	5.5	
Sheep	dig coef %		4.	4.	
Cattle	dig prot %	*	1.6	1.7	
Goats	dig prot %	*	1.6	1.7	
Horses	dig prot %	*	2.1	2.2	
Rabbits	dig prot %	*	2.7	2.9	
Sheep	dig prot %		0.2	0.2	
Energy	GE Mcal/kg				
Cattle	DE Mcal/kg	*	1.47	1.58	
Sheep	DE Mcal/kg	*	1.33	1.42	
Cattle	ME Mcal/kg	*	1.21	1.30	
Sheep	ME Mcal/kg	*	1.09	1.17	
Cattle	TDN %	*	33.4	35.8	
Sheep	TDN %		30.1	32.3	

Muhly, bush, aerial part, fresh, over ripe, (2)

Ref No 2 03 140 United States

Feed Name or Analyses			As Fed	Dry	C.V. ± %
Dry matter	%			100.0	
Ash	%			5.9	8
Crude fiber	%			36.9	2
Ether extract	%			1.8	7
N-free extract	%			48.1	
Protein (N x 6.25)	%			7.3	4
Cattle	dig prot %	*		4.1	
Goats	dig prot %	*		3.4	
Horses	dig prot %	*		3.7	
Rabbits	dig prot %	*		4.3	
Sheep	dig prot %	*		3.8	
Calcium	%			0.27	9
Phosphorus	%			0.09	21

Feed Name or Analyses			Mean As Fed	Mean Dry	C.V. ± %

Muhly, bush, aerial part wo lower stems, fresh, early leaf, (2)

Ref No 2 08 714 United States

			As Fed	Dry	C.V.
Dry matter	%			100.0	
Organic matter	%			93.2	
Ash	%			6.8	
Crude fiber	%			35.3	
Ether extract	%			2.1	
N-free extract	%			48.1	
Protein (N x 6.25)	%			7.7	
Cattle	dig prot %	*		4.4	
Goats	dig prot %	*		3.7	
Horses	dig prot %	*		4.0	
Rabbits	dig prot %	*		4.6	
Sheep	dig prot %	*		4.1	
Calcium	%			0.32	
Phosphorus	%			0.06	
Carotene	mg/kg			50.5	
Vitamin A equivalent	IU/g			84.2	

Muhly, bush, aerial part wo lower stems, fresh, mid-bloom, (2)

Ref No 2 08 715 United States

			As Fed	Dry	C.V.
Dry matter	%			100.0	
Organic matter	%			94.5	
Ash	%			5.5	
Crude fiber	%			37.7	
Ether extract	%			1.7	
N-free extract	%			48.6	
Protein (N x 6.25)	%			6.5	
Cattle	dig prot %	*		3.4	
Goats	dig prot %	*		2.6	
Horses	dig prot %	*		3.0	
Rabbits	dig prot %	*		3.7	
Sheep	dig prot %	*		3.0	
Calcium	%			0.30	
Phosphorus	%			0.07	

Muhly, bush, aerial part wo lower stems, fresh, mature, (2)

Ref No 2 08 779 United States

			As Fed	Dry	C.V.
Dry matter	%			100.0	
Organic matter	%			93.4	1
Ash	%			6.6	12
Crude fiber	%			37.3	7
Ether extract	%			2.1	31
N-free extract	%			46.8	4
Protein (N x 6.25)	%			7.2	28
Cattle	dig prot %	*		4.0	
Goats	dig prot %	*		3.2	
Horses	dig prot %	*		3.6	
Rabbits	dig prot %	*		4.2	
Sheep	dig prot %	*		3.7	
Calcium	%			0.39	29
Phosphorus	%			0.10	24

(1) dry forages and roughages (2) pasture, range plants, and forages fed green (3) silages (4) energy feeds (5) protein supplements (6) minerals (7) vitamins (8) additives

Muhly, bush, aerial part wo lower stems, fresh, dormant, (2)

Ref No 2 08 713 United States

			As Fed	Dry	C.V.
Dry matter	%			100.0	
Organic matter	%			94.3	1
Ash	%			5.7	9
Crude fiber	%			39.0	4
Ether extract	%			1.9	44
N-free extract	%			46.4	4
Protein (N x 6.25)	%			6.9	16
Cattle	dig prot %	*		3.8	
Goats	dig prot %	*		3.0	
Horses	dig prot %	*		3.4	
Rabbits	dig prot %	*		4.0	
Sheep	dig prot %	*		3.5	
Calcium	%			0.34	35
Phosphorus	%			0.07	42
Carotene	mg/kg			13.0	90
Vitamin A equivalent	IU/g			21.7	

MUHLY, EAR. Muhlenbergia arenacea

Muhly, ear, aerial part, fresh, immature, (2)

Ref No 2 03 141 United States

			As Fed	Dry	C.V.
Dry matter	%			100.0	
Ash	%			14.4	
Crude fiber	%			30.5	
Ether extract	%			2.0	
N-free extract	%			39.8	
Protein (N x 6.25)	%			13.3	
Cattle	dig prot %	*		9.2	
Goats	dig prot %	*		9.0	
Horses	dig prot %	*		8.8	
Rabbits	dig prot %	*		8.9	
Sheep	dig prot %	*		9.4	
Calcium	%			1.04	
Phosphorus	%			0.33	

Muhly, ear, aerial part, fresh, mid-bloom, (2)

Ref No 2 03 142 United States

			As Fed	Dry	C.V.
Dry matter	%			100.0	
Ash	%			15.4	
Crude fiber	%			33.6	
Ether extract	%			1.3	
N-free extract	%			42.5	
Protein (N x 6.25)	%			7.2	
Cattle	dig prot %	*		4.0	
Goats	dig prot %	*		3.3	
Horses	dig prot %	*		3.6	
Rabbits	dig prot %	*		4.2	
Sheep	dig prot %	*		3.7	

MUHLY, HAIRAWN. Muhlenbergia capillaris

Muhly, hairawn, aerial part, fresh, immature, (2)

Ref No 2 03 143 United States

			As Fed	Dry	C.V.
Dry matter	%			100.0	
Calcium	%			0.43	9
Phosphorus	%			0.28	14

Muhly, hairawn, aerial part, fresh, early bloom, (2)

Ref No 2 03 144 United States

			As Fed	Dry	C.V.
Dry matter	%			100.0	
Ash	%			9.8	
Crude fiber	%			35.0	
Ether extract	%			1.9	
N-free extract	%			43.4	
Protein (N x 6.25)	%			9.9	
Cattle	dig prot %	*		6.3	
Goats	dig prot %	*		5.8	
Horses	dig prot %	*		5.9	
Rabbits	dig prot %	*		6.3	
Sheep	dig prot %	*		6.2	

Muhly, hairawn, aerial part, fresh, mature, (2)

Ref No 2 03 145 United States

			As Fed	Dry	C.V.
Dry matter	%			100.0	
Ash	%			7.4	
Crude fiber	%			37.1	
Ether extract	%			1.7	
N-free extract	%			49.9	
Protein (N x 6.25)	%			3.9	
Cattle	dig prot %	*		1.2	
Goats	dig prot %	*		0.2	
Horses	dig prot %	*		0.8	
Rabbits	dig prot %	*		1.7	
Sheep	dig prot %	*		0.6	

Muhly, hairawn, aerial part, fresh, over ripe, (2)

Ref No 2 03 146 United States

			As Fed	Dry	C.V.
Dry matter	%			100.0	
Calcium	%			0.26	
Phosphorus	%			0.17	

MUHLY, RING. Muhlenbergia torreyi

Muhly, ring, aerial part wo lower stems, fresh, early leaf, (2)

Ref No 2 08 724 United States

			As Fed	Dry	C.V.
Dry matter	%			100.0	
Organic matter	%			86.1	10
Ash	%			13.9	63
Crude fiber	%			30.9	5
Ether extract	%			1.4	7
N-free extract	%			47.5	17
Protein (N x 6.25)	%			6.2	33
Cattle	dig prot %	*		3.2	
Goats	dig prot %	*		2.3	
Horses	dig prot %	*		2.8	
Rabbits	dig prot %	*		3.5	
Sheep	dig prot %	*		2.8	
Calcium	%			1.06	96
Phosphorus	%			0.09	34
Carotene	mg/kg			36.3	
Vitamin A equivalent	IU/g			60.5	

Feed Name or Analyses		Mean As Fed	Mean Dry	C.V. ± %

Muhly, ring, aerial part wo lower stems, fresh, mid-bloom, (2)

Ref No 2 08 725 United States

Feed Name or Analyses		As Fed	Dry	C.V. ± %
Dry matter	%		100.0	
Organic matter	%		91.1	
Ash	%		8.9	
Crude fiber	%		34.0	
Ether extract	%		1.7	
N-free extract	%		47.1	
Protein (N x 6.25)	%		8.3	
Cattle	dig prot % *		4.9	
Goats	dig prot % *		4.3	
Horses	dig prot % *		4.5	
Rabbits	dig prot % *		5.0	
Sheep	dig prot % *		4.7	
Calcium	%		0.44	
Phosphorus	%		0.10	

Muhly, ring, aerial part wo lower stems, fresh, mature, (2)

Ref No 2 08 726 United States

Feed Name or Analyses		As Fed	Dry	C.V. ± %
Dry matter	%		100.0	
Organic matter	%		90.1	5
Ash	%		9.9	48
Crude fiber	%		32.6	5
Ether extract	%		1.7	25
N-free extract	%		48.3	6
Protein (N x 6.25)	%		7.6	23
Cattle	dig prot % *		4.3	
Goats	dig prot % *		3.6	
Horses	dig prot % *		4.0	
Rabbits	dig prot % *		4.5	
Sheep	dig prot % *		4.1	
Calcium	%		0.77	64
Phosphorus	%		0.10	24

Muhly, ring, aerial part wo lower stems, fresh, dormant, (2)

Ref No 2 08 723 United States

Feed Name or Analyses		As Fed	Dry	C.V. ± %
Dry matter	%		100.0	
Organic matter	%		90.2	6
Ash	%		10.6	48
Crude fiber	%		35.2	5
Ether extract	%		1.2	18
N-free extract	%		47.3	9
Protein (N x 6.25)	%		5.8	18
Cattle	dig prot % *		2.8	
Goats	dig prot % *		1.9	
Horses	dig prot % *		2.4	
Rabbits	dig prot % *		3.1	
Sheep	dig prot % *		2.3	
Calcium	%		0.61	78
Phosphorus	%		0.07	39
Carotene	mg/kg		6.6	41
Vitamin A equivalent	IU/g		11.0	

MUHLY, SANDHILL. Muhlenbergia pungens

Muhly, sandhill, aerial part, fresh, immature, (2)

Ref No 2 03 147 United States

Feed Name or Analyses		As Fed	Dry	C.V. ± %
Dry matter	%		100.0	
Ash	%		11.1	
Crude fiber	%		26.9	
Ether extract	%		1.7	
N-free extract	%		42.0	
Protein (N x 6.25)	%		18.3	
Cattle	dig prot % *		13.4	
Goats	dig prot % *		13.6	
Horses	dig prot % *		13.1	
Rabbits	dig prot % *		12.8	
Sheep	dig prot % *		14.0	
Calcium	%		0.98	
Phosphorus	%		0.45	

MUHLY, STONYHILLS. Muhlenbergia cuspidata

Muhly, stonyhills, aerial part, fresh, full bloom, (2)

Ref No 2 03 148 United States

Feed Name or Analyses		As Fed	Dry	C.V. ± %
Dry matter	%		100.0	
Ash	%		10.5	
Crude fiber	%		33.5	
Ether extract	%		1.9	
N-free extract	%		46.7	
Protein (N x 6.25)	%		7.4	
Cattle	dig prot % *		4.2	
Goats	dig prot % *		3.5	
Horses	dig prot % *		3.8	
Rabbits	dig prot % *		4.4	
Sheep	dig prot % *		3.9	

MULBERRY. Morus spp

Mulberry, browse, (2)

Ref No 2 03 150 United States

Feed Name or Analyses		As Fed	Dry	C.V. ± %
Dry matter	%		100.0	
Ash	%		15.0	
Crude fiber	%		11.2	
Ether extract	%		5.7	
N-free extract	%		50.1	
Protein (N x 6.25)	%		18.0	
Cattle	dig prot % *		13.2	
Goats	dig prot % *		13.4	
Horses	dig prot % *		12.8	
Rabbits	dig prot % *		12.6	
Sheep	dig prot % *		13.8	

Mulberry, browse, immature, (2)

Ref No 2 03 149 United States

Feed Name or Analyses		As Fed	Dry	C.V. ± %
Dry matter	%		100.0	
Calcium	%		4.64	
Magnesium	%		0.53	
Phosphorus	%		0.17	
Potassium	%		1.98	

MUSSELS. Mytilus spp

Mussels, meat w liquid, raw, (5)

Ref No 5 08 119 United States

Feed Name or Analyses		As Fed	Dry	C.V. ± %
Dry matter	%	16.2	100.0	
Ash	%	2.1	13.0	
Ether extract	%	1.4	8.6	
N-free extract	%	3.1	19.1	
Protein (N x 6.25)	%	9.6	59.3	
Energy	GE Mcal/kg	0.66	4.07	
Cattle	DE Mcal/kg *	0.59	3.63	
Sheep	DE Mcal/kg *	0.57	3.51	
Swine	DE kcal/kg *	606.	3742.	
Cattle	ME Mcal/kg *	0.48	2.98	
Sheep	ME Mcal/kg *	0.47	2.88	
Swine	ME kcal/kg *	509.	3144.	
Cattle	TDN % *	13.3	82.4	
Sheep	TDN % *	12.9	79.7	
Swine	TDN % *	13.7	84.9	

MUSTARD. Brassica spp

Mustard, hay, s-c, (1)

Ref No 1 03 153 United States

Feed Name or Analyses		As Fed	Dry	C.V. ± %
Dry matter	%		100.0	
Ash	%		6.1	
Crude fiber	%		33.0	
Ether extract	%		11.5	
N-free extract	%		31.7	
Protein (N x 6.25)	%		17.7	
Cattle	dig prot % *		12.3	
Goats	dig prot % *		13.1	
Horses	dig prot % *		12.6	
Rabbits	dig prot % *		12.3	
Sheep	dig prot % *		12.4	

Mustard, seeds, extn unspecified grnd, (5)

Mustard seed meal (CFA)

Mustard seed oil meal, extraction unspecified

Ref No 5 03 154 United States

Feed Name or Analyses		As Fed	Dry	C.V. ± %
Dry matter	%	92.7	100.0	2
Ash	%	7.0	7.6	5
Crude fiber	%	10.8	11.6	
Cattle	dig coef % *	24.	24.	
Sheep	dig coef %	53.	53.	
Ether extract	%	5.1	5.5	43
Cattle	dig coef % *	89.	89.	
Sheep	dig coef %	89.	89.	
N-free extract	%	37.8	40.8	
Cattle	dig coef % *	72.	72.	
Sheep	dig coef %	74.	74.	
Protein (N x 6.25)	%	31.9	34.5	14
Cattle	dig coef % *	86.	86.	
Sheep	dig coef %	86.	86.	
Cattle	dig prot %	27.5	29.6	
Sheep	dig prot %	27.5	29.6	
Energy	GE Mcal/kg			
Cattle	DE Mcal/kg *	2.97	3.21	
Sheep	DE Mcal/kg *	3.15	3.40	
Swine	DE kcal/kg *	3504.	3782.	
Cattle	ME Mcal/kg *	2.44	2.63	
Sheep	ME Mcal/kg *	2.58	2.78	
Swine	ME kcal/kg *	3120.	3367.	
Cattle	TDN %	67.5	72.8	
Sheep	TDN %	71.4	77.0	
Swine	TDN % *	79.5	85.8	

Feed Name or Analyses		Mean As Fed	Dry	C.V. ± %

MUSTARD, WILD YELLOW. Brassica spp

Mustard, wild yellow, seeds, (5)

Ref No 5 08 461 United States

Analyses		As Fed	Dry	C.V.
Dry matter	%	95.9	100.0	
Ash	%	5.5	5.7	
Crude fiber	%	5.0	5.2	
Ether extract	%	38.8	40.5	
N-free extract	%	23.6	24.6	
Protein (N x 6.25)	%	23.0	24.0	
Energy	GE Mcal/kg			
Cattle	DE Mcal/kg	* 3.96	4.13	
Sheep	DE Mcal/kg	* 3.72	3.88	
Swine	DE kcal/kg	*4232.	4413.	
Cattle	ME Mcal/kg	* 3.25	3.39	
Sheep	ME Mcal/kg	* 3.05	3.19	
Swine	ME kcal/kg	*3860.	4025.	
Cattle	TDN %	* 89.9	93.7	
Sheep	TDN %	* 84.5	88.1	
Swine	TDN %	* 96.0	100.1	

Mustard seed meal (CFA) —
see Mustard, seeds, extn unspecified grnd, (5)

Mustard seed oil meal, extraction unspecified —
see Mustard, seeds, extn unspecified grnd, (5)

NAMA. Nama spp

Nama, browse, (2)

Ref No 2 03 155 United States

Analyses		As Fed	Dry	C.V.
Dry matter	%		100.0	
Ash	%		18.3	
Crude fiber	%		19.2	
Ether extract	%		2.4	
N-free extract	%		46.0	
Protein (N x 6.25)	%		14.1	
Cattle	dig prot %	*	9.9	
Goats	dig prot %	*	9.7	
Horses	dig prot %	*	9.5	
Rabbits	dig prot %	*	9.6	
Sheep	dig prot %	*	10.1	
Calcium	%		5.02	
Magnesium	%		0.46	
Phosphorus	%		0.12	
Potassium	%		2.57	

NANDINA. Nandina domestica

Nandina, seeds w hulls, (5)

Ref No 5 09 017 United States

Analyses		As Fed	Dry	C.V.
Dry matter	%		100.0	
Protein (N x 6.25)	%		20.0	
Alanine	%		0.80	
Arginine	%		1.16	
Aspartic acid	%		1.76	
Glutamic acid	%		3.18	
Glycine	%		0.88	
Histidine	%		0.40	
Hydroxyproline	%		0.84	
Isoleucine	%		0.64	
Leucine	%		0.86	
Lysine	%		0.86	
Methionine	%		0.28	
Phenylalanine	%		0.64	
Proline	%		1.14	
Serine	%		0.84	
Threonine	%		0.82	
Tyrosine	%		0.72	
Valine	%		1.10	

NAPIERGRASS. Pennisetum purpureum

Napiergrass, hay, s-c, (1)

Ref No 1 08 462 United States

Analyses		As Fed	Dry	C.V.
Dry matter	%	89.1	100.0	
Ash	%	10.5	11.8	
Crude fiber	%	34.0	38.2	
Ether extract	%	1.8	2.0	
N-free extract	%	34.6	38.8	
Protein (N x 6.25)	%	8.2	9.2	
Cattle	dig prot %	* 4.4	4.9	
Goats	dig prot %	* 4.6	5.1	
Horses	dig prot %	* 4.8	5.3	
Rabbits	dig prot %	* 5.1	5.8	
Sheep	dig prot %	* 4.3	4.8	
Energy	GE Mcal/kg			
Cattle	DE Mcal/kg	* 2.09	2.35	
Sheep	DE Mcal/kg	* 2.08	2.34	
Cattle	ME Mcal/kg	* 1.72	1.93	
Sheep	ME Mcal/kg	* 1.71	1.92	
Cattle	TDN %	* 47.4	53.2	
Sheep	TDN %	* 47.2	53.0	

Napiergrass, aerial part, fresh, (2)

Ref No 2 03 166 United States

Analyses		As Fed	Dry	C.V.
Dry matter	%	36.9	100.0	87
Ash	%	3.7	10.0	
Crude fiber	%	12.2	33.2	9
Cattle	dig coef %	68.	68.	
Sheep	dig coef %	59.	59.	
Ether extract	%	1.0	2.6	29
Cattle	dig coef %	59.	59.	
Sheep	dig coef %	57.	57.	
N-free extract	%	16.7	45.4	
Cattle	dig coef %	68.	68.	
Sheep	dig coef %	57.	57.	
Protein (N x 6.25)	%	3.2	8.8	30
Cattle	dig coef %	63.	63.	
Sheep	dig coef %	61.	61.	
Cattle	dig prot %	2.0	5.5	
Goats	dig prot %	* 1.7	4.7	
Horses	dig prot %	* 1.8	5.0	
Rabbits	dig prot %	* 2.0	5.4	
Sheep	dig prot %	2.0	5.3	
Energy	GE Mcal/kg			
Cattle	DE Mcal/kg	* 1.02	2.76	
Sheep	DE Mcal/kg	* 0.88	2.39	
Cattle	ME Mcal/kg	* 0.83	2.26	
Sheep	ME Mcal/kg	* 0.72	1.96	
Cattle	TDN %	23.0	62.5	
Sheep	TDN %	20.0	54.2	
Calcium	%	0.18	0.48	
Phosphorus	%	0.14	0.37	

Napiergrass, aerial part, fresh, early leaf, (2)

Ref No 2 03 156 United States

Analyses		As Fed	Dry	C.V.
Dry matter	%	20.8	100.0	
Ash	%	1.7	8.2	
Crude fiber	%	6.0	28.9	
Cattle	dig coef %	68.	68.	
Ether extract	%	0.9	4.4	
Cattle	dig coef %	58.	58.	
N-free extract	%	9.5	45.9	
Cattle	dig coef %	70.	70.	
Protein (N x 6.25)	%	2.6	12.6	
Cattle	dig coef %	64.	64.	
Cattle	dig prot %	1.7	8.1	
Goats	dig prot %	* 1.7	8.3	
Horses	dig prot %	* 1.7	8.2	
Rabbits	dig prot %	* 1.7	8.4	
Sheep	dig prot %	* 1.8	8.7	
Energy	GE Mcal/kg			
Cattle	DE Mcal/kg	* 0.60	2.89	
Sheep	DE Mcal/kg	* 0.59	2.82	
Cattle	ME Mcal/kg	* 0.49	2.37	
Sheep	ME Mcal/kg	* 0.48	2.31	
Cattle	TDN %	13.6	65.6	
Sheep	TDN %	* 13.3	64.0	

Napiergrass, aerial part, fresh, immature, (2)

Ref No 2 03 157 United States

Analyses		As Fed	Dry	C.V.
Dry matter	%	22.2	100.0	15
Ash	%	2.6	11.6	17
Crude fiber	%	8.7	39.3	8
Ether extract	%	0.4	1.7	32
N-free extract	%	9.3	41.9	
Protein (N x 6.25)	%	1.2	5.5	43
Cattle	dig prot %	* 0.6	2.6	
Goats	dig prot %	* 0.4	1.7	
Horses	dig prot %	* 0.5	2.2	
Rabbits	dig prot %	* 0.6	2.9	
Sheep	dig prot %	* 0.5	2.1	
Energy	GE Mcal/kg			
Cattle	DE Mcal/kg	* 0.57	2.57	
Sheep	DE Mcal/kg	* 0.56	2.50	
Cattle	ME Mcal/kg	* 0.47	2.11	
Sheep	ME Mcal/kg	* 0.46	2.05	
Cattle	TDN %	* 12.9	58.3	
Sheep	TDN %	* 12.6	56.7	
Calcium	%	0.11	0.48	
Magnesium	%	0.06	0.26	
Phosphorus	%	0.08	0.36	
Potassium	%	0.29	1.31	

Napiergrass, aerial part, fresh, pre-bloom, (2)

Ref No 2 03 158 United States

Analyses		As Fed	Dry	C.V.
Dry matter	%	25.6	100.0	
Ash	%	2.2	8.6	
Crude fiber	%	8.8	34.5	
Sheep	dig coef %	57.	57.	
Ether extract	%	0.8	3.0	
Sheep	dig coef %	61.	61.	
N-free extract	%	12.2	47.5	
Sheep	dig coef %	55.	55.	
Protein (N x 6.25)	%	1.6	6.4	
Sheep	dig coef %	58.	58.	
Cattle	dig prot %	* 0.9	3.3	
Goats	dig prot %	* 0.6	2.5	

(1) dry forages and roughages (2) pasture, range plants, and forages fed green (3) silages (4) energy feeds (5) protein supplements (6) minerals (7) vitamins (8) additives

Feed Name or Analyses			Mean As Fed	Mean Dry	C.V. ± %
Horses	dig prot %	*	0.8	3.0	
Rabbits	dig prot %	*	0.9	3.6	
Sheep	dig prot %		1.0	3.7	
Energy	GE Mcal/kg				
Cattle	DE Mcal/kg	*	0.63	2.44	
Sheep	DE Mcal/kg	*	0.61	2.36	
Cattle	ME Mcal/kg	*	0.51	2.00	
Sheep	ME Mcal/kg	*	0.50	1.94	
Cattle	NE_m Mcal/kg	*	0.35	1.36	
Cattle	NE_{gain} Mcal/kg	*	0.19	0.76	
Cattle	$NE_{lactating\ cows}$ Mcal/kg	*	0.40	1.57	
Cattle	TDN %	*	14.2	55.4	
Sheep	TDN %		13.7	53.6	

Napiergrass, aerial part, fresh, early bloom, (2)

Ref No 2 03 159 United States

Feed Name or Analyses			Mean As Fed	Mean Dry	C.V. ± %
Dry matter	%		14.9	100.0	
Ash	%		1.4	9.2	
Crude fiber	%		4.7	31.5	
Cattle	dig coef %	*	24.	24.	
Sheep	dig coef %	*	59.	59.	
Ether extract	%		0.4	3.0	
Cattle	dig coef %	*	89.	89.	
Sheep	dig coef %	*	57.	57.	
N-free extract	%		6.7	45.3	
Cattle	dig coef %	*	72.	72.	
Sheep	dig coef %	*	57.	57.	
Protein (N x 6.25)	%		1.6	11.0	
Cattle	dig coef %	*	86.	86.	
Sheep	dig coef %	*	61.	61.	
Cattle	dig prot %		1.4	9.5	
Goats	dig prot %	*	1.0	6.8	
Horses	dig prot %	*	1.0	6.9	
Rabbits	dig prot %	*	1.1	7.2	
Sheep	dig prot %		1.0	6.7	
Energy	GE Mcal/kg				
Cattle	DE Mcal/kg	*	0.37	2.45	
Sheep	DE Mcal/kg	*	0.36	2.42	
Cattle	ME Mcal/kg	*	0.30	2.01	
Sheep	ME Mcal/kg	*	0.30	1.99	
Cattle	TDN %	*	8.3	55.6	
Sheep	TDN %	*	8.2	55.0	

Napiergrass, aerial part, fresh, mid-bloom, (2)

Ref No 2 03 160 United States

Feed Name or Analyses			Mean As Fed	Mean Dry	C.V. ± %
Dry matter	%		23.0	100.0	11
Ash	%		2.4	10.3	16
Crude fiber	%		8.0	34.9	5
Cattle	dig coef %	*	68.	68.	
Sheep	dig coef %	*	59.	59.	
Ether extract	%		0.6	2.5	21
Cattle	dig coef %	*	44.	44.	
Sheep	dig coef %	*	57.	57.	
N-free extract	%		9.9	43.1	
Cattle	dig coef %	*	59.	59.	
Sheep	dig coef %	*	57.	57.	
Protein (N x 6.25)	%		2.1	9.2	5
Cattle	dig coef %	*	63.	63.	
Sheep	dig coef %	*	61.	61.	
Cattle	dig prot %		1.3	5.8	
Goats	dig prot %	*	1.2	5.1	
Horses	dig prot %	*	1.2	5.3	
Rabbits	dig prot %	*	1.3	5.8	
Sheep	dig prot %		1.3	5.6	
Energy	GE Mcal/kg				
Cattle	DE Mcal/kg	*	0.58	2.53	
Sheep	DE Mcal/kg	*	0.55	2.38	
Cattle	ME Mcal/kg	*	0.48	2.08	
Sheep	ME Mcal/kg	*	0.45	1.95	
Cattle	TDN %	*	13.2	57.4	
Sheep	TDN %	*	12.4	54.0	

Napiergrass, aerial part, fresh, full bloom, (2)

Ref No 2 03 161 United States

Feed Name or Analyses			Mean As Fed	Mean Dry	C.V. ± %
Dry matter	%		22.5	100.0	9
Ash	%		2.3	10.1	15
Crude fiber	%		8.1	36.1	7
Cattle	dig coef %	*	68.	68.	
Sheep	dig coef %	*	59.	59.	
Ether extract	%		0.6	2.5	26
Cattle	dig coef %	*	59.	59.	
Sheep	dig coef %	*	57.	57.	
N-free extract	%		9.8	43.7	
Cattle	dig coef %	*	68.	68.	
Sheep	dig coef %	*	57.	57.	
Protein (N x 6.25)	%		1.7	7.6	37
Cattle	dig coef %	*	63.	63.	
Sheep	dig coef %	*	61.	61.	
Cattle	dig prot %		1.1	4.8	
Goats	dig prot %	*	0.8	3.7	
Horses	dig prot %	*	0.9	4.0	
Rabbits	dig prot %	*	1.0	4.5	
Sheep	dig prot %		1.0	4.6	
Energy	GE Mcal/kg				
Cattle	DE Mcal/kg	*	0.62	2.75	
Sheep	DE Mcal/kg	*	0.54	2.38	
Cattle	ME Mcal/kg	*	0.51	2.26	
Sheep	ME Mcal/kg	*	0.44	1.95	
Cattle	TDN %	*	14.0	62.4	
Sheep	TDN %	*	12.2	54.1	

Napiergrass, aerial part, fresh, late bloom, (2)

Ref No 2 03 162 United States

Feed Name or Analyses			Mean As Fed	Mean Dry	C.V. ± %
Dry matter	%		23.0	100.0	
Ash	%		1.2	5.3	
Crude fiber	%		9.0	39.0	
Sheep	dig coef %		56.	56.	
Ether extract	%		0.3	1.1	
Sheep	dig coef %		30.	30.	
N-free extract	%		10.8	46.8	
Sheep	dig coef %		49.	49.	
Protein (N x 6.25)	%		1.8	7.8	
Sheep	dig coef %		46.	46.	
Cattle	dig prot %	*	1.0	4.5	
Goats	dig prot %	*	0.9	3.8	
Horses	dig prot %	*	1.0	4.2	
Rabbits	dig prot %	*	1.1	4.7	
Sheep	dig prot %		0.8	3.6	
Energy	GE Mcal/kg				
Cattle	DE Mcal/kg	*	0.58	2.53	
Sheep	DE Mcal/kg	*	0.50	2.17	
Cattle	ME Mcal/kg	*	0.48	2.07	
Sheep	ME Mcal/kg	*	0.41	1.78	
Cattle	NE_m Mcal/kg	*	0.25	1.10	
Cattle	NE_{gain} Mcal/kg	*	0.08	0.35	
Cattle	$NE_{lactating\ cows}$ Mcal/kg	*	0.26	1.15	
Cattle	TDN %	*	13.2	57.3	
Sheep	TDN %		11.3	49.1	

Napiergrass, aerial part, fresh, milk stage, (2)

Ref No 2 03 163 United States

Feed Name or Analyses			Mean As Fed	Mean Dry	C.V. ± %
Dry matter	%		23.7	100.0	
Ash	%		2.7	11.4	
Crude fiber	%		8.1	34.2	
Sheep	dig coef %		59.	59.	
Ether extract	%		0.7	3.0	
Sheep	dig coef %		56.	56.	
N-free extract	%		9.9	41.8	
Sheep	dig coef %		57.	57.	
Protein (N x 6.25)	%		2.3	9.6	
Sheep	dig coef %		65.	65.	
Cattle	dig prot %	*	1.4	6.1	
Goats	dig prot %	*	1.3	5.5	
Horses	dig prot %	*	1.3	5.7	
Rabbits	dig prot %	*	1.4	6.1	
Sheep	dig prot %		1.5	6.2	
Energy	GE Mcal/kg				
Cattle	DE Mcal/kg	*	0.63	2.66	
Sheep	DE Mcal/kg	*	0.56	2.38	
Cattle	ME Mcal/kg	*	0.52	2.18	
Sheep	ME Mcal/kg	*	0.46	1.95	
Cattle	TDN %	*	14.3	60.2	
Sheep	TDN %		12.8	54.0	

Napiergrass, aerial part, fresh, mature, (2)

Ref No 2 03 164 United States

Feed Name or Analyses			Mean As Fed	Mean Dry	C.V. ± %
Dry matter	%		26.7	100.0	12
Ash	%		2.3	8.6	20
Crude fiber	%		10.0	37.6	4
Cattle	dig coef %	*	68.	68.	
Sheep	dig coef %	*	59.	59.	
Ether extract	%		0.5	2.0	11
Cattle	dig coef %	*	59.	59.	
Sheep	dig coef %	*	57.	57.	
N-free extract	%		12.4	46.4	
Cattle	dig coef %	*	68.	68.	
Sheep	dig coef %	*	57.	57.	
Protein (N x 6.25)	%		1.4	5.4	10
Cattle	dig coef %	*	63.	63.	
Sheep	dig coef %	*	61.	61.	
Cattle	dig prot %		0.9	3.4	
Goats	dig prot %	*	0.4	1.6	
Horses	dig prot %	*	0.6	2.1	
Rabbits	dig prot %	*	0.8	2.8	
Sheep	dig prot %		0.9	3.3	
Energy	GE Mcal/kg				
Cattle	DE Mcal/kg	*	0.74	2.79	
Sheep	DE Mcal/kg	*	0.64	2.40	
Cattle	ME Mcal/kg	*	0.61	2.28	
Sheep	ME Mcal/kg	*	0.53	1.97	
Cattle	TDN %	*	16.9	63.2	
Sheep	TDN %	*	14.5	54.5	

Napiergrass, aerial part, fresh, over ripe, (2)

Ref No 2 03 165 United States

Feed Name or Analyses			Mean As Fed	Mean Dry	C.V. ± %
Dry matter	%		30.4	100.0	
Ash	%		2.7	9.0	
Crude fiber	%		12.0	39.6	
Cattle	dig coef %	*	68.	68.	
Sheep	dig coef %	*	59.	59.	
Ether extract	%		0.5	1.6	
Cattle	dig coef %	*	59.	59.	

Continued

Feed Name or Analyses			Mean As Fed	Mean Dry	C.V. ± %
Sheep	dig coef %	*	57.	57.	
N-free extract	%		13.9	45.7	
Cattle	dig coef %	*	68.	68.	
Sheep	dig coef %	*	57.	57.	
Protein (N x 6.25)	%		1.2	4.1	
Cattle	dig coef %	*	63.	63.	
Sheep	dig coef %	*	61.	61.	
Cattle	dig prot %		0.8	2.6	
Goats	dig prot %	*	0.1	0.4	
Horses	dig prot %	*	0.3	1.0	
Rabbits	dig prot %	*	0.6	1.8	
Sheep	dig prot %		0.8	2.5	
Energy	GE Mcal/kg				
Cattle	DE Mcal/kg	*	0.84	2.77	
Sheep	DE Mcal/kg	*	0.72	2.38	
Cattle	ME Mcal/kg	*	0.69	2.27	
Sheep	ME Mcal/kg	*	0.59	1.95	
Cattle	TDN %	*	19.1	62.7	
Sheep	TDN %	*	16.4	54.0	

Napiergrass, leaves, fresh, (2)
Ref No 2 03 167 United States

Feed Name or Analyses			Mean As Fed	Mean Dry	C.V. ± %
Dry matter	%		19.8	100.0	12
Ash	%		1.3	6.6	10
Crude fiber	%		6.4	32.5	5
Ether extract	%		0.6	2.9	18
N-free extract	%		9.0	45.7	
Protein (N x 6.25)	%		2.4	12.3	12
Cattle	dig prot %	*	1.7	8.3	
Goats	dig prot %	*	1.6	8.0	
Horses	dig prot %	*	1.6	8.0	
Rabbits	dig prot %	*	1.6	8.2	
Sheep	dig prot %	*	1.7	8.5	
Energy	GE Mcal/kg				
Cattle	DE Mcal/kg	*	0.56	2.82	
Sheep	DE Mcal/kg	*	0.55	2.79	
Cattle	ME Mcal/kg	*	0.46	2.31	
Sheep	ME Mcal/kg	*	0.45	2.29	
Cattle	TDN %	*	12.7	63.9	
Sheep	TDN %	*	12.5	63.3	
Calcium	%		0.10	0.48	12
Magnesium	%		0.07	0.33	
Phosphorus	%		0.07	0.33	25

Napiergrass, aerial part, ensiled, (3)
Ref No 3 03 170 United States

Feed Name or Analyses			Mean As Fed	Mean Dry	C.V. ± %
Dry matter	%		25.3	100.0	15
Ash	%		2.3	9.0	22
Crude fiber	%		9.9	39.1	9
Ether extract	%		0.6	2.2	21
N-free extract	%		11.4	44.9	
Protein (N x 6.25)	%		1.2	4.8	23
Cattle	dig prot %	*	0.1	0.5	
Goats	dig prot %	*	0.1	0.5	
Horses	dig prot %	*	0.1	0.5	
Sheep	dig prot %	*	0.1	0.5	
Energy	GE Mcal/kg				
Cattle	DE Mcal/kg	*	0.53	2.10	
Sheep	DE Mcal/kg	*	0.48	1.91	
Cattle	ME Mcal/kg	*	0.44	1.72	
Sheep	ME Mcal/kg	*	0.40	1.57	

(1) dry forages and roughages (2) pasture, range plants, and forages fed green (3) silages (4) energy feeds (5) protein supplements (6) minerals (7) vitamins (8) additives

Feed Name or Analyses			Mean As Fed	Mean Dry	C.V. ± %
Cattle	TDN %	*	12.0	47.6	
Sheep	TDN %	*	11.0	43.3	
Calcium	%		0.08	0.31	
Phosphorus	%		0.08	0.31	
Carotene	mg/kg		24.4	96.6	35
Vitamin A equivalent	IU/g		40.7	161.0	

Napiergrass, aerial part, ensiled, immature, (3)
Ref No 3 08 463 United States

Feed Name or Analyses			Mean As Fed	Mean Dry	C.V. ± %
Dry matter	%		19.9	100.0	
Ash	%		2.6	13.1	
Crude fiber	%		7.3	36.7	
Ether extract	%		0.5	2.5	
N-free extract	%		8.4	42.2	
Protein (N x 6.25)	%		1.1	5.5	
Cattle	dig prot %	*	0.2	1.2	
Goats	dig prot %	*	0.2	1.2	
Horses	dig prot %	*	0.2	1.2	
Sheep	dig prot %	*	0.2	1.2	
Energy	GE Mcal/kg				
Cattle	DE Mcal/kg	*	0.41	2.05	
Sheep	DE Mcal/kg	*	0.35	1.77	
Cattle	ME Mcal/kg	*	0.33	1.68	
Sheep	ME Mcal/kg	*	29.04	1.45	
Cattle	TDN %	*	9.3	46.6	
Sheep	TDN %	*	8.0	40.1	

Napiergrass, aerial part, ensiled, mature, (3)
Ref No 3 03 169 United States

Feed Name or Analyses			Mean As Fed	Mean Dry	C.V. ± %
Dry matter	%		25.5	100.0	20
Ash	%		2.3	9.1	33
Crude fiber	%		10.4	40.9	8
Ether extract	%		0.6	2.2	16
N-free extract	%		11.1	43.5	
Protein (N x 6.25)	%		1.1	4.3	14
Cattle	dig prot %	*	0.0	0.1	
Goats	dig prot %	*	0.0	0.1	
Horses	dig prot %	*	0.0	0.1	
Sheep	dig prot %	*	0.0	0.1	

Napiergrass, aerial part w molasses added, ensiled, (3)
Ref No 3 03 172 United States

Feed Name or Analyses			Mean As Fed	Mean Dry	C.V. ± %
Dry matter	%		25.4	100.0	14
Ash	%		2.8	11.1	11
Crude fiber	%		8.9	35.2	11
Ether extract	%		0.7	2.6	17
N-free extract	%		11.6	45.5	
Protein (N x 6.25)	%		1.4	5.6	21
Cattle	dig prot %	*	0.3	1.3	
Goats	dig prot %	*	0.3	1.3	
Horses	dig prot %	*	0.3	1.3	
Sheep	dig prot %	*	0.3	1.3	

Napiergrass, aerial part w molasses added, ensiled, milk stage, (3)
Ref No 3 03 171 United States

Feed Name or Analyses			Mean As Fed	Mean Dry	C.V. ± %
Dry matter	%		30.1	100.0	15
Ash	%		3.2	10.7	10
Crude fiber	%		10.1	33.6	5
Ether extract	%		0.5	1.7	14
N-free extract	%		14.0	46.6	

Feed Name or Analyses			Mean As Fed	Mean Dry	C.V. ± %
Protein (N x 6.25)	%		2.2	7.4	11
Cattle	dig prot %	*	0.9	2.9	
Goats	dig prot %	*	0.9	2.9	
Horses	dig prot %	*	0.9	2.9	
Sheep	dig prot %	*	0.9	2.9	

NAPIERGRASS-KUDZU. Pennisetum purpureum, Pueraria spp

Napiergrass-Kudzu, aerial part, fresh, (2)
Ref No 2 03 175 United States

Feed Name or Analyses			Mean As Fed	Mean Dry	C.V. ± %
Dry matter	%		20.5	100.0	11
Ash	%		1.8	8.7	8
Crude fiber	%		7.6	36.9	6
Ether extract	%		0.3	1.6	15
N-free extract	%		9.3	45.3	
Protein (N x 6.25)	%		1.5	7.5	24
Cattle	dig prot %	*	0.9	4.3	
Goats	dig prot %	*	0.7	3.6	
Horses	dig prot %	*	0.8	3.9	
Rabbits	dig prot %	*	0.9	4.5	
Sheep	dig prot %	*	0.8	4.0	

Napiergrass-Kudzu, aerial part, ensiled, (3)
Ref No 3 03 177 United States

Feed Name or Analyses			Mean As Fed	Mean Dry	C.V. ± %
Dry matter	%		28.5	100.0	8
Ash	%		2.5	8.9	15
Crude fiber	%		10.3	36.2	7
Ether extract	%		0.4	1.4	17
N-free extract	%		13.9	48.6	
Protein (N x 6.25)	%		1.4	4.9	13
Cattle	dig prot %	*	0.2	0.7	
Goats	dig prot %	*	0.2	0.7	
Horses	dig prot %	*	0.2	0.7	
Sheep	dig prot %	*	0.2	0.7	

Napiergrass-Kudzu, aerial part, ensiled, milk stage, (3)
Ref No 3 03 176 United States

Feed Name or Analyses			Mean As Fed	Mean Dry	C.V. ± %
Dry matter	%		30.1	100.0	7
Ash	%		2.6	8.7	17
Crude fiber	%		11.0	36.5	7
Ether extract	%		0.4	1.4	15
N-free extract	%		14.6	48.4	
Protein (N x 6.25)	%		1.5	5.0	14
Cattle	dig prot %	*	0.2	0.8	
Goats	dig prot %	*	0.2	0.8	
Horses	dig prot %	*	0.2	0.8	
Sheep	dig prot %	*	0.2	0.8	

Napiergrass-Kudzu, aerial part w molasses added, ensiled, (3)
Ref No 3 03 178 United States

Feed Name or Analyses			Mean As Fed	Mean Dry	C.V. ± %
Dry matter	%		30.6	100.0	8
Ash	%		2.3	7.4	12
Crude fiber	%		10.7	34.9	7
Ether extract	%		0.4	1.4	13
N-free extract	%		15.6	51.0	
Protein (N x 6.25)	%		1.6	5.3	11
Cattle	dig prot %	*	0.3	1.0	
Goats	dig prot %	*	0.3	1.0	

Feed Name or Analyses			Mean As Fed	Mean Dry	C.V. ± %
Horses	dig prot %	*	0.3	1.0	
Sheep	dig prot %	*	0.3	1.0	

NAPIERGRASS-LEADTREE, WHITEPOPINAC. Pennisetum purpureum, Leucaena glauca

Napiergrass-Leadtree, whitepopinac, aerial part, fresh, (2)

Ref No 2 03 173 United States

Feed Name or Analyses			Mean As Fed	Mean Dry	C.V. ± %
Dry matter	%		33.4	100.0	
Ash	%		3.1	9.2	
Crude fiber	%		12.6	37.6	
Ether extract	%		0.7	2.1	
N-free extract	%		14.5	43.4	
Protein (N x 6.25)	%		2.6	7.7	
Cattle	dig prot %	*	1.5	4.4	
Goats	dig prot %	*	1.3	3.7	
Horses	dig prot %	*	1.4	4.1	
Rabbits	dig prot %	*	1.5	4.6	
Sheep	dig prot %	*	1.4	4.2	

Napiergrass-Leadtree, whitepopinac, aerial part, ensiled, (3)

Ref No 3 03 174 United States

Feed Name or Analyses			Mean As Fed	Mean Dry	C.V. ± %
Dry matter	%		28.9	100.0	
Ash	%		2.6	8.9	
Crude fiber	%		13.2	45.6	
Ether extract	%		0.4	1.4	
N-free extract	%		10.1	35.1	
Protein (N x 6.25)	%		2.6	9.0	
Cattle	dig prot %	*	1.3	4.4	
Goats	dig prot %	*	1.3	4.4	
Horses	dig prot %	*	1.3	4.4	
Sheep	dig prot %	*	1.3	4.4	

NATALGRASS. Rhynchelytrum roseum

Natalgrass, hay, s-c, (1)

Ref No 1 03 180 United States

Feed Name or Analyses			Mean As Fed	Mean Dry	C.V. ± %
Dry matter	%		91.4	100.0	2
Ash	%		5.5	6.0	40
Crude fiber	%		37.2	40.6	8
Ether extract	%		1.7	1.9	15
N-free extract	%		40.4	44.2	
Protein (N x 6.25)	%		6.6	7.3	38
Cattle	dig prot %	*	2.9	3.2	
Goats	dig prot %	*	3.0	3.3	
Horses	dig prot %	*	3.4	3.7	
Rabbits	dig prot %	*	3.9	4.3	
Sheep	dig prot %	*	2.8	3.1	
Fatty acids	%		1.8	2.0	
Energy	GE Mcal/kg				
Cattle	DE Mcal/kg	*	2.20	2.41	
Sheep	DE Mcal/kg	*	2.09	2.29	
Cattle	ME Mcal/kg	*	1.81	1.98	
Sheep	ME Mcal/kg	*	1.72	1.88	
Cattle	TDN %	*	50.5	54.7	
Sheep	TDN %	*	47.5	52.0	
Calcium	%		0.47	0.51	
Phosphorus	%		0.31	0.34	

Natalgrass, hay, s-c, mature, (1)

Ref No 1 03 179 United States

Feed Name or Analyses			Mean As Fed	Mean Dry	C.V. ± %
Dry matter	%		92.4	100.0	
Ash	%		4.8	5.2	
Crude fiber	%		39.5	42.8	
Ether extract	%		1.4	1.5	
N-free extract	%		43.1	46.6	
Protein (N x 6.25)	%		3.6	3.9	
Cattle	dig prot %	*	0.3	0.3	
Goats	dig prot %	*	0.2	0.2	
Horses	dig prot %	*	0.8	0.8	
Rabbits	dig prot %	*	1.6	1.7	
Sheep	dig prot %	*	0.1	0.1	
Energy	GE Mcal/kg				
Cattle	DE Mcal/kg	*	1.80	1.95	
Sheep	DE Mcal/kg	*	2.03	2.20	
Cattle	ME Mcal/kg	*	1.48	1.60	
Sheep	ME Mcal/kg	*	1.67	1.80	
Cattle	TDN %	*	40.8	44.2	
Sheep	TDN %	*	46.1	49.9	

NATIVE PLANTS, CE 8 PLUS YEARS POST SEEDING BROAD LEAF. Scientific name not used

Native plants, CE 8 plus years post seeding broad leaf, hay, s-c, mn 75% broad leaf, (1)

Ref No 1 09 084 Canada

Feed Name or Analyses			Mean As Fed	Mean Dry	C.V. ± %
Dry matter	%			100.0	
Ash	%			5.5	
Crude fiber	%			32.6	
Ether extract	%			2.5	
N-free extract	%			52.1	
Protein (N x 6.25)	%			7.3	
Sheep	dig coef %			51.	
Cattle	dig prot %	*		3.3	
Goats	dig prot %	*		3.4	
Horses	dig prot %	*		3.7	
Rabbits	dig prot %	*		4.3	
Sheep	dig prot %			3.7	
Energy	GE Mcal/kg				
Cattle	DE Mcal/kg	*		2.09	
Sheep	DE Mcal/kg	*		1.98	
Cattle	ME Mcal/kg	*		1.71	
Sheep	ME Mcal/kg	*		1.63	
Cattle	TDN %	*		47.4	
Sheep	TDN %			45.0	

NATIVE PLANTS, CE 8 PLUS YEARS POST SEEDING NARROW LEAF. Scientific name not used

Native plants, CE 8 plus years post seeding narrow leaf, hay, s-c, mn 75% narrow leaf, (1)

Ref No 1 09 083 Canada

Feed Name or Analyses			Mean As Fed	Mean Dry	C.V. ± %
Dry matter	%			100.0	
Ash	%			7.0	
Crude fiber	%			31.7	
Ether extract	%			2.7	
N-free extract	%			50.0	
Protein (N x 6.25)	%			8.6	
Sheep	dig coef %			55.	
Cattle	dig prot %	*		4.4	
Goats	dig prot %	*		4.6	
Horses	dig prot %	*		4.8	
Rabbits	dig prot %	*		5.3	
Sheep	dig prot %			4.7	
Energy	GE Mcal/kg				
Cattle	DE Mcal/kg	*		1.93	
Sheep	DE Mcal/kg	*		1.76	
Cattle	ME Mcal/kg	*		1.58	
Sheep	ME Mcal/kg	*		1.45	
Cattle	TDN %	*		43.7	
Sheep	TDN %			40.0	

NATIVE PLANTS, INTERMOUNTAIN. Scientific name not used

Native plants, intermountain, aerial part, s-c, immature, (1)

Ref No 1 08 464 United States

Feed Name or Analyses			Mean As Fed	Mean Dry	C.V. ± %
Dry matter	%		90.0	100.0	
Ash	%		6.8	7.6	
Crude fiber	%		14.0	15.6	
Ether extract	%		3.1	3.4	
N-free extract	%		49.1	54.6	
Protein (N x 6.25)	%		17.0	18.9	
Cattle	dig prot %	*	12.0	13.3	
Goats	dig prot %	*	12.8	14.2	
Horses	dig prot %	*	12.2	13.6	
Rabbits	dig prot %	*	11.9	13.3	
Sheep	dig prot %	*	12.2	13.5	
Energy	GE Mcal/kg				
Cattle	DE Mcal/kg	*	2.72	3.02	
Sheep	DE Mcal/kg	*	2.70	3.00	
Cattle	ME Mcal/kg	*	2.23	2.48	
Sheep	ME Mcal/kg	*	2.21	2.46	
Cattle	TDN %	*	61.7	68.6	
Sheep	TDN %	*	61.2	68.0	
Calcium	%		1.21	1.34	
Phosphorus	%		0.38	0.42	

Native plants, intermountain, aerial part, s-c, mature, (1)

Ref No 1 08 465 United States

Feed Name or Analyses			Mean As Fed	Mean Dry	C.V. ± %
Dry matter	%		90.0	100.0	
Ash	%		8.1	9.0	
Crude fiber	%		17.4	19.3	
Ether extract	%		4.3	4.8	
N-free extract	%		51.4	57.1	
Protein (N x 6.25)	%		8.8	9.8	
Cattle	dig prot %	*	4.9	5.4	
Goats	dig prot %	*	5.1	5.7	
Horses	dig prot %	*	5.2	5.8	
Rabbits	dig prot %	*	5.6	6.2	
Sheep	dig prot %	*	4.8	5.3	
Energy	GE Mcal/kg				
Cattle	DE Mcal/kg	*	3.35	3.72	
Sheep	DE Mcal/kg	*	2.98	2.31	
Cattle	ME Mcal/kg	*	2.75	3.05	
Sheep	ME Mcal/kg	*	2.44	2.71	
Cattle	TDN %	*	76.0	84.5	
Sheep	TDN %	*	67.6	75.1	

Feed Name or Analyses			Mean As Fed	Mean Dry	C.V. ± %
Native plants, intermountain, hay, s-c, (1)					
Meadow hay					
Ref No 1 03 181			United States		
Dry matter	%		93.5	100.0	1
Ash	%		8.0	8.6	6
Crude fiber	%		30.5	32.7	4
Ether extract	%		2.3	2.5	19
N-free extract	%		44.8	47.9	
Protein (N x 6.25)	%		7.8	8.3	12
Cattle	dig prot %	*	3.9	4.1	
Goats	dig prot %	*	4.0	4.3	
Horses	dig prot %	*	4.3	4.6	
Rabbits	dig prot %	*	4.8	5.1	
Sheep	dig prot %	*	3.8	4.0	
Lignin (Ellis)	%		7.3	7.8	
Energy	GE Mcal/kg				
Cattle	DE Mcal/kg	*	2.47	2.65	
Sheep	DE Mcal/kg	*	2.30	2.46	
Cattle	ME Mcal/kg	*	2.03	2.17	
Sheep	ME Mcal/kg	*	1.88	2.01	
Cattle	NE_m Mcal/kg	*	0.94	1.00	
Cattle	NE_{gain} Mcal/kg	*	0.10	0.11	
Cattle	TDN %	*	56.1	60.0	
Sheep	TDN %	*	52.1	55.7	
Calcium	%		0.55	0.58	24
Phosphorus	%		0.15	0.16	12
Carotene	mg/kg		40.0	42.8	
Vitamin A equivalent	IU/g		66.6	71.3	

Feed Name or Analyses			Mean As Fed	Mean Dry	C.V. ± %
Native plants, intermountain, hay, s-c, weathered, mature, (1)					
Ref No 1 08 466			United States		
Dry matter	%		90.0	100.0	
Ash	%		7.5	8.3	
Crude fiber	%		33.6	37.3	
Ether extract	%		1.4	1.6	
N-free extract	%		43.6	48.4	
Protein (N x 6.25)	%		3.9	4.3	
Cattle	dig prot %	*	0.6	0.7	
Goats	dig prot %	*	0.5	0.6	
Horses	dig prot %	*	1.1	1.2	
Rabbits	dig prot %	*	1.8	2.0	
Sheep	dig prot %	*	0.4	0.5	
Energy	GE Mcal/kg				
Cattle	DE Mcal/kg	*	1.74	1.93	
Sheep	DE Mcal/kg	*	1.59	1.77	
Cattle	ME Mcal/kg	*	1.42	1.58	
Sheep	ME Mcal/kg	*	1.30	1.45	
Cattle	TDN %	*	39.3	43.7	
Sheep	TDN %	*	36.1	40.1	
Calcium	%		0.53	0.59	
Phosphorus	%		0.07	0.08	

(1) dry forages and roughages (2) pasture, range plants, and forages fed green (3) silages (4) energy feeds (5) protein supplements (6) minerals (7) vitamins (8) additives

NATIVE PLANTS, MARSH. Scientific name not used

Feed Name or Analyses			Mean As Fed	Mean Dry	C.V. ± %
Native plants, marsh, hay, s-c, (1)					
Marsh hay					
Ref No 1 03 182			United States		
Dry matter	%		88.2	100.0	1
Ash	%		6.8	7.7	7
Crude fiber	%		28.5	32.3	8
Ether extract	%		2.1	2.4	7
N-free extract	%		42.1	47.7	
Protein (N x 6.25)	%		8.7	9.9	9
Cattle	dig prot %	*	4.8	5.5	
Goats	dig prot %	*	5.1	5.8	
Horses	dig prot %	*	5.2	5.9	
Rabbits	dig prot %	*	5.5	6.3	
Sheep	dig prot %	*	4.8	5.4	
Fatty acids	%		2.2	2.5	
Energy	GE Mcal/kg				
Cattle	DE Mcal/kg	*	2.15	2.44	
Sheep	DE Mcal/kg	*	2.04	2.31	
Cattle	ME Mcal/kg	*	1.76	2.00	
Sheep	ME Mcal/kg	*	1.67	1.89	
Cattle	TDN %	*	48.9	55.4	
Sheep	TDN %	*	46.1	52.3	
Calcium	%		0.31	0.35	
Iron	%		0.010	0.011	
Magnesium	%		0.25	0.29	
Phosphorus	%		0.10	0.11	
Potassium	%		0.67	0.77	

NATIVE PLANTS, MIDWEST. Scientific name not used

Feed Name or Analyses			Mean As Fed	Mean Dry	C.V. ± %
Native plants, midwest, hay, s-c, (1)					
Prairie hay					
Ref No 1 03 191			United States		
Dry matter	%		91.0	100.0	2
Organic matter	%		84.7	93.1	
Ash	%		7.2	8.0	
Crude fiber	%		30.7	33.7	6
Cattle	dig coef %	*	61.	61.	
Ether extract	%		2.1	2.3	28
Cattle	dig coef %	*	33.	33.	
N-free extract	%		45.2	49.6	5
Cattle	dig coef %	*	58.	58.	
Protein (N x 6.25)	%		5.8	6.4	26
Cattle	dig coef %	*	28.	28.	
Cattle	dig prot %		1.6	1.8	
Goats	dig prot %	*	2.3	2.5	
Horses	dig prot %	*	2.7	2.9	
Rabbits	dig prot %	*	3.3	3.6	
Sheep	dig prot %	*	2.1	2.3	
Energy	GE Mcal/kg		3.80	4.17	
Cattle	DE Mcal/kg	*	2.12	2.33	
Sheep	DE Mcal/kg	*	2.00	2.20	
Cattle	ME Mcal/kg	*	1.74	1.91	
Sheep	ME Mcal/kg	*	1.64	1.80	
Cattle	TDN %		48.1	52.9	
Sheep	TDN %	*	45.3	49.8	
Calcium	%		0.32	0.35	
Iron	%		0.008	0.009	
Magnesium	%		0.22	0.24	
Phosphorus	%		0.12	0.14	
Potassium	%		0.98	1.08	
Vitamin D_2	IU/g		0.9	1.0	38

Feed Name or Analyses			Mean As Fed	Mean Dry	C.V. ± %
Ref No 1 03 191			Canada		
Dry matter	%		91.3	100.0	
Ash	%		8.6	9.4	
Crude fiber	%		29.7	32.5	
Ether extract	%		3.0	3.3	
N-free extract	%		44.0	48.2	
Protein (N x 6.25)	%		6.0	6.6	
Cattle	dig prot %	*	2.4	2.6	
Goats	dig prot %	*	2.5	2.7	
Horses	dig prot %	*	2.8	3.1	
Rabbits	dig prot %	*	3.4	3.7	
Sheep	dig prot %	*	2.3	2.5	
Energy	GE Mcal/kg		3.81	4.17	
Cattle	DE Mcal/kg	*	2.12	2.32	
Sheep	DE Mcal/kg	*	2.01	2.20	
Cattle	ME Mcal/kg	*	1.74	1.91	
Sheep	ME Mcal/kg	*	1.64	1.80	
Cattle	TDN %		48.1	52.7	
Sheep	TDN %	*	45.5	49.8	
Calcium	%		0.33	0.36	
Magnesium	%		0.25	0.27	
Phosphorus	%		0.12	0.13	

Feed Name or Analyses			Mean As Fed	Mean Dry	C.V. ± %
Native plants, midwest, hay, s-c, immature, (1)					
Prairie hay, immature					
Ref No 1 03 183			United States		
Dry matter	%		90.2	100.0	4
Ash	%		8.0	8.9	17
Crude fiber	%		28.3	31.4	4
Cattle	dig coef %	*	61.	61.	
Ether extract	%		2.4	2.7	24
Cattle	dig coef %	*	33.	33.	
N-free extract	%		41.6	46.2	
Cattle	dig coef %	*	58.	58.	
Protein (N x 6.25)	%		9.7	10.8	7
Cattle	dig coef %	*	28.	28.	
Cattle	dig prot %		2.7	3.0	
Goats	dig prot %	*	6.0	6.6	
Horses	dig prot %	*	6.0	6.7	
Rabbits	dig prot %	*	6.3	7.0	
Sheep	dig prot %	*	5.6	6.3	
Energy	GE Mcal/kg				
Cattle	DE Mcal/kg	*	2.03	2.25	
Sheep	DE Mcal/kg	*	2.27	2.52	
Cattle	ME Mcal/kg	*	1.66	1.84	
Sheep	ME Mcal/kg	*	1.86	2.07	
Cattle	NE_m Mcal/kg	*	0.98	1.09	
Cattle	NE_{gain} Mcal/kg	*	0.28	0.31	
Cattle	TDN %		46.0	51.0	
Sheep	TDN %	*	51.5	57.2	
Calcium	%		0.44	0.49	18
Iron	%		0.008	0.009	19
Magnesium	%		0.22	0.24	18
Phosphorus	%		0.21	0.23	17
Potassium	%		0.97	1.08	7

Feed Name or Analyses			Mean As Fed	Mean Dry	C.V. ± %
Native plants, midwest, hay, s-c, full bloom, (1)					
Prairie hay, full bloom					
Ref No 1 03 184			United States		
Dry matter	%		83.3	100.0	9
Ash	%		8.5	10.2	8
Crude fiber	%		27.5	33.0	2
Cattle	dig coef %	*	61.	61.	
Ether extract	%		2.7	3.2	15
Cattle	dig coef %	*	33.	33.	

Feed Name or Analyses			Mean As Fed	Mean Dry	C.V. ± %
N-free extract	%		38.3	46.0	
Cattle	dig coef %	*	58.	58.	
Protein (N x 6.25)	%		6.3	7.6	25
Cattle	dig coef %	*	28.	28.	
Cattle	dig prot %		1.8	2.1	
Goats	dig prot %	*	3.0	3.7	
Horses	dig prot %	*	3.3	4.0	
Rabbits	dig prot %	*	3.8	4.5	
Sheep	dig prot %	*	2.8	3.4	
Energy	GE Mcal/kg				
Cattle	DE Mcal/kg	*	1.88	2.26	
Sheep	DE Mcal/kg	*	2.02	2.42	
Cattle	ME Mcal/kg	*	1.55	1.86	
Sheep	ME Mcal/kg	*	1.65	1.99	
Cattle	NE_m Mcal/kg	*	0.91	1.09	
Cattle	NE_{gain} Mcal/kg	*	0.26	0.31	
Cattle	TDN %	*	42.7	51.3	
Sheep	TDN %	*	45.8	54.9	

Native plants, midwest, hay, s-c, milk stage, (1)
Prairie hay, milk stage
Ref No 1 03 185 United States

Feed Name or Analyses			Mean As Fed	Mean Dry	C.V. ± %
Dry matter	%		89.6	100.0	4
Energy	GE Mcal/kg				
Cattle	NE_m Mcal/kg	*	0.92	1.03	
Cattle	NE_{gain} Mcal/kg	*	0.17	0.19	
Calcium	%		0.47	0.52	24
Magnesium	%		0.35	0.39	10
Phosphorus	%		0.07	0.08	18

Native plants, midwest, hay, s-c, dough stage, (1)
Prairie hay, dough stage
Ref No 1 03 186 United States

Feed Name or Analyses			Mean As Fed	Mean Dry	C.V. ± %
Dry matter	%		88.6	100.0	
Ash	%		8.6	9.7	6
Crude fiber	%		29.4	33.2	4
Ether extract	%		2.2	2.5	29
N-free extract	%		43.1	48.6	
Protein (N x 6.25)	%		5.3	6.0	20
Cattle	dig prot %	*	1.9	2.1	
Goats	dig prot %	*	1.9	2.2	
Horses	dig prot %	*	2.3	2.6	
Rabbits	dig prot %	*	2.9	3.3	
Sheep	dig prot %	*	1.7	2.0	
Energy	GE Mcal/kg				
Cattle	DE Mcal/kg	*	2.16	2.44	
Sheep	DE Mcal/kg	*	2.13	2.40	
Cattle	ME Mcal/kg	*	1.77	2.00	
Sheep	ME Mcal/kg	*	1.74	1.97	
Cattle	TDN %	*	49.0	55.3	
Sheep	TDN %	*	48.2	54.4	

Native plants, midwest, hay, s-c, mature, (1)
Prairie hay, mature
Ref No 1 03 187 United States

Feed Name or Analyses			Mean As Fed	Mean Dry	C.V. ± %
Dry matter	%		91.4	100.0	2
Ash	%		6.7	7.4	
Crude fiber	%		31.6	34.5	
Cattle	dig coef %	*	61.	61.	
Ether extract	%		2.5	2.7	
Cattle	dig coef %	*	33.	33.	
N-free extract	%		46.1	50.5	
Cattle	dig coef %	*	58.	58.	
Protein (N x 6.25)	%		4.5	4.9	
Cattle	dig coef %	*	28.	28.	
Cattle	dig prot %		1.3	1.4	
Goats	dig prot %	*	1.1	1.2	
Horses	dig prot %	*	1.6	1.7	
Rabbits	dig prot %	*	2.3	2.5	
Sheep	dig prot %	*	0.9	1.0	
Energy	GE Mcal/kg				
Cattle	DE Mcal/kg	*	2.16	2.37	
Sheep	DE Mcal/kg	*	2.17	2.37	
Cattle	ME Mcal/kg	*	1.78	1.94	
Sheep	ME Mcal/kg	*	1.78	1.94	
Cattle	NE_m Mcal/kg	*	1.05	1.15	
Cattle	NE_{gain} Mcal/kg	*	0.40	0.44	
Cattle	TDN %		49.1	53.7	
Sheep	TDN %	*	49.1	53.7	
Calcium	%		0.35	0.38	
Copper	mg/kg		20.8	22.8	
Iron	%		0.010	0.011	
Magnesium	%		0.26	0.28	
Manganese	mg/kg		44.3	48.5	
Phosphorus	%		0.11	0.12	
Potassium	%		0.73	0.79	
Carotene	mg/kg		9.5	10.4	52
Vitamin A equivalent	IU/g		15.8	17.3	

Native plants, midwest, hay, s-c, over ripe, (1)
Prairie hay, over ripe
Ref No 1 03 188 United States

Feed Name or Analyses			Mean As Fed	Mean Dry	C.V. ± %
Dry matter	%		91.5	100.0	1
Ash	%		7.2	7.9	11
Crude fiber	%		31.5	34.4	6
Cattle	dig coef %	*	61.	61.	
Ether extract	%		2.7	2.9	19
Cattle	dig coef %	*	33.	33.	
N-free extract	%		46.5	50.8	
Cattle	dig coef %	*	58.	58.	
Protein (N x 6.25)	%		3.7	4.0	31
Cattle	dig coef %	*	28.	28.	
Cattle	dig prot %		1.0	1.1	
Goats	dig prot %	*	0.3	0.3	
Horses	dig prot %	*	0.8	0.9	
Rabbits	dig prot %	*	1.6	1.8	
Sheep	dig prot %	*	0.1	0.2	
Energy	GE Mcal/kg				
Cattle	DE Mcal/kg	*	2.17	2.37	
Sheep	DE Mcal/kg	*	2.15	2.35	
Cattle	ME Mcal/kg	*	1.78	1.94	
Sheep	ME Mcal/kg	*	1.76	1.93	
Cattle	NE_m Mcal/kg	*	1.05	1.15	
Cattle	NE_{gain} Mcal/kg	*	0.40	0.44	
Cattle	TDN %	*	49.2	53.7	
Sheep	TDN %	*	48.8	53.3	

Native plants, midwest, hay, s-c, cut 1, (1)
Prairie hay, cut 1
Ref No 1 03 189 United States

Feed Name or Analyses			Mean As Fed	Mean Dry	C.V. ± %
Dry matter	%		91.6	100.0	1
Cobalt	mg/kg		0.109	0.119	
Sodium	%		0.04	0.04	
Carotene	mg/kg		22.4	24.5	
Vitamin A equivalent	IU/g		37.4	40.8	

Native plants, midwest, hay, s-c, cut 2, (1)
Prairie hay, cut 2
Ref No 1 03 190 United States

Feed Name or Analyses			Mean As Fed	Mean Dry	C.V. ± %
Dry matter	%			100.0	
Calcium	%			0.45	
Phosphorus	%			0.09	

Native plants, midwest, hay, s-c, gr 2 US, (1)
Prairie hay, gr 2
Ref No 1 03 193 United States

Feed Name or Analyses			Mean As Fed	Mean Dry	C.V. ± %
Dry matter	%			100.0	
Vitamin D_2	IU/g			1.7	

Native plants, midwest, hay, s-c, gr 3 US, (1)
Prairie hay, gr 3
Ref No 1 03 194 United States

Feed Name or Analyses			Mean As Fed	Mean Dry	C.V. ± %
Dry matter	%			100.0	
Vitamin D_2	IU/g			1.4	

Native plants, midwest, hay, s-c, gr sample US, (1)
Prairie hay, gr sample
Ref No 1 03 192 United States

Feed Name or Analyses			Mean As Fed	Mean Dry	C.V. ± %
Dry matter	%			100.0	
Vitamin D_2	IU/g			1.3	

Native plants, midwest, hay, s-c weathered, mature, (1)
Ref No 1 08 467 United States

Feed Name or Analyses			Mean As Fed	Mean Dry	C.V. ± %
Dry matter	%		91.9	100.0	
Ash	%		6.5	7.1	
Crude fiber	%		31.5	34.3	
Ether extract	%		2.3	2.5	
N-free extract	%		48.7	53.0	
Protein (N x 6.25)	%		2.9	3.2	
Cattle	dig prot %	*	-0.2	-0.2	
Goats	dig prot %	*	-0.4	-0.4	
Horses	dig prot %	*	0.2	0.2	
Rabbits	dig prot %	*	1.0	1.1	
Sheep	dig prot %	*	-0.5	-0.5	
Energy	GE Mcal/kg				
Cattle	DE Mcal/kg	*	1.81	1.97	
Sheep	DE Mcal/kg	*	1.76	1.91	
Cattle	ME Mcal/kg	*	1.49	1.62	
Sheep	ME Mcal/kg	*	1.44	1.57	
Cattle	TDN %	*	41.1	44.7	
Sheep	TDN %	*	39.9	43.3	
Calcium	%		0.41	0.45	
Phosphorus	%		0.03	0.03	

Native plants, midwest, aerial part, fresh, dormant, (2)
Ref No 2 07 853 United States

Feed Name or Analyses			Mean As Fed	Mean Dry	C.V. ± %
Dry matter	%		94.2	100.0	
Ash	%		7.2	7.6	
Crude fiber	%		33.5	35.6	
Ether extract	%		2.4	2.6	

Continued

Feed Name or Analyses		Mean As Fed	Mean Dry	C.V. ± %
N-free extract	%	47.9	50.8	
Protein (N x 6.25)	%	3.2	3.4	
Cattle	dig prot % *	0.7	0.8	
Goats	dig prot % *	-0.2	-0.2	
Horses	dig prot % *	0.4	0.4	
Rabbits	dig prot % *	1.2	1.3	
Sheep	dig prot % *	0.1	0.2	
Energy	GE Mcal/kg			
Cattle	DE Mcal/kg *	2.30	2.44	
Sheep	DE Mcal/kg *	2.19	2.32	
Cattle	ME Mcal/kg *	1.89	2.00	
Sheep	ME Mcal/kg *	1.79	1.90	
Cattle	TDN % *	52.2	55.4	
Sheep	TDN % *	49.6	52.6	

NEEDLEANDTHREAD. Stipa comata

Needleandthread, aerial part, fresh, immature, (2)

Ref No 2 03 195 United States

Feed Name or Analyses		Mean As Fed	Mean Dry	C.V. ± %
Dry matter	%		100.0	
Ash	%		12.3	16
Crude fiber	%		29.0	5
Ether extract	%		2.6	11
N-free extract	%		44.1	
Protein (N x 6.25)	%		12.0	5
Cattle	dig prot % *		8.1	
Goats	dig prot % *		7.8	
Horses	dig prot % *		7.7	
Rabbits	dig prot % *		7.9	
Sheep	dig prot % *		8.2	
Energy	GE Mcal/kg			
Cattle	DE Mcal/kg *		2.70	
Sheep	DE Mcal/kg *		2.81	
Cattle	ME Mcal/kg *		2.21	
Sheep	ME Mcal/kg *		2.30	
Cattle	TDN % *		61.2	
Sheep	TDN % *		63.7	
Calcium	%		0.93	
Cobalt	mg/kg		0.051	
Phosphorus	%		0.16	
Carotene	mg/kg		88.2	
Vitamin A equivalent	IU/g		147.0	

Needleandthread, aerial part, fresh, early bloom, (2)

Ref No 2 03 196 United States

Feed Name or Analyses		Mean As Fed	Mean Dry	C.V. ± %
Dry matter	%		100.0	
Ash	%		10.6	
Crude fiber	%		31.1	
Ether extract	%		3.6	
N-free extract	%		43.3	
Protein (N x 6.25)	%		11.4	
Cattle	dig prot % *		7.6	
Goats	dig prot % *		7.2	
Horses	dig prot % *		7.2	
Rabbits	dig prot % *		7.5	
Sheep	dig prot % *		7.6	
Energy	GE Mcal/kg			
Cattle	DE Mcal/kg *		2.73	
Sheep	DE Mcal/kg *		2.80	
Cattle	ME Mcal/kg *		2.24	
Sheep	ME Mcal/kg *		2.29	
Cattle	TDN % *		61.9	
Sheep	TDN % *		63.4	

Needleandthread, aerial part, fresh, mid-bloom, (2)

Ref No 2 03 197 United States

Feed Name or Analyses		Mean As Fed	Mean Dry	C.V. ± %
Dry matter	%		100.0	
Ash	%		11.2	
Crude fiber	%		32.9	
Ether extract	%		3.1	
N-free extract	%		43.9	
Protein (N x 6.25)	%		8.9	
Cattle	dig prot % *		5.5	
Goats	dig prot % *		4.9	
Horses	dig prot % *		5.1	
Rabbits	dig prot % *		5.5	
Sheep	dig prot % *		5.3	
Energy	GE Mcal/kg			
Cattle	DE Mcal/kg *		2.62	
Sheep	DE Mcal/kg *		2.84	
Cattle	ME Mcal/kg *		2.15	
Sheep	ME Mcal/kg *		2.33	
Cattle	TDN % *		59.3	
Sheep	TDN % *		64.5	

Needleandthread, aerial part, fresh, full bloom, (2)

Ref No 2 03 198 United States

Feed Name or Analyses		Mean As Fed	Mean Dry	C.V. ± %
Dry matter	%		100.0	
Ash	%		9.1	
Crude fiber	%		37.3	
Ether extract	%		2.2	
N-free extract	%		44.8	
Protein (N x 6.25)	%		6.6	
Cattle	dig prot % *		3.5	
Goats	dig prot % *		2.7	
Horses	dig prot % *		3.1	
Rabbits	dig prot % *		3.8	
Sheep	dig prot % *		3.1	
Energy	GE Mcal/kg			
Cattle	DE Mcal/kg *		2.74	
Sheep	DE Mcal/kg *		2.87	
Cattle	ME Mcal/kg *		2.24	
Sheep	ME Mcal/kg *		2.35	
Cattle	TDN % *		62.1	
Sheep	TDN % *		65.1	
Calcium	%		0.34	
Phosphorus	%		0.11	

Needleandthread, aerial part, fresh, mature, (2)

Ref No 2 03 199 United States

Feed Name or Analyses		Mean As Fed	Mean Dry	C.V. ± %
Dry matter	%		100.0	
Ash	%		11.6	48
Crude fiber	%		36.2	8
Ether extract	%		3.3	46
N-free extract	%		44.2	
Protein (N x 6.25)	%		4.7	19
Cattle	dig prot % *		1.9	
Goats	dig prot % *		0.9	
Horses	dig prot % *		1.5	
Rabbits	dig prot % *		2.3	
Sheep	dig prot % *		1.4	
Energy	GE Mcal/kg			
Cattle	DE Mcal/kg *		2.40	
Sheep	DE Mcal/kg *		2.47	
Cattle	ME Mcal/kg *		1.97	
Sheep	ME Mcal/kg *		2.03	
Cattle	TDN % *		54.4	
Sheep	TDN % *		56.1	
Calcium	%		0.30	
Phosphorus	%		0.09	
Carotene	mg/kg		0.4	
Vitamin A equivalent	IU/g		0.7	

Needleandthread, aerial part, fresh, over ripe, (2)

Ref No 2 03 200 United States

Feed Name or Analyses		Mean As Fed	Mean Dry	C.V. ± %
Dry matter	%		100.0	
Ash	%		9.2	
Crude fiber	%		39.4	
Ether extract	%		2.2	
N-free extract	%		46.3	
Protein (N x 6.25)	%		2.9	
Cattle	dig prot % *		0.4	
Goats	dig prot % *		-0.6	
Horses	dig prot % *		0.0	
Rabbits	dig prot % *		0.9	
Sheep	dig prot % *		-0.2	
Energy	GE Mcal/kg			
Cattle	DE Mcal/kg *		2.56	
Sheep	DE Mcal/kg *		2.51	
Cattle	ME Mcal/kg *		2.10	
Sheep	ME Mcal/kg *		2.06	
Cattle	TDN % *		58.2	
Sheep	TDN % *		57.0	
Calcium	%		0.26	
Phosphorus	%		0.07	

Needleandthread, aerial part wo lower stems, fresh, early leaf, (2)

Ref No 2 08 702 United States

Feed Name or Analyses		Mean As Fed	Mean Dry	C.V. ± %
Dry matter	%		100.0	
Organic matter	%		90.7	
Ash	%		9.3	
Crude fiber	%		32.1	
Ether extract	%		2.0	
N-free extract	%		49.8	
Protein (N x 6.25)	%		6.8	
Cattle	dig prot % *		3.7	
Goats	dig prot % *		2.9	
Horses	dig prot % *		3.3	
Rabbits	dig prot % *		3.9	
Sheep	dig prot % *		3.4	
Energy	GE Mcal/kg			
Cattle	DE Mcal/kg *		2.70	
Sheep	DE Mcal/kg *		2.75	
Cattle	ME Mcal/kg *		2.22	
Sheep	ME Mcal/kg *		2.26	
Cattle	TDN % *		61.3	
Sheep	TDN % *		62.3	
Calcium	%		0.45	
Phosphorus	%		0.09	
Carotene	mg/kg		0.0	

Needleandthread, aerial part wo lower stems, fresh, mature, (2)

Ref No 2 08 704 United States

Feed Name or Analyses		Mean As Fed	Mean Dry	C.V. ± %
Dry matter	%		100.0	
Organic matter	%		90.2	
Ash	%		9.8	
Crude fiber	%		33.9	

(1) dry forages and roughages (2) pasture, range plants, and forages fed green (3) silages (4) energy feeds (5) protein supplements (6) minerals (7) vitamins (8) additives

Feed Name or Analyses			Mean As Fed	Mean Dry	C.V. ± %
Ether extract	%			2.1	
N-free extract	%			49.0	
Protein (N x 6.25)	%			5.3	
Cattle	dig prot %	*		2.4	
Goats	dig prot %	*		1.5	
Horses	dig prot %	*		2.0	
Rabbits	dig prot %	*		2.7	
Sheep	dig prot %	*		1.9	
Energy	GE Mcal/kg				
Cattle	DE Mcal/kg	*		2.60	
Sheep	DE Mcal/kg	*		2.54	
Cattle	ME Mcal/kg	*		2.13	
Sheep	ME Mcal/kg	*		2.08	
Cattle	TDN %	*		59.0	
Sheep	TDN %	*		65.77	
Calcium	%			0.63	
Phosphorus	%			0.07	

Needleandthread, aerial part wo lower stems, fresh, dormant, (2)

Ref No 2 08 701 United States

Feed Name or Analyses			Mean As Fed	Mean Dry	C.V. ± %
Dry matter	%			100.0	
Organic matter	%			89.2	
Ash	%			10.8	
Crude fiber	%			33.8	
Ether extract	%			2.1	
N-free extract	%			48.5	
Protein (N x 6.25)	%			4.9	
Cattle	dig prot %	*		2.0	
Goats	dig prot %	*		1.1	
Horses	dig prot %	*		1.6	
Rabbits	dig prot %	*		2.4	
Sheep	dig prot %	*		1.5	
Energy	GE Mcal/kg				
Cattle	DE Mcal/kg	*		2.52	
Sheep	DE Mcal/kg	*		2.63	
Cattle	ME Mcal/kg	*		2.06	
Sheep	ME Mcal/kg	*		2.18	
Cattle	TDN %	*		57.1	
Sheep	TDN %	*		59.7	
Calcium	%			0.45	
Phosphorus	%			0.04	
Carotene	mg/kg			9.0	
Vitamin A equivalent	IU/g			15.0	

NEEDLEGRASS. Stipa spp

Needlegrass, hay, s-c, (1)

Ref No 1 03 202 United States

Feed Name or Analyses			Mean As Fed	Mean Dry	C.V. ± %
Dry matter	%		88.1	100.0	2
Ash	%		6.2	7.1	3
Crude fiber	%		30.0	34.0	5
Ether extract	%		2.2	2.5	15
N-free extract	%		41.9	47.5	
Protein (N x 6.25)	%		7.8	8.8	24
Cattle	dig prot %	*	4.0	4.6	
Goats	dig prot %	*	4.2	4.8	
Horses	dig prot %	*	4.4	5.0	
Rabbits	dig prot %	*	4.8	5.5	
Sheep	dig prot %	*	4.0	4.5	
Fatty acids	%		2.0	2.3	
Energy	GE Mcal/kg				
Cattle	DE Mcal/kg	*	1.88	2.13	
Sheep	DE Mcal/kg	*	1.81	2.06	
Cattle	ME Mcal/kg	*	1.54	1.75	
Sheep	ME Mcal/kg	*	1.49	1.69	

Feed Name or Analyses			Mean As Fed	Mean Dry	C.V. ± %
Cattle	TDN %	*	42.6	48.3	
Sheep	TDN %	*	41.1	46.7	

Needlegrass, aerial part, fresh, full bloom, (2)

Ref No 2 03 203 United States

Feed Name or Analyses			Mean As Fed	Mean Dry	C.V. ± %
Dry matter	%			100.0	
Ash	%			9.9	
Crude fiber	%			36.7	
Ether extract	%			1.9	
N-free extract	%			43.2	
Protein (N x 6.25)	%			8.3	
Cattle	dig prot %	*		4.9	
Goats	dig prot %	*		4.3	
Horses	dig prot %	*		4.6	
Rabbits	dig prot %	*		5.1	
Sheep	dig prot %	*		4.7	

Needlegrass, aerial part, fresh, over ripe, (2)

Ref No 2 03 204 United States

Feed Name or Analyses			Mean As Fed	Mean Dry	C.V. ± %
Dry matter	%			100.0	
Carotene	mg/kg			5.3	
Vitamin A equivalent	IU/g			8.8	

Needlegrass, aerial part, dormant, (2)

Ref No 2 03 201 United States

Feed Name or Analyses			Mean As Fed	Mean Dry	C.V. ± %
Dry matter	%			100.0	
Carotene	mg/kg			11.0	
Vitamin A equivalent	IU/g			18.4	

NEEDLEGRASS, CALIFORNIA. Stipa pulchra

Needlegrass, California, aerial part, fresh, (2)

Purple needlegrass

Ref No 2 08 565 United States

Feed Name or Analyses			Mean As Fed	Mean Dry	C.V. ± %
Dry matter	%			100.0	
Ash	%			9.4	
Ether extract	%			1.7	
Protein (N x 6.25)	%			3.4	
Cattle	dig prot %	*		0.7	
Goats	dig prot %	*		-0.2	
Horses	dig prot %	*		0.4	
Rabbits	dig prot %	*		1.3	
Sheep	dig prot %	*		0.1	
Other carbohydrates	%			35.7	
Lignin (Ellis)	%			11.1	
Silicon	%			7.20	

NEEDLEGRASS, GREEN. Stipa spp

Needlegrass, green, aerial part, fresh, mature, (2)

Ref No 2 03 205 United States

Feed Name or Analyses			Mean As Fed	Mean Dry	C.V. ± %
Dry matter	%			100.0	
Ash	%			6.9	
Crude fiber	%			40.7	
Ether extract	%			1.1	
N-free extract	%			43.7	
Protein (N x 6.25)	%			7.6	
Cattle	dig prot %	*		4.4	
Goats	dig prot %	*		3.7	
Horses	dig prot %	*		4.0	

Feed Name or Analyses			Mean As Fed	Mean Dry	C.V. ± %
Rabbits	dig prot %	*		4.5	
Sheep	dig prot %	*		4.1	
Energy	GE Mcal/kg				

NEEDLEGRASS, LETTERMAN. Stipa lettermanii

Needlegrass, letterman, hay, s-c, (1)

Ref No 1 03 208 United States

Feed Name or Analyses			Mean As Fed	Mean Dry	C.V. ± %
Dry matter	%			100.0	
Ash	%			7.1	4
Crude fiber	%			31.3	6
Ether extract	%			3.2	12
N-free extract	%			47.5	
Protein (N x 6.25)	%			10.9	23
Cattle	dig prot %	*		6.4	
Goats	dig prot %	*		6.7	
Horses	dig prot %	*		6.8	
Rabbits	dig prot %	*		7.1	
Sheep	dig prot %	*		6.3	

Needlegrass, letterman, hay, s-c, immature, (1)

Ref No 1 03 206 United States

Feed Name or Analyses			Mean As Fed	Mean Dry	C.V. ± %
Dry matter	%			100.0	
Calcium	%			0.73	
Magnesium	%			0.15	
Phosphorus	%			0.26	
Sulphur	%			0.19	

Needlegrass, letterman, hay, s-c, mature, (1)

Ref No 1 03 207 United States

Feed Name or Analyses			Mean As Fed	Mean Dry	C.V. ± %
Dry matter	%			100.0	
Calcium	%			0.96	
Magnesium	%			0.12	
Phosphorus	%			0.16	
Sulphur	%			0.12	

NEEDLEGRASS, TEXAS. Stipa leucotricha

Needlegrass, Texas, aerial part, fresh, (2)

Ref No 2 03 212 United States

Feed Name or Analyses			Mean As Fed	Mean Dry	C.V. ± %
Dry matter	%			100.0	
Ash	%			11.9	19
Crude fiber	%			31.2	7
Ether extract	%			2.7	21
N-free extract	%			45.0	
Protein (N x 6.25)	%			9.2	26
Cattle	dig prot %	*		5.7	
Goats	dig prot %	*		5.1	
Horses	dig prot %	*		5.3	
Rabbits	dig prot %	*		5.8	
Sheep	dig prot %	*		5.6	
Calcium	%			0.56	26
Magnesium	%			0.14	
Phosphorus	%			0.15	21
Potassium	%			1.35	

Feed Name or Analyses			Mean As Fed	Mean Dry	C.V. ± %

Needlegrass, Texas, aerial part, fresh, immature, (2)
Ref No 2 03 209 United States

Analyses			As Fed	Dry	C.V. ± %
Dry matter		%		100.0	
Ash		%		10.3	8
Crude fiber		%		30.3	4
Ether extract		%		2.9	7
N-free extract		%		45.1	
Protein (N x 6.25)		%		11.4	19
Cattle	dig prot	%	*	7.6	
Goats	dig prot	%	*	7.2	
Horses	dig prot	%	*	7.2	
Rabbits	dig prot	%	*	7.5	
Sheep	dig prot	%	*	7.6	
Calcium		%		0.71	
Magnesium		%		0.19	
Phosphorus		%		0.17	11
Potassium		%		1.51	

Needlegrass, Texas, aerial part, fresh, full bloom, (2)
Ref No 2 03 210 United States

Analyses			As Fed	Dry	C.V. ± %
Dry matter		%		100.0	
Calcium		%		0.52	
Phosphorus		%		0.14	

Needlegrass, Texas, aerial part, fresh, mature, (2)
Ref No 2 03 211 United States

Analyses			As Fed	Dry	C.V. ± %
Dry matter		%		100.0	
Ash		%		12.4	
Crude fiber		%		31.9	
Ether extract		%		2.8	
N-free extract		%		46.0	
Protein (N x 6.25)		%		6.9	
Cattle	dig prot	%	*	3.8	
Goats	dig prot	%	*	3.0	
Horses	dig prot	%	*	3.4	
Rabbits	dig prot	%	*	4.0	
Sheep	dig prot	%	*	3.4	
Calcium		%		0.26	
Phosphorus		%		0.10	

NEMESIA, POUCH SUTTON. Nemesia suttonii

Nemesia, pouch sutton, seeds w hulls, (5)
Ref No 5 09 015 United States

Analyses			As Fed	Dry	C.V. ± %
Dry matter		%		100.0	
Protein (N x 6.25)		%		29.0	
Alanine		%		1.19	
Arginine		%		2.93	
Aspartic acid		%		2.41	
Glutamic acid		%		4.64	
Glycine		%		1.22	
Histidine		%		0.61	
Hydroxyproline		%		0.00	
Isoleucine		%		0.99	
Leucine		%		1.51	
Lysine		%		1.13	
Methionine		%		0.46	
Phenylalanine		%		1.16	
Proline		%		1.02	
Serine		%		1.13	
Threonine		%		0.87	
Tyrosine		%		0.78	
Valine		%		1.31	

NETTLE. Urtica spp

Nettle, aerial part, grnd, (1)
Ref No 1 03 214 United States

Analyses			As Fed	Dry	C.V. ± %
Dry matter		%	92.5	100.0	
Ash		%	5.6	6.1	
Crude fiber		%	45.5	49.2	
Sheep	dig coef	%	28.	28.	
Ether extract		%	0.7	0.8	
Sheep	dig coef	%	62.	62.	
N-free extract		%	33.9	36.6	
Sheep	dig coef	%	43.	43.	
Protein (N x 6.25)		%	6.8	7.3	
Sheep	dig coef	%	55.	55.	
Cattle	dig prot	%	* 3.0	3.3	
Goats	dig prot	%	* 3.1	3.4	
Horses	dig prot	%	* 3.4	3.7	
Rabbits	dig prot	%	* 4.0	4.3	
Sheep	dig prot	%	3.7	4.0	
Energy	GE Mcal/kg				
Cattle	DE Mcal/kg		* 1.17	1.27	
Sheep	DE Mcal/kg		* 1.41	1.53	
Cattle	ME Mcal/kg		* 0.96	1.04	
Sheep	ME Mcal/kg		* 1.16	1.25	
Cattle	TDN	%	* 26.6	28.7	
Sheep	TDN	%	32.0	34.6	

NETTLESPURGE. Jatropha macrorhiza

Nettlespurge, seeds w hulls, (5)
Ref No 5 09 033 United States

Analyses			As Fed	Dry	C.V. ± %
Dry matter		%		100.0	
Protein (N x 6.25)		%		36.0	
Alanine		%		1.62	
Arginine		%		4.46	
Aspartic acid		%		3.35	
Glutamic acid		%		5.36	
Glycine		%		1.48	
Histidine		%		0.86	
Hydroxyproline		%		0.14	
Isoleucine		%		1.69	
Leucine		%		2.48	
Lysine		%		1.15	
Methionine		%		0.58	
Phenylalanine		%		1.55	
Proline		%		1.66	
Serine		%		1.62	
Threonine		%		1.30	
Tyrosine		%		1.01	
Valine		%		1.66	

NIGERSEED. Guizotia spp

Nigerseed, seeds, extn unspecified grnd, (5)
Ref No 5 03 220 United States

Analyses			As Fed	Dry	C.V. ± %
Dry matter		%	90.0	100.0	
Ash		%	10.8	12.0	
Crude fiber		%	19.7	21.9	
Sheep	dig coef	%	42.	42.	
Ether extract		%	5.6	6.2	
Sheep	dig coef	%	92.	92.	
N-free extract		%	20.9	23.2	
Sheep	dig coef	%	73.	73.	
Protein (N x 6.25)		%	33.0	36.7	
Sheep	dig coef	%	91.	91.	
Sheep	dig prot	%	30.1	33.4	
Energy	GE Mcal/kg				
Cattle	DE Mcal/kg		* 2.74	3.05	
Sheep	DE Mcal/kg		* 2.87	3.19	
Swine	DE kcal/kg		*2816.	3129.	
Cattle	ME Mcal/kg		* 2.25	2.50	
Sheep	ME Mcal/kg		* 2.35	2.62	
Swine	ME kcal/kg		*2495.	2772.	
Cattle	TDN	%	* 62.3	69.2	
Sheep	TDN	%	65.1	72.4	
Swine	TDN	%	* 63.9	71.0	

NIGGERHEAD. Rudbeckia occidentalis

Niggerhead, aerial part, fresh, (2)
Ref No 2 03 223 United States

Analyses			As Fed	Dry	C.V. ± %
Dry matter		%		100.0	
Ash		%		13.3	7
Crude fiber		%		11.4	1
Ether extract		%		6.0	17
N-free extract		%		47.6	
Protein (N x 6.25)		%		21.7	17
Cattle	dig prot	%	*	16.3	
Goats	dig prot	%	*	16.8	
Horses	dig prot	%	*	16.0	
Rabbits	dig prot	%	*	15.4	
Sheep	dig prot	%	*	17.2	

Niggerhead, aerial part, fresh, immature, (2)
Ref No 2 03 221 United States

Analyses			As Fed	Dry	C.V. ± %
Dry matter		%		100.0	
Ash		%		11.5	
Crude fiber		%		11.2	
Ether extract		%		4.1	
N-free extract		%		44.8	
Protein (N x 6.25)		%		28.4	
Cattle	dig prot	%	*	22.0	
Goats	dig prot	%	*	23.1	
Horses	dig prot	%	*	21.6	
Rabbits	dig prot	%	*	20.6	
Sheep	dig prot	%	*	23.5	
Calcium		%		1.57	
Magnesium		%		0.32	
Phosphorus		%		0.62	
Sulphur		%		0.39	

Niggerhead, aerial part, fresh, full bloom, (2)
Ref No 2 03 222 United States

Analyses			As Fed	Dry	C.V. ± %
Dry matter		%		100.0	
Calcium		%		2.93	
Magnesium		%		0.55	
Phosphorus		%		0.44	
Sulphur		%		0.32	

(1) dry forages and roughages (2) pasture, range plants, and forages fed green (3) silages (4) energy feeds (5) protein supplements (6) minerals (7) vitamins (8) additives

Feed Name or Analyses			Mean As Fed	Mean Dry	C.V. ± %

Niggerhead, leaves, fresh, (2)
Ref No 2 03 224 United States

Dry matter		%		100.0	
Ash		%		12.0	
Crude fiber		%		8.7	
Ether extract		%		2.2	
N-free extract		%		42.7	
Protein (N x 6.25)		%		34.4	
Cattle	dig prot	%	*	27.1	
Goats	dig prot	%	*	28.7	
Horses	dig prot	%	*	26.7	
Rabbits	dig prot	%	*	25.2	
Sheep	dig prot	%	*	29.1	

NIGHTSHADE. Solanum spp

Nightshade, aerial part, fresh, (2)
Ref No 2 03 227 United States

Dry matter		%		100.0	
Ash		%		9.1	33
Crude fiber		%		24.4	15
Ether extract		%		3.3	69
N-free extract		%		42.1	
Protein (N x 6.25)		%		21.1	22
Cattle	dig prot	%	*	15.8	
Goats	dig prot	%	*	16.2	
Horses	dig prot	%	*	15.4	
Rabbits	dig prot	%	*	15.0	
Sheep	dig prot	%	*	16.7	
Calcium		%		1.79	38
Magnesium		%		0.30	61
Phosphorus		%		0.27	29
Potassium		%		2.74	15

Nightshade, aerial part, fresh, immature, (2)
Ref No 2 03 225 United States

Dry matter		%		100.0	
Ash		%		10.0	29
Crude fiber		%		24.2	6
Ether extract		%		2.4	53
N-free extract		%		39.9	
Protein (N x 6.25)		%		23.5	13
Cattle	dig prot	%	*	17.9	
Goats	dig prot	%	*	18.5	
Horses	dig prot	%	*	17.5	
Rabbits	dig prot	%	*	16.8	
Sheep	dig prot	%	*	18.9	

Nightshade, aerial part, fresh, full bloom, (2)
Ref No 2 03 226 United States

Dry matter		%		100.0	
Ash		%		7.4	
Crude fiber		%		24.7	
Ether extract		%		5.1	
N-free extract		%		46.5	
Protein (N x 6.25)		%		16.3	
Cattle	dig prot	%	*	11.7	
Goats	dig prot	%	*	11.8	
Horses	dig prot	%	*	11.4	
Rabbits	dig prot	%	*	11.3	
Sheep	dig prot	%	*	12.2	
Calcium		%		1.94	
Magnesium		%		0.30	
Phosphorus		%		0.22	
Potassium		%		2.55	

NIGHTSHADE, BLACK. Solanum nigrun

Nightshade, black, seeds w hulls, (4)
Ref No 4 09 133 United States

Dry matter		%		100.0	
Protein (N x 6.25)		%		17.0	
Alanine		%		0.87	
Arginine		%		1.53	
Aspartic acid		%		1.51	
Glutamic acid		%		3.13	
Glycine		%		0.80	
Histidine		%		0.39	
Hydroxyproline		%		0.24	
Isoleucine		%		0.68	
Leucine		%		1.09	
Lysine		%		0.73	
Methionine		%		0.32	
Phenylalanine		%		0.77	
Proline		%		0.70	
Serine		%		0.80	
Threonine		%		0.66	
Tyrosine		%		0.44	
Valine		%		0.80	

NIGHTSHADE, SILVERLEAF. Solanum elaeagnifolium

Nightshade, silverleaf, aerial part, fresh, (2)
Ref No 2 03 230 United States

Dry matter		%		100.0	
Ash		%		9.0	40
Crude fiber		%		25.6	13
Ether extract		%		3.2	88
N-free extract		%		40.4	
Protein (N x 6.25)		%		21.8	26
Cattle	dig prot	%	*	16.4	
Goats	dig prot	%	*	16.9	
Horses	dig prot	%	*	16.0	
Rabbits	dig prot	%	*	15.5	
Sheep	dig prot	%	*	17.3	

Nightshade, silverleaf, aerial part, fresh, immature, (2)
Ref No 2 03 228 United States

Dry matter		%		100.0	
Ash		%		10.1	
Crude fiber		%		24.3	
Ether extract		%		1.8	
N-free extract		%		40.0	
Protein (N x 6.25)		%		23.8	
Cattle	dig prot	%	*	18.1	
Goats	dig prot	%	*	18.8	
Horses	dig prot	%	*	17.7	
Rabbits	dig prot	%	*	17.0	
Sheep	dig prot	%	*	19.2	
Calcium		%		1.64	
Magnesium		%		0.31	
Phosphorus		%		0.31	
Potassium		%		2.94	

Nightshade, silverleaf, aerial part, fresh, full bloom, (2)
Ref No 2 03 229 United States

Dry matter		%		100.0	
Ash		%		5.6	
Crude fiber		%		29.4	
Ether extract		%		7.5	
N-free extract		%		41.8	
Protein (N x 6.25)		%		15.7	
Cattle	dig prot	%	*	11.2	
Goats	dig prot	%	*	11.2	
Horses	dig prot	%	*	10.9	
Rabbits	dig prot	%	*	10.8	
Sheep	dig prot	%	*	11.6	
Calcium		%		1.07	
Magnesium		%		0.00	
Phosphorus		%		0.23	
Potassium		%		2.14	

NIGHTSHADE, TEXAS. Solanum triquetrum

Nightshade, Texas, aerial part, fresh, immature, (2)
Ref No 2 03 231 United States

Dry matter		%		100.0	
Ash		%		9.6	
Crude fiber		%		23.9	
Ether extract		%		4.3	
N-free extract		%		39.8	
Protein (N x 6.25)		%		22.4	
Cattle	dig prot	%	*	16.9	
Goats	dig prot	%	*	17.5	
Horses	dig prot	%	*	16.5	
Rabbits	dig prot	%	*	16.0	
Sheep	dig prot	%	*	17.9	

Nightshade, Texas, aerial part, fresh, full bloom, (2)
Ref No 2 03 232 United States

Dry matter		%		100.0	
Calcium		%		2.77	
Magnesium		%		0.46	
Phosphorus		%		0.26	
Potassium		%		2.76	

NIGHTSHADE, TORREY. Solanum torreyi

Nightshade, torrey, aerial part, fresh, full bloom, (2)
Ref No 2 03 233 United States

Dry matter		%		100.0	
Ash		%		9.1	
Crude fiber		%		19.9	
Ether extract		%		2.7	
N-free extract		%		51.5	
Protein (N x 6.25)		%		16.8	
Cattle	dig prot	%	*	12.2	
Goats	dig prot	%	*	12.2	
Horses	dig prot	%	*	11.8	
Rabbits	dig prot	%	*	11.6	
Sheep	dig prot	%	*	12.6	

Continued

Feed Name or Analyses		Mean As Fed	Mean Dry	C.V. ± %
Calcium	%		1.97	
Magnesium	%		0.43	
Phosphorus	%		0.16	
Potassium	%		2.74	

NINEBARK. Physocarpus spp

Ninebark, twigs, fresh, (2)

Ref No 2 03 235 United States

Feed Name or Analyses		Mean As Fed	Mean Dry	C.V. ± %
Dry matter	%		100.0	
Ash	%		3.3	
Crude fiber	%		43.2	
Ether extract	%		2.8	
N-free extract	%		46.9	
Protein (N x 6.25)	%		3.8	
Cattle	dig prot % *		1.1	
Goats	dig prot % *		0.1	
Horses	dig prot % *		0.8	
Rabbits	dig prot % *		1.6	
Sheep	dig prot % *		0.5	

NITGRASS. Gastridium ventricosum

Nitgrass, aerial part, fresh, (2)

Ref No 2 08 570 United States

Feed Name or Analyses		Mean As Fed	Mean Dry	C.V. ± %
Dry matter	%	92.9	100.0	
Ash	%	5.8	6.3	
Ether extract	%	1.3	1.4	
Protein (N x 6.25)	%	3.8	4.1	
Cattle	dig prot % *	1.3	1.4	
Goats	dig prot % *	0.4	0.4	
Horses	dig prot % *	1.0	1.0	
Rabbits	dig prot % *	1.7	1.9	
Sheep	dig prot % *	0.8	0.8	
Cellulose (Matrone)	%	36.0	38.7	
Other carbohydrates	%	37.7	40.6	
Lignin (Ellis)	%	8.9	9.5	
Energy	GE Mcal/kg	4.00	4.31	
Silicon	%	3.19	3.43	

NOLINA. Nolina spp

Nolina, browse, (2)

Ref No 2 03 239 United States

Feed Name or Analyses		Mean As Fed	Mean Dry	C.V. ± %
Dry matter	%		100.0	
Ash	%		4.1	43
Crude fiber	%		42.7	6
Ether extract	%		3.1	23
N-free extract	%		43.8	
Protein (N x 6.25)	%		6.3	16
Cattle	dig prot % *		3.2	
Goats	dig prot % *		2.4	
Horses	dig prot % *		2.9	
Rabbits	dig prot % *		3.5	
Sheep	dig prot % *		2.9	
Calcium	%		0.92	38
Magnesium	%		0.16	61
Phosphorus	%		0.09	22
Potassium	%		0.70	17

Nolina, buds, fresh, (2)

Ref No 2 03 240 United States

Feed Name or Analyses		Mean As Fed	Mean Dry	C.V. ± %
Dry matter	%		100.0	
Ash	%		6.3	
Crude fiber	%		17.9	
Ether extract	%		2.9	
N-free extract	%		48.7	
Protein (N x 6.25)	%		24.2	
Cattle	dig prot % *		18.5	
Goats	dig prot % *		19.1	
Horses	dig prot % *		18.1	
Rabbits	dig prot % *		17.4	
Sheep	dig prot % *		19.5	
Calcium	%		0.42	
Magnesium	%		0.08	
Phosphorus	%		0.46	
Potassium	%		2.86	

NOSEBURN. Tragia spp

Noseburn, browse, immature, (2)

Ref No 2 03 242 United States

Feed Name or Analyses		Mean As Fed	Mean Dry	C.V. ± %
Dry matter	%		100.0	
Ash	%		7.6	
Crude fiber	%		19.7	
Ether extract	%		3.2	
N-free extract	%		53.3	
Protein (N x 6.25)	%		16.2	
Cattle	dig prot % *		11.7	
Goats	dig prot % *		11.7	
Horses	dig prot % *		11.3	
Rabbits	dig prot % *		11.2	
Sheep	dig prot % *		12.1	
Calcium	%		2.00	
Magnesium	%		0.36	
Phosphorus	%		0.22	
Potassium	%		1.54	

No. 1 feed screenings (CFA) –
see Grains, screenings, gr 1 mn 35% grain mx 7% fiber mx 6% foreign material mx 8% wild oats, (4)

No 2 feed screenings (CFA) –
see Grains, screenings, gr 2 mx 11% fiber mx 10% foreign material mx 49% wild oats, (4)

OAK. Quercus spp

Oak, acorn hulls, (1)

Ref No 1 07 756 United States

Feed Name or Analyses		Mean As Fed	Mean Dry	C.V. ± %
Dry matter	%	91.6	100.0	
Organic matter	%	88.2	96.3	
Ash	%	3.4	3.7	
Crude fiber	%	32.2	35.1	
Ether extract	%	1.7	1.9	
N-free extract	%	50.8	55.5	
Protein (N x 6.25)	%	3.5	3.8	
Cattle	dig prot % *	0.2	0.2	
Goats	dig prot % *	0.1	0.1	
Horses	dig prot % *	0.7	0.8	
Rabbits	dig prot % *	1.5	1.6	
Sheep	dig prot % *	0.0	0.0	

Oak, leaves, (1)

Ref No 1 03 245 United States

Feed Name or Analyses		Mean As Fed	Mean Dry	C.V. ± %
Dry matter	%	92.7	100.0	1
Ash	%	5.2	5.6	18
Crude fiber	%	25.4	27.4	11
Ether extract	%	2.3	2.5	22
N-free extract	%	50.3	54.3	
Protein (N x 6.25)	%	9.5	10.2	29
Cattle	dig prot % *	5.4	5.8	
Goats	dig prot % *	5.6	6.1	
Horses	dig prot % *	5.7	6.2	
Rabbits	dig prot % *	6.1	6.5	
Sheep	dig prot % *	5.3	5.7	

Oak, acorn meats, (4)

Ref No 4 07 757 United States

Feed Name or Analyses		Mean As Fed	Mean Dry	C.V. ± %
Dry matter	%	91.3	100.0	
Organic matter	%	89.5	98.0	
Ash	%	1.8	2.0	
Crude fiber	%	1.8	2.0	
Ether extract	%	8.1	8.9	
N-free extract	%	73.7	80.7	
Protein (N x 6.25)	%	5.8	6.4	
Cattle	dig prot % *	1.7	1.9	
Goats	dig prot % *	2.8	3.1	
Horses	dig prot % *	2.8	3.1	
Sheep	dig prot % *	2.8	3.1	

Oak, acorns, (4)

Ref No 4 07 755 United States

Feed Name or Analyses		Mean As Fed	Mean Dry	C.V. ± %
Dry matter	%	70.7	100.0	
Organic matter	%	68.9	97.5	
Ash	%	1.7	2.5	
Crude fiber	%	9.8	13.9	
Ether extract	%	3.8	5.4	
N-free extract	%	51.9	73.5	
Protein (N x 6.25)	%	3.4	4.8	
Cattle	dig prot % *	0.3	0.5	
Goats	dig prot % *	1.2	1.7	
Horses	dig prot % *	1.2	1.7	
Sheep	dig prot % *	1.2	1.7	
Other carbohydrates	%	44.5	63.0	
Lignin (Ellis)	%	6.2	8.8	
Energy	GE Mcal/kg			
Cattle	DE Mcal/kg *	1.48	2.09	
Sheep	DE Mcal/kg *	1.46	2.06	
Cattle	ME Mcal/kg *	1.21	1.71	
Sheep	ME Mcal/kg *	1.19	1.69	
Cattle	TDN % *	33.6	47.5	
Sheep	TDN % *	33.1	46.8	

OAK, BIGELOW. Quercus durandii breviloba

Oak, bigelow, leaves, (1)

Ref No 1 03 247 United States

Feed Name or Analyses		Mean As Fed	Mean Dry	C.V. ± %
Dry matter	%	92.6	100.0	1
Ash	%	5.0	5.4	24
Crude fiber	%	23.4	25.3	10
Ether extract	%	2.3	2.5	20
N-free extract	%	53.0	57.2	
Protein (N x 6.25)	%	8.9	9.6	16
Cattle	dig prot % *	4.9	5.3	

(1) dry forages and roughages (2) pasture, range plants, and forages fed green (3) silages (4) energy feeds (5) protein supplements (6) minerals (7) vitamins (8) additives

Feed Name or Analyses				Mean As Fed	Mean Dry	C.V. ± %
Goats	dig prot	%	*	5.1	5.5	
Horses	dig prot	%	*	5.3	5.7	
Rabbits	dig prot	%	*	5.6	6.1	
Sheep	dig prot	%	*	4.8	5.2	
Calcium		%		1.33	1.44	45
Magnesium		%		0.19	0.21	56
Phosphorus		%		0.10	0.11	47
Potassium		%		0.70	0.76	50

Oak, bigelow, leaves, mature, (1)
Ref No 1 03 246 United States

Dry matter		%		92.7	100.0	1
Calcium		%		1.21	1.30	50
Magnesium		%		0.16	0.17	53
Phosphorus		%		0.10	0.11	24
Potassium		%		0.73	0.79	54

OAK, BLACK. Quercus velutina

Oak, black, acorn hulls, dehy grnd, (1)
Ref No 1 08 302 United States

Dry matter		%		92.1	100.0	
Organic matter		%		90.3	98.0	
Ash		%		1.8	2.0	
Crude fiber		%		35.9	39.0	
Ether extract		%		2.8	3.0	
N-free extract		%		49.6	53.9	
Protein (N x 6.25)		%		1.9	2.1	
Cattle	dig prot	%	*	-1.0	-1.1	
Goats	dig prot	%	*	-1.3	-1.4	
Horses	dig prot	%	*	-0.5	-0.6	
Rabbits	dig prot	%	*	0.3	0.3	
Sheep	dig prot	%	*	-1.3	-1.4	
Silicon		%		0.01	0.01	

Oak, black, acorn meats, dehy grnd, (4)
Ref No 4 08 303 United States

Dry matter		%		94.3	100.0	
Organic matter		%		92.8	98.4	
Ash		%		1.5	1.6	
Crude fiber		%		3.4	3.6	
Ether extract		%		24.4	25.9	
N-free extract		%		59.7	63.3	
Protein (N x 6.25)		%		5.3	5.6	
Cattle	dig prot	%	*	1.1	1.2	
Goats	dig prot	%	*	2.2	2.4	
Horses	dig prot	%	*	2.2	2.4	
Sheep	dig prot	%	*	2.2	2.4	
Silicon		%		0.00	0.00	

Oak, black, acorns, dehy grnd, (4)
Ref No 4 08 301 United States

Dry matter		%		93.9	100.0	
Organic matter		%		92.4	98.4	
Ash		%		1.6	1.7	
Crude fiber		%		12.5	13.3	
Ether extract		%		18.3	19.5	
N-free extract		%		57.3	61.0	
Protein (N x 6.25)		%		4.2	4.5	
Cattle	dig prot	%	*	0.1	0.2	
Goats	dig prot	%	*	1.3	1.4	
Horses	dig prot	%	*	1.3	1.4	

Feed Name or Analyses				Mean As Fed	Mean Dry	C.V. ± %
Sheep	dig prot	%	*	1.3	1.4	
Silicon		%		0.00	0.00	

OAK, BLUE. Quercus douglasi

Oak, blue, acorn hulls, dehy grnd, (1)
Ref No 1 08 305 United States

Dry matter		%		91.7	100.0	
Organic matter		%		88.7	96.7	
Ash		%		3.0	3.3	
Crude fiber		%		29.3	32.0	
Ether extract		%		1.2	1.3	
N-free extract		%		54.8	59.8	
Protein (N x 6.25)		%		3.3	3.6	
Cattle	dig prot	%	*	0.1	0.1	
Goats	dig prot	%	*	0.0	0.0	
Horses	dig prot	%	*	0.5	0.6	
Rabbits	dig prot	%	*	1.3	1.4	
Sheep	dig prot	%	*	-0.1	-0.1	
Silicon		%		0.01	0.01	

Oak, blue, acorn meats, dehy grnd, (4)
Ref No 4 08 306 United States

Dry matter		%		91.3	100.0	
Organic matter		%		89.9	98.5	
Ash		%		1.4	1.5	
Crude fiber		%		2.2	2.4	
Ether extract		%		9.3	10.2	
N-free extract		%		73.6	80.6	
Protein (N x 6.25)		%		4.8	5.3	
Cattle	dig prot	%	*	0.8	0.9	
Goats	dig prot	%	*	1.9	2.1	
Horses	dig prot	%	*	1.9	2.1	
Sheep	dig prot	%	*	1.9	2.1	
Silicon		%		0.03	0.03	

Oak, blue, acorns, dehy grnd, (4)
Ref No 4 08 304 United States

Dry matter		%		91.6	100.0	
Organic matter		%		89.7	97.9	
Ash		%		1.9	2.1	
Crude fiber		%		9.5	10.4	
Ether extract		%		7.4	8.1	
N-free extract		%		68.1	74.3	
Protein (N x 6.25)		%		4.7	5.1	
Cattle	dig prot	%	*	0.6	0.7	
Goats	dig prot	%	*	1.8	1.9	
Horses	dig prot	%	*	1.8	1.9	
Sheep	dig prot	%	*	1.8	1.9	
Silicon		%		0.00	0.00	

OAK, CALIFORNIA SCRUB. Quercus dumosa

Oak, California scrub, acorn hulls, dehy grnd, (1)
Ref No 1 08 308 United States

Dry matter		%		92.6	100.0	
Organic matter		%		90.4	97.6	
Ash		%		2.2	2.4	
Crude fiber		%		32.4	35.0	
Ether extract		%		1.0	1.1	
N-free extract		%		53.9	58.2	

Feed Name or Analyses				Mean As Fed	Mean Dry	C.V. ± %
Protein (N x 6.25)		%		3.1	3.3	
Cattle	dig prot	%	*	-0.1	-0.1	
Goats	dig prot	%	*	-0.2	-0.3	
Horses	dig prot	%	*	0.3	0.3	
Rabbits	dig prot	%	*	1.1	1.2	
Sheep	dig prot	%	*	-0.3	-0.4	
Silicon		%		0.03	0.03	

Oak, California scrub, acorn meats, dehy grnd, (4)
Ref No 4 08 309 United States

Dry matter		%		91.1	100.0	
Organic matter		%		89.7	98.5	
Ash		%		1.4	1.5	
Crude fiber		%		1.5	1.7	
Ether extract		%		4.6	5.1	
N-free extract		%		80.4	88.3	
Protein (N x 6.25)		%		3.1	3.4	
Cattle	dig prot	%	*	-0.7	-0.8	
Goats	dig prot	%	*	0.3	0.4	
Horses	dig prot	%	*	0.3	0.4	
Sheep	dig prot	%	*	0.3	0.4	
Silicon		%		0.01	0.01	

Oak, California scrub, acorns, dehy grnd, (4)
Ref No 4 08 307 United States

Dry matter		%		91.7	100.0	
Organic matter		%		89.9	98.0	
Ash		%		1.8	2.0	
Crude fiber		%		13.6	14.8	
Ether extract		%		3.2	3.5	
N-free extract		%		69.9	76.2	
Protein (N x 6.25)		%		3.2	3.5	
Cattle	dig prot	%	*	-0.6	-0.7	
Goats	dig prot	%	*	0.4	0.4	
Horses	dig prot	%	*	0.4	0.4	
Sheep	dig prot	%	*	0.4	0.4	
Silicon		%		0.00	0.00	

OAK, LIVE. Quercus virginiana

Oak, live, acorn hulls, dehy grnd, (1)
Ref No 1 08 311 United States

Dry matter		%		92.9	100.0	
Organic matter		%		91.0	98.0	
Ash		%		2.0	2.1	
Crude fiber		%		41.0	44.1	
Ether extract		%		1.6	1.7	
N-free extract		%		45.6	49.1	
Protein (N x 6.25)		%		2.8	3.0	
Cattle	dig prot	%	*	-0.3	-0.4	
Goats	dig prot	%	*	-0.5	-0.5	
Horses	dig prot	%	*	0.1	0.1	
Rabbits	dig prot	%	*	0.9	1.0	
Sheep	dig prot	%	*	-0.6	-0.6	
Silicon		%		0.00	0.00	

Oak, live, browse, fan air dried grnd pelleted, early leaf, (1)
Ref No 1 07 859 United States

Dry matter		%		90.0	100.0	
Ash		%		2.8	3.1	

Continued

Feed Name or Analyses			Mean As Fed	Mean Dry	C.V. ± %
Crude fiber	%		24.7	27.4	
Sheep	dig coef %		20.	20.	
Ether extract	%		1.3	1.4	
Sheep	dig coef %		46.	46.	
N-free extract	%		54.6	60.7	
Sheep	dig coef %		56.	56.	
Protein (N x 6.25)	%		6.7	7.4	
Sheep	dig coef %		-10.	-10.	
Cattle	dig prot %	*	3.0	3.3	
Goats	dig prot %	*	3.1	3.5	
Horses	dig prot %	*	3.4	3.8	
Rabbits	dig prot %	*	3.9	4.4	
Sheep	dig prot %		-0.6	-0.7	
Energy	GE Mcal/kg				
Cattle	DE Mcal/kg	*	1.72	1.91	
Sheep	DE Mcal/kg	*	1.59	1.77	
Cattle	ME Mcal/kg	*	1.41	1.57	
Sheep	ME Mcal/kg	*	1.31	1.45	
Cattle	TDN %	*	39.1	43.4	
Sheep	TDN %		36.1	40.1	

Oak, live, leaves, (1)

Ref No 1 03 249 United States

Feed Name or Analyses			Mean As Fed	Mean Dry	C.V. ± %
Dry matter	%		93.4	100.0	1
Ash	%		6.1	6.6	15
Crude fiber	%		28.8	30.8	8
Sheep	dig coef %		10.	10.	
Ether extract	%		2.5	2.7	23
Sheep	dig coef %		26.	26.	
N-free extract	%		46.5	49.8	
Sheep	dig coef %		27.	27.	
Protein (N x 6.25)	%		9.5	10.1	29
Sheep	dig coef %		0.	0.	29
Cattle	dig prot %	*	5.3	5.7	
Goats	dig prot %	*	5.6	6.0	
Horses	dig prot %	*	5.7	6.1	
Rabbits	dig prot %	*	6.1	6.5	
Fatty acids	%		2.7	2.9	
Energy	GE Mcal/kg				
Cattle	DE Mcal/kg	*	0.82	0.88	
Sheep	DE Mcal/kg	*	0.75	0.80	
Cattle	ME Mcal/kg	*	0.67	0.72	
Sheep	ME Mcal/kg	*	0.61	0.65	
Cattle	TDN %	*	18.7	20.0	
Sheep	TDN %		16.9	18.1	
Carotene	mg/kg		87.7	93.9	
Vitamin A equivalent	IU/g		146.3	156.6	

Oak, live, leaves, mature, (1)

Ref No 1 03 248 United States

Feed Name or Analyses			Mean As Fed	Mean Dry	C.V. ± %
Dry matter	%		92.8	100.0	1
Ash	%		5.4	5.8	19
Crude fiber	%		27.1	29.2	6
Ether extract	%		2.2	2.4	22
N-free extract	%		48.7	52.5	
Protein (N x 6.25)	%		9.4	10.1	12
Cattle	dig prot %	*	5.3	5.7	
Goats	dig prot %	*	5.6	6.0	
Horses	dig prot %	*	5.7	6.1	
Rabbits	dig prot %	*	6.0	6.5	
Sheep	dig prot %	*	5.2	5.6	

(1) dry forages and roughages (2) pasture, range plants, and forages fed green (3) silages (4) energy feeds (5) protein supplements (6) minerals (7) vitamins (8) additives

Feed Name or Analyses			Mean As Fed	Mean Dry	C.V. ± %
Energy	GE Mcal/kg				
Cattle	DE Mcal/kg	*	0.93	1.00	
Sheep	DE Mcal/kg	*	0.76	0.82	
Cattle	ME Mcal/kg	*	0.76	0.82	
Sheep	ME Mcal/kg	*	0.62	0.67	
Cattle	TDN %	*	21.1	22.7	
Sheep	TDN %	*	17.2	18.5	

Oak, live, acorn meats, dehy grnd, (4)

Ref No 4 08 312 United States

Feed Name or Analyses			Mean As Fed	Mean Dry	C.V. ± %
Dry matter	%		94.1	100.0	
Organic matter	%		93.0	98.8	
Ash	%		1.1	1.2	
Crude fiber	%		1.9	2.0	
Ether extract	%		23.1	24.6	
N-free extract	%		63.3	67.3	
Protein (N x 6.25)	%		4.6	4.9	
Cattle	dig prot %	*	0.5	0.5	
Goats	dig prot %	*	1.6	1.7	
Horses	dig prot %	*	1.6	1.7	
Sheep	dig prot %	*	1.6	1.7	
Silicon	%		0.04	0.04	

Oak, live, acorns, dehy grnd, (4)

Ref No 4 08 310 United States

Feed Name or Analyses			Mean As Fed	Mean Dry	C.V. ± %
Dry matter	%		93.6	100.0	
Organic matter	%		92.3	98.6	
Ash	%		1.3	1.4	
Crude fiber	%		13.7	14.6	
Ether extract	%		15.0	16.0	
N-free extract	%		59.6	63.7	
Protein (N x 6.25)	%		4.0	4.3	
Cattle	dig prot %	*	0.0	0.0	
Goats	dig prot %	*	1.1	1.2	
Horses	dig prot %	*	1.1	1.2	
Sheep	dig prot %	*	1.1	1.2	
Silicon	%		0.05	0.05	

OAK, OREGON WHITE. Quercus garryana

Oak, Oregon white, acorn hulls, dehy grnd, (1)

Ref No 1 08 314 United States

Feed Name or Analyses			Mean As Fed	Mean Dry	C.V. ± %
Dry matter	%		91.6	100.0	
Organic matter	%		88.4	96.5	
Ash	%		3.2	3.5	
Crude fiber	%		25.6	27.9	
Ether extract	%		1.0	1.1	
N-free extract	%		57.1	62.3	
Protein (N x 6.25)	%		4.8	5.2	
Cattle	dig prot %	*	1.3	1.4	
Goats	dig prot %	*	1.3	1.4	
Horses	dig prot %	*	1.8	1.9	
Rabbits	dig prot %	*	2.5	2.7	
Sheep	dig prot %	*	1.1	1.2	
Silicon	%		0.00	0.00	

Oak, Oregon white, acorn meats, dehy grnd, (4)

Ref No 4 08 315 United States

Feed Name or Analyses			Mean As Fed	Mean Dry	C.V. ± %
Dry matter	%		91.6	100.0	
Organic matter	%		90.1	98.4	
Ash	%		1.5	1.6	
Crude fiber	%		1.7	1.9	

Feed Name or Analyses			Mean As Fed	Mean Dry	C.V. ± %
Ether extract	%		9.5	10.4	
N-free extract	%		73.2	79.9	
Protein (N x 6.25)	%		5.7	6.2	
Cattle	dig prot %	*	1.6	1.7	
Goats	dig prot %	*	2.7	2.9	
Horses	dig prot %	*	2.7	2.9	
Sheep	dig prot %	*	2.7	2.9	
Silicon	%		0.03	0.03	

Oak, Oregon white, acorns, dehy grnd, (4)

Ref No 4 08 313 United States

Feed Name or Analyses			Mean As Fed	Mean Dry	C.V. ± %
Dry matter	%		91.5	100.0	
Organic matter	%		89.7	98.0	
Ash	%		2.0	2.2	
Crude fiber	%		8.0	8.7	
Ether extract	%		6.7	7.3	
N-free extract	%		69.3	75.7	
Protein (N x 6.25)	%		5.6	6.1	
Cattle	dig prot %	*	1.5	1.6	
Goats	dig prot %	*	2.6	2.8	
Horses	dig prot %	*	2.6	2.8	
Sheep	dig prot %	*	2.6	2.8	
Silicon	%		0.00	0.00	

OAK, POST. Quercus stellata

Oak, post, leaves, (1)

Ref No 1 03 250 United States

Feed Name or Analyses			Mean As Fed	Mean Dry	C.V. ± %
Dry matter	%			100.0	
Carotene	mg/kg			112.4	
Vitamin A equivalent	IU/g			187.4	

OAK, RED. Quercus spp

Oak, red, acorns, (4)

Ref No 4 08 469 United States

Feed Name or Analyses			Mean As Fed	Mean Dry	C.V. ± %
Dry matter	%		50.0	100.0	
Ash	%		1.2	2.4	
Crude fiber	%		9.9	19.8	
Ether extract	%		10.7	21.4	
N-free extract	%		25.0	50.0	
Protein (N x 6.25)	%		3.2	6.4	
Cattle	dig prot %	*	0.9	1.9	
Goats	dig prot %	*	1.6	3.1	
Horses	dig prot %	*	1.6	3.1	
Sheep	dig prot %	*	1.6	3.1	
Energy	GE Mcal/kg				
Cattle	DE Mcal/kg	*	1.48	2.97	
Sheep	DE Mcal/kg	*	1.40	2.80	
Swine	DE kcal/kg	*	1365.	2729.	
Cattle	ME Mcal/kg	*	1.22	2.44	
Sheep	ME Mcal/kg	*	1.15	2.30	
Swine	ME kcal/kg	*	1292.	2584.	
Cattle	TDN %	*	33.7	67.4	
Sheep	TDN %	*	31.8	63.5	
Swine	TDN %	*	31.0	61.9	

Feed Name or Analyses			Mean As Fed	Mean Dry	C.V. ± %

OAK, WHITE. Quercus alba

Oak, white, buds, fresh, (2)

Ref No 2 03 251 United States

Analyses	Unit		As Fed	Dry	C.V. ± %
Dry matter	%		37.4	100.0	
Biotin	mg/kg		0.04	0.11	
Folic acid	mg/kg		0.09	0.24	
Niacin	mg/kg		3.6	9.7	
Pantothenic acid	mg/kg		2.3	6.2	
Riboflavin	mg/kg		1.2	3.1	
Thiamine	mg/kg		1.2	3.1	
Vitamin B6	mg/kg		1.40	3.75	

Oat feed (CFA) –

see Oats, groats by-product, mx 22% fiber, (1)

OATGRASS. Arrhenatherum spp

Oatgrass, aerial part, fresh, immature, (2)

Ref No 2 03 252 United States

Analyses	Unit		As Fed	Dry	C.V. ± %
Dry matter	%		14.0	100.0	
Carotene	mg/kg		74.7	533.3	44
Vitamin A equivalent	IU/g		124.5	889.0	

Oatgrass, aerial part, ensiled, immature, (3)

Ref No 3 03 253 United States

Analyses	Unit		As Fed	Dry	C.V. ± %
Dry matter	%		35.4	100.0	
Carotene	mg/kg		154.5	436.5	
Vitamin A equivalent	IU/g		257.6	727.7	

OATGRASS, TALL. Arrhenatherum elatius

Oatgrass, tall, hay, dehy, (1)

Ref No 1 03 254 United States

Analyses	Unit		As Fed	Dry	C.V. ± %
Dry matter	%			100.0	
Arginine	%			0.50	
Histidine	%			0.20	
Isoleucine	%			0.60	
Leucine	%			0.90	
Lysine	%			0.50	
Methionine	%			0.20	
Phenylalanine	%			0.50	
Threonine	%			0.60	
Tryptophan	%			0.20	
Valine	%			0.80	

Oatgrass, tall, hay, s-c, (1)

Ref No 1 03 259 United States

Analyses	Unit		As Fed	Dry	C.V. ± %
Dry matter	%		87.3	100.0	2
Ash	%		6.7	7.7	15
Crude fiber	%		29.5	33.8	4
Sheep	dig coef %		58.	58.	
Ether extract	%		2.0	2.3	20
Sheep	dig coef %		48.	48.	
N-free extract	%		41.9	48.0	
Sheep	dig coef %		56.	56.	
Protein (N x 6.25)	%		7.2	8.2	29
Sheep	dig coef %		46.	46.	
Cattle	dig prot %	*	3.5	4.1	
Goats	dig prot %	*	3.7	4.2	
Horses	dig prot %	*	3.9	4.5	
Rabbits	dig prot %	*	4.4	5.0	
Sheep	dig prot %		3.3	3.8	
Cellulose (Matrone)	%		25.0	28.7	
Lignin (Ellis)	%		5.1	5.8	
Fatty acids	%		2.4	2.7	
Energy	GE Mcal/kg				
Cattle	DE Mcal/kg	*	2.20	2.52	
Sheep	DE Mcal/kg	*	2.03	2.33	
Cattle	ME Mcal/kg	*	1.80	2.06	
Sheep	ME Mcal/kg	*	1.67	1.91	
Cattle	TDN %	*	49.8	57.1	
Sheep	TDN %		46.1	52.8	
Calcium	%		0.40	0.46	
Magnesium	%		0.29	0.33	
Manganese	mg/kg		32.1	36.8	
Phosphorus	%		0.30	0.35	
Potassium	%		1.98	2.27	

Oatgrass, tall, hay, s-c, immature, (1)

Ref No 1 03 255 United States

Analyses	Unit		As Fed	Dry	C.V. ± %
Dry matter	%			100.0	
Calcium	%			0.51	
Phosphorus	%			0.60	

Oatgrass, tall, hay, s-c, pre-bloom, (1)

Ref No 1 03 256 United States

Analyses	Unit		As Fed	Dry	C.V. ± %
Dry matter	%		87.0	100.0	
Ash	%		10.3	11.8	
Crude fiber	%		33.8	38.8	
Sheep	dig coef %		61.	61.	
Ether extract	%		1.3	1.5	
Sheep	dig coef %		55.	55.	
N-free extract	%		39.0	44.8	
Sheep	dig coef %		54.	54.	
Protein (N x 6.25)	%		2.7	3.1	
Sheep	dig coef %		47.	47.	
Cattle	dig prot %	*	-0.2	-0.3	
Goats	dig prot %	*	-0.4	-0.4	
Horses	dig prot %	*	0.1	0.2	
Rabbits	dig prot %	*	0.9	1.1	
Sheep	dig prot %		1.3	1.5	
Energy	GE Mcal/kg				
Cattle	DE Mcal/kg	*	2.04	2.34	
Sheep	DE Mcal/kg	*	1.96	2.26	
Cattle	ME Mcal/kg	*	1.67	1.92	
Sheep	ME Mcal/kg	*	1.61	1.85	
Cattle	TDN %	*	46.2	53.1	
Sheep	TDN %		44.5	51.2	

Oatgrass, tall, hay, s-c, early bloom, (1)

Ref No 1 05 678 United States

Analyses	Unit		As Fed	Dry	C.V. ± %
Dry matter	%			100.0	
Ash	%			4.9	
Crude fiber	%			32.9	
Ether extract	%			2.8	
N-free extract	%			51.7	
Protein (N x 6.25)	%			7.7	
Cattle	dig prot %	*		3.6	
Goats	dig prot %	*		3.7	
Horses	dig prot %	*		4.1	
Rabbits	dig prot %	*		4.6	
Sheep	dig prot %	*		3.5	
Cellulose (Matrone)	%			31.4	
Lignin (Ellis)	%			6.7	
Energy	GE Mcal/kg				
Cattle	DE Mcal/kg	*		2.49	
Sheep	DE Mcal/kg	*		2.47	
Cattle	ME Mcal/kg	*		2.05	
Sheep	ME Mcal/kg	*		2.02	
Cattle	TDN %	*		56.6	
Sheep	TDN %	*		55.9	

Oatgrass, tall, hay, s-c, full bloom, (1)

Ref No 1 03 257 United States

Analyses	Unit		As Fed	Dry	C.V. ± %
Dry matter	%		85.9	100.0	
Ash	%		8.8	10.3	
Crude fiber	%		32.2	37.5	
Sheep	dig coef %		66.	66.	
Ether extract	%		1.4	1.6	
Sheep	dig coef %		49.	49.	
N-free extract	%		38.0	44.2	
Sheep	dig coef %		48.	48.	
Protein (N x 6.25)	%		5.5	6.4	
Sheep	dig coef %		44.	44.	
Cattle	dig prot %	*	2.1	2.5	
Goats	dig prot %	*	2.2	2.5	
Horses	dig prot %	*	2.5	3.0	
Rabbits	dig prot %	*	3.1	3.6	
Sheep	dig prot %		2.4	2.8	
Energy	GE Mcal/kg				
Cattle	DE Mcal/kg	*	2.01	2.35	
Sheep	DE Mcal/kg	*	1.91	2.23	
Cattle	ME Mcal/kg	*	1.65	1.92	
Sheep	ME Mcal/kg	*	1.57	1.83	
Cattle	TDN %	*	45.7	53.2	
Sheep	TDN %		43.4	50.5	

Oatgrass, tall, hay, s-c, mature, (1)

Ref No 1 03 258 United States

Analyses	Unit		As Fed	Dry	C.V. ± %
Dry matter	%		83.9	100.0	
Ash	%		5.9	7.0	
Crude fiber	%		27.5	32.8	
Sheep	dig coef %		44.	44.	
Ether extract	%		0.7	0.8	
Sheep	dig coef %		60.	60.	
N-free extract	%		45.4	54.1	
Sheep	dig coef %		62.	62.	
Protein (N x 6.25)	%		4.4	5.3	
Sheep	dig coef %		37.	37.	
Cattle	dig prot %	*	1.3	1.5	
Goats	dig prot %	*	1.3	1.5	
Horses	dig prot %	*	1.7	2.0	
Rabbits	dig prot %	*	2.3	2.8	
Sheep	dig prot %		1.6	2.0	
Energy	GE Mcal/kg				
Cattle	DE Mcal/kg	*	2.12	2.53	
Sheep	DE Mcal/kg	*	1.89	2.25	
Cattle	ME Mcal/kg	*	1.74	2.07	
Sheep	ME Mcal/kg	*	1.55	1.84	
Cattle	TDN %	*	48.1	57.4	
Sheep	TDN %		42.8	51.0	

Oatgrass, tall, aerial part, fresh, (2)

Ref No 2 03 267 United States

Analyses	Unit		As Fed	Dry	C.V. ± %
Dry matter	%		30.3	100.0	
Ash	%		2.0	6.6	

Continued

Feed Name or Analyses		Mean As Fed	Mean Dry	C.V. ± %
Crude fiber	%	10.5	34.7	
Ether extract	%	0.9	3.0	
N-free extract	%	14.3	47.2	
Protein (N x 6.25)	%	2.6	8.6	
Cattle	dig prot % *	1.6	5.2	
Goats	dig prot % *	1.4	4.6	
Horses	dig prot % *	1.5	4.8	
Rabbits	dig prot % *	1.6	5.3	
Sheep	dig prot % *	1.5	5.0	
Hemicellulose	%	5.7	18.7	6
Energy	GE Mcal/kg			
Cattle	DE Mcal/kg *	0.78	2.56	
Sheep	DE Mcal/kg *	0.75	2.48	
Cattle	ME Mcal/kg *	0.64	2.10	
Sheep	ME Mcal/kg *	0.62	2.04	
Cattle	TDN % *	17.6	58.1	
Sheep	TDN % *	17.1	56.5	
Calcium	%	0.12	0.40	
Phosphorus	%	0.14	0.46	
Potassium	%	0.91	3.00	

Oatgrass, tall, aerial part, fresh, immature, (2)

Ref No 2 03 261 United States

		As Fed	Dry	C.V.
Dry matter	%	21.6	100.0	
Ash	%	2.2	10.1	20
Crude fiber	%	4.1	19.0	7
Ether extract	%	1.1	4.9	11
N-free extract	%	10.0	46.3	
Protein (N x 6.25)	%	4.3	19.7	5
Cattle	dig prot % *	3.2	14.6	
Goats	dig prot % *	3.2	14.9	
Horses	dig prot % *	3.1	14.3	
Rabbits	dig prot % *	3.0	13.9	
Sheep	dig prot % *	3.3	15.4	
Energy	GE Mcal/kg			
Cattle	DE Mcal/kg *	0.55	2.54	
Sheep	DE Mcal/kg *	0.57	2.64	
Cattle	ME Mcal/kg *	0.45	2.08	
Sheep	ME Mcal/kg *	0.42	2.16	
Cattle	TDN % *	12.5	57.7	
Sheep	TDN % *	12.9	59.9	

Oatgrass, tall, aerial part, fresh, immature, pure stand, (2)

Ref No 2 03 260 United States

		As Fed	Dry	C.V.
Dry matter	%		100.0	
Carotene	mg/kg		420.0	
Vitamin A equivalent	IU/g		700.1	

Oatgrass, tall, aerial part, fresh, early bloom, (2)

Ref No 2 03 262 United States

		As Fed	Dry	C.V.
Dry matter	%		100.0	
Ash	%		8.2	10
Crude fiber	%		30.7	5
Ether extract	%		3.5	8
N-free extract	%		45.4	
Protein (N x 6.25)	%		12.2	8
Cattle	dig prot % *		8.3	
Goats	dig prot % *		7.9	

(1) dry forages and roughages (2) pasture, range plants, and forages fed green (3) silages (4) energy feeds (5) protein supplements (6) minerals (7) vitamins (8) additives

Feed Name or Analyses		Mean As Fed	Mean Dry	C.V. ± %
Horses	dig prot % *		7.9	
Rabbits	dig prot % *		8.1	
Sheep	dig prot % *		8.4	
Energy	GE Mcal/kg			
Cattle	DE Mcal/kg *		2.65	
Sheep	DE Mcal/kg *		2.57	
Cattle	ME Mcal/kg *		2.17	
Sheep	ME Mcal/kg *		2.11	
Cattle	TDN % *		60.2	
Sheep	TDN % *		58.3	

Oatgrass, tall, aerial part, fresh, mid-bloom, (2)

Ref No 2 03 263 United States

		As Fed	Dry	C.V.
Dry matter	%		100.0	
Ash	%		6.9	7
Crude fiber	%		33.2	5
Ether extract	%		2.9	7
N-free extract	%		48.4	
Protein (N x 6.25)	%		8.6	8
Cattle	dig prot % *		5.2	
Goats	dig prot % *		4.6	
Horses	dig prot % *		4.8	
Rabbits	dig prot % *		5.3	
Sheep	dig prot % *		5.0	
Energy	GE Mcal/kg			
Cattle	DE Mcal/kg *		2.71	
Sheep	DE Mcal/kg *		2.62	
Cattle	ME Mcal/kg *		2.22	
Sheep	ME Mcal/kg *		2.15	
Cattle	TDN % *		61.4	
Sheep	TDN % *		59.5	

Oatgrass, tall, aerial part, fresh, milk stage, (2)

Ref No 2 03 264 United States

		As Fed	Dry	C.V.
Dry matter	%		100.0	
Ash	%		6.3	6
Crude fiber	%		34.1	3
Ether extract	%		2.7	6
N-free extract	%		49.6	
Protein (N x 6.25)	%		7.3	5
Cattle	dig prot % *		4.1	
Goats	dig prot % *		3.4	
Horses	dig prot % *		3.7	
Rabbits	dig prot % *		4.3	
Sheep	dig prot % *		3.8	
Energy	GE Mcal/kg			
Cattle	DE Mcal/kg *		2.69	
Sheep	DE Mcal/kg *		2.79	
Cattle	ME Mcal/kg *		2.21	
Sheep	ME Mcal/kg *		2.29	
Cattle	TDN % *		61.0	
Sheep	TDN % *		63.2	

Outgrass, tall, aerial part, fresh, dough stage, (2)

Ref No 2 03 265 United States

		As Fed	Dry	C.V.
Dry matter	%		100.0	
Ash	%		6.0	4
Crude fiber	%		34.3	4
Ether extract	%		2.6	4
N-free extract	%		51.1	
Protein (N x 6.25)	%		6.0	12
Cattle	dig prot % *		3.0	
Goats	dig prot % *		2.2	
Horses	dig prot % *		2.6	

Feed Name or Analyses		Mean As Fed	Mean Dry	C.V. ± %
Rabbits	dig prot % *		3.3	
Sheep	dig prot % *		2.6	
Energy	GE Mcal/kg			
Cattle	DE Mcal/kg *		2.64	
Sheep	DE Mcal/kg *		2.69	
Cattle	ME Mcal/kg *		2.16	
Sheep	ME Mcal/kg *		2.21	
Cattle	TDN % *		59.8	
Sheep	TDN % *		61.0	

Oat groats (AAFCO) (CFA) –
see Oats, groats, (4)

Oat huller feed –
see Oats, huller by-product, (1)

Oat meal (CFA) –
see Oats, cereal by-product, grnd, mx 2% Fiber, (4)

Oat meal –
see Oats, cereal by-product, mx 4% fiber, (4)

Oat middlings (CFA) –
see Oats, cereal by-product, mx 4% fiber, (4)

Oat mill by-product (AAFCO) –
see Oats, groats by-product, mx 22% fiber, (1)

Oat mill feed –
see Oats, huller by-product, (1)

OATS. Avena sativa

Oats, chaff, (1)

Ref No 1 03 305 United States

		As Fed	Dry	C.V.
Dry matter	%	90.4	100.0	
Ash	%	10.1	11.2	
Crude fiber	%	26.0	28.8	
Sheep	dig coef %	39.	39.	
Ether extract	%	1.9	2.1	
Sheep	dig coef %	42.	42.	
N-free extract	%	47.4	52.5	
Sheep	dig coef %	43.	43.	
Protein (N x 6.25)	%	5.0	5.5	
Sheep	dig coef %	16.	16.	
Cattle	dig prot % *	1.5	1.7	
Goats	dig prot % *	1.5	1.7	
Horses	dig prot % *	2.0	2.2	
Rabbits	dig prot % *	2.6	2.9	
Sheep	dig prot %	0.8	0.9	
Fatty acids	%	2.4	2.6	
Energy	GE Mcal/kg			
Cattle	DE Mcal/kg *	1.52	1.68	
Sheep	DE Mcal/kg *	1.46	1.61	
Cattle	ME Mcal/kg *	1.25	1.38	
Sheep	ME Mcal/kg *	1.20	1.32	
Cattle	TDN % *	34.4	38.1	
Sheep	TDN %	33.1	36.6	
Calcium	%	0.79	0.87	
Phosphorus	%	0.30	0.33	
Potassium	%	0.85	0.94	

Ref No 1 03 305 Canada

		As Fed	Dry	C.V.
Dry matter	%	91.5	100.0	2
Crude fiber	%	25.8	28.2	9
Sheep	dig coef %	39.	39.	

Feed Name or Analyses			Mean As Fed	Mean Dry	C.V. ± %
Ether extract		%			
Sheep	dig coef	%	42.	42.	
N-free extract		%			
Sheep	dig coef	%	43.	43.	
Protein (N x 6.25)		%	7.6	8.3	29
Sheep	dig coef	%	16.	16.	
Cattle	dig prot	%	* 3.7	4.1	
Goats	dig prot	%	* 3.9	4.3	
Horses	dig prot	%	* 4.2	4.5	
Rabbits	dig prot	%	* 4.6	5.0	
Sheep	dig prot	%	1.2	1.3	
Energy	GE Mcal/kg				
Sheep	DE Mcal/kg		* 1.45	1.59	
Sheep	ME Mcal/kg		* 1.19	1.30	
Sheep	TDN	%	33.0	36.0	
Calcium		%	0.46	0.51	40
Phosphorus		%	0.12	0.13	98
Carotene		mg/kg	27.4	29.9	
Vitamin A equivalent		IU/g	45.7	49.9	

Oats, grain clippings, (1)

Clipped oat by-product (AAFCO)

Ref No 1 03 269 United States

Feed Name or Analyses			Mean As Fed	Mean Dry	C.V. ± %
Dry matter		%	92.2	100.0	
Ash		%	10.8	11.8	
Crude fiber		%	25.2	27.4	
Sheep	dig coef	%	58.	58.	
Ether extract		%	2.3	2.5	
Sheep	dig coef	%	70.	70.	
N-free extract		%	45.0	48.8	
Sheep	dig coef	%	62.	62.	
Protein (N x 6.25)		%	8.8	9.6	
Sheep	dig coef	%	40.	40.	
Cattle	dig prot	%	* 4.8	5.2	
Goats	dig prot	%	* 5.1	5.5	
Horses	dig prot	%	* 5.2	5.7	
Rabbits	dig prot	%	* 5.6	6.1	
Sheep	dig prot	%	3.5	3.8	
Energy	GE Mcal/kg				
Cattle	DE Mcal/kg		* 2.29	2.48	
Sheep	DE Mcal/kg		* 2.19	2.38	
Cattle	ME Mcal/kg		* 1.87	2.03	
Sheep	ME Mcal/kg		* 1.80	1.95	
Cattle	TDN	%	* 51.9	56.3	
Sheep	TDN	%	49.7	53.9	

Oats, groats by-product, mx 22% fiber, (1)

Oat mill by-product (AAFCO)

Oat feed (CFA)

Ref No 1 03 332 United States

Feed Name or Analyses			Mean As Fed	Mean Dry	C.V. ± %
Dry matter		%	92.2	100.0	1
Ash		%	5.9	6.4	6
Crude fiber		%	27.1	29.4	13
Cattle	dig coef	%	26.	26.	
Horses	dig coef	%	42.	42.	
Sheep	dig coef	%	38.	38.	
Swine	dig coef	%	8.	8.	
Ether extract		%	1.9	2.1	40
Cattle	dig coef	%	39.	39.	
Horses	dig coef	%	62.	62.	
Sheep	dig coef	%	79.	79.	
Swine	dig coef	%	73.	73.	
N-free extract		%	51.6	56.0	3
Cattle	dig coef	%	33.	33.	
Horses	dig coef	%	38.	38.	
Sheep	dig coef	%	41.	41.	
Swine	dig coef	%	28.	28.	
Protein (N x 6.25)		%	5.7	6.1	34
Cattle	dig coef	%	55.	55.	
Horses	dig coef	%	65.	65.	
Sheep	dig coef	%	58.	58.	
Swine	dig coef	%	59.	59.	
Cattle	dig prot	%	3.1	3.3	
Goats	dig prot	%	* 2.1	2.3	
Horses	dig prot	%	3.6	4.0	
Rabbits	dig prot	%	* 3.1	3.4	
Sheep	dig prot	%	3.3	3.6	
Swine	dig prot	%	3.3	3.6	
Cellulose (Matrone)		%	10.3	11.2	
Lignin (Ellis)		%	4.5	4.9	
Energy	GE Mcal/kg		4.00	4.34	
Cattle	DE Mcal/kg		* 1.27	1.38	
Horses	DE Mcal/kg		* 1.65	1.78	
Sheep	DE Mcal/kg		* 1.68	1.83	
Swine	DE kcal/kg		*1019.	1105.	
Cattle	ME Mcal/kg		* 1.04	1.13	
Chickens	ME_n kcal/kg		391.	424.	
Horses	ME Mcal/kg		* 1.35	1.46	
Sheep	ME Mcal/kg		* 1.38	1.50	
Swine	ME kcal/kg		* 966.	1048.	
Cattle	TDN	%	28.9	31.3	
Horses	TDN	%	37.3	40.5	
Sheep	TDN	%	38.2	41.4	
Swine	TDN	%	23.1	25.1	
Calcium		%	0.14	0.16	
Chlorine		%	0.10	0.11	
Copper		mg/kg	2.3	2.5	
Iron		%	0.005	0.005	
Magnesium		%	0.07	0.08	
Phosphorus		%	0.20	0.22	64
Potassium		%	0.56	0.60	
Sodium		%	0.04	0.04	
Sulphur		%	0.13	0.15	
Choline		mg/kg	326.	353.	
Niacin		mg/kg	9.1	9.9	
Pantothenic acid		mg/kg	3.3	3.5	
Riboflavin		mg/kg	4.6	4.9	
Arginine		%	0.20	0.21	
Cystine		%	0.10	0.11	
Glycine		%	0.10	0.11	
Lysine		%	0.20	0.21	
Methionine		%	0.10	0.11	
Tryptophan		%	0.10	0.11	

Ref No 1 03 332 Canada

Feed Name or Analyses			Mean As Fed	Mean Dry	C.V. ± %
Dry matter		%	90.5	100.0	2
Ash		%	5.2	5.7	
Crude fiber		%	29.7	32.9	8
Cattle	dig coef	%	26.	26.	
Horses	dig coef	%	42.	42.	
Sheep	dig coef	%	38.	38.	
Swine	dig coef	%	8.	8.	
Ether extract		%	1.5	1.6	25
Cattle	dig coef	%	39.	39.	
Horses	dig coef	%	62.	62.	
Sheep	dig coef	%	79.	79.	
Swine	dig coef	%	73.	73.	
N-free extract		%	50.0	55.3	
Cattle	dig coef	%	33.	33.	
Horses	dig coef	%	38.	38.	
Sheep	dig coef	%	41.	41.	
Swine	dig coef	%	28.	28.	
Protein (N x 6.25)		%	4.1	4.5	17
Cattle	dig coef	%	55.	55.	
Horses	dig coef	%	65.	65.	
Sheep	dig coef	%	58.	58.	
Swine	dig coef	%	59.	59.	
Cattle	dig prot	%	2.2	2.5	
Goats	dig prot	%	* 0.7	0.8	
Horses	dig prot	%	2.6	2.9	
Rabbits	dig prot	%	* 2.0	2.2	
Sheep	dig prot	%	2.4	2.6	
Swine	dig prot	%	2.4	2.7	
Energy	GE Mcal/kg		3.93	4.34	
Cattle	DE Mcal/kg		* 1.22	1.35	
Horses	DE Mcal/kg		* 1.60	1.76	
Sheep	DE Mcal/kg		* 1.62	1.79	
Swine	DE kcal/kg		* 937.	1035.	
Cattle	ME Mcal/kg		* 1.00	1.11	
Chickens	ME_n kcal/kg		1122.	1240.	
Horses	ME Mcal/kg		* 1.31	1.45	
Sheep	ME Mcal/kg		* 1.33	1.47	
Swine	ME kcal/kg		* 891.	984.	
Cattle	TDN	%	27.8	30.7	
Horses	TDN	%	36.2	40.0	
Sheep	TDN	%	36.8	40.7	
Swine	TDN	%	21.2	23.5	
Niacin		mg/kg	9.4	10.3	
Pantothenic acid		mg/kg	3.5	3.8	
Riboflavin		mg/kg	2.5	2.7	
Thiamine		mg/kg	1.2	1.3	
Tryptophan		%	0.16	0.18	

Oats, hay, dehy, (1)

Ref No 1 03 271 United States

Feed Name or Analyses			Mean As Fed	Mean Dry	C.V. ± %
Dry matter		%	87.1	100.0	0
Ash		%	8.3	9.5	
Crude fiber		%	20.3	23.3	
Cattle	dig coef	%	76.	76.	
Ether extract		%	3.0	3.4	
Cattle	dig coef	%	68.	68.	
N-free extract		%	41.6	47.8	
Cattle	dig coef	%	78.	78.	
Protein (N x 6.25)		%	13.9	16.0	
Cattle	dig coef	%	66.	66.	
Cattle	dig prot	%	9.2	10.6	
Goats	dig prot	%	* 10.0	11.5	
Horses	dig prot	%	* 9.7	11.1	
Rabbits	dig prot	%	* 9.6	11.0	
Sheep	dig prot	%	* 9.5	10.9	
Energy	GE Mcal/kg				
Cattle	DE Mcal/kg		* 2.72	3.12	
Sheep	DE Mcal/kg		* 2.52	2.89	
Cattle	ME Mcal/kg		* 2.23	2.56	
Sheep	ME Mcal/kg		* 2.06	2.37	
Cattle	TDN	%	61.6	70.8	
Sheep	TDN	%	* 57.0	65.5	
Carotene		mg/kg	155.9	179.0	46
Vitamin A equivalent		IU/g	259.9	298.4	

Oats, hay, fan air dried, immature, (1)

Ref No 1 03 270 United States

Feed Name or Analyses			Mean As Fed	Mean Dry	C.V. ± %
Dry matter		%		100.0	
Riboflavin		mg/kg		26.0	

Oats, hay, s-c, (1)

Ref No 1 03 280 — United States

Feed Name or Analyses				Mean As Fed	Mean Dry	C.V. ± %
Dry matter		%		90.7	100.0	3
Ash		%		7.5	8.2	10
Crude fiber		%		27.9	30.8	7
Cattle	dig coef	%		65.	65.	
Sheep	dig coef	%		52.	52.	
Ether extract		%		1.9	2.1	44
Cattle	dig coef	%		66.	66.	
Sheep	dig coef	%		63.	63.	
N-free extract		%		45.7	50.4	
Cattle	dig coef	%		65.	65.	
Sheep	dig coef	%		57.	57.	
Protein (N x 6.25)		%		7.7	8.5	22
Cattle	dig coef	%		48.	48.	
Sheep	dig coef	%		55.	55.	
Cattle	dig prot	%		3.7	4.1	
Goats	dig prot	%	*	4.1	4.5	
Horses	dig prot	%	*	4.3	4.7	
Rabbits	dig prot	%	*	4.7	5.2	
Sheep	dig prot	%		4.2	4.7	
Cellulose (Matrone)		%		27.9	30.8	
Fatty acids		%		2.8	3.1	
Energy	GE	Mcal/kg				
Cattle	DE	Mcal/kg	*	2.40	2.64	
Sheep	DE	Mcal/kg	*	2.10	2.31	
Cattle	ME	Mcal/kg	*	1.97	2.17	
Sheep	ME	Mcal/kg	*	1.72	1.89	
Cattle	NE_m	Mcal/kg	*	1.19	1.31	
Cattle	NE_{gain}	Mcal/kg	*	0.65	0.70	
Cattle	$NE_{lactating\ cows}$	Mcal/kg	*	1.35	1.49	
Cattle	TDN	%		54.4	60.0	
Sheep	TDN	%		47.5	52.4	
Calcium		%		0.22	0.24	
Chlorine		%		0.47	0.52	
Iron		%		0.050	0.056	
Magnesium		%		0.16	0.18	
Manganese		mg/kg		83.1	91.6	
Phosphorus		%		0.20	0.22	
Potassium		%		0.85	0.94	
Silicon		%		3.32	3.66	
Sodium		%		0.15	0.17	

Ref No 1 03 280 — Canada

Feed Name or Analyses				Mean As Fed	Mean Dry	C.V. ± %
Dry matter		%		90.7	100.0	
Crude fiber		%		35.1	38.7	
Cattle	dig coef	%		65.	65.	
Sheep	dig coef	%		52.	52.	
Ether extract		%				
Cattle	dig coef	%		66.	66.	
Sheep	dig coef	%		63.	63.	
N-free extract		%				
Cattle	dig coef	%		65.	65.	
Sheep	dig coef	%		57.	57.	
Protein (N x 6.25)		%		8.8	9.7	
Cattle	dig coef	%		48.	48.	
Sheep	dig coef	%		55.	55.	
Cattle	dig prot	%		4.2	4.6	
Goats	dig prot	%	*	5.0	5.6	
Horses	dig prot	%	*	5.2	5.7	
Rabbits	dig prot	%	*	5.6	6.1	
Sheep	dig prot	%		4.8	5.3	
Energy	GE	Mcal/kg				
Cattle	DE	Mcal/kg	*	2.40	2.64	
Sheep	DE	Mcal/kg	*	2.09	2.31	
Cattle	ME	Mcal/kg	*	1.96	2.17	
Sheep	ME	Mcal/kg	*	1.72	1.89	
Cattle	TDN	%		54.3	59.9	
Sheep	TDN	%		47.5	52.3	
Calcium		%		0.28	0.31	
Phosphorus		%		0.04	0.04	
Potassium		%		0.36	0.40	
Biotin		mg/kg		6.41	7.07	
Carotene		mg/kg		47.5	52.4	
Vitamin A equivalent		IU/g		79.3	87.4	

Oats, hay, s-c, immature, (1)

Ref No 1 03 272 — United States

Feed Name or Analyses				Mean As Fed	Mean Dry	C.V. ± %
Dry matter		%		89.2	100.0	3
Ash		%		8.0	9.0	15
Crude fiber		%		23.2	26.0	17
Cattle	dig coef	%	*	65.	65.	
Sheep	dig coef	%	*	52.	52.	
Ether extract		%		3.2	3.6	29
Cattle	dig coef	%	*	66.	66.	
Sheep	dig coef	%	*	63.	63.	
N-free extract		%		42.5	47.6	
Cattle	dig coef	%	*	65.	65.	
Sheep	dig coef	%	*	57.	57.	
Protein (N x 6.25)		%		12.3	13.8	15
Cattle	dig coef	%	*	48.	48.	
Sheep	dig coef	%	*	55.	55.	
Cattle	dig prot	%		5.9	6.6	
Goats	dig prot	%	*	8.4	9.4	
Horses	dig prot	%	*	8.2	9.2	
Rabbits	dig prot	%	*	8.3	9.3	
Sheep	dig prot	%		6.8	7.6	
Energy	GE	Mcal/kg				
Cattle	DE	Mcal/kg	*	2.35	2.64	
Sheep	DE	Mcal/kg	*	2.10	2.35	
Cattle	ME	Mcal/kg	*	1.93	2.16	
Sheep	ME	Mcal/kg	*	1.72	1.93	
Cattle	TDN	%		53.4	59.8	
Sheep	TDN	%		47.6	53.3	
Calcium		%		0.29	0.33	
Cobalt		mg/kg		0.041	0.046	
Phosphorus		%		0.38	0.43	
Carotene		mg/kg		233.8	262.1	
Riboflavin		mg/kg		20.1	22.5	
Thiamine		mg/kg		3.5	4.0	
Vitamin A equivalent		IU/g		389.8	437.0	

Oats, hay, s-c, early bloom, (1)

Ref No 1 09 099 — United States

Feed Name or Analyses				Mean As Fed	Mean Dry	C.V. ± %
Dry matter		%		85.5	100.0	
Cattle	dig coef	%		67.	67.	
Ash		%		6.9	8.1	
Crude fiber		%		24.2	28.3	
Ether extract		%		2.9	3.4	
N-free extract		%		43.3	50.6	
Protein (N x 6.25)		%		8.2	9.6	
Cattle	dig coef	%		49.	49.	
Cattle	dig prot	%		4.0	4.7	
Goats	dig prot	%	*	4.7	5.5	
Horses	dig prot	%	*	4.9	5.7	
Rabbits	dig prot	%	*	5.2	6.1	
Sheep	dig prot	%	*	4.4	5.2	
Cell walls (Van Soest)		%		49.1	57.4	
Cellulose (Matrone)		%		22.7	26.6	
Cattle	dig coef	%		64.	64.	
Fiber, acid detergent (VS)		%		26.4	30.9	
Energy	GE	Mcal/kg		3.56	4.17	
Cattle	GE dig coef	%		67.	67.	
Cattle	DE	Mcal/kg		2.39	2.79	
Sheep	DE	Mcal/kg	*	2.38	2.78	
Cattle	ME	Mcal/kg	*	1.96	2.29	
Sheep	ME	Mcal/kg	*	1.95	2.28	
Cattle	TDN	%	*	57.6	67.4	
Sheep	TDN	%	*	54.0	63.1	

Oats, hay, s-c, full bloom, (1)

Ref No 1 03 274 — United States

Feed Name or Analyses				Mean As Fed	Mean Dry	C.V. ± %
Dry matter		%		85.0	100.0	6
Cattle	dig coef	%		68.	68.	
Ash		%		6.1	7.2	10
Crude fiber		%		29.4	34.6	8
Cattle	dig coef	%	*	65.	65.	
Sheep	dig coef	%		60.	60.	
Ether extract		%		1.9	2.3	25
Cattle	dig coef	%	*	66.	66.	
Sheep	dig coef	%		48.	48.	
N-free extract		%		39.7	46.7	5
Cattle	dig coef	%	*	65.	65.	
Sheep	dig coef	%		51.	51.	
Protein (N x 6.25)		%		7.8	9.2	11
Cattle	dig coef	%	*	55.	55.	
Sheep	dig coef	%		54.	54.	
Cattle	dig prot	%		4.3	5.0	
Goats	dig prot	%	*	4.4	5.1	
Horses	dig prot	%	*	4.5	5.3	
Rabbits	dig prot	%	*	4.9	5.8	
Sheep	dig prot	%		4.2	5.0	
Cellulose (Matrone)		%		30.1	35.5	
Cattle	dig coef	%		75.	75.	
Energy	GE	Mcal/kg		3.63	4.27	
Cattle	GE dig coef	%		65.	65.	
Cattle	DE	Mcal/kg		2.37	2.79	
Sheep	DE	Mcal/kg	*	1.95	2.29	
Cattle	ME	Mcal/kg	*	1.94	2.29	
Sheep	ME	Mcal/kg	*	1.60	1.88	
Cattle	TDN	%		52.1	61.3	
Sheep	TDN	%		44.2	52.0	
Calcium		%		0.25	0.30	
Cobalt		mg/kg		0.064	0.075	
Copper		mg/kg		3.7	4.4	
Iron		%		0.015	0.018	
Magnesium		%		1.26	1.48	
Manganese		mg/kg		143.1	168.4	
Phosphorus		%		0.30	0.35	
Potassium		%		2.05	2.41	
Carotene		mg/kg		85.8	101.0	
Riboflavin		mg/kg		4.5	5.3	
Thiamine		mg/kg		2.8	3.3	
Vitamin A equivalent		IU/g		143.0	168.3	

Oats, hay, s-c, full bloom, gr 3 US, (1)

Ref No 1 03 273 — United States

Feed Name or Analyses				Mean As Fed	Mean Dry	C.V. ± %
Dry matter		%		88.2	100.0	5
Cellulose (Matrone)		%		26.9	30.5	
Hemicellulose		%		23.1	26.2	16
Sugars, total		%		3.4	3.9	
Lignin (Ellis)		%		7.5	8.5	
Energy	GE	Mcal/kg		4.00	4.54	
Chlorine		%		0.46	0.52	9

(1) dry forages and roughages (2) pasture, range plants, and forages fed green (3) silages (4) energy feeds (5) protein supplements (6) minerals (7) vitamins (8) additives

Feed Name or Analyses			Mean As Fed	Mean Dry	C.V. ± %
Cobalt	mg/kg		0.060	0.068	
Sodium	%		0.15	0.17	18

Oats, hay, s-c, milk stage, (1)

Ref No 1 03 275 United States

Feed Name or Analyses			Mean As Fed	Mean Dry	C.V. ± %
Dry matter	%		87.9	100.0	6
Cattle	dig coef %		59.	59.	
Ash	%		6.0	6.8	3
Crude fiber	%		29.5	33.6	4
Cattle	dig coef %		64.	64.	
Sheep	dig coef %		49.	49.	
Ether extract	%		2.5	2.8	5
Cattle	dig coef %		71.	71.	
Sheep	dig coef %		63.	63.	
N-free extract	%		42.6	48.5	
Cattle	dig coef %		65.	65.	
Sheep	dig coef %		54.	54.	
Protein (N x 6.25)	%		7.3	8.4	9
Cattle	dig coef %		48.	48.	
Sheep	dig coef %		58.	58.	
Cattle	dig prot %		3.5	4.0	
Goats	dig prot %	*	3.8	4.4	
Horses	dig prot %	*	4.1	4.6	
Rabbits	dig prot %	*	4.5	5.1	
Sheep	dig prot %		4.3	4.8	
Cellulose (Matrone)	%		31.8	36.2	
Cattle	dig coef %		65.	65.	
Energy	GE Mcal/kg		3.69	4.19	
Cattle	GE dig coef %		58.	58.	
Cattle	DE Mcal/kg		2.14	2.43	
Sheep	DE Mcal/kg	*	1.99	2.27	
Cattle	ME Mcal/kg	*	1.75	1.99	
Sheep	ME Mcal/kg	*	1.64	1.86	
Cattle	TDN %		54.1	61.5	
Sheep	TDN %		45.2	51.4	
Calcium	%		0.22	0.25	
Phosphorus	%		0.18	0.20	
Carotene	mg/kg		17.4	19.8	
Vitamin A equivalent	IU/g		29.1	33.1	

Oats, hay, s-c, dough stage, (1)

Ref No 1 03 276 United States

Feed Name or Analyses			Mean As Fed	Mean Dry	C.V. ± %
Dry matter	%		88.7	100.0	3
Ash	%		6.0	6.8	28
Crude fiber	%		28.7	32.3	6
Sheep	dig coef %		49.	49.	
Ether extract	%		2.9	3.3	19
Sheep	dig coef %		64.	64.	
N-free extract	%		43.7	49.2	
Sheep	dig coef %		59.	59.	
Protein (N x 6.25)	%		7.5	8.5	18
Sheep	dig coef %		52.	52.	
Cattle	dig prot %	*	3.8	4.3	
Goats	dig prot %	*	3.9	4.4	
Horses	dig prot %	*	4.2	4.7	
Rabbits	dig prot %	*	4.6	5.2	
Sheep	dig prot %		3.9	4.4	
Energy	GE Mcal/kg				
Cattle	DE Mcal/kg	*	2.33	2.63	
Sheep	DE Mcal/kg	*	2.11	2.38	
Cattle	ME Mcal/kg	*	1.91	2.16	
Sheep	ME Mcal/kg	*	1.73	1.95	
Cattle	TDN %	*	52.9	59.6	
Sheep	TDN %		47.9	54.0	

Feed Name or Analyses			Mean As Fed	Mean Dry	C.V. ± %
Calcium	%		0.29	0.33	
Phosphorus	%		0.15	0.17	

Oats, hay, s-c, mature, (1)

Ref No 1 03 277 United States

Feed Name or Analyses			Mean As Fed	Mean Dry	C.V. ± %
Dry matter	%		91.7	100.0	1
Ash	%		6.3	6.9	16
Crude fiber	%		28.4	30.9	3
Cattle	dig coef %	*	65.	65.	
Sheep	dig coef %		52.	52.	
Ether extract	%		3.2	3.5	13
Cattle	dig coef %	*	66.	66.	
Sheep	dig coef %		74.	74.	
N-free extract	%		46.5	50.7	
Cattle	dig coef %	*	65.	65.	
Sheep	dig coef %		63.	63.	
Protein (N x 6.25)	%		7.3	8.0	16
Cattle	dig coef %	*	48.	48.	
Sheep	dig coef %		70.	70.	
Cattle	dig prot %		3.5	3.8	
Goats	dig prot %	*	3.7	4.0	
Horses	dig prot %	*	3.9	4.3	
Rabbits	dig prot %	*	4.4	4.8	
Sheep	dig prot %		5.1	5.6	
Energy	GE Mcal/kg				
Cattle	DE Mcal/kg	*	2.51	2.74	
Sheep	DE Mcal/kg	*	2.40	2.62	
Cattle	ME Mcal/kg	*	2.06	2.25	
Sheep	ME Mcal/kg	*	1.97	2.15	
Cattle	TDN %		57.0	62.1	
Sheep	TDN %		54.5	59.5	
Calcium	%		0.32	0.35	
Cobalt	mg/kg		0.079	0.086	
Phosphorus	%		0.21	0.23	
Carotene	mg/kg		5.7	6.2	
Vitamin A equivalent	IU/g		9.4	10.3	

Oats, hay, s-c, cut 1, (1)

Ref No 1 03 278 United States

Feed Name or Analyses			Mean As Fed	Mean Dry	C.V. ± %
Dry matter	%		87.9	100.0	
Ash	%		6.0	6.8	
Crude fiber	%		32.3	36.7	
Ether extract	%		2.1	2.4	
N-free extract	%		40.2	45.7	
Protein (N x 6.25)	%		7.4	8.4	
Cattle	dig prot %	*	3.7	4.2	
Goats	dig prot %	*	3.9	4.4	
Horses	dig prot %	*	4.1	4.7	
Rabbits	dig prot %	*	4.5	5.2	
Sheep	dig prot %	*	3.6	4.1	
Energy	GE Mcal/kg				
Cattle	DE Mcal/kg	*	2.00	2.27	
Sheep	DE Mcal/kg	*	2.10	2.39	
Cattle	ME Mcal/kg	*	1.64	1.86	
Sheep	ME Mcal/kg	*	1.72	1.96	
Cattle	TDN %	*	45.3	51.5	
Sheep	TDN %	*	47.6	54.1	
Calcium	%		0.21	0.24	12
Chlorine	%		0.46	0.52	9
Cobalt	mg/kg		0.066	0.075	
Copper	mg/kg		3.9	4.4	
Iron	%		0.047	0.053	22
Magnesium	%		0.26	0.30	98
Manganese	mg/kg		86.6	98.5	25
Phosphorus	%		0.20	0.23	22
Potassium	%		0.94	1.07	43

Feed Name or Analyses			Mean As Fed	Mean Dry	C.V. ± %
Sodium	%		0.15	0.17	18
Carotene	mg/kg		88.8	101.0	
Riboflavin	mg/kg		4.7	5.3	
Thiamine	mg/kg		2.9	3.3	
Vitamin A equivalent	IU/g		148.0	168.3	

Oats, hay, s-c, gr 3 US, (1)

Ref No 1 03 279 United States

Feed Name or Analyses			Mean As Fed	Mean Dry	C.V. ± %
Dry matter	%		87.8	100.0	
Ash	%		6.0	6.8	
Crude fiber	%		32.2	36.7	
Ether extract	%		2.1	2.4	
N-free extract	%		40.1	45.7	
Protein (N x 6.25)	%		7.4	8.4	
Cattle	dig prot %	*	3.7	4.2	
Goats	dig prot %	*	3.9	4.4	
Horses	dig prot %	*	4.1	4.7	
Rabbits	dig prot %	*	4.5	5.2	
Sheep	dig prot %	*	3.6	4.1	
Energy	GE Mcal/kg				
Cattle	DE Mcal/kg	*	1.99	2.27	
Sheep	DE Mcal/kg	*	2.09	2.39	
Cattle	ME Mcal/kg	*	1.63	1.86	
Sheep	ME Mcal/kg	*	1.72	1.96	
Cattle	TDN %	*	45.2	51.5	
Sheep	TDN %	*	47.5	54.1	
Calcium	%		0.26	0.30	
Cobalt	mg/kg		0.066	0.075	
Copper	mg/kg		3.9	4.4	
Iron	%		0.016	0.018	
Magnesium	%		1.30	1.48	
Manganese	mg/kg		147.9	168.4	
Phosphorus	%		0.31	0.35	
Potassium	%		2.12	2.41	
Carotene	mg/kg		47.6	54.2	
Riboflavin	mg/kg		4.6	5.3	
Thiamine	mg/kg		2.9	3.3	
Vitamin A equivalent	IU/g		79.4	90.4	

Oats, huller by-product, (1)

Oat huller feed

Oat mill feed

Ref No 1 08 316 United States

Feed Name or Analyses			Mean As Fed	Mean Dry	C.V. ± %
Dry matter	%		92.9	100.0	
Ash	%		5.9	6.4	
Crude fiber	%		32.6	35.1	
Ether extract	%		1.0	1.1	
N-free extract	%		50.5	54.4	
Protein (N x 6.25)	%		2.9	3.1	
Cattle	dig prot %	*	-0.2	-0.3	
Goats	dig prot %	*	-0.4	-0.4	
Horses	dig prot %	*	0.2	0.2	
Rabbits	dig prot %	*	1.0	1.1	
Sheep	dig prot %	*	-0.5	-0.5	
Energy	GE Mcal/kg				
Cattle	DE Mcal/kg	*	1.38	1.49	
Sheep	DE Mcal/kg	*	1.36	1.46	
Cattle	ME Mcal/kg	*	1.13	1.22	
Chickens	ME_n kcal/kg		331.	356.	
Sheep	ME Mcal/kg	*	1.11	1.20	
Cattle	TDN %	*	31.3	33.7	
Sheep	TDN %	*	31.0	33.2	
Calcium	%		0.10	0.11	
Phosphorus	%		0.05	0.05	
Choline	mg/kg		331.	356.	

Continued

Feed Name or Analyses			Mean As Fed	Mean Dry	C.V. ± %
Niacin	mg/kg		9.3	10.0	
Pantothenic acid	mg/kg		3.3	3.6	
Riboflavin	mg/kg		1.8	1.9	

Oats, hulls, (1)
Oat hulls (AAFCO)
Oat hulls (CFA)

Ref No 1 03 281 United States

Feed Name or Analyses			Mean As Fed	Mean Dry	C.V. ± %
Dry matter	%		92.7	100.0	0
Ash	%		6.0	6.5	12
Crude fiber	%		29.8	32.2	5
Cattle	dig coef %		37.	37.	
Horses	dig coef %		25.	25.	
Sheep	dig coef %	*	40.	40.	
Swine	dig coef %		2.	2.	
Ether extract	%		1.4	1.5	57
Cattle	dig coef %		64.	64.	
Horses	dig coef %		74.	74.	
Sheep	dig coef %	*	74.	74.	
Swine	dig coef %		73.	73.	
N-free extract	%		51.8	55.9	2
Cattle	dig coef %		35.	35.	
Horses	dig coef %		21.	21.	
Sheep	dig coef %	*	33.	33.	
Swine	dig coef %		37.	37.	
Protein (N x 6.25)	%		3.6	3.8	29
Cattle	dig coef %		33.	33.	
Horses	dig coef %		91.	91.	
Sheep	dig coef %	*	28.	28.	
Swine	dig coef %		57.	57.	
Cattle	dig prot %		1.2	1.3	
Goats	dig prot %	*	0.1	0.1	
Horses	dig prot %		3.2	3.5	
Rabbits	dig prot %	*	1.5	1.6	
Sheep	dig prot %		1.0	1.1	
Swine	dig prot %		2.0	2.2	
Cellulose (Matrone)	%		47.4	51.1	
Pentosans	%		36.0	38.8	
Lignin (Ellis)	%		13.2	14.2	
Fatty acids	%		1.4	1.5	
Energy	GE Mcal/kg				
Cattle	DE Mcal/kg	*	1.42	1.53	
Horses	DE Mcal/kg	*	1.05	1.14	
Sheep	DE Mcal/kg	*	1.43	1.54	
Swine	DE kcal/kg	*	1065.	1149.	
Cattle	ME Mcal/kg	*	1.17	1.26	
Chickens	ME_n kcal/kg		734.	792.	
Horses	ME Mcal/kg	*	0.86	0.93	
Sheep	ME Mcal/kg	*	1.17	1.26	
Swine	ME kcal/kg	*	1014.	1094.	
Cattle	TDN %		32.2	34.8	
Horses	TDN %		23.9	25.8	
Sheep	TDN %		32.4	35.0	
Swine	TDN %		24.1	26.1	
Calcium	%		0.15	0.16	40
Copper	mg/kg		3.1	3.3	
Iron	%		0.010	0.011	54
Manganese	mg/kg		18.6	20.1	10
Phosphorus	%		0.10	0.11	52
Potassium	%		0.48	0.52	
Choline	mg/kg		192.	207.	31
Folic acid	mg/kg		0.96	1.04	
Niacin	mg/kg		7.6	8.2	34
Riboflavin	mg/kg		1.4	1.5	58
Thiamine	mg/kg		0.6	0.7	32

Ref No 1 03 281 Canada

Feed Name or Analyses			Mean As Fed	Mean Dry	C.V. ± %
Dry matter	%		90.1	100.0	
Ash	%		3.7	4.1	
Crude fiber	%		19.3	21.4	
Cattle	dig coef %		37.	37.	
Horses	dig coef %		25.	25.	
Swine	dig coef %		2.	2.	
Ether extract	%		3.4	3.8	
Cattle	dig coef %		64.	64.	
Horses	dig coef %		74.	74.	
Swine	dig coef %		73.	73.	
N-free extract	%		52.4	58.1	
Cattle	dig coef %		35.	35.	
Horses	dig coef %		21.	21.	
Swine	dig coef %		37.	37.	
Protein (N x 6.25)	%		11.3	12.6	
Cattle	dig coef %		33.	33.	
Horses	dig coef %		91.	91.	
Swine	dig coef %		57.	57.	
Cattle	dig prot %		3.7	4.2	
Goats	dig prot %	*	7.5	8.3	
Horses	dig prot %		10.3	11.4	
Rabbits	dig prot %	*	7.6	8.4	
Sheep	dig prot %	*	7.1	7.9	
Swine	dig prot %		6.5	7.2	
Energy	GE Mcal/kg				
Cattle	DE Mcal/kg	*	1.50	1.66	
Horses	DE Mcal/kg	*	1.40	1.55	
Sheep	DE Mcal/kg	*	1.41	1.56	
Swine	DE kcal/kg	*	1403.	1557.	
Cattle	ME Mcal/kg	*	1.23	1.36	
Horses	ME Mcal/kg	*	1.15	1.27	
Sheep	ME Mcal/kg	*	1.16	1.28	
Swine	ME kcal/kg	*	1311.	1456.	
Cattle	TDN %		34.0	37.7	
Horses	TDN %		31.7	35.2	
Sheep	TDN %		32.0	35.5	
Swine	TDN %		31.8	35.3	

Oats, straw, (1)

Ref No 1 03 283 United States

Feed Name or Analyses			Mean As Fed	Mean Dry	C.V. ± %
Dry matter	%		88.6	100.0	4
Ash	%		6.8	7.6	9
Crude fiber	%		36.3	41.0	5
Cattle	dig coef %		58.	58.	
Horses	dig coef %		58.	58.	
Sheep	dig coef %		57.	57.	
Ether extract	%		2.1	2.3	22
Cattle	dig coef %		44.	44.	
Horses	dig coef %		29.	29.	
Sheep	dig coef %		36.	36.	
N-free extract	%		39.6	44.7	3
Cattle	dig coef %		56.	56.	
Horses	dig coef %		46.	46.	
Sheep	dig coef %		47.	47.	
Protein (N x 6.25)	%		3.8	4.3	13
Cattle	dig coef %		33.	33.	
Horses	dig coef %		58.	58.	
Sheep	dig coef %		3.	3.	
Cattle	dig prot %		1.2	1.4	
Goats	dig prot %	*	0.5	0.6	
Horses	dig prot %		2.2	2.5	
Rabbits	dig prot %	*	1.8	2.0	
Sheep	dig prot %		0.1	0.1	
Cellulose (Matrone)	%		35.5	40.1	
Lignin (Ellis)	%		12.9	14.6	
Fatty acids	%		2.2	2.4	
Energy	GE Mcal/kg		4.15	4.69	
Sheep	GE dig coef %		58.	58.	
Cattle	DE Mcal/kg	*	2.05	2.32	
Horses	DE Mcal/kg	*	1.88	2.12	
Sheep	DE Mcal/kg		2.41	2.72	
Cattle	ME Mcal/kg	*	1.68	1.90	
Horses	ME Mcal/kg	*	1.54	1.74	
Sheep	ME Mcal/kg	*	1.97	2.23	
Cattle	NE_m Mcal/kg	*	0.98	1.11	
Cattle	NE_{gain} Mcal/kg	*	0.32	0.35	
Cattle	$NE_{lactating\ cows}$ Mcal/kg	*	1.02	1.15	
Cattle	TDN %		46.5	52.5	
Horses	TDN %		42.6	48.1	
Sheep	TDN %		40.9	46.1	
Calcium	%		0.24	0.27	
Chlorine	%		0.69	0.78	8
Copper	mg/kg		9.8	11.0	
Iron	%		0.018	0.020	
Magnesium	%		0.18	0.20	
Manganese	mg/kg		29.3	33.1	
Phosphorus	%		0.09	0.10	
Potassium	%		1.97	2.23	
Sodium	%		0.40	0.45	21
Sulphur	%		0.19	0.22	12

Ref No 1 03 283 Canada

Feed Name or Analyses			Mean As Fed	Mean Dry	C.V. ± %
Dry matter	%		92.8	100.0	2
Sheep	dig coef %		62.	62.	
Ash	%		5.6	6.0	
Crude fiber	%		36.5	39.4	10
Cattle	dig coef %		58.	58.	
Horses	dig coef %		58.	58.	
Sheep	dig coef %		57.	57.	
Ether extract	%		1.7	1.8	
Cattle	dig coef %		44.	44.	
Horses	dig coef %		29.	29.	
Sheep	dig coef %		36.	36.	
N-free extract	%		44.5	48.0	
Cattle	dig coef %		56.	56.	
Horses	dig coef %		46.	46.	
Sheep	dig coef %		63.	63.	
Protein (N x 6.25)	%		4.4	4.8	44
Cattle	dig coef %		33.	33.	
Horses	dig coef %		58.	58.	
Sheep	dig coef %		3.	3.	
Cattle	dig prot %		1.4	1.6	
Goats	dig prot %	*	1.0	1.0	
Horses	dig prot %		2.6	2.8	
Rabbits	dig prot %	*	2.2	2.4	
Sheep	dig prot %		0.1	0.1	
Energy	GE Mcal/kg		4.19	4.51	
Sheep	GE dig coef %		58.	58.	
Cattle	DE Mcal/kg	*	2.17	2.34	
Horses	DE Mcal/kg	*	1.99	2.14	
Sheep	DE Mcal/kg		2.43	2.62	
Cattle	ME Mcal/kg	*	1.78	1.92	
Horses	ME Mcal/kg	*	1.63	1.76	
Sheep	ME Mcal/kg	*	1.99	2.15	
Cattle	TDN %		49.2	53.0	
Horses	TDN %		45.1	48.6	
Sheep	TDN %		50.3	54.2	
Calcium	%		0.24	0.26	44
Phosphorus	%		0.41	0.45	98
Carotene	mg/kg		8.2	8.9	98
Vitamin A equivalent	IU/g		13.7	14.8	

(1) dry forages and roughages (2) pasture, range plants, and forages fed green (3) silages (4) energy feeds (5) protein supplements (6) minerals (7) vitamins (8) additives

Oats, straw, treated w calcium hydroxide, (1)

Ref No 1 03 284 United States

Feed Name or Analyses			Mean As Fed	Mean Dry	C.V. ± %
Dry matter	%		21.2	100.0	
Ash	%		1.6	7.5	
Crude fiber	%		10.7	50.7	
Sheep	dig coef %		81.	81.	
Ether extract	%		0.3	1.2	
Sheep	dig coef %		56.	56.	
N-free extract	%		8.3	39.1	
Sheep	dig coef %		42.	42.	
Protein (N x 6.25)	%		0.3	1.5	
Sheep	dig coef %		-169.	-169.	
Cattle	dig prot %	*	-0.3	-1.7	
Goats	dig prot %	*	-0.3	-1.9	
Horses	dig prot %	*	-0.2	-1.1	
Rabbits	dig prot %	*	-0.0	-0.1	
Sheep	dig prot %		-0.4	-2.5	
Energy	GE Mcal/kg				
Cattle	DE Mcal/kg	*	0.46	2.17	
Sheep	DE Mcal/kg	*	0.53	2.49	
Cattle	ME Mcal/kg	*	0.38	1.78	
Sheep	ME Mcal/kg	*	0.43	2.04	
Cattle	TDN %	*	10.5	49.3	
Sheep	TDN %		12.0	56.5	

Oats, straw, treated w sodium hydroxide, (1)

Ref No 1 03 285 United States

Feed Name or Analyses			Mean As Fed	Mean Dry	C.V. ± %
Dry matter	%		28.7	100.0	
Ash	%		1.4	5.0	
Crude fiber	%		17.8	62.0	
Sheep	dig coef %	*	74.	74.	
Ether extract	%		0.4	1.4	
Sheep	dig coef %	*	39.	39.	
N-free extract	%		8.6	29.8	
Sheep	dig coef %	*	63.	63.	
Protein (N x 6.25)	%		0.5	1.8	
Sheep	dig coef %	*	-36.	-36.	
Cattle	dig prot %	*	-0.3	-1.4	
Goats	dig prot %	*	-0.4	-1.7	
Horses	dig prot %	*	-0.2	-0.8	
Rabbits	dig prot %	*	0.0	0.1	
Sheep	dig prot %		-0.1	-0.6	
Energy	GE Mcal/kg				
Cattle	DE Mcal/kg	*	0.74	2.58	
Sheep	DE Mcal/kg	*	0.82	2.86	
Cattle	ME Mcal/kg	*	0.61	2.12	
Sheep	ME Mcal/kg	*	0.67	2.35	
Cattle	TDN %	*	16.8	58.6	
Sheep	TDN %		18.6	64.9	

Oats, straw, good quality, (1)

Ref No 1 03 282 United States

Feed Name or Analyses			Mean As Fed	Mean Dry	C.V. ± %
Dry matter	%			100.0	
Choline	mg/kg			234.	

Oats, straw pulp, treated w sodium hydroxide wet, (1)

Ref No 1 08 472 United States

Feed Name or Analyses			Mean As Fed	Mean Dry	C.V. ± %
Dry matter	%		28.7	100.0	
Ash	%		1.4	4.9	
Crude fiber	%		17.8	62.0	
Ether extract	%		0.4	1.4	
N-free extract	%		8.6	30.0	
Protein (N x 6.25)	%		0.5	1.7	
Cattle	dig prot %	*	-0.3	-1.5	
Goats	dig prot %	*	-0.4	-1.7	
Horses	dig prot %	*	-0.2	-0.9	
Rabbits	dig prot %	*	0.0	0.0	
Sheep	dig prot %	*	-0.4	-1.8	

Oats, aerial part, fresh, (2)

Ref No 2 03 292 United States

Feed Name or Analyses			Mean As Fed	Mean Dry	C.V. ± %
Dry matter	%		20.4	100.0	15
Ash	%		2.0	9.7	
Crude fiber	%		5.7	28.0	
Cattle	dig coef %		76.	76.	
Sheep	dig coef %		55.	55.	
Ether extract	%		0.7	3.5	
Cattle	dig coef %		50.	50.	
Sheep	dig coef %		70.	70.	
N-free extract	%		9.8	48.2	
Cattle	dig coef %		79.	79.	
Sheep	dig coef %		63.	63.	
Protein (N x 6.25)	%		2.2	10.8	
Cattle	dig coef %		72.	72.	
Sheep	dig coef %		73.	73.	
Cattle	dig prot %		1.6	7.7	
Goats	dig prot %	*	1.3	6.6	
Horses	dig prot %	*	1.4	6.7	
Rabbits	dig prot %	*	1.4	7.0	
Sheep	dig prot %		1.6	7.8	
Energy	GE Mcal/kg				
Cattle	DE Mcal/kg	*	0.64	3.13	
Sheep	DE Mcal/kg	*	0.53	2.60	
Cattle	ME Mcal/kg	*	0.52	2.56	
Sheep	ME Mcal/kg	*	0.43	2.13	
Cattle	TDN %		14.4	70.9	
Sheep	TDN %		12.0	59.0	
Biotin	mg/kg		0.05	0.26	58
Carotene	mg/kg		87.0	427.7	31
Folic acid	mg/kg		0.04	0.18	27
Niacin	mg/kg		10.0	49.4	45
Pantothenic acid	mg/kg		4.0	19.6	69
Riboflavin	mg/kg		2.2	11.0	49
Thiamine	mg/kg		0.3	1.3	29
Vitamin B6	mg/kg		0.49	2.43	23
Vitamin A equivalent	IU/g		145.1	713.0	
Glutamic acid	%		0.39	1.90	
Isoleucine	%		0.26	1.30	
Leucine	%		0.37	1.80	
Lysine	%		0.28	1.40	
Methionine	%		0.04	0.20	
Threonine	%		0.33	1.60	
Tryptophan	%		0.04	0.20	
Valine	%		0.24	1.20	

Oats, aerial part, fresh, immature, (2)

Ref No 2 03 286 United States

Feed Name or Analyses			Mean As Fed	Mean Dry	C.V. ± %
Dry matter	%		14.2	100.0	25
Crude fiber	%				
Cattle	dig coef %	*	76.	76.	
Sheep	dig coef %	*	55.	55.	
Ether extract	%				
Cattle	dig coef %	*	50.	50.	
Sheep	dig coef %	*	36.	36.	
N-free extract	%				
Cattle	dig coef %	*	79.	79.	
Sheep	dig coef %	*	62.	62.	
Protein (N x 6.25)	%				
Cattle	dig coef %	*	72.	72.	
Sheep	dig coef %	*	46.	46.	
Energy	GE Mcal/kg				
Cattle	DE Mcal/kg	*	0.44	3.07	
Sheep	DE Mcal/kg	*	0.32	2.28	
Cattle	ME Mcal/kg	*	0.36	2.52	
Sheep	ME Mcal/kg	*	0.27	1.87	
Cattle	TDN %		9.9	69.7	
Sheep	TDN %		7.3	51.7	
Chlorine	%		0.01	0.10	
Sodium	%		0.02	0.11	
Sulphur	%		0.01	0.08	
Carotene	mg/kg		79.6	560.6	18
Vitamin A equivalent	IU/g		132.7	934.6	

Ref No 2 03 286 Canada

Feed Name or Analyses			Mean As Fed	Mean Dry	C.V. ± %
Dry matter	%		17.9	100.0	
Ash	%		1.9	10.6	
Crude fiber	%		4.5	24.9	
Ether extract	%		0.5	2.6	
N-free extract	%		8.3	46.3	
Protein (N x 6.25)	%		2.8	15.6	
Cattle	dig prot %	*	2.0	11.2	
Goats	dig prot %	*	2.0	11.1	
Horses	dig prot %	*	1.9	10.8	
Rabbits	dig prot %	*	1.9	10.7	
Sheep	dig prot %	*	2.1	11.5	
Energy	GE Mcal/kg				
Cattle	DE Mcal/kg	*	0.55	3.07	
Sheep	DE Mcal/kg	*	0.41	2.28	
Cattle	ME Mcal/kg	*	0.45	2.52	
Sheep	ME Mcal/kg	*	0.33	1.87	
Cattle	TDN %		12.5	69.7	
Sheep	TDN %		9.3	51.7	

Oats, aerial part, fresh, pre-bloom, (2)

Ref No 2 08 470 United States

Feed Name or Analyses			Mean As Fed	Mean Dry	C.V. ± %
Dry matter	%		14.1	100.0	
Ash	%		2.0	14.2	
Crude fiber	%		2.8	19.9	
Ether extract	%		0.6	4.3	
N-free extract	%		5.5	39.0	
Protein (N x 6.25)	%		3.2	22.7	
Cattle	dig prot %	*	2.4	17.2	
Goats	dig prot %	*	2.5	17.7	
Horses	dig prot %	*	2.4	16.8	
Rabbits	dig prot %	*	2.3	16.2	
Sheep	dig prot %	*	2.6	18.1	
Energy	GE Mcal/kg				
Cattle	DE Mcal/kg	*	0.41	2.92	
Sheep	DE Mcal/kg	*	0.44	3.13	
Cattle	ME Mcal/kg	*	0.34	2.39	
Sheep	ME Mcal/kg	*	0.36	2.56	
Cattle	TDN %	*	9.3	66.1	
Sheep	TDN %	*	10.0	70.9	
Calcium	%		0.06	0.43	
Magnesium	%		0.06	0.43	
Phosphorus	%		0.09	0.64	

Oats, aerial part, fresh, early bloom, (2)

Ref No 2 03 287 United States

Feed Name or Analyses			Mean As Fed	Mean Dry	C.V. ± %
Dry matter	%		26.6	100.0	
Ash	%		2.2	8.2	23

Continued

Feed Name or Analyses			Mean As Fed	Mean Dry	C.V. ± %
Crude fiber	%		7.6	28.7	15
Cattle	dig coef %	*	76.	76.	
Sheep	dig coef %	*	55.	55.	
Ether extract	%		0.9	3.4	8
Cattle	dig coef %	*	50.	50.	
Sheep	dig coef %	*	70.	70.	
N-free extract	%		12.6	47.4	
Cattle	dig coef %	*	79.	79.	
Sheep	dig coef %	*	63.	63.	
Protein (N x 6.25)	%		3.3	12.3	38
Cattle	dig coef %	*	72.	72.	
Sheep	dig coef %	*	73.	73.	
Cattle	dig prot %		2.4	8.9	
Goats	dig prot %	*	2.1	8.1	
Horses	dig prot %	*	2.1	8.0	
Rabbits	dig prot %	*	2.2	8.2	
Sheep	dig prot %		2.4	9.0	
Energy	GE Mcal/kg				
Cattle	DE Mcal/kg	*	0.84	3.17	
Sheep	DE Mcal/kg	*	0.70	2.65	
Cattle	ME Mcal/kg	*	0.69	2.60	
Sheep	ME Mcal/kg	*	0.58	2.17	
Cattle	TDN %		19.1	72.0	
Sheep	TDN %		16.0	60.0	
Calcium	%		0.09	0.34	
Phosphorus	%		0.09	0.34	
Potassium	%		0.50	1.88	

Ref No 2 03 287 Canada

Feed Name or Analyses			Mean As Fed	Mean Dry	C.V. ± %
Dry matter	%		19.7	100.0	
Ash	%		1.8	9.1	
Crude fiber	%		5.0	25.2	
Ether extract	%		0.5	2.5	
N-free extract	%		10.4	52.6	
Protein (N x 6.25)	%		2.1	10.6	
Cattle	dig prot %	*	1.4	6.9	
Goats	dig prot %	*	1.3	6.4	
Horses	dig prot %	*	1.3	6.5	
Rabbits	dig prot %	*	1.4	6.9	
Sheep	dig prot %	*	1.4	6.9	
Energy	GE Mcal/kg				
Cattle	DE Mcal/kg	*	0.62	3.16	
Sheep	DE Mcal/kg	*	0.52	2.63	
Cattle	ME Mcal/kg	*	0.51	2.59	
Sheep	ME Mcal/kg	*	0.42	2.15	
Cattle	TDN %		14.1	71.7	
Sheep	TDN %		11.7	59.6	

Oats, aerial part, fresh, full bloom, (2)
Ref No 2 03 288 United States

Feed Name or Analyses			Mean As Fed	Mean Dry	C.V. ± %
Dry matter	%		32.2	100.0	31
Crude fiber	%				
Cattle	dig coef %	*	76.	76.	
Sheep	dig coef %	*	55.	55.	
Ether extract	%				
Cattle	dig coef %	*	50.	50.	
Sheep	dig coef %	*	70.	70.	
N-free extract	%				
Cattle	dig coef %	*	79.	79.	
Sheep	dig coef %	*	63.	63.	
Protein (N x 6.25)	%				
Cattle	dig coef %	*	72.	72.	
Sheep	dig coef %	*	73.	73.	
Calcium	%		0.09	0.28	16
Phosphorus	%		0.10	0.31	15
Potassium	%		0.67	2.07	

Oats, aerial part, fresh, milk stage, (2)
Ref No 2 03 268 United States

Feed Name or Analyses			Mean As Fed	Mean Dry	C.V. ± %
Dry matter	%		28.2	100.0	
Ash	%		2.2	7.8	
Crude fiber	%		9.2	32.7	
Horses	dig coef %		49.	49.	
Sheep	dig coef %		51.	51.	
Ether extract	%		1.0	3.5	
Horses	dig coef %		30.	30.	
Sheep	dig coef %		69.	69.	
N-free extract	%		13.3	47.0	
Horses	dig coef %		66.	66.	
Sheep	dig coef %		63.	63.	
Protein (N x 6.25)	%		2.5	9.0	
Horses	dig coef %		72.	72.	
Sheep	dig coef %		71.	71.	
Cattle	dig prot %	*	1.6	5.5	
Goats	dig prot %	*	1.4	5.0	
Horses	dig prot %		1.8	6.5	
Rabbits	dig prot %	*	1.6	5.6	
Sheep	dig prot %		1.8	6.4	
Energy	GE Mcal/kg				
Cattle	DE Mcal/kg	*	0.79	2.82	
Horses	DE Mcal/kg	*	0.69	2.46	
Sheep	DE Mcal/kg	*	0.72	2.56	
Cattle	ME Mcal/kg	*	0.65	2.31	
Horses	ME Mcal/kg	*	0.57	2.02	
Sheep	ME Mcal/kg	*	0.59	2.10	
Cattle	TDN %	*	18.0	63.9	
Horses	TDN %		15.8	55.9	
Sheep	TDN %		16.4	58.1	

Oats, aerial part, fresh, dough stage, (2)
Ref No 2 03 289 United States

Feed Name or Analyses			Mean As Fed	Mean Dry	C.V. ± %
Dry matter	%		25.6	100.0	
Ash	%		2.0	8.0	8
Crude fiber	%		8.1	31.7	12
Cattle	dig coef %		47.	47.	
Ether extract	%		0.9	3.6	14
Cattle	dig coef %		71.	71.	
N-free extract	%		11.9	46.6	
Protein (N x 6.25)	%		2.6	10.1	28
Cattle	dig coef %		66.	66.	
Cattle	dig prot %		1.7	6.7	
Goats	dig prot %	*	1.5	6.0	
Horses	dig prot %	*	1.6	6.1	
Rabbits	dig prot %	*	1.7	6.5	
Sheep	dig prot %	*	1.6	6.4	
Energy	GE Mcal/kg		1.11	4.33	
Cattle	GE dig coef %		55.	55.	
Cattle	DE Mcal/kg		0.60	2.36	
Sheep	DE Mcal/kg	*	0.73	2.86	
Cattle	ME Mcal/kg	*	0.50	1.94	
Sheep	ME Mcal/kg	*	0.60	2.35	
Cattle	TDN %	*	16.5	64.5	
Sheep	TDN %	*	16.6	64.9	
Calcium	%		0.08	0.30	52
Phosphorus	%		0.08	0.30	26
Potassium	%		0.60	2.34	

Ref No 2 03 289 Canada

Feed Name or Analyses			Mean As Fed	Mean Dry	C.V. ± %
Dry matter	%		90.5	100.0	10
Ash	%		7.2	8.0	
Crude fiber	%		27.1	30.0	15
Cattle	dig coef %		47.	47.	
Ether extract	%		3.2	3.5	
Cattle	dig coef %		71.	71.	
N-free extract	%		-9.9	-11.0	
Protein (N x 6.25)	%		6.3	7.0	98
Cattle	dig coef %		66.	66.	
Cattle	dig prot %		41.6	46.0	
Goats	dig prot %	*	55.7	61.6	
Horses	dig prot %	*	51.3	56.7	
Rabbits	dig prot %	*	47.5	52.5	
Sheep	dig prot %	*	56.1	61.9	
Cellulose (Matrone)	%		24.3	26.9	
Cattle	dig coef %		48.	48.	
Energy	GE Mcal/kg		3.92	4.33	
Cattle	GE dig coef %		55.	55.	
Cattle	DE Mcal/kg		2.14	2.36	
Sheep	DE Mcal/kg	*	2.59	2.86	
Cattle	ME Mcal/kg	*	1.75	1.94	
Sheep	ME Mcal/kg	*	2.12	2.35	
Cattle	TDN %	*	48.4	53.5	
Sheep	TDN %	*	58.8	64.9	
Calcium	%		0.23	0.26	40
Phosphorus	%		0.21	0.23	40
Carotene	mg/kg		22.2	24.5	82
Vitamin A equivalent	IU/g		37.0	40.8	

Oats, aerial part, fresh, mature, (2)
Ref No 2 03 290 United States

Feed Name or Analyses			Mean As Fed	Mean Dry	C.V. ± %
Dry matter	%			100.0	
Ash	%			7.9	18
Crude fiber	%			31.9	11
Cattle	dig coef %	*		65.	
Sheep	dig coef %	*		55.	
Ether extract	%			3.6	7
Cattle	dig coef %	*		66.	
Sheep	dig coef %	*		70.	
N-free extract	%			48.3	
Cattle	dig coef %	*		65.	
Sheep	dig coef %	*		63.	
Protein (N x 6.25)	%			8.3	42
Cattle	dig coef %	*		48.	
Sheep	dig coef %	*		77.	
Cattle	dig prot %			4.0	
Goats	dig prot %	*		4.3	
Horses	dig prot %	*		4.6	
Rabbits	dig prot %	*		5.1	
Sheep	dig prot %			6.4	
Energy	GE Mcal/kg				
Cattle	DE Mcal/kg	*		2.71	
Sheep	DE Mcal/kg	*		2.65	
Cattle	ME Mcal/kg	*		2.22	
Sheep	ME Mcal/kg	*		2.17	
Cattle	TDN %			61.5	
Sheep	TDN %			60.0	
Calcium	%			0.27	25
Iron	%			0.054	
Phosphorus	%			0.28	12

(1) dry forages and roughages (2) pasture, range plants, and forages fed green (3) silages (4) energy feeds (5) protein supplements (6) minerals (7) vitamins (8) additives

Feed Name or Analyses		Mean As Fed	Mean Dry	C.V. ± %

Oats, aerial part, fresh, cut 1, (2)

Ref No 2 03 291 — United States

Feed Name or Analyses		As Fed	Dry	C.V. ± %
Dry matter	%	26.8	100.0	44
Ash	%	2.9	10.8	20
Crude fiber	%	4.4	16.5	21
Ether extract	%	1.2	4.5	22
N-free extract	%	10.5	39.0	
Protein (N x 6.25)	%	7.8	29.2	28
Cattle	dig prot % *	6.1	22.7	
Goats	dig prot % *	6.4	23.8	
Horses	dig prot % *	6.0	22.3	
Rabbits	dig prot % *	5.7	21.2	
Sheep	dig prot % *	6.5	24.2	
Energy	GE Mcal/kg			
Cattle	DE Mcal/kg *	0.76	2.84	
Sheep	DE Mcal/kg *	0.73	2.74	
Cattle	ME Mcal/kg *	0.62	2.33	
Sheep	ME Mcal/kg *	0.60	2.25	
Cattle	TDN % *	17.3	64.5	
Sheep	TDN % *	16.6	62.1	
Calcium	%	0.18	0.69	37
Phosphorus	%	0.16	0.59	36
Potassium	%	0.69	2.57	55

Oats, leaves, fresh, immature, (2)

Ref No 2 03 293 — United States

Feed Name or Analyses		As Fed	Dry	C.V. ± %
Dry matter	%		100.0	
N-free extract	%		55.5	
Protein (N x 6.25)	%		35.1	
Cattle	dig prot % *		27.7	
Goats	dig prot % *		29.3	
Horses	dig prot % *		27.3	
Rabbits	dig prot % *		25.8	
Sheep	dig prot % *		29.7	
Calcium	%		0.89	
Magnesium	%		0.40	
Phosphorus	%		0.29	
Potassium	%		6.67	
Riboflavin	mg/kg		21.6	

Oats, stems, fresh, immature, (2)

Ref No 2 03 294 — United States

Feed Name or Analyses		As Fed	Dry	C.V. ± %
Dry matter	%		100.0	
Calcium	%		0.69	
Magnesium	%		0.40	
Phosphorus	%		0.27	
Riboflavin	mg/kg		5.3	

Oats, aerial part, ensiled, (3)

Ref No 3 03 298 — United States

Feed Name or Analyses		As Fed	Dry	C.V. ± %
Dry matter	%	32.4	100.0	
Ash	%	2.7	8.4	
Crude fiber	%	11.5	35.5	
Cattle	dig coef %	65.	65.	
Ether extract	%	1.0	3.1	
Cattle	dig coef %	57.	57.	
N-free extract	%	14.3	44.1	
Cattle	dig coef %	60.	60.	
Protein (N x 6.25)	%	2.9	8.9	
Cattle	dig coef %	57.	57.	
Cattle	dig prot %	1.6	5.1	
Goats	dig prot % *	1.4	4.3	
Horses	dig prot % *	1.4	4.3	
Sheep	dig prot % *	1.4	4.3	
Energy	GE Mcal/kg			
Cattle	DE Mcal/kg *	0.84	2.58	
Sheep	DE Mcal/kg *	0.90	2.78	
Cattle	ME Mcal/kg *	0.69	2.12	
Sheep	ME Mcal/kg *	0.74	2.28	
Cattle	NE_m Mcal/kg *	0.41	1.27	
Cattle	NE_{gain} Mcal/kg *	0.21	0.64	
Cattle	$NE_{lactating\ cows}$ Mcal/kg *	0.46	1.41	
Cattle	TDN %	19.0	58.6	
Sheep	TDN % *	20.4	63.0	

Ref No 3 03 298 — Canada

Feed Name or Analyses		As Fed	Dry	C.V. ± %
Dry matter	%	31.0	100.0	26
Crude fiber	%	9.6	30.9	14
Cattle	dig coef %	65.	65.	
Ether extract	%			
Cattle	dig coef %	57.	57.	
N-free extract	%			
Cattle	dig coef %	60.	60.	
Protein (N x 6.25)	%	3.1	10.1	50
Cattle	dig coef %	57.	57.	
Cattle	dig prot %	1.8	5.7	
Goats	dig prot % *	1.7	5.4	
Horses	dig prot % *	1.7	5.4	
Sheep	dig prot % *	1.7	5.4	
Energy	GE Mcal/kg			
Cattle	DE Mcal/kg *	0.80	2.57	
Cattle	ME Mcal/kg *	0.65	2.11	
Cattle	TDN %	18.1	58.3	
Calcium	%	0.11	0.36	44
Phosphorus	%	0.08	0.25	34
Carotene	mg/kg	15.7	50.8	97
Vitamin A equivalent	IU/g	26.2	84.7	

Oats, aerial part, ensiled, immature, (3)

Ref No 3 03 295 — United States

Feed Name or Analyses		As Fed	Dry	C.V. ± %
Dry matter	%	29.4	100.0	
Calcium	%	0.16	0.54	
Chlorine	%	0.46	1.56	
Cobalt	mg/kg	0.013	0.044	
Copper	mg/kg	1.6	5.5	
Iron	%	0.003	0.011	
Manganese	mg/kg	11.9	40.3	
Phosphorus	%	0.09	0.32	
Potassium	%	1.00	3.41	
Sulphur	%	0.03	0.10	
Carotene	mg/kg	98.3	334.4	38
Vitamin A equivalent	IU/g	163.9	557.5	

Oats, aerial part, ensiled, full bloom, (3)

Ref No 3 07 893 — United States

Feed Name or Analyses		As Fed	Dry	C.V. ± %
Dry matter	%	44.4	100.0	
Cattle	dig coef %	60.	60.	
Ash	%	3.7	8.3	
Crude fiber	%	15.0	33.7	
Cattle	dig coef %	72.	72.	
Ether extract	%	1.4	3.2	
Cattle	dig coef %	44.	44.	
N-free extract	%	20.0	45.1	
Cattle	dig coef %	64.	64.	
Protein (N x 6.25)	%	4.3	9.6	
Cattle	dig coef %	43.	43.	
Cattle	dig prot %	1.8	4.1	
Goats	dig prot % *	2.2	5.0	
Horses	dig prot % *	2.2	5.0	
Sheep	dig prot % *	2.2	5.0	
Cellulose (Matrone)	%	14.9	33.5	
Cattle	dig coef %	75.	75.	
Energy	GE Mcal/kg	1.97	4.44	
Cattle	GE dig coef %	60.	60.	
Cattle	DE Mcal/kg	1.18	2.66	
Sheep	DE Mcal/kg *	1.24	2.79	
Cattle	ME Mcal/kg *	0.97	2.18	
Sheep	ME Mcal/kg *	1.01	2.29	
Cattle	TDN %	26.7	60.1	
Sheep	TDN % *	28.1	63.2	

Oats, aerial part, ensiled, dough stage, (3)

Ref No 3 03 296 — United States

Feed Name or Analyses		As Fed	Dry	C.V. ± %
Dry matter	%	36.6	100.0	14
Cattle	dig coef %	55.	55.	
Ash	%	2.7	7.3	8
Crude fiber	%	12.6	34.4	6
Cattle	dig coef %	54.	54.	
Ether extract	%	1.3	3.6	8
Cattle	dig coef %	67.	67.	
N-free extract	%	16.5	45.0	
Cattle	dig coef %	58.	58.	
Protein (N x 6.25)	%	3.6	9.7	7
Cattle	dig coef %	55.	55.	
Cattle	dig prot %	2.0	5.3	
Goats	dig prot % *	1.9	5.1	
Horses	dig prot % *	1.9	5.1	
Sheep	dig prot % *	1.9	5.1	
Cellulose (Matrone)	%	14.1	38.4	
Cattle	dig coef %	65.	65.	
Energy	GE Mcal/kg	1.61	4.41	
Cattle	GE dig coef %	55.	55.	
Cattle	DE Mcal/kg	0.89	2.43	
Sheep	DE Mcal/kg *	0.91	2.48	
Cattle	ME Mcal/kg *	0.73	1.99	
Sheep	ME Mcal/kg *	0.74	2.03	
Cattle	TDN %	20.2	55.2	
Sheep	TDN % *	30.6	56.3	
Calcium	%	0.17	0.47	
Phosphorus	%	0.12	0.33	21

Oats, aerial part, ensiled, mature, (3)

Ref No 3 03 297 — United States

Feed Name or Analyses		As Fed	Dry	C.V. ± %
Dry matter	%	25.8	100.0	
Ash	%	1.2	4.7	
Crude fiber	%	10.4	40.3	
Ether extract	%	0.6	2.2	
N-free extract	%	12.4	48.2	
Protein (N x 6.25)	%	1.2	4.6	
Cattle	dig prot % *	0.1	0.4	
Goats	dig prot % *	0.1	0.4	
Horses	dig prot % *	0.1	0.4	
Sheep	dig prot % *	0.1	0.4	
Energy	GE Mcal/kg			
Cattle	DE Mcal/kg *	0.61	2.35	
Sheep	DE Mcal/kg *	0.63	2.45	
Cattle	ME Mcal/kg *	0.50	1.93	
Sheep	ME Mcal/kg *	0.52	2.01	
Cattle	TDN % *	13.8	53.4	
Sheep	TDN % *	14.3	55.5	

Oats, aerial part w molasses added, ensiled, (3)

Ref No 3 03 300 United States

Feed Name or Analyses		Mean As Fed	Mean Dry	C.V. ± %
Dry matter	%	33.0	100.0	8
Ash	%	2.5	7.5	8
Crude fiber	%	10.5	31.7	5
Ether extract	%	1.2	3.8	14
N-free extract	%	15.9	48.3	
Protein (N x 6.25)	%	2.9	8.8	7
Cattle	dig prot %	* 1.4	4.2	
Goats	dig prot %	* 1.4	4.2	
Horses	dig prot %	* 1.4	4.2	
Sheep	dig prot %	* 1.4	4.2	
Energy	GE Mcal/kg			
Cattle	DE Mcal/kg	* 0.76	2.31	
Sheep	DE Mcal/kg	* 0.79	2.38	
Cattle	ME Mcal/kg	* 0.62	1.89	
Sheep	ME Mcal/kg	* 0.64	1.95	
Cattle	TDN %	* 17.3	52.4	
Sheep	TDN %	* 17.8	54.0	
Calcium	%	0.10	0.31	
Phosphorus	%	0.09	0.28	

Oats, aerial part w molasses added, ensiled, dough stage, (3)

Ref No 3 03 299 United States

Feed Name or Analyses		Mean As Fed	Mean Dry	C.V. ± %
Dry matter	%	34.8	100.0	8
Ash	%	2.6	7.5	9
Crude fiber	%	10.4	30.0	6
Cattle	dig coef %	48.	48.	
Ether extract	%	1.3	3.9	13
Cattle	dig coef %	64.	64.	
N-free extract	%	17.1	49.0	
Cattle	dig coef %	60.	60.	
Protein (N x 6.25)	%	3.4	9.7	7
Cattle	dig coef %	57.	57.	
Cattle	dig prot %	1.9	5.5	
Goats	dig prot %	* 1.8	5.0	
Horses	dig prot %	* 1.8	5.0	
Sheep	dig prot %	* 1.8	5.0	
Energy	GE Mcal/kg			
Cattle	DE Mcal/kg	* 0.84	2.42	
Sheep	DE Mcal/kg	* 0.83	2.39	
Cattle	ME Mcal/kg	* 0.69	1.98	
Sheep	ME Mcal/kg	* 0.68	1.96	
Cattle	TDN %	19.1	54.8	
Sheep	TDN %	* 18.8	54.1	
Calcium	%	0.18	0.53	
Phosphorus	%	0.13	0.36	

Oats, aerial part w sulfur dioxide added, ensiled, mature, (3)

Ref No 3 03 301 United States

Feed Name or Analyses		Mean As Fed	Mean Dry	C.V. ± %
Dry matter	%	34.9	100.0	10
Ash	%	2.2	6.4	9
Crude fiber	%	9.6	27.5	10
Ether extract	%	2.1	6.1	11
N-free extract	%	17.4	49.8	
Protein (N x 6.25)	%	3.6	10.2	5
Cattle	dig prot %	* 1.9	5.5	
Goats	dig prot %	* 1.9	5.5	
Horses	dig prot %	* 1.9	5.5	
Sheep	dig prot %	* 1.9	5.5	
Energy	GE Mcal/kg			
Cattle	DE Mcal/kg	* 0.87	2.48	
Sheep	DE Mcal/kg	* 0.83	2.38	
Cattle	ME Mcal/kg	* 0.71	2.03	
Sheep	ME Mcal/kg	* 0.68	1.95	
Cattle	TDN %	* 19.6	56.3	
Sheep	TDN %	* 18.8	53.9	
Calcium	%	0.14	0.39	12
Phosphorus	%	0.09	0.26	8
Carotene	mg/kg	22.7	65.0	14
Vitamin A equivalent	IU/g	37.8	108.4	

Oats, cereal by-product, grnd, mx 2% fiber, (4)

Oat meal (CFA)

Rolled oats (CFA)

Ref No 4 03 302 United States

Feed Name or Analyses		Mean As Fed	Mean Dry	C.V. ± %
Dry matter	%	90.8	100.0	1
Ash	%	2.3	2.5	15
Crude fiber	%	2.8	3.1	26
Sheep	dig coef %	* 80.	80.	
Ether extract	%	5.9	6.5	12
Sheep	dig coef %	* 96.	96.	
N-free extract	%	63.6	70.1	
Sheep	dig coef %	* 98.	98.	
Protein (N x 6.25)	%	16.1	17.7	10
Sheep	dig coef %	* 90.	90.	
Cattle	dig prot %	* 11.2	12.3	
Goats	dig prot %	* 12.2	13.5	
Horses	dig prot %	* 12.2	13.5	
Sheep	dig prot %	14.5	16.0	
Pentosans	%	4.6	5.1	19
Energy	GE Mcal/kg			
Cattle	DE Mcal/kg	* 4.00	4.40	
Sheep	DE Mcal/kg	* 4.05	4.46	
Swine	DE kcal/kg	*3649.	4021.	
Cattle	ME Mcal/kg	* 3.28	3.61	
Chickens	ME_n kcal/kg	3547.	3909.	
Sheep	ME Mcal/kg	* 3.32	3.66	
Swine	ME kcal/kg	*3372.	3716.	
Cattle	TDN %	* 90.6	99.8	
Sheep	TDN %	91.9	101.2	
Swine	TDN %	* 82.8	91.2	
Calcium	%	0.07	0.08	52
Cobalt	mg/kg	0.044	0.049	
Copper	mg/kg	3.5	3.9	
Iron	%	0.006	0.007	8
Manganese	mg/kg	41.2	45.5	18
Phosphorus	%	0.45	0.50	21
Choline	mg/kg	1113.	1227.	14
Folic acid	mg/kg	0.60	0.66	14
Niacin	mg/kg	11.7	12.9	46
Pantothenic acid	mg/kg	12.9	14.2	21
Riboflavin	mg/kg	1.7	1.8	30
Thiamine	mg/kg	7.1	7.8	31
Arginine	%	1.00	1.10	13
Cystine	%	0.26	0.29	
Glycine	%	0.60	0.66	
Histidine	%	0.30	0.33	15
Isoleucine	%	0.50	0.55	26
Leucine	%	1.10	1.21	5
Lysine	%	0.60	0.66	15
Methionine	%	0.20	0.22	
Phenylalanine	%	0.70	0.77	13
Threonine	%	0.50	0.55	9
Tryptophan	%	0.20	0.22	
Tyrosine	%	0.70	0.77	8
Valine	%	0.70	0.77	25

Oats, cereal by-product, mx 4% fiber, (4)

Feeding oat meal (AAFCO)

Oat middlings (CFA)

Oat meal

Ref No 4 03 303 United States

Feed Name or Analyses		Mean As Fed	Mean Dry	C.V. ± %
Dry matter	%	91.2	100.0	1
Ash	%	2.4	2.6	16
Crude fiber	%	2.9	3.2	26
Sheep	dig coef %	49.	49.	
Ether extract	%	5.9	6.4	19
Sheep	dig coef %	93.	93.	
N-free extract	%	64.0	70.2	
Sheep	dig coef %	95.	95.	
Protein (N x 6.25)	%	16.0	17.5	10
Sheep	dig coef %	80.	80.	
Cattle	dig prot %	* 11.0	12.1	
Goats	dig prot %	* 12.1	13.3	
Horses	dig prot %	* 12.1	13.3	
Sheep	dig prot %	12.8	14.0	
Linoleic	%	2.000	2.193	
Energy	GE Mcal/kg			
Cattle	DE Mcal/kg	* 3.98	4.36	
Sheep	DE Mcal/kg	* 3.85	4.22	
Swine	DE kcal/kg	*3642.	3995.	
Cattle	ME Mcal/kg	* 3.26	3.58	
Chickens	ME_n kcal/kg	3120.	3421.	
Sheep	ME Mcal/kg	* 3.16	3.46	
Swine	ME kcal/kg	*3367.	3694.	
Cattle	NE_m Mcal/kg	* 2.12	2.32	
Cattle	NE_{gain} Mcal/kg	* 1.37	1.50	
Cattle	$NE_{lactating\ cows}$ Mcal/kg	* 2.41	2.64	
Cattle	TDN %	* 90.2	98.9	
Sheep	TDN %	87.3	95.8	
Swine	TDN %	* 82.6	90.6	
Calcium	%	0.08	0.08	10
Chlorine	%	0.02	0.02	
Copper	mg/kg	5.3	5.8	
Iron	%	0.028	0.030	69
Magnesium	%	0.16	0.18	
Manganese	mg/kg	43.1	47.3	25
Phosphorus	%	0.47	0.51	8
Potassium	%	0.50	0.55	
Sodium	%	0.14	0.15	
Sulphur	%	0.26	0.29	
Niacin	mg/kg	28.2	30.9	18
Pantothenic acid	mg/kg	23.1	25.4	8
Riboflavin	mg/kg	1.8	1.9	15
Thiamine	mg/kg	7.0	7.7	19
α-tocopherol	mg/kg	24.0	26.3	
Arginine	%	0.70	0.77	
Histidine	%	0.30	0.33	
Lysine	%	0.10	0.11	
Tyrosine	%	0.90	0.99	

Ref No 4 03 303 Canada

Feed Name or Analyses		Mean As Fed	Mean Dry	C.V. ± %
Dry matter	%	90.0	100.0	0
Crude fiber	%	5.3	5.9	31
Sheep	dig coef %	49.	49.	
Ether extract	%	7.1	7.8	6
Sheep	dig coef %	93.	93.	
N-free extract	%			
Sheep	dig coef %	95.	95.	
Protein (N x 6.25)	%	13.6	15.1	5
Sheep	dig coef %	80.	80.	

(1) dry forages and roughages (2) pasture, range plants, and forages fed green (3) silages (4) energy feeds (5) protein supplements (6) minerals (7) vitamins (8) additives

Feed Name or Analyses		Mean As Fed	Mean Dry	C.V. ± %
Cattle	dig prot %	* 8.9	9.8	
Goats	dig prot %	* 9.9	11.0	
Horses	dig prot %	* 9.9	11.0	
Sheep	dig prot %	10.8	12.0	
Energy	GE Mcal/kg			
Sheep	DE Mcal/kg	* 3.82	4.25	
Sheep	ME Mcal/kg	* 3.13	3.48	
Sheep	TDN %	86.7	96.3	

Oats, cereal by-product, mx 7% fiber, (4)

Oat shorts (CFA)

Ref No 4 03 304 United States

Feed Name or Analyses		Mean As Fed	Mean Dry	C.V. ± %
Dry matter	%	91.2	100.0	
Ash	%	4.8	5.3	
Crude fiber	%	13.5	14.8	
Cattle	dig coef %	29.	29.	
Sheep	dig coef %	31.	31.	
Swine	dig coef %	1.	1.	
Ether extract	%	5.6	6.1	
Cattle	dig coef %	74.	74.	
Sheep	dig coef %	85.	85.	
Swine	dig coef %	88.	88.	
N-free extract	%	54.5	59.8	
Cattle	dig coef %	61.	61.	
Sheep	dig coef %	66.	66.	
Swine	dig coef %	66.	66.	
Protein (N x 6.25)	%	12.8	14.1	
Cattle	dig coef %	66.	66.	
Sheep	dig coef %	73.	73.	
Swine	dig coef %	72.	72.	
Cattle	dig prot %	8.5	9.3	
Goats	dig prot %	* 9.2	10.1	
Horses	dig prot %	* 9.2	10.1	
Sheep	dig prot %	9.3	10.2	
Swine	dig prot %	9.2	10.1	
Linoleic	%	.700	.768	
Energy	GE Mcal/kg			
Cattle	DE Mcal/kg	* 2.42	2.65	
Sheep	DE Mcal/kg	* 2.64	2.90	
Swine	DE kcal/kg	*2485.	2725.	
Cattle	ME Mcal/kg	* 1.98	2.18	
Sheep	ME Mcal/kg	* 2.17	2.38	
Swine	ME kcal/kg	*2315.	2539.	
Cattle	TDN %	54.9	60.2	
Sheep	TDN %	60.0	65.8	
Swine	TDN %	56.4	61.8	

Oats, flour, coarse bolt, (4)

Ref No 4 03 306 United States

Feed Name or Analyses		Mean As Fed	Mean Dry	C.V. ± %
Dry matter	%	91.0	100.0	1
Ash	%	1.9	2.1	14
Crude fiber	%	1.3	1.4	33
Ether extract	%	6.1	6.7	7
N-free extract	%	64.7	71.1	
Protein (N x 6.25)	%	17.0	18.7	5
Cattle	dig prot %	* 12.0	13.2	
Goats	dig prot %	* 13.1	14.4	
Horses	dig prot %	* 13.1	14.4	
Sheep	dig prot %	* 13.1	14.4	
Energy	GE Mcal/kg			
Cattle	DE Mcal/kg	* 3.33	3.66	
Sheep	DE Mcal/kg	* 3.38	3.71	
Swine	DE kcal/kg	*3877.	4261.	
Cattle	ME Mcal/kg	* 2.73	3.00	
Sheep	ME Mcal/kg	* 2.77	3.05	
Swine	ME kcal/kg	*3576.	3929.	

Feed Name or Analyses		Mean As Fed	Mean Dry	C.V. ± %
Cattle	TDN %	* 75.6	83.1	
Sheep	TDN %	* 76.7	84.3	
Swine	TDN %	* 87.9	96.6	
Niacin	mg/kg	6.8	7.5	
Arginine	%	0.90	0.99	18
Histidine	%	0.30	0.33	
Isoleucine	%	0.80	0.88	13
Leucine	%	1.20	1.32	
Lysine	%	0.80	0.88	7
Methionine	%	0.20	0.22	
Phenylalanine	%	0.70	0.77	8
Threonine	%	0.50	0.55	
Tryptophan	%	0.20	0.22	
Tyrosine	%	0.60	0.66	13
Valine	%	0.70	0.77	15

Oats, grain, (4)

Ref No 4 03 309 United States

Feed Name or Analyses		Mean As Fed	Mean Dry	C.V. ± %
Dry matter	%	88.9	100.0	3
Ash	%	3.4	3.8	10
Crude fiber	%	10.6	11.9	12
Cattle	dig coef %	* 24.	24.	
Horses	dig coef %	27.	27.	
Sheep	dig coef %	35.	35.	
Swine	dig coef %	-1.	-1.	
Ether extract	%	4.5	5.1	16
Cattle	dig coef %	* 83.	83.	
Horses	dig coef %	73.	73.	
Sheep	dig coef %	81.	81.	
Swine	dig coef %	79.	79.	
N-free extract	%	58.7	66.0	2
Cattle	dig coef %	* 79.	79.	
Horses	dig coef %	77.	77.	
Sheep	dig coef %	81.	81.	
Swine	dig coef %	76.	76.	
Protein (N x 6.25)	%	11.7	13.2	6
Cattle	dig coef %	* 75.	75.	
Horses	dig coef %	78.	78.	
Sheep	dig coef %	78.	78.	
Swine	dig coef %	81.	81.	
Cattle	dig prot %	8.7	9.8	
Goats	dig prot %	* 8.3	9.3	
Horses	dig prot %	9.1	10.3	
Sheep	dig prot %	9.1	10.3	
Swine	dig prot %	9.5	10.7	
Energy	GE Mcal/kg			
Cattle	DE Mcal/kg	* 2.92	3.29	
Horses	DE Mcal/kg	* 2.83	3.19	
Sheep	DE Mcal/kg	* 3.01	3.38	
Swine	DE kcal/kg	*2731.	3072.	
Cattle	ME Mcal/kg	* 2.40	2.70	
Chickens	ME_n kcal/kg	2498.	2810.	
Horses	ME Mcal/kg	* 2.32	2.61	
Sheep	ME Mcal/kg	* 2.47	2.78	
Swine	ME kcal/kg	*2549.	2868.	
Cattle	NE_m Mcal/kg	* 1.54	1.73	
Cattle	NE_{gain} Mcal/kg	* 1.01	1.14	
Cattle	$NE_{lactating\ cows}$ Mcal/kg	* 1.81	2.04	
Cattle	TDN %	66.3	74.5	
Horses	TDN %	64.3	72.3	
Sheep	TDN %	68.2	76.8	
Swine	TDN %	61.9	69.7	
Calcium	%	0.09	0.10	
Chlorine	%	0.11	0.12	31
Cobalt	mg/kg	0.057	0.064	
Copper	mg/kg	8.3	9.3	
Iron	%	0.007	0.008	
Magnesium	%	0.16	0.18	

Feed Name or Analyses		Mean As Fed	Mean Dry	C.V. ± %
Manganese	mg/kg	43.2	48.6	
Phosphorus	%	0.33	0.37	
Potassium	%	0.42	0.48	
Sodium	%	0.08	0.08	49
Sulphur	%	0.21	0.23	20
Choline	mg/kg	945.	1063.	
Niacin	mg/kg	13.7	15.4	
Pantothenic acid	mg/kg	13.0	14.7	
Riboflavin	mg/kg	1.1	1.2	
Arginine	%	0.57	0.64	
Cystine	%	0.15	0.17	
Glycine	%	0.49	0.55	
Histidine	%	0.09	0.10	
Lysine	%	0.34	0.38	
Methionine	%	0.18	0.20	
Tryptophan	%	0.12	0.14	
Tyrosine	%	1.07	1.20	

Ref No 4 03 309 Canada

Feed Name or Analyses		Mean As Fed	Mean Dry	C.V. ± %
Dry matter	%	90.0	100.0	
Ash	%	2.4	2.7	
Crude fiber	%	10.0	11.1	
Horses	dig coef %	27.	27.	
Sheep	dig coef %	35.	35.	
Swine	dig coef %	-1.	-1.	
Ether extract	%	5.6	6.2	
Horses	dig coef %	73.	73.	
Sheep	dig coef %	81.	81.	
Swine	dig coef %	79.	79.	
N-free extract	%	60.6	67.4	
Horses	dig coef %	77.	77.	
Sheep	dig coef %	81.	81.	
Swine	dig coef %	76.	76.	
Protein (N x 6.25)	%	11.4	12.6	
Horses	dig coef %	78.	78.	
Sheep	dig coef %	78.	78.	
Swine	dig coef %	81.	81.	
Cattle	dig prot %	* 6.9	7.6	
Goats	dig prot %	* 7.9	8.8	
Horses	dig prot %	8.9	9.9	
Sheep	dig prot %	8.9	9.9	
Swine	dig prot %	9.2	10.2	
Energy	GE Mcal/kg			
Cattle	DE Mcal/kg	* 2.97	3.30	
Horses	DE Mcal/kg	* 2.96	3.29	
Sheep	DE Mcal/kg	* 3.15	3.49	
Swine	DE kcal/kg	*2870.	3189.	
Cattle	ME Mcal/kg	* 2.43	2.70	
Chickens	ME_n kcal/kg	2529.	2810.	
Horses	ME Mcal/kg	* 2.43	2.70	
Sheep	ME Mcal/kg	* 2.58	2.87	
Swine	ME kcal/kg	*2682.	2980.	
Cattle	TDN %	67.3	74.8	
Horses	TDN %	67.1	74.6	
Sheep	TDN %	71.3	79.3	
Swine	TDN %	65.1	72.3	
Calcium	%	0.08	0.09	
Phosphorus	%	0.32	0.36	
Potassium	%	0.41	0.46	
Niacin	mg/kg	11.8	13.1	
Pantothenic acid	mg/kg	13.4	14.9	
Riboflavin	mg/kg	1.1	1.3	
Thiamine	mg/kg	5.1	5.6	
Tryptophan	%	0.17	0.19	

Feed Name or Analyses		Mean As Fed	Mean Dry	C.V. ± %

Oats, grain, grnd, (4)

Ref No 4 08 471 United States

Analyses		As Fed	Dry	C.V. ± %
Dry matter	%	89.8	100.0	
Ash	%	4.3	4.8	
Crude fiber	%	12.1	13.5	
Ether extract	%	4.1	4.6	
N-free extract	%	57.7	64.3	
Protein (N x 6.25)	%	11.6	12.9	
Cattle	dig prot %	* 7.1	7.9	
Goats	dig prot %	* 8.1	9.1	
Horses	dig prot %	* 8.1	9.1	
Sheep	dig prot %	* 8.1	9.1	
Cattle	DE Mcal/kg	* 3.27	3.64	
Sheep	DE Mcal/kg	* 3.02	3.36	
Swine	DE kcal/kg	*3092.	3443.	
Cattle	ME Mcal/kg	* 2.68	2.99	
Chickens	ME_n kcal/kg	2640.	2940.	
Sheep	ME Mcal/kg	* 2.48	2.76	
Swine	ME kcal/kg	*2889.	3216.	
Cattle	TDN %	* 74.2	82.6	
Sheep	TDN %	* 68.5	76.3	
Swine	TDN %	* 70.1	78.1	

Ref No 4 08 471 Canada

Analyses		As Fed	Dry	C.V. ± %
Dry matter	%	87.8	100.0	
Ash	%	2.9	3.3	
Crude fiber	%	14.5	16.5	
Ether extract	%	2.6	2.9	
N-free extract	%	55.2	62.9	
Protein (N x 6.25)	%	12.6	14.4	
Cattle	dig prot %	* 8.1	9.2	
Goats	dig prot %	* 9.1	10.4	
Horses	dig prot %	* 9.1	10.4	
Sheep	dig prot %	* 9.1	10.4	
Energy	GE Mcal/kg			
Cattle	DE Mcal/kg	* 2.94	3.35	
Sheep	DE Mcal/kg	* 2.87	3.27	
Swine	DE kcal/kg	*3012.	3430.	
Cattle	ME Mcal/kg	* 2.41	2.75	
Sheep	ME Mcal/kg	* 2.35	2.68	
Swine	ME kcal/kg	*2803.	3193.	
Cattle	TDN %	* 66.7	76.0	
Sheep	TDN %	* 65.1	74.1	
Swine	TDN %	* 68.3	77.8	

Oats, grain, rolled, (4)

Ref No 4 03 307 United States

Analyses		As Fed	Dry	C.V. ± %
Dry matter	%	90.7	100.0	1
Ash	%	2.1	2.3	17
Crude fiber	%	2.2	2.5	39
Sheep	dig coef %	80.	80.	
Ether extract	%	5.8	6.5	18
Sheep	dig coef %	96.	96.	
N-free extract	%	64.1	70.7	
Sheep	dig coef %	98.	98.	
Protein (N x 6.25)	%	16.4	18.1	10
Sheep	dig coef %	90.	90.	
Cattle	dig prot %	* 11.4	12.6	
Goats	dig prot %	* 12.5	13.8	
Horses	dig prot %	* 12.5	13.8	
Sheep	dig prot %	14.7	16.2	
Energy	GE Mcal/kg	4.30	4.74	
Cattle	DE Mcal/kg	* 3.90	4.30	
Sheep	DE Mcal/kg	* 4.06	4.47	
Swine	DE kcal/kg	3826.	4220.	
Cattle	ME Mcal/kg	* 3.20	3.53	
Chickens	ME_n kcal/kg	3212.	3542.	
Sheep	ME Mcal/kg	* 3.33	3.67	
Swine	ME kcal/kg	3617.	3990.	
Cattle	TDN %	* 88.5	97.6	
Sheep	TDN %	92.0	101.5	
Swine	TDN %	* 84.6	93.3	
Calcium	%	0.08	0.09	60
Chlorine	%	0.09	0.10	
Cobalt	mg/kg	0.042	0.046	
Copper	mg/kg	5.8	6.4	34
Iron	%	0.005	0.006	18
Magnesium	%	0.17	0.19	
Manganese	mg/kg	31.4	34.6	56
Phosphorus	%	0.43	0.47	32
Potassium	%	0.25	0.28	
Sodium	%	0.15	0.16	
Sulphur	%	0.24	0.26	
Choline	mg/kg	1149.	1268.	19
Folic acid	mg/kg	0.40	0.44	34
Niacin	mg/kg	9.8	10.8	39
Pantothenic acid	mg/kg	13.4	14.8	35
Riboflavin	mg/kg	1.8	2.0	64
Thiamine	mg/kg	6.4	7.1	16
Vitamin B_6	mg/kg	1.20	1.32	35
Alanine	%	0.36	0.40	
Arginine	%	0.91	1.00	16
Aspartic acid	%	0.63	0.70	
Cystine	%	0.18	0.20	20
Glutamic acid	%	2.81	3.10	
Glycine	%	0.18	0.20	
Histidine	%	0.27	0.30	38
Isoleucine	%	0.54	0.60	15
Leucine	%	0.91	1.00	10
Lysine	%	0.45	0.50	27
Methionine	%	0.18	0.20	28
Phenylalanine	%	0.54	0.60	15
Threonine	%	0.45	0.50	12
Tryptophan	%	0.18	0.20	27
Tyrosine	%	0.54	0.60	20
Valine	%	0.63	0.70	17

Oats, grain, thresher-run, mn wt 34 lb per bushel mx 10% foreign material, (4)

Ref No 4 08 162 Canada

Analyses		As Fed	Dry	C.V. ± %
Dry matter	%		100.0	
Sheep	dig coef %		72.	
Ash	%		3.8	
Crude fiber	%		11.6	
Ether extract	%		2.7	
N-free extract	%		68.1	
Protein (N x 6.25)	%		13.8	
Cattle	dig prot %	*	8.7	
Goats	dig prot %	*	9.9	
Horses	dig prot %	*	9.9	
Sheep	dig prot %	*	9.9	
Energy	GE Mcal/kg		4.55	
Cattle	DE Mcal/kg	*	3.31	
Sheep	DE Mcal/kg	*	3.40	
Swine	DE kcal/kg	*	2877.	
Cattle	ME Mcal/kg	*	2.71	
Sheep	ME Mcal/kg	*	2.79	
Swine	ME kcal/kg	*	2681.	
Cattle	TDN %	*	75.0	
Sheep	TDN %	*	77.2	
Swine	TDN %	*	65.2	

Oats, grain, thresher-run, mx wt 34 lb per bushel mn 10% mx 20% foreign material, (4)

Ref No 4 08 161 Canada

Analyses		As Fed	Dry	C.V. ± %
Dry matter	%	90.0	100.0	0
Ash	%	3.5	3.9	6
Crude fiber	%	10.2	11.3	9
Ether extract	%	1.7	1.9	31
N-free extract	%	63.2	70.2	1
Protein (N x 6.25)	%	11.4	12.7	8
Cattle	dig prot %	* 6.9	7.7	
Goats	dig prot %	* 8.0	8.9	
Horses	dig prot %	* 8.0	8.9	
Sheep	dig prot %	* 8.0	8.9	
Energy	GE Mcal/kg	4.32	4.80	0
Cattle	DE Mcal/kg	* 3.10	3.44	
Sheep	DE Mcal/kg	* 3.07	3.41	
Swine	DE kcal/kg	*2577.	2863.	
Cattle	ME Mcal/kg	* 2.54	2.82	
Sheep	ME Mcal/kg	* 2.52	2.80	
Swine	ME kcal/kg	*2408.	2676.	
Cattle	TDN %	* 70.2	78.0	
Sheep	TDN %	* 69.7	77.4	
Swine	TDN %	* 58.4	64.9	

Oats, grain, thresher-run, mx wt 34 lb per bushel mx 10% foreign material, (4)

Ref No 4 08 160 Canada

Analyses		As Fed	Dry	C.V. ± %
Dry matter	%	90.0	100.0	0
Ash	%	3.4	3.7	8
Crude fiber	%	10.3	11.5	9
Ether extract	%	1.6	1.8	23
N-free extract	%	63.3	70.4	2
Protein (N x 6.25)	%	11.4	12.7	13
Cattle	dig prot %	* 6.9	7.6	
Goats	dig prot %	* 8.0	8.8	
Horses	dig prot %	* 8.0	8.8	
Sheep	dig prot %	* 8.0	8.8	
Energy	GE Mcal/kg	4.41	4.90	5
Cattle	DE Mcal/kg	* 3.10	3.44	
Sheep	DE Mcal/kg	* 3.07	3.42	
Swine	DE kcal/kg	*2583.	2870.	
Cattle	ME Mcal/kg	* 2.54	2.82	
Sheep	ME Mcal/kg	* 2.52	2.80	
Swine	ME kcal/kg	*2414.	2682.	
Cattle	TDN %	* 70.2	78.0	
Sheep	TDN %	* 69.7	77.5	
Swine	TDN %	* 58.6	65.1	

Oats, grain, germinated, (4)

Ref No 4 03 308 United States

Analyses		As Fed	Dry	C.V. ± %
Dry matter	%		100.0	
Biotin	mg/kg		1.41	
Niacin	mg/kg		44.1	
Pantothenic acid	mg/kg		21.8	
Riboflavin	mg/kg		11.7	
Thiamine	mg/kg		12.1	
Vitamin B_6	mg/kg		1.76	

(1) dry forages and roughages (2) pasture, range plants, and forages fed green (3) silages (4) energy feeds (5) protein supplements (6) minerals (7) vitamins (8) additives

Oats, grain, Can extra 1 feed mn wt 38 lb per bushel, (4)

Ref No 4 08 181 Canada

Feed Name or Analyses		Mean As Fed	Mean Dry	C.V. ± %
Dry matter	%	88.3	100.0	2
Ash	%	3.0	3.4	8
Crude fiber	%	10.5	11.9	9
Ether extract	%	4.9	5.6	10
N-free extract	%	59.0	66.8	
Protein (N x 6.25)	%	10.9	12.3	13
Cattle	dig prot %	* 6.5	7.3	
Goats	dig prot %	* 7.5	8.5	
Horses	dig prot %	* 7.5	8.5	
Sheep	dig prot %	* 7.5	8.5	
Energy	GE Mcal/kg			
Cattle	DE Mcal/kg	* 3.28	3.71	
Sheep	DE Mcal/kg	* 3.07	3.47	
Swine	DE kcal/kg	*2951.	3342.	
Cattle	ME Mcal/kg	* 2.69	3.04	
Sheep	ME Mcal/kg	* 2.52	2.85	
Swine	ME kcal/kg	*2759.	3125.	
Cattle	TDN %	* 74.3	84.2	
Sheep	TDN %	* 69.6	78.8	
Swine	TDN %	* 66.9	75.8	
Calcium	%	0.07	0.08	15
Phosphorus	%	0.34	0.39	15

Oats, grain, Can extra 3 CW mn wt 40 lb per bushel, (4)

Ref No 4 08 182 Canada

Feed Name or Analyses		Mean As Fed	Mean Dry	C.V. ± %
Dry matter	%	86.5	100.0	0
Ash	%	2.9	3.4	5
Crude fiber	%	10.5	12.1	10
Ether extract	%	4.7	5.4	16
N-free extract	%	57.3	66.3	
Protein (N x 6.25)	%	11.1	12.8	5
Cattle	dig prot %	* 6.7	7.8	
Goats	dig prot %	* 7.8	9.0	
Horses	dig prot %	* 7.8	9.0	
Sheep	dig prot %	* 7.8	9.0	
Energy	GE Mcal/kg			
Cattle	DE Mcal/kg	* 3.13	3.62	
Sheep	DE Mcal/kg	* 2.99	3.46	
Swine	DE kcal/kg	*2880.	3329.	
Cattle	ME Mcal/kg	* 2.57	2.97	
Sheep	ME Mcal/kg	* 2.45	2.83	
Swine	ME kcal/kg	*2689.	3109.	
Cattle	TDN %	* 71.1	82.2	
Sheep	TDN %	* 67.8	78.4	
Swine	TDN %	* 65.3	75.5	
Calcium	%	0.04	0.05	
Phosphorus	%	0.31	0.36	

Oats, grain, Can 1 CW mn wt 38 lb per bushel mx 1% foreign material, (4)

Ref No 4 09 076 Canada

Feed Name or Analyses		Mean As Fed	Mean Dry	C.V. ± %
Dry matter	%	90.0	100.0	
Ether extract	%	5.2	5.8	
Protein (N x 6.25)	%	11.8	13.1	
Cattle	dig prot %	* 7.3	8.1	
Goats	dig prot %	* 8.3	9.2	
Horses	dig prot %	* 8.3	9.2	
Sheep	dig prot %	* 8.3	9.2	
Calcium	%	0.08	0.09	
Phosphorus	%	0.28	0.31	

Feed Name or Analyses		Mean As Fed	Mean Dry	C.V. ± %
Pantothenic acid	mg/kg	6.3	7.0	
Riboflavin	mg/kg	0.5	0.6	
Lysine	%	0.40	0.44	
Methionine	%	0.20	0.22	
Tryptophan	%	0.20	0.22	

Oats, grain, gr 1 heavy US mn wt 36 lb per bushel mx 2% foreign material, (4)

Ref No 4 03 312 United States

Feed Name or Analyses		Mean As Fed	Mean Dry	C.V. ± %
Dry matter	%	89.0	100.0	
Ash	%	2.7	3.0	
Crude fiber	%	8.9	10.0	
Cattle	dig coef %	* 30.	30.	
Sheep	dig coef %	24.	24.	
Swine	dig coef %	* -16.	-16.	
Ether extract	%	5.2	5.8	
Cattle	dig coef %	* 82.	82.	
Sheep	dig coef %	92.	92.	
Swine	dig coef %	* 82.	82.	
N-free extract	%	59.6	67.0	
Cattle	dig coef %	* 81.	81.	
Sheep	dig coef %	81.	81.	
Swine	dig coef %	* 78.	78.	
Protein (N x 6.25)	%	12.6	14.2	
Cattle	dig coef %	* 75.	75.	
Sheep	dig coef %	88.	88.	
Swine	dig coef %	* 84.	84.	
Cattle	dig prot %	9.5	10.7	
Goats	dig prot %	* 9.1	10.2	
Horses	dig prot %	* 9.1	10.2	
Sheep	dig prot %	11.1	12.5	
Swine	dig prot %	10.6	11.9	
Energy	GE Mcal/kg			
Cattle	DE Mcal/kg	* 3.09	3.47	
Sheep	DE Mcal/kg	* 3.19	3.58	
Swine	DE kcal/kg	*2872.	3227.	
Cattle	ME Mcal/kg	* 2.53	2.84	
Sheep	ME Mcal/kg	* 2.61	2.93	
Swine	ME kcal/kg	*2675.	3005.	
Cattle	TDN %	70.0	78.6	
Sheep	TDN %	72.2	81.2	
Swine	TDN %	65.1	73.2	

Oats, grain, gr 1 US mn wt 34 lb per bushel mx 2% foreign material, (4)

Ref No 4 03 313 United States

Feed Name or Analyses		Mean As Fed	Mean Dry	C.V. ± %
Dry matter	%	90.2	100.0	3
Crude fiber	%			
Cattle	dig coef %	* 30.	30.	
Sheep	dig coef %	* 37.	37.	
Ether extract	%			
Cattle	dig coef %	* 82.	82.	
Sheep	dig coef %	* 82.	82.	
N-free extract	%			
Cattle	dig coef %	* 81.	81.	
Sheep	dig coef %	* 83.	83.	
Protein (N x 6.25)	%			
Cattle	dig coef %	* 75.	75.	
Sheep	dig coef %	* 78.	78.	
Starch	%	37.2	41.2	7
Sugars, total	%	1.4	1.5	27
Linoleic	%	1.500	1.663	
Energy	GE Mcal/kg			
Chickens	ME_n kcal/kg	2620.	2905.	

Feed Name or Analyses		Mean As Fed	Mean Dry	C.V. ± %
Calcium	%	0.08	0.09	
Phosphorus	%	0.30	0.33	

Oats, grain, gr 2 heavy US mn wt 36 lb per bushel mx 3% foreign material, (4)

Oats, grain, heavy

Ref No 4 03 315 United States

Feed Name or Analyses		Mean As Fed	Mean Dry	C.V. ± %
Dry matter	%	89.7	100.0	1
Ash	%	3.2	3.6	8
Crude fiber	%	9.4	10.5	13
Cattle	dig coef %	* 30.	30.	
Sheep	dig coef %	* 37.	37.	
Ether extract	%	4.3	4.8	6
Cattle	dig coef %	* 82.	82.	
Sheep	dig coef %	* 82.	82.	
N-free extract	%	60.1	67.1	
Cattle	dig coef %	* 81.	81.	
Sheep	dig coef %	* 83.	83.	
Protein (N x 6.25)	%	12.7	14.2	3
Cattle	dig coef %	* 75.	75.	
Sheep	dig coef %	* 78.	78.	
Cattle	dig prot %	9.5	10.6	
Goats	dig prot %	* 9.1	10.2	
Horses	dig prot %	* 9.1	10.2	
Sheep	dig prot %	9.9	11.0	
Energy	GE Mcal/kg			
Cattle	DE Mcal/kg	* 3.04	3.39	
Sheep	DE Mcal/kg	* 3.14	3.50	
Swine	DE kcal/kg	*2736.	3052.	
Cattle	ME Mcal/kg	* 2.49	2.78	
Sheep	ME Mcal/kg	* 2.57	2.87	
Swine	ME kcal/kg	*2548.	2842.	
Cattle	TDN %	* 68.9	76.8	
Sheep	TDN %	* 71.1	79.3	
Swine	TDN %	* 62.0	69.2	
α-tocopherol	mg/kg	20.0	22.3	

Oats, grain, gr 2 US mn wt 32 lb per bushel mx 3% foreign material, (4)

Ref No 4 03 316 United States

Feed Name or Analyses		Mean As Fed	Mean Dry	C.V. ± %
Dry matter	%	88.8	100.0	2
Ash	%	3.3	3.7	13
Crude fiber	%	10.8	12.2	13
Cattle	dig coef %	* 30.	30.	
Sheep	dig coef %	* 37.	37.	
Ether extract	%	4.2	4.7	11
Cattle	dig coef %	* 82.	82.	
Sheep	dig coef %	* 82.	82.	
N-free extract	%	59.1	66.6	
Cattle	dig coef %	* 81.	81.	
Sheep	dig coef %	* 83.	83.	
Protein (N x 6.25)	%	11.4	12.8	9
Cattle	dig coef %	* 75.	75.	
Sheep	dig coef %	* 78.	78.	
Cattle	dig prot %	8.5	9.6	
Goats	dig prot %	* 8.0	9.0	
Horses	dig prot %	* 8.0	9.0	
Sheep	dig prot %	8.9	10.0	
Energy	GE Mcal/kg			
Cattle	DE Mcal/kg	* 2.97	3.35	
Sheep	DE Mcal/kg	* 3.07	3.46	
Swine	DE kcal/kg	*2517.	2834.	
Cattle	ME Mcal/kg	* 2.44	2.74	
Sheep	ME Mcal/kg	* 2.52	2.84	
Swine	ME kcal/kg	*2351.	2648.	

Continued

Feed Name or Analyses			Mean As Fed	Mean Dry	C.V. ± %
Cattle	TDN %		67.4	75.9	
Sheep	TDN %		69.7	78.4	
Swine	TDN %	*	57.1	64.3	
Calcium	%		0.06	0.07	
Phosphorus	%		0.27	0.30	

Oats, grain, gr 3 US mn wt 30 lb per bushel mx 4% foreign material, (4)

Ref No 4 03 317 — United States

Feed Name or Analyses			Mean As Fed	Mean Dry	C.V. ± %
Dry matter	%		91.4	100.0	
Ash	%		3.7	4.1	
Crude fiber	%		12.8	14.0	
Cattle	dig coef %	*	30.	30.	
Sheep	dig coef %	*	37.	37.	
Ether extract	%		4.6	5.0	
Cattle	dig coef %	*	82.	82.	
Sheep	dig coef %	*	82.	82.	
N-free extract	%		58.2	63.7	
Cattle	dig coef %	*	81.	81.	
Sheep	dig coef %	*	83.	83.	
Protein (N x 6.25)	%		12.1	13.2	
Cattle	dig coef %	*	75.	75.	
Sheep	dig coef %	*	78.	78.	
Cattle	dig prot %		9.0	9.9	
Goats	dig prot %	*	8.5	9.3	
Horses	dig prot %	*	8.5	9.3	
Sheep	dig prot %		9.4	10.3	
Energy	GE Mcal/kg				
Cattle	DE Mcal/kg	*	3.02	3.30	
Sheep	DE Mcal/kg	*	3.13	3.42	
Swine	DE kcal/kg	*	2387.	2612.	
Cattle	ME Mcal/kg	*	2.48	2.71	
Sheep	ME Mcal/kg	*	2.56	2.80	
Swine	ME kcal/kg	*	2228.	2438.	
Cattle	TDN %	*	68.5	74.9	
Sheep	TDN %	*	70.9	77.6	
Swine	TDN %	*	54.1	59.2	

Oats, grain, gr 4 US mn wt 27 lb per bushel mx 5% foreign material, (4)

Oats, grain, light

Ref No 4 03 318 — United States

Feed Name or Analyses			Mean As Fed	Mean Dry	C.V. ± %
Dry matter	%		90.0	100.0	2
Ash	%		4.4	4.9	11
Crude fiber	%		17.2	19.1	21
Cattle	dig coef %	*	30.	30.	
Swine	dig coef %		6.	6.	
Ether extract	%		3.6	4.0	26
Cattle	dig coef %	*	82.	82.	
Swine	dig coef %		35.	35.	
N-free extract	%		54.3	60.3	
Cattle	dig coef %	*	81.	81.	
Swine	dig coef %		27.	27.	
Protein (N x 6.25)	%		10.5	11.7	13
Cattle	dig coef %	*	75.	75.	
Swine	dig coef %		66.	66.	
Cattle	dig prot %		7.9	8.8	
Goats	dig prot %	*	7.2	7.9	
Horses	dig prot %	*	7.2	7.9	
Sheep	dig prot %	*	7.2	7.9	
Swine	dig prot %		6.9	7.7	

(1) dry forages and roughages (2) pasture, range plants, and forages fed green (3) silages (4) energy feeds (5) protein supplements (6) minerals (7) vitamins (8) additives

Feed Name or Analyses			Mean As Fed	Mean Dry	C.V. ± %
Cellulose (Matrone)	%		24.0	26.6	
Lignin (Ellis)	%		9.3	10.3	
Energy	GE Mcal/kg				
Cattle	DE Mcal/kg	*	2.81	3.12	
Sheep	DE Mcal/kg	*	2.89	3.21	
Swine	DE kcal/kg	*	1122.	1246.	
Cattle	ME Mcal/kg	*	2.30	2.55	
Sheep	ME Mcal/kg	*	2.37	2.63	
Swine	ME kcal/kg	*	1051.	1167.	
Cattle	TDN %		63.6	70.7	
Sheep	TDN %	*	65.5	72.8	
Swine	TDN %		25.4	28.3	

Oats, grain, high fat, (4)

Ref No 4 03 319 — United States

Feed Name or Analyses			Mean As Fed	Mean Dry	C.V. ± %
Dry matter	%		93.1	100.0	
Ash	%		3.2	3.4	
Crude fiber	%		12.4	13.3	
Swine	dig coef %		22.	22.	
Ether extract	%		7.0	7.5	
Swine	dig coef %		86.	86.	
N-free extract	%		58.2	62.5	
Swine	dig coef %		78.	78.	
Protein (N x 6.25)	%		12.4	13.3	
Swine	dig coef %		78.	78.	
Cattle	dig prot %	*	7.7	8.2	
Goats	dig prot %	*	8.8	9.4	
Horses	dig prot %	*	8.8	9.4	
Sheep	dig prot %	*	8.8	9.4	
Swine	dig prot %		9.7	10.4	
Energy	GE Mcal/kg				
Cattle	DE Mcal/kg	*	3.44	3.69	
Sheep	DE Mcal/kg	*	3.20	3.44	
Swine	DE kcal/kg	*	3143.	3376.	
Cattle	ME Mcal/kg	*	2.82	3.03	
Sheep	ME Mcal/kg	*	2.63	2.82	
Swine	ME kcal/kg	*	2933.	3150.	
Cattle	TDN %	*	78.0	83.8	
Sheep	TDN %	*	72.7	78.1	
Swine	TDN %		71.3	76.6	

Oats, grain, low fat, (4)

Ref No 4 03 320 — United States

Feed Name or Analyses			Mean As Fed	Mean Dry	C.V. ± %
Dry matter	%		89.1	100.0	
Ash	%		2.7	3.0	
Crude fiber	%		7.0	7.9	
Swine	dig coef %		-26.	-26.	
Ether extract	%		3.5	3.9	
Swine	dig coef %		81.	81.	
N-free extract	%		62.3	69.9	
Swine	dig coef %		78.	78.	
Protein (N x 6.25)	%		13.6	15.3	
Swine	dig coef %		85.	85.	
Cattle	dig prot %	*	9.0	10.1	
Goats	dig prot %	*	10.0	11.3	
Horses	dig prot %	*	10.0	11.3	
Sheep	dig prot %	*	10.0	11.3	
Swine	dig prot %		11.6	13.0	
Energy	GE Mcal/kg				
Cattle	DE Mcal/kg	*	3.06	3.43	
Sheep	DE Mcal/kg	*	3.14	3.52	
Swine	DE kcal/kg	*	2848.	3197.	
Cattle	ME Mcal/kg	*	2.50	2.81	
Sheep	ME Mcal/kg	*	2.57	2.89	
Swine	ME kcal/kg	*	2646.	2970.	
Cattle	TDN %	*	69.2	77.7	

Feed Name or Analyses			Mean As Fed	Mean Dry	C.V. ± %
Sheep	TDN %	*	71.2	79.9	
Swine	TDN %		64.6	72.5	

Oats, grain, Pacific coast, (4)

Ref No 4 07 999 — United States

Feed Name or Analyses			Mean As Fed	Mean Dry	C.V. ± %
Dry matter	%		91.2	100.0	
Ash	%		3.7	4.1	
Crude fiber	%		11.0	12.1	
Ether extract	%		5.4	5.9	
N-free extract	%		62.1	68.1	
Protein (N x 6.25)	%		9.0	9.9	
Cattle	dig prot %	*	4.6	5.1	
Goats	dig prot %	*	5.7	6.3	
Horses	dig prot %	*	5.7	6.3	
Sheep	dig prot %	*	5.7	6.3	
Linoleic	%		1.400	1.535	
Energy	GE Mcal/kg				
Cattle	DE Mcal/kg	*	3.80	4.17	
Sheep	DE Mcal/kg	*	3.19	3.50	
Swine	DE kcal/kg	*	2939.	3223.	
Cattle	ME Mcal/kg	*	3.12	3.42	
Chickens	ME_n kcal/kg		2611.	2863.	
Sheep	ME Mcal/kg	*	2.62	2.87	
Swine	ME kcal/kg	*	2763.	3030.	
Cattle	NE_m Mcal/kg	*	1.61	1.76	
Cattle	NE_{gain} Mcal/kg	*	1.06	1.16	
Cattle	$NE_{lactating\ cows}$ Mcal/kg	*	1.91	2.09	
Cattle	TDN %	*	86.2	94.5	
Sheep	TDN %	*	72.4	79.3	
Swine	TDN %	*	66.7	73.1	
Calcium	%		0.08	0.09	
Phosphorus	%		0.30	0.33	
Choline	mg/kg		959.	1052.	
Niacin	mg/kg		13.9	15.2	
Pantothenic acid	mg/kg		13.2	14.5	
Riboflavin	mg/kg		1.1	1.2	
a-tocopherol	mg/kg		20.0	21.9	
Arginine	%		0.60	0.66	
Cystine	%		0.17	0.19	
Glycine	%		0.40	0.44	
Lysine	%		0.40	0.44	
Methionine	%		0.13	0.14	
Tryptophan	%		0.12	0.13	

Oats, grain, gr sample US, (4)

Ref No 4 03 310 — United States

Feed Name or Analyses			Mean As Fed	Mean Dry	C.V. ± %
Dry matter	%		87.3	100.0	4
Ash	%		4.6	5.3	18
Crude fiber	%		12.0	13.7	14
Cattle	dig coef %	*	30.	30.	
Sheep	dig coef %	*	37.	37.	
Ether extract	%		4.2	4.8	9
Cattle	dig coef %	*	82.	82.	
Sheep	dig coef %	*	82.	82.	
N-free extract	%		55.6	63.7	
Cattle	dig coef %	*	81.	81.	
Sheep	dig coef %	*	83.	83.	
Protein (N x 6.25)	%		10.9	12.5	5
Cattle	dig coef %	*	75.	75.	
Sheep	dig coef %	*	78.	78.	
Cattle	dig prot %		8.2	9.4	
Goats	dig prot %	*	7.6	8.7	
Horses	dig prot %	*	7.6	8.7	
Sheep	dig prot %		8.5	9.8	
Energy	GE Mcal/kg				
Cattle	DE Mcal/kg	*	2.84	3.26	

Feed Name or Analyses			Mean As Fed	Mean Dry	C.V. ± %
Sheep	DE Mcal/kg	*	2.94	3.38	
Swine	DE kcal/kg	*	2571.	2945.	
Cattle	ME Mcal/kg	*	2.33	2.67	
Sheep	ME Mcal/kg	*	2.41	2.77	
Swine	ME kcal/kg	*	2403.	2751.	
Cattle	TDN %	*	64.5	73.9	
Sheep	TDN %	*	66.8	76.5	
Swine	TDN %	*	58.3	66.8	

Oats, grain screenings, (4)

Ref No 4 03 329 United States

Feed Name or Analyses			Mean As Fed	Mean Dry	C.V. ± %
Dry matter	%		90.2	100.0	
Ash	%		4.9	5.5	
Crude fiber	%		14.6	16.2	
Sheep	dig coef %		18.	18.	
Ether extract	%		4.3	4.7	
Sheep	dig coef %		92.	92.	
N-free extract	%		55.4	61.5	
Sheep	dig coef %		81.	81.	
Protein (N x 6.25)	%		10.9	12.1	
Sheep	dig coef %		88.	88.	
Cattle	dig prot %	*	6.4	7.1	
Goats	dig prot %	*	7.5	8.3	
Horses	dig prot %	*	7.5	8.3	
Sheep	dig prot %		9.6	10.6	
Energy	GE Mcal/kg				
Cattle	DE Mcal/kg	*	3.51	3.89	
Sheep	DE Mcal/kg	*	2.91	3.23	
Swine	DE kcal/kg	*	2394.	2654.	
Cattle	ME Mcal/kg	*	2.88	3.19	
Sheep	ME Mcal/kg	*	2.39	2.65	
Swine	ME kcal/kg	*	2241.	2484.	
Cattle	TDN %	*	79.6	88.3	
Sheep	TDN %		66.0	73.2	
Swine	TDN %	*	54.3	60.2	

Ref No 4 03 329 Canada

Feed Name or Analyses			Mean As Fed	Mean Dry	C.V. ± %
Dry matter	%		89.7	100.0	
Crude fiber	%		12.7	14.2	
Sheep	dig coef %		18.	18.	
Ether extract	%				
Sheep	dig coef %		92.	92.	
N-free extract	%				
Sheep	dig coef %		81.	81.	
Protein (N x 6.25)	%		14.2	15.8	
Sheep	dig coef %		88.	88.	
Cattle	dig prot %	*	9.4	10.5	
Goats	dig prot %	*	10.5	11.7	
Horses	dig prot %	*	10.5	11.7	
Sheep	dig prot %		12.5	13.9	
Energy	GE Mcal/kg				
Sheep	DE Mcal/kg	*	2.92	3.26	
Sheep	ME Mcal/kg	*	2.40	2.67	
Sheep	TDN %		66.3	74.0	
Calcium	%		0.18	0.20	
Phosphorus	%		0.32	0.36	

Oats, groats, (4)

Oat groats (AAFCO)

Oat groats (CFA)

Hulled oats (CFA)

Ref No 4 03 331 United States

Feed Name or Analyses			Mean As Fed	Mean Dry	C.V. ± %
Dry matter	%		90.7	100.0	1
Ash	%		2.2	2.4	18
Crude fiber	%		2.6	2.8	40
Sheep	dig coef %	*	80.	80.	
Ether extract	%		6.0	6.6	19
Sheep	dig coef %	*	96.	96.	
N-free extract	%		63.5	70.0	
Sheep	dig coef %	*	98.	98.	
Protein (N x 6.25)	%		16.5	18.2	10
Sheep	dig coef %	*	90.	90.	
Cattle	dig prot %	*	11.5	12.7	
Goats	dig prot %	*	12.6	13.9	
Horses	dig prot %	*	12.6	13.9	
Sheep	dig prot %		14.8	16.3	
Starch	%		56.5	62.3	7
Sugars, total	%		1.4	1.5	13
Energy	GE Mcal/kg				
Cattle	DE Mcal/kg	*	3.99	4.40	
Sheep	DE Mcal/kg	*	4.06	4.47	
Swine	DE kcal/kg	*	3690.	4068.	
Cattle	ME Mcal/kg	*	3.27	3.61	
Sheep	ME Mcal/kg	*	3.33	3.67	
Swine	ME kcal/kg	*	3407.	3756.	
Cattle	NE_m Mcal/kg	*	2.14	2.36	
Cattle	NE_{gain} Mcal/kg	*	1.38	1.52	
Cattle	$NE_{lactating\ cows}$ Mcal/kg	*	2.42	2.67	
Cattle	TDN %	*	90.5	99.8	
Sheep	TDN %		92.0	101.4	
Swine	TDN %	*	83.7	92.3	
Calcium	%		0.08	0.08	
Chlorine	%		0.09	0.10	
Copper	mg/kg		6.4	7.1	36
Iron	%		0.007	0.008	0
Magnesium	%		0.12	0.13	
Manganese	mg/kg		28.7	31.6	14
Phosphorus	%		0.45	0.49	6
Potassium	%		0.37	0.40	
Sodium	%		0.05	0.06	
Sulphur	%		0.20	0.22	
Niacin	mg/kg		8.1	9.0	11
Pantothenic acid	mg/kg		14.8	16.3	14
Riboflavin	mg/kg		1.3	1.5	8
Thiamine	mg/kg		6.8	7.5	10
Vitamin B_6	mg/kg		1.10	1.21	18

Ref No 4 03 331 Canada

Feed Name or Analyses			Mean As Fed	Mean Dry	C.V. ± %
Dry matter	%		88.5	100.0	1
Ash	%		2.2	2.5	
Crude fiber	%		2.6	2.9	45
Ether extract	%		6.7	7.6	13
N-free extract	%		61.4	69.4	
Protein (N x 6.25)	%		15.6	17.6	12
Cattle	dig prot %	*	10.8	12.2	
Goats	dig prot %	*	11.9	13.4	
Horses	dig prot %	*	11.9	13.4	
Sheep	dig prot %	*	11.9	13.4	
Energy	GE Mcal/kg				
Cattle	DE Mcal/kg	*	3.81	4.30	
Sheep	DE Mcal/kg	*	4.00	4.52	
Swine	DE kcal/kg	*	3596.	4062.	
Cattle	ME Mcal/kg	*	3.12	3.52	
Sheep	ME Mcal/kg	*	3.28	3.71	
Swine	ME kcal/kg	*	3324.	3755.	
Cattle	TDN %	*	86.4	97.6	
Sheep	TDN %		90.7	102.5	
Swine	TDN %	*	81.6	92.1	
Calcium	%		0.10	0.11	
Copper	mg/kg		3.8	4.3	
Phosphorus	%		0.42	0.48	
Salt (NaCl)	%		0.10	0.11	
Pantothenic acid	mg/kg		18.3	20.7	

OATS, NORTEX 107. *Avena byzantina*

Oats, nortex 107, hay, s-c, prebloom, (1)

Ref No 1 09 241 United States

Feed Name or Analyses			Mean As Fed	Mean Dry	C.V. ± %
Dry matter	%		85.5	100.0	
Cattle	dig coef %		67.	67.	
Ash	%		6.9	8.1	
Crude fiber	%		24.2	28.3	
Ether extract	%		2.9	3.4	
N-free extract	%		43.3	50.6	
Protein (N x 6.25)	%		8.2	9.6	
Cattle	dig coef %		49.	49.	
Cattle	dig prot %		4.0	4.7	
Goats	dig prot %	*	4.7	5.5	
Horses	dig prot %	*	4.9	5.7	
Rabbits	dig prot %	*	5.2	6.1	
Sheep	dig prot %	*	4.4	5.2	
Cellulose (Matrone)	%		22.7	26.6	
Cattle	dig coef %		64.	64.	
Energy	GE Mcal/kg		3.56	4.17	
Cattle	GE dig coef %		67.	67.	
Cattle	DE Mcal/kg		2.39	2.79	
Sheep	DE Mcal/kg	*	2.20	2.57	
Cattle	ME Mcal/kg	*	1.96	2.29	
Sheep	ME Mcal/kg	*	1.80	2.11	
Cattle	TDN %	*	54.5	63.7	
Sheep	TDN %	*	49.8	58.2	

Oats, nortex 107, hay, s-c, full bloom, (1)

Ref No 1 09 243 United States

Feed Name or Analyses			Mean As Fed	Mean Dry	C.V. ± %
Dry matter	%		85.5	100.0	
Cattle	dig coef %		72.	72.	
Ash	%		6.2	7.2	
Crude fiber	%		26.1	30.5	
Ether extract	%		2.0	2.3	
N-free extract	%		42.4	49.6	
Protein (N x 6.25)	%		8.9	10.4	
Cattle	dig coef %		62.	62.	
Cattle	dig prot %		5.5	6.4	
Goats	dig prot %	*	5.4	6.3	
Horses	dig prot %	*	5.4	6.4	
Rabbits	dig prot %	*	5.7	6.7	
Sheep	dig prot %	*	5.0	5.9	
Cellulose (Matrone)	%		26.4	30.9	
Cattle	dig coef %		79.	79.	
Lignin (Ellis)	%		6.7	7.8	
Energy	GE Mcal/kg		3.72	4.35	
Cattle	GE dig coef %		72.	72.	
Cattle	DE Mcal/kg		2.68	3.13	
Sheep	DE Mcal/kg	*	2.46	2.88	
Cattle	ME Mcal/kg	*	2.20	2.57	
Sheep	ME Mcal/kg	*	2.02	2.36	
Cattle	TDN %	*	61.2	71.6	
Sheep	TDN %	*	55.9	65.4	

Oats, nortex 107, hay, s-c, late bloom, (1)

Ref No 1 09 100 United States

Feed Name or Analyses			Mean As Fed	Mean Dry	C.V. ± %
Dry matter	%		87.3	100.0	
Cattle	dig coef %		63.	63.	
Ash	%		6.4	7.3	

Continued

Feed Name or Analyses		Mean As Fed	Mean Dry	C.V. ± %
Crude fiber	%	31.6	36.2	
Ether extract	%	2.2	2.5	
N-free extract	%	40.6	46.5	
Protein (N x 6.25)	%	6.6	7.6	
Cattle	dig coef %	56.	56.	
Cattle	dig prot %	3.7	4.3	
Goats	dig prot %	* 3.2	3.7	
Horses	dig prot %	* 3.5	4.0	
Rabbits	dig prot %	* 4.0	4.5	
Sheep	dig prot %	* 3.0	3.4	
Cell walls (Van Soest)	%	63.6	72.8	
Cellulose (Matrone)	%	30.9	35.4	
Cattle	dig coef %	68.	68.	
Fiber, acid detergent (VS)	%	35.9	41.1	
Energy	GE Mcal/kg	3.71	4.25	
Cattle	GE dig coef %	60.	60.	
Cattle	DE Mcal/kg	2.21	2.53	
Sheep	DE Mcal/kg	* 2.08	2.38	
Cattle	ME Mcal/kg	* 1.81	2.08	
Sheep	ME Mcal/kg	* 1.70	1.95	
Cattle	TDN %	* 50.2	57.5	
Sheep	TDN %	* 47.1	54.0	

Oats, nortex 107, hay, s-c, milk stage, (1)

Ref No 1 09 242 United States

Feed Name or Analyses		Mean As Fed	Mean Dry	C.V. ± %
Dry matter	%	86.6	100.0	
Cattle	dig coef %	62.	62.	
Ash	%	5.6	6.4	
Crude fiber	%	31.7	36.6	
Ether extract	%	2.0	2.4	
N-free extract	%	40.9	47.2	
Protein (N x 6.25)	%	6.4	7.4	
Cattle	dig coef %	53.	53.	
Cattle	dig prot %	3.4	3.9	
Goats	dig prot %	* 3.0	3.5	
Horses	dig prot %	* 3.3	3.8	
Rabbits	dig prot %	* 3.8	4.4	
Sheep	dig prot %	* 2.8	3.2	
Cellulose (Matrone)	%	30.3	34.9	
Cattle	dig coef %	67.	67.	
Lignin (Ellis)	%	7.7	8.9	
Energy	GE Mcal/kg	3.63	4.19	
Cattle	GE dig coef %	60.	60.	
Cattle	DE Mcal/kg	2.17	2.50	
Sheep	DE Mcal/kg	* 2.06	2.38	
Cattle	ME Mcal/kg	* 1.78	2.05	
Sheep	ME Mcal/kg	* 1.69	1.95	
Cattle	TDN %	* 49.5	57.2	
Sheep	TDN %	* 46.6	53.9	

Oats, nortex 107, aerial part, ensiled, milk stage, (3)

Ref No 3 09 107 United States

Feed Name or Analyses		Mean As Fed	Mean Dry	C.V. ± %
Dry matter	%	28.2	100.0	
Cattle	dig coef %	61.	61.	
Ash	%	2.3	8.1	
Crude fiber	%	10.4	37.0	
Ether extract	%	0.9	3.1	
N-free extract	%	11.7	41.6	
Protein (N x 6.25)	%	2.9	10.2	
Cattle	dig coef %	61.	61.	
Cattle	dig prot %	1.8	6.2	

(1) dry forages and roughages (2) pasture, range plants, and forages fed green (3) silages (4) energy feeds (5) protein supplements (6) minerals (7) vitamins (8) additives

Feed Name or Analyses		Mean As Fed	Mean Dry	C.V. ± %
Goats	dig prot %	* 1.5	5.5	
Horses	dig prot %	* 1.5	5.5	
Sheep	dig prot %	* 1.5	5.5	
Cellulose (Matrone)	%	12.0	42.5	
Cattle	dig coef %	70.	70.	
Energy	GE Mcal/kg	1.24	4.39	
Cattle	GE dig coef %	60.	60.	
Cattle	DE Mcal/kg	0.74	2.63	
Sheep	DE Mcal/kg	* 0.78	2.76	
Cattle	ME Mcal/kg	* 0.61	2.16	
Sheep	ME Mcal/kg	* 0.64	2.26	
Cattle	TDN %	* 17.5	59.9	
Sheep	TDN %	* 17.7	62.6	

OATS, RED. Avena byzantina

Oats, red, grain, (4)

Ref No 4 03 369 United States

Feed Name or Analyses		Mean As Fed	Mean Dry	C.V. ± %
Dry matter	%		100.0	
Crude fiber	%			
Cattle	dig coef %	*	30.	
Sheep	dig coef %	*	37.	
Swine	dig coef %	*	-16.	
Ether extract	%			
Cattle	dig coef %	*	82.	
Sheep	dig coef %	*	82.	
Swine	dig coef %	*	82.	
N-free extract	%			
Cattle	dig coef %	*	81.	
Sheep	dig coef %	*	83.	
Swine	dig coef %	*	78.	
Protein (N x 6.25)	%			
Cattle	dig coef %	*	75.	
Sheep	dig coef %	*	78.	
Swine	dig coef %	*	84.	

Oats, red, grain, gr 1 heavy US mn wt 36 lb per bushel mx 2% foreign material, (4)

Ref No 4 03 362 United States

Feed Name or Analyses		Mean As Fed	Mean Dry	C.V. ± %
Dry matter	%	91.6	100.0	
Ash	%	3.7	4.0	
Crude fiber	%	8.6	9.4	
Cattle	dig coef %	* 30.	30.	
Sheep	dig coef %	* 37.	37.	
Swine	dig coef %	* -16.	-16.	
Ether extract	%	5.0	5.5	
Cattle	dig coef %	* 82.	82.	
Sheep	dig coef %	* 82.	82.	
Swine	dig coef %	* 82.	82.	
N-free extract	%	59.2	64.6	
Cattle	dig coef %	* 81.	81.	
Sheep	dig coef %	* 83.	83.	
Swine	dig coef %	* 78.	78.	
Protein (N x 6.25)	%	15.1	16.5	
Cattle	dig coef %	* 75.	75.	
Sheep	dig coef %	* 78.	78.	
Swine	dig coef %	* 84.	84.	
Cattle	dig prot %	11.3	12.4	
Goats	dig prot %	* 11.3	12.4	
Horses	dig prot %	* 11.3	12.4	
Sheep	dig prot %	11.8	12.9	
Swine	dig prot %	12.7	13.9	
Energy	GE Mcal/kg			
Cattle	DE Mcal/kg	* 3.14	3.42	
Sheep	DE Mcal/kg	* 3.24	3.53	
Swine	DE kcal/kg	*2940.	3210.	

Feed Name or Analyses		Mean As Fed	Mean Dry	C.V. ± %
Cattle	ME Mcal/kg	* 2.57	2.81	
Sheep	ME Mcal/kg	* 2.65	2.90	
Swine	ME kcal/kg	*2725.	2974.	
Cattle	TDN %	* 71.1	77.7	
Sheep	TDN %	* 73.4	80.1	
Swine	TDN %	* 66.7	72.8	

Oats, red, grain, gr 1 US mn wt 34 lb per bushel mx 2% foreign material, (4)

Ref No 4 03 363 United States

Feed Name or Analyses		Mean As Fed	Mean Dry	C.V. ± %
Dry matter	%	92.0	100.0	
Ash	%	3.7	4.0	15
Crude fiber	%	13.1	14.2	
Cattle	dig coef %	* 30.	30.	
Sheep	dig coef %	* 37.	37.	
Swine	dig coef %	* -16.	-16.	
Ether extract	%	5.6	6.1	7
Cattle	dig coef %	* 82.	82.	
Sheep	dig coef %	* 82.	82.	
Swine	dig coef %	* 82.	82.	
N-free extract	%	57.4	62.4	
Cattle	dig coef %	* 81.	81.	
Sheep	dig coef %	* 83.	83.	
Swine	dig coef %	* 78.	78.	
Protein (N x 6.25)	%	12.2	13.3	6
Cattle	dig coef %	* 75.	75.	
Sheep	dig coef %	* 78.	78.	
Swine	dig coef %	* 84.	84.	
Cattle	dig prot %	9.2	10.0	
Goats	dig prot %	* 8.7	9.4	
Horses	dig prot %	* 8.7	9.4	
Sheep	dig prot %	9.5	10.4	
Swine	dig prot %	10.3	11.2	
Energy	GE Mcal/kg			
Cattle	DE Mcal/kg	* 3.08	3.35	
Sheep	DE Mcal/kg	* 3.19	3.47	
Swine	DE kcal/kg	*2786.	3028.	
Cattle	ME Mcal/kg	* 2.53	2.75	
Sheep	ME Mcal/kg	* 2.62	2.84	
Swine	ME kcal/kg	*2600.	2826.	
Cattle	TDN %	* 70.0	76.0	
Sheep	TDN %	* 72.4	78.7	
Swine	TDN %	* 63.2	68.7	

Oats, red, grain, gr 2 US mn wt 32 lb per bushel mx 3% foreign material, (4)

Ref No 4 03 366 United States

Feed Name or Analyses		Mean As Fed	Mean Dry	C.V. ± %
Dry matter	%	91.8	100.0	
Ash	%	3.6	3.9	
Crude fiber	%	13.3	14.5	
Cattle	dig coef %	* 30.	30.	
Sheep	dig coef %	* 37.	37.	
Swine	dig coef %	* -16.	-16.	
Ether extract	%	5.1	5.6	
Cattle	dig coef %	* 82.	82.	
Sheep	dig coef %	* 82.	82.	
Swine	dig coef %	* 82.	82.	
N-free extract	%	58.6	63.8	
Cattle	dig coef %	* 81.	81.	
Sheep	dig coef %	* 83.	83.	
Swine	dig coef %	* 78.	78.	
Protein (N x 6.25)	%	11.2	12.2	
Cattle	dig coef %	* 75.	75.	
Sheep	dig coef %	* 78.	78.	
Swine	dig coef %	* 84.	84.	
Cattle	dig prot %	8.4	9.2	

Feed Name or Analyses		Mean As Fed	Mean Dry	C.V. ± %
Goats	dig prot %	* 7.7	8.4	
Horses	dig prot %	* 7.7	8.4	
Sheep	dig prot %	8.7	9.5	
Swine	dig prot %	9.4	10.2	
Starch	%	35.5	38.7	
Sugars, total	%	1.1	1.2	
Energy	GE Mcal/kg			
Cattle	DE Mcal/kg	* 3.06	3.33	
Sheep	DE Mcal/kg	* 3.16	3.45	
Swine	DE kcal/kg	*2748.	2993.	
Cattle	ME Mcal/kg	* 2.51	2.73	
Sheep	ME Mcal/kg	* 2.59	2.83	
Swine	ME kcal/kg	*2570.	2799.	
Cattle	TDN %	69.3	75.5	
Sheep	TDN %	71.8	78.2	
Swine	TDN %	62.3	67.9	

Oats, red, grain, gr 3 US mn wt 30 lb per bushel mx 4% foreign material, (4)

Ref No 4 03 367 United States

Feed Name or Analyses		Mean As Fed	Mean Dry	C.V. ± %
Dry matter	%	91.8	100.0	
Ash	%	3.8	4.1	
Crude fiber	%	13.7	14.9	
Cattle	dig coef %	* 30.	30.	
Sheep	dig coef %	* 52.	52.	
Swine	dig coef %	* -16.	-16.	
Ether extract	%	5.0	5.4	
Cattle	dig coef %	* 82.	82.	
Sheep	dig coef %	* 67.	67.	
Swine	dig coef %	* 82.	82.	
N-free extract	%	57.9	63.1	
Cattle	dig coef %	* 81.	81.	
Sheep	dig coef %	* 71.	71.	
Swine	dig coef %	* 78.	78.	
Protein (N x 6.25)	%	11.5	12.5	
Cattle	dig coef %	* 75.	75.	
Sheep	dig coef %	* 67.	67.	
Swine	dig coef %	* 84.	84.	
Cattle	dig prot %	8.6	9.4	
Goats	dig prot %	* 8.0	8.7	
Horses	dig prot %	* 8.0	8.7	
Sheep	dig prot %	7.7	8.4	
Swine	dig prot %	9.6	10.5	
Energy	GE Mcal/kg			
Cattle	DE Mcal/kg	* 3.03	3.30	
Sheep	DE Mcal/kg	* 2.80	3.05	
Swine	DE kcal/kg	*2718.	2961.	
Cattle	ME Mcal/kg	* 2.49	2.71	
Sheep	ME Mcal/kg	* 2.29	2.50	
Swine	ME kcal/kg	*2541.	2768.	
Cattle	TDN %	* 68.8	74.9	
Sheep	TDN %	* 63.4	69.1	
Swine	TDN %	* 61.6	67.1	

Oats, red, grain, gr 4 US mn wt 27 lb per bushel mx 5% foreign material, (4)

Ref No 4 03 368 United States

Feed Name or Analyses		Mean As Fed	Mean Dry	C.V. ± %
Dry matter	%	92.2	100.0	
Ash	%	3.6	3.9	
Crude fiber	%	14.3	15.5	
Cattle	dig coef %	* 30.	30.	
Sheep	dig coef %	* 37.	37.	
Swine	dig coef %	* -16.	-16.	
Ether extract	%	4.8	5.2	
Cattle	dig coef %	* 82.	82.	
Sheep	dig coef %	* 82.	82.	
Swine	dig coef %	* 82.	82.	
N-free extract	%	58.1	63.0	
Cattle	dig coef %	* 81.	81.	
Sheep	dig coef %	* 83.	83.	
Swine	dig coef %	* 78.	78.	
Protein (N x 6.25)	%	11.4	12.4	
Cattle	dig coef %	* 75.	75.	
Sheep	dig coef %	* 78.	78.	
Swine	dig coef %	* 84.	84.	
Cattle	dig prot %	8.6	9.3	
Goats	dig prot %	* 7.9	8.6	
Horses	dig prot %	* 7.9	8.6	
Sheep	dig prot %	8.9	9.7	
Swine	dig prot %	9.6	10.4	
Energy	GE Mcal/kg			
Cattle	DE Mcal/kg	* 3.03	3.29	
Sheep	DE Mcal/kg	* 3.14	3.41	
Swine	DE kcal/kg	*2704.	2933.	
Cattle	ME Mcal/kg	* 2.49	2.70	
Sheep	ME Mcal/kg	* 2.58	2.79	
Swine	ME kcal/kg	*2528.	2742.	
Cattle	TDN %	* 68.8	74.6	
Sheep	TDN %	* 71.3	77.3	
Swine	TDN %	* 61.3	66.5	

Oats, red, grain, gr sample US, (4)

Ref No 4 03 360 United States

Feed Name or Analyses		Mean As Fed	Mean Dry	C.V. ± %
Dry matter	%	92.3	100.0	
Ash	%	4.2	4.6	
Crude fiber	%	13.6	14.8	
Ether extract	%	5.0	5.4	
N-free extract	%	56.8	61.6	
Protein (N x 6.25)	%	12.6	13.7	
Cattle	dig prot %	* 7.9	8.6	
Goats	dig prot %	* 9.0	9.8	
Horses	dig prot %	* 9.0	9.8	
Sheep	dig prot %	* 9.0	9.8	
Energy	GE Mcal/kg			
Cattle	DE Mcal/kg	* 3.31	3.59	
Sheep	DE Mcal/kg	* 3.07	3.33	
Swine	DE kcal/kg	*2898.	3227.	
Cattle	ME Mcal/kg	* 2.72	2.94	
Sheep	ME Mcal/kg	* 2.52	2.73	
Swine	ME kcal/kg	*2862.	3101.	
Cattle	TDN %	* 75.1	81.4	
Sheep	TDN %	* 69.6	75.5	
Swine	TDN %	* 67.6	73.2	

OATS, RUSSIAN. Avena spp

Oats, Russian, grain, (4)

Ref No 4 03 370 United States

Feed Name or Analyses		Mean As Fed	Mean Dry	C.V. ± %
Dry matter	%	86.0	100.0	
Ash	%	3.2	3.7	
Crude fiber	%	8.9	10.3	
Horses	dig coef %	22.	22.	
Ether extract	%	4.0	4.7	
Horses	dig coef %	80.	80.	
N-free extract	%	59.8	69.5	
Horses	dig coef %	80.	80.	
Protein (N x 6.25)	%	10.1	11.8	
Horses	dig coef %	77.	77.	
Cattle	dig prot %	* 5.9	6.9	
Goats	dig prot %	* 6.9	8.0	
Horses	dig prot %	7.8	9.1	
Sheep	dig prot %	* 6.9	8.0	
Energy	GE Mcal/kg			
Cattle	DE Mcal/kg	* 3.19	3.71	
Horses	DE Mcal/kg	* 2.86	3.33	
Sheep	DE Mcal/kg	* 3.02	3.51	
Swine	DE kcal/kg	*2596.	3019.	
Cattle	ME Mcal/kg	* 2.62	3.04	
Horses	ME Mcal/kg	* 2.34	2.73	
Sheep	ME Mcal/kg	* 2.47	2.88	
Swine	ME kcal/kg	*2431.	2826.	
Cattle	TDN %	* 72.4	84.2	
Horses	TDN %	64.9	75.4	
Sheep	TDN %	* 68.4	79.5	
Swine	TDN %	* 58.9	68.5	

OATS, SLENDER. Avena barbata

Oats, slender, aerial part, dehy grnd, (1)

Ref No 1 08 269 United States

Feed Name or Analyses		Mean As Fed	Mean Dry	C.V. ± %
Dry matter	%	92.8	100.0	
Organic matter	%	87.3	94.2	
Ash	%	5.5	5.9	
Crude fiber	%	31.5	34.0	
Ether extract	%	1.8	1.9	
N-free extract	%	50.5	54.4	
Protein (N x 6.25)	%	3.6	3.9	
Cattle	dig prot %	* 0.3	0.3	
Goats	dig prot %	* 0.1	0.2	
Horses	dig prot %	* 0.7	0.8	
Rabbits	dig prot %	* 1.5	1.6	
Sheep	dig prot %	* 0.0	0.0	
Energy	GE Mcal/kg			
Cattle	DE Mcal/kg	* 2.28	2.46	
Sheep	DE Mcal/kg	* 2.21	2.38	
Cattle	ME Mcal/kg	* 1.87	2.02	
Sheep	ME Mcal/kg	* 1.81	1.95	
Cattle	TDN %	* 51.8	55.8	
Sheep	TDN %	* 50.0	54.0	
Silicon	%	1.89	2.04	

Oats, slender, aerial part, dehy grnd, mid-bloom, (1)

Ref No 1 08 266 United States

Feed Name or Analyses		Mean As Fed	Mean Dry	C.V. ± %
Dry matter	%	92.9	100.0	
Organic matter	%	82.9	89.2	
Ash	%	10.0	10.8	
Crude fiber	%	15.0	16.1	
Ether extract	%	4.3	4.6	
N-free extract	%	45.1	48.6	
Protein (N x 6.25)	%	18.5	19.9	
Cattle	dig prot %	* 13.2	14.2	
Goats	dig prot %	* 14.1	15.1	
Horses	dig prot %	* 13.4	14.4	
Rabbits	dig prot %	* 13.0	14.0	
Sheep	dig prot %	* 13.4	14.4	
Energy	GE Mcal/kg			
Cattle	DE Mcal/kg	* 2.15	2.31	
Sheep	DE Mcal/kg	* 2.21	2.38	
Cattle	ME Mcal/kg	* 1.76	1.89	
Sheep	ME Mcal/kg	* 1.81	1.95	
Cattle	TDN %	* 48.6	52.3	
Sheep	TDN %	* 50.2	54.0	
Silicon	%	2.05	2.21	

Oats, slender, aerial part, dehy grnd, over ripe, (1)

Ref No 1 08 267 United States

Feed Name or Analyses		Mean As Fed	Mean Dry	C.V. ± %
Dry matter	%	94.2	100.0	
Organic matter	%	88.5	94.0	
Ash	%	5.7	6.0	
Crude fiber	%	33.8	35.9	
Ether extract	%	1.9	2.0	
N-free extract	%	50.3	53.4	
Protein (N x 6.25)	%	2.5	2.7	
Cattle	dig prot % *	-0.6	-0.6	
Goats	dig prot % *	-0.8	-0.8	
Horses	dig prot % *	-0.1	-0.1	
Rabbits	dig prot % *	0.7	0.8	
Sheep	dig prot % *	-0.8	-0.9	
Energy	GE Mcal/kg			
Cattle	DE Mcal/kg *	2.20	2.34	
Sheep	DE Mcal/kg *	2.18	2.32	
Cattle	ME Mcal/kg *	1.81	1.92	
Sheep	ME Mcal/kg *	1.79	1.90	
Cattle	TDN % *	49.9	53.0	
Sheep	TDN % *	49.5	52.6	
Silicon	%	1.79	1.90	

Oats, slender, aerial part, dehy grnd, dormant, (1)

Ref No 1 08 268 United States

Feed Name or Analyses		Mean As Fed	Mean Dry	C.V. ± %
Dry matter	%	93.0	100.0	0
Organic matter	%	88.8	95.5	1
Ash	%	4.2	4.5	25
Crude fiber	%	37.6	40.4	8
Ether extract	%	1.0	1.1	32
N-free extract	%	47.4	51.0	3
Protein (N x 6.25)	%	2.7	2.9	31
Cattle	dig prot % *	-0.4	-0.4	
Goats	dig prot % *	-0.5	-0.6	
Horses	dig prot % *	0.0	0.0	
Rabbits	dig prot % *	0.9	0.9	
Sheep	dig prot % *	-0.6	-0.7	
Energy	GE Mcal/kg			
Cattle	DE Mcal/kg *	1.77	1.91	
Sheep	DE Mcal/kg *	2.08	2.24	
Cattle	ME Mcal/kg *	1.45	1.56	
Sheep	ME Mcal/kg *	1.71	1.84	
Cattle	TDN % *	40.2	43.2	
Sheep	TDN % *	47.3	50.9	
Silicon	%	1.69	1.81	22

Oats, slender, leaves, dehy grnd, (1)

Ref No 1 08 275 United States

Feed Name or Analyses		Mean As Fed	Mean Dry	C.V. ± %
Dry matter	%	93.2	100.0	
Organic matter	%	84.1	90.2	
Ash	%	9.1	9.8	
Crude fiber	%	19.3	20.7	
Ether extract	%	5.8	6.2	
N-free extract	%	51.4	55.1	
Protein (N x 6.25)	%	7.6	8.2	
Cattle	dig prot % *	3.8	4.0	
Goats	dig prot % *	3.9	4.2	
Horses	dig prot % *	4.2	4.5	
Rabbits	dig prot % *	4.7	5.0	
Sheep	dig prot % *	3.7	3.9	
Silicon	%	3.50	3.76	

Oats, slender, leaves, dehy grnd, early leaf, (1)

Ref No 1 08 270 United States

Feed Name or Analyses		Mean As Fed	Mean Dry	C.V. ± %
Dry matter	%	93.7	100.0	
Organic matter	%	83.2	88.8	
Ash	%	10.5	11.2	
Crude fiber	%	12.6	13.4	
Ether extract	%	4.7	5.0	
N-free extract	%	40.9	43.6	
Protein (N x 6.25)	%	25.1	26.8	
Cattle	dig prot % *	18.9	20.1	
Goats	dig prot % *	20.2	21.6	
Horses	dig prot % *	19.0	20.3	
Rabbits	dig prot % *	18.1	19.4	
Sheep	dig prot % *	19.3	20.6	
Energy	GE Mcal/kg			
Cattle	DE Mcal/kg *	2.83	2.80	
Sheep	DE Mcal/kg *	2.73	2.91	
Cattle	ME Mcal/kg *	2.31	2.30	
Sheep	ME Mcal/kg *	2.24	2.39	
Cattle	TDN % *	64.1	63.5	
Sheep	TDN % *	61.9	66.1	
Silicon	%	1.05	1.12	

Oats, slender, leaves, dehy grnd, immature, (1)

Ref No 1 08 271 United States

Feed Name or Analyses		Mean As Fed	Mean Dry	C.V. ± %
Dry matter	%	93.1	100.0	
Organic matter	%	83.7	89.9	
Ash	%	9.4	10.1	
Crude fiber	%	16.6	17.8	
Ether extract	%	4.5	4.8	
N-free extract	%	46.9	50.4	
Protein (N x 6.25)	%	15.7	16.9	
Cattle	dig prot % *	10.8	11.6	
Goats	dig prot % *	11.5	12.3	
Horses	dig prot % *	11.1	11.9	
Rabbits	dig prot % *	10.9	11.7	
Sheep	dig prot % *	10.9	11.7	
Energy	GE Mcal/kg			
Cattle	DE Mcal/kg *	2.92	3.14	
Sheep	DE Mcal/kg *	2.58	2.77	
Cattle	ME Mcal/kg *	2.40	2.57	
Sheep	ME Mcal/kg *	2.12	2.27	
Cattle	TDN % *	66.3	71.2	
Sheep	TDN % *	58.5	62.8	
Silicon	%	2.67	2.87	

Oats, slender, leaves, dehy grnd, pre-bloom, (1)

Ref No 1 08 272 United States

Feed Name or Analyses		Mean As Fed	Mean Dry	C.V. ± %
Dry matter	%	93.2	100.0	
Organic matter	%	82.5	88.5	
Ash	%	10.7	11.5	
Protein (N x 6.25)	%	15.8	16.9	
Cattle	dig prot % *	10.8	11.6	
Goats	dig prot % *	11.5	12.3	
Horses	dig prot % *	11.1	11.9	
Rabbits	dig prot % *	10.9	11.7	
Sheep	dig prot % *	10.9	11.7	
Silicon	%	2.90	3.11	

Oats, slender, leaves, dehy grnd, mid-bloom, (1)

Ref No 1 08 273 United States

Feed Name or Analyses		Mean As Fed	Mean Dry	C.V. ± %
Dry matter	%	92.7	100.0	
Organic matter	%	83.3	89.9	
Ash	%	9.5	10.2	
Crude fiber	%	14.6	15.8	
Ether extract	%	3.9	4.2	
N-free extract	%	46.9	50.6	
Protein (N x 6.25)	%	17.8	19.2	
Cattle	dig prot % *	12.6	13.6	
Goats	dig prot % *	13.4	14.5	
Horses	dig prot % *	12.8	13.8	
Rabbits	dig prot % *	12.5	13.5	
Sheep	dig prot % *	12.8	13.8	
Energy	GE Mcal/kg			
Cattle	DE Mcal/kg *	2.54	2.74	
Sheep	DE Mcal/kg *	2.65	2.86	
Cattle	ME Mcal/kg *	2.09	2.25	
Sheep	ME Mcal/kg *	2.18	2.35	
Cattle	TDN % *	57.6	62.1	
Sheep	TDN % *	60.2	64.9	
Silicon	%	2.50	2.70	

Oats, slender, leaves, dehy grnd, dormant, (1)

Ref No 1 08 274 United States

Feed Name or Analyses		Mean As Fed	Mean Dry	C.V. ± %
Dry matter	%	93.1	100.0	
Organic matter	%	85.0	91.3	
Ash	%	8.1	8.7	
Crude fiber	%	23.0	24.7	
Ether extract	%	3.8	4.1	
N-free extract	%	54.2	58.2	
Protein (N x 6.25)	%	4.0	4.3	
Cattle	dig prot % *	0.6	0.7	
Goats	dig prot % *	0.5	0.6	
Horses	dig prot % *	1.1	1.2	
Rabbits	dig prot % *	1.9	2.0	
Sheep	dig prot % *	0.4	0.4	
Silicon	%	4.05	4.35	

Oats, slender, leaves w some stems, dehy grnd, mid-bloom, (1)

Ref No 1 08 276 United States

Feed Name or Analyses		Mean As Fed	Mean Dry	C.V. ± %
Dry matter	%	93.3	100.0	
Organic matter	%	86.4	92.6	
Ash	%	6.9	7.4	
Crude fiber	%	14.2	15.2	
Ether extract	%	4.0	4.3	
N-free extract	%	51.5	55.2	
Protein (N x 6.25)	%	16.7	17.9	
Cattle	dig prot % *	11.6	12.4	
Goats	dig prot % *	12.4	13.3	
Horses	dig prot % *	11.9	12.7	
Rabbits	dig prot % *	11.7	12.5	
Sheep	dig prot % *	11.8	12.6	
Silicon	%	1.60	1.71	

Oats, slender, aerial part, fresh, (2)

Ref No 2 03 375 United States

Feed Name or Analyses		Mean As Fed	Mean Dry	C.V. ± %
Dry matter	%		100.0	
Ash	%		7.5	47
Crude fiber	%		34.1	17
Ether extract	%		1.3	

(1) dry forages and roughages (2) pasture, range plants, and forages fed green (3) silages (4) energy feeds (5) protein supplements (6) minerals (7) vitamins (8) additives

Feed Name or Analyses				Mean As Fed	Mean Dry	C.V. ± %
N-free extract		%			53.2	
Protein (N x 6.25)		%			3.9	98
Cattle	dig prot	%	*		1.2	
Goats	dig prot	%	*		0.2	
Horses	dig prot	%	*		0.9	
Rabbits	dig prot	%	*		1.7	
Sheep	dig prot	%	*		0.6	
Other carbohydrates		%			38.7	
Lignin (Ellis)		%			9.8	
Calcium		%			0.32	45
Phosphorus		%			0.31	34
Potassium		%			2.62	32
Silicon		%			2.43	

Oats, slender, aerial part, fresh, immature, (2)

Ref No 2 03 371 United States

Feed Name or Analyses				Mean As Fed	Mean Dry	C.V. ± %
Dry matter		%			100.0	
Ash		%			11.7	34
Crude fiber		%			22.3	10
Protein (N x 6.25)		%			19.1	11
Cattle	dig prot	%	*		14.1	
Goats	dig prot	%	*		14.4	
Horses	dig prot	%	*		13.7	
Rabbits	dig prot	%	*		13.4	
Sheep	dig prot	%	*		14.8	
Calcium		%			0.46	30
Phosphorus		%			0.43	23
Potassium		%			3.60	22

Oats, slender, aerial part, fresh, full bloom, (2)

Ref No 2 03 372 United States

Feed Name or Analyses				Mean As Fed	Mean Dry	C.V. ± %
Dry matter		%			100.0	
Ash		%			7.6	
Crude fiber		%			33.3	
Protein (N x 6.25)		%			7.4	
Cattle	dig prot	%	*		4.2	
Goats	dig prot	%	*		3.5	
Horses	dig prot	%	*		3.8	
Rabbits	dig prot	%	*		4.4	
Sheep	dig prot	%	*		3.9	
Calcium		%			0.28	
Phosphorus		%			0.35	
Potassium		%			2.07	

Oats, slender, aerial part, fresh, dough stage, (2)

Ref No 2 03 373 United States

Feed Name or Analyses				Mean As Fed	Mean Dry	C.V. ± %
Dry matter		%			100.0	
Ash		%			7.4	
Crude fiber		%			36.4	
Protein (N x 6.25)		%			6.6	
Cattle	dig prot	%	*		3.5	
Goats	dig prot	%	*		2.7	
Horses	dig prot	%	*		3.1	
Rabbits	dig prot	%	*		3.8	
Sheep	dig prot	%	*		3.1	
Calcium		%			0.22	
Phosphorus		%			0.27	
Potassium		%			2.06	

Oats, slender, aerial part, fresh, over ripe, (2)

Ref No 2 03 374 United States

Feed Name or Analyses				Mean As Fed	Mean Dry	C.V. ± %
Dry matter		%			100.0	
Ash		%			7.1	8
Crude fiber		%			40.1	4
Protein (N x 6.25)		%			2.6	41
Cattle	dig prot	%	*		0.1	
Goats	dig prot	%	*		-0.9	
Horses	dig prot	%	*		-0.2	
Rabbits	dig prot	%	*		0.7	
Sheep	dig prot	%	*		-0.5	
Calcium		%			0.21	25
Phosphorus		%			0.23	21
Potassium		%			1.84	

OATS, SWEDISH. Avena spp

Oats, Swedish, grain, (4)

Ref No 4 03 376 United States

Feed Name or Analyses				Mean As Fed	Mean Dry	C.V. ± %
Dry matter		%		83.2	100.0	
Ash		%		3.2	3.9	
Crude fiber		%		8.1	9.7	
Horses	dig coef	%		38.	38.	
Ether extract		%		5.2	6.3	
Horses	dig coef	%		84.	84.	
N-free extract		%		57.7	69.4	
Horses	dig coef	%		78.	78.	
Protein (N x 6.25)		%		8.9	10.7	
Horses	dig coef	%		75.	75.	
Cattle	dig prot	%	*	4.9	5.8	
Goats	dig prot	%	*	5.9	7.0	
Horses	dig prot	%		6.7	8.0	
Sheep	dig prot	%	*	5.9	7.0	
Energy		GE Mcal/kg				
Cattle		DE Mcal/kg	*	3.02	3.63	
Horses		DE Mcal/kg	*	2.85	3.43	
Sheep		DE Mcal/kg	*	2.96	3.56	
Swine		DE kcal/kg	*	2550.	3065.	
Cattle		ME Mcal/kg	*	2.48	2.98	
Horses		ME Mcal/kg	*	2.34	2.81	
Sheep		ME Mcal/kg	*	2.43	2.92	
Swine		ME kcal/kg	*	2393.	2876.	
Cattle		TDN %	*	68.5	82.3	
Horses		TDN %		64.7	77.8	
Sheep		TDN %	*	67.2	80.8	
Swine		TDN %	*	57.8	69.5	

Ref No 4 03 376 Canada

Feed Name or Analyses				Mean As Fed	Mean Dry	C.V. ± %
Dry matter		%		94.0	100.0	
Crude fiber		%				
Horses	dig coef	%		38.	38.	
Ether extract		%				
Horses	dig coef	%		84.	84.	
N-free extract		%				
Horses	dig coef	%		78.	78.	
Protein (N x 6.25)		%		12.3	13.1	
Horses	dig coef	%		75.	75.	
Cattle	dig prot	%	*	7.6	8.0	
Goats	dig prot	%	*	8.7	9.2	
Horses	dig prot	%		9.2	9.8	
Sheep	dig prot	%	*	8.7	9.2	
Energy		GE Mcal/kg				
Horses		DE Mcal/kg	*	3.22	3.43	
Horses		ME Mcal/kg	*	2.64	2.81	
Horses		TDN %		73.1	77.7	

Feed Name or Analyses				Mean As Fed	Mean Dry	C.V. ± %
Alanine		%		0.48	0.51	
Arginine		%		0.58	0.62	
Aspartic acid		%		0.80	0.85	
Glutamic acid		%		2.14	2.28	
Glycine		%		0.52	0.55	
Histidine		%		0.22	0.23	
Isoleucine		%		0.37	0.39	
Leucine		%		0.74	0.79	
Lysine		%		0.42	0.45	
Methionine		%		0.09	0.10	
Phenylalanine		%		0.52	0.55	
Proline		%		0.59	0.63	
Tyrosine		%		0.17	0.18	
Valine		%		0.59	0.63	

OATS, WHITE. Avena sativa

Oats, white, grain, (4)

Ref No 4 03 391 United States

Feed Name or Analyses				Mean As Fed	Mean Dry	C.V. ± %
Dry matter		%		90.2	100.0	1
Ash		%		3.1	3.5	21
Crude fiber		%		10.5	11.7	11
Cattle	dig coef	%	*	30.	30.	
Sheep	dig coef	%		22.	22.	
Swine	dig coef	%	*	-16.	-16.	
Ether extract		%		4.8	5.3	11
Cattle	dig coef	%	*	82.	82.	
Sheep	dig coef	%		90.	90.	
Swine	dig coef	%	*	82.	82.	
N-free extract		%		60.3	66.9	
Cattle	dig coef	%	*	81.	81.	
Sheep	dig coef	%		79.	79.	
Swine	dig coef	%	*	78.	78.	
Protein (N x 6.25)		%		11.4	12.7	8
Cattle	dig coef	%	*	75.	75.	
Sheep	dig coef	%		78.	78.	
Swine	dig coef	%	*	84.	84.	
Cattle	dig prot	%		8.6	9.5	
Goats	dig prot	%	*	8.0	8.8	
Horses	dig prot	%	*	8.0	8.8	
Sheep	dig prot	%		8.9	9.9	
Swine	dig prot	%		9.6	10.6	
Energy		GE Mcal/kg				
Cattle		DE Mcal/kg	*	3.06	3.39	
Sheep		DE Mcal/kg	*	3.02	3.35	
Swine		DE kcal/kg	*	2807.	3113.	
Cattle		ME Mcal/kg	*	2.51	2.78	
Sheep		ME Mcal/kg	*	2.48	2.75	
Swine		ME kcal/kg	*	2623.	2909.	
Cattle		TDN %		69.4	77.0	
Sheep		TDN %		68.5	76.0	
Swine		TDN %		63.7	70.6	
Folic acid		mg/kg		0.38	0.42	46

Oats, white, grain, Can 1 feed mn wt 34 lb per bushel mx 12% foreign material, (4)

Ref No 4 03 377 United States

Feed Name or Analyses				Mean As Fed	Mean Dry	C.V. ± %
Dry matter		%		87.9	100.0	
Crude fiber		%		8.4	9.6	
Cattle	dig coef	%	*	30.	30.	
Sheep	dig coef	%	*	37.	37.	
Swine	dig coef	%	*	-16.	-16.	
Ether extract		%		5.6	6.4	
Cattle	dig coef	%	*	82.	82.	
Sheep	dig coef	%	*	82.	82.	

Continued

Feed Name or Analyses		Mean As Fed	Mean Dry	C.V. ± %
Swine	dig coef %	* 82.	82.	
N-free extract	%			
Cattle	dig coef %	* 81.	81.	
Sheep	dig coef %	* 83.	83.	
Swine	dig coef %	* 78.	78.	
Protein (N x 6.25)	%	8.7	9.9	
Cattle	dig coef %	* 75.	75.	
Sheep	dig coef %	* 78.	78.	
Swine	dig coef %	* 84.	84.	
Cattle	dig prot %	6.5	7.4	
Goats	dig prot %	* 5.5	6.3	
Horses	dig prot %	* 5.5	6.3	
Sheep	dig prot %	6.8	7.7	
Swine	dig prot %	7.3	8.3	
Energy	GE Mcal/kg	2.84	3.23	
Cattle	DE Mcal/kg	* 2.99	3.40	
Sheep	DE Mcal/kg	* 3.09	3.51	
Swine	DE kcal/kg	*2743.	3121.	
Cattle	ME Mcal/kg	* 2.45	2.79	
Chickens	ME_n kcal/kg	2510.	2855.	
Sheep	ME Mcal/kg	* 2.53	2.88	
Swine	ME kcal/kg	*2579.	2933.	
Cattle	TDN %	67.8	77.2	
Sheep	TDN %	70.1	79.7	
Swine	TDN %	62.2	70.8	
Calcium	%	0.03	0.03	
Phosphorus	%	0.36	0.41	

Ref No 4 03 377 — Canada

Feed Name or Analyses		Mean As Fed	Mean Dry	C.V. ± %
Dry matter	%	88.5	100.0	2
Ash	%	3.0	3.4	7
Crude fiber	%	10.5	11.8	9
Ether extract	%	4.8	5.4	17
N-free extract	%	59.2	66.9	
Protein (N x 6.25)	%	11.1	12.6	9
Cattle	dig prot %	* 6.7	7.5	
Goats	dig prot %	* 7.7	8.7	
Horses	dig prot %	* 7.7	8.7	
Sheep	dig prot %	* 7.7	8.7	
Cellulose (Matrone)	%	9.0	10.2	
Energy	GE Mcal/kg	2.86	3.23	
Cattle	DE Mcal/kg	* 3.01	3.40	
Sheep	DE Mcal/kg	* 3.11	3.51	
Swine	DE kcal/kg	*2763.	3121.	
Cattle	ME Mcal/kg	* 2.47	2.79	
Chickens	ME_n kcal/kg	2527.	2855.	15
Sheep	ME Mcal/kg	* 2.55	2.88	
Swine	ME kcal/kg	*2582.	2917.	
Cattle	TDN %	68.3	77.2	
Sheep	TDN %	70.6	79.7	
Swine	TDN %	62.7	70.8	
Calcium	%	0.07	0.08	22
Copper	mg/kg	4.7	5.3	
Magnesium	%	0.12	0.14	
Manganese	mg/kg	29.8	33.7	
Phosphorus	%	0.33	0.37	31
Selenium	mg/kg	0.244	0.276	
Zinc	mg/kg	30.5	34.5	
Niacin	mg/kg	15.0	17.0	
Pantothenic acid	mg/kg	29.2	33.0	
Riboflavin	mg/kg	1.2	1.4	
Thiamine	mg/kg	1.0	1.1	
Vitamin B6	mg/kg	3.61	4.08	
Alanine	%	0.48	0.54	
Arginine	%	0.62	0.70	
Aspartic acid	%	0.83	0.94	
Glutamic acid	%	2.18	2.47	
Glycine	%	0.52	0.59	
Histidine	%	0.21	0.23	
Isoleucine	%	0.38	0.43	
Leucine	%	0.75	0.85	
Lysine	%	0.39	0.44	
Methionine	%	0.08	0.10	
Phenylalanine	%	0.54	0.61	
Proline	%	0.52	0.59	
Serine	%	0.45	0.51	
Threonine	%	0.33	0.37	
Tyrosine	%	0.28	0.32	
Valine	%	0.57	0.65	

(1) dry forages and roughages (2) pasture, range plants, and forages fed green (3) silages (4) energy feeds (5) protein supplements (6) minerals (7) vitamins (8) additives

Oats, white, grain, Can 2 CW mn wt 36 lb per bushel mx 3% foreign material, (4)

Ref No 4 03 378 — United States

Feed Name or Analyses		Mean As Fed	Mean Dry	C.V. ± %
Dry matter	%		100.0	
Crude fiber	%			
Cattle	dig coef %	*	30.	
Sheep	dig coef %	*	37.	
Swine	dig coef %	*	-16.	
Ether extract	%			
Cattle	dig coef %	*	82.	
Sheep	dig coef %	*	82.	
Swine	dig coef %	*	82.	
N-free extract	%			
Cattle	dig coef %	*	81.	
Sheep	dig coef %	*	83.	
Swine	dig coef %	*	78.	
Protein (N x 6.25)	%			
Cattle	dig coef %	*	75.	
Sheep	dig coef %	*	78.	
Swine	dig coef %	*	84.	
Energy	GE Mcal/kg			
Cattle	DE Mcal/kg	*	3.37	
Sheep	DE Mcal/kg	*	3.49	
Swine	DE kcal/kg	*	3082.	
Cattle	ME Mcal/kg	*	2.77	
Sheep	ME Mcal/kg	*	2.86	
Cattle	TDN %	*	76.5	
Sheep	TDN %	*	79.1	
Swine	TDN %	*	69.9	

Ref No 4 03 378 — Canada

Feed Name or Analyses		Mean As Fed	Mean Dry	C.V. ± %
Dry matter	%	88.5	100.0	2
Ash	%	3.1	3.5	5
Crude fiber	%	10.8	12.2	12
Ether extract	%	4.5	5.1	9
N-free extract	%	58.9	66.5	
Protein (N x 6.25)	%	11.2	12.7	13
Cattle	dig prot %	* 6.8	7.7	
Goats	dig prot %	* 7.8	8.9	
Horses	dig prot %	* 7.8	8.9	
Sheep	dig prot %	* 7.8	8.9	
Cellulose (Matrone)	%	12.0	13.6	
Energy	GE Mcal/kg			
Cattle	DE Mcal/kg	* 2.99	3.37	
Sheep	DE Mcal/kg	* 3.09	3.49	
Swine	DE kcal/kg	*2728.	3082.	
Cattle	ME Mcal/kg	* 2.45	2.77	
Sheep	ME Mcal/kg	* 2.53	2.86	
Swine	ME kcal/kg	*2549.	2880.	
Cattle	TDN %	67.7	76.5	
Sheep	TDN %	70.0	79.1	
Swine	TDN %	61.9	69.9	
Calcium	%	0.08	0.09	27
Copper	mg/kg	11.2	12.6	
Iron	%	0.004	0.005	
Manganese	mg/kg	32.2	36.3	
Phosphorus	%	0.34	0.38	16
Selenium	mg/kg	0.263	0.297	
Niacin	mg/kg	15.0	16.9	
Pantothenic acid	mg/kg	26.4	29.8	
Riboflavin	mg/kg	1.9	2.1	
Thiamine	mg/kg	1.2	1.3	
Vitamin B6	mg/kg	1.84	2.08	

Oats, white, grain, Can 2 feed mn wt 28 lb per bushel mx 22% foreign material, (4)

Ref No 4 03 379 — United States

Feed Name or Analyses		Mean As Fed	Mean Dry	C.V. ± %
Dry matter	%		100.0	
Crude fiber	%			
Cattle	dig coef %	*	30.	
Sheep	dig coef %	*	37.	
Swine	dig coef %	*	-16.	
Ether extract	%			
Cattle	dig coef %	*	82.	
Sheep	dig coef %	*	82.	
Swine	dig coef %	*	82.	
N-free extract	%			
Cattle	dig coef %	*	81.	
Sheep	dig coef %	*	83.	
Swine	dig coef %	*	78.	
Protein (N x 6.25)	%			
Cattle	dig coef %	*	75.	
Sheep	dig coef %	*	78.	
Swine	dig coef %	*	84.	
Energy	GE Mcal/kg	*	4.75	
Cattle	DE Mcal/kg	*	3.36	
Sheep	DE Mcal/kg	*	3.47	
Swine	DE kcal/kg	*	2682.	
Cattle	ME Mcal/kg	*	2.75	
Sheep	ME Mcal/kg	*	2.85	
Cattle	TDN %	*	76.2	
Sheep	TDN %	*	78.8	
Swine	TDN %	*	67.2	
Phosphorus	%	*	57.	

Ref No 4 03 379 — Canada

Feed Name or Analyses		Mean As Fed	Mean Dry	C.V. ± %
Dry matter	%	88.0	100.0	3
Swine	dig coef %	57.	57.	
Ash	%	3.1	3.5	19
Crude fiber	%	10.9	12.4	17
Swine	dig coef %	24.	24.	
Ether extract	%	4.4	5.0	19
Swine	dig coef %	88.	88.	
N-free extract	%	58.7	66.7	
Swine	dig coef %	65.	65.	
Protein (N x 6.25)	%	10.9	12.4	11
Swine	dig coef %	80.	80.	
Cattle	dig prot %	* 6.5	7.4	
Goats	dig prot %	* 7.6	8.6	
Horses	dig prot %	* 7.6	8.6	
Sheep	dig prot %	* 7.6	8.6	
Swine	dig prot %	8.7	9.9	
Cellulose (Matrone)	%	9.1	10.3	
Energy	GE Mcal/kg	4.18	4.75	
Cattle	DE Mcal/kg	* 2.96	3.36	
Sheep	DE Mcal/kg	* 3.06	3.47	
Swine	DE kcal/kg	2361.	2682.	
Cattle	ME Mcal/kg	* 2.42	2.75	

Feed Name or Analyses		Mean As Fed	Mean Dry	C.V. ± %
Sheep	ME Mcal/kg	* 2.51	2.85	
Swine	ME kcal/kg	*2207.	2508.	
Cattle	TDN %	67.1	76.2	
Sheep	TDN %	69.3	78.8	
Swine	TDN %	58.1	66.1	
Calcium	%	0.09	0.10	23
Copper	mg/kg	4.5	5.1	
Iron	%	0.004	0.005	
Magnesium	%	0.12	0.14	
Manganese	mg/kg	29.8	33.9	
Phosphorus	%	57.	57.	
Phosphorus	%	0.33	0.37	15
Selenium	mg/kg	0.289	0.329	
Zinc	mg/kg	29.4	33.4	
Niacin	mg/kg	22.4	25.5	
Pantothenic acid	mg/kg	25.9	29.4	
Riboflavin	mg/kg	1.4	1.6	
Thiamine	mg/kg	1.0	1.2	
Vitamin B6	mg/kg	2.73	3.10	
Alanine	%	0.45	0.51	
Arginine	%	0.51	0.57	
Aspartic acid	%	0.73	0.83	
Glutamic acid	%	2.10	2.38	
Glycine	%	0.45	0.51	
Histidine	%	0.16	0.18	
Isoleucine	%	0.24	0.28	
Leucine	%	0.64	0.72	
Lysine	%	0.29	0.33	
Methionine	%	0.11	0.13	
Phenylalanine	%	0.43	0.49	
Proline	%	0.58	0.66	
Serine	%	0.49	0.55	
Threonine	%	0.31	0.35	
Tyrosine	%	0.25	0.29	
Valine	%	0.34	0.38	

Oats, white, grain, Can 3 CW mn wt 34 lb per bushel mx 6% foreign material, (4)

Ref No 4 03 380 United States

Feed Name or Analyses		Mean As Fed	Mean Dry	C.V. ± %
Dry matter	%		100.0	
Crude fiber	%			
Cattle	dig coef %	*	30.	
Sheep	dig coef %	*	37.	
Swine	dig coef %	*	-16.	
Ether extract	%			
Cattle	dig coef %	*	82.	
Sheep	dig coef %	*	82.	
Swine	dig coef %	*	82.	
N-free extract	%			
Cattle	dig coef %	*	81.	
Sheep	dig coef %	*	83.	
Swine	dig coef %	*	78.	
Protein (N x 6.25)	%			
Cattle	dig coef %	*	75.	
Sheep	dig coef %	*	78.	
Swine	dig coef %	*	84.	
Energy	GE Mcal/kg	*	4.74	
Sheep	GE dig coef %	*	69.	
Swine	GE dig coef %	*	65.	
Cattle	DE Mcal/kg	*	3.38	
Sheep	DE Mcal/kg	*	3.27	
Swine	DE kcal/kg	*	3096.	
Cattle	ME Mcal/kg	*	2.77	
Sheep	ME Mcal/kg	*	2.68	
Cattle	TDN %	*	76.7	
Sheep	TDN %	*	75.2	
Swine	TDN %	*	73.1	

Ref No 4 03 380 Canada

Feed Name or Analyses		Mean As Fed	Mean Dry	C.V. ± %
Dry matter	%	88.5	100.0	3
Sheep	dig coef %	69.	69.	
Swine	dig coef %	63.	63.	
Ash	%	3.0	3.4	8
Crude fiber	%	11.1	12.5	17
Sheep	dig coef %	40.	40.	
Swine	dig coef %	18.	18.	
Ether extract	%	4.8	5.4	12
Sheep	dig coef %	96.	96.	
Swine	dig coef %	81.	81.	
N-free extract	%	58.1	65.7	
Sheep	dig coef %	71.	71.	
Swine	dig coef %	78.	78.	
Protein (N x 6.25)	%	11.4	12.9	10
Sheep	dig coef %	77.	77.	
Swine	dig coef %	82.	82.	
Cattle	dig prot %	* 7.0	7.9	
Goats	dig prot %	* 8.0	9.1	
Horses	dig prot %	* 8.0	9.1	
Sheep	dig prot %	8.7	9.9	
Swine	dig prot %	9.3	10.5	
Cellulose (Matrone)	%	11.2	12.7	
Energy	GE Mcal/kg	4.20	4.74	0
Sheep	GE dig coef %	69.	69.	
Swine	GE dig coef %	65.	65.	
Cattle	DE Mcal/kg	* 2.99	3.38	
Sheep	DE Mcal/kg	2.90	3.27	
Swine	DE kcal/kg	2741.	3097.	
Cattle	ME Mcal/kg	* 2.45	2.77	
Sheep	ME Mcal/kg	* 2.37	2.68	
Swine	ME kcal/kg	*2560.	2893.	
Cattle	TDN %	67.9	76.7	
Sheep	TDN %	64.8	73.2	
Swine	TDN %	65.7	74.2	
Calcium	%	0.10	0.12	45
Copper	mg/kg	4.4	5.0	
Iron	%	.004	.004	
Magnesium	%	.122	.138	
Manganese	mg/kg	29.5	33.3	
Phosphorus	%	0.34	0.38	15
Salt (NaCl)	%	0.10	0.11	
Selenium	mg/kg	0.339	0.383	
Zinc	mg/kg	27.8	31.4	
Niacin	mg/kg	14.2	16.1	
Pantothenic acid	mg/kg	21.5	24.3	
Riboflavin	mg/kg	1.5	1.7	
Thiamine	mg/kg	2.4	2.7	
Vitamin B6	mg/kg	1.87	2.12	
Alanine	%	0.44	0.50	
Arginine	%	0.56	0.63	
Aspartic acid	%	0.78	0.88	
Glutamic acid	%	1.98	2.24	
Glycine	%	0.47	0.53	
Histidine	%	0.17	0.19	
Isoleucine	%	0.26	0.30	
Leucine	%	0.66	0.75	
Lysine	%	0.32	0.36	
Methionine	%	0.10	0.12	
Phenylalanine	%	0.46	0.52	
Proline	%	0.51	0.58	
Serine	%	0.49	0.55	
Threonine	%	0.32	0.36	
Tyrosine	%	0.25	0.29	
Valine	%	0.40	0.45	

Oats, white, grain, Can 3 feed mx 33% foreign material, (4)

Ref No 4 03 381 United States

Feed Name or Analyses		Mean As Fed	Mean Dry	C.V. ± %
Dry matter	%		100.0	
Crude fiber	%			
Cattle	dig coef %	*	30.	
Sheep	dig coef %	*	37.	
Swine	dig coef %	*	-16.	
Ether extract	%			
Cattle	dig coef %	*	82.	
Sheep	dig coef %	*	82.	
Swine	dig coef %	*	82.	
N-free extract	%			
Cattle	dig coef %	*	81.	
Sheep	dig coef %	*	83.	
Swine	dig coef %	*	78.	
Protein (N x 6.25)	%			
Cattle	dig coef %	*	75.	
Sheep	dig coef %	*	78.	
Swine	dig coef %	*	84.	
Energy	GE Mcal/kg	*	4.84	
Swine	GE dig coef %	*	68.	
Cattle	DE Mcal/kg	*	3.33	
Sheep	DE Mcal/kg	*	3.45	
Swine	DE kcal/kg	*	3288.	
Cattle	ME Mcal/kg	*	2.73	
Sheep	ME Mcal/kg	*	2.83	
Cattle	TDN %	*	75.6	
Sheep	TDN %	*	78.2	
Swine	TDN %	*	68.4	

Ref No 4 03 381 Canada

Feed Name or Analyses		Mean As Fed	Mean Dry	C.V. ± %
Dry matter	%	88.7	100.0	4
Ash	%	3.2	3.6	
Crude fiber	%	11.9	13.4	
Ether extract	%	4.3	4.9	
N-free extract	%	58.5	66.0	
Protein (N x 6.25)	%	10.7	12.1	13
Cattle	dig prot %	* 6.3	7.1	
Goats	dig prot %	* 7.4	8.3	
Horses	dig prot %	* 7.4	8.3	
Sheep	dig prot %	* 7.4	8.3	
Cellulose (Matrone)	%	16.2	18.3	
Energy	GE Mcal/kg	4.29	4.84	
Swine	GE dig coef %	68.	68.	
Cattle	DE Mcal/kg	* 2.96	3.33	
Sheep	DE Mcal/kg	* 3.06	3.45	
Swine	DE kcal/kg	2916.	3288.	
Cattle	ME Mcal/kg	* 2.42	2.73	
Sheep	ME Mcal/kg	* 2.51	2.83	
Swine	ME kcal/kg	*2728.	3076.	
Cattle	TDN %	67.0	75.6	
Sheep	TDN %	69.3	78.2	
Swine	TDN %	60.7	68.4	
Calcium	%	0.09	0.10	38
Copper	mg/kg	5.1	5.8	
Iron	%	0.005	0.006	
Magnesium	%	0.14	0.15	
Manganese	mg/kg	37.1	41.8	
Phosphorus	%	0.37	0.41	14
Selenium	mg/kg	0.168	0.189	
Zinc	mg/kg	27.0	30.5	
Niacin	mg/kg	16.3	18.4	
Pantothenic acid	mg/kg	19.9	22.4	
Riboflavin	mg/kg	1.8	2.0	
Thiamine	mg/kg	2.9	3.3	

Continued

Feed Name or Analyses		Mean As Fed	Mean Dry	C.V. ± %
Vitamin B_6	mg/kg	2.61	2.94	
Alanine	%	0.38	0.43	
Arginine	%	0.39	0.44	
Aspartic acid	%	0.64	0.73	
Glutamic acid	%	1.34	1.51	
Glycine	%	0.37	0.42	
Histidine	%	0.13	0.15	
Isoleucine	%	0.21	0.24	
Leucine	%	0.49	0.56	
Lysine	%	0.27	0.31	
Methionine	%	0.07	0.08	
Phenylalanine	%	0.27	0.31	
Proline	%	0.49	0.55	
Serine	%	0.35	0.40	
Threonine	%	0.26	0.29	
Tyrosine	%	0.20	0.22	
Valine	%	0.30	0.34	

Oats, white, grain, gr 1 heavy US mn wt 36 lb per bushel mx 2% foreign material, (4)

Ref No 4 03 384 United States

Feed Name or Analyses		Mean As Fed	Mean Dry	C.V. ± %
Dry matter	%	89.4	100.0	1
Ash	%	3.0	3.4	26
Crude fiber	%	10.2	11.4	10
Cattle	dig coef %	* 30.	30.	
Sheep	dig coef %	25.	25.	
Swine	dig coef %	* -16.	-16.	
Ether extract	%	4.3	4.9	10
Cattle	dig coef %	* 82.	82.	
Sheep	dig coef %	90.	90.	
Swine	dig coef %	* 82.	82.	
N-free extract	%	59.8	66.9	
Cattle	dig coef %	* 81.	81.	
Sheep	dig coef %	81.	81.	
Swine	dig coef %	* 78.	78.	
Protein (N x 6.25)	%	12.1	13.5	10
Cattle	dig coef %	* 75.	75.	
Sheep	dig coef %	84.	84.	
Swine	dig coef %	* 84.	84.	
Cattle	dig prot %	9.1	10.1	
Goats	dig prot %	* 8.6	9.6	
Horses	dig prot %	* 8.6	9.6	
Sheep	dig prot %	10.1	11.3	
Swine	dig prot %	10.1	11.3	
Energy	GE Mcal/kg			
Cattle	DE Mcal/kg	* 3.02	3.38	
Sheep	DE Mcal/kg	* 3.08	3.45	
Swine	DE kcal/kg	*2779.	3108.	
Cattle	ME Mcal/kg	* 2.48	2.77	
Sheep	ME Mcal/kg	* 2.53	2.83	
Swine	ME kcal/kg	*2592.	2899.	
Cattle	NE_m Mcal/kg	* 1.55	1.73	
Cattle	NE_{gain} Mcal/kg	* 1.02	1.14	
Cattle	$NE_{lactating\ cows}$ Mcal/kg	* 1.82	2.04	
Cattle	TDN %	68.5	76.6	
Sheep	TDN %	69.9	78.2	
Swine	TDN %	63.0	70.5	

Oats, white, grain, gr 1 US mn wt 34 lb per bushel mx 2% foreign material, (4)

Ref No 4 03 385 United States

Feed Name or Analyses		Mean As Fed	Mean Dry	C.V. ± %
Dry matter	%	89.5	100.0	
Ash	%	3.3	3.7	12
Crude fiber	%	11.7	13.1	
Cattle	dig coef %	* 30.	30.	
Sheep	dig coef %	* 37.	37.	
Swine	dig coef %	* -16.	-16.	
Ether extract	%	4.4	4.9	9
Cattle	dig coef %	* 82.	82.	
Sheep	dig coef %	* 82.	82.	
Swine	dig coef %	* 82.	82.	
N-free extract	%	57.6	64.4	
Cattle	dig coef %	* 81.	81.	
Sheep	dig coef %	* 83.	83.	
Swine	dig coef %	* 78.	78.	
Protein (N x 6.25)	%	12.4	13.9	6
Cattle	dig coef %	* 75.	75.	
Sheep	dig coef %	* 78.	78.	
Swine	dig coef %	* 84.	84.	
Cattle	dig prot %	9.3	10.4	
Goats	dig prot %	* 8.9	10.0	
Horses	dig prot %	* 8.9	10.0	
Sheep	dig prot %	9.7	10.8	
Swine	dig prot %	10.5	11.7	
Starch	%	37.1	41.4	7
Sugars, total	%	1.3	1.5	27
Energy	GE Mcal/kg			
Cattle	DE Mcal/kg	* 2.98	3.33	
Sheep	DE Mcal/kg	* 3.09	3.45	
Swine	DE kcal/kg	*2712.	3030.	
Cattle	ME Mcal/kg	* 2.44	2.73	
Sheep	ME Mcal/kg	* 2.53	2.83	
Swine	ME kcal/kg	*2527.	2824.	
Cattle	TDN %	67.6	75.6	
Sheep	TDN %	70.0	78.2	
Swine	TDN %	61.5	68.7	

Oats, white, grain, gr 2 US mn wt 32 lb per bushel mx 3% foreign material, (4)

Ref No 4 03 388 United States

Feed Name or Analyses		Mean As Fed	Mean Dry	C.V. ± %
Dry matter	%	89.9	100.0	1
Ash	%	3.2	3.6	7
Crude fiber	%	10.8	12.0	11
Cattle	dig coef %	* 30.	30.	
Sheep	dig coef %	* 37.	37.	
Swine	dig coef %	* -16.	-16.	
Ether extract	%	4.1	4.6	7
Cattle	dig coef %	* 82.	82.	
Sheep	dig coef %	* 82.	82.	
Swine	dig coef %	* 82.	82.	
N-free extract	%	60.5	67.3	
Cattle	dig coef %	* 81.	81.	
Sheep	dig coef %	* 83.	83.	
Swine	dig coef %	* 78.	78.	
Protein (N x 6.25)	%	11.2	12.5	3
Cattle	dig coef %	* 75.	75.	
Sheep	dig coef %	* 78.	78.	
Swine	dig coef %	* 84.	84.	
Cattle	dig prot %	8.4	9.4	
Goats	dig prot %	* 7.8	8.7	
Horses	dig prot %	* 7.8	8.7	
Sheep	dig prot %	8.8	9.8	
Swine	dig prot %	9.4	10.5	
Energy	GE Mcal/kg			
Cattle	DE Mcal/kg	* 3.01	3.35	
Sheep	DE Mcal/kg	* 3.11	3.46	
Swine	DE kcal/kg	*2753.	3062.	
Cattle	ME Mcal/kg	* 2.47	2.75	
Sheep	ME Mcal/kg	* 2.55	2.84	
Swine	ME kcal/kg	*2573.	2862.	
Cattle	NE_m Mcal/kg	* 1.56	1.73	
Cattle	NE_{gain} Mcal/kg	* 1.02	1.14	
Cattle	$NE_{lactating\ cows}$ Mcal/kg	* 1.83	2.04	
Cattle	TDN %	* 68.3	76.0	
Sheep	TDN %	* 70.6	78.5	
Swine	TDN %	* 62.4	69.4	

Oats, white, grain, gr 3 US mn wt 30 lb per bushel mx 4% foreign material, (4)

Ref No 4 03 389 United States

Feed Name or Analyses		Mean As Fed	Mean Dry	C.V. ± %
Dry matter	%	90.9	100.0	
Ash	%	3.6	4.0	
Crude fiber	%	11.9	13.1	
Cattle	dig coef %	* 30.	30.	
Sheep	dig coef %	* 37.	37.	
Swine	dig coef %	* -16.	-16.	
Ether extract	%	4.1	4.5	
Cattle	dig coef %	* 82.	82.	
Sheep	dig coef %	* 82.	82.	
Swine	dig coef %	* 82.	82.	
N-free extract	%	58.7	64.6	
Cattle	dig coef %	* 81.	81.	
Sheep	dig coef %	* 83.	83.	
Swine	dig coef %	* 78.	78.	
Protein (N x 6.25)	%	12.5	13.8	
Cattle	dig coef %	* 75.	75.	
Sheep	dig coef %	* 78.	78.	
Swine	dig coef %	* 84.	84.	
Cattle	dig prot %	9.4	10.4	
Goats	dig prot %	* 9.0	9.9	
Horses	dig prot %	* 9.0	9.9	
Sheep	dig prot %	9.8	10.8	
Swine	dig prot %	10.5	11.6	
Energy	GE Mcal/kg			
Cattle	DE Mcal/kg	* 3.00	3.30	
Sheep	DE Mcal/kg	* 3.11	3.42	
Swine	DE kcal/kg	*2728.	3001.	
Cattle	ME Mcal/kg	* 2.46	2.71	
Sheep	ME Mcal/kg	* 2.55	2.80	
Swine	ME kcal/kg	*2542.	2797.	
Cattle	NE_m Mcal/kg	* 1.57	1.73	
Cattle	NE_{gain} Mcal/kg	* 1.04	1.14	
Cattle	$NE_{lactating\ cows}$ Mcal/kg	* 1.85	2.04	
Cattle	TDN %	* 68.1	74.9	
Sheep	TDN %	* 70.5	77.5	
Swine	TDN %	* 61.9	68.1	

Oats, white, grain, gr 4 US mn wt 27 lb per bushel mx 5% foreign material, (4)

Ref No 4 03 390 United States

Feed Name or Analyses		Mean As Fed	Mean Dry	C.V. ± %
Dry matter	%	91.9	100.0	
Ash	%	3.3	3.6	
Crude fiber	%	11.6	12.6	
Cattle	dig coef %	* 30.	30.	
Sheep	dig coef %	* 37.	37.	
Swine	dig coef %	* -16.	-16.	
Ether extract	%	4.4	4.8	
Cattle	dig coef %	* 82.	82.	
Sheep	dig coef %	* 82.	82.	

(1) dry forages and roughages (2) pasture, range plants, and forages fed green (3) silages (4) energy feeds (5) protein supplements (6) minerals (7) vitamins (8) additives

Feed Name or Analyses		Mean As Fed	Mean Dry	C.V. ± %
Swine	dig coef %	* 82.	82.	
N-free extract	%	59.9	65.2	
Cattle	dig coef %	* 81.	81.	
Sheep	dig coef %	* 83.	83.	
Swine	dig coef %	* 78.	78.	
Protein (N x 6.25)	%	12.7	13.8	
Cattle	dig coef %	* 75.	75.	
Sheep	dig coef %	* 78.	78.	
Swine	dig coef %	* 84.	84.	
Cattle	dig prot %	9.5	10.4	
Goats	dig prot %	* 9.1	9.9	
Horses	dig prot %	* 9.1	9.9	
Sheep	dig prot %	9.9	10.8	
Swine	dig prot %	10.7	11.6	
Energy	GE Mcal/kg			
Cattle	DE Mcal/kg	* 3.07	3.34	
Sheep	DE Mcal/kg	* 3.18	3.46	
Swine	DE kcal/kg	*2802.	3049.	
Cattle	ME Mcal/kg	* 2.52	2.74	
Sheep	ME Mcal/kg	* 2.60	2.83	
Swine	ME kcal/kg	*2612.	2843.	
Cattle	NE_m Mcal/kg	* 1.59	1.73	
Cattle	NE_{gain} Mcal/kg	* 1.05	1.14	
Cattle	$NE_{lactating\ cows}$ Mcal/kg	* 1.87	2.04	
Cattle	TDN %	* 69.7	75.8	
Sheep	TDN %	* 72.0	78.4	
Swine	TDN %	* 63.6	69.2	

Oats, white, grain, gr sample US, (4)

Ref No 4 03 382 United States

Feed Name or Analyses		Mean As Fed	Mean Dry	C.V. ± %
Dry matter	%	91.1	100.0	
Ash	%	2.9	3.2	
Crude fiber	%	11.0	12.1	
Ether extract	%	3.9	4.3	
N-free extract	%	60.7	66.6	
Protein (N x 6.25)	%	12.6	13.8	
Cattle	dig prot %	* 7.9	8.7	
Goats	dig prot %	* 9.0	9.9	
Horses	dig prot %	* 9.0	9.9	
Sheep	dig prot %	* 9.0	9.9	
Energy	GE Mcal/kg			
Cattle	DE Mcal/kg	* 3.09	3.39	
Sheep	DE Mcal/kg	* 3.12	3.43	
Swine	DE kcal/kg	*2649.	2908.	
Cattle	ME Mcal/kg	* 2.53	2.78	
Sheep	ME Mcal/kg	* 2.56	2.81	
Swine	ME kcal/kg	*2469.	2711.	
Cattle	TDN %	* 70.1	76.9	
Sheep	TDN %	* 70.8	77.7	
Swine	TDN %	* 60.1	66.0	

OATS, WILD. Avena fatua

Oats, wild, hay, s-c, (1)

Ref No 1 03 392 United States

Feed Name or Analyses		Mean As Fed	Mean Dry	C.V. ± %
Dry matter	%	92.1	100.0	1
Ash	%	6.9	7.5	13
Crude fiber	%	32.4	35.2	4
Ether extract	%	2.6	2.9	10
N-free extract	%	43.4	47.2	
Protein (N x 6.25)	%	6.7	7.3	9
Cattle	dig prot %	* 3.0	3.3	
Goats	dig prot %	* 3.1	3.4	
Horses	dig prot %	* 3.4	3.7	
Rabbits	dig prot %	* 4.0	4.3	
Sheep	dig prot %	* 2.9	3.1	

Feed Name or Analyses		Mean As Fed	Mean Dry	C.V. ± %
Fatty acids	%	2.6	2.8	
Energy	GE Mcal/kg			
Cattle	DE Mcal/kg	* 2.25	2.44	
Sheep	DE Mcal/kg	* 2.20	2.39	
Cattle	ME Mcal/kg	* 1.84	2.00	
Sheep	ME Mcal/kg	* 1.81	1.96	
Cattle	TDN %	* 51.0	55.4	
Sheep	TDN %	* 50.0	54.3	
Calcium	%	0.22	0.24	
Phosphorus	%	0.25	0.27	

Oats, wild, aerial part, fresh, (2)

Ref No 2 03 393 United States

Feed Name or Analyses		Mean As Fed	Mean Dry	C.V. ± %
Dry matter	%	35.4	100.0	
Ash	%	2.6	7.4	
Crude fiber	%	8.3	23.5	
Ether extract	%	1.3	3.8	
N-free extract	%	20.6	58.2	
Protein (N x 6.25)	%	2.5	7.1	
Cattle	dig prot %	* 1.4	3.9	
Goats	dig prot %	* 1.1	3.2	
Horses	dig prot %	* 1.3	3.6	
Rabbits	dig prot %	* 1.5	4.2	
Sheep	dig prot %	* 1.3	3.6	
Energy	GE Mcal/kg			
Cattle	DE Mcal/kg	* 0.94	2.65	
Sheep	DE Mcal/kg	* 1.12	3.15	
Cattle	ME Mcal/kg	* 0.77	2.17	
Sheep	ME Mcal/kg	* 0.91	2.58	
Cattle	TDN %	* 21.3	60.1	
Sheep	TDN %	* 25.3	71.5	
Calcium	%	0.09	0.25	
Phosphorus	%	0.10	0.27	

Oats, wild, grain, (4)

Ref No 4 03 394 United States

Feed Name or Analyses		Mean As Fed	Mean Dry	C.V. ± %
Dry matter	%	89.2	100.0	1
Ash	%	4.6	5.2	10
Crude fiber	%	15.0	16.8	6
Cattle	dig coef %	* 30.	30.	
Sheep	dig coef %	* 37.	37.	
Ether extract	%	5.6	6.2	12
Cattle	dig coef %	* 82.	82.	
Sheep	dig coef %	* 82.	82.	
N-free extract	%	51.8	58.1	
Cattle	dig coef %	* 81.	81.	
Sheep	dig coef %	* 83.	83.	
Protein (N x 6.25)	%	12.2	13.7	9
Cattle	dig coef %	* 75.	75.	
Sheep	dig coef %	* 78.	78.	
Cattle	dig prot %	9.1	10.3	
Goats	dig prot %	* 8.7	9.8	
Horses	dig prot %	* 8.7	9.8	
Sheep	dig prot %	9.5	10.7	
Energy	GE Mcal/kg			
Cattle	DE Mcal/kg	* 2.90	3.26	
Sheep	DE Mcal/kg	* 3.01	3.38	
Swine	DE kcal/kg	*1985.	2226.	
Cattle	ME Mcal/kg	* 2.38	2.67	
Sheep	ME Mcal/kg	* 2.47	2.77	
Swine	ME kcal/kg	*1850.	2076.	
Cattle	TDN %	65.9	73.9	
Sheep	TDN %	68.3	76.6	
Swine	TDN %	* 45.0	50.5	

Ref No 4 03 394 Canada

Feed Name or Analyses		Mean As Fed	Mean Dry	C.V. ± %
Dry matter	%	92.9	100.0	
Ash	%	3.9	4.2	
Crude fiber	%	14.8	15.9	
Ether extract	%	5.4	5.9	
N-free extract	%	55.6	59.8	
Protein (N x 6.25)	%	13.3	14.3	
Cattle	dig prot %	* 8.5	9.1	
Goats	dig prot %	* 9.6	10.3	
Horses	dig prot %	* 9.6	10.3	
Sheep	dig prot %	* 9.6	10.3	
Cellulose (Matrone)	%	15.4	16.5	
Energy	GE Mcal/kg			
Cattle	DE Mcal/kg	* 3.04	3.27	
Sheep	DE Mcal/kg	* 3.15	3.39	
Swine	DE kcal/kg	*2263.	2435.	
Cattle	ME Mcal/kg	* 2.49	2.68	
Sheep	ME Mcal/kg	* 2.58	2.78	
Swine	ME kcal/kg	*2107.	2267.	
Cattle	TDN %	68.9	74.2	
Sheep	TDN %	71.5	76.9	
Swine	TDN %	* 51.3	55.2	
Copper	mg/kg	6.3	6.8	
Iron	%	0.005	0.006	
Magnesium	%	0.15	0.16	
Manganese	mg/kg	30.3	32.6	
Selenium	mg/kg	0.594	0.640	
Zinc	mg/kg	34.0	36.6	
Niacin	mg/kg	27.7	29.9	
Pantothenic acid	mg/kg	36.1	38.8	
Riboflavin	mg/kg	1.6	1.7	
Thiamine	mg/kg	1.0	1.0	
Vitamin B_6	mg/kg	2.10	2.26	
Alanine	%	0.53	0.57	
Arginine	%	0.83	0.90	
Aspartic acid	%	0.82	0.88	
Glutamic acid	%	1.84	1.98	
Glycine	%	0.54	0.58	
Histidine	%	0.24	0.26	
Isoleucine	%	0.40	0.43	
Leucine	%	0.73	0.78	
Lysine	%	0.41	0.44	
Methionine	%	0.11	0.12	
Phenylalanine	%	0.49	0.53	
Proline	%	0.48	0.52	
Serine	%	0.42	0.46	
Threonine	%	0.33	0.35	
Tyrosine	%	0.29	0.31	
Valine	%	0.50	0.53	

OATS-CLOVER. Avena sativa, Trifolium spp

Oats-Clover, aerial part, ensiled, (3)

Ref No 3 03 395 United States

Feed Name or Analyses		Mean As Fed	Mean Dry	C.V. ± %
Dry matter	%		100.0	
Calcium	%		0.48	
Phosphorus	%		0.17	

OATS-CLOVER, LADINO. Avena sativa, Trifolium repens

Oats-Clover, ladino, aerial part, fresh, (2)
Ref No 2 03 396 United States

Feed Name or Analyses	Unit		Mean As Fed	Mean Dry	C.V. ± %
Dry matter	%		34.2	100.0	8
Ash	%		2.5	7.4	7
Crude fiber	%		8.7	25.5	10
Ether extract	%		0.9	2.7	14
N-free extract	%		16.7	48.9	
Protein (N x 6.25)	%		5.3	15.5	15
Cattle	dig prot %	*	3.8	11.1	
Goats	dig prot %	*	3.8	11.0	
Horses	dig prot %	*	3.7	10.7	
Rabbits	dig prot %	*	3.6	10.6	
Sheep	dig prot %	*	3.9	11.4	
Energy	GE Mcal/kg				
Cattle	DE Mcal/kg	*	1.02	2.97	
Sheep	DE Mcal/kg	*	0.97	2.82	
Cattle	ME Mcal/kg	*	0.83	2.44	
Sheep	ME Mcal/kg	*	0.79	2.31	
Cattle	TDN %	*	23.0	67.3	
Sheep	TDN %	*	21.9	64.0	

OATS-COWPEA. Avena sativa, Vigna spp

Oats-Cowpea, aerial part, ensiled, (3)
Ref No 3 03 397 United States

Feed Name or Analyses	Unit		Mean As Fed	Mean Dry	C.V. ± %
Dry matter	%			100.0	
Ash	%			11.0	
Crude fiber	%			28.8	
Ether extract	%			4.2	
N-free extract	%			41.2	
Protein (N x 6.25)	%			14.8	
Cattle	dig prot %	*		9.7	
Goats	dig prot %	*		9.7	
Horses	dig prot %	*		9.7	
Sheep	dig prot %	*		9.7	
Energy	GE Mcal/kg				
Cattle	DE Mcal/kg	*		2.79	
Sheep	DE Mcal/kg	*		2.70	
Cattle	ME Mcal/kg	*		2.29	
Sheep	ME Mcal/kg	*		2.21	
Cattle	TDN %	*		63.4	
Sheep	TDN %	*		61.2	

Oats, grain, heavy —
see Oats, grain, gr 2 heavy US mn wt 36 lb per bushel mx 3% foreign material, (4)

Oats, grain, light —
see Oats, grain, gr 4 US mn wt 27 lb per bushel mx 5% foreign material, (4)

Oat shorts (CFA) —
see Oats, cereal by-product, mx 7% fiber, (4)

(1) dry forages and roughages
(2) pasture, range plants, and forages fed green
(3) silages
(4) energy feeds
(5) protein supplements
(6) minerals
(7) vitamins
(8) additives

OATS-PEA. Avena sativa, Pisum spp

Oats-Pea, hay, s-c, (1)
Ref No 1 03 398 United States

Feed Name or Analyses	Unit		Mean As Fed	Mean Dry	C.V. ± %
Dry matter	%		87.3	100.0	3
Ash	%		7.4	8.5	3
Crude fiber	%		26.8	30.6	6
Sheep	dig coef %		60.	60.	
Ether extract	%		3.0	3.4	20
Sheep	dig coef %		58.	58.	
N-free extract	%		38.6	44.2	6
Sheep	dig coef %		62.	62.	
Protein (N x 6.25)	%		11.6	13.3	11
Sheep	dig coef %		71.	71.	
Cattle	dig prot %	*	7.4	8.5	
Goats	dig prot %	*	7.8	9.0	
Horses	dig prot %	*	7.7	8.8	
Rabbits	dig prot %	*	7.8	8.9	
Sheep	dig prot %		8.3	9.5	
Fatty acids	%		2.8	3.3	
Energy	GE Mcal/kg				
Cattle	DE Mcal/kg	*	2.29	2.62	
Sheep	DE Mcal/kg	*	2.30	2.63	
Cattle	ME Mcal/kg	*	1.87	2.15	
Sheep	ME Mcal/kg	*	1.88	2.16	
Cattle	TDN %	*	51.8	59.4	
Sheep	TDN %		52.1	59.6	
Calcium	%		0.71	0.81	
Phosphorus	%		0.22	0.25	
Potassium	%		1.02	1.17	

Oats-Pea, aerial part, fresh, early bloom, (2)
Ref No 2 03 399 United States

Feed Name or Analyses	Unit		Mean As Fed	Mean Dry	C.V. ± %
Dry matter	%		22.9	100.0	22
Ash	%		2.0	8.8	19
Crude fiber	%		7.0	30.5	6
Ether extract	%		0.9	3.9	16
N-free extract	%		9.5	41.5	
Protein (N x 6.25)	%		3.5	15.3	12
Cattle	dig prot %	*	2.5	10.9	
Goats	dig prot %	*	2.5	10.8	
Horses	dig prot %	*	2.4	10.5	
Rabbits	dig prot %	*	2.4	10.5	
Sheep	dig prot %	*	2.6	11.2	
Energy	GE Mcal/kg				
Cattle	DE Mcal/kg	*	0.69	3.02	
Sheep	DE Mcal/kg	*	0.62	2.69	
Cattle	ME Mcal/kg	*	0.57	2.47	
Sheep	ME Mcal/kg	*	0.50	2.20	
Cattle	TDN %	*	15.7	68.4	
Sheep	TDN %	*	14.0	61.0	

Oats-Pea, aerial part, fresh, dough stage, (2)
Ref No 2 03 400 United States

Feed Name or Analyses	Unit		Mean As Fed	Mean Dry	C.V. ± %
Dry matter	%		21.8	100.0	17
Ash	%		1.8	8.2	10
Crude fiber	%		6.1	27.9	6
Ether extract	%		0.8	3.8	10
N-free extract	%		10.1	46.4	
Protein (N x 6.25)	%		3.0	13.7	10
Cattle	dig prot %	*	2.1	9.5	
Goats	dig prot %	*	2.0	9.3	
Horses	dig prot %	*	2.0	9.2	
Rabbits	dig prot %	*	2.0	9.2	
Sheep	dig prot %	*	2.1	9.8	
Energy	GE Mcal/kg				
Cattle	DE Mcal/kg	*	0.65	2.96	
Sheep	DE Mcal/kg	*	0.61	2.81	
Cattle	ME Mcal/kg	*	0.53	2.43	
Sheep	ME Mcal/kg	*	0.50	2.30	
Cattle	TDN %	*	14.6	67.2	
Sheep	TDN %	*	13.9	63.7	
Calcium	%		0.08	0.35	
Phosphorus	%		0.10	0.45	

Oats-Pea, aerial part, ensiled, (3)
Ref No 3 03 402 United States

Feed Name or Analyses	Unit		Mean As Fed	Mean Dry	C.V. ± %
Dry matter	%		26.4	100.0	16
Ash	%		2.6	9.8	11
Crude fiber	%		8.6	32.4	9
Sheep	dig coef %		54.	54.	
Ether extract	%		0.9	3.4	15
Sheep	dig coef %		70.	70.	
N-free extract	%		11.5	43.4	8
Sheep	dig coef %		63.	63.	
Protein (N x 6.25)	%		2.9	11.0	7
Sheep	dig coef %		50.	50.	
Cattle	dig prot %	*	1.6	6.2	
Goats	dig prot %	*	1.6	6.2	
Horses	dig prot %	*	1.6	6.2	
Sheep	dig prot %		1.5	5.5	
Energy	GE Mcal/kg				
Cattle	DE Mcal/kg	*	0.72	2.72	
Sheep	DE Mcal/kg	*	0.65	2.46	
Cattle	ME Mcal/kg	*	0.59	2.23	
Sheep	ME Mcal/kg	*	0.53	2.01	
Cattle	TDN %	*	16.3	61.6	
Sheep	TDN %		14.7	55.7	
Calcium	%		0.16	0.62	18
Chlorine	%		0.31	1.19	
Cobalt	mg/kg		0.064	0.240	
Copper	mg/kg		4.9	18.5	
Iron	%		0.016	0.062	
Magnesium	%		0.12	0.44	
Manganese	mg/kg		8.7	33.1	
Phosphorus	%		0.09	0.32	20
Potassium	%		0.49	1.86	
Carotene	mg/kg		20.5	77.7	
Vitamin A equivalent	IU/g		34.2	129.5	

Oats-Pea, aerial part, ensiled, full bloom, (3)
Ref No 3 03 401 United States

Feed Name or Analyses	Unit		Mean As Fed	Mean Dry	C.V. ± %
Dry matter	%		26.2	100.0	13
Ash	%		2.0	7.8	20
Crude fiber	%		8.8	33.4	7
Ether extract	%		1.6	6.1	13
N-free extract	%		10.5	40.0	
Protein (N x 6.25)	%		3.3	12.7	13
Cattle	dig prot %	*	2.0	7.8	
Goats	dig prot %	*	2.0	7.8	
Horses	dig prot %	*	2.0	7.8	
Sheep	dig prot %	*	2.0	7.8	
Energy	GE Mcal/kg				
Cattle	DE Mcal/kg	*	0.69	2.63	
Sheep	DE Mcal/kg	*	0.71	2.70	
Cattle	ME Mcal/kg	*	0.57	2.16	
Sheep	ME Mcal/kg	*	0.58	2.22	
Cattle	TDN %	*	15.6	59.7	
Sheep	TDN %	*	16.1	61.3	

Feed Name or Analyses			Mean As Fed	Mean Dry	C.V. ± %

OATS-VETCH. Avena sativa, Vicia spp

Oats-Vetch, hay, s-c, (1)
Ref No 1 03 404 United States

Analyses			As Fed	Dry	C.V. ± %
Dry matter	%		85.4	100.0	3
Ash	%		6.7	7.8	19
Crude fiber	%		27.5	32.2	4
Cattle	dig coef %		63.	63.	
Sheep	dig coef %		51.	51.	
Ether extract	%		2.5	2.9	23
Cattle	dig coef %		72.	72.	
Sheep	dig coef %		52.	52.	
N-free extract	%		38.7	45.3	
Cattle	dig coef %		63.	63.	
Sheep	dig coef %		67.	67.	
Protein (N x 6.25)	%		10.0	11.7	20
Cattle	dig coef %		88.	88.	
Sheep	dig coef %		71.	71.	
Cattle	dig prot %		8.8	10.3	
Goats	dig prot %	*	6.4	7.5	
Horses	dig prot %	*	6.4	7.5	
Rabbits	dig prot %	*	6.6	7.7	
Sheep	dig prot %		7.1	8.3	
Energy	GE Mcal/kg				
Cattle	DE Mcal/kg	*	2.41	2.82	
Sheep	DE Mcal/kg	*	2.21	2.58	
Cattle	ME Mcal/kg	*	1.97	2.31	
Sheep	ME Mcal/kg	*	1.81	2.12	
Cattle	TDN %		54.6	63.9	
Sheep	TDN %		50.0	58.6	
Vitamin D_2	IU/g		0.8	1.0	

Oats-Vetch, hay, s-c, milk stage, (1)
Ref No 1 03 403 United States

Analyses			As Fed	Dry	C.V. ± %
Dry matter	%		80.8	100.0	3
Ash	%		6.4	8.0	14
Crude fiber	%		26.6	33.0	1
Sheep	dig coef %		49.	49.	
Ether extract	%		2.2	2.8	7
Sheep	dig coef %		63.	63.	
N-free extract	%		36.6	45.3	
Sheep	dig coef %		59.	59.	
Protein (N x 6.25)	%		8.9	11.1	14
Sheep	dig coef %		65.	65.	
Cattle	dig prot %	*	5.3	6.5	
Goats	dig prot %	*	5.6	6.9	
Horses	dig prot %	*	5.6	6.9	
Rabbits	dig prot %	*	5.8	7.2	
Sheep	dig prot %		5.8	7.2	
Energy	GE Mcal/kg				
Cattle	DE Mcal/kg	*	2.09	2.58	
Sheep	DE Mcal/kg	*	1.92	2.38	
Cattle	ME Mcal/kg	*	1.71	2.12	
Sheep	ME Mcal/kg	*	1.58	1.95	
Cattle	TDN %	*	47.3	58.6	
Sheep	TDN %		43.6	54.0	

Oats-Vetch, hay, s-c, gr 2 US, (1)
Ref No 1 03 405 United States

Analyses			As Fed	Dry	C.V. ± %
Dry matter	%		89.7	100.0	
Ash	%		5.2	5.8	
Crude fiber	%		30.1	33.6	
Ether extract	%		2.4	2.7	
N-free extract	%		45.5	50.7	
Protein (N x 6.25)	%		6.5	7.2	
Cattle	dig prot %	*	2.8	3.2	
Goats	dig prot %	*	2.9	3.3	
Horses	dig prot %	*	3.3	3.6	
Rabbits	dig prot %	*	3.8	4.2	
Sheep	dig prot %	*	2.7	3.0	
Energy	GE Mcal/kg				
Cattle	DE Mcal/kg	*	2.23	2.48	
Sheep	DE Mcal/kg	*	2.19	2.44	
Cattle	ME Mcal/kg	*	1.83	2.03	
Sheep	ME Mcal/kg	*	1.79	2.00	
Cattle	TDN %	*	50.5	56.3	
Sheep	TDN %	*	49.6	55.3	
Calcium	%		0.39	0.44	
Phosphorus	%		0.20	0.22	
Carotene	mg/kg		19.8	22.0	
Vitamin A equivalent	IU/g		33.0	36.8	

Oats-Vetch, aerial part, fresh, milk stage, (2)
Ref No 2 03 407 United States

Analyses			As Fed	Dry	C.V. ± %
Dry matter	%		32.5	100.0	
Ash	%		2.5	7.8	
Crude fiber	%		9.1	28.1	
Ether extract	%		1.0	3.0	
N-free extract	%		16.3	50.3	
Protein (N x 6.25)	%		3.5	10.8	
Cattle	dig prot %	*	2.3	7.1	
Goats	dig prot %	*	2.2	6.6	
Horses	dig prot %	*	2.2	6.7	
Rabbits	dig prot %	*	2.3	7.0	
Sheep	dig prot %	*	2.3	7.1	
Energy	GE Mcal/kg				
Cattle	DE Mcal/kg	*	0.94	2.90	
Sheep	DE Mcal/kg	*	0.95	2.92	
Cattle	ME Mcal/kg	*	0.77	2.38	
Sheep	ME Mcal/kg	*	0.78	2.40	
Cattle	TDN %	*	21.4	65.7	
Sheep	TDN %	*	21.6	66.3	

Oats-Vetch, aerial part, ensiled, (3)
Ref No 3 03 409 United States

Analyses			As Fed	Dry	C.V. ± %
Dry matter	%		31.0	100.0	11
Ash	%		2.7	8.7	6
Crude fiber	%		9.7	31.2	6
Ether extract	%		1.3	4.3	1
N-free extract	%		12.8	41.4	
Protein (N x 6.25)	%		4.5	14.4	12
Cattle	dig prot %	*	2.9	9.3	
Goats	dig prot %	*	2.9	9.3	
Horses	dig prot %	*	2.9	9.3	
Sheep	dig prot %	*	2.9	9.3	
Energy	GE Mcal/kg				
Cattle	DE Mcal/kg	*	0.80	2.59	
Sheep	DE Mcal/kg	*	0.85	2.73	
Cattle	ME Mcal/kg	*	0.66	2.13	
Sheep	ME Mcal/kg	*	0.69	2.24	
Cattle	TDN %	*	18.2	58.8	
Sheep	TDN %	*	19.2	62.0	
Calcium	%		0.17	0.56	
Phosphorus	%		0.07	0.21	

Oats-Vetch, aerial part, ensiled, milk stage, (3)
Ref No 3 03 408 United States

Analyses			As Fed	Dry	C.V. ± %
Dry matter	%		27.3	100.0	
Ash	%		2.2	8.1	
Crude fiber	%		8.0	29.4	
Ether extract	%		1.2	4.3	
N-free extract	%		12.4	45.6	
Protein (N x 6.25)	%		3.4	12.6	
Cattle	dig prot %	*	2.1	7.7	
Goats	dig prot %	*	2.1	7.7	
Horses	dig prot %	*	2.1	7.7	
Sheep	dig prot %	*	2.1	7.7	
Energy	GE Mcal/kg				
Cattle	DE Mcal/kg	*	0.74	2.72	
Sheep	DE Mcal/kg	*	0.76	2.78	
Cattle	ME Mcal/kg	*	0.61	2.23	
Sheep	ME Mcal/kg	*	0.62	2.28	
Cattle	TDN %	*	16.9	61.7	
Sheep	TDN %	*	17.2	63.1	

Ocean perch, whole, raw –
see Fish, redfish, whole, raw, (5)

OCOTILLO. Fouquieria splendens

Ocotillo, aerial part, fresh, mature, (2)
Ref No 2 03 410 United States

Analyses			As Fed	Dry	C.V. ± %
Dry matter	%		36.2	100.0	
Ash	%		1.3	3.5	
Protein (N x 6.25)	%		1.3	3.7	
Cattle	dig prot %	*	0.4	1.0	
Goats	dig prot %	*	0.0	0.0	
Horses	dig prot %	*	0.2	0.7	
Rabbits	dig prot %	*	0.6	1.5	
Sheep	dig prot %	*	0.2	0.4	
Calcium	%		0.23	0.63	
Chlorine	%		0.04	0.10	
Iron	%		0.006	0.017	
Magnesium	%		0.05	0.15	
Manganese	mg/kg		28.1	77.6	
Phosphorus	%		0.03	0.07	
Potassium	%		0.18	0.50	
Sodium	%		0.01	0.04	
Sulphur	%		0.27	0.75	

Ocotillo, seeds w hulls, (5)
Ref No 5 09 008 United States

Analyses			As Fed	Dry	C.V. ± %
Dry matter	%			100.0	
Protein (N x 6.25)	%			43.0	
Alanine	%			1.63	
Arginine	%			3.96	
Aspartic acid	%			3.23	
Glutamic acid	%			8.00	
Glycine	%			1.51	
Histidine	%			0.95	
Hydroxyproline	%			0.43	
Isoleucine	%			1.42	
Leucine	%			2.54	
Lysine	%			1.76	
Methionine	%			0.65	
Phenylalanine	%			1.42	
Proline	%			1.55	
Serine	%			1.68	

Continued

Feed Name or Analyses	Mean As Fed	Mean Dry	C.V. ± %
Threonine %		1.25	
Tyrosine %		1.03	
Valine %		1.72	

OIL FUNGUS. Endomyces vernalis

Oil fungus, entire plant, (5)

Ref No 5 02 074 United States

Feed Name or Analyses	Mean As Fed	Mean Dry	C.V. ± %
Dry matter %	93.8	100.0	
Ash %	4.9	5.2	
Ether extract %	27.9	29.7	
Sheep dig coef %	80.	80.	
N-free extract %	38.6	41.1	
Sheep dig coef %	75.	75.	
Protein (N x 6.25) %	22.5	24.0	
Sheep dig coef %	65.	65.	
Sheep dig prot %	14.6	15.6	
Energy GE Mcal/kg			
Cattle DE Mcal/kg	* 4.23	4.51	
Sheep DE Mcal/kg	* 4.13	4.40	
Swine DE kcal/kg	*4867.	5189.	
Cattle ME Mcal/kg	* 3.47	3.70	
Sheep ME Mcal/kg	* 3.39	3.61	
Swine ME kcal/kg	*4439.	4732.	
Cattle TDN %	* 96.0	102.3	
Sheep TDN %	93.7	99.9	
Swine TDN %	* 110.4	117.7	

Oily fish meal (CFA) –
see Fish, whole or cuttings, cooked dehy grnd, mn 9% oil salt declared above 4%, (5)

OKRA. Hibiscus esculentus

Okra, seed hulls, (1)

Ref No 1 03 411 United States

Feed Name or Analyses	Mean As Fed	Mean Dry	C.V. ± %
Dry matter %	91.3	100.0	1
Ether extract %	1.1	1.2	44
Protein (N x 6.25) %	3.8	4.2	8
Cattle dig prot %	* 0.5	0.6	
Goats dig prot %	* 0.4	0.5	
Horses dig prot %	* 1.0	1.1	
Rabbits dig prot %	* 1.7	1.9	
Sheep dig prot %	* 0.3	0.3	

OLIVES. Olea europaea

Olives, fruit wo pits, extn unspecified grnd, (1)

Ref No 1 03 413 United States

Feed Name or Analyses	Mean As Fed	Mean Dry	C.V. ± %
Dry matter %	92.0	100.0	
Ash %	2.5	2.7	
Crude fiber %	36.4	39.6	
Sheep dig coef %	-19.	-19.	
Ether extract %	15.6	17.0	
Sheep dig coef %	86.	86.	
N-free extract %	31.6	34.3	
Sheep dig coef %	20.	20.	
Protein (N x 6.25) %	5.9	6.4	
Sheep dig coef %	-10.	-10.	

(1) dry forages and roughages (2) pasture, range plants, and forages fed green (3) silages (4) energy feeds (5) protein supplements (6) minerals (7) vitamins (8) additives

Feed Name or Analyses	Mean As Fed	Mean Dry	C.V. ± %
Cattle dig prot %	* 2.3	2.5	
Goats dig prot %	* 2.3	2.5	
Horses dig prot %	* 2.7	3.0	
Rabbits dig prot %	* 3.3	3.6	
Sheep dig prot %	-0.5	-0.6	
Energy GE Mcal/kg			
Cattle DE Mcal/kg	* 1.69	1.84	
Sheep DE Mcal/kg	* 1.26	1.37	
Cattle ME Mcal/kg	* 1.39	1.51	
Sheep ME Mcal/kg	* 1.04	1.13	
Cattle TDN %	* 38.5	41.8	
Sheep TDN %	28.6	31.1	

Olives, fruit wo pits, dehy, (1)

Ref No 1 08 474 United States

Feed Name or Analyses	Mean As Fed	Mean Dry	C.V. ± %
Dry matter %	95.1	100.0	
Ash %	3.4	3.6	
Crude fiber %	19.3	20.3	
Ether extract %	27.4	28.8	
N-free extract %	31.0	32.6	
Protein (N x 6.25) %	14.0	14.7	
Cattle dig prot %	* 9.2	9.7	
Goats dig prot %	* 9.8	10.3	
Horses dig prot %	* 9.5	10.0	
Rabbits dig prot %	* 9.5	10.0	
Sheep dig prot %	* 9.3	9.8	
Energy GE Mcal/kg			
Cattle DE Mcal/kg	* 3.41	3.59	
Sheep DE Mcal/kg	* 3.19	3.35	
Cattle ME Mcal/kg	* 2.80	2.94	
Sheep ME Mcal/kg	* 2.62	2.75	
Cattle TDN %	* 77.4	81.4	
Sheep TDN %	* 72.2	75.9	

Olives, fruit wo pits, solv-extd dehy, (1)

Ref No 1 08 475 United States

Feed Name or Analyses	Mean As Fed	Mean Dry	C.V. ± %
Dry matter %	91.3	100.0	
Ash %	8.5	9.3	
Crude fiber %	24.6	26.9	
Ether extract %	3.6	3.9	
N-free extract %	41.6	45.6	
Protein (N x 6.25) %	13.0	14.2	
Cattle dig prot %	* 8.5	9.3	
Goats dig prot %	* 9.0	9.8	
Horses dig prot %	* 8.8	9.6	
Rabbits dig prot %	* 8.8	9.7	
Sheep dig prot %	* 8.5	9.3	
Energy GE Mcal/kg			
Cattle DE Mcal/kg	* 1.57	1.72	
Sheep DE Mcal/kg	* 1.40	1.53	
Cattle ME Mcal/kg	* 1.29	1.41	
Sheep ME Mcal/kg	* 1.14	1.25	
Cattle TDN %	* 35.5	38.9	
Sheep TDN %	* 31.7	34.7	

Olives, fruit w pits, dehy, (1)

Ref No 1 08 473 United States

Feed Name or Analyses	Mean As Fed	Mean Dry	C.V. ± %
Dry matter %	92.0	100.0	
Ash %	2.5	2.7	
Crude fiber %	36.5	39.7	
Ether extract %	15.6	17.0	
N-free extract %	31.5	34.2	
Protein (N x 6.25) %	5.9	6.4	
Cattle dig prot %	* 2.3	2.5	

Feed Name or Analyses	Mean As Fed	Mean Dry	C.V. ± %
Goats dig prot %	* 2.3	2.5	
Horses dig prot %	* 2.7	3.0	
Rabbits dig prot %	* 3.3	3.6	
Sheep dig prot %	* 2.1	2.3	
Energy GE Mcal/kg			
Cattle DE Mcal/kg	* 1.71	1.86	
Sheep DE Mcal/kg	* 1.56	1.70	
Cattle ME Mcal/kg	* 1.41	1.53	
Sheep ME Mcal/kg	* 1.28	1.39	
Cattle TDN %	* 38.8	42.2	
Sheep TDN %	* 35.5	38.6	

One-flowered sunflower-
See Helianthella, oneflower, aerial part, fresh, dormant, (2)

One-flowered sunflower-
See Helianthella, oneflower, aerial part, fresh, mature, (2)

One-flowered sunflower -
See Helianthella, oneflower, aerial part, fresh, early leaf, (2)

ONION. Allium spp

Onion, aerial part, fresh, (2)

Ref No 2 03 417 United States

Feed Name or Analyses	Mean As Fed	Mean Dry	C.V. ± %
Dry matter %	91.4	100.0	
Ash %	7.3	8.0	
Crude fiber %	20.7	22.6	
Ether extract %	1.8	2.0	
N-free extract %	50.1	54.8	
Protein (N x 6.25) %	11.5	12.6	
Cattle dig prot %	* 7.9	8.6	
Goats dig prot %	* 7.6	8.3	
Horses dig prot %	* 7.5	8.2	
Rabbits dig prot %	* 7.7	8.4	
Sheep dig prot %	* 8.0	8.7	
Calcium %	1.65	1.80	
Magnesium %	0.15	0.16	
Phosphorus %	0.19	0.21	
Potassium %	1.61	1.76	

ORACH, SILVERY. Atriplex argentea

Orach, silvery, leaves, fresh, mature, (2)

Ref No 2 03 419 United States

Feed Name or Analyses	Mean As Fed	Mean Dry	C.V. ± %
Dry matter %		100.0	
Ash %		20.4	
Crude fiber %		28.9	
Ether extract %		1.5	
N-free extract %		38.9	
Protein (N x 6.25) %		10.3	
Cattle dig prot %	*	6.6	
Goats dig prot %	*	6.2	
Horses dig prot %	*	6.3	
Rabbits dig prot %	*	6.6	
Sheep dig prot %	*	6.6	

Feed Name or Analyses	Mean As Fed	Mean Dry	C.V. ± %

ORACH, TUMBLING. Atriplex rosea

Orach, tumbling, leaves, fresh, (2)
Ref No 2 03 421 United States

Feed Name or Analyses	As Fed	Dry	C.V. ± %
Dry matter %		100.0	
Ash %		23.0	27
Crude fiber %		14.5	27
Ether extract %		1.8	19
N-free extract %		45.2	
Protein (N x 6.25) %		15.5	31
Cattle dig prot % *		11.1	
Goats dig prot % *		11.0	
Horses dig prot % *		10.7	
Rabbits dig prot % *		10.6	
Sheep dig prot % *		11.4	

Orach, tumbling, leaves, fresh, immature, (2)
Ref No 2 03 420 United States

Feed Name or Analyses	As Fed	Dry	C.V. ± %
Dry matter %		100.0	
Ash %		31.7	
Crude fiber %		8.1	
Ether extract %		2.1	
N-free extract %		34.4	
Protein (N x 6.25) %		23.7	
Cattle dig prot % *		18.0	
Goats dig prot % *		18.7	
Horses dig prot % *		17.7	
Rabbits dig prot % *		17.0	
Sheep dig prot % *		19.1	

Orange pulp, ammoniated –
see Citrus, orange, pulp wo fines, ammoniated shredded dehy, (4)

Orange pulp, dried –
see Citrus, orange, pulp wo fines, shredded dehy, (4)

Orange, pulp, wet –
see Citrus, orange, pulp, shredded wet, (4)

ORCHARDGRASS. Dactylis glomerata

Orchardgrass, hay, fan air dried, early leaf, (1)
Ref No 1 08 754 United States

Feed Name or Analyses	As Fed	Dry	C.V. ± %
Dry matter %		100.0	
Lignin (Ellis) %		4.0	

Orchardgrass, hay, fan air dried, immature, (1)
Ref No 1 08 752 United States

Feed Name or Analyses	As Fed	Dry	C.V. ± %
Dry matter %		100.0	
Lignin (Ellis) %		4.4	

Orchardgrass, hay, fan air dried, pre-bloom, (1)
Ref No 1 08 750 United States

Feed Name or Analyses	As Fed	Dry	C.V. ± %
Dry matter %		100.0	
Lignin (Ellis) %		4.0	

Orchardgrass, hay, fan air dried, early bloom, (1)
Ref No 1 08 751 United States

Feed Name or Analyses	As Fed	Dry	C.V. ± %
Dry matter %		100.0	
Lignin (Ellis) %		6.1	

Orchardgrass, hay, fan air dried, mid-bloom, (1)
Ref No 1 08 753 United States

Feed Name or Analyses	As Fed	Dry	C.V. ± %
Dry matter %		100.0	
Lignin (Ellis) %		7.5	

Orchardgrass, hay, fan air dried chopped, pre-bloom, (1)
Ref No 1 07 715 United States

Feed Name or Analyses	As Fed	Dry	C.V. ± %
Dry matter %		100.0	
Sheep dig coef %		64.	
Ash %		9.1	
Protein (N x 6.25) %		17.5	
Sheep dig coef %		72.	
Cattle dig prot % *		12.1	
Goats dig prot % *		12.9	
Horses dig prot % *		12.4	
Rabbits dig prot % *		12.2	
Sheep dig prot %		12.6	
Cellulose (Matrone) %		31.1	
Sheep dig coef %		70.	
Energy GE Mcal/kg		4.40	
Sheep GE dig coef %		62.	
Sheep DE Mcal/kg		2.73	
Sheep ME Mcal/kg *		2.24	
Sheep TDN % *		62.0	

Orchardgrass, hay, fan air dried chopped, mid-bloom, (1)
Ref No 1 07 716 United States

Feed Name or Analyses	As Fed	Dry	C.V. ± %
Dry matter %		100.0	
Sheep dig coef %		58.	
Ash %		7.6	
Protein (N x 6.25) %		13.8	
Sheep dig coef %		68.	
Cattle dig prot % *		8.9	
Goats dig prot % *		9.4	
Horses dig prot % *		9.2	
Rabbits dig prot % *		9.3	
Sheep dig prot %		9.4	
Cellulose (Matrone) %		33.5	
Sheep dig coef %		60.	
Energy GE Mcal/kg		4.39	
Sheep GE dig coef %		55.	
Sheep DE Mcal/kg		2.41	
Sheep ME Mcal/kg *		1.98	
Sheep TDN % *		54.8	

Orchardgrass, hay, fan air dried w heat, (1)
Ref No 1 05 575 United States

Feed Name or Analyses	As Fed	Dry	C.V. ± %
Dry matter %		100.0	
Rabbits dig coef %		51.	
Sheep dig coef %		65.	
Ash %		8.7	16
Crude fiber %		25.5	8
Rabbits dig coef %		37.	
Sheep dig coef %		70.	
Ether extract %		6.4	15
Rabbits dig coef %		42.	
Sheep dig coef %		45.	
N-free extract %		35.3	11
Rabbits dig coef %		48.	
Sheep dig coef %		63.	
Protein (N x 6.25) %		24.2	14
Rabbits dig coef %		75.	
Sheep dig coef %		76.	
Cattle dig prot % *		17.9	
Goats dig prot % *		19.1	
Horses dig prot % *		18.0	
Rabbits dig prot %		18.1	
Sheep dig prot %		18.4	
Soluble carbohydrates %		9.5	
Lignin (Ellis) %		5.2	
Rabbits dig coef %		2.	
Energy GE Mcal/kg			
Cattle DE Mcal/kg *		2.90	
Rabbits DE kcal/kg *		2227.	
Sheep DE Mcal/kg *		2.85	
Cattle ME Mcal/kg *		2.38	
Sheep ME Mcal/kg *		2.34	
Cattle TDN % *		65.8	
Rabbits TDN %		50.5	
Sheep TDN %		64.7	

Orchardgrass, hay, fan air dried w heat, cut 1, (1)
Ref No 1 05 567 United States

Feed Name or Analyses	As Fed	Dry	C.V. ± %
Dry matter %	91.6	100.0	
Sheep dig coef %	71.	71.	
Ash %	6.0	6.6	
Crude fiber %	24.9	27.2	
Sheep dig coef %	75.	75.	
Ether extract %	4.5	4.9	
Sheep dig coef %	52.	52.	
N-free extract %	39.8	43.4	
Sheep dig coef %	74.	74.	
Protein (N x 6.25) %	16.4	17.9	
Sheep dig coef %	72.	72.	
Cattle dig prot % *	11.4	12.4	
Goats dig prot % *	12.1	13.3	
Horses dig prot % *	11.7	12.7	
Rabbits dig prot % *	11.4	12.5	
Sheep dig prot %	11.8	12.9	
Energy GE Mcal/kg	4.26	4.65	
Sheep GE dig coef %	68.	68.	
Cattle DE Mcal/kg *	2.77	3.02	
Sheep DE Mcal/kg	2.89	3.16	
Cattle ME Mcal/kg *	2.27	2.48	
Sheep ME Mcal/kg	2.36	2.57	
Cattle TDN % *	62.7	68.5	
Sheep TDN %	65.2	71.1	

Orchardgrass, hay, fan air dried w heat, cut 3, (1)
Ref No 1 05 571 United States

Feed Name or Analyses	As Fed	Dry	C.V. ± %
Dry matter %	91.3	100.0	
Sheep dig coef %	70.	70.	
Ash %	7.2	7.9	
Crude fiber %	24.2	26.6	
Sheep dig coef %	75.	75.	
Ether extract %	5.2	5.7	
Sheep dig coef %	44.	44.	

Continued

Feed Name or Analyses				Mean As Fed	Mean Dry	C.V. ± %
N-free extract		%		35.8	39.3	
Sheep	dig coef	%		69.	69.	
Protein (N x 6.25)		%		18.9	20.7	
Sheep	dig coef	%		77.	77.	
Cattle	dig prot	%	*	13.6	14.9	
Goats	dig prot	%	*	14.5	15.9	
Horses	dig prot	%	*	13.8	15.1	
Rabbits	dig prot	%	*	13.4	14.7	
Sheep	dig prot	%		14.6	15.9	
Energy	GE	Mcal/kg		4.36	4.78	
Sheep	GE dig coef	%		66.	66.	
Cattle	DE	Mcal/kg	*	2.58	2.82	
Sheep	DE	Mcal/kg		2.88	3.15	
Cattle	ME	Mcal/kg	*	2.11	2.32	
Sheep	ME	Mcal/kg		2.30	2.52	
Cattle	TDN	%	*	58.5	64.1	
Sheep	TDN	%		62.6	68.5	

Orchardgrass, hay, s-c, (1)
Ref No 1 03 438 United States

Feed Name or Analyses				Mean As Fed	Mean Dry	C.V. ± %
Dry matter		%		88.6	100.0	2
Cattle	dig coef	%		62.	62.	
Horses	dig coef	%		52.	52.	25
Rabbits	dig coef	%		41.	41.	
Sheep	dig coef	%		65.	65.	
Organic matter		%		81.6	92.1	0
Ash		%		6.6	7.4	13
Crude fiber		%		30.3	34.2	12
Cattle	dig coef	%		64.	64.	
Horses	dig coef	%		46.	46.	7
Sheep	dig coef	%		67.	67.	
Ether extract		%		2.9	3.3	22
Cattle	dig coef	%		56.	56.	
Sheep	dig coef	%		40.	40.	
N-free extract		%		39.0	44.0	6
Cattle	dig coef	%		62.	62.	
Horses	dig coef	%		48.	48.	8
Sheep	dig coef	%		63.	63.	
Protein (N x 6.25)		%		9.8	11.1	34
Cattle	dig coef	%		60.	60.	
Rabbits	dig coef	%		58.	58.	
Sheep	dig coef	%		59.	59.	
Cattle	dig prot	%		5.8	6.6	
Goats	dig prot	%	*	6.1	6.9	
Horses	dig prot	%	*	6.1	6.9	
Rabbits	dig prot	%		5.7	6.4	
Sheep	dig prot	%		5.8	6.5	
Cell contents (Van Soest)		%		25.0	28.2	6
Horses	dig coef	%		55.	55.	10
Cell walls (Van Soest)		%		63.6	71.8	2
Horses	dig coef	%		44.	44.	5
Cellulose (Matrone)		%		31.2	35.2	7
Horses	dig coef	%		45.	45.	7
Fiber, acid detergent (VS)		%		35.7	40.3	2
Horses	dig coef	%		40.	40.	8
Hemicellulose		%		24.1	27.2	4
Horses	dig coef	%		50.	50.	5
Holocellulose		%		56.1	63.3	3
Soluble carbohydrates		%		10.8	12.2	15
Lignin (Ellis)		%		4.0	4.6	20
Horses	dig coef	%		-6.	-6.	98
Fatty acids		%		3.3	3.7	
Energy	GE	Mcal/kg		3.93	4.44	1
Horses	GE dig coef	%		43.	43.	9
Sheep	GE dig coef	%		68.	68.	
Cattle	DE	Mcal/kg	*	2.33	2.63	
Horses	DE	Mcal/kg		1.67	1.89	9
Sheep	DE	Mcal/kg		2.68	3.02	
Cattle	ME	Mcal/kg	*	1.91	2.16	
Horses	ME	Mcal/kg	*	1.37	1.55	
Sheep	ME	Mcal/kg	*	2.19	2.47	
Cattle	NE_m	Mcal/kg	*	1.08	1.22	
Cattle	NE_{gain}	Mcal/kg	*	0.49	0.55	
Cattle	$NE_{lactating\ cows}$	Mcal/kg	*	1.19	1.34	
Cattle	TDN	%		52.9	59.7	
Sheep	TDN	%		53.4	60.2	
Calcium		%		0.28	0.31	17
Chlorine		%		0.37	0.42	
Cobalt		mg/kg		0.020	0.022	
Copper		mg/kg		14.8	16.7	
Iron		%		0.012	0.014	
Magnesium		%		0.15	0.17	20
Manganese		mg/kg		69.4	78.3	
Phosphorus		%		0.23	0.26	25
Potassium		%		2.91	3.28	15
Sulphur		%		0.23	0.26	
Zinc		mg/kg		16.0	18.1	
Carotene		mg/kg		17.2	19.4	37
Riboflavin		mg/kg		6.1	6.8	
Thiamine		mg/kg		2.5	2.9	
α-tocopherol		mg/kg		219.6	247.8	
Vitamin A equivalent		IU/g		28.7	32.4	

Orchardgrass, hay, s-c, immature, (1)
Ref No 1 03 423 United States

Feed Name or Analyses				Mean As Fed	Mean Dry	C.V. ± %
Dry matter		%		89.3	100.0	3
Crude fiber		%				
Cattle	dig coef	%	*	63.	63.	
Sheep	dig coef	%	*	60.	60.	
Ether extract		%				
Cattle	dig coef	%	*	56.	56.	
Sheep	dig coef	%	*	40.	40.	
N-free extract		%				
Cattle	dig coef	%	*	56.	56.	
Sheep	dig coef	%	*	56.	56.	
Protein (N x 6.25)		%				
Cattle	dig coef	%	*	60.	60.	
Sheep	dig coef	%	*	50.	50.	
Calcium		%		0.32	0.36	
Phosphorus		%		0.28	0.31	
Potassium		%		3.34	3.74	

Orchardgrass, hay, s-c, pre-bloom, (1)
Ref No 1 03 424 United States

Feed Name or Analyses				Mean As Fed	Mean Dry	C.V. ± %
Dry matter		%		84.1	100.0	
Sheep	dig coef	%		57.	57.	
Ash		%		6.6	7.8	
Crude fiber		%		28.2	33.6	
Sheep	dig coef	%		61.	61.	
Ether extract		%		2.5	3.0	
Sheep	dig coef	%		56.	56.	
N-free extract		%		37.6	44.7	
Sheep	dig coef	%		60.	60.	
Protein (N x 6.25)		%		9.2	11.0	
Sheep	dig coef	%		59.	59.	
Cattle	dig prot	%	*	5.4	6.4	
Goats	dig prot	%	*	5.7	6.8	
Horses	dig prot	%	*	5.7	6.8	
Rabbits	dig prot	%	*	6.0	7.1	
Sheep	dig prot	%		5.4	6.4	
Energy	GE	Mcal/kg				
Cattle	DE	Mcal/kg	*	2.14	2.54	
Sheep	DE	Mcal/kg	*	2.12	2.53	
Cattle	ME	Mcal/kg	*	1.75	2.09	
Sheep	ME	Mcal/kg	*	1.74	2.07	
Cattle	TDN	%	*	48.5	57.7	
Sheep	TDN	%		48.2	57.3	

Orchardgrass, hay, s-c, early bloom, (1)
Ref No 1 03 425 United States

Feed Name or Analyses				Mean As Fed	Mean Dry	C.V. ± %
Dry matter		%		86.9	100.0	4
Ash		%		7.4	8.5	15
Crude fiber		%		29.8	34.4	10
Cattle	dig coef	%	*	63.	63.	
Sheep	dig coef	%		65.	65.	
Ether extract		%		2.3	2.6	26
Cattle	dig coef	%	*	56.	56.	
Sheep	dig coef	%		40.	40.	
N-free extract		%		38.6	44.4	
Cattle	dig coef	%	*	56.	56.	
Sheep	dig coef	%		53.	53.	
Protein (N x 6.25)		%		8.8	10.2	23
Cattle	dig coef	%	*	60.	60.	
Sheep	dig coef	%		48.	48.	
Cattle	dig prot	%		5.3	6.1	
Goats	dig prot	%	*	5.2	6.0	
Horses	dig prot	%	*	5.3	6.1	
Rabbits	dig prot	%	*	5.7	6.5	
Sheep	dig prot	%		4.2	4.9	
Cellulose (Matrone)		%		29.8	34.3	
Sugars, total		%		5.0	5.7	
Lignin (Ellis)		%		5.6	6.5	
Energy	GE	Mcal/kg				
Cattle	DE	Mcal/kg	*	2.14	2.46	
Sheep	DE	Mcal/kg	*	2.03	2.34	
Cattle	ME	Mcal/kg	*	1.76	2.02	
Sheep	ME	Mcal/kg	*	1.67	1.92	
Cattle	TDN	%		48.5	55.9	
Sheep	TDN	%		46.1	53.1	

Orchardgrass, hay, s-c, mid-bloom, (1)
Ref No 1 03 426 United States

Feed Name or Analyses				Mean As Fed	Mean Dry	C.V. ± %
Dry matter		%		94.6	100.0	
Calcium		%		0.38	0.40	
Phosphorus		%		0.43	0.45	
Potassium		%		1.94	2.05	

Orchardgrass, hay, s-c, full bloom, (1)
Ref No 1 03 427 United States

Feed Name or Analyses				Mean As Fed	Mean Dry	C.V. ± %
Dry matter		%		89.2	100.0	
Ash		%		6.9	7.7	11
Crude fiber		%		29.5	33.1	10
Cattle	dig coef	%	*	63.	63.	
Sheep	dig coef	%	*	60.	60.	
Ether extract		%		3.2	3.6	17
Cattle	dig coef	%	*	56.	56.	
Sheep	dig coef	%	*	40.	40.	
N-free extract		%		38.5	43.2	
Cattle	dig coef	%	*	56.	56.	
Sheep	dig coef	%	*	56.	56.	
Protein (N x 6.25)		%		11.1	12.4	4
Cattle	dig coef	%	*	60.	60.	

(1) dry forages and roughages (2) pasture, range plants, and forages fed green (3) silages (4) energy feeds (5) protein supplements (6) minerals (7) vitamins (8) additives

Feed Name or Analyses			Mean As Fed	Mean Dry	C.V. ± %
Sheep	dig coef %	*	50.	50.	
Cattle	dig prot %		6.6	7.4	
Goats	dig prot %	*	7.3	8.1	
Horses	dig prot %	*	7.2	8.1	
Rabbits	dig prot %	*	7.4	8.2	
Sheep	dig prot %		5.5	6.2	
Energy	GE Mcal/kg				
Cattle	DE Mcal/kg	*	2.24	2.51	
Sheep	DE Mcal/kg	*	2.10	2.36	
Cattle	ME Mcal/kg	*	1.84	2.06	
Sheep	ME Mcal/kg	*	1.73	1.93	
Cattle	TDN %	*	50.9	57.0	
Sheep	TDN %	*	47.7	53.5	

Orchardgrass, hay, s-c, late bloom, (1)

Ref No 1 03 428 United States

Feed Name or Analyses			Mean As Fed	Mean Dry	C.V. ± %
Dry matter	%		89.0	100.0	
Ash	%		6.2	7.0	
Crude fiber	%		33.0	37.1	
Sheep	dig coef %		58.	58.	
Ether extract	%		3.0	3.4	
Sheep	dig coef %		51.	51.	
N-free extract	%		39.2	44.1	
Sheep	dig coef %		54.	54.	
Protein (N x 6.25)	%		7.5	8.4	
Sheep	dig coef %		58.	58.	
Cattle	dig prot %	*	3.8	4.2	
Goats	dig prot %	*	3.9	4.4	
Horses	dig prot %	*	4.1	4.7	
Rabbits	dig prot %	*	4.6	5.2	
Sheep	dig prot %		4.3	4.9	
Energy	GE Mcal/kg				
Cattle	DE Mcal/kg	*	2.04	2.29	
Sheep	DE Mcal/kg	*	2.12	2.39	
Cattle	ME Mcal/kg	*	1.67	1.88	
Sheep	ME Mcal/kg	*	1.74	1.96	
Cattle	TDN %	*	46.3	52.0	
Sheep	TDN %		48.2	54.1	

Orchardgrass, hay, s-c, milk stage, (1)

Ref No 1 03 429 United States

Feed Name or Analyses			Mean As Fed	Mean Dry	C.V. ± %
Dry matter	%		89.8	100.0	
Ash	%		7.9	8.9	
Crude fiber	%		32.2	35.9	
Ether extract	%		2.6	2.9	
N-free extract	%		38.0	42.4	
Protein (N x 6.25)	%		9.0	10.1	
Cattle	dig prot %	*	5.1	5.6	
Goats	dig prot %	*	5.3	5.9	
Horses	dig prot %	*	5.4	6.1	
Rabbits	dig prot %	*	5.8	6.4	
Sheep	dig prot %	*	5.0	5.6	
Cellulose (Matrone)	%		34.1	38.0	
Energy	GE Mcal/kg				
Cattle	DE Mcal/kg	*	2.17	2.42	
Sheep	DE Mcal/kg	*	2.17	2.41	
Cattle	ME Mcal/kg	*	1.78	1.98	
Sheep	ME Mcal/kg	*	1.78	1.98	
Cattle	TDN %	*	49.2	54.8	
Sheep	TDN %	*	49.1	54.7	

Orchardgrass, hay, s-c, dough stage, (1)

Ref No 1 03 430 United States

Feed Name or Analyses			Mean As Fed	Mean Dry	C.V. ± %
Dry matter	%		91.4	100.0	3
Ash	%		5.5	6.0	20
Crude fiber	%		33.3	36.4	5
Ether extract	%		2.3	2.5	38
N-free extract	%		44.0	48.1	
Protein (N x 6.25)	%		6.4	7.0	13
Cattle	dig prot %	*	2.7	3.0	
Goats	dig prot %	*	2.8	3.1	
Horses	dig prot %	*	3.2	3.5	
Rabbits	dig prot %	*	3.7	4.1	
Sheep	dig prot %	*	2.6	2.8	
Energy	GE Mcal/kg				
Cattle	DE Mcal/kg	*	2.08	2.28	
Sheep	DE Mcal/kg	*	2.17	2.37	
Cattle	ME Mcal/kg	*	1.71	1.87	
Sheep	ME Mcal/kg	*	1.78	1.95	
Cattle	TDN %	*	47.2	51.7	
Sheep	TDN %	*	49.2	53.8	

Orchardgrass, hay, s-c, mature, (1)

Ref No 1 03 431 United States

Feed Name or Analyses			Mean As Fed	Mean Dry	C.V. ± %
Dry matter	%		90.4	100.0	
Ash	%		5.8	6.4	19
Crude fiber	%		34.3	38.0	8
Cattle	dig coef %	*	63.	63.	
Sheep	dig coef %		42.	42.	
Ether extract	%		2.4	2.7	19
Cattle	dig coef %	*	56.	56.	
Sheep	dig coef %		99.	99.	
N-free extract	%		40.7	45.0	
Cattle	dig coef %	*	56.	56.	
Sheep	dig coef %		35.	35.	
Protein (N x 6.25)	%		7.1	7.9	20
Cattle	dig coef %	*	60.	60.	
Sheep	dig coef %		50.	50.	
Cattle	dig prot %		4.3	4.7	
Goats	dig prot %	*	3.6	3.9	
Horses	dig prot %	*	3.8	4.2	
Rabbits	dig prot %	*	4.3	4.8	
Sheep	dig prot %		3.6	4.0	
Lignin (Ellis)	%		8.4	9.3	
Energy	GE Mcal/kg				
Cattle	DE Mcal/kg	*	2.28	2.52	
Sheep	DE Mcal/kg	*	1.66	1.83	
Cattle	ME Mcal/kg	*	1.87	2.07	
Sheep	ME Mcal/kg	*	1.36	1.50	
Cattle	TDN %		51.7	57.2	
Sheep	TDN %		37.6	41.6	

Orchardgrass, hay, s-c, over ripe, (1)

Ref No 1 03 432 United States

Feed Name or Analyses			Mean As Fed	Mean Dry	C.V. ± %
Dry matter	%		90.0	100.0	
Ash	%		4.9	5.4	
Crude fiber	%		39.6	44.0	
Ether extract	%		2.1	2.3	
N-free extract	%		38.1	42.3	
Protein (N x 6.25)	%		5.4	6.0	11
Cattle	dig prot %	*	1.9	2.1	
Goats	dig prot %	*	1.9	2.2	
Horses	dig prot %	*	2.4	2.6	
Rabbits	dig prot %	*	3.0	3.3	
Sheep	dig prot %	*	1.8	2.0	
Energy	GE Mcal/kg				
Cattle	DE Mcal/kg	*	1.81	2.01	
Sheep	DE Mcal/kg	*	1.98	2.20	
Cattle	ME Mcal/kg	*	1.48	1.65	
Sheep	ME Mcal/kg	*	1.63	1.81	
Cattle	TDN %	*	41.0	45.6	
Sheep	TDN %	*	44.9	49.9	

Orchardgrass, hay, s-c, cut 1, (1)

Ref No 1 03 433 United States

Feed Name or Analyses			Mean As Fed	Mean Dry	C.V. ± %
Dry matter	%		51.9	100.0	4
Crude fiber	%				
Cattle	dig coef %	*	63.	63.	
Sheep	dig coef %	*	60.	60.	
Ether extract	%				
Cattle	dig coef %	*	56.	56.	
Sheep	dig coef %	*	40.	40.	
N-free extract	%				
Cattle	dig coef %	*	56.	56.	
Sheep	dig coef %	*	56.	56.	
Protein (N x 6.25)	%				
Cattle	dig coef %	*	60.	60.	
Sheep	dig coef %	*	50.	50.	
Sugars, total	%		3.0	5.7	
Energy	GE Mcal/kg		2.35	4.53	
Calcium	%		0.21	0.40	
Phosphorus	%		0.23	0.45	
Potassium	%		1.06	2.05	

Orchardgrass, hay, s-c, cut 2, (1)

Ref No 1 03 434 United States

Feed Name or Analyses			Mean As Fed	Mean Dry	C.V. ± %
Dry matter	%			100.0	
Ash	%			7.5	
Crude fiber	%			35.6	
Cattle	dig coef %	*		63.	
Sheep	dig coef %	*		68.	
Ether extract	%			4.0	
Cattle	dig coef %	*		56.	
Sheep	dig coef %	*		33.	
N-free extract	%			36.5	
Cattle	dig coef %	*		56.	
Sheep	dig coef %	*		66.	
Protein (N x 6.25)	%			16.5	
Cattle	dig coef %	*		60.	
Sheep	dig coef %	*		66.	
Cattle	dig prot %			9.9	
Goats	dig prot %	*		11.9	
Horses	dig prot %	*		11.5	
Rabbits	dig prot %	*		11.4	
Sheep	dig prot %			10.9	
Energy	GE Mcal/kg			4.66	
Sheep	GE dig coef %			64.	
Cattle	DE Mcal/kg	*		2.55	
Sheep	DE Mcal/kg			2.97	
Cattle	ME Mcal/kg	*		2.09	
Sheep	ME Mcal/kg	*		2.43	
Cattle	TDN %			57.8	
Sheep	TDN %			62.0	

Orchardgrass, hay, s-c, cut 3, (1)

Ref No 1 03 435 United States

Feed Name or Analyses			Mean As Fed	Mean Dry	C.V. ± %
Dry matter	%		91.3	100.0	
Ash	%		6.8	7.5	5

Continued

Feed Name or Analyses			Mean As Fed	Mean Dry	C.V. ± %
Crude fiber	%		28.1	30.8	14
Cattle	dig coef %	*	63.	63.	
Sheep	dig coef %	*	60.	60.	
Ether extract	%		4.5	4.9	18
Cattle	dig coef %	*	56.	56.	
Sheep	dig coef %	*	40.	40.	
N-free extract	%		36.7	40.2	
Cattle	dig coef %	*	56.	56.	
Sheep	dig coef %	*	56.	56.	
Protein (N x 6.25)	%		15.2	16.6	25
Cattle	dig coef %	*	60.	60.	
Sheep	dig coef %	*	50.	50.	
Cattle	dig prot %		9.1	10.0	
Goats	dig prot %	*	11.0	12.0	
Horses	dig prot %	*	10.6	11.6	
Rabbits	dig prot %	*	10.5	11.5	
Sheep	dig prot %		7.6	8.3	
Energy	GE Mcal/kg		4.36	4.78	
Cattle	DE Mcal/kg	*	2.34	2.56	
Sheep	DE Mcal/kg	*	2.16	2.37	
Cattle	ME Mcal/kg	*	1.92	2.10	
Sheep	ME Mcal/kg	*	1.77	1.94	
Cattle	TDN %		53.0	58.1	
Sheep	TDN %		49.0	53.7	

Orchardgrass, hay, s-c, gr 1 US, (1)
Ref No 1 03 436 United States

Feed Name or Analyses			Mean As Fed	Mean Dry	C.V. ± %
Dry matter	%		90.0	100.0	
Ash	%		4.5	5.0	
Crude fiber	%		33.4	37.1	
Ether extract	%		1.4	1.6	
N-free extract	%		44.9	49.9	
Protein (N x 6.25)	%		5.8	6.4	
Cattle	dig prot %	*	2.2	2.5	
Goats	dig prot %	*	2.3	2.5	
Horses	dig prot %	*	2.7	3.0	
Rabbits	dig prot %	*	3.2	3.6	
Sheep	dig prot %	*	2.1	2.3	
Energy	GE Mcal/kg				
Cattle	DE Mcal/kg	*	1.94	2.15	
Sheep	DE Mcal/kg	*	2.13	2.36	
Cattle	ME Mcal/kg	*	1.59	1.77	
Sheep	ME Mcal/kg	*	1.74	1.94	
Cattle	TDN %	*	43.9	48.8	
Sheep	TDN %	*	48.2	53.6	

Orchardgrass, hay, s-c, gr 2 US, (1)
Ref No 1 03 437 United States

Feed Name or Analyses			Mean As Fed	Mean Dry	C.V. ± %
Dry matter	%		90.0	100.0	
Ash	%		4.9	5.4	
Crude fiber	%		33.8	37.6	
Ether extract	%		1.4	1.6	
N-free extract	%		43.8	48.7	
Protein (N x 6.25)	%		6.0	6.7	
Cattle	dig prot %	*	2.5	2.7	
Goats	dig prot %	*	2.5	2.8	
Horses	dig prot %	*	2.9	3.2	
Rabbits	dig prot %	*	3.5	3.8	
Sheep	dig prot %	*	2.3	2.6	
Energy	GE Mcal/kg				
Cattle	DE Mcal/kg	*	1.92	2.13	
Sheep	DE Mcal/kg	*	2.12	2.35	
Cattle	ME Mcal/kg	*	1.57	1.75	
Sheep	ME Mcal/kg	*	1.74	1.93	
Cattle	TDN %	*	43.5	48.3	
Sheep	TDN %	*	48.1	53.4	

(1) dry forages and roughages (2) pasture, range plants, and forages fed green (3) silages (4) energy feeds (5) protein supplements (6) minerals (7) vitamins (8) additives

Orchardgrass, aerial part, fresh, (2)
Ref No 2 03 451 United States

Feed Name or Analyses			Mean As Fed	Mean Dry	C.V. ± %
Dry matter	%		24.9	100.0	10
Organic matter	%		22.8	91.6	1
Crude fiber	%		6.2	25.0	16
Cattle	dig coef %		63.	63.	
Sheep	dig coef %		73.	73.	9
Ether extract	%		1.6	6.5	18
Cattle	dig coef %		74.	74.	
Sheep	dig coef %		48.	48.	37
N-free extract	%		11.2	45.0	4
Cattle	dig coef %		74.	74.	
Sheep	dig coef %		70.	70.	12
Protein (N x 6.25)	%		3.8	15.4	24
Cattle	dig coef %		73.	73.	
Sheep	dig coef %		67.	67.	16
Cattle	dig prot %		2.8	11.2	
Goats	dig prot %	*	2.7	10.9	
Horses	dig prot %	*	2.6	10.6	
Rabbits	dig prot %	*	2.6	10.5	
Sheep	dig prot %		2.6	10.3	
Lignin (Ellis)	%		1.5	5.9	25
Energy	GE Mcal/kg				
Cattle	DE Mcal/kg	*	0.78	3.12	
Sheep	DE Mcal/kg	*	0.74	2.96	
Cattle	ME Mcal/kg	*	0.64	2.56	
Sheep	ME Mcal/kg	*	0.60	2.42	
Cattle	TDN %		17.6	70.7	
Sheep	TDN %		16.7	67.1	
Carotene	mg/kg		79.4	319.0	49
Thiamine	mg/kg		1.8	7.3	
α-tocopherol	mg/kg		108.5	435.6	
Vitamin A equivalent	IU/g		132.4	531.8	

Orchardgrass, aerial part, fresh, immature, (2)
Ref No 2 03 440 United States

Feed Name or Analyses			Mean As Fed	Mean Dry	C.V. ± %
Dry matter	%		23.9	100.0	11
Ash	%		2.7	11.3	
Crude fiber	%		5.6	23.4	
Ether extract	%		1.2	5.0	
N-free extract	%		10.0	41.8	
Protein (N x 6.25)	%		4.4	18.4	
Cattle	dig prot %	*	3.2	13.5	
Goats	dig prot %	*	3.3	13.7	
Horses	dig prot %	*	3.1	13.2	
Rabbits	dig prot %	*	3.1	12.9	
Sheep	dig prot %	*	3.4	14.1	
Energy	GE Mcal/kg				
Cattle	DE Mcal/kg	*	0.69	2.87	
Sheep	DE Mcal/kg	*	0.71	2.97	
Cattle	ME Mcal/kg	*	0.56	2.36	
Sheep	ME Mcal/kg	*	0.58	2.43	
Cattle	NE_m Mcal/kg	*	0.34	1.41	
Cattle	NE_{gain} Mcal/kg	*	0.20	0.82	
Cattle	$NE_{lactating\ cows}$ Mcal/kg	*	0.39	1.64	
Cattle	TDN %	*	15.5	65.2	
Sheep	TDN %	*	16.1	67.3	
Calcium	%		0.13	0.54	
Iron	%		0.004	0.017	
Magnesium	%		0.08	0.33	
Manganese	mg/kg		29.5	123.6	
Phosphorus	%		0.12	0.50	
Potassium	%		0.63	2.64	
Sodium	%		0.01	0.04	
Sulphur	%		0.05	0.21	
Carotene	mg/kg		80.4	337.3	44
Vitamin A equivalent	IU/g		134.1	562.3	

Orchardgrass, aerial part, fresh, immature, pure stand, (2)
Ref No 2 03 439 United States

Feed Name or Analyses			Mean As Fed	Mean Dry	C.V. ± %
Dry matter	%		23.0	100.0	4
Calcium	%		0.14	0.60	5
Magnesium	%		0.06	0.27	13
Phosphorus	%		0.13	0.58	29
Potassium	%		0.87	3.78	
Carotene	mg/kg		144.2	626.8	20
Vitamin A equivalent	IU/g		240.3	1044.8	

Orchardgrass, aerial part, fresh, pre-bloom, (2)
Ref No 2 08 476 United States

Feed Name or Analyses			Mean As Fed	Mean Dry	C.V. ± %
Dry matter	%		27.5	100.0	
Ash	%		2.2	8.0	
Crude fiber	%		8.1	29.5	
Ether extract	%		1.3	4.7	
N-free extract	%		12.4	45.1	
Protein (N x 6.25)	%		3.5	12.7	
Cattle	dig prot %	*	2.4	8.7	
Goats	dig prot %	*	2.3	8.4	
Horses	dig prot %	*	2.3	8.3	
Rabbits	dig prot %	*	2.3	8.5	
Sheep	dig prot %	*	2.4	8.9	
Energy	GE Mcal/kg				
Cattle	DE Mcal/kg	*	0.79	2.88	
Sheep	DE Mcal/kg	*	0.77	2.80	
Cattle	ME Mcal/kg	*	0.65	2.36	
Sheep	ME Mcal/kg	*	0.63	2.30	
Cattle	TDN %	*	18.0	65.3	
Sheep	TDN %	*	17.5	63.6	
Calcium	%		0.07	0.25	
Phosphorus	%		0.08	0.29	

Orchardgrass, aerial part, fresh, pre-bloom, pure stand, (2)
Ref No 2 03 441 United States

Feed Name or Analyses			Mean As Fed	Mean Dry	C.V. ± %
Dry matter	%			100.0	
Carotene	mg/kg			520.7	23
Vitamin A equivalent	IU/g			868.1	

Orchardgrass, aerial part, fresh, early bloom, (2)
Ref No 2 03 442 United States

Feed Name or Analyses			Mean As Fed	Mean Dry	C.V. ± %
Dry matter	%		24.2	100.0	12
Ash	%		1.7	7.2	20
Crude fiber	%		7.9	32.8	7
Ether extract	%		0.9	3.5	14
N-free extract	%		10.9	45.0	
Protein (N x 6.25)	%		2.8	11.5	18
Cattle	dig prot %	*	1.9	7.7	
Goats	dig prot %	*	1.8	7.3	
Horses	dig prot %	*	1.8	7.3	
Rabbits	dig prot %	*	1.8	7.6	
Sheep	dig prot %	*	1.9	7.7	

Feed Name or Analyses			Mean As Fed	Mean Dry	C.V. ± %
Hemicellulose	%		4.6	18.9	
Energy	GE Mcal/kg		1.08	4.45	
Cattle	DE Mcal/kg	*	0.72	2.98	
Sheep	DE Mcal/kg	*	0.68	2.80	
Cattle	ME Mcal/kg	*	0.59	2.45	
Sheep	ME Mcal/kg	*	0.56	2.30	
Cattle	NE_m Mcal/kg	*	0.44	1.60	
Cattle	NE_{gain} Mcal/kg	*	0.25	1.03	
Cattle	$NE_{lactating\ cows}$ Mcal/kg	*	0.40	1.67	
Cattle	TDN %	*	16.4	67.6	
Sheep	TDN %	*	15.4	63.5	
Calcium	%		0.06	0.25	
Copper	mg/kg		8.0	33.1	
Iron	%		0.019	0.077	
Manganese	mg/kg		25.2	104.1	
Phosphorus	%		0.09	0.39	50

Orchardgrass, aerial part, fresh, mid-bloom, (2)
Ref No 2 03 443 United States

Feed Name or Analyses			Mean As Fed	Mean Dry	C.V. ± %
Dry matter	%		30.5	100.0	
Ash	%		2.3	7.5	7
Crude fiber	%		10.2	33.5	5
Ether extract	%		1.1	3.5	7
N-free extract	%		14.2	46.7	
Protein (N x 6.25)	%		2.7	8.8	4
Cattle	dig prot %	*	1.6	5.4	
Goats	dig prot %	*	1.5	4.8	
Horses	dig prot %	*	1.5	5.0	
Rabbits	dig prot %	*	1.7	5.5	
Sheep	dig prot %	*	1.6	5.2	
Hemicellulose	%		6.6	21.6	
Energy	GE Mcal/kg				
Cattle	DE Mcal/kg	*	0.86	2.83	
Sheep	DE Mcal/kg	*	0.88	2.88	
Cattle	ME Mcal/kg	*	0.71	2.32	
Sheep	ME Mcal/kg	*	0.72	2.36	
Cattle	TDN %	*	19.6	64.2	
Sheep	TDN %	*	19.9	65.2	
Calcium	%		0.07	0.23	
Phosphorus	%		0.07	0.23	

Orchardgrass, aerial part, fresh, full bloom, (2)
Ref No 2 03 445 United States

Feed Name or Analyses			Mean As Fed	Mean Dry	C.V. ± %
Dry matter	%		29.9	100.0	5
Ash	%		2.4	7.9	7
Crude fiber	%		9.9	33.1	5
Ether extract	%		1.0	3.3	8
N-free extract	%		14.1	47.2	
Protein (N x 6.25)	%		2.5	8.5	6
Cattle	dig prot %	*	1.5	5.1	
Goats	dig prot %	*	1.3	4.5	
Horses	dig prot %	*	1.4	4.7	
Rabbits	dig prot %	*	1.6	5.2	
Sheep	dig prot %	*	1.5	4.9	
Energy	GE Mcal/kg				
Cattle	DE Mcal/kg	*	0.84	2.80	
Sheep	DE Mcal/kg	*	0.87	2.89	
Cattle	ME Mcal/kg	*	0.69	2.30	
Sheep	ME Mcal/kg	*	0.71	2.37	
Cattle	TDN %	*	19.0	63.5	
Sheep	TDN %	*	19.6	65.6	
Calcium	%		0.07	0.23	7
Phosphorus	%		0.07	0.22	11
Carotene	mg/kg		73.2	244.8	34
Vitamin A equivalent	IU/g		122.0	408.1	

Orchardgrass, aerial part, fresh, milk stage, (2)
Ref No 2 03 446 United States

Feed Name or Analyses			Mean As Fed	Mean Dry	C.V. ± %
Dry matter	%			100.0	
Ash	%			6.0	5
Crude fiber	%			35.2	4
Ether extract	%			3.7	7
N-free extract	%			46.7	
Protein (N x 6.25)	%			8.4	7
Cattle	dig prot %	*		5.0	
Goats	dig prot %	*		4.4	
Horses	dig prot %	*		4.7	
Rabbits	dig prot %	*		5.2	
Sheep	dig prot %	*		4.8	
Hemicellulose	%			21.3	
Energy	GE Mcal/kg				
Cattle	DE Mcal/kg	*		2.91	
Sheep	DE Mcal/kg	*		2.87	
Cattle	ME Mcal/kg	*		2.39	
Sheep	ME Mcal/kg	*		2.35	
Cattle	TDN %	*		66.0	
Sheep	TDN %	*		65.1	

Orchardgrass, aerial part, fresh, dough stage, (2)
Ref No 2 03 447 United States

Feed Name or Analyses			Mean As Fed	Mean Dry	C.V. ± %
Dry matter	%			100.0	
Ash	%			6.3	5
Crude fiber	%			36.2	2
Ether extract	%			3.0	6
N-free extract	%			48.4	
Protein (N x 6.25)	%			6.1	2
Cattle	dig prot %	*		3.1	
Goats	dig prot %	*		2.3	
Horses	dig prot %	*		2.7	
Rabbits	dig prot %	*		3.4	
Sheep	dig prot %	*		2.7	
Energy	GE Mcal/kg				
Cattle	DE Mcal/kg	*		2.83	
Sheep	DE Mcal/kg	*		2.94	
Cattle	ME Mcal/kg	*		2.32	
Sheep	ME Mcal/kg	*		2.41	
Cattle	TDN %	*		64.3	
Sheep	TDN %	*		66.6	

Orchardgrass, aerial part, fresh, mature, (2)
Ref No 2 03 449 United States

Feed Name or Analyses			Mean As Fed	Mean Dry	C.V. ± %
Dry matter	%			100.0	
Ash	%			6.1	7
Crude fiber	%			36.5	2
Ether extract	%			3.0	9
N-free extract	%			48.7	
Protein (N x 6.25)	%			5.7	4
Cattle	dig prot %	*		2.7	
Goats	dig prot %	*		1.9	
Horses	dig prot %	*		2.4	
Rabbits	dig prot %	*		3.1	
Sheep	dig prot %	*		2.3	
Hemicellulose	%			22.5	
Energy	GE Mcal/kg				
Cattle	DE Mcal/kg	*		2.83	
Sheep	DE Mcal/kg	*		2.95	
Cattle	ME Mcal/kg	*		2.32	
Sheep	ME Mcal/kg	*		2.42	

Feed Name or Analyses			Mean As Fed	Mean Dry	C.V. ± %
Cattle	TDN %	*		64.2	
Sheep	TDN %	*		66.8	

Orchardgrass, aerial part, fresh, mature, pure stand, (2)
Ref No 2 03 448 United States

Feed Name or Analyses			Mean As Fed	Mean Dry	C.V. ± %
Dry matter	%			100.0	
Carotene	mg/kg			125.0	30
Vitamin A equivalent	IU/g			208.4	

Orchardgrass, aerial part, fresh, cut 1, (2)
Ref No 2 03 450 United States

Feed Name or Analyses			Mean As Fed	Mean Dry	C.V. ± %
Dry matter	%			100.0	
Copper	mg/kg			33.1	
Iron	%			0.077	
Manganese	mg/kg			104.1	
Phosphorus	%			0.69	

Orchardgrass, aerial part, fresh, cut 2, (2)
Ref No 2 07 723 Canada

Feed Name or Analyses			Mean As Fed	Mean Dry	C.V. ± %
Dry matter	%		25.5	100.0	
Ash	%		2.6	10.1	
Crude fiber	%		7.3	28.7	
Ether extract	%		0.7	2.7	
N-free extract	%		10.6	41.8	
Protein (N x 6.25)	%		4.3	16.8	
Cattle	dig prot %	*	3.1	12.1	
Goats	dig prot %	*	3.1	12.2	
Horses	dig prot %	*	3.0	11.8	
Rabbits	dig prot %	*	3.0	11.6	
Sheep	dig prot %	*	3.2	12.6	
Energy	GE Mcal/kg				
Cattle	DE Mcal/kg	*	0.71	2.78	
Sheep	DE Mcal/kg	*	0.68	2.67	
Cattle	ME Mcal/kg	*	0.58	2.28	
Sheep	ME Mcal/kg	*	0.56	2.19	
Cattle	TDN %	*	16.1	63.1	
Sheep	TDN %	*	15.5	60.7	

Orchardgrass, aerial part, fresh, pure stand, (2)
Ref No 2 03 452 United States

Feed Name or Analyses			Mean As Fed	Mean Dry	C.V. ± %
Dry matter	%		23.0	100.0	3
Ash	%		3.1	13.5	9
Crude fiber	%		5.4	23.6	8
Ether extract	%		1.1	4.6	12
N-free extract	%		9.3	40.5	
Protein (N x 6.25)	%		4.1	17.8	3
Cattle	dig prot %	*	3.0	13.0	
Goats	dig prot %	*	3.0	13.2	
Horses	dig prot %	*	2.9	12.6	
Rabbits	dig prot %	*	2.9	12.4	
Sheep	dig prot %	*	3.1	13.6	
Energy	GE Mcal/kg				
Cattle	DE Mcal/kg	*	0.63	2.73	
Sheep	DE Mcal/kg	*	0.64	2.78	
Cattle	ME Mcal/kg	*	0.51	2.24	
Sheep	ME Mcal/kg	*	0.52	2.28	
Cattle	TDN %	*	14.2	61.9	
Sheep	TDN %	*	14.5	63.1	
Carotene	mg/kg		92.2	401.0	43
Vitamin A equivalent	IU/g		153.8	668.5	

Feed Name or Analyses			As Fed (Mean)	Dry (Mean)	C.V. ± %

Orchardgrass, leaves, fresh, (2)
Ref No 2 03 453 United States

Feed Name or Analyses			As Fed	Dry	C.V. ± %
Dry matter	%			100.0	
Alanine	%			1.50	
Arginine	%			5.10	
Aspartic acid	%			1.60	
Cystine	%			0.50	
Glutamic acid	%			2.70	
Glycine	%			1.30	
Histidine	%			0.80	
Leucine	%			2.90	
Lysine	%			2.00	
Methionine	%			0.40	
Phenylalanine	%			0.80	
Proline	%			0.70	
Tryptophan	%			0.60	
Tyrosine	%			0.80	
Valine	%			1.40	

Orchardgrass, aerial part, ensiled, (3)
Ref No 3 03 457 United States

Feed Name or Analyses			As Fed	Dry	C.V. ± %
Dry matter	%		29.5	100.0	2
Cattle	dig coef %		66.	66.	
Sheep	dig coef %		64.	64.	
Ash	%		2.7	9.1	
Crude fiber	%		10.5	35.4	
Cattle	dig coef %		75.	75.	
Sheep	dig coef %		75.	75.	
Ether extract	%		1.2	4.2	
N-free extract	%		10.9	37.0	
Cattle	dig coef %		56.	56.	
Sheep	dig coef %		55.	55.	
Protein (N x 6.25)	%		4.2	14.3	
Cattle	dig coef %		68.	68.	
Sheep	dig coef %		69.	69.	
Cattle	dig prot %		2.9	9.7	
Goats	dig prot %	*	2.7	9.2	
Horses	dig prot %	*	2.7	9.2	
Sheep	dig prot %		2.9	9.8	
Energy	GE Mcal/kg				
Cattle	DE Mcal/kg	*	0.86	2.93	
Sheep	DE Mcal/kg	*	0.79	2.69	
Cattle	ME Mcal/kg	*	0.71	2.40	
Sheep	ME Mcal/kg	*	0.65	2.20	
Cattle	TDN %	*	19.6	66.5	
Sheep	TDN %	*	18.0	60.9	
Sulphur	%		0.24	0.82	48
Carotene	mg/kg		49.9	169.1	49
Vitamin A equivalent	IU/g		83.2	281.9	

Orchardgrass, aerial part, ensiled, immature, (3)
Ref No 3 03 454 United States

Feed Name or Analyses			As Fed	Dry	C.V. ± %
Dry matter	%		22.4	100.0	13
Ash	%		1.3	5.9	3
Crude fiber	%		6.2	27.7	4
Ether extract	%		1.1	4.8	11
N-free extract	%		9.3	41.4	
Protein (N x 6.25)	%		4.5	20.2	7
Cattle	dig prot %	*	3.3	14.6	
Goats	dig prot %	*	3.3	14.6	
Horses	dig prot %	*	3.3	14.6	
Sheep	dig prot %	*	3.3	14.6	
Energy	GE Mcal/kg				
Cattle	DE Mcal/kg	*	0.52	2.33	
Sheep	DE Mcal/kg	*	0.53	2.37	
Cattle	ME Mcal/kg	*	0.43	1.91	
Sheep	ME Mcal/kg	*	0.44	1.94	
Cattle	TDN %	*	11.8	52.8	
Sheep	TDN %	*	12.0	53.8	

(1) dry forages and roughages (2) pasture, range plants, and forages fed green (3) silages (4) energy feeds (5) protein supplements (6) minerals (7) vitamins (8) additives

Orchardgrass, aerial part, ensiled, early bloom, (3)
Ref No 3 07 906 United States

Feed Name or Analyses			As Fed	Dry	C.V. ± %
Dry matter	%		19.1	100.0	
Sheep	dig coef %		65.	65.	
Ash	%		1.3	7.1	
Crude fiber	%		7.0	36.6	
Sheep	dig coef %		75.	75.	
Ether extract	%		0.8	4.0	
Sheep	dig coef %		61.	61.	
N-free extract	%		7.9	41.3	
Sheep	dig coef %		60.	60.	
Protein (N x 6.25)	%		2.1	11.1	
Sheep	dig coef %		58.	58.	
Cattle	dig prot %	*	1.2	6.3	
Goats	dig prot %	*	1.2	6.3	
Horses	dig prot %	*	1.2	6.3	
Sheep	dig prot %		1.2	6.4	
Energy	GE Mcal/kg		0.86	4.48	
Sheep	GE dig coef %		62.	62.	
Cattle	DE Mcal/kg	*	0.52	2.71	
Sheep	DE Mcal/kg		0.53	2.78	
Cattle	ME Mcal/kg	*	0.42	2.22	
Sheep	ME Mcal/kg		0.42	2.22	
Cattle	TDN %	*	11.7	61.5	
Sheep	TDN %		12.2	64.1	
Carotene	mg/kg		28.7	150.4	
Vitamin A equivalent	IU/g		47.9	250.6	

Orchardgrass, aerial part, ensiled, mature, (3)
Ref No 3 03 455 United States

Feed Name or Analyses			As Fed	Dry	C.V. ± %
Dry matter	%		30.0	100.0	
Ash	%		1.8	6.0	
Crude fiber	%		13.0	43.2	
Ether extract	%		1.3	4.3	
N-free extract	%		11.7	38.9	
Protein (N x 6.25)	%		2.3	7.6	
Cattle	dig prot %	*	0.9	3.1	
Goats	dig prot %	*	0.9	3.1	
Horses	dig prot %	*	0.9	3.1	
Sheep	dig prot %	*	0.9	3.1	
Energy	GE Mcal/kg				
Cattle	DE Mcal/kg	*	0.79	2.64	
Sheep	DE Mcal/kg	*	0.82	2.75	
Cattle	ME Mcal/kg	*	0.65	2.16	
Sheep	ME Mcal/kg	*	0.68	2.25	
Cattle	TDN %	*	17.9	59.8	
Sheep	TDN %	*	18.7	62.3	

Orchardgrass, aerial part, ensiled, cut 1, (3)
Ref No 3 03 456 United States

Feed Name or Analyses			As Fed	Dry	C.V. ± %
Dry matter	%		28.5	100.0	
Ash	%		2.0	6.9	
Crude fiber	%		9.9	34.7	
Ether extract	%		1.0	3.5	
N-free extract	%		12.7	44.4	
Protein (N x 6.25)	%		3.0	10.5	
Cattle	dig prot %	*	1.6	5.8	
Goats	dig prot %	*	1.6	5.8	
Horses	dig prot %	*	1.6	5.8	
Sheep	dig prot %	*	1.6	5.8	
Energy	GE Mcal/kg				
Cattle	DE Mcal/kg	*	0.75	2.62	
Sheep	DE Mcal/kg	*	0.80	2.80	
Cattle	ME Mcal/kg	*	0.61	2.15	
Sheep	ME Mcal/kg	*	0.65	2.30	
Cattle	TDN %	*	16.9	59.4	
Sheep	TDN %	*	18.1	63.5	

Orchardgrass, aerial part w corn grits by-product added, ensiled, (3)
Ref No 3 03 458 United States

Feed Name or Analyses			As Fed	Dry	C.V. ± %
Dry matter	%			100.0	
Ash	%			7.6	
Crude fiber	%			34.4	
Ether extract	%			3.9	
N-free extract	%			42.9	
Protein (N x 6.25)	%			11.2	
Cattle	dig prot %	*		6.4	
Goats	dig prot %	*		6.4	
Horses	dig prot %	*		6.4	
Sheep	dig prot %	*		6.4	
Energy	GE Mcal/kg				
Cattle	DE Mcal/kg	*		2.53	
Sheep	DE Mcal/kg	*		2.77	
Cattle	ME Mcal/kg	*		2.08	
Sheep	ME Mcal/kg	*		2.27	
Cattle	TDN %	*		57.4	
Sheep	TDN %	*		62.8	

Orchardgrass, aerial part w molasses added, ensiled, (3)
Ref No 3 03 459 United States

Feed Name or Analyses			As Fed	Dry	C.V. ± %
Dry matter	%		28.5	100.0	
Ash	%		2.0	7.0	6
Crude fiber	%		9.5	33.5	4
Ether extract	%		1.1	3.7	12
N-free extract	%		12.9	45.3	
Protein (N x 6.25)	%		3.0	10.5	7
Cattle	dig prot %	*	1.6	5.8	
Goats	dig prot %	*	1.6	5.8	
Horses	dig prot %	*	1.6	5.8	
Sheep	dig prot %	*	1.6	5.8	
Energy	GE Mcal/kg				
Cattle	DE Mcal/kg	*	0.73	2.56	
Sheep	DE Mcal/kg	*	0.80	2.80	
Cattle	ME Mcal/kg	*	0.60	2.10	
Sheep	ME Mcal/kg	*	0.66	2.30	
Cattle	TDN %	*	16.5	58.1	
Sheep	TDN %	*	18.1	63.6	
Carotene	mg/kg		32.0	112.2	
Vitamin A equivalent	IU/g		53.3	187.1	

Orchardgrass, aerial part w molasses added, ensiled, early bloom, (3)
Ref No 3 08 896 United States

Feed Name or Analyses			As Fed	Dry	C.V. ± %
Dry matter	%			100.0	
Sheep	dig coef %			68.	
Ash	%			7.0	

Feed Name or Analyses			Mean As Fed	Mean Dry	C.V. ± %
Crude fiber	%			32.2	
Sheep	dig coef %			75.	
Ether extract	%			3.8	
Sheep	dig coef %			62.	
N-free extract	%			46.6	
Sheep	dig coef %			68.	
Protein (N x 6.25)	%			10.4	
Sheep	dig coef %			60.	
Cattle	dig prot %	*		5.7	
Goats	dig prot %	*		5.7	
Horses	dig prot %	*		5.7	
Sheep	dig prot %			6.2	
Energy	GE Mcal/kg			4.36	
Sheep	GE dig coef %			65.	
Cattle	DE Mcal/kg	*		2.97	
Sheep	DE Mcal/kg			2.83	
Cattle	ME Mcal/kg	*		2.44	
Sheep	ME Mcal/kg			2.29	
Cattle	TDN %	*		67.3	
Sheep	TDN %			67.4	
Carotene	mg/kg			178.1	
Vitamin A equivalent	IU/g			296.9	

Orchardgrass, aerial part w sodium bisulfite preservative added, ensiled, early bloom, (3)

Ref No 3 07 907 United States

Feed Name or Analyses			Mean As Fed	Mean Dry	C.V. ± %
Dry matter	%			100.0	
Sheep	dig coef %			66.	
Ash	%			7.7	
Crude fiber	%			33.5	
Sheep	dig coef %			74.	
Ether extract	%			4.4	
Sheep	dig coef %			63.	
N-free extract	%			43.3	
Sheep	dig coef %			64.	
Protein (N x 6.25)	%			11.1	
Sheep	dig coef %			62.	
Cattle	dig prot %	*		6.3	
Goats	dig prot %	*		6.3	
Horses	dig prot %	*		6.3	
Sheep	dig prot %			6.9	
Energy	GE Mcal/kg			4.40	
Sheep	GE dig coef %			63.	
Cattle	DE Mcal/kg	*		2.61	
Sheep	DE Mcal/kg			2.77	
Cattle	ME Mcal/kg	*		2.14	
Sheep	ME Mcal/kg			2.22	
Cattle	TDN %	*		59.1	
Sheep	TDN %			65.6	
Carotene	mg/kg			239.2	
Vitamin A equivalent	IU/g			398.7	

Orchardgrass, aerial part w sulfur dioxide preservative added, ensiled, (3)

Ref No 3 03 460 United States

Feed Name or Analyses			Mean As Fed	Mean Dry	C.V. ± %
Dry matter	%			100.0	
Ash	%			8.9	15
Crude fiber	%			34.6	7
Ether extract	%			3.0	25
N-free extract	%			44.2	
Protein (N x 6.25)	%			9.3	5
Cattle	dig prot %	*		4.7	
Goats	dig prot %	*		4.7	
Horses	dig prot %	*		4.7	
Sheep	dig prot %	*		4.7	
Energy	GE Mcal/kg				
Cattle	DE Mcal/kg	*		2.64	
Sheep	DE Mcal/kg	*		2.77	
Cattle	ME Mcal/kg	*		2.16	
Sheep	ME Mcal/kg	*		2.27	
Cattle	TDN %	*		59.8	
Sheep	TDN %	*		62.9	
Sulphur	%			0.93	24

ORCHARDGRASS, LATOR. Dactylis glomerata

Orchardgrass, lator, hay, s-c, full bloom, (1)

Ref No 1 08 780 United States

Feed Name or Analyses			Mean As Fed	Mean Dry	C.V. ± %
Dry matter	%			100.0	
Sheep	dig coef %			62.	
Ash	%			7.2	
Crude fiber	%			35.8	
Ether extract	%			1.2	
N-free extract	%			49.1	
Protein (N x 6.25)	%			6.7	
Cattle	dig prot %	*		2.7	
Goats	dig prot %	*		2.8	
Horses	dig prot %	*		3.2	
Rabbits	dig prot %	*		3.8	
Sheep	dig prot %	*		2.6	
Lignin (Ellis)	%			7.2	
Energy	GE Mcal/kg			4.25	
Sheep	GE dig coef %			60.	
Cattle	DE Mcal/kg	*		2.32	
Sheep	DE Mcal/kg			2.55	
Cattle	ME Mcal/kg	*		1.90	
Sheep	ME Mcal/kg	*		2.09	
Cattle	TDN %	*		52.6	
Sheep	TDN %	*		54.0	

Orchardgrass, lator, hay, s-c chopped, (1)

Ref No 1 08 661 United States

Feed Name or Analyses			Mean As Fed	Mean Dry	C.V. ± %
Dry matter	%			100.0	
Sheep	dig coef %			57.	
Ash	%			8.3	
Crude fiber	%			38.2	
Ether extract	%			1.9	
N-free extract	%			44.0	
Protein (N x 6.25)	%			7.6	
Cattle	dig prot %	*		3.5	
Goats	dig prot %	*		3.7	
Horses	dig prot %	*		4.0	
Rabbits	dig prot %	*		4.5	
Sheep	dig prot %	*		3.4	
Lignin (Ellis)	%			8.4	
Energy	GE Mcal/kg			4.32	
Cattle	DE Mcal/kg	*		2.21	
Sheep	DE Mcal/kg	*		2.33	
Cattle	ME Mcal/kg	*		1.81	
Sheep	ME Mcal/kg	*		1.91	
Cattle	TDN %	*		50.1	
Sheep	TDN %	*		52.9	

ORCHARDGRASS, POTOMAC. Dactylis glomerata

Orchardgrass, potomac, aerial part, ensiled, pre-bloom, (3)

Ref No 3 09 260 United States

Feed Name or Analyses			Mean As Fed	Mean Dry	C.V. ± %
Dry matter	%			100.0	
Sheep	dig coef %			67.	
Ash	%			8.3	
Crude fiber	%			30.7	
Ether extract	%			4.2	
N-free extract	%			36.5	
Protein (N x 6.25)	%			20.4	
Cattle	dig prot %	*		14.7	
Goats	dig prot %	*		14.7	
Horses	dig prot %	*		14.7	
Sheep	dig prot %	*		14.7	
Energy	GE Mcal/kg				
Cattle	DE Mcal/kg	*		2.30	
Sheep	DE Mcal/kg	*		2.44	
Cattle	ME Mcal/kg	*		1.88	
Sheep	ME Mcal/kg	*		2.00	
Cattle	TDN %	*		52.1	
Sheep	TDN %	*		55.4	

Orchardgrass, potomac, aerial part, ensiled, early bloom, (3)

Ref No 3 09 259 United States

Feed Name or Analyses			Mean As Fed	Mean Dry	C.V. ± %
Dry matter	%			100.0	
Sheep	dig coef %			63.	
Ash	%			7.2	
Crude fiber	%			33.3	
Ether extract	%			3.4	
N-free extract	%			40.7	
Protein (N x 6.25)	%			15.5	
Cattle	dig prot %	*		10.3	
Goats	dig prot %	*		10.3	
Horses	dig prot %	*		10.3	
Sheep	dig prot %	*		10.3	
Energy	GE Mcal/kg				
Cattle	DE Mcal/kg	*		2.32	
Sheep	DE Mcal/kg	*		2.43	
Cattle	ME Mcal/kg	*		1.90	
Sheep	ME Mcal/kg	*		1.99	
Cattle	TDN %	*		52.5	
Sheep	TDN %	*		55.1	

ORCHARDGRASS, STERLING. Dactylus glomerata

Orchardgrass, sterling, hay, fan-air dried w heat, midbloom, cut 1, (1)

Ref No 1 08 890 United States

Feed Name or Analyses			Mean As Fed	Mean Dry	C.V. ± %
Dry matter	%		96.6	100.0	
Goats	dig coef %		74.	74.	
Protein (N x 6.25)	%		22.2	23.0	
Goats	dig coef %		83.	83.	
Cattle	dig prot %	*	16.3	16.9	
Goats	dig prot %		18.4	19.1	
Horses	dig prot %	*	16.5	17.1	
Rabbits	dig prot %	*	15.9	16.4	
Sheep	dig prot %	*	16.6	17.2	
Fiber, acid detergent (VS)	%		31.1	32.2	
Goats	dig coef %		69.	69.	

Continued

Feed Name or Analyses			Mean As Fed	Mean Dry	C.V. ± %
Lignin (Van Soest)	%		3.1	3.2	
Dry matter intake	g/W^0.75		59.795	61.900	

ORCHARDGRASS, S-143. Dactylis glomerata

Orchardgrass, S-143, hay, s-c, full bloom, (1)

Ref No 1 08 665 United States

Feed Name or Analyses			As Fed	Dry	C.V. ± %
Dry matter	%			100.0	
Sheep	dig coef %			57.	
Ash	%			7.3	
Crude fiber	%			35.6	
Ether extract	%			1.3	
N-free extract	%			49.1	
Protein (N x 6.25)	%			6.7	
Cattle	dig prot %	*		2.7	
Goats	dig prot %	*		2.8	
Horses	dig prot %	*		3.2	
Rabbits	dig prot %	*		3.8	
Sheep	dig prot %	*		2.6	
Lignin (Ellis)	%			7.7	
Energy	GE Mcal/kg			4.29	
Sheep	GE dig coef %			55.	
Cattle	DE Mcal/kg	*		2.34	
Sheep	DE Mcal/kg			2.36	
Cattle	ME Mcal/kg	*		1.92	
Sheep	ME Mcal/kg	*		1.93	
Cattle	TDN %	*		53.2	
Sheep	TDN %	*		54.1	

Orchardgrass, S-143, hay, s-c chopped, full bloom, (1)

Ref No 1 08 781 United States

Feed Name or Analyses			As Fed	Dry	C.V. ± %
Dry matter	%			100.0	
Sheep	dig coef %			48.	
Ash	%			9.0	
Crude fiber	%			39.6	
Ether extract	%			1.7	
N-free extract	%			42.0	
Protein (N x 6.25)	%			7.7	
Cattle	dig prot %	*		3.6	
Goats	dig prot %	*		3.7	
Horses	dig prot %	*		4.1	
Rabbits	dig prot %	*		4.6	
Sheep	dig prot %	*		3.5	
Lignin (Ellis)	%			9.9	
Energy	GE Mcal/kg			4.27	
Sheep	GE dig coef %			47.	
Cattle	DE Mcal/kg	*		2.13	
Sheep	DE Mcal/kg			2.01	
Cattle	ME Mcal/kg	*		1.74	
Sheep	ME Mcal/kg	*		1.65	
Cattle	TDN %	*		48.2	
Sheep	TDN %	*		52.2	

ORCHARDGRASS, S-37. Dactylis glomerata

Orchardgrass, S-37, aerial part, fresh, cut 2, (2)

Ref No 2 07 913 United States

Feed Name or Analyses			As Fed	Dry	C.V. ± %
Dry matter	%			100.0	
Calcium	%			0.53	

(1) dry forages and roughages (2) pasture, range plants, and forages fed green (3) silages (4) energy feeds (5) protein supplements (6) minerals (7) vitamins (8) additives

Feed Name or Analyses			As Fed	Dry	C.V. ± %
Magnesium	%			0.27	
Phosphorus	%			0.39	
Potassium	%			3.43	
Carotene	mg/kg			191.8	
Vitamin A equivalent	IU/g			319.7	
Vitamin D3	ICU/g			0.1	

ORCHARDGRASS-CLOVER, LADINO. Dactylis glomerata, Trifolium repens

Orchard-Clover, ladino, aerial part, fresh, (2)

Ref No 2 03 463 United States

Feed Name or Analyses			As Fed	Dry	C.V. ± %
Dry matter	%			100.0	
Cattle	dig coef %			78.	
Calcium	%			0.53	17
Magnesium	%			0.31	22
Phosphorus	%			0.48	
Potassium	%			3.53	
Sodium	%			0.06	41

Orchardgrass-Clover, ladino, aerial part, fresh, early bloom, (2)

Ref No 2 03 462 United States

Feed Name or Analyses			As Fed	Dry	C.V. ± %
Dry matter	%			100.0	
Ash	%			11.4	16
Protein (N x 6.25)	%			22.9	12
Cattle	dig prot %	*		17.4	
Goats	dig prot %	*		17.9	
Horses	dig prot %	*		17.0	
Rabbits	dig prot %	*		16.3	
Sheep	dig prot %	*		18.3	

Orchardgrass-Clover, ladino, aerial part, ensiled, (3)

Ref No 3 03 466 United States

Feed Name or Analyses			As Fed	Dry	C.V. ± %
Dry matter	%		31.4	100.0	35
Cattle	dig coef %		53.	53.	
Ash	%		2.9	9.2	19
Crude fiber	%		9.8	31.1	22
Cattle	dig coef %		57.	57.	
Ether extract	%		1.4	4.5	11
Cattle	dig coef %		58.	58.	
N-free extract	%		11.5	36.7	
Cattle	dig coef %		49.	49.	
Protein (N x 6.25)	%		5.9	18.7	19
Cattle	dig coef %		62.	62.	
Cattle	dig prot %		3.6	11.6	
Goats	dig prot %	*	4.1	13.2	
Horses	dig prot %	*	4.1	13.2	
Sheep	dig prot %	*	4.1	13.2	
Energy	GE Mcal/kg				
Cattle	DE Mcal/kg	*	0.74	2.34	
Sheep	DE Mcal/kg	*	0.78	2.48	
Cattle	ME Mcal/kg	*	0.60	1.92	
Sheep	ME Mcal/kg	*	0.64	2.04	
Cattle	TDN %		16.7	53.1	
Sheep	TDN %	*	17.7	56.3	
Calcium	%		0.23	0.72	60
Phosphorus	%		0.07	0.21	43
Carotene	mg/kg		56.2	178.8	71
Vitamin A equivalent	IU/g		93.7	298.0	

Orchardgrass-Clover, ladino, aerial part, ensiled, immature, (3)

Ref No 3 03 464 United States

Feed Name or Analyses			As Fed	Dry	C.V. ± %
Dry matter	%			100.0	
Ash	%			11.8	7
Crude fiber	%			25.3	7
Ether extract	%			4.9	6
N-free extract	%			38.4	
Protein (N x 6.25)	%			19.6	1
Cattle	dig prot %	*		14.0	
Goats	dig prot %	*		14.0	
Horses	dig prot %	*		14.0	
Sheep	dig prot %	*		14.0	
Energy	GE Mcal/kg				
Cattle	DE Mcal/kg	*		2.65	
Sheep	DE Mcal/kg	*		2.51	
Cattle	ME Mcal/kg	*		2.17	
Sheep	ME Mcal/kg	*		2.06	
Cattle	TDN %	*		63.3	
Sheep	TDN %	*		60.1	
Carotene	mg/kg			381.0	
Vitamin A equivalent	IU/g			635.1	

Orchardgrass-Clover, ladino, aerial part, ensiled, mature, (3)

Ref No 3 03 465 United States

Feed Name or Analyses			As Fed	Dry	C.V. ± %
Dry matter	%		23.3	100.0	28
Ash	%		1.8	7.6	
Crude fiber	%		10.2	43.8	9
Ether extract	%		0.9	4.0	
N-free extract	%		8.9	38.3	
Protein (N x 6.25)	%		1.5	6.3	5
Cattle	dig prot %	*	0.5	2.0	
Goats	dig prot %	*	0.5	2.0	
Horses	dig prot %	*	0.5	2.0	
Sheep	dig prot %	*	0.5	2.0	
Energy	GE Mcal/kg				
Cattle	DE Mcal/kg	*	0.55	2.38	
Sheep	DE Mcal/kg	*	0.59	2.53	
Cattle	ME Mcal/kg	*	0.46	1.95	
Sheep	ME Mcal/kg	*	0.48	2.07	
Cattle	TDN %	*	12.6	54.0	
Sheep	TDN %	*	13.4	57.4	

Orchardgrass-Clover, ladino, aerial part w sulfur dioxide preservative added, ensiled, (3)

Ref No 3 03 469 United States

Feed Name or Analyses			As Fed	Dry	C.V. ± %
Dry matter	%		32.3	100.0	29
Ash	%		3.0	9.3	20
Crude fiber	%		9.5	29.3	15
Ether extract	%		1.4	4.3	11
N-free extract	%		13.5	41.9	
Protein (N x 6.25)	%		4.9	15.2	25
Cattle	dig prot %	*	3.2	10.0	
Goats	dig prot %	*	3.2	10.0	
Horses	dig prot %	*	3.2	10.0	
Sheep	dig prot %	*	3.2	10.0	
Energy	GE Mcal/kg				
Cattle	DE Mcal/kg	*	0.86	2.67	
Sheep	DE Mcal/kg	*	0.80	2.48	
Cattle	ME Mcal/kg	*	0.71	2.19	
Sheep	ME Mcal/kg	*	0.66	2.03	
Cattle	TDN %	*	19.5	60.5	
Sheep	TDN %	*	18.2	56.2	

Feed Name or Analyses			Mean As Fed	Mean Dry	C.V. ± %
Carotene	mg/kg		68.1	211.0	68
Vitamin A equivalent	IU/g		113.6	351.7	

Orchardgrass-Clover, ladino, aerial part w sulfur dioxide preservative added, ensiled, milk stage, (3)

Ref No 3 03 467 United States

Feed Name or Analyses			Mean As Fed	Mean Dry	C.V. ± %
Dry matter	%		33.5	100.0	
Calcium	%		0.21	0.62	
Phosphorus	%		0.08	0.24	
Carotene	mg/kg		34.8	103.8	
Vitamin A equivalent	IU/g		58.0	173.1	

Orchardgrass-Clover, ladino, aerial part, w sulfur dioxide preservative added, ensiled, mature, (3)

Ref No 3 03 468 United States

Feed Name or Analyses			Mean As Fed	Mean Dry	C.V. ± %
Dry matter	%		30.1	100.0	
Calcium	%		0.12	0.40	
Phosphorus	%		0.04	0.12	
Carotene	mg/kg		6.7	22.3	
Vitamin A equivalent	IU/g		11.2	37.1	

ORCHARDGRASS, PENNLATE-CLOVER, NEW ZEALAND WHITE. Dactylis glomerata, Trifolium repens

Orchardgrass, pennlate-Clover, New Zealand white, aerial part, ensiled, pre-bloom, (3)

Ref No 3 08 821 United States

Feed Name or Analyses			Mean As Fed	Mean Dry	C.V. ± %
Dry matter	%		20.6	100.0	
Cattle dig coef	%		72.	72.	
Organic matter	%		19.1	92.6	
Ash	%		1.5	7.4	
Crude fiber	%		6.8	33.2	
Cattle dig coef	%		79.	79.	
Ether extract	%		1.4	7.0	
Cattle dig coef	%		76.	76.	
N-free extract	%		6.9	33.5	
Cattle dig coef	%		71.	71.	
Protein (N x 6.25)	%		3.9	18.9	
Cattle dig coef	%		77.	77.	
Cattle dig prot	%		3.0	14.5	
Goats dig prot	%	*	2.8	13.4	
Horses dig prot	%	*	2.8	13.4	
Sheep dig prot	%	*	2.8	13.4	
Cell walls (Van Soest)	%		10.8	52.5	
Fiber, acid detergent (VS)	%		8.9	43.3	
Lignin (Van Soest)	%		0.9	4.3	
Energy	GE Mcal/kg		0.97	4.71	
Cattle GE dig coef	%		71.	71.	
Cattle	DE Mcal/kg		0.69	3.33	
Sheep	DE Mcal/kg	*	0.62	3.02	
Cattle	ME Mcal/kg	*	0.56	2.73	
Sheep	ME Mcal/kg	*	0.51	2.48	
Cattle	TDN %		15.7	76.4	
Sheep	TDN %	*	14.1	68.4	

Orchardgrass, pennlate-Clover, New Zealand white, aerial part, ensiled, pre-bloom, cut 1, (3)

Ref No 3 08 823 United States

Feed Name or Analyses			Mean As Fed	Mean Dry	C.V. ± %
Dry matter	%		21.2	100.0	
Cattle dig coef	%		70.	70.	
Organic matter	%		19.7	92.9	
Ash	%		1.5	7.1	
Crude fiber	%		7.3	34.2	
Cattle dig coef	%		77.	77.	
Ether extract	%		1.5	7.0	
Cattle dig coef	%		71.	71.	
N-free extract	%		7.3	34.5	
Cattle dig coef	%		73.	73.	
Protein (N x 6.25)	%		3.6	17.2	
Cattle dig coef	%		73.	73.	
Cattle dig prot	%		2.7	12.6	
Goats dig prot	%	*	2.5	11.8	
Horses dig prot	%	*	2.5	11.8	
Sheep dig prot	%	*	2.5	11.8	
Cell walls (Van Soest)	%		11.7	55.4	
Fiber, acid detergent (VS)	%		9.4	44.3	
Lignin (Van Soest)	%		0.9	4.3	
Energy	GE Mcal/kg		0.99	4.69	
Cattle GE dig coef	%		68.	68.	
Cattle	DE Mcal/kg		0.68	3.19	
Sheep	DE Mcal/kg	*	0.66	3.10	
Cattle	ME Mcal/kg	*	0.55	2.61	
Sheep	ME Mcal/kg	*	0.54	2.54	
Cattle	TDN %		16.0	75.3	
Sheep	TDN %	*	14.9	70.2	

Orchardgrass, pennlate-Clover, New Zealand white, aerial part, wilted ensiled, pre-bloom, (3)

Ref No 3 08 822 United States

Feed Name or Analyses			Mean As Fed	Mean Dry	C.V. ± %
Dry matter	%		28.5	100.0	
Cattle dig coef	%		71.	71.	
Organic matter	%		25.9	91.0	
Ash	%		2.6	9.0	
Crude fiber	%		8.8	30.9	
Cattle dig coef	%		76.	76.	
Ether extract	%		1.8	6.3	
Cattle dig coef	%		80.	80.	
N-free extract	%		10.5	36.8	
Cattle dig coef	%		71.	71.	
Protein (N x 6.25)	%		4.8	17.0	
Cattle dig coef	%		72.	72.	
Cattle dig prot	%		3.5	12.2	
Goats dig prot	%	*	3.3	11.7	
Horses dig prot	%	*	3.3	11.7	
Sheep dig prot	%	*	3.3	11.7	
Cell walls (Van Soest)	%		13.7	48.0	
Fiber, acid detergent (VS)	%		11.5	40.2	
Lignin (Van Soest)	%		1.2	4.1	
Energy	GE Mcal/kg		1.32	4.64	
Cattle GE dig coef	%		71.	71.	
Cattle	DE Mcal/kg		0.94	3.29	
Sheep	DE Mcal/kg	*	0.88	3.08	
Cattle	ME Mcal/kg	*	0.77	2.70	
Sheep	ME Mcal/kg	*	0.72	2.53	
Cattle	TDN %		20.8	73.0	
Sheep	TDN %	*	19.9	69.8	
pH	pH units		3.93	3.93	

Orchardgrass, pennlate-Clover, New Zealand white, aerial part, wilted ensiled, pre-bloom, cut 1, (3)

Ref No 3 08 824 United States

Feed Name or Analyses			Mean As Fed	Mean Dry	C.V. ± %
Dry matter	%		30.7	100.0	
Cattle dig coef	%		70.	70.	
Organic matter	%		27.8	90.6	
Ash	%		2.9	9.4	
Crude fiber	%		11.1	36.3	
Cattle dig coef	%		77.	77.	
Ether extract	%		1.5	5.0	
Cattle dig coef	%		74.	74.	
N-free extract	%		10.0	32.6	
Cattle dig coef	%		64.	64.	
Protein (N x 6.25)	%		5.1	16.7	
Cattle dig coef	%		75.	75.	
Cattle dig prot	%		3.8	12.5	
Goats dig prot	%	*	3.5	11.4	
Horses dig prot	%	*	3.5	11.4	
Sheep dig prot	%	*	3.5	11.4	
Cell walls (Van Soest)	%		15.1	49.3	
Fiber, acid detergent (VS)	%		12.3	40.2	
Lignin (Van Soest)	%		1.4	4.6	
Energy	GE Mcal/kg		1.41	4.59	
Cattle GE dig coef	%		69.	69.	
Cattle	DE Mcal/kg		0.97	3.17	
Sheep	DE Mcal/kg	*	0.92	2.99	
Cattle	ME Mcal/kg	*	0.80	2.60	
Sheep	ME Mcal/kg	*	0.75	2.45	
Cattle	TDN %		21.4	69.7	
Sheep	TDN %	*	20.8	67.9	

ORCHARDGRASS-RYEGRASS. Dactylis glomerata, Lolium spp

Orchardgrass-Ryegrass, hay, s-c, (1)

Ref No 1 03 471 United States

Feed Name or Analyses			Mean As Fed	Mean Dry	C.V. ± %
Dry matter	%		85.5	100.0	
Ash	%		6.6	7.7	
Crude fiber	%		28.8	33.8	
Sheep dig coef	%		68.	68.	
Ether extract	%		1.5	1.7	
Sheep dig coef	%		56.	56.	
N-free extract	%		42.4	49.6	
Sheep dig coef	%		66.	66.	
Protein (N x 6.25)	%		6.2	7.3	
Sheep dig coef	%		44.	44.	
Cattle dig prot	%	*	2.8	3.2	
Goats dig prot	%	*	2.8	3.3	
Horses dig prot	%	*	3.1	3.7	
Rabbits dig prot	%	*	3.6	4.3	
Sheep dig prot	%		2.7	3.2	
Energy	GE Mcal/kg				
Cattle	DE Mcal/kg	*	2.14	2.51	
Sheep	DE Mcal/kg	*	2.30	2.69	
Cattle	ME Mcal/kg	*	1.76	2.06	
Sheep	ME Mcal/kg	*	1.89	2.21	
Cattle	TDN %	*	48.6	56.9	
Sheep	TDN %		52.1	61.0	

Orchardgrass-Ryegrass, hay, s-c, milk stage, (1)

Ref No 1 03 470 United States

Feed Name or Analyses			Mean As Fed	Mean Dry	C.V. ± %
Dry matter	%		85.7	100.0	
Ash	%		6.3	7.3	
Crude fiber	%		27.8	32.4	
Ether extract	%		1.4	1.6	
N-free extract	%		45.2	52.8	
Protein (N x 6.25)	%		5.1	5.9	
Cattle dig prot	%	*	1.8	2.0	
Goats dig prot	%	*	1.8	2.1	
Horses dig prot	%	*	2.2	2.5	

Continued

Feed Name or Analyses			Mean As Fed	Mean Dry	C.V. ± %
Rabbits	dig prot %	*	2.8	3.2	
Sheep	dig prot %	*	1.6	1.9	
Energy	GE Mcal/kg				
Cattle	DE Mcal/kg	*	2.23	2.60	
Sheep	DE Mcal/kg	*	2.09	2.44	
Cattle	ME Mcal/kg	*	1.83	2.13	
Sheep	ME Mcal/kg	*	1.71	2.00	
Cattle	TDN %	*	50.5	59.0	
Sheep	TDN %	*	47.4	55.3	

Orchardgrass-Ryegrass, aerial part, fresh, immature, (2)

Ref No 2 03 472 — United States

Feed Name or Analyses			Mean As Fed	Mean Dry	C.V. ± %
Dry matter	%		19.4	100.0	
Ash	%		1.4	7.4	
Crude fiber	%		5.4	27.8	
Ether extract	%		0.5	2.5	
N-free extract	%		9.5	49.2	
Protein (N x 6.25)	%		2.5	13.1	
Cattle	dig prot %	*	1.8	9.0	
Goats	dig prot %	*	1.7	8.8	
Horses	dig prot %	*	1.7	8.7	
Rabbits	dig prot %	*	1.7	8.8	
Sheep	dig prot %	*	1.8	9.2	
Energy	GE Mcal/kg				
Cattle	DE Mcal/kg	*	0.60	3.07	
Sheep	DE Mcal/kg	*	0.55	2.86	
Cattle	ME Mcal/kg	*	0.49	2.52	
Sheep	ME Mcal/kg	*	0.46	2.35	
Cattle	TDN %	*	13.5	69.6	
Sheep	TDN %	*	12.6	64.9	

Orchardgrass-Ryegrass, aerial part, ensiled, immature, (3)

Ref No 3 03 473 — United States

Feed Name or Analyses			Mean As Fed	Mean Dry	C.V. ± %
Dry matter	%		21.0	100.0	
Ash	%		1.7	8.1	
Crude fiber	%		6.7	31.8	
Ether extract	%		0.8	3.6	
N-free extract	%		9.0	43.0	
Protein (N x 6.25)	%		2.8	13.5	
Cattle	dig prot %	*	1.8	8.5	
Goats	dig prot %	*	1.8	8.5	
Horses	dig prot %	*	1.8	8.5	
Sheep	dig prot %	*	1.8	8.5	
Energy	GE Mcal/kg				
Cattle	DE Mcal/kg	*	0.53	2.54	
Sheep	DE Mcal/kg	*	0.58	2.77	
Cattle	ME Mcal/kg	*	0.44	2.08	
Sheep	ME Mcal/kg	*	0.48	2.27	
Cattle	TDN %	*	12.1	57.6	
Sheep	TDN %	*	13.2	62.8	

OSAGEORANGE. Maclura pomifera

Osageorange, seeds w hulls, (5)

Ref No 5 09 266 — United States

Feed Name or Analyses			Mean As Fed	Mean Dry	C.V. ± %
Dry matter	%			100.0	
Protein (N x 6.25)	%			38.0	

(1) dry forages and roughages (2) pasture, range plants, and forages fed green (3) silages (4) energy feeds (5) protein supplements (6) minerals (7) vitamins (8) additives

Feed Name or Analyses			Mean As Fed	Mean Dry	C.V. ± %
Alanine	%			1.44	
Arginine	%			5.32	
Aspartic acid	%			3.88	
Glutamic acid	%			6.73	
Glycine	%			1.60	
Histidine	%			0.80	
Isoleucine	%			1.41	
Leucine	%			2.36	
Lysine	%			1.14	
Methionine	%			0.46	
Phenylalanine	%			1.71	
Proline	%			1.25	
Serine	%			1.90	
Threonine	%			1.18	
Tyrosine	%			1.06	
Valine	%			2.32	

OSTEOPERMUM, ECKLONIS. Osteospermum ecklonis

Osteopermum, ecklonis, seeds w hulls, (5)

Ref No 5 09 294 — United States

Feed Name or Analyses			Mean As Fed	Mean Dry	C.V. ± %
Dry matter	%			100.0	
Protein (N x 6.25)	%			31.0	
Alanine	%			1.09	
Arginine	%			2.36	
Aspartic acid	%			2.76	
Glutamic acid	%			5.12	
Glycine	%			1.43	
Histidine	%			0.62	
Hydroxyproline	%			0.00	
Isoleucine	%			1.21	
Leucine	%			1.80	
Lysine	%			1.05	
Methionine	%			0.34	
Phenylalanine	%			1.05	
Proline	%			1.09	
Serine	%			1.09	
Threonine	%			0.93	
Tyrosine	%			0.78	
Valine	%			1.36	

OSTEOSPERMUM, SPINESCENS. Osteopermum spinescens

Osteospermum, spinescens, seeds w hulls, (5)

Ref No 5 09 295 — United States

Feed Name or Analyses			Mean As Fed	Mean Dry	C.V. ± %
Dry matter	%			100.0	
Protein (N x 6.25)	%			40.0	
Alanine	%			1.32	
Arginine	%			3.20	
Aspartic acid	%			3.32	
Glutamic acid	%			6.80	
Glycine	%			1.64	
Histidine	%			0.84	
Hydroxyproline	%			0.00	
Isoleucine	%			1.48	
Leucine	%			2.08	
Lysine	%			1.32	
Methionine	%			0.60	
Phenylalanine	%			1.44	
Proline	%			1.32	
Serine	%			1.44	
Threonine	%			1.12	

Feed Name or Analyses			Mean As Fed	Mean Dry	C.V. ± %
Tyrosine	%			0.92	
Valine	%			1.64	

OWLCLOVER, YELLOW. Orthocarpus luteus

Owlclover, yellow, aerial part, fresh, (2)

Ref No 2 03 474 — United States

Feed Name or Analyses			Mean As Fed	Mean Dry	C.V. ± %
Dry matter	%			100.0	
Ash	%			9.2	
Ether extract	%			2.1	
Protein (N x 6.25)	%			6.7	
Cattle	dig prot %	*		3.6	
Goats	dig prot %	*		2.8	
Horses	dig prot %	*		3.2	
Rabbits	dig prot %	*		3.8	
Sheep	dig prot %	*		3.2	
Calcium	%			1.03	
Phosphorus	%			0.49	

Owlclover, yellow, leaves, fresh, (2)

Ref No 2 03 475 — United States

Feed Name or Analyses			Mean As Fed	Mean Dry	C.V. ± %
Dry matter	%			100.0	
Ash	%			11.6	
Ether extract	%			3.9	
Protein (N x 6.25)	%			10.9	
Cattle	dig prot %	*		7.2	
Goats	dig prot %	*		6.7	
Horses	dig prot %	*		6.8	
Rabbits	dig prot %	*		7.1	
Sheep	dig prot %	*		7.1	
Calcium	%			1.42	
Phosphorus	%			0.52	

Owlclover, yellow, stems, fresh, (2)

Ref No 2 03 477 — United States

Feed Name or Analyses			Mean As Fed	Mean Dry	C.V. ± %
Dry matter	%			100.0	
Calcium	%			0.74	
Phosphorus	%			0.44	

Owlclover, yellow, stems, fresh, mature, (2)

Ref No 2 03 476 — United States

Feed Name or Analyses			Mean As Fed	Mean Dry	C.V. ± %
Dry matter	%			100.0	
Ash	%			7.7	
Ether extract	%			0.9	
Protein (N x 6.25)	%			3.9	
Cattle	dig prot %	*		1.2	
Goats	dig prot %	*		0.2	
Horses	dig prot %	*		0.8	
Rabbits	dig prot %	*		1.7	
Sheep	dig prot %	*		0.6	

OXEYEDAISY. Chrysanthemum leucanthemum

Oxeyedaisy, aerial part, s-c, full bloom, (1)

Ref No 1 03 478 — United States

Feed Name or Analyses			Mean As Fed	Mean Dry	C.V. ± %
Dry matter	%		90.4	100.0	
Ash	%		6.9	7.6	
Crude fiber	%		29.0	32.1	
Sheep	dig coef %		46.	46.	

Feed Name or Analyses			Mean As Fed	Mean Dry	C.V. ± %
Ether extract	%		4.3	4.8	
Sheep	dig coef %		62.	62.	
N-free extract	%		41.8	46.2	
Sheep	dig coef %		67.	67.	
Protein (N x 6.25)	%		8.4	9.3	
Sheep	dig coef %		58.	58.	
Cattle	dig prot %	*	4.5	5.0	
Goats	dig prot %	*	4.7	5.2	
Horses	dig prot %	*	4.9	5.4	
Rabbits	dig prot %	*	5.3	5.8	
Sheep	dig prot %		4.9	5.4	
Energy	GE Mcal/kg				
Cattle	DE Mcal/kg	*	2.47	2.73	
Sheep	DE Mcal/kg	*	2.30	2.55	
Cattle	ME Mcal/kg	*	2.03	2.24	
Sheep	ME Mcal/kg	*	1.89	2.09	
Cattle	TDN %	*	56.0	62.0	
Sheep	TDN %		52.3	57.8	

OYSTERS. Crassostrea spp, Ostrea spp

Oysters, meat, raw, (5)

Ref No 5 03 480 — United States

Analyses			As Fed	Dry	C.V. ± %
Dry matter	%		18.2	100.0	
Ash	%		1.8	9.9	
Ether extract	%		2.0	11.1	
N-free extract	%		4.8	26.3	
Protein (N x 6.25)	%		9.6	52.6	
Energy	GE Mcal/kg		0.78	4.32	
Cattle	DE Mcal/kg	*	0.71	3.93	
Sheep	DE Mcal/kg	*	0.68	3.77	
Swine	DE kcal/kg	*	762.	4197.	
Cattle	ME Mcal/kg	*	0.59	3.23	
Sheep	ME Mcal/kg	*	0.56	3.09	
Swine	ME kcal/kg	*	650.	3583.	
Cattle	TDN %	*	16.2	89.2	
Sheep	TDN %	*	15.5	85.6	
Swine	TDN %	*	17.3	95.2	
Calcium	%		0.09	0.51	
Iron	%		0.006	0.035	
Phosphorus	%		0.15	0.83	
Potassium	%		0.14	0.79	
Sodium	%		0.09	0.47	
Ascorbic acid	mg/kg		260.5	1435.4	
Niacin	mg/kg		20.4	112.3	
Riboflavin	mg/kg		2.1	11.7	
Thiamine	mg/kg		1.3	7.4	
Vitamin A	IU/g		3.7	20.1	

Oysters, shells, fine grnd, mn 33% calcium, (6)

Oyster shell flour (AAFCO)

Ref No 6 03 481 — United States

Analyses			As Fed	Dry	C.V. ± %
Dry matter	%		99.6	100.0	98
Ash	%		90.2	90.6	
Protein (N x 6.25)	%		1.0	1.0	35
Calcium	%		37.95	38.10	2
Chlorine	%		0.01	0.01	
Iron	%		0.286	0.287	
Magnesium	%		0.30	0.30	98
Manganese	mg/kg		133.6	134.1	62
Phosphorus	%		0.07	0.07	76
Potassium	%		0.10	0.10	
Sodium	%		0.21	0.21	

Ref No 6 03 481 — Canada

Analyses			As Fed	Dry	C.V. ± %
Dry matter	%		95.0	100.0	
Calcium	%		36.60	38.53	

Oyster shell flour (AAFCO) –
see Oysters, shells, fine grnd, mn 33% calcium, (6)

OZARKGRASS. Limnodea arkansana

Ozarkgrass, aerial part, fresh, (2)

Ref No 2 03 482 — United States

Analyses			As Fed	Dry	C.V. ± %
Dry matter	%			100.0	
Ash	%			6.6	
Crude fiber	%			29.2	
Ether extract	%			2.4	
N-free extract	%			50.4	
Protein (N x 6.25)	%			11.4	
Cattle	dig prot %	*		7.6	
Goats	dig prot %	*		7.2	
Horses	dig prot %	*		7.2	
Rabbits	dig prot %	*		7.5	
Sheep	dig prot %	*		7.6	
Energy	GE Mcal/kg				
Cattle	DE Mcal/kg	*		3.06	
Sheep	DE Mcal/kg	*		2.90	
Cattle	ME Mcal/kg	*		2.51	
Sheep	ME Mcal/kg	*		2.38	
Cattle	TDN %	*		69.3	
Sheep	TDN %	*		65.8	
Calcium	%			0.34	
Magnesium	%			0.14	
Phosphorus	%			0.19	
Potassium	%			1.77	

Paddy rice –
see Rice, grain w hulls, (4)

PAINTEDCUP. Castilleja spp

Paintedcup, aerial part, fresh, (2)

Ref No 2 03 484 — United States

Analyses			As Fed	Dry	C.V. ± %
Dry matter	%			100.0	
Ash	%			12.5	
Crude fiber	%			14.4	
Ether extract	%			2.6	
N-free extract	%			60.6	
Protein (N x 6.25)	%			9.9	
Cattle	dig prot %	*		6.3	
Goats	dig prot %	*		5.8	
Horses	dig prot %	*		5.9	
Rabbits	dig prot %	*		6.3	
Sheep	dig prot %	*		6.2	
Calcium	%			0.91	
Phosphorus	%			0.49	
Carotene	mg/kg			23.6	
Vitamin A equivalent	IU/g			39.3	

PAINTEDCUP, SCARLET. Castilleja miniata

Paintedcup, scarlet, aerial part, s-c, (1)

Ref No 1 03 485 — United States

Analyses			As Fed	Dry	C.V. ± %
Dry matter	%		93.3	100.0	
Ash	%		12.5	13.4	
Crude fiber	%		21.2	22.7	
Sheep	dig coef %		49.	49.	
Ether extract	%		4.8	5.1	
Sheep	dig coef %		77.	77.	
N-free extract	%		45.4	48.7	
Sheep	dig coef %		80.	80.	
Protein (N x 6.25)	%		9.4	10.1	
Sheep	dig coef %		65.	65.	
Cattle	dig prot %	*	5.3	5.7	
Goats	dig prot %	*	5.6	6.0	
Horses	dig prot %	*	5.7	6.1	
Rabbits	dig prot %	*	6.0	6.5	
Sheep	dig prot %		6.1	6.6	
Energy	GE Mcal/kg				
Cattle	DE Mcal/kg	*	2.73	2.93	
Sheep	DE Mcal/kg	*	2.69	2.89	
Cattle	ME Mcal/kg	*	2.24	2.40	
Sheep	ME Mcal/kg	*	2.21	2.37	
Cattle	TDN %	*	62.0	66.4	
Sheep	TDN %		61.1	65.5	

PALM. Elaeis spp

Palm, kernels, solv-extd grnd, (5)

Ref No 5 03 486 — United States

Analyses			As Fed	Dry	C.V. ± %
Dry matter	%		90.3	100.0	
Ash	%		3.9	4.3	
Crude fiber	%		13.0	14.4	
Sheep	dig coef %		60.	60.	
Ether extract	%		3.5	3.9	
N-free extract	%		51.7	57.2	
Sheep	dig coef %		88.	88.	
Protein (N x 6.25)	%		18.2	20.2	
Sheep	dig coef %		76.	76.	
Sheep	dig prot %		13.9	15.4	
Energy	GE Mcal/kg				
Cattle	DE Mcal/kg	*	3.02	3.35	
Sheep	DE Mcal/kg	*	3.07	3.40	
Swine	DE kcal/kg	*	3168.	3470.	
Cattle	ME Mcal/kg	*	2.48	2.74	
Sheep	ME Mcal/kg	*	2.52	2.79	
Swine	ME kcal/kg	*	2886.	3161.	
Cattle	TDN %	*	68.5	75.9	
Sheep	TDN %		69.6	77.1	
Swine	TDN %	*	71.9	78.7	
Copper	mg/kg		42.0	46.6	
Iron	%		0.017	0.019	
Manganese	mg/kg		221.2	245.1	
Phosphorus	%		0.68	0.75	
Potassium	%		0.41	0.46	

Palm, seeds, extn unspecified grnd, (5)

Ref No 5 03 487 — United States

Analyses			As Fed	Dry	C.V. ± %
Dry matter	%		91.3	100.0	1
Ash	%		3.9	4.3	7
Crude fiber	%		11.9	13.0	15
Ether extract	%		7.1	7.8	48
N-free extract	%		50.5	55.3	
Protein (N x 6.25)	%		17.9	19.6	11
Energy	GE Mcal/kg		5.28	5.79	
Cattle	DE Mcal/kg	*	3.29	3.60	
Sheep	DE Mcal/kg	*	3.03	3.32	
Swine	DE kcal/kg	*	2587.	2833.	
Cattle	ME Mcal/kg	*	2.69	2.95	
Sheep	ME Mcal/kg	*	2.48	2.72	

Continued

Feed Name or Analyses			Mean As Fed	Mean Dry	C.V. ± %
Swine	ME kcal/kg	*	2381.	2608.	
Cattle	TDN %	*	74.5	81.6	
Sheep	TDN %	*	68.6	75.2	
Swine	TDN %	*	58.7	64.3	
Copper	mg/kg		30.0	32.8	
Iron	%		0.017	0.019	4
Manganese	mg/kg		211.4	231.6	19
Phosphorus	%		0.68	0.74	
Potassium	%		0.42	0.46	
Niacin	mg/kg		44.1	48.3	1

PALM, ROYAL. Roystonea spp

Palm, royal, seeds, (1)
Ref No 1 08 477 — United States

Feed Name or Analyses			Mean As Fed	Mean Dry	C.V. ± %
Dry matter	%		86.5	100.0	
Ash	%		5.5	6.4	
Crude fiber	%		22.8	26.4	
Ether extract	%		8.3	9.6	
N-free extract	%		43.8	50.6	
Protein (N x 6.25)	%		6.1	7.1	
Cattle	dig prot %	*	2.6	3.0	
Goats	dig prot %	*	2.7	3.1	
Horses	dig prot %	*	3.0	3.5	
Rabbits	dig prot %	*	3.6	4.1	
Sheep	dig prot %	*	2.5	2.9	
Energy	GE Mcal/kg				
Cattle	DE Mcal/kg	*	2.88	3.33	
Sheep	DE Mcal/kg	*	2.19	2.54	
Cattle	ME Mcal/kg	*	2.36	2.73	
Sheep	ME Mcal/kg	*	1.80	2.08	
Cattle	TDN %	*	65.3	75.5	
Sheep	TDN %	*	49.7	57.5	

Palmo middlings –
see Wheat-Palm, flour by-product w palm oil added, mx 9.5% fiber, (4)

PANGOLAGRASS. Digitaria decumbens

Pangolagrass, aerial part, weathered post frost, (1)
Ref No 1 03 488 — United States

Feed Name or Analyses			Mean As Fed	Mean Dry	C.V. ± %
Dry matter	%			100.0	
Vitamin D_2	IU/g			0.9	

Pangolagrass, hay, s-c, early leaf, (1)
Ref No 1 03 489 — United States

Feed Name or Analyses			Mean As Fed	Mean Dry	C.V. ± %
Dry matter	%			100.0	
Vitamin D_2	IU/g			1.2	

Pangolagrass, hay, s-c, milk stage, (1)
Ref No 1 03 490 — United States

Feed Name or Analyses			Mean As Fed	Mean Dry	C.V. ± %
Dry matter	%		87.5	100.0	4
Ash	%		11.6	13.3	8
Crude fiber	%		24.0	27.4	4
Ether extract	%		1.5	1.7	7
N-free extract	%		42.0	48.0	
Protein (N x 6.25)	%		8.4	9.6	7
Cattle	dig prot %	*	4.6	5.3	
Goats	dig prot %	*	4.8	5.5	
Horses	dig prot %	*	5.0	5.7	
Rabbits	dig prot %	*	5.3	6.1	
Sheep	dig prot %	*	4.5	5.2	

Pangolagrass, aerial part, fresh, (2)
Ref No 2 03 493 — United States

Feed Name or Analyses			Mean As Fed	Mean Dry	C.V. ± %
Dry matter	%		20.0	100.0	8
Ash	%		2.0	9.8	4
Crude fiber	%		6.0	29.8	6
Ether extract	%		0.4	2.0	24
N-free extract	%		9.5	47.6	
Protein (N x 6.25)	%		2.2	10.8	22
Cattle	dig prot %	*	1.4	7.1	
Goats	dig prot %	*	1.3	6.6	
Horses	dig prot %	*	1.3	6.7	
Rabbits	dig prot %	*	1.4	7.0	
Sheep	dig prot %	*	1.4	7.1	
Calcium	%		0.09	0.45	68
Magnesium	%		0.03	0.14	88
Phosphorus	%		0.07	0.35	60

Pangolagrass, aerial part, fresh, early bloom, (2)
Ref No 2 03 491 — United States

Feed Name or Analyses			Mean As Fed	Mean Dry	C.V. ± %
Dry matter	%		16.7	100.0	
Ash	%		1.8	10.7	
Crude fiber	%		4.5	27.2	
Ether extract	%		0.4	2.4	
N-free extract	%		7.2	43.1	
Protein (N x 6.25)	%		2.8	16.6	
Cattle	dig prot %	*	2.0	12.0	
Goats	dig prot %	*	2.0	12.0	
Horses	dig prot %	*	1.9	11.6	
Rabbits	dig prot %	*	1.9	11.5	
Sheep	dig prot %	*	2.1	12.5	

Pangolagrass, aerial part, fresh, full bloom, (2)
Ref No 2 03 492 — United States

Feed Name or Analyses			Mean As Fed	Mean Dry	C.V. ± %
Dry matter	%		20.4	100.0	7
Ash	%		2.0	9.7	2
Crude fiber	%		6.1	30.1	2
Ether extract	%		0.4	2.0	15
N-free extract	%		9.8	48.1	
Protein (N x 6.25)	%		2.1	10.1	10
Cattle	dig prot %	*	1.3	6.5	
Goats	dig prot %	*	1.2	6.0	
Horses	dig prot %	*	1.2	6.1	
Rabbits	dig prot %	*	1.3	6.5	
Sheep	dig prot %	*	1.3	6.4	

Pangolagrass, aerial part, ensiled, (3)
Ref No 3 03 495 — United States

Feed Name or Analyses			Mean As Fed	Mean Dry	C.V. ± %
Dry matter	%		32.3	100.0	12
Ash	%		2.3	7.0	19
Crude fiber	%		12.3	38.0	6
Ether extract	%		0.6	2.0	16
N-free extract	%		14.2	44.1	
Protein (N x 6.25)	%		2.9	8.9	40
Cattle	dig prot %	*	1.4	4.3	
Goats	dig prot %	*	1.4	4.3	
Horses	dig prot %	*	1.4	4.3	
Sheep	dig prot %	*	1.4	4.3	
Calcium	%		0.16	0.50	18
Magnesium	%		0.01	0.03	25
Phosphorus	%		0.07	0.23	35
Carotene	mg/kg		13.4	41.4	
Vitamin A equivalent	IU/g		22.3	69.1	

Pangolagrass, aerial part, ensiled, mature, (3)
Ref No 3 03 494 — United States

Feed Name or Analyses			Mean As Fed	Mean Dry	C.V. ± %
Dry matter	%		32.1	100.0	
Calcium	%		0.13	0.39	
Magnesium	%		0.01	0.03	
Phosphorus	%		0.03	0.08	

Panic grass hay –
see Panicum, hay, s-c, (1)

PANICUM. Panicum spp

Panicum, hay, s-c, (1)
Panic grass hay
Ref No 1 03 496 — United States

Feed Name or Analyses			Mean As Fed	Mean Dry	C.V. ± %
Dry matter	%		92.6	100.0	
Ash	%		6.0	6.5	
Crude fiber	%		30.8	33.2	
Sheep	dig coef %		51.	51.	
Ether extract	%		2.0	2.2	
Sheep	dig coef %		44.	44.	
N-free extract	%		45.4	49.1	
Sheep	dig coef %		54.	54.	
Protein (N x 6.25)	%		8.3	9.0	
Sheep	dig coef %		52.	52.	
Cattle	dig prot %	*	4.4	4.7	
Goats	dig prot %	*	4.6	5.0	
Horses	dig prot %	*	4.8	5.2	
Rabbits	dig prot %	*	5.2	5.6	
Sheep	dig prot %		4.3	4.7	
Fatty acids	%		2.3	2.5	
Energy	GE Mcal/kg				
Cattle	DE Mcal/kg	*	2.17	2.34	
Sheep	DE Mcal/kg		2.05	2.22	
Cattle	ME Mcal/kg	*	1.78	1.92	
Sheep	ME Mcal/kg	*	1.68	1.82	
Cattle	TDN %	*	49.2	53.1	
Sheep	TDN %		46.6	50.3	

Panicum, aerial part, fresh, (2)
Ref No 2 03 499 — United States

Feed Name or Analyses			Mean As Fed	Mean Dry	C.V. ± %
Dry matter	%			100.0	
Ash	%			14.2	18
Crude fiber	%			29.6	15
Ether extract	%			2.3	27
N-free extract	%			41.2	
Protein (N x 6.25)	%			12.7	30
Cattle	dig prot %	*		8.7	
Goats	dig prot %	*		8.4	
Horses	dig prot %	*		8.3	
Rabbits	dig prot %	*		8.5	
Sheep	dig prot %	*		8.8	
Energy	GE Mcal/kg				
Cattle	DE Mcal/kg	*		2.63	
Sheep	DE Mcal/kg	*		2.49	

(1) dry forages and roughages (2) pasture, range plants, and forages fed green (3) silages (4) energy feeds (5) protein supplements (6) minerals (7) vitamins (8) additives

Feed Name or Analyses			Mean As Fed	Mean Dry	C.V. ± %
Cattle	ME Mcal/kg	*		2.16	
Sheep	ME Mcal/kg	*		2.04	
Cattle	TDN %	*		59.7	
Sheep	TDN %	*		56.4	
Calcium	%			0.47	43
Magnesium	%			0.36	
Phosphorus	%			0.18	31
Potassium	%			3.19	

Panicum, aerial part, fresh, immature, (2)
Ref No 2 03 497 United States

Feed Name or Analyses			Mean As Fed	Mean Dry	C.V. ± %
Dry matter	%			100.0	
Ash	%			12.9	10
Crude fiber	%			24.8	13
Ether extract	%			2.6	17
N-free extract	%			41.3	
Protein (N x 6.25)	%			18.4	16
Cattle	dig prot %	*		13.5	
Goats	dig prot %	*		13.7	
Horses	dig prot %	*		13.2	
Rabbits	dig prot %	*		12.9	
Sheep	dig prot %	*		14.1	
Calcium	%			0.61	25
Magnesium	%			0.39	
Phosphorus	%			0.22	24
Potassium	%			4.02	

Panicum, aerial part, fresh, full bloom, (2)
Ref No 2 03 498 United States

Feed Name or Analyses			Mean As Fed	Mean Dry	C.V. ± %
Dry matter	%			100.0	
Ash	%			13.1	
Crude fiber	%			38.3	
Ether extract	%			1.3	
N-free extract	%			38.8	
Protein (N x 6.25)	%			8.5	
Cattle	dig prot %	*		5.1	
Goats	dig prot %	*		4.5	
Horses	dig prot %	*		4.7	
Rabbits	dig prot %	*		5.2	
Sheep	dig prot %	*		4.9	
Energy	GE Mcal/kg				
Cattle	DE Mcal/kg	*		2.64	
Sheep	DE Mcal/kg	*		2.74	
Cattle	ME Mcal/kg	*		2.16	
Sheep	ME Mcal/kg	*		2.25	
Cattle	TDN %	*		59.8	
Sheep	TDN %	*		62.1	

PANICUM, ARIZONA. Panicum arizonicum

Panicum, Arizona, aerial part, fresh, (2)
Ref No 2 03 502 United States

Feed Name or Analyses			Mean As Fed	Mean Dry	C.V. ± %
Dry matter	%			100.0	
Ash	%			12.4	9
Crude fiber	%			36.9	9
Ether extract	%			1.6	22
N-free extract	%			41.1	
Protein (N x 6.25)	%			8.0	8
Cattle	dig prot %	*		4.7	
Goats	dig prot %	*		4.0	
Horses	dig prot %	*		4.3	
Rabbits	dig prot %	*		4.8	
Sheep	dig prot %	*		4.4	

Feed Name or Analyses			Mean As Fed	Mean Dry	C.V. ± %
Energy	GE Mcal/kg				
Cattle	DE Mcal/kg	*		2.63	
Sheep	DE Mcal/kg	*		2.79	
Cattle	ME Mcal/kg	*		2.15	
Sheep	ME Mcal/kg	*		2.29	
Cattle	TDN %	*		59.5	
Sheep	TDN %	*		63.4	
Calcium	%			0.54	16
Phosphorus	%			0.13	67

Panicum, Arizona, aerial part, fresh, mid-bloom, (2)
Ref No 2 03 500 United States

Feed Name or Analyses			Mean As Fed	Mean Dry	C.V. ± %
Dry matter	%			100.0	
Ash	%			10.2	
Crude fiber	%			29.7	
Ether extract	%			2.2	
N-free extract	%			48.9	
Protein (N x 6.25)	%			9.0	
Cattle	dig prot %	*		5.5	
Goats	dig prot %	*		5.0	
Horses	dig prot %	*		5.2	
Rabbits	dig prot %	*		5.6	
Sheep	dig prot %	*		5.4	
Energy	GE Mcal/kg				
Cattle	DE Mcal/kg	*		2.72	
Sheep	DE Mcal/kg	*		2.93	
Cattle	ME Mcal/kg	*		2.23	
Sheep	ME Mcal/kg	*		2.41	
Cattle	TDN %	*		61.6	
Sheep	TDN %	*		66.5	

Panicum, Arizona, aerial part, fresh, full bloom, (2)
Ref No 2 03 501 United States

Feed Name or Analyses			Mean As Fed	Mean Dry	C.V. ± %
Dry matter	%			100.0	
Ash	%			13.1	
Crude fiber	%			38.3	
Ether extract	%			1.3	
N-free extract	%			38.8	
Protein (N x 6.25)	%			8.5	
Cattle	dig prot %	*		5.1	
Goats	dig prot %	*		4.5	
Horses	dig prot %	*		4.7	
Rabbits	dig prot %	*		5.2	
Sheep	dig prot %	*		4.9	
Energy	GE Mcal/kg				
Cattle	DE Mcal/kg	*		2.64	
Sheep	DE Mcal/kg	*		2.74	
Cattle	ME Mcal/kg	*		2.16	
Sheep	ME Mcal/kg	*		2.25	
Cattle	TDN %	*		59.8	
Sheep	TDN %	*		62.1	

PANICUM, BROWNTOP. Panicum fasciculatum

Panicum, browntop, aerial part, fresh, immature, (2)
Ref No 2 03 503 United States

Feed Name or Analyses			Mean As Fed	Mean Dry	C.V. ± %
Dry matter	%			100.0	
Ash	%			13.7	
Crude fiber	%			27.6	
Ether extract	%			2.5	
N-free extract	%			37.9	

Feed Name or Analyses			Mean As Fed	Mean Dry	C.V. ± %
Protein (N x 6.25)	%			18.3	
Cattle	dig prot %	*		13.4	
Goats	dig prot %	*		13.6	
Horses	dig prot %	*		13.1	
Rabbits	dig prot %	*		12.8	
Sheep	dig prot %	*		14.0	
Energy	GE Mcal/kg				
Cattle	DE Mcal/kg	*		2.91	
Sheep	DE Mcal/kg	*		2.80	
Cattle	ME Mcal/kg	*		2.39	
Sheep	ME Mcal/kg	*		2.30	
Cattle	TDN %	*		66.0	
Sheep	TDN %	*		63.5	

PANICUM, BULB. Panicum bulbosum

Panicum, bulb, aerial part, fresh, immature, (2)
Ref No 2 03 504 United States

Feed Name or Analyses			Mean As Fed	Mean Dry	C.V. ± %
Dry matter	%			100.0	
Ash	%			13.1	
Crude fiber	%			24.5	
Ether extract	%			3.4	
N-free extract	%			38.2	
Protein (N x 6.25)	%			20.8	
Cattle	dig prot %	*		15.6	
Goats	dig prot %	*		16.0	
Horses	dig prot %	*		15.2	
Rabbits	dig prot %	*		14.7	
Sheep	dig prot %	*		16.4	
Energy	GE Mcal/kg				
Cattle	DE Mcal/kg	*		2.99	
Sheep	DE Mcal/kg	*		2.90	
Cattle	ME Mcal/kg	*		2.45	
Sheep	ME Mcal/kg	*		2.38	
Cattle	TDN %	*		67.8	
Sheep	TDN %	*		65.8	
Calcium	%			0.62	
Magnesium	%			0.39	
Phosphorus	%			0.28	
Potassium	%			4.02	

PANICUM, EGYPTIAN. Panicum geminatum

Panicum, Egyptian, aerial part, fresh, (2)
Ref No 2 03 506 United States

Feed Name or Analyses			Mean As Fed	Mean Dry	C.V. ± %
Dry matter	%			100.0	
Ash	%			14.9	
Crude fiber	%			31.1	
Ether extract	%			2.1	
N-free extract	%			42.0	
Protein (N x 6.25)	%			9.9	
Cattle	dig prot %	*		6.3	
Goats	dig prot %	*		5.8	
Horses	dig prot %	*		5.9	
Rabbits	dig prot %	*		6.3	
Sheep	dig prot %	*		6.2	

Panicum, Egyptian, aerial part, fresh, immature, (2)
Ref No 2 03 505 United States

Feed Name or Analyses			Mean As Fed	Mean Dry	C.V. ± %
Dry matter	%			100.0	
Calcium	%			0.49	
Phosphorus	%			0.23	

Feed Name or Analyses				Mean As Fed	Mean Dry	C.V. ± %

PANICUM, GIANT. Panicum antidotale

Panicum, giant, aerial part, fresh, (2)
Ref No 2 03 508 — United States

Feed Name or Analyses				As Fed	Dry	C.V. ± %
Dry matter		%			100.0	
Ash		%			13.1	3
Crude fiber		%			23.5	18
Ether extract		%			2.3	15
N-free extract		%			42.2	
Protein (N x 6.25)		%			18.9	21
Cattle	dig prot	%	*		14.0	
Goats	dig prot	%	*		14.2	
Horses	dig prot	%	*		13.6	
Rabbits	dig prot	%	*		13.3	
Sheep	dig prot	%	*		14.6	

Panicum, giant, aerial part, fresh immature, (2)
Ref No 2 03 507 — United States

Feed Name or Analyses				As Fed	Dry	C.V. ± %
Dry matter		%			100.0	
Calcium		%			0.61	
Phosphorus		%			0.26	

PANICUM, HALL. Panicum hallii

Panicum, hall, aerial part, fresh, (2)
Ref No 2 03 509 — United States

Feed Name or Analyses				As Fed	Dry	C.V. ± %
Dry matter		%			100.0	
Ash		%			9.7	
Crude fiber		%			29.3	
Ether extract		%			2.0	
N-free extract		%			48.0	
Protein (N x 6.25)		%			11.0	
Cattle	dig prot	%	*		7.2	
Goats	dig prot	%	*		6.8	
Horses	dig prot	%	*		6.9	
Rabbits	dig prot	%	*		7.2	
Sheep	dig prot	%	*		7.2	
Calcium		%			0.52	
Magnesium		%			0.32	
Phosphorus		%			0.17	
Potassium		%			1.52	

PANICUM, HELLER. Panicum hallii

Panicum, heller, aerial part, fresh, (2)
Ref No 2 03 510 — United States

Feed Name or Analyses				As Fed	Dry	C.V. ± %
Dry matter		%			100.0	
Ash		%			14.6	10
Crude fiber		%			28.1	2
Ether extract		%			2.5	9
N-free extract		%			44.5	
Protein (N x 6.25)		%			10.3	6
Cattle	dig prot	%	*		6.6	
Goats	dig prot	%	*		6.2	
Horses	dig prot	%	*		6.3	
Rabbits	dig prot	%	*		6.6	
Sheep	dig prot	%	*		6.6	

(1) dry forages and roughages (2) pasture, range plants, and forages fed green (3) silages (4) energy feeds (5) protein supplements (6) minerals (7) vitamins (8) additives

Feed Name or Analyses				As Fed	Dry	C.V. ± %
Calcium		%			0.32	
Phosphorus		%			0.10	

PANICUM, LINDHEIMER. Panicum lindheimeri

Panicum, lindheimer, aerial part, fresh, immature, (2)
Ref No 2 03 511 — United States

Feed Name or Analyses				As Fed	Dry	C.V. ± %
Dry matter		%			100.0	
Calcium		%			0.57	12
Phosphorus		%			0.13	11

PANICUM, TEXAS. Panicum texanum

Panicum, Texas, aerial part, fresh, (2)
Ref No 2 03 512 — United States

Feed Name or Analyses				As Fed	Dry	C.V. ± %
Dry matter		%			100.0	
Ash		%			17.6	
Crude fiber		%			28.3	
Ether extract		%			2.7	
N-free extract		%			38.0	
Protein (N x 6.25)		%			13.4	
Cattle	dig prot	%	*		9.3	
Goats	dig prot	%	*		9.1	
Horses	dig prot	%	*		8.9	
Rabbits	dig prot	%	*		9.0	
Sheep	dig prot	%	*		9.5	
Calcium		%			0.23	74
Phosphorus		%			0.23	21

PARAGRASS. Panicum purpurascens

Paragrass, hay, dehy, (1)
Ref No 1 03 514 — United States

Feed Name or Analyses				As Fed	Dry	C.V. ± %
Dry matter		%		85.3	100.0	
Ash		%		5.6	6.6	
Crude fiber		%		27.3	32.0	
Ether extract		%		2.3	2.7	
N-free extract		%		43.9	51.5	
Protein (N x 6.25)		%		6.1	7.2	
Cattle	dig prot	%	*	2.7	3.2	
Goats	dig prot	%	*	2.8	3.3	
Horses	dig prot	%	*	3.1	3.6	
Rabbits	dig prot	%	*	3.6	4.2	
Sheep	dig prot	%	*	2.6	3.0	
Energy	GE Mcal/kg					
Cattle	DE Mcal/kg		*	2.25	2.63	
Sheep	DE Mcal/kg		*	2.10	2.47	
Cattle	ME Mcal/kg		*	1.84	2.16	
Sheep	ME Mcal/kg		*	1.72	2.02	
Cattle	TDN	%	*	50.9	59.7	
Sheep	TDN	%	*	47.7	55.9	
Calcium		%		0.53	0.62	
Magnesium		%		0.19	0.22	
Phosphorus		%		0.28	0.33	

Paragrass, hay, s-c, (1)
Ref No 1 03 517 — United States

Feed Name or Analyses				As Fed	Dry	C.V. ± %
Dry matter		%		90.9	100.0	
Ash		%		6.7	7.4	
Crude fiber		%		33.7	37.1	
Sheep	dig coef	%		53.	53.	
Ether extract		%		0.9	1.0	
Sheep	dig coef	%		45.	45.	
N-free extract		%		45.6	50.2	
Sheep	dig coef	%		47.	47.	
Protein (N x 6.25)		%		3.9	4.3	
Sheep	dig coef	%		10.	10.	
Cattle	dig prot	%	*	0.6	0.7	
Goats	dig prot	%	*	0.5	0.6	
Horses	dig prot	%	*	1.1	1.2	
Rabbits	dig prot	%	*	1.8	2.0	
Sheep	dig prot	%		0.4	0.4	
Fatty acids		%		0.9	1.0	
Energy	GE Mcal/kg					
Cattle	DE Mcal/kg		*	1.96	2.16	
Sheep	DE Mcal/kg		*	1.79	1.97	
Cattle	ME Mcal/kg		*	1.61	1.77	
Sheep	ME Mcal/kg		*	1.47	1.62	
Cattle	TDN	%	*	44.6	49.1	
Sheep	TDN	%		40.6	44.7	
Calcium		%		0.35	0.39	
Phosphorus		%		0.35	0.39	
Potassium		%		1.45	1.60	

Paragrass, hay, s-c, milk stage, (1)
Ref No 1 03 515 — United States

Feed Name or Analyses				As Fed	Dry	C.V. ± %
Dry matter		%		52.5	100.0	
Ash		%		4.9	9.3	
Crude fiber		%		16.2	30.8	
Sheep	dig coef	%		56.	56.	
Ether extract		%		1.4	2.7	
Sheep	dig coef	%		51.	51.	
N-free extract		%		25.6	48.9	
Sheep	dig coef	%		61.	61.	
Protein (N x 6.25)		%		4.4	8.4	
Sheep	dig coef	%		65.	65.	
Cattle	dig prot	%	*	2.2	4.2	
Goats	dig prot	%	*	2.3	4.4	
Horses	dig prot	%	*	2.4	4.7	
Rabbits	dig prot	%	*	2.7	5.2	
Sheep	dig prot	%		2.9	5.5	
Energy	GE Mcal/kg					
Cattle	DE Mcal/kg		*	1.33	2.53	
Sheep	DE Mcal/kg		*	1.29	2.45	
Cattle	ME Mcal/kg		*	1.09	2.07	
Sheep	ME Mcal/kg		*	1.05	2.01	
Cattle	TDN	%	*	30.1	57.3	
Sheep	TDN	%		29.2	55.6	

Paragrass, hay, s-c, mature, (1)
Ref No 1 03 516 — United States

Feed Name or Analyses				As Fed	Dry	C.V. ± %
Dry matter		%		90.4	100.0	1
Ash		%		5.9	6.5	19
Crude fiber		%		34.2	37.8	2
Ether extract		%		1.0	1.1	4
N-free extract		%		46.4	51.3	
Protein (N x 6.25)		%		3.0	3.3	10
Cattle	dig prot	%	*	-0.1	-0.1	
Goats	dig prot	%	*	-0.2	-0.3	
Horses	dig prot	%	*	0.3	0.3	
Rabbits	dig prot	%	*	1.1	1.2	
Sheep	dig prot	%	*	-0.3	-0.4	
Energy	GE Mcal/kg					
Cattle	DE Mcal/kg		*	1.97	2.18	
Sheep	DE Mcal/kg		*	2.07	2.29	
Cattle	ME Mcal/kg		*	1.61	1.78	
Sheep	ME Mcal/kg		*	1.70	1.88	

Feed Name or Analyses			As Fed	Dry	C.V. ± %
Cattle	TDN %	*	44.6	49.3	
Sheep	TDN %	*	47.0	51.9	

Paragrass, aerial part, fresh, (2)

Ref No 2 03 525 United States

Feed Name or Analyses			As Fed	Dry	C.V. ± %
Dry matter	%		25.6	100.0	
Ash	%		2.9	11.3	
Crude fiber	%		8.9	34.6	
Cattle	dig coef %		59.	59.	
Sheep	dig coef %	*	56.	56.	
Ether extract	%		0.4	1.5	
Cattle	dig coef %		62.	62.	
Sheep	dig coef %	*	51.	51.	
N-free extract	%		11.6	45.1	
Cattle	dig coef %		60.	60.	
Sheep	dig coef %	*	61.	61.	
Protein (N x 6.25)	%		1.9	7.4	
Cattle	dig coef %		65.	65.	
Sheep	dig coef %	*	65.	65.	
Cattle	dig prot %		1.2	4.8	
Goats	dig prot %	*	0.9	3.5	
Horses	dig prot %	*	1.0	3.8	
Rabbits	dig prot %	*	1.1	4.4	
Sheep	dig prot %		1.2	4.8	
Energy	GE Mcal/kg				
Cattle	DE Mcal/kg	*	0.61	2.40	
Sheep	DE Mcal/kg	*	0.60	2.36	
Cattle	ME Mcal/kg	*	0.50	1.97	
Sheep	ME Mcal/kg	*	0.50	1.93	
Cattle	TDN %		13.9	54.5	
Sheep	TDN %		13.7	53.5	
Calcium	%		0.10	0.40	
Phosphorus	%		0.10	0.40	
Potassium	%		0.41	1.58	

Paragrass, aerial part, fresh, immature, (2)

Ref No 2 03 518 United States

Feed Name or Analyses			As Fed	Dry	C.V. ± %
Dry matter	%		24.6	100.0	18
Ash	%		2.6	10.4	20
Crude fiber	%		8.0	32.6	10
Ether extract	%		0.4	1.7	23
N-free extract	%		11.7	47.6	
Protein (N x 6.25)	%		1.9	7.7	25
Cattle	dig prot %	*	1.1	4.4	
Goats	dig prot %	*	0.9	3.7	
Horses	dig prot %	*	1.0	4.1	
Rabbits	dig prot %	*	1.1	4.6	
Sheep	dig prot %	*	1.0	4.2	
Energy	GE Mcal/kg				
Cattle	DE Mcal/kg	*	0.66	2.70	
Sheep	DE Mcal/kg	*	0.61	2.48	
Cattle	ME Mcal/kg	*	0.54	2.21	
Sheep	ME Mcal/kg	*	0.50	2.03	
Cattle	TDN %	*	15.1	61.2	
Sheep	TDN %	*	13.8	56.3	
Calcium	%		0.15	0.60	
Iron	%		0.004	0.015	
Magnesium	%		0.09	0.38	
Phosphorus	%		0.08	0.34	
Potassium	%		0.38	1.55	

Paragrass, aerial part, fresh, pre-bloom, cut 1, (2)

Ref No 2 03 526 United States

Feed Name or Analyses			As Fed	Dry	C.V. ± %
Dry matter	%		26.2	100.0	
Ash	%		2.4	9.1	
Crude fiber	%		8.1	31.0	
Sheep	dig coef %		64.	64.	
Ether extract	%		0.7	2.5	
Sheep	dig coef %		61.	61.	
N-free extract	%		13.3	50.8	
Sheep	dig coef %		65.	65.	
Protein (N x 6.25)	%		1.7	6.6	
Sheep	dig coef %		66.	66.	
Cattle	dig prot %	*	0.9	3.5	
Goats	dig prot %	*	0.7	2.7	
Horses	dig prot %	*	0.8	3.1	
Rabbits	dig prot %	*	1.0	3.8	
Sheep	dig prot %		1.1	4.4	
Energy	GE Mcal/kg				
Cattle	DE Mcal/kg	*	0.70	2.66	
Sheep	DE Mcal/kg	*	0.70	2.67	
Cattle	ME Mcal/kg	*	0.57	2.18	
Sheep	ME Mcal/kg	*	0.57	2.19	
Cattle	TDN %	*	15.8	60.4	
Sheep	TDN %		15.9	60.6	

Paragrass, aerial part, fresh, early bloom, (2)

Ref No 2 03 519 United States

Feed Name or Analyses			As Fed	Dry	C.V. ± %
Dry matter	%		21.1	100.0	16
Ash	%		2.2	10.5	8
Crude fiber	%		6.2	29.2	8
Cattle	dig coef %	*	59.	59.	
Sheep	dig coef %	*	56.	56.	
Ether extract	%		0.5	2.2	6
Cattle	dig coef %	*	62.	62.	
Sheep	dig coef %	*	51.	51.	
N-free extract	%		9.9	46.9	
Cattle	dig coef %	*	60.	60.	
Sheep	dig coef %	*	61.	61.	
Protein (N x 6.25)	%		2.4	11.2	8
Cattle	dig coef %	*	65.	65.	
Sheep	dig coef %	*	65.	65.	
Cattle	dig prot %		1.5	7.3	
Goats	dig prot %	*	1.5	7.0	
Horses	dig prot %	*	1.5	7.0	
Rabbits	dig prot %	*	1.5	7.3	
Sheep	dig prot %		1.5	7.3	
Energy	GE Mcal/kg				
Cattle	DE Mcal/kg	*	0.52	2.46	
Sheep	DE Mcal/kg	*	0.51	2.41	
Cattle	ME Mcal/kg	*	0.43	2.01	
Sheep	ME Mcal/kg	*	0.42	1.98	
Cattle	TDN %		11.8	55.7	
Sheep	TDN %		11.6	54.8	
Calcium	%		0.18	0.83	
Iron	%		0.003	0.015	
Magnesium	%		0.11	0.53	
Phosphorus	%		0.10	0.49	

Paragrass, aerial part, fresh, full bloom, (2)

Ref No 2 03 520 United States

Feed Name or Analyses			As Fed	Dry	C.V. ± %
Dry matter	%		26.7	100.0	12
Ash	%		2.3	8.7	9
Crude fiber	%		8.9	33.4	5
Cattle	dig coef %	*	56.	56.	
Sheep	dig coef %	*	56.	56.	
Ether extract	%		0.5	1.7	16
Cattle	dig coef %	*	62.	62.	
Sheep	dig coef %	*	51.	51.	
N-free extract	%		13.2	49.5	
Cattle	dig coef %	*	60.	60.	
Sheep	dig coef %	*	61.	61.	
Protein (N x 6.25)	%		1.8	6.7	23
Cattle	dig coef %	*	65.	65.	
Sheep	dig coef %	*	65.	65.	
Cattle	dig prot %		1.2	4.4	
Goats	dig prot %	*	0.8	2.8	
Horses	dig prot %	*	0.9	3.2	
Rabbits	dig prot %	*	1.0	3.8	
Sheep	dig prot %		1.2	4.4	
Energy	GE Mcal/kg				
Cattle	DE Mcal/kg	*	0.65	2.43	
Sheep	DE Mcal/kg	*	0.65	2.43	
Cattle	ME Mcal/kg	*	0.53	1.99	
Sheep	ME Mcal/kg	*	0.53	2.00	
Cattle	TDN %	*	14.7	55.1	
Sheep	TDN %	*	14.7	55.2	

Paragrass, aerial part, fresh, full bloom, cut 2, (2)

Ref No 2 03 527 United States

Feed Name or Analyses			As Fed	Dry	C.V. ± %
Dry matter	%		33.6	100.0	
Ash	%		2.4	7.0	
Crude fiber	%		11.0	32.7	
Sheep	dig coef %		47.	47.	
Ether extract	%		0.6	1.8	
Sheep	dig coef %		64.	64.	
N-free extract	%		18.5	55.2	
Sheep	dig coef %		61.	61.	
Protein (N x 6.25)	%		1.1	3.3	
Sheep	dig coef %		35.	35.	
Cattle	dig prot %	*	0.2	0.7	
Goats	dig prot %	*	0.0	-0.3	
Horses	dig prot %	*	0.1	0.3	
Rabbits	dig prot %	*	0.4	1.2	
Sheep	dig prot %		0.4	1.2	
Energy	GE Mcal/kg				
Cattle	DE Mcal/kg	*	0.81	2.41	
Sheep	DE Mcal/kg	*	0.78	2.33	
Cattle	ME Mcal/kg	*	0.67	1.98	
Sheep	ME Mcal/kg	*	0.64	1.91	
Cattle	TDN %	*	18.4	54.7	
Sheep	TDN %		17.7	52.8	

Paragrass, aerial part, fresh, late bloom, (2)

Ref No 2 03 521 United States

Feed Name or Analyses			As Fed	Dry	C.V. ± %
Dry matter	%		25.6	100.0	
Ash	%		3.2	12.6	
Crude fiber	%		8.8	34.2	
Cattle	dig coef %		67.	67.	
Ether extract	%		0.5	1.8	
Cattle	dig coef %		52.	52.	
N-free extract	%		11.1	43.3	
Cattle	dig coef %		64.	64.	
Protein (N x 6.25)	%		2.1	8.1	
Cattle	dig coef %		60.	60.	
Cattle	dig prot %		1.2	4.9	
Goats	dig prot %	*	1.1	4.1	
Horses	dig prot %	*	1.1	4.4	
Rabbits	dig prot %	*	1.3	4.9	
Sheep	dig prot %	*	1.2	4.5	

Continued

Feed Name or Analyses			Mean As Fed	Mean Dry	C.V. ± %
Energy	GE Mcal/kg				
Cattle	DE Mcal/kg	*	0.65	2.54	
Sheep	DE Mcal/kg	*	0.68	2.66	
Cattle	ME Mcal/kg	*	0.53	2.08	
Sheep	ME Mcal/kg	*	0.56	2.18	
Cattle	TDN %		14.7	57.6	
Sheep	TDN %	*	15.4	60.3	

Paragrass, aerial part, fresh, milk stage, (2)
Ref No 2 03 522 United States

Feed Name or Analyses			Mean As Fed	Mean Dry	C.V. ± %
Dry matter	%		21.9	100.0	15
Ash	%		2.9	13.2	11
Crude fiber	%		7.1	32.6	7
Ether extract	%		0.4	1.8	22
N-free extract	%		9.6	44.0	
Protein (N x 6.25)	%		1.8	8.4	16
Cattle	dig prot %	*	1.1	5.0	
Goats	dig prot %	*	1.0	4.4	
Horses	dig prot %	*	1.0	4.7	
Rabbits	dig prot %	*	1.1	5.2	
Sheep	dig prot %	*	1.1	4.8	
Energy	GE Mcal/kg				
Cattle	DE Mcal/kg	*	0.56	2.54	
Sheep	DE Mcal/kg	*	0.57	2.60	
Cattle	ME Mcal/kg	*	0.46	2.09	
Sheep	ME Mcal/kg	*	0.47	2.13	
Cattle	TDN %	*	12.6	57.7	
Sheep	TDN %	*	12.9	58.9	
Carotene	mg/kg		51.2	233.7	22
Vitamin A equivalent	IU/g		85.3	389.6	

Paragrass, aerial part, fresh, mature, (2)
Ref No 2 03 523 United States

Feed Name or Analyses			Mean As Fed	Mean Dry	C.V. ± %
Dry matter	%		22.0	100.0	
Ash	%		1.9	8.6	
Crude fiber	%		7.4	33.8	
Cattle	dig coef %	*	59.	59.	
Sheep	dig coef %	*	56.	56.	
Ether extract	%		0.3	1.5	
Cattle	dig coef %	*	62.	62.	
Sheep	dig coef %	*	51.	51.	
N-free extract	%		11.3	51.2	
Cattle	dig coef %	*	60.	60.	
Sheep	dig coef %	*	61.	61.	
Protein (N x 6.25)	%		1.1	4.9	
Cattle	dig coef %	*	65.	65.	
Sheep	dig coef %	*	65.	65.	
Cattle	dig prot %		0.7	3.2	
Goats	dig prot %	*	0.2	1.1	
Horses	dig prot %	*	0.4	1.7	
Rabbits	dig prot %	*	0.5	2.5	
Sheep	dig prot %		0.7	3.2	
Energy	GE Mcal/kg				
Cattle	DE Mcal/kg	*	0.54	2.47	
Sheep	DE Mcal/kg	*	0.53	2.43	
Cattle	ME Mcal/kg	*	0.44	2.02	
Sheep	ME Mcal/kg	*	0.44	1.99	
Cattle	TDN %	*	12.3	55.9	
Sheep	TDN %	*	12.1	55.1	

(1) dry forages and roughages (2) pasture, range plants, and forages fed green (3) silages (4) energy feeds (5) protein supplements (6) minerals (7) vitamins (8) additives

Paragrass, aerial part, fresh, cut 1, (2)
Ref No 2 03 524 United States

Feed Name or Analyses			Mean As Fed	Mean Dry	C.V. ± %
Dry matter	%		26.2	100.0	
Ash	%		2.4	9.1	
Crude fiber	%		8.1	31.0	
Ether extract	%		0.7	2.5	
N-free extract	%		13.3	50.8	
Protein (N x 6.25)	%		1.7	6.6	
Cattle	dig prot %	*	0.9	3.5	
Goats	dig prot %	*	0.7	2.7	
Horses	dig prot %	*	0.8	3.1	
Rabbits	dig prot %	*	1.0	3.8	
Sheep	dig prot %	*	0.8	3.1	
Energy	GE Mcal/kg				
Cattle	DE Mcal/kg	*	0.70	2.66	
Sheep	DE Mcal/kg	*	0.72	2.74	
Cattle	ME Mcal/kg	*	0.57	2.18	
Sheep	ME Mcal/kg	*	0.59	2.25	
Cattle	TDN %	*	15.8	60.4	
Sheep	TDN %	*	16.3	62.1	

Paragrass, aerial part, ensiled, (3)
Ref No 3 03 529 United States

Feed Name or Analyses			Mean As Fed	Mean Dry	C.V. ± %
Dry matter	%		25.6	100.0	9
Ash	%		4.4	17.2	4
Crude fiber	%		7.0	27.3	6
Ether extract	%		0.3	1.3	17
N-free extract	%		11.9	46.4	
Protein (N x 6.25)	%		2.0	7.8	7
Cattle	dig prot %	*	0.8	3.3	
Goats	dig prot %	*	0.8	3.3	
Horses	dig prot %	*	0.8	3.3	
Sheep	dig prot %	*	0.8	3.3	
Carotene	mg/kg		21.1	82.5	56
Vitamin A equivalent	IU/g		35.2	137.4	

Paragrass, aerial part w molasses added, ensiled, (3)
Ref No 3 03 530 United States

Feed Name or Analyses			Mean As Fed	Mean Dry	C.V. ± %
Dry matter	%		25.6	100.0	10
Ash	%		4.3	16.9	5
Crude fiber	%		7.0	27.5	7
Ether extract	%		0.4	1.4	19
N-free extract	%		11.9	46.4	
Protein (N x 6.25)	%		2.0	7.8	5
Cattle	dig prot %	*	0.8	3.3	
Goats	dig prot %	*	0.8	3.3	
Horses	dig prot %	*	0.8	3.3	
Sheep	dig prot %	*	0.8	3.3	

PARAGRASS-KUDZU. Panicum purpurascens, Pueraria spp

Paragrass-Kudzu, aerial part, fresh, (2)
Ref No 2 03 532 United States

Feed Name or Analyses			Mean As Fed	Mean Dry	C.V. ± %
Dry matter	%		23.9	100.0	9
Ash	%		2.1	8.6	13
Crude fiber	%		7.7	32.2	7
Ether extract	%		0.3	1.3	22
N-free extract	%		11.5	48.0	
Protein (N x 6.25)	%		2.4	9.9	12
Cattle	dig prot %	*	1.5	6.3	
Goats	dig prot %	*	1.4	5.8	
Horses	dig prot %	*	1.4	5.9	
Rabbits	dig prot %	*	1.5	6.3	
Sheep	dig prot %	*	1.5	6.2	

Paragrass-Kudzu, aerial part, fresh, milk stage, (2)
Ref No 2 03 531 United States

Feed Name or Analyses			Mean As Fed	Mean Dry	C.V. ± %
Dry matter	%		20.2	100.0	
Ash	%		1.8	9.0	
Crude fiber	%		6.7	33.3	
Ether extract	%		0.3	1.4	
N-free extract	%		9.0	44.5	
Protein (N x 6.25)	%		2.4	11.8	
Cattle	dig prot %	*	1.6	7.9	
Goats	dig prot %	*	1.5	7.6	
Horses	dig prot %	*	1.5	7.5	
Rabbits	dig prot %	*	1.6	7.8	
Sheep	dig prot %	*	1.6	8.0	

PARSNIP. Pastinaca sativa

Parsnip, leaves, fresh, (2)
Ref No 2 03 534 United States

Feed Name or Analyses			Mean As Fed	Mean Dry	C.V. ± %
Dry matter	%			100.0	
Crude fiber	%			8.0	
Ether extract	%			5.0	
Protein (N x 6.25)	%			22.9	
Cattle	dig prot %	*		17.4	
Goats	dig prot %	*		17.9	
Horses	dig prot %	*		17.0	
Rabbits	dig prot %	*		16.3	
Sheep	dig prot %	*		18.3	
Carotene	mg/kg			231.9	
Riboflavin	mg/kg			11.9	
Vitamin A equivalent	IU/g			386.6	

Parsnip, stems, fresh, (2)
Ref No 2 03 535 United States

Feed Name or Analyses			Mean As Fed	Mean Dry	C.V. ± %
Dry matter	%			100.0	
Crude fiber	%			17.2	
Protein (N x 6.25)	%			6.0	
Cattle	dig prot %	*		3.0	
Goats	dig prot %	*		2.2	
Horses	dig prot %	*		2.6	
Rabbits	dig prot %	*		3.3	
Sheep	dig prot %	*		2.6	
Carotene	mg/kg			4.0	
Riboflavin	mg/kg			4.4	
Vitamin A equivalent	IU/g			6.6	

Parsnip, roots, fresh, (4)
Ref No 2 03 536 United States

Feed Name or Analyses			Mean As Fed	Mean Dry	C.V. ± %
Dry matter	%		18.8	100.0	
Ash	%		1.3	6.8	
Crude fiber	%		1.6	8.7	
Ether extract	%		0.5	2.4	
N-free extract	%		13.7	72.9	
Protein (N x 6.25)	%		1.7	9.2	
Cattle	dig prot %	*	0.8	4.5	
Goats	dig prot %	*	1.1	5.7	
Horses	dig prot %	*	1.1	5.7	
Sheep	dig prot %	*	1.1	5.7	

Feed Name or Analyses			Mean As Fed	Mean Dry	C.V. ± %
Energy	GE Mcal/kg		0.68	3.64	
Cattle	DE Mcal/kg	*	0.76	4.07	
Sheep	DE Mcal/kg	*	0.74	3.92	
Swine	DE kcal/kg	*	523.	2791.	
Cattle	ME Mcal/kg	*	0.63	3.34	
Sheep	ME Mcal/kg	*	0.60	3.21	
Swine	ME kcal/kg	*	493.	2628.	
Cattle	TDN %	*	17.3	92.3	
Sheep	TDN %	*	16.7	88.9	
Swine	TDN %	*	11.9	63.3	
Calcium	%		0.06	0.30	
Iron	%		0.001	0.003	
Phosphorus	%		0.08	0.43	
Potassium	%		0.54	2.86	
Sodium	%		0.01	0.06	
Ascorbic acid	mg/kg		143.5	765.6	
Niacin	mg/kg		1.8	9.6	
Riboflavin	mg/kg		0.8	4.3	
Thiamine	mg/kg		0.7	3.8	
Vitamin A	IU/g		0.3	1.4	

PARSNIP, GARDEN. Pastinaca sativa

Parsnip, garden, seeds w hulls, (4)
Ref No 4 09 139 — United States

Feed Name or Analyses			Mean As Fed	Mean Dry	C.V. ± %
Dry matter	%			100.0	
Protein (N x 6.25)	%			16.0	
Alanine	%			0.66	
Arginine	%			0.78	
Aspartic acid	%			1.60	
Glutamic acid	%			2.77	
Glycine	%			1.04	
Histidine	%			0.37	
Hydroxyproline	%			0.03	
Isoleucine	%			0.61	
Leucine	%			0.90	
Lysine	%			0.78	
Methionine	%			0.27	
Phenylalanine	%			0.66	
Proline	%			0.74	
Serine	%			0.67	
Threonine	%			0.56	
Tyrosine	%			0.38	
Valine	%			0.78	

PARTHENIUM. Parthenium spp

Parthenium, browse, immature, (2)
Ref No 2 03 537 — United States

Feed Name or Analyses			Mean As Fed	Mean Dry	C.V. ± %
Dry matter	%			100.0	
Calcium	%			2.67	
Magnesium	%			0.26	
Phosphorus	%			0.34	
Potassium	%			5.78	

Parthenium, browse, early bloom, (2)
Ref No 2 03 538 — United States

Feed Name or Analyses			Mean As Fed	Mean Dry	C.V. ± %
Dry matter	%			100.0	
Calcium	%			2.42	
Magnesium	%			0.25	
Phosphorus	%			0.24	
Potassium	%			1.57	

PARTHENIUM, MARIOLA. Parthenium incanum

Parthenium, mariola, browse, (2)
Ref No 2 03 539 — United States

Feed Name or Analyses			Mean As Fed	Mean Dry	C.V. ± %
Dry matter	%			100.0	
Ash	%			10.6	
Crude fiber	%			19.4	
Ether extract	%			6.8	
N-free extract	%			42.0	
Protein (N x 6.25)	%			21.2	
Cattle	dig prot %	*		15.9	
Goats	dig prot %	*		16.3	
Horses	dig prot %	*		15.5	
Rabbits	dig prot %	*		15.0	
Sheep	dig prot %	*		16.7	

PASPALUM. Paspalum spp

Paspalum, aerial part, fresh, (2)
Ref No 2 03 543 — United States

Feed Name or Analyses			Mean As Fed	Mean Dry	C.V. ± %
Dry matter	%		26.8	100.0	56
Ash	%		3.5	12.9	21
Crude fiber	%		8.9	33.4	13
Cattle	dig coef %		61.	61.	
Ether extract	%		0.5	1.7	31
Cattle	dig coef %		81.	81.	
N-free extract	%		12.1	45.2	
Cattle	dig coef %		52.	52.	
Protein (N x 6.25)	%		1.8	6.9	46
Cattle	dig coef %		50.	50.	
Cattle	dig prot %		0.9	3.4	
Goats	dig prot %	*	0.8	3.0	
Horses	dig prot %	*	0.9	3.3	
Rabbits	dig prot %	*	1.1	4.0	
Sheep	dig prot %	*	0.9	3.4	
Energy	GE Mcal/kg				
Cattle	DE Mcal/kg	*	0.59	2.22	
Sheep	DE Mcal/kg	*	0.62	2.31	
Cattle	ME Mcal/kg	*	0.49	1.82	
Sheep	ME Mcal/kg	*	0.51	1.89	
Cattle	TDN %		13.5	50.4	
Sheep	TDN %	*	14.0	52.3	
Iodine	mg/kg		0.046	0.172	

Paspalum, aerial part, fresh, immature, (2)
Ref No 2 03 540 — United States

Feed Name or Analyses			Mean As Fed	Mean Dry	C.V. ± %
Dry matter	%			100.0	
Ash	%			13.2	14
Crude fiber	%			28.8	5
Ether extract	%			2.3	14
N-free extract	%			44.2	
Protein (N x 6.25)	%			11.5	16
Cattle	dig prot %	*		7.7	
Goats	dig prot %	*		7.3	
Horses	dig prot %	*		7.3	
Rabbits	dig prot %	*		7.5	
Sheep	dig prot %	*		7.7	
Energy	GE Mcal/kg				
Cattle	DE Mcal/kg	*		2.23	
Sheep	DE Mcal/kg	*		2.31	
Cattle	ME Mcal/kg	*		1.83	
Sheep	ME Mcal/kg	*		1.89	
Cattle	TDN %	*		50.6	
Sheep	TDN %	*		52.4	

Feed Name or Analyses			Mean As Fed	Mean Dry	C.V. ± %
Calcium	%			0.59	14
Phosphorus	%			0.17	18

Paspalum, aerial part, fresh, full bloom, (2)
Ref No 2 03 541 — United States

Feed Name or Analyses			Mean As Fed	Mean Dry	C.V. ± %
Dry matter	%			100.0	
Ash	%			10.9	20
Crude fiber	%			32.3	4
Ether extract	%			1.8	11
N-free extract	%			48.7	
Protein (N x 6.25)	%			6.3	11
Cattle	dig prot %	*		3.2	
Goats	dig prot %	*		2.4	
Horses	dig prot %	*		2.9	
Rabbits	dig prot %	*		3.5	
Sheep	dig prot %	*		2.9	
Energy	GE Mcal/kg				
Cattle	DE Mcal/kg	*		2.18	
Sheep	DE Mcal/kg	*		2.26	
Cattle	ME Mcal/kg	*		1.79	
Sheep	ME Mcal/kg	*		1.85	
Cattle	TDN %	*		49.5	
Sheep	TDN %	*		51.3	
Calcium	%			0.55	8
Phosphorus	%			0.11	32

Paspalum, aerial part, fresh, mature,(2)
Ref No 2 03 542 — United States

Feed Name or Analyses			Mean As Fed	Mean Dry	C.V. ± %
Dry matter	%		24.9	100.0	
Ash	%		2.4	9.8	17
Crude fiber	%		8.4	33.8	10
Ether extract	%		0.5	1.9	29
N-free extract	%		12.2	49.1	
Protein (N x 6.25)	%		1.3	5.4	18
Cattle	dig prot %	*	0.6	2.5	
Goats	dig prot %	*	0.4	1.6	
Horses	dig prot %	*	0.5	2.1	
Rabbits	dig prot %	*	0.7	2.8	
Sheep	dig prot %	*	0.5	2.0	
Energy	GE Mcal/kg				
Cattle	DE Mcal/kg	*	0.56	2.26	
Sheep	DE Mcal/kg	*	0.59	2.36	
Cattle	ME Mcal/kg	*	0.46	1.85	
Sheep	ME Mcal/kg	*	0.48	1.94	
Cattle	TDN %	*	12.7	51.2	
Sheep	TDN %	*	13.3	53.5	
Calcium	%		0.13	0.51	7
Phosphorus	%		0.02	0.07	21

PASPALUM, BROWNSEED. Paspalum plicatulum

Paspalum, brownseed, aerial part, fresh, (2)
Ref No 2 03 547 — United States

Feed Name or Analyses			Mean As Fed	Mean Dry	C.V. ± %
Dry matter	%			100.0	
Ash	%			10.0	12
Crude fiber	%			31.4	7
Ether extract	%			2.0	16
N-free extract	%			50.2	
Protein (N x 6.25)	%			6.4	29
Cattle	dig prot %	*		3.3	
Goats	dig prot %	*		2.5	
Horses	dig prot %	*		3.0	
Rabbits	dig prot %	*		3.6	

Continued

Feed Name or Analyses			Mean As Fed	Mean Dry	C.V. ± %
Sheep	dig prot	%	*	3.0	
Calcium		%		0.53	8
Phosphorus		%		0.09	29

Paspalum, brownseed, aerial part, fresh, immature, (2)
Ref No 2 03 544 United States

Feed Name or Analyses			Mean As Fed	Mean Dry	C.V. ± %
Dry matter		%		100.0	
Ash		%		11.5	
Crude fiber		%		28.6	
Ether extract		%		2.2	
N-free extract		%		47.9	
Protein (N x 6.25)		%		9.8	
Cattle	dig prot	%	*	6.2	
Goats	dig prot	%	*	5.7	
Horses	dig prot	%	*	5.9	
Rabbits	dig prot	%	*	6.2	
Sheep	dig prot	%	*	6.1	
Calcium		%		0.57	5
Phosphorus		%		0.13	13

Paspalum, brownseed, aerial part, fresh, full bloom, (2)
Ref No 2 03 545 United States

Feed Name or Analyses			Mean As Fed	Mean Dry	C.V. ± %
Dry matter		%		100.0	
Ash		%		10.2	
Crude fiber		%		31.8	
Ether extract		%		1.7	
N-free extract		%		50.0	
Protein (N x 6.25)		%		6.3	
Cattle	dig prot	%	*	3.2	
Goats	dig prot	%	*	2.4	
Horses	dig prot	%	*	2.9	
Rabbits	dig prot	%	*	3.5	
Sheep	dig prot	%	*	2.9	
Calcium		%		0.54	4
Phosphorus		%		0.09	13

Paspalum, brownseed, aerial part, fresh, mature, (2)
Ref No 2 03 546 United States

Feed Name or Analyses			Mean As Fed	Mean Dry	C.V. ± %
Dry matter		%		100.0	
Ash		%		9.2	
Crude fiber		%		32.6	
Ether extract		%		2.0	
N-free extract		%		51.5	
Protein (N x 6.25)		%		4.7	
Cattle	dig prot	%	*	1.9	
Goats	dig prot	%	*	0.9	
Horses	dig prot	%	*	1.5	
Rabbits	dig prot	%	*	2.3	
Sheep	dig prot	%	*	1.4	
Calcium		%		0.53	7
Phosphorus		%		0.07	19

PASPALUM, COMBS. Paspalum almum

Paspalum, combs, aerial part, fresh, (2)
Ref No 2 03 548 United States

Feed Name or Analyses			Mean As Fed	Mean Dry	C.V. ± %
Dry matter		%		100.0	
Ash		%		9.0	
Crude fiber		%		30.7	
Ether extract		%		2.2	
N-free extract		%		51.3	
Protein (N x 6.25)		%		6.8	
Cattle	dig prot	%	*	3.7	
Goats	dig prot	%	*	2.9	
Horses	dig prot	%	*	3.3	
Rabbits	dig prot	%	*	3.9	
Sheep	dig prot	%	*	3.3	
Calcium		%		0.80	
Phosphorus		%		0.12	

(1) dry forages and roughages (2) pasture, range plants, and forages fed green (3) silages (4) energy feeds (5) protein supplements (6) minerals (7) vitamins (8) additives

PASPALUM, FLORIDA. Paspalum floridanum

Paspalum, Florida, aerial part, fresh, (2)
Ref No 2 03 551 United States

Feed Name or Analyses			Mean As Fed	Mean Dry	C.V. ± %
Dry matter		%		100.0	
Ash		%		9.5	6
Crude fiber		%		34.0	3
Ether extract		%		1.6	12
N-free extract		%		49.3	
Protein (N x 6.25)		%		5.6	7
Cattle	dig prot	%	*	2.7	
Goats	dig prot	%	*	1.8	
Horses	dig prot	%	*	2.3	
Rabbits	dig prot	%	*	3.0	
Sheep	dig prot	%	*	2.2	
Calcium		%		0.47	7
Phosphorus		%		0.09	16

Paspalum, Florida, aerial part, fresh, full bloom, (2)
Ref No 2 03 549 United States

Feed Name or Analyses		Mean As Fed	Mean Dry	C.V. ± %
Dry matter	%		100.0	
Calcium	%		0.52	
Phosphorus	%		0.11	

Paspalum, Florida, aerial part, fresh, mature, (2)
Ref No 2 03 550 United States

Feed Name or Analyses		Mean As Fed	Mean Dry	C.V. ± %
Dry matter	%		100.0	
Calcium	%		0.47	
Phosphorus	%		0.07	

PASPALUM, FRINGELEAF. Paspalum ciliatifolium

Paspalum, fringeleaf, aerial part, fresh, full bloom, (2)
Ref No 2 03 552 United States

Feed Name or Analyses			Mean As Fed	Mean Dry	C.V. ± %
Dry matter		%		100.0	
Ash		%		14.7	
Crude fiber		%		33.3	
Ether extract		%		1.8	
N-free extract		%		42.9	
Protein (N x 6.25)		%		7.3	
Cattle	dig prot	%	*	4.1	
Goats	dig prot	%	*	3.4	
Horses	dig prot	%	*	3.7	
Rabbits	dig prot	%	*	4.3	
Sheep	dig prot	%	*	3.8	
Calcium		%		0.64	
Phosphorus		%		0.15	

PASPALUM, HAIRYSEED. Paspalum pubiflorum

Paspalum, hairyseed, aerial part, fresh, (2)
Ref No 2 03 553 United States

Feed Name or Analyses			Mean As Fed	Mean Dry	C.V. ± %
Dry matter		%		100.0	
Ash		%		14.5	
Crude fiber		%		31.5	
Ether extract		%		1.5	
N-free extract		%		43.5	
Protein (N x 6.25)		%		9.0	
Cattle	dig prot	%	*	5.5	
Goats	dig prot	%	*	5.0	
Horses	dig prot	%	*	5.2	
Rabbits	dig prot	%	*	5.6	
Sheep	dig prot	%	*	5.4	
Calcium		%		0.49	
Phosphorus		%		0.10	

PASPALUM, HARTWEG. Paspalum hartwegianum

Paspalum, hartweg, aerial part, fresh, immature, (2)
Ref No 2 03 554 United States

Feed Name or Analyses			Mean As Fed	Mean Dry	C.V. ± %
Dry matter		%		100.0	
Ash		%		12.2	
Crude fiber		%		26.7	
Ether extract		%		2.6	
N-free extract		%		46.0	
Protein (N x 6.25)		%		12.5	
Cattle	dig prot	%	*	8.5	
Goats	dig prot	%	*	8.2	
Horses	dig prot	%	*	8.1	
Rabbits	dig prot	%	*	8.3	
Sheep	dig prot	%	*	8.6	
Calcium		%		0.57	
Phosphorus		%		0.17	

Paspalum, hartweg, aerial part, fresh, mature, (2)
Ref No 2 03 555 United States

Feed Name or Analyses			Mean As Fed	Mean Dry	C.V. ± %
Dry matter		%		100.0	
Ash		%		8.6	
Crude fiber		%		35.3	
Ether extract		%		1.4	
N-free extract		%		51.0	
Protein (N x 6.25)		%		3.7	
Cattle	dig prot	%	*	1.0	
Goats	dig prot	%	*	0.0	
Horses	dig prot	%	*	0.7	
Rabbits	dig prot	%	*	1.5	
Sheep	dig prot	%	*	0.4	
Calcium		%		0.45	
Phosphorus		%		0.06	

PASPALUM, ONESPIKE. Paspalum unispicatum

Paspalum, onespike, aerial part, fresh, immature, (2)
Ref No 2 03 556 United States

Feed Name or Analyses		Mean As Fed	Mean Dry	C.V. ± %
Dry matter	%		100.0	
Ash	%		15.9	
Crude fiber	%		26.5	
Ether extract	%		1.8	
N-free extract	%		41.4	

Feed Name or Analyses		Mean As Fed	Mean Dry	C.V. ± %
Protein (N x 6.25)	%		14.4	
Cattle	dig prot % *		10.1	
Goats	dig prot % *		10.0	
Horses	dig prot % *		9.8	
Rabbits	dig prot % *		9.8	
Sheep	dig prot % *		10.4	
Calcium	%		0.71	
Phosphorus	%		0.18	

PASPALUM, RUSTYSEED. Paspalum langei

Paspalum, rustyseed, aerial part, fresh, (2)

Ref No 2 03 557 United States

Feed Name or Analyses		Mean As Fed	Mean Dry	C.V. ± %
Dry matter	%		100.0	
Ash	%		13.9	
Crude fiber	%		27.9	
Ether extract	%		2.9	
N-free extract	%		39.3	
Protein (N x 6.25)	%		16.0	
Cattle	dig prot % *		11.5	
Goats	dig prot % *		11.5	
Horses	dig prot % *		11.1	
Rabbits	dig prot % *		11.0	
Sheep	dig prot % *		11.9	
Calcium	%		0.48	
Phosphorus	%		0.22	

PASPALUM, SAND. Paspalum stramineum

Paspalum, sand, aerial part, fresh, (2)

Ref No 2 03 561 United States

Feed Name or Analyses		Mean As Fed	Mean Dry	C.V. ± %
Dry matter	%	36.7	100.0	
Ash	%	4.1	11.3	20
Crude fiber	%	12.6	34.3	11
Ether extract	%	0.8	2.3	25
N-free extract	%	16.1	43.8	
Protein (N x 6.25)	%	3.0	8.3	35
Cattle	dig prot % *	1.8	4.9	
Goats	dig prot % *	1.6	4.3	
Horses	dig prot % *	1.7	4.6	
Rabbits	dig prot % *	1.9	5.1	
Sheep	dig prot % *	1.7	4.7	
Calcium	%	0.22	0.60	16
Phosphorus	%	0.06	0.15	32

Paspalum, sand, aerial part, fresh, immature, (2)

Ref No 2 03 558 United States

Feed Name or Analyses		Mean As Fed	Mean Dry	C.V. ± %
Dry matter	%		100.0	
Ash	%		13.6	
Crude fiber	%		31.1	
Ether extract	%		2.2	
N-free extract	%		41.9	
Protein (N x 6.25)	%		11.2	
Cattle	dig prot % *		7.4	
Goats	dig prot % *		7.0	
Horses	dig prot % *		7.0	
Rabbits	dig prot % *		7.3	
Sheep	dig prot % *		7.4	
Calcium	%		0.65	14
Phosphorus	%		0.18	11

Paspalum, sand, aerial part, fresh, dough stage, (2)

Ref No 2 03 559 United States

Feed Name or Analyses		Mean As Fed	Mean Dry	C.V. ± %
Dry matter	%	33.7	100.0	
Calcium	%	0.19	0.56	
Phosphorus	%	0.04	0.13	

Paspalum, sand, aerial part, fresh, over ripe, (2)

Ref No 2 03 560 United States

Feed Name or Analyses		Mean As Fed	Mean Dry	C.V. ± %
Dry matter	%	51.4	100.0	
Ash	%	4.0	7.7	
Crude fiber	%	18.9	36.7	
Ether extract	%	1.2	2.4	
N-free extract	%	25.4	49.4	
Protein (N x 6.25)	%	2.0	3.8	
Cattle	dig prot % *	0.6	1.1	
Goats	dig prot % *	0.1	0.1	
Horses	dig prot % *	0.4	0.8	
Rabbits	dig prot % *	0.8	1.6	
Sheep	dig prot % *	0.3	0.5	
Phosphorus	%	0.05	0.09	38

PASPALUM, THIN. Paspalum setaceum

Paspalum, thin, aerial part, fresh, (2)

Ref No 2 03 565 United States

Feed Name or Analyses		Mean As Fed	Mean Dry	C.V. ± %
Dry matter	%		100.0	
Ash	%		14.4	
Crude fiber	%		30.6	
Ether extract	%		1.9	
N-free extract	%		44.8	
Protein (N x 6.25)	%		8.3	
Cattle	dig prot % *		4.9	
Goats	dig prot % *		4.3	
Horses	dig prot % *		4.6	
Rabbits	dig prot % *		5.1	
Sheep	dig prot % *		4.7	
Calcium	%		0.68	
Phosphorus	%		0.16	

Paspalum, thin, aerial part, fresh, immature, (2)

Ref No 2 03 562 United States

Feed Name or Analyses		Mean As Fed	Mean Dry	C.V. ± %
Dry matter	%		100.0	
Ash	%		17.8	
Crude fiber	%		25.7	
Ether extract	%		2.1	
N-free extract	%		42.4	
Protein (N x 6.25)	%		12.0	
Cattle	dig prot % *		8.1	
Goats	dig prot % *		7.8	
Horses	dig prot % *		7.7	
Rabbits	dig prot % *		7.9	
Sheep	dig prot % *		8.2	

Paspalum, thin, aerial part, fresh, mid-bloom, (2)

Ref No 2 03 563 United States

Feed Name or Analyses		Mean As Fed	Mean Dry	C.V. ± %
Dry matter	%		100.0	
Ash	%		10.2	
Crude fiber	%		32.8	
Ether extract	%		1.9	
N-free extract	%		47.5	
Protein (N x 6.25)	%		7.6	
Cattle	dig prot % *		4.4	
Goats	dig prot % *		3.7	
Horses	dig prot % *		4.0	
Rabbits	dig prot % *		4.5	
Sheep	dig prot % *		4.1	

Paspalum, thin, aerial part, fresh, full bloom, (2)

Ref No 2 03 564 United States

Feed Name or Analyses		Mean As Fed	Mean Dry	C.V. ± %
Dry matter	%		100.0	
Ash	%		15.2	
Crude fiber	%		33.3	
Ether extract	%		1.6	
N-free extract	%		44.7	
Protein (N x 6.25)	%		5.2	
Cattle	dig prot % *		2.3	
Goats	dig prot % *		1.4	
Horses	dig prot % *		1.9	
Rabbits	dig prot % *		2.7	
Sheep	dig prot % *		1.8	

PEA. Pisum spp

Pea, aerial part wo pods, dehy, (1)

Ref No 1 03 566 United States

Feed Name or Analyses		Mean As Fed	Mean Dry	C.V. ± %
Dry matter	%	85.8	100.0	
Ash	%	7.7	9.0	
Crude fiber	%	22.3	26.0	
Sheep	dig coef %	51.	51.	
Ether extract	%	2.3	2.7	
Sheep	dig coef %	55.	55.	
N-free extract	%	43.4	50.6	
Sheep	dig coef %	73.	73.	
Protein (N x 6.25)	%	10.0	11.7	
Sheep	dig coef %	59.	59.	
Cattle	dig prot % *	6.1	7.1	
Goats	dig prot % *	6.4	7.5	
Horses	dig prot % *	6.4	7.5	
Rabbits	dig prot % *	6.6	7.7	
Sheep	dig prot %	5.9	6.9	
Energy	GE Mcal/kg			
Cattle	DE Mcal/kg *	2.40	2.80	
Sheep	DE Mcal/kg *	2.29	2.67	
Cattle	ME Mcal/kg *	1.97	2.30	
Sheep	ME Mcal/kg *	1.87	2.19	
Cattle	TDN % *	54.4	63.4	
Sheep	TDN %	51.9	60.4	

Pea, aerial part wo pods, s-c, (1)

Ref No 1 03 567 United States

Feed Name or Analyses		Mean As Fed	Mean Dry	C.V. ± %
Dry matter	%	86.8	100.0	
Ash	%	5.9	6.8	
Crude fiber	%	23.3	26.9	
Sheep	dig coef %	52.	52.	
Ether extract	%	2.4	2.8	
Sheep	dig coef %	54.	54.	
N-free extract	%	41.2	47.5	
Sheep	dig coef %	73.	73.	
Protein (N x 6.25)	%	13.9	16.0	
Sheep	dig coef %	72.	72.	
Cattle	dig prot % *	9.4	10.8	
Goats	dig prot % *	10.0	11.5	
Horses	dig prot % *	9.6	11.1	

Continued

Feed Name or Analyses			Mean As Fed	Mean Dry	C.V. ± %
Rabbits	dig prot %	*	9.6	11.0	
Sheep	dig prot %		10.0	11.5	
Energy	GE Mcal/kg				
Cattle	DE Mcal/kg	*	2.26	2.61	
Sheep	DE Mcal/kg	*	2.43	2.80	
Cattle	ME Mcal/kg	*	1.86	2.14	
Sheep	ME Mcal/kg	*	2.00	2.30	
Cattle	TDN %	*	51.3	59.1	
Sheep	TDN %		55.2	63.6	

Pea, aerial part wo seeds, (1)

Ref No 1 03 568 — United States

Feed Name or Analyses			Mean As Fed	Mean Dry	C.V. ± %
Dry matter	%		89.9	100.0	
Ash	%		3.1	3.4	
Crude fiber	%		23.0	25.6	
Cattle	dig coef %		87.	87.	
Ether extract	%		1.6	1.8	
Cattle	dig coef %		68.	68.	
N-free extract	%		44.2	49.2	
Cattle	dig coef %		93.	93.	
Protein (N x 6.25)	%		18.0	20.0	
Cattle	dig coef %		82.	82.	
Cattle	dig prot %		14.7	16.4	
Goats	dig prot %	*	13.7	15.2	
Horses	dig prot %	*	13.0	14.5	
Rabbits	dig prot %	*	12.7	14.1	
Sheep	dig prot %	*	13.0	14.5	
Energy	GE Mcal/kg				
Cattle	DE Mcal/kg	*	3.46	3.84	
Sheep	DE Mcal/kg	*	3.02	3.36	
Cattle	ME Mcal/kg	*	2.83	3.15	
Sheep	ME Mcal/kg	*	2.48	2.76	
Cattle	TDN %		78.4	87.2	
Sheep	TDN %	*	68.6	76.3	

Ref No 1 03 568 — Canada

Feed Name or Analyses			Mean As Fed	Mean Dry	C.V. ± %
Dry matter	%		92.9	100.0	
Crude fiber	%		24.1	25.9	
Cattle	dig coef %		87.	87.	
Ether extract	%				
Cattle	dig coef %		68.	68.	
N-free extract	%				
Cattle	dig coef %		93.	93.	
Protein (N x 6.25)	%		18.6	20.0	
Cattle	dig coef %		82.	82.	
Cattle	dig prot %		15.3	16.4	
Goats	dig prot %	*	14.2	15.2	
Horses	dig prot %	*	13.5	14.5	
Rabbits	dig prot %	*	13.1	14.1	
Sheep	dig prot %	*	13.5	14.5	
Energy	GE Mcal/kg				
Cattle	DE Mcal/kg	*	3.57	3.84	
Cattle	ME Mcal/kg	*	2.93	3.15	
Cattle	TDN %		81.0	87.2	
Calcium	%		0.69	0.74	
Phosphorus	%		0.35	0.38	

Pea, hay, dehy, (1)

Ref No 1 03 569 — United States

Feed Name or Analyses			Mean As Fed	Mean Dry	C.V. ± %
Dry matter	%		86.8	100.0	
Ash	%		5.9	6.8	
Crude fiber	%		23.3	26.9	
Ether extract	%		2.4	2.8	
N-free extract	%		41.2	47.5	
Protein (N x 6.25)	%		13.9	16.0	
Cattle	dig prot %	*	9.4	10.8	
Goats	dig prot %	*	10.0	11.5	
Horses	dig prot %	*	9.6	11.1	
Rabbits	dig prot %	*	9.6	11.0	
Sheep	dig prot %	*	9.5	10.9	
Energy	GE Mcal/kg				
Cattle	DE Mcal/kg	*	2.26	2.61	
Sheep	DE Mcal/kg	*	2.34	2.69	
Cattle	ME Mcal/kg	*	1.86	2.14	
Sheep	ME Mcal/kg	*	1.92	2.21	
Cattle	TDN %	*	51.3	59.1	
Sheep	TDN %	*	53.1	61.1	

Pea, hay, s-c, (1)

Ref No 1 03 572 — United States

Feed Name or Analyses			Mean As Fed	Mean Dry	C.V. ± %
Dry matter	%		88.0	100.0	3
Ash	%		6.7	7.6	10
Crude fiber	%		26.5	30.2	14
Sheep	dig coef %	*	51.	51.	
Ether extract	%		2.2	2.5	20
Sheep	dig coef %	*	48.	48.	
N-free extract	%		40.5	46.0	
Sheep	dig coef %	*	72.	72.	
Protein (N x 6.25)	%		12.0	13.6	17
Sheep	dig coef %	*	71.	71.	
Cattle	dig prot %	*	7.7	8.7	
Goats	dig prot %	*	8.2	9.3	
Horses	dig prot %	*	8.0	9.1	
Rabbits	dig prot %	*	8.1	9.2	
Sheep	dig prot %		8.5	9.7	
Energy	GE Mcal/kg				
Cattle	DE Mcal/kg	*	2.25	2.56	
Sheep	DE Mcal/kg	*	2.36	2.69	
Cattle	ME Mcal/kg	*	1.84	2.10	
Sheep	ME Mcal/kg	*	1.94	2.20	
Cattle	TDN %	*	51.0	58.0	
Sheep	TDN %		53.6	60.9	

Pea, hay, s-c, dough stage, (1)

Ref No 1 03 570 — United States

Feed Name or Analyses			Mean As Fed	Mean Dry	C.V. ± %
Dry matter	%		84.1	100.0	
Ash	%		6.1	7.3	
Crude fiber	%		26.8	31.9	
Sheep	dig coef %		52.	52.	
Ether extract	%		2.0	2.4	
Sheep	dig coef %		46.	46.	
N-free extract	%		37.3	44.4	
Sheep	dig coef %		64.	64.	
Protein (N x 6.25)	%		11.8	14.0	
Sheep	dig coef %		60.	60.	
Cattle	dig prot %	*	7.6	9.1	
Goats	dig prot %	*	8.1	9.6	
Horses	dig prot %	*	7.9	9.4	
Rabbits	dig prot %	*	8.0	9.5	
Sheep	dig prot %		7.1	8.4	
Energy	GE Mcal/kg				
Cattle	DE Mcal/kg	*	2.09	2.49	
Sheep	DE Mcal/kg	*	2.07	2.46	
Cattle	ME Mcal/kg	*	1.72	2.04	
Sheep	ME Mcal/kg	*	1.70	2.02	
Cattle	TDN %	*	47.5	56.4	
Sheep	TDN %		47.0	55.9	

Pea, hay, s-c, over ripe, (1)

Ref No 1 03 571 — United States

Feed Name or Analyses			Mean As Fed	Mean Dry	C.V. ± %
Dry matter	%		93.2	100.0	
Ash	%		5.8	6.2	
Crude fiber	%		28.6	30.7	
Sheep	dig coef %		51.	51.	
Ether extract	%		1.4	1.5	
Sheep	dig coef %		50.	50.	
N-free extract	%		43.1	46.2	
Sheep	dig coef %		79.	79.	
Protein (N x 6.25)	%		14.4	15.4	
Sheep	dig coef %		78.	78.	
Cattle	dig prot %	*	9.6	10.3	
Goats	dig prot %	*	10.2	10.9	
Horses	dig prot %	*	9.9	10.6	
Rabbits	dig prot %	*	9.8	10.6	
Sheep	dig prot %		11.2	12.0	
Energy	GE Mcal/kg				
Cattle	DE Mcal/kg	*	2.59	2.78	
Sheep	DE Mcal/kg	*	2.71	2.90	
Cattle	ME Mcal/kg	*	2.12	2.28	
Sheep	ME Mcal/kg	*	2.22	2.38	
Cattle	TDN %	*	58.8	63.1	
Sheep	TDN %		61.4	65.9	

Pea, pods, s-c, (1)

Ref No 1 03 575 — United States

Feed Name or Analyses			Mean As Fed	Mean Dry	C.V. ± %
Dry matter	%		88.4	100.0	
Ash	%		4.7	5.3	
Crude fiber	%		31.5	35.6	
Sheep	dig coef %	*	79.	79.	
Ether extract	%		1.0	1.1	
Sheep	dig coef %	*	57.	57.	
N-free extract	%		41.7	47.2	
Sheep	dig coef %	*	83.	83.	
Protein (N x 6.25)	%		9.5	10.8	
Sheep	dig coef %	*	53.	53.	
Cattle	dig prot %	*	5.6	6.3	
Goats	dig prot %	*	5.9	6.6	
Horses	dig prot %	*	5.9	6.7	
Rabbits	dig prot %	*	6.2	7.0	
Sheep	dig prot %		5.1	5.7	
Energy	GE Mcal/kg				
Cattle	DE Mcal/kg	*	2.64	2.99	
Sheep	DE Mcal/kg	*	2.90	3.28	
Cattle	ME Mcal/kg	*	2.17	2.45	
Sheep	ME Mcal/kg	*	2.38	2.69	
Cattle	TDN %	*	59.9	67.8	
Sheep	TDN %		65.8	74.4	

Pea, seed skins, (1)

Pea bran

Pea hulls

Ref No 1 03 602 — United States

Feed Name or Analyses			Mean As Fed	Mean Dry	C.V. ± %
Dry matter	%		89.8	100.0	4
Ash	%		3.1	3.5	15
Crude fiber	%		37.1	41.3	46
Sheep	dig coef %		83.	83.	
Swine	dig coef %		72.	72.	
Ether extract	%		0.9	1.0	57
Sheep	dig coef %		83.	83.	

(1) dry forages and roughages (2) pasture, range plants, and forages fed green (3) silages (4) energy feeds (5) protein supplements (6) minerals (7) vitamins (8) additives

Feed Name or Analyses			Mean As Fed	Mean Dry	C.V. ± %
Swine	dig coef %		42.	42.	
N-free extract	%		38.2	42.5	25
Sheep	dig coef %		80.	80.	
Swine	dig coef %		98.	98.	
Protein (N x 6.25)	%		10.4	11.6	73
Sheep	dig coef %		41.	41.	
Swine	dig coef %		91.	91.	
Cattle	dig prot %	*	6.3	7.0	
Goats	dig prot %	*	6.7	7.4	
Horses	dig prot %	*	6.7	7.4	
Rabbits	dig prot %	*	6.9	7.7	
Sheep	dig prot %		4.3	4.8	
Swine	dig prot %		9.5	10.6	
Energy	GE Mcal/kg				
Cattle	DE Mcal/kg	*	3.02	3.36	
Sheep	DE Mcal/kg	*	2.95	3.29	
Swine	DE kcal/kg	*	3285.	3658.	
Cattle	ME Mcal/kg	*	2.48	2.76	
Sheep	ME Mcal/kg	*	2.42	2.69	
Swine	ME kcal/kg	*	3076.	3426.	
Cattle	TDN %	*	68.4	76.2	
Sheep	TDN %		66.9	74.5	
Swine	TDN %		74.5	83.0	
Calcium	%		0.67	0.75	
Phosphorus	%		0.13	0.14	

Ref No 1 03 602 Canada

Feed Name or Analyses			Mean As Fed	Mean Dry	C.V. ± %
Dry matter	%		88.3	100.0	
Crude fiber	%		38.5	43.6	
Sheep	dig coef %		83.	83.	
Swine	dig coef %		72.	72.	
Ether extract	%		0.7	0.8	
Sheep	dig coef %		83.	83.	
Swine	dig coef %		42.	42.	
N-free extract	%				
Sheep	dig coef %		80.	80.	
Swine	dig coef %		98.	98.	
Protein (N x 6.25)	%		11.6	13.2	
Sheep	dig coef %		41.	41.	
Swine	dig coef %		91.	91.	
Cattle	dig prot %	*	7.4	8.4	
Goats	dig prot %	*	7.8	8.9	
Horses	dig prot %	*	7.7	8.7	
Rabbits	dig prot %	*	7.8	8.9	
Sheep	dig prot %		4.8	5.4	
Swine	dig prot %		10.6	12.0	
Energy	GE Mcal/kg				
Sheep	DE Mcal/kg	*	2.89	3.28	
Swine	DE kcal/kg	*	3225.	3653.	
Sheep	ME Mcal/kg	*	2.37	2.69	
Swine	ME kcal/kg	*	3010.	3409.	
Sheep	TDN %		65.6	74.3	
Swine	TDN %		73.2	82.8	

Pea, split pea by-product, grnd, (1)
Pea feed
Pea meal
Ref No 1 08 478 United States

Feed Name or Analyses			Mean As Fed	Mean Dry	C.V. ± %
Dry matter	%		90.0	100.0	
Ash	%		3.5	3.9	
Crude fiber	%		23.7	26.3	
Ether extract	%		1.4	1.6	
N-free extract	%		43.7	48.6	
Protein (N x 6.25)	%		17.7	19.7	
Cattle	dig prot %	*	12.6	14.0	
Goats	dig prot %	*	13.4	14.9	
Horses	dig prot %	*	12.8	14.2	

Feed Name or Analyses			Mean As Fed	Mean Dry	C.V. ± %
Rabbits	dig prot %	*	12.5	13.9	
Sheep	dig prot %	*	12.8	14.2	
Energy	GE Mcal/kg				
Cattle	DE Mcal/kg	*	3.35	3.72	
Sheep	DE Mcal/kg	*	3.39	3.77	
Cattle	ME Mcal/kg	*	2.75	3.05	
Sheep	ME Mcal/kg	*	2.78	3.09	
Cattle	TDN %	*	75.9	84.3	
Sheep	TDN %	*	77.0	85.6	

Pea, straw, (1)
Ref No 1 03 577 United States

Feed Name or Analyses			Mean As Fed	Mean Dry	C.V. ± %
Dry matter	%		84.7	100.0	4
Ash	%		5.5	6.5	9
Crude fiber	%		33.4	39.5	10
Cattle	dig coef %		53.	53.	
Sheep	dig coef %		39.	39.	
Ether extract	%		1.5	1.8	38
Cattle	dig coef %		56.	56.	
Sheep	dig coef %		34.	34.	
N-free extract	%		36.7	43.4	
Cattle	dig coef %		66.	66.	
Sheep	dig coef %		46.	46.	
Protein (N x 6.25)	%		7.6	8.9	19
Cattle	dig coef %		53.	53.	
Sheep	dig coef %		51.	51.	
Cattle	dig prot %		4.0	4.7	
Goats	dig prot %	*	4.1	4.9	
Horses	dig prot %	*	4.3	5.1	
Rabbits	dig prot %	*	4.7	5.6	
Sheep	dig prot %		3.9	4.6	
Energy	GE Mcal/kg				
Cattle	DE Mcal/kg	*	2.11	2.49	
Sheep	DE Mcal/kg	*	1.52	1.80	
Cattle	ME Mcal/kg	*	1.73	2.04	
Sheep	ME Mcal/kg	*	1.25	1.48	
Cattle	TDN %		47.8	56.5	
Sheep	TDN %		34.6	40.8	

Pea, straw, treated w sodium hydroxide wet, (1)
Ref No 2 03 578 United States

Feed Name or Analyses			Mean As Fed	Mean Dry	C.V. ± %
Dry matter	%		21.7	100.0	
Ash	%		1.3	5.9	
Crude fiber	%		13.8	63.6	
Sheep	dig coef %		54.	54.	
Ether extract	%		0.3	1.2	
Sheep	dig coef %		46.	46.	
N-free extract	%		5.3	24.2	
Sheep	dig coef %		59.	59.	
Protein (N x 6.25)	%		1.1	5.1	
Sheep	dig coef %		-6.	-6.	
Cattle	dig prot %	*	0.3	1.4	
Goats	dig prot %	*	0.3	1.3	
Horses	dig prot %	*	0.4	1.9	
Rabbits	dig prot %	*	0.6	2.6	
Sheep	dig prot %		0.0	-0.3	
Energy	GE Mcal/kg				
Cattle	DE Mcal/kg	*	0.41	1.90	
Sheep	DE Mcal/kg	*	0.47	2.18	
Cattle	ME Mcal/kg	*	0.34	1.56	
Sheep	ME Mcal/kg	*	0.39	1.79	
Cattle	TDN %	*	9.3	43.0	
Sheep	TDN %		10.7	49.5	

Pea, aerial part, fresh, (2)
Ref No 2 03 582 United States

Feed Name or Analyses			Mean As Fed	Mean Dry	C.V. ± %
Dry matter	%		14.5	100.0	20
Ash	%		2.1	14.7	18
Crude fiber	%		3.7	25.3	9
Sheep	dig coef %	*	62.	62.	
Ether extract	%		0.6	3.9	17
Sheep	dig coef %	*	58.	58.	
N-free extract	%		4.9	33.6	
Sheep	dig coef %	*	80.	80.	
Protein (N x 6.25)	%		3.3	22.6	10
Sheep	dig coef %	*	84.	84.	
Cattle	dig prot %	*	2.5	17.1	
Goats	dig prot %	*	2.5	17.6	
Horses	dig prot %	*	2.4	16.7	
Rabbits	dig prot %	*	2.3	16.1	
Sheep	dig prot %		2.7	18.9	
Energy	GE Mcal/kg				
Cattle	DE Mcal/kg	*	0.43	2.95	
Sheep	DE Mcal/kg	*	0.42	2.94	
Cattle	ME Mcal/kg	*	0.35	2.42	
Sheep	ME Mcal/kg	*	0.35	2.41	
Cattle	TDN %	*	9.7	66.9	
Sheep	TDN %		9.6	66.6	

Pea, aerial part, fresh, immature, (2)
Ref No 2 03 579 United States

Feed Name or Analyses			Mean As Fed	Mean Dry	C.V. ± %
Dry matter	%		13.5	100.0	
Ash	%		1.4	10.4	
Crude fiber	%		3.0	22.2	
Ether extract	%		0.6	4.4	
N-free extract	%		4.8	35.6	
Protein (N x 6.25)	%		3.7	27.4	
Cattle	dig prot %	*	2.9	21.2	
Goats	dig prot %	*	3.0	22.1	
Horses	dig prot %	*	2.8	20.8	
Rabbits	dig prot %	*	2.7	19.8	
Sheep	dig prot %	*	3.0	22.5	
Energy	GE Mcal/kg				
Cattle	DE Mcal/kg	*	0.40	2.97	
Sheep	DE Mcal/kg	*	0.38	2.85	
Cattle	ME Mcal/kg	*	0.33	2.44	
Sheep	ME Mcal/kg	*	0.32	2.33	
Cattle	TDN %	*	9.1	67.3	
Sheep	TDN %	*	8.7	64.6	

Pea, aerial part, fresh, pre-bloom, (2)
Ref No 2 03 580 United States

Feed Name or Analyses			Mean As Fed	Mean Dry	C.V. ± %
Dry matter	%		13.5	100.0	
Ash	%		1.4	10.4	
Crude fiber	%		3.0	22.2	
Sheep	dig coef %		62.	62.	
Ether extract	%		0.6	4.4	
Sheep	dig coef %		52.	52.	
N-free extract	%		4.8	35.6	
Sheep	dig coef %		71.	71.	
Protein (N x 6.25)	%		3.7	27.4	
Sheep	dig coef %		82.	82.	
Cattle	dig prot %	*	2.9	21.2	
Goats	dig prot %	*	3.0	22.1	
Horses	dig prot %	*	2.8	20.8	
Rabbits	dig prot %	*	2.7	19.8	
Sheep	dig prot %		3.0	22.5	

Continued

Feed Name or Analyses			Mean As Fed	Mean Dry	C.V. ± %
Energy	GE Mcal/kg				
Cattle	DE Mcal/kg	*	0.40	2.97	
Sheep	DE Mcal/kg	*	0.40	2.94	
Cattle	ME Mcal/kg	*	0.33	2.44	
Sheep	ME Mcal/kg	*	0.33	2.41	
Cattle	TDN %	*	9.1	67.3	
Sheep	TDN %		9.0	66.7	

Pea, aerial part, fresh, early bloom, (2)

Ref No 2 03 581 United States

Feed Name or Analyses			As Fed	Dry	C.V. ± %
Dry matter	%		12.2	100.0	11
Ash	%		2.1	17.1	9
Crude fiber	%		3.2	25.9	7
Sheep	dig coef %		62.	62.	
Ether extract	%		0.5	4.0	6
Sheep	dig coef %		59.	59.	
N-free extract	%		3.7	30.6	
Sheep	dig coef %		82.	82.	
Protein (N x 6.25)	%		2.7	22.4	6
Sheep	dig coef %		84.	84.	
Cattle	dig prot %	*	2.1	16.9	
Goats	dig prot %	*	2.1	17.5	
Horses	dig prot %	*	2.0	16.5	
Rabbits	dig prot %	*	1.9	16.0	
Sheep	dig prot %		2.3	18.8	
Energy	GE Mcal/kg				
Cattle	DE Mcal/kg	*	0.34	2.78	
Sheep	DE Mcal/kg	*	0.35	2.88	
Cattle	ME Mcal/kg	*	0.28	2.28	
Sheep	ME Mcal/kg	*	0.29	2.36	
Cattle	TDN %	*	7.7	63.0	
Sheep	TDN %		8.0	65.3	

Pea, pods, fresh, (2)

Ref No 2 03 588 United States

Feed Name or Analyses			As Fed	Dry	C.V. ± %
Dry matter	%		12.9	100.0	
Ash	%		0.9	7.0	
Crude fiber	%		3.0	23.3	
Ether extract	%		0.3	2.3	
N-free extract	%		6.9	53.4	
Protein (N x 6.25)	%		1.8	14.0	
Cattle	dig prot %	*	1.3	9.8	
Goats	dig prot %	*	1.2	9.6	
Horses	dig prot %	*	1.2	9.4	
Rabbits	dig prot %	*	1.2	9.5	
Sheep	dig prot %	*	1.3	10.0	
Energy	GE Mcal/kg				
Cattle	DE Mcal/kg	*	0.40	3.11	
Sheep	DE Mcal/kg	*	0.41	3.21	
Cattle	ME Mcal/kg	*	0.33	2.55	
Sheep	ME Mcal/kg	*	0.34	2.63	
Cattle	TDN %	*	9.1	70.6	
Sheep	TDN %	*	9.4	72.9	
Calcium	%		0.19	1.47	
Phosphorus	%		0.03	0.23	
Biotin	mg/kg		0.03	0.22	
Niacin	mg/kg		4.2	32.2	22
Pantothenic acid	mg/kg		1.3	10.1	7
Riboflavin	mg/kg		0.9	6.8	6
Thiamine	mg/kg		0.3	2.6	14

Pea, aerial part, ensiled, (3)

Ref No 3 03 590 United States

Feed Name or Analyses			As Fed	Dry	C.V. ± %
Dry matter	%		24.4	100.0	16
Ash	%		1.8	7.4	
Crude fiber	%		7.9	32.2	
Ether extract	%		0.8	3.3	
N-free extract	%		10.8	44.1	
Protein (N x 6.25)	%		3.2	13.0	
Cattle	dig prot %	*	2.0	8.0	
Goats	dig prot %	*	2.0	8.0	
Horses	dig prot %	*	2.0	8.0	
Sheep	dig prot %	*	2.0	8.0	
Energy	GE Mcal/kg				
Cattle	DE Mcal/kg	*	0.61	2.49	
Sheep	DE Mcal/kg	*	0.57	2.35	
Cattle	ME Mcal/kg	*	0.50	2.04	
Sheep	ME Mcal/kg	*	0.47	1.93	
Cattle	TDN %	*	13.8	56.5	
Sheep	TDN %	*	13.0	53.4	
Calcium	%		0.42	1.71	38
Phosphorus	%		0.07	0.30	98
Carotene	mg/kg		31.4	128.7	65
Riboflavin	mg/kg		1.6	6.4	
Vitamin A equivalent	IU/g		52.4	214.6	

Pea, aerial part, ensiled, immature, (3)

Ref No 3 03 589 United States

Feed Name or Analyses			As Fed	Dry	C.V. ± %
Dry matter	%		25.7	100.0	
Ash	%		3.9	15.3	
Crude fiber	%		6.5	25.3	
Ether extract	%		0.7	2.6	
N-free extract	%		10.4	40.3	
Protein (N x 6.25)	%		4.2	16.5	
Cattle	dig prot %	*	2.9	11.2	
Goats	dig prot %	*	2.9	11.2	
Horses	dig prot %	*	2.9	11.2	
Sheep	dig prot %	*	2.9	11.2	
Energy	GE Mcal/kg				
Cattle	DE Mcal/kg	*	0.70	2.74	
Sheep	DE Mcal/kg	*	0.66	2.56	
Cattle	ME Mcal/kg	*	0.58	2.25	
Sheep	ME Mcal/kg	*	0.54	2.10	
Cattle	TDN %	*	16.0	62.1	
Sheep	TDN %	*	14.9	58.0	
Carotene	mg/kg		53.8	209.4	
Vitamin A equivalent	IU/g		89.7	349.1	

Pea, aerial part, molded ensiled, (3)

Ref No 3 03 592 United States

Feed Name or Analyses			As Fed	Dry	C.V. ± %
Dry matter	%		38.0	100.0	
Ash	%		3.2	8.4	
Crude fiber	%		17.5	46.0	
Ether extract	%		0.3	0.8	
N-free extract	%		14.2	37.4	
Protein (N x 6.25)	%		2.8	7.4	
Cattle	dig prot %	*	1.1	2.9	
Goats	dig prot %	*	1.1	2.9	
Horses	dig prot %	*	1.1	2.9	
Sheep	dig prot %	*	1.1	2.9	
Carotene	mg/kg		8.0	20.9	
Vitamin A equivalent	IU/g		13.3	34.9	

Pea, aerial part w phosphoric acid preservative added, ensiled, (3)

Ref No 3 03 593 United States

Feed Name or Analyses			As Fed	Dry	C.V. ± %
Dry matter	%		28.7	100.0	
Ash	%		5.3	18.5	
Crude fiber	%		7.1	24.7	
Ether extract	%		1.1	3.8	
N-free extract	%		10.4	36.3	
Protein (N x 6.25)	%		4.8	16.7	
Cattle	dig prot %	*	3.3	11.4	
Goats	dig prot %	*	3.3	11.4	
Horses	dig prot %	*	3.3	11.4	
Sheep	dig prot %	*	3.3	11.4	
Energy	GE Mcal/kg				
Cattle	DE Mcal/kg	*	0.78	2.73	
Sheep	DE Mcal/kg	*	0.77	2.67	
Cattle	ME Mcal/kg	*	0.64	2.24	
Sheep	ME Mcal/kg	*	0.63	2.19	
Cattle	TDN %	*	17.8	62.0	
Sheep	TDN %	*	17.4	60.5	
Phosphorus	%		0.30	1.05	
Carotene	mg/kg		38.8	135.1	27
Vitamin A equivalent	IU/g		64.7	225.3	

Pea, aerial part wo seeds, ensiled, (3)

Pea vine silage

Ref No 3 03 596 United States

Feed Name or Analyses			As Fed	Dry	C.V. ± %
Dry matter	%		24.0	100.0	
Ash	%		2.0	8.2	
Crude fiber	%		7.4	31.0	
Sheep	dig coef %		50.	50.	
Ether extract	%		0.8	3.3	
Sheep	dig coef %		58.	58.	
N-free extract	%		10.7	44.5	
Sheep	dig coef %		68.	68.	
Protein (N x 6.25)	%		3.1	13.0	
Sheep	dig coef %		59.	59.	
Cattle	dig prot %	*	1.9	8.1	
Goats	dig prot %	*	1.9	8.1	
Horses	dig prot %	*	1.9	8.1	
Sheep	dig prot %		1.8	7.7	
Energy	GE Mcal/kg				
Cattle	DE Mcal/kg	*	0.62	2.58	
Sheep	DE Mcal/kg	*	0.61	2.55	
Cattle	ME Mcal/kg	*	0.51	2.12	
Sheep	ME Mcal/kg	*	0.50	2.09	
Cattle	NE_m Mcal/kg	*	0.29	1.20	
Cattle	NE_{gain} Mcal/kg	*	0.13	0.53	
Cattle	$NE_{lactating\ cows}$ Mcal/kg	*	0.31	1.30	
Cattle	TDN %	*	14.1	58.6	
Sheep	TDN %		13.9	57.7	
Calcium	%		0.31	1.31	
Phosphorus	%		0.06	0.24	
Carotene	mg/kg		45.3	188.9	
Vitamin A equivalent	IU/g		75.6	315.0	

Pea, aerial part w sulfur dioxide preservative added, ensiled, (3)

Ref No 3 03 594 United States

Feed Name or Analyses			As Fed	Dry	C.V. ± %
Dry matter	%			100.0	
Ash	%			16.5	
Crude fiber	%			22.0	
Ether extract	%			2.3	
N-free extract	%			43.9	

(1) dry forages and roughages (3) silages (6) minerals
(2) pasture, range plants, and forages fed green (4) energy feeds (7) vitamins
(5) protein supplements (8) additives

Feed Name or Analyses		Mean As Fed	Mean Dry	C.V. ± %
Protein (N x 6.25)	%		15.3	
Cattle	dig prot %	*	10.1	
Goats	dig prot %	*	10.1	
Horses	dig prot %	*	10.1	
Sheep	dig prot %	*	10.1	
Energy	GE Mcal/kg			
Cattle	DE Mcal/kg	*	2.62	
Sheep	DE Mcal/kg	*	2.56	
Cattle	ME Mcal/kg	*	2.15	
Sheep	ME Mcal/kg	*	2.10	
Cattle	TDN %	*	69.4	
Sheep	TDN %	*	58.0	

Pea, seed skins w molasses added, ensiled, (3)

Pea bran, molasses added

Ref No 3 03 595 United States

Feed Name or Analyses		Mean As Fed	Mean Dry	C.V. ± %
Dry matter	%	54.6	100.0	
Ash	%	8.0	14.7	
Crude fiber	%	10.8	19.8	
Sheep	dig coef %	83.	83.	
Ether extract	%	1.1	2.0	
Sheep	dig coef %	47.	47.	
N-free extract	%	25.3	46.3	
Sheep	dig coef %	86.	86.	
Protein (N x 6.25)	%	9.4	17.3	
Sheep	dig coef %	79.	79.	
Cattle	dig prot %	* 6.5	11.9	
Goats	dig prot %	* 6.5	11.9	
Horses	dig prot %	* 6.5	11.9	
Sheep	dig prot %	7.5	13.7	
Energy	GE Mcal/kg			
Cattle	DE Mcal/kg	* 1.71	3.13	
Sheep	DE Mcal/kg	* 1.73	3.17	
Cattle	ME Mcal/kg	* 1.40	2.57	
Sheep	ME Mcal/kg	* 1.42	2.60	
Cattle	TDN %	* 38.7	71.0	
Sheep	TDN %	39.3	72.0	

Pea, seeds, (5)

Ref No 5 03 600 United States

Feed Name or Analyses		Mean As Fed	Mean Dry	C.V. ± %
Dry matter	%	89.5	100.0	
Ash	%	2.8	3.1	
Crude fiber	%	5.5	6.1	
Ether extract	%	1.3	1.4	
N-free extract	%	56.2	62.8	
Protein (N x 6.25)	%	23.8	26.5	
Energy	GE Mcal/kg	3.45	3.85	
Cattle	DE Mcal/kg	* 3.28	3.66	
Sheep	DE Mcal/kg	* 2.98	3.33	
Swine	DE kcal/kg	*3679.	4111.	
Cattle	ME Mcal/kg	* 2.69	3.00	
Chickens	ME_n kcal/kg	2567.	2868.	
Sheep	ME Mcal/kg	* 2.45	2.73	
Swine	ME kcal/kg	*3335.	3726.	
Cattle	TDN %	* 74.3	83.0	
Sheep	TDN %	* 67.6	75.6	
Swine	TDN %	* 83.4	93.2	
Calcium	%	0.11	0.13	
Iron	%	0.005	0.006	
Phosphorus	%	0.42	0.47	
Potassium	%	1.02	1.14	
Sodium	%	0.04	0.05	
Choline	mg/kg	642.	717.	
Niacin	mg/kg	33.9	37.9	
Pantothenic acid	mg/kg	10.0	11.2	
Riboflavin	mg/kg	2.3	2.6	
Thiamine	mg/kg	7.5	8.4	
Vitamin A	IU/g	1.2	1.4	
Arginine	%	1.38	1.54	
Cystine	%	0.17	0.19	
Lysine	%	1.58	1.76	
Methionine	%	0.31	0.34	
Tryptophan	%	0.24	0.26	

Ref No 5 03 600 Canada

Feed Name or Analyses		Mean As Fed	Mean Dry	C.V. ± %
Dry matter	%	88.9	100.0	
Crude fiber	%	5.3	5.9	
Ether extract	%	0.8	0.9	
Protein (N x 6.25)	%	21.6	24.3	
Energy	GE Mcal/kg	3.42	3.85	
Chickens	ME_n kcal/kg	2550.	2868.	
Calcium	%	0.11	0.12	
Phosphorus	%	0.47	0.53	

Pea, seeds, cooked, (5)

Ref No 5 03 597 United States

Feed Name or Analyses		Mean As Fed	Mean Dry	C.V. ± %
Dry matter	%	82.5	100.0	
Ash	%	2.7	3.3	
Crude fiber	%	6.3	7.6	
Swine	dig coef %	55.	55.	
Ether extract	%	1.5	1.8	
Swine	dig coef %	36.	36.	
N-free extract	%	50.2	60.8	
Swine	dig coef %	95.	95.	
Protein (N x 6.25)	%	21.9	26.5	
Swine	dig coef %	87.	87.	
Swine	dig prot %	19.0	23.1	
Energy	GE Mcal/kg			
Cattle	DE Mcal/kg	* 2.97	3.60	
Sheep	DE Mcal/kg	* 2.72	3.29	
Swine	DE kcal/kg	*3145.	3812.	
Cattle	ME Mcal/kg	* 2.43	2.95	
Sheep	ME Mcal/kg	* 2.23	2.70	
Swine	ME kcal/kg	*2851.	3455.	
Cattle	TDN %	* 67.3	81.6	
Sheep	TDN %	* 61.6	74.7	
Swine	TDN %	71.3	86.5	

Pea, seeds, grnd, (5)

Ref No 5 03 598 United States

Feed Name or Analyses		Mean As Fed	Mean Dry	C.V. ± %
Dry matter	%	87.7	100.0	3
Ash	%	3.0	3.4	12
Crude fiber	%	6.8	7.7	26
Horses	dig coef %	8.	8.	
Sheep	dig coef %	50.	50.	
Swine	dig coef %	61.	61.	
Ether extract	%	1.4	1.6	23
Horses	dig coef %	7.	7.	
Sheep	dig coef %	64.	64.	
Swine	dig coef %	32.	32.	
N-free extract	%	53.3	60.7	3
Horses	dig coef %	90.	90.	
Sheep	dig coef %	93.	93.	
Swine	dig coef %	93.	93.	
Protein (N x 6.25)	%	23.2	26.5	9
Horses	dig coef %	83.	83.	
Sheep	dig coef %	86.	86.	
Swine	dig coef %	88.	88.	
Horses	dig prot %	19.3	22.0	
Sheep	dig prot %	20.0	22.8	
Swine	dig prot %	20.3	23.2	
Energy	GE Mcal/kg			
Cattle	DE Mcal/kg	* 3.14	3.58	
Horses	DE Mcal/kg	* 3.00	3.42	
Sheep	DE Mcal/kg	* 3.31	3.77	
Swine	DE kcal/kg	*3295.	3756.	
Cattle	ME Mcal/kg	* 2.57	2.94	
Horses	ME Mcal/kg	* 2.46	2.80	
Sheep	ME Mcal/kg	* 2.71	3.09	
Swine	ME kcal/kg	*2986.	3405.	
Cattle	TDN %	* 71.2	81.2	
Horses	TDN %	68.0	77.5	
Sheep	TDN %	75.0	85.5	
Swine	TDN %	74.7	85.2	
Choline	mg/kg	624.	712.	
Thiamine	mg/kg	193.4	220.5	0

Ref No 5 03 598 Canada

Feed Name or Analyses		Mean As Fed	Mean Dry	C.V. ± %
Dry matter	%	90.8	100.0	
Crude fiber	%	28.7	31.6	
Horses	dig coef %	8.	8.	
Sheep	dig coef %	50.	50.	
Swine	dig coef %	61.	61.	
Ether extract	%	1.0	1.1	
Horses	dig coef %	7.	7.	
Sheep	dig coef %	64.	64.	
Swine	dig coef %	32.	32.	
N-free extract	%			
Horses	dig coef %	90.	90.	
Sheep	dig coef %	93.	93.	
Swine	dig coef %	93.	93.	
Protein (N x 6.25)	%	15.9	17.5	
Horses	dig coef %	83.	83.	
Sheep	dig coef %	86.	86.	
Swine	dig coef %	88.	88.	
Horses	dig prot %	13.2	14.5	
Sheep	dig prot %	13.7	15.1	
Swine	dig prot %	13.9	15.3	
Energy	GE Mcal/kg			
Horses	DE Mcal/kg	* 2.98	3.28	
Sheep	DE Mcal/kg	* 3.36	3.70	
Swine	DE kcal/kg	*3363.	3704.	
Horses	ME Mcal/kg	* 2.44	2.69	
Sheep	ME Mcal/kg	* 2.75	3.03	
Swine	ME kcal/kg	*3109.	3425.	
Horses	TDN %	67.6	74.4	
Sheep	TDN %	76.1	83.8	
Swine	TDN %	76.3	84.0	

Pea, seed screenings, (5)

Ref No 5 03 601 Canada

Feed Name or Analyses		Mean As Fed	Mean Dry	C.V. ± %
Dry matter	%	91.9	100.0	
Crude fiber	%	7.7	8.4	
Ether extract	%	2.4	2.6	
Protein (N x 6.25)	%	17.6	19.2	
Calcium	%	0.28	0.30	
Phosphorus	%	0.34	0.37	

PEA, AUSTRALIAN FIELD. Pisum spp

Pea, Australian field, aerial part, fresh, (2)

Ref No 2 03 603 United States

Feed Name or Analyses		Mean As Fed	Mean Dry	C.V. ± %
Dry matter	%		100.0	
Cobalt	mg/kg		0.150	

PEA, AUSTRIAN FIELD. Pisum spp

Pea, Austrian field, aerial part, fresh, full bloom, (2)

Ref No 2 03 604 United States

Feed Name or Analyses			Mean As Fed	Mean Dry	C.V. ± %
Dry matter	%			100.0	
Ash	%			8.2	
Protein (N x 6.25)	%			21.4	
Cattle	dig prot %	*		16.1	
Goats	dig prot %	*		16.5	
Horses	dig prot %	*		15.7	
Rabbits	dig prot %	*		15.2	
Sheep	dig prot %	*		16.9	
Calcium	%			1.02	
Iron	%			0.040	
Magnesium	%			0.22	
Manganese	mg/kg			85.1	
Phosphorus	%			0.64	
Potassium	%			2.52	

PEA, FIELD. Pisum sativum arvense

Pea, field, hay, s-c, (1)

Ref No 1 03 605 United States

Feed Name or Analyses			Mean As Fed	Mean Dry	C.V. ± %
Dry matter	%		89.3	100.0	2
Ash	%		7.7	8.6	6
Crude fiber	%		24.3	27.2	15
Ether extract	%		3.3	3.7	5
N-free extract	%		39.1	43.7	
Protein (N x 6.25)	%		15.0	16.8	6
Cattle	dig prot %	*	10.3	11.5	
Goats	dig prot %	*	10.9	12.2	
Horses	dig prot %	*	10.5	11.8	
Rabbits	dig prot %	*	10.4	11.6	
Sheep	dig prot %	*	10.4	11.6	
Fatty acids	%		3.3	3.7	
Energy	GE Mcal/kg				
Cattle	DE Mcal/kg	*	2.42	2.71	
Sheep	DE Mcal/kg	*	2.31	2.59	
Cattle	ME Mcal/kg	*	1.98	2.22	
Sheep	ME Mcal/kg	*	1.90	2.12	
Cattle	TDN %	*	54.9	61.5	
Sheep	TDN %	*	52.5	58.7	
Calcium	%		1.22	1.37	
Magnesium	%		0.33	0.37	
Phosphorus	%		0.25	0.28	
Potassium	%		1.25	1.40	
Sulphur	%		0.21	0.23	

Pea, field, hay wo seeds, s-c, (1)

Ref No 1 03 607 United States

Feed Name or Analyses			Mean As Fed	Mean Dry	C.V. ± %
Dry matter	%		87.8	100.0	3
Ash	%		6.8	7.8	17
Crude fiber	%		23.4	26.7	10
Ether extract	%		2.3	2.6	23
N-free extract	%		42.5	48.4	
Protein (N x 6.25)	%		12.7	14.5	26
Cattle	dig prot %	*	8.3	9.5	
Goats	dig prot %	*	8.9	10.1	
Horses	dig prot %	*	8.6	9.8	
Rabbits	dig prot %	*	8.7	9.9	
Sheep	dig prot %	*	8.4	9.6	
Energy	GE Mcal/kg				
Cattle	DE Mcal/kg	*	2.33	2.65	
Sheep	DE Mcal/kg	*	2.34	2.66	
Cattle	ME Mcal/kg	*	1.91	2.18	
Sheep	ME Mcal/kg	*	1.92	2.18	
Cattle	TDN %	*	52.8	60.2	
Sheep	TDN %	*	53.0	60.4	

Pea, field, hay wo seeds, s-c, immature, (1)

Ref No 1 03 606 United States

Feed Name or Analyses			Mean As Fed	Mean Dry	C.V. ± %
Dry matter	%		91.7	100.0	
Ash	%		8.1	8.8	
Crude fiber	%		21.0	22.9	
Ether extract	%		2.0	2.2	
N-free extract	%		40.9	44.6	
Protein (N x 6.25)	%		19.7	21.5	
Cattle	dig prot %	*	14.3	15.6	
Goats	dig prot %	*	15.2	16.6	
Horses	dig prot %	*	14.5	15.8	
Rabbits	dig prot %	*	14.0	15.3	
Sheep	dig prot %	*	14.5	15.9	
Energy	GE Mcal/kg				
Cattle	DE Mcal/kg	*	2.52	2.75	
Sheep	DE Mcal/kg	*	2.56	2.79	
Cattle	ME Mcal/kg	*	2.06	2.25	
Sheep	ME Mcal/kg	*	2.10	2.29	
Cattle	TDN %	*	57.1	62.3	
Sheep	TDN %	*	58.1	63.4	

Pea, field, straw, s-c, (1)

Ref No 1 08 479 United States

Feed Name or Analyses			Mean As Fed	Mean Dry	C.V. ± %
Dry matter	%		90.2	100.0	
Ash	%		5.4	6.0	
Crude fiber	%		33.1	36.7	
Ether extract	%		1.6	1.8	
N-free extract	%		44.0	48.8	
Protein (N x 6.25)	%		6.1	6.8	
Cattle	dig prot %	*	2.5	2.8	
Goats	dig prot %	*	2.6	2.9	
Horses	dig prot %	*	3.0	3.3	
Rabbits	dig prot %	*	3.5	3.9	
Sheep	dig prot %	*	2.4	2.6	
Energy	GE Mcal/kg				
Cattle	DE Mcal/kg	*	2.01	2.23	
Sheep	DE Mcal/kg	*	2.14	2.37	
Cattle	ME Mcal/kg	*	1.65	1.83	
Sheep	ME Mcal/kg	*	1.75	1.94	
Cattle	TDN %	*	45.6	50.5	
Sheep	TDN %	*	48.5	53.7	
Phosphorus	%		0.10	0.11	
Potassium	%		1.08	1.20	

Pea, field, aerial part, fresh, (2)

Ref No 2 03 608 United States

Feed Name or Analyses			Mean As Fed	Mean Dry	C.V. ± %
Dry matter	%		18.4	100.0	9
Ash	%		1.7	9.1	12
Crude fiber	%		4.7	25.4	7
Ether extract	%		0.6	3.4	9
N-free extract	%		7.7	42.1	
Protein (N x 6.25)	%		3.7	20.0	9
Cattle	dig prot %	*	2.7	14.9	
Goats	dig prot %	*	2.8	15.2	
Horses	dig prot %	*	2.7	14.5	
Rabbits	dig prot %	*	2.6	14.1	
Sheep	dig prot %	*	2.9	15.6	
Energy	GE Mcal/kg				
Cattle	DE Mcal/kg	*	0.59	3.22	
Sheep	DE Mcal/kg	*	0.55	3.01	
Cattle	ME Mcal/kg	*	0.48	2.64	
Sheep	ME Mcal/kg	*	0.45	2.47	
Cattle	TDN %	*	13.4	72.9	
Sheep	TDN %	*	12.7	68.3	
Calcium	%		0.22	1.21	14
Iron	%		0.007	0.040	
Magnesium	%		0.04	0.22	
Manganese	mg/kg		15.6	85.1	
Phosphorus	%		0.04	0.23	86
Potassium	%		0.28	1.50	40

Pea, field, aerial part, ensiled, (3)

Ref No 3 03 609 United States

Feed Name or Analyses			Mean As Fed	Mean Dry	C.V. ± %
Dry matter	%		27.0	100.0	
Ash	%		2.5	9.3	
Crude fiber	%		7.4	27.5	
Sheep	dig coef %	*	50.	50.	
Ether extract	%		1.2	4.3	
Sheep	dig coef %	*	58.	58.	
N-free extract	%		12.1	44.8	
Sheep	dig coef %	*	68.	68.	
Protein (N x 6.25)	%		3.8	14.1	
Sheep	dig coef %	*	59.	59.	
Cattle	dig prot %	*	2.4	9.0	
Goats	dig prot %	*	2.4	9.0	
Horses	dig prot %	*	2.4	9.0	
Sheep	dig prot %		2.2	8.3	
Energy	GE Mcal/kg				
Cattle	DE Mcal/kg	*	0.75	2.80	
Sheep	DE Mcal/kg	*	0.69	2.56	
Cattle	ME Mcal/kg	*	0.62	2.29	
Sheep	ME Mcal/kg	*	0.57	2.10	
Cattle	TDN %	*	17.1	63.4	
Sheep	TDN %		15.7	58.1	
Calcium	%		0.37	1.36	7
Magnesium	%		0.11	0.39	13
Phosphorus	%		0.08	0.29	21
Potassium	%		0.38	1.40	7
Sulphur	%		0.07	0.25	19

Pea, field, seeds, (5)

Ref No 5 08 481 United States

Feed Name or Analyses			Mean As Fed	Mean Dry	C.V. ± %
Dry matter	%		90.7	100.0	
Ash	%		3.0	3.3	
Crude fiber	%		6.1	6.7	
Ether extract	%		1.2	1.3	
N-free extract	%		57.0	62.8	
Protein (N x 6.25)	%		23.4	25.8	
Energy	GE Mcal/kg				
Cattle	DE Mcal/kg	*	3.29	3.62	
Sheep	DE Mcal/kg	*	3.01	3.32	
Swine	DE kcal/kg	*	3707.	4087.	
Cattle	ME Mcal/kg	*	2.70	2.97	
Chickens	ME_n kcal/kg		2200.	2426.	
Sheep	ME Mcal/kg	*	2.47	2.72	
Swine	ME kcal/kg	*	3366.	3711.	
Cattle	TDN %	*	74.5	82.2	
Sheep	TDN %	*	68.3	75.4	
Swine	TDN %	*	84.1	92.7	
Calcium	%		0.17	0.19	

(1) dry forages and roughages (2) pasture, range plants, and forages fed green (3) silages (4) energy feeds (5) protein supplements (6) minerals (7) vitamins (8) additives

Feed Name or Analyses		Mean As Fed	Mean Dry	C.V. ± %
Phosphorus	%	0.50	0.55	
Potassium	%	1.03	1.14	

Pea, field, seeds, cull, (5)
Ref No 5 08 480 United States

Feed Name or Analyses		Mean As Fed	Mean Dry	C.V. ± %
Dry matter	%	91.6	100.0	
Ash	%	2.8	3.1	
Crude fiber	%	5.9	6.4	
Ether extract	%	1.1	1.2	
N-free extract	%	59.8	65.3	
Protein (N x 6.25)	%	22.0	24.0	
Energy	GE Mcal/kg			
Cattle	DE Mcal/kg	* 3.35	3.65	
Sheep	DE Mcal/kg	* 3.09	3.37	
Swine	DE kcal/kg	*3794.	4141.	
Cattle	ME Mcal/kg	* 2.74	3.00	
Sheep	ME Mcal/kg	* 2.53	2.76	
Swine	ME kcal/kg	*3458.	3775.	
Cattle	TDN %	* 75.9	82.9	
Sheep	TDN %	* 70.0	76.4	
Swine	TDN %	* 86.0	93.9	
Calcium	%	0.17	0.19	
Phosphorus	%	0.32	0.35	

PEA, GARDEN. Pisum sativum

Pea, garden, seeds, (5)
Ref No 5 08 482 United States

Feed Name or Analyses		Mean As Fed	Mean Dry	C.V. ± %
Dry matter	%	89.2	100.0	
Ash	%	2.9	3.3	
Crude fiber	%	5.7	6.4	
Ether extract	%	1.7	1.9	
N-free extract	%	53.6	60.1	
Protein (N x 6.25)	%	25.3	28.4	
Energy	GE Mcal/kg			
Cattle	DE Mcal/kg	* 3.26	3.66	
Sheep	DE Mcal/kg	* 3.55	3.98	
Swine	DE kcal/kg	*3648.	4090.	
Cattle	ME Mcal/kg	* 2.67	3.00	
Sheep	ME Mcal/kg	* 2.91	3.26	
Swine	ME kcal/kg	*3293.	3692.	
Cattle	TDN %	* 74.0	82.9	
Sheep	TDN %	* 80.5	90.3	
Swine	TDN %	* 82.7	92.8	
Calcium	%	0.08	0.09	
Phosphorus	%	0.40	0.45	
Potassium	%	0.90	1.01	

PEA, YELLOW. Pisum spp

Pea, yellow, seeds, grnd, (5)
Ref No 5 03 610 United States

Feed Name or Analyses		Mean As Fed	Mean Dry	C.V. ± %
Dry matter	%	90.0	100.0	
Niacin	mg/kg	31.7	35.3	16

PEA-BARLEY. Pisum spp, Hordeum vulgare

Pea-Barley, aerial part, fresh, (2)
Ref No 2 03 611 United States

Feed Name or Analyses		Mean As Fed	Mean Dry	C.V. ± %
Dry matter	%	21.0	100.0	13
Ash	%	1.7	8.3	9
Crude fiber	%	5.3	25.2	7
Ether extract	%	0.9	4.1	8
N-free extract	%	9.4	45.1	
Protein (N x 6.25)	%	3.6	17.3	9
Cattle	dig prot %	* 2.6	12.6	
Goats	dig prot %	* 2.7	12.7	
Horses	dig prot %	* 2.6	12.2	
Rabbits	dig prot %	* 2.5	12.0	
Sheep	dig prot %	* 2.7	13.1	
Energy	GE Mcal/kg			
Cattle	DE Mcal/kg	* 0.57	2.73	
Sheep	DE Mcal/kg	* 0.58	2.78	
Cattle	ME Mcal/kg	* 0.46	2.24	
Sheep	ME Mcal/kg	* 0.48	2.28	
Cattle	TDN %	* 13.0	61.9	
Sheep	TDN %	* 13.2	63.0	
Calcium	%	0.17	0.79	
Phosphorus	%	0.07	0.35	
Potassium	%	0.31	1.49	

Pea bran —
see Pea, seed skins, (1)

Pea bran, molasses added —
see Pea, seed skins w molasses added, ensiled, (3)

PEACHBRUSH, SMALLFLOWER. Prunus minutiflora

Peachbrush, smallflower, browse, immature, (2)
Ref No 2 03 613 United States

Feed Name or Analyses		Mean As Fed	Mean Dry	C.V. ± %
Dry matter	%		100.0	
Ash	%		7.0	
Crude fiber	%		10.8	
Ether extract	%		4.2	
N-free extract	%		54.6	
Protein (N x 6.25)	%		23.4	
Cattle	dig prot %	*	17.8	
Goats	dig prot %	*	18.4	
Horses	dig prot %	*	17.4	
Rabbits	dig prot %	*	16.7	
Sheep	dig prot %	*	18.8	
Calcium	%		2.01	
Magnesium	%		0.30	
Phosphorus	%		0.27	
Potassium	%		1.67	

Pea feed —
see Pea, split pea by-product, grnd, (1)

Pea hulls —
see Pea, seed skins, (1)

Pea meal —
see Pea, split pea by-product, grnd, (1)

PEANUT. Arachis hypogaea

Peanut, hay, s-c, (1)
Ref No 1 03 619 United States

Feed Name or Analyses		Mean As Fed	Mean Dry	C.V. ± %
Dry matter	%	91.2	100.0	2
Ash	%	9.7	10.6	
Crude fiber	%	23.7	26.0	
Sheep	dig coef %	* 51.	51.	
Ether extract	%	5.1	5.6	
Sheep	dig coef %	* 62.	62.	
N-free extract	%	42.1	46.2	
Sheep	dig coef %	* 77.	77.	
Protein (N x 6.25)	%	10.6	11.6	
Sheep	dig coef %	* 65.	65.	
Cattle	dig prot %	* 6.4	7.0	
Goats	dig prot %	* 6.7	7.4	
Horses	dig prot %	* 6.7	7.4	
Rabbits	dig prot %	* 6.9	7.6	
Sheep	dig prot %	6.9	7.5	
Fatty acids	%	5.1	5.6	
Energy	GE Mcal/kg			
Cattle	DE Mcal/kg	* 2.57	2.82	
Sheep	DE Mcal/kg	* 2.58	2.83	
Cattle	ME Mcal/kg	* 2.11	2.31	
Sheep	ME Mcal/kg	* 2.11	2.32	
Cattle	TDN %	* 58.3	63.9	
Sheep	TDN %	58.5	64.2	
Cobalt	mg/kg	0.072	0.079	
Carotene	mg/kg	45.8	50.3	82
Riboflavin	mg/kg	8.8	9.7	
Vitamin A equivalent	IU/g	76.4	83.8	

Peanut, hay, s-c, dough stage, (1)
Ref No 1 03 615 United States

Feed Name or Analyses		Mean As Fed	Mean Dry	C.V. ± %
Dry matter	%	91.3	100.0	1
Ash	%	9.9	10.8	20
Crude fiber	%	26.2	28.7	16
Sheep	dig coef %	* 51.	51.	
Ether extract	%	2.6	2.8	29
Sheep	dig coef %	* 62.	62.	
N-free extract	%	41.7	45.7	
Sheep	dig coef %	* 77.	77.	
Protein (N x 6.25)	%	11.0	12.0	16
Sheep	dig coef %	* 65.	65.	
Cattle	dig prot %	* 6.7	7.3	
Goats	dig prot %	* 7.1	7.8	
Horses	dig prot %	* 7.0	7.7	
Rabbits	dig prot %	* 7.2	7.9	
Sheep	dig prot %	7.1	7.8	
Energy	GE Mcal/kg			
Cattle	DE Mcal/kg	* 2.56	2.80	
Sheep	DE Mcal/kg	* 2.48	2.71	
Cattle	ME Mcal/kg	* 2.10	2.30	
Sheep	ME Mcal/kg	* 2.03	2.22	
Cattle	TDN %	* 57.9	63.4	
Sheep	TDN %	* 56.2	61.5	

Peanut, hay, s-c, gr 1 US, (1)
Ref No 1 03 616 United States

Feed Name or Analyses		Mean As Fed	Mean Dry	C.V. ± %
Dry matter	%		100.0	
Energy	GE Mcal/kg		4.01	

Peanut, hay, s-c, gr 2 US, (1)
Ref No 1 03 617 United States

Feed Name or Analyses		Mean As Fed	Mean Dry	C.V. ± %
Dry matter	%	89.0	100.0	
Ash	%	6.2	7.0	
Crude fiber	%	38.8	43.6	
Ether extract	%	2.5	2.8	
N-free extract	%	32.8	36.9	
Protein (N x 6.25)	%	8.6	9.7	
Cattle	dig prot %	* 4.8	5.3	
Goats	dig prot %	* 5.0	5.6	
Horses	dig prot %	* 5.1	5.8	
Rabbits	dig prot %	* 5.5	6.2	
Sheep	dig prot %	* 4.7	5.3	

Continued

Feed Name or Analyses			Mean As Fed	Mean Dry	C.V. ± %
Energy	GE Mcal/kg		3.95	4.43	
Cattle	DE Mcal/kg	*	2.10	2.36	
Sheep	DE Mcal/kg	*	2.01	2.26	
Cattle	ME Mcal/kg	*	1.73	1.94	
Sheep	ME Mcal/kg	*	1.65	1.85	
Cattle	TDN %	*	47.6	53.5	
Sheep	TDN %	*	45.5	51.2	
Vitamin D_2	IU/g		3.2	3.6	

Peanut, hay, s-c, stemmy, (1)
Ref No 1 03 618 United States

Feed Name or Analyses			Mean As Fed	Mean Dry	C.V. ± %
Dry matter	%		90.0	100.0	
Ash	%		8.6	9.6	
Crude fiber	%		34.2	38.0	
Sheep	dig coef %	*	51.	51.	
Ether extract	%		3.2	3.6	
Sheep	dig coef %	*	62.	62.	
N-free extract	%		33.9	37.7	
Sheep	dig coef %	*	77.	77.	
Protein (N x 6.25)	%		10.0	11.1	
Sheep	dig coef %	*	65.	65.	
Cattle	dig prot %	*	5.9	6.6	
Goats	dig prot %	*	6.2	6.9	
Horses	dig prot %	*	6.3	7.0	
Rabbits	dig prot %	*	6.5	7.2	
Sheep	dig prot %		6.5	7.2	
Energy	GE Mcal/kg				
Cattle	DE Mcal/kg	*	2.08	2.31	
Sheep	DE Mcal/kg	*	2.41	2.67	
Cattle	ME Mcal/kg	*	1.71	1.90	
Sheep	ME Mcal/kg	*	1.97	2.19	
Cattle	TDN %	*	47.2	52.5	
Sheep	TDN %		54.6	60.6	

Peanut, hay w nuts, s-c, (1)
Ref No 1 03 620 United States

Feed Name or Analyses			Mean As Fed	Mean Dry	C.V. ± %
Dry matter	%		91.5	100.0	1
Ash	%		8.0	8.7	17
Crude fiber	%		22.7	24.8	7
Sheep	dig coef %		48.	48.	
Ether extract	%		12.6	13.7	6
Sheep	dig coef %		92.	92.	
N-free extract	%		34.9	38.1	
Sheep	dig coef %		68.	68.	
Protein (N x 6.25)	%		13.3	14.6	6
Sheep	dig coef %		76.	76.	
Cattle	dig prot %	*	8.7	9.5	
Goats	dig prot %	*	9.3	10.1	
Horses	dig prot %	*	9.0	9.9	
Rabbits	dig prot %	*	9.1	9.9	
Sheep	dig prot %		10.1	11.1	
Fatty acids	%		12.5	13.7	
Energy	GE Mcal/kg				
Cattle	DE Mcal/kg	*	3.20	3.50	
Sheep	DE Mcal/kg	*	3.12	3.41	
Cattle	ME Mcal/kg	*	2.63	2.87	
Sheep	ME Mcal/kg	*	2.56	2.80	
Cattle	TDN %	*	72.7	79.4	
Sheep	TDN %		70.8	77.3	
Calcium	%		1.13	1.23	
Magnesium	%		0.33	0.36	

(1) dry forages and roughages (2) pasture, range plants, and forages fed green (3) silages (4) energy feeds (5) protein supplements (6) minerals (7) vitamins (8) additives

Feed Name or Analyses			Mean As Fed	Mean Dry	C.V. ± %
Phosphorus	%		0.14	0.15	
Potassium	%		0.85	0.92	

Peanut, hay wo nuts, s-c, (1)
Ref No 1 03 626 United States

Feed Name or Analyses			Mean As Fed	Mean Dry	C.V. ± %
Dry matter	%		91.1	100.0	2
Ash	%		9.5	10.4	27
Crude fiber	%		24.2	26.5	18
Sheep	dig coef %		51.	51.	
Ether extract	%		3.1	3.4	26
Sheep	dig coef %		62.	62.	
N-free extract	%		44.4	48.8	
Sheep	dig coef %		77.	77.	
Protein (N x 6.25)	%		9.8	10.8	14
Sheep	dig coef %		65.	65.	
Cattle	dig prot %	*	5.7	6.3	
Goats	dig prot %	*	6.1	6.6	
Horses	dig prot %	*	6.1	6.7	
Rabbits	dig prot %	*	6.4	7.0	
Sheep	dig prot %		6.4	7.0	
Fatty acids	%		3.2	3.5	
Energy	GE Mcal/kg				
Cattle	DE Mcal/kg	*	2.36	2.59	
Sheep	DE Mcal/kg	*	2.53	2.78	
Cattle	ME Mcal/kg	*	1.93	2.12	
Sheep	ME Mcal/kg	*	2.07	2.28	
Cattle	TDN %	*	53.5	58.7	
Sheep	TDN %		57.3	62.9	
Calcium	%		1.16	1.28	15
Magnesium	%		0.43	0.47	14
Phosphorus	%		0.21	0.23	98
Potassium	%		1.26	1.38	12
Sulphur	%		0.21	0.23	

Peanut, hay wo nuts, s-c, dough stage,(1)
Ref No 1 03 622 United States

Feed Name or Analyses			Mean As Fed	Mean Dry	C.V. ± %
Dry matter	%		91.2	100.0	1
Ash	%		9.7	10.6	19
Crude fiber	%		25.9	28.4	16
Ether extract	%		2.6	2.9	28
N-free extract	%		42.3	46.4	
Protein (N x 6.25)	%		10.7	11.7	16
Cattle	dig prot %	*	6.4	7.1	
Goats	dig prot %	*	6.8	7.5	
Horses	dig prot %	*	6.8	7.5	
Rabbits	dig prot %	*	7.0	7.7	
Sheep	dig prot %	*	6.4	7.1	
Energy	GE Mcal/kg				
Cattle	DE Mcal/kg	*	2.40	2.63	
Sheep	DE Mcal/kg	*	2.36	2.59	
Cattle	ME Mcal/kg	*	1.96	2.16	
Sheep	ME Mcal/kg	*	1.93	2.12	
Cattle	TDN %	*	54.4	59.6	
Sheep	TDN %	*	53.5	58.6	

Peanut, hay wo nuts, s-c, mature, (1)
Ref No 1 03 623 United States

Feed Name or Analyses			Mean As Fed	Mean Dry	C.V. ± %
Dry matter	%		90.5	100.0	
Ash	%		8.1	8.9	25
Crude fiber	%		31.0	34.2	26
Ether extract	%		2.2	2.4	40
N-free extract	%		40.7	45.0	
Protein (N x 6.25)	%		8.6	9.5	17
Cattle	dig prot %	*	4.7	5.2	

Feed Name or Analyses			Mean As Fed	Mean Dry	C.V. ± %
Goats	dig prot %	*	4.9	5.4	
Horses	dig prot %	*	5.1	5.6	
Rabbits	dig prot %	*	5.4	6.0	
Sheep	dig prot %	*	4.6	5.1	
Energy	GE Mcal/kg				
Cattle	DE Mcal/kg	*	2.29	2.53	
Sheep	DE Mcal/kg	*	2.21	2.44	
Cattle	ME Mcal/kg	*	1.88	2.08	
Sheep	ME Mcal/kg	*	1.81	2.00	
Cattle	TDN %	*	51.9	57.4	
Sheep	TDN %	*	50.1	55.4	

Peanut, hay wo nuts, s-c, gr 1 US, (1)
Ref No 1 03 624 United States

Feed Name or Analyses			Mean As Fed	Mean Dry	C.V. ± %
Dry matter	%			100.0	
Ash	%			6.3	
Crude fiber	%			34.2	
Ether extract	%			2.5	
N-free extract	%			44.6	
Protein (N x 6.25)	%			12.4	
Cattle	dig prot %	*		7.7	
Goats	dig prot %	*		8.1	
Horses	dig prot %	*		8.1	
Rabbits	dig prot %	*		8.2	
Sheep	dig prot %	*		7.7	
Energy	GE Mcal/kg				
Cattle	DE Mcal/kg	*		2.40	
Sheep	DE Mcal/kg	*		2.51	
Cattle	ME Mcal/kg	*		1.97	
Sheep	ME Mcal/kg	*		2.06	
Cattle	TDN %	*		54.4	
Sheep	TDN %	*		56.9	

Peanut, hay wo nuts, s-c, gr 2 US, (1)
Ref No 1 03 625 United States

Feed Name or Analyses			Mean As Fed	Mean Dry	C.V. ± %
Dry matter	%		89.0	100.0	
Ash	%		6.1	6.9	
Crude fiber	%		38.8	43.6	
Ether extract	%		2.5	2.8	
N-free extract	%		33.0	37.1	
Protein (N x 6.25)	%		8.5	9.6	
Cattle	dig prot %	*	4.7	5.3	
Goats	dig prot %	*	4.9	5.5	
Horses	dig prot %	*	5.1	5.7	
Rabbits	dig prot %	*	5.4	6.1	
Sheep	dig prot %	*	4.6	5.2	
Energy	GE Mcal/kg				
Cattle	DE Mcal/kg	*	1.90	2.13	
Sheep	DE Mcal/kg	*	2.01	2.26	
Cattle	ME Mcal/kg	*	1.56	1.75	
Sheep	ME Mcal/kg	*	1.65	1.85	
Cattle	TDN %	*	42.9	48.2	
Sheep	TDN %	*	45.5	51.2	

Peanut, hay wo nuts, s-c, stemmy, (1)
Ref No 1 08 484 United States

Feed Name or Analyses			Mean As Fed	Mean Dry	C.V. ± %
Dry matter	%		90.0	100.0	
Ash	%		8.6	9.6	
Crude fiber	%		34.2	38.0	
Ether extract	%		3.2	3.6	
N-free extract	%		34.0	37.8	
Protein (N x 6.25)	%		10.0	11.1	
Cattle	dig prot %	*	5.9	6.6	
Goats	dig prot %	*	6.2	6.9	

Feed Name or Analyses			Mean As Fed	Mean Dry	C.V. ± %
Horses	dig prot %	*	6.3	7.0	
Rabbits	dig prot %	*	6.5	7.2	
Sheep	dig prot %	*	5.9	6.5	
Energy	GE Mcal/kg				
Cattle	DE Mcal/kg	*	1.68	1.87	
Sheep	DE Mcal/kg	*	1.81	2.01	
Cattle	ME Mcal/kg	*	1.38	1.53	
Sheep	ME Mcal/kg	*	1.48	1.65	
Cattle	TDN %	*	38.2	42.4	
Sheep	TDN %	*	41.1	45.7	

Peanut, hay w some nuts, (1)

Ref No 1 03 621 United States

Feed Name or Analyses			Mean As Fed	Mean Dry	C.V. ± %
Dry matter	%		92.6	100.0	
Ash	%		4.2	4.5	
Crude fiber	%		53.4	57.7	
Sheep	dig coef %		16.	16.	
Ether extract	%		3.7	4.0	
Sheep	dig coef %		84.	84.	
N-free extract	%		22.1	23.9	
Sheep	dig coef %		57.	57.	
Protein (N x 6.25)	%		9.2	9.9	
Sheep	dig coef %		52.	52.	
Cattle	dig prot %	*	5.1	5.5	
Goats	dig prot %	*	5.4	5.8	
Horses	dig prot %	*	5.5	5.9	
Rabbits	dig prot %	*	5.8	6.3	
Sheep	dig prot %		4.8	5.1	
Energy	GE Mcal/kg				
Cattle	DE Mcal/kg	*	1.40	1.51	
Sheep	DE Mcal/kg	*	1.45	1.57	
Cattle	ME Mcal/kg	*	1.15	1.24	
Sheep	ME Mcal/kg	*	1.19	1.29	
Cattle	TDN %	*	31.7	34.2	
Sheep	TDN %		32.9	35.6	

Peanut, hulls, (1)

Peanut hulls (AAFCO)

Ref No 1 08 028 United States

Feed Name or Analyses			Mean As Fed	Mean Dry	C.V. ± %
Dry matter	%		91.5	100.0	2
Ash	%		4.4	4.8	
Crude fiber	%		59.8	65.4	
Ether extract	%		1.2	1.3	
N-free extract	%		19.4	21.2	
Protein (N x 6.25)	%		6.6	7.3	
Cattle	dig prot %	*	3.0	3.2	
Goats	dig prot %	*	3.0	3.3	
Horses	dig prot %	*	3.4	3.7	
Rabbits	dig prot %	*	3.9	4.3	
Sheep	dig prot %	*	2.8	3.1	
Cellulose (Matrone)	%		44.4	48.5	
Pentosans	%		17.6	19.2	6
Starch	%		0.7	0.8	34
Sugars, total	%		2.2	2.4	20
Lignin (Ellis)	%		26.2	28.7	
Fatty acids	%		1.2	1.3	
Energy	GE Mcal/kg				
Cattle	DE Mcal/kg	*	0.74	0.81	
Sheep	DE Mcal/kg	*	0.95	1.04	
Cattle	ME Mcal/kg	*	0.60	0.66	
Sheep	ME Mcal/kg	*	0.78	0.85	
Cattle	TDN %	*	16.7	18.3	
Sheep	TDN %	*	21.6	23.6	
Calcium	%		0.25	0.27	
Cobalt	mg/kg		0.107	0.117	
Magnesium	%		0.17	0.18	

Feed Name or Analyses			Mean As Fed	Mean Dry	C.V. ± %
Phosphorus	%		0.06	0.07	
Potassium	%		0.81	0.89	
Carotene	mg/kg		0.8	0.9	
Vitamin A equivalent	IU/g		1.3	1.5	

Peanut, hulls, grnd, (1)

Ref No 1 03 629 United States

Feed Name or Analyses			Mean As Fed	Mean Dry	C.V. ± %
Dry matter	%		94.4	100.0	
Ash	%		4.2	4.4	
Crude fiber	%		41.8	44.3	
Ether extract	%		0.8	0.8	
N-free extract	%		42.7	45.2	
Protein (N x 6.25)	%		4.9	5.2	
Cattle	dig prot %	*	1.4	1.4	
Goats	dig prot %	*	1.3	1.4	
Horses	dig prot %	*	1.8	1.9	
Rabbits	dig prot %	*	2.5	2.7	
Sheep	dig prot %	*	1.2	1.2	
Energy	GE Mcal/kg				
Cattle	DE Mcal/kg	*	1.49	1.58	
Sheep	DE Mcal/kg	*	1.66	1.76	
Cattle	ME Mcal/kg	*	1.23	1.30	
Sheep	ME Mcal/kg	*	1.36	1.44	
Cattle	TDN %	*	33.9	35.9	
Sheep	TDN %	*	37.7	39.9	

Peanut, hulls w skins, grnd, (1)

Ref No 1 03 630 United States

Feed Name or Analyses			Mean As Fed	Mean Dry	C.V. ± %
Dry matter	%		94.2	100.0	
Ash	%		13.3	14.1	
Crude fiber	%		13.3	14.1	
Ether extract	%		13.8	14.6	
N-free extract	%		41.0	43.5	
Protein (N x 6.25)	%		12.8	13.6	
Cattle	dig prot %	*	8.2	8.7	
Goats	dig prot %	*	8.7	9.2	
Horses	dig prot %	*	8.5	9.1	
Rabbits	dig prot %	*	8.6	9.2	
Sheep	dig prot %	*	8.3	8.8	
Energy	GE Mcal/kg				
Cattle	DE Mcal/kg	*	2.34	2.48	
Sheep	DE Mcal/kg	*	2.20	2.34	
Cattle	ME Mcal/kg	*	1.91	2.03	
Sheep	ME Mcal/kg	*	1.81	1.92	
Cattle	TDN %	*	53.0	56.3	
Sheep	TDN %	*	49.9	53.0	

Peanut, seed screenings, (1)

Ref No 1 08 485 United States

Feed Name or Analyses			Mean As Fed	Mean Dry	C.V. ± %
Dry matter	%		90.6	100.0	
Ash	%		6.9	7.6	
Crude fiber	%		23.8	26.3	
Ether extract	%		8.8	9.7	
N-free extract	%		37.3	41.2	
Protein (N x 6.25)	%		13.8	15.2	
Cattle	dig prot %	*	9.2	10.1	
Goats	dig prot %	*	9.8	10.8	
Horses	dig prot %	*	9.5	10.5	
Rabbits	dig prot %	*	9.4	10.4	
Sheep	dig prot %	*	9.3	10.2	
Energy	GE Mcal/kg				
Cattle	DE Mcal/kg	*	1.75	1.93	
Sheep	DE Mcal/kg	*	1.88	2.08	
Cattle	ME Mcal/kg	*	1.43	1.58	

Feed Name or Analyses			Mean As Fed	Mean Dry	C.V. ± %
Sheep	ME Mcal/kg	*	1.55	1.71	
Cattle	TDN %	*	39.7	43.8	
Sheep	TDN %	*	42.8	47.2	

Peanut, aerial part, fresh, (2)

Ref No 2 03 638 United States

Feed Name or Analyses			Mean As Fed	Mean Dry	C.V. ± %
Dry matter	%		24.6	100.0	26
Ash	%		2.4	9.9	
Crude fiber	%		7.8	31.7	
Sheep	dig coef %	*	46.	46.	
Ether extract	%		0.2	1.0	
Sheep	dig coef %	*	31.	31.	
N-free extract	%		12.2	49.6	
Sheep	dig coef %	*	79.	79.	
Protein (N x 6.25)	%		1.9	7.8	
Sheep	dig coef %	*	47.	47.	
Cattle	dig prot %	*	1.1	4.5	
Goats	dig prot %	*	0.9	3.8	
Horses	dig prot %	*	1.0	4.2	
Rabbits	dig prot %	*	1.2	4.7	
Sheep	dig prot %		0.9	3.7	
Energy	GE Mcal/kg				
Cattle	DE Mcal/kg	*	0.68	2.77	
Sheep	DE Mcal/kg	*	0.63	2.56	
Cattle	ME Mcal/kg	*	0.56	2.27	
Sheep	ME Mcal/kg	*	0.52	2.10	
Cattle	TDN %	*	15.5	62.9	
Sheep	TDN %		14.3	58.1	
Calcium	%		0.42	1.71	16
Magnesium	%		0.12	0.50	25
Phosphorus	%		0.07	0.29	19
Potassium	%		0.24	0.97	44

Peanut, aerial part, fresh, immature, (2)

Ref No 2 03 634 United States

Feed Name or Analyses			Mean As Fed	Mean Dry	C.V. ± %
Dry matter	%		20.1	100.0	
Calcium	%		0.40	1.99	
Magnesium	%		0.08	0.42	
Phosphorus	%		0.05	0.25	
Potassium	%		0.29	1.44	

Peanut, aerial part, fresh, full bloom, (2)

Ref No 2 03 635 United States

Feed Name or Analyses			Mean As Fed	Mean Dry	C.V. ± %
Dry matter	%		17.7	100.0	14
Calcium	%		0.30	1.71	8
Magnesium	%		0.10	0.59	25
Phosphorus	%		0.06	0.36	13
Potassium	%		0.17	0.97	38

Peanut, aerial part, fresh, milk stage, (2)

Ref No 2 03 636 United States

Feed Name or Analyses			Mean As Fed	Mean Dry	C.V. ± %
Dry matter	%		22.3	100.0	6
Calcium	%		0.36	1.61	3
Magnesium	%		0.11	0.49	11
Phosphorus	%		0.06	0.28	7
Potassium	%		0.17	0.77	38

Peanut, aerial part, fresh, mature, (2)

Ref No 2 03 637 United States

Feed Name or Analyses			Mean As Fed	Mean Dry	C.V. ± %
Dry matter	%		37.2	100.0	
Ash	%		3.1	8.4	
Crude fiber	%		8.6	23.1	
Ether extract	%		1.5	4.0	
N-free extract	%		19.6	52.7	
Protein (N x 6.25)	%		4.4	11.9	
Cattle	dig prot %	*	3.0	8.0	
Goats	dig prot %	*	2.9	7.7	
Horses	dig prot %	*	2.8	7.6	
Rabbits	dig prot %	*	2.9	7.9	
Sheep	dig prot %	*	3.0	8.1	
Energy	GE Mcal/kg				
Cattle	DE Mcal/kg	*	1.04	2.81	
Sheep	DE Mcal/kg	*	1.11	2.97	
Cattle	ME Mcal/kg	*	0.86	2.30	
Sheep	ME Mcal/kg	*	0.91	2.44	
Cattle	TDN %	*	23.7	63.7	
Sheep	TDN %	*	25.1	67.5	
Calcium	%		0.46	1.24	
Magnesium	%		0.14	0.38	
Phosphorus	%		0.09	0.24	
Potassium	%		0.14	0.37	

Peanut, aerial part w nuts, fresh, (2)

Ref No 2 03 642 United States

Feed Name or Analyses			Mean As Fed	Mean Dry	C.V. ± %
Dry matter	%		18.7	100.0	14
Ash	%		1.2	6.4	18
Crude fiber	%		4.4	23.6	9
Ether extract	%		2.4	13.0	35
N-free extract	%		6.8	36.1	
Protein (N x 6.25)	%		3.9	20.9	16
Cattle	dig prot %	*	2.9	15.6	
Goats	dig prot %	*	3.0	16.0	
Horses	dig prot %	*	2.9	15.2	
Rabbits	dig prot %	*	2.8	14.8	
Sheep	dig prot %	*	3.1	16.4	
Energy	GE Mcal/kg				
Cattle	DE Mcal/kg	*	0.52	2.79	
Sheep	DE Mcal/kg	*	0.57	3.04	
Cattle	ME Mcal/kg	*	0.43	2.29	
Sheep	ME Mcal/kg	*	0.47	2.49	
Cattle	TDN %	*	11.9	63.4	
Sheep	TDN %	*	12.9	68.8	

Peanut, aerial part w nuts, fresh, milk stage, (2)

Ref No 2 03 640 United States

Feed Name or Analyses			Mean As Fed	Mean Dry	C.V. ± %
Dry matter	%		21.9	100.0	
Protein (N x 6.25)	%		3.9	17.7	
Cattle	dig prot %	*	2.8	12.9	
Goats	dig prot %	*	2.9	13.1	
Horses	dig prot %	*	2.8	12.6	
Rabbits	dig prot %	*	2.7	12.3	
Sheep	dig prot %	*	3.0	13.5	

(1) dry forages and roughages (2) pasture, range plants, and forages fed green (3) silages (4) energy feeds (5) protein supplements (6) minerals (7) vitamins (8) additives

Peanut, aerial part w nuts, fresh, mature, (2)

Ref No 2 03 641 United States

Feed Name or Analyses			Mean As Fed	Mean Dry	C.V. ± %
Dry matter	%			100.0	
Ash	%			7.0	
Crude fiber	%			23.4	
Ether extract	%			10.5	
N-free extract	%			45.2	
Protein (N x 6.25)	%			13.9	
Cattle	dig prot %	*		9.7	
Goats	dig prot %	*		9.5	
Horses	dig prot %	*		9.3	
Rabbits	dig prot %	*		9.4	
Sheep	dig prot %	*		9.9	
Energy	GE Mcal/kg				
Cattle	DE Mcal/kg	*		2.57	
Sheep	DE Mcal/kg	*		2.83	
Cattle	ME Mcal/kg	*		2.11	
Sheep	ME Mcal/kg	*		2.32	
Cattle	TDN %	*		58.3	
Sheep	TDN %	*		64.2	

Peanut, leaves w stems, fresh, (2)

Ref No 2 03 643 United States

Feed Name or Analyses			Mean As Fed	Mean Dry	C.V. ± %
Dry matter	%			100.0	
Ash	%			8.4	14
Crude fiber	%			25.3	11
Ether extract	%			1.7	9
N-free extract	%			55.0	
Protein (N x 6.25)	%			9.6	26
Cattle	dig prot %	*		6.1	
Goats	dig prot %	*		5.5	
Horses	dig prot %	*		5.7	
Rabbits	dig prot %	*		6.1	
Sheep	dig prot %	*		5.9	
Energy	GE Mcal/kg				
Cattle	DE Mcal/kg	*		2.86	
Sheep	DE Mcal/kg	*		3.04	
Cattle	ME Mcal/kg	*		2.35	
Sheep	ME Mcal/kg	*		2.49	
Cattle	TDN %	*		64.9	
Sheep	TDN %	*		68.9	

Peanut, oil, (4)

Ref No 4 03 658 United States

Feed Name or Analyses			Mean As Fed	Mean Dry	C.V. ± %
Dry matter	%		99.9	100.0	
Ash	%		0.1	0.1	
Ether extract	%		99.8	99.9	
Swine	dig coef %		98.	98.	
Energy	GE Mcal/kg				
Swine	DE kcal/kg	*	9704.	9713.	
Swine	TDN %		220.1	220.3	

Peanut, skins, grnd, (4)

Peanut skins (AAFCO)

Peanut skins (CFA)

Ref No 4 03 631 United States

Feed Name or Analyses			Mean As Fed	Mean Dry	C.V. ± %
Dry matter	%		94.3	100.0	1
Ash	%		2.9	3.0	18
Crude fiber	%		10.1	10.7	30
Sheep	dig coef %		0.	0.	30
Ether extract	%		26.6	28.2	4
Sheep	dig coef %		92.	92.	
N-free extract	%		37.5	39.7	
Sheep	dig coef %		16.	16.	
Protein (N x 6.25)	%		17.3	18.3	5
Sheep	dig coef %		25.	25.	
Cattle	dig prot %	*	12.1	12.9	
Goats	dig prot %	*	13.2	14.0	
Horses	dig prot %	*	13.2	14.0	
Sheep	dig prot %		4.3	4.6	
Energy	GE Mcal/kg				
Cattle	DE Mcal/kg	*	2.69	2.85	
Sheep	DE Mcal/kg	*	2.88	3.05	
Cattle	ME Mcal/kg	*	2.21	2.34	
Sheep	ME Mcal/kg	*	2.36	2.50	
Cattle	TDN %	*	61.0	64.7	
Sheep	TDN %		65.3	69.3	
Calcium	%		0.18	0.19	
Phosphorus	%		0.19	0.20	

Peanut, flour, coarse bolt, (5)

Ref No 5 03 645 United States

Feed Name or Analyses			Mean As Fed	Mean Dry	C.V. ± %
Dry matter	%		91.4	100.0	
Ash	%		4.0	4.4	
Crude fiber	%		2.7	2.9	
Ether extract	%		9.1	9.9	
N-free extract	%		23.2	25.3	
Protein (N x 6.25)	%		52.4	57.4	4
Energy	GE Mcal/kg		3.66	4.00	
Cattle	DE Mcal/kg	*	3.63	3.98	
Sheep	DE Mcal/kg	*	3.47	3.80	
Swine	DE kcal/kg	*	3767.	4124.	
Cattle	ME Mcal/kg	*	2.98	3.26	
Sheep	ME Mcal/kg	*	2.85	3.12	
Swine	ME kcal/kg	*	3180.	3481.	
Cattle	TDN %	*	82.4	90.2	
Sheep	TDN %	*	78.7	86.2	
Swine	TDN %	*	85.4	93.5	
Calcium	%		0.10	0.11	
Iron	%		0.003	0.004	
Phosphorus	%		0.71	0.78	
Potassium	%		1.17	1.28	
Sodium	%		0.01	0.01	
Niacin	mg/kg		274.0	299.9	
Riboflavin	mg/kg		2.2	2.4	
Thiamine	mg/kg		7.4	8.1	
Arginine	%		7.00	7.67	2
Histidine	%		1.22	1.33	4
Isoleucine	%		2.33	2.56	11
Leucine	%		3.96	4.33	4
Lysine	%		2.33	2.56	22
Methionine	%		0.41	0.44	29
Phenylalanine	%		3.04	3.33	1
Threonine	%		1.62	1.78	5
Tryptophan	%		0.61	0.67	22
Tyrosine	%		2.03	2.22	16
Valine	%		2.94	3.22	8

Peanut, kernels, extn unspecified grnd, 39% protein, (5)

Ref No 5 03 646 United States

Feed Name or Analyses			Mean As Fed	Mean Dry	C.V. ± %
Dry matter	%		90.2	100.0	
Ash	%		5.3	5.9	
Crude fiber	%		5.3	5.9	
Cattle	dig coef %		55.	55.	
Ether extract	%		6.2	6.9	
Cattle	dig coef %		97.	97.	

Feed Name or Analyses		As Fed	Dry	C.V. ± %
N-free extract	%	34.3	38.0	
Cattle	dig coef %	79.	79.	
Protein (N x 6.25)	%	39.1	43.3	
Cattle	dig coef %	94.	94.	
Cattle	dig prot %	36.7	40.7	
Energy	GE Mcal/kg			
Cattle	DE Mcal/kg	* 3.54	3.93	
Sheep	DE Mcal/kg	* 3.39	3.76	
Swine	DE kcal/kg	*3626.	4020.	
Cattle	ME Mcal/kg	* 2.90	3.22	
Sheep	ME Mcal/kg	* 2.78	3.08	
Swine	ME kcal/kg	*3163.	3507.	
Cattle	TDN %	80.3	89.0	
Sheep	TDN %	* 76.9	85.2	
Swine	TDN %	* 82.2	91.2	

Peanut, kernels, extn unspecified grnd, 45% protein, (5)

Ref No 5 03 647 United States

Feed Name or Analyses		As Fed	Dry	C.V. ± %
Dry matter	%	87.0	100.0	
Ash	%	5.1	5.9	
Crude fiber	%	5.1	5.9	
Cattle	dig coef %	32.	32.	
Ether extract	%	6.8	7.8	
Cattle	dig coef %	91.	91.	
N-free extract	%	23.6	27.1	
Cattle	dig coef %	83.	83.	
Protein (N x 6.25)	%	46.4	53.3	
Cattle	dig coef %	90.	90.	
Cattle	dig prot %	41.7	48.0	
Energy	GE Mcal/kg			
Cattle	DE Mcal/kg	* 3.39	3.89	
Sheep	DE Mcal/kg	* 3.16	3.63	
Swine	DE kcal/kg	*3362.	3865.	
Cattle	ME Mcal/kg	* 2.78	3.19	
Sheep	ME Mcal/kg	* 2.59	2.98	
Swine	ME kcal/kg	*2866.	3294.	
Cattle	TDN %	76.8	88.3	
Sheep	TDN %	* 71.7	82.4	
Swine	TDN %	* 76.3	87.7	

Peanut, kernels, mech-extd grnd, mx 7% fiber, (5)

Peanut meal (AAFCO)

Peanut meal (CFA)

Peanut oil meal, expeller extracted

Ref No 5 03 649 United States

Feed Name or Analyses		As Fed	Dry	C.V. ± %
Dry matter	%	89.2	100.0	2
Ash	%	5.3	5.9	13
Crude fiber	%	6.8	7.6	47
Cattle	dig coef %	* 32.	32.	
Sheep	dig coef %	56.	56.	
Swine	dig coef %	70.	70.	
Ether extract	%	7.3	8.2	23
Cattle	dig coef %	* 91.	91.	
Sheep	dig coef %	91.	91.	
Swine	dig coef %	84.	84.	
N-free extract	%	24.7	27.7	
Cattle	dig coef %	* 83.	83.	
Sheep	dig coef %	90.	90.	
Swine	dig coef %	84.	84.	
Protein (N x 6.25)	%	45.1	50.6	9
Cattle	dig coef %	* 90.	90.	
Sheep	dig coef %	91.	91.	
Swine	dig coef %	94.	94.	
Cattle	dig prot %	40.6	45.5	
Sheep	dig prot %	41.1	46.0	
Swine	dig prot %	42.4	47.5	
Energy	GE Mcal/kg			
Cattle	DE Mcal/kg	* 3.45	3.87	
Sheep	DE Mcal/kg	* 3.62	4.05	
Swine	DE kcal/kg	*3603.	4038.	
Cattle	ME Mcal/kg	* 2.83	3.17	
Chickens	ME_n kcal/kg	2610.	2926.	
Sheep	ME Mcal/kg	* 2.97	3.32	
Swine	ME kcal/kg	*3091.	3464.	
Cattle	NE_m Mcal/kg	* 1.75	1.96	
Cattle	NE_{gain} Mcal/kg	* 1.17	1.31	
Cattle	$NE_{lactating\ cows}$ Mcal/kg	* 2.05	2.30	
Cattle	TDN %	78.2	87.7	
Sheep	TDN %	82.0	91.9	
Swine	TDN %	81.7	91.6	
Calcium	%	0.16	0.18	35
Chlorine	%	0.03	0.03	
Magnesium	%	0.32	0.36	
Manganese	mg/kg	24.8	27.8	
Phosphorus	%	0.55	0.62	17
Potassium	%	1.12	1.25	
Sulphur	%	0.28	0.32	
Carotene	mg/kg	0.2	0.2	
Choline	mg/kg	1636.	1833.	
Niacin	mg/kg	164.2	184.0	
Pantothenic acid	mg/kg	46.8	52.5	5
Riboflavin	mg/kg	5.1	5.8	
Thiamine	mg/kg	7.1	7.9	
Vitamin A equivalent	IU/g	0.4	0.4	
Lysine	%	1.26	1.41	
Methionine	%	0.58	0.65	

Ref No 5 03 649 Canada

Feed Name or Analyses		As Fed	Dry	C.V. ± %
Dry matter	%	94.4	100.0	
Crude fiber	%	5.3	5.6	
Sheep	dig coef %	56.	56.	
Swine	dig coef %	70.	70.	
Ether extract	%	33.1	35.1	
Sheep	dig coef %	91.	91.	
Swine	dig coef %	84.	84.	
N-free extract	%			
Sheep	dig coef %	90.	90.	
Swine	dig coef %	84.	84.	
Protein (N x 6.25)	%	23.3	24.7	
Sheep	dig coef %	91.	91.	
Swine	dig coef %	94.	94.	
Sheep	dig prot %	21.2	22.5	
Swine	dig prot %	21.9	23.2	
Energy	GE Mcal/kg			
Cattle	DE Mcal/kg	* 3.98	4.22	
Sheep	DE Mcal/kg	* 4.15	4.40	
Swine	DE kcal/kg	*4082.	4324.	
Cattle	ME Mcal/kg	* 3.26	3.46	
Sheep	ME Mcal/kg	* 3.41	3.61	
Swine	ME kcal/kg	*3715.	3935.	
Cattle	TDN %	90.3	95.7	
Sheep	TDN %	94.2	99.8	
Swine	TDN %	92.6	98.1	

Peanut, kernels, mech-extd grnd, 45% protein, (5)

Ref No 5 08 604 United States

Feed Name or Analyses		As Fed	Dry	C.V. ± %
Dry matter	%	94.0	100.0	
Ash	%	5.4	5.7	
Crude fiber	%	5.5	5.9	
Ether extract	%	6.3	6.7	
N-free extract	%	30.2	32.1	
Protein (N x 6.25)	%	46.6	49.6	
Energy	GE Mcal/kg			
Cattle	DE Mcal/kg	* 3.44	3.66	
Sheep	DE Mcal/kg	* 3.46	3.68	
Swine	DE kcal/kg	*3645.	3878.	
Cattle	ME Mcal/kg	* 2.82	3.00	
Sheep	ME Mcal/kg	* 2.83	3.01	
Swine	ME kcal/kg	*3134.	3334.	
Cattle	TDN %	* 78.1	83.1	
Sheep	TDN %	* 78.4	83.4	
Swine	TDN %	* 82.7	87.9	
Calcium	%	0.16	0.17	
Chlorine	%	0.03	0.03	
Magnesium	%	0.24	0.26	
Phosphorus	%	0.54	0.57	
Potassium	%	1.15	1.22	
Sodium	%	0.42	0.45	
Sulphur	%	0.18	0.19	

Peanut, kernels, solv-extd grnd, mx 7% fiber, (5)

Solvent extracted peanut meal (AAFCO)

Groundnut oil meal, solvent extracted

Peanut oil meal, solvent extracted

Ref No 5 03 650 United States

Feed Name or Analyses		As Fed	Dry	C.V. ± %
Dry matter	%	91.5	100.0	1
Ash	%	4.5	4.9	7
Crude fiber	%	13.1	14.3	33
Cattle	dig coef %	* 32.	32.	
Sheep	dig coef %	* 56.	56.	
Ether extract	%	1.2	1.3	50
Cattle	dig coef %	* 91.	91.	
Sheep	dig coef %	* 91.	91.	
N-free extract	%	25.3	27.7	
Cattle	dig coef %	* 83.	83.	
Sheep	dig coef %	* 90.	90.	
Protein (N x 6.25)	%	47.4	51.8	10
Cattle	dig coef %	* 90.	90.	
Sheep	dig coef %	* 91.	91.	
Cattle	dig prot %	42.7	46.6	
Sheep	dig prot %	43.1	47.1	
Linoleic	%	.380	.415	
Energy	GE Mcal/kg			
Cattle	DE Mcal/kg	* 3.10	3.39	
Sheep	DE Mcal/kg	* 3.34	3.65	
Swine	DE kcal/kg	*2845.	3109.	
Cattle	ME Mcal/kg	* 2.54	2.78	
Chickens	ME_n kcal/kg	2200.	2404.	
Sheep	ME Mcal/kg	* 2.74	2.99	
Swine	ME kcal/kg	*2433.	2659.	
Cattle	NE_m Mcal/kg	* 1.61	1.76	
Cattle	NE_{gain} Mcal/kg	* 1.06	1.16	
Cattle	$NE_{lactating\ cows}$ Mcal/kg	* 1.91	2.09	
Cattle	TDN %	70.3	76.8	
Sheep	TDN %	75.7	82.7	
Swine	TDN %	* 64.5	70.5	
Calcium	%	0.20	0.22	
Magnesium	%	0.04	0.04	
Phosphorus	%	0.65	0.71	13
Riboflavin	mg/kg	11.0	12.0	
a-tocopherol	mg/kg	3.0	3.3	
Arginine	%	5.90	6.45	2
Histidine	%	1.20	1.31	
Isoleucine	%	2.00	2.19	5
Leucine	%	3.70	4.04	4
Lysine	%	2.30	2.51	9
Methionine	%	0.40	0.44	
Phenylalanine	%	2.70	2.95	2

Continued

Feed Name or Analyses		Mean As Fed	Mean Dry	C.V. ± %
Threonine	%	1.50	1.64	4
Tryptophan	%	0.50	0.55	
Tyrosine	%	1.80	1.97	9
Valine	%	2.80	3.06	2

Ref No 5 03 650 India

Feed Name or Analyses		Mean As Fed	Mean Dry	C.V. ± %
Dry matter	%	92.0	100.0	0
Protein (N x 6.25)	%	67.0	72.9	1
Energy	GE Mcal/kg			
Cattle	DE Mcal/kg	* 3.17	3.45	
Sheep	DE Mcal/kg	* 3.36	3.66	
Cattle	ME Mcal/kg	* 2.60	2.83	
Sheep	ME Mcal/kg	* 2.76	3.00	
Cattle	TDN %	71.9	78.2	
Sheep	TDN %	76.3	82.9	
Alanine	%	1.98	2.15	3
Arginine	%	5.19	5.64	1
Aspartic acid	%	5.52	6.00	1
Cystine	%	0.76	0.82	5
Glutamic acid	%	8.72	9.48	1
Glycine	%	2.78	3.02	1
Histidine	%	1.33	1.45	3
Isoleucine	%	1.59	1.72	2
Leucine	%	3.19	3.47	2
Lysine	%	1.88	2.04	2
Methionine	%	0.41	0.45	9
Phenylalanine	%	2.49	2.71	3
Proline	%	1.89	2.05	4
Serine	%	2.42	2.63	3
Threonine	%	1.20	1.30	3
Tryptophan	%	0.44	0.47	2
Tyrosine	%	1.93	2.10	2
Valine	%	1.82	1.98	1

Peanut, kernels, solv-extd grnd, 50% protein, (5)

Ref No 5 08 605 United States

Feed Name or Analyses		Mean As Fed	Mean Dry	C.V. ± %
Dry matter	%	93.0	100.0	
Ash	%	5.9	6.3	
Crude fiber	%	6.9	7.4	
Ether extract	%	1.6	1.7	
N-free extract	%	26.3	28.3	
Protein (N x 6.25)	%	52.3	56.2	
Linoleic	%	.320	.344	
Energy	GE Mcal/kg			
Cattle	DE Mcal/kg	* 3.06	3.29	
Sheep	DE Mcal/kg	* 3.17	3.41	
Swine	DE kcal/kg	*3055.	3285.	
Cattle	ME Mcal/kg	* 2.51	2.70	
Chickens	ME_n kcal/kg	2750.	2957.	
Sheep	ME Mcal/kg	* 2.60	2.80	
Swine	ME kcal/kg	*2586.	2780.	
Cattle	TDN %	* 69.3	74.6	
Sheep	TDN %	* 72.0	77.4	
Swine	TDN %	* 69.3	74.5	

Peanut, kernels wo shells, (5)

Ref No 5 03 657 United States

Feed Name or Analyses		Mean As Fed	Mean Dry	C.V. ± %
Dry matter	%	94.8	100.0	
Ash	%	2.4	2.5	
Crude fiber	%	2.8	3.0	
Sheep	dig coef %	39.	39.	
Ether extract	%	47.7	50.3	
Sheep	dig coef %	89.	89.	
N-free extract	%	13.6	14.3	
Sheep	dig coef %	-16.	-16.	
Protein (N x 6.25)	%	28.4	29.9	
Sheep	dig coef %	83.	83.	
Sheep	dig prot %	23.6	24.9	
Energy	GE Mcal/kg	5.69	6.00	
Cattle	DE Mcal/kg	* 5.78	6.10	
Sheep	DE Mcal/kg	* 5.20	5.48	
Swine	DE kcal/kg	*7254.	7652.	
Cattle	ME Mcal/kg	* 4.74	5.00	
Sheep	ME Mcal/kg	* 4.26	4.49	
Swine	ME kcal/kg	*6525.	6883.	
Cattle	TDN %	* 131.1	138.3	
Sheep	TDN %	117.8	124.3	
Swine	TDN %	* 164.5	173.6	
Calcium	%	0.06	0.06	
Chlorine	%	0.02	0.02	
Iron	%	0.002	0.002	
Magnesium	%	0.18	0.19	
Phosphorus	%	0.43	0.45	
Potassium	%	0.61	0.64	
Sodium	%	0.29	0.30	
Sulphur	%	0.25	0.26	
Niacin	mg/kg	158.3	167.0	
Riboflavin	mg/kg	1.3	1.4	
Thiamine	mg/kg	9.9	10.5	

Peanut, kernels w skins, fresh, (5)

Ref No 2 08 120 United States

Feed Name or Analyses		Mean As Fed	Mean Dry	C.V. ± %
Dry matter	%	94.4	100.0	
Ash	%	2.3	2.4	
Crude fiber	%	2.4	2.5	
Ether extract	%	47.5	50.3	
N-free extract	%	16.2	17.2	
Protein (N x 6.25)	%	26.0	27.5	
Energy	GE Mcal/kg	5.64	5.97	
Cattle	DE Mcal/kg	* 5.23	5.54	
Sheep	DE Mcal/kg	* 3.90	4.13	
Swine	DE kcal/kg	*7294.	7727.	
Cattle	ME Mcal/kg	* 4.29	4.54	
Sheep	ME Mcal/kg	* 3.20	3.39	
Swine	ME kcal/kg	*6596.	6988.	
Cattle	TDN %	* 118.6	125.6	
Sheep	TDN %	* 88.5	93.7	
Swine	TDN %	* 165.4	175.2	
Calcium	%	0.07	0.07	
Iron	%	0.002	0.002	
Phosphorus	%	0.40	0.42	
Potassium	%	0.67	0.71	
Sodium	%	0.09	0.10	
Niacin	mg/kg	172.0	182.2	
Riboflavin	mg/kg	1.3	1.4	
Thiamine	mg/kg	11.4	12.1	

Peanut, kernels w skins w hulls, (5)

Ref No 5 03 653 United States

Feed Name or Analyses		Mean As Fed	Mean Dry	C.V. ± %
Dry matter	%	93.1	100.0	
Ash	%	3.0	3.2	
Crude fiber	%	19.6	21.1	
Sheep	dig coef %	25.	25.	
Ether extract	%	35.3	37.9	
Sheep	dig coef %	98.	98.	
N-free extract	%	13.2	14.1	
Sheep	dig coef %	19.	19.	
Protein (N x 6.25)	%	22.0	23.6	
Sheep	dig coef %	79.	79.	
Sheep	dig prot %	17.4	18.7	
Energy	GE Mcal/kg			
Cattle	DE Mcal/kg	* 4.31	4.63	
Sheep	DE Mcal/kg	* 4.52	4.86	
Swine	DE kcal/kg	*5656.	6079.	
Cattle	ME Mcal/kg	* 3.53	3.80	
Sheep	ME Mcal/kg	* 3.71	3.99	
Swine	ME kcal/kg	*5160.	5545.	
Cattle	TDN %	* 97.7	105.0	
Sheep	TDN %	102.6	110.3	
Swine	TDN %	* 128.3	137.9	
Phosphorus	%	0.33	0.35	
Potassium	%	0.52	0.56	

Peanut, kernels w skins w hulls, mech-extd grnd, (5)

Peanut meal and hulls (AAFCO)

Peanut oil meal w shells, mechanical extracted

Ref No 5 03 655 United States

Feed Name or Analyses		Mean As Fed	Mean Dry	C.V. ± %
Dry matter	%	92.3	100.0	
Ash	%	5.2	5.6	
Crude fiber	%	12.9	14.0	
Ether extract	%	6.8	7.3	
N-free extract	%	24.1	26.1	
Protein (N x 6.25)	%	43.3	46.9	
Energy	GE Mcal/kg			
Cattle	DE Mcal/kg	* 3.05	3.31	
Sheep	DE Mcal/kg	* 3.12	3.38	
Swine	DE kcal/kg	*3342.	3620.	
Cattle	ME Mcal/kg	* 2.50	2.71	
Chickens	ME_n kcal/kg	2505.	2714.	
Sheep	ME Mcal/kg	* 2.56	2.77	
Swine	ME kcal/kg	*2892.	3132.	
Cattle	TDN %	* 69.3	75.0	
Sheep	TDN %	* 70.8	76.6	
Swine	TDN %	* 75.8	82.1	
Calcium	%	0.10	0.11	
Phosphorus	%	0.60	0.65	
Choline	mg/kg	2073.	2245.	
Niacin	mg/kg	174.7	189.2	
Pantothenic acid	mg/kg	52.1	56.4	
Riboflavin	mg/kg	4.7	5.0	
Arginine	%	4.53	4.90	
Cystine	%	0.65	0.71	
Glycine	%	2.21	2.40	
Lysine	%	1.61	1.74	
Methionine	%	0.40	0.44	
Tryptophan	%	0.45	0.49	

Peanut, kernels w skins w hulls, solv-extd grnd, (5)

Solvent extracted peanut meal and hulls (AAFCO)

Peanut oil meal w shells, solvent extracted

Ref No 5 03 656 United States

Feed Name or Analyses		Mean As Fed	Mean Dry	C.V. ± %
Dry matter	%	92.1	100.0	
Ash	%	4.6	5.0	
Crude fiber	%	15.6	16.9	
Ether extract	%	1.6	1.8	
N-free extract	%	25.1	27.3	
Protein (N x 6.25)	%	45.1	49.0	
Energy	GE Mcal/kg			
Cattle	DE Mcal/kg	* 2.69	2.93	
Sheep	DE Mcal/kg	* 2.91	3.17	
Swine	DE kcal/kg	*2838.	3082.	

(1) dry forages and roughages (2) pasture, range plants, and forages fed green (3) silages (4) energy feeds (5) protein supplements (6) minerals (7) vitamins (8) additives

Feed Name or Analyses		Mean As Fed	Mean Dry	C.V. ± %
Cattle	ME Mcal/kg	* 2.21	2.40	
Chickens	ME_n kcal/kg	2221.	2412.	
Sheep	ME Mcal/kg	* 2.39	2.60	
Swine	ME kcal/kg	*2443.	2654.	
Cattle	TDN %	* 61.1	66.4	
Sheep	TDN %	* 66.1	71.8	
Swine	TDN %	* 64.4	69.9	
Calcium	%	0.10	0.11	
Phosphorus	%	0.60	0.66	
Choline	mg/kg	2076.	2255.	
Niacin	mg/kg	175.0	190.1	
Pantothenic acid	mg/kg	52.2	56.7	
Riboflavin	mg/kg	4.7	5.1	
Arginine	%	4.83	5.25	
Cystine	%	0.68	0.74	
Glycine	%	2.32	2.52	
Lysine	%	1.61	1.75	
Methionine	%	0.42	0.46	
Tryptophan	%	0.46	0.50	

Peanut hulls (AAFCO) –
see Peanut, hulls, (1)

Peanut meal (AAFCO) (CFA) –
see Peanut, kernels, mech-extd grnd, mx 7% fiber, (5)

Peanut meal and hulls (AAFCO) –
see Peanut, kernels w skins w hulls, mech-extd grnd, (5)

Peanut oil meal, expeller extracted –
see Peanut, kernels, mech-extd grnd, mx 7% fiber, (5)

Peanut oil meal, solvent extracted –
see Peanut, kernels, solv-extd grnd, mx 7% fiber, (5)

Peanut oil meal w shells, mechanical extracted –
see Peanut, kernels w skins w hulls, mech-extd grnd, (5)

Peanut oil meal w shells, solvent extracted –
see Peanut, kernels w skins w hulls, solv-extd grnd, (5)

PEA-OATS. Pisum spp, Aneva sativa

Pea-Oats, aerial part, fresh, (2)
Ref No 2 08 483 — United States

Analyses		As Fed	Dry	C.V. ± %
Dry matter	%	22.7	100.0	
Ash	%	1.9	8.4	
Crude fiber	%	6.4	28.2	
Ether extract	%	0.9	4.0	
N-free extract	%	10.3	45.4	
Protein (N x 6.25)	%	3.2	14.1	
Cattle	dig prot %	* 2.2	9.9	
Goats	dig prot %	* 2.2	9.7	
Horses	dig prot %	* 2.2	9.5	
Rabbits	dig prot %	* 2.2	9.6	
Sheep	dig prot %	* 2.3	10.1	
Energy	GE Mcal/kg			
Cattle	DE Mcal/kg	* 0.67	2.96	
Sheep	DE Mcal/kg	* 0.63	2.79	
Cattle	ME Mcal/kg	* 0.55	2.43	
Sheep	ME Mcal/kg	* 0.52	2.28	
Cattle	TDN %	* 15.2	67.2	

Feed Name or Analyses		Mean As Fed	Mean Dry	C.V. ± %
Sheep	TDN %	* 14.3	63.2	
Calcium	%	0.17	0.75	
Phosphorus	%	0.07	0.31	
Potassium	%	0.38	1.67	

Pearl barley by-product (AAFCO) –
see Barley, pearl by-product, (4)

PEARLBRUSH, COMMON. Exochorda racemosa

Pearlbrush, common, seeds w hulls, (5)
Ref No 5 09 028 — United States

Analyses		As Fed	Dry	C.V. ± %
Dry matter	%		100.0	
Protein (N x 6.25)	%		48.0	
Alanine	%		1.44	
Arginine	%		5.62	
Aspartic acid	%		4.27	
Glutamic acid	%		10.70	
Glycine	%		1.92	
Histidine	%		0.91	
Hydroxyproline	%		0.10	
Isoleucine	%		0.91	
Leucine	%		2.78	
Lysine	%		0.82	
Methionine	%		0.29	
Phenylalanine	%		0.96	
Proline	%		1.34	
Serine	%		1.49	
Threonine	%		1.06	
Tyrosine	%		1.49	
Valine	%		1.63	

PEARS, Pyrus spp

Pears, pulp, dehy grnd, (1)
Ref No 1 03 661 — United States

Analyses		As Fed	Dry	C.V. ± %
Dry matter	%	91.5	100.0	4
Ash	%	3.7	4.0	76
Crude fiber	%	21.8	23.8	40
Ether extract	%	2.0	2.1	38
N-free extract	%	58.5	63.9	
Protein (N x 6.25)	%	5.6	6.1	54
Cattle	dig prot %	* 2.0	2.2	
Goats	dig prot %	* 2.0	2.2	
Horses	dig prot %	* 2.5	2.7	
Rabbits	dig prot %	* 3.1	3.4	
Sheep	dig prot %	* 1.9	2.0	
Cellulose (Matrone)	%	23.7	25.8	
Lignin (Ellis)	%	24.8	27.1	
Energy	GE Mcal/kg			
Cattle	DE Mcal/kg	* 2.85	3.11	
Sheep	DE Mcal/kg	* 2.77	3.03	
Cattle	ME Mcal/kg	* 2.34	2.55	
Sheep	ME Mcal/kg	* 2.27	2.48	
Cattle	TDN %	* 64.6	70.6	
Sheep	TDN %	* 62.9	68.7	
Calcium	%	2.18	2.38	
Phosphorus	%	0.11	0.12	

Pears, cannery residue, fresh, (2)
Ref No 2 08 486 — United States

Analyses		As Fed	Dry	C.V. ± %
Dry matter	%	15.2	100.0	
Ash	%	0.3	2.0	
Crude fiber	%	2.6	17.1	

Feed Name or Analyses		Mean As Fed	Mean Dry	C.V. ± %
Ether extract	%	0.2	1.3	
N-free extract	%	11.5	75.7	
Protein (N x 6.25)	%	0.6	3.9	
Cattle	dig prot %	* 0.2	1.2	
Goats	dig prot %	* 0.0	0.2	
Horses	dig prot %	* 0.1	0.9	
Rabbits	dig prot %	* 0.3	1.7	
Sheep	dig prot %	* 0.1	0.7	
Energy	GE Mcal/kg			
Cattle	DE Mcal/kg	* 0.47	3.06	
Sheep	DE Mcal/kg	* 0.50	3.28	
Cattle	ME Mcal/kg	* 0.38	2.51	
Sheep	ME Mcal/kg	* 0.41	2.69	
Cattle	TDN %	* 10.5	69.3	
Sheep	TDN %	* 11.3	74.4	

Pears, cannery residue, ensiled, (3)
Ref No 3 03 659 — United States

Analyses		As Fed	Dry	C.V. ± %
Dry matter	%	27.6	100.0	
Ash	%	2.2	8.0	
Crude fiber	%	9.0	32.6	
Ether extract	%	0.9	3.3	
N-free extract	%	12.2	44.1	
Protein (N x 6.25)	%	3.3	12.0	
Cattle	dig prot %	* 2.0	7.1	
Goats	dig prot %	* 2.0	7.1	
Horses	dig prot %	* 2.0	7.1	
Sheep	dig prot %	* 2.0	7.1	
Energy	GE Mcal/kg			
Cattle	DE Mcal/kg	* 0.71	2.56	
Sheep	DE Mcal/kg	* 0.77	2.78	
Cattle	ME Mcal/kg	* 0.58	2.10	
Sheep	ME Mcal/kg	* 0.63	2.28	
Cattle	TDN %	* 16.0	58.0	
Sheep	TDN %	* 17.4	63.2	

Pears, molasses, (4)
Ref No 4 08 487 — United States

Analyses		As Fed	Dry	C.V. ± %
Dry matter	%	76.4	100.0	
Ash	%	5.3	6.9	
Crude fiber	%	0.0	0.0	
Ether extract	%	0.0	0.0	
N-free extract	%	69.9	91.5	
Protein (N x 6.25)	%	1.2	1.6	
Cattle	dig prot %	* -1.8	-2.4	
Goats	dig prot %	* -0.9	-1.2	
Horses	dig prot %	* -0.9	-1.2	
Sheep	dig prot %	* -0.9	-1.2	
Energy	GE Mcal/kg			
Cattle	DE Mcal/kg	* 2.62	3.43	
Sheep	DE Mcal/kg	* 2.92	3.83	
Swine	DE kcal/kg	*2707.	3543.	
Cattle	ME Mcal/kg	* 2.15	2.81	
Sheep	ME Mcal/kg	* 2.40	3.14	
Swine	ME kcal/kg	*2590.	3390.	
Cattle	TDN %	* 59.4	77.8	
Sheep	TDN %	* 66.3	86.8	
Swine	TDN %	* 61.4	80.3	

Pears, pulp, (4)
Ref No 4 03 662 — United States

Analyses		As Fed	Dry	C.V. ± %
Dry matter	%	88.2	100.0	
Ash	%	1.4	1.6	

Continued

Feed Name or Analyses				Mean As Fed	Mean Dry	C.V. ± %
Crude fiber		%		28.8	32.7	
Swine	dig coef	%		14.	14.	
Ether extract		%		1.9	2.2	
Swine	dig coef	%		48.	48.	
N-free extract		%		52.3	59.3	
Swine	dig coef	%		56.	56.	
Protein (N x 6.25)		%		3.7	4.2	
Swine	dig coef	%		0.	0.	
Cattle	dig prot	%	*	0.0	0.0	
Goats	dig prot	%	*	1.0	1.1	
Horses	dig prot	%	*	1.0	1.1	
Sheep	dig prot	%	*	1.0	1.1	
Energy		GE Mcal/kg				
Swine		DE kcal/kg	*	1562.	1771.	
Swine		ME kcal/kg	*	1486.	1685.	
Swine		TDN %		35.4	40.2	

PEA-VETCH. Pisum spp, Vicia spp

Pea-vetch, aerial part, ensiled, (3)

Ref No 3 03 612 United States

Feed Name or Analyses				Mean As Fed	Mean Dry	C.V. ± %
Dry matter		%		10.0	100.0	
Ash		%		1.4	14.4	
Crude fiber		%		4.6	45.8	
Ether extract		%		0.7	6.6	
N-free extract		%		1.5	14.8	
Protein (N x 6.25)		%		1.8	18.4	
Cattle	dig prot	%	*	1.3	12.9	
Goats	dig prot	%	*	1.3	12.9	
Horses	dig prot	%	*	1.3	12.9	
Sheep	dig prot	%	*	1.3	12.9	
Energy		GE Mcal/kg				
Cattle		DE Mcal/kg	*	0.22	2.22	
Sheep		DE Mcal/kg	*	0.24	2.36	
Cattle		ME Mcal/kg	*	0.18	1.82	
Sheep		ME Mcal/kg	*	0.19	1.94	
Cattle		TDN %	*	5.0	50.4	
Sheep		TDN %	*	5.4	53.5	

PEAVINE. Lathyrus spp

Peavine, hay, dehy, (1)

Ref No 1 03 663 United States

Feed Name or Analyses				Mean As Fed	Mean Dry	C.V. ± %
Dry matter		%			100.0	
Carotene		mg/kg			72.8	
Riboflavin		mg/kg			32.2	
Vitamin A equivalent		IU/g			121.3	

Peavine, hay, s-c, (1)

Ref No 1 03 666 United States

Feed Name or Analyses				Mean As Fed	Mean Dry	C.V. ± %
Dry matter		%		89.2	100.0	
Ash		%		6.9	7.7	
Crude fiber		%		24.2	27.1	
Sheep	dig coef	%		44.	44.	
Ether extract		%		2.5	2.8	
Sheep	dig coef	%		53.	53.	
N-free extract		%		39.8	44.6	
Sheep	dig coef	%		70.	70.	
Protein (N x 6.25)		%		15.9	17.8	
Sheep	dig coef	%		78.	78.	
Cattle	dig prot	%	*	11.0	12.4	
Goats	dig prot	%	*	11.7	13.2	
Horses	dig prot	%	*	11.3	12.6	
Rabbits	dig prot	%	*	11.1	12.4	
Sheep	dig prot	%		12.4	13.9	
Fatty acids		%		2.5	2.8	
Energy		GE Mcal/kg				
Cattle		DE Mcal/kg	*	2.35	2.63	
Sheep		DE Mcal/kg	*	2.38	2.66	
Cattle		ME Mcal/kg	*	1.93	2.16	
Sheep		ME Mcal/kg	*	1.95	2.18	
Cattle		TDN %	*	53.3	59.7	
Sheep		TDN %		53.9	60.4	
Calcium		%		1.53	1.71	
Phosphorus		%		0.17	0.19	
Carotene		mg/kg		28.3	31.7	
Riboflavin		mg/kg		8.3	9.3	
Vitamin A equivalent		IU/g		47.2	52.9	

(1) dry forages and roughages (2) pasture, range plants, and forages fed green (3) silages (4) energy feeds (5) protein supplements (6) minerals (7) vitamins (8) additives

Peavine, hay, s-c, immature, (1)

Ref No 1 03 664 United States

Feed Name or Analyses				Mean As Fed	Mean Dry	C.V. ± %
Dry matter		%		91.7	100.0	
Ash		%		8.8	9.6	4
Crude fiber		%		24.7	26.9	5
Ether extract		%		2.7	2.9	9
N-free extract		%		35.2	38.4	
Protein (N x 6.25)		%		20.4	22.2	7
Cattle	dig prot	%	*	14.8	16.2	
Goats	dig prot	%	*	15.8	17.3	
Horses	dig prot	%	*	15.0	16.4	
Rabbits	dig prot	%	*	14.5	15.8	
Sheep	dig prot	%	*	15.1	16.5	
Energy		GE Mcal/kg				
Cattle		DE Mcal/kg	*	2.47	2.70	
Sheep		DE Mcal/kg	*	2.41	2.62	
Cattle		ME Mcal/kg	*	2.03	2.21	
Sheep		ME Mcal/kg	*	1.97	2.15	
Cattle		TDN %	*	56.1	61.2	
Sheep		TDN %	*	54.6	59.5	

Peavine, hay, s-c, dough stage, (1)

Ref No 1 03 665 United States

Feed Name or Analyses				Mean As Fed	Mean Dry	C.V. ± %
Dry matter		%			100.0	
Ash		%			7.9	8
Crude fiber		%			36.3	4
Ether extract		%			2.5	13
N-free extract		%			40.0	
Protein (N x 6.25)		%			13.3	10
Cattle	dig prot	%	*		8.5	
Goats	dig prot	%	*		9.0	
Horses	dig prot	%	*		8.8	
Rabbits	dig prot	%	*		8.9	
Sheep	dig prot	%	*		8.5	
Energy		GE Mcal/kg				
Cattle		DE Mcal/kg	*		2.40	
Sheep		DE Mcal/kg	*		2.37	
Cattle		ME Mcal/kg	*		1.97	
Sheep		ME Mcal/kg	*		1.94	
Cattle		TDN %	*		54.5	
Sheep		TDN %	*		53.7	

Peavine, aerial part, fresh, (2)

Ref No 2 03 669 United States

Feed Name or Analyses				Mean As Fed	Mean Dry	C.V. ± %
Dry matter		%		21.0	100.0	22
Ash		%		5.2	24.9	22
Crude fiber		%		5.5	26.2	11
Ether extract		%		0.7	3.4	13
N-free extract		%		5.9	28.3	
Protein (N x 6.25)		%		3.6	17.2	9
Cattle	dig prot	%	*	2.6	12.5	
Goats	dig prot	%	*	2.6	12.6	
Horses	dig prot	%	*	2.5	12.1	
Rabbits	dig prot	%	*	2.5	11.9	
Sheep	dig prot	%	*	2.7	13.0	
Energy		GE Mcal/kg				
Cattle		DE Mcal/kg	*	0.56	2.66	
Sheep		DE Mcal/kg	*	0.60	2.84	
Cattle		ME Mcal/kg	*	0.47	2.18	
Sheep		ME Mcal/kg	*	0.49	2.33	
Cattle		TDN %	*	12.7	60.3	
Sheep		TDN %	*	13.5	64.5	
Carotene		mg/kg		21.4	102.1	
Riboflavin		mg/kg		2.6	12.6	
Vitamin A equivalent		IU/g		35.6	170.2	

Peavine, aerial part, fresh, winter cut, (2)

Ref No 2 03 667 United States

Feed Name or Analyses				Mean As Fed	Mean Dry	C.V. ± %
Dry matter		%		10.5	100.0	
Ash		%		3.3	31.8	
Crude fiber		%		2.3	22.2	
Sheep	dig coef	%		63.	63.	
Ether extract		%		0.4	4.0	
Sheep	dig coef	%		52.	52.	
N-free extract		%		2.4	22.6	
Sheep	dig coef	%		88.	88.	
Protein (N x 6.25)		%		2.0	19.4	
Sheep	dig coef	%		79.	79.	
Cattle	dig prot	%	*	1.5	14.4	
Goats	dig prot	%	*	1.5	14.7	
Horses	dig prot	%	*	1.5	14.0	
Rabbits	dig prot	%	*	1.4	13.6	
Sheep	dig prot	%		1.6	15.3	
Energy		GE Mcal/kg				
Cattle		DE Mcal/kg	*	0.23	2.16	
Sheep		DE Mcal/kg	*	0.25	2.38	
Cattle		ME Mcal/kg	*	0.19	2.77	
Sheep		ME Mcal/kg	*	0.20	1.95	
Cattle		TDN %	*	5.1	48.9	
Sheep		TDN %		5.7	53.9	

Peavine, aerial part, fresh, cut 1, (2)

Ref No 2 03 668 United States

Feed Name or Analyses				Mean As Fed	Mean Dry	C.V. ± %
Dry matter		%		25.3	100.0	
Ash		%		15.4	60.7	
Crude fiber		%		3.9	15.4	
Sheep	dig coef	%		55.	55.	
Ether extract		%		9.1	35.9	
Sheep	dig coef	%		48.	48.	
N-free extract		%		6.6	26.5	
Sheep	dig coef	%		81.	81.	
Protein (N x 6.25)		%		3.7	14.6	
Sheep	dig coef	%		74.	74.	
Cattle	dig prot	%	*	2.6	10.3	
Goats	dig prot	%	*	2.6	10.2	
Horses	dig prot	%	*	2.5	9.9	

Feed Name or Analyses		Mean As Fed	Mean Dry	C.V. ± %
Rabbits	dig prot %	* 2.5	9.9	
Sheep	dig prot %	2.7	10.8	
Energy	GE Mcal/kg			
Sheep	DE Mcal/kg	* 0.41	1.61	
Sheep	ME Mcal/kg	* 0.33	1.32	
Sheep	TDN %	9.2	36.5	

Peavine, leaves, fresh, (2)
Ref No 2 03 670 United States

Feed Name or Analyses		Mean As Fed	Mean Dry	C.V. ± %
Dry matter	%	27.5	100.0	31
Ash	%	2.8	10.2	
Crude fiber	%	4.1	14.8	23
Ether extract	%	1.8	6.6	60
N-free extract	%	13.0	47.1	
Protein (N x 6.25)	%	5.9	21.3	15
Cattle	dig prot %	* 4.4	16.0	
Goats	dig prot %	* 4.5	16.4	
Horses	dig prot %	* 4.3	15.6	
Rabbits	dig prot %	* 4.2	15.1	
Sheep	dig prot %	* 4.6	16.8	
Energy	GE Mcal/kg			
Cattle	DE Mcal/kg	* 0.80	2.91	
Sheep	DE Mcal/kg	* 0.90	3.29	
Cattle	ME Mcal/kg	* 0.66	2.39	
Sheep	ME Mcal/kg	* 0.74	2.70	
Cattle	TDN %	* 18.2	66.0	
Sheep	TDN %	* 20.5	74.6	
Calcium	%	0.51	1.87	
Phosphorus	%	0.07	0.27	
Carotene	mg/kg	63.4	230.4	66
Riboflavin	mg/kg	5.5	20.1	17
Vitamin A equivalent	IU/g	105.6	384.0	
Arginine	%	0.33	1.20	
Histidine	%	0.11	0.40	
Isoleucine	%	0.30	1.10	
Leucine	%	0.55	2.00	
Lysine	%	0.33	1.20	
Methionine	%	0.06	0.20	
Phenylalanine	%	0.41	1.50	
Threonine	%	0.30	1.10	
Tryptophan	%	0.11	0.40	
Valine	%	0.39	1.40	

Peavine, pods, fresh, (2)
Ref No 2 03 671 United States

Feed Name or Analyses		Mean As Fed	Mean Dry	C.V. ± %
Dry matter	%	18.8	100.0	
Carotene	mg/kg	3.5	18.5	
Riboflavin	mg/kg	1.6	8.6	
Vitamin A equivalent	IU/g	5.8	30.9	

Peavine, stems, fresh, (2)
Ref No 2 03 672 United States

Feed Name or Analyses		Mean As Fed	Mean Dry	C.V. ± %
Dry matter	%	44.0	100.0	
Ash	%	3.4	7.8	
Crude fiber	%	12.9	29.4	22
Ether extract	%	1.1	2.4	33
N-free extract	%	21.5	48.9	
Protein (N x 6.25)	%	5.1	11.5	12
Cattle	dig prot %	* 3.4	7.7	
Goats	dig prot %	* 3.2	7.3	
Horses	dig prot %	* 3.2	7.3	
Rabbits	dig prot %	* 3.3	7.5	
Sheep	dig prot %	* 3.4	7.7	
Calcium	%	0.26	0.60	

Feed Name or Analyses		Mean As Fed	Mean Dry	C.V. ± %
Phosphorus	%	0.08	0.18	
Carotene	mg/kg	21.2	48.3	23
Riboflavin	mg/kg	5.3	12.1	19
Vitamin A equivalent	IU/g	35.4	80.5	

PEAVINE, ASPEN. Lathyrus leucanthus

Peavine, aspen, hay, s-c, (1)
Ref No 1 03 675 United States

Feed Name or Analyses		Mean As Fed	Mean Dry	C.V. ± %
Dry matter	%	87.5	100.0	3
Cellulose (Matrone)	%	25.6	29.2	
Sugars, total	%	15.7	17.9	
Lignin (Ellis)	%	10.4	11.9	
Calcium	%	1.20	1.37	2
Magnesium	%	0.21	0.24	18
Phosphorus	%	0.27	0.31	16

Peavine, aspen, hay, s-c, immature, (1)
Ref No 1 03 673 United States

Feed Name or Analyses		Mean As Fed	Mean Dry	C.V. ± %
Dry matter	%		100.0	
Sulphur	%		0.23	

Peavine, aspen, hay, s-c, full bloom, (1)
Ref No 1 03 674 United States

Feed Name or Analyses		Mean As Fed	Mean Dry	C.V. ± %
Dry matter	%		100.0	
Ash	%		9.1	
Ether extract	%		2.8	9
Protein (N x 6.25)	%		16.1	8
Cattle	dig prot %	*	10.9	
Goats	dig prot %	*	11.6	
Horses	dig prot %	*	11.2	
Rabbits	dig prot %	*	11.1	
Sheep	dig prot %	*	11.0	
Sulphur	%		0.15	

PEAVINE, GRASS. Lathyrus sativus

Peavine, grass, seeds, (5)
Ref No 5 03 676 United States

Feed Name or Analyses		Mean As Fed	Mean Dry	C.V. ± %
Dry matter	%	88.6	100.0	
Ash	%	6.1	6.9	
Crude fiber	%	7.3	8.2	
Sheep	dig coef %	83.	83.	
Ether extract	%	0.8	0.9	
Sheep	dig coef %	106.	106.	
N-free extract	%	49.9	56.3	
Sheep	dig coef %	92.	92.	
Protein (N x 6.25)	%	24.5	27.7	
Sheep	dig coef %	88.	88.	
Sheep	dig prot %	21.6	24.4	
Energy	GE Mcal/kg			
Cattle	DE Mcal/kg	* 2.96	3.34	
Sheep	DE Mcal/kg	* 3.33	3.75	
Swine	DE kcal/kg	*3358.	3790.	
Cattle	ME Mcal/kg	* 2.43	2.74	
Sheep	ME Mcal/kg	* 2.73	3.08	
Swine	ME kcal/kg	*3036.	3426.	
Cattle	TDN %	* 67.2	75.8	
Sheep	TDN %	75.4	85.1	
Swine	TDN %	* 76.2	86.0	

PEAVINE, SHOWY. Lathyrus ornatus

Peavine, showy, seeds w hulls, (5)
Ref No 5 09 059 United States

Feed Name or Analyses		Mean As Fed	Mean Dry	C.V. ± %
Dry matter	%		100.0	
Protein (N x 6.25)	%		33.0	
Alanine	%		1.02	
Arginine	%		2.21	
Aspartic acid	%		2.77	
Glutamic acid	%		2.77	
Glycine	%		1.16	
Histidine	%		0.76	
Hydroxyproline	%		0.07	
Isoleucine	%		1.06	
Leucine	%		1.82	
Lysine	%		1.85	
Methionine	%		0.23	
Phenylalanine	%		1.12	
Proline	%		1.12	
Serine	%		1.19	
Threonine	%		0.99	
Tyrosine	%		0.83	
Valine	%		1.12	

PEAVINE, THICKLEAF. Lathyrus lanszwerti

Peavine, thickleaf, aerial part, s-c, (1)
Ref No 1 03 677 United States

Feed Name or Analyses		Mean As Fed	Mean Dry	C.V. ± %
Dry matter	%	93.1	100.0	
Ash	%	7.4	7.9	
Crude fiber	%	27.7	29.7	
Sheep	dig coef %	36.	36.	
Ether extract	%	4.0	4.3	
Sheep	dig coef %	32.	32.	
N-free extract	%	44.8	48.1	
Sheep	dig coef %	65.	65.	
Protein (N x 6.25)	%	9.3	10.0	
Sheep	dig coef %	48.	48.	
Cattle	dig prot %	* 5.2	5.6	
Goats	dig prot %	* 5.5	5.9	
Horses	dig prot %	* 5.6	6.0	
Rabbits	dig prot %	* 5.9	6.4	
Sheep	dig prot %	4.5	4.8	
Energy	GE Mcal/kg			
Cattle	DE Mcal/kg	* 2.19	2.35	
Sheep	DE Mcal/kg	* 2.05	2.20	
Cattle	ME Mcal/kg	* 1.80	1.93	
Sheep	ME Mcal/kg	* 1.68	1.80	
Cattle	TDN %	* 49.7	53.4	
Sheep	TDN %	46.4	49.9	

Pea vine silage –
see Pea, aerial part wo seeds, ensiled, (3)

PECAN. Carya illinoensis

Pecan, hulls, (1)
Ref No 1 03 678 United States

Feed Name or Analyses		Mean As Fed	Mean Dry	C.V. ± %
Dry matter	%	91.9	100.0	
Ash	%	2.2	2.4	
Crude fiber	%	54.6	59.4	
Ether extract	%	0.6	0.7	
N-free extract	%	32.5	35.4	

Continued

Feed Name or Analyses		As Fed	Dry	C.V. ± %
Protein (N x 6.25)	%	1.9	2.1	
Cattle	dig prot % *	-1.0	-1.1	
Goats	dig prot % *	-1.3	-1.4	
Horses	dig prot % *	-0.5	-0.6	
Rabbits	dig prot % *	0.3	0.3	
Sheep	dig prot % *	-1.3	-1.4	

PEDICULARIS, Sickletop. Pedicularis racemosa

Pedicularis, sickletop, aerial part, fresh, (2)
Ref No 2 03 680 United States

Feed Name or Analyses		As Fed	Dry	C.V. ± %
Dry matter	%		100.0	
Ash	%		11.6	
Crude fiber	%		17.7	
Ether extract	%		1.8	
N-free extract	%		53.9	
Protein (N x 6.25)	%		15.0	
Cattle	dig prot % *		10.6	
Goats	dig prot % *		10.6	
Horses	dig prot % *		10.3	
Rabbits	dig prot % *		10.3	
Sheep	dig prot % *		11.0	

PENSTEMON, SHOWY. Penstemon spectabilis

Penstemon, showy, seed w hulls, (4)
Ref No 4 09 134 United States

Feed Name or Analyses		As Fed	Dry	C.V. ± %
Dry matter	%		100.0	
Protein (N x 6.25)	%		11.0	
Alanine	%		0.50	
Arginine	%		0.72	
Aspartic acid	%		0.86	
Glutamic acid	%		1.35	
Glycine	%		0.55	
Histidine	%		0.22	
Hydroxyproline	%		0.22	
Isoleucine	%		0.44	
Leucine	%		0.67	
Lysine	%		0.44	
Methionine	%		0.17	
Phenylalanine	%		0.42	
Proline	%		0.46	
Serine	%		0.52	
Threonine	%		0.41	
Tyrosine	%		0.35	
Valine	%		0.51	

PEPPERWEED. Lepidium spp

Pepperweed, aerial part, fresh, pre-bloom, (2)
Ref No 2 03 686 United States

Feed Name or Analyses		As Fed	Dry	C.V. ± %
Dry matter	%		100.0	
Ash	%		13.5	
Crude fiber	%		11.9	
Ether extract	%		1.9	
N-free extract	%		33.8	
Protein (N x 6.25)	%		38.9	
Cattle	dig prot % *		31.0	
Goats	dig prot % *		32.9	
Horses	dig prot % *		30.6	
Rabbits	dig prot % *		28.7	
Sheep	dig prot % *		33.2	
Calcium	%		1.95	
Magnesium	%		0.49	
Phosphorus	%		0.60	
Potassium	%		3.18	

Pepperweed, aerial part, fresh, mid-bloom, (2)
Ref No 2 03 687 United States

Feed Name or Analyses		As Fed	Dry	C.V. ± %
Dry matter	%		100.0	
Ash	%		8.6	
Crude fiber	%		22.3	
Ether extract	%		2.0	
N-free extract	%		42.7	
Protein (N x 6.25)	%		24.4	
Cattle	dig prot % *		18.6	
Goats	dig prot % *		19.3	
Horses	dig prot % *		18.2	
Rabbits	dig prot % *		17.5	
Sheep	dig prot % *		19.7	
Calcium	%		1.55	
Magnesium	%		0.36	
Phosphorus	%		0.38	
Potassium	%		2.05	

PEPPERWEED, VIRGINIA. Lepidium virginicum

Pepperweed, Virginia, aerial part, fresh, (2)
Ref No 2 03 688 United States

Feed Name or Analyses		As Fed	Dry	C.V. ± %
Dry matter	%	21.9	100.0	
Niacin	mg/kg	25.4	116.0	
Riboflavin	mg/kg	6.2	28.4	
Thiamine	mg/kg	0.4	2.0	

PERILLA. Perilla spp

Perilla, seeds, extn unspecified grnd, (5)
Ref No 5 03 690 United States

Feed Name or Analyses		As Fed	Dry	C.V. ± %
Dry matter	%	91.4	100.0	
Ash	%	8.2	9.0	
Crude fiber	%	20.5	22.4	
Sheep	dig coef %	18.	18.	
Ether extract	%	8.6	9.4	
Sheep	dig coef %	92.	92.	
N-free extract	%	16.2	17.7	
Sheep	dig coef %	38.	38.	
Protein (N x 6.25)	%	38.0	41.5	
Sheep	dig coef %	89.	89.	
Sheep	dig prot %	33.8	37.0	
Energy	GE Mcal/kg			
Cattle	DE Mcal/kg *	2.61	2.86	
Sheep	DE Mcal/kg *	2.71	2.96	
Swine	DE kcal/kg	*3070.	3359.	
Cattle	ME Mcal/kg *	2.14	2.34	
Sheep	ME Mcal/kg *	2.22	2.43	
Swine	ME kcal/kg	*2690.	2943.	
Cattle	TDN % *	59.2	64.8	
Sheep	TDN %	61.4	67.1	
Swine	TDN % *	69.6	76.2	
Calcium	%	0.56	0.61	
Phosphorus	%	0.47	0.51	

PERSIMMON, COMMON. Diospyros virginiana

Persimmon, common, seeds w hulls, (4)
Ref No 4 09 138 United States

Feed Name or Analyses		As Fed	Dry	C.V. ± %
Dry matter	%		100.0	
Protein (N x 6.25)	%		10.0	
Alanine	%		0.36	
Arginine	%		0.66	
Aspartic acid	%		0.68	
Glutamic acid	%		1.23	
Glycine	%		0.39	
Histidine	%		0.18	
Hydroxyproline	%		0.04	
Isoleucine	%		0.31	
Leucine	%		0.48	
Lysine	%		0.62	
Methionine	%		0.12	
Phenylalanine	%		0.34	
Proline	%		0.48	
Serine	%		0.36	
Threonine	%		0.32	
Tyrosine	%		0.20	
Valine	%		0.36	

PERSIMMON, TEXAS. Diospyros texana

Persimmon, Texas, browse, (2)
Ref No 2 03 692 United States

Feed Name or Analyses		As Fed	Dry	C.V. ± %
Dry matter	%		100.0	
Ash	%		10.4	
Crude fiber	%		21.0	
Ether extract	%		7.4	
N-free extract	%		48.2	
Protein (N x 6.25)	%		13.0	
Cattle	dig prot % *		8.9	
Goats	dig prot % *		8.7	
Horses	dig prot % *		8.6	
Rabbits	dig prot % *		8.7	
Sheep	dig prot % *		9.1	
Calcium	%		4.11	
Magnesium	%		0.48	
Phosphorus	%		0.12	
Potassium	%		0.53	

Persimmon, Texas, browse, immature, (2)
Ref No 2 03 691 United States

Feed Name or Analyses		As Fed	Dry	C.V. ± %
Dry matter	%		100.0	
Ash	%		7.9	
Crude fiber	%		21.9	
Ether extract	%		4.3	
N-free extract	%		49.6	
Protein (N x 6.25)	%		16.3	
Cattle	dig prot % *		11.7	
Goats	dig prot % *		11.8	
Horses	dig prot % *		11.4	
Rabbits	dig prot % *		11.3	
Sheep	dig prot % *		12.2	

(1) dry forages and roughages (2) pasture, range plants, and forages fed green (3) silages (4) energy feeds (5) protein supplements (6) minerals (7) vitamins (8) additives

PHACELIA. Phacelia spp

Phacelia, aerial part, fresh, (2)
Ref No 2 03 695 United States

Feed Name or Analyses				Mean As Fed	Mean Dry	C.V. ± %
Dry matter		%		13.1	100.0	16
Ash		%		2.1	16.1	16
Crude fiber		%		3.0	22.9	40
Ether extract		%		0.3	2.1	21
N-free extract		%		5.6	42.5	
Protein (N x 6.25)		%		2.1	16.4	31
Cattle	dig prot	%	*	1.5	11.8	
Goats	dig prot	%	*	1.6	11.9	
Horses	dig prot	%	*	1.5	11.5	
Rabbits	dig prot	%	*	1.5	11.3	
Sheep	dig prot	%	*	1.6	12.3	

Phacelia, aerial part, fresh, immature, (2)
Ref No 2 03 693 United States

Feed Name or Analyses				Mean As Fed	Mean Dry	C.V. ± %
Dry matter		%			100.0	
Ash		%			16.3	
Crude fiber		%			7.9	
Ether extract		%			1.6	
N-free extract		%			47.2	
Protein (N x 6.25)		%			27.0	
Cattle	dig prot	%	*		20.8	
Goats	dig prot	%	*		21.8	
Horses	dig prot	%	*		20.5	
Rabbits	dig prot	%	*		19.5	
Sheep	dig prot	%	*		22.2	
Calcium		%			3.87	
Magnesium		%			0.35	
Phosphorus		%			0.32	
Potassium		%			3.01	

Phacelia, aerial part, fresh, full bloom, (2)
Ref No 2 03 694 United States

Feed Name or Analyses				Mean As Fed	Mean Dry	C.V. ± %
Dry matter		%		13.1	100.0	16
Ash		%		2.3	17.6	8
Crude fiber		%		3.6	27.7	26
Sheep	dig coef	%		42.	42.	
Ether extract		%		0.3	2.4	12
Sheep	dig coef	%		47.	47.	
N-free extract		%		4.7	35.6	
Sheep	dig coef	%		58.	58.	
Protein (N x 6.25)		%		2.2	16.7	8
Sheep	dig coef	%		64.	64.	
Cattle	dig prot	%	*	1.6	12.1	
Goats	dig prot	%	*	1.6	12.1	
Horses	dig prot	%	*	1.5	11.7	
Rabbits	dig prot	%	*	1.5	11.5	
Sheep	dig prot	%		1.4	10.7	
Energy	GE	Mcal/kg				
Cattle	DE	Mcal/kg	*	0.28	2.14	
Sheep	DE	Mcal/kg	*	0.26	1.99	
Cattle	ME	Mcal/kg	*	0.23	1.75	
Sheep	ME	Mcal/kg	*	0.21	1.63	
Cattle	TDN	%	*	6.4	48.6	
Sheep	TDN	%		5.9	45.2	

Phacelia, aerial part, ensiled, (3)
Ref No 3 03 697 United States

Feed Name or Analyses				Mean As Fed	Mean Dry	C.V. ± %
Dry matter		%		12.0	100.0	15
Ash		%		2.1	17.3	13
Crude fiber		%		3.3	27.3	8
Sheep	dig coef	%		54.	54.	
Ether extract		%		0.4	3.7	15
Sheep	dig coef	%		70.	70.	
N-free extract		%		4.4	37.0	
Sheep	dig coef	%		70.	70.	
Protein (N x 6.25)		%		1.8	14.8	15
Sheep	dig coef	%		67.	67.	
Cattle	dig prot	%	*	1.2	9.7	
Goats	dig prot	%	*	1.2	9.7	
Horses	dig prot	%	*	1.2	9.7	
Sheep	dig prot	%		1.2	9.9	
Energy	GE	Mcal/kg				
Cattle	DE	Mcal/kg	*	0.31	2.57	
Sheep	DE	Mcal/kg	*	0.30	2.48	
Cattle	ME	Mcal/kg	*	0.25	2.11	
Sheep	ME	Mcal/kg	*	0.24	2.04	
Cattle	TDN	%	*	7.0	58.4	
Sheep	TDN	%		6.8	56.3	

Phacelia, aerial part, ensiled, full bloom, (3)
Ref No 3 03 696 United States

Feed Name or Analyses				Mean As Fed	Mean Dry	C.V. ± %
Dry matter		%		12.2	100.0	18
Ash		%		2.2	17.9	17
Crude fiber		%		3.2	26.2	8
Sheep	dig coef	%		58.	58.	
Ether extract		%		0.5	4.1	14
Sheep	dig coef	%		69.	69.	
N-free extract		%		4.4	36.4	
Sheep	dig coef	%		73.	73.	
Protein (N x 6.25)		%		1.9	15.5	17
Sheep	dig coef	%		66.	66.	
Cattle	dig prot	%	*	1.3	10.3	
Goats	dig prot	%	*	1.3	10.3	
Horses	dig prot	%	*	1.3	10.3	
Sheep	dig prot	%		1.2	10.2	
Energy	GE	Mcal/kg				
Cattle	DE	Mcal/kg	*	0.32	2.66	
Sheep	DE	Mcal/kg	*	0.31	2.57	
Cattle	ME	Mcal/kg	*	0.27	2.18	
Sheep	ME	Mcal/kg	*	0.26	2.11	
Cattle	TDN	%	*	7.4	60.3	
Sheep	TDN	%		7.1	58.3	

PHACELIA, TANSY. Phacelia tanacetifolia

Phacelia, tansy, seeds w hulls, (5)
Ref No 5 09 136 United States

Feed Name or Analyses				Mean As Fed	Mean Dry	C.V. ± %
Dry matter		%			100.0	
Protein (N x 6.25)		%			18.0	
Alanine		%			0.81	
Arginine		%			1.42	
Aspartic acid		%			1.67	
Glutamic acid		%			2.75	
Glycine		%			1.10	
Histidine		%			0.40	
Hydroxyproline		%			0.00	
Isoleucine		%			0.76	
Leucine		%			1.04	
Lysine		%			0.88	
Methionine		%			0.29	
Phenylalanine		%			0.70	
Proline		%			0.65	
Serine		%			0.86	
Threonine		%			0.68	
Tyrosine		%			0.63	
Valine		%			0.83	

PHACELIA, TANSY. Phacelia tanacetifolia

Phacelia, tansy, aerial part, s-c, (1)
Ref No 1 03 699 United States

Feed Name or Analyses				Mean As Fed	Mean Dry	C.V. ± %
Dry matter		%		85.4	100.0	
Ash		%		9.7	11.4	
Crude fiber		%		26.7	31.3	
Sheep	dig coef	%		19.	19.	
Ether extract		%		1.5	1.8	
Sheep	dig coef	%		52.	52.	
N-free extract		%		37.1	43.4	
Sheep	dig coef	%		48.	48.	
Protein (N x 6.25)		%		10.3	12.1	
Sheep	dig coef	%		56.	56.	
Cattle	dig prot	%	*	6.3	7.4	
Goats	dig prot	%	*	6.7	7.8	
Horses	dig prot	%	*	6.7	7.8	
Rabbits	dig prot	%	*	6.8	8.0	
Sheep	dig prot	%		5.8	6.8	
Energy	GE	Mcal/kg				
Cattle	DE	Mcal/kg	*	1.63	1.91	
Sheep	DE	Mcal/kg	*	1.34	1.57	
Cattle	ME	Mcal/kg	*	1.34	1.57	
Sheep	ME	Mcal/kg	*	1.10	1.29	
Cattle	TDN	%	*	37.1	43.4	
Sheep	TDN	%		30.5	35.7	

PHACELIA, VARILEAF. Phacelia heterophylla

Phacelia, varileaf, aerial part, fresh, (2)
Ref No 2 03 700 United States

Feed Name or Analyses				Mean As Fed	Mean Dry	C.V. ± %
Dry matter		%			100.0	
Ash		%			17.6	
Crude fiber		%			12.2	
Ether extract		%			2.0	
N-free extract		%			53.1	
Protein (N x 6.25)		%			15.1	
Cattle	dig prot	%	*		10.7	
Goats	dig prot	%	*		10.6	
Horses	dig prot	%	*		10.3	
Rabbits	dig prot	%	*		10.3	
Sheep	dig prot	%	*		11.1	

Phacelia, varileaf, flowers, fresh, (2)
Ref No 2 03 701 United States

Feed Name or Analyses				Mean As Fed	Mean Dry	C.V. ± %
Dry matter		%			100.0	
Ash		%			14.6	
Crude fiber		%			15.3	
Ether extract		%			2.3	
N-free extract		%			55.0	
Protein (N x 6.25)		%			12.8	
Cattle	dig prot	%	*		8.8	
Goats	dig prot	%	*		8.5	
Horses	dig prot	%	*		8.4	

Continued

Feed Name or Analyses				Mean As Fed	Mean Dry	C.V. ± %
Rabbits	dig prot	%	*		8.6	
Sheep	dig prot	%	*		8.9	

PHLOX. Phlox spp

Phlox, aerial part, fresh, (2)
Ref No 2 03 704 — United States

Feed Name or Analyses				Mean As Fed	Mean Dry	C.V. ± %
Dry matter		%			100.0	
Ash		%			24.3	38
Crude fiber		%			22.4	9
Ether extract		%			1.9	20
N-free extract		%			42.9	
Protein (N x 6.25)		%			8.5	31
Cattle	dig prot	%	*		5.1	
Goats	dig prot	%	*		4.5	
Horses	dig prot	%	*		4.7	
Rabbits	dig prot	%	*		5.2	
Sheep	dig prot	%	*		4.9	
Calcium		%			1.60	
Magnesium		%			0.18	
Phosphorus		%			0.21	
Potassium		%			3.17	

Phlox, aerial part, fresh, mid-bloom, (2)
Ref No 2 03 703 — United States

Feed Name or Analyses				Mean As Fed	Mean Dry	C.V. ± %
Dry matter		%			100.0	
Ash		%			10.2	
Crude fiber		%			25.1	
Ether extract		%			2.4	
N-free extract		%			49.8	
Protein (N x 6.25)		%			12.5	
Cattle	dig prot	%	*		8.5	
Goats	dig prot	%	*		8.2	
Horses	dig prot	%	*		8.1	
Rabbits	dig prot	%	*		8.3	
Sheep	dig prot	%	*		8.6	

PHLOX, ROEMER. Phlox roemeriana

Phlox, roemer, aerial part, fresh, full bloom, (2)
Ref No 2 03 706 — United States

Feed Name or Analyses				Mean As Fed	Mean Dry	C.V. ± %
Dry matter		%			100.0	
Calcium		%			1.45	
Magnesium		%			0.18	
Phosphorus		%			0.31	
Potassium		%			3.17	

PHOSPHATE ROCK

Phosphate rock, defluorinated grnd, mx 1 part fluorine per 100 parts phosphorus, (6)
Phosphate, defluorinated (AAFCO)
Defluorinated phosphate (CFA)
Ref No 6 01 780 — United States

Feed Name or Analyses				Mean As Fed	Mean Dry	C.V. ± %
Dry matter		%			100.0	
Ash		%			100.0	
Calcium		%			32.00	
Phosphorus		%			16.25	

CURACAO PHOSPHATE

Phosphate rock, curacao, (6)
Curacao phosphate
Ref No 6 05 586 — United States

Feed Name or Analyses				Mean As Fed	Mean Dry	C.V. ± %
Dry matter		%			100.0	
Ash		%			100.0	
Calcium		%			34.00	
Phosphorus		%			15.00	

PHOSPHORIC ACID

Phosphoric acid, H_3PO_4, cp, (6)
Ref No 6 03 708 — United States

Feed Name or Analyses				Mean As Fed	Mean Dry	C.V. ± %
Dry matter		%			100.0	
Iron		%			0.003	
Phosphorus		%			31.61	
Potassium		%			0.01	
Sodium		%			0.03	
Sulphur		%			0.07	

PIANOGRASS. Themeda arguens

Pianograss, hay, s-c, mature, (1)
Ref No 1 03 711 — United States

Feed Name or Analyses				Mean As Fed	Mean Dry	C.V. ± %
Dry matter		%			100.0	
Ash		%			3.6	
Crude fiber		%			31.2	
Protein (N x 6.25)		%			4.0	
Cattle	dig prot	%	*		0.4	
Goats	dig prot	%	*		0.3	
Horses	dig prot	%	*		0.9	
Rabbits	dig prot	%	*		1.8	
Sheep	dig prot	%	*		0.2	
Calcium		%			0.72	
Phosphorus		%			0.21	
Potassium		%			0.49	

PIGEONPEA. Cajanus cajan

Pigeonpea, aerial part, fresh, (2)
Ref No 2 03 715 — United States

Feed Name or Analyses				Mean As Fed	Mean Dry	C.V. ± %
Dry matter		%		40.0	100.0	15
Ash		%		1.9	4.8	10
Crude fiber		%		15.6	39.2	7
Ether extract		%		1.7	4.2	13
N-free extract		%		15.3	38.3	
Protein (N x 6.25)		%		5.4	13.5	16
Cattle	dig prot	%	*	3.7	9.4	
Goats	dig prot	%	*	3.7	9.1	
Horses	dig prot	%	*	3.6	9.0	
Rabbits	dig prot	%	*	3.6	9.1	
Sheep	dig prot	%	*	3.8	9.6	
Energy		GE Mcal/kg				
Cattle		DE Mcal/kg	*	1.08	2.69	
Sheep		DE Mcal/kg	*	1.04	2.61	
Cattle		ME Mcal/kg	*	0.88	2.21	
Sheep		ME Mcal/kg	*	0.86	2.14	
Cattle		TDN %	*	24.4	61.0	
Sheep		TDN %	*	23.7	59.2	

Pigeonpea, aerial part, fresh, pre-bloom, (2)
Ref No 2 03 713 — United States

Feed Name or Analyses				Mean As Fed	Mean Dry	C.V. ± %
Dry matter		%		25.2	100.0	
Ash		%		1.5	5.8	
Crude fiber		%		7.7	30.7	
Sheep	dig coef	%		38.	38.	
Ether extract		%		1.4	5.7	
Sheep	dig coef	%		61.	61.	
N-free extract		%		9.5	37.7	
Sheep	dig coef	%		69.	69.	
Protein (N x 6.25)		%		5.1	20.1	
Sheep	dig coef	%		62.	62.	
Cattle	dig prot	%	*	3.8	15.0	
Goats	dig prot	%	*	3.9	15.3	
Horses	dig prot	%	*	3.7	14.6	
Rabbits	dig prot	%	*	3.6	14.2	
Sheep	dig prot	%		3.1	12.5	
Energy		GE Mcal/kg				
Cattle		DE Mcal/kg	*	0.67	2.66	
Sheep		DE Mcal/kg	*	0.64	2.56	
Cattle		ME Mcal/kg	*	0.55	2.18	
Sheep		ME Mcal/kg	*	0.53	2.10	
Cattle		TDN %	*	15.2	60.3	
Sheep		TDN %		14.6	58.0	

Pigeonpea, aerial part, fresh, milk stage, (2)
Ref No 2 03 714 — United States

Feed Name or Analyses				Mean As Fed	Mean Dry	C.V. ± %
Dry matter		%		49.7	100.0	
Ash		%		2.8	5.7	
Crude fiber		%		14.8	29.7	
Cattle	dig coef	%		50.	50.	
Ether extract		%		2.6	5.3	
Cattle	dig coef	%		69.	69.	
N-free extract		%		20.1	40.4	
Cattle	dig coef	%		78.	78.	
Protein (N x 6.25)		%		9.4	18.9	
Cattle	dig coef	%		69.	69.	
Cattle	dig prot	%		6.5	13.0	
Goats	dig prot	%	*	7.1	14.2	
Horses	dig prot	%	*	6.7	13.6	
Rabbits	dig prot	%	*	6.6	13.3	
Sheep	dig prot	%	*	7.3	14.6	
Energy		GE Mcal/kg				
Cattle		DE Mcal/kg	*	1.48	2.98	
Sheep		DE Mcal/kg	*	1.38	2.78	
Cattle		ME Mcal/kg	*	1.22	2.45	
Sheep		ME Mcal/kg	*	1.13	2.28	
Cattle		TDN %		33.6	67.6	
Sheep		TDN %	*	31.4	63.1	

Pigeonpea, seeds, (4)
Ref No 5 03 716 — United States

Feed Name or Analyses				Mean As Fed	Mean Dry	C.V. ± %
Dry matter		%		88.8	100.0	
Ash		%		4.5	5.1	
Crude fiber		%		10.5	11.8	
Swine	dig coef	%		86.	86.	
Ether extract		%		3.3	3.7	
Swine	dig coef	%		36.	36.	
N-free extract		%		52.2	58.8	
Swine	dig coef	%		86.	86.	

(1) dry forages and roughages (2) pasture, range plants, and forages fed green (3) silages (4) energy feeds (5) protein supplements (6) minerals (7) vitamins (8) additives

Feed Name or Analyses			Mean As Fed	Mean Dry	C.V. ± %
Protein (N x 6.25)	%		18.3	20.6	
Swine	dig coef %		92.	92.	
Cattle	dig prot %	*	13.3	14.9	
Goats	dig prot %	*	14.3	16.1	
Horses	dig prot %	*	14.3	16.1	
Sheep	dig prot %	*	14.3	16.1	
Swine	dig prot %		16.8	19.0	
Energy	GE Mcal/kg		3.41	3.83	
Cattle	DE Mcal/kg	*	2.34	2.66	
Sheep	DE Mcal/kg	*	2.59	2.94	
Swine	DE kcal/kg	*	3237.	3644.	
Cattle	ME Mcal/kg	*	1.92	2.18	
Sheep	ME Mcal/kg	*	2.12	2.41	
Swine	ME kcal/kg	*	2973.	3347.	
Cattle	TDN %	*	53.2	60.4	
Sheep	TDN %	*	58.7	66.7	
Swine	TDN %		73.4	82.6	
Calcium	%		0.11	0.12	
Iron	%		0.008	0.009	
Phosphorus	%		0.32	0.36	
Potassium	%		0.98	1.10	
Sodium	%		0.03	0.03	
Niacin	mg/kg		29.9	33.6	
Riboflavin	mg/kg		1.6	1.8	
Thiamine	mg/kg		3.2	3.6	
Vitamin A	IU/g		0.8	0.9	

PIGWEED. Axyris spp

Pigweed, seeds (4)

Ref No 4 08 580 United States

Analyses			As Fed	Dry	C.V. ± %
Dry matter	%		90.0	100.0	
Ash	%		3.3	3.7	
Crude fiber	%		15.9	17.7	
Ether extract	%		6.2	6.9	
N-free extract	%		47.8	53.1	
Protein (N x 6.25)	%		16.8	18.7	
Cattle	dig prot %	*	11.8	13.2	
Goats	dig prot %	*	12.9	14.3	
Horses	dig prot %	*	12.9	14.3	
Sheep	dig prot %	*	12.9	14.3	
Energy	GE Mcal/kg				
Cattle	DE Mcal/kg	*	2.61	2.90	
Sheep	DE Mcal/kg	*	2.88	3.20	
Swine	DE kcal/kg	*	2706.	3007.	
Cattle	ME Mcal/kg	*	2.14	2.37	
Sheep	ME Mcal/kg	*	2.36	2.63	
Swine	ME kcal/kg	*	2495.	2772.	
Cattle	TDN %	*	59.1	65.7	
Sheep	TDN %	*	65.4	72.6	
Swine	TDN %	*	61.4	68.2	

PINE. Pinus spp

Pine, needles, collected from ground alcohol extd, (1)

Ref No 1 03 718 United States

Analyses			As Fed	Dry	C.V. ± %
Dry matter	%		94.6	100.0	
Ash	%		8.5	9.0	
Crude fiber	%		42.4	44.8	
Sheep	dig coef %		3.	3.	
Ether extract	%		0.2	0.2	
Sheep	dig coef %		-97.	-97.	
N-free extract	%		38.5	40.7	
Sheep	dig coef %		24.	24.	
Protein (N x 6.25)	%		5.0	5.3	
Sheep	dig coef %		-33.	-33.	
Cattle	dig prot %	*	1.4	1.5	
Goats	dig prot %	*	1.4	1.5	
Horses	dig prot %	*	1.9	2.0	
Rabbits	dig prot %	*	2.6	2.8	
Sheep	dig prot %		-1.6	-1.7	
Energy	GE Mcal/kg				
Sheep	DE Mcal/kg	*	0.37	0.39	
Sheep	ME Mcal/kg	*	0.30	0.32	
Sheep	TDN %		8.4	8.9	

Pine, needles, collected from tree alcohol extd air dried, (1)

Ref No 1 03 719 United States

Analyses			As Fed	Dry	C.V. ± %
Dry matter	%		95.3	100.0	
Ash	%		2.1	2.2	
Crude fiber	%		45.0	47.2	
Sheep	dig coef %		33.	33.	
Ether extract	%		2.4	2.5	
Sheep	dig coef %		18.	18.	
N-free extract	%		36.9	38.7	
Sheep	dig coef %		50.	50.	
Protein (N x 6.25)	%		9.0	9.4	
Sheep	dig coef %		-5.	-5.	
Cattle	dig prot %	*	4.8	5.1	
Goats	dig prot %	*	5.1	5.3	
Horses	dig prot %	*	5.3	5.5	
Rabbits	dig prot %	*	5.6	5.9	
Sheep	dig prot %		-0.4	-0.5	
Energy	GE Mcal/kg				
Cattle	DE Mcal/kg	*	1.24	1.30	
Sheep	DE Mcal/kg	*	1.49	1.56	
Cattle	ME Mcal/kg	*	1.02	1.07	
Sheep	ME Mcal/kg	*	1.22	1.28	
Cattle	TDN %	*	28.1	29.5	
Sheep	TDN %		33.7	35.4	

Pine needles, fresh, (2)

Ref No 2 03 717 United States

Analyses			As Fed	Dry	C.V. ± %
Dry matter	%		94.6	100.0	
Ash	%		10.3	10.9	
Crude fiber	%		39.3	41.5	
Sheep	dig coef %		25.	25.	
Ether extract	%		10.8	11.4	
Sheep	dig coef %		65.	65.	
N-free extract	%		28.9	30.6	
Sheep	dig coef %		18.	18.	
Protein (N x 6.25)	%		5.3	5.6	
Sheep	dig coef %		-14.	-14.	
Cattle	dig prot %	*	2.5	2.7	
Goats	dig prot %	*	1.7	1.8	
Horses	dig prot %	*	2.2	2.3	
Rabbits	dig prot %	*	2.8	3.0	
Sheep	dig prot %		-0.7	-0.7	
Energy	GE Mcal/kg				
Cattle	DE Mcal/kg	*	1.52	1.61	
Sheep	DE Mcal/kg	*	1.32	1.40	
Cattle	ME Mcal/kg	*	1.25	1.32	
Sheep	ME Mcal/kg	*	1.08	1.15	
Cattle	TDN %	*	34.6	36.6	
Sheep	TDN %		30.0	31.7	

PINEAPPLE. Ananas comosus

Pineapple, tops, fresh, (2)

Ref No 2 03 720 United States

Analyses			As Fed	Dry	C.V. ± %
Dry matter	%		16.0	100.0	
Ash	%		1.5	9.4	
Crude fiber	%		3.7	23.1	
Ether extract	%		0.5	3.1	
N-free extract	%		8.7	54.4	
Protein (N x 6.25)	%		1.6	10.0	
Cattle	dig prot %	*	1.0	6.4	
Goats	dig prot %	*	0.9	5.9	
Horses	dig prot %	*	1.0	6.0	
Rabbits	dig prot %	*	1.0	6.4	
Sheep	dig prot %	*	1.0	6.3	
Fatty acids	%		0.5	3.1	
Energy	GE Mcal/kg				
Cattle	DE Mcal/kg	*	0.40	1.89	
Sheep	DE Mcal/kg	*	0.37	1.73	
Cattle	ME Mcal/kg	*	0.33	1.55	
Sheep	ME Mcal/kg	*	0.30	1.42	
Cattle	TDN %	*	9.1	42.8	
Sheep	TDN %	*	8.4	39.3	

Pineapple, tops, ensiled, (3)

Ref No 3 08 488 United States

Analyses			As Fed	Dry	C.V. ± %
Dry matter	%		21.3	100.0	
Ash	%		1.5	7.0	
Crude fiber	%		4.8	22.5	
Ether extract	%		0.6	2.8	
N-free extract	%		12.8	60.1	
Protein (N x 6.25)	%		1.6	7.5	
Cattle	dig prot %	*	0.6	3.1	
Goats	dig prot %	*	0.6	3.1	
Horses	dig prot %	*	0.6	3.1	
Sheep	dig prot %	*	0.6	3.1	
Energy	GE Mcal/kg				
Cattle	DE Mcal/kg	*	0.66	3.12	
Sheep	DE Mcal/kg	*	0.63	2.96	
Cattle	ME Mcal/kg	*	0.54	2.56	
Sheep	ME Mcal/kg	*	0.52	2.42	
Cattle	TDN %	*	15.1	70.7	
Sheep	TDN %	*	14.3	67.0	

Pineapple, cannery residue, dehy, (4)

Pineapple bran dried

Pineapple pulp dried

Ref No 4 03 722 United States

Analyses			As Fed	Dry	C.V. ± %
Dry matter	%		87.1	100.0	3
Ash	%		3.0	3.4	14
Crude fiber	%		17.0	19.6	12
Cattle	dig coef %		79.	79.	
Sheep	dig coef %		70.	70.	
Ether extract	%		1.4	1.6	29
Cattle	dig coef %		53.	53.	
Sheep	dig coef %		-6.	-6.	
N-free extract	%		61.7	70.8	3
Cattle	dig coef %		79.	79.	
Sheep	dig coef %		80.	80.	
Protein (N x 6.25)	%		4.0	4.6	3
Cattle	dig coef %		15.	15.	
Sheep	dig coef %		21.	21.	
Cattle	dig prot %		0.6	0.7	

Continued

Feed Name or Analyses			Mean As Fed	Mean Dry	C.V. ± %
Goats	dig prot %	*	1.2	1.4	
Horses	dig prot %	*	1.2	1.4	
Sheep	dig prot %		0.8	1.0	
Energy	GE Mcal/kg				
Cattle	DE Mcal/kg	*	2.84	3.26	
Sheep	DE Mcal/kg	*	2.73	3.13	
Cattle	ME Mcal/kg	*	2.33	2.68	
Sheep	ME Mcal/kg	*	2.24	2.57	
Cattle	TDN %		64.4	74.0	
Sheep	TDN %		61.9	71.1	
Calcium	%		0.21	0.24	21
Iron	%		0.049	0.056	98
Phosphorus	%		0.10	0.12	26

Pineapple, cannery residue w molasses added, dehy, (4)
Pineapple bran w molasses, dried
Pineapple pulp w molasses, dried
Ref No 4 08 489 United States

Feed Name or Analyses			Mean As Fed	Mean Dry	C.V. ± %
Dry matter	%		87.4	100.0	
Ash	%		3.2	3.7	
Crude fiber	%		15.9	18.2	
Ether extract	%		1.0	1.1	
N-free extract	%		63.4	72.5	
Protein (N x 6.25)	%		3.9	4.5	
Cattle	dig prot %	*	0.1	0.1	
Goats	dig prot %	*	1.2	1.3	
Horses	dig prot %	*	1.2	1.3	
Sheep	dig prot %	*	1.2	1.3	
Energy	GE Mcal/kg				
Cattle	DE Mcal/kg	*	2.69	3.08	
Sheep	DE Mcal/kg	*	2.94	3.36	
Cattle	ME Mcal/kg	*	2.21	2.53	
Sheep	ME Mcal/kg	*	2.41	2.76	
Cattle	TDN %	*	61.0	69.8	
Sheep	TDN %	*	66.6	76.2	

Pineapple, fruit, fresh, (4)
Ref No 2 08 122 United States

Feed Name or Analyses			Mean As Fed	Mean Dry	C.V. ± %
Dry matter	%		14.7	100.0	
Ash	%		0.4	2.7	
Crude fiber	%		0.4	2.7	
Ether extract	%		0.2	1.4	
N-free extract	%		13.3	90.5	
Protein (N x 6.25)	%		0.4	2.7	
Cattle	dig prot %	*	-0.1	-1.4	
Goats	dig prot %	*	0.0	-0.2	
Horses	dig prot %	*	0.0	-0.2	
Sheep	dig prot %	*	0.0	-0.2	
Energy	GE Mcal/kg		0.52	3.54	
Cattle	DE Mcal/kg	*	0.52	3.53	
Sheep	DE Mcal/kg	*	0.51	3.50	
Swine	DE kcal/kg	*	537.	3655.	
Cattle	ME Mcal/kg	*	0.42	2.89	
Sheep	ME Mcal/kg	*	0.42	2.87	
Swine	ME kcal/kg	*	513.	3491.	
Cattle	TDN %	*	11.8	80.1	
Sheep	TDN %	*	11.7	79.4	
Swine	TDN %	*	12.2	82.8	
Calcium	%		0.02	0.14	
Iron	%		0.001	0.007	
Phosphorus	%		0.01	0.07	

(1) dry forages and roughages (2) pasture, range plants, and forages fed green (3) silages (4) energy feeds (5) protein supplements (6) minerals (7) vitamins (8) additives

Feed Name or Analyses			Mean As Fed	Mean Dry	C.V. ± %
Potassium	%		0.15	1.02	
Sodium	%		0.00	0.00	
Ascorbic acid	mg/kg		170.0	1156.5	
Niacin	mg/kg		2.0	13.6	
Riboflavin	mg/kg		0.3	2.0	
Thiamine	mg/kg		0.9	6.1	
Vitamin A	IU/g		0.7	4.8	

Pineapple bran dried –
see Pineapple, cannery residue, dehy, (4)

Pineapple bran w molasses, dried –
see Pineapple, cannery residue w molasses added, dehy, (4)

Pineapple pulp dried –
see Pineapple, cannery residue, dehy, (4)

Pineapple pulp w molasses, dried –
see Pineapple, cannery residue w molasses added, dehy, (4)

PLANT. Scientific name not used

Plant, charcoal, (6)
Charcoal, vegetable
Ref No 6 03 727 United States

Feed Name or Analyses			Mean As Fed	Mean Dry	C.V. ± %
Dry matter	%		90.0	100.0	
Ash	%		9.6	10.7	18
Calcium	%		4.70	5.22	1
Cobalt	mg/kg		0.441	0.490	98
Copper	mg/kg		0.2	0.2	
Iron	%		0.041	0.046	6
Manganese	mg/kg		0.2	0.2	80
Phosphorus	%		0.03	0.03	26

PLANTAIN. Plantago spp

Plantain, aerial part, fresh, (2)
Ref No 2 03 729 United States

Feed Name or Analyses			Mean As Fed	Mean Dry	C.V. ± %
Dry matter	%			100.0	
Ash	%			9.8	46
Crude fiber	%			15.9	42
Ether extract	%			1.9	41
N-free extract	%			59.1	
Protein (N x 6.25)	%			13.3	33
Cattle	dig prot %	*		9.2	
Goats	dig prot %	*		9.0	
Horses	dig prot %	*		8.8	
Rabbits	dig prot %	*		8.9	
Sheep	dig prot %	*		9.4	
Calcium	%			4.16	69
Magnesium	%			0.23	13
Phosphorus	%			0.19	26
Potassium	%			2.35	39

Plantain, leaves, fresh, (2)
Ref No 2 03 730 United States

Feed Name or Analyses			Mean As Fed	Mean Dry	C.V. ± %
Dry matter	%			100.0	
Carotene	mg/kg			100.1	
Vitamin A equivalent	IU/g			166.8	

PLANTAIN, BLOND. Plantago ovata

Plantain, blond, seeds w hulls, (4)
Ref No 4 09 130 United States

Feed Name or Analyses			Mean As Fed	Mean Dry	C.V. ± %
Dry matter	%			100.0	
Protein (N x 6.25)	%			16.0	
Alanine	%			0.86	
Arginine	%			1.17	
Aspartic acid	%			1.30	
Glutamic acid	%			3.81	
Glycine	%			1.12	
Histidine	%			0.37	
Hydroxyproline	%			0.10	
Isoleucine	%			0.66	
Leucine	%			1.01	
Lysine	%			0.69	
Methionine	%			0.30	
Phenylalanine	%			0.62	
Proline	%			0.61	
Serine	%			0.67	
Threonine	%			0.61	
Tyrosine	%			0.45	
Valine	%			0.75	

PLANTAIN, REDSEED. Plantago rhodosperma

Plantain, redseed, aerial part, fresh, (2)
Ref No 2 03 733 United States

Feed Name or Analyses			Mean As Fed	Mean Dry	C.V. ± %
Dry matter	%			100.0	
Ash	%			12.9	
Crude fiber	%			13.1	
Ether extract	%			1.5	
N-free extract	%			58.0	
Protein (N x 6.25)	%			14.5	
Cattle	dig prot %	*		10.2	
Goats	dig prot %	*		10.1	
Horses	dig prot %	*		9.8	
Rabbits	dig prot %	*		9.9	
Sheep	dig prot %	*		10.5	
Calcium	%			5.24	
Magnesium	%			0.24	
Phosphorus	%			0.19	
Potassium	%			2.70	

Plantain, redseed, aerial part, fresh, immature, (2)
Ref No 2 03 731 United States

Feed Name or Analyses			Mean As Fed	Mean Dry	C.V. ± %
Dry matter	%			100.0	
Ash	%			9.1	
Crude fiber	%			9.3	
Ether extract	%			1.1	
N-free extract	%			60.7	
Protein (N x 6.25)	%			19.8	
Cattle	dig prot %	*		14.7	
Goats	dig prot %	*		15.0	
Horses	dig prot %	*		14.3	
Rabbits	dig prot %	*		14.0	
Sheep	dig prot %	*		15.4	
Calcium	%			6.84	
Magnesium	%			0.25	
Phosphorus	%			0.24	
Potassium	%			3.00	

Feed Name or Analyses				Mean As Fed	Mean Dry	C.V. ± %

Plantain, redseed, aerial part, fresh, mature, (2)
Ref No 2 03 732 United States

Feed Name or Analyses				As Fed	Dry	C.V. ± %
Dry matter		%			100.0	
Ash		%			13.8	
Crude fiber		%			21.7	
Ether extract		%			1.3	
N-free extract		%			53.6	
Protein (N x 6.25)		%			9.6	
Cattle	dig prot	%	*		6.1	
Goats	dig prot	%	*		5.5	
Horses	dig prot	%	*		5.7	
Rabbits	dig prot	%	*		6.1	
Sheep	dig prot	%	*		5.9	
Calcium		%			3.79	
Magnesium		%			0.24	
Phosphorus		%			0.18	
Potassium		%			1.90	

PLUCHEA, ARROWWEED. Pluchea sericea

Pluchea, arrowweed, aerial part, fresh, (2)
Ref No 2 03 734 United States

Feed Name or Analyses				As Fed	Dry	C.V. ± %
Dry matter		%		26.0	100.0	
Calcium		%		0.08	0.30	
Chlorine		%		0.11	0.42	
Iron		%		0.002	0.008	
Magnesium		%		0.10	0.37	
Manganese		mg/kg		20.2	77.6	
Phosphorus		%		0.01	0.03	
Potassium		%		0.07	0.27	
Sodium		%		0.15	0.56	
Sulphur		%		0.26	1.00	

PLUMEGRASS. Erianthus spp

Plumegrass, aerial part, fresh, mature, (2)
Ref No 2 03 735 United States

Feed Name or Analyses				As Fed	Dry	C.V. ± %
Dry matter		%			100.0	
Ash		%			1.9	
Crude fiber		%			48.9	
Ether extract		%			3.5	
N-free extract		%			39.6	
Protein (N x 6.25)		%			6.1	
Cattle	dig prot	%	*		3.1	
Goats	dig prot	%	*		2.3	
Horses	dig prot	%	*		2.7	
Rabbits	dig prot	%	*		3.4	
Sheep	dig prot	%	*		2.7	
Calcium		%			0.61	
Cobalt		mg/kg			0.020	
Copper		mg/kg			36.6	
Manganese		mg/kg			19.6	
Phosphorus		%			0.10	
Carotene		mg/kg			5.3	
Vitamin A equivalent		IU/g			8.8	

POKEBERRY. Phytolacca spp

Pokeberry, aerial part, fresh, (2)
Ref No 2 03 737 United States

Feed Name or Analyses				As Fed	Dry	C.V. ± %
Dry matter		%		14.2	100.0	
Ash		%		0.5	3.4	
Crude fiber		%		1.9	13.4	
Ether extract		%		1.3	9.3	
N-free extract		%		8.4	59.2	
Protein (N x 6.25)		%		2.1	14.7	
Cattle	dig prot	%	*	1.5	10.4	
Goats	dig prot	%	*	1.5	10.3	
Horses	dig prot	%	*	1.4	10.0	
Rabbits	dig prot	%	*	1.4	10.0	
Sheep	dig prot	%	*	1.5	10.7	
Niacin		mg/kg		11.9	83.8	
Riboflavin		mg/kg		3.3	23.1	
Thiamine		mg/kg		0.8	5.7	

POLEMONIUM, WHITE. Polemonium albiflorum

Polemonium, white, aerial part, fresh, (2)
Ref No 2 03 740 United States

Feed Name or Analyses				As Fed	Dry	C.V. ± %
Dry matter		%			100.0	
Ash		%			11.7	
Ether extract		%			2.3	
Protein (N x 6.25)		%			10.6	
Cattle	dig prot	%	*		6.9	
Goats	dig prot	%	*		6.4	
Horses	dig prot	%	*		6.5	
Rabbits	dig prot	%	*		6.9	
Sheep	dig prot	%	*		6.9	
Calcium		%			1.35	
Phosphorus		%			0.43	

Polemonium, white, aerial part, fresh, immature, (2)
Ref No 2 03 738 United States

Feed Name or Analyses				As Fed	Dry	C.V. ± %
Dry matter		%			100.0	
Ash		%			14.0	
Ether extract		%			3.9	
Protein (N x 6.25)		%			16.4	
Cattle	dig prot	%	*		11.8	
Goats	dig prot	%	*		11.9	
Horses	dig prot	%	*		11.5	
Rabbits	dig prot	%	*		11.3	
Sheep	dig prot	%	*		12.3	
Calcium		%			1.96	
Phosphorus		%			0.50	

Polemonium, white, aerial part, fresh, mature, (2)
Ref No 2 03 739 United States

Feed Name or Analyses				As Fed	Dry	C.V. ± %
Dry matter		%			100.0	
Ash		%			9.3	
Ether extract		%			0.8	
Protein (N x 6.25)		%			4.6	
Cattle	dig prot	%	*		1.8	
Goats	dig prot	%	*		0.9	
Horses	dig prot	%	*		1.4	
Rabbits	dig prot	%	*		2.2	
Sheep	dig prot	%	*		1.3	
Calcium		%			0.75	
Phosphorus		%			0.35	

POLYPOGON, RABBITFOOT. Polypogon monspeliensis

Polypogon, rabbitfoot, aerial part, fresh, immature, (2)
Ref No 2 03 741 United States

Feed Name or Analyses				As Fed	Dry	C.V. ± %
Dry matter		%			100.0	
Ash		%			14.6	
Crude fiber		%			28.3	
Ether extract		%			2.7	
N-free extract		%			40.8	
Protein (N x 6.25)		%			13.6	
Cattle	dig prot	%	*		9.5	
Goats	dig prot	%	*		9.2	
Horses	dig prot	%	*		9.1	
Rabbits	dig prot	%	*		9.2	
Sheep	dig prot	%	*		9.7	
Calcium		%			0.34	
Phosphorus		%			0.33	

POPCORNFLOWER, REDSTAIN. Plagiobothrys Nothofulvus

Popcornflower, redstain, aerial part, fresh, (2)
Ref No 2 03 748 United States

Feed Name or Analyses				As Fed	Dry	C.V. ± %
Dry matter		%			100.0	
Ash		%			14.7	20
Crude fiber		%			22.0	27
Protein (N x 6.25)		%			14.2	56
Cattle	dig prot	%	*		10.0	
Goats	dig prot	%	*		9.8	
Horses	dig prot	%	*		9.6	
Rabbits	dig prot	%	*		9.6	
Sheep	dig prot	%	*		10.2	
Calcium		%			1.94	21
Phosphorus		%			0.52	35
Potassium		%			3.94	43

Popcornflower, redstain, aerial part, fresh, immature, (2)
Ref No 2 03 742 United States

Feed Name or Analyses				As Fed	Dry	C.V. ± %
Dry matter		%			100.0	
Ash		%			18.0	13
Crude fiber		%			11.3	
Protein (N x 6.25)		%			26.6	12
Cattle	dig prot	%	*		20.5	
Goats	dig prot	%	*		21.4	
Horses	dig prot	%	*		20.1	
Rabbits	dig prot	%	*		19.2	
Sheep	dig prot	%	*		21.8	
Calcium		%			2.17	12
Phosphorus		%			0.77	8
Potassium		%			6.26	15

Popcornflower, redstain, aerial part, fresh, early bloom, (2)
Ref No 2 03 743 United States

Feed Name or Analyses				As Fed	Dry	C.V. ± %
Dry matter		%			100.0	
Ash		%			15.4	
Crude fiber		%			14.6	
Protein (N x 6.25)		%			20.4	
Cattle	dig prot	%	*		15.2	
Goats	dig prot	%	*		15.6	

Continued

Feed Name or Analyses			Mean As Fed	Mean Dry	C.V. ± %
Horses	dig prot	% *		14.8	
Rabbits	dig prot	% *		14.4	
Sheep	dig prot	% *		16.0	
Calcium		%		2.07	
Phosphorus		%		0.57	
Potassium		%		4.79	

Popcornflower, redstain, aerial part, fresh, full bloom, (2)
Ref No 2 03 744 United States

Feed Name or Analyses			Mean As Fed	Mean Dry	C.V. ± %
Dry matter		%		100.0	
Ash		%		13.7	5
Crude fiber		%		20.6	14
Protein (N x 6.25)		%		12.0	8
Cattle	dig prot	% *		8.1	
Goats	dig prot	% *		7.8	
Horses	dig prot	% *		7.7	
Rabbits	dig prot	% *		7.9	
Sheep	dig prot	% *		8.2	
Calcium		%		1.98	21
Phosphorus		%		0.51	24
Potassium		%		3.47	29

Popcornflower, redstain, aerial part, fresh, milk stage, (2)
Ref No 2 03 745 United States

Feed Name or Analyses			Mean As Fed	Mean Dry	C.V. ± %
Dry matter		%		100.0	
Ash		%		12.8	
Crude fiber		%		24.6	
Protein (N x 6.25)		%		9.6	
Cattle	dig prot	% *		6.1	
Goats	dig prot	% *		5.5	
Horses	dig prot	% *		5.7	
Rabbits	dig prot	% *		6.1	
Sheep	dig prot	% *		5.9	
Calcium		%		1.82	
Phosphorus		%		0.49	
Potassium		%		2.83	

Popcornflower, redstain, aerial part, fresh, mature, (2)
Ref No 2 03 746 United States

Feed Name or Analyses			Mean As Fed	Mean Dry	C.V. ± %
Dry matter		%		100.0	
Ash		%		13.5	
Crude fiber		%		28.5	
Protein (N x 6.25)		%		5.2	
Cattle	dig prot	% *		2.3	
Goats	dig prot	% *		1.4	
Horses	dig prot	% *		1.9	
Rabbits	dig prot	% *		2.7	
Sheep	dig prot	% *		1.8	
Calcium		%		2.12	
Phosphorus		%		0.23	
Potassium		%		2.44	

(1) dry forages and roughages
(2) pasture, range plants, and forages fed green
(3) silages
(4) energy feeds
(5) protein supplements
(6) minerals
(7) vitamins
(8) additives

Popcornflower, redstain, aerial part, fresh, dormant, (2)
Ref No 2 03 747 United States

Feed Name or Analyses			Mean As Fed	Mean Dry	C.V. ± %
Dry matter		%		100.0	
Ash		%		12.2	
Crude fiber		%		30.5	
Protein (N x 6.25)		%		2.4	
Cattle	dig prot	% *		0.0	
Goats	dig prot	% *		-1.1	
Horses	dig prot	% *		-0.3	
Rabbits	dig prot	% *		0.5	
Sheep	dig prot	% *		-0.7	
Calcium		%		1.36	
Phosphorus		%		0.29	
Potassium		%		2.02	

POPLAR. Populus spp

Poplar, twigs, s-c, (1)
Ref No 1 03 749 United States

Feed Name or Analyses			Mean As Fed	Mean Dry	C.V. ± %
Dry matter		%	76.4	100.0	
Ash		%	3.0	3.9	
Crude fiber		%	30.3	39.7	
Sheep	dig coef	%	28.	28.	
Ether extract		%	2.6	3.4	
Sheep	dig coef	%	39.	39.	
N-free extract		%	34.5	45.2	
Sheep	dig coef	%	51.	51.	
Protein (N x 6.25)		%	6.0	7.8	
Sheep	dig coef	%	39.	39.	
Cattle	dig prot	% *	2.8	3.7	
Goats	dig prot	% *	2.9	3.8	
Horses	dig prot	% *	3.2	4.2	
Rabbits	dig prot	% *	3.6	4.7	
Sheep	dig prot	%	2.3	3.0	
Energy	GE Mcal/kg				
Cattle	DE Mcal/kg	*	1.51	1.98	
Sheep	DE Mcal/kg	*	1.35	1.77	
Cattle	ME Mcal/kg	*	1.24	1.62	
Sheep	ME Mcal/kg	*	1.11	1.45	
Cattle	TDN	% *	34.3	44.8	
Sheep	TDN	%	30.7	40.2	

POPLAR, BLACK. Populus nigra

Poplar, black, leaves, s-c, (1)
Ref No 1 03 750 United States

Feed Name or Analyses			Mean As Fed	Mean Dry	C.V. ± %
Dry matter		%	85.4	100.0	
Ash		%	5.9	6.9	
Crude fiber		%	20.1	23.5	
Sheep	dig coef	%	32.	32.	
Ether extract		%	2.7	3.2	
Sheep	dig coef	%	65.	65.	
N-free extract		%	40.7	47.6	
Sheep	dig coef	%	72.	72.	
Protein (N x 6.25)		%	16.1	18.8	
Sheep	dig coef	%	80.	80.	
Cattle	dig prot	% *	11.3	13.2	
Goats	dig prot	% *	12.0	14.1	
Horses	dig prot	% *	11.5	13.5	
Rabbits	dig prot	% *	11.3	13.2	
Sheep	dig prot	%	12.8	15.0	
Energy	GE Mcal/kg				
Cattle	DE Mcal/kg	*	2.31	2.70	

Feed Name or Analyses			Mean As Fed	Mean Dry	C.V. ± %
Sheep	DE Mcal/kg	*	2.32	2.71	
Cattle	ME Mcal/kg	*	1.89	2.22	
Sheep	ME Mcal/kg	*	1.90	2.22	
Cattle	TDN	% *	52.4	61.3	
Sheep	TDN	%	52.5	61.5	

POPPY. Papaver spp

Poppy, seeds, extn unspecified grnd, (5)
Ref No 5 03 751 United States

Feed Name or Analyses			Mean As Fed	Mean Dry	C.V. ± %
Dry matter		%	89.2	100.0	
Ash		%	12.4	13.9	
Crude fiber		%	11.6	13.0	
Cattle	dig coef	%	61.	61.	
Sheep	dig coef	%	28.	28.	
Ether extract		%	7.9	8.8	
Cattle	dig coef	%	92.	92.	
Sheep	dig coef	%	94.	94.	
N-free extract		%	20.7	23.2	
Cattle	dig coef	%	64.	64.	
Sheep	dig coef	%	52.	52.	
Protein (N x 6.25)		%	36.6	41.0	
Cattle	dig coef	%	79.	79.	
Sheep	dig coef	%	84.	84.	
Cattle	dig prot	%	28.9	32.4	
Sheep	dig prot	%	30.7	34.4	
Energy	GE Mcal/kg				
Cattle	DE Mcal/kg	*	2.89	3.24	
Sheep	DE Mcal/kg	*	2.71	3.03	
Swine	DE kcal/kg	*	3147.	3529.	
Cattle	ME Mcal/kg	*	2.37	2.66	
Sheep	ME Mcal/kg	*	2.22	2.49	
Swine	ME kcal/kg	*	2761.	3095.	
Cattle	TDN	%	65.5	73.5	
Sheep	TDN	%	61.4	68.8	
Swine	TDN	% *	71.4	80.0	

POPPY, CORN. Papaver rhoeas

Poppy, corn, seeds w hulls, (5)
Ref No 5 09 052 United States

Feed Name or Analyses			Mean As Fed	Mean Dry	C.V. ± %
Dry matter		%		100.0	
Protein (N x 6.25)		%		22.0	
Alanine		%		0.99	
Arginine		%		2.13	
Aspartic acid		%		2.11	
Glutamic acid		%		4.53	
Glycine		%		1.03	
Histidine		%		0.53	
Isoleucine		%		0.88	
Leucine		%		1.39	
Lysine		%		0.92	
Methionine		%		0.51	
Phenylalanine		%		0.86	
Proline		%		0.86	
Serine		%		1.01	
Threonine		%		0.81	
Tyrosine		%		0.79	
Valine		%		1.14	

POPPYMALLOW. Callirrhoe pedata

Poppymallow, aerial part, fresh, (2)

Ref No 2 03 752 — United States

Feed Name or Analyses			Mean As Fed	Mean Dry	C.V. ± %
Dry matter	%			100.0	
Ash	%			9.3	
Crude fiber	%			30.2	
Ether extract	%			2.7	
N-free extract	%			47.4	
Protein (N x 6.25)	%			10.4	
Cattle	dig prot %	*		6.7	
Goats	dig prot %	*		6.3	
Horses	dig prot %	*		6.4	
Rabbits	dig prot %	*		6.7	
Sheep	dig prot %	*		6.7	
Energy	GE Mcal/kg				
Cattle	DE Mcal/kg	*		2.82	
Sheep	DE Mcal/kg	*		2.88	
Cattle	ME Mcal/kg	*		2.31	
Sheep	ME Mcal/kg	*		2.36	
Cattle	TDN %	*		63.9	
Sheep	TDN %	*		65.3	
Calcium	%			2.46	
Magnesium	%			0.25	
Phosphorus	%			0.45	
Potassium	%			2.23	

PORCUPINEGRASS. Stipa spartea

Porcupinegrass, aerial part, fresh, full bloom, (2)

Ref No 2 03 753 — United States

Feed Name or Analyses			Mean As Fed	Mean Dry	C.V. ± %
Dry matter	%			100.0	
Ash	%			8.5	
Crude fiber	%			41.8	
Ether extract	%			0.9	
N-free extract	%			43.2	
Protein (N x 6.25)	%			5.6	
Cattle	dig prot %	*		2.7	
Goats	dig prot %	*		1.8	
Horses	dig prot %	*		2.3	
Rabbits	dig prot %	*		3.0	
Sheep	dig prot %	*		2.2	
Energy	GE Mcal/kg				
Cattle	DE Mcal/kg	*		2.85	
Sheep	DE Mcal/kg	*		2.83	
Cattle	ME Mcal/kg	*		2.34	
Sheep	ME Mcal/kg	*		2.32	
Cattle	TDN %	*		64.6	
Sheep	TDN %	*		64.2	

Pork cracklings (CFA) —
see Swine, cracklings, dry rendered dehy, (5)

PORTULACA. Portulaca spp

Portulaca, aerial part, fresh, (2)

Purslane

Ref No 2 08 490 — United States

Feed Name or Analyses			Mean As Fed	Mean Dry	C.V. ± %
Dry matter	%		10.3	100.0	
Ash	%		1.9	18.4	
Crude fiber	%		1.5	14.6	
Ether extract	%		0.3	2.9	
N-free extract	%		4.4	42.7	
Protein (N x 6.25)	%		2.2	21.4	
Cattle	dig prot %	*	1.7	16.0	
Goats	dig prot %	*	1.7	16.5	
Horses	dig prot %	*	1.6	15.7	
Rabbits	dig prot %	*	1.6	15.2	
Sheep	dig prot %	*	1.7	16.9	
Energy	GE Mcal/kg				
Cattle	DE Mcal/kg	*	0.29	2.79	
Sheep	DE Mcal/kg	*	0.31	2.98	
Cattle	ME Mcal/kg	*	0.24	2.29	
Sheep	ME Mcal/kg	*	0.25	2.44	
Cattle	TDN %	*	6.5	63.2	
Sheep	TDN %	*	7.0	67.5	
Phosphorus	%		0.04	0.39	
Potassium	%		0.94	9.13	

Portulaca, aerial part, fresh, mid-bloom, (2)

Ref No 2 03 754 — United States

Feed Name or Analyses			Mean As Fed	Mean Dry	C.V. ± %
Dry matter	%			100.0	
Ash	%			25.2	
Crude fiber	%			17.9	
Ether extract	%			5.0	
N-free extract	%			42.0	
Protein (N x 6.25)	%			9.9	
Cattle	dig prot %	*		6.3	
Goats	dig prot %	*		5.8	
Horses	dig prot %	*		5.9	
Rabbits	dig prot %	*		6.3	
Sheep	dig prot %	*		6.2	
Energy	GE Mcal/kg				
Cattle	DE Mcal/kg	*		2.75	
Sheep	DE Mcal/kg	*		2.93	
Cattle	ME Mcal/kg	*		2.26	
Sheep	ME Mcal/kg	*		2.40	
Cattle	TDN %	*		62.4	
Sheep	TDN %	*		66.5	
Calcium	%			2.84	
Magnesium	%			0.85	
Phosphorus	%			0.32	
Potassium	%			5.39	

POTASSIUM CHLORIDE

Potassium chloride, KCl, cp, (6)

Ref No 6 03 756 — United States

Feed Name or Analyses			Mean As Fed	Mean Dry	C.V. ± %
Dry matter	%			100.0	
Chlorine	%			47.55	
Iodine	%			0.0	
Iron	%			0.000	
Potassium	%			52.44	
Sodium	%			0.01	

POTASSIUM IODATE

Potassium iodate, KIO_3, cp, (6)

Ref No 6 03 758 — United States

Feed Name or Analyses			Mean As Fed	Mean Dry	C.V. ± %
Dry matter	%			100.0	
Chlorine	%			0.01	
Iodine	%			59.3	
Iron	%			0.001	
Potassium	%			18.27	
Sodium	%			0.01	

POTASSIUM IODIDE

Potassium iodide, KI, cp, (6)

Ref No 6 03 760 — United States

Feed Name or Analyses			Mean As Fed	Mean Dry	C.V. ± %
Dry matter	%			100.0	
Chlorine	%			0.01	
Iodine	%			76.5	
Iron	%			0.000	
Potassium	%			23.56	
Sodium	%			0.01	

POTATO. Solanum tuberosum

Potato, aerial part, s-c, (1)

Ref No 1 03 762 — United States

Feed Name or Analyses			Mean As Fed	Mean Dry	C.V. ± %
Dry matter	%		87.2	100.0	
Ash	%		15.8	18.1	
Crude fiber	%		22.6	25.9	
Sheep	dig coef %		61.	61.	
Ether extract	%		2.4	2.8	
Sheep	dig coef %		54.	54.	
N-free extract	%		35.6	40.8	
Sheep	dig coef %		64.	64.	
Protein (N x 6.25)	%		10.8	12.4	
Sheep	dig coef %		52.	52.	
Cattle	dig prot %	*	6.7	7.7	
Goats	dig prot %	*	7.1	8.1	
Horses	dig prot %	*	7.0	8.1	
Rabbits	dig prot %	*	7.2	8.2	
Sheep	dig prot %		5.6	6.4	
Energy	GE Mcal/kg				
Cattle	DE Mcal/kg	*	2.04	2.34	
Sheep	DE Mcal/kg	*	1.99	2.28	
Cattle	ME Mcal/kg	*	1.67	1.92	
Sheep	ME Mcal/kg	*	1.63	1.87	
Cattle	TDN %	*	46.3	53.1	
Sheep	TDN %		45.1	51.8	

Potato, aerial part w seedballs, s-c, (1)

Ref No 1 03 763 — United States

Feed Name or Analyses			Mean As Fed	Mean Dry	C.V. ± %
Dry matter	%		82.6	100.0	
Ash	%		10.2	12.4	
Crude fiber	%		26.0	31.5	
Sheep	dig coef %		60.	60.	
Ether extract	%		4.3	5.2	
Sheep	dig coef %		47.	47.	
N-free extract	%		30.6	37.1	
Sheep	dig coef %		49.	49.	
Protein (N x 6.25)	%		11.4	13.8	
Sheep	dig coef %		50.	50.	
Cattle	dig prot %	*	7.3	8.9	
Goats	dig prot %	*	7.8	9.4	
Horses	dig prot %	*	7.6	9.2	
Rabbits	dig prot %	*	7.7	9.3	
Sheep	dig prot %		5.7	6.9	
Energy	GE Mcal/kg				
Cattle	DE Mcal/kg	*	1.92	2.32	
Sheep	DE Mcal/kg	*	1.80	2.18	
Cattle	ME Mcal/kg	*	1.57	1.90	
Sheep	ME Mcal/kg	*	1.48	1.79	
Cattle	TDN %	*	43.4	52.6	
Sheep	TDN %		40.9	49.5	

Feed Name or Analyses			Mean As Fed	Mean Dry	C.V. ± %

Potato, aerial part, ensiled, (3)

Ref No 3 03 765 United States

Feed Name or Analyses			As Fed	Dry	C.V. ± %
Dry matter	%		14.8	100.0	3
Ash	%		2.8	19.1	
Crude fiber	%		3.4	23.0	
Sheep	dig coef %		53.	53.	
Ether extract	%		0.5	3.7	
Sheep	dig coef %		63.	63.	
N-free extract	%		5.7	38.6	
Sheep	dig coef %		74.	74.	
Protein (N x 6.25)	%		2.3	15.6	
Sheep	dig coef %		72.	72.	
Cattle	dig prot %	*	1.5	10.4	
Goats	dig prot %	*	1.5	10.4	
Horses	dig prot %	*	1.5	10.4	
Sheep	dig prot %		1.7	11.2	
Energy	GE Mcal/kg				
Cattle	DE Mcal/kg	*	0.38	2.58	
Sheep	DE Mcal/kg	*	0.37	2.52	
Cattle	ME Mcal/kg	*	0.31	2.12	
Sheep	ME Mcal/kg	*	0.31	2.07	
Cattle	TDN %	*	8.7	58.5	
Sheep	TDN %		8.5	57.2	
Calcium	%		0.31	2.12	5
Chlorine	%		0.06	0.38	10
Magnesium	%		0.02	0.14	25
Phosphorus	%		0.03	0.20	15
Potassium	%		0.59	3.95	11
Sulphur	%		0.05	0.37	7

Potato, aerial part w sugar added, ensiled, (3)

Ref No 3 03 766 United States

Feed Name or Analyses			As Fed	Dry	C.V. ± %
Dry matter	%		17.3	100.0	
Ash	%		5.1	29.4	
Crude fiber	%		4.5	26.0	
Sheep	dig coef %		44.	44.	
Ether extract	%		0.5	2.8	
Sheep	dig coef %		58.	58.	
N-free extract	%		5.1	29.2	
Sheep	dig coef %		52.	52.	
Protein (N x 6.25)	%		2.2	12.6	
Sheep	dig coef %		48.	48.	
Cattle	dig prot %	*	1.3	7.7	
Goats	dig prot %	*	1.3	7.7	
Horses	dig prot %	*	1.3	7.7	
Sheep	dig prot %		1.0	6.0	
Energy	GE Mcal/kg				
Sheep	DE Mcal/kg	*	0.28	1.60	
Sheep	ME Mcal/kg	*	0.23	1.31	
Sheep	TDN %		6.3	36.3	

Potato, tubers, cooked ensiled stored at 34 degrees centigrade, (3)

Ref No 3 03 767 United States

Feed Name or Analyses			As Fed	Dry	C.V. ± %
Dry matter	%		25.1	100.0	21
Ash	%		1.5	6.1	26
Crude fiber	%		2.1	8.5	55
Ether extract	%		0.2	0.8	46
N-free extract	%		18.7	74.6	
Protein (N x 6.25)	%		2.5	10.0	21
Cattle	dig prot %	*	1.3	5.3	
Goats	dig prot %	*	1.3	5.3	
Horses	dig prot %	*	1.3	5.3	
Sheep	dig prot %	*	1.3	5.3	
Energy	GE Mcal/kg				
Cattle	DE Mcal/kg	*	0.86	3.42	
Sheep	DE Mcal/kg	*	0.79	3.14	
Cattle	ME Mcal/kg	*	0.70	2.81	
Sheep	ME Mcal/kg	*	0.65	2.58	
Cattle	TDN %	*	19.5	77.6	
Sheep	TDN %	*	17.9	71.3	

Potato, tubers, ensiled, (3)

Ref No 3 03 768 United States

Feed Name or Analyses			As Fed	Dry	C.V. ± %
Dry matter	%		24.4	100.0	
Ash	%		1.3	5.5	
Crude fiber	%		0.7	3.0	
Sheep	dig coef %		15.	15.	
Swine	dig coef %		43.	43.	
Ether extract	%		0.1	0.4	
Sheep	dig coef %		64.	64.	
Swine	dig coef %		0.	0.	
N-free extract	%		20.5	84.0	
Sheep	dig coef %		93.	93.	
Swine	dig coef %		97.	97.	
Protein (N x 6.25)	%		1.7	7.1	
Sheep	dig coef %		54.	54.	
Swine	dig coef %		75.	75.	
Cattle	dig prot %	*	0.7	2.7	
Goats	dig prot %	*	0.7	2.7	
Horses	dig prot %	*	0.7	2.7	
Sheep	dig prot %		0.9	3.8	
Swine	dig prot %		1.3	5.3	
Energy	GE Mcal/kg				
Cattle	DE Mcal/kg	*	0.91	3.74	
Sheep	DE Mcal/kg	*	0.89	3.64	
Swine	DE kcal/kg	*	948.	3886.	
Cattle	ME Mcal/kg	*	0.75	3.07	
Sheep	ME Mcal/kg	*	0.73	2.98	
Swine	ME kcal/kg	*	897.	3674.	
Cattle	NE_m Mcal/kg	*	0.45	1.83	
Cattle	NE_{gain} Mcal/kg	*	0.30	1.22	
Cattle	$NE_{lactating\ cows}$ Mcal/kg	*	0.52	2.15	
Cattle	TDN %	*	20.7	84.8	
Sheep	TDN %		20.1	82.5	
Swine	TDN %		21.5	88.1	

Potato, tubers, steamed ensiled, (3)

Ref No 3 03 769 United States

Feed Name or Analyses			As Fed	Dry	C.V. ± %
Dry matter	%		23.9	100.0	11
Ash	%		1.5	6.2	35
Crude fiber	%		0.8	3.2	9
Cattle	dig coef %		0.	0.	9
Sheep	dig coef %		62.	62.	
Swine	dig coef %		79.	79.	
Ether extract	%		0.1	0.2	37
Cattle	dig coef %		0.	0.	37
Sheep	dig coef %		81.	81.	
Swine	dig coef %		0.	0.	
N-free extract	%		19.8	82.6	3
Cattle	dig coef %		83.	83.	
Sheep	dig coef %		96.	96.	
Swine	dig coef %		98.	98.	
Protein (N x 6.25)	%		1.9	7.8	7
Cattle	dig coef %		-9.	-9.	
Sheep	dig coef %		53.	53.	
Swine	dig coef %		62.	62.	
Cattle	dig prot %		-0.1	-0.7	
Goats	dig prot %	*	0.8	3.3	
Horses	dig prot %	*	0.8	3.3	
Sheep	dig prot %		1.0	4.1	
Swine	dig prot %		1.2	4.8	
Energy	GE Mcal/kg				
Cattle	DE Mcal/kg	*	0.72	2.99	
Sheep	DE Mcal/kg	*	0.91	3.78	
Swine	DE kcal/kg	*	927.	3873.	
Cattle	ME Mcal/kg	*	0.59	2.45	
Sheep	ME Mcal/kg	*	0.74	3.10	
Swine	ME kcal/kg	*	875.	3657.	
Cattle	TDN %		16.2	67.7	
Sheep	TDN %		20.5	85.8	
Swine	TDN %		21.0	87.8	

Potato, tubers w alfalfa hay added, ensiled, (3)

Ref No 3 03 770 United States

Feed Name or Analyses			As Fed	Dry	C.V. ± %
Dry matter	%		33.7	100.0	12
Ash	%		2.0	6.1	3
Crude fiber	%		6.8	20.2	25
Ether extract	%		0.5	1.4	26
N-free extract	%		19.8	58.9	
Protein (N x 6.25)	%		4.5	13.5	16
Cattle	dig prot %	*	2.9	8.5	
Goats	dig prot %	*	2.9	8.5	
Horses	dig prot %	*	2.9	8.5	
Sheep	dig prot %	*	2.9	8.5	
Energy	GE Mcal/kg				
Cattle	DE Mcal/kg	*	0.84	2.49	
Sheep	DE Mcal/kg	*	0.89	2.64	
Cattle	ME Mcal/kg	*	0.69	2.04	
Sheep	ME Mcal/kg	*	0.73	2.16	
Cattle	TDN %	*	19.0	56.4	
Sheep	TDN %	*	20.2	59.8	

Potato, tubers w grnd corn added, ensiled, (3)

Ref No 3 03 771 United States

Feed Name or Analyses			As Fed	Dry	C.V. ± %
Dry matter	%		31.7	100.0	
Ash	%		1.0	3.2	
Crude fiber	%		0.9	2.8	
Ether extract	%		0.3	0.9	
N-free extract	%		27.5	86.8	
Protein (N x 6.25)	%		2.0	6.3	
Cattle	dig prot %	*	0.6	2.0	
Goats	dig prot %	*	0.6	2.0	
Horses	dig prot %	*	0.6	2.0	
Sheep	dig prot %	*	0.6	2.0	
Energy	GE Mcal/kg				
Cattle	DE Mcal/kg	*	1.17	3.69	
Sheep	DE Mcal/kg	*	1.16	3.66	
Cattle	ME Mcal/kg	*	0.96	3.02	
Sheep	ME Mcal/kg	*	0.95	3.00	
Cattle	TDN %	*	26.5	83.7	
Sheep	TDN %	*	26.3	83.1	

Potato, tubers w oat hay added, ensiled, (3)

Ref No 3 03 772 United States

Feed Name or Analyses			As Fed	Dry	C.V. ± %
Dry matter	%		35.2	100.0	5
Ash	%		2.4	6.7	15
Crude fiber	%		7.4	21.0	8
Ether extract	%		0.5	1.4	14

(1) dry forages and roughages (3) silages (6) minerals
(2) pasture, range plants, and forages fed green (4) energy feeds (7) vitamins
(5) protein supplements (8) additives

Feed Name or Analyses			Mean As Fed	Mean Dry	C.V. ± %
N-free extract	%		21.5	61.2	
Protein (N x 6.25)	%		3.4	9.7	13
Cattle	dig prot %	*	1.8	5.0	
Goats	dig prot %	*	1.8	5.0	
Horses	dig prot %	*	1.8	5.0	
Sheep	dig prot %	*	1.8	5.0	
Energy	GE Mcal/kg				
Cattle	DE Mcal/kg	*	1.05	2.98	
Sheep	DE Mcal/kg	*	1.05	3.00	
Cattle	ME Mcal/kg	*	0.86	2.44	
Sheep	ME Mcal/kg	*	0.87	2.46	
Cattle	TDN %	*	23.8	67.6	
Sheep	TDN %	*	23.9	68.0	

Potato, peelings, (4)

Ref No 4 03 774 — United States

Feed Name or Analyses			Mean As Fed	Mean Dry	C.V. ± %
Dry matter	%		23.0	100.0	
Ash	%		1.4	6.0	
Crude fiber	%		0.8	3.4	
Swine	dig coef %		67.	67.	
Ether extract	%		0.1	0.4	
N-free extract	%		18.5	80.3	
Swine	dig coef %		96.	96.	
Protein (N x 6.25)	%		2.3	9.9	
Swine	dig coef %		74.	74.	
Cattle	dig prot %	*	1.2	5.1	
Goats	dig prot %	*	1.5	6.3	
Horses	dig prot %	*	1.5	6.3	
Sheep	dig prot %	*	1.5	6.3	
Swine	dig prot %		1.7	7.3	
Energy	GE Mcal/kg				
Cattle	DE Mcal/kg	*	0.84	3.65	
Sheep	DE Mcal/kg	*	0.83	3.61	
Swine	DE kcal/kg	*	880.	3827.	
Cattle	ME Mcal/kg	*	0.69	3.00	
Sheep	ME Mcal/kg	*	0.68	2.96	
Swine	ME kcal/kg	*	827.	3598.	
Cattle	TDN %	*	19.1	82.9	
Sheep	TDN %	*	18.9	82.0	
Swine	TDN %		20.0	86.8	
Calcium	%		0.03	0.14	
Phosphorus	%		0.04	0.19	

Ref No 4 03 774 — Canada

Feed Name or Analyses			Mean As Fed	Mean Dry	C.V. ± %
Dry matter	%		25.0	100.0	
Ash	%		0.8	3.1	
Crude fiber	%		0.6	2.5	
Swine	dig coef %		67.	67.	
Ether extract	%		0.0	0.1	
N-free extract	%		21.9	87.4	
Swine	dig coef %		96.	96.	
Protein (N x 6.25)	%		1.7	6.9	
Swine	dig coef %		74.	74.	
Cattle	dig prot %	*	0.6	2.4	
Goats	dig prot %	*	0.9	3.6	
Horses	dig prot %	*	0.9	3.6	
Sheep	dig prot %	*	0.9	3.6	
Swine	dig prot %		1.3	5.1	
Energy	GE Mcal/kg				
Cattle	DE Mcal/kg	*	0.90	3.59	
Sheep	DE Mcal/kg	*	0.94	3.77	
Swine	DE kcal/kg	*	957.	3827.	
Cattle	ME Mcal/kg	*	0.74	2.94	
Sheep	ME Mcal/kg	*	0.77	3.09	
Swine	ME kcal/kg	*	905.	3621.	
Cattle	TDN %	*	20.4	81.4	

Feed Name or Analyses			Mean As Fed	Mean Dry	C.V. ± %
Sheep	TDN %	*	21.4	85.6	
Swine	TDN %		21.7	86.8	

Potato, process residue, dehy, (4)

Potato by-product, dried

Potato pomace, dried

Potato pulp, dried

Potato waste, dried

Ref No 4 03 775 — United States

Feed Name or Analyses			Mean As Fed	Mean Dry	C.V. ± %
Dry matter	%		88.6	100.0	2
Ash	%		4.2	4.7	33
Crude fiber	%		6.1	6.9	47
Sheep	dig coef %	*	-129.	-129.	
Swine	dig coef %		75.	75.	
Ether extract	%		0.4	0.5	33
Sheep	dig coef %	*	47.	47.	
N-free extract	%		70.2	79.2	
Sheep	dig coef %	*	96.	96.	
Protein (N x 6.25)	%		7.7	8.7	19
Sheep	dig coef %	*	27.	27.	
Swine	dig coef %	*	26.	26.	
Cattle	dig prot %	*	3.5	4.0	
Goats	dig prot %	*	4.6	5.2	
Horses	dig prot %	*	4.6	5.2	
Sheep	dig prot %		2.1	2.3	
Swine	dig prot %		2.0	2.3	
Energy	GE Mcal/kg				
Swine	GE dig coef %	*	76.	76.	
Cattle	DE Mcal/kg	*	3.48	3.93	
Sheep	DE Mcal/kg	*	2.73	3.08	
Swine	DE kcal/kg	*	3496.	3946.	
Cattle	ME Mcal/kg	*	2.85	3.22	
Sheep	ME Mcal/kg	*	2.24	2.53	
Swine	ME kcal/kg	*	3293.	3717.	
Cattle	TDN %	*	78.9	89.1	
Sheep	TDN %		62.0	69.9	
Sheep	TDN %	*	62.0	69.9	
Swine	TDN %	*	79.3	89.5	

Ref No 4 03 775 — Canada

Feed Name or Analyses			Mean As Fed	Mean Dry	C.V. ± %
Dry matter	%		90.0	100.0	
Ash	%		2.4	2.7	
Crude fiber	%		7.0	7.8	
Swine	dig coef %		75.	75.	
Ether extract	%		0.1	0.1	
N-free extract	%		74.1	82.4	
Protein (N x 6.25)	%		6.3	7.0	
Swine	dig coef %		26.	26.	
Cattle	dig prot %	*	2.2	2.4	
Goats	dig prot %	*	3.3	3.7	
Horses	dig prot %	*	3.3	3.7	
Sheep	dig prot %	*	3.3	3.7	
Swine	dig prot %		1.6	1.8	
Energy	GE Mcal/kg				
Swine	GE dig coef %		76.	76.	
Cattle	DE Mcal/kg	*	3.56	3.95	
Sheep	DE Mcal/kg	*	2.80	3.11	
Swine	DE kcal/kg	*	2782.	3091.	
Cattle	ME Mcal/kg	*	2.92	3.24	
Sheep	ME Mcal/kg	*	2.29	2.55	
Swine	ME kcal/kg	*	2632.	2924.	
Cattle	TDN %	*	80.6	89.5	
Swine	TDN %	*	63.1	90.1	

Potato, process residue, wet, (4)

Potato by-product, wet

Potato pulp, wet

Potato waste, wet

Ref No 2 03 777 — United States

Feed Name or Analyses			Mean As Fed	Mean Dry	C.V. ± %
Dry matter	%		13.1	100.0	
Ash	%		0.4	2.9	
Crude fiber	%		1.6	12.4	
Sheep	dig coef %	*	-129.	-129.	
Swine	dig coef %	*	85.	85.	
Ether extract	%		0.1	0.9	
Sheep	dig coef %	*	47.	47.	
N-free extract	%		9.9	75.9	
Sheep	dig coef %	*	96.	96.	
Swine	dig coef %	*	98.	98.	
Protein (N x 6.25)	%		1.0	7.9	
Sheep	dig coef %	*	27.	27.	
Swine	dig coef %	*	27.	27.	
Cattle	dig prot %	*	0.4	3.2	
Goats	dig prot %	*	0.6	4.4	
Horses	dig prot %	*	0.6	4.4	
Sheep	dig prot %		0.3	2.1	
Swine	dig prot %		0.3	2.1	
Energy	GE Mcal/kg				
Cattle	DE Mcal/kg	*	0.43	3.82	
Sheep	DE Mcal/kg	*	0.34	2.63	
Swine	DE kcal/kg	*	360.	2762.	
Cattle	ME Mcal/kg	*	0.36	2.72	
Sheep	ME Mcal/kg	*	0.28	2.16	
Swine	ME kcal/kg	*	340.	2608.	
Cattle	TDN %	*	9.9	75.3	
Sheep	TDN %		7.8	59.7	
Swine	TDN %	*	8.2	62.6	
Calcium	%		0.03	0.23	
Phosphorus	%		0.03	0.24	
Potassium	%		0.17	1.33	

Potato, process residue w lime added, dehy, (4)

Ref No 4 08 491 — United States

Feed Name or Analyses			Mean As Fed	Mean Dry	C.V. ± %
Dry matter	%		88.0	100.0	
Ash	%		9.5	10.8	
Crude fiber	%		10.4	11.8	
Ether extract	%		0.3	0.3	
N-free extract	%		64.2	73.0	
Protein (N x 6.25)	%		3.6	4.1	
Cattle	dig prot %	*	-0.1	-0.1	
Goats	dig prot %	*	0.9	1.0	
Horses	dig prot %	*	0.9	1.0	
Sheep	dig prot %	*	0.9	1.0	
Energy	GE Mcal/kg				
Cattle	DE Mcal/kg	*	3.40	3.86	
Sheep	DE Mcal/kg	*	2.93	3.33	
Swine	DE kcal/kg	*	2798.	3178.	
Cattle	ME Mcal/kg	*	2.79	3.17	
Sheep	ME Mcal/kg	*	2.41	2.73	
Swine	ME kcal/kg	*	2636.	2995.	
Cattle	TDN %	*	77.0	87.5	
Sheep	TDN %	*	66.6	75.6	
Swine	TDN %	*	63.4	72.1	
Calcium	%		3.69	4.19	
Phosphorus	%		0.16	0.18	

Feed Name or Analyses			Mean As Fed	Mean Dry	C.V. ± %
Potato, pulp, dehy, (4)					
Ref No 4 03 778			United States		
Dry matter		%	85.7	100.0	
Ash		%	1.5	1.7	
Crude fiber		%	3.2	3.7	
Swine	dig coef	%	73.	73.	
Ether extract		%	0.2	0.2	
Swine	dig coef	%	0.	0.	
N-free extract		%	77.1	90.0	
Swine	dig coef	%	97.	97.	
Protein (N x 6.25)		%	3.8	4.4	
Swine	dig coef	%	0.	0.	
Cattle	dig prot	%	* 0.1	0.1	
Goats	dig prot	%	* 1.1	1.3	
Horses	dig prot	%	* 1.1	1.3	
Sheep	dig prot	%	* 1.1	1.3	
Energy	GE Mcal/kg				
Cattle	DE Mcal/kg		* 3.34	3.90	
Sheep	DE Mcal/kg		* 3.28	3.82	
Swine	DE kcal/kg		*3401.	3968.	
Cattle	ME Mcal/kg		* 2.74	3.20	
Sheep	ME Mcal/kg		* 2.69	3.14	
Swine	ME kcal/kg		*3235.	3774.	
Cattle	TDN	%	* 75.8	88.4	
Sheep	TDN	%	* 74.3	86.7	
Swine	TDN	%	77.1	90.0	
Potato, seedballs, s-c, (4)					
Ref No 4 03 780			United States		
Dry matter		%	92.3	100.0	
Ash		%	8.4	9.1	
Crude fiber		%	21.7	23.5	
Sheep	dig coef	%	26.	26.	
Ether extract		%	7.1	7.7	
Sheep	dig coef	%	68.	68.	
N-free extract		%	38.9	42.1	
Sheep	dig coef	%	54.	54.	
Protein (N x 6.25)		%	16.2	17.6	
Sheep	dig coef	%	33.	33.	
Cattle	dig prot	%	* 11.2	12.2	
Goats	dig prot	%	* 12.3	13.4	
Horses	dig prot	%	* 12.3	13.4	
Sheep	dig prot	%	5.4	5.8	
Energy	GE Mcal/kg				
Cattle	DE Mcal/kg		* 2.10	2.27	
Sheep	DE Mcal/kg		* 1.89	2.05	
Swine	DE kcal/kg		*1094.	1185.	
Cattle	ME Mcal/kg		* 1.72	1.86	
Sheep	ME Mcal/kg		* 1.55	1.68	
Swine	ME kcal/kg		*1011.	1096.	
Cattle	TDN	%	* 47.5	51.5	
Sheep	TDN	%	42.9	46.4	
Swine	TDN	%	* 24.8	26.9	
Potato, starch, (4)					
Ref No 4 03 783			United States		
Dry matter		%	82.5	100.0	
Ash		%	0.2	0.2	
N-free extract		%	81.4	98.7	
Swine	dig coef	%	100.	100.	
Protein (N x 6.25)		%	0.9	1.1	
Swine	dig coef	%	-8.	-8.	
Cattle	dig prot	%	* -2.4	-2.9	
Goats	dig prot	%	* -1.3	-1.7	
Horses	dig prot	%	* -1.3	-1.7	
Sheep	dig prot	%	* -1.3	-1.7	
Swine	dig prot	%	0.0	0.0	
Energy	GE Mcal/kg				
Swine	DE kcal/kg		*3587.	4347.	
Swine	ME kcal/kg		*3435.	4164.	
Swine	TDN	%	81.3	98.6	
Potato, tubers, cooked, (4)					
Ref No 4 03 784			United States		
Dry matter		%	24.3	100.0	
Ash		%	1.3	5.3	
Crude fiber		%	0.7	3.0	
Cattle	dig coef	%	0.	0.	
Sheep	dig coef	%	-256.	-256.	
Swine	dig coef	%	70.	70.	
Ether extract		%	0.1	0.3	
Cattle	dig coef	%	0.	0.	
Sheep	dig coef	%	-42.	-42.	
Swine	dig coef	%	-42.	-42.	
N-free extract		%	20.0	82.3	
Cattle	dig coef	%	80.	80.	
Sheep	dig coef	%	92.	92.	
Swine	dig coef	%	98.	98.	
Protein (N x 6.25)		%	2.2	9.2	
Cattle	dig coef	%	-19.	-19.	
Sheep	dig coef	%	43.	43.	
Swine	dig coef	%	66.	66.	
Cattle	dig prot	%	-0.3	-1.7	
Goats	dig prot	%	* 1.4	5.6	
Horses	dig prot	%	* 1.4	5.6	
Sheep	dig prot	%	1.0	3.9	
Swine	dig prot	%	1.5	6.0	
Energy	GE Mcal/kg				
Cattle	DE Mcal/kg		* 0.68	2.82	
Sheep	DE Mcal/kg		* 0.77	3.16	
Swine	DE kcal/kg		* 941.	3881.	
Cattle	ME Mcal/kg		* 0.56	2.31	
Sheep	ME Mcal/kg		* 0.63	2.59	
Swine	ME kcal/kg		* 886.	3654.	
Cattle	TDN	%	15.5	64.0	
Sheep	TDN	%	17.4	71.6	
Swine	TDN	%	21.3	88.0	
Potato, tubers, dehy grnd, (4)					
Potato meal					
Ref No 4 07 850			United States		
Dry matter		%	91.4	100.0	
Ash		%	4.3	4.7	
Crude fiber		%	2.1	2.3	
Ether extract		%	0.3	0.3	
N-free extract		%	75.0	82.1	
Protein (N x 6.25)		%	9.7	10.6	
Cattle	dig prot	%	* 5.3	5.8	
Goats	dig prot	%	* 6.4	7.0	
Horses	dig prot	%	* 6.4	7.0	
Sheep	dig prot	%	* 6.4	7.0	
Energy	GE Mcal/kg				
Cattle	DE Mcal/kg		* 3.16	3.46	
Sheep	DE Mcal/kg		* 3.35	3.67	
Swine	DE kcal/kg		*3366.	3682.	
Cattle	ME Mcal/kg		* 2.59	2.84	
Chickens	MEn kcal/kg		3527.	3859.	
Sheep	ME Mcal/kg		* 2.75	3.01	
Swine	ME kcal/kg		*3159.	3456.	
Cattle	TDN	%	* 71.7	78.4	
Sheep	TDN	%	* 76.0	83.2	
Swine	TDN	%	* 76.3	83.5	
Calcium		%	0.07	0.08	
Chlorine		%	0.36	0.39	
Manganese		mg/kg	2.9	3.1	
Phosphorus		%	0.20	0.22	
Potassium		%	1.97	2.16	
Potato, tubers, flaked dehy, (4)					
Ref No 4 03 785			United States		
Dry matter		%	86.6	100.0	
Ash		%	3.8	4.4	
Crude fiber		%	1.9	2.2	
Cattle	dig coef	%	-7.	-7.	
Sheep	dig coef	%	-71.	-71.	
Swine	dig coef	%	67.	67.	
Ether extract		%	0.2	0.3	
Cattle	dig coef	%	96.	96.	
Sheep	dig coef	%	-9.	-9.	
Swine	dig coef	%	-8.	-8.	
N-free extract		%	73.6	85.0	
Cattle	dig coef	%	94.	94.	
Sheep	dig coef	%	86.	86.	
Swine	dig coef	%	98.	98.	
Protein (N x 6.25)		%	7.1	8.2	
Cattle	dig coef	%	39.	39.	
Sheep	dig coef	%	34.	34.	
Swine	dig coef	%	72.	72.	
Cattle	dig prot	%	2.8	3.2	
Goats	dig prot	%	* 4.1	4.8	
Horses	dig prot	%	* 4.1	4.8	
Sheep	dig prot	%	2.4	2.8	
Swine	dig prot	%	5.1	5.9	
Energy	GE Mcal/kg				
Cattle	DE Mcal/kg		* 3.19	3.68	
Sheep	DE Mcal/kg		* 2.83	3.27	
Swine	DE kcal/kg		*3441.	3974.	
Cattle	ME Mcal/kg		* 2.61	3.02	
Sheep	ME Mcal/kg		* 2.32	2.68	
Swine	ME kcal/kg		*3247.	3749.	
Cattle	TDN	%	72.2	83.4	
Sheep	TDN	%	64.3	74.2	
Swine	TDN	%	78.0	90.1	
Potato, tubers, fresh, (4)					
Ref No 2 03 787			United States		
Dry matter		%	23.1	100.0	6
Ash		%	1.1	4.8	14
Crude fiber		%	0.6	2.4	21
Cattle	dig coef	%	-22.	-22.	
Sheep	dig coef	%	-52.	-52.	
Swine	dig coef	%	79.	79.	
Ether extract		%	0.1	0.3	39
Cattle	dig coef	%	1.	1.	
Sheep	dig coef	%	55.	55.	
Swine	dig coef	%	0.	0.	
N-free extract		%	19.1	82.9	2
Cattle	dig coef	%	91.	91.	
Sheep	dig coef	%	93.	93.	
Swine	dig coef	%	96.	96.	

(1) dry forages and roughages (2) pasture, range plants, and forages fed green (3) silages (4) energy feeds (5) protein supplements (6) minerals (7) vitamins (8) additives

Feed Name or Analyses			Mean As Fed	Mean Dry	C.V. ± %
Protein (N x 6.25)	%		2.2	9.6	15
Cattle	dig coef %		55.	55.	
Sheep	dig coef %		63.	63.	
Swine	dig coef %		33.	33.	
Cattle	dig prot %		1.2	5.3	
Goats	dig prot %	*	1.4	6.0	
Horses	dig prot %	*	1.4	6.0	
Sheep	dig prot %		1.4	6.0	
Swine	dig prot %		0.7	3.2	
Energy	GE Mcal/kg				
Cattle	DE Mcal/kg	*	0.82	3.53	
Sheep	DE Mcal/kg	*	0.84	3.63	
Swine	DE kcal/kg	*	857.	3715.	
Cattle	ME Mcal/kg	*	0.67	2.90	
Sheep	ME Mcal/kg	*	0.69	2.97	
Swine	ME kcal/kg	*	806.	3494.	
Cattle	NE_m Mcal/kg	*	0.45	1.95	
Cattle	NE_{gain} Mcal/kg	*	0.30	1.30	
Cattle	$NE_{lactating\ cows}$ Mcal/kg	*	0.50	2.15	
Cattle	TDN %		18.5	80.2	
Sheep	TDN %		19.0	82.2	
Swine	TDN %		19.4	84.2	
Calcium	%		0.01	0.05	
Chlorine	%		0.07	0.28	
Copper	mg/kg		4.1	17.7	
Iron	%		0.002	0.009	
Magnesium	%		0.03	0.14	
Manganese	mg/kg		9.6	41.6	
Phosphorus	%		0.05	0.24	
Potassium	%		0.52	2.26	
Sodium	%		0.02	0.09	
Sulphur	%		0.02	0.09	

Potato, distillers residue, dehy, (5)
Potato distillers dried residue (AAFCO)
Ref No 5 03 773 United States

Feed Name or Analyses			Mean As Fed	Mean Dry	C.V. ± %
Dry matter	%		95.7	100.0	1
Ash	%		6.7	7.0	18
Crude fiber	%		20.6	21.5	48
Ether extract	%		3.1	3.2	40
N-free extract	%		42.4	44.3	
Protein (N x 6.25)	%		22.9	23.9	17
Energy	GE Mcal/kg				
Cattle	DE Mcal/kg	*	2.69	2.81	
Sheep	DE Mcal/kg	*	2.73	2.85	
Swine	DE kcal/kg	*	3291.	3439.	
Cattle	ME Mcal/kg	*	2.21	2.30	
Sheep	ME Mcal/kg	*	2.24	2.34	
Swine	ME kcal/kg	*	3000.	3135.	
Cattle	TDN %	*	61.0	63.7	
Sheep	TDN %	*	61.8	64.6	
Swine	TDN %	*	74.6	78.0	

Potato, pulp w protein, dehy flaked, (5)
Ref No 5 03 779 United States

Feed Name or Analyses			Mean As Fed	Mean Dry	C.V. ± %
Dry matter	%		89.0	100.0	
Ash	%		6.3	7.1	
Crude fiber	%		4.9	5.5	
Swine	dig coef %		-177.	-177.	
Ether extract	%		0.6	0.7	
Swine	dig coef %		-47.	-47.	
N-free extract	%		39.8	44.7	
Swine	dig coef %		69.	69.	
Protein (N x 6.25)	%		37.4	42.0	
Swine	dig coef %		68.	68.	
Swine	dig prot %		25.4	28.6	
Energy	GE Mcal/kg				
Swine	DE kcal/kg	*	1917.	2154.	
Swine	ME kcal/kg	*	1678.	1885.	
Swine	TDN %		43.5	48.9	

Potato, spent residue liquid, (5)
Ref No 5 03 782 United States

Feed Name or Analyses			Mean As Fed	Mean Dry	C.V. ± %
Dry matter	%		4.2	100.0	
Ash	%		0.6	14.3	
Crude fiber	%		0.6	14.3	
Horses	dig coef %		103.	103.	
Ether extract	%		0.1	2.4	
Horses	dig coef %		149.	149.	
N-free extract	%		1.8	42.8	
Horses	dig coef %		104.	104.	
Protein (N x 6.25)	%		1.1	26.2	
Horses	dig coef %		77.	77.	
Horses	dig prot %		0.8	20.2	
Energy	GE Mcal/kg				
Cattle	DE Mcal/kg	*	0.12	2.76	
Horses	DE Mcal/kg	*	0.16	3.86	
Sheep	DE Mcal/kg	*	0.12	2.79	
Swine	DE kcal/kg	*	140.	3327.	
Cattle	ME Mcal/kg	*	0.09	2.26	
Horses	ME Mcal/kg	*	0.13	3.16	
Sheep	ME Mcal/kg	*	0.10	2.29	
Swine	ME kcal/kg	*	127.	3018.	
Cattle	TDN %	*	2.6	62.5	
Horses	TDN %		3.7	87.5	
Sheep	TDN %	*	2.7	63.4	
Swine	TDN %	*	3.2	75.5	

Potato, tubers w 15% urea added, flaked dehy, (5)
Ref No 5 03 788 United States

Feed Name or Analyses			Mean As Fed	Mean Dry	C.V. ± %
Dry matter	%		94.1	100.0	
Ash	%		4.0	4.2	
Crude fiber	%		2.1	2.2	
Sheep	dig coef %		50.	50.	
Ether extract	%		0.4	0.4	
Sheep	dig coef %		16.	16.	
N-free extract	%		37.5	39.8	
Sheep	dig coef %		93.	93.	
Protein (N x 6.25)	%		50.2	53.4	
Sheep	dig coef %		92.	92.	
Sheep	dig prot %		46.2	49.1	
Energy	GE Mcal/kg				
Cattle	DE Mcal/kg	*	3.39	3.61	
Sheep	DE Mcal/kg	*	3.63	3.85	
Swine	DE kcal/kg	*	3371.	3582.	
Cattle	ME Mcal/kg	*	2.78	2.96	
Sheep	ME Mcal/kg	*	2.97	3.16	
Swine	ME kcal/kg	*	2872.	3052.	
Cattle	TDN %	*	77.0	81.8	
Sheep	TDN %		82.2	87.4	
Swine	TDN %	*	76.4	81.2	

POTATO, WHITE. Solanum tuberosum

Potato, white, tubers, cooked ensiled trench silo, (3)
Ref No 3 03 789 United States

Feed Name or Analyses			Mean As Fed	Mean Dry	C.V. ± %
Dry matter	%		19.1	100.0	5
Ash	%		1.6	8.6	19
Crude fiber	%		0.6	3.3	12
Ether extract	%		0.1	0.5	32
N-free extract	%		14.1	73.9	
Protein (N x 6.25)	%		2.6	13.7	11
Cattle	dig prot %	*	1.7	8.7	
Goats	dig prot %	*	1.7	8.7	
Horses	dig prot %	*	1.7	8.7	
Sheep	dig prot %	*	1.7	8.7	
Energy	GE Mcal/kg				
Cattle	DE Mcal/kg	*	0.68	3.58	
Sheep	DE Mcal/kg	*	0.59	3.11	
Cattle	ME Mcal/kg	*	0.56	2.94	
Sheep	ME Mcal/kg	*	0.49	2.55	
Cattle	TDN %	*	15.5	81.2	
Sheep	TDN %	*	13.5	70.4	

Potato, white, tubers, fresh, (4)
Ref No 2 03 790 United States

Feed Name or Analyses			Mean As Fed	Mean Dry	C.V. ± %
Dry matter	%		20.2	100.0	
Ash	%		0.9	4.5	
Crude fiber	%		0.5	2.5	
Ether extract	%		0.1	0.5	
N-free extract	%		16.6	82.2	
Protein (N x 6.25)	%		2.1	10.4	
Cattle	dig prot %	*	1.1	5.6	
Goats	dig prot %	*	1.4	6.8	
Horses	dig prot %	*	1.4	6.8	
Sheep	dig prot %	*	1.4	6.8	
Energy	GE Mcal/kg		0.76	3.76	
Cattle	DE Mcal/kg	*	0.71	3.50	
Sheep	DE Mcal/kg	*	0.74	3.68	
Swine	DE kcal/kg	*	745.	3689.	
Cattle	ME Mcal/kg	*	0.58	2.87	
Sheep	ME Mcal/kg	*	0.61	3.01	
Swine	ME kcal/kg	*	700.	3464.	
Cattle	TDN %	*	16.0	79.4	
Sheep	TDN %	*	16.8	83.4	
Swine	TDN %	*	16.9	83.7	
Calcium	%		0.01	0.05	
Iron	%		0.001	0.005	
Phosphorus	%		0.05	0.25	
Potassium	%		0.41	2.03	
Sodium	%		0.00	0.00	
Ascorbic acid	mg/kg		200.0	990.1	
Niacin	mg/kg		15.0	74.3	
Riboflavin	mg/kg		0.4	2.0	
Thiamine	mg/kg		1.0	5.0	

Ref No 2 03 790 Canada

Feed Name or Analyses			Mean As Fed	Mean Dry	C.V. ± %
Dry matter	%		30.0	100.0	
Ash	%		1.6	5.3	
Crude fiber	%		0.8	2.7	
Ether extract	%		0.2	0.7	
N-free extract	%		24.6	82.0	
Protein (N x 6.25)	%		2.8	9.3	
Cattle	dig prot %	*	1.4	4.6	
Goats	dig prot %	*	1.7	5.8	
Horses	dig prot %	*	1.7	5.8	
Sheep	dig prot %	*	1.7	5.8	
Energy	GE Mcal/kg		1.13	3.76	
Cattle	DE Mcal/kg	*	1.12	3.72	
Sheep	DE Mcal/kg	*	1.10	3.67	
Swine	DE kcal/kg	*	1068.	3561.	
Cattle	ME Mcal/kg	*	0.92	3.05	
Sheep	ME Mcal/kg	*	0.90	3.01	
Swine	ME kcal/kg	*	1006.	3352.	
Cattle	TDN %	*	25.3	84.4	

Continued

Feed Name or Analyses		Mean As Fed	Mean Dry	C.V. ± %
Sheep	TDN %	* 25.0	83.2	
Swine	TDN %	* 24.2	80.8	

Potato by-product, dried –
see Potato, process residue, dehy, (4)

Potato by-product, wet –
see Potato, process residue, wet, (4)

Potato distillers dried residue (AAFCO) –
see Potato, distillers residue, dehy, (5)

Potato meal –
see Potato, tubers, dehy grnd, (4)

Potato pomace, dried –
see Potato, process residue, dehy, (4)

Potato pulp, dried –
see Potato, process residue, dehy, (4)

Potato pulp, wet –
see Potato, process residue, wet, (4)

Potato, waste, dried –
see Potato, process residue, dehy, (4)

Potato waste, wet –
see Potato, process residue, wet, (4)

POULTRY. Scientific name not used

Poultry, feathers, hydrolyzed dehy grnd, mn 70% of protein digestible, (5)
Hydrolyzed poultry feathers (CFA)

Ref No 5 03 794 United States

Feed Name or Analyses		Mean As Fed	Mean Dry	C.V. ± %
Dry matter	%		100.0	
Protein (N x 6.25)	%		89.6	
Energy	GE Mcal/kg			
Chickens	ME_n kcal/kg		2778.	
Niacin	mg/kg		17.3	
Pantothenic acid	mg/kg		8.1	
Riboflavin	mg/kg		1.9	
Vitamin B_{12}	µg/kg		71.0	
Arginine	%		5.62	
Cystine	%		3.09	
Glutamic acid	%		8.04	
Histidine	%		0.40	
Isoleucine	%		3.79	
Leucine	%		7.43	
Lysine	%		1.55	
Methionine	%		0.51	
Phenylalanine	%		3.77	
Threonine	%		3.96	
Tryptophan	%		0.57	
Tyrosine	%		2.43	
Valine	%		6.53	

Ref No 5 03 794 Canada

Feed Name or Analyses		Mean As Fed	Mean Dry	C.V. ± %
Dry matter	%	92.8	100.0	3
Ash	%	4.7	5.1	
Ether extract	%	7.5	8.0	30
Protein (N x 6.25)	%	82.9	89.3	6
Energy	GE Mcal/kg			
Chickens	ME_n kcal/kg	2579.	2778.	
Phosphorus	%	0.47	0.51	

Poultry, feathers, hydrolyzed dehy grnd, mn 75% of protein digestible, (5)
Hydrolyzed poultry feathers (AAFCO)
Feather meal

Ref No 5 03 795 United States

Feed Name or Analyses		Mean As Fed	Mean Dry	C.V. ± %
Dry matter	%	94.6	100.0	
Ash	%	3.7	3.9	
Crude fiber	%	0.6	0.6	
Ether extract	%	2.9	3.1	
N-free extract	%	0.0	0.0	
Protein (N x 6.25)	%	87.4	92.4	
Energy	GE Mcal/kg			
Cattle	DE Mcal/kg	* 2.75	2.90	
Sheep	DE Mcal/kg	* 2.63	2.78	
Swine	DE kcal/kg	*2839.	3001.	
Cattle	ME Mcal/kg	* 2.26	2.39	
Chickens	ME_n kcal/kg	2423.	2561.	
Sheep	ME Mcal/kg	* 2.16	2.28	
Swine	ME kcal/kg	*2196.	2321.	
Cattle	TDN %	* 62.2	65.7	
Sheep	TDN %	* 59.7	63.1	
Swine	TDN %	* 64.4	68.1	
Choline	mg/kg	882.	932.	
Niacin	mg/kg	30.9	32.6	
Pantothenic acid	mg/kg	11.0	11.7	
Riboflavin	mg/kg	2.2	2.3	
Arginine	%	5.90	6.24	
Cystine	%	3.00	3.17	
Glycine	%	6.80	7.19	
Lysine	%	2.00	2.11	
Methionine	%	0.60	0.63	
Tryptophan	%	0.50	0.53	

Poultry, viscera w feet w heads, dry or wet rendered dehy grnd, (5)
Poultry by-product meal (CFA)

Ref No 5 03 799 Canada

Feed Name or Analyses		Mean As Fed	Mean Dry	C.V. ± %
Dry matter	%	92.8	100.0	2
Ash	%	18.3	19.7	
Ether extract	%	12.0	12.9	10
Protein (N x 6.25)	%	61.0	65.7	4
Calcium	%	6.82	7.36	
Phosphorus	%	2.96	3.19	
Sodium	%	1.38	1.49	

Poultry, viscera w feet w heads, dry or wet rendered dehy grnd, mx 16% ash 4% acid insoluble ash, (5)
Poultry by-product meal (AAFCO)

Ref No 5 03 798 United States

Feed Name or Analyses		Mean As Fed	Mean Dry	C.V. ± %
Dry matter	%	93.4	100.0	
Ash	%	18.7	20.0	
Crude fiber	%	1.6	1.7	
Ether extract	%	13.1	14.0	
N-free extract	%	4.6	4.9	
Protein (N x 6.25)	%	55.4	59.3	
Linoleic	%	2.181	2.355	
Energy	GE Mcal/kg			
Cattle	DE Mcal/kg	* 3.22	3.45	
Sheep	DE Mcal/kg	* 3.01	3.22	
Swine	DE kcal/kg	*3483.	3729.	
Cattle	ME Mcal/kg	* 2.64	2.83	
Chickens	ME_n kcal/kg	2668.	2856.	
Sheep	ME Mcal/kg	* 2.47	2.64	
Swine	ME kcal/kg	*2926.	3133.	
Cattle	TDN %	* 73.0	78.2	
Sheep	TDN %	* 68.3	73.1	
Swine	TDN %	* 79.0	84.6	
Calcium	%	3.00	3.21	
Phosphorus	%	1.70	1.82	
Choline	mg/kg	5952.	6373.	
Niacin	mg/kg	39.7	42.5	
Pantothenic acid	mg/kg	8.8	9.4	
Riboflavin	mg/kg	11.0	11.8	
α-tocopherol	mg/kg	2.0	2.1	
Arginine	%	3.50	3.75	
Cystine	%	1.00	1.07	
Glycine	%	7.10	7.60	
Lysine	%	3.70	3.96	
Methionine	%	1.00	1.07	
Tryptophan	%	0.45	0.48	

Poultry by-product meal (AAFCO) –
see Poultry, viscera w feet w heads, dry or wet rendered dehy grnd, mx 16% ash 4% acid insoluble ash, (5)

Poultry by-product meal (CFA) –
see Poultry, viscera w feet w heads, dry or wet rendered dehy grnd, (5)

PRAIRIECLOVER. Petalostemon spp

Prairieclover, aerial part, fresh, (2)

Ref No 2 03 804 United States

Feed Name or Analyses		Mean As Fed	Mean Dry	C.V. ± %
Dry matter	%	44.4	100.0	
Ash	%	3.1	7.0	
Crude fiber	%	11.4	25.7	
Ether extract	%	2.5	5.6	
N-free extract	%	22.8	51.3	
Protein (N x 6.25)	%	4.6	10.4	
Cattle	dig prot %	* 3.0	6.7	
Goats	dig prot %	* 2.8	6.3	
Horses	dig prot %	* 2.8	6.4	
Rabbits	dig prot %	* 3.0	6.7	
Sheep	dig prot %	* 3.0	6.7	

Prairieclover, aerial part, fresh, immature, (2)

Ref No 2 03 803 United States

Feed Name or Analyses		Mean As Fed	Mean Dry	C.V. ± %
Dry matter	%		100.0	
Ash	%		5.8	
Crude fiber	%		25.1	
Ether extract	%		2.9	
N-free extract	%		52.1	
Protein (N x 6.25)	%		14.1	
Cattle	dig prot %	*	9.9	
Goats	dig prot %	*	9.7	
Horses	dig prot %	*	9.5	
Rabbits	dig prot %	*	9.6	
Sheep	dig prot %	*	10.1	

(1) dry forages and roughages (2) pasture, range plants, and forages fed green (3) silages (4) energy feeds (5) protein supplements (6) minerals (7) vitamins (8) additives

Feed Name or Analyses			Mean As Fed	Mean Dry	C.V. ± %

PRAIRIECLOVER, PURPLE. Petalostemon purpureus

Prairieclover, purple, aerial part, fresh, (2)
Ref No 2 03 805 United States

Analyses	Unit		As Fed	Dry	C.V. ± %
Dry matter	%		44.4	100.0	
Calcium	%		1.24	2.80	
Phosphorus	%		0.19	0.43	

PRAIRIECONEFLOWER. Ratibida spp

Prairieconeflower, browse, immature, (2)
Ref No 2 03 801 United States

Analyses	Unit		As Fed	Dry	C.V. ± %
Dry matter	%			100.0	
Ash	%			16.1	
Crude fiber	%			14.0	
Ether extract	%			4.1	
N-free extract	%			46.1	
Protein (N x 6.25)	%			19.7	
Cattle	dig prot %	*		14.6	
Goats	dig prot %	*		14.9	
Horses	dig prot %	*		14.3	
Rabbits	dig prot %	*		13.9	
Sheep	dig prot %	*		15.4	
Calcium	%			3.28	
Magnesium	%			0.50	
Phosphorus	%			0.26	
Potassium	%			4.53	

Prairieconeflower, heads, fresh, mature, (2)
Ref No 2 03 802 United States

Analyses	Unit		As Fed	Dry	C.V. ± %
Dry matter	%			100.0	
Ash	%			7.2	
Crude fiber	%			37.2	
Ether extract	%			6.1	
N-free extract	%			35.9	
Protein (N x 6.25)	%			13.6	
Cattle	dig prot %	*		9.5	
Goats	dig prot %	*		9.2	
Horses	dig prot %	*		9.1	
Rabbits	dig prot %	*		9.2	
Sheep	dig prot %	*		9.7	
Calcium	%			1.83	
Magnesium	%			0.00	
Phosphorus	%			0.31	
Potassium	%			1.30	

Prairie hay –
see Native plants, midwest, hay, s-c, (1)

Prairie hay, cut 1 –
see Native plants, midwest, hay, s-c, cut 1, (1)

Prairie hay, cut 2 –
see Native plants, midwest, hay, s-c, cut 2, (1)

Prairie hay, dough stage –
see Native plants, midwest, hay, s-c, dough stage, (1)

Prairie hay, full bloom –
see Native plants, midwest, hay, s-c, full bloom, (1)

Prairie hay, gr sample –
see Native plants, midwest, hay, s-c, gr sample US, (1)

Prairie hay, gr 2 –
see Native plants, midwest, hay, s-c, gr 2 US, (1)

Prairie hay, gr 3 –
see Native plants, midwest, hay, s-c, gr 3 US, (1)

Prairie hay, immature -
see Native plants, midwest, hay, s-c, immature, (1)

Prairie hay, mature –
see Native plants, midwest, hay, s-c, mature, (1)

Prairie hay, milk stage –
see Native plants, midwest, hay, s-c, milk stage, (1)

Prairie hay, over ripe –
see Native plants, midwest, hay, s-c, over ripe, (1)

PRICKLEPOPPY. Argemone intermedia

Pricklepoppy, seeds w hulls, (4)
Ref No 4 09 051 United States

Analyses	Unit		As Fed	Dry	C.V. ± %
Dry matter	%			100.0	
Protein (N x 6.25)	%			19.0	
Cattle	dig prot %	*		13.5	
Goats	dig prot %	*		14.6	
Horses	dig prot %	*		14.6	
Sheep	dig prot %	*		14.6	
Alanine	%			0.59	
Arginine	%			1.67	
Aspartic acid	%			2.11	
Glutamic acid	%			2.81	
Glycine	%			1.22	
Histidine	%			0.36	
Hydroxyproline	%			0.19	
Isoleucine	%			0.59	
Leucine	%			0.91	
Lysine	%			0.72	
Methionine	%			0.27	
Phenylalanine	%			0.48	
Proline	%			0.57	
Serine	%			0.72	
Threonine	%			0.53	
Tyrosine	%			0.76	
Valine	%			0.74	

Primary dried yeast (AAFCO) –
see Yeast, primary Saccharomyces, dehy, mn 40% protein, (7)

Proso millet grain –
see Millet, proso, grain, (4)

Proso, millet, hay, s-c –
see Millet, proso, hay, s-c, (1)

PUMPKINS. Cucurbita pepo

Pumpkins, seed hulls, (1)
Ref No 1 03 819 United States

Analyses	Unit		As Fed	Dry	C.V. ± %
Dry matter	%		89.8	100.0	
Ash	%		2.5	2.8	
Crude fiber	%		64.7	72.0	
Sheep	dig coef %		13.	13.	
Ether extract	%		5.3	5.9	
Sheep	dig coef %		77.	77.	
N-free extract	%		0.3	0.3	
Sheep	dig coef %		40.	40.	
Protein (N x 6.25)	%		17.1	19.0	
Sheep	dig coef %		38.	38.	
Cattle	dig prot %	*	12.0	13.4	
Goats	dig prot %	*	12.8	14.3	
Horses	dig prot %	*	12.3	13.7	
Rabbits	dig prot %	*	12.0	13.3	
Sheep	dig prot %		6.5	7.2	
Energy	GE Mcal/kg				
Cattle	DE Mcal/kg	*	1.21	1.35	
Sheep	DE Mcal/kg	*	1.07	1.19	
Cattle	ME Mcal/kg	*	0.99	1.10	
Sheep	ME Mcal/kg	*	0.87	0.97	
Cattle	TDN %	*	27.4	30.5	
Sheep	TDN %		24.2	26.9	

Pumpkins, fruit, fresh, (4)
Ref No 2 03 815 United States

Analyses	Unit		As Fed	Dry	C.V. ± %
Dry matter	%		8.7	100.0	32
Ash	%		0.8	8.9	8
Crude fiber	%		1.2	14.2	7
Cattle	dig coef %		68.	68.	
Sheep	dig coef %		61.	61.	
Swine	dig coef %		68.	68.	
Ether extract	%		0.8	8.9	50
Cattle	dig coef %		90.	90.	
Sheep	dig coef %		92.	92.	
Swine	dig coef %		57.	57.	
N-free extract	%		4.5	51.8	14
Cattle	dig coef %		89.	89.	
Sheep	dig coef %		89.	89.	
Swine	dig coef %		93.	93.	
Protein (N x 6.25)	%		1.4	16.2	17
Cattle	dig coef %		70.	70.	
Sheep	dig coef %		77.	77.	
Swine	dig coef %		72.	72.	
Cattle	dig prot %		1.0	11.3	
Goats	dig prot %	*	1.1	12.1	
Horses	dig prot %	*	1.1	12.1	
Sheep	dig prot %		1.1	12.5	
Swine	dig prot %		1.0	11.7	
Energy	GE Mcal/kg		0.27	3.10	
Cattle	DE Mcal/kg	*	0.33	3.75	
Sheep	DE Mcal/kg	*	0.33	3.78	
Swine	DE kcal/kg	*	311.	3568.	
Cattle	ME Mcal/kg	*	0.27	3.08	
Sheep	ME Mcal/kg	*	0.27	3.10	
Swine	ME kcal/kg	*	289.	3309.	
Cattle	TDN %		7.4	85.2	
Sheep	TDN %		7.5	85.7	
Swine	TDN %		7.1	80.9	
Calcium	%		0.02	0.24	
Iron	%		0.001	0.012	
Phosphorus	%		0.04	0.43	

Continued

Feed Name or Analyses			Mean As Fed	Mean Dry	C.V. ± %
Potassium		%	0.29	3.32	
Sodium		%	0.00	0.00	
Ascorbic acid		mg/kg	93.4	1071.4	
Niacin		mg/kg	6.2	71.4	
Riboflavin		mg/kg	1.1	13.1	
Thiamine		mg/kg	0.5	6.0	
Vitamin A		IU/g	16.6	190.5	

Pumpkins, fruit wo seeds, (4)
Ref No 4 03 816 United States

Feed Name or Analyses			Mean As Fed	Mean Dry	C.V. ± %
Dry matter		%	6.4	100.0	
Ash		%	0.5	7.8	
Crude fiber		%	1.0	16.2	
Sheep	dig coef	%	116.	116.	
Ether extract		%	0.2	2.7	
Sheep	dig coef	%	93.	93.	
N-free extract		%	3.7	58.3	
Sheep	dig coef	%	106.	106.	
Protein (N x 6.25)		%	1.0	15.1	
Sheep	dig coef	%	93.	93.	
Cattle	dig prot	%	* 0.6	9.9	
Goats	dig prot	%	* 0.7	11.0	
Horses	dig prot	%	* 0.7	11.0	
Sheep	dig prot	%	0.9	14.0	
Energy	GE Mcal/kg				
Cattle	DE Mcal/kg		* 0.25	3.95	
Sheep	DE Mcal/kg		* 0.28	4.41	
Swine	DE kcal/kg		* 241.	3765.	
Cattle	ME Mcal/kg		* 0.21	3.24	
Sheep	ME Mcal/kg		* 0.23	3.62	
Swine	ME kcal/kg		* 222.	3464.	
Cattle	TDN	%	* 5.7	89.6	
Sheep	TDN	%	6.4	100.1	
Swine	TDN	%	* 5.5	85.4	
Calcium		%	0.02	0.27	
Phosphorus		%	0.03	0.55	
Potassium		%	0.23	3.56	

Pumpkins, seeds, fresh, (5)
Ref No 5 08 492 United States

Feed Name or Analyses			Mean As Fed	Mean Dry	C.V. ± %
Dry matter		%	55.0	100.0	
Ash		%	1.9	3.5	
Crude fiber		%	10.8	19.6	
Ether extract		%	20.6	37.5	
N-free extract		%	4.1	7.5	
Protein (N x 6.25)		%	17.6	32.0	
Energy	GE Mcal/kg				
Cattle	DE Mcal/kg		* 2.54	4.62	
Sheep	DE Mcal/kg		* 2.47	4.49	
Swine	DE kcal/kg		*3254.	5916.	
Cattle	ME Mcal/kg		* 2.08	3.79	
Sheep	ME Mcal/kg		* 2.02	3.68	
Swine	ME kcal/kg		*2913.	5297.	
Cattle	TDN	%	* 57.6	104.8	
Sheep	TDN	%	* 56.0	101.8	
Swine	TDN	%	* 73.8	134.2	

(1) dry forages and roughages (2) pasture, range plants, and forages fed green (3) silages (4) energy feeds (5) protein supplements (6) minerals (7) vitamins (8) additives

Pumpkins, seeds, grnd, (5)
Ref No 5 03 817 United States

Feed Name or Analyses			Mean As Fed	Mean Dry	C.V. ± %
Dry matter		%	91.4	100.0	
Ash		%	7.4	8.1	
Crude fiber		%	12.7	13.9	
Sheep	dig coef	%	61.	61.	
Ether extract		%	16.7	18.3	
Sheep	dig coef	%	96.	96.	
N-free extract		%	9.3	10.2	
Sheep	dig coef	%	68.	68.	
Protein (N x 6.25)		%	45.2	49.5	
Sheep	dig coef	%	84.	84.	
Sheep	dig prot	%	38.0	41.6	
Energy	GE Mcal/kg				
Cattle	DE Mcal/kg		* 3.37	3.69	
Sheep	DE Mcal/kg		* 3.89	4.26	
Swine	DE kcal/kg		*3869.	4233.	
Cattle	ME Mcal/kg		* 2.76	3.02	
Sheep	ME Mcal/kg		* 3.19	3.49	
Swine	ME kcal/kg		*3327.	3640.	
Cattle	TDN	%	* 76.4	83.6	
Sheep	TDN	%	88.2	96.5	
Swine	TDN	%	* 87.7	96.0	

Pumpkins, seeds, solv-extd grnd, (5)
Ref No 5 03 818 United States

Feed Name or Analyses			Mean As Fed	Mean Dry	C.V. ± %
Dry matter		%	89.5	100.0	
Ash		%	7.6	8.5	
Crude fiber		%	30.5	34.1	
Sheep	dig coef	%	40.	40.	
Ether extract		%	1.2	1.3	
Sheep	dig coef	%	48.	48.	
N-free extract		%	8.1	9.1	
Sheep	dig coef	%	96.	96.	
Protein (N x 6.25)		%	42.1	47.0	
Sheep	dig coef	%	81.	81.	
Sheep	dig prot	%	34.1	38.1	
Energy	GE Mcal/kg				
Cattle	DE Mcal/kg		* 2.32	2.59	
Sheep	DE Mcal/kg		* 2.44	2.73	
Swine	DE kcal/kg		*1943.	2171.	
Cattle	ME Mcal/kg		* 1.90	2.12	
Sheep	ME Mcal/kg		* 2.00	2.24	
Swine	ME kcal/kg		*1681.	1878.	
Cattle	TDN	%	* 52.5	58.7	
Sheep	TDN	%	55.4	61.9	
Swine	TDN	%	* 44.1	49.2	

PUNCTUREVINE. Tribulus terrestris

Puncturevine, aerial part, fresh, immature, (2)
Ref No 2 03 820 United States

Feed Name or Analyses			Mean As Fed	Mean Dry	C.V. ± %
Dry matter		%		100.0	
Ash		%		11.8	
Crude fiber		%		25.7	
Ether extract		%		2.7	
N-free extract		%		40.9	
Protein (N x 6.25)		%		18.9	
Cattle	dig prot	%	*	14.0	
Goats	dig prot	%	*	14.2	
Horses	dig prot	%	*	13.6	
Rabbits	dig prot	%	*	13.3	
Sheep	dig prot	%	*	14.6	
Calcium		%		3.35	

Feed Name or Analyses			Mean As Fed	Mean Dry	C.V. ± %
Magnesium		%		0.32	
Phosphorus		%		0.24	
Potassium		%		2.84	

Purple needlegrass –
see Needlegrass, California aerial part, fresh, (2)

PURSLANE, COMMON. Portulaca oleracea

Purselane, common, seeds w hulls, (5)
Ref No 5 09 045 United States

Feed Name or Analyses			Mean As Fed	Mean Dry	C.V. ± %
Dry matter		%		100.0	
Protein (N x 6.25)		%		21.0	
Alanine		%		0.59	
Arginine		%		1.70	
Aspartic acid		%		1.37	
Glutamic acid		%		2.73	
Glycine		%		1.34	
Histidine		%		0.44	
Hydroxyproline		%		0.06	
Isoleucine		%		0.57	
Leucine		%		0.97	
Lysine		%		0.59	
Methionine		%		0.40	
Phenylalanine		%		0.69	
Proline		%		0.69	
Serine		%		0.61	
Threonine		%		0.55	
Tyrosine		%		0.80	
Valine		%		0.69	

Purslane –
see Portulaca, aerial part, fresh, (2)

QUACKGRASS. Agropyron repens

Quackgrass, hay, s-c, (1)
Ref No 1 03 827 United States

Feed Name or Analyses			Mean As Fed	Mean Dry	C.V. ± %
Dry matter		%	88.8	100.0	3
Ash		%	6.2	7.0	13
Crude fiber		%	30.1	33.9	14
Sheep	dig coef	%	* 61.	61.	
Ether extract		%	2.4	2.8	18
Sheep	dig coef	%	* 56.	56.	
N-free extract		%	40.2	45.3	
Sheep	dig coef	%	* 67.	67.	
Protein (N x 6.25)		%	9.8	11.1	33
Sheep	dig coef	%	* 57.	57.	
Cattle	dig prot	%	* 5.8	6.5	
Goats	dig prot	%	* 6.1	6.9	
Horses	dig prot	%	* 6.1	6.9	
Rabbits	dig prot	%	* 6.4	7.2	
Sheep	dig prot	%	5.6	6.3	
Fatty acids		%	1.8	2.1	
Energy	GE Mcal/kg				
Cattle	DE Mcal/kg		* 2.20	2.48	
Sheep	DE Mcal/kg		* 2.38	2.68	
Cattle	ME Mcal/kg		* 1.80	2.03	
Sheep	ME Mcal/kg		* 1.95	2.20	
Cattle	TDN	%	* 49.8	56.2	
Sheep	TDN	%	54.0	60.8	
Manganese		mg/kg	40.9	46.1	32

Feed Name or Analyses			As Fed	Dry	C.V. ± %
Ref No 1 03 827			Canada		
Dry matter	%		92.8	100.0	1
Crude fiber	%		29.2	31.5	9
Protein (N x 6.25)	%		10.0	10.8	48
Cattle	dig prot %	*	5.8	6.3	
Goats	dig prot %	*	6.1	6.6	
Horses	dig prot %	*	6.2	6.7	
Rabbits	dig prot %	*	6.5	7.0	
Sheep	dig prot %	*	5.8	6.2	
Energy	GE Mcal/kg				
Sheep	DE Mcal/kg	*	2.49	2.69	
Sheep	ME Mcal/kg	*	2.04	2.20	
Sheep	TDN %		56.5	60.9	
Calcium	%		0.31	0.33	38
Phosphorus	%		0.11	0.12	44
Carotene	mg/kg		34.5	37.2	
Vitamin A equivalent	IU/g		57.5	62.0	

Quackgrass, hay, s-c, immature, (1)

Ref No 1 03 825			United States		
Dry matter	%		86.4	100.0	
Ash	%		6.0	7.0	
Crude fiber	%		24.3	28.1	
Ether extract	%		3.0	3.5	
N-free extract	%		38.9	45.0	
Protein (N x 6.25)	%		14.2	16.4	
Cattle	dig prot %	*	9.6	11.1	
Goats	dig prot %	*	10.2	11.9	
Horses	dig prot %	*	9.9	11.5	
Rabbits	dig prot %	*	9.8	11.3	
Sheep	dig prot %	*	9.7	11.3	
Energy	GE Mcal/kg				
Cattle	DE Mcal/kg	*	2.26	2.61	
Sheep	DE Mcal/kg	*	2.27	2.63	
Cattle	ME Mcal/kg	*	1.85	2.14	
Sheep	ME Mcal/kg	*	1.86	2.15	
Cattle	TDN %	*	51.2	59.2	
Sheep	TDN %	*	51.5	59.6	

Quackgrass, hay, s-c, full bloom, (1)

Ref No 1 03 826			United States		
Dry matter	%		92.1	100.0	
Ash	%		5.1	5.5	
Crude fiber	%		30.6	33.2	
Sheep	dig coef %		61.	61.	
Ether extract	%		2.9	3.2	
Sheep	dig coef %		56.	56.	
N-free extract	%		45.8	49.7	
Sheep	dig coef %		67.	67.	
Protein (N x 6.25)	%		7.7	8.4	
Sheep	dig coef %		57.	57.	
Cattle	dig prot %	*	3.9	4.2	
Goats	dig prot %	*	4.0	4.4	
Horses	dig prot %	*	4.3	4.7	
Rabbits	dig prot %	*	4.7	5.2	
Sheep	dig prot %		4.4	4.8	
Energy	GE Mcal/kg				
Cattle	DE Mcal/kg	*	2.31	2.51	
Sheep	DE Mcal/kg	*	2.53	2.75	
Cattle	ME Mcal/kg	*	1.89	2.06	
Sheep	ME Mcal/kg	*	2.08	2.26	
Cattle	TDN %	*	52.4	56.9	
Sheep	TDN %		57.4	62.4	

Feed Name or Analyses			As Fed	Dry	C.V. ± %
Ref No 1 03 826			Canada		
Dry matter	%			100.0	
Ash	%			4.3	15
Crude fiber	%			32.2	4
Sheep	dig coef %			61.	
Ether extract	%			2.5	8
Sheep	dig coef %			56.	
N-free extract	%			49.3	
Sheep	dig coef %			67.	
Protein (N x 6.25)	%			11.7	19
Sheep	dig coef %			57.	
Cattle	dig prot %	*		7.1	
Goats	dig prot %	*		7.5	
Horses	dig prot %	*		7.5	
Rabbits	dig prot %	*		7.7	
Sheep	dig prot %			6.7	
Energy	GE Mcal/kg				
Cattle	DE Mcal/kg	*		2.47	
Sheep	DE Mcal/kg	*		2.76	
Cattle	ME Mcal/kg	*		2.02	
Sheep	ME Mcal/kg	*		2.26	
Cattle	TDN %	*		56.0	
Sheep	TDN %			62.5	
Calcium	%			0.39	
Copper	mg/kg			13.4	27
Iron	%			.007	
Manganese	mg/kg			28.9	17
Phosphorus	%			0.17	
Potassium	%			1.23	
Sulphur	%			0.04	

Quackgrass, hay, s-c, grnd, (1)

Ref No 1 03 824			United States		
Dry matter	%		94.0	100.0	
Ash	%		8.3	8.8	
Crude fiber	%		29.2	31.1	
Sheep	dig coef %		27.	27.	
Ether extract	%		0.7	0.7	
Sheep	dig coef %		5.	5.	
N-free extract	%		48.3	51.4	
Sheep	dig coef %		63.	63.	
Protein (N x 6.25)	%		7.5	8.0	
Sheep	dig coef %		45.	45.	
Cattle	dig prot %	*	3.6	3.9	
Goats	dig prot %	*	3.8	4.0	
Horses	dig prot %	*	4.1	4.3	
Rabbits	dig prot %	*	4.6	4.8	
Sheep	dig prot %		3.4	3.6	
Energy	GE Mcal/kg				
Cattle	DE Mcal/kg	*	2.54	2.70	
Sheep	DE Mcal/kg	*	1.84	1.96	
Cattle	ME Mcal/kg	*	2.08	2.21	
Sheep	ME Mcal/kg	*	1.51	1.61	
Cattle	TDN %	*	57.5	61.2	
Sheep	TDN %		41.8	44.5	

Quackgrass, aerial part, fresh, (2)

Coach grass

Ref No 2 03 829			United States		
Dry matter	%		27.6	100.0	13
Ash	%		2.4	8.6	19
Crude fiber	%		8.3	30.2	10
Ether extract	%		1.1	4.1	17
N-free extract	%		12.0	43.6	

Feed Name or Analyses			As Fed	Dry	C.V. ± %
Protein (N x 6.25)	%		3.7	13.5	23
Cattle	dig prot %	*	2.6	9.4	
Goats	dig prot %	*	2.5	9.2	
Horses	dig prot %	*	2.5	9.0	
Rabbits	dig prot %	*	2.5	9.1	
Sheep	dig prot %	*	2.6	9.6	
Energy	GE Mcal/kg				
Cattle	DE Mcal/kg	*	0.81	2.93	
Sheep	DE Mcal/kg	*	0.76	2.76	
Cattle	ME Mcal/kg	*	0.66	2.40	
Sheep	ME Mcal/kg	*	0.62	2.26	
Cattle	TDN %	*	18.3	66.4	
Sheep	TDN %	*	17.2	62.6	
Calcium	%		0.10	0.35	12
Phosphorus	%		0.08	0.30	16
Carotene	mg/kg		33.8	122.6	
Vitamin A equivalent	IU/g		56.3	204.3	

Quackgrass, aerial part, fresh, early bloom, (2)

Ref No 2 08 493			United States		
Dry matter	%		32.0	100.0	
Ash	%		1.8	5.6	
Crude fiber	%		11.3	35.3	
Ether extract	%		0.9	2.8	
N-free extract	%		15.7	49.1	
Protein (N x 6.25)	%		2.3	7.2	
Cattle	dig prot %	*	1.3	4.0	
Goats	dig prot %	*	1.0	3.3	
Horses	dig prot %	*	1.2	3.6	
Rabbits	dig prot %	*	1.3	4.2	
Sheep	dig prot %	*	1.2	3.7	
Energy	GE Mcal/kg				
Cattle	DE Mcal/kg	*	0.94	2.94	
Sheep	DE Mcal/kg	*	0.94	2.93	
Cattle	ME Mcal/kg	*	0.77	2.41	
Sheep	ME Mcal/kg	*	0.77	2.40	
Cattle	TDN %	*	21.3	66.6	
Sheep	TDN %	*	21.2	66.4	
Calcium	%		0.09	0.28	
Phosphorus	%		0.07	0.22	

Quackgrass, aerial part, fresh, mature, (2)

Ref No 2 03 828			United States		
Dry matter	%		31.7	100.0	7
Ash	%		1.7	5.4	5
Crude fiber	%		11.4	35.9	4
Ether extract	%		0.9	2.7	4
N-free extract	%		15.5	49.0	
Protein (N x 6.25)	%		2.2	7.0	12
Cattle	dig prot %	*	1.2	3.8	
Goats	dig prot %	*	1.0	3.1	
Horses	dig prot %	*	1.1	3.5	
Rabbits	dig prot %	*	1.3	4.1	
Sheep	dig prot %	*	1.1	3.5	
Energy	GE Mcal/kg				
Cattle	DE Mcal/kg	*	0.94	2.95	
Sheep	DE Mcal/kg	*	0.93	2.93	
Cattle	ME Mcal/kg	*	0.77	2.42	
Sheep	ME Mcal/kg	*	0.76	2.40	
Cattle	TDN %	*	21.2	67.0	
Sheep	TDN %	*	21.0	66.3	
Calcium	%		0.09	0.28	
Phosphorus	%		0.07	0.22	

Feed Name or Analyses		Mean As Fed	Mean Dry	C.V. ± %

QUACKGRASS-BLUEGRASS. Agropyron repens, Poa spp

Quackgrass-Bluegrass, hay, s-c, immature, (1)
Ref No 1 03 830 United States

Analyses		As Fed	Dry	C.V. ± %
Dry matter	%		100.0	
Manganese	mg/kg		44.5	

Quackgrass-Bluegrass, hay, s-c, mature, (1)
Ref No 1 03 831 United States

Analyses		As Fed	Dry	C.V. ± %
Dry matter	%		100.0	
Manganese	mg/kg		36.6	

RABBIT. Oryclolagus cuniculus

Rabbit, bone, dehy grnd, (6)
Ref No 6 07 857 United States

Analyses		As Fed	Dry	C.V. ± %
Dry matter	%		100.0	
Ash	%		47.8	
Ether extract	%		17.4	
Calcium	%		7.14	
Phosphorus	%		7.30	

RABBITBRUSH. Chrysothamnus spp

Rabbitbrush, aerial part, fresh, (2)
Ref No 2 03 835 United States

Analyses		As Fed	Dry	C.V. ± %
Dry matter	%		100.0	
Ash	%		8.0	25
Crude fiber	%		23.8	29
Ether extract	%		5.4	50
N-free extract	%		49.4	
Protein (N x 6.25)	%		13.4	34
Cattle	dig prot % *		9.3	
Goats	dig prot % *		9.1	
Horses	dig prot % *		8.9	
Rabbits	dig prot % *		9.0	
Sheep	dig prot % *		9.5	
Cobalt	mg/kg		0.110	
Sulphur	%		0.32	32

Rabbitbrush, aerial part, fresh, immature, (2)
Ref No 2 03 832 United States

Analyses		As Fed	Dry	C.V. ± %
Dry matter	%		100.0	
Ash	%		9.3	12
Crude fiber	%		16.9	13
Ether extract	%		3.1	24
N-free extract	%		49.0	
Protein (N x 6.25)	%		21.7	7
Cattle	dig prot % *		16.3	
Goats	dig prot % *		16.8	
Horses	dig prot % *		16.0	
Rabbits	dig prot % *		15.4	
Sheep	dig prot % *		17.2	
Sulphur	%		0.33	21

(1) dry forages and roughages (2) pasture, range plants, and forages fed green (3) silages (4) energy feeds (5) protein supplements (6) minerals (7) vitamins (8) additives

Rabbitbrush, aerial part, fresh, mid-bloom, (2)
Ref No 2 03 833 United States

Analyses		As Fed	Dry	C.V. ± %
Dry matter	%		100.0	
Ash	%		8.4	9
Crude fiber	%		24.2	8
Ether extract	%		4.0	17
N-free extract	%		50.9	
Protein (N x 6.25)	%		12.5	10
Cattle	dig prot % *		8.5	
Goats	dig prot % *		8.2	
Horses	dig prot % *		8.1	
Rabbits	dig prot % *		8.3	
Sheep	dig prot % *		8.6	
Sulphur	%		0.28	13

Rabbitbrush, aerial part, fresh, mature, (2)
Ref No 2 03 834 United States

Analyses		As Fed	Dry	C.V. ± %
Dry matter	%		100.0	
Ash	%		6.1	29
Crude fiber	%		32.1	14
Ether extract	%		9.2	30
N-free extract	%		46.8	
Protein (N x 6.25)	%		5.8	20
Cattle	dig prot % *		2.8	
Goats	dig prot % *		2.0	
Horses	dig prot % *		2.5	
Rabbits	dig prot % *		3.1	
Sheep	dig prot % *		2.4	

Rabbitbrush, stems, fresh, (2)
Ref No 2 03 836 United States

Analyses		As Fed	Dry	C.V. ± %
Dry matter	%		100.0	
Ash	%		4.8	
Crude fiber	%		27.4	
Ether extract	%		13.4	
N-free extract	%		47.3	
Protein (N x 6.25)	%		7.1	
Cattle	dig prot % *		3.9	
Goats	dig prot % *		3.2	
Horses	dig prot % *		3.6	
Rabbits	dig prot % *		4.2	
Sheep	dig prot % *		3.6	

RABBITBRUSH, DOUGLAS. Chrysothamnus viscidiflorus

Rabbitbrush, douglas, aerial part, fresh, (2)
Ref No 2 03 840 United States

Analyses		As Fed	Dry	C.V. ± %
Dry matter	%		100.0	
Ash	%		8.2	21
Crude fiber	%		22.8	26
Ether extract	%		5.1	28
N-free extract	%		49.2	
Protein (N x 6.25)	%		14.7	35
Cattle	dig prot % *		10.4	
Goats	dig prot % *		10.3	
Horses	dig prot % *		10.0	
Rabbits	dig prot % *		10.0	
Sheep	dig prot % *		10.7	
Calcium	%		1.40	31
Magnesium	%		0.33	28
Phosphorus	%		0.34	36
Sulphur	%		0.36	22

Rabbitbrush, douglas, aerial part, fresh, immature, (2)
Ref No 2 03 837 United States

Analyses		As Fed	Dry	C.V. ± %
Dry matter	%		100.0	
Ash	%		9.8	
Crude fiber	%		13.8	
Ether extract	%		3.8	
N-free extract	%		50.0	
Protein (N x 6.25)	%		22.6	
Cattle	dig prot % *		17.1	
Goats	dig prot % *		17.6	
Horses	dig prot % *		16.7	
Rabbits	dig prot % *		16.1	
Sheep	dig prot % *		18.1	
Calcium	%		2.00	9
Magnesium	%		0.45	16
Phosphorus	%		0.46	13
Sulphur	%		0.36	22

Rabbitbrush, douglas, aerial part, fresh, mid-bloom, (2)
Ref No 2 03 838 United States

Analyses		As Fed	Dry	C.V. ± %
Dry matter	%		100.0	
Ash	%		8.0	
Crude fiber	%		25.8	
Ether extract	%		4.7	
N-free extract	%		49.4	
Protein (N x 6.25)	%		12.1	
Cattle	dig prot % *		8.2	
Goats	dig prot % *		7.8	
Horses	dig prot % *		7.8	
Rabbits	dig prot % *		8.0	
Sheep	dig prot % *		8.3	
Calcium	%		1.46	10
Magnesium	%		0.28	14
Phosphorus	%		0.35	18
Sulphur	%		0.27	16

Rabbitbrush, douglas, aerial part, fresh, mature, (2)
Ref No 2 03 839 United States

Analyses		As Fed	Dry	C.V. ± %
Dry matter	%		100.0	
Ash	%		5.9	
Crude fiber	%		32.6	
Ether extract	%		7.8	
N-free extract	%		47.8	
Protein (N x 6.25)	%		5.9	
Cattle	dig prot % *		2.9	
Goats	dig prot % *		2.1	
Horses	dig prot % *		2.5	
Rabbits	dig prot % *		3.2	
Sheep	dig prot % *		2.5	
Calcium	%		0.99	24
Magnesium	%		0.23	17
Phosphorus	%		0.11	29

Feed Name or Analyses		Mean As Fed	Mean Dry	C.V. ± %

RABBITBRUSH, LANCELEAF. Chrysothamnus lanceolatus

Rabbitbrush, lanceleaf, aerial part, fresh, mature, (2)

Ref No 2 03 841 United States

Analyses		As Fed	Dry	C.V. ± %
Dry matter	%		100.0	
Ash	%		6.3	
Crude fiber	%		28.6	
Ether extract	%		5.0	
N-free extract	%		54.0	
Protein (N x 6.25)	%		6.1	
Cattle	dig prot % *		3.1	
Goats	dig prot % *		2.3	
Horses	dig prot % *		2.7	
Rabbits	dig prot % *		3.4	
Sheep	dig prot % *		2.7	

RABBITBRUSH, RUBBER. Chrysothamnus nauseosus

Rabbitbrush, rubber, aerial part, fresh, (2)

Ref No 2 03 845 United States

Analyses		As Fed	Dry	C.V. ± %
Dry matter	%		100.0	
Ash	%		8.2	18
Crude fiber	%		22.5	14
Ether extract	%		4.6	69
N-free extract	%		49.8	
Protein (N x 6.25)	%		14.9	30
Cattle	dig prot % *		10.6	
Goats	dig prot % *		10.5	
Horses	dig prot % *		10.2	
Rabbits	dig prot % *		10.2	
Sheep	dig prot % *		10.9	
Calcium	%		1.09	20
Magnesium	%		0.32	22
Phosphorus	%		0.37	33
Sulphur	%		0.27	22

Rabbitbrush, rubber, aerial part, fresh, immature, (2)

Ref No 2 03 842 United States

Analyses		As Fed	Dry	C.V. ± %
Dry matter	%		100.0	
Ash	%		8.8	
Crude fiber	%		20.0	
Ether extract	%		2.4	
N-free extract	%		48.1	
Protein (N x 6.25)	%		20.7	
Cattle	dig prot % *		15.5	
Goats	dig prot % *		15.9	
Horses	dig prot % *		15.1	
Rabbits	dig prot % *		14.7	
Sheep	dig prot % *		16.3	
Calcium	%		1.38	6
Magnesium	%		0.40	13
Phosphorus	%		0.45	14
Sulphur	%		0.30	11

Rabbitbrush, rubber, aerial part, fresh, mid-bloom, (2)

Ref No 2 03 843 United States

Analyses		As Fed	Dry	C.V. ± %
Dry matter	%		100.0	
Ash	%		8.8	
Crude fiber	%		22.6	
Ether extract	%		3.2	
N-free extract	%		52.6	
Protein (N x 6.25)	%		12.8	
Cattle	dig prot % *		8.8	
Goats	dig prot % *		8.5	
Horses	dig prot % *		8.4	
Rabbits	dig prot % *		8.6	
Sheep	dig prot % *		8.9	
Calcium	%		0.95	8
Magnesium	%		0.29	10
Phosphorus	%		0.38	16
Sulphur	%		0.29	13

Rabbitbrush, rubber, aerial part, fresh, mature, (2)

Ref No 2 03 844 United States

Analyses		As Fed	Dry	C.V. ± %
Dry matter	%		100.0	
Ash	%		5.4	
Crude fiber	%		28.2	
Ether extract	%		13.3	
N-free extract	%		47.2	
Protein (N x 6.25)	%		5.9	
Cattle	dig prot % *		2.9	
Goats	dig prot % *		2.1	
Horses	dig prot % *		2.5	
Rabbits	dig prot % *		3.2	
Sheep	dig prot % *		2.5	
Calcium	%		0.74	7
Magnesium	%		0.22	11
Phosphorus	%		0.14	28
Sulphur	%		0.15	13

RABBITBRUSH, SMALL. Chrysothamnus stenophyllus

Rabbitbrush, small, aerial part, mature, (1)

Ref No 1 03 846 United States

Analyses		As Fed	Dry	C.V. ± %
Dry matter	%		100.0	
Ash	%		7.9	17
Crude fiber	%		29.8	14
Ether extract	%		8.9	38
N-free extract	%		46.8	
Protein (N x 6.25)	%		6.6	14
Cattle	dig prot % *		2.7	
Goats	dig prot % *		2.7	
Horses	dig prot % *		3.1	
Rabbits	dig prot % *		3.8	
Sheep	dig prot % *		2.5	
Cellulose (Matrone)	%		19.0	
Lignin (Ellis)	%		11.8	
Calcium	%		2.03	16
Phosphorus	%		0.08	38

Rabbitbrush, small, aerial part, fresh, (2)

Ref No 2 03 849 United States

Analyses		As Fed	Dry	C.V. ± %
Dry matter	%		100.0	
Calcium	%		2.21	29
Phosphorus	%		0.11	61

Rabbitbrush, small, aerial part, fresh, milk stage, (2)

Ref No 2 03 847 United States

Analyses		As Fed	Dry	C.V. ± %
Dry matter	%		100.0	
Calcium	%		2.85	
Phosphorus	%		0.10	

RAGWEED. Ambrosia spp

Ragweed, aerial part, fresh, (2)

Ref No 2 03 853 United States

Analyses		As Fed	Dry	C.V. ± %
Dry matter	%	26.6	100.0	
Ash	%	2.4	9.2	59
Crude fiber	%	6.7	25.1	56
Ether extract	%	1.9	7.3	38
N-free extract	%	10.3	38.9	
Protein (N x 6.25)	%	5.2	19.5	18
Cattle	dig prot % *	3.8	14.5	
Goats	dig prot % *	3.9	14.8	
Horses	dig prot % *	3.7	14.1	
Rabbits	dig prot % *	3.7	13.7	
Sheep	dig prot % *	4.0	15.2	

Ragweed, aerial part, fresh, immature, (2)

Ref No 2 03 852 United States

Analyses		As Fed	Dry	C.V. ± %
Dry matter	%	26.6	100.0	
Ash	%	2.8	10.7	
Crude fiber	%	5.3	19.9	
Ether extract	%	1.7	6.4	
N-free extract	%	11.3	42.6	
Protein (N x 6.25)	%	5.4	20.4	
Cattle	dig prot % *	4.1	15.2	
Goats	dig prot % *	4.1	15.6	
Horses	dig prot % *	3.9	14.8	
Rabbits	dig prot % *	3.8	14.4	
Sheep	dig prot % *	4.3	16.0	

Ragweed, aerial part, fresh, immature, pure stand, (2)

Ref No 2 03 851 United States

Analyses		As Fed	Dry	C.V. ± %
Dry matter	%	26.6	100.0	
Ash	%	2.8	10.7	
Crude fiber	%	5.3	19.8	
Ether extract	%	1.1	4.2	
N-free extract	%	12.9	48.6	
Protein (N x 6.25)	%	4.4	16.7	
Cattle	dig prot % *	3.2	12.1	
Goats	dig prot % *	3.2	12.1	
Horses	dig prot % *	3.1	11.7	
Rabbits	dig prot % *	3.1	11.6	
Sheep	dig prot % *	3.3	12.6	
Calcium	%	0.45	1.68	
Phosphorus	%	0.04	0.16	

RAGWEED, COMMON. Ambrosia artemisiaefolia

Ragweed, common, aerial part, fresh, (2)

Ref No 2 03 855 United States

Analyses		As Fed	Dry	C.V. ± %
Dry matter	%		100.0	
Ash	%		5.1	
Crude fiber	%		34.4	
Ether extract	%		9.8	
N-free extract	%		32.0	
Protein (N x 6.25)	%		18.7	
Cattle	dig prot % *		13.8	
Goats	dig prot % *		14.0	
Horses	dig prot % *		13.4	
Rabbits	dig prot % *		13.1	
Sheep	dig prot % *		14.4	

Ragweed, common, aerial part, fresh, immature, (2)
Ref No 2 03 854 — United States

Feed Name or Analyses			Mean As Fed	Mean Dry	C.V. ± %
Dry matter	%			100.0	
Ash	%			5.6	
Crude fiber	%			28.0	
Ether extract	%			9.6	
N-free extract	%			36.2	
Protein (N x 6.25)	%			20.6	
Cattle	dig prot %	*		15.4	
Goats	dig prot %	*		15.8	
Horses	dig prot %	*		15.0	
Rabbits	dig prot %	*		14.6	
Sheep	dig prot %	*		16.2	

RAGWEED, GIANT. Ambrosia trifida

Ragweed, giant, seeds w hulls, (5)
Ref No 5 09 118 — United States

Feed Name or Analyses			Mean As Fed	Mean Dry	C.V. ± %
Dry matter	%			100.0	
Protein (N x 6.25)	%			22.0	
Alanine	%			0.90	
Arginine	%			2.11	
Aspartic acid	%			2.07	
Glutamic acid	%			4.84	
Glycine	%			0.92	
Histidine	%			0.57	
Hydroxyproline	%			0.13	
Isoleucine	%			0.88	
Leucine	%			1.34	
Lysine	%			0.68	
Methionine	%			0.37	
Phenylalanine	%			0.99	
Proline	%			1.14	
Serine	%			0.90	
Threonine	%			0.66	
Tyrosine	%			0.59	
Valine	%			1.08	

RAGWEED, WESTERN. Ambrosia psilostachya

Ragweed, western, aerial part, fresh, immature, (2)
Ref No 2 03 856 — United States

Feed Name or Analyses			Mean As Fed	Mean Dry	C.V. ± %
Dry matter	%			100.0	
Ash	%			15.8	
Crude fiber	%			12.0	
Ether extract	%			5.5	
N-free extract	%			42.9	
Protein (N x 6.25)	%			23.8	
Cattle	dig prot %	*		18.1	
Goats	dig prot %	*		18.8	
Horses	dig prot %	*		17.7	
Rabbits	dig prot %	*		17.0	
Sheep	dig prot %	*		19.2	
Calcium	%			3.33	
Magnesium	%			0.49	
Phosphorus	%			0.33	
Potassium	%			3.99	

(1) dry forages and roughages (2) pasture, range plants, and forages fed green (3) silages (4) energy feeds (5) protein supplements (6) minerals (7) vitamins (8) additives

Raisin pulp dried –
see Grapes, raisin syrup by-product, (4)

Raisins, cull –
see Grapes, fruit, dehy, cull, (4)

RAMIE. Boehmeria nivea

Ramie, leaves, (1)
Ref No 1 03 861 — United States

Feed Name or Analyses			Mean As Fed	Mean Dry	C.V. ± %
Dry matter	%		91.3	100.0	
Ash	%		17.6	19.3	
Crude fiber	%		13.5	14.8	
Ether extract	%		4.1	4.5	
N-free extract	%		36.1	39.5	
Protein (N x 6.25)	%		20.0	21.9	
Cattle	dig prot %	*	14.5	15.9	
Goats	dig prot %	*	15.5	17.0	
Horses	dig prot %	*	14.7	16.1	
Rabbits	dig prot %	*	14.2	15.6	
Sheep	dig prot %	*	14.8	16.2	

Ramie, leaves, dehy grnd, (1)
Ref No 1 03 857 — United States

Feed Name or Analyses			Mean As Fed	Mean Dry	C.V. ± %
Dry matter	%		92.3	100.0	2
Calcium	%		4.24	4.59	14
Iron	%		0.071	0.077	
Magnesium	%		0.79	0.86	16
Phosphorus	%		0.22	0.24	20
Carotene	mg/kg		123.5	133.8	
Niacin	mg/kg		51.7	56.0	
Pantothenic acid	mg/kg		27.1	29.3	
Riboflavin	mg/kg		10.0	10.8	
Thiamine	mg/kg		2.8	3.1	
Vitamin A equivalent	IU/g		205.9	223.1	

Ramie, leaves, s-c grnd, (1)
Ref No 1 03 858 — United States

Feed Name or Analyses			Mean As Fed	Mean Dry	C.V. ± %
Dry matter	%		92.1	100.0	
Ash	%		13.4	14.5	
Crude fiber	%		19.7	21.4	
Ether extract	%		3.8	4.1	
N-free extract	%		35.8	38.9	
Protein (N x 6.25)	%		19.4	21.1	
Cattle	dig prot %	*	14.0	15.2	
Goats	dig prot %	*	14.9	16.2	
Horses	dig prot %	*	14.2	15.4	
Rabbits	dig prot %	*	13.8	14.9	
Sheep	dig prot %	*	14.2	15.5	
Energy	GE Mcal/kg				
Cattle	DE Mcal/kg	*	2.38	2.58	
Sheep	DE Mcal/kg	*	2.35	2.55	
Cattle	ME Mcal/kg	*	1.95	2.12	
Sheep	ME Mcal/kg	*	1.92	2.09	
Cattle	TDN %	*	53.9	58.5	
Sheep	TDN %	*	53.2	57.8	
Calcium	%		4.52	4.91	
Phosphorus	%		0.26	0.28	

Ramie, aerial part, fresh, (2)
Ref No 2 03 859 — United States

Feed Name or Analyses			Mean As Fed	Mean Dry	C.V. ± %
Dry matter	%		8.9	100.0	
Ash	%		1.2	13.4	
Crude fiber	%		2.3	25.9	
Sheep	dig coef %		70.	70.	
Ether extract	%		0.4	4.1	
Sheep	dig coef %		53.	53.	
N-free extract	%		3.1	34.7	
Sheep	dig coef %		86.	86.	
Protein (N x 6.25)	%		1.9	21.9	
Sheep	dig coef %		84.	84.	
Cattle	dig prot %	*	1.5	16.5	
Goats	dig prot %	*	1.5	17.0	
Horses	dig prot %	*	1.4	16.1	
Rabbits	dig prot %	*	1.4	15.6	
Sheep	dig prot %		1.6	18.4	
Energy	GE Mcal/kg				
Cattle	DE Mcal/kg	*	0.27	2.99	
Sheep	DE Mcal/kg	*	0.28	3.14	
Cattle	ME Mcal/kg	*	0.22	2.45	
Sheep	ME Mcal/kg	*	0.23	2.57	
Cattle	TDN %	*	6.0	67.8	
Sheep	TDN %		6.3	71.2	
Calcium	%		0.41	4.60	
Phosphorus	%		0.02	0.24	
Carotene	mg/kg		27.1	304.9	
Choline	mg/kg		169.	1900.	
Niacin	mg/kg		6.0	67.0	
Pantothenic acid	mg/kg		2.4	26.7	
Riboflavin	mg/kg		1.5	16.5	
Thiamine	mg/kg		0.6	7.3	
Vitamin A equivalent	IU/g		45.2	508.3	

Ref No 2 03 859 — Guatemala

Feed Name or Analyses			Mean As Fed	Mean Dry	C.V. ± %
Dry matter	%			100.0	
Ash	%			16.2	
Crude fiber	%			27.4	
Sheep	dig coef %			70.	
Ether extract	%			2.6	
Sheep	dig coef %			53.	
N-free extract	%			33.8	
Sheep	dig coef %			86.	
Protein (N x 6.25)	%			20.1	
Sheep	dig coef %			84.	
Cattle	dig prot %	*		14.9	
Goats	dig prot %	*		15.3	
Horses	dig prot %	*		14.6	
Rabbits	dig prot %	*		14.1	
Sheep	dig prot %			16.8	
Energy	GE Mcal/kg				
Cattle	DE Mcal/kg	*		2.83	
Sheep	DE Mcal/kg	*		3.00	
Cattle	ME Mcal/kg	*		2.32	
Sheep	ME Mcal/kg	*		2.46	
Cattle	TDN %	*		64.1	
Sheep	TDN %			68.1	

Ramie, stems, fresh, (2)
Ref No 2 03 860 — United States

Feed Name or Analyses			Mean As Fed	Mean Dry	C.V. ± %
Dry matter	%		8.9	100.0	
Carotene	mg/kg		1.4	16.1	
Riboflavin	mg/kg		0.6	7.1	
Vitamin A equivalent	IU/g		2.4	26.8	

RAPE. Brassica spp

Rape, aerial part, dehy, early bloom, (1)

Ref No 1 03 862 United States

Feed Name or Analyses		Unit		Mean As Fed	Mean Dry	C.V. ± %
Dry matter		%		82.5	100.0	
Ash		%		11.3	13.8	
Crude fiber		%		19.4	23.5	
Sheep	dig coef	%		56.	56.	
Swine	dig coef	%		51.	51.	
Ether extract		%		2.3	2.8	
Sheep	dig coef	%		61.	61.	
Swine	dig coef	%		64.	64.	
N-free extract		%		37.7	45.7	
Sheep	dig coef	%		83.	83.	
Swine	dig coef	%		78.	78.	
Protein (N x 6.25)		%		11.8	14.3	
Sheep	dig coef	%		59.	59.	
Swine	dig coef	%		43.	43.	
Cattle	dig prot	%	*	7.7	9.3	
Goats	dig prot	%	*	8.1	9.9	
Horses	dig prot	%	*	7.9	9.6	
Rabbits	dig prot	%	*	8.0	9.7	
Sheep	dig prot	%		6.9	8.4	
Swine	dig prot	%		5.1	6.1	
Energy	GE Mcal/kg					
Cattle	DE Mcal/kg		*	2.48	3.00	
Sheep	DE Mcal/kg		*	2.30	2.79	
Swine	DE kcal/kg		*	2102.	2548.	
Cattle	ME Mcal/kg		*	2.03	2.46	
Sheep	ME Mcal/kg		*	1.89	2.29	
Swine	ME kcal/kg		*	1958.	2373.	
Cattle	TDN	%	*	56.2	68.1	
Sheep	TDN	%		52.3	63.3	
Swine	TDN	%		47.7	57.8	

Rape, pods w seeds, grnd, (1)

Ref No 1 03 868 United States

Feed Name or Analyses		Unit		Mean As Fed	Mean Dry	C.V. ± %
Dry matter		%		88.7	100.0	
Ash		%		14.0	15.8	
Crude fiber		%		22.9	25.8	
Sheep	dig coef	%		55.	55.	
Ether extract		%		7.5	8.5	
Sheep	dig coef	%		86.	86.	
N-free extract		%		34.5	38.9	
Sheep	dig coef	%		53.	53.	
Protein (N x 6.25)		%		9.8	11.0	
Sheep	dig coef	%		52.	52.	
Cattle	dig prot	%	*	5.7	6.5	
Goats	dig prot	%	*	6.1	6.8	
Horses	dig prot	%	*	6.1	6.9	
Rabbits	dig prot	%	*	6.4	7.2	
Sheep	dig prot	%		5.1	5.7	
Energy	GE Mcal/kg					
Cattle	DE Mcal/kg		*	2.48	2.80	
Sheep	DE Mcal/kg		*	2.23	2.51	
Cattle	ME Mcal/kg		*	2.04	2.30	
Sheep	ME Mcal/kg		*	1.83	2.06	
Cattle	TDN	%	*	56.2	63.4	
Sheep	TDN	%		50.5	57.0	

Rape, straw, (1)

Ref No 1 03 863 United States

Feed Name or Analyses		Unit		Mean As Fed	Mean Dry	C.V. ± %
Dry matter		%		79.7	100.0	
Ash		%		3.4	4.3	
Crude fiber		%		43.7	54.8	
Sheep	dig coef	%		25.	25.	
Ether extract		%		0.9	1.1	
Sheep	dig coef	%		25.	25.	
N-free extract		%		28.9	36.3	
Sheep	dig coef	%		38.	38.	
Protein (N x 6.25)		%		2.8	3.5	
Sheep	dig coef	%		24.	24.	
Cattle	dig prot	%	*	0.0	0.0	
Goats	dig prot	%	*	0.0	-0.1	
Horses	dig prot	%	*	0.4	0.5	
Rabbits	dig prot	%	*	1.1	1.4	
Sheep	dig prot	%		0.7	0.8	
Energy	GE Mcal/kg					
Cattle	DE Mcal/kg		*	0.66	0.83	
Sheep	DE Mcal/kg		*	1.02	1.28	
Cattle	ME Mcal/kg		*	0.54	0.68	
Sheep	ME Mcal/kg		*	0.83	1.05	
Cattle	TDN	%	*	15.0	18.9	
Sheep	TDN	%		23.1	29.0	

Rape, straw, treated w sodium hydroxide wet, (1)

Ref No 2 03 864 United States

Feed Name or Analyses		Unit		Mean As Fed	Mean Dry	C.V. ± %
Dry matter		%		25.7	100.0	
Ash		%		1.4	5.5	
Crude fiber		%		16.0	62.1	
Sheep	dig coef	%		44.	44.	
Ether extract		%		0.3	1.3	
Sheep	dig coef	%		41.	41.	
N-free extract		%		7.0	27.4	
Sheep	dig coef	%		51.	51.	
Protein (N x 6.25)		%		1.0	3.7	
Sheep	dig coef	%		-35.	-35.	
Cattle	dig prot	%	*	0.0	0.1	
Goats	dig prot	%	*	0.0	0.0	
Horses	dig prot	%	*	0.2	0.7	
Rabbits	dig prot	%	*	0.4	1.5	
Sheep	dig prot	%		-0.2	-1.2	
Energy	GE Mcal/kg					
Cattle	DE Mcal/kg		*	0.40	1.57	
Sheep	DE Mcal/kg		*	0.47	1.82	
Cattle	ME Mcal/kg		*	0.33	1.29	
Sheep	ME Mcal/kg		*	0.38	1.49	
Cattle	TDN	%	*	9.1	35.6	
Sheep	TDN	%		10.6	41.2	

Rape, aerial part, fresh, (2)

Ref No 2 03 867 United States

Feed Name or Analyses		Unit		Mean As Fed	Mean Dry	C.V. ± %
Dry matter		%		16.9	100.0	
Ash		%		2.1	12.6	
Crude fiber		%		2.5	14.7	
Sheep	dig coef	%		87.	87.	
Ether extract		%		0.6	3.8	
Sheep	dig coef	%		50.	50.	
N-free extract		%		8.6	51.2	
Sheep	dig coef	%		93.	93.	
Protein (N x 6.25)		%		3.0	17.6	
Sheep	dig coef	%		82.	82.	
Cattle	dig prot	%	*	2.2	12.9	
Goats	dig prot	%	*	2.2	13.0	
Horses	dig prot	%	*	2.1	12.5	
Rabbits	dig prot	%	*	2.1	12.3	
Sheep	dig prot	%		2.4	14.5	
Energy	GE Mcal/kg					
Cattle	DE Mcal/kg		*	0.57	3.37	
Sheep	DE Mcal/kg		*	0.59	3.49	
Cattle	ME Mcal/kg		*	0.47	2.76	
Sheep	ME Mcal/kg		*	0.48	2.86	
Cattle	TDN	%	*	12.9	76.4	
Sheep	TDN	%		13.3	79.2	
Calcium		%		0.25	1.47	
Copper		mg/kg		1.4	8.1	
Iron		%		0.003	0.018	
Magnesium		%		0.01	0.06	
Manganese		mg/kg		7.7	46.0	
Phosphorus		%		0.07	0.43	
Potassium		%		0.57	3.37	
Sulphur		%		0.11	0.67	

Rape, aerial part, fresh, early leaf, (2)

Ref No 2 03 865 United States

Feed Name or Analyses		Unit		Mean As Fed	Mean Dry	C.V. ± %
Dry matter		%		18.5	100.0	
Ash		%		2.1	11.4	
Crude fiber		%		2.4	13.0	
Sheep	dig coef	%		89.	89.	
Ether extract		%		0.7	4.0	
Sheep	dig coef	%		49.	49.	
N-free extract		%		10.2	55.2	
Sheep	dig coef	%		94.	94.	
Protein (N x 6.25)		%		3.0	16.4	
Sheep	dig coef	%		81.	81.	
Cattle	dig prot	%	*	2.2	11.8	
Goats	dig prot	%	*	2.2	11.9	
Horses	dig prot	%	*	2.1	11.5	
Rabbits	dig prot	%	*	2.1	11.3	
Sheep	dig prot	%		2.5	13.3	
Energy	GE Mcal/kg					
Cattle	DE Mcal/kg		*	0.63	3.39	
Sheep	DE Mcal/kg		*	0.66	3.58	
Cattle	ME Mcal/kg		*	0.51	2.78	
Sheep	ME Mcal/kg		*	0.54	2.93	
Cattle	TDN	%	*	14.2	76.9	
Sheep	TDN	%		15.0	81.2	

Rape, aerial part, fresh, early bloom, (2)

Ref No 2 03 866 United States

Feed Name or Analyses		Unit		Mean As Fed	Mean Dry	C.V. ± %
Dry matter		%		11.3	100.0	
Ash		%		1.6	14.0	
Crude fiber		%		1.8	15.8	
Sheep	dig coef	%		77.	77.	
Ether extract		%		0.4	3.8	
Sheep	dig coef	%		56.	56.	
N-free extract		%		4.8	42.9	
Sheep	dig coef	%		88.	88.	
Protein (N x 6.25)		%		2.7	23.5	
Sheep	dig coef	%		86.	86.	
Cattle	dig prot	%	*	2.0	17.9	
Goats	dig prot	%	*	2.1	18.5	
Horses	dig prot	%	*	2.0	17.5	
Rabbits	dig prot	%	*	1.9	16.8	
Sheep	dig prot	%		2.3	20.2	
Energy	GE Mcal/kg					
Cattle	DE Mcal/kg		*	0.33	2.96	
Sheep	DE Mcal/kg		*	0.37	3.30	
Cattle	ME Mcal/kg		*	0.27	2.43	
Sheep	ME Mcal/kg		*	0.31	2.71	
Cattle	TDN	%	*	7.6	67.2	
Sheep	TDN	%		8.5	74.9	

Rape, leaves w stems, fresh, (2)

Ref No 2 08 494 United States

Feed Name or Analyses			Mean As Fed	Mean Dry	C.V. ± %
Dry matter		%	15.4	100.0	
Ash		%	1.6	10.4	
Crude fiber		%	1.0	6.5	
Ether extract		%	0.6	3.9	
N-free extract		%	7.9	51.3	
Protein (N x 6.25)		%	4.3	27.9	
Cattle	dig prot	%	* 3.3	21.6	
Goats	dig prot	%	* 3.5	22.6	
Horses	dig prot	%	* 3.3	21.2	
Rabbits	dig prot	%	* 3.1	20.2	
Sheep	dig prot	%	* 3.5	23.0	
Energy	GE	Mcal/kg			
Cattle	DE	Mcal/kg	* 0.51	3.33	
Sheep	DE	Mcal/kg	* 0.53	3.44	
Cattle	ME	Mcal/kg	* 0.42	2.73	
Sheep	ME	Mcal/kg	* 0.43	2.82	
Cattle	TDN	%	* 11.6	75.6	
Sheep	TDN	%	* 12.0	78.1	

Rape, seeds, (5)

Ref No 5 08 109 United States

Feed Name or Analyses			Mean As Fed	Mean Dry	C.V. ± %
Dry matter		%	90.5	100.0	
Ash		%	4.2	4.6	
Crude fiber		%	6.6	7.3	
Ether extract		%	43.6	48.2	
N-free extract		%	15.7	17.3	
Protein (N x 6.25)		%	20.4	22.5	
Energy	GE	Mcal/kg			
Cattle	DE	Mcal/kg	* 5.18	5.72	
Sheep	DE	Mcal/kg	* 4.69	5.18	
Swine	DE	kcal/kg	*6021.	6653.	
Cattle	ME	Mcal/kg	* 4.24	4.69	
Sheep	ME	Mcal/kg	* 3.84	4.24	
Swine	ME	kcal/kg	*5509.	6087.	
Cattle	TDN	%	* 117.4	129.7	
Sheep	TDN	%	* 106.4	117.6	
Swine	TDN	%	* 136.6	150.9	

Ref No 5 08 109 Canada

Feed Name or Analyses			Mean As Fed	Mean Dry	C.V. ± %
Dry matter		%	92.4	100.0	
Protein (N x 6.25)		%	22.9	24.8	
Calcium		%	0.29	0.31	
Phosphorus		%	0.56	0.61	

Rape, seeds, extn unspecified grnd, (5)

Rapeseed meal (CFA)

Colza oil meal, extraction unspecified

Ref No 5 03 869 United States

Feed Name or Analyses			Mean As Fed	Mean Dry	C.V. ± %
Dry matter		%	89.3	100.0	2
Ash		%	7.0	7.8	8
Crude fiber		%	11.0	12.3	17
Cattle	dig coef	%	24.	24.	
Sheep	dig coef	%	30.	30.	
Ether extract		%	7.8	8.7	30
Cattle	dig coef	%	89.	89.	
Sheep	dig coef	%	86.	86.	
N-free extract		%	29.8	33.3	8
Cattle	dig coef	%	72.	72.	
Sheep	dig coef	%	73.	73.	
Protein (N x 6.25)		%	33.8	37.9	5
Cattle	dig coef	%	86.	86.	
Sheep	dig coef	%	85.	85.	
Cattle	dig prot	%	29.1	32.6	
Sheep	dig prot	%	28.6	32.0	
Linoleic		%	1.000	1.120	
Energy	GE	Mcal/kg			
Cattle	DE	Mcal/kg	* 3.03	3.39	
Sheep	DE	Mcal/kg	* 3.02	3.38	
Swine	DE	kcal/kg	*3478.	3896.	
Cattle	ME	Mcal/kg	* 2.48	2.78	
Chickens	ME_n	kcal/kg	679.	760.	
Chickens	ME_n	kcal/kg	2200.	2464.	
Chickens	ME_n	kcal/kg	1440.	1612.	
Sheep	ME	Mcal/kg	* 2.47	2.77	
Swine	ME	kcal/kg	*3073.	3442.	
Cattle	TDN	%	68.7	76.9	
Sheep	TDN	%	68.4	76.6	
Swine	TDN	%	* 78.9	88.4	
Calcium		%	0.51	0.57	39
Manganese		mg/kg	40.5	45.4	0
Phosphorus		%	0.94	1.06	22
Riboflavin		mg/kg	1.7	1.9	

Ref No 5 03 869 Canada

Feed Name or Analyses			Mean As Fed	Mean Dry	C.V. ± %
Dry matter		%	91.2	100.0	
Crude fiber		%	13.9	15.3	
Cattle	dig coef	%	24.	24.	
Sheep	dig coef	%	30.	30.	
Ether extract		%	2.4	2.7	
Cattle	dig coef	%	89.	89.	
Sheep	dig coef	%	86.	86.	
N-free extract		%			
Cattle	dig coef	%	72.	72.	
Sheep	dig coef	%	73.	73.	
Protein (N x 6.25)		%	39.8	43.6	
Cattle	dig coef	%	86.	86.	
Sheep	dig coef	%	85.	85.	
Cattle	dig prot	%	34.2	37.5	
Sheep	dig prot	%	33.6	36.9	
Energy	GE	Mcal/kg			
Cattle	DE	Mcal/kg	* 3.04	3.33	
Sheep	DE	Mcal/kg	* 3.03	3.32	
Cattle	ME	Mcal/kg	* 2.49	2.73	
Sheep	ME	Mcal/kg	* 2.48	2.72	
Cattle	TDN	%	68.9	75.6	
Sheep	TDN	%	68.7	75.3	
Calcium		%	0.40	0.44	
Phosphorus		%	0.91	1.00	

Rape, seeds, mech-extd grnd, (5)

Rapeseed oil meal, expeller extracted

Rapeseed meal, expeller extracted

Ref No 5 03 870 United States

Feed Name or Analyses			Mean As Fed	Mean Dry	C.V. ± %
Dry matter		%	92.0	100.0	1
Ash		%	7.2	7.8	9
Crude fiber		%	12.1	13.2	21
Cattle	dig coef	%	* 24.	24.	
Sheep	dig coef	%	* 24.	24.	
Ether extract		%	8.6	9.3	
Cattle	dig coef	%	* 89.	89.	
Sheep	dig coef	%	* 90.	90.	
N-free extract		%	32.6	35.4	
Cattle	dig coef	%	* 72.	72.	
Sheep	dig coef	%	* 63.	63.	
Protein (N x 6.25)		%	31.5	34.2	20
Cattle	dig coef	%	* 86.	86.	
Sheep	dig coef	%	* 82.	82.	
Cattle	dig prot	%	27.1	29.4	
Sheep	dig prot	%	25.8	28.1	
Energy	GE	Mcal/kg			
Cattle	DE	Mcal/kg	* 3.12	3.39	
Sheep	DE	Mcal/kg	* 2.94	3.20	
Swine	DE	kcal/kg	*3662.	3981.	
Cattle	ME	Mcal/kg	* 2.56	2.78	
Sheep	ME	Mcal/kg	* 2.41	2.62	
Swine	ME	kcal/kg	*3262.	3546.	
Cattle	NE_m	Mcal/kg	* 1.54	1.67	
Cattle	NE_{gain}	Mcal/kg	* 1.00	1.09	
Cattle	$NE_{lactating\ cows}$	Mcal/kg	* 1.81	1.97	
Cattle	TDN	%	70.7	76.8	
Sheep	TDN	%	66.7	72.5	
Swine	TDN	%	* 83.1	90.3	
Calcium		%	0.92	1.00	
Manganese		mg/kg	60.8	66.1	
Phosphorus		%	1.29	1.40	
a-tocopherol		mg/kg	19.0	20.2	

Ref No 5 03 870 Canada

Feed Name or Analyses			Mean As Fed	Mean Dry	C.V. ± %
Dry matter		%	94.2	100.0	
Crude fiber		%	11.6	12.3	
Ether extract		%	7.0	7.4	
Protein (N x 6.25)		%	36.2	38.4	
Energy	GE	Mcal/kg			
Cattle	DE	Mcal/kg	* 3.15	3.34	
Sheep	DE	Mcal/kg	* 2.96	3.15	
Cattle	ME	Mcal/kg	* 2.58	2.74	
Sheep	ME	Mcal/kg	* 2.43	2.58	
Cattle	TDN	%	71.4	75.7	
Sheep	TDN	%	67.2	71.3	
Calcium		%	0.60	0.63	
Phosphorus		%	1.01	1.07	
Alanine		%	1.65	1.75	
Arginine		%	1.99	2.11	
Aspartic acid		%	2.59	2.75	
Cystine		%	0.29	0.31	
Glutamic acid		%	6.27	6.65	
Glycine		%	1.83	1.94	
Histidine		%	0.92	0.97	
Isoleucine		%	1.43	1.52	
Leucine		%	2.49	2.65	
Lysine		%	1.74	1.84	
Methionine		%	0.73	0.77	
Phenylalanine		%	1.49	1.58	
Proline		%	2.21	2.35	
Serine		%	1.58	1.67	
Threonine		%	1.62	1.71	
Tyrosine		%	0.84	0.90	
Valine		%	1.86	1.98	
Isothiocyanate		%	2.572	2.729	
Oxazoladinethione		%	2.75	2.92	

Rape, seeds, solv-extd grnd, (5)

Rapeseed oil meal, solvent extracted

Rapeseed meal, solvent extracted

Ref No 5 03 871 United States

Feed Name or Analyses			Mean As Fed	Mean Dry	C.V. ± %
Dry matter		%		100.0	
Crude fiber		%			
Cattle	dig coef	%	*	24.	
Sheep	dig coef	%	*	24.	

(1) dry forages and roughages (2) pasture, range plants, and forages fed green (3) silages (4) energy feeds (5) protein supplements (6) minerals (7) vitamins (8) additives

Feed Name or Analyses			Mean As Fed	Mean Dry	C.V. ± %
Ether extract	%				
Cattle	dig coef %	*		89.	
Sheep	dig coef %	*		90.	
N-free extract	%				
Cattle	dig coef %	*		72.	
Sheep	dig coef %	*		63.	
Protein (N x 6.25)	%				
Cattle	dig coef %	*		86.	
Sheep	dig coef %	*		82.	
Energy	GE Mcal/kg				
Cattle	DE Mcal/kg	*		3.00	
Sheep	DE Mcal/kg	*		2.79	
Cattle	ME Mcal/kg	*		2.46	
Sheep	ME Mcal/kg	*		2.29	
Cattle	NE_m Mcal/kg	*		1.51	
Cattle	NE_{gain} Mcal/kg	*		0.94	
Cattle	$NE_{lactating\ cows}$ Mcal/kg	*		1.78	
Cattle	TDN %	*		68.1	
Sheep	TDN %	*		63.2	

Ref No 5 03 871 Canada

Feed Name or Analyses			Mean As Fed	Mean Dry	C.V. ± %
Dry matter	%		91.2	100.0	2
Ash	%		7.1	7.8	
Crude fiber	%		11.7	12.9	8
Ether extract	%		1.6	1.7	36
N-free extract	%		33.4	36.6	
Protein (N x 6.25)	%		37.4	41.0	4
Energy	GE Mcal/kg				
Cattle	DE Mcal/kg	*	2.74	3.00	
Sheep	DE Mcal/kg	*	2.54	2.79	
Swine	DE kcal/kg	*	3020.	3310.	
Cattle	ME Mcal/kg	*	2.25	2.46	
Sheep	ME Mcal/kg	*	2.09	2.29	
Swine	ME kcal/kg	*	2649.	2903.	
Cattle	TDN %		62.2	68.1	
Sheep	TDN %		57.7	63.2	
Swine	TDN %	*	68.5	75.1	
Calcium	%		0.61	0.67	15
Phosphorus	%		0.95	1.04	11
Alanine	%		1.76	1.93	5
Arginine	%		2.29	2.51	7
Aspartic acid	%		2.75	3.02	6
Cystine	%		0.32	0.35	14
Glutamic acid	%		6.93	7.60	7
Glycine	%		1.96	2.15	6
Histidine	%		1.14	1.25	13
Isoleucine	%		1.51	1.65	7
Leucine	%		2.74	3.00	6
Lysine	%		2.21	2.42	8
Methionine	%		0.77	0.85	8
Phenylalanine	%		1.56	1.71	6
Proline	%		2.51	2.75	9
Serine	%		1.72	1.89	5
Threonine	%		1.73	1.89	6
Tryptophan	%		0.57	0.62	
Tyrosine	%		0.87	0.95	7
Valine	%		2.08	2.28	20
Isothiocyanate	%		3.034	3.326	67
Oxazoladinethione	%		2.47	2.71	80

RAPE, ARGENTINE. Brassica napus

Rape, Argentine, seeds, mech-extd grnd, (5)

Ref No 5 07 869 Canada

Feed Name or Analyses			Mean As Fed	Mean Dry	C.V. ± %
Dry matter	%		93.2	100.0	
Ash	%		5.9	6.3	
Crude fiber	%		12.8	13.7	
Ether extract	%		6.7	7.1	
N-free extract	%		27.0	29.0	
Protein (N x 6.25)	%		40.8	43.8	
Energy	GE Mcal/kg				
Cattle	DE Mcal/kg	*	3.07	3.30	
Sheep	DE Mcal/kg	*	3.17	3.40	
Swine	DE kcal/kg	*	3406.	3657.	
Cattle	ME Mcal/kg	*	2.52	2.70	
Sheep	ME Mcal/kg	*	2.60	2.79	
Swine	ME kcal/kg	*	2969.	3187.	
Cattle	TDN %	*	69.7	74.8	
Sheep	TDN %	*	71.9	77.2	
Swine	TDN %	*	77.3	82.9	
Calcium	%		0.57	0.61	
Phosphorus	%		1.07	1.15	
Choline	mg/kg		6996.	7511.	
Niacin	mg/kg		166.9	179.2	
Pantothenic acid	mg/kg		9.9	10.6	
Riboflavin	mg/kg		4.2	4.5	
Thiamine	mg/kg		1.9	2.0	
Alanine	%		1.68	1.80	
Arginine	%		2.40	2.58	
Aspartic acid	%		2.55	2.74	
Cystine	%		0.29	0.32	
Glutamic acid	%		7.94	8.53	
Glycine	%		2.32	2.49	
Histidine	%		1.11	1.19	
Isoleucine	%		1.79	1.93	
Leucine	%		2.93	3.15	
Lysine	%		1.68	1.80	
Methionine	%		0.76	0.82	
Phenylalanine	%		1.65	1.77	
Proline	%		2.31	2.48	
Serine	%		1.58	1.70	
Threonine	%		1.75	1.88	
Tryptophan	%		0.61	0.65	
Tyrosine	%		1.02	1.09	
Valine	%		2.12	2.28	
Isothiocyanate	%		2.301	2.471	
Oxazoladinethione	%		2.71	2.91	

Rape, Argentine, seeds, solv-extd grnd, (5)

Ref No 5 07 868 Canada

Feed Name or Analyses			Mean As Fed	Mean Dry	C.V. ± %
Dry matter	%		93.2	100.0	
Crude fiber	%		12.5	13.4	
Ether extract	%		1.7	1.8	
Protein (N x 6.25)	%		40.1	43.0	
Calcium	%		0.49	0.53	
Phosphorus	%		0.94	1.01	
Alanine	%		1.72	1.85	
Arginine	%		2.20	2.36	
Aspartic acid	%		2.83	3.04	
Cystine	%		0.26	0.28	
Glutamic acid	%		6.69	7.17	
Glycine	%		1.90	2.04	
Histidine	%		1.02	1.10	
Isoleucine	%		1.50	1.60	
Leucine	%		2.68	2.87	
Lysine	%		2.13	2.29	
Methionine	%		0.79	0.85	
Phenylalanine	%		1.50	1.61	
Proline	%		2.52	2.71	
Serine	%		1.65	1.77	
Threonine	%		1.65	1.77	
Tyrosine	%		0.83	0.89	
Valine	%		2.00	2.15	
Isothiocyanate	%		3.120	3.348	
Oxazoladinethione	%		5.34	5.73	

RAPE, BIRD. Brassica campestris

Rape, bird, aerial part, fresh, (2)

Ref No 2 03 872 United States

Feed Name or Analyses			Mean As Fed	Mean Dry	C.V. ± %
Dry matter	%		14.9	100.0	
Ash	%		2.3	15.6	
Crude fiber	%		4.0	26.6	
Cattle	dig coef %		28.	28.	
Ether extract	%		0.4	2.8	
Cattle	dig coef %		73.	73.	
N-free extract	%		6.9	46.2	
Cattle	dig coef %		54.	54.	
Protein (N x 6.25)	%		1.3	8.8	
Cattle	dig coef %		65.	65.	
Cattle	dig prot %		0.9	5.7	
Goats	dig prot %	*	0.7	4.8	
Horses	dig prot %	*	0.7	5.0	
Rabbits	dig prot %	*	0.8	5.5	
Sheep	dig prot %	*	0.8	5.2	
Energy	GE Mcal/kg				
Cattle	DE Mcal/kg	*	0.28	1.88	
Sheep	DE Mcal/kg	*	0.32	2.18	
Cattle	ME Mcal/kg	*	0.23	1.54	
Sheep	ME Mcal/kg	*	0.27	1.79	
Cattle	TDN %		6.4	42.7	
Sheep	TDN %	*	7.4	49.5	

RAPE, CANADA. Brassica napus var

Rape, Canada, seeds, cooked mech-extd grnd, Can 1 mx 6% fat, (5)

Canada rapeseed meal (CFA)

Ref No 5 08 136 Canada

Feed Name or Analyses			Mean As Fed	Mean Dry	C.V. ± %
Dry matter	%		94.0	100.0	
Ash	%		6.8	7.2	
Crude fiber	%		15.5	16.5	
Ether extract	%		7.0	7.4	
N-free extract	%		29.5	31.4	
Protein (N x 6.25)	%		35.2	37.4	
Energy	GE Mcal/kg				
Cattle	DE Mcal/kg	*	2.98	3.17	
Sheep	DE Mcal/kg	*	3.14	3.34	
Swine	DE kcal/kg	*	3435.	3654.	
Cattle	ME Mcal/kg	*	2.44	2.60	
Sheep	ME Mcal/kg	*	2.57	2.74	
Swine	ME kcal/kg	*	3038.	3232.	
Cattle	TDN %	*	67.5	71.8	
Sheep	TDN %	*	71.2	75.7	
Swine	TDN %	*	77.9	82.9	
Calcium	%		0.71	0.76	
Phosphorus	%		1.00	1.06	
Choline	mg/kg		70.	74.	
Niacin	mg/kg		167.0	177.7	
Pantothenic acid	mg/kg		9.9	10.5	
Riboflavin	mg/kg		4.2	4.5	
Thiamine	mg/kg		1.9	2.0	
Alanine	%		1.58	1.68	
Arginine	%		1.90	2.03	
Aspartic acid	%		2.46	2.62	
Glutamic acid	%		6.05	6.44	
Glycine	%		1.75	1.86	
Histidine	%		0.90	0.95	

Continued

Feed Name or Analyses		Mean As Fed	Mean Dry	C.V. ± %
Isoleucine	%	1.39	1.48	
Leucine	%	2.41	2.57	
Lysine	%	1.64	1.75	
Methionine	%	0.70	0.75	
Phenylalanine	%	1.40	1.49	
Proline	%	2.14	2.27	
Serine	%	1.51	1.61	
Threonine	%	1.53	1.63	
Tryptophan	%	0.35	0.37	
Tyrosine	%	0.81	0.86	
Valine	%	1.78	1.89	

Rape, Canada, seeds, cooked pre-press solv-extd grnd, Can 1 mx 1% fat, (5)
Canada rapeseed pre-press solvent extracted meal (CFA)
Ref No 5 08 135 — Canada

Feed Name or Analyses		Mean As Fed	Mean Dry	C.V. ± %
Dry matter	%	92.0	100.0	
Ash	%	7.2	7.8	
Crude fiber	%	9.3	10.1	
Ether extract	%	1.1	1.2	
N-free extract	%	33.9	36.8	
Protein (N x 6.25)	%	40.5	44.0	
Energy	GE Mcal/kg			
Cattle	DE Mcal/kg	* 2.88	3.13	
Sheep	DE Mcal/kg	* 3.12	3.40	
Swine	DE kcal/kg	*3062.	3328.	
Cattle	ME Mcal/kg	* 2.36	2.56	
Sheep	ME Mcal/kg	* 2.56	2.79	
Swine	ME kcal/kg	*2667.	2899.	
Cattle	TDN %	* 65.2	70.9	
Sheep	TDN %	* 70.9	77.0	
Swine	TDN %	* 69.4	75.5	
Calcium	%	0.66	0.72	
Phosphorus	%	0.93	1.01	
Alanine	%	1.89	2.06	
Arginine	%	2.42	2.63	
Aspartic acid	%	2.95	3.20	
Glutamic acid	%	7.40	8.04	
Glycine	%	2.11	2.30	
Histidine	%	1.19	1.29	
Isoleucine	%	1.58	1.72	
Leucine	%	2.95	3.20	
Lysine	%	2.33	2.54	
Methionine	%	0.84	0.91	
Phenylalanine	%	1.67	1.82	
Proline	%	2.69	2.92	
Serine	%	1.85	2.01	
Threonine	%	1.85	2.01	
Tryptophan	%	0.53	0.57	
Tyrosine	%	0.92	1.00	
Valine	%	2.11	2.30	

RAPE, POLISH. Brassica campestris

Rape, Polish, seeds, mech-extd grnd, (5)
Ref No 5 07 871 — Canada

Feed Name or Analyses		Mean As Fed	Mean Dry	C.V. ± %
Dry matter	%	93.5	100.0	
Ash	%	6.8	7.2	
Crude fiber	%	13.8	14.8	
Ether extract	%	6.6	7.1	
N-free extract	%	32.0	34.2	
Protein (N x 6.25)	%	34.2	36.6	
Energy	GE Mcal/kg			
Cattle	DE Mcal/kg	* 3.03	3.24	
Sheep	DE Mcal/kg	* 3.20	3.42	
Swine	DE kcal/kg	*3482.	3726.	
Cattle	ME Mcal/kg	* 2.48	2.66	
Sheep	ME Mcal/kg	* 2.62	2.80	
Swine	ME kcal/kg	*3085.	3301.	
Cattle	TDN %	* 68.7	73.5	
Sheep	TDN %	* 72.5	77.6	
Swine	TDN %	* 79.0	84.5	
Calcium	%	0.71	0.76	
Phosphorus	%	1.01	1.08	
Choline	mg/kg	6402.	6851.	
Niacin	mg/kg	151.1	161.7	
Pantothenic acid	mg/kg	8.5	9.1	
Riboflavin	mg/kg	3.3	3.5	
Thiamine	mg/kg	1.7	1.8	
Alanine	%	1.45	1.55	
Arginine	%	1.93	2.06	
Aspartic acid	%	2.28	2.44	
Cystine	%	0.22	0.24	
Glutamic acid	%	6.16	6.59	
Glycine	%	1.91	2.04	
Histidine	%	0.91	0.98	
Isoleucine	%	1.51	1.62	
Leucine	%	2.39	2.56	
Lysine	%	1.61	1.72	
Methionine	%	0.65	0.70	
Phenylalanine	%	1.35	1.44	
Proline	%	2.01	2.15	
Serine	%	1.42	1.52	
Threonine	%	1.52	1.63	
Tryptophan	%	0.48	0.52	
Tyrosine	%	0.86	0.92	
Valine	%	1.78	1.91	
Isothiocyanate	%	2.213	2.368	
Oxazoladinethione	%	1.05	1.12	

(1) dry forages and roughages (2) pasture, range plants, and forages fed green (3) silages (4) energy feeds (5) protein supplements (6) minerals (7) vitamins (8) additives

Rape, Polish, seeds, solv-extd grnd, (5)
Ref No 5 07 870 — Canada

Feed Name or Analyses		Mean As Fed	Mean Dry	C.V. ± %
Dry matter	%	92.8	100.0	
Crude fiber	%	12.6	13.6	
Ether extract	%	1.3	1.4	
Protein (N x 6.25)	%	37.6	40.5	
Calcium	%	0.66	0.71	
Phosphorus	%	0.93	1.00	
Alanine	%	1.67	1.80	
Arginine	%	1.97	2.12	
Aspartic acid	%	2.64	2.85	
Cystine	%	0.26	0.28	
Glutamic acid	%	6.56	7.07	
Glycine	%	1.89	2.03	
Histidine	%	0.97	1.04	
Isoleucine	%	1.43	1.54	
Leucine	%	2.60	2.80	
Lysine	%	2.04	2.20	
Methionine	%	0.81	0.87	
Phenylalanine	%	1.42	1.53	
Proline	%	2.52	2.72	
Serine	%	1.65	1.78	
Threonine	%	1.68	1.81	
Tyrosine	%	0.78	0.85	
Valine	%	1.90	2.04	
Isothiocyanate	%	4.350	4.688	
Oxazoladinethione	%	2.15	2.32	

RAPE, WINTER. Brassica napus

Rape, winter, hay, dehy, early bloom, (1)
Ref No 1 03 873 — United States

Feed Name or Analyses		Mean As Fed	Mean Dry	C.V. ± %
Dry matter	%	82.5	100.0	
Ash	%	11.4	13.8	
Crude fiber	%	19.4	23.5	
Ether extract	%	2.3	2.8	
N-free extract	%	37.6	45.6	
Protein (N x 6.25)	%	11.8	14.3	
Cattle	dig prot %	* 7.7	9.3	
Goats	dig prot %	* 8.2	9.9	
Horses	dig prot %	* 8.0	9.7	
Rabbits	dig prot %	* 8.0	9.7	
Sheep	dig prot %	* 7.8	9.4	

Rape, winter, hay, s-c, (1)
Ref No 1 03 874 — United States

Feed Name or Analyses		Mean As Fed	Mean Dry	C.V. ± %
Dry matter	%		100.0	
Copper	mg/kg		7.7	
Iron	%		0.024	
Manganese	mg/kg		51.1	

Rape, winter, straw, (1)
Ref No 1 03 875 — United States

Feed Name or Analyses		Mean As Fed	Mean Dry	C.V. ± %
Dry matter	%	79.7	100.0	
Ash	%	3.4	4.3	
Crude fiber	%	43.7	54.8	
Ether extract	%	0.9	1.1	
N-free extract	%	28.9	36.3	
Protein (N x 6.25)	%	2.8	3.5	
Cattle	dig prot %	* 0.0	0.0	
Goats	dig prot %	* 0.0	-0.1	
Horses	dig prot %	* 0.4	0.5	
Rabbits	dig prot %	* 1.1	1.4	
Sheep	dig prot %	* -0.1	-0.2	

Rape, winter, aerial part, fresh, immature, (2)
Ref No 2 03 876 — United States

Feed Name or Analyses		Mean As Fed	Mean Dry	C.V. ± %
Dry matter	%	16.0	100.0	5
Ash	%	2.1	12.8	10
Crude fiber	%	2.5	15.6	14
Ether extract	%	0.6	3.8	12
N-free extract	%	8.1	50.6	
Protein (N x 6.25)	%	2.8	17.2	22
Cattle	dig prot %	* 2.0	12.5	
Goats	dig prot %	* 2.0	12.6	
Horses	dig prot %	* 1.9	12.1	
Rabbits	dig prot %	* 1.9	11.9	
Sheep	dig prot %	* 2.1	13.0	
Energy	GE Mcal/kg			
Cattle	DE Mcal/kg	* 0.44	2.73	
Sheep	DE Mcal/kg	* 0.43	2.69	
Cattle	ME Mcal/kg	* 0.36	2.24	
Sheep	ME Mcal/kg	* 0.35	2.21	
Cattle	TDN %	* 9.9	61.9	
Sheep	TDN %	* 9.7	60.9	
Calcium	%	0.19	1.19	48
Chlorine	%	0.07	0.45	
Copper	mg/kg	1.3	8.2	13
Iron	%	0.003	0.018	26
Magnesium	%	0.01	0.08	18

Feed Name or Analyses			Mean As Fed	Mean Dry	C.V. ± %
Manganese	mg/kg		7.4	46.1	19
Phosphorus	%		0.05	0.34	40
Potassium	%		0.41	2.58	29
Sodium	%		0.01	0.05	
Sulphur	%		0.08	0.49	34
Choline	mg/kg		458.	2859.	
Niacin	mg/kg		13.1	82.0	

Rape, winter, aerial part, ensiled, early bloom, (3)
Ref No 3 03 877 United States

Feed Name or Analyses			Mean As Fed	Mean Dry	C.V. ± %
Dry matter	%		13.8	100.0	19
Ash	%		2.0	14.7	13
Crude fiber	%		2.6	19.2	13
Sheep	dig coef %		56.	56.	
Ether extract	%		1.0	7.0	15
Sheep	dig coef %		70.	70.	
N-free extract	%		5.2	38.0	
Sheep	dig coef %		71.	71.	
Protein (N x 6.25)	%		2.9	21.2	13
Sheep	dig coef %		74.	74.	
Cattle	dig prot %	*	2.1	15.5	
Goats	dig prot %	*	2.1	15.5	
Horses	dig prot %	*	2.1	15.5	
Sheep	dig prot %		2.2	15.7	
Energy	GE Mcal/kg				
Cattle	DE Mcal/kg	*	0.41	3.00	
Sheep	DE Mcal/kg	*	0.39	2.84	
Cattle	ME Mcal/kg	*	0.34	2.46	
Sheep	ME Mcal/kg	*	0.32	2.33	
Cattle	TDN %	*	9.4	68.1	
Sheep	TDN %		8.9	64.4	

Rapeseed meal (CFA) –
see Rape, seeds, extn unspecified grnd, (5)

Rapeseed meal, expeller extracted –
see Rape, seeds, mech-extd grnd, (5)

Rapeseed meal, solvent extracted –
see Rape, seeds, solv-extd grnd, (5)

Rapeseed oil meal, expeller extracted –
see Rape, seeds, mech-extd grnd, (5)

Rapeseed oil meal, solvent extracted –
see Rape, seeds, solv-extd grnd, (5)

Rattlesnake weed –
see Carrot, southwestern, aerial part, fresh, (2)

RATTLEWEED. Rhinanthus spp

Rattleweed, aerial part, fresh, immature, (2)
Ref No 2 03 879 United States

Feed Name or Analyses			Mean As Fed	Mean Dry	C.V. ± %
Dry matter	%			100.0	
Ash	%			6.6	
Crude fiber	%			23.4	
Ether extract	%			1.6	
N-free extract	%			43.7	
Protein (N x 6.25)	%			24.7	
Cattle	dig prot %	*		18.9	
Goats	dig prot %	*		19.6	
Horses	dig prot %	*		18.5	
Rabbits	dig prot %	*		17.7	
Sheep	dig prot %	*		20.0	
Calcium	%			0.42	

Feed Name or Analyses			Mean As Fed	Mean Dry	C.V. ± %
Cobalt	mg/kg			0.190	
Copper	mg/kg			14.6	
Manganese	mg/kg			158.1	
Phosphorus	%			0.13	
Carotene	mg/kg			24.3	
Vitamin A equivalent	IU/g			40.4	

REDPEPPER, BUSH. Capsicum frutescens

Redpepper, bush, seeds w hulls, (8)
Ref No 8 09 014 United States

Feed Name or Analyses			Mean As Fed	Mean Dry	C.V. ± %
Dry matter	%			100.0	
Protein (N x 6.25)	%			16.0	
Alanine	%			0.62	
Arginine	%			1.15	
Aspartic acid	%			2.32	
Glutamic acid	%			2.54	
Glycine	%			0.70	
Histidine	%			0.32	
Hydroxyproline	%			0.34	
Isoleucine	%			0.54	
Leucine	%			0.91	
Lysine	%			0.77	
Methionine	%			0.24	
Phenylalanine	%			0.59	
Proline	%			0.77	
Serine	%			0.66	
Threonine	%			0.59	
Tyrosine	%			0.40	
Valine	%			0.70	

Red stem filaree–
see Alfileria, redstem, hay,s-c, (1)

REDTOP. Agrostis alba

Redtop, hay, s-c, (1)
Ref No 1 03 885 United States

Feed Name or Analyses			Mean As Fed	Mean Dry	C.V. ± %
Dry matter	%		92.3	100.0	1
Ash	%		6.0	6.6	24
Crude fiber	%		28.5	30.9	13
Sheep	dig coef %	*	61.	61.	
Ether extract	%		2.8	3.1	30
Sheep	dig coef %	*	53.	53.	
N-free extract	%		47.5	51.5	
Sheep	dig coef %	*	63.	63.	
Protein (N x 6.25)	%		7.4	8.1	11
Sheep	dig coef %	*	62.	62.	
Cattle	dig prot %	*	3.6	3.9	
Goats	dig prot %	*	3.8	4.1	
Horses	dig prot %	*	4.0	4.4	
Rabbits	dig prot %	*	4.5	4.9	
Sheep	dig prot %		4.6	5.0	
Cellulose (Matrone)	%		24.5	26.6	
Lignin (Ellis)	%		10.0	10.8	11
Fatty acids	%		2.3	2.5	
Energy	GE Mcal/kg				
Cattle	DE Mcal/kg	*	2.51	2.72	
Sheep	DE Mcal/kg	*	2.44	2.64	
Cattle	ME Mcal/kg	*	2.06	2.23	
Sheep	ME Mcal/kg	*	2.00	2.17	
Cattle	TDN %	*	57.0	61.7	
Sheep	TDN %		55.3	59.9	
Calcium	%		0.39	0.43	
Chlorine	%		0.06	0.07	43
Cobalt	mg/kg		0.134	0.146	12

Feed Name or Analyses			Mean As Fed	Mean Dry	C.V. ± %
Copper	mg/kg		3.6	3.9	
Iodine	mg/kg		0.092	0.099	
Iron	%		0.014	0.015	
Magnesium	%		0.20	0.22	
Manganese	mg/kg		208.1	225.5	
Phosphorus	%		0.20	0.22	
Potassium	%		1.74	1.89	
Sodium	%		0.06	0.07	25
Sulphur	%		0.23	0.25	12
Zinc	mg/kg		16.5	17.9	10
Carotene	mg/kg		3.7	4.0	46
Vitamin A equivalent	IU/g		6.1	6.6	

Redtop, hay, s-c, immature, (1)
Ref No 1 03 880 United States

Feed Name or Analyses			Mean As Fed	Mean Dry	C.V. ± %
Dry matter	%		92.0	100.0	2
Ash	%		5.9	6.4	14
Crude fiber	%		29.3	31.9	9
Sheep	dig coef %	*	61.	61.	
Ether extract	%		2.7	2.9	51
Sheep	dig coef %	*	53.	53.	
N-free extract	%		41.4	45.0	
Sheep	dig coef %	*	63.	63.	
Protein (N x 6.25)	%		12.7	13.8	14
Sheep	dig coef %	*	62.	62.	
Cattle	dig prot %	*	8.2	8.9	
Goats	dig prot %	*	8.7	9.4	
Horses	dig prot %	*	8.5	9.2	
Rabbits	dig prot %	*	8.6	9.3	
Sheep	dig prot %		7.9	8.6	
Energy	GE Mcal/kg				
Cattle	DE Mcal/kg	*	2.27	2.47	
Sheep	DE Mcal/kg	*	2.43	2.64	
Cattle	ME Mcal/kg	*	1.86	2.03	
Sheep	ME Mcal/kg	*	1.99	2.16	
Cattle	TDN %	*	51.6	56.1	
Sheep	TDN %	*	55.0	59.8	

Redtop, hay, s-c, early bloom, (1)
Ref No 1 05 683 United States

Feed Name or Analyses			Mean As Fed	Mean Dry	C.V. ± %
Dry matter	%			100.0	
Ash	%			4.1	
Crude fiber	%			27.5	
Ether extract	%			2.8	
N-free extract	%			60.2	
Protein (N x 6.25)	%			5.4	
Cattle	dig prot %	*		1.6	
Goats	dig prot %	*		1.6	
Horses	dig prot %	*		2.1	
Rabbits	dig prot %	*		2.8	
Sheep	dig prot %	*		1.4	
Cellulose (Matrone)	%			26.8	
Lignin (Ellis)	%			5.5	
Energy	GE Mcal/kg				
Cattle	DE Mcal/kg	*		2.65	
Sheep	DE Mcal/kg	*		2.55	
Cattle	ME Mcal/kg	*		2.17	
Sheep	ME Mcal/kg	*		2.09	
Cattle	TDN %	*		60.1	
Sheep	TDN %	*		57.8	

Feed Name or Analyses			Mean As Fed	Mean Dry	C.V. ± %

Redtop, hay, s-c, mid-bloom, (1)

Ref No 1 03 886 United States

Feed Name or Analyses			As Fed	Dry	C.V. ± %
Dry matter	%		92.8	100.0	1
Ash	%		6.0	6.5	9
Crude fiber	%		29.0	31.2	11
Sheep	dig coef %	*	61.	61.	
Ether extract	%		2.4	2.6	27
Sheep	dig coef %	*	53.	53.	
N-free extract	%		44.3	47.7	
Sheep	dig coef %	*	63.	63.	
Protein (N x 6.25)	%		11.1	12.0	20
Sheep	dig coef %	*	62.	62.	
Cattle	dig prot %	*	6.8	7.3	
Goats	dig prot %	*	7.2	7.8	
Horses	dig prot %	*	7.2	7.7	
Rabbits	dig prot %	*	7.4	7.9	
Sheep	dig prot %		6.9	7.4	
Energy	GE Mcal/kg				
Cattle	DE Mcal/kg	*	2.44	2.63	
Sheep	DE Mcal/kg	*	2.44	2.63	
Cattle	ME Mcal/kg	*	2.00	2.16	
Sheep	ME Mcal/kg	*	2.00	2.16	
Cattle	TDN %	*	55.4	59.7	
Sheep	TDN %	*	55.3	59.6	

Redtop, hay, s-c, mid-bloom, cut 1, (1)

Ref No 1 03 881 United States

Feed Name or Analyses			As Fed	Dry	C.V. ± %
Dry matter	%		95.3	100.0	1
Ash	%		6.2	6.5	9
Crude fiber	%		28.8	30.2	9
Ether extract	%		2.5	2.6	29
N-free extract	%		47.1	49.4	
Protein (N x 6.25)	%		10.8	11.3	17
Cattle	dig prot %	*	6.4	6.7	
Goats	dig prot %	*	6.8	7.1	
Horses	dig prot %	*	6.8	7.1	
Rabbits	dig prot %	*	7.0	7.4	
Sheep	dig prot %	*	6.4	6.7	
Energy	GE Mcal/kg				
Cattle	DE Mcal/kg	*	2.59	2.71	
Sheep	DE Mcal/kg	*	2.45	2.57	
Cattle	ME Mcal/kg	*	2.12	2.23	
Sheep	ME Mcal/kg	*	2.01	2.11	
Cattle	TDN %	*	58.6	61.6	
Sheep	TDN %	*	55.6	58.3	
Calcium	%		0.60	0.63	
Phosphorus	%		0.33	0.35	
Potassium	%		1.61	1.69	

Redtop, hay, s-c, full bloom, (1)

Ref No 1 03 882 United States

Feed Name or Analyses			As Fed	Dry	C.V. ± %
Dry matter	%		90.7	100.0	
Ash	%		4.3	4.7	
Crude fiber	%		28.2	31.1	
Sheep	dig coef %		61.	61.	
Ether extract	%		3.4	3.8	
Sheep	dig coef %		53.	53.	
N-free extract	%		46.3	51.0	
Sheep	dig coef %		63.	63.	
Protein (N x 6.25)	%		8.5	9.4	
Sheep	dig coef %		62.	62.	
Cattle	dig prot %	*	4.6	5.1	
Goats	dig prot %	*	4.8	5.3	
Horses	dig prot %	*	5.0	5.5	
Rabbits	dig prot %	*	5.4	5.9	
Sheep	dig prot %		5.3	5.8	
Energy	GE Mcal/kg				
Cattle	DE Mcal/kg	*	2.40	2.64	
Sheep	DE Mcal/kg	*	2.46	2.71	
Cattle	ME Mcal/kg	*	1.97	2.17	
Sheep	ME Mcal/kg	*	2.02	2.22	
Cattle	TDN %	*	54.4	59.9	
Sheep	TDN %		55.7	61.5	

Redtop, hay, s-c, mature, (1)

Ref No 1 03 883 United States

Feed Name or Analyses			As Fed	Dry	C.V. ± %
Dry matter	%		91.7	100.0	2
Ash	%		6.4	7.0	45
Crude fiber	%		33.6	36.6	12
Sheep	dig coef %	*	51.	61.	
Ether extract	%		2.2	2.4	27
Sheep	dig coef %	*	53.	53.	
N-free extract	%		44.2	48.3	
Sheep	dig coef %	*	63.	63.	
Protein (N x 6.25)	%		5.4	5.9	5
Sheep	dig coef %	*	62.	62.	
Cattle	dig prot %	*	1.8	2.0	
Goats	dig prot %	*	1.9	2.0	
Horses	dig prot %	*	2.3	2.5	
Rabbits	dig prot %	*	2.9	3.2	
Sheep	dig prot %		3.3	3.6	
Energy	GE Mcal/kg				
Cattle	DE Mcal/kg	*	2.32	2.53	
Sheep	DE Mcal/kg	*	2.39	2.61	
Cattle	ME Mcal/kg	*	1.90	2.07	
Sheep	ME Mcal/kg	*	1.96	2.14	
Cattle	TDN %	*	52.5	57.3	
Sheep	TDN %	*	54.2	59.2	

Redtop, hay, s-c, cut 2, (1)

Ref No 1 03 884 United States

Feed Name or Analyses			As Fed	Dry	C.V. ± %
Dry matter	%		91.1	100.0	
Ash	%		7.7	8.5	
Crude fiber	%		21.6	23.7	
Sheep	dig coef %	*	61.	61.	
Ether extract	%		2.4	2.6	
Sheep	dig coef %	*	53.	53.	
N-free extract	%		51.1	56.1	
Sheep	dig coef %	*	63.	63.	
Protein (N x 6.25)	%		8.3	9.1	
Sheep	dig coef %	*	62.	62.	
Cattle	dig prot %	*	4.4	4.8	
Goats	dig prot %	*	4.6	5.1	
Horses	dig prot %	*	4.8	5.3	
Rabbits	dig prot %	*	5.2	5.7	
Sheep	dig prot %		5.1	5.6	
Energy	GE Mcal/kg				
Cattle	DE Mcal/kg	*	2.30	2.52	
Sheep	DE Mcal/kg	*	2.35	2.58	
Cattle	ME Mcal/kg	*	1.89	2.07	
Sheep	ME Mcal/kg	*	1.93	2.12	
Cattle	TDN %	*	52.1	57.2	
Sheep	TDN %	*	53.3	58.5	

Redtop, aerial part, fresh, (2)

Ref No 2 03 897 United States

Feed Name or Analyses			As Fed	Dry	C.V. ± %
Dry matter	%		26.7	100.0	17
Organic matter	%		24.6	92.1	
Ash	%		2.8	10.4	
Crude fiber	%		6.6	24.8	
Sheep	dig coef %		74.	74.	
Ether extract	%		1.3	5.0	
Sheep	dig coef %		50.	50.	
N-free extract	%		12.0	44.7	
Sheep	dig coef %		72.	72.	
Protein (N x 6.25)	%		4.0	15.0	
Sheep	dig coef %		64.	64.	
Cattle	dig prot %	*	2.8	10.7	
Goats	dig prot %	*	2.8	10.6	
Horses	dig prot %	*	2.7	10.3	
Rabbits	dig prot %	*	2.7	10.3	
Sheep	dig prot %		2.6	9.6	
Hemicellulose	%		5.0	18.8	5
Lignin (Ellis)	%		1.3	4.9	
Energy	GE Mcal/kg				
Cattle	DE Mcal/kg	*	0.74	2.78	
Sheep	DE Mcal/kg	*	0.77	2.90	
Cattle	ME Mcal/kg	*	0.61	2.28	
Sheep	ME Mcal/kg	*	0.63	2.37	
Cattle	TDN %	*	16.8	63.0	
Sheep	TDN %		17.5	65.7	
Calcium	%		0.17	0.65	
Chlorine	%		0.02	0.09	
Magnesium	%		0.08	0.31	
Phosphorus	%		0.10	0.38	
Potassium	%		0.57	2.12	
Sodium	%		0.01	0.05	
Sulphur	%		0.04	0.16	67
Carotene	mg/kg		53.1	198.9	50
Vitamin A equivalent	IU/g		88.6	331.5	

Redtop, aerial part, fresh, immature, (2)

Ref No 2 03 888 United States

Feed Name or Analyses			As Fed	Dry	C.V. ± %
Dry matter	%		25.3	100.0	4
Calcium	%		0.16	0.64	15
Magnesium	%		0.06	0.23	12
Phosphorus	%		0.11	0.42	14
Potassium	%		0.67	2.64	9
Carotene	mg/kg		92.3	364.9	25
Vitamin A equivalent	IU/g		153.9	608.2	

Redtop, aerial part, fresh, immature, pure stand, (2)

Ref No 2 03 887 United States

Feed Name or Analyses			As Fed	Dry	C.V. ± %
Dry matter	%		25.0	100.0	7
Ash	%		2.7	10.9	8
Crude fiber	%		6.2	24.9	10
Ether extract	%		1.0	3.8	19
N-free extract	%		10.8	43.1	
Protein (N x 6.25)	%		4.3	17.3	8
Cattle	dig prot %	*	3.1	12.6	
Goats	dig prot %	*	3.2	12.7	
Horses	dig prot %	*	3.1	12.2	
Rabbits	dig prot %	*	3.0	12.0	
Sheep	dig prot %	*	3.3	13.1	
Energy	GE Mcal/kg				
Cattle	DE Mcal/kg	*	0.73	2.94	
Sheep	DE Mcal/kg	*	0.72	2.87	
Cattle	ME Mcal/kg	*	0.60	2.41	

(1) dry forages and roughages (2) pasture, range plants, and forages fed green (3) silages (4) energy feeds (5) protein supplements (6) minerals (7) vitamins (8) additives

Feed Name or Analyses			Mean As Fed	Mean Dry	C.V. ± %
Sheep	ME Mcal/kg	*	0.59	2.35	
Cattle	TDN %	*	16.7	66.6	
Sheep	TDN %	*	16.3	65.1	
Calcium	%		0.15	0.61	14
Magnesium	%		0.06	0.23	12
Phosphorus	%		0.11	0.42	17
Potassium	%		0.69	2.76	

Redtop, aerial part, fresh, early bloom, (2)
Ref No 2 03 889 United States

Feed Name or Analyses			Mean As Fed	Mean Dry	C.V. ± %
Dry matter	%		26.3	100.0	
Ash	%		1.8	7.0	13
Crude fiber	%		6.6	25.1	9
Ether extract	%		0.9	3.5	21
N-free extract	%		14.8	56.3	
Protein (N x 6.25)	%		2.1	8.1	43
Cattle	dig prot %	*	1.3	4.8	
Goats	dig prot %	*	1.1	4.1	
Horses	dig prot %	*	1.2	4.4	
Rabbits	dig prot %	*	1.3	4.9	
Sheep	dig prot %	*	1.2	4.5	
Energy	GE Mcal/kg				
Cattle	DE Mcal/kg	*	0.73	2.76	
Sheep	DE Mcal/kg	*	0.77	2.92	
Cattle	ME Mcal/kg	*	0.59	2.26	
Sheep	ME Mcal/kg	*	0.63	2.37	
Cattle	TDN %	*	16.5	62.6	
Sheep	TDN %	*	17.4	66.2	
Carotene	mg/kg		40.1	152.6	16
Vitamin A equivalent	IU/g		66.9	254.3	

Redtop, aerial part, fresh, mid-bloom, (2)
Ref No 2 03 890 United States

Feed Name or Analyses			Mean As Fed	Mean Dry	C.V. ± %
Dry matter	%		39.0	100.0	
Ash	%		2.6	6.7	
Crude fiber	%		12.6	32.3	
Ether extract	%		1.1	2.8	
N-free extract	%		19.8	50.8	
Protein (N x 6.25)	%		2.9	7.4	
Cattle	dig prot %	*	1.6	4.2	
Goats	dig prot %	*	1.4	3.5	
Horses	dig prot %	*	1.5	3.8	
Rabbits	dig prot %	*	1.7	4.4	
Sheep	dig prot %	*	1.5	3.9	
Energy	GE Mcal/kg				
Cattle	DE Mcal/kg	*	1.00	2.56	
Sheep	DE Mcal/kg	*	1.06	2.71	
Cattle	ME Mcal/kg	*	0.82	2.10	
Sheep	ME Mcal/kg	*	0.87	2.22	
Cattle	TDN %	*	22.7	58.1	
Sheep	TDN %	*	24.0	61.5	
Calcium	%		0.13	0.33	
Magnesium	%		0.07	0.18	
Phosphorus	%		0.09	0.23	
Potassium	%		0.83	2.13	
Sulphur	%		0.10	0.26	

Redtop, aerial part, fresh, milk stage, (2)
Ref No 2 03 892 United States

Feed Name or Analyses			Mean As Fed	Mean Dry	C.V. ± %
Dry matter	%		39.0	100.0	8
Ash	%		2.6	6.6	3
Crude fiber	%		12.4	31.8	4
Ether extract	%		1.1	2.8	1
N-free extract	%		20.2	51.7	
Protein (N x 6.25)	%		2.8	7.1	11
Cattle	dig prot %	*	1.5	3.9	
Goats	dig prot %	*	1.2	3.2	
Horses	dig prot %	*	1.4	3.6	
Rabbits	dig prot %	*	1.6	4.2	
Sheep	dig prot %	*	1.4	3.6	
Energy	GE Mcal/kg				
Cattle	DE Mcal/kg	*	1.01	2.58	
Sheep	DE Mcal/kg	*	1.06	2.71	
Cattle	ME Mcal/kg	*	0.83	2.12	
Sheep	ME Mcal/kg	*	0.87	2.22	
Cattle	TDN %	*	22.9	58.6	
Sheep	TDN %	*	24.0	61.5	
Calcium	%		0.15	0.39	8
Magnesium	%		0.07	0.18	
Phosphorus	%		0.10	0.25	9
Potassium	%		0.83	2.13	

Redtop, aerial part, fresh, dough stage, (2)
Ref No 2 03 893 United States

Feed Name or Analyses			Mean As Fed	Mean Dry	C.V. ± %
Dry matter	%			100.0	
Calcium	%			0.23	
Phosphorus	%			0.14	

Redtop, aerial part, fresh, mature, (2)
Ref No 2 03 894 United States

Feed Name or Analyses			Mean As Fed	Mean Dry	C.V. ± %
Dry matter	%			100.0	
Ash	%			5.1	20
Crude fiber	%			28.5	8
Ether extract	%			2.8	15
N-free extract	%			57.9	
Protein (N x 6.25)	%			5.7	7
Cattle	dig prot %	*		2.7	
Goats	dig prot %	*		1.9	
Horses	dig prot %	*		2.4	
Rabbits	dig prot %	*		3.1	
Sheep	dig prot %	*		2.3	
Energy	GE Mcal/kg				
Cattle	DE Mcal/kg	*		2.75	
Sheep	DE Mcal/kg	*		2.81	
Cattle	ME Mcal/kg	*		2.26	
Sheep	ME Mcal/kg	*		2.30	
Cattle	TDN %	*		62.3	
Sheep	TDN %	*		63.8	

Redtop, aerial part, fresh, cut 1, (2)
Ref No 2 03 895 United States

Feed Name or Analyses			Mean As Fed	Mean Dry	C.V. ± %
Dry matter	%			100.0	
Copper	mg/kg			26.0	
Phosphorus	%			0.71	

Redtop, aerial part, fresh, cut 2, (2)
Ref No 2 03 896 United States

Feed Name or Analyses			Mean As Fed	Mean Dry	C.V. ± %
Dry matter	%			100.0	
Calcium	%			0.23	
Phosphorus	%			0.14	

REED. Phragmites spp

Reed, aerial part, ensiled, (3)
Ref No 3 03 898 United States

Feed Name or Analyses			Mean As Fed	Mean Dry	C.V. ± %
Dry matter	%		25.0	100.0	
Ash	%		2.9	11.7	
Crude fiber	%		11.0	43.9	
Sheep	dig coef %		67.	67.	
Ether extract	%		0.9	3.5	
Sheep	dig coef %		62.	62.	
N-free extract	%		7.9	31.6	
Sheep	dig coef %		28.	28.	
Protein (N x 6.25)	%		2.3	9.3	
Sheep	dig coef %		50.	50.	
Cattle	dig prot %	*	1.2	4.7	
Goats	dig prot %	*	1.2	4.7	
Horses	dig prot %	*	1.2	4.7	
Sheep	dig prot %		1.2	4.7	
Energy	GE Mcal/kg				
Cattle	DE Mcal/kg	*	0.58	2.13	
Sheep	DE Mcal/kg	*	0.53	2.11	
Cattle	ME Mcal/kg	*	0.47	1.89	
Sheep	ME Mcal/kg	*	0.43	1.73	
Cattle	TDN %	*	13.0	52.1	
Sheep	TDN %		11.9	47.8	

REED, COMMON. Phragmites communis

Reed, common, aerial part, fresh, (2)
Ref No 2 03 899 United States

Feed Name or Analyses			Mean As Fed	Mean Dry	C.V. ± %
Dry matter	%			100.0	
Ash	%			14.6	
Crude fiber	%			31.9	
Ether extract	%			2.1	
N-free extract	%			40.8	
Protein (N x 6.25)	%			10.6	
Cattle	dig prot %	*		6.9	
Goats	dig prot %	*		6.4	
Horses	dig prot %	*		6.5	
Rabbits	dig prot %	*		6.9	
Sheep	dig prot %	*		6.9	
Energy	GE Mcal/kg				
Cattle	DE Mcal/kg	*		2.53	
Sheep	DE Mcal/kg	*		2.78	
Cattle	ME Mcal/kg	*		2.08	
Sheep	ME Mcal/kg	*		2.28	
Cattle	TDN %	*		57.5	
Sheep	TDN %	*		62.9	
Calcium	%			0.48	
Magnesium	%			0.13	
Phosphorus	%			0.06	

Reed, common, leaves, fresh, (2)
Ref No 2 03 900 United States

Feed Name or Analyses			Mean As Fed	Mean Dry	C.V. ± %
Dry matter	%			100.0	
Ash	%			15.7	
Crude fiber	%			27.4	
Ether extract	%			3.5	
N-free extract	%			36.3	
Protein (N x 6.25)	%			17.1	
Cattle	dig prot %	*		12.4	
Goats	dig prot %	*		12.5	
Horses	dig prot %	*		12.0	

Continued

Feed Name or Analyses			Mean As Fed	Mean Dry	C.V. ± %
Rabbits	dig prot %	*		11.9	
Sheep	dig prot %	*		12.9	
Energy	GE Mcal/kg				
Cattle	DE Mcal/kg	*		2.65	
Sheep	DE Mcal/kg	*		2.93	
Cattle	ME Mcal/kg	*		2.18	
Sheep	ME Mcal/kg	*		2.40	
Cattle	TDN %	*		60.2	
Sheep	TDN %	*		66.4	

Reed, common, stems, fresh, (2)
Ref No 2 03 901 United States

Feed Name or Analyses			Mean As Fed	Mean Dry	C.V. ± %
Dry matter	%			100.0	
Ash	%			4.4	
Crude fiber	%			41.2	
Ether extract	%			0.8	
N-free extract	%			48.8	
Protein (N x 6.25)	%			4.8	
Cattle	dig prot %	*		2.0	
Goats	dig prot %	*		1.0	
Horses	dig prot %	*		1.6	
Rabbits	dig prot %	*		2.4	
Sheep	dig prot %	*		1.5	
Energy	GE Mcal/kg				
Cattle	DE Mcal/kg	*		3.08	
Sheep	DE Mcal/kg	*		2.93	
Cattle	ME Mcal/kg	*		2.53	
Sheep	ME Mcal/kg	*		2.40	
Cattle	TDN %	*		69.9	
Sheep	TDN %	*		66.4	

REEDGRASS. Calamagrostis spp

Reedgrass, hay, s-c, full bloom, (1)
Ref No 1 03 902 United States

Feed Name or Analyses			Mean As Fed	Mean Dry	C.V. ± %
Dry matter	%		93.2	100.0	
Ash	%		5.3	5.7	
Crude fiber	%		33.8	36.3	
Sheep	dig coef %		72.	72.	
Ether extract	%		2.3	2.5	
Sheep	dig coef %		52.	52.	
N-free extract	%		42.2	45.3	
Sheep	dig coef %		69.	69.	
Protein (N x 6.25)	%		9.5	10.2	
Sheep	dig coef %		70.	70.	
Cattle	dig prot %	*	5.4	5.8	
Goats	dig prot %	*	5.7	6.1	
Horses	dig prot %	*	5.8	6.2	
Rabbits	dig prot %	*	6.1	6.5	
Sheep	dig prot %		6.7	7.1	
Energy	GE Mcal/kg				
Cattle	DE Mcal/kg	*	2.59	2.78	
Sheep	DE Mcal/kg	*	2.77	2.97	
Cattle	ME Mcal/kg	*	2.12	2.23	
Sheep	ME Mcal/kg	*	2.27	2.44	
Cattle	TDN %	*	58.8	63.1	
Sheep	TDN %		62.9	67.5	

(1) dry forages and roughages (2) pasture, range plants, and forages fed green (3) silages (4) energy feeds (5) protein supplements (6) minerals (7) vitamins (8) additives

REEDGRASS, BLUEJOINT. Calamagrostis canadensis

Reedgrass, bluejoint, hay, s-c, (1)
Ref No 1 03 903 United States

Feed Name or Analyses			Mean As Fed	Mean Dry	C.V. ± %
Dry matter	%		90.2	100.0	1
Ash	%		6.3	7.0	20
Crude fiber	%		32.7	36.2	6
Sheep	dig coef %		60.	60.	
Ether extract	%		2.4	2.7	25
Sheep	dig coef %		47.	47.	
N-free extract	%		40.9	45.3	
Sheep	dig coef %		60.	60.	
Protein (N x 6.25)	%		7.9	8.7	12
Sheep	dig coef %		66.	66.	
Cattle	dig prot %	*	4.1	4.5	
Goats	dig prot %	*	4.3	4.7	
Horses	dig prot %	*	4.5	5.0	
Rabbits	dig prot %	*	4.9	5.4	
Sheep	dig prot %		5.2	5.8	
Fatty acids	%		2.3	2.6	
Energy	GE Mcal/kg				
Cattle	DE Mcal/kg	*	2 09	2.32	
Sheep	DE Mcal/kg	*	2.29	2.54	
Cattle	ME Mcal/kg	*	1.72	1.90	
Sheep	ME Mcal/kg	*	1.88	2.08	
Cattle	TDN %	*	47.5	52.7	
Sheep	TDN %		51.9	57.6	

Ref No 1 03 903 Canada

Feed Name or Analyses			Mean As Fed	Mean Dry	C.V. ± %
Dry matter	%		95.3	100.0	1
Crude fiber	%		31.6	33.2	8
Sheep	dig coef %		60.	60.	
Ether extract	%		1.4	1.5	
Sheep	dig coef %		47.	47.	
N-free extract	%				
Sheep	dig coef %		60.	60.	
Protein (N x 6.25)	%		9.6	10.1	38
Sheep	dig coef %		66.	66.	
Cattle	dig prot %	*	5.4	5.7	
Goats	dig prot %	*	5.7	6.0	
Horses	dig prot %	*	5.8	6.1	
Rabbits	dig prot %	*	6.1	6.4	
Sheep	dig prot %		6.3	6.6	
Energy	GE Mcal/kg				
Sheep	DE Mcal/kg	*	2.42	2.53	
Sheep	ME Mcal/kg	*	1.98	2.08	
Sheep	TDN %		54.8	57.5	
Calcium	%		0.40	0.42	7
Phosphorus	%		0.15	0.15	54
Potassium	%		0.10	0.11	
Carotene	mg/kg		19.5	20.4	98
Vitamin A equivalent	IU/g		32.4	34.0	

Reedgrass, bluejoint, aerial part, fresh, (2)
Ref No 2 03 904 United States

Feed Name or Analyses			Mean As Fed	Mean Dry	C.V. ± %
Dry matter	%		44.6	100.0	
Ash	%		4.1	9.2	
Crude fiber	%		15.2	34.1	
Ether extract	%		1.2	2.7	
N-free extract	%		20.0	44.8	
Protein (N x 6.25)	%		4.1	9.2	
Cattle	dig prot %	*	2.5	5.7	
Goats	dig prot %	*	2.3	5.1	
Horses	dig prot %	*	2.4	5.3	
Rabbits	dig prot %	*	2.6	5.8	
Sheep	dig prot %	*	2.5	5.6	
Energy	GE Mcal/kg				
Cattle	DE Mcal/kg	*	1.09	2.44	
Sheep	DE Mcal/kg	*	1.12	2.52	
Cattle	ME Mcal/kg	*	0.89	2.00	
Sheep	ME Mcal/kg	*	0.92	2.07	
Cattle	TDN %	*	24.7	55.4	
Sheep	TDN %	*	25.5	57.2	
Phosphorus	%		0.10	0.22	

Refuse screenings (CFA) –
see Grains, screenings, refuse mx 100% small weed seeds chaff hulls dust scourings noxious seeds, (4)

REINDEERMOSS. Cladonia rangiferina

Reindeermoss, aerial part, fresh, immature, (2)
Ref No 2 03 905 United States

Feed Name or Analyses			Mean As Fed	Mean Dry	C.V. ± %
Dry matter	%		80.8	100.0	
Ash	%		1.1	1.3	
Crude fiber	%		26.6	32.9	
Sheep	dig coef %		29.	29.	
Ether extract	%		1.7	2.1	
Sheep	dig coef %		46.	46.	
N-free extract	%		47.0	58.2	
Sheep	dig coef %		22.	22.	
Protein (N x 6.25)	%		4.4	5.5	
Sheep	dig coef %		-1.	-1.	
Cattle	dig prot %	*	2.1	2.6	
Goats	dig prot %	*	1.4	1.7	
Horses	dig prot %	*	1.8	2.2	
Rabbits	dig prot %	*	2.4	2.9	
Sheep	dig prot %		0.0	0.0	
Energy	GE Mcal/kg				
Cattle	DE Mcal/kg	*	1.04	1.29	
Sheep	DE Mcal/kg	*	0.87	1.08	
Cattle	ME Mcal/kg	*	0.86	1.06	
Sheep	ME Mcal/kg	*	0.71	0.88	
Cattle	TDN %	*	23.6	29.2	
Sheep	TDN %		19.7	24.4	

Reindeermoss, aerial part, fresh, dormant, (2)
Ref No 2 03 907 United States

Feed Name or Analyses			Mean As Fed	Mean Dry	C.V. ± %
Dry matter	%		91.4	100.0	
Ash	%		1.4	1.5	
Crude fiber	%		32.8	35.9	
Sheep	dig coef %		26.	26.	
Ether extract	%		2.6	2.8	
Sheep	dig coef %		33.	33.	
N-free extract	%		50.4	55.1	
Sheep	dig coef %		40.	40.	
Protein (N x 6.25)	%		4.3	4.7	
Sheep	dig coef %		-40.	-40.	
Cattle	dig prot %	*	1.7	1.9	
Goats	dig prot %	*	0.9	0.9	
Horses	dig prot %	*	1.4	1.5	
Rabbits	dig prot %	*	2.1	2.3	
Sheep	dig prot %		-1.7	-1.8	
Energy	GE Mcal/kg				
Cattle	DE Mcal/kg	*	1.49	1.63	
Sheep	DE Mcal/kg	*	1.27	1.39	
Cattle	ME Mcal/kg	*	1.22	1.34	
Sheep	ME Mcal/kg	*	1.04	1.14	

Feed Name or Analyses			Mean As Fed	Mean Dry	C.V. ± %
Cattle	TDN %	*	33.9	37.1	
Sheep	TDN %		28.8	31.5	

RHODESGRASS. Chloris gayana

Rhodesgrass, hay, s-c, (1)

Ref No 1 03 913 United States

Feed Name or Analyses			Mean As Fed	Mean Dry	C.V. ± %
Dry matter	%		90.7	100.0	
Ash	%		8.2	9.1	
Crude fiber	%		32.4	35.7	
Sheep	dig coef %		69.	69.	
Ether extract	%		1.5	1.7	
Sheep	dig coef %		49.	49.	
N-free extract	%		42.8	47.2	
Sheep	dig coef %		61.	61.	
Protein (N x 6.25)	%		5.7	6.3	
Sheep	dig coef %		45.	45.	
Cattle	dig prot %	*	2.2	2.4	
Goats	dig prot %	*	2.2	2.4	
Horses	dig prot %	*	2.6	2.9	
Rabbits	dig prot %	*	3.2	3.5	
Sheep	dig prot %		2.6	2.8	
Fatty acids	%		1.3	1.5	
Energy	GE Mcal/kg				
Cattle	DE Mcal/kg	*	2.20	2.43	
Sheep	DE Mcal/kg	*	2.33	2.56	
Cattle	ME Mcal/kg	*	1.81	1.99	
Sheep	ME Mcal/kg	*	1.91	2.10	
Cattle	TDN %	*	50.0	55.1	
Sheep	TDN %		52.7	58.1	
Calcium	%		0.36	0.39	
Phosphorus	%		0.28	0.30	
Potassium	%		1.20	1.33	

Rhodesgrass, hay, s-c, immature, (1)

Ref No 1 03 910 United States

Feed Name or Analyses			Mean As Fed	Mean Dry	C.V. ± %
Dry matter	%			100.0	
Calcium	%			0.59	
Magnesium	%			0.35	0
Phosphorus	%			0.41	
Potassium	%			2.27	

Rhodesgrass, hay, s-c, mature, (1)

Ref No 1 03 911 United States

Feed Name or Analyses			Mean As Fed	Mean Dry	C.V. ± %
Dry matter	%		92.8	100.0	
Ash	%		10.5	11.3	
Crude fiber	%		32.1	34.6	
Sheep	dig coef %		68.	68.	
Ether extract	%		1.7	1.8	
Sheep	dig coef %		45.	45.	
N-free extract	%		43.1	46.4	
Sheep	dig coef %		58.	58.	
Protein (N x 6.25)	%		5.5	5.9	
Sheep	dig coef %		44.	44.	
Cattle	dig prot %	*	1.9	2.0	
Goats	dig prot %	*	1.9	2.1	
Horses	dig prot %	*	2.4	2.5	
Rabbits	dig prot %	*	3.0	3.2	
Sheep	dig prot %		2.4	2.6	
Energy	GE Mcal/kg				
Cattle	DE Mcal/kg	*	2.43	2.61	
Sheep	DE Mcal/kg	*	2.24	2.42	
Cattle	ME Mcal/kg	*	1.99	2.14	
Sheep	ME Mcal/kg	*	1.84	1.98	
Cattle	TDN %	*	55.0	59.3	
Sheep	TDN %		50.9	54.9	

Rhodesgrass, hay, s-c, over ripe, (1)

Ref No 1 03 912 United States

Feed Name or Analyses			Mean As Fed	Mean Dry	C.V. ± %
Dry matter	%		90.7	100.0	1
Ash	%		8.6	9.5	7
Crude fiber	%		32.5	35.8	5
Ether extract	%		1.4	1.5	14
N-free extract	%		41.7	46.0	
Protein (N x 6.25)	%		6.5	7.2	24
Cattle	dig prot %	*	2.9	3.2	
Goats	dig prot %	*	3.0	3.3	
Horses	dig prot %	*	3.3	3.6	
Rabbits	dig prot %	*	3.8	4.2	
Sheep	dig prot %	*	2.7	3.0	
Energy	GE Mcal/kg				
Cattle	DE Mcal/kg	*	2.20	2.42	
Sheep	DE Mcal/kg	*	2.15	2.37	
Cattle	ME Mcal/kg	*	1.80	1.99	
Sheep	ME Mcal/kg	*	1.76	1.95	
Cattle	TDN %	*	49.9	55.0	
Sheep	TDN %	*	48.8	53.8	
Carotene	mg/kg		1.8	2.0	
Vitamin A equivalent	IU/g		3.0	3.3	

Rhodesgrass, aerial part, fresh, (2)

Ref No 2 03 916 United States

Feed Name or Analyses			Mean As Fed	Mean Dry	C.V. ± %
Dry matter	%		27.1	100.0	
Ash	%		3.3	12.1	
Crude fiber	%		10.1	37.4	
Cattle	dig coef %		70.	70.	
Sheep	dig coef %	*	75.	75.	
Ether extract	%		0.5	1.8	
Cattle	dig coef %		50.	50.	
Sheep	dig coef %	*	36.	36.	
N-free extract	%		11.1	41.2	
Cattle	dig coef %		60.	60.	
Sheep	dig coef %	*	67.	67.	
Protein (N x 6.25)	%		2.0	7.6	
Cattle	dig coef %		58.	58.	
Sheep	dig coef %	*	62.	62.	
Cattle	dig prot %		1.2	4.4	
Goats	dig prot %	*	1.0	3.6	
Horses	dig prot %	*	1.1	3.9	
Rabbits	dig prot %	*	1.2	4.5	
Sheep	dig prot %		1.3	4.7	
Energy	GE Mcal/kg				
Cattle	DE Mcal/kg	*	0.68	2.53	
Sheep	DE Mcal/kg	*	0.74	2.72	
Cattle	ME Mcal/kg	*	0.56	2.07	
Sheep	ME Mcal/kg	*	0.60	2.23	
Cattle	TDN %		15.5	57.3	
Sheep	TDN %		16.7	61.8	
Calcium	%		0.17	0.63	
Phosphorus	%		0.11	0.40	
Potassium	%		0.62	2.29	

Rhodesgrass, aerial part, fresh, immature, (2)

Ref No 2 03 914 United States

Feed Name or Analyses			Mean As Fed	Mean Dry	C.V. ± %
Dry matter	%		26.4	100.0	13
Ash	%		2.8	10.5	17
Crude fiber	%		9.9	37.5	6
Ether extract	%		0.4	1.6	17
N-free extract	%		11.4	43.0	
Protein (N x 6.25)	%		2.0	7.4	9
Cattle	dig prot %	*	1.1	4.2	
Goats	dig prot %	*	0.9	3.5	
Horses	dig prot %	*	1.0	3.8	
Rabbits	dig prot %	*	1.2	4.4	
Sheep	dig prot %	*	1.0	3.9	
Energy	GE Mcal/kg				
Cattle	DE Mcal/kg	*	0.72	2.73	
Sheep	DE Mcal/kg	*	0.75	2.83	
Cattle	ME Mcal/kg	*	0.59	2.23	
Sheep	ME Mcal/kg	*	0.61	2.32	
Cattle	TDN %	*	16.3	61.8	
Sheep	TDN %	*	16.9	64.1	
Carotene	mg/kg		74.7	283.1	
Vitamin A equivalent	IU/g		124.6	471.9	

Rhodesgrass, aerial part, fresh, early bloom, (2)

Ref No 2 03 915 United States

Feed Name or Analyses			Mean As Fed	Mean Dry	C.V. ± %
Dry matter	%		28.8	100.0	
Ash	%		1.7	5.8	
Crude fiber	%		10.9	37.7	
Sheep	dig coef %		75.	75.	
Ether extract	%		0.4	1.4	
Sheep	dig coef %		36.	36.	
N-free extract	%		13.3	46.3	
Sheep	dig coef %		67.	67.	
Protein (N x 6.25)	%		2.5	8.8	
Sheep	dig coef %		62.	62.	
Cattle	dig prot %	*	1.5	5.4	
Goats	dig prot %	*	1.4	4.8	
Horses	dig prot %	*	1.4	5.0	
Rabbits	dig prot %	*	1.6	5.5	
Sheep	dig prot %		1.6	5.5	
Energy	GE Mcal/kg				
Cattle	DE Mcal/kg	*	0.90	3.12	
Sheep	DE Mcal/kg	*	0.84	2.91	
Cattle	ME Mcal/kg	*	0.74	2.56	
Sheep	ME Mcal/kg	*	0.69	2.38	
Cattle	TDN %	*	20.4	70.7	
Sheep	TDN %		19.0	65.9	

RHODODENDRON. Rhododendron spp

Rhododendron, browse, mature, (2)

Ref No 2 03 917 United States

Feed Name or Analyses			Mean As Fed	Mean Dry	C.V. ± %
Dry matter	%			100.0	
Ash	%			4.8	11
Crude fiber	%			21.3	3
Ether extract	%			6.0	10
N-free extract	%			61.6	
Protein (N x 6.25)	%			6.3	4
Cattle	dig prot %	*		3.2	
Goats	dig prot %	*		2.4	
Horses	dig prot %	*		2.9	
Rabbits	dig prot %	*		3.5	
Sheep	dig prot %	*		2.9	
Calcium	%			1.42	11
Cobalt	mg/kg			0.053	27
Copper	mg/kg			4.4	24
Iron	%			0.037	15
Manganese	mg/kg			646.2	37
Phosphorus	%			0.08	30

Feed Name or Analyses			Mean As Fed	Mean Dry	C.V. ± %

RHUBARB. Rheun rhaponticum

Rhubarb, leaves, fresh, (2)
Ref No 2 03 918 United States

Feed Name or Analyses			As Fed	Dry	C.V. ± %
Dry matter	%		11.4	100.0	
Crude fiber	%		0.9	8.0	
Ether extract	%		0.7	6.5	
Protein (N x 6.25)	%		3.1	27.6	
Cattle	dig prot %	*	2.4	21.4	
Goats	dig prot %	*	2.5	22.3	
Horses	dig prot %	*	2.4	21.0	
Rabbits	dig prot %	*	2.3	20.0	
Sheep	dig prot %	*	2.6	22.7	
Carotene	mg/kg		45.9	403.0	
Riboflavin	mg/kg		0.8	7.1	
α-tocopherol	mg/kg		139.1	1220.0	
Vitamin A equivalent	IU/g		76.6	671.8	
Arginine	%		0.15	1.30	
Histidine	%		0.06	0.50	
Isoleucine	%		0.13	1.10	
Leucine	%		0.27	2.40	
Lysine	%		0.17	1.50	
Methionine	%		0.03	0.30	
Phenylalanine	%		0.19	1.70	
Threonine	%		0.13	1.10	
Tryptophan	%		0.05	0.40	
Valine	%		0.17	1.50	

RHYNCHOSIA, TEXAS. Rhynchosia texana

Rhynchosia, Texas, aerial part, fresh, mid-bloom, (2)
Ref No 2 03 919 United States

Feed Name or Analyses			As Fed	Dry	C.V. ± %
Dry matter	%		90.2	100.0	2
Ash	%		7.5	8.3	33
Crude fiber	%		23.2	25.8	22
Ether extract	%		3.1	3.4	22
N-free extract	%		42.6	47.3	
Protein (N x 6.25)	%		13.8	15.3	8
Cattle	dig prot %	*	9.8	10.9	
Goats	dig prot %	*	9.8	10.8	
Horses	dig prot %	*	9.5	10.5	
Rabbits	dig prot %	*	9.5	10.5	
Sheep	dig prot %	*	10.1	11.2	
Calcium	%		1.49	1.65	
Magnesium	%		0.23	0.25	
Phosphorus	%		0.21	0.23	
Potassium	%		1.24	1.37	

RICE. Oryza sativa

Rice, bran w germ w hulls, (1)
Ref No 1 03 931 United States

Feed Name or Analyses			As Fed	Dry	C.V. ± %
Dry matter	%		91.2	100.0	
Ash	%		15.0	16.5	
Crude fiber	%		25.4	27.8	
Sheep	dig coef %		7.	7.	
Ether extract	%		4.5	4.9	
Sheep	dig coef %		75.	75.	
N-free extract	%		39.6	43.4	
Sheep	dig coef %		53.	53.	
Protein (N x 6.25)	%		6.7	7.4	
Sheep	dig coef %		46.	46.	
Cattle	dig prot %	*	3.1	3.3	
Goats	dig prot %	*	3.2	3.5	
Horses	dig prot %	*	3.5	3.8	
Rabbits	dig prot %	*	4.0	4.4	
Sheep	dig prot %		3.1	3.4	
Energy	GE Mcal/kg				
Sheep	DE Mcal/kg	*	.47	1.61	
Sheep	ME Mcal/kg	*	1.21	1.32	
Sheep	TDN %		33.4	36.6	

Rice, hay, s-c, dough stage, (1)
Ref No 1 03 922 United States

Feed Name or Analyses			As Fed	Dry	C.V. ± %
Dry matter	%		93.0	100.0	
Ash	%		15.1	16.2	
Crude fiber	%		31.0	33.3	
Sheep	dig coef %		51.	51.	
Ether extract	%		1.4	1.5	
Sheep	dig coef %		56.	56.	
N-free extract	%		39.9	42.9	
Sheep	dig coef %		48.	48.	
Protein (N x 6.25)	%		5.7	6.1	
Sheep	dig coef %		38.	38.	
Cattle	dig prot %	*	2.1	2.2	
Goats	dig prot %	*	2.1	2.3	
Horses	dig prot %	*	2.5	2.7	
Rabbits	dig prot %	*	3.1	3.4	
Sheep	dig prot %		2.2	2.3	
Energy	GE Mcal/kg				
Cattle	DE Mcal/kg	*	1.88	2.02	
Sheep	DE Mcal/kg	*	1.71	1.84	
Cattle	ME Mcal/kg	*	1.54	1.66	
Sheep	ME Mcal/kg	*	1.40	1.51	
Cattle	TDN %	*	42.6	45.8	
Sheep	TDN %		38.9	41.8	

Rice, hulls, (1)
Rice hulls (AAFCO)
Ref No 1 08 075 United States

Feed Name or Analyses			As Fed	Dry	C.V. ± %
Dry matter	%		92.4	100.0	1
Ash	%		18.4	19.9	8
Crude fiber	%		41.1	44.5	2
Sheep	dig coef %		12.	12.	
Ether extract	%		0.8	0.9	43
Sheep	dig coef %		31.	31.	
N-free extract	%		29.2	31.6	
Sheep	dig coef %		29.	29.	
Protein (N x 6.25)	%		2.8	3.1	8
Sheep	dig coef %		7.	7.	
Cattle	dig prot %	*	-0.3	-0.3	
Goats	dig prot %	*	-0.4	-0.5	
Horses	dig prot %	*	0.1	0.1	
Rabbits	dig prot %	*	1.0	1.0	
Sheep	dig prot %		0.2	0.2	
Cellulose (Matrone)	%		39.0	42.2	
Pentosans	%		19.8	21.4	14
Lignin (Ellis)	%		19.8	21.4	
Fatty acids	%		0.8	0.9	
Energy	GE Mcal/kg				
Cattle	DE Mcal/kg	*	0.44	0.48	
Sheep	DE Mcal/kg	*	0.63	0.68	
Cattle	ME Mcal/kg	*	0.36	0.39	
Sheep	ME Mcal/kg	*	0.51	0.56	
Cattle	TDN %	*	10.0	10.8	
Sheep	TDN %		14.2	15.4	
Calcium	%		0.08	0.09	
Manganese	mg/kg		308.0	333.3	
Phosphorus	%		0.07	0.08	
Potassium	%		0.31	0.34	
Niacin	mg/kg		36.5	39.5	35
Riboflavin	mg/kg		0.6	0.7	
Thiamine	mg/kg		2.2	2.4	43

Rice, straw, (1)
Ref No 1 03 925 United States

Feed Name or Analyses			As Fed	Dry	C.V. ± %
Dry matter	%		90.5	100.0	2
Ash	%		15.4	17.0	14
Crude fiber	%		31.8	35.1	7
Cattle	dig coef %		64.	64.	
Sheep	dig coef %		60.	60.	
Ether extract	%		1.3	1.4	28
Cattle	dig coef %		44.	44.	
Sheep	dig coef %		34.	34.	
N-free extract	%		38.0	42.0	13
Cattle	dig coef %		50.	50.	
Sheep	dig coef %		42.	42.	
Protein (N x 6.25)	%		4.0	4.5	30
Cattle	dig coef %		6.	6.	
Sheep	dig coef %		37.	37.	
Cattle	dig prot %		0.2	0.2	
Goats	dig prot %	*	0.7	0.7	
Horses	dig prot %	*	1.2	1.3	
Rabbits	dig prot %	*	1.9	2.1	
Sheep	dig prot %		1.5	1.7	
Pentosans	%		16.0	17.7	
Fatty acids	%		1.4	1.5	
Energy	GE Mcal/kg				
Cattle	DE Mcal/kg	*	1.80	1.99	
Sheep	DE Mcal/kg	*	1.65	1.82	
Cattle	ME Mcal/kg	*	1.48	1.63	
Sheep	ME Mcal/kg	*	1.35	1.49	
Cattle	TDN %		40.9	45.2	
Sheep	TDN %		37.4	41.3	
Calcium	%		0.19	0.21	
Magnesium	%		0.10	0.11	
Manganese	mg/kg		313.1	345.8	
Phosphorus	%		0.07	0.08	
Potassium	%		1.19	1.32	
Sodium	%		0.28	0.31	

Rice, straw, boiled in water, (1)
Ref No 1 03 924 United States

Feed Name or Analyses			As Fed	Dry	C.V. ± %
Dry matter	%		91.7	100.0	
Ash	%		17.4	19.0	
Crude fiber	%		33.5	36.5	
Sheep	dig coef %		62.	62.	
Ether extract	%		1.7	1.8	
Sheep	dig coef %		47.	47.	
N-free extract	%		33.7	36.7	
Sheep	dig coef %		47.	47.	
Protein (N x 6.25)	%		5.5	6.0	
Sheep	dig coef %		37.	37.	
Cattle	dig prot %	*	2.0	2.1	
Goats	dig prot %	*	2.0	2.2	
Horses	dig prot %	*	2.4	2.6	
Rabbits	dig prot %	*	3.0	3.3	
Sheep	dig prot %		2.0	2.2	
Energy	GE Mcal/kg				
Cattle	DE Mcal/kg	*	1.84	2.01	
Sheep	DE Mcal/kg	*	1.78	1.94	
Cattle	ME Mcal/kg	*	1.51	1.65	

(1) dry forages and roughages (2) pasture, range plants, and forages fed green (3) silages (4) energy feeds (5) protein supplements (6) minerals (7) vitamins (8) additives

Feed Name or Analyses			Mean As Fed	Mean Dry	C.V. ± %
Sheep	ME Mcal/kg	*	1.46	1.59	
Cattle	TDN %	*	41.9	45.7	
Sheep	TDN %		40.4	44.0	

Rice, straw, treated w calcium hydroxide, (1)

Ref No 1 03 926 United States

Feed Name or Analyses			Mean As Fed	Mean Dry	C.V. ± %
Dry matter	%		91.2	100.0	
Ash	%		19.2	21.0	
Crude fiber	%		39.7	43.5	
Sheep	dig coef %		86.	86.	
Ether extract	%		1.6	1.8	
Sheep	dig coef %		27.	27.	
N-free extract	%		26.4	28.9	
Sheep	dig coef %		62.	62.	
Protein (N x 6.25)	%		4.4	4.8	
Sheep	dig coef %		0.	0.	
Cattle	dig prot %	*	1.0	1.1	
Goats	dig prot %	*	0.9	1.0	
Horses	dig prot %	*	1.5	1.6	
Rabbits	dig prot %	*	2.2	2.4	
Energy	GE Mcal/kg				
Cattle	DE Mcal/kg	*	2.17	2.38	
Sheep	DE Mcal/kg	*	2.27	2.49	
Cattle	ME Mcal/kg	*	1.78	1.95	
Sheep	ME Mcal/kg	*	1.86	2.04	
Cattle	TDN %	*	49.2	53.9	
Sheep	TDN %		51.5	56.4	

Rice, straw, treated w sodium hydroxide dehy, (1)

Ref No 1 03 927 United States

Feed Name or Analyses			Mean As Fed	Mean Dry	C.V. ± %
Dry matter	%		91.4	100.0	
Ash	%		14.3	15.7	
Crude fiber	%		36.6	40.0	
Sheep	dig coef %		78.	78.	
Ether extract	%		1.5	1.6	
Sheep	dig coef %		31.	31.	
N-free extract	%		33.6	36.8	
Sheep	dig coef %		59.	59.	
Protein (N x 6.25)	%		5.4	5.9	
Sheep	dig coef %		20.	20.	
Cattle	dig prot %	*	1.9	2.0	
Goats	dig prot %	*	1.9	2.1	
Horses	dig prot %	*	2.3	2.5	
Rabbits	dig prot %	*	2.9	3.2	
Sheep	dig prot %		1.1	1.2	
Energy	GE Mcal/kg				
Cattle	DE Mcal/kg	*	2.19	2.39	
Sheep	DE Mcal/kg	*	2.22	2.43	
Cattle	ME Mcal/kg	*	1.79	1.96	
Sheep	ME Mcal/kg	*	1.82	2.00	
Cattle	TDN %	*	49.6	54.3	
Sheep	TDN %		50.5	55.2	

Rice, bran w germ, dry milled, (4)

Rice bran (CFA)

Ref No 4 03 929 United States

Feed Name or Analyses			Mean As Fed	Mean Dry	C.V. ± %
Dry matter	%		90.8	100.0	1
Ash	%		10.8	11.9	14
Crude fiber	%		12.5	13.7	14
Cattle	dig coef %		13.	13.	
Sheep	dig coef %		30.	30.	
Swine	dig coef %		20.	20.	
Ether extract	%		13.1	14.4	16
Cattle	dig coef %		55.	55.	
Sheep	dig coef %		86.	86.	
Swine	dig coef %		83.	83.	
N-free extract	%		42.0	46.3	7
Cattle	dig coef %		78.	78.	
Sheep	dig coef %		76.	76.	
Swine	dig coef %		73.	73.	
Protein (N x 6.25)	%		12.5	13.8	10
Cattle	dig coef %		65.	65.	
Sheep	dig coef %		68.	68.	
Swine	dig coef %		72.	72.	
Cattle	dig prot %		8.1	9.0	
Goats	dig prot %	*	8.9	9.9	
Horses	dig prot %	*	8.9	9.9	
Sheep	dig prot %		8.5	9.4	
Swine	dig prot %		8.9	9.8	
Energy	GE Mcal/kg		2.78	3.06	
Cattle	DE Mcal/kg	*	2.59	2.85	
Sheep	DE Mcal/kg	*	3.06	3.37	
Swine	DE kcal/kg	*	2932.	3228.	
Cattle	ME Mcal/kg	*	2.12	2.34	
Sheep	ME Mcal/kg	*	2.51	2.76	
Swine	ME kcal/kg	*	2733.	3009.	
Cattle	TDN %		58.7	64.6	
Sheep	TDN %		69.4	76.5	
Swine	TDN %		66.5	73.2	
Calcium	%		0.07	0.08	
Iron	%		0.019	0.021	
Phosphorus	%		1.61	1.77	
Potassium	%		1.51	1.66	
Choline	mg/kg		972.	1071.	
Niacin	mg/kg		282.4	311.0	
Pantothenic acid	mg/kg		21.0	23.1	
Riboflavin	mg/kg		2.4	2.6	
Thiamine	mg/kg		22.7	25.0	
Arginine	%		1.00	1.10	
Cystine	%		0.09	0.10	
Lysine	%		0.50	0.55	
Methionine	%		0.20	0.22	
Tryptophan	%		0.20	0.22	

Rice, bran w germ, dry milled, mx 13% fiber calcium carbonate declared above 3% mn, (4)

Rice bran (AAFCO)

Ref No 4 03 928 United States

Feed Name or Analyses			Mean As Fed	Mean Dry	C.V. ± %
Dry matter	%		90.7	100.0	2
Ash	%		12.1	13.3	9
Crude fiber	%		11.2	12.4	15
Cattle	dig coef %	*	13.	13.	
Sheep	dig coef %	*	30.	30.	
Ether extract	%		14.4	15.8	18
Cattle	dig coef %	*	55.	55.	
Sheep	dig coef %	*	86.	86.	
N-free extract	%		40.1	44.2	
Cattle	dig coef %	*	78.	78.	
Sheep	dig coef %	*	76.	76.	
Protein (N x 6.25)	%		13.0	14.3	8
Cattle	dig coef %	*	65.	65.	
Sheep	dig coef %	*	68.	68.	
Cattle	dig prot %		8.4	9.3	
Goats	dig prot %	*	9.4	10.3	
Horses	dig prot %	*	9.4	10.3	
Sheep	dig prot %		8.8	9.7	
Pentosans	%		10.4	11.5	10
Linoleic	%		3.601	3.970	
Energy	GE Mcal/kg				
Cattle	DE Mcal/kg	*	2.60	2.86	
Sheep	DE Mcal/kg	*	3.10	3.42	
Swine	DE kcal/kg	*	2907.	3205.	
Cattle	ME Mcal/kg	*	2.13	2.35	
Chickens	ME_n kcal/kg		1630.	1797.	
Sheep	ME Mcal/kg	*	2.54	2.81	
Swine	ME kcal/kg	*	2706.	2984.	
Cattle	NE_m Mcal/kg	*	1.30	1.43	
Cattle	NE_{gain} Mcal/kg	*	0.77	0.85	
Cattle	$NE_{lactating\ cows}$ Mcal/kg	*	1.51	1.67	
Cattle	TDN %		58.9	64.9	
Sheep	TDN %		70.4	77.6	
Swine	TDN %	*	65.9	72.7	
Calcium	%		0.07	0.08	41
Chlorine	%		0.07	0.08	9
Copper	mg/kg		13.0	14.3	22
Iron	%		0.019	0.021	18
Magnesium	%		0.95	1.05	11
Manganese	mg/kg		418.7	461.6	37
Phosphorus	%		1.59	1.75	17
Potassium	%		1.74	1.92	4
Sodium	%		0.00	0.00	4
Sulphur	%		0.18	0.20	4
Zinc	mg/kg		30.0	33.1	
Biotin	mg/kg		0.42	0.46	23
Niacin	mg/kg		304.1	335.3	17
Pantothenic acid	mg/kg		23.6	26.0	38
Riboflavin	mg/kg		2.6	2.9	27
Thiamine	mg/kg		22.5	24.8	37
a-tocopherol	mg/kg		60.0	66.4	
Vitamin B_6	mg/kg		29.13	32.12	
Arginine	%		0.50	0.55	64
Cystine	%		0.10	0.11	80
Histidine	%		0.20	0.22	80
Isoleucine	%		0.40	0.44	
Leucine	%		0.60	0.66	
Lysine	%		0.50	0.55	21
Phenylalanine	%		0.40	0.44	
Threonine	%		0.40	0.44	
Tryptophan	%		0.10	0.11	
Valine	%		0.60	0.66	

Rice, bran w germ, solv-extd, mn 14% protein mx 14% fiber, (4)

Solvent extracted rice bran (AAFCO)

Ref No 4 03 930 United States

Feed Name or Analyses			Mean As Fed	Mean Dry	C.V. ± %
Dry matter	%		90.9	100.0	
Ash	%		13.6	15.0	
Crude fiber	%		12.0	13.2	
Cattle	dig coef %	*	13.	13.	
Sheep	dig coef %	*	30.	30.	
Ether extract	%		3.1	3.4	
Cattle	dig coef %	*	55.	55.	
Sheep	dig coef %	*	86.	86.	
N-free extract	%		47.9	52.7	
Cattle	dig coef %	*	78.	78.	
Sheep	dig coef %	*	76.	76.	
Protein (N x 6.25)	%		14.3	15.7	
Cattle	dig coef %	*	65.	65.	
Sheep	dig coef %	*	68.	68.	
Cattle	dig prot %		9.3	10.2	
Goats	dig prot %	*	10.6	11.7	
Horses	dig prot %	*	10.6	11.7	
Sheep	dig prot %		9.7	10.7	
Energy	GE Mcal/kg				
Cattle	DE Mcal/kg	*	2.30	2.52	
Sheep	DE Mcal/kg	*	2.46	2.70	
Swine	DE kcal/kg	*	2381.	2619.	
Cattle	ME Mcal/kg	*	1.88	2.07	
Chickens	ME_n kcal/kg		1181.	1299.	

Continued

Feed Name or Analyses		Mean As Fed	Mean Dry	C.V. ± %
Sheep	ME Mcal/kg	* 2.01	2.22	
Swine	ME kcal/kg	*2209.	2430.	
Cattle	TDN %	52.1	57.3	
Sheep	TDN %	55.7	61.3	
Swine	TDN %	* 54.0	59.4	
Calcium	%	0.08	0.09	
Phosphorus	%	1.36	1.50	
Choline	mg/kg	1080.	1188.	
Niacin	mg/kg	303.8	334.2	
Pantothenic acid	mg/kg	23.6	26.0	
Riboflavin	mg/kg	2.6	2.9	
Arginine	%	1.10	1.21	
Cystine	%	0.10	0.11	
Lysine	%	0.60	0.66	
Methionine	%	0.24	0.26	
Tryptophan	%	0.24	0.26	

Rice, bran w germ w broken grain w polish, dehy, (4)

Rice feed (CFA)

Ref No 4 03 937 United States

Feed Name or Analyses		Mean As Fed	Mean Dry	C.V. ± %
Dry matter	%	88.9	100.0	
Ash	%	9.2	10.3	
Crude fiber	%	6.4	7.3	
Cattle	dig coef %	-18.	-18.	
Sheep	dig coef %	5.	5.	
Ether extract	%	13.8	15.6	
Cattle	dig coef %	84.	84.	
Sheep	dig coef %	78.	78.	
N-free extract	%	46.4	52.3	
Cattle	dig coef %	82.	82.	
Sheep	dig coef %	77.	77.	
Protein (N x 6.25)	%	13.0	14.7	
Cattle	dig coef %	65.	65.	
Sheep	dig coef %	58.	58.	
Cattle	dig prot %	8.5	9.5	
Goats	dig prot %	* 9.5	10.7	
Horses	dig prot %	* 9.5	10.7	
Sheep	dig prot %	7.5	8.5	
Energy	GE Mcal/kg			
Cattle	DE Mcal/kg	* 3.15	3.54	
Sheep	DE Mcal/kg	* 2.99	3.37	
Swine	DE kcal/kg	*2515.	2831.	
Cattle	ME Mcal/kg	* 2.58	2.91	
Sheep	ME Mcal/kg	* 2.45	2.76	
Swine	ME kcal/kg	*2340.	2634.	
Cattle	TDN %	71.4	80.4	
Sheep	TDN %	67.9	76.4	
Swine	TDN %	* 57.0	64.2	

Rice, bran w germ w calcium carbonate, added, (4)

Ref No 4 08 495 United States

Feed Name or Analyses		Mean As Fed	Mean Dry	C.V. ± %
Dry matter	%	90.1	100.0	
Ash	%	13.3	14.8	
Crude fiber	%	11.6	12.9	
Ether extract	%	13.6	15.1	
N-free extract	%	40.3	44.7	
Protein (N x 6.25)	%	11.3	12.5	
Cattle	dig prot %	* 6.8	7.5	
Goats	dig prot %	* 7.9	8.7	
Horses	dig prot %	* 7.9	8.7	
Sheep	dig prot %	* 7.9	8.7	

(1) dry forages and roughages (2) pasture, range plants, and forages fed green (3) silages (4) energy feeds (5) protein supplements (6) minerals (7) vitamins (8) additives

Feed Name or Analyses		Mean As Fed	Mean Dry	C.V. ± %
Energy	GE Mcal/kg			
Cattle	DE Mcal/kg	* 3.00	3.33	
Sheep	DE Mcal/kg	* 2.93	3.25	
Swine	DE kcal/kg	*2820.	3130.	
Cattle	ME Mcal/kg	* 2.46	2.73	
Sheep	ME Mcal/kg	* 2.40	2.66	
Swine	ME kcal/kg	*2637.	2927.	
Cattle	TDN %	* 68.1	75.6	
Sheep	TDN %	* 66.4	73.7	
Swine	TDN %	* 64.0	71.0	
Calcium	%	8.82	9.79	

Rice, grain, polished cooked, (4)

Ref No 4 03 933 United States

Feed Name or Analyses		Mean As Fed	Mean Dry	C.V. ± %
Dry matter	%	87.0	100.0	
Ash	%	0.3	0.4	
Crude fiber	%	0.1	0.1	
Swine	dig coef %	9.	9.	
Ether extract	%	0.6	0.7	
Swine	dig coef %	70.	70.	
N-free extract	%	79.6	91.5	
Swine	dig coef %	100.	100.	
Protein (N x 6.25)	%	6.4	7.3	
Swine	dig coef %	86.	86.	
Cattle	dig prot %	* 2.4	2.7	
Goats	dig prot %	* 3.4	3.9	
Horses	dig prot %	* 3.4	3.9	
Sheep	dig prot %	* 3.4	3.9	
Swine	dig prot %	5.5	6.3	
Energy	GE Mcal/kg			
Cattle	DE Mcal/kg	* 3.34	3.84	
Sheep	DE Mcal/kg	* 3.41	3.92	
Swine	DE kcal/kg	*3793.	4360.	
Cattle	ME Mcal/kg	* 2.74	3.15	
Sheep	ME Mcal/kg	* 2.80	3.22	
Swine	ME kcal/kg	*3586.	4122.	
Cattle	TDN %	* 75.8	87.1	
Sheep	TDN %	* 77.4	88.9	
Swine	TDN %	86.0	98.9	

Rice, grain w hulls, (4)

Rough rice

Paddy rice

Ref No 4 03 939 United States

Feed Name or Analyses		Mean As Fed	Mean Dry	C.V. ± %
Dry matter	%	88.8	100.0	0
Ash	%	4.8	5.4	11
Crude fiber	%	8.8	9.9	4
Sheep	dig coef %	23.	23.	
Ether extract	%	1.7	1.9	14
Sheep	dig coef %	76.	76.	
N-free extract	%	65.6	73.9	1
Sheep	dig coef %	91.	91.	
Protein (N x 6.25)	%	7.9	8.9	5
Sheep	dig coef %	76.	76.	
Cattle	dig prot %	* 3.7	4.2	
Goats	dig prot %	* 4.8	5.4	
Horses	dig prot %	* 4.8	5.4	
Sheep	dig prot %	6.0	6.7	
Pentosans	%	3.8	4.3	23
Sugars, total	%	0.5	0.6	46
Linoleic	%	.650	.732	
Energy	GE Mcal/kg	3.31	3.73	6
Cattle	DE Mcal/kg	* 3.04	3.42	
Sheep	DE Mcal/kg	* 3.11	3.51	
Swine	DE kcal/kg	*3270.	3682.	
Cattle	ME Mcal/kg	* 2.49	2.80	

Feed Name or Analyses		Mean As Fed	Mean Dry	C.V. ± %
Chickens	ME_n kcal/kg	2667.	3005.	
Sheep	ME Mcal/kg	* 2.55	2.88	
Swine	ME kcal/kg	*3080.	3468.	
Cattle	TDN %	* 68.8	77.5	
Sheep	TDN %	70.6	79.6	
Swine	TDN %	* 74.1	83.5	
Calcium	%	0.06	0.07	37
Chlorine	%	0.08	0.09	32
Copper	mg/kg	3.3	3.7	22
Iron	%	0.004	0.004	88
Magnesium	%	0.12	0.14	25
Manganese	mg/kg	17.6	19.8	77
Phosphorus	%	0.29	0.33	15
Potassium	%	0.29	0.33	14
Sodium	%	0.04	0.05	29
Sulphur	%	0.04	0.05	86
Zinc	mg/kg	1.8	2.0	
Biotin	mg/kg	0.08	0.09	
Choline	mg/kg	1014.	1142.	
Niacin	mg/kg	34.0	38.3	15
Pantothenic acid	mg/kg	11.0	12.4	
Riboflavin	mg/kg	0.9	1.0	37
Thiamine	mg/kg	2.7	3.1	23
Alanine	%	0.27	0.30	
Arginine	%	0.56	0.63	33
Cystine	%	0.09	0.11	26
Glycine	%	0.80	0.90	
Histidine	%	0.09	0.10	30
Hydroxyproline	%	0.09	0.10	
Isoleucine	%	0.31	0.35	25
Leucine	%	0.53	0.60	29
Lysine	%	0.28	0.31	25
Methionine	%	0.17	0.20	28
Phenylalanine	%	0.31	0.35	45
Threonine	%	0.22	0.25	30
Tryptophan	%	0.10	0.12	22
Tyrosine	%	0.62	0.70	
Valine	%	0.44	0.50	16

Rice, grain w hulls, grnd, (4)

Ground rough rice (AAFCO)

Ground paddy rice (AAFCO)

Ref No 4 03 938 United States

Feed Name or Analyses		Mean As Fed	Mean Dry	C.V. ± %
Dry matter	%	88.1	100.0	1
Ash	%	5.2	5.8	
Crude fiber	%	7.8	8.8	
Sheep	dig coef %	* 23.	23.	
Ether extract	%	1.5	1.7	
Sheep	dig coef %	* 76.	76.	
N-free extract	%	66.6	75.6	
Sheep	dig coef %	* 91.	91.	
Protein (N x 6.25)	%	7.1	8.0	
Sheep	dig coef %	* 76.	76.	
Cattle	dig prot %	* 3.0	3.4	
Goats	dig prot %	* 4.0	4.6	
Horses	dig prot %	* 4.0	4.6	
Sheep	dig prot %	5.4	6.1	
Pentosans	%	5.4	6.1	
Energy	GE Mcal/kg			
Cattle	DE Mcal/kg	* 3.31	3.76	
Sheep	DE Mcal/kg	* 3.10	3.52	
Swine	DE kcal/kg	*3286.	3730.	
Cattle	ME Mcal/kg	* 2.71	3.08	
Sheep	ME Mcal/kg	* 2.54	2.89	
Swine	ME kcal/kg	*3102.	3521.	
Cattle	TDN %	* 75.1	85.3	
Sheep	TDN %	70.3	79.8	
Swine	TDN %	* 74.5	84.6	

Feed Name or Analyses				Mean As Fed	Mean Dry	C.V. ± %
Calcium		%		0.05	0.06	
Chlorine		%		0.05	0.06	
Magnesium		%		0.10	0.11	
Phosphorus		%		0.39	0.45	
Potassium		%		0.22	0.25	
Sulphur		%		0.05	0.06	
Biotin		mg/kg		0.08	0.09	
Folic acid		mg/kg		0.29	0.33	14
Niacin		mg/kg		49.0	55.6	
α-tocopherol		mg/kg		14.0	15.9	

Rice, groats, (4)

Rice grain without hulls

Rice grain, brown

Ref No 4 03 936 United States

Feed Name or Analyses				Mean As Fed	Mean Dry	C.V. ± %
Dry matter		%		88.2	100.0	0
Ash		%		1.0	1.2	27
Crude fiber		%		0.9	1.0	20
Ether extract		%		1.7	1.9	23
N-free extract		%		76.3	86.5	
Protein (N x 6.25)		%		8.4	9.5	9
Cattle	dig prot	%	*	4.2	4.7	
Goats	dig prot	%	*	5.2	5.9	
Horses	dig prot	%	*	5.2	5.9	
Sheep	dig prot	%	*	5.2	5.9	
Pentosans		%		2.1	2.4	9
Sugars, total		%		0.9	1.0	19
Energy	GE	Mcal/kg		3.59	4.07	1
Cattle	DE	Mcal/kg	*	3.55	4.03	
Sheep	DE	Mcal/kg	*	3.40	3.85	
Swine	DE	kcal/kg	*	3704.	4200.	
Cattle	ME	Mcal/kg	*	2.91	3.30	
Sheep	ME	Mcal/kg	*	2.79	3.16	
Swine	ME	kcal/kg	*	3485.	3952.	
Cattle	TDN	%	*	80.6	91.4	
Sheep	TDN	%	*	77.0	87.4	
Swine	TDN	%	*	84.0	95.3	
Calcium		%		0.04	0.05	39
Chlorine		%		0.13	0.15	16
Copper		mg/kg		3.4	3.9	16
Iron		%		0.003	0.003	98
Magnesium		%		0.09	0.10	16
Manganese		mg/kg		13.3	15.1	2
Phosphorus		%		0.25	0.28	14
Potassium		%		0.21	0.23	10
Sodium		%		0.03	0.04	40
Sulphur		%		0.02	0.03	98
Biotin		mg/kg		0.09	0.10	37
Folic acid		mg/kg		0.19	0.21	17
Niacin		mg/kg		45.3	51.4	33
Pantothenic acid		mg/kg		10.7	12.1	31
Riboflavin		mg/kg		0.6	0.7	98
Thiamine		mg/kg		3.2	3.6	28
α-tocopherol		mg/kg		8.7	9.9	
Vitamin B6		mg/kg		7.00	7.94	26

Rice, groats, grnd, (4)

Ground brown rice (AAFCO)

Rice grain without hulls, ground

Ref No 4 03 935 United States

Feed Name or Analyses				Mean As Fed	Mean Dry	C.V. ± %
Dry matter		%			100.0	
Crude fiber		%				
Sheep	dig coef	%	*		23.	
Ether extract		%				
Sheep	dig coef	%	*		76.	
N-free extract		%				
Sheep	dig coef	%	*		91.	
Protein (N x 6.25)		%				
Sheep	dig coef	%	*		76.	

Rice, groats, polished, (4)

Rice, white, polished

Ref No 4 03 942 United States

Feed Name or Analyses				Mean As Fed	Mean Dry	C.V. ± %
Dry matter		%		88.5	100.0	1
Ash		%		0.5	0.6	52
Crude fiber		%		0.4	0.4	46
Sheep	dig coef	%	*	23.	23.	
Ether extract		%		0.4	0.5	34
Sheep	dig coef	%	*	76.	76.	
N-free extract		%		80.0	90.4	
Sheep	dig coef	%	*	91.	91.	
Protein (N x 6.25)		%		7.1	8.1	10
Sheep	dig coef	%	*	76.	76.	
Cattle	dig prot	%	*	3.0	3.4	
Goats	dig prot	%	*	4.1	4.6	
Horses	dig prot	%	*	4.1	4.6	
Sheep	dig prot	%		5.4	6.1	
Pentosans		%		1.4	1.6	
Sugars, total		%		0.3	0.3	
Linoleic		%		.200	.226	
Energy	GE	Mcal/kg		3.62	4.09	
Cattle	DE	Mcal/kg	*	3.44	3.89	
Sheep	DE	Mcal/kg	*	3.48	3.94	
Swine	DE	kcal/kg	*	3783.	4277.	
Cattle	ME	Mcal/kg	*	2.82	3.19	
Chickens	ME_n	kcal/kg		3100.	3503.	
Sheep	ME	Mcal/kg	*	2.86	3.23	
Swine	ME	kcal/kg	*	3570.	4036.	
Cattle	NE_m	Mcal/kg	*	1.77	2.00	
Cattle	NE_{gain}	Mcal/kg	*	1.18	1.33	
Cattle	$NE_{lactating\ cows}$	Mcal/kg	*	2.07	2.34	
Cattle	TDN	%	*	78.1	88.3	
Sheep	TDN	%		79.0	89.3	
Swine	TDN	%	*	85.8	97.0	
Calcium		%		0.02	0.02	57
Chlorine		%		0.02	0.02	
Copper		mg/kg		2.9	3.3	
Iron		%		0.001	0.002	93
Magnesium		%		0.02	0.02	
Manganese		mg/kg		10.9	12.3	66
Phosphorus		%		0.10	0.11	61
Potassium		%		0.09	0.10	
Sodium		%		0.01	0.01	
Sulphur		%		0.08	0.09	
Zinc		mg/kg		1.8	2.0	
Choline		mg/kg		903.	1021.	
Niacin		mg/kg		16.5	18.7	24
Pantothenic acid		mg/kg		3.3	3.7	52
Riboflavin		mg/kg		0.4	0.5	42
Thiamine		mg/kg		0.8	0.9	36
α-tocopherol		mg/kg		3.5	4.0	
Vitamin B6		mg/kg		0.39	0.44	
Vitamin A		IU/g		0.0	0.0	
Arginine		%		0.35	0.40	38
Cystine		%		0.09	0.10	
Glycine		%		0.71	0.80	
Histidine		%		0.18	0.20	
Isoleucine		%		0.44	0.50	
Leucine		%		0.71	0.80	
Lysine		%		0.27	0.30	20
Methionine		%		0.22	0.25	13
Phenylalanine		%		0.53	0.60	
Threonine		%		0.35	0.40	
Tryptophan		%		0.09	0.10	
Tyrosine		%		0.62	0.70	
Valine		%		0.49	0.55	

Rice, groats, polished and broken, (4)

Chipped rice (AAFCO)

Broken rice (AAFCO)

Brewers rice (AAFCO)

Ref No 4 03 932 United States

Feed Name or Analyses				Mean As Fed	Mean Dry	C.V. ± %
Dry matter		%		88.3	100.0	0
Ash		%		0.8	0.9	22
Crude fiber		%		0.6	0.7	23
Sheep	dig coef	%	*	23.	23.	
Ether extract		%		0.7	0.8	30
Sheep	dig coef	%	*	76.	76.	
N-free extract		%		78.4	88.8	
Sheep	dig coef	%	*	91.	91.	
Protein (N x 6.25)		%		7.8	8.8	7
Sheep	dig coef	%	*	76.	76.	
Cattle	dig prot	%	*	3.7	4.1	
Goats	dig prot	%	*	4.7	5.3	
Horses	dig prot	%	*	4.7	5.3	
Sheep	dig prot	%		5.9	6.7	
Pentosans		%		1.7	1.9	10
Sugars, total		%		0.4	0.5	27
Energy	GE	Mcal/kg				
Cattle	DE	Mcal/kg	*	3.41	3.86	
Sheep	DE	Mcal/kg	*	3.47	3.92	
Swine	DE	kcal/kg	*	3738.	4232.	
Cattle	ME	Mcal/kg	*	2.80	3.17	
Sheep	ME	Mcal/kg	*	2.84	3.22	
Swine	ME	kcal/kg	*	3521.	3987.	
Cattle	TDN	%	*	77.4	87.6	
Sheep	TDN	%		78.6	89.0	
Swine	TDN	%	*	84.8	96.0	
Calcium		%		0.03	0.04	23
Magnesium		%		0.05	0.06	22
Phosphorus		%		0.12	0.13	31
Potassium		%		0.13	0.15	22
Niacin		mg/kg		31.9	36.2	
Thiamine		mg/kg		1.4	1.5	

Rice, polishings, dehy, (4)

Rice polishings (AAFCO)

Rice polish (CFA)

Ref No 4 03 943 United States

Feed Name or Analyses				Mean As Fed	Mean Dry	C.V. ± %
Dry matter		%		90.4	100.0	1
Ash		%		6.9	7.6	30
Crude fiber		%		3.2	3.6	26
Cattle	dig coef	%		22.	22.	
Sheep	dig coef	%		37.	37.	
Swine	dig coef	%		44.	44.	
Ether extract		%		11.8	13.1	22
Cattle	dig coef	%		73.	73.	
Sheep	dig coef	%		88.	88.	
Swine	dig coef	%		88.	88.	
N-free extract		%		56.1	62.0	8
Cattle	dig coef	%		93.	93.	
Sheep	dig coef	%		87.	87.	
Swine	dig coef	%		93.	93.	
Protein (N x 6.25)		%		12.5	13.8	7
Cattle	dig coef	%		66.	66.	
Sheep	dig coef	%		78.	78.	
Swine	dig coef	%		83.	83.	
Cattle	dig prot	%		8.2	9.1	

Continued

Feed Name or Analyses		Mean As Fed	Mean Dry	C.V. ± %
Goats	dig prot %	* 8.9	9.9	
Horses	dig prot %	* 8.9	9.9	
Sheep	dig prot %	9.7	10.7	
Swine	dig prot %	10.3	11.4	
Pentosans	%	3.8	4.2	3
Linoleic	%	3.600	3.982	
Energy	GE Mcal/kg	2.66	2.94	
Cattle	DE Mcal/kg	* 3.55	3.92	
Sheep	DE Mcal/kg	* 3.66	4.05	
Swine	DE kcal/kg	*3837.	4242.	
Cattle	ME Mcal/kg	* 2.91	3.22	
Chickens	ME_n kcal/kg	2860.	3164.	
Sheep	ME Mcal/kg	* 3.00	3.32	
Swine	ME kcal/kg	*3577.	3955.	
Cattle	TDN %	80.5	89.0	
Sheep	TDN %	83.1	91.9	
Swine	TDN %	87.0	96.2	
Calcium	%	0.05	0.06	33
Chlorine	%	0.13	0.14	
Iron	%	0.016	0.018	
Magnesium	%	0.65	0.72	
Phosphorus	%	1.24	1.37	18
Potassium	%	1.02	1.13	
Sodium	%	0.11	0.12	
Sulphur	%	0.17	0.19	
Biotin	mg/kg	0.62	0.69	25
Choline	mg/kg	1317.	1457.	7
Niacin	mg/kg	409.4	452.7	51
Pantothenic acid	mg/kg	58.8	65.0	66
Riboflavin	mg/kg	1.8	2.0	30
Thiamine	mg/kg	19.6	21.7	40
α-tocopherol	mg/kg	90.0	99.6	
Vitamin B_6	mg/kg	27.95	30.90	
Arginine	%	0.50	0.56	95
Cystine	%	0.10	0.11	
Histidine	%	0.10	0.11	79
Isoleucine	%	0.30	0.33	
Leucine	%	0.50	0.56	
Lysine	%	0.50	0.56	32
Phenylalanine	%	0.30	0.33	
Threonine	%	0.30	0.33	
Tryptophan	%	0.10	0.11	

Rice, polishings w calcium carbonate added, dehy, (4)
Ref No 4 08 496 United States

Feed Name or Analyses		Mean As Fed	Mean Dry	C.V. ± %
Dry matter	%	89.9	100.0	
Ash	%	10.7	11.9	
Crude fiber	%	4.1	4.6	
Ether extract	%	14.9	16.6	
N-free extract	%	48.1	53.5	
Protein (N x 6.25)	%	12.1	13.5	
Cattle	dig prot %	* 7.5	8.4	
Goats	dig prot %	* 8.6	9.6	
Horses	dig prot %	* 8.6	9.6	
Sheep	dig prot %	* 8.6	9.6	
Energy	GE Mcal/kg			
Cattle	DE Mcal/kg	* 3.74	4.17	
Sheep	DE Mcal/kg	* 3.21	3.57	
Swine	DE kcal/kg	*3460.	3849.	
Cattle	ME Mcal/kg	* 3.07	3.42	
Sheep	ME Mcal/kg	* 2.64	2.93	
Swine	ME kcal/kg	*3228.	3591.	

(1) dry forages and roughages
(2) pasture, range plants, and forages fed green
(3) silages
(4) energy feeds
(5) protein supplements
(6) minerals
(7) vitamins
(8) additives

Feed Name or Analyses		Mean As Fed	Mean Dry	C.V. ± %
Cattle	TDN %	* 84.9	94.5	
Sheep	TDN %	* 72.9	81.1	
Swine	TDN %	* 78.5	87.3	

Rice bran (AAFCO) –
see Rice, bran w germ, dry milled, mx 13% fiber calcium carbonate declared above 3% mn, (4)

Rice bran (CFA) –
see Rice, bran w germ, dry milled, (4)

Rice feed (CFA) –
see Rice, bran w germ w broken grain w polish, dehy, (4)

Rice grain, brown –
see Rice, groats, (4)

Rice grain without hulls –
see Rice, groats, (4)

Rice grain without hulls, ground –
see Rice, groats, grnd, (4)

RICEGRASS, INDIAN. Oryzopsis hymenoides

Ricegrass, Indian, aerial part, fresh, over ripe, (2)
Ref No 2 03 944 United States

Feed Name or Analyses		Mean As Fed	Mean Dry	C.V. ± %
Dry matter	%		100.0	
Calcium	%		0.58	50
Magnesium	%		0.15	21
Phosphorus	%		0.05	35
Sulphur	%		0.14	20
Carotene	mg/kg		0.4	
Vitamin A equivalent	IU/g		0.7	

Rice polish (CFA) –
see Rice, polishings, dehy, (4)

Rice, white, polished –
see Rice, groats, polished, (4)

ROCKETSALAD. Eruca sativa

Rocketsalad, aerial part, dehy, (1)
Ref No 1 03 948 United States

Feed Name or Analyses		Mean As Fed	Mean Dry	C.V. ± %
Dry matter	%	88.3	100.0	
Ash	%	9.3	10.5	
Crude fiber	%	29.1	33.0	
Sheep	dig coef %	53.	53.	
Ether extract	%	2.0	2.3	
Sheep	dig coef %	71.	71.	
N-free extract	%	35.8	40.6	
Sheep	dig coef %	72.	72.	
Protein (N x 6.25)	%	12.0	13.6	
Sheep	dig coef %	79.	79.	
Cattle	dig prot %	* 7.7	8.7	
Goats	dig prot %	* 8.2	9.2	
Horses	dig prot %	* 8.0	9.1	
Rabbits	dig prot %	* 8.1	9.2	
Sheep	dig prot %	9.5	10.7	
Energy	GE Mcal/kg			
Cattle	DE Mcal/kg	* 2.29	2.59	
Sheep	DE Mcal/kg	* 2.38	2.70	
Cattle	ME Mcal/kg	* 1.88	2.13	

Feed Name or Analyses		Mean As Fed	Mean Dry	C.V. ± %
Sheep	ME Mcal/kg	* 1.95	2.21	
Cattle	TDN %	* 51.9	58.8	
Sheep	TDN %	54.0	61.1	

Rocketsalad, aerial part, fresh, (2)
Ref No 2 03 949 United States

Feed Name or Analyses		Mean As Fed	Mean Dry	C.V. ± %
Dry matter	%	15.8	100.0	
Ash	%	1.5	9.3	
Crude fiber	%	5.8	36.9	
Sheep	dig coef %	61.	61.	
Ether extract	%	0.3	2.2	
Sheep	dig coef %	52.	52.	
N-free extract	%	6.4	40.2	
Sheep	dig coef %	71.	71.	
Protein (N x 6.25)	%	1.8	11.4	
Sheep	dig coef %	57.	57.	
Cattle	dig prot %	* 1.2	7.6	
Goats	dig prot %	* 1.1	7.2	
Horses	dig prot %	* 1.1	7.2	
Rabbits	dig prot %	* 1.2	7.5	
Sheep	dig prot %	1.0	6.5	
Energy	GE Mcal/kg			
Cattle	DE Mcal/kg	* 0.44	2.76	
Sheep	DE Mcal/kg	* 0.42	2.65	
Cattle	ME Mcal/kg	* 0.36	2.26	
Sheep	ME Mcal/kg	* 0.34	2.17	
Cattle	TDN %	* 9.9	62.5	
Sheep	TDN %	9.5	60.1	

Rocketsalad, aerial part, ensiled, (3)
Ref No 3 03 950 United States

Feed Name or Analyses		Mean As Fed	Mean Dry	C.V. ± %
Dry matter	%	14.2	100.0	
Ash	%	2.5	17.7	
Crude fiber	%	2.6	18.6	
Sheep	dig coef %	66.	66.	
Ether extract	%	0.6	4.3	
Sheep	dig coef %	86.	86.	
N-free extract	%	5.9	41.6	
Sheep	dig coef %	87.	87.	
Protein (N x 6.25)	%	2.5	17.8	
Sheep	dig coef %	82.	82.	
Cattle	dig prot %	* 1.8	12.4	
Goats	dig prot %	* 1.8	12.4	
Horses	dig prot %	* 1.8	12.4	
Sheep	dig prot %	2.1	14.6	
Energy	GE Mcal/kg			
Cattle	DE Mcal/kg	* 0.45	3.17	
Sheep	DE Mcal/kg	* 0.45	3.15	
Cattle	ME Mcal/kg	* 0.37	2.60	
Sheep	ME Mcal/kg	* 0.37	2.58	
Cattle	TDN %	* 10.2	72.0	
Sheep	TDN %	10.1	71.4	

ROCK PHOSPHATE. Scientific name not applicable

Rock phosphate, grnd, (6)
Rock phosphate, ground (AAFCO)
Ref No 6 03 945 United States

Feed Name or Analyses		Mean As Fed	Mean Dry	C.V. ± %
Dry matter	%		100.0	
Ash	%		100.0	
Calcium	%		30.00	
Phosphorus	%		12.00	

Feed Name or Analyses			Mean As Fed	Mean Dry	C.V. ± %

Rock phosphate, soft, mn 9% phosphorus mn 15% calcium mx 1.5% fluorine mx 30% clay, (6)

Rock phosphate, soft (AAFCO)

Ref No 6 03 947 United States

Analyses			As Fed	Dry	C.V. ± %
Dry matter	%			100.0	
Ash	%			100.0	
Calcium	%			18.00	
Phosphorus	%			9.00	

Rockweed —

see Rockweed, entire plant, dehy, (1)

ROCKWEED. Fucus spp

Rockweed, entire plant, dehy, (1)

Rockweed

Ref No 1 08 508 United States

Analyses			As Fed	Dry	C.V. ± %
Dry matter	%		88.7	100.0	
Ash	%		16.3	18.4	
Crude fiber	%		9.4	10.6	
Ether extract	%		4.2	4.7	
N-free extract	%		53.6	60.4	
Protein (N x 6.25)	%		5.2	5.9	
Cattle	dig prot %	*	1.8	2.0	
Goats	dig prot %	*	1.8	2.0	
Horses	dig prot %	*	2.2	2.5	
Rabbits	dig prot %	*	2.8	3.2	
Sheep	dig prot %	*	1.6	1.8	
Energy	GE Mcal/kg				
Cattle	DE Mcal/kg	*	1.89	2.13	
Sheep	DE Mcal/kg	*	1.73	1.95	
Cattle	ME Mcal/kg	*	1.55	1.75	
Sheep	ME Mcal/kg	*	1.42	1.60	
Cattle	TDN %	*	42.9	48.4	
Sheep	TDN %	*	39.1	44.1	

Rockweed, aerial part, fresh, (2)

Ref No 2 03 951 United States

Analyses			As Fed	Dry	C.V. ± %
Dry matter	%		88.2	100.0	
Ash	%		17.8	20.2	
Crude fiber	%		8.4	9.6	
Sheep	dig coef %		-41.	-41.	
Swine	dig coef %		0.	0.	
Ether extract	%		3.2	3.7	
Sheep	dig coef %		97.	97.	
Swine	dig coef %		83.	83.	
N-free extract	%		49.0	55.6	
Sheep	dig coef %		55.	55.	
Swine	dig coef %		55.	55.	
Protein (N x 6.25)	%		9.7	11.1	
Sheep	dig coef %		33.	33.	
Swine	dig coef %		0.	0.	
Cattle	dig prot %	*	6.4	7.3	
Goats	dig prot %	*	6.1	6.9	
Horses	dig prot %	*	6.1	6.9	
Rabbits	dig prot %	*	6.4	7.2	
Sheep	dig prot %		3.2	3.6	
Energy	GE Mcal/kg				
Cattle	DE Mcal/kg	*	1.62	1.84	
Sheep	DE Mcal/kg	*	1.48	1.68	
Swine	DE kcal/kg	*	1454.	1649.	
Cattle	ME Mcal/kg	*	1.33	1.51	
Sheep	ME Mcal/kg	*	1.22	1.38	
Swine	ME kcal/kg	*	1364.	1546.	
Cattle	TDN %	*	36.8	41.7	
Sheep	TDN %		33.7	38.2	
Swine	TDN %		33.0	37.4	

ROCKWEEP, DRUG. Ascophyllum nodosum

Rockweep, drug, aerial part, fresh, (2)

Ref No 2 03 952 United States

Analyses			As Fed	Dry	C.V. ± %
Dry matter	%			100.0	
Ash	%			20.9	
Ether extract	%			3.2	
Protein (N x 6.25)	%			8.4	
Cattle	dig prot %	*		5.0	
Goats	dig prot %	*		4.4	
Horses	dig prot %	*		4.7	
Rabbits	dig prot %	*		5.2	
Sheep	dig prot %	*		4.8	
Calcium	%			1.16	
Phosphorus	%			0.11	

Rolled oats (CFA) —

see Oats, cereal by-product, grnd, mx 2% fiber, (4)

ROSE, FENDLER. Rosa woodsii fendleri

Rose, fendler, twigs, fresh, (2)

Ref No 2 03 954 United States

Analyses			As Fed	Dry	C.V. ± %
Dry matter	%			100.0	
Ash	%			3.2	
Crude fiber	%			32.7	
Ether extract	%			3.0	
N-free extract	%			55.1	
Protein (N x 6.25)	%			6.0	
Cattle	dig prot %	*		3.0	
Goats	dig prot %	*		2.2	
Horses	dig prot %	*		2.6	
Rabbits	dig prot %	*		3.3	
Sheep	dig prot %	*		2.6	

ROSE, SPALDING. Rosa spaldingii

Rose, spalding, browse, immature, (2)

Ref No 2 03 955 United States

Analyses			As Fed	Dry	C.V. ± %
Dry matter	%			100.0	
Ash	%			7.6	
Crude fiber	%			11.7	
Ether extract	%			3.7	
N-free extract	%			58.4	
Protein (N x 6.25)	%			18.6	
Cattle	dig prot %	*		13.7	
Goats	dig prot %	*		13.9	
Horses	dig prot %	*		13.3	
Rabbits	dig prot %	*		13.0	
Sheep	dig prot %	*		14.3	
Calcium	%			1.50	
Magnesium	%			0.45	
Phosphorus	%			0.53	
Sulphur	%			0.26	

Rose, spalding, browse, mid-bloom, (2)

Ref No 2 03 956 United States

Analyses			As Fed	Dry	C.V. ± %
Dry matter	%			100.0	
Ash	%			8.3	
Crude fiber	%			11.4	
Ether extract	%			6.2	
N-free extract	%			62.5	
Protein (N x 6.25)	%			11.6	
Cattle	dig prot %	*		7.8	
Goats	dig prot %	*		7.4	
Horses	dig prot %	*		7.4	
Rabbits	dig prot %	*		7.6	
Sheep	dig prot %	*		7.8	
Calcium	%			2.34	
Magnesium	%			0.57	
Phosphorus	%			0.65	
Sulphur	%			0.24	

ROSEMALLOW. Hibiscus moscheutos

Rosemallow, seeds w hulls, (5)

Ref No 5 09 301 United States

Analyses			As Fed	Dry	C.V. ± %
Dry matter	%			100.0	
Protein (N x 6.25)	%			22.0	
Alanine	%			0.92	
Arginine	%			1.91	
Aspartic acid	%			1.98	
Glutamic acid	%			3.39	
Glycine	%			1.10	
Histidine	%			0.48	
Hydroxyproline	%			0.02	
Isoleucine	%			0.66	
Leucine	%			1.19	
Lysine	%			0.90	
Methionine	%			0.31	
Phenylalanine	%			0.90	
Proline	%			0.79	
Serine	%			0.92	
Threonine	%			0.68	
Tyrosine	%			0.55	
Valine	%			0.90	

ROSEWOOD. Dalbergia spp

Rosewood, leaves, fresh, (2)

Ref No 2 03 957 United States

Analyses			As Fed	Dry	C.V. ± %
Dry matter	%		28.4	100.0	
Ash	%		2.9	10.2	
Crude fiber	%		5.6	19.7	
Cattle	dig coef %		91.	91.	
Ether extract	%		1.1	4.0	
Cattle	dig coef %		122.	122.	
N-free extract	%		13.7	48.1	
Cattle	dig coef %		127.	127.	
Protein (N x 6.25)	%		5.1	18.0	
Cattle	dig coef %		93.	93.	
Cattle	dig prot %		4.8	16.7	
Goats	dig prot %	*	3.8	13.4	
Horses	dig prot %	*	3.6	12.8	
Rabbits	dig prot %	*	3.6	12.6	
Sheep	dig prot %	*	3.9	13.8	

Rosewood, leaves, ensiled, (3)

Ref No 3 03 958 — United States

Feed Name or Analyses				Mean As Fed	Mean Dry	C.V. ± %
Dry matter		%		24.6	100.0	
Ash		%		4.5	18.2	
Crude fiber		%		7.4	30.1	
Cattle	dig coef	%		34.	34.	
Ether extract		%		0.9	3.6	
Cattle	dig coef	%		28.	28.	
N-free extract		%		8.4	34.1	
Cattle	dig coef	%		50.	50.	
Protein (N x 6.25)		%		3.4	14.0	
Cattle	dig coef	%		52.	52.	
Cattle	dig prot	%		1.8	7.3	
Goats	dig prot	%	*	2.2	8.9	
Horses	dig prot	%	*	2.2	8.9	
Sheep	dig prot	%	*	2.2	8.9	
Energy	GE Mcal/kg					
Cattle	DE Mcal/kg		*	0.40	1.62	
Sheep	DE Mcal/kg		*	0.42	1.70	
Cattle	ME Mcal/kg		*	0.33	1.33	
Sheep	ME Mcal/kg		*	0.34	1.39	
Cattle	TDN	%		9.1	36.8	
Sheep	TDN	%	*	9.5	38.5	

Rough rice –
see Rice, grain w hulls, (4)

Rubber-seed meal –
see Rubbertree, para, seeds, mech-extd grnd, (5)

RUBBERTREE, PARA. Hevea brasiliensis

Rubbertree, para, seeds, mech-extd grnd, (5)

Ref No 5 03 959 — United States

Feed Name or Analyses				Mean As Fed	Mean Dry	C.V. ± %
Dry matter		%		90.9	100.0	
Ash		%		5.1	5.6	
Crude fiber		%		6.6	7.2	
Sheep	dig coef	%		121.	121.	
Ether extract		%		14.7	16.1	
Sheep	dig coef	%		97.	97.	
N-free extract		%		35.3	38.8	
Sheep	dig coef	%		95.	95.	
Protein (N x 6.25)		%		29.3	32.3	
Sheep	dig coef	%		91.	91.	
Sheep	dig prot	%		26.7	29.4	
Energy	GE Mcal/kg					
Cattle	DE Mcal/kg		*	3.76	4.14	
Sheep	DE Mcal/kg		*	4.42	4.86	
Swine	DE kcal/kg		*	4424.	4867.	
Cattle	ME Mcal/kg		*	3.08	3.39	
Sheep	ME Mcal/kg		*	3.62	3.98	
Swine	ME kcal/kg		*	3959.	4356.	
Cattle	TDN	%	*	85.3	93.8	
Sheep	TDN	%		100.2	110.2	
Swine	TDN	%	*	100.3	110.4	

Rumen contents, dried –
see Animal, rumen contents, dehy grnd, (7)

(1) dry forages and roughages
(2) pasture, range plants, and forages fed green
(3) silages
(4) energy feeds
(5) protein supplements
(6) minerals
(7) vitamins
(8) additives

RUSH. Juncus spp

Rush, hay, s-c, (1)

Ref No 1 03 960 — United States

Feed Name or Analyses				Mean As Fed	Mean Dry	C.V. ± %
Dry matter		%		89.8	100.0	2
Ash		%		5.8	6.5	12
Crude fiber		%		28.4	31.6	8
Ether extract		%		1.9	2.2	29
N-free extract		%		44.7	49.8	
Protein (N x 6.25)		%		9.0	10.0	10
Cattle	dig prot	%	*	5.0	5.6	
Goats	dig prot	%	*	5.3	5.9	
Horses	dig prot	%	*	5.4	6.0	
Rabbits	dig prot	%	*	5.8	6.4	
Sheep	dig prot	%	*	5.0	5.6	
Energy	GE Mcal/kg					
Cattle	DE Mcal/kg		*	2.34	2.61	
Sheep	DE Mcal/kg		*	2.27	2.53	
Cattle	ME Mcal/kg		*	1.92	2.14	
Sheep	ME Mcal/kg		*	1.86	2.07	
Cattle	TDN	%	*	53.1	59.1	
Sheep	TDN	%	*	51.4	57.3	
Calcium		%		0.31	0.35	
Iron		%		0.010	0.011	
Magnesium		%		0.26	0.29	
Phosphorus		%		0.10	0.11	
Potassium		%		0.68	0.76	

Rush, aerial part, fresh, (2)

Ref No 2 03 965 — United States

Feed Name or Analyses				Mean As Fed	Mean Dry	C.V. ± %
Dry matter		%		31.1	100.0	
Ash		%		2.2	7.1	
Crude fiber		%		9.8	31.5	
Ether extract		%		0.6	1.9	
N-free extract		%		15.1	48.6	
Protein (N x 6.25)		%		3.4	10.9	
Cattle	dig prot	%	*	2.2	7.2	
Goats	dig prot	%	*	2.1	6.8	
Horses	dig prot	%	*	2.1	6.8	
Rabbits	dig prot	%	*	2.2	7.1	
Sheep	dig prot	%	*	2.2	7.2	
Energy	GE Mcal/kg					
Cattle	DE Mcal/kg		*	0.83	2.68	
Sheep	DE Mcal/kg		*	0.87	2.79	
Cattle	ME Mcal/kg		*	0.68	2.20	
Sheep	ME Mcal/kg		*	0.71	2.29	
Cattle	TDN	%	*	18.9	60.8	
Sheep	TDN	%	*	19.7	63.3	
Calcium		%		0.09	0.30	16
Phosphorus		%		0.08	0.25	44
Potassium		%		0.73	2.36	31

Rush, aerial part, fresh, immature, (2)

Ref No 2 03 961 — United States

Feed Name or Analyses				Mean As Fed	Mean Dry	C.V. ± %
Dry matter		%			100.0	
Ash		%			7.7	30
Crude fiber		%			26.9	18
Ether extract		%			1.7	
N-free extract		%			44.4	
Protein (N x 6.25)		%			19.3	26
Cattle	dig prot	%	*		14.3	
Goats	dig prot	%	*		14.6	
Horses	dig prot	%	*		13.9	
Rabbits	dig prot	%	*		13.6	
Sheep	dig prot	%	*		15.0	
Energy	GE Mcal/kg					
Cattle	DE Mcal/kg		*		2.79	
Sheep	DE Mcal/kg		*		2.58	
Cattle	ME Mcal/kg		*		2.29	
Sheep	ME Mcal/kg		*		2.12	
Cattle	TDN	%	*		63.3	
Sheep	TDN	%	*		58.6	

Rush, aerial part, fresh, early bloom, (2)

Ref No 2 03 962 — United States

Feed Name or Analyses				Mean As Fed	Mean Dry	C.V. ± %
Dry matter		%			100.0	
Ash		%			5.3	23
Crude fiber		%			31.6	7
Ether extract		%			2.1	29
N-free extract		%			50.3	
Protein (N x 6.25)		%			10.7	10
Cattle	dig prot	%	*		7.0	
Goats	dig prot	%	*		6.5	
Horses	dig prot	%	*		6.6	
Rabbits	dig prot	%	*		6.9	
Sheep	dig prot	%	*		7.0	
Energy	GE Mcal/kg					
Cattle	DE Mcal/kg		*		2.83	
Sheep	DE Mcal/kg		*		2.90	
Cattle	ME Mcal/kg		*		2.32	
Sheep	ME Mcal/kg		*		2.37	
Cattle	TDN	%	*		64.1	
Sheep	TDN	%	*		65.7	

Rush, aerial part, fresh, full bloom, (2)

Ref No 2 03 963 — United States

Feed Name or Analyses				Mean As Fed	Mean Dry	C.V. ± %
Dry matter		%			100.0	
Ash		%			6.6	15
Crude fiber		%			31.3	14
Protein (N x 6.25)		%			9.4	9
Cattle	dig prot	%	*		5.9	
Goats	dig prot	%	*		5.3	
Horses	dig prot	%	*		5.5	
Rabbits	dig prot	%	*		5.9	
Sheep	dig prot	%	*		5.8	

Rush, aerial part, fresh, over ripe, (2)

Ref No 2 03 964 — United States

Feed Name or Analyses				Mean As Fed	Mean Dry	C.V. ± %
Dry matter		%			100.0	
Ash		%			6.3	25
Crude fiber		%			33.8	7
Protein (N x 6.25)		%			4.3	12
Cattle	dig prot	%	*		1.5	
Goats	dig prot	%	*		0.6	
Horses	dig prot	%	*		1.2	
Rabbits	dig prot	%	*		2.0	
Sheep	dig prot	%	*		1.0	

RUSH, BALTIC. Juncus balticus

Rush, baltic, aerial part, fresh, (2)

Ref No 2 03 968 — United States

Feed Name or Analyses				Mean As Fed	Mean Dry	C.V. ± %
Dry matter		%			100.0	
Ash		%			5.7	17
Crude fiber		%			33.0	6
Ether extract		%			1.8	19

Feed Name or Analyses				Mean As Fed	Mean Dry	C.V. ± %
N-free extract		%			49.5	
Protein (N x 6.25)		%			10.0	27
Cattle	dig prot	%	*		6.4	
Goats	dig prot	%	*		5.9	
Horses	dig prot	%	*		6.0	
Rabbits	dig prot	%	*		6.4	
Sheep	dig prot	%	*		6.3	

Rush, baltic, aerial part, fresh, immature, (2)
Ref No 2 03 966 United States

Analyses				As Fed	Dry	C.V. ± %
Dry matter		%			100.0	
Ash		%			5.8	
Crude fiber		%			31.5	
Ether extract		%			1.7	
N-free extract		%			46.5	
Protein (N x 6.25)		%			14.5	
Cattle	dig prot	%	*		10.2	
Goats	dig prot	%	*		10.1	
Horses	dig prot	%	*		9.8	
Rabbits	dig prot	%	*		9.9	
Sheep	dig prot	%	*		10.5	

Rush, baltic, aerial part, fresh, early bloom, (2)
Ref No 2 03 967 United States

Analyses				As Fed	Dry	C.V. ± %
Dry matter		%			100.0	
Ash		%			5.2	26
Crude fiber		%			32.4	3
Ether extract		%			2.0	27
N-free extract		%			49.4	
Protein (N x 6.25)		%			11.0	7
Cattle	dig prot	%	*		7.2	
Goats	dig prot	%	*		6.8	
Horses	dig prot	%	*		6.9	
Rabbits	dig prot	%	*		7.2	
Sheep	dig prot	%	*		7.2	

RUSH, LONGSTYLE. Juncus longistylis

Rush, longstyle, hay, s-c, (1)
Ref No 1 03 969 United States

Analyses				As Fed	Dry	C.V. ± %
Dry matter		%			100.0	
Ash		%			6.3	
Crude fiber		%			33.8	
Ether extract		%			1.4	
N-free extract		%			47.7	
Protein (N x 6.25)		%			10.8	
Cattle	dig prot	%	*		6.3	
Goats	dig prot	%	*		6.6	
Horses	dig prot	%	*		6.7	
Rabbits	dig prot	%	*		7.0	
Sheep	dig prot	%	*		6.3	
Energy	GE Mcal/kg					
Cattle	DE Mcal/kg		*		2.40	
Sheep	DE Mcal/kg		*		2.50	
Cattle	ME Mcal/kg		*		1.96	
Sheep	ME Mcal/kg		*		2.05	
Cattle	TDN	%	*		54.3	
Sheep	TDN	%	*		56.6	

RUSH, SALTMEADOW. Juncus gerardii

Rush, saltmeadow, hay, s-c, (1)
Black grass hay
Ref No 1 03 970 United States

Analyses				As Fed	Dry	C.V. ± %
Dry matter		%		87.7	100.0	2
Ash		%		7.1	8.0	10
Crude fiber		%		24.8	28.2	7
Sheep	dig coef	%		57.	57.	
Ether extract		%		2.4	2.7	18
Sheep	dig coef	%		46.	46.	
N-free extract		%		46.1	52.5	
Sheep	dig coef	%		49.	49.	
Protein (N x 6.25)		%		7.4	8.5	8
Sheep	dig coef	%		54.	54.	
Cattle	dig prot	%	*	3.8	4.3	
Goats	dig prot	%	*	3.9	4.5	
Horses	dig prot	%	*	4.2	4.7	
Rabbits	dig prot	%	*	4.6	5.2	
Sheep	dig prot	%		4.0	4.6	
Fatty acids		%		2.4	2.8	
Energy	GE Mcal/kg					
Cattle	DE Mcal/kg		*	2.37	2.70	
Sheep	DE Mcal/kg		*	1.90	2.17	
Cattle	ME Mcal/kg		*	1.94	2.21	
Sheep	ME Mcal/kg		*	1.56	1.78	
Cattle	TDN	%	*	53.8	61.3	
Sheep	TDN	%		43.2	49.2	
Phosphorus		%		0.09	0.10	
Potassium		%		1.53	1.74	

RUSH, SLIMPOD. Juncus oxymeris

Rush, slimpod, aerial part, fresh, (2)
Ref No 2 03 976 United States

Analyses				As Fed	Dry	C.V. ± %
Dry matter		%			100.0	
Ash		%			8.1	10
Crude fiber		%			31.3	15
Protein (N x 6.25)		%			10.8	64
Cattle	dig prot	%	*		7.1	
Goats	dig prot	%	*		6.6	
Horses	dig prot	%	*		6.7	
Rabbits	dig prot	%	*		7.0	
Sheep	dig prot	%	*		7.1	
Calcium		%			0.29	19
Phosphorus		%			0.21	57
Potassium		%			2.79	15

Rush, slimpod, aerial part, fresh, immature, (2)
Ref No 2 03 971 United States

Analyses				As Fed	Dry	C.V. ± %
Dry matter		%			100.0	
Ash		%			9.7	
Crude fiber		%			22.3	
Protein (N x 6.25)		%			24.1	
Cattle	dig prot	%	*		18.4	
Goats	dig prot	%	*		19.0	
Horses	dig prot	%	*		18.0	
Rabbits	dig prot	%	*		17.3	
Sheep	dig prot	%	*		19.5	
Calcium		%			0.38	
Phosphorus		%			0.42	
Potassium		%			3.66	

Rush, slimpod, aerial part, fresh, mid-bloom, (2)
Ref No 2 03 972 United States

Analyses				As Fed	Dry	C.V. ± %
Dry matter		%			100.0	
Ash		%			7.4	
Crude fiber		%			33.7	
Protein (N x 6.25)		%			10.4	
Cattle	dig prot	%	*		6.7	
Goats	dig prot	%	*		6.3	
Horses	dig prot	%	*		6.4	
Rabbits	dig prot	%	*		6.7	
Sheep	dig prot	%	*		6.7	
Calcium		%			0.25	
Phosphorus		%			0.26	
Potassium		%			2.77	

Rush, slimpod, aerial part, fresh, full bloom, (2)
Ref No 2 03 973 United States

Analyses				As Fed	Dry	C.V. ± %
Dry matter		%			100.0	
Ash		%			7.6	
Crude fiber		%			34.8	
Protein (N x 6.25)		%			8.9	
Cattle	dig prot	%	*		5.5	
Goats	dig prot	%	*		4.9	
Horses	dig prot	%	*		5.1	
Rabbits	dig prot	%	*		5.5	
Sheep	dig prot	%	*		5.3	

Rush, slimpod, aerial part, fresh, dough stage, (2)
Ref No 2 03 974 United States

Analyses				As Fed	Dry	C.V. ± %
Dry matter		%			100.0	
Calcium		%			0.31	
Phosphorus		%			0.15	
Potassium		%			2.67	

Rush, slimrod, aerial part, fresh, dormant, (2)
Ref No 2 03 975 United States

Analyses				As Fed	Dry	C.V. ± %
Dry matter		%			100.0	
Ash		%			7.8	
Crude fiber		%			35.3	
Protein (N x 6.25)		%			4.1	
Cattle	dig prot	%	*		1.4	
Goats	dig prot	%	*		0.4	
Horses	dig prot	%	*		1.0	
Rabbits	dig prot	%	*		1.8	
Sheep	dig prot	%	*		0.8	
Calcium		%			0.29	
Phosphorus		%			0.09	
Potassium		%			2.52	

RUSH, TOAD. Juncus bufonius

Rush, toad, aerial part, fresh, (2)
Ref No 2 03 979 United States

Analyses				As Fed	Dry	C.V. ± %
Dry matter		%			100.0	
Ash		%			5.5	9
Crude fiber		%			30.8	12
Protein (N x 6.25)		%			6.6	39
Cattle	dig prot	%	*		3.5	
Goats	dig prot	%	*		2.7	

Continued

Feed Name or Analyses		Mean As Fed	Mean Dry	C.V. ± %
Horses	dig prot % *		3.1	
Rabbits	dig prot % *		3.8	
Sheep	dig prot % *		3.1	

Rush, toad, aerial part, fresh, full bloom, (2)

Ref No 2 03 977 United States

Feed Name or Analyses		Mean As Fed	Mean Dry	C.V. ± %
Dry matter	%		100.0	
Ash	%		5.7	
Crude fiber	%		27.7	
Protein (N x 6.25)	%		10.0	
Cattle	dig prot % *		6.4	
Goats	dig prot % *		5.9	
Horses	dig prot % *		6.0	
Rabbits	dig prot % *		6.4	
Sheep	dig prot % *		6.3	
Calcium	%		0.28	
Phosphorus	%		0.39	
Potassium	%		1.91	

Rush, toad, aerial part, fresh, dormant, (2)

Ref No 2 03 978 United States

Feed Name or Analyses		Mean As Fed	Mean Dry	C.V. ± %
Dry matter	%		100.0	
Ash	%		5.3	
Crude fiber	%		32.9	
Protein (N x 6.25)	%		4.4	
Cattle	dig prot % *		1.6	
Goats	dig prot % *		0.7	
Horses	dig prot % *		1.3	
Rabbits	dig prot % *		2.1	
Sheep	dig prot % *		1.1	
Calcium	%		0.25	
Phosphorus	%		0.18	
Potassium	%		1.30	

RUSHPEA. Hoffmannseggia spp

Rushpea, aerial part, fresh, mature, (2)

Ref No 2 03 980 United States

Feed Name or Analyses		Mean As Fed	Mean Dry	C.V. ± %
Dry matter	%		100.0	
Ash	%		7.9	
Crude fiber	%		17.1	
Ether extract	%		2.5	
N-free extract	%		57.8	
Protein (N x 6.25)	%		14.7	
Cattle	dig prot % *		10.4	
Goats	dig prot % *		10.3	
Horses	dig prot % *		10.0	
Rabbits	dig prot % *		10.0	
Sheep	dig prot % *		10.7	
Calcium	%		3.18	
Magnesium	%		0.21	
Phosphorus	%		0.18	
Potassium	%		1.21	

(1) dry forages and roughages (2) pasture, range plants, and forages fed green (3) silages (4) energy feeds (5) protein supplements (6) minerals (7) vitamins (8) additives

RUSSIANTHISTLE. Salsola spp

Russianthistle, hay, s-c, (1)

Ref No 1 08 497 United States

Feed Name or Analyses		Mean As Fed	Mean Dry	C.V. ± %
Dry matter	%	87.5	100.0	
Ash	%	12.7	14.5	
Crude fiber	%	26.9	30.7	
Ether extract	%	1.6	1.8	
N-free extract	%	37.4	42.7	
Protein (N x 6.25)	%	8.9	10.2	
Cattle	dig prot % *	5.0	5.7	
Goats	dig prot % *	5.3	6.0	
Horses	dig prot % *	5.4	6.2	
Rabbits	dig prot % *	5.7	6.5	
Sheep	dig prot % *	5.0	5.7	
Energy	GE Mcal/kg			
Cattle	DE Mcal/kg *	1.72	1.97	
Sheep	DE Mcal/kg *	1.58	1.80	
Cattle	ME Mcal/kg *	1.42	1.62	
Sheep	ME Mcal/kg *	1.30	1.48	
Cattle	TDN % *	39.0	44.6	
Sheep	TDN % *	35.7	40.8	

Russianthistle, aerial part, fresh, (2)

Ref No 2 03 983 United States

Feed Name or Analyses		Mean As Fed	Mean Dry	C.V. ± %
Dry matter	%	34.7	100.0	46
Ash	%	5.2	15.0	
Crude fiber	%	10.4	30.0	
Ether extract	%	0.7	2.0	
N-free extract	%	14.5	41.7	
Protein (N x 6.25)	%	3.9	11.3	
Cattle	dig prot % *	2.6	7.5	
Goats	dig prot % *	2.5	7.1	
Horses	dig prot % *	2.5	7.2	
Rabbits	dig prot % *	2.6	7.4	
Sheep	dig prot % *	2.6	7.6	
Energy	GE Mcal/kg			
Cattle	DE Mcal/kg *	0.65	1.86	
Sheep	DE Mcal/kg *	0.67	1.94	
Cattle	ME Mcal/kg *	0.53	1.53	
Sheep	ME Mcal/kg *	0.55	1.59	
Cattle	TDN % *	14.7	42.3	
Sheep	TDN % *	15.3	44.1	
Calcium	%	0.76	2.20	
Cobalt	mg/kg	0.059	0.170	
Magnesium	%	0.32	0.93	
Phosphorus	%	0.08	0.23	
Potassium	%	2.37	6.83	
Sulphur	%	0.06	0.17	15
Carotene	mg/kg	30.9	89.1	
Vitamin A equivalent	IU/g	51.5	148.5	

Russianthistle, aerial part, fresh, immature, (2)

Ref No 2 03 981 United States

Feed Name or Analyses		Mean As Fed	Mean Dry	C.V. ± %
Dry matter	%	61.6	100.0	36
Ash	%	12.0	19.5	13
Crude fiber	%	12.4	20.2	
Ether extract	%	1.5	2.5	
N-free extract	%	24.1	39.1	
Protein (N x 6.25)	%	11.5	18.7	19
Cattle	dig prot % *	8.5	13.8	
Goats	dig prot % *	8.6	14.0	
Horses	dig prot % *	8.3	13.4	
Rabbits	dig prot % *	8.1	13.1	
Sheep	dig prot % *	8.9	14.4	
Energy	GE Mcal/kg			
Cattle	DE Mcal/kg *	1.17	1.90	
Sheep	DE Mcal/kg *	1.26	2.04	
Cattle	ME Mcal/kg *	0.96	1.56	
Sheep	ME Mcal/kg *	1.03	1.67	
Cattle	TDN % *	26.6	43.2	
Sheep	TDN % *	28.5	46.3	
Calcium	%	1.67	2.71	19
Magnesium	%	0.51	0.82	9
Phosphorus	%	0.12	0.20	18
Potassium	%	2.85	4.63	
Carotene	mg/kg	65.9	106.9	98
Vitamin A equivalent	IU/g	109.8	178.2	

Russianthistle, aerial part, fresh, early bloom, (2)

Ref No 2 03 982 United States

Feed Name or Analyses		Mean As Fed	Mean Dry	C.V. ± %
Dry matter	%	51.4	100.0	
Ash	%	9.7	18.9	
Ether extract	%	1.1	2.2	
Protein (N x 6.25)	%	6.9	13.5	
Cattle	dig prot % *	4.8	9.4	
Goats	dig prot % *	4.7	9.2	
Horses	dig prot % *	4.6	9.0	
Rabbits	dig prot % *	4.7	9.1	
Sheep	dig prot % *	4.9	9.6	
Carotene	mg/kg	36.3	70.5	
Vitamin A equivalent	IU/g	60.4	117.6	

Russianthistle, aerial part, ensiled, (3)

Ref No 3 03 984 United States

Feed Name or Analyses		Mean As Fed	Mean Dry	C.V. ± %
Dry matter	%	34.4	100.0	
Ash	%	6.2	17.9	
Crude fiber	%	10.3	29.8	
Ether extract	%	0.9	2.8	
N-free extract	%	14.5	42.2	
Protein (N x 6.25)	%	2.5	7.4	
Cattle	dig prot % *	1.0	2.9	
Goats	dig prot % *	1.0	2.9	
Horses	dig prot % *	1.0	2.9	
Sheep	dig prot % *	1.0	2.9	
Energy	GE Mcal/kg			
Cattle	DE Mcal/kg *	0.65	1.89	
Sheep	DE Mcal/kg *	0.59	1.72	
Cattle	ME Mcal/kg *	0.53	1.55	
Sheep	ME Mcal/kg *	0.49	1.41	
Cattle	TDN % *	14.7	42.8	
Sheep	TDN % *	13.4	38.9	

Russianthistle, aerial part w molasses added, ensiled, immature, (3)

Ref No 3 03 985 United States

Feed Name or Analyses		Mean As Fed	Mean Dry	C.V. ± %
Dry matter	%	34.5	100.0	
Calcium	%	1.01	2.92	
Magnesium	%	0.55	1.60	
Phosphorus	%	0.07	0.21	

Feed Name or Analyses			Mean As Fed	Mean Dry	C.V. ± %

RUSSIANTHISTLE, TUMBLING. Salsola kali Tenuifolia

Russianthistle, tumbling, hay, s-c, (1)

Ref No 1 03 988 United States

Feed Name or Analyses			As Fed	Dry	C.V. ± %
Dry matter	%		85.9	100.0	1
Ash	%		13.2	15.4	18
Crude fiber	%		24.4	28.4	17
Cattle	dig coef %		45.	45.	
Sheep	dig coef %		35.	35.	
Ether extract	%		1.8	2.1	38
Cattle	dig coef %		40.	40.	
Sheep	dig coef %		71.	71.	
N-free extract	%		35.9	41.8	
Cattle	dig coef %		62.	62.	
Sheep	dig coef %		51.	51.	
Protein (N x 6.25)	%		10.7	12.4	16
Cattle	dig coef %		63.	63.	
Sheep	dig coef %		66.	66.	
Cattle	dig prot %		6.7	7.8	
Goats	dig prot %	*	7.0	8.1	
Horses	dig prot %	*	6.9	8.1	
Rabbits	dig prot %	*	7.1	8.2	
Sheep	dig prot %		7.0	8.2	
Energy	GE Mcal/kg				
Cattle	DE Mcal/kg	*	1.83	2.13	
Sheep	DE Mcal/kg	*	1.62	1.88	
Cattle	ME Mcal/kg	*	1.50	1.75	
Sheep	ME Mcal/kg	*	1.33	1.54	
Cattle	NE_m Mcal/kg	*	0.90	1.05	
Cattle	NE_{gain} Mcal/kg	*	0.18	0.21	
Cattle	TDN %		41.5	48.3	
Sheep	TDN %		36.7	42.7	
Calcium	%		1.30	1.51	
Magnesium	%		0.64	0.75	
Phosphorus	%		0.21	0.25	
Potassium	%		5.88	6.85	

Russianthistle, tumbling, hay, s-c, immature, (1)

Ref No 1 03 986 United States

Feed Name or Analyses			As Fed	Dry	C.V. ± %
Dry matter	%		89.8	100.0	3
Ash	%		16.2	18.1	
Crude fiber	%		19.7	22.0	
Cattle	dig coef %	*	45.	45.	
Sheep	dig coef %	*	35.	35.	
Ether extract	%		1.4	1.6	
Cattle	dig coef %	*	40.	40.	
Sheep	dig coef %	*	71.	71.	
N-free extract	%		38.6	43.0	
Cattle	dig coef %	*	62.	62.	
Sheep	dig coef %	*	51.	51.	
Protein (N x 6.25)	%		13.7	15.3	
Cattle	dig coef %	*	63.	63.	
Sheep	dig coef %	*	66.	66.	
Cattle	dig prot %		8.7	9.6	
Goats	dig prot %	*	9.7	10.8	
Horses	dig prot %	*	9.4	10.5	
Rabbits	dig prot %	*	9.4	10.5	
Sheep	dig prot %		9.1	10.1	
Energy	GE Mcal/kg		3.60	4.01	
Cattle	DE Mcal/kg	*	1.89	2.10	
Sheep	DE Mcal/kg	*	1.67	1.86	
Cattle	ME Mcal/kg	*	1.55	1.72	
Sheep	ME Mcal/kg	*	1.37	1.53	
Cattle	TDN %		42.8	47.6	
Sheep	TDN %		38.0	42.3	
Sulphur	%		0.38	0.42	
Carotene	mg/kg		4.2	4.6	
Vitamin A equivalent	IU/g		6.9	7.7	

Russianthistle, tumbling, hay, s-c, mature, (1)

Ref No 1 03 987 United States

Feed Name or Analyses			As Fed	Dry	C.V. ± %
Dry matter	%		91.2	100.0	3
Ash	%		11.8	12.9	32
Crude fiber	%		29.4	32.2	
Cattle	dig coef %	*	45.	45.	
Sheep	dig coef %	*	35.	35.	
Ether extract	%		0.9	1.0	
Cattle	dig coef %	*	40.	40.	
Sheep	dig coef %	*	71.	71.	
N-free extract	%		43.9	48.1	
Cattle	dig coef %	*	62.	62.	
Sheep	dig coef %	*	51.	51.	
Protein (N x 6.25)	%		5.3	5.8	22
Cattle	dig coef %	*	63.	63.	
Sheep	dig coef %	*	66.	66.	
Cattle	dig prot %		3.3	3.7	
Goats	dig prot %	*	1.8	2.0	
Horses	dig prot %	*	2.2	2.5	
Rabbits	dig prot %	*	2.9	3.1	
Sheep	dig prot %		3.5	3.8	
Energy	GE Mcal/kg				
Cattle	DE Mcal/kg	*	1.96	2.15	
Sheep	DE Mcal/kg	*	1.66	1.82	
Cattle	ME Mcal/kg	*	1.61	1.77	
Sheep	ME Mcal/kg	*	1.36	1.49	
Cattle	TDN %		44.6	48.9	
Sheep	TDN %		37.6	41.2	
Calcium	%		1.55	1.70	
Magnesium	%		0.94	1.03	
Phosphorus	%		0.18	0.20	
Potassium	%		6.25	6.85	

Russianthistle, tumbling, aerial part, fresh, (2)

Ref No 2 03 990 United States

Feed Name or Analyses			As Fed	Dry	C.V. ± %
Dry matter	%			100.0	
Ash	%			15.5	54
Crude fiber	%			31.7	
Ether extract	%			2.0	78
N-free extract	%			37.7	
Protein (N x 6.25)	%			13.1	53
Cattle	dig prot %	*		9.0	
Goats	dig prot %	*		8.8	
Horses	dig prot %	*		8.7	
Rabbits	dig prot %	*		8.8	
Sheep	dig prot %	*		9.2	
Energy	GE Mcal/kg				
Cattle	DE Mcal/kg	*		2.37	
Sheep	DE Mcal/kg	*		2.28	
Cattle	ME Mcal/kg	*		1.94	
Sheep	ME Mcal/kg	*		1.87	
Cattle	TDN %	*		53.7	
Sheep	TDN %	*		51.8	
Calcium	%			2.73	20
Copper	mg/kg			19.2	
Magnesium	%			0.82	9
Manganese	mg/kg			33.3	
Phosphorus	%			0.13	41
Potassium	%			4.63	
Carotene	mg/kg			4.6	
Vitamin A equivalent	IU/g			7.7	

Russianthistle, tumbling, aerial part, immature, (2)

Ref No 2 03 991 United States

Feed Name or Analyses			As Fed	Dry	C.V. ± %
Dry matter	%			100.0	
Ash	%			20.0	
Crude fiber	%			20.2	
Ether extract	%			2.5	
N-free extract	%			40.1	
Protein (N x 6.25)	%			17.2	
Cattle	dig prot %	*		12.5	
Goats	dig prot %	*		12.6	
Horses	dig prot %	*		12.1	
Rabbits	dig prot %	*		11.9	
Sheep	dig prot %	*		13.0	
Energy	GE Mcal/kg				
Cattle	DE Mcal/kg	*		2.38	
Sheep	DE Mcal/kg	*		2.48	
Cattle	ME Mcal/kg	*		1.95	
Sheep	ME Mcal/kg	*		2.03	
Cattle	TDN %	*		54.0	
Sheep	TDN %	*		56.2	
Calcium	%			2.87	19
Magnesium	%			0.82	9
Phosphorus	%			0.20	15
Potassium	%			4.63	

Russianthistle, tumbling, aerial part, early bloom, (2)

Ref No 2 03 992 United States

Feed Name or Analyses			As Fed	Dry	C.V. ± %
Dry matter	%			100.0	
Ash	%			16.5	
Ether extract	%			2.2	
Protein (N x 6.25)	%			13.2	
Cattle	dig prot %	*		9.1	
Goats	dig prot %	*		8.9	
Horses	dig prot %	*		8.7	
Rabbits	dig prot %	*		8.9	
Sheep	dig prot %	*		9.3	

Russianthistle, tumbling, aerial part, over ripe, (2)

Ref No 2 03 993 United States

Feed Name or Analyses			As Fed	Dry	C.V. ± %
Dry matter	%			100.0	
Ash	%			5.4	
Crude fiber	%			43.1	
Ether extract	%			0.6	
N-free extract	%			46.0	
Protein (N x 6.25)	%			4.9	
Cattle	dig prot %	*		2.1	
Goats	dig prot %	*		1.1	
Horses	dig prot %	*		1.7	
Rabbits	dig prot %	*		2.5	
Sheep	dig prot %	*		1.6	
Calcium	%			2.31	10
Cobalt	mg/kg			0.170	
Copper	mg/kg			19.2	
Manganese	mg/kg			33.3	
Phosphorus	%			0.07	22
Carotene	mg/kg			0.0	

Feed Name or Analyses		Mean As Fed	Mean Dry	C.V. ± %
RUTABAGA. Brassica napobrassica				
Rutabaga, aerial part w crowns, dehy, (1)				
Ref No 1 03 995			United States	
Dry matter	%	87.3	100.0	
Ash	%	34.4	39.4	
Crude fiber	%	8.6	9.8	
Sheep	dig coef %	96.	96.	
Ether extract	%	1.9	2.2	
Sheep	dig coef %	65.	65.	
N-free extract	%	32.7	37.4	
Sheep	dig coef %	85.	85.	
Protein (N x 6.25)	%	9.8	11.2	
Sheep	dig coef %	78.	78.	
Cattle	dig prot % *	5.8	6.6	
Goats	dig prot % *	6.1	7.0	
Horses	dig prot % *	6.1	7.0	
Rabbits	dig prot % *	6.4	7.3	
Sheep	dig prot %	7.6	8.7	
Energy	GE Mcal/kg			
Cattle	DE Mcal/kg *	2.12	2.43	
Sheep	DE Mcal/kg *	2.05	2.34	
Cattle	ME Mcal/kg *	1.74	1.99	
Sheep	ME Mcal/kg *	1.68	1.92	
Cattle	TDN % *	48.1	55.1	
Sheep	TDN %	46.4	53.2	
Rutabaga, aerial part w crowns, fresh, (2)				
Ref No 2 03 996			United States	
Dry matter	%	10.9	100.0	
Ash	%	2.2	19.9	
Crude fiber	%	1.5	14.1	
Sheep	dig coef %	85.	85.	
Ether extract	%	0.5	4.6	
Sheep	dig coef %	68.	68.	
N-free extract	%	4.7	42.9	
Sheep	dig coef %	86.	86.	
Protein (N x 6.25)	%	2.0	18.6	
Sheep	dig coef %	86.	86.	
Cattle	dig prot % *	1.5	13.7	
Goats	dig prot % *	1.5	13.9	
Horses	dig prot % *	1.5	13.3	
Rabbits	dig prot % *	1.4	13.0	
Sheep	dig prot %	1.7	15.9	
Energy	GE Mcal/kg			
Cattle	DE Mcal/kg *	0.33	3.02	
Sheep	DE Mcal/kg *	0.34	3.15	
Cattle	ME Mcal/kg *	0.27	2.48	
Sheep	ME Mcal/kg *	0.28	2.59	
Cattle	TDN % *	7.5	68.4	
Sheep	TDN %	7.8	71.5	
Rutabaga, leaves, fresh, (2)				
Ref No 2 03 997			United States	
Dry matter	%	17.8	100.0	
Crude fiber	%	1.1	6.3	
Ether extract	%	1.2	6.5	
Protein (N x 6.25)	%	5.6	31.5	
Cattle	dig prot % *	4.4	24.7	
Goats	dig prot % *	4.6	25.9	
Horses	dig prot % *	4.3	24.3	
Rabbits	dig prot % *	4.1	23.0	
Sheep	dig prot % *	4.7	26.3	
Carotene	mg/kg	45.8	257.1	
Riboflavin	mg/kg	3.7	20.9	
Vitamin A equivalent	IU/g	76.3	428.5	
Rutabaga, stems, fresh, (2)				
Ref No 2 03 998			United States	
Dry matter	%	9.5	100.0	
Crude fiber	%	1.4	14.9	
Protein (N x 6.25)	%	1.8	18.5	
Cattle	dig prot % *	1.3	13.6	
Goats	dig prot % *	1.3	13.8	
Horses	dig prot % *	1.3	13.2	
Rabbits	dig prot % *	1.2	13.0	
Sheep	dig prot % *	1.4	14.2	
Carotene	mg/kg	1.2	13.0	
Riboflavin	mg/kg	0.8	8.6	
Vitamin A equivalent	IU/g	2.1	21.7	
Rutabaga, aerial part, ensiled, (3)				
Ref No 3 03 999			United States	
Dry matter	%	14.7	100.0	
Ash	%	4.5	30.8	
Crude fiber	%	1.8	12.3	
Sheep	dig coef %	85.	85.	
Ether extract	%	0.5	3.5	
Sheep	dig coef %	68.	68.	
N-free extract	%	5.4	37.0	
Sheep	dig coef %	85.	85.	
Protein (N x 6.25)	%	2.4	16.4	
Sheep	dig coef %	79.	79.	
Cattle	dig prot % *	1.6	11.1	
Goats	dig prot % *	1.6	11.1	
Horses	dig prot % *	1.6	11.1	
Sheep	dig prot %	1.9	13.0	
Energy	GE Mcal/kg			
Cattle	DE Mcal/kg *	0.38	2.61	
Sheep	DE Mcal/kg *	0.39	2.66	
Cattle	ME Mcal/kg *	0.31	2.14	
Sheep	ME Mcal/kg *	0.32	2.18	
Cattle	TDN % *	8.7	59.3	
Sheep	TDN %	8.9	60.2	
Rutabaga, aerial part w crowns, ensiled, (3)				
Ref No 3 04 000			United States	
Dry matter	%	17.0	100.0	
Ash	%	5.5	32.5	
Crude fiber	%	2.1	12.4	
Sheep	dig coef %	94.	94.	
Ether extract	%	0.7	4.3	
Sheep	dig coef %	70.	70.	
N-free extract	%	6.3	37.1	
Sheep	dig coef %	86.	86.	
Protein (N x 6.25)	%	2.3	13.8	
Sheep	dig coef %	82.	82.	
Cattle	dig prot % *	1.5	8.8	
Goats	dig prot % *	1.5	8.8	
Horses	dig prot % *	1.5	8.8	
Sheep	dig prot %	1.9	11.3	
Energy	GE Mcal/kg			
Cattle	DE Mcal/kg *	0.44	2.59	
Sheep	DE Mcal/kg *	0.46	2.71	
Cattle	ME Mcal/kg *	0.36	2.12	
Sheep	ME Mcal/kg *	0.38	2.23	
Cattle	TDN % *	10.0	58.7	
Sheep	TDN %	10.4	61.6	
Rutabaga, roots, fresh, (4)				
Ref No 2 04 001			United States	
Dry matter	%	12.4	100.0	9
Ash	%	0.9	7.3	19
Crude fiber	%	1.2	10.1	23
Horses	dig coef %	-74.	74.	
Sheep	dig coef %	89.	89.	
Swine	dig coef %	41.	41.	
Ether extract	%	0.2	1.4	61
Horses	dig coef %	-11.	11.	
Sheep	dig coef %	79.	79.	
Swine	dig coef %	0.	0.	
N-free extract	%	9.0	72.5	9
Horses	dig coef %	86.	86.	
Sheep	dig coef %	96.	96.	
Swine	dig coef %	93.	93.	
Protein (N x 6.25)	%	1.1	8.7	26
Horses	dig coef %	83.	83.	
Sheep	dig coef %	75.	75.	
Swine	dig coef %	84.	84.	
Cattle	dig prot % *	0.5	4.0	
Goats	dig prot % *	0.6	5.2	
Horses	dig prot %	0.9	7.2	
Sheep	dig prot %	0.8	6.6	
Swine	dig prot %	0.9	7.3	
Energy	GE Mcal/kg	0.44	3.54	
Cattle	DE Mcal/kg *	0.51	4.14	
Horses	DE Mcal/kg *	0.34	2.72	
Sheep	DE Mcal/kg *	0.48	3.86	
Swine	DE kcal/kg *	431.	3479.	
Cattle	ME Mcal/kg *	0.42	3.40	
Horses	ME Mcal/kg *	0.28	2.23	
Sheep	ME Mcal/kg *	0.39	3.17	
Swine	ME kcal/kg *	407.	3279.	
Cattle	TDN % *	11.7	94.0	
Horses	TDN %	7.7	61.7	
Sheep	TDN %	10.9	87.6	
Swine	TDN %	9.8	78.9	
Calcium	%	0.06	0.49	
Copper	mg/kg	1.0	7.9	
Iron	%	0.002	0.014	
Magnesium	%	0.02	0.18	
Manganese	mg/kg	1.2	9.9	
Phosphorus	%	0.04	0.29	
Potassium	%	0.23	1.87	
Sodium	%	0.01	0.08	
Sulphur	%	0.03	0.27	
Ascorbic acid	mg/kg	410.2	3307.7	
Niacin	mg/kg	10.5	84.6	
Riboflavin	mg/kg	0.7	5.4	
Thiamine	mg/kg	0.7	5.4	
Vitamin A	IU/g	5.5	44.6	
RYE. Secale cereale				
Rye, hay, s-c, (1)				
Ref No 1 04 004			United States	
Dry matter	%	91.8	100.0	2
Ash	%	5.0	5.5	
Crude fiber	%	36.7	40.0	

(1) dry forages and roughages (2) pasture, range plants, and forages fed green (3) silages (4) energy feeds (5) protein supplements (6) minerals (7) vitamins (8) additives

Feed Name or Analyses			Mean As Fed	Mean Dry	C.V. ± %
Ether extract	%		2.1	2.3	
N-free extract	%		41.2	44.9	
Protein (N x 6.25)	%		6.7	7.3	
Cattle	dig prot %	*	3.0	3.3	
Goats	dig prot %	*	3.1	3.4	
Horses	dig prot %	*	3.5	3.8	
Rabbits	dig prot %	*	4.0	4.3	
Sheep	dig prot %	*	2.9	3.2	
Fatty acids	%		2.1	2.3	
Energy	GE Mcal/kg				
Cattle	DE Mcal/kg	*	1.82	1.98	
Sheep	DE Mcal/kg	*	1.91	2.08	
Cattle	ME Mcal/kg	*	1.49	1.62	
Sheep	ME Mcal/kg	*	1.57	1.71	
Cattle	TDN %	*	41.2	44.9	
Sheep	TDN %	*	43.2	47.1	
Calcium	%		0.32	0.35	43
Phosphorus	%		0.26	0.28	98
Potassium	%		1.01	1.10	

Ref No 1 04 004 Canada

Feed Name or Analyses			Mean As Fed	Mean Dry	C.V. ± %
Dry matter	%		92.2	100.0	
Ash	%		3.7	4.0	
Crude fiber	%		26.6	28.9	
Ether extract	%		1.7	1.8	
N-free extract	%		52.0	56.5	
Protein (N x 6.25)	%		8.2	8.8	
Cattle	dig prot %	*	4.2	4.6	
Goats	dig prot %	*	4.4	4.8	
Horses	dig prot %	*	4.6	5.0	
Rabbits	dig prot %	*	5.1	5.5	
Sheep	dig prot %	*	4.2	4.5	
Energy	GE Mcal/kg				
Cattle	DE Mcal/kg	*	1.92	2.08	
Sheep	DE Mcal/kg	*	1.82	1.97	
Cattle	ME Mcal/kg	*	1.58	1.71	
Sheep	ME Mcal/kg	*	1.49	1.62	
Cattle	TDN %	*	43.5	47.2	
Sheep	TDN %	*	41.2	44.7	
Calcium	%		0.20	0.22	
Phosphorus	%		0.15	0.16	
Zinc	mg/kg		31.2	33.9	
Carotene	mg/kg		6.0	6.5	
Vitamin A equivalent	IU/g		10.0	10.9	

Rye, hay, s-c, dough stage, (1)

Ref No 1 04 002 United States

Feed Name or Analyses			Mean As Fed	Mean Dry	C.V. ± %
Dry matter	%		93.1	100.0	1
Ash	%		5.4	5.8	12
Crude fiber	%		34.4	37.0	5
Ether extract	%		2.1	2.3	7
N-free extract	%		41.3	44.4	
Protein (N x 6.25)	%		9.8	10.5	11
Cattle	dig prot %	*	5.6	6.0	
Goats	dig prot %	*	5.9	6.4	
Horses	dig prot %	*	6.0	6.4	
Rabbits	dig prot %	*	6.3	6.8	
Sheep	dig prot %	*	5.6	6.0	
Energy	GE Mcal/kg				
Cattle	DE Mcal/kg	*	2.03	2.18	
Sheep	DE Mcal/kg	*	2.25	2.42	
Cattle	ME Mcal/kg	*	1.66	1.79	
Sheep	ME Mcal/kg	*	1.85	1.99	
Cattle	TDN %	*	46.0	49.5	
Sheep	TDN %	*	51.1	54.9	

Rye, hay, s-c, mature, (1)

Ref No 1 04 003 United States

Feed Name or Analyses			Mean As Fed	Mean Dry	C.V. ± %
Dry matter	%		90.9	100.0	2
Ash	%		6.7	7.4	26
Crude fiber	%		34.4	37.8	5
Ether extract	%		2.0	2.2	4
N-free extract	%		40.5	44.5	
Protein (N x 6.25)	%		7.4	8.1	17
Cattle	dig prot %	*	3.6	4.0	
Goats	dig prot %	*	3.7	4.1	
Horses	dig prot %	*	4.0	4.4	
Rabbits	dig prot %	*	4.5	4.9	
Sheep	dig prot %	*	3.5	3.8	
Energy	GE Mcal/kg				
Cattle	DE Mcal/kg	*	2.01	2.21	
Sheep	DE Mcal/kg	*	2.14	2.35	
Cattle	ME Mcal/kg	*	1.65	1.81	
Sheep	ME Mcal/kg	*	1.76	1.93	
Cattle	TDN %	*	45.6	50.1	
Sheep	TDN %	*	48.5	53.4	

Rye, straw, (1)

Ref No 1 04 007 United States

Feed Name or Analyses			Mean As Fed	Mean Dry	C.V. ± %
Dry matter	%		89.4	100.0	2
Ash	%		3.7	4.2	24
Crude fiber	%		41.5	46.4	9
Sheep	dig coef %		55.	55.	
Ether extract	%		1.3	1.5	15
Sheep	dig coef %		44.	44.	
N-free extract	%		40.1	44.9	
Sheep	dig coef %		43.	43.	
Protein (N x 6.25)	%		2.8	3.1	44
Sheep	dig coef %		-52.	-52.	
Cattle	dig prot %	*	-0.2	-0.3	
Goats	dig prot %	*	-0.4	-0.4	
Horses	dig prot %	*	0.2	0.2	
Rabbits	dig prot %	*	1.0	1.1	
Sheep	dig prot %		-1.4	-1.6	
Fatty acids	%		1.2	1.3	
Energy	GE Mcal/kg				
Cattle	DE Mcal/kg	*	1.84	2.06	
Sheep	DE Mcal/kg	*	1.76	1.97	
Cattle	ME Mcal/kg	*	1.51	1.69	
Sheep	ME Mcal/kg	*	1.44	1.61	
Cattle	NE_m Mcal/kg	*	0.67	0.75	
Cattle	NE_{gain} Mcal/kg	*	0.00	0.00	
Cattle	TDN %	*	41.7	46.7	
Sheep	TDN %		39.9	44.6	
Calcium	%		0.25	0.28	16
Chlorine	%		0.21	0.24	12
Copper	mg/kg		3.6	4.0	10
Magnesium	%		0.07	0.08	28
Manganese	mg/kg		5.9	6.6	8
Phosphorus	%		0.09	0.10	29
Potassium	%		0.87	0.97	11
Sodium	%		0.12	0.13	20
Sulphur	%		0.10	0.11	22

Ref No 1 04 007 Canada

Feed Name or Analyses			Mean As Fed	Mean Dry	C.V. ± %
Dry matter	%		94.1	100.0	
Crude fiber	%		33.6	35.7	
Sheep	dig coef %		55.	55.	
Ether extract	%				
Sheep	dig coef %		44.	44.	
N-free extract	%				
Sheep	dig coef %		43.	43.	
Protein (N x 6.25)	%		5.6	5.9	
Sheep	dig coef %		-52.	-52.	
Cattle	dig prot %	*	1.9	2.1	
Goats	dig prot %	*	2.0	2.1	
Horses	dig prot %	*	2.4	2.6	
Rabbits	dig prot %	*	3.0	3.2	
Sheep	dig prot %		-2.9	-3.0	
Energy	GE Mcal/kg				
Sheep	DE Mcal/kg	*	1.78	1.90	
Sheep	ME Mcal/kg	*	1.46	1.56	
Sheep	TDN %		40.5	43.0	
Calcium	%		0.19	0.20	
Phosphorus	%		0.08	0.09	

Rye, straw, steamed, (1)

Ref No 1 04 005 United States

Feed Name or Analyses			Mean As Fed	Mean Dry	C.V. ± %
Dry matter	%		91.4	100.0	
Ash	%		2.0	2.2	
Crude fiber	%		66.9	73.2	
Sheep	dig coef %		86.	86.	
Ether extract	%		0.6	0.7	
Sheep	dig coef %		130.	130.	
N-free extract	%		21.7	23.7	
Sheep	dig coef %		78.	78.	
Protein (N x 6.25)	%		0.2	0.2	
Sheep	dig coef %		-192.	-192.	
Cattle	dig prot %	*	-2.5	-2.8	
Goats	dig prot %	*	-2.9	-3.2	
Horses	dig prot %	*	-2.0	-2.2	
Rabbits	dig prot %	*	-1.0	-1.1	
Sheep	dig prot %		-0.3	-0.3	
Energy	GE Mcal/kg				
Sheep	DE Mcal/kg	*	3.35	3.66	
Sheep	ME Mcal/kg	*	2.75	3.00	
Sheep	TDN %		76.0	83.1	

Rye, straw, steamed dehy, (1)

Ref No 1 04 006 United States

Feed Name or Analyses			Mean As Fed	Mean Dry	C.V. ± %
Dry matter	%		87.8	100.0	
Ash	%		2.5	2.8	
Crude fiber	%		39.8	45.3	
Sheep	dig coef %		66.	66.	
Ether extract	%		1.9	2.2	
Sheep	dig coef %		68.	68.	
N-free extract	%		41.3	47.0	
Sheep	dig coef %		49.	49.	
Protein (N x 6.25)	%		2.4	2.7	
Sheep	dig coef %		-57.	-57.	
Cattle	dig prot %	*	-0.5	-0.6	
Goats	dig prot %	*	-0.7	-0.8	
Horses	dig prot %	*	-0.1	-0.1	
Rabbits	dig prot %	*	0.7	0.8	
Sheep	dig prot %		-1.3	-1.5	
Energy	GE Mcal/kg				
Cattle	DE Mcal/kg	*	1.90	2.16	
Sheep	DE Mcal/kg	*	2.12	2.41	
Cattle	ME Mcal/kg	*	1.55	1.77	
Sheep	ME Mcal/kg	*	1.74	1.98	
Cattle	TDN %	*	42.9	48.9	
Sheep	TDN %		48.1	54.7	

Feed Name or Analyses			Mean As Fed	Mean Dry	C.V. ± %

Rye, straw, treated w calcium hydroxide, (1)
Ref No 1 04 009 United States

Analyses			As Fed	Dry	C.V. ± %
Dry matter	%		30.4	100.0	
Ash	%		2.9	9.6	
Crude fiber	%		17.3	57.0	
Sheep	dig coef %		80.	80.	
Ether extract	%		0.2	0.8	
Sheep	dig coef %		32.	32.	
N-free extract	%		9.3	30.5	
Sheep	dig coef %		25.	25.	
Protein (N x 6.25)	%		0.6	2.1	
Sheep	dig coef %		-184.	-184.	
Cattle	dig prot %	*	-0.3	-1.1	
Goats	dig prot %	*	-0.4	-1.4	
Horses	dig prot %	*	-0.1	-0.6	
Rabbits	dig prot %	*	0.1	0.3	
Sheep	dig prot %		-1.1	-3.8	
Energy	GE Mcal/kg				
Cattle	DE Mcal/kg	*	0.59	1.93	
Sheep	DE Mcal/kg	*	0.67	2.20	
Cattle	ME Mcal/kg	*	0.48	1.58	
Sheep	ME Mcal/kg	*	0.55	1.80	
Cattle	TDN %	*	13.3	43.7	
Sheep	TDN %		15.2	49.9	

Rye, straw, treated w hydrochloric acid dehy, (1)
Ref No 1 04 010 United States

Analyses			As Fed	Dry	C.V. ± %
Dry matter	%		89.9	100.0	
Ash	%		7.2	8.0	
Crude fiber	%		37.3	41.5	
Sheep	dig coef %		52.	52.	
Ether extract	%		1.3	1.4	
Sheep	dig coef %		59.	59.	
N-free extract	%		41.4	46.1	
Sheep	dig coef %		52.	52.	
Protein (N x 6.25)	%		2.7	3.0	
Sheep	dig coef %		11.	11.	
Cattle	dig prot %	*	-0.3	-0.4	
Goats	dig prot %	*	-0.5	-0.5	
Horses	dig prot %	*	0.1	0.1	
Rabbits	dig prot %	*	0.9	1.0	
Sheep	dig prot %		0.3	0.3	
Energy	GE Mcal/kg				
Cattle	DE Mcal/kg	*	1.78	1.98	
Sheep	DE Mcal/kg	*	1.89	2.10	
Cattle	ME Mcal/kg	*	1.46	1.62	
Sheep	ME Mcal/kg	*	1.55	1.73	
Cattle	TDN %	*	40.4	44.9	
Sheep	TDN %		42.9	47.7	

Rye, straw, treated w sodium hydroxide dehy, (1)
Ref No 1 04 011 United States

Analyses			As Fed	Dry	C.V. ± %
Dry matter	%		89.6	100.0	
Ash	%		4.4	4.9	
Crude fiber	%		51.2	57.1	
Sheep	dig coef %		78.	78.	
Ether extract	%		1.3	1.5	
Sheep	dig coef %		44.	44.	
N-free extract	%		30.9	34.5	
Sheep	dig coef %		54.	54.	
Protein (N x 6.25)	%		1.8	2.0	
Sheep	dig coef %		-102.	-102.	
Cattle	dig prot %	*	-1.1	-1.2	
Goats	dig prot %	*	-1.3	-1.5	
Horses	dig prot %	*	-0.6	-0.7	
Rabbits	dig prot %	*	0.2	0.2	
Sheep	dig prot %		-1.7	-2.0	
Energy	GE Mcal/kg				
Cattle	DE Mcal/kg	*	2.60	2.90	
Sheep	DE Mcal/kg	*	2.47	2.76	
Cattle	ME Mcal/kg	*	1.95	2.38	
Sheep	ME Mcal/kg	*	2.03	2.26	
Cattle	TDN %	*	59.0	65.8	
Sheep	TDN %		56.1	62.6	

Rye, straw, treated w sodium hydroxide wet, (1)
Ref No 2 04 012 United States

Analyses			As Fed	Dry	C.V. ± %
Dry matter	%		16.5	100.0	
Ash	%		0.5	2.8	
Crude fiber	%		8.9	53.8	
Cattle	dig coef %		79.	79.	
Ether extract	%		0.3	1.6	
Cattle	dig coef %		49.	49.	
N-free extract	%		6.6	39.8	
Cattle	dig coef %		56.	56.	
Protein (N x 6.25)	%		0.3	2.0	
Cattle	dig coef %		59.	59.	
Cattle	dig prot %		0.2	1.2	
Goats	dig prot %	*	-0.2	-1.5	
Horses	dig prot %	*	0.0	-0.7	
Rabbits	dig prot %	*	0.0	0.2	
Sheep	dig prot %	*	-0.2	-1.5	
Energy	GE Mcal/kg				
Cattle	DE Mcal/kg	*	0.49	2.99	
Sheep	DE Mcal/kg	*	0.47	2.82	
Cattle	ME Mcal/kg	*	0.40	2.45	
Sheep	ME Mcal/kg	*	0.38	2.31	
Cattle	TDN %		11.2	67.7	
Sheep	TDN %	*	10.5	63.9	

Rye, straw, treated w sodium sulfide dehy, (1)
Ref No 1 04 008 United States

Analyses			As Fed	Dry	C.V. ± %
Dry matter	%		91.0	100.0	
Ash	%		3.9	4.3	
Crude fiber	%		54.1	59.5	
Sheep	dig coef %		94.	94.	
Ether extract	%		1.2	1.3	
Sheep	dig coef %		80.	80.	
N-free extract	%		29.8	32.7	
Sheep	dig coef %		63.	63.	
Protein (N x 6.25)	%		2.0	2.2	
Sheep	dig coef %		-37.	-37.	
Cattle	dig prot %	*	-1.0	-1.1	
Goats	dig prot %	*	-1.2	-1.3	
Horses	dig prot %	*	-0.4	-0.5	
Rabbits	dig prot %	*	0.3	0.4	
Sheep	dig prot %		-0.7	-0.7	
Energy	GE Mcal/kg				
Cattle	DE Mcal/kg	*	2.93	3.22	
Sheep	DE Mcal/kg	*	3.13	3.44	
Cattle	ME Mcal/kg	*	2.40	2.64	
Sheep	ME Mcal/kg	*	2.57	2.82	
Cattle	TDN %	*	66.5	73.1	
Sheep	TDN %		71.0	78.0	

Rye, aerial part, fresh, (2)
Ref No 2 04 018 United States

Analyses			As Fed	Dry	C.V. ± %
Dry matter	%		20.6	100.0	34
Ash	%		2.1	10.3	35
Crude fiber	%		5.2	25.2	27
Cattle	dig coef %	*	80.	80.	
Ether extract	%		0.9	4.1	24
Cattle	dig coef %	*	74.	74.	
N-free extract	%		8.2	39.7	
Cattle	dig coef %	*	71.	71.	
Protein (N x 6.25)	%		4.3	20.7	36
Cattle	dig coef %	*	80.	80.	
Cattle	dig prot %		3.4	16.5	
Goats	dig prot %	*	3.3	15.9	
Horses	dig prot %	*	3.1	15.1	
Rabbits	dig prot %	*	3.0	14.6	
Sheep	dig prot %	*	3.4	16.3	
Energy	GE Mcal/kg				
Cattle	DE Mcal/kg	*	0.65	3.16	
Sheep	DE Mcal/kg	*	0.62	3.02	
Cattle	ME Mcal/kg	*	0.53	2.59	
Sheep	ME Mcal/kg	*	0.51	2.48	
Cattle	TDN %		14.8	71.8	
Sheep	TDN %	*	14.1	68.5	
Calcium	%		0.11	0.51	
Magnesium	%		0.07	0.36	
Phosphorus	%		0.09	0.41	
Carotene	mg/kg		70.6	342.6	4
Vitamin A equivalent	IU/g		117.6	571.1	
Glutamic acid	%		0.47	2.30	
Isoleucine	%		0.35	1.70	
Leucine	%		0.54	2.60	
Lysine	%		0.29	1.40	
Methionine	%		0.06	0.30	
Phenylalanine	%		0.23	1.10	
Threonine	%		0.47	2.30	
Tryptophan	%		0.04	0.20	
Valine	%		0.29	1.40	

Ref No 2 04 018 Canada

Analyses			As Fed	Dry	C.V. ± %
Dry matter	%		92.0	100.0	3
Crude fiber	%		27.1	29.5	8
Protein (N x 6.25)	%		8.0	8.7	25
Cattle	dig prot %	*	4.9	5.3	
Goats	dig prot %	*	4.3	4.7	
Horses	dig prot %	*	4.5	4.9	
Rabbits	dig prot %	*	5.0	5.4	
Sheep	dig prot %	*	4.7	5.1	
Energy	GE Mcal/kg				
Cattle	DE Mcal/kg	*	2.89	3.14	
Cattle	ME Mcal/kg	*	2.37	2.58	
Cattle	TDN %		65.6	71.3	
Calcium	%		0.24	0.26	46
Phosphorus	%		0.15	0.17	44
Carotene	mg/kg		15.6	16.9	
Vitamin A equivalent	IU/g		26.0	28.2	

Rye, aerial part, fresh, immature, (2)
Ref No 2 04 013 United States

Analyses			As Fed	Dry	C.V. ± %
Dry matter	%		15.9	100.0	21
Crude fiber	%				
Cattle	dig coef %	*	80.	80.	

(1) dry forages and roughages (2) pasture, range plants, and forages fed green (3) silages (4) energy feeds (5) protein supplements (6) minerals (7) vitamins (8) additives

Feed Name or Analyses		As Fed	Dry	C.V. ± %
Ether extract	%			
Cattle	dig coef %	* 74.	74.	
N-free extract	%			
Cattle	dig coef %	* 71.	71.	
Protein (N x 6.25)	%			
Cattle	dig coef %	* 80.	80.	
Sodium	%	0.01	0.07	79
Carotene	mg/kg	88.6	558.9	28
Vitamin A equivalent	IU/g	147.7	931.6	

Rye, aerial part, fresh, full bloom, (2)

Ref No 2 04 014 United States

Feed Name or Analyses		As Fed	Dry	C.V. ± %
Dry matter	%	23.1	100.0	
Calcium	%	0.10	0.44	
Phosphorus	%	0.07	0.30	

Rye, aerial part, fresh, milk stage, (2)

Ref No 2 04 015 United States

Feed Name or Analyses		As Fed	Dry	C.V. ± %
Dry matter	%		100.0	
Calcium	%		0.30	
Phosphorus	%		0.30	

Rye, aerial part, fresh, dough stage, (2)

Ref No 2 04 016 United States

Feed Name or Analyses		As Fed	Dry	C.V. ± %
Dry matter	%	33.4	100.0	
Ash	%	2.3	6.9	
Crude fiber	%	9.2	27.4	
Cattle	dig coef %	80.	80.	
Ether extract	%	1.2	3.5	
Cattle	dig coef %	74.	74.	
N-free extract	%	16.3	48.9	
Cattle	dig coef %	71.	71.	
Protein (N x 6.25)	%	4.5	13.4	
Cattle	dig coef %	80.	80.	
Cattle	dig prot %	3.6	10.7	
Goats	dig prot %	* 3.0	9.0	
Horses	dig prot %	* 3.0	8.9	
Rabbits	dig prot %	* 3.0	9.0	
Sheep	dig prot %	* 3.2	9.4	
Energy	GE Mcal/kg			
Cattle	DE Mcal/kg	* 1.08	3.23	
Sheep	DE Mcal/kg	* 1.02	3.06	
Cattle	ME Mcal/kg	* 0.88	2.64	
Sheep	ME Mcal/kg	* 0.84	2.51	
Cattle	TDN %	24.4	73.1	
Sheep	TDN %	* 23.2	69.5	
Calcium	%	0.09	0.28	
Phosphorus	%	0.10	0.30	

Rye, aerial part, fresh, mature, (2)

Ref No 2 04 017 United States

Feed Name or Analyses		As Fed	Dry	C.V. ± %
Dry matter	%		100.0	
Ash	%		5.0	
Crude fiber	%		43.7	
Ether extract	%		2.7	
N-free extract	%		42.3	
Protein (N x 6.25)	%		6.3	
Cattle	dig prot %	*	3.2	
Goats	dig prot %	*	2.4	
Horses	dig prot %	*	2.9	
Rabbits	dig prot %	*	3.5	
Sheep	dig prot %	*	2.9	

Feed Name or Analyses		As Fed	Dry	C.V. ± %
Energy	GE Mcal/kg			
Cattle	DE Mcal/kg	*	3.01	
Sheep	DE Mcal/kg	*	2.79	
Cattle	ME Mcal/kg	*	2.47	
Sheep	ME Mcal/kg	*	2.28	
Cattle	TDN %	*	68.3	
Sheep	TDN %	*	63.2	

Rye, aerial part, fresh nitrogen fertilized immature, (2)

Ref No 2 04 019 United States

Feed Name or Analyses		As Fed	Dry	C.V. ± %
Dry matter	%	15.0	100.0	25
Ash	%	1.9	12.5	11
Protein (N x 6.25)	%	5.6	37.3	11
Cattle	dig prot %	* 4.4	29.6	
Goats	dig prot %	* 4.7	31.4	
Horses	dig prot %	* 4.4	29.2	
Rabbits	dig prot %	* 4.1	27.5	
Sheep	dig prot %	* 4.8	31.8	

Rye, aerial part, ensiled, (3)

Ref No 3 04 020 United States

Feed Name or Analyses		As Fed	Dry	C.V. ± %
Dry matter	%	27.6	100.0	10
Ash	%	2.2	7.9	6
Crude fiber	%	10.2	36.8	7
Ether extract	%	0.9	3.3	9
N-free extract	%	11.5	41.5	
Protein (N x 6.25)	%	2.9	10.5	12
Cattle	dig prot %	* 1.6	5.8	
Goats	dig prot %	* 1.6	5.8	
Horses	dig prot %	* 1.6	5.8	
Sheep	dig prot %	* 1.6	5.8	
Energy	GE Mcal/kg			
Cattle	DE Mcal/kg	* 0.68	2.45	
Sheep	DE Mcal/kg	* 0.65	2.34	
Cattle	ME Mcal/kg	* 0.55	2.01	
Sheep	ME Mcal/kg	* 0.53	1.92	
Cattle	TDN %	* 15.3	55.5	
Sheep	TDN %	* 14.7	53.1	

Ref No 3 04 020 Canada

Feed Name or Analyses		As Fed	Dry	C.V. ± %
Dry matter	%	29.3	100.0	
Crude fiber	%	9.6	32.8	
Protein (N x 6.25)	%	3.9	13.3	
Cattle	dig prot %	* 2.4	8.3	
Goats	dig prot %	* 2.4	8.3	
Horses	dig prot %	* 2.4	8.3	
Sheep	dig prot %	* 2.4	8.3	
Calcium	%	0.11	0.39	
Phosphorus	%	0.09	0.32	
Carotene	mg/kg	16.9	57.7	
Vitamin A equivalent	IU/g	28.2	96.3	

Rye, aerial part, wilted ensiled, (3)

Ref No 3 08 601 United States

Feed Name or Analyses		As Fed	Dry	C.V. ± %
Dry matter	%	30.3	100.0	
Ash	%	2.5	8.3	
Crude fiber	%	10.8	35.6	
Ether extract	%	1.0	3.3	
N-free extract	%	12.5	41.3	
Protein (N x 6.25)	%	3.5	11.6	
Cattle	dig prot %	* 2.0	6.7	
Goats	dig prot %	* 2.0	6.7	
Horses	dig prot %	* 2.0	6.7	
Sheep	dig prot %	* 2.0	6.7	
Energy	GE Mcal/kg			
Cattle	DE Mcal/kg	* 0.66	2.17	
Sheep	DE Mcal/kg	* 0.61	2.01	
Cattle	ME Mcal/kg	* 0.54	1.78	
Sheep	ME Mcal/kg	* 0.50	1.65	
Cattle	TDN %	* 14.9	49.3	
Sheep	TDN %	* 13.8	45.7	
Phosphorus	%	0.07	0.23	
Potassium	%	0.56	1.85	

Rye, aerial part w molasses added, ensiled, (3)

Ref No 3 04 021 United States

Feed Name or Analyses		As Fed	Dry	C.V. ± %
Dry matter	%	23.7	100.0	
Ash	%	1.8	7.6	
Crude fiber	%	8.4	35.4	
Ether extract	%	0.7	3.0	
N-free extract	%	10.4	43.9	
Protein (N x 6.25)	%	2.4	10.1	
Cattle	dig prot %	* 1.3	5.4	
Goats	dig prot %	* 1.3	5.4	
Horses	dig prot %	* 1.3	5.4	
Sheep	dig prot %	* 1.3	5.4	
Energy	GE Mcal/kg			
Cattle	DE Mcal/kg	* 0.53	2.20	
Sheep	DE Mcal/kg	* 0.48	2.04	
Cattle	ME Mcal/kg	* 0.43	1.80	
Sheep	ME Mcal/kg	* 0.40	1.67	
Cattle	TDN %	* 11.8	49.8	
Sheep	TDN %	* 10.9	46.2	

Rye, bran, dry milled dehy, (4)

Rye bran (CFA)

Ref No 4 04 022 United States

Feed Name or Analyses		As Fed	Dry	C.V. ± %
Dry matter	%	90.7	100.0	
Ash	%	4.7	5.2	
Crude fiber	%	6.9	7.6	
Cattle	dig coef %	* 59.	59.	
Sheep	dig coef %	23.	23.	
Swine	dig coef %	25.	25.	
Ether extract	%	3.1	3.4	
Cattle	dig coef %	* 84.	84.	
Sheep	dig coef %	68.	68.	
Swine	dig coef %	78.	78.	
N-free extract	%	60.2	66.4	
Cattle	dig coef %	* 90.	90.	
Sheep	dig coef %	64.	64.	
Swine	dig coef %	74.	74.	
Protein (N x 6.25)	%	15.9	17.5	
Cattle	dig coef %	* 84.	84.	
Sheep	dig coef %	67.	67.	
Swine	dig coef %	70.	70.	
Cattle	dig prot %	13.3	14.7	
Goats	dig prot %	* 12.0	13.2	
Horses	dig prot %	* 12.0	13.2	
Sheep	dig prot %	10.6	11.7	
Swine	dig prot %	11.1	12.2	
Energy	GE Mcal/kg			
Cattle	DE Mcal/kg	* 3.41	3.76	
Sheep	DE Mcal/kg	* 2.44	2.70	
Swine	DE kcal/kg	*2768.	3051.	
Cattle	ME Mcal/kg	* 2.80	3.08	
Sheep	ME Mcal/kg	* 2.00	2.21	
Swine	ME kcal/kg	*2559.	2822.	

Continued

Feed Name or Analyses			Mean As Fed	Mean Dry	C.V. ± %
Cattle	TDN %		77.4	85.3	
Sheep	TDN %		55.4	61.1	
Swine	TDN %		62.8	69.2	

Rye, flour, coarse bolt, (4)

Ref No 4 04 028 United States

Feed Name or Analyses			Mean As Fed	Mean Dry	C.V. ± %
Dry matter	%		87.1	100.0	1
Ash	%		1.1	1.3	
Crude fiber	%		0.4	0.4	
Sheep	dig coef %		-471.	-471.	
Swine	dig coef %		-343.	-343.	
Ether extract	%		1.5	1.8	
Sheep	dig coef %		75.	75.	
Swine	dig coef %		71.	71.	
N-free extract	%		73.4	84.2	
Sheep	dig coef %		96.	96.	
Swine	dig coef %		96.	96.	
Protein (N x 6.25)	%		10.7	12.2	
Sheep	dig coef %		76.	76.	
Swine	dig coef %		84.	84.	
Cattle	dig prot %	*	6.3	7.3	
Goats	dig prot %	*	7.4	8.5	
Horses	dig prot %	*	7.4	8.5	
Sheep	dig prot %		8.1	9.3	
Swine	dig prot %		9.0	10.3	
Energy	GE Mcal/kg				
Cattle	DE Mcal/kg	*	3.23	3.71	
Sheep	DE Mcal/kg	*	3.50	4.02	
Swine	DE kcal/kg	*	3551.	4076.	
Cattle	ME Mcal/kg	*	2.65	3.04	
Sheep	ME Mcal/kg	*	2.87	3.29	
Swine	ME kcal/kg	*	3322.	3812.	
Cattle	TDN %	*	73.3	84.1	
Sheep	TDN %		79.3	91.1	
Swine	TDN %		80.5	92.5	
Calcium	%		0.02	0.02	
Phosphorus	%		0.28	0.32	
Potassium	%		0.46	0.52	
Niacin	mg/kg		8.2	9.4	46
Pantothenic acid	mg/kg		10.1	11.6	54
Riboflavin	mg/kg		0.9	1.0	
Thiamine	mg/kg		2.4	2.7	58

Rye, flour by-product, coarse sift, mx 8.5% fiber, (4)

Rye middlings (AAFCO)

Ref No 4 04 031 United States

Feed Name or Analyses			Mean As Fed	Mean Dry	C.V. ± %
Dry matter	%		87.8	100.0	2
Ash	%		3.5	4.0	12
Crude fiber	%		4.5	5.1	18
Cattle	dig coef %	*	59.	59.	
Sheep	dig coef %		19.	19.	
Swine	dig coef %		26.	26.	
Ether extract	%		3.1	3.5	9
Cattle	dig coef %	*	84.	84.	
Sheep	dig coef %		81.	81.	
Swine	dig coef %		75.	75.	
N-free extract	%		61.7	70.3	6
Cattle	dig coef %	*	90.	90.	
Sheep	dig coef %		85.	85.	
Swine	dig coef %		84.	84.	
Protein (N x 6.25)	%		15.1	17.2	14
Cattle	dig coef %	*	84.	84.	
Sheep	dig coef %		74.	74.	
Swine	dig coef %		72.	72.	
Cattle	dig prot %		12.7	14.4	
Goats	dig prot %	*	11.4	13.0	
Horses	dig prot %	*	11.4	13.0	
Sheep	dig prot %		11.1	12.6	
Swine	dig prot %		10.9	12.4	
Energy	GE Mcal/kg				
Cattle	DE Mcal/kg	*	3.38	3.85	
Sheep	DE Mcal/kg	*	3.08	3.51	
Swine	DE kcal/kg	*	3042.	3465.	
Cattle	ME Mcal/kg	*	2.77	3.15	
Sheep	ME Mcal/kg	*	2.53	2.88	
Swine	ME kcal/kg	*	2815.	3206.	
Cattle	NE_m Mcal/kg	*	1.85	2.11	
Cattle	NE_{gain} Mcal/kg	*	1.23	1.40	
Cattle	TDN %		76.6	87.2	
Sheep	TDN %		69.9	79.6	
Swine	TDN %		69.0	78.6	
Calcium	%		0.06	0.07	
Manganese	mg/kg		43.1	49.1	1
Phosphorus	%		0.61	0.70	1
Potassium	%		0.61	0.70	
Niacin	mg/kg		16.6	18.9	2
Pantothenic acid	mg/kg		22.6	25.8	1
Riboflavin	mg/kg		2.4	2.7	6
Thiamine	mg/kg		3.2	3.7	3

(1) dry forages and roughages (2) pasture, range plants, and forages fed green (3) silages (4) energy feeds (5) protein supplements (6) minerals (7) vitamins (8) additives

Rye, flour by-product, fine sift, mx 4.5% fiber, (4)

Rye middlings (CFA)

Ref No 4 04 032 United States

Feed Name or Analyses			Mean As Fed	Mean Dry	C.V. ± %
Dry matter	%		90.6	100.0	
Ash	%		3.3	3.6	
Crude fiber	%		4.2	4.6	
Ether extract	%		3.5	3.9	
N-free extract	%		63.1	69.6	
Protein (N x 6.25)	%		16.5	18.2	
Cattle	dig prot %	*	11.5	12.7	
Goats	dig prot %	*	12.6	13.9	
Horses	dig prot %	*	12.6	13.9	
Sheep	dig prot %	*	12.6	13.9	
Energy	GE Mcal/kg				
Cattle	DE Mcal/kg	*	3.18	3.51	
Sheep	DE Mcal/kg	*	3.21	3.55	
Swine	DE kcal/kg	*	3364.	3713.	
Cattle	ME Mcal/kg	*	2.61	2.88	
Sheep	ME Mcal/kg	*	2.63	2.91	
Swine	ME kcal/kg	*	3106.	3428.	
Cattle	TDN %	*	72.1	79.6	
Sheep	TDN %	*	72.9	80.4	
Swine	TDN %	*	76.3	84.2	

Rye, flour by-product mill run, (4)

Rye feed (CFA)

Ref No 4 04 035 United States

Feed Name or Analyses			Mean As Fed	Mean Dry	C.V. ± %
Dry matter	%		86.2	100.0	
Ash	%		5.3	6.2	
Crude fiber	%		5.0	5.8	
Cattle	dig coef %		-22.	-22.	
Sheep	dig coef %		35.	35.	
Swine	dig coef %		44.	44.	
Ether extract	%		3.0	3.4	
Cattle	dig coef %		64.	64.	
Sheep	dig coef %		59.	59.	
Swine	dig coef %		57.	57.	
N-free extract	%		58.2	67.5	
Cattle	dig coef %		82.	82.	
Sheep	dig coef %		80.	80.	
Swine	dig coef %		84.	84.	
Protein (N x 6.25)	%		14.7	17.1	
Cattle	dig coef %		78.	78.	
Sheep	dig coef %		74.	74.	
Swine	dig coef %		72.	72.	
Cattle	dig prot %		11.5	13.3	
Goats	dig prot %	*	11.1	12.9	
Horses	dig prot %	*	11.1	12.9	
Sheep	dig prot %		10.9	12.6	
Swine	dig prot %		10.6	12.3	
Energy	GE Mcal/kg				
Cattle	DE Mcal/kg	*	2.75	3.19	
Sheep	DE Mcal/kg	*	2.78	3.23	
Swine	DE kcal/kg	*	2887.	3350.	
Cattle	ME Mcal/kg	*	2.25	2.61	
Sheep	ME Mcal/kg	*	2.28	2.65	
Swine	ME kcal/kg	*	2672.	3100.	
Cattle	TDN %		62.3	72.3	
Sheep	TDN %		63.1	73.2	
Swine	TDN %		65.5	76.0	

Rye, flour by-product mill run, mx 9.5% fiber, (4)

Rye mill run (AAFCO)

Ref No 4 04 034 United States

Feed Name or Analyses			Mean As Fed	Mean Dry	C.V. ± %
Dry matter	%		90.1	100.0	1
Ash	%		3.8	4.2	8
Crude fiber	%		4.6	5.1	9
Cattle	dig coef %	*	-22.	-22.	
Sheep	dig coef %	*	35.	35.	
Ether extract	%		3.4	3.7	9
Cattle	dig coef %	*	64.	64.	
Sheep	dig coef %	*	59.	59.	
N-free extract	%		61.7	68.5	
Cattle	dig coef %	*	82.	82.	
Sheep	dig coef %	*	80.	80.	
Protein (N x 6.25)	%		16.7	18.5	11
Cattle	dig coef %	*	78.	78.	
Sheep	dig coef %	*	74.	74.	
Cattle	dig prot %		13.0	14.4	
Goats	dig prot %	*	12.8	14.2	
Horses	dig prot %	*	12.8	14.2	
Sheep	dig prot %		12.3	13.7	
Energy	GE Mcal/kg				
Cattle	DE Mcal/kg	*	2.97	3.30	
Sheep	DE Mcal/kg	*	2.99	3.32	
Swine	DE kcal/kg	*	3252.	3611.	
Cattle	ME Mcal/kg	*	2.43	2.70	
Sheep	ME Mcal/kg	*	2.45	2.72	
Swine	ME kcal/kg	*	3000.	3332.	
Cattle	NE_m Mcal/kg	*	1.50	1.67	
Cattle	NE_{gain} Mcal/kg	*	0.98	1.09	
Cattle	TDN %		67.3	74.8	
Sheep	TDN %		67.7	75.2	
Swine	TDN %	*	73.7	81.9	
Calcium	%		0.07	0.08	
Magnesium	%		0.23	0.26	
Phosphorus	%		0.64	0.71	
Potassium	%		0.83	0.92	
Sulphur	%		0.04	0.04	

Feed Name or Analyses			Mean As Fed	Mean Dry	C.V. ± %
Rye, grain, (4)					
Ref No 4 04 047			United States		
Dry matter		%	88.2	100.0	2
Ash		%	1.7	2.0	9
Crude fiber		%	2.0	2.3	17
Sheep	dig coef	%	-26.	-26.	
Swine	dig coef	%	30.	30.	
Ether extract		%	1.5	1.7	15
Sheep	dig coef	%	53.	53.	
Swine	dig coef	%	24.	24.	
N-free extract		%	71.6	81.2	3
Sheep	dig coef	%	90.	90.	
Swine	dig coef	%	94.	94.	
Protein (N x 6.25)		%	11.3	12.8	14
Sheep	dig coef	%	79.	79.	
Swine	dig coef	%	81.	81.	
Cattle	dig prot	%	* 6.8	7.7	
Goats	dig prot	%	* 7.9	8.9	
Horses	dig prot	%	* 7.9	8.9	
Sheep	dig prot	%	8.9	10.1	
Swine	dig prot	%	9.1	10.3	
Energy	GE Mcal/kg		3.31	3.75	
Cattle	DE Mcal/kg		* 3.15	3.57	
Sheep	DE Mcal/kg		* 3.29	3.73	
Swine	DE kcal/kg		*3434.	3894.	
Cattle	ME Mcal/kg		* 2.58	2.93	
Chickens	ME_n kcal/kg		2549.	2890.	
Sheep	ME Mcal/kg		* 2.70	3.06	
Swine	ME kcal/kg		*3208.	3638.	
Cattle	NE_m Mcal/kg		* 1.80	2.04	
Cattle	NE_{gain} Mcal/kg		* 1.20	1.36	
Cattle	TDN	%	* 71.4	80.9	
Sheep	TDN	%	74.6	84.6	
Swine	TDN	%	77.9	88.3	
Calcium		%	0.07	0.08	30
Chlorine		%	0.03	0.03	59
Copper		mg/kg	7.6	8.6	17
Iron		%	0.007	0.007	27
Magnesium		%	0.12	0.13	19
Manganese		mg/kg	73.4	83.3	41
Phosphorus		%	0.34	0.39	21
Potassium		%	0.46	0.52	15
Sodium		%	0.02	0.02	
Sulphur		%	0.15	0.17	14
Zinc		mg/kg	30.3	34.4	44
Biotin		mg/kg	0.06	0.07	22
Carotene		mg/kg	0.0	0.0	22
Choline		mg/kg	434.	493.	
Folic acid		mg/kg	0.64	0.73	65
Niacin		mg/kg	14.7	16.7	20
Pantothenic acid		mg/kg	8.0	9.0	30
Riboflavin		mg/kg	1.8	2.0	35
Thiamine		mg/kg	4.1	4.6	30
α-tocopherol		mg/kg	15.4	17.4	
Arginine		%	0.56	0.64	18
Cystine		%	0.14	0.16	
Glutamic acid		%	3.35	3.80	8
Histidine		%	0.26	0.30	29
Isoleucine		%	0.53	0.60	24
Leucine		%	0.71	0.80	24
Lysine		%	0.47	0.53	22
Methionine		%	0.18	0.21	14
Phenylalanine		%	0.62	0.70	18
Serine		%	0.62	0.70	
Threonine		%	0.35	0.40	20
Tryptophan		%	0.12	0.14	21
Tyrosine		%	0.26	0.30	
Valine		%	0.62	0.70	12

Feed Name or Analyses			Mean As Fed	Mean Dry	C.V. ± %
Rye, grain, Can 1 CW mn wt 58 lb per bushel mx 0% foreign material, (4)					
Ref No 4 04 036			United States		
Dry matter		%		100.0	
Crude fiber		%			
Sheep	dig coef	%	*	-26.	
Swine	dig coef	%	*	30.	
Ether extract		%			
Sheep	dig coef	%	*	53.	
Swine	dig coef	%	*	24.	
N-free extract		%			
Sheep	dig coef	%	*	90.	
Swine	dig coef	%	*	94.	
Protein (N x 6.25)		%			
Sheep	dig coef	%	*	79.	
Swine	dig coef	%	*	81.	
Energy	GE Mcal/kg				
Cattle	DE Mcal/kg		* 3.09	3.54	
Sheep	DE Mcal/kg		*	3.71	
Swine	DE kcal/kg		*	3889.	
Cattle	ME Mcal/kg		* 2.53	2.90	
Chickens	ME_n kcal/kg		*	2570.	
Sheep	ME Mcal/kg		*	3.04	
Cattle	TDN	%	* 69.9	80.2	
Sheep	TDN	%	*	84.1	
Swine	TDN	%	*	88.2	
Ref No 4 04 036			Canada		
Dry matter		%	87.2	100.0	2
Ash		%	1.5	1.8	6
Crude fiber		%	2.4	2.8	28
Ether extract		%	1.4	1.6	
N-free extract		%	70.1	80.4	
Protein (N x 6.25)		%	11.8	13.5	11
Cattle	dig prot	%	* 7.3	8.4	
Goats	dig prot	%	* 8.4	9.6	
Horses	dig prot	%	* 8.4	9.6	
Sheep	dig prot	%	* 8.4	9.6	
Energy	GE Mcal/kg				
Cattle	DE Mcal/kg		* 3.09	3.54	
Sheep	DE Mcal/kg		* 3.24	3.71	
Swine	DE kcal/kg		*3391.	3889.	
Cattle	ME Mcal/kg		* 2.53	2.90	
Chickens	ME_n kcal/kg		2241.	2570.	
Sheep	ME Mcal/kg		* 2.65	3.04	
Swine	ME kcal/kg		*3163.	3627.	
Cattle	TDN	%	* 69.9	80.2	
Sheep	TDN	%	73.4	84.1	
Swine	TDN	%	76.9	88.2	
Calcium		%	0.05	0.06	
Phosphorus		%	0.29	0.34	
Rye, grain, Can 2 CE, (4)					
Ref No 4 04 037			United States		
Dry matter		%		100.0	
Crude fiber		%			
Sheep	dig coef	%	*	-26.	
Swine	dig coef	%	*	30.	
Ether extract		%			
Sheep	dig coef	%	*	53.	
Swine	dig coef	%	*	24.	
N-free extract		%			
Sheep	dig coef	%	*	90.	
Swine	dig coef	%	*	94.	

Feed Name or Analyses			Mean As Fed	Mean Dry	C.V. ± %
Protein (N x 6.25)		%			
Sheep	dig coef	%	*	79.	
Swine	dig coef	%	*	81.	
Energy	GE Mcal/kg				
Cattle	DE Mcal/kg		* 3.04	3.52	
Sheep	DE Mcal/kg		*	3.66	
Swine	DE kcal/kg		*	3871.	
Cattle	ME Mcal/kg		* 2.50	2.89	
Sheep	ME Mcal/kg		*	3.00	
Cattle	TDN	%	* 69.1	79.9	
Sheep	TDN	%	*	83.1	
Swine	TDN	%	*	87.8	
Ref No 4 04 037			Canada		
Dry matter		%	86.5	100.0	
Ash		%	1.7	2.0	
Crude fiber		%	3.1	3.6	
Ether extract		%	1.1	1.3	
N-free extract		%	70.2	81.2	
Protein (N x 6.25)		%	10.4	12.0	
Cattle	dig prot	%	* 6.1	7.1	
Goats	dig prot	%	* 7.1	8.3	
Horses	dig prot	%	* 7.1	8.3	
Sheep	dig prot	%	* 7.1	8.3	
Energy	GE Mcal/kg				
Cattle	DE Mcal/kg		* 3.04	3.52	
Sheep	DE Mcal/kg		* 3.17	3.66	
Swine	DE kcal/kg		*3348.	3871.	
Cattle	ME Mcal/kg		* 2.50	2.89	
Sheep	ME Mcal/kg		* 2.60	3.00	
Swine	ME kcal/kg		*3133.	3622.	
Cattle	TDN	%	* 69.1	79.9	
Sheep	TDN	%	71.9	83.1	
Swine	TDN	%	75.9	87.8	
Rye, grain, Can 2 CW mn wt 56 lb per bushel mx 2% foreign material, (4)					
Ref No 4 04 038			United States		
Dry matter		%		100.0	
Crude fiber		%			
Sheep	dig coef	%	*	-26.	
Swine	dig coef	%	*	30.	
Ether extract		%			
Sheep	dig coef	%	*	53.	
Swine	dig coef	%	*	24.	
N-free extract		%			
Sheep	dig coef	%	*	90.	
Swine	dig coef	%	*	94.	
Protein (N x 6.25)		%			
Sheep	dig coef	%	*	79.	
Swine	dig coef	%	*	81.	
Energy	GE Mcal/kg				
Cattle	DE Mcal/kg		* 3.11	3.58	
Sheep	DE Mcal/kg		*	3.72	
Swine	DE kcal/kg		*	3885.	
Cattle	ME Mcal/kg		* 2.56	2.94	
Sheep	ME Mcal/kg		*	3.05	
Cattle	TDN	%	* 70.7	81.3	
Sheep	TDN	%	*	84.4	
Swine	TDN	%	*	88.1	
Ref No 4 04 038			Canada		
Dry matter		%	87.0	100.0	2
Ash		%	1.6	1.8	7
Crude fiber		%	2.2	2.5	30

Continued

Feed Name or Analyses			Mean As Fed	Mean Dry	C.V. ± %
Ether extract	%		1.6	1.8	17
N-free extract	%		69.6	80.0	
Protein (N x 6.25)	%		12.0	13.8	5
Cattle	dig prot %	*	7.6	8.7	
Goats	dig prot %	*	8.6	9.9	
Horses	dig prot %	*	8.6	9.9	
Sheep	dig prot %	*	8.6	9.9	
Cellulose (Matrone)	%		2.0	2.3	
Energy	GE Mcal/kg				
Cattle	DE Mcal/kg	*	3.11	3.58	
Sheep	DE Mcal/kg	*	3.24	3.72	
Swine	DE kcal/kg	*	3381.	3885.	
Cattle	ME Mcal/kg	*	2.56	2.94	
Sheep	ME Mcal/kg	*	2.65	3.05	
Swine	ME kcal/kg	*	3152.	3621.	
Cattle	TDN %	*	70.7	81.3	
Sheep	TDN %		73.4	84.4	
Swine	TDN %		76.7	88.1	
Calcium	%		0.05	0.06	
Copper	mg/kg		5.3	6.0	
Iron	%		0.003	0.003	
Magnesium	%		0.13	0.15	
Manganese	mg/kg		21.9	25.1	
Phosphorus	%		0.30	0.35	
Selenium	mg/kg		0.382	0.439	
Zinc	mg/kg		32.3	37.1	
Niacin	mg/kg		30.6	35.1	
Pantothenic acid	mg/kg		9.2	10.5	
Riboflavin	mg/kg		1.3	1.5	
Thiamine	mg/kg		0.7	0.8	
Vitamin B6	mg/kg		2.37	2.72	
Alanine	%		0.46	0.53	
Arginine	%		0.56	0.65	
Aspartic acid	%		0.82	0.95	
Glutamic acid	%		3.11	3.57	
Glycine	%		0.51	0.58	
Histidine	%		0.26	0.30	
Isoleucine	%		0.42	0.48	
Leucine	%		0.75	0.86	
Lysine	%		0.41	0.47	
Methionine	%		0.15	0.18	
Phenylalanine	%		0.55	0.64	
Proline	%		1.09	1.25	
Serine	%		0.47	0.54	
Threonine	%		0.42	0.48	
Tyrosine	%		0.28	0.32	
Valine	%		0.56	0.65	

Rye, grain, Can 3 CE, (4)

Ref No 4 04 039 United States

Feed Name or Analyses			Mean As Fed	Mean Dry	C.V. ± %
Dry matter	%			100.0	
Crude fiber	%				
Sheep	dig coef %	*		-26.	
Swine	dig coef %	*		30.	
Ether extract	%				
Sheep	dig coef %	*		53.	
Swine	dig coef %	*		24.	
N-free extract	%				
Sheep	dig coef %	*		90.	
Swine	dig coef %	*		94.	
Protein (N x 6.25)	%				
Sheep	dig coef %	*		79.	
Swine	dig coef %	*		81.	

(1) dry forages and roughages (2) pasture, range plants, and forages fed green (3) silages (4) energy feeds (5) protein supplements (6) minerals (7) vitamins (8) additives

Feed Name or Analyses			Mean As Fed	Mean Dry	C.V. ± %
Energy	GE Mcal/kg				
Cattle	DE Mcal/kg	*	3.06	3.54	
Sheep	DE Mcal/kg	*		3.69	
Swine	DE kcal/kg	*		3884.	
Cattle	ME Mcal/kg	*	2.51	2.90	
Sheep	ME Mcal/kg	*		3.03	
Cattle	TDN %	*	69.5	80.4	
Sheep	TDN %	*		83.8	
Swine	TDN %	*		88.1	

Ref No 4 04 039 Canada

Feed Name or Analyses			Mean As Fed	Mean Dry	C.V. ± %
Dry matter	%		86.5	100.0	
Ash	%		1.8	2.1	
Crude fiber	%		2.5	2.9	
Ether extract	%		1.1	1.3	
N-free extract	%		70.5	81.5	
Protein (N x 6.25)	%		10.6	12.3	
Cattle	dig prot %	*	6.3	7.3	
Goats	dig prot %	*	7.3	8.5	
Horses	dig prot %	*	7.3	8.5	
Sheep	dig prot %	*	7.3	8.5	
Energy	GE Mcal/kg				
Cattle	DE Mcal/kg	*	3.06	3.54	
Sheep	DE Mcal/kg	*	3.19	3.69	
Cattle	ME Mcal/kg	*	3360.	3884.	
Cattle	ME Mcal/kg	*	2.51	2.90	
Sheep	ME Mcal/kg	*	2.62	3.03	
Swine	ME kcal/kg	*	3142.	3633.	
Cattle	TDN %	*	69.5	80.4	
Sheep	TDN %		72.5	83.8	
Swine	TDN %		76.2	88.1	

Rye, grain, Can 3 CW mn wt 54 lb per bushel mx 5% foreign material, (4)

Ref No 4 04 040 United States

Feed Name or Analyses			Mean As Fed	Mean Dry	C.V. ± %
Dry matter	%			100.0	
Crude fiber	%				
Sheep	dig coef %	*		-26.	
Swine	dig coef %	*		30.	
Ether extract	%				
Sheep	dig coef %	*		53.	
Swine	dig coef %	*		24.	
N-free extract	%				
Sheep	dig coef %	*		90.	
Swine	dig coef %	*		94.	
Protein (N x 6.25)	%				
Sheep	dig coef %	*		79.	
Swine	dig coef %	*		81.	
Energy	GE Mcal/kg				
Cattle	DE Mcal/kg	*	3.18	3.66	
Sheep	DE Mcal/kg	*		3.71	
Swine	DE kcal/kg	*		3883.	
Cattle	ME Mcal/kg	*	2.61	3.00	
Sheep	ME Mcal/kg	*		3.05	
Cattle	TDN %	*	72.1	82.9	
Sheep	TDN %	*		84.3	
Swine	TDN %	*		88.1	

Ref No 4 04 040 Canada

Feed Name or Analyses			Mean As Fed	Mean Dry	C.V. ± %
Dry matter	%		87.0	100.0	2
Ash	%		1.6	1.8	5
Crude fiber	%		2.3	2.6	34
Ether extract	%		1.5	1.8	16
N-free extract	%		69.4	79.8	
Protein (N x 6.25)	%		12.2	14.0	3
Cattle	dig prot %	*	7.7	8.9	

Feed Name or Analyses			Mean As Fed	Mean Dry	C.V. ± %
Goats	dig prot %	*	8.8	10.1	
Horses	dig prot %	*	8.8	10.1	
Sheep	dig prot %	*	8.8	10.1	
Cellulose (Matrone)	%		1.8	2.1	
Energy	GE Mcal/kg				
Cattle	DE Mcal/kg	*	3.18	3.66	
Sheep	DE Mcal/kg	*	3.23	3.71	
Swine	DE kcal/kg	*	3379.	3883.	
Cattle	ME Mcal/kg	*	2.61	3.00	
Sheep	ME Mcal/kg	*	2.65	3.05	
Swine	ME kcal/kg	*	3149.	3618.	
Cattle	TDN %	*	72.1	82.9	
Sheep	TDN %		73.3	84.3	
Swine	TDN %		76.6	88.1	
Calcium	%		0.06	0.07	
Copper	mg/kg		4.7	5.4	
Iron	%		0.003	0.004	
Manganese	mg/kg		21.0	24.1	
Phosphorus	%		0.32	0.37	
Selenium	mg/kg		0.382	0.439	
Zinc	mg/kg		34.6	39.7	
Niacin	mg/kg		32.5	37.3	
Pantothenic acid	mg/kg		10.3	11.9	
Riboflavin	mg/kg		1.5	1.7	
Thiamine	mg/kg		0.6	0.7	
Vitamin B6	mg/kg		2.74	3.15	
Alanine	%		0.47	0.54	
Arginine	%		0.56	0.65	
Aspartic acid	%		0.80	0.92	
Glutamic acid	%		3.23	3.71	
Glycine	%		0.51	0.58	
Histidine	%		0.27	0.31	
Isoleucine	%		0.43	0.49	
Leucine	%		0.75	0.86	
Lysine	%		0.41	0.47	
Methionine	%		0.14	0.16	
Phenylalanine	%		0.57	0.66	
Proline	%		1.12	1.29	
Serine	%		0.48	0.55	
Threonine	%		0.35	0.41	
Tyrosine	%		0.26	0.30	
Valine	%		0.46	0.53	

Rye, grain, Can 4 CW mx 0.3% ergot mx 10% foreign material, (4)

Ref No 4 04 041 United States

Feed Name or Analyses			Mean As Fed	Mean Dry	C.V. ± %
Dry matter	%			100.0	
Crude fiber	%				
Sheep	dig coef %	*		-26.	
Swine	dig coef %	*		30.	
Ether extract	%				
Sheep	dig coef %	*		53.	
Swine	dig coef %	*		24.	
N-free extract	%				
Sheep	dig coef %	*		90.	
Swine	dig coef %	*		94.	
Protein (N x 6.25)	%				
Sheep	dig coef %	*		79.	
Swine	dig coef %	*		81.	
Energy	GE Mcal/kg				
Cattle	DE Mcal/kg	*	3.12	3.60	
Sheep	DE Mcal/kg	*		3.71	
Swine	DE kcal/kg	*		3877.	
Cattle	ME Mcal/kg	*	2.56	2.95	
Sheep	ME Mcal/kg	*		3.04	
Cattle	TDN %	*	70.8	81.7	

Feed Name or Analyses	Mean As Fed	Mean Dry	C.V. ±%
Sheep TDN %	*	84.1	
Swine TDN %	*	87.9	

Ref No 4 04 041 Canada

Feed Name or Analyses	Mean As Fed	Mean Dry	C.V. ±%
Dry matter %	86.7	100.0	1
Ash %	1.6	1.9	4
Crude fiber %	2.3	2.7	29
Ether extract %	1.6	1.8	14
N-free extract %	68.7	79.2	
Protein (N x 6.25) %	12.5	14.4	8
Cattle dig prot %	* 8.0	9.2	
Goats dig prot %	* 9.1	10.4	
Horses dig prot %	* 9.1	10.4	
Sheep dig prot %	* 9.1	10.4	
Energy GE Mcal/kg			
Cattle DE Mcal/kg	* 3.12	3.60	
Sheep DE Mcal/kg	* 3.22	3.71	
Swine DE kcal/kg	*3362.	3877.	
Cattle ME Mcal/kg	* 2.56	2.95	
Sheep ME Mcal/kg	* 2.64	3.04	
Swine ME kcal/kg	*3129.	3609.	
Cattle TDN %	* 70.8	81.7	
Sheep TDN %	72.9	84.1	
Swine TDN %	76.2	87.9	
Calcium %	0.06	0.07	
Phosphorus %	0.29	0.33	

Rye, grain, gr 2 US mn wt 54 lb per bushel mx 2% foreign material, (4)

Ref No 4 04 044 United States

Feed Name or Analyses	Mean As Fed	Mean Dry	C.V. ±%
Dry matter %	87.3	100.0	2
Ether extract %	1.7	1.9	12
Protein (N x 6.25) %	12.5	14.3	5
Cattle dig prot %	* 8.0	9.1	
Goats dig prot %	* 9.0	10.3	
Horses dig prot %	* 9.0	10.3	
Sheep dig prot %	* 9.0	10.3	
Starch %	57.7	66.1	3

Rye, distillers grains, dehy, (5)

Rye distillers dried grains (AAFCO)

Rye distillers dried grains (CFA)

Ref No 5 04 023 United States

Feed Name or Analyses	Mean As Fed	Mean Dry	C.V. ±%
Dry matter %	92.2	100.0	3
Ash %	2.3	2.5	50
Crude fiber %	13.3	14.4	19
Cattle dig coef %	4.	4.	
Sheep dig coef %	56.	56.	
Ether extract %	7.5	8.1	18
Cattle dig coef %	70.	70.	
Sheep dig coef %	72.	72.	
N-free extract %	48.7	52.8	
Cattle dig coef %	47.	47.	
Sheep dig coef %	60.	60.	
Protein (N x 6.25) %	20.4	22.1	16
Cattle dig coef %	43.	43.	
Sheep dig coef %	60.	60.	
Cattle dig prot %	8.8	9.5	
Sheep dig prot %	12.2	13.3	
Energy GE Mcal/kg			
Cattle DE Mcal/kg	* 1.94	2.10	
Sheep DE Mcal/kg	* 2.69	2.92	
Swine DE kcal/kg	*2736.	2967.	
Cattle ME Mcal/kg	* 1.59	1.73	
Sheep ME Mcal/kg	* 2.21	2.39	
Swine ME kcal/kg	*2503.	2715.	
Cattle NE_m Mcal/kg	* 0.91	0.99	
Cattle NE_{gain} Mcal/kg	* 0.10	0.11	
Cattle TDN %	44.0	47.7	
Sheep TDN %	61.1	66.2	
Swine TDN %	* 62.1	67.3	
Calcium %	0.13	0.14	14
Chlorine %	0.05	0.05	65
Magnesium %	0.17	0.18	14
Manganese mg/kg	18.4	20.0	
Phosphorus %	0.42	0.45	15
Potassium %	0.07	0.08	98
Sodium %	0.17	0.18	96
Sulphur %	0.44	0.47	21
Niacin mg/kg	16.8	18.3	8
Pantothenic acid mg/kg	5.2	5.7	24
Riboflavin mg/kg	3.3	3.6	9
Thiamine mg/kg	1.3	1.4	46

Rye, distillers grains w solubles, dehy, (5)

Rye distillers dried grains with solubles (CFA)

Ref No 5 04 025 United States

Feed Name or Analyses	Mean As Fed	Mean Dry	C.V. ±%
Dry matter %	90.5	100.0	3
Ash %	6.4	7.1	13
Crude fiber %	8.1	9.0	48
Ether extract %	4.1	4.5	30
N-free extract %	44.7	49.4	
Protein (N x 6.25) %	27.2	30.1	5
Energy GE Mcal/kg			
Cattle DE Mcal/kg	* 2.48	2.74	
Sheep DE Mcal/kg	* 2.59	2.86	
Swine DE kcal/kg	*2650.	2928.	
Cattle ME Mcal/kg	* 2.04	2.25	
Sheep ME Mcal/kg	* 2.13	2.35	
Swine ME kcal/kg	*2173.	2401.	
Cattle TDN %	* 56.2	62.1	
Sheep TDN %	* 58.6	64.8	
Swine TDN %	* 60.1	66.4	
Niacin mg/kg	62.8	69.4	
Pantothenic acid mg/kg	17.4	19.2	
Riboflavin mg/kg	8.2	9.0	
Thiamine mg/kg	3.1	3.4	
Arginine %	1.00	1.11	5
Glutamic acid %	5.50	6.08	10
Histidine %	0.70	0.77	8
Isoleucine %	1.50	1.66	7
Leucine %	2.10	2.32	8
Lysine %	1.00	1.11	5
Methionine %	0.40	0.44	
Phenylalanine %	1.30	1.44	8
Serine %	1.20	1.33	
Threonine %	1.10	1.22	19
Tryptophan %	0.30	0.33	
Tyrosine %	0.50	0.55	11
Valine %	1.60	1.77	7

Rye, distillers grains w solubles, dehy, mn 75% original solids, (5)

Rye distillers dried grains with solubles (AAFCO)

Ref No 5 04 024 United States

Feed Name or Analyses	Mean As Fed	Mean Dry	C.V. ±%
Dry matter %		100.0	
Crude fiber %			
Cattle dig coef %	*	4.	
Sheep dig coef %	*	56.	
Ether extract %			
Cattle dig coef %	*	70.	
Sheep dig coef %	*	72.	
N-free extract %			
Cattle dig coef %	*	47.	
Sheep dig coef %	*	60.	
Protein (N x 6.25) %			
Cattle dig coef %	*	43.	
Sheep dig coef %	*	60.	

Rye, distillers solubles, dehy, (5)

Rye distillers dried solubles (AAFCO)

Rye distillers dried solubles (CFA)

Ref No 5 04 026 United States

Feed Name or Analyses	Mean As Fed	Mean Dry	C.V. ±%
Dry matter %	94.4	100.0	1
Ash %	7.2	7.6	14
Crude fiber %	3.4	3.6	48
Ether extract %	1.2	1.3	58
N-free extract %	47.5	50.3	
Protein (N x 6.25) %	35.1	37.2	6
Calcium %	0.35	0.37	38
Phosphorus %	1.19	1.26	27
Niacin mg/kg	66.1	70.1	41
Pantothenic acid mg/kg	28.7	30.4	22
Riboflavin mg/kg	12.8	13.5	9
Thiamine mg/kg	3.1	3.3	14
Arginine %	1.00	1.06	
Histidine %	0.70	0.74	
Isoleucine %	1.80	1.91	
Leucine %	1.80	1.91	
Lysine %	0.60	0.64	
Methionine %	0.50	0.53	
Phenylalanine %	1.70	1.80	
Serine %	1.60	1.69	
Threonine %	1.10	1.17	
Tryptophan %	0.20	0.21	
Tyrosine %	0.60	0.64	
Valine %	1.90	2.01	

Ref No 5 04 026 Canada

Feed Name or Analyses	Mean As Fed	Mean Dry	C.V. ±%
Dry matter %	92.0	100.0	
Pantothenic acid mg/kg	15.2	16.5	
Riboflavin mg/kg	14.8	16.1	

Rye, distillers stillage, wet, (5)

Ref No 2 08 498 United States

Feed Name or Analyses	Mean As Fed	Mean Dry	C.V. ±%
Dry matter %	5.9	100.0	
Ash %	0.3	5.1	
Crude fiber %	0.5	8.5	
Ether extract %	0.3	5.1	
N-free extract %	2.9	49.2	
Protein (N x 6.25) %	1.9	32.2	
Energy GE Mcal/kg			
Cattle DE Mcal/kg	* 0.16	2.66	
Sheep DE Mcal/kg	* 0.16	2.78	
Swine DE kcal/kg	* 172.	2910.	
Cattle ME Mcal/kg	* 0.13	218.	
Sheep ME Mcal/kg	* 0.13	228.	
Swine ME kcal/kg	* 141.	2386.	
Cattle TDN %	* 3.6	60.3	
Sheep TDN %	* 3.7	63.1	
Swine TDN %	* 3.9	66.0	

Rye, germ, grnd, (5)
Ref No 5 04 049 United States

Feed Name or Analyses				Mean As Fed	Mean Dry	C.V. ± %
Dry matter		%		89.5	100.0	
Ash		%		5.1	5.8	
Crude fiber		%		3.4	3.9	
Sheep	dig coef	%		66.	66.	
Swine	dig coef	%		95.	95.	
Ether extract		%		7.7	8.6	
Sheep	dig coef	%		86.	86.	
Swine	dig coef	%		67.	67.	
N-free extract		%		39.8	44.5	
Sheep	dig coef	%		93.	93.	
Swine	dig coef	%		98.	98.	
Protein (N x 6.25)		%		33.4	37.4	
Sheep	dig coef	%		91.	91.	
Swine	dig coef	%		87.	87.	
Sheep	dig prot	%		30.4	34.0	
Swine	dig prot	%		29.1	32.5	
Energy	GE Mcal/kg					
Cattle	DE Mcal/kg		*	3.51	3.92	
Sheep	DE Mcal/kg		*	3.73	4.17	
Swine	DE kcal/kg		*	3657.	4086.	
Cattle	ME Mcal/kg		*	2.88	3.21	
Sheep	ME Mcal/kg		*	3.06	3.42	
Swine	ME kcal/kg		*	3235.	3615.	
Cattle	TDN	%	*	79.5	88.9	
Sheep	TDN	%		84.6	94.5	
Swine	TDN	%		82.9	92.7	
Niacin	mg/kg			26.5	29.6	3
Pantothenic acid	mg/kg			13.6	15.2	3
Riboflavin	mg/kg			6.6	7.3	37
Thiamine	mg/kg			12.3	13.7	52

RYE, WINTER. Secale cereale

Rye, winter, straw, (1)
Ref No 1 04 050 United States

Feed Name or Analyses				Mean As Fed	Mean Dry	C.V. ± %
Dry matter		%		85.7	100.0	
Ash		%		4.2	4.9	
Crude fiber		%		39.1	45.6	
Sheep	dig coef	%		54.	54.	
Ether extract		%		1.4	1.6	
Sheep	dig coef	%		48.	48.	
N-free extract		%		38.2	44.6	
Sheep	dig coef	%		39.	39.	
Protein (N x 6.25)		%		2.8	3.3	
Sheep	dig coef	%		4.	4.	
Cattle	dig prot	%	*	-0.1	-0.1	
Goats	dig prot	%	*	-0.2	-0.3	
Horses	dig prot	%	*	0.3	0.3	
Rabbits	dig prot	%	*	1.0	1.2	
Sheep	dig prot	%		0.1	0.1	
Energy	GE Mcal/kg					
Cattle	DE Mcal/kg		*	1.33	1.56	
Sheep	DE Mcal/kg		*	1.66	1.93	
Cattle	ME Mcal/kg		*	1.09	1.28	
Sheep	ME Mcal/kg		*	1.36	1.59	
Cattle	TDN	%	*	30.2	35.3	
Sheep	TDN	%		37.6	43.9	

(1) dry forages and roughages (2) pasture, range plants, and forages fed green (3) silages (4) energy feeds (5) protein supplements (6) minerals (7) vitamins (8) additives

Rye, winter, straw, treated w sodium bicarbonate, dehy, (1)
Ref No 1 04 051 United States

Feed Name or Analyses				Mean As Fed	Mean Dry	C.V. ± %
Dry matter		%		89.6	100.0	
Ash		%		3.0	3.4	
Crude fiber		%		51.3	57.3	
Sheep	dig coef	%		80.	80.	
Ether extract		%		1.0	1.1	
Sheep	dig coef	%		78.	78.	
N-free extract		%		31.7	35.4	
Sheep	dig coef	%		41.	41.	
Protein (N x 6.25)		%		2.5	2.8	
Sheep	dig coef	%		-94.	-94.	
Cattle	dig prot	%	*	-0.5	-0.5	
Goats	dig prot	%	*	-0.6	-0.7	
Horses	dig prot	%	*	0.0	0.0	
Rabbits	dig prot	%	*	0.7	0.8	
Sheep	dig prot	%		-2.3	-2.6	
Energy	GE Mcal/kg					
Cattle	DE Mcal/kg		*	2.20	2.46	
Sheep	DE Mcal/kg		*	2.36	2.63	
Cattle	ME Mcal/kg		*	1.81	2.02	
Sheep	ME Mcal/kg		*	1.93	2.16	
Cattle	TDN	%	*	50.0	55.8	
Sheep	TDN	%		53.4	59.6	

Rye, winter, straw, treated w sodium hydroxide wet, (1)
Ref No 2 04 052 United States

Feed Name or Analyses				Mean As Fed	Mean Dry	C.V. ± %
Dry matter		%		21.9	100.0	
Ash		%		1.3	6.0	
Crude fiber		%		13.9	63.3	
Sheep	dig coef	%		85.	85.	
Ether extract		%		0.2	0.8	
Sheep	dig coef	%		12.	12.	
N-free extract		%		6.1	27.9	
Sheep	dig coef	%		39.	39.	
Protein (N x 6.25)		%		0.4	2.0	
Sheep	dig coef	%		-223.	-223.	
Cattle	dig prot	%	*	-0.2	-1.2	
Goats	dig prot	%	*	-0.2	-1.5	
Horses	dig prot	%	*	-0.1	-0.7	
Rabbits	dig prot	%	*	0.0	0.2	
Sheep	dig prot	%		-0.9	-4.4	
Energy	GE Mcal/kg					
Cattle	DE Mcal/kg		*	0.55	2.49	
Sheep	DE Mcal/kg		*	0.58	2.66	
Cattle	ME Mcal/kg		*	0.45	2.04	
Sheep	ME Mcal/kg		*	0.48	2.18	
Cattle	TDN	%	*	12.3	56.3	
Sheep	TDN	%		13.2	60.4	

Rye distillers dried grains (AAFCO) (CFA) – see Rye, distillers grains, dehy, (5)

Rye distillers dried grains with solubles (AAFCO) – see Rye, distillers grains w solubles, dehy, mn 75% original solids, (5)

Rye distillers dried grains with solubles (CFA) – see Rye, distillers grains w solubles, dehy, (5)

Rye distillers dried solubles (AAFCO) (CFA) – see Rye, distillers solubles, dehy, (5)

Rye feed (CFA) – see Rye, flour by-product mill run, (4)

RYEGRASS. Lolium spp

Ryegrass, hay, s-c, (1)
Ref No 1 04 057 United States

Feed Name or Analyses				Mean As Fed	Mean Dry	C.V. ± %
Dry matter		%		83.9	100.0	
Ash		%		6.8	8.1	
Crude fiber		%		28.6	34.1	
Cattle	dig coef	%		68.	68.	
Sheep	dig coef	%	*	66.	66.	
Ether extract		%		1.8	2.1	
Cattle	dig coef	%		55.	55.	
Sheep	dig coef	%	*	46.	46.	
N-free extract		%		39.2	46.7	
Cattle	dig coef	%		64.	64.	
Sheep	dig coef	%	*	66.	66.	
Protein (N x 6.25)		%		7.6	9.0	
Cattle	dig coef	%		48.	48.	
Sheep	dig coef	%	*	51.	51.	
Cattle	dig prot	%		3.6	4.3	
Goats	dig prot	%	*	4.2	5.0	
Horses	dig prot	%	*	4.3	5.2	
Rabbits	dig prot	%	*	4.7	5.6	
Sheep	dig prot	%		3.9	4.6	
Energy	GE Mcal/kg					
Cattle	DE Mcal/kg		*	2.22	2.65	
Sheep	DE Mcal/kg		*	2.22	2.65	
Cattle	ME Mcal/kg		*	1.82	2.17	
Sheep	ME Mcal/kg		*	1.82	2.17	
Cattle	TDN	%		50.3	60.0	
Sheep	TDN	%		50.4	60.1	
Carotene	mg/kg			100.6	119.9	
Vitamin A equivalent	IU/g			167.7	199.9	

Ryegrass, hay, s-c, immature, (1)
Ref No 1 04 055 United States

Feed Name or Analyses				Mean As Fed	Mean Dry	C.V. ± %
Dry matter		%		91.4	100.0	4
Ash		%		11.1	12.1	10
Crude fiber		%		20.7	22.7	16
Cattle	dig coef	%	*	68.	68.	
Sheep	dig coef	%	*	66.	66.	
Ether extract		%		2.7	3.0	12
Cattle	dig coef	%	*	55.	55.	
Sheep	dig coef	%	*	46.	46.	
N-free extract		%		44.1	48.2	
Cattle	dig coef	%	*	64.	64.	
Sheep	dig coef	%	*	66.	66.	
Protein (N x 6.25)		%		12.8	14.0	17
Cattle	dig coef	%	*	48.	48.	
Sheep	dig coef	%	*	51.	51.	
Cattle	dig prot	%		6.1	6.7	
Goats	dig prot	%	*	8.8	9.6	
Horses	dig prot	%	*	8.6	9.4	
Rabbits	dig prot	%	*	8.7	9.5	
Sheep	dig prot	%		6.5	7.1	
Energy	GE Mcal/kg					
Cattle	DE Mcal/kg		*	2.29	2.50	
Sheep	DE Mcal/kg		*	2.30	2.51	
Cattle	ME Mcal/kg		*	1.87	2.05	
Sheep	ME Mcal/kg		*	1.88	2.06	
Cattle	TDN	%		51.8	56.7	
Sheep	TDN	%		52.1	57.0	
Cobalt	mg/kg			0.127	0.139	

Feed Name or Analyses			Mean As Fed	Mean Dry	C.V. ± %
Carotene	mg/kg		265.0	289.9	
Vitamin A equivalent	IU/g		441.7	483.3	

Ryegrass, hay, s-c, mature, (1)
Ref No 1 04 056 United States

Analyses			As Fed	Dry	C.V. ± %
Dry matter	%		83.5	100.0	1
Ash	%		7.0	8.4	13
Crude fiber	%		30.3	36.3	5
Ether extract	%		0.8	0.9	18
N-free extract	%		40.8	48.9	
Protein (N x 6.25)	%		4.6	5.5	4
Cattle	dig prot %	*	1.4	1.7	
Goats	dig prot %	*	1.4	1.7	
Horses	dig prot %	*	1.8	2.2	
Rabbits	dig prot %	*	2.4	2.9	
Sheep	dig prot %	*	1.3	1.5	
Energy	GE Mcal/kg				
Cattle	DE Mcal/kg	*	1.95	2.33	
Sheep	DE Mcal/kg	*	1.96	2.35	
Cattle	ME Mcal/kg	*	1.60	1.91	
Sheep	ME Mcal/kg	*	1.61	1.92	
Cattle	TDN %	*	44.2	52.9	
Sheep	TDN %	*	44.4	53.2	

Ref No 1 04 056 Canada

Analyses			As Fed	Dry	C.V. ± %
Dry matter	%		92.4	100.0	
Crude fiber	%		33.7	36.5	
Protein (N x 6.25)	%		5.2	5.6	
Cattle	dig prot %	*	1.7	1.8	
Goats	dig prot %	*	1.7	1.8	
Horses	dig prot %	*	2.1	2.3	
Rabbits	dig prot %	*	2.8	3.0	
Sheep	dig prot %	*	1.5	1.6	
Calcium	%		0.28	0.30	
Phosphorus	%		0.06	0.06	
Carotene	mg/kg		6.6	7.2	
Vitamin A equivalent	IU/g		11.0	11.9	

Ryegrass, straw, (1)
Ref No 1 04 059 United States

Analyses			As Fed	Dry	C.V. ± %
Dry matter	%		89.1	100.0	2
Ash	%		5.3	5.9	18
Crude fiber	%		32.8	36.9	7
Sheep	dig coef %		60.	60.	
Ether extract	%		1.0	1.2	16
Sheep	dig coef %		44.	44.	
N-free extract	%		46.2	51.8	
Sheep	dig coef %		65.	65.	
Protein (N x 6.25)	%		3.8	4.3	6
Sheep	dig coef %		22.	22.	
Cattle	dig prot %	*	0.6	0.7	
Goats	dig prot %	*	0.5	0.6	
Horses	dig prot %	*	1.1	1.2	
Rabbits	dig prot %	*	1.8	2.0	
Sheep	dig prot %		0.8	0.9	
Energy	GE Mcal/kg				
Cattle	DE Mcal/kg	*	2.09	2.35	
Sheep	DE Mcal/kg	*	2.27	2.55	
Cattle	ME Mcal/kg	*	1.72	1.93	
Sheep	ME Mcal/kg	*	1.86	2.09	
Cattle	TDN %	*	47.5	53.3	
Sheep	TDN %		51.6	57.9	

Ryegrass, aerial part, fresh, (2)
Ref No 2 04 062 United States

Analyses			As Fed	Dry	C.V. ± %
Dry matter	%		21.5	100.0	
Ash	%		1.6	7.6	
Crude fiber	%		6.6	30.6	
Cattle	dig coef %		70.	70.	
Ether extract	%		0.6	2.8	
Cattle	dig coef %		62.	62.	
N-free extract	%		10.6	49.5	
Cattle	dig coef %		73.	73.	
Protein (N x 6.25)	%		2.0	9.5	
Cattle	dig coef %		62.	62.	
Cattle	dig prot %		1.3	5.9	
Goats	dig prot %	*	1.2	5.4	
Horses	dig prot %	*	1.2	5.6	
Rabbits	dig prot %	*	1.3	6.0	
Sheep	dig prot %	*	1.3	5.8	
Energy	GE Mcal/kg				
Cattle	DE Mcal/kg	*	0.64	2.97	
Sheep	DE Mcal/kg	*	0.63	2.92	
Cattle	ME Mcal/kg	*	0.52	2.44	
Sheep	ME Mcal/kg	*	0.52	2.40	
Cattle	TDN %		14.5	67.4	
Sheep	TDN %	*	14.2	66.3	

Ryegrass, aerial part, fresh, immature, (2)
Ref No 2 04 060 United States

Analyses			As Fed	Dry	C.V. ± %
Dry matter	%		21.3	100.0	8
Ash	%		3.7	17.3	19
Crude fiber	%		4.0	18.8	11
Ether extract	%		0.9	4.1	11
N-free extract	%		7.2	33.9	
Protein (N x 6.25)	%		5.5	25.9	13
Cattle	dig prot %	*	4.2	19.9	
Goats	dig prot %	*	4.4	20.7	
Horses	dig prot %	*	4.2	19.5	
Rabbits	dig prot %	*	4.0	18.7	
Sheep	dig prot %	*	4.5	21.1	
Energy	GE Mcal/kg				
Cattle	DE Mcal/kg	*	0.61	2.87	
Sheep	DE Mcal/kg	*	0.62	2.91	
Cattle	ME Mcal/kg	*	0.50	2.35	
Sheep	ME Mcal/kg	*	0.51	2.39	
Cattle	TDN %	*	13.9	65.1	
Sheep	TDN %	*	14.5	66.0	

Ryegrass, aerial part, fresh, mid-bloom, (2)
Ref No 2 04 061 United States

Analyses			As Fed	Dry	C.V. ± %
Dry matter	%		35.3	100.0	
Ash	%		3.2	9.0	9
Crude fiber	%		8.3	23.6	17
Ether extract	%		0.6	1.6	40
N-free extract	%		19.5	55.2	
Protein (N x 6.25)	%		3.7	10.6	32
Cattle	dig prot %	*	2.4	6.9	
Goats	dig prot %	*	2.3	6.4	
Horses	dig prot %	*	2.3	6.5	
Rabbits	dig prot %	*	2.4	6.9	
Sheep	dig prot %	*	2.4	6.9	
Energy	GE Mcal/kg				
Cattle	DE Mcal/kg	*	1.01	2.86	
Sheep	DE Mcal/kg	*	1.07	3.03	
Cattle	ME Mcal/kg	*	0.83	2.35	
Sheep	ME Mcal/kg	*	0.88	2.49	

Analyses			As Fed	Dry	C.V. ± %
Cattle	TDN %	*	22.9	64.9	
Sheep	TDN %	*	24.3	68.8	

RYEGRASS, ANNUAL. Lolium spp

Ryegrass, annual, hay, s-c, (1)
Ref No 1 08 642 United States

Analyses			As Fed	Dry	C.V. ± %
Dry matter	%		93.5	100.0	
Ash	%		7.4	7.9	
Crude fiber	%		12.8	13.7	
Ether extract	%		1.6	1.7	
Silicon	%		1.77	1.89	

Ryegrass, annual, aerial part, fresh, (2)
Ref No 2 04 063 United States

Analyses			As Fed	Dry	C.V. ± %
Dry matter	%		24.1	100.0	18
Ash	%		3.2	13.2	32
Crude fiber	%		5.6	23.2	14
Ether extract	%		0.9	3.9	22
N-free extract	%		10.5	43.5	
Protein (N x 6.25)	%		3.9	16.2	30
Cattle	dig prot %	*	2.8	11.7	
Goats	dig prot %	*	2.8	11.7	
Horses	dig prot %	*	2.7	11.3	
Rabbits	dig prot %	*	2.7	11.2	
Sheep	dig prot %	*	2.9	12.1	
Energy	GE Mcal/kg				
Cattle	DE Mcal/kg	*	0.65	2.71	
Sheep	DE Mcal/kg	*	0.66	2.76	
Cattle	ME Mcal/kg	*	0.54	2.22	
Sheep	ME Mcal/kg	*	0.55	2.26	
Cattle	TDN %	*	14.8	61.5	
Sheep	TDN %	*	15.1	62.6	
Carotene	mg/kg		96.6	401.0	39
Vitamin A equivalent	IU/g		161.1	668.5	
Glutamic acid	%		0.55	2.30	
Isoleucine	%		0.39	1.60	
Leucine	%		0.55	2.30	
Lysine	%		0.36	1.50	
Methionine	%		0.07	0.30	
Phenylalanine	%		0.29	1.20	
Threonine	%		0.51	2.10	
Tryptophan	%		0.07	0.30	
Valine	%		0.31	1.30	

RYEGRASS, GULF. Lolium multiflorum

Ryegrass, gulf, hay, s-c grnd pelleted, (1)
Ref No 1 07 912 United States

Analyses			As Fed	Dry	C.V. ± %
Dry matter	%			100.0	
Protein (N x 6.25)	%			9.4	36
Cattle	dig prot %	*		5.1	
Goats	dig prot %	*		5.4	
Horses	dig prot %	*		5.5	
Rabbits	dig prot %	*		6.0	
Sheep	dig prot %	*		5.0	
Soluble carbohydrates	%			6.8	68
Lignin (Ellis)	%			6.9	32
Energy	GE Mcal/kg				
Sheep	DE Mcal/kg			2.50	16
Sheep	ME Mcal/kg	*		2.05	

Feed Name or Analyses				Mean As Fed	Mean Dry	C.V. ± %

Ryegrass, gulf, hay, s-c grnd pelleted, immature, (1)
Ref No 1 09 162 United States

Analyses		Unit		As Fed	Dry	C.V. ± %
Dry matter		%			100.0	
Sheep	dig coef	%			71.	
Ash		%			11.1	
Crude fiber		%			20.9	
Ether extract		%			4.6	
N-free extract		%			48.3	
Protein (N x 6.25)		%			15.2	
Cattle	dig prot	%	*		10.1	
Goats	dig prot	%	*		10.7	
Horses	dig prot	%	*		10.4	
Rabbits	dig prot	%	*		10.4	
Sheep	dig prot	%	*		10.2	
Lignin (Ellis)		%			3.5	
Energy		GE Mcal/kg				
Cattle		DE Mcal/kg	*		3.10	
Sheep		DE Mcal/kg			3.10	
Cattle		ME Mcal/kg	*		2.54	
Sheep		ME Mcal/kg	*		2.54	
Cattle		TDN %	*		70.2	
Sheep		TDN %	*		70.2	

Ryegrass, gulf, hay, s-c grnd pelleted, prebloom, (1)
Ref No 1 09 163 United States

Analyses		Unit		As Fed	Dry	C.V. ± %
Dry matter		%			100.0	
Sheep	dig coef	%			65.	
Ash		%			11.2	
Crude fiber		%			27.6	
Ether extract		%			3.7	
N-free extract		%			46.3	
Protein (N x 6.25)		%			11.2	
Cattle	dig prot	%	*		6.6	
Goats	dig prot	%	*		7.0	
Horses	dig prot	%	*		7.0	
Rabbits	dig prot	%	*		7.3	
Sheep	dig prot	%	*		6.6	
Lignin (Ellis)		%			4.2	
Energy		GE Mcal/kg				
Cattle		DE Mcal/kg	*		2.75	
Sheep		DE Mcal/kg			2.85	
Cattle		ME Mcal/kg	*		2.26	
Sheep		ME Mcal/kg	*		2.34	
Cattle		TDN %	*		62.3	
Sheep		TDN %	*		64.8	

Ryegrass, gulf, hay, s-c grnd pelleted, early bloom, (1)
Ref No 1 09 164 United States

Analyses		Unit		As Fed	Dry	C.V. ± %
Dry matter		%			100.0	
Sheep	dig coef	%			55.	
Ash		%			10.6	
Crude fiber		%			33.7	
Ether extract		%			2.9	
N-free extract		%			44.9	
Protein (N x 6.25)		%			7.9	
Cattle	dig prot	%	*		3.8	
Goats	dig prot	%	*		3.9	
Horses	dig prot	%	*		4.2	
Rabbits	dig prot	%	*		4.8	
Sheep	dig prot	%	*		3.7	
Lignin (Ellis)		%			10.6	
Energy		GE Mcal/kg				
Cattle		DE Mcal/kg	*		2.31	
Sheep		DE Mcal/kg			2.27	
Cattle		ME Mcal/kg	*		1.89	
Sheep		ME Mcal/kg	*		1.86	
Cattle		TDN %	*		52.3	
Sheep		TDN %	*		51.6	

Ryegrass, gulf, hay, s-c grnd pelleted, midbloom, (1)
Ref No 1 09 165 United States

Analyses		Unit		As Fed	Dry	C.V. ± %
Dry matter		%			100.0	
Sheep	dig coef	%			58.	
Ash		%			9.5	
Crude fiber		%			31.1	
Ether extract		%			2.1	
N-free extract		%			49.3	
Protein (N x 6.25)		%			8.0	
Cattle	dig prot	%	*		3.9	
Goats	dig prot	%	*		4.0	
Horses	dig prot	%	*		4.3	
Rabbits	dig prot	%	*		4.8	
Sheep	dig prot	%	*		3.7	
Energy		GE Mcal/kg				
Cattle		DE Mcal/kg	*		2.61	
Sheep		DE Mcal/kg			2.51	
Cattle		ME Mcal/kg	*		2.74	
Sheep		ME Mcal/kg	*		2.05	
Cattle		TDN %	*		59.2	
Sheep		TDN %	*		57.0	

Ryegrass, gulf, hay, s-c grnd pelleted, milk stage, (1)
Ref No 1 09 166 United States

Analyses		Unit		As Fed	Dry	C.V. ± %
Dry matter		%			100.0	
Sheep	dig coef	%			50.	
Ash		%			9.5	
Crude fiber		%			34.2	
Ether extract		%			2.1	
N-free extract		%			46.9	
Protein (N x 6.25)		%			7.4	
Cattle	dig prot	%	*		3.3	
Goats	dig prot	%	*		3.4	
Horses	dig prot	%	*		3.8	
Rabbits	dig prot	%	*		4.3	
Sheep	dig prot	%	*		3.2	
Lignin (Ellis)		%			7.0	
Energy		GE Mcal/kg				
Cattle		DE Mcal/kg	*		2.30	
Sheep		DE Mcal/kg			2.19	
Cattle		ME Mcal/kg	*		1.89	
Sheep		ME Mcal/kg	*		1.80	
Cattle		TDN %	*		52.1	
Sheep		TDN %	*		49.8	

Ryegrass, gulf, hay, s-c grnd pelleted, dough stage, (1)
Ref No 1 09 167 United States

Analyses		Unit		As Fed	Dry	C.V. ± %
Dry matter		%			100.0	
Sheep	dig coef	%			52.	
Ash		%			8.8	
Crude fiber		%			34.2	
Ether extract		%			1.9	
N-free extract		%			48.0	
Protein (N x 6.25)		%			7.2	
Cattle	dig prot	%	*		3.2	
Goats	dig prot	%	*		3.3	
Horses	dig prot	%	*		3.6	
Rabbits	dig prot	%	*		4.2	
Sheep	dig prot	%	*		3.0	
Lignin (Ellis)		%			8.8	
Energy		GE Mcal/kg				
Cattle		DE Mcal/kg	*		2.39	
Sheep		DE Mcal/kg			2.25	
Cattle		ME Mcal/kg	*		1.96	
Sheep		ME Mcal/kg	*		1.85	
Cattle		TDN %	*		54.2	
Sheep		TDN %	*		51.1	

RYEGRASS, ITALIAN. Lolium multiflorum

Ryegrass, Italian, aerial part, dehy grnd, early bloom, (1)
Ref No 1 08 277 United States

Analyses		Unit		As Fed	Dry	C.V. ± %
Dry matter		%		92.5	100.0	
Organic matter		%		87.9	95.0	
Ash		%		4.7	5.1	
Crude fiber		%		21.1	22.8	
Ether extract		%		1.5	1.6	
N-free extract		%		59.8	64.6	
Protein (N x 6.25)		%		5.5	5.9	
Cattle	dig prot	%	*	1.9	2.0	
Goats	dig prot	%	*	1.9	2.1	
Horses	dig prot	%	*	2.3	2.5	
Rabbits	dig prot	%	*	3.0	3.2	
Sheep	dig prot	%	*	1.7	1.9	
Silicon		%		0.90	0.97	

Ryegrass, Italian, aerial part, dehy grnd, over ripe, (1)
Ref No 1 08 278 United States

Analyses		Unit		As Fed	Dry	C.V. ± %
Dry matter		%		93.5	100.0	
Organic matter		%		88.2	94.3	
Ash		%		5.3	5.7	
Crude fiber		%		28.6	30.6	
Ether extract		%		1.5	1.6	
N-free extract		%		54.7	58.5	
Protein (N x 6.25)		%		3.4	3.6	
Cattle	dig prot	%	*	0.1	0.1	
Goats	dig prot	%	*	0.0	0.0	
Horses	dig prot	%	*	0.5	0.6	
Rabbits	dig prot	%	*	1.4	1.4	
Sheep	dig prot	%	*	-0.1	-0.1	
Silicon		%		1.12	1.20	

Ryegrass, Italian, aerial part, dehy grnd, dormant, (1)
Ref No 1 08 279 United States

Analyses		Unit		As Fed	Dry	C.V. ± %
Dry matter		%		92.5	100.0	0
Organic matter		%		87.7	94.8	1
Ash		%		4.9	5.3	14
Crude fiber		%		29.1	31.4	13
Ether extract		%		1.2	1.3	10
N-free extract		%		54.2	58.6	7
Protein (N x 6.25)		%		3.2	3.4	11
Cattle	dig prot	%	*	0.0	0.0	
Goats	dig prot	%	*	-0.1	-0.1	
Horses	dig prot	%	*	0.4	0.4	
Rabbits	dig prot	%	*	1.2	1.3	

(1) dry forages and roughages (2) pasture, range plants, and forages fed green (3) silages (4) energy feeds (5) protein supplements (6) minerals (7) vitamins (8) additives

Feed Name or Analyses			Mean As Fed	Mean Dry	C.V. ± %
Sheep	dig prot %	*	-0.2	-0.3	
Silicon	%		2.37	2.56	10

Ryegrass, Italian, hay, s-c, (1)

Ref No 1 04 069 United States

Feed Name or Analyses			As Fed	Dry	C.V. ± %
Dry matter	%		86.5	100.0	3
Ash	%		7.2	8.4	19
Crude fiber	%		27.3	31.5	16
Sheep	dig coef %		65.	65.	
Ether extract	%		1.6	1.9	31
Sheep	dig coef %		42.	42.	
N-free extract	%		43.5	50.2	
Sheep	dig coef %		67.	67.	
Protein (N x 6.25)	%		7.0	8.0	32
Sheep	dig coef %		42.	42.	
Cattle	dig prot %	*	3.4	3.9	
Goats	dig prot %	*	3.5	4.1	
Horses	dig prot %	*	3.8	4.4	
Rabbits	dig prot %	*	4.2	4.9	
Sheep	dig prot %		2.9	3.4	
Fatty acids	%		1.9	2.1	
Energy	GE Mcal/kg				
Cattle	DE Mcal/kg	*	2.34	2.70	
Sheep	DE Mcal/kg	*	2.26	2.61	
Cattle	ME Mcal/kg	*	1.92	2.21	
Sheep	ME Mcal/kg	*	1.85	2.14	
Cattle	TDN %	*	53.0	61.2	
Sheep	TDN %		51.3	59.3	
Calcium	%		0.54	0.62	9
Magnesium	%		0.28	0.32	12
Phosphorus	%		0.29	0.34	11
Potassium	%		1.35	1.56	5

Ryegrass, Italian, hay, s-c, immature, (1)

Ref No 1 04 064 United States

Feed Name or Analyses			As Fed	Dry	C.V. ± %
Dry matter	%		89.3	100.0	4
Ash	%		11.6	13.0	9
Crude fiber	%		17.6	19.7	12
Sheep	dig coef %	*	74.	74.	
Ether extract	%		2.9	3.2	13
Sheep	dig coef %	*	53.	53.	
N-free extract	%		43.7	48.9	
Sheep	dig coef %	*	79.	79.	
Protein (N x 6.25)	%		13.6	15.2	18
Sheep	dig coef %	*	56.	56.	
Cattle	dig prot %	*	9.0	10.1	
Goats	dig prot %	*	9.6	10.7	
Horses	dig prot %	*	9.3	10.4	
Rabbits	dig prot %	*	9.3	10.4	
Sheep	dig prot %		7.6	8.5	
Energy	GE Mcal/kg				
Cattle	DE Mcal/kg	*	2.75	3.08	
Sheep	DE Mcal/kg	*	2.58	2.89	
Cattle	ME Mcal/kg	*	2.26	2.53	
Sheep	ME Mcal/kg	*	2.12	2.37	
Cattle	TDN %	*	62.4	69.9	
Sheep	TDN %	*	58.5	65.5	

Ryegrass, Italian, hay, s-c, pre-bloom, (1)

Ref No 1 04 065 United States

Feed Name or Analyses			As Fed	Dry	C.V. ± %
Dry matter	%		82.2	100.0	
Ash	%		9.0	11.0	
Crude fiber	%		19.6	23.8	
Sheep	dig coef %		74.	74.	
Ether extract	%		2.0	2.4	
Sheep	dig coef %		53.	53.	
N-free extract	%		43.2	52.5	
Sheep	dig coef %		79.	79.	
Protein (N x 6.25)	%		8.5	10.3	
Sheep	dig coef %		56.	56.	
Cattle	dig prot %	*	4.8	5.9	
Goats	dig prot %	*	5.1	6.2	
Horses	dig prot %	*	5.2	6.3	
Rabbits	dig prot %	*	5.4	6.6	
Sheep	dig prot %		4.7	5.8	
Energy	GE Mcal/kg				
Cattle	DE Mcal/kg	*	2.52	3.06	
Sheep	DE Mcal/kg	*	2.45	2.99	
Cattle	ME Mcal/kg	*	2.06	2.51	
Sheep	ME Mcal/kg	*	2.01	2.45	
Cattle	TDN %	*	57.1	69.5	
Sheep	TDN %		55.7	67.7	

Ryegrass, Italian, hay, s-c, early bloom, (1)

Ref No 1 04 066 United States

Feed Name or Analyses			As Fed	Dry	C.V. ± %
Dry matter	%		83.4	100.0	
Ash	%		7.0	8.4	
Crude fiber	%		30.3	36.3	
Sheep	dig coef %		64.	64.	
Ether extract	%		0.8	0.9	
Sheep	dig coef %		35.	35.	
N-free extract	%		40.8	48.9	
Sheep	dig coef %		58.	58.	
Protein (N x 6.25)	%		4.6	5.5	
Sheep	dig coef %		38.	38.	
Cattle	dig prot %	*	1.4	1.7	
Goats	dig prot %	*	1.4	1.7	
Horses	dig prot %	*	1.8	2.2	
Rabbits	dig prot %	*	2.4	2.9	
Sheep	dig prot %		1.7	2.1	
Energy	GE Mcal/kg				
Cattle	DE Mcal/kg	*	1.95	2.33	
Sheep	DE Mcal/kg	*	2.00	2.40	
Cattle	ME Mcal/kg	*	1.60	1.91	
Sheep	ME Mcal/kg	*	1.64	1.97	
Cattle	TDN %	*	44.1	52.9	
Sheep	TDN %		45.4	54.4	

Ryegrass, Italian, hay, s-c, full bloom, (1)

Ref No 1 04 067 United States

Feed Name or Analyses			As Fed	Dry	C.V. ± %
Dry matter	%		85.9	100.0	2
Ash	%		6.1	7.1	12
Crude fiber	%		26.5	30.9	3
Sheep	dig coef %		46.	46.	
Ether extract	%		1.3	1.6	24
Sheep	dig coef %		36.	36.	
N-free extract	%		46.3	53.9	
Sheep	dig coef %		68.	68.	
Protein (N x 6.25)	%		5.7	6.6	11
Sheep	dig coef %		26.	26.	
Cattle	dig prot %	*	2.3	2.7	
Goats	dig prot %	*	2.3	2.7	
Horses	dig prot %	*	2.7	3.1	
Rabbits	dig prot %	*	3.2	3.8	
Sheep	dig prot %		1.5	1.7	
Energy	GE Mcal/kg				
Cattle	DE Mcal/kg	*	2.12	2.47	
Sheep	DE Mcal/kg	*	2.04	2.37	
Cattle	ME Mcal/kg	*	1.74	2.03	
Sheep	ME Mcal/kg	*	1.67	1.95	
Cattle	TDN %	*	48.1	56.0	
Sheep	TDN %		46.2	53.8	

Ryegrass, Italian, hay, s-c, mature, (1)

Ref No 1 04 068 United States

Feed Name or Analyses			As Fed	Dry	C.V. ± %
Dry matter	%		83.2	100.0	1
Ash	%		6.6	7.9	8
Crude fiber	%		30.6	36.8	5
Sheep	dig coef %	*	65.	65.	
Ether extract	%		0.9	1.1	16
Sheep	dig coef %	*	42.	42.	
N-free extract	%		40.4	48.5	
Sheep	dig coef %	*	67.	67.	
Protein (N x 6.25)	%		4.7	5.7	7
Sheep	dig coef %	*	42.	42.	
Cattle	dig prot %	*	1.6	1.9	
Goats	dig prot %	*	1.6	1.9	
Horses	dig prot %	*	2.0	2.4	
Rabbits	dig prot %	*	2.6	3.1	
Sheep	dig prot %		2.0	2.4	
Energy	GE Mcal/kg				
Cattle	DE Mcal/kg	*	2.10	2.53	
Sheep	DE Mcal/kg	*	2.20	2.64	
Cattle	ME Mcal/kg	*	1.72	2.07	
Sheep	ME Mcal/kg	*	1.80	2.16	
Cattle	TDN %	*	47.7	57.3	
Sheep	TDN %	*	49.8	59.8	

Ryegrass, Italian, leaves, dehy grnd, early leaf, (1)

Ref No 1 08 280 United States

Feed Name or Analyses			As Fed	Dry	C.V. ± %
Dry matter	%		93.0	100.0	
Organic matter	%		84.5	90.9	
Ash	%		8.5	9.1	
Crude fiber	%		13.3	14.3	
Ether extract	%		3.8	4.1	
N-free extract	%		50.6	54.4	
Protein (N x 6.25)	%		16.8	18.1	
Cattle	dig prot %	*	11.7	12.6	
Goats	dig prot %	*	12.5	13.4	
Horses	dig prot %	*	12.0	12.9	
Rabbits	dig prot %	*	11.8	12.6	
Sheep	dig prot %	*	11.9	12.8	
Silicon	%		0.85	0.91	

Ryegrass, Italian, leaves, dehy grnd, immature, (1)

Ref No 1 08 281 United States

Feed Name or Analyses			As Fed	Dry	C.V. ± %
Dry matter	%		93.0	100.0	
Organic matter	%		83.5	89.8	
Ash	%		9.5	10.2	
Crude fiber	%		13.5	14.6	
Ether extract	%		2.6	2.8	
N-free extract	%		53.2	57.2	
Protein (N x 6.25)	%		14.3	15.4	
Cattle	dig prot %	*	9.5	10.2	
Goats	dig prot %	*	10.1	10.9	
Horses	dig prot %	*	9.8	10.6	
Rabbits	dig prot %	*	9.8	10.5	
Sheep	dig prot %	*	9.6	10.4	
Silicon	%		1.86	2.00	

Continued

Ryegrass, Italian, leaves, dehy grnd, early bloom, (1)

Ref No 1 08 282 United States

Feed Name or Analyses				Mean As Fed	Mean Dry	C.V. ± %
Dry matter		%		93.2	100.0	
Organic matter		%		85.2	91.4	
Ash		%		8.0	8.6	
Crude fiber		%		16.7	17.9	
Ether extract		%		3.7	4.0	
N-free extract		%		57.4	61.6	
Protein (N x 6.25)		%		7.4	7.9	
Cattle	dig prot	%	*	3.5	3.8	
Goats	dig prot	%	*	3.7	3.9	
Horses	dig prot	%	*	3.9	4.2	
Rabbits	dig prot	%	*	4.4	4.8	
Sheep	dig prot	%	*	3.4	3.7	
Silicon		%		1.49	1.60	

Ryegrass, Italian, leaves, dehy grnd, dormant, (1)

Ref No 1 08 283 United States

Feed Name or Analyses				Mean As Fed	Mean Dry	C.V. ± %
Dry matter		%		92.0	100.0	
Organic matter		%		82.1	89.2	
Ash		%		9.9	10.8	
Protein (N x 6.25)		%		5.6	6.1	
Cattle	dig prot	%	*	2.0	2.2	
Goats	dig prot	%	*	2.1	2.3	
Horses	dig prot	%	*	2.5	2.7	
Rabbits	dig prot	%	*	3.1	3.4	
Sheep	dig prot	%	*	1.9	2.0	
Silicon		%		3.00	3.26	

Ryegrass, Italian, leaves w some stems, dehy grnd, pre-bloom, (1)

Ref No 1 08 284 United States

Feed Name or Analyses				Mean As Fed	Mean Dry	C.V. ± %
Dry matter		%		93.0	100.0	
Organic matter		%		86.5	93.0	
Ash		%		6.5	7.0	
Protein (N x 6.25)		%		7.5	8.1	
Cattle	dig prot	%	*	3.7	4.0	
Goats	dig prot	%	*	3.8	4.1	
Horses	dig prot	%	*	4.1	4.4	
Rabbits	dig prot	%	*	4.6	4.9	
Sheep	dig prot	%	*	3.6	3.8	
Silicon		%		1.10	1.18	

Ryegrass, Italian, aerial part, fresh, (2)

Ref No 2 04 073 United States

Feed Name or Analyses				Mean As Fed	Mean Dry	C.V. ± %
Dry matter		%		23.8	100.0	15
Ash		%		3.4	14.1	
Crude fiber		%		5.4	22.5	
Ether extract		%		1.0	4.4	
N-free extract		%		10.3	43.5	
Protein (N x 6.25)		%		3.7	15.5	
Cattle	dig prot	%	*	2.6	11.0	
Goats	dig prot	%	*	2.6	11.0	
Horses	dig prot	%	*	2.5	10.7	
Rabbits	dig prot	%	*	2.5	10.6	
Sheep	dig prot	%	*	2.7	11.4	
Energy	GE Mcal/kg					
Cattle	DE Mcal/kg		*	0.61	2.58	
Sheep	DE Mcal/kg		*	0.66	2.78	
Cattle	ME Mcal/kg		*	0.50	2.11	
Sheep	ME Mcal/kg		*	0.54	2.28	
Cattle	NEm Mcal/kg		*	0.32	1.33	
Cattle	NEgain Mcal/kg		*	0.17	0.72	
Cattle	NElactating cows Mcal/kg		*	0.36	1.52	
Cattle	TDN	%	*	13.9	58.4	
Sheep	TDN	%	*	15.0	63.1	
Calcium		%		0.15	0.65	8
Magnesium		%		0.08	0.35	12
Phosphorus		%		0.10	0.41	17
Potassium		%		0.48	2.00	4

Ryegrass, Italian, aerial part, fresh, immature, (2)

Ref No 2 04 070 United States

Feed Name or Analyses				Mean As Fed	Mean Dry	C.V. ± %
Dry matter		%			100.0	
Ash		%			18.4	14
Crude fiber		%			18.8	6
Ether extract		%			4.2	12
N-free extract		%			34.4	
Protein (N x 6.25)		%			24.2	9
Cattle	dig prot	%	*		18.5	
Goats	dig prot	%	*		19.1	
Horses	dig prot	%	*		18.1	
Rabbits	dig prot	%	*		17.4	
Sheep	dig prot	%	*		19.5	
Energy	GE Mcal/kg					
Cattle	DE Mcal/kg		*		2.71	
Sheep	DE Mcal/kg		*		2.88	
Cattle	ME Mcal/kg		*		2.22	
Sheep	ME Mcal/kg		*		2.36	
Cattle	TDN	%	*		61.4	
Sheep	TDN	%	*		65.3	

Ryegrass, Italian, aerial part, fresh, early bloom, (2)

Ref No 2 04 071 United States

Feed Name or Analyses				Mean As Fed	Mean Dry	C.V. ± %
Dry matter		%		35.3	100.0	
Ash		%		2.9	8.1	
Crude fiber		%		10.6	30.1	
Sheep	dig coef	%		64.	64.	
Ether extract		%		0.3	0.8	
Sheep	dig coef	%		46.	46.	
N-free extract		%		19.5	55.2	
Sheep	dig coef	%		62.	62.	
Protein (N x 6.25)		%		2.0	5.8	
Sheep	dig coef	%		43.	43.	
Cattle	dig prot	%	*	1.0	2.8	
Goats	dig prot	%	*	0.7	2.0	
Horses	dig prot	%	*	0.9	2.5	
Rabbits	dig prot	%	*	1.1	3.1	
Sheep	dig prot	%		0.9	2.5	
Energy	GE Mcal/kg					
Cattle	DE Mcal/kg		*	0.91	2.58	
Sheep	DE Mcal/kg		*	0.88	2.50	
Cattle	ME Mcal/kg		*	0.75	2.12	
Sheep	ME Mcal/kg		*	0.73	2.05	
Cattle	TDN	%	*	20.7	58.5	
Sheep	TDN	%		20.1	56.8	

Ryegrass, Italian, aerial part, fresh, mature, (2)

Ref No 2 04 072 United States

Feed Name or Analyses				Mean As Fed	Mean Dry	C.V. ± %
Dry matter		%		35.3	100.0	
Ash		%		2.9	8.1	
Crude fiber		%		10.6	30.1	
Ether extract		%		0.3	0.8	
N-free extract		%		19.5	55.2	
Protein (N x 6.25)		%		2.0	5.8	
Cattle	dig prot	%	*	1.0	2.8	
Goats	dig prot	%	*	0.7	2.0	
Horses	dig prot	%	*	0.9	2.5	
Rabbits	dig prot	%	*	1.1	3.1	
Sheep	dig prot	%	*	0.8	2.4	
Energy	GE Mcal/kg					
Cattle	DE Mcal/kg		*	1.03	2.94	
Sheep	DE Mcal/kg		*	0.99	2.80	
Cattle	ME Mcal/kg		*	0.85	2.41	
Sheep	ME Mcal/kg		*	0.81	2.30	
Cattle	TDN	%	*	23.5	66.7	
Sheep	TDN	%	*	22.4	63.4	

RYEGRASS, PERENNIAL. Lolium perenne

Ryegrass, perennial, hay, s-c, (1)

Ref No 1 04 077 United States

Feed Name or Analyses				Mean As Fed	Mean Dry	C.V. ± %
Dry matter		%		86.1	100.0	2
Ash		%		7.8	9.1	
Crude fiber		%		24.9	29.0	
Sheep	dig coef	%		66.	66.	
Ether extract		%		2.4	2.8	
Sheep	dig coef	%		46.	46.	
N-free extract		%		42.6	49.5	
Sheep	dig coef	%		66.	66.	
Protein (N x 6.25)		%		8.3	9.7	
Sheep	dig coef	%		51.	51.	
Cattle	dig prot	%	*	4.6	5.3	
Goats	dig prot	%	*	4.8	5.6	
Horses	dig prot	%	*	4.9	5.7	
Rabbits	dig prot	%	*	5.3	6.1	
Sheep	dig prot	%		4.3	4.9	
Fatty acids		%		3.0	3.5	
Energy	GE Mcal/kg					
Cattle	DE Mcal/kg		*	2.36	2.74	
Sheep	DE Mcal/kg		*	2.26	2.63	
Cattle	ME Mcal/kg		*	1.94	2.25	
Sheep	ME Mcal/kg		*	1.86	2.15	
Cattle	TDN	%	*	53.5	62.1	
Sheep	TDN	%		51.3	59.6	
Phosphorus		%		0.23	0.27	
Potassium		%		1.22	1.42	
α-tocopherol		mg/kg		181.7	211.0	

Ryegrass, perennial, hay, s-c, pre-bloom, (1)

Ref No 1 04 074 United States

Feed Name or Analyses				Mean As Fed	Mean Dry	C.V. ± %
Dry matter		%		83.1	100.0	
Ash		%		8.5	10.2	
Crude fiber		%		23.8	28.7	
Sheep	dig coef	%		75.	75.	
Ether extract		%		2.2	2.6	
Sheep	dig coef	%		49.	49.	
N-free extract		%		39.1	47.0	
Sheep	dig coef	%		78.	78.	
Protein (N x 6.25)		%		9.6	11.5	
Sheep	dig coef	%		62.	62.	

(1) dry forages and roughages (2) pasture, range plants, and forages fed green (3) silages (4) energy feeds (5) protein supplements (6) minerals (7) vitamins (8) additives

Feed Name or Analyses			Mean As Fed	Mean Dry	C.V. ± %
Cattle	dig prot %	*	5.7	6.9	
Goats	dig prot %	*	6.1	7.3	
Horses	dig prot %	*	6.1	7.3	
Rabbits	dig prot %	*	6.3	7.5	
Sheep	dig prot %		5.9	7.1	
Energy	GE Mcal/kg				
Cattle	DE Mcal/kg	*	2.47	2.98	
Sheep	DE Mcal/kg	*	2.50	3.01	
Cattle	ME Mcal/kg	*	2.03	2.44	
Sheep	ME Mcal/kg	*	2.05	2.47	
Cattle	TDN %	*	56.1	67.5	
Sheep	TDN %		56.7	68.2	

Ryegrass, perennial, hay, s-c, early bloom, (1)
Ref No 1 04 075 United States

Feed Name or Analyses			Mean As Fed	Mean Dry	C.V. ± %
Dry matter	%		84.5	100.0	
Ash	%		8.8	10.4	
Crude fiber	%		29.2	34.6	
Sheep	dig coef %		65.	65.	
Ether extract	%		1.1	1.3	
Sheep	dig coef %		42.	42.	
N-free extract	%		40.3	47.7	
Sheep	dig coef %		56.	56.	
Protein (N x 6.25)	%		5.1	6.0	
Sheep	dig coef %		44.	44.	
Cattle	dig prot %	*	1.8	2.1	
Goats	dig prot %	*	1.8	2.2	
Horses	dig prot %	*	2.2	2.6	
Rabbits	dig prot %	*	2.8	3.3	
Sheep	dig prot %		2.2	2.6	
Energy	GE Mcal/kg				
Cattle	DE Mcal/kg	*	2.16	2.55	
Sheep	DE Mcal/kg	*	1.98	2.34	
Cattle	ME Mcal/kg	*	1.77	2.09	
Sheep	ME Mcal/kg	*	1.62	1.92	
Cattle	TDN %	*	48.9	57.9	
Sheep	TDN %		44.8	53.1	

Ryegrass, perennial, hay, s-c, full bloom, (1)
Ref No 1 04 076 United States

Feed Name or Analyses			Mean As Fed	Mean Dry	C.V. ± %
Dry matter	%		80.9	100.0	
Ash	%		6.0	7.4	
Crude fiber	%		24.5	30.3	
Sheep	dig coef %		55.	55.	
Ether extract	%		1.1	1.4	
Sheep	dig coef %		49.	49.	
N-free extract	%		44.3	54.7	
Sheep	dig coef %		65.	65.	
Protein (N x 6.25)	%		5.0	6.2	
Sheep	dig coef %		41.	41.	
Cattle	dig prot %	*	1.9	2.3	
Goats	dig prot %	*	1.9	2.3	
Horses	dig prot %	*	2.3	2.8	
Rabbits	dig prot %	*	2.8	3.5	
Sheep	dig prot %		2.1	2.5	
Energy	GE Mcal/kg				
Cattle	DE Mcal/kg	*	2.10	2.60	
Sheep	DE Mcal/kg	*	2.01	2.48	
Cattle	ME Mcal/kg	*	1.72	2.13	
Sheep	ME Mcal/kg	*	1.65	2.04	
Cattle	TDN %	*	47.7	58.9	
Sheep	TDN %		45.6	56.3	

Ryegrass, perennial, aerial part, fresh, (2)
Ref No 2 04 086 United States

Feed Name or Analyses			Mean As Fed	Mean Dry	C.V. ± %
Dry matter	%		26.1	100.0	
Ash	%		2.4	9.0	
Crude fiber	%		6.6	25.2	
Ether extract	%		1.3	4.9	
N-free extract	%		12.9	49.6	
Protein (N x 6.25)	%		2.9	11.3	
Cattle	dig prot %	*	1.9	7.5	
Goats	dig prot %	*	1.8	7.1	
Horses	dig prot %	*	1.9	7.1	
Rabbits	dig prot %	*	1.9	7.4	
Sheep	dig prot %	*	2.0	7.5	
Energy	GE Mcal/kg				
Cattle	DE Mcal/kg	*	0.73	2.98	
Sheep	DE Mcal/kg	*	0.76	2.93	
Cattle	ME Mcal/kg	*	0.64	2.44	
Sheep	ME Mcal/kg	*	0.63	2.40	
Cattle	TDN %	*	17.7	67.7	
Sheep	TDN %	*	17.3	66.5	
Calcium	%		0.14	0.53	
Cobalt	mg/kg		0.016	0.060	98
Copper	mg/kg		1.1	4.4	44
Phosphorus	%		0.10	0.37	
Potassium	%		0.50	1.92	
Carotene	mg/kg		69.2	265.7	40
α-tocopherol	mg/kg		45.2	173.5	64
Vitamin A equivalent	IU/g		115.4	442.8	

Ryegrass, perennial, aerial part, fresh, immature, (2)
Ref No 2 04 079 United States

Feed Name or Analyses			Mean As Fed	Mean Dry	C.V. ± %
Dry matter	%		20.9	100.0	8
Ash	%		3.4	16.5	10
Crude fiber	%		3.8	18.4	8
Ether extract	%		0.8	4.0	9
N-free extract	%		7.4	35.4	
Protein (N x 6.25)	%		5.4	25.9	15
Cattle	dig prot %	*	4.1	19.9	
Goats	dig prot %	*	4.3	20.7	
Horses	dig prot %	*	4.1	19.5	
Rabbits	dig prot %	*	3.9	18.6	
Sheep	dig prot %	*	4.4	21.1	
Energy	GE Mcal/kg				
Cattle	DE Mcal/kg	*	0.61	2.93	
Sheep	DE Mcal/kg	*	0.61	2.91	
Cattle	ME Mcal/kg	*	0.50	2.40	
Sheep	ME Mcal/kg	*	0.50	2.39	
Cattle	TDN %	*	13.9	66.4	
Sheep	TDN %	*	14.0	66.9	
Calcium	%		0.14	0.65	
Phosphorus	%		0.08	0.40	
Potassium	%		0.40	1.90	
α-tocopherol	mg/kg		73.5	352.1	

Ryegrass, perennial, aerial part, fresh, immature, pure stand, (2)
Ref No 2 04 078 United States

Feed Name or Analyses			Mean As Fed	Mean Dry	C.V. ± %
Dry matter	%			100.0	
Carotene	mg/kg			470.0	
Vitamin A equivalent	IU/g			783.5	

Ryegrass, perennial, aerial part, fresh, early bloom, (2)
Ref No 2 04 080 United States

Feed Name or Analyses			Mean As Fed	Mean Dry	C.V. ± %
Dry matter	%			100.0	
Ash	%			9.5	8
Crude fiber	%			20.3	7
Ether extract	%			2.1	16
N-free extract	%			55.1	
Protein (N x 6.25)	%			13.0	10
Cattle	dig prot %	*		8.9	
Goats	dig prot %	*		8.7	
Horses	dig prot %	*		8.6	
Rabbits	dig prot %	*		8.7	
Sheep	dig prot %	*		9.1	
α-tocopherol	mg/kg			269.0	

Ryegrass, perennial, aerial part, fresh, mid-bloom, (2)
Ref No 2 04 081 United States

Feed Name or Analyses			Mean As Fed	Mean Dry	C.V. ± %
Dry matter	%			100.0	
α-tocopherol	mg/kg			145.9	

Ryegrass, perennial, aerial part, fresh, full bloom, (2)
Ref No 2 04 082 United States

Feed Name or Analyses			Mean As Fed	Mean Dry	C.V. ± %
Dry matter	%			100.0	
Ash	%			7.2	
Crude fiber	%			24.2	
Ether extract	%			1.1	
N-free extract	%			60.5	
Protein (N x 6.25)	%			7.0	
Cattle	dig prot %	*		3.8	
Goats	dig prot %	*		3.1	
Horses	dig prot %	*		3.5	
Rabbits	dig prot %	*		4.1	
Sheep	dig prot %	*		3.5	
α-tocopherol	mg/kg			123.2	

Ryegrass, perennial, aerial part, fresh, milk stage, (2)
Ref No 2 04 083 United States

Feed Name or Analyses			Mean As Fed	Mean Dry	C.V. ± %
Dry matter	%			100.0	
α-tocopherol	mg/kg			81.1	

Ryegrass, perennial, aerial part, fresh, mature, (2)
Ref No 2 04 084 United States

Feed Name or Analyses			Mean As Fed	Mean Dry	C.V. ± %
Dry matter	%			100.0	
Ash	%			6.4	
Crude fiber	%			30.8	
Ether extract	%			1.5	
N-free extract	%			54.7	
Protein (N x 6.25)	%			6.6	
Cattle	dig prot %	*		3.5	
Goats	dig prot %	*		2.7	
Horses	dig prot %	*		3.1	
Rabbits	dig prot %	*		3.8	
Sheep	dig prot %	*		3.1	

Ryegrass, perennial, aerial part, fresh, over ripe, (2)
Ref No 2 04 085 United States

Feed Name or Analyses			Mean As Fed	Mean Dry	C.V. ± %
Dry matter	%			100.0	
Ash	%			7.0	
Crude fiber	%			32.7	
Ether extract	%			1.5	
N-free extract	%			53.6	
Protein (N x 6.25)	%			5.2	
Cattle	dig prot %	*		2.3	
Goats	dig prot %	*		1.4	
Horses	dig prot %	*		1.9	
Rabbits	dig prot %	*		2.7	
Sheep	dig prot %	*		1.8	
α-tocopherol	mg/kg			22.0	

RYEGRASS, ANNUAL-CLOVER, CRIMSON. Lolium spp, Trifolium incarnatum

Ryegrass, annual-Clover, crimson, aerial part, fresh, (2)
Ref No 2 04 090 United States

Feed Name or Analyses			Mean As Fed	Mean Dry	C.V. ± %
Dry matter	%		19.7	100.0	20
Ash	%		1.8	9.1	6
Crude fiber	%		3.3	17.0	21
Ether extract	%		1.0	5.1	26
N-free extract	%		9.1	46.2	
Protein (N x 6.25)	%		4.5	22.6	15
Cattle	dig prot %	*	3.4	17.1	
Goats	dig prot %	*	3.5	17.6	
Horses	dig prot %	*	3.3	16.7	
Rabbits	dig prot %	*	3.2	16.1	
Sheep	dig prot %	*	3.6	18.1	

Ryegrass, annual-Clover, crimson, aerial part, fresh, immature, (2)
Ref No 2 04 088 United States

Feed Name or Analyses			Mean As Fed	Mean Dry	C.V. ± %
Dry matter	%		15.8	100.0	11
Ash	%		1.5	9.4	4
Crude fiber	%		2.3	14.7	5
Ether extract	%		0.9	5.8	30
N-free extract	%		7.1	45.1	
Protein (N x 6.25)	%		4.0	25.0	8
Cattle	dig prot %	*	3.0	19.1	
Goats	dig prot %	*	3.1	19.9	
Horses	dig prot %	*	3.0	18.8	
Rabbits	dig prot %	*	2.8	18.0	
Sheep	dig prot %	*	3.2	20.3	
Calcium	%		0.11	0.68	23
Phosphorus	%		0.08	0.48	7
Carotene	mg/kg		83.7	529.8	23
Vitamin A equivalent	IU/g		139.5	883.1	

Ryegrass, annual-Clover, crimson, aerial part, fresh, mid-bloom, (2)
Ref No 2 04 089 United States

Feed Name or Analyses			Mean As Fed	Mean Dry	C.V. ± %
Dry matter	%		23.6	100.0	14
Ash	%		2.1	8.8	5
Crude fiber	%		4.5	19.2	20
Ether extract	%		1.0	4.3	23
N-free extract	%		11.2	47.5	
Protein (N x 6.25)	%		4.8	20.2	11
Cattle	dig prot %	*	3.6	15.1	
Goats	dig prot %	*	3.6	15.4	
Horses	dig prot %	*	3.5	14.7	
Rabbits	dig prot %	*	3.4	14.3	
Sheep	dig prot %	*	3.7	15.8	
Calcium	%		0.13	0.56	11
Phosphorus	%		0.10	0.42	13
Carotene	mg/kg		69.0	292.3	23
Vitamin A equivalent	IU/g		115.0	487.3	

(1) dry forages and roughages (2) pasture, range plants, and forages fed green (3) silages (4) energy feeds (5) protein supplements (6) minerals (7) vitamins (8) additives

RYEGRASS-FESCUE. Lolium spp, Festuca spp

Ryegrass-Fescue, hay, s-c, (1)
Ref No 1 04 087 United States

Feed Name or Analyses			Mean As Fed	Mean Dry	C.V. ± %
Dry matter	%		93.0	100.0	
Calcium	%		0.72	0.77	
Magnesium	%		0.35	0.38	
Phosphorus	%		0.44	0.47	

RYEGRASS, PERENNIAL-ALFALFA. Lolium perenne, Medicago sativa

Ryegrass, perennial-Alfalfa, aerial part, fresh, (2)
Ref No 2 04 091 United States

Feed Name or Analyses			Mean As Fed	Mean Dry	C.V. ± %
Dry matter	%			100.0	
Calcium	%			0.66	
Magnesium	%			0.29	
Potassium	%			3.19	
Sodium	%			0.05	

RYEGRASS, PERENNIAL-CLOVER, LADINO. Lolium Perenne, Trifolium repens

Ryegrass, perennial-Clover, ladino, aerial part, fresh, (2)
Ref No 2 04 092 United States

Feed Name or Analyses			Mean As Fed	Mean Dry	C.V. ± %
Dry matter	%			100.0	
Calcium	%			0.83	11
Magnesium	%			0.35	14
Potassium	%			3.10	11
Sodium	%			0.08	28

Rye middlings (AAFCO) –
see Rye, flour by-product, coarse sift, mx 8.5% fiber, (4)

Rye middlings (CFA) –
see Rye, flour by-product, fine sift, mx 4.5% fiber, (4)

Rye mill run (AAFCO) –
see Rye, flour by-product mill, run, mx 9.5% fiber, (4)

RYE-VETCH. Secale cereale, Vicia spp

Rye-Vetch, aerial part, fresh, (2)
Ref No 2 04 053 United States

Feed Name or Analyses			Mean As Fed	Mean Dry	C.V. ± %
Dry matter	%			100.0	
Calcium	%			0.28	
Iodine	mg/kg			0.476	
Iron	%			0.011	
Magnesium	%			0.18	
Phosphorus	%			0.28	

Rye-Vetch, aerial part, ensiled, full bloom, (3)
Ref No 3 04 054 United States

Feed Name or Analyses			Mean As Fed	Mean Dry	C.V. ± %
Dry matter	%		20.6	100.0	
Ash	%		2.1	10.2	
Crude fiber	%		6.8	33.0	
Ether extract	%		1.0	4.9	
N-free extract	%		8.3	40.2	
Protein (N x 6.25)	%		2.4	11.7	
Cattle	dig prot %	*	1.4	6.9	
Goats	dig prot %	*	1.4	6.9	
Horses	dig prot %	*	1.4	6.9	
Sheep	dig prot %	*	1.4	6.9	
Energy	GE Mcal/kg				
Cattle	DE Mcal/kg	*	0.57	2.76	
Sheep	DE Mcal/kg	*	0.55	2.69	
Cattle	ME Mcal/kg	*	0.47	2.27	
Sheep	ME Mcal/kg	*	0.45	2.20	
Cattle	TDN %	*	12.9	62.7	
Sheep	TDN %	*	12.5	60.9	

SACAHUISTA. Nolina microcarpa

Sacahuista, hay, s-c, (1)
Ref No 1 04 093 United States

Feed Name or Analyses			Mean As Fed	Mean Dry	C.V. ± %
Dry matter	%		92.1	100.0	
Ash	%		3.0	3.3	
Crude fiber	%		43.9	47.7	
Sheep	dig coef %		4.	4.	
Ether extract	%		2.2	2.4	
Sheep	dig coef %		0.	0.	
N-free extract	%		37.3	40.5	
Sheep	dig coef %		9.	9.	
Protein (N x 6.25)	%		5.6	6.1	
Sheep	dig coef %		0.	0.	
Cattle	dig prot %	*	2.0	2.2	
Goats	dig prot %	*	2.1	2.3	
Horses	dig prot %	*	2.5	2.7	
Rabbits	dig prot %	*	3.1	3.4	
Energy	GE Mcal/kg				
Cattle	DE Mcal/kg	*	1.23	1.34	
Sheep	DE Mcal/kg	*	0.23	0.24	
Cattle	ME Mcal/kg	*	1.01	1.10	
Sheep	ME Mcal/kg	*	0.18	0.20	
Cattle	TDN %	*	28.0	30.4	
Sheep	TDN %		5.1	5.6	

Sacahuista, browse, mature, (2)
Ref No 2 04 094 United States

Feed Name or Analyses			Mean As Fed	Mean Dry	C.V. ± %
Dry matter	%			100.0	
Ash	%			3.9	
Crude fiber	%			41.9	

Feed Name or Analyses			Mean As Fed	Mean Dry	C.V. ± %
Ether extract		%		3.0	
N-free extract		%		44.9	
Protein (N x 6.25)		%		6.3	
Cattle	dig prot	%	*	3.2	
Goats	dig prot	%	*	2.4	
Horses	dig prot	%	*	2.9	
Rabbits	dig prot	%	*	3.5	
Sheep	dig prot	%	*	2.9	

Sacahuista, buds, fresh, (2)

Ref No 2 04 095 United States

Analyses			As Fed	Dry	C.V. ± %
Dry matter		%		100.0	
Ash		%		6.0	
Crude fiber		%		16.5	
Ether extract		%		2.7	
N-free extract		%		48.8	
Protein (N x 6.25)		%		26.0	
Cattle	dig prot	%	*	20.0	
Goats	dig prot	%	*	20.8	
Horses	dig prot	%	*	19.6	
Rabbits	dig prot	%	*	18.7	
Sheep	dig prot	%	*	21.2	

SAFFLOWER. Carthamus tinctorius

Safflower, hay, s-c, mature, (1)

Ref No 1 04 104 United States

Analyses			As Fed	Dry	C.V. ± %
Dry matter		%	91.3	100.0	
Ash		%	1.4	1.5	
Crude fiber		%	53.1	58.2	
Ether extract		%	4.7	5.1	
N-free extract		%	28.3	31.0	
Protein (N x 6.25)		%	3.8	4.2	
Cattle	dig prot	%	* 0.5	0.6	
Goats	dig prot	%	* 0.4	0.5	
Horses	dig prot	%	* 1.0	1.1	
Rabbits	dig prot	%	* 1.7	1.9	
Sheep	dig prot	%	* 0.3	0.3	
Energy	GE Mcal/kg				
Cattle	DE Mcal/kg		* 0.57	0.62	
Sheep	DE Mcal/kg		* 0.88	0.96	
Cattle	ME Mcal/kg		* 0.47	0.51	
Sheep	ME Mcal/kg		* 0.72	0.79	
Cattle	TDN	%	* 12.9	14.2	
Sheep	TDN	%	* 19.9	21.8	

Safflower, hulls, (1)

Ref No 1 04 105 United States

Analyses			As Fed	Dry	C.V. ± %
Dry matter		%	91.3	100.0	
Ash		%	1.6	1.8	13
Crude fiber		%	53.1	58.2	4
Ether extract		%	3.4	3.7	36
N-free extract		%	29.9	32.7	
Protein (N x 6.25)		%	3.3	3.6	14
Cattle	dig prot	%	* 0.1	0.1	
Goats	dig prot	%	* 0.0	0.0	
Horses	dig prot	%	* 0.6	0.6	
Rabbits	dig prot	%	* 1.3	1.5	
Sheep	dig prot	%	* -0.1	-0.1	
Fatty acids		%	4.7	5.1	
Energy	GE Mcal/kg				
Cattle	DE Mcal/kg		* 0.53	0.59	
Sheep	DE Mcal/kg		* 0.70	0.77	
Cattle	ME Mcal/kg		* 0.44	0.48	

Feed Name or Analyses			Mean As Fed	Mean Dry	C.V. ± %
Chickens	ME_n kcal/kg		924.	1012.	
Sheep	ME Mcal/kg		* 0.34	0.37	
Cattle	TDN	%	* 12.1	13.3	
Sheep	TDN	%	* 16.0	17.5	

Safflower, aerial part, fresh, pre-bloom, (2)

Ref No 2 04 106 United States

Analyses			As Fed	Dry	C.V. ± %
Dry matter		%	16.8	100.0	
Ash		%	1.6	9.5	
Crude fiber		%	6.3	37.4	
Sheep	dig coef	%	59.	59.	
Ether extract		%	0.4	2.5	
Sheep	dig coef	%	60.	60.	
N-free extract		%	6.4	38.2	
Sheep	dig coef	%	66.	66.	
Protein (N x 6.25)		%	2.1	12.4	
Sheep	dig coef	%	64.	64.	
Cattle	dig prot	%	* 1.4	8.4	
Goats	dig prot	%	* 1.4	8.1	
Horses	dig prot	%	* 1.4	8.1	
Rabbits	dig prot	%	* 1.4	8.2	
Sheep	dig prot	%	1.3	7.9	
Energy	GE Mcal/kg				
Cattle	DE Mcal/kg		* 0.44	2.62	
Sheep	DE Mcal/kg		* 0.43	2.58	
Cattle	ME Mcal/kg		* 0.36	2.15	
Sheep	ME Mcal/kg		* 0.36	2.12	
Cattle	TDN	%	* 10.0	59.4	
Sheep	TDN	%	9.8	58.6	

Safflower, aerial part, ensiled, (3)

Ref No 3 04 107 United States

Analyses			As Fed	Dry	C.V. ± %
Dry matter		%	16.6	100.0	
Ash		%	1.5	8.9	
Crude fiber		%	5.2	31.4	
Sheep	dig coef	%	61.	61.	
Ether extract		%	0.7	4.4	
Sheep	dig coef	%	82.	82.	
N-free extract		%	7.1	42.7	
Sheep	dig coef	%	74.	74.	
Protein (N x 6.25)		%	2.1	12.6	
Sheep	dig coef	%	76.	76.	
Cattle	dig prot	%	* 1.3	7.7	
Goats	dig prot	%	* 1.3	7.7	
Horses	dig prot	%	* 1.3	7.7	
Sheep	dig prot	%	1.6	9.5	
Energy	GE Mcal/kg				
Cattle	DE Mcal/kg		* 0.45	2.69	
Sheep	DE Mcal/kg		* 0.50	3.02	
Cattle	ME Mcal/kg		* 0.37	2.20	
Sheep	ME Mcal/kg		* 0.41	2.47	
Cattle	TDN	%	* 10.1	61.0	
Sheep	TDN	%	11.4	68.4	

Safflower, seeds, (4)

Ref No 4 07 958 United States

Analyses			As Fed	Dry	C.V. ± %
Dry matter		%	93.1	100.0	
Ash		%	2.9	3.1	
Crude fiber		%	26.6	28.6	
Ether extract		%	29.8	32.0	
N-free extract		%	17.5	18.8	
Protein (N x 6.25)		%	16.3	17.5	
Cattle	dig prot	%	* 11.3	12.1	
Goats	dig prot	%	* 12.4	13.3	

Feed Name or Analyses			Mean As Fed	Mean Dry	C.V. ± %
Horses	dig prot	%	* 12.4	13.3	
Sheep	dig prot	%	* 12.4	13.3	
Energy	GE Mcal/kg				
Cattle	DE Mcal/kg		* 3.74	4.02	
Sheep	DE Mcal/kg		* 2.00	2.15	
Swine	DE kcal/kg		*3551.	3814.	
Cattle	ME Mcal/kg		* 3.07	3.30	
Sheep	ME Mcal/kg		* 1.64	1.76	
Swine	ME kcal/kg		*2911.	3127.	
Cattle	NE_m Mcal/kg		* 2.05	2.20	
Cattle	NE_{gain} Mcal/kg		* 1.34	1.44	
Cattle	$NE_{lactating\ cows}$ Mcal/kg		* 1.42	1.52	
Cattle	TDN	%	* 84.9	91.2	
Sheep	TDN	%	* 82.3	88.4	
Swine	TDN	%	* 80.5	86.5	
α-tocopherol		mg/kg	1.0	1.1	

Safflower, seeds, extn unspecified grnd, (5)

Ref No 5 04 108 United States

Analyses			As Fed	Dry	C.V. ± %
Dry matter		%	92.2	100.0	
Ash		%	6.9	7.5	
Crude fiber		%	21.6	23.4	
Sheep	dig coef	%	-6.	-6.	
Ether extract		%	9.1	9.9	
Sheep	dig coef	%	96.	96.	
N-free extract		%	21.7	23.5	
Sheep	dig coef	%	48.	48.	
Protein (N x 6.25)		%	32.9	35.7	
Sheep	dig coef	%	86.	86.	
Sheep	dig prot	%	28.3	30.7	
Linoleic		%	4.400	4.772	
Energy	GE Mcal/kg				
Cattle	DE Mcal/kg		* 2.71	2.94	
Sheep	DE Mcal/kg		* 2.51	2.72	
Swine	DE kcal/kg		*3269.	3545.	
Cattle	ME Mcal/kg		* 2.22	2.41	
Chickens	ME_n kcal/kg		810.	878.	
Sheep	ME Mcal/kg		* 2.06	2.23	
Swine	ME kcal/kg		*2902.	3148.	
Cattle	TDN	%	* 61.4	66.6	
Sheep	TDN	%	56.9	61.7	
Swine	TDN	%	* 74.1	80.4	

Safflower, seeds, mech-extd grnd, (5)

Safflower seed, mechanical extracted (AAFCO)

Safflower oil meal, expeller extracted

Safflower oil meal, hydraulic extracted

Ref No 5 04 109 United States

Analyses			As Fed	Dry	C.V. ± %
Dry matter		%	91.9	100.0	1
Ash		%	3.8	4.1	8
Crude fiber		%	32.2	35.0	5
Sheep	dig coef	%	* -6.	-6.	
Ether extract		%	6.6	7.2	11
Sheep	dig coef	%	* 96.	96.	
N-free extract		%	27.8	30.2	
Sheep	dig coef	%	* 48.	48.	
Protein (N x 6.25)		%	21.6	23.5	2
Sheep	dig coef	%	* 86.	86.	
Sheep	dig prot	%	18.6	20.2	
Energy	GE Mcal/kg				
Cattle	DE Mcal/kg		* 2.27	2.47	
Sheep	DE Mcal/kg		* 1.94	2.11	
Swine	DE kcal/kg		*2435.	2650.	
Cattle	ME Mcal/kg		* 1.87	2.03	
Sheep	ME Mcal/kg		* 1.59	1.73	

Continued

Feed Name or Analyses		Mean As Fed	Mean Dry	C.V. ± %
Swine	ME kcal/kg	*1997.	2173.	
Cattle	NE_m Mcal/kg	* 1.12	1.22	
Cattle	NE_{gain} Mcal/kg	* 0.51	0.56	
Cattle $NE_{lactating\ cows}$	Mcal/kg	* 1.23	1.34	
Cattle	TDN %	* 51.6	56.1	
Sheep	TDN %	43.9	47.8	
Swine	TDN %	* 55.2	60.1	
Calcium	%	0.24	0.26	
Copper	mg/kg	9.8	10.7	
Iron	%	0.046	0.050	
Magnesium	%	0.33	0.36	
Manganese	mg/kg	18.1	19.7	
Phosphorus	%	0.61	0.66	
Potassium	%	0.73	0.79	
Sodium	%	0.05	0.06	
Sulphur	%	0.05	0.06	
Zinc	mg/kg	40.5	44.0	
Choline	mg/kg	1514.	1647.	
Niacin	mg/kg	13.0	14.1	
Pantothenic acid	mg/kg	51.9	56.5	
Riboflavin	mg/kg	2.4	2.6	
Arginine	%	1.44	1.57	
Cystine	%	0.53	0.57	
Glycine	%	1.47	1.60	
Lysine	%	0.65	0.71	
Methionine	%	0.37	0.40	
Tryptophan	%	0.28	0.30	

Safflower, seeds wo hulls, mech-extd grnd, (5)
Ref No 5 08 499 United States

Feed Name or Analyses		Mean As Fed	Mean Dry	C.V. ± %
Dry matter	%	88.5	100.0	
Ash	%	6.4	7.2	
Crude fiber	%	8.5	9.6	
Ether extract	%	6.7	7.6	
N-free extract	%	26.4	29.9	
Protein (N x 6.25)	%	40.4	45.7	
Linoleic	%	1.100	1.243	
Energy	GE Mcal/kg			
Cattle	DE Mcal/kg	* 3.07	3.47	
Sheep	DE Mcal/kg	* 3.11	3.52	
Swine	DE kcal/kg	*3352.	3790.	
Cattle	ME Mcal/kg	* 2.51	2.84	
Chickens	ME_n kcal/kg	1690.	1910.	
Sheep	ME Mcal/kg	* 2.55	2.89	
Swine	ME kcal/kg	*2908.	3288.	
Cattle	TDN %	* 69.5	78.6	
Sheep	TDN %	* 70.6	79.8	
Swine	TDN %	* 76.0	85.9	
Calcium	%	0.32	0.36	
Phosphorus	%	0.59	0.67	
Choline	mg/kg	2641.	2985.	
Niacin	mg/kg	22.6	25.5	
Pantothenic acid	mg/kg	88.0	99.5	
Riboflavin	mg/kg	4.1	4.6	
Arginine	%	3.07	3.47	
Cystine	%	0.51	0.58	
Glycine	%	2.56	2.89	
Lysine	%	1.33	1.50	
Methionine	%	0.72	0.81	
Tryptophan	%	0.51	0.58	

(1) dry forages and roughages (2) pasture, range plants, and forages fed green (3) silages (4) energy feeds (5) protein supplements (6) minerals (7) vitamins (8) additives

Safflower, seeds w some hulls, mech-extd grnd, (5)
Ref No 5 08 500 United States

Feed Name or Analyses		Mean As Fed	Mean Dry	C.V. ± %
Dry matter	%	94.0	100.0	
Ash	%	6.2	6.6	
Crude fiber	%	18.4	19.6	
Ether extract	%	6.2	6.6	
N-free extract	%	28.8	30.6	
Protein (N x 6.25)	%	34.4	36.6	
Energy	GE Mcal/kg			
Cattle	DE Mcal/kg	* 2.83	3.01	
Sheep	DE Mcal/kg	* 3.04	3.23	
Swine	DE kcal/kg	*3294.	3504.	
Cattle	ME Mcal/kg	* 2.32	2.47	
Sheep	ME Mcal/kg	* 2.49	2.65	
Swine	ME kcal/kg	*2918.	3105.	
Cattle	TDN %	* 64.2	68.3	
Sheep	TDN %	* 68.9	73.3	
Swine	TDN %	* 74.7	79.5	

Safflower, seeds w some hulls, solv-extd grnd, (5)
Ref No 5 08 501 United States

Feed Name or Analyses		Mean As Fed	Mean Dry	C.V. ± %
Dry matter	%	93.8	100.0	
Ash	%	4.0	4.3	
Crude fiber	%	26.9	28.7	
Ether extract	%	1.0	1.1	
N-free extract	%	40.1	42.8	
Protein (N x 6.25)	%	21.8	23.2	
Energy	GE Mcal/kg			
Cattle	DE Mcal/kg	* 2.34	2.50	
Sheep	DE Mcal/kg	* 2.54	2.70	
Swine	DE kcal/kg	*2577.	2747.	
Cattle	ME Mcal/kg	* 1.92	2.05	
Sheep	ME Mcal/kg	* 2.08	2.22	
Swine	ME kcal/kg	*2098.	2236.	
Cattle	NE_m Mcal/kg	* 1.09	1.17	
Cattle	NE_{gain} Mcal/kg	* 0.45	0.48	
Cattle $NE_{lactating\ cows}$	Mcal/kg	* 1.18	1.26	
Cattle	TDN %	* 53.1	56.6	
Sheep	TDN %	* 57.5	61.3	
Swine	TDN %	* 58.4	62.3	
α-tocopherol	mg/kg	1.0	1.1	

Safflower oil meal, expeller extracted – see Safflower, seeds, mech-extd grnd, (5)

Safflower oil meal, hydraulic extracted – see Safflower, seeds, mech-extd grnd, (5)

Safflower seed, mechanical extracted (AAFCO) – see Safflower, seeds, mech-extd grnd, (5)

SAGE. Salvia spp

Sage, browse, fresh, (2)
Ref No 2 04 112 United States

Feed Name or Analyses		Mean As Fed	Mean Dry	C.V. ± %
Dry matter	%		100.0	
Calcium	%		1.49	24
Magnesium	%		0.52	11
Phosphorus	%		0.36	26
Potassium	%		4.63	

Sage, browse, fresh, immature, (2)
Ref No 2 04 111 United States

Feed Name or Analyses		Mean As Fed	Mean Dry	C.V. ± %
Dry matter	%		100.0	
Ash	%		14.0	
Crude fiber	%		12.2	
Ether extract	%		3.9	
N-free extract	%		45.5	
Protein (N x 6.25)	%		24.4	
Cattle	dig prot %	*	18.6	
Goats	dig prot %	*	19.3	
Horses	dig prot %	*	18.2	
Rabbits	dig prot %	*	17.5	
Sheep	dig prot %	*	19.7	
Calcium	%		1.61	6
Magnesium	%		0.52	11
Phosphorus	%		0.39	16
Potassium	%		4.63	

SAGE, BLACK. Salvia mellifera

Sage, black, browse, fresh, dormant, (2)
Ref No 2 05 564 United States

Feed Name or Analyses		Mean As Fed	Mean Dry	C.V. ± %
Dry matter	%		100.0	
Ash	%		5.5	
Ether extract	%		10.8	
Sheep	dig coef %		57.	
Protein (N x 6.25)	%		8.5	
Sheep	dig coef %		54.	
Cattle	dig prot %	*	5.1	
Goats	dig prot %	*	4.4	
Horses	dig prot %	*	4.7	
Rabbits	dig prot %	*	5.2	
Sheep	dig prot %		4.5	
Cellulose (Matrone)	%		21.5	
Sheep	dig coef %		29.	
Other carbohydrates	%		40.1	
Lignin (Ellis)	%		15.6	
Energy	GE Mcal/kg		5.10	
Sheep	GE dig coef %		42.	
Sheep	DE Mcal/kg		2.12	
Sheep	ME Mcal/kg		1.04	
Sheep	TDN %		49.5	
Calcium	%		0.81	
Phosphorus	%		0.17	

SAGEBRUSH. Artemisia spp

Sagebrush, browse, fresh, (2)
Ref No 2 04 116 United States

Feed Name or Analyses		Mean As Fed	Mean Dry	C.V. ± %
Dry matter	%	50.5	100.0	20
Ash	%	4.9	9.7	
Crude fiber	%	12.5	24.8	
Ether extract	%	4.6	9.2	
N-free extract	%	22.0	43.5	
Protein (N x 6.25)	%	6.5	12.9	
Cattle	dig prot %	* 4.5	8.8	
Goats	dig prot %	* 4.3	8.6	
Horses	dig prot %	* 4.3	8.5	
Rabbits	dig prot %	* 4.3	8.6	
Sheep	dig prot %	* 4.5	9.0	
Energy	GE Mcal/kg			
Cattle	DE Mcal/kg	* 1.11	2.20	
Sheep	DE Mcal/kg	* 1.22	2.41	
Cattle	ME Mcal/kg	* 0.91	1.80	

Feed Name or Analyses			Mean As Fed	Mean Dry	C.V. ± %
Sheep	ME Mcal/kg	*	1.00	1.98	
Cattle	TDN %	*	25.2	49.9	
Sheep	TDN %	*	27.6	54.6	
Calcium	%		0.51	1.01	
Cobalt	mg/kg		0.101	0.201	
Phosphorus	%		0.13	0.25	
Sulphur	%		0.11	0.22	19

Sagebrush, browse, fresh, immature, (2)

Ref No 2 04 113 United States

Feed Name or Analyses			Mean As Fed	Mean Dry	C.V. ± %
Dry matter	%		28.9	100.0	
Ash	%		5.3	18.5	35
Crude fiber	%		6.8	23.4	2
Ether extract	%		0.9	3.1	51
N-free extract	%		11.7	40.6	
Protein (N x 6.25)	%		4.2	14.4	22
Cattle	dig prot %	*	2.9	10.1	
Goats	dig prot %	*	2.9	10.0	
Horses	dig prot %	*	2.8	9.8	
Rabbits	dig prot %	*	2.8	9.8	
Sheep	dig prot %	*	3.0	10.4	
Energy	GE Mcal/kg				
Cattle	DE Mcal/kg	*	0.67	2.33	
Sheep	DE Mcal/kg	*	0.67	2.33	
Cattle	ME Mcal/kg	*	0.55	1.91	
Sheep	ME Mcal/kg	*	0.55	1.91	
Cattle	TDN %	*	15.3	52.8	
Sheep	TDN %	*	15.3	52.9	
Calcium	%		0.31	1.06	35
Magnesium	%		0.14	0.49	
Phosphorus	%		0.08	0.29	73

Sagebrush, browse, fresh, mid-bloom, (2)

Ref No 2 04 114 United States

Feed Name or Analyses			Mean As Fed	Mean Dry	C.V. ± %
Dry matter	%		43.2	100.0	
Ash	%		2.8	6.4	
Crude fiber	%		11.4	26.5	
Ether extract	%		2.2	5.2	
N-free extract	%		23.1	53.4	
Protein (N x 6.25)	%		3.7	8.5	
Cattle	dig prot %	*	2.2	5.1	
Goats	dig prot %	*	1.9	4.5	
Horses	dig prot %	*	2.1	4.7	
Rabbits	dig prot %	*	2.3	5.2	
Sheep	dig prot %	*	2.1	4.9	
Energy	GE Mcal/kg				
Cattle	DE Mcal/kg	*	0.98	2.28	
Sheep	DE Mcal/kg	*	1.03	2.39	
Cattle	ME Mcal/kg	*	0.81	1.87	
Sheep	ME Mcal/kg	*	0.85	1.96	
Cattle	TDN %	*	22.3	51.7	
Sheep	TDN %	*	23.5	54.3	

Sagebrush, browse, fresh, mature, (2)

Ref No 2 04 115 United States

Feed Name or Analyses			Mean As Fed	Mean Dry	C.V. ± %
Dry matter	%			100.0	
Ash	%			12.8	
Crude fiber	%			31.7	
Ether extract	%			3.5	
N-free extract	%			46.0	
Protein (N x 6.25)	%			6.0	
Cattle	dig prot %	*		3.0	
Goats	dig prot %	*		2.2	
Horses	dig prot %	*		2.6	

Feed Name or Analyses			Mean As Fed	Mean Dry	C.V. ± %
Rabbits	dig prot %	*		3.3	
Sheep	dig prot %	*		2.6	
Energy	GE Mcal/kg				
Cattle	DE Mcal/kg	*		2.33	
Sheep	DE Mcal/kg	*		2.39	
Cattle	ME Mcal/kg	*		1.91	
Sheep	ME Mcal/kg	*		1.96	
Cattle	TDN %	*		52.9	
Sheep	TDN %	*		54.1	

Sagebrush, buds, fresh, (2)

Ref No 2 04 117 United States

Feed Name or Analyses			Mean As Fed	Mean Dry	C.V. ± %
Dry matter	%			100.0	
Ash	%			19.6	58
Crude fiber	%			29.4	22
Ether extract	%			3.4	50
N-free extract	%			38.7	
Protein (N x 6.25)	%			8.9	27
Cattle	dig prot %	*		5.5	
Goats	dig prot %	*		4.9	
Horses	dig prot %	*		5.1	
Rabbits	dig prot %	*		5.5	
Sheep	dig prot %	*		5.3	
Energy	GE Mcal/kg				
Cattle	DE Mcal/kg	*		2.59	
Sheep	DE Mcal/kg	*		2.81	
Cattle	ME Mcal/kg	*		2.12	
Sheep	ME Mcal/kg	*		2.30	
Cattle	TDN %	*		58.3	
Sheep	TDN %	*		63.7	

Sagebrush, leaves, fresh, (2)

Ref No 2 04 118 United States

Feed Name or Analyses			Mean As Fed	Mean Dry	C.V. ± %
Dry matter	%		50.0	100.0	
Ash	%		3.4	6.8	
Crude fiber	%		6.3	12.6	
Ether extract	%		7.9	15.8	
N-free extract	%		24.0	48.0	
Protein (N x 6.25)	%		8.4	16.8	
Cattle	dig prot %	*	6.1	12.2	
Goats	dig prot %	*	6.1	12.2	
Horses	dig prot %	*	5.9	11.8	
Rabbits	dig prot %	*	5.8	11.6	
Sheep	dig prot %	*	6.3	12.6	
Energy	GE Mcal/kg				
Cattle	DE Mcal/kg	*	1.37	2.73	
Sheep	DE Mcal/kg	*	1.45	2.90	
Cattle	ME Mcal/kg	*	1.12	2.24	
Sheep	ME Mcal/kg	*	1.19	2.38	
Cattle	TDN %	*	31.0	61.9	
Sheep	TDN %	*	32.9	65.7	

SAGEBRUSH, BIG. Artemisia tridentata

Sagebrush, big, browse, fresh, (2)

Ref No 2 04 119 United States

Feed Name or Analyses			Mean As Fed	Mean Dry	C.V. ± %
Dry matter	%		52.0	100.0	
Ash	%		3.1	6.0	25
Crude fiber	%		13.2	25.3	16
Ether extract	%		6.8	13.0	32
N-free extract	%		23.8	45.8	
Protein (N x 6.25)	%		5.2	9.9	15
Cattle	dig prot %	*	3.3	6.3	
Goats	dig prot %	*	3.0	5.8	

Feed Name or Analyses			Mean As Fed	Mean Dry	C.V. ± %
Horses	dig prot %	*	3.1	5.9	
Rabbits	dig prot %	*	3.3	6.3	
Sheep	dig prot %	*	3.2	6.2	
Calcium	%		0.37	0.72	24
Cobalt	mg/kg		0.073	0.141	
Copper	mg/kg		7.2	13.9	
Magnesium	%		0.11	0.21	
Manganese	mg/kg		16.1	30.9	
Phosphorus	%		0.09	0.18	22
Sulphur	%		0.11	0.21	
Carotene	mg/kg		10.4	20.1	42
Vitamin A equivalent	IU/g		17.4	33.4	

Sagebrush, big, browse, fresh, dormant, (2)

Ref No 2 07 992 United States

Feed Name or Analyses			Mean As Fed	Mean Dry	C.V. ± %
Dry matter	%			100.0	
Ash	%			9.6	
Ether extract	%			8.2	
Sheep	dig coef %			66.	
Protein (N x 6.25)	%			9.0	
Sheep	dig coef %			42.	
Cattle	dig prot %	*		5.5	
Goats	dig prot %	*		5.0	
Horses	dig prot %	*		5.2	
Rabbits	dig prot %	*		5.6	
Sheep	dig prot %			3.8	
Cellulose (Matrone)	%			18.5	
Sheep	dig coef %			44.	
Other carbohydrates	%			38.1	
Lignin (Ellis)	%			16.6	
Energy	GE Mcal/kg			4.83	
Sheep	GE dig coef %			41.	
Sheep	DE Mcal/kg			1.98	
Sheep	ME Mcal/kg			1.13	
Sheep	TDN %			43.4	
Calcium	%			0.98	
Phosphorus	%			0.21	

Sagebrush, big, leaves, fresh, (2)

Ref No 2 04 120 United States

Feed Name or Analyses			Mean As Fed	Mean Dry	C.V. ± %
Dry matter	%			100.0	
Ash	%			6.2	41
Crude fiber	%			15.8	30
Ether extract	%			14.9	16
N-free extract	%			49.3	
Protein (N x 6.25)	%			13.8	24
Cattle	dig prot %	*		9.6	
Goats	dig prot %	*		9.4	
Horses	dig prot %	*		9.2	
Rabbits	dig prot %	*		9.3	
Sheep	dig prot %	*		9.9	
Energy	GE Mcal/kg				
Cattle	DE Mcal/kg	*		2.27	
Sheep	DE Mcal/kg	*		2.44	
Cattle	ME Mcal/kg	*		1.86	
Sheep	ME Mcal/kg	*		2.00	
Cattle	TDN %	*		51.4	
Sheep	TDN %	*		55.4	

SAGEBRUSH, BIRDFOOT. Artemisia pedatifida

Sagebrush, birdfoot, buds, fresh, (2)

Ref No 2 04 121 United States

Feed Name or Analyses			Mean As Fed	Mean Dry	C.V. ± %
Dry matter	%			100.0	
Ash	%			27.8	
Crude fiber	%			23.7	
Ether extract	%			2.4	
N-free extract	%			39.1	
Protein (N x 6.25)	%			7.0	
Cattle	dig prot %	*		3.8	
Goats	dig prot %	*		3.1	
Horses	dig prot %	*		3.5	
Rabbits	dig prot %	*		4.1	
Sheep	dig prot %	*		3.5	

SAGEBRUSH, BLACK. Artemisia nova

Sagebrush, black, browse, fresh, (2)

Ref No 2 04 122 United States

Feed Name or Analyses			Mean As Fed	Mean Dry	C.V. ± %
Dry matter	%			100.0	
Ash	%			6.1	18
Crude fiber	%			27.5	10
Ether extract	%			10.0	19
N-free extract	%			47.8	
Protein (N x 6.25)	%			8.6	8
Cattle	dig prot %	*		5.2	
Goats	dig prot %	*		4.6	
Horses	dig prot %	*		4.8	
Rabbits	dig prot %	*		5.3	
Sheep	dig prot %	*		5.0	
Calcium	%			0.65	50
Magnesium	%			0.29	13
Phosphorus	%			0.20	95
Sulphur	%			0.20	17
Carotene	mg/kg			17.6	47
Vitamin A equivalent	IU/g			29.4	

SAGEBRUSH, BROWN. Artemisia spp

Sagebrush, brown, browse, fresh, (2)

Ref No 2 04 123 United States

Feed Name or Analyses			Mean As Fed	Mean Dry	C.V. ± %
Dry matter	%			100.0	
Calcium	%			0.68	
Phosphorus	%			0.28	

SAGEBRUSH, BUD. Artemisia spinescens

Sagebrush, bud, browse, fresh, (2)

Ref No 2 04 125 United States

Feed Name or Analyses			Mean As Fed	Mean Dry	C.V. ± %
Dry matter	%			100.0	
Calcium	%			1.57	24
Magnesium	%			0.49	
Phosphorus	%			0.42	64
Sulphur	%			0.26	
Carotene	mg/kg			23.8	
Vitamin A equivalent	IU/g			39.7	

(1) dry forages and roughages (2) pasture, range plants, and forages fed green (3) silages (4) energy feeds (5) protein supplements (6) minerals (7) vitamins (8) additives

Sagebrush, bud, browse, fresh, immature, (2)

Ref No 2 04 124 United States

Feed Name or Analyses			Mean As Fed	Mean Dry	C.V. ± %
Dry matter	%			100.0	
Ash	%			21.6	5
Crude fiber	%			22.7	
Ether extract	%			2.5	62
N-free extract	%			35.7	
Protein (N x 6.25)	%			17.5	1
Cattle	dig prot %	*		12.8	
Goats	dig prot %	*		12.9	
Horses	dig prot %	*		12.4	
Rabbits	dig prot %	*		12.2	
Sheep	dig prot %	*		13.3	

Sagebrush, bud, buds, fresh, (2)

Ref No 2 04 126 United States

Feed Name or Analyses			Mean As Fed	Mean Dry	C.V. ± %
Dry matter	%			100.0	
Ash	%			15.6	51
Crude fiber	%			32.3	11
Ether extract	%			4.0	44
N-free extract	%			38.3	
Protein (N x 6.25)	%			9.8	22
Cattle	dig prot %	*		6.2	
Goats	dig prot %	*		5.7	
Horses	dig prot %	*		5.9	
Rabbits	dig prot %	*		6.2	
Sheep	dig prot %	*		6.1	

Sagebrush, bud, leaves, fresh, (2)

Ref No 2 04 127 United States

Feed Name or Analyses			Mean As Fed	Mean Dry	C.V. ± %
Dry matter	%			100.0	
Ash	%			16.4	
Crude fiber	%			20.1	
Ether extract	%			3.3	
N-free extract	%			41.7	
Protein (N x 6.25)	%			18.5	
Cattle	dig prot %	*		13.6	
Goats	dig prot %	*		13.8	
Horses	dig prot %	*		13.2	
Rabbits	dig prot %	*		13.0	
Sheep	dig prot %	*		14.2	

SAGEBRUSH, CUDWEED. Artemisia gnaphalodes

Sagebrush, cudweed, browse, fresh, (2)

Ref No 2 04 128 United States

Feed Name or Analyses			Mean As Fed	Mean Dry	C.V. ± %
Dry matter	%			100.0	
Ash	%			8.1	
Crude fiber	%			26.9	
Ether extract	%			3.1	
N-free extract	%			50.8	
Protein (N x 6.25)	%			11.1	
Cattle	dig prot %	*		7.3	
Goats	dig prot %	*		6.9	
Horses	dig prot %	*		7.0	
Rabbits	dig prot %	*		7.2	
Sheep	dig prot %	*		7.3	

SAGEBRUSH, FRINGED. Artemisia frigida

Sagebrush, fringed, browse, fresh, (2)

Ref No 2 04 131 United States

Feed Name or Analyses			Mean As Fed	Mean Dry	C.V. ± %
Dry matter	%			100.0	
Ash	%			14.3	53
Crude fiber	%			30.8	15
Ether extract	%			4.0	34
N-free extract	%			41.5	
Protein (N x 6.25)	%			9.4	37
Cattle	dig prot %	*		5.9	
Goats	dig prot %	*		5.3	
Horses	dig prot %	*		5.5	
Rabbits	dig prot %	*		5.9	
Sheep	dig prot %	*		5.8	
Calcium	%			0.60	
Phosphorus	%			0.29	

Sagebrush, fringed, browse, fresh, mid-bloom, (2)

Ref No 2 04 129 United States

Feed Name or Analyses			Mean As Fed	Mean Dry	C.V. ± %
Dry matter	%			100.0	
Ash	%			6.5	
Crude fiber	%			33.2	
Ether extract	%			2.0	
N-free extract	%			48.9	
Protein (N x 6.25)	%			9.4	
Cattle	dig prot %	*		5.9	
Goats	dig prot %	*		5.3	
Horses	dig prot %	*		5.5	
Rabbits	dig prot %	*		5.9	
Sheep	dig prot %	*		5.8	

Sagebrush, fringed, browse, fresh, mature, (2)

Ref No 2 04 130 United States

Feed Name or Analyses			Mean As Fed	Mean Dry	C.V. ± %
Dry matter	%			100.0	
Ash	%			17.1	
Crude fiber	%			31.8	
Ether extract	%			3.4	
N-free extract	%			42.3	
Protein (N x 6.25)	%			5.4	
Cattle	dig prot %	*		2.5	
Goats	dig prot %	*		1.6	
Horses	dig prot %	*		2.1	
Rabbits	dig prot %	*		2.8	
Sheep	dig prot %	*		2.0	

SAGEBRUSH, LOW. Artemisia arbuscula

Sagebrush, low, leaves, fresh, (2)

Ref No 2 04 132 United States

Feed Name or Analyses			Mean As Fed	Mean Dry	C.V. ± %
Dry matter	%			100.0	
Ash	%			6.8	
Crude fiber	%			20.2	
Ether extract	%			7.3	
N-free extract	%			57.1	
Protein (N x 6.25)	%			8.6	
Cattle	dig prot %	*		5.2	
Goats	dig prot %	*		4.6	
Horses	dig prot %	*		4.8	
Rabbits	dig prot %	*		5.3	
Sheep	dig prot %	*		5.0	

SAGEBRUSH, SAND. Artemisia filifolia

Sagebrush, sand, browse, fresh, immature, (2)
Ref No 2 04 133 — United States

Feed Name or Analyses	Mean As Fed	Mean Dry	C.V. ± %
Dry matter %	28.9	100.0	
Ash %	1.8	6.4	
Crude fiber %	6.5	22.6	
Ether extract %	1.6	5.4	
N-free extract %	15.4	53.4	
Protein (N x 6.25) %	3.5	12.2	
Cattle dig prot % *	2.4	8.3	
Goats dig prot % *	2.3	7.9	
Horses dig prot % *	2.3	7.9	
Rabbits dig prot % *	2.3	8.1	
Sheep dig prot % *	2.4	8.4	

Sagebrush, sand, browse, fresh, mid-bloom, (2)
Ref No 2 04 134 — United States

Feed Name or Analyses	Mean As Fed	Mean Dry	C.V. ± %
Dry matter %		100.0	
Ash %		6.8	
Crude fiber %		22.2	
Ether extract %		10.0	
N-free extract %		51.1	
Protein (N x 6.25) %		9.9	
Cattle dig prot % *		6.3	
Goats dig prot % *		5.8	
Horses dig prot % *		5.9	
Rabbits dig prot % *		6.3	
Sheep dig prot % *		6.2	
Calcium %		0.97	
Phosphorus %		0.21	

Sagebrush, sand, browse, fresh, mature, (2)
Ref No 2 04 135 — United States

Feed Name or Analyses	Mean As Fed	Mean Dry	C.V. ± %
Dry matter %		100.0	
Ash %		4.2	
Crude fiber %		31.7	
Ether extract %		3.8	
N-free extract %		53.1	
Protein (N x 6.25) %		7.2	
Cattle dig prot % *		4.0	
Goats dig prot % *		3.3	
Horses dig prot % *		3.6	
Rabbits dig prot % *		4.2	
Sheep dig prot % *		3.7	
Calcium %		0.48	
Phosphorus %		0.12	

SAGEBRUSH, STIFF. Artemisia rigida

Sagebrush, stiff, browse, fresh, immature, (2)
Ref No 2 04 137 — United States

Feed Name or Analyses	Mean As Fed	Mean Dry	C.V. ± %
Dry matter %	65.0	100.0	
Ash %	4.3	6.6	
Crude fiber %	14.3	22.0	
Ether extract %	2.4	3.7	
N-free extract %	30.3	46.6	
Protein (N x 6.25) %	13.7	21.1	
Cattle dig prot % *	10.3	15.8	
Goats dig prot % *	10.6	16.2	
Horses dig prot % *	10.0	15.4	
Rabbits dig prot % *	9.7	15.0	
Sheep dig prot % *	10.8	16.7	

SAGEBRUSH, THREETIP. Artemisia tripartita

Sagebrush, threetip, browse, fresh, immature, (2)
Ref No 2 04 138 — United States

Feed Name or Analyses	Mean As Fed	Mean Dry	C.V. ± %
Dry matter %		100.0	
N-free extract %		46.6	
Protein (N x 6.25) %		13.1	25
Cattle dig prot % *		9.0	
Goats dig prot % *		8.8	
Horses dig prot % *		8.7	
Rabbits dig prot % *		8.8	
Sheep dig prot % *		9.2	
Calcium %		0.79	11
Phosphorus %		0.21	25

Sagebrush, threetip, browse, fresh, over ripe, (2)
Ref No 2 04 139 — United States

Feed Name or Analyses	Mean As Fed	Mean Dry	C.V. ± %
Dry matter %		100.0	
Calcium %		0.49	
Phosphorus %		0.11	

SAGOPALM, SMOOTH. Metroxylon sagu

Sagopalm, smooth, pith, grnd, (4)
Ref No 4 04 140 — United States

Feed Name or Analyses	Mean As Fed	Mean Dry	C.V. ± %
Dry matter %	86.9	100.0	
Ash %	3.4	3.9	
Crude fiber %	4.3	5.0	
Swine dig coef %	-10.	-10.	
Ether extract %	0.3	0.4	
Swine dig coef %	46.	46.	
N-free extract %	77.2	88.8	
Swine dig coef %	86.	86.	
Protein (N x 6.25) %	1.7	1.9	
Swine dig coef %	-275.	-275.	
Cattle dig prot % *	-1.8	-2.1	
Goats dig prot % *	-0.8	-0.9	
Horses dig prot % *	-0.8	-0.9	
Sheep dig prot % *	-0.8	-0.9	
Swine dig prot %	-4.5	-5.1	
Energy GE Mcal/kg			
Swine DE kcal/kg	*2720.	3130.	
Swine ME kcal/kg	*2601.	2993.	
Swine TDN %	61.7	71.0	

SAINFOIN. Onobrychis spp

Sainfoin, hay, s-c, (1)
Ref No 1 04 143 — United States

Feed Name or Analyses	Mean As Fed	Mean Dry	C.V. ± %
Dry matter %	85.7	100.0	2
Ash %	6.5	7.6	11
Crude fiber %	22.7	26.5	14
Sheep dig coef % *	36.	36.	
Ether extract %	3.2	3.7	19
Sheep dig coef % *	66.	66.	
N-free extract %	39.1	45.6	
Sheep dig coef % *	74.	74.	
Protein (N x 6.25) %	14.2	16.6	23
Sheep dig coef % *	70.	70.	
Cattle dig prot % *	9.7	11.3	
Goats dig prot % *	10.3	12.0	
Horses dig prot % *	10.0	11.6	
Rabbits dig prot % *	9.8	11.5	
Sheep dig prot %	10.0	11.6	
Energy GE Mcal/kg			
Cattle DE Mcal/kg *	2.30	2.69	
Sheep DE Mcal/kg *	2.28	2.66	
Cattle ME Mcal/kg *	1.89	2.20	
Sheep ME Mcal/kg *	1.87	2.18	
Cattle TDN % *	52.2	60.9	
Sheep TDN % *	51.8	60.4	

Sainfoin, hay, s-c, early bloom, (1)
Ref No 1 04 141 — United States

Feed Name or Analyses	Mean As Fed	Mean Dry	C.V. ± %
Dry matter %	88.4	100.0	
Ash %	5.7	6.4	
Crude fiber %	24.9	28.2	
Sheep dig coef % *	36.	36.	
Ether extract %	3.4	3.8	
Sheep dig coef % *	66.	66.	
N-free extract %	34.5	39.0	
Sheep dig coef % *	74.	74.	
Protein (N x 6.25) %	20.0	22.6	
Sheep dig coef % *	70.	70.	
Cattle dig prot % *	14.6	16.5	
Goats dig prot % *	15.6	17.6	
Horses dig prot % *	14.8	16.7	
Rabbits dig prot % *	14.2	16.1	
Sheep dig prot %	14.0	15.8	
Energy GE Mcal/kg			
Cattle DE Mcal/kg *	2.27	2.56	
Sheep DE Mcal/kg *	2.36	2.67	
Cattle ME Mcal/kg *	1.86	2.10	
Sheep ME Mcal/kg *	1.93	2.19	
Cattle TDN % *	51.4	58.1	
Sheep TDN % *	53.5	60.5	

Sainfoin, hay, s-c, mid-bloom, (1)
Ref No 1 04 142 — United States

Feed Name or Analyses	Mean As Fed	Mean Dry	C.V. ± %
Dry matter %	87.0	100.0	
Ash %	6.1	7.0	
Crude fiber %	28.2	32.4	
Sheep dig coef % *	36.	36.	
Ether extract %	4.3	4.9	
Sheep dig coef % *	66.	66.	
N-free extract %	30.5	35.0	
Sheep dig coef % *	74.	74.	
Protein (N x 6.25) %	18.0	20.7	
Sheep dig coef % *	70.	70.	
Cattle dig prot % *	12.9	14.9	
Goats dig prot % *	13.8	15.9	
Horses dig prot % *	13.1	15.1	
Rabbits dig prot % *	12.7	14.7	
Sheep dig prot %	12.6	14.5	
Energy GE Mcal/kg			
Cattle DE Mcal/kg *	2.20	2.53	
Sheep DE Mcal/kg *	2.28	2.62	
Cattle ME Mcal/kg *	1.81	2.08	
Sheep ME Mcal/kg *	1.87	2.15	
Cattle TDN % *	50.0	57.4	
Sheep TDN % *	51.6	59.3	

Sainfoin, hay, s-c, brown, mid-bloom, (1)

Ref No 1 04 144 United States

Feed Name or Analyses				Mean As Fed	Mean Dry	C.V. ± %
Dry matter		%		87.0	100.0	
Ash		%		6.1	7.0	
Crude fiber		%		28.2	32.4	
Sheep	dig coef	%		45.	45.	
Ether extract		%		4.3	4.9	
Sheep	dig coef	%		76.	76.	
N-free extract		%		30.5	35.0	
Sheep	dig coef	%		67.	67.	
Protein (N x 6.25)		%		18.0	20.7	
Sheep	dig coef	%		64.	64.	
Cattle	dig prot	%	*	12.9	14.9	
Goats	dig prot	%	*	13.8	15.9	
Horses	dig prot	%	*	13.1	15.1	
Rabbits	dig prot	%	*	12.7	14.7	
Sheep	dig prot	%		11.5	13.2	
Energy	GE	Mcal/kg				
Cattle	DE	Mcal/kg	*	2.20	2.53	
Sheep	DE	Mcal/kg	*	2.29	2.63	
Cattle	ME	Mcal/kg	*	1.81	2.08	
Sheep	ME	Mcal/kg	*	1.88	2.16	
Cattle	TDN	%	*	50.0	57.4	
Sheep	TDN	%		51.9	59.7	

Sainfoin, aerial part, fresh, (2)

Ref No 2 04 146 United States

Feed Name or Analyses				Mean As Fed	Mean Dry	C.V. ± %
Dry matter		%		22.7	100.0	15
Ash		%		1.9	8.4	14
Crude fiber		%		5.8	25.5	6
Ether extract		%		0.7	3.3	8
N-free extract		%		10.3	45.4	
Protein (N x 6.25)		%		3.9	17.4	18
Cattle	dig prot	%	*	2.9	12.7	
Goats	dig prot	%	*	2.9	12.8	
Horses	dig prot	%	*	2.8	12.3	
Rabbits	dig prot	%	*	2.7	12.1	
Sheep	dig prot	%		3.0	13.2	
Energy	GE	Mcal/kg				
Cattle	DE	Mcal/kg	*	0.65	2.88	
Sheep	DE	Mcal/kg	*	0.62	2.74	
Cattle	ME	Mcal/kg	*	0.54	2.36	
Sheep	ME	Mcal/kg	*	0.51	2.25	
Cattle	TDN	%	*	14.8	65.4	
Sheep	TDN	%	*	14.1	62.2	

Sainfoin, aerial part, fresh, early bloom, (2)

Ref No 2 04 145 United States

Feed Name or Analyses				Mean As Fed	Mean Dry	C.V. ± %
Dry matter		%		16.8	100.0	
Ash		%		1.1	6.4	
Crude fiber		%		4.7	28.2	
Sheep	dig coef	%		42.	42.	
Ether extract		%		0.6	3.8	
Sheep	dig coef	%		67.	67.	
N-free extract		%		6.6	39.0	
Sheep	dig coef	%		78.	78.	
Protein (N x 6.25)		%		3.8	22.6	
Sheep	dig coef	%		72.	72.	
Cattle	dig prot	%	*	2.9	17.1	
Goats	dig prot	%	*	3.0	17.6	
Horses	dig prot	%	*	2.8	16.7	
Rabbits	dig prot	%	*	2.7	16.1	
Sheep	dig prot	%		2.7	16.3	
Energy	GE	Mcal/kg				
Cattle	DE	Mcal/kg	*	0.49	2.92	
Sheep	DE	Mcal/kg	*	0.48	2.83	
Cattle	ME	Mcal/kg	*	0.40	2.39	
Sheep	ME	Mcal/kg	*	0.39	2.32	
Cattle	TDN	%	*	11.1	66.2	
Sheep	TDN	%		10.8	64.3	

(1) dry forages and roughages (2) pasture, range plants, and forages fed green (3) silages (4) energy feeds (5) protein supplements (6) minerals (7) vitamins (8) additives

Sainfoin, aerial part, ensiled, mid-bloom, (3)

Ref No 3 04 147 United States

Feed Name or Analyses				Mean As Fed	Mean Dry	C.V. ± %
Dry matter		%		24.0	100.0	
Ash		%		1.8	7.5	
Crude fiber		%		8.4	35.2	
Sheep	dig coef	%		29.	29.	
Ether extract		%		1.4	6.0	
Sheep	dig coef	%		74.	74.	
N-free extract		%		7.4	30.9	
Sheep	dig coef	%		53.	53.	
Protein (N x 6.25)		%		4.9	20.4	
Sheep	dig coef	%		50.	50.	
Cattle	dig prot	%	*	3.5	14.8	
Goats	dig prot	%	*	3.5	14.8	
Horses	dig prot	%	*	3.5	14.8	
Sheep	dig prot	%		2.4	10.2	
Energy	GE	Mcal/kg				
Cattle	DE	Mcal/kg	*	0.52	2.16	
Sheep	DE	Mcal/kg	*	0.49	2.06	
Cattle	ME	Mcal/kg	*	0.42	1.77	
Sheep	ME	Mcal/kg	*	0.41	1.69	
Cattle	TDN	%	*	11.8	49.0	
Sheep	TDN	%		11.2	46.8	

SAINFOIN, COMMON. Onobrychis viciaefolia

Sainfoin, common, hay, s-c, (1)

Ref No 1 08 502 United States

Feed Name or Analyses				Mean As Fed	Mean Dry	C.V. ± %
Dry matter		%		84.1	100.0	
Ash		%		7.1	8.4	
Crude fiber		%		19.7	23.4	
Ether extract		%		2.6	3.1	
N-free extract		%		44.2	52.6	
Protein (N x 6.25)		%		10.5	12.5	
Cattle	dig prot	%	*	6.5	7.8	
Goats	dig prot	%	*	6.9	8.2	
Horses	dig prot	%	*	6.8	8.1	
Rabbits	dig prot	%	*	7.0	8.3	
Sheep	dig prot	%	*	6.5	7.8	
Energy	GE	Mcal/kg				
Cattle	DE	Mcal/kg	*	2.24	2.66	
Sheep	DE	Mcal/kg	*	2.14	2.54	
Cattle	ME	Mcal/kg	*	1.83	2.18	
Sheep	ME	Mcal/kg	*	1.75	2.08	
Cattle	TDN	%	*	50.7	60.3	
Sheep	TDN	%	*	48.4	57.5	

Sainfoin, common, hay, s-c, early bloom, (1)

Ref No 1 04 148 United States

Feed Name or Analyses				Mean As Fed	Mean Dry	C.V. ± %
Dry matter		%		88.4	100.0	
Ash		%		5.7	6.4	
Crude fiber		%		24.9	28.2	
Sheep	dig coef	%		36.	36.	
Ether extract		%		3.4	3.8	
Sheep	dig coef	%		66.	66.	
N-free extract		%		34.5	39.0	
Sheep	dig coef	%		74.	74.	
Protein (N x 6.25)		%		20.0	22.6	
Sheep	dig coef	%		70.	70.	
Cattle	dig prot	%	*	14.6	16.5	
Goats	dig prot	%	*	15.6	17.6	
Horses	dig prot	%	*	14.8	16.7	
Rabbits	dig prot	%	*	14.2	16.1	
Sheep	dig prot	%		14.0	15.8	
Energy	GE	Mcal/kg				
Cattle	DE	Mcal/kg	*	2.27	2.56	
Sheep	DE	Mcal/kg	*	2.36	2.67	
Cattle	ME	Mcal/kg	*	1.86	2.10	
Sheep	ME	Mcal/kg	*	1.93	2.19	
Cattle	TDN	%	*	51.4	58.1	
Sheep	TDN	%		53.5	60.5	

Sainfoin, common, aerial part, fresh, (2)

Ref No 2 08 503 United States

Feed Name or Analyses				Mean As Fed	Mean Dry	C.V. ± %
Dry matter		%		25.6	100.0	
Ash		%		2.4	9.4	
Crude fiber		%		6.2	24.2	
Ether extract		%		0.8	3.1	
N-free extract		%		12.4	48.4	
Protein (N x 6.25)		%		3.8	14.8	
Cattle	dig prot	%	*	2.7	10.5	
Goats	dig prot	%	*	2.7	10.4	
Horses	dig prot	%	*	2.6	10.1	
Rabbits	dig prot	%	*	2.6	10.1	
Sheep	dig prot	%	*	2.8	10.8	
Energy	GE	Mcal/kg				
Cattle	DE	Mcal/kg	*	0.76	2.95	
Sheep	DE	Mcal/kg	*	0.73	2.84	
Cattle	ME	Mcal/kg	*	0.62	2.42	
Sheep	ME	Mcal/kg	*	0.60	2.33	
Cattle	TDN	%	*	17.1	67.0	
Sheep	TDN	%	*	16.5	64.5	

SAINT AUGUSTINEGRASS. Stenotaphrum secundatum

Saint Augustinegrass, hay, s-c, (1)

Ref No 1 04 150 United States

Feed Name or Analyses				Mean As Fed	Mean Dry	C.V. ± %
Dry matter		%		90.5	100.0	
Ash		%		7.0	7.7	
Crude fiber		%		25.4	28.1	
Ether extract		%		3.3	3.7	
N-free extract		%		43.9	48.5	
Protein (N x 6.25)		%		10.9	12.0	
Cattle	dig prot	%	*	6.6	7.3	
Goats	dig prot	%	*	7.0	7.8	
Horses	dig prot	%	*	7.0	7.7	
Rabbits	dig prot	%	*	7.2	7.9	
Sheep	dig prot	%	*	6.6	7.3	
Calcium		%		0.64	0.71	
Magnesium		%		0.39	0.43	
Phosphorus		%		0.20	0.22	

Feed Name or Analyses			Mean As Fed	Mean Dry	C.V. ± %
Saint Augustinegrass, hay, s-c, immature, (1)					
Ref No 1 04 149				United States	
Dry matter	%		90.0	100.0	
Calcium	%		0.67	0.74	
Magnesium	%		0.39	0.43	
Phosphorus	%		0.23	0.26	

Feed Name or Analyses			Mean As Fed	Mean Dry	C.V. ± %
Saint Augustinegrass, aerial part, fresh, (2)					
Ref No 2 08 090				United States	
Dry matter	%		18.1	100.0	
Ash	%		1.3	7.2	
Crude fiber	%		5.4	29.8	
Ether extract	%		0.5	2.8	
N-free extract	%		8.2	45.3	
Protein (N x 6.25)	%		2.7	14.9	
Cattle	dig prot %	*	1.9	10.6	
Goats	dig prot %	*	1.9	10.5	
Horses	dig prot %	*	1.8	10.2	
Rabbits	dig prot %	*	1.8	10.2	
Sheep	dig prot %	*	2.0	10.9	
Energy	GE Mcal/kg				
Cattle	DE Mcal/kg	*	0.52	2.87	
Sheep	DE Mcal/kg	*	0.50	2.75	
Cattle	ME Mcal/kg	*	0.43	23.5	
Sheep	ME Mcal/kg	*	0.41	2.26	
Cattle	TDN %	*	11.8	65.2	
Sheep	TDN %	*	11.3	62.4	

Feed Name or Analyses			Mean As Fed	Mean Dry	C.V. ± %
Saint Augustinegrass, aerial part, fresh, early bloom, (2)					
Ref No 2 04 676				United States	
Dry matter	%		28.0	100.0	27
Ash	%		2.2	7.7	17
Crude fiber	%		7.6	27.1	10
Ether extract	%		0.7	2.6	26
N-free extract	%		13.6	48.7	
Protein (N x 6.25)	%		3.9	13.9	10
Cattle	dig prot %	*	2.7	9.7	
Goats	dig prot %	*	2.7	9.5	
Horses	dig prot %	*	2.6	9.3	
Rabbits	dig prot %	*	2.6	9.4	
Sheep	dig prot %	*	2.8	9.9	
Energy	GE Mcal/kg				
Cattle	DE Mcal/kg	*	0.79	2.78	
Sheep	DE Mcal/kg	*	0.76	2.70	
Cattle	ME Mcal/kg	*	0.64	2.28	
Sheep	ME Mcal/kg	*	0.62	2.21	
Cattle	TDN %	*	17.6	63.0	
Sheep	TDN %	*	17.1	61.2	
Calcium	%		0.14	0.51	25
Iron	%		0.006	0.020	
Magnesium	%		0.11	0.39	35
Phosphorus	%		0.06	0.21	29
Potassium	%		0.18	0.63	

Feed Name or Analyses			Mean As Fed	Mean Dry	C.V. ± %
Saint Augustinegrass, aerial part, fresh, mature, (2)					
Ref No 2 04 677				United States	
Dry matter	%		30.5	100.0	13
Ash	%		2.3	7.6	20
Crude fiber	%		8.8	28.9	9
Ether extract	%		0.3	1.1	25
N-free extract	%		17.6	57.6	
Protein (N x 6.25)	%		1.5	4.8	22
Cattle	dig prot %	*	0.6	2.0	
Goats	dig prot %	*	0.3	1.0	
Horses	dig prot %	*	0.5	1.6	
Rabbits	dig prot %	*	0.7	2.4	
Sheep	dig prot %	*	0.4	1.5	
Energy	GE Mcal/kg				
Cattle	DE Mcal/kg	*	0.87	2.84	
Sheep	DE Mcal/kg	*	0.84	2.75	
Cattle	ME Mcal/kg	*	0.71	2.33	
Sheep	ME Mcal/kg	*	0.69	2.26	
Cattle	TDN %	*	19.7	64.5	
Sheep	TDN %	*	19.0	62.3	

SALTBUSH. Atriplex spp

Feed Name or Analyses			Mean As Fed	Mean Dry	C.V. ± %
Saltbush, browse, s-c, (1)					
Ref No 1 08 504				United States	
Dry matter	%		93.5	100.0	
Ash	%		17.2	18.4	
Crude fiber	%		22.1	23.6	
Ether extract	%		1.6	1.7	
N-free extract	%		38.8	41.5	
Protein (N x 6.25)	%		13.8	14.8	
Cattle	dig prot %	*	9.1	9.7	
Goats	dig prot %	*	9.7	10.3	
Horses	dig prot %	*	9.4	10.1	
Rabbits	dig prot %	*	9.4	10.1	
Sheep	dig prot %	*	9.2	9.8	
Energy	GE Mcal/kg				
Cattle	DE Mcal/kg	*	1.68	1.80	
Sheep	DE Mcal/kg	*	1.51	1.62	
Cattle	ME Mcal/kg	*	1.38	1.48	
Sheep	ME Mcal/kg	*	1.24	1.33	
Cattle	TDN %	*	38.1	40.8	
Sheep	TDN %	*	34.3	36.7	
Calcium	%		1.88	2.01	
Phosphorus	%		0.11	0.12	
Potassium	%		4.69	5.02	

Feed Name or Analyses			Mean As Fed	Mean Dry	C.V. ± %
Saltbush, browse, s-c, dormant, (1)					
Ref No 1 08 505				United States	
Dry matter	%		90.0	100.0	
Ash	%		16.0	17.8	
Crude fiber	%		25.8	28.7	
Ether extract	%		4.2	4.7	
N-free extract	%		36.8	40.9	
Protein (N x 6.25)	%		7.2	8.0	
Cattle	dig prot %	*	3.5	3.9	
Goats	dig prot %	*	3.6	4.0	
Horses	dig prot %	*	3.9	4.3	
Rabbits	dig prot %	*	4.4	4.8	
Sheep	dig prot %	*	3.4	3.7	
Energy	GE Mcal/kg				
Cattle	DE Mcal/kg	*	1.91	2.12	
Sheep	DE Mcal/kg	*	1.72	1.91	
Cattle	ME Mcal/kg	*	1.57	1.74	
Sheep	ME Mcal/kg	*	1.42	1.57	
Cattle	TDN %	*	43.3	48.1	
Sheep	TDN %	*	39.0	43.3	
Calcium	%		2.13	2.37	
Phosphorus	%		0.09	0.10	

Feed Name or Analyses			Mean As Fed	Mean Dry	C.V. ± %
Saltbush, browse, fresh, (2)					
Ref No 2 04 155				United States	
Dry matter	%		30.7	100.0	52
Ash	%		5.6	18.3	
Crude fiber	%		8.3	27.0	
Ether extract	%		0.5	1.7	
N-free extract	%		13.3	43.3	
Protein (N x 6.25)	%		3.0	9.7	
Cattle	dig prot %	*	1.9	6.1	
Goats	dig prot %	*	1.7	5.6	
Horses	dig prot %	*	1.8	5.8	
Rabbits	dig prot %	*	1.9	6.1	
Sheep	dig prot %	*	1.8	6.0	
Energy	GE Mcal/kg				
Cattle	DE Mcal/kg	*	0.45	1.47	
Sheep	DE Mcal/kg	*	0.40	1.31	
Cattle	ME Mcal/kg	*	0.37	1.21	
Sheep	ME Mcal/kg	*	0.33	1.07	
Cattle	TDN %	*	10.3	33.4	
Sheep	TDN %	*	9.1	29.6	
Calcium	%		0.62	2.01	
Chlorine	%		0.02	0.08	
Cobalt	mg/kg		0.112	0.364	
Phosphorus	%		0.03	0.10	
Sodium	%		0.06	0.18	
Sulphur	%		0.10	0.32	26
Carotene	mg/kg		7.2	23.4	38
Vitamin A equivalent	IU/g		12.0	39.0	

Feed Name or Analyses			Mean As Fed	Mean Dry	C.V. ± %
Saltbush, browse, fresh, over ripe, (2)					
Ref No 2 04 154				United States	
Dry matter	%			100.0	
Calcium	%			1.39	
Phosphorus	%			0.09	

Feed Name or Analyses			Mean As Fed	Mean Dry	C.V. ± %
Saltbush, browse, fresh, dormant, (2)					
Ref No 2 08 506				United States	
Dry matter	%		85.0	100.0	
Ash	%		15.1	17.8	
Crude fiber	%		24.4	28.7	
Ether extract	%		4.0	4.7	
N-free extract	%		34.8	40.9	
Protein (N x 6.25)	%		6.7	7.9	
Cattle	dig prot %	*	3.9	4.6	
Goats	dig prot %	*	3.3	3.9	
Horses	dig prot %	*	3.6	4.2	
Rabbits	dig prot %	*	4.0	4.8	
Sheep	dig prot %	*	3.7	4.3	
Energy	GE Mcal/kg				
Cattle	DE Mcal/kg	*	1.78	2.09	
Sheep	DE Mcal/kg	*	1.71	2.01	
Cattle	ME Mcal/kg	*	1.45	1.71	
Sheep	ME Mcal/kg	*	1.40	1.65	
Cattle	TDN %	*	40.4	47.5	
Sheep	TDN %	*	38.8	45.7	
Calcium	%		2.01	2.36	
Phosphorus	%		0.08	0.09	

Saltbush, leaves, fresh, (2)
Ref No 2 04 156 United States

Feed Name or Analyses			Mean As Fed	Mean Dry	C.V. ± %
Dry matter	%		28.0	100.0	12
Ash	%		7.6	27.0	12
Crude fiber	%		3.3	11.9	17
Ether extract	%		0.6	2.2	19
N-free extract	%		12.3	43.8	
Protein (N x 6.25)	%		4.3	15.2	31
Cattle	dig prot %	*	3.0	10.8	
Goats	dig prot %	*	3.0	10.7	
Horses	dig prot %	*	2.9	10.4	
Rabbits	dig prot %	*	2.9	10.4	
Sheep	dig prot %	*	3.1	11.2	
Energy	GE Mcal/kg				
Cattle	DE Mcal/kg	*	0.50	1.77	
Sheep	DE Mcal/kg	*	0.56	2.01	
Cattle	ME Mcal/kg	*	0.41	1.46	
Sheep	ME Mcal/kg	*	0.46	1.65	
Cattle	TDN %	*	11.3	40.2	
Sheep	TDN %	*	12.7	45.5	

SALTBUSH, AUSTRALIAN. Atriplex semibaccata

Saltbush, Australian, browse, s-c, (1)
Ref No 1 04 157 United States

Feed Name or Analyses			Mean As Fed	Mean Dry	C.V. ± %
Dry matter	%		96.4	100.0	
Ash	%		18.6	19.3	
Crude fiber	%		16.4	17.0	
Sheep	dig coef %		27.	27.	
Ether extract	%		1.3	1.4	
Sheep	dig coef %		16.	16.	
N-free extract	%		39.4	40.9	
Sheep	dig coef %		61.	61.	
Protein (N x 6.25)	%		20.6	21.4	
Sheep	dig coef %		83.	83.	
Cattle	dig prot %	*	14.9	15.5	
Goats	dig prot %	*	15.9	16.5	
Horses	dig prot %	*	15.1	15.7	
Rabbits	dig prot %	*	14.6	15.2	
Sheep	dig prot %		17.1	17.8	
Energy	GE Mcal/kg				
Cattle	DE Mcal/kg	*	1.92	1.99	
Sheep	DE Mcal/kg	*	2.03	2.11	
Cattle	ME Mcal/kg	*	1.57	1.63	
Sheep	ME Mcal/kg	*	1.67	1.73	
Cattle	TDN %	*	43.6	45.2	
Sheep	TDN %		46.1	47.8	

Saltbush, Australian, browse, fresh, (2)
Ref No 2 04 158 United States

Feed Name or Analyses			Mean As Fed	Mean Dry	C.V. ± %
Dry matter	%		23.3	100.0	
Ash	%		5.4	23.2	
Crude fiber	%		4.4	18.9	
Ether extract	%		0.4	1.7	
N-free extract	%		9.4	40.3	
Protein (N x 6.25)	%		3.7	15.9	
Cattle	dig prot %	*	2.7	11.4	
Goats	dig prot %	*	2.7	11.4	
Horses	dig prot %	*	2.6	11.0	
Rabbits	dig prot %	*	2.5	10.9	
Sheep	dig prot %	*	2.7	11.8	
Energy	GE Mcal/kg				
Cattle	DE Mcal/kg	*	0.50	2.15	
Sheep	DE Mcal/kg	*	0.50	2.16	
Cattle	ME Mcal/kg	*	0.41	1.76	
Sheep	ME Mcal/kg	*	0.41	1.77	
Cattle	TDN %	*	11.4	48.7	
Sheep	TDN %	*	11.4	49.0	
Calcium	%		0.19	0.82	
Phosphorus	%		0.03	0.13	

SALTBUSH, FOURWING. Atriplex canescens

Saltbush, fourwing, browse, fresh, (2)
Ref No 2 04 160 United States

Feed Name or Analyses			Mean As Fed	Mean Dry	C.V. ± %
Dry matter	%		40.9	100.0	
Ash	%		7.6	18.6	30
Crude fiber	%		5.9	14.5	18
Cattle	dig coef %		4.	4.	
Ether extract	%		1.1	2.8	16
Cattle	dig coef %		-69.	-69.	
N-free extract	%		18.3	44.8	
Cattle	dig coef %		62.	62.	
Protein (N x 6.25)	%		7.9	19.4	22
Cattle	dig coef %		77.	77.	
Cattle	dig prot %		6.1	14.9	
Goats	dig prot %	*	6.0	14.6	
Horses	dig prot %	*	5.7	14.0	
Rabbits	dig prot %	*	5.6	13.6	
Sheep	dig prot %	*	6.1	15.0	
Energy	GE Mcal/kg				
Cattle	DE Mcal/kg	*	0.70	1.71	
Sheep	DE Mcal/kg	*	0.76	1.86	
Cattle	ME Mcal/kg	*	0.57	1.40	
Sheep	ME Mcal/kg	*	0.63	1.53	
Cattle	TDN %		15.9	38.8	
Sheep	TDN %	*	17.2	42.1	
Calcium	%		0.49	1.19	26
Chlorine	%		0.03	0.08	
Iron	%		0.004	0.010	
Magnesium	%		0.24	0.58	25
Manganese	mg/kg		63.3	154.8	
Phosphorus	%		0.06	0.15	18
Potassium	%		0.33	0.81	
Sodium	%		0.07	0.18	
Sulphur	%		0.17	0.41	11

Saltbush, fourwing, browse, fresh, immature, (2)
Ref No 2 04 159 United States

Feed Name or Analyses			Mean As Fed	Mean Dry	C.V. ± %
Dry matter	%			100.0	
Ash	%			10.5	6
Crude fiber	%			16.3	17
Ether extract	%			1.9	3
N-free extract	%			57.2	
Protein (N x 6.25)	%			14.1	24
Cattle	dig prot %	*		9.9	
Goats	dig prot %	*		9.7	
Horses	dig prot %	*		9.5	
Rabbits	dig prot %	*		9.6	
Sheep	dig prot %	*		10.1	

SALTBUSH, GARDNER. Atriplex gardneri

Saltbush, gardner, browse, fresh, mature, (2)
Ref No 2 04 161 United States

Feed Name or Analyses			Mean As Fed	Mean Dry	C.V. ± %
Dry matter	%		64.6	100.0	
Ash	%		10.1	15.6	26
Crude fiber	%		19.8	30.6	17
Ether extract	%		1.9	3.0	67
N-free extract	%		27.4	42.4	
Protein (N x 6.25)	%		5.4	8.4	17
Cattle	dig prot %	*	3.2	5.0	
Goats	dig prot %	*	2.8	4.4	
Horses	dig prot %	*	3.0	4.7	
Rabbits	dig prot %	*	3.3	5.2	
Sheep	dig prot %	*	3.1	4.8	

Saltbush, gardner, leaves, fresh, (2)
Ref No 2 04 162 United States

Feed Name or Analyses			Mean As Fed	Mean Dry	C.V. ± %
Dry matter	%			100.0	
Ash	%			24.1	16
Crude fiber	%			11.3	15
Ether extract	%			2.7	18
N-free extract	%			45.0	
Protein (N x 6.25)	%			16.9	9
Cattle	dig prot %	*		12.3	
Goats	dig prot %	*		12.3	
Horses	dig prot %	*		11.9	
Rabbits	dig prot %	*		11.7	
Sheep	dig prot %	*		12.7	

Saltbush, gardner, stems, fresh, (2)
Ref No 2 04 163 United States

Feed Name or Analyses			Mean As Fed	Mean Dry	C.V. ± %
Dry matter	%			100.0	
Ash	%			7.7	
Crude fiber	%			35.7	
Ether extract	%			5.0	
N-free extract	%			43.9	
Protein (N x 6.25)	%			7.7	
Cattle	dig prot %	*		4.4	
Goats	dig prot %	*		3.7	
Horses	dig prot %	*		4.1	
Rabbits	dig prot %	*		4.6	
Sheep	dig prot %	*		4.2	

SALTBUSH, NUTTALL. Atriplex nuttallii

Saltbush, nuttall, browse, fresh, (2)
Ref No 2 04 164 United States

Feed Name or Analyses			Mean As Fed	Mean Dry	C.V. ± %
Dry matter	%			100.0	
Calcium	%			3.27	
Phosphorus	%			0.13	

SALTBUSH, SHADSCALE. Atriplex confertifolia

Saltbush, shadscale, browse, fresh, (2)
Ref No 2 04 165 United States

Feed Name or Analyses			Mean As Fed	Mean Dry	C.V. ± %
Dry matter	%			100.0	
Carotene	mg/kg			24.3	41
Vitamin A equivalent	IU/g			40.4	

(1) dry forages and roughages (3) silages (6) minerals
(2) pasture, range plants, and forages fed green (4) energy feeds (7) vitamins
(5) protein supplements (8) additives

Feed Name or Analyses			Mean As Fed	Mean Dry	C.V. ± %

Saltbush, shadscale, browse, fresh, dormant, (2)
Ref No 2 05 565 — United States

Feed Name or Analyses			As Fed	Dry	C.V. ± %
Dry matter	%			100.0	
Ash	%			25.4	
Ether extract	%			2.6	
Sheep	dig coef %			23.	
Protein (N x 6.25)	%			7.1	
Sheep	dig coef %			51.	
Cattle	dig prot %	*		3.9	
Goats	dig prot %	*		3.2	
Horses	dig prot %	*		3.6	
Rabbits	dig prot %	*		4.2	
Sheep	dig prot %			3.6	
Cellulose (Matrone)	%			15.1	
Sheep	dig coef %			33.	
Other carbohydrates	%			37.1	
Lignin (Ellis)	%			12.9	
Energy	GE Mcal/kg			3.50	
Sheep	GE dig coef %			39.	
Sheep	DE Mcal/kg			1.35	
Sheep	ME Mcal/kg			0.85	
Cattle	NE_m Mcal/kg	*		0.66	
Cattle	NE_{gain} Mcal/kg	*		0.00	
Sheep	TDN %			33.4	
Calcium	%			2.41	
Phosphorus	%			0.08	

Saltbush, shadscale, leaves, fresh, (2)
Ref No 2 04 166 — United States

Feed Name or Analyses			As Fed	Dry	C.V. ± %
Dry matter	%			100.0	
Ash	%			31.1	9
Crude fiber	%			10.8	28
Ether extract	%			1.9	20
N-free extract	%			41.3	
Protein (N x 6.25)	%			14.9	46
Cattle	dig prot %	*		10.6	
Goats	dig prot %	*		10.5	
Horses	dig prot %	*		10.2	
Rabbits	dig prot %	*		10.2	
Sheep	dig prot %	*		10.9	
Energy	GE Mcal/kg				
Cattle	DE Mcal/kg	*		1.50	
Sheep	DE Mcal/kg	*		1.60	
Cattle	ME Mcal/kg	*		1.23	
Sheep	ME Mcal/kg	*		1.31	
Cattle	TDN %	*		34.0	
Sheep	TDN %	*		36.3	

SALTBUSH, SILVERY. Atriplex argentea

Saltbush, silvery, browse, s-c, (1)
Ref No 1 04 167 — United States

Feed Name or Analyses			As Fed	Dry	C.V. ± %
Dry matter	%		94.7	100.0	
Ash	%		19.3	20.4	
Crude fiber	%		27.4	28.9	
Sheep	dig coef %		8.	8.	
Ether extract	%		1.4	1.5	
Sheep	dig coef %		52.	52.	
N-free extract	%		36.8	38.9	
Sheep	dig coef %		49.	49.	
Protein (N x 6.25)	%		9.8	10.3	
Sheep	dig coef %		66.	66.	
Cattle	dig prot %	*	5.5	5.9	
Goats	dig prot %	*	5.8	6.2	
Horses	dig prot %	*	5.9	6.3	
Rabbits	dig prot %	*	6.3	6.6	
Sheep	dig prot %		6.4	6.8	
Energy	GE Mcal/kg				
Cattle	DE Mcal/kg	*	1.39	1.47	
Sheep	DE Mcal/kg	*	1.25	1.32	
Cattle	ME Mcal/kg	*	1.15	1.21	
Sheep	ME Mcal/kg	*	1.02	1.08	
Cattle	TDN %	*	31.6	33.4	
Sheep	TDN %		28.3	29.9	

SALTGRASS. Distichlis spp

Saltgrass, hay, s-c, (1)
Ref No 1 04 168 — United States

Feed Name or Analyses			As Fed	Dry	C.V. ± %
Dry matter	%		89.2	100.0	5
Ash	%		11.3	12.7	15
Crude fiber	%		28.2	31.6	6
Ether extract	%		1.8	2.1	14
N-free extract	%		39.9	44.7	
Protein (N x 6.25)	%		7.9	8.9	5
Cattle	dig prot %	*	4.1	4.6	
Goats	dig prot %	*	4.3	4.9	
Horses	dig prot %	*	4.5	5.1	
Rabbits	dig prot %	*	4.9	5.5	
Sheep	dig prot %	*	4.1	4.6	
Fatty acids	%		1.8	2.0	
Energy	GE Mcal/kg				
Cattle	DE Mcal/kg	*	2.11	2.36	
Sheep	DE Mcal/kg	*	1.93	2.16	
Cattle	ME Mcal/kg	*	1.73	1.94	
Sheep	ME Mcal/kg	*	1.58	1.77	
Cattle	NE_m Mcal/kg	*	1.29	1.45	
Cattle	NE_{gain} Mcal/kg	*	0.78	0.87	
Cattle	TDN %	*	47.7	53.5	
Sheep	TDN %	*	43.6	48.7	
Calcium	%		2.11	2.37	
Phosphorus	%		0.09	0.10	

Saltgrass, aerial part, fresh, over ripe, (2)
Ref No 2 04 169 — United States

Feed Name or Analyses			As Fed	Dry	C.V. ± %
Dry matter	%		74.4	100.0	
Ash	%		5.4	7.3	34
Crude fiber	%		26.0	34.9	5
Ether extract	%		1.9	2.6	13
N-free extract	%		37.9	51.0	
Protein (N x 6.25)	%		3.1	4.2	22
Cattle	dig prot %	*	1.1	1.5	
Goats	dig prot %	*	0.4	0.5	
Horses	dig prot %	*	0.8	1.1	
Rabbits	dig prot %	*	1.4	1.9	
Sheep	dig prot %	*	0.7	0.9	
Energy	GE Mcal/kg				
Cattle	DE Mcal/kg	*	1.69	2.27	
Sheep	DE Mcal/kg	*	1.61	2.16	
Cattle	ME Mcal/kg	*	1.38	1.86	
Sheep	ME Mcal/kg	*	1.32	1.77	
Cattle	NE_m Mcal/kg	*	0.97	1.31	
Cattle	NE_{gain} Mcal/kg	*	0.52	0.70	
Cattle	TDN %	*	38.2	51.4	
Sheep	TDN %	*	36.4	48.9	
Calcium	%		0.17	0.23	19
Magnesium	%		0.22	0.30	
Phosphorus	%		0.05	0.07	32

SALTGRASS, DESERT. Distichlis stricta

Saltgrass, desert, aerial part, fresh, (2)
Ref No 2 04 171 — United States

Feed Name or Analyses			As Fed	Dry	C.V. ± %
Dry matter	%			100.0	
Ash	%			6.8	9
Crude fiber	%			29.7	1
Ether extract	%			1.7	5
N-free extract	%			55.9	
Protein (N x 6.25)	%			5.9	9
Cattle	dig prot %	*		2.9	
Goats	dig prot %	*		2.1	
Horses	dig prot %	*		2.5	
Rabbits	dig prot %	*		3.2	
Sheep	dig prot %	*		2.5	
Calcium	%			0.16	24
Phosphorus	%			0.09	25

SALTGRASS, SEASHORE. Distichlis spicata

Saltgrass, seashore, hay, s-c, (1)
Ref No 1 04 172 — United States

Feed Name or Analyses			As Fed	Dry	C.V. ± %
Dry matter	%		79.0	100.0	
Ash	%		6.2	7.8	
Crude fiber	%		20.9	26.5	
Sheep	dig coef %		25.	25.	
Ether extract	%		2.2	2.8	
Sheep	dig coef %		37.	37.	
N-free extract	%		43.5	55.0	
Sheep	dig coef %		46.	46.	
Protein (N x 6.25)	%		6.2	7.9	
Sheep	dig coef %		52.	52.	
Cattle	dig prot %	*	3.0	3.8	
Goats	dig prot %	*	3.1	3.9	
Horses	dig prot %	*	3.3	4.2	
Rabbits	dig prot %	*	3.8	4.8	
Sheep	dig prot %		3.2	4.1	
Energy	GE Mcal/kg				
Cattle	DE Mcal/kg	*	1.48	1.87	
Sheep	DE Mcal/kg	*	1.34	1.69	
Cattle	ME Mcal/kg	*	1.21	1.53	
Sheep	ME Mcal/kg	*	1.10	1.39	
Cattle	TDN %	*	33.6	42.5	
Sheep	TDN %		30.3	38.4	

Saltgrass, seashore, aerial part, fresh, (2)
Ref No 2 04 175 — United States

Feed Name or Analyses			As Fed	Dry	C.V. ± %
Dry matter	%		74.4	100.0	
Ash	%		6.9	9.3	13
Crude fiber	%		22.2	29.8	9
Ether extract	%		1.3	1.7	4
N-free extract	%		38.9	52.3	
Protein (N x 6.25)	%		5.1	6.9	34
Cattle	dig prot %	*	2.8	3.8	
Goats	dig prot %	*	2.2	3.0	
Horses	dig prot %	*	2.5	3.4	
Rabbits	dig prot %	*	3.0	4.0	
Sheep	dig prot %	*	2.5	3.4	
Energy	GE Mcal/kg				
Cattle	DE Mcal/kg	*	1.39	1.87	
Sheep	DE Mcal/kg	*	1.32	1.78	
Cattle	ME Mcal/kg	*	1.14	1.53	
Sheep	ME Mcal/kg	*	1.09	1.46	

Continued

Feed Name or Analyses			Mean As Fed	Mean Dry	C.V. ± %
Cattle	TDN %	*	31.6	42.5	
Sheep	TDN %	*	30.0	40.3	
Calcium	%		0.25	0.33	49
Chlorine	%		0.13	0.17	
Iron	%		0.014	0.019	
Magnesium	%		0.16	0.22	
Manganese	mg/kg		115.1	154.8	
Phosphorus	%		0.09	0.12	43
Potassium	%		0.18	0.24	
Sodium	%		0.16	0.22	
Sulphur	%		0.18	0.24	

Saltgrass, seashore, aerial part, fresh, full bloom, (2)
Ref No 2 04 173 United States

Feed Name or Analyses			As Fed	Dry	C.V. ± %
Dry matter	%			100.0	
Ash	%			9.7	
Crude fiber	%			27.7	
Ether extract	%			1.8	
N-free extract	%			53.0	
Protein (N x 6.25)	%			7.8	
Cattle	dig prot %	*		4.5	
Goats	dig prot %	*		3.8	
Horses	dig prot %	*		4.2	
Rabbits	dig prot %	*		4.7	
Sheep	dig prot %	*		4.3	
Calcium	%			0.63	
Phosphorus	%			0.17	

Saltgrass, seashore, aerial part, fresh, over ripe, (2)
Ref No 2 04 174 United States

Feed Name or Analyses			As Fed	Dry	C.V. ± %
Dry matter	%		74.4	100.0	
Ash	%		8.6	11.6	
Protein (N x 6.25)	%		2.1	2.8	
Cattle	dig prot %	*	0.2	0.3	
Goats	dig prot %	*	-0.5	-0.7	
Horses	dig prot %	*	0.0	0.0	
Rabbits	dig prot %	*	0.6	0.8	
Sheep	dig prot %	*	-0.2	-0.3	

SALTMARSHMALLOW. Kosteletzkya spp

Saltmarshmallow, hay, s-c, (1)
Ref No 1 04 176 United States

Feed Name or Analyses			As Fed	Dry	C.V. ± %
Dry matter	%		90.3	100.0	
Ash	%		5.1	5.7	
Crude fiber	%		32.8	36.3	
Ether extract	%		2.0	2.2	
N-free extract	%		38.8	43.0	
Protein (N x 6.25)	%		11.6	12.8	
Cattle	dig prot %	*	7.2	8.0	
Goats	dig prot %	*	7.7	8.5	
Horses	dig prot %	*	7.6	8.4	
Rabbits	dig prot %	*	7.7	8.6	
Sheep	dig prot %	*	7.3	8.1	
Calcium	%		0.23	0.26	
Magnesium	%		0.17	0.19	
Phosphorus	%		0.19	0.21	
Potassium	%		0.81	0.90	
Sulphur	%		0.12	0.13	

(1) dry forages and roughages (2) pasture, range plants, and forages fed green (3) sitages (4) energy feeds (5) protein supplements (6) minerals (7) vitamins (8) additives

SANDBUR. Cenchrus spp

Sandbur, aerial part, fresh, (2)
Ref No 2 04 181 United States

Feed Name or Analyses			As Fed	Dry	C.V. ± %
Dry matter	%		20.0	100.0	
Ash	%		1.4	7.0	11
Crude fiber	%		7.3	36.4	7
Sheep	dig coef %		74.	74.	
Ether extract	%		0.5	2.3	16
Sheep	dig coef %		69.	69.	
N-free extract	%		8.7	43.7	
Sheep	dig coef %		69.	69.	
Protein (N x 6.25)	%		2.1	10.7	9
Sheep	dig coef %		70.	70.	
Cattle	dig prot %	*	1.4	7.0	
Goats	dig prot %	*	1.3	6.5	
Horses	dig prot %	*	1.3	6.6	
Rabbits	dig prot %	*	1.4	6.9	
Sheep	dig prot %		1.5	7.5	
Energy	GE Mcal/kg				
Cattle	DE Mcal/kg	*	0.61	3.07	
Sheep	DE Mcal/kg	*	0.60	3.00	
Cattle	ME Mcal/kg	*	0.50	2.51	
Sheep	ME Mcal/kg	*	0.49	2.46	
Cattle	TDN %	*	13.9	69.5	
Sheep	TDN %		13.6	68.0	

Sandbur, aerial part, fresh, early-bloom, (2)
Ref No 2 04 177 United States

Feed Name or Analyses			As Fed	Dry	C.V. ± %
Dry matter	%		20.0	100.0	
Ash	%		1.3	6.7	
Crude fiber	%		7.6	37.8	
Sheep	dig coef %		75.	75.	
Ether extract	%		0.4	2.0	
Sheep	dig coef %		64.	64.	
N-free extract	%		8.6	43.2	
Sheep	dig coef %		69.	69.	
Protein (N x 6.25)	%		2.1	10.3	
Sheep	dig coef %		71.	71.	
Cattle	dig prot %	*	1.3	6.6	
Goats	dig prot %	*	1.2	6.2	
Horses	dig prot %	*	1.3	6.3	
Rabbits	dig prot %	*	1.3	6.6	
Sheep	dig prot %		1.5	7.3	
Energy	GE Mcal/kg				
Cattle	DE Mcal/kg	*	0.62	3.09	
Sheep	DE Mcal/kg	*	0.60	3.01	
Cattle	ME Mcal/kg	*	0.51	2.54	
Sheep	ME Mcal/kg	*	0.49	2.47	
Cattle	TDN %	*	14.0	70.2	
Sheep	TDN %		13.7	68.4	

Sandbur, aerial part, fresh, full bloom, (2)
Ref No 2 04 178 United States

Feed Name or Analyses			As Fed	Dry	C.V. ± %
Dry matter	%			100.0	
Ash	%			7.7	
Crude fiber	%			33.9	
Ether extract	%			2.8	
N-free extract	%			44.0	
Protein (N x 6.25)	%			11.6	
Cattle	dig prot %	*		7.8	
Goats	dig prot %	*		7.4	
Horses	dig prot %	*		7.4	
Rabbits	dig prot %	*		7.6	
Sheep	dig prot %	*		7.8	
Energy	GE Mcal/kg				
Cattle	DE Mcal/kg	*		3.01	
Sheep	DE Mcal/kg	*		2.77	
Cattle	ME Mcal/kg	*		2.47	
Sheep	ME Mcal/kg	*		2.28	
Cattle	TDN %	*		68.2	
Sheep	TDN %	*		62.9	

Sandbur, aerial part, fresh, late bloom, (2)
Ref No 2 04 179 United States

Feed Name or Analyses			As Fed	Dry	C.V. ± %
Dry matter	%		20.0	100.0	
Ash	%		1.5	7.7	
Crude fiber	%		6.8	33.9	
Sheep	dig coef %		72.	72.	
Ether extract	%		0.6	2.8	
Sheep	dig coef %		78.	78.	
N-free extract	%		8.8	44.0	
Sheep	dig coef %		68.	68.	
Protein (N x 6.25)	%		2.3	11.6	
Sheep	dig coef %		70.	70.	
Cattle	dig prot %	*	1.6	7.8	
Goats	dig prot %	*	1.5	7.4	
Horses	dig prot %	*	1.5	7.4	
Rabbits	dig prot %	*	1.5	7.6	
Sheep	dig prot %		1.6	8.1	
Energy	GE Mcal/kg				
Cattle	DE Mcal/kg	*	0.60	3.01	
Sheep	DE Mcal/kg	*	0.59	2.97	
Cattle	ME Mcal/kg	*	0.49	2.47	
Sheep	ME Mcal/kg	*	0.49	2.44	
Cattle	TDN %	*	13.6	68.2	
Sheep	TDN %		13.5	67.4	

Sandbur, aerial part, fresh, milk stage, (2)
Ref No 2 04 180 United States

Feed Name or Analyses			As Fed	Dry	C.V. ± %
Dry matter	%			100.0	
Ash	%			6.7	
Crude fiber	%			37.8	
Ether extract	%			2.0	
N-free extract	%			43.2	
Protein (N x 6.25)	%			10.3	
Cattle	dig prot %	*		6.6	
Goats	dig prot %	*		6.2	
Horses	dig prot %	*		6.3	
Rabbits	dig prot %	*		6.6	
Sheep	dig prot %	*		6.6	
Energy	GE Mcal/kg				
Cattle	DE Mcal/kg	*		3.02	
Sheep	DE Mcal/kg	*		2.83	
Cattle	ME Mcal/kg	*		2.48	
Sheep	ME Mcal/kg	*		2.32	
Cattle	TDN %	*		68.5	
Sheep	TDN %	*		64.3	

SANDBUR, INDIA. Cenchrus biflorus

Sandbur, India, hay, s-c, early bloom, (1)
Ref No 1 04 182 United States

Feed Name or Analyses			As Fed	Dry	C.V. ± %
Dry matter	%		94.8	100.0	
Ash	%		10.9	11.5	
Crude fiber	%		31.2	32.9	
Cattle	dig coef %		58.	58.	

Feed Name or Analyses				Mean As Fed	Mean Dry	C.V. ± %
Ether extract		%		0.9	1.0	
Cattle	dig coef	%		37.	37.	
N-free extract		%		43.2	45.6	
Cattle	dig coef	%		50.	50.	
Protein (N x 6.25)		%		8.5	9.0	
Cattle	dig coef	%		58.	58.	
Cattle	dig prot	%		4.9	5.2	
Goats	dig prot	%	*	4.7	5.0	
Horses	dig prot	%	*	4.9	5.2	
Rabbits	dig prot	%	*	5.3	5.6	
Sheep	dig prot	%	*	4.4	4.6	
Energy	GE Mcal/kg					
Cattle	DE Mcal/kg		*	2.00	2.11	
Sheep	DE Mcal/kg		*	2.33	2.45	
Cattle	ME Mcal/kg		*	1.64	1.73	
Sheep	ME Mcal/kg		*	1.91	2.01	
Cattle	TDN	%		45.4	47.9	
Sheep	TDN	%	*	52.7	55.6	

Sand dropseed –

see Dropseed, sand, aerial part, fresh, mature, (2)

Sand dropseed –

see Dropseed, sand, aerial part, fresh, early bloom, (2)

Sand dropseed –

see Dropseed, sand, aerial part, fresh, over ripe, (2)

SANDPEA. Tephrosia spp

Sandpea, leaves w stems, fresh, (2)

Ref No 2 04 185 United States

Feed Name or Analyses				As Fed	Dry	C.V. ± %
Dry matter		%		38.0	100.0	
Ash		%		1.9	5.1	29
Crude fiber		%		16.2	42.6	26
Ether extract		%		0.9	2.3	43
N-free extract		%		15.7	41.4	
Protein (N x 6.25)		%		3.3	8.6	49
Cattle	dig prot	%	*	2.0	5.2	
Goats	dig prot	%	*	1.7	4.6	
Horses	dig prot	%	*	1.8	4.8	
Rabbits	dig prot	%	*	2.0	5.3	
Sheep	dig prot	%	*	1.9	5.0	
Calcium		%		0.92	2.42	98
Phosphorus		%		0.14	0.37	98

Sandpea, leaves w stems, fresh, immature, (2)

Ref No 2 04 183 United States

Feed Name or Analyses				As Fed	Dry	C.V. ± %
Dry matter		%		26.0	100.0	
Ash		%		1.3	5.1	
Crude fiber		%		7.2	27.5	
Ether extract		%		0.7	2.5	
N-free extract		%		13.1	50.4	
Protein (N x 6.25)		%		3.8	14.5	
Cattle	dig prot	%	*	2.7	10.2	
Goats	dig prot	%	*	2.6	10.1	
Horses	dig prot	%	*	2.6	9.8	
Rabbits	dig prot	%	*	2.6	9.9	
Sheep	dig prot	%	*	2.7	10.5	

Sandpea, leaves w stems, fresh, over ripe, (2)

Ref No 2 04 184 United States

Feed Name or Analyses				As Fed	Dry	C.V. ± %
Dry matter		%			100.0	
Ash		%			3.8	
Crude fiber		%			54.2	
Ether extract		%			1.2	
N-free extract		%			34.0	
Protein (N x 6.25)		%			6.8	
Cattle	dig prot	%	*		3.7	
Goats	dig prot	%	*		2.9	
Horses	dig prot	%	*		3.3	
Rabbits	dig prot	%	*		3.9	
Sheep	dig prot	%	*		3.3	

SANDREED, PRAIRIE. Calamovilfa longifolia

Sandreed, prairie, aerial part, fresh, full bloom, (2)

Ref No 2 04 186 United States

Feed Name or Analyses				As Fed	Dry	C.V. ± %
Dry matter		%			100.0	
Ash		%			5.4	
Crude fiber		%			37.8	
Ether extract		%			1.9	
N-free extract		%			49.0	
Protein (N x 6.25)		%			5.9	
Cattle	dig prot	%	*		2.9	
Goats	dig prot	%	*		2.1	
Horses	dig prot	%	*		2.5	
Rabbits	dig prot	%	*		3.2	
Sheep	dig prot	%	*		2.5	

SATINTAIL, BRAZIL. Imperata brasiliensis

Satintail, Brazil, aerial part, dehy, early leaf, (1)

Ref No 1 04 187 United States

Feed Name or Analyses				As Fed	Dry	C.V. ± %
Dry matter		%		83.8	100.0	
Ash		%		7.0	8.3	
Crude fiber		%		35.5	42.4	
Sheep	dig coef	%		56.	56.	
Ether extract		%		2.3	2.8	
Sheep	dig coef	%		38.	38.	
N-free extract		%		29.9	35.7	
Sheep	dig coef	%		42.	42.	
Protein (N x 6.25)		%		9.1	10.8	
Sheep	dig coef	%		52.	52.	
Cattle	dig prot	%	*	5.3	6.3	
Goats	dig prot	%	*	5.6	6.6	
Horses	dig prot	%	*	5.6	6.7	
Rabbits	dig prot	%	*	5.9	7.0	
Sheep	dig prot	%		4.7	5.6	
Energy	GE Mcal/kg					
Cattle	DE Mcal/kg		*	1.60	1.91	
Sheep	DE Mcal/kg		*	1.73	2.06	
Cattle	ME Mcal/kg		*	1.31	1.56	
Sheep	ME Mcal/kg		*	1.42	1.69	
Cattle	TDN	%	*	36.2	43.2	
Sheep	TDN	%		39.2	46.7	

SEAL. Pagophilus groenlandica

Seal, blubber oil, steam rendered, (4)

Ref No 4 09 282 Canada

Feed Name or Analyses			As Fed	Dry	C.V. ± %
Dry matter		%	100.0	100.0	
Ether extract		%	100.0	100.0	
Fatty acids		%			
Docosahexaenoic		%	6.000	6.000	
Docosapentaenoic		%	3.500	3.500	
Eicosapentaenoic		%	6.800	6.800	
Eicosenoic		%	11.200	11.200	
Linoleic		%	1.300	1.300	
Myristic		%	4.000	4.000	
Octadecatetraenoic		%	1.100	1.100	
Oleic		%	22.200	22.200	
Palmitic		%	5.800	5.800	
Palmitoleic		%	16.200	16.200	
Stearic		%	0.900	0.900	

SEAL, HARBOUR. Phoca vitulina concolor dekay

Seal, harbour, blubber oil, solv-extd, (4)

Ref No 4 09 283 Canada

Feed Name or Analyses			As Fed	Dry	C.V. ± %
Dry matter		%		100.0	
Docosahexaenoic		%		6.750	
Docosapentaenoic		%		3.780	
Eicosapentaenoic		%		4.950	
Eicosenoic		%		12.240	
Myristic		%		2.970	
Myristoleic		%		0.990	
Oleic		%		27.180	
Palmitic		%		5.850	
Palmitoleic		%		15.840	

SEAMOSS. Scientific name not used

Seamoss, entire plant, fresh, (2)

Ref No 2 04 189 United States

Feed Name or Analyses				As Fed	Dry	C.V. ± %
Dry matter		%			100.0	
Ash		%			19.7	
Crude fiber		%			3.0	
Protein (N x 6.25)		%			16.0	
Cattle	dig prot	%	*		11.5	
Goats	dig prot	%	*		11.5	
Horses	dig prot	%	*		11.1	
Rabbits	dig prot	%	*		11.0	
Sheep	dig prot	%	*		11.9	
Calcium		%			1.20	
Chlorine		%			1.12	
Phosphorus		%			0.11	

SEAWEED. Laminariales (order), Fucales (order)

Seaweed, entire plant, s-c grnd, (1)

Ref No 1 04 190 United States

Feed Name or Analyses			As Fed	Dry	C.V. ± %
Dry matter		%	89.4	100.0	5
Ash		%	20.0	22.4	14
Crude fiber		%	6.9	7.7	33
Ether extract		%	2.2	2.5	54
N-free extract		%	51.7	57.8	

Continued

Feed Name or Analyses				Mean As Fed	Mean Dry	C.V. ± %
Protein (N x 6.25)		%		8.6	9.6	27
Cattle	dig prot	%	*	4.7	5.3	
Goats	dig prot	%	*	4.9	5.5	
Horses	dig prot	%	*	5.1	5.7	
Rabbits	dig prot	%	*	5.4	6.1	
Sheep	dig prot	%	*	4.6	5.2	
Calcium		%		1.64	1.83	34
Magnesium		%		5.69	6.37	31
Phosphorus		%		0.16	0.18	40

SEAWEED, BLADEKELP. Laminaria spp

Seaweed, bladekelp, entire plant, dehy, (1)
Bladekelp
Ref No 1 08 507 United States

Feed Name or Analyses				Mean As Fed	Mean Dry	C.V. ± %
Dry matter		%		87.3	100.0	3
Ash		%		21.1	24.2	16
Crude fiber		%		8.0	9.2	24
Sheep	dig coef	%		0.	0.	24
Swine	dig coef	%		41.	41.	
Ether extract		%		0.9	1.0	61
Sheep	dig coef	%		0.	0.	61
Swine	dig coef	%		53.	53.	
N-free extract		%		47.8	54.7	
Sheep	dig coef	%		74.	74.	
Swine	dig coef	%		69.	69.	
Protein (N x 6.25)		%		9.5	10.9	17
Sheep	dig coef	%		53.	53.	
Swine	dig coef	%		18.	18.	
Cattle	dig prot	%	*	5.5	6.3	
Goats	dig prot	%	*	5.8	6.7	
Horses	dig prot	%	*	5.9	6.7	
Rabbits	dig prot	%	*	6.2	7.1	
Sheep	dig prot	%		5.0	5.8	
Swine	dig prot	%		1.7	2.0	
Energy	GE Mcal/kg					
Cattle	DE Mcal/kg		*	1.55	1.77	
Sheep	DE Mcal/kg		*	1.78	2.04	
Cattle	ME Mcal/kg		*	1720.	1971.	
Cattle	ME Mcal/kg		*	1.27	1.45	
Sheep	ME Mcal/kg		*	1.46	1.67	
Swine	ME kcal/kg		*	1614.	1849.	
Cattle	TDN	%	*	35.0	40.1	
Sheep	TDN	%		40.4	46.3	
Swine	TDN	%		39.0	44.7	
Calcium		%		1.31	1.50	11
Copper		mg/kg		7.9	9.0	
Iodine		mg/kg		0.029	0.033	
Iron		%		0.001	0.001	
Magnesium		%		0.62	0.71	
Manganese		mg/kg		6.4	7.3	98
Phosphorus		%		0.24	0.27	28
Potassium		%		3.84	4.40	

(1) dry forages and roughages
(2) pasture, range plants, and forages fed green
(3) silages
(4) energy feeds
(5) protein supplements
(6) minerals
(7) vitamins
(8) additives

SEAWEED, KELP. Laminariales (order), Fucales (order)

Seaweed, kelp, entire plant, dehy, (1)
Dried kelp (AAFCO)
Kelp, dried
Ref No 1 08 073 United States

Feed Name or Analyses				Mean As Fed	Mean Dry	C.V. ± %
Dry matter		%		91.3	100.0	
Ash		%		35.2	38.6	
Crude fiber		%		6.5	7.1	
Ether extract		%		0.5	0.5	
N-free extract		%		42.6	46.7	
Protein (N x 6.25)		%		6.5	7.1	
Cattle	dig prot	%	*	2.8	3.1	
Goats	dig prot	%	*	2.9	3.2	
Horses	dig prot	%	*	3.3	3.6	
Rabbits	dig prot	%	*	3.8	4.2	
Sheep	dig prot	%	*	2.7	3.0	
Energy	GE Mcal/kg					
Cattle	DE Mcal/kg		*	1.31	1.43	
Sheep	DE Mcal/kg		*	1.24	1.36	
Cattle	ME Mcal/kg		*	1.07	1.17	
Sheep	ME Mcal/kg		*	1.02	1.12	
Cattle	TDN	%	*	29.7	32.5	
Sheep	TDN	%	*	28.1	30.8	
Calcium		%		2.48	2.72	
Magnesium		%		0.85	0.93	
Phosphorus		%		0.28	0.31	

SEDGE. Carex spp

Sedge, hay, s-c, (1)
Ref No 1 04 193 United States

Feed Name or Analyses				Mean As Fed	Mean Dry	C.V. ± %
Dry matter		%		89.0	100.0	
Ash		%		6.3	7.1	
Crude fiber		%		27.6	31.0	
Sheep	dig coef	%		58.	58.	
Ether extract		%		2.1	2.4	
Sheep	dig coef	%		23.	23.	
N-free extract		%		44.3	49.7	
Sheep	dig coef	%		53.	53.	
Protein (N x 6.25)		%		8.6	9.7	
Sheep	dig coef	%		50.	50.	
Cattle	dig prot	%	*	4.7	5.3	
Goats	dig prot	%	*	5.0	5.6	
Horses	dig prot	%	*	5.1	5.8	
Rabbits	dig prot	%	*	5.5	6.2	
Sheep	dig prot	%		4.3	4.8	
Energy	GE Mcal/kg					
Cattle	DE Mcal/kg		*	2.36	2.54	
Sheep	DE Mcal/kg		*	1.98	2.22	
Cattle	ME Mcal/kg		*	1.85	2.08	
Sheep	ME Mcal/kg		*	1.62	1.82	
Cattle	NE_m Mcal/kg		*	0.95	1.07	
Cattle	NE_{gain} Mcal/kg		*	0.25	0.28	
Cattle	TDN	%	*	51.3	57.6	
Sheep	TDN	%		44.9	50.5	
Calcium		%		0.59	0.66	
Phosphorus		%		0.24	0.26	

Sedge, aerial part, fresh, (2)
Ref No 2 04 195 United States

Feed Name or Analyses				Mean As Fed	Mean Dry	C.V. ± %
Dry matter		%			100.0	
Chlorine		%			0.24	
Sodium		%			0.21	
Sulphur		%			0.23	20

Sedge, aerial part, fresh, immature, (2)
Ref No 2 04 194 United States

Feed Name or Analyses				Mean As Fed	Mean Dry	C.V. ± %
Dry matter		%			100.0	
Ash		%			7.6	9
Crude fiber		%			25.6	12
Ether extract		%			4.3	18
N-free extract		%			45.3	
Protein (N x 6.25)		%			17.2	15
Cattle	dig prot	%	*		12.5	
Goats	dig prot	%	*		12.6	
Horses	dig prot	%	*		12.1	
Rabbits	dig prot	%	*		11.9	
Sheep	dig prot	%	*		13.0	
Energy	GE Mcal/kg					
Cattle	DE Mcal/kg		*		2.67	
Sheep	DE Mcal/kg		*		2.75	
Cattle	ME Mcal/kg		*		2.19	
Sheep	ME Mcal/kg		*		2.25	
Cattle	TDN	%	*		60.5	
Sheep	TDN	%	*		62.3	

SEDGE, BEAKED. Carex rostrata

Sedge, beaked, aerial part, fresh, (2)
Ref No 2 04 196 United States

Feed Name or Analyses				Mean As Fed	Mean Dry	C.V. ± %
Dry matter		%			100.0	
Carotene		mg/kg			166.9	40
Vitamin A equivalent		IU/g			278.2	

SEDGE, CROWFOOT. Carex cruscorvi

Sedge, crowfoot, seeds w hulls, (4)
Ref No 4 09 269 United States

Feed Name or Analyses				Mean As Fed	Mean Dry	C.V. ± %
Dry matter		%			100.0	
Protein (N x 6.25)		%			10.0	
Alanine		%			0.44	
Arginine		%			0.93	
Aspartic acid		%			0.87	
Glutamic acid		%			1.54	
Glycine		%			0.38	
Histidine		%			0.20	
Hydroxyproline		%			0.01	
Isoleucine		%			0.39	
Leucine		%			0.71	
Lysine		%			0.31	
Methionine		%			0.24	
Phenylalanine		%			0.63	
Proline		%			0.37	
Serine		%			0.42	
Threonine		%			0.37	
Tyrosine		%			0.28	
Valine		%			0.48	

SEDGE, DOUGLAS. Carex douglasii

Sedge, douglas, aerial part, fresh, (2)
Ref No 2 04 197 United States

Feed Name or Analyses				Mean As Fed	Mean Dry	C.V. ± %
Dry matter		%			100.0	
Ash		%			6.8	

Feed Name or Analyses		Mean As Fed	Mean Dry	C.V. ± %
Crude fiber	%		29.7	
Ether extract	%		4.0	
N-free extract	%		49.4	
Protein (N x 6.25)	%		10.1	
Cattle	dig prot % *		6.5	
Goats	dig prot % *		6.0	
Horses	dig prot % *		6.1	
Rabbits	dig prot % *		6.5	
Sheep	dig prot % *		6.4	

SEDGE, NEBRASKA. Carex nebraskensis

Sedge, Nebraska, aerial part, fresh, (2)
Ref No 2 04 198 United States

		As Fed	Dry	C.V. ± %
Dry matter	%		100.0	
Ash	%		7.7	12
Crude fiber	%		30.2	5
Ether extract	%		3.2	17
N-free extract	%		49.3	
Protein (N x 6.25)	%		9.6	15
Cattle	dig prot % *		6.1	
Goats	dig prot % *		5.5	
Horses	dig prot % *		5.7	
Rabbits	dig prot % *		6.1	
Sheep	dig prot % *		5.9	

SEDGE, OVALHEAD. Carex festivella

Sedge, ovalhead, aerial part, fresh, (2)
Ref No 2 04 199 United States

		As Fed	Dry	C.V. ± %
Dry matter	%		100.0	
Ash	%		4.9	
Crude fiber	%		29.6	
Ether extract	%		2.5	
N-free extract	%		50.1	
Protein (N x 6.25)	%		12.9	
Cattle	dig prot % *		8.9	
Goats	dig prot % *		8.6	
Horses	dig prot % *		8.5	
Rabbits	dig prot % *		8.6	
Sheep	dig prot % *		9.0	

SEDGE, PRESL. Carex preslii

Sedge, presl, aerial part, fresh, immature, (2)
Ref No 2 04 200 United States

		As Fed	Dry	C.V. ± %
Dry matter	%		100.0	
Ash	%		8.7	
Crude fiber	%		20.9	
Ether extract	%		2.9	
N-free extract	%		44.7	
Protein (N x 6.25)	%		22.8	
Cattle	dig prot % *		17.3	
Goats	dig prot % *		17.8	
Horses	dig prot % *		16.9	
Rabbits	dig prot % *		16.3	
Sheep	dig prot % *		18.2	

SEDGE, SILVERTOP. Carex siccata

Sedge, silvertop, aerial part, fresh, (2)
Ref No 2 04 201 United States

		As Fed	Dry	C.V. ± %
Dry matter	%		100.0	
Ash	%		6.9	15
Crude fiber	%		30.6	5
Ether extract	%		2.8	27
N-free extract	%		50.8	
Protein (N x 6.25)	%		8.9	9
Cattle	dig prot % *		5.5	
Goats	dig prot % *		4.9	
Horses	dig prot % *		5.1	
Rabbits	dig prot % *		5.5	
Sheep	dig prot % *		5.3	

SEDGE, SLENDERBEAK. Carex athrostachya

Sedge, slenderbeak, aerial part, fresh, (2)
Ref No 2 04 202 United States

		As Fed	Dry	C.V. ± %
Dry matter	%		100.0	
Ash	%		6.3	
Crude fiber	%		32.8	
Ether extract	%		3.1	
N-free extract	%		49.8	
Protein (N x 6.25)	%		8.0	
Cattle	dig prot % *		4.7	
Goats	dig prot % *		4.0	
Horses	dig prot % *		4.3	
Rabbits	dig prot % *		4.8	
Sheep	dig prot % *		4.4	

SEDGE, SUN. Carex heliophila

Sedge, sun, aerial part, fresh, early bloom, (2)
Ref No 2 04 203 United States

		As Fed	Dry	C.V. ± %
Dry matter	%		100.0	
Ash	%		7.8	
Crude fiber	%		22.5	
Ether extract	%		2.4	
N-free extract	%		51.4	
Protein (N x 6.25)	%		15.9	
Cattle	dig prot % *		11.4	
Goats	dig prot % *		11.4	
Horses	dig prot % *		11.0	
Rabbits	dig prot % *		10.9	
Sheep	dig prot % *		11.8	

SEDGE, THREADLEAF. Carex filifolia

Sedge, threadleaf, aerial part, fresh, (2)
Ref No 2 04 205 United States

		As Fed	Dry	C.V. ± %
Dry matter	%		100.0	
Ash	%		10.8	25
Crude fiber	%		32.9	23
Ether extract	%		3.2	23
N-free extract	%		45.6	
Protein (N x 6.25)	%		7.5	72
Cattle	dig prot % *		4.3	
Goats	dig prot % *		3.6	
Horses	dig prot % *		3.9	
Rabbits	dig prot % *		4.5	
Sheep	dig prot % *		4.0	

Sedge, threadleaf, aerial part, fresh, over ripe, (2)
Ref No 2 04 204 United States

		As Fed	Dry	C.V. ± %
Dry matter	%		100.0	
Ash	%		11.8	
Crude fiber	%		37.5	
Ether extract	%		3.2	
N-free extract	%		43.4	
Protein (N x 6.25)	%		4.1	
Cattle	dig prot % *		1.4	
Goats	dig prot % *		0.4	
Horses	dig prot % *		1.0	
Rabbits	dig prot % *		1.8	
Sheep	dig prot % *		0.8	

SEDGE, WATER. Carex aquatilis

Sedge, water, aerial part, fresh, (2)
Ref No 2 04 207 United States

		As Fed	Dry	C.V. ± %
Dry matter	%		100.0	
Ash	%		8.4	24
Crude fiber	%		30.2	11
Ether extract	%		4.0	19
N-free extract	%		44.2	
Protein (N x 6.25)	%		13.2	26
Cattle	dig prot % *		9.1	
Goats	dig prot % *		8.9	
Horses	dig prot % *		8.7	
Rabbits	dig prot % *		8.9	
Sheep	dig prot % *		9.3	
Calcium	%		0.40	16
Phosphorus	%		0.31	21
Carotene	mg/kg		186.5	
Vitamin A equivalent	IU/g		310.9	

Sedge, water, aerial part, fresh, immature, (2)
Ref No 2 04 206 United States

		As Fed	Dry	C.V. ± %
Dry matter	%		100.0	
Ash	%		7.5	8
Crude fiber	%		26.2	7
Ether extract	%		4.4	15
N-free extract	%		45.2	
Protein (N x 6.25)	%		16.7	10
Cattle	dig prot % *		12.1	
Goats	dig prot % *		12.1	
Horses	dig prot % *		11.7	
Rabbits	dig prot % *		11.6	
Sheep	dig prot % *		12.6	

SEDGE, WOOLLY. Carex lanuginosa

Sedge, woolly, aerial part, fresh, (2)
Ref No 2 04 208 United States

		As Fed	Dry	C.V. ± %
Dry matter	%		100.0	
Ash	%		7.3	
Crude fiber	%		32.3	
Ether extract	%		2.3	
N-free extract	%		48.7	

Continued

Feed Name or Analyses				Mean As Fed	Mean Dry	C.V. ± %
Protein (N x 6.25)		%			9.4	
Cattle	dig prot	%	*		5.9	
Goats	dig prot	%	*		5.3	
Horses	dig prot	%	*		5.5	
Rabbits	dig prot	%	*		5.9	
Sheep	dig prot	%	*		5.8	

SENNA, MARGINATA. Cassia marginata

Senna, marginata, seeds w hulls, (5)
Ref No 5 09 289 United States

Feed Name or Analyses				Mean As Fed	Mean Dry	C.V. ± %
Dry matter		%			100.0	
Protein (N x 6.25)		%			25.0	
Alanine		%			0.90	
Arginine		%			2.55	
Aspartic acid		%			2.05	
Glutamic acid		%			5.00	
Glycine		%			1.28	
Histidine		%			0.70	
Hydroxyproline		%			0.25	
Isoleucine		%			0.83	
Leucine		%			1.15	
Lysine		%			1.33	
Methionine		%			0.40	
Phenylalanine		%			0.95	
Proline		%			1.05	
Serine		%			1.05	
Threonine		%			0.78	
Tyrosine		%			0.70	
Valine		%			1.05	

SENSITIVEBRIER, NARROWPOD. Schrankia Angustisiliqua

Sensitivebrier, narrowpod, aerial part, fresh, (2)
Ref No 2 04 209 United States

Feed Name or Analyses				Mean As Fed	Mean Dry	C.V. ± %
Dry matter		%			100.0	
Calcium		%			1.04	
Magnesium		%			0.31	
Phosphorus		%			0.23	
Potassium		%			1.32	

SENSITIVEBRIER, ROEMER. Schrankia roemeriana

Sansitivebrier, roemer, leaves, fresh, (2)
Ref No 2 04 210 United States

Feed Name or Analyses				Mean As Fed	Mean Dry	C.V. ± %
Dry matter		%			100.0	
Ash		%			10.7	
Crude fiber		%			14.2	
Ether extract		%			2.4	
N-free extract		%			54.0	
Protein (N x 6.25)		%			18.7	
Cattle	dig prot	%	*		13.8	
Goats	dig prot	%	*		14.0	
Horses	dig prot	%	*		13.4	
Rabbits	dig prot	%	*		13.1	
Sheep	dig prot	%	*		14.4	
Calcium		%			4.37	
Magnesium		%			0.30	

(1) dry forages and roughages (2) pasture, range plants, and forages fed green (3) silages (4) energy feeds (5) protein supplements (6) minerals (7) vitamins (8) additives

Feed Name or Analyses				Mean As Fed	Mean Dry	C.V. ± %
Phosphorus		%			0.20	
Potassium		%			1.38	

SERRADELLA. Ornithopus spp

Serradella, hay, s-c, (1)
Ref No 1 04 213 United States

Feed Name or Analyses				Mean As Fed	Mean Dry	C.V. ± %
Dry matter		%		85.2	100.0	
Ash		%		7.3	8.6	
Crude fiber		%		28.5	33.4	
Sheep	dig coef	%		43.	43.	
Ether extract		%		3.1	3.6	
Sheep	dig coef	%		64.	64.	
N-free extract		%		30.6	35.9	
Sheep	dig coef	%		58.	58.	
Protein (N x 6.25)		%		15.7	18.5	
Sheep	dig coef	%		70.	70.	
Cattle	dig prot	%	*	11.0	12.9	
Goats	dig prot	%	*	11.7	13.8	
Horses	dig prot	%	*	11.2	13.2	
Rabbits	dig prot	%	*	11.0	12.9	
Sheep	dig prot	%		11.0	12.9	
Fatty acids		%		3.1	3.6	
Energy		GE Mcal/kg				
Cattle		DE Mcal/kg	*	2.15	2.53	
Sheep		DE Mcal/kg	*	2.00	2.35	
Cattle		ME Mcal/kg	*	1.77	2.07	
Sheep		ME Mcal/kg	*	1.64	1.93	
Cattle		TDN %	*	48.8	57.3	
Sheep		TDN %		45.4	53.3	
Phosphorus		%		0.32	0.37	
Potassium		%		1.20	1.40	

Serradella, hay, s-c, full bloom, (1)
Ref No 1 04 211 United States

Feed Name or Analyses				Mean As Fed	Mean Dry	C.V. ± %
Dry matter		%		85.6	100.0	
Ash		%		9.9	11.6	
Crude fiber		%		25.3	29.6	
Sheep	dig coef	%		50.	50.	
Ether extract		%		4.5	5.2	
Sheep	dig coef	%		65.	65.	
N-free extract		%		26.5	31.0	
Sheep	dig coef	%		63.	63.	
Protein (N x 6.25)		%		19.3	22.6	
Sheep	dig coef	%		74.	74.	
Cattle	dig prot	%	*	14.1	16.5	
Goats	dig prot	%	*	15.1	17.6	
Horses	dig prot	%	*	14.3	16.7	
Rabbits	dig prot	%	*	13.8	16.1	
Sheep	dig prot	%		14.3	16.7	
Energy		GE Mcal/kg				
Cattle		DE Mcal/kg	*	2.40	2.81	
Sheep		DE Mcal/kg	*	2.21	2.59	
Cattle		ME Mcal/kg	*	1.97	2.30	
Sheep		ME Mcal/kg	*	1.82	2.12	
Cattle		TDN %	*	54.5	63.6	
Sheep		TDN %		50.2	58.7	

Serradella, hay, s-c, late bloom, (1)
Ref No 1 04 212 United States

Feed Name or Analyses				Mean As Fed	Mean Dry	C.V. ± %
Dry matter		%		80.7	100.0	
Ash		%		6.0	7.4	
Crude fiber		%		29.3	36.3	
Sheep	dig coef	%		37.	37.	
Ether extract		%		3.2	4.0	
Sheep	dig coef	%		66.	66.	
N-free extract		%		26.6	32.9	
Sheep	dig coef	%		48.	48.	
Protein (N x 6.25)		%		15.7	19.4	
Sheep	dig coef	%		63.	63.	
Cattle	dig prot	%	*	11.1	13.7	
Goats	dig prot	%	*	11.8	14.7	
Horses	dig prot	%	*	11.3	14.0	
Rabbits	dig prot	%	*	11.0	13.6	
Sheep	dig prot	%		9.9	12.2	
Energy		GE Mcal/kg				
Cattle		DE Mcal/kg	*	1.86	2.30	
Sheep		DE Mcal/kg	*	1.69	2.09	
Cattle		ME Mcal/kg	*	1.53	1.89	
Sheep		ME Mcal/kg	*	1.38	1.71	
Cattle		TDN %	*	42.0	52.1	
Sheep		TDN %		38.2	47.4	

Serradella, hay, s-c, brown, (1)
Ref No 1 04 214 United States

Feed Name or Analyses				Mean As Fed	Mean Dry	C.V. ± %
Dry matter		%		81.7	100.0	
Ash		%		7.0	8.6	
Crude fiber		%		25.8	31.6	
Sheep	dig coef	%		45.	45.	
Ether extract		%		2.4	2.9	
Sheep	dig coef	%		76.	76.	
N-free extract		%		33.5	41.0	
Sheep	dig coef	%		62.	62.	
Protein (N x 6.25)		%		13.0	15.9	
Sheep	dig coef	%		59.	59.	
Cattle	dig prot	%	*	8.7	10.7	
Goats	dig prot	%	*	9.3	11.4	
Horses	dig prot	%	*	9.0	11.0	
Rabbits	dig prot	%	*	8.9	10.9	
Sheep	dig prot	%		7.7	9.4	
Energy		GE Mcal/kg				
Cattle		DE Mcal/kg	*	2.09	2.56	
Sheep		DE Mcal/kg	*	1.94	2.38	
Cattle		ME Mcal/kg	*	1.72	2.10	
Sheep		ME Mcal/kg	*	1.59	1.95	
Cattle		TDN %	*	47.5	58.1	
Sheep		TDN %		44.1	54.0	

SERRADELLA, COMMON. Ornithopus sativus

Serradella, common, hay, s-c, (1)
Ref No 1 04 215 United States

Feed Name or Analyses				Mean As Fed	Mean Dry	C.V. ± %
Dry matter		%		89.0	100.0	
Ash		%		7.6	8.5	
Crude fiber		%		29.8	33.5	
Ether extract		%		3.2	3.6	
N-free extract		%		32.0	36.0	
Protein (N x 6.25)		%		16.4	18.4	
Cattle	dig prot	%	*	11.5	12.9	
Goats	dig prot	%	*	12.2	13.7	
Horses	dig prot	%	*	11.7	13.2	
Rabbits	dig prot	%	*	11.5	12.9	
Sheep	dig prot	%	*	11.6	13.1	
Energy		GE Mcal/kg				
Cattle		DE Mcal/kg	*	2.25	2.52	
Sheep		DE Mcal/kg	*	2.15	2.42	
Cattle		ME Mcal/kg	*	1.84	2.07	
Sheep		ME Mcal/kg	*	1.77	1.98	

Feed Name or Analyses			Mean As Fed	Mean Dry	C.V. ± %
Cattle	TDN %	*	50.9	57.2	
Sheep	TDN %	*	48.8	54.9	

Serradella, common, aerial part, fresh, (2)
Ref No 2 04 216 United States

Feed Name or Analyses			Mean As Fed	Mean Dry	C.V. ± %
Dry matter	%		20.8	100.0	13
Ash	%		3.1	15.1	16
Crude fiber	%		4.9	23.4	7
Ether extract	%		0.7	3.6	11
N-free extract	%		9.0	43.4	
Protein (N x 6.25)	%		3.0	14.6	13
Cattle	dig prot %	*	2.1	10.3	
Goats	dig prot %	*	2.1	10.2	
Horses	dig prot %	*	2.1	9.9	
Rabbits	dig prot %	*	2.1	9.9	
Sheep	dig prot %	*	2.2	10.6	
Energy	GE Mcal/kg				
Cattle	DE Mcal/kg	*	0.53	2.53	
Sheep	DE Mcal/kg	*	0.58	2.79	
Cattle	ME Mcal/kg	*	0.43	2.07	
Sheep	ME Mcal/kg	*	0.48	2.29	
Cattle	TDN %	*	11.9	57.4	
Sheep	TDN %	*	13.1	63.3	
Calcium	%		0.29	1.39	
Phosphorus	%		0.09	0.45	
Potassium	%		0.44	2.13	

SERVICEBERRY, SASKATOON. Amelanchier alnifolia

Serviceberry, saskatoon, leaves, fresh, immature, (2)
Ref No 2 04 217 United States

Feed Name or Analyses			Mean As Fed	Mean Dry	C.V. ± %
Dry matter	%			100.0	
Ash	%			6.7	
Crude fiber	%			13.4	
Ether extract	%			3.9	
N-free extract	%			58.0	
Protein (N x 6.25)	%			18.0	
Cattle	dig prot %	*		13.2	
Goats	dig prot %	*		13.4	
Horses	dig prot %	*		12.8	
Rabbits	dig prot %	*		12.6	
Sheep	dig prot %	*		13.8	
Calcium	%			1.60	
Magnesium	%			0.43	0
Phosphorus	%			0.48	
Sulphur	%			0.25	

Serviceberry, saskatoon, leaves, fresh, mid-bloom, (2)
Ref No 2 04 218 United States

Feed Name or Analyses			Mean As Fed	Mean Dry	C.V. ± %
Dry matter	%			100.0	
Ash	%			8.5	
Crude fiber	%			14.8	
Ether extract	%			6.2	
N-free extract	%			59.6	
Protein (N x 6.25)	%			10.9	
Cattle	dig prot %	*		7.2	
Goats	dig prot %	*		6.7	
Horses	dig prot %	*		6.8	
Rabbits	dig prot %	*		7.1	
Sheep	dig prot %	*		7.1	
Calcium	%			2.32	
Magnesium	%			0.47	0

Feed Name or Analyses			Mean As Fed	Mean Dry	C.V. ± %
Phosphorus	%			0.62	
Sulphur	%			0.14	

Serviceberry, saskatoon, twigs, fresh, (2)
Ref No 2 04 219 United States

Feed Name or Analyses			Mean As Fed	Mean Dry	C.V. ± %
Dry matter	%			100.0	
Ash	%			2.8	
Crude fiber	%			34.8	
Ether extract	%			3.4	
N-free extract	%			53.5	
Protein (N x 6.25)	%			5.5	
Cattle	dig prot %	*		2.6	
Goats	dig prot %	*		1.7	
Horses	dig prot %	*		2.2	
Rabbits	dig prot %	*		2.9	
Sheep	dig prot %	*		2.1	

SESAME. Sesamum indicum

Sesame, seeds, (5)
Ref No 5 08 509 United States

Feed Name or Analyses			Mean As Fed	Mean Dry	C.V. ± %
Dry matter	%		92.0	100.0	
Ash	%		5.6	6.1	
Crude fiber	%		10.3	11.2	
Ether extract	%		42.9	46.6	
N-free extract	%		10.9	11.8	
Protein (N x 6.25)	%		22.3	24.2	
Energy	GE Mcal/kg				
Cattle	DE Mcal/kg	*	4.48	4.87	
Sheep	DE Mcal/kg	*	4.29	4.66	
Swine	DE kcal/kg	*	4685.	5092.	
Cattle	ME Mcal/kg	*	3.67	3.99	
Sheep	ME Mcal/kg	*	3.51	3.82	
Swine	ME kcal/kg	*	3841.	4175.	
Cattle	TDN %	*	101.6	110.4	
Sheep	TDN %	*	97.2	105.6	
Swine	TDN %	*	106.3	115.5	
Calcium	%		0.94	1.02	
Phosphorus	%		0.70	0.76	

Sesame, seeds, extn unspecified grnd, (5)
Ref No 5 04 221 United States

Feed Name or Analyses			Mean As Fed	Mean Dry	C.V. ± %
Dry matter	%		92.4	100.0	2
Ash	%		11.3	12.2	11
Crude fiber	%		6.1	6.6	11
Sheep	dig coef %		55.	55.	
Ether extract	%		9.6	10.4	32
Sheep	dig coef %		65.	65.	
N-free extract	%		23.8	25.8	
Sheep	dig coef %		65.	65.	
Protein (N x 6.25)	%		41.6	45.1	6
Sheep	dig coef %		91.	91.	
Sheep	dig prot %		37.9	41.0	
Linoleic	%		1.900	2.056	
Energy	GE Mcal/kg		4.07	4.40	
Cattle	DE Mcal/kg	*	3.24	3.50	
Sheep	DE Mcal/kg	*	3.12	3.37	
Swine	DE kcal/kg	*	3616.	3914.	
Cattle	ME Mcal/kg	*	2.65	2.87	
Chickens	ME_n kcal/kg		1910.	2067.	
Sheep	ME Mcal/kg	*	2.56	2.77	
Swine	ME kcal/kg	*	3143.	3401.	
Cattle	TDN %	*	73.4	79.4	
Sheep	TDN %		70.7	76.5	

Feed Name or Analyses			Mean As Fed	Mean Dry	C.V. ± %
Swine	TDN %	*	82.0	88.8	
Calcium	%		2.02	2.19	10
Manganese	mg/kg		106.5	115.3	17
Phosphorus	%		1.48	1.60	14
Potassium	%		1.33	1.44	
Carotene	mg/kg		0.4	0.5	
Choline	mg/kg		1384.	1498.	23
Pantothenic acid	mg/kg		6.3	6.9	24
Riboflavin	mg/kg		3.7	4.0	25
Thiamine	mg/kg		2.8	3.1	
Vitamin A equivalent	IU/g		0.7	0.8	
Arginine	%		4.26	4.61	15
Cystine	%		0.59	0.64	
Glycine	%		3.96	4.29	4
Histidine	%		1.09	1.18	5
Isoleucine	%		1.58	1.71	3
Leucine	%		2.77	3.00	4
Lysine	%		1.19	1.29	
Methionine	%		1.19	1.29	15
Phenylalanine	%		1.98	2.14	
Threonine	%		1.58	1.71	
Tryptophan	%		0.59	0.64	7
Tyrosine	%		1.98	2.14	14
Valine	%		2.18	2.36	

Sesame, seeds, mech-extd grnd, (5)
Sesame oil meal, expeller extracted
Ref No 5 04 220 United States

Feed Name or Analyses			Mean As Fed	Mean Dry	C.V. ± %
Dry matter	%		92.2	100.0	1
Organic matter	%		80.9	87.8	
Ash	%		10.3	11.2	
Crude fiber	%		5.4	5.8	14
Cattle	dig coef %	*	20.	20.	
Sheep	dig coef %	*	55.	55.	
Ether extract	%		8.6	9.4	22
Cattle	dig coef %	*	96.	96.	
Sheep	dig coef %	*	65.	65.	
N-free extract	%		23.6	25.6	
Cattle	dig coef %	*	74.	74.	
Sheep	dig coef %	*	65.	65.	
Protein (N x 6.25)	%		44.3	48.0	4
Cattle	dig coef %	*	80.	80.	
Sheep	dig coef %	*	91.	91.	
Cattle	dig prot %		35.4	38.4	
Sheep	dig prot %		40.3	43.7	
Energy	GE Mcal/kg				
Cattle	DE Mcal/kg	*	3.20	3.47	
Sheep	DE Mcal/kg	*	3.14	3.41	
Swine	DE kcal/kg	*	3556.	3858.	
Cattle	ME Mcal/kg	*	2.62	2.85	
Chickens	ME_n kcal/kg		2603.	2824.	
Sheep	ME Mcal/kg	*	2.57	2.79	
Swine	ME kcal/kg	*	3069.	3329.	
Cattle	NE_m Mcal/kg	*	1.56	1.69	
Cattle	NE_{gain} Mcal/kg	*	1.02	1.11	
Cattle	$NE_{lactating\ cows}$ Mcal/kg	*	1.85	2.01	
Cattle	TDN %		72.6	78.8	
Sheep	TDN %		71.2	77.2	
Swine	TDN %	*	80.6	87.5	
Calcium	%		1.99	2.16	
Chlorine	%		0.07	0.08	
Manganese	mg/kg		47.7	51.8	
Phosphorus	%		1.33	1.44	
Choline	mg/kg		1520.	1649.	3
Pantothenic acid	mg/kg		6.3	6.8	
Riboflavin	mg/kg		3.7	4.0	
Arginine	%		4.72	5.12	

Continued

Feed Name or Analyses		Mean As Fed	Mean Dry	C.V. ± %
Cystine	%	0.58	0.63	
Glycine	%	4.13	4.48	
Lysine	%	1.28	1.39	
Methionine	%	1.38	1.49	
Tryptophan	%	0.77	0.83	

Sesame oil meal, expeller extracted –
see Sesame, seeds, mech-extd grnd, (5)

SESBANIA, HEMP. Sesbania macrocarpa

Sesbania, hemp, seeds w hulls, (5)

Ref No 5 09 291 United States

Feed Name or Analyses		Mean As Fed	Mean Dry	C.V. ± %
Dry matter	%		100.0	
Protein (N x 6.25)	%		38.0	
Alanine	%		1.22	
Arginine	%		3.12	
Aspartic acid	%		3.00	
Glutamic acid	%		5.43	
Glycine	%		1.63	
Histidine	%		0.99	
Hydroxyproline	%		0.04	
Isoleucine	%		1.22	
Leucine	%		2.17	
Lysine	%		1.86	
Methionine	%		0.34	
Phenylalanine	%		1.41	
Proline	%		1.44	
Serine	%		1.52	
Threonine	%		1.10	
Tyrosine	%		1.03	
Valine	%		1.22	

SHEEP. Ovis aries

Sheep, livers, raw, (5)

Ref No 5 08 116 United States

Feed Name or Analyses		Mean As Fed	Mean Dry	C.V. ± %
Dry matter	%	29.2	100.0	
Ash	%	1.4	4.8	
Ether extract	%	3.9	13.4	
N-free extract	%	2.9	9.9	
Protein (N x 6.25)	%	21.0	71.9	
Energy	GE Mcal/kg	1.36	4.66	
Cattle	DE Mcal/kg	* 1.22	4.18	
Sheep	DE Mcal/kg	* 1.10	3.76	
Swine	DE kcal/kg	*1215.	4162.	
Cattle	ME Mcal/kg	* 1.00	3.43	
Sheep	ME Mcal/kg	* 0.90	3.08	
Swine	ME kcal/kg	* 990.	3391.	
Cattle	TDN %	* 27.7	94.8	
Sheep	TDN %	* 24.9	85.2	
Swine	TDN %	* 27.6	94.4	
Calcium	%	0.01	0.03	
Iron	%	0.011	0.038	
Phosphorus	%	0.35	1.20	
Potassium	%	0.20	0.68	
Sodium	%	0.05	0.17	
Ascorbic acid	mg/kg	330.0	1130.1	
Niacin	mg/kg	169.0	578.8	
Riboflavin	mg/kg	32.8	112.3	

(1) dry forages and roughages (2) pasture, range plants, and forages fed green (3) silages (4) energy feeds (5) protein supplements (6) minerals (7) vitamins (8) additives

Feed Name or Analyses		Mean As Fed	Mean Dry	C.V. ± %
Thiamine	mg/kg	4.0	13.7	
Vitamin A	IU/g	505.0	1729.5	

Sheep, milk, fresh, (5)

Ref No 2 08 510 United States

Feed Name or Analyses		Mean As Fed	Mean Dry	C.V. ± %
Dry matter	%	19.2	100.0	
Ash	%	0.9	4.7	
Crude fiber	%	0.0	0.0	
Ether extract	%	6.9	35.9	
N-free extract	%	4.9	25.5	
Protein (N x 6.25)	%	6.5	33.9	
Energy	GE Mcal/kg			
Cattle	DE Mcal/kg	* 1.05	5.45	
Sheep	DE Mcal/kg	* 0.89	4.65	
Swine	DE kcal/kg	*1265.	6589.	
Cattle	ME Mcal/kg	* 0.86	4.47	
Sheep	ME Mcal/kg	* 0.73	3.81	
Swine	ME kcal/kg	*1128.	5875.	
Cattle	TDN %	* 23.7	123.5	
Sheep	TDN %	* 20.2	105.5	
Swine	TDN %	* 28.7	149.4	
Calcium	%	0.21	1.09	
Phosphorus	%	0.12	0.63	
Potassium	%	0.19	0.99	

SHEEPBUSH, AUSTRALIAN. Pentzia incana

Sheepbush, Australian, browse, s-c, early bloom, (1)

Ref No 1 04 222 United States

Feed Name or Analyses		Mean As Fed	Mean Dry	C.V. ± %
Dry matter	%	87.0	100.0	
Ash	%	11.3	13.0	
Crude fiber	%	22.2	25.5	
Sheep	dig coef %	25.	25.	
Ether extract	%	2.3	2.6	
Sheep	dig coef %	66.	66.	
N-free extract	%	39.2	45.0	
Sheep	dig coef %	71.	71.	
Protein (N x 6.25)	%	12.1	13.9	
Sheep	dig coef %	74.	74.	
Cattle	dig prot %	* 7.8	9.0	
Goats	dig prot %	* 8.3	9.5	
Horses	dig prot %	* 8.1	9.3	
Rabbits	dig prot %	* 8.2	9.4	
Sheep	dig prot %	8.9	10.3	
Energy	GE Mcal/kg			
Cattle	DE Mcal/kg	* 2.08	2.39	
Sheep	DE Mcal/kg	* 2.01	2.31	
Cattle	ME Mcal/kg	* 1.71	1.96	
Sheep	ME Mcal/kg	* 1.65	1.90	
Cattle	TDN %	* 47.2	54.3	
Sheep	TDN %	45.7	52.5	

Sheepbush, Australian, browse, s-c, mature, (1)

Ref No 1 04 223 United States

Feed Name or Analyses		Mean As Fed	Mean Dry	C.V. ± %
Dry matter	%	87.0	100.0	
Ash	%	12.2	14.0	
Crude fiber	%	22.0	25.3	
Sheep	dig coef %	17.	17.	
Ether extract	%	4.5	5.2	
Sheep	dig coef %	77.	77.	
N-free extract	%	34.7	39.9	
Sheep	dig coef %	70.	70.	
Protein (N x 6.25)	%	13.6	15.6	
Sheep	dig coef %	74.	74.	
Cattle	dig prot %	* 9.1	10.4	
Goats	dig prot %	* 9.7	11.1	
Horses	dig prot %	* 9.4	10.8	
Rabbits	dig prot %	* 9.3	10.7	
Sheep	dig prot %	10.0	11.5	
Energy	GE Mcal/kg			
Cattle	DE Mcal/kg	* 2.13	2.45	
Sheep	DE Mcal/kg	* 2.02	2.33	
Cattle	ME Mcal/kg	* 1.75	2.01	
Sheep	ME Mcal/kg	* 1.66	1.91	
Cattle	TDN %	* 48.3	55.5	
Sheep	TDN %	45.9	52.8	

SHEEPBUSH, BALLHEAD. Pentzia spp

Sheepbush, ballhead, browse, s-c, (1)

Ref No 1 04 224 United States

Feed Name or Analyses		Mean As Fed	Mean Dry	C.V. ± %
Dry matter	%	87.0	100.0	
Ash	%	8.4	9.6	
Crude fiber	%	21.3	24.5	
Sheep	dig coef %	48.	48.	
Ether extract	%	5.0	5.7	
Sheep	dig coef %	39.	39.	
N-free extract	%	46.5	53.4	
Sheep	dig coef %	78.	78.	
Protein (N x 6.25)	%	5.9	6.8	
Sheep	dig coef %	48.	48.	
Cattle	dig prot %	* 2.5	2.8	
Goats	dig prot %	* 2.5	2.9	
Horses	dig prot %	* 2.9	3.3	
Rabbits	dig prot %	* 3.4	3.9	
Sheep	dig prot %	2.8	3.3	
Energy	GE Mcal/kg			
Cattle	DE Mcal/kg	* 2.43	2.79	
Sheep	DE Mcal/kg	* 2.37	2.72	
Cattle	ME Mcal/kg	* 1.99	2.29	
Sheep	ME Mcal/kg	* 1.94	2.23	
Cattle	TDN %	* 55.0	63.2	
Sheep	TDN %	53.7	61.7	

SHOREGRASS. Monanthochloe littoralis

Shoregrass, aerial part, fresh, (2)

Ref No 2 04 225 United States

Feed Name or Analyses		Mean As Fed	Mean Dry	C.V. ± %
Dry matter	%		100.0	
Ash	%		12.8	9
Crude fiber	%		29.1	3
Ether extract	%		1.5	13
N-free extract	%		50.0	
Protein (N x 6.25)	%		6.6	5
Cattle	dig prot %	*	3.5	
Goats	dig prot %	*	2.7	
Horses	dig prot %	*	3.1	
Rabbits	dig prot %	*	3.8	
Sheep	dig prot %	*	3.1	
Calcium	%		0.31	5
Phosphorus	%		0.18	35

Shorts, mx 8 fbr (CFA) –
see Wheat, flour by-product, coarse sift, mx 7% fiber, (4)

Feed Name or Analyses		Mean As Fed	Mean Dry	C.V. ± %

SHRIMP. Pandalus spp, Penaeus spp

Shrimp, process residue, dehy grnd, salt declared above 3% mx 7%, (5)
Shrimp meal (AAFCO)
Ref No 5 04 226 United States

Analyses		As Fed	Dry	C.V. ± %
Dry matter	%	89.9	100.0	5
Ash	%	27.4	30.5	14
Crude fiber	%	11.1	12.4	20
Ether extract	%	2.9	3.2	39
N-free extract	%	1.5	1.7	
Protein (N x 6.25)	%	46.9	52.2	13
Energy	GE Mcal/kg			
Cattle	DE Mcal/kg *	1.75	1.94	
Sheep	DE Mcal/kg *	2.02	2.25	
Swine	DE kcal/kg *	1960.	2180.	
Cattle	ME Mcal/kg *	1.43	1.59	
Chickens	ME_n kcal/kg	1678.	1862.	
Sheep	ME Mcal/kg *	1.66	1.85	
Swine	ME kcal/kg *	1674.	1863.	
Cattle	TDN % *	39.6	44.1	
Sheep	TDN % *	45.9	51.0	
Swine	TDN % *	44.4	49.5	
Calcium	%	7.34	8.17	17
Chlorine	%	1.04	1.15	
Iron	%	0.011	0.012	
Magnesium	%	0.54	0.60	
Manganese	mg/kg	30.1	33.5	33
Phosphorus	%	1.59	1.77	13
Choline	mg/kg	5836.	6494.	
Riboflavin	mg/kg	4.0	4.4	75
Arginine	%	2.30	2.56	
Cystine	%	0.45	0.50	
Lysine	%	2.00	2.23	
Methionine	%	0.72	0.80	
Tryptophan	%	0.39	0.43	

Shrimp meal (AAFCO) –
see Shrimp, process residue, dehy grnd, salt declared above 3% mx 7%, (5)

SHRUBALTHEA. Hibiscus syriacus

Shrubalthea, seeds w hulls, (5)
Ref No 5 09 145 United States

Analyses		As Fed	Dry	C.V. ± %
Dry matter	%		100.0	
Protein (N x 6.25)	%		29.0	
Alanine	%		1.19	
Arginine	%		3.07	
Aspartic acid	%		2.81	
Glutamic acid	%		4.18	
Glycine	%		1.13	
Histidine	%		0.52	
Hydroxyproline	%		0.03	
Isoleucine	%		0.90	
Leucine	%		1.62	
Lysine	%		1.31	
Methionine	%		0.35	
Phenylalanine	%		0.96	
Proline	%		0.87	
Serine	%		1.19	
Threonine	%		1.02	
Tyrosine	%		0.70	
Valine	%		1.13	

SIDA. Sida spp

Sida, aerial part, fresh, mid-bloom, (2)
Ref No 2 04 227 United States

Analyses		As Fed	Dry	C.V. ± %
Dry matter	%		100.0	
Ash	%		9.9	
Crude fiber	%		17.5	
Ether extract	%		1.8	
N-free extract	%		51.7	
Protein (N x 6.25)	%		19.1	
Cattle	dig prot % *		14.1	
Goats	dig prot % *		14.4	
Horses	dig prot % *		13.7	
Rabbits	dig prot % *		13.4	
Sheep	dig prot % *		14.8	
Calcium	%		2.63	
Magnesium	%		0.63	
Phosphorus	%		0.26	
Potassium	%		1.80	

SILENE, FRENCH. Silene gallica

Silene, French, aerial part, fresh, (2)
Ref No 2 04 232 United States

Analyses		As Fed	Dry	C.V. ± %
Dry matter	%		100.0	
Ash	%		12.8	34
Crude fiber	%		21.1	22
Protein (N x 6.25)	%		10.7	59
Cattle	dig prot % *		7.0	
Goats	dig prot % *		6.5	
Horses	dig prot % *		6.6	
Rabbits	dig prot % *		6.9	
Sheep	dig prot % *		7.0	
Calcium	%		1.53	18
Phosphorus	%		0.48	36
Potassium	%		3.60	55

Silene, French, aerial part, fresh, immature, (2)
Ref No 2 04 228 United States

Analyses		As Fed	Dry	C.V. ± %
Dry matter	%		100.0	
Ash	%		19.1	
Crude fiber	%		14.8	
Protein (N x 6.25)	%		20.1	
Cattle	dig prot % *		15.0	
Goats	dig prot % *		15.3	
Horses	dig prot % *		14.6	
Rabbits	dig prot % *		14.2	
Sheep	dig prot % *		15.7	
Calcium	%		1.95	9
Phosphorus	%		0.78	9
Potassium	%		6.15	

Silene, French, aerial part, fresh, full bloom, (2)
Ref No 2 04 229 United States

Analyses		As Fed	Dry	C.V. ± %
Dry matter	%		100.0	
Ash	%		16.7	
Crude fiber	%		19.2	
Protein (N x 6.25)	%		10.4	
Cattle	dig prot % *		6.7	
Goats	dig prot % *		6.3	
Horses	dig prot % *		6.4	
Rabbits	dig prot % *		6.7	
Sheep	dig prot % *		6.7	
Calcium	%		1.38	8
Phosphorus	%		0.43	20
Potassium	%		2.63	

Silene, French, aerial part, fresh, mature, (2)
Ref No 2 04 230 United States

Analyses		As Fed	Dry	C.V. ± %
Dry matter	%		100.0	
Ash	%		8.9	
Crude fiber	%		23.5	
Protein (N x 6.25)	%		7.4	
Cattle	dig prot % *		4.1	
Goats	dig prot % *		3.4	
Horses	dig prot % *		3.8	
Rabbits	dig prot % *		4.3	
Sheep	dig prot % *		3.8	
Calcium	%		1.23	7
Phosphorus	%		0.24	19
Potassium	%		2.31	

Silene, French, aerial part, fresh, dormant, (2)
Ref No 2 04 231 United States

Analyses		As Fed	Dry	C.V. ± %
Dry matter	%		100.0	
Ash	%		9.1	
Crude fiber	%		27.2	
Protein (N x 6.25)	%		4.1	
Cattle	dig prot % *		1.4	
Goats	dig prot % *		0.4	
Horses	dig prot % *		1.0	
Rabbits	dig prot % *		1.8	
Sheep	dig prot % *		0.8	
Calcium	%		1.05	
Phosphorus	%		0.21	

SILVERLEAF. Leucophyllum spp

Silverleaf, aerial part, fresh, mid-bloom, (2)
Ref No 2 04 233 United States

Analyses		As Fed	Dry	C.V. ± %
Dry matter	%		100.0	
Ash	%		3.7	
Crude fiber	%		25.8	
Ether extract	%		5.8	
N-free extract	%		57.7	
Protein (N x 6.25)	%		7.0	
Cattle	dig prot % *		3.8	
Goats	dig prot % *		3.1	
Horses	dig prot % *		3.5	
Rabbits	dig prot % *		4.1	
Sheep	dig prot % *		3.5	
Calcium	%		1.14	
Magnesium	%		0.14	
Phosphorus	%		0.10	
Potassium	%		0.70	

Skimmilk, centrifugal –
see Cattle, milk, skimmed centrifugal, (5)

Skimmilk, dried–
see Cattle, milk, skimmed dehy, mx 8% moisture, (5)

Feed Name or Analyses	Mean As Fed	Mean Dry	C.V. ± %

Skimmilk, gravity –
see Cattle, milk, skimmed, (5)

SKULLCAP. Scutellaria spp

Skullcap, aerial part, fresh, (2)
Ref No 2 04 235 United States

Feed Name or Analyses		As Fed	Dry	C.V. ± %
Dry matter	%	41.7	100.0	
Ash	%	3.9	9.4	
Crude fiber	%	10.6	25.3	
Ether extract	%	2.7	6.5	
N-free extract	%	21.6	51.8	
Protein (N x 6.25)	%	2.9	7.0	
Cattle	dig prot % *	1.6	3.8	
Goats	dig prot % *	1.3	3.1	
Horses	dig prot % *	1.4	3.5	
Rabbits	dig prot % *	1.7	4.1	
Sheep	dig prot % *	1.5	3.5	

Skullcap, aerial part, fresh, mid-bloom, (2)
Ref No 2 04 234 United States

Feed Name or Analyses		As Fed	Dry	C.V. ± %
Dry matter	%		100.0	
Ash	%		7.8	
Crude fiber	%		31.0	
Ether extract	%		2.8	
N-free extract	%		48.6	
Protein (N x 6.25)	%		9.8	
Cattle	dig prot % *		6.2	
Goats	dig prot % *		5.7	
Horses	dig prot % *		5.9	
Rabbits	dig prot % *		6.2	
Sheep	dig prot % *		6.1	
Calcium	%		1.53	
Magnesium	%		0.36	
Phosphorus	%		0.17	
Potassium	%		1.72	

SKULLCAP, RESINOUS. Scutellaria resinosa

Skullcap, resinous, aerial part, fresh, (2)
Ref No 2 04 236 United States

Feed Name or Analyses		As Fed	Dry	C.V. ± %
Dry matter	%	41.7	100.0	
Ash	%	4.6	11.0	
Crude fiber	%	8.1	19.5	
Ether extract	%	4.3	10.2	
N-free extract	%	23.0	55.1	
Protein (N x 6.25)	%	1.8	4.2	
Cattle	dig prot % *	0.6	1.5	
Goats	dig prot % *	0.2	0.5	
Horses	dig prot % *	0.5	1.1	
Rabbits	dig prot % *	0.8	1.9	
Sheep	dig prot % *	0.4	0.9	
Calcium	%	0.50	1.20	
Phosphorus	%	0.19	0.45	

(1) dry forages and roughages (2) pasture, range plants, and forages fed green (3) silages (4) energy feeds (5) protein supplements (6) minerals (7) vitamins (8) additives

SLOUGHGRASS, AMERICAN. Beckmannia syzigachne

Sloughgrass, American, hay, s-c, mature, (1)
Ref No 1 04 237 United States

Feed Name or Analyses		As Fed	Dry	C.V. ± %
Dry matter	%	94.6	100.0	1
Ash	%	11.0	11.6	14
Crude fiber	%	35.9	37.9	7
Cattle	dig coef %	58.	58.	
Ether extract	%	1.2	1.3	26
Cattle	dig coef %	6.	6.	
N-free extract	%	41.8	44.2	
Protein (N x 6.25)	%	4.7	5.0	35
Cattle	dig coef %	62.	62.	
Sheep	dig coef %	60.	60.	
Cattle	dig prot %	2.9	3.1	
Goats	dig prot % *	1.2	1.2	
Horses	dig prot % *	1.7	1.8	
Rabbits	dig prot % *	2.4	2.5	
Sheep	dig prot %	2.8	3.0	
Energy	GE Mcal/kg	4.10	4.34	
Cattle	GE dig coef %	54.	54.	
Sheep	GE dig coef %	52.	52.	
Cattle	DE Mcal/kg	2.21	2.34	
Sheep	DE Mcal/kg	2.13	2.25	
Cattle	ME Mcal/kg *	1.82	1.92	
Sheep	ME Mcal/kg *	1.75	1.85	
Cattle	TDN % *	50.3	53.2	
Sheep	TDN % *	48.3	51.1	

Ref No 1 04 237 Canada

Feed Name or Analyses		As Fed	Dry	C.V. ± %
Dry matter	%	93.3	100.0	2
Ash	%	8.0	8.6	
Crude fiber	%	29.6	31.7	14
Cattle	dig coef %	58.	58.	
Ether extract	%	1.4	1.5	
Cattle	dig coef %	6.	6.	
N-free extract	%	46.5	49.9	
Protein (N x 6.25)	%	7.8	8.4	37
Cattle	dig coef %	62.	62.	
Sheep	dig coef %	60.	60.	
Cattle	dig prot %	4.8	5.2	
Goats	dig prot % *	4.1	4.4	
Horses	dig prot % *	4.3	4.6	
Rabbits	dig prot % *	4.8	5.1	
Sheep	dig prot %	4.7	5.0	
Cellulose (Matrone)	%	32.1	34.4	
Cattle	dig coef %	58.	58.	
Energy	GE Mcal/kg	4.04	4.34	
Cattle	GE dig coef %	54.	54.	
Sheep	GE dig coef %	52.	52.	
Cattle	DE Mcal/kg	2.18	2.34	
Sheep	DE Mcal/kg	2.10	2.25	
Cattle	ME Mcal/kg *	1.79	1.92	
Sheep	ME Mcal/kg *	1.72	1.85	
Cattle	TDN % *	52.8	53.2	
Sheep	TDN % *	47.7	51.1	
Calcium	%	0.39	0.42	46
Phosphorus	%	0.13	0.14	51
Carotene	mg/kg	50.4	54.0	59
Vitamin A equivalent	IU/g	84.0	90.0	

Sloughgrass, American, hay, s-c, over ripe, (1)
Ref No 1 04 238 United States

Feed Name or Analyses		As Fed	Dry	C.V. ± %
Dry matter	%	94.4	100.0	
Ash	%	11.7	12.4	
Crude fiber	%	37.0	39.2	
Ether extract	%	1.0	1.1	
N-free extract	%	41.5	44.0	
Protein (N x 6.25)	%	3.1	3.3	
Cattle	dig prot % *	-0.1	-0.1	
Goats	dig prot % *	-0.2	-0.3	
Horses	dig prot % *	0.3	0.3	
Rabbits	dig prot % *	1.1	1.2	
Sheep	dig prot % *	-0.3	-0.4	
Energy	GE Mcal/kg			
Cattle	DE Mcal/kg *	2.19	2.32	
Sheep	DE Mcal/kg *	2.09	2.22	
Cattle	ME Mcal/kg *	1.80	1.90	
Sheep	ME Mcal/kg *	1.72	1.82	
Cattle	TDN % *	49.7	52.6	
Sheep	TDN % *	47.5	50.3	

SMARTWEED, PENNSYLVANIA. Polygonium pennsylvanicum

Smartweed, Pennsylvania, seeds w hulls, (4)
Ref No 4 09 039 United States

Feed Name or Analyses		As Fed	Dry	C.V. ± %
Dry matter	%		100.0	
Protein (N x 6.25)	%		10.0	
Cattle	dig prot % *		5.2	
Goats	dig prot % *		6.4	
Horses	dig prot % *		6.4	
Sheep	dig prot % *		6.4	
Alanine	%		0.40	
Arginine	%		0.68	
Aspartic acid	%		0.73	
Glutamic acid	%		1.24	
Glycine	%		0.41	
Histidine	%		0.22	
Hydroxyproline	%		0.01	
Isoleucine	%		0.36	
Leucine	%		0.60	
Lysine	%		0.46	
Methionine	%		0.18	
Phenylalanine	%		0.37	
Proline	%		0.34	
Serine	%		0.40	
Threonine	%		0.33	
Tyrosine	%		0.23	
Valine	%		0.46	

SMUTGRASS, RATTAIL. Sporobolus poiretii

Smutgrass, rattail, aerial part, fresh, mature, (2)
Ref No 2 04 239 United States

Feed Name or Analyses		As Fed	Dry	C.V. ± %
Dry matter	%		100.0	
Ash	%		8.3	5
Crude fiber	%		32.7	3
Ether extract	%		1.8	19
N-free extract	%		50.5	
Protein (N x 6.25)	%		6.7	8
Cattle	dig prot % *		3.6	
Goats	dig prot % *		2.8	
Horses	dig prot % *		3.2	
Rabbits	dig prot % *		3.8	
Sheep	dig prot % *		3.2	

Feed Name or Analyses			As Fed	Dry	C.V. ± %

SNAIL. Helix spp, Achatina spp

Snail, meat w shell, dehy grnd, (5)
Ref No 5 04 241 United States

			As Fed	Dry	C.V.
Dry matter		%	90.0	100.0	
Protein (N x 6.25)		%	67.4	74.9	7
Arginine		%	6.40	7.11	42
Histidine		%	1.00	1.11	
Isoleucine		%	3.20	3.56	17
Leucine		%	3.90	4.33	18
Lysine		%	5.30	5.89	34
Methionine		%	0.70	0.78	8
Phenylalanine		%	2.40	2.67	11
Threonine		%	2.60	2.89	
Tryptophan		%	0.40	0.44	
Valine		%	3.20	3.56	5

SNAILSEED, CAROLINA. Cocculus carolinus

Snailseed, Carolina, seeds w hulls, (4)
Ref No 4 09 049 United States

			As Fed	Dry	C.V.
Dry matter		%		100.0	
Protein (N x 6.25)		%		17.0	
Cattle	dig prot	% *		11.6	
Goats	dig prot	% *		12.8	
Horses	dig prot	% *		12.8	
Sheep	dig prot	% *		12.8	
Alanine		%		0.63	
Arginine		%		1.48	
Aspartic acid		%		1.24	
Glutamic acid		%		2.64	
Glycine		%		0.78	
Histidine		%		0.37	
Hydroxyproline		%		0.02	
Isoleucine		%		0.53	
Leucine		%		0.90	
Lysine		%		0.58	
Methionine		%		0.22	
Phenylalanine		%		0.54	
Proline		%		0.60	
Serine		%		0.63	
Threonine		%		0.44	
Tyrosine		%		0.46	
Valine		%		0.73	

SNAKEWEED. Gutierrezia spp

Snakeweed, aerial part, fresh, (2)
Ref No 2 04 243 United States

			As Fed	Dry	C.V.
Dry matter		%		100.0	
Chlorine		%		0.04	
Cobalt		mg/kg		0.306	
Sodium		%		0.04	
Sulphur		%		0.20	
Carotene		mg/kg		15.2	
Vitamin A equivalent		IU/g		25.4	

Snakeweed, aerial part, fresh, over ripe, (2)
Ref No 2 04 242 United States

			As Fed	Dry	C.V.
Dry matter		%		100.0	
Ash		%		4.7	
Crude fiber		%		43.2	
Ether extract		%		2.9	
N-free extract		%		44.3	
Protein (N x 6.25)		%		4.9	
Cattle	dig prot	% *		2.1	
Goats	dig prot	% *		1.1	
Horses	dig prot	% *		1.7	
Rabbits	dig prot	% *		2.5	
Sheep	dig prot	% *		1.6	

SNAKEWEED, BROOM. Gutierrezia sarothrae

Snakeweed, broom, aerial part, fresh, (2)
Ref No 2 04 244 United States

			As Fed	Dry	C.V.
Dry matter		%		100.0	
Ash		%		5.3	8
Crude fiber		%		29.3	7
Ether extract		%		11.2	24
N-free extract		%		47.5	
Protein (N x 6.25)		%		6.7	8
Cattle	dig prot	% *		3.6	
Goats	dig prot	% *		2.8	
Horses	dig prot	% *		3.2	
Rabbits	dig prot	% *		3.8	
Sheep	dig prot	% *		3.2	
Calcium		%		1.32	12
Magnesium		%		0.28	
Phosphorus		%		0.20	98
Sulphur		%		0.21	

SNAKEWEED, TEXAS. Gutierrezia texana

Snakeweed, Texas, aerial part, fresh, pre-bloom, (2)
Ref No 2 04 245 United States

			As Fed	Dry	C.V.
Dry matter		%		100.0	
Ash		%		6.0	
Crude fiber		%		19.0	
Ether extract		%		7.2	
N-free extract		%		53.1	
Protein (N x 6.25)		%		14.7	
Cattle	dig prot	% *		10.4	
Goats	dig prot	% *		10.3	
Horses	dig prot	% *		10.0	
Rabbits	dig prot	% *		10.0	
Sheep	dig prot	% *		10.7	
Calcium		%		0.65	
Magnesium		%		0.21	
Phosphorus		%		0.17	
Potassium		%		2.98	

SNAKEWEED, THREADLEAF. Gutierrezia microcephala

Snakeweed, threadleaf, aerial part, fresh, mid-bloom, (2)
Ref No 2 04 246 United States

			As Fed	Dry	C.V.
Dry matter		%		100.0	
Ash		%		4.2	
Crude fiber		%		22.5	
Ether extract		%		15.8	
N-free extract		%		47.1	
Protein (N x 6.25)		%		10.4	
Cattle	dig prot	% *		6.7	
Goats	dig prot	% *		6.3	
Horses	dig prot	% *		6.4	
Rabbits	dig prot	% *		6.7	
Sheep	dig prot	% *		6.7	
Calcium		%		0.74	
Magnesium		%		0.16	
Phosphorus		%		0.31	
Potassium		%		1.35	

SNEEZEWEED. Helenium spp

Sneezeweed, aerial part, fresh, immature, (2)
Ref No 2 04 247 United States

			As Fed	Dry	C.V.
Dry matter		%		100.0	
Ash		%		12.3	
Crude fiber		%		11.0	
Ether extract		%		4.8	
N-free extract		%		53.2	
Protein (N x 6.25)		%		18.7	
Cattle	dig prot	% *		13.8	
Goats	dig prot	% *		14.0	
Horses	dig prot	% *		13.4	
Rabbits	dig prot	% *		13.1	
Sheep	dig prot	% *		14.4	
Calcium		%		2.44	
Magnesium		%		0.18	
Phosphorus		%		0.61	
Potassium		%		3.95	

SNOWBERRY. Symphoricarpos spp

Snowberry, browse, (2)
Ref No 2 04 253 United States

			As Fed	Dry	C.V.
Dry matter		%	43.9	100.0	11
Ash		%	3.3	7.5	15
Crude fiber		%	6.4	14.6	46
Ether extract		%	2.3	5.3	17
N-free extract		%	27.2	62.0	
Protein (N x 6.25)		%	4.7	10.6	30
Cattle	dig prot	% *	3.0	6.9	
Goats	dig prot	% *	2.8	6.4	
Horses	dig prot	% *	2.9	6.5	
Rabbits	dig prot	% *	3.0	6.9	
Sheep	dig prot	% *	3.0	6.9	
Calcium		%	0.74	1.68	26
Magnesium		%	0.21	0.47	29
Phosphorus		%	0.17	0.39	26

Snowberry, browse, immature, (2)
Ref No 2 04 248 United States

			As Fed	Dry	C.V.
Dry matter		%	37.2	100.0	8
Ash		%	2.6	7.1	7
Crude fiber		%	4.5	12.2	42
Ether extract		%	1.8	4.8	13
N-free extract		%	22.8	61.3	
Protein (N x 6.25)		%	5.4	14.6	11
Cattle	dig prot	% *	3.8	10.3	
Goats	dig prot	% *	3.8	10.2	
Horses	dig prot	% *	3.7	9.9	
Rabbits	dig prot	% *	3.7	9.9	
Sheep	dig prot	% *	3.9	10.6	

Feed Name or Analyses				Mean As Fed	Mean Dry	C.V. ± %

Snowberry, browse, mid-bloom, (2)
Ref No 2 04 250 United States

Feed Name or Analyses				As Fed	Dry	C.V. ± %
Dry matter		%		43.9	100.0	3
Ash		%		3.4	7.8	11
Crude fiber		%		6.0	13.7	12
Ether extract		%		2.4	5.5	10
N-free extract		%		27.8	63.4	
Protein (N x 6.25)		%		4.2	9.6	8
Cattle	dig prot	%	*	2.7	6.1	
Goats	dig prot	%	*	2.4	5.5	
Horses	dig prot	%	*	2.5	5.7	
Rabbits	dig prot	%	*	2.7	6.1	
Sheep	dig prot	%	*	2.6	5.9	
Calcium		%		0.91	2.08	16
Magnesium		%		0.26	0.60	16
Phosphorus		%		0.18	0.41	21

Snowberry, browse, mature, (2)
Ref No 2 04 251 United States

Feed Name or Analyses				As Fed	Dry	C.V. ± %
Dry matter		%		49.1	100.0	4
Ash		%		4.1	8.3	16
Crude fiber		%		7.8	15.8	48
Ether extract		%		2.9	5.9	21
N-free extract		%		30.7	62.6	
Protein (N x 6.25)		%		3.6	7.4	19
Cattle	dig prot	%	*	2.1	4.2	
Goats	dig prot	%	*	1.7	3.5	
Horses	dig prot	%	*	1.9	3.8	
Rabbits	dig prot	%	*	2.2	4.4	
Sheep	dig prot	%	*	1.9	3.9	
Calcium		%		0.92	1.88	21
Phosphorus		%		0.14	0.28	29

Snowberry, browse, over ripe, (2)
Ref No 2 04 252 United States

Feed Name or Analyses				As Fed	Dry	C.V. ± %
Dry matter		%		52.9	100.0	
Ash		%		3.2	6.0	
Crude fiber		%		13.8	26.1	45
Ether extract		%		2.3	4.3	36
N-free extract		%		30.8	58.2	
Protein (N x 6.25)		%		2.9	5.4	19
Cattle	dig prot	%	*	1.3	2.5	
Goats	dig prot	%	*	0.8	1.6	
Horses	dig prot	%	*	1.1	2.1	
Rabbits	dig prot	%	*	1.5	2.8	
Sheep	dig prot	%	*	1.1	2.0	
Calcium		%		0.38	0.72	
Phosphorus		%		0.13	0.24	

Snowberry, stems, fresh, (2)
Ref No 2 04 258 United States

Feed Name or Analyses				As Fed	Dry	C.V. ± %
Dry matter		%		54.7	100.0	10
Ash		%		2.5	4.5	19
Crude fiber		%		21.5	39.3	17
Ether extract		%		1.2	2.2	50
N-free extract		%		27.2	49.7	
Protein (N x 6.25)		%		2.4	4.3	38
Cattle	dig prot	%	*	0.8	1.5	
Goats	dig prot	%	*	0.3	0.6	
Horses	dig prot	%	*	0.6	1.2	
Rabbits	dig prot	%	*	1.1	2.0	
Sheep	dig prot	%	*	0.5	1.0	
Calcium		%		0.45	0.82	29
Phosphorus		%		0.09	0.16	47

(1) dry forages and roughages (3) silages (6) minerals
(2) pasture, range plants, and forages fed green (4) energy feeds (7) vitamins
(5) protein supplements (8) additives

Snowberry, stems, fresh, immature, (2)
Ref No 2 04 254 United States

Feed Name or Analyses				As Fed	Dry	C.V. ± %
Dry matter		%		39.3	100.0	
Ash		%		3.1	7.9	
Crude fiber		%		4.9	12.4	
Ether extract		%		2.5	6.3	
N-free extract		%		24.4	62.1	
Protein (N x 6.25)		%		4.4	11.3	
Cattle	dig prot	%	*	2.9	7.5	
Goats	dig prot	%	*	2.8	7.1	
Horses	dig prot	%	*	2.8	7.1	
Rabbits	dig prot	%	*	2.9	7.4	
Sheep	dig prot	%	*	3.0	7.5	
Calcium		%		0.66	1.69	
Phosphorus		%		0.17	0.44	

Snowberry, stems, fresh, mid-bloom, (2)
Ref No 2 04 255 United States

Feed Name or Analyses				As Fed	Dry	C.V. ± %
Dry matter		%		52.1	100.0	7
Calcium		%		0.42	0.81	7
Phosphorus		%		0.08	0.16	25

Snowberry, stems, fresh, mature, (2)
Ref No 2 04 257 United States

Feed Name or Analyses				As Fed	Dry	C.V. ± %
Dry matter		%		61.3	100.0	8
Ash		%		2.7	4.4	5
Crude fiber		%		24.3	39.7	4
Ether extract		%		1.1	1.8	13
N-free extract		%		30.6	49.9	
Protein (N x 6.25)		%		2.6	4.2	9
Cattle	dig prot	%	*	0.9	1.5	
Goats	dig prot	%	*	0.3	0.5	
Horses	dig prot	%	*	0.7	1.1	
Rabbits	dig prot	%	*	1.2	1.9	
Sheep	dig prot	%	*	0.6	0.9	
Calcium		%		0.47	0.77	10
Phosphorus		%		0.09	0.15	20

SNOWBERRY, ROUNDLEAF. Symphoricarpos Rotundifolius

Snowberry, roundleaf, browse, immature, (2)
Ref No 2 04 259 United States

Feed Name or Analyses				As Fed	Dry	C.V. ± %
Dry matter		%			100.0	
Ash		%			6.9	
Crude fiber		%			11.7	
Ether extract		%			3.9	
N-free extract		%			61.5	
Protein (N x 6.25)		%			16.0	
Cattle	dig prot	%	*		11.5	
Goats	dig prot	%	*		11.5	
Horses	dig prot	%	*		11.1	
Rabbits	dig prot	%	*		11.0	
Sheep	dig prot	%	*		11.9	
Calcium		%			1.22	19
Magnesium		%			0.34	26
Phosphorus		%			0.38	18
Sulphur		%			0.24	

Snowberry, roundleaf, browse, mid-bloom, (2)
Ref No 2 04 260 United States

Feed Name or Analyses				As Fed	Dry	C.V. ± %
Dry matter		%			100.0	
Ash		%			9.7	
Crude fiber		%			13.6	
Ether extract		%			4.6	
N-free extract		%			63.4	
Protein (N x 6.25)		%			8.7	
Cattle	dig prot	%	*		5.3	
Goats	dig prot	%	*		4.7	
Horses	dig prot	%	*		4.9	
Rabbits	dig prot	%	*		5.4	
Sheep	dig prot	%	*		5.1	
Calcium		%			2.58	8
Magnesium		%			0.60	16
Phosphorus		%			0.52	12
Sulphur		%			0.32	

Snowberry, roundleaf, twigs, fresh, (2)
Ref No 2 04 261 United States

Feed Name or Analyses				As Fed	Dry	C.V. ± %
Dry matter		%			100.0	
Ash		%			3.8	
Crude fiber		%			41.1	
Ether extract		%			2.0	
N-free extract		%			48.9	
Protein (N x 6.25)		%			4.2	
Cattle	dig prot	%	*		1.5	
Goats	dig prot	%	*		0.5	
Horses	dig prot	%	*		1.1	
Rabbits	dig prot	%	*		1.9	
Sheep	dig prot	%	*		0.9	

SNOWBERRY, WHORTLELEAF. Symphoricarpos vaccinioides

Snowberry, whortleleaf, browse, mature, (2)
Ref No 2 04 262 United States

Feed Name or Analyses				As Fed	Dry	C.V. ± %
Dry matter		%		55.6	100.0	
Ash		%		2.3	4.1	
Crude fiber		%		21.6	38.8	
Ether extract		%		1.2	2.2	
N-free extract		%		28.4	51.0	
Protein (N x 6.25)		%		2.2	3.9	
Cattle	dig prot	%	*	0.7	1.2	
Goats	dig prot	%	*	0.1	0.2	
Horses	dig prot	%	*	0.5	0.8	
Rabbits	dig prot	%	*	0.9	1.7	
Sheep	dig prot	%	*	0.3	0.6	
Calcium		%		0.43	0.77	
Phosphorus		%		0.08	0.14	

SOAPBERRY, CHINESE. Sapindus mukorossi

Soapberry, Chinese, seeds w hulls, (5)
Ref No 5 09 120 — United States

Feed Name or Analyses				Mean As Fed	Mean Dry	C.V. ± %
Dry matter		%			100.0	
Protein (N x 6.25)		%			31.0	
Alanine		%			1.21	
Arginine		%			2.98	
Aspartic acid		%			2.02	
Glutamic acid		%			5.33	
Glycine		%			1.33	
Histidine		%			0.53	
Hydroxyproline		%			0.00	
Isoleucine		%			0.96	
Leucine		%			1.64	
Lysine		%			1.77	
Methionine		%			0.43	
Phenylalanine		%			1.27	
Proline		%			1.33	
Serine		%			1.12	
Threonine		%			0.84	
Tyrosine		%			0.81	
Valine		%			1.33	

SOAPWEED. Yucca spp

Soapweed, aerial part, s-c, (1)
Ref No 1 04 263 — United States

Feed Name or Analyses				Mean As Fed	Mean Dry	C.V. ± %
Dry matter		%		47.9	100.0	
Ash		%		3.6	7.5	
Crude fiber		%		15.3	31.9	
Cattle	dig coef	%		34.	34.	
Ether extract		%		0.9	1.8	
Cattle	dig coef	%		-11.	-11.	
N-free extract		%		25.3	52.9	
Cattle	dig coef	%		73.	73.	
Protein (N x 6.25)		%		2.8	5.9	
Cattle	dig coef	%		43.	43.	
Cattle	dig prot	%		1.2	2.5	
Goats	dig prot	%	*	1.0	2.1	
Horses	dig prot	%	*	1.2	2.5	
Rabbits	dig prot	%	*	1.5	3.2	
Sheep	dig prot	%	*	0.9	1.9	
Energy	GE Mcal/kg					
Cattle	DE Mcal/kg		*	1.09	2.27	
Sheep	DE Mcal/kg		*	1.17	2.45	
Cattle	ME Mcal/kg		*	0.89	1.86	
Sheep	ME Mcal/kg		*	0.96	2.01	
Cattle	TDN	%		24.7	51.5	
Sheep	TDN	%	*	26.6	55.5	

Soapweed, stems, s-c, (1)
Ref No 1 04 264 — United States

Feed Name or Analyses				Mean As Fed	Mean Dry	C.V. ± %
Dry matter		%		91.4	100.0	
Ash		%		5.6	6.1	
Crude fiber		%		35.4	38.7	
Sheep	dig coef	%		33.	33.	
Ether extract		%		1.0	1.1	
Sheep	dig coef	%		-67.	-67.	
N-free extract		%		45.1	49.3	
Sheep	dig coef	%		70.	70.	
Protein (N x 6.25)		%		4.4	4.8	
Sheep	dig coef	%		14.	14.	
Cattle	dig prot	%	*	1.0	1.1	
Goats	dig prot	%	*	0.9	1.0	
Horses	dig prot	%	*	1.5	1.6	
Rabbits	dig prot	%	*	2.2	2.4	
Sheep	dig prot	%		0.6	0.7	
Energy	GE Mcal/kg					
Cattle	DE Mcal/kg		*	1.90	2.08	
Sheep	DE Mcal/kg		*	1.86	2.04	
Cattle	ME Mcal/kg		*	1.56	1.70	
Sheep	ME Mcal/kg		*	1.53	1.67	
Cattle	TDN	%	*	43.1	47.1	
Sheep	TDN	%		42.3	46.3	

Soapweed, leaves, fresh, (2)
Ref No 2 04 265 — United States

Feed Name or Analyses				Mean As Fed	Mean Dry	C.V. ± %
Dry matter		%		30.3	100.0	41
Ash		%		2.2	7.1	21
Crude fiber		%		10.1	33.3	5
Ether extract		%		0.9	3.0	23
N-free extract		%		14.8	49.0	
Protein (N x 6.25)		%		2.3	7.6	8
Cattle	dig prot	%	*	1.3	4.4	
Goats	dig prot	%	*	1.1	3.7	
Horses	dig prot	%	*	1.2	4.0	
Rabbits	dig prot	%	*	1.4	4.5	
Sheep	dig prot	%	*	1.2	4.1	
Energy	GE Mcal/kg					
Cattle	DE Mcal/kg		*	0.86	2.83	
Sheep	DE Mcal/kg		*	0.89	2.94	
Cattle	ME Mcal/kg		*	0.70	2.32	
Sheep	ME Mcal/kg		*	0.73	2.41	
Cattle	TDN	%	*	19.4	64.2	
Sheep	TDN	%	*	20.2	66.6	
Calcium		%		0.50	1.64	
Chlorine		%		0.12	0.40	
Iron		%		0.028	0.091	
Magnesium		%		0.39	1.28	
Manganese		mg/kg		70.4	232.4	
Phosphorus		%		0.08	0.27	
Potassium		%		0.35	1.16	
Sodium		%		0.09	0.30	
Sulphur		%		0.08	0.26	

Soapweed, stems, fresh, (2)
Ref No 2 04 266 — United States

Feed Name or Analyses				Mean As Fed	Mean Dry	C.V. ± %
Dry matter		%		42.6	100.0	
Ash		%		3.5	8.1	20
Crude fiber		%		17.0	39.8	13
Ether extract		%		0.9	2.1	81
N-free extract		%		19.2	45.1	
Protein (N x 6.25)		%		2.1	4.9	39
Cattle	dig prot	%	*	0.9	2.1	
Goats	dig prot	%	*	0.5	1.1	
Horses	dig prot	%	*	0.7	1.7	
Rabbits	dig prot	%	*	1.0	2.5	
Sheep	dig prot	%	*	0.7	1.6	
Calcium		%		0.78	1.82	
Chlorine		%		0.02	0.05	
Iron		%		0.003	0.008	
Magnesium		%		0.27	0.64	
Manganese		mg/kg		0.0	0.0	
Phosphorus		%		0.03	0.07	
Potassium		%		0.26	0.61	
Sodium		%		0.07	0.16	
Sulphur		%		0.09	0.21	

SOAPWEED, SMALL. Yucca glauca

Soapweed, small, hay, s-c, (1)
Ref No 1 04 267 — United States

Feed Name or Analyses				Mean As Fed	Mean Dry	C.V. ± %
Dry matter		%		92.8	100.0	
Ash		%		7.4	8.0	
Crude fiber		%		38.6	41.6	
Sheep	dig coef	%		71.	71.	
Ether extract		%		2.1	2.3	
Sheep	dig coef	%		-8.	-8.	
N-free extract		%		37.9	40.8	
Sheep	dig coef	%		64.	64.	
Protein (N x 6.25)		%		6.8	7.3	
Sheep	dig coef	%		24.	24.	
Cattle	dig prot	%	*	3.0	3.3	
Goats	dig prot	%	*	3.1	3.4	
Horses	dig prot	%	*	3.5	3.7	
Rabbits	dig prot	%	*	4.0	4.3	
Sheep	dig prot	%		1.6	1.8	
Energy	GE Mcal/kg					
Cattle	DE Mcal/kg		*	2.20	2.37	
Sheep	DE Mcal/kg		*	2.33	2.51	
Cattle	ME Mcal/kg		*	1.80	1.94	
Sheep	ME Mcal/kg		*	1.91	2.06	
Cattle	TDN	%	*	49.8	53.7	
Sheep	TDN	%		52.8	56.9	

Soapweed, small, aerial part, fresh, (2)
Ref No 2 04 268 — United States

Feed Name or Analyses				Mean As Fed	Mean Dry	C.V. ± %
Dry matter		%		45.2	100.0	4
Ash		%		3.4	7.5	8
Crude fiber		%		14.9	32.9	8
Ether extract		%		0.9	1.9	12
N-free extract		%		23.5	52.0	
Protein (N x 6.25)		%		2.6	5.7	9
Cattle	dig prot	%	*	1.2	2.7	
Goats	dig prot	%	*	0.8	1.9	
Horses	dig prot	%	*	1.1	2.4	
Rabbits	dig prot	%	*	1.4	3.1	
Sheep	dig prot	%	*	1.0	2.3	
Energy	GE Mcal/kg					
Cattle	DE Mcal/kg		*	1.26	2.78	
Sheep	DE Mcal/kg		*	1.37	3.02	
Cattle	ME Mcal/kg		*	1.03	2.28	
Sheep	ME Mcal/kg		*	1.12	2.48	
Cattle	TDN	%	*	28.5	63.0	
Sheep	TDN	%	*	31.0	68.5	

Soapweed, small, heads, fresh, (2)
Ref No 2 04 269 — United States

Feed Name or Analyses				Mean As Fed	Mean Dry	C.V. ± %
Dry matter		%			100.0	
Ash		%			11.0	
Crude fiber		%			22.5	
Ether extract		%			1.5	
N-free extract		%			59.3	
Protein (N x 6.25)		%			5.7	
Cattle	dig prot	%	*		2.7	
Goats	dig prot	%	*		1.9	
Horses	dig prot	%	*		2.4	
Rabbits	dig prot	%	*		3.1	
Sheep	dig prot	%	*		2.3	
Energy	GE Mcal/kg					
Cattle	DE Mcal/kg		*		2.49	

Continued

Feed Name or Analyses		Mean As Fed	Mean Dry	C.V. ± %
Sheep	DE Mcal/kg *		2.60	
Cattle	ME Mcal/kg *		2.04	
Sheep	ME Mcal/kg *		2.21	
Cattle	TDN % *		56.4	
Sheep	TDN % *		58.9	

SODIUM BICARBONATE

Sodium bicarbonate, $NaHCO_3$, cp, (6)

Ref No 6 04 273 — United States

Analyses		As Fed	Dry	C.V. ± %
Dry matter	%		100.0	
Chlorine	%		0.00	
Iron	%		0.001	
Potassium	%		0.01	
Sodium	%		27.36	

SODIUM FLUORIDE

Sodium fluoride, NaF, cp, (6)

Ref No 6 04 276 — United States

Analyses		As Fed	Dry	C.V. ± %
Dry matter	%		100.0	
Chlorine	%		0.00	
Fluorine	%		45.2	
Iron	%		0.003	
Potassium	%		0.02	
Sodium	%		54.75	

SODIUM IODIDE

Sodium iodide, NaI, cp, (6)

Ref No 6 04 280 — United States

Analyses		As Fed	Dry	C.V. ± %
Dry matter	%		100.0	
Chlorine	%		0.01	
Iodine	%		84.7	
Iron	%		0.001	
Sodium	%		15.33	

SODIUM MOLYBDATE

Sodium molybdate, $Na_2MoO_4 \cdot 2H_2O$, cp, (6)

Ref No 6 04 282 — United States

Analyses		As Fed	Dry	C.V. ± %
Dry matter	%	85.1	100.0	
Chlorine	%	0.01	0.01	
Iron	%	0.001	0.001	
Molybdenum	%	39.7	46.6	
Sodium	%	19.01	22.34	

SODIUM PHOSPHATE, DIBASIC

Sodium phosphate, dibasic, anhydrous, Na_2HPO_4, cp, (6)

Ref No 6 04 285 — United States

Analyses		As Fed	Dry	C.V. ± %
Dry matter	%		100.0	
Chlorine	%		0.00	
Iron	%		0.001	
Phosphorus	%		21.81	
Sodium	%		32.38	

SODIUM PHOSPHATE, MONOBASIC

Sodium phosphate, monobasic, $NaH_2PO_4 \cdot H_2O$, cp, (6)

Ref No 6 04 287 — United States

Analyses		As Fed	Dry	C.V. ± %
Dry matter	%	87.0	100.0	
Iron	%	0.001	0.001	
Phosphorus	%	22.45	25.80	
Sodium	%	16.66	19.15	

SODIUM SULFATE

Sodium sulfate, $Na_2SO_4 \cdot 10H_2O$, cp, (6)

Ref No 6 04 292 — United States

Analyses		As Fed	Dry	C.V. ± %
Dry matter	%	44.1	100.0	
Chlorine	%	0.00	0.00	
Iron	%	0.000	0.001	
Sodium	%	14.27	32.36	
Sulphur	%	9.96	22.59	

Solvent extracted linseed meal (CFA) –
see Flax, seeds, solv-extd grnd, mx 0.5% acid insoluble ash, (5)

Solvent extracted peanut meal (AAFCO) –
see Peanut, kernels, solv-extd grnd, mx 7% fiber, (5)

Solvent extracted peanut meal and hulls (AAFCO) –
see Peanut, kernels w skins w hulls, solv-extd grnd, (5)

Solvent extracted rice bran (AAFCO) –
see Rice, bran w germ, solv-extd, mn 14% protein mx 14% fiber, (4)

Solvent extracted soy flour (AAFCO) –
see Soybean, flour, solv-extd fine sift, mx 3% fiber, (5)

SOPHORA, MESCALBEAN. Sophora secundiflora

Sophora, mescalbean, browse, immature, (2)

Ref No 2 04 295 — United States

Analyses		As Fed	Dry	C.V. ± %
Dry matter	%		100.0	
Ash	%		5.2	
Crude fiber	%		31.1	
Ether extract	%		2.6	
N-free extract	%		42.0	
Protein (N x 6.25)	%		19.1	
Cattle	dig prot % *		14.1	
Goats	dig prot % *		14.4	
Horses	dig prot % *		13.7	
Rabbits	dig prot % *		13.4	
Sheep	dig prot % *		14.8	
Calcium	%		1.99	
Magnesium	%		0.21	
Phosphorus	%		0.14	
Potassium	%		0.89	

Sophora, mescalbean, pods, fresh, immature, (2)

Ref No 2 04 296 — United States

Analyses		As Fed	Dry	C.V. ± %
Dry matter	%		100.0	
Ash	%		3.2	
Crude fiber	%		15.1	
Ether extract	%		0.6	
N-free extract	%		62.9	
Protein (N x 6.25)	%		18.2	
Cattle	dig prot % *		13.4	
Goats	dig prot % *		13.5	
Horses	dig prot % *		13.0	
Rabbits	dig prot % *		12.7	
Sheep	dig prot % *		14.0	
Calcium	%		0.50	
Magnesium	%		0.17	
Phosphorus	%		0.14	
Potassium	%		1.15	

SORGHUM. Sorghum vulgare

Sorghum, aerial part, dehy, (1)

Ref No 1 08 511 — United States

Analyses		As Fed	Dry	C.V. ± %
Dry matter	%	90.3	100.0	
Ash	%	7.8	8.6	
Crude fiber	%	18.5	20.5	
Ether extract	%	1.7	1.9	
N-free extract	%	56.7	62.8	
Protein (N x 6.25)	%	5.6	6.2	
Cattle	dig prot % *	2.1	2.3	
Goats	dig prot % *	2.1	2.3	
Horses	dig prot % *	2.5	2.8	
Rabbits	dig prot % *	3.1	3.5	
Sheep	dig prot % *	1.9	2.1	
Energy	GE Mcal/kg			
Cattle	DE Mcal/kg *	2.24	3.03	
Sheep	DE Mcal/kg *	2.42	2.68	
Cattle	ME Mcal/kg *	2.24	2.48	
Sheep	ME Mcal/kg *	1.99	2.20	
Cattle	TDN % *	60.9	67.4	
Sheep	TDN % *	55.0	60.9	
Calcium	%	0.39	0.43	
Phosphorus	%	0.16	0.18	

Sorghum, aerial part, s-c, early leaf, (1)

Ref No 1 04 298 — United States

Analyses		As Fed	Dry	C.V. ± %
Dry matter	%	87.9	100.0	
Ash	%	8.7	9.9	
Crude fiber	%	29.1	33.1	
Sheep	dig coef %	66.	66.	
Ether extract	%	1.3	1.5	
Sheep	dig coef %	18.	18.	
N-free extract	%	36.0	41.0	
Sheep	dig coef %	50.	50.	
Protein (N x 6.25)	%	12.7	14.5	
Sheep	dig coef %	56.	56.	
Cattle	dig prot % *	8.3	9.5	
Goats	dig prot % *	8.9	10.1	
Horses	dig prot % *	8.6	9.8	
Rabbits	dig prot % *	8.7	9.9	
Sheep	dig prot %	7.1	8.1	
Energy	GE Mcal/kg			
Cattle	DE Mcal/kg *	2.22	2.53	
Sheep	DE Mcal/kg *	1.98	2.25	
Cattle	ME Mcal/kg *	1.82	2.07	

(1) dry forages and roughages (2) pasture, range plants, and forages fed green (3) silages (4) energy feeds (5) protein supplements (6) minerals (7) vitamins (8) additives

Feed Name or Analyses			Mean As Fed	Mean Dry	C.V. ± %
Sheep	ME Mcal/kg	*	1.62	1.85	
Cattle	TDN %	*	50.3	57.3	
Sheep	TDN %		44.9	51.1	

Sorghum, aerial part, s-c, immature, (1)
Sorghum fodder, sun-cured, immature
Ref No 1 04 299 United States

Feed Name or Analyses			As Fed	Dry	C.V. ± %
Dry matter	%			100.0	
Crude fiber	%				
Cattle	dig coef %	*		60.	
Sheep	dig coef %	*		63.	
Ether extract	%				
Cattle	dig coef %	*		61.	
Sheep	dig coef %	*		53.	
N-free extract	%				
Cattle	dig coef %	*		66.	
Sheep	dig coef %	*		61.	
Protein (N x 6.25)	%				
Cattle	dig coef %	*		38.	
Sheep	dig coef %	*		38.	

Sorghum, aerial part, s-c, mature, (1)
Sorghum fodder, sun-cured, mature
Ref No 1 04 301 United States

Feed Name or Analyses			As Fed	Dry	C.V. ± %
Dry matter	%			100.0	
Crude fiber	%				
Cattle	dig coef %	*		60.	
Sheep	dig coef %	*		63.	
Ether extract	%				
Cattle	dig coef %	*		61.	
Sheep	dig coef %	*		53.	
N-free extract	%				
Cattle	dig coef %	*		66.	
Sheep	dig coef %	*		61.	
Protein (N x 6.25)	%				
Cattle	dig coef %	*		38.	
Sheep	dig coef %	*		38.	

Ref No 1 04 301 Canada

Feed Name or Analyses			As Fed	Dry	C.V. ± %
Dry matter	%		48.6	100.0	
Protein (N x 6.25)	%		6.9	14.2	
Cattle	dig prot %	*	4.5	9.2	
Goats	dig prot %	*	4.8	9.8	
Horses	dig prot %	*	4.7	9.6	
Rabbits	dig prot %	*	4.7	9.6	
Sheep	dig prot %	*	4.5	9.3	

Sorghum, aerial part wo heads, s-c, (1)
Sorghum stover, sun-cured
Ref No 1 04 302 United States

Feed Name or Analyses			As Fed	Dry	C.V. ± %
Dry matter	%			100.0	
Crude fiber	%				
Cattle	dig coef %	*		67.	
Sheep	dig coef %	*		66.	
Ether extract	%				
Cattle	dig coef %	*		75.	
Sheep	dig coef %	*		58.	
N-free extract	%				
Cattle	dig coef %	*		60.	
Sheep	dig coef %	*		49.	

Feed Name or Analyses			Mean As Fed	Mean Dry	C.V. ± %
Protein (N x 6.25)	%				
Cattle	dig coef %	*		34.	
Sheep	dig coef %	*		-17.	

Sorghum, heads wo seeds, (1)
Ref No 1 04 303 United States

Feed Name or Analyses			As Fed	Dry	C.V. ± %
Dry matter	%		92.0	100.0	1
Ash	%		7.3	7.9	11
Crude fiber	%		21.1	22.9	7
Ether extract	%		1.3	1.4	34
N-free extract	%		56.2	61.1	
Protein (N x 6.25)	%		6.2	6.7	12
Cattle	dig prot %	*	2.5	2.7	
Goats	dig prot %	*	2.6	2.8	
Horses	dig prot %	*	3.0	3.2	
Rabbits	dig prot %	*	3.5	3.8	
Sheep	dig prot %	*	2.4	2.6	

Sorghum, hulls, mature, (1)
Ref No 1 04 305 United States

Feed Name or Analyses			As Fed	Dry	C.V. ± %
Dry matter	%		92.7	100.0	
Cellulose (Matrone)	%		47.6	51.4	
Pentosans	%		28.0	30.2	
Starch	%		2.6	2.8	
Sugars, total	%		0.6	0.6	
Lignin (Ellis)	%		15.4	16.6	

Sorghum, leaves, (1)
Ref No 1 04 308 United States

Feed Name or Analyses			As Fed	Dry	C.V. ± %
Dry matter	%		89.5	100.0	
Ash	%		4.8	5.4	
Crude fiber	%		20.6	23.0	
Ether extract	%		5.2	5.8	
N-free extract	%		49.1	54.9	
Protein (N x 6.25)	%		9.8	10.9	
Cattle	dig prot %	*	5.7	6.4	
Goats	dig prot %	*	6.0	6.7	
Horses	dig prot %	*	6.1	6.8	
Rabbits	dig prot %	*	6.3	7.1	
Sheep	dig prot %	*	5.7	6.3	
Energy	GE Mcal/kg				
Cattle	DE Mcal/kg	*	2.59	2.89	
Sheep	DE Mcal/kg	*	2.42	2.70	
Cattle	ME Mcal/kg	*	2.12	2.37	
Sheep	ME Mcal/kg	*	1.98	2.22	
Cattle	TDN %	*	58.6	65.5	
Sheep	TDN %	*	54.9	61.3	

Sorghum, leaves, s-c, dough stage, (1)
Ref No 1 04 306 United States

Feed Name or Analyses			As Fed	Dry	C.V. ± %
Dry matter	%		87.6	100.0	
Ash	%		4.5	5.1	
Crude fiber	%		24.0	27.4	
Cattle	dig coef %		76.	76.	
Ether extract	%		4.6	5.2	
Cattle	dig coef %		46.	46.	
N-free extract	%		44.9	51.3	
Cattle	dig coef %		67.	67.	
Protein (N x 6.25)	%		9.6	11.0	
Cattle	dig coef %		62.	62.	
Cattle	dig prot %		6.0	6.8	
Goats	dig prot %	*	6.0	6.8	

Feed Name or Analyses			Mean As Fed	Mean Dry	C.V. ± %
Horses	dig prot %	*	6.0	6.9	
Rabbits	dig prot %	*	6.3	7.2	
Sheep	dig prot %	*	5.6	6.4	
Energy	GE Mcal/kg				
Cattle	DE Mcal/kg	*	2.60	2.97	
Sheep	DE Mcal/kg	*	2.29	2.62	
Cattle	ME Mcal/kg	*	2.13	2.44	
Sheep	ME Mcal/kg	*	1.88	2.15	
Cattle	TDN %		59.0	67.4	
Sheep	TDN %	*	52.0	59.4	

Sorghum, leaves, dough stage, (1)
Ref No 1 04 307 United States

Feed Name or Analyses			As Fed	Dry	C.V. ± %
Dry matter	%		87.6	100.0	
Ash	%		4.3	4.9	
Crude fiber	%		24.2	27.6	
Ether extract	%		4.6	5.3	
N-free extract	%		44.9	51.2	
Protein (N x 6.25)	%		9.6	11.0	
Cattle	dig prot %	*	5.7	6.5	
Goats	dig prot %	*	6.0	6.8	
Horses	dig prot %	*	6.0	6.9	
Rabbits	dig prot %	*	6.3	7.2	
Sheep	dig prot %	*	5.6	6.4	
Energy	GE Mcal/kg				
Cattle	DE Mcal/kg	*	2.59	2.95	
Sheep	DE Mcal/kg	*	2.29	2.61	
Cattle	ME Mcal/kg	*	2.12	2.42	
Sheep	ME Mcal/kg	*	1.88	2.14	
Cattle	TDN %	*	58.7	67.0	
Sheep	TDN %	*	51.9	59.3	

Sorghum, pulp, (1)
Ref No 1 04 330 United States

Feed Name or Analyses			As Fed	Dry	C.V. ± %
Dry matter	%		89.3	100.0	
Ash	%		3.5	3.9	
Crude fiber	%		31.3	35.0	
Ether extract	%		1.4	1.6	
N-free extract	%		50.0	56.0	
Protein (N x 6.25)	%		3.1	3.5	
Cattle	dig prot %	*	0.0	0.0	
Goats	dig prot %	*	-0.1	-0.1	
Horses	dig prot %	*	0.4	0.5	
Rabbits	dig prot %	*	1.2	1.4	
Sheep	dig prot %	*	-0.2	-0.2	
Energy	GE Mcal/kg				
Cattle	DE Mcal/kg	*	2.05	2.29	
Sheep	DE Mcal/kg	*	2.11	2.37	
Cattle	ME Mcal/kg	*	1.68	1.88	
Sheep	ME Mcal/kg	*	1.73	1.94	
Cattle	TDN %	*	46.4	52.0	
Sheep	TDN %	*	47.9	53.7	

Sorghum, pulp, dehy, (1)
Sorghum bagasse, dried
Ref No 1 04 328 United States

Feed Name or Analyses			As Fed	Dry	C.V. ± %
Dry matter	%		90.9	100.0	2
Ash	%		3.3	3.6	21
Crude fiber	%		29.9	32.9	2
Ether extract	%		1.4	1.5	6
N-free extract	%		53.1	58.4	
Protein (N x 6.25)	%		3.3	3.6	27
Cattle	dig prot %	*	0.1	0.1	

Continued

Feed Name or Analyses	Mean As Fed	Mean Dry	C.V. ± %
Goats dig prot % *	0.0	0.0	
Horses dig prot % *	0.6	0.6	
Rabbits dig prot % *	1.3	1.5	
Sheep dig prot % *	-0.1	-0.1	
Cellulose (Matrone) %	46.1	50.7	1
Pentosans %	24.9	27.4	
Lignin (Ellis) %	10.7	11.8	
Fatty acids %	1.4	1.6	
Energy GE Mcal/kg			
Cattle DE Mcal/kg *	1.84	2.02	
Sheep DE Mcal/kg *	1.80	1.98	
Cattle ME Mcal/kg *	1.51	1.66	
Sheep ME Mcal/kg *	1.47	1.62	
Cattle TDN % *	41.6	46.8	
Sheep TDN % *	40.7	44.8	

Sorghum, stems, (1)
Ref No 1 04 309 United States

Feed Name or Analyses	Mean As Fed	Mean Dry	C.V. ± %
Dry matter %	90.1	100.0	2
Ash %	7.7	8.5	18
Crude fiber %	31.1	34.5	8
Ether extract %	1.5	1.7	9
N-free extract %	43.2	47.9	
Protein (N x 6.25) %	6.7	7.4	27
Cattle dig prot % *	3.0	3.3	
Goats dig prot % *	3.1	3.5	
Horses dig prot % *	3.4	3.8	
Rabbits dig prot % *	3.9	4.4	
Sheep dig prot % *	2.9	3.2	
Calcium %	0.20	0.22	
Magnesium %	0.69	0.77	
Phosphorus %	0.17	0.19	
Potassium %	1.11	1.23	
Sulphur %	0.05	0.05	

Sorghum, aerial part, fresh, (2)
Sorghum fodder, fresh
Ref No 2 04 317 United States

Feed Name or Analyses	Mean As Fed	Mean Dry	C.V. ± %
Dry matter %	82.5	100.0	10
Crude fiber %			
Cattle dig coef % *	61.	61.	
Sheep dig coef % *	56.	56.	
Ether extract %			
Cattle dig coef % *	39.	39.	
Sheep dig coef % *	65.	65.	
N-free extract %			
Cattle dig coef % *	61.	61.	
Sheep dig coef % *	72.	72.	
Protein (N x 6.25) %			
Cattle dig coef % *	37.	37.	
Sheep dig coef % *	44.	44.	
Biotin mg/kg	0.45	0.55	25
Vitamin D2 IU/g	0.5	0.6	

Sorghum, aerial part, fresh, immature, (2)
Sorghum fodder, fresh, immature
Ref No 2 04 310 United States

Feed Name or Analyses	Mean As Fed	Mean Dry	C.V. ± %
Dry matter %		100.0	
Crude fiber %			
Cattle dig coef % *		60.	
Sheep dig coef % *		56.	
Ether extract %			
Cattle dig coef % *		61.	
Sheep dig coef % *		65.	
N-free extract %			
Cattle dig coef % *		66.	
Sheep dig coef % *		72.	
Protein (N x 6.25) %			
Cattle dig coef % *		38.	
Sheep dig coef % *		44.	

(1) dry forages and roughages (2) pasture, range plants, and forages fed green (3) silages (4) energy feeds (5) protein supplements (6) minerals (7) vitamins (8) additives

Sorghum, aerial part, fresh, mid-bloom, (2)
Sorghum fodder, fresh, mid-bloom
Ref No 2 04 311 United States

Feed Name or Analyses	Mean As Fed	Mean Dry	C.V. ± %
Dry matter %	21.5	100.0	11
Ash %	2.2	10.3	8
Crude fiber %	7.4	34.3	11
Cattle dig coef % *	60.	60.	
Sheep dig coef % *	56.	56.	
Ether extract %	0.4	1.7	27
Cattle dig coef % *	61.	61.	
Sheep dig coef % *	65.	65.	
N-free extract %	9.5	44.0	
Cattle dig coef % *	66.	66.	
Sheep dig coef % *	72.	72.	
Protein (N x 6.25) %	2.1	9.7	19
Cattle dig coef % *	38.	38.	
Sheep dig coef % *	44.	44.	
Cattle dig prot %	0.8	3.7	
Goats dig prot % *	1.2	5.6	
Horses dig prot % *	1.2	5.8	
Rabbits dig prot % *	1.3	6.2	
Sheep dig prot %	0.9	4.3	
Energy GE Mcal/kg			
Cattle DE Mcal/kg *	0.53	2.45	
Sheep DE Mcal/kg *	0.55	2.54	
Cattle ME Mcal/kg *	0.43	2.01	
Sheep ME Mcal/kg *	0.45	2.08	
Cattle TDN % *	12.0	55.6	
Sheep TDN % *	12.4	57.6	

Sorghum, aerial part, fresh, late bloom, (2)
Ref No 2 04 312 United States

Feed Name or Analyses	Mean As Fed	Mean Dry	C.V. ± %
Dry matter %	33.0	100.0	
Ash %	2.9	8.7	
Crude fiber %	11.5	34.8	
Cattle dig coef %	61.	61.	
Ether extract %	0.5	1.5	
Cattle dig coef %	39.	39.	
N-free extract %	17.0	51.6	
Cattle dig coef %	61.	61.	
Protein (N x 6.25) %	1.1	3.4	
Cattle dig coef %	37.	37.	
Cattle dig prot %	0.4	1.3	
Goats dig prot % *	0.0	-0.2	
Horses dig prot % *	0.1	0.4	
Rabbits dig prot % *	0.4	1.3	
Sheep dig prot % *	0.1	0.2	
Energy GE Mcal/kg			
Cattle DE Mcal/kg *	0.80	2.44	
Sheep DE Mcal/kg *	0.84	2.54	
Cattle ME Mcal/kg *	0.66	2.00	
Sheep ME Mcal/kg *	0.69	2.08	
Cattle TDN %	18.2	55.3	
Sheep TDN % *	19.0	57.6	

Sorghum, aerial part, fresh, milk stage, (2)
Sorghum fodder, fresh, milk stage
Ref No 2 04 313 United States

Feed Name or Analyses	Mean As Fed	Mean Dry	C.V. ± %
Dry matter %	22.5	100.0	8
Ash %	2.3	10.0	12
Crude fiber %	7.9	35.3	5
Cattle dig coef % *	60.	60.	
Sheep dig coef % *	56.	56.	
Ether extract %	0.4	1.8	9
Cattle dig coef % *	61.	61.	
Sheep dig coef % *	65.	65.	
N-free extract %	10.1	44.9	
Cattle dig coef % *	66.	66.	
Sheep dig coef % *	72.	72.	
Protein (N x 6.25) %	1.8	8.0	6
Cattle dig coef % *	38.	38.	
Sheep dig coef % *	44.	44.	
Cattle dig prot %	0.7	3.0	
Goats dig prot % *	0.9	4.0	
Horses dig prot % *	1.0	4.3	
Rabbits dig prot % *	1.1	4.8	
Sheep dig prot %	0.8	3.5	
Energy GE Mcal/kg			
Cattle DE Mcal/kg *	0.56	2.48	
Sheep DE Mcal/kg *	0.58	2.57	
Cattle ME Mcal/kg *	0.46	2.04	
Sheep ME Mcal/kg *	0.47	2.11	
Cattle TDN % *	12.7	56.3	
Sheep TDN % *	13.1	58.2	

Sorghum, aerial part, fresh, dough stage, (2)
Ref No 2 04 314 United States

Feed Name or Analyses	Mean As Fed	Mean Dry	C.V. ± %
Dry matter %	25.5	100.0	
Ash %	2.9	11.3	
Crude fiber %	6.6	25.8	
Ether extract %	0.5	2.2	
N-free extract %	14.0	54.9	
Protein (N x 6.25) %	1.5	5.9	
Cattle dig prot % *	0.7	2.9	
Goats dig prot % *	0.5	2.0	
Horses dig prot % *	0.6	2.5	
Rabbits dig prot % *	0.8	3.2	
Sheep dig prot % *	0.6	2.4	
Lignin (Ellis) %	1.8	7.0	
Energy GE Mcal/kg			
Cattle DE Mcal/kg *	0.63	2.46	
Sheep DE Mcal/kg *	0.66	2.57	
Cattle ME Mcal/kg *	0.51	2.02	
Sheep ME Mcal/kg *	0.54	2.11	
Cattle TDN % *	14.2	55.8	
Sheep TDN % *	14.9	58.3	

Sorghum, aerial part, fresh, mature, (2)
Sorghum fodder, fresh, mature
Ref No 2 04 315 United States

Feed Name or Analyses	Mean As Fed	Mean Dry	C.V. ± %
Dry matter %	39.2	100.0	
Ash %	3.3	8.4	
Crude fiber %	10.7	27.3	
Cattle dig coef %	57.	57.	
Sheep dig coef % *	56.	56.	
Ether extract %	0.5	1.4	
Cattle dig coef %	28.	28.	
Sheep dig coef % *	65.	65.	

Feed Name or Analyses			Mean As Fed	Mean Dry	C.V. ± %
N-free extract	%		23.2	59.3	
Cattle	dig coef %		65.	65.	
Sheep	dig coef %	*	72.	72.	
Protein (N x 6.25)	%		1.4	3.6	
Cattle	dig coef %		15.	15.	
Sheep	dig coef %	*	44.	44.	
Cattle	dig prot %		0.2	0.5	
Goats	dig prot %	*	0.0	0.0	
Horses	dig prot %	*	0.2	0.6	
Rabbits	dig prot %	*	0.6	1.4	
Sheep	dig prot %		0.6	1.6	
Energy	GE Mcal/kg				
Cattle	DE Mcal/kg	*	0.96	2.45	
Sheep	DE Mcal/kg	*	1.06	2.72	
Cattle	ME Mcal/kg	*	0.79	2.01	
Sheep	ME Mcal/kg	*	0.87	2.23	
Cattle	TDN %		21.8	55.5	
Sheep	TDN %		24.2	61.6	

Sorghum, aerial part, fresh, over ripe, (2)
Ref No 2 04 316 United States

Feed Name or Analyses			Mean As Fed	Mean Dry	C.V. ± %
Dry matter	%		28.6	100.0	19
Ash	%		1.6	5.5	26
Crude fiber	%		10.8	37.7	8
Ether extract	%		0.4	1.4	16
N-free extract	%		15.3	53.4	
Protein (N x 6.25)	%		0.6	2.0	22
Cattle	dig prot %	*	0.0	-0.3	
Goats	dig prot %	*	-0.4	-1.5	
Horses	dig prot %	*	-0.1	-0.7	
Rabbits	dig prot %	*	0.1	0.2	
Sheep	dig prot %	*	-0.2	-1.0	
Energy	GE Mcal/kg				
Cattle	DE Mcal/kg	*	0.67	2.36	
Sheep	DE Mcal/kg	*	0.76	2.64	
Cattle	ME Mcal/kg	*	0.55	1.94	
Sheep	ME Mcal/kg	*	0.62	2.16	
Cattle	TDN %	*	15.3	53.5	
Sheep	TDN %	*	17.1	59.9	

Sorghum, heads, fresh, (2)
Ref No 2 04 318 United States

Feed Name or Analyses			Mean As Fed	Mean Dry	C.V. ± %
Dry matter	%			100.0	
Calcium	%			0.22	
Magnesium	%			0.11	
Phosphorus	%			0.51	
Potassium	%			0.55	
Sulphur	%			0.06	

Sorghum, mill residue, fresh, (2)
Ref No 2 04 319 United States

Feed Name or Analyses			Mean As Fed	Mean Dry	C.V. ± %
Dry matter	%		26.9	100.0	
Ash	%		1.3	5.0	
Crude fiber	%		11.5	42.7	
Cattle	dig coef %		47.	47.	
Ether extract	%		0.5	1.7	
Cattle	dig coef %		45.	45.	
N-free extract	%		13.0	48.3	
Cattle	dig coef %		34.	34.	
Protein (N x 6.25)	%		0.6	2.3	
Cattle	dig coef %		-122.	-122.	
Cattle	dig prot %		-0.7	-2.7	
Goats	dig prot %	*	-0.2	-1.2	
Horses	dig prot %	*	0.0	-0.4	
Rabbits	dig prot %	*	0.1	0.4	
Sheep	dig prot %	*	-0.1	-0.8	
Energy	GE Mcal/kg				
Cattle	DE Mcal/kg	*	0.42	1.56	
Sheep	DE Mcal/kg	*	0.47	1.74	
Cattle	ME Mcal/kg	*	0.34	1.28	
Sheep	ME Mcal/kg	*	0.38	1.43	
Cattle	TDN %		9.5	35.4	
Sheep	TDN %	*	10.6	39.4	

Sorghum, aerial part, ensiled, (3)
Sorghum fodder silage
Ref No 3 04 323 United States

Feed Name or Analyses			Mean As Fed	Mean Dry	C.V. ± %
Dry matter	%		28.9	100.0	11
Ash	%		3.1	10.8	
Crude fiber	%		7.1	24.7	
Cattle	dig coef %	*	61.	61.	
Sheep	dig coef %	*	56.	56.	
Ether extract	%		0.5	1.8	
Cattle	dig coef %	*	39.	39.	
Sheep	dig coef %	*	63.	63.	
N-free extract	%				
Cattle	dig coef %	*	61.	61.	
Sheep	dig coef %	*	67.	67.	
Protein (N x 6.25)	%				
Cattle	dig coef %	*	37.	37.	
Sheep	dig coef %	*	23.	23.	
Lignin (Ellis)	%		2.3	7.9	
Chlorine	%		0.03	0.12	51
Cobalt	mg/kg		0.087	0.300	51
Sodium	%		0.01	0.04	51
Carotene	mg/kg		3.7	12.9	98
Vitamin A equivalent	IU/g		6.2	21.6	

Sorghum, aerial part, ensiled, dough stage, (3)
Ref No 3 04 321 United States

Feed Name or Analyses			Mean As Fed	Mean Dry	C.V. ± %
Dry matter	%		25.5	100.0	13
Cattle	dig coef %		53.	53.	
Ash	%		2.4	9.3	15
Crude fiber	%		7.3	28.8	14
Ether extract	%		0.6	2.3	11
N-free extract	%		13.4	52.8	
Protein (N x 6.25)	%		1.8	6.9	13
Cattle	dig coef %		39.	39.	
Cattle	dig prot %		0.7	2.7	
Goats	dig prot %	*	0.6	2.5	
Horses	dig prot %	*	0.6	2.5	
Sheep	dig prot %	*	0.6	2.5	
Cellulose (Matrone)	%		8.5	33.2	
Cattle	dig coef %		58.	58.	
Lignin (Ellis)	%		2.0	7.9	
Energy	GE Mcal/kg		1.11	4.34	
Cattle	GE dig coef %		54.	54.	
Cattle	DE Mcal/kg		0.60	2.35	
Sheep	DE Mcal/kg	*	0.63	2.49	
Cattle	ME Mcal/kg	*	0.49	1.92	
Sheep	ME Mcal/kg	*	0.52	2.04	
Cattle	TDN %	*	13.6	53.3	
Sheep	TDN %	*	14.4	56.4	

Sorghum, aerial part, ensiled, mature, (3)
Ref No 3 04 322 United States

Feed Name or Analyses			Mean As Fed	Mean Dry	C.V. ± %
Dry matter	%		33.9	100.0	14
Cattle	dig coef %		51.	51.	
Ash	%		2.6	7.6	33
Crude fiber	%		8.2	24.2	13
Cattle	dig coef %		53.	53.	
Ether extract	%		0.9	2.6	15
Cattle	dig coef %		61.	61.	
N-free extract	%		19.7	58.0	8
Cattle	dig coef %		61.	61.	
Protein (N x 6.25)	%		2.6	7.5	12
Cattle	dig coef %		29.	29.	
Cattle	dig prot %		0.7	2.2	
Goats	dig prot %	*	1.0	3.1	
Horses	dig prot %	*	1.0	3.1	
Sheep	dig prot %	*	1.0	3.1	
Cell walls (Van Soest)	%		14.7	43.3	
Cellulose (Matrone)	%		8.6	25.2	
Cattle	dig coef %		53.	53.	
Fiber, acid detergent (VS)	%		7.3	21.4	
Lignin (Ellis)	%		1.6	4.7	
Energy	GE Mcal/kg		1.44	4.23	1
Cattle	GE dig coef %		50.	50.	
Cattle	DE Mcal/kg		0.72	2.12	
Sheep	DE Mcal/kg	*	0.85	2.52	
Cattle	ME Mcal/kg	*	0.59	1.74	
Sheep	ME Mcal/kg	*	0.70	2.07	
Cattle	TDN %		18.3	54.0	
Sheep	TDN %	*	19.4	57.2	

Sorghum, aerial part w molasses added, ensiled, (3)
Ref No 3 04 325 United States

Feed Name or Analyses			Mean As Fed	Mean Dry	C.V. ± %
Dry matter	%		20.7	100.0	4
Ash	%		1.7	8.2	12
Crude fiber	%		7.0	33.6	7
Ether extract	%		0.8	4.0	35
N-free extract	%		9.0	43.7	
Protein (N x 6.25)	%		2.2	10.5	30
Cattle	dig prot %	*	1.2	5.8	
Goats	dig prot %	*	1.2	5.8	
Horses	dig prot %	*	1.2	5.8	
Sheep	dig prot %	*	1.2	5.8	
Energy	GE Mcal/kg				
Cattle	DE Mcal/kg	*	0.46	2.20	
Sheep	DE Mcal/kg	*	0.48	2.32	
Cattle	ME Mcal/kg	*	0.37	1.80	
Sheep	ME Mcal/kg	*	0.39	1.90	
Cattle	TDN %	*	10.3	49.8	
Sheep	TDN %	*	10.9	52.7	

Sorghum, aerial part wo heads, ensiled, (3)
Sorghum stover silage
Ref No 3 04 326 United States

Feed Name or Analyses			Mean As Fed	Mean Dry	C.V. ± %
Dry matter	%		85.1	100.0	7
Crude fiber	%				
Cattle	dig coef %	*	50.	50.	
Sheep	dig coef %	*	56.	56.	
Ether extract	%				
Cattle	dig coef %	*	52.	52.	
Sheep	dig coef %	*	63.	63.	
N-free extract	%				
Cattle	dig coef %	*	68.	68.	
Sheep	dig coef %	*	67.	67.	
Protein (N x 6.25)	%				
Cattle	dig coef %	*	27.	27.	
Sheep	dig coef %	*	23.	23.	
Calcium	%		0.34	0.40	10
Manganese	mg/kg		125.9	147.9	31

Continued

Feed Name or Analyses		Mean As Fed	Mean Dry	C.V. ± %
Phosphorus	%	0.09	0.11	32
Carotene	mg/kg	5.8	6.8	98
Vitamin A equivalent	IU/g	9.7	11.4	

Sorghum, grain, (4)

Ref No 4 04 327 United States

Feed Name or Analyses		Mean As Fed	Mean Dry	C.V. ± %
Dry matter	%	89.0	100.0	0
Ash	%	1.9	2.1	9
Crude fiber	%	2.4	2.7	26
Sheep	dig coef %	33.	33.	
Ether extract	%	3.0	3.4	5
Sheep	dig coef %	76.	76.	
N-free extract	%	70.8	79.7	
Sheep	dig coef %	89.	89.	
Protein (N x 6.25)	%	10.9	12.2	4
Sheep	dig coef %	67.	67.	
Cattle	dig prot %	* 6.4	7.2	
Goats	dig prot %	* 7.5	8.4	
Horses	dig prot %	* 7.5	8.4	
Sheep	dig prot %	7.3	8.2	
Pentosans	%	2.1	2.4	34
Starch	%	60.8	68.3	4
Sugars, total	%	1.2	1.3	13
Energy	GE Mcal/kg			
Cattle	DE Mcal/kg	* 3.28	3.69	
Sheep	DE Mcal/kg	* 3.36	3.78	
Swine	DE kcal/kg	*3545.	3985.	
Cattle	ME Mcal/kg	* 2.70	3.03	
Sheep	ME Mcal/kg	* 2.76	3.10	
Swine	ME kcal/kg	*3315.	3727.	
Cattle	TDN %	* 74.4	83.6	
Sheep	TDN %	76.3	85.7	
Swine	TDN %	* 80.4	90.4	
Calcium	%	0.02	0.02	
Phosphorus	%	0.32	0.36	
Biotin	mg/kg	0.18	0.20	36

Sorghum, heads, chopped, (4)

Ref No 4 04 304 United States

Feed Name or Analyses		Mean As Fed	Mean Dry	C.V. ± %
Dry matter	%	89.2	100.0	4
Ash	%	5.0	5.6	24
Crude fiber	%	6.0	6.7	21
Ether extract	%	2.8	3.1	16
N-free extract	%	65.1	73.0	
Protein (N x 6.25)	%	10.3	11.6	12
Cattle	dig prot %	* 5.9	6.7	
Goats	dig prot %	* 7.0	7.9	
Horses	dig prot %	* 7.0	7.9	
Sheep	dig prot %	* 7.0	7.9	
Energy	GE Mcal/kg			
Cattle	DE Mcal/kg	* 3.21	3.60	
Sheep	DE Mcal/kg	* 3.15	3.54	
Cattle	ME Mcal/kg	*2830.	3175.	
Cattle	ME Mcal/kg	* 2.64	2.96	
Sheep	ME Mcal/kg	* 2.59	2.90	
Swine	ME kcal/kg	*2651.	2973.	
Cattle	TDN %	* 72.9	81.8	
Sheep	TDN %	* 71.5	80.2	
Swine	TDN %	* 64.2	72.0	
Carotene	mg/kg	1.4	1.5	70

(1) dry forages and roughages (2) pasture, range plants, and forages fed green (3) silages (4) energy feeds (5) protein supplements (6) minerals (7) vitamins (8) additives

Feed Name or Analyses		Mean As Fed	Mean Dry	C.V. ± %
Riboflavin	mg/kg	2.0	2.2	21
Vitamin A equivalent	IU/g	2.3	2.6	

Sorghum, distillers grains, dehy, (5)

Ref No 5 08 512 United States

Feed Name or Analyses		Mean As Fed	Mean Dry	C.V. ± %
Dry matter	%	94.0	100.0	
Ash	%	4.7	5.0	
Crude fiber	%	13.9	14.8	
Ether extract	%	7.4	7.9	
N-free extract	%	40.0	42.6	
Protein (N x 6.25)	%	28.0	29.8	
Energy	GE Mcal/kg			
Cattle	DE Mcal/kg	* 3.22	3.42	
Sheep	DE Mcal/kg	* 3.40	3.62	
Swine	DE kcal/kg	*3785.	4027.	
Cattle	ME Mcal/kg	* 2.64	2.81	
Sheep	ME Mcal/kg	* 2.79	2.97	
Swine	ME kcal/kg	*3406.	3624.	
Cattle	TDN %	* 73.0	77.6	
Sheep	TDN %	* 77.2	82.1	
Swine	TDN %	* 85.9	91.3	
Calcium	%	0.15	0.16	
Phosphorus	%	0.77	0.82	

Sorghum, gluten, wet milled dehy, (5)

Sorghum gluten meal

Ref No 5 04 329 United States

Feed Name or Analyses		Mean As Fed	Mean Dry	C.V. ± %
Dry matter	%	90.1	100.0	1
Ash	%	1.2	1.4	81
Crude fiber	%	2.9	3.3	9
Ether extract	%	4.3	4.8	14
N-free extract	%	40.0	44.4	
Protein (N x 6.25)	%	41.6	46.2	3
Energy	GE Mcal/kg			
Cattle	DE Mcal/kg	* 3.54	3.93	
Sheep	DE Mcal/kg	* 3.59	3.99	
Swine	DE kcal/kg	*3714.	4125.	
Cattle	ME Mcal/kg	* 2.91	3.23	
Sheep	ME Mcal/kg	* 2.94	3.27	
Swine	ME kcal/kg	*3219.	3575.	
Cattle	TDN %	* 80.4	89.2	
Sheep	TDN %	* 81.5	90.5	
Swine	TDN %	* 84.2	93.5	
Calcium	%	0.03	0.03	89
Magnesium	%	0.16	0.18	
Manganese	mg/kg	16.1	17.9	
Phosphorus	%	0.24	0.27	98
Potassium	%	0.50	0.56	
Choline	mg/kg	714.	793.	
Niacin	mg/kg	34.2	37.9	
Pantothenic acid	mg/kg	10.1	11.3	
Riboflavin	mg/kg	1.5	1.7	
Arginine	%	1.20	1.33	
Histidine	%	0.80	0.89	
Isoleucine	%	2.30	2.55	
Leucine	%	7.40	8.21	
Lysine	%	0.80	0.89	30
Methionine	%	0.60	0.67	40
Phenylalanine	%	2.60	2.89	
Threonine	%	1.40	1.55	
Tryptophan	%	0.40	0.44	
Valine	%	2.50	2.77	

Sorghum, gluten w bran, wet milled dehy grnd, (5)

Sorghum gluten feed

Ref No 5 08 086 United States

Feed Name or Analyses		Mean As Fed	Mean Dry	C.V. ± %
Dry matter	%	90.7	100.0	
Ash	%	7.3	8.0	
Crude fiber	%	6.5	7.2	
Ether extract	%	3.6	4.0	
N-free extract	%	48.4	53.4	
Protein (N x 6.25)	%	24.9	27.5	
Energy	GE Mcal/kg			
Cattle	DE Mcal/kg	* 3.16	3.48	
Sheep	DE Mcal/kg	* 3.28	3.62	
Swine	DE kcal/kg	*3634.	4007.	
Cattle	ME Mcal/kg	* 2.59	2.85	
Sheep	ME Mcal/kg	* 2.69	2.97	
Swine	ME kcal/kg	*3287.	3624.	
Cattle	TDN %	* 71.6	79.0	
Sheep	TDN %	* 74.5	82.1	
Swine	TDN %	* 82.4	90.9	
Calcium	%	0.09	0.10	
Magnesium	%	0.44	0.49	
Manganese	mg/kg	48.5	53.5	
Phosphorus	%	0.59	0.65	
Potassium	%	1.45	1.60	

SORGHUM, ATLAS. Sorghum vulgare

Sorghum, atlas, aerial part, dehy, mature, (1)

Sorghum, atlas, fodder, dehy, mature

Ref No 1 04 332 United States

Feed Name or Analyses		Mean As Fed	Mean Dry	C.V. ± %
Dry matter	%	90.8	100.0	
Carotene	mg/kg	17.6	19.4	
Vitamin A equivalent	IU/g	29.4	32.3	

Sorghum, atlas, aerial part, fan air dried, immature, (1)

Sorghum, atlas, fodder, barn-cured, immature

Ref No 1 04 331 United States

Feed Name or Analyses		Mean As Fed	Mean Dry	C.V. ± %
Dry matter	%		100.0	
Carotene	mg/kg		71.2	
Vitamin A equivalent	IU/g		118.7	

Sorghum, atlas, aerial part, s-c, (1)

Sorghum, atlas, fodder

Ref No 1 04 334 United States

Feed Name or Analyses		Mean As Fed	Mean Dry	C.V. ± %
Dry matter	%	57.6	100.0	16
Ash	%	4.1	7.2	
Crude fiber	%	12.8	22.2	
Ether extract	%	1.8	3.1	
N-free extract	%	34.8	60.5	
Protein (N x 6.25)	%	4.1	7.1	
Cattle	dig prot %	* 1.8	3.1	
Goats	dig prot %	* 1.8	3.1	
Horses	dig prot %	* 2.0	3.5	
Rabbits	dig prot %	* 2.4	4.1	
Sheep	dig prot %	* 1.7	2.9	
Sugars, total	%	9.3	16.2	
Energy	GE Mcal/kg			
Cattle	DE Mcal/kg	* 1.49	2.59	
Sheep	DE Mcal/kg	* 1.53	2.66	
Cattle	ME Mcal/kg	* 1.22	2.12	
Sheep	ME Mcal/kg	* 1.26	2.18	

Feed Name or Analyses			Mean As Fed	Mean Dry	C.V. ± %
Cattle	TDN %	*	33.8	58.7	
Sheep	TDN %	*	34.8	60.4	
Calcium	%		0.46	0.80	
Phosphorus	%		0.17	0.30	
Carotene	mg/kg		29.8	51.8	
Vitamin A equivalent	IU/g		49.7	86.4	

Sorghum, atlas, aerial part, s-c, mature, (1)
Sorghum, atlas, fodder, mature
Ref No 1 04 333 United States

Feed Name or Analyses			Mean As Fed	Mean Dry	C.V. ± %
Dry matter	%		69.9	100.0	15
Ash	%		6.6	9.4	26
Crude fiber	%		19.6	28.0	12
Ether extract	%		1.3	1.9	11
N-free extract	%		37.7	54.0	
Protein (N x 6.25)	%		4.7	6.7	18
Cattle	dig prot %	*	1.9	2.7	
Goats	dig prot %	*	2.0	2.8	
Horses	dig prot %	*	2.2	3.2	
Rabbits	dig prot %	*	2.7	3.8	
Sheep	dig prot %	*	1.8	2.6	
Energy	GE Mcal/kg				
Cattle	DE Mcal/kg	*	1.80	2.57	
Sheep	DE Mcal/kg	*	1.77	2.53	
Cattle	ME Mcal/kg	*	1.47	2.11	
Sheep	ME Mcal/kg	*	1.45	2.07	
Cattle	TDN %	*	40.8	58.4	
Sheep	TDN %	*	40.1	57.3	
Carotene	mg/kg		3.1	4.4	
Vitamin A equivalent	IU/g		5.1	7.4	

Sorghum, atlas, aerial part, wo heads, s-c, (1)
Sorghum, atlas, stover
Ref No 1 04 336 United States

Feed Name or Analyses			Mean As Fed	Mean Dry	C.V. ± %
Dry matter	%		75.0	100.0	
Ash	%		5.9	7.9	
Crude fiber	%		23.2	30.9	
Ether extract	%		1.7	2.3	
N-free extract	%		40.7	54.3	
Protein (N x 6.25)	%		3.5	4.7	
Cattle	dig prot %	*	0.7	1.0	
Goats	dig prot %	*	0.7	0.9	
Horses	dig prot %	*	1.1	1.5	
Rabbits	dig prot %	*	1.7	2.3	
Sheep	dig prot %	*	0.6	0.8	
Energy	GE Mcal/kg				
Cattle	DE Mcal/kg	*	1.98	2.64	
Sheep	DE Mcal/kg	*	1.83	2.44	
Cattle	ME Mcal/kg	*	1.62	2.16	
Sheep	ME Mcal/kg	*	1.50	2.00	
Cattle	TDN %	*	44.9	59.8	
Sheep	TDN %	*	41.5	55.3	

Sorghum, atlas, aerial part wo heads, s-c, over ripe, (1)
Sorghum, atlas, stover, over ripe
Ref No 1 04 335 United States

Feed Name or Analyses			Mean As Fed	Mean Dry	C.V. ± %
Dry matter	%			100.0	
Ash	%			7.9	
Crude fiber	%			32.0	
Ether extract	%			2.4	
N-free extract	%			54.1	
Protein (N x 6.25)	%			3.6	
Cattle	dig prot %	*		0.1	
Goats	dig prot %	*		0.0	
Horses	dig prot %	*		0.6	
Rabbits	dig prot %	*		1.4	
Sheep	dig prot %	*		-0.1	
Energy	GE Mcal/kg				
Cattle	DE Mcal/kg	*		2.48	
Sheep	DE Mcal/kg	*		2.40	
Cattle	ME Mcal/kg	*		2.03	
Sheep	ME Mcal/kg	*		1.97	
Cattle	TDN %	*		56.3	
Sheep	TDN %	*		54.4	

Sorghum, atlas, aerial part, fresh, dough stage, (2)
Ref No 2 04 337 United States

Feed Name or Analyses			Mean As Fed	Mean Dry	C.V. ± %
Dry matter	%		66.6	100.0	
Carotene	mg/kg		15.4	23.1	
Vitamin A equivalent	IU/g		25.7	38.6	

Sorghum, atlas, aerial part, ensiled, (3)
Ref No 3 08 513 United States

Feed Name or Analyses			Mean As Fed	Mean Dry	C.V. ± %
Dry matter	%		29.7	100.0	
Ash	%		2.3	7.7	
Crude fiber	%		7.4	24.9	
Ether extract	%		0.8	2.7	
N-free extract	%		16.7	56.2	
Protein (N x 6.25)	%		2.5	8.4	
Cattle	dig prot %	*	1.2	3.9	
Goats	dig prot %	*	1.2	3.9	
Horses	dig prot %	*	1.2	3.9	
Sheep	dig prot %	*	1.2	3.9	
Energy	GE Mcal/kg				
Cattle	DE Mcal/kg	*	0.84	2.83	
Sheep	DE Mcal/kg	*	0.79	2.65	
Cattle	ME Mcal/kg	*	0.69	2.32	
Sheep	ME Mcal/kg	*	0.64	2.17	
Cattle	TDN %	*	19.1	64.3	
Sheep	TDN %	*	17.8	60.1	
Calcium	%		0.12	0.40	
Chlorine	%		0.04	0.13	
Copper	mg/kg		10.4	34.9	
Iron	%		0.008	0.027	
Magnesium	%		0.10	0.34	
Manganese	mg/kg		14.3	48.2	
Phosphorus	%		0.06	0.20	
Potassium	%		0.33	1.11	
Sodium	%		0.08	0.27	

Sorghum, atlas, aerial part wo heads, ensiled, (3)
Ref No 3 04 340 United States

Feed Name or Analyses			Mean As Fed	Mean Dry	C.V. ± %
Dry matter	%		61.5	100.0	7
Ash	%		3.9	6.3	
Crude fiber	%		15.4	25.0	
Ether extract	%		1.1	1.9	
N-free extract	%		37.6	61.2	
Protein (N x 6.25)	%		3.4	5.6	
Cattle	dig prot %	*	0.8	1.3	
Goats	dig prot %	*	0.8	1.3	
Horses	dig prot %	*	0.8	1.3	
Sheep	dig prot %	*	0.8	1.3	
Energy	GE Mcal/kg				
Cattle	DE Mcal/kg	*	1.66	2.70	
Sheep	DE Mcal/kg	*	1.60	2.60	
Cattle	ME Mcal/kg	*	1.36	2.21	
Sheep	ME Mcal/kg	*	1.31	2.13	
Cattle	TDN %	*	37.6	61.2	
Sheep	TDN %	*	36.2	58.9	
Calcium	%		0.24	0.39	14
Phosphorus	%		0.07	0.11	31
Carotene	mg/kg		14.0	22.7	
Vitamin A equivalent	IU/g		23.3	37.9	

Sorghum, atlas, heads, ensiled trench silo, (3)
Ref No 3 04 341 United States

Feed Name or Analyses			Mean As Fed	Mean Dry	C.V. ± %
Dry matter	%		29.3	100.0	18
Calcium	%		0.09	0.30	66
Chlorine	%		0.04	0.14	45
Cobalt	mg/kg		0.076	0.260	62
Copper	mg/kg		10.3	35.1	41
Iron	%		0.008	0.028	17
Magnesium	%		0.10	0.35	21
Manganese	mg/kg		14.1	48.1	21
Phosphorus	%		0.07	0.24	25
Potassium	%		0.32	1.10	15
Sodium	%		0.01	0.04	56
Carotene	mg/kg		7.0	23.8	10
Vitamin A equivalent	IU/g		11.6	39.7	

Sorghum, atlas, grain, (4)
Ref No 4 04 342 United States

Feed Name or Analyses			Mean As Fed	Mean Dry	C.V. ± %
Dry matter	%		89.1	100.0	2
Ash	%		1.9	2.1	14
Crude fiber	%		2.0	2.3	23
Cattle	dig coef %	*	-96.	-96.	
Sheep	dig coef %	*	33.	33.	
Ether extract	%		3.3	3.8	8
Cattle	dig coef %	*	52.	52.	
Sheep	dig coef %	*	76.	76.	
N-free extract	%		70.5	79.1	
Cattle	dig coef %	*	79.	79.	
Sheep	dig coef %	*	89.	89.	
Protein (N x 6.25)	%		11.4	12.7	9
Cattle	dig coef %	*	58.	58.	
Sheep	dig coef %	*	67.	67.	
Cattle	dig prot %		6.6	7.4	
Goats	dig prot %	*	7.9	8.9	
Horses	dig prot %	*	7.9	8.9	
Sheep	dig prot %		7.6	8.5	
Pentosans	%		2.1	2.4	
Starch	%		61.7	69.2	
Sugars, total	%		1.2	1.3	
Energy	GE Mcal/kg				
Cattle	DE Mcal/kg	*	2.83	3.18	
Sheep	DE Mcal/kg	*	3.38	3.80	
Swine	DE kcal/kg	*	3597.	4037.	
Cattle	ME Mcal/kg	*	2.32	2.61	
Sheep	ME Mcal/kg	*	2.77	3.11	
Swine	ME kcal/kg	*	3360.	3771.	
Cattle	TDN %		64.2	72.1	
Sheep	TDN %		76.7	86.1	
Swine	TDN %	*	81.6	91.5	
Biotin	mg/kg		0.26	0.29	22
Niacin	mg/kg		34.4	38.6	
Pantothenic acid	mg/kg		13.4	15.0	
Riboflavin	mg/kg		2.0	2.2	

Sorghum, atlas, heads, chopped, (4)
Ref No 4 04 343 United States

Feed Name or Analyses			Mean As Fed	Mean Dry	C.V. ± %
Dry matter	%		85.6	100.0	5
Ash	%		4.5	5.3	14
Crude fiber	%		10.0	11.7	17
Cattle	dig coef %	*	40.	40.	
Sheep	dig coef %	*	61.	61.	
Ether extract	%		2.7	3.2	12
Cattle	dig coef %	*	80.	80.	
Sheep	dig coef %	*	74.	74.	
N-free extract	%		59.1	69.1	
Cattle	dig coef %	*	83.	83.	
Sheep	dig coef %	*	80.	80.	
Protein (N x 6.25)	%		9.2	10.7	7
Cattle	dig coef %	*	57.	57.	
Sheep	dig coef %	*	63.	63.	
Cattle	dig prot %		5.2	6.1	
Goats	dig prot %	*	6.1	7.1	
Horses	dig prot %	*	6.1	7.1	
Sheep	dig prot %		5.8	6.8	
Energy	GE Mcal/kg				
Cattle	DE Mcal/kg	*	2.79	3.26	
Sheep	DE Mcal/kg	*	2.81	3.28	
Swine	DE kcal/kg	*	2660.	3108.	
Cattle	ME Mcal/kg	*	2.29	2.67	
Sheep	ME Mcal/kg	*	2.30	2.69	
Swine	ME kcal/kg	*	2182.	2549.	
Cattle	TDN %		63.2	73.9	
Sheep	TDN %		63.7	74.5	
Swine	TDN %	*	60.3	70.5	

SORGHUM, BRAWLEY. Sorghum vulgare

Sorghum, brawley, aerial part, fresh, dough stage, cut 1, (2)
Ref No 2 07 892 United States

Feed Name or Analyses			Mean As Fed	Mean Dry	C.V. ± %
Dry matter	%		28.6	100.0	
Ash	%		2.6	9.1	
Crude fiber	%		6.3	22.2	
Ether extract	%		0.7	2.3	
N-free extract	%		17.5	61.2	
Protein (N x 6.25)	%		1.5	5.3	
Cattle	dig prot %	*	0.7	2.4	
Goats	dig prot %	*	0.4	1.5	
Horses	dig prot %	*	0.6	2.0	
Rabbits	dig prot %	*	0.8	2.7	
Sheep	dig prot %	*	0.5	1.9	
Lignin (Ellis)	%		1.7	6.0	
Energy	GE Mcal/kg				
Cattle	DE Mcal/kg	*	0.72	2.54	
Sheep	DE Mcal/kg	*	0.71	2.47	
Cattle	ME Mcal/kg	*	0.59	2.08	
Sheep	ME Mcal/kg	*	0.58	2.03	
Cattle	TDN %	*	16.4	57.5	
Sheep	TDN %	*	15.8	55.1	

(1) dry forages and roughages (2) pasture, range plants, and forages fed green (3) silages (4) energy feeds (5) protein supplements (6) minerals (7) vitamins (8) additives

Sorghum, brawley, aerial part, ensiled, dough stage, cut 1, (3)
Ref No 3 07 891 United States

Feed Name or Analyses			Mean As Fed	Mean Dry	C.V. ± %
Dry matter	%		26.7	100.0	
Ash	%		2.7	10.1	
Crude fiber	%		6.3	23.7	
Ether extract	%		0.5	2.0	
N-free extract	%		15.8	59.0	
Protein (N x 6.25)	%		1.4	5.2	
Cattle	dig prot %	*	0.3	1.0	
Goats	dig prot %	*	0.3	1.0	
Horses	dig prot %	*	0.3	1.0	
Sheep	dig prot %	*	0.3	1.0	
Lignin (Ellis)	%		2.0	7.7	
Energy	GE Mcal/kg				
Cattle	DE Mcal/kg		0.65	2.42	
Sheep	DE Mcal/kg	*	0.62	2.32	
Cattle	ME Mcal/kg	*	0.53	1.98	
Sheep	ME Mcal/kg	*	0.51	1.90	
Cattle	TDN %	*	14.7	54.9	
Sheep	TDN %	*	14.1	52.7	

SORGHUM, BROOMCORN. Sorghum vulgare technicum

Sorghum, broomcorn, aerial part wo heads, s-c, (1)
Sorghum, broomcorn, stover
Ref No 1 04 344 United States

Feed Name or Analyses			Mean As Fed	Mean Dry	C.V. ± %
Dry matter	%		90.6	100.0	
Ash	%		5.7	6.3	
Crude fiber	%		36.8	40.6	
Ether extract	%		1.8	2.0	
N-free extract	%		42.4	46.8	
Protein (N x 6.25)	%		3.9	4.3	
Cattle	dig prot %	*	0.6	0.7	
Goats	dig prot %	*	0.5	0.6	
Horses	dig prot %	*	1.1	1.2	
Rabbits	dig prot %	*	1.8	2.0	
Sheep	dig prot %	*	0.4	0.4	
Fatty acids	%		1.8	2.0	
Energy	GE Mcal/kg				
Cattle	DE Mcal/kg	*	1.80	1.99	
Sheep	DE Mcal/kg	*	2.03	2.24	
Cattle	ME Mcal/kg	*	1.48	1.63	
Sheep	ME Mcal/kg	*	1.67	1.84	
Cattle	TDN %	*	40.8	45.0	
Sheep	TDN %	*	46.1	50.8	

Sorghum, broomcorn, aerial part, ensiled, (3)
Ref No 3 04 345 United States

Feed Name or Analyses			Mean As Fed	Mean Dry	C.V. ± %
Dry matter	%		20.0	100.0	
Ash	%		1.9	9.3	
Crude fiber	%		6.8	34.2	
Sheep	dig coef %		50.	50.	
Ether extract	%		0.2	1.2	
Sheep	dig coef %		69.	69.	
N-free extract	%		9.8	48.8	
Sheep	dig coef %		55.	55.	
Protein (N x 6.25)	%		1.3	6.5	
Sheep	dig coef %		36.	36.	
Cattle	dig prot %	*	0.4	2.1	
Goats	dig prot %	*	0.4	2.1	
Horses	dig prot %	*	0.4	2.1	
Sheep	dig prot %		0.5	2.3	
Energy	GE Mcal/kg				
Cattle	DE Mcal/kg	*	0.46	2.28	
Sheep	DE Mcal/kg	*	0.42	2.12	
Cattle	ME Mcal/kg	*	0.37	1.87	
Sheep	ME Mcal/kg	*	0.35	1.74	
Cattle	TDN %	*	10.3	51.6	
Sheep	TDN %		9.6	48.1	

Sorghum, broomcorn, aerial part wo heads, ensiled, (3)
Ref No 3 04 346 United States

Feed Name or Analyses			Mean As Fed	Mean Dry	C.V. ± %
Dry matter	%		30.0	100.0	
Ash	%		2.3	7.7	
Crude fiber	%		10.9	36.2	
Ether extract	%		0.4	1.4	
N-free extract	%		15.1	50.2	
Protein (N x 6.25)	%		1.4	4.5	
Cattle	dig prot %	*	0.1	0.3	
Goats	dig prot %	*	0.1	0.3	
Horses	dig prot %	*	0.1	0.3	
Sheep	dig prot %	*	0.1	0.3	

Sorghum, broomcorn, grain, (4)
Ref No 4 04 349 United States

Feed Name or Analyses			Mean As Fed	Mean Dry	C.V. ± %
Dry matter	%		87.5	100.0	2
Ash	%		2.9	3.3	
Crude fiber	%		5.5	6.3	
Cattle	dig coef %	*	46.	46.	
Horses	dig coef %		-10.	-10.	
Sheep	dig coef %	*	28.	28.	
Ether extract	%		4.0	4.6	
Cattle	dig coef %	*	63.	63.	
Horses	dig coef %		61.	61.	
Sheep	dig coef %	*	78.	78.	
N-free extract	%		65.3	74.6	
Cattle	dig coef %	*	79.	79.	
Horses	dig coef %		74.	74.	
Sheep	dig coef %	*	75.	75.	
Protein (N x 6.25)	%		9.8	11.2	
Cattle	dig coef %	*	42.	42.	
Horses	dig coef %		42.	42.	
Sheep	dig coef %	*	48.	48.	
Cattle	dig prot %		4.1	4.7	
Goats	dig prot %	*	6.6	7.5	
Horses	dig prot %		4.1	4.7	
Sheep	dig prot %		4.7	5.4	
Energy	GE Mcal/kg		3.97	4.53	2
Cattle	DE Mcal/kg	*	2.82	3.22	
Horses	DE Mcal/kg	*	2.53	2.89	
Sheep	DE Mcal/kg	*	2.75	3.14	
Swine	DE kcal/kg	*	3044.	3480.	
Cattle	ME Mcal/kg	*	2.31	2.64	
Horses	ME Mcal/kg	*	2.07	2.37	
Sheep	ME Mcal/kg	*	2.25	2.57	
Swine	ME kcal/kg	*	2854.	3262.	
Cattle	TDN %		63.9	73.1	
Horses	TDN %		57.4	65.6	
Sheep	TDN %		62.3	71.2	
Swine	TDN %	*	69.0	78.9	
Calcium	%		0.47	0.54	
Phosphorus	%		0.34	0.39	

Feed Name or Analyses		Mean As Fed	Mean Dry	C.V. ± %

Sorghum, broomcorn, grain, cracked, (4)
Ref No 4 04 347 United States

Analyses	Unit	As Fed	Dry	C.V. ± %
Dry matter	%	84.9	100.0	
Ash	%	2.7	3.2	
Crude fiber	%	5.3	6.3	
Cattle	dig coef %	46.	46.	
Ether extract	%	3.6	4.2	
Cattle	dig coef %	63.	63.	
N-free extract	%	62.7	73.9	
Cattle	dig coef %	79.	79.	
Protein (N x 6.25)	%	10.5	12.4	
Cattle	dig coef %	42.	42.	
Cattle	dig prot %	4.4	5.2	
Goats	dig prot %	* 7.3	8.6	
Horses	dig prot %	* 7.3	8.6	
Sheep	dig prot %	* 7.3	8.6	
Energy	GE Mcal/kg			
Cattle	DE Mcal/kg	* 2.71	3.19	
Sheep	DE Mcal/kg	* 2.63	3.10	
Swine	DE kcal/kg	*2969.	3497.	
Cattle	ME Mcal/kg	* 2.22	2.62	
Sheep	ME Mcal/kg	* 2.18	2.57	
Swine	ME kcal/kg	*2776.	3270.	
Cattle	TDN %	61.5	72.4	
Sheep	TDN %	* 59.7	70.3	
Swine	TDN %	* 67.3	79.3	

Sorghum, broomcorn, grain, grnd, (4)
Ref No 4 04 348 United States

Analyses	Unit	As Fed	Dry	C.V. ± %
Dry matter	%	85.1	100.0	
Ash	%	2.5	3.0	
Crude fiber	%	3.8	4.5	
Swine	dig coef %	36.	36.	
Ether extract	%	4.1	4.8	
Swine	dig coef %	73.	73.	
N-free extract	%	64.2	75.5	
Swine	dig coef %	82.	82.	
Protein (N x 6.25)	%	10.5	12.4	
Swine	dig coef %	52.	52.	
Cattle	dig prot %	* 6.3	7.4	
Goats	dig prot %	* 7.3	8.6	
Horses	dig prot %	* 7.3	8.6	
Sheep	dig prot %	* 7.3	8.6	
Swine	dig prot %	5.5	6.4	
Energy	GE Mcal/kg			
Cattle	DE Mcal/kg	* 2.91	3.43	
Sheep	DE Mcal/kg	* 2.77	3.25	
Swine	DE kcal/kg	*2914.	3426.	
Cattle	ME Mcal/kg	* 2.39	2.81	
Sheep	ME Mcal/kg	* 2.27	2.67	
Swine	ME kcal/kg	*2725.	3203.	
Cattle	TDN %	* 66.1	77.7	
Sheep	TDN %	* 62.7	73.7	
Swine	TDN %	66.1	77.7	

SORGHUM, DARSO. Sorghum vulgare

Sorghum, darso, aerial part, fresh, (2)
Ref No 2 04 355 United States

Analyses	Unit	As Fed	Dry	C.V. ± %
Dry matter	%	29.0	100.0	
Ash	%	2.1	7.2	
Crude fiber	%	7.3	25.2	
Ether extract	%	0.6	2.1	
N-free extract	%	17.7	61.0	
Protein (N x 6.25)	%	1.3	4.5	
Cattle	dig prot %	* 0.5	1.7	
Goats	dig prot %	* 0.2	0.8	
Horses	dig prot %	* 0.4	1.3	
Rabbits	dig prot %	* 0.6	2.1	
Sheep	dig prot %	* 0.3	1.2	
Energy	GE Mcal/kg			
Cattle	DE Mcal/kg	* 0.83	2.87	
Sheep	DE Mcal/kg	* 0.86	3.02	
Cattle	ME Mcal/kg	* 0.68	2.35	
Sheep	ME Mcal/kg	* 0.72	2.48	
Cattle	TDN %	* 18.9	65.2	
Sheep	TDN %	* 19.8	68.4	
Phosphorus	%	0.04	0.14	

Sorghum, darso, aerial part, ensiled, (3)
Ref No 3 04 356 United States

Analyses	Unit	As Fed	Dry	C.V. ± %
Dry matter	%	27.0	100.0	
Ash	%	1.5	5.7	
Crude fiber	%	6.5	24.0	
Sheep	dig coef %	39.	39.	
Ether extract	%	0.4	1.4	
Sheep	dig coef %	60.	60.	
N-free extract	%	16.7	61.9	
Sheep	dig coef %	70.	70.	
Protein (N x 6.25)	%	1.9	7.0	
Sheep	dig coef %	3.	3.	
Cattle	dig prot %	* 0.7	2.6	
Goats	dig prot %	* 0.7	2.6	
Horses	dig prot %	* 0.7	2.6	
Sheep	dig prot %	0.1	0.2	
Energy	GE Mcal/kg			
Cattle	DE Mcal/kg	* 0.67	2.49	
Sheep	DE Mcal/kg	* 0.65	2.42	
Cattle	ME Mcal/kg	* 0.55	2.04	
Sheep	ME Mcal/kg	* 0.54	1.98	
Cattle	TDN %	* 15.2	56.4	
Sheep	TDN %	14.8	54.8	

Sorghum, darso, grain, (4)
Ref No 4 04 357 United States

Analyses	Unit	As Fed	Dry	C.V. ± %
Dry matter	%	89.7	100.0	1
Ash	%	1.5	1.6	24
Crude fiber	%	2.1	2.3	21
Cattle	dig coef %	* 40.	40.	
Sheep	dig coef %	* 10.	10.	
Ether extract	%	3.1	3.5	7
Cattle	dig coef %	* 80.	80.	
Sheep	dig coef %	* 78.	78.	
N-free extract	%	72.6	80.9	
Cattle	dig coef %	* 83.	83.	
Sheep	dig coef %	* 88.	88.	
Protein (N x 6.25)	%	10.4	11.6	9
Cattle	dig coef %	* 57.	57.	
Sheep	dig coef %	* 64.	64.	
Cattle	dig prot %	5.9	6.6	
Goats	dig prot %	* 7.0	7.8	
Horses	dig prot %	* 7.0	7.8	
Sheep	dig prot %	6.6	7.4	
Energy	GE Mcal/kg			
Cattle	DE Mcal/kg	* 3.20	3.57	
Sheep	DE Mcal/kg	* 3.36	3.75	
Swine	DE kcal/kg	*3642.	4060.	
Cattle	ME Mcal/kg	* 2.63	2.93	
Sheep	ME Mcal/kg	* 2.76	3.07	
Swine	ME kcal/kg	*3411.	3802.	
Cattle	TDN %	72.7	81.0	
Sheep	TDN %	76.3	85.0	
Swine	TDN %	* 82.6	92.1	
Calcium	%	0.08	0.09	92
Phosphorus	%	0.30	0.34	19
Carotene	mg/kg	2.7	3.0	29
Niacin	mg/kg	28.1	31.3	
Pantothenic acid	mg/kg	13.2	14.8	
Riboflavin	mg/kg	1.4	1.5	
Vitamin A equivalent	IU/g	4.5	5.0	
Arginine	%	0.36	0.40	
Histidine	%	0.18	0.20	
Isoleucine	%	0.45	0.50	
Leucine	%	1.17	1.30	
Lysine	%	0.18	0.20	
Methionine	%	0.09	0.10	
Phenylalanine	%	0.45	0.50	
Threonine	%	0.27	0.30	
Tryptophan	%	0.09	0.10	
Valine	%	0.45	0.50	

Sorghum, darso, heads, chopped, (4)
Ref No 4 04 358 United States

Analyses	Unit	As Fed	Dry	C.V. ± %
Dry matter	%	90.3	100.0	1
Ash	%	2.1	2.3	22
Crude fiber	%	8.2	9.1	
Cattle	dig coef %	* 40.	40.	
Sheep	dig coef %	* 61.	61.	
Ether extract	%	2.8	3.1	10
Cattle	dig coef %	* 80.	80.	
Sheep	dig coef %	* 74.	74.	
N-free extract	%	67.5	74.8	
Cattle	dig coef %	* 83.	83.	
Sheep	dig coef %	* 80.	80.	
Protein (N x 6.25)	%	9.7	10.7	5
Cattle	dig coef %	* 57.	57.	
Sheep	dig coef %	* 63.	63.	
Cattle	dig prot %	5.5	6.1	
Goats	dig prot %	* 6.4	7.0	
Horses	dig prot %	* 6.4	7.0	
Sheep	dig prot %	6.1	6.7	
Energy	GE Mcal/kg			
Cattle	DE Mcal/kg	* 3.08	3.41	
Sheep	DE Mcal/kg	* 3.08	3.41	
Swine	DE kcal/kg	*2940.	3256.	
Cattle	ME Mcal/kg	* 2.53	2.80	
Sheep	ME Mcal/kg	* 2.52	2.79	
Swine	ME kcal/kg	*2759.	3055.	
Cattle	TDN %	69.9	77.4	
Sheep	TDN %	69.8	77.3	
Swine	TDN %	* 66.7	73.8	

SORGHUM, DEKALB FS1A. Sorghum vulgare

Sorghum, DeKalb FS1A, aerial part, ensiled, milk stage, (3)
Ref No 3 09 245 United States

Analyses	Unit	As Fed	Dry	C.V. ± %
Dry matter	%	26.4	100.0	
Cattle	dig coef %	51.	51.	
Ash	%	1.5	5.8	
Crude fiber	%	8.1	30.8	
Ether extract	%	0.5	1.9	
N-free extract	%	14.6	55.3	
Protein (N x 6.25)	%	1.6	6.2	
Cattle	dig coef %	23.	23.	

Continued

Feed Name or Analyses		Mean As Fed	Mean Dry	C.V. ± %
Cattle	dig prot %	0.4	1.4	
Goats	dig prot % *	0.5	1.9	
Horses	dig prot % *	0.5	1.9	
Sheep	dig prot % *	0.5	1.9	
Cellulose (Matrone)	%	7.5	28.4	
Cattle	dig coef %	51.	51.	
Energy	GE Mcal/kg			
Cattle	DE Mcal/kg *	0.72	2.72	
Sheep	DE Mcal/kg *	0.78	2.94	
Cattle	ME Mcal/kg *	0.59	2.23	
Sheep	ME Mcal/kg *	0.64	2.41	
Cattle	TDN % *	16.3	61.6	
Sheep	TDN % *	17.6	66.8	

SORGHUM, DE KALB, FS 22. Sorghum vulgare

Sorghum, FS 22, aerial part, ensiled, early bloom, (3)

Ref No 3 09 249 United States

Feed Name or Analyses		Mean As Fed	Mean Dry	C.V. ± %
Dry matter	%		100.0	
Cattle	dig coef %		52.0	
Ash	%		5.2	
Crude fiber	%		36.1	
Cattle	dig coef %		30.0	
Ether extract	%		1.8	
N-free extract	%		50.7	
Cattle	dig coef %		75.0	
Protein (N x 6.25)	%		6.2	
Cattle	dig coef %		20.0	
Cattle	dig prot %		1.2	
Goats	dig prot % *		1.9	
Horses	dig prot % *		1.9	
Sheep	dig prot % *		1.9	
Energy	GE Mcal/kg			
Cattle	DE Mcal/kg *		2.31	
Sheep	DE Mcal/kg *		2.47	
Cattle	ME Mcal/kg *		1.90	
Sheep	ME Mcal/kg *		2.03	
Cattle	TDN %		52.5	
Sheep	TDN % *		56.0	

SORGHUM, DURRA. Sorghum vulgare

Sorghum, durra, aerial part, s-c, (1)

Sorghum, durra, fodder

Ref No 1 04 359 United States

Feed Name or Analyses		Mean As Fed	Mean Dry	C.V. ± %
Dry matter	%	89.9	100.0	
Ash	%	5.2	5.8	
Crude fiber	%	24.1	26.8	
Ether extract	%	2.8	3.1	
N-free extract	%	51.4	57.2	
Protein (N x 6.25)	%	6.4	7.1	
Cattle	dig prot % *	2.8	3.1	
Goats	dig prot % *	2.9	3.2	
Horses	dig prot % *	3.2	3.6	
Rabbits	dig prot % *	3.7	4.2	
Sheep	dig prot % *	2.7	3.0	
Fatty acids	%	2.8	3.1	
Energy	GE Mcal/kg			
Cattle	DE Mcal/kg *	2.39	2.66	
Sheep	DE Mcal/kg *	2.32	2.58	

(1) dry forages and roughages (2) pasture, range plants, and forages fed green (3) silages (4) energy feeds (5) protein supplements (6) minerals (7) vitamins (8) additives

Feed Name or Analyses		Mean As Fed	Mean Dry	C.V. ± %
Cattle	ME Mcal/kg *	1.96	2.18	
Sheep	ME Mcal/kg *	1.90	2.11	
Cattle	TDN % *	54.2	60.3	
Sheep	TDN % *	52.5	58.4	

Sorghum, durra, aerial part, fresh, (2)

Ref No 2 04 363 United States

Feed Name or Analyses		Mean As Fed	Mean Dry	C.V. ± %
Dry matter	%	22.4	100.0	
Ash	%	1.8	8.0	
Crude fiber	%	6.2	27.7	
Ether extract	%	0.6	2.7	
N-free extract	%	11.8	52.7	
Protein (N x 6.25)	%	2.0	8.9	
Cattle	dig prot % *	1.2	5.5	
Goats	dig prot % *	1.1	4.9	
Horses	dig prot % *	1.1	5.1	
Rabbits	dig prot % *	1.2	5.6	
Sheep	dig prot % *	1.2	5.3	
Energy	GE Mcal/kg			
Cattle	DE Mcal/kg *	0.63	2.81	
Sheep	DE Mcal/kg *	0.67	3.00	
Cattle	ME Mcal/kg *	0.52	2.30	
Sheep	ME Mcal/kg *	0.55	2.46	
Cattle	TDN % *	14.3	63.7	
Sheep	TDN % *	15.3	68.1	

Sorghum, durra, aerial part, fresh, immature, (2)

Ref No 2 04 360 United States

Feed Name or Analyses		Mean As Fed	Mean Dry	C.V. ± %
Dry matter	%	15.0	100.0	4
Ash	%	2.3	15.2	30
Crude fiber	%	3.7	24.9	25
Ether extract	%	0.2	1.2	24
N-free extract	%	5.7	38.3	
Protein (N x 6.25)	%	3.1	20.4	12
Cattle	dig prot % *	2.3	15.2	
Goats	dig prot % *	2.3	15.6	
Horses	dig prot % *	2.2	14.8	
Rabbits	dig prot % *	2.2	14.4	
Sheep	dig prot % *	2.4	16.0	
Energy	GE Mcal/kg			
Cattle	DE Mcal/kg *	0.45	2.98	
Sheep	DE Mcal/kg *	0.43	2.89	
Cattle	ME Mcal/kg *	0.37	2.44	
Sheep	ME Mcal/kg *	0.36	2.37	
Cattle	TDN % *	10.1	67.6	
Sheep	TDN % *	9.8	65.5	

Sorghum, durra, aerial part, fresh, mid-bloom, (2)

Ref No 2 04 361 United States

Feed Name or Analyses		Mean As Fed	Mean Dry	C.V. ± %
Dry matter	%	15.2	100.0	8
Ash	%	1.4	9.5	11
Crude fiber	%	3.8	24.8	8
Ether extract	%	0.2	1.4	45
N-free extract	%	7.5	49.3	
Protein (N x 6.25)	%	2.3	15.0	2
Cattle	dig prot % *	1.6	10.6	
Goats	dig prot % *	1.6	10.6	
Horses	dig prot % *	1.6	10.3	
Rabbits	dig prot % *	1.6	10.3	
Sheep	dig prot % *	1.7	11.0	
Energy	GE Mcal/kg			
Cattle	DE Mcal/kg *	0.47	3.07	
Sheep	DE Mcal/kg *	0.43	2.85	
Cattle	ME Mcal/kg *	0.38	2.52	

Feed Name or Analyses		Mean As Fed	Mean Dry	C.V. ± %
Sheep	ME Mcal/kg *	0.35	2.33	
Cattle	TDN % *	10.6	69.7	
Sheep	TDN % *	9.8	64.5	

Sorghum, durra, aerial part, fresh, full bloom, (2)

Ref No 2 04 362 United States

Feed Name or Analyses		Mean As Fed	Mean Dry	C.V. ± %
Dry matter	%	19.7	100.0	
Ash	%	1.3	6.5	
Crude fiber	%	4.9	24.7	
Ether extract	%	0.0	0.2	
N-free extract	%	11.5	58.5	
Protein (N x 6.25)	%	2.0	10.1	
Cattle	dig prot % *	1.3	6.5	
Goats	dig prot % *	1.2	6.0	
Horses	dig prot % *	1.2	6.1	
Rabbits	dig prot % *	1.3	6.5	
Sheep	dig prot % *	1.3	6.4	
Energy	GE Mcal/kg			
Cattle	DE Mcal/kg *	0.59	2.97	
Sheep	DE Mcal/kg *	0.57	2.88	
Cattle	ME Mcal/kg *	0.48	2.44	
Sheep	ME Mcal/kg *	0.46	2.36	
Cattle	TDN % *	13.3	67.3	
Sheep	TDN % *	12.9	65.4	

Sorghum, durra, aerial part, ensiled, (3)

Ref No 3 04 364 United States

Feed Name or Analyses		Mean As Fed	Mean Dry	C.V. ± %
Dry matter	%	29.3	100.0	
Ash	%	2.7	9.4	
Crude fiber	%	10.1	34.4	
Ether extract	%	1.0	3.3	
N-free extract	%	13.7	46.9	
Protein (N x 6.25)	%	1.8	6.0	
Cattle	dig prot % *	0.5	1.7	
Goats	dig prot % *	0.5	1.7	
Horses	dig prot % *	0.5	1.7	
Sheep	dig prot % *	0.5	1.7	
Energy	GE Mcal/kg			
Cattle	DE Mcal/kg *	0.74	2.52	
Sheep	DE Mcal/kg *	0.73	2.48	
Cattle	ME Mcal/kg *	0.61	2.07	
Sheep	ME Mcal/kg *	0.59	2.03	
Cattle	TDN % *	16.7	57.1	
Sheep	TDN % *	16.5	56.2	

Sorghum, durra, grain, (4)

Ref No 4 04 365 United States

Feed Name or Analyses		Mean As Fed	Mean Dry	C.V. ± %
Dry matter	%	88.9	100.0	1
Ash	%	3.0	3.4	20
Crude fiber	%	2.1	2.4	15
Cattle	dig coef % *	40.	40.	
Sheep	dig coef %	60.	60.	
Ether extract	%	3.6	4.1	5
Cattle	dig coef % *	80.	80.	
Sheep	dig coef %	79.	79.	
N-free extract	%	69.9	78.7	
Cattle	dig coef % *	83.	83.	
Sheep	dig coef %	85.	85.	
Protein (N x 6.25)	%	10.2	11.5	8
Cattle	dig coef % *	57.	57.	
Sheep	dig coef %	56.	56.	
Cattle	dig prot %	5.8	6.5	
Goats	dig prot % *	6.9	7.8	
Horses	dig prot % *	6.9	7.8	

Feed Name or Analyses			Mean As Fed	Mean Dry	C.V. ± %
Sheep	dig prot %		5.7	6.4	
Energy	GE Mcal/kg				
Cattle	DE Mcal/kg	*	3.14	3.53	
Sheep	DE Mcal/kg	*	3.21	3.61	
Swine	DE kcal/kg	*	3444.	3875.	
Cattle	ME Mcal/kg	*	2.57	2.90	
Sheep	ME Mcal/kg	*	2.63	2.96	
Swine	ME kcal/kg	*	3226.	3630.	
Cattle	TDN %		71.2	80.1	
Sheep	TDN %		72.8	82.0	
Swine	TDN %	*	78.1	87.9	
Biotin	mg/kg		0.29	0.33	
Niacin	mg/kg		47.8	53.8	
Pantothenic acid	mg/kg		8.6	9.7	
Riboflavin	mg/kg		0.8	0.9	
Vitamin B6	mg/kg		1.96	2.20	

SORGHUM, ELLIS. Sorghum vulgare

Sorghum, ellis, aerial part, ensiled, (3)

Ref No 3 08 602 United States

Dry matter	%		32.2	100.0	
Ash	%		2.4	7.5	
Crude fiber	%		7.6	23.6	
Ether extract	%		0.9	2.8	
N-free extract	%		18.6	57.8	
Protein (N x 6.25)	%		2.7	8.4	
Cattle	dig prot %	*	1.2	3.8	
Goats	dig prot %	*	1.2	3.8	
Horses	dig prot %	*	1.2	3.8	
Sheep	dig prot %	*	1.2	3.8	
Energy	GE Mcal/kg				
Cattle	DE Mcal/kg	*	0.86	2.66	
Sheep	DE Mcal/kg	*	0.82	2.56	
Cattle	ME Mcal/kg	*	0.70	2.18	
Sheep	ME Mcal/kg	*	0.68	2.10	
Cattle	TDN %	*	19.4	60.3	
Sheep	TDN %	*	18.7	58.1	

SORGHUM, FETERITA. Sorghum vulgare

Sorghum, feterita, aerial part, s-c, (1)

Sorghum, feterita, fodder

Ref No 1 04 366 United States

Dry matter	%		87.2	100.0	1
Ash	%		8.0	9.2	5
Crude fiber	%		23.6	27.1	20
Sheep	dig coef %		66.	66.	
Ether extract	%		1.9	2.2	10
Sheep	dig coef %		61.	61.	
N-free extract	%		47.0	53.9	
Sheep	dig coef %		61.	61.	
Protein (N x 6.25)	%		6.7	7.7	17
Sheep	dig coef %		50.	50.	
Cattle	dig prot %	*	3.1	3.6	
Goats	dig prot %	*	3.2	3.7	
Horses	dig prot %	*	3.5	4.0	
Rabbits	dig prot %	*	4.0	4.6	
Sheep	dig prot %		3.3	3.8	
Energy	GE Mcal/kg				
Cattle	DE Mcal/kg	*	2.29	2.63	
Sheep	DE Mcal/kg	*	2.21	2.54	
Cattle	ME Mcal/kg	*	1.88	2.16	
Sheep	ME Mcal/kg	*	1.82	2.08	
Cattle	TDN %	*	52.1	59.7	
Sheep	TDN %		50.2	57.6	
Calcium	%		0.28	0.33	
Phosphorus	%		0.20	0.23	

Sorghum, feterita, aerial part wo heads, s-c, (1)

Sorghum, feterita, stover

Ref No 1 04 367 United States

Dry matter	%		86.3	100.0	
Ash	%		8.3	9.6	
Crude fiber	%		29.2	33.8	
Ether extract	%		1.7	2.0	
N-free extract	%		41.9	48.6	
Protein (N x 6.25)	%		5.2	6.0	
Cattle	dig prot %	*	1.9	2.1	
Goats	dig prot %	*	1.9	2.2	
Horses	dig prot %	*	2.3	2.6	
Rabbits	dig prot %	*	2.9	3.3	
Sheep	dig prot %	*	1.7	2.0	
Fatty acids	%		1.7	2.0	
Energy	GE Mcal/kg				
Cattle	DE Mcal/kg	*	2.25	2.61	
Sheep	DE Mcal/kg	*	2.06	2.39	
Cattle	ME Mcal/kg	*	1.84	2.14	
Sheep	ME Mcal/kg	*	1.69	1.96	
Cattle	TDN %	*	51.0	59.1	
Sheep	TDN %	*	46.8	54.2	

Sorghum, feterita, aerial part, ensiled, (3)

Ref No 3 04 368 United States

Dry matter	%		33.2	100.0	
Ash	%		2.4	7.4	
Crude fiber	%		7.1	21.4	
Ether extract	%		0.8	2.3	
N-free extract	%		20.1	60.7	
Protein (N x 6.25)	%		2.7	8.3	
Cattle	dig prot %	*	1.2	3.8	
Goats	dig prot %	*	1.2	3.8	
Horses	dig prot %	*	1.2	3.8	
Sheep	dig prot %	*	1.2	3.8	
Energy	GE Mcal/kg				
Cattle	DE Mcal/kg	*	0.89	2.68	
Sheep	DE Mcal/kg	*	0.82	2.48	
Cattle	ME Mcal/kg	*	0.73	2.20	
Sheep	ME Mcal/kg	*	0.67	2.03	
Cattle	TDN %	*	20.2	60.7	
Sheep	TDN %	*	18.7	56.3	
Calcium	%		0.12	0.37	
Phosphorus	%		0.09	0.27	

Sorghum, feterita, grain, (4)

Ref No 4 04 369 United States

Dry matter	%		89.6	100.0	0
Ash	%		1.7	1.9	
Crude fiber	%		1.9	2.1	
Cattle	dig coef %	*	40.	40.	
Sheep	dig coef %	*	-65.	-65.	
Ether extract	%		3.1	3.5	
Cattle	dig coef %	*	80.	80.	
Sheep	dig coef %	*	74.	74.	
N-free extract	%		70.4	78.6	
Cattle	dig coef %	*	83.	83.	
Sheep	dig coef %	*	91.	91.	
Protein (N x 6.25)	%		12.5	14.0	
Cattle	dig coef %	*	57.	57.	
Sheep	dig coef %	*	76.	76.	
Cattle	dig prot %		7.1	8.0	
Goats	dig prot %	*	9.0	10.0	
Horses	dig prot %	*	9.0	10.0	
Sheep	dig prot %		9.5	10.6	
Pentosans	%		2.9	3.2	24
Starch	%		59.0	65.9	6
Sugars, total	%		1.3	1.5	10
Energy	GE Mcal/kg				
Cattle	DE Mcal/kg	*	3.17	3.54	
Sheep	DE Mcal/kg	*	3.42	3.81	
Swine	DE kcal/kg	*	3670.	4096.	
Cattle	ME Mcal/kg	*	2.60	2.90	
Sheep	ME Mcal/kg	*	2.80	3.13	
Swine	ME kcal/kg	*	3419.	3816.	
Cattle	TDN %		71.9	80.3	
Sheep	TDN %		77.5	86.5	
Swine	TDN %	*	83.2	92.9	
Calcium	%		0.02	0.02	33
Phosphorus	%		0.30	0.34	25
Carotene	mg/kg		0.6	0.7	
Niacin	mg/kg		50.2	56.0	9
Pantothenic acid	mg/kg		14.0	15.7	11
Riboflavin	mg/kg		1.6	1.8	18
Vitamin A equivalent	IU/g		1.0	1.1	
Arginine	%		0.45	0.50	
Histidine	%		0.27	0.30	
Isoleucine	%		0.54	0.60	
Leucine	%		1.70	1.90	
Lysine	%		0.18	0.20	
Methionine	%		0.18	0.20	
Phenylalanine	%		0.63	0.70	
Threonine	%		0.45	0.50	
Tryptophan	%		0.18	0.20	
Valine	%		0.63	0.70	

Sorghum, feterita, heads, chopped, (4)

Ref No 4 04 370 United States

Dry matter	%		89.6	100.0	2
Ash	%		3.2	3.6	13
Crude fiber	%		7.4	8.3	15
Cattle	dig coef %	*	40.	40.	
Sheep	dig coef %	*	61.	61.	
Ether extract	%		2.6	2.9	12
Cattle	dig coef %	*	80.	80.	
Sheep	dig coef %	*	74.	74.	
N-free extract	%		65.7	73.3	
Cattle	dig coef %	*	83.	83.	
Sheep	dig coef %	*	80.	80.	
Protein (N x 6.25)	%		10.7	11.9	12
Cattle	dig coef %	*	57.	57.	
Sheep	dig coef %	*	63.	63.	
Cattle	dig prot %		6.1	6.8	
Goats	dig prot %	*	7.3	8.2	
Horses	dig prot %	*	7.3	8.2	
Sheep	dig prot %		6.7	7.5	
Energy	GE Mcal/kg				
Cattle	DE Mcal/kg	*	3.01	3.36	
Sheep	DE Mcal/kg	*	3.00	3.35	
Swine	DE kcal/kg	*	2888.	3223.	
Cattle	ME Mcal/kg	*	2.47	2.75	
Sheep	ME Mcal/kg	*	2.46	2.75	
Swine	ME kcal/kg	*	2703.	3016.	
Cattle	TDN %		68.3	76.2	
Sheep	TDN %		68.1	76.0	
Swine	TDN %	*	65.5	73.1	

Feed Name or Analyses			Mean As Fed	Mean Dry	C.V. ± %

SORGHUM. GA 609. Sorghum vulgare

Sorghum, GA 609, aerial part, ensiled, dough stage, (3)

Ref No 3 09 244 United States

Analyses	Unit		As Fed	Dry	C.V. ± %
Dry matter	%			100.0	
Cattle	dig coef %			56.	
Ash	%			5.7	
Crude fiber	%			41.0	
Cattle	dig coef %			64.	
Ether extract	%			2.3	
N-free extract	%			40.0	
Cattle	dig coef %			51.	
Protein (N x 6.25)	%			11.0	
Cattle	dig coef %			51.	
Cattle	dig prot %			5.6	
Goats	dig prot %	*		6.2	
Horses	dig prot %	*		6.2	
Sheep	dig prot %	*		6.2	
Energy	GE Mcal/kg				
Cattle	DE Mcal/kg	*		2.46	
Sheep	DE Mcal/kg	*		2.50	
Cattle	ME Mcal/kg	*		2.01	
Sheep	ME Mcal/kg	*		2.08	
Cattle	TDN %			55.7	
Sheep	TDN %	*		57.5	

SORGHUM, GRAIN VARIETY. Sorghum vulgare

Sorghum, grain variety, aerial part, s-c, (1)

Grain sorghum fodder, sun-cured

Ref No 1 04 372 United States

Analyses	Unit		As Fed	Dry	C.V. ± %
Dry matter	%		90.2	100.0	1
Ash	%		8.9	9.9	14
Crude fiber	%		26.1	28.9	7
Sheep	dig coef %		63.	63.	
Ether extract	%		1.7	1.9	8
Sheep	dig coef %		53.	53.	
N-free extract	%		47.2	52.4	
Sheep	dig coef %		61.	61.	
Protein (N x 6.25)	%		6.2	6.9	6
Sheep	dig coef %		38.	38.	
Cattle	dig prot %	*	2.6	2.9	
Goats	dig prot %	*	2.7	3.0	
Horses	dig prot %	*	3.1	3.4	
Rabbits	dig prot %	*	3.6	4.0	
Sheep	dig prot %		2.4	2.6	
Energy	GE Mcal/kg				
Cattle	DE Mcal/kg	*	2.28	2.53	
Sheep	DE Mcal/kg	*	2.19	2.43	
Cattle	ME Mcal/kg	*	1.87	2.07	
Sheep	ME Mcal/kg	*	1.79	1.99	
Cattle	NE_m Mcal/kg	*	1.12	1.24	
Cattle	NE_{gain} Mcal/kg	*	0.52	0.58	
Cattle	$NE_{lactating\ cows}$ Mcal/kg	*	1.24	1.38	
Cattle	TDN %	*	51.7	57.3	
Sheep	TDN %		49.6	55.1	

(1) dry forages and roughages (2) pasture, range plants, and forages fed green (3) silages (4) energy feeds (5) protein supplements (6) minerals (7) vitamins (8) additives

Sorghum, grain variety, aerial part, s-c, full bloom, (1)

Sorghum, grain, fodder, full bloom

Ref No 1 04 371 United States

Analyses	Unit		As Fed	Dry	C.V. ± %
Dry matter	%		89.5	100.0	
Ash	%		5.0	5.6	
Crude fiber	%		21.3	23.8	
Ether extract	%		1.7	1.9	
N-free extract	%		55.8	62.3	
Protein (N x 6.25)	%		5.7	6.4	
Cattle	dig prot %	*	2.2	2.5	
Goats	dig prot %	*	2.3	2.5	
Horses	dig prot %	*	2.7	3.0	
Rabbits	dig prot %	*	3.2	3.6	
Sheep	dig prot %	*	2.1	2.3	
Energy	GE Mcal/kg				
Cattle	DE Mcal/kg	*	2.24	2.50	
Sheep	DE Mcal/kg	*	2.18	2.44	
Cattle	ME Mcal/kg	*	1.83	2.05	
Sheep	ME Mcal/kg	*	1.79	2.00	
Cattle	TDN %	*	50.7	56.7	
Sheep	TDN %	*	49.6	55.4	
Calcium	%		0.55	0.62	22
Magnesium	%		0.27	0.30	25
Phosphorus	%		0.17	0.19	39
Potassium	%		1.11	1.24	25
Sodium	%		0.02	0.02	40

Sorghum, grain variety, aerial part, ensiled, milk stage, (3)

Ref No 3 09 092 United States

Analyses	Unit		As Fed	Dry	C.V. ± %
Dry matter	%		26.8	100.0	4
Cattle	dig coef %		58.	58.	3
Ash	%		1.9	6.9	7
Crude fiber	%		7.5	28.1	9
Cattle	dig coef %		58.	58.	
Ether extract	%		0.8	2.8	15
Cattle	dig coef %		61.	61.	
N-free extract	%		14.0	52.1	4
Cattle	dig coef %		62.	62.	
Protein (N x 6.25)	%		2.7	10.1	12
Cattle	dig coef %		54.	54.	7
Cattle	dig prot %		1.5	5.4	
Goats	dig prot %	*	1.5	5.4	
Horses	dig prot %	*	1.5	5.4	
Sheep	dig prot %	*	1.5	5.4	
Cellulose (Matrone)	%		7.0	26.2	9
Cattle	dig coef %		57.	57.	5
Energy	GE Mcal/kg		1.16	4.33	1
Cattle	GE dig coef %		59.	59.	4
Cattle	DE Mcal/kg		0.69	2.55	
Sheep	DE Mcal/kg	*	0.69	2.56	
Cattle	ME Mcal/kg	*	0.56	2.09	
Sheep	ME Mcal/kg	*	0.55	2.10	
Cattle	TDN %		15.5	57.9	
Sheep	TDN %	*	15.6	58.1	

Sorghum, grain variety, aerial part, ensiled, dough stage, (3)

Ref No 3 09 091 United States

Analyses	Unit		As Fed	Dry	C.V. ± %
Dry matter	%		26.9	100.0	3
Cattle	dig coef %		59.	59.	7
Ash	%		1.8	6.7	8
Crude fiber	%		7.5	28.1	9
Cattle	dig coef %		55.	55.	
Ether extract	%		0.7	2.7	17
Cattle	dig coef %		63.	63.	
N-free extract	%		14.3	53.4	5
Cattle	dig coef %		67.	67.	
Protein (N x 6.25)	%		2.5	9.2	17
Cattle	dig coef %		48.	48.	28
Cattle	dig prot %		1.2	4.5	
Goats	dig prot %	*	1.2	4.6	
Horses	dig prot %	*	1.2	4.6	
Sheep	dig prot %	*	1.2	4.6	
Cellulose (Matrone)	%		7.1	26.3	8
Cattle	dig coef %		56.	56.	8
Energy	GE Mcal/kg		1.16	4.32	1
Cattle	GE dig coef %		61.	61.	5
Cattle	DE Mcal/kg		0.70	2.62	
Sheep	DE Mcal/kg	*	0.68	2.54	
Cattle	ME Mcal/kg	*	0.58	2.15	
Sheep	ME Mcal/kg	*	1.56	2.08	
Cattle	TDN %		16.0	59.5	
Sheep	TDN %	*	15.5	57.7	

Sorghum, grain variety, aerial part, ensiled, mature, (3)

Ref No 3 09 087 United States

Analyses	Unit		As Fed	Dry	C.V. ± %
Dry matter	%		30.7	100.0	12
Cattle	dig coef %		56.	56.	6
Ash	%		2.3	7.3	21
Crude fiber	%		7.3	23.8	13
Cattle	dig coef %		50.	50.	8
Ether extract	%		0.9	2.8	16
Cattle	dig coef %		63.	63.	12
N-free extract	%		17.3	56.4	8
Cattle	dig coef %		66.	66.	4
Protein (N x 6.25)	%		3.0	9.6	13
Cattle	dig coef %		45.	45.	16
Cattle	dig prot %		1.3	4.3	
Goats	dig prot %	*	1.5	5.0	
Horses	dig prot %	*	1.5	5.0	
Sheep	dig prot %	*	1.5	5.0	
Cell walls (Van Soest)	%		12.1	39.3	
Cellulose (Matrone)	%		7.1	23.1	15
Cattle	dig coef %		53.	53.	9
Fiber, acid detergent (VS)	%		6.5	21.2	
Lignin (Ellis)	%		1.4	4.6	
Energy	GE Mcal/kg		1.33	4.33	0
Cattle	GE dig coef %		57.	57.	5
Cattle	DE Mcal/kg		0.76	2.46	
Sheep	DE Mcal/kg	*	0.79	2.57	
Cattle	ME Mcal/kg	*	0.62	2.02	
Sheep	ME Mcal/kg	*	0.65	2.11	
Cattle	TDN %		17.6	57.4	
Sheep	TDN %	*	17.9	58.4	

Sorghum, grain variety, aerial part, ensiled rolled, mature, (3)

Ref No 3 09 093 United States

Analyses	Unit		As Fed	Dry	C.V. ± %
Dry matter	%		31.3	100.0	
Cattle	dig coef %		59.	59.	
Ash	%		2.1	6.8	
Crude fiber	%		7.5	24.1	
Cattle	dig coef %		48.	48.	
Ether extract	%		0.9	3.0	
Cattle	dig coef %		65.	65.	

Feed Name or Analyses		Mean As Fed	Mean Dry	C.V. ± %
N-free extract	%	17.7	56.5	
Cattle	dig coef %	71.	71.	
Protein (N x 6.25)	%	3.0	9.6	
Cattle	dig coef %	47.	47.	
Cattle	dig prot %	1.4	4.5	
Goats	dig prot %	* 1.5	4.9	
Horses	dig prot %	* 1.5	4.9	
Sheep	dig prot %	* 1.5	4.9	
Cellulose (Matrone)	%	6.9	21.9	
Cattle	dig coef %	48.	48.	
Energy	GE Mcal/kg	1.35	4.32	
Cattle	GE dig coef %	59.	59.	
Cattle	DE Mcal/kg	0.80	2.55	
Sheep	DE Mcal/kg	* 0.82	2.61	
Cattle	ME Mcal/kg	* 0.65	2.09	
Sheep	ME Mcal/kg	* 0.67	2.14	
Cattle	TDN %	19.0	60.6	
Sheep	TDN %	* 18.6	59.3	

Sorghum, grain variety, grain, (4)

Ref No 4 04 383 United States

Feed Name or Analyses		Mean As Fed	Mean Dry	C.V. ± %
Dry matter	%	88.5	100.0	4
Ash	%	2.1	2.4	
Crude fiber	%	2.3	2.6	
Cattle	dig coef %	* 40.	40.	
Sheep	dig coef %	* 33.	33.	
Ether extract	%	3.1	3.5	
Cattle	dig coef %	* 80.	80.	
Sheep	dig coef %	* 76.	76.	
N-free extract	%	72.0	81.5	
Cattle	dig coef %	* 83.	83.	
Sheep	dig coef %	* 89.	89.	
Protein (N x 6.25)	%	8.9	10.0	
Cattle	dig coef %	* 57.	57.	
Sheep	dig coef %	* 67.	67.	
Cattle	dig prot %	5.1	5.7	
Goats	dig prot %	* 5.7	6.4	
Horses	dig prot %	* 5.7	6.4	
Sheep	dig prot %	5.9	6.7	
Energy	GE Mcal/kg			
Cattle	DE Mcal/kg	* 3.14	3.55	
Sheep	DE Mcal/kg	* 3.35	3.79	
Swine	DE kcal/kg	*3466.	3918.	
Cattle	ME Mcal/kg	* 2.58	2.91	
Chickens	ME_n kcal/kg	3370.	3810.	
Sheep	ME Mcal/kg	* 2.75	3.11	
Swine	ME kcal/kg	*3257.	3682.	
Cattle	NE_m Mcal/kg	* 1.73	1.96	
Cattle	NE_{gain} Mcal/kg	* 1.16	1.31	
Cattle	$NE_{lactating\ cows}$ Mcal/kg	* 2.04	2.30	
Cattle	TDN %	71.3	80.6	
Sheep	TDN %	76.1	86.0	
Swine	TDN %	* 78.6	88.9	
Calcium	%	0.03	0.03	
Chlorine	%	0.09	0.10	26
Cobalt	mg/kg	0.269	0.304	60
Phosphorus	%	0.29	0.32	
Sodium	%	0.04	0.05	54
Sulphur	%	0.16	0.18	13
Zinc	mg/kg	13.6	15.4	
Choline	mg/kg	449.	508.	
Niacin	mg/kg	41.1	46.5	
Pantothenic acid	mg/kg	11.7	13.2	
Riboflavin	mg/kg	1.1	1.3	
Arginine	%	0.38	0.43	12
Cystine	%	0.15	0.17	19
Glutamic acid	%	2.48	2.80	
Glycine	%	0.31	0.35	

Feed Name or Analyses		Mean As Fed	Mean Dry	C.V. ± %
Histidine	%	0.27	0.30	17
Isoleucine	%	0.53	0.60	14
Leucine	%	1.42	1.60	14
Lysine	%	0.29	0.32	24
Methionine	%	0.12	0.14	
Phenylalanine	%	0.44	0.50	23
Serine	%	0.53	0.60	
Threonine	%	0.27	0.30	35
Tryptophan	%	0.10	0.11	26
Tyrosine	%	0.35	0.40	26
Valine	%	0.53	0.60	10
Xanthophylls	mg/kg	1.12	1.27	

Sorghum, grain variety, grain, grnd, (4)

Ground grain sorghum (AAFCO)

Ref No 4 04 379 United States

Feed Name or Analyses		Mean As Fed	Mean Dry	C.V. ± %
Dry matter	%		100.0	
Crude fiber	%			
Cattle	dig coef %	*	-96.	
Sheep	dig coef %	*	33.	
Ether extract	%			
Cattle	dig coef %	*	52.	
Sheep	dig coef %	*	76.	
N-free extract	%			
Cattle	dig coef %	*	79.	
Sheep	dig coef %	*	89.	
Protein (N x 6.25)	%			
Cattle	dig coef %	*	58.	
Sheep	dig coef %	*	67.	

Sorghum, grain variety, heads, chopped, (4)

Ref No 4 04 387 United States

Feed Name or Analyses		Mean As Fed	Mean Dry	C.V. ± %
Dry matter	%	89.3	100.0	
Ash	%	5.9	6.6	
Crude fiber	%	7.2	8.1	
Cattle	dig coef %	* 40.	40.	
Sheep	dig coef %	* 61.	61.	
Ether extract	%	2.1	2.4	
Cattle	dig coef %	* 80.	80.	
Sheep	dig coef %	* 74.	74.	
N-free extract	%	64.0	71.6	
Cattle	dig coef %	* 83.	83.	
Sheep	dig coef %	* 80.	80.	
Protein (N x 6.25)	%	10.1	11.3	
Cattle	dig coef %	* 57.	57.	
Sheep	dig coef %	* 63.	63.	
Cattle	dig prot %	5.8	6.5	
Goats	dig prot %	* 6.8	7.6	
Horses	dig prot %	* 6.8	7.6	
Sheep	dig prot %	6.4	7.1	
Energy	GE Mcal/kg			
Cattle	DE Mcal/kg	* 2.89	3.24	
Sheep	DE Mcal/kg	* 2.89	3.23	
Swine	DE kcal/kg	*2604.	2916.	
Cattle	ME Mcal/kg	* 2.37	2.66	
Sheep	ME Mcal/kg	* 2.37	2.65	
Swine	ME kcal/kg	*2440.	2733.	
Cattle	TDN %	65.6	73.5	
Sheep	TDN %	65.5	73.4	
Swine	TDN %	* 59.1	66.1	
Calcium	%	0.14	0.16	
Phosphorus	%	0.26	0.29	
Riboflavin	mg/kg	2.0	2.2	

Sorghum, grain variety, distillers grains, dehy, (5)

Grain sorghum distillers dried grains (AAFCO)

Ref No 5 04 374 United States

Feed Name or Analyses		Mean As Fed	Mean Dry	C.V. ± %
Dry matter	%	93.8	100.0	2
Ash	%	3.9	4.2	31
Crude fiber	%	12.3	13.1	34
Ether extract	%	8.4	9.0	29
N-free extract	%	38.0	40.5	
Protein (N x 6.25)	%	31.2	33.3	20
Energy	GE Mcal/kg			
Cattle	DE Mcal/kg	* 3.36	3.58	
Sheep	DE Mcal/kg	* 3.47	3.70	
Swine	DE kcal/kg	*3883.	4140.	
Cattle	ME Mcal/kg	* 2.75	2.93	
Sheep	ME Mcal/kg	* 2.84	3.03	
Swine	ME kcal/kg	*3467.	3696.	
Cattle	NE_m Mcal/kg	* 1.81	1.93	
Cattle	NE_{gain} Mcal/kg	* 1.21	1.29	
Cattle	$NE_{lactating\ cows}$ Mcal/kg	* 2.13	2.27	
Cattle	TDN %	* 76.1	81.2	
Sheep	TDN %	* 78.7	83.9	
Swine	TDN %	* 88.1	93.9	
Calcium	%	0.14	0.15	20
Phosphorus	%	0.59	0.63	33

Sorghum, grain variety, distillers grains w solubles, dehy, mn 75% original solids, (5)

Grain sorghum distillers dried grains with solubles (AAFCO)

Ref No 5 04 375 United States

Feed Name or Analyses		Mean As Fed	Mean Dry	C.V. ± %
Dry matter	%	94.8	100.0	2
Ash	%	4.2	4.4	26
Crude fiber	%	10.1	10.7	19
Ether extract	%	9.4	9.9	6
N-free extract	%	38.0	40.1	
Protein (N x 6.25)	%	33.1	34.9	12
Energy	GE Mcal/kg			
Cattle	DE Mcal/kg	* 3.53	3.72	
Sheep	DE Mcal/kg	* 3.59	3.78	
Swine	DE kcal/kg	*4047.	4269.	
Cattle	ME Mcal/kg	* 2.89	3.05	
Sheep	ME Mcal/kg	* 2.94	3.10	
Swine	ME kcal/kg	*3599.	3797.	
Cattle	TDN %	* 80.0	84.4	
Sheep	TDN %	* 81.3	85.8	
Swine	TDN %	* 91.8	96.8	
Calcium	%	0.17	0.18	
Manganese	mg/kg	104.5	110.2	
Phosphorus	%	0.92	0.97	
Choline	mg/kg	844.	891.	51
Niacin	mg/kg	61.1	64.4	
Pantothenic acid	mg/kg	12.3	13.0	
Riboflavin	mg/kg	4.2	4.4	9
Thiamine	mg/kg	1.3	1.4	

Sorghum, grain variety, distillers solubles, dehy, (5)

Grain sorghum distillers dried solubles (AAFCO)

Ref No 5 04 376 United States

Feed Name or Analyses		Mean As Fed	Mean Dry	C.V. ± %
Dry matter	%	92.6	100.0	2
Ash	%	8.4	9.1	9
Crude fiber	%	3.9	4.2	23
Ether extract	%	5.5	5.9	64

Continued

Feed Name or Analyses		Mean As Fed	Mean Dry	C.V. ± %
N-free extract	%	48.4	52.3	
Protein (N x 6.25)	%	26.4	28.5	11
Energy	GE Mcal/kg			
Cattle	DE Mcal/kg	* 3.39	3.66	
Sheep	DE Mcal/kg	* 3.60	3.89	
Swine	DE kcal/kg	*3897.	4209.	
Cattle	ME Mcal/kg	* 2.78	3.00	
Sheep	ME Mcal/kg	* 2.95	3.19	
Swine	ME kcal/kg	*3517.	3798.	
Cattle	TDN %	* 76.9	83.1	
Sheep	TDN %	* 81.6	88.1	
Swine	TDN %	* 88.4	95.4	
Calcium	%	0.68	0.73	49
Phosphorus	%	1.37	1.48	24
Niacin	mg/kg	142.0	153.3	9
Riboflavin	mg/kg	16.1	17.4	
Thiamine	mg/kg	4.4	4.8	45

Sorghum, grain variety, gluten, wet milled dehy, (5)
Grain sorghum gluten meal (AAFCO)
Ref No 5 04 388 United States

Feed Name or Analyses		Mean As Fed	Mean Dry	C.V. ± %
Dry matter	%	91.4	100.0	
Ash	%	1.4	1.5	
Crude fiber	%	6.5	7.2	
Ether extract	%	4.5	4.9	
N-free extract	%	35.0	38.3	
Protein (N x 6.25)	%	43.9	48.1	
Energy	GE Mcal/kg			
Cattle	DE Mcal/kg	* 3.41	3.73	
Sheep	DE Mcal/kg	* 3.47	3.79	
Swine	DE kcal/kg	*3578.	3915.	
Cattle	ME Mcal/kg	* 2.80	3.06	
Chickens	ME_n kcal/kg	2574.	2817.	
Sheep	ME Mcal/kg	* 2.84	3.11	
Swine	ME kcal/kg	*3088.	3378.	
Cattle	TDN %	* 77.4	84.6	
Sheep	TDN %	* 78.6	86.0	
Swine	TDN %	* 81.2	88.8	
Calcium	%	0.02	0.02	
Phosphorus	%	0.17	0.19	
Choline	mg/kg	728.	796.	
Niacin	mg/kg	34.7	38.0	
Pantothenic acid	mg/kg	10.3	11.3	
Riboflavin	mg/kg	1.6	1.7	
Arginine	%	1.22	1.33	
Cystine	%	0.51	0.56	
Lysine	%	0.71	0.78	
Methionine	%	0.71	0.78	
Tryptophan	%	0.41	0.44	

Sorghum, grain variety, gluten w bran, wet milled dehy, (5)
Grain sorghum gluten feed (AAFCO)
Ref No 5 04 389 United States

Feed Name or Analyses		Mean As Fed	Mean Dry	C.V. ± %
Dry matter	%	91.3	100.0	
Ash	%	8.6	9.4	
Crude fiber	%	7.1	7.8	
Ether extract	%	3.3	3.6	
N-free extract	%	46.1	50.5	
Protein (N x 6.25)	%	26.1	28.6	
Energy	GE Mcal/kg			
Cattle	DE Mcal/kg	* 3.06	3.36	
Sheep	DE Mcal/kg	* 3.36	3.68	
Swine	DE kcal/kg	*3527.	3866.	
Cattle	ME Mcal/kg	* 2.51	2.75	
Chickens	ME_n kcal/kg	1697.	1859.	
Sheep	ME Mcal/kg	* 2.75	3.02	
Swine	ME kcal/kg	*3182.	3487.	
Cattle	TDN %	* 69.5	76.2	
Sheep	TDN %	* 76.1	83.4	
Swine	TDN %	* 80.0	87.7	
Calcium	%	0.09	0.10	
Phosphorus	%	0.61	0.66	
Choline	mg/kg	1810.	1984.	
Niacin	mg/kg	55.9	61.3	
Pantothenic acid	mg/kg	19.9	21.8	
Riboflavin	mg/kg	2.9	3.2	
Arginine	%	1.03	1.12	
Cystine	%	0.21	0.23	
Lysine	%	0.82	0.90	
Methionine	%	0.31	0.34	
Tryptophan	%	0.21	0.23	

(1) dry forages and roughages (2) pasture, range plants, and forages fed green (3) silages (4) energy feeds (5) protein supplements (6) minerals (7) vitamins (8) additives

SORGHUM, HEGARI. Sorghum vulgare

Sorghum, hegari, aerial part, dehy, (1)
Ref No 1 08 514 United States

Feed Name or Analyses		Mean As Fed	Mean Dry	C.V. ± %
Dry matter	%	90.1	100.0	
Ash	%	8.2	9.1	
Crude fiber	%	17.1	19.0	
Ether extract	%	1.7	1.9	
N-free extract	%	57.7	64.0	
Protein (N x 6.25)	%	5.4	6.0	
Cattle	dig prot %	* 1.9	2.1	
Goats	dig prot %	* 1.9	2.2	
Horses	dig prot %	* 2.4	2.6	
Rabbits	dig prot %	* 3.0	3.3	
Sheep	dig prot %	* 1.8	1.9	
Energy	GE Mcal/kg			
Cattle	DE Mcal/kg	* 2.51	2.79	
Sheep	DE Mcal/kg	* 2.44	2.71	
Cattle	ME Mcal/kg	* 2.06	2.29	
Sheep	ME Mcal/kg	* 2.00	2.22	
Cattle	TDN %	* 56.9	63.2	
Sheep	TDN %	* 55.4	61.5	

Sorghum, hegari, aerial part, s-c, (1)
Ref No 1 08 088 United States

Feed Name or Analyses		Mean As Fed	Mean Dry	C.V. ± %
Dry matter	%	86.3	100.0	
Ash	%	7.5	8.7	
Crude fiber	%	18.2	21.1	
Ether extract	%	1.7	2.0	
N-free extract	%	52.8	61.2	
Protein (N x 6.25)	%	6.1	7.1	
Cattle	dig prot %	* 2.6	3.1	
Goats	dig prot %	* 2.7	3.2	
Horses	dig prot %	* 3.0	3.5	
Rabbits	dig prot %	* 3.6	4.1	
Sheep	dig prot %	* 2.5	2.9	
Energy	GE Mcal/kg			
Cattle	DE Mcal/kg	* 2.32	2.69	
Sheep	DE Mcal/kg	* 2.32	2.68	
Cattle	ME Mcal/kg	* 1.91	2.21	
Sheep	ME Mcal/kg	* 1.90	2.20	
Cattle	TDN %	* 52.6	61.0	
Sheep	TDN %	* 52.5	60.9	
Calcium	%	0.27	0.31	
Phosphorus	%	0.16	0.19	

Sorghum, hegari, aerial part, s-c, dough stage, (1)
Sorghum, hegari, fodder, dough stage
Ref No 1 04 390 United States

Feed Name or Analyses		Mean As Fed	Mean Dry	C.V. ± %
Dry matter	%	78.4	100.0	2
Ash	%	6.7	8.5	3
Crude fiber	%	16.5	21.0	7
Ether extract	%	1.8	2.3	8
N-free extract	%	47.0	60.0	
Protein (N x 6.25)	%	6.4	8.2	7
Cattle	dig prot %	* 3.2	4.0	
Goats	dig prot %	* 3.3	4.2	
Horses	dig prot %	* 3.5	4.5	
Rabbits	dig prot %	* 3.9	5.0	
Sheep	dig prot %	* 3.1	3.9	
Energy	GE Mcal/kg			
Cattle	DE Mcal/kg	* 2.16	2.76	
Sheep	DE Mcal/kg	* 2.12	2.70	
Cattle	ME Mcal/kg	* 1.77	2.26	
Sheep	ME Mcal/kg	* 1.74	2.22	
Cattle	TDN %	* 49.2	62.7	
Sheep	TDN %	* 48.1	61.3	

Sorghum, hegari, aerial part, s-c, mature, (1)
Sorghum, hegari, fodder, sun-cured, mature
Ref No 1 04 391 United States

Feed Name or Analyses		Mean As Fed	Mean Dry	C.V. ± %
Dry matter	%		100.0	
Crude fiber	%			
Cattle	dig coef %	*	60.	
Sheep	dig coef %	*	63.	
Ether extract	%			
Cattle	dig coef %	*	61.	
Sheep	dig coef %	*	53.	
N-free extract	%			
Cattle	dig coef %	*	66.	
Sheep	dig coef %	*	61.	
Protein (N x 6.25)	%			
Cattle	dig coef %	*	38.	
Sheep	dig coef %	*	38.	

Sorghum, hegari, aerial part wo heads, s-c, (1)
Ref No 1 08 515 United States

Feed Name or Analyses		Mean As Fed	Mean Dry	C.V. ± %
Dry matter	%	87.0	100.0	
Ash	%	9.9	11.4	
Crude fiber	%	28.0	32.2	
Ether extract	%	1.8	2.1	
N-free extract	%	41.7	47.9	
Protein (N x 6.25)	%	5.6	6.4	
Cattle	dig prot %	* 2.2	2.5	
Goats	dig prot %	* 2.2	2.6	
Horses	dig prot %	* 2.6	3.0	
Rabbits	dig prot %	* 3.2	3.6	
Sheep	dig prot %	* 2.0	2.3	
Energy	GE Mcal/kg			
Cattle	DE Mcal/kg	* 2.21	2.54	
Sheep	DE Mcal/kg	* 2.10	2.42	
Cattle	ME Mcal/kg	* 1.81	2.08	
Sheep	ME Mcal/kg	* 1.73	1.98	
Cattle	TDN %	* 50.0	57.5	
Sheep	TDN %	* 47.7	54.9	

Feed Name or Analyses		Mean As Fed	Mean Dry	C.V. ± %
Calcium	%	0.33	0.38	
Phosphorus	%	0.08	0.09	

Sorghum, hegari, aerial part wo heads, s-c, mature, (1)

Sorghum, hegari, stover, sun-cured, mature

Ref No 1 04 392 United States

Feed Name or Analyses		Mean As Fed	Mean Dry	C.V. ± %
Dry matter	%	85.7	100.0	
Ash	%	6.9	8.1	
Crude fiber	%	20.3	23.7	
Cattle	dig coef %	* 67.	67.	
Sheep	dig coef %	* 66.	66.	
Ether extract	%	2.2	2.6	
Cattle	dig coef %	* 75.	75.	
Sheep	dig coef %	* 58.	58.	
N-free extract	%	51.7	60.3	
Cattle	dig coef %	* 60.	60.	
Sheep	dig coef %	* 49.	49.	
Protein (N x 6.25)	%	4.5	5.3	
Cattle	dig coef %	* 34.	34.	
Sheep	dig coef %	* -17.	-17.	
Cattle	dig prot %	1.5	1.8	
Goats	dig prot %	* 1.3	1.5	
Horses	dig prot %	* 1.7	2.0	
Rabbits	dig prot %	* 2.4	2.8	
Sheep	dig prot %	-0.7	-0.9	
Energy	GE Mcal/kg			
Cattle	DE Mcal/kg	* 2.20	2.57	
Sheep	DE Mcal/kg	* 1.80	2.10	
Cattle	ME Mcal/kg	* 1.80	2.11	
Sheep	ME Mcal/kg	* 1.48	1.72	
Cattle	TDN %	* 49.9	58.2	
Sheep	TDN %	* 40.8	47.6	

Sorghum, hegari, aerial part, ensiled, (3)

Ref No 3 04 395 United States

Feed Name or Analyses		Mean As Fed	Mean Dry	C.V. ± %
Dry matter	%	34.9	100.0	5
Ash	%	3.6	10.3	15
Crude fiber	%	7.7	21.9	6
Ether extract	%	0.7	2.0	28
N-free extract	%	21.4	61.2	
Protein (N x 6.25)	%	1.6	4.5	19
Cattle	dig prot %	* 0.1	0.4	
Goats	dig prot %	* 0.1	0.4	
Horses	dig prot %	* 0.1	0.4	
Sheep	dig prot %	* 0.1	0.4	
Energy	GE Mcal/kg			
Cattle	DE Mcal/kg	* 0.89	2.54	
Sheep	DE Mcal/kg	* 0.87	2.48	
Cattle	ME Mcal/kg	* 0.73	2.08	
Sheep	ME Mcal/kg	* 0.71	2.03	
Cattle	TDN %	* 20.1	57.7	
Sheep	TDN %	* 19.6	56.3	
Carotene	mg/kg	6.5	18.7	
Vitamin A equivalent	IU/g	10.9	31.2	
Vitamin D_2	IU/g	0.2	0.7	

Sorghum, hegari, aerial part, ensiled, milk stage, (3)

Ref No 3 09 104 United States

Feed Name or Analyses		Mean As Fed	Mean Dry	C.V. ± %
Dry matter	%	23.3	100.0	
Cattle	dig coef %	56.	56.	
Ash	%	1.7	7.1	
Crude fiber	%	7.3	31.5	
Ether extract	%	0.6	2.7	
N-free extract	%	11.7	50.4	
Protein (N x 6.25)	%	2.0	8.4	
Cattle	dig coef %	43.	43.	
Cattle	dig prot %	0.8	3.6	
Goats	dig prot %	* 0.9	3.9	
Horses	dig prot %	* 0.9	3.9	
Sheep	dig prot %	* 0.9	3.9	
Cellulose (Matrone)	%	7.7	33.0	
Cattle	dig coef %	60.	60.	
Energy	GE Mcal/kg	1.04	4.45	
Cattle	GE dig coef %	57.	57.	
Cattle	DE Mcal/kg	0.59	2.51	
Sheep	DE Mcal/kg	* 0.56	2.42	
Cattle	ME Mcal/kg	* 0.48	2.06	
Sheep	ME Mcal/kg	* 0.46	1.98	
Cattle	TDN %	* 13.3	56.9	
Sheep	TDN %	* 12.8	55.0	

Sorghum, hegari, aerial part, ensiled, dough stage, (3)

Ref No 3 09 161 United States

Feed Name or Analyses		Mean As Fed	Mean Dry	C.V. ± %
Dry matter	%	22.0	100.0	
Cattle	dig coef %	55.	55.	
Ash	%	1.7	7.9	
Crude fiber	%	6.8	30.9	
Ether extract	%	0.7	3.0	
N-free extract	%	10.8	48.9	
Protein (N x 6.25)	%	2.0	9.3	
Cattle	dig coef %	47.	47.	
Cattle	dig prot %	1.0	4.4	
Goats	dig prot %	* 1.0	4.7	
Horses	dig prot %	* 1.0	4.7	
Sheep	dig prot %	* 1.0	4.7	
Cellulose (Matrone)	%	7.3	33.1	
Cattle	dig coef %	61.	61.	
Energy	GE Mcal/kg	0.99	4.48	
Cattle	GE dig coef %	57.	57.	
Cattle	DE Mcal/kg	0.56	2.55	
Sheep	DE Mcal/kg	* 0.54	2.44	
Cattle	ME Mcal/kg	* 0.46	2.09	
Sheep	ME Mcal/kg	* 0.44	2.00	
Cattle	TDN %	* 12.7	57.8	
Sheep	TDN %	* 12.2	55.4	

Sorghum, hegari, aerial part wo heads, ensiled, (3)

Ref No 3 04 396 United States

Feed Name or Analyses		Mean As Fed	Mean Dry	C.V. ± %
Dry matter	%	69.2	100.0	4
Ash	%	3.3	4.8	
Crude fiber	%	13.5	19.6	
Ether extract	%	1.2	1.7	
N-free extract	%	43.3	62.5	
Protein (N x 6.25)	%	7.8	11.3	
Cattle	dig prot %	* 4.5	6.5	
Goats	dig prot %	* 4.5	6.5	
Horses	dig prot %	* 4.5	6.5	
Sheep	dig prot %	* 4.5	6.5	
Energy	GE Mcal/kg			
Cattle	DE Mcal/kg	* 1.59	2.30	
Sheep	DE Mcal/kg	* 1.63	2.36	
Cattle	ME Mcal/kg	* 1.31	1.89	
Sheep	ME Mcal/kg	* 1.34	1.94	
Cattle	TDN %	* 36.1	52.1	
Sheep	TDN %	* 37.1	53.6	
Calcium	%	0.30	0.43	
Phosphorus	%	0.07	0.10	
Carotene	mg/kg	2.0	2.9	63
Vitamin A equivalent	IU/g	3.3	4.8	

Sorghum, hegari, grain, (4)

Ref No 4 04 398 United States

Feed Name or Analyses		Mean As Fed	Mean Dry	C.V. ± %
Dry matter	%	89.4	100.0	0
Ash	%	1.6	1.8	
Crude fiber	%	2.0	2.2	
Cattle	dig coef %	* 40.	40.	
Sheep	dig coef %	* 55.	55.	
Ether extract	%	2.6	2.9	
Cattle	dig coef %	* 80.	80.	
Sheep	dig coef %	* 76.	76.	
N-free extract	%	73.7	82.4	
Cattle	dig coef %	* 83.	83.	
Sheep	dig coef %	* 92.	92.	
Protein (N x 6.25)	%	9.6	10.7	
Cattle	dig coef %	* 57.	57.	
Sheep	dig coef %	* 81.	81.	
Cattle	dig prot %	5.5	6.1	
Goats	dig prot %	* 6.3	7.0	
Horses	dig prot %	* 6.3	7.0	
Sheep	dig prot %	7.7	8.7	
Pentosans	%	1.6	1.8	
Starch	%	64.5	72.2	
Sugars, total	%	1.0	1.1	
Energy	GE Mcal/kg			
Cattle	DE Mcal/kg	* 3.18	3.55	
Sheep	DE Mcal/kg	* 3.57	4.00	
Swine	DE kcal/kg	*3605.	4033.	
Cattle	ME Mcal/kg	* 2.60	2.91	
Sheep	ME Mcal/kg	* 2.93	3.28	
Swine	ME kcal/kg	*3383.	3784.	
Cattle	TDN %	72.0	80.6	
Sheep	TDN %	81.0	90.6	
Swine	TDN %	* 81.8	91.5	
Calcium	%	0.10	0.11	
Phosphorus	%	0.27	0.31	
Arginine	%	0.27	0.30	
Histidine	%	0.18	0.20	
Isoleucine	%	0.45	0.50	
Leucine	%	1.34	1.50	
Lysine	%	0.18	0.20	
Methionine	%	0.09	0.10	
Phenylalanine	%	0.54	0.60	
Threonine	%	0.36	0.40	
Tryptophan	%	0.09	0.10	
Valine	%	0.54	0.60	

Sorghum, hegari, grain, gr 2 US mn wt 55 lb per bushel, (4)

Ref No 4 04 397 United States

Feed Name or Analyses		Mean As Fed	Mean Dry	C.V. ± %
Dry matter	%	89.2	100.0	
Ash	%	1.4	1.6	
Crude fiber	%	2.5	2.8	
Ether extract	%	2.6	2.9	
N-free extract	%	71.4	80.0	
Protein (N x 6.25)	%	11.3	12.7	
Cattle	dig prot %	* 6.8	7.7	
Goats	dig prot %	* 7.9	8.9	
Horses	dig prot %	* 7.9	8.9	
Sheep	dig prot %	* 7.9	8.9	
Energy	GE Mcal/kg			
Cattle	DE Mcal/kg	* 3.22	3.61	
Sheep	DE Mcal/kg	* 3.34	3.74	

Continued

Feed Name or Analyses		Mean As Fed	Mean Dry	C.V. ± %
Swine	DE kcal/kg	*3589.	4024.	
Cattle	ME Mcal/kg	* 2.64	2.96	
Sheep	ME Mcal/kg	* 2.74	3.07	
Swine	ME kcal/kg	*3353.	3759.	
Cattle	TDN %	* 73.1	81.9	
Sheep	TDN %	* 75.7	84.9	
Swine	TDN %	* 81.4	91.3	

Sorghum, hegari, heads, chopped, (4)

Ref No 4 04 399 United States

Feed Name or Analyses		Mean As Fed	Mean Dry	C.V. ± %
Dry matter	%	89.9	100.0	1
Ash	%	5.0	5.6	
Crude fiber	%	11.9	13.3	
Cattle	dig coef %	* 40.	40.	
Sheep	dig coef %	* 61.	61.	
Ether extract	%	2.1	2.3	
Cattle	dig coef %	* 80.	80.	
Sheep	dig coef %	* 74.	74.	
N-free extract	%	60.8	67.6	
Cattle	dig coef %	* 83.	83.	
Sheep	dig coef %	* 80.	80.	
Protein (N x 6.25)	%	10.0	11.2	
Cattle	dig coef %	* 57.	57.	
Sheep	dig coef %	* 63.	63.	
Cattle	dig prot %	5.7	6.4	
Goats	dig prot %	* 6.7	7.5	
Horses	dig prot %	* 6.7	7.5	
Sheep	dig prot %	6.3	7.0	
Energy	GE Mcal/kg			
Cattle	DE Mcal/kg	* 2.85	3.18	
Sheep	DE Mcal/kg	* 2.90	3.22	
Swine	DE kcal/kg	*2787.	3100.	
Cattle	ME Mcal/kg	* 2.34	2.60	
Sheep	ME Mcal/kg	* 2.38	2.64	
Swine	ME kcal/kg	*2285.	2542.	
Cattle	TDN %	64.7	72.0	
Sheep	TDN %	65.7	73.1	
Swine	TDN %	* 63.2	70.3	
Calcium	%	0.10	0.11	
Magnesium	%	0.10	0.11	
Manganese	mg/kg	13.1	14.6	
Phosphorus	%	0.19	0.21	

Sorghum, hegari, heads, dehy chopped, (4)

Ref No 4 04 393 United States

Feed Name or Analyses		Mean As Fed	Mean Dry	C.V. ± %
Dry matter	%	90.1	100.0	1
Ash	%	8.2	9.1	9
Crude fiber	%	17.1	19.0	8
Ether extract	%	1.7	1.9	12
N-free extract	%	57.7	64.0	
Protein (N x 6.25)	%	5.4	6.0	14
Cattle	dig prot %	* 1.4	1.5	
Goats	dig prot %	* 2.5	2.7	
Horses	dig prot %	* 2.5	2.7	
Sheep	dig prot %	* 2.5	2.7	
Energy	GE Mcal/kg			
Cattle	DE Mcal/kg	* 2.79	3.10	
Sheep	DE Mcal/kg	* 2.85	3.16	
Cattle	ME Mcal/kg	* 2.29	2.54	
Sheep	ME Mcal/kg	* 2.34	2.59	

(1) dry forages and roughages (2) pasture, range plants, and forages fed green (3) silages (4) energy feeds (5) protein supplements (6) minerals (7) vitamins (8) additives

Feed Name or Analyses		Mean As Fed	Mean Dry	C.V. ± %
Cattle	TDN %	* 63.3	70.3	
Sheep	TDN %	* 64.7	71.8	

SORGHUM, JOHNSONGRASS. Sorghum halepense

Sorghum, johnsongrass, hay, fan air dried, (1)

Johnsongrass, hay, fan air dried

Ref No 1 04 400 United States

Feed Name or Analyses		Mean As Fed	Mean Dry	C.V. ± %
Dry matter	%		100.0	
Carotene	mg/kg		57.5	
Vitamin A equivalent	IU/g		95.9	

Sorghum, johnsongrass, hay, s-c, (1)

Ref No 1 04 407 United States

Feed Name or Analyses		Mean As Fed	Mean Dry	C.V. ± %
Dry matter	%	90.6	100.0	0
Ash	%	8.1	8.9	15
Crude fiber	%	30.0	33.1	7
Sheep	dig coef %	* 68.	68.	
Ether extract	%	1.8	2.0	20
Sheep	dig coef %	* 47.	47.	
N-free extract	%	43.8	48.4	
Sheep	dig coef %	* 57.	57.	
Protein (N x 6.25)	%	6.9	7.6	30
Sheep	dig coef %	* 44.	44.	
Cattle	dig prot %	* 3.2	3.6	
Goats	dig prot %	* 3.3	3.7	
Horses	dig prot %	* 3.6	4.0	
Rabbits	dig prot %	* 4.1	4.6	
Sheep	dig prot %	3.0	3.4	
Energy	GE Mcal/kg			
Cattle	DE Mcal/kg	* 2.37	2.62	
Sheep	DE Mcal/kg	* 2.22	2.45	
Cattle	ME Mcal/kg	* 1.94	2.15	
Sheep	ME Mcal/kg	* 1.82	2.01	
Cattle	NE_m Mcal/kg	* 1.08	1.19	
Cattle	NE_{gain} Mcal/kg	* 0.46	0.51	
Cattle	$NE_{lactating\ cows}$ Mcal/kg	* 1.18	1.30	
Cattle	TDN %	* 53.8	59.3	
Sheep	TDN %	50.3	55.5	
Calcium	%	0.80	0.89	29
Iron	%	0.053	0.058	
Magnesium	%	0.31	0.35	14
Phosphorus	%	0.27	0.30	25
Potassium	%	1.22	1.35	10
Carotene	mg/kg	38.6	42.7	21
Vitamin A equivalent	IU/g	64.4	71.1	

Sorghum, johnsongrass, hay, s-c, immature, (1)

Ref No 1 04 401 United States

Feed Name or Analyses		Mean As Fed	Mean Dry	C.V. ± %
Dry matter	%	91.8	100.0	
Ash	%	10.0	10.9	
Crude fiber	%	26.6	29.0	
Ether extract	%	2.6	2.8	
N-free extract	%	38.8	42.3	
Protein (N x 6.25)	%	13.8	15.0	
Cattle	dig prot %	* 9.1	9.9	
Goats	dig prot %	* 9.7	10.6	
Horses	dig prot %	* 9.4	10.3	
Rabbits	dig prot %	* 9.4	10.3	
Sheep	dig prot %	* 9.2	10.0	
Energy	GE Mcal/kg			
Cattle	DE Mcal/kg	* 2.36	2.57	
Sheep	DE Mcal/kg	* 2.27	2.47	
Cattle	ME Mcal/kg	* 1.94	2.11	

Feed Name or Analyses		Mean As Fed	Mean Dry	C.V. ± %
Sheep	ME Mcal/kg	* 1.86	2.03	
Cattle	TDN %	* 53.5	58.3	
Sheep	TDN %	* 51.5	56.1	
Calcium	%	0.63	0.69	
Phosphorus	%	0.45	0.49	

Sorghum, johnsongrass, hay, s-c, prebloom, cut 1, (1)

Ref No 1 09 090 United States

Feed Name or Analyses		Mean As Fed	Mean Dry	C.V. ± %
Dry matter	%	88.0	100.0	
Cattle	dig coef %	63.	63.	
Ash	%	7.2	8.2	
Crude fiber	%	30.4	34.6	
Ether extract	%	1.4	1.6	
N-free extract	%	41.4	47.0	
Protein (N x 6.25)	%	7.6	8.6	
Cattle	dig coef %	49.	49.	
Cattle	dig prot %	3.7	4.2	
Goats	dig prot %	* 4.0	4.6	
Horses	dig prot %	* 4.3	4.8	
Rabbits	dig prot %	* 4.7	5.3	
Sheep	dig prot %	* 3.8	4.3	
Cellulose (Matrone)	%	31.2	35.4	
Cattle	dig coef %	70.	70.	
Energy	GE Mcal/kg	3.70	4.21	
Cattle	GE dig coef %	61.	61.	
Cattle	DE Mcal/kg	2.26	2.57	
Sheep	DE Mcal/kg	* 2.14	2.43	
Cattle	ME Mcal/kg	* 1.85	2.10	
Sheep	ME Mcal/kg	* 1.75	1.99	
Cattle	TDN %	* 51.3	58.3	
Sheep	TDN %	* 48.5	55.1	

Sorghum, johnsongrass, hay, s-c, prebloom, cut 2, (1)

Ref No 1 09 105 United States

Feed Name or Analyses		Mean As Fed	Mean Dry	C.V. ± %
Dry matter	%	88.5	100.0	1
Cattle	dig coef %	62.	62.	5
Ash	%	7.0	8.0	21
Crude fiber	%	29.4	33.2	7
Cattle	dig coef %	68.	68.	
Ether extract	%	2.2	2.5	25
Cattle	dig coef %	61.	61.	
N-free extract	%	41.4	46.8	11
Cattle	dig coef %	62.	62.	
Protein (N x 6.25)	%	8.4	9.5	28
Cattle	dig coef %	54.	54.	18
Cattle	dig prot %	4.5	5.1	
Goats	dig prot %	* 4.8	5.4	
Horses	dig prot %	* 4.9	5.6	
Rabbits	dig prot %	* 5.3	6.0	
Sheep	dig prot %	* 4.5	5.1	
Cellulose (Matrone)	%	32.6	36.8	5
Cattle	dig coef %	72.	72.	5
Energy	GE Mcal/kg	3.84	4.34	0
Cattle	GE dig coef %	63.	63.	3
Cattle	DE Mcal/kg	2.42	2.74	
Sheep	DE Mcal/kg	* 2.18	2.47	
Cattle	ME Mcal/kg	* 1.99	2.24	
Sheep	ME Mcal/kg	* 1.79	2.02	
Cattle	TDN %	53.3	60.3	
Sheep	TDN %	* 49.5	56.0	

Feed Name or Analyses			Mean As Fed	Mean Dry	C.V. ± %

Sorghum, johnsongrass, hay, s-c, early bloom, cut 1, (1)
Ref No 1 09 089 United States

Feed Name or Analyses			As Fed	Dry	C.V. ± %
Dry matter	%		89.0	100.0	
Cattle	dig coef %		59.	59.	
Ash	%		7.5	8.4	
Crude fiber	%		30.1	33.8	
Ether extract	%		1.4	1.6	
N-free extract	%		42.9	48.2	
Protein (N x 6.25)	%		7.1	8.0	
Cattle	dig coef %		48.	48.	
Cattle	dig prot %		3.4	3.8	
Goats	dig prot %	*	3.6	4.0	
Horses	dig prot %	*	3.8	4.3	
Rabbits	dig prot %	*	4.3	4.8	
Sheep	dig prot %	*	3.3	3.7	
Cellulose (Matrone)	%		31.7	35.6	
Cattle	dig coef %		69.	69.	
Energy	GE Mcal/kg		3.82	4.30	
Cattle	GE dig coef %		59.	59.	
Cattle	DE Mcal/kg		2.26	2.53	
Sheep	DE Mcal/kg	*	2.17	2.44	
Cattle	ME Mcal/kg	*	1.85	2.08	
Sheep	ME Mcal/kg	*	1.78	2.00	
Cattle	TDN %	*	50.9	57.2	
Sheep	TDN %	*	49.2	55.2	

Sorghum, johnsongrass, hay, s-c, mid-bloom, (1)
Ref No 1 04 403 United States

Feed Name or Analyses			As Fed	Dry	C.V. ± %
Dry matter	%		91.8	100.0	
Ash	%		9.4	10.2	
Crude fiber	%		29.9	32.6	
Sheep	dig coef %	*	68.	68.	
Ether extract	%		2.1	2.3	
Sheep	dig coef %	*	47.	47.	
N-free extract	%		42.5	46.3	
Sheep	dig coef %	*	57.	57.	
Protein (N x 6.25)	%		7.9	8.6	
Sheep	dig coef %	*	44.	44.	
Cattle	dig prot %	*	4.0	4.4	
Goats	dig prot %	*	4.2	4.6	
Horses	dig prot %	*	4.4	4.8	
Rabbits	dig prot %	*	4.9	5.3	
Sheep	dig prot %		3.5	3.8	
Energy	GE Mcal/kg				
Cattle	DE Mcal/kg	*	2.33	2.54	
Sheep	DE Mcal/kg	*	2.22	2.42	
Cattle	ME Mcal/kg	*	1.91	2.08	
Sheep	ME Mcal/kg	*	1.82	1.98	
Cattle	TDN %	*	52.9	57.6	
Sheep	TDN %		50.3	54.8	
Calcium	%		0.58	0.63	
Phosphorus	%		0.21	0.23	

Sorghum, johnsongrass, hay, s-c, mature, (1)
Ref No 1 04 404 United States

Feed Name or Analyses			As Fed	Dry	C.V. ± %
Dry matter	%		93.1	100.0	
Calcium	%		0.17	0.18	
Phosphorus	%		0.13	0.14	

Sorghum, johnsongrass, hay, s-c, over ripe, (1)
Ref No 1 04 405 United States

Feed Name or Analyses			As Fed	Dry	C.V. ± %
Dry matter	%		93.1	100.0	1
Ash	%		8.6	9.2	6
Crude fiber	%		31.8	34.2	2
Sheep	dig coef %	*	68.	68.	
Ether extract	%		1.3	1.4	7
Sheep	dig coef %	*	47.	47.	
N-free extract	%		46.1	49.5	
Sheep	dig coef %	*	57.	57.	
Protein (N x 6.25)	%		5.3	5.7	3
Sheep	dig coef %	*	44.	44.	
Cattle	dig prot %	*	1.7	1.9	
Goats	dig prot %	*	1.7	1.9	
Horses	dig prot %	*	2.2	2.4	
Rabbits	dig prot %	*	2.9	3.1	
Sheep	dig prot %		2.3	2.5	
Energy	GE Mcal/kg				
Cattle	DE Mcal/kg	*	2.37	2.54	
Sheep	DE Mcal/kg	*	2.28	2.45	
Cattle	ME Mcal/kg	*	1.94	2.08	
Sheep	ME Mcal/kg	*	1.87	2.01	
Cattle	TDN %	*	53.6	57.6	
Sheep	TDN %	*	51.6	55.5	

Sorghum, johnsongrass, hay, s-c, cut 1, (1)
Ref No 1 04 406 United States

Feed Name or Analyses			As Fed	Dry	C.V. ± %
Dry matter	%		90.1	100.0	
Calcium	%		0.86	0.96	19
Magnesium	%		0.31	0.34	15
Phosphorus	%		0.26	0.29	18
Potassium	%		1.22	1.35	10

Sorghum, johnsongrass, hay, 180 to 200 lb nitrogen per acre spring s-c, prebloom, cut 2, (1)
Ref No 1 09 258 United States

Feed Name or Analyses			As Fed	Dry	C.V. ± %
Dry matter	%		87.6	100.0	
Cattle	dig coef %		65.	65.	
Ash	%		8.1	9.2	
Crude fiber	%		30.0	34.2	
Ether extract	%		2.2	2.5	
N-free extract	%		36.4	41.5	
Protein (N x 6.25)	%		11.0	12.5	
Cattle	dig coef %		64.	64.	
Cattle	dig prot %		7.0	8.0	
Goats	dig prot %	*	7.2	8.2	
Horses	dig prot %	*	7.1	8.1	
Rabbits	dig prot %	*	7.3	8.3	
Sheep	dig prot %	*	6.8	7.8	
Cellulose (Matrone)	%		32.5	37.1	
Cattle	dig coef %		74.	74.	
Energy	GE Mcal/kg		3.84	4.39	
Cattle	GE dig coef %		64.	64.	
Cattle	DE Mcal/kg		2.45	2.79	
Sheep	DE Mcal/kg	*	2.30	2.62	
Cattle	ME Mcal/kg	*	2.01	2.29	
Sheep	ME Mcal/kg	*	1.88	2.15	
Cattle	TDN %	*	55.4	63.2	
Sheep	TDN %	*	52.1	59.5	

Sorghum, johnsongrass, hay, 40 to 60 lb nitrogen per acre spring s-c, prebloom, (1)
Ref No 1 09 257 United States

Feed Name or Analyses			As Fed	Dry	C.V. ± %
Dry matter	%		88.0	100.0	
Cattle	dig coef %		63.	63.	
Ash	%		7.2	8.2	
Crude fiber	%		30.4	34.6	
Ether extract	%		1.4	1.6	
N-free extract	%		41.4	47.0	
Protein (N x 6.25)	%		7.6	8.6	
Cattle	dig coef %		49.	49.	
Cattle	dig prot %		3.7	4.2	
Goats	dig prot %	*	4.0	4.6	
Horses	dig prot %	*	4.3	4.8	
Rabbits	dig prot %	*	4.7	5.3	
Sheep	dig prot %	*	3.8	4.3	
Cellulose (Matrone)	%		31.2	35.4	
Cattle	dig coef %		70.	70.	
Energy	GE Mcal/kg		3.70	4.21	
Cattle	GE dig coef %		61.	61.	
Cattle	DE Mcal/kg		2.26	2.57	
Sheep	DE Mcal/kg	*	2.14	2.43	
Cattle	ME Mcal/kg	*	1.85	2.10	
Sheep	ME Mcal/kg	*	1.75	1.99	
Cattle	TDN %	*	51.3	58.3	
Sheep	TDN %	*	48.5	55.1	

Sorghum, johnsongrass, hay, 80 to 100 lb nitrogen per acre spring s-c, prebloom, cut 2, (1)
Ref No 1 09 253 United States

Feed Name or Analyses			As Fed	Dry	C.V. ± %
Dry matter	%			100.0	
Cattle	dig coef %			63.	
Ash	%			7.7	
Crude fiber	%			32.0	
Cattle	dig coef %			71.	
Ether extract	%			3.7	
N-free extract	%			47.8	
Cattle	dig coef %			62.	
Protein (N x 6.25)	%			8.8	
Cattle	dig coef %			53.	
Cattle	dig prot %			4.7	
Goats	dig prot %	*		4.8	
Horses	dig prot %	*		5.0	
Rabbits	dig prot %	*		5.5	
Sheep	dig prot %	*		4.5	
Cellulose (Matrone)	%			36.0	
Cattle	dig coef %			71.	
Energy	GE Mcal/kg				
Cattle	DE Mcal/kg	*		2.83	
Sheep	DE Mcal/kg	*		2.63	
Cattle	ME Mcal/kg	*		2.32	
Sheep	ME Mcal/kg	*		2.16	
Cattle	TDN %			64.1	
Sheep	TDN %	*		59.7	

Sorghum, johnsongrass, aerial part, fresh, (2)
Ref No 2 04 412 United States

Feed Name or Analyses			As Fed	Dry	C.V. ± %
Dry matter	%		24.8	100.0	20
Ash	%		2.8	11.2	
Crude fiber	%		7.4	29.6	
Ether extract	%		0.7	2.8	
N-free extract	%		10.4	42.0	

Continued

Feed Name or Analyses			Mean As Fed	Mean Dry	C.V. ± %
Protein (N x 6.25)	%		3.6	14.4	
Cattle	dig prot %	*	2.5	10.1	
Goats	dig prot %	*	2.5	10.0	
Horses	dig prot %	*	2.4	9.8	
Rabbits	dig prot %	*	2.4	9.8	
Sheep	dig prot %	*	2.6	10.4	
Energy	GE Mcal/kg				
Cattle	DE Mcal/kg	*	0.72	2.88	
Sheep	DE Mcal/kg	*	0.68	2.72	
Cattle	ME Mcal/kg	*	0.59	2.36	
Sheep	ME Mcal/kg	*	0.55	2.23	
Cattle	TDN %	*	16.2	65.3	
Sheep	TDN %	*	15.3	61.8	
Calcium	%		0.23	0.91	18
Magnesium	%		0.06	0.25	
Phosphorus	%		0.06	0.26	26
Potassium	%		0.77	3.12	
Carotene	mg/kg		49.3	198.4	64
Vitamin A equivalent	IU/g		82.1	330.8	

Sorghum, johnsongrass, aerial part, fresh, immature, (2)

Ref No 2 04 409 United States

Feed Name or Analyses			Mean As Fed	Mean Dry	C.V. ± %
Dry matter	%		19.8	100.0	
Ash	%		2.1	10.6	4
Crude fiber	%		5.6	28.5	5
Ether extract	%		0.6	3.2	33
N-free extract	%		8.4	42.2	
Protein (N x 6.25)	%		3.1	15.5	10
Cattle	dig prot %	*	2.2	11.1	
Goats	dig prot %	*	2.2	11.0	
Horses	dig prot %	*	2.1	10.7	
Rabbits	dig prot %	*	2.1	10.6	
Sheep	dig prot %	*	2.3	11.4	
Energy	GE Mcal/kg				
Cattle	DE Mcal/kg	*	0.51	2.58	
Sheep	DE Mcal/kg	*	0.54	2.71	
Cattle	ME Mcal/kg	*	0.42	2.12	
Sheep	ME Mcal/kg	*	0.44	2.22	
Cattle	TDN %	*	11.6	58.5	
Sheep	TDN %	*	12.2	61.5	
Calcium	%		0.18	0.93	13
Phosphorus	%		0.06	0.31	17

Sorghum, johnsongrass, aerial part, fresh, mid-bloom, (2)

Ref No 2 08 516 United States

Feed Name or Analyses			Mean As Fed	Mean Dry	C.V. ± %
Dry matter	%		35.0	100.0	
Ash	%		3.5	10.0	
Crude fiber	%		11.4	32.6	
Ether extract	%		0.8	2.3	
N-free extract	%		16.5	47.1	
Protein (N x 6.25)	%		2.8	8.0	
Cattle	dig prot %	*	1.6	4.7	
Goats	dig prot %	*	1.4	4.0	
Horses	dig prot %	*	1.5	4.3	
Rabbits	dig prot %	*	1.7	4.8	
Sheep	dig prot %	*	1.6	4.4	
Energy	GE Mcal/kg				
Cattle	DE Mcal/kg	*	0.95	2.70	
Sheep	DE Mcal/kg	*	0.90	2.57	
Cattle	ME Mcal/kg	*	0.78	2.21	
Sheep	ME Mcal/kg	*	0.74	2.11	
Cattle	TDN %	*	21.4	61.3	
Sheep	TDN %	*	20.4	58.4	
Calcium	%		0.29	0.83	
Phosphorus	%		0.06	0.17	

Sorghum, johnsongrass, aerial part, fresh, full bloom, (2)

Ref No 2 04 410 United States

Feed Name or Analyses			Mean As Fed	Mean Dry	C.V. ± %
Dry matter	%		35.0	100.0	7
Ash	%		3.5	10.0	4
Crude fiber	%		11.4	32.7	2
Ether extract	%		0.8	2.3	5
N-free extract	%		16.4	46.9	
Protein (N x 6.25)	%		2.8	8.1	4
Cattle	dig prot %	*	1.7	4.8	
Goats	dig prot %	*	1.4	4.1	
Horses	dig prot %	*	1.5	4.4	
Rabbits	dig prot %	*	1.7	4.9	
Sheep	dig prot %	*	1.6	4.5	
Energy	GE Mcal/kg				
Cattle	DE Mcal/kg	*	0.95	2.71	
Sheep	DE Mcal/kg	*	0.92	2.64	
Cattle	ME Mcal/kg	*	0.78	2.22	
Sheep	ME Mcal/kg	*	0.76	2.16	
Cattle	TDN %	*	21.5	61.4	
Sheep	TDN %	*	20.9	59.8	
Calcium	%		0.29	0.83	22
Phosphorus	%		0.06	0.17	13

Sorghum, johnsongrass, aerial part, fresh, dough stage, (2)

Ref No 2 04 411 United States

Feed Name or Analyses			Mean As Fed	Mean Dry	C.V. ± %
Dry matter	%			100.0	
Carotene	mg/kg			86.9	
Vitamin A equivalent	IU/g			144.8	

Sorghum, johnsongrass, aerial part, ensiled, (3)

Johnsongrass, aerial part, ensiled

Ref No 3 04 413 United States

Feed Name or Analyses			Mean As Fed	Mean Dry	C.V. ± %
Dry matter	%		38.8	100.0	53
Ash	%		3.8	9.9	16
Crude fiber	%		12.6	32.5	6
Ether extract	%		0.9	2.4	23
N-free extract	%		18.7	48.2	
Protein (N x 6.25)	%		2.7	7.0	19
Cattle	dig prot %	*	1.0	2.6	
Goats	dig prot %	*	1.0	2.6	
Horses	dig prot %	*	1.0	2.6	
Sheep	dig prot %	*	1.0	2.6	
Energy	GE Mcal/kg				
Cattle	DE Mcal/kg	*	0.94	2.42	
Sheep	DE Mcal/kg	*	0.92	2.36	
Cattle	ME Mcal/kg	*	0.77	1.98	
Sheep	ME Mcal/kg	*	0.75	1.94	
Cattle	TDN %	*	21.3	54.9	
Sheep	TDN %	*	20.8	53.6	

Sorghum, johnsongrass, aerial part, ensiled, prebloom, (3)

Ref No 3 09 170 United States

Feed Name or Analyses			Mean As Fed	Mean Dry	C.V. ± %
Dry matter	%		27.6	100.0	
Cattle	dig coef %		56.	56.	
Ash	%		3.0	11.0	
Crude fiber	%		8.9	32.1	
Ether extract	%		0.8	3.0	
N-free extract	%		12.3	44.4	
Protein (N x 6.25)	%		2.6	9.5	
Cattle	dig coef %		52.	52.	
Cattle	dig prot %		1.4	4.9	
Goats	dig prot %	*	1.3	4.9	
Horses	dig prot %	*	1.3	4.9	
Sheep	dig prot %	*	1.3	4.9	
Cellulose (Matrone)	%		9.7	35.0	
Cattle	dig coef %		69.	69.	
Energy	GE Mcal/kg				
Cattle	GE dig coef %		59.	59.	
Cattle	DE Mcal/kg	*	0.67	2.41	
Sheep	DE Mcal/kg	*	0.63	2.30	
Cattle	ME Mcal/kg	*	0.55	1.98	
Sheep	ME Mcal/kg	*	0.52	1.89	
Cattle	TDN %	*	15.1	54.6	
Sheep	TDN %	*	14.4	52.2	

Sorghum, johnsongrass, aerial part, w molasses added, ensiled, (3)

Ref No 3 04 414 United States

Feed Name or Analyses			Mean As Fed	Mean Dry	C.V. ± %
Dry matter	%		32.9	100.0	3
Ash	%		3.5	10.8	13
Crude fiber	%		10.5	31.8	7
Ether extract	%		0.7	2.2	17
N-free extract	%		16.0	48.8	
Protein (N x 6.25)	%		2.1	6.4	9
Cattle	dig prot %	*	0.7	2.1	
Goats	dig prot %	*	0.7	2.1	
Horses	dig prot %	*	0.7	2.1	
Sheep	dig prot %	*	0.7	2.1	
Energy	GE Mcal/kg				
Cattle	DE Mcal/kg	*	0.83	2.51	
Sheep	DE Mcal/kg	*	0.77	2.35	
Cattle	ME Mcal/kg	*	0.68	2.06	
Sheep	ME Mcal/kg	*	0.63	1.93	
Cattle	TDN %	*	18.7	56.9	
Sheep	TDN %	*	17.6	53.4	

Sorghum, johnsongrass, aerial part w 40 lb molasses added per ton, ensiled, prebloom, (3)

Ref No 3 09 255 United States

Feed Name or Analyses			Mean As Fed	Mean Dry	C.V. ± %
Dry matter	%		29.9	100.0	
Cattle	dig coef %		63.	63.	
Ash	%		3.3	11.2	
Crude fiber	%		9.0	30.0	
Ether extract	%		1.0	3.2	
N-free extract	%		13.6	45.6	
Protein (N x 6.25)	%		3.0	10.0	
Cattle	dig coef %		56.	56.	
Cattle	dig prot %		1.7	5.6	
Goats	dig prot %	*	1.6	5.3	
Horses	dig prot %	*	1.6	5.3	
Sheep	dig prot %	*	1.6	5.3	
Cellulose (Matrone)	%		9.8	32.9	
Cattle	dig coef %		72.	72.	

(1) dry forages and roughages (2) pasture, range plants, and forages fed green (3) silages (4) energy feeds (5) protein supplements (6) minerals (7) vitamins (8) additives

Feed Name or Analyses			Mean As Fed	Mean Dry	C.V. ± %
Energy	GE Mcal/kg				
Cattle	GE dig coef %		63.	63.	
Cattle	DE Mcal/kg	*	0.74	2.49	
Sheep	DE Mcal/kg	*	0.76	2.53	
Cattle	ME Mcal/kg	*	0.61	2.04	
Sheep	ME Mcal/kg	*	0.62	2.07	
Cattle	TDN %	*	16.9	56.5	
Sheep	TDN %	*	17.1	52.3	

Sorghum, johnsongrass, aerial part w 5 g zinc bacitracin preservative added per ton, ensiled, prebloom, (3)

Ref No 3 09 256 United States

Feed Name or Analyses			Mean As Fed	Mean Dry	C.V. ± %
Dry matter	%		27.9	100.0	
Cattle	dig coef %		56.	56.	
Ash	%		3.7	13.4	
Crude fiber	%		8.7	31.2	
Ether extract	%		0.8	3.0	
N-free extract	%		12.0	43.0	
Protein (N x 6.25)	%		2.6	9.4	
Cattle	dig coef %		50.	50.	
Cattle	dig prot %		1.3	4.7	
Goats	dig prot %	*	1.3	4.8	
Horses	dig prot %	*	1.3	4.8	
Sheep	dig prot %	*	1.3	4.8	
Cellulose (Matrone)	%		9.5	34.2	
Cattle	dig coef %		71.	71.	
Energy	GE Mcal/kg				
Cattle	GE dig coef %		59.	59.	
Cattle	DE Mcal/kg	*	0.69	2.48	
Sheep	DE Mcal/kg	*	0.64	2.30	
Cattle	ME Mcal/kg	*	0.57	2.03	
Sheep	ME Mcal/kg	*	0.53	1.89	
Cattle	TDN %	*	15.7	56.3	
Sheep	TDN %	*	14.5	52.1	

Sorghum, johnsongrass, aerial part w 8 lb sodium metabisulfite preservative added per ton, ensiled, prebloom, (3)

Ref No 3 09 254 United States

Feed Name or Analyses			Mean As Fed	Mean Dry	C.V. ± %
Dry matter	%		30.6	100.0	
Cattle	dig coef %		54.	54.	
Ash	%		3.9	12.6	
Crude fiber	%		9.4	30.7	
Ether extract	%		0.8	2.6	
N-free extract	%		13.8	45.1	
Protein (N x 6.25)	%		2.8	9.0	
Cattle	dig coef %		48.	48.	
Cattle	dig prot %		1.3	4.3	
Goats	dig prot %	*	1.3	4.4	
Horses	dig prot %	*	1.3	4.4	
Sheep	dig prot %	*	1.3	4.4	
Cellulose (Matrone)	%		10.5	34.4	
Cattle	dig coef %		69.	69.	
Energy	GE Mcal/kg				
Cattle	GE dig coef %		56.	56.	
Cattle	DE Mcal/kg	*	0.75	2.44	
Sheep	DE Mcal/kg	*	0.72	2.34	
Cattle	ME Mcal/kg	*	0.61	2.00	
Sheep	ME Mcal/kg	*	0.59	1.92	
Cattle	TDN %	*	16.9	55.3	
Sheep	TDN %	*	16.2	53.0	

SORGHUM, JUAR. Sorghum vulgare

Sorghum, juar, aerial part, fresh, (2)

Ref No 2 04 415 United States

Feed Name or Analyses			Mean As Fed	Mean Dry	C.V. ± %
Dry matter	%		34.8	100.0	
Ash	%		2.9	8.2	
Crude fiber	%		10.8	30.9	
Cattle	dig coef %		58.	58.	
Ether extract	%		0.6	1.6	
Cattle	dig coef %		40.	40.	
N-free extract	%		19.3	55.6	
Cattle	dig coef %		59.	59.	
Protein (N x 6.25)	%		1.3	3.7	
Cattle	dig coef %		26.	26.	
Cattle	dig prot %		0.3	1.0	
Goats	dig prot %	*	0.0	0.0	
Horses	dig prot %	*	0.2	0.7	
Rabbits	dig prot %	*	0.5	1.5	
Sheep	dig prot %	*	0.2	0.4	
Energy	GE Mcal/kg				
Cattle	DE Mcal/kg	*	0.82	2.34	
Sheep	DE Mcal/kg	*	0.84	2.42	
Cattle	ME Mcal/kg	*	0.67	1.92	
Sheep	ME Mcal/kg	*	0.69	1.98	
Cattle	TDN %		18.5	53.1	
Sheep	TDN %	*	19.1	54.8	

SORGHUM, KAFIR. Sorghum vulgare

Sorghum, kafir, aerial part, s-c, (1)

Sorghum, kafir, fodder, sun-cured

Ref No 1 04 418 United States

Feed Name or Analyses			Mean As Fed	Mean Dry	C.V. ± %
Dry matter	%		83.9	100.0	
Ash	%		7.1	8.5	
Crude fiber	%		24.4	29.1	
Cattle	dig coef %	*	60.	60.	
Sheep	dig coef %		61.	61.	
Ether extract	%		2.4	2.9	
Cattle	dig coef %	*	61.	61.	
Sheep	dig coef %		52.	52.	
N-free extract	%		42.3	50.4	
Cattle	dig coef %	*	66.	66.	
Sheep	dig coef %		63.	63.	
Protein (N x 6.25)	%		7.7	9.1	
Cattle	dig coef %	*	38.	38.	
Sheep	dig coef %		50.	50.	
Cattle	dig prot %		2.9	3.5	
Goats	dig prot %	*	4.3	5.1	
Horses	dig prot %	*	4.4	5.3	
Rabbits	dig prot %	*	4.8	5.7	
Sheep	dig prot %		3.8	4.6	
Energy	GE Mcal/kg				
Cattle	DE Mcal/kg	*	2.15	2.56	
Sheep	DE Mcal/kg	*	2.13	2.53	
Cattle	ME Mcal/kg	*	1.76	2.10	
Sheep	ME Mcal/kg	*	1.74	2.08	
Cattle	TDN %		48.8	58.2	
Sheep	TDN %		48.2	57.5	
Calcium	%		0.33	0.39	
Magnesium	%		0.24	0.29	
Phosphorus	%		0.17	0.20	
Potassium	%		1.43	1.71	

Sorghum, kafir, aerial part, s-c, late bloom, (1)

Ref No 1 04 417 United States

Feed Name or Analyses			Mean As Fed	Mean Dry	C.V. ± %
Dry matter	%		91.4	100.0	
Ash	%		11.2	12.3	
Crude fiber	%		23.8	26.0	
Sheep	dig coef %		64.	64.	
Ether extract	%		2.1	2.3	
Sheep	dig coef %		53.	53.	
N-free extract	%		44.3	48.5	
Sheep	dig coef %		69.	69.	
Protein (N x 6.25)	%		10.0	10.9	
Sheep	dig coef %		63.	63.	
Cattle	dig prot %	*	5.8	6.4	
Goats	dig prot %	*	6.2	6.7	
Horses	dig prot %	*	6.2	6.8	
Rabbits	dig prot %	*	6.5	7.1	
Sheep	dig prot %		6.3	6.9	
Energy	GE Mcal/kg				
Cattle	DE Mcal/kg	*	2.51	2.75	
Sheep	DE Mcal/kg	*	2.41	2.63	
Cattle	ME Mcal/kg	*	2.07	2.26	
Sheep	ME Mcal/kg	*	1.97	2.16	
Cattle	TDN %	*	56.9	62.3	
Sheep	TDN %		54.6	59.7	

Sorghum, kafir, aerial part, s-c, mature, (1)

Ref No 1 04 416 United States

Feed Name or Analyses			Mean As Fed	Mean Dry	C.V. ± %
Dry matter	%		88.2	100.0	
Ash	%		4.1	4.6	
Crude fiber	%		20.8	23.6	
Cattle	dig coef %		60.	60.	
Ether extract	%		2.4	2.7	
Cattle	dig coef %		61.	61.	
N-free extract	%		55.2	62.6	
Cattle	dig coef %		66.	66.	
Protein (N x 6.25)	%		5.7	6.5	
Cattle	dig coef %		38.	38.	
Cattle	dig prot %		2.2	2.5	
Goats	dig prot %	*	2.3	2.6	
Horses	dig prot %	*	2.7	3.0	
Rabbits	dig prot %	*	3.3	3.7	
Sheep	dig prot %	*	2.1	2.4	
Energy	GE Mcal/kg				
Cattle	DE Mcal/kg	*	2.40	2.72	
Sheep	DE Mcal/kg	*	2.33	2.65	
Cattle	ME Mcal/kg	*	1.97	2.23	
Sheep	ME Mcal/kg	*	1.91	2.17	
Cattle	TDN %		54.4	61.7	
Sheep	TDN %	*	52.9	60.0	

Sorghum, kafir, aerial part wo heads, s-c, (1)

Sorghum, Kafir, stover, sun-cured

Ref No 1 04 419 United States

Feed Name or Analyses			Mean As Fed	Mean Dry	C.V. ± %
Dry matter	%		81.6	100.0	8
Ash	%		8.1	9.9	6
Crude fiber	%		26.8	32.9	3
Cattle	dig coef %		67.	67.	
Ether extract	%		1.5	1.9	11
Cattle	dig coef %		75.	75.	
N-free extract	%		40.4	49.6	
Cattle	dig coef %		60.	60.	
Protein (N x 6.25)	%		4.7	5.8	7
Cattle	dig coef %		34.	34.	

Continued

Feed Name or Analyses			Mean As Fed	Mean Dry	C.V. ± %
Cattle	dig prot %		1.6	2.0	
Goats	dig prot %	*	1.6	2.0	
Horses	dig prot %	*	2.0	2.4	
Rabbits	dig prot %	*	2.6	3.1	
Sheep	dig prot %	*	1.4	1.8	
Energy	GE Mcal/kg				
Cattle	DE Mcal/kg	*	2.05	2.51	
Sheep	DE Mcal/kg	*	1.96	2.41	
Cattle	ME Mcal/kg	*	1.68	2.06	
Sheep	ME Mcal/kg	*	1.61	1.97	
Cattle	TDN %		46.5	56.9	
Sheep	TDN %	*	44.5	54.6	
Calcium	%		0.49	0.60	
Phosphorus	%		0.08	0.10	

Sorghum, kafir, heads wo seeds, (1)
Ref No 1 04 421 United States

Feed Name or Analyses			Mean As Fed	Mean Dry	C.V. ± %
Dry matter	%		92.0	100.0	
Ash	%		7.9	8.6	
Crude fiber	%		19.6	21.3	
Ether extract	%		1.0	1.1	
N-free extract	%		56.8	61.7	
Protein (N x 6.25)	%		6.7	7.3	
Cattle	dig prot %	*	3.0	3.3	
Goats	dig prot %	*	3.1	3.4	
Horses	dig prot %	*	3.4	3.7	
Rabbits	dig prot %	*	4.0	4.3	
Sheep	dig prot %	*	2.9	3.1	

Sorghum, kafir, heads wo seeds, s-c, (1)
Ref No 1 04 420 United States

Feed Name or Analyses			Mean As Fed	Mean Dry	C.V. ± %
Dry matter	%		92.0	100.0	
Ash	%		7.9	8.6	
Crude fiber	%		19.6	21.3	
Sheep	dig coef %		33.	33.	
Ether extract	%		1.0	1.1	
Sheep	dig coef %		53.	53.	
N-free extract	%		56.8	61.7	
Sheep	dig coef %		58.	58.	
Protein (N x 6.25)	%		6.7	7.3	
Sheep	dig coef %		20.	20.	
Cattle	dig prot %	*	3.0	3.3	
Goats	dig prot %	*	3.1	3.4	
Horses	dig prot %	*	3.4	3.7	
Rabbits	dig prot %	*	4.0	4.3	
Sheep	dig prot %		1.3	1.5	
Energy	GE Mcal/kg				
Cattle	DE Mcal/kg	*	1.94	2.11	
Sheep	DE Mcal/kg	*	1.85	2.01	
Cattle	ME Mcal/kg	*	1.59	1.73	
Sheep	ME Mcal/kg	*	1.52	1.65	
Cattle	TDN %	*	44.1	47.9	
Sheep	TDN %		41.9	45.6	

Sorghum, kafir, aerial part, fresh, (2)
Ref No 2 04 424 United States

Feed Name or Analyses			Mean As Fed	Mean Dry	C.V. ± %
Dry matter	%		57.1	100.0	
Ash	%		4.6	8.1	
Crude fiber	%		16.0	28.0	
Ether extract	%		1.7	3.0	
N-free extract	%		29.0	50.8	
Protein (N x 6.25)	%		5.8	10.2	
Cattle	dig prot %	*	3.7	6.5	
Goats	dig prot %	*	3.5	6.0	
Horses	dig prot %	*	3.5	6.2	
Rabbits	dig prot %	*	3.7	6.5	
Sheep	dig prot %	*	3.7	6.5	
Sugars, total	%		3.4	5.9	42
Energy	GE Mcal/kg				
Cattle	DE Mcal/kg	*	1.63	2.85	
Sheep	DE Mcal/kg	*	1.54	2.69	
Cattle	ME Mcal/kg	*	1.33	2.34	
Sheep	ME Mcal/kg	*	1.26	2.31	
Cattle	TDN %	*	36.9	64.7	
Sheep	TDN %	*	34.8	60.9	
Calcium	%		0.22	0.38	
Magnesium	%		0.17	0.30	
Phosphorus	%		0.10	0.17	
Potassium	%		0.97	1.69	
Carotene	mg/kg		10.1	17.6	
Niacin	mg/kg		22.4	39.2	44
Pantothenic acid	mg/kg		8.0	14.1	46
Riboflavin	mg/kg		2.4	4.2	20
Vitamin B6	mg/kg		3.40	5.95	
Vitamin A equivalent	IU/g		16.8	29.4	

(1) dry forages and roughages (2) pasture, range plants, and forages fed green (3) silages (4) energy feeds (5) protein supplements (6) minerals (7) vitamins (8) additives

Sorghum, kafir, aerial part, fresh, immature, (2)
Ref No 2 04 422 United States

Feed Name or Analyses			Mean As Fed	Mean Dry	C.V. ± %
Dry matter	%		16.1	100.0	
Ash	%		1.9	12.1	
Crude fiber	%		3.3	20.2	
Ether extract	%		0.4	2.6	
N-free extract	%		7.1	43.8	
Protein (N x 6.25)	%		3.4	21.3	
Cattle	dig prot %	*	2.6	16.0	
Goats	dig prot %	*	2.6	16.4	
Horses	dig prot %	*	2.5	15.6	
Rabbits	dig prot %	*	2.4	15.1	
Sheep	dig prot %	*	2.7	16.8	
Energy	GE Mcal/kg				
Cattle	DE Mcal/kg	*	0.43	2.65	
Sheep	DE Mcal/kg	*	0.41	2.56	
Cattle	ME Mcal/kg	*	0.35	2.17	
Sheep	ME Mcal/kg	*	0.34	2.10	
Cattle	TDN %	*	9.7	60.2	
Sheep	TDN %	*	9.4	58.1	

Sorghum, kafir, aerial part, fresh, early bloom, (2)
Ref No 2 08 517 United States

Feed Name or Analyses			Mean As Fed	Mean Dry	C.V. ± %
Dry matter	%		19.9	100.0	
Ash	%		1.3	6.5	
Crude fiber	%		6.5	32.7	
Ether extract	%		0.4	2.0	
N-free extract	%		10.1	50.8	
Protein (N x 6.25)	%		1.6	8.0	
Cattle	dig prot %	*	0.9	4.7	
Goats	dig prot %	*	0.8	4.1	
Horses	dig prot %	*	0.9	4.4	
Rabbits	dig prot %	*	1.0	4.9	
Sheep	dig prot %	*	0.9	4.5	
Energy	GE Mcal/kg				
Cattle	DE Mcal/kg	*	0.53	2.68	
Sheep	DE Mcal/kg	*	0.55	2.74	
Cattle	ME Mcal/kg	*	0.44	2.20	
Sheep	ME Mcal/kg	*	0.45	2.25	
Cattle	TDN %	*	12.1	60.7	
Sheep	TDN %	*	12.4	62.1	

Sorghum, kafir, aerial part, fresh, dough stage, (2)
Ref No 2 04 423 United States

Feed Name or Analyses			Mean As Fed	Mean Dry	C.V. ± %
Dry matter	%		19.9	100.0	
Ash	%		1.3	6.5	
Crude fiber	%		6.5	32.7	
Ether extract	%		0.4	2.0	
N-free extract	%		10.1	50.8	
Protein (N x 6.25)	%		1.6	8.0	
Cattle	dig prot %	*	0.9	4.7	
Goats	dig prot %	*	0.8	4.0	
Horses	dig prot %	*	0.9	4.3	
Rabbits	dig prot %	*	1.0	4.8	
Sheep	dig prot %	*	0.9	4.4	
Energy	GE Mcal/kg				
Cattle	DE Mcal/kg	*	0.54	2.73	
Sheep	DE Mcal/kg	*	0.56	2.79	
Cattle	ME Mcal/kg	*	0.45	2.24	
Sheep	ME Mcal/kg	*	0.46	2.29	
Cattle	TDN %	*	12.3	62.0	
Sheep	TDN %	*	12.6	63.3	

Sorghum, kafir, aerial part, ensiled, (3)
Ref No 3 04 425 United States

Feed Name or Analyses			Mean As Fed	Mean Dry	C.V. ± %
Dry matter	%		29.5	100.0	14
Ash	%		2.2	7.6	18
Crude fiber	%		8.0	27.3	13
Ether extract	%		1.0	3.3	21
N-free extract	%		16.1	54.6	
Protein (N x 6.25)	%		2.1	7.2	23
Cattle	dig prot %	*	0.8	2.8	
Goats	dig prot %	*	0.8	2.8	
Horses	dig prot %	*	0.8	2.8	
Sheep	dig prot %	*	0.8	2.8	
Energy	GE Mcal/kg				
Cattle	DE Mcal/kg	*	0.66	2.55	
Sheep	DE Mcal/kg	*	0.63	2.45	
Cattle	ME Mcal/kg	*	0.54	2.09	
Sheep	ME Mcal/kg	*	0.52	2.01	
Cattle	TDN %	*	15.0	57.9	
Sheep	TDN %	*	14.4	55.5	
Calcium	%		0.07	0.24	21
Magnesium	%		0.08	0.27	18
Phosphorus	%		0.05	0.17	36
Potassium	%		0.50	1.68	5
Carotene	mg/kg		3.2	10.8	94
Vitamin A equivalent	IU/g		5.3	18.0	

Sorghum, kafir, grain, (4)
Ref No 4 04 428 United States

Feed Name or Analyses			Mean As Fed	Mean Dry	C.V. ± %
Dry matter	%		88.6	100.0	1
Ash	%		1.6	1.8	18
Crude fiber	%		1.9	2.2	23
Cattle	dig coef %	*	61.	61.	
Sheep	dig coef %		55.	55.	
Swine	dig coef %		67.	67.	
Ether extract	%		2.9	3.3	12
Cattle	dig coef %	*	21.	21.	
Sheep	dig coef %		76.	76.	
Swine	dig coef %		62.	62.	
N-free extract	%		71.5	80.7	1
Cattle	dig coef %	*	53.	53.	

Feed Name or Analyses		Mean As Fed	Mean Dry	C.V. ± %
Sheep	dig coef %	92.	92.	
Swine	dig coef %	96.	96.	
Protein (N x 6.25)	%	10.7	12.1	6
Cattle	dig coef %	* 47.	47.	
Sheep	dig coef %	81.	81.	
Swine	dig coef %	77.	77.	
Cattle	dig prot %	5.0	5.6	
Goats	dig prot %	* 7.3	8.3	
Horses	dig prot %	* 7.3	8.3	
Sheep	dig prot %	8.7	9.8	
Swine	dig prot %	8.2	9.3	
Starch	%	64.1	72.4	2
Sugars, total	%	1.3	1.5	9
Energy	GE Mcal/kg			
Cattle	DE Mcal/kg	* 2.00	2.26	
Sheep	DE Mcal/kg	* 3.55	4.01	
Swine	DE kcal/kg	*3624.	4092.	
Cattle	ME Mcal/kg	* 1.64	1.85	
Sheep	ME Mcal/kg	* 2.91	3.28	
Swine	ME kcal/kg	*3391.	3829.	
Cattle	NE_m Mcal/kg	* 1.42	1.60	
Cattle	NE_{gain} Mcal/kg	* 0.91	1.03	
Cattle	TDN %	45.4	51.3	
Sheep	TDN %	80.5	90.8	
Swine	TDN %	82.2	92.8	
Calcium	%	0.03	0.03	
Chlorine	%	0.10	0.11	19
Cobalt	mg/kg	0.385	0.434	54
Copper	mg/kg	3.0	3.3	
Iron	%	0.001	0.001	
Magnesium	%	0.15	0.17	
Manganese	mg/kg	7.3	8.2	
Phosphorus	%	0.31	0.35	
Potassium	%	0.34	0.38	
Sodium	%	0.06	0.07	25
Sulphur	%	0.16	0.18	13
Biotin	mg/kg	0.23	0.26	43
Folic acid	mg/kg	0.20	0.22	23
Arginine	%	0.35	0.40	
Histidine	%	0.27	0.30	
Isoleucine	%	0.53	0.60	
Leucine	%	1.59	1.80	
Lysine	%	0.27	0.30	
Methionine	%	0.18	0.20	
Phenylalanine	%	0.62	0.70	
Threonine	%	0.44	0.50	
Tryptophan	%	0.18	0.20	
Valine	%	0.62	0.70	

Sorghum, kafir, grain, grnd, (4)
Ref No 4 04 426 United States

Feed Name or Analyses		Mean As Fed	Mean Dry	C.V. ± %
Dry matter	%	86.8	100.0	
Ash	%	1.6	1.8	
Crude fiber	%	2.2	2.5	
Cattle	dig coef %	-96.	-96.	
Ether extract	%	2.5	2.9	
Cattle	dig coef %	52.	52.	
N-free extract	%	70.7	81.5	
Cattle	dig coef %	79.	79.	
Protein (N x 6.25)	%	9.8	11.3	
Cattle	dig coef %	58.	58.	
Cattle	dig prot %	5.7	6.6	
Goats	dig prot %	* 6.6	7.6	
Horses	dig prot %	* 6.6	7.6	
Sheep	dig prot %	* 6.6	7.6	
Energy	GE Mcal/kg			
Cattle	DE Mcal/kg	* 2.75	3.17	
Sheep	DE Mcal/kg	* 3.27	3.77	

Feed Name or Analyses		Mean As Fed	Mean Dry	C.V. ± %
Swine	DE kcal/kg	*3483.	4012.	
Cattle	ME Mcal/kg	* 2.26	2.60	
Sheep	ME Mcal/kg	* 2.68	3.09	
Swine	ME kcal/kg	*3264.	3760.	
Cattle	TDN %	62.4	71.9	
Sheep	TDN %	* 74.2	85.5	
Swine	TDN %	* 79.0	91.0	

Sorghum, kafir, heads, chopped, (4)
Ref No 4 04 429 United States

Feed Name or Analyses		Mean As Fed	Mean Dry	C.V. ± %
Dry matter	%	86.9	100.0	1
Ash	%	2.9	3.4	
Crude fiber	%	7.3	8.4	
Cattle	dig coef %	27.	27.	
Sheep	dig coef %	61.	61.	
Ether extract	%	2.5	2.9	
Cattle	dig coef %	31.	31.	
Sheep	dig coef %	74.	74.	
N-free extract	%	64.7	74.5	
Cattle	dig coef %	31.	31.	
Sheep	dig coef %	80.	80.	
Protein (N x 6.25)	%	9.4	10.8	
Cattle	dig coef %	10.	10.	
Sheep	dig coef %	63.	63.	
Cattle	dig prot %	0.9	1.1	
Goats	dig prot %	* 6.2	7.2	
Horses	dig prot %	* 6.2	7.2	
Sheep	dig prot %	5.9	6.8	
Energy	GE Mcal/kg			
Cattle	DE Mcal/kg	* 1.09	1.25	
Sheep	DE Mcal/kg	* 2.93	3.37	
Swine	DE kcal/kg	*2791.	3211.	
Cattle	ME Mcal/kg	* 0.89	1.03	
Sheep	ME Mcal/kg	* 2.40	2.76	
Swine	ME kcal/kg	*2618.	3012.	
Cattle	TDN %	24.7	28.5	
Sheep	TDN %	66.3	76.3	
Swine	TDN %	* 63.3	72.8	
Calcium	%	0.08	0.09	
Magnesium	%	0.23	0.27	
Phosphorus	%	0.24	0.28	

SORGHUM, KALO. Sorghum vulgare

Sorghum, kalo, grain, (4)
Ref No 4 04 430 United States

Feed Name or Analyses		Mean As Fed	Mean Dry	C.V. ± %
Dry matter	%	89.2	100.0	
Ash	%	1.7	1.9	
Crude fiber	%	1.6	1.8	
Cattle	dig coef %	* 40.	40.	
Sheep	dig coef %	* 33.	33.	
Ether extract	%	3.2	3.6	
Cattle	dig coef %	* 80.	80.	
Sheep	dig coef %	* 76.	76.	
N-free extract	%	70.9	79.5	
Cattle	dig coef %	* 83.	83.	
Sheep	dig coef %	* 89.	89.	
Protein (N x 6.25)	%	11.8	13.2	
Cattle	dig coef %	* 57.	57.	
Sheep	dig coef %	* 67.	67.	
Cattle	dig prot %	6.7	7.5	
Goats	dig prot %	* 8.3	9.3	
Horses	dig prot %	* 8.3	9.3	
Sheep	dig prot %	7.9	8.9	
Energy	GE Mcal/kg			
Cattle	DE Mcal/kg	* 3.17	3.56	

Feed Name or Analyses		Mean As Fed	Mean Dry	C.V. ± %
Sheep	DE Mcal/kg	* 3.40	3.81	
Swine	DE kcal/kg	*3670.	4115.	
Cattle	ME Mcal/kg	* 2.60	2.92	
Sheep	ME Mcal/kg	* 2.78	3.12	
Swine	ME kcal/kg	*3426.	3840.	
Cattle	TDN %	72.0	80.7	
Sheep	TDN %	77.0	86.3	
Swine	TDN %	* 83.2	93.3	
Biotin	mg/kg	0.26	0.29	
Niacin	mg/kg	33.4	37.5	
Pantothenic acid	mg/kg	4.9	5.5	
Riboflavin	mg/kg	1.2	1.3	
Vitamin B_6	mg/kg	5.11	5.73	

SORGHUM, KAOLIANG. Sorghum vulgare

Sorghum, kaoliang, grain, (4)
Ref No 4 04 431 United States

Feed Name or Analyses		Mean As Fed	Mean Dry	C.V. ± %
Dry matter	%	89.6	100.0	3
Ash	%	1.9	2.1	27
Crude fiber	%	1.6	1.8	50
Cattle	dig coef %	* 40.	40.	
Sheep	dig coef %	* 33.	33.	
Ether extract	%	4.1	4.6	14
Cattle	dig coef %	* 80.	80.	
Sheep	dig coef %	* 76.	76.	
N-free extract	%	71.6	79.9	
Cattle	dig coef %	* 83.	83.	
Sheep	dig coef %	* 89.	89.	
Protein (N x 6.25)	%	10.3	11.5	10
Cattle	dig coef %	* 57.	57.	
Sheep	dig coef %	* 67.	67.	
Cattle	dig prot %	5.9	6.6	
Goats	dig prot %	* 7.0	7.8	
Horses	dig prot %	* 7.0	7.8	
Sheep	dig prot %	6.9	7.7	
Energy	GE Mcal/kg	3.34	3.73	
Cattle	DE Mcal/kg	* 3.23	3.61	
Sheep	DE Mcal/kg	* 3.45	3.85	
Swine	DE kcal/kg	*3649.	4075.	
Cattle	ME Mcal/kg	* 2.65	2.96	
Sheep	ME Mcal/kg	* 2.83	3.16	
Swine	ME kcal/kg	*3418.	3817.	
Cattle	TDN %	73.3	81.9	
Sheep	TDN %	78.2	87.3	
Swine	TDN %	* 82.8	92.4	
Calcium	%	0.03	0.03	
Phosphorus	%	0.29	0.32	

SORGHUM, MILO. Sorghum vulgare

Sorghum, milo, aerial part, s-c, (1)
Sorghum, milo, fodder, sun-cured
Ref No 1 04 433 United States

Feed Name or Analyses		Mean As Fed	Mean Dry	C.V. ± %
Dry matter	%	89.0	100.0	
Ash	%	7.2	8.1	
Crude fiber	%	21.9	24.7	
Cattle	dig coef %	* 60.	60.	
Sheep	dig coef %	72.	72.	
Ether extract	%	2.5	2.8	
Cattle	dig coef %	* 61.	61.	
Sheep	dig coef %	71.	71.	
N-free extract	%	50.8	57.1	
Cattle	dig coef %	* 66.	66.	
Sheep	dig coef %	78.	78.	

Continued

Feed Name or Analyses			Mean As Fed	Mean Dry	C.V. ± %
Protein (N x 6.25)	%		6.5	7.3	
Cattle	dig coef %	*	38.	38.	
Sheep	dig coef %		38.	38.	
Cattle	dig prot %		2.5	2.8	
Goats	dig prot %	*	3.0	3.3	
Horses	dig prot %	*	3.3	3.7	
Rabbits	dig prot %	*	3.8	4.3	
Sheep	dig prot %		2.5	2.8	
Fatty acids	%		3.3	3.7	
Energy	GE Mcal/kg				
Cattle	DE Mcal/kg	*	2.32	2.61	
Sheep	DE Mcal/kg	*	2.73	3.07	
Cattle	ME Mcal/kg	*	1.90	2.14	
Sheep	ME Mcal/kg	*	2.24	2.51	
Cattle	TDN %		52.6	59.1	
Sheep	TDN %		61.9	69.6	
Calcium	%		0.35	0.40	
Phosphorus	%		0.18	0.20	

Sorghum, milo, aerial part wo heads, s-c, (1)
Sorghum, milo, stover, sun-cured
Ref No 1 04 434 United States

Feed Name or Analyses			Mean As Fed	Mean Dry	C.V. ± %
Dry matter	%		92.5	100.0	1
Ash	%		9.9	10.7	2
Crude fiber	%		31.7	34.2	5
Cattle	dig coef %	*	67.	67.	
Sheep	dig coef %		66.	66.	
Ether extract	%		1.4	1.5	16
Cattle	dig coef %	*	75.	75.	
Sheep	dig coef %		58.	58.	
N-free extract	%		46.3	50.0	
Cattle	dig coef %	*	60.	60.	
Sheep	dig coef %		49.	49.	
Protein (N x 6.25)	%		3.3	3.6	7
Cattle	dig coef %	*	34.	34.	
Sheep	dig coef %		-17.	-17.	
Cattle	dig prot %		1.1	1.2	
Goats	dig prot %	*	0.0	0.0	
Horses	dig prot %	*	0.5	0.6	
Rabbits	dig prot %	*	1.3	1.4	
Sheep	dig prot %		-0.5	-0.5	
Energy	GE Mcal/kg				
Cattle	DE Mcal/kg	*	2.31	2.50	
Sheep	DE Mcal/kg	*	1.97	2.13	
Cattle	ME Mcal/kg	*	1.90	2.05	
Sheep	ME Mcal/kg	*	1.62	1.75	
Cattle	TDN %		52.4	56.7	
Sheep	TDN %		44.8	48.4	
Calcium	%		0.59	0.64	
Phosphorus	%		0.11	0.12	

Sorghum, milo, stalks, (1)
Ref No 1 04 435 United States

Feed Name or Analyses			Mean As Fed	Mean Dry	C.V. ± %
Dry matter	%		91.6	100.0	
Ash	%		10.4	11.4	
Crude fiber	%		31.9	34.8	
Ether extract	%		1.3	1.4	
N-free extract	%		45.1	49.2	
Protein (N x 6.25)	%		2.9	3.2	
Cattle	dig prot %	*	-0.2	-0.2	
Goats	dig prot %	*	-0.3	-0.4	
Horses	dig prot %	*	0.2	0.2	
Rabbits	dig prot %	*	1.0	1.1	
Sheep	dig prot %	*	-0.4	-0.5	
Energy	GE Mcal/kg				
Cattle	DE Mcal/kg	*	2.26	2.47	
Sheep	DE Mcal/kg	*	2.12	2.32	
Cattle	ME Mcal/kg	*	1.86	2.03	
Sheep	ME Mcal/kg	*	1.74	1.90	
Cattle	TDN %	*	51.3	56.0	
Sheep	TDN %	*	48.1	52.5	
Calcium	%		0.47	0.51	
Magnesium	%		0.18	0.20	
Phosphorus	%		0.10	0.11	

Sorghum, milo, aerial part, fresh, (2)
Ref No 2 04 436 United States

Feed Name or Analyses			Mean As Fed	Mean Dry	C.V. ± %
Dry matter	%		67.0	100.0	3
Ash	%		4.1	6.2	
Crude fiber	%		20.7	30.8	
Ether extract	%		1.2	1.8	
N-free extract	%		35.7	53.3	
Protein (N x 6.25)	%		5.3	7.9	
Cattle	dig prot %	*	3.1	4.6	
Goats	dig prot %	*	2.7	4.0	
Horses	dig prot %	*	2.9	4.3	
Rabbits	dig prot %	*	3.2	4.8	
Sheep	dig prot %	*	2.9	4.4	
Energy	GE Mcal/kg				
Cattle	DE Mcal/kg	*	1.70	2.53	
Sheep	DE Mcal/kg	*	1.72	2.56	
Cattle	ME Mcal/kg	*	1.39	2.09	
Sheep	ME Mcal/kg	*	1.41	2.10	
Cattle	TDN %	*	38.4	57.3	
Sheep	TDN %	*	38.9	58.1	
Calcium	%		0.27	0.40	
Phosphorus	%		0.14	0.21	
Potassium	%		1.83	2.73	
Carotene	mg/kg		1.3	2.0	
Vitamin A equivalent	IU/g		2.2	3.3	

Sorghum, milo, aerial part, ensiled, (3)
Ref No 3 04 437 United States

Feed Name or Analyses			Mean As Fed	Mean Dry	C.V. ± %
Dry matter	%		32.3	100.0	
Ash	%		2.0	6.2	
Crude fiber	%		6.5	20.1	
Ether extract	%		0.5	1.5	
N-free extract	%		21.0	65.1	
Protein (N x 6.25)	%		2.3	7.1	
Cattle	dig prot %	*	0.9	2.7	
Goats	dig prot %	*	0.9	2.7	
Horses	dig prot %	*	0.9	2.7	
Sheep	dig prot %	*	0.9	2.7	
Energy	GE Mcal/kg				
Cattle	DE Mcal/kg	*	0.87	2.68	
Sheep	DE Mcal/kg	*	0.84	2.60	
Cattle	ME Mcal/kg	*	0.71	2.20	
Sheep	ME Mcal/kg	*	0.69	2.13	
Cattle	TDN %	*	19.6	60.8	
Sheep	TDN %	*	19.0	58.9	

Sorghum, milo, grain, (4)
Ref No 4 04 444 United States

Feed Name or Analyses			Mean As Fed	Mean Dry	C.V. ± %
Dry matter	%		88.9	100.0	1
Ash	%		1.8	2.0	10
Crude fiber	%		2.3	2.5	6
Cattle	dig coef %		40.	40.	
Sheep	dig coef %		66.	66.	
Swine	dig coef %		66.	66.	
Ether extract	%		2.9	3.2	7
Cattle	dig coef %		80.	80.	
Sheep	dig coef %		88.	88.	
Swine	dig coef %		60.	60.	
N-free extract	%		71.1	80.1	1
Cattle	dig coef %		83.	83.	
Sheep	dig coef %		96.	96.	
Swine	dig coef %		90.	90.	
Protein (N x 6.25)	%		10.9	12.2	5
Cattle	dig coef %		57.	57.	
Sheep	dig coef %		78.	78.	
Swine	dig coef %		71.	71.	
Cattle	dig prot %		6.2	7.0	
Goats	dig prot %	*	7.5	8.4	
Horses	dig prot %	*	7.5	8.4	
Sheep	dig prot %		8.5	9.5	
Swine	dig prot %		7.7	8.7	
Starch	%		63.7	71.6	3
Sugars, total	%		1.4	1.5	9
Linoleic	%		1.100	1.237	
Energy	GE Mcal/kg				
Cattle	DE Mcal/kg	*	3.14	3.54	
Sheep	DE Mcal/kg	*	3.70	4.16	
Swine	DE kcal/kg	*	3399.	3825.	
Cattle	ME Mcal/kg	*	2.58	2.90	
Chickens	ME_n kcal/kg		3250.	3656.	
Sheep	ME Mcal/kg	*	3.03	3.41	
Swine	ME kcal/kg	*	3180.	3577.	
Cattle	NE_m Mcal/kg	*	1.64	1.85	
Cattle	NE_{gain} Mcal/kg	*	1.09	1.23	
Cattle	TDN %		71.3	80.2	
Sheep	TDN %		83.9	94.4	
Swine	TDN %		77.1	86.7	
Calcium	%		0.03	0.03	
Chlorine	%		0.08	0.09	
Cobalt	mg/kg		0.055	0.062	66
Copper	mg/kg		16.1	18.1	
Iron	%		0.004	0.005	
Magnesium	%		0.13	0.15	
Manganese	mg/kg		13.2	14.9	
Phosphorus	%		0.28	0.31	
Potassium	%		0.35	0.39	
Sodium	%		0.01	0.01	
Biotin	mg/kg		0.25	0.29	
Folic acid	mg/kg		0.22	0.24	13
α-tocopherol	mg/kg		12.0	13.5	
Arginine	%		0.36	0.40	14
Cystine	%		0.18	0.20	16
Glutamic acid	%		2.49	2.80	
Histidine	%		0.27	0.30	18
Isoleucine	%		0.53	0.60	11
Leucine	%		1.42	1.60	13
Lysine	%		0.27	0.30	27
Methionine	%		0.09	0.10	
Phenylalanine	%		0.44	0.50	27
Serine	%		0.53	0.60	
Threonine	%		0.27	0.30	38
Tryptophan	%		0.09	0.10	31
Tyrosine	%		0.36	0.40	26
Valine	%		0.53	0.60	5

(1) dry forages and roughages (2) pasture, range plants, and forages fed green (3) silages (4) energy feeds (5) protein supplements (6) minerals (7) vitamins (8) additives

Sorghum, milo, grain, chopped, (4)

Ref No 4 04 441 United States

Feed Name or Analyses		Mean As Fed	Mean Dry	C.V. ± %
Dry matter	%	88.1	100.0	0
Ash	%	2.2	2.5	7
Crude fiber	%	2.7	3.1	14
Ether extract	%	2.7	3.1	2
N-free extract	%	69.3	78.7	
Protein (N x 6.25)	%	11.2	12.7	5
Cattle	dig prot %	* 6.8	7.7	
Goats	dig prot %	* 7.8	8.9	
Horses	dig prot %	* 7.8	8.9	
Sheep	dig prot %	* 7.8	8.9	
Energy	GE Mcal/kg			
Cattle	DE Mcal/kg	* 2.83	3.21	
Sheep	DE Mcal/kg	* 3.27	3.71	
Cattle	ME Mcal/kg	*3439.	3903.	
Cattle	ME Mcal/kg	* 2.32	2.64	
Sheep	ME Mcal/kg	* 2.68	3.04	
Swine	ME kcal/kg	*3213.	3647.	
Cattle	TDN %	* 64.2	72.9	
Sheep	TDN %	* 74.2	84.2	
Swine	TDN %	* 78.0	88.5	
Calcium	%	0.04	0.05	
Magnesium	%	0.12	0.14	
Manganese	mg/kg	22.0	25.0	
Phosphorus	%	0.27	0.31	

Sorghum, milo, grain, grnd, (4)

Ref No 4 05 643 United States

Feed Name or Analyses		Mean As Fed	Mean Dry	C.V. ± %
Dry matter	%	89.5	100.0	
Ash	%	1.7	1.9	
Crude fiber	%	2.5	2.8	
Ether extract	%	2.7	3.0	
N-free extract	%	71.9	80.4	
Protein (N x 6.25)	%	10.6	11.9	
Cattle	dig prot %	* 6.2	6.9	
Goats	dig prot %	* 7.3	8.1	
Horses	dig prot %	* 7.3	8.1	
Sheep	dig prot %	* 7.3	8.1	
Lignin (Ellis)	%	0.8	0.9	
Energy	GE Mcal/kg	4.05	4.53	
Cattle	DE Mcal/kg	* 2.97	3.32	
Sheep	DE Mcal/kg	* 3.35	3.75	
Swine	DE kcal/kg	3363.	3760.	
Cattle	ME Mcal/kg	* 2.44	2.72	
Chickens	ME_n kcal/kg	3318.	3707.	
Sheep	ME Mcal/kg	* 2.75	3.07	
Swine	ME kcal/kg	3283.	3670.	
Cattle	TDN %	* 67.4	75.3	
Sheep	TDN %	* 76.0	85.0	
Swine	TDN %	* 76.3	85.3	
Phosphorus	%	0.31	0.35	

Sorghum, milo, grain, rolled, (4)

Ref No 4 05 642 United States

Feed Name or Analyses		Mean As Fed	Mean Dry	C.V. ± %
Dry matter	%	90.4	100.0	
Ash	%	2.0	2.2	
Crude fiber	%	1.9	2.2	
Ether extract	%	1.9	2.1	
N-free extract	%	73.9	81.8	
Protein (N x 6.25)	%	10.7	11.8	
Cattle	dig prot %	* 6.2	6.9	
Goats	dig prot %	* 7.3	8.0	
Horses	dig prot %	* 7.3	8.0	
Sheep	dig prot %	* 7.3	8.0	
Lignin (Ellis)	%	1.4	1.5	
Energy	GE Mcal/kg	3.96	4.38	
Cattle	DE Mcal/kg	* 2.96	3.27	
Sheep	DE Mcal/kg	* 3.39	3.75	
Swine	DE kcal/kg	*3619.	4004.	
Cattle	ME Mcal/kg	* 2.43	2.69	
Sheep	ME Mcal/kg	* 2.78	3.07	
Swine	ME kcal/kg	*3388.	3748.	
Cattle	NE_{gain} Mcal/kg	1.30	1.44	
Cattle	TDN %	* 67.1	74.3	
Sheep	TDN %	* 76.8	85.0	
Swine	TDN %	* 82.1	90.8	

Sorghum, milo, grain, gr 1 US mn wt 57 lb per bushel, (4)

Ref No 4 04 442 United States

Feed Name or Analyses		Mean As Fed	Mean Dry	C.V. ± %
Dry matter	%	89.0	100.0	
Ash	%	1.6	1.8	7
Crude fiber	%	2.4	2.7	
Ether extract	%	3.0	3.4	5
N-free extract	%	71.6	80.4	
Protein (N x 6.25)	%	10.4	11.7	5
Cattle	dig prot %	* 6.0	6.8	
Goats	dig prot %	* 7.1	8.0	
Horses	dig prot %	* 7.1	8.0	
Sheep	dig prot %	* 7.1	8.0	
Starch	%	64.7	72.7	1
Sugars, total	%	1.3	1.5	9
Energy	GE Mcal/kg			
Cattle	DE Mcal/kg	* 3.00	3.37	
Sheep	DE Mcal/kg	* 3.35	3.76	
Swine	DE kcal/kg	*3563.	4004.	
Cattle	ME Mcal/kg	* 2.46	2.76	
Sheep	ME Mcal/kg	* 2.75	3.09	
Swine	ME kcal/kg	*3337.	3749.	
Cattle	TDN %	* 68.0	76.5	
Sheep	TDN %	* 76.0	85.4	
Swine	TDN %	* 80.8	90.8	
Calcium	%	0.04	0.04	
Phosphorus	%	0.34	0.38	

Sorghum, milo, grain, gr 2 US mn wt 55 lb per bushel, (4)

Ref No 4 04 443 United States

Feed Name or Analyses		Mean As Fed	Mean Dry	C.V. ± %
Dry matter	%	88.3	100.0	2
Ash	%	1.5	1.8	12
Crude fiber	%	2.1	2.4	4
Ether extract	%	2.7	3.1	23
N-free extract	%	71.7	81.2	
Protein (N x 6.25)	%	10.2	11.6	2
Cattle	dig prot %	* 5.9	6.7	
Goats	dig prot %	* 6.9	7.9	
Horses	dig prot %	* 6.9	7.9	
Sheep	dig prot %	* 6.9	7.9	
Starch	%	64.0	72.5	1
Sugars, total	%	1.3	1.5	9
Energy	GE Mcal/kg			
Cattle	DE Mcal/kg	* 2.96	3.36	
Sheep	DE Mcal/kg	* 3.33	3.77	
Swine	DE kcal/kg	*3561.	4035.	
Cattle	ME Mcal/kg	* 2.43	2.75	
Sheep	ME Mcal/kg	* 2.73	3.09	
Swine	ME kcal/kg	*3335.	3779.	
Cattle	TDN %	* 67.2	76.2	
Sheep	TDN %	* 75.5	85.5	
Swine	TDN %	* 80.8	91.5	
Calcium	%	0.03	0.03	
Phosphorus	%	0.31	0.35	

Sorghum, milo, grits, cracked fine screened, (4)

Milo grits

Ref No 4 04 445 United States

Feed Name or Analyses		Mean As Fed	Mean Dry	C.V. ± %
Dry matter	%	87.8	100.0	4
Ash	%	0.7	0.8	45
Crude fiber	%	0.8	0.9	26
Cattle	dig coef %	* 40.	40.	
Sheep	dig coef %	* 66.	66.	
Ether extract	%	1.5	1.7	76
Cattle	dig coef %	* 80.	80.	
Sheep	dig coef %	* 88.	88.	
N-free extract	%	73.7	83.9	
Cattle	dig coef %	* 83.	83.	
Sheep	dig coef %	* 96.	96.	
Protein (N x 6.25)	%	11.1	12.6	7
Cattle	dig coef %	* 57.	57.	
Sheep	dig coef %	* 78.	78.	
Cattle	dig prot %	6.3	7.2	
Goats	dig prot %	* 7.7	8.8	
Horses	dig prot %	* 7.7	8.8	
Sheep	dig prot %	8.7	9.9	
Energy	GE Mcal/kg			
Cattle	DE Mcal/kg	* 3.11	3.54	
Sheep	DE Mcal/kg	* 3.66	4.16	
Swine	DE kcal/kg	*3768.	4291.	
Cattle	ME Mcal/kg	* 2.55	2.90	
Sheep	ME Mcal/kg	* 3.00	3.41	
Swine	ME kcal/kg	*3521.	4010.	
Cattle	TDN %	* 70.5	80.3	
Sheep	TDN %	* 82.9	94.4	
Swine	TDN %	* 85.5	97.3	

Sorghum, milo, heads, chopped, (4)

Ref No 4 04 446 United States

Feed Name or Analyses		Mean As Fed	Mean Dry	C.V. ± %
Dry matter	%	90.0	100.0	17
Ash	%	4.8	5.3	
Crude fiber	%	7.0	7.8	
Cattle	dig coef %	* 40.	40.	
Sheep	dig coef %	52.	52.	
Ether extract	%	2.5	2.7	
Cattle	dig coef %	* 80.	80.	
Sheep	dig coef %	87.	87.	
N-free extract	%	66.5	73.9	
Cattle	dig coef %	* 83.	83.	
Sheep	dig coef %	91.	91.	
Protein (N x 6.25)	%	9.2	10.2	
Cattle	dig coef %	* 57.	57.	
Sheep	dig coef %	76.	76.	
Cattle	dig prot %	5.2	5.8	
Goats	dig prot %	* 6.0	6.6	
Horses	dig prot %	* 6.0	6.6	
Sheep	dig prot %	7.0	7.8	
Energy	GE Mcal/kg			
Cattle	DE Mcal/kg	* 2.99	3.32	
Sheep	DE Mcal/kg	* 3.35	3.72	
Swine	DE kcal/kg	*2756.	3062.	
Cattle	ME Mcal/kg	* 2.45	2.72	
Sheep	ME Mcal/kg	* 2.75	3.05	
Swine	ME kcal/kg	*2588.	2876.	
Cattle	NE_m Mcal/kg	* 1.56	1.73	
Cattle	NE_{gain} Mcal/kg	* 1.03	1.14	

Continued

Feed Name or Analyses		Mean As Fed	Mean Dry	C.V. ± %
Cattle	TDN %	67.7	75.2	
Sheep	TDN %	76.0	84.5	
Swine	TDN %	* 62.5	69.4	
Calcium	%	0.12	0.13	25
Phosphorus	%	0.25	0.28	12
Carotene	mg/kg	0.6	0.7	37
Riboflavin	mg/kg	2.0	2.2	22
Vitamin A equivalent	IU/g	1.0	1.1	

Sorghum, milo, distillers grains, dehy, (5)
Ref No 5 04 438 United States

Feed Name or Analyses		Mean As Fed	Mean Dry	C.V. ± %
Dry matter	%	93.5	100.0	1
Ash	%	2.2	2.4	64
Crude fiber	%	11.1	11.9	16
Ether extract	%	11.9	12.7	24
N-free extract	%	29.2	31.2	
Protein (N x 6.25)	%	39.1	41.8	11
Energy	GE Mcal/kg			
Cattle	DE Mcal/kg	* 3.61	3.86	
Sheep	DE Mcal/kg	* 3.52	3.77	
Swine	DE kcal/kg	*4071.	4354.	
Cattle	ME Mcal/kg	* 2.96	3.16	
Sheep	ME Mcal/kg	* 2.89	3.09	
Swine	ME kcal/kg	*3565.	3812.	
Cattle	TDN %	* 81.8	87.4	
Sheep	TDN %	* 79.9	85.4	
Swine	TDN %	* 92.3	98.8	
Biotin	mg/kg	0.37	0.40	29
Choline	mg/kg	769.	823.	6
Niacin	mg/kg	54.2	58.0	4
Pantothenic acid	mg/kg	5.7	6.1	4
Riboflavin	mg/kg	2.9	3.1	4
Thiamine	mg/kg	0.7	0.7	

Sorghum, milo, distillers grains w solubles, dehy, mn 75% original solids, (5)
Ref No 5 04 439 United States

Feed Name or Analyses		Mean As Fed	Mean Dry	C.V. ± %
Dry matter	%	89.0	100.0	1
Ash	%	4.2	4.7	6
Crude fiber	%	10.1	11.3	13
Ether extract	%	8.0	9.0	13
N-free extract	%	35.3	39.7	
Protein (N x 6.25)	%	31.4	35.3	5
Energy	GE Mcal/kg			
Cattle	DE Mcal/kg	* 3.23	3.63	
Sheep	DE Mcal/kg	* 3.31	3.72	
Swine	DE kcal/kg	*3695.	4152.	
Cattle	ME Mcal/kg	* 2.65	2.98	
Sheep	ME Mcal/kg	* 2.72	3.05	
Swine	ME kcal/kg	*3284.	3690.	
Cattle	TDN %	* 73.2	82.3	
Sheep	TDN %	* 75.1	84.4	
Swine	TDN %	* 83.8	94.2	
Biotin	mg/kg	0.20	0.22	
Folic acid	mg/kg	0.20	0.22	
Niacin	mg/kg	64.8	72.8	
Pantothenic acid	mg/kg	7.9	8.9	
Riboflavin	mg/kg	6.0	6.7	
Glutamic acid	%	6.70	7.53	8

(1) dry forages and roughages (2) pasture, range plants, and forages fed green (3) silages (4) energy feeds (5) protein supplements (6) minerals (7) vitamins (8) additives

Sorghum, milo, distillers solubles, dehy, (5)
Ref No 5 04 440 United States

Feed Name or Analyses		Mean As Fed	Mean Dry	C.V. ± %
Dry matter	%	95.0	100.0	2
Ash	%	11.9	12.5	19
Crude fiber	%	2.6	2.7	20
Ether extract	%	7.2	7.6	19
N-free extract	%	48.3	50.8	
Protein (N x 6.25)	%	25.0	26.3	15
Energy	GE Mcal/kg			
Cattle	DE Mcal/kg	* 3.47	3.65	
Sheep	DE Mcal/kg	* 3.09	3.25	
Swine	DE kcal/kg	*4070.	4284.	
Cattle	ME Mcal/kg	* 2.85	3.00	
Sheep	ME Mcal/kg	* 2.53	2.67	
Swine	ME kcal/kg	*3690.	3885.	
Cattle	TDN %	* 78.7	82.8	
Sheep	TDN %	* 70.0	73.7	
Swine	TDN %	* 92.3	97.2	
Biotin	mg/kg	1.46	1.53	
Choline	mg/kg	2227.	2344.	52
Folic acid	mg/kg	1.26	1.32	
Niacin	mg/kg	141.1	148.5	10
Pantothenic acid	mg/kg	26.5	27.8	6
Riboflavin	mg/kg	14.8	15.5	18
Thiamine	mg/kg	4.9	5.1	4
Arginine	%	1.00	1.05	
Glutamic acid	%	7.20	7.58	
Histidine	%	0.90	0.95	
Isoleucine	%	0.90	0.95	
Leucine	%	1.70	1.79	
Lysine	%	0.90	0.95	
Methionine	%	0.50	0.53	
Phenylalanine	%	1.80	1.89	
Serine	%	2.10	2.21	
Threonine	%	1.00	1.05	
Tryptophan	%	0.30	0.32	
Tyrosine	%	0.90	0.95	
Valine	%	2.00	2.11	

Sorghum, milo, gluten, wet milled dehy grnd, (5)
Milo gluten meal
Ref No 5 08 087 United States

Feed Name or Analyses		Mean As Fed	Mean Dry	C.V. ± %
Dry matter	%	90.1	100.0	
Ash	%	1.5	1.7	
Crude fiber	%	3.3	3.7	
Ether extract	%	4.3	4.8	
N-free extract	%	38.5	42.7	
Protein (N x 6.25)	%	42.5	47.2	
Energy	GE Mcal/kg			
Cattle	DE Mcal/kg	* 3.51	3.90	
Sheep	DE Mcal/kg	* 3.56	3.95	
Swine	DE kcal/kg	*3670.	4074.	
Cattle	ME Mcal/kg	* 2.88	3.19	
Sheep	ME Mcal/kg	* 2.92	3.24	
Swine	ME kcal/kg	*3174.	3523.	
Cattle	TDN %	* 79.6	88.4	
Sheep	TDN %	* 80.6	89.5	
Swine	TDN %	* 83.2	92.4	

Sorghum, milo, gluten w bran, wet milled dehy grnd, (5)
Milo gluten feed
Ref No 5 08 089 United States

Feed Name or Analyses		Mean As Fed	Mean Dry	C.V. ± %
Dry matter	%	88.9	100.0	
Ash	%	6.8	7.6	
Crude fiber	%	6.5	7.3	
Ether extract	%	3.4	3.8	
N-free extract	%	47.7	53.7	
Protein (N x 6.25)	%	24.5	27.6	
Energy	GE Mcal/kg			
Cattle	DE Mcal/kg	* 3.10	3.49	
Sheep	DE Mcal/kg	* 2.83	3.18	
Swine	DE kcal/kg	*3561.	4006.	
Cattle	ME Mcal/kg	* 2.54	2.86	
Sheep	ME Mcal/kg	* 2.32	2.61	
Swine	ME kcal/kg	*3220.	3622.	
Cattle	TDN %	* 70.3	79.1	
Sheep	TDN %	* 64.2	72.2	
Swine	TDN %	* 80.8	90.8	

SORGHUM, MILO RED. Sorghum vulgare

Sorghum, milo red, grain, grnd, (4)
Ref No 4 08 983 United States

Feed Name or Analyses		Mean As Fed	Mean Dry	C.V. ± %
Dry matter	%	92.2	100.0	
Ash	%	2.3	2.5	
Crude fiber	%	2.4	2.6	
Ether extract	%	3.0	3.2	
Lignin (Ellis)	%	1.1	1.2	

Sorghum, milo red, grain, rolled, (4)
Ref No 4 08 984 United States

Feed Name or Analyses		Mean As Fed	Mean Dry	C.V. ± %
Dry matter	%	92.1	100.0	
Ash	%	2.0	2.2	
Crude fiber	%	2.1	2.3	
Ether extract	%	1.9	2.1	
Lignin (Ellis)	%	1.2	1.3	

SORGHUM, NK 300. Sorghum vulgare

Sorghum, NK 300, aerial part, ensiled, milk stage, (3)
Ref No 3 09 250 United States

Feed Name or Analyses		Mean As Fed	Mean Dry	C.V. ± %
Dry matter	%	25.0	100.0	
Cattle	dig coef %	53.	53.	
Ash	%	1.5	5.9	
Crude fiber	%	9.0	35.8	
Ether extract	%	0.8	3.0	
N-free extract	%	11.8	47.2	
Protein (N x 6.25)	%	2.0	8.1	
Cattle	dig coef %	39.	39.	
Cattle	dig prot %	0.8	3.2	
Goats	dig prot %	* 0.9	3.6	
Horses	dig prot %	* 0.9	3.6	
Sheep	dig prot %	* 0.9	3.6	
Cellulose (Matrone)	%	8.3	33.2	
Cattle	dig coef %	58.	58.	
Lignin (Ellis)	%	2.7	10.9	
Energy	GE Mcal/kg	1.09	4.34	
Cattle	GE dig coef %	54.	54.	
Cattle	DE Mcal/kg	0.59	2.35	

Feed Name or Analyses			Mean As Fed	Mean Dry	C.V. ± %
Sheep	DE Mcal/kg	*	0.61	2.43	
Cattle	ME Mcal/kg	*	0.48	1.92	
Sheep	ME Mcal/kg	*	0.50	1.99	
Cattle	TDN %	*	13.3	53.3	
Sheep	TDN %	*	13.9	55.7	

Sorghum, NK 300, aerial part, ensiled, dough stage, (3)
Ref No 3 09 251 United States

Feed Name or Analyses			Mean As Fed	Mean Dry	C.V. ± %
Dry matter	%			100.0	
Cattle	dig coef %			58.	
Ash	%			4.7	
Crude fiber	%			32.2	
Cattle	dig coef %			34.	
Ether extract	%			2.2	
N-free extract	%			53.4	
Cattle	dig coef %			78.	
Protein (N x 6.25)	%			7.7	
Cattle	dig coef %			34.	
Cattle	dig prot %			2.6	
Goats	dig prot %	*		3.2	
Horses	dig prot %	*		3.2	
Sheep	dig prot %	*		3.2	
Energy	GE Mcal/kg				
Cattle	DE Mcal/kg	*		2.55	
Sheep	DE Mcal/kg	*		2.50	
Cattle	ME Mcal/kg	*		2.09	
Sheep	ME Mcal/kg	*		2.05	
Cattle	TDN %			57.9	
Sheep	TDN %	*		56.7	

SORGHUM, NK 320. Sorghum vulgare

Sorghum, NK 320, aerial part, ensiled, dough stage, (3)
Ref No 3 09 248 United States

Feed Name or Analyses			Mean As Fed	Mean Dry	C.V. ± %
Dry matter	%			100.0	
Cattle	dig coef %			47.	
Ash	%			3.8	
Crude fiber	%			33.3	
Cattle	dig coef %			44.	
Ether extract	%			2.0	
N-free extract	%			54.8	
Cattle	dig coef %			54.	
Protein (N x 6.25)	%			6.1	
Cattle	dig coef %			27.	
Cattle	dig prot %			1.6	
Goats	dig prot %	*		1.8	
Horses	dig prot %	*		1.8	
Sheep	dig prot %	*		1.8	
Energy	GE Mcal/kg				
Cattle	DE Mcal/kg	*		2.22	
Sheep	DE Mcal/kg	*		2.34	
Cattle	ME Mcal/kg	*		1.82	
Sheep	ME Mcal/kg	*		1.92	
Cattle	TDN %			50.3	
Sheep	TDN %	*		53.1	

SORGHUM, NORGHUM. Sorghum vulgare

Sorghum, norghum, grain, (4)
Ref No 4 05 563 United States

Feed Name or Analyses			Mean As Fed	Mean Dry	C.V. ± %
Dry matter	%		89.7	100.0	1
Sheep	dig coef %		87.	87.	
Ash	%		1.5	1.7	1
Crude fiber	%		1.8	2.0	1
Sheep	dig coef %		39.	39.	
Ether extract	%		2.8	3.2	4
Sheep	dig coef %		80.	80.	
N-free extract	%		71.6	79.9	0
Sheep	dig coef %		92.	92.	
Protein (N x 6.25)	%		11.9	13.3	1
Sheep	dig coef %		74.	74.	
Cattle	dig prot %	*	7.4	8.2	
Goats	dig prot %	*	8.4	9.4	
Horses	dig prot %	*	8.4	9.4	
Sheep	dig prot %		8.8	9.9	
Energy	GE Mcal/kg				
Cattle	DE Mcal/kg	*	3.38	3.77	
Sheep	DE Mcal/kg	*	3.55	3.96	
Swine	DE kcal/kg	*	3688.	4112.	
Cattle	ME Mcal/kg	*	2.77	3.09	
Sheep	ME Mcal/kg	*	2.91	3.25	
Swine	ME kcal/kg	*	3441.	3837.	
Cattle	TDN %	*	76.8	85.6	
Sheep	TDN %		80.6	89.9	
Swine	TDN %	*	83.6	93.3	

SORGHUM, RS 610. Sorghum vulgare

Sorghum, RS 610, aerial part, ensiled, milk stage, (3)
Ref No 3 09 246 United States

Feed Name or Analyses			Mean As Fed	Mean Dry	C.V. ± %
Dry matter	%		26.7	100.0	
Cattle	dig coef %		59.	59.	
Ash	%		1.7	6.4	
Crude fiber	%		7.6	28.6	
Ether extract	%		0.8	3.0	
N-free extract	%		14.1	52.7	
Protein (N x 6.25)	%		2.5	9.3	
Cattle	dig coef %		53.	53.	
Cattle	dig prot %		1.3	4.9	
Goats	dig prot %	*	1.3	4.7	
Horses	dig prot %	*	1.3	4.7	
Sheep	dig prot %	*	1.3	4.7	
Cellulose (Matrone)	%		7.4	27.9	
Cattle	dig coef %		57.	57.	
Lignin (Ellis)	%		2.1	8.0	
Energy	GE Mcal/kg		1.14	4.29	1
Cattle	GE dig coef %		60.	60.	
Cattle	DE Mcal/kg		0.69	2.58	
Sheep	DE Mcal/kg	*	0.70	2.63	
Cattle	ME Mcal/kg	*	0.57	2.12	
Sheep	ME Mcal/kg	*	0.58	2.16	
Cattle	TDN %	*	15.6	58.5	
Sheep	TDN %	*	15.9	59.7	

Sorghum, RS 610, aerial part, ensiled, dough stage, (3)
Ref No 3 09 247 United States

Feed Name or Analyses			Mean As Fed	Mean Dry	C.V. ± %
Dry matter	%		22.7	100.0	
Cattle	dig coef %		52.	52.	
Ash	%		2.6	11.4	
Crude fiber	%		6.5	28.6	
Ether extract	%		0.8	3.7	
N-free extract	%		10.2	44.9	
Protein (N x 6.25)	%		2.6	11.4	
Cattle	dig coef %		52.	52.	
Cattle	dig prot %		1.3	5.9	
Goats	dig prot %	*	1.5	6.6	
Horses	dig prot %	*	1.5	6.6	
Sheep	dig prot %	*	1.5	6.6	
Cellulose (Matrone)	%		6.8	30.0	
Cattle	dig coef %		59.	59.	
Energy	GE Mcal/kg		0.97	4.28	
Cattle	GE dig coef %		55.	55.	
Cattle	DE Mcal/kg		0.53	2.35	
Sheep	DE Mcal/kg	*	0.52	2.27	
Cattle	ME Mcal/kg	*	0.44	1.93	
Sheep	ME Mcal/kg	*	0.42	1.86	
Cattle	TDN %	*	12.1	53.3	
Sheep	TDN %	*	11.7	51.4	

SORGHUM, SAGRAIN. Sorghum vulgare

Sorghum, sagrain, aerial part, ensiled, mature, (3)
Ref No 3 08 518 United States

Feed Name or Analyses			Mean As Fed	Mean Dry	C.V. ± %
Dry matter	%		38.1	100.0	
Ash	%		2.3	6.0	
Crude fiber	%		9.1	23.9	
Ether extract	%		1.2	3.1	
N-free extract	%		22.7	59.6	
Protein (N x 6.25)	%		2.8	7.3	
Cattle	dig prot %	*	1.1	2.9	
Goats	dig prot %	*	1.1	2.9	
Horses	dig prot %	*	1.1	2.9	
Sheep	dig prot %	*	1.1	2.9	
Energy	GE Mcal/kg				
Cattle	DE Mcal/kg	*	0.99	2.59	
Sheep	DE Mcal/kg	*	0.96	2.52	
Cattle	ME Mcal/kg	*	0.81	2.12	
Sheep	ME Mcal/kg	*	0.79	2.07	
Cattle	TDN %	*	22.4	58.8	
Sheep	TDN %	*	21.8	57.2	

Sorghum, sagrain, grain, (4)
Ref No 4 08 603 United States

Feed Name or Analyses			Mean As Fed	Mean Dry	C.V. ± %
Dry matter	%		90.0	100.0	
Ash	%		1.5	1.7	
Crude fiber	%		2.1	2.3	
Ether extract	%		3.5	3.9	
N-free extract	%		73.4	81.6	
Protein (N x 6.25)	%		9.5	10.6	
Cattle	dig prot %	*	5.1	5.7	
Goats	dig prot %	*	6.2	6.9	
Horses	dig prot %	*	6.2	6.9	
Sheep	dig prot %	*	6.2	6.9	
Energy	GE Mcal/kg				
Cattle	DE Mcal/kg	*	3.22	3.58	
Sheep	DE Mcal/kg	*	3.43	3.81	
Swine	DE kcal/kg	*	3640.	4045.	
Cattle	ME Mcal/kg	*	2.64	2.93	
Sheep	ME Mcal/kg	*	2.81	3.12	
Swine	ME kcal/kg	*	3417.	3797.	
Cattle	TDN %	*	73.0	81.1	
Sheep	TDN %	*	77.7	86.3	
Swine	TDN %	*	82.6	91.7	
Calcium	%		0.02	0.02	
Phosphorus	%		0.27	0.30	

Feed Name or Analyses			Mean As Fed	Mean Dry	C.V. ± %

SORGHUM, SCHROCK. Sorghum vulgare

Sorghum, schrock, grain, (4)
Ref No 4 04 452 United States

Analyses	Unit		As Fed	Dry	C.V. ± %
Dry matter	%		88.7	100.0	2
Ash	%		1.6	1.9	12
Crude fiber	%		3.1	3.5	16
Cattle	dig coef %	*	40.	40.	
Sheep	dig coef %	*	33.	33.	
Ether extract	%		3.1	3.5	9
Cattle	dig coef %	*	80.	80.	
Sheep	dig coef %	*	76.	76.	
N-free extract	%		70.8	79.8	
Cattle	dig coef %	*	83.	83.	
Sheep	dig coef %	*	89.	89.	
Protein (N x 6.25)	%		10.0	11.3	6
Cattle	dig coef %	*	57.	57.	
Sheep	dig coef %	*	67.	67.	
Cattle	dig prot %		5.7	6.5	
Goats	dig prot %	*	6.7	7.6	
Horses	dig prot %	*	6.7	7.6	
Sheep	dig prot %		6.7	7.6	
Starch	%		65.2	73.5	
Sugars, total	%		0.7	0.8	
Energy	GE Mcal/kg				
Cattle	DE Mcal/kg	*	3.14	3.55	
Sheep	DE Mcal/kg	*	3.35	3.78	
Swine	DE kcal/kg	*	3470.	3914.	
Cattle	ME Mcal/kg	*	2.58	2.91	
Sheep	ME Mcal/kg	*	2.75	3.10	
Swine	ME kcal/kg	*	3252.	3668.	
Cattle	TDN %		71.3	80.5	
Sheep	TDN %		76.1	85.8	
Swine	TDN %	*	78.7	88.8	
Calcium	%		0.02	0.02	
Phosphorus	%		0.30	0.34	
Biotin	mg/kg		0.25	0.29	

SORGHUM, SHALLU. Sorghum vulgare

Sorghum, shallu, aerial part, s-c, (1)
Ref No 1 04 454 United States

Analyses	Unit		As Fed	Dry	C.V. ± %
Dry matter	%		93.0	100.0	
Ash	%		7.9	8.5	
Crude fiber	%		35.4	38.1	
Sheep	dig coef %		65.	65.	
Ether extract	%		1.3	1.4	
Sheep	dig coef %		32.	32.	
N-free extract	%		45.6	49.0	
Sheep	dig coef %		50.	50.	
Protein (N x 6.25)	%		2.8	3.0	
Sheep	dig coef %		-8.	-8.	
Cattle	dig prot %	*	-0.3	-0.4	
Goats	dig prot %	*	-0.5	-0.5	
Horses	dig prot %	*	0.1	0.1	
Rabbits	dig prot %	*	0.9	1.0	
Sheep	dig prot %		-0.2	-0.2	
Energy	GE Mcal/kg				
Cattle	DE Mcal/kg	*	2.09	2.25	
Sheep	DE Mcal/kg	*	2.05	2.20	
Cattle	ME Mcal/kg	*	1.72	1.85	
Sheep	ME Mcal/kg	*	1.68	1.81	
Cattle	TDN %	*	47.5	51.1	
Sheep	TDN %		46.5	50.0	

Sorghum, shallu, aerial part, s-c, mature, (1)
Sorghum, shallu, fodder, sun-cured, mature
Ref No 1 04 453 United States

Analyses	Unit		As Fed	Dry	C.V. ± %
Dry matter	%		93.0	100.0	
Ash	%		7.9	8.5	
Crude fiber	%		35.4	38.1	
Cattle	dig coef %	*	60.	60.	
Sheep	dig coef %	*	65.	65.	
Ether extract	%		1.3	1.4	
Cattle	dig coef %	*	61.	61.	
Sheep	dig coef %	*	32.	32.	
N-free extract	%		45.6	49.0	
Cattle	dig coef %	*	66.	66.	
Sheep	dig coef %	*	50.	50.	
Protein (N x 6.25)	%		2.8	3.0	
Cattle	dig coef %	*	38.	38.	
Sheep	dig coef %	*	-8.	-8.	
Cattle	dig prot %		1.1	1.1	
Goats	dig prot %	*	-0.5	-0.5	
Horses	dig prot %	*	0.1	0.1	
Rabbits	dig prot %	*	0.9	1.0	
Sheep	dig prot %		-0.2	-0.2	
Energy	GE Mcal/kg				
Cattle	DE Mcal/kg	*	2.39	2.57	
Sheep	DE Mcal/kg	*	2.05	2.20	
Cattle	ME Mcal/kg	*	1.96	2.11	
Sheep	ME Mcal/kg	*	1.68	1.81	
Cattle	TDN %	*	54.2	58.3	
Sheep	TDN %	*	46.5	50.0	

Sorghum, shallu, aerial part wo heads, s-c, (1)
Sorghum, shallu, stover, sun-cured
Ref No 1 04 455 United States

Analyses	Unit		As Fed	Dry	C.V. ± %
Dry matter	%		93.0	100.0	
Ash	%		7.9	8.5	
Crude fiber	%		35.5	38.2	
Cattle	dig coef %	*	67.	67.	
Sheep	dig coef %	*	65.	65.	
Ether extract	%		1.4	1.5	
Cattle	dig coef %	*	75.	75.	
Sheep	dig coef %	*	32.	32.	
N-free extract	%		45.4	48.8	
Cattle	dig coef %	*	60.	60.	
Sheep	dig coef %	*	50.	50.	
Protein (N x 6.25)	%		2.8	3.0	
Cattle	dig coef %	*	34.	34.	
Sheep	dig coef %	*	-8.	-8.	
Cattle	dig prot %		0.9	1.0	
Goats	dig prot %	*	-0.5	-0.5	
Horses	dig prot %	*	0.1	0.1	
Rabbits	dig prot %	*	0.9	1.0	
Sheep	dig prot %		-0.2	-0.2	
Energy	GE Mcal/kg				
Cattle	DE Mcal/kg	*	2.40	2.58	
Sheep	DE Mcal/kg	*	2.05	2.21	
Cattle	ME Mcal/kg	*	1.96	2.11	
Sheep	ME Mcal/kg	*	1.68	1.81	
Cattle	TDN %	*	54.3	58.4	
Sheep	TDN %	*	46.5	50.0	

Sorghum, shallu, grain, (4)
Ref No 4 04 456 United States

Analyses	Unit		As Fed	Dry	C.V. ± %
Dry matter	%		90.2	100.0	1
Ash	%		1.9	2.1	19
Crude fiber	%		2.0	2.2	10
Cattle	dig coef %	*	40.	40.	
Sheep	dig coef %	*	33.	33.	
Ether extract	%		3.7	4.1	8
Cattle	dig coef %	*	80.	80.	
Sheep	dig coef %	*	76.	76.	
N-free extract	%		69.4	77.0	
Cattle	dig coef %	*	83.	83.	
Sheep	dig coef %	*	89.	89.	
Protein (N x 6.25)	%		13.1	14.6	6
Cattle	dig coef %	*	57.	57.	
Sheep	dig coef %	*	67.	67.	
Cattle	dig prot %		7.5	8.3	
Goats	dig prot %	*	9.5	10.6	
Horses	dig prot %	*	9.5	10.6	
Sheep	dig prot %		8.8	9.8	
Energy	GE Mcal/kg				
Cattle	DE Mcal/kg	*	3.20	3.55	
Sheep	DE Mcal/kg	*	3.42	3.79	
Swine	DE kcal/kg	*	3677.	4078.	
Cattle	ME Mcal/kg	*	2.62	2.91	
Sheep	ME Mcal/kg	*	2.81	3.11	
Swine	ME kcal/kg	*	3422.	3795.	
Cattle	TDN %		72.6	80.5	
Sheep	TDN %		77.6	86.1	
Swine	TDN %	*	83.4	92.5	
Arginine	%		0.27	0.30	
Histidine	%		0.18	0.20	
Isoleucine	%		0.36	0.40	
Leucine	%		0.90	1.00	
Lysine	%		0.18	0.20	
Methionine	%		0.18	0.20	
Phenylalanine	%		0.36	0.40	
Threonine	%		0.27	0.30	
Tryptophan	%		0.09	0.10	
Valine	%		0.45	0.50	

Sorghum, shallu, heads, chopped, (4)
Ref No 4 04 457 United States

Analyses	Unit		As Fed	Dry	C.V. ± %
Dry matter	%		90.5	100.0	
Ash	%		3.2	3.5	
Crude fiber	%		9.2	10.2	
Cattle	dig coef %	*	40.	40.	
Sheep	dig coef %	*	61.	61.	
Ether extract	%		3.5	3.9	
Cattle	dig coef %	*	80.	80.	
Sheep	dig coef %	*	74.	74.	
N-free extract	%		61.9	68.4	
Cattle	dig coef %	*	83.	83.	
Sheep	dig coef %	*	80.	80.	
Protein (N x 6.25)	%		12.7	14.0	
Cattle	dig coef %	*	57.	57.	
Sheep	dig coef %	*	63.	63.	
Cattle	dig prot %		7.2	8.0	
Goats	dig prot %	*	9.1	10.1	
Horses	dig prot %	*	9.1	10.1	
Sheep	dig prot %		8.0	8.8	
Energy	GE Mcal/kg				
Cattle	DE Mcal/kg	*	3.02	3.34	
Sheep	DE Mcal/kg	*	3.04	3.36	
Swine	DE kcal/kg	*	2783.	3075.	
Cattle	ME Mcal/kg	*	2.48	2.74	

(1) dry forages and roughages (2) pasture, range plants, and forages fed green (3) silages (4) energy feeds (5) protein supplements (6) minerals (7) vitamins (8) additives

Feed Name or Analyses			Mean As Fed	Mean Dry	C.V. ± %
Sheep	ME Mcal/kg	*	2.49	2.76	
Swine	ME kcal/kg	*	2593.	2865.	
Cattle	TDN %		68.6	75.8	
Sheep	TDN %		69.0	76.2	
Swine	TDN %	*	63.1	69.7	

SORGHUM X SORGHUM SUDANGRASS. Sorghum vulgare X Sorghum vulgare sudanese

Sorghum x Sorghum sudangrass, hay, s-c, prebloom, cut 1, (1)

Ref No 1 09 094 United States

Feed Name or Analyses			As Fed	Dry	C.V. ± %
Dry matter	%		66.9	100.0	
Cattle	dig coef %		61.	61.	
Ash	%		4.9	7.3	
Crude fiber	%		21.3	31.9	
Cattle	dig coef %		70.	70.	
Ether extract	%		1.6	2.5	
Cattle	dig coef %		64.	64.	
N-free extract	%		32.1	48.0	
Cattle	dig coef %		61.	61.	
Protein (N x 6.25)	%		7.0	10.5	
Cattle	dig coef %		47.	47.	
Cattle	dig prot %		3.3	4.9	
Goats	dig prot %	*	4.2	6.3	
Horses	dig prot %	*	4.3	6.4	
Rabbits	dig prot %	*	4.5	6.7	
Sheep	dig prot %	*	4.0	5.9	
Cellulose (Matrone)	%		19.9	29.8	
Cattle	dig coef %		73.	73.	
Energy	GE Mcal/kg		2.86	4.28	
Cattle	GE dig coef %		60.	60.	
Cattle	DE Mcal/kg		1.72	2.57	
Sheep	DE Mcal/kg	*	1.69	2.52	
Cattle	ME Mcal/kg	*	1.41	2.11	
Sheep	ME Mcal/kg	*	1.38	2.07	
Cattle	TDN %		40.1	60.0	
Sheep	TDN %	*	38.2	57.1	

Sorghum x Sorghum sudangrass, hay, s-c, prebloom, cut 2, (1)

Ref No 1 09 095 United States

Feed Name or Analyses			As Fed	Dry	C.V. ± %
Dry matter	%		87.9	100.0	
Cattle	dig coef %		69.	69.	
Ash	%		5.1	5.8	
Crude fiber	%		26.7	30.4	
Cattle	dig coef %		74.	74.	
Ether extract	%		2.0	2.3	
Cattle	dig coef %		72.	72.	
N-free extract	%		47.5	54.0	
Cattle	dig coef %		70.	70.	
Protein (N x 6.25)	%		6.6	7.5	
Cattle	dig coef %		61.	61.	
Cattle	dig prot %		4.0	4.6	
Goats	dig prot %	*	3.1	3.6	
Horses	dig prot %	*	3.4	3.9	
Rabbits	dig prot %	*	3.9	4.5	
Sheep	dig prot %	*	2.9	3.3	
Cellulose (Matrone)	%		24.7	28.1	
Cattle	dig coef %		75.	75.	
Energy	GE Mcal/kg		3.60	4.09	
Cattle	GE dig coef %		66.	66.	
Cattle	DE Mcal/kg		2.37	2.70	
Sheep	DE Mcal/kg	*	2.44	2.78	
Cattle	ME Mcal/kg	*	1.95	2.21	

Feed Name or Analyses			Mean As Fed	Mean Dry	C.V. ± %
Sheep	ME Mcal/kg	*	2.00	2.28	
Cattle	TDN %		60.3	68.6	
Sheep	TDN %	*	55.5	63.1	

SORGHUM, SORGO. Sorghum vulgare saccharatum

Sorghum, sorgo, aerial part, s-c, (1)

Sorghum, sorgo, fodder, sun-cured

Ref No 1 04 460 United States

Feed Name or Analyses			As Fed	Dry	C.V. ± %
Dry matter	%		82.3	100.0	12
Ash	%		6.0	7.3	
Crude fiber	%		23.4	28.4	
Cattle	dig coef %	*	60.	60.	
Sheep	dig coef %		63.	63.	
Ether extract	%		2.3	2.8	
Cattle	dig coef %	*	61.	61.	
Sheep	dig coef %		65.	65.	
N-free extract	%		45.0	54.6	
Cattle	dig coef %	*	66.	66.	
Sheep	dig coef %		63.	63.	
Protein (N x 6.25)	%		5.7	6.9	
Cattle	dig coef %	*	38.	38.	
Sheep	dig coef %		39.	39.	
Cattle	dig prot %		2.2	2.6	
Goats	dig prot %	*	2.4	3.0	
Horses	dig prot %	*	2.8	3.4	
Rabbits	dig prot %	*	3.3	4.0	
Sheep	dig prot %		2.2	2.7	
Sugars, total	%		17.4	21.1	46
Energy	GE Mcal/kg				
Cattle	DE Mcal/kg	*	2.16	2.62	
Sheep	DE Mcal/kg	*	2.14	2.60	
Cattle	ME Mcal/kg	*	1.77	2.15	
Sheep	ME Mcal/kg	*	1.76	2.13	
Cattle	TDN %		49.0	59.5	
Sheep	TDN %		48.6	59.1	
Calcium	%		0.31	0.38	
Chlorine	%		0.32	0.39	
Magnesium	%		0.29	0.35	
Manganese	mg/kg		107.6	130.7	
Phosphorus	%		0.13	0.16	
Potassium	%		1.20	1.46	
Carotene	mg/kg		2.2	2.6	39
Vitamin A equivalent	IU/g		3.6	4.4	

Sorghum, sorgo, aerial part, s-c, dough stage, (1)

Ref No 1 04 458 United States

Feed Name or Analyses			As Fed	Dry	C.V. ± %
Dry matter	%		89.4	100.0	
Ash	%		7.1	7.9	
Crude fiber	%		27.8	31.1	
Sheep	dig coef %		65.	65.	
Ether extract	%		1.9	2.1	
Sheep	dig coef %		57.	57.	
N-free extract	%		46.8	52.4	
Sheep	dig coef %		64.	64.	
Protein (N x 6.25)	%		5.8	6.5	
Sheep	dig coef %		34.	34.	
Cattle	dig prot %	*	2.3	2.6	
Goats	dig prot %	*	2.3	2.6	
Horses	dig prot %	*	2.7	3.0	
Rabbits	dig prot %	*	3.3	3.7	
Sheep	dig prot %		2.0	2.2	
Energy	GE Mcal/kg				
Cattle	DE Mcal/kg	*	2.45	2.74	
Sheep	DE Mcal/kg	*	2.31	2.59	
Cattle	ME Mcal/kg	*	2.00	2.24	

Feed Name or Analyses			Mean As Fed	Mean Dry	C.V. ± %
Sheep	ME Mcal/kg	*	1.90	2.12	
Cattle	TDN %	*	55.5	62.0	
Sheep	TDN %		52.4	58.7	

Sorghum, sorgo, aerial part, s-c, mature, (1)

Sorghum, sorgo, fodder, mature

Ref No 1 04 459 United States

Feed Name or Analyses			As Fed	Dry	C.V. ± %
Dry matter	%		84.5	100.0	10
Ash	%		6.9	8.2	27
Crude fiber	%		25.0	29.6	11
Ether extract	%		1.9	2.3	16
N-free extract	%		45.2	53.5	
Protein (N x 6.25)	%		5.4	6.4	15
Cattle	dig prot %	*	2.1	2.5	
Goats	dig prot %	*	2.1	2.5	
Horses	dig prot %	*	2.5	3.0	
Rabbits	dig prot %	*	3.1	3.6	
Sheep	dig prot %	*	2.0	2.3	
Energy	GE Mcal/kg				
Cattle	DE Mcal/kg	*	2.26	2.68	
Sheep	DE Mcal/kg	*	2.20	2.60	
Cattle	ME Mcal/kg	*	1.86	2.20	
Sheep	ME Mcal/kg	*	1.80	2.13	
Cattle	TDN %	*	51.4	60.8	
Sheep	TDN %	*	49.8	58.9	

Sorghum, sorgo, aerial part wo heads, s-c, (1)

Sorghum, sorgo, stover

Ref No 1 04 461 United States

Feed Name or Analyses			As Fed	Dry	C.V. ± %
Dry matter	%		93.0	100.0	
Sugars, total	%		32.4	34.8	21

Sorghum, sorgo, aerial part, fresh, (2)

Ref No 2 04 465 United States

Feed Name or Analyses			As Fed	Dry	C.V. ± %
Dry matter	%		22.1	100.0	
Ash	%		1.2	5.5	
Crude fiber	%		6.0	27.2	
Sheep	dig coef %		56.	56.	
Ether extract	%		0.7	3.4	
Sheep	dig coef %		65.	65.	
N-free extract	%		12.8	57.9	
Sheep	dig coef %		72.	72.	
Protein (N x 6.25)	%		1.3	6.1	
Sheep	dig coef %		44.	44.	
Cattle	dig prot %	*	0.7	3.0	
Goats	dig prot %	*	0.5	2.2	
Horses	dig prot %	*	0.6	2.7	
Rabbits	dig prot %	*	0.7	3.3	
Sheep	dig prot %		0.6	2.7	
Energy	GE Mcal/kg				
Cattle	DE Mcal/kg	*	0.62	2.79	
Sheep	DE Mcal/kg	*	0.63	2.84	
Cattle	ME Mcal/kg	*	0.50	2.28	
Sheep	ME Mcal/kg	*	0.52	2.33	
Cattle	TDN %	*	14.0	63.2	
Sheep	TDN %		14.3	64.5	
Calcium	%		0.08	0.36	
Manganese	mg/kg		29.0	131.0	
Phosphorus	%		0.03	0.12	
Potassium	%		0.32	1.45	
Sodium	%		0.08	0.36	

Sorghum, sorgo, aerial part, fresh, early bloom, (2)

Ref No 2 04 462 United States

Feed Name or Analyses				Mean As Fed	Mean Dry	C.V. ± %
Dry matter		%		16.4	100.0	
Ash		%		1.0	6.1	
Crude fiber		%		4.8	29.3	
Sheep	dig coef	%		68.	68.	
Ether extract		%		0.4	2.4	
Sheep	dig coef	%		63.	63.	
N-free extract		%		9.2	56.0	
Sheep	dig coef	%		74.	74.	
Protein (N x 6.25)		%		1.0	6.2	
Sheep	dig coef	%		48.	48.	
Cattle	dig prot	%	*	0.5	3.2	
Goats	dig prot	%	*	0.4	2.3	
Horses	dig prot	%	*	0.5	2.8	
Rabbits	dig prot	%	*	0.6	3.5	
Sheep	dig prot	%		0.5	3.0	
Energy	GE Mcal/kg					
Cattle	DE Mcal/kg		*	0.46	2.83	
Sheep	DE Mcal/kg		*	0.49	2.99	
Cattle	ME Mcal/kg		*	0.38	2.32	
Sheep	ME Mcal/kg		*	0.40	2.45	
Cattle	TDN	%	*	10.5	64.2	
Sheep	TDN	%		11.1	67.7	

Sorghum, sorgo, aerial part, fresh, dough stage, (2)

Ref No 2 04 463 United States

Feed Name or Analyses				Mean As Fed	Mean Dry	C.V. ± %
Dry matter		%		29.4	100.0	
Ash		%		2.0	6.7	
Crude fiber		%		8.6	29.1	
Cattle	dig coef	%		74.	74.	
Ether extract		%		1.1	3.8	
Cattle	dig coef	%		81.	81.	
N-free extract		%		15.9	54.2	
Cattle	dig coef	%		78.	78.	
Protein (N x 6.25)		%		1.8	6.2	
Cattle	dig coef	%		53.	53.	
Cattle	dig prot	%		1.0	3.3	
Goats	dig prot	%	*	0.7	2.3	
Horses	dig prot	%	*	0.8	2.8	
Rabbits	dig prot	%	*	1.0	3.5	
Sheep	dig prot	%	*	0.8	2.8	
Energy	GE Mcal/kg					
Cattle	DE Mcal/kg		*	0.96	3.26	
Sheep	DE Mcal/kg		*	0.90	3.07	
Cattle	ME Mcal/kg		*	0.79	2.68	
Sheep	ME Mcal/kg		*	0.74	2.52	
Cattle	TDN	%		21.8	74.0	
Sheep	TDN	%	*	20.5	69.7	

Sorghum, sorgo, aerial part, fresh, over ripe, (2)

Ref No 2 04 464 United States

Feed Name or Analyses				Mean As Fed	Mean Dry	C.V. ± %
Dry matter		%		31.1	100.0	12
Ash		%		1.9	6.2	24
Crude fiber		%		10.7	34.5	8
Ether extract		%		0.4	1.4	18
N-free extract		%		17.4	55.8	
Protein (N x 6.25)		%		0.7	2.1	23
Cattle	dig prot	%	*	0.0	-0.2	
Goats	dig prot	%	*	-0.4	-1.4	
Horses	dig prot	%	*	-0.1	-0.6	
Rabbits	dig prot	%	*	0.1	0.3	
Sheep	dig prot	%	*	-0.2	-1.0	
Energy	GE Mcal/kg					
Cattle	DE Mcal/kg		*	0.85	2.73	
Sheep	DE Mcal/kg		*	0.90	2.88	
Cattle	ME Mcal/kg		*	0.70	2.24	
Sheep	ME Mcal/kg		*	0.73	2.36	
Cattle	TDN	%	*	19.3	62.0	
Sheep	TDN	%	*	20.3	65.3	

Sorghum, sorgo, aerial part, ensiled, (3)

Sorghum, sorgo, fodder, silage

Ref No 3 04 468 United States

Feed Name or Analyses				Mean As Fed	Mean Dry	C.V. ± %
Dry matter		%		28.2	100.0	10
Ash		%		2.2	7.7	16
Crude fiber		%		6.6	23.3	11
Cattle	dig coef	%		50.	50.	
Sheep	dig coef	%		56.	56.	
Ether extract		%		0.7	2.5	20
Cattle	dig coef	%		52.	52.	
Sheep	dig coef	%		63.	63.	
N-free extract		%		16.9	59.7	
Cattle	dig coef	%		68.	68.	
Sheep	dig coef	%		67.	67.	
Protein (N x 6.25)		%		1.9	6.8	19
Cattle	dig coef	%		27.	27.	
Sheep	dig coef	%		23.	23.	
Cattle	dig prot	%		0.5	1.8	
Goats	dig prot	%	*	0.7	2.4	
Horses	dig prot	%	*	0.7	2.4	
Sheep	dig prot	%		0.4	1.6	
Lignin (Ellis)		%		2.2	7.7	
Energy	GE Mcal/kg					
Cattle	DE Mcal/kg		*	0.71	2.51	
Sheep	DE Mcal/kg		*	0.72	2.56	
Cattle	ME Mcal/kg		*	0.58	2.06	
Sheep	ME Mcal/kg		*	0.59	2.10	
Cattle	NE_m Mcal/kg		*	0.35	1.25	
Cattle	NE_{gain} Mcal/kg		*	0.17	0.61	
Cattle	$NE_{lactating\ cows}$ Mcal/kg		*	0.39	1.38	
Cattle	TDN	%		16.1	57.0	
Sheep	TDN	%		16.4	58.1	
Calcium		%		0.09	0.33	83
Chlorine		%		0.02	0.06	
Copper		mg/kg		8.8	31.3	15
Iron		%		0.006	0.020	9
Magnesium		%		0.08	0.27	12
Manganese		mg/kg		19.0	67.3	43
Phosphorus		%		0.06	0.20	62
Potassium		%		0.32	1.12	12
Sodium		%		0.01	0.03	
Carotene		mg/kg		7.3	25.9	98
Vitamin A equivalent		IU/g		12.2	43.1	

Sorghum, sorgo, aerial part, ensiled, milk stage, (3)

Ref No 3 09 106 United States

Feed Name or Analyses				Mean As Fed	Mean Dry	C.V. ± %
Dry matter		%		25.6	100.0	
Cattle	dig coef	%		55.	55.	
Ash		%		1.4	5.3	
Crude fiber		%		7.5	29.3	
Cattle	dig coef	%		48.	48.	
Ether extract		%		0.8	3.0	
N-free extract		%		14.3	55.9	
Cattle	dig coef	%		66.	66.	
Protein (N x 6.25)		%		1.7	6.5	
Cattle	dig coef	%		29.	29.	
Cattle	dig prot	%		0.5	1.9	
Goats	dig prot	%	*	0.6	2.2	
Horses	dig prot	%	*	0.6	2.2	
Sheep	dig prot	%	*	0.6	2.2	
Cellulose (Matrone)		%		9.1	35.5	
Cattle	dig coef	%		56.	56.	
Energy	GE Mcal/kg			1.15	4.48	
Cattle	GE dig coef	%		55.	55.	
Cattle	DE Mcal/kg			0.63	2.45	
Sheep	DE Mcal/kg		*	0.67	2.61	
Cattle	ME Mcal/kg		*	0.51	2.01	
Sheep	ME Mcal/kg		*	0.55	2.14	
Cattle	TDN	%		14.8	57.6	
Sheep	TDN	%	*	15.2	59.3	
Calcium		%		0.13	0.50	
Phosphorus		%		0.03	0.14	

Sorghum, sorgo, aerial part, ensiled, dough stage, (3)

Ref No 3 04 466 United States

Feed Name or Analyses				Mean As Fed	Mean Dry	C.V. ± %
Dry matter		%		25.7	100.0	8
Cattle	dig coef	%		53.	53.	6
Ash		%		1.4	5.4	16
Crude fiber		%		8.0	30.9	9
Cattle	dig coef	%		58.	58.	10
Ether extract		%		0.7	2.7	26
Cattle	dig coef	%		42.	42.	
N-free extract		%		14.2	55.2	6
Cattle	dig coef	%		59.	59.	8
Protein (N x 6.25)		%		1.5	5.8	13
Cattle	dig coef	%		20.	20.	45
Cattle	dig prot	%		0.3	1.1	
Goats	dig prot	%	*	0.4	1.5	
Horses	dig prot	%	*	0.4	1.5	
Sheep	dig prot	%	*	0.4	1.5	
Cellulose (Matrone)		%		9.1	35.4	9
Cattle	dig coef	%		56.	56.	9
Energy	GE Mcal/kg			1.15	4.45	0
Cattle	GE dig coef	%		55.	55.	8
Cattle	DE Mcal/kg			0.63	2.45	
Sheep	DE Mcal/kg		*	0.64	2.50	
Cattle	ME Mcal/kg		*	0.52	2.01	
Sheep	ME Mcal/kg		*	0.53	2.05	
Cattle	TDN	%		13.9	53.9	4
Sheep	TDN	%	*	14.5	56.6	
Calcium		%		0.07	0.27	22
Phosphorus		%		0.04	0.15	21

Sorghum, sorgo, aerial part, ensiled, mature, (3)

Ref No 3 04 467 United States

Feed Name or Analyses				Mean As Fed	Mean Dry	C.V. ± %
Dry matter		%		26.5	100.0	5
Cattle	dig coef	%		59.	59.	
Ash		%		2.2	8.5	
Crude fiber		%		7.3	27.6	
Cattle	dig coef	%		57.	57.	
Ether extract		%		0.8	2.9	
N-free extract		%		14.4	54.5	
Cattle	dig coef	%		66.	66.	
Protein (N x 6.25)		%		1.7	6.6	
Cattle	dig coef	%		38.	38.	
Cattle	dig prot	%		0.7	2.5	
Goats	dig prot	%	*	0.6	2.2	
Horses	dig prot	%	*	0.6	2.2	
Sheep	dig prot	%	*	0.6	2.2	

(1) dry forages and roughages (2) pasture, range plants, and forages fed green (3) silages (4) energy feeds (5) protein supplements (6) minerals (7) vitamins (8) additives

Feed Name or Analyses			Mean As Fed	Mean Dry	C.V. ± %
Energy	GE Mcal/kg		1.17	4.43	
Cattle	GE dig coef %		58.	58.	
Cattle	DE Mcal/kg		0.68	2.57	
Sheep	DE Mcal/kg	*	0.76	2.88	
Cattle	ME Mcal/kg	*	0.56	2.11	
Sheep	ME Mcal/kg	*	0.63	2.36	
Cattle	TDN %		16.2	61.0	
Sheep	TDN %	*	17.3	65.2	
Calcium	%		0.07	0.26	
Phosphorus	%		0.04	0.14	
Carotene	mg/kg		0.8	2.9	
Vitamin A equivalent	IU/g		1.3	4.8	

Sorghum, sorgo, grain, (4)

Ref No 4 04 469 United States

Feed Name or Analyses			Mean As Fed	Mean Dry	C.V. ± %
Dry matter	%		89.4	100.0	0
Ash	%		1.7	1.9	17
Crude fiber	%		3.7	4.1	85
Cattle	dig coef %	*	40.	40.	
Sheep	dig coef %		100.	100.	
Ether extract	%		3.0	3.4	10
Cattle	dig coef %	*	80.	80.	
Sheep	dig coef %		66.	66.	
N-free extract	%		71.0	79.4	5
Cattle	dig coef %	*	83.	83.	
Sheep	dig coef %		89.	89.	
Protein (N x 6.25)	%		10.0	11.2	6
Cattle	dig coef %	*	57.	57.	
Sheep	dig coef %		61.	61.	
Cattle	dig prot %		5.7	6.4	
Goats	dig prot %	*	6.7	7.5	
Horses	dig prot %	*	6.7	7.5	
Sheep	dig prot %		6.1	6.8	
Energy	GE Mcal/kg				
Cattle	DE Mcal/kg	*	3.15	3.53	
Sheep	DE Mcal/kg	*	3.42	3.82	
Swine	DE kcal/kg	*	3427.	3836.	
Cattle	ME Mcal/kg	*	2.59	2.90	
Sheep	ME Mcal/kg	*	2.80	3.13	
Swine	ME kcal/kg	*	3213.	3596.	
Cattle	TDN %		71.5	80.1	
Sheep	TDN %		77.5	86.7	
Swine	TDN %	*	77.7	87.0	
Calcium	%		0.02	0.03	37
Iron	%		0.007	0.008	
Manganese	mg/kg		16.9	19.0	
Phosphorus	%		0.30	0.33	15
Potassium	%		0.37	0.41	
Niacin	mg/kg		18.1	20.3	4
Thiamine	mg/kg		3.5	4.0	10

SORGHUM, SUDANGRASS. Sorghum vulgare sudanense

Sorghum, sudangrass, aerial part, dehy, (1)

Ref No 1 04 472 United States

Feed Name or Analyses			Mean As Fed	Mean Dry	C.V. ± %
Dry matter	%		87.7	100.0	
Ash	%		9.3	10.6	
Crude fiber	%		19.4	22.1	
Cattle	dig coef %		72.	72.	
Ether extract	%		2.2	2.5	
Cattle	dig coef %		66.	66.	
N-free extract	%		41.6	47.4	
Cattle	dig coef %		76.	76.	
Protein (N x 6.25)	%		15.3	17.4	
Cattle	dig coef %		64.	64.	
Cattle	dig prot %		9.8	11.1	
Goats	dig prot %	*	11.2	12.8	
Horses	dig prot %	*	10.8	12.3	
Rabbits	dig prot %	*	10.6	12.1	
Sheep	dig prot %	*	10.7	12.2	
Energy	GE Mcal/kg				
Cattle	DE Mcal/kg	*	2.58	2.94	
Sheep	DE Mcal/kg	*	2.38	2.71	
Cattle	ME Mcal/kg	*	2.12	2.41	
Sheep	ME Mcal/kg	*	1.95	2.23	
Cattle	TDN %		58.6	66.8	
Sheep	TDN %	*	54.0	61.6	

Sorghum, sudangrass, aerial part, dehy, immature, (1)

Ref No 1 04 470 United States

Feed Name or Analyses			Mean As Fed	Mean Dry	C.V. ± %
Dry matter	%		87.9	100.0	2
Ash	%		9.3	10.6	6
Crude fiber	%		20.2	22.9	8
Cattle	dig coef %	*	72.	72.	
Ether extract	%		2.4	2.8	34
Cattle	dig coef %	*	66.	66.	
N-free extract	%		41.3	47.1	
Cattle	dig coef %	*	76.	76.	
Protein (N x 6.25)	%		14.6	16.6	10
Cattle	dig coef %	*	64.	64.	
Cattle	dig prot %		9.3	10.6	
Goats	dig prot %	*	10.6	12.0	
Horses	dig prot %	*	10.2	11.6	
Rabbits	dig prot %	*	10.1	11.5	
Sheep	dig prot %	*	10.1	11.4	
Fatty acids	%		2.5	2.8	
Energy	GE Mcal/kg				
Cattle	DE Mcal/kg	*	2.60	2.95	
Sheep	DE Mcal/kg	*	2.46	2.80	
Cattle	ME Mcal/kg	*	2.13	2.42	
Sheep	ME Mcal/kg	*	2.02	2.30	
Cattle	TDN %		58.9	67.0	
Sheep	TDN %	*	55.9	63.6	
Calcium	%		0.52	0.59	
Phosphorus	%		0.39	0.44	

Sorghum, sudangrass, aerial part, dehy, pre-bloom, (1)

Ref No 1 04 471 United States

Feed Name or Analyses			Mean As Fed	Mean Dry	C.V. ± %
Dry matter	%		86.4	100.0	
Ash	%		8.6	10.0	
Crude fiber	%		18.7	21.7	
Cattle	dig coef %		71.	71.	
Ether extract	%		2.4	2.8	
Cattle	dig coef %		71.	71.	
N-free extract	%		42.4	49.1	
Cattle	dig coef %		77.	77.	
Protein (N x 6.25)	%		14.2	16.4	
Cattle	dig coef %		64.	64.	
Cattle	dig prot %		9.1	10.5	
Goats	dig prot %	*	10.2	11.9	
Horses	dig prot %	*	9.9	11.5	
Rabbits	dig prot %	*	9.8	11.3	
Sheep	dig prot %	*	9.7	11.3	
Energy	GE Mcal/kg				
Cattle	DE Mcal/kg	*	2.60	3.01	
Sheep	DE Mcal/kg	*	2.47	2.86	
Cattle	ME Mcal/kg	*	2.13	2.47	
Sheep	ME Mcal/kg	*	2.03	2.35	
Cattle	TDN %		58.9	68.2	
Sheep	TDN %	*	56.1	64.9	

Sorghum, sudangrass, hay, s-c, (1)

Ref No 1 04 480 United States

Feed Name or Analyses			Mean As Fed	Mean Dry	C.V. ± %
Dry matter	%		89.6	100.0	2
Ash	%		8.6	9.6	20
Crude fiber	%		27.5	30.7	8
Cattle	dig coef %	*	69.	69.	
Sheep	dig coef %	*	62.	62.	
Ether extract	%		1.6	1.8	25
Cattle	dig coef %	*	53.	53.	
Sheep	dig coef %	*	51.	51.	
N-free extract	%		43.1	48.1	
Cattle	dig coef %	*	66.	66.	
Sheep	dig coef %	*	54.	54.	
Protein (N x 6.25)	%		8.7	9.7	
Cattle	dig coef %	*	43.	43.	
Sheep	dig coef %	*	51.	51.	
Cattle	dig prot %		3.8	4.2	
Goats	dig prot %	*	5.1	5.6	
Horses	dig prot %	*	5.2	5.8	
Rabbits	dig prot %	*	5.5	6.2	
Sheep	dig prot %		4.5	5.0	
Cellulose (Matrone)	%		28.6	31.9	
Hemicellulose	%		12.2	13.6	
Sugars, total	%		1.2	1.3	
Lignin (Ellis)	%		10.4	11.6	
Energy	GE Mcal/kg		3.79	4.22	
Cattle	DE Mcal/kg	*	2.34	2.61	
Sheep	DE Mcal/kg	*	2.06	2.30	
Cattle	ME Mcal/kg	*	1.92	2.14	
Sheep	ME Mcal/kg	*	1.69	1.88	
Cattle	NE_m Mcal/kg	*	1.13	1.26	
Cattle	NE_{gain} Mcal/kg	*	0.56	0.63	
Cattle	$NE_{lactating\ cows}$ Mcal/kg	*	1.26	1.41	
Cattle	TDN %		53.1	59.3	
Sheep	TDN %		46.7	52.1	
Calcium	%		0.36	0.40	
Cobalt	mg/kg		0.111	0.124	
Iron	%		0.017	0.019	
Magnesium	%		0.31	0.35	
Manganese	mg/kg		82.0	91.5	
Phosphorus	%		0.27	0.30	
Potassium	%		1.89	2.10	
Silicon	%		4.17	4.65	
Sodium	%		0.02	0.02	40
Sulphur	%		0.05	0.06	30
Carotene	mg/kg		4.7	5.3	54
Vitamin A equivalent	IU/g		7.9	8.8	

Sorghum, sudangrass, hay, s-c, immature, (1)

Ref No 1 04 473 United States

Feed Name or Analyses			Mean As Fed	Mean Dry	C.V. ± %
Dry matter	%		88.0	100.0	4
Ash	%		9.2	10.5	7
Crude fiber	%		22.5	25.6	10
Cattle	dig coef %	*	69.	69.	
Ether extract	%		2.3	2.6	44
Cattle	dig coef %	*	53.	53.	
N-free extract	%		40.2	45.7	
Cattle	dig coef %	*	66.	66.	
Protein (N x 6.25)	%		13.7	15.6	15
Cattle	dig coef %	*	43.	43.	
Cattle	dig prot %		5.9	6.7	
Goats	dig prot %	*	9.8	11.1	
Horses	dig prot %	*	9.5	10.8	
Rabbits	dig prot %	*	9.4	10.7	
Sheep	dig prot %	*	9.3	10.6	

Continued

Feed Name or Analyses			Mean As Fed	Mean Dry	C.V. ± %
Energy	GE Mcal/kg				
Cattle	DE Mcal/kg	*	2.24	2.54	
Sheep	DE Mcal/kg	*	2.29	2.60	
Cattle	ME Mcal/kg	*	1.83	2.08	
Sheep	ME Mcal/kg	*	1.88	2.13	
Cattle	TDN %		50.7	57.6	
Sheep	TDN %	*	51.9	59.0	
Calcium	%		0.75	0.85	17
Magnesium	%		0.35	0.40	27
Phosphorus	%		0.32	0.36	62
Potassium	%		0.92	1.04	14

Sorghum, sudangrass, hay, s-c, pre-bloom, (1)
Ref No 1 04 474 United States

Feed Name or Analyses			Mean As Fed	Mean Dry	C.V. ± %
Dry matter	%		85.3	100.0	
Ash	%		8.2	9.6	
Crude fiber	%		27.2	31.9	
Sheep	dig coef %		73.	73.	
Ether extract	%		1.4	1.6	
Sheep	dig coef %		35.	35.	
N-free extract	%		36.6	43.0	
Sheep	dig coef %		54.	54.	
Protein (N x 6.25)	%		11.8	13.9	
Sheep	dig coef %		62.	62.	
Cattle	dig prot %	*	7.6	8.9	
Goats	dig prot %	*	8.1	9.5	
Horses	dig prot %	*	7.9	9.3	
Rabbits	dig prot %	*	8.0	9.4	
Sheep	dig prot %		7.3	8.6	
Fatty acids	%		1.4	1.7	
Energy	GE Mcal/kg				
Cattle	DE Mcal/kg	*	2.18	2.55	
Sheep	DE Mcal/kg	*	2.12	2.49	
Cattle	ME Mcal/kg	*	1.79	2.09	
Sheep	ME Mcal/kg	*	1.74	2.04	
Cattle	TDN %	*	49.4	57.9	
Sheep	TDN %		48.1	56.4	
Calcium	%		0.39	0.46	
Phosphorus	%		0.27	0.31	

Sorghum, sudangrass, hay, s-c, mid-bloom, (1)
Ref No 1 04 475 United States

Feed Name or Analyses			Mean As Fed	Mean Dry	C.V. ± %
Dry matter	%		90.7	100.0	1
Ash	%		8.9	9.8	12
Crude fiber	%		28.3	31.1	4
Cattle	dig coef %	*	69.	69.	
Sheep	dig coef %		60.	60.	
Ether extract	%		1.9	2.1	25
Cattle	dig coef %	*	53.	53.	
Sheep	dig coef %		61.	61.	
N-free extract	%		41.1	45.3	
Cattle	dig coef %	*	66.	66.	
Sheep	dig coef %		53.	53.	
Protein (N x 6.25)	%		10.6	11.7	11
Cattle	dig coef %	*	43.	43.	
Sheep	dig coef %		64.	64.	
Cattle	dig prot %		4.6	5.0	
Goats	dig prot %	*	6.8	7.5	
Horses	dig prot %	*	6.8	7.5	
Rabbits	dig prot %	*	7.0	7.7	
Sheep	dig prot %		6.8	7.5	

(1) dry forages and roughages
(2) pasture, range plants, and forages fed green
(3) silages
(4) energy feeds
(5) protein supplements
(6) minerals
(7) vitamins
(8) additives

Feed Name or Analyses			Mean As Fed	Mean Dry	C.V. ± %
Cellulose (Matrone)	%		30.8	33.9	
Lignin (Ellis)	%		13.5	14.9	
Energy	GE Mcal/kg				
Cattle	DE Mcal/kg	*	2.36	2.60	
Sheep	DE Mcal/kg	*	2.12	2.34	
Cattle	ME Mcal/kg	*	1.93	2.13	
Sheep	ME Mcal/kg	*	1.74	1.92	
Cattle	TDN %		53.4	58.9	
Sheep	TDN %		48.1	53.0	

Sorghum, sudangrass, hay, s-c, full bloom, (1)
Ref No 1 04 476 United States

Feed Name or Analyses			Mean As Fed	Mean Dry	C.V. ± %
Dry matter	%		87.9	100.0	1
Ash	%		7.5	8.6	5
Crude fiber	%		28.8	32.7	6
Cattle	dig coef %		71.	71.	
Sheep	dig coef %		62.	62.	
Ether extract	%		1.6	1.9	11
Cattle	dig coef %		58.	58.	
Sheep	dig coef %		42.	42.	
N-free extract	%		41.1	46.7	
Cattle	dig coef %		68.	68.	
Sheep	dig coef %		46.	46.	
Protein (N x 6.25)	%		8.9	10.1	13
Cattle	dig coef %		47.	47.	
Sheep	dig coef %		57.	57.	
Cattle	dig prot %		4.2	4.7	
Goats	dig prot %	*	5.3	6.0	
Horses	dig prot %	*	5.4	6.1	
Rabbits	dig prot %	*	5.7	6.5	
Sheep	dig prot %		5.1	5.8	
Energy	GE Mcal/kg				
Cattle	DE Mcal/kg	*	2.41	2.74	
Sheep	DE Mcal/kg	*	1.91	2.17	
Cattle	ME Mcal/kg	*	1.98	2.25	
Sheep	ME Mcal/kg	*	1.57	1.78	
Cattle	TDN %		54.7	62.2	
Sheep	TDN %		43.4	49.3	

Sorghum, sudangrass, hay, s-c, milk stage, (1)
Ref No 1 04 477 United States

Feed Name or Analyses			Mean As Fed	Mean Dry	C.V. ± %
Dry matter	%		90.1	100.0	
Ash	%		4.3	4.8	
Crude fiber	%		31.4	34.9	
Cattle	dig coef %		67.	67.	
Ether extract	%		1.7	1.9	
Cattle	dig coef %		41.	41.	
N-free extract	%		46.8	51.9	
Cattle	dig coef %		63.	63.	
Protein (N x 6.25)	%		5.9	6.5	
Cattle	dig coef %		35.	35.	
Cattle	dig prot %		2.0	2.3	
Goats	dig prot %	*	2.4	2.6	
Horses	dig prot %	*	2.7	3.0	
Rabbits	dig prot %	*	3.3	3.7	
Sheep	dig prot %	*	2.2	2.4	
Energy	GE Mcal/kg				
Cattle	DE Mcal/kg	*	2.39	2.65	
Sheep	DE Mcal/kg	*	2.17	2.41	
Cattle	ME Mcal/kg	*	1.96	2.17	
Sheep	ME Mcal/kg	*	1.78	1.98	
Cattle	TDN %		54.2	60.1	
Sheep	TDN %	*	49.3	54.7	

Sorghum, sudangrass, hay, s-c, dough stage, (1)
Ref No 1 04 478 United States

Feed Name or Analyses			Mean As Fed	Mean Dry	C.V. ± %
Dry matter	%		89.6	100.0	2
Ash	%		6.9	7.7	13
Crude fiber	%		30.3	33.9	6
Ether extract	%		1.6	1.8	32
N-free extract	%		43.9	49.0	
Protein (N x 6.25)	%		6.8	7.6	11
Cattle	dig prot %	*	3.2	3.5	
Goats	dig prot %	*	3.3	3.6	
Horses	dig prot %	*	3.6	4.0	
Rabbits	dig prot %	*	4.1	4.5	
Sheep	dig prot %	*	3.0	3.4	
Fatty acids	%		1.6	1.8	
Energy	GE Mcal/kg				
Cattle	DE Mcal/kg	*	2.24	2.50	
Sheep	DE Mcal/kg	*	2.18	2.43	
Cattle	ME Mcal/kg	*	1.84	2.05	
Sheep	ME Mcal/kg	*	1.78	1.99	
Cattle	TDN %	*	50.8	56.8	
Sheep	TDN %	*	49.4	55.1	
Carotene	mg/kg		52.7	58.9	
Vitamin A equivalent	IU/g		87.9	98.1	

Sorghum, sudangrass, hay, s-c, cut 2, (1)
Ref No 1 04 479 United States

Feed Name or Analyses			Mean As Fed	Mean Dry	C.V. ± %
Dry matter	%		89.6	100.0	
Calcium	%		0.58	0.65	
Copper	mg/kg		33.0	36.8	
Iron	%		0.019	0.021	
Magnesium	%		0.51	0.57	
Manganese	mg/kg		78.0	87.1	
Phosphorus	%		0.26	0.29	
Potassium	%		2.34	2.61	

Sorghum, sudangrass, hay, s-c, gr sample US, (1)
Ref No 1 04 481 United States

Feed Name or Analyses			Mean As Fed	Mean Dry	C.V. ± %
Dry matter	%		89.6	100.0	
Ash	%		8.1	9.0	
Crude fiber	%		28.4	31.7	
Ether extract	%		2.4	2.7	
N-free extract	%		38.3	42.8	
Protein (N x 6.25)	%		12.4	13.8	
Cattle	dig prot %	*	8.0	8.9	
Goats	dig prot %	*	8.5	9.4	
Horses	dig prot %	*	8.3	9.2	
Rabbits	dig prot %	*	8.4	9.3	
Sheep	dig prot %	*	8.0	8.9	
Sugars, total	%		1.2	1.3	
Energy	GE Mcal/kg				
Cattle	DE Mcal/kg	*	2.31	2.58	
Sheep	DE Mcal/kg	*	2.20	2.46	
Cattle	ME Mcal/kg	*	1.89	2.11	
Sheep	ME Mcal/kg	*	1.81	2.02	
Cattle	TDN %	*	52.4	58.5	
Sheep	TDN %	*	50.0	55.8	
Calcium	%		0.58	0.65	
Copper	mg/kg		33.0	36.8	
Iron	%		0.019	0.021	
Magnesium	%		0.51	0.57	
Manganese	mg/kg		78.0	87.1	
Phosphorus	%		0.26	0.29	
Potassium	%		2.34	2.61	

Feed Name or Analyses			Mean As Fed	Mean Dry	C.V. ± %
Carotene	mg/kg		4.9	5.5	
Vitamin A equivalent	IU/g		8.2	9.2	

Sorghum, sudangrass, straw, (1)

Ref No 1 04 483 United States

Feed Name or Analyses			Mean As Fed	Mean Dry	C.V. ± %
Dry matter	%		90.4	100.0	
Ash	%		7.0	7.8	
Crude fiber	%		31.8	35.2	
Sheep	dig coef %		60.	60.	
Ether extract	%		1.5	1.7	
Sheep	dig coef %		34.	34.	
N-free extract	%		42.6	47.1	
Sheep	dig coef %		48.	48.	
Protein (N x 6.25)	%		7.4	8.2	
Sheep	dig coef %		46.	46.	
Cattle	dig prot %	*	3.7	4.1	
Goats	dig prot %	*	3.8	4.2	
Horses	dig prot %	*	4.1	4.5	
Rabbits	dig prot %	*	4.5	5.0	
Sheep	dig prot %		3.4	3.8	
Energy	GE Mcal/kg				
Cattle	DE Mcal/kg	*	2.17	2.40	
Sheep	DE Mcal/kg	*	1.94	2.15	
Cattle	ME Mcal/kg	*	1.78	1.97	
Sheep	ME Mcal/kg	*	1.59	1.76	
Cattle	TDN %	*	49.2	54.4	
Sheep	TDN %		44.1	48.8	

Sorghum, sudangrass, aerial part, fresh, (2)

Ref No 2 04 489 United States

Feed Name or Analyses			Mean As Fed	Mean Dry	C.V. ± %
Dry matter	%		20.8	100.0	6
Cattle	dig coef %		72.	72.	
Ash	%		1.6	7.8	
Crude fiber	%		5.7	27.5	
Sheep	dig coef %	*	73.	73.	
Ether extract	%		0.7	3.3	
Sheep	dig coef %	*	68.	68.	
N-free extract	%		9.8	47.2	
Sheep	dig coef %	*	67.	67.	
Protein (N x 6.25)	%		2.9	14.1	
Sheep	dig coef %	*	71.	71.	
Cattle	dig prot %	*	2.1	9.9	
Goats	dig prot %	*	2.0	9.8	
Horses	dig prot %	*	2.0	9.5	
Rabbits	dig prot %	*	2.0	9.6	
Sheep	dig prot %		2.1	10.0	
Energy	GE Mcal/kg				
Cattle	DE Mcal/kg	*	0.63	3.04	
Sheep	DE Mcal/kg	*	0.61	2.94	
Cattle	ME Mcal/kg	*	0.52	2.49	
Sheep	ME Mcal/kg	*	0.50	2.41	
Cattle	TDN %	*	14.3	68.9	
Sheep	TDN %		13.9	66.8	
Calcium	%		0.10	0.49	20
Cobalt	mg/kg		0.027	0.132	
Copper	mg/kg		7.5	35.9	
Iron	%		0.004	0.021	88
Magnesium	%		0.07	0.35	10
Manganese	mg/kg		16.9	81.3	32
Phosphorus	%		0.09	0.44	38
Potassium	%		0.44	2.14	24
Sulphur	%		0.02	0.11	52
Carotene	mg/kg		38.0	182.8	27
Vitamin A equivalent	IU/g		63.3	304.7	

Sorghum, sudangrass, aerial part, fresh, early leaf, (2)

Ref No 2 04 491 United States

Feed Name or Analyses			Mean As Fed	Mean Dry	C.V. ± %
Dry matter	%		19.0	100.0	
Ash	%		2.1	11.0	
Crude fiber	%		8.1	42.6	
Sheep	dig coef %		70.	70.	
Ether extract	%		0.3	1.8	
Sheep	dig coef %		63.	63.	
N-free extract	%		6.7	35.5	
Sheep	dig coef %		66.	66.	
Protein (N x 6.25)	%		1.7	9.1	
Sheep	dig coef %		54.	54.	
Cattle	dig prot %	*	1.1	5.6	
Goats	dig prot %	*	1.0	5.0	
Horses	dig prot %	*	1.0	5.2	
Rabbits	dig prot %	*	1.1	5.7	
Sheep	dig prot %		0.9	4.9	
Lignin (Ellis)	%		1.2	6.5	
Energy	GE Mcal/kg				
Cattle	DE Mcal/kg	*	0.53	2.81	
Sheep	DE Mcal/kg	*	0.51	2.68	
Cattle	ME Mcal/kg	*	0.44	2.30	
Sheep	ME Mcal/kg	*	0.42	2.19	
Cattle	TDN %	*	12.1	63.6	
Sheep	TDN %		11.5	60.7	

Sorghum, sudangrass, aerial part, fresh, immature, (2)

Ref No 2 04 484 United States

Feed Name or Analyses			Mean As Fed	Mean Dry	C.V. ± %
Dry matter	%		17.6	100.0	9
Ash	%		1.6	9.0	17
Crude fiber	%		5.4	30.9	12
Sheep	dig coef %	*	70.	70.	
Ether extract	%		0.7	3.9	18
Sheep	dig coef %	*	63.	63.	
N-free extract	%		6.9	39.4	
Sheep	dig coef %	*	66.	66.	
Protein (N x 6.25)	%		3.0	16.8	20
Sheep	dig coef %	*	69.	69.	
Cattle	dig prot %	*	2.1	12.2	
Goats	dig prot %	*	2.2	12.2	
Horses	dig prot %	*	2.1	11.8	
Rabbits	dig prot %	*	2.0	11.6	
Sheep	dig prot %		2.0	11.6	
Energy	GE Mcal/kg				
Cattle	DE Mcal/kg	*	0.54	3.08	
Sheep	DE Mcal/kg	*	0.50	2.86	
Cattle	ME Mcal/kg	*	0.44	2.53	
Sheep	ME Mcal/kg	*	0.41	2.34	
Cattle	NE_m Mcal/kg	*	0.27	1.56	
Cattle	NE_{gain} Mcal/kg	*	0.18	1.00	
Cattle	$NE_{lactating\ cows}$ Mcal/kg	*	0.32	1.83	
Cattle	TDN %	*	12.3	69.9	
Sheep	TDN %	*	11.4	64.8	

Sorghum, sudangrass, aerial part, fresh, pre-bloom, (2)

Ref No 2 04 492 United States

Feed Name or Analyses			Mean As Fed	Mean Dry	C.V. ± %
Dry matter	%		19.6	100.0	
Ash	%		1.4	7.2	
Crude fiber	%		5.9	30.3	
Sheep	dig coef %		82.	82.	
Ether extract	%		0.9	4.7	
Sheep	dig coef %		83.	83.	
N-free extract	%		8.5	43.6	
Sheep	dig coef %		69.	69.	
Protein (N x 6.25)	%		2.8	14.2	
Sheep	dig coef %		79.	79.	
Cattle	dig prot %	*	2.0	10.0	
Goats	dig prot %	*	1.9	9.8	
Horses	dig prot %	*	1.9	9.6	
Rabbits	dig prot %	*	1.9	9.6	
Sheep	dig prot %		2.2	11.2	
Energy	GE Mcal/kg				
Cattle	DE Mcal/kg	*	0.59	3.01	
Sheep	DE Mcal/kg	*	0.65	3.30	
Cattle	ME Mcal/kg	*	0.48	2.47	
Sheep	ME Mcal/kg	*	0.53	2.71	
Cattle	TDN %	*	13.4	68.3	
Sheep	TDN %		14.7	74.9	

Sorghum, sudangrass, aerial part, fresh, early bloom, (2)

Ref No 2 08 519 United States

Feed Name or Analyses			Mean As Fed	Mean Dry	C.V. ± %
Dry matter	%		23.4	100.0	
Ash	%		2.4	10.3	
Crude fiber	%		8.4	35.9	
Ether extract	%		0.4	1.7	
N-free extract	%		10.3	44.0	
Protein (N x 6.25)	%		1.9	8.1	
Cattle	dig prot %	*	1.1	4.8	
Goats	dig prot %	*	1.0	4.1	
Horses	dig prot %	*	1.0	4.4	
Rabbits	dig prot %	*	1.2	4.9	
Sheep	dig prot %	*	1.1	4.6	
Energy	GE Mcal/kg				
Cattle	DE Mcal/kg	*	0.64	2.76	
Sheep	DE Mcal/kg	*	0.66	2.84	
Cattle	ME Mcal/kg	*	0.53	2.26	
Sheep	ME Mcal/kg	*	0.54	2.33	
Cattle	TDN %	*	14.6	62.5	
Sheep	TDN %	*	15.1	64.3	
Calcium	%		0.09	0.38	
Iron	%		0.003	0.013	
Magnesium	%		0.08	0.34	
Manganese	mg/kg		21.2	90.4	
Phosphorus	%		0.07	0.30	
Potassium	%		0.34	1.45	
Sulphur	%		0.01	0.04	

Sorghum, sudangrass, aerial part, fresh, mid-bloom, (2)

Ref No 2 04 485 United States

Feed Name or Analyses			Mean As Fed	Mean Dry	C.V. ± %
Dry matter	%		22.7	100.0	9
Ash	%		2.4	10.4	6
Crude fiber	%		8.2	36.1	3
Sheep	dig coef %	*	70.	70.	
Ether extract	%		0.4	1.8	12
Sheep	dig coef %	*	63.	63.	
N-free extract	%		9.8	43.0	
Sheep	dig coef %	*	66.	66.	
Protein (N x 6.25)	%		2.0	8.7	18
Sheep	dig coef %	*	69.	69.	
Cattle	dig prot %	*	1.2	5.3	
Goats	dig prot %	*	1.1	4.7	
Horses	dig prot %	*	1.1	4.9	
Rabbits	dig prot %	*	1.2	5.4	

Continued

Feed Name or Analyses			Mean As Fed	Mean Dry	C.V. ± %
Sheep	dig prot %		1.4	6.0	
Energy	GE Mcal/kg				
Cattle	DE Mcal/kg	*	0.63	2.77	
Sheep	DE Mcal/kg	*	0.62	2.74	
Cattle	ME Mcal/kg	*	0.52	2.27	
Sheep	ME Mcal/kg	*	0.51	2.25	
Cattle	NE_m Mcal/kg	*	0.31	1.37	
Cattle	NE_{gain} Mcal/kg	*	0.17	0.77	
Cattle	$NE_{lactating\ cows}$ Mcal/kg	*	0.36	1.57	
Cattle	TDN %	*	14.3	62.8	
Sheep	TDN %	*	14.1	62.2	

Sorghum, sudangrass, aerial part, fresh, full bloom, (2)

Ref No 2 04 486 United States

Analyses			As Fed	Dry	C.V. ± %
Dry matter	%		23.6	100.0	8
Ash	%		2.6	11.0	9
Crude fiber	%		8.6	36.4	7
Sheep	dig coef %	*	70.	70.	
Ether extract	%		0.4	1.7	14
Sheep	dig coef %	*	63.	63.	
N-free extract	%		10.1	42.8	
Sheep	dig coef %	*	66.	66.	
Protein (N x 6.25)	%		1.9	8.1	9
Sheep	dig coef %	*	69.	69.	
Cattle	dig prot %	*	1.1	4.8	
Goats	dig prot %	*	1.0	4.1	
Horses	dig prot %	*	1.0	4.4	
Rabbits	dig prot %	*	1.2	4.9	
Sheep	dig prot %		1.3	5.6	
Energy	GE Mcal/kg				
Cattle	DE Mcal/kg	*	0.64	2.71	
Sheep	DE Mcal/kg	*	0.64	2.72	
Cattle	ME Mcal/kg	*	0.52	2.22	
Sheep	ME Mcal/kg	*	0.53	2.23	
Cattle	TDN %	*	14.5	61.5	
Sheep	TDN %	*	14.6	61.7	

Sorghum, sudangrass, aerial part, fresh, dough stage, (2)

Ref No 2 04 493 United States

Analyses			As Fed	Dry	C.V. ± %
Dry matter	%		35.4	100.0	
Ash	%		3.6	10.2	
Crude fiber	%		11.3	32.0	
Cattle	dig coef %		58.	58.	
Ether extract	%		0.6	1.6	
Cattle	dig coef %		32.	32.	
N-free extract	%		18.4	52.1	
Cattle	dig coef %		51.	51.	
Protein (N x 6.25)	%		1.5	4.1	
Cattle	dig coef %		28.	28.	
Cattle	dig prot %		0.4	1.1	
Goats	dig prot %	*	0.1	0.4	
Horses	dig prot %	*	0.4	1.0	
Rabbits	dig prot %	*	0.6	1.8	
Sheep	dig prot %	*	0.3	0.8	
Energy	GE Mcal/kg				
Cattle	DE Mcal/kg	*	0.74	2.09	
Sheep	DE Mcal/kg	*	0.93	2.63	
Cattle	ME Mcal/kg	*	0.61	1.71	
Sheep	ME Mcal/kg	*	0.76	2.16	

(1) dry forages and roughages (2) pasture, range plants, and forages fed green (3) silages (4) energy feeds (5) protein supplements (6) minerals (7) vitamins (8) additives

Feed Name or Analyses			Mean As Fed	Mean Dry	C.V. ± %
Cattle	TDN %		16.8	47.4	
Sheep	TDN %	*	21.1	59.6	

Sorghum, sudangrass, aerial part, fresh, mature, (2)

Ref No 2 04 487 United States

Analyses			As Fed	Dry	C.V. ± %
Dry matter	%		30.3	100.0	9
Ash	%		2.4	8.1	14
Crude fiber	%		10.8	35.8	12
Sheep	dig coef %	*	70.	70.	
Ether extract	%		0.5	1.7	5
Sheep	dig coef %	*	63.	63.	
N-free extract	%		14.8	48.9	
Sheep	dig coef %	*	66.	66.	
Protein (N x 6.25)	%		1.7	5.5	14
Sheep	dig coef %	*	69.	69.	
Cattle	dig prot %	*	0.8	2.6	
Goats	dig prot %	*	0.5	1.7	
Horses	dig prot %	*	0.7	2.2	
Rabbits	dig prot %	*	0.9	2.9	
Sheep	dig prot %		1.2	3.8	
Energy	GE Mcal/kg				
Cattle	DE Mcal/kg	*	0.84	2.77	
Sheep	DE Mcal/kg	*	0.85	2.80	
Cattle	ME Mcal/kg	*	0.69	2.27	
Sheep	ME Mcal/kg	*	0.70	2.30	
Cattle	TDN %	*	19.0	62.8	
Sheep	TDN %		19.2	63.6	
Calcium	%		0.10	0.32	
Phosphorus	%		0.06	0.21	

Sorghum, sudangrass, aerial part, fresh, cut 1, (2)

Ref No 2 04 488 United States

Analyses			As Fed	Dry	C.V. ± %
Dry matter	%		14.3	100.0	
Ash	%		1.4	10.0	
Crude fiber	%		3.5	24.2	
Ether extract	%		0.7	5.0	
N-free extract	%		5.7	40.0	
Protein (N x 6.25)	%		3.0	20.8	
Cattle	dig prot %	*	2.2	15.6	
Goats	dig prot %	*	2.3	16.0	
Horses	dig prot %	*	2.2	15.2	
Rabbits	dig prot %	*	2.1	14.7	
Sheep	dig prot %	*	2.3	16.4	
Energy	GE Mcal/kg		0.64	4.46	
Sheep	GE dig coef %		66.	66.	
Cattle	DE Mcal/kg	*	0.40	2.80	
Sheep	DE Mcal/kg		0.42	2.93	
Cattle	ME Mcal/kg	*	0.33	2.30	
Sheep	ME Mcal/kg	*	0.34	2.40	
Cattle	TDN %	*	9.1	63.4	
Sheep	TDN %	*	9.5	66.4	
Calcium	%		0.07	0.48	
Copper	mg/kg		5.1	35.9	
Iron	%		0.009	0.060	
Magnesium	%		0.05	0.36	
Manganese	mg/kg		5.1	35.9	
Phosphorus	%		0.03	0.18	
Potassium	%		0.28	1.93	

Ref No 2 04 488 Canada

Analyses			As Fed	Dry	C.V. ± %
Dry matter	%			100.0	
Sheep	dig coef %			66.	
Ash	%			8.2	
Crude fiber	%			27.3	

Feed Name or Analyses			Mean As Fed	Mean Dry	C.V. ± %
Energy	GE Mcal/kg			4.46	1
Sheep	GE dig coef %			66.	
Sheep	DE Mcal/kg			2.93	
Sheep	ME Mcal/kg	*		2.40	

Sorghum, sudangrass, aerial part, ensiled, (3)

Ref No 3 04 499 United States

Analyses			As Fed	Dry	C.V. ± %
Dry matter	%		22.1	100.0	19
Ash	%		2.1	9.7	37
Crude fiber	%		8.1	36.6	9
Sheep	dig coef %		67.	67.	
Ether extract	%		0.7	3.1	29
Sheep	dig coef %		71.	71.	
N-free extract	%		8.7	39.4	
Sheep	dig coef %		49.	49.	
Protein (N x 6.25)	%		2.5	11.3	24
Sheep	dig coef %		69.	69.	
Cattle	dig prot %	*	1.4	6.5	
Goats	dig prot %	*	1.4	6.5	
Horses	dig prot %	*	1.4	6.5	
Sheep	dig prot %		1.7	7.8	
Energy	GE Mcal/kg				
Cattle	DE Mcal/kg	*	0.55	2.51	
Sheep	DE Mcal/kg	*	0.55	2.49	
Cattle	ME Mcal/kg	*	0.45	2.05	
Sheep	ME Mcal/kg	*	0.45	2.04	
Cattle	NE_m Mcal/kg	*	0.28	1.27	
Cattle	NE_{gain} Mcal/kg	*	0.14	0.64	
Cattle	$NE_{lactating\ cows}$ Mcal/kg	*	0.31	1.41	
Cattle	TDN %	*	12.6	56.8	
Sheep	TDN %		12.5	56.6	
Calcium	%		0.12	0.53	
Copper	mg/kg		8.1	36.6	
Iron	%		0.003	0.014	
Magnesium	%		0.11	0.49	
Manganese	mg/kg		21.8	98.8	
Phosphorus	%		0.04	0.19	
Potassium	%		0.68	3.07	
Carotene	mg/kg		5.7	26.0	
Vitamin A equivalent	IU/g		9.6	43.4	

Sorghum, sudangrass, aerial part, ensiled, prebloom, (3)

Ref No 3 09 086 United States

Analyses			As Fed	Dry	C.V. ± %
Dry matter	%		42.2	100.0	
Cattle	dig coef %		56.	56.	
Ash	%		5.7	13.5	
Crude fiber	%		12.0	28.5	
Ether extract	%		3.9	9.3	
N-free extract	%		15.6	37.0	
Protein (N x 6.25)	%		5.0	11.8	
Cattle	dig coef %		47.	47.	
Cattle	dig prot %		2.4	5.6	
Goats	dig prot %	*	2.9	6.9	
Horses	dig prot %	*	2.9	6.9	
Sheep	dig prot %	*	2.9	6.9	
Cellulose (Matrone)	%		10.9	25.8	
Cattle	dig coef %		74.	74.	
Energy	GE Mcal/kg		1.64	3.89	10
Cattle	GE dig coef %		62.	62.	
Cattle	DE Mcal/kg		1.02	2.42	
Sheep	DE Mcal/kg	*	1.07	2.53	
Cattle	ME Mcal/kg	*	0.84	1.98	
Sheep	ME Mcal/kg	*	0.88	2.08	

1848

Feed Name or Analyses			As Fed (Mean)	Dry (Mean)	C.V. ± %
Cattle	TDN %		20.2	47.9	
Sheep	TDN %	*	24.2	57.4	

Sorghum, sudangrass, aerial part, ensiled, early bloom, (3)

Ref No 3 09 169 United States

Analyses			As Fed	Dry	C.V. ± %
Dry matter	%			100.0	
Cattle	dig coef %			56.	
Ash	%			5.4	
Crude fiber	%			39.1	
Cattle	dig coef %			66.	
Ether extract	%			2.1	
N-free extract	%			43.6	
Cattle	dig coef %			53.	
Protein (N x 6.25)	%			10.0	
Cattle	dig coef %			44.	
Cattle	dig prot %			4.3	
Goats	dig prot %	*		5.3	
Horses	dig prot %	*		5.3	
Sheep	dig prot %	*		5.3	
Energy	GE Mcal/kg				
Cattle	DE Mcal/kg	*		2.49	
Sheep	DE Mcal/kg	*		2.84	
Cattle	ME Mcal/kg	*		2.04	
Sheep	ME Mcal/kg	*		2.33	
Cattle	TDN %			56.4	
Sheep	TDN %	*		64.4	

Sorghum, sudangrass, aerial part, ensiled, mid-bloom, (3)

Ref No 3 04 495 United States

Analyses			As Fed	Dry	C.V. ± %
Dry matter	%		25.5	100.0	8
Ash	%		1.9	7.4	21
Crude fiber	%		8.3	32.7	3
Ether extract	%		0.8	3.3	31
N-free extract	%		11.8	46.4	
Protein (N x 6.25)	%		2.6	10.2	18
Cattle	dig prot %	*	1.4	5.5	
Goats	dig prot %	*	1.4	5.5	
Horses	dig prot %	*	1.4	5.5	
Sheep	dig prot %	*	1.4	5.5	
Energy	GE Mcal/kg				
Cattle	DE Mcal/kg	*	0.66	2.61	
Sheep	DE Mcal/kg	*	0.72	2.81	
Cattle	ME Mcal/kg	*	0.55	2.14	
Sheep	ME Mcal/kg	*	0.59	2.31	
Cattle	TDN %	*	15.1	59.1	
Sheep	TDN %	*	16.3	63.8	

Sorghum, sudangrass, aerial part, ensiled, full bloom, (3)

Ref No 3 04 496 United States

Analyses			As Fed	Dry	C.V. ± %
Dry matter	%		25.8	100.0	5
Ash	%		1.5	5.8	14
Crude fiber	%		8.8	34.1	2
Ether extract	%		0.7	2.9	10
N-free extract	%		12.4	47.9	
Protein (N x 6.25)	%		2.4	9.3	11
Cattle	dig prot %	*	1.2	4.7	
Goats	dig prot %	*	1.2	4.7	
Horses	dig prot %	*	1.2	4.7	
Sheep	dig prot %	*	1.2	4.7	
Energy	GE Mcal/kg				
Cattle	DE Mcal/kg	*	0.64	2.48	

Feed Name or Analyses			As Fed (Mean)	Dry (Mean)	C.V. ± %
Sheep	DE Mcal/kg	*	0.69	2.68	
Cattle	ME Mcal/kg	*	0.53	2.04	
Sheep	ME Mcal/kg	*	0.57	2.20	
Cattle	TDN %	*	14.5	56.3	
Sheep	TDN %	*	15.7	60.8	

Sorghum, sudangrass, aerial part, ensiled, dough stage, (3)

Ref No 3 04 497 United States

Analyses			As Fed	Dry	C.V. ± %
Dry matter	%		24.0	100.0	
Ash	%		2.3	9.6	
Crude fiber	%		8.4	35.0	
Ether extract	%		0.5	2.1	
N-free extract	%		10.9	45.4	
Protein (N x 6.25)	%		1.9	7.9	
Cattle	dig prot %	*	0.8	3.4	
Goats	dig prot %	*	0.8	3.4	
Horses	dig prot %	*	0.8	3.4	
Sheep	dig prot %	*	0.8	3.4	
Energy	GE Mcal/kg				
Cattle	DE Mcal/kg	*	0.64	2.68	
Sheep	DE Mcal/kg	*	0.67	2.79	
Cattle	ME Mcal/kg	*	0.53	2.20	
Sheep	ME Mcal/kg	*	0.55	2.28	
Cattle	TDN %	*	14.6	60.7	
Sheep	TDN %	*	15.2	63.2	

Sorghum, sudangrass, aerial part, ensiled, cut 2, (3)

Ref No 3 04 498 United States

Analyses			As Fed	Dry	C.V. ± %
Dry matter	%		16.3	100.0	
Ash	%		3.0	18.4	
Crude fiber	%		4.6	28.2	
Ether extract	%		0.6	3.7	
N-free extract	%		5.4	33.1	
Protein (N x 6.25)	%		2.7	16.6	
Cattle	dig prot %	*	1.8	11.3	
Goats	dig prot %	*	1.8	11.3	
Horses	dig prot %	*	1.8	11.3	
Sheep	dig prot %	*	1.8	11.3	
Energy	GE Mcal/kg				
Cattle	DE Mcal/kg	*	0.43	2.61	
Sheep	DE Mcal/kg	*	0.44	2.69	
Cattle	ME Mcal/kg	*	0.35	2.14	
Sheep	ME Mcal/kg	*	0.36	2.21	
Cattle	TDN %	*	9.7	59.3	
Sheep	TDN %	*	10.0	61.1	
Cobalt	mg/kg		0.088	0.540	

Sorghum, sudangrass, aerial part, wilted ensiled, immature, (3)

Ref No 3 09 168 United States

Analyses			As Fed	Dry	C.V. ± %
Dry matter	%		41.5	100.0	
Cattle	dig coef %		51.	51.	
Ash	%		8.0	19.3	
Crude fiber	%		11.5	27.8	
Ether extract	%		1.2	3.0	
N-free extract	%		16.2	39.2	
Protein (N x 6.25)	%		4.5	10.8	
Cattle	dig coef %		43.	43.	
Cattle	dig prot %		1.9	4.6	
Goats	dig prot %	*	2.5	6.0	
Horses	dig prot %	*	2.5	6.0	
Sheep	dig prot %	*	2.5	6.0	

Feed Name or Analyses			As Fed (Mean)	Dry (Mean)	C.V. ± %
Cellulose (Matrone)	%		9.9	23.8	
Cattle	dig coef %		70.	70.	
Energy	GE Mcal/kg		1.49	3.59	
Cattle	GE dig coef %		59.	59.	
Cattle	DE Mcal/kg		0.87	2.10	
Sheep	DE Mcal/kg	*	1.06	2.56	
Cattle	ME Mcal/kg	*	0.72	1.72	
Sheep	ME Mcal/kg	*	0.87	2.10	
Cattle	TDN %	*	19.8	47.6	
Sheep	TDN %	*	24.1	58.0	

Sorghum, sudangrass, aerial part w molasses added, ensiled, (3)

Ref No 3 04 500 United States

Analyses			As Fed	Dry	C.V. ± %
Dry matter	%		20.7	100.0	4
Ash	%		1.7	8.2	14
Crude fiber	%		7.1	34.4	5
Ether extract	%		0.9	4.5	25
N-free extract	%		8.6	41.4	
Protein (N x 6.25)	%		2.4	11.5	21
Cattle	dig prot %	*	1.4	6.7	
Goats	dig prot %	*	1.4	6.7	
Horses	dig prot %	*	1.4	6.7	
Sheep	dig prot %	*	1.4	6.7	
Energy	GE Mcal/kg				
Cattle	DE Mcal/kg	*	0.53	2.58	
Sheep	DE Mcal/kg	*	0.57	2.74	
Cattle	ME Mcal/kg	*	0.44	2.12	
Sheep	ME Mcal/kg	*	0.46	2.24	
Cattle	TDN %	*	12.1	58.6	
Sheep	TDN %	*	12.8	62.0	

Sorghum, sudangrass, aerial part w 100 lb grnd corn ears added per ton, wilted ensiled, early bloom, (3)

Ref No 3 09 252 United States

Analyses			As Fed	Dry	C.V. ± %
Dry matter	%			100.0	
Cattle	dig coef %			61.	
Ash	%			7.7	
Crude fiber	%			34.8	
Cattle	dig coef %			71.	
Ether extract	%			2.6	
N-free extract	%			44.1	
Cattle	dig coef %			60.	
Protein (N x 6.25)	%			10.8	
Cattle	dig coef %			53.	
Cattle	dig prot %			5.7	
Goats	dig prot %	*		6.0	
Horses	dig prot %	*		6.0	
Sheep	dig prot %	*		6.0	
Cellulose (Matrone)	%			38.2	
Cattle	dig coef %			70.	
Energy	GE Mcal/kg				
Cattle	DE Mcal/kg	*		2.68	
Sheep	DE Mcal/kg	*		2.80	
Cattle	ME Mcal/kg	*		2.19	
Sheep	ME Mcal/kg	*		2.30	
Cattle	TDN %			60.7	
Sheep	TDN %	*		63.5	

Sorghum, sudangrass, grain, (4)
Ref No 4 08 520 United States

Feed Name or Analyses				Mean As Fed	Mean Dry	C.V. ± %
Dry matter		%		92.4	100.0	
Ash		%		12.0	13.0	
Crude fiber		%		25.4	27.5	
Ether extract		%		2.4	2.6	
N-free extract		%		38.4	41.6	
Protein (N x 6.25)		%		14.2	15.4	
Cattle	dig prot	%	*	9.4	10.1	
Goats	dig prot	%	*	10.5	11.3	
Horses	dig prot	%	*	10.5	11.3	
Sheep	dig prot	%	*	10.5	11.3	
Energy	GE	Mcal/kg				
Cattle	DE	Mcal/kg	*	2.17	2.35	
Sheep	DE	Mcal/kg	*	2.01	2.31	
Cattle	ME	Mcal/kg	*	1.78	1.93	
Sheep	ME	Mcal/kg	*	1.67	1.81	
Cattle	TDN	%	*	49.2	53.2	
Sheep	TDN	%	*	46.3	50.1	

SORGHUM, SUDANGRASS PIPER. Sorghum vulgare

Sorghum, sudangrass piper, hay, fan-air dried w heat, immature, cut 1, (1)
Ref No 1 08 891 United States

Feed Name or Analyses				Mean As Fed	Mean Dry	C.V. ± %
Dry matter		%			100.0	
Goats	dig coef	%			68.	
Protein (N x 6.25)		%			18.2	
Cattle	dig prot	%	*		12.7	
Goats	dig prot	%	*		13.5	
Horses	dig prot	%	*		13.0	
Rabbits	dig prot	%	*		12.7	
Sheep	dig prot	%	*		12.9	
Cell walls (Van Soest)		%			54.7	
Cellulose (Matrone)		%			26.3	
Fiber, acid detergent (VS)		%			34.5	
Lignin (Van Soest)		%			4.5	
Energy	GE	Mcal/kg				
Goats	GE dig coef	%			69.	
Nutritive value index (NVI)		%			38.	
Relative intake		%			55.	

Sorghum, sudangrass piper, hay, fan-air dried w heat, prebloom, cut 1, (1)
Ref No 1 08 892 United States

Feed Name or Analyses				Mean As Fed	Mean Dry	C.V. ± %
Dry matter		%			100.0	
Goats	dig coef	%			67.	
Protein (N x 6.25)		%			11.1	
Cattle	dig prot	%	*		6.6	
Goats	dig prot	%	*		6.9	
Horses	dig prot	%	*		7.0	
Rabbits	dig prot	%	*		7.2	
Sheep	dig prot	%	*		6.5	
Cell walls (Van Soest)		%			60.6	
Cellulose (Matrone)		%			32.1	
Fiber, acid detergent (VS)		%			36.3	
Lignin (Van Soest)		%			3.9	
Energy	GE	Mcal/kg				
Goats	GE dig coef	%			65.	

(1) dry forages and roughages (2) pasture, range plants, and forages fed green (3) silages (4) energy feeds (5) protein supplements (6) minerals (7) vitamins (8) additives

Feed Name or Analyses				Mean As Fed	Mean Dry	C.V. ± %
Nutritive value index (NVI)		%			35.	
Relative intake		%			54.	

Sorghum, sudangrass piper, hay, s-c (1)
Ref No 1 08 980 United States

Feed Name or Analyses				Mean As Fed	Mean Dry	C.V. ± %
Dry matter		%		93.4	100.0	
Ash		%		12.8	13.7	
Crude fiber		%		24.2	25.9	
Ether extract		%		1.4	1.6	
Lignin (Ellis)		%		5.8	6.3	
Silicon		%		6.27	6.71	

Sorghum, sudangrass piper, hay, s-c early bloom, (1)
Ref No 1 08 981 United States

Feed Name or Analyses				Mean As Fed	Mean Dry	C.V. ± %
Dry matter		%		93.8	100.0	
Ash		%		10.7	11.4	
Crude fiber		%		26.1	27.8	
Ether extract		%		1.2	1.3	
Lignin (Ellis)		%		6.2	6.6	
Silicon		%		4.65	4.96	

SORGHUM, SUDANGRASS SWEET SIOUX. Sorghum vulgare

Sorghum, sudangrass sweet Sioux, hay, fan-air dried w heat, immature, cut 1, (1)
Ref No 1 08 893 United States

Feed Name or Analyses				Mean As Fed	Mean Dry	C.V. ± %
Dry matter		%			100.0	
Goats	dig coef	%			68.	
Protein (N x 6.25)		%			14.0	
Cattle	dig prot	%	*		9.1	
Goats	dig prot	%	*		9.6	
Horses	dig prot	%	*		9.4	
Rabbits	dig prot	%	*		9.5	
Sheep	dig prot	%	*		9.1	
Cell walls (Van Soest)		%			59.6	
Cellulose (Matrone)		%			31.2	
Fiber, acid detergent (VS)		%			35.2	
Lignin (Van Soest)		%			3.7	
Energy	GE	Mcal/kg				
Goats	GE dig coef	%			67.	
Goats	DE	Mcal/kg			0.07	
Nutritive value index (NVI)		%			32.	
Relative intake		%			48.	

SORGHUM, SUDANGRASS TIFT. Sorghum vulgare sudanense

Sorghum, sudangrass tift, aerial part, ensiled, (3)
Ref No 3 09 108 United States

Feed Name or Analyses				Mean As Fed	Mean Dry	C.V. ± %
Dry matter		%		24.1	100.0	
Cattle	dig coef	%		61.	61.	
Ash		%		1.9	7.7	
Crude fiber		%		8.4	34.8	
Cattle	dig coef	%		71.	71.	
Ether extract		%		0.6	2.6	
Cattle	dig coef	%		67.	67.	
N-free extract		%		10.6	44.1	
Cattle	dig coef	%		60.	60.	
Protein (N x 6.25)		%		2.6	10.8	
Cattle	dig coef	%		53.	53.	
Cattle	dig prot	%		1.4	5.7	

Feed Name or Analyses				Mean As Fed	Mean Dry	C.V. ± %
Goats	dig prot	%	*	1.5	6.0	
Horses	dig prot	%	*	1.5	6.0	
Sheep	dig prot	%	*	1.5	6.0	
Cellulose (Matrone)		%		9.2	38.2	
Cattle	dig coef	%		70.	70.	
Energy	GE	Mcal/kg				
Cattle	DE	Mcal/kg	*	0.65	2.68	
Sheep	DE	Mcal/kg	*	0.67	2.80	
Cattle	ME	Mcal/kg	*	0.53	2.20	
Sheep	ME	Mcal/kg	*	0.55	2.30	
Cattle	TDN	%		14.7	60.8	
Sheep	TDN	%	*	15.3	63.5	

SORGHUM, SUMAC. Sorghum vulgare

Sorghum, sumac, aerial part, s-c, immature, (1)
Sorghum sumac, fodder, immature
Ref No 1 04 501 United States

Feed Name or Analyses				Mean As Fed	Mean Dry	C.V. ± %
Dry matter		%		87.8	100.0	5
Calcium		%		0.31	0.35	20
Phosphorus		%		0.18	0.20	28

Sorghum, sumac, aerial part, s-c, mature, (1)
Sorghum, sumac, fodder, sun-cured, mature
Ref No 1 04 502 United States

Feed Name or Analyses				Mean As Fed	Mean Dry	C.V. ± %
Dry matter		%		81.2	100.0	6
Ash		%		6.7	8.3	12
Crude fiber		%		24.8	30.6	5
Cattle	dig coef	%	*	60.	60.	
Sheep	dig coef	%	*	48.	48.	
Ether extract		%		2.1	2.6	9
Cattle	dig coef	%	*	61.	61.	
Sheep	dig coef	%	*	47.	47.	
N-free extract		%		42.4	52.2	
Cattle	dig coef	%	*	66.	66.	
Sheep	dig coef	%	*	70.	70.	
Protein (N x 6.25)		%		5.1	6.3	6
Cattle	dig coef	%	*	38.	38.	
Sheep	dig coef	%	*	74.	74.	
Cattle	dig prot	%		1.9	2.4	
Goats	dig prot	%	*	2.0	2.4	
Horses	dig prot	%	*	2.3	2.9	
Rabbits	dig prot	%	*	2.9	3.5	
Sheep	dig prot	%		3.8	4.7	
Energy	GE	Mcal/kg				
Cattle	DE	Mcal/kg	*	2.10	2.59	
Sheep	DE	Mcal/kg	*	2.10	2.59	
Cattle	ME	Mcal/kg	*	1.73	2.13	
Sheep	ME	Mcal/kg	*	1.72	2.12	
Cattle	TDN	%	*	47.7	58.8	
Sheep	TDN	%	*	47.6	58.6	

Sorghum, sumac, aerial part wo heads, s-c, mature, (1)
Sorghum, sumac, stover, mature
Ref No 1 04 503 United States

Feed Name or Analyses				Mean As Fed	Mean Dry	C.V. ± %
Dry matter		%			100.0	
Ash		%			8.0	4
Crude fiber		%			33.2	7
Ether extract		%			2.3	14
N-free extract		%			52.2	
Protein (N x 6.25)		%			4.3	26
Cattle	dig prot	%	*		0.7	
Goats	dig prot	%	*		0.6	

Feed Name or Analyses	Mean As Fed	Mean Dry	C.V. ± %
Horses dig prot % *		1.2	
Rabbits dig prot % *		2.0	
Sheep dig prot % *		0.4	
Energy GE Mcal/kg			
Cattle DE Mcal/kg *		2.62	
Sheep DE Mcal/kg *		2.39	
Cattle ME Mcal/kg *		2.15	
Sheep ME Mcal/kg *		1.96	
Cattle TDN % *		59.3	
Sheep TDN % *		54.1	
Calcium %		0.39	14
Phosphorus %		0.11	9
Carotene mg/kg		6.8	52
Vitamin A equivalent IU/g		11.4	

Sorghum, sumac, leaves, (1)

Ref No 1 04 504 United States

Feed Name or Analyses	Mean As Fed	Mean Dry	C.V. ± %
Dry matter %	93.4	100.0	
Ash %	6.0	6.4	
Crude fiber %	13.0	13.9	
Ether extract %	6.5	7.0	
N-free extract %	57.8	61.9	
Protein (N x 6.25) %	10.1	10.8	
Cattle dig prot % *	5.9	6.3	
Goats dig prot % *	6.2	6.6	
Horses dig prot % *	6.3	6.7	
Rabbits dig prot % *	6.5	7.0	
Sheep dig prot % *	5.8	6.3	
Energy GE Mcal/kg			
Cattle DE Mcal/kg *	2.59	2.77	
Sheep DE Mcal/kg *	2.69	2.88	
Cattle ME Mcal/kg *	2.12	2.27	
Sheep ME Mcal/kg *	2.21	2.36	
Cattle TDN % *	58.7	62.9	
Sheep TDN % *	61.0	65.3	

Sorghum, sumac, aerial part, ensiled, (3)

Ref No 3 04 507 United States

Feed Name or Analyses	Mean As Fed	Mean Dry	C.V. ± %
Dry matter %	37.4	100.0	45
Ash %	3.0	7.9	26
Crude fiber %	8.9	23.9	21
Ether extract %	1.1	2.9	16
N-free extract %	21.2	56.6	
Protein (N x 6.25) %	3.3	8.7	23
Cattle dig prot % *	1.5	4.1	
Goats dig prot % *	1.5	4.1	
Horses dig prot % *	1.5	4.1	
Sheep dig prot % *	1.5	4.1	
Calcium %	0.14	0.38	46
Phosphorus %	0.08	0.22	33
Carotene mg/kg	6.8	18.1	71
Vitamin A equivalent IU/g	11.3	30.1	

Sorghum, sumac, aerial part, ensiled, dough stage, (3)

Ref No 3 04 506 United States

Feed Name or Analyses	Mean As Fed	Mean Dry	C.V. ± %
Dry matter %	29.8	100.0	4
Ash %	2.2	7.3	15
Crude fiber %	7.2	24.0	3
Ether extract %	0.7	2.5	12
N-free extract %	17.3	57.9	
Protein (N x 6.25) %	2.5	8.3	3
Cattle dig prot % *	1.1	3.8	
Goats dig prot % *	1.1	3.8	
Horses dig prot % *	1.1	3.8	
Sheep dig prot % *	1.1	3.8	

Sorghum, sumac, aerial part, ensiled trench silo, (3)

Ref No 3 04 508 United States

Feed Name or Analyses	Mean As Fed	Mean Dry	C.V. ± %
Dry matter %	57.4	100.0	28
Ash %	5.3	9.2	23
Crude fiber %	7.9	13.7	11
Ether extract %	2.0	3.5	14
N-free extract %	36.2	63.1	
Protein (N x 6.25) %	6.0	10.5	9
Cattle dig prot % *	3.3	5.8	
Goats dig prot % *	3.3	5.8	
Horses dig prot % *	3.3	5.8	
Sheep dig prot % *	3.3	5.8	
Carotene mg/kg	1.1	2.0	53
Vitamin A equivalent IU/g	1.9	3.3	

Sorghum, sumac, grain, (4)

Ref No 4 04 509 United States

Feed Name or Analyses	Mean As Fed	Mean Dry	C.V. ± %
Dry matter %	81.9	100.0	9
Ash %	2.1	2.6	60
Crude fiber %	1.8	2.2	29
Cattle dig coef % *	40.	40.	
Sheep dig coef % *	33.	33.	
Ether extract %	3.4	4.2	7
Cattle dig coef % *	80.	80.	
Sheep dig coef % *	76.	76.	
N-free extract %	64.5	78.7	
Cattle dig coef % *	83.	83.	
Sheep dig coef % *	89.	89.	
Protein (N x 6.25) %	10.1	12.3	8
Cattle dig coef % *	57.	57.	
Sheep dig coef % *	67.	67.	
Cattle dig prot %	5.7	7.0	
Goats dig prot % *	7.0	8.5	
Horses dig prot % *	7.0	8.5	
Sheep dig prot %	6.7	8.2	
Energy GE Mcal/kg			
Cattle DE Mcal/kg *	2.92	3.56	
Sheep DE Mcal/kg *	3.11	3.80	
Swine DE kcal/kg *	3269.	3992.	
Cattle ME Mcal/kg *	2.39	2.92	
Sheep ME Mcal/kg *	2.55	3.12	
Swine ME kcal/kg *	3057.	3733.	
Cattle TDN %	66.2	80.8	
Sheep TDN %	70.6	86.2	
Swine TDN % *	74.2	90.5	
Carotene mg/kg	1.4	1.8	42
Niacin mg/kg	26.2	32.0	17
Pantothenic acid mg/kg	12.5	15.2	8
Riboflavin mg/kg	1.4	1.8	27
Vitamin A equivalent IU/g	2.4	2.9	

Sorghum, sumac, heads, chopped, (4)

Ref No 4 04 510 United States

Feed Name or Analyses	Mean As Fed	Mean Dry	C.V. ± %
Dry matter %	79.8	100.0	9
Ash %	3.7	4.7	
Crude fiber %	8.3	10.4	
Cattle dig coef % *	40.	40.	
Sheep dig coef % *	61.	61.	
Ether extract %	2.8	3.5	
Cattle dig coef % *	80.	80.	
Sheep dig coef % *	74.	74.	
N-free extract %	55.8	70.0	
Cattle dig coef % *	83.	83.	
Sheep dig coef % *	80.	80.	
Protein (N x 6.25) %	9.2	11.5	
Cattle dig coef % *	57.	57.	
Sheep dig coef % *	63.	63.	
Cattle dig prot %	5.2	6.6	
Goats dig prot % *	6.2	7.8	
Horses dig prot % *	6.2	7.8	
Sheep dig prot %	5.8	7.2	
Sugars, total %	2.5	3.1	
Energy GE Mcal/kg			
Cattle DE Mcal/kg *	2.64	3.31	
Sheep DE Mcal/kg *	2.65	3.32	
Swine DE kcal/kg *	2306.	2888.	
Cattle ME Mcal/kg *	2.17	2.71	
Sheep ME Mcal/kg *	2.18	2.73	
Swine ME kcal/kg *	2160.	2705.	
Cattle TDN %	59.9	75.1	
Sheep TDN %	60.2	75.4	
Swine TDN % *	52.3	65.5	
Calcium %	0.36	0.45	
Phosphorus %	0.09	0.11	
Carotene mg/kg	2.6	3.3	
Vitamin A equivalent IU/g	4.4	5.5	

Sorghum, atlas, fodder –
see Sorghum, atlas, aerial part, s-c, (1)

Sorghum, atlas, fodder, barn-cured, immature –
see Sorghum, atlas, aerial part, fan air dried, immature, (1)

Sorghum, atlas, fodder, dehy, mature –
see Sorghum, atlas, aerial part, dehy, mature, (1)

Sorghum, atlas, fodder, mature –
see Sorghum, atlas, aerial part, s-c, mature, (1)

Sorghum, atlas, stover –
see Sorghum, atlas, aerial part wo heads, s-c, (1)

Sorghum, atlas, stover, over ripe –
see Sorghum, atlas, aerial part wo heads, s-c, over ripe, (1)

Sorghum, bagasse, dried –
see Sorghum, pulp, dehy, (1)

Sorghum, broomcorn, stover –
see Sorghum, broomcorn, aerial part wo heads, s-c, (1)

SORGHUM-COWPEA. Sorghum vulgare, Vigna spp

Sorghum-cowpea, aerial part, ensiled, (3)

Ref No 3 04 521 United States

Feed Name or Analyses	Mean As Fed	Mean Dry	C.V. ± %
Dry matter %	31.9	100.0	
Ash %	2.2	6.8	
Crude fiber %	8.4	26.3	
Ether extract %	1.0	3.1	
N-free extract %	18.0	56.4	
Protein (N x 6.25) %	2.4	7.4	
Cattle dig prot % *	0.9	3.0	
Goats dig prot % *	0.9	3.0	
Horses dig prot % *	0.9	3.0	

Continued

Feed Name or Analyses			Mean As Fed	Mean Dry	C.V. ± %
Sheep	dig prot %	*	0.9	3.0	
Energy	GE Mcal/kg				
Cattle	DE Mcal/kg	*	0.81	2.53	
Sheep	DE Mcal/kg	*	0.79	2.48	
Cattle	ME Mcal/kg	*	0.66	2.07	
Sheep	ME Mcal/kg	*	0.65	2.03	
Cattle	TDN %	*	18.3	57.3	
Sheep	TDN %	*	17.9	56.2	
Calcium	%		0.14	0.43	
Phosphorus	%		0.04	0.12	
Potassium	%		0.30	0.93	

Sorghum, durra, fodder –
see Sorghum, durra, aerial part, s-c, (1)

Sorghum, feterita, fodder –
see Sorghum, feterita, aerial part, s-c, (1)

Sorghum, feterita, stover –
see Sorghum, feterita, aerial part wo heads, s-c, (1)

Sorghum fodder, fresh –
see Sorghum, aerial part, fresh, (2)

Sorghum fodder, fresh, immature –
see Sorghum, aerial part, fresh, immature, (2)

Sorghum fodder, fresh, mature –
see Sorghum, aerial part, fresh, mature, (2)

Sorghum fodder, fresh, mid-bloom, –
see Sorghum, aerial part, fresh, mid-bloom, (2)

Sorghum fodder, fresh, milk stage –
see Sorghum, aerial part, fresh, milk stage, (2)

Sorghum fodder silage –
see Sorghum, aerial part, ensiled, (3)

Sorghum fodder, sun-cured, immature –
see Sorghum, aerial part, s-c, immature, (1)

Sorghum fodder, sun-cured, mature –
see Sorghum, aerial part, s-c, mature, (1)

Sorghum gluten feed –
see Sorghum, gluten w bran, wet milled dehy grnd, (5)

Sorghum gluten meal –
see Sorghum, gluten, wet milled, dehy, (5)

Sorghum, grain, fodder, full bloom –
see Sorghum, grain variety, aerial part, s-c, full bloom, (1)

Sorghum, hegari, fodder, dough stage –
see Sorghum, hegari, aerial part, s-c, dough stage, (1)

Sorghum, hegari, fodder, sun-cured, mature –
see Sorghum, hegari, aerial part, s-c, mature, (1)

(1) dry forages and roughages (2) pasture, range plants, and forages fed green (3) silages (4) energy feeds (5) protein supplements (6) minerals (7) vitamins (8) additives

Sorghum, hegari, stover, sun-cured, mature –
see Sorghum, hegari, aerial part wo heads, s-c, mature, (1)

Sorghum, kafir, fodder, sun-cured –
see Sorghum, kafir, aerial part, s-c, (1)

Sorghum, kafir, stover, sun-cured –
see Sorghum, kafir, aerial part wo heads, s-c, (1)

SORGHUM-MILLET. Sorghum vulgare, Setaria spp

Sorghum-Millet, aerial part w molasses added, ensiled, mature, (3)
Ref No 3 04 522 United States

Analyses			As Fed	Dry	C.V. ± %
Dry matter	%		20.3	100.0	
Ash	%		1.6	7.9	
Crude fiber	%		9.0	44.5	
Ether extract	%		0.4	2.1	
N-free extract	%		8.3	40.9	
Protein (N x 6.25)	%		0.9	4.6	
Cattle	dig prot %	*	0.1	0.4	
Goats	dig prot %	*	0.1	0.4	
Horses	dig prot %	*	0.1	0.4	
Sheep	dig prot %	*	0.1	0.4	
Carotene	mg/kg		5.8	28.7	
Vitamin A equivalent	IU/g		9.7	47.8	

Sorghum-Millet, aerial part w molasses added, ensiled, cut 1, (3)
Ref No 3 04 523 United States

Analyses			As Fed	Dry	C.V. ± %
Dry matter	%		20.3	100.0	
Chlorine	%		0.29	1.43	

Sorghum, milo, fodder, sun-cured –
see Sorghum, milo, aerial part, s-c, (1)

Sorghum, milo, stover, sun-cured –
see Sorghum, milo, aerial part wo heads, s-c, (1)

SORGHUM-REDTOP. Sorghum vulgare, Agrostis alba

Sorghum-Redtop, aerial part, ensiled, (3)
Ref No 3 04 524 United States

Analyses			As Fed	Dry	C.V. ± %
Dry matter	%		28.9	100.0	19
Ash	%		2.1	7.1	11
Crude fiber	%		6.5	22.4	14
Ether extract	%		0.8	2.9	12
N-free extract	%		17.3	60.0	
Protein (N x 6.25)	%		2.2	7.6	7
Cattle	dig prot %	*	0.9	3.1	
Goats	dig prot %	*	0.9	3.1	
Horses	dig prot %	*	0.9	3.1	
Sheep	dig prot %	*	0.9	3.1	
Calcium	%		0.09	0.32	10
Phosphorus	%		0.05	0.16	11

Sorghum, shallu, fodder, sun-cured, mature –
see Sorghum, shallu, aerial part, s-c, mature, (1)

Sorghum, shallu, stover, sun-cured –
see Sorghum, shallu, aerial part wo heads, s-c, (1)

Sorghum, sorgo, fodder, mature –
see Sorghum, sorgo, aerial part, s-c, mature, (1)

Sorghum, sorgo, fodder silage –
see Sorghum, sorgo, aerial part, ensiled, (3)

Sorghum, sorgo, fodder, sun-cured –
see Sorghum, sorgo, aerial part, s-c, (1)

Sorghum, sorgo, stover –
see Sorghum, sorgo, aerial part wo heads, s-c, (1)

Sorghum stover silage –
see Sorghum, aerial part wo heads, ensiled, (3)

Sorghum stover, sun-cured –
see Sorghum, aerial part wo heads, s-c, (1)

SORGHUM, SUDANGRASS-SOYBEAN. Sorghum vulgare sudanense, Glycine max

Sorghum, Sudangrass-Soybean, hay, s-c, (1)
Ref No 1 08 521 United States

Analyses			As Fed	Dry	C.V. ± %
Dry matter	%		89.0	100.0	
Ash	%		4.9	5.5	
Crude fiber	%		31.1	34.9	
Ether extract	%		2.2	2.5	
N-free extract	%		43.4	48.8	
Protein (N x 6.25)	%		7.4	8.3	
Cattle	dig prot %	*	3.7	4.1	
Goats	dig prot %	*	3.8	4.3	
Horses	dig prot %	*	4.1	4.6	
Rabbits	dig prot %	*	4.5	5.1	
Sheep	dig prot %	*	3.6	4.0	
Energy	GE Mcal/kg				
Cattle	DE Mcal/kg	*	2.09	2.35	
Sheep	DE Mcal/kg	*	2.16	2.43	
Cattle	ME Mcal/kg	*	1.71	1.93	
Sheep	ME Mcal/kg	*	1.77	1.99	
Cattle	TDN %	*	47.4	53.3	
Sheep	TDN %	*	49.1	55.1	

Sorghum sumac, fodder, immature –
see Sorghum, sumac, aerial part, s-c, immature, (1)

Sorghum, sumac, fodder, sun-cured, mature –
see Sorghum, sumac, aerial part, s-c, mature, (1)

Sorghum, sumac, stover, mature –
see Sorghum, sumac, aerial part wo heads, s-c, mature, (1)

SOTOL. Dasylirion spp

Sotol, bulbs, fresh, (2)
Ref No 2 04 525 United States

Analyses			As Fed	Dry	C.V. ± %
Dry matter	%		43.7	100.0	
Ash	%		2.2	5.1	
Crude fiber	%		11.6	26.6	
Ether extract	%		0.8	1.8	
N-free extract	%		26.1	59.7	
Protein (N x 6.25)	%		3.0	6.8	
Cattle	dig prot %	*	1.6	3.7	
Goats	dig prot %	*	1.3	2.9	
Horses	dig prot %	*	1.4	3.3	

Feed Name or Analyses			Mean As Fed	Mean Dry	C.V. ± %
Rabbits	dig prot %	*	1.7	3.9	
Sheep	dig prot %	*	1.5	3.3	
Carotene	mg/kg		0.9	2.0	
Vitamin A equivalent	IU/g		1.4	3.3	

Sotol, heads, fresh, (2)

Ref No 2 04 526 United States

Feed Name or Analyses			Mean As Fed	Mean Dry	C.V. ± %
Dry matter	%		38.4	100.0	6
Ash	%		1.7	4.4	4
Crude fiber	%		9.9	25.7	6
Ether extract	%		0.6	1.6	13
N-free extract	%		24.0	62.6	
Protein (N x 6.25)	%		2.2	5.7	13
Cattle	dig prot %	*	1.0	2.7	
Goats	dig prot %	*	0.7	1.9	
Horses	dig prot %	*	0.9	2.3	
Rabbits	dig prot %	*	1.2	3.0	
Sheep	dig prot %	*	0.9	2.3	
Energy	GE Mcal/kg				
Cattle	DE Mcal/kg	*	1.05	2.74	
Sheep	DE Mcal/kg	*	1.08	2.80	
Cattle	ME Mcal/kg	*	0.86	2.25	
Sheep	ME Mcal/kg	*	0.88	2.30	
Cattle	TDN %	*	23.8	62.1	
Sheep	TDN %	*	24.4	63.5	

Sotol, leaves, fresh, (2)

Ref No 2 04 527 United States

Feed Name or Analyses			Mean As Fed	Mean Dry	C.V. ± %
Dry matter	%		57.2	100.0	
Ash	%		2.7	4.7	43
Crude fiber	%		24.9	43.6	12
Ether extract	%		1.3	2.2	26
N-free extract	%		25.1	43.9	
Protein (N x 6.25)	%		3.2	5.6	23
Cattle	dig prot %	*	1.5	2.7	
Goats	dig prot %	*	1.0	1.8	
Horses	dig prot %	*	1.3	2.3	
Rabbits	dig prot %	*	1.7	3.0	
Sheep	dig prot %	*	1.3	2.2	
Carotene	mg/kg		24.1	42.1	
Vitamin A equivalent	IU/g		40.2	70.2	

SOTOL, TEXAS. Dasylirion texanum

Sotol, Texas, leaves, fresh, mature, (2)

Ref No 2 04 528 United States

Feed Name or Analyses			Mean As Fed	Mean Dry	C.V. ± %
Dry matter	%			100.0	
Ash	%			3.0	
Crude fiber	%			39.0	
Ether extract	%			2.2	
N-free extract	%			48.7	
Protein (N x 6.25)	%			7.1	
Cattle	dig prot %	*		3.9	
Goats	dig prot %	*		3.2	
Horses	dig prot %	*		3.6	
Rabbits	dig prot %	*		4.2	
Sheep	dig prot %	*		3.6	
Calcium	%			0.69	
Magnesium	%			0.11	
Phosphorus	%			0.08	
Potassium	%			0.76	

SOTOL, WHEELER. Dasylirion wheeleri

Sotol, wheeler, leaves, fresh, mature, (2)

Ref No 2 04 529 United States

Feed Name or Analyses			Mean As Fed	Mean Dry	C.V. ± %
Dry matter	%		60.9	100.0	
Ash	%		4.7	7.7	
Protein (N x 6.25)	%		2.5	4.1	
Cattle	dig prot %	*	0.8	1.4	
Goats	dig prot %	*	0.2	0.4	
Horses	dig prot %	*	0.6	1.0	
Rabbits	dig prot %	*	1.1	1.8	
Sheep	dig prot %	*	0.5	0.8	
Calcium	%		0.88	1.44	
Chlorine	%		0.30	0.50	
Iron	%		0.008	0.013	
Magnesium	%		0.43	0.71	
Manganese	mg/kg		141.5	232.4	
Phosphorus	%		0.16	0.26	
Potassium	%		0.99	1.63	
Sodium	%		0.04	0.07	
Sulphur	%		0.17	0.28	

Sotol, wheeler, stems, fresh, (2)

Ref No 2 04 530 United States

Feed Name or Analyses			Mean As Fed	Mean Dry	C.V. ± %
Dry matter	%		32.8	100.0	
Ash	%		3.8	11.6	
Protein (N x 6.25)	%		1.2	3.7	
Cattle	dig prot %	*	0.3	1.0	
Goats	dig prot %	*	0.0	0.0	
Horses	dig prot %	*	0.2	0.7	
Rabbits	dig prot %	*	0.5	1.5	
Sheep	dig prot %	*	0.1	0.4	
Calcium	%		0.29	0.89	
Chlorine	%		0.04	0.11	
Iron	%		0.006	0.019	
Magnesium	%		0.31	0.94	
Manganese	mg/kg		101.6	309.7	
Phosphorus	%		0.07	0.21	
Potassium	%		0.69	2.10	
Sodium	%		0.03	0.09	
Sulphur	%		0.04	0.13	

Sotol, Wheeler, seeds w hulls, (5)

Ref No 5 09 019 United States

Feed Name or Analyses			Mean As Fed	Mean Dry	C.V. ± %
Dry matter	%			100.0	
Protein (N x 6.25)	%			50.0	
Alanine	%			1.85	
Arginine	%			5.75	
Aspartic acid	%			5.35	
Glutamic acid	%			7.50	
Glycine	%			1.90	
Histidine	%			1.50	
Hydroxyproline	%			0.05	
Isoleucine	%			1.50	
Leucine	%			2.30	
Lysine	%			1.80	
Methionine	%			0.80	
Phenylalanine	%			1.55	
Proline	%			1.50	
Serine	%			1.90	
Threonine	%			1.45	
Tyrosine	%			1.30	
Valine	%			1.90	

SOYBEAN. Glycine max

Soybean, hay, dehy, (1)

Ref No 1 04 535 United States

Feed Name or Analyses			Mean As Fed	Mean Dry	C.V. ± %
Dry matter	%		89.2	100.0	2
Ash	%		7.3	8.2	12
Crude fiber	%		25.3	28.4	7
Cattle	dig coef %		40.	40.	
Ether extract	%		2.0	2.3	18
Cattle	dig coef %		55.	55.	
N-free extract	%		40.6	45.6	
Cattle	dig coef %		72.	72.	
Protein (N x 6.25)	%		13.9	15.6	8
Cattle	dig coef %		59.	59.	
Cattle	dig prot %		8.2	9.2	
Goats	dig prot %	*	9.9	11.1	
Horses	dig prot %	*	9.6	10.8	
Rabbits	dig prot %	*	9.6	10.7	
Sheep	dig prot %	*	9.4	10.6	
Energy	GE Mcal/kg				
Cattle	DE Mcal/kg	*	2.21	2.48	
Sheep	DE Mcal/kg	*	2.33	2.62	
Cattle	ME Mcal/kg	*	1.81	2.03	
Sheep	ME Mcal/kg	*	1.91	2.15	
Cattle	TDN %		50.1	56.1	
Sheep	TDN %	*	52.9	59.3	

Soybean, hay, dehy, dough stage, (1)

Ref No 1 04 534 United States

Feed Name or Analyses			Mean As Fed	Mean Dry	C.V. ± %
Dry matter	%		86.4	100.0	
Ash	%		7.0	8.1	
Crude fiber	%		18.7	21.7	
Cattle	dig coef %		45.	45.	
Ether extract	%		1.7	2.0	
Cattle	dig coef %		51.	51.	
N-free extract	%		43.2	50.0	
Cattle	dig coef %		80.	80.	
Protein (N x 6.25)	%		15.7	18.2	
Cattle	dig coef %		71.	71.	
Cattle	dig prot %		11.2	12.9	
Goats	dig prot %	*	11.7	13.5	
Horses	dig prot %	*	11.2	13.0	
Rabbits	dig prot %	*	11.0	12.7	
Sheep	dig prot %	*	11.1	12.9	
Energy	GE Mcal/kg				
Cattle	DE Mcal/kg	*	2.48	2.87	
Sheep	DE Mcal/kg	*	2.46	2.84	
Cattle	ME Mcal/kg	*	2.03	2.35	
Sheep	ME Mcal/kg	*	2.01	2.33	
Cattle	TDN %		56.1	65.0	
Sheep	TDN %	*	55.7	64.5	

Soybean, hay, fan air dried, (1)

Ref No 1 04 533 United States

Feed Name or Analyses			Mean As Fed	Mean Dry	C.V. ± %
Dry matter	%			100.0	
Carotene	mg/kg			61.3	
Vitamin A equivalent	IU/g			102.2	

Feed Name or Analyses			Mean As Fed	Mean Dry	C.V. ± %

Soybean, hay, fan air dried, gr 1 US,(1)
Ref No 1 04 532 United States

Feed Name or Analyses			As Fed	Dry	C.V. ± %
Dry matter	%			100.0	
Vitamin D_2	IU/g			0.6	

Soybean, hay, s-c, (1)
Ref No 1 04 558 United States

Feed Name or Analyses			As Fed	Dry	C.V. ± %
Dry matter	%		88.9	100.0	1
Ash	%		7.1	8.0	
Crude fiber	%		33.2	37.4	
Cattle	dig coef %		47.	47.	
Sheep	dig coef %		46.	46.	
Ether extract	%		2.0	2.2	
Cattle	dig coef %		36.	36.	
Sheep	dig coef %		57.	57.	
N-free extract	%		33.6	37.7	
Cattle	dig coef %		61.	61.	
Sheep	dig coef %		63.	63.	
Protein (N x 6.25)	%		13.1	14.7	
Cattle	dig coef %		62.	62.	
Sheep	dig coef %		72.	72.	
Cattle	dig prot %		8.1	9.1	
Goats	dig prot %	*	9.1	10.3	
Horses	dig prot %	*	8.9	10.0	
Rabbits	dig prot %	*	8.9	10.0	
Sheep	dig prot %		9.4	10.6	
Hemicellulose	%		11.7	13.2	
Lignin (Ellis)	%		12.2	13.7	
Energy	GE Mcal/kg				
Cattle	DE Mcal/kg	*	2.02	2.27	
Sheep	DE Mcal/kg	*	2.12	2.39	
Cattle	ME Mcal/kg	*	1.66	1.86	
Sheep	ME Mcal/kg	*	1.74	1.96	
Cattle	NE_m Mcal/kg	*	0.97	1.11	
Cattle	NE_{gain} Mcal/kg	*	0.32	0.36	
Cattle	$NE_{lactating\ cows}$ Mcal/kg	*	1.02	1.15	
Cattle	TDN %		45.8	51.5	
Sheep	TDN %		48.2	54.2	
Carotene	mg/kg		31.8	35.7	90
α-tocopherol	mg/kg		26.3	29.5	
Vitamin A equivalent	IU/g		52.9	59.5	
Vitamin D_2	IU/g		0.7	0.8	63

Soybean, hay, s-c, immature, (1)
Ref No 1 04 536 United States

Feed Name or Analyses			As Fed	Dry	C.V. ± %
Dry matter	%		91.6	100.0	1
Ash	%		8.4	9.2	24
Crude fiber	%		24.5	26.8	24
Cattle	dig coef %	*	47.	47.	
Sheep	dig coef %	*	39.	39.	
Ether extract	%		2.7	3.0	59
Cattle	dig coef %	*	36.	36.	
Sheep	dig coef %	*	54.	54.	
N-free extract	%		37.3	40.7	
Cattle	dig coef %	*	61.	61.	
Sheep	dig coef %	*	65.	65.	
Protein (N x 6.25)	%		18.6	20.3	15
Cattle	dig coef %	*	62.	62.	
Sheep	dig coef %	*	69.	69.	
Cattle	dig prot %		11.5	12.6	
Goats	dig prot %	*	14.2	15.5	
Horses	dig prot %	*	13.5	14.8	
Rabbits	dig prot %	*	13.1	14.3	
Sheep	dig prot %		12.8	14.0	
Energy	GE Mcal/kg				
Cattle	DE Mcal/kg	*	2.12	2.31	
Sheep	DE Mcal/kg	*	2.20	2.41	
Cattle	ME Mcal/kg	*	1.74	1.90	
Sheep	ME Mcal/kg	*	1.81	1.97	
Cattle	TDN %		48.0	52.4	
Sheep	TDN %		50.0	54.6	
Calcium	%		1.02	1.11	13
Phosphorus	%		0.21	0.23	34
Carotene	mg/kg		127.6	139.3	
Vitamin A equivalent	IU/g		212.8	232.3	

Soybean, hay, s-c, early bloom, (1)
Ref No 1 04 537 United States

Feed Name or Analyses			As Fed	Dry	C.V. ± %
Dry matter	%		89.4	100.0	1
Ash	%		8.9	10.0	22
Crude fiber	%		23.1	25.9	24
Sheep	dig coef %	*	39.	39.	
Ether extract	%		3.3	3.7	47
Sheep	dig coef %	*	54.	54.	
N-free extract	%		37.2	41.7	
Sheep	dig coef %	*	65.	65.	
Protein (N x 6.25)	%		16.8	18.8	9
Sheep	dig coef %	*	69.	69.	
Cattle	dig prot %	*	11.8	13.3	
Goats	dig prot %	*	12.6	14.1	
Horses	dig prot %	*	12.1	13.5	
Rabbits	dig prot %	*	11.8	13.2	
Sheep	dig prot %		11.6	13.0	
Fatty acids	%		3.4	3.8	
Energy	GE Mcal/kg				
Cattle	DE Mcal/kg	*	2.49	2.79	
Sheep	DE Mcal/kg	*	2.15	2.41	
Cattle	ME Mcal/kg	*	2.04	2.29	
Sheep	ME Mcal/kg	*	1.77	1.98	
Cattle	TDN %	*	56.6	63.3	
Sheep	TDN %		48.8	54.6	
Calcium	%		1.31	1.47	
Magnesium	%		0.65	0.73	
Phosphorus	%		0.35	0.39	
Sulphur	%		0.16	0.18	

Soybean, hay, s-c, mid-bloom, (1)
Ref No 1 04 538 United States

Feed Name or Analyses			As Fed	Dry	C.V. ± %
Dry matter	%		93.6	100.0	3
Ash	%		8.2	8.8	28
Crude fiber	%		27.9	29.8	8
Cattle	dig coef %	*	47.	47.	
Sheep	dig coef %	*	39.	39.	
Ether extract	%		5.1	5.4	40
Cattle	dig coef %	*	36.	36.	
Sheep	dig coef %	*	54.	54.	
N-free extract	%	*	35.8	38.2	
Cattle	dig coef %	*	61.	61.	
Sheep	dig coef %	*	65.	65.	
Protein (N x 6.25)	%		16.7	17.8	11
Cattle	dig coef %	*	62.	62.	
Sheep	dig coef %	*	69.	69.	
Cattle	dig prot %		10.3	11.0	
Goats	dig prot %	*	12.3	13.2	
Horses	dig prot %	*	11.8	12.6	
Rabbits	dig prot %	*	11.6	12.4	
Sheep	dig prot %		11.5	12.3	
Energy	GE Mcal/kg				
Cattle	DE Mcal/kg	*	2.18	2.32	
Sheep	DE Mcal/kg	*	2.28	2.44	
Cattle	ME Mcal/kg	*	1.78	1.91	
Sheep	ME Mcal/kg	*	1.87	2.00	
Cattle	TDN %		49.3	52.7	
Sheep	TDN %		51.8	55.3	
Carotene	mg/kg		31.2	33.3	31
Vitamin A equivalent	IU/g		51.9	55.5	

Soybean, hay, s-c, full bloom, (1)
Ref No 1 04 539 United States

Feed Name or Analyses			As Fed	Dry	C.V. ± %
Dry matter	%		88.3	100.0	4
Cattle	dig coef %		57.	57.	
Ash	%		7.6	8.6	
Crude fiber	%		33.0	37.4	
Cattle	dig coef %	*	47.	47.	
Sheep	dig coef %		48.	48.	
Ether extract	%		2.1	2.3	
Cattle	dig coef %	*	36.	36.	
Sheep	dig coef %		47.	47.	
N-free extract	%		33.2	37.6	
Cattle	dig coef %	*	61.	61.	
Sheep	dig coef %		70.	70.	
Protein (N x 6.25)	%		12.4	14.0	
Cattle	dig coef %	*	66.	66.	
Sheep	dig coef %		74.	74.	
Cattle	dig prot %		8.2	9.2	
Goats	dig prot %	*	8.5	9.6	
Horses	dig prot %	*	8.3	9.4	
Rabbits	dig prot %	*	8.4	9.5	
Sheep	dig prot %		9.1	10.4	
Cellulose (Matrone)	%		33.4	37.8	
Cattle	dig coef %		55.	55.	
Energy	GE Mcal/kg		3.73	4.22	
Cattle	GE dig coef %		58.	58.	
Cattle	DE Mcal/kg		2.16	2.45	
Sheep	DE Mcal/kg	*	2.22	2.52	
Cattle	ME Mcal/kg	*	1.77	2.01	
Sheep	ME Mcal/kg	*	1.82	2.07	
Cattle	TDN %		45.6	51.7	
Sheep	TDN %		50.4	57.1	
Sodium	%		0.02	0.02	24
Sulphur	%		0.16	0.18	

Soybean, hay, s-c, late bloom, (1)
Ref No 1 04 540 United States

Feed Name or Analyses			As Fed	Dry	C.V. ± %
Dry matter	%		85.3	100.0	
Ash	%		6.8	8.0	
Crude fiber	%		26.4	30.9	
Cattle	dig coef %		60.	60.	
Ether extract	%		2.5	3.0	
Cattle	dig coef %		40.	40.	
N-free extract	%		34.5	40.4	
Cattle	dig coef %		66.	66.	
Protein (N x 6.25)	%		15.1	17.7	
Cattle	dig coef %		72.	72.	
Cattle	dig prot %		10.9	12.7	
Goats	dig prot %	*	11.1	13.1	
Horses	dig prot %	*	10.7	12.6	
Rabbits	dig prot %	*	10.5	12.3	
Sheep	dig prot %	*	10.6	12.4	
Fatty acids	%		2.3	2.7	

(1) dry forages and roughages (2) pasture, range plants, and forages fed green (3) silages (4) energy feeds (5) protein supplements (6) minerals (7) vitamins (8) additives

Feed Name or Analyses			Mean As Fed	Mean Dry	C.V. ± %
Energy	GE Mcal/kg				
Cattle	DE Mcal/kg	*	2.28	2.67	
Sheep	DE Mcal/kg	*	2.16	2.54	
Cattle	ME Mcal/kg	*	1.87	2.19	
Sheep	ME Mcal/kg	*	1.77	2.08	
Cattle	TDN %		51.7	60.6	
Sheep	TDN %	*	49.0	57.5	
Calcium	%		1.20	1.41	
Phosphorus	%		0.24	0.28	
Potassium	%		0.87	1.02	

Soybean, hay, s-c, milk stage, (1)
Ref No 1 04 541 United States

Feed Name or Analyses			Mean As Fed	Mean Dry	C.V. ± %
Dry matter	%		87.4	100.0	4
Ash	%		7.7	8.8	16
Crude fiber	%		32.4	37.1	3
Sheep	dig coef %		58.	58.	
Ether extract	%		2.4	2.8	10
Sheep	dig coef %		14.	14.	
N-free extract	%		30.2	34.6	
Sheep	dig coef %		61.	61.	
Protein (N x 6.25)	%		14.6	16.8	6
Sheep	dig coef %		64.	64.	
Cattle	dig prot %	*	10.0	11.4	
Goats	dig prot %	*	10.7	12.2	
Horses	dig prot %	*	10.3	11.8	
Rabbits	dig prot %	*	10.1	11.6	
Sheep	dig prot %		9.4	10.7	
Energy	GE Mcal/kg				
Cattle	DE Mcal/kg	*	2.11	2.42	
Sheep	DE Mcal/kg	*	2.09	2.39	
Cattle	ME Mcal/kg	*	1.73	1.98	
Sheep	ME Mcal/kg	*	1.71	1.96	
Cattle	TDN %	*	47.9	54.8	
Sheep	TDN %		47.4	54.2	

Soybean, hay, s-c, dough stage, (1)
Ref No 1 04 542 United States

Feed Name or Analyses			Mean As Fed	Mean Dry	C.V. ± %
Dry matter	%		87.6	100.0	2
Ash	%		6.0	6.8	20
Crude fiber	%		25.0	28.5	8
Cattle	dig coef %		49.	49.	
Ether extract	%		3.6	4.1	32
Cattle	dig coef %		26.	26.	
N-free extract	%		38.3	43.8	
Cattle	dig coef %		73.	73.	
Protein (N x 6.25)	%		14.7	16.8	8
Cattle	dig coef %		72.	72.	
Cattle	dig prot %		10.6	12.1	
Goats	dig prot %	*	10.7	12.3	
Horses	dig prot %	*	10.3	11.8	
Rabbits	dig prot %	*	10.2	11.7	
Sheep	dig prot %	*	10.2	11.7	
Fatty acids	%		6.6	7.5	
Energy	GE Mcal/kg				
Cattle	DE Mcal/kg	*	2.33	2.66	
Sheep	DE Mcal/kg	*	2.28	2.60	
Cattle	ME Mcal/kg	*	1.91	2.18	
Sheep	ME Mcal/kg	*	1.87	2.13	
Cattle	TDN %		52.9	60.4	
Sheep	TDN %	*	51.7	59.0	
Calcium	%		1.07	1.22	6
Phosphorus	%		0.29	0.33	13
Potassium	%		0.85	0.97	

Soybean, hay, s-c, mature, (1)
Ref No 1 04 543 United States

Feed Name or Analyses			Mean As Fed	Mean Dry	C.V. ± %
Dry matter	%		91.1	100.0	2
Ash	%		6.6	7.3	31
Crude fiber	%		32.6	35.8	20
Cattle	dig coef %	*	47.	47.	
Sheep	dig coef %	*	39.	39.	
Ether extract	%		3.5	3.9	40
Cattle	dig coef %	*	36.	36.	
Sheep	dig coef %	*	54.	54.	
N-free extract	%		35.2	38.7	
Cattle	dig coef %	*	61.	61.	
Sheep	dig coef %	*	65.	65.	
Protein (N x 6.25)	%		13.1	14.4	18
Cattle	dig coef %	*	62.	62.	
Sheep	dig coef %	*	69.	69.	
Cattle	dig prot %		8.1	8.9	
Goats	dig prot %	*	9.1	10.0	
Horses	dig prot %	*	8.9	9.7	
Rabbits	dig prot %	*	8.9	9.8	
Sheep	dig prot %		9.0	9.9	
Fatty acids	%		4.9	5.3	
Energy	GE Mcal/kg				
Cattle	DE Mcal/kg	*	2.11	2.31	
Sheep	DE Mcal/kg	*	2.16	2.37	
Cattle	ME Mcal/kg	*	1.73	1.90	
Sheep	ME Mcal/kg	*	1.77	1.94	
Cattle	TDN %		47.8	52.5	
Sheep	TDN %		48.9	53.7	
Calcium	%		0.95	1.04	
Phosphorus	%		0.25	0.28	
Potassium	%		0.73	0.81	
Carotene	mg/kg		6.6	7.3	42
Vitamin A equivalent	IU/g		11.0	12.1	

Soybean, hay, s-c, over ripe, (1)
Ref No 1 04 544 United States

Feed Name or Analyses			Mean As Fed	Mean Dry	C.V. ± %
Dry matter	%		89.0	100.0	1
Ash	%		7.2	8.1	12
Crude fiber	%		41.0	46.1	7
Cattle	dig coef %	*	47.	47.	
Sheep	dig coef %	*	39.	39.	
Ether extract	%		1.2	1.3	13
Cattle	dig coef %	*	36.	36.	
Sheep	dig coef %	*	54.	54.	
N-free extract	%		30.4	34.2	
Cattle	dig coef %	*	61.	61.	
Sheep	dig coef %	*	65.	65.	
Protein (N x 6.25)	%		9.2	10.3	18
Cattle	dig coef %	*	62.	62.	
Sheep	dig coef %	*	69.	69.	
Cattle	dig prot %		5.7	6.4	
Goats	dig prot %	*	5.5	6.2	
Horses	dig prot %	*	5.6	6.3	
Rabbits	dig prot %	*	5.9	6.6	
Sheep	dig prot %		6.3	7.1	
Energy	GE Mcal/kg				
Cattle	DE Mcal/kg	*	1.96	2.20	
Sheep	DE Mcal/kg	*	1.92	2.16	
Cattle	ME Mcal/kg	*	1.61	1.81	
Sheep	ME Mcal/kg	*	1.57	1.77	
Cattle	TDN %	*	44.5	50.0	
Sheep	TDN %	*	43.5	48.9	

Soybean, hay, s-c, cut 1, (1)
Ref No 1 04 545 United States

Feed Name or Analyses			Mean As Fed	Mean Dry	C.V. ± %
Dry matter	%		92.9	100.0	
Ash	%		6.6	7.1	
Crude fiber	%		33.5	36.1	
Ether extract	%		2.0	2.2	
N-free extract	%		37.7	40.6	
Protein (N x 6.25)	%		13.0	14.0	
Cattle	dig prot %	*	8.4	9.1	
Goats	dig prot %	*	8.9	9.6	
Horses	dig prot %	*	8.7	9.4	
Rabbits	dig prot %	*	8.8	9.5	
Sheep	dig prot %	*	8.5	9.1	
Energy	GE Mcal/kg				
Cattle	DE Mcal/kg	*	2.19	2.36	
Sheep	DE Mcal/kg	*	2.25	2.42	
Cattle	ME Mcal/kg	*	1.80	1.93	
Sheep	ME Mcal/kg	*	1.84	1.99	
Cattle	TDN %	*	49.7	53.5	
Sheep	TDN %	*	51.0	54.9	
Calcium	%		1.00	1.08	
Phosphorus	%		0.24	0.26	
Potassium	%		1.15	1.24	

Soybean, hay, s-c, cut 2, (1)
Ref No 1 04 546 United States

Feed Name or Analyses			Mean As Fed	Mean Dry	C.V. ± %
Dry matter	%		91.1	100.0	2
Ash	%		6.9	7.6	50
Crude fiber	%		30.0	32.9	9
Ether extract	%		2.7	3.0	68
N-free extract	%		38.9	42.7	
Protein (N x 6.25)	%		12.6	13.8	36
Cattle	dig prot %	*	8.1	8.9	
Goats	dig prot %	*	8.6	9.4	
Horses	dig prot %	*	8.4	9.2	
Rabbits	dig prot %	*	8.5	9.3	
Sheep	dig prot %	*	8.2	8.9	
Energy	GE Mcal/kg				
Cattle	DE Mcal/kg	*	2.28	2.50	
Sheep	DE Mcal/kg	*	2.25	2.47	
Cattle	ME Mcal/kg	*	1.87	2.05	
Sheep	ME Mcal/kg	*	1.84	2.02	
Cattle	TDN %	*	51.7	56.7	
Sheep	TDN %	*	50.9	55.9	

Soybean, hay, s-c, gr 1 US, (1)
Ref No 1 04 548 United States

Feed Name or Analyses			Mean As Fed	Mean Dry	C.V. ± %
Dry matter	%		90.0	100.0	
Ash	%		8.3	9.2	
Crude fiber	%		33.3	37.0	
Ether extract	%		2.5	2.8	
N-free extract	%		31.6	35.1	
Protein (N x 6.25)	%		14.3	15.9	
Cattle	dig prot %	*	9.6	10.7	
Goats	dig prot %	*	10.3	11.4	
Horses	dig prot %	*	9.9	11.0	
Rabbits	dig prot %	*	9.9	10.9	
Sheep	dig prot %	*	9.7	10.8	
Energy	GE Mcal/kg				
Cattle	DE Mcal/kg	*	2.19	2.44	
Sheep	DE Mcal/kg	*	2.07	2.30	
Cattle	ME Mcal/kg	*	1.80	2.00	
Sheep	ME Mcal/kg	*	1.70	1.89	

Continued

Feed Name or Analyses			Mean As Fed	Mean Dry	C.V. ± %
Cattle	TDN %	*	49.8	55.3	
Sheep	TDN %	*	47.0	52.3	
Carotene	mg/kg		70.8	78.7	
Vitamin A equivalent	IU/g		118.1	131.2	
Vitamin D2	IU/g		1.5	1.7	

Soybean, hay, s-c, gr 2 US, (1)
Ref No 1 04 549 United States

Feed Name or Analyses			Mean As Fed	Mean Dry	C.V. ± %
Dry matter	%			100.0	
Ash	%			6.8	
Crude fiber	%			34.0	
Ether extract	%			1.6	
N-free extract	%			43.9	
Protein (N x 6.25)	%			13.7	
Cattle	dig prot %	*		8.8	
Goats	dig prot %	*		9.3	
Horses	dig prot %	*		9.2	
Rabbits	dig prot %	*		9.2	
Sheep	dig prot %	*		8.9	
Energy	GE Mcal/kg			4.23	
Cattle	DE Mcal/kg	*		2.38	
Sheep	DE Mcal/kg	*		2.52	
Cattle	ME Mcal/kg	*		1.95	
Sheep	ME Mcal/kg	*		2.06	
Cattle	TDN %	*		53.9	
Sheep	TDN %	*		57.0	

Soybean, hay, s-c, gr 3 US, (1)
Ref No 1 04 550 United States

Feed Name or Analyses			Mean As Fed	Mean Dry	C.V. ± %
Dry matter	%			100.0	
Ash	%			8.6	
Crude fiber	%			29.7	
Ether extract	%			2.7	
N-free extract	%			42.9	
Protein (N x 6.25)	%			16.1	
Cattle	dig prot %	*		10.9	
Goats	dig prot %	*		11.6	
Horses	dig prot %	*		11.2	
Rabbits	dig prot %	*		11.1	
Sheep	dig prot %	*		11.0	
Energy	GE Mcal/kg				
Cattle	DE Mcal/kg	*		2.60	
Sheep	DE Mcal/kg	*		2.55	
Cattle	ME Mcal/kg	*		2.13	
Sheep	ME Mcal/kg	*		2.09	
Cattle	TDN %	*		59.0	
Sheep	TDN %	*		57.8	

Soybean, hay, s-c, mn 29% fiber, (1)
Ref No 1 04 551 United States

Feed Name or Analyses			Mean As Fed	Mean Dry	C.V. ± %
Dry matter	%		84.8	100.0	
Ash	%		5.8	6.8	
Crude fiber	%		31.7	37.4	
Sheep	dig coef %		53.	53.	
Ether extract	%		2.8	3.3	
Sheep	dig coef %		30.	30.	
N-free extract	%		28.5	33.6	
Sheep	dig coef %		66.	66.	
Protein (N x 6.25)	%		16.0	18.9	
Sheep	dig coef %		69.	69.	
Cattle	dig prot %	*	11.3	13.3	
Goats	dig prot %	*	12.0	14.2	
Horses	dig prot %	*	11.5	13.6	
Rabbits	dig prot %	*	11.2	13.3	
Sheep	dig prot %		11.1	13.0	
Energy	GE Mcal/kg				
Cattle	DE Mcal/kg	*	1.98	2.33	
Sheep	DE Mcal/kg	*	2.14	2.52	
Cattle	ME Mcal/kg	*	1.62	1.91	
Sheep	ME Mcal/kg	*	1.76	2.07	
Cattle	TDN %	*	44.9	52.9	
Sheep	TDN %		48.6	57.3	

(1) dry forages and roughages (2) pasture, range plants, and forages fed green (3) silages (4) energy feeds (5) protein supplements (6) minerals (7) vitamins (8) additives

Soybean, hay, s-c, mn 35% fiber, (1)
Ref No 1 04 552 United States

Feed Name or Analyses			Mean As Fed	Mean Dry	C.V. ± %
Dry matter	%		93.3	100.0	
Ash	%		7.7	8.2	
Crude fiber	%		42.9	46.0	
Cattle	dig coef %		36.	36.	
Ether extract	%		1.3	1.4	
Cattle	dig coef %		41.	41.	
N-free extract	%		30.5	32.7	
Cattle	dig coef %		58.	58.	
Protein (N x 6.25)	%		10.9	11.7	
Cattle	dig coef %		49.	49.	
Cattle	dig prot %		5.3	5.7	
Goats	dig prot %	*	7.0	7.5	
Horses	dig prot %	*	7.0	7.5	
Rabbits	dig prot %	*	7.2	7.7	
Sheep	dig prot %	*	6.6	7.1	
Energy	GE Mcal/kg				
Cattle	DE Mcal/kg	*	1.75	1.88	
Sheep	DE Mcal/kg	*	2.09	2.24	
Cattle	ME Mcal/kg	*	1.44	1.54	
Sheep	ME Mcal/kg	*	1.71	1.84	
Cattle	TDN %		39.7	42.6	
Sheep	TDN %	*	47.4	50.8	

Soybean, hay, s-c, mx 13% protein, (1)
Ref No 1 04 553 United States

Feed Name or Analyses			Mean As Fed	Mean Dry	C.V. ± %
Dry matter	%		91.7	100.0	
Ash	%		7.7	8.4	
Crude fiber	%		36.1	39.4	
Cattle	dig coef %		45.	45.	
Ether extract	%		1.5	1.6	
Cattle	dig coef %		43.	43.	
N-free extract	%		35.6	38.8	
Cattle	dig coef %		64.	64.	
Protein (N x 6.25)	%		10.8	11.8	
Cattle	dig coef %		52.	52.	
Cattle	dig prot %		5.6	6.1	
Goats	dig prot %	*	6.9	7.6	
Horses	dig prot %	*	6.9	7.5	
Rabbits	dig prot %	*	7.1	7.8	
Sheep	dig prot %	*	6.6	7.2	
Energy	GE Mcal/kg				
Cattle	DE Mcal/kg	*	2.03	2.22	
Sheep	DE Mcal/kg	*	2.18	2.38	
Cattle	ME Mcal/kg	*	1.67	1.82	
Sheep	ME Mcal/kg	*	1.79	1.95	
Cattle	TDN %		46.1	50.2	
Sheep	TDN %	*	49.5	53.9	

Soybean, hay, s-c, 13% protein, (1)
Ref No 1 04 554 United States

Feed Name or Analyses			Mean As Fed	Mean Dry	C.V. ± %
Dry matter	%		92.9	100.0	
Ash	%		7.4	8.0	
Crude fiber	%		35.5	38.2	
Cattle	dig coef %		31.	31.	
Ether extract	%		2.8	3.0	
Cattle	dig coef %		60.	60.	
N-free extract	%		33.1	35.6	
Cattle	dig coef %		59.	59.	
Protein (N x 6.25)	%		14.1	15.2	
Cattle	dig coef %		63.	63.	
Cattle	dig prot %		8.9	9.6	
Goats	dig prot %	*	10.0	10.7	
Horses	dig prot %	*	9.7	10.4	
Rabbits	dig prot %	*	9.7	10.4	
Sheep	dig prot %	*	9.5	10.2	
Energy	GE Mcal/kg				
Cattle	DE Mcal/kg	*	1.90	2.05	
Sheep	DE Mcal/kg	*	2.14	2.30	
Cattle	ME Mcal/kg	*	1.56	1.68	
Sheep	ME Mcal/kg	*	1.75	1.89	
Cattle	TDN %		43.2	46.5	
Sheep	TDN %	*	48.5	52.2	

Soybean, hay, s-c, 15% protein, (1)
Ref No 1 04 555 United States

Feed Name or Analyses			Mean As Fed	Mean Dry	C.V. ± %
Dry matter	%		86.2	100.0	
Ash	%		7.0	8.1	
Crude fiber	%		23.2	26.9	
Cattle	dig coef %		51.	51.	
Ether extract	%		2.2	2.5	
Cattle	dig coef %		43.	43.	
N-free extract	%		37.7	43.7	
Cattle	dig coef %		71.	71.	
Protein (N x 6.25)	%		16.2	18.8	
Cattle	dig coef %		75.	75.	
Cattle	dig prot %		12.2	14.1	
Goats	dig prot %	*	12.2	14.1	
Horses	dig prot %	*	11.6	13.5	
Rabbits	dig prot %	*	11.4	13.2	
Sheep	dig prot %	*	11.6	13.4	
Energy	GE Mcal/kg				
Cattle	DE Mcal/kg	*	2.33	2.70	
Sheep	DE Mcal/kg	*	2.31	2.67	
Cattle	ME Mcal/kg	*	1.91	2.22	
Sheep	ME Mcal/kg	*	1.89	2.19	
Cattle	TDN %		52.8	61.3	
Sheep	TDN %	*	52.3	60.7	

Soybean, hay, s-c, 25% fiber, (1)
Ref No 1 04 556 United States

Feed Name or Analyses			Mean As Fed	Mean Dry	C.V. ± %
Dry matter	%		87.4	100.0	
Ash	%		6.4	7.4	
Crude fiber	%		22.6	25.9	
Cattle	dig coef %		49.	49.	
Sheep	dig coef %		35.	35.	
Ether extract	%		2.8	3.2	
Cattle	dig coef %		41.	41.	
Sheep	dig coef %		62.	62.	
N-free extract	%		39.9	45.7	
Cattle	dig coef %		77.	77.	
Sheep	dig coef %		64.	64.	

Feed Name or Analyses			Mean As Fed	Mean Dry	C.V. ± %
Protein (N x 6.25)	%		15.7	18.0	
Cattle	dig coef %		68.	68.	
Sheep	dig coef %		69.	69.	
Cattle	dig prot %		10.7	12.2	
Goats	dig prot %	*	11.6	13.3	
Horses	dig prot %	*	11.2	12.8	
Rabbits	dig prot %	*	10.9	12.5	
Sheep	dig prot %		10.8	12.4	
Energy	GE Mcal/kg				
Cattle	DE Mcal/kg	*	2.43	2.78	
Sheep	DE Mcal/kg	*	2.12	2.43	
Cattle	ME Mcal/kg	*	1.99	2.28	
Sheep	ME Mcal/kg	*	1.74	1.99	
Cattle	TDN %		55.0	63.0	
Sheep	TDN %		48.2	55.1	

Soybean, hay, s-c, 25-35% fiber, (1)

Ref No 1 04 557 United States

Feed Name or Analyses			Mean As Fed	Mean Dry	C.V. ± %
Dry matter	%		89.0	100.0	
Ash	%		7.2	8.1	
Crude fiber	%		29.4	33.0	
Cattle	dig coef %		47.	47.	
Ether extract	%		3.0	3.4	
Cattle	dig coef %		57.	57.	
N-free extract	%		34.6	38.9	
Cattle	dig coef %		61.	61.	
Protein (N x 6.25)	%		14.8	16.6	
Cattle	dig coef %		71.	71.	
Cattle	dig prot %		10.5	11.8	
Goats	dig prot %	*	10.7	12.0	
Horses	dig prot %	*	10.3	11.6	
Rabbits	dig prot %	*	10.2	11.5	
Sheep	dig prot %	*	10.2	11.4	
Energy	GE Mcal/kg				
Cattle	DE Mcal/kg	*	2.17	2.44	
Sheep	DE Mcal/kg	*	2.18	2.45	
Cattle	ME Mcal/kg	*	1.78	2.00	
Sheep	ME Mcal/kg	*	1.79	2.01	
Cattle	TDN %		49.3	55.4	
Sheep	TDN %	*	49.4	55.5	
Calcium	%		1.11	1.25	
Copper	mg/kg		8.0	9.0	
Iron	%		0.021	0.024	
Magnesium	%		0.60	0.67	
Manganese	mg/kg		106.0	119.1	
Phosphorus	%		0.22	0.25	
Potassium	%		1.10	1.24	
Sodium	%		0.09	0.10	
Sulphur	%		0.26	0.30	

Soybean, hay, s-c, gr sample US, (1)

Ref No 1 04 547 United States

Feed Name or Analyses			Mean As Fed	Mean Dry	C.V. ± %
Dry matter	%			100.0	
Ash	%			11.5	
Crude fiber	%			34.7	
Ether extract	%			1.6	
N-free extract	%			37.7	
Protein (N x 6.25)	%			14.5	
Cattle	dig prot %	*		9.5	
Goats	dig prot %	*		10.1	
Horses	dig prot %	*		9.8	
Rabbits	dig prot %	*		9.9	
Sheep	dig prot %	*		9.6	
Energy	GE Mcal/kg				
Cattle	DE Mcal/kg	*		2.55	
Sheep	DE Mcal/kg	*		2.32	

Feed Name or Analyses			Mean As Fed	Mean Dry	C.V. ± %
Cattle	ME Mcal/kg	*		2.09	
Sheep	ME Mcal/kg	*		1.90	
Cattle	TDN %	*		57.9	
Sheep	TDN %	*		52.7	

Soybean, hay, s-c, grnd, mx 33% fiber, (1)

Ground soybean hay (AAFCO)

Ref No 1 04 559 United States

Feed Name or Analyses			Mean As Fed	Mean Dry	C.V. ± %
Dry matter	%		90.3	100.0	
Ash	%		6.9	7.6	
Crude fiber	%		31.7	35.1	
Cattle	dig coef %		55.	55.	
Ether extract	%		4.7	5.2	
Cattle	dig coef %		69.	69.	
N-free extract	%		31.4	34.8	
Cattle	dig coef %		47.	47.	
Protein (N x 6.25)	%		15.6	17.3	
Cattle	dig coef %		94.	94.	
Cattle	dig prot %		14.7	16.3	
Goats	dig prot %	*	11.5	12.7	
Horses	dig prot %	*	11.0	12.2	
Rabbits	dig prot %	*	10.9	12.0	
Sheep	dig prot %	*	10.9	12.1	
Energy	GE Mcal/kg				
Cattle	DE Mcal/kg	*	2.39	2.65	
Sheep	DE Mcal/kg	*	2.32	2.57	
Cattle	ME Mcal/kg	*	1.96	2.17	
Sheep	ME Mcal/kg	*	1.91	2.11	
Cattle	TDN %		54.2	60.0	
Sheep	TDN %	*	52.6	58.3	

Soybean, hay, s-c, weathered, mature, (1)

Ref No 1 08 522 United States

Feed Name or Analyses			Mean As Fed	Mean Dry	C.V. ± %
Dry matter	%		89.0	100.0	
Ash	%		7.2	8.1	
Crude fiber	%		41.0	46.1	
Ether extract	%		1.2	1.3	
N-free extract	%		30.4	34.2	
Protein (N x 6.25)	%		9.2	10.3	
Cattle	dig prot %	*	5.2	5.9	
Goats	dig prot %	*	5.5	6.2	
Horses	dig prot %	*	5.6	6.3	
Rabbits	dig prot %	*	5.9	6.7	
Sheep	dig prot %	*	5.2	5.8	
Energy	GE Mcal/kg				
Cattle	DE Mcal/kg	*	1.58	1.78	
Sheep	DE Mcal/kg	*	1.80	2.02	
Cattle	ME Mcal/kg	*	1.30	1.46	
Sheep	ME Mcal/kg	*	1.48	1.66	
Cattle	TDN %	*	36.0	40.4	
Sheep	TDN %	*	40.6	45.6	

Soybean, hulls, (1)

Soybean hulls (AAFCO)

Soybran flakes

Ref No 1 04 560 United States

Feed Name or Analyses			Mean As Fed	Mean Dry	C.V. ± %
Dry matter	%		91.6	100.0	1
Organic matter	%		88.0	96.1	
Ash	%		3.8	4.2	17
Crude fiber	%		33.1	36.1	3
Sheep	dig coef %		72.	72.	
Ether extract	%		2.5	2.8	23
Sheep	dig coef %		76.	76.	
N-free extract	%		40.9	44.6	
Sheep	dig coef %		75.	75.	
Protein (N x 6.25)	%		11.3	12.4	6
Sheep	dig coef %		48.	48.	
Cattle	dig prot %	*	7.0	7.6	
Goats	dig prot %	*	7.4	8.1	
Horses	dig prot %	*	7.3	8.0	
Rabbits	dig prot %	*	7.5	8.2	
Sheep	dig prot %		5.5	6.0	
Cellulose (Crampton)	%				
Sheep	dig coef %		71.	71.	
Cellulose (Matrone)	%		47.7	52.1	
Pentosans	%		21.2	23.1	
Lignin (Ellis)	%		6.0	6.5	
Energy	GE Mcal/kg				
Cattle	DE Mcal/kg	*	2.59	2.83	
Sheep	DE Mcal/kg	*	2.83	3.09	
Cattle	ME Mcal/kg	*	2.13	2.32	
Chickens	ME_n kcal/kg		12.	13.	
Sheep	ME Mcal/kg	*	2.32	2.53	
Cattle	NE_m Mcal/kg	*	0.90	0.98	
Cattle	NE_{gain} Mcal/kg	*	0.06	0.07	
Cattle	$NE_{lactating\ cows}$ Mcal/kg	*	0.82	0.89	
Cattle	TDN %	*	58.8	64.2	
Sheep	TDN %		64.1	70.0	
Calcium	%		0.54	0.59	11
Manganese	mg/kg		12.7	13.9	18
Phosphorus	%		0.16	0.17	23
Choline	mg/kg		632.	690.	
Thiamine	mg/kg		1.6	1.8	

Ref No 1 04 560 Canada

Feed Name or Analyses			Mean As Fed	Mean Dry	C.V. ± %
Dry matter	%		92.4	100.0	
Ash	%		4.3	4.7	
Crude fiber	%		37.4	40.5	
Sheep	dig coef %		72.	72.	
Ether extract	%		1.8	1.9	
Sheep	dig coef %		76.	76.	
N-free extract	%		38.9	42.1	
Sheep	dig coef %		75.	75.	
Protein (N x 6.25)	%		10.0	10.8	
Sheep	dig coef %		48.	48.	
Cattle	dig prot %	*	5.8	6.3	
Goats	dig prot %	*	6.2	6.7	
Horses	dig prot %	*	6.2	6.7	
Rabbits	dig prot %	*	6.5	7.0	
Sheep	dig prot %		4.8	5.2	
Energy	GE Mcal/kg				
Cattle	DE Mcal/kg	*	2.57	2.78	
Sheep	DE Mcal/kg	*	2.82	3.05	
Cattle	ME Mcal/kg	*	2.11	2.28	
Sheep	ME Mcal/kg	*	2.31	2.50	
Cattle	TDN %	*	58.3	63.1	
Sheep	TDN %		63.9	69.1	

Soybean, leaves, (1)

Ref No 1 04 561 United States

Feed Name or Analyses			Mean As Fed	Mean Dry	C.V. ± %
Dry matter	%			100.0	
N-free extract	%			40.1	
Protein (N x 6.25)	%			16.7	11
Cattle	dig prot %	*		11.4	
Goats	dig prot %	*		12.1	
Horses	dig prot %	*		11.7	
Rabbits	dig prot %	*		11.6	
Sheep	dig prot %	*		11.5	

Soybean, pods, (1)
Ref No 1 04 564 — United States

Feed Name or Analyses			Mean As Fed	Mean Dry	C.V. ± %
Dry matter	%			100.0	
Calcium	%			0.99	6
Magnesium	%			0.82	6
Phosphorus	%			0.20	25
Potassium	%			0.93	7

Soybean, pods, s-c, (1)
Ref No 1 04 563 — United States

Feed Name or Analyses			Mean As Fed	Mean Dry	C.V. ± %
Dry matter	%		83.2	100.0	
Ash	%		7.8	9.4	
Crude fiber	%		28.0	33.7	
Sheep	dig coef %		51.	51.	
Ether extract	%		1.2	1.5	
Sheep	dig coef %		57.	57.	
N-free extract	%		41.2	49.5	
Sheep	dig coef %		73.	73.	
Protein (N x 6.25)	%		4.9	5.9	
Sheep	dig coef %		44.	44.	
Cattle	dig prot %	*	1.7	2.0	
Goats	dig prot %	*	1.7	2.1	
Horses	dig prot %	*	2.1	2.5	
Rabbits	dig prot %	*	2.7	3.2	
Sheep	dig prot %		2.2	2.6	
Energy	GE Mcal/kg				
Cattle	DE Mcal/kg	*	2.15	2.59	
Sheep	DE Mcal/kg	*	2.12	2.55	
Cattle	ME Mcal/kg	*	1.77	2.12	
Sheep	ME Mcal/kg	*	1.74	2.09	
Cattle	TDN %	*	48.8	58.7	
Sheep	TDN %		48.1	57.8	

Soybean, stems, (1)
Ref No 1 04 565 — United States

Feed Name or Analyses			Mean As Fed	Mean Dry	C.V. ± %
Dry matter	%			100.0	
Niacin	mg/kg			7.9	
Riboflavin	mg/kg			2.6	
Thiamine	mg/kg			1.5	

Soybean, straw, (1)
Ref No 1 04 567 — United States

Feed Name or Analyses			Mean As Fed	Mean Dry	C.V. ± %
Dry matter	%		87.5	100.0	2
Ash	%		5.6	6.4	9
Crude fiber	%		38.8	44.3	5
Sheep	dig coef %		36.	36.	
Ether extract	%		1.3	1.4	35
Sheep	dig coef %		31.	31.	
N-free extract	%		37.3	42.7	
Sheep	dig coef %		58.	58.	
Protein (N x 6.25)	%		4.5	5.2	24
Sheep	dig coef %		29.	29.	
Cattle	dig prot %	*	1.3	1.4	
Goats	dig prot %	*	1.2	1.4	
Horses	dig prot %	*	1.7	1.9	
Rabbits	dig prot %	*	2.3	2.7	
Sheep	dig prot %		1.3	1.5	
Energy	GE Mcal/kg				
Cattle	DE Mcal/kg	*	1.63	1.86	
Sheep	DE Mcal/kg	*	1.67	1.91	
Cattle	ME Mcal/kg	*	1.34	1.53	
Sheep	ME Mcal/kg	*	1.37	1.56	
Cattle	NE_m Mcal/kg	*	0.74	0.85	
Cattle	NE_{gain} Mcal/kg	*	0.00	0.00	
Cattle	$NE_{lactating\ cows}$ Mcal/kg	*	0.56	0.64	
Cattle	TDN %	*		42.1	
Sheep	TDN %		37.8	43.2	
Calcium	%		1.39	1.59	12
Magnesium	%		0.80	0.92	15
Manganese	mg/kg		44.7	51.1	
Phosphorus	%		0.05	0.06	98
Potassium	%		0.49	0.56	14

Soybean, straw, grnd, (1)
Ref No 1 04 566 — United States

Feed Name or Analyses			Mean As Fed	Mean Dry	C.V. ± %
Dry matter	%		89.8	100.0	
Ash	%		5.4	6.0	
Crude fiber	%		39.3	43.8	
Sheep	dig coef %		40.	40.	
Ether extract	%		0.7	0.8	
Sheep	dig coef %		3.	3.	
N-free extract	%		39.2	43.7	
Sheep	dig coef %		66.	66.	
Protein (N x 6.25)	%		5.1	5.7	
Sheep	dig coef %		41.	41.	
Cattle	dig prot %	*	1.7	1.9	
Goats	dig prot %	*	1.7	1.9	
Horses	dig prot %	*	2.1	2.4	
Rabbits	dig prot %	*	2.8	3.1	
Sheep	dig prot %		2.1	2.3	
Energy	GE Mcal/kg				
Cattle	DE Mcal/kg	*	1.75	1.95	
Sheep	DE Mcal/kg	*	1.93	2.15	
Cattle	ME Mcal/kg	*	1.44	1.60	
Sheep	ME Mcal/kg	*	1.58	1.76	
Cattle	TDN %	*	39.8	44.3	
Sheep	TDN %		43.8	48.8	

Soybean, aerial part, fresh, (2)
Ref No 2 04 574 — United States

Feed Name or Analyses			Mean As Fed	Mean Dry	C.V. ± %
Dry matter	%		22.7	100.0	5
Ash	%		2.4	10.5	
Crude fiber	%		6.2	27.3	
Sheep	dig coef %		48.	48.	
Ether extract	%		0.9	4.0	
Sheep	dig coef %		64.	64.	
N-free extract	%		9.1	40.3	
Sheep	dig coef %		75.	75.	
Protein (N x 6.25)	%		4.1	17.9	
Sheep	dig coef %		79.	79.	
Cattle	dig prot %	*	3.0	13.1	
Goats	dig prot %	*	3.0	13.3	
Horses	dig prot %	*	2.9	12.8	
Rabbits	dig prot %	*	2.8	12.5	
Sheep	dig prot %		3.2	14.2	
Energy	GE Mcal/kg				
Cattle	DE Mcal/kg	*	0.68	3.00	
Sheep	DE Mcal/kg	*	0.63	2.78	
Cattle	ME Mcal/kg	*	0.56	2.46	
Sheep	ME Mcal/kg	*	0.52	2.28	
Cattle	TDN %	*	15.5	68.1	
Sheep	TDN %		14.3	63.0	
Calcium	%		0.25	1.08	
Copper	mg/kg		2.1	9.2	
Iron	%		0.005	0.021	
Magnesium	%		0.12	0.54	
Manganese	mg/kg		27.1	119.4	
Phosphorus	%		0.07	0.29	
Potassium	%		0.21	0.92	
Carotene	mg/kg		65.8	290.1	24
α-tocopherol	mg/kg		63.6	280.4	
Vitamin A equivalent	IU/g		109.7	483.6	

Soybean, aerial part, fresh, immature, (2)
Ref No 2 04 568 — United States

Feed Name or Analyses			Mean As Fed	Mean Dry	C.V. ± %
Dry matter	%		20.0	100.0	7
Ash	%		2.0	9.9	24
Crude fiber	%		5.4	27.2	5
Ether extract	%		0.9	4.6	20
N-free extract	%		8.1	40.6	
Protein (N x 6.25)	%		3.5	17.7	9
Cattle	dig prot %	*	2.6	12.9	
Goats	dig prot %	*	2.6	13.1	
Horses	dig prot %	*	2.5	12.6	
Rabbits	dig prot %	*	2.5	12.3	
Sheep	dig prot %	*	2.7	13.5	
Energy	GE Mcal/kg				
Cattle	DE Mcal/kg	*	0.60	2.99	
Sheep	DE Mcal/kg	*	0.55	2.77	
Cattle	ME Mcal/kg	*	0.49	2.45	
Sheep	ME Mcal/kg	*	0.45	2.27	
Cattle	TDN %	*	13.6	67.8	
Sheep	TDN %	*	12.5	62.7	

Soybean, aerial part, fresh, pre-bloom, (2)
Ref No 2 08 523 — United States

Feed Name or Analyses			Mean As Fed	Mean Dry	C.V. ± %
Dry matter	%		20.0	100.0	
Ash	%		1.8	9.0	
Crude fiber	%		5.9	29.5	
Ether extract	%		1.0	5.0	
N-free extract	%		8.1	40.5	
Protein (N x 6.25)	%		3.2	16.0	
Cattle	dig prot %	*	2.3	11.5	
Goats	dig prot %	*	2.3	11.5	
Horses	dig prot %	*	2.2	11.1	
Rabbits	dig prot %	*	2.2	11.0	
Sheep	dig prot %	*	2.4	11.9	
Energy	GE Mcal/kg				
Cattle	DE Mcal/kg	*	0.59	2.96	
Sheep	DE Mcal/kg	*	0.53	2.67	
Cattle	ME Mcal/kg	*	0.49	2.43	
Sheep	ME Mcal/kg	*	0.44	2.19	
Cattle	TDN %	*	13.4	67.1	
Sheep	TDN %	*	12.1	60.5	

Soybean, aerial part, fresh, mid-bloom, (2)
Ref No 2 04 569 — United States

Feed Name or Analyses			Mean As Fed	Mean Dry	C.V. ± %
Dry matter	%		21.3	100.0	
Ash	%		2.3	10.8	
Crude fiber	%		6.4	30.1	
Sheep	dig coef %		47.	47.	
Ether extract	%		0.7	3.3	
Sheep	dig coef %		50.	50.	
N-free extract	%		8.3	39.1	
Sheep	dig coef %		71.	71.	
Protein (N x 6.25)	%		3.6	16.7	
Sheep	dig coef %		77.	77.	

(1) dry forages and roughages (2) pasture, range plants, and forages fed green (3) silages (4) energy feeds (5) protein supplements (6) minerals (7) vitamins (8) additives

Feed Name or Analyses			Mean As Fed	Mean Dry	C.V. ± %
Cattle	dig prot %	*	2.6	12.1	
Goats	dig prot %	*	2.6	12.2	
Horses	dig prot %	*	2.5	11.7	
Rabbits	dig prot %	*	2.5	11.6	
Sheep	dig prot %		2.7	12.9	
Energy	GE Mcal/kg				
Cattle	DE Mcal/kg	*	0.62	2.83	
Sheep	DE Mcal/kg	*	0.55	2.58	
Cattle	ME Mcal/kg	*	0.49	2.32	
Sheep	ME Mcal/kg	*	0.45	2.11	
Cattle	TDN %	*	13.7	64.2	
Sheep	TDN %		12.4	58.5	
Calcium	%		0.37	1.73	
Magnesium	%		0.15	0.72	
Phosphorus	%		0.06	0.29	
Potassium	%		0.22	1.06	
Sulphur	%		0.04	0.19	

Soybean, aerial part, fresh, full bloom, (2)

Ref No 2 04 570 United States

Feed Name or Analyses			Mean As Fed	Mean Dry	C.V. ± %
Dry matter	%		16.2	100.0	
Ash	%		1.8	11.4	
Crude fiber	%		4.4	27.2	
Sheep	dig coef %		58.	58.	
Ether extract	%		0.6	3.9	
Sheep	dig coef %		56.	56.	
N-free extract	%		6.4	39.3	
Sheep	dig coef %		73.	73.	
Protein (N x 6.25)	%		2.9	18.2	
Sheep	dig coef %		75.	75.	
Cattle	dig prot %	*	2.2	13.4	
Goats	dig prot %	*	2.2	13.5	
Horses	dig prot %	*	2.1	13.0	
Rabbits	dig prot %	*	2.1	12.7	
Sheep	dig prot %		2.2	13.7	
Energy	GE Mcal/kg				
Cattle	DE Mcal/kg	*	0.48	2.96	
Sheep	DE Mcal/kg	*	0.45	2.78	
Cattle	ME Mcal/kg	*	0.39	2.43	
Sheep	ME Mcal/kg	*	0.37	2.28	
Cattle	TDN %	*	10.9	67.2	
Sheep	TDN %		10.2	63.0	

Soybean, aerial part, fresh, late bloom, (2)

Ref No 2 04 571 United States

Feed Name or Analyses			Mean As Fed	Mean Dry	C.V. ± %
Dry matter	%		23.3	100.0	
Ash	%		2.4	10.4	
Crude fiber	%		6.2	26.7	
Sheep	dig coef %		50.	50.	
Ether extract	%		1.0	4.2	
Sheep	dig coef %		56.	56.	
N-free extract	%		9.9	42.4	
Sheep	dig coef %		74.	74.	
Protein (N x 6.25)	%		3.8	16.4	
Sheep	dig coef %		77.	77.	
Cattle	dig prot %	*	2.7	11.8	
Goats	dig prot %	*	2.7	11.8	
Horses	dig prot %	*	2.7	11.4	
Rabbits	dig prot %	*	2.6	11.3	
Sheep	dig prot %		2.9	12.6	
Energy	GE Mcal/kg				
Cattle	DE Mcal/kg	*	0.68	2.92	
Sheep	DE Mcal/kg	*	0.64	2.76	
Cattle	ME Mcal/kg	*	0.56	2.39	
Sheep	ME Mcal/kg	*	0.53	2.26	
Cattle	TDN %	*	15.4	66.2	

Feed Name or Analyses			Mean As Fed	Mean Dry	C.V. ± %
Sheep	TDN %		14.5	62.6	
Calcium	%		0.36	1.53	
Phosphorus	%		0.07	0.29	
Potassium	%		0.20	0.87	

Soybean, aerial part, fresh, milk stage, (2)

Ref No 2 04 572 United States

Feed Name or Analyses			Mean As Fed	Mean Dry	C.V. ± %
Dry matter	%		20.8	100.0	10
Ash	%		2.0	9.8	7
Crude fiber	%		5.7	27.5	5
Sheep	dig coef %		43.	43.	
Ether extract	%		0.7	3.4	8
Sheep	dig coef %		42.	42.	
N-free extract	%		9.0	43.2	
Sheep	dig coef %		77.	77.	
Protein (N x 6.25)	%		3.4	16.2	6
Sheep	dig coef %		74.	74.	
Cattle	dig prot %	*	2.4	11.6	
Goats	dig prot %	*	2.4	11.6	
Horses	dig prot %	*	2.3	11.2	
Rabbits	dig prot %	*	2.3	11.1	
Sheep	dig prot %		2.5	12.0	
Energy	GE Mcal/kg				
Cattle	DE Mcal/kg	*	0.59	2.82	
Sheep	DE Mcal/kg	*	0.55	2.66	
Cattle	ME Mcal/kg	*	0.48	2.31	
Sheep	ME Mcal/kg	*	0.45	2.18	
Cattle	TDN %	*	13.3	63.9	
Sheep	TDN %		12.5	60.2	
Calcium	%		0.35	1.71	6
Magnesium	%		0.15	0.72	5
Phosphorus	%		0.06	0.30	16
Potassium	%		0.25	1.19	45
Sulphur	%		0.04	0.19	23
Carotene	mg/kg		22.2	106.9	
Vitamin A equivalent	IU/g		37.0	178.2	

Soybean, aerial part, fresh, dough stage, (2)

Ref No 2 04 573 United States

Feed Name or Analyses			Mean As Fed	Mean Dry	C.V. ± %
Dry matter	%		25.0	100.0	12
Ash	%		2.5	9.8	10
Crude fiber	%		7.3	29.0	5
Sheep	dig coef %		52.	52.	
Ether extract	%		1.3	5.1	10
Sheep	dig coef %		75.	75.	
N-free extract	%		9.6	38.4	
Sheep	dig coef %		73.	73.	
Protein (N x 6.25)	%		4.4	17.7	7
Sheep	dig coef %		80.	80.	
Cattle	dig prot %	*	3.2	12.9	
Goats	dig prot %	*	3.3	13.1	
Horses	dig prot %	*	3.1	12.6	
Rabbits	dig prot %	*	3.1	12.3	
Sheep	dig prot %		3.5	14.1	
Energy	GE Mcal/kg				
Cattle	DE Mcal/kg	*	0.75	2.98	
Sheep	DE Mcal/kg	*	0.72	2.89	
Cattle	ME Mcal/kg	*	0.61	2.44	
Sheep	ME Mcal/kg	*	0.59	2.37	
Cattle	TDN %	*	16.9	67.5	
Sheep	TDN %		16.4	65.6	
Calcium	%		0.33	1.31	6
Magnesium	%		0.21	0.83	6
Phosphorus	%		0.08	0.31	21
Potassium	%		0.20	0.79	10

Soybean, stems, fresh, (2)

Ref No 2 04 577 United States

Feed Name or Analyses			Mean As Fed	Mean Dry	C.V. ± %
Dry matter	%			100.0	
N-free extract	%			40.1	
Protein (N x 6.25)	%			8.9	13
Cattle	dig prot %	*		5.5	
Goats	dig prot %	*		4.9	
Horses	dig prot %	*		5.1	
Rabbits	dig prot %	*		5.5	
Sheep	dig prot %	*		5.3	
Calcium	%			0.83	9
Magnesium	%			0.81	8
Phosphorus	%			0.25	16
Potassium	%			0.51	25

Soybean, aerial part, ensiled, (3)

Ref No 3 04 581 United States

Feed Name or Analyses			Mean As Fed	Mean Dry	C.V. ± %
Dry matter	%		27.5	100.0	10
Sheep	dig coef %		52.	52.	
Ash	%		2.8	10.0	13
Crude fiber	%		7.9	28.6	9
Cattle	dig coef %		45.	45.	
Sheep	dig coef %	*	48.	48.	
Ether extract	%		0.7	2.6	12
Cattle	dig coef %		51.	51.	
Sheep	dig coef %	*	72.	72.	
N-free extract	%		11.4	41.3	4
Cattle	dig coef %		66.	66.	
Sheep	dig coef %	*	59.	59.	
Protein (N x 6.25)	%		4.8	17.5	10
Cattle	dig coef %		63.	63.	
Sheep	dig coef %	*	68.	68.	
Cattle	dig prot %		3.0	11.0	
Goats	dig prot %	*	3.3	12.1	
Horses	dig prot %	*	3.3	12.1	
Sheep	dig prot %		3.3	11.8	
Energy	GE Mcal/kg				
Cattle	DE Mcal/kg	*	0.66	2.39	
Sheep	DE Mcal/kg	*	0.65	2.38	
Cattle	ME Mcal/kg	*	0.54	1.96	
Sheep	ME Mcal/kg	*	0.54	1.95	
Cattle	NE_m Mcal/kg	*	0.32	1.16	
Cattle	NE_{gain} Mcal/kg	*	0.12	0.45	
Cattle	$NE_{lactating\ cows}$ Mcal/kg	*	0.34	1.23	
Cattle	TDN %		14.9	54.1	
Sheep	TDN %		14.9	53.9	
Calcium	%		0.38	1.39	6
Phosphorus	%		0.13	0.46	63
Potassium	%		0.26	0.93	
Carotene	mg/kg		21.4	77.6	46
Vitamin A equivalent	IU/g		35.6	129.4	

Soybean, aerial part, ensiled, immature, (3)

Ref No 3 04 578 United States

Feed Name or Analyses			Mean As Fed	Mean Dry	C.V. ± %
Dry matter	%		24.6	100.0	
Carotene	mg/kg		31.7	129.0	
Vitamin A equivalent	IU/g		52.9	215.0	

Feed Name or Analyses				Mean As Fed	Mean Dry	C.V. ± %

Soybean, aerial part, ensiled, full bloom, (3)
Ref No 3 04 579 United States

Feed Name or Analyses				As Fed	Dry	C.V. ± %
Dry matter		%		31.2	100.0	10
Ash		%		2.6	8.4	8
Crude fiber		%		7.6	24.3	7
Ether extract		%		0.9	2.9	17
N-free extract		%		14.1	45.3	
Protein (N x 6.25)		%		6.0	19.1	6
Cattle	dig prot	%	*	4.2	13.6	
Goats	dig prot	%	*	4.2	13.6	
Horses	dig prot	%	*	4.2	13.6	
Sheep	dig prot	%	*	4.2	13.6	
Energy	GE Mcal/kg					
Cattle	DE Mcal/kg		*	0.80	2.56	
Sheep	DE Mcal/kg		*	0.74	2.36	
Cattle	ME Mcal/kg		*	0.65	2.10	
Sheep	ME Mcal/kg		*	0.60	1.94	
Cattle	TDN	%	*	18.1	58.0	
Sheep	TDN	%	*	16.7	53.6	

Soybean, aerial part, ensiled, dough stage, (3)
Ref No 3 04 580 United States

Feed Name or Analyses				As Fed	Dry	C.V. ± %
Dry matter		%		26.0	100.0	7
Ash		%		2.5	9.7	10
Crude fiber		%		7.4	28.4	15
Cattle	dig coef	%		48.	48.	
Ether extract		%		0.6	2.4	14
Cattle	dig coef	%		53.	53.	
N-free extract		%		10.9	42.1	
Cattle	dig coef	%		70.	70.	
Protein (N x 6.25)		%		4.5	17.5	14
Cattle	dig coef	%		71.	71.	
Cattle	dig prot	%		3.2	12.4	
Goats	dig prot	%	*	3.1	12.1	
Horses	dig prot	%	*	3.1	12.1	
Sheep	dig prot	%	*	3.1	12.1	
Energy	GE Mcal/kg					
Cattle	DE Mcal/kg		*	0.67	2.57	
Sheep	DE Mcal/kg		*	0.63	2.42	
Cattle	ME Mcal/kg		*	0.55	2.11	
Sheep	ME Mcal/kg		*	0.52	1.99	
Cattle	TDN	%		15.2	58.3	
Sheep	TDN	%	*	14.3	55.0	

Soybean, aerial part, wilted ensiled, (3)
Ref No 3 04 584 United States

Feed Name or Analyses				As Fed	Dry	C.V. ± %
Dry matter		%		34.9	100.0	10
Ash		%		3.2	9.2	13
Crude fiber		%		9.6	27.5	7
Cattle	dig coef	%	*	45.	45.	
Sheep	dig coef	%	*	55.	55.	
Ether extract		%		0.9	2.6	33
Cattle	dig coef	%	*	51.	51.	
Sheep	dig coef	%	*	72.	72.	
N-free extract		%		14.9	42.7	
Cattle	dig coef	%	*	66.	66.	
Sheep	dig coef	%	*	52.	52.	
Protein (N x 6.25)		%		6.3	18.0	11
Cattle	dig coef	%	*	63.	63.	
Sheep	dig coef	%	*	79.	79.	
Cattle	dig prot	%		3.9	11.3	
Goats	dig prot	%	*	4.4	12.5	
Horses	dig prot	%	*	4.4	12.5	
Sheep	dig prot	%		5.0	14.2	
Energy	GE Mcal/kg					
Cattle	DE Mcal/kg		*	0.84	2.42	
Sheep	DE Mcal/kg		*	0.86	2.46	
Cattle	ME Mcal/kg		*	0.69	1.98	
Sheep	ME Mcal/kg		*	0.70	2.01	
Cattle	TDN	%		19.1	54.8	
Sheep	TDN	%		19.4	55.7	
Calcium		%		0.48	1.38	4
Copper		mg/kg		3.2	9.2	7
Iron		%		0.006	0.018	34
Magnesium		%		0.16	0.45	11
Manganese		mg/kg		45.8	131.1	9
Phosphorus		%		0.12	0.36	14
Potassium		%		0.32	0.92	12
Sodium		%		0.03	0.09	
Sulphur		%		0.10	0.30	16

(1) dry forages and roughages
(2) pasture, range plants, and forages fed green
(3) silages
(4) energy feeds
(5) protein supplements
(6) minerals
(7) vitamins
(8) additives

Soybean, aerial part, wilted ensiled, full bloom, (3)
Ref No 3 04 582 United States

Feed Name or Analyses				As Fed	Dry	C.V. ± %
Dry matter		%		37.1	100.0	6
Ash		%		2.9	7.7	11
Crude fiber		%		8.6	23.3	5
Ether extract		%		0.9	2.4	16
N-free extract		%		17.3	46.7	
Protein (N x 6.25)		%		7.4	19.9	5
Cattle	dig prot	%	*	5.3	14.3	
Goats	dig prot	%	*	5.3	14.3	
Horses	dig prot	%	*	5.3	14.3	
Sheep	dig prot	%	*	5.3	14.3	
Energy	GE Mcal/kg					
Cattle	DE Mcal/kg		*	0.93	2.50	
Sheep	DE Mcal/kg		*	0.86	2.32	
Cattle	ME Mcal/kg		*	0.76	2.05	
Sheep	ME Mcal/kg		*	0.71	1.90	
Cattle	TDN	%	*	21.0	56.6	
Sheep	TDN	%	*	19.5	52.6	

Soybean, aerial part, wilted ensiled, dough stage, (3)
Ref No 3 04 583 United States

Feed Name or Analyses				As Fed	Dry	C.V. ± %
Dry matter		%		34.6	100.0	
Ash		%		2.6	7.6	
Crude fiber		%		8.5	24.5	
Cattle	dig coef	%		46.	46.	
Ether extract		%		0.8	2.4	
Cattle	dig coef	%		56.	56.	
N-free extract		%		15.7	45.5	
Cattle	dig coef	%		55.	55.	
Protein (N x 6.25)		%		6.9	20.0	
Cattle	dig coef	%		70.	70.	
Cattle	dig prot	%		4.8	14.0	
Goats	dig prot	%	*	5.0	14.4	
Horses	dig prot	%	*	5.0	14.4	
Sheep	dig prot	%	*	5.0	14.4	
Energy	GE Mcal/kg					
Cattle	DE Mcal/kg		*	0.81	2.35	
Sheep	DE Mcal/kg		*	0.81	2.33	
Cattle	ME Mcal/kg		*	0.67	1.93	
Sheep	ME Mcal/kg		*	0.66	1.91	
Cattle	TDN	%		18.4	53.3	
Sheep	TDN	%	*	18.3	52.8	

Soybean, aerial part w grnd corn grain added, ensiled, (3)
Ref No 3 04 585 United States

Feed Name or Analyses				As Fed	Dry	C.V. ± %
Dry matter		%		30.6	100.0	
Ash		%		2.3	7.4	
Crude fiber		%		7.2	23.6	
Ether extract		%		1.1	3.7	
N-free extract		%		15.5	50.7	
Protein (N x 6.25)		%		4.5	14.6	
Cattle	dig prot	%	*	2.9	9.5	
Goats	dig prot	%	*	2.9	9.5	
Horses	dig prot	%	*	2.9	9.5	
Sheep	dig prot	%	*	2.9	9.5	
Energy	GE Mcal/kg					
Cattle	DE Mcal/kg		*	0.86	2.80	
Sheep	DE Mcal/kg		*	0.87	2.85	
Cattle	ME Mcal/kg		*	0.70	2.29	
Sheep	ME Mcal/kg		*	0.72	2.34	
Cattle	TDN	%	*	19.4	63.4	
Sheep	TDN	%	*	19.8	64.7	
Calcium		%		0.33	1.09	
Phosphorus		%		0.13	0.43	

Soybean, aerial part w phosphoric acid preservative added, ensiled, (3)
Ref No 3 04 587 United States

Feed Name or Analyses				As Fed	Dry	C.V. ± %
Dry matter		%		22.8	100.0	7
Ash		%		2.6	11.6	15
Crude fiber		%		7.4	32.5	10
Ether extract		%		0.8	3.4	25
N-free extract		%		8.2	35.8	
Protein (N x 6.25)		%		3.8	16.7	8
Cattle	dig prot	%	*	2.6	11.4	
Goats	dig prot	%	*	2.6	11.4	
Horses	dig prot	%	*	2.6	11.4	
Sheep	dig prot	%	*	2.6	11.4	
Energy	GE Mcal/kg					
Cattle	DE Mcal/kg		*	0.58	2.54	
Sheep	DE Mcal/kg		*	0.58	2.54	
Cattle	ME Mcal/kg		*	0.47	2.08	
Sheep	ME Mcal/kg		*	0.47	2.08	
Cattle	TDN	%	*	13.1	57.5	
Sheep	TDN	%	*	13.1	57.6	
Phosphorus		%		0.31	1.36	4

Soybean, aerial part w phosphoric acid preservative added, ensiled, dough stage, (3)
Ref No 3 04 586 United States

Feed Name or Analyses				As Fed	Dry	C.V. ± %
Dry matter		%		25.7	100.0	
Ash		%		2.5	9.9	
Crude fiber		%		7.0	27.2	
Cattle	dig coef	%		43.	43.	
Ether extract		%		0.9	3.5	
Cattle	dig coef	%		66.	66.	
N-free extract		%		10.8	42.0	
Cattle	dig coef	%		73.	73.	
Protein (N x 6.25)		%		4.5	17.6	
Cattle	dig coef	%		71.	71.	
Cattle	dig prot	%		3.2	12.5	
Goats	dig prot	%	*	3.1	12.2	
Horses	dig prot	%	*	3.1	12.2	

Feed Name or Analyses				Mean As Fed	Mean Dry	C.V. ± %
Sheep	dig prot	%	*	3.1	12.2	
Energy	GE	Mcal/kg				
Cattle	DE	Mcal/kg	*	0.67	2.63	
Sheep	DE	Mcal/kg	*	0.63	2.45	
Cattle	ME	Mcal/kg	*	0.55	2.15	
Sheep	ME	Mcal/kg	*	0.51	2.01	
Cattle	TDN	%		15.3	59.6	
Sheep	TDN	%	*	14.2	55.5	

Soybean, aerial part w molasses added, ensiled, (3)

Ref No 3 04 590 United States

Feed Name or Analyses				Mean As Fed	Mean Dry	C.V. ± %
Dry matter		%		28.5	100.0	13
Ash		%		2.6	9.2	14
Crude fiber		%		7.2	25.1	13
Cattle	dig coef	%		45.	45.	
Ether extract		%		1.2	4.4	19
Cattle	dig coef	%		72.	72.	
N-free extract		%		12.6	44.4	
Cattle	dig coef	%		71.	71.	
Protein (N x 6.25)		%		4.8	17.0	14
Cattle	dig coef	%		70.	70.	
Cattle	dig prot	%		3.4	11.9	
Goats	dig prot	%	*	3.3	11.7	
Horses	dig prot	%	*	3.3	11.7	
Sheep	dig prot	%	*	3.3	11.7	
Energy	GE	Mcal/kg				
Cattle	DE	Mcal/kg	*	0.78	2.72	
Sheep	DE	Mcal/kg	*	0.69	2.44	
Cattle	ME	Mcal/kg	*	0.64	2.23	
Sheep	ME	Mcal/kg	*	0.57	2.00	
Cattle	TDN	%		17.6	61.7	
Sheep	TDN	%	*	15.8	55.3	

Soybean, aerial part w molasses added, ensiled, full bloom, (3)

Ref No 3 04 588 United States

Feed Name or Analyses				Mean As Fed	Mean Dry	C.V. ± %
Dry matter		%		29.9	100.0	11
Ash		%		2.5	8.3	5
Crude fiber		%		7.1	23.6	9
Ether extract		%		0.9	3.1	15
N-free extract		%		13.9	46.6	
Protein (N x 6.25)		%		5.5	18.4	3
Cattle	dig prot	%	*	3.9	12.9	
Goats	dig prot	%	*	3.9	12.9	
Horses	dig prot	%	*	3.9	12.9	
Sheep	dig prot	%	*	3.9	12.9	
Energy	GE	Mcal/kg				
Cattle	DE	Mcal/kg	*	0.79	2.63	
Sheep	DE	Mcal/kg	*	0.71	2.36	
Cattle	ME	Mcal/kg	*	0.64	2.15	
Sheep	ME	Mcal/kg	*	0.58	1.94	
Cattle	TDN	%	*	17.8	59.6	
Sheep	TDN	%	*	16.0	53.6	
Calcium		%		0.41	1.36	8
Phosphorus		%		0.11	0.36	15
Carotene		mg/kg		5.3	17.9	
Vitamin A equivalent		IU/g		8.9	29.8	

Soybean, aerial part w molasses added, ensiled, dough stage, (3)

Ref No 3 04 589 United States

Feed Name or Analyses				Mean As Fed	Mean Dry	C.V. ± %
Dry matter		%		27.0	100.0	
Ash		%		2.4	8.8	
Crude fiber		%		6.6	24.5	
Cattle	dig coef	%		48.	48.	
Ether extract		%		1.1	3.9	
Cattle	dig coef	%		66.	66.	
N-free extract		%		12.3	45.6	
Cattle	dig coef	%		72.	72.	
Protein (N x 6.25)		%		4.6	17.2	
Cattle	dig coef	%		68.	68.	
Cattle	dig prot	%		3.2	11.7	
Goats	dig prot	%	*	3.2	11.8	
Horses	dig prot	%	*	3.2	11.8	
Sheep	dig prot	%	*	3.2	11.8	
Energy	GE	Mcal/kg				
Cattle	DE	Mcal/kg	*	0.74	2.74	
Sheep	DE	Mcal/kg	*	0.69	2.57	
Cattle	ME	Mcal/kg	*	0.61	2.24	
Sheep	ME	Mcal/kg	*	0.57	2.11	
Cattle	TDN	%		16.8	62.1	
Sheep	TDN	%	*	15.7	58.3	

Soybean, aerial part w molasses added, wilted ensiled, dough stage, (3)

Ref No 3 04 591 United States

Feed Name or Analyses				Mean As Fed	Mean Dry	C.V. ± %
Dry matter		%		39.6	100.0	
Ash		%		3.0	7.7	
Crude fiber		%		8.8	22.1	
Cattle	dig coef	%		48.	48.	
Ether extract		%		1.0	2.4	
Cattle	dig coef	%		54.	54.	
N-free extract		%		19.0	48.1	
Cattle	dig coef	%		73.	73.	
Protein (N x 6.25)		%		7.8	19.7	
Cattle	dig coef	%		69.	69.	
Cattle	dig prot	%		5.4	13.6	
Goats	dig prot	%	*	5.6	14.1	
Horses	dig prot	%	*	5.6	14.1	
Sheep	dig prot	%	*	5.6	14.1	
Energy	GE	Mcal/kg				
Cattle	DE	Mcal/kg	*	1.09	2.74	
Sheep	DE	Mcal/kg	*	1.03	2.60	
Cattle	ME	Mcal/kg	*	0.89	2.25	
Sheep	ME	Mcal/kg	*	0.84	2.13	
Cattle	TDN	%		24.6	62.2	
Sheep	TDN	%	*	23.3	58.9	

Soybean, flour, mech-extd fine sift, mx 3% fiber, (5)

Soy flour (AAFCO)

Soy grits (AAFCO)

Soy flour, expeller extracted

Soy grits, expeller extracted

Ref No 5 04 592 United States

Feed Name or Analyses				Mean As Fed	Mean Dry	C.V. ± %
Dry matter		%		92.9	100.0	
Ash		%		6.0	6.5	
Crude fiber		%		2.4	2.6	
Ether extract		%		6.7	7.2	
N-free extract		%		29.9	32.2	
Protein (N x 6.25)		%		47.9	51.6	
Energy	GE	Mcal/kg				
Cattle	DE	Mcal/kg	*	3.53	3.80	
Sheep	DE	Mcal/kg	*	3.50	3.76	
Swine	DE	kcal/kg	*	3701.	3984.	
Cattle	ME	Mcal/kg	*	2.90	3.12	
Sheep	ME	Mcal/kg	*	2.87	3.09	
Swine	ME	kcal/kg	*	3168.	3410.	
Cattle	TDN	%	*	80.1	86.2	
Sheep	TDN	%	*	79.3	85.4	
Swine	TDN	%	*	83.9	90.4	

Soybean, flour, solv-extd fine sift, mx 3% fiber, (5)

Solvent extracted soy flour (AAFCO)

Solvent extracted soy grits (AAFCO)

Ref No 5 04 593 United States

Feed Name or Analyses				Mean As Fed	Mean Dry	C.V. ± %
Dry matter		%		92.3	100.0	
Ash		%		6.1	6.6	
Crude fiber		%		2.3	2.5	
Cattle	dig coef	%	*	57.	57.	
Sheep	dig coef	%	*	101.	101.	
Swine	dig coef	%	*	83.	83.	
Ether extract		%		0.8	0.9	
Cattle	dig coef	%	*	38.	38.	
Sheep	dig coef	%	*	52.	52.	
Swine	dig coef	%	*	58.	58.	
N-free extract		%		32.8	35.6	
Cattle	dig coef	%	*	91.	91.	
Sheep	dig coef	%	*	96.	96.	
Swine	dig coef	%	*	91.	91.	
Protein (N x 6.25)		%		50.2	54.5	
Cattle	dig coef	%	*	90.	90.	
Sheep	dig coef	%	*	94.	94.	
Swine	dig coef	%	*	91.	91.	
Cattle	dig prot	%		45.2	49.0	
Sheep	dig prot			47.2	51.2	
Swine	dig prot	%		45.7	49.6	
Energy	GE	Mcal/kg				
Cattle	DE	Mcal/kg	*	3.40	3.68	
Sheep	DE	Mcal/kg	*	3.61	3.92	
Swine	DE	kcal/kg	*	3462.	3753.	
Cattle	ME	Mcal/kg	*	2.79	3.02	
Sheep	ME	Mcal/kg	*	2.96	3.21	
Swine	ME	kcal/kg	*	2943.	3190.	
Cattle	TDN	%		77.1	83.5	
Sheep	TDN	%		82.0	88.9	
Swine	TDN	%		78.5	85.1	
Calcium		%		0.33	0.35	
Chlorine		%		0.20	0.22	
Copper		mg/kg		16.0	17.3	
Iron		%		0.020	0.022	
Magnesium		%		0.20	0.22	
Manganese		mg/kg		31.7	34.4	
Phosphorus		%		0.62	0.67	
Potassium		%		1.89	2.05	
Sodium		%		0.34	0.37	
Sulphur		%		0.41	0.44	
Zinc		mg/kg		19.9	21.6	
Biotin		mg/kg		0.72	0.78	
Carotene		mg/kg		0.4	0.5	
Choline		mg/kg		2233.	2420.	
Niacin		mg/kg		59.5	64.5	
Pantothenic acid		mg/kg		14.0	15.2	
Riboflavin		mg/kg		3.9	4.3	
Thiamine		mg/kg		1.5	1.7	
Vitamin A equivalent		IU/g		0.7	0.8	
Arginine		%		3.07	3.33	
Cystine		%		0.60	0.65	
Histidine		%		0.69	0.75	
Lysine		%		4.17	4.52	
Methionine		%		0.89	0.97	
Phenylalanine		%		1.79	1.94	
Tryptophan		%		0.99	1.08	
Valine		%		0.30	0.32	

Soybean, flour by-product, grnd, mn 13% protein mx 32% fiber, (5)

Soybean mill feed (AAFCO)

Soybean mill feed (CFA)

Ref No 5 04 594 United States

Feed Name or Analyses		Mean As Fed	Mean Dry	C.V. ± %
Dry matter	%	89.4	100.0	1
Ash	%	4.5	5.0	8
Crude fiber	%	33.0	36.9	10
Ether extract	%	1.6	1.8	58
N-free extract	%	37.0	41.4	
Protein (N x 6.25)	%	13.3	14.9	18
Energy	GE Mcal/kg			
Cattle	DE Mcal/kg	* 1.91	2.14	
Sheep	DE Mcal/kg	* 1.96	2.19	
Cattle	ME Mcal/kg	* 1.56	1.75	
Chickens	ME_n kcal/kg	864.	967.	
Sheep	ME Mcal/kg	* 1.61	1.80	
Cattle	TDN %	* 43.3	48.4	
Sheep	TDN %	* 44.4	49.7	
Calcium	%	0.37	0.42	61
Manganese	mg/kg	28.5	31.8	
Phosphorus	%	0.19	0.21	25

Soybean, groats by-product, grnd, mn 11% protein mx 35% fiber, (5)

Soybean mill run (AAFCO)

Ref No 5 04 595 United States

Feed Name or Analyses		Mean As Fed	Mean Dry	C.V. ± %
Dry matter	%	88.0	100.0	
Ash	%	4.5	5.1	
Crude fiber	%	35.8	40.7	
Ether extract	%	1.2	1.4	
N-free extract	%	34.5	39.2	
Protein (N x 6.25)	%	12.0	13.6	
Energy	GE Mcal/kg			
Cattle	DE Mcal/kg	* 1.70	1.93	
Sheep	DE Mcal/kg	* 1.70	1.93	
Cattle	ME Mcal/kg	* 1.39	1.58	
Sheep	ME Mcal/kg	* 1.39	1.58	
Cattle	TDN %	* 38.5	43.8	
Sheep	TDN %	* 38.4	43.7	

Soybean, seeds, (5)

Ref No 5 04 610 United States

Feed Name or Analyses		Mean As Fed	Mean Dry	C.V. ± %
Dry matter	%	90.9	100.0	1
Ash	%	4.9	5.4	5
Crude fiber	%	5.3	5.8	12
Cattle	dig coef %	-125.	-125.	
Sheep	dig coef %	-10.	-10.	
Swine	dig coef %	30.	30.	
Ether extract	%	17.4	19.2	6
Cattle	dig coef %	98.	98.	
Sheep	dig coef %	89.	89.	
Swine	dig coef %	84.	84.	
N-free extract	%	25.4	27.9	9
Cattle	dig coef %	68.	68.	
Sheep	dig coef %	59.	59.	
Swine	dig coef %	102.	102.	
Protein (N x 6.25)	%	37.9	41.7	6
Cattle	dig coef %	90.	90.	
Sheep	dig coef %	90.	90.	
Swine	dig coef %	82.	82.	
Cattle	dig prot %	34.1	37.5	
Sheep	dig prot %	34.1	37.5	
Swine	dig prot %	31.1	34.2	
Energy	GE Mcal/kg	4.07	4.48	
Cattle	DE Mcal/kg	* 3.66	4.03	
Sheep	DE Mcal/kg	* 3.68	4.04	
Swine	DE kcal/kg	*4032.	4436.	
Cattle	ME Mcal/kg	* 3.00	3.30	
Sheep	ME Mcal/kg	* 3.01	3.32	
Swine	ME kcal/kg	*3531.	3885.	
Cattle	NE_m Mcal/kg	* 2.19	2.41	
Cattle	NE_{gain} Mcal/kg	* 1.39	1.53	
Cattle	$NE_{lactating\ cows}$ Mcal/kg	* 2.46	2.71	
Cattle	TDN %	83.1	91.4	
Sheep	TDN %	83.4	91.7	
Swine	TDN %	91.4	100.6	
Calcium	%	0.24	0.27	
Chlorine	%	0.03	0.03	
Copper	mg/kg	15.8	17.4	
Iron	%	0.008	0.009	
Magnesium	%	0.28	0.31	
Manganese	mg/kg	29.8	32.8	
Phosphorus	%	0.58	0.63	
Potassium	%	1.61	1.77	
Sodium	%	0.12	0.13	
Sulphur	%	0.22	0.24	
Niacin	mg/kg	22.2	24.4	
Riboflavin	mg/kg	3.1	3.4	
Thiamine	mg/kg	11.1	12.2	
Vitamin A	IU/g	0.8	0.9	

Soybean, seeds, mech-extd grnd, (5)

Soybean meal (CFA)

Ref No 5 04 601 United States

Feed Name or Analyses		Mean As Fed	Mean Dry	C.V. ± %
Dry matter	%	92.6	100.0	
Ash	%	6.3	6.8	
Crude fiber	%	3.4	3.7	
Ether extract	%	1.1	1.2	
N-free extract	%	34.2	36.9	
Protein (N x 6.25)	%	47.6	51.4	
Lignin (Ellis)	%	0.3	0.3	
Energy	GE Mcal/kg			
Cattle	DE Mcal/kg	* 3.19	3.45	
Sheep	DE Mcal/kg	* 3.34	3.61	
Swine	DE kcal/kg	*3237.	3496.	
Cattle	ME Mcal/kg	* 2.62	2.83	
Sheep	ME Mcal/kg	* 2.74	2.96	
Swine	ME kcal/kg	*2771.	2993.	
Cattle	TDN %	* 72.5	78.2	
Sheep	TDN %	* 75.8	81.9	
Swine	TDN %	* 73.4	79.3	
Calcium	%	0.28	0.30	
Phosphorus	%	2.14	2.31	
Silicon	%	0.03	0.03	

Soybean, seeds, mech-extd grnd, mx 7% fiber, (5)

Soybean meal, mechanical extracted (AAFCO)

Soybean meal, expeller extracted

Soybean meal, hydraulic extracted

Soybean oil meal, expeller extracted

Soybean oil meal, hydraulic extracted

Ref No 5 04 600 United States

Feed Name or Analyses		Mean As Fed	Mean Dry	C.V. ± %
Dry matter	%	90.7	100.0	1
Organic matter	%	84.5	93.1	
Ash	%	6.1	6.7	4
Crude fiber	%	6.1	6.8	17
Cattle	dig coef %	72.	72.	
Sheep	dig coef %	65.	65.	
Ether extract	%	4.7	5.2	24
Cattle	dig coef %	84.	84.	
Sheep	dig coef %	80.	80.	
N-free extract	%	31.4	34.6	7
Cattle	dig coef %	88.	88.	
Sheep	dig coef %	89.	89.	
Protein (N x 6.25)	%	42.4	46.7	4
Cattle	dig coef %	85.	85.	
Sheep	dig coef %	87.	87.	
Cattle	dig prot %	36.0	39.7	
Sheep	dig prot %	36.7	40.4	
Pentosans	%	5.7	6.2	9
Energy	GE Mcal/kg	4.39	4.84	1
Cattle	DE Mcal/kg	* 3.40	3.74	
Sheep	DE Mcal/kg	* 3.39	3.74	
Swine	DE kcal/kg	*3405.	3753.	
Cattle	ME Mcal/kg	* 2.78	3.07	
Chickens	ME_n kcal/kg	2379.	2623.	
Sheep	ME Mcal/kg	* 2.78	3.07	
Swine	ME kcal/kg	*2947.	3249.	
Cattle	NE_m Mcal/kg	* 1.87	2.06	
Cattle	NE_{gain} Mcal/kg	* 1.24	1.37	
Cattle	$NE_{lactating\ cows}$ Mcal/kg	* 2.16	2.38	
Cattle	TDN %	77.0	84.9	
Sheep	TDN %	76.9	84.8	
Swine	TDN %	* 77.2	85.1	
Calcium	%	0.28	0.31	4
Chlorine	%	0.07	0.08	
Cobalt	mg/kg	0.156	0.172	19
Copper	mg/kg	18.2	20.0	28
Iron	%	0.016	0.018	25
Magnesium	%	0.25	0.28	11
Manganese	mg/kg	32.5	35.9	20
Phosphorus	%	0.59	0.65	8
Potassium	%	1.75	1.93	0
Sodium	%	0.24	0.27	75
Sulphur	%	0.33	0.37	
Carotene	mg/kg	0.2	0.2	53
Choline	mg/kg	2653.	2924.	12
Folic acid	mg/kg	6.67	7.35	29
Niacin	mg/kg	31.0	34.2	15
Pantothenic acid	mg/kg	15.0	16.6	34
Riboflavin	mg/kg	3.5	3.9	12
Thiamine	mg/kg	4.0	4.4	84
α-tocopherol	mg/kg	6.6	7.3	15
Vitamin A equivalent	IU/g	0.4	0.4	
Arginine	%	2.90	3.20	10
Cysteine	%	1.01	1.11	5
Cystine	%	0.63	0.70	
Glutamic acid	%	7.59	8.36	9
Glycine	%	2.41	2.65	
Histidine	%	1.11	1.23	6
Isoleucine	%	2.83	3.12	17
Leucine	%	3.64	4.01	5
Lysine	%	2.81	3.09	9
Methionine	%	0.71	0.79	60
Phenylalanine	%	2.12	2.34	5
Serine	%	2.02	2.23	
Threonine	%	1.72	1.90	13
Tryptophan	%	0.59	0.65	11
Tyrosine	%	1.42	1.56	23
Valine	%	2.23	2.45	5

(1) dry forages and roughages (2) pasture, range plants, and forages fed green (3) silages (4) energy feeds (5) protein supplements (6) minerals (7) vitamins (8) additives

Soybean, seeds, solv-extd flaked, (5)
Soybean flakes (CFA)
Ref No 5 04 603 Canada

Feed Name or Analyses		Mean As Fed	Mean Dry	C.V. ± %
Dry matter	%	88.0	100.0	
Crude fiber	%	6.3	7.2	
Protein (N x 6.25)	%	46.7	53.1	

Soybean, seeds, solv-extd grnd, (5)
Soybean meal (CFA)
Ref No 5 04 605 Canada

Feed Name or Analyses		Mean As Fed	Mean Dry	C.V. ± %
Dry matter	%	89.1	100.0	1
Ash	%	6.0	6.7	
Crude fiber	%	6.4	7.1	14
Ether extract	%	1.1	1.2	
N-free extract	%	29.1	32.7	
Protein (N x 6.25)	%	46.6	52.3	4
Energy	GE Mcal/kg	4.22	4.73	
Cattle	DE Mcal/kg	* 2.92	3.28	
Sheep	DE Mcal/kg	* 3.08	3.45	
Swine	DE kcal/kg	*2968.	3330.	
Cattle	ME Mcal/kg	* 2.40	2.69	
Chickens	ME_n kcal/kg	1884.	2114.	
Sheep	ME Mcal/kg	* 2.52	2.83	
Swine	ME kcal/kg	*2536.	2845.	
Cattle	TDN %	* 66.3	74.4	
Sheep	TDN %	* 69.8	78.4	
Swine	TDN %	* 67.3	75.5	
Choline	mg/kg	2734.	3068.	
Niacin	mg/kg	26.7	30.0	
Pantothenic acid	mg/kg	14.5	16.2	
Riboflavin	mg/kg	3.3	3.7	
Thiamine	mg/kg	6.6	7.4	
Alanine	%	2.02	2.26	
Arginine	%	3.29	3.69	
Aspartic acid	%	5.20	5.84	
Cystine	%	0.68	0.76	
Glutamic acid	%	8.22	9.22	
Glycine	%	1.96	2.20	
Histidine	%	1.28	1.43	
Isoleucine	%	2.44	2.74	
Leucine	%	3.77	4.23	
Lysine	%	3.06	3.44	
Methionine	%	0.53	0.59	
Phenylalanine	%	2.47	2.77	
Proline	%	2.47	2.77	
Serine	%	2.27	2.55	
Threonine	%	1.83	2.06	
Tryptophan	%	0.72	0.81	
Tyrosine	%	1.47	1.65	
Valine	%	2.43	2.72	

Soybean, seeds, solv-extd grnd, mx 7% fiber, (5)
Soybean meal, solvent extracted (AAFCO)
Soybean meal, solvent extracted
Soybean oil meal, solvent extracted
Ref No 5 04 604 United States

Feed Name or Analyses		Mean As Fed	Mean Dry	C.V. ± %
Dry matter	%	89.1	100.0	1
Organic matter	%	82.8	93.0	
Ash	%	5.9	6.6	5
Crude fiber	%	5.2	5.9	19
Cattle	dig coef %	57.	57.	
Sheep	dig coef %	101.	101.	
Swine	dig coef %	83.	83.	
Ether extract	%	1.2	1.3	13
Cattle	dig coef %	38.	38.	
Sheep	dig coef %	52.	52.	
Swine	dig coef %	58.	58.	
N-free extract	%	30.1	33.8	2
Cattle	dig coef %	91.	91.	
Sheep	dig coef %	96.	96.	
Swine	dig coef %	91.	91.	
Protein (N x 6.25)	%	46.7	52.4	5
Cattle	dig coef %	90.	90.	
Sheep	dig coef %	94.	94.	
Swine	dig coef %	91.	91.	
Cattle	dig prot %	42.0	47.2	
Sheep	dig prot %	43.9	49.2	
Swine	dig prot %	42.5	47.7	
Energy	GE Mcal/kg	4.17	4.69	
Cattle	DE Mcal/kg	* 3.23	3.63	
Sheep	DE Mcal/kg	* 3.50	3.93	
Swine	DE kcal/kg	*3338.	3748.	
Cattle	ME Mcal/kg	* 2.65	2.98	
Chickens	ME_n kcal/kg	2219.	2491.	
Sheep	ME Mcal/kg	* 2.87	3.22	
Swine	ME kcal/kg	*2851.	3201.	
Cattle	NE_m Mcal/kg	* 1.72	1.93	
Cattle	NE_{gain} Mcal/kg	* 1.15	1.29	
Cattle	$NE_{lactating\ cows}$ Mcal/kg	* 1.99	2.23	
Cattle	TDN %	73.3	82.4	
Sheep	TDN %	79.4	89.1	
Swine	TDN %	75.7	85.0	
Calcium	%	0.30	0.33	9
Chlorine	%	0.00	0.00	9
Cobalt	mg/kg	0.092	0.104	55
Copper	mg/kg	14.2	15.9	10
Iron	%	0.013	0.015	30
Magnesium	%	0.27	0.30	53
Manganese	mg/kg	27.3	30.7	11
Phosphorus	%	0.65	0.73	3
Potassium	%	1.92	2.15	6
Sodium	%	0.34	0.38	61
Sulphur	%	0.43	0.48	
Carotene	mg/kg	0.2	0.2	
Choline	mg/kg	2796.	3139.	8
Folic acid	mg/kg	0.59	0.67	3
Niacin	mg/kg	26.8	30.1	18
Pantothenic acid	mg/kg	15.8	17.7	8
Riboflavin	mg/kg	3.1	3.5	8
Thiamine	mg/kg	6.6	7.4	50
α-tocopherol	mg/kg	1.2	1.3	36
Vitamin A equivalent	IU/g	0.4	0.4	
Alanine	%	2.46	2.76	
Arginine	%	3.39	3.81	11
Aspartic acid	%	6.53	7.34	
Cystine	%	0.70	0.79	
Glutamic acid	%	9.31	10.46	6
Glycine	%	2.44	2.74	
Histidine	%	1.29	1.44	10
Isoleucine	%	2.47	2.77	15
Leucine	%	3.84	4.31	10
Lysine	%	3.05	3.43	11
Methionine	%	0.60	0.67	38
Phenylalanine	%	2.50	2.80	5
Proline	%	2.92	3.28	
Serine	%	2.59	2.91	
Threonine	%	1.98	2.22	3
Tryptophan	%	0.58	0.66	7
Tyrosine	%	1.54	1.73	11
Valine	%	2.46	2.76	5

Ref No 5 04 604 Canada

Feed Name or Analyses		Mean As Fed	Mean Dry	C.V. ± %
Dry matter	%	89.2	100.0	0
Crude fiber	%	5.0	5.6	
Cattle	dig coef %	57.	57.	
Sheep	dig coef %	101.	101.	
Swine	dig coef %	83.	83.	
Ether extract	%	1.5	1.7	
Cattle	dig coef %	38.	38.	
Sheep	dig coef %	52.	52.	
Swine	dig coef %	58.	58.	
N-free extract	%			
Cattle	dig coef %	91.	91.	
Sheep	dig coef %	96.	96.	
Swine	dig coef %	91.	91.	
Protein (N x 6.25)	%	46.0	51.6	6
Cattle	dig coef %	90.	90.	
Sheep	dig coef %	94.	94.	
Swine	dig coef %	91.	91.	
Cattle	dig prot %	41.4	46.5	
Sheep	dig prot %	43.3	48.5	
Swine	dig prot %	41.9	47.0	
Energy	GE Mcal/kg	4.14	4.64	
Cattle	DE Mcal/kg	* 3.24	3.63	
Sheep	DE Mcal/kg	* 3.51	3.93	
Swine	DE kcal/kg	*3344.	3750.	
Cattle	ME Mcal/kg	* 2.66	2.98	
Chickens	ME_n kcal/kg	2222.	2491.	
Sheep	ME Mcal/kg	* 2.87	3.22	
Swine	ME kcal/kg	*2861.	3209.	
Cattle	TDN %	73.5	82.4	
Sheep	TDN %	79.5	89.2	
Swine	TDN %	75.8	85.0	
Niacin	mg/kg	26.6	29.9	
Pantothenic acid	mg/kg	16.4	18.4	20
Riboflavin	mg/kg	2.8	3.1	14
Thiamine	mg/kg	4.7	5.3	
Alanine	%	2.31	2.59	
Arginine	%	3.49	3.91	
Aspartic acid	%	6.18	6.93	
Glutamic acid	%	9.94	11.15	
Glycine	%	2.26	2.53	
Histidine	%	1.40	1.57	
Isoleucine	%	2.31	2.59	
Leucine	%	4.08	4.58	
Lysine	%	3.33	3.74	
Methionine	%	0.48	0.54	
Phenylalanine	%	2.63	2.95	
Proline	%	2.74	3.07	
Serine	%	2.79	3.13	
Threonine	%	2.15	2.41	
Tyrosine	%	1.61	1.81	
Valine	%	2.42	2.71	

Soybean, seeds, solv-extd grnd pelleted, (5)
Soybean meal, solvent extracted, pelleted
Soybean oil meal, solvent extracted, pelleted
Ref No 5 04 606 United States

Feed Name or Analyses		Mean As Fed	Mean Dry	C.V. ± %
Dry matter	%	90.0	100.0	
Ash	%	5.9	6.6	
Crude fiber	%	5.3	5.9	
Ether extract	%	0.7	0.8	
N-free extract	%	31.7	35.2	
Protein (N x 6.25)	%	46.4	51.6	
Energy	GE Mcal/kg			
Cattle	DE Mcal/kg	* 3.00	3.34	
Sheep	DE Mcal/kg	* 3.17	3.52	

Continued

Feed Name or Analyses		Mean As Fed	Mean Dry	C.V. ± %
Swine	DE kcal/kg	*3041.	3379.	
Cattle	ME Mcal/kg	* 2.46	2.73	
Sheep	ME Mcal/kg	* 2.60	2.89	
Swine	ME kcal/kg	*2603.	2892.	
Cattle	TDN %	* 68.1	75.6	
Sheep	TDN %	* 71.8	79.8	
Swine	TDN %	* 69.0	76.6	

Soybean, seeds, solv-extd toasted grnd, (5)
Soybean meal, solvent extracted, toasted
Soybean oil meal, solvent extracted, toasted
Ref No 5 04 607 United States

Feed Name or Analyses		Mean As Fed	Mean Dry	C.V. ± %
Dry matter	%	91.3	100.0	2
Ash	%	5.9	6.5	9
Crude fiber	%	5.5	6.0	18
Cattle	dig coef %	* 57.	57.	
Sheep	dig coef %	* 101.	101.	
Swine	dig coef %	* 83.	83.	
Ether extract	%	1.0	1.1	34
Cattle	dig coef %	* 39.	39.	
Sheep	dig coef %	* 52.	52.	
Swine	dig coef %	* 58.	58.	
N-free extract	%	32.8	35.9	
Cattle	dig coef %	* 91.	91.	
Sheep	dig coef %	* 96.	96.	
Swine	dig coef %	* 91.	91.	
Protein (N x 6.25)	%	46.1	50.5	4
Cattle	dig coef %	* 90.	90.	
Sheep	dig coef %	* 94.	94.	
Swine	dig coef %	* 91.	91.	
Cattle	dig prot %	41.5	45.4	
Sheep	dig prot %	43.3	47.5	
Swine	dig prot %	42.0	45.9	
Energy	GE Mcal/kg			
Cattle	DE Mcal/kg	* 3.32	3.64	
Sheep	DE Mcal/kg	* 3.60	3.94	
Swine	DE kcal/kg	*3425.	3751.	
Cattle	ME Mcal/kg	* 2.72	2.98	
Sheep	ME Mcal/kg	* 2.95	3.23	
Swine	ME kcal/kg	*2938.	3218.	
Cattle	TDN %	75.4	82.5	
Sheep	TDN %	81.5	89.3	
Swine	TDN %	77.7	85.1	
Arginine	%	3.60	3.94	3
Glutamic acid	%	8.30	9.09	1
Histidine	%	1.10	1.20	4
Isoleucine	%	2.40	2.63	2
Leucine	%	3.50	3.83	1
Lysine	%	2.90	3.18	3
Methionine	%	0.70	0.77	6
Phenylalanine	%	2.30	2.52	5
Threonine	%	1.80	1.97	4
Tryptophan	%	0.70	0.77	
Tyrosine	%	0.70	0.77	
Valine	%	2.40	2.63	2

Soybean, seeds, steamed, (5)
Ref No 5 04 608 United States

Feed Name or Analyses		Mean As Fed	Mean Dry	C.V. ± %
Dry matter	%	64.6	100.0	
Ash	%	3.4	5.3	
Crude fiber	%	4.2	6.5	
Sheep	dig coef %	178.	178.	
Ether extract	%	12.1	18.8	
Sheep	dig coef %	95.	95.	
N-free extract	%	19.6	30.3	
Sheep	dig coef %	120.	120.	
Protein (N x 6.25)	%	25.3	39.2	
Sheep	dig coef %	91.	91.	
Sheep	dig prot %	23.0	35.7	
Energy	GE Mcal/kg	2.89	4.48	
Cattle	DE Mcal/kg	* 3.38	5.23	
Sheep	DE Mcal/kg	* 3.52	5.45	
Swine	DE kcal/kg	*3194.	4947.	
Cattle	ME Mcal/kg	* 2.77	4.29	
Chickens	ME_n kcal/kg	2689.	4164.	
Sheep	ME Mcal/kg	* 2.89	4.47	
Swine	ME kcal/kg	*2813.	4357.	
Cattle	TDN %	* 76.7	118.7	
Sheep	TDN %	79.8	123.6	
Swine	TDN %	* 72.4	112.2	
Calcium	%	0.17	0.26	
Iron	%	0.007	0.010	
Phosphorus	%	0.41	0.64	
Potassium	%	1.20	1.86	
Sodium	%	0.00	0.00	
Choline	mg/kg	1740.	2695.	
Niacin	mg/kg	14.6	22.6	
Pantothenic acid	mg/kg	11.2	17.4	
Riboflavin	mg/kg	2.0	3.0	
Thiamine	mg/kg	4.7	7.2	
Vitamin A	IU/g	0.7	1.0	
Arginine	%	1.94	3.00	
Cystine	%	0.43	0.67	
Glycine	%	1.51	2.33	
Lysine	%	1.72	2.67	
Methionine	%	0.39	0.60	
Tryptophan	%	0.34	0.52	

(1) dry forages and roughages (2) pasture, range plants, and forages fed green (3) silages (4) energy feeds (5) protein supplements (6) minerals (7) vitamins (8) additives

Soybean, seeds, gr sample US mn 18% moisture, (5)
Ref No 5 04 609 United States

Feed Name or Analyses		Mean As Fed	Mean Dry	C.V. ± %
Dry matter	%	74.7	100.0	
Ash	%	4.2	5.6	
Crude fiber	%	6.2	8.3	
Sheep	dig coef %	178.	178.	
Ether extract	%	12.4	16.6	
Sheep	dig coef %	95.	95.	
N-free extract	%	23.9	32.0	
Sheep	dig coef %	120.	120.	
Protein (N x 6.25)	%	28.0	37.5	
Sheep	dig coef %	91.	91.	
Sheep	dig prot %	25.5	34.1	
Energy	GE Mcal/kg			
Cattle	DE Mcal/kg	* 3.80	5.09	
Sheep	DE Mcal/kg	* 4.04	5.41	
Swine	DE kcal/kg	*3538.	4736.	
Cattle	ME Mcal/kg	* 3.11	4.17	
Sheep	ME Mcal/kg	* 3.32	4.44	
Swine	ME kcal/kg	*3128.	4188.	
Cattle	TDN %	* 86.2	115.4	
Sheep	TDN %	91.7	122.8	
Swine	TDN %	* 80.2	107.4	

Soybean, seeds w calcium carbonate added, mech-extd grnd, (5)
Ref No 5 08 526 United States

Feed Name or Analyses		Mean As Fed	Mean Dry	C.V. ± %
Dry matter	%	91.0	100.0	
Ash	%	7.7	8.5	
Crude fiber	%	5.9	6.5	
Ether extract	%	5.3	5.8	
N-free extract	%	27.9	30.7	
Protein (N x 6.25)	%	44.2	48.6	
Energy	GE Mcal/kg			
Cattle	DE Mcal/kg	* 3.15	3.46	
Sheep	DE Mcal/kg	* 3.22	3.53	
Swine	DE kcal/kg	*3357.	3689.	
Cattle	ME Mcal/kg	* 2.58	2.84	
Sheep	ME Mcal/kg	* 2.64	2.90	
Swine	ME kcal/kg	*2893.	3179.	
Cattle	TDN %	* 71.5	78.5	
Sheep	TDN %	* 72.9	80.1	
Swine	TDN %	* 76.1	83.7	

Soybean, seeds wo hulls, solv-extd grnd, mx 3% fiber, (5)
Soybean meal, dehulled, solvent extracted (AAFCO)
Soybean oil meal, dehulled, solvent extracted
Ref No 5 04 612 United States

Feed Name or Analyses		Mean As Fed	Mean Dry	C.V. ± %
Dry matter	%	91.3	100.0	1
Ash	%	5.9	6.5	6
Crude fiber	%	2.9	3.2	13
Sheep	dig coef %	* 101.	101.	
Ether extract	%	1.1	1.2	33
Sheep	dig coef %	* 52.	52.	
N-free extract	%	30.5	33.5	
Sheep	dig coef %	* 96.	96.	
Protein (N x 6.25)	%	50.8	55.6	1
Sheep	dig coef %	* 94.	94.	
Sheep	dig prot %	47.7	52.3	
Linoleic	%	.405	.444	
Energy	GE Mcal/kg	4.25	4.66	
Cattle	DE Mcal/kg	* 3.37	3.69	
Sheep	DE Mcal/kg	* 3.58	3.93	
Swine	DE kcal/kg	4007.	4390.	
Cattle	ME Mcal/kg	* 2.77	3.03	
Chickens	ME_n kcal/kg	2489.	2726.	
Sheep	ME Mcal/kg	* 2.94	3.22	
Swine	ME kcal/kg	3542.	3880.	
Cattle	TDN %	* 76.3	83.6	
Sheep	TDN %	81.3	89.1	
Swine	TDN %	* 71.2	78.0	
Calcium	%	0.29	0.32	31
Chlorine	%	0.07	0.08	
Manganese	mg/kg	46.4	50.8	53
Phosphorus	%	0.65	0.71	8
Potassium	%	2.05	2.25	
Sodium	%	0.51	0.56	
Choline	mg/kg	2783.	3049.	8
Niacin	mg/kg	21.7	23.8	3
Pantothenic acid	mg/kg	13.6	14.9	
Riboflavin	mg/kg	3.1	3.4	29
Thiamine	mg/kg	2.5	2.7	7
α-tocopherol	mg/kg	3.3	3.6	
Arginine	%	3.58	3.93	
Cystine	%	0.83	0.91	
Glycine	%	2.89	3.16	
Lysine	%	3.19	3.49	

Feed Name or Analyses		Mean As Fed	Mean Dry	C.V. ± %
Methionine	%	0.74	0.81	
Tryptophan	%	0.64	0.70	

Ref No 5 04 612 Canada

Feed Name or Analyses			Mean As Fed	Mean Dry	C.V. ± %
Dry matter	%		90.1	100.0	2
Crude fiber	%		4.2	4.6	37
Protein (N x 6.25)	%		49.1	54.6	4
Energy	GE Mcal/kg		4.20	4.66	
Sheep	DE Mcal/kg	*	3.54	3.93	
Swine	DE kcal/kg		3953.	4390.	
Chickens	ME_n kcal/kg		2381.	2644.	
Sheep	ME Mcal/kg	*	2.90	3.22	
Swine	ME kcal/kg		3494.	3880.	
Sheep	TDN %		80.3	89.1	
Calcium	%		0.30	0.33	
Copper	mg/kg		14.9	16.6	
Phosphorus	%		0.55	0.62	
Niacin	mg/kg		21.8	24.3	
Pantothenic acid	mg/kg		17.6	19.6	
Riboflavin	mg/kg		2.6	2.8	
Thiamine	mg/kg		6.6	7.3	
Tryptophan	%		0.66	0.73	

Soybean flakes (CFA) –
see Soybean, seeds, solv-extd flaked, (5)

Soybean meal (CFA) –
see Soybean, seeds, mech-extd grnd, (5)

Soybean meal (CFA) –
see Soybean, seeds, solv-extd grnd, (5)

Soybean meal, dehulled, solvent extracted (AAFCO) –
see Soybean, seeds wo hulls, solv-extd grnd, mx 3% fiber, (5)

Soybean meal, expeller extracted –
see Soybean, seeds, mech-extd grnd, mx 7% fiber, (5)

Soybean meal, hydraulic extracted –
see Soybean, seeds, mech-extd grnd, mx 7% fiber, (5)

Soybean meal, mechanical extracted (AAFCO) –
see Soybean, seeds, mech-extd grnd, mx 7% fiber, (5)

Soybean meal, solvent extracted (AAFCO) –
see Soybean, seeds, solv-extd grnd, mx 7% fiber, (5)

Soybean meal, solvent extracted –
see Soybean, seeds, solv-extd grnd, mx 7% fiber, (5)

Soybean meal, solvent extracted, pelleted –
see Soybean, seeds, solv-extd grnd pelleted, (5)

Soybean meal, solvent extracted, toasted –
see Soybean, seeds, solv-extd toasted grnd, (5)

SOYBEAN-MILLET. Glycine max, Setaria spp

Soybean-Millet, aerial part, fresh, (2)

Ref No 2 04 639 United States

Feed Name or Analyses			Mean As Fed	Mean Dry	C.V. ± %
Dry matter	%		23.5	100.0	
Ash	%		1.8	7.7	
Crude fiber	%		7.5	31.9	
Ether extract	%		0.5	2.1	
N-free extract	%		11.4	48.5	
Protein (N x 6.25)	%		2.3	9.8	
Cattle	dig prot %	*	1.5	6.2	
Goats	dig prot %	*	1.3	5.7	
Horses	dig prot %	*	1.4	5.8	
Rabbits	dig prot %	*	1.5	6.2	
Sheep	dig prot %	*	1.4	6.1	
Energy	GE Mcal/kg				
Cattle	DE Mcal/kg	*	0.69	2.95	
Sheep	DE Mcal/kg	*	0.68	2.89	
Cattle	ME Mcal/kg	*	0.57	2.42	
Sheep	ME Mcal/kg	*	0.56	2.37	
Cattle	TDN %	*	15.7	66.8	
Sheep	TDN %	*	15.4	65.6	

Soybean-Millet, aerial part, ensiled, (3)

Ref No 3 04 640 United States

Feed Name or Analyses			Mean As Fed	Mean Dry	C.V. ± %
Dry matter	%		22.6	100.0	13
Ash	%		2.9	12.7	12
Crude fiber	%		7.6	33.8	3
Ether extract	%		1.0	4.5	13
N-free extract	%		8.1	36.0	
Protein (N x 6.25)	%		2.9	13.0	5
Cattle	dig prot %	*	1.8	8.0	
Goats	dig prot %	*	1.8	8.0	
Horses	dig prot %	*	1.8	8.0	
Sheep	dig prot %	*	1.8	8.0	
Energy	GE Mcal/kg				
Cattle	DE Mcal/kg	*	0.63	2.78	
Sheep	DE Mcal/kg	*	0.59	2.61	
Cattle	ME Mcal/kg	*	0.52	2.28	
Sheep	ME Mcal/kg	*	0.48	2.14	
Cattle	TDN %	*	14.3	63.1	
Sheep	TDN %	*	13.4	59.3	

Soybean-Millet, aerial part w molasses added, ensiled, (3)

Ref No 3 04 641 United States

Feed Name or Analyses			Mean As Fed	Mean Dry	C.V. ± %
Dry matter	%		29.6	100.0	
Ash	%		2.9	9.8	
Crude fiber	%		9.4	31.7	
Ether extract	%		1.0	3.3	
N-free extract	%		12.8	43.3	
Protein (N x 6.25)	%		3.5	11.9	
Cattle	dig prot %	*	2.1	7.0	
Goats	dig prot %	*	2.1	7.0	
Horses	dig prot %	*	2.1	7.0	
Sheep	dig prot %	*	2.1	7.0	
Energy	GE Mcal/kg				
Cattle	DE Mcal/kg	*	0.80	2.70	
Sheep	DE Mcal/kg	*	0.81	2.75	
Cattle	ME Mcal/kg	*	0.65	2.21	
Sheep	ME Mcal/kg	*	0.67	2.25	
Cattle	TDN %	*	18.1	61.2	
Sheep	TDN %	*	18.4	62.3	

SOYBEAN-MILLET, PEARL. Glycine max, Pennisetum Glaucum

Soybean-Millet, pearl, aerial part, fresh, (2)

Ref No 2 08 524 United States

Feed Name or Analyses			Mean As Fed	Mean Dry	C.V. ± %
Dry matter	%		24.5	100.0	
Ash	%		1.9	7.8	
Crude fiber	%		6.4	26.1	
Ether extract	%		0.9	3.7	
N-free extract	%		11.1	45.3	
Protein (N x 6.25)	%		4.2	17.1	
Cattle	dig prot %	*	3.1	12.5	
Goats	dig prot %	*	3.1	12.6	
Horses	dig prot %	*	3.0	12.1	
Rabbits	dig prot %	*	2.9	11.9	
Sheep	dig prot %	*	3.2	13.0	
Energy	GE Mcal/kg				
Cattle	DE Mcal/kg	*	0.72	2.93	
Sheep	DE Mcal/kg	*	0.59	2.40	
Cattle	ME Mcal/kg	*	0.59	2.40	
Sheep	ME Mcal/kg	*	0.48	1.97	
Cattle	TDN %	*	16.3	66.4	
Sheep	TDN %	*	15.3	62.3	

Soybean-Millet, pearl, aerial part, fresh, immature, (2)

Ref No 2 04 642 United States

Feed Name or Analyses			Mean As Fed	Mean Dry	C.V. ± %
Dry matter	%		24.5	100.0	
Ash	%		1.9	7.8	
Crude fiber	%		6.4	26.1	
Ether extract	%		0.9	3.7	
N-free extract	%		11.1	45.3	
Protein (N x 6.25)	%		4.2	17.1	
Cattle	dig prot %	*	3.0	12.4	
Goats	dig prot %	*	3.1	12.5	
Horses	dig prot %	*	3.0	12.0	
Rabbits	dig prot %	*	2.9	11.9	
Sheep	dig prot %	*	3.2	12.9	
Energy	GE Mcal/kg				
Cattle	DE Mcal/kg	*	0.71	2.90	
Sheep	DE Mcal/kg	*	0.68	2.78	
Cattle	ME Mcal/kg	*	0.58	2.38	
Sheep	ME Mcal/kg	*	0.56	2.28	
Cattle	TDN %	*	16.1	65.7	
Sheep	TDN %	*	15.5	63.1	

Soybean mill feed (CFA) –
see Soybean, flour by-product, grnd, mn 13% protein mx 32% fiber, (5)

Soybean mill run (AAFCO) –
see Soybean, groats by-product, grnd, mn 11% protein mx 35% fiber, (5)

Soybean oil meal, dehulled, solvent extracted –
see Soybean, seeds wo hulls, solv-extd grnd, mx 3% fiber, (5)

Soybean oil meal, expeller extracted –
see Soybean, seeds, mech-extd grnd, mx 7% fiber, (5)

Soybean oil meal, hydraulic extracted –
see Soybean, seeds, mech-extd grnd, mx 7% fiber, (5)

Feed Name or Analyses			Mean As Fed	Dry	C.V. ± %

Soybean oil meal, solvent extracted –
see Soybean, seeds, solv-extd grnd, mx 7% fiber, (5)

Soybean oil meal, solvent extracted, pelleted –
see Soybean, seeds, solv-extd grnd pelleted, (5)

Soybean oil meal, solvent extracted, toasted –
see Soybean, seeds, solv-extd toasted grnd, (5)

SOYBEAN-SORGHUM. Glycine max, Sorghum vulgare

Soybean-Sorghum, aerial part, ensiled, (3)
Ref No 3 04 643 United States

Analyses	Unit		As Fed	Dry	C.V. ± %
Dry matter	%		25.5	100.0	
Ash	%		1.2	4.8	
Crude fiber	%		7.1	28.0	
Cattle	dig coef %		53.	53.	
Ether extract	%		2.5	9.7	
Cattle	dig coef %		94.	94.	
N-free extract	%		12.2	47.8	
Cattle	dig coef %		58.	58.	
Protein (N x 6.25)	%		2.5	9.7	
Cattle	dig coef %		50.	50.	
Cattle	dig prot %		1.2	4.9	
Goats	dig prot %	*	1.3	5.0	
Horses	dig prot %	*	1.3	5.0	
Sheep	dig prot %	*	1.3	5.0	
Energy	GE Mcal/kg				
Cattle	DE Mcal/kg	*	0.76	3.00	
Sheep	DE Mcal/kg	*	0.71	2.77	
Cattle	ME Mcal/kg	*	0.63	2.46	
Sheep	ME Mcal/kg	*	0.58	2.27	
Cattle	TDN %		17.3	67.9	
Sheep	TDN %	*	16.0	62.7	

SOYBEAN-SORGHUM, SUDANGRASS. Glycine max, Sorghum vulgare sudanese

Soybean-Sorghum, Sudangrass, aerial part, fresh, (2)
Ref No 2 04 646 United States

Analyses	Unit		As Fed	Dry	C.V. ± %
Dry matter	%		24.2	100.0	
Ash	%		1.6	6.6	
Crude fiber	%		8.3	34.3	
Ether extract	%		0.5	2.1	
N-free extract	%		11.1	45.8	
Protein (N x 6.25)	%		2.7	11.2	
Cattle	dig prot %	*	1.8	7.4	
Goats	dig prot %	*	1.7	7.0	
Horses	dig prot %	*	1.7	7.0	
Rabbits	dig prot %	*	1.8	7.3	
Sheep	dig prot %	*	1.8	7.4	
Energy	GE Mcal/kg				
Cattle	DE Mcal/kg	*	0.72	2.97	
Sheep	DE Mcal/kg	*	0.68	2.80	
Cattle	ME Mcal/kg	*	0.59	2.44	
Sheep	ME Mcal/kg	*	0.56	2.30	
Cattle	TDN %	*	16.3	67.3	
Sheep	TDN %	*	15.4	63.6	

Soybean-Sorghum, Sudangrass, aerial part, ensiled, (3)
Ref No 3 04 644 United States

Analyses	Unit		As Fed	Dry	C.V. ± %
Dry matter	%		25.4	100.0	10
Ash	%		2.2	8.6	9
Crude fiber	%		8.7	34.1	5
Ether extract	%		0.5	2.1	11
N-free extract	%		10.6	41.9	
Protein (N x 6.25)	%		3.4	13.3	10
Cattle	dig prot %	*	2.1	8.3	
Goats	dig prot %	*	2.1	8.3	
Horses	dig prot %	*	2.1	8.3	
Sheep	dig prot %	*	2.1	8.3	
Energy	GE Mcal/kg				
Cattle	DE Mcal/kg	*	0.63	2.48	
Sheep	DE Mcal/kg	*	0.68	2.68	
Cattle	ME Mcal/kg	*	0.52	2.03	
Sheep	ME Mcal/kg	*	0.56	2.20	
Cattle	TDN %	*	14.3	56.3	
Sheep	TDN %	*	15.4	60.7	
Carotene	mg/kg		9.5	37.5	
Vitamin A equivalent	IU/g		15.9	62.5	

SOYBEAN-SUGARCANE. Glycine max, Saccharum Officinarum

Soybean-Sugarcane, aerial part, ensiled, (3)
Ref No 3 04 645 United States

Analyses	Unit		As Fed	Dry	C.V. ± %
Dry matter	%		29.8	100.0	
Ash	%		1.5	5.0	
Crude fiber	%		7.4	24.8	
Ether extract	%		0.9	3.0	
N-free extract	%		17.2	57.8	
Protein (N x 6.25)	%		2.8	9.4	
Cattle	dig prot %	*	1.4	4.8	
Goats	dig prot %	*	1.4	4.8	
Horses	dig prot %	*	1.4	4.8	
Sheep	dig prot %	*	1.4	4.8	

SOYBEAN-SWEETPOTATO. Glycine max, Ipomoea batatas

Soybean-Sweetpotato, aerial part, ensiled, (3)
Ref No 3 04 647 United States

Analyses	Unit		As Fed	Dry	C.V. ± %
Dry matter	%		27.7	100.0	
Ash	%		2.3	8.3	
Crude fiber	%		7.9	28.7	
Ether extract	%		1.7	6.1	
N-free extract	%		10.7	38.5	
Protein (N x 6.25)	%		5.1	18.4	
Cattle	dig prot %	*	3.6	12.9	
Goats	dig prot %	*	3.6	12.9	
Horses	dig prot %	*	3.6	12.9	
Sheep	dig prot %	*	3.6	12.9	

Soybran flakes –
see Soybean, hulls, (1)

Soy flour (AAFCO) –
see Soybean, flour, mech-extd fine sift, mx 3% fiber, (5)

Soy flour, expeller extracted –
see Soybean, flour, mech-extd fine sift, mx 3% fiber, (5)

Soy grits, expeller extracted –
see Soybean, flour, mech-extd fine sift, mx 3% fiber, (5)

Soy flour, expeller extracted –
see Soybean, flour, mech-extd fine sift, mx 3% fiber, (5)

Soy grits (AAFCO) –
see Soybean, flour, mech-extd fine sift, mx 3% fiber, (5)

Soy grits, expeller extracted –
see Soybean, flour, mech-extd fine sift, mx 3% fiber, (5)

SPANISHMOSS. Tillandsia usneoides

Spanishmoss, entire plant, s-c, (1)
Ref No 1 04 868 United States

Analyses	Unit		As Fed	Dry	C.V. ± %
Dry matter	%		90.3	100.0	
Ash	%		8.3	9.2	
Crude fiber	%		26.8	29.7	
Sheep	dig coef %		52.	52.	
Ether extract	%		2.4	2.6	
Sheep	dig coef %		16.	16.	
N-free extract	%		47.8	53.0	
Sheep	dig coef %		77.	77.	
Protein (N x 6.25)	%		4.9	5.5	
Sheep	dig coef %		-5.	-5.	
Cattle	dig prot %	*	1.5	1.7	
Goats	dig prot %	*	1.5	1.6	
Horses	dig prot %	*	1.9	2.2	
Rabbits	dig prot %	*	2.6	2.9	
Sheep	dig prot %		-0.2	-0.2	
Energy	GE Mcal/kg				
Cattle	DE Mcal/kg	*	2.36	2.61	
Sheep	DE Mcal/kg	*	2.26	2.51	
Cattle	ME Mcal/kg	*	1.93	2.14	
Sheep	ME Mcal/kg	*	1.86	2.06	
Cattle	TDN %	*	53.4	59.1	
Sheep	TDN %		51.4	56.9	
Phosphorus	%		0.04	0.04	
Potassium	%		0.47	0.52	

Spanishmoss, entire plant, fresh, (2)
Tillandsia, treebeard
Ref No 2 04 648 United States

Analyses	Unit		As Fed	Dry	C.V. ± %
Dry matter	%		41.1	100.0	
Ash	%		3.1	7.6	18
Crude fiber	%		12.3	29.9	8
Ether extract	%		0.9	2.3	16
N-free extract	%		22.7	55.3	
Protein (N x 6.25)	%		2.0	4.9	10
Cattle	dig prot %	*	0.8	2.1	
Goats	dig prot %	*	0.5	1.1	
Horses	dig prot %	*	0.7	1.7	
Rabbits	dig prot %	*	1.0	2.5	
Sheep	dig prot %	*	0.6	1.6	
Energy	GE Mcal/kg				
Cattle	DE Mcal/kg	*	1.10	2.68	
Sheep	DE Mcal/kg	*	1.13	2.75	

(1) dry forages and roughages (2) pasture, range plants, and forages fed green (3) silages (4) energy feeds (5) protein supplements (6) minerals (7) vitamins (8) additives

Feed Name or Analyses			Mean As Fed	Mean Dry	C.V. ± %
Cattle	ME Mcal/kg	*	0.90	2.20	
Sheep	ME Mcal/kg	*	0.93	2.26	
Cattle	TDN %	*	25.0	60.8	
Sheep	TDN %	*	25.6	62.3	
Carotene	mg/kg		6.2	15.0	
Vitamin A equivalent	IU/g		10.3	25.0	

SPELT. Triticum spelta

Spelt, straw, (1)

Ref No 1 04 649 United States

Feed Name or Analyses			As Fed	Dry	C.V. ± %
Dry matter	%		82.8	100.0	
Ash	%		5.8	7.0	
Crude fiber	%		39.9	48.2	
Horses	dig coef %		30.	30.	
Sheep	dig coef %		46.	46.	
Ether extract	%		1.2	1.5	
Horses	dig coef %		20.	20.	
Sheep	dig coef %		60.	60.	
N-free extract	%		33.2	40.1	
Horses	dig coef %		18.	18.	
Sheep	dig coef %		35.	35.	
Protein (N x 6.25)	%		2.7	3.3	
Horses	dig coef %		23.	23.	
Sheep	dig coef %		42.	42.	
Cattle	dig prot %	*	-0.1	-0.1	
Goats	dig prot %	*	-0.2	-0.3	
Horses	dig prot %		0.6	0.8	
Rabbits	dig prot %	*	1.0	1.2	
Sheep	dig prot %		1.1	1.4	
Energy	GE Mcal/kg				
Cattle	DE Mcal/kg	*	1.19	1.44	
Horses	DE Mcal/kg	*	0.84	1.02	
Sheep	DE Mcal/kg	*	1.44	1.74	
Cattle	ME Mcal/kg	*	0.98	1.18	
Horses	ME Mcal/kg	*	0.69	0.83	
Sheep	ME Mcal/kg	*	1.18	1.43	
Cattle	TDN %	*	27.1	32.8	
Horses	TDN %		19.1	23.1	
Sheep	TDN %		32.7	39.6	

Spelt, straw, pulp, (1)

Ref No 1 04 650 United States

Feed Name or Analyses			As Fed	Dry	C.V. ± %
Dry matter	%		93.9	100.0	
Ash	%		5.3	5.6	
Crude fiber	%		71.5	76.1	
Sheep	dig coef %		79.	79.	
Ether extract	%		0.6	0.6	
Sheep	dig coef %		0.	0.	
N-free extract	%		15.0	16.0	
Sheep	dig coef %		24.	24.	
Protein (N x 6.25)	%		1.6	1.7	
Sheep	dig coef %		0.	0.	
Cattle	dig prot %	*	-1.4	-1.5	
Goats	dig prot %	*	-1.6	-1.8	
Horses	dig prot %	*	-0.9	-0.9	
Rabbits	dig prot %	*	0.0	0.0	
Energy	GE Mcal/kg				
Sheep	DE Mcal/kg	*	2.65	2.82	
Sheep	ME Mcal/kg	*	2.17	2.31	
Sheep	TDN %		60.1	64.0	

Spelt, grain, (4)

Ref No 4 04 651 United States

Feed Name or Analyses			As Fed	Dry	C.V. ± %
Dry matter	%		89.6	100.0	2
Ash	%		3.5	3.9	6
Crude fiber	%		8.2	9.1	14
Cattle	dig coef %	*	51.	51.	
Sheep	dig coef %	*	37.	37.	
Ether extract	%		2.0	2.2	11
Cattle	dig coef %	*	43.	43.	
Sheep	dig coef %	*	82.	82.	
N-free extract	%		64.1	71.5	
Cattle	dig coef %	*	71.	71.	
Sheep	dig coef %	*	83.	83.	
Protein (N x 6.25)	%		11.9	13.3	5
Cattle	dig coef %	*	75.	75.	
Sheep	dig coef %	*	78.	78.	
Cattle	dig prot %		8.9	10.0	
Goats	dig prot %	*	8.4	9.4	
Horses	dig prot %	*	8.4	9.4	
Sheep	dig prot %		9.3	10.4	
Energy	GE Mcal/kg				
Cattle	DE Mcal/kg	*	2.67	2.98	
Sheep	DE Mcal/kg	*	3.05	3.40	
Swine	DE kcal/kg	*	2792.	3116.	
Cattle	ME Mcal/kg	*	2.19	2.44	
Sheep	ME Mcal/kg	*	2.50	2.79	
Swine	ME kcal/kg	*	2605.	2907.	
Cattle	TDN %		60.5	67.5	
Sheep	TDN %		69.1	77.1	
Swine	TDN %	*	63.3	70.7	
Niacin	mg/kg		47.6	53.1	
Arginine	%		0.45	0.50	
Histidine	%		0.18	0.20	
Isoleucine	%		0.36	0.40	
Leucine	%		0.63	0.70	
Lysine	%		0.27	0.30	
Methionine	%		0.18	0.20	
Phenylalanine	%		0.45	0.50	
Threonine	%		0.36	0.40	
Tryptophan	%		0.09	0.10	
Valine	%		0.45	0.50	

SPIDERWORT, WANDERINGJEW. Tradescantia

Spiderwort, wanderingjew, aerial part, fresh, (2)

Ref No 2 04 652 United States

Feed Name or Analyses			As Fed	Dry	C.V. ± %
Dry matter	%		10.0	100.0	
Carotene	mg/kg		5.7	57.1	
Niacin	mg/kg		2.4	24.0	
Riboflavin	mg/kg		1.3	13.0	
Thiamine	mg/kg		0.4	4.2	
Vitamin A equivalent	IU/g		9.5	95.2	

Fluminensis

Spikefescue, aerial part, fresh, (2)

Ref No 2 04 657 United States

Feed Name or Analyses			As Fed	Dry	C.V. ± %
Dry matter	%			100.0	
Ash	%			7.4	11
Crude fiber	%			33.9	9
Ether extract	%			4.2	15
N-free extract	%			43.0	
Protein (N x 6.25)	%			11.5	38
Cattle	dig prot %	*		7.7	
Goats	dig prot %	*		7.3	
Horses	dig prot %	*		7.3	
Rabbits	dig prot %	*		7.5	
Sheep	dig prot %	*		7.7	
Calcium	%			0.44	25
Phosphorus	%			0.30	31
Carotene	mg/kg			182.8	57
Vitamin A equivalent	IU/g			304.7	

Spikefescue, aerial part, fresh, immature, (2)

Ref No 2 04 653 United States

Feed Name or Analyses			As Fed	Dry	C.V. ± %
Dry matter	%			100.0	
Ash	%			7.6	9
Crude fiber	%			29.2	6
Ether extract	%			4.4	10
N-free extract	%			40.0	
Protein (N x 6.25)	%			18.8	9
Cattle	dig prot %	*		13.9	
Goats	dig prot %	*		14.1	
Horses	dig prot %	*		13.5	
Rabbits	dig prot %	*		13.2	
Sheep	dig prot %	*		14.5	
Carotene	mg/kg			288.4	
Vitamin A equivalent	IU/g			480.7	

Spikefescue, aerial part, fresh, early bloom, (2)

Ref No 2 04 654 United States

Feed Name or Analyses			As Fed	Dry	C.V. ± %
Dry matter	%			100.0	
Ash	%			7.8	7
Crude fiber	%			33.6	6
Ether extract	%			4.3	16
N-free extract	%			39.8	
Protein (N x 6.25)	%			14.5	9
Cattle	dig prot %	*		10.2	
Goats	dig prot %	*		10.1	
Horses	dig prot %	*		9.8	
Rabbits	dig prot %	*		9.9	
Sheep	dig prot %	*		10.5	
Calcium	%			0.49	16
Phosphorus	%			0.38	13
Carotene	mg/kg			233.2	35
Vitamin A equivalent	IU/g			388.8	

Spikefescue, aerial part, fresh, mature, (2)

Ref No 2 04 655 United States

Feed Name or Analyses			As Fed	Dry	C.V. ± %
Dry matter	%			100.0	
Ash	%			7.5	13
Crude fiber	%			37.2	5
Ether extract	%			3.8	17
N-free extract	%			44.9	
Protein (N x 6.25)	%			6.6	10
Cattle	dig prot %	*		3.5	
Goats	dig prot %	*		2.7	
Horses	dig prot %	*		3.1	
Rabbits	dig prot %	*		3.8	
Sheep	dig prot %	*		3.1	
Calcium	%			0.44	15
Phosphorus	%			0.20	37
Carotene	mg/kg			78.5	45
Vitamin A equivalent	IU/g			130.8	

Feed Name or Analyses		Mean As Fed	Mean Dry	C.V. ± %

Spikefescue, aerial part, fresh, over ripe, (2)
Ref No 2 04 656 United States

Analyses		As Fed	Dry	C.V. ± %
Dry matter	%		100.0	
Ash	%		6.4	
Crude fiber	%		39.0	
Ether extract	%		5.3	
N-free extract	%		45.5	
Protein (N x 6.25)	%		3.8	
Cattle	dig prot % *		1.1	
Goats	dig prot % *		0.1	
Horses	dig prot % *		0.8	
Rabbits	dig prot % *		1.6	
Sheep	dig prot % *		0.5	
Calcium	%		0.25	
Phosphorus	%		0.07	
Carotene	mg/kg		18.5	
Vitamin A equivalent	IU/g		30.9	

SPIKESEDGE, COMMON. Eleocharis palustris

Spikesedge, common, aerial part, fresh, (2)
Ref No 2 04 662 United States

Analyses		As Fed	Dry	C.V. ± %
Dry matter	%		100.0	
Ash	%		13.9	17
Crude fiber	%		29.0	7
Protein (N x 6.25)	%		10.3	47
Cattle	dig prot % *		6.6	
Goats	dig prot % *		6.2	
Horses	dig prot % *		6.3	
Rabbits	dig prot % *		6.6	
Sheep	dig prot % *		6.6	
Calcium	%		0.45	25
Phosphorus	%		0.21	23
Potassium	%		2.72	11

Spikesedge, common, aerial part, fresh, immature, (2)
Ref No 2 04 658 United States

Analyses		As Fed	Dry	C.V. ± %
Dry matter	%		100.0	
Ash	%		11.6	
Crude fiber	%		27.8	
Protein (N x 6.25)	%		17.0	
Cattle	dig prot % *		12.3	
Goats	dig prot % *		12.4	
Horses	dig prot % *		12.0	
Rabbits	dig prot % *		11.8	
Sheep	dig prot % *		12.8	
Calcium	%		0.50	11
Phosphorus	%		0.32	13
Potassium	%		3.07	

Spikesedge, common, aerial part, fresh, full bloom, (2)
Ref No 2 04 659 United States

Analyses		As Fed	Dry	C.V. ± %
Dry matter	%		100.0	
Ash	%		13.7	
Crude fiber	%		29.1	
Protein (N x 6.25)	%		8.6	
Cattle	dig prot % *		5.2	
Goats	dig prot % *		4.6	
Horses	dig prot % *		4.8	
Rabbits	dig prot % *		5.3	
Sheep	dig prot % *		5.0	
Calcium	%		0.36	10
Phosphorus	%		0.19	11
Potassium	%		2.71	

Spikesedge, common, aerial part, fresh, mature, (2)
Ref No 2 04 660 United States

Analyses		As Fed	Dry	C.V. ± %
Dry matter	%		100.0	
Ash	%		14.0	
Crude fiber	%		29.2	
Protein (N x 6.25)	%		6.9	
Cattle	dig prot % *		3.8	
Goats	dig prot % *		3.0	
Horses	dig prot % *		3.4	
Rabbits	dig prot % *		4.0	
Sheep	dig prot % *		3.4	
Calcium	%		0.25	14
Phosphorus	%		0.14	21
Potassium	%		2.65	

Spikesedge, common, aerial part, fresh, over ripe, (2)
Ref No 2 04 661 United States

Analyses		As Fed	Dry	C.V. ± %
Dry matter	%		100.0	
Ash	%		17.3	
Crude fiber	%		32.4	
Protein (N x 6.25)	%		5.0	
Cattle	dig prot % *		2.1	
Goats	dig prot % *		1.2	
Horses	dig prot % *		1.8	
Rabbits	dig prot % *		2.5	
Sheep	dig prot % *		1.7	

SPINACH. Spinacia oleracea

Spinach, leaves, fresh, (2)
Ref No 2 08 125 United States

Analyses		As Fed	Dry	C.V. ± %
Dry matter	%	9.4	100.0	
Ash	%	2.2	23.6	
Crude fiber	%	0.7	7.5	15
Ether extract	%	0.4	4.0	16
N-free extract	%	3.0	32.2	
Protein (N x 6.25)	%	3.1	32.7	5
Cattle	dig prot % *	2.4	25.7	
Goats	dig prot % *	2.5	27.1	
Horses	dig prot % *	2.4	25.3	
Rabbits	dig prot % *	2.2	23.9	
Sheep	dig prot % *	2.6	27.5	
Calcium	%	0.09	0.97	
Iron	%	0.003	0.032	
Phosphorus	%	0.05	0.54	
Potassium	%	0.48	5.05	
Sodium	%	0.07	0.75	
Ascorbic acid	mg/kg	515.5	5483.9	
Carotene	mg/kg	33.8	360.0	
Niacin	mg/kg	6.1	64.5	
Riboflavin	mg/kg	2.0	21.1	
Thiamine	mg/kg	1.0	10.8	
α-tocopherol	mg/kg	38.5	410.1	
Vitamin A	IU/g	81.9	871.0	
Vitamin A equivalent	IU/g	56.4	600.1	
Arginine	%	0.11	1.20	
Histidine	%	0.04	0.40	
Isoleucine	%	0.09	1.00	
Leucine	%	0.18	1.90	
Lysine	%	0.12	1.30	
Methionine	%	0.06	0.60	
Phenylalanine	%	0.12	1.30	
Threonine	%	0.10	1.10	
Tryptophan	%	0.03	0.30	
Valine	%	0.13	1.40	

Spinach, leaves w stems, fresh, (2)
Ref No 2 04 664 United States

Analyses		As Fed	Dry	C.V. ± %
Dry matter	%	7.9	100.0	
Carotene	mg/kg	13.7	173.3	
Folic acid	mg/kg	0.87	11.07	
Riboflavin	mg/kg	0.5	6.2	95
Vitamin A equivalent	IU/g	22.8	288.9	

Spinach, stems, fresh, (2)
Ref No 2 04 665 United States

Analyses		As Fed	Dry	C.V. ± %
Dry matter	%	6.5	100.0	
Crude fiber	%	0.6	9.3	
Protein (N x 6.25)	%	1.5	22.5	
Cattle	dig prot % *	1.1	17.0	
Goats	dig prot % *	1.1	17.6	
Horses	dig prot % *	1.1	16.6	
Rabbits	dig prot % *	1.0	16.0	
Sheep	dig prot % *	1.2	18.0	

SPRANGLETOP. Leptochloa spp

Sprangletop, aerial part, fresh, (2)
Ref No 2 04 666 United States

Analyses		As Fed	Dry	C.V. ± %
Dry matter	%		100.0	
Calcium	%		0.40	
Magnesium	%		0.17	
Phosphorus	%		0.23	
Potassium	%		1.85	

SPRANGLETOP, GREEN. Leptochloa dubia

Sprangletop, green, aerial part, fresh, (2)
Ref No 2 04 667 United States

Analyses		As Fed	Dry	C.V. ± %
Dry matter	%		100.0	
Ash	%		9.5	
Crude fiber	%		32.4	
Ether extract	%		2.1	
N-free extract	%		44.4	
Protein (N x 6.25)	%		11.6	
Cattle	dig prot % *		7.8	
Goats	dig prot % *		7.4	
Horses	dig prot % *		7.4	
Rabbits	dig prot % *		7.6	
Sheep	dig prot % *		7.8	

(1) dry forages and roughages (2) pasture, range plants, and forages fed green (3) silages (4) energy feeds (5) protein supplements (6) minerals (7) vitamins (8) additives

SQUID. Illex illecebrosus

Squid, liver oil, solv-extd, iodine value 190, (4)

Ref No 4 09 285 Canada

Feed Name or Analyses		Mean As Fed	Mean Dry	C.V. ± %
Dry matter	%	100.0	100.0	
Fatty acids	%			
Docosahexaenoic	%	14.400	14.400	
Docosapentaenoic	%	1.100	1.100	
Eicosapentaenoic	%	11.800	11.800	
Eicosenoic	%	10.500	10.500	
Myristic	%	3.700	3.700	
Oleic	%	13.900	13.900	
Palmitic	%	11.600	11.600	
Palmitoleic	%	7.700	7.700	
Stearic	%	1.500	1.500	

Squid, flesh, fresh, (5)

Ref No 5 09 287 Canada

Feed Name or Analyses		Mean As Fed	Mean Dry	C.V. ± %
Dry matter	%		100.0	
Ether extract	%		8.0	
Fatty acids	%			
Arachidonic	%		0.040	
Docosahexaenoic	%		1.780	
Eicosapentaenoic	%		0.760	
Myristic	%		0.100	
Oleic	%		0.240	
Palmitic	%		1.330	
Palmitoleic	%		0.020	
Stearic	%		0.210	

Squid, liver, (5)

Ref No 5 09 286 Canada

Feed Name or Analyses		Mean As Fed	Mean Dry	C.V. ± %
Dry matter	%		100.0	
Ether extract	%		22.0	
Fatty acids	%			
Docosahexaenoic	%		3.720	
Docosapentaenoic	%		0.290	
Eicosapentaenoic	%		3.060	
Eicosenoic	%		2.730	
Myristic	%		0.970	
Oleic	%		3.590	
Palmitic	%		3.010	
Palmitoleic	%		1.980	
Stearic	%		0.400	

SQUID. Ommastrephes spp, Loligo spp

Squid, whole or cuttings, cooked dehy grnd, (5)

Ref No 5 04 671 United States

Feed Name or Analyses		Mean As Fed	Mean Dry	C.V. ± %
Dry matter	%	90.0	100.0	
Ash	%	6.5	7.2	5
Crude fiber	%	0.0	0.0	5
Ether extract	%	12.6	14.0	40
N-free extract	%	3.6	4.0	
Protein (N x 6.25)	%	67.3	74.8	6
Energy	GE Mcal/kg			
Cattle	DE Mcal/kg	* 3.67	4.08	
Sheep	DE Mcal/kg	* 3.26	3.62	
Swine	DE kcal/kg	*3633.	4037.	
Cattle	ME Mcal/kg	* 3.01	3.34	
Sheep	ME Mcal/kg	* 2.68	2.97	
Swine	ME kcal/kg	*2939.	3266.	
Cattle	TDN %	* 83.2	92.4	
Sheep	TDN %	* 74.0	82.2	
Swine	TDN %	* 82.4	91.6	
Calcium	%	0.09	0.10	
Phosphorus	%	0.99	1.10	

SQUIRRELTAIL, BOTTLEBRUSH. Sitanion hystrix

Squirreltail, bottlebrush, aerial part, fresh, (2)

Ref No 2 04 674 United States

Feed Name or Analyses		Mean As Fed	Mean Dry	C.V. ± %
Dry matter	%		100.0	
Ash	%		14.6	30
Crude fiber	%		36.5	10
Ether extract	%		2.5	27
N-free extract	%		39.5	
Protein (N x 6.25)	%		6.9	53
Cattle	dig prot %	*	3.8	
Goats	dig prot %	*	3.0	
Horses	dig prot %	*	3.4	
Rabbits	dig prot %	*	4.0	
Sheep	dig prot %	*	3.4	

Squirreltail, bottlebrush, aerial part, fresh, mature, (2)

Ref No 2 04 672 United States

Feed Name or Analyses		Mean As Fed	Mean Dry	C.V. ± %
Dry matter	%		100.0	
Ash	%		15.6	
Crude fiber	%		34.0	
Ether extract	%		2.8	
N-free extract	%		41.6	
Protein (N x 6.25)	%		6.0	
Cattle	dig prot %	*	3.0	
Goats	dig prot %	*	2.2	
Horses	dig prot %	*	2.6	
Rabbits	dig prot %	*	3.3	
Sheep	dig prot %	*	2.6	
Carotene	mg/kg		1.1	
Vitamin A equivalent	IU/g		1.8	

Squirreltail, bottlebrush, aerial part, fresh, over ripe, (2)

Ref No 2 04 673 United States

Feed Name or Analyses		Mean As Fed	Mean Dry	C.V. ± %
Dry matter	%		100.0	
Ash	%		17.0	
Ether extract	%		2.2	
Protein (N x 6.25)	%		3.1	
Cattle	dig prot %	*	0.5	
Goats	dig prot %	*	-0.4	
Horses	dig prot %	*	0.2	
Rabbits	dig prot %	*	1.1	
Sheep	dig prot %	*	0.0	

Squirreltail, bottlebrush, browse, mature, (2)

Ref No 2 04 675 United States

Feed Name or Analyses		Mean As Fed	Mean Dry	C.V. ± %
Dry matter	%		100.0	
Calcium	%		1.01	45
Magnesium	%		0.14	
Phosphorus	%		0.10	55
Potassium	%		1.94	

SQUIRRELTAIL. Sitanion spp

Squirreltail, aerial part, fresh, dormant, (2)

Ref No 2 05 566 United States

Feed Name or Analyses		Mean As Fed	Mean Dry	C.V. ± %
Dry matter	%		100.0	
Ash	%		17.0	
Ether extract	%		2.2	
Sheep	dig coef %		35.	
Protein (N x 6.25)	%		3.1	
Sheep	dig coef %		1.	
Cattle	dig prot %	*	0.5	
Goats	dig prot %	*	-0.4	
Horses	dig prot %	*	0.2	
Rabbits	dig prot %	*	1.1	
Sheep	dig prot %		0.0	
Cellulose (Matrone)	%		40.4	
Sheep	dig coef %		76.	
Other carbohydrates	%		29.0	
Lignin (Ellis)	%		8.3	
Energy	GE Mcal/kg		3.79	
Sheep	GE dig coef %		53.	
Sheep	DE Mcal/kg		2.01	
Sheep	ME Mcal/kg		1.70	
Sheep	TDN %		50.4	
Calcium	%		0.37	
Phosphorus	%		0.06	

STARFISH. Asteroidea (class)

Starfish, entire animal, dehy grnd, (5)

Ref No 5 08 527 United States

Feed Name or Analyses		Mean As Fed	Mean Dry	C.V. ± %
Dry matter	%	96.5	100.0	
Ash	%	43.9	45.5	
Crude fiber	%	1.9	2.0	
Ether extract	%	5.8	6.0	
N-free extract	%	14.3	14.8	
Protein (N x 6.25)	%	30.6	31.7	
Energy	GE Mcal/kg			
Cattle	DE Mcal/kg	* 1.52	1.57	
Sheep	DE Mcal/kg	* 1.69	1.75	
Swine	DE kcal/kg	*1842.	1909.	
Cattle	ME Mcal/kg	* 1.24	1.29	
Chickens	ME_n kcal/kg	150.	156.	
Sheep	ME Mcal/kg	* 1.39	1.44	
Swine	ME kcal/kg	*1650.	1710.	
Cattle	TDN %	* 34.5	35.7	
Sheep	TDN %	* 38.2	39.6	
Swine	TDN %	* 41.8	43.3	
Calcium	%	15.70	16.27	
Phosphorus	%	0.45	0.47	

STERCULIA, HAZEL. Sterculia foetida

Sterculia, hazel, seeds w hulls, (5)

Ref No 5 09 007 United States

Feed Name or Analyses		Mean As Fed	Mean Dry	C.V. ± %
Dry matter	%		100.0	
Protein (N x 6.25)	%		20.0	
Alanine	%		0.94	
Arginine	%		1.74	
Aspartic acid	%		2.00	
Glutamic acid	%		3.96	
Glycine	%		0.82	
Histidine	%		0.58	

Continued

Feed Name or Analyses				Mean As Fed	Mean Dry	C.V. ± %
Hydroxyproline		%			0.02	
Isoleucine		%			0.82	
Leucine		%			1.22	
Lysine		%			1.20	
Methionine		%			0.40	
Phenylalanine		%			0.80	
Proline		%			0.64	
Serine		%			1.00	
Threonine		%			0.70	
Tyrosine		%			0.58	
Valine		%			1.00	

STICKSEED. Lappula spp

Stickseed, leaves, fresh, (2)

Ref No 2 04 678 United States

Feed Name or Analyses				Mean As Fed	Mean Dry	C.V. ± %
Dry matter		%			100.0	
Ash		%			11.6	
Crude fiber		%			10.6	
Ether extract		%			3.3	
N-free extract		%			51.7	
Protein (N x 6.25)		%			22.8	
Cattle	dig prot	%	*		17.3	
Goats	dig prot	%	*		17.8	
Horses	dig prot	%	*		16.9	
Rabbits	dig prot	%	*		16.3	
Sheep	dig prot	%	*		18.2	

STILLINGIA. Stillingia spp

Stillingia, aerial part, fresh, immature, (2)

Ref No 2 04 679 United States

Feed Name or Analyses				Mean As Fed	Mean Dry	C.V. ± %
Dry matter		%			100.0	
Ash		%			11.7	
Crude fiber		%			13.9	
Ether extract		%			3.2	
N-free extract		%			49.5	
Protein (N x 6.25)		%			21.7	
Cattle	dig prot	%	*		16.3	
Goats	dig prot	%	*		16.8	
Horses	dig prot	%	*		16.0	
Rabbits	dig prot	%	*		15.4	
Sheep	dig prot	%	*		17.2	
Calcium		%			0.31	
Magnesium		%			0.42	
Phosphorus		%			0.24	
Potassium		%			2.59	

STINKGRASS. Eragrostis cilianensis

Stinkgrass, aerial part, fresh, (2)

Ref No 2 04 680 United States

Feed Name or Analyses				Mean As Fed	Mean Dry	C.V. ± %
Dry matter		%			100.0	
Ash		%			10.9	
Crude fiber		%			27.8	
Ether extract		%			1.6	
N-free extract		%			47.5	
Protein (N x 6.25)		%			12.2	
Cattle	dig prot	%	*		8.3	

(1) dry forages and roughages (2) pasture, range plants, and forages fed green (3) silages (4) energy feeds (5) protein supplements (6) minerals (7) vitamins (8) additives

Feed Name or Analyses				Mean As Fed	Mean Dry	C.V. ± %
Goats	dig prot	%	*		7.9	
Horses	dig prot	%	*		7.9	
Rabbits	dig prot	%	*		8.1	
Sheep	dig prot	%	*		8.4	
Calcium		%			0.95	
Magnesium		%			0.26	
Phosphorus		%			0.24	
Potassium		%			1.61	

Stock melons –
see Melons, pie, fruit w seeds, fresh, (4)

STONECROP. Sedum spp

Stonecrop, aerial part, fresh, (2)

Ref No 2 04 681 United States

Feed Name or Analyses				Mean As Fed	Mean Dry	C.V. ± %
Dry matter		%			100.0	
Ash		%			20.7	
Crude fiber		%			11.7	
Ether extract		%			6.1	
N-free extract		%			53.0	
Protein (N x 6.25)		%			8.5	
Cattle	dig prot	%	*		5.1	
Goats	dig prot	%	*		4.5	
Horses	dig prot	%	*		4.7	
Rabbits	dig prot	%	*		5.2	
Sheep	dig prot	%	*		4.9	
Calcium		%			7.51	
Magnesium		%			0.21	
Phosphorus		%			0.40	
Potassium		%			2.64	

STRAWFLOWER. Helichrysum bracteatum

Strawflower, seeds w hulls, (5)

Ref No 5 09 038 United States

Feed Name or Analyses				Mean As Fed	Mean Dry	C.V. ± %
Dry matter		%			100.0	
Protein (N x 6.25)		%			22.0	
Alanine		%			0.95	
Arginine		%			1.47	
Aspartic acid		%			1.78	
Glutamic acid		%			3.72	
Glycine		%			1.25	
Histidine		%			0.33	
Hydroxyproline		%			0.15	
Isoleucine		%			0.70	
Leucine		%			1.17	
Lysine		%			0.68	
Methionine		%			0.40	
Phenylalanine		%			0.73	
Proline		%			0.75	
Serine		%			1.06	
Threonine		%			0.75	
Tyrosine		%			0.48	
Valine		%			0.88	

Strip cane –
see Sugarcane, leaves w top of aerial part, fresh, (2)

Sucrose –
see Sugarcane, sugar, (4)

SUGARCANE. Saccharum officinarum

Sugarcane, aerial part, s-c, (1)

Ref No 1 04 685 United States

Feed Name or Analyses				Mean As Fed	Mean Dry	C.V. ± %
Dry matter		%		92.7	100.0	4
Ash		%		3.8	4.1	34
Crude fiber		%		40.0	43.1	20
Ether extract		%		1.1	1.2	98
N-free extract		%		45.4	49.0	
Protein (N x 6.25)		%		2.4	2.6	72
Cattle	dig prot	%	*	-0.6	-0.7	
Goats	dig prot	%	*	-0.8	-0.9	
Horses	dig prot	%	*	-0.1	-0.2	
Rabbits	dig prot	%	*	0.6	0.7	
Sheep	dig prot	%	*	-0.9	-1.0	
Energy	GE Mcal/kg					
Cattle	DE Mcal/kg		*	1.96	2.11	
Sheep	DE Mcal/kg		*	2.02	2.18	
Cattle	ME Mcal/kg		*	1.60	1.73	
Sheep	ME Mcal/kg		*	1.66	1.79	
Cattle	TDN	%	*	44.3	47.8	
Sheep	TDN	%	*	45.9	49.5	

Sugarcane, pulp, dehy, (1)
Cane bagasse, dried

Ref No 1 04 686 United States

Feed Name or Analyses				Mean As Fed	Mean Dry	C.V. ± %
Dry matter		%		91.5	100.0	6
Ash		%		2.9	3.1	14
Crude fiber		%		44.5	48.6	11
Cattle	dig coef	%	*	58.	58.	
Ether extract		%		0.7	0.7	29
Cattle	dig coef	%	*	53.	53.	
N-free extract		%		42.0	45.9	
Cattle	dig coef	%	*	41.	41.	
Protein (N x 6.25)		%		1.5	1.7	20
Cattle	dig prot	%	*	-1.4	-1.5	
Goats	dig prot	%	*	-1.6	-1.8	
Horses	dig prot	%	*	-0.9	-1.0	
Rabbits	dig prot	%	*	0.0	0.0	
Sheep	dig prot	%	*	-1.7	-1.8	
Fatty acids		%		0.5	0.5	
Energy	GE Mcal/kg					
Cattle	DE Mcal/kg		*	1.13	1.24	
Sheep	DE Mcal/kg		*	1.88	2.06	
Cattle	ME Mcal/kg		*	0.93	1.02	
Sheep	ME Mcal/kg		*	1.55	1.69	
Cattle	TDN	%	*	25.7	28.1	
Sheep	TDN	%	*	42.7	46.7	

Sugarcane, aerial part, fresh, (2)

Ref No 2 04 689 United States

Feed Name or Analyses				Mean As Fed	Mean Dry	C.V. ± %
Dry matter		%		25.1	100.0	9
Ash		%		1.7	6.8	
Crude fiber		%		7.9	31.3	
Cattle	dig coef	%		61.	61.	
Ether extract		%		0.6	2.2	
Cattle	dig coef	%		52.	52.	
N-free extract		%		13.4	53.5	
Cattle	dig coef	%		61.	61.	
Protein (N x 6.25)		%		1.5	6.2	
Cattle	dig coef	%		53.	53.	
Cattle	dig prot	%		0.8	3.3	
Goats	dig prot	%	*	0.6	2.3	
Horses	dig prot	%	*	0.7	2.8	

Feed Name or Analyses			Mean As Fed	Mean Dry	C.V. ± %
Rabbits	dig prot %	*	0.9	3.4	
Sheep	dig prot %	*	0.7	2.7	
Lignin (Ellis)	%		2.1	8.2	
Energy	GE Mcal/kg				
Cattle	DE Mcal/kg	*	0.63	2.52	
Sheep	DE Mcal/kg	*	0.69	2.75	
Cattle	ME Mcal/kg	*	0.52	2.07	
Sheep	ME Mcal/kg	*	0.57	2.26	
Cattle	TDN %		14.3	57.2	
Sheep	TDN %	*	15.6	62.3	
Calcium	%		0.12	0.47	
Iron	%		0.002	0.008	
Magnesium	%		0.16	0.62	
Phosphorus	%		0.04	0.17	
Potassium	%		0.40	1.59	

Sugarcane, aerial part, fresh, mature, (2)

Ref No 2 04 687 United States

Feed Name or Analyses			Mean As Fed	Mean Dry	C.V. ± %
Dry matter	%		27.2	100.0	
Ash	%		1.6	6.0	
Crude fiber	%		8.7	32.0	
Cattle	dig coef %		59.	59.	
Ether extract	%		0.6	2.1	
Cattle	dig coef %		50.	50.	
N-free extract	%		14.3	52.7	
Cattle	dig coef %		63.	63.	
Protein (N x 6.25)	%		2.0	7.3	
Cattle	dig coef %		52.	52.	
Cattle	dig prot %		1.0	3.8	
Goats	dig prot %	*	0.9	3.3	
Horses	dig prot %	*	1.0	3.7	
Rabbits	dig prot %	*	1.2	4.3	
Sheep	dig prot %	*	1.0	3.7	
Energy	GE Mcal/kg				
Cattle	DE Mcal/kg	*	0.69	2.56	
Sheep	DE Mcal/kg	*	0.72	2.66	
Cattle	ME Mcal/kg	*	0.57	2.10	
Sheep	ME Mcal/kg	*	0.59	2.18	
Cattle	TDN %		15.7	58.0	
Sheep	TDN %	*	16.4	60.3	

Sugarcane, leaves, fresh, (2)

Ref No 2 04 691 United States

Feed Name or Analyses			Mean As Fed	Mean Dry	C.V. ± %
Dry matter	%		35.5	100.0	
Ash	%		2.8	8.0	
Crude fiber	%		13.1	36.8	
Ether extract	%		0.9	2.5	
N-free extract	%		16.3	46.0	
Protein (N x 6.25)	%		2.4	6.7	
Cattle	dig prot %	*	1.3	3.6	
Goats	dig prot %	*	1.0	2.8	
Horses	dig prot %	*	1.1	3.2	
Rabbits	dig prot %	*	1.4	3.8	
Sheep	dig prot %	*	1.1	3.2	
Energy	GE Mcal/kg				
Cattle	DE Mcal/kg	*	0.99	2.79	
Sheep	DE Mcal/kg	*	1.02	2.89	
Cattle	ME Mcal/kg	*	0.81	2.29	
Sheep	ME Mcal/kg	*	0.84	2.37	
Cattle	TDN %	*	22.5	63.3	
Sheep	TDN %	*	23.2	65.5	

Sugarcane, leaves w top of aerial part, fresh, (2)

Strip cane

Ref No 2 04 692 United States

Feed Name or Analyses			Mean As Fed	Mean Dry	C.V. ± %
Dry matter	%		25.5	100.0	1
Ash	%		3.0	11.8	15
Crude fiber	%		8.5	33.1	2
Ether extract	%		0.4	1.7	9
N-free extract	%		12.3	48.1	
Protein (N x 6.25)	%		1.3	5.2	4
Cattle	dig prot %	*	0.6	2.3	
Goats	dig prot %	*	0.4	1.4	
Horses	dig prot %	*	0.5	2.0	
Rabbits	dig prot %	*	0.7	2.7	
Sheep	dig prot %	*	0.5	1.9	
Energy	GE Mcal/kg				
Cattle	DE Mcal/kg	*	0.53	2.09	
Sheep	DE Mcal/kg	*	0.58	2.27	
Cattle	ME Mcal/kg	*	0.44	1.71	
Sheep	ME Mcal/kg	*	0.47	1.86	
Cattle	TDN %	*	12.1	47.3	
Sheep	TDN %	*	12.9	50.5	
Calcium	%		0.09	0.35	
Phosphorus	%		0.07	0.27	
Potassium	%		0.75	2.96	

Sugarcane, aerial part, ensiled, (3)

Ref No 3 04 693 United States

Feed Name or Analyses			Mean As Fed	Mean Dry	C.V. ± %
Dry matter	%		22.1	100.0	10
Ash	%		0.9	4.1	
Crude fiber	%		8.7	39.3	
Ether extract	%		0.6	2.7	
N-free extract	%		11.0	49.8	
Protein (N x 6.25)	%		0.9	4.1	
Cattle	dig prot %	*	0.0	0.0	
Goats	dig prot %	*	0.0	0.0	
Horses	dig prot %	*	0.0	0.0	
Sheep	dig prot %	*	0.0	0.0	
Energy	GE Mcal/kg				
Cattle	DE Mcal/kg	*	0.58	2.62	
Sheep	DE Mcal/kg	*	0.61	2.78	
Cattle	ME Mcal/kg	*	0.48	2.15	
Sheep	ME Mcal/kg	*	0.50	2.28	
Cattle	TDN %	*	13.1	59.4	
Sheep	TDN %	*	13.9	63.1	
Calcium	%		0.08	0.35	
Magnesium	%		0.05	0.22	
Phosphorus	%		0.04	0.18	

Sugarcane, top of aerial part, ensiled, (3)

Ref No 3 08 528 United States

Feed Name or Analyses			Mean As Fed	Mean Dry	C.V. ± %
Dry matter	%		29.6	100.0	
Ash	%		2.8	9.5	
Crude fiber	%		10.6	35.8	
Ether extract	%		0.6	2.0	
N-free extract	%		14.1	47.6	
Protein (N x 6.25)	%		1.5	5.1	
Cattle	dig prot %	*	0.2	0.8	
Goats	dig prot %	*	0.2	0.8	
Horses	dig prot %	*	0.2	0.8	
Sheep	dig prot %	*	0.2	0.8	
Energy	GE Mcal/kg				
Cattle	DE Mcal/kg	*	0.66	2.24	
Sheep	DE Mcal/kg	*	0.70	2.38	
Cattle	ME Mcal/kg	*	0.54	1.84	
Sheep	ME Mcal/kg	*	0.58	1.95	
Cattle	TDN %	*	15.0	50.8	
Sheep	TDN %	*	16.0	53.9	

Sugarcane, molasses, dehy, (4)

Cane molasses, dried

Molasses, cane, dried

Ref No 4 04 695 United States

Feed Name or Analyses			Mean As Fed	Mean Dry	C.V. ± %
Dry matter	%		93.5	100.0	2
Ash	%		13.4	14.4	6
Crude fiber	%		2.7	2.9	15
Ether extract	%		0.5	0.6	98
N-free extract	%		66.8	71.5	
Protein (N x 6.25)	%		10.0	10.7	6
Cattle	dig prot %	*	5.5	5.9	
Goats	dig prot %	*	6.6	7.1	
Horses	dig prot %	*	6.6	7.1	
Sheep	dig prot %	*	6.6	7.1	
Energy	GE Mcal/kg		3.83	4.10	
Cattle	DE Mcal/kg	*	3.52	3.76	
Sheep	DE Mcal/kg	*	3.15	3.37	
Swine	DE kcal/kg	*	2997.	3205.	
Cattle	ME Mcal/kg	*	2.88	3.08	
Chickens	ME_n kcal/kg		2345.	2508.	
Sheep	ME Mcal/kg	*	2.59	2.77	
Swine	ME kcal/kg	*	2811.	3006.	
Cattle	TDN %	*	79.8	85.3	
Sheep	TDN %	*	71.5	76.5	
Swine	TDN %	*	68.0	72.7	
Calcium	%		1.15	1.23	
Phosphorus	%		0.14	0.15	
Silicon	%		5.32	5.69	

Sugarcane, molasses, mn 48% invert sugar mn 79.5 degrees brix, (4)

Cane molasses (AAFCO)

Molasses, cane

Ref No 4 04 696 United States

Feed Name or Analyses			Mean As Fed	Mean Dry	C.V. ± %
Dry matter	%		77.0	100.0	8
Ash	%		7.8	10.1	24
Crude fiber	%		0.0	0.0	24
Ether extract	%		0.0	0.0	98
N-free extract	%		64.7	84.0	6
Cattle	dig coef %		105.	105.	
Sheep	dig coef %		83.	83.	
Protein (N x 6.25)	%		4.5	5.9	46
Cattle	dig coef %		57.	57.	
Sheep	dig coef %		28.	28.	
Cattle	dig prot %		2.6	3.4	
Goats	dig prot %	*	2.0	2.6	
Horses	dig prot %	*	2.0	2.6	
Sheep	dig prot %		1.2	1.6	
Energy	GE Mcal/kg		2.68	3.48	11
Cattle	DE Mcal/kg	*	3.26	4.23	
Sheep	DE Mcal/kg	*	2.41	3.13	
Swine	DE kcal/kg	*	2524.	3277.	
Cattle	ME Mcal/kg	*	2.67	3.47	
Chickens	ME_n kcal/kg		1007.	2607.	
Sheep	ME Mcal/kg	*	1.98	2.57	
Swine	ME kcal/kg	*	2393.	3107.	
Cattle	NE_m Mcal/kg	*	1.75	2.27	
Cattle	NE_{gain} Mcal/kg	*	1.14	1.48	
Cattle	$NE_{lactating\ cows}$ Mcal/kg	*	2.00	2.60	
Cattle	TDN %		73.9	96.0	
Sheep	TDN %		54.8	71.1	

Continued

Feed Name or Analyses		Mean As Fed	Mean Dry	C.V. ± %
Swine	TDN %	*57.2	74.3	
Calcium	%	0.81	1.05	42
Chlorine	%	2.86	3.72	23
Cobalt	mg/kg	0.937	1.216	98
Copper	mg/kg	61.8	80.2	45
Iron	%	0.019	0.024	42
Magnesium	%	0.36	0.47	41
Manganese	mg/kg	44.1	57.2	42
Phosphorus	%	0.08	0.11	53
Potassium	%	3.09	4.02	12
Sodium	%	0.15	0.20	94
Sulphur	%	0.35	0.46	55
Choline	mg/kg	915.	1189.	66
Niacin	mg/kg	30.6	39.8	37
Pantothenic acid	mg/kg	39.9	51.8	24
Riboflavin	mg/kg	2.9	3.8	87
Thiamine	mg/kg	1.0	1.3	41

Sugarcane, molasses distillation solubles, condensed, (4)

Molasses distillers condensed solubles (AAFCO)

Ref No 4 04 697 United States

Feed Name or Analyses		Mean As Fed	Mean Dry	C.V. ± %
Dry matter	%	95.5	100.0	2
Ash	%	27.4	28.7	17
Crude fiber	%	1.3	1.4	70
Ether extract	%	0.6	0.6	65
N-free extract	%	54.2	56.8	
Protein (N x 6.25)	%	12.0	12.6	27
Cattle	dig prot %	*7.2	7.6	
Goats	dig prot %	*8.4	8.7	
Horses	dig prot %	*8.4	8.7	
Sheep	dig prot %	*8.4	8.7	
Energy	GE Mcal/kg			
Cattle	DE Mcal/kg	*3.66	3.84	
Sheep	DE Mcal/kg	*2.84	2.97	
Swine	DE kcal/kg	*2960.	3100.	
Cattle	ME Mcal/kg	*3.00	3.15	
Sheep	ME Mcal/kg	*2.33	2.44	
Swine	ME kcal/kg	*2768.	2898.	
Cattle	TDN %	*83.1	87.0	
Sheep	TDN %	*64.4	67.4	
Swine	TDN %	*67.1	70.3	
Choline	mg/kg	1799.	1884.	66
Folic acid	mg/kg	0.18	0.18	
Niacin	mg/kg	65.0	68.1	49
Pantothenic acid	mg/kg	47.8	50.1	15
Riboflavin	mg/kg	43.2	45.2	98

Sugarcane, molasses distillation solubles, dehy, (4)

Molasses distillers dried solubles (AAFCO)

Molasses dried fermentation solubles

Ref No 4 04 698 United States

Feed Name or Analyses		Mean As Fed	Mean Dry	C.V. ± %
Dry matter	%	94.9	100.0	
Ash	%	24.7	26.0	
Crude fiber	%	0.7	0.7	
Ether extract	%	0.6	0.6	
N-free extract	%	56.0	59.0	
Protein (N x 6.25)	%	12.9	13.6	
Cattle	dig prot %	*8.1	8.5	
Goats	dig prot %	*9.2	9.7	
Horses	dig prot %	*9.2	9.7	

(1) dry forages and roughages (2) pasture, range plants, and forages fed green (3) silages (4) energy feeds (5) protein supplements (6) minerals (7) vitamins (8) additives

Feed Name or Analyses		Mean As Fed	Mean Dry	C.V. ± %
Sheep	dig prot %	*9.2	9.7	
Energy	GE Mcal/kg			
Cattle	DE Mcal/kg	*3.38	3.57	
Sheep	DE Mcal/kg	*2.89	3.05	
Swine	DE kcal/kg	*3147.	3316.	
Cattle	ME Mcal/kg	*2.77	2.92	
Sheep	ME Mcal/kg	*2.37	2.50	
Swine	ME kcal/kg	*2936.	3094.	
Cattle	TDN %	*76.7	80.9	
Sheep	TDN %	*65.6	69.1	
Swine	TDN %	*71.4	75.2	

Sugarcane, pulp, sifted, (4)

Ref No 4 04 700 United States

Feed Name or Analyses		Mean As Fed	Mean Dry	C.V. ± %
Dry matter	%	90.3	100.0	
Ash	%	7.1	7.9	
Crude fiber	%	40.6	45.0	
Cattle	dig coef %	58.	58.	
Ether extract	%	0.9	1.0	
Cattle	dig coef %	53.	53.	
N-free extract	%	39.9	44.2	
Cattle	dig coef %	41.	41.	
Protein (N x 6.25)	%	1.7	1.9	
Cattle	dig prot %	*-1.9	-2.1	
Goats	dig prot %	*-0.8	-0.9	
Horses	dig prot %	*-0.8	-0.9	
Sheep	dig prot %	*-0.8	-0.9	
Energy	GE Mcal/kg			
Cattle	DE Mcal/kg	*1.81	2.00	
Sheep	DE Mcal/kg	*2.00	2.22	
Cattle	ME Mcal/kg	*1.48	1.64	
Sheep	ME Mcal/kg	*1.64	1.82	
Cattle	TDN %	41.0	45.4	
Sheep	TDN %	*45.5	50.4	

Sugarcane, sugar, (4)

Ref No 4 04 701 United States

Feed Name or Analyses		Mean As Fed	Mean Dry	C.V. ± %
Dry matter	%	99.8	100.0	
N-free extract	%	99.8	100.0	
Swine	dig coef %	99.	99.	
Energy	GE Mcal/kg			
Swine	DE kcal/kg	*4352.	4361.	
Chickens	ME_n kcal/kg	3720.	3727.	
Swine	TDN %	98.7	98.9	

Sugarcane, molasses, ammoniated, (5)

Molasses, cane, ammoniated

Cane, molasses, ammoniated

Ref No 5 04 702 United States

Feed Name or Analyses		Mean As Fed	Mean Dry	C.V. ± %
Dry matter	%	66.4	100.0	4
Cattle	dig coef %	74.	74.	
Ash	%	6.0	9.1	17
Crude fiber	%	0.0	0.0	17
Ether extract	%	0.0	0.0	17
N-free extract	%	35.8	53.8	
Cattle	dig coef %	89.	89.	
Protein (N x 6.25)	%	24.6	37.1	36
Cattle	dig coef %	59.	59.	
Cattle	dig prot %	14.5	21.9	
Energy	GE Mcal/kg			
Cattle	DE Mcal/kg	*2.06	3.10	
Sheep	DE Mcal/kg	*2.26	3.41	
Swine	DE kcal/kg	*2225.	3351.	

Feed Name or Analyses		Mean As Fed	Mean Dry	C.V. ± %
Cattle	ME Mcal/kg	*1.69	2.54	
Sheep	ME Mcal/kg	*1.86	2.80	
Swine	ME kcal/kg	*1969.	2966.	
Cattle	TDN %	*46.7	70.4	
Sheep	TDN %	*51.4	77.4	
Swine	TDN %	*50.5	76.0	
Calcium	%	0.81	1.22	11
Phosphorus	%	0.13	0.20	13

Sugarcane, molasses fermentation solubles, dehy, (5)

Dried molasses fermentation solubles (CFA)

Molasses fermentation solubles, dried

Ref No 5 04 699 United States

Feed Name or Analyses		Mean As Fed	Mean Dry	C.V. ± %
Dry matter	%	95.5	100.0	0
Ash	%	15.0	15.7	52
Crude fiber	%	1.2	1.3	62
Ether extract	%	0.7	0.7	60
N-free extract	%	55.6	58.2	
Protein (N x 6.25)	%	23.0	24.1	34
Calcium	%	1.73	1.81	
Manganese	mg/kg	69.7	72.9	
Phosphorus	%	0.34	0.36	
Biotin	mg/kg	2.16	2.26	
Pantothenic acid	mg/kg	188.3	197.1	
Riboflavin	mg/kg	72.3	75.7	7

SUGARCANE, JAPANESE. Saccharum officinarum

Sugarcane, Japanese, aerial part, s-c, (1)

Ref No 1 08 529 United States

Feed Name or Analyses		Mean As Fed	Mean Dry	C.V. ± %
Dry matter	%	89.0	100.0	
Ash	%	1.9	2.1	
Crude fiber	%	19.7	22.1	
Ether extract	%	1.8	2.0	
N-free extract	%	64.3	72.2	
Protein (N x 6.25)	%	1.3	1.5	
Cattle	dig prot %	*-1.5	-1.7	
Goats	dig prot %	*-1.7	-2.0	
Horses	dig prot %	*-1.0	-1.1	
Rabbits	dig prot %	*-0.1	-0.1	
Sheep	dig prot %	*-1.8	-2.0	
Energy	GE Mcal/kg			
Cattle	DE Mcal/kg	*2.50	2.81	
Sheep	DE Mcal/kg	*2.33	2.62	
Cattle	ME Mcal/kg	*2.05	2.30	
Sheep	ME Mcal/kg	*1.91	2.15	
Cattle	TDN %	*56.7	63.7	
Sheep	TDN %	*52.8	59.3	
Calcium	%	0.32	0.36	
Phosphorus	%	0.14	0.16	
Potassium	%	0.58	0.65	

Sugarcane, Japanese, aerial part, fresh, (2)

Ref No 2 08 575 United States

Feed Name or Analyses		Mean As Fed	Mean Dry	C.V. ± %
Dry matter	%	28.2	100.0	
Ash	%	1.1	3.9	
Crude fiber	%	7.7	27.3	
Ether extract	%	0.5	1.8	
N-free extract	%	18.2	64.5	
Protein (N x 6.25)	%	0.7	2.5	
Cattle	dig prot %	*0.0	0.0	
Goats	dig prot %	*-0.2	-1.0	
Horses	dig prot %	*0.0	-0.3	
Rabbits	dig prot %	*0.2	0.6	

Feed Name or Analyses	Mean As Fed	Mean Dry	C.V. ± %
Sheep dig prot % *	-0.1	-0.6	
Energy GE Mcal/kg			
Cattle DE Mcal/kg *	0.72	2.57	
Sheep DE Mcal/kg *	0.78	2.78	
Cattle ME Mcal/kg *	0.60	2.11	
Sheep ME Mcal/kg *	0.64	2.28	
Cattle TDN % *	16.5	58.4	
Sheep TDN % *	17.8	63.1	
Calcium %	0.10	0.35	
Phosphorus %	0.04	0.14	
Potassium %	0.18	0.64	

Sugar, sugarcane –
see Sugarcane, sugar, (4)

SUMAC. Rhus spp

Sumac, browse, (2)
Ref No 2 04 707 United States

Feed Name or Analyses	Mean As Fed	Mean Dry	C.V. ± %
Dry matter %	48.1	100.0	23
Ash %	2.9	6.0	24
Crude fiber %	6.5	13.5	37
Ether extract %	2.3	4.7	31
N-free extract %	31.6	65.8	
Protein (N x 6.25) %	4.8	10.0	30
Cattle dig prot % *	3.1	6.4	
Goats dig prot % *	2.8	5.9	
Horses dig prot % *	2.9	6.0	
Rabbits dig prot % *	3.1	6.4	
Sheep dig prot % *	3.0	6.3	
Calcium %	0.93	1.94	28
Magnesium %	0.14	0.30	40
Phosphorus %	0.07	0.15	30
Potassium %	0.84	1.74	26

SUMAC, EVERGREEN. Rhus sempervirens

Sumac, evergreen, browse, (2)
Ref No 2 04 709 United States

Feed Name or Analyses	Mean As Fed	Mean Dry	C.V. ± %
Dry matter %		100.0	
Ash %		4.8	
Crude fiber %		18.7	
Ether extract %		6.3	
N-free extract %		60.3	
Protein (N x 6.25) %		9.9	
Cattle dig prot % *		6.3	
Goats dig prot % *		5.8	
Horses dig prot % *		5.9	
Rabbits dig prot % *		6.3	
Sheep dig prot % *		6.2	
Calcium %		1.67	
Magnesium %		0.12	
Phosphorus %		0.05	
Potassium %		0.73	

Sumac, evergreen, browse, mature, (2)
Ref No 2 04 708 United States

Feed Name or Analyses	Mean As Fed	Mean Dry	C.V. ± %
Dry matter %		100.0	
Calcium %		1.50	
Magnesium %		0.23	
Potassium %		0.70	

SUMAC, LITTLELEAF. Rhus microphylla

Sumac, littleleaf, browse, immature, (2)
Ref No 2 04 710 United States

Feed Name or Analyses	Mean As Fed	Mean Dry	C.V. ± %
Dry matter %		100.0	
Ash %		7.6	24
Crude fiber %		11.4	6
Ether extract %		4.7	37
N-free extract %		61.4	
Protein (N x 6.25) %		14.9	11
Cattle dig prot % *		10.6	
Goats dig prot % *		10.5	
Horses dig prot % *		10.2	
Rabbits dig prot % *		10.2	
Sheep dig prot % *		10.9	

SUMAC, SKUNKBUSH. Rhus trilobata

Sumac, skunkbush, browse, (2)
Ref No 2 04 711 United States

Feed Name or Analyses	Mean As Fed	Mean Dry	C.V. ± %
Dry matter %	48.1	100.0	23
Ash %	2.6	5.5	15
Crude fiber %	6.6	13.7	41
Ether extract %	2.1	4.4	35
N-free extract %	32.9	68.4	
Protein (N x 6.25) %	3.8	8.0	20
Cattle dig prot % *	2.3	4.7	
Goats dig prot % *	1.9	4.0	
Horses dig prot % *	2.1	4.3	
Rabbits dig prot % *	2.3	4.8	
Sheep dig prot % *	2.1	4.4	
Calcium %	0.93	1.93	35
Magnesium %	0.13	0.28	48
Phosphorus %	0.05	0.11	22
Potassium %	0.81	1.69	44

Sumac, skunkbush, stems, fresh, (2)
Ref No 2 04 712 United States

Feed Name or Analyses	Mean As Fed	Mean Dry	C.V. ± %
Dry matter %	56.7	100.0	17
Calcium %	0.46	0.81	22
Phosphorus %	0.09	0.16	24

SUMMERCYPRESS. Kochia spp

Summercypress, hay, s-c, (1)
Ref No 1 04 713 United States

Feed Name or Analyses	Mean As Fed	Mean Dry	C.V. ± %
Dry matter %	91.8	100.0	3
Ash %	12.4	13.5	22
Crude fiber %	23.6	25.7	20
Ether extract %	1.5	1.6	29
N-free extract %	41.8	45.5	
Protein (N x 6.25) %	12.6	13.7	26
Cattle dig prot % *	8.1	8.8	
Goats dig prot % *	8.6	9.3	
Horses dig prot % *	8.4	9.2	
Rabbits dig prot % *	8.5	9.2	
Sheep dig prot % *	8.1	8.9	
Energy GE Mcal/kg			
Cattle DE Mcal/kg *	2.34	2.55	
Sheep DE Mcal/kg *	2.30	2.51	
Cattle ME Mcal/kg *	1.92	2.09	
Sheep ME Mcal/kg *	1.89	2.06	
Cattle TDN % *	53.1	57.8	
Sheep TDN % *	52.2	56.9	

SUMMERCYPRESS, BELVEDERE. Kochia scoparia

Summercypress, belvedere, hay, s-c, (1)
Kochia scoparia hay
Ref No 1 04 717 United States

Feed Name or Analyses	Mean As Fed	Mean Dry	C.V. ± %
Dry matter %	91.6	100.0	3
Ash %	12.7	13.8	22
Crude fiber %	23.4	25.6	20
Ether extract %	1.5	1.6	25
N-free extract %	42.0	45.8	
Protein (N x 6.25) %	12.1	13.2	28
Cattle dig prot % *	7.7	8.4	
Goats dig prot % *	8.2	8.9	
Horses dig prot % *	8.0	8.8	
Rabbits dig prot % *	8.1	8.9	
Sheep dig prot % *	7.7	8.4	
Fatty acids %	1.5	1.7	
Energy GE Mcal/kg			
Cattle DE Mcal/kg *	2.31	2.52	
Sheep DE Mcal/kg *	2.29	2.50	
Cattle ME Mcal/kg *	1.90	2.07	
Sheep ME Mcal/kg *	1.88	2.05	
Cattle TDN % *	52.3	57.1	
Sheep TDN % *	51.9	56.7	
Carotene mg/kg	26.5	28.9	46
Vitamin A equivalent IU/g	44.1	48.1	

Summercypress, belvedere, hay, s-c, immature, (1)
Ref No 1 04 714 United States

Feed Name or Analyses	Mean As Fed	Mean Dry	C.V. ± %
Dry matter %	95.0	100.0	1
Ash %	13.0	13.7	17
Crude fiber %	18.7	19.7	15
Ether extract %	1.3	1.4	17
N-free extract %	44.1	46.4	
Protein (N x 6.25) %	17.9	18.8	15
Cattle dig prot % *	12.6	13.2	
Goats dig prot % *	13.4	14.1	
Horses dig prot % *	12.8	13.5	
Rabbits dig prot % *	12.5	13.2	
Sheep dig prot % *	12.8	13.4	
Energy GE Mcal/kg			
Cattle DE Mcal/kg *	2.45	2.58	
Sheep DE Mcal/kg *	2.32	2.44	
Cattle ME Mcal/kg *	2.01	2.12	
Sheep ME Mcal/kg *	1.90	2.00	
Cattle TDN % *	55.6	58.5	
Sheep TDN % *	52.6	55.4	
Carotene mg/kg	41.9	44.1	
Vitamin A equivalent IU/g	69.8	73.5	

Summercypress, belvedere, hay, s-c, early bloom, (1)
Ref No 1 04 715 United States

Feed Name or Analyses	Mean As Fed	Mean Dry	C.V. ± %
Dry matter %	95.6	100.0	
Ash %	13.3	13.9	27
Crude fiber %	26.4	27.6	22
Ether extract %	1.5	1.6	30
N-free extract %	41.9	43.8	
Protein (N x 6.25) %	12.5	13.1	13
Cattle dig prot % *	7.9	8.3	
Goats dig prot % *	8.4	8.8	

Continued

Feed Name or Analyses			Mean As Fed	Mean Dry	C.V. ± %
Horses	dig prot %	*	8.3	8.7	
Rabbits	dig prot %	*	8.4	8.8	
Sheep	dig prot %	*	8.0	8.3	
Energy	GE Mcal/kg				
Cattle	DE Mcal/kg	*	2.43	2.54	
Sheep	DE Mcal/kg	*	2.33	2.43	
Cattle	ME Mcal/kg	*	1.99	2.08	
Sheep	ME Mcal/kg	*	1.91	2.00	
Cattle	TDN %	*	55.2	57.7	
Sheep	TDN %	*	52.8	55.2	

Summercypress, belvedere, hay, s-c, mature, (1)
Ref No 1 04 716 United States

Feed Name or Analyses			Mean As Fed	Mean Dry	C.V. ± %
Dry matter	%		87.4	100.0	
Ash	%		10.7	12.2	
Crude fiber	%		31.1	35.6	
Ether extract	%		1.3	1.5	
N-free extract	%		34.5	39.5	
Protein (N x 6.25)	%		9.8	11.2	
Cattle	dig prot %	*	5.8	6.6	
Goats	dig prot %	*	6.1	7.0	
Horses	dig prot %	*	6.2	7.0	
Rabbits	dig prot %	*	6.4	7.3	
Sheep	dig prot %	*	5.8	6.6	
Energy	GE Mcal/kg				
Cattle	DE Mcal/kg	*	2.19	2.51	
Sheep	DE Mcal/kg	*	2.12	2.42	
Cattle	ME Mcal/kg	*	1.80	2.06	
Sheep	ME Mcal/kg	*	1.74	1.99	
Cattle	TDN %	*	49.8	56.9	
Sheep	TDN %	*	48.0	55.0	
Carotene	mg/kg		8.9	10.1	
Vitamin A equivalent	IU/g		14.8	16.9	

Summercypress, belvedere, aerial part, ensiled, (3)
Ref No 3 04 718 United States

Feed Name or Analyses			Mean As Fed	Mean Dry	C.V. ± %
Dry matter	%		35.7	100.0	65
Ash	%		4.5	12.6	4
Crude fiber	%		11.2	31.3	98
Ether extract	%		0.8	2.3	9
N-free extract	%		13.4	37.4	
Protein (N x 6.25)	%		5.8	16.4	10
Cattle	dig prot %	*	4.0	11.1	
Goats	dig prot %	*	4.0	11.1	
Horses	dig prot %	*	4.0	11.1	
Sheep	dig prot %	*	4.0	11.1	
Energy	GE Mcal/kg				
Cattle	DE Mcal/kg	*	0.93	2.60	
Sheep	DE Mcal/kg	*	0.90	2.53	
Cattle	ME Mcal/kg	*	0.76	2.13	
Sheep	ME Mcal/kg	*	0.74	2.07	
Cattle	TDN %	*	21.0	58.9	
Sheep	TDN %	*	20.5	57.3	

Summercypress, Belvedere, seeds w hulls, (5)
Ref No 5 09 042 United States

Feed Name or Analyses			Mean As Fed	Mean Dry	C.V. ± %
Dry matter	%			100.0	
Protein (N x 6.25)	%			28.0	
Alanine	%			1.09	
Arginine	%			1.99	
Aspartic acid	%			2.24	
Glutamic acid	%			3.44	
Glycine	%			1.43	
Histidine	%			0.67	
Hydroxyproline	%			0.00	
Isoleucine	%			0.98	
Leucine	%			1.65	
Lysine	%			1.54	
Methionine	%			0.48	
Phenylalanine	%			1.06	
Proline	%			0.95	
Serine	%			1.12	
Threonine	%			0.92	
Tyrosine	%			1.01	
Valine	%			1.29	

(1) dry forages and roughages (2) pasture, range plants, and forages fed green (3) silages (4) energy feeds (5) protein supplements (6) minerals (7) vitamins (8) additives

SUMMERCYPRESS, GRAY. Kochia vestita

Summercypress, gray, aerial part, fresh, mature, (2)
Gray-molly
Ref No 2 08 844 United States

Feed Name or Analyses			Mean As Fed	Mean Dry	C.V. ± %
Dry matter	%			100.0	
Crude fiber	%			24.0	
Ether extract	%			2.0	
N-free extract	%			45.5	
Protein (N x 6.25)	%			9.0	
Cattle	dig prot %	*		5.5	
Goats	dig prot %	*		5.0	
Horses	dig prot %	*		5.2	
Rabbits	dig prot %	*		5.6	
Sheep	dig prot %	*		5.4	

Summercypress, gray, aerial part, fresh, dormant, (2)
Gray molly
Ref No 2 08 843 United States

Feed Name or Analyses			Mean As Fed	Mean Dry	C.V. ± %
Dry matter	%			100.0	
Crude fiber	%			22.0	
Ether extract	%			2.5	
N-free extract	%			48.0	
Protein (N x 6.25)	%			9.0	
Cattle	dig prot %	*		5.5	
Goats	dig prot %	*		5.0	
Horses	dig prot %	*		5.2	
Rabbits	dig prot %	*		5.6	
Sheep	dig prot %	*		5.4	

Summercypress, gray, browse, (2)
Ref No 2 04 719 United States

Feed Name or Analyses			Mean As Fed	Mean Dry	C.V. ± %
Dry matter	%			100.0	
Ash	%			24.4	7
Crude fiber	%			15.5	
Ether extract	%			3.7	41
N-free extract	%			47.4	
Protein (N x 6.25)	%			9.0	31
Cattle	dig prot %	*		5.5	
Goats	dig prot %	*		5.0	
Horses	dig prot %	*		5.2	
Rabbits	dig prot %	*		5.6	
Sheep	dig prot %	*		5.4	
Calcium	%			2.37	
Phosphorus	%			0.12	
Carotene	mg/kg			18.1	
Vitamin A equivalent	IU/g			30.1	

Sun-cured alfalfa meal (AAFCO) –
see Alfalfa, hay, s-c, grnd, (1)

Suncured alfalfa stem meal (AAFCO) –
see Alfalfa, stems, s-c grnd, (1)

Suncured chopped alfalfa (AAFCO)–
see Alfalfa, hay, s-c chopped, (1)

Suncured chopped alfalfa (AAFCO) –
see Alfalfa, hay, s-c, chopped, (1)

Suncured chopped alfalfa (AAFCO) –
see Alfalfa, hay, s-c chopped, (1)

SUNFLOWER. Helianthus spp

Sunflower, hulls, (1)
Sunflower hulls (AAFCO)
Ref No 1 04 720 United States

Feed Name or Analyses			Mean As Fed	Mean Dry	C.V. ± %
Dry matter	%		89.5	100.0	
Ash	%		2.1	2.4	11
Crude fiber	%		22.1	24.7	
Ether extract	%		3.4	3.8	
N-free extract	%		58.4	65.2	
Protein (N x 6.25)	%		3.5	3.9	13
Sheep	dig coef %		28.	28.	
Cattle	dig prot %	*	0.3	0.3	
Goats	dig prot %	*	0.2	0.2	
Horses	dig prot %	*	0.8	0.8	
Rabbits	dig prot %	*	1.5	1.7	
Sheep	dig prot %		1.0	1.1	
Energy	GE Mcal/kg		4.13	4.61	
Sheep	GE dig coef %		27.	27.	
Cattle	DE Mcal/kg	*	2.75	3.07	
Sheep	DE Mcal/kg		1.11	1.25	
Cattle	ME Mcal/kg	*	2.26	2.52	
Sheep	ME Mcal/kg	*	0.91	1.02	
Cattle	TDN %	*	62.4	69.7	
Sheep	TDN %	*	52.5	58.7	

Ref No 1 04 720 Canada

Feed Name or Analyses			Mean As Fed	Mean Dry	C.V. ± %
Dry matter	%		92.8	100.0	
Sheep	dig coef %		30.	30.	
Crude fiber	%		56.7	61.1	
Ether extract	%		3.2	3.5	
Protein (N x 6.25)	%		4.0	4.3	
Sheep	dig coef %		28.	28.	
Cattle	dig prot %	*	0.6	0.6	
Goats	dig prot %	*	0.5	0.5	
Horses	dig prot %	*	1.1	1.1	
Rabbits	dig prot %	*	1.8	2.0	
Sheep	dig prot %		1.1	1.2	
Lignin (Ellis)	%		20.7	22.4	
Energy	GE Mcal/kg		4.28	4.61	
Sheep	GE dig coef %		27.	27.	
Sheep	DE Mcal/kg		1.16	1.25	
Sheep	ME Mcal/kg	*	0.95	1.02	
Sheep	TDN %	*	26.4	28.4	

Feed Name or Analyses			Mean As Fed	Mean Dry	C.V. ± %

Sunflower, aerial part, fresh, (2)

Ref No 2 04 723 United States

Feed Name or Analyses			As Fed	Dry	C.V. ± %
Dry matter	%		15.4	100.0	29
Ash	%		1.9	12.2	13
Crude fiber	%		4.5	29.3	18
Cattle	dig coef %		20.	20.	
Sheep	dig coef %		52.	52.	
Ether extract	%		0.4	2.4	50
Cattle	dig coef %		44.	44.	
Sheep	dig coef %		66.	66.	
N-free extract	%		7.2	46.9	
Cattle	dig coef %		71.	71.	
Sheep	dig coef %		77.	77.	
Protein (N x 6.25)	%		1.4	9.2	32
Cattle	dig coef %		72.	72.	
Sheep	dig coef %		56.	56.	
Cattle	dig prot %		1.0	6.6	
Goats	dig prot %	*	0.8	5.2	
Horses	dig prot %	*	0.8	5.4	
Rabbits	dig prot %	*	0.9	5.8	
Sheep	dig prot %		0.8	5.1	
Energy	GE Mcal/kg				
Cattle	DE Mcal/kg	*	0.33	2.12	
Sheep	DE Mcal/kg	*	0.41	2.65	
Cattle	ME Mcal/kg	*	0.27	1.74	
Sheep	ME Mcal/kg	*	0.33	2.17	
Cattle	TDN %		7.4	48.2	
Sheep	TDN %		9.2	60.0	

Sunflower, aerial part, fresh, immature, (2)

Ref No 2 04 721 United States

Feed Name or Analyses			As Fed	Dry	C.V. ± %
Dry matter	%			100.0	
Ash	%			13.5	
Crude fiber	%			16.1	
Ether extract	%			5.4	
N-free extract	%			49.1	
Protein (N x 6.25)	%			15.9	
Cattle	dig prot %	*		11.4	
Goats	dig prot %	*		11.4	
Horses	dig prot %	*		11.0	
Rabbits	dig prot %	*		10.9	
Sheep	dig prot %	*		11.8	
Energy	GE Mcal/kg				
Cattle	DE Mcal/kg	*		2.52	
Sheep	DE Mcal/kg	*		2.90	
Cattle	ME Mcal/kg	*		2.06	
Sheep	ME Mcal/kg	*		2.38	
Cattle	TDN %	*		57.1	
Sheep	TDN %	*		65.8	
Calcium	%			2.13	
Magnesium	%			0.68	
Phosphorus	%			0.22	
Potassium	%			4.47	

Sunflower, aerial part, fresh, pre-bloom, (2)

Ref No 2 04 722 United States

Feed Name or Analyses			As Fed	Dry	C.V. ± %
Dry matter	%		11.5	100.0	
Ash	%		1.2	10.5	
Crude fiber	%		3.9	34.2	
Sheep	dig coef %		56.	56.	
Ether extract	%		0.2	1.5	
Sheep	dig coef %		65.	65.	
N-free extract	%		5.2	45.5	
Sheep	dig coef %		82.	82.	
Protein (N x 6.25)	%		1.0	8.3	
Sheep	dig coef %		59.	59.	
Cattle	dig prot %	*	0.6	4.9	
Goats	dig prot %	*	0.5	4.3	
Horses	dig prot %	*	0.5	4.6	
Rabbits	dig prot %	*	0.6	5.1	
Sheep	dig prot %		0.6	4.9	
Energy	GE Mcal/kg				
Cattle	DE Mcal/kg	*	0.32	2.75	
Sheep	DE Mcal/kg	*	0.32	2.80	
Cattle	ME Mcal/kg	*	0.26	2.25	
Sheep	ME Mcal/kg	*	0.26	2.30	
Cattle	TDN %	*	7.2	62.3	
Sheep	TDN %		7.3	63.6	

Sunflower, leaves, fresh, over ripe, (2)

Ref No 2 04 726 United States

Feed Name or Analyses			As Fed	Dry	C.V. ± %
Dry matter	%			100.0	
Ash	%			8.5	
Crude fiber	%			24.5	
Ether extract	%			2.2	
N-free extract	%			61.0	
Protein (N x 6.25)	%			3.8	
Cattle	dig prot %	*		1.1	
Goats	dig prot %	*		0.1	
Horses	dig prot %	*		0.8	
Rabbits	dig prot %	*		1.6	
Sheep	dig prot %	*		0.5	
Energy	GE Mcal/kg				
Cattle	DE Mcal/kg	*		2.53	
Sheep	DE Mcal/kg	*		2.56	
Cattle	ME Mcal/kg	*		2.07	
Sheep	ME Mcal/kg	*		2.10	
Cattle	TDN %	*		57.3	
Sheep	TDN %	*		58.1	

Sunflower, stems, fresh, over ripe, (2)

Ref No 2 04 727 United States

Feed Name or Analyses			As Fed	Dry	C.V. ± %
Dry matter	%			100.0	
Ash	%			4.2	
Crude fiber	%			49.0	
Ether extract	%			0.8	
N-free extract	%			44.5	
Protein (N x 6.25)	%			1.5	
Cattle	dig prot %	*		-0.7	
Goats	dig prot %	*		-1.9	
Horses	dig prot %	*		-1.1	
Rabbits	dig prot %	*		-0.1	
Sheep	dig prot %	*		-1.5	

Sunflower, aerial part, ensiled, (3)

Ref No 3 04 736 United States

Feed Name or Analyses			As Fed	Dry	C.V. ± %
Dry matter	%		22.5	100.0	5
Ash	%		2.4	10.8	17
Crude fiber	%		7.5	33.2	9
Cattle	dig coef %		42.	42.	
Ether extract	%		0.8	3.6	35
Cattle	dig coef %		76.	76.	
N-free extract	%		9.7	43.2	
Cattle	dig coef %		59.	59.	
Protein (N x 6.25)	%		2.1	9.1	15
Cattle	dig coef %		49.	49.	
Cattle	dig prot %		1.0	4.4	
Goats	dig prot %	*	1.0	4.5	
Horses	dig prot %	*	1.0	4.5	
Sheep	dig prot %	*	1.0	4.5	
Energy	GE Mcal/kg				
Cattle	DE Mcal/kg	*	0.50	2.20	
Sheep	DE Mcal/kg	*	0.55	2.44	
Cattle	ME Mcal/kg	*	0.41	1.80	
Sheep	ME Mcal/kg	*	0.45	2.00	
Cattle	TDN %		11.2	49.8	
Sheep	TDN %	*	12.4	55.3	
Calcium	%		0.39	1.72	6
Magnesium	%		0.02	0.09	
Manganese	mg/kg		2387.	1060.8	
Phosphorus	%		0.05	0.20	41
Potassium	%		0.66	2.92	
Sulphur	%		0.01	0.04	46

Sunflower, aerial part, ensiled, pre-bloom, (3)

Ref No 3 04 728 United States

Feed Name or Analyses			As Fed	Dry	C.V. ± %
Dry matter	%		26.2	100.0	
Ash	%		2.6	10.0	
Crude fiber	%		8.5	32.6	
Sheep	dig coef %		42.	42.	
Ether extract	%		0.7	2.8	
Sheep	dig coef %		66.	66.	
N-free extract	%		11.6	44.4	
Sheep	dig coef %		60.	60.	
Protein (N x 6.25)	%		2.7	10.2	
Sheep	dig coef %		51.	51.	
Cattle	dig prot %	*	1.4	5.5	
Goats	dig prot %	*	1.4	5.5	
Horses	dig prot %	*	1.4	5.5	
Sheep	dig prot %		1.4	5.2	
Energy	GE Mcal/kg				
Cattle	DE Mcal/kg	*	0.63	2.39	
Sheep	DE Mcal/kg	*	0.57	2.19	
Cattle	ME Mcal/kg	*	0.51	1.96	
Sheep	ME Mcal/kg	*	0.47	1.80	
Cattle	TDN %	*	14.2	54.1	
Sheep	TDN %		13.0	49.7	

Sunflower, aerial part, ensiled, early bloom, (3)

Ref No 3 04 729 United States

Feed Name or Analyses			As Fed	Dry	C.V. ± %
Dry matter	%		21.5	100.0	5
Ash	%		2.0	9.4	36
Crude fiber	%		7.1	33.0	6
Cattle	dig coef %		42.	42.	
Sheep	dig coef %		43.	43.	
Ether extract	%		0.5	2.5	12
Cattle	dig coef %		75.	75.	
Sheep	dig coef %		48.	48.	
N-free extract	%		9.7	45.4	
Cattle	dig coef %		71.	71.	
Sheep	dig coef %		67.	67.	
Protein (N x 6.25)	%		2.1	9.7	7
Cattle	dig coef %		58.	58.	
Sheep	dig coef %		59.	59.	
Cattle	dig prot %		1.2	5.6	
Goats	dig prot %	*	1.1	5.0	
Horses	dig prot %	*	1.1	5.0	
Sheep	dig prot %		1.2	5.7	
Energy	GE Mcal/kg				
Cattle	DE Mcal/kg	*	0.53	2.46	
Sheep	DE Mcal/kg	*	0.50	2.34	
Cattle	ME Mcal/kg	*	0.43	2.01	
Sheep	ME Mcal/kg	*	0.41	1.92	

Continued

Feed Name or Analyses				Mean As Fed	Mean Dry	C.V. ± %
Cattle	TDN	%		12.0	55.7	
Sheep	TDN	%		11.4	53.0	

Sunflower, aerial part, ensiled, mid-bloom, (3)
Ref No 3 04 730 United States

Feed Name or Analyses				Mean As Fed	Mean Dry	C.V. ± %
Dry matter		%		21.2	100.0	18
Ash		%		2.3	10.8	26
Crude fiber		%		6.9	32.8	11
Cattle	dig coef	%		45.	45.	
Sheep	dig coef	%		53.	53.	
Ether extract		%		0.8	3.9	36
Cattle	dig coef	%		74.	74.	
Sheep	dig coef	%		64.	64.	
N-free extract		%		8.7	41.1	8
Cattle	dig coef	%		57.	57.	
Sheep	dig coef	%		70.	70.	
Protein (N x 6.25)		%		2.4	11.3	41
Cattle	dig coef	%		46.	46.	
Sheep	dig coef	%		66.	66.	
Cattle	dig prot	%		1.1	5.2	
Goats	dig prot	%	*	1.4	6.5	
Horses	dig prot	%	*	1.4	6.5	
Sheep	dig prot	%		1.6	7.5	
Energy	GE Mcal/kg					
Cattle	DE Mcal/kg		*	0.47	2.20	
Sheep	DE Mcal/kg		*	0.55	2.60	
Cattle	ME Mcal/kg		*	0.38	1.81	
Sheep	ME Mcal/kg		*	0.45	2.13	
Cattle	TDN	%		10.6	49.9	
Sheep	TDN	%		12.5	58.9	

Sunflower, aerial part, ensiled, late bloom, (3)
Ref No 3 04 732 United States

Feed Name or Analyses				Mean As Fed	Mean Dry	C.V. ± %
Dry matter		%		26.0	100.0	
Ash		%		3.9	15.0	
Crude fiber		%		9.9	38.2	
Cattle	dig coef	%		40.	40.	
Ether extract		%		0.5	2.1	
Cattle	dig coef	%		72.	72.	
N-free extract		%		9.5	36.5	
Cattle	dig coef	%		57.	57.	
Protein (N x 6.25)		%		2.1	8.2	
Cattle	dig coef	%		44.	44.	
Cattle	dig prot	%		0.9	3.6	
Goats	dig prot	%	*	1.0	3.7	
Horses	dig prot	%	*	1.0	3.7	
Sheep	dig prot	%	*	1.0	3.7	
Energy	GE Mcal/kg					
Cattle	DE Mcal/kg		*	0.49	1.90	
Sheep	DE Mcal/kg		*	0.54	2.09	
Cattle	ME Mcal/kg		*	0.41	1.56	
Sheep	ME Mcal/kg		*	0.44	1.71	
Cattle	TDN	%		11.2	43.1	
Sheep	TDN	%	*	12.3	47.3	

Sunflower, aerial part, ensiled, milk stage, (3)
Ref No 3 04 733 United States

Feed Name or Analyses				Mean As Fed	Mean Dry	C.V. ± %
Dry matter		%		21.2	100.0	
Ash		%		2.1	10.0	
Crude fiber		%		6.2	29.4	
Cattle	dig coef	%		37.	37.	
Ether extract		%		1.3	5.9	
Cattle	dig coef	%		74.	74.	
N-free extract		%		9.5	45.0	
Cattle	dig coef	%		56.	56.	
Protein (N x 6.25)		%		2.1	9.7	
Cattle	dig coef	%		48.	48.	
Cattle	dig prot	%		1.0	4.7	
Goats	dig prot	%	*	1.1	5.0	
Horses	dig prot	%	*	1.1	5.0	
Sheep	dig prot	%	*	1.1	5.0	
Energy	GE Mcal/kg					
Cattle	DE Mcal/kg		*	0.47	2.23	
Sheep	DE Mcal/kg		*	0.50	2.35	
Cattle	ME Mcal/kg		*	0.39	1.83	
Sheep	ME Mcal/kg		*	0.41	1.93	
Cattle	TDN	%		10.7	50.6	
Sheep	TDN	%	*	11.3	53.4	

Sunflower, aerial part, ensiled, mature, (3)
Ref No 3 04 735 United States

Feed Name or Analyses				Mean As Fed	Mean Dry	C.V. ± %
Dry matter		%		22.8	100.0	
Ash		%		2.2	9.7	
Crude fiber		%		8.9	39.1	
Cattle	dig coef	%		35.	35.	
Sheep	dig coef	%	*	51.	51.	
Ether extract		%		1.1	4.9	
Cattle	dig coef	%		65.	65.	
Sheep	dig coef	%	*	66.	66.	
N-free extract		%		8.9	39.1	
Cattle	dig coef	%		40.	40.	
Sheep	dig coef	%	*	68.	68.	
Protein (N x 6.25)		%		1.6	7.2	
Cattle	dig coef	%		22.	22.	
Sheep	dig coef	%	*	45.	45.	
Cattle	dig prot	%		0.4	1.6	
Goats	dig prot	%	*	0.6	2.8	
Horses	dig prot	%	*	0.6	2.8	
Sheep	dig prot	%		0.7	3.2	
Energy	GE Mcal/kg					
Cattle	DE Mcal/kg		*	0.38	1.68	
Sheep	DE Mcal/kg		*	0.57	2.52	
Cattle	ME Mcal/kg		*	0.31	1.38	
Sheep	ME Mcal/kg		*	0.47	2.06	
Cattle	TDN	%		8.7	38.1	
Sheep	TDN	%		13.0	57.0	

Sunflower, seeds, (4)
Ref No 4 08 530 United States

Feed Name or Analyses				Mean As Fed	Mean Dry	C.V. ± %
Dry matter		%		93.6	100.0	
Ash		%		3.1	3.3	
Crude fiber		%		29.0	31.0	
Ether extract		%		25.9	27.7	
N-free extract		%		18.8	20.1	
Protein (N x 6.25)		%		16.8	17.9	
Energy	GE Mcal/kg					
Cattle	DE Mcal/kg		*	3.46	3.69	
Sheep	DE Mcal/kg		*	3.24	3.46	
Swine	DE kcal/kg		*	3363.	3593.	
Cattle	ME Mcal/kg		*	2.83	3.03	
Sheep	ME Mcal/kg		*	2.66	2.84	
Swine	ME kcal/kg		*	3108.	3320.	
Cattle	TDN	%	*	78.4	83.7	
Sheep	TDN	%	*	73.4	78.4	
Swine	TDN	%	*	76.3	81.5	
Calcium		%		0.17	0.18	
Iron		%		0.003	0.003	
Manganese		mg/kg		21.6	23.1	
Phosphorus		%		0.52	0.56	
Potassium		%		0.66	0.71	

Sunflower, seeds, extn unspecified grnd, (5)
Whole pressed sunflower meal (AAFCO)
Sunflower seed meal (CFA)
Sunflower seed oil meal, extraction unspecified
Ref No 5 04 737 United States

Feed Name or Analyses				Mean As Fed	Mean Dry	C.V. ± %
Dry matter		%		89.7	100.0	2
Ash		%		6.0	6.7	14
Crude fiber		%		23.7	26.5	8
Cattle	dig coef	%		22.	22.	
Ether extract		%		1.9	2.1	46
Cattle	dig coef	%		86.	86.	
N-free extract		%		25.7	28.7	
Cattle	dig coef	%		36.	36.	
Protein (N x 6.25)		%		32.3	36.0	12
Cattle	dig coef	%		83.	83.	
Cattle	dig prot	%		26.8	29.9	
Linoleic		%		2.899	3.232	
Energy	GE Mcal/kg			3.87	4.31	11
Cattle	DE Mcal/kg		*	1.98	2.21	
Sheep	DE Mcal/kg		*	2.13	2.37	
Swine	DE kcal/kg		*	2590.	2888.	
Cattle	ME Mcal/kg		*	1.63	1.81	
Chickens	ME_n kcal/kg			2262.	2522.	
Sheep	ME Mcal/kg		*	1.74	1.94	
Swine	ME kcal/kg		*	2298.	2562.	
Cattle	TDN	%		45.0	50.1	
Sheep	TDN	%	*	48.2	53.7	
Swine	TDN	%	*	58.8	65.5	
Calcium		%		0.38	0.43	39
Magnesium		%		0.68	0.75	
Manganese		mg/kg		5.0	5.5	
Phosphorus		%		0.97	1.08	39
Carotene		mg/kg		8.6	9.6	
Choline		mg/kg		4227.	4712.	
Niacin		mg/kg		285.8	318.7	4
Pantothenic acid		mg/kg		40.2	44.8	
Riboflavin		mg/kg		3.2	3.6	24
Thiamine		mg/kg		33.9	37.8	
Vitamin B_6		mg/kg		11.23	12.52	
Vitamin A equivalent		IU/g		14.4	16.0	
Arginine		%		3.72	4.15	
Cystine		%		0.69	0.76	
Glycine		%		1.76	1.97	
Lysine		%		1.66	1.86	29
Methionine		%		1.57	1.75	
Tryptophan		%		0.59	0.66	

Ref No 5 04 737 Canada

Feed Name or Analyses				Mean As Fed	Mean Dry	C.V. ± %
Dry matter		%		91.1	100.0	
Crude fiber		%		13.2	14.5	
Cattle	dig coef	%		22.	22.	
Ether extract		%		0.6	0.6	
Cattle	dig coef	%		86.	86.	
N-free extract		%				
Cattle	dig coef	%		36.	36.	
Protein (N x 6.25)		%		47.0	51.6	
Cattle	dig coef	%		83.	83.	
Cattle	dig prot	%		39.0	42.8	
Energy	GE Mcal/kg			4.16	4.57	
Cattle	DE Mcal/kg		*	2.17	2.39	
Cattle	ME Mcal/kg		*	1.78	1.96	

(1) dry forages and roughages (2) pasture, range plants, and forages fed green (3) silages (4) energy feeds (5) protein supplements (6) minerals (7) vitamins (8) additives

Feed Name or Analyses		Mean As Fed	Mean Dry	C.V. ± %
Chickens	ME_n kcal/kg	2249.	2469.	
Cattle	TDN %	49.3	54.2	
Calcium	%	0.28	0.31	
Phosphorus	%	1.17	1.28	

Sunflower, seeds, solv-extd caked, (5)

Ref No 5 08 531 United States

Feed Name or Analyses		As Fed	Dry	C.V. ± %
Dry matter	%	89.2	100.0	
Ash	%	5.6	6.3	
Crude fiber	%	35.9	40.2	
Ether extract	%	1.1	1.2	
N-free extract	%	27.0	30.3	
Protein (N x 6.25)	%	19.6	22.0	
Energy	GE Mcal/kg			
Cattle	DE Mcal/kg	* 1.64	1.84	
Sheep	DE Mcal/kg	* 1.69	1.90	
Cattle	ME Mcal/kg	* 1.35	1.51	
Sheep	ME Mcal/kg	* 1.39	1.56	
Cattle	TDN %	* 37.2	41.7	
Sheep	TDN %	* 38.4	43.0	

Sunflower, seeds wo hulls, dehy, (5)

Ref No 5 08 128 United States

Feed Name or Analyses		As Fed	Dry	C.V. ± %
Dry matter	%	95.4	100.0	
Ash	%	3.9	4.1	
Crude fiber	%	5.0	5.3	
Ether extract	%	44.4	46.5	
N-free extract	%	16.2	17.0	
Protein (N x 6.25)	%	25.8	27.1	
Energy	GE Mcal/kg	5.61	5.88	
Cattle	DE Mcal/kg	* 5.11	5.36	
Sheep	DE Mcal/kg	* 4.68	4.91	
Cattle	ME Mcal/kg	* 4.20	4.40	
Sheep	ME Mcal/kg	* 3.84	4.03	
Cattle	TDN %	* 116.0	121.6	
Sheep	TDN %	* 106.2	110.3	
Calcium	%	0.16	0.17	
Chlorine	%	0.01	0.01	
Iron	%	0.007	0.007	
Magnesium	%	0.38	0.40	
Phosphorus	%	0.90	0.94	
Potassium	%	0.92	0.96	
Sodium	%	0.03	0.03	
Sulphur	%	0.02	0.02	
Niacin	mg/kg	54.1	56.7	
Riboflavin	mg/kg	2.3	2.4	
Thiamine	mg/kg	19.6	20.6	
Vitamin A	IU/g	0.5	0.5	

Sunflower, seeds wo hulls, extn unspecified grnd, (5)

Sunflower oil meal, without hulls, extraction unspecified

Ref No 5 04 740 United States

Feed Name or Analyses		As Fed	Dry	C.V. ± %
Dry matter	%	92.5	100.0	3
Ash	%	6.1	6.6	15
Crude fiber	%	9.7	10.4	39
Cattle	dig coef %	16.	16.	
Ether extract	%	8.1	8.8	14
Cattle	dig coef %	86.	86.	
N-free extract	%	25.1	27.1	
Cattle	dig coef %	47.	47.	
Protein (N x 6.25)	%	43.5	47.0	14
Cattle	dig coef %	89.	89.	
Cattle	dig prot %	38.7	41.8	

Feed Name or Analyses		As Fed	Dry	C.V. ± %
Energy	GE Mcal/kg			
Cattle	DE Mcal/kg	* 2.99	3.23	
Sheep	DE Mcal/kg	* 3.25	3.51	
Swine	DE kcal/kg	*3549.	3835.	
Cattle	ME Mcal/kg	* 2.45	2.65	
Sheep	ME Mcal/kg	* 2.66	2.88	
Swine	ME kcal/kg	*3070.	3317.	
Cattle	TDN %	67.8	73.2	
Sheep	TDN %	* 73.7	79.6	
Swine	TDN %	* 80.5	87.0	
Calcium	%	0.35	0.38	26
Chlorine	%	0.19	0.20	
Manganese	mg/kg	22.5	24.3	
Phosphorus	%	1.12	1.21	8
Potassium	%	1.06	1.15	

Ref No 5 04 740 Canada

Feed Name or Analyses		As Fed	Dry	C.V. ± %
Dry matter	%	94.0	100.0	
Crude fiber	%	2.8	3.0	
Cattle	dig coef %	16.	16.	
Ether extract	%	47.9	51.0	
Cattle	dig coef %	86.	86.	
N-free extract	%			
Cattle	dig coef %	47.	47.	
Protein (N x 6.25)	%	23.9	25.5	
Cattle	dig coef %	89.	89.	
Cattle	dig prot %	21.3	22.7	
Energy	GE Mcal/kg			
Cattle	DE Mcal/kg	* 3.95	4.20	
Cattle	ME Mcal/kg	* 3.24	3.44	
Cattle	TDN %	89.5	95.3	

Sunflower, seeds wo hulls, mech-extd grnd, (5)

Sunflower meal (AAFCO)

Sunflower oil meal, without hulls, expeller extracted

Ref No 5 04 738 United States

Feed Name or Analyses		As Fed	Dry	C.V. ± %
Dry matter	%	92.9	100.0	2
Ash	%	6.8	7.3	10
Crude fiber	%	13.3	14.3	17
Cattle	dig coef %	* 16.	16.	
Sheep	dig coef %	* 28.	28.	
Ether extract	%	7.6	8.2	56
Cattle	dig coef %	* 86.	86.	
Sheep	dig coef %	* 90.	90.	
N-free extract	%	24.2	26.0	
Cattle	dig coef %	* 47.	47.	
Sheep	dig coef %	* 71.	71.	
Protein (N x 6.25)	%	41.0	44.1	13
Cattle	dig coef %	* 89.	89.	
Sheep	dig coef %	* 91.	91.	
Cattle	dig prot %	36.5	39.3	
Sheep	dig prot %	37.3	40.2	
Energy	GE Mcal/kg			
Cattle	DE Mcal/kg	* 2.85	3.07	
Sheep	DE Mcal/kg	* 3.25	3.49	
Swine	DE kcal/kg	*3394.	3653.	
Cattle	ME Mcal/kg	* 2.34	2.52	
Sheep	ME Mcal/kg	* 2.66	2.86	
Swine	ME kcal/kg	*2955.	3181.	
Cattle	NE_m Mcal/kg	* 1.43	1.54	
Cattle	NE_{gain} Mcal/kg	* 0.90	0.97	
Cattle	$NE_{lactating\ cows}$ Mcal/kg	* 1.70	1.83	
Cattle	TDN %	64.7	69.6	
Sheep	TDN %	73.6	79.2	
Swine	TDN %	* 77.0	82.9	
Calcium	%	0.43	0.46	

Feed Name or Analyses		As Fed	Dry	C.V. ± %
Chlorine	%	0.19	0.20	
Manganese	mg/kg	22.9	24.7	
Phosphorus	%	1.04	1.12	
Potassium	%	1.08	1.16	
Lysine	%	2.00	2.15	
Methionine	%	1.60	1.72	

Sunflower, seeds wo hulls, solv-extd grnd, (5)

Sunflower meal (AAFCO)

Sunflower oil meal, without hulls, solvent extracted

Ref No 5 04 739 United States

Feed Name or Analyses		As Fed	Dry	C.V. ± %
Dry matter	%	93.0	100.0	4
Ash	%	7.7	8.3	4
Crude fiber	%	10.8	11.6	18
Cattle	dig coef %	* 16.	16.	
Sheep	dig coef %	* 28.	28.	
Ether extract	%	2.9	3.1	62
Cattle	dig coef %	* 86.	86.	
Sheep	dig coef %	* 90.	90.	
N-free extract	%	24.8	26.7	
Cattle	dig coef %	* 47.	47.	
Sheep	dig coef %	* 71.	71.	
Protein (N x 6.25)	%	46.8	50.3	3
Cattle	dig coef %	* 89.	89.	
Sheep	dig coef %	* 91.	91.	
Cattle	dig prot %	41.7	44.8	
Sheep	dig prot %	42.6	45.8	
Linoleic	%	1.600	1.720	
Energy	GE Mcal/kg			
Cattle	DE Mcal/kg	* 2.67	2.88	
Sheep	DE Mcal/kg	* 3.05	3.28	
Swine	DE kcal/kg	*3012.	3239.	
Cattle	ME Mcal/kg	* 2.19	2.36	
Chickens	ME_n kcal/kg	2289.	2464.	
Chickens	ME_n kcal/kg	1760.	1892.	
Sheep	ME Mcal/kg	* 2.50	2.69	
Swine	ME kcal/kg	*2586.	2780.	
Cattle	NE_m Mcal/kg	* 1.31	1.41	
Cattle	NE_{gain} Mcal/kg	* 0.77	0.83	
Cattle	$NE_{lactating\ cows}$ Mcal/kg	* 1.53	1.64	
Cattle	TDN %	60.6	65.2	
Sheep	TDN %	69.1	74.3	
Swine	TDN %	* 68.3	73.5	
Riboflavin	mg/kg	3.1	3.3	
α-tocopherol	mg/kg	11.0	11.8	

Ref No 5 04 739 Canada

Feed Name or Analyses		As Fed	Dry	C.V. ± %
Dry matter	%	92.6	100.0	
Ash	%	6.8	7.3	
Crude fiber	%	11.8	12.7	
Ether extract	%	2.4	2.6	
N-free extract	%	25.5	27.6	
Protein (N x 6.25)	%	46.1	49.8	
Energy	GE Mcal/kg			
Cattle	DE Mcal/kg	* 2.64	2.85	
Sheep	DE Mcal/kg	* 3.02	3.26	
Swine	DE kcal/kg	*2968.	3205.	
Cattle	ME Mcal/kg	* 2.17	2.34	
Sheep	ME Mcal/kg	* 2.48	2.67	
Swine	ME kcal/kg	*2551.	2755.	
Cattle	TDN %	59.9	64.7	
Sheep	TDN %	68.5	74.0	
Swine	TDN %	* 67.3	72.7	
Calcium	%	0.53	0.57	
Phosphorus	%	0.50	0.54	

Continued

Feed Name or Analyses			Mean As Fed	Mean Dry	C.V. ± %
Pantothenic acid	mg/kg		14.9	16.1	
Arginine	%		9.45	10.20	
Histidine	%		2.18	2.35	
Isoleucine	%		4.05	4.37	
Leucine	%		6.23	6.73	
Lysine	%		3.11	3.36	
Methionine	%		1.56	1.68	
Phenylalanine	%		4.36	4.71	
Threonine	%		3.32	3.59	
Tryptophan	%		1.04	1.12	
Valine	%		5.09	5.49	

SUNFLOWER, COMMON. Helianthus annuus

Sunflower, common, aerial part, fresh, immature, (2)
Ref No 2 04 741 United States

Feed Name or Analyses			Mean As Fed	Mean Dry	C.V. ± %
Dry matter	%			100.0	
Ash	%			16.9	
Crude fiber	%			10.7	
Ether extract	%			6.9	
N-free extract	%			45.0	
Protein (N x 6.25)	%			20.5	
Cattle	dig prot %	*		15.3	
Goats	dig prot %	*		15.7	
Horses	dig prot %	*		14.9	
Rabbits	dig prot %	*		14.5	
Sheep	dig prot %	*		16.1	
Energy	GE Mcal/kg				
Cattle	DE Mcal/kg	*		2.38	
Sheep	DE Mcal/kg	*		3.70	
Cattle	ME Mcal/kg	*		1.95	
Sheep	ME Mcal/kg	*		3.04	
Cattle	TDN %	*		53.9	
Sheep	TDN %	*		83.9	
Calcium	%			4.04	
Magnesium	%			0.46	
Phosphorus	%			0.20	
Potassium	%			4.14	

SUNFLOWER. MAXIMILIAN. Helianthus maximilianii

Sunflower, maximilian, seeds w hulls, (5)
Ref No 5 09 128 United States

Feed Name or Analyses		Mean As Fed	Mean Dry	C.V. ± %
Dry matter	%		100.0	
Protein (N x 6.25)	%		31.0	
Alanine	%		1.24	
Arginine	%		3.04	
Aspartic acid	%		2.73	
Glutamic acid	%		7.32	
Glycine	%		1.36	
Histidine	%		0.68	
Isoleucine	%		1.30	
Leucine	%		2.05	
Lysine	%		0.99	
Methionine	%		0.56	
Phenylalanine	%		1.43	
Proline	%		1.49	
Serine	%		1.30	
Threonine	%		0.93	

(1) dry forages and roughages (2) pasture, range plants, and forages fed green (3) silages (4) energy feeds (5) protein supplements (6) minerals (7) vitamins (8) additives

Feed Name or Analyses		Mean As Fed	Mean Dry	C.V. ± %
Tyrosine	%		0.65	
Valine	%		1.77	

SUNFLOWER, PRAIRIE. Helianthus petiolaris

Sunflower, prairie, aerial part, fresh, (2)
Ref No 2 04 742 United States

Feed Name or Analyses			Mean As Fed	Mean Dry	C.V. ± %
Dry matter	%			100.0	
Ash	%			11.5	
Crude fiber	%			35.2	
Ether extract	%			6.9	
N-free extract	%			36.2	
Protein (N x 6.25)	%			10.2	
Cattle	dig prot %	*		6.6	
Goats	dig prot %	*		6.1	
Horses	dig prot %	*		6.2	
Rabbits	dig prot %	*		6.5	
Sheep	dig prot %	*		6.5	
Energy	GE Mcal/kg				
Cattle	DE Mcal/kg	*		2.43	
Sheep	DE Mcal/kg	*		2.70	
Cattle	ME Mcal/kg	*		1.99	
Sheep	ME Mcal/kg	*		2.21	
Cattle	TDN %	*		55.1	
Sheep	TDN %	*		61.1	

SUNFLOWER, RUSSIAN. Helianthus annuus

Sunflower, Russian, aerial part, fresh, (2)
Ref No 2 08 576 United States

Feed Name or Analyses			Mean As Fed	Mean Dry	C.V. ± %
Dry matter	%		16.9	100.0	
Ash	%		1.7	10.1	
Crude fiber	%		5.2	30.8	
Ether extract	%		0.7	4.1	
N-free extract	%		7.9	46.7	
Protein (N x 6.25)	%		1.4	8.3	
Cattle	dig prot %	*	0.8	4.9	
Goats	dig prot %	*	0.7	4.3	
Horses	dig prot %	*	0.8	4.6	
Rabbits	dig prot %	*	0.9	5.1	
Sheep	dig prot %	*	0.8	4.7	
Energy	GE Mcal/kg				
Cattle	DE Mcal/kg	*	0.43	2.57	
Sheep	DE Mcal/kg	*	0.49	2.91	
Cattle	ME Mcal/kg	*	0.36	2.11	
Sheep	ME Mcal/kg	*	0.40	2.39	
Cattle	TDN %	*	9.9	58.4	
Sheep	TDN %	*	11.2	66.1	
Calcium	%		0.29	1.72	
Phosphorus	%		0.04	0.24	
Potassium	%		0.63	3.73	

Sunflower meal (AAFCO) –
see Sunflower, seeds wo hulls, solv-extd grnd, (5)

Sunflower meal (AAFCO) –
see Sunflower, seeds wo hulls, mech-extd grnd, (5)

Sunflower oil meal, without hulls, expeller extracted –
see Sunflower, seeds wo hulls, mech-extd grnd, (5)

Sunflower oil meal, without hulls, extraction unspecified –
see Sunflower, seeds wo hulls, extn unspecified grnd, (5)

Sunflower oil meal, without hulls, solvent extracted –
see Sunflower, seeds wo hulls, solv-extd grnd, (5)

Sunflower seed meal (CFA) –
see Sunflower, seeds, extn unspecified grnd, (5)

Sunflower seed oil meal, extraction unspecified –
see Sunflower, seeds, extn unspecified grnd, (5)

SWEETANISE. Osmorhiza occidentalis

Sweetanise, aerial part, fresh, (2)
Ref No 2 04 743 United States

Feed Name or Analyses			Mean As Fed	Mean Dry	C.V. ± %
Dry matter	%			100.0	
Ash	%			16.5	
Ether extract	%			2.9	
Protein (N x 6.25)	%			10.5	
Cattle	dig prot %	*		6.8	
Goats	dig prot %	*		6.4	
Horses	dig prot %	*		6.4	
Rabbits	dig prot %	*		6.8	
Sheep	dig prot %	*		6.8	
Calcium	%			2.94	
Phosphorus	%			0.45	

SWEETCLOVER. Melilotus spp

Sweetclover, hay, fan air dried, (1)
Ref No 1 04 744 United States

Feed Name or Analyses		Mean As Fed	Mean Dry	C.V. ± %
Dry matter	%		100.0	
Carotene	mg/kg		139.8	
Vitamin A equivalent	IU/g		233.0	

Sweetclover, hay, s-c, (1)
Ref No 1 04 754 United States

Feed Name or Analyses			Mean As Fed	Mean Dry	C.V. ± %
Dry matter	%		91.3	100.0	
Ash	%		8.0	8.8	
Crude fiber	%		27.4	30.0	
Sheep	dig coef %	*	42.	42.	
Ether extract	%		2.2	2.4	
Sheep	dig coef %	*	32.	32.	
N-free extract	%		38.6	42.4	
Sheep	dig coef %	*	65.	65.	
Protein (N x 6.25)	%		15.0	16.4	
Sheep	dig coef %	*	71.	71.	
Cattle	dig prot %	*	10.2	11.2	
Goats	dig prot %	*	10.8	11.9	
Horses	dig prot %	*	10.5	11.5	
Rabbits	dig prot %	*	10.4	11.4	
Sheep	dig prot %		10.6	11.7	
Fatty acids	%		2.2	2.4	
Energy	GE Mcal/kg				
Cattle	DE Mcal/kg	*	2.36	2.59	
Sheep	DE Mcal/kg	*	2.15	2.36	
Cattle	ME Mcal/kg	*	1.94	2.12	
Sheep	ME Mcal/kg	*	1.77	1.94	
Cattle	NE_m Mcal/kg	*	1.11	1.22	
Cattle	NE_{gain} Mcal/kg	*	0.50	0.55	
Cattle	$NE_{lactating\ cows}$ Mcal/kg	*	1.22	1.34	
Cattle	TDN %	*	53.5	58.7	
Sheep	TDN %		48.9	53.5	
Calcium	%		1.31	1.44	
Chlorine	%		0.34	0.37	
Copper	mg/kg		9.1	10.0	

Feed Name or Analyses	Unit		Mean As Fed	Mean Dry	C.V. ± %
Iron	%		0.012	0.013	
Magnesium	%		0.23	0.25	
Manganese	mg/kg		107.6	117.9	
Phosphorus	%		0.24	0.27	
Potassium	%		1.68	1.84	
Sodium	%		0.08	0.09	
Sulphur	%		0.41	0.45	
Carotene	mg/kg		90.1	98.8	
Vitamin A equivalent	IU/g		150.2	164.6	

Sweetclover, hay, s-c, immature, (1)

Ref No 1 04 746 United States

Feed Name or Analyses	Unit		Mean As Fed	Mean Dry	C.V. ± %
Dry matter	%		90.2	100.0	3
Crude fiber	%				
Sheep dig coef	%	*	42.	42.	
Ether extract	%				
Sheep dig coef	%	*	32.	32.	
N-free extract	%				
Sheep dig coef	%	*	65.	65.	
Protein (N x 6.25)	%				
Sheep dig coef	%	*	71.	71.	
Energy	GE Mcal/kg		4.02	4.46	
Calcium	%		1.73	1.92	13
Magnesium	%		0.51	0.57	16
Phosphorus	%		0.29	0.32	17
Potassium	%		1.44	1.60	9

Sweetclover, hay, s-c, immature, cut 2, (1)

Ref No 1 04 745 United States

Feed Name or Analyses	Unit		Mean As Fed	Mean Dry	C.V. ± %
Dry matter	%		86.1	100.0	5
Ash	%		7.3	8.5	9
Crude fiber	%		28.6	33.2	12
Ether extract	%		1.7	2.0	22
N-free extract	%		34.6	40.2	
Protein (N x 6.25)	%		13.9	16.1	8
Cattle dig prot	%	*	9.4	10.9	
Goats dig prot	%	*	10.0	11.6	
Horses dig prot	%	*	9.6	11.2	
Rabbits dig prot	%	*	9.6	11.1	
Sheep dig prot	%	*	9.5	11.0	
Energy	GE Mcal/kg				
Cattle	DE Mcal/kg	*	2.13	2.48	
Sheep	DE Mcal/kg	*	2.13	2.47	
Cattle	ME Mcal/kg	*	1.75	2.03	
Sheep	ME Mcal/kg	*	1.75	2.03	
Cattle	TDN %	*	48.4	56.2	
Sheep	TDN %	*	48.3	56.1	
Calcium	%		1.28	1.49	24
Chlorine	%		0.32	0.37	
Copper	mg/kg		8.5	9.9	10
Iron	%		0.011	0.013	25
Magnesium	%		0.22	0.25	21
Manganese	mg/kg		101.4	117.7	11
Phosphorus	%		0.23	0.27	15
Potassium	%		1.69	1.96	6
Sodium	%		0.08	0.09	
Sulphur	%		0.39	0.45	
Carotene	mg/kg		130.0	151.0	
Vitamin A equivalent	IU/g		216.7	251.7	

Sweetclover, hay, s-c, early bloom, (1)

Ref No 1 04 750 United States

Feed Name or Analyses	Unit		Mean As Fed	Mean Dry	C.V. ± %
Dry matter	%		82.6	100.0	
Ash	%		6.9	8.4	
Crude fiber	%		27.3	33.1	
Cattle dig coef	%		46.	46.	
Sheep dig coef	%	*	42.	42.	
Ether extract	%		1.4	1.7	
Cattle dig coef	%		11.	11.	
Sheep dig coef	%	*	32.	32.	
N-free extract	%		32.2	39.0	
Sheep dig coef	%	*	65.	65.	
Protein (N x 6.25)	%		14.7	17.8	
Cattle dig coef	%	*	75.	75.	
Sheep dig coef	%	*	71.	71.	
Cattle dig prot	%		11.0	13.4	
Goats dig prot	%	*	10.9	13.2	
Horses dig prot	%	*	10.4	12.6	
Rabbits dig prot	%	*	10.3	12.4	
Sheep dig prot	%		10.4	12.6	
Energy	GE Mcal/kg	*	3.62	4.39	
Cattle	GE dig coef %	*	63.	63.	
Cattle	DE Mcal/kg	*	2.28	2.76	
Sheep	DE Mcal/kg	*	1.93	2.34	
Cattle	ME Mcal/kg	*	1.87	2.27	
Sheep	ME Mcal/kg	*	1.59	1.92	
Cattle	TDN %	*	46.0	55.6	
Sheep	TDN %	*	43.9	53.1	

Ref No 1 04 750 Canada

Feed Name or Analyses	Unit		Mean As Fed	Mean Dry	C.V. ± %
Dry matter	%			100.0	
Ash	%			8.6	
Crude fiber	%			29.4	
Cattle dig coef	%			46.	
Ether extract	%			1.0	
Cattle dig coef	%			11.	
N-free extract	%			45.1	
Protein (N x 6.25)	%			15.9	
Cattle dig coef	%			75.	
Cattle dig prot	%			11.9	
Goats dig prot	%	*		11.4	
Horses dig prot	%	*		11.0	
Rabbits dig prot	%	*		10.9	
Sheep dig prot	%	*		10.8	
Energy	GE Mcal/kg			4.39	
Cattle	GE dig coef %			63.	
Cattle	DE Mcal/kg			2.76	
Sheep	DE Mcal/kg	*		2.35	
Cattle	ME Mcal/kg	*		2.27	
Sheep	ME Mcal/kg	*		1.93	
Cattle	TDN %	*		57.6	
Sheep	TDN %			53.4	

Sweetclover, hay, s-c, full bloom, (1)

Ref No 1 04 751 United States

Feed Name or Analyses	Unit		Mean As Fed	Mean Dry	C.V. ± %
Dry matter	%		88.9	100.0	1
Ash	%		8.4	9.5	3
Crude fiber	%		30.1	33.9	12
Sheep dig coef	%	*	42.	42.	
Ether extract	%		1.8	2.0	26
Sheep dig coef	%	*	32.	32.	
N-free extract	%		37.1	41.7	
Sheep dig coef	%	*	65.	65.	
Protein (N x 6.25)	%		11.5	12.9	11
Sheep dig coef	%	*	71.	71.	
Cattle dig prot	%	*	7.2	8.1	
Goats dig prot	%	*	7.6	8.6	
Horses dig prot	%	*	7.5	8.5	
Rabbits dig prot	%	*	7.7	8.6	
Sheep dig prot	%		8.1	9.2	
Energy	GE Mcal/kg				
Cattle	DE Mcal/kg	*	2.11	2.37	
Sheep	DE Mcal/kg	*	2.04	2.29	
Cattle	ME Mcal/kg	*	1.72	1.94	
Sheep	ME Mcal/kg	*	1.67	1.88	
Cattle	TDN %	*	47.8	53.8	
Sheep	TDN %	*	46.2	51.9	

Sweetclover, hay, s-c, mature, (1)

Ref No 1 04 752 United States

Feed Name or Analyses	Unit		Mean As Fed	Mean Dry	C.V. ± %
Dry matter	%		90.6	100.0	1
Ash	%		5.5	6.1	21
Crude fiber	%		35.2	38.8	18
Sheep dig coef	%	*	42.	42.	
Ether extract	%		1.2	1.3	15
Sheep dig coef	%	*	32.	32.	
N-free extract	%		39.0	43.1	
Sheep dig coef	%	*	65.	65.	
Protein (N x 6.25)	%		9.7	10.7	5
Sheep dig coef	%	*	71.	71.	
Cattle dig prot	%	*	5.6	6.2	
Goats dig prot	%	*	5.9	6.5	
Horses dig prot	%	*	6.0	6.6	
Rabbits dig prot	%	*	6.3	6.9	
Sheep dig prot	%		6.9	7.6	
Energy	GE Mcal/kg				
Cattle	DE Mcal/kg	*	1.96	2.16	
Sheep	DE Mcal/kg	*	2.11	2.33	
Cattle	ME Mcal/kg	*	1.60	1.77	
Sheep	ME Mcal/kg	*	1.73	1.91	
Cattle	TDN %	*	44.3	48.9	
Sheep	TDN %		47.9	52.8	
Calcium	%		1.20	1.32	5
Iron	%		0.014	0.015	17
Magnesium	%		0.56	0.62	18
Manganese	mg/kg		90.7	100.1	19
Phosphorus	%		0.17	0.19	29
Potassium	%		0.72	0.80	36

Ref No 1 04 752 Canada

Feed Name or Analyses	Unit		Mean As Fed	Mean Dry	C.V. ± %
Dry matter	%		90.4	100.0	6
Crude fiber	%		33.4	37.0	23
Ether extract	%		0.9	1.0	
Protein (N x 6.25)	%		11.8	13.1	27
Cattle dig prot	%	*	7.4	8.2	
Goats dig prot	%	*	7.9	8.7	
Horses dig prot	%	*	7.8	8.6	
Rabbits dig prot	%	*	7.9	8.7	
Sheep dig prot	%	*	7.5	8.3	
Energy	GE Mcal/kg				
Sheep	DE Mcal/kg	*	2.13	2.35	
Sheep	ME Mcal/kg	*	1.74	1.93	
Sheep	TDN %		48.2	53.4	
Calcium	%		0.98	1.09	33
Phosphorus	%		0.17	0.19	51
Carotene	mg/kg		21.1	23.4	66
Vitamin A equivalent	IU/g		35.2	39.0	

Sweetclover, hay, s-c, cut 1, (1)

Ref No 1 04 753 United States

Feed Name or Analyses	Unit		Mean As Fed	Mean Dry	C.V. ± %
Dry matter	%		87.5	100.0	4
Ash	%		7.9	9.0	23
Crude fiber	%		29.8	34.0	17
Ether extract	%		2.1	2.4	19

Continued

Feed Name or Analyses			Mean As Fed	Mean Dry	C.V. ± %
N-free extract	%		33.2	37.9	
Protein (N x 6.25)	%		14.6	16.7	22
Cattle	dig prot %	*	10.0	11.4	
Goats	dig prot %	*	10.6	12.1	
Horses	dig prot %	*	10.2	11.7	
Rabbits	dig prot %	*	10.1	11.6	
Sheep	dig prot %	*	10.1	11.5	
Energy	GE Mcal/kg				
Cattle	DE Mcal/kg	*	2.18	2.49	
Sheep	DE Mcal/kg	*	2.12	2.42	
Cattle	ME Mcal/kg	*	1.79	2.04	
Sheep	ME Mcal/kg	*	1.74	1.99	
Cattle	TDN %	*	49.4	56.5	
Sheep	TDN %	*	48.0	54.9	
Calcium	%		1.40	1.60	16
Phosphorus	%		0.29	0.33	6

Sweetclover, hay, s-c, gr 3 US, (1)
Ref No 1 04 755 United States

Feed Name or Analyses			Mean As Fed	Mean Dry	C.V. ± %
Dry matter	%		91.3	100.0	
Ash	%		4.3	4.7	
Crude fiber	%		43.2	47.3	
Ether extract	%		1.1	1.2	
N-free extract	%		33.4	36.6	
Protein (N x 6.25)	%		9.3	10.2	
Cattle	dig prot %	*	5.3	5.8	
Goats	dig prot %	*	5.5	6.1	
Horses	dig prot %	*	5.7	6.2	
Rabbits	dig prot %	*	6.0	6.5	
Sheep	dig prot %	*	5.2	5.7	
Energy	GE Mcal/kg				
Cattle	DE Mcal/kg	*	1.84	2.02	
Sheep	DE Mcal/kg	*	2.02	2.21	
Cattle	ME Mcal/kg	*	1.52	1.66	
Sheep	ME Mcal/kg	*	1.66	1.82	
Cattle	TDN %	*	41.8	45.8	
Sheep	TDN %	*	45.8	50.2	

Sweetclover, leaves, (1)
Ref No 1 04 757 United States

Feed Name or Analyses			Mean As Fed	Mean Dry	C.V. ± %
Dry matter	%		90.1	100.0	3
Ash	%		10.8	12.0	6
Crude fiber	%		9.7	10.8	16
Ether extract	%		3.1	3.5	8
N-free extract	%		41.3	45.8	
Protein (N x 6.25)	%		25.2	28.0	11
Cattle	dig prot %	*	19.1	21.2	
Goats	dig prot %	*	20.4	22.7	
Horses	dig prot %	*	19.2	21.3	
Rabbits	dig prot %	*	18.3	20.3	
Sheep	dig prot %	*	19.5	21.7	
Fatty acids	%		3.1	3.5	
Energy	GE Mcal/kg				
Cattle	DE Mcal/kg	*	2.50	2.78	
Sheep	DE Mcal/kg	*	2.33	2.59	
Cattle	ME Mcal/kg	*	2.05	2.28	
Sheep	ME Mcal/kg	*	1.95	2.12	
Cattle	TDN %	*	56.9	63.1	
Sheep	TDN %	*	52.9	58.7	
Calcium	%		2.81	3.12	
Phosphorus	%		0.29	0.32	

(1) dry forages and roughages (2) pasture, range plants, and forages fed green (3) silages (4) energy feeds (5) protein supplements (6) minerals (7) vitamins (8) additives

Sweetclover, stems, (1)
Ref No 1 04 758 United States

Feed Name or Analyses			Mean As Fed	Mean Dry	C.V. ± %
Dry matter	%		91.7	100.0	2
Ash	%		6.7	7.3	17
Crude fiber	%		39.0	42.5	7
Ether extract	%		1.1	1.2	16
N-free extract	%		35.2	38.5	
Protein (N x 6.25)	%		9.7	10.6	16
Cattle	dig prot %	*	5.6	6.1	
Goats	dig prot %	*	5.9	6.4	
Horses	dig prot %	*	6.0	6.5	
Rabbits	dig prot %	*	6.3	6.8	
Sheep	dig prot %	*	5.5	6.0	
Fatty acids	%		1.1	1.2	
Energy	GE Mcal/kg				
Cattle	DE Mcal/kg	*	1.93	2.10	
Sheep	DE Mcal/kg	*	2.11	2.30	
Cattle	ME Mcal/kg	*	1.58	1.22	
Sheep	ME Mcal/kg	*	1.73	1.89	
Cattle	TDN %	*	43.6	47.6	
Sheep	TDN %	*	47.9	52.2	
Calcium	%		0.57	0.62	
Phosphorus	%		0.30	0.33	

Sweetclover, aerial part, fresh, (2)
Ref No 2 04 766 United States

Feed Name or Analyses			Mean As Fed	Mean Dry	C.V. ± %
Dry matter	%		25.0	100.0	18
Ash	%		2.1	8.2	23
Crude fiber	%		7.4	29.4	23
Cattle	dig coef %	*	58.	58.	
Ether extract	%		0.7	2.8	36
Cattle	dig coef %	*	43.	43.	
N-free extract	%		10.5	41.8	
Cattle	dig coef %	*	76.	76.	
Protein (N x 6.25)	%		4.5	17.8	24
Cattle	dig coef %	*	82.	82.	
Cattle	dig prot %		3.6	14.6	
Goats	dig prot %	*	3.3	13.2	
Horses	dig prot %	*	3.2	12.6	
Rabbits	dig prot %	*	3.1	12.4	
Sheep	dig prot %	*	3.4	13.6	
Energy	GE Mcal/kg				
Cattle	DE Mcal/kg	*	0.73	2.92	
Sheep	DE Mcal/kg	*	0.68	2.74	
Cattle	ME Mcal/kg	*	0.60	2.39	
Sheep	ME Mcal/kg	*	0.56	2.25	
Cattle	TDN %		16.5	66.1	
Sheep	TDN %	*	15.5	62.1	
Calcium	%		0.33	1.32	16
Chlorine	%		0.10	0.38	14
Copper	mg/kg		2.5	9.9	10
Iron	%		0.004	0.014	14
Magnesium	%		0.08	0.33	8
Manganese	mg/kg		31.4	125.4	24
Phosphorus	%		0.07	0.27	41
Potassium	%		0.41	1.65	33
Sodium	%		0.03	0.10	26
Sulphur	%		0.12	0.49	23
Zinc	mg/kg		12.5	50.0	
Carotene	mg/kg		66.5	266.1	
Niacin	mg/kg		9.0	36.2	
Riboflavin	mg/kg		21.0	84.0	
Thiamine	mg/kg		1.3	5.3	
Vitamin A equivalent	IU/g		110.9	443.6	

Sweetclover, aerial part, fresh, immature, (2)
Ref No 2 04 759 United States

Feed Name or Analyses			Mean As Fed	Mean Dry	C.V. ± %
Dry matter	%		19.7	100.0	28
Ash	%		1.9	9.6	22
Crude fiber	%		4.5	22.9	25
Cattle	dig coef %	*	58.	58.	
Sheep	dig coef %	*	58.	58.	
Ether extract	%		0.6	3.2	21
Cattle	dig coef %	*	43.	43.	
Sheep	dig coef %	*	51.	51.	
N-free extract	%		8.4	42.6	
Cattle	dig coef %	*	76.	76.	
Sheep	dig coef %	*	69.	69.	
Protein (N x 6.25)	%		4.3	21.7	12
Cattle	dig coef %	*	82.	82.	
Sheep	dig coef %	*	78.	78.	
Cattle	dig prot %		3.5	17.8	
Goats	dig prot %	*	3.3	16.8	
Horses	dig prot %	*	3.1	16.0	
Rabbits	dig prot %	*	3.0	15.4	
Sheep	dig prot %		3.3	16.9	
Energy	GE Mcal/kg				
Cattle	DE Mcal/kg	*	0.58	2.93	
Sheep	DE Mcal/kg	*	0.55	2.79	
Cattle	ME Mcal/kg	*	0.47	2.41	
Sheep	ME Mcal/kg	*	0.45	2.29	
Cattle	TDN %		13.1	66.5	
Sheep	TDN %		12.5	63.3	
Calcium	%		0.34	1.73	
Magnesium	%		0.06	0.30	
Phosphorus	%		0.09	0.44	
Potassium	%		0.58	2.95	

Sweetclover, aerial part, fresh, pre-bloom, (2)
Ref No 2 04 760 United States

Feed Name or Analyses			Mean As Fed	Mean Dry	C.V. ± %
Dry matter	%		20.8	100.0	
Ash	%		1.9	9.1	
Crude fiber	%		4.9	23.6	
Ether extract	%		0.7	3.4	
N-free extract	%		9.2	44.2	
Protein (N x 6.25)	%		4.1	19.7	
Cattle	dig prot %	*	3.0	14.6	
Goats	dig prot %	*	3.1	15.0	
Horses	dig prot %	*	3.0	14.3	
Rabbits	dig prot %	*	2.9	13.9	
Sheep	dig prot %	*	3.2	15.4	
Energy	GE Mcal/kg				
Cattle	DE Mcal/kg	*	0.59	2.83	
Sheep	DE Mcal/kg	*	0.55	2.66	
Cattle	ME Mcal/kg	*	0.48	2.32	
Sheep	ME Mcal/kg	*	0.45	2.18	
Cattle	TDN %	*	13.4	64.3	
Sheep	TDN %	*	12.6	60.4	
Calcium	%		0.34	1.63	
Phosphorus	%		0.10	0.48	

Sweetclover, aerial part, fresh, early bloom, (2)
Ref No 2 04 761 United States

Feed Name or Analyses			Mean As Fed	Mean Dry	C.V. ± %
Dry matter	%		25.0	100.0	24
Ash	%		2.4	9.5	4
Crude fiber	%		7.2	28.7	14
Ether extract	%		0.6	2.5	15
N-free extract	%		10.4	41.7	

Feed Name or Analyses			Mean As Fed	Mean Dry	C.V. ± %
Protein (N x 6.25)	%		4.4	17.6	9
Cattle	dig prot %	*	3.2	12.9	
Goats	dig prot %	*	3.2	13.0	
Horses	dig prot %	*	3.1	12.5	
Rabbits	dig prot %	*	3.1	12.3	
Sheep	dig prot %	*	3.3	13.4	
Energy	GE Mcal/kg				
Cattle	DE Mcal/kg	*	0.70	2.78	
Sheep	DE Mcal/kg	*	0.66	2.62	
Cattle	ME Mcal/kg	*	0.57	2.28	
Sheep	ME Mcal/kg	*	0.54	2.15	
Cattle	TDN %	*	15.8	63.0	
Sheep	TDN %	*	14.9	59.5	

Sweetclover, aerial part, fresh, mid-bloom, (2)

Ref No 2 08 532 United States

Analyses			As Fed	Dry	C.V. ± %
Dry matter	%		29.2	100.0	
Ash	%		2.2	7.5	
Crude fiber	%		9.5	32.5	
Ether extract	%		0.8	2.7	
N-free extract	%		11.8	40.4	
Protein (N x 6.25)	%		4.9	16.8	
Cattle	dig prot %	*	3.5	12.2	
Goats	dig prot %	*	3.6	12.2	
Horses	dig prot %	*	3.4	11.8	
Rabbits	dig prot %	*	3.4	11.6	
Sheep	dig prot %	*	3.7	12.6	
Energy	GE Mcal/kg				
Cattle	DE Mcal/kg	*	0.87	2.98	
Sheep	DE Mcal/kg	*	0.83	2.83	
Cattle	ME Mcal/kg	*	0.71	2.44	
Sheep	ME Mcal/kg	*	0.68	2.32	
Cattle	TDN %	*	19.7	67.5	
Sheep	TDN %	*	18.7	64.1	
Calcium	%		0.36	1.23	
Chlorine	%		0.11	0.38	
Copper	mg/kg		2.9	9.8	
Iron	%		0.004	0.014	
Magnesium	%		0.10	0.34	
Manganese	mg/kg		34.4	117.8	
Phosphorus	%		0.07	0.24	
Potassium	%		0.42	1.44	
Sodium	%		0.03	0.10	
Sulphur	%		0.13	0.45	

Sweetclover, aerial part, fresh, full bloom, (2)

Ref No 2 04 762 United States

Analyses			As Fed	Dry	C.V. ± %
Dry matter	%		32.7	100.0	14
Ash	%		2.3	7.0	17
Crude fiber	%		10.5	32.2	7
Cattle	dig coef %	*	58.	58.	
Sheep	dig coef %	*	58.	58.	
Ether extract	%		0.8	2.5	6
Cattle	dig coef %	*	43.	43.	
Sheep	dig coef %	*	51.	51.	
N-free extract	%		13.6	41.5	
Cattle	dig coef %	*	76.	76.	
Sheep	dig coef %	*	69.	69.	
Protein (N x 6.25)	%		5.5	16.8	11
Cattle	dig coef %	*	82.	82.	
Sheep	dig coef %	*	78.	78.	
Cattle	dig prot %		4.5	13.8	
Goats	dig prot %	*	4.0	12.2	
Horses	dig prot %	*	3.9	11.8	
Rabbits	dig prot %	*	3.8	11.6	
Sheep	dig prot %		4.3	13.1	

Feed Name or Analyses			Mean As Fed	Mean Dry	C.V. ± %
Energy	GE Mcal/kg				
Cattle	DE Mcal/kg	*	0.96	2.93	
Sheep	DE Mcal/kg	*	0.91	2.79	
Cattle	ME Mcal/kg	*	0.79	2.40	
Sheep	ME Mcal/kg	*	0.75	2.29	
Cattle	TDN %		21.7	66.4	
Sheep	TDN %		20.7	63.3	
Calcium	%		0.43	1.30	9
Copper	mg/kg		3.2	9.9	10
Iron	%		0.005	0.014	15
Magnesium	%		0.11	0.34	8
Manganese	mg/kg		38.5	117.7	27
Phosphorus	%		0.07	0.22	23
Potassium	%		0.47	1.44	14

Sweetclover, aerial part, fresh, milk stage, (2)

Ref No 2 04 763 United States

Analyses			As Fed	Dry	C.V. ± %
Dry matter	%		38.9	100.0	15
Ash	%		2.1	5.3	9
Crude fiber	%		15.4	39.7	6
Ether extract	%		0.8	2.1	9
N-free extract	%		17.1	43.9	
Protein (N x 6.25)	%		3.5	9.0	5
Cattle	dig prot %	*	2.2	5.5	
Goats	dig prot %	*	1.9	5.0	
Horses	dig prot %	*	2.0	5.2	
Rabbits	dig prot %	*	2.2	5.6	
Sheep	dig prot %	*	2.1	5.4	
Energy	GE Mcal/kg				
Cattle	DE Mcal/kg	*	1.11	2.86	
Sheep	DE Mcal/kg	*	1.08	2.78	
Cattle	ME Mcal/kg	*	0.91	2.35	
Sheep	ME Mcal/kg	*	0.89	2.28	
Cattle	TDN %	*	25.2	64.8	
Sheep	TDN %	*	24.5	63.1	

Sweetclover, aerial part, fresh, cut 1, (2)

Ref No 2 04 764 United States

Analyses			As Fed	Dry	C.V. ± %
Dry matter	%		15.1	100.0	9
Ash	%		1.4	9.3	26
Crude fiber	%		5.5	36.3	17
Ether extract	%		0.5	3.2	56
N-free extract	%		4.8	32.0	
Protein (N x 6.25)	%		2.9	19.2	30
Cattle	dig prot %	*	2.1	14.2	
Goats	dig prot %	*	2.2	14.5	
Horses	dig prot %	*	2.1	13.8	
Rabbits	dig prot %	*	2.0	13.5	
Sheep	dig prot %	*	2.2	14.9	
Energy	GE Mcal/kg				
Cattle	DE Mcal/kg	*	0.43	2.83	
Sheep	DE Mcal/kg	*	0.42	2.76	
Cattle	ME Mcal/kg	*	0.35	2.32	
Sheep	ME Mcal/kg	*	0.34	2.26	
Cattle	TDN %	*	9.7	64.2	
Sheep	TDN %	*	9.5	62.6	

Sweetclover, aerial part, fresh, cut 2, (2)

Ref No 2 04 765 United States

Analyses			As Fed	Dry	C.V. ± %
Dry matter	%		28.3	100.0	32
Ash	%		1.8	6.5	22
Crude fiber	%		10.2	35.9	10
Ether extract	%		0.6	2.1	49
N-free extract	%		11.7	41.5	

Feed Name or Analyses			Mean As Fed	Mean Dry	C.V. ± %
Protein (N x 6.25)	%		4.0	14.0	35
Cattle	dig prot %	*	2.8	9.8	
Goats	dig prot %	*	2.7	9.6	
Horses	dig prot %	*	2.7	9.4	
Rabbits	dig prot %	*	2.7	9.5	
Sheep	dig prot %	*	2.8	10.0	
Energy	GE Mcal/kg				
Cattle	DE Mcal/kg	*	0.80	2.83	
Sheep	DE Mcal/kg	*	0.76	2.67	
Cattle	ME Mcal/kg	*	0.67	2.32	
Sheep	ME Mcal/kg	*	0.62	2.19	
Cattle	TDN %	*	18.1	64.1	
Sheep	TDN %	*	17.1	60.5	

Sweetclover, aerial part, ensiled, (3)

Ref No 3 04 771 United States

Analyses			As Fed	Dry	C.V. ± %
Dry matter	%		30.1	100.0	19
Ash	%		2.8	9.4	
Crude fiber	%		9.8	32.7	
Cattle	dig coef %	*	42.	42.	
Sheep	dig coef %	*	37.	37.	
Ether extract	%		1.0	3.4	
Cattle	dig coef %	*	62.	62.	
Sheep	dig coef %	*	63.	63.	
N-free extract	%		10.7	35.7	
Cattle	dig coef %	*	72.	72.	
Sheep	dig coef %	*	57.	57.	
Protein (N x 6.25)	%		5.7	18.8	
Cattle	dig coef %	*	77.	77.	
Sheep	dig coef %	*	76.	76.	
Cattle	dig prot %		4.4	14.5	
Goats	dig prot %	*	4.0	13.3	
Horses	dig prot %	*	4.0	13.3	
Sheep	dig prot %		4.3	14.3	
Energy	GE Mcal/kg		1.10	3.87	
Energy	GE Mcal/kg		1.16	3.87	
Cattle	DE Mcal/kg	*	0.72	2.54	
Cattle	DE Mcal/kg	*	0.78	2.58	
Sheep	DE Mcal/kg	*	0.62	2.20	
Sheep	DE Mcal/kg	*	0.68	2.27	
Cattle	ME Mcal/kg	*	0.59	2.08	
Cattle	ME Mcal/kg	*	0.64	2.12	
Sheep	ME Mcal/kg	*	0.51	1.80	
Sheep	ME Mcal/kg	*	0.56	1.86	
Cattle	TDN %		16.3	57.6	
Cattle	TDN %		17.6	58.6	
Sheep	TDN %		14.1	49.8	
Sheep	TDN %		15.5	51.5	
Calcium	%		0.39	1.38	26
Calcium	%		0.40	1.34	15
Magnesium	%		0.19	0.62	
Manganese	mg/kg		7.2	24.0	
Phosphorus	%		0.06	0.21	31
Phosphorus	%		0.06	0.21	29
Potassium	%		0.59	1.96	9
Carotene	mg/kg		20.2	71.1	63
Carotene	mg/kg		8.6	28.4	57
Vitamin A equivalent	IU/g		33.6	118.6	
Vitamin A equivalent	IU/g		14.3	47.4	

Ref No 3 04 771 Canada

Analyses			As Fed	Dry	C.V. ± %
Dry matter	%		28.4	100.0	27
Ash	%		3.8	13.2	
Crude fiber	%		8.9	31.4	15
Ether extract	%		0.6	2.0	
N-free extract	%		11.1	39.1	

Continued

Feed Name or Analyses				Mean As Fed	Mean Dry	C.V. ± %
Protein (N x 6.25)		%		4.0	14.2	16
Cattle	dig prot	%	*	2.6	9.2	
Goats	dig prot	%	*	2.6	9.2	
Horses	dig prot	%	*	2.6	9.2	
Sheep	dig prot	%	*	2.6	9.2	

Sweetclover, aerial part, ensiled, pre-bloom, (3)

Ref No 3 04 767 United States

Feed Name or Analyses				As Fed	Dry	C.V. ± %
Dry matter		%		34.8	100.0	
Ash		%		3.4	9.7	
Crude fiber		%		12.4	35.7	
Sheep	dig coef	%		37.	37.	
Ether extract		%		1.1	3.3	
Sheep	dig coef	%		63.	63.	
N-free extract		%		10.4	29.8	
Sheep	dig coef	%		57.	57.	
Protein (N x 6.25)		%		7.5	21.5	
Sheep	dig coef	%		76.	76.	
Cattle	dig prot	%	*	5.5	15.8	
Goats	dig prot	%	*	5.5	15.8	
Horses	dig prot	%	*	5.5	15.8	
Sheep	dig prot	%		5.7	16.3	
Energy		GE Mcal/kg				
Cattle		DE Mcal/kg	*	0.76	2.17	
Sheep		DE Mcal/kg	*	0.79	2.26	
Cattle		ME Mcal/kg	*	0.62	1.78	
Sheep		ME Mcal/kg	*	0.64	1.85	
Cattle		TDN %	*	17.0	48.9	
Sheep		TDN %		17.8	51.2	

Sweetclover, aerial part, ensiled, early bloom, (3)

Ref No 3 04 768 United States

Feed Name or Analyses				As Fed	Dry	C.V. ± %
Dry matter		%		27.2	100.0	10
Ash		%		2.7	9.9	3
Crude fiber		%		8.4	31.0	3
Cattle	dig coef	%		42.	42.	
Sheep	dig coef	%	*	37.	37.	
Ether extract		%		1.0	3.7	9
Cattle	dig coef	%		65.	65.	
Sheep	dig coef	%	*	63.	63.	
N-free extract		%		9.2	34.1	
Cattle	dig coef	%		72.	72.	
Sheep	dig coef	%	*	57.	57.	
Protein (N x 6.25)		%		5.8	21.4	6
Cattle	dig coef	%		78.	78.	
Sheep	dig coef	%	*	76.	76.	
Cattle	dig prot	%		4.5	16.7	
Goats	dig prot	%	*	4.3	15.7	
Horses	dig prot	%	*	4.3	15.7	
Sheep	dig prot	%		4.4	16.3	
Energy		GE Mcal/kg				
Cattle		DE Mcal/kg	*	0.71	2.63	
Sheep		DE Mcal/kg	*	0.63	2.31	
Cattle		ME Mcal/kg	*	0.58	2.15	
Sheep		ME Mcal/kg	*	0.51	1.89	
Cattle		TDN %		16.2	59.6	
Sheep		TDN %		14.2	52.3	

(1) dry forages and roughages (2) pasture, range plants, and forages fed green (3) silages (4) energy feeds (5) protein supplements (6) minerals (7) vitamins (8) additives

Sweetclover, aerial part, ensiled, full bloom, (3)

Ref No 3 04 769 United States

Feed Name or Analyses				As Fed	Dry	C.V. ± %
Dry matter		%		30.1	100.0	9
Ash		%		2.9	9.5	6
Crude fiber		%		10.3	34.1	7
Cattle	dig coef	%		44.	44.	
Sheep	dig coef	%	*	37.	37.	
Ether extract		%		1.2	3.9	33
Cattle	dig coef	%		49.	49.	
Sheep	dig coef	%	*	63.	63.	
N-free extract		%		10.2	33.9	
Cattle	dig coef	%		67.	67.	
Sheep	dig coef	%	*	57.	57.	
Protein (N x 6.25)		%		5.6	18.7	15
Cattle	dig coef	%		73.	73.	
Sheep	dig coef	%	*	76.	76.	
Cattle	dig prot	%		4.1	13.6	
Goats	dig prot	%	*	4.0	13.2	
Horses	dig prot	%	*	4.0	13.2	
Sheep	dig prot	%		4.3	14.2	
Energy		GE Mcal/kg				
Cattle		DE Mcal/kg	*	0.74	2.45	
Sheep		DE Mcal/kg	*	0.68	2.28	
Cattle		ME Mcal/kg	*	0.61	2.01	
Sheep		ME Mcal/kg	*	0.56	1.87	
Cattle		TDN %		16.7	55.6	
Sheep		TDN %		15.5	51.6	

Sweetclover, aerial part, ensiled, cut 1, (3)

Ref No 3 04 770 United States

Feed Name or Analyses				As Fed	Dry	C.V. ± %
Dry matter		%		42.2	100.0	
Calcium		%		0.52	1.24	19
Magnesium		%		0.26	0.62	
Phosphorus		%		0.05	0.12	59

Sweetclover, aerial part, wilted ensiled, (3)

Ref No 3 04 773 United States

Feed Name or Analyses				As Fed	Dry	C.V. ± %
Dry matter		%		38.8	100.0	5
Ash		%		3.8	9.9	4
Crude fiber		%		12.4	31.9	3
Ether extract		%		1.3	3.4	20
N-free extract		%		13.6	35.1	
Protein (N x 6.25)		%		7.6	19.7	8
Cattle	dig prot	%	*	5.5	14.1	
Goats	dig prot	%	*	5.5	14.1	
Horses	dig prot	%	*	5.5	14.1	
Sheep	dig prot	%	*	5.5	14.1	
Energy		GE Mcal/kg				
Cattle		DE Mcal/kg	*	0.90	2.32	
Sheep		DE Mcal/kg	*	0.96	2.48	
Cattle		ME Mcal/kg	*	0.74	1.90	
Sheep		ME Mcal/kg	*	0.79	2.03	
Cattle		TDN %	*	20.4	52.7	
Sheep		TDN %	*	21.8	56.2	
Calcium		%		0.53	1.38	
Phosphorus		%		0.10	0.25	
Potassium		%		0.76	1.95	

Sweetclover, aerial part w grnd corn grain added, ensiled, cut 1, (3)

Ref No 3 04 774 United States

Feed Name or Analyses				As Fed	Dry	C.V. ± %
Dry matter		%		42.2	100.0	
Calcium		%		0.49	1.16	13
Phosphorus		%		0.04	0.10	24
Carotene		mg/kg		20.8	49.4	18
Vitamin A equivalent		IU/g		34.7	82.3	

Sweetclover, aerial part w molasses added, ensiled, (3)

Ref No 3 04 775 United States

Feed Name or Analyses				As Fed	Dry	C.V. ± %
Dry matter		%			100.0	
Manganese		mg/kg			24.0	

Sweetclover, seeds, (5)

Ref No 5 08 533 United States

Feed Name or Analyses				As Fed	Dry	C.V. ± %
Dry matter		%		92.2	100.0	
Ash		%		3.5	3.8	
Crude fiber		%		11.3	12.3	
Ether extract		%		4.2	4.6	
N-free extract		%		35.8	38.8	
Protein (N x 6.25)		%		37.4	40.6	
Energy		GE Mcal/kg				
Cattle		DE Mcal/kg	*	2.72	2.95	
Sheep		DE Mcal/kg	*	2.89	3.13	
Swine		DE kcal/kg	*	3029.	3285.	
Cattle		ME Mcal/kg	*	2.23	2.42	
Sheep		ME Mcal/kg	*	2.37	2.57	
Swine		ME kcal/kg	*	2659.	2884.	
Cattle		TDN %	*	61.8	67.0	
Sheep		TDN %	*	65.6	71.1	
Swine		TDN %	*	68.9	74.5	

Sweetclover, seed screenings, (5)

Ref No 5 08 534 United States

Feed Name or Analyses				As Fed	Dry	C.V. ± %
Dry matter		%		90.1	100.0	
Ash		%		8.9	9.9	
Crude fiber		%		14.7	16.3	
Ether extract		%		3.7	4.1	
N-free extract		%		41.1	45.6	
Protein (N x 6.25)		%		21.7	24.1	
Energy		GE Mcal/kg				
Cattle		DE Mcal/kg	*	2.67	2.97	
Sheep		DE Mcal/kg	*	2.64	2.93	
Swine		DE kcal/kg	*	3244.	3601.	
Cattle		ME Mcal/kg	*	2.19	2.43	
Sheep		ME Mcal/kg	*	2.16	2.40	
Swine		ME kcal/kg	*	2956.	3281.	
Cattle		TDN %	*	60.6	67.2	
Sheep		TDN %	*	59.8	66.4	
Swine		TDN %	*	73.6	81.7	

SWEETCLOVER, ANNUAL YELLOW. Melilotus indica

Sweetclover, annual yellow, aerial part, fresh, (2)

Ref No 2 04 776 United States

Feed Name or Analyses				As Fed	Dry	C.V. ± %
Dry matter		%		21.6	100.0	
Ash		%		2.8	13.1	

Feed Name or Analyses				Mean As Fed	Mean Dry	C.V. ± %
Crude fiber		%		6.4	29.4	
Cattle	dig coef	%		58.	58.	
Ether extract		%		0.4	1.7	
Cattle	dig coef	%		43.	43.	
N-free extract		%		8.7	40.5	
Cattle	dig coef	%		76.	76.	
Protein (N x 6.25)		%		3.3	15.3	
Cattle	dig coef	%		82.	82.	
Cattle	dig prot	%		2.7	12.5	
Goats	dig prot	%	*	2.3	10.8	
Horses	dig prot	%	*	2.3	10.5	
Rabbits	dig prot	%	*	2.3	10.5	
Sheep	dig prot	%	*	2.4	11.2	
Energy	GE Mcal/kg					
Cattle	DE Mcal/kg		*	0.59	2.73	
Sheep	DE Mcal/kg		*	0.58	2.69	
Cattle	ME Mcal/kg		*	0.48	2.24	
Sheep	ME Mcal/kg		*	0.48	2.20	
Cattle	TDN	%		13.4	62.0	
Sheep	TDN	%	*	13.2	60.9	

Sweetclover, annual yellow, seeds w hulls, (5)
Ref No 5 09 056 United States

Feed Name or Analyses				Mean As Fed	Mean Dry	C.V. ± %
Dry matter		%			100.0	
Protein (N x 6.25)		%			33.0	
Alanine		%			1.25	
Arginine		%			2.61	
Aspartic acid		%			3.40	
Glutamic acid		%			4.69	
Glycine		%			1.52	
Histidine		%			0.92	
Hydroxyproline		%			0.00	
Isoleucine		%			1.29	
Leucine		%			2.05	
Lysine		%			1.82	
Methionine		%			0.40	
Phenylalanine		%			1.49	
Proline		%			1.22	
Serine		%			1.39	
Threonine		%			1.06	
Tyrosine		%			0.96	
Valine		%			1.29	

SWEETCLOVER, WHITE. Melilotus alba

Sweetclover, white, aerial part, s-c, cut 1, (1)
Ref No 1 04 777 United States

Feed Name or Analyses				Mean As Fed	Mean Dry	C.V. ± %
Dry matter		%		82.2	100.0	
Ash		%		7.7	9.4	
Crude fiber		%		25.7	31.3	
Sheep	dig coef	%		42.	42.	
Ether extract		%		2.1	2.6	
Sheep	dig coef	%		41.	41.	
N-free extract		%		34.3	41.7	
Sheep	dig coef	%		65.	65.	
Protein (N x 6.25)		%		12.3	15.0	
Sheep	dig coef	%		72.	72.	
Cattle	dig prot	%	*	8.2	9.9	
Goats	dig prot	%	*	8.7	10.6	
Horses	dig prot	%	*	8.4	10.3	
Rabbits	dig prot	%	*	8.4	10.3	
Sheep	dig prot	%		8.9	10.8	
Energy	GE Mcal/kg					
Cattle	DE Mcal/kg		*	2.13	2.60	
Sheep	DE Mcal/kg		*	1.94	2.36	
Cattle	ME Mcal/kg		*	1.75	2.13	

Feed Name or Analyses				Mean As Fed	Mean Dry	C.V. ± %
Sheep	ME Mcal/kg		*	1.59	1.93	
Cattle	TDN	%	*	48.4	58.9	
Sheep	TDN	%		43.9	53.4	

SWEETCLOVER, YELLOW. Melilotus officinalis

Sweetclover, yellow, seed screenings, (5)
Ref No 5 08 007 Canada

Feed Name or Analyses				Mean As Fed	Mean Dry	C.V. ± %
Dry matter		%		87.3	100.0	
Crude fiber		%		15.5	17.8	
Protein (N x 6.25)		%		18.5	21.2	
Calcium		%		0.82	0.94	
Phosphorus		%		0.41	0.47	

SWEETGRASS. Hierochloe odorata

Sweetgrass, hay, s-c, (1)
Ref No 1 04 778 United States

Feed Name or Analyses				Mean As Fed	Mean Dry	C.V. ± %
Dry matter		%		90.0	100.0	
Ash		%		8.4	9.3	
Crude fiber		%		32.1	35.7	
Sheep	dig coef	%		68.	68.	
Ether extract		%		2.2	2.4	
Sheep	dig coef	%		46.	46.	
N-free extract		%		36.7	40.8	
Sheep	dig coef	%		62.	62.	
Protein (N x 6.25)		%		10.6	11.8	
Sheep	dig coef	%		68.	68.	
Cattle	dig prot	%	*	6.4	7.2	
Goats	dig prot	%	*	6.8	7.6	
Horses	dig prot	%	*	6.8	7.5	
Rabbits	dig prot	%	*	7.0	7.8	
Sheep	dig prot	%		7.2	8.0	
Energy	GE Mcal/kg					
Cattle	DE Mcal/kg		*	2.29	2.54	
Sheep	DE Mcal/kg		*	2.38	2.65	
Cattle	ME Mcal/kg		*	1.87	2.08	
Sheep	ME Mcal/kg		*	1.95	2.17	
Cattle	TDN	%	*	51.8	57.5	
Sheep	TDN	%		54.1	60.1	

SWEETGUM, AMERICAN. Liquidamber styraciflua

Sweetgum, American, seeds w hulls, (5)
Ref No 5 09 053 United States

Feed Name or Analyses				Mean As Fed	Mean Dry	C.V. ± %
Dry matter		%			100.0	
Protein (N x 6.25)		%			33.0	
Alanine		%			1.25	
Arginine		%			4.29	
Aspartic acid		%			2.94	
Glutamic acid		%			7.82	
Glycine		%			1.52	
Histidine		%			0.79	
Hydroxyproline		%			0.03	
Isoleucine		%			1.02	
Leucine		%			2.01	
Lysine		%			1.06	
Methionine		%			0.59	
Phenylalanine		%			1.39	
Proline		%			1.35	
Serine		%			1.22	
Threonine		%			0.86	

Feed Name or Analyses				Mean As Fed	Mean Dry	C.V. ± %
Tyrosine		%			0.83	
Valine		%			1.55	

SWEETPOTATO. Ipomoea batatas

Sweetpotato, aerial part, s-c, (1)
Ref No 1 04 779 United States

Feed Name or Analyses				Mean As Fed	Mean Dry	C.V. ± %
Dry matter		%		89.2	100.0	
Ash		%		10.1	11.3	
Crude fiber		%		22.2	24.9	
Sheep	dig coef	%		56.	56.	
Ether extract		%		3.2	3.6	
Sheep	dig coef	%		57.	57.	
N-free extract		%		42.3	47.4	
Sheep	dig coef	%		61.	61.	
Protein (N x 6.25)		%		11.5	12.8	
Sheep	dig coef	%		39.	39.	
Cattle	dig prot	%	*	7.2	8.1	
Goats	dig prot	%	*	7.6	8.5	
Horses	dig prot	%	*	7.5	8.4	
Rabbits	dig prot	%	*	7.7	8.6	
Sheep	dig prot	%		4.5	5.0	
Fatty acids		%		3.2	3.6	
Energy	GE Mcal/kg					
Cattle	DE Mcal/kg		*	2.29	2.57	
Sheep	DE Mcal/kg		*	2.06	2.31	
Cattle	ME Mcal/kg		*	1.88	2.11	
Sheep	ME Mcal/kg		*	1.69	1.90	
Cattle	TDN	%	*	52.1	58.4	
Sheep	TDN	%		46.8	52.5	

Sweetpotato, aerial part, immature, (1)
Ref No 1 04 781 United States

Feed Name or Analyses				Mean As Fed	Mean Dry	C.V. ± %
Dry matter		%		91.6	100.0	
Ash		%		10.8	11.8	
Crude fiber		%		12.2	13.3	
Ether extract		%		5.0	5.5	
N-free extract		%		48.7	53.2	
Protein (N x 6.25)		%		14.8	16.2	
Cattle	dig prot	%	*	10.0	11.0	
Goats	dig prot	%	*	10.7	11.7	
Horses	dig prot	%	*	10.3	11.3	
Rabbits	dig prot	%	*	10.2	11.2	
Sheep	dig prot	%	*	10.2	11.1	

Sweetpotato, hay, s-c, (1)
Ref No 1 04 780 United States

Feed Name or Analyses				Mean As Fed	Mean Dry	C.V. ± %
Dry matter		%			100.0	
Calcium		%			0.09	
Phosphorus		%			0.36	

Sweetpotato, aerial part, fresh, (2)
Ref No 2 04 784 United States

Feed Name or Analyses				Mean As Fed	Mean Dry	C.V. ± %
Dry matter		%		21.9	100.0	
Ash		%		2.4	10.9	
Crude fiber		%		5.9	26.9	
Cattle	dig coef	%		36.	36.	
Ether extract		%		0.6	2.8	
Cattle	dig coef	%		75.	75.	
N-free extract		%		10.2	46.6	
Cattle	dig coef	%		72.	72.	

Continued

Feed Name or Analyses		Mean As Fed	Mean Dry	C.V. ± %
Protein (N x 6.25)	%	2.8	12.8	
Cattle	dig coef %	65.	65.	
Cattle	dig prot %	1.8	8.3	
Goats	dig prot %	* 1.9	8.5	
Horses	dig prot %	* 1.8	8.4	
Rabbits	dig prot %	* 1.9	8.6	
Sheep	dig prot %	* 2.0	8.9	
Energy	GE Mcal/kg			
Cattle	DE Mcal/kg	* 0.54	2.48	
Sheep	DE Mcal/kg	* 0.53	2.40	
Cattle	ME Mcal/kg	* 0.45	2.03	
Sheep	ME Mcal/kg	* 0.43	1.97	
Cattle	TDN %	12.3	56.3	
Sheep	TDN %	* 11.9	54.5	

Sweetpotato, aerial part, ensiled, (3)
Ref No 3 04 785 United States

Feed Name or Analyses		Mean As Fed	Mean Dry	C.V. ± %
Dry matter	%	12.1	100.0	
Ash	%	1.4	11.5	
Crude fiber	%	3.5	29.0	
Sheep	dig coef %	56.	56.	
Ether extract	%	0.5	4.3	
Sheep	dig coef %	60.	60.	
N-free extract	%	5.1	42.0	
Sheep	dig coef %	50.	50.	
Protein (N x 6.25)	%	1.6	13.2	
Sheep	dig coef %	40.	40.	
Cattle	dig prot %	* 1.0	8.2	
Goats	dig prot %	* 1.0	8.2	
Horses	dig prot %	* 1.0	8.2	
Sheep	dig prot %	0.6	5.3	
Energy	GE Mcal/kg			
Cattle	DE Mcal/kg	* 0.28	2.30	
Sheep	DE Mcal/kg	* 0.26	2.13	
Cattle	ME Mcal/kg	* 0.23	1.89	
Sheep	ME Mcal/kg	* 0.21	1.75	
Cattle	TDN %	* 6.3	52.1	
Sheep	TDN %	5.8	48.3	

Sweetpotato, aerial part w molasses added, ensiled, (3)
Ref No 3 04 786 United States

Feed Name or Analyses		Mean As Fed	Mean Dry	C.V. ± %
Dry matter	%	27.9	100.0	16
Ash	%	3.3	12.0	4
Crude fiber	%	4.8	17.2	2
Ether extract	%	1.0	3.6	5
N-free extract	%	15.7	56.2	
Protein (N x 6.25)	%	3.1	11.0	6
Cattle	dig prot %	* 1.7	6.2	
Goats	dig prot %	* 1.7	6.2	
Horses	dig prot %	* 1.7	6.2	
Sheep	dig prot %	* 1.7	6.2	
Energy	GE Mcal/kg			
Cattle	DE Mcal/kg	* 0.68	2.44	
Sheep	DE Mcal/kg	* 0.66	2.36	
Cattle	ME Mcal/kg	* 0.56	2.00	
Sheep	ME Mcal/kg	* 0.54	1.96	
Cattle	TDN %	* 15.5	55.4	
Sheep	TDN %	* 14.9	53.5	

(1) dry forages and roughages (2) pasture, range plants, and forages fed green (3) silages (4) energy feeds (5) protein supplements (6) minerals (7) vitamins (8) additives

Sweetpotato, roots, ensiled, (3)
Ref No 3 04 787 United States

Feed Name or Analyses		Mean As Fed	Mean Dry	C.V. ± %
Dry matter	%	14.2	100.0	38
Ash	%	1.5	10.6	23
Crude fiber	%	3.7	25.8	32
Ether extract	%	0.6	3.9	27
N-free extract	%	6.7	47.1	
Protein (N x 6.25)	%	1.8	12.6	12
Cattle	dig prot %	* 1.1	7.7	
Goats	dig prot %	* 1.1	7.7	
Horses	dig prot %	* 1.1	7.7	
Sheep	dig prot %	* 1.1	7.7	
Energy	GE Mcal/kg			
Cattle	DE Mcal/kg	* 0.42	2.99	
Sheep	DE Mcal/kg	* 0.39	2.76	
Cattle	ME Mcal/kg	* 0.35	2.45	
Sheep	ME Mcal/kg	* 0.32	2.26	
Cattle	TDN %	* 9.6	67.8	
Sheep	TDN %	* 8.9	62.6	

Sweetpotato, process residue, dehy, (4)
Sweetpotato by-product, dried
Sweetpotato pomace, dried
Sweetpotato pulp, dried
Sweetpotato waste, dried
Ref No 4 08 535 United States

Feed Name or Analyses		Mean As Fed	Mean Dry	C.V. ± %
Dry matter	%	90.2	100.0	
Ash	%	6.0	6.7	
Crude fiber	%	9.6	10.6	
Ether extract	%	0.3	0.3	
N-free extract	%	71.8	79.6	
Protein (N x 6.25)	%	2.5	2.8	
Cattle	dig prot %	* -1.2	-1.3	
Goats	dig prot %	* -0.1	-0.1	
Horses	dig prot %	* -0.1	-0.1	
Sheep	dig prot %	* -0.1	-0.1	
Energy	GE Mcal/kg			
Cattle	DE Mcal/kg	* 2.98	3.30	
Sheep	DE Mcal/kg	* 3.17	3.51	
Swine	DE kcal/kg	*3006.	3333.	
Cattle	ME Mcal/kg	* 2.44	2.71	
Sheep	ME Mcal/kg	* 2.60	2.88	
Swine	ME kcal/kg	*2868.	3180.	
Cattle	TDN %	* 67.6	74.9	
Sheep	TDN %	* 71.8	79.7	
Swine	TDN %	* 68.2	75.6	

Sweetpotato, roots, dehy grnd, (4)
Sweetpotato meal
Ref No 4 08 536 United States

Feed Name or Analyses		Mean As Fed	Mean Dry	C.V. ± %
Dry matter	%	90.2	100.0	
Ash	%	4.1	4.5	
Crude fiber	%	3.3	3.7	
Ether extract	%	0.9	1.0	
N-free extract	%	77.0	85.4	
Protein (N x 6.25)	%	4.9	5.4	
Cattle	dig prot %	* 0.9	1.0	
Goats	dig prot %	* 2.0	2.2	
Horses	dig prot %	* 2.0	2.2	
Sheep	dig prot %	* 2.0	2.2	
Energy	GE Mcal/kg			
Cattle	DE Mcal/kg	* 3.20	3.55	
Sheep	DE Mcal/kg	* 3.37	3.74	
Swine	DE kcal/kg	*3141.	3483.	

Feed Name or Analyses		Mean As Fed	Mean Dry	C.V. ± %
Cattle	ME Mcal/kg	* 2.62	2.91	
Sheep	ME Mcal/kg	* 2.76	3.06	
Swine	ME kcal/kg	*2981.	3305.	
Cattle	TDN %	* 72.7	80.6	
Sheep	TDN %	* 76.4	84.7	
Swine	TDN %	* 71.2	79.0	
Calcium	%	0.15	0.17	
Phosphorus	%	0.14	0.16	

Sweetpotato, roots, fresh, (4)
Ref No 2 04 788 United States

Feed Name or Analyses		Mean As Fed	Mean Dry	C.V. ± %
Dry matter	%	30.6	100.0	
Ash	%	1.1	3.6	
Crude fiber	%	1.3	4.2	
Ether extract	%	0.4	1.3	
N-free extract	%	26.2	85.5	
Protein (N x 6.25)	%	1.7	5.4	
Cattle	dig prot %	* 0.3	1.0	
Goats	dig prot %	* 0.7	2.2	
Horses	dig prot %	* 0.7	2.2	
Sheep	dig prot %	* 0.7	2.2	
Energy	GE Mcal/kg	1.19	3.88	
Cattle	DE Mcal/kg	* 1.12	3.66	
Sheep	DE Mcal/kg	* 1.15	3.75	
Swine	DE kcal/kg	*1081.	3533.	
Cattle	ME Mcal/kg	* 0.92	3.00	
Sheep	ME Mcal/kg	* 0.94	3.08	
Swine	ME kcal/kg	*1026.	3353.	
Cattle	TDN %	* 25.4	83.1	
Sheep	TDN %	* 26.1	85.1	
Swine	TDN %	* 24.5	80.1	
Calcium	%	0.03	0.10	
Chlorine	%	0.02	0.06	
Copper	mg/kg	1.3	4.2	
Iron	%	0.001	0.005	
Magnesium	%	0.05	0.16	
Manganese	mg/kg	3.4	11.1	
Phosphorus	%	0.05	0.15	
Potassium	%	0.31	1.01	
Sodium	%	0.01	0.05	
Sulphur	%	0.04	0.13	
Ascorbic acid	mg/kg	218.6	714.3	
Niacin	mg/kg	6.2	20.4	
Riboflavin	mg/kg	0.6	2.0	
Thiamine	mg/kg	1.0	3.4	
Vitamin A	IU/g	91.6	299.3	

Sweetpotato by-product, dried –
see Sweetpotato, process residue, dehy, (4)

Sweetpotato meal –
see Sweetpotato, roots, dehy grnd, (4)

Sweetpotato pomace, dried –
see Sweetpotato, process residue, dehy, (4)

Sweetpotato pulp, dried –
see Sweetpotato, process residue, dehy, (4)

Sweetpotato waste, dried –
see Sweetpotato, process residue, dehy, (4)

SWEETSHRUB, COMMON. Calycanthus floridus

Sweetshrub, common, seeds w hulls, (5)
Carolina Allspice
Ref No 5 09 050 — United States

Feed Name or Analyses		Mean As Fed	Mean Dry	C.V. ± %
Dry matter	%		100.0	
Protein (N x 6.25)	%		24.0	
Alanine	%		0.98	
Arginine	%		1.85	
Aspartic acid	%		1.94	
Glutamic acid	%		3.14	
Glycine	%		0.84	
Histidine	%		0.50	
Hydroxyproline	%		0.02	
Isoleucine	%		0.82	
Leucine	%		1.46	
Lysine	%		1.25	
Methionine	%		0.36	
Phenylalanine	%		1.15	
Proline	%		0.86	
Serine	%		0.96	
Threonine	%		0.84	
Tyrosine	%		0.72	
Valine	%		1.03	

SWEETVETCH. Hedysarum varium

Sweetvetch, seeds w hulls, (5)
Ref No 5 09 298 — United States

Feed Name or Analyses		Mean As Fed	Mean Dry	C.V. ± %
Dry matter	%		100.0	
Protein (N x 6.25)	%		50.0	
Alanine	%		1.55	
Arginine	%		2.70	
Aspartic acid	%		3.65	
Glutamic acid	%		4.55	
Glycine	%		1.95	
Histidine	%		1.00	
Hydroxyproline	%		0.05	
Isoleucine	%		1.30	
Leucine	%		2.15	
Lysine	%		1.80	
Methionine	%		0.40	
Phenylalanine	%		1.20	
Proline	%		1.35	
Serine	%		1.50	
Threonine	%		1.35	
Tyrosine	%		1.10	
Valine	%		1.45	

SWINE. Sus scrofa

Swine, lard, (4)
Ref No 4 04 790 — United States

Feed Name or Analyses		Mean As Fed	Mean Dry	C.V. ± %
Dry matter	%		100.0	
Ether extract	%		100.0	
Energy	GE Mcal/kg		9.20	
Swine	DE kcal/kg		7760.	
Chickens	ME_n kcal/kg		8600.	
Swine	ME kcal/kg		7700.	

Swine, carcass, raw, (5)
Ref No 5 05 703 — United States

Feed Name or Analyses		Mean As Fed	Mean Dry	C.V. ± %
Dry matter	%	53.7	100.0	
Ash	%	2.8	5.1	
Ether extract	%	36.0	67.1	
Protein (N x 6.25)	%	13.7	25.6	
Potassium	%	0.21	0.38	
Sodium	%	0.08	0.14	

Swine, cracklings, dry rendered dehy, (5)
Pork cracklings (CFA)
Cracklings
Ref No 5 04 791 — United States

Feed Name or Analyses		Mean As Fed	Mean Dry	C.V. ± %
Dry matter	%	92.5	100.0	2
Ash	%	22.8	24.6	13
Crude fiber	%	1.4	1.5	46
Ether extract	%	10.9	11.8	33
N-free extract	%	0.5	0.5	
Protein (N x 6.25)	%	56.9	61.5	15
Energy	GE Mcal/kg			
Cattle	DE Mcal/kg	* 2.88	3.11	
Sheep	DE Mcal/kg	* 2.75	2.97	
Swine	DE kcal/kg	*3076.	3326.	
Cattle	ME Mcal/kg	* 2.36	2.55	
Sheep	ME Mcal/kg	* 2.25	2.43	
Swine	ME kcal/kg	*2571.	2780.	
Cattle	TDN %	* 65.3	70.6	
Sheep	TDN %	* 62.3	67.3	
Swine	TDN %	* 69.8	75.4	
Calcium	%	7.40	8.00	14
Manganese	mg/kg	14.3	15.5	79
Phosphorus	%	4.13	4.46	30
Riboflavin	mg/kg	4.0	4.3	12

Swine, cracklings, grnd, (5)
Ref No 5 04 789 — United States

Feed Name or Analyses		Mean As Fed	Mean Dry	C.V. ± %
Dry matter	%	93.7	100.0	
Ash	%	2.2	2.4	
Ether extract	%	30.7	32.8	
Swine	dig coef %	98.	98.	
Protein (N x 6.25)	%	59.1	63.1	
Swine	dig coef %	95.	95.	
Swine	dig prot %	56.2	59.9	
Energy	GE Mcal/kg			
Swine	DE kcal/kg	*5466.	5833.	
Swine	ME kcal/kg	*4551.	4857.	
Swine	TDN %	124.0	132.3	

Swine, livers, raw, (5)
Ref No 5 04 792 — United States

Feed Name or Analyses		Mean As Fed	Mean Dry	C.V. ± %
Dry matter	%	28.4	100.0	
Ash	%	1.5	5.3	
Crude fiber	%	0.0	0.0	
Ether extract	%	3.7	13.0	
N-free extract	%	2.6	9.2	
Protein (N x 6.25)	%	20.6	72.5	
Energy	GE Mcal/kg	1.31	4.61	
Cattle	DE Mcal/kg	* 1.18	4.14	
Sheep	DE Mcal/kg	* 1.06	3.72	
Swine	DE kcal/kg	*1165.	4104.	
Cattle	ME Mcal/kg	* 0.96	3.39	
Sheep	ME Mcal/kg	* 0.87	3.05	
Swine	ME kcal/kg	* 948.	3338.	
Cattle	TDN %	* 26.7	93.9	
Sheep	TDN %	* 24.0	84.5	
Swine	TDN %	* 26.4	93.1	
Calcium	%	0.01	0.04	
Iron	%	0.019	0.068	
Phosphorus	%	0.36	1.25	
Potassium	%	0.26	0.92	
Sodium	%	0.07	0.26	
Ascorbic acid	mg/kg	230.0	809.9	
Niacin	mg/kg	164.0	577.5	
Riboflavin	mg/kg	30.3	106.7	
Thiamine	mg/kg	3.0	10.6	
Vitamin A	IU/g	109.0	383.8	

Swine, milk, fresh, (5)
Ref No 2 08 537 — United States

Feed Name or Analyses		Mean As Fed	Mean Dry	C.V. ± %
Dry matter	%	20.1	100.0	
Ash	%	1.0	5.0	
Crude fiber	%	0.0	0.0	
Ether extract	%	6.7	33.3	
N-free extract	%	5.1	25.4	
Protein (N x 6.25)	%	7.3	36.3	
Energy	GE Mcal/kg			
Cattle	DE Mcal/kg	* 1.06	5.30	
Sheep	DE Mcal/kg	* 0.92	4.56	
Swine	DE kcal/kg	*1274.	6339.	
Cattle	ME Mcal/kg	* 0.87	4.34	
Sheep	ME Mcal/kg	* 0.75	3.74	
Swine	ME kcal/kg	*1130.	5620.	
Cattle	TDN %	* 24.2	120.2	
Sheep	TDN %	* 20.8	103.5	
Swine	TDN %	* 28.9	143.8	

SWITCHGRASS. Panicum virgatum

Switchgrass, hay, fan air dried, (1)
Ref No 1 04 793 — United States

Feed Name or Analyses		Mean As Fed	Mean Dry	C.V. ± %
Dry matter	%		100.0	
Carotene	mg/kg		47.8	
Vitamin A equivalent	IU/g		79.7	

Switchgrass, hay, s-c, (1)
Ref No 1 04 794 — United States

Feed Name or Analyses		Mean As Fed	Mean Dry	C.V. ± %
Dry matter	%		100.0	
Carotene	mg/kg		39.2	
Vitamin A equivalent	IU/g		65.4	

Switchgrass, aerial part, fresh, (2)
Ref No 2 04 800 — United States

Feed Name or Analyses		Mean As Fed	Mean Dry	C.V. ± %
Dry matter	%	55.3	100.0	18
Ash	%	3.2	5.9	9
Crude fiber	%	19.2	34.7	6
Ether extract	%	1.3	2.3	19
N-free extract	%	28.6	51.7	3
Protein (N x 6.25)	%	3.0	5.4	21
Cattle	dig prot %	* 1.4	2.5	
Goats	dig prot %	* 0.9	1.6	
Horses	dig prot %	* 1.2	2.1	
Rabbits	dig prot %	* 1.6	2.9	
Sheep	dig prot %	* 1.1	2.0	

Continued

Feed Name or Analyses			Mean As Fed	Mean Dry	C.V. ± %
Energy	GE Mcal/kg				
Cattle	DE Mcal/kg	*	1.58	2.86	
Sheep	DE Mcal/kg	*	1.66	3.01	
Cattle	ME Mcal/kg	*	1.30	2.34	
Sheep	ME Mcal/kg	*	1.36	2.47	
Cattle	TDN %	*	35.8	64.8	
Sheep	TDN %	*	37.7	68.2	
Calcium	%		0.16	0.29	10
Phosphorus	%		0.04	0.08	20
Carotene	mg/kg		49.8	90.1	64
Vitamin A equivalent	IU/g		83.0	150.1	

Switchgrass, aerial part, fresh, immature, (2)
Ref No 2 04 795 United States

Feed Name or Analyses			Mean As Fed	Mean Dry	C.V. ± %
Dry matter	%		33.2	100.0	
Ash	%		2.6	7.9	24
Crude fiber	%		9.8	29.7	4
Ether extract	%		0.8	2.4	21
N-free extract	%		16.3	49.2	
Protein (N x 6.25)	%		3.6	10.8	12
Cattle	dig prot %	*	2.3	7.1	
Goats	dig prot %	*	2.2	6.6	
Horses	dig prot %	*	2.2	6.7	
Rabbits	dig prot %	*	2.3	7.0	
Sheep	dig prot %	*	2.3	7.0	
Energy	GE Mcal/kg				
Cattle	DE Mcal/kg	*	0.90	2.72	
Sheep	DE Mcal/kg	*	0.89	2.68	
Cattle	ME Mcal/kg	*	0.74	2.23	
Sheep	ME Mcal/kg	*	0.73	2.20	
Cattle	TDN %	*	20.5	61.7	
Sheep	TDN %	*	20.2	60.7	
Calcium	%		0.15	0.46	18
Phosphorus	%		0.07	0.20	13

Switchgrass, aerial part, fresh, full bloom, (2)
Ref No 2 04 796 United States

Feed Name or Analyses			Mean As Fed	Mean Dry	C.V. ± %
Dry matter	%			100.0	
Ash	%			6.2	15
Crude fiber	%			33.9	7
Ether extract	%			1.8	20
N-free extract	%			51.8	
Protein (N x 6.25)	%			6.3	25
Cattle	dig prot %	*		3.2	
Goats	dig prot %	*		2.4	
Horses	dig prot %	*		2.9	
Rabbits	dig prot %	*		3.5	
Sheep	dig prot %	*		2.9	
Energy	GE Mcal/kg				
Cattle	DE Mcal/kg	*		2.68	
Sheep	DE Mcal/kg	*		2.77	
Cattle	ME Mcal/kg	*		2.21	
Sheep	ME Mcal/kg	*		2.27	
Cattle	TDN %	*		61.0	
Sheep	TDN %	*		62.9	
Calcium	%			0.33	12
Phosphorus	%			0.12	25

(1) dry forages and roughages (2) pasture, range plants, and forages fed green (3) silages (4) energy feeds (5) protein supplements (6) minerals (7) vitamins (8) additives

Switchgrass, aerial part, fresh, milk stage, (2)
Ref No 2 04 797 United States

Feed Name or Analyses			Mean As Fed	Mean Dry	C.V. ± %
Dry matter	%			100.0	
Ash	%			4.1	20
Crude fiber	%			35.1	6
Ether extract	%			2.6	41
N-free extract	%			52.8	
Protein (N x 6.25)	%			5.4	8
Cattle	dig prot %	*		2.5	
Goats	dig prot %	*		1.6	
Horses	dig prot %	*		2.1	
Rabbits	dig prot %	*		2.8	
Sheep	dig prot %	*		2.0	
Energy	GE Mcal/kg				
Cattle	DE Mcal/kg	*		2.74	
Sheep	DE Mcal/kg	*		2.80	
Cattle	ME Mcal/kg	*		2.25	
Sheep	ME Mcal/kg	*		2.30	
Cattle	TDN %	*		62.1	
Sheep	TDN %	*		63.4	
Calcium	%			0.26	9
Phosphorus	%			0.11	18

Switchgrass, aerial part, fresh, mature, (2)
Ref No 2 04 798 United States

Feed Name or Analyses			Mean As Fed	Mean Dry	C.V. ± %
Dry matter	%		42.0	100.0	
Ash	%		2.3	5.5	
Crude fiber	%		18.2	43.3	
Ether extract	%		1.1	2.6	
N-free extract	%		18.1	43.1	
Protein (N x 6.25)	%		2.3	5.5	
Cattle	dig prot %	*	1.1	2.5	
Goats	dig prot %	*	0.7	1.7	
Horses	dig prot %	*	0.9	2.2	
Rabbits	dig prot %	*	1.2	2.9	
Sheep	dig prot %	*	0.9	2.1	
Energy	GE Mcal/kg				
Cattle	DE Mcal/kg	*	0.97	2.31	
Sheep	DE Mcal/kg	*	0.89	2.11	
Cattle	ME Mcal/kg	*	0.79	1.89	
Sheep	ME Mcal/kg	*	0.73	1.73	
Cattle	TDN %	*	22.0	52.5	
Sheep	TDN %	*	20.1	47.8	

Switchgrass, aerial part, fresh, over ripe, (2)
Ref No 2 04 799 United States

Feed Name or Analyses			Mean As Fed	Mean Dry	C.V. ± %
Dry matter	%		76.7	100.0	
Ash	%		3.6	4.7	9
Crude fiber	%		31.6	41.2	9
Ether extract	%		1.5	2.0	28
N-free extract	%		38.5	50.2	
Protein (N x 6.25)	%		1.5	1.9	15
Cattle	dig prot %	*	-0.3	-0.4	
Goats	dig prot %	*	-1.2	-1.6	
Horses	dig prot %	*	-0.6	-0.8	
Rabbits	dig prot %	*	0.1	0.1	
Sheep	dig prot %	*	-0.9	-1.1	
Energy	GE Mcal/kg				
Cattle	DE Mcal/kg	*	1.53	2.00	
Sheep	DE Mcal/kg	*	1.69	2.20	
Cattle	ME Mcal/kg	*	1.26	1.64	
Sheep	ME Mcal/kg	*	1.38	1.81	
Cattle	TDN %	*	34.7	45.3	
Sheep	TDN %	*	38.3	49.9	

Feed Name or Analyses			Mean As Fed	Mean Dry	C.V. ± %
Calcium	%		0.18	0.24	12
Phosphorus	%		0.04	0.05	39
Carotene	mg/kg		0.5	0.7	
Vitamin A equivalent	IU/g		0.8	1.1	

Switchgrass, aerial part, weathered, mature, (1)
Ref No 1 08 577 United States

Feed Name or Analyses			Mean As Fed	Mean Dry	C.V. ± %
Dry matter	%		90.0	100.0	
Ash	%		4.4	4.9	
Crude fiber	%		37.2	41.3	
Ether extract	%		1.6	1.8	
N-free extract	%		45.1	50.1	
Protein (N x 6.25)	%		1.7	1.9	
Cattle	dig prot %	*	-0.4	-0.4	
Goats	dig prot %	*	-1.4	-1.6	
Horses	dig prot %	*	-0.7	-0.8	
Rabbits	dig prot %	*	0.1	0.1	
Sheep	dig prot %	*	-1.0	-1.1	
Energy	GE Mcal/kg				
Cattle	DE Mcal/kg	*	1.84	2.04	
Sheep	DE Mcal/kg	*	1.98	2.20	
Cattle	ME Mcal/kg	*	1.50	1.67	
Sheep	ME Mcal/kg	*	1.62	1.80	
Cattle	TDN %	*	41.7	46.3	
Sheep	TDN %	*	44.9	49.9	
Calcium	%		0.25	0.28	
Phosphorus	%		0.03	0.03	

Tangerine pulp, dried –
see Citrus, tangerine, pulp wo fines, shredded dehy, (4)

TANGLEHEAD. Heteropogon contortus

Tanglehead, hay, s-c, milk stage, (1)
Ref No 1 04 801 United States

Feed Name or Analyses			Mean As Fed	Mean Dry	C.V. ± %
Dry matter	%		92.6	100.0	
Ash	%		7.6	8.2	
Crude fiber	%		37.1	40.1	
Cattle	dig coef %		65.	65.	
Ether extract	%		0.9	1.0	
Cattle	dig coef %		34.	34.	
N-free extract	%		43.7	47.2	
Cattle	dig coef %		44.	44.	
Protein (N x 6.25)	%		3.2	3.5	
Cattle	dig coef %		25.	25.	
Cattle	dig prot %		0.8	0.9	
Goats	dig prot %	*	-0.1	-0.1	
Horses	dig prot %	*	0.5	0.5	
Rabbits	dig prot %	*	1.3	1.4	
Sheep	dig prot %	*	-0.2	-0.2	
Energy	GE Mcal/kg				
Cattle	DE Mcal/kg	*	1.98	2.14	
Sheep	DE Mcal/kg	*	2.07	2.23	
Cattle	ME Mcal/kg	*	1.62	1.75	
Sheep	ME Mcal/kg	*	1.70	1.83	
Cattle	TDN %		44.9	48.5	
Sheep	TDN %	*	46.9	50.7	

Tanglehead, aerial part, fresh, (2)
Ref No 2 04 805 United States

Feed Name or Analyses			Mean As Fed	Mean Dry	C.V. ± %
Dry matter	%			100.0	
Ash	%			7.6	24
Crude fiber	%			33.5	6

Feed Name or Analyses		Mean As Fed	Mean Dry	C.V. ± %
Ether extract	%		2.6	26
N-free extract	%		49.9	
Protein (N x 6.25)	%		6.4	42
Cattle	dig prot % *		3.3	
Goats	dig prot % *		2.5	
Horses	dig prot % *		3.0	
Rabbits	dig prot % *		3.6	
Sheep	dig prot % *		3.0	

Tanglehead, aerial part, fresh, immature, (2)
Ref No 2 04 802 United States

Analyses		As Fed	Dry	C.V. ± %
Dry matter	%		100.0	
Ash	%		10.0	
Crude fiber	%		31.0	
Ether extract	%		2.0	
N-free extract	%		47.4	
Protein (N x 6.25)	%		9.6	
Cattle	dig prot % *		6.1	
Goats	dig prot % *		5.5	
Horses	dig prot % *		5.7	
Rabbits	dig prot % *		6.1	
Sheep	dig prot % *		5.9	
Calcium	%		0.47	
Phosphorus	%		0.16	

Tanglehead, aerial part, fresh, full bloom, (2)
Ref No 2 04 803 United States

Analyses		As Fed	Dry	C.V. ± %
Dry matter	%		100.0	
Calcium	%		0.37	
Phosphorus	%		0.07	

Tanglehead, aerial part, fresh, mature, (2)
Ref No 2 04 804 United States

Analyses		As Fed	Dry	C.V. ± %
Dry matter	%		100.0	
Ash	%		6.6	
Crude fiber	%		34.4	
Ether extract	%		1.9	
N-free extract	%		53.7	
Protein (N x 6.25)	%		3.4	
Cattle	dig prot % *		0.8	
Goats	dig prot % *		-0.2	
Horses	dig prot % *		0.4	
Rabbits	dig prot % *		1.3	
Sheep	dig prot % *		0.2	
Calcium	%		0.22	
Phosphorus	%		0.05	

TANSY, ROCK. Tanacetum capitatum

Tansy, rock, aerial part, fresh, mature, (2)
Ref No 2 04 806 United States

Analyses		As Fed	Dry	C.V. ± %
Dry matter	%		100.0	
Ash	%		10.4	
Crude fiber	%		26.8	
Ether extract	%		7.6	
N-free extract	%		51.1	
Protein (N x 6.25)	%		4.1	
Cattle	dig prot % *		1.4	
Goats	dig prot % *		0.4	
Horses	dig prot % *		1.0	

Feed Name or Analyses		Mean As Fed	Mean Dry	C.V. ± %
Rabbits	dig prot % *		1.8	
Sheep	dig prot % *		0.8	

TANSYMUSTARD. Descurainia spp

Tansymustard, aerial part, fresh, mature, (2)
Ref No 2 04 807 United States

Analyses		As Fed	Dry	C.V. ± %
Dry matter	%		100.0	
Ash	%		5.6	
Ether extract	%		2.4	
Protein (N x 6.25)	%		7.6	
Cattle	dig prot % *		4.4	
Goats	dig prot % *		3.7	
Horses	dig prot % *		4.0	
Rabbits	dig prot % *		4.5	
Sheep	dig prot % *		4.1	
Calcium	%		1.49	
Magnesium	%		0.32	
Phosphorus	%		0.49	
Potassium	%		4.45	

Tansymustard, leaves, fresh, (2)
Ref No 2 04 808 United States

Analyses		As Fed	Dry	C.V. ± %
Dry matter	%		100.0	
Ash	%		9.7	
Ether extract	%		7.6	
Protein (N x 6.25)	%		20.2	
Cattle	dig prot % *		15.1	
Goats	dig prot % *		15.4	
Horses	dig prot % *		14.7	
Rabbits	dig prot % *		14.3	
Sheep	dig prot % *		15.8	
Calcium	%		2.07	
Phosphorus	%		0.59	

Tansymustard, stems, fresh, (2)
Ref No 2 04 809 United States

Analyses		As Fed	Dry	C.V. ± %
Dry matter	%		100.0	
Ash	%		4.5	
Ether extract	%		1.3	
Protein (N x 6.25)	%		4.3	
Cattle	dig prot % *		1.5	
Goats	dig prot % *		0.6	
Horses	dig prot % *		1.2	
Rabbits	dig prot % *		2.0	
Sheep	dig prot % *		1.0	
Calcium	%		0.76	
Phosphorus	%		0.20	

TANSYMUSTARD, PINNATE. Descurainia spp

Tansymustard, pinnate, aerial part, fresh, immature, (2)
Ref No 2 04 810 United States

Analyses		As Fed	Dry	C.V. ± %
Dry matter	%		100.0	
Calcium	%		1.97	
Magnesium	%		0.32	
Phosphorus	%		0.69	
Potassium	%		4.45	

Tapioca flour –
see Cassava, common, root flour, fine grnd, (4)

Tapioca meal –
see Cassava, starch by-product, dehy, (4)

TARWEED. Madia spp

Tarweed, aerial part, fresh, (2)
Ref No 2 04 816 United States

Analyses		As Fed	Dry	C.V. ± %
Dry matter	%		100.0	
Ash	%		11.1	29
Crude fiber	%		18.0	26
Protein (N x 6.25)	%		13.0	55
Cattle	dig prot % *		8.9	
Goats	dig prot % *		8.7	
Horses	dig prot % *		8.6	
Rabbits	dig prot % *		8.7	
Sheep	dig prot % *		9.1	
Calcium	%		1.54	28
Phosphorus	%		0.34	69
Potassium	%		3.27	57

Tarweed, aerial part, fresh, immature, (2)
Ref No 2 04 811 United States

Analyses		As Fed	Dry	C.V. ± %
Dry matter	%		100.0	
Ash	%		15.7	11
Crude fiber	%		12.8	25
Protein (N x 6.25)	%		22.9	24
Cattle	dig prot % *		17.4	
Goats	dig prot % *		17.9	
Horses	dig prot % *		17.0	
Rabbits	dig prot % *		16.3	
Sheep	dig prot % *		18.3	
Calcium	%		2.05	14
Phosphorus	%		0.52	49
Potassium	%		5.23	27

Tarweed, aerial part, fresh, mid-bloom, (2)
Ref No 2 04 812 United States

Analyses		As Fed	Dry	C.V. ± %
Dry matter	%		100.0	
Calcium	%		1.64	
Phosphorus	%		0.47	
Potassium	%		4.93	

Tarweed, aerial part, fresh, full bloom, (2)
Ref No 2 04 813 United States

Analyses		As Fed	Dry	C.V. ± %
Dry matter	%		100.0	
Ash	%		6.9	
Crude fiber	%		19.3	
Protein (N x 6.25)	%		6.7	
Cattle	dig prot % *		3.6	
Goats	dig prot % *		2.8	
Horses	dig prot % *		3.2	
Rabbits	dig prot % *		3.8	
Sheep	dig prot % *		3.2	

Tarweed, aerial part, fresh, milk stage, (2)
Ref No 2 04 814 United States

Analyses		As Fed	Dry	C.V. ± %
Dry matter	%		100.0	
Calcium	%		1.25	

Continued

Feed Name or Analyses				Mean As Fed	Mean Dry	C.V. ± %
Phosphorus		%			0.39	
Potassium		%			2.91	

Tarweed, aerial part, fresh, mature, (2)
Ref No 2 04 815 United States

Feed Name or Analyses				As Fed	Dry	C.V. ± %
Dry matter		%			100.0	
Ash		%			7.0	
Crude fiber		%			22.2	
Protein (N x 6.25)		%			5.5	
Cattle	dig prot	%	*		2.6	
Goats	dig prot	%	*		1.7	
Horses	dig prot	%	*		2.2	
Rabbits	dig prot	%	*		2.9	
Sheep	dig prot	%	*		2.1	
Calcium		%			1.04	
Phosphorus		%			0.11	
Potassium		%			1.09	

TEASEL, VENUSCUP. Dipsacus sylvestris

Teasel, venuscup, leaves, fresh, (2)
Ref No 2 04 817 United States

Feed Name or Analyses				As Fed	Dry	C.V. ± %
Dry matter		%			100.0	
Calcium		%			2.09	
Iron		%			0.019	
Manganese		mg/kg			50.0	
Phosphorus		%			0.54	
Potassium		%			2.74	

TEFF. Eragrostis abyssinica

Teff, hay, s-c, (1)
Ref No 1 04 821 United States

Feed Name or Analyses				As Fed	Dry	C.V. ± %
Dry matter		%		92.6	100.0	
Ash		%		6.2	6.7	
Crude fiber		%		31.3	33.8	
Sheep	dig coef	%		70.	70.	
Ether extract		%		1.0	1.1	
Sheep	dig coef	%		32.	32.	
N-free extract		%		45.4	49.0	
Sheep	dig coef	%		58.	58.	
Protein (N x 6.25)		%		8.7	9.4	
Sheep	dig coef	%		58.	58.	
Cattle	dig prot	%	*	4.7	5.1	
Goats	dig prot	%	*	4.9	5.3	
Horses	dig prot	%	*	5.1	5.5	
Rabbits	dig prot	%	*	5.5	5.9	
Sheep	dig prot	%		5.0	5.5	
Energy	GE Mcal/kg					
Cattle	DE Mcal/kg		*	2.24	2.41	
Sheep	DE Mcal/kg		*	2.38	2.57	
Cattle	ME Mcal/kg		*	1.83	1.98	
Sheep	ME Mcal/kg		*	1.95	2.11	
Cattle	TDN	%	*	50.7	54.8	
Sheep	TDN	%		54.0	58.3	

(1) dry forages and roughages (2) pasture, range plants, and forages fed green (3) silages (4) energy feeds (5) protein supplements (6) minerals (7) vitamins (8) additives

Teff, hay, s-c, milk stage, (1)
Ref No 1 04 818 United States

Feed Name or Analyses				As Fed	Dry	C.V. ± %
Dry matter		%		93.2	100.0	
Ash		%		4.9	5.3	
Crude fiber		%		31.9	34.2	
Sheep	dig coef	%		67.	67.	
Ether extract		%		1.0	1.1	
Sheep	dig coef	%		23.	23.	
N-free extract		%		45.6	48.9	
Sheep	dig coef	%		58.	58.	
Protein (N x 6.25)		%		9.8	10.5	
Sheep	dig coef	%		61.	61.	
Cattle	dig prot	%	*	5.6	6.0	
Goats	dig prot	%	*	5.9	6.4	
Horses	dig prot	%	*	6.0	6.4	
Rabbits	dig prot	%	*	6.3	6.8	
Sheep	dig prot	%		6.0	6.4	
Energy	GE Mcal/kg					
Cattle	DE Mcal/kg		*	2.16	2.32	
Sheep	DE Mcal/kg		*	2.39	2.57	
Cattle	ME Mcal/kg		*	1.77	1.90	
Sheep	ME Mcal/kg		*	1.96	2.11	
Cattle	TDN	%	*	48.9	52.5	
Sheep	TDN	%		54.3	58.3	

Teff, hay, s-c, mature, (1)
Ref No 1 04 819 United States

Feed Name or Analyses				As Fed	Dry	C.V. ± %
Dry matter		%		89.8	100.0	
Ash		%		4.3	4.8	
Crude fiber		%		27.3	30.4	
Cattle	dig coef	%		73.	73.	
Ether extract		%		1.6	1.8	
Cattle	dig coef	%		60.	60.	
N-free extract		%		48.1	53.6	
Cattle	dig coef	%		69.	69.	
Protein (N x 6.25)		%		8.4	9.4	
Cattle	dig coef	%		59.	59.	
Cattle	dig prot	%		5.0	5.5	
Goats	dig prot	%	*	4.8	5.3	
Horses	dig prot	%	*	4.9	5.5	
Rabbits	dig prot	%	*	5.3	5.9	
Sheep	dig prot	%	*	4.5	5.0	
Energy	GE Mcal/kg					
Cattle	DE Mcal/kg		*	2.66	2.96	
Sheep	DE Mcal/kg		*	2.51	2.80	
Cattle	ME Mcal/kg		*	2.18	2.43	
Sheep	ME Mcal/kg		*	2.07	2.30	
Cattle	TDN	%		60.3	67.2	
Sheep	TDN	%	*	56.9	63.4	

Teff, hay, s-c, over ripe, (1)
Ref No 1 04 820 United States

Feed Name or Analyses				As Fed	Dry	C.V. ± %
Dry matter		%		91.4	100.0	
Ash		%		6.8	7.4	
Crude fiber		%		30.3	33.2	
Cattle	dig coef	%		72.	72.	
Sheep	dig coef	%		71.	71.	
Ether extract		%		1.0	1.1	
Cattle	dig coef	%		43.	43.	
Sheep	dig coef	%		38.	38.	
N-free extract		%		44.3	48.5	
Cattle	dig coef	%		59.	59.	
Sheep	dig coef	%		58.	58.	
Protein (N x 6.25)		%		9.0	9.9	
Cattle	dig coef	%		52.	52.	
Sheep	dig coef	%		56.	56.	
Cattle	dig prot	%		4.7	5.1	
Goats	dig prot	%	*	5.3	5.8	
Horses	dig prot	%	*	5.4	5.9	
Rabbits	dig prot	%	*	5.7	6.3	
Sheep	dig prot	%		5.0	5.5	
Energy	GE Mcal/kg					
Cattle	DE Mcal/kg		*	2.36	2.59	
Sheep	DE Mcal/kg		*	2.34	2.56	
Cattle	ME Mcal/kg		*	1.94	2.12	
Sheep	ME Mcal/kg		*	1.92	2.10	
Cattle	TDN	%		53.6	58.7	
Sheep	TDN	%		53.1	58.1	

Teff, aerial part, fresh, (2)
Ref No 2 04 822 United States

Feed Name or Analyses				As Fed	Dry	C.V. ± %
Dry matter		%		34.2	100.0	
Ash		%		3.3	9.6	
Crude fiber		%		14.1	41.3	
Sheep	dig coef	%		66.	66.	
Ether extract		%		0.6	1.8	
Sheep	dig coef	%		50.	50.	
N-free extract		%		14.3	41.9	
Sheep	dig coef	%		62.	62.	
Protein (N x 6.25)		%		1.8	5.4	
Sheep	dig coef	%		44.	44.	
Cattle	dig prot	%	*	0.8	2.5	
Goats	dig prot	%	*	0.5	1.6	
Horses	dig prot	%	*	0.7	2.1	
Rabbits	dig prot	%	*	1.0	2.8	
Sheep	dig prot	%		0.8	2.4	
Energy	GE Mcal/kg					
Cattle	DE Mcal/kg		*	0.92	2.70	
Sheep	DE Mcal/kg		*	0.87	2.54	
Cattle	ME Mcal/kg		*	0.76	2.22	
Sheep	ME Mcal/kg		*	0.71	2.08	
Cattle	TDN	%	*	21.0	61.3	
Sheep	TDN	%		19.7	57.6	

TEOSINTE. Euchlaena spp

Teosinte, hay, s-c, (1)
Ref No 1 08 538 United States

Feed Name or Analyses				As Fed	Dry	C.V. ± %
Dry matter		%		89.4	100.0	
Ash		%		10.3	11.5	
Crude fiber		%		26.4	29.5	
Ether extract		%		1.9	2.1	
N-free extract		%		41.7	46.6	
Protein (N x 6.25)		%		9.1	10.2	
Cattle	dig prot	%	*	5.1	5.8	
Goats	dig prot	%	*	5.4	6.1	
Horses	dig prot	%	*	5.5	6.2	
Rabbits	dig prot	%	*	5.8	6.5	
Sheep	dig prot	%	*	5.1	5.7	
Energy	GE Mcal/kg					
Cattle	DE Mcal/kg		*	2.20	2.46	
Sheep	DE Mcal/kg		*	2.27	2.54	
Cattle	ME Mcal/kg		*	1.81	2.02	
Sheep	ME Mcal/kg		*	1.86	2.08	
Cattle	TDN	%	*	50.0	55.9	
Sheep	TDN	%	*	51.4	57.5	
Phosphorus		%		0.17	0.19	
Potassium		%		0.88	0.98	

Feed Name or Analyses			Mean As Fed	Mean Dry	C.V. ± %
Teosinte, aerial part, fresh, (2)					
Ref No 2 04 823				United States	
Dry matter	%		21.3	100.0	10
Ash	%		2.0	9.6	11
Crude fiber	%		6.7	31.3	7
Ether extract	%		0.5	2.3	16
N-free extract	%		10.4	48.6	
Protein (N x 6.25)	%		1.7	8.2	14
Cattle	dig prot %	*	1.0	4.9	
Goats	dig prot %	*	0.9	4.2	
Horses	dig prot %	*	1.0	4.5	
Rabbits	dig prot %	*	1.1	5.0	
Sheep	dig prot %	*	1.0	4.6	
Energy	GE Mcal/kg				
Cattle	DE Mcal/kg	*	0.58	2.72	
Sheep	DE Mcal/kg	*	0.63	2.93	
Cattle	ME Mcal/kg	*	0.48	2.23	
Sheep	ME Mcal/kg	*	0.51	2.41	
Cattle	TDN %	*	13.2	61.8	
Sheep	TDN %	*	14.2	66.6	
Calcium	%		0.25	1.17	
Phosphorus	%		0.03	0.14	
Teosinte, grain, (4)					
Ref No 4 04 824				United States	
Dry matter	%			100.0	
Protein (N x 6.25)	%			10.0	
Cattle	dig prot %	*		5.2	
Goats	dig prot %	*		6.4	
Horses	dig prot %	*		6.4	
Sheep	dig prot %	*		6.4	
Alanine	%			0.81	
Arginine	%			0.33	
Aspartic acid	%			0.58	
Glutamic acid	%			1.92	
Glycine	%			0.26	
Histidine	%			0.20	
Hydroxyproline	%			0.00	
Isoleucine	%			0.33	
Leucine	%			1.37	
Lysine	%			0.17	
Methionine	%			0.23	
Phenylalanine	%			0.49	
Proline	%			0.85	
Serine	%			0.47	
Threonine	%			0.33	
Tyrosine	%			0.36	
Valine	%			0.43	
Teosinte, groats, (5)					
Ref No 5 04 825				United States	
Dry matter	%		89.8	100.0	
Ether extract	%		3.5	3.9	29
Protein (N x 6.25)	%		21.5	23.9	6
Cattle	dig prot %	*	16.1	18.0	
Goats	dig prot %	*	17.2	19.1	
Horses	dig prot %	*	17.2	19.1	
Sheep	dig prot %	*	17.2	19.1	
Niacin	mg/kg		7.7	8.6	57
Arginine	%		0.63	0.70	
Histidine	%		0.45	0.50	
Isoleucine	%		0.99	1.10	
Lysine	%		0.36	0.40	32
Methionine	%		0.45	0.50	26

Feed Name or Analyses			Mean As Fed	Mean Dry	C.V. ± %
Phenylalanine	%		1.62	1.80	
Threonine	%		0.81	0.90	
Tryptophan	%		0.09	0.10	
Valine	%		0.99	1.10	

TETRAGONIA. Tetragonia arbuscula

Feed Name or Analyses			Mean As Fed	Mean Dry	C.V. ± %
Tetragonia, bush, browse, fresh, (2)					
Ref No 2 04 826				United States	
Dry matter	%		20.0	100.0	
Ash	%		3.2	16.0	
Crude fiber	%		5.5	27.6	
Sheep	dig coef %		20.	20.	
Ether extract	%		0.6	2.8	
Sheep	dig coef %		59.	59.	
N-free extract	%		7.0	35.0	
Sheep	dig coef %		74.	74.	
Protein (N x 6.25)	%		3.7	18.6	
Sheep	dig coef %		76.	76.	
Cattle	dig prot %	*	2.7	13.7	
Goats	dig prot %	*	2.8	13.9	
Horses	dig prot %	*	2.7	13.3	
Rabbits	dig prot %	*	2.6	13.0	
Sheep	dig prot %		2.8	14.1	
Energy	GE Mcal/kg				
Cattle	DE Mcal/kg	*	0.46	2.31	
Sheep	DE Mcal/kg	*	0.43	2.17	
Cattle	ME Mcal/kg	*	0.38	1.89	
Sheep	ME Mcal/kg	*	0.36	1.78	
Cattle	TDN %	*	10.5	52.5	
Sheep	TDN %		9.9	49.3	

THISTLE. Cirsium spp

Feed Name or Analyses			Mean As Fed	Mean Dry	C.V. ± %
Thistle, aerial part, fresh, (2)					
Ref No 2 04 832				United States	
Dry matter	%			100.0	
Carotene	mg/kg			22.0	
Vitamin A equivalent	IU/g			36.8	
Thistle, aerial part, fresh, immature, (2)					
Ref No 2 04 831				United States	
Dry matter	%			100.0	
Ash	%			14.8	
Crude fiber	%			14.4	
Ether extract	%			2.7	
N-free extract	%			43.2	
Protein (N x 6.25)	%			24.9	
Cattle	dig prot %	*		19.1	
Goats	dig prot %	*		19.8	
Horses	dig prot %	*		18.7	
Rabbits	dig prot %	*		17.9	
Sheep	dig prot %	*		20.2	
Calcium	%			4.26	
Magnesium	%			0.41	
Phosphorus	%			0.21	
Potassium	%			3.18	

THREEAWN. Aristida spp

Feed Name or Analyses			Mean As Fed	Mean Dry	C.V. ± %
Threeawn, hay, s-c, (1)					
Wiregrass hay					
Ref No 1 08 539				United States	
Dry matter	%		90.0	100.0	
Ash	%		5.3	5.9	
Crude fiber	%		32.9	36.6	
Ether extract	%		1.4	1.5	
N-free extract	%		44.5	49.4	
Protein (N x 6.25)	%		5.9	6.6	
Cattle	dig prot %	*	2.4	2.7	
Goats	dig prot %	*	2.5	2.7	
Horses	dig prot %	*	2.8	3.1	
Rabbits	dig coef %	*	3.4	3.8	
Sheep	dig prot %	*	2.3	2.5	
Energy	GE Mcal/kg				
Cattle	DE Mcal/kg	*	2.00	2.22	
Sheep	DE Mcal/kg	*	2.13	2.37	
Cattle	ME Mcal/kg	*	1.64	1.82	
Sheep	ME Mcal/kg	*	1.75	1.94	
Cattle	TDN %	*	45.3	50.3	
Sheep	TDN %	*	48.4	53.8	
Calcium	%		0.15	0.17	
Phosphorus	%		0.14	0.16	
Threeawn, aerial part, fresh, (2)					
Wiregrass					
Ref No 2 04 838				United States	
Dry matter	%		38.9	100.0	
Ash	%		2.3	5.9	
Crude fiber	%		13.3	34.2	
Ether extract	%		0.9	2.3	
N-free extract	%		18.6	47.8	
Protein (N x 6.25)	%		3.8	9.8	
Cattle	dig prot %	*	2.4	6.2	
Goats	dig prot %	*	2.2	5.7	
Horses	dig prot %	*	2.3	5.8	
Rabbits	dig coef %	*	2.4	6.2	
Sheep	dig prot %	*	2.4	6.1	
Energy	GE Mcal/kg				
Cattle	DE Mcal/kg	*	1.09	2.80	
Sheep	DE Mcal/kg	*	1.03	2.65	
Cattle	ME Mcal/kg	*	0.89	2.30	
Sheep	ME Mcal/kg	*	0.84	2.17	
Cattle	TDN %	*	24.7	63.4	
Sheep	TDN %	*	23.4	60.1	
Threeawn, aerial part, fresh, immature, (2)					
Ref No 2 04 833				United States	
Dry matter	%			100.0	
Ash	%			9.9	8
Crude fiber	%			29.8	8
Ether extract	%			1.7	9
N-free extract	%			49.4	
Protein (N x 6.25)	%			9.2	8
Cattle	dig prot %	*		5.7	
Goats	dig prot %	*		5.1	
Horses	dig prot %	*		5.3	
Rabbits	dig coef %	*		5.8	
Sheep	dig prot %	*		5.6	
Energy	GE Mcal/kg				
Cattle	DE Mcal/kg	*		2.78	

Continued

Feed Name or Analyses		Mean As Fed	Mean Dry	C.V. ± %
Sheep	DE Mcal/kg *		2.94	
Cattle	ME Mcal/kg *		2.28	
Sheep	ME Mcal/kg *		2.41	
Cattle	TDN % *		63.1	
Sheep	TDN % *		66.6	
Calcium	%		0.85	22
Phosphorus	%		0.14	14

Threeawn, aerial part, fresh, mid-bloom, (2)
Ref No 2 04 834 United States

Feed Name or Analyses		Mean As Fed	Mean Dry	C.V. ± %
Dry matter	%		100.0	
Calcium	%		0.80	15
Phosphorus	%		0.10	18

Threeawn, aerial part, fresh, full bloom, (2)
Ref No 2 04 835 United States

Feed Name or Analyses		Mean As Fed	Mean Dry	C.V. ± %
Dry matter	%		100.0	
Ash	%		10.3	14
Crude fiber	%		33.0	8
Ether extract	%		1.3	10
N-free extract	%		49.3	
Protein (N x 6.25)	%		6.1	5
Cattle	dig prot % *		3.1	
Goats	dig prot % *		2.3	
Horses	dig prot % *		2.7	
Rabbits	dig coef % *		3.4	
Sheep	dig prot % *		2.7	
Energy	GE Mcal/kg			
Cattle	DE Mcal/kg *		2.66	
Sheep	DE Mcal/kg *		2.71	
Cattle	ME Mcal/kg *		2.18	
Sheep	ME Mcal/kg *		2.22	
Cattle	TDN % *		60.2	
Sheep	TDN % *		61.5	
Calcium	%		0.48	21
Phosphorus	%		0.08	21

Threeawn, aerial part, fresh, mature, (2)
Ref No 2 04 836 United States

Feed Name or Analyses		Mean As Fed	Mean Dry	C.V. ± %
Dry matter	%		100.0	
Ash	%		12.6	14
Crude fiber	%		34.6	5
Ether extract	%		1.4	29
N-free extract	%		46.5	
Protein (N x 6.25)	%		4.9	10
Cattle	dig prot % *		2.1	
Goats	dig prot % *		1.1	
Horses	dig prot % *		1.7	
Rabbits	dig coef % *		2.5	
Sheep	dig prot % *		1.6	
Energy	GE Mcal/kg			
Cattle	DE Mcal/kg *		2.45	
Sheep	DE Mcal/kg *		2.61	
Cattle	ME Mcal/kg *		2.01	
Sheep	ME Mcal/kg *		2.14	
Cattle	TDN % *		55.6	
Sheep	TDN % *		59.1	
Calcium	%		0.37	14
Magnesium	%		0.10	29

(1) dry forages and roughages (2) pasture, range plants, and forages fed green (3) silages (4) energy feeds (5) protein supplements (6) minerals (7) vitamins (8) additives

Feed Name or Analyses		Mean As Fed	Mean Dry	C.V. ± %
Phosphorus	%		0.08	35
Potassium	%		0.40	28

Threeawn, aerial part, fresh, over ripe, (2)
Ref No 2 04 837 United States

Feed Name or Analyses		Mean As Fed	Mean Dry	C.V. ± %
Dry matter	%		100.0	
Ash	%		8.3	
Crude fiber	%		38.5	
Ether extract	%		1.3	
N-free extract	%		48.7	
Protein (N x 6.25)	%		3.2	
Cattle	dig prot % *		0.6	
Goats	dig prot % *		-0.4	
Horses	dig prot % *		0.2	
Rabbits	dig coef % *		1.1	
Sheep	dig prot % *		0.0	
Calcium	%		0.36	39
Iron	%		0.014	
Magnesium	%		0.26	
Phosphorus	%		0.06	32
Carotene	mg/kg		0.2	
Vitamin A equivalent	IU/g		0.4	

THREEAWN, ARROWFEATHER. ***Aristida purpurascens***

Threeawn, arrowfeather, aerial part, fresh, (2)
Ref No 2 04 840 United States

Feed Name or Analyses		Mean As Fed	Mean Dry	C.V. ± %
Dry matter	%		100.0	
Calcium	%		0.71	32
Phosphorus	%		0.06	26

Threeawn, arrowfeather, aerial part, fresh, mature, (2)
Ref No 2 04 839 United States

Feed Name or Analyses		Mean As Fed	Mean Dry	C.V. ± %
Dry matter	%		100.0	
Ash	%		11.6	19
Crude fiber	%		35.9	6
Ether extract	%		1.2	10
N-free extract	%		46.1	
Protein (N x 6.25)	%		5.2	18
Cattle	dig prot % *		2.3	
Goats	dig prot % *		1.4	
Horses	dig prot % *		1.9	
Rabbits	dig coef % *		2.7	
Sheep	dig prot % *		1.8	

THREEAWN, PRAIRIE. ***Aristida oligantha***

Threeawn, prairie, aerial part, fresh, mature, (2)
Ref No 2 04 841 United States

Feed Name or Analyses		Mean As Fed	Mean Dry	C.V. ± %
Dry matter	%		100.0	
Ash	%		9.8	
Crude fiber	%		31.4	
Ether extract	%		2.2	
N-free extract	%		51.4	
Protein (N x 6.25)	%		5.2	
Cattle	dig prot % *		2.3	
Goats	dig prot % *		1.4	
Horses	dig prot % *		1.9	
Rabbits	dig coef % *		2.7	
Sheep	dig prot % *		1.8	

THREEAWN, PURPLE. ***Aristida purpurea***

Threeawn, purple, aerial part, fresh, (2)
Ref No 2 04 846 United States

Feed Name or Analyses		Mean As Fed	Mean Dry	C.V. ± %
Dry matter	%		100.0	
Ash	%		11.8	26
Crude fiber	%		33.9	5
Ether extract	%		1.4	16
N-free extract	%		46.1	
Protein (N x 6.25)	%		6.8	26
Cattle	dig prot % *		3.7	
Goats	dig prot % *		2.9	
Horses	dig prot % *		3.3	
Rabbits	dig coef % *		3.9	
Sheep	dig prot % *		3.3	
Calcium	%		0.47	24
Magnesium	%		0.12	16
Phosphorus	%		0.10	26
Potassium	%		0.51	34

Threeawn, purple, aerial part, fresh, immature, (2)
Ref No 2 04 842 United States

Feed Name or Analyses		Mean As Fed	Mean Dry	C.V. ± %
Dry matter	%		100.0	
Ash	%		9.8	
Crude fiber	%		32.4	
Ether extract	%		1.7	
N-free extract	%		46.1	
Protein (N x 6.25)	%		10.0	
Cattle	dig prot % *		6.4	
Goats	dig prot % *		5.9	
Horses	dig prot % *		6.0	
Rabbits	dig coef % *		6.4	
Sheep	dig prot % *		6.3	
Calcium	%		0.68	
Phosphorus	%		0.15	

Threeawn, purple, aerial part, fresh, mid-bloom, (2)
Ref No 2 04 843 United States

Feed Name or Analyses		Mean As Fed	Mean Dry	C.V. ± %
Dry matter	%		100.0	
Calcium	%		0.63	
Phosphorus	%		0.09	

Threeawn, purple, aerial part, fresh, full bloom, (2)
Ref No 2 04 844 United States

Feed Name or Analyses		Mean As Fed	Mean Dry	C.V. ± %
Dry matter	%		100.0	
Calcium	%		0.45	
Phosphorus	%		0.08	

Threeawn, purple, aerial part, fresh, mature, (2)
Ref No 2 04 845 United States

Feed Name or Analyses		Mean As Fed	Mean Dry	C.V. ± %
Dry matter	%		100.0	
Ash	%		14.8	
Crude fiber	%		34.6	
Ether extract	%		1.4	
N-free extract	%		44.2	
Protein (N x 6.25)	%		5.0	
Cattle	dig prot % *		2.1	
Goats	dig prot % *		1.2	
Horses	dig prot % *		1.8	
Rabbits	dig coef % *		2.5	

Feed Name or Analyses			Mean As Fed	Mean Dry	C.V. ± %
Sheep	dig prot %	*		1.7	
Calcium	%			0.39	
Phosphorus	%			0.08	

THREEAWN, RED. Aristida longiseta

Threeawn, red, aerial part, fresh, (2)
Ref No 2 04 852 United States

Feed Name or Analyses			As Fed	Dry	C.V. ± %
Dry matter	%			100.0	
Asn	%			11.1	18
Crude fiber	%			36.1	8
Ether extract	%			1.4	16
N-free extract	%			45.8	
Protein (N x 6.25)	%			5.6	29
Cattle	dig prot %	*		2.7	
Goats	dig prot %	*		1.8	
Horses	dig prot %	*		2.3	
Rabbits	dig coef %	*		3.0	
Sheep	dig prot %	*		2.2	
Calcium	%			0.76	30
Phosphorus	%			0.09	27

Threeawn, red, aerial part, fresh, immature, (2)
Ref No 2 04 847 United States

Feed Name or Analyses		As Fed	Dry	C.V. ± %
Dry matter	%		100.0	
Calcium	%		1.02	
Phosphorus	%		0.12	

Threeawn, red, aerial part, fresh, mid-bloom, (2)
Ref No 2 04 848 United States

Feed Name or Analyses		As Fed	Dry	C.V. ± %
Dry matter	%		100.0	
Calcium	%		0.94	4
Phosphorus	%		0.10	15

Threeawn, red, aerial part, fresh, full bloom, (2)
Ref No 2 04 849 United States

Feed Name or Analyses		As Fed	Dry	C.V. ± %
Dry matter	%		100.0	
Calcium	%		0.41	
Phosphorus	%		0.08	

Threeawn, red, aerial part, fresh, mature, (2)
Ref No 2 04 850 United States

Feed Name or Analyses			As Fed	Dry	C.V. ± %
Dry matter	%			100.0	
Ash	%			12.8	10
Crude fiber	%			34.7	4
Ether extract	%			1.3	14
N-free extract	%			46.3	
Protein (N x 6.25)	%			4.9	10
Cattle	dig prot %	*		2.1	
Goats	dig prot %	*		1.1	
Horses	dig prot %	*		1.7	
Rabbits	dig coef %	*		2.5	
Sheep	dig prot %	*		1.6	

Threeawn, red, aerial part, fresh, over ripe, (2)
Ref No 2 04 851 United States

Feed Name or Analyses			As Fed	Dry	C.V. ± %
Dry matter	%			100.0	
Ash	%			10.7	
Crude fiber	%			40.2	
Ether extract	%			1.3	
N-free extract	%			44.7	
Protein (N x 6.25)	%			3.1	
Cattle	dig prot %	*		0.5	
Goats	dig prot %	*		-0.4	
Horses	dig prot %	*		0.2	
Rabbits	dig coef %	*		1.1	
Sheep	dig prot %	*		-0.0	
Calcium	%			0.42	
Phosphorus	%			0.07	

THREEAWN, SLIMSPIKE. Aristida longspica

Threeawn, slimspike, aerial part, fresh, (2)
Ref No 2 04 854 United States

Feed Name or Analyses			As Fed	Dry	C.V. ± %
Dry matter	%			100.0	
Ash	%			10.6	13
Crude fiber	%			31.9	6
Ether extract	%			2.0	17
N-free extract	%			48.7	
Protein (N x 6.25)	%			6.8	9
Cattle	dig prot %	*		3.7	
Goats	dig prot %	*		2.9	
Horses	dig prot %	*		3.3	
Rabbits	dig coef %	*		3.9	
Sheep	dig prot %	*		3.3	
Calcium	%			0.50	25
Phosphorus	%			0.09	19

Threeawn, slimspike, aerial part, fresh, mature, (2)
Ref No 2 04 853 United States

Feed Name or Analyses		As Fed	Dry	C.V. ± %
Dry matter	%		100.0	
Calcium	%		0.34	
Phosphorus	%		0.08	

THREEAWN, WRIGHT. Aristida wrightii

Threeawn, wright, aerial part, fresh, (2)
Ref No 2 04 856 United States

Feed Name or Analyses			As Fed	Dry	C.V. ± %
Dry matter	%			100.0	
Ash	%			11.4	19
Crude fiber	%			32.4	12
Ether extract	%			1.7	13
N-free extract	%			46.7	
Protein (N x 6.25)	%			7.8	17
Cattle	dig prot %	*		4.5	
Goats	dig prot %	*		3.8	
Horses	dig prot %	*		4.2	
Rabbits	dig coef %	*		4.7	
Sheep	dig prot %	*		4.3	
Calcium	%			0.48	39
Magnesium	%			0.10	33
Phosphorus	%			0.12	22
Potassium	%			0.47	30

Threeawn, wright, aerial part, fresh, mature, (2)
Ref No 2 04 855 United States

Feed Name or Analyses		As Fed	Dry	C.V. ± %
Dry matter	%		100.0	
Calcium	%		0.38	18
Magnesium	%		0.10	29
Phosphorus	%		0.11	18
Potassium	%		0.40	28

TICKCLOVER. Desmodium spp

Tickclover, hay, s-c, (1)
Ref No 1 04 859 United States

Feed Name or Analyses			As Fed	Dry	C.V. ± %
Dry matter	%		90.9	100.0	2
Ash	%		7.8	8.6	11
Crude fiber	%		28.4	31.2	7
Ether extract	%		2.3	2.5	12
N-free extract	%		37.3	41.0	
Protein (N x 6.25)	%		15.2	16.7	21
Cattle	dig prot %	*	10.4	11.4	
Goats	dig prot %	*	11.0	12.1	
Horses	dig prot %	*	10.6	11.7	
Rabbits	dig coef %	*	10.5	11.6	
Sheep	dig prot %	*	10.5	11.5	
Energy	GE Mcal/kg				
Cattle	DE Mcal/kg	*	2.32	2.55	
Sheep	DE Mcal/kg	*	2.29	2.52	
Cattle	ME Mcal/kg	*	1.90	2.09	
Sheep	ME Mcal/kg	*	1.87	2.06	
Cattle	TDN %	*	52.6	57.9	
Sheep	TDN %	*	51.9	57.0	

Tickclover, leaves, dehy grnd, (1)
Ref No 1 04 860 United States

Feed Name or Analyses			As Fed	Dry	C.V. ± %
Dry matter	%		27.1	100.0	
Ash	%		2.9	10.8	19
Crude fiber	%		7.1	26.1	12
Ether extract	%		0.7	2.5	54
N-free extract	%		12.4	45.6	
Protein (N x 6.25)	%		4.1	15.0	7
Cattle	dig prot %	*	2.7	9.9	
Goats	dig prot %	*	2.9	10.6	
Horses	dig prot %	*	2.8	10.3	
Rabbits	dig coef %	*	2.8	10.3	
Sheep	dig prot %	*	2.7	10.0	
Energy	GE Mcal/kg				
Cattle	DE Mcal/kg	*	0.76	2.79	
Sheep	DE Mcal/kg	*	0.70	2.57	
Cattle	ME Mcal/kg	*	0.62	2.29	
Sheep	ME Mcal/kg	*	0.57	2.11	
Cattle	TDN %	*	17.1	63.3	
Sheep	TDN %	*	15.8	58.4	
Calcium	%		0.34	1.24	
Iron	%		0.013	0.047	
Phosphorus	%		0.08	0.28	
Carotene	mg/kg		49.9	184.3	
Niacin	mg/kg		14.2	52.5	
Riboflavin	mg/kg		2.1	7.7	
Thiamine	mg/kg		1.0	3.7	
Vitamin A equivalent	IU/g		83.3	307.2	

TICKCLOVER, CHEROKEE. Desmodium tortuosum

Tickclover, cherokee, hay, s-c, (1)
Beggarweed hay
Ref No 1 08 540 United States

Feed Name or Analyses		As Fed	Dry	C.V. ± %
Dry matter	%	90.9	100.0	
Ash	%	7.8	8.6	
Crude fiber	%	28.4	31.2	

Continued

Feed Name or Analyses				Mean As Fed	Mean Dry	C.V. ± %
Ether extract		%		2.3	2.5	
N-free extract		%		37.2	40.9	
Protein (N x 6.25)		%		15.2	16.7	
Cattle	dig prot	%	*	10.4	11.4	
Goats	dig prot	%	*	11.1	12.2	
Horses	dig prot	%	*	10.7	11.7	
Rabbits	dig coef	%	*	10.5	11.6	
Sheep	dig prot	%	*	10.5	11.6	
Energy	GE Mcal/kg					
Cattle	DE Mcal/kg		*	2.16	2.38	
Sheep	DE Mcal/kg		*	2.06	2.27	
Cattle	ME Mcal/kg		*	1.77	1.95	
Sheep	ME Mcal/kg		*	1.69	1.86	
Cattle	TDN	%	*	49.0	53.9	
Sheep	TDN	%	*	46.8	51.5	
Calcium		%		1.05	1.16	
Phosphorus		%		0.27	0.30	
Potassium		%		2.32	2.55	

Tickclover, cherokee, aerial part, fresh, (2)
Beggarweed
Ref No 2 08 541 United States

				As Fed	Dry	C.V.
Dry matter		%		27.1	100.0	
Ash		%		3.2	11.8	
Crude fiber		%		7.5	27.7	
Ether extract		%		0.5	1.8	
N-free extract		%		11.7	43.2	
Protein (N x 6.25)		%		4.2	15.5	
Cattle	dig prot	%	*	3.0	11.1	
Goats	dig prot	%	*	3.0	11.0	
Horses	dig prot	%	*	2.9	10.7	
Rabbits	dig coef	%	*	2.9	10.6	
Sheep	dig prot	%	*	3.1	11.4	
Energy	GE Mcal/kg					
Cattle	DE Mcal/kg		*	0.65	2.41	
Sheep	DE Mcal/kg		*	0.62	2.29	
Cattle	ME Mcal/kg		*	0.54	1.98	
Sheep	ME Mcal/kg		*	0.51	1.88	
Cattle	TDN	%	*	14.8	54.6	
Sheep	TDN	%	*	14.1	51.9	
Phosphorus		%		0.12	0.44	
Potassium		%		0.47	1.73	

TICKCLOVER. SPIKE. Desmodium cinerascens

Tickclover, spike, seeds w hulls, (5)
Ref No 5 09 061 United States

		As Fed	Dry	C.V.
Dry matter	%		100.0	
Protein (N x 6.25)	%		38.0	
Alanine	%		1.29	
Arginine	%		2.55	
Aspartic acid	%		3.69	
Glutamic acid	%		4.94	
Glycine	%		1.22	
Histidine	%		0.91	
Hydroxyproline	%		0.08	
Isoleucine	%		1.29	
Leucine	%		2.36	
Lysine	%		2.36	
Methionine	%		0.46	
Phenylalanine	%		1.22	
Proline	%		1.48	
Serine	%		1.48	
Threonine	%		1.18	
Tyrosine	%		0.95	
Valine	%		1.41	

(1) dry forages and roughages (2) pasture, range plants, and forages fed green (3) silages (4) energy feeds (5) protein supplements (6) minerals (7) vitamins (8) additives

TIDESTROMIA. Tidestromia spp

Tidestromia, aerial part, fresh, mid-bloom, (2)
Ref No 2 04 861 United States

				As Fed	Dry	C.V.
Dry matter		%			100.0	
Ash		%			11.1	
Crude fiber		%			22.5	
Ether extract		%			1.3	
N-free extract		%			48.9	
Protein (N x 6.25)		%			16.2	
Cattle	dig prot	%	*		11.7	
Goats	dig prot	%	*		11.7	
Horses	dig prot	%	*		11.3	
Rabbits	dig coef	%	*		11.2	
Sheep	dig prot	%	*		12.1	
Calcium		%			2.40	
Magnesium		%			0.54	
Phosphorus		%			0.13	
Potassium		%			3.31	

TIDYTIPS. Layia spp

Tidytips, aerial part, fresh, (2)
Ref No 2 04 867 United States

				As Fed	Dry	C.V.
Dry matter		%			100.0	
Ash		%			10.8	28
Crude fiber		%			24.0	23
Protein (N x 6.25)		%			10.0	70
Cattle	dig prot	%	*		6.4	
Goats	dig prot	%	*		5.9	
Horses	dig prot	%	*		6.0	
Rabbits	dig coef	%	*		6.4	
Sheep	dig prot	%	*		6.3	
Calcium		%			1.36	15
Phosphorus		%			0.41	36
Potassium		%			2.90	37

Tidytips, aerial part, fresh, immature, (2)
Ref No 2 04 862 United States

				As Fed	Dry	C.V.
Dry matter		%			100.0	
Ash		%			16.2	
Crude fiber		%			15.3	
Protein (N x 6.25)		%			21.3	
Cattle	dig prot	%	*		16.0	
Goats	dig prot	%	*		16.4	
Horses	dig prot	%	*		15.6	
Rabbits	dig coef	%	*		15.1	
Sheep	dig prot	%	*		16.8	
Calcium		%			1.49	
Phosphorus		%			0.60	
Potassium		%			5.34	

Tidytips, aerial part, fresh, early bloom, (2)
Ref No 2 04 863 United States

				As Fed	Dry	C.V.
Dry matter		%			100.0	
Ash		%			13.4	
Crude fiber		%			15.4	
Protein (N x 6.25)		%			12.0	
Cattle	dig prot	%	*		8.1	
Goats	dig prot	%	*		7.8	
Horses	dig prot	%	*		7.7	
Rabbits	dig coef	%	*		7.9	
Sheep	dig prot	%	*		8.2	
Calcium		%			1.44	
Phosphorus		%			0.34	
Potassium		%			3.70	

Tidytips, aerial part, fresh, full bloom, (2)
Ref No 2 04 864 United States

				As Fed	Dry	C.V.
Dry matter		%			100.0	
Ash		%			10.4	7
Crude fiber		%			22.6	12
Protein (N x 6.25)		%			9.7	11
Cattle	dig prot	%	*		6.1	
Goats	dig prot	%	*		5.6	
Horses	dig prot	%	*		5.8	
Rabbits	dig coef	%	*		6.2	
Sheep	dig prot	%	*		6.0	
Calcium		%			1.32	8
Phosphorus		%			0.54	13
Potassium		%			3.21	7

Tidytips, aerial part, fresh, mature, (2)
Ref No 2 04 865 United States

				As Fed	Dry	C.V.
Dry matter		%			100.0	
Ash		%			10.2	
Crude fiber		%			27.9	
Protein (N x 6.25)		%			6.3	
Cattle	dig prot	%	*		3.2	
Goats	dig prot	%	*		2.4	
Horses	dig prot	%	*		2.9	
Rabbits	dig coef	%	*		3.5	
Sheep	dig prot	%	*		2.9	

Tidytips, aerial part, fresh, over ripe, (2)
Ref No 2 04 866 United States

				As Fed	Dry	C.V.
Dry matter		%			100.0	
Ash		%			8.9	
Crude fiber		%			30.4	
Protein (N x 6.25)		%			3.7	
Cattle	dig prot	%	*		1.0	
Goats	dig prot	%	*		0.0	
Horses	dig prot	%	*		0.7	
Rabbits	dig coef	%	*		1.5	
Sheep	dig prot	%	*		0.4	
Calcium		%			1.24	
Phosphorus		%			0.29	
Potassium		%			2.31	

Tillandsia, treebeard –
see Spanishmoss, entire plant, fresh, (2)

TIMOTHY. Phleum pratense

Timothy, aerial part, dehy, (1)
Ref No 1 04 879 United States

		As Fed	Dry	C.V.
Dry matter	%	88.2	100.0	1
Ash	%	5.4	6.1	9

Feed Name or Analyses			Mean As Fed	Mean Dry	C.V. ± %
Crude fiber	%		27.8	31.5	5
Cattle	dig coef %		60.	60.	
Ether extract	%		2.4	2.8	12
Cattle	dig coef %		54.	54.	
N-free extract	%		44.5	50.5	
Cattle	dig coef %		66.	66.	
Protein (N x 6.25)	%		8.1	9.2	13
Cattle	dig coef %		51.	51.	
Cattle	dig prot %		4.1	4.7	
Goats	dig prot %	*	4.5	5.1	
Horses	dig prot %	*	4.7	5.3	
Rabbits	dig coef %	*	5.1	5.8	
Sheep	dig prot %	*	4.3	4.8	
Energy	GE Mcal/kg				
Cattle	DE Mcal/kg	*	2.34	2.66	
Sheep	DE Mcal/kg	*	2.21	2.51	
Cattle	ME Mcal/kg	*	1.92	2.18	
Sheep	ME Mcal/kg	*	1.82	2.06	
Cattle	TDN %		53.1	60.2	
Sheep	TDN %	*	50.2	57.0	

Timothy, aerial part, dehy, pre-bloom, (1)

Ref No 1 04 874 United States

Feed Name or Analyses			Mean As Fed	Mean Dry	C.V. ± %
Dry matter	%		86.7	100.0	
Ash	%		6.1	7.0	
Crude fiber	%		25.7	29.6	
Cattle	dig coef %		68.	68.	
Sheep	dig coef %		73.	73.	
Ether extract	%		2.8	3.2	
Cattle	dig coef %		59.	59.	
N-free extract	%		42.0	48.4	
Cattle	dig coef %		69.	69.	
Protein (N x 6.25)	%		10.2	11.8	
Cattle	dig coef %		59.	59.	
Sheep	dig coef %		70.	70.	
Cattle	dig prot %		6.0	7.0	
Goats	dig prot %	*	6.6	7.6	
Horses	dig prot %	*	6.5	7.5	
Rabbits	dig coef %	*	6.7	7.8	
Sheep	dig prot %		7.1	8.2	
Energy	GE Mcal/kg		3.91	4.51	
Cattle	GE dig coef %		52.	52.	
Sheep	GE dig coef %		68.	68.	
Cattle	DE Mcal/kg		2.03	2.34	
Sheep	DE Mcal/kg		2.66	3.06	
Cattle	ME Mcal/kg	*	1.67	1.92	
Sheep	ME Mcal/kg	*	2.18	2.51	
Cattle	TDN %		56.1	64.7	
Sheep	TDN %		59.0	68.1	

Ref No 1 04 874 Canada

Feed Name or Analyses			Mean As Fed	Mean Dry	C.V. ± %
Dry matter	%		86.3	100.0	
Sheep	dig coef %		70.	70.	7
Organic matter	%		79.7	92.3	
Ash	%		6.9	8.0	13
Sheep	dig coef %		36.	41.	
Crude fiber	%		23.2	26.9	15
Cattle	dig coef %		50.	50.	
Sheep	dig coef %		73.	73.	4
Ether extract	%		2.2	2.5	25
Cattle	dig coef %		19.	19.	
N-free extract	%		40.0	46.3	
Cattle	dig coef %		69.	69.	
Protein (N x 6.25)	%		14.1	16.3	36
Cattle	dig coef %		32.	32.	
Sheep	dig coef %		70.	70.	7
Cattle	dig prot %		4.5	5.2	

Feed Name or Analyses			Mean As Fed	Mean Dry	C.V. ± %
Goats	dig prot %	*	10.2	11.8	
Horses	dig prot %	*	9.8	11.4	
Rabbits	dig coef %	*	9.7	11.3	
Sheep	dig prot %		9.8	11.4	
Cellulose (Matrone)	%		29.6	34.3	
Cattle	dig coef %		53.	53.	
Energy	GE Mcal/kg		3.89	4.51	2
Cattle	GE dig coef %		52.	52.	
Sheep	GE dig coef %		68.	68.	6
Cattle	DE Mcal/kg		2.02	2.34	
Sheep	DE Mcal/kg		2.65	3.06	7
Cattle	ME Mcal/kg	*	1.66	1.92	
Sheep	ME Mcal/kg	*	2.17	2.51	
Cattle	TDN %		44.6	51.7	
Sheep	TDN %		58.8	68.1	6
Potassium	%		0.99	1.15	51

Timothy, aerial part, dehy, early bloom, (1)

Ref No 1 04 875 United States

Feed Name or Analyses			Mean As Fed	Mean Dry	C.V. ± %
Dry matter	%		89.0	100.0	
Ash	%		6.0	6.7	
Crude fiber	%		26.6	29.9	
Cattle	dig coef %		71.	71.	
Ether extract	%		2.3	2.6	
Cattle	dig coef %		45.	45.	
N-free extract	%		45.9	51.6	
Cattle	dig coef %		72.	72.	
Protein (N x 6.25)	%		8.2	9.2	
Cattle	dig coef %		50.	50.	
Cattle	dig prot %		4.1	4.6	
Goats	dig prot %	*	4.6	5.1	
Horses	dig prot %	*	4.8	5.3	
Rabbits	dig coef %	*	5.1	5.8	
Sheep	dig prot %	*	4.3	4.8	
Energy	GE Mcal/kg		4.08	4.58	
Sheep	GE dig coef %		58.	58.	
Cattle	DE Mcal/kg	*	2.57	2.89	
Sheep	DE Mcal/kg		2.36	2.65	
Cattle	ME Mcal/kg	*	2.11	2.37	
Sheep	ME Mcal/kg	*	1.93	2.17	
Cattle	TDN %		58.4	65.6	
Sheep	TDN %	*	53.5	60.1	
Sheep	TDN %	*	51.3	57.7	

Ref No 1 04 875 Canada

Feed Name or Analyses			Mean As Fed	Mean Dry	C.V. ± %
Dry matter	%			100.0	
Sheep	dig coef %			61.	5
Ash	%			6.1	6
Crude fiber	%			29.7	7
Cattle	dig coef %			71.	
Ether extract	%			2.4	12
Cattle	dig coef %			45.	
N-free extract	%			48.7	
Cattle	dig coef %			72.	
Protein (N x 6.25)	%			13.1	13
Cattle	dig coef %			50.	
Cattle	dig prot %			6.5	
Goats	dig prot %	*		8.7	
Horses	dig prot %	*		8.6	
Rabbits	dig coef %	*		8.7	
Sheep	dig prot %	*		6.3	
Energy	GE Mcal/kg			4.58	0
Sheep	GE dig coef %			58.	5
Cattle	DE Mcal/kg	*		2.87	
Sheep	DE Mcal/kg			2.65	6
Cattle	ME Mcal/kg	*		2.36	

Feed Name or Analyses			Mean As Fed	Mean Dry	C.V. ± %
Sheep	ME Mcal/kg	*		2.17	
Cattle	TDN %			65.2	

Timothy, aerial part, dehy, mid-bloom, (1)

Ref No 1 08 542 United States

Feed Name or Analyses			Mean As Fed	Mean Dry	C.V. ± %
Dry matter	%		89.0	100.0	
Ash	%		5.2	5.8	
Crude fiber	%		28.3	31.8	
Ether extract	%		2.3	2.6	
N-free extract	%		45.5	51.1	
Protein (N x 6.25)	%		7.7	8.7	
Cattle	dig prot %	*	3.9	4.4	
Goats	dig prot %	*	4.1	4.6	
Horses	dig prot %	*	4.3	4.9	
Rabbits	dig coef %	*	4.8	5.3	
Sheep	dig prot %	*	3.9	4.3	
Energy	GE Mcal/kg				
Cattle	DE Mcal/kg	*	2.31	2.60	
Sheep	DE Mcal/kg	*	2.22	2.50	
Cattle	ME Mcal/kg	*	1.89	2.13	
Sheep	ME Mcal/kg	*	1.82	2.05	
Cattle	TDN %	*	52.4	58.9	
Sheep	TDN %	*	50.4	56.7	

Timothy, aerial part, dehy, full bloom, (1)

Ref No 1 04 876 United States

Feed Name or Analyses			Mean As Fed	Mean Dry	C.V. ± %
Dry matter	%		87.6	100.0	
Ash	%		5.0	5.7	
Crude fiber	%		28.6	32.6	
Cattle	dig coef %		62.	62.	
Ether extract	%		2.3	2.6	
Cattle	dig coef %		58.	58.	
N-free extract	%		44.2	50.4	
Cattle	dig coef %		68.	68.	
Protein (N x 6.25)	%		7.6	8.7	
Cattle	dig coef %		53.	53.	
Cattle	dig prot %		4.0	4.6	
Goats	dig prot %	*	4.1	4.7	
Horses	dig prot %	*	4.3	4.9	
Rabbits	dig coef %	*	4.7	5.4	
Sheep	dig prot %	*	3.8	4.4	
Energy	GE Mcal/kg		4.01	4.58	
Sheep	GE dig coef %		52.	52.	
Cattle	DE Mcal/kg	*	2.41	2.76	
Sheep	DE Mcal/kg		2.06	2.36	
Cattle	ME Mcal/kg	*	1.98	2.26	
Sheep	ME Mcal/kg	*	1.69	1.93	
Cattle	TDN %		54.7	62.5	
Sheep	TDN %	*	47.7	53.5	

Ref No 1 04 876 Canada

Feed Name or Analyses			Mean As Fed	Mean Dry	C.V. ± %
Dry matter	%		89.1	100.0	
Sheep	dig coef %		53.	53.	
Organic matter	%		84.0	94.3	
Ash	%		4.6	5.2	
Sheep	dig coef %		27.	30.	
Crude fiber	%		26.9	30.2	
Cattle	dig coef %		62.	62.	
Ether extract	%		2.0	2.3	
Cattle	dig coef %		58.	58.	
N-free extract	%		45.5	51.1	
Cattle	dig coef %		68.	68.	
Protein (N x 6.25)	%		10.1	11.3	
Cattle	dig coef %		53.	53.	

Continued

Feed Name or Analyses			Mean As Fed	Mean Dry	C.V. ± %
Cattle	dig prot %		5.3	6.0	
Goats	dig prot %	*	6.3	7.1	
Horses	dig prot %	*	6.3	7.1	
Rabbits	dig coef %	*	6.6	7.4	
Sheep	dig prot %	*	6.0	6.7	
Energy	GE Mcal/kg		4.08	4.58	
Sheep	GE dig coef %		52.	52.	
Cattle	DE Mcal/kg	*	2.45	2.75	
Sheep	DE Mcal/kg		2.10	2.36	
Cattle	ME Mcal/kg	*	2.01	2.26	
Sheep	ME Mcal/kg	*	1.72	1.93	
Cattle	TDN %		55.6	62.4	
Sheep	TDN %	*	47.7	53.5	

Timothy, aerial part, dehy, late bloom, (1)

Ref No 1 04 877 United States

Feed Name or Analyses			Mean As Fed	Mean Dry	C.V. ± %
Dry matter	%		88.9	100.0	
Ash	%		6.0	6.7	
Crude fiber	%		28.7	32.3	
Cattle	dig coef %		64.	64.	
Ether extract	%		2.5	2.8	
Cattle	dig coef %		48.	48.	
N-free extract	%		43.7	49.2	
Cattle	dig coef %		68.	68.	
Protein (N x 6.25)	%		8.0	9.0	
Cattle	dig coef %		53.	53.	
Cattle	dig prot %		4.2	4.8	
Goats	dig prot %	*	4.4	5.0	
Horses	dig prot %	*	4.6	5.2	
Rabbits	dig coef %	*	5.0	5.6	
Sheep	dig prot %	*	4.1	4.6	
Energy	GE Mcal/kg				
Cattle	DE Mcal/kg	*	2.43	2.73	
Sheep	DE Mcal/kg	*	2.21	2.49	
Cattle	ME Mcal/kg	*	1.99	2.24	
Sheep	ME Mcal/kg	*	1.81	2.04	
Cattle	TDN %		55.0	61.9	
Sheep	TDN %	*	50.1	56.4	

Timothy, aerial part, dehy, milk stage, (1)

Ref No 1 04 878 United States

Feed Name or Analyses			Mean As Fed	Mean Dry	C.V. ± %
Dry matter	%		89.2	100.0	
Ash	%		4.9	5.5	
Crude fiber	%		30.9	34.6	
Cattle	dig coef %		53.	53.	
Ether extract	%		2.1	2.4	
Cattle	dig coef %		61.	61.	
N-free extract	%		45.5	51.0	
Cattle	dig coef %		62.	62.	
Protein (N x 6.25)	%		5.8	6.5	
Cattle	dig coef %		39.	39.	
Cattle	dig prot %		2.3	2.5	
Goats	dig prot %	*	2.3	2.6	
Horses	dig prot %	*	2.7	3.0	
Rabbits	dig coef %	*	3.3	3.7	
Sheep	dig prot %	*	2.1	2.4	
Energy	GE Mcal/kg				
Cattle	DE Mcal/kg	*	2.19	2.46	
Sheep	DE Mcal/kg	*	2.15	2.41	
Cattle	ME Mcal/kg	*	1.80	2.02	
Sheep	ME Mcal/kg	*	1.76	1.97	
Cattle	TDN %		49.8	55.8	
Sheep	TDN %	*	48.7	54.6	

Ref No 1 04 878 Canada

Feed Name or Analyses			Mean As Fed	Mean Dry	C.V. ± %
Dry matter	%		86.3	100.0	
Sheep	dig coef %		49.	49.	
Organic matter	%		81.0	93.9	
Ash	%		5.6	6.5	
Sheep	dig coef %		8.	9.	
Crude fiber	%				
Cattle	dig coef %		53.	53.	
Ether extract	%				
Cattle	dig coef %		61.	61.	
N-free extract	%				
Cattle	dig coef %		62.	62.	
Protein (N x 6.25)	%				
Cattle	dig coef %		39.	39.	
Energy	GE Mcal/kg				
Cattle	DE Mcal/kg	*	2.11	2.45	
Cattle	ME Mcal/kg	*	1.73	2.01	
Cattle	TDN %		47.9	55.5	

Timothy, hay, bleached s-c, cut 1, (1)

Ref No 1 04 873 United States

Feed Name or Analyses			Mean As Fed	Mean Dry	C.V. ± %
Dry matter	%		93.3	100.0	
Sugars, total	%		4.1	4.4	
Carotene	mg/kg		6.2	6.6	
Vitamin A equivalent	IU/g		10.3	11.0	

Timothy, hay, fan air dried, (1)

Ref No 1 04 872 United States

Feed Name or Analyses			Mean As Fed	Mean Dry	C.V. ± %
Dry matter	%		88.2	100.0	2
Ash	%		4.7	5.3	10
Crude fiber	%		28.8	32.6	7
Ether extract	%		2.1	2.4	11
N-free extract	%		44.7	50.7	
Protein (N x 6.25)	%		7.9	9.0	14
Cattle	dig prot %	*	4.2	4.7	
Goats	dig prot %	*	4.4	5.0	
Horses	dig prot %	*	4.6	5.2	
Rabbits	dig coef %	*	5.0	5.6	
Sheep	dig prot %	*	4.1	4.6	
Energy	GE Mcal/kg				
Cattle	DE Mcal/kg	*	2.21	2.50	
Sheep	DE Mcal/kg	*	2.20	2.49	
Cattle	ME Mcal/kg	*	1.81	2.05	
Sheep	ME Mcal/kg	*	1.80	2.04	
Cattle	TDN %	*	50.1	56.8	
Sheep	TDN %	*	49.9	56.5	
Carotene	mg/kg		41.8	47.4	
Riboflavin	mg/kg		12.8	14.6	
Vitamin A equivalent	IU/g		69.7	79.0	
Vitamin D_2	IU/g		0.9	1.0	

Timothy, hay, fan air dried, immature, (1)

Ref No 1 04 869 United States

Feed Name or Analyses			Mean As Fed	Mean Dry	C.V. ± %
Dry matter	%			100.0	
Sheep	dig coef %			60.	
Ash	%			7.8	
Protein (N x 6.25)	%			25.4	
Sheep	dig coef %			80.	
Cattle	dig prot %	*		18.9	
Goats	dig prot %	*		20.2	
Horses	dig prot %	*		19.1	
Rabbits	dig coef %	*		18.2	
Sheep	dig prot %			20.3	
Cellulose (Matrone)	%			26.7	
Sheep	dig coef %			51.	
Energy	GE Mcal/kg			4.56	
Sheep	GE dig coef %			69.	
Sheep	DE Mcal/kg			3.12	
Sheep	ME Mcal/kg	*		2.56	
Sheep	TDN %			79.7	
Riboflavin	mg/kg			17.0	

Timothy, hay, fan air dried, early bloom, (1)

Ref No 1 04 870 United States

Feed Name or Analyses			Mean As Fed	Mean Dry	C.V. ± %
Dry matter	%		88.9	100.0	
Ash	%		4.7	5.3	
Crude fiber	%		29.0	32.6	
Sheep	dig coef %		65.	65.	
Ether extract	%		2.1	2.4	
Sheep	dig coef %		48.	48.	
N-free extract	%		44.1	49.7	
Sheep	dig coef %		70.	70.	
Protein (N x 6.25)	%		9.0	10.1	
Sheep	dig coef %		66.	66.	
Cattle	dig prot %	*	5.0	5.7	
Goats	dig prot %	*	5.3	6.0	
Horses	dig prot %	*	5.4	6.1	
Rabbits	dig coef %	*	5.7	6.5	
Sheep	dig prot %		5.9	6.7	
Energy	GE Mcal/kg		4.03	4.54	
Sheep	GE dig coef %		60.	60.	
Cattle	DE Mcal/kg	*	2.21	2.49	
Sheep	DE Mcal/kg		2.42	2.72	
Cattle	ME Mcal/kg	*	1.81	2.04	
Sheep	ME Mcal/kg	*	1.98	2.23	
Cattle	TDN %	*	50.2	56.5	
Sheep	TDN %		57.9	65.2	
Calcium	%		0.55	0.62	20
Phosphorus	%		0.21	0.24	31
Riboflavin	mg/kg		10.6	11.9	

Ref No 1 04 870 Canada

Feed Name or Analyses			Mean As Fed	Mean Dry	C.V. ± %
Dry matter	%			100.0	
Sheep	dig coef %			65.	
Ash	%			6.6	
Crude fiber	%			27.8	
Sheep	dig coef %			65.	
Ether extract	%			1.9	
Sheep	dig coef %			48.	
N-free extract	%			53.7	
Sheep	dig coef %			70.	
Protein (N x 6.25)	%			10.0	
Sheep	dig coef %			61.	
Cattle	dig prot %	*		5.6	
Goats	dig prot %	*		5.9	
Horses	dig prot %	*		6.0	
Rabbits	dig coef %	*		6.4	
Sheep	dig prot %			6.1	
Cellulose (Matrone)	%			31.0	
Sheep	dig coef %			65.	
Lignin (Ellis)	%			5.3	
Energy	GE Mcal/kg			4.45	
Sheep	GE dig coef %			65.	
Cattle	DE Mcal/kg	*		2.88	
Sheep	DE Mcal/kg			2.89	
Cattle	ME Mcal/kg	*		2.36	
Sheep	ME Mcal/kg	*		2.37	

(1) dry forages and roughages (2) pasture, range plants, and forages fed green (3) silages (4) energy feeds (5) protein supplements (6) minerals (7) vitamins (8) additives

Feed Name or Analyses		Mean As Fed	Mean Dry	C.V. ± %
Cattle	TDN % *		65.3	
Sheep	TDN %		63.8	

Timothy, hay, fan air dried, full bloom, (1)

Ref No 1 07 719 — Canada

Feed Name or Analyses		Mean As Fed	Mean Dry	C.V. ± %
Dry matter	%		100.0	
Sheep	dig coef %		51.	
Ash	%		5.3	
Crude fiber	%		31.5	
Sheep	dig coef %		46.	
Ether extract	%		1.3	
Sheep	dig coef %		45.	
N-free extract	%		55.6	
Sheep	dig coef %		57.	
Protein (N x 6.25)	%		6.3	
Sheep	dig coef %		44.	
Cattle	dig prot % *		2.4	
Goats	dig prot % *		2.4	
Horses	dig prot % *		2.9	
Rabbits	dig coef % *		3.5	
Sheep	dig prot %		2.8	
Cellulose (Matrone)	%		32.8	
Sheep	dig coef %		48.	
Lignin (Ellis)	%		9.7	
Energy	GE Mcal/kg		4.44	
Sheep	GE dig coef %		51.	
Cattle	DE Mcal/kg *		2.39	
Sheep	DE Mcal/kg		2.26	
Cattle	ME Mcal/kg *		1.96	
Sheep	ME Mcal/kg *		1.86	
Cattle	TDN % *		54.3	
Sheep	TDN %		50.3	

Timothy, hay, fan air dried, cut 2, (1)

Ref No 1 04 871 — United States

Feed Name or Analyses		Mean As Fed	Mean Dry	C.V. ± %
Dry matter	%	88.7	100.0	
Calcium	%	0.58	0.65	
Phosphorus	%	0.19	0.21	

Timothy, hay, fan air dried chopped, pre-bloom, cut 1, (1)

Ref No 1 07 708 — United States

Feed Name or Analyses		Mean As Fed	Mean Dry	C.V. ± %
Dry matter	%	92.1	100.0	
Sheep	dig coef %	66.	66.	8
Organic matter	%	86.6	94.1	
Ash	%	5.6	6.1	14
Crude fiber	%	28.3	30.8	
Sheep	dig coef %	69.	69.	
Ether extract	%	2.8	3.0	
Sheep	dig coef %	48.	48.	
N-free extract	%	41.6	45.2	
Sheep	dig coef %	70.	70.	
Protein (N x 6.25)	%	13.7	14.9	27
Sheep	dig coef %	70.	70.	7
Cattle	dig prot % *	9.1	9.9	
Goats	dig prot % *	9.6	10.5	
Horses	dig prot % *	9.4	10.2	
Rabbits	dig coef % *	9.4	10.2	
Sheep	dig prot %	9.6	10.5	
Cellulose (Crampton)	%			
Sheep	dig coef %	72.	72.	
Cellulose (Matrone)	%	29.5	32.1	
Sheep	dig coef %	69.	69.	
Energy	GE Mcal/kg	4.19	4.55	1
Sheep	GE dig coef %	66.	66.	9
Cattle	DE Mcal/kg *	2.54	2.76	
Sheep	DE Mcal/kg	2.75	2.99	
Cattle	ME Mcal/kg *	2.08	2.26	
Sheep	ME Mcal/kg *	2.25	2.45	
Cattle	TDN % *	57.6	62.5	
Sheep	TDN %	61.0	66.3	8

Timothy, hay, fan air dried chopped, mid-bloom, cut 1, (1)

Ref No 1 07 709 — United States

Feed Name or Analyses		Mean As Fed	Mean Dry	C.V. ± %
Dry matter	%		100.0	
Crude fiber	%			
Sheep	dig coef %		52.	
Ether extract	%			
Sheep	dig coef %		46.	
N-free extract	%			
Sheep	dig coef %		64.	
Protein (N x 6.25)	%		10.5	
Sheep	dig coef %		66.	
Cattle	dig prot % *		6.0	
Goats	dig prot % *		6.4	
Horses	dig prot % *		6.4	
Rabbits	dig coef % *		6.8	
Sheep	dig prot %		6.9	
Energy	GE Mcal/kg		4.58	
Sheep	GE dig coef %		60.	
Sheep	DE Mcal/kg		2.75	
Sheep	ME Mcal/kg *		2.25	
Sheep	TDN %		61.1	

Ref No 1 07 709 — Canada

Feed Name or Analyses		Mean As Fed	Mean Dry	C.V. ± %
Dry matter	%		100.0	
Sheep	dig coef %		57.	
Ash	%		5.8	
Crude fiber	%		30.3	
Sheep	dig coef %		52.	
Ether extract	%		1.4	
Sheep	dig coef %		46.	
N-free extract	%		54.4	
Sheep	dig coef %		64.	
Protein (N x 6.25)	%		8.1	
Sheep	dig coef %		55.	
Cattle	dig prot % *		4.0	
Goats	dig prot % *		4.1	
Horses	dig prot % *		4.4	
Rabbits	dig coef % *		4.9	
Sheep	dig prot %		4.5	
Cellulose (Matrone)	%		33.5	
Sheep	dig coef %		56.	
Lignin (Ellis)	%		7.4	
Energy	GE Mcal/kg		4.45	
Sheep	GE dig coef %		56.	
Cattle	DE Mcal/kg *		2.66	
Sheep	DE Mcal/kg		2.49	
Cattle	ME Mcal/kg *		2.18	
Sheep	ME Mcal/kg *		2.04	
Cattle	TDN % *		60.3	
Sheep	TDN %		56.5	

Timothy, hay, fan air dried chopped, late bloom, cut 1, (1)

Ref No 1 07 710 — United States

Feed Name or Analyses		Mean As Fed	Mean Dry	C.V. ± %
Dry matter	%	90.6	100.0	
Cattle	dig coef %	50.	50.	
Ash	%	3.4	3.7	
Crude fiber	%	36.4	40.2	
Cattle	dig coef %	53.	53.	
Ether extract	%	1.8	2.0	
Cattle	dig coef %	48.	48.	
N-free extract	%	42.3	46.7	
Cattle	dig coef %	52.	52.	
Protein (N x 6.25)	%	6.8	7.5	
Cattle	dig coef %	48.	48.	
Sheep	dig coef %	53.	53.	
Cattle	dig prot %	3.2	3.6	
Goats	dig prot % *	3.2	3.5	
Horses	dig prot % *	3.5	3.9	
Rabbits	dig coef % *	4.0	4.4	
Sheep	dig prot %	3.6	3.9	
Cellulose (Crampton)	%			
Cattle	dig coef %	55.	55.	
Energy	GE Mcal/kg	4.19	4.62	
Sheep	GE dig coef %	52.	52.	
Cattle	DE Mcal/kg *	2.04	2.25	
Sheep	DE Mcal/kg	2.18	2.40	
Cattle	ME Mcal/kg *	1.67	1.84	
Sheep	ME Mcal/kg *	1.79	1.97	
Cattle	TDN %	46.2	51.0	
Sheep	TDN %	50.1	55.3	

Timothy, hay, fan air dried chopped, milk stage, cut 1, (1)

Ref No 1 07 711 — United States

Feed Name or Analyses		Mean As Fed	Mean Dry	C.V. ± %
Dry matter	%		100.0	
Crude fiber	%			
Sheep	dig coef %		46.	
Ether extract	%			
Sheep	dig coef %		27.	
N-free extract	%			
Sheep	dig coef %		55.	
Protein (N x 6.25)	%		8.1	
Sheep	dig coef %		50.	
Cattle	dig prot % *		4.0	
Goats	dig prot % *		4.1	
Horses	dig prot % *		4.4	
Rabbits	dig coef % *		4.9	
Sheep	dig prot %		4.1	
Energy	GE Mcal/kg		4.60	
Sheep	GE dig coef %		49.	
Sheep	DE Mcal/kg		2.25	
Sheep	ME Mcal/kg *		1.85	
Sheep	TDN %		51.7	

Ref No 1 07 711 — Canada

Feed Name or Analyses		Mean As Fed	Mean Dry	C.V. ± %
Dry matter	%		100.0	
Sheep	dig coef %		48.	
Ash	%		5.7	
Crude fiber	%		33.2	
Sheep	dig coef %		46.	
Ether extract	%		1.1	
Sheep	dig coef %		27.	
N-free extract	%		54.1	
Sheep	dig coef %		55.	

Continued

Feed Name or Analyses			Mean As Fed	Mean Dry	C.V. ± %
Protein (N x 6.25)	%			5.9	
Sheep	dig coef %			32.	
Cattle	dig prot %	*		2.0	
Goats	dig prot %	*		2.1	
Horses	dig prot %	*		2.5	
Rabbits	dig coef %	*		3.2	
Sheep	dig prot %			1.9	
Cellulose (Matrone)	%			34.0	
Sheep	dig coef %			47.	
Lignin (Ellis)	%			10.1	
Energy	GE Mcal/kg			4.47	
Sheep	GE dig coef %			47.	
Cattle	DE Mcal/kg	*		2.27	
Sheep	DE Mcal/kg			2.10	
Cattle	ME Mcal/kg	*		1.86	
Sheep	ME Mcal/kg	*		1.72	
Cattle	TDN %	*		51.4	
Sheep	TDN %			47.6	

Timothy, hay, fan air dried chopped, dough stage, cut 1, (1)
Ref No 1 07 712 United States

Feed Name or Analyses			Mean As Fed	Mean Dry	C.V. ± %
Dry matter	%			100.0	
Protein (N x 6.25)	%			8.1	
Sheep	dig coef %			52.	
Cattle	dig prot %	*		4.0	
Goats	dig prot %	*		4.1	
Horses	dig prot %	*		4.4	
Rabbits	dig coef %	*		4.9	
Sheep	dig prot %			4.2	
Energy	GE Mcal/kg			4.66	
Sheep	GE dig coef %			51.	
Sheep	DE Mcal/kg			2.38	
Sheep	ME Mcal/kg	*		1.95	
Sheep	TDN %			51.8	

Timothy, hay, fan air dried chopped, mature, cut 1, (1)
Ref No 1 07 713 United States

Feed Name or Analyses			Mean As Fed	Mean Dry	C.V. ± %
Dry matter	%			100.0	
Protein (N x 6.25)	%			8.4	
Sheep	dig coef %			52.	
Cattle	dig prot %	*		4.2	
Goats	dig prot %	*		4.4	
Horses	dig prot %	*		4.7	
Rabbits	dig coef %	*		5.2	
Sheep	dig prot %			4.4	
Energy	GE Mcal/kg			4.69	
Sheep	GE dig coef %			50.	
Sheep	DE Mcal/kg			2.35	
Sheep	ME Mcal/kg	*		1.92	
Sheep	TDN %			51.1	

Timothy, hay, fan air dried w heat, pre-bloom, cut 1, (1)
Ref No 1 05 577 United States

Feed Name or Analyses			Mean As Fed	Mean Dry	C.V. ± %
Dry matter	%		91.7	100.0	1
Sheep	dig coef %		62.	62.	4
Ash	%		4.8	5.2	17
Crude fiber	%		31.3	34.2	6
Sheep	dig coef %		65.	65.	5
Ether extract	%		3.0	3.3	11
Sheep	dig coef %		53.	53.	4
N-free extract	%		42.5	46.4	15
Sheep	dig coef %		62.	62.	4
Protein (N x 6.25)	%		10.1	11.0	61
Sheep	dig coef %		54.	54.	26
Cattle	dig prot %	*	6.0	6.5	
Goats	dig prot %	*	6.3	6.9	
Horses	dig prot %	*	6.3	6.9	
Rabbits	dig coef %	*	6.6	7.2	
Sheep	dig prot %		5.4	5.9	
Energy	GE Mcal/kg		4.15	4.53	98
Sheep	GE dig coef %		59.	59.	5
Cattle	DE Mcal/kg	*	2.37	2.58	
Sheep	DE Mcal/kg		2.46	2.68	5
Cattle	ME Mcal/kg	*	1.94	2.12	
Sheep	ME Mcal/kg		2.02	2.20	5
Cattle	TDN %	*	53.6	58.5	
Sheep	TDN %		55.8	60.8	4

Timothy, hay, fan air dried w heat, early bloom, cut 2, (1)
Ref No 1 05 578 United States

Feed Name or Analyses			Mean As Fed	Mean Dry	C.V. ± %
Dry matter	%		92.1	100.0	
Sheep	dig coef %		59.	59.	
Ash	%		4.3	4.7	
Crude fiber	%		33.4	36.3	
Sheep	dig coef %		60.	60.	
Ether extract	%		2.5	2.7	
Sheep	dig coef %		47.	47.	
N-free extract	%		44.1	47.9	
Sheep	dig coef %		61.	61.	
Protein (N x 6.25)	%		7.7	8.4	
Sheep	dig coef %		52.	52.	
Cattle	dig prot %	*	3.9	4.2	
Goats	dig prot %	*	4.0	4.4	
Horses	dig prot %	*	4.3	4.7	
Rabbits	dig coef %	*	4.7	5.2	
Sheep	dig prot %		4.0	4.4	
Energy	GE Mcal/kg		4.20	4.56	
Sheep	GE dig coef %		56.	56.	
Cattle	DE Mcal/kg	*	2.22	2.41	
Sheep	DE Mcal/kg		2.35	2.56	
Cattle	ME Mcal/kg	*	1.82	1.98	
Sheep	ME Mcal/kg		1.95	2.12	
Cattle	TDN %	*	50.3	54.6	
Sheep	TDN %		53.6	58.2	

Timothy, hay, fan air dried w heat, late bloom, cut 3, (1)
Ref No 1 05 579 United States

Feed Name or Analyses			Mean As Fed	Mean Dry	C.V. ± %
Dry matter	%		92.5	100.0	
Sheep	dig coef %		50.	50.	
Ash	%		3.9	4.2	
Crude fiber	%		35.3	38.2	
Sheep	dig coef %		52.	52.	
Ether extract	%		1.9	2.0	
Sheep	dig coef %		31.	31.	
N-free extract	%		45.5	49.2	
Sheep	dig coef %		54.	54.	
Protein (N x 6.25)	%		5.9	6.4	
Sheep	dig coef %		35.	35.	
Cattle	dig prot %	*	2.3	2.5	
Goats	dig prot %	*	2.3	2.5	
Horses	dig prot %	*	2.7	3.0	
Rabbits	dig coef %	*	3.3	3.6	
Sheep	dig prot %		2.1	2.2	
Energy	GE Mcal/kg		4.19	4.53	
Sheep	GE dig coef %		47.	47.	
Cattle	DE Mcal/kg	*	1.90	2.06	
Sheep	DE Mcal/kg		1.97	2.13	
Cattle	ME Mcal/kg	*	1.56	1.69	
Sheep	ME Mcal/kg		1.59	1.72	
Cattle	TDN %	*	43.1	46.6	
Sheep	TDN %		46.3	50.1	

Timothy, hay, s-c, (1)
Ref No 1 04 893 United States

Feed Name or Analyses			Mean As Fed	Mean Dry	C.V. ± %
Dry matter	%		88.6	100.0	3
Rabbits	dig coef %		38.	38.	
Ash	%		4.5	5.1	10
Crude fiber	%		30.2	34.0	7
Cattle	dig coef %		55.	55.	
Horses	dig coef %		43.	43.	
Sheep	dig coef %		50.	50.	
Ether extract	%		2.3	2.5	16
Cattle	dig coef %		55.	55.	
Horses	dig coef %		13.	13.	
Sheep	dig coef %		51.	51.	
N-free extract	%		45.4	51.2	5
Cattle	dig coef %		60.	60.	
Horses	dig coef %		53.	53.	
Sheep	dig coef %		63.	63.	
Protein (N x 6.25)	%		6.3	7.1	10
Cattle	dig coef %		41.	41.	
Horses	dig coef %		43.	43.	
Rabbits	dig coef %		55.	55.	
Sheep	dig coef %		35.	35.	
Cattle	dig prot %		2.6	2.9	
Goats	dig prot %	*	2.8	3.2	
Horses	dig prot %		2.7	3.0	
Rabbits	dig coef %		3.5	3.9	
Sheep	dig prot %		2.2	2.5	
Cellulose (Matrone)	%		25.0	28.2	
Hemicellulose	%		15.8	17.8	8
Lignin (Ellis)	%		7.7	8.7	
Energy	GE Mcal/kg				
Cattle	DE Mcal/kg	*	2.15	2.43	
Horses	DE Mcal/kg	*	1.78	2.01	
Sheep	DE Mcal/kg	*	2.13	2.40	
Cattle	ME Mcal/kg	*	1.76	1.99	
Horses	ME Mcal/kg	*	1.46	1.65	
Sheep	ME Mcal/kg	*	1.75	1.97	
Cattle	TDN %		48.8	55.1	
Horses	TDN %		40.4	45.6	
Sheep	TDN %		48.3	54.5	
Calcium	%		0.36	0.40	39
Chlorine	%		0.55	0.62	
Copper	mg/kg		4.4	5.0	
Iron	%		0.012	0.014	
Magnesium	%		0.16	0.18	20
Manganese	mg/kg		41.8	47.1	41
Phosphorus	%		0.15	0.16	63
Potassium	%		1.41	1.59	29
Sodium	%		0.16	0.18	
Sulphur	%		0.12	0.13	
Biotin	mg/kg		0.06	0.07	32
Folic acid	mg/kg		2.03	2.29	
α-tocopherol	mg/kg		55.9	63.1	98
Vitamin D_2	IU/g		1.8	2.0	32

(1) dry forages and roughages (2) pasture, range plants, and forages fed green (3) silages (4) energy feeds (5) protein supplements (6) minerals (7) vitamins (8) additives

Feed Name or Analyses			Mean As Fed	Mean Dry	C.V. ± %
Ref No 1 04 893				Canada	
Dry matter	%		93.7	100.0	1
Crude fiber	%		29.5	31.5	8
Cattle	dig coef %		55.	55.	
Horses	dig coef %		43.	43.	
Sheep	dig coef %		50.	50.	
Ether extract	%				
Cattle	dig coef %		55.	55.	
Horses	dig coef %		13.	13.	
Sheep	dig coef %		51.	51.	
N-free extract	%				
Cattle	dig coef %		60.	60.	
Horses	dig coef %		53.	53.	
Sheep	dig coef %		63.	63.	
Protein (N x 6.25)	%		7.5	8.0	55
Cattle	dig coef %		41.	41.	
Horses	dig coef %		43.	43.	
Rabbits	dig coef %		55.	55.	
Sheep	dig coef %		35.	35.	
Cattle	dig prot %		3.1	3.3	
Goats	dig prot %	*	3.8	4.0	
Horses	dig prot %		3.2	3.4	
Rabbits	dig coef %		4.1	4.4	
Sheep	dig prot %		2.6	2.8	
Energy	GE Mcal/kg				
Cattle	DE Mcal/kg	*	2.27	2.43	
Horses	DE Mcal/kg	*	1.89	2.01	
Sheep	DE Mcal/kg	*	2.26	2.41	
Cattle	ME Mcal/kg	*	1.87	1.99	
Horses	ME Mcal/kg	*	1.55	1.65	
Sheep	ME Mcal/kg	*	1.85	1.97	
Cattle	TDN %		51.6	55.0	
Horses	TDN %		42.8	45.7	
Sheep	TDN %		51.2	54.6	
Calcium	%		0.33	0.36	66
Phosphorus	%		0.19	0.21	72
Carotene	mg/kg		42.4	45.2	98
Vitamin A equivalent	IU/g		70.7	75.4	

Timothy, hay, s-c, immature, (1)

Feed Name or Analyses			Mean As Fed	Mean Dry	C.V. ± %
Ref No 1 04 880				United States	
Dry matter	%		87.7	100.0	2
Crude fiber	%				
Cattle	dig coef %	*	75.	75.	
Sheep	dig coef %	*	75.	75.	
Ether extract	%				
Cattle	dig coef %	*	34.	34.	
Sheep	dig coef %	*	44.	44.	
N-free extract	%				
Cattle	dig coef %	*	65.	65.	
Sheep	dig coef %	*	71.	71.	
Protein (N x 6.25)	%				
Cattle	dig coef %	*	54.	54.	
Sheep	dig coef %	*	64.	64.	
Cellulose (Matrone)	%		24.5	27.9	
Energy	GE Mcal/kg		3.96	4.52	
Calcium	%		0.58	0.66	30
Magnesium	%		0.26	0.30	19
Phosphorus	%		0.30	0.34	9
Potassium	%		1.67	1.90	8
Carotene	mg/kg		108.4	123.7	53
Riboflavin	mg/kg		14.9	17.0	
Vitamin A equivalent	IU/g		180.7	206.2	

Timothy, hay, s-c, pre-bloom, (1)

Feed Name or Analyses			Mean As Fed	Mean Dry	C.V. ± %
Ref No 1 04 881				United States	
Dry matter	%		86.3	100.0	
Ash	%		6.1	7.1	
Crude fiber	%		28.2	32.6	
Cattle	dig coef %		75.	75.	
Sheep	dig coef %		75.	75.	
Ether extract	%		2.1	2.4	
Cattle	dig coef %		34.	34.	
Sheep	dig coef %		44.	44.	
N-free extract	%		40.0	46.4	
Cattle	dig coef %		65.	65.	
Sheep	dig coef %		71.	71.	
Protein (N x 6.25)	%		9.9	11.4	
Cattle	dig coef %		54.	54.	
Sheep	dig coef %		64.	64.	
Cattle	dig prot %		5.3	6.2	
Goats	dig prot %	*	6.2	7.2	
Horses	dig prot %	*	6.2	7.2	
Rabbits	dig coef %	*	6.5	7.5	
Sheep	dig prot %		6.3	7.3	
Cellulose (Matrone)	%		28.0	32.4	
Energy	GE Mcal/kg				
Cattle	DE Mcal/kg	*	2.38	2.76	
Sheep	DE Mcal/kg	*	2.55	2.96	
Cattle	ME Mcal/kg	*	1.96	2.27	
Sheep	ME Mcal/kg	*	2.10	2.43	
Cattle	NE_m Mcal/kg	*	1.16	1.34	
Cattle	NE_{gain} Mcal/kg	*	0.63	0.73	
Cattle	$NE_{lactating\ cows}$ Mcal/kg	*	1.31	1.52	
Cattle	TDN %		54.1	62.7	
Sheep	TDN %		57.9	67.2	

Timothy, hay, s-c, early bloom, (1)

Feed Name or Analyses			Mean As Fed	Mean Dry	C.V. ± %
Ref No 1 04 882				United States	
Dry matter	%		86.4	100.0	5
Ash	%		4.8	5.5	
Crude fiber	%		29.6	34.3	
Cattle	dig coef %		68.	68.	
Sheep	dig coef %		57.	57.	
Ether extract	%		2.2	2.6	
Cattle	dig coef %		33.	33.	
Sheep	dig coef %		40.	40.	
N-free extract	%		43.2	50.0	
Cattle	dig coef %		60.	60.	
Sheep	dig coef %		60.	60.	
Protein (N x 6.25)	%		6.5	7.6	
Cattle	dig coef %		58.	58.	
Sheep	dig coef %		54.	54.	
Cattle	dig prot %		3.8	4.4	
Goats	dig prot %	*	3.1	3.6	
Horses	dig prot %	*	3.4	3.9	
Rabbits	dig coef %	*	3.9	4.5	
Sheep	dig prot %		3.5	4.1	
Cellulose (Matrone)	%		25.6	29.6	
Lignin (Ellis)	%		6.6	7.6	
Energy	GE Mcal/kg		3.83	4.44	
Cattle	DE Mcal/kg	*	2.27	2.63	
Sheep	DE Mcal/kg	*	2.13	2.47	
Cattle	ME Mcal/kg	*	1.86	2.16	
Sheep	ME Mcal/kg	*	1.75	2.02	
Cattle	NE_m Mcal/kg	*	1.09	1.26	
Cattle	NE_{gain} Mcal/kg	*	0.54	0.62	
Cattle	$NE_{lactating\ cows}$ Mcal/kg	*	1.22	1.41	
Cattle	TDN %		51.5	59.6	
Sheep	TDN %		48.3	56.0	
Calcium	%		0.46	0.53	22
Phosphorus	%		0.21	0.25	31
Potassium	%		0.79	0.92	

Timothy, hay, s-c, early bloom, cut 1, (1)

Feed Name or Analyses			Mean As Fed	Mean Dry	C.V. ± %
Ref No 1 09 003				United States	
Dry matter	%		92.4	100.0	1
Horses	dig coef %		50.	50.	2
Organic matter	%		86.7	93.9	0
Ash	%		5.7	6.1	3
Crude fiber	%		36.3	39.3	1
Horses	dig coef %		45.	45.	6
Ether extract	%		2.4	2.6	5
Horses	dig coef %		39.	39.	22
N-free extract	%		40.4	43.7	1
Horses	dig coef %		62.	62.	5
Protein (N x 6.25)	%		7.7	8.3	3
Horses	dig coef %		54.	54.	7
Cattle	dig prot %	*	3.8	4.1	
Goats	dig prot %	*	4.0	4.3	
Horses	dig prot %		4.2	4.5	
Sheep	dig prot %	*	3.7	4.0	
Cell contents (Van Soest)	%		26.5	28.6	7
Horses	dig coef %		62.	62.	3
Cell walls (Van Soest)	%		66.4	71.8	3
Horses	dig coef %		46.	46.	4
Cellulose (Matrone)	%		34.4	37.2	1
Horses	dig coef %		48.	48.	5
Fiber, acid detergent (VS)	%		39.1	42.3	1
Horses	dig coef %		40.	40.	6
Hemicellulose	%		24.0	26.0	8
Horses	dig coef %		53.	53.	3
Holocellulose	%		58.0	62.7	3
Soluble carbohydrates	%		13.9	15.1	13
Lignin (Ellis)	%		4.7	5.1	4
Horses	dig coef %		24.	24.	13
Energy	GE Mcal/kg		4.12	4.46	0
Horses	GE dig coef %		44.	44.	7
Cattle	DE Mcal/kg	*	1.90	2.06	
Horses	DE Mcal/kg		1.79	1.94	
Sheep	DE Mcal/kg	*	2.16	2.33	
Cattle	ME Mcal/kg	*	1.56	1.69	
Horses	ME Mcal/kg	*	1.47	1.59	
Sheep	ME Mcal/kg	*	1.77	1.91	
Cattle	TDN %	*	43.2	46.7	
Horses	TDN %		47.8	51.7	
Sheep	TDN %	*	48.9	52.9	
Calcium	%		0.33	0.36	14
Magnesium	%		0.18	0.20	7
Phosphorus	%		0.19	0.21	6
Potassium	%		0.23	0.25	10
Carotene	mg/kg		17.7	19.2	14
Vitamin A equivalent	IU/g		29.5	31.9	

Timothy, hay, s-c, mid-bloom, (1)

Feed Name or Analyses			Mean As Fed	Mean Dry	C.V. ± %
Ref No 1 04 883				United States	
Dry matter	%		88.1	100.0	2
Ash	%		6.0	6.9	
Crude fiber	%		30.6	34.7	
Cattle	dig coef %		68.	68.	
Sheep	dig coef %		57.	57.	
Ether extract	%		2.2	2.5	
Cattle	dig coef %		52.	52.	
Sheep	dig coef %		62.	62.	

Continued

Feed Name or Analyses		Mean As Fed	Mean Dry	C.V. ± %
N-free extract	%	40.1	45.5	
Cattle dig coef	%	62.	62.	
Sheep dig coef	%	65.	65.	
Protein (N x 6.25)	%	9.2	10.5	
Cattle dig coef	%	54.	54.	
Sheep dig coef	%	61.	61.	
Cattle dig prot	%	5.0	5.7	
Goats dig prot	% *	5.6	6.4	
Horses dig prot	% *	5.7	6.4	
Rabbits dig coef	% *	6.0	6.8	
Sheep dig prot	%	5.6	6.4	
Cellulose (Matrone)	%	30.0	34.1	
Energy	GE Mcal/kg	3.84	4.37	
Cattle	DE Mcal/kg *	2.34	2.66	
Sheep	DE Mcal/kg *	2.30	2.61	
Cattle	ME Mcal/kg *	1.92	2.18	
Sheep	ME Mcal/kg *	1.88	2.14	
Cattle	NE_m Mcal/kg *	1.15	1.31	
Cattle	NE_{gain} Mcal/kg *	0.61	0.69	
Cattle	$NE_{lactating\ cows}$ Mcal/kg *	1.31	1.49	
Cattle	TDN %	53.1	60.3	
Sheep	TDN %	52.1	59.2	
Calcium	%	0.36	0.41	
Magnesium	%	0.14	0.16	
Phosphorus	%	0.17	0.19	
Carotene	mg/kg	47.0	53.4	
Vitamin A equivalent	IU/g	78.3	88.9	

Timothy, hay, s-c, full bloom, (1)
Ref No 1 04 884 United States

Feed Name or Analyses		Mean As Fed	Mean Dry	C.V. ± %
Dry matter	%	86.6	100.0	6
Ash	%	4.5	5.2	10
Crude fiber	%	30.6	35.4	4
Cattle dig coef	%	62.	62.	
Horses dig coef	%	43.	43.	
Sheep dig coef	%	58.	58.	
Ether extract	%	2.7	3.1	22
Cattle dig coef	%	42.	42.	
Horses dig coef	%	47.	47.	
Sheep dig coef	%	58.	58.	
N-free extract	%	42.2	48.7	2
Cattle dig coef	%	59.	59.	
Horses dig coef	%	47.	47.	
Sheep dig coef	%	62.	62.	
Protein (N x 6.25)	%	6.6	7.6	23
Cattle dig coef	%	49.	49.	
Horses dig coef	%	21.	21.	
Sheep dig coef	%	54.	54.	
Cattle dig prot	%	3.2	3.7	
Goats dig prot	% *	3.2	3.7	
Horses dig prot	%	1.4	1.6	
Rabbits dig coef	% *	4.0	4.6	
Sheep dig prot	%	3.6	4.1	
Energy	GE Mcal/kg			
Cattle	DE Mcal/kg *	2.19	2.53	
Horses	DE Mcal/kg *	1.64	1.89	
Sheep	DE Mcal/kg *	2.25	2.60	
Cattle	ME Mcal/kg *	1.79	2.07	
Horses	ME Mcal/kg *	1.35	1.55	
Sheep	ME Mcal/kg *	1.84	2.13	
Cattle	TDN %	49.6	57.3	
Horses	TDN %	37.2	43.0	
Sheep	TDN %	51.0	58.9	

(1) dry forages and roughages (2) pasture, range plants, and forages fed green (3) silages (4) energy feeds (5) protein supplements (6) minerals (7) vitamins (8) additives

Feed Name or Analyses		Mean As Fed	Mean Dry	C.V. ± %
Calcium	%	0.39	0.45	
Chlorine	%	0.54	0.62	9
Copper	mg/kg	4.3	5.0	
Iron	%	0.014	0.016	
Magnesium	%	0.12	0.13	
Manganese	mg/kg	70.1	81.0	
Phosphorus	%	0.20	0.23	
Potassium	%	1.42	1.63	
Sodium	%	0.16	0.18	18
Sulphur	%	0.11	0.13	24

Timothy, hay, s-c, late bloom, (1)
Ref No 1 04 885 United States

Feed Name or Analyses		Mean As Fed	Mean Dry	C.V. ± %
Dry matter	%	88.3	100.0	
Ash	%	4.8	5.4	
Crude fiber	%	28.7	32.5	
Cattle dig coef	%	62.	62.	
Sheep dig coef	%	52.	52.	
Ether extract	%	2.6	2.9	
Cattle dig coef	%	36.	36.	
Sheep dig coef	%	48.	48.	
N-free extract	%	45.4	51.5	
Cattle dig coef	%	63.	63.	
Sheep dig coef	%	63.	63.	
Protein (N x 6.25)	%	6.8	7.7	
Cattle dig coef	%	49.	49.	
Sheep dig coef	%	43.	43.	
Cattle dig prot	%	3.3	3.7	
Goats dig prot	% *	3.3	3.7	
Horses dig prot	% *	3.6	4.0	
Rabbits dig coef	% *	4.0	4.6	
Sheep dig prot	%	2.9	3.3	
Energy	GE Mcal/kg			
Cattle	DE Mcal/kg *	2.29	2.59	
Sheep	DE Mcal/kg *	2.17	2.46	
Cattle	ME Mcal/kg *	1.87	2.12	
Sheep	ME Mcal/kg *	1.78	2.02	
Cattle	NE_m Mcal/kg *	1.09	1.24	
Cattle	NE_{gain} Mcal/kg *	0.52	0.59	
Cattle	$NE_{lactating\ cows}$ Mcal/kg *	1.22	1.38	
Cattle	TDN %	51.9	58.7	
Sheep	TDN %	49.3	55.8	

Timothy, hay, s-c, milk stage, (1)
Ref No 1 04 886 United States

Feed Name or Analyses		Mean As Fed	Mean Dry	C.V. ± %
Dry matter	%	82.9	100.0	
Ash	%	4.5	5.4	
Crude fiber	%	29.5	35.7	
Cattle dig coef	%	50.	50.	
Sheep dig coef	%	50.	50.	
Ether extract	%	2.2	2.7	
Cattle dig coef	%	47.	47.	
Sheep dig coef	%	54.	54.	
N-free extract	%	40.6	49.1	
Cattle dig coef	%	51.	51.	
Sheep dig coef	%	59.	59.	
Protein (N x 6.25)	%	6.0	7.2	
Cattle dig coef	%	41.	41.	
Sheep dig coef	%	48.	48.	
Cattle dig prot	%	2.4	3.0	
Goats dig prot	% *	2.7	3.3	
Horses dig prot	% *	3.0	3.6	
Rabbits dig coef	% *	3.5	4.2	
Sheep dig prot	%	2.9	3.5	
Energy	GE Mcal/kg	3.67	4.43	
Sheep	GE dig coef %	53.	53.	
Cattle	DE Mcal/kg *	1.78	2.14	
Sheep	DE Mcal/kg	1.93	2.33	
Cattle	ME Mcal/kg *	1.46	1.76	
Sheep	ME Mcal/kg *	1.58	1.91	
Cattle	TDN %	40.3	48.6	
Sheep	TDN %	44.3	53.5	
Calcium	%	0.31	0.38	
Magnesium	%	0.15	0.18	
Phosphorus	%	0.15	0.18	

Ref No 1 04 886 Canada

Feed Name or Analyses		Mean As Fed	Mean Dry	C.V. ± %
Dry matter	%		100.0	
Sheep dig coef	%		54.	
Ash	%		7.1	
Crude fiber	%		33.4	
Cattle dig coef	%		50.	
Sheep dig coef	%		49.	
Ether extract	%		1.8	
Cattle dig coef	%		47.	
Sheep dig coef	%		54.	
N-free extract	%		47.5	
Cattle dig coef	%		51.	
Sheep dig coef	%		59.	
Protein (N x 6.25)	%		10.2	
Cattle dig coef	%		41.	
Sheep dig coef	%		56.	
Cattle dig prot	%		4.2	
Goats dig prot	% *		6.1	
Horses dig prot	% *		6.2	
Rabbits dig coef	% *		6.5	
Sheep dig prot	%		5.7	
Energy	GE Mcal/kg		4.43	
Sheep	GE dig coef %		53.	
Cattle	DE Mcal/kg *		2.07	
Sheep	DE Mcal/kg		2.33	
Cattle	ME Mcal/kg *		1.70	
Sheep	ME Mcal/kg *		1.91	
Cattle	TDN %		47.0	
Sheep	TDN %		52.4	
Calcium	%		0.18	
Magnesium	%		0.06	
Potassium	%		1.00	
Sodium	%		0.01	

Timothy, hay, s-c, dough stage, (1)
Ref No 1 04 887 United States

Feed Name or Analyses		Mean As Fed	Mean Dry	C.V. ± %
Dry matter	%	94.1	100.0	
Ash	%	3.4	3.6	
Crude fiber	%	36.5	38.8	
Ether extract	%	1.1	1.2	
N-free extract	%	47.2	50.2	
Protein (N x 6.25)	%	5.9	6.3	12
Cattle dig prot	% *	2.2	2.4	
Goats dig prot	% *	2.3	2.4	
Horses dig prot	% *	2.7	2.8	
Rabbits dig coef	% *	3.3	3.5	
Sheep dig prot	% *	2.0	2.2	
Cellulose (Matrone)	%	30.0	31.9	
Energy	GE Mcal/kg	4.48	4.76	
Sheep	GE dig coef %	48.	48.	
Cattle	DE Mcal/kg *	2.06	2.19	
Sheep	DE Mcal/kg	2.13	2.26	
Cattle	ME Mcal/kg *	1.69	1.80	
Sheep	ME Mcal/kg *	1.74	1.85	
Cattle	TDN % *	46.8	49.7	
Sheep	TDN % *	48.2	51.2	

Feed Name or Analyses			Mean As Fed	Mean Dry	C.V. ± %
Ref No 1 04 887				**Canada**	
Dry matter	%			100.0	
Sheep	dig coef %			51.	
Ash	%			4.0	15
Crude fiber	%			34.7	4
Ether extract	%			2.3	12
N-free extract	%			51.0	
Protein (N x 6.25)	%			8.0	22
Cattle	dig prot %	*		3.9	
Goats	dig prot %	*		4.0	
Horses	dig prot %	*		4.3	
Rabbits	dig coef %	*		4.8	
Sheep	dig prot %	*		3.7	
Energy	GE Mcal/kg			4.59	
Sheep	GE dig coef %			48.	
Cattle	DE Mcal/kg	*		2.30	
Sheep	DE Mcal/kg			2.18	
Cattle	ME Mcal/kg	*		1.89	
Sheep	ME Mcal/kg	*		1.79	
Cattle	TDN %	*		52.1	
Sheep	TDN %	*		55.4	
Calcium	%			0.33	
Copper	mg/kg			13.6	23
Iron	%			.000	8
Manganese	mg/kg			35.0	21
Phosphorus	%			0.15	
Potassium	%			1.26	
Sulphur	%			0.03	

Timothy, hay, s-c, mature, (1)

Feed Name or Analyses			Mean As Fed	Mean Dry	C.V. ± %
Ref No 1 04 888				**United States**	
Dry matter	%		83.8	100.0	7
Ash	%		4.6	5.5	
Crude fiber	%		30.7	36.6	
Cattle	dig coef %		51.	51.	
Sheep	dig coef %	*	55.	55.	
Ether extract	%		2.2	2.6	
Cattle	dig coef %		43.	43.	
Sheep	dig coef %	*	47.	47.	
N-free extract	%		41.7	49.8	
Cattle	dig coef %		48.	48.	
Sheep	dig coef %	*	61.	61.	
Protein (N x 6.25)	%		4.6	5.5	8
Cattle	dig coef %		42.	42.	
Sheep	dig coef %	*	47.	47.	
Cattle	dig prot %		1.9	2.3	
Goats	dig prot %	*	1.4	1.7	
Horses	dig prot %	*	1.8	2.2	
Rabbits	dig coef %	*	2.4	2.9	
Sheep	dig prot %		2.2	2.6	
Energy	GE Mcal/kg		3.70	4.42	
Cattle	DE Mcal/kg	*	1.75	2.09	
Sheep	DE Mcal/kg	*	2.06	2.46	
Cattle	ME Mcal/kg	*	1.44	1.71	
Sheep	ME Mcal/kg	*	1.69	2.02	
Cattle	TDN %		39.7	47.4	
Sheep	TDN %		46.8	55.8	
Calcium	%		0.17	0.20	32
Chlorine	%		0.49	0.58	8
Iron	%		0.023	0.027	10
Magnesium	%		0.06	0.07	31
Manganese	mg/kg		85.5	102.1	12
Phosphorus	%		0.14	0.17	22
Potassium	%		1.32	1.58	21
Sodium	%		0.06	0.07	74
Sulphur	%		0.13	0.16	22

Feed Name or Analyses			Mean As Fed	Mean Dry	C.V. ± %
Carotene	mg/kg		4.4	5.3	67
Vitamin A equivalent	IU/g		7.4	8.8	
Ref No 1 04 888				**Canada**	
Dry matter	%		93.9	100.0	
Crude fiber	%		33.2	35.4	
Cattle	dig coef %		51.	51.	
Ether extract	%				
Cattle	dig coef %		43.	43.	
N-free extract	%				
Cattle	dig coef %		48.	48.	
Protein (N x 6.25)	%		6.3	6.7	
Cattle	dig coef %		42.	42.	
Cattle	dig prot %		2.6	2.8	
Goats	dig prot %	*	2.6	2.8	
Horses	dig prot %	*	3.0	3.2	
Rabbits	dig coef %	*	3.6	3.8	
Sheep	dig prot %	*	2.4	2.6	
Energy	GE Mcal/kg		4.15	4.42	
Cattle	DE Mcal/kg	*	1.96	2.09	
Sheep	DE Mcal/kg	*	2.31	2.46	
Cattle	ME Mcal/kg	*	1.61	1.71	
Sheep	ME Mcal/kg	*	1.89	2.02	
Cattle	TDN %		44.4	47.3	
Sheep	TDN %		52.4	55.8	
Calcium	%		0.34	0.37	
Phosphorus	%		0.15	0.16	
Carotene	mg/kg		24.9	26.5	
Vitamin A equivalent	IU/g		41.5	44.2	

Timothy, hay, s-c, over ripe, (1)

Feed Name or Analyses			Mean As Fed	Mean Dry	C.V. ± %
Ref No 1 04 889				**United States**	
Dry matter	%		85.9	100.0	
Ash	%		4.6	5.4	
Crude fiber	%		27.6	32.2	
Cattle	dig coef %		55.	55.	
Sheep	dig coef %		45.	45.	
Ether extract	%		2.3	2.7	
Cattle	dig coef %		36.	36.	
Sheep	dig coef %		27.	27.	
N-free extract	%		46.1	53.7	
Cattle	dig coef %		59.	59.	
Sheep	dig coef %		56.	56.	
Protein (N x 6.25)	%		5.3	6.1	
Cattle	dig coef %		35.	35.	
Sheep	dig coef %		33.	33.	
Cattle	dig prot %		1.8	2.1	
Goats	dig prot %	*	1.9	2.3	
Horses	dig prot %	*	2.3	2.7	
Rabbits	dig coef %	*	2.9	3.4	
Sheep	dig prot %		1.7	2.0	
Lignin (Ellis)	%		11.5	13.4	
Energy	GE Mcal/kg				
Cattle	DE Mcal/kg	*	2.03	2.37	
Sheep	DE Mcal/kg	*	1.82	2.12	
Cattle	ME Mcal/kg	*	1.67	1.94	
Sheep	ME Mcal/kg	*	1.50	1.74	
Cattle	TDN %		46.1	53.7	
Sheep	TDN %		41.4	48.2	
Calcium	%		0.15	0.17	28
Chlorine	%		0.50	0.58	
Cobalt	mg/kg		0.112	0.130	
Iron	%		0.023	0.027	
Magnesium	%		0.06	0.07	
Manganese	mg/kg		88.9	103.6	
Phosphorus	%		0.14	0.16	28
Potassium	%		1.23	1.43	

Feed Name or Analyses			Mean As Fed	Mean Dry	C.V. ± %
Sodium	%		0.06	0.07	
Sulphur	%		0.14	0.16	
Carotene	mg/kg		2.9	3.4	55
Vitamin A equivalent	IU/g		4.9	5.7	

Timothy, hay, s-c, cut 1, (1)

Feed Name or Analyses			Mean As Fed	Mean Dry	C.V. ± %
Ref No 1 04 890				**United States**	
Dry matter	%		90.0	100.0	1
Sheep	dig coef %		63.	63.	
Ash	%		4.9	5.5	18
Crude fiber	%		30.7	34.1	9
Cattle	dig coef %	*	31.	31.	
Sheep	dig coef %	*	55.	55.	
Ether extract	%		2.7	3.0	10
Cattle	dig coef %	*	60.	60.	
Sheep	dig coef %	*	47.	47.	
N-free extract	%		41.8	46.5	
Cattle	dig coef %	*	59.	59.	
Sheep	dig coef %	*	61.	61.	
Protein (N x 6.25)	%		9.9	11.0	22
Cattle	dig coef %	*	63.	63.	
Sheep	dig coef %	*	47.	47.	
Cattle	dig prot %		6.3	6.9	
Goats	dig prot %	*	6.2	6.8	
Horses	dig prot %	*	6.2	6.9	
Rabbits	dig coef %	*	6.5	7.2	
Sheep	dig prot %		4.7	5.2	
Fiber, acid detergent (VS)	%		30.2	33.6	
Sugars, total	%		4.0	4.4	
Lignin (Ellis)	%		6.8	7.5	18
Energy	GE Mcal/kg		4.16	4.62	1
Cattle	DE Mcal/kg	*	1.94	2.16	
Sheep	DE Mcal/kg	*	2.20	2.44	
Cattle	ME Mcal/kg	*	1.59	1.77	
Sheep	ME Mcal/kg	*	1.80	2.00	
Cattle	TDN %		44.1	48.9	
Sheep	TDN %		49.9	55.4	
Calcium	%		0.36	0.40	24
Cobalt	mg/kg		0.073	0.082	62
Magnesium	%		0.14	0.16	18
Manganese	mg/kg		40.5	45.0	43
Phosphorus	%		0.21	0.23	29
Potassium	%		1.28	1.42	36
Sulphur	%		0.16	0.18	24
Carotene	mg/kg		9.7	10.8	
Vitamin A equivalent	IU/g		16.2	18.0	
Vitamin D_2	IU/g		2.0	2.3	16
Dry matter intake	g/$w^{0.75}$		72.415	80.450	
Gain, sheep	kg/day		0.004	0.004	

Timothy, hay, s-c, cut 2, (1)

Feed Name or Analyses			Mean As Fed	Mean Dry	C.V. ± %
Ref No 1 04 891				**United States**	
Dry matter	%		87.8	100.0	2
Sheep	dig coef %		61.	61.	
Ash	%		6.4	7.2	26
Crude fiber	%		27.0	30.7	13
Cattle	dig coef %	*	58.	58.	
Sheep	dig coef %	*	55.	55.	
Ether extract	%		3.6	4.1	30
Cattle	dig coef %	*	46.	46.	
Sheep	dig coef %	*	47.	47.	
N-free extract	%		36.1	41.1	
Cattle	dig coef %	*	57.	57.	
Sheep	dig coef %	*	61.	61.	

Continued

Feed Name or Analyses			Mean As Fed	Mean Dry	C.V. ± %
Protein (N x 6.25)	%		14.8	16.8	21
Cattle	dig coef %	*	46.	46.	
Sheep	dig coef %	*	47.	47.	
Cattle	dig prot %		6.8	7.7	
Goats	dig prot %	*	10.8	12.3	
Horses	dig prot %	*	10.4	11.8	
Rabbits	dig coef %	*	10.2	11.7	
Sheep	dig prot %		6.9	7.9	
Fiber, acid detergent (VS)	%		26.2	29.8	
Lignin (Ellis)	%		3.2	3.6	
Energy	GE Mcal/kg		4.14	4.71	
Sheep	GE dig coef %		59.	59.	
Cattle	DE Mcal/kg	*	2.06	2.35	
Sheep	DE Mcal/kg		2.44	2.78	
Cattle	ME Mcal/kg	*	1.69	1.92	
Sheep	ME Mcal/kg	*	2.00	2.28	
Cattle	TDN %		46.7	53.2	
Sheep	TDN %		47.6	54.2	
Calcium	%		0.60	0.68	
Manganese	mg/kg		41.0	46.7	
Phosphorus	%		0.24	0.28	
Sulphur	%		0.42	0.48	
Dry matter intake	$g/w^{.75}$		65.825	75.000	
Gain, sheep	kg/day		0.047	0.054	

Timothy, hay, s-c, cut 3, (1)
Ref No 1 04 892 United States

Feed Name or Analyses			Mean As Fed	Mean Dry	C.V. ± %
Dry matter	%		90.2	100.0	
Ash	%		4.4	4.9	17
Crude fiber	%		33.4	37.0	11
Ether extract	%		2.3	2.5	52
N-free extract	%		44.7	49.6	
Protein (N x 6.25)	%		5.4	6.0	16
Cattle	dig prot %	*	1.9	2.1	
Goats	dig prot %	*	1.9	2.2	
Horses	dig prot %	*	2.4	2.6	
Rabbits	dig coef %	*	3.0	3.3	
Sheep	dig prot %	*	1.8	2.0	
Energy	GE Mcal/kg				
Cattle	DE Mcal/kg	*	1.98	2.20	
Sheep	DE Mcal/kg	*	2.12	2.35	
Cattle	ME Mcal/kg	*	1.63	1.80	
Sheep	ME Mcal/kg	*	1.74	1.93	
Cattle	TDN %	*	45.0	49.8	
Sheep	TDN %	*	48.1	53.4	

Timothy, hay, s-c, good quality, (1)
Ref No 1 04 894 United States

Feed Name or Analyses			Mean As Fed	Mean Dry	C.V. ± %
Dry matter	%			100.0	
Chlorine	%			0.13	
Sodium	%			0.01	
Choline	mg/kg			811.	

Timothy, hay, s-c, gr 1 US, (1)
Ref No 1 04 896 United States

Feed Name or Analyses			Mean As Fed	Mean Dry	C.V. ± %
Dry matter	%		91.7	100.0	2
N-free extract	%		47.5	51.8	
Protein (N x 6.25)	%		6.7	7.3	16
Cattle	dig prot %	*	3.0	3.3	

(1) dry forages and roughages (2) pasture, range plants, and forages fed green (3) silages (4) energy feeds (5) protein supplements (6) minerals (7) vitamins (8) additives

Feed Name or Analyses			Mean As Fed	Mean Dry	C.V. ± %
Goats	dig prot %	*	3.1	3.4	
Horses	dig prot %	*	3.4	3.7	
Rabbits	dig coef %	*	3.9	4.3	
Sheep	dig prot %	*	2.9	3.1	
Lignin (Ellis)	%		9.0	9.8	
Calcium	%		0.37	0.40	27
Cobalt	mg/kg		0.065	0.071	41
Manganese	mg/kg		35.8	39.0	43
Phosphorus	%		0.19	0.21	42
Carotene	mg/kg		16.3	17.7	30
Vitamin A equivalent	IU/g		27.1	29.6	
Vitamin D_2	IU/g		2.1	2.3	5

Timothy, hay, s-c, gr 2 US, (1)
Ref No 1 04 897 United States

Feed Name or Analyses			Mean As Fed	Mean Dry	C.V. ± %
Dry matter	%			100.0	
Ash	%			3.8	
Crude fiber	%			38.3	
Ether extract	%			3.3	
N-free extract	%			47.9	
Protein (N x 6.25)	%			6.7	22
Cattle	dig prot %	*		2.7	
Goats	dig prot %	*		2.8	
Horses	dig prot %	*		3.2	
Rabbits	dig coef %	*		3.8	
Sheep	dig prot %	*		2.6	
Lignin (Ellis)	%			10.3	10
Energy	GE Mcal/kg				
Cattle	DE Mcal/kg	*		2.17	
Sheep	DE Mcal/kg	*		2.34	
Cattle	ME Mcal/kg	*		2.04	
Sheep	ME Mcal/kg	*		1.92	
Cattle	TDN %	*		49.3	
Sheep	TDN %	*		53.0	
Calcium	%			0.42	24
Cobalt	mg/kg			0.084	65
Manganese	mg/kg			45.0	53
Phosphorus	%			0.16	25
Carotene	mg/kg			10.0	31
Vitamin A equivalent	IU/g			16.7	

Timothy, hay, s-c, gr 3 US, (1)
Ref No 1 04 898 United States

Feed Name or Analyses			Mean As Fed	Mean Dry	C.V. ± %
Dry matter	%			100.0	
Ash	%			2.8	
Crude fiber	%			36.7	
Ether extract	%			3.1	
N-free extract	%			50.8	
Protein (N x 6.25)	%			6.6	14
Cattle	dig prot %	*		2.7	
Goats	dig prot %	*		2.7	
Horses	dig prot %	*		3.1	
Rabbits	dig coef %	*		3.8	
Sheep	dig prot %	*		2.5	
Lignin (Ellis)	%			10.1	13
Energy	GE Mcal/kg				
Cattle	DE Mcal/kg	*		2.15	
Sheep	DE Mcal/kg	*		2.38	
Cattle	ME Mcal/kg	*		1.76	
Sheep	ME Mcal/kg	*		1.95	
Cattle	TDN %	*		48.8	
Sheep	TDN %	*		54.0	
Calcium	%			0.41	29
Cobalt	mg/kg			0.079	65
Manganese	mg/kg			45.9	47
Phosphorus	%			0.16	15

Feed Name or Analyses			Mean As Fed	Mean Dry	C.V. ± %
Carotene	mg/kg			6.6	
Niacin	mg/kg			20.1	14
Vitamin A equivalent	IU/g			11.0	
Vitamin D_2	IU/g			2.1	12

Timothy, hay, s-c, gr sample US, (1)
Ref No 1 04 895 United States

Feed Name or Analyses			Mean As Fed	Mean Dry	C.V. ± %
Dry matter	%			100.0	
N-free extract	%			51.8	
Protein (N x 6.25)	%			6.7	19
Cattle	dig prot %	*		2.7	
Goats	dig prot %	*		2.8	
Horses	dig prot %	*		3.2	
Rabbits	dig coef %	*		3.8	
Sheep	dig prot %	*		2.5	
Lignin (Ellis)	%			12.4	13
Calcium	%			0.40	26
Cobalt	mg/kg			0.086	40
Manganese	mg/kg			52.5	51
Phosphorus	%			0.14	13
Carotene	mg/kg			7.8	77
Vitamin A equivalent	IU/g			13.0	
Vitamin D_2	IU/g			2.8	5

Timothy, hay, s-c grnd, (1)
Ref No 1 04 900 United States

Feed Name or Analyses			Mean As Fed	Mean Dry	C.V. ± %
Dry matter	%		90.7	100.0	
Ash	%		4.4	4.9	
Crude fiber	%		28.7	31.7	
Cattle	dig coef %		56.	56.	
Sheep	dig coef %		56.	56.	
Ether extract	%		2.3	2.6	
Cattle	dig coef %		31.	31.	
Sheep	dig coef %		32.	32.	
N-free extract	%		49.3	54.4	
Cattle	dig coef %		57.	57.	
Sheep	dig coef %		59.	59.	
Protein (N x 6.25)	%		5.9	6.5	
Cattle	dig coef %		42.	42.	
Sheep	dig coef %		43.	43.	
Cattle	dig prot %		2.5	2.7	
Goats	dig prot %	*	2.4	2.6	
Horses	dig prot %	*	2.8	3.0	
Rabbits	dig coef %	*	3.3	3.7	
Sheep	dig prot %		2.5	2.8	
Energy	GE Mcal/kg				
Cattle	DE Mcal/kg	*	2.13	2.35	
Sheep	DE Mcal/kg	*	2.18	2.40	
Cattle	ME Mcal/kg	*	1.75	1.93	
Sheep	ME Mcal/kg	*	1.79	1.97	
Cattle	TDN %		48.3	53.3	
Sheep	TDN %		49.4	54.5	

Timothy, hay, s-c grnd pelleted, (1)
Ref No 1 07 717 United States

Feed Name or Analyses			Mean As Fed	Mean Dry	C.V. ± %
Dry matter	%		93.1	100.0	
Sheep	dig coef %		47.	47.	
Organic matter	%		85.7	92.1	
Ash	%		7.4	7.9	
Crude fiber	%		24.4	26.2	
Sheep	dig coef %		23.	23.	
Ether extract	%		2.2	2.4	
Sheep	dig coef %		44.	44.	

Feed Name or Analyses				Mean As Fed	Mean Dry	C.V. ± %
N-free extract		%		50.7	54.5	
Sheep	dig coef	%		60.	60.	
Protein (N x 6.25)		%		8.4	9.0	
Sheep	dig coef	%		53.	53.	
Cattle	dig prot	%	*	4.4	4.7	
Goats	dig prot	%	*	4.6	5.0	
Horses	dig prot	%	*	4.8	5.2	
Rabbits	dig coef	%	*	5.2	5.6	
Sheep	dig prot	%		4.4	4.8	
Energy		GE Mcal/kg				
Cattle		DE Mcal/kg	*	2.29	2.46	
Sheep		DE Mcal/kg	*	1.88	2.02	
Cattle		ME Mcal/kg	*	2.15	2.31	
Sheep		ME Mcal/kg	*	1.54	1.66	
Cattle		TDN %	*	52.0	55.9	
Sheep		TDN %		42.7	45.9	

Timothy, aerial part, fresh, (2)

Ref No 2 04 912 United States

Feed Name or Analyses				Mean As Fed	Mean Dry	C.V. ± %
Dry matter		%		27.6	100.0	6
Organic matter		%		25.7	93.3	
Ash		%		2.2	8.0	
Crude fiber		%		7.4	26.8	21
Cattle	dig coef	%		68.	68.	
Sheep	dig coef	%		68.	68.	
Ether extract		%		1.2	4.2	25
Cattle	dig coef	%		57.	57.	
Sheep	dig coef	%		56.	56.	
N-free extract		%		13.3	48.2	6
Cattle	dig coef	%		72.	72.	
Sheep	dig coef	%		70.	70.	
Protein (N x 6.25)		%		3.5	12.8	42
Cattle	dig coef	%		59.	59.	
Sheep	dig coef	%		60.	60.	
Cattle	dig prot	%		2.1	7.5	
Goats	dig prot	%	*	2.3	8.5	
Horses	dig prot	%	*	2.3	8.4	
Rabbits	dig coef	%	*	2.4	8.5	
Sheep	dig prot	%		2.1	7.7	
Lignin (Ellis)		%		1.3	4.6	
Energy		GE Mcal/kg				
Cattle		DE Mcal/kg	*	0.80	2.91	
Sheep		DE Mcal/kg	*	0.79	2.87	
Cattle		ME Mcal/kg	*	0.66	2.38	
Sheep		ME Mcal/kg	*	0.65	2.35	
Cattle		TDN %		18.2	65.9	
Sheep		TDN %		17.9	65.0	
Calcium		%		0.16	0.59	
Chlorine		%		0.14	0.51	36
Cobalt		mg/kg		0.011	0.040	27
Magnesium		%		0.07	0.25	
Phosphorus		%		0.10	0.38	
Potassium		%		0.58	2.09	
Sodium		%		0.03	0.11	56
Sulphur		%		0.04	0.13	19
Carotene		mg/kg		61.8	224.0	37
Riboflavin		mg/kg		3.2	11.5	45
Thiamine		mg/kg		0.8	2.9	30
α-tocopherol		mg/kg		42.4	153.9	47
Vitamin A equivalent		IU/g		103.0	373.4	

Timothy, aerial part, fresh, immature, (2)

Ref No 2 04 902 United States

Feed Name or Analyses				Mean As Fed	Mean Dry	C.V. ± %
Dry matter		%		25.5	100.0	24
Crude fiber		%				
Cattle	dig coef	%	*	68.	68.	
Sheep	dig coef	%	*	51.	51.	
Ether extract		%				
Cattle	dig coef	%	*	57.	57.	
Sheep	dig coef	%	*	59.	59.	
N-free extract		%				
Cattle	dig coef	%	*	72.	72.	
Sheep	dig coef	%	*	57.	57.	
Protein (N x 6.25)		%				
Cattle	dig coef	%	*	59.	59.	
Sheep	dig coef	%	*	39.	39.	
Carotene		mg/kg		59.6	233.9	33
Riboflavin		mg/kg		4.8	19.0	
α-tocopherol		mg/kg		54.2	212.7	6
Vitamin A equivalent		IU/g		99.4	389.9	

Timothy, aerial part, fresh, immature, pure stand, (2)

Ref No 2 04 901 United States

Feed Name or Analyses				Mean As Fed	Mean Dry	C.V. ± %
Dry matter		%		26.1	100.0	2
Ash		%		2.6	9.9	16
Crude fiber		%		5.9	22.7	3
Ether extract		%		0.9	3.6	7
N-free extract		%		12.6	48.1	
Protein (N x 6.25)		%		4.1	15.7	11
Cattle	dig prot	%	*	2.9	11.2	
Goats	dig prot	%	*	2.9	11.2	
Horses	dig prot	%	*	2.8	10.9	
Rabbits	dig coef	%	*	2.8	10.8	
Sheep	dig prot	%	*	3.0	11.6	
Energy		GE Mcal/kg				
Cattle		DE Mcal/kg	*	0.76	2.92	
Sheep		DE Mcal/kg	*	0.74	2.83	
Cattle		ME Mcal/kg	*	0.62	2.39	
Sheep		ME Mcal/kg	*	0.61	2.32	
Cattle		TDN %	*	17.3	66.2	
Sheep		TDN %	*	16.8	64.3	
Calcium		%		0.12	0.47	
Magnesium		%		0.05	0.18	
Phosphorus		%		0.11	0.43	
Potassium		%		0.66	2.54	

Timothy, aerial part, fresh, pre-bloom, (2)

Ref No 2 04 903 United States

Feed Name or Analyses				Mean As Fed	Mean Dry	C.V. ± %
Dry matter		%		26.5	100.0	
Ash		%		1.8	6.6	
Crude fiber		%		8.5	32.1	
Sheep	dig coef	%		59.	59.	
Ether extract		%		1.0	3.8	
Sheep	dig coef	%		65.	65.	
N-free extract		%		12.6	47.5	
Sheep	dig coef	%		59.	59.	
Protein (N x 6.25)		%		2.6	9.9	
Sheep	dig coef	%		54.	54.	
Cattle	dig prot	%	*	1.7	6.3	
Goats	dig prot	%	*	1.5	5.8	
Horses	dig prot	%	*	1.6	5.9	
Rabbits	dig prot	%	*	1.7	6.3	
Sheep	dig prot	%		1.4	5.4	
Energy		GE Mcal/kg				
Cattle		DE Mcal/kg	*	0.71	2.68	
Sheep		DE Mcal/kg	*	0.68	2.55	
Cattle		ME Mcal/kg	*	0.58	2.20	
Sheep		ME Mcal/kg	*	0.55	2.09	
Cattle		NE_m Mcal/kg	*	0.39	1.47	
Cattle		NE_{gain} Mcal/kg	*	0.24	0.90	
Cattle		$NE_{lactating\ cows}$ Mcal/kg	*	0.45	1.71	
Cattle		TDN %	*	16.1	60.8	
Sheep		TDN %		15.3	57.9	
Calcium		%		0.08	0.28	
Phosphorus		%		0.08	0.28	

Timothy, aerial part, fresh, early bloom, (2)

Ref No 2 04 904 United States

Feed Name or Analyses				Mean As Fed	Mean Dry	C.V. ± %
Dry matter		%		27.9	100.0	17
Crude fiber		%				
Cattle	dig coef	%	*	68.	68.	
Sheep	dig coef	%	*	51.	51.	
Ether extract		%				
Cattle	dig coef	%	*	57.	57.	
Sheep	dig coef	%	*	59.	59.	
N-free extract		%				
Cattle	dig coef	%	*	72.	72.	
Sheep	dig coef	%	*	57.	57.	
Protein (N x 6.25)		%				
Cattle	dig coef	%	*	59.	59.	
Sheep	dig coef	%	*	39.	39.	
Calcium		%		0.14	0.50	10
Magnesium		%		0.04	0.15	16
Phosphorus		%		0.10	0.35	12
Potassium		%		0.67	2.40	7
Carotene		mg/kg		54.1	194.0	
α-tocopherol		mg/kg		48.0	172.0	
Vitamin A equivalent		IU/g		90.2	323.4	

Timothy, aerial part, fresh, mid-bloom, (2)

Ref No 2 04 905 United States

Feed Name or Analyses				Mean As Fed	Mean Dry	C.V. ± %
Dry matter		%		29.8	100.0	
Ash		%		2.0	6.6	2
Crude fiber		%		10.0	33.5	5
Cattle	dig coef	%	*	68.	68.	
Sheep	dig coef	%	*	51.	51.	
Ether extract		%		0.9	3.0	5
Cattle	dig coef	%	*	57.	57.	
Sheep	dig coef	%	*	59.	59.	
N-free extract		%		14.3	47.9	
Cattle	dig coef	%	*	72.	72.	
Sheep	dig coef	%	*	57.	57.	
Protein (N x 6.25)		%		2.7	9.1	26
Cattle	dig coef	%	*	59.	59.	
Sheep	dig coef	%	*	39.	39.	
Cattle	dig prot	%		1.6	5.4	
Goats	dig prot	%	*	1.5	5.0	
Horses	dig prot	%	*	1.6	5.2	
Rabbits	dig prot	%	*	1.7	5.7	
Sheep	dig prot	%		1.1	3.5	
Energy		GE Mcal/kg				
Cattle		DE Mcal/kg	*	0.87	2.93	
Sheep		DE Mcal/kg	*	0.68	2.29	
Cattle		ME Mcal/kg	*	0.72	2.40	
Sheep		ME Mcal/kg	*	0.56	1.88	
Cattle		NE_m Mcal/kg	*	0.43	1.43	
Cattle		NE_{gain} Mcal/kg	*	0.25	0.85	
Cattle		$NE_{lactating\ cows}$ Mcal/kg	*	0.50	1.67	
Cattle		TDN %		19.8	66.4	
Sheep		TDN %		15.5	51.9	
Calcium		%		0.08	0.25	
Chlorine		%		0.19	0.63	
Copper		mg/kg		3.3	11.2	
Iron		%		0.005	0.016	
Magnesium		%		0.04	0.13	
Manganese		mg/kg		57.4	192.5	

Continued

Feed Name or Analyses			Mean As Fed	Mean Dry	C.V. ± %
Phosphorus	%		0.08	0.25	
Potassium	%		0.51	1.71	
Sodium	%		0.06	0.19	
Sulphur	%		0.04	0.13	

Timothy, aerial part, fresh, full bloom, (2)
Ref No 2 04 906 United States

Feed Name or Analyses			Mean As Fed	Mean Dry	C.V. ± %
Dry matter	%		31.8	100.0	8
Crude fiber	%				
Cattle	dig coef %	*	68.	68.	
Sheep	dig coef %	*	51.	51.	
Ether extract	%				
Cattle	dig coef %	*	57.	57.	
Sheep	dig coef %	*	59.	59.	
N-free extract	%				
Cattle	dig coef %	*	72.	72.	
Sheep	dig coef %	*	57.	57.	
Protein (N x 6.25)	%				
Cattle	dig coef %	*	59.	59.	
Sheep	dig coef %	*	39.	39.	
Calcium	%		0.08	0.25	17
Chlorine	%		0.19	0.61	15
Copper	mg/kg		3.5	11.0	46
Iron	%		0.005	0.016	46
Magnesium	%		0.04	0.13	21
Manganese	mg/kg		55.6	174.8	25
Phosphorus	%		0.07	0.21	22
Potassium	%		0.53	1.68	5
Sodium	%		0.06	0.19	11
Sulphur	%		0.04	0.13	18
Carotene	mg/kg		33.1	104.1	
α-tocopherol	mg/kg		22.7	71.4	
Vitamin A equivalent	IU/g		55.2	173.5	

Timothy, aerial part, fresh, late bloom, (2)
Ref No 2 04 907 United States

Feed Name or Analyses			Mean As Fed	Mean Dry	C.V. ± %
Dry matter	%		27.7	100.0	
Ash	%		1.6	5.9	
Crude fiber	%		9.8	35.3	
Sheep	dig coef %		54.	54.	
Ether extract	%		0.9	3.3	
Sheep	dig coef %		48.	48.	
N-free extract	%		13.9	50.0	
Sheep	dig coef %		58.	58.	
Protein (N x 6.25)	%		1.5	5.5	
Sheep	dig coef %		40.	40.	
Cattle	dig prot %	*	0.7	2.6	
Goats	dig prot %	*	0.5	1.7	
Horses	dig prot %	*	0.6	2.2	
Rabbits	dig prot %	*	0.8	2.9	
Sheep	dig prot %		0.6	2.2	
Energy	GE Mcal/kg				
Cattle	DE Mcal/kg	*	0.69	2.49	
Sheep	DE Mcal/kg	*	0.66	2.37	
Cattle	ME Mcal/kg	*	0.57	2.04	
Sheep	ME Mcal/kg	*	0.54	1.95	
Cattle	TDN %	*	15.6	56.4	
Sheep	TDN %		14.9	53.8	

(1) dry forages and roughages (2) pasture, range plants, and forages fed green (3) silages (4) energy feeds (5) protein supplements (6) minerals (7) vitamins (8) additives

Timothy, aerial part, fresh, milk stage, (2)
Ref No 2 04 908 United States

Feed Name or Analyses			Mean As Fed	Mean Dry	C.V. ± %
Dry matter	%		28.3	100.0	
Ash	%		1.4	5.1	
Crude fiber	%		9.7	34.3	
Sheep	dig coef %		39.	39.	
Ether extract	%		1.0	3.5	
Sheep	dig coef %		65.	65.	
N-free extract	%		14.9	52.7	
Sheep	dig coef %		53.	53.	
Protein (N x 6.25)	%		1.2	4.4	
Sheep	dig coef %		23.	23.	
Cattle	dig prot %	*	0.5	1.6	
Goats	dig prot %	*	0.2	0.7	
Horses	dig prot %	*	0.4	1.3	
Rabbits	dig prot %	*	0.6	2.1	
Sheep	dig prot %		0.3	1.0	
Energy	GE Mcal/kg				
Cattle	DE Mcal/kg	*	0.66	2.34	
Sheep	DE Mcal/kg	*	0.59	2.09	
Cattle	ME Mcal/kg	*	0.54	1.92	
Sheep	ME Mcal/kg	*	0.49	1.72	
Cattle	TDN %	*	15.0	53.1	
Sheep	TDN %		13.4	47.4	

Timothy, aerial part, fresh, dough stage, (2)
Ref No 2 04 909 United States

Feed Name or Analyses			Mean As Fed	Mean Dry	C.V. ± %
Dry matter	%		46.4	100.0	5
Ash	%		2.3	5.0	6
Crude fiber	%		15.3	33.0	4
Ether extract	%		1.3	2.8	6
N-free extract	%		24.4	52.5	
Protein (N x 6.25)	%		3.1	6.7	4
Cattle	dig prot %	*	1.7	3.6	
Goats	dig prot %	*	1.3	2.8	
Horses	dig prot %	*	1.5	3.2	
Rabbits	dig prot %	*	1.8	3.8	
Sheep	dig prot %	*	1.5	3.2	
Energy	GE Mcal/kg				
Cattle	DE Mcal/kg	*	1.12	2.41	
Sheep	DE Mcal/kg	*	1.09	2.34	
Cattle	ME Mcal/kg	*	0.92	1.98	
Sheep	ME Mcal/kg	*	0.89	1.92	
Cattle	TDN %	*	25.3	54.6	
Sheep	TDN %	*	24.0	53.1	
Carotene	mg/kg		34.8	75.0	
α-tocopherol	mg/kg		22.3	48.1	
Vitamin A equivalent	IU/g		58.0	125.0	

Timothy, aerial part, fresh, mature, (2)
Ref No 2 04 910 United States

Feed Name or Analyses			Mean As Fed	Mean Dry	C.V. ± %
Dry matter	%		35.8	100.0	
Ash	%		1.7	4.7	11
Crude fiber	%		12.0	33.4	2
Cattle	dig coef %	*	68.	68.	
Sheep	dig coef %	*	51.	51.	
Ether extract	%		1.0	2.9	6
Cattle	dig coef %	*	57.	57.	
Sheep	dig coef %	*	59.	59.	
N-free extract	%		18.9	52.9	
Cattle	dig coef %	*	72.	72.	
Sheep	dig coef %	*	57.	57.	
Protein (N x 6.25)	%		2.2	6.1	19
Cattle	dig coef %	*	59.	59.	
Sheep	dig coef %	*	39.	39.	
Cattle	dig prot %		1.3	3.6	
Goats	dig prot %	*	0.8	2.3	
Horses	dig prot %	*	1.0	2.7	
Rabbits	dig prot %	*	1.2	3.4	
Sheep	dig prot %		0.9	2.4	
Energy	GE Mcal/kg				
Cattle	DE Mcal/kg	*	1.07	3.00	
Sheep	DE Mcal/kg	*	0.84	2.35	
Cattle	ME Mcal/kg	*	0.88	2.46	
Sheep	ME Mcal/kg	*	0.69	1.93	
Cattle	TDN %		24.4	68.1	
Sheep	TDN %		19.1	53.4	
Calcium	%		0.06	0.16	30
Chlorine	%		0.19	0.52	
Iron	%		0.009	0.026	24
Magnesium	%		0.02	0.06	34
Manganese	mg/kg		37.3	104.1	13
Phosphorus	%		0.07	0.18	14
Potassium	%		0.56	1.57	6
Sodium	%		0.02	0.06	
Sulphur	%		0.05	0.13	

Timothy, aerial part, fresh, over ripe, (2)
Ref No 2 04 911 United States

Feed Name or Analyses			Mean As Fed	Mean Dry	C.V. ± %
Dry matter	%			100.0	
α-tocopherol	mg/kg			9.0	

Timothy, heads, fresh, (2)
Ref No 2 04 913 United States

Feed Name or Analyses			Mean As Fed	Mean Dry	C.V. ± %
Dry matter	%		32.9	100.0	
Protein (N x 6.25)	%		4.6	14.0	
Cattle	dig prot %	*	3.2	9.8	
Goats	dig prot %	*	3.2	9.6	
Horses	dig prot %	*	3.1	9.4	
Rabbits	dig prot %	*	3.1	9.5	
Sheep	dig prot %	*	3.3	10.0	

Timothy, leaves, fresh, (2)
Ref No 2 04 914 United States

Feed Name or Analyses			Mean As Fed	Mean Dry	C.V. ± %
Dry matter	%		32.3	100.0	
Protein (N x 6.25)	%		5.5	16.9	
Cattle	dig prot %	*	4.0	12.3	
Goats	dig prot %	*	4.0	12.3	
Horses	dig prot %	*	3.8	11.9	
Rabbits	dig prot %	*	3.8	11.7	
Sheep	dig prot %	*	4.1	12.7	

Timothy, stems, fresh, (2)
Ref No 2 04 915 United States

Feed Name or Analyses			Mean As Fed	Mean Dry	C.V. ± %
Dry matter	%		29.3	100.0	7
Protein (N x 6.25)	%		1.6	5.6	33
Cattle	dig prot %	*	0.8	2.7	
Goats	dig prot %	*	0.5	1.8	
Horses	dig prot %	*	0.7	2.3	
Rabbits	dig prot %	*	0.9	3.0	
Sheep	dig prot %	*	0.6	2.2	

Timothy, aerial part, ensiled, (3)

Ref No 3 04 922 United States

Feed Name or Analyses			Mean As Fed	Mean Dry	C.V. ± %
Dry matter	%		34.0	100.0	11
Ash	%		2.4	7.1	10
Crude fiber	%		12.0	35.2	7
Cattle	dig coef %		66.	66.	
Ether extract	%		1.2	3.4	11
Cattle	dig coef %		59.	59.	
N-free extract	%		14.9	43.8	
Cattle	dig coef %		59.	59.	
Protein (N x 6.25)	%		3.6	10.6	9
Cattle	dig coef %		56.	56.	
Cattle	dig prot %		2.0	5.9	
Goats	dig prot %	*	2.0	5.8	
Horses	dig prot %	*	2.0	5.8	
Sheep	dig prot %	*	2.0	5.8	
Energy	GE Mcal/kg				
Cattle	DE Mcal/kg	*	0.89	2.62	
Sheep	DE Mcal/kg	*	0.95	2.79	
Cattle	ME Mcal/kg	*	0.73	2.15	
Sheep	ME Mcal/kg	*	0.78	2.29	
Cattle	NE_m Mcal/kg	*	0.43	1.26	
Cattle	NE_{gain} Mcal/kg	*	0.21	0.62	
Cattle	$NE_{lactating\ cows}$ Mcal/kg	*	0.48	1.41	
Cattle	TDN %		20.2	59.5	
Sheep	TDN %	*	21.5	63.3	
Calcium	%		0.20	0.58	
Phosphorus	%		0.10	0.29	
Potassium	%		0.57	1.68	
Carotene	mg/kg		26.8	78.9	30
Vitamin A equivalent	IU/g		44.7	131.6	

Timothy, aerial part, ensiled, immature, (3)

Ref No 3 04 916 United States

Feed Name or Analyses			Mean As Fed	Mean Dry	C.V. ± %
Dry matter	%		29.7	100.0	25
Ash	%		2.3	7.8	6
Crude fiber	%		9.9	33.5	3
Cattle	dig coef %	*	66.	66.	
Ether extract	%		1.1	3.8	1
Cattle	dig coef %	*	59.	59.	
N-free extract	%		12.9	43.4	
Cattle	dig coef %	*	59.	59.	
Protein (N x 6.25)	%		3.5	11.7	12
Cattle	dig coef %	*	56.	56.	
Cattle	dig prot %		1.9	6.5	
Goats	dig prot %	*	2.0	6.8	
Horses	dig prot %	*	2.0	6.8	
Sheep	dig prot %	*	2.0	6.8	
Energy	GE Mcal/kg				
Cattle	DE Mcal/kg	*	0.78	2.61	
Sheep	DE Mcal/kg	*	0.82	2.77	
Cattle	ME Mcal/kg	*	0.64	2.14	
Sheep	ME Mcal/kg	*	0.68	2.27	
Cattle	TDN %		17.6	59.2	
Sheep	TDN %	*	18.7	62.9	
Calcium	%		0.16	0.54	11
Phosphorus	%		0.10	0.32	10

Timothy, aerial part, ensiled, pre-bloom, (3)

Ref No 3 04 917 United States

Feed Name or Analyses			Mean As Fed	Mean Dry	C.V. ± %
Dry matter	%		26.9	100.0	
Ash	%		2.1	7.9	
Crude fiber	%		9.1	33.8	
Cattle	dig coef %		70.	70.	
Ether extract	%		0.9	3.5	
Cattle	dig coef %		67.	67.	
N-free extract	%		11.5	42.8	
Cattle	dig coef %		63.	63.	
Protein (N x 6.25)	%		3.2	12.0	
Cattle	dig coef %		60.	60.	
Cattle	dig prot %		1.9	7.2	
Goats	dig prot %	*	1.9	7.1	
Horses	dig prot %	*	1.9	7.1	
Sheep	dig prot %	*	1.9	7.1	
Energy	GE Mcal/kg				
Cattle	DE Mcal/kg	*	0.75	2.78	
Sheep	DE Mcal/kg	*	0.75	2.77	
Cattle	ME Mcal/kg	*	0.61	2.28	
Sheep	ME Mcal/kg	*	0.61	2.27	
Cattle	TDN %		17.0	63.1	
Sheep	TDN %	*	16.9	62.8	

Timothy, aerial part, ensiled, early bloom, (3)

Ref No 3 04 918 United States

Feed Name or Analyses			Mean As Fed	Mean Dry	C.V. ± %
Dry matter	%		36.3	100.0	22
Ash	%		2.5	6.8	9
Crude fiber	%		12.8	35.3	4
Cattle	dig coef %		64.	64.	
Ether extract	%		1.2	3.2	13
Cattle	dig coef %		50.	50.	
N-free extract	%		16.2	44.6	
Cattle	dig coef %		54.	54.	
Protein (N x 6.25)	%		3.7	10.2	13
Cattle	dig coef %		50.	50.	
Cattle	dig prot %		1.9	5.1	
Goats	dig prot %	*	2.0	5.5	
Horses	dig prot %	*	2.0	5.5	
Sheep	dig prot %	*	2.0	5.5	
Energy	GE Mcal/kg				
Cattle	DE Mcal/kg	*	0.89	2.44	
Sheep	DE Mcal/kg	*	0.91	2.52	
Cattle	ME Mcal/kg	*	0.73	2.00	
Sheep	ME Mcal/kg	*	0.75	2.07	
Cattle	TDN %		20.1	55.3	
Sheep	TDN %	*	20.7	57.1	

Timothy, aerial part, ensiled, mid-bloom, cut 2, (3)

Ref No 3 07 917 Canada

Feed Name or Analyses			Mean As Fed	Mean Dry	C.V. ± %
Dry matter	%		18.4	100.0	
Sheep	dig coef %		56.	56.	
Ash	%		1.4	7.8	
Sheep	dig coef %		6.	34.	
pH	pH units		4.22	4.22	

Timothy, aerial part, ensiled, full bloom, (3)

Ref No 3 04 920 United States

Feed Name or Analyses			Mean As Fed	Mean Dry	C.V. ± %
Dry matter	%		35.7	100.0	9
Ash	%		2.5	6.9	7
Crude fiber	%		12.9	36.3	5
Cattle	dig coef %		63.	63.	
Ether extract	%		1.2	3.2	9
Cattle	dig coef %		60.	60.	
N-free extract	%		15.6	43.8	
Cattle	dig coef %		60.	60.	
Protein (N x 6.25)	%		3.5	9.7	9
Cattle	dig coef %		57.	57.	
Cattle	dig prot %		2.0	5.5	
Goats	dig prot %	*	1.8	5.1	
Horses	dig prot %	*	1.8	5.1	
Sheep	dig prot %	*	1.8	5.1	
Energy	GE Mcal/kg				
Cattle	DE Mcal/kg	*	0.93	2.60	
Sheep	DE Mcal/kg	*	1.00	2.80	
Cattle	ME Mcal/kg	*	0.76	2.14	
Sheep	ME Mcal/kg	*	0.82	2.29	
Cattle	TDN %		21.1	59.1	
Sheep	TDN %	*	22.6	63.5	
Calcium	%		0.19	0.54	16
Phosphorus	%		0.10	0.29	11

Timothy, aerial part, ensiled, milk stage, (3)

Ref No 3 04 921 United States

Feed Name or Analyses			Mean As Fed	Mean Dry	C.V. ± %
Dry matter	%		41.5	100.0	7
Ash	%		2.7	6.6	7
Crude fiber	%		15.9	38.4	3
Ether extract	%		1.3	3.1	7
N-free extract	%		18.1	43.5	
Protein (N x 6.25)	%		3.5	8.4	5
Cattle	dig prot %	*	1.6	3.9	
Goats	dig prot %	*	1.6	3.9	
Horses	dig prot %	*	1.6	3.9	
Sheep	dig prot %	*	1.6	3.9	
Energy	GE Mcal/kg				
Cattle	DE Mcal/kg	*	1.07	2.59	
Sheep	DE Mcal/kg	*	1.10	2.66	
Cattle	ME Mcal/kg	*	0.88	2.12	
Sheep	ME Mcal/kg	*	0.90	2.18	
Cattle	TDN %	*	24.4	58.8	
Sheep	TDN %	*	25.1	60.4	

Timothy, aerial part, wilted ensiled, (3)

Ref No 3 04 930 United States

Feed Name or Analyses			Mean As Fed	Mean Dry	C.V. ± %
Dry matter	%		41.7	100.0	1
Ash	%		2.9	7.0	9
Crude fiber	%		14.3	34.2	5
Cattle	dig coef %		64.	64.	
Ether extract	%		1.3	3.2	9
Cattle	dig coef %		58.	58.	
N-free extract	%		18.8	45.0	
Cattle	dig coef %		61.	61.	
Protein (N x 6.25)	%		4.4	10.7	10
Cattle	dig coef %		51.	51.	
Cattle	dig prot %		2.3	5.4	
Goats	dig prot %	*	2.5	5.9	
Horses	dig prot %	*	2.5	5.9	
Sheep	dig prot %	*	2.5	5.9	
Energy	GE Mcal/kg				
Cattle	DE Mcal/kg	*	1.08	2.60	
Sheep	DE Mcal/kg	*	1.06	2.54	
Cattle	ME Mcal/kg	*	0.89	2.13	
Sheep	ME Mcal/kg	*	0.87	2.08	
Cattle	TDN %		24.6	59.0	
Sheep	TDN %	*	24.0	57.6	
Calcium	%		0.23	0.56	9
Chlorine	%		0.28	0.66	
Copper	mg/kg		2.3	5.5	17
Iron	%		0.005	0.012	17
Magnesium	%		0.06	0.15	23
Manganese	mg/kg		37.6	90.2	19
Phosphorus	%		0.12	0.29	17
Potassium	%		0.71	1.69	8
Sodium	%		0.08	0.20	
Sulphur	%		0.06	0.15	20

Feed Name or Analyses			Mean As Fed	Mean Dry	C.V. ± %

Timothy, aerial part, wilted ensiled, immature, (3)

Ref No 3 04 923 United States

Feed Name or Analyses			As Fed	Dry	C.V. ± %
Dry matter	%		43.0	100.0	10
Ash	%		3.2	7.4	5
Crude fiber	%		13.5	31.4	5
Ether extract	%		1.6	3.7	7
N-free extract	%		19.4	45.1	
Protein (N x 6.25)	%		5.3	12.4	8
Cattle	dig prot %	*	3.2	7.5	
Goats	dig prot %	*	3.2	7.5	
Horses	dig prot %	*	3.2	7.5	
Sheep	dig prot %	*	3.2	7.5	
Energy	GE Mcal/kg				
Cattle	DE Mcal/kg	*	1.14	2.66	
Sheep	DE Mcal/kg	*	1.20	2.80	
Cattle	ME Mcal/kg	*	0.94	2.18	
Sheep	ME Mcal/kg	*	0.99	2.29	
Cattle	TDN %	*	25.9	60.3	
Sheep	TDN %	*	27.3	63.5	

Timothy, aerial part, wilted ensiled, pre-bloom, (3)

Ref No 3 04 924 United States

Feed Name or Analyses			As Fed	Dry	C.V. ± %
Dry matter	%		44.0	100.0	
Ash	%		3.1	7.0	
Crude fiber	%		14.3	32.6	
Cattle	dig coef %		71.	71.	
Ether extract	%		1.7	3.8	
Cattle	dig coef %		58.	58.	
N-free extract	%		19.4	44.0	
Cattle	dig coef %		66.	66.	
Protein (N x 6.25)	%		5.5	12.6	
Cattle	dig coef %		55.	55.	
Cattle	dig prot %		3.0	6.9	
Goats	dig prot %	*	3.4	7.7	
Horses	dig prot %	*	3.4	7.7	
Sheep	dig prot %	*	3.4	7.7	
Energy	GE Mcal/kg				
Cattle	DE Mcal/kg	*	1.24	2.83	
Sheep	DE Mcal/kg	*	1.23	2.79	
Cattle	ME Mcal/kg	*	1.02	2.32	
Sheep	ME Mcal/kg	*	1.01	2.29	
Cattle	TDN %		28.2	64.1	
Sheep	TDN %	*	27.9	63.3	

Timothy, aerial part, wilted ensiled, early bloom, (3)

Ref No 3 04 925 United States

Feed Name or Analyses			As Fed	Dry	C.V. ± %
Dry matter	%		39.4	100.0	17
Ash	%		2.8	7.0	10
Crude fiber	%		13.7	34.8	5
Cattle	dig coef %		66.	66.	
Ether extract	%		1.2	3.1	9
Cattle	dig coef %		53.	53.	
N-free extract	%		18.0	45.7	
Cattle	dig coef %		58.	58.	
Protein (N x 6.25)	%		3.7	9.5	13
Cattle	dig coef %		47.	47.	
Cattle	dig prot %		1.8	4.5	
Goats	dig prot %	*	1.9	4.9	
Horses	dig prot %	*	1.9	4.9	
Sheep	dig prot %	*	1.9	4.9	
Energy	GE Mcal/kg				
Cattle	DE Mcal/kg	*	1.00	2.54	
Sheep	DE Mcal/kg	*	1.06	2.69	
Cattle	ME Mcal/kg	*	0.82	2.08	
Sheep	ME Mcal/kg	*	0.87	2.21	
Cattle	TDN %		22.7	57.6	
Sheep	TDN %	*	24.0	61.0	

(1) dry forages and roughages (2) pasture, range plants, and forages fed green (3) silages (4) energy feeds (5) protein supplements (6) minerals (7) vitamins (8) additives

Timothy, aerial part, wilted ensiled, early bloom, cut 1, (3)

Ref No 3 07 918 Canada

Feed Name or Analyses			As Fed	Dry	C.V. ± %
Dry matter	%		37.8	100.0	
Sheep	dig coef %		59.	59.	
Ash	%		2.8	7.5	
Sheep	dig coef %		17.	46.	
pH	pH units		4.93	4.93	

Timothy, aerial part, wilted ensiled, mid-bloom, (3)

Ref No 3 04 926 United States

Feed Name or Analyses			As Fed	Dry	C.V. ± %
Dry matter	%		39.7	100.0	24
Ash	%		2.9	7.2	16
Crude fiber	%		13.3	33.5	4
Ether extract	%		1.3	3.3	14
N-free extract	%		18.3	46.1	
Protein (N x 6.25)	%		3.9	9.9	14
Cattle	dig prot %	*	2.1	5.2	
Goats	dig prot %	*	2.1	5.2	
Horses	dig prot %	*	2.1	5.2	
Sheep	dig prot %	*	2.1	5.2	
Energy	GE Mcal/kg				
Cattle	DE Mcal/kg	*	1.02	2.58	
Sheep	DE Mcal/kg	*	1.08	2.71	
Cattle	ME Mcal/kg	*	0.84	2.11	
Sheep	ME Mcal/kg	*	0.88	2.22	
Cattle	TDN %	*	23.2	58.5	
Sheep	TDN %	*	24.4	61.4	

Timothy, aerial part, wilted ensiled, full bloom, (3)

Ref No 3 04 927 United States

Feed Name or Analyses			As Fed	Dry	C.V. ± %
Dry matter	%		45.0	100.0	
Ash	%		3.0	6.6	
Crude fiber	%		15.3	34.0	
Cattle	dig coef %		54.	54.	
Ether extract	%		1.4	3.1	
Cattle	dig coef %		62.	62.	
N-free extract	%		20.5	45.5	
Cattle	dig coef %		60.	60.	
Protein (N x 6.25)	%		4.9	10.8	
Cattle	dig coef %		50.	50.	
Cattle	dig prot %		2.4	5.4	
Goats	dig prot %	*	2.7	6.0	
Horses	dig prot %	*	2.7	6.0	
Sheep	dig prot %	*	2.7	6.0	
Energy	GE Mcal/kg				
Cattle	DE Mcal/kg	*	1.10	2.44	
Sheep	DE Mcal/kg	*	1.15	2.56	
Cattle	ME Mcal/kg	*	0.90	2.00	
Sheep	ME Mcal/kg	*	0.95	2.10	
Cattle	TDN %		24.9	55.4	
Sheep	TDN %	*	26.1	58.1	

Timothy, aerial part, wilted ensiled, late bloom, (3)

Ref No 3 04 928 United States

Feed Name or Analyses			As Fed	Dry	C.V. ± %
Dry matter	%		39.4	100.0	6
Ash	%		2.6	6.6	6
Crude fiber	%		14.3	36.3	1
Ether extract	%		1.2	3.0	6
N-free extract	%		17.6	44.6	
Protein (N x 6.25)	%		3.7	9.5	3
Cattle	dig prot %	*	1.9	4.9	
Goats	dig prot %	*	1.9	4.9	
Horses	dig prot %	*	1.9	4.9	
Sheep	dig prot %	*	1.9	4.9	
Energy	GE Mcal/kg				
Cattle	DE Mcal/kg	*	0.96	2.43	
Sheep	DE Mcal/kg	*	1.00	2.55	
Cattle	ME Mcal/kg	*	0.79	1.99	
Sheep	ME Mcal/kg	*	0.82	2.09	
Cattle	TDN %	*	21.7	55.1	
Sheep	TDN %	*	22.8	57.9	

Timothy, aerial part, wilted ensiled, milk stage, (3)

Ref No 3 04 929 United States

Feed Name or Analyses			As Fed	Dry	C.V. ± %
Dry matter	%		43.3	100.0	8
Ash	%		2.9	6.8	8
Crude fiber	%		16.7	38.5	2
Ether extract	%		1.3	3.0	8
N-free extract	%		18.7	43.3	
Protein (N x 6.25)	%		3.6	8.4	6
Cattle	dig prot %	*	1.7	3.9	
Goats	dig prot %	*	1.7	3.9	
Horses	dig prot %	*	1.7	3.9	
Sheep	dig prot %	*	1.7	3.9	
Energy	GE Mcal/kg				
Cattle	DE Mcal/kg	*	1.04	2.40	
Sheep	DE Mcal/kg	*	1.11	2.57	
Cattle	ME Mcal/kg	*	0.85	1.97	
Sheep	ME Mcal/kg	*	0.91	2.11	
Cattle	TDN %	*	23.6	54.5	
Sheep	TDN %	*	25.3	58.4	

Timothy, aerial part w AIV preservative added, ensiled, (3)

Ref No 3 04 934 United States

Feed Name or Analyses			As Fed	Dry	C.V. ± %
Dry matter	%		34.8	100.0	8
Ash	%		2.3	6.7	4
Crude fiber	%		12.3	35.4	6
Cattle	dig coef %		60.	60.	
Ether extract	%		1.0	2.8	6
Cattle	dig coef %		60.	60.	
N-free extract	%		15.8	45.5	
Cattle	dig coef %		56.	56.	
Protein (N x 6.25)	%		3.4	9.8	10
Cattle	dig coef %		53.	53.	
Cattle	dig prot %		1.8	5.2	
Goats	dig prot %	*	1.8	5.1	
Horses	dig prot %	*	1.8	5.1	
Sheep	dig prot %	*	1.8	5.1	
Energy	GE Mcal/kg				
Cattle	DE Mcal/kg	*	0.85	2.45	
Sheep	DE Mcal/kg	*	0.91	2.61	
Cattle	ME Mcal/kg	*	0.70	2.01	
Sheep	ME Mcal/kg	*	0.74	2.14	

Feed Name or Analyses				Mean As Fed	Mean Dry	C.V. ± %
Cattle	TDN	%		19.3	55.5	
Sheep	TDN	%	*	20.6	59.3	

Timothy, aerial part w AIV preservative added, ensiled, early bloom, (3)

Ref No 3 04 931 United States

Feed Name or Analyses				Mean As Fed	Mean Dry	C.V. ± %
Dry matter		%		34.8	100.0	
Ash		%		2.3	6.7	
Crude fiber		%		12.3	35.3	
Cattle	dig coef	%		61.	61.	
Ether extract		%		1.0	2.9	
Cattle	dig coef	%		57.	57.	
N-free extract		%		15.8	45.4	
Cattle	dig coef	%		53.	53.	
Protein (N x 6.25)		%		3.4	9.9	
Cattle	dig coef	%		49.	49.	
Cattle	dig prot	%		1.7	4.8	
Goats	dig prot	%	*	1.8	5.2	
Horses	dig prot	%	*	1.8	5.2	
Sheep	dig prot	%	*	1.8	5.2	
Energy	GE	Mcal/kg				
Cattle	DE	Mcal/kg	*	0.83	2.38	
Sheep	DE	Mcal/kg	*	0.89	2.56	
Cattle	ME	Mcal/kg	*	0.68	1.95	
Sheep	ME	Mcal/kg	*	0.73	2.10	
Cattle	TDN	%		18.8	54.0	
Sheep	TDN	%	*	20.2	58.1	

Timothy, aerial part w AIV preservative added, ensiled, mid-bloom, (3)

Ref No 3 04 932 United States

Feed Name or Analyses				Mean As Fed	Mean Dry	C.V. ± %
Dry matter		%		39.7	100.0	
Ash		%		2.5	6.4	
Crude fiber		%		12.9	32.5	
Cattle	dig coef	%		59.	59.	
Ether extract		%		1.0	2.6	
Cattle	dig coef	%		64.	64.	
N-free extract		%		18.8	47.3	
Cattle	dig coef	%		59.	59.	
Protein (N x 6.25)		%		4.4	11.2	
Cattle	dig coef	%		57.	57.	
Cattle	dig prot	%		2.5	6.4	
Goats	dig prot	%	*	2.5	6.4	
Horses	dig prot	%	*	2.5	6.4	
Sheep	dig prot	%	*	2.5	6.4	
Energy	GE	Mcal/kg				
Cattle	DE	Mcal/kg	*	1.00	2.52	
Sheep	DE	Mcal/kg	*	1.04	2.63	
Cattle	ME	Mcal/kg	*	0.82	2.07	
Sheep	ME	Mcal/kg	*	0.86	2.16	
Cattle	TDN	%		22.7	57.2	
Sheep	TDN	%	*	23.7	59.6	

Timothy, aerial part w AIV preservative added, ensiled, full bloom, (3)

Ref No 3 04 933 United States

Feed Name or Analyses				Mean As Fed	Mean Dry	C.V. ± %
Dry matter		%		30.4	100.0	1
Ash		%		2.1	7.0	2
Crude fiber		%		11.4	37.6	3
Ether extract		%		0.9	2.8	6
N-free extract		%		13.5	44.3	
Protein (N x 6.25)		%		2.5	8.3	2
Cattle	dig prot	%	*	1.1	3.8	
Goats	dig prot	%	*	1.1	3.8	
Horses	dig prot	%	*	1.1	3.8	
Sheep	dig prot	%	*	1.1	3.8	
Energy	GE	Mcal/kg				
Cattle	DE	Mcal/kg	*	0.74	2.45	
Sheep	DE	Mcal/kg	*	0.78	2.57	
Cattle	ME	Mcal/kg	*	0.61	2.01	
Sheep	ME	Mcal/kg	*	0.64	2.11	
Cattle	TDN	%	*	16.9	55.5	
Sheep	TDN	%	*	17.8	58.4	

Timothy, aerial part w barley added, ensiled, full bloom, (3)

Ref No 3 04 935 United States

Feed Name or Analyses				Mean As Fed	Mean Dry	C.V. ± %
Dry matter		%		34.0	100.0	2
Ash		%		2.1	6.1	7
Crude fiber		%		11.6	34.1	2
Cattle	dig coef	%		56.	56.	
Ether extract		%		1.1	3.3	3
Cattle	dig coef	%		64.	64.	
N-free extract		%		15.8	46.6	
Cattle	dig coef	%		57.	57.	
Protein (N x 6.25)		%		3.4	10.1	6
Cattle	dig coef	%		40.	40.	
Cattle	dig prot	%		1.4	4.0	
Goats	dig prot	%	*	1.8	5.4	
Horses	dig prot	%	*	1.8	5.4	
Sheep	dig prot	%	*	1.8	5.4	
Energy	GE	Mcal/kg				
Cattle	DE	Mcal/kg	*	0.82	2.40	
Sheep	DE	Mcal/kg	*	0.86	2.52	
Cattle	ME	Mcal/kg	*	0.67	1.97	
Sheep	ME	Mcal/kg	*	0.70	2.07	
Cattle	TDN	%		18.5	54.4	
Sheep	TDN	%	*	19.4	57.1	

Timothy, aerial part w grain added, ensiled, (3)

Ref No 3 08 543 United States

Feed Name or Analyses				Mean As Fed	Mean Dry	C.V. ± %
Dry matter		%		33.1	100.0	
Ash		%		2.0	6.0	
Crude fiber		%		10.4	31.4	
Ether extract		%		1.1	3.3	
N-free extract		%		15.6	47.1	
Protein (N x 6.25)		%		4.0	12.1	
Cattle	dig prot	%	*	2.4	7.2	
Goats	dig prot	%	*	2.4	7.2	
Horses	dig prot	%	*	2.4	7.2	
Sheep	dig prot	%	*	2.4	7.2	
Energy	GE	Mcal/kg				
Cattle	DE	Mcal/kg	*	0.83	2.50	
Sheep	DE	Mcal/kg	*	0.87	2.63	
Cattle	ME	Mcal/kg	*	0.68	2.05	
Sheep	ME	Mcal/kg	*	0.71	2.16	
Cattle	TDN	%	*	18.7	56.6	
Sheep	TDN	%	*	19.7	59.6	
Calcium		%		0.19	0.57	
Phosphorus		%		0.11	0.33	

Timothy, aerial part w phosphoric acid preservative added, ensiled, (3)

Ref No 3 04 940 United States

Feed Name or Analyses				Mean As Fed	Mean Dry	C.V. ± %
Dry matter		%		34.1	100.0	12
Ash		%		2.5	7.3	8
Crude fiber		%		12.5	36.8	4
Cattle	dig coef	%		59.	59.	
Ether extract		%		1.0	2.8	6
Cattle	dig coef	%		60.	60.	
N-free extract		%		15.3	44.8	
Cattle	dig coef	%		57.	57.	
Protein (N x 6.25)		%		2.8	8.3	6
Cattle	dig coef	%		44.	44.	
Cattle	dig prot	%		1.2	3.7	
Goats	dig prot	%	*	1.3	3.8	
Horses	dig prot	%	*	1.3	3.8	
Sheep	dig prot	%	*	1.3	3.8	
Energy	GE	Mcal/kg				
Cattle	DE	Mcal/kg	*	0.82	2.41	
Sheep	DE	Mcal/kg	*	0.89	2.60	
Cattle	ME	Mcal/kg	*	0.67	1.98	
Sheep	ME	Mcal/kg	*	0.73	2.13	
Cattle	TDN	%		18.6	54.7	
Sheep	TDN	%	*	20.1	58.7	

Timothy, aerial part w phosphoric acid preservative added, ensiled, early bloom, (3)

Ref No 3 04 936 United States

Feed Name or Analyses				Mean As Fed	Mean Dry	C.V. ± %
Dry matter		%		34.1	100.0	11
Ash		%		2.3	6.8	6
Crude fiber		%		12.6	37.1	3
Cattle	dig coef	%		63.	63.	
Ether extract		%		1.0	2.9	6
Cattle	dig coef	%		59.	59.	
N-free extract		%		15.4	45.1	
Cattle	dig coef	%		54.	54.	
Protein (N x 6.25)		%		2.8	8.2	8
Cattle	dig coef	%		46.	46.	
Cattle	dig prot	%		1.3	3.8	
Goats	dig prot	%	*	1.3	3.7	
Horses	dig prot	%	*	1.3	3.7	
Sheep	dig prot	%	*	1.3	3.7	
Energy	GE	Mcal/kg				
Cattle	DE	Mcal/kg	*	0.83	2.44	
Sheep	DE	Mcal/kg	*	0.88	2.58	
Cattle	ME	Mcal/kg	*	0.68	2.00	
Sheep	ME	Mcal/kg	*	0.72	2.12	
Cattle	TDN	%		18.8	55.3	
Sheep	TDN	%	*	20.0	58.6	

Timothy, aerial part w phosphoric acid preservative added, ensiled, full bloom, (3)

Ref No 3 04 937 United States

Feed Name or Analyses				Mean As Fed	Mean Dry	C.V. ± %
Dry matter		%		31.0	100.0	4
Ash		%		2.3	7.4	9
Crude fiber		%		12.1	39.0	3
Ether extract		%		0.8	2.7	3
N-free extract		%		13.3	43.0	
Protein (N x 6.25)		%		2.4	7.9	3
Cattle	dig prot	%	*	1.1	3.4	
Goats	dig prot	%	*	1.1	3.4	
Horses	dig prot	%	*	1.1	3.4	
Sheep	dig prot	%	*	1.1	3.4	
Energy	GE	Mcal/kg				
Cattle	DE	Mcal/kg	*	0.75	2.42	
Sheep	DE	Mcal/kg	*	0.79	2.55	
Cattle	ME	Mcal/kg	*	0.62	1.99	
Sheep	ME	Mcal/kg	*	0.65	2.09	
Cattle	TDN	%	*	17.0	54.9	
Sheep	TDN	%	*	17.9	57.8	

Feed Name or Analyses		Mean As Fed	Mean Dry	C.V. ± %

Timothy, aerial part w phosphoric acid preservative added, ensiled, late bloom, (3)

Ref No 3 04 938 United States

Feed Name or Analyses		As Fed	Dry	C.V. ± %
Dry matter %		43.1	100.0	
Ash %		3.0	7.0	
Crude fiber %		14.4	33.3	
Cattle dig coef %		64.	64.	
Ether extract %		1.4	3.2	
Cattle dig coef %		52.	52.	
N-free extract %		20.7	48.1	
Cattle dig coef %		64.	64.	
Protein (N x 6.25) %		3.6	8.4	
Cattle dig coef %		46.	46.	
Cattle dig prot %		1.7	3.9	
Goats dig prot %	*	1.7	3.9	
Horses dig prot %	*	1.7	3.9	
Sheep dig prot %	*	1.7	3.9	
Energy GE Mcal/kg				
Cattle DE Mcal/kg	*	1.13	2.63	
Sheep DE Mcal/kg	*	1.16	2.70	
Cattle ME Mcal/kg	*	0.93	2.16	
Sheep ME Mcal/kg	*	0.95	2.21	
Cattle TDN %		25.7	59.7	
Sheep TDN %	*	26.4	61.3	

Timothy, aerial part w phosphoric acid preservative added, ensiled, milk stage, (3)

Ref No 3 04 939 United States

Feed Name or Analyses		As Fed	Dry	C.V. ± %
Dry matter %		37.0	100.0	8
Ash %		2.8	7.6	4
Crude fiber %		13.7	37.0	4
Cattle dig coef %		54.	54.	
Ether extract %		1.0	2.8	9
Cattle dig coef %		64.	64.	
N-free extract %		16.2	44.0	
Cattle dig coef %		54.	54.	
Protein (N x 6.25) %		3.3	8.8	3
Cattle dig coef %		44.	44.	
Cattle dig prot %		1.4	3.9	
Goats dig prot %	*	1.6	4.2	
Horses dig prot %	*	1.6	4.2	
Sheep dig prot %	*	1.6	4.2	
Energy GE Mcal/kg				
Cattle DE Mcal/kg	*	0.84	2.27	
Sheep DE Mcal/kg	*	0.92	2.49	
Cattle ME Mcal/kg	*	0.69	1.86	
Sheep ME Mcal/kg	*	0.75	2.04	
Cattle TDN %		19.0	51.5	
Sheep TDN %	*	20.9	56.4	

Timothy, aerial part w phosphoric acid preservative added, wilted ensiled, (3)

Ref No 3 08 544 United States

Feed Name or Analyses		As Fed	Dry	C.V. ± %
Dry matter %		36.6	100.0	
Ash %		2.6	7.1	
Crude fiber %		12.6	34.4	
Ether extract %		1.1	3.0	
N-free extract %		17.1	46.7	
Protein (N x 6.25) %		3.2	8.7	
Cattle dig prot %	*	1.5	4.2	
Goats dig prot %	*	1.5	4.2	
Horses dig prot %	*	1.5	4.2	
Sheep dig prot %	*	1.5	4.2	
Energy GE Mcal/kg				
Cattle DE Mcal/kg	*	0.94	2.57	
Sheep DE Mcal/kg	*	0.94	2.56	
Cattle ME Mcal/kg	*	0.77	2.11	
Sheep ME Mcal/kg	*	0.77	2.10	
Cattle TDN %	*	21.4	58.4	
Sheep TDN %	*	21.3	58.4	

Timothy, aerial part w limestone preservative added, wilted ensiled, early bloom, cut 1, (3)

Ref No 3 07 915 Canada

Feed Name or Analyses		As Fed	Dry	C.V. ± %
Dry matter %		48.9	100.0	
Sheep dig coef %		59.	59.	
Ash %		3.9	8.0	
Sheep dig coef %		21.	42.	

Timothy, aerial part w molasses added, ensiled, (3)

Ref No 3 04 948 United States

Feed Name or Analyses		As Fed	Dry	C.V. ± %
Dry matter %		33.9	100.0	16
Ash %		2.4	7.1	9
Crude fiber %		11.3	33.4	7
Cattle dig coef %		65.	65.	
Ether extract %		1.1	3.4	7
Cattle dig coef %		58.	58.	
N-free extract %		15.5	45.7	
Cattle dig coef %		60.	60.	
Protein (N x 6.25) %		3.5	10.4	10
Cattle dig coef %		52.	52.	
Cattle dig prot %		1.8	5.4	
Goats dig prot %	*	1.9	5.7	
Horses dig prot %	*	1.9	5.7	
Sheep dig prot %	*	1.9	5.7	
Energy GE Mcal/kg				
Cattle DE Mcal/kg	*	0.88	2.60	
Sheep DE Mcal/kg	*	0.95	2.81	
Cattle ME Mcal/kg	*	0.72	2.13	
Sheep ME Mcal/kg	*	0.78	2.31	
Cattle TDN %		20.0	59.0	
Sheep TDN %	*	21.6	63.8	
Calcium %		0.18	0.53	
Phosphorus %		0.09	0.27	

Timothy, aerial part w molasses added, ensiled, immature, (3)

Ref No 3 04 941 United States

Feed Name or Analyses		As Fed	Dry	C.V. ± %
Dry matter %		36.8	100.0	15
Ash %		2.9	7.9	6
Crude fiber %		11.5	31.4	3
Ether extract %		1.4	3.8	4
N-free extract %		16.3	44.4	
Protein (N x 6.25) %		4.6	12.5	6
Cattle dig prot %	*	2.8	7.6	
Goats dig prot %	*	2.8	7.6	
Horses dig prot %	*	2.8	7.6	
Sheep dig prot %	*	2.8	7.6	
Energy GE Mcal/kg				
Cattle DE Mcal/kg	*	0.96	2.60	
Sheep DE Mcal/kg	*	1.02	2.78	
Cattle ME Mcal/kg	*	0.78	2.14	
Sheep ME Mcal/kg	*	0.84	2.28	
Cattle TDN %	*	21.7	59.1	
Sheep TDN %	*	23.2	63.1	
Calcium %		0.20	0.54	12
Phosphorus %		0.12	0.32	5

Timothy, aerial part w molasses added, ensiled, pre-bloom, (3)

Ref No 3 04 942 United States

Feed Name or Analyses		As Fed	Dry	C.V. ± %
Dry matter %		25.3	100.0	
Ash %		2.1	8.3	
Crude fiber %		8.2	32.4	
Cattle dig coef %		72.	72.	
Ether extract %		1.0	3.8	
Cattle dig coef %		58.	58.	
N-free extract %		10.9	43.2	
Cattle dig coef %		62.	62.	
Protein (N x 6.25) %		3.1	12.3	
Cattle dig coef %		55.	55.	
Cattle dig prot %		1.7	6.8	
Goats dig prot %	*	1.9	7.4	
Horses dig prot %	*	1.9	7.4	
Sheep dig prot %	*	1.9	7.4	
Energy GE Mcal/kg				
Cattle DE Mcal/kg	*	0.69	2.73	
Sheep DE Mcal/kg	*	0.70	2.76	
Cattle ME Mcal/kg	*	0.57	2.24	
Sheep ME Mcal/kg	*	0.57	2.27	
Cattle TDN %		15.6	61.8	
Sheep TDN %	*	15.9	62.7	

Timothy, aerial part w molasses added, ensiled, early bloom, (3)

Ref No 3 04 943 United States

Feed Name or Analyses		As Fed	Dry	C.V. ± %
Dry matter %		34.6	100.0	17
Ash %		2.4	7.0	8
Crude fiber %		12.2	35.2	5
Cattle dig coef %		63.	63.	
Ether extract %		1.1	3.2	9
Cattle dig coef %		59.	59.	
N-free extract %		15.5	44.9	
Cattle dig coef %		54.	54.	
Protein (N x 6.25) %		3.4	9.9	10
Cattle dig coef %		53.	53.	
Cattle dig prot %		1.8	5.2	
Goats dig prot %	*	1.8	5.2	
Horses dig prot %	*	1.8	5.2	
Sheep dig prot %	*	1.8	5.2	
Energy GE Mcal/kg				
Cattle DE Mcal/kg	*	0.85	2.46	
Sheep DE Mcal/kg	*	0.90	2.59	
Cattle ME Mcal/kg	*	0.70	2.02	
Sheep ME Mcal/kg	*	0.73	2.12	
Cattle TDN %		19.3	55.8	
Sheep TDN %	*	20.3	58.7	

Timothy, aerial part w molasses added, ensiled, mid-bloom, (3)

Ref No 3 04 944 United States

Feed Name or Analyses		As Fed	Dry	C.V. ± %
Dry matter %		39.7	100.0	24
Ash %		2.9	7.2	16
Crude fiber %		13.3	33.5	4
Ether extract %		1.3	3.3	14
N-free extract %		18.3	46.1	

(1) dry forages and roughages (2) pasture, range plants, and forages fed green (3) silages (4) energy feeds (5) protein supplements (6) minerals (7) vitamins (8) additives

Feed Name or Analyses			Mean As Fed	Mean Dry	C.V. ± %
Protein (N x 6.25)	%		3.9	9.9	14
Cattle	dig prot %	*	2.1	5.2	
Goats	dig prot %	*	2.1	5.2	
Horses	dig prot %	*	2.1	5.2	
Sheep	dig prot %	*	2.1	5.2	
Energy	GE Mcal/kg				
Cattle	DE Mcal/kg	*	1.02	2.58	
Sheep	DE Mcal/kg	*	1.06	2.66	
Cattle	ME Mcal/kg	*	0.84	2.11	
Sheep	ME Mcal/kg	*	0.87	2.18	
Cattle	TDN %	*	23.2	58.5	
Sheep	TDN %	*	24.0	60.4	

Timothy, aerial part w molasses added, ensiled, full bloom, (3)

Ref No 3 04 945 United States

Feed Name or Analyses			Mean As Fed	Mean Dry	C.V. ± %
Dry matter	%		38.4	100.0	24
Ash	%		2.6	6.8	11
Crude fiber	%		13.2	34.5	7
Cattle	dig coef %		63.	63.	
Ether extract	%		1.3	3.3	4
Cattle	dig coef %		59.	59.	
N-free extract	%		17.5	45.5	
Cattle	dig coef %		58.	58.	
Protein (N x 6.25)	%		3.8	10.0	3
Cattle	dig coef %		54.	54.	
Cattle	dig prot %		2.1	5.4	
Goats	dig prot %	*	2.0	5.3	
Horses	dig prot %	*	2.0	5.3	
Sheep	dig prot %	*	2.0	5.3	
Energy	GE Mcal/kg				
Cattle	DE Mcal/kg	*	0.98	2.55	
Sheep	DE Mcal/kg	*	1.02	2.66	
Cattle	ME Mcal/kg	*	0.80	2.09	
Sheep	ME Mcal/kg	*	0.84	2.18	
Cattle	TDN %		22.2	57.8	
Sheep	TDN %	*	23.2	60.3	
Calcium	%		0.19	0.49	23
Phosphorus	%		0.10	0.26	7

Timothy, aerial part w molasses added, ensiled, late bloom, (3)

Ref No 3 04 946 United States

Feed Name or Analyses			Mean As Fed	Mean Dry	C.V. ± %
Dry matter	%		39.8	100.0	
Ash	%		2.4	6.1	
Crude fiber	%		13.1	32.8	
Cattle	dig coef %		61.	61.	
Ether extract	%		1.2	3.1	
Cattle	dig coef %		58.	58.	
N-free extract	%		19.8	49.8	
Cattle	dig coef %		62.	62.	
Protein (N x 6.25)	%		3.3	8.2	
Cattle	dig coef %		42.	42.	
Cattle	dig prot %		1.4	3.4	
Goats	dig prot %	*	1.5	3.7	
Horses	dig prot %	*	1.5	3.7	
Sheep	dig prot %	*	1.5	3.7	
Energy	GE Mcal/kg				
Cattle	DE Mcal/kg	*	1.02	2.57	
Sheep	DE Mcal/kg	*	1.07	2.69	
Cattle	ME Mcal/kg	*	0.84	2.11	
Sheep	ME Mcal/kg	*	0.88	2.21	
Cattle	TDN %		23.2	58.4	
Sheep	TDN %	*	24.3	61.0	

Timothy, aerial part w molasses added, ensiled, milk stage, (3)

Ref No 3 04 947 United States

Feed Name or Analyses			Mean As Fed	Mean Dry	C.V. ± %
Dry matter	%		42.9	100.0	4
Ash	%		2.7	6.3	3
Crude fiber	%		16.3	38.0	3
Ether extract	%		1.3	3.1	5
N-free extract	%		19.0	44.4	
Protein (N x 6.25)	%		3.5	8.2	4
Cattle	dig prot %	*	1.6	3.7	
Goats	dig prot %	*	1.6	3.7	
Horses	dig prot %	*	1.6	3.7	
Sheep	dig prot %	*	1.6	3.7	
Energy	GE Mcal/kg				
Cattle	DE Mcal/kg	*	1.03	2.41	
Sheep	DE Mcal/kg	*	1.10	2.67	
Cattle	ME Mcal/kg	*	0.85	1.98	
Sheep	ME Mcal/kg	*	0.91	2.11	
Cattle	TDN %	*	23.5	54.7	
Sheep	TDN %	*	25.0	58.2	

Timothy, aerial part w molasses added, wilted ensiled, (3)

Ref No 3 04 954 United States

Feed Name or Analyses			Mean As Fed	Mean Dry	C.V. ± %
Dry matter	%		26.8	100.0	
Ash	%		1.8	6.8	
Crude fiber	%		8.9	33.2	
Cattle	dig coef %		66.	66.	
Ether extract	%		0.9	3.2	
Cattle	dig coef %		59.	59.	
N-free extract	%		12.5	46.7	
Cattle	dig coef %		63.	63.	
Protein (N x 6.25)	%		2.7	10.1	
Cattle	dig coef %		52.	52.	
Cattle	dig prot %		1.4	5.3	
Goats	dig prot %	*	1.5	5.4	
Horses	dig prot %	*	1.5	5.4	
Sheep	dig prot %	*	1.5	5.4	
Energy	GE Mcal/kg				
Cattle	DE Mcal/kg	*	0.72	2.68	
Sheep	DE Mcal/kg	*	0.76	2.83	
Cattle	ME Mcal/kg	*	0.59	2.20	
Sheep	ME Mcal/kg	*	0.62	2.32	
Cattle	TDN %		16.3	60.9	
Sheep	TDN %	*	17.2	64.1	
Calcium	%		0.14	0.52	
Phosphorus	%		0.10	0.38	

Timothy, aerial part w molasses added, wilted ensiled, pre-bloom, (3)

Ref No 3 04 949 United States

Feed Name or Analyses			Mean As Fed	Mean Dry	C.V. ± %
Dry matter	%		44.7	100.0	
Ash	%		3.4	7.6	
Crude fiber	%		13.6	30.5	
Cattle	dig coef %		72.	72.	
Ether extract	%		1.7	3.8	
Cattle	dig coef %		58.	58.	
N-free extract	%		20.2	45.3	
Cattle	dig coef %		68.	68.	
Protein (N x 6.25)	%		5.7	12.8	
Cattle	dig coef %		56.	56.	
Cattle	dig prot %		3.2	7.2	
Goats	dig prot %	*	3.5	7.9	
Horses	dig prot %	*	3.5	7.9	

Feed Name or Analyses			Mean As Fed	Mean Dry	C.V. ± %
Sheep	dig prot %	*	3.5	7.9	
Energy	GE Mcal/kg				
Cattle	DE Mcal/kg	*	1.28	2.86	
Sheep	DE Mcal/kg	*	1.25	2.80	
Cattle	ME Mcal/kg	*	1.05	2.35	
Sheep	ME Mcal/kg	*	1.02	2.29	
Cattle	TDN %		29.0	64.9	
Sheep	TDN %	*	28.3	63.4	

Timothy, aerial part w molasses added, wilted ensiled, early bloom, (3)

Ref No 3 04 950 United States

Feed Name or Analyses			Mean As Fed	Mean Dry	C.V. ± %
Dry matter	%		39.6	100.0	
Ash	%		2.3	5.8	
Crude fiber	%		14.4	36.4	
Cattle	dig coef %		62.	62.	
Ether extract	%		1.1	2.9	
Cattle	dig coef %		58.	58.	
N-free extract	%		18.0	45.5	
Cattle	dig coef %		56.	56.	
Protein (N x 6.25)	%		3.7	9.4	
Cattle	dig coef %		48.	48.	
Cattle	dig prot %		1.8	4.5	
Goats	dig prot %	*	1.9	4.8	
Horses	dig prot %	*	1.9	4.8	
Sheep	dig prot %	*	1.9	4.8	
Energy	GE Mcal/kg				
Cattle	DE Mcal/kg	*	0.98	2.48	
Sheep	DE Mcal/kg	*	1.03	2.61	
Cattle	ME Mcal/kg	*	0.81	2.04	
Sheep	ME Mcal/kg	*	0.85	2.14	
Cattle	TDN %		22.3	56.3	
Sheep	TDN %	*	23.4	59.1	

Timothy, aerial part w molasses added, wilted ensiled, mid-bloom, (3)

Ref No 3 04 951 United States

Feed Name or Analyses			Mean As Fed	Mean Dry	C.V. ± %
Dry matter	%		26.4	100.0	
Ash	%		2.3	8.8	
Crude fiber	%		9.0	34.0	
Cattle	dig coef %		70.	70.	
Ether extract	%		1.0	3.8	
Cattle	dig coef %		68.	68.	
N-free extract	%		11.2	42.4	
Cattle	dig coef %		64.	64.	
Protein (N x 6.25)	%		2.9	11.0	
Cattle	dig coef %		60.	60.	
Cattle	dig prot %		1.7	6.6	
Goats	dig prot %	*	1.6	6.2	
Horses	dig prot %	*	1.6	6.2	
Sheep	dig prot %	*	1.6	6.2	
Energy	GE Mcal/kg				
Cattle	DE Mcal/kg	*	0.74	2.79	
Sheep	DE Mcal/kg	*	0.72	2.75	
Cattle	ME Mcal/kg	*	0.60	2.29	
Sheep	ME Mcal/kg	*	0.59	2.25	
Cattle	TDN %		16.7	63.4	
Sheep	TDN %	*	16.4	62.3	

Feed Name or Analyses			Mean As Fed	Mean Dry	C.V. ± %

Timothy, aerial part w molasses added, wilted ensiled, full bloom, (3)

Ref No 3 04 952 — United States

Analyses			As Fed	Dry	C.V. ± %
Dry matter	%		46.5	100.0	
Ash	%		3.1	6.7	
Crude fiber	%		15.7	33.8	
Cattle	dig coef %		62.	62.	
Ether extract	%		1.5	3.2	
Cattle	dig coef %		60.	60.	
N-free extract	%		21.7	46.6	
Cattle	dig coef %		60.	60.	
Protein (N x 6.25)	%		4.5	9.7	
Cattle	dig coef %		51.	51.	
Cattle	dig prot %		2.3	4.9	
Goats	dig prot %	*	2.3	5.0	
Horses	dig prot %	*	2.3	5.0	
Sheep	dig prot %	*	2.3	5.0	
Energy	GE Mcal/kg				
Cattle	DE Mcal/kg	*	1.19	2.57	
Sheep	DE Mcal/kg	*	1.26	2.22	
Cattle	ME Mcal/kg	*	0.98	2.10	
Sheep	ME Mcal/kg	*	1.04	2.23	
Cattle	TDN %		27.1	58.2	
Sheep	TDN %	*	28.7	61.7	

Timothy, aerial part w molasses added, wilted ensiled, late bloom, (3)

Ref No 3 04 953 — United States

Analyses			As Fed	Dry	C.V. ± %
Dry matter	%		46.6	100.0	
Ash	%		3.1	6.6	
Crude fiber	%		15.5	33.2	
Cattle	dig coef %		66.	66.	
Ether extract	%		1.5	3.2	
Cattle	dig coef %		53.	53.	
N-free extract	%		22.6	48.6	
Cattle	dig coef %		63.	63.	
Protein (N x 6.25)	%		3.9	8.4	
Cattle	dig coef %		46.	46.	
Cattle	dig prot %		1.8	3.9	
Goats	dig prot %	*	1.8	3.9	
Horses	dig prot %	*	1.8	3.9	
Sheep	dig prot %	*	1.8	3.9	
Energy	GE Mcal/kg				
Cattle	DE Mcal/kg	*	1.24	2.65	
Sheep	DE Mcal/kg	*	1.33	2.85	
Cattle	ME Mcal/kg	*	1.01	2.18	
Sheep	ME Mcal/kg	*	1.09	2.34	
Cattle	TDN %		28.1	60.2	
Sheep	TDN %	*	30.1	64.6	

Timothy, aerial part w newspaper added, ensiled, mid-bloom, cut 2, (3)

Ref No 3 07 916 — Canada

Analyses			As Fed	Dry	C.V. ± %
Dry matter	%		21.0	100.0	
Sheep	dig coef %		56.	56.	
Ash	%		1.4	6.7	
Sheep	dig coef %		7.	33.	

(1) dry forages and roughages (2) pasture, range plants, and forages fed green (3) silages (4) energy feeds (5) protein supplements (6) minerals (7) vitamins (8) additives

Timothy, aerial part w 10% molasses added, ensiled, full bloom, (3)

Ref No 3 04 955 — United States

Analyses			As Fed	Dry	C.V. ± %
Dry matter	%		40.0	100.0	
Ash	%		2.7	6.8	
Crude fiber	%		11.7	29.2	
Cattle	dig coef %		55.	55.	
Ether extract	%		1.0	2.5	
Cattle	dig coef %		59.	59.	
N-free extract	%		20.9	52.3	
Cattle	dig coef %		64.	64.	
Protein (N x 6.25)	%		3.7	9.2	
Cattle	dig coef %		46.	46.	
Cattle	dig prot %		1.7	4.2	
Goats	dig prot %	*	1.8	4.6	
Horses	dig prot %	*	1.8	4.6	
Sheep	dig prot %	*	1.8	4.6	
Energy	GE Mcal/kg				
Cattle	DE Mcal/kg	*	1.01	2.52	
Sheep	DE Mcal/kg	*	1.06	2.66	
Cattle	ME Mcal/kg	*	0.83	2.06	
Sheep	ME Mcal/kg	*	0.87	2.18	
Cattle	TDN %		22.8	57.1	
Sheep	TDN %	*	24.1	60.3	

Timothy, aerial part w 8% molasses added, ensiled, early bloom, (3)

Ref No 3 04 956 — United States

Analyses			As Fed	Dry	C.V. ± %
Dry matter	%		32.1	100.0	
Ash	%		2.7	8.3	
Crude fiber	%		9.5	29.6	
Cattle	dig coef %		73.	73.	
Ether extract	%		1.0	3.2	
Cattle	dig coef %		57.	57.	
N-free extract	%		15.6	48.5	
Cattle	dig coef %		70.	70.	
Protein (N x 6.25)	%		3.3	10.4	
Cattle	dig coef %		49.	49.	
Cattle	dig prot %		1.6	5.1	
Goats	dig prot %	*	1.8	5.7	
Horses	dig prot %	*	1.8	5.7	
Sheep	dig prot %	*	1.8	5.7	
Energy	GE Mcal/kg				
Cattle	DE Mcal/kg	*	0.92	2.86	
Sheep	DE Mcal/kg	*	0.91	2.82	
Cattle	ME Mcal/kg	*	0.75	2.34	
Sheep	ME Mcal/kg	*	0.74	2.31	
Cattle	TDN %		20.8	64.8	
Sheep	TDN %	*	20.5	64.0	

Timothy, aerial part w 8-10% molasses added, ensiled, (3)

Ref No 3 04 957 — United States

Analyses			As Fed	Dry	C.V. ± %
Dry matter	%		37.3	100.0	
Ash	%		2.7	7.3	
Crude fiber	%		10.9	29.3	
Cattle	dig coef %		61.	61.	
Ether extract	%		1.0	2.7	
Cattle	dig coef %		59.	59.	
N-free extract	%		19.1	51.1	
Cattle	dig coef %		66.	66.	
Protein (N x 6.25)	%		3.6	9.6	
Cattle	dig coef %		47.	47.	
Cattle	dig prot %		1.7	4.5	
Goats	dig prot %	*	1.8	4.9	
Horses	dig prot %	*	1.8	4.9	
Sheep	dig prot %	*	1.8	4.9	
Energy	GE Mcal/kg				
Cattle	DE Mcal/kg	*	0.98	2.63	
Sheep	DE Mcal/kg	*	1.02	2.24	
Cattle	ME Mcal/kg	*	0.81	2.16	
Sheep	ME Mcal/kg	*	0.84	2.25	
Cattle	TDN %		22.3	59.7	
Sheep	TDN %	*	23.2	62.2	

TIMOTHY, CLIMAX. Phleum pratense

Timothy, climax, hay, fan air dried, immature, (1)

Ref No 1 07 923 — Canada

Analyses			As Fed	Dry	C.V. ± %
Dry matter	%			100.0	
Organic matter	%			90.9	
Ash	%			9.0	
Crude fiber	%			23.5	
Ether extract	%			3.1	
N-free extract	%			41.7	
Protein (N x 6.25)	%			22.8	
Cattle	dig prot %	*		16.7	
Goats	dig prot %	*		17.8	
Horses	dig prot %	*		16.9	
Rabbits	dig prot %	*		16.3	
Sheep	dig prot %	*		17.0	
Energy	GE Mcal/kg			4.58	
Sheep	GE dig coef %			74.	
Cattle	DE Mcal/kg	*		2.84	
Sheep	DE Mcal/kg			3.39	
Cattle	ME Mcal/kg	*		2.33	
Sheep	ME Mcal/kg	*		2.78	
Cattle	TDN %	*		64.3	
Sheep	TDN %	*		62.3	

Timothy, climax, hay, fan air dried, pre-bloom, (1)

Ref No 1 07 925 — Canada

Analyses			As Fed	Dry	C.V. ± %
Dry matter	%			100.0	
Organic matter	%			93.2	
Ash	%			6.8	
Crude fiber	%			32.4	
Ether extract	%			2.3	
N-free extract	%			45.9	
Protein (N x 6.25)	%			12.7	
Cattle	dig prot %	*		7.9	
Goats	dig prot %	*		8.4	
Horses	dig prot %	*		8.3	
Rabbits	dig prot %	*		8.5	
Sheep	dig prot %	*		8.0	
Energy	GE Mcal/kg			4.53	
Sheep	GE dig coef %			63.	
Cattle	DE Mcal/kg	*		2.54	
Sheep	DE Mcal/kg			2.83	
Cattle	ME Mcal/kg	*		2.08	
Sheep	ME Mcal/kg	*		2.32	
Cattle	TDN %	*		57.6	
Sheep	TDN %	*		57.8	

Timothy, climax, hay, fan air dried, early bloom, (1)

Ref No 1 07 922 — Canada

Analyses			As Fed	Dry	C.V. ± %
Dry matter	%			100.0	
Organic matter	%			94.1	
Ash	%			5.9	

Feed Name or Analyses			Mean As Fed	Mean Dry	C.V. ± %
Crude fiber	%			31.5	
Ether extract	%			2.5	
N-free extract	%			47.7	
Protein (N x 6.25)	%			12.4	
Cattle	dig prot %	*		7.7	
Goats	dig prot %	*		8.1	
Horses	dig prot %	*		8.1	
Rabbits	dig prot %	*		8.2	
Sheep	dig prot %	*		7.7	
Energy	GE Mcal/kg			4.57	
Sheep	GE dig coef %			56.	
Cattle	DE Mcal/kg	*		2.58	
Sheep	DE Mcal/kg			2.56	
Cattle	ME Mcal/kg	*		2.11	
Sheep	ME Mcal/kg	*		2.10	
Cattle	TDN %	*		58.5	
Sheep	TDN %	*		58.2	

Timothy, climax, hay, fan air dried, late bloom, (1)

Ref No 1 07 924 Canada

Feed Name or Analyses			Mean As Fed	Mean Dry	C.V. ± %
Dry matter	%			100.0	
Organic matter	%			95.5	
Ash	%			4.6	
Crude fiber	%			35.1	
Ether extract	%			2.1	
N-free extract	%			50.7	
Protein (N x 6.25)	%			7.7	
Cattle	dig prot %	*		3.6	
Goats	dig prot %	*		3.7	
Horses	dig prot %	*		4.0	
Rabbits	dig prot %	*		4.6	
Sheep	dig prot %	*		3.4	
Energy	GE Mcal/kg			4.57	
Sheep	GE dig coef %			50.	
Cattle	DE Mcal/kg	*		2.29	
Sheep	DE Mcal/kg			2.28	
Cattle	ME Mcal/kg	*		1.88	
Sheep	ME Mcal/kg	*		1.87	
Cattle	TDN %	*		51.9	
Sheep	TDN %	*		55.0	

Timothy, climax, hay, fan-air dried w heat, midbloom, cut 1, (1)

Ref No 1 08 895 United States

Feed Name or Analyses			Mean As Fed	Mean Dry	C.V. ± %
Dry matter	%		95.6	100.0	
Goats	dig coef %		67.	67.	
Protein (N x 6.25)	%		13.7	14.3	
Goats	dig coef %		72.	72.	
Cattle	dig prot %	*	8.9	9.3	
Goats	dig prot %		9.8	10.3	
Horses	dig prot %	*	9.2	9.7	
Rabbits	dig prot %	*	9.3	9.7	
Sheep	dig prot %	*	9.0	9.4	
Fiber, acid detergent (VS)	%		35.4	37.0	
Goats	dig coef %		63.	63.	
Lignin (Van Soest)	%		4.5	4.7	
Dry matter intake	$g/W^{.75}$		48.852	51.100	

Timothy, climax, hay, s-c chopped, cut 2, (1)

Ref No 1 07 802 United States

Feed Name or Analyses			Mean As Fed	Mean Dry	C.V. ± %
Dry matter	%			100.0	
Sheep	dig coef %			63.	
Ash	%			9.7	

Feed Name or Analyses			Mean As Fed	Mean Dry	C.V. ± %
Crude fiber	%			27.5	
Sheep	dig coef %			71.	
Ether extract	%			3.5	
Sheep	dig coef %			50.	
N-free extract	%			42.0	
Sheep	dig coef %			63.	
Protein (N x 6.25)	%			17.4	
Sheep	dig coef %			67.	
Cattle	dig prot %	*		12.0	
Goats	dig prot %	*		12.8	
Horses	dig prot %	*		12.3	
Rabbits	dig prot %	*		12.1	
Sheep	dig prot %			11.6	
Energy	GE Mcal/kg			4.45	1
Cattle	DE Mcal/kg	*		2.80	
Sheep	DE Mcal/kg			2.73	
Cattle	ME Mcal/kg	*		2.30	
Sheep	ME Mcal/kg	*		2.23	
Cattle	TDN %	*		63.6	
Sheep	TDN %			61.3	

TIMOTHY, COARSE. Phleum pratense

Timothy, coarse, hay, s-c, (1)

Ref No 1 04 958 United States

Feed Name or Analyses			Mean As Fed	Mean Dry	C.V. ± %
Dry matter	%		89.0	100.0	
Ash	%		4.5	5.1	
Crude fiber	%		29.9	33.6	
Horses	dig coef %		42.	42.	
Ether extract	%		2.0	2.2	
Horses	dig coef %		7.	7.	
N-free extract	%		46.9	52.7	
Horses	dig coef %		53.	53.	
Protein (N x 6.25)	%		5.7	6.4	
Horses	dig coef %		44.	44.	
Cattle	dig prot %	*	2.2	2.5	
Goats	dig prot %	*	2.3	2.5	
Horses	dig prot %		2.5	2.8	
Rabbits	dig prot %	*	3.2	3.6	
Sheep	dig prot %	*	2.1	2.3	
Energy	GE Mcal/kg				
Cattle	DE Mcal/kg	*	2.17	2.44	
Horses	DE Mcal/kg	*	1.77	1.99	
Sheep	DE Mcal/kg	*	2.16	2.43	
Cattle	ME Mcal/kg	*	1.78	2.00	
Horses	ME Mcal/kg	*	1.45	1.63	
Sheep	ME Mcal/kg	*	1.77	1.99	
Cattle	TDN %	*	49.2	55.3	
Horses	TDN %		40.2	45.2	
Sheep	TDN %	*	49.1	55.2	

TIMOTHY, DRUMMOND. Phleum pratense

Timothy, drummond, hay, fan air dried, immature, (1)

Ref No 1 07 928 Canada

Feed Name or Analyses			Mean As Fed	Mean Dry	C.V. ± %
Dry matter	%			100.0	
Organic matter	%			91.2	
Ash	%			8.8	
Crude fiber	%			23.3	
Ether extract	%			3.1	
N-free extract	%			42.5	
Protein (N x 6.25)	%			22.3	
Cattle	dig prot %	*		16.3	
Goats	dig prot %	*		17.4	
Horses	dig prot %	*		16.5	

Feed Name or Analyses			Mean As Fed	Mean Dry	C.V. ± %
Rabbits	dig prot %	*		15.9	
Sheep	dig prot %	*		16.6	
Energy	GE Mcal/kg			4.52	
Sheep	GE dig coef. %			75.	
Cattle	DE Mcal/kg	*		2.84	
Sheep	DE Mcal/kg			3.37	
Cattle	ME Mcal/kg	*		2.33	
Sheep	ME Mcal/kg	*		2.76	
Cattle	TDN %	*		64.3	
Sheep	TDN %	*		62.5	

Timothy, drummond, hay, fan air dried, pre-bloom, (1)

Ref No 1 07 929 Canada

Feed Name or Analyses			Mean As Fed	Mean Dry	C.V. ± %
Dry matter	%			100.0	
Organic matter	%			93.2	
Ash	%			6.9	
Crude fiber	%			31.2	
Ether extract	%			2.6	
N-free extract	%			47.2	
Protein (N x 6.25)	%			12.2	
Cattle	dig prot %	*		7.5	
Goats	dig prot %	*		7.9	
Horses	dig prot %	*		7.8	
Rabbits	dig prot %	*		8.0	
Sheep	dig prot %	*		7.5	
Energy	GE Mcal/kg			4.51	
Sheep	GE dig coef %			64.	
Cattle	DE Mcal/kg	*		2.65	
Sheep	DE Mcal/kg			2.88	
Cattle	ME Mcal/kg	*		2.17	
Sheep	ME Mcal/kg	*		2.36	
Cattle	TDN %	*		60.0	
Sheep	TDN %	*		58.1	

Timothy, drummond, hay, fan air dried, early bloom, (1)

Ref No 1 07 926 Canada

Feed Name or Analyses			Mean As Fed	Mean Dry	C.V. ± %
Dry matter	%			100.0	
Organic matter	%			93.7	
Ash	%			6.3	
Crude fiber	%			30.8	
Ether extract	%			2.6	
N-free extract	%			47.2	
Protein (N x 6.25)	%			13.1	
Cattle	dig prot %	*		8.3	
Goats	dig prot %	*		8.8	
Horses	dig prot %	*		8.7	
Rabbits	dig prot %	*		8.8	
Sheep	dig prot %	*		8.3	
Energy	GE Mcal/kg			4.58	
Sheep	GE dig coef %			60.	
Cattle	DE Mcal/kg	*		2.56	
Sheep	DE Mcal/kg			2.75	
Cattle	ME Mcal/kg	*		2.10	
Sheep	ME Mcal/kg	*		2.25	
Cattle	TDN %	*		58.0	
Sheep	TDN %	*		58.7	

Timothy, drummond, hay, fan air dried, full bloom, (1)

Ref No 1 07 927 Canada

Feed Name or Analyses			Mean As Fed	Mean Dry	C.V. ± %
Dry matter	%			100.0	
Organic matter	%			95.3	
Ash	%			4.7	
Crude fiber	%			35.6	
Ether extract	%			2.2	
N-free extract	%			49.6	
Protein (N x 6.25)	%			8.0	
Cattle	dig prot %	*		3.9	
Goats	dig prot %	*		4.0	
Horses	dig prot %	*		4.3	
Rabbits	dig prot %	*		4.8	
Sheep	dig prot %	*		3.7	
Energy	GE Mcal/kg			4.57	
Sheep	GE dig coef %			52.	
Cattle	DE Mcal/kg	*		2.26	
Sheep	DE Mcal/kg			2.37	
Cattle	ME Mcal/kg	*		1.86	
Sheep	ME Mcal/kg	*		1.95	
Cattle	TDN %	*		51.3	
Sheep	TDN %	*		54.9	

TIMOTHY, FINE. Phleum pratense

Timothy, fine, hay, s-c, (1)

Ref No 1 04 959 United States

Feed Name or Analyses			Mean As Fed	Mean Dry	C.V. ± %
Dry matter	%		88.0	100.0	
Ash	%		4.8	5.4	
Crude fiber	%		28.3	32.2	
Horses	dig coef %		47.	47.	
Ether extract	%		2.5	2.8	
Horses	dig coef %		8.	8.	
N-free extract	%		44.7	50.8	
Horses	dig coef %		56.	56.	
Protein (N x 6.25)	%		7.7	8.8	
Horses	dig coef %		58.	58.	
Cattle	dig prot %	*	4.0	4.6	
Goats	dig prot %	*	4.2	4.8	
Horses	dig prot %		4.5	5.1	
Rabbits	dig prot %	*	4.8	5.5	
Sheep	dig prot %	*	3.9	4.5	
Energy	GE Mcal/kg				
Cattle	DE Mcal/kg	*	2.25	2.56	
Horses	DE Mcal/kg	*	1.91	2.17	
Sheep	DE Mcal/kg	*	2.20	2.49	
Cattle	ME Mcal/kg	*	1.84	2.10	
Horses	ME Mcal/kg	*	1.56	1.78	
Sheep	ME Mcal/kg	*	1.80	2.05	
Cattle	TDN %	*	51.0	58.0	
Horses	TDN %		43.3	49.2	
Sheep	TDN %	*	49.8	56.6	

(1) dry forages and roughages (2) pasture, range plants, and forages fed green (3) silages (4) energy feeds (5) protein supplements (6) minerals (7) vitamins (8) additives

TIMOTHY, QUEBEC COMMON. Phleum pratense

Timothy, Quebec common, hay, fan air dried, immature, (1)

Ref No 1 07 931 Canada

Feed Name or Analyses			Mean As Fed	Mean Dry	C.V. ± %
Dry matter	%			100.0	
Organic matter	%			91.5	
Ash	%			8.5	
Crude fiber	%			23.5	
Ether extract	%			3.1	
N-free extract	%			43.4	
Protein (N x 6.25)	%			21.6	
Cattle	dig prot %	*		15.6	
Goats	dig prot %	*		16.7	
Horses	dig prot %	*		15.9	
Rabbits	dig prot %	*		15.3	
Sheep	dig prot %	*		15.9	
Energy	GE Mcal/kg			4.56	
Sheep	GE dig coef %			73.	
Cattle	DE Mcal/kg	*		2.82	
Sheep	DE Mcal/kg			3.30	
Cattle	ME Mcal/kg	*		2.31	
Sheep	ME Mcal/kg	*		2.71	
Cattle	TDN %	*		64.0	
Sheep	TDN %	*		62.5	

Timothy, Quebec common, hay, fan air dried, pre-bloom, (1)

Ref No 1 07 933 Canada

Feed Name or Analyses			Mean As Fed	Mean Dry	C.V. ± %
Dry matter	%			100.0	
Organic matter	%			93.7	
Ash	%			6.3	
Crude fiber	%			32.0	
Ether extract	%			2.3	
N-free extract	%			48.4	
Protein (N x 6.25)	%			11.1	
Cattle	dig prot %	*		6.5	
Goats	dig prot %	*		6.9	
Horses	dig prot %	*		6.9	
Rabbits	dig prot %	*		7.2	
Sheep	dig prot %	*		6.5	
Energy	GE Mcal/kg			4.50	
Sheep	GE dig coef %			64.	
Cattle	DE Mcal/kg	*		2.57	
Sheep	DE Mcal/kg			2.86	
Cattle	ME Mcal/kg	*		2.11	
Sheep	ME Mcal/kg	*		2.34	
Cattle	TDN %	*		58.2	
Sheep	TDN %	*		57.5	

Timothy, Quebec common, hay, fan air dried, early bloom, (1)

Ref No 1 07 930 Canada

Feed Name or Analyses			Mean As Fed	Mean Dry	C.V. ± %
Dry matter	%			100.0	
Organic matter	%			94.3	
Ash	%			5.7	
Crude fiber	%			32.5	
Ether extract	%			2.3	
N-free extract	%			48.5	
Protein (N x 6.25)	%			11.0	
Cattle	dig prot %	*		6.5	
Goats	dig prot %	*		6.8	
Horses	dig prot %	*		6.9	
Rabbits	dig prot %	*		7.2	
Sheep	dig prot %	*		6.4	
Energy	GE Mcal/kg			4.56	
Sheep	GE dig coef %			54.	
Cattle	DE Mcal/kg	*		2.50	
Sheep	DE Mcal/kg			2.46	
Cattle	ME Mcal/kg	*		2.05	
Sheep	ME Mcal/kg	*		2.02	
Cattle	TDN %	*		56.8	
Sheep	TDN %	*		57.3	

Timothy, Quebec common, hay, fan air dried, late bloom, (1)

Ref No 1 07 932 Canada

Feed Name or Analyses			Mean As Fed	Mean Dry	C.V. ± %
Dry matter	%			100.0	
Organic matter	%			95.6	
Ash	%			4.5	
Crude fiber	%			35.2	
Ether extract	%			2.1	
N-free extract	%			51.4	
Protein (N x 6.25)	%			7.0	
Cattle	dig prot %	*		3.0	
Goats	dig prot %	*		3.0	
Horses	dig prot %	*		3.4	
Rabbits	dig prot %	*		4.0	
Sheep	dig prot %	*		2.8	
Energy	GE Mcal/kg			4.56	
Sheep	GE dig coef %			49.	
Cattle	DE Mcal/kg	*		2.29	
Sheep	DE Mcal/kg			2.21	
Cattle	ME Mcal/kg	*		1.87	
Sheep	ME Mcal/kg	*		1.81	
Cattle	TDN %	*		51.9	
Sheep	TDN %	*		54.7	

TIMOTHY, S-50. Phleum nodosum

Timothy, S-50, hay, fan air dried, immature, (1)

Ref No 1 07 936 Canada

Feed Name or Analyses			Mean As Fed	Mean Dry	C.V. ± %
Dry matter	%			100.0	
Organic matter	%			91.3	
Ash	%			8.7	
Crude fiber	%			23.8	
Ether extract	%			3.3	
N-free extract	%			38.1	
Protein (N x 6.25)	%			26.1	
Cattle	dig prot %	*		19.5	
Goats	dig prot %	*		20.9	
Horses	dig prot %	*		19.7	
Rabbits	dig prot %	*		18.8	
Sheep	dig prot %	*		20.0	
Energy	GE Mcal/kg			4.72	
Sheep	GE dig coef %			72.	
Cattle	DE Mcal/kg	*		2.81	
Sheep	DE Mcal/kg			3.40	
Cattle	ME Mcal/kg	*		2.30	
Sheep	ME Mcal/kg	*		2.79	
Cattle	TDN %	*		63.7	
Sheep	TDN %	*		62.6	

Timothy, S-50, hay, fan air dried, pre-bloom, (1)

Ref No 1 07 937 Canada

Feed Name or Analyses			Mean As Fed	Mean Dry	C.V. ± %
Dry matter	%			100.0	
Organic matter	%			92.7	
Ash	%			7.3	

Feed Name or Analyses			Mean As Fed	Mean Dry	C.V. ± %
Crude fiber	%			25.5	
Ether extract	%			3.1	
N-free extract	%			46.5	
Protein (N x 6.25)	%			17.6	
Cattle	dig prot %	*		12.2	
Goats	dig prot %	*		13.0	
Horses	dig prot %	*		12.5	
Rabbits	dig prot %	*		12.3	
Sheep	dig prot %	*		12.4	
Energy	GE Mcal/kg			4.63	
Sheep	GE dig coef %			65.	
Cattle	DE Mcal/kg	*		2.74	
Sheep	DE Mcal/kg			3.01	
Cattle	ME Mcal/kg	*		2.25	
Sheep	ME Mcal/kg	*		2.47	
Cattle	TDN %	*		62.1	
Sheep	TDN %	*		61.6	

Timothy, S-50, hay, fan air dried, early bloom, (1)
Ref No 1 07 934 Canada

			As Fed	Dry	C.V. ± %
Dry matter	%			100.0	
Organic matter	%			93.6	
Ash	%			6.4	
Crude fiber	%			26.7	
Ether extract	%			3.0	
N-free extract	%			47.8	
Protein (N x 6.25)	%			16.1	
Cattle	dig prot %	*		10.9	
Goats	dig prot %	*		11.6	
Horses	dig prot %	*		11.2	
Rabbits	dig prot %	*		11.1	
Sheep	dig prot %	*		11.0	
Energy	GE Mcal/kg			4.67	
Sheep	GE dig coef %			61.	
Cattle	DE Mcal/kg	*		2.67	
Sheep	DE Mcal/kg			2.85	
Cattle	ME Mcal/kg	*		2.19	
Sheep	ME Mcal/kg	*		2.34	
Cattle	TDN %	*		60.6	
Sheep	TDN %	*		61.4	

Timothy, S-50, hay, fan air dried, full bloom, (1)
Ref No 1 07 935 Canada

			As Fed	Dry	C.V. ± %
Dry matter	%			100.0	
Organic matter	%			95.0	
Ash	%			5.0	
Crude fiber	%			27.7	
Ether extract	%			2.4	
N-free extract	%			53.6	
Protein (N x 6.25)	%			11.3	
Cattle	dig prot %	*		6.7	
Goats	dig prot %	*		7.1	
Horses	dig prot %	*		7.1	
Rabbits	dig prot %	*		7.4	
Sheep	dig prot %	*		6.7	
Energy	GE Mcal/kg			4.58	
Sheep	GE dig coef %			53.	
Cattle	DE Mcal/kg	*		2.83	
Sheep	DE Mcal/kg			2.43	
Cattle	ME Mcal/kg	*		2.32	
Sheep	ME Mcal/kg	*		1.99	
Cattle	TDN %	*		64.1	
Sheep	TDN %	*		59.8	

TIMOTHY, WILD. Phleum pratense

Timothy, wild, seeds, (4)
Ref No 5 04 960 United States

			As Fed	Dry	C.V. ± %
Dry matter	%		86.0	100.0	
Ash	%		3.1	3.6	
Crude fiber	%		6.3	7.3	
Sheep	dig coef %		17.	17.	
Ether extract	%		3.1	3.6	
Sheep	dig coef %		48.	48.	
N-free extract	%		55.0	63.9	
Sheep	dig coef %		64.	64.	
Protein (N x 6.25)	%		18.6	21.6	
Sheep	dig coef %		59.	59.	
Cattle	dig prot %	*	13.6	15.8	
Goats	dig prot %	*	14.6	17.0	
Horses	dig prot %	*	14.6	17.0	
Sheep	dig prot %		11.0	12.7	
Energy	GE Mcal/kg				
Cattle	DE Mcal/kg	*	2.17	2.52	
Sheep	DE Mcal/kg	*	2.23	2.59	
Cattle	ME Mcal/kg	*	1.78	2.07	
Sheep	ME Mcal/kg	*	1.83	2.12	
Cattle	TDN %	*	49.1	57.1	
Sheep	TDN %		50.5	58.8	

TIMOTHY-ALFALFA. Phleum pratense, Medicago sativa

Timothy-Alfalfa, hay, s-c, (1)
Ref No 1 04 961 United States

			As Fed	Dry	C.V. ± %
Dry matter	%			100.0	
Vitamin D2	IU/g			1.5	

Timothy-Alfalfa, hay, s-c, 25% alfalfa, (1)
Ref No 1 08 739 United States

			As Fed	Dry	C.V. ± %
Dry matter	%		90.0	100.0	
Ash	%		5.4	6.0	
Crude fiber	%		35.7	39.7	
Ether extract	%		2.0	2.2	
N-free extract	%		38.1	42.3	
Protein (N x 6.25)	%		8.8	9.8	
Cattle	dig prot %	*	4.9	5.4	
Goats	dig prot %	*	5.1	5.7	
Horses	dig prot %	*	5.2	5.8	
Rabbits	dig prot %	*	5.6	6.2	
Sheep	dig prot %	*	4.8	5.3	
Energy	GE Mcal/kg				
Cattle	DE Mcal/kg	*	2.02	2.24	
Sheep	DE Mcal/kg	*	2.12	2.35	
Cattle	ME Mcal/kg	*	1.66	1.84	
Sheep	ME Mcal/kg	*	1.74	1.93	
Cattle	TDN %	*	45.6	50.7	
Sheep	TDN %	*	48.0	53.4	

TIMOTHY-ALFALFA, LIGHT. Phleum pratense, Medicago sativa

Timothy-Alfalfa, light, hay, s-c, (1)
Ref No 1 04 965 United States

			As Fed	Dry	C.V. ± %
Dry matter	%		91.7	100.0	
Ash	%		4.7	5.1	
Crude fiber	%		32.0	34.9	
Ether extract	%		1.8	2.0	
N-free extract	%		43.8	47.8	
Protein (N x 6.25)	%		9.4	10.2	23
Cattle	dig prot %	*	5.3	5.8	
Goats	dig prot %	*	5.6	6.1	
Horses	dig prot %	*	5.7	6.2	
Rabbits	dig prot %	*	6.0	6.5	
Sheep	dig prot %	*	5.2	5.7	
Energy	GE Mcal/kg				
Cattle	DE Mcal/kg	*	2.11	2.30	
Sheep	DE Mcal/kg	*	2.26	2.47	
Cattle	ME Mcal/kg	*	1.73	1.88	
Sheep	ME Mcal/kg	*	1.86	2.02	
Cattle	TDN %	*	47.8	52.1	
Sheep	TDN %	*	51.3	56.0	
Calcium	%		0.49	0.53	20
Cobalt	mg/kg		0.081	0.088	98
Copper	mg/kg		10.9	11.9	
Iron	%		0.012	0.013	
Magnesium	%		0.28	0.31	
Manganese	mg/kg		30.7	33.5	40
Phosphorus	%		0.17	0.18	26
Potassium	%		1.29	1.41	
Carotene	mg/kg		15.0	16.3	73
Vitamin A equivalent	IU/g		24.9	27.2	

Timothy-Alfalfa, light, hay, s-c, immature, (1)
Ref No 1 04 962 United States

			As Fed	Dry	C.V. ± %
Dry matter	%			100.0	
N-free extract	%			49.1	
Protein (N x 6.25)	%			15.2	9
Cattle	dig prot %	*		10.1	
Goats	dig prot %	*		10.7	
Horses	dig prot %	*		10.4	
Rabbits	dig prot %	*		10.4	
Sheep	dig prot %	*		10.2	

Timothy-Alfalfa, light, hay, s-c, full bloom, (1)
Ref No 1 04 963 United States

			As Fed	Dry	C.V. ± %
Dry matter	%		89.4	100.0	
Calcium	%		0.70	0.78	
Cobalt	mg/kg		0.061	0.068	
Copper	mg/kg		10.6	11.9	
Iron	%		0.012	0.013	
Magnesium	%		0.28	0.31	
Manganese	mg/kg		15.6	17.4	
Phosphorus	%		0.17	0.19	
Potassium	%		1.26	1.41	
Carotene	mg/kg		10.2	11.5	
Vitamin A equivalent	IU/g		17.1	19.1	

Timothy-Alfalfa, light, hay, s-c, cut 1, (1)
Ref No 1 04 964 United States

			As Fed	Dry	C.V. ± %
Dry matter	%		89.4	100.0	
N-free extract	%		42.7	47.8	
Protein (N x 6.25)	%		7.7	8.6	20
Cattle	dig prot %	*	3.9	4.4	
Goats	dig prot %	*	4.1	4.6	
Horses	dig prot %	*	4.3	4.8	
Rabbits	dig prot %	*	4.7	5.3	
Sheep	dig prot %	*	3.8	4.3	
Calcium	%		0.49	0.55	23

Continued

Feed Name or Analyses		Mean As Fed	Mean Dry	C.V. ± %
Cobalt	mg/kg	0.057	0.064	24
Copper	mg/kg	10.6	11.9	
Iron	%	0.012	0.013	
Magnesium	%	0.28	0.31	
Manganese	mg/kg	26.2	29.3	37
Phosphorus	%	0.16	0.18	23
Potassium	%	1.26	1.41	
Carotene	mg/kg	14.4	16.1	70
Vitamin A equivalent	IU/g	24.0	26.8	
Vitamin D2	IU/g	1.2	1.4	21

Timothy-Alfalfa, light, hay, s-c, gr 1 US, (1)
Ref No 1 04 967 United States

Feed Name or Analyses		As Fed	Dry	C.V. ± %
Dry matter	%		100.0	
N-free extract	%		47.8	
Protein (N x 6.25)	%		10.5	18
Cattle	dig prot % *		6.0	
Goats	dig prot % *		6.4	
Horses	dig prot % *		6.4	
Rabbits	dig prot % *		6.8	
Sheep	dig prot % *		6.0	
Calcium	%		0.47	15
Cobalt	mg/kg		0.053	
Manganese	mg/kg		35.6	
Phosphorus	%		0.21	13
Carotene	mg/kg		20.1	73
Vitamin A equivalent	IU/g		33.4	
Vitamin D2	IU/g		1.6	10

Timothy-Alfalfa, light, hay, s-c, gr 2 US, (1)
Ref No 1 04 968 United States

Feed Name or Analyses		As Fed	Dry	C.V. ± %
Dry matter	%	91.0	100.0	2
Ash	%	4.6	5.1	
Crude fiber	%	31.7	34.9	
Ether extract	%	1.8	2.0	
N-free extract	%	44.7	49.1	
Protein (N x 6.25)	%	8.1	8.9	23
Cattle	dig prot % *	4.2	4.6	
Goats	dig prot % *	4.4	4.9	
Horses	dig prot % *	4.6	5.1	
Rabbits	dig prot % *	5.0	5.5	
Sheep	dig prot % *	4.1	4.6	
Energy	GE Mcal/kg			
Cattle	DE Mcal/kg *	2.10	2.31	
Sheep	DE Mcal/kg *	2.23	2.45	
Cattle	ME Mcal/kg *	1.72	1.89	
Sheep	ME Mcal/kg *	1.83	2.01	
Cattle	TDN % *	47.7	52.4	
Sheep	TDN % *	50.5	55.5	
Calcium	%	0.58	0.64	21
Cobalt	mg/kg	0.112	0.124	98
Copper	mg/kg	10.8	11.9	
Iron	%	0.012	0.013	
Magnesium	%	0.28	0.31	
Manganese	mg/kg	22.6	24.8	77
Phosphorus	%	0.15	0.17	34
Potassium	%	1.28	1.41	
Carotene	mg/kg	12.4	13.7	49
Vitamin A equivalent	IU/g	20.7	22.8	

(1) dry forages and roughages (2) pasture, range plants, and forages fed green (3) silages (4) energy feeds (5) protein supplements (6) minerals (7) vitamins (8) additives

Timothy-Alfalfa, light, hay, s-c, gr 3 US, (1)
Ref No 1 04 969 United States

Feed Name or Analyses		As Fed	Dry	C.V. ± %
Dry matter	%		100.0	
N-free extract	%		49.1	
Protein (N x 6.25)	%		10.5	32
Cattle	dig prot % *		6.0	
Goats	dig prot % *		6.4	
Horses	dig prot % *		6.4	
Rabbits	dig prot % *		6.8	
Sheep	dig prot % *		6.0	
Calcium	%		0.45	17
Cobalt	mg/kg		0.090	
Manganese	mg/kg		30.4	
Phosphorus	%		0.18	19
Carotene	mg/kg		7.7	97
Vitamin A equivalent	IU/g		12.9	
Vitamin D2	IU/g		1.0	

Timothy-Alfalfa, light, hay, s-c, gr sample US, (1)
Ref No 1 04 966 United States

Feed Name or Analyses		As Fed	Dry	C.V. ± %
Dry matter	%		100.0	
N-free extract	%		49.1	
Protein (N x 6.25)	%		11.3	26
Cattle	dig prot % *		6.7	
Goats	dig prot % *		7.1	
Horses	dig prot % *		7.1	
Rabbits	dig prot % *		7.4	
Sheep	dig prot % *		6.7	
Calcium	%		0.52	
Cobalt	mg/kg		0.051	
Manganese	mg/kg		43.2	
Phosphorus	%		0.17	
Carotene	mg/kg		5.7	
Vitamin A equivalent	IU/g		9.6	
Vitamin D2	IU/g		1.1	

TIMOTHY-CLOVER. Phleum pratense, Trifolium spp

Timothy-Clover, hay, s-c, (1)
Ref No 1 04 973 United States

Feed Name or Analyses		As Fed	Dry	C.V. ± %
Dry matter	%	88.9	100.0	
Ash	%	5.5	6.2	
Crude fiber	%	31.2	35.1	
Ether extract	%	2.2	2.5	
N-free extract	%	41.8	47.0	
Protein (N x 6.25)	%	8.3	9.3	
Cattle	dig prot % *	4.4	5.0	
Goats	dig prot % *	4.7	5.2	
Horses	dig prot % *	4.8	5.4	
Rabbits	dig prot % *	5.2	5.8	
Sheep	dig prot % *	4.4	4.9	
Energy	GE Mcal/kg			
Cattle	DE Mcal/kg *	2.10	2.36	
Sheep	DE Mcal/kg *	2.17	2.44	
Cattle	ME Mcal/kg *	1.72	1.94	
Sheep	ME Mcal/kg *	1.78	2.00	
Cattle	TDN % *	47.6	53.5	
Sheep	TDN % *	49.2	55.3	
Calcium	%	0.62	0.70	
Chlorine	%	0.53	0.59	
Copper	mg/kg	6.2	7.0	
Iron	%	0.014	0.015	
Magnesium	%	0.23	0.26	
Manganese	mg/kg	47.4	53.3	
Phosphorus	%	0.15	0.17	
Potassium	%	1.61	1.81	
Sodium	%	0.17	0.19	
Sulphur	%	0.13	0.14	
Carotene	mg/kg	23.7	26.7	58
Vitamin A equivalent	IU/g	39.5	44.5	

Timothy-Clover, hay, s-c, immature, (1)
Ref No 1 04 971 United States

Feed Name or Analyses		As Fed	Dry	C.V. ± %
Dry matter	%		100.0	
Vitamin D2	IU/g		2.1	11

Timothy-Clover, hay, s-c, cut 1, (1)
Ref No 1 04 972 United States

Feed Name or Analyses		As Fed	Dry	C.V. ± %
Dry matter	%		100.0	
Carotene	mg/kg		29.5	
Vitamin A equivalent	IU/g		49.2	

Timothy-Clover, hay, s-c, gr 1 US, (1)
Ref No 1 04 974 United States

Feed Name or Analyses		As Fed	Dry	C.V. ± %
Dry matter	%		100.0	
Vitamin D2	IU/g		2.1	

Timothy-Clover, hay, s-c, gr 2 US, (1)
Ref No 1 04 975 United States

Feed Name or Analyses		As Fed	Dry	C.V. ± %
Dry matter	%		100.0	
Carotene	mg/kg		48.5	
Vitamin A equivalent	IU/g		80.9	
Vitamin D2	IU/g		2.2	

Timothy-Clover, hay, s-c, gr 3 US, (1)
Ref No 1 04 976 United States

Feed Name or Analyses		As Fed	Dry	C.V. ± %
Dry matter	%		100.0	
Vitamin D2	IU/g		1.8	

Timothy-Clover, hay, s-c, 40% clover, (1)
Ref No 1 08 740 United States

Feed Name or Analyses		As Fed	Dry	C.V. ± %
Dry matter	%	90.0	100.0	
Ash	%	5.6	6.2	
Crude fiber	%	32.6	36.2	
Ether extract	%	2.1	2.3	
N-free extract	%	40.4	44.9	
Protein (N x 6.25)	%	9.3	10.3	
Cattle	dig prot % *	5.3	5.9	
Goats	dig prot % *	5.6	6.2	
Horses	dig prot % *	5.7	6.3	
Rabbits	dig prot % *	6.0	6.6	
Sheep	dig prot % *	5.3	5.8	
Energy	GE Mcal/kg			
Cattle	DE Mcal/kg *	2.03	2.26	
Sheep	DE Mcal/kg *	2.19	2.43	
Cattle	ME Mcal/kg *	1.67	1.85	
Sheep	ME Mcal/kg *	1.79	1.99	
Cattle	TDN % *	46.1	51.2	
Sheep	TDN % *	49.6	55.2	

Feed Name or Analyses			Mean As Fed	Mean Dry	C.V. ± %
Timothy-Clover, aerial part, ensiled, (3)					
Ref No 3 04 980				United States	
Dry matter	%		27.1	100.0	7
Ash	%		2.1	7.9	23
Crude fiber	%		9.3	34.4	10
Ether extract	%		1.0	3.7	30
N-free extract	%		11.6	42.6	
Protein (N x 6.25)	%		3.1	11.4	19
Cattle	dig prot %	*	1.8	6.6	
Goats	dig prot %	*	1.8	6.6	
Horses	dig prot %	*	1.8	6.6	
Sheep	dig prot %	*	1.8	6.6	
Energy	GE Mcal/kg				
Cattle	DE Mcal/kg	*	0.69	2.53	
Sheep	DE Mcal/kg	*	0.72	2.66	
Cattle	ME Mcal/kg	*	0.56	2.07	
Sheep	ME Mcal/kg	*	0.59	2.18	
Cattle	TDN %	*	15.6	57.3	
Sheep	TDN %	*	16.3	60.3	
Calcium	%		0.25	0.91	31
Phosphorus	%		0.07	0.26	25
Carotene	mg/kg		39.3	144.8	58
Vitamin A equivalent	IU/g		65.5	241.5	
Timothy-Clover, aerial part, ensiled, immature, (3)					
Ref No 3 04 977				United States	
Dry matter	%		34.1	100.0	
Ash	%		4.3	12.5	
Crude fiber	%		11.4	33.6	
Ether extract	%		0.8	2.3	
N-free extract	%		12.9	37.8	
Protein (N x 6.25)	%		4.7	13.8	
Cattle	dig prot %	*	3.0	8.8	
Goats	dig prot %	*	3.0	8.8	
Horses	dig prot %	*	3.0	8.8	
Sheep	dig prot %	*	3.0	8.8	
Energy	GE Mcal/kg				
Cattle	DE Mcal/kg	*	0.89	2.62	
Sheep	DE Mcal/kg	*	0.91	2.67	
Cattle	ME Mcal/kg	*	0.73	2.15	
Sheep	ME Mcal/kg	*	0.75	2.19	
Cattle	TDN %	*	20.2	59.4	
Sheep	TDN %	*	20.6	60.5	
Chlorine	%		0.43	1.26	
Carotene	mg/kg		70.5	207.0	33
Vitamin A equivalent	IU/g		117.6	345.1	
Timothy-Clover, aerial part, ensiled, mature, (3)					
Ref No 3 04 978				United States	
Dry matter	%		27.6	100.0	
Carotene	mg/kg		2.3	8.4	
Vitamin A equivalent	IU/g		3.9	14.0	
Timothy-Clover, aerial part, ensiled, cut 1, (3)					
Ref No 3 04 979				United States	
Dry matter	%		26.5	100.0	6
Ash	%		2.0	7.6	19
Crude fiber	%		8.9	33.6	7
Ether extract	%		1.0	3.6	18
N-free extract	%		11.7	44.2	
Protein (N x 6.25)	%		2.9	11.0	16
Cattle	dig prot %	*	1.6	6.2	
Goats	dig prot %	*	1.6	6.2	
Horses	dig prot %	*	1.6	6.2	
Sheep	dig prot %	*	1.6	6.2	
Energy	GE Mcal/kg				
Cattle	DE Mcal/kg	*	0.68	2.56	
Sheep	DE Mcal/kg	*	0.72	2.70	
Cattle	ME Mcal/kg	*	0.56	2.10	
Sheep	ME Mcal/kg	*	0.58	2.21	
Cattle	TDN %	*	15.4	58.0	
Sheep	TDN %	*	16.2	61.2	
Carotene	mg/kg		45.1	170.2	49
Vitamin A equivalent	IU/g		75.2	283.7	
Timothy-Clover, aerial part, w phosphoric acid preservative added, ensiled, (3)					
Ref No 3 04 982				United States	
Dry matter	%		22.4	100.0	
Ash	%		1.7	7.8	
Crude fiber	%		9.4	42.0	
Ether extract	%		1.4	6.4	
N-free extract	%		7.8	34.6	
Protein (N x 6.25)	%		2.1	9.2	
Cattle	dig prot %	*	1.0	4.6	
Goats	dig prot %	*	1.0	4.6	
Horses	dig prot %	*	1.0	4.6	
Sheep	dig prot %	*	1.0	4.6	
Energy	GE Mcal/kg				
Cattle	DE Mcal/kg	*	0.55	2.46	
Sheep	DE Mcal/kg	*	0.59	2.64	
Cattle	ME Mcal/kg	*	0.45	2.02	
Sheep	ME Mcal/kg	*	0.49	2.17	
Cattle	TDN %	*	12.5	55.8	
Sheep	TDN %	*	13.4	60.0	
Timothy-Clover, aerial part w sodium bisulfite preservative added, ensiled, immature, (3)					
Ref No 3 04 981				United States	
Dry matter	%		25.7	100.0	3
Carotene	mg/kg		58.5	227.7	22
Vitamin A equivalent	IU/g		97.6	379.6	
Timothy-Clover, aerial part w sodium metabisulfite preservative added, ensiled, (3)					
Ref No 3 04 984				United States	
Dry matter	%		25.7	100.0	3
Ash	%		1.9	7.2	9
Crude fiber	%		8.5	32.9	4
Ether extract	%		1.1	4.1	11
N-free extract	%		11.6	45.3	
Protein (N x 6.25)	%		2.7	10.5	8
Cattle	dig prot %	*	1.5	5.8	
Goats	dig prot %	*	1.5	5.8	
Horses	dig prot %	*	1.5	5.8	
Sheep	dig prot %	*	1.5	5.8	
Energy	GE Mcal/kg				
Cattle	DE Mcal/kg	*	0.67	2.62	
Sheep	DE Mcal/kg	*	0.72	2.79	
Cattle	ME Mcal/kg	*	0.55	2.15	
Sheep	ME Mcal/kg	*	0.59	2.29	
Cattle	TDN %	*	15.3	59.4	
Sheep	TDN %	*	16.3	63.4	
TIMOTHY-CLOVER, HEAVY. Phleum pratense, Trifolium spp					
Timothy-Clover, heavy, hay, s-c, (1)					
Ref No 1 04 987				United States	
Dry matter	%			100.0	
Calcium	%			0.83	19
Cobalt	mg/kg			0.130	77
Manganese	mg/kg			54.7	48
Phosphorus	%			0.21	26
Carotene	mg/kg			18.6	55
Vitamin A equivalent	IU/g			31.1	
Timothy-Clover, heavy, hay, s-c, cut 1, (1)					
Ref No 1 04 985				United States	
Dry matter	%			100.0	
Calcium	%			0.82	22
Cobalt	mg/kg			0.135	77
Manganese	mg/kg			56.9	46
Phosphorus	%			0.20	28
Carotene	mg/kg			16.1	53
Vitamin A equivalent	IU/g			26.8	
Vitamin D_2	IU/g			2.1	
Timothy-Clover, heavy, hay, s-c, cut 2, (1)					
Ref No 1 04 986				United States	
Dry matter	%			100.0	
Calcium	%			0.96	
Cobalt	mg/kg			0.071	
Manganese	mg/kg			41.2	
Phosphorus	%			0.29	
Timothy-Clover, heavy, hay, s-c, gr 1 US, (1)					
Ref No 1 04 989				United States	
Dry matter	%			100.0	
Calcium	%			0.91	22
Cobalt	mg/kg			0.085	70
Manganese	mg/kg			41.1	41
Phosphorus	%			0.21	25
Carotene	mg/kg			19.8	55
Vitamin A equivalent	IU/g			33.1	
Vitamin D_2	IU/g			2.2	
Timothy-Clover, heavy, hay, s-c, gr 2 US, (1)					
Ref No 1 04 990				United States	
Dry matter	%			100.0	
Calcium	%			0.74	21
Cobalt	mg/kg			0.155	50
Manganese	mg/kg			74.5	32
Phosphorus	%			0.19	19
Carotene	mg/kg			15.7	38
Vitamin A equivalent	IU/g			26.1	
Timothy-Clover, heavy, hay, s-c, gr 3 US, (1)					
Ref No 1 04 991				United States	
Dry matter	%			100.0	
Calcium	%			0.95	

Continued

Feed Name or Analyses			Mean As Fed	Mean Dry	C.V. ± %
Cobalt	mg/kg			0.115	
Manganese	mg/kg			46.1	
Phosphorus	%			0.26	

Timothy-Clover, heavy, hay, s-c, gr sample US, (1)
Ref No 1 04 988 United States

Feed Name or Analyses			Mean As Fed	Mean Dry	C.V. ± %
Dry matter	%			100.0	
Calcium	%			0.87	24
Cobalt	mg/kg			0.130	88
Manganese	mg/kg			40.2	55
Phosphorus	%			0.22	24
Carotene	mg/kg			16.5	87
Vitamin A equivalent	IU/g			27.6	

TIMOTHY-CLOVER, LIGHT. Phleum pratense, Trifolium spp

Timothy-Clover, light, hay, s-c, (1)
Ref No 1 04 996 United States

Feed Name or Analyses			Mean As Fed	Mean Dry	C.V. ± %
Dry matter	%		88.5	100.0	2
Ash	%		4.7	5.4	7
Crude fiber	%		31.4	35.5	3
Ether extract	%		1.9	2.1	10
N-free extract	%		43.4	49.0	
Protein (N x 6.25)	%		7.2	8.1	21
Cattle	dig prot %	*	3.5	4.0	
Goats	dig prot %	*	3.6	4.1	
Horses	dig prot %	*	3.9	4.4	
Rabbits	dig prot %	*	4.4	4.9	
Sheep	dig prot %	*	3.4	3.8	
Energy	GE Mcal/kg				
Cattle	DE Mcal/kg	*	2.03	2.29	
Sheep	DE Mcal/kg	*	2.14	2.42	
Cattle	ME Mcal/kg	*	1.66	1.88	
Sheep	ME Mcal/kg	*	1.76	1.98	
Cattle	TDN %	*	46.0	52.0	
Sheep	TDN %	*	48.6	54.9	
Calcium	%		0.51	0.58	19
Cobalt	mg/kg		0.072	0.082	42
Manganese	mg/kg		46.4	52.5	31
Phosphorus	%		0.15	0.17	25
Carotene	mg/kg		10.5	11.9	28
Vitamin A equivalent	IU/g		17.6	19.8	
Vitamin D_2	IU/g		1.9	2.2	15

Timothy-Clover, light, hay, s-c, immature, (1)
Ref No 1 04 993 United States

Feed Name or Analyses			Mean As Fed	Mean Dry	C.V. ± %
Dry matter	%		90.1	100.0	
Ash	%		5.0	5.5	
Crude fiber	%		33.1	36.7	
Ether extract	%		1.6	1.8	
N-free extract	%		40.4	44.8	
Protein (N x 6.25)	%		10.1	11.2	10
Cattle	dig prot %	*	6.0	6.6	
Goats	dig prot %	*	6.3	7.0	
Horses	dig prot %	*	6.3	7.0	
Rabbits	dig prot %	*	6.6	7.3	
Sheep	dig prot %	*	6.0	6.6	
Energy	GE Mcal/kg				
Cattle	DE Mcal/kg	*	1.95	2.16	
Sheep	DE Mcal/kg	*	2.13	2.36	
Cattle	ME Mcal/kg	*	1.60	1.77	
Sheep	ME Mcal/kg	*	1.75	1.94	
Cattle	TDN %	*	44.2	49.0	
Sheep	TDN %	*	48.2	53.5	

(1) dry forages and roughages (2) pasture, range plants, and forages fed green (3) silages (4) energy feeds (5) protein supplements (6) minerals (7) vitamins (8) additives

Timothy-Clover, light, hay, s-c, mature, (1)
Ref No 1 04 994 United States

Feed Name or Analyses			Mean As Fed	Mean Dry	C.V. ± %
Dry matter	%			100.0	
N-free extract	%			44.8	
Protein (N x 6.25)	%			6.0	10
Cattle	dig prot %	*		2.1	
Goats	dig prot %	*		2.2	
Horses	dig prot %	*		2.6	
Rabbits	dig prot %	*		3.3	
Sheep	dig prot %	*		2.0	

Timothy-Clover, light, hay, s-c, cut 1, (1)
Ref No 1 04 995 United States

Feed Name or Analyses			Mean As Fed	Mean Dry	C.V. ± %
Dry matter	%			100.0	
N-free extract	%			47.8	
Protein (N x 6.25)	%			8.0	23
Cattle	dig prot %	*		3.9	
Goats	dig prot %	*		4.0	
Horses	dig prot %	*		4.3	
Rabbits	dig prot %	*		4.8	
Sheep	dig prot %	*		3.7	
Calcium	%			0.55	13
Cobalt	mg/kg			0.073	40
Manganese	mg/kg			52.2	35
Phosphorus	%			0.16	9
Carotene	mg/kg			13.9	27
Vitamin A equivalent	IU/g			23.2	
Vitamin D_2	IU/g			1.8	

Timothy-Clover, light, hay, s-c, gr 1 US, (1)
Ref No 1 04 999 United States

Feed Name or Analyses			Mean As Fed	Mean Dry	C.V. ± %
Dry matter	%			100.0	
Calcium	%			0.59	22
Cobalt	mg/kg			0.090	
Magnesium	%			0.18	
Manganese	mg/kg			71.2	
Phosphorus	%			0.17	15
Potassium	%			1.34	
Sulphur	%			0.18	
Carotene	mg/kg			13.0	
Vitamin A equivalent	IU/g			21.7	
Vitamin D_2	IU/g			2.4	

Timothy-Clover, light, hay, s-c, gr 2 US, (1)
Ref No 1 05 000 United States

Feed Name or Analyses			Mean As Fed	Mean Dry	C.V. ± %
Dry matter	%		90.1	100.0	
Ash	%		5.0	5.5	
Crude fiber	%		33.1	36.7	
Ether extract	%		1.6	1.8	
N-free extract	%		42.5	47.2	
Protein (N x 6.25)	%		7.9	8.8	22
Cattle	dig prot %	*	4.1	4.6	
Goats	dig prot %	*	4.3	4.8	
Horses	dig prot %	*	4.5	5.0	
Rabbits	dig coef %	*	4.9	5.5	
Sheep	dig prot %	*	4.0	4.5	
Energy	GE Mcal/kg				
Cattle	DE Mcal/kg	*	1.97	2.19	
Sheep	DE Mcal/kg	*	2.17	2.41	
Cattle	ME Mcal/kg	*	1.62	1.79	
Sheep	ME Mcal/kg	*	1.78	1.97	
Cattle	TDN %	*	44.7	49.6	
Sheep	TDN %	*	49.2	54.6	
Calcium	%		0.51	0.57	30
Cobalt	mg/kg		0.099	0.110	66
Manganese	mg/kg		45.2	50.2	40
Phosphorus	%		0.15	0.17	25
Carotene	mg/kg		9.3	10.4	42
Vitamin A equivalent	IU/g		15.6	17.3	

Timothy-Clover, light, hay, s-c, gr 3 US, (1)
Ref No 1 05 001 United States

Feed Name or Analyses			Mean As Fed	Mean Dry	C.V. ± %
Dry matter	%			100.0	
N-free extract	%			47.2	
Protein (N x 6.25)	%			7.2	18
Cattle	dig prot %	*		3.2	
Goats	dig prot %	*		3.3	
Horses	dig prot %	*		3.6	
Rabbits	dig coef %	*		4.2	
Sheep	dig prot %	*		3.0	
Calcium	%			0.54	22
Cobalt	mg/kg			0.064	15
Manganese	mg/kg			41.3	39
Phosphorus	%			0.17	17
Carotene	mg/kg			10.1	41
Vitamin A equivalent	IU/g			16.9	
Vitamin D_2	IU/g			1.9	

Timothy-Clover, light, hay, s-c, gr sample US, (1)
Ref No 1 04 997 United States

Feed Name or Analyses			Mean As Fed	Mean Dry	C.V. ± %
Dry matter	%			100.0	
N-free extract	%			47.2	
Protein (N x 6.25)	%			8.3	26
Cattle	dig prot %	*		4.1	
Goats	dig prot %	*		4.3	
Horses	dig prot %	*		4.6	
Rabbits	dig coef %	*		5.1	
Sheep	dig prot %	*		4.0	
Calcium	%			0.65	42
Cobalt	mg/kg			0.095	30
Manganese	mg/kg			48.4	24
Phosphorus	%			0.21	24
Carotene	mg/kg			9.3	
Vitamin A equivalent	IU/g			15.4	

TIMOTHY-CLOVER, RED. Phleum pratense, Trifolium Pratense

Timothy-Clover, red, hay, s-c, (1)
Ref No 1 05 006 United States

Feed Name or Analyses			Mean As Fed	Mean Dry	C.V. ± %
Dry matter	%		90.4	100.0	
Calcium	%		0.57	0.63	
Cobalt	mg/kg		0.078	0.086	
Copper	mg/kg		14.9	16.5	
Iron	%		0.012	0.013	
Magnesium	%		0.28	0.31	
Manganese	mg/kg		40.5	44.8	
Phosphorus	%		0.15	0.17	

Feed Name or Analyses			Mean As Fed	Mean Dry	C.V. ± %
Potassium	%		1.37	1.52	
Carotene	mg/kg		15.3	17.0	
Vitamin A equivalent	IU/g		25.6	28.3	

Timothy-Clover, red, hay, s-c, full bloom, (1)

Ref No 1 05 003 United States

Feed Name or Analyses			Mean As Fed	Mean Dry	C.V. ± %
Dry matter	%		89.5	100.0	
Calcium	%		0.86	0.96	
Copper	mg/kg		15.0	16.8	
Iron	%		0.013	0.014	
Magnesium	%		0.44	0.49	
Manganese	mg/kg		19.9	22.3	
Phosphorus	%		0.17	0.19	
Potassium	%		1.36	1.52	
Carotene	mg/kg		20.3	22.7	
Vitamin A equivalent	IU/g		33.9	37.9	

Timothy-Clover, red, hay, s-c, cut 1, (1)

Ref No 1 05 004 United States

Feed Name or Analyses			Mean As Fed	Mean Dry	C.V. ± %
Dry matter	%		90.2	100.0	
Calcium	%		0.54	0.60	
Copper	mg/kg		12.1	13.4	
Iron	%		0.013	0.014	
Magnesium	%		0.29	0.32	
Manganese	mg/kg		46.7	51.8	
Phosphorus	%		0.15	0.17	
Potassium	%		1.27	1.41	
Carotene	mg/kg		13.1	14.6	
Vitamin A equivalent	IU/g		21.9	24.3	

Timothy-Clover, red, hay, s-c, gr 2 US, (1)

Ref No 1 05 005 United States

Feed Name or Analyses			Mean As Fed	Mean Dry	C.V. ± %
Dry matter	%		90.2	100.0	
Calcium	%		0.54	0.60	
Copper	mg/kg		12.1	13.4	
Iron	%		0.013	0.014	
Magnesium	%		0.29	0.32	
Manganese	mg/kg		46.7	51.8	
Phosphorus	%		0.15	0.17	
Potassium	%		1.27	1.41	
Carotene	mg/kg		13.1	14.6	
Vitamin A equivalent	IU/g		21.9	24.3	

Timothy-Clover, red, aerial part, ensiled, (3)

Ref No 3 05 008 United States

Feed Name or Analyses			Mean As Fed	Mean Dry	C.V. ± %
Dry matter	%			100.0	
Ash	%			7.4	19
Crude fiber	%			30.5	12
Ether extract	%			3.1	10
N-free extract	%			46.0	
Protein (N x 6.25)	%			13.2	13
Cattle	dig prot %	*		8.2	
Goats	dig prot %	*		8.2	
Horses	dig prot %	*		8.2	
Sheep	dig prot %	*		8.2	
Energy	GE Mcal/kg				
Cattle	DE Mcal/kg	*		2.54	
Sheep	DE Mcal/kg	*		2.72	
Cattle	ME Mcal/kg	*		2.08	
Sheep	ME Mcal/kg	*		2.23	
Cattle	TDN %	*		57.6	
Sheep	TDN %	*		61.8	

Feed Name or Analyses			Mean As Fed	Mean Dry	C.V. ± %
Sulphur	%			0.63	47
Carotene	mg/kg			73.9	53
Vitamin A equivalent	IU/g			123.1	

Timothy-Clover, red, aerial part, ensiled, dough stage, (3)

Ref No 3 05 007 United States

Feed Name or Analyses			Mean As Fed	Mean Dry	C.V. ± %
Dry matter	%			100.0	
Carotene	mg/kg			29.1	
Vitamin A equivalent	IU/g			48.5	

Timothy-Clover, red, aerial part w corn grits by-product added, ensiled, (3)

Ref No 3 05 009 United States

Feed Name or Analyses			Mean As Fed	Mean Dry	C.V. ± %
Dry matter	%			100.0	
Ash	%			7.7	
Crude fiber	%			27.3	
Ether extract	%			3.5	
N-free extract	%			47.9	
Protein (N x 6.25)	%			13.6	
Cattle	dig prot %	*		8.6	
Goats	dig prot %	*		8.6	
Horses	dig prot %	*		8.6	
Sheep	dig prot %	*		8.6	
Energy	GE Mcal/kg				
Cattle	DE Mcal/kg	*		2.70	
Sheep	DE Mcal/kg	*		2.82	
Cattle	ME Mcal/kg	*		2.21	
Sheep	ME Mcal/kg	*		2.32	
Cattle	TDN %	*		61.1	
Sheep	TDN %	*		64.1	
Sulphur	%			0.20	

Timothy-Clover, red, aerial part w sulfur dioxide preservative added, ensiled, (3)

Ref No 3 05 010 United States

Feed Name or Analyses			Mean As Fed	Mean Dry	C.V. ± %
Dry matter	%			100.0	
Ash	%			8.0	8
Crude fiber	%			29.4	6
Ether extract	%			3.3	10
N-free extract	%			44.7	
Protein (N x 6.25)	%			14.6	7
Cattle	dig prot %	*		9.5	
Goats	dig prot %	*		9.5	
Horses	dig prot %	*		9.5	
Sheep	dig prot %	*		9.5	
Energy	GE Mcal/kg				
Cattle	DE Mcal/kg	*		2.56	
Sheep	DE Mcal/kg	*		2.79	
Cattle	ME Mcal/kg	*		2.10	
Sheep	ME Mcal/kg	*		2.29	
Cattle	TDN %	*		58.2	
Sheep	TDN %	*		63.3	
Sulphur	%			0.66	
Carotene	mg/kg			98.5	
Vitamin A equivalent	IU/g			164.3	

TIMOTHY-GRASS-ALFALFA. Phleum pratense, Scientific name not used, Medicago sativa

Timothy-Grass-Alfalfa, hay, (1)

Ref No 1 05 013 United States

Feed Name or Analyses			Mean As Fed	Mean Dry	C.V. ± %
Dry matter	%		90.9	100.0	1
Vitamin D_2	IU/g		0.9	1.0	67

Timothy-Grass-Alfalfa, hay, fan air dried, (1)

Ref No 1 05 011 United States

Feed Name or Analyses			Mean As Fed	Mean Dry	C.V. ± %
Dry matter	%		91.1	100.0	1
Vitamin D_2	IU/g		0.9	0.9	49

TIMOTHY-GRASS, HEAVY. Phleum pratense, Scientific name not used

Timothy-Grass, heavy, hay, s-c, (1)

Ref No 1 05 016 United States

Feed Name or Analyses			Mean As Fed	Mean Dry	C.V. ± %
Dry matter	%			100.0	
N-free extract	%			42.8	
Protein (N x 6.25)	%			7.0	20
Cattle	dig prot %	*		3.0	
Goats	dig prot %	*		3.1	
Horses	dig prot %	*		3.5	
Rabbits	dig coef %	*		4.1	
Sheep	dig prot %	*		2.8	
Calcium	%			0.49	26
Chlorine	%			0.79	
Cobalt	mg/kg			0.090	68
Copper	mg/kg			8.8	
Iron	%			0.001	
Magnesium	%			0.19	
Manganese	mg/kg			42.8	54
Phosphorus	%			0.16	20
Potassium	%			1.80	

Timothy-Grass, heavy, hay, s-c, mature, (1)

Ref No 1 05 014 United States

Feed Name or Analyses			Mean As Fed	Mean Dry	C.V. ± %
Dry matter	%			100.0	
N-free extract	%			42.8	
Protein (N x 6.25)	%			5.4	15
Cattle	dig prot %	*		1.6	
Goats	dig prot %	*		1.6	
Horses	dig prot %	*		2.1	
Rabbits	dig coef %	*		2.8	
Sheep	dig prot %	*		1.4	

Timothy-Grass, heavy, hay, s-c, cut 1, (1)

Ref No 1 05 015 United States

Feed Name or Analyses			Mean As Fed	Mean Dry	C.V. ± %
Dry matter	%			100.0	
N-free extract	%			42.8	
Protein (N x 6.25)	%			7.0	17
Cattle	dig prot %	*		3.0	
Goats	dig prot %	*		3.1	
Horses	dig prot %	*		3.5	
Rabbits	dig coef %	*		4.1	
Sheep	dig prot %	*		2.8	

Timothy-Grass, heavy, hay, s-c, gr 1 US, (1)
Ref No 1 05 018 United States

Feed Name or Analyses		As Fed	Dry	C.V. ± %
Dry matter	%		100.0	
N-free extract	%		42.8	
Protein (N x 6.25)	%		7.4	
Cattle	dig prot % *		3.3	
Goats	dig prot % *		3.5	
Horses	dig prot % *		3.8	
Rabbits	dig coef % *		4.4	
Sheep	dig prot % *		3.2	
Calcium	%		0.46	
Cobalt	mg/kg		0.057	
Manganese	mg/kg		32.4	
Phosphorus	%		0.15	

Timothy-Grass, heavy, hay, s-c, gr 2 US, (1)
Ref No 1 05 019 United States

Feed Name or Analyses		As Fed	Dry	C.V. ± %
Dry matter	%		100.0	
N-free extract	%		42.8	
Protein (N x 6.25)	%		7.0	23
Cattle	dig prot % *		3.0	
Goats	dig prot % *		3.1	
Horses	dig prot % *		3.5	
Rabbits	dig coef % *		4.1	
Sheep	dig prot % *		2.8	
Calcium	%		0.48	30
Cobalt	mg/kg		0.104	68
Manganese	mg/kg		49.4	24
Phosphorus	%		0.17	15
Carotene	mg/kg		13.4	64
Vitamin A equivalent	IU/g		22.4	

Timothy-Grass, heavy, hay, s-c, gr 3 US, (1)
Ref No 1 05 020 United States

Feed Name or Analyses		As Fed	Dry	C.V. ± %
Dry matter	%		100.0	
N-free extract	%		42.8	
Protein (N x 6.25)	%		7.0	18
Cattle	dig prot % *		3.0	
Goats	dig prot % *		3.1	
Horses	dig prot % *		3.5	
Rabbits	dig coef % *		4.1	
Sheep	dig prot % *		2.8	
Calcium	%		0.51	17
Cobalt	mg/kg		0.095	88
Manganese	mg/kg		36.8	77
Phosphorus	%		0.15	21
Carotene	mg/kg		8.6	34
Vitamin A equivalent	IU/g		14.3	

Timothy-Grass, heavy, hay, s-c, gr sample US, (1)
Ref No 1 05 017 United States

Feed Name or Analyses		As Fed	Dry	C.V. ± %
Dry matter	%		100.0	
N-free extract	%		42.8	
Protein (N x 6.25)	%		6.9	13
Cattle	dig prot % *		2.9	
Goats	dig prot % *		3.0	
Horses	dig prot % *		3.4	
Rabbits	dig coef % *		4.0	
Sheep	dig prot % *		2.8	
Calcium	%		0.47	38
Cobalt	mg/kg		0.066	
Manganese	mg/kg		64.2	
Phosphorus	%		0.17	16
Carotene	mg/kg		11.2	74
Vitamin A equivalent	IU/g		18.7	

Timothy-Grass, heavy, aerial part w molasses, added, ensiled, mid-bloom, (3)
Ref No 3 05 021 United States

Feed Name or Analyses		As Fed	Dry	C.V. ± %
Dry matter	%	29.4	100.0	
Ash	%	2.5	8.4	
Crude fiber	%	9.8	33.4	
Ether extract	%	1.0	3.3	
N-free extract	%	12.6	42.8	
Protein (N x 6.25)	%	3.6	12.1	
Cattle	dig prot % *	2.1	7.2	
Goats	dig prot % *	2.1	7.2	
Horses	dig prot % *	2.1	7.2	
Sheep	dig prot % *	2.1	7.2	
Energy	GE Mcal/kg			
Cattle	DE Mcal/kg *	0.75	2.54	
Sheep	DE Mcal/kg *	0.81	2.77	
Cattle	ME Mcal/kg *	0.61	2.08	
Sheep	ME Mcal/kg *	0.67	2.27	
Cattle	TDN % *	16.9	57.6	
Sheep	TDN % *	18.4	62.7	

TIMOTHY-GRASS, LIGHT. Phleum pratense,
Scientific name not used

Timothy-Grass, light, hay, s-c, (1)
Ref No 1 05 025 United States

Feed Name or Analyses		As Fed	Dry	C.V. ± %
Dry matter	%	89.9	100.0	1
Ash	%	5.5	6.1	21
Crude fiber	%	31.6	35.2	12
Ether extract	%	1.7	1.9	15
N-free extract	%	44.1	49.1	
Protein (N x 6.25)	%	6.9	7.7	28
Cattle	dig prot % *	3.2	3.6	
Goats	dig prot % *	3.4	3.7	
Horses	dig prot % *	3.7	4.1	
Rabbits	dig coef % *	4.1	4.6	
Sheep	dig prot % *	3.1	3.5	
Energy	GE Mcal/kg			
Cattle	DE Mcal/kg *	2.10	2.34	
Sheep	DE Mcal/kg *	2.17	2.41	
Cattle	ME Mcal/kg *	1.72	1.92	
Sheep	ME Mcal/kg *	1.78	1.98	
Cattle	TDN % *	47.7	53.0	
Sheep	TDN % *	49.2	54.8	
Calcium	%	0.44	0.49	32
Cobalt	mg/kg	0.065	0.073	60
Copper	mg/kg	17.0	19.0	
Iron	%	0.022	0.025	
Magnesium	%	0.17	0.19	15
Manganese	mg/kg	43.6	48.5	50
Phosphorus	%	0.17	0.19	37
Potassium	%	1.46	1.62	14
Sulphur	%	0.14	0.16	14

Timothy-Grass, light, hay, s-c, full bloom, (1)
Ref No 1 05 022 United States

Feed Name or Analyses		As Fed	Dry	C.V. ± %
Dry matter	%	89.4	100.0	
Ash	%	5.6	6.3	
Crude fiber	%	31.6	35.4	
Ether extract	%	1.7	1.9	
N-free extract	%	40.2	45.0	
Protein (N x 6.25)	%	10.2	11.4	22
Cattle	dig prot % *	6.1	6.8	
Goats	dig prot % *	6.4	7.2	
Horses	dig prot % *	6.4	7.2	
Rabbits	dig coef % *	6.7	7.5	
Sheep	dig prot % *	6.1	6.8	
Energy	GE Mcal/kg			
Cattle	DE Mcal/kg *	2.05	2.29	
Sheep	DE Mcal/kg *	2.21	2.47	
Cattle	ME Mcal/kg *	1.68	1.88	
Sheep	ME Mcal/kg *	1.81	2.03	
Cattle	TDN % *	46.5	52.0	
Sheep	TDN % *	50.1	56.0	

Timothy-Grass, light, hay, s-c, mature, (1)
Ref No 1 05 023 United States

Feed Name or Analyses		As Fed	Dry	C.V. ± %
Dry matter	%		100.0	
N-free extract	%		45.0	
Protein (N x 6.25)	%		5.7	11
Cattle	dig prot % *		1.9	
Goats	dig prot % *		1.9	
Horses	dig prot % *		2.4	
Rabbits	dig coef % *		3.1	
Sheep	dig prot % *		1.7	

Timothy-Grass, light, hay, s-c, cut 1, (1)
Ref No 1 05 024 United States

Feed Name or Analyses		As Fed	Dry	C.V. ± %
Dry matter	%	89.6	100.0	1
Ash	%	5.4	6.0	24
Crude fiber	%	31.7	35.4	13
Ether extract	%	1.6	1.8	17
N-free extract	%	41.7	46.5	
Protein (N x 6.25)	%	9.2	10.3	32
Cattle	dig prot % *	5.3	5.9	
Goats	dig prot % *	5.5	6.2	
Horses	dig prot % *	5.6	6.3	
Rabbits	dig coef % *	5.9	6.6	
Sheep	dig prot % *	5.2	5.8	
Energy	GE Mcal/kg			
Cattle	DE Mcal/kg *	2.05	2.29	
Sheep	DE Mcal/kg *	2.20	2.45	
Cattle	ME Mcal/kg *	1.68	1.88	
Sheep	ME Mcal/kg *	1.80	2.01	
Cattle	TDN % *	46.5	51.9	
Sheep	TDN % *	49.9	55.7	
Calcium	%	0.36	0.40	
Copper	mg/kg	17.0	19.0	
Iron	%	0.022	0.025	
Magnesium	%	0.19	0.21	
Manganese	mg/kg	24.1	26.9	
Phosphorus	%	0.23	0.26	
Potassium	%	2.08	2.32	
Carotene	mg/kg	6.3	7.1	36
Vitamin A equivalent	IU/g	10.5	11.8	

(1) dry forages and roughages (2) pasture, range plants, and forages fed green (3) silages (4) energy feeds (5) protein supplements (6) minerals (7) vitamins (8) additives

Feed Name or Analyses			Mean As Fed	Mean Dry	C.V. ± %

Timothy-Grass, light, hay, s-c, gr 1 US, (1)

Ref No 1 05 027 United States

Analyses	Unit		As Fed	Dry	C.V. ± %
Dry matter	%			100.0	
Calcium	%			0.31	
Phosphorus	%			0.18	
Carotene	mg/kg			10.6	
Vitamin A equivalent	IU/g			17.6	

Timothy-Grass, light, hay, s-c, gr 2 US, (1)

Ref No 1 05 028 United States

Analyses	Unit		As Fed	Dry	C.V. ± %
Dry matter	%		89.4	100.0	
Ash	%		5.6	6.3	
Crude fiber	%		31.6	35.4	
Ether extract	%		1.7	1.9	
N-free extract	%		43.2	48.3	
Protein (N x 6.25)	%		7.2	8.1	30
Cattle	dig prot %	*	3.5	4.0	
Goats	dig prot %	*	3.7	4.1	
Horses	dig prot %	*	3.9	4.4	
Rabbits	dig coef %	*	4.4	4.9	
Sheep	dig prot %	*	3.4	3.8	
Energy	GE Mcal/kg				
Cattle	DE Mcal/kg	*	2.08	2.33	
Sheep	DE Mcal/kg	*	2.16	2.42	
Cattle	ME Mcal/kg	*	1.71	1.91	
Sheep	ME Mcal/kg	*	1.77	1.98	
Cattle	TDN %	*	47.2	52.8	
Sheep	TDN %	*	49.0	54.8	
Calcium	%		0.41	0.46	40
Cobalt	mg/kg		0.069	0.077	71
Manganese	mg/kg		41.0	45.9	55
Phosphorus	%		0.15	0.17	17
Carotene	mg/kg		13.0	14.6	
Vitamin A equivalent	IU/g		21.7	24.3	

Timothy-Grass, light, hay, s-c, gr 3 US, (1)

Ref No 1 05 029 United States

Analyses	Unit		As Fed	Dry	C.V. ± %
Dry matter	%		89.5	100.0	
Ash	%		6.4	7.2	
Crude fiber	%		32.0	35.8	
Ether extract	%		1.5	1.7	
N-free extract	%		43.0	48.1	
Protein (N x 6.25)	%		6.4	7.2	22
Cattle	dig prot %	*	2.8	3.2	
Goats	dig prot %	*	2.9	3.3	
Horses	dig prot %	*	3.3	3.6	
Rabbits	dig coef %	*	3.8	4.2	
Sheep	dig prot %	*	2.7	3.0	
Energy	GE Mcal/kg				
Cattle	DE Mcal/kg	*	2.09	2.34	
Sheep	DE Mcal/kg	*	2.14	2.39	
Cattle	ME Mcal/kg	*	1.71	1.92	
Sheep	ME Mcal/kg	*	1.75	1.96	
Cattle	TDN %	*	47.4	53.0	
Sheep	TDN %	*	48.5	54.1	
Calcium	%		0.42	0.47	39
Cobalt	mg/kg		0.061	0.068	39
Copper	mg/kg		17.0	19.0	
Iron	%		0.022	0.025	
Magnesium	%		0.19	0.21	
Manganese	mg/kg		48.3	54.0	37
Phosphorus	%		0.17	0.19	42
Potassium	%		2.08	2.32	
Sulphur	%		0.14	0.16	
Carotene	mg/kg		12.2	13.7	
Vitamin A equivalent	IU/g		20.4	22.8	

Timothy-Grass, light, hay, s-c, gr sample US, (1)

Ref No 1 05 026 United States

Analyses	Unit		As Fed	Dry	C.V. ± %
Dry matter	%			100.0	
N-free extract	%			48.1	
Protein (N x 6.25)	%			7.1	26
Cattle	dig prot %	*		3.1	
Goats	dig prot %	*		3.2	
Horses	dig prot %	*		3.6	
Rabbits	dig coef %	*		4.2	
Sheep	dig prot %	*		2.9	
Calcium	%			0.49	32
Cobalt	mg/kg			0.071	27
Manganese	mg/kg			45.6	77
Phosphorus	%			0.16	34
Carotene	mg/kg			10.8	72
Vitamin A equivalent	IU/g			18.0	

TIMOTHY-REDTOP. Phleum pratense, Agrostis alba

Timothy-Redtop, hay, fan air dried, cut 1, (1)

Ref No 1 05 030 United States

Analyses	Unit		As Fed	Dry	C.V. ± %
Dry matter	%		92.0	100.0	
Carotene	mg/kg		7.1	7.7	
Vitamin A equivalent	IU/g		11.8	12.9	

Timothy-Redtop, hay, s-c, (1)

Ref No 1 05 032 United States

Analyses	Unit		As Fed	Dry	C.V. ± %
Dry matter	%		89.7	100.0	1
Ash	%		5.7	6.3	11
Crude fiber	%		31.0	34.6	6
Ether extract	%		1.9	2.1	23
N-free extract	%		44.9	50.0	
Protein (N x 6.25)	%		6.3	7.0	14
Cattle	dig prot %	*	2.7	3.0	
Goats	dig prot %	*	2.8	3.1	
Horses	dig prot %	*	3.1	3.5	
Rabbits	dig coef %	*	3.7	4.1	
Sheep	dig prot %	*	2.6	2.8	
Energy	GE Mcal/kg				
Cattle	DE Mcal/kg	*	2.16	2.41	
Sheep	DE Mcal/kg	*	2.16	2.41	
Cattle	ME Mcal/kg	*	1.77	1.97	
Sheep	ME Mcal/kg	*	1.77	1.98	
Cattle	TDN %	*	48.9	54.6	
Sheep	TDN %	*	49.1	54.7	
Calcium	%		0.22	0.24	
Iron	%		0.007	0.008	
Magnesium	%		0.13	0.14	
Phosphorus	%		0.16	0.18	
Potassium	%		1.09	1.22	
Sulphur	%		0.05	0.06	
Carotene	mg/kg		5.5	6.2	
Vitamin A equivalent	IU/g		9.2	10.3	

TIMOTHY-SWEETCLOVER. Phleum pratense, Melilotus spp

Timothy-Sweetclover, aerial part, ensiled, (3)

Ref No 3 05 033 United States

Analyses	Unit		As Fed	Dry	C.V. ± %
Dry matter	%		22.3	100.0	
Calcium	%		0.14	0.63	
Phosphorus	%		0.01	0.04	

TOBOSA, Hilaria mutica

Tobosa, hay, s-c, (1)

Ref No 1 05 034 United States

Analyses	Unit		As Fed	Dry	C.V. ± %
Dry matter	%		93.4	100.0	
Ash	%		10.0	10.7	
Crude fiber	%		32.8	35.1	
Sheep	dig coef %		54.	54.	
Ether extract	%		1.1	1.2	
Sheep	dig coef %		37.	37.	
N-free extract	%		45.9	49.1	
Sheep	dig coef %		47.	47.	
Protein (N x 6.25)	%		3.6	3.9	
Sheep	dig coef %		20.	20.	
Cattle	dig prot %	*	0.3	0.3	
Goats	dig prot %	*	0.2	0.2	
Horses	dig prot %	*	0.8	0.8	
Rabbits	dig coef %	*	1.6	1.7	
Sheep	dig prot %		0.7	0.8	
Energy	GE Mcal/kg				
Cattle	DE Mcal/kg	*	1.97	2.11	
Sheep	DE Mcal/kg	*	1.80	1.93	
Cattle	ME Mcal/kg	*	1.62	1.73	
Sheep	ME Mcal/kg	*	1.48	1.58	
Cattle	TDN %	*	44.6	47.8	
Sheep	TDN %		40.9	43.8	

Tobosa, aerial part, fresh, immature, (2)

Ref No 2 08 578 United States

Analyses	Unit		As Fed	Dry	C.V. ± %
Dry matter	%		39.6	100.0	
Ash	%		5.2	13.1	
Crude fiber	%		12.3	31.1	
Ether extract	%		0.6	1.5	
N-free extract	%		17.3	43.7	
Protein (N x 6.25)	%		4.2	10.6	
Cattle	dig prot %	*	2.7	6.9	
Goats	dig prot %	*	2.6	6.5	
Horses	dig prot %	*	2.6	6.5	
Rabbits	dig coef %	*	2.7	6.9	
Sheep	dig prot %	*	2.7	6.9	
Energy	GE Mcal/kg				
Cattle	DE Mcal/kg	*	0.96	2.42	
Sheep	DE Mcal/kg	*	0.94	2.38	
Cattle	ME Mcal/kg	*	0.78	1.98	
Sheep	ME Mcal/kg	*	0.77	1.95	
Cattle	TDN %	*	21.7	54.8	
Sheep	TDN %	*	21.3	53.9	
Calcium	%		0.12	0.30	
Phosphorus	%		0.08	0.20	

Feed Name or Analyses			Mean As Fed	Mean Dry	C.V. ± %

Tobosa, aerial part, fresh, full bloom, (2)

Ref No 2 05 035 — United States

Analyses			As Fed	Dry	C.V. ± %
Dry matter	%		50.0	100.0	
Ash	%		5.9	11.8	
Crude fiber	%		17.0	34.0	
Ether extract	%		1.0	2.0	
N-free extract	%		22.9	45.8	
Protein (N x 6.25)	%		3.2	6.4	
Cattle	dig prot %	*	1.7	3.3	
Goats	dig prot %	*	1.3	2.5	
Horses	dig prot %	*	1.5	3.0	
Rabbits	dig coef %	*	1.8	3.6	
Sheep	dig prot %	*	1.5	3.0	

Tobosa, aerial part, fresh, mature, (2)

Ref No 2 05 036 — United States

Analyses			As Fed	Dry	C.V. ± %
Dry matter	%		50.0	100.0	
Ash	%		5.4	10.8	12
Crude fiber	%		17.2	34.4	3
Ether extract	%		0.8	1.7	10
N-free extract	%		23.7	47.4	
Protein (N x 6.25)	%		2.9	5.8	9
Cattle	dig prot %	*	1.4	2.8	
Goats	dig prot %	*	1.0	1.9	
Horses	dig prot %	*	1.2	2.4	
Rabbits	dig coef %	*	1.6	3.1	
Sheep	dig prot %	*	1.2	2.3	
Energy	GE Mcal/kg				
Cattle	DE Mcal/kg	*	0.95	1.90	
Sheep	DE Mcal/kg	*	1.03	2.06	
Cattle	ME Mcal/kg	*	0.78	1.56	
Sheep	ME Mcal/kg	*	0.85	1.69	
Cattle	TDN %	*	21.6	43.1	
Sheep	TDN %	*	23.4	46.8	
Calcium	%		0.12	0.24	18
Cobalt	mg/kg		0.165	0.331	
Phosphorus	%		0.06	0.12	18
Carotene	mg/kg		1.2	2.4	
Vitamin A equivalent	IU/g		2.0	4.0	

Tobosa, aerial part, fresh, over ripe, (2)

Ref No 2 05 037 — United States

Analyses			As Fed	Dry	C.V. ± %
Dry matter	%			100.0	
Ash	%			9.2	12
Crude fiber	%			35.4	3
Ether extract	%			1.5	15
N-free extract	%			50.0	
Protein (N x 6.25)	%			3.9	16
Cattle	dig prot %	*		1.2	
Goats	dig prot %	*		0.2	
Horses	dig prot %	*		0.8	
Rabbits	dig coef %	*		1.7	
Sheep	dig prot %	*		0.6	

(1) dry forages and roughages (2) pasture, range plants, and forages fed green (3) silages (4) energy feeds (5) protein supplements (6) minerals (7) vitamins (8) additives

Feed Name or Analyses			Mean As Fed	Mean Dry	C.V. ± %

Tobosa, aerial part wo lower stems, fresh, early leaf, (2)

Ref No 2 08 710 — United States

Analyses			As Fed	Dry	C.V. ± %
Dry matter	%			100.0	
Organic matter	%			88.2	4
Ash	%			12.0	27
Crude fiber	%			30.9	5
Ether extract	%			1.5	29
N-free extract	%			49.4	6
Protein (N x 6.25)	%			6.2	29
Cattle	dig prot %	*		3.1	
Goats	dig prot %	*		2.3	
Horses	dig prot %	*		2.8	
Rabbits	dig coef %	*		3.4	
Sheep	dig prot %	*		2.8	
Calcium	%			0.41	45
Phosphorus	%			0.10	34
Carotene	mg/kg			34.0	98
Vitamin A equivalent	IU/g			56.6	

Tobosa, aerial part wo lower stems, fresh, mid-bloom, (2)

Ref No 2 08 711 — United States

Analyses			As Fed	Dry	C.V. ± %
Dry matter	%			100.0	
Organic matter	%			88.9	2
Ash	%			11.1	19
Crude fiber	%			33.1	5
Ether extract	%			1.7	28
N-free extract	%			47.9	25
Protein (N x 6.25)	%			6.2	20
Cattle	dig prot %	*		3.1	
Goats	dig prot %	*		2.3	
Horses	dig prot %	*		2.8	
Rabbits	dig coef %	*		3.4	
Sheep	dig prot %	*		2.7	
Calcium	%			0.33	46
Phosphorus	%			0.11	31
Carotene	mg/kg			20.6	98
Vitamin A equivalent	IU/g			34.4	

Tobosa, aerial part wo lower stems, fresh, mature, (2)

Ref No 2 08 712 — United States

Analyses			As Fed	Dry	C.V. ± %
Dry matter	%			100.0	
Organic matter	%			88.2	2
Ash	%			12.0	17
Crude fiber	%			32.6	5
Ether extract	%			1.3	25
N-free extract	%			48.0	5
Protein (N x 6.25)	%			6.1	27
Cattle	dig prot %	*		3.1	
Goats	dig prot %	*		2.3	
Horses	dig prot %	*		2.7	
Rabbits	dig coef %	*		3.4	
Sheep	dig prot %	*		2.7	
Calcium	%			0.37	41
Phosphorus	%			0.10	33
Carotene	mg/kg			12.9	
Vitamin A equivalent	IU/g			21.4	

Feed Name or Analyses			Mean As Fed	Mean Dry	C.V. ± %

Tobosa, aerial part wo lower stems, fresh, dormant, (2)

Ref No 2 08 709 — United States

Analyses			As Fed	Dry	C.V. ± %
Dry matter	%			100.0	
Organic matter	%			88.4	3
Ash	%			11.6	22
Crude fiber	%			34.3	8
Ether extract	%			1.2	26
N-free extract	%			48.1	5
Protein (N x 6.25)	%			4.9	25
Cattle	dig prot %	*		2.0	
Goats	dig prot %	*		1.1	
Horses	dig prot %	*		1.7	
Rabbits	dig coef %	*		2.4	
Sheep	dig prot %	*		1.5	
Calcium	%			0.33	59
Phosphorus	%			0.09	59
Carotene	mg/kg			9.9	98
Vitamin A equivalent	IU/g			16.5	

TOMATO. Lycopersicon esculentum

Tomato, leaves w stems, fresh, (2)

Ref No 2 05 039 — United States

Analyses			As Fed	Dry	C.V. ± %
Dry matter	%			100.0	
Ash	%			26.6	
Crude fiber	%			15.4	
Ether extract	%			1.7	
N-free extract	%			29.9	
Protein (N x 6.25)	%			26.4	
Cattle	dig prot %	*		20.3	
Goats	dig prot %	*		21.2	
Horses	dig prot %	*		19.9	
Rabbits	dig coef %	*		19.1	
Sheep	dig prot %	*		21.6	

Tomato, fruit, fresh, (4)

Ref No 2 05 040 — United States

Analyses			As Fed	Dry	C.V. ± %
Dry matter	%		6.1	100.0	
Ash	%		0.5	8.2	
Crude fiber	%		0.6	9.1	
Ether extract	%		0.3	5.0	
N-free extract	%		3.7	61.3	
Protein (N x 6.25)	%		1.0	16.4	
Cattle	dig prot %	*	0.7	11.0	
Goats	dig prot %	*	0.7	12.2	
Horses	dig prot %	*	0.7	12.2	
Sheep	dig prot %	*	0.7	12.2	
Energy	GE Mcal/kg		0.21	3.38	
Cattle	DE Mcal/kg	*	0.19	3.06	
Sheep	DE Mcal/kg	*	0.20	3.33	
Swine	DE kcal/kg	*	167.	2743.	
Cattle	ME Mcal/kg	*	0.15	2.51	
Sheep	ME Mcal/kg	*	0.17	2.73	
Swine	ME kcal/kg	*	155.	2542.	
Cattle	TDN %	*	4.2	69.5	
Sheep	TDN %	*	4.6	75.6	
Swine	TDN %	*	3.8	62.2	
Calcium	%		0.01	0.16	
Iron	%		0.001	0.015	
Phosphorus	%		0.03	0.49	
Potassium	%		0.26	4.21	
Sodium	%		0.00	0.00	
Ascorbic acid	mg/kg		215.8	3538.5	

Feed Name or Analyses		Mean As Fed	Mean Dry	C.V. ± %
Niacin	mg/kg	6.6	107.7	
Riboflavin	mg/kg	0.4	6.2	
Thiamine	mg/kg	0.6	9.2	
Vitamin A	IU/g	8.4	138.5	

Tomato, pulp, dehy, (5)
Dried tomato pomace (AAFCO)
Ref No 5 05 041 United States

Feed Name or Analyses		Mean As Fed	Mean Dry	C.V. ± %
Dry matter	%	91.9	100.0	3
Ash	%	6.0	6.5	57
Crude fiber	%	24.2	26.3	30
Cattle	dig coef %	46.	46.	
Ether extract	%	9.8	10.6	55
Cattle	dig coef %	91.	91.	
N-free extract	%	30.0	32.7	30
Cattle	dig coef %	79.	79.	
Protein (N x 6.25)	%	21.9	23.9	1
Cattle	dig coef %	57.	57.	
Cattle	dig prot %	12.4	13.5	
Pectic substances	%	20.7	22.5	
Energy	GE Mcal/kg	4.05	4.41	
Cattle	DE Mcal/kg *	2.96	3.22	
Sheep	DE Mcal/kg *	2.60	2.83	
Cattle	ME Mcal/kg *	2.43	2.64	
Chickens	ME_n kcal/kg	1760.	1915.	
Sheep	ME Mcal/kg *	2.13	2.32	
Cattle	TDN %	67.1	73.0	
Sheep	TDN % *	59.0	64.2	
Calcium	%	0.41	0.44	25
Copper	mg/kg	30.0	32.6	
Iron	%	0.423	0.460	
Magnesium	%	0.18	0.20	
Manganese	mg/kg	47.4	51.5	26
Phosphorus	%	0.54	0.59	16
Potassium	%	3.34	3.63	
Riboflavin	mg/kg	6.1	6.6	
Thiamine	mg/kg	11.9	12.9	
Arginine	%	1.20	1.30	5
Histidine	%	0.40	0.43	14
Isoleucine	%	0.70	0.76	6
Leucine	%	1.70	1.85	6
Lysine	%	1.60	1.74	7
Methionine	%	0.10	0.11	
Phenylalanine	%	0.90	0.98	6
Threonine	%	0.70	0.76	
Tryptophan	%	0.20	0.22	
Tyrosine	%	0.90	0.98	10
Valine	%	1.00	1.09	4

Tomato, pulp, wet, (5)
Ref No 2 05 042 United States

Feed Name or Analyses		Mean As Fed	Mean Dry	C.V. ± %
Dry matter	%	13.9	100.0	
Ash	%	0.6	4.3	
Crude fiber	%	3.8	27.3	
Ether extract	%	1.7	12.3	
N-free extract	%	4.8	34.3	
Protein (N x 6.25)	%	3.0	21.7	
Energy	GE Mcal/kg			
Cattle	DE Mcal/kg *	0.40	2.87	
Sheep	DE Mcal/kg *	0.38	2.71	
Cattle	ME Mcal/kg *	0.33	2.35	
Sheep	ME Mcal/kg *	0.31	2.22	
Cattle	TDN % *	9.0	65.1	
Sheep	TDN % *	8.5	61.4	

Tomato, pulp wo skin, dehy screened, (5)
Tomato fines
Ref No 5 07 744 United States

Feed Name or Analyses		Mean As Fed	Mean Dry	C.V. ± %
Dry matter	%	90.9	100.0	
Ash	%	8.2	9.1	
Crude fiber	%	16.4	18.1	
Ether extract	%	5.7	6.3	
N-free extract	%	41.2	45.3	
Protein (N x 6.25)	%	19.3	21.3	
Pectic substances	%	16.6	18.3	
Energy	GE Mcal/kg	4.03	4.43	
Cattle	DE Mcal/kg *	2.76	3.04	
Sheep	DE Mcal/kg *	2.71	2.98	
Cattle	ME Mcal/kg *	2.26	2.49	
Sheep	ME Mcal/kg *	2.22	2.44	
Cattle	TDN % *	62.6	68.9	
Sheep	TDN % *	61.5	67.6	
Calcium	%	0.52	0.57	
Copper	mg/kg	24.1	26.5	
Iron	%	0.436	0.480	
Magnesium	%	0.30	0.33	
Phosphorus	%	0.43	0.48	
Potassium	%	2.38	2.62	

Tomato, skins w juice, dehy, (5)
Tomato flakes
Ref No 5 07 743 United States

Feed Name or Analyses		Mean As Fed	Mean Dry	C.V. ± %
Dry matter	%	89.6	100.0	
Ash	%	8.8	9.8	
Crude fiber	%	17.1	19.1	
Ether extract	%	2.2	2.5	
N-free extract	%	43.0	48.0	
Protein (N x 6.25)	%	18.5	20.7	
Pectic substances	%	16.7	18.6	
Energy	GE Mcal/kg	4.09	4.56	
Cattle	DE Mcal/kg *	2.49	2.78	
Sheep	DE Mcal/kg *	2.59	2.89	
Cattle	ME Mcal/kg *	2.04	2.28	
Sheep	ME Mcal/kg *	2.12	2.37	
Cattle	TDN % *	56.4	63.0	
Sheep	TDN % *	58.7	65.5	
Calcium	%	0.55	0.61	
Copper	mg/kg	21.2	23.7	
Iron	%	0.332	0.370	
Magnesium	%	0.27	0.30	
Phosphorus	%	0.41	0.46	
Potassium	%	2.77	3.09	

Tomato fines –
see Tomato, pulp wo skin, dehy screened, (5)

Tomato flakes –
see Tomato, skins w juice, dehy, (5)

TREADSOFTLY. Cnidoscolus angustidens

Treadsoftly, seeds w hulls, (5)
Ref No 5 09 264 United States

Feed Name or Analyses		Mean As Fed	Mean Dry	C.V. ± %
Dry matter	%		100.0	
Protein (N x 6.25)	%		42.0	
Alanine	%		1.72	
Arginine	%		3.19	
Aspartic acid	%		3.99	
Glutamic acid	%		6.26	
Glycine	%		1.60	
Histidine	%		0.80	
Hydroxyproline	%		0.04	
Isoleucine	%		1.34	
Leucine	%		2.52	
Lysine	%		1.39	
Methionine	%		0.63	
Phenylalanine	%		1.64	
Proline	%		1.81	
Serine	%		1.72	
Threonine	%		1.43	
Tyrosine	%		1.05	
Valine	%		2.69	

TREFOIL, BIG. Lotus uliginosus

Trefoil, big, hay, s-c, (1)
Ref No 1 08 545 United States

Feed Name or Analyses		Mean As Fed	Mean Dry	C.V. ± %
Dry matter	%	89.3	100.0	
Ash	%	7.3	8.2	
Crude fiber	%	21.9	24.5	
Ether extract	%	2.7	3.0	
N-free extract	%	45.4	50.8	
Protein (N x 6.25)	%	12.0	13.4	
Cattle	dig prot % *	7.7	8.6	
Goats	dig prot % *	8.1	9.1	
Horses	dig prot % *	8.0	8.9	
Rabbits	dig coef % *	8.1	9.0	
Sheep	dig prot % *	7.7	8.6	
Energy	GE Mcal/kg			
Cattle	DE Mcal/kg *	2.52	2.82	
Sheep	DE Mcal/kg *	2.40	2.69	
Cattle	ME Mcal/kg *	2.06	2.31	
Sheep	ME Mcal/kg *	1.97	2.21	
Cattle	TDN % *	57.1	63.9	
Sheep	TDN % *	54.5	61.0	

TREFOIL, BIRDSFOOT. Lotus corniculatus

Trefoil, birdsfoot, aerial part, dehy grnd, (1)
Ref No 1 08 765 United States

Feed Name or Analyses		Mean As Fed	Mean Dry	C.V. ± %
Dry matter	%	92.0	100.0	
Crude fiber	%	20.7	22.5	
Ether extract	%	4.0	4.3	
Protein (N x 6.25)	%	15.7	17.1	
Cattle	dig prot % *	10.8	11.7	
Goats	dig prot % *	11.5	12.5	
Horses	dig prot % *	11.1	12.0	
Rabbits	dig coef % *	10.9	11.8	
Sheep	dig prot % *	10.9	11.9	
Calcium	%	1.00	1.09	
Cobalt	mg/kg	0.100	0.109	
Copper	mg/kg	8.6	9.3	
Iron	%	0.037	0.040	
Magnesium	%	0.26	0.28	
Manganese	mg/kg	26.0	28.3	
Molybdenum	mg/kg	0.87	0.95	
Phosphorus	%	0.29	0.32	
Potassium	%	1.85	2.01	
Sodium	%	0.07	0.07	
Zinc	mg/kg	70.0	76.1	
Carotene	mg/kg	132.1	143.5	
Vitamin A equivalent	IU/g	220.1	239.3	

Trefoil, birdsfoot, hay, s-c, (1)

Ref No 1 05 044 United States

Feed Name or Analyses				Mean As Fed	Mean Dry	C.V. ± %
Dry matter		%		89.0	100.0	
Ash		%		6.8	7.6	
Crude fiber		%		26.3	29.6	
Sheep	dig coef	%		48.	48.	
Ether extract		%		1.9	2.2	
Sheep	dig coef	%		43.	43.	
N-free extract		%		39.7	44.6	
Sheep	dig coef	%		64.	64.	
Protein (N x 6.25)		%		14.2	16.0	
Sheep	dig coef	%		69.	69.	
Cattle	dig prot	%	*	9.6	10.8	
Goats	dig prot	%	*	10.2	11.5	
Horses	dig prot	%	*	9.9	11.1	
Rabbits	dig coef	%	*	9.8	11.0	
Sheep	dig prot	%		9.8	11.0	
Fatty acids		%		2.0	2.3	
Energy	GE	Mcal/kg		3.99	4.49	
Sheep	GE dig coef	%		50.	50.	
Cattle	DE	Mcal/kg	*	2.26	2.54	
Sheep	DE	Mcal/kg		1.98	2.22	
Cattle	ME	Mcal/kg	*	1.86	2.09	
Sheep	ME	Mcal/kg	*	1.62	1.82	
Cattle	NE$_m$	Mcal/kg	*	1.17	1.31	
Cattle	NE$_{gain}$	Mcal/kg	*	0.61	0.69	
Cattle	NE$_{lactating\ cows}$	Mcal/kg	*	1.33	1.49	
Cattle	TDN	%	*	51.3	57.7	
Sheep	TDN	%		49.7	55.9	
Calcium		%		1.56	1.75	
Phosphorus		%		0.20	0.22	
Potassium		%		1.62	1.82	

Ref No 1 05 044 Canada

Feed Name or Analyses				Mean As Fed	Mean Dry	C.V. ± %
Dry matter		%			100.0	
Sheep	dig coef	%			52.	
Ash		%			7.2	
Crude fiber		%			40.1	
Sheep	dig coef	%			48.	
Ether extract		%			1.5	
Sheep	dig coef	%			43.	
N-free extract		%			38.2	
Sheep	dig coef	%			64.	
Protein (N x 6.25)		%			13.1	
Sheep	dig coef	%			69.	
Cattle	dig prot	%	*		8.3	
Goats	dig prot	%	*		8.8	
Horses	dig prot	%	*		8.7	
Rabbits	dig coef	%	*		8.8	
Sheep	dig prot	%			9.0	
Energy	GE	Mcal/kg			4.49	
Sheep	GE dig coef	%			50.	
Cattle	DE	Mcal/kg	*		2.66	
Sheep	DE	Mcal/kg			2.22	
Cattle	ME	Mcal/kg	*		2.18	
Sheep	ME	Mcal/kg	*		1.82	
Cattle	TDN	%	*		60.3	
Sheep	TDN	%			54.1	

Trefoil, birdsfoot, aerial part, fresh, (2)

Ref No 2 07 998 United States

Feed Name or Analyses				Mean As Fed	Mean Dry	C.V. ± %
Dry matter		%		22.5	100.0	
Ash		%		1.5	6.6	
Crude fiber		%		4.7	20.7	
Ether extract		%		1.1	4.7	
N-free extract		%		10.5	46.7	
Protein (N x 6.25)		%		4.8	21.4	
Cattle	dig prot	%	*	3.6	16.1	
Goats	dig prot	%	*	3.7	16.5	
Horses	dig prot	%	*	3.5	15.7	
Rabbits	dig coef	%	*	3.4	15.2	
Sheep	dig prot	%	*	3.8	16.9	
Energy	GE	Mcal/kg				
Cattle	DE	Mcal/kg	*	0.75	3.33	
Sheep	DE	Mcal/kg	*	0.64	2.86	
Cattle	ME	Mcal/kg	*	0.61	2.73	
Sheep	ME	Mcal/kg	*	0.53	2.35	
Cattle	TDN	%	*	17.0	75.5	
Sheep	TDN	%	*	14.6	64.9	
Calcium		%		0.40	1.76	
Phosphorus		%		0.05	0.20	
Potassium		%		0.41	1.84	

Trefoil, birdsfoot, aerial part, fresh, full bloom, (2)

Ref No 2 07 722 Canada

Feed Name or Analyses				Mean As Fed	Mean Dry	C.V. ± %
Dry matter		%		26.1	100.0	
Ash		%		2.1	8.2	
Crude fiber		%		7.9	30.2	
Ether extract		%		0.5	2.0	
N-free extract		%		12.2	46.8	
Protein (N x 6.25)		%		3.4	13.0	
Cattle	dig prot	%	*	2.3	8.9	
Goats	dig prot	%	*	2.3	8.6	
Horses	dig prot	%	*	2.2	8.5	
Rabbits	dig coef	%	*	2.3	8.7	
Sheep	dig prot	%	*	2.4	9.1	
Energy	GE	Mcal/kg				
Cattle	DE	Mcal/kg	*	0.80	3.06	
Sheep	DE	Mcal/kg	*	0.73	2.81	
Cattle	ME	Mcal/kg	*	0.66	2.51	
Sheep	ME	Mcal/kg	*	0.60	2.31	
Cattle	TDN	%	*	18.1	69.5	
Sheep	TDN	%	*	16.6	63.8	

Trefoil, birdsfoot, aerial part, fresh, late bloom, (2)

Ref No 2 07 834 United States

Feed Name or Analyses				Mean As Fed	Mean Dry	C.V. ± %
Dry matter		%			100.0	
Ash		%			12.0	
Ether extract		%			2.6	
Protein (N x 6.25)		%			27.0	
Sheep	dig coef	%			78.	
Cattle	dig prot	%	*		20.8	
Goats	dig prot	%	*		21.8	
Horses	dig prot	%	*		20.5	
Rabbits	dig coef	%	*		19.5	
Sheep	dig prot	%			21.0	
Lignin (Ellis)		%			6.6	
Energy	GE	Mcal/kg				
Sheep	DE	Mcal/kg	*		2.84	
Sheep	ME	Mcal/kg	*		2.33	
Sheep	TDN	%			64.3	

Trefoil, birdsfoot, aerial part, fresh, milk stage, (2)

Ref No 2 07 721 Canada

Feed Name or Analyses				Mean As Fed	Mean Dry	C.V. ± %
Dry matter		%		25.9	100.0	
Ash		%		1.7	6.5	
Crude fiber		%		7.6	29.5	
Ether extract		%		0.6	2.2	
N-free extract		%		12.5	48.2	
Protein (N x 6.25)		%		3.5	13.6	
Cattle	dig prot	%	*	2.4	9.5	
Goats	dig prot	%	*	2.4	9.2	
Horses	dig prot	%	*	2.4	9.1	
Rabbits	dig coef	%	*	2.4	9.2	
Sheep	dig prot	%	*	2.5	9.7	
Energy	GE	Mcal/kg				
Cattle	DE	Mcal/kg	*	0.83	3.19	
Sheep	DE	Mcal/kg	*	0.73	2.82	
Cattle	ME	Mcal/kg	*	0.68	2.61	
Sheep	ME	Mcal/kg	*	0.60	2.31	
Cattle	TDN	%	*	18.7	72.3	
Sheep	TDN	%	*	16.6	64.0	

TREFOIL, BIRDSFOOT VIKING. Lotus corniculatus

Trefoil, birdsfoot viking, hay, s-c, cut 2, (1)

Ref No 1 08 853 United States

Feed Name or Analyses				Mean As Fed	Mean Dry	C.V. ± %
Dry matter		%			100.0	
Sheep	dig coef	%			61.	
Ash		%			7.2	
Ether extract		%			2.8	
Protein (N x 6.25)		%			17.8	
Cattle	dig prot	%	*		12.4	
Goats	dig prot	%	*		13.2	
Horses	dig prot	%	*		12.6	
Rabbits	dig coef	%	*		12.4	
Sheep	dig prot	%	*		12.5	
Fiber, acid detergent (VS)		%			31.4	
Lignin (Ellis)		%			9.1	
Energy	GE	Mcal/kg			4.63	
Sheep	DE	Mcal/kg	*		2.65	
Sheep	ME	Mcal/kg	*		2.17	
Sheep	TDN	%			60.2	
Phosphorus		%			0.21	
Sulphur		%			0.25	
Dry matter intake		g/W$^{0.75}$			110.200	
Gain, sheep		kg/day			0.118	

Trefoil, birdsfoot viking, hay, s-c, cut 3, (1)

Ref No 1 08 854 United States

Feed Name or Analyses				Mean As Fed	Mean Dry	C.V. ± %
Dry matter		%			100.0	
Sheep	dig coef	%			64.	
Ash		%			10.6	
Ether extract		%			3.6	
Protein (N x 6.25)		%			23.2	
Cattle	dig prot	%	*		17.0	
Goats	dig prot	%	*		18.2	
Horses	dig prot	%	*		17.2	
Rabbits	dig coef	%	*		16.6	
Sheep	dig prot	%	*		17.4	
Fiber, acid detergent (VS)		%			29.2	
Lignin (Ellis)		%			7.4	
Energy	GE	Mcal/kg			4.50	
Sheep	DE	Mcal/kg	*		2.78	
Sheep	ME	Mcal/kg	*		2.28	
Sheep	TDN	%			63.0	

(1) dry forages and roughages (2) pasture, range plants, and forages fed green (3) sitages (4) energy feeds (5) protein supplements (6) minerals (7) vitamins (8) additives

Feed Name or Analyses		Mean As Fed	Mean Dry	C.V. ± %
Phosphorus	%		0.30	
Sulphur	%		0.28	
Dry matter intake	g/W0.75		93.300	
Gain, sheep	kg/day		0.136	

Trefoil, birdsfoot viking, hay, s-c baled fan-air dried w heat, cut 1, (1)

Ref No 1 08 852 United States

Feed Name or Analyses		Mean As Fed	Mean Dry	C.V. ± %
Dry matter	%		100.0	
Sheep	dig coef %		62.	
Ash	%		6.5	
Ether extract	%		2.9	
Protein (N x 6.25)	%		16.2	
Cattle	dig prot % *		11.0	
Goats	dig prot % *		11.7	
Horses	dig prot % *		11.3	
Rabbits	dig coef % *		11.2	
Sheep	dig prot % *		11.1	
Fiber, acid detergent (VS)	%		35.1	
Lignin (Ellis)	%		9.3	
Energy	GE Mcal/kg		4.62	
Sheep	DE Mcal/kg *		2.70	
Sheep	ME Mcal/kg *		2.21	
Sheep	TDN %		61.3	
Phosphorus	%		0.19	
Sulphur	%		0.23	
Dry matter intake	g/W0.75		102.450	
Gain, sheep	kg/day		0.100	

TREFOIL, BIRDSFOOT EMPIRE BFT-BLUEGRASS, KENTUCKY COMMON Lotus corniculatus, poa pratensis

Trefoil, birdsfoot empire bft-bluegrass, Kentucky common, hay, fan-air dried w/heat, prebloom, cut 2, (1)

Ref No 1 08 910 United States

Feed Name or Analyses		Mean As Fed	Mean Dry	C.V. ± %
Dry matter	%		100.0	
Sheep	dig coef %		44.	
Crude fiber	%		32.2	
Protein (N x 6.25)	%		15.5	
Cattle	dig prot % *		10.4	
Goats	dig prot % *		11.0	
Horses	dig prot % *		10.7	
Rabbits	dig coef % *		10.6	
Sheep	dig prot % *		10.5	
Dry matter intake	g/W0.75		46.860	

Trefoil, birdsfoot empire bft-bluegrass, Kentucky common, hay, s-c baled, prebloom after frost, cut 2, (1)

Ref No 1 08 909 United States

Feed Name or Analyses		Mean As Fed	Mean Dry	C.V. ± %
Dry matter	%		100.0	
Sheep	dig coef %		50.	
Crude fiber	%		31.2	
Protein (N x 6.25)	%		12.8	
Cattle	dig prot % *		8.0	
Goats	dig prot % *		8.5	
Horses	dig prot % *		8.4	
Rabbits	dig coef % *		8.6	
Sheep	dig prot % *		8.1	
Dry matter intake	g/W0.75		46.860	

TRIODIA. Triodia spp

Triodia, aerial part, fresh, (2)

Ref No 2 05 047 United States

Feed Name or Analyses		Mean As Fed	Mean Dry	C.V. ± %
Dry matter	%		100.0	
Ash	%		10.9	30
Crude fiber	%		34.1	7
Ether extract	%		1.8	16
N-free extract	%		44.1	
Protein (N x 6.25)	%		9.1	19
Cattle	dig prot % *		5.6	
Goats	dig prot % *		5.1	
Horses	dig prot % *		5.3	
Rabbits	dig coef % *		5.7	
Sheep	dig prot % *		5.5	
Calcium	%		0.61	45
Magnesium	%		0.11	
Phosphorus	%		0.17	33
Potassium	%		1.61	

Triodia, aerial part, fresh, immature, (2)

Ref No 2 05 045 United States

Feed Name or Analyses		Mean As Fed	Mean Dry	C.V. ± %
Dry matter	%		100.0	
Ash	%		10.6	4
Crude fiber	%		29.8	7
Ether extract	%		1.9	7
N-free extract	%		45.8	
Protein (N x 6.25)	%		11.9	6
Cattle	dig prot % *		8.0	
Goats	dig prot % *		7.7	
Horses	dig prot % *		7.6	
Rabbits	dig coef % *		7.9	
Sheep	dig prot % *		8.1	

Triodia, aerial part, fresh, full bloom, (2)

Ref No 2 05 046 United States

Feed Name or Analyses		Mean As Fed	Mean Dry	C.V. ± %
Dry matter	%		100.0	
Ash	%		14.7	35
Crude fiber	%		31.4	9
Ether extract	%		1.5	10
N-free extract	%		45.0	
Protein (N x 6.25)	%		7.4	15
Cattle	dig prot % *		4.2	
Goats	dig prot % *		3.5	
Horses	dig prot % *		3.8	
Rabbits	dig coef % *		4.4	
Sheep	dig prot % *		3.9	
Calcium	%		0.80	30
Phosphorus	%		0.12	32

TRIODIA, SHORTLEAF. Triodia grandiflora

Triodia, shortleaf, aerial part, fresh, full bloom, (2)

Ref No 2 05 048 United States

Feed Name or Analyses		Mean As Fed	Mean Dry	C.V. ± %
Dry matter	%		100.0	
Calcium	%		0.91	
Phosphorus	%		0.10	

TRIODIA, SLIM. Triodia mutica

Triodia, slim, aerial part, fresh, (2)

Ref No 2 05 049 United States

Feed Name or Analyses		Mean As Fed	Mean Dry	C.V. ± %
Dry matter	%		100.0	
Calcium	%		1.01	
Phosphorus	%		0.08	

TRIODIA, TEXAS. Triodia texana

Triodia, Texas, aerial part, fresh, (2)

Ref No 2 05 050 United States

Feed Name or Analyses		Mean As Fed	Mean Dry	C.V. ± %
Dry matter	%		100.0	
Ash	%		10.7	
Crude fiber	%		33.4	
Ether extract	%		2.0	
N-free extract	%		41.7	
Protein (N x 6.25)	%		12.2	
Cattle	dig prot % *		8.3	
Goats	dig prot % *		7.9	
Horses	dig prot % *		7.9	
Rabbits	dig coef % *		8.1	
Sheep	dig prot % *		8.4	
Calcium	%		0.44	
Phosphorus	%		0.17	

TRIODIA, WHITE. Triodia albescens

Triodia, white, aerial part, fresh, (2)

Ref No 2 05 053 United States

Feed Name or Analyses		Mean As Fed	Mean Dry	C.V. ± %
Dry matter	%		100.0	
Ash	%		10.2	9 1
Crude fiber	%		33.7	5 1
Ether extract	%		2.0	10 1
N-free extract	%		44.4	
Protein (N x 6.25)	%		9.7	20 1
Cattle	dig prot % *		6.1	
Goats	dig prot % *		5.6	
Horses	dig prot % *		5.8	
Rabbits	dig coef % *		6.2	
Sheep	dig prot % *		6.0	
Calcium	%		0.52	16 1
Magnesium	%		0.11	
Phosphorus	%		0.20	16 1
Potassium	%		1.61	

Triodia, white, aerial part, fresh, full bloom, (2)

Ref No 2 05 052 United States

Feed Name or Analyses		Mean As Fed	Mean Dry	C.V. ± %
Dry matter	%		100.0	
Calcium	%		0.48	
Phosphorus	%		0.16	

TRIPTERIS, THICKWING. Tripteris pachypteris

Tripteris, thickwing, aerial part, s-c, (1)

Ref No 1 05 054 United States

Feed Name or Analyses		Mean As Fed	Mean Dry	C.V. ± %
Dry matter	%	87.0	100.0	
Ash	%	14.5	16.7	

Continued

Feed Name or Analyses			Mean As Fed	Mean Dry	C.V. ± %
Crude fiber	%		15.7	18.1	
Sheep	dig coef %		4.	4.	
Ether extract	%		3.0	3.4	
Sheep	dig coef %		0.	0.	
N-free extract	%		39.5	45.4	
Sheep	dig coef %		93.	93.	
Protein (N x 6.25)	%		14.3	16.4	
Sheep	dig coef %		77.	77.	
Cattle	dig prot %	*	9.7	11.1	
Goats	dig prot %	*	10.3	11.9	
Horses	dig prot %	*	10.0	11.5	
Rabbits	dig coef %	*	9.9	11.3	
Sheep	dig prot %		11.0	12.6	
Energy	GE Mcal/kg				
Cattle	DE Mcal/kg	*	2.17	2.49	
Sheep	DE Mcal/kg	*	2.13	2.45	
Cattle	ME Mcal/kg	*	1.77	2.04	
Sheep	ME Mcal/kg	*	1.75	2.01	
Cattle	TDN %	*	49.2	56.5	
Sheep	TDN %		48.3	55.6	

TRISETUM. Trisetum spp

Trisetum, aerial part, fresh, (2)
Ref No 2 05 055 United States

Analyses			As Fed	Dry	C.V.
Dry matter	%			100.0	
Ash	%			10.7	
Crude fiber	%			31.3	
Ether extract	%			1.8	
N-free extract	%			45.0	
Protein (N x 6.25)	%			11.2	
Cattle	dig prot %	*		7.4	
Goats	dig prot %	*		7.0	
Horses	dig prot %	*		7.0	
Rabbits	dig coef %	*		7.3	
Sheep	dig prot %	*		7.4	

TRISETUM, PRAIRIE. Trisetum interruptum

Trisetum, prairie, aerial part, fresh, (2)
Ref No 2 05 056 United States

Analyses		As Fed	Dry	C.V.
Dry matter	%		100.0	
Calcium	%		0.92	
Magnesium	%		0.15	
Phosphorus	%		0.22	
Potassium	%		1.40	

TUCUM. Astrocaryum spp

Tucum, seeds, extn unspecified grnd, (4)
Tucum oil meal, extraction unspecified
Ref No 4 05 057 United States

Analyses			As Fed	Dry	C.V.
Dry matter	%		91.3	100.0	
Ash	%		2.7	3.0	
Crude fiber	%		10.5	11.5	16
Ether extract	%		8.0	8.8	1
N-free extract	%		58.4	64.0	
Protein (N x 6.25)	%		11.7	12.8	7
Cattle	dig prot %	*	7.1	7.8	
Goats	dig prot %	*	8.2	9.0	
Horses	dig prot %	*	8.2	9.0	
Sheep	dig prot %	*	8.2	9.0	
Arginine	%		1.10	1.20	11
Histidine	%		0.20	0.22	
Isoleucine	%		0.40	0.44	10
Leucine	%		0.70	0.77	21
Lysine	%		0.50	0.55	16
Methionine	%		0.10	0.11	39
Phenylalanine	%		0.40	0.44	10
Threonine	%		0.30	0.33	
Tryptophan	%		0.10	0.11	
Tyrosine	%		0.30	0.33	13
Valine	%		0.40	0.44	10

Tucum oil meal, extraction unspecified — see Tucum, seeds, extn unspecified grnd, (5)

TUMBLEGRASS. Schedonnardus paniculatus

Tumblegrass, aerial part, fresh, early bloom, (2)
Ref No 2 05 058 United States

Analyses			As Fed	Dry	C.V.
Dry matter	%			100.0	
Ash	%			10.8	
Crude fiber	%			30.5	
Ether extract	%			2.1	
N-free extract	%			46.9	
Protein (N x 6.25)	%			9.7	
Cattle	dig prot %	*		6.1	
Goats	dig prot %	*		5.6	
Horses	dig prot %	*		5.8	
Rabbits	dig coef %	*		6.2	
Sheep	dig prot %	*		6.0	
Calcium	%			0.68	
Phosphorus	%			0.15	

Tumblegrass, aerial part, fresh, full bloom, (2)
Ref No 2 05 059 United States

Analyses			As Fed	Dry	C.V.
Dry matter	%			100.0	
Ash	%			10.2	
Crude fiber	%			32.8	
Ether extract	%			1.8	
N-free extract	%			47.7	
Protein (N x 6.25)	%			7.5	
Cattle	dig prot %	*		4.3	
Goats	dig prot %	*		3.6	
Horses	dig prot %	*		3.9	
Rabbits	dig coef %	*		4.5	
Sheep	dig prot %	*		4.0	
Calcium	%			0.56	
Phosphorus	%			0.16	

TURKEY, Meleagris gallopavo

Turkey, carcass, raw, (5)
Ref No 5 08 133 United States

Analyses		As Fed	Dry	C.V.
Dry matter	%	36.7	100.0	
Ash	%	1.0	2.7	
Ether extract	%	15.8	43.1	
Protein (N x 6.25)	%	19.9	54.2	
Energy	GE Mcal/kg	2.27	6.19	

Turkey, carcass, roasted, (5)
Ref No 5 08 131 United States

Analyses		As Fed	Dry	C.V.
Dry matter	%	42.7	100.0	
Ash	%	1.2	2.8	
Ether extract	%	9.6	22.5	
Protein (N x 6.25)	%	31.9	74.7	
Energy	GE Mcal/kg	2.23	5.22	
Niacin	mg/kg	81.0	189.7	
Riboflavin	mg/kg	1.6	3.7	
Thiamine	mg/kg	0.7	1.6	

Turkey, carcass young birds, raw, (5)
Ref No 5 08 132 United States

Analyses		As Fed	Dry	C.V.
Dry matter	%	28.5	100.0	
Ash	%	1.1	3.9	
Ether extract	%	6.0	21.1	
Protein (N x 6.25)	%	21.4	75.1	
Energy	GE Mcal/kg	1.45	5.09	

Turkey, gizzards, raw, (5)
Ref No 5 08 115 United States

Analyses			As Fed	Dry	C.V.
Dry matter	%		29.7	100.0	
Ash	%		1.0	3.4	
Ether extract	%		7.3	24.6	
N-free extract	%		1.1	3.7	
Protein (N x 6.25)	%		20.3	68.4	
Energy	GE Mcal/kg		1.57	5.29	
Cattle	DE Mcal/kg	*	1.42	4.79	
Sheep	DE Mcal/kg	*	1.33	4.49	
Swine	DE kcal/kg	*	1513.	5094.	
Cattle	ME Mcal/kg	*	1.17	3.93	
Sheep	ME Mcal/kg	*	1.09	3.68	
Swine	ME kcal/kg	*	1244.	4187.	
Cattle	TDN %	*	32.3	108.7	
Sheep	TDN %	*	30.2	101.8	
Swine	TDN %	*	34.3	115.5	
Potassium	%		0.17	0.57	
Sodium	%		0.06	0.20	
Niacin	mg/kg		50.0	168.4	
Riboflavin	mg/kg		1.3	4.4	
Thiamine	mg/kg		0.5	1.7	

TURKEY, MATURE BIRDS. Meleagris gallapavo

Turkey, mature birds, viscera w feet, raw, (5)
Ref No 5 08 618 United States

Analyses		As Fed	Dry	C.V.
Dry matter	%	35.5	100.0	
Ash	%	4.6	13.0	
Ether extract	%	18.3	51.5	
Protein (N x 6.25)	%	12.1	34.0	

Turkey, mature birds, viscera, raw, (5)
Ref No 5 08 616 United States

Analyses		As Fed	Dry	C.V.
Dry matter	%	27.5	100.0	
Ash	%	1.7	6.0	
Ether extract	%	11.1	40.5	
Protein (N x 6.25)	%	12.1	44.0	

(1) dry forages and roughages (2) pasture, range plants, and forages fed green (3) silages (4) energy feeds (5) protein supplements (6) minerals (7) vitamins (8) additives

TURKEY, YOUNG BIRDS. *Meleagris gallapavo*

Turkey, young birds, viscera, raw, (5)

Ref No 5 08 617 United States

Feed Name or Analyses	Unit		As Fed	Dry	C.V. ± %
Dry matter	%		22.5	100.0	
Ash	%		1.9	8.5	
Ether extract	%		5.1	22.5	
Protein (N x 6.25)	%		12.2	54.0	

TURNIP. *Brassica rapa*

Turnip, aerial part, s-c, (1)

Ref No 1 05 061 United States

Feed Name or Analyses	Unit		As Fed	Dry	C.V. ± %
Dry matter	%		87.7	100.0	
Ash	%		21.4	24.4	
Crude fiber	%		10.2	11.6	
Sheep	dig coef %		72.	72.	
Ether extract	%		1.4	1.6	
Sheep	dig coef %		34.	34.	
N-free extract	%		41.5	47.3	
Sheep	dig coef %		81.	81.	
Protein (N x 6.25)	%		13.2	15.1	
Sheep	dig coef %		75.	75.	
Cattle	dig prot %	*	8.8	10.0	
Goats	dig prot %	*	9.3	10.6	
Horses	dig prot %	*	9.1	10.3	
Rabbits	dig coef %	*	9.1	10.3	
Sheep	dig prot %		9.9	11.3	
Energy	GE Mcal/kg				
Cattle	DE Mcal/kg	*	2.41	2.75	
Sheep	DE Mcal/kg	*	2.29	2.61	
Cattle	ME Mcal/kg	*	1.98	2.26	
Sheep	ME Mcal/kg	*	1.88	2.14	
Cattle	TDN %	*	54.6	62.3	
Sheep	TDN %		51.9	59.2	

Turnip, aerial part, fresh, (2)

Ref No 2 05 063 United States

Feed Name or Analyses	Unit		As Fed	Dry	C.V. ± %
Dry matter	%		12.7	100.0	
Ash	%		2.1	16.8	
Crude fiber	%		1.3	10.3	
Sheep	dig coef %		96.	96.	
Ether extract	%		0.3	2.6	
Sheep	dig coef %		64.	64.	
N-free extract	%		6.2	48.5	
Sheep	dig coef %		93.	93.	
Protein (N x 6.25)	%		2.8	21.8	
Sheep	dig coef %		38.	38.	
Cattle	dig prot %	*	2.1	16.4	
Goats	dig prot %	*	2.1	16.9	
Horses	dig prot %	*	2.0	16.1	
Rabbits	dig coef %	*	2.0	15.5	
Sheep	dig prot %		1.1	8.3	
Energy	GE Mcal/kg		0.37	2.89	
Cattle	DE Mcal/kg	*	0.39	3.10	
Sheep	DE Mcal/kg	*	0.38	2.95	
Cattle	ME Mcal/kg	*	0.32	2.54	
Sheep	ME Mcal/kg	*	0.31	2.42	
Cattle	TDN %	*	8.9	70.2	
Sheep	TDN %		8.5	67.0	
Calcium	%		0.37	2.92	
Chlorine	%		0.25	1.93	
Copper	mg/kg		2.2	17.6	
Iron	%		0.005	0.040	
Magnesium	%		0.10	0.80	
Manganese	mg/kg		51.9	408.6	
Phosphorus	%		0.06	0.51	
Potassium	%		0.38	3.00	
Sodium	%		0.14	1.13	
Sulphur	%		0.03	0.27	
Niacin	mg/kg		10.5	82.5	
Riboflavin	mg/kg		5.1	40.2	
Thiamine	mg/kg		2.7	21.6	
Vitamin A	IU/g		99.5	783.5	

Turnip, leaves, fresh, (2)

Ref No 2 05 064 United States

Feed Name or Analyses	Unit		As Fed	Dry	C.V. ± %
Dry matter	%		11.2	100.0	
Crude fiber	%		0.9	8.1	4
Ether extract	%		0.5	4.4	
Protein (N x 6.25)	%		3.1	28.1	11
Cattle	dig prot %	*	2.4	21.8	
Goats	dig prot %	*	2.6	22.8	
Horses	dig prot %	*	2.4	21.4	
Rabbits	dig coef %	*	2.3	20.4	
Sheep	dig prot %	*	2.6	23.2	
Carotene	mg/kg		39.1	349.4	37
Riboflavin	mg/kg		2.2	19.4	19
α-tocopherol	mg/kg		33.6	300.0	
Vitamin A equivalent	IU/g		65.2	582.5	
Arginine	%		0.13	1.20	
Histidine	%		0.04	0.40	
Leucine	%		0.11	1.00	
Lysine	%		0.09	0.80	
Methionine	%		0.07	0.60	
Phenylalanine	%		0.16	1.40	
Threonine	%		0.12	1.10	
Tryptophan	%		0.03	0.30	
Valine	%		0.15	1.30	

Turnip, stems, fresh, (2)

Ref No 2 05 065 United States

Feed Name or Analyses	Unit		As Fed	Dry	C.V. ± %
Dry matter	%		6.9	100.0	
Crude fiber	%		0.7	10.3	
Protein (N x 6.25)	%		1.2	18.0	
Cattle	dig prot %	*	0.9	13.2	
Goats	dig prot %	*	0.9	13.4	
Horses	dig prot %	*	0.9	12.8	
Rabbits	dig coef %	*	0.9	12.6	
Sheep	dig prot %	*	0.9	13.8	
Carotene	mg/kg		3.7	54.0	
Riboflavin	mg/kg		0.8	11.7	
Vitamin A equivalent	IU/g		6.2	90.0	

Turnip, aerial part, ensiled, (3)

Ref No 3 05 066 United States

Feed Name or Analyses	Unit		As Fed	Dry	C.V. ± %
Dry matter	%		14.3	100.0	4
Ash	%		3.4	24.0	
Crude fiber	%		2.4	17.0	
Sheep	dig coef %		98.	98.	
Ether extract	%		0.4	3.1	
Sheep	dig coef %		100.	100.	
N-free extract	%		5.9	41.1	
Sheep	dig coef %		88.	88.	
Protein (N x 6.25)	%		2.1	14.8	
Sheep	dig coef %		72.	72.	
Cattle	dig prot %	*	1.4	9.7	
Goats	dig prot %	*	1.4	9.7	
Horses	dig prot %	*	1.4	9.7	
Sheep	dig prot %		1.5	10.7	
Energy	GE Mcal/kg				
Cattle	DE Mcal/kg	*	0.46	3.24	
Sheep	DE Mcal/kg	*	0.44	3.11	
Cattle	ME Mcal/kg	*	0.38	2.66	
Sheep	ME Mcal/kg	*	0.36	2.55	
Cattle	TDN %	*	10.5	73.4	
Sheep	TDN %		10.1	70.5	
Calcium	%		0.41	2.87	31
Copper	mg/kg		2.5	17.6	
Iron	%		0.005	0.038	41
Magnesium	%		0.07	0.47	28
Manganese	mg/kg		58.3	408.5	
Phosphorus	%		0.08	0.58	20
Potassium	%		0.54	3.79	29
Carotene	mg/kg		71.5	500.7	30
Riboflavin	mg/kg		3.7	26.2	21
Thiamine	mg/kg		2.0	14.3	28
α-tocopherol	mg/kg		3.2	22.5	
Vitamin A equivalent	IU/g		119.1	834.6	

Turnip, roots, fresh, (4)

Ref No 2 05 067 United States

Feed Name or Analyses	Unit		As Fed	Dry	C.V. ± %
Dry matter	%		9.7	100.0	
Ash	%		0.8	8.6	
Crude fiber	%		1.1	11.1	
Sheep	dig coef %		91.	91.	
Swine	dig coef %		78.	78.	
Ether extract	%		0.1	1.5	
Sheep	dig coef %		54.	54.	
Swine	dig coef %		0.	0.	
N-free extract	%		6.5	67.4	
Sheep	dig coef %		94.	94.	
Swine	dig coef %		94.	94.	
Protein (N x 6.25)	%		1.1	11.4	
Sheep	dig coef %		64.	64.	
Swine	dig coef %		34.	34.	
Cattle	dig prot %	*	0.6	6.4	
Goats	dig prot %	*	0.7	7.6	
Horses	dig prot %	*	0.7	7.6	
Sheep	dig prot %		0.7	7.3	
Swine	dig prot %		0.4	3.9	
Energy	GE Mcal/kg		0.34	3.53	
Cattle	DE Mcal/kg	*	0.37	3.78	
Sheep	DE Mcal/kg	*	0.35	3.64	
Swine	DE kcal/kg	*	324.	3347.	
Cattle	ME Mcal/kg	*	0.30	3.10	
Sheep	ME Mcal/kg	*	0.29	2.98	
Swine	ME kcal/kg	*	303.	3136.	
Cattle	NE_m Mcal/kg	*	0.19	1.97	
Cattle	NE_{gain} Mcal/kg	*	0.13	1.31	
Cattle	$NE_{lactating\ cows}$ Mcal/kg	*	0.23	2.34	
Cattle	TDN %	*	8.3	85.7	
Sheep	TDN %		8.0	82.6	
Swine	TDN %		7.3	75.9	
Calcium	%		0.05	0.56	
Chlorine	%		0.06	0.65	
Copper	mg/kg		2.1	21.3	
Iron	%		0.001	0.011	
Magnesium	%		0.02	0.22	
Manganese	mg/kg		4.1	42.7	
Phosphorus	%		0.03	0.28	
Potassium	%		0.29	2.99	
Sodium	%		0.10	1.05	
Sulphur	%		0.04	0.43	

Continued

Feed Name or Analyses			Mean As Fed	Mean Dry	C.V. ± %
Ascorbic acid	mg/kg		409.8	4235.3	
Niacin	mg/kg		6.8	70.6	
Riboflavin	mg/kg		0.8	8.2	
Thiamine	mg/kg		0.5	4.7	

TURNIP, ENGLISH FLAT. Brassica rapa

Turnip, English flat, roots, (4)

Ref No 4 05 068 United States

Feed Name or Analyses			Mean As Fed	Mean Dry	C.V. ± %
Dry matter	%		9.8	100.0	
Ash	%		0.9	9.4	
Crude fiber	%		1.0	10.4	
Sheep	dig coef %		103.	103.	
Ether extract	%		0.4	4.1	
Sheep	dig coef %		88.	88.	
N-free extract	%		6.4	64.8	
Sheep	dig coef %		96.	96.	
Protein (N x 6.25)	%		1.1	11.3	
Sheep	dig coef %		90.	90.	
Cattle	dig prot %	*	0.6	6.4	
Goats	dig prot %	*	0.7	7.6	
Horses	dig prot %	*	0.7	7.6	
Sheep	dig prot %		1.0	10.2	
Energy	GE Mcal/kg				
Cattle	DE Mcal/kg	*	0.39	3.94	
Sheep	DE Mcal/kg	*	0.39	4.02	
Swine	DE kcal/kg	*	234.	2391.	
Cattle	ME Mcal/kg	*	0.32	3.23	
Sheep	ME Mcal/kg	*	0.32	3.30	
Swine	ME kcal/kg	*	220.	2241.	
Cattle	TDN %	*	8.7	89.3	
Sheep	TDN %		8.9	91.2	
Swine	TDN %	*	5.3	54.2	

Uncleaned screenings (CFA) —
see Grains, screenings, uncleaned, mn 12% grain mx 3% wild oats mx 17% buckwheat and large seeds mx 68% small weed seeds chaff hulls dust scourings noxious seeds, (4)

VANILLA. Vanilla spp

Vanilla, hulls, (1)

Ref No 1 05 071 United States

Feed Name or Analyses			Mean As Fed	Mean Dry	C.V. ± %
Dry matter	%			100.0	
Ash	%			6.0	
Crude fiber	%			26.3	
Ether extract	%			21.2	
N-free extract	%			36.5	
Protein (N x 6.25)	%			10.0	
Cattle	dig prot %	*		5.6	
Goats	dig prot %	*		5.9	
Horses	dig prot %	*		6.0	
Rabbits	dig coef %	*		6.4	
Sheep	dig prot %	*		5.5	
Calcium	%			1.90	
Manganese	mg/kg			30.0	
Phosphorus	%			0.07	

(1) dry forages and roughages (2) pasture, range plants, and forages fed green (3) silages (4) energy feeds (5) protein supplements (6) minerals (7) vitamins (8) additives

VASEYGRASS. Paspalum urvillei

Vaseygrass, aerial part, fresh, mature, (2)

Ref No 2 05 072 United States

Feed Name or Analyses			Mean As Fed	Mean Dry	C.V. ± %
Dry matter	%			100.0	
Ash	%			10.2	8
Crude fiber	%			37.7	8
Ether extract	%			2.1	29
N-free extract	%			42.5	
Protein (N x 6.25)	%			7.6	13
Cattle	dig prot %	*		4.4	
Goats	dig prot %	*		3.7	
Horses	dig prot %	*		4.0	
Rabbits	dig coef %	*		4.5	
Sheep	dig prot %	*		4.1	
Calcium	%			0.54	6
Phosphorus	%			0.13	17

VELVETBEAN. Stizolobium spp

Velvetbean, hay, s-c, (1)

Ref No 1 05 080 United States

Feed Name or Analyses			Mean As Fed	Mean Dry	C.V. ± %
Dry matter	%		92.8	100.0	
Ash	%		7.4	8.0	
Crude fiber	%		27.5	29.6	
Ether extract	%		3.1	3.3	
N-free extract	%		38.4	41.4	
Protein (N x 6.25)	%		16.4	17.7	
Cattle	dig prot %	*	11.4	12.3	
Goats	dig prot %	*	12.1	13.1	
Horses	dig prot %	*	11.6	12.5	
Rabbits	dig coef %	*	11.4	12.3	
Sheep	dig prot %	*	11.5	12.4	
Fatty acids	%		3.1	3.3	
Energy	GE Mcal/kg				
Cattle	DE Mcal/kg	*	2.41	2.60	
Sheep	DE Mcal/kg	*	2.38	2.56	
Cattle	ME Mcal/kg	*	1.98	2.13	
Sheep	ME Mcal/kg	*	1.95	2.10	
Cattle	TDN %	*	54.7	58.9	
Sheep	TDN %	*	53.9	58.1	
Phosphorus	%		0.24	0.26	
Potassium	%		2.20	2.37	

Velvetbean, hay, s-c, milk stage, (1)

Ref No 1 05 079 United States

Feed Name or Analyses			Mean As Fed	Mean Dry	C.V. ± %
Dry matter	%		77.5	100.0	
Ash	%		5.4	7.0	
Crude fiber	%		32.3	41.7	
Cattle	dig coef %		78.	78.	
Sheep	dig coef %		68.	68.	
Ether extract	%		1.5	2.0	
Cattle	dig coef %		79.	79.	
Sheep	dig coef %		66.	66.	
N-free extract	%		29.0	37.4	
Cattle	dig coef %		76.	76.	
Sheep	dig coef %		61.	61.	
Protein (N x 6.25)	%		9.3	12.0	
Cattle	dig coef %		69.	69.	
Sheep	dig coef %		58.	58.	
Cattle	dig prot %		6.4	8.3	
Goats	dig prot %	*	6.0	7.8	
Horses	dig prot %	*	6.0	7.7	
Rabbits	dig coef %	*	6.1	7.9	

Feed Name or Analyses			Mean As Fed	Mean Dry	C.V. ± %
Sheep	dig prot %		5.4	7.0	
Energy	GE Mcal/kg				
Cattle	DE Mcal/kg	*	2.48	3.20	
Sheep	DE Mcal/kg	*	2.08	2.69	
Cattle	ME Mcal/kg	*	2.04	2.63	
Sheep	ME Mcal/kg	*	1.71	2.21	
Cattle	TDN %		56.3	72.7	
Sheep	TDN %		47.3	61.0	

Velvetbean, hulls, (1)

Ref No 1 05 081 United States

Feed Name or Analyses			Mean As Fed	Mean Dry	C.V. ± %
Dry matter	%		86.8	100.0	
Ash	%		6.1	7.0	
Crude fiber	%		24.5	28.2	
Ether extract	%		1.9	2.2	
N-free extract	%		42.7	49.2	
Protein (N x 6.25)	%		11.6	13.4	
Cattle	dig prot %	*	7.4	8.5	
Goats	dig prot %	*	7.9	9.1	
Horses	dig prot %	*	7.7	8.9	
Rabbits	dig coef %	*	7.8	9.0	
Sheep	dig prot %	*	7.5	8.6	
Energy	GE Mcal/kg				
Cattle	DE Mcal/kg	*	2.23	2.57	
Sheep	DE Mcal/kg	*	2.30	2.66	
Cattle	ME Mcal/kg	*	1.83	2.11	
Sheep	ME Mcal/kg	*	1.89	2.18	
Cattle	TDN %	*	50.6	58.3	
Sheep	TDN %	*	52.3	60.2	

Velvetbean, pods, (1)

Ref No 1 05 082 United States

Feed Name or Analyses			Mean As Fed	Mean Dry	C.V. ± %
Dry matter	%		88.6	100.0	
Ash	%		7.0	7.9	
Crude fiber	%		28.0	31.6	
Ether extract	%		0.8	0.9	
N-free extract	%		48.2	54.4	
Protein (N x 6.25)	%		4.6	5.2	
Cattle	dig prot %	*	1.3	1.4	
Goats	dig prot %	*	1.3	1.4	
Horses	dig prot %	*	1.7	1.9	
Rabbits	dig coef %	*	2.4	2.7	
Sheep	dig prot %	*	1.1	1.2	
Energy	GE Mcal/kg				
Cattle	DE Mcal/kg	*	2.36	2.66	
Sheep	DE Mcal/kg	*	2.17	2.44	
Cattle	ME Mcal/kg	*	1.93	2.18	
Sheep	ME Mcal/kg	*	1.78	2.00	
Cattle	TDN %	*	53.5	60.4	
Sheep	TDN %	*	49.1	55.4	

Velvetbean, aerial part, fresh, (2)

Ref No 2 05 085 United States

Feed Name or Analyses			Mean As Fed	Mean Dry	C.V. ± %
Dry matter	%		19.9	100.0	
Ash	%		2.0	10.0	
Crude fiber	%		5.8	28.9	
Cattle	dig coef %		58.	58.	
Sheep	dig coef %		58.	58.	
Ether extract	%		0.7	3.3	
Cattle	dig coef %		63.	63.	
Sheep	dig coef %		68.	68.	
N-free extract	%		8.0	40.1	
Cattle	dig coef %		74.	74.	
Sheep	dig coef %		80.	80.	

Feed Name or Analyses			Mean As Fed	Mean Dry	C.V. ± %
Protein (N x 6.25)	%		3.5	17.7	
Cattle	dig coef %		69.	69.	
Sheep	dig coef %		71.	71.	
Cattle	dig prot %		2.4	12.2	
Goats	dig prot %	*	2.6	13.0	
Horses	dig prot %	*	2.5	12.5	
Rabbits	dig coef %	*	2.5	12.3	
Sheep	dig prot %		2.5	12.5	
Energy	GE Mcal/kg				
Cattle	DE Mcal/kg	*	0.56	2.79	
Sheep	DE Mcal/kg	*	0.58	2.93	
Cattle	ME Mcal/kg	*	0.46	2.29	
Sheep	ME Mcal/kg	*	0.48	2.40	
Cattle	TDN %		12.6	63.3	
Sheep	TDN %		13.2	66.5	
Phosphorus	%		0.07	0.34	
Potassium	%		0.41	2.07	

Velvetbean, aerial part, fresh, full bloom, (2)

Ref No 2 05 083 United States

Feed Name or Analyses			Mean As Fed	Mean Dry	C.V. ± %
Dry matter	%		19.2	100.0	
Ash	%		1.1	5.8	
Crude fiber	%		6.6	34.5	
Sheep	dig coef %		57.	57.	
Ether extract	%		0.8	4.2	
Sheep	dig coef %		67.	67.	
N-free extract	%		7.7	39.9	
Sheep	dig coef %		77.	77.	
Protein (N x 6.25)	%		3.0	15.6	
Sheep	dig coef %		71.	71.	
Cattle	dig prot %	*	2.1	11.2	
Goats	dig prot %	*	2.1	11.1	
Horses	dig prot %	*	2.1	10.8	
Rabbits	dig coef %	*	2.1	10.7	
Sheep	dig prot %		2.1	11.1	
Energy	GE Mcal/kg				
Cattle	DE Mcal/kg	*	0.62	3.24	
Sheep	DE Mcal/kg	*	0.57	2.99	
Cattle	ME Mcal/kg	*	0.51	2.66	
Sheep	ME Mcal/kg	*	0.47	2.45	
Cattle	TDN %	*	14.1	73.5	
Sheep	TDN %		13.0	67.8	

Velvetbean, aerial part, fresh, dough stage, (2)

Ref No 2 05 084 United States

Feed Name or Analyses			Mean As Fed	Mean Dry	C.V. ± %
Dry matter	%		22.5	100.0	
Ash	%		2.4	10.7	
Crude fiber	%		5.3	23.7	
Cattle	dig coef %		59.	59.	
Sheep	dig coef %		59.	59.	
Ether extract	%		0.5	2.3	
Cattle	dig coef %		64.	64.	
Sheep	dig coef %		70.	70.	
N-free extract	%		10.7	47.7	
Cattle	dig coef %		80.	80.	
Sheep	dig coef %		85.	85.	
Protein (N x 6.25)	%		3.5	15.8	
Cattle	dig coef %		70.	70.	
Sheep	dig coef %		70.	70.	
Cattle	dig prot %		2.5	11.0	
Goats	dig prot %	*	2.5	11.3	
Horses	dig prot %	*	2.4	10.9	
Rabbits	dig coef %	*	2.4	10.8	
Sheep	dig prot %		2.5	11.0	
Energy	GE Mcal/kg				
Cattle	DE Mcal/kg	*	0.66	2.92	
Sheep	DE Mcal/kg	*	0.68	3.04	
Cattle	ME Mcal/kg	*	0.54	2.40	
Sheep	ME Mcal/kg	*	0.56	2.50	
Cattle	TDN %		14.9	66.3	
Sheep	TDN %		15.5	69.0	

Velvetbean, pods w seeds, (4)

Velvet bean feed

Ref No 4 05 087 United States

Feed Name or Analyses			Mean As Fed	Mean Dry	C.V. ± %
Dry matter	%		88.7	100.0	
Ash	%		4.7	5.3	
Crude fiber	%		12.8	14.5	
Horses	dig coef %		63.	63.	
Sheep	dig coef %		74.	74.	
Ether extract	%		4.3	4.8	
Horses	dig coef %		81.	81.	
Sheep	dig coef %		86.	86.	
N-free extract	%		49.6	55.9	
Horses	dig coef %		85.	85.	
Sheep	dig coef %		89.	89.	
Protein (N x 6.25)	%		17.2	19.4	
Horses	dig coef %		75.	75.	
Sheep	dig coef %		75.	75.	
Cattle	dig prot %	*	12.3	13.9	
Goats	dig prot %	*	13.3	15.0	
Horses	dig prot %		12.9	14.6	
Sheep	dig prot %		12.9	14.6	
Energy	GE Mcal/kg				
Cattle	DE Mcal/kg	*	2.31	2.61	
Horses	DE Mcal/kg	*	3.13	3.53	
Sheep	DE Mcal/kg	*	3.30	3.72	
Cattle	ME Mcal/kg	*	1.90	2.14	
Horses	ME Mcal/kg	*	2.57	2.89	
Sheep	ME Mcal/kg	*	2.71	3.05	
Cattle	TDN %	*	52.4	59.1	
Horses	TDN %		71.0	80.0	
Sheep	TDN %		74.9	84.4	
Calcium	%		0.24	0.27	
Chlorine	%		0.22	0.24	
Iron	%		0.013	0.014	
Magnesium	%		0.21	0.23	
Phosphorus	%		0.37	0.42	
Potassium	%		1.18	1.33	
Sodium	%		0.14	0.16	
Sulphur	%		0.15	0.17	

Velvetbean, seeds, (5)

Ref No 5 05 089 United States

Feed Name or Analyses			Mean As Fed	Mean Dry	C.V. ± %
Dry matter	%		89.3	100.0	
Ash	%		3.3	3.7	
Crude fiber	%		9.6	10.8	
Cattle	dig coef %		79.	79.	
Sheep	dig coef %		72.	72.	
Ether extract	%		4.8	5.4	
Cattle	dig coef %		76.	76.	
Sheep	dig coef %		64.	64.	
N-free extract	%		50.4	56.4	
Cattle	dig coef %		91.	91.	
Sheep	dig coef %		97.	97.	
Protein (N x 6.25)	%		21.2	23.8	
Cattle	dig coef %		76.	76.	
Sheep	dig coef %		81.	81.	
Cattle	dig prot %		16.1	18.1	
Sheep	dig prot %		17.2	19.3	
Energy	GE Mcal/kg				
Cattle	DE Mcal/kg	*	3.43	3.84	
Sheep	DE Mcal/kg	*	3.52	3.94	
Swine	DE kcal/kg	*	3766.	4217.	
Cattle	ME Mcal/kg	*	2.81	3.15	
Sheep	ME Mcal/kg	*	2.89	3.23	
Swine	ME kcal/kg	*	3435.	3846.	
Cattle	TDN %		77.8	87.1	
Sheep	TDN %		79.9	89.4	
Swine	TDN %	*	85.4	95.6	

Velvetbean, seeds, grnd, (5)

Ref No 5 05 088 United States

Feed Name or Analyses			Mean As Fed	Mean Dry	C.V. ± %
Dry matter	%		88.5	100.0	0
Ash	%		3.5	4.0	18
Crude fiber	%		12.3	13.9	2
Ether extract	%		3.9	4.4	4
N-free extract	%		50.5	57.1	
Protein (N x 6.25)	%		18.3	20.7	3
Energy	GE Mcal/kg				
Cattle	DE Mcal/kg	*	3.02	3.41	
Sheep	DE Mcal/kg	*	2.86	3.23	
Swine	DE kcal/kg	*	2432.	2748.	
Cattle	ME Mcal/kg	*	2.47	2.80	
Sheep	ME Mcal/kg	*	2.35	2.65	
Swine	ME kcal/kg	*	2233.	2523.	
Cattle	TDN %	*	68.4	77.3	
Sheep	TDN %	*	64.9	73.3	
Swine	TDN %	*	55.2	62.3	
Calcium	%		0.24	0.27	4
Iron	%		0.010	0.011	
Magnesium	%		0.18	0.20	
Phosphorus	%		0.32	0.36	2

Velvet bean feed —
see Velvetbean, pods w seeds, (4)

VERBENA. Verbena spp

Verbena, browse, immature, (2)

Ref No 2 05 090 United States

Feed Name or Analyses			Mean As Fed	Mean Dry	C.V. ± %
Dry matter	%			100.0	
Ash	%			9.8	
Crude fiber	%			15.6	
Ether extract	%			3.0	
N-free extract	%			60.1	
Protein (N x 6.25)	%			11.5	
Cattle	dig prot %	*		7.7	
Goats	dig prot %	*		7.3	
Horses	dig prot %	*		7.3	
Rabbits	dig coef %	*		7.5	
Sheep	dig prot %	*		7.7	
Calcium	%			2.46	
Magnesium	%			0.37	
Phosphorus	%			0.17	
Potassium	%			1.59	

VERBENA, DAKOTA. Verbena bipinnatifida

Verbena, Dakota, browse, immature, (2)

Ref No 2 05 091 United States

Feed Name or Analyses			Mean As Fed	Mean Dry	C.V. ± %
Dry matter	%			100.0	
Ash	%			8.4	
Crude fiber	%			18.1	
Ether extract	%			1.8	

Continued

Feed Name or Analyses			Mean As Fed	Mean Dry	C.V. ± %
N-free extract	%			60.2	
Protein (N x 6.25)	%			11.5	
Cattle	dig prot %	*		7.7	
Goats	dig prot %	*		7.3	
Horses	dig prot %	*		7.3	
Rabbits	dig coef %	*		7.5	
Sheep	dig prot %	*		7.7	
Calcium	%			1.99	
Magnesium	%			0.34	
Phosphorus	%			0.18	
Potassium	%			1.60	

VERNALGRASS, SWEET. Anthoxanthum odoratum

Vernalgrass, sweet, aerial part, fresh, (2)
Ref No 2 05 094 United States

Feed Name or Analyses			Mean As Fed	Mean Dry	C.V. ± %
Dry matter	%		22.2	100.0	14
Ash	%		2.2	10.0	10
Crude fiber	%		3.9	17.8	9
Ether extract	%		1.0	4.6	17
N-free extract	%		10.6	47.7	
Protein (N x 6.25)	%		4.4	20.0	16
Cattle	dig prot %	*	3.3	14.9	
Goats	dig prot %	*	3.4	15.2	
Horses	dig prot %	*	3.2	14.5	
Rabbits	dig coef %	*	3.1	14.1	
Sheep	dig prot %	*	3.5	15.6	
Energy	GE Mcal/kg				
Cattle	DE Mcal/kg	*	0.59	2.66	
Sheep	DE Mcal/kg	*	0.61	2.76	
Cattle	ME Mcal/kg	*	0.48	2.18	
Sheep	ME Mcal/kg	*	0.50	2.26	
Cattle	TDN %	*	13.4	60.4	
Sheep	TDN %	*	13.9	62.6	
Calcium	%		0.14	0.64	
Phosphorus	%		0.08	0.36	

Vernalgrass, sweet, aerial part, fresh, immature, pure stand, (2)
Ref No 2 05 092 United States

Feed Name or Analyses			Mean As Fed	Mean Dry	C.V. ± %
Dry matter	%		21.9	100.0	9
Ash	%		2.2	10.2	11
Crude fiber	%		3.9	17.8	10
Ether extract	%		1.0	4.6	19
N-free extract	%		10.2	46.6	
Protein (N x 6.25)	%		4.6	20.8	13
Cattle	dig prot %	*	3.4	15.6	
Goats	dig prot %	*	3.5	16.0	
Horses	dig prot %	*	3.3	15.2	
Rabbits	dig coef %	*	3.2	14.7	
Sheep	dig prot %	*	3.6	16.4	
Energy	GE Mcal/kg				
Cattle	DE Mcal/kg	*	0.57	2.60	
Sheep	DE Mcal/kg	*	0.58	2.67	
Cattle	ME Mcal/kg	*	0.47	2.13	
Sheep	ME Mcal/kg	*	0.48	2.19	
Cattle	TDN %	*	12.9	59.0	
Sheep	TDN %	*	13.2	60.5	
Calcium	%		0.14	0.64	7
Copper	mg/kg		2.6	11.9	
Iron	%		0.013	0.061	

(1) dry forages and roughages
(2) pasture, range plants, and forages fed green
(3) silages
(4) energy feeds
(5) protein supplements
(6) minerals
(7) vitamins
(8) additives

Feed Name or Analyses			Mean As Fed	Mean Dry	C.V. ± %
Magnesium	%		0.06	0.29	18
Manganese	mg/kg		16.7	76.1	
Phosphorus	%		0.09	0.39	11
Carotene	mg/kg		25.8	117.9	23
Vitamin A equivalent	IU/g		43.1	196.6	

Vernalgrass, sweet, aerial part, fresh, mid-bloom, pure stand, (2)
Ref No 2 05 093 United States

Feed Name or Analyses			Mean As Fed	Mean Dry	C.V. ± %
Dry matter	%		26.0	100.0	
Ash	%		2.4	9.4	
Crude fiber	%		4.8	18.6	
Ether extract	%		1.3	5.0	
N-free extract	%		13.7	52.5	
Protein (N x 6.25)	%		3.8	14.5	
Cattle	dig prot %	*	2.7	10.2	
Goats	dig prot %	*	2.6	10.1	
Horses	dig prot %	*	2.6	9.8	
Rabbits	dig coef %	*	2.6	9.9	
Sheep	dig prot %	*	2.7	10.5	
Energy	GE Mcal/kg				
Cattle	DE Mcal/kg	*	0.72	2.76	
Sheep	DE Mcal/kg	*	0.77	2.95	
Cattle	ME Mcal/kg	*	0.59	2.27	
Sheep	ME Mcal/kg	*	0.63	2.42	
Cattle	TDN %	*	16.3	62.7	
Sheep	TDN %	*	17.4	67.0	
Calcium	%		0.16	0.63	
Magnesium	%		0.06	0.22	
Phosphorus	%		0.11	0.41	

Vernalgrass, sweet, aerial part, fresh, pure stand, (2)
Ref No 2 05 095 United States

Feed Name or Analyses			Mean As Fed	Mean Dry	C.V. ± %
Dry matter	%		22.4	100.0	15
Ash	%		2.3	10.1	10
Crude fiber	%		4.0	17.9	10
Ether extract	%		1.1	4.7	18
N-free extract	%		10.6	47.3	
Protein (N x 6.25)	%		4.5	20.0	18
Cattle	dig prot %	*	3.3	14.9	
Goats	dig prot %	*	3.4	15.2	
Horses	dig prot %	*	3.3	14.5	
Rabbits	dig coef %	*	3.2	14.1	
Sheep	dig prot %	*	3.5	15.6	
Energy	GE Mcal/kg				
Cattle	DE Mcal/kg	*	0.63	2.83	
Sheep	DE Mcal/kg	*	0.65	2.88	
Cattle	ME Mcal/kg	*	0.52	2.32	
Sheep	ME Mcal/kg	*	0.53	2.36	
Cattle	TDN %	*	14.4	64.1	
Sheep	TDN %	*	14.6	65.3	

VETCH. Vicia spp

Vetch, aerial part, dehy, immature, (1)
Ref No 1 05 097 United States

Feed Name or Analyses			Mean As Fed	Mean Dry	C.V. ± %
Dry matter	%		93.4	100.0	
Carotene	mg/kg		307.8	329.6	
Vitamin A equivalent	IU/g		513.2	549.4	

Vetch, hay, s-c, (1)
Ref No 1 05 106 United States

Feed Name or Analyses			Mean As Fed	Mean Dry	C.V. ± %
Dry matter	%		88.0	100.0	3
Ash	%		3.3	3.8	
Crude fiber	%		3.1	3.5	
Cattle	dig coef %	*	58.	58.	
Sheep	dig coef %	*	20.	20.	
Ether extract	%		1.2	1.4	
Cattle	dig coef %	*	70.	70.	
Sheep	dig coef %	*	67.	67.	
N-free extract	%		52.0	59.1	
Cattle	dig coef %	*	71.	71.	
Sheep	dig coef %	*	86.	86.	
Protein (N x 6.25)	%		28.3	32.2	
Cattle	dig coef %	*	66.	66.	
Sheep	dig coef %	*	83.	83.	
Cattle	dig prot %		18.7	21.3	
Goats	dig prot %	*	23.4	26.6	
Horses	dig prot %	*	21.9	24.9	
Rabbits	dig coef %	*	20.7	23.5	
Sheep	dig prot %		23.5	26.7	
Energy	GE Mcal/kg				
Cattle	DE Mcal/kg	*	2.62	2.97	
Sheep	DE Mcal/kg	*	3.12	3.54	
Cattle	ME Mcal/kg	*	2.15	2.44	
Sheep	ME Mcal/kg	*	2.56	2.90	
Cattle	NE_m Mcal/kg	*	1.17	1.33	
Cattle	NE_{gain} Mcal/kg	*	0.64	0.73	
Cattle	$NE_{lactating\ cows}$ Mcal/kg	*	1.34	1.52	
Cattle	TDN %		59.4	67.4	
Sheep	TDN %		70.7	80.3	
Cobalt	mg/kg		0.312	0.355	
Iodine	mg/kg		0.433	0.492	22
Sodium	%		0.46	0.52	11
Sulphur	%		0.13	0.15	26

Vetch, hay, s-c, immature, (1)
Ref No 1 05 098 United States

Feed Name or Analyses			Mean As Fed	Mean Dry	C.V. ± %
Dry matter	%		91.9	100.0	3
Ash	%		7.7	8.4	20
Crude fiber	%		24.0	26.1	9
Cattle	dig coef %	*	58.	58.	
Sheep	dig coef %	*	50.	50.	
Ether extract	%		3.0	3.3	25
Cattle	dig coef %	*	70.	70.	
Sheep	dig coef %	*	46.	46.	
N-free extract	%		34.0	37.0	
Cattle	dig coef %	*	71.	71.	
Sheep	dig coef %	*	72.	72.	
Protein (N x 6.25)	%		23.2	25.2	9
Cattle	dig coef %	*	66.	66.	
Sheep	dig coef %	*	78.	78.	
Cattle	dig prot %		15.3	16.6	
Goats	dig prot %	*	18.4	20.1	
Horses	dig prot %	*	17.4	18.9	
Rabbits	dig coef %	*	16.7	18.1	
Sheep	dig prot %		18.1	19.7	
Energy	GE Mcal/kg				
Cattle	DE Mcal/kg	*	2.56	2.79	
Sheep	DE Mcal/kg	*	2.54	2.77	
Cattle	ME Mcal/kg	*	2.10	2.29	
Sheep	ME Mcal/kg	*	2.09	2.27	
Cattle	TDN %		58.1	63.2	
Sheep	TDN %		57.7	62.8	

Feed Name or Analyses			Mean As Fed	Mean Dry	C.V. ± %
Carotene	mg/kg		355.0	386.2	29
Vitamin A equivalent	IU/g		591.7	643.9	

Vetch, hay, s-c, early bloom, (1)

Ref No 1 05 099 United States

Feed Name or Analyses			As Fed	Dry	C.V. ± %
Dry matter	%		83.0	100.0	
Ash	%		9.3	11.2	
Crude fiber	%		23.7	28.6	
Sheep	dig coef %		55.	55.	
Ether extract	%		2.2	2.7	
Sheep	dig coef %		52.	52.	
N-free extract	%		28.4	34.2	
Sheep	dig coef %		65.	65.	
Protein (N x 6.25)	%		19.3	23.3	
Sheep	dig coef %		76.	76.	
Cattle	dig prot %	*	14.2	17.1	
Goats	dig prot %	*	15.2	18.3	
Horses	dig prot %	*	14.4	17.3	
Rabbits	dig prot %	*	13.8	16.7	
Sheep	dig prot %		14.7	17.7	
Energy	GE Mcal/kg				
Cattle	DE Mcal/kg	*	2.25	2.71	
Sheep	DE Mcal/kg	*	2.15	2.59	
Cattle	ME Mcal/kg	*	1.84	2.22	
Sheep	ME Mcal/kg	*	1.77	2.13	
Cattle	TDN %	*	50.9	61.4	
Sheep	TDN %		48.8	58.8	

Vetch, hay, s-c, mid-bloom, (1)

Ref No 1 05 100 United States

Feed Name or Analyses			As Fed	Dry	C.V. ± %
Dry matter	%			100.0	
Ash	%			6.8	5
Crude fiber	%			29.3	3
Ether extract	%			4.8	5
N-free extract	%			38.3	
Protein (N x 6.25)	%			20.8	7
Cattle	dig prot %	*		15.0	
Goats	dig prot %	*		16.0	
Horses	dig prot %	*		15.2	
Rabbits	dig prot %	*		14.7	
Sheep	dig prot %	*		15.2	
Energy	GE Mcal/kg				
Cattle	DE Mcal/kg	*		2.60	
Sheep	DE Mcal/kg	*		2.58	
Cattle	ME Mcal/kg	*		2.13	
Sheep	ME Mcal/kg	*		2.11	
Cattle	TDN %	*		59.0	
Sheep	TDN %	*		58.5	

Vetch, hay, s-c, full bloom, (1)

Ref No 1 05 101 United States

Feed Name or Analyses			As Fed	Dry	C.V. ± %
Dry matter	%		85.6	100.0	
Ash	%		8.1	9.5	12
Crude fiber	%		29.9	34.9	4
Cattle	dig coef %	*	58.	58.	
Sheep	dig coef %		47.	47.	
Ether extract	%		1.9	2.2	24
Cattle	dig coef %	*	70.	70.	
Sheep	dig coef %		-3.	-3.	
N-free extract	%		29.1	34.0	
Cattle	dig coef %	*	71.	71.	
Sheep	dig coef %		60.	60.	
Protein (N x 6.25)	%		16.7	19.5	7
Cattle	dig coef %	*	66.	66.	
Sheep	dig coef %		71.	71.	
Cattle	dig prot %		11.0	12.9	
Goats	dig prot %	*	12.6	14.8	
Horses	dig prot %	*	12.1	14.1	
Rabbits	dig prot %	*	11.7	13.7	
Sheep	dig prot %		11.9	13.8	
Energy	GE Mcal/kg				
Cattle	DE Mcal/kg	*	2.29	2.68	
Sheep	DE Mcal/kg	*	1.90	2.22	
Cattle	ME Mcal/kg	*	1.88	2.19	
Sheep	ME Mcal/kg	*	1.56	1.82	
Cattle	TDN %		51.9	60.7	
Sheep	TDN %		43.2	50.4	
Calcium	%		0.87	1.02	
Iron	%		0.024	0.028	
Magnesium	%		0.19	0.22	
Manganese	mg/kg		72.8	85.1	
Phosphorus	%		0.24	0.28	
Potassium	%		2.16	2.52	

Vetch, hay, s-c, dough stage, (1)

Ref No 1 05 102 United States

Feed Name or Analyses			As Fed	Dry	C.V. ± %
Dry matter	%		85.1	100.0	
Ash	%		7.4	8.8	8
Crude fiber	%		30.0	35.2	5
Cattle	dig coef %	*	58.	58.	
Sheep	dig coef %		45.	45.	
Ether extract	%		1.8	2.1	26
Cattle	dig coef %	*	70.	70.	
Sheep	dig coef %		7.	7.	
N-free extract	%		34.3	40.3	
Cattle	dig coef %	*	71.	71.	
Sheep	dig coef %		62.	62.	
Protein (N x 6.25)	%		11.7	13.7	11
Cattle	dig coef %	*	66.	66.	
Sheep	dig coef %		62.	62.	
Cattle	dig prot %		7.7	9.0	
Goats	dig prot %	*	8.0	9.3	
Horses	dig prot %	*	7.8	9.2	
Rabbits	dig prot %	*	7.9	9.2	
Sheep	dig prot %		7.2	8.5	
Energy	GE Mcal/kg				
Cattle	DE Mcal/kg	*	2.30	2.70	
Sheep	DE Mcal/kg	*	1.86	2.19	
Cattle	ME Mcal/kg	*	1.89	2.22	
Sheep	ME Mcal/kg	*	1.53	1.79	
Cattle	TDN %		52.2	61.3	
Sheep	TDN %		42.2	49.6	

Vetch, hay, s-c, over ripe, (1)

Ref No 1 05 103 United States

Feed Name or Analyses			As Fed	Dry	C.V. ± %
Dry matter	%		81.8	100.0	
Ash	%		7.0	8.6	
Crude fiber	%		21.6	26.4	
Cattle	dig coef %		58.	58.	
Ether extract	%		2.6	3.2	
Cattle	dig coef %		71.	71.	
N-free extract	%		37.9	46.3	
Cattle	dig coef %		72.	72.	
Protein (N x 6.25)	%		12.7	15.5	
Cattle	dig coef %		70.	70.	
Cattle	dig prot %		8.9	10.9	
Goats	dig prot %	*	9.0	11.0	
Horses	dig prot %	*	8.7	10.7	
Rabbits	dig prot %	*	8.7	10.6	
Sheep	dig prot %	*	8.6	10.5	
Energy	GE Mcal/kg				
Cattle	DE Mcal/kg	*	2.33	2.85	
Sheep	DE Mcal/kg	*	2.15	2.62	
Cattle	ME Mcal/kg	*	1.91	2.34	
Sheep	ME Mcal/kg	*	1.76	2.15	
Cattle	TDN %		52.9	64.6	
Sheep	TDN %	*	48.7	59.5	

Vetch, hay, s-c, cut 2, (1)

Ref No 1 05 104 United States

Feed Name or Analyses			As Fed	Dry	C.V. ± %
Dry matter	%			100.0	
Calcium	%			1.26	
Phosphorus	%			0.37	
Carotene	mg/kg			587.1	
Vitamin A equivalent	IU/g			978.7	

Vetch, hay, s-c, cut 3, (1)

Ref No 1 05 105 United States

Feed Name or Analyses			As Fed	Dry	C.V. ± %
Dry matter	%			100.0	
Calcium	%			1.09	
Phosphorus	%			0.30	
Carotene	mg/kg			334.0	
Vitamin A equivalent	IU/g			556.8	

Vetch, aerial part, fresh, (2)

Ref No 2 05 111 United States

Feed Name or Analyses			As Fed	Dry	C.V. ± %
Dry matter	%		22.0	100.0	12
Ash	%		2.1	9.4	23
Crude fiber	%		6.1	27.6	10
Sheep	dig coef %		40.	40.	
Ether extract	%		0.5	2.3	23
Sheep	dig coef %		49.	49.	
N-free extract	%		8.7	39.7	
Sheep	dig coef %		76.	76.	
Protein (N x 6.25)	%		4.6	20.9	15
Sheep	dig coef %		75.	75.	
Cattle	dig prot %	*	3.4	15.6	
Goats	dig prot %	*	3.5	16.0	
Horses	dig prot %	*	3.3	15.3	
Rabbits	dig prot %	*	3.2	14.8	
Sheep	dig prot %		3.4	15.7	
Energy	GE Mcal/kg				
Cattle	DE Mcal/kg	*	0.73	3.33	
Sheep	DE Mcal/kg	*	0.58	2.62	
Cattle	ME Mcal/kg	*	0.60	2.73	
Sheep	ME Mcal/kg	*	0.47	2.15	
Cattle	TDN %	*	16.6	75.5	
Sheep	TDN %		13.1	59.5	
Chlorine	%		0.41	1.85	
Cobalt	mg/kg		0.066	0.300	
Sodium	%		0.11	0.49	
Sulphur	%		0.03	0.15	40

Vetch, aerial part, fresh, pre-bloom, (2)

Ref No 2 05 108 United States

Feed Name or Analyses			As Fed	Dry	C.V. ± %
Dry matter	%		22.3	100.0	
Ash	%		2.3	10.2	
Crude fiber	%		6.2	27.7	
Sheep	dig coef %		40.	40.	
Ether extract	%		0.5	2.2	
Sheep	dig coef %		47.	47.	

Continued

Feed Name or Analyses			Mean As Fed	Mean Dry	C.V. ± %
N-free extract	%		8.7	39.1	
Sheep	dig coef %		76.	76.	
Protein (N x 6.25)	%		4.6	20.8	
Sheep	dig coef %		76.	76.	
Cattle	dig prot %	*	3.5	15.6	
Goats	dig prot %	*	3.6	16.0	
Horses	dig prot %	*	3.4	15.2	
Rabbits	dig prot %	*	3.3	14.7	
Sheep	dig prot %		3.5	15.8	
Energy	GE Mcal/kg				
Cattle	DE Mcal/kg	*	0.59	2.66	
Sheep	DE Mcal/kg	*	0.58	2.60	
Cattle	ME Mcal/kg	*	0.49	2.18	
Sheep	ME Mcal/kg	*	0.48	2.13	
Cattle	TDN %	*	13.5	60.4	
Sheep	TDN %		13.1	58.9	

Vetch, aerial part, fresh, early bloom, (2)

Ref No 2 05 109 United States

Feed Name or Analyses			Mean As Fed	Mean Dry	C.V. ± %
Dry matter	%			100.0	
Calcium	%			1.62	
Magnesium	%			0.26	
Phosphorus	%			0.26	
Potassium	%			3.15	

Vetch, aerial part, fresh, full bloom, (2)

Ref No 2 05 110 United States

Feed Name or Analyses			Mean As Fed	Mean Dry	C.V. ± %
Dry matter	%			100.0	
Calcium	%			1.07	
Magnesium	%			0.31	
Manganese	mg/kg			144.2	
Phosphorus	%			0.35	
Potassium	%			2.51	

Vetch, aerial part, ensiled, (3)

Ref No 3 05 112 United States

Feed Name or Analyses			Mean As Fed	Mean Dry	C.V. ± %
Dry matter	%		30.1	100.0	9
Ash	%		2.4	7.9	7
Crude fiber	%		9.8	32.7	5
Cattle	dig coef %		63.	63.	
Ether extract	%		1.0	3.3	15
Cattle	dig coef %		77.	77.	
N-free extract	%		13.4	44.4	
Cattle	dig coef %		67.	67.	
Protein (N x 6.25)	%		3.5	11.6	15
Cattle	dig coef %		56.	56.	
Cattle	dig prot %		2.0	6.5	
Goats	dig prot %	*	2.0	6.8	
Horses	dig prot %	*	2.0	6.8	
Sheep	dig prot %	*	2.0	6.8	
Energy	GE Mcal/kg				
Cattle	DE Mcal/kg	*	0.83	2.76	
Sheep	DE Mcal/kg	*	0.84	2.79	
Cattle	ME Mcal/kg	*	0.68	2.27	
Sheep	ME Mcal/kg	*	0.69	2.29	
Cattle	TDN %		18.9	62.7	
Sheep	TDN %	*	19.1	63.2	

(1) dry forages and roughages (2) pasture, range plants, and forages fed green (3) silages (4) energy feeds (5) protein supplements (6) minerals (7) vitamins (8) additives

Vetch, aerial part, steamed ensiled, (3)

Ref No 3 05 113 United States

Feed Name or Analyses			Mean As Fed	Mean Dry	C.V. ± %
Dry matter	%		30.5	100.0	
Ash	%		2.3	7.6	
Crude fiber	%		10.1	33.2	
Cattle	dig coef %		51.	51.	
Ether extract	%		0.8	2.6	
Cattle	dig coef %		63.	63.	
N-free extract	%		13.5	44.2	
Cattle	dig coef %		63.	63.	
Protein (N x 6.25)	%		3.8	12.4	
Cattle	dig coef %		15.	15.	
Cattle	dig prot %		0.6	1.9	
Goats	dig prot %	*	2.3	7.5	
Horses	dig prot %	*	2.3	7.5	
Sheep	dig prot %	*	2.3	7.5	
Energy	GE Mcal/kg				
Cattle	DE Mcal/kg	*	0.68	2.22	
Sheep	DE Mcal/kg	*	0.72	2.36	
Cattle	ME Mcal/kg	*	0.55	1.82	
Sheep	ME Mcal/kg	*	0.59	1.94	
Cattle	TDN %		15.3	50.3	
Sheep	TDN %	*	16.3	53.6	

Vetch, seeds, (5)

Ref No 5 08 546 United States

Feed Name or Analyses			Mean As Fed	Mean Dry	C.V. ± %
Dry matter	%		90.7	100.0	
Ash	%		3.1	3.4	
Crude fiber	%		5.7	6.3	
Ether extract	%		0.8	0.9	
N-free extract	%		51.5	56.8	
Protein (N x 6.25)	%		29.6	32.6	
Energy	GE Mcal/kg				
Cattle	DE Mcal/kg	*	2.72	3.00	
Sheep	DE Mcal/kg	*	2.89	3.19	
Swine	DE kcal/kg	*	2923.	3223.	
Cattle	ME Mcal/kg	*	2.23	2.46	
Sheep	ME Mcal/kg	*	2.38	2.62	
Swine	ME kcal/kg	*	2613.	2881.	
Cattle	TDN %	*	61.8	68.1	
Sheep	TDN %	*	65.6	72.3	
Swine	TDN %	*	66.3	73.1	

VETCH, AMERICAN. Vicia americana

Vetch, American, hay, s-c, immature, (1)

Ref No 1 05 114 United States

Feed Name or Analyses			Mean As Fed	Mean Dry	C.V. ± %
Dry matter	%			100.0	
Ash	%			8.5	15
Crude fiber	%			24.2	6
Ether extract	%			2.3	19
N-free extract	%			37.7	
Protein (N x 6.25)	%			27.3	9
Cattle	dig prot %	*		20.6	
Goats	dig prot %	*		22.0	
Horses	dig prot %	*		20.7	
Rabbits	dig prot %	*		19.7	
Sheep	dig prot %	*		21.1	
Energy	GE Mcal/kg				
Cattle	DE Mcal/kg	*		2.67	
Sheep	DE Mcal/kg	*		2.80	
Cattle	ME Mcal/kg	*		2.19	
Sheep	ME Mcal/kg	*		2.30	
Cattle	TDN %	*		60.6	
Sheep	TDN %	*		63.5	
Calcium	%			1.30	15
Magnesium	%			0.28	13
Phosphorus	%			0.39	12
Sulphur	%			0.20	12

Vetch, American, hay, s-c, dough stage, (1)

Ref No 1 05 115 United States

Feed Name or Analyses			Mean As Fed	Mean Dry	C.V. ± %
Dry matter	%			100.0	
Ash	%			8.1	8
Crude fiber	%			33.5	5
Ether extract	%			2.5	20
N-free extract	%			43.9	
Protein (N x 6.25)	%			12.0	12
Cattle	dig prot %	*		7.3	
Goats	dig prot %	*		7.8	
Horses	dig prot %	*		7.7	
Rabbits	dig prot %	*		7.9	
Sheep	dig prot %	*		7.3	
Energy	GE Mcal/kg				
Cattle	DE Mcal/kg	*		2.53	
Sheep	DE Mcal/kg	*		2.50	
Cattle	ME Mcal/kg	*		2.07	
Sheep	ME Mcal/kg	*		2.05	
Cattle	TDN %	*		57.3	
Sheep	TDN %	*		56.8	
Calcium	%			2.10	4
Magnesium	%			0.31	16
Phosphorus	%			0.22	28
Sulphur	%			0.13	16

VETCH, BITTER. Vicia ervilia

Vetch, bitter, aerial part, fresh, early bloom, (2)

Ref No 2 05 116 United States

Feed Name or Analyses			Mean As Fed	Mean Dry	C.V. ± %
Dry matter	%		17.8	100.0	
Ash	%		4.6	26.0	
Crude fiber	%		3.7	20.7	
Sheep	dig coef %		55.	55.	
Ether extract	%		0.6	3.2	
Sheep	dig coef %		71.	71.	
N-free extract	%		5.4	30.5	
Sheep	dig coef %		86.	86.	
Protein (N x 6.25)	%		3.5	19.6	
Sheep	dig coef %		82.	82.	
Cattle	dig prot %	*	2.6	14.6	
Goats	dig prot %	*	2.6	14.8	
Horses	dig prot %	*	2.5	14.2	
Rabbits	dig prot %	*	2.5	13.8	
Sheep	dig prot %		2.9	16.1	
Energy	GE Mcal/kg				
Cattle	DE Mcal/kg	*	0.45	2.50	
Sheep	DE Mcal/kg	*	0.46	2.59	
Cattle	ME Mcal/kg	*	0.36	2.05	
Sheep	ME Mcal/kg	*	0.38	2.13	
Cattle	TDN %	*	10.1	56.8	
Sheep	TDN %		10.5	58.8	

VETCH, BLACK BITTER. Vicia ervilia

Vetch, black bitter, seeds, (5)

Ref No 5 05 117 United States

Feed Name or Analyses			Mean As Fed	Mean Dry	C.V. ± %
Dry matter	%		90.1	100.0	
Ash	%		3.8	4.2	

Feed Name or Analyses			Mean As Fed	Mean Dry	C.V. ± %
Crude fiber	%		9.0	10.0	
Sheep	dig coef %		94.	94.	
Ether extract	%		0.8	0.9	
Sheep	dig coef %		76.	76.	
N-free extract	%		54.5	60.5	
Sheep	dig coef %		95.	95.	
Protein (N x 6.25)	%		22.0	24.4	
Sheep	dig coef %		82.	82.	
Sheep	dig prot %		18.0	20.0	
Energy	GE Mcal/kg				
Cattle	DE Mcal/kg	*	3.34	3.71	
Sheep	DE Mcal/kg	*	3.51	3.90	
Swine	DE kcal/kg	*	3516.	3902.	
Cattle	ME Mcal/kg	*	2.74	3.04	
Sheep	ME Mcal/kg	*	2.88	3.20	
Swine	ME kcal/kg	*	3202.	3554.	
Cattle	TDN %	*	75.8	84.1	
Sheep	TDN %		79.7	88.4	
Swine	TDN %	*	79.7	88.5	

VETCH, COMMON. Vicia sativa

Vetch, common, hay, s-c, (1)

Ref No 1 05 122 United States

Feed Name or Analyses			Mean As Fed	Mean Dry	C.V. ± %
Dry matter	%		85.4	100.0	
Ash	%		7.2	8.4	
Crude fiber	%		24.0	28.1	
Cattle	dig coef %		58.	58.	
Sheep	dig coef %		50.	50.	
Ether extract	%		1.7	2.0	
Cattle	dig coef %		70.	70.	
Sheep	dig coef %		46.	46.	
N-free extract	%		37.6	44.1	
Cattle	dig coef %		71.	71.	
Sheep	dig coef %		72.	72.	
Protein (N x 6.25)	%		14.9	17.5	
Cattle	dig coef %		66.	66.	
Sheep	dig coef %		78.	78.	
Cattle	dig prot %		9.9	11.5	
Goats	dig prot %	*	11.0	12.9	
Horses	dig prot %	*	10.6	12.4	
Rabbits	dig prot %	*	10.4	12.2	
Sheep	dig prot %		11.6	13.6	
Fatty acids	%		1.1	1.2	
Energy	GE Mcal/kg				
Cattle	DE Mcal/kg	*	2.34	2.74	
Sheep	DE Mcal/kg	*	2.31	2.71	
Cattle	ME Mcal/kg	*	1.92	2.25	
Sheep	ME Mcal/kg	*	1.90	2.22	
Cattle	TDN %		53.1	62.2	
Sheep	TDN %		52.5	61.5	
Calcium	%		1.13	1.33	
Copper	mg/kg		8.5	9.9	
Iron	%		0.027	0.032	
Magnesium	%		0.17	0.20	
Manganese	mg/kg		45.5	53.3	
Phosphorus	%		0.31	0.36	
Potassium	%		2.13	2.49	
Sodium	%		0.44	0.52	
Sulphur	%		0.09	0.10	

Vetch, common, hay, s-c, immature, (1)

Ref No 1 05 118 United States

Feed Name or Analyses			Mean As Fed	Mean Dry	C.V. ± %
Dry matter	%		86.9	100.0	2
Ash	%		7.3	8.4	23
Crude fiber	%		21.1	24.3	8
Ether extract	%		2.1	2.4	17
N-free extract	%		33.4	38.4	
Protein (N x 6.25)	%		23.0	26.5	15
Cattle	dig prot %	*	17.3	19.9	
Goats	dig prot %	*	18.5	21.3	
Horses	dig prot %	*	17.4	20.0	
Rabbits	dig prot %	*	16.6	19.1	
Sheep	dig prot %	*	17.7	20.3	
Energy	GE Mcal/kg				
Cattle	DE Mcal/kg	*	2.32	2.67	
Sheep	DE Mcal/kg	*	2.43	2.79	
Cattle	ME Mcal/kg	*	1.90	2.19	
Sheep	ME Mcal/kg	*	1.99	2.29	
Cattle	TDN %	*	52.7	60.6	
Sheep	TDN %	*	55.0	63.3	

Vetch, common, hay, s-c, early bloom, (1)

Ref No 1 05 119 United States

Feed Name or Analyses			Mean As Fed	Mean Dry	C.V. ± %
Dry matter	%		83.0	100.0	2
Ash	%		8.9	10.7	9
Crude fiber	%		25.0	30.1	4
Ether extract	%		2.2	2.6	9
N-free extract	%		28.8	34.7	
Protein (N x 6.25)	%		18.2	21.9	5
Cattle	dig prot %	*	13.2	15.9	
Goats	dig prot %	*	14.1	17.0	
Horses	dig prot %	*	13.4	16.1	
Rabbits	dig prot %	*	12.9	15.6	
Sheep	dig prot %	*	13.5	16.2	
Energy	GE Mcal/kg				
Cattle	DE Mcal/kg	*	2.20	2.65	
Sheep	DE Mcal/kg	*	2.07	2.50	
Cattle	ME Mcal/kg	*	1.80	2.17	
Sheep	ME Mcal/kg	*	1.70	2.05	
Cattle	TDN %	*	49.8	60.0	
Sheep	TDN %	*	47.0	56.6	

Vetch, common, hay, s-c, full bloom, (1)

Ref No 1 05 120 United States

Feed Name or Analyses			Mean As Fed	Mean Dry	C.V. ± %
Dry matter	%		85.6	100.0	
Ash	%		7.9	9.2	
Crude fiber	%		30.8	36.0	
Ether extract	%		1.5	1.8	
N-free extract	%		29.4	34.4	
Protein (N x 6.25)	%		15.9	18.6	
Cattle	dig prot %	*	11.2	13.0	
Goats	dig prot %	*	11.9	13.9	
Horses	dig prot %	*	11.4	13.3	
Rabbits	dig prot %	*	11.2	13.0	
Sheep	dig prot %	*	11.3	13.3	
Energy	GE Mcal/kg				
Cattle	DE Mcal/kg	*	2.06	2.41	
Sheep	DE Mcal/kg	*	2.04	2.39	
Cattle	ME Mcal/kg	*	1.69	1.98	
Sheep	ME Mcal/kg	*	1.68	1.96	
Cattle	TDN %	*	46.8	54.6	
Sheep	TDN %	*	46.4	54.2	

Vetch, common, hay, s-c, dough stage, (1)

Ref No 1 05 121 United States

Feed Name or Analyses			Mean As Fed	Mean Dry	C.V. ± %
Dry matter	%		85.1	100.0	
Ash	%		7.9	9.3	
Crude fiber	%		30.4	35.7	
Ether extract	%		1.5	1.8	
N-free extract	%		32.4	38.1	
Protein (N x 6.25)	%		12.9	15.1	
Cattle	dig prot %	*	8.5	10.0	
Goats	dig prot %	*	9.1	10.6	
Horses	dig prot %	*	8.8	10.3	
Rabbits	dig prot %	*	8.8	10.3	
Sheep	dig prot %	*	8.6	10.1	
Energy	GE Mcal/kg				
Cattle	DE Mcal/kg	*	2.08	2.44	
Sheep	DE Mcal/kg	*	2.02	2.37	
Cattle	ME Mcal/kg	*	1.70	2.00	
Sheep	ME Mcal/kg	*	1.66	1.94	
Cattle	TDN %	*	47.1	55.3	
Sheep	TDN %	*	45.8	53.8	

Vetch, common, aerial part, fresh, (2)

Ref No 2 05 123 United States

Feed Name or Analyses			Mean As Fed	Mean Dry	C.V. ± %
Dry matter	%		20.4	100.0	8
Ash	%		2.1	10.3	17
Crude fiber	%		5.5	27.0	11
Sheep	dig coef %	*	40.	40.	
Ether extract	%		0.5	2.5	13
Sheep	dig coef %	*	49.	49.	
N-free extract	%		8.5	41.6	
Sheep	dig coef %	*	76.	76.	
Protein (N x 6.25)	%		3.8	18.6	16
Sheep	dig coef %	*	75.	75.	
Cattle	dig prot %	*	2.8	13.7	
Goats	dig prot %	*	2.8	13.9	
Horses	dig prot %	*	2.7	13.3	
Rabbits	dig prot %	*	2.7	13.0	
Sheep	dig prot %		2.8	14.0	
Energy	GE Mcal/kg				
Cattle	DE Mcal/kg	*	0.50	2.43	
Sheep	DE Mcal/kg	*	0.53	2.61	
Cattle	ME Mcal/kg	*	0.41	1.99	
Sheep	ME Mcal/kg	*	0.44	2.14	
Cattle	TDN %	*	12.3	60.5	
Sheep	TDN %		12.1	59.1	
Calcium	%		0.27	1.32	
Copper	mg/kg		2.0	9.7	
Iron	%		0.008	0.039	
Magnesium	%		0.04	0.20	
Manganese	mg/kg		24.5	120.0	
Phosphorus	%		0.07	0.34	
Potassium	%		0.51	2.50	
Sulphur	%		0.02	0.10	

VETCH, HAIRY. Vicia villosa

Vetch, hairy, hay, s-c, (1)

Ref No 1 05 126 United States

Feed Name or Analyses			Mean As Fed	Mean Dry	C.V. ± %
Dry matter	%		88.8	100.0	
Ash	%		8.8	9.9	
Crude fiber	%		25.9	29.2	
Sheep	dig coef %		59.	59.	
Ether extract	%		3.0	3.3	
Sheep	dig coef %		67.	67.	
N-free extract	%		30.7	34.6	
Sheep	dig coef %		71.	71.	
Protein (N x 6.25)	%		20.4	23.0	
Sheep	dig coef %		79.	79.	
Cattle	dig prot %	*	14.9	16.8	
Goats	dig prot %	*	16.0	18.0	
Horses	dig prot %	*	15.1	17.0	

Continued

Feed Name or Analyses			Mean As Fed	Mean Dry	C.V. ± %
Rabbits	dig prot %	*	14.6	16.4	
Sheep	dig prot %		16.1	18.1	
Fatty acids	%		2.6	3.0	
Energy	GE Mcal/kg				
Cattle	DE Mcal/kg	*	2.37	2.66	
Sheep	DE Mcal/kg	*	2.54	2.86	
Cattle	ME Mcal/kg	*	1.94	2.18	
Sheep	ME Mcal/kg	*	2.08	2.35	
Cattle	TDN %	*	53.6	60.4	
Sheep	TDN %		57.7	64.9	
Calcium	%		1.14	1.28	
Phosphorus	%		0.32	0.36	
Potassium	%		1.98	2.23	

Vetch, hairy, hay, s-c, immature, (1)
Ref No 1 05 125 United States

Feed Name or Analyses			Mean As Fed	Mean Dry	C.V. ± %
Dry matter	%		88.5	100.0	2
Ash	%		8.8	9.9	6
Crude fiber	%		25.4	28.7	4
Ether extract	%		2.8	3.2	14
N-free extract	%		31.5	35.6	
Protein (N x 6.25)	%		20.0	22.6	5
Cattle	dig prot %	*	14.6	16.5	
Goats	dig prot %	*	15.6	17.6	
Horses	dig prot %	*	14.8	16.7	
Rabbits	dig prot %	*	14.3	16.1	
Sheep	dig prot %	*	14.9	16.8	
Energy	GE Mcal/kg				
Cattle	DE Mcal/kg	*	2.37	2.67	
Sheep	DE Mcal/kg	*	2.26	2.55	
Cattle	ME Mcal/kg	*	1.94	2.19	
Sheep	ME Mcal/kg	*	1.85	2.09	
Cattle	TDN %	*	53.7	60.6	
Sheep	TDN %	*	51.2	57.8	
Calcium	%		1.09	1.23	29
Copper	mg/kg		8.8	9.9	8
Iodine	mg/kg		0.484	0.547	
Iron	%		0.032	0.036	27
Magnesium	%		0.20	0.23	41
Manganese	mg/kg		47.6	53.8	36
Phosphorus	%		0.32	0.36	15
Potassium	%		1.85	2.09	24
Sodium	%		0.46	0.52	11
Sulphur	%		0.09	0.10	21

Vetch, hairy, aerial part, fresh, (2)
Ref No 2 05 124 United States

Feed Name or Analyses			Mean As Fed	Mean Dry	C.V. ± %
Dry matter	%		18.5	100.0	
Ash	%		2.0	10.7	
Crude fiber	%		5.2	28.3	
Sheep	dig coef %		63.	63.	
Ether extract	%		0.6	3.5	
Sheep	dig coef %		73.	73.	
N-free extract	%		6.2	33.7	
Sheep	dig coef %		76.	76.	
Protein (N x 6.25)	%		4.4	23.8	
Sheep	dig coef %		83.	83.	
Cattle	dig prot %	*	3.3	18.1	
Goats	dig prot %	*	3.5	18.8	
Horses	dig prot %	*	3.3	17.7	
Rabbits	dig prot %	*	3.1	17.0	
Sheep	dig prot %		3.6	19.7	
Energy	GE Mcal/kg				
Cattle	DE Mcal/kg	*	0.55	2.98	
Sheep	DE Mcal/kg	*	0.56	3.04	
Cattle	ME Mcal/kg	*	0.45	2.44	
Sheep	ME Mcal/kg	*	0.46	2.49	
Cattle	TDN %	*	12.5	67.6	
Sheep	TDN %		12.7	68.9	
Calcium	%		0.20	1.10	
Phosphorus	%		0.06	0.33	
Potassium	%		0.42	2.25	

Vetch, hairy, aerial part, fresh, mature, (2)
Ref No 2 05 127 United States

Feed Name or Analyses			Mean As Fed	Mean Dry	C.V. ± %
Dry matter	%			100.0	
Calcium	%			1.10	
Phosphorus	%			0.33	
Potassium	%			2.25	

VETCH, NARBONNE. Vicia narbonensis

Vetch, narbonne, aerial part, fresh, (2)
Ref No 2 05 129 United States

Feed Name or Analyses			Mean As Fed	Mean Dry	C.V. ± %
Dry matter	%		16.4	100.0	
Ash	%		2.3	13.8	
Crude fiber	%		4.3	26.1	
Sheep	dig coef %		58.	58.	
Ether extract	%		0.5	3.2	
Sheep	dig coef %		74.	74.	
N-free extract	%		6.4	38.9	
Sheep	dig coef %		74.	74.	
Protein (N x 6.25)	%		3.0	18.0	
Sheep	dig coef %		80.	80.	
Cattle	dig prot %	*	2.2	13.2	
Goats	dig prot %	*	2.2	13.4	
Horses	dig prot %	*	2.1	12.8	
Rabbits	dig prot %	*	2.1	12.6	
Sheep	dig prot %		2.4	14.4	
Energy	GE Mcal/kg				
Cattle	DE Mcal/kg	*	0.46	2.83	
Sheep	DE Mcal/kg	*	0.46	2.81	
Cattle	ME Mcal/kg	*	0.38	2.32	
Sheep	ME Mcal/kg	*	0.38	2.30	
Cattle	TDN %	*	10.5	64.2	
Sheep	TDN %		10.4	63.7	

VETCH-GRASS. Vicia spp, Scientific name not used

Vetch-Grass, aerial part, dehy, (1)
Ref No 1 05 130 United States

Feed Name or Analyses			Mean As Fed	Mean Dry	C.V. ± %
Dry matter	%		91.8	100.0	
Ash	%		13.5	14.7	
Crude fiber	%		20.5	22.3	
Ether extract	%		3.3	3.6	
N-free extract	%		32.4	35.3	
Protein (N x 6.25)	%		22.1	24.1	
Cattle	dig prot %	*	16.4	17.8	
Goats	dig prot %	*	17.5	19.0	
Horses	dig prot %	*	16.5	18.0	
Rabbits	dig prot %	*	15.9	17.3	
Sheep	dig prot %	*	16.7	18.2	

Vetch-Grass, hay, s-c, (1)
Ref No 1 05 131 United States

Feed Name or Analyses			Mean As Fed	Mean Dry	C.V. ± %
Dry matter	%		88.6	100.0	4
Ash	%		9.7	11.0	33
Crude fiber	%		18.9	21.3	5
Ether extract	%		3.5	4.0	10
N-free extract	%		39.2	44.3	
Protein (N x 6.25)	%		17.2	19.4	24
Cattle	dig prot %	*	12.2	13.7	
Goats	dig prot %	*	13.0	14.7	
Horses	dig prot %	*	12.4	14.0	
Rabbits	dig prot %	*	12.1	13.6	
Sheep	dig prot %	*	12.4	14.0	

VETCH-OATS. Vicia spp, Avena sativa

Vetch-Oats, hay, s-c, (1)
Ref No 1 05 132 United States

Feed Name or Analyses			Mean As Fed	Mean Dry	C.V. ± %
Dry matter	%		87.6	100.0	
Ash	%		8.2	9.4	
Crude fiber	%		27.3	31.2	
Ether extract	%		2.7	3.1	
N-free extract	%		37.5	42.8	
Protein (N x 6.25)	%		11.9	13.6	
Cattle	dig prot %	*	7.6	8.7	
Goats	dig prot %	*	8.1	9.2	
Horses	dig prot %	*	7.9	9.1	
Rabbits	dig prot %	*	8.0	9.2	
Sheep	dig prot %	*	7.7	8.8	
Fatty acids	%		2.7	3.1	
Energy	GE Mcal/kg				
Cattle	DE Mcal/kg	*	2.30	2.63	
Sheep	DE Mcal/kg	*	2.14	2.45	
Cattle	ME Mcal/kg	*	1.89	2.15	
Sheep	ME Mcal/kg	*	1.76	2.00	
Cattle	TDN %	*	52.2	59.6	
Sheep	TDN %	*	48.6	55.5	
Calcium	%		0.76	0.87	
Phosphorus	%		0.27	0.31	
Potassium	%		1.51	1.72	

Vetch-Oats, aerial part, fresh, (2)
Ref No 2 05 133 United States

Feed Name or Analyses			Mean As Fed	Mean Dry	C.V. ± %
Dry matter	%		25.9	100.0	6
Ash	%		3.2	12.3	20
Crude fiber	%		6.3	24.3	8
Ether extract	%		0.9	3.5	8
N-free extract	%		10.8	41.8	
Protein (N x 6.25)	%		4.7	18.1	20
Cattle	dig prot %	*	3.4	13.3	
Goats	dig prot %	*	3.5	13.5	
Horses	dig prot %	*	3.4	12.9	
Rabbits	dig prot %	*	3.3	12.7	
Sheep	dig prot %	*	3.6	13.9	
Energy	GE Mcal/kg				
Cattle	DE Mcal/kg	*	0.75	2.90	
Sheep	DE Mcal/kg	*	0.76	2.92	
Cattle	ME Mcal/kg	*	0.62	2.38	
Sheep	ME Mcal/kg	*	0.62	2.40	
Cattle	TDN %	*	17.0	65.8	
Sheep	TDN %	*	17.2	66.3	
Calcium	%		0.14	0.54	
Magnesium	%		0.06	0.24	

(1) dry forages and roughages (3) silages (6) minerals
(2) pasture, range plants, and forages fed green (4) energy feeds (7) vitamins
(5) protein supplements (8) additives

Feed Name or Analyses			Mean As Fed	Mean Dry	C.V. ± %
Phosphorus	%		0.11	0.41	
Potassium	%		0.44	1.70	

Vetch-Oats, aerial part, ensiled, (3)

Ref No 3 08 547 United States

Feed Name or Analyses			As Fed	Dry	C.V. ± %
Dry matter	%		26.4	100.0	
Ash	%		1.9	7.2	
Crude fiber	%		8.8	33.3	
Ether extract	%		0.6	2.3	
N-free extract	%		12.9	48.9	
Protein (N x 6.25)	%		2.2	8.3	
Cattle	dig prot %	*	1.0	3.8	
Goats	dig prot %	*	1.0	3.8	
Horses	dig prot %	*	1.0	3.8	
Sheep	dig prot %	*	1.0	3.8	
Energy	GE Mcal/kg				
Cattle	DE Mcal/kg	*	0.69	2.60	
Sheep	DE Mcal/kg	*	0.75	2.86	
Cattle	ME Mcal/kg	*	0.56	2.14	
Sheep	ME Mcal/kg	*	0.62	2.34	
Cattle	TDN %	*	15.6	59.1	
Sheep	TDN %	*	17.1	64.8	

Vetch-Oats, aerial part, ensiled, mature, (3)

Ref No 3 05 134 United States

Feed Name or Analyses			As Fed	Dry	C.V. ± %
Dry matter	%		25.6	100.0	
Ash	%		1.8	7.2	
Crude fiber	%		8.5	33.3	
Ether extract	%		0.6	2.3	
N-free extract	%		12.5	48.9	
Protein (N x 6.25)	%		2.1	8.3	
Cattle	dig prot %	*	1.0	3.8	
Goats	dig prot %	*	1.0	3.8	
Horses	dig prot %	*	1.0	3.8	
Sheep	dig prot %	*	1.0	3.8	
Energy	GE Mcal/kg				
Cattle	DE Mcal/kg	*	0.67	2.61	
Sheep	DE Mcal/kg	*	0.73	2.86	
Cattle	ME Mcal/kg	*	0.55	2.14	
Sheep	ME Mcal/kg	*	0.60	2.34	
Cattle	TDN %	*	15.1	59.2	
Sheep	TDN %	*	16.6	64.8	

Vetch-Oats, aerial part w molasses added, ensiled, (3)

Ref No 3 05 135 United States

Feed Name or Analyses			As Fed	Dry	C.V. ± %
Dry matter	%		28.5	100.0	
Ash	%		2.1	7.5	
Crude fiber	%		7.7	27.1	
Ether extract	%		1.0	3.4	
N-free extract	%		14.8	52.1	
Protein (N x 6.25)	%		2.8	9.9	
Cattle	dig prot %	*	1.5	5.2	
Goats	dig prot %	*	1.5	5.2	
Horses	dig prot %	*	1.5	5.2	
Sheep	dig prot %	*	1.5	5.2	
Energy	GE Mcal/kg				
Cattle	DE Mcal/kg	*	0.82	2.87	
Sheep	DE Mcal/kg	*	0.82	2.87	
Cattle	ME Mcal/kg	*	0.67	2.35	
Sheep	ME Mcal/kg	*	0.67	2.35	
Cattle	TDN %	*	18.5	65.0	
Sheep	TDN %	*	18.5	65.0	

VETCH-WHEAT. Vicia spp, Triticum aestivum

Vetch-Wheat, hay, s-c, (1)

Ref No 1 05 136 United States

Feed Name or Analyses			As Fed	Dry	C.V. ± %
Dry matter	%		90.0	100.0	
Ash	%		7.2	8.0	
Crude fiber	%		28.8	32.0	
Ether extract	%		2.2	2.4	
N-free extract	%		36.4	40.5	
Protein (N x 6.25)	%		15.4	17.1	
Cattle	dig prot %	*	10.6	11.8	
Goats	dig prot %	*	11.3	12.5	
Horses	dig prot %	*	10.8	12.1	
Rabbits	dig prot %	*	10.7	11.9	
Sheep	dig prot %	*	10.7	11.9	
Fatty acids	%		2.2	2.4	
Energy	GE Mcal/kg				
Cattle	DE Mcal/kg	*	2.53	2.81	
Sheep	DE Mcal/kg	*	2.57	2.86	
Cattle	ME Mcal/kg	*	2.07	2.30	
Sheep	ME Mcal/kg	*	2.12	2.35	
Cattle	TDN %	*	57.3	63.7	
Sheep	TDN %	*	58.4	64.9	

Vetch-Wheat, aerial part, fresh, (2)

Ref No 2 05 137 United States

Feed Name or Analyses			As Fed	Dry	C.V. ± %
Dry matter	%		28.3	100.0	15
Ash	%		2.4	8.6	19
Crude fiber	%		7.4	26.3	7
Ether extract	%		0.7	2.5	20
N-free extract	%		13.8	48.9	
Protein (N x 6.25)	%		3.9	13.7	26
Cattle	dig prot %	*	2.7	9.5	
Goats	dig prot %	*	2.6	9.3	
Horses	dig prot %	*	2.6	9.1	
Rabbits	dig prot %	*	2.6	9.2	
Sheep	dig prot %	*	2.7	9.7	
Energy	GE Mcal/kg				
Cattle	DE Mcal/kg	*	0.85	3.01	
Sheep	DE Mcal/kg	*	0.81	2.86	
Cattle	ME Mcal/kg	*	0.70	2.46	
Sheep	ME Mcal/kg	*	0.66	2.34	
Cattle	TDN %	*	19.3	68.2	
Sheep	TDN %	*	18.3	64.8	

VETCH-WHEAT. Vicia spp, Triticum spp

Vetch-Wheat, aerial part w molasses added, ensiled, (3)

Ref No 3 08 548 United States

Feed Name or Analyses			As Fed	Dry	C.V. ± %
Dry matter	%		29.2	100.0	
Ash	%		2.2	7.5	
Crude fiber	%		7.9	27.1	
Ether extract	%		1.0	3.4	
N-free extract	%		15.2	52.1	
Protein (N x 6.25)	%		2.9	9.9	
Cattle	dig prot %	*	1.5	5.2	
Goats	dig prot %	*	1.5	5.2	
Horses	dig prot %	*	1.5	5.2	
Sheep	dig prot %	*	1.5	5.2	
Energy	GE Mcal/kg				
Cattle	DE Mcal/kg	*	0.84	2.87	
Sheep	DE Mcal/kg	*	0.84	2.86	
Cattle	ME Mcal/kg	*	0.69	2.35	
Sheep	ME Mcal/kg	*	0.69	2.35	
Cattle	TDN %	*	19.0	65.1	
Sheep	TDN %	*	19.0	64.9	

VINE-MESQUITE. Panicum obtusum

Vine-mesquite, aerial part, fresh, immature, (2)

Ref No 2 05 138 United States

Feed Name or Analyses			As Fed	Dry	C.V. ± %
Dry matter	%			100.0	
Ash	%			11.7	13
Crude fiber	%			30.3	11
Ether extract	%			2.4	9
N-free extract	%			43.0	
Protein (N x 6.25)	%			12.6	16
Cattle	dig prot %	*		8.6	
Goats	dig prot %	*		8.3	
Horses	dig prot %	*		8.2	
Rabbits	dig prot %	*		8.4	
Sheep	dig prot %	*		8.7	
Energy	GE Mcal/kg				
Cattle	DE Mcal/kg	*		2.79	
Sheep	DE Mcal/kg	*		2.77	
Cattle	ME Mcal/kg	*		2.29	
Sheep	ME Mcal/kg	*		2.27	
Cattle	TDN %	*		63.3	
Sheep	TDN %	*		62.9	
Calcium	%			0.56	19
Magnesium	%			0.26	
Phosphorus	%			0.25	16
Potassium	%			2.48	

Vine-mesquite, aerial part, fresh, mature, (2)

Ref No 2 05 139 United States

Feed Name or Analyses			As Fed	Dry	C.V. ± %
Dry matter	%			100.0	
Ash	%			10.5	7
Crude fiber	%			35.8	5
Ether extract	%			1.8	8
N-free extract	%			46.5	
Protein (N x 6.25)	%			5.4	18
Cattle	dig prot %	*		2.5	
Goats	dig prot %	*		1.6	
Horses	dig prot %	*		2.1	
Rabbits	dig prot %	*		2.8	
Sheep	dig prot %	*		2.0	
Energy	GE Mcal/kg				
Cattle	DE Mcal/kg	*		2.60	
Sheep	DE Mcal/kg	*		2.93	
Cattle	ME Mcal/kg	*		2.13	
Sheep	ME Mcal/kg	*		2.40	
Cattle	TDN %	*		58.9	
Sheep	TDN %	*		66.5	
Calcium	%			0.32	8
Phosphorus	%			0.19	30

VITAMIN D

Vitamin D (for prevention of milk fever), (7)

Vitamin D (for prevention of milk fever) (AAFCO)

Ref No 7 05 148 United States

Feed Name or Analyses			As Fed	Dry	C.V. ± %
Dry matter	%			100.0	
Crude fiber	%				
Sheep	dig coef %	*		33.	

Continued

Feed Name or Analyses			Mean As Fed	Mean Dry	C.V. ± %
Ether extract	%				
Sheep	dig coef %	*		72.	
N-free extract	%				
Sheep	dig coef %	*		92.	
Protein (N x 6.25)	%				
Sheep	dig coef %	*		78.	

WALNUTS. Juglans spp

Walnuts, meats, grnd, (5)

Ref No 5 05 152 United States

Feed Name or Analyses			Mean As Fed	Mean Dry	C.V. ± %
Dry matter	%		91.2	100.0	
Ash	%		5.3	5.8	
Crude fiber	%		3.5	3.8	
Sheep	dig coef %		37.	37.	
Ether extract	%		8.8	9.6	
Sheep	dig coef %		98.	98.	
N-free extract	%		31.9	35.0	
Sheep	dig coef %		100.	100.	
Protein (N x 6.25)	%		41.8	45.8	
Sheep	dig coef %		91.	91.	
Cattle	dig prot %	*	34.7	38.1	
Goats	dig prot %	*	35.7	39.2	
Horses	dig prot %	*	35.7	39.2	
Sheep	dig prot %		38.0	41.7	
Energy	GE Mcal/kg				
Cattle	DE Mcal/kg	*	3.77	4.13	
Sheep	DE Mcal/kg	*	3.99	4.38	
Swine	DE kcal/kg	*	3879.	4253.	
Cattle	ME Mcal/kg	*	3.09	3.39	
Sheep	ME Mcal/kg	*	3.27	3.59	
Swine	ME kcal/kg	*	3365.	3690.	
Cattle	TDN %	*	85.5	93.7	
Sheep	TDN %		90.5	99.3	
Swine	TDN %	*	88.0	96.5	

Walnuts, meats w hulls, grnd, (5)

Ref No 5 05 153 United States

Feed Name or Analyses			Mean As Fed	Mean Dry	C.V. ± %
Dry matter	%		91.7	100.0	
Ash	%		4.6	5.0	
Crude fiber	%		28.3	30.9	
Sheep	dig coef %		14.	14.	
Ether extract	%		8.3	9.0	
Sheep	dig coef %		98.	98.	
N-free extract	%		29.3	31.9	
Sheep	dig coef %		59.	59.	
Protein (N x 6.25)	%		21.3	23.2	
Sheep	dig coef %		84.	84.	
Cattle	dig prot %	*	15.9	17.3	
Goats	dig prot %	*	17.0	18.5	
Horses	dig prot %	*	17.0	18.5	
Sheep	dig prot %		17.9	19.5	
Energy	GE Mcal/kg				
Cattle	DE Mcal/kg	*	2.46	2.69	
Sheep	DE Mcal/kg	*	2.53	2.75	
Swine	DE kcal/kg	*	3239.	3532.	
Cattle	ME Mcal/kg	*	2.02	2.20	
Sheep	ME Mcal/kg	*	2.07	2.26	
Swine	ME kcal/kg	*	2957.	3225.	
Cattle	TDN %	*	55.9	60.9	

(1) dry forages and roughages (2) pasture, range plants, and forages fed green (3) silages (4) energy feeds (5) protein supplements (6) minerals (7) vitamins (8) additives

Feed Name or Analyses			Mean As Fed	Mean Dry	C.V. ± %
Sheep	TDN %		57.3	62.5	
Swine	TDN %	*	73.5	80.1	

WATERGRASS. Hydrochloa caroliniensis

Watergrass, hay, s-c, (1)

Ref No 1 05 154 United States

Feed Name or Analyses			Mean As Fed	Mean Dry	C.V. ± %
Dry matter	%		94.2	100.0	
Ash	%		18.0	19.1	
Crude fiber	%		31.2	33.1	
Sheep	dig coef %		58.	58.	
Ether extract	%		1.1	1.2	
Sheep	dig coef %		27.	27.	
N-free extract	%		36.3	38.5	
Sheep	dig coef %		46.	46.	
Protein (N x 6.25)	%		7.6	8.1	
Sheep	dig coef %		50.	50.	
Cattle	dig prot %	*	3.7	4.0	
Goats	dig prot %	*	3.9	4.1	
Horses	dig prot %	*	4.2	4.4	
Rabbits	dig prot %	*	4.6	4.9	
Sheep	dig prot %		3.8	4.1	
Energy	GE Mcal/kg				
Cattle	DE Mcal/kg	*	1.89	2.01	
Sheep	DE Mcal/kg	*	1.73	1.84	
Cattle	ME Mcal/kg	*	1.55	1.65	
Sheep	ME Mcal/kg	*	1.42	1.51	
Cattle	TDN %	*	43.0	45.7	
Sheep	TDN %		39.3	41.7	

Watermelon, seeds w hulls, (5)

Ref No 5 09 131 United States

Feed Name or Analyses			Mean As Fed	Mean Dry	C.V. ± %
Dry matter	%			100.0	
Protein (N x 6.25)	%			38.0	
Alanine	%			1.60	
Arginine	%			5.93	
Aspartic acid	%			3.08	
Glutamic acid	%			6.57	
Glycine	%			1.79	
Histidine	%			0.91	
Hydroxyproline	%			0.00	
Isoleucine	%			1.37	
Leucine	%			2.32	
Lysine	%			1.06	
Methionine	%			0.91	
Phenylalanine	%			1.86	
Proline	%			1.25	
Serine	%			1.79	
Threonine	%			1.14	
Tyrosine	%			1.10	
Valine	%			1.52	

WEDGESCALE, PRAIRIE. Sphenopholis obtusata

Wedgescale, prairie, aerial part, fresh, milk stage, (2)

Ref No 2 05 156 United States

Feed Name or Analyses			Mean As Fed	Mean Dry	C.V. ± %
Dry matter	%			100.0	
Ash	%			9.9	
Crude fiber	%			34.5	
Ether extract	%			5.1	
N-free extract	%			43.8	
Protein (N x 6.25)	%			6.7	
Cattle	dig prot %	*		3.6	
Goats	dig prot %	*		2.8	

Feed Name or Analyses			Mean As Fed	Mean Dry	C.V. ± %
Horses	dig prot %	*		3.2	
Rabbits	dig prot %	*		3.8	
Sheep	dig prot %	*		3.2	

WHALE. Balaenoptera physalus

Whale, blubber oil, solv-extd, (4)

Ref No 4 09 284 Canada

Feed Name or Analyses			Mean As Fed	Mean Dry	C.V. ± %
Dry matter	%		100.0	100.0	
Fatty acids	%				
Docosahexaenoic	%		2.500	2.500	
Eicosapentaenoic	%		2.100	2.100	
Eicosenoic	%		16.200	16.200	
Linoleic	%		1.700	1.700	
Myristic	%		4.900	4.900	
Oleic	%		25.500	25.500	
Palmitic	%		8.000	8.000	
Palmitoleic	%		6.500	6.500	
Stearic	%		1.700	1.700	

WHALE. Balaena glacialis, Balaenoptera spp, Physeter catadon

Whale, livers, dehy grnd, (5)

Ref No 5 05 157 United States

Feed Name or Analyses			Mean As Fed	Mean Dry	C.V. ± %
Dry matter	%		93.4	100.0	5
Chlorine	%		1.99	2.13	
Choline	mg/kg		3351.	3588.	
Niacin	mg/kg		200.2	214.3	36
Pantothenic acid	mg/kg		36.4	38.9	33
Riboflavin	mg/kg		79.1	84.7	14
Thiamine	mg/kg		2.6	2.8	55
Vitamin B_6	mg/kg		9.04	9.68	75
Vitamin B_{12}	µg/kg		499.1	534.4	15

Whale, meat, dehy grnd, high fat, (5)

Ref No 5 05 158 United States

Feed Name or Analyses			Mean As Fed	Mean Dry	C.V. ± %
Dry matter	%		91.8	100.0	
Ash	%		2.7	2.9	
Ether extract	%		21.8	23.7	
Sheep	dig coef %		100.	100.	
N-free extract	%		2.1	2.3	
Sheep	dig coef %		36.	36.	
Protein (N x 6.25)	%		65.3	71.1	
Sheep	dig coef %		75.	75.	
Sheep	dig prot %		49.0	53.3	
Energy	GE Mcal/kg				
Cattle	DE Mcal/kg	*	4.37	4.76	
Sheep	DE Mcal/kg	*	4.35	4.74	
Swine	DE kcal/kg	*	4584.	4993.	
Cattle	ME Mcal/kg	*	3.58	3.90	
Sheep	ME Mcal/kg	*	3.57	3.89	
Swine	ME kcal/kg	*	3742.	4076.	
Cattle	TDN %	*	99.1	107.9	
Sheep	TDN %		98.7	107.5	
Swine	TDN %	*	104.0	113.2	

Whale, meat, extn unspecified grnd, (5)

Ref No 5 05 159 United States

Feed Name or Analyses			Mean As Fed	Mean Dry	C.V. ± %
Dry matter	%		89.7	100.0	
Ash	%		3.9	4.3	

Feed Name or Analyses			Mean As Fed	Mean Dry	C.V. ± %
Ether extract		%	1.6	1.8	
Swine	dig coef	%	69.	69.	
N-free extract		%	1.3	1.5	
Swine	dig coef	%	49.	49.	
Protein (N x 6.25)		%	82.9	92.4	
Swine	dig coef	%	91.	91.	
Swine	dig prot	%	75.4	84.1	
Energy	GE Mcal/kg				
Cattle	DE Mcal/kg	*	3.19	3.56	
Sheep	DE Mcal/kg	*	2.97	3.31	
Swine	DE kcal/kg		*3465.	3862.	
Cattle	ME Mcal/kg	*	2.62	2.92	
Sheep	ME Mcal/kg	*	2.44	2.72	
Swine	ME kcal/kg		*2679.	2987.	
Cattle	TDN %	*	72.3	80.6	
Sheep	TDN %	*	67.4	75.1	
Swine	TDN %		78.6	87.6	

Whale, meat, heat rendered dehy grnd, salt declared above 3% mx 7%, (5)

Whale meal (AAFCO)

Ref No 5 05 160 United States

Feed Name or Analyses			Mean As Fed	Mean Dry	C.V. ± %
Dry matter		%	91.3	100.0	1
Ash		%	4.0	4.4	38
Crude fiber		%	0.2	0.2	98
Swine	dig coef	%	275.	275.	
Ether extract		%	7.6	8.4	16
Sheep	dig coef	% *	98.	98.	
Swine	dig coef	%	92.	92.	
N-free extract		%	1.9	2.1	
Sheep	dig coef	% *	-37.	-37.	
Swine	dig coef	%	79.	79.	
Protein (N x 6.25)		%	77.5	84.9	4
Sheep	dig coef	% *	92.	92.	
Swine	dig coef	%	95.	95.	
Sheep	dig prot	%	71.3	78.1	
Swine	dig prot	%	73.6	80.6	
Energy	GE Mcal/kg				
Cattle	DE Mcal/kg	*	3.55	3.89	
Sheep	DE Mcal/kg	*	3.36	3.68	
Swine	DE kcal/kg		*4034.	4419.	
Cattle	ME Mcal/kg	*	2.91	3.19	
Sheep	ME Mcal/kg	*	2.76	3.02	
Swine	ME kcal/kg		*3181.	3484.	
Cattle	TDN %	*	80.5	88.1	
Sheep	TDN %	*	76.1	83.4	
Swine	TDN %		91.5	100.2	
Calcium		%	0.40	0.44	56
Phosphorus		%	0.56	0.61	8
Niacin		mg/kg	104.0	113.9	18
Pantothenic acid		mg/kg	2.6	2.9	20
Riboflavin		mg/kg	8.3	9.1	27
Thiamine		mg/kg	1.3	1.4	27
Vitamin B6		mg/kg	8.30	9.10	69
Vitamin B12		μg/kg	87.9	96.2	79
Lysine		%	5.65	6.19	
Methionine		%	2.28	2.50	

Whale, meat, partially extd dehy grnd, (5)

Ref No 5 05 162 United States

Feed Name or Analyses			Mean As Fed	Mean Dry	C.V. ± %
Dry matter		%	91.0	100.0	
Ash		%	3.5	3.9	
Ether extract		%	4.9	5.4	
Sheep	dig coef	%	98.	98.	
N-free extract		%	1.9	2.1	
Sheep	dig coef	%	-37.	-37.	
Protein (N x 6.25)		%	80.6	88.6	
Sheep	dig coef	%	92.	92.	
Sheep	dig prot	%	74.2	81.5	
Energy	GE Mcal/kg				
Cattle	DE Mcal/kg	*	3.42	3.76	
Sheep	DE Mcal/kg	*	3.72	4.08	
Swine	DE kcal/kg		*2982.	3277.	
Cattle	ME Mcal/kg	*	2.81	3.09	
Sheep	ME Mcal/kg	*	3.05	3.35	
Swine	ME kcal/kg		*2329.	2560.	
Cattle	TDN %	*	77.7	85.4	
Sheep	TDN %		84.3	92.6	
Swine	TDN %	*	67.6	74.3	

Whale, meat, partially extd grnd, (5)

Ref No 5 05 163 United States

Feed Name or Analyses			Mean As Fed	Mean Dry	C.V. ± %
Dry matter		%	86.9	100.0	
Ash		%	3.2	3.7	
Ether extract		%	4.7	5.4	
Swine	dig coef	%	81.	81.	
N-free extract		%	1.9	2.2	
Swine	dig coef	%	65.	65.	
Protein (N x 6.25)		%	77.1	88.7	
Swine	dig coef	%	93.	93.	
Swine	dig prot	%	71.7	82.5	
Energy	GE Mcal/kg				
Cattle	DE Mcal/kg	*	3.28	3.77	
Sheep	DE Mcal/kg	*	2.99	3.44	
Swine	DE kcal/kg		*3594.	4136.	
Cattle	ME Mcal/kg	*	2.69	3.09	
Sheep	ME Mcal/kg	*	2.45	2.82	
Swine	ME kcal/kg		*2806.	3229.	
Cattle	TDN %	*	74.4	85.6	
Sheep	TDN %	*	67.9	78.1	
Swine	TDN %		81.5	93.8	

Whale, meat, raw, (5)

Ref No 5 07 986 United States

Feed Name or Analyses			Mean As Fed	Mean Dry	C.V. ± %
Dry matter		%	29.1	100.0	
Ash		%	1.0	3.4	
Ether extract		%	7.5	25.8	
Protein (N x 6.25)		%	20.6	70.8	
Energy	GE Mcal/kg		1.56	5.36	
Calcium		%	0.01	0.03	
Phosphorus		%	0.14	0.48	
Potassium		%	0.02	0.07	
Sodium		%	0.08	0.27	
Ascorbic acid		mg/kg	60.0	206.2	
Riboflavin		mg/kg	0.8	2.7	
Thiamine		mg/kg	0.9	3.1	
Vitamin A		IU/g	18.6	63.9	

Whale, meat, solv-extd dehy grnd, (5)

Ref No 5 05 164 United States

Feed Name or Analyses			Mean As Fed	Mean Dry	C.V. ± %
Dry matter		%	90.2	100.0	
Ash		%	3.9	4.3	
Ether extract		%	1.6	1.8	
Sheep	dig coef	%	103.	103.	
N-free extract		%	1.4	1.5	
Sheep	dig coef	%	-4.	-4.	
Protein (N x 6.25)		%	83.3	92.4	
Sheep	dig coef	%	92.	92.	
Sheep	dig prot	%	76.7	85.0	
Energy	GE Mcal/kg				
Cattle	DE Mcal/kg	*	3.21	3.56	
Sheep	DE Mcal/kg	*	3.54	3.93	
Swine	DE kcal/kg		*2636.	2923.	
Cattle	ME Mcal/kg	*	2.63	2.92	
Sheep	ME Mcal/kg	*	2.91	3.22	
Swine	ME kcal/kg		*2039.	2260.	
Cattle	TDN %	*	72.7	80.6	
Sheep	TDN %		80.4	89.1	
Swine	TDN %	*	59.8	66.3	

Whale, meat w bone, heat rendered dehy grnd, (5)

Ref No 5 05 165 United States

Feed Name or Analyses			Mean As Fed	Mean Dry	C.V. ± %
Dry matter		%	92.3	100.0	3
Ash		%	27.0	29.2	24
Crude fiber		%	1.4	1.5	67
Ether extract		%	11.0	11.9	29
Sheep	dig coef	%	95.	95.	
Swine	dig coef	%	106.	106.	
N-free extract		%	1.1	1.2	
Protein (N x 6.25)		%	51.8	56.1	12
Sheep	dig coef	%	65.	65.	
Swine	dig coef	%	76.	76.	
Sheep	dig prot	%	33.7	36.5	
Swine	dig prot	%	39.4	42.7	
Energy	GE Mcal/kg				
Cattle	DE Mcal/kg	*	2.70	2.92	
Sheep	DE Mcal/kg	*	2.70	2.92	
Swine	DE kcal/kg		*2744.	2972.	
Cattle	ME Mcal/kg	*	2.21	2.40	
Sheep	ME Mcal/kg	*	2.21	2.40	
Swine	ME kcal/kg		*2323.	2516.	
Cattle	TDN %	*	61.2	66.3	
Sheep	TDN %		61.2	66.3	
Swine	TDN %		62.2	67.4	
Calcium		%	8.28	8.97	2
Phosphorus		%	4.16	4.50	1

Whale, stickwater solubles, cooked condensed, (5)

Ref No 5 05 166 United States

Feed Name or Analyses			Mean As Fed	Mean Dry	C.V. ± %
Dry matter		%	51.2	100.0	1
Ash		%	3.2	6.2	
Crude fiber		%	0.0	0.0	
Ether extract		%	0.7	1.4	
N-free extract		%	9.6	18.8	
Protein (N x 6.25)		%	37.6	73.6	16
Energy	GE Mcal/kg				
Cattle	DE Mcal/kg	*	1.82	3.55	
Sheep	DE Mcal/kg	*	1.78	3.49	
Swine	DE kcal/kg		*1649.	3223.	
Cattle	ME Mcal/kg	*	1.49	2.91	
Sheep	ME Mcal/kg	*	1.46	2.86	
Swine	ME kcal/kg		*1338.	2615.	
Cattle	TDN %	*	41.2	80.5	
Sheep	TDN %	*	40.5	79.1	
Swine	TDN %	*	37.4	73.1	
Calcium		%	0.03	0.06	
Iron		%	0.001	0.002	
Magnesium		%	0.02	0.04	
Phosphorus		%	0.15	0.30	
Potassium		%	0.02	0.04	
Folic acid		mg/kg	0.14	0.27	
Niacin		mg/kg	22.7	44.4	
Pantothenic acid		mg/kg	10.9	21.3	
Riboflavin		mg/kg	1.6	3.1	

Continued

Feed Name or Analyses			Mean		C.V.
			As Fed	Dry	± %
Vitamin B_{12}	µg/kg		65.1	127.3	98
Arginine	%		1.65	3.22	9
Cystine	%		0.10	0.20	
Glutamic acid	%		1.96	3.82	
Histidine	%		0.31	0.60	20
Isoleucine	%		0.51	1.01	
Leucine	%		1.03	2.01	12
Lysine	%		1.34	2.62	25
Methionine	%		0.41	0.80	7
Phenylalanine	%		0.62	1.21	
Threonine	%		0.62	1.21	7
Tryptophan	%		0.10	0.20	59
Tyrosine	%		0.41	0.80	
Valine	%		1.03	2.01	3

Whale, bone, grnd, (6)

Ref No 6 05 167 United States

Analyses	Units		As Fed	Dry	C.V. ± %
Dry matter	%		97.1	100.0	
Ash	%		64.7	66.6	
Ether extract	%		10.1	10.4	
Swine	dig coef %		90.	90.	
N-free extract	%		3.5	3.6	
Protein (N x 6.25)	%		18.8	19.4	
Swine	dig coef %		88.	88.	
Swine	dig prot %		16.6	17.1	
Energy	GE Mcal/kg				
Cattle	DE Mcal/kg	*	1.73	1.78	
Sheep	DE Mcal/kg	*	1.88	1.93	
Swine	DE kcal/kg	*	1631.	1680.	
Cattle	ME Mcal/kg	*	1.42	1.46	
Sheep	ME Mcal/kg	*	1.54	1.58	
Swine	ME kcal/kg	*	1502.	1547.	
Cattle	TDN %	*	39.1	40.3	
Sheep	TDN %	*	42.6	43.8	
Swine	TDN %		37.0	38.1	

Whale meal (AAFCO) —
see Whale, meat, heat rendered dehy grnd, salt declared above 3% mx 7%, (5)

WHEAT. Triticum spp

Wheat, chaff, s-c, (1)

Ref No 1 05 192 United States

Analyses	Units		As Fed	Dry	C.V. ± %
Dry matter	%		92.6	100.0	
Ash	%		16.7	18.0	
Crude fiber	%		29.8	32.2	
Sheep	dig coef %		33.	33.	
Swine	dig coef %		3.	3.	
Ether extract	%		1.5	1.7	
Sheep	dig coef %		121.	121.	
Swine	dig coef %		238.	238.	
N-free extract	%		39.4	42.6	
Sheep	dig coef %		44.	44.	
Swine	dig coef %		28.	28.	
Protein (N x 6.25)	%		5.1	5.5	
Sheep	dig coef %		46.	46.	
Swine	dig coef %		-20.	-20.	
Cattle	dig prot %	*	1.6	1.7	
Goats	dig prot %	*	1.6	1.7	
Horses	dig prot %	*	2.1	2.2	
Rabbits	dig prot %	*	2.7	2.9	
Sheep	dig prot %		2.4	2.5	
Swine	dig prot %		-1.0	-1.1	
Fatty acids	%		1.5	1.7	
Energy	GE Mcal/kg				
Cattle	DE Mcal/kg	*	1.62	1.75	
Sheep	DE Mcal/kg	*	1.49	1.61	
Swine	DE kcal/kg	*	841.	908.	
Cattle	ME Mcal/kg	*	1.33	1.44	
Sheep	ME Mcal/kg	*	1.22	1.32	
Swine	ME kcal/kg	*	798.	862.	
Cattle	TDN %	*	36.9	39.8	
Sheep	TDN %		33.7	36.4	
Swine	TDN %		19.1	20.6	
Calcium	%		0.18	0.20	
Phosphorus	%		0.14	0.15	
Potassium	%		0.51	0.56	

Ref No 1 05 192 Canada

Analyses	Units		As Fed	Dry	C.V. ± %
Dry matter	%		93.0	100.0	2
Crude fiber	%		30.3	32.5	17
Sheep	dig coef %		33.	33.	
Swine	dig coef %		3.	3.	
Ether extract	%				
Sheep	dig coef %		121.	121.	
Swine	dig coef %		238.	238.	
N-free extract	%				
Sheep	dig coef %		44.	44.	
Swine	dig coef %		28.	28.	
Protein (N x 6.25)	%		5.9	6.4	33
Sheep	dig coef %		46.	46.	
Swine	dig coef %		-20.	-20.	
Cattle	dig prot %	*	2.3	2.5	
Goats	dig prot %	*	2.3	2.5	
Horses	dig prot %	*	2.7	2.9	
Rabbits	dig prot %	*	3.3	3.6	
Sheep	dig prot %		2.7	2.9	
Swine	dig prot %		-1.1	-1.2	
Energy	GE Mcal/kg				
Sheep	DE Mcal/kg	*	1.49	1.61	
Swine	DE kcal/kg	*	829.	891.	
Sheep	ME Mcal/kg	*	1.22	1.32	
Swine	ME kcal/kg	*	785.	844.	
Sheep	TDN %		33.9	36.4	
Swine	TDN %		18.8	20.2	
Calcium	%		0.20	0.21	29
Phosphorus	%		0.08	0.08	78

Wheat, hay, fan air dried, early bloom, (1)

Ref No 1 05 168 United States

Analyses	Units		As Fed	Dry	C.V. ± %
Dry matter	%			100.0	
Riboflavin	mg/kg			17.0	

Wheat, hay, s-c, (1)

Ref No 1 05 172 United States

Analyses	Units		As Fed	Dry	C.V. ± %
Dry matter	%		86.4	100.0	3
Ash	%		5.9	6.8	
Crude fiber	%		24.0	27.8	
Sheep	dig coef %	*	41.	41.	
Ether extract	%		1.7	2.0	
Sheep	dig coef %	*	42.	42.	
N-free extract	%		48.2	55.9	
Sheep	dig coef %	*	62.	62.	
Protein (N x 6.25)	%		6.5	7.5	
Sheep	dig coef %	*	54.	54.	
Cattle	dig prot %	*	2.9	3.4	
Goats	dig prot %	*	3.1	3.5	
Horses	dig prot %	*	3.3	3.9	
Rabbits	dig prot %	*	3.8	4.4	
Sheep	dig prot %		3.5	4.0	
Fatty acids	%		1.7	2.0	
Energy	GE Mcal/kg				
Cattle	DE Mcal/kg	*	2.04	2.36	
Sheep	DE Mcal/kg	*	1.98	2.29	
Cattle	ME Mcal/kg	*	1.68	1.94	
Sheep	ME Mcal/kg	*	1.62	1.88	
Cattle	NE_m Mcal/kg	*	1.24	1.44	
Cattle	NE_{gain} Mcal/kg	*	0.74	0.86	
Cattle	$NE_{lactating\ cows}$ Mcal/kg	*	1.44	1.67	
Cattle	TDN %	*	46.3	53.6	
Sheep	TDN %		44.9	52.0	
Calcium	%		0.12	0.14	
Phosphorus	%		0.15	0.18	
Potassium	%		0.87	1.00	
Carotene	mg/kg		96.3	111.6	
Riboflavin	mg/kg		14.7	17.0	
Vitamin A equivalent	IU/g		160.6	186.0	

Ref No 1 05 172 Canada

Analyses	Units		As Fed	Dry	C.V. ± %
Dry matter	%		91.7	100.0	
Crude fiber	%		28.7	31.3	
Protein (N x 6.25)	%		10.1	11.1	
Cattle	dig prot %	*	6.0	6.5	
Goats	dig prot %	*	6.3	6.9	
Horses	dig prot %	*	6.4	6.9	
Rabbits	dig prot %	*	6.6	7.2	
Sheep	dig prot %	*	6.0	6.5	
Energy	GE Mcal/kg				
Sheep	DE Mcal/kg	*	2.08	2.26	
Sheep	ME Mcal/kg	*	1.70	1.86	
Sheep	TDN %		47.1	51.4	
Calcium	%		0.14	0.16	
Phosphorus	%		0.22	0.24	
Carotene	mg/kg		54.3	59.2	
Vitamin A equivalent	IU/g		90.5	98.7	

Wheat, hay, s-c, milk stage, (1)

Ref No 1 05 169 United States

Analyses	Units		As Fed	Dry	C.V. ± %
Dry matter	%		84.3	100.0	2
Ash	%		5.4	6.4	6
Crude fiber	%		20.0	23.7	6
Sheep	dig coef %		34.	34.	
Ether extract	%		1.6	1.9	12
Sheep	dig coef %		40.	40.	
N-free extract	%		49.7	58.9	
Sheep	dig coef %		62.	62.	
Protein (N x 6.25)	%		7.7	9.1	4
Sheep	dig coef %		58.	58.	
Cattle	dig prot %	*	4.1	4.8	
Goats	dig prot %	*	4.3	5.1	
Horses	dig prot %	*	4.4	5.3	
Rabbits	dig prot %	*	4.8	5.7	
Sheep	dig prot %		4.4	5.3	
Energy	GE Mcal/kg				
Cattle	DE Mcal/kg	*	1.98	2.35	
Sheep	DE Mcal/kg	*	1.92	2.27	
Cattle	ME Mcal/kg	*	1.63	1.93	
Sheep	ME Mcal/kg	*	1.57	1.86	
Cattle	TDN %	*	45.0	53.4	
Sheep	TDN %		43.5	51.6	

(1) dry forages and roughages (2) pasture, range plants, and forages fed green (3) silages (4) energy feeds (5) protein supplements (6) minerals (7) vitamins (8) additives

Feed Name or Analyses			Mean As Fed	Mean Dry	C.V. ± %
Wheat, hay, s-c, dough stage, (1)					
Ref No 1 05 170			United States		
Dry matter	%		82.6	100.0	0
Ash	%		5.4	6.6	12
Crude fiber	%		23.4	28.3	6
Sheep	dig coef %		42.	42.	
Ether extract	%		1.9	2.3	13
Sheep	dig coef %		44.	44.	
N-free extract	%		45.3	54.9	
Sheep	dig coef %		61.	61.	
Protein (N x 6.25)	%		6.6	8.0	5
Sheep	dig coef %		51.	51.	
Cattle	dig prot %	*	3.2	3.9	
Goats	dig prot %	*	3.3	4.0	
Horses	dig prot %	*	3.6	4.3	
Rabbits	dig prot %	*	4.0	4.8	
Sheep	dig prot %		3.4	4.1	
Energy	GE Mcal/kg				
Cattle	DE Mcal/kg	*	1.92	2.33	
Sheep	DE Mcal/kg	*	1.88	2.28	
Cattle	ME Mcal/kg	*	1.58	1.91	
Sheep	ME Mcal/kg	*	1.54	1.87	
Cattle	TDN %	*	43.6	52.8	
Sheep	TDN %		42.7	51.7	
Wheat, hay, s-c, mature, (1)					
Ref No 1 05 171			United States		
Dry matter	%		85.2	100.0	4
Ash	%		5.9	6.9	5
Crude fiber	%		25.9	30.4	15
Sheep	dig coef %		47.	47.	
Ether extract	%		1.4	1.7	10
Sheep	dig coef %		43.	43.	
N-free extract	%		45.7	53.7	
Sheep	dig coef %		65.	65.	
Protein (N x 6.25)	%		6.3	7.5	12
Sheep	dig coef %		55.	55.	
Cattle	dig prot %	*	2.9	3.4	
Goats	dig prot %	*	3.0	3.5	
Horses	dig prot %	*	3.3	3.9	
Rabbits	dig prot %	*	3.8	4.4	
Sheep	dig prot %		3.5	4.1	
Energy	GE Mcal/kg				
Cattle	DE Mcal/kg	*	2.14	2.51	
Sheep	DE Mcal/kg	*	2.06	2.42	
Cattle	ME Mcal/kg	*	1.76	2.06	
Sheep	ME Mcal/kg	*	1.69	1.98	
Cattle	TDN %	*	48.5	56.9	
Sheep	TDN %		46.7	54.8	
Ref No 1 05 171			Canada		
Dry matter	%		93.5	100.0	
Crude fiber	%		31.5	33.7	
Sheep	dig coef %		47.	47.	
Ether extract	%				
Sheep	dig coef %		43.	43.	
N-free extract	%				
Sheep	dig coef %		65.	65.	
Protein (N x 6.25)	%		8.1	8.6	
Sheep	dig coef %		55.	55.	
Cattle	dig prot %	*	4.1	4.4	
Goats	dig prot %	*	4.3	4.6	
Horses	dig prot %	*	4.6	4.9	
Rabbits	dig prot %	*	5.0	5.3	
Sheep	dig prot %		4.4	4.8	
Energy	GE Mcal/kg				
Sheep	DE Mcal/kg	*	2.24	2.40	
Sheep	ME Mcal/kg	*	1.84	1.97	
Sheep	TDN %		50.9	54.5	
Calcium	%		0.13	0.14	
Phosphorus	%		0.14	0.16	
Wheat, leaves, s-c, (1)					
Ref No 1 05 174			United States		
Dry matter	%			100.0	
Carotene	mg/kg			111.6	
Vitamin A equivalent	IU/g			186.0	
Wheat, straw, (1)					
Ref No 1 05 175			United States		
Dry matter	%		87.8	100.0	5
Ash	%		6.3	7.2	33
Crude fiber	%		38.3	43.6	9
Cattle	dig coef %		58.	58.	
Horses	dig coef %		48.	48.	
Sheep	dig coef %		51.	51.	
Ether extract	%		1.4	1.5	17
Cattle	dig coef %		42.	42.	
Horses	dig coef %		82.	82.	
Sheep	dig coef %		21.	21.	
N-free extract	%		38.6	44.0	7
Cattle	dig coef %		49.	49.	
Horses	dig coef %		37.	37.	
Sheep	dig coef %		40.	40.	
Protein (N x 6.25)	%		3.2	3.7	15
Cattle	dig coef %		14.	14.	
Horses	dig coef %		19.	19.	
Sheep	dig coef %		-16.	-16.	
Cattle	dig prot %		0.4	0.5	
Goats	dig prot %	*	0.0	0.0	
Horses	dig prot %		0.6	0.7	
Rabbits	dig prot %	*	1.3	1.5	
Sheep	dig prot %		-0.4	-0.5	
Cellulose (Matrone)	%		44.0	50.1	23
Pentosans	%		10.1	11.5	
Starch	%		1.0	1.1	
Lignin (Ellis)	%		11.2	12.8	
Fatty acids	%		1.4	1.6	
Energy	GE Mcal/kg				
Cattle	DE Mcal/kg	*	1.89	2.15	
Horses	DE Mcal/kg	*	1.57	1.79	
Sheep	DE Mcal/kg	*	1.53	1.74	
Cattle	ME Mcal/kg	*	1.55	1.76	
Horses	ME Mcal/kg	*	1.29	1.47	
Sheep	ME Mcal/kg	*	1.25	1.43	
Cattle	NE_m Mcal/kg	*	0.90	1.03	
Cattle	NE_{gain} Mcal/kg	*	0.17	0.19	
Cattle	$NE_{lactating\ cows}$ Mcal/kg	*	0.87	1.01	
Cattle	TDN %		42.8	48.8	
Horses	TDN %		35.6	40.5	
Sheep	TDN %		34.7	39.5	
Calcium	%		0.14	0.16	
Chlorine	%		0.29	0.33	
Copper	mg/kg		2.7	3.1	
Iron	%		0.015	0.017	
Magnesium	%		0.10	0.12	
Manganese	mg/kg		48.3	55.0	
Phosphorus	%		0.07	0.08	
Potassium	%		0.59	0.67	
Sodium	%		0.12	0.14	
Sulphur	%		0.16	0.18	
Carotene	mg/kg		1.9	2.2	
Riboflavin	mg/kg		2.1	2.4	
Vitamin A equivalent	IU/g		3.2	3.7	
Ref No 1 05 175			Canada		
Dry matter	%		93.2	100.0	2
Crude fiber	%		35.6	38.2	13
Cattle	dig coef %		58.	58.	
Horses	dig coef %		48.	48.	
Sheep	dig coef %		51.	51.	
Ether extract	%				
Cattle	dig coef %		42.	42.	
Horses	dig coef %		82.	82.	
Sheep	dig coef %		21.	21.	
N-free extract	%				
Cattle	dig coef %		49.	49.	
Horses	dig coef %		37.	37.	
Sheep	dig coef %		40.	40.	
Protein (N x 6.25)	%		4.7	5.1	35
Cattle	dig coef %		14.	14.	
Horses	dig coef %		19.	19.	
Sheep	dig coef %		-16.	-16.	
Cattle	dig prot %		0.7	0.7	
Goats	dig prot %	*	1.2	1.3	
Horses	dig prot %		0.9	1.0	
Rabbits	dig prot %	*	2.4	2.6	
Sheep	dig prot %		-0.7	-0.7	
Energy	GE Mcal/kg				
Cattle	DE Mcal/kg	*	1.98	2.12	
Horses	DE Mcal/kg	*	1.64	1.76	
Sheep	DE Mcal/kg	*	1.59	1.70	
Cattle	ME Mcal/kg	*	1.62	1.74	
Horses	ME Mcal/kg	*	1.35	1.45	
Sheep	ME Mcal/kg	*	1.30	1.40	
Cattle	TDN %		44.9	48.1	
Horses	TDN %		37.3	40.0	
Sheep	TDN %		36.0	38.7	
Calcium	%		0.21	0.22	56
Phosphorus	%		0.07	0.07	91
Ref No 1 05 175			Egypt		
Dry matter	%		89.6	100.0	
Ash	%		8.9	9.9	
Crude fiber	%		44.3	49.4	
Cattle	dig coef %		58.	58.	
Horses	dig coef %		48.	48.	
Sheep	dig coef %		51.	51.	
Ether extract	%		1.4	1.6	
Cattle	dig coef %		42.	42.	
Horses	dig coef %		82.	82.	
Sheep	dig coef %		21.	21.	
N-free extract	%		33.0	36.8	
Cattle	dig coef %		49.	49.	
Horses	dig coef %		37.	37.	
Sheep	dig coef %		40.	40.	
Protein (N x 6.25)	%		2.0	2.2	
Cattle	dig coef %		14.	14.	
Horses	dig coef %		19.	19.	
Sheep	dig coef %		-16.	-16.	
Cattle	dig prot %		0.3	0.3	
Goats	dig prot %	*	-1.1	-1.3	
Horses	dig prot %		0.4	0.4	
Rabbits	dig prot %	*	0.4	0.4	
Sheep	dig prot %		-0.2	-0.3	
Energy	GE Mcal/kg				
Cattle	DE Mcal/kg	*	1.92	2.14	
Horses	DE Mcal/kg	*	1.60	1.78	

Continued

Feed Name or Analyses			Mean As Fed	Mean Dry	C.V. ± %
Sheep	DE Mcal/kg	*	1.57	1.76	
Cattle	ME Mcal/kg	*	1.57	1.75	
Horses	ME Mcal/kg	*	1.31	1.46	
Sheep	ME Mcal/kg	*	1.29	1.44	
Cattle	TDN %		43.5	48.5	
Horses	TDN %		36.2	40.4	
Sheep	TDN %		35.7	39.9	

Wheat, aerial part, fresh, (2)
Ref No 2 08 078 United States

Feed Name or Analyses			Mean As Fed	Mean Dry	C.V. ± %
Dry matter	%		25.9	100.0	
Cattle	dig coef %		81.	81.	
Ash	%		2.4	9.5	
Crude fiber	%		5.9	22.9	
Cattle	dig coef %		88.	88.	
Ether extract	%		0.9	3.3	
N-free extract	%		12.5	48.3	
Cattle	dig coef %		88.	88.	
Protein (N x 6.25)	%		4.1	16.0	
Cattle	dig coef %		73.	73.	
Cattle	dig prot %		3.0	11.7	
Goats	dig prot %	*	3.0	11.5	
Horses	dig prot %	*	2.9	11.1	
Rabbits	dig prot %	*	2.8	11.0	
Sheep	dig prot %	*	3.1	11.9	
Energy	GE Mcal/kg		1.15	4.44	
Cattle	GE dig coef %		79.	79.	
Cattle	DE Mcal/kg		0.91	3.50	
Sheep	DE Mcal/kg	*	0.66	2.54	
Cattle	ME Mcal/kg	*	0.74	2.87	
Sheep	ME Mcal/kg	*	0.54	2.08	
Cattle	TDN %	*	20.5	79.0	
Sheep	TDN %	*	14.9	57.6	
Calcium	%		0.08	0.31	
Magnesium	%		0.08	0.30	
Phosphorus	%		0.08	0.31	

Wheat, aerial part, fresh, immature, (2)
Ref No 2 05 176 United States

Feed Name or Analyses			Mean As Fed	Mean Dry	C.V. ± %
Dry matter	%		22.9	100.0	24
Energy	GE Mcal/kg				
Cattle	NE_m Mcal/kg	*	0.38	1.64	
Cattle	NE_{gain} Mcal/kg	*	0.25	1.07	
Cattle	$NE_{lactating\ cows}$ Mcal/kg	*	0.44	1.93	
Carotene	mg/kg		119.1	520.1	39
Niacin	mg/kg		13.0	56.9	18
Pantothenic acid	mg/kg		4.8	21.2	48
Riboflavin	mg/kg		6.3	27.6	30
Vitamin A equivalent	IU/g		198.5	866.9	

Wheat, aerial part, fresh, early bloom, (2)
Ref No 2 05 177 United States

Feed Name or Analyses			Mean As Fed	Mean Dry	C.V. ± %
Dry matter	%		21.6	100.0	
Calcium	%		0.08	0.39	
Phosphorus	%		0.11	0.50	
Potassium	%		0.59	2.72	

(1) dry forages and roughages (2) pasture, range plants, and forages fed green (3) silages (4) energy feeds (5) protein supplements (6) minerals (7) vitamins (8) additives

Wheat, aerial part, fresh, milk stage, (2)
Ref No 2 05 178 United States

Feed Name or Analyses			Mean As Fed	Mean Dry	C.V. ± %
Dry matter	%		26.6	100.0	
Ash	%		1.7	6.5	22
Crude fiber	%		7.4	27.9	13
Ether extract	%		1.0	3.8	9
N-free extract	%		13.2	49.8	
Protein (N x 6.25)	%		3.2	12.0	32
Cattle	dig prot %	*	2.2	8.1	
Goats	dig prot %	*	2.1	7.8	
Horses	dig prot %	*	2.1	7.7	
Rabbits	dig prot %	*	2.1	7.9	
Sheep	dig prot %	*	2.2	8.2	
Energy	GE Mcal/kg				
Cattle	DE Mcal/kg	*	0.79	2.99	
Sheep	DE Mcal/kg	*	0.77	2.89	
Cattle	ME Mcal/kg	*	0.65	2.45	
Sheep	ME Mcal/kg	*	0.63	2.37	
Cattle	TDN %	*	18.0	67.8	
Sheep	TDN %	*	17.5	65.6	
Chlorine	%		0.18	0.67	
Sodium	%		0.07	0.28	
Sulphur	%		0.06	0.24	
Niacin	mg/kg		9.6	35.9	
Riboflavin	mg/kg		4.1	15.4	
Thiamine	mg/kg		0.9	3.5	

Wheat, aerial part, fresh, dough stage, (2)
Ref No 2 05 179 United States

Feed Name or Analyses			Mean As Fed	Mean Dry	C.V. ± %
Dry matter	%		41.6	100.0	
Calcium	%		0.06	0.14	
Phosphorus	%		0.09	0.22	

Wheat, aerial part, fresh, mature, (2)
Ref No 2 05 180 Canada

Feed Name or Analyses			Mean As Fed	Mean Dry	C.V. ± %
Dry matter	%		93.1	100.0	2
Crude fiber	%		26.9	28.9	23
Protein (N x 6.25)	%		9.5	10.3	29
Cattle	dig prot %	*	6.2	6.6	
Goats	dig prot %	*	5.7	6.1	
Horses	dig prot %	*	5.8	6.2	
Rabbits	dig prot %	*	6.1	6.6	
Sheep	dig prot %	*	6.1	6.5	
Calcium	%		0.14	0.15	35
Phosphorus	%		0.19	0.21	40
Carotene	mg/kg		17.6	18.9	
Vitamin A equivalent	IU/g		29.4	31.6	

Wheat, aerial part, fresh, over ripe, (2)
Ref No 2 05 181 United States

Feed Name or Analyses			Mean As Fed	Mean Dry	C.V. ± %
Dry matter	%		39.9	100.0	
Calcium	%		0.06	0.15	
Magnesium	%		0.04	0.10	
Phosphorus	%		0.07	0.17	
Potassium	%		0.80	2.00	
Sodium	%		0.01	0.02	

Ref No 2 05 181 Canada

Feed Name or Analyses			Mean As Fed	Mean Dry	C.V. ± %
Dry matter	%		94.1	100.0	
Crude fiber	%		25.2	26.8	
Protein (N x 6.25)	%		17.2	18.3	
Cattle	dig prot %	*	12.6	13.4	
Goats	dig prot %	*	12.8	13.6	
Horses	dig prot %	*	12.3	13.0	
Rabbits	dig prot %	*	12.0	12.8	
Sheep	dig prot %	*	13.2	14.0	
Calcium	%		0.69	0.73	
Phosphorus	%		0.12	0.13	

Wheat, aerial part, fresh, cut 1, (2)
Ref No 2 05 182 United States

Feed Name or Analyses			Mean As Fed	Mean Dry	C.V. ± %
Dry matter	%		23.8	100.0	20
Ash	%		2.8	11.6	24
Crude fiber	%		5.0	21.1	25
Ether extract	%		0.9	3.9	32
N-free extract	%		9.2	38.6	
Protein (N x 6.25)	%		5.9	24.8	30
Cattle	dig prot %	*	4.5	19.0	
Goats	dig prot %	*	4.7	19.7	
Horses	dig prot %	*	4.4	18.6	
Rabbits	dig prot %	*	4.2	17.8	
Sheep	dig prot %	*	4.8	20.1	
Energy	GE Mcal/kg				
Cattle	DE Mcal/kg	*	0.77	3.22	
Sheep	DE Mcal/kg	*	0.70	2.95	
Cattle	ME Mcal/kg	*	0.63	2.64	
Sheep	ME Mcal/kg	*	0.58	2.42	
Cattle	TDN %	*	17.4	73.1	
Sheep	TDN %	*	15.9	66.9	

Wheat, leaves, fresh, mature, (2)
Ref No 2 05 183 United States

Feed Name or Analyses			Mean As Fed	Mean Dry	C.V. ± %
Dry matter	%			100.0	
Carotene	mg/kg			239.9	
Vitamin A equivalent	IU/g			399.8	

Wheat, aerial part, ensiled, (3)
Ref No 3 05 186 United States

Feed Name or Analyses			Mean As Fed	Mean Dry	C.V. ± %
Dry matter	%		25.1	100.0	4
Ash	%		2.2	8.8	29
Crude fiber	%		7.9	31.3	18
Ether extract	%		0.8	3.1	12
N-free extract	%		12.1	48.1	
Protein (N x 6.25)	%		2.2	8.7	17
Cattle	dig prot %	*	1.0	4.1	
Goats	dig prot %	*	1.0	4.1	
Horses	dig prot %	*	1.0	4.1	
Sheep	dig prot %	*	1.0	4.1	
Energy	GE Mcal/kg				
Cattle	DE Mcal/kg	*	0.70	2.81	
Sheep	DE Mcal/kg	*	0.71	2.81	
Cattle	ME Mcal/kg	*	0.58	2.30	
Sheep	ME Mcal/kg	*	0.58	2.30	
Cattle	TDN %	*	16.0	63.6	
Sheep	TDN %	*	16.0	63.7	
Carotene	mg/kg		44.0	175.5	
Vitamin A equivalent	IU/g		73.4	292.5	

Ref No 3 05 186 Canada

Feed Name or Analyses			Mean As Fed	Mean Dry	C.V. ± %
Dry matter	%		29.2	100.0	
Crude fiber	%		8.6	29.7	
Protein (N x 6.25)	%		6.7	23.0	
Cattle	dig prot %	*	5.0	17.1	
Goats	dig prot %	*	5.0	17.1	
Horses	dig prot %	*	5.0	17.1	
Sheep	dig prot %	*	5.0	17.1	

Feed Name or Analyses			Mean As Fed	Mean Dry	C.V. ± %
Calcium	%		0.13	0.44	
Phosphorus	%		0.08	0.26	
Carotene	mg/kg		12.3	42.1	
Vitamin A equivalent	IU/g		20.4	70.1	

Wheat, aerial part, ensiled, immature, (3)

Ref No 3 05 184 United States

Feed Name or Analyses			Mean As Fed	Mean Dry	C.V. ± %
Dry matter	%		26.1	100.0	
Crude fiber	%				
Cattle	dig coef %		38.	38.	
Ether extract	%				
Cattle	dig coef %		66.	66.	
Protein (N x 6.25)	%				
Cattle	dig coef %		57.	57.	
Energy	GE Mcal/kg		1.15	4.39	
Cattle	GE dig coef %		50.	50.	
Cattle	DE Mcal/kg		0.57	2.19	
Cattle	ME Mcal/kg	*	0.47	1.80	
Carotene	mg/kg		67.0	256.8	
Vitamin A equivalent	IU/g		111.7	428.1	

Ref No 3 05 184 Canada

Feed Name or Analyses			Mean As Fed	Mean Dry	C.V. ± %
Dry matter	%		37.6	100.0	
Ash	%		2.8	7.5	
Crude fiber	%		10.1	26.9	
Cattle	dig coef %		38.	38.	
Ether extract	%		0.9	2.5	
Cattle	dig coef %		66.	66.	
N-free extract	%		19.2	51.2	
Protein (N x 6.25)	%		4.5	11.9	
Cattle	dig coef %		57.	57.	
Cattle	dig prot %		2.5	6.8	
Goats	dig prot %	*	2.6	7.0	
Horses	dig prot %	*	2.6	7.0	
Sheep	dig prot %	*	2.6	7.0	
Cellulose (Matrone)	%		10.2	27.2	
Cattle	dig coef %		33.	33.	
Energy	GE Mcal/kg		1.65	4.39	
Cattle	GE dig coef %		50.	50.	
Cattle	DE Mcal/kg		0.83	2.19	
Sheep	DE Mcal/kg	*	1.08	2.87	
Cattle	ME Mcal/kg	*	0.68	1.80	
Sheep	ME Mcal/kg	*	0.89	2.36	
Cattle	TDN %	*	23.3	61.9	
Sheep	TDN %	*	24.5	65.2	
Calcium	%		0.10	0.27	
Phosphorus	%		0.10	0.27	
Carotene	mg/kg		13.7	36.4	
Vitamin A equivalent	IU/g		22.8	60.6	

Wheat, aerial part, ensiled, full bloom, (3)

Ref No 3 05 185 United States

Feed Name or Analyses			Mean As Fed	Mean Dry	C.V. ± %
Dry matter	%		25.1	100.0	4
Ash	%		2.1	8.4	32
Crude fiber	%		7.8	30.9	19
Ether extract	%		0.8	3.0	10
N-free extract	%		12.4	49.6	
Protein (N x 6.25)	%		2.0	8.1	12
Cattle	dig prot %	*	0.9	3.6	
Goats	dig prot %	*	0.9	3.6	
Horses	dig prot %	*	0.9	3.6	
Sheep	dig prot %	*	0.9	3.6	
Energy	GE Mcal/kg				
Cattle	DE Mcal/kg	*	0.71	2.82	
Sheep	DE Mcal/kg	*	0.71	2.83	

Feed Name or Analyses			Mean As Fed	Mean Dry	C.V. ± %
Cattle	ME Mcal/kg	*	0.58	2.32	
Sheep	ME Mcal/kg	*	0.58	2.32	
Cattle	TDN %	*	16.1	64.0	
Sheep	TDN %	*	16.1	64.2	

Wheat, aerial part w AIV preservative added, ensiled, (3)

Ref No 3 05 187 United States

Feed Name or Analyses			Mean As Fed	Mean Dry	C.V. ± %
Dry matter	%		23.8	100.0	5
Ash	%		2.3	9.7	14
Crude fiber	%		7.1	30.0	7
Ether extract	%		0.7	3.0	6
N-free extract	%		11.3	47.3	
Protein (N x 6.25)	%		2.4	10.0	11
Cattle	dig prot %	*	1.3	5.3	
Goats	dig prot %	*	1.3	5.3	
Horses	dig prot %	*	1.3	5.3	
Sheep	dig prot %	*	1.3	5.3	
Energy	GE Mcal/kg				
Cattle	DE Mcal/kg	*	0.68	2.84	
Sheep	DE Mcal/kg	*	0.66	2.79	
Cattle	ME Mcal/kg	*	0.55	2.33	
Sheep	ME Mcal/kg	*	0.54	2.29	
Cattle	TDN %	*	15.3	64.4	
Sheep	TDN %	*	15.1	63.3	
Phosphorus	%		0.06	0.25	23

Wheat, aerial part w phosphoric acid preservative added, ensiled, (3)

Ref No 3 05 188 United States

Feed Name or Analyses			Mean As Fed	Mean Dry	C.V. ± %
Dry matter	%		25.9	100.0	
Carotene	mg/kg		3.3	12.6	
Vitamin A equivalent	IU/g		5.4	20.9	

Wheat, aerial part w salt added, ensiled, immature, (3)

Ref No 3 05 189 United States

Feed Name or Analyses			Mean As Fed	Mean Dry	C.V. ± %
Dry matter	%		26.1	100.0	
Carotene	mg/kg		66.7	255.7	
Vitamin A equivalent	IU/g		111.3	426.3	

Wheat, bran, (4)

Ref No 4 05 191 United States

Feed Name or Analyses			Mean As Fed	Mean Dry	C.V. ± %
Dry matter	%		88.1	100.0	2
Ash	%		5.9	6.7	4
Crude fiber	%		9.9	11.3	8
Cattle	dig coef %		30.	30.	
Sheep	dig coef %		36.	36.	
Ether extract	%		4.3	4.8	6
Cattle	dig coef %		74.	74.	
Sheep	dig coef %		55.	55.	
N-free extract	%		52.9	60.0	
Cattle	dig coef %		76.	76.	
Sheep	dig coef %		72.	72.	
Protein (N x 6.25)	%		15.2	17.2	
Cattle	dig coef %		76.	76.	
Sheep	dig coef %		75.	75.	
Cattle	dig prot %		11.5	13.1	
Goats	dig prot %	*	11.5	13.0	
Horses	dig prot %	*	11.5	13.0	
Sheep	dig prot %		11.4	12.9	
Energy	GE Mcal/kg		4.03	4.57	
Cattle	DE Mcal/kg	*	2.72	3.09	

Feed Name or Analyses			Mean As Fed	Mean Dry	C.V. ± %
Sheep	DE Mcal/kg	*	2.57	2.92	
Swine	DE kcal/kg		2353.	2670.	
Cattle	ME Mcal/kg	*	2.23	2.54	
Chickens	ME_n kcal/kg		1139.	1292.	
Sheep	ME Mcal/kg	*	2.11	2.39	
Swine	ME kcal/kg		2230.	2530.	
Cattle	TDN %		61.8	70.1	
Sheep	TDN %		58.3	66.2	
Swine	TDN %	*	53.8	61.0	
Calcium	%		0.13	0.15	
Chlorine	%		0.04	0.04	
Copper	mg/kg		11.4	13.0	
Iron	%		0.017	0.019	
Magnesium	%		0.58	0.65	
Manganese	mg/kg		113.4	128.7	
Phosphorus	%		1.21	1.38	
Potassium	%		1.20	1.37	
Sodium	%		0.06	0.07	
Sulphur	%		0.22	0.24	
Choline	mg/kg		1073.	1218.	
Niacin	mg/kg		208.3	236.4	
Pantothenic acid	mg/kg		28.9	32.8	
Riboflavin	mg/kg		3.1	3.5	
Arginine	%		0.99	1.13	
Cystine	%		0.24	0.27	
Glycine	%		0.89	1.01	
Lysine	%		0.60	0.68	
Methionine	%		0.17	0.19	
Tryptophan	%		0.27	0.30	

Wheat, bran, dry milled, (4)

Wheat bran (AAFCO)

Bran (CFA)

Ref No 4 05 190 United States

Feed Name or Analyses			Mean As Fed	Mean Dry	C.V. ± %
Dry matter	%		88.7	100.0	1
Ash	%		5.9	6.7	7
Crude fiber	%		8.6	9.7	27
Cattle	dig coef %		44.	44.	
Horses	dig coef %		50.	50.	
Sheep	dig coef %	*	36.	36.	
Swine	dig coef %		23.	23.	
Ether extract	%		4.4	5.0	12
Cattle	dig coef %		62.	62.	
Horses	dig coef %		57.	57.	
Sheep	dig coef %	*	55.	55.	
Swine	dig coef %		58.	58.	
N-free extract	%		54.0	60.9	5
Cattle	dig coef %		75.	75.	
Horses	dig coef %		74.	74.	
Sheep	dig coef %	*	72.	72.	
Swine	dig coef %		69.	69.	
Protein (N x 6.25)	%		15.7	17.8	3
Cattle	dig coef %		78.	78.	
Horses	dig coef %		85.	85.	
Sheep	dig coef %	*	75.	75.	
Swine	dig coef %		76.	76.	
Cattle	dig prot %		12.3	13.8	
Goats	dig prot %	*	12.0	13.5	
Horses	dig prot %		13.3	15.0	
Sheep	dig prot %		11.8	13.3	
Swine	dig prot %		12.0	13.5	
Cellulose (Matrone)	%		13.6	15.4	
Pentosans	%		29.3	33.0	
Linoleic	%		1.699	1.916	
Energy	GE Mcal/kg		3.77	4.26	2
Cattle	DE Mcal/kg	*	2.76	3.12	
Horses	DE Mcal/kg	*	2.77	3.13	

Continued

Feed Name or Analyses		Mean As Fed	Mean Dry	C.V. ± %
Sheep	DE Mcal/kg	* 2.61	2.95	
Swine	DE kcal/kg	*2511.	2832.	
Cattle	ME Mcal/kg	* 2.27	2.56	
Chickens	ME_n kcal/kg	1392.	1570.	
Horses	ME Mcal/kg	* 2.27	2.56	
Sheep	ME Mcal/kg	* 2.14	2.42	
Swine	ME kcal/kg	*2320.	2617.	
Cattle	NE_m Mcal/kg	* 1.36	1.53	
Cattle	NE_{gain} Mcal/kg	* 0.84	0.96	
Cattle	$NE_{lactating\ cows}$ Mcal/kg	* 1.62	1.83	
Cattle	TDN %	62.7	70.7	
Horses	TDN %	62.9	70.9	
Sheep	TDN %	59.2	66.8	
Swine	TDN %	56.9	64.2	
Calcium	%	0.14	0.16	64
Chlorine	%	0.04	0.04	66
Cobalt	mg/kg	0.097	0.109	35
Copper	mg/kg	12.3	13.9	45
Iron	%	0.017	0.019	23
Magnesium	%	0.55	0.62	16
Manganese	mg/kg	115.4	130.1	13
Phosphorus	%	1.16	1.31	28
Potassium	%	1.23	1.39	19
Sodium	%	0.06	0.07	50
Sulphur	%	0.22	0.25	23
Carotene	mg/kg	2.6	3.0	65
Choline	mg/kg	1077.	1215.	31
Folic acid	mg/kg	1.65	1.86	85
Niacin	mg/kg	208.6	235.3	38
Pantothenic acid	mg/kg	29.0	32.7	22
Riboflavin	mg/kg	3.1	3.5	33
Thiamine	mg/kg	7.9	8.9	38
α-tocopherol	mg/kg	9.8	11.1	45
Vitamin A equivalent	IU/g	4.4	4.9	
Arginine	%	1.00	1.12	12
Cystine	%	0.30	0.34	
Glycine	%	0.90	1.01	
Histidine	%	0.30	0.34	21
Isoleucine	%	0.60	0.67	21
Leucine	%	0.90	1.01	7
Lysine	%	0.60	0.67	19
Methionine	%	0.10	0.11	53
Phenylalanine	%	0.50	0.56	13
Serine	%	0.90	1.01	
Threonine	%	0.40	0.45	16
Tryptophan	%	0.30	0.34	10
Tyrosine	%	0.40	0.45	35
Valine	%	0.70	0.79	14

Ref No 4 05 190 Canada

Feed Name or Analyses		Mean As Fed	Mean Dry	C.V. ± %
Dry matter	%	89.8	100.0	1
Ash	%	5.6	6.2	
Crude fiber	%	12.0	13.3	12
Cattle	dig coef %	44.	44.	
Horses	dig coef %	50.	50.	
Swine	dig coef %	23.	23.	
Ether extract	%	4.4	4.9	16
Cattle	dig coef %	62.	62.	
Horses	dig coef %	57.	57.	
Swine	dig coef %	58.	58.	
N-free extract	%	52.3	58.2	
Cattle	dig coef %	75.	75.	
Horses	dig coef %	74.	74.	
Swine	dig coef %	69.	69.	
Protein (N x 6.25)	%	15.6	17.4	10
Cattle	dig coef %	78.	78.	
Horses	dig coef %	85.	85.	
Swine	dig coef %	76.	76.	
Cattle	dig prot %	12.2	13.5	
Goats	dig prot %	* 11.8	13.1	
Horses	dig prot %	13.2	14.7	
Sheep	dig prot %	* 11.8	13.1	
Swine	dig prot %	11.9	13.2	
Energy	GE Mcal/kg	3.42	3.81	
Cattle	DE Mcal/kg	* 2.77	3.08	
Horses	DE Mcal/kg	* 2.79	3.10	
Sheep	DE Mcal/kg	* 2.61	2.91	
Swine	DE kcal/kg	*2489.	2771.	
Cattle	ME Mcal/kg	* 2.27	2.53	
Chickens	ME_n kcal/kg	1411.	1570.	
Horses	ME Mcal/kg	* 2.29	2.54	
Sheep	ME Mcal/kg	* 2.14	2.38	
Swine	ME kcal/kg	*2302.	2563.	
Cattle	TDN %	62.8	69.9	
Horses	TDN %	63.2	70.4	
Sheep	TDN %	59.2	65.9	
Swine	TDN %	56.5	62.8	
Calcium	%	0.17	0.18	
Copper	mg/kg	15.8	17.6	
Magnesium	%	0.44	0.49	
Phosphorus	%	1.06	1.18	
Potassium	%	1.09	1.21	
Salt (NaCl)	%	0.13	0.14	
Sodium	%	0.01	0.01	
Choline	mg/kg	5.	5.	
Niacin	mg/kg	123.3	137.2	
Pantothenic acid	mg/kg	32.5	36.1	
Riboflavin	mg/kg	5.9	6.5	39

(1) dry forages and roughages (2) pasture, range plants, and forages fed green (3) silages (4) energy feeds (5) protein supplements (6) minerals (7) vitamins (8) additives

Wheat, endosperm, (4)

Ref No 4 05 197 United States

Feed Name or Analyses		Mean As Fed	Mean Dry	C.V. ± %
Dry matter	%	87.9	100.0	0
Ash	%	1.2	1.4	98
Crude fiber	%	0.3	0.3	
Ether extract	%	1.1	1.3	
N-free extract	%	74.2	84.4	
Protein (N x 6.25)	%	11.1	12.6	15
Cattle	dig prot %	* 6.7	7.6	
Goats	dig prot %	* 7.7	8.8	
Horses	dig prot %	* 7.7	8.8	
Sheep	dig prot %	* 7.7	8.8	
Energy	GE Mcal/kg			
Cattle	DE Mcal/kg	* 2.65	3.01	
Sheep	DE Mcal/kg	* 3.34	3.80	
Swine	DE kcal/kg	*3767.	4285.	
Cattle	ME Mcal/kg	* 2.17	2.47	
Sheep	ME Mcal/kg	* 2.74	3.11	
Swine	ME kcal/kg	*3520.	4005.	
Cattle	TDN %	* 60.0	68.3	
Sheep	TDN %	* 75.7	86.1	
Swine	TDN %	* 85.4	97.2	
Arginine	%	0.60	0.68	52
Cystine	%	0.30	0.34	52
Histidine	%	0.30	0.34	
Isoleucine	%	1.10	1.25	14
Leucine	%	1.70	1.93	46
Lysine	%	0.40	0.46	39
Methionine	%	0.20	0.23	39
Phenylalanine	%	0.60	0.68	13
Threonine	%	0.40	0.46	19
Tryptophan	%	0.30	0.34	78
Valine	%	0.60	0.68	26

Wheat, flour, coarse bolt, feed gr mx 2% fiber, (4)

Wheat feed flour, mx 1.5% fiber (AAFCO)

Feed flour, mx 2.0% fiber (CFA)

Ref No 4 05 199 United States

Feed Name or Analyses		Mean As Fed	Mean Dry	C.V. ± %
Dry matter	%	88.4	100.0	
Ash	%	0.9	1.0	
Crude fiber	%	0.5	0.6	
Sheep	dig coef %	* 15.	15.	
Ether extract	%	1.9	2.1	
Sheep	dig coef %	* 85.	85.	
N-free extract	%	69.7	78.8	
Sheep	dig coef %	* 90.	90.	
Protein (N x 6.25)	%	15.4	17.4	
Sheep	dig coef %	* 71.	71.	
Cattle	dig prot %	* 10.6	12.0	
Goats	dig prot %	* 11.7	13.2	
Horses	dig prot %	* 11.7	13.2	
Sheep	dig prot %	10.9	12.4	
Energy	GE Mcal/kg			
Cattle	DE Mcal/kg	* 3.77	4.27	
Sheep	DE Mcal/kg	* 3.41	3.86	
Swine	DE kcal/kg	*3877.	4385.	
Cattle	ME Mcal/kg	* 3.09	3.50	
Chickens	ME_n kcal/kg	3041.	3440.	
Sheep	ME Mcal/kg	* 2.80	3.16	
Swine	ME kcal/kg	*3585.	4056.	
Cattle	TDN %	* 85.6	96.8	
Sheep	TDN %	77.4	87.5	
Swine	TDN %	* 87.9	99.5	
Calcium	%	0.04	0.05	
Phosphorus	%	0.08	0.09	
Choline	mg/kg	441.	499.	
Niacin	mg/kg	26.0	29.4	
Pantothenic acid	mg/kg	5.5	6.2	
Riboflavin	mg/kg	0.7	0.7	
Arginine	%	1.00	1.13	
Cystine	%	0.20	0.23	
Glycine	%	0.40	0.45	
Lysine	%	0.60	0.68	
Methionine	%	0.10	0.11	
Tryptophan	%	0.20	0.23	

Wheat, flour, human gr, (4)

Ref No 4 05 200 United States

Feed Name or Analyses		Mean As Fed	Mean Dry	C.V. ± %
Dry matter	%	88.0	100.0	
Ash	%	0.4	0.5	
Crude fiber	%	0.3	0.3	
Ether extract	%	0.9	1.0	
N-free extract	%	75.6	85.9	
Protein (N x 6.25)	%	10.8	12.3	
Cattle	dig prot %	* 6.4	7.3	
Goats	dig prot %	* 7.5	8.5	
Horses	dig prot %	* 7.5	8.5	
Sheep	dig prot %	* 7.5	8.5	
Energy	GE Mcal/kg			
Cattle	DE Mcal/kg	* 3.43	3.90	
Sheep	DE Mcal/kg	* 3.37	3.83	
Swine	DE kcal/kg	*3848.	4373.	
Cattle	ME Mcal/kg	* 2.82	3.20	
Sheep	ME Mcal/kg	* 2.76	3.14	
Swine	ME kcal/kg	*3599.	4090.	
Cattle	TDN %	* 77.8	88.4	
Sheep	TDN %	* 76.4	86.8	

Feed Name or Analyses		Mean As Fed	Mean Dry	C.V. ± %
Swine	TDN %	* 87.3	99.2	
Calcium	%	0.02	0.02	
Phosphorus	%	0.09	0.10	
Potassium	%	0.05	0.06	

Wheat, flour by-product, coarse sift, mx 7% fiber, (4)

Wheat shorts, mx 7% fiber (AAFCO)

Shorts, mx 8% fiber (CFA)

Ref No 4 05 201 United States

Feed Name or Analyses		Mean As Fed	Mean Dry	C.V. ± %
Dry matter	%	87.1	100.0	2
Ash	%	4.0	4.5	8
Crude fiber	%	6.3	7.3	9
Sheep	dig coef %	* 60.	60.	
Ether extract	%	4.8	5.5	16
Sheep	dig coef %	* 85.	85.	
N-free extract	%	56.2	64.6	
Sheep	dig coef %	* 85.	85.	
Protein (N x 6.25)	%	15.8	18.1	8
Sheep	dig coef %	* 85.	85.	
Cattle	dig prot %	* 11.0	12.6	
Goats	dig prot %	* 12.0	13.8	
Horses	dig prot %	* 12.0	13.8	
Sheep	dig prot %	13.4	15.4	
Starch	%	19.0	21.8	11
Linoleic	%	1.644	1.888	
Energy	GE Mcal/kg	4.02	4.62	
Cattle	DE Mcal/kg	* 3.31	3.68	
Sheep	DE Mcal/kg	* 3.27	3.76	
Swine	DE kcal/kg	*2924.	3358.	
Cattle	ME Mcal/kg	* 2.63	3.02	
Chickens	ME_n kcal/kg	2603.	2990.	
Sheep	ME Mcal/kg	* 2.68	3.08	
Swine	ME kcal/kg	*2700.	3101.	
Cattle	NE_m Mcal/kg	* 1.81	2.08	
Cattle	NE_{gain} Mcal/kg	* 1.20	1.38	
Cattle	$NE_{lactating\ cows}$ Mcal/kg	* 2.10	2.41	
Cattle	TDN %	* 72.7	83.5	
Sheep	TDN %	74.1	85.2	
Swine	TDN %	* 66.3	76.2	
Calcium	%	0.09	0.10	27
Cobalt	mg/kg	0.105	0.120	18
Copper	mg/kg	11.6	13.3	19
Iron	%	0.007	0.008	80
Magnesium	%	0.25	0.29	12
Manganese	mg/kg	115.8	133.0	13
Phosphorus	%	0.81	0.93	12
Potassium	%	0.93	1.07	10
Selenium	mg/kg	0.388	0.445	66
Sodium	%	0.02	0.02	46
Zinc	mg/kg	107.3	123.3	24
Choline	mg/kg	1905.	2188.	20
Folic acid	mg/kg	1.59	1.82	17
Niacin	mg/kg	105.5	121.1	14
Pantothenic acid	mg/kg	22.7	26.1	19
Riboflavin	mg/kg	4.4	5.0	24
Thiamine	mg/kg	19.6	22.5	10
α-tocopherol	mg/kg	57.0	65.4	
Alanine	%	0.94	1.08	7
Arginine	%	1.20	1.38	12
Aspartic acid	%	1.33	1.52	6
Cystine	%	0.36	0.41	19
Glutamic acid	%	3.11	3.57	6
Glycine	%	0.90	1.04	21
Histidine	%	0.45	0.52	9
Isoleucine	%	0.56	0.64	10
Leucine	%	1.08	1.24	6
Lysine	%	0.81	0.93	9
Methionine	%	0.27	0.31	16
Phenylalanine	%	0.67	0.77	8
Proline	%	0.98	1.12	5
Serine	%	0.77	0.88	5
Threonine	%	0.61	0.70	6
Tryptophan	%	0.19	0.21	
Tyrosine	%	0.49	0.57	6
Valine	%	0.82	0.95	12

Ref No 4 05 201 Canada

Feed Name or Analyses		Mean As Fed	Mean Dry	C.V. ± %
Dry matter	%	90.0	100.0	2
Ash	%	4.3	4.8	
Crude fiber	%	7.3	8.1	22
Ether extract	%	4.9	5.5	25
N-free extract	%	55.6	61.8	
Protein (N x 6.25)	%	17.9	19.9	7
Cattle	dig prot %	* 12.8	14.2	
Goats	dig prot %	* 13.9	15.4	
Horses	dig prot %	* 13.9	15.4	
Sheep	dig prot %	* 13.9	15.4	
Energy	GE Mcal/kg	4.16	4.62	
Cattle	DE Mcal/kg	* 3.29	3.66	
Sheep	DE Mcal/kg	* 3.38	3.75	
Swine	DE kcal/kg	*2948.	3276.	
Cattle	ME Mcal/kg	* 2.70	3.00	
Chickens	ME_n kcal/kg	1818.	2020.	
Sheep	ME Mcal/kg	* 2.77	3.08	
Swine	ME kcal/kg	*2712.	3013.	
Cattle	TDN %	* 74.6	82.9	
Sheep	TDN %	76.6	85.1	
Swine	TDN %	* 66.9	74.3	
Calcium	%	0.22	0.25	
Phosphorus	%	0.96	1.06	
Salt (NaCl)	%	0.14	0.15	

Wheat, flour by-product, fine sift, mx 4% fiber, (4)

Wheat red dog, mx 4% fiber (AAFCO)

Middlings, mx 4.5% fiber (CFA)

Ref No 4 05 203 United States

Feed Name or Analyses		Mean As Fed	Mean Dry	C.V. ± %
Dry matter	%	87.4	100.0	2
Ash	%	2.5	2.8	30
Crude fiber	%	2.5	2.8	33
Sheep	dig coef %	43.	43.	
Swine	dig coef %	-33.	-33.	
Ether extract	%	3.4	3.9	22
Sheep	dig coef %	87.	87.	
Swine	dig coef %	26.	26.	
N-free extract	%	63.5	72.7	5
Sheep	dig coef %	95.	95.	
Swine	dig coef %	91.	91.	
Protein (N x 6.25)	%	15.5	17.7	12
Sheep	dig coef %	81.	81.	
Swine	dig coef %	89.	89.	
Cattle	dig prot %	* 10.7	12.3	
Goats	dig prot %	* 11.8	13.5	
Horses	dig prot %	* 11.8	13.5	
Sheep	dig prot %	12.5	14.3	
Swine	dig prot %	13.8	15.8	
Starch	%	43.6	49.9	10
Energy	GE Mcal/kg			
Cattle	DE Mcal/kg	* 3.45	3.95	
Sheep	DE Mcal/kg	* 3.54	4.05	
Swine	DE kcal/kg	*3208.	3673.	
Cattle	ME Mcal/kg	* 2.83	3.24	
Chickens	ME_n kcal/kg	2633.	3013.	
Sheep	ME Mcal/kg	* 2.90	3.32	
Swine	ME kcal/kg	*2965.	3394.	
Cattle	TDN %	* 78.3	89.6	
Sheep	TDN %	80.2	91.9	
Swine	TDN %	72.8	83.3	
Calcium	%	0.08	0.09	98
Chlorine	%	0.10	0.12	
Cobalt	mg/kg	0.112	0.128	37
Copper	mg/kg	6.3	7.2	30
Iron	%	0.004	0.005	98
Magnesium	%	0.17	0.19	48
Manganese	mg/kg	53.2	60.8	27
Phosphorus	%	0.50	0.57	19
Potassium	%	0.54	0.62	27
Selenium	mg/kg	0.299	0.342	54
Sodium	%	0.12	0.14	98
Sulphur	%	0.23	0.27	
Zinc	mg/kg	63.7	72.9	41
Choline	mg/kg	1560.	1786.	18
Folic acid	mg/kg	0.75	0.86	35
Niacin	mg/kg	44.3	50.7	29
Pantothenic acid	mg/kg	13.2	15.1	19
Riboflavin	mg/kg	2.3	2.6	26
Thiamine	mg/kg	23.1	26.5	22
a-tocopherol	mg/kg	44.0	50.3	
Alanine	%	0.69	0.79	10
Arginine	%	0.93	1.06	12
Aspartic acid	%	0.95	1.09	10
Cystine	%	0.35	0.41	19
Glutamic acid	%	3.97	4.54	13
Glycine	%	0.69	0.79	18
Histidine	%	0.38	0.43	8
Isoleucine	%	0.53	0.60	14
Leucine	%	1.03	1.18	9
Lysine	%	0.56	0.64	10
Methionine	%	0.23	0.26	23
Phenylalanine	%	0.64	0.73	12
Proline	%	1.20	1.37	15
Serine	%	0.74	0.85	14
Threonine	%	0.49	0.56	9
Tryptophan	%	0.21	0.24	17
Tyrosine	%	0.45	0.52	9
Valine	%	0.71	0.81	9

Wheat, flour by-product, mx 9.5% fiber, (4)

Wheat middlings (AAFCO)

Wheat standard middlings

Ref No 4 05 205 United States

Feed Name or Analyses		Mean As Fed	Mean Dry	C.V. ± %
Dry matter	%	88.9	100.0	2
Ash	%	4.3	4.8	21
Crude fiber	%	6.5	7.3	22
Sheep	dig coef %	39.	39.	
Swine	dig coef %	23.	23.	16
Ether extract	%	4.8	5.4	10
Sheep	dig coef %	88.	88.	
Swine	dig coef %	75.	75.	15
N-free extract	%	55.9	62.9	3
Sheep	dig coef %	83.	83.	
Swine	dig coef %	80.	80.	4
Protein (N x 6.25)	%	17.4	19.6	6
Sheep	dig coef %	81.	81.	
Swine	dig coef %	82.	82.	3
Cattle	dig prot %	* 12.5	14.0	
Goats	dig prot %	* 13.5	15.2	
Horses	dig prot %	* 13.5	15.2	
Sheep	dig prot %	14.1	15.9	
Swine	dig prot %	14.2	16.0	
Linoleic	%	1.877	2.111	

Continued

Feed Name or Analyses		Mean As Fed	Mean Dry	C.V. ± %
Energy	GE Mcal/kg	4.16	4.68	
Cattle	DE Mcal/kg	* 3.12	3.51	
Sheep	DE Mcal/kg	* 3.20	3.59	
Swine	DE kcal/kg	*3028.	3406.	
Cattle	ME Mcal/kg	* 2.56	2.88	
Chickens	ME_n kcal/kg	1820.	2048.	
Sheep	ME Mcal/kg	* 2.62	2.95	
Swine	ME kcal/kg	*2787.	3135.	
Cattle	NE_m Mcal/kg	* 1.74	1.96	
Cattle	NE_{gain} Mcal/kg	* 1.16	1.31	
Cattle	$NE_{lactating\ cows}$ Mcal/kg	* 2.04	2.30	
Cattle	TDN %	* 70.8	79.6	
Sheep	TDN %	72.5	81.5	
Swine	TDN %	68.7	77.3	5
Calcium	%	0.10	0.11	7
Chlorine	%	0.03	0.03	18
Cobalt	mg/kg	0.083	0.093	18
Copper	mg/kg	16.0	18.0	11
Iron	%	0.007	0.008	40
Magnesium	%	0.35	0.40	15
Manganese	mg/kg	108.3	121.8	6
Phosphorus	%	0.88	0.99	14
Potassium	%	0.99	1.11	7
Sodium	%	0.18	0.20	23
Sulphur	%	0.16	0.19	9
Carotene	mg/kg	3.1	3.4	22
Choline	mg/kg	1154.	1298.	12
Folic acid	mg/kg	0.97	1.10	73
Niacin	mg/kg	98.6	110.9	21
Pantothenic acid	mg/kg	16.6	18.6	31
Riboflavin	mg/kg	2.0	2.2	24
Thiamine	mg/kg	16.5	18.6	52
α-tocopherol	mg/kg	20.8	23.4	11
Vitamin A equivalent	IU/g	5.1	5.7	
Arginine	%	0.90	1.01	22
Cystine	%	0.19	0.22	
Glutamic acid	%	4.06	4.57	6
Glycine	%	0.40	0.45	
Histidine	%	0.40	0.45	27
Isoleucine	%	0.79	0.89	25
Leucine	%	1.19	1.34	18
Lysine	%	0.70	0.79	24
Methionine	%	0.18	0.20	35
Phenylalanine	%	0.69	0.78	14
Serine	%	0.79	0.89	
Threonine	%	0.59	0.67	6
Tryptophan	%	0.19	0.22	17
Tyrosine	%	0.40	0.45	25
Valine	%	0.79	0.89	21

Ref No 4 05 205 Canada

Feed Name or Analyses		Mean As Fed	Mean Dry	C.V. ± %
Dry matter	%	90.0	100.0	
Ash	%	2.7	3.1	
Crude fiber	%	4.3	4.8	
Sheep	dig coef %	39.	39.	
Swine	dig coef %	23.	23.	
Ether extract	%	4.2	4.6	
Sheep	dig coef %	88.	88.	
Swine	dig coef %	75.	75.	
N-free extract	%	61.7	68.5	
Sheep	dig coef %	83.	83.	
Swine	dig coef %	80.	80.	
Protein (N x 6.25)	%	17.2	19.1	
Sheep	dig coef %	81.	81.	
Swine	dig coef %	82.	82.	
Cattle	dig prot %	* 12.2	13.5	
Goats	dig prot %	* 13.2	14.7	
Horses	dig prot %	* 13.2	14.7	
Sheep	dig prot %	13.9	15.5	
Swine	dig prot %	14.0	15.6	
Energy	GE Mcal/kg	4.21	4.68	
Cattle	DE Mcal/kg	* 3.18	3.53	
Sheep	DE Mcal/kg	* 3.30	3.67	
Swine	DE kcal/kg	*3154.	3503.	
Cattle	ME Mcal/kg	* 2.60	2.89	
Chickens	ME_n kcal/kg	2453.	2725.	
Sheep	ME Mcal/kg	* 2.71	3.01	
Swine	ME kcal/kg	*2906.	3228.	
Cattle	TDN %	* 72.1	80.1	
Sheep	TDN %	75.0	83.3	
Swine	TDN %	71.5	79.4	
Calcium	%	0.12	0.14	
Phosphorus	%	0.60	0.67	
Salt (NaCl)	%	0.20	0.23	

Wheat, flour by-product mill run, mx 9.5% fiber, (4)

Wheat mill run (AAFCO)

Ref No 4 05 206 United States

Feed Name or Analyses		Mean As Fed	Mean Dry	C.V. ± %
Dry matter	%	90.1	100.0	2
Ash	%	5.1	5.7	10
Crude fiber	%	8.1	9.0	13
Sheep	dig coef %	* 34.	34.	
Ether extract	%	4.3	4.7	14
Sheep	dig coef %	* 86.	86.	
N-free extract	%	57.0	63.3	
Sheep	dig coef %	* 78.	78.	
Protein (N x 6.25)	%	15.6	17.3	9
Sheep	dig coef %	* 83.	83.	
Cattle	dig prot %	* 10.7	11.9	
Goats	dig prot %	* 11.8	13.1	
Horses	dig prot %	* 11.8	13.1	
Sheep	dig prot %	12.9	14.3	
Energy	GE Mcal/kg	3.97	4.40	
Cattle	DE Mcal/kg	* 2.94	3.26	
Sheep	DE Mcal/kg	* 3.02	3.35	
Swine	DE kcal/kg	*2734.	3035.	
Cattle	ME Mcal/kg	* 2.41	2.67	
Chickens	ME_n kcal/kg	1752.	1945.	
Sheep	ME Mcal/kg	* 2.47	2.75	
Swine	ME kcal/kg	*2529.	2808.	
Cattle	NE_m Mcal/kg	* 1.70	1.89	
Cattle	NE_{gain} Mcal/kg	* 1.14	1.26	
Cattle	$NE_{lactating\ cows}$ Mcal/kg	* 2.01	2.23	
Cattle	TDN %	* 66.6	73.9	
Sheep	TDN %	68.4	76.0	
Swine	TDN %	* 62.0	68.8	
Calcium	%	0.10	0.11	24
Cobalt	mg/kg	0.225	0.250	
Copper	mg/kg	18.7	20.8	
Iron	%	0.010	0.011	
Magnesium	%	0.51	0.56	
Manganese	mg/kg	102.7	114.0	30
Phosphorus	%	1.06	1.18	8
Potassium	%	1.28	1.42	
Sodium	%	0.22	0.24	
Choline	mg/kg	1029.	1142.	6
Niacin	mg/kg	132.8	147.5	11
Pantothenic acid	mg/kg	18.7	20.7	
Riboflavin	mg/kg	2.4	2.7	11
Thiamine	mg/kg	15.2	16.9	
Arginine	%	0.89	0.99	
Cysteine	%	0.40	0.44	
Cystine	%	0.20	0.22	
Glycine	%	0.60	0.66	
Lysine	%	0.60	0.66	
Methionine	%	0.28	0.31	
Tryptophan	%	0.20	0.22	

Wheat, graham flour, coarse sift, (4)

Ref No 4 05 208 United States

Feed Name or Analyses		Mean As Fed	Mean Dry	C.V. ± %
Dry matter	%	88.1	100.0	0
Ash	%	1.5	1.7	6
Crude fiber	%	1.8	2.1	16
Ether extract	%	1.9	2.2	8
N-free extract	%	70.3	79.7	
Protein (N x 6.25)	%	12.6	14.3	4
Cattle	dig prot %	* 8.1	9.1	
Goats	dig prot %	* 9.1	10.3	
Horses	dig prot %	* 9.1	10.3	
Sheep	dig prot %	* 9.1	10.3	
Energy	GE Mcal/kg			
Cattle	DE Mcal/kg	* 3.23	3.67	
Sheep	DE Mcal/kg	* 3.28	3.72	
Swine	DE kcal/kg	*3614.	4102.	
Cattle	ME Mcal/kg	* 2.65	3.01	
Sheep	ME Mcal/kg	* 2.69	3.05	
Swine	ME kcal/kg	*3365.	3820.	
Cattle	TDN %	* 73.3	83.2	
Sheep	TDN %	* 74.3	84.4	
Swine	TDN %	* 82.0	93.0	
Calcium	%	0.04	0.05	
Copper	mg/kg	19.4	22.0	
Iron	%	0.009	0.010	
Manganese	mg/kg	16.1	18.3	
Phosphorus	%	0.38	0.43	
Potassium	%	0.46	0.52	

Wheat, grain, (4)

Ref No 4 05 211 United States

Feed Name or Analyses		Mean As Fed	Mean Dry	C.V. ± %
Dry matter	%	88.9	100.0	1
Ash	%	1.8	2.1	6
Crude fiber	%	2.5	2.8	12
Sheep	dig coef %	33.	33.	
Swine	dig coef %	19.	19.	
Ether extract	%	1.9	2.1	6
Sheep	dig coef %	72.	72.	
Swine	dig coef %	74.	74.	
N-free extract	%	70.8	79.6	2
Sheep	dig coef %	92.	92.	
Swine	dig coef %	90.	90.	
Protein (N x 6.25)	%	11.9	13.4	16
Cattle	dig coef %	78.	78.	
Sheep	dig coef %	78.	78.	
Swine	dig coef %	88.	88.	
Cattle	dig prot %	9.3	10.5	
Goats	dig prot %	* 8.5	9.5	
Horses	dig prot %	* 8.5	9.5	
Sheep	dig prot %	9.3	10.5	
Swine	dig prot %	10.5	11.8	
Starch	%	68.5	77.0	
Energy	GE Mcal/kg			
Cattle	DE Mcal/kg	3.45	3.88	
Sheep	DE Mcal/kg	3.45	3.88	
Swine	DE kcal/kg	3607.	4056.	
Cattle	ME Mcal/kg	2.83	3.18	
Chickens	ME_n kcal/kg	3012.	3388.	
Sheep	ME Mcal/kg	2.83	3.18	

(1) dry forages and roughages (2) pasture, range plants, and forages fed green (3) silages (4) energy feeds (5) protein supplements (6) minerals (7) vitamins (8) additives

Feed Name or Analyses		Mean As Fed	Mean Dry	C.V. ± %
Swine	ME kcal/kg	3378.	3798.	
Cattle	NEm Mcal/kg	* 1.91	2.15	
Cattle	NEgain Mcal/kg	* 1.26	1.42	
Cattle	NE lactating cows Mcal/kg	* 2.21	2.49	
Cattle	TDN %	78.3	88.0	
Sheep	TDN %	78.3	88.0	
Swine	TDN %	77.8	87.5	
Calcium	%	0.08	0.09	
Chlorine	%	0.08	0.09	
Copper	mg/kg	8.1	9.1	
Iron	%	0.006	0.007	
Magnesium	%	0.14	0.16	
Manganese	mg/kg	43.6	49.0	
Phosphorus	%	0.34	0.39	
Potassium	%	0.42	0.47	
Sodium	%	0.06	0.07	
Sulphur	%	0.20	0.22	
Biotin	mg/kg	0.10	0.11	84
Folic acid	mg/kg	0.43	0.49	
Niacin	mg/kg	59.0	66.3	
Pantothenic acid	mg/kg	11.5	12.9	
Riboflavin	mg/kg	1.1	1.2	
Thiamine	mg/kg	4.9	5.5	
α-tocopherol	mg/kg	19.8	22.3	27
Alanine	%	0.62	0.70	
Arginine	%	0.71	0.80	23
Aspartic acid	%	0.18	0.20	
Cysteine	%	0.27	0.30	
Glycine	%	0.89	1.00	
Histidine	%	0.27	0.30	36
Isoleucine	%	0.53	0.60	32
Leucine	%	0.89	1.00	25
Lysine	%	0.44	0.50	32
Phenylalanine	%	0.62	0.70	22
Proline	%	1.25	1.40	
Threonine	%	0.36	0.40	18

Ref No 4 05 211 Canada

Feed Name or Analyses		Mean As Fed	Mean Dry	C.V. ± %
Dry matter	%	85.1	100.0	9
Crude fiber	%			
Sheep	dig coef %	33.	33.	
Swine	dig coef %	19.	19.	
Ether extract	%			
Sheep	dig coef %	72.	72.	
Swine	dig coef %	74.	74.	
N-free extract	%			
Sheep	dig coef %	92.	92.	
Swine	dig coef %	90.	90.	
Protein (N x 6.25)	%	12.6	14.8	
Cattle	dig coef %	78.	78.	
Sheep	dig coef %	78.	78.	
Swine	dig coef %	88.	88.	
Cattle	dig prot %	9.8	11.6	
Goats	dig prot %	* 9.2	10.8	
Horses	dig prot %	* 9.2	10.8	
Sheep	dig prot %	9.8	11.6	
Swine	dig prot %	11.1	13.1	
Energy	GE Mcal/kg			
Cattle	DE Mcal/kg	3.30	3.88	
Sheep	DE Mcal/kg	3.30	3.88	
Swine	DE kcal/kg	3450.	4056.	
Cattle	ME Mcal/kg	2.71	3.18	
Sheep	ME Mcal/kg	2.71	3.18	
Swine	ME kcal/kg	3231.	3798.	
Cattle	TDN %	74.9	88.0	
Sheep	TDN %	74.8	88.0	
Swine	TDN %	74.4	87.5	
Calcium	%	0.22	0.26	
Phosphorus	%	0.28	0.33	

Feed Name or Analyses		Mean As Fed	Mean Dry	C.V. ± %
Alanine	%	0.51	0.60	
Arginine	%	0.62	0.73	
Aspartic acid	%	0.69	0.82	
Cystine	%	0.31	0.36	
Glutamic acid	%	3.31	3.89	
Glycine	%	0.58	0.69	
Histidine	%	0.33	0.39	
Isoleucine	%	0.51	0.60	
Leucine	%	0.97	1.14	
Lysine	%	0.38	0.45	
Methionine	%	0.16	0.18	
Phenylalanine	%	0.66	0.78	
Proline	%	1.41	1.66	
Serine	%	0.65	0.76	
Threonine	%	0.42	0.49	
Tyrosine	%	0.31	0.36	
Valine	%	0.63	0.74	

Wheat, grain, cooked, (4)

Ref No 5 05 209 United States

Feed Name or Analyses		Mean As Fed	Mean Dry	C.V. ± %
Dry matter	%	87.7	100.0	
Ash	%	2.1	2.4	
Crude fiber	%	3.6	4.1	
Sheep	dig coef %	40.	40.	
Ether extract	%	2.4	2.7	
Sheep	dig coef %	78.	78.	
N-free extract	%	61.2	69.8	
Sheep	dig coef %	84.	84.	
Protein (N x 6.25)	%	18.4	21.0	
Sheep	dig coef %	80.	80.	
Cattle	dig prot %	* 13.4	15.3	
Goats	dig prot %	* 14.4	16.5	
Horses	dig prot %	* 14.4	16.5	
Sheep	dig prot %	14.7	16.8	
Energy	GE Mcal/kg			
Cattle	DE Mcal/kg	* 3.11	3.55	
Sheep	DE Mcal/kg	* 3.16	3.61	
Swine	DE kcal/kg	*3900.	4447.	
Cattle	ME Mcal/kg	* 2.55	2.91	
Sheep	ME Mcal/kg	* 2.59	2.96	
Swine	ME kcal/kg	*3578.	4080.	
Cattle	TDN %	* 70.7	80.6	
Sheep	TDN %	71.7	81.8	
Swine	TDN %	* 88.4	100.8	

Wheat, grain, flaked, (4)

Ref No 4 08 319 United States

Feed Name or Analyses		Mean As Fed	Mean Dry	C.V. ± %
Dry matter	%	99.8	100.0	
Ash	%	3.6	3.6	
Crude fiber	%	3.6	3.6	
Ether extract	%	1.6	1.6	
N-free extract	%	80.2	80.4	
Protein (N x 6.25)	%	10.8	10.8	
Cattle	dig prot %	* 5.9	6.0	
Goats	dig prot %	* 7.1	7.2	
Horses	dig prot %	* 7.1	7.2	
Sheep	dig prot %	* 7.1	7.2	
Energy	GE Mcal/kg			
Cattle	DE Mcal/kg	* 3.49	3.50	
Sheep	DE Mcal/kg	* 3.67	3.68	
Swine	DE kcal/kg	*3673.	3680.	
Cattle	ME Mcal/kg	* 2.86	2.87	
Chickens	MEn kcal/kg	3527.	3534.	
Sheep	ME Mcal/kg	* 3.01	3.01	
Swine	ME kcal/kg	*3445.	3452.	
Cattle	TDN %	* 79.2	79.4	

Feed Name or Analyses		Mean As Fed	Mean Dry	C.V. ± %
Sheep	TDN %	* 83.2	83.4	
Swine	TDN %	* 83.3	83.5	
Calcium	%	0.01	0.01	
Phosphorus	%	0.30	0.30	
Choline	mg/kg	441.	442.	
Niacin	mg/kg	48.1	48.2	
Pantothenic acid	mg/kg	4.4	4.4	
Riboflavin	mg/kg	1.8	1.8	
Arginine	%	0.60	0.60	
Cystine	%	0.10	0.10	
Lysine	%	0.40	0.40	
Methionine	%	0.10	0.10	
Tryptophan	%	0.15	0.15	

Wheat, grain, grnd, (4)

Ref No 4 09 074 Canada

Feed Name or Analyses		Mean As Fed	Mean Dry	C.V. ± %
Dry matter	%	89.0	100.0	
Ash	%	1.6	1.8	
Crude fiber	%	2.6	2.9	
Ether extract	%	1.8	2.0	
N-free extract	%	67.6	76.0	
Protein (N x 6.25)	%	15.4	17.3	
Cattle	dig prot %	* 10.6	11.9	
Goats	dig prot %	* 11.6	13.1	
Horses	dig prot %	* 11.6	13.1	
Sheep	dig prot %	* 11.6	13.1	
Energy	GE Mcal/kg			
Cattle	DE Mcal/kg	* 3.13	3.52	
Sheep	DE Mcal/kg	* 3.24	3.64	
Swine	DE kcal/kg	*3603.	4048.	
Cattle	ME Mcal/kg	* 2.57	2.89	
Sheep	ME Mcal/kg	* 2.66	2.98	
Swine	ME kcal/kg	*3333.	3745.	
Cattle	TDN %	* 71.0	79.8	
Sheep	TDN %	* 73.4	82.5	
Swine	TDN %	* 81.7	91.8	

Wheat, grain, thresher-run, mn wt 60 lb per bushel mx 5% foreign material, (4)

Ref No 4 08 164 Canada

Feed Name or Analyses		Mean As Fed	Mean Dry	C.V. ± %
Dry matter	%	88.0	100.0	0
Ash	%	1.8	2.0	4
Crude fiber	%	2.2	2.5	8
Ether extract	%	1.4	1.6	13
N-free extract	%	69.0	78.5	2
Protein (N x 6.25)	%	13.6	15.4	8
Cattle	dig prot %	* 8.9	10.2	
Goats	dig prot %	* 10.0	11.4	
Horses	dig prot %	* 10.0	11.4	
Sheep	dig prot %	* 10.0	11.4	
Energy	GE Mcal/kg	4.17	4.74	1
Cattle	DE Mcal/kg	* 3.12	3.55	
Sheep	DE Mcal/kg	* 3.23	3.67	
Swine	DE kcal/kg	*3547.	4031.	
Cattle	ME Mcal/kg	* 2.56	2.91	
Sheep	ME Mcal/kg	* 2.65	3.01	
Swine	ME kcal/kg	*3295.	3744.	
Cattle	TDN %	* 70.9	80.6	
Sheep	TDN %	* 73.2	83.2	
Swine	TDN %	* 80.4	91.4	

Feed Name or Analyses			Mean As Fed	Mean Dry	C.V. ± %

Wheat, grain, thresher-run, mn wt 55 lb mx wt 60 lb per bushel mx 5% foreign material, (4)

Ref No 4 08 165 Canada

Analyses			As Fed	Dry	C.V. ± %
Dry matter	%		88.0	100.0	
Ash	%		1.6	1.8	
Crude fiber	%		2.6	2.9	
Ether extract	%		1.2	1.4	
N-free extract	%		66.8	75.9	
Protein (N x 6.25)	%		15.8	18.0	
Cattle	dig prot %	*	11.0	12.5	
Goats	dig prot %	*	12.1	13.7	
Horses	dig prot %	*	12.1	13.7	
Sheep	dig prot %	*	12.1	13.7	
Energy	GE Mcal/kg		4.22	4.80	
Cattle	DE Mcal/kg	*	3.16	3.59	
Sheep	DE Mcal/kg	*	3.18	3.62	
Swine	DE kcal/kg	*	3569.	4056.	
Cattle	ME Mcal/kg	*	2.59	2.94	
Sheep	ME Mcal/kg	*	2.61	2.97	
Swine	ME kcal/kg	*	3297.	3746.	
Cattle	TDN %	*	71.7	81.5	
Sheep	TDN %	*	72.2	82.0	
Swine	TDN %	*	80.9	92.0	

Wheat, grain, germinated, (4)

Ref No 4 05 210 United States

Analyses			As Fed	Dry	C.V. ± %
Dry matter	%			100.0	
Biotin	mg/kg			0.35	
Niacin	mg/kg			103.0	
Pantothenic acid	mg/kg			12.6	
Riboflavin	mg/kg			5.3	
Thiamine	mg/kg			9.0	
Vitamin B6	mg/kg			4.63	

Wheat, grain, Can feed gr mx 13% foreign material, (4)

Ref No 4 08 175 Canada

Analyses			As Fed	Dry	C.V. ± %
Dry matter	%		87.6	100.0	2
Swine	dig coef %		83.	83.	
Ash	%		1.8	2.0	8
Crude fiber	%		4.0	4.6	84
Swine	dig coef %		16.	16.	
Ether extract	%		2.0	2.3	37
Swine	dig coef %		40.	40.	
N-free extract	%		65.5	74.8	
Swine	dig coef %		91.	91.	
Protein (N x 6.25)	%		14.3	16.3	11
Swine	dig coef %		78.	78.	
Cattle	dig prot %	*	9.6	11.0	
Goats	dig prot %	*	10.7	12.2	
Horses	dig prot %	*	10.7	12.2	
Sheep	dig prot %	*	10.7	12.2	
Swine	dig prot %		11.1	12.6	
Cellulose (Matrone)	%		2.8	3.2	
Energy	GE Mcal/kg		3.95	4.51	
Swine	GE dig coef %		82.	82.	
Cattle	DE Mcal/kg	*	3.09	3.53	
Sheep	DE Mcal/kg	*	3.16	3.61	
Swine	DE kcal/kg		3220.	3677.	
Cattle	ME Mcal/kg	*	2.53	2.89	
Sheep	ME Mcal/kg	*	2.59	2.96	
Swine	ME kcal/kg	*	2985.	3409.	
Cattle	TDN %	*	70.2	80.1	
Sheep	TDN %	*	71.6	81.8	
Swine	TDN %		73.1	83.5	
Calcium	%		0.04	0.05	
Copper	mg/kg		5.3	6.1	
Iron	%		0.004	0.004	
Magnesium	%		0.20	0.23	
Manganese	mg/kg		24.6	28.0	
Phosphorus	%		0.39	0.45	
Potassium	%		0.18	0.20	
Selenium	mg/kg		0.330	0.377	
Zinc	mg/kg		39.0	44.6	
Niacin	mg/kg		73.8	84.3	
Pantothenic acid	mg/kg		11.4	13.0	
Riboflavin	mg/kg		1.8	2.0	
Thiamine	mg/kg		0.6	0.7	
Vitamin B6	mg/kg		3.22	3.68	
Alanine	%		0.50	0.57	
Arginine	%		0.54	0.61	
Aspartic acid	%		0.72	0.82	
Glutamic acid	%		3.43	3.91	
Glycine	%		0.51	0.58	
Histidine	%		0.23	0.27	
Isoleucine	%		0.28	0.32	
Leucine	%		0.74	0.84	
Lysine	%		0.34	0.39	
Methionine	%		0.17	0.19	
Phenylalanine	%		0.53	0.60	
Proline	%		1.12	1.28	
Serine	%		0.59	0.68	
Threonine	%		0.36	0.41	
Tyrosine	%		0.30	0.34	
Valine	%		0.37	0.42	

Wheat, grain, Can 1 CW mn wt 60 lb per bushel mx 1% foreign material, (4)

Ref No 4 08 173 Canada

Analyses			As Fed	Dry	C.V. ± %
Dry matter	%		91.0	100.0	
Ash	%		1.7	1.9	
Crude fiber	%		2.5	2.7	
Ether extract	%		2.3	2.5	
N-free extract	%		67.1	73.8	
Protein (N x 6.25)	%		17.3	19.1	
Cattle	dig prot %	*	12.3	13.5	
Goats	dig prot %	*	13.4	14.7	
Horses	dig prot %	*	13.4	14.7	
Sheep	dig prot %	*	13.4	14.7	
Cellulose (Matrone)	%		2.5	2.7	
Energy	GE Mcal/kg				
Cattle	DE Mcal/kg	*	3.30	3.63	
Sheep	DE Mcal/kg	*	3.29	3.62	
Swine	DE kcal/kg	*	3728.	4096.	
Cattle	ME Mcal/kg	*	2.71	2.98	
Sheep	ME Mcal/kg	*	2.70	2.97	
Swine	ME kcal/kg	*	3435.	3775.	
Cattle	TDN %	*	74.9	82.3	
Sheep	TDN %	*	74.6	82.0	
Swine	TDN %	*	84.5	92.9	
Copper	mg/kg		6.3	6.9	
Iron	%		0.003	0.004	
Magnesium	%		0.16	0.18	
Manganese	mg/kg		35.8	39.3	
Selenium	mg/kg		1.200	1.319	
Zinc	mg/kg		33.5	36.8	
Niacin	mg/kg		62.0	68.1	
Pantothenic acid	mg/kg		11.5	12.6	
Riboflavin	mg/kg		1.2	1.3	
Thiamine	mg/kg		0.9	1.0	
Vitamin B6	mg/kg		2.87	3.15	
Alanine	%		0.54	0.59	
Arginine	%		0.67	0.74	
Aspartic acid	%		0.75	0.82	
Glutamic acid	%		4.83	5.31	
Glycine	%		0.62	0.68	
Histidine	%		0.34	0.37	
Isoleucine	%		0.48	0.53	
Leucine	%		0.98	1.08	
Lysine	%		0.45	0.49	
Methionine	%		0.17	0.19	
Phenylalanine	%		0.72	0.79	
Proline	%		1.57	1.73	
Serine	%		0.59	0.65	
Threonine	%		0.38	0.42	
Tyrosine	%		0.41	0.45	
Valine	%		0.60	0.66	

Wheat, grain, Can 5 CW mn wt 53 lb per bushel mx 3% foreign material, (4)

Ref No 4 08 177 Canada

Analyses			As Fed	Dry	C.V. ± %
Dry matter	%		91.2	100.0	
Ash	%		1.7	1.9	
Crude fiber	%		2.7	3.0	
Ether extract	%		2.4	2.6	
N-free extract	%		68.6	75.2	
Protein (N x 6.25)	%		15.8	17.3	
Cattle	dig prot %	*	10.9	11.9	
Goats	dig prot %	*	12.0	13.1	
Horses	dig prot %	*	12.0	13.1	
Sheep	dig prot %	*	12.0	13.1	
Cellulose (Matrone)	%		2.6	2.8	
Energy	GE Mcal/kg				
Cattle	DE Mcal/kg	*	3.36	3.68	
Sheep	DE Mcal/kg	*	3.32	3.64	
Swine	DE kcal/kg	*	3690.	4049.	
Cattle	ME Mcal/kg	*	2.75	3.02	
Sheep	ME Mcal/kg	*	2.72	2.99	
Swine	ME kcal/kg	*	3414.	3745.	
Cattle	TDN %	*	76.1	83.4	
Sheep	TDN %	*	75.3	82.6	
Swine	TDN %	*	83.7	91.8	
Copper	mg/kg		6.3	6.9	
Iron	%		0.004	0.005	
Magnesium	%		0.17	0.18	
Manganese	mg/kg		30.7	33.7	
Selenium	mg/kg		0.399	0.438	
Zinc	mg/kg		37.3	41.0	
Niacin	mg/kg		45.9	50.4	
Pantothenic acid	mg/kg		11.5	12.6	
Riboflavin	mg/kg		1.3	1.5	
Thiamine	mg/kg		0.8	0.9	
Vitamin B6	mg/kg		2.80	3.07	
Alanine	%		0.52	0.57	
Arginine	%		0.67	0.74	
Aspartic acid	%		0.76	0.84	
Glutamic acid	%		4.94	5.42	
Glycine	%		0.60	0.66	
Histidine	%		0.33	0.36	
Isoleucine	%		0.56	0.62	
Leucine	%		1.03	1.13	
Lysine	%		0.40	0.44	
Methionine	%		0.19	0.21	
Phenylalanine	%		0.73	0.80	
Proline	%		1.45	1.59	

(1) dry forages and roughages (2) pasture, range plants, and forages fed green (3) silages (4) energy feeds (5) protein supplements (6) minerals (7) vitamins (8) additives

Feed Name or Analyses		Mean As Fed	Dry	C.V. ± %
Serine	%	0.65	0.71	
Threonine	%	0.42	0.46	
Tyrosine	%	0.39	0.43	
Valine	%	0.68	0.75	

Wheat, grain, Can 5 mx 0.33% foreign material, (4)
Ref No 4 08 179 Canada

Feed Name or Analyses		Mean As Fed	Dry	C.V. ± %
Dry matter	%	86.8	100.0	1
Ash	%	1.5	1.8	7
Crude fiber	%	2.4	2.8	25
Ether extract	%	1.8	2.1	16
N-free extract	%	67.0	77.2	
Protein (N x 6.25)	%	14.0	16.1	4
Cattle	dig prot %	* 9.4	10.8	
Goats	dig prot %	* 10.4	12.0	
Horses	dig prot %	* 10.4	12.0	
Sheep	dig prot %	* 10.4	12.0	
Energy	GE Mcal/kg			
Cattle	DE Mcal/kg	* 3.12	3.60	
Sheep	DE Mcal/kg	* 3.18	3.67	
Swine	DE kcal/kg	*3514.	4050.	
Cattle	ME Mcal/kg	* 2.56	2.95	
Sheep	ME Mcal/kg	* 2.61	3.01	
Swine	ME kcal/kg	*3259.	3756.	
Cattle	TDN %	* 70.8	81.6	
Sheep	TDN %	* 72.1	83.1	
Swine	TDN %	* 79.7	91.9	
Calcium	%	0.06	0.07	
Phosphorus	%	0.27	0.31	

Wheat, grain, Can 6 CW mn wt 51 lb per bushel mx 3% foreign material, (4)
Ref No 4 08 174 Canada

Feed Name or Analyses		Mean As Fed	Dry	C.V. ± %
Dry matter	%	90.1	100.0	
Ash	%	1.8	2.0	
Crude fiber	%	2.8	3.1	
Ether extract	%	2.5	2.8	
N-free extract	%	67.7	75.2	
Protein (N x 6.25)	%	15.3	16.9	
Cattle	dig prot %	* 10.4	11.6	
Goats	dig prot %	* 11.5	12.8	
Horses	dig prot %	* 11.5	12.8	
Sheep	dig prot %	* 11.5	12.8	
Cellulose (Matrone)	%	2.5	2.8	
Energy	GE Mcal/kg			
Cattle	DE Mcal/kg	* 3.24	3.60	
Sheep	DE Mcal/kg	* 3.28	3.64	
Swine	DE kcal/kg	*3615.	4014.	
Cattle	ME Mcal/kg	* 2.66	2.95	
Sheep	ME Mcal/kg	* 2.69	2.99	
Swine	ME kcal/kg	*3346.	3716.	
Cattle	TDN %	* 22.2	80.1	
Sheep	TDN %	* 74.4	82.7	
Swine	TDN %	* 82.0	91.0	
Copper	mg/kg	4.9	5.4	
Iron	%	0.004	0.004	
Magnesium	%	0.17	0.19	
Manganese	mg/kg	27.6	30.6	
Selenium	mg/kg	0.400	0.444	
Zinc	mg/kg	42.3	46.9	
Niacin	mg/kg	62.0	68.8	
Pantothenic acid	mg/kg	12.2	13.5	
Riboflavin	mg/kg	1.1	1.2	
Thiamine	mg/kg	0.7	0.8	
Vitamin B6	mg/kg	3.00	3.33	
Alanine	%	0.54	0.60	
Arginine	%	0.68	0.76	
Aspartic acid	%	0.76	0.84	
Glutamic acid	%	4.54	5.04	
Glycine	%	0.60	0.67	
Histidine	%	0.33	0.37	
Isoleucine	%	0.50	0.56	
Leucine	%	1.00	1.11	
Lysine	%	0.41	0.46	
Methionine	%	0.18	0.20	
Phenylalanine	%	0.72	0.80	
Proline	%	1.51	1.68	
Serine	%	0.74	0.82	
Threonine	%	0.44	0.49	
Tyrosine	%	0.39	0.43	
Valine	%	0.61	0.68	

Wheat, grain, Can 6 mx 0.5% foreign material, (4)
Ref No 4 08 180 Canada

Feed Name or Analyses		Mean As Fed	Dry	C.V. ± %
Dry matter	%	86.5	100.0	0
Ash	%	1.6	1.8	5
Crude fiber	%	2.5	2.9	31
Ether extract	%	1.7	2.0	13
N-free extract	%	66.9	77.3	
Protein (N x 6.25)	%	13.8	15.9	5
Cattle	dig prot %	* 9.2	10.6	
Goats	dig prot %	* 10.2	11.8	
Horses	dig prot %	* 10.2	11.8	
Sheep	dig prot %	* 10.2	11.8	
Energy	GE Mcal/kg			
Cattle	DE Mcal/kg	* 3.14	3.63	
Sheep	DE Mcal/kg	* 3.17	3.66	
Cattle	ME Mcal/kg	*3486.	4030.	
Cattle	ME Mcal/kg	* 2.58	2.98	
Sheep	ME Mcal/kg	* 2.60	3.00	
Swine	ME kcal/kg	*3234.	3739.	
Cattle	TDN %	* 71.2	82.3	
Sheep	TDN %	* 71.9	83.1	
Swine	TDN %	* 79.1	91.4	
Calcium	%	0.10	0.12	
Phosphorus	%	0.25	0.29	

Wheat, grain, high fiber, (4)
Ref No 4 05 213 United States

Feed Name or Analyses		Mean As Fed	Dry	C.V. ± %
Dry matter	%	87.0	100.0	
Ash	%	2.2	2.5	
Crude fiber	%	9.4	10.8	
Swine	dig coef %	76.	76.	
Ether extract	%	2.0	2.3	
Swine	dig coef %	99.	99.	
N-free extract	%	58.8	67.6	
Swine	dig coef %	93.	93.	
Protein (N x 6.25)	%	14.6	16.8	
Swine	dig coef %	94.	94.	
Cattle	dig prot %	* 10.0	11.4	
Goats	dig prot %	* 11.0	12.6	
Horses	dig prot %	* 11.0	12.6	
Sheep	dig prot %	* 11.0	12.6	
Swine	dig prot %	13.7	15.8	
Energy	GE Mcal/kg			
Cattle	DE Mcal/kg	* 2.89	3.32	
Sheep	DE Mcal/kg	* 2.96	3.40	
Swine	DE kcal/kg	*3529.	4056.	
Cattle	ME Mcal/kg	* 2.37	2.72	
Sheep	ME Mcal/kg	* 2.43	2.79	
Swine	ME kcal/kg	*3268.	3756.	
Cattle	TDN %	* 65.5	75.3	
Sheep	TDN %	* 67.2	77.2	
Swine	TDN %	80.0	92.0	

Wheat, grain screenings, (4)
Ref No 4 05 216 United States

Feed Name or Analyses		Mean As Fed	Dry	C.V. ± %
Dry matter	%	88.8	100.0	2
Ash	%	4.0	4.5	29
Crude fiber	%	7.3	8.2	58
Sheep	dig coef %	6.	6.	
Swine	dig coef %	2.	2.	
Ether extract	%	3.3	3.8	22
Sheep	dig coef %	88.	88.	
Swine	dig coef %	27.	27.	
N-free extract	%	59.6	67.2	
Sheep	dig coef %	84.	84.	
Swine	dig coef %	80.	80.	
Protein (N x 6.25)	%	14.5	16.3	15
Sheep	dig coef %	72.	72.	
Swine	dig coef %	80.	80.	
Cattle	dig prot %	* 9.8	11.0	
Goats	dig prot %	* 10.8	12.2	
Horses	dig prot %	* 10.8	12.2	
Sheep	dig prot %	10.4	11.8	
Swine	dig prot %	11.6	13.1	
Cellulose (Matrone)	%	4.6	5.2	
Lignin (Ellis)	%	6.6	7.4	
Energy	GE Mcal/kg			
Cattle	DE Mcal/kg	* 2.53	2.86	
Sheep	DE Mcal/kg	* 2.98	3.36	
Swine	DE kcal/kg	*2710.	3054.	
Cattle	ME Mcal/kg	* 2.08	2.34	
Sheep	ME Mcal/kg	* 2.44	2.75	
Swine	ME kcal/kg	*2512.	2831.	
Cattle	NE_m Mcal/kg	* 1.56	1.76	
Cattle	NE_{gain} Mcal/kg	* 1.04	1.17	
Cattle	$NE_{lactating\ cows}$ Mcal/kg	* 1.86	2.09	
Cattle	TDN %	* 57.5	64.8	
Sheep	TDN %	67.6	76.1	
Swine	TDN %	61.5	69.3	
Calcium	%	0.26	0.29	
Manganese	mg/kg	28.5	32.1	
Phosphorus	%	0.37	0.42	
Thiamine	mg/kg	6.4	7.2	

Ref No 4 05 216 Canada

Feed Name or Analyses		Mean As Fed	Dry	C.V. ± %
Dry matter	%	88.7	100.0	4
Ash	%	11.3	12.7	
Crude fiber	%	8.4	9.4	
Sheep	dig coef %	6.	6.	
Swine	dig coef %	2.	2.	
Ether extract	%	3.8	4.3	
Sheep	dig coef %	88.	88.	
Swine	dig coef %	27.	27.	
N-free extract	%	51.3	57.9	
Sheep	dig coef %	84.	84.	
Swine	dig coef %	80.	80.	
Protein (N x 6.25)	%	13.9	15.6	18
Sheep	dig coef %	72.	72.	
Swine	dig coef %	80.	80.	
Cattle	dig prot %	* 9.2	10.4	
Goats	dig prot %	* 10.3	11.6	
Horses	dig prot %	* 10.3	11.6	
Sheep	dig prot %	10.0	11.3	
Swine	dig prot %	11.1	12.5	
Energy	GE Mcal/kg			
Cattle	DE Mcal/kg	* 2.91	3.29	

Continued

Feed Name or Analyses		Mean As Fed	Mean Dry	C.V. ± %
Sheep	DE Mcal/kg	* 2.70	3.04	
Swine	DE kcal/kg	*2410.	2717.	
Cattle	ME Mcal/kg	* 2.39	2.69	
Sheep	ME Mcal/kg	* 2.21	2.49	
Swine	ME kcal/kg	*2237.	2522.	
Cattle	TDN %	* 66.1	74.5	
Sheep	TDN %	61.2	69.0	
Swine	TDN %	54.6	61.6	
Calcium	%	0.11	0.12	
Phosphorus	%	0.36	0.41	

Wheat, grits, cracked fine screened, (4)
Farina
Wheat endosperm
Ref No 4 07 852 United States

Feed Name or Analyses		Mean As Fed	Mean Dry	C.V. ± %
Dry matter	%	89.7	100.0	
Ash	%	0.4	0.4	
Crude fiber	%	0.4	0.4	
Ether extract	%	0.9	1.0	
N-free extract	%	76.6	85.4	
Protein (N x 6.25)	%	11.4	12.7	
Cattle	dig prot %	* 6.9	7.7	
Goats	dig prot %	* 8.0	8.9	
Horses	dig prot %	* 8.0	8.9	
Sheep	dig prot %	* 8.0	8.9	
Energy	GE Mcal/kg	3.71	4.14	
Cattle	DE Mcal/kg	* 3.48	3.58	
Sheep	DE Mcal/kg	* 3.42	3.82	
Swine	DE kcal/kg	*3919.	4369.	
Cattle	ME Mcal/kg	* 2.64	2.94	
Sheep	ME Mcal/kg	* 2.81	3.13	
Swine	ME kcal/kg	*3662.	4082.	
Cattle	TDN %	* 72.9	81.3	
Sheep	TDN %	* 77.6	86.6	
Swine	TDN %	* 88.9	99.1	
Calcium	%	0.03	0.03	
Iron	%	0.002	0.002	
Phosphorus	%	0.11	0.12	
Potassium	%	0.08	0.09	
Sodium	%	0.00	0.00	
Niacin	mg/kg	7.0	7.8	
Riboflavin	mg/kg	1.0	1.1	
Thiamine	mg/kg	0.6	0.7	

Wheat, distillers grains, dehy, (5)
Wheat distillers dried grains (AAFCO)
Wheat distillers dried grains (CFA)
Ref No 5 05 193 United States

Feed Name or Analyses		Mean As Fed	Mean Dry	C.V. ± %
Dry matter	%	93.4	100.0	2
Ash	%	3.0	3.3	25
Crude fiber	%	12.7	13.6	16
Ether extract	%	5.9	6.4	23
N-free extract	%	40.4	43.3	
Protein (N x 6.25)	%	31.3	33.5	21
Energy	GE Mcal/kg			
Cattle	DE Mcal/kg	* 3.25	3.48	
Sheep	DE Mcal/kg	* 3.43	3.67	
Cattle	ME Mcal/kg	*3707.	3971.	
Cattle	ME Mcal/kg	* 2.66	2.85	
Sheep	ME Mcal/kg	* 2.81	3.01	
Swine	ME kcal/kg	*3308.	3543.	
Cattle	TDN %	* 73.6	78.9	
Sheep	TDN %	* 77.7	83.2	
Swine	TDN %	* 84.1	90.1	
Calcium	%	0.08	0.08	62
Manganese	mg/kg	15.0	16.1	
Phosphorus	%	0.52	0.56	29
Carotene	mg/kg	1.1	1.2	
Niacin	mg/kg	56.2	60.2	28
Pantothenic acid	mg/kg	8.2	8.8	78
Riboflavin	mg/kg	3.8	4.0	54
Thiamine	mg/kg	2.0	2.1	72
Vitamin A equivalent	IU/g	1.8	2.0	
Arginine	%	1.10	1.18	
Glutamic acid	%	8.13	8.71	
Histidine	%	0.80	0.86	
Isoleucine	%	2.01	2.15	
Leucine	%	1.71	1.83	
Lysine	%	0.70	0.75	
Phenylalanine	%	1.71	1.83	
Threonine	%	0.90	0.97	
Tyrosine	%	0.50	0.54	
Valine	%	1.71	1.83	

Wheat, distillers grains w solubles, dehy, mn 75% original solids, (5)
Wheat distillers dried grains with solubles (AAFCO)
Ref No 5 05 194 United States

Feed Name or Analyses		Mean As Fed	Mean Dry	C.V. ± %
Dry matter	%	92.5	100.0	2
Ash	%	4.1	4.4	11
Crude fiber	%	9.8	10.6	22
Ether extract	%	6.3	6.8	26
N-free extract	%	40.3	43.6	
Protein (N x 6.25)	%	32.0	34.6	12
Energy	GE Mcal/kg			
Cattle	DE Mcal/kg	* 3.31	3.58	
Sheep	DE Mcal/kg	* 3.46	3.74	
Cattle	ME Mcal/kg	*3755.	4059.	
Cattle	ME Mcal/kg	* 2.72	2.94	
Sheep	ME Mcal/kg	* 2.84	3.07	
Swine	ME kcal/kg	*3342.	3613.	
Cattle	TDN %	* 75.2	81.2	
Sheep	TDN %	* 78.5	84.8	
Swine	TDN %	* 85.2	92.1	
Calcium	%	0.19	0.21	28
Phosphorus	%	0.71	0.77	17
Carotene	mg/kg	2.4	2.6	
Niacin	mg/kg	75.0	81.0	28
Pantothenic acid	mg/kg	12.1	13.1	15
Riboflavin	mg/kg	10.6	11.4	25
Vitamin A equivalent	IU/g	4.0	4.4	
Arginine	%	1.10	1.19	7
Glutamic acid	%	8.60	9.30	11
Histidine	%	0.80	0.86	5
Isoleucine	%	1.90	2.05	13
Leucine	%	2.00	2.16	38
Lysine	%	0.80	0.86	9
Methionine	%	0.50	0.54	
Phenylalanine	%	1.90	2.05	10
Serine	%	1.70	1.84	
Threonine	%	1.00	1.08	7
Tryptophan	%	0.40	0.43	55
Tyrosine	%	0.60	0.65	13
Valine	%	1.90	2.05	6

Wheat, distillers solubles, dehy, (5)
Wheat distillers dried solubles (AAFCO)
Ref No 5 05 195 United States

Feed Name or Analyses		Mean As Fed	Mean Dry	C.V. ± %
Dry matter	%	93.9	100.0	2
Ash	%	6.8	7.2	26
Crude fiber	%	3.4	3.6	47
Ether extract	%	2.2	2.3	73
N-free extract	%	50.4	53.7	
Protein (N x 6.25)	%	31.1	33.1	11
Energy	GE Mcal/kg			
Cattle	DE Mcal/kg	* 3.37	3.59	
Sheep	DE Mcal/kg	* 3.63	3.87	
Swine	DE kcal/kg	*3721.	3962.	
Cattle	ME Mcal/kg	* 2.76	2.94	
Sheep	ME Mcal/kg	* 2.98	3.17	
Swine	ME kcal/kg	*3323.	3539.	
Cattle	TDN %	* 76.4	81.4	
Sheep	TDN %	* 82.4	87.7	
Swine	TDN %	* 84.4	89.9	
Calcium	%	0.36	0.38	20
Phosphorus	%	1.51	1.61	13
Carotene	mg/kg	2.2	2.3	
Niacin	mg/kg	191.8	204.3	11
Pantothenic acid	mg/kg	33.7	35.9	23
Riboflavin	mg/kg	15.0	16.0	21
Thiamine	mg/kg	5.1	5.4	38
Vitamin A equivalent	IU/g	3.7	3.9	
Arginine	%	1.00	1.07	12
Glutamic acid	%	9.20	9.80	14
Histidine	%	0.80	0.85	10
Isoleucine	%	1.60	1.70	5
Leucine	%	1.50	1.60	11
Lysine	%	0.70	0.75	6
Methionine	%	0.40	0.43	
Phenylalanine	%	1.70	1.81	17
Serine	%	1.90	2.02	
Threonine	%	1.00	1.07	4
Tryptophan	%	0.50	0.53	98
Tyrosine	%	0.70	0.75	20
Valine	%	1.50	1.60	8

Wheat, distillers stillage, wet, (5)
Ref No 2 05 196 United States

Feed Name or Analyses		Mean As Fed	Mean Dry	C.V. ± %
Dry matter	%	7.2	100.0	18
Ash	%	0.3	4.2	26
Crude fiber	%	0.7	9.7	20
Ether extract	%	0.3	4.2	32
N-free extract	%	3.3	45.8	
Protein (N x 6.25)	%	2.6	36.1	10
Energy	GE Mcal/kg	0.35	4.87	0
Cattle	DE Mcal/kg	* 0.25	3.51	
Sheep	DE Mcal/kg	* 0.27	3.73	
Swine	DE kcal/kg	* 280.	3891.	
Cattle	ME Mcal/kg	* 0.21	2.88	
Sheep	ME Mcal/kg	* 0.22	3.06	
Swine	ME kcal/kg	* 248.	3451.	
Cattle	TDN %	* 5.7	79.6	
Sheep	TDN %	* 6.1	84.5	
Swine	TDN %	* 6.4	88.2	
Carotene	mg/kg	0.1	1.4	98
Vitamin A equivalent	IU/g	0.2	2.3	

(1) dry forages and roughages (2) pasture, range plants, and forages fed green (3) silages (4) energy feeds (5) protein supplements (6) minerals (7) vitamins (8) additives

Wheat, germ, extn unspecified grnd, mn 30% protein, (5)

Defatted wheat germ meal (AAFCO)

Ref No 5 05 217 United States

Feed Name or Analyses			Mean As Fed	Mean Dry	C.V. ± %
Dry matter	%		88.6	100.0	4
Ash	%		4.8	5.4	5
Crude fiber	%		2.2	2.5	15
Sheep	dig coef %	*	-23.	-23.	
Swine	dig coef %		41.	41.	
Ether extract	%		6.0	6.8	40
Sheep	dig coef %	*	89.	89.	
Swine	dig coef %		85.	85.	
N-free extract	%		45.5	51.3	4
Sheep	dig coef %	*	91.	91.	
Swine	dig coef %		88.	88.	
Protein (N x 6.25)	%		30.1	33.9	12
Sheep	dig coef %	*	94.	94.	
Swine	dig coef %		90.	90.	
Sheep	dig prot %		28.3	31.9	
Swine	dig prot %		27.1	30.5	
Energy	GE Mcal/kg				
Cattle	DE Mcal/kg	*	3.49	3.94	
Sheep	DE Mcal/kg	*	3.58	4.04	
Swine	DE kcal/kg	*	3507.	3959.	
Cattle	ME Mcal/kg	*	2.86	3.23	
Sheep	ME Mcal/kg	*	2.94	3.32	
Swine	ME kcal/kg	*	3127.	3530.	
Cattle	TDN %	*	79.1	89.3	
Sheep	TDN %		81.2	91.7	
Swine	TDN %		79.5	89.8	
Calcium	%		0.06	0.07	46
Cobalt	mg/kg		0.041	0.047	20
Copper	mg/kg		9.1	10.3	24
Iron	%		0.011	0.012	13
Manganese	mg/kg		163.4	184.4	23
Phosphorus	%		1.10	1.24	12
Carotene	mg/kg		6.5	7.4	
Choline	mg/kg		3861.	4359.	
Niacin	mg/kg		50.7	57.3	13
Pantothenic acid	mg/kg		12.2	13.8	31
Riboflavin	mg/kg		5.0	5.7	15
Thiamine	mg/kg		25.0	28.3	18
α-tocopherol	mg/kg		73.1	82.6	17
Vitamin A equivalent	IU/g		10.9	12.3	

Wheat, germ, grnd, mn 25% protein mn 7% fat, (5)

Wheat germ meal (AAFCO)

Ref No 5 05 218 United States

Feed Name or Analyses			Mean As Fed	Mean Dry	C.V. ± %
Dry matter	%		87.4	100.0	2
Ash	%		4.4	5.0	8
Crude fiber	%		3.2	3.7	13
Sheep	dig coef %	*	-23.	-23.	
Ether extract	%		8.6	9.8	18
Sheep	dig coef %	*	89.	89.	
N-free extract	%		47.3	54.1	
Sheep	dig coef %	*	91.	91.	
Protein (N x 6.25)	%		23.9	27.3	8
Sheep	dig coef %	*	94.	94.	
Sheep	dig prot %		22.4	25.7	
Starch	%		18.9	21.6	18
Linoleic	%		3.799	4.347	
Energy	GE Mcal/kg		4.11	4.70	
Cattle	DE Mcal/kg	*	3.56	4.07	
Sheep	DE Mcal/kg	*	3.61	4.13	
Swine	DE kcal/kg	*	4117.	4712.	
Cattle	ME Mcal/kg	*	2.92	3.34	
Chickens	ME_n kcal/kg		2850.	3261.	
Sheep	ME Mcal/kg	*	2.96	3.39	
Swine	ME kcal/kg	*	3725.	4264.	
Cattle	NE_m Mcal/kg	*	2.13	2.44	
Cattle	NE_{gain} Mcal/kg	*	1.35	1.55	
Cattle	$NE_{lactating\ cows}$ Mcal/kg	*	2.40	2.75	
Cattle	TDN %	*	80.7	92.4	
Sheep	TDN %		81.9	93.7	
Swine	TDN %	*	93.4	106.9	
Calcium	%		0.06	0.07	26
Chlorine	%		0.07	0.08	
Cobalt	mg/kg		0.118	0.135	20
Copper	mg/kg		9.1	10.4	14
Iron	%		0.006	0.007	85
Magnesium	%		0.25	0.29	17
Manganese	mg/kg		123.4	141.2	16
Phosphorus	%		0.95	1.08	10
Potassium	%		0.86	0.98	40
Selenium	mg/kg		0.359	0.411	66
Sodium	%		0.16	0.18	98
Sulphur	%		0.31	0.35	
Zinc	mg/kg		119.6	136.9	14
Choline	mg/kg		3199.	3662.	13
Folic acid	mg/kg		2.01	2.30	26
Niacin	mg/kg		68.6	78.5	23
Pantothenic acid	mg/kg		25.0	28.7	77
Riboflavin	mg/kg		5.6	6.4	9
Thiamine	mg/kg		22.8	26.1	11
α-tocopherol	mg/kg		130.0	148.7	
Vitamin B_6	mg/kg		7.09	8.11	34
Alanine	%		1.52	1.74	9
Arginine	%		1.83	2.10	10
Aspartic acid	%		2.12	2.42	6
Cystine	%		0.48	0.55	11
Glutamic acid	%		4.12	4.72	7
Glycine	%		1.46	1.67	7
Histidine	%		0.65	0.75	15
Isoleucine	%		0.87	1.00	15
Leucine	%		1.55	1.78	13
Lysine	%		1.51	1.73	10
Methionine	%		0.42	0.48	21
Phenylalanine	%		0.94	1.07	8
Proline	%		1.28	1.47	10
Serine	%		1.13	1.29	7
Threonine	%		0.96	1.10	10
Tryptophan	%		0.29	0.33	13
Tyrosine	%		0.73	0.84	7
Valine	%		1.15	1.32	9

Wheat, germ, mn 75% embryo mx 2.5% fiber, (5)

Wheat germ (CFA)

Ref No 5 05 219 United States

Feed Name or Analyses			Mean As Fed	Mean Dry	C.V. ± %
Dry matter	%			100.0	
Crude fiber	%				
Sheep	dig coef %	*		-23.	
Swine	dig coef %	*		41.	
Ether extract	%				
Sheep	dig coef %	*		89.	
Swine	dig coef %	*		85.	
N-free extract	%				
Sheep	dig coef %	*		91.	
Swine	dig coef %	*		88.	
Protein (N x 6.25)	%				
Sheep	dig coef %	*		94.	
Swine	dig coef %	*		90.	

Ref No 5 05 219 Canada

Feed Name or Analyses		Mean As Fed	Mean Dry	C.V. ± %
Dry matter	%	90.0	100.0	
Pantothenic acid	mg/kg	19.9	22.1	
Riboflavin	mg/kg	8.4	9.4	

Wheat, Gluten, (5)

Ref No 5 05 221 United States

Feed Name or Analyses		Mean As Fed	Mean Dry	C.V. ± %
Dry matter	%	88.0	100.0	4
Protein (N x 6.25)	%	80.6	91.6	4
Copper	mg/kg	16.3	18.5	
Iron	%	0.006	0.007	
Manganese	mg/kg	23.1	26.3	
Sulphur	%	0.93	1.06	
Niacin	mg/kg	23.6	26.8	
Alanine	%	1.70	1.93	23
Arginine	%	2.90	3.30	17
Aspartic acid	%	2.70	3.07	29
Cystine	%	1.70	1.93	9
Glutamic acid	%	27.20	30.91	20
Glycine	%	2.70	3.07	20
Histidine	%	1.60	1.82	10
Isoleucine	%	3.30	3.75	9
Leucine	%	5.40	6.14	11
Lysine	%	1.50	1.70	12
Methionine	%	1.20	1.36	15
Phenylalanine	%	4.10	4.66	8
Proline	%	9.90	11.25	16
Serine	%	4.00	4.55	
Threonine	%	2.10	2.39	7
Tryptophan	%	0.70	0.80	18
Tyrosine	%	2.30	2.61	20
Valine	%	3.80	4.32	11

Wheat, gluten, grnd, (5)

Ref No 5 05 220 United States

Feed Name or Analyses			Mean As Fed	Mean Dry	C.V. ± %
Dry matter	%		89.0	100.0	
Ash	%		3.6	4.1	
Crude fiber	%		6.0	6.7	
Sheep	dig coef %		120.	120.	
Ether extract	%		4.4	4.9	
Sheep	dig coef %		46.	46.	
N-free extract	%		56.9	63.9	
Sheep	dig coef %		92.	92.	
Protein (N x 6.25)	%		18.2	20.4	
Sheep	dig coef %		88.	88.	
Sheep	dig prot %		16.0	18.0	
Energy	GE Mcal/kg				
Cattle	DE Mcal/kg	*	3.37	3.78	
Sheep	DE Mcal/kg	*	3.53	3.96	
Swine	DE kcal/kg	*	3111.	3495.	
Cattle	ME Mcal/kg	*	2.76	3.10	
Sheep	ME Mcal/kg	*	2.89	3.25	
Swine	ME kcal/kg	*	2858.	3211.	
Cattle	TDN %	*	76.3	85.8	
Sheep	TDN %		80.0	89.9	
Swine	TDN %	*	70.5	79.3	

Feed Name or Analyses			Mean As Fed	Mean Dry	C.V. ± %

WHEAT, AMBER DURUM. Triticum durum

Wheat, amber durum, grain, Can 1 CW mn wt 62 lb per bushel mx 0.5% foreign material, (4)

Ref No 5 08 176 Canada

Analyses	Unit		As Fed	Dry	C.V. ± %
Dry matter	%		90.2	100.0	
Ash	%		1.6	1.8	
Crude fiber	%		2.3	2.5	
Ether extract	%		2.6	2.9	
N-free extract	%		65.3	72.4	
Protein (N x 6.25)	%		18.4	20.4	
Cattle	dig prot %	*	13.3	14.8	
Goats	dig prot %	*	14.4	15.9	
Horses	dig prot %	*	14.4	15.9	
Sheep	dig prot %	*	14.4	15.9	
Cellulose (Matrone)	%		2.5	2.8	
Energy	GE Mcal/kg				
Cattle	DE Mcal/kg	*	3.17	3.51	
Sheep	DE Mcal/kg	*	3.25	3.61	
Swine	DE kcal/kg	*	3744.	4153.	
Cattle	ME Mcal/kg	*	2.60	2.88	
Sheep	ME Mcal/kg	*	2.67	2.96	
Swine	ME kcal/kg	*	3440.	3816.	
Cattle	TDN %	*	71.8	79.6	
Sheep	TDN %	*	73.7	81.8	
Swine	TDN %	*	84.9	94.2	
Copper	mg/kg		5.5	6.1	
Iron	%		0.004	0.005	
Magnesium	%		0.15	0.16	
Manganese	mg/kg		29.0	32.1	
Selenium	mg/kg		0.918	1.019	
Zinc	mg/kg		33.7	37.4	
Niacin	mg/kg		61.9	68.7	
Pantothenic acid	mg/kg		9.1	10.1	
Riboflavin	mg/kg		0.9	1.0	
Thiamine	mg/kg		1.2	1.3	
Vitamin B6	mg/kg		3.08	3.42	
Alanine	%		0.55	0.61	
Arginine	%		0.68	0.76	
Aspartic acid	%		0.78	0.87	
Glutamic acid	%		5.35	5.93	
Glycine	%		0.60	0.67	
Histidine	%		0.35	0.39	
Isoleucine	%		0.60	0.67	
Leucine	%		1.08	1.20	
Lysine	%		0.39	0.43	
Methionine	%		0.19	0.21	
Phenylalanine	%		0.83	0.92	
Proline	%		1.66	1.84	
Serine	%		0.73	0.81	
Threonine	%		0.43	0.48	
Tyrosine	%		0.40	0.44	
Valine	%		0.71	0.79	

Wheat, amber durum, grain, Can 6 CW mn wt 51 lb per bushel mx 3% foreign material, (4)

Ref No 4 08 183 Canada

Analyses	Unit		As Fed	Dry	C.V. ± %
Dry matter	%		86.5	100.0	
Ash	%		1.8	2.1	
Crude fiber	%		3.0	3.4	
Ether extract	%		2.1	2.4	
N-free extract	%		66.3	76.6	
Protein (N x 6.25)	%		13.3	15.4	
Cattle	dig prot %	*	8.8	10.1	
Goats	dig prot %	*	9.8	11.3	
Horses	dig prot %	*	9.8	11.3	
Sheep	dig prot %	*	9.8	11.3	
Energy	GE Mcal/kg				
Cattle	DE Mcal/kg	*	2.99	3.46	
Sheep	DE Mcal/kg	*	3.16	3.65	
Swine	DE kcal/kg	*	3407.	3938.	
Cattle	ME Mcal/kg	*	2.46	2.84	
Sheep	ME Mcal/kg	*	2.59	3.00	
Swine	ME kcal/kg	*	3164.	3658.	
Cattle	TDN %	*	67.8	78.4	
Sheep	TDN %	*	71.7	82.9	
Swine	TDN %	*	77.3	89.3	

(1) dry forages and roughages (2) pasture, range plants, and forages fed green (3) silages (4) energy feeds (5) protein supplements (6) minerals (7) vitamins (8) additives

WHEAT, BLEDSO. Triticum vulgare

Wheat, bledso, aerial part, ensiled, prebloom, (3)

Ref No 3 09 240 United States

Analyses	Unit		As Fed	Dry	C.V. ± %
Dry matter	%			100.0	
Cattle	dig coef %			69.	
Ash	%			5.3	
Crude fiber	%			31.5	
Cattle	dig coef %			73.	
Ether extract	%			3.4	
N-free extract	%			48.7	
Cattle	dig coef %			71.	
Protein (N x 6.25)	%			11.1	
Cattle	dig coef %			63.	
Cattle	dig prot %			7.0	
Goats	dig prot %	*		6.3	
Horses	dig prot %	*		6.3	
Sheep	dig prot %	*		6.3	
Energy	GE Mcal/kg				
Cattle	DE Mcal/kg	*		3.04	
Sheep	DE Mcal/kg	*		2.87	
Cattle	ME Mcal/kg	*		2.49	
Sheep	ME Mcal/kg	*		2.35	
Cattle	TDN %			69.0	
Sheep	TDN %	*		65.1	

WHEAT, DURUM. Triticum durum

Wheat, durum, bran, dry milled, (4)

Ref No 4 05 222 United States

Analyses	Unit		As Fed	Dry	C.V. ± %
Dry matter	%		90.3	100.0	
Ash	%		3.8	4.2	
Crude fiber	%		8.6	9.5	
Ether extract	%		5.0	5.5	
N-free extract	%		55.3	61.2	
Protein (N x 6.25)	%		17.6	19.5	
Cattle	dig prot %	*	12.6	13.9	
Goats	dig prot %	*	13.6	15.1	
Horses	dig prot %	*	13.6	15.1	
Sheep	dig prot %	*	13.6	15.1	
Energy	GE Mcal/kg				
Cattle	DE Mcal/kg	*	2.95	3.27	
Sheep	DE Mcal/kg	*	3.06	3.39	
Swine	DE kcal/kg	*	2871.	3179.	
Cattle	ME Mcal/kg	*	2.42	2.68	
Sheep	ME Mcal/kg	*	2.51	2.78	
Swine	ME kcal/kg	*	2643.	2927.	
Cattle	TDN %	*	66.9	74.1	
Sheep	TDN %	*	69.4	76.9	
Swine	TDN %	*	65.1	72.1	

Wheat, durum, grain, (4)

Ref No 4 05 224 United States

Analyses	Unit		As Fed	Dry	C.V. ± %
Dry matter	%		88.8	100.0	1
Ash	%		1.7	1.9	
Crude fiber	%		2.5	2.8	
Sheep	dig coef %	*	24.	24.	
Ether extract	%		2.6	2.9	
Sheep	dig coef %	*	65.	65.	
N-free extract	%		69.3	78.1	
Sheep	dig coef %	*	92.	92.	
Protein (N x 6.25)	%		12.7	14.3	
Sheep	dig coef %	*	78.	78.	
Cattle	dig prot %	*	8.2	9.2	
Goats	dig prot %	*	9.2	10.4	
Horses	dig prot %	*	9.2	10.4	
Sheep	dig prot %		9.9	11.2	
Starch	%		57.0	64.2	3
Sugars, total	%		2.9	3.3	12
Energy	GE Mcal/kg		3.39	3.82	
Cattle	DE Mcal/kg	*	3.33	3.75	
Sheep	DE Mcal/kg	*	3.44	3.88	
Swine	DE kcal/kg	*	3568.	4016.	
Cattle	ME Mcal/kg	*	2.74	3.08	
Sheep	ME Mcal/kg	*	2.82	3.18	
Swine	ME kcal/kg	*	3322.	3739.	
Cattle	TDN %	*	75.5	85.0	
Sheep	TDN %		78.1	87.9	
Swine	TDN %	*	80.9	91.1	
Calcium	%		0.10	0.11	
Copper	mg/kg		7.6	8.6	
Iron	%		0.004	0.005	
Manganese	mg/kg		28.4	32.0	
Phosphorus	%		0.40	0.45	4
Potassium	%		0.45	0.51	
Sodium	%		0.00	0.00	
Folic acid	mg/kg		0.39	0.44	30
Niacin	mg/kg		44.9	50.6	
Riboflavin	mg/kg		1.2	1.4	
Thiamine	mg/kg		6.5	7.3	10

Wheat, durum, grain, Can 4 CW mn wt 56 lb per bushel mx 2.5% foreign material, (4)

Ref No 4 05 225 United States

Analyses	Unit		As Fed	Dry	C.V. ± %
Dry matter	%			100.0	
Crude fiber	%				
Sheep	dig coef %	*		24.	
Ether extract	%				
Sheep	dig coef %	*		65.	
N-free extract	%				
Sheep	dig coef %	*		92.	
Protein (N x 6.25)	%				
Sheep	dig coef %	*		78.	
Energy	GE Mcal/kg				
Sheep	DE Mcal/kg	*		3.85	
Sheep	ME Mcal/kg	*		3.16	
Sheep	TDN %	*		87.4	

Ref No 4 05 225 Canada

Analyses	Unit		As Fed	Dry	C.V. ± %
Dry matter	%		86.5	100.0	0
Ash	%		1.6	1.8	7
Crude fiber	%		2.2	2.5	21
Ether extract	%		1.6	1.9	14

Feed Name or Analyses		Mean As Fed	Mean Dry	C.V. ± %
N-free extract	%	67.2	77.7	
Protein (N x 6.25)	%	14.0	16.2	9
Cattle	dig prot %	* 9.4	10.9	
Goats	dig prot %	* 10.4	12.0	
Horses	dig prot %	* 10.4	12.0	
Sheep	dig prot %	* 10.4	12.0	
Energy	GE Mcal/kg			
Cattle	DE Mcal/kg	* 3.30	3.82	
Sheep	DE Mcal/kg	* 3.33	3.85	
Swine	DE kcal/kg	*3525.	4075.	
Cattle	ME Mcal/kg	* 2.71	3.13	
Sheep	ME Mcal/kg	* 2.73	3.16	
Swine	ME kcal/kg	*3269.	3779.	
Cattle	TDN %	* 75.0	86.7	
Sheep	TDN %	75.6	87.4	
Swine	TDN %	* 79.9	92.4	
Calcium	%	0.07	0.09	
Phosphorus	%	0.31	0.36	

Wheat, durum, grain, Can 5 CW mn wt 54 lb per bushel mx 3% foreign material, (4)

Ref No 4 05 226 United States

Feed Name or Analyses		Mean As Fed	Mean Dry	C.V. ± %
Dry matter	%		100.0	
Crude fiber	%			
Sheep	dig coef %	*	24.	
Ether extract	%			
Sheep	dig coef %	*	65.	
N-free extract	%			
Sheep	dig coef %	*	92.	
Protein (N x 6.25)	%			
Sheep	dig coef %	*	78.	
Energy	GE Mcal/kg			
Sheep	DE Mcal/kg	*	3.86	
Sheep	ME Mcal/kg	*	3.16	
Sheep	TDN %	*	87.4	

Ref No 4 05 226 Canada

Feed Name or Analyses		Mean As Fed	Mean Dry	C.V. ± %
Dry matter	%	86.5	100.0	0
Ash	%	1.6	1.8	11
Crude fiber	%	2.3	2.7	20
Ether extract	%	1.8	2.1	21
N-free extract	%	67.2	77.7	
Protein (N x 6.25)	%	13.6	15.7	11
Cattle	dig prot %	* 9.0	10.4	
Goats	dig prot %	* 10.0	11.6	
Horses	dig prot %	* 10.0	11.6	
Sheep	dig prot %	* 10.0	11.6	
Energy	GE Mcal/kg			
Cattle	DE Mcal/kg	* 3.28	3.79	
Sheep	DE Mcal/kg	* 3.33	3.86	
Swine	DE kcal/kg	*3502.	4049.	
Cattle	ME Mcal/kg	* 2.69	3.11	
Sheep	ME Mcal/kg	* 2.73	3.16	
Swine	ME kcal/kg	*3251.	3758.	
Cattle	TDN %	* 74.3	85.9	
Sheep	TDN %	75.6	87.4	
Swine	TDN %	* 79.4	91.8	
Calcium	%	0.08	0.09	
Phosphorus	%	0.28	0.32	

Wheat, durum, grain, gr 1 US mn wt 60 lb per bushel mx 0.5% foreign material, (4)

Ref No 4 05 229 United States

Feed Name or Analyses		Mean As Fed	Mean Dry	C.V. ± %
Dry matter	%		100.0	
Ash	%		1.8	11
Ether extract	%		2.1	10
Protein (N x 6.25)	%		13.8	17
Cattle	dig prot %	*	8.7	
Goats	dig prot %	*	9.9	
Horses	dig prot %	*	9.9	
Sheep	dig prot %	*	9.9	
Starch	%		65.2	2
Sugars, total	%		3.2	10

Wheat, durum, grain, gr 2 US mn wt 58 lb per bushel mx 1% foreign material, (4)

Ref No 4 05 231 United States

Feed Name or Analyses		Mean As Fed	Mean Dry	C.V. ± %
Dry matter	%		100.0	
Ash	%		1.9	10
Ether extract	%		2.1	9
Protein (N x 6.25)	%		13.9	6
Cattle	dig prot %	*	8.8	
Goats	dig prot %	*	10.0	
Horses	dig prot %	*	10.0	
Sheep	dig prot %	*	10.0	
Starch	%		64.6	3
Sugars, total	%		3.3	6

Wheat, durum, grain, gr 3 US mn wt 56 lb per bushel mx 2% foreign material, (4)

Ref No 4 05 233 United States

Feed Name or Analyses		Mean As Fed	Mean Dry	C.V. ± %
Dry matter	%		100.0	
Copper	mg/kg		9.0	
Iron	%		0.005	
Manganese	mg/kg		32.0	
Phosphorus	%		0.48	

Wheat, durum, germ, (5)

Ref No 5 05 223 United States

Feed Name or Analyses		Mean As Fed	Mean Dry	C.V. ± %
Dry matter	%	91.9	100.0	
Ash	%	4.6	5.0	
Crude fiber	%	2.6	2.8	
Ether extract	%	13.5	14.7	
N-free extract	%	40.0	43.5	
Protein (N x 6.25)	%	31.2	33.9	
Energy	GE Mcal/kg			
Cattle	DE Mcal/kg	* 3.96	4.31	
Sheep	DE Mcal/kg	* 3.84	4.18	
Swine	DE kcal/kg	*4023.	4378.	
Cattle	ME Mcal/kg	* 3.24	3.53	
Sheep	ME Mcal/kg	* 3.15	3.43	
Swine	ME kcal/kg	*3589.	3905.	
Cattle	TDN %	* 89.7	97.7	
Sheep	TDN %	* 87.1	94.8	
Swine	TDN %	* 91.3	99.3	
Calcium	%	0.06	0.07	
Chlorine	%	0.08	0.09	
Copper	mg/kg	26.2	28.5	
Iron	%	0.006	0.007	
Magnesium	%	0.38	0.41	
Manganese	mg/kg	181.2	197.2	
Phosphorus	%	1.14	1.24	
Potassium	%	1.01	1.10	
Sulphur	%	0.26	0.28	

WHEAT, GAINES. Triticum

Wheat, gaines, grain, grnd, (4)

Ref No 4 08 664 United States

Feed Name or Analyses		Mean As Fed	Mean Dry	C.V. ± %
Dry matter	%	86.0	100.0	
Ash	%	1.2	1.4	
Crude fiber	%	2.1	2.4	
Ether extract	%	1.6	1.9	
N-free extract	%	71.9	83.6	
Protein (N x 6.25)	%	9.2	10.7	
Cattle	dig prot %	* 5.0	5.8	
Goats	dig prot %	* 6.1	7.0	
Horses	dig prot %	* 6.1	7.0	
Sheep	dig prot %	* 6.1	7.0	
Starch	%	59.2	68.8	
Energy	GE Mcal/kg			
Cattle	DE Mcal/kg	* 2.95	3.43	
Sheep	DE Mcal/kg	* 3.25	3.78	
Swine	DE kcal/kg	*3473.	4038.	
Cattle	ME Mcal/kg	* 2.42	2.82	
Sheep	ME Mcal/kg	* 2.67	3.10	
Swine	ME kcal/kg	*3259.	3789.	
Cattle	TDN %	* 67.0	77.9	
Sheep	TDN %	* 73.7	85.7	
Swine	TDN %	* 78.8	91.6	
Calcium	%	0.02	0.02	
Cobalt	mg/kg	0.130	0.151	
Copper	mg/kg	4.2	4.9	
Iron	%	0.003	0.004	
Magnesium	%	0.09	0.10	
Manganese	mg/kg	24.0	27.9	
Phosphorus	%	0.28	0.33	
Potassium	%	0.37	0.43	
Selenium	mg/kg	0.040	0.047	
Sodium	%	0.01	0.01	
Zinc	mg/kg	21.0	24.4	
Choline	mg/kg	1140.	1325.	
Folic acid	mg/kg	0.37	0.43	
Niacin	mg/kg	46.6	54.2	
Pantothenic acid	mg/kg	8.4	9.8	
Riboflavin	mg/kg	1.3	1.5	
Thiamine	mg/kg	4.1	4.8	
Alanine	%	0.40	0.47	
Arginine	%	0.55	0.64	
Aspartic acid	%	0.57	0.66	
Cystine	%	0.29	0.34	
Glutamic acid	%	3.45	4.01	
Glycine	%	0.47	0.55	
Histidine	%	0.26	0.30	
Isoleucine	%	0.38	0.44	
Leucine	%	0.74	0.86	
Lysine	%	0.32	0.37	
Methionine	%	0.16	0.19	
Phenylalanine	%	0.49	0.57	
Proline	%	1.06	1.23	
Serine	%	0.52	0.60	
Threonine	%	0.32	0.37	
Tyrosine	%	0.32	0.37	
Valine	%	0.48	0.56	

Feed Name or Analyses		Mean As Fed	Mean Dry	C.V. ± %

WHEAT, HARD. Triticum aestivum

Wheat, hard, bran, dry milled, (4)

Ref No 4 05 246 United States

Feed Name or Analyses		As Fed	Dry	C.V. ± %
Dry matter	%	86.5	100.0	2
Ash	%	6.3	7.3	12
Crude fiber	%	10.4	12.0	7
Ether extract	%	3.6	4.2	18
N-free extract	%	52.2	60.3	
Protein (N x 6.25)	%	14.1	16.3	9
Cattle	dig prot %	* 9.5	10.9	
Goats	dig prot %	* 10.5	12.1	
Horses	dig prot %	* 10.5	12.1	
Sheep	dig prot %	* 10.5	12.1	
Starch	%	5.9	6.8	17
Energy	GE Mcal/kg			
Cattle	DE Mcal/kg	* 2.67	3.09	
Sheep	DE Mcal/kg	* 2.83	3.27	
Swine	DE kcal/kg	*2605.	3011.	
Cattle	ME Mcal/kg	* 2.19	2.53	
Sheep	ME Mcal/kg	* 2.32	2.68	
Swine	ME kcal/kg	*2443.	2824.	
Cattle	TDN %	* 60.6	70.0	
Sheep	TDN %	* 64.1	74.1	
Swine	TDN %	* 59.1	68.3	
Calcium	%	0.09	0.10	29
Cobalt	mg/kg	0.108	0.125	15
Copper	mg/kg	12.5	14.4	26
Iron	%	0.009	0.010	78
Magnesium	%	0.50	0.58	14
Manganese	mg/kg	107.8	124.7	23
Molybdenum	mg/kg	0.34	0.39	
Phosphorus	%	1.28	1.48	14
Potassium	%	1.48	1.72	16
Selenium	mg/kg	0.370	0.428	71
Sodium	%	0.02	0.02	56
Zinc	mg/kg	109.7	126.9	26
Choline	mg/kg	2122.	2454.	15
Folic acid	mg/kg	1.15	1.33	14
Niacin	mg/kg	300.4	347.4	12
Pantothenic acid	mg/kg	32.9	38.1	16
Riboflavin	mg/kg	4.9	5.7	9
Thiamine	mg/kg	6.2	7.1	11
Alanine	%	0.75	0.87	4
Arginine	%	1.07	1.24	4
Aspartic acid	%	1.12	1.30	4
Cystine	%	0.36	0.42	9
Glutamic acid	%	2.77	3.21	12
Glycine	%	0.89	1.02	4
Histidine	%	0.43	0.50	8
Isoleucine	%	0.46	0.53	9
Leucine	%	0.90	1.04	6
Lysine	%	0.59	0.68	4
Methionine	%	0.22	0.26	11
Phenylalanine	%	0.57	0.66	6
Proline	%	0.85	0.98	10
Serine	%	0.65	0.76	7
Threonine	%	0.48	0.56	11
Tyrosine	%	0.43	0.49	6
Valine	%	0.69	0.80	12

Wheat, hard, flour by-product, mx 9.5% fiber, (4)

Wheat, hard, standard middlings

Ref No 4 08 549 United States

Feed Name or Analyses		As Fed	Dry	C.V. ± %
Dry matter	%	90.5	100.0	
Ash	%	4.9	5.4	
Crude fiber	%	8.0	8.8	
Ether extract	%	5.2	5.7	
N-free extract	%	54.9	60.7	
Protein (N x 6.25)	%	17.5	19.3	
Cattle	dig prot %	* 12.5	13.8	
Goats	dig prot %	* 13.5	15.0	
Horses	dig prot %	* 13.5	15.0	
Sheep	dig prot %	* 13.5	15.0	
Energy	GE Mcal/kg			
Cattle	DE Mcal/kg	* 3.40	3.76	
Sheep	DE Mcal/kg	* 3.47	3.83	
Swine	DE kcal/kg	*3228.	3567.	
Cattle	ME Mcal/kg	* 2.79	3.08	
Sheep	ME Mcal/kg	* 2.83	3.13	
Swine	ME kcal/kg	*2901.	3205.	
Cattle	TDN %	* 77.1	85.2	
Sheep	TDN %	* 78.4	86.6	
Swine	TDN %	* 73.2	80.9	
Calcium	%	0.08	0.09	
Phosphorus	%	0.82	0.91	

Wheat, hard, flour by-product mill run, mx 9.5% fiber, (4)

Wheat, hard, mill run

Ref No 4 08 550 United States

Feed Name or Analyses		As Fed	Dry	C.V. ± %
Dry matter	%	91.0	100.0	
Ash	%	5.1	5.6	
Crude fiber	%	8.9	9.8	
Ether extract	%	4.9	5.4	
N-free extract	%	55.3	60.8	
Protein (N x 6.25)	%	16.8	18.5	
Cattle	dig prot %	* 11.8	13.0	
Goats	dig prot %	* 12.9	14.2	
Horses	dig prot %	* 12.9	14.2	
Sheep	dig prot %	* 12.9	14.2	
Energy	GE Mcal/kg			
Cattle	DE Mcal/kg	* 3.05	3.35	
Sheep	DE Mcal/kg	* 3.06	3.36	
Swine	DE kcal/kg	*3118.	3426.	
Cattle	ME Mcal/kg	* 2.50	2.75	
Sheep	ME Mcal/kg	* 2.51	2.75	
Swine	ME kcal/kg	*2877. 3	3162.	
Cattle	TDN %	* 69.1	75.9	
Sheep	TDN %	* 69.3	76.1	
Swine	TDN %	* 70.7	77.7	
Calcium	%	0.11	0.12	
Phosphorus	%	1.09	1.20	

Wheat, hard, grain, (4)

Ref No 4 05 247 United States

Feed Name or Analyses		As Fed	Dry	C.V. ± %
Dry matter	%	89.6	100.0	2
Linoleic	%	.599	.669	
Energy	GE Mcal/kg			
Chickens	MEn kcal/kg	3250.	3627.	
Carotene	mg/kg	0.0	0.0	2
Choline	mg/kg	770.	860.	18
Niacin	mg/kg	54.1	60.4	30
Pantothenic acid	mg/kg	13.2	14.8	14
Riboflavin	mg/kg	1.2	1.3	43
Thiamine	mg/kg	5.9	6.6	20
Vitamin B6	mg/kg	4.15	4.63	18

Ref No 4 05 247 Canada

Feed Name or Analyses		As Fed	Dry	C.V. ± %
Dry matter	%	89.5	100.0	
Niacin	mg/kg	55.1	61.6	
Pantothenic acid	mg/kg	13.7	15.3	
Riboflavin	mg/kg	0.9	1.0	
Thiamine	mg/kg	5.1	5.7	

WHEAT, HARD RED SPRING. Triticum aestivum

Wheat, hard red spring, bran, (4)

Ref No 4 08 009 United States

Feed Name or Analyses		As Fed	Dry	C.V. ± %
Dry matter	%	88.0	100.0	
Ash	%	6.5	7.4	
Crude fiber	%	10.8	12.3	
Ether extract	%	4.0	4.5	
N-free extract	%	52.9	60.1	
Protein (N x 6.25)	%	13.8	15.7	
Cattle	dig prot %	* 9.2	10.4	
Goats	dig prot %	* 10.2	11.6	
Horses	dig prot %	* 10.2	11.6	
Sheep	dig prot %	* 10.2	11.6	
Starch	%	5.0	5.6	
Energy	GE Mcal/kg	4.04	4.60	
Cattle	DE Mcal/kg	* 2.83	3.22	
Sheep	DE Mcal/kg	* 2.88	3.27	
Swine	DE kcal/kg	*2654.	3016.	
Cattle	ME Mcal/kg	* 2.32	2.64	
Sheep	ME Mcal/kg	* 2.36	2.68	
Swine	ME kcal/kg	*2463.	2799.	
Cattle	TDN %	* 64.2	73.0	
Sheep	TDN %	* 65.3	74.2	
Swine	TDN %	* 60.2	68.4	
Aluminum	mg/kg	5.12	5.81	
Calcium	%	0.06	0.07	
Cobalt	mg/kg	0.113	0.128	
Copper	mg/kg	15.1	17.2	
Iron	%	0.008	0.009	
Magnesium	%	0.57	0.65	
Manganese	mg/kg	144.3	164.0	
Phosphorus	%	1.23	1.40	
Potassium	%	1.43	1.63	
Selenium	mg/kg	0.358	0.407	
Sodium	%	0.02	0.02	
Zinc	mg/kg	110.5	125.6	
Choline	mg/kg	2034.	2312.	
Folic acid	mg/kg	1.31	1.49	
Niacin	mg/kg	329.1	374.0	
Pantothenic acid	mg/kg	38.5	43.7	
Riboflavin	mg/kg	4.8	5.5	
Thiamine	mg/kg	6.7	7.6	
α-tocopherol	mg/kg	22.9	26.0	
Vitamin B6	mg/kg	9.88	11.23	
Alanine	%	0.76	0.86	
Arginine	%	1.09	1.24	
Aspartic acid	%	1.13	1.28	
Cystine	%	0.35	0.40	
Glutamic acid	%	2.66	3.02	
Glycine	%	0.90	1.02	
Histidine	%	0.44	0.50	
Isoleucine	%	0.44	0.50	
Leucine	%	0.90	1.02	
Lysine	%	0.61	0.70	
Methionine	%	0.24	0.27	
Phenylalanine	%	0.57	0.65	

(1) dry forages and roughages (2) pasture, range plants, and forages fed green (3) silages (4) energy feeds (5) protein supplements (6) minerals (7) vitamins (8) additives

Feed Name or Analyses			Mean As Fed	Mean Dry	C.V. ± %
Proline	%		0.82	0.93	
Serine	%		0.65	0.74	
Threonine	%		0.53	0.60	
Tyrosine	%		0.42	0.48	
Valine	%		0.76	0.86	

Wheat, hard red spring, flour, (4)

Ref No 4 08 113 United States

Feed Name or Analyses			Mean As Fed	Mean Dry	C.V. ± %
Dry matter	%		87.6	100.0	
Ash	%		0.4	0.5	
Ether extract	%		1.0	1.1	
Protein (N x 6.25)	%		11.7	13.4	
Cattle	dig prot %	*	7.3	8.3	
Goats	dig prot %	*	8.3	9.5	
Horses	dig prot %	*	8.3	9.5	
Sheep	dig prot %	*	8.3	9.5	
Starch	%		68.2	77.8	
Energy	GE Mcal/kg		3.86	4.40	
Aluminum	mg/kg		5.09	5.81	
Calcium	%		0.03	0.03	
Cobalt	mg/kg		0.061	0.070	
Copper	mg/kg		0.6	0.7	
Iron	%		0.000	0.001	
Magnesium	%		0.02	0.02	
Manganese	mg/kg		3.6	4.1	
Phosphorus	%		0.10	0.11	
Potassium	%		0.08	0.09	
Selenium	mg/kg		0.295	0.337	
Sodium	%		0.01	0.01	
Zinc	mg/kg		5.9	6.7	
Choline	mg/kg		927.	1058.	
Folic acid	mg/kg		0.13	0.14	
Niacin	mg/kg		10.2	11.6	
Pantothenic acid	mg/kg		4.1	4.7	
Riboflavin	mg/kg		0.4	0.4	
Thiamine	mg/kg		1.4	1.6	
α-tocopherol	mg/kg		2.2	2.6	
Vitamin B6	mg/kg		0.72	0.83	
Alanine	%		0.39	0.44	
Arginine	%		0.43	0.49	
Aspartic acid	%		0.50	0.57	
Cystine	%		0.35	0.40	
Glutamic acid	%		4.28	4.88	
Glycine	%		0.47	0.53	
Histidine	%		0.24	0.28	
Isoleucine	%		0.45	0.51	
Leucine	%		0.90	1.02	
Lysine	%		0.23	0.27	
Methionine	%		0.21	0.24	
Phenylalanine	%		0.65	0.74	
Proline	%		1.57	1.79	
Serine	%		0.63	0.72	
Threonine	%		0.35	0.40	
Tyrosine	%		0.36	0.41	
Valine	%		0.48	0.55	

Wheat, hard red spring, flour by-product, coarse sift, mx 7% fiber, (4)

Ref No 4 08 015 United States

Feed Name or Analyses			Mean As Fed	Mean Dry	C.V. ± %
Dry matter	%		86.3	100.0	
Ash	%		4.0	4.6	
Crude fiber	%		6.9	8.0	
Ether extract	%		6.0	7.0	
N-free extract	%		53.7	62.2	
Protein (N x 6.25)	%		15.7	18.2	
Cattle	dig prot %	*	11.0	12.7	
Goats	dig prot %	*	12.0	13.9	
Horses	dig prot %	*	12.0	13.9	
Sheep	dig prot %	*	12.0	13.9	
Starch	%		15.9	18.4	
Energy	GE Mcal/kg		4.32	5.00	
Cattle	DE Mcal/kg	*	2.93	3.39	
Sheep	DE Mcal/kg	*	2.99	3.46	
Swine	DE kcal/kg	*	2843.	3294.	
Cattle	ME Mcal/kg	*	2.40	2.78	
Sheep	ME Mcal/kg	*	2.45	2.84	
Swine	ME kcal/kg	*	2624.	3041.	
Cattle	TDN %	*	66.4	76.9	
Sheep	TDN %	*	67.8	78.6	
Swine	TDN %	*	64.5	74.7	
Aluminum	mg/kg		5.02	5.81	
Calcium	%		0.06	0.07	
Cobalt	mg/kg		0.100	0.116	
Copper	mg/kg		12.4	14.4	
Iron	%		0.006	0.007	
Magnesium	%		0.26	0.30	
Manganese	mg/kg		134.5	155.8	
Phosphorus	%		0.80	0.93	
Potassium	%		0.88	1.02	
Selenium	mg/kg		0.642	0.744	
Sodium	%		0.02	0.02	
Zinc	mg/kg		105.9	122.7	
Choline	mg/kg		1995.	2312.	
Folic acid	mg/kg		1.97	2.28	
Niacin	mg/kg		122.7	142.2	
Pantothenic acid	mg/kg		25.9	30.0	
Riboflavin	mg/kg		4.8	5.6	
Thiamine	mg/kg		21.8	25.2	
α-tocopherol	mg/kg		62.9	72.9	
Vitamin B6	mg/kg		6.54	7.58	
Alanine	%		0.96	1.12	
Arginine	%		1.17	1.36	
Aspartic acid	%		1.26	1.47	
Cystine	%		0.34	0.40	
Glutamic acid	%		3.07	3.56	
Glycine	%		0.92	1.07	
Histidine	%		0.43	0.50	
Isoleucine	%		0.52	0.60	
Leucine	%		1.04	1.21	
Lysine	%		0.80	0.93	
Methionine	%		0.28	0.33	
Phenylalanine	%		0.64	0.74	
Proline	%		0.96	1.12	
Serine	%		0.76	0.88	
Threonine	%		0.60	0.70	
Tyrosine	%		0.48	0.56	
Valine	%		0.77	0.90	

Wheat, hard red spring, flour by-product, fine sift, mx 4% fiber, (4)

Ref No 4 08 019 United States

Feed Name or Analyses			Mean As Fed	Mean Dry	C.V. ± %
Dry matter	%		88.7	100.0	
Ash	%		2.6	2.9	
Crude fiber	%		2.6	2.9	
Ether extract	%		4.3	4.8	
N-free extract	%		64.6	72.8	
Protein (N x 6.25)	%		14.6	16.5	
Cattle	dig prot %	*	9.9	11.1	
Goats	dig prot %	*	10.9	12.3	
Horses	dig prot %	*	10.9	12.3	
Sheep	dig prot %	*	10.9	12.3	
Starch	%		40.6	45.8	
Energy	GE Mcal/kg		4.19	4.73	
Cattle	DE Mcal/kg	*	3.25	3.66	
Sheep	DE Mcal/kg	*	3.25	3.66	
Swine	DE kcal/kg	*	3507.	3954.	
Cattle	ME Mcal/kg	*	2.66	3.00	
Sheep	ME Mcal/kg	*	2.66	3.00	
Swine	ME kcal/kg	*	3250.	3664.	
Cattle	TDN %	*	73.5	82.9	
Sheep	TDN %	*	73.6	83.0	
Swine	TDN %	*	79.5	89.7	
Aluminum	mg/kg		5.16	5.81	
Calcium	%		0.04	0.05	
Cobalt	mg/kg		0.113	0.128	
Copper	mg/kg		8.9	10.0	
Iron	%		0.005	0.005	
Magnesium	%		0.19	0.21	
Manganese	mg/kg		73.2	82.6	
Phosphorus	%		0.59	0.66	
Potassium	%		0.62	0.70	
Selenium	mg/kg		0.299	0.337	
Sodium	%		0.01	0.01	
Zinc	mg/kg		68.1	76.7	
Choline	mg/kg		1860.	2097.	
Folic acid	mg/kg		1.12	1.27	
Niacin	mg/kg		63.3	71.4	
Pantothenic acid	mg/kg		17.8	20.1	
Riboflavin	mg/kg		3.2	3.6	
Thiamine	mg/kg		29.7	33.5	
α-tocopherol	mg/kg		38.4	43.3	
Vitamin B6	mg/kg		5.07	5.72	
Alanine	%		0.79	0.90	
Arginine	%		1.07	1.21	
Aspartic acid	%		1.05	1.19	
Cystine	%		0.38	0.43	
Glutamic acid	%		4.12	4.64	
Glycine	%		0.83	0.93	
Histidine	%		0.42	0.48	
Isoleucine	%		0.54	0.60	
Leucine	%		1.07	1.21	
Lysine	%		0.64	0.72	
Methionine	%		0.27	0.30	
Phenylalanine	%		0.70	0.79	
Proline	%		1.22	1.37	
Serine	%		0.78	0.88	
Threonine	%		0.56	0.63	
Tyrosine	%		0.47	0.53	
Valine	%		0.74	0.84	

Wheat, hard red spring, grain, (4)

Ref No 4 05 258 United States

Feed Name or Analyses			Mean As Fed	Mean Dry	C.V. ± %
Dry matter	%		89.2	100.0	2 1
Ash	%		1.7	1.9	
Crude fiber	%		2.4	2.7	
Sheep	dig coef %	*	33.	33.	
Ether extract	%		2.0	2.3	
Sheep	dig coef %	*	72.	72.	
N-free extract	%		69.5	78.0	
Sheep	dig coef %	*	92.	92.	
Protein (N x 6.25)	%		13.5	15.1	
Sheep	dig coef %	*	78.	78.	
Cattle	dig prot %	*	8.8	9.9	
Goats	dig prot %	*	9.9	11.1	
Horses	dig prot %	*	9.9	11.1	
Sheep	dig prot %		10.5	11.8	
Starch	%		59.1	66.3	
Energy	GE Mcal/kg		3.67	4.11	
Cattle	DE Mcal/kg	*	3.43	3.85	
Sheep	DE Mcal/kg	*	3.47	3.89	
Swine	DE kcal/kg	*	3598.	4034.	

Continued

Feed Name or Analyses			Mean As Fed	Mean Dry	C.V. ± %
Cattle	ME Mcal/kg	*	2.82	3.16	
Sheep	ME Mcal/kg	*	2.84	3.19	
Swine	ME kcal/kg	*	3344.	3750.	
Cattle	NEm Mcal/kg	*	1.92	2.15	
Cattle	NEgain Mcal/kg	*	1.27	1.42	
Cattle	NElactating cows Mcal/kg	*	2.22	2.49	
Cattle	TDN %	*	77.9	87.3	
Sheep	TDN %		78.6	88.2	
Swine	TDN %	*	81.6	91.5	
Aluminum	mg/kg		5.18	5.81	
Calcium	%		0.04	0.04	31
Cobalt	mg/kg		0.124	0.140	
Copper	mg/kg		8.2	9.2	28
Iron	%		0.003	0.004	36
Magnesium	%		0.11	0.13	
Manganese	mg/kg		50.7	56.9	27
Phosphorus	%		0.39	0.44	20
Potassium	%		0.36	0.40	
Selenium	mg/kg		0.518	0.581	
Sodium	%		0.01	0.01	
Zinc	mg/kg		38.4	43.0	
Carotene	mg/kg		0.0	0.0	
Choline	mg/kg		1026.	1151.	10
Folic acid	mg/kg		0.44	0.49	25
Niacin	mg/kg		56.2	63.1	34
Pantothenic acid	mg/kg		12.5	14.0	13
Riboflavin	mg/kg		1.3	1.4	18
Thiamine	mg/kg		5.1	5.8	13
α-tocopherol	mg/kg		14.4	16.2	
Vitamin B6	mg/kg		3.64	4.08	17
Alanine	%		0.47	0.52	
Arginine	%		0.58	0.65	22
Aspartic acid	%		0.64	0.72	
Cystine	%		0.26	0.29	17
Glutamic acid	%		4.43	4.97	
Glycine	%		0.72	0.80	
Histidine	%		0.22	0.25	19
Isoleucine	%		0.58	0.66	
Leucine	%		0.94	1.05	
Lysine	%		0.34	0.38	26
Methionine	%		0.19	0.22	
Phenylalanine	%		0.71	0.79	
Proline	%		1.40	1.57	
Serine	%		0.63	0.71	
Threonine	%		0.37	0.41	
Tryptophan	%		0.18	0.20	24
Tyrosine	%		0.60	0.67	
Valine	%		0.63	0.71	

Wheat, hard red spring, grain, grnd, (4)
Ref No 4 08 667 United States

Feed Name or Analyses			Mean As Fed	Mean Dry	C.V. ± %
Dry matter	%		86.0	100.0	
Ash	%		1.5	1.7	
Crude fiber	%		2.1	2.4	
Ether extract	%		1.8	2.1	
N-free extract	%		69.5	80.8	
Protein (N x 6.25)	%		11.1	12.9	
Cattle	dig prot %	*	6.8	7.9	
Goats	dig prot %	*	7.8	9.1	
Horses	dig prot %	*	7.8	9.1	
Sheep	dig prot %	*	7.8	9.1	
Starch	%		56.8	66.0	
Energy	GE Mcal/kg				
Cattle	DE Mcal/kg	*	3.10	3.60	
Sheep	DE Mcal/kg	*	3.21	3.73	
Swine	DE kcal/kg	*	3473.	4039.	
Cattle	ME Mcal/kg	*	2.54	2.95	
Sheep	ME Mcal/kg	*	2.63	3.06	
Swine	ME kcal/kg	*	3244.	3772.	
Cattle	TDN %	*	70.3	81.7	
Sheep	TDN %	*	72.8	84.7	
Swine	TDN %	*	78.8	91.6	
Calcium	%		0.02	0.02	
Cobalt	mg/kg		0.110	0.128	
Copper	mg/kg		4.2	4.9	
Iron	%		0.002	0.003	
Magnesium	%		0.12	0.14	
Manganese	mg/kg		34.0	39.5	
Phosphorus	%		0.36	0.42	
Potassium	%		0.32	0.37	
Selenium	mg/kg		0.700	0.814	
Sodium	%		0.01	0.01	
Zinc	mg/kg		25.0	29.1	
Choline	mg/kg		1153.	1340.	
Folic acid	mg/kg		0.40	0.47	
Niacin	mg/kg		49.7	57.8	
Pantothenic acid	mg/kg		8.3	9.7	
Riboflavin	mg/kg		1.5	1.7	
Thiamine	mg/kg		4.2	4.9	
Alanine	%		0.40	0.47	
Arginine	%		0.50	0.58	
Aspartic acid	%		0.55	0.64	
Cystine	%		0.31	0.36	
Glutamic acid	%		3.72	4.33	
Glycine	%		0.46	0.53	
Histidine	%		0.24	0.28	
Isoleucine	%		0.38	0.44	
Leucine	%		0.74	0.86	
Lysine	%		0.28	0.33	
Methionine	%		0.18	0.21	
Phenylalanine	%		0.51	0.59	
Proline	%		1.15	1.34	
Serine	%		0.53	0.62	
Threonine	%		0.33	0.38	
Tyrosine	%		0.33	0.38	
Valine	%		0.50	0.58	

Wheat, hard red spring, grain, Can 3 Northern mn wt 57 lb per bushel mx 2% foreign material, (4)
Ref No 4 09 075 Canada

Feed Name or Analyses			Mean As Fed	Mean Dry	C.V. ± %
Dry matter	%		90.5	100.0	
Ash	%		1.6	1.7	
Ether extract	%		2.0	2.2	
Protein (N x 6.25)	%		16.7	18.4	
Cattle	dig prot %	*	11.7	12.9	
Goats	dig prot %	*	12.8	14.1	
Horses	dig prot %	*	12.8	14.1	
Sheep	dig prot %	*	12.8	14.1	
Calcium	%		0.03	0.04	
Magnesium	%		0.10	0.11	
Phosphorus	%		0.41	0.45	
Potassium	%		0.31	0.34	
Sodium	%		0.01	0.01	
Choline	mg/kg		2.	2.	
Niacin	mg/kg		24.8	27.4	
Pantothenic acid	mg/kg		7.1	7.8	
Riboflavin	mg/kg		0.5	0.5	

Wheat, hard red spring, grain, feed gr, (4)
Ref No 4 05 248 United States

Feed Name or Analyses			Mean As Fed	Mean Dry	C.V. ± %
Dry matter	%			100.0	
Crude fiber	%				
Swine	dig coef %	*		24.	
Ether extract	%				
Swine	dig coef %	*		80.	
N-free extract	%				
Swine	dig coef %	*		94.	
Protein (N x 6.25)	%				
Swine	dig coef %	*		92.	

Wheat, hard red spring, grain, gr 1 US mn wt 58 lb per bushel mx 0.5% foreign material, (4)
Ref No 4 05 251 United States

Feed Name or Analyses			Mean As Fed	Mean Dry	C.V. ± %
Dry matter	%		86.0	100.0	
Ash	%		1.8	2.1	9
Ether extract	%		1.7	2.0	9
Protein (N x 6.25)	%		13.2	15.4	8
Cattle	dig prot %	*	8.7	10.2	
Goats	dig prot %	*	9.8	11.3	
Horses	dig prot %	*	9.8	11.3	
Sheep	dig prot %	*	9.8	11.3	
Starch	%		54.3	63.1	2
Sugars, total	%		2.4	2.8	7
Thiamine	mg/kg		5.3	6.2	

Wheat, hard red spring, grain, gr 2 US mn wt 57 lb per bushel mx 1% foreign material, (4)
Ref No 4 05 253 United States

Feed Name or Analyses			Mean As Fed	Mean Dry	C.V. ± %
Dry matter	%		88.2	100.0	2
Ash	%		1.8	2.1	11
Crude fiber	%		2.4	2.7	
Ether extract	%		1.6	1.9	10
N-free extract	%		69.7	79.1	
Protein (N x 6.25)	%		12.6	14.3	13
Cattle	dig prot %	*	8.1	9.2	
Goats	dig prot %	*	9.1	10.4	
Horses	dig prot %	*	9.1	10.4	
Sheep	dig prot %	*	9.1	10.4	
Starch	%		55.5	62.9	2
Sugars, total	%		2.5	2.8	5
Energy	GE Mcal/kg				
Cattle	DE Mcal/kg	*	3.26	3.70	
Sheep	DE Mcal/kg	*	3.25	3.69	
Swine	DE kcal/kg	*	3524.	3996.	
Cattle	ME Mcal/kg	*	2.67	3.03	
Sheep	ME Mcal/kg	*	2.67	3.02	
Swine	ME kcal/kg	*	3281.	3720.	
Cattle	TDN %	*	74.1	84.0	
Sheep	TDN %	*	73.8	83.6	
Swine	TDN %	*	79.9	90.6	
Calcium	%		0.05	0.06	
Phosphorus	%		0.41	0.46	

Wheat, hard red spring, grain, gr 3 US mn wt 55 lb per bushel mx 2% foreign material, (4)
Ref No 4 05 255 United States

Feed Name or Analyses			Mean As Fed	Mean Dry	C.V. ± %
Dry matter	%		88.7	100.0	1
Ash	%		1.8	2.0	16
Crude fiber	%		2.4	2.7	16
Ether extract	%		1.8	2.0	15

(1) dry forages and roughages (2) pasture, range plants, and forages fed green (3) silages (4) energy feeds (5) protein supplements (6) minerals (7) vitamins (8) additives

Feed Name or Analyses			Mean As Fed	Mean Dry	C.V. ± %
N-free extract	%		68.4	77.1	
Protein (N x 6.25)	%		14.4	16.2	14
Cattle	dig prot %	*	9.7	10.9	
Goats	dig prot %	*	10.7	12.1	
Horses	dig prot %	*	10.7	12.1	
Sheep	dig prot %	*	10.7	12.1	
Starch	%		56.2	63.3	3
Sugars, total	%		2.5	2.8	6
Energy	GE Mcal/kg				
Cattle	DE Mcal/kg	*	3.23	3.64	
Sheep	DE Mcal/kg	*	3.25	3.66	
Swine	DE kcal/kg	*	3580.	4035.	
Cattle	ME Mcal/kg	*	2.64	2.98	
Sheep	ME Mcal/kg	*	2.66	3.00	
Swine	ME kcal/kg	*	3320.	3742.	
Cattle	TDN %	*	73.3	82.6	
Sheep	TDN %	*	73.6	83.0	
Swine	TDN %	*	81.2	91.5	
Calcium	%		0.08	0.09	
Copper	mg/kg		8.0	9.0	
Iron	%		0.004	0.005	
Manganese	mg/kg		28.4	32.0	
Phosphorus	%		0.42	0.47	

Wheat, hard red spring, grain, gr 4 US mn wt 53 lb per bushel mx 3% foreign material, (4)

Ref No 4 05 256 United States

Feed Name or Analyses			Mean As Fed	Mean Dry	C.V. ± %
Dry matter	%		89.1	100.0	
Ash	%		2.0	2.3	10
Crude fiber	%		2.5	2.8	
Ether extract	%		1.8	2.0	8
N-free extract	%		69.2	77.6	
Protein (N x 6.25)	%		13.7	15.3	8
Cattle	dig prot %	*	9.0	10.1	
Goats	dig prot %	*	10.1	11.3	
Horses	dig prot %	*	10.1	11.3	
Sheep	dig prot %	*	10.1	11.3	
Starch	%		55.6	62.4	3
Sugars, total	%		2.6	2.9	5
Energy	GE Mcal/kg				
Cattle	DE Mcal/kg	*	3.24	3.64	
Sheep	DE Mcal/kg	*	3.26	3.66	
Swine	DE kcal/kg	*	3548.	3982.	
Cattle	ME Mcal/kg	*	2.66	2.98	
Sheep	ME Mcal/kg	*	2.68	3.00	
Swine	ME kcal/kg	*	3297.	3700.	
Cattle	TDN %	*	73.5	82.5	
Sheep	TDN %	*	74.0	83.1	
Swine	TDN %	*	80.5	90.3	
Calcium	%		0.07	0.08	
Phosphorus	%		0.37	0.42	

Wheat, hard red spring, grain, gr 5 US mn wt 50 lb per bushel mx 5% foreign material, (4)

Ref No 4 05 257 United States

Feed Name or Analyses			Mean As Fed	Mean Dry	C.V. ± %
Dry matter	%		89.3	100.0	
Starch	%		55.5	62.1	3
Sugars, total	%		2.5	2.8	7
Calcium	%		0.05	0.06	
Phosphorus	%		0.45	0.50	

Wheat, hard red spring, grain, gr sample US, (4)

Ref No 4 05 249 United States

Feed Name or Analyses			Mean As Fed	Mean Dry	C.V. ± %
Dry matter	%			100.0	
Ash	%			2.2	7
Ether extract	%			2.1	12
Protein (N x 6.25)	%			15.0	11
Cattle	dig prot %	*		9.8	
Goats	dig prot %	*		11.0	
Horses	dig prot %	*		11.0	
Sheep	dig prot %	*		11.0	
Starch	%			62.4	2
Sugars, total	%			2.9	3

Wheat, hard red spring, germ, (5)

Ref No 5 08 013 United States

Feed Name or Analyses			Mean As Fed	Mean Dry	C.V. ± %
Dry matter	%		88.3	100.0	
Ash	%		4.2	4.8	
Crude fiber	%		2.8	3.2	
Ether extract	%		10.4	11.8	
N-free extract	%		46.9	53.1	
Protein (N x 6.25)	%		24.0	27.2	
Starch	%		19.2	21.7	
Energy	GE Mcal/kg		4.52	5.12	
Cattle	DE Mcal/kg	*	3.71	4.20	
Sheep	DE Mcal/kg	*	3.58	4.06	
Swine	DE kcal/kg	*	3842.	4352.	
Cattle	ME Mcal/kg	*	3.04	3.45	
Sheep	ME Mcal/kg	*	2.94	3.33	
Swine	ME kcal/kg	*	3478.	3939.	
Cattle	TDN %	*	84.2	95.4	
Sheep	TDN %	*	81.3	92.1	
Swine	TDN %	*	87.2	98.7	
Aluminum	mg/kg		5.13	5.81	
Calcium	%		0.04	0.05	
Cobalt	mg/kg		0.144	0.163	
Copper	mg/kg		8.8	10.0	
Iron	%		0.004	0.005	
Magnesium	%		0.23	0.26	
Manganese	mg/kg		131.4	148.8	
Phosphorus	%		0.88	1.00	
Potassium	%		0.92	1.05	
Selenium	mg/kg		0.472	0.535	
Sodium	%		0.02	0.02	
Zinc	mg/kg		127.3	144.2	
Choline	mg/kg		3261.	3693.	
Folic acid	mg/kg		2.87	3.26	
Niacin	mg/kg		72.3	81.9	
Pantothenic acid	mg/kg		23.8	27.0	
Riboflavin	mg/kg		6.3	7.1	
Thiamine	mg/kg		22.5	25.5	
α-tocopherol	mg/kg		180.0	203.8	
Vitamin B6	mg/kg		11.41	12.92	
Alanine	%		1.62	1.84	
Arginine	%		2.06	2.34	
Aspartic acid	%		2.25	2.55	
Cystine	%		0.48	0.55	
Glutamic acid	%		4.07	4.60	
Glycine	%		1.58	1.79	
Histidine	%		0.78	0.88	
Isoleucine	%		0.92	1.05	
Leucine	%		1.72	1.95	
Lysine	%		1.66	1.88	
Methionine	%		0.49	0.56	
Phenylalanine	%		1.01	1.14	
Proline	%		1.23	1.40	
Serine	%		1.14	1.29	
Threonine	%		1.09	1.23	
Tyrosine	%		0.80	0.91	
Valine	%		1.31	1.49	

WHEAT, HARD RED WINTER. *Triticum aestivum*

Wheat, hard red winter, bran, (4)

Ref No 4 08 008 United States

Feed Name or Analyses			Mean As Fed	Mean Dry	C.V. ± %
Dry matter	%		87.9	100.0	
Ash	%		6.8	7.7	
Crude fiber	%		10.1	11.5	
Ether extract	%		3.3	3.8	
N-free extract	%		54.1	61.5	
Protein (N x 6.25)	%		13.6	15.5	
Cattle	dig prot %	*	9.0	10.2	
Goats	dig prot %	*	10.0	11.4	
Horses	dig prot %	*	10.0	11.4	
Sheep	dig prot %	*	10.0	11.4	
Starch	%		5.7	6.5	
Energy	GE Mcal/kg		3.99	4.54	
Cattle	DE Mcal/kg	*	2.81	3.20	
Sheep	DE Mcal/kg	*	2.88	3.28	
Swine	DE kcal/kg	*	2624.	2985.	
Cattle	ME Mcal/kg	*	2.30	2.62	
Sheep	ME Mcal/kg	*	2.36	2.69	
Swine	ME kcal/kg	*	2437.	2773.	
Cattle	TDN %	*	63.7	72.5	
Sheep	TDN %	*	65.3	74.3	
Swine	TDN %	*	59.5	67.7	
Aluminum	mg/kg		6.13	6.98	
Calcium	%		0.10	0.12	
Cobalt	mg/kg		0.112	0.128	
Copper	mg/kg		12.3	14.0	
Iron	%		0.009	0.010	
Magnesium	%		0.51	0.58	
Manganese	mg/kg		96.5	109.8	
Phosphorus	%		1.33	1.51	
Potassium	%		1.53	1.74	
Selenium	mg/kg		0.409	0.465	
Sodium	%		0.02	0.02	
Zinc	mg/kg		123.1	140.0	
Choline	mg/kg		2285.	2600.	
Folic acid	mg/kg		1.19	1.35	
Niacin	mg/kg		307.3	349.7	
Pantothenic acid	mg/kg		31.7	36.0	
Riboflavin	mg/kg		5.3	6.0	
Thiamine	mg/kg		6.2	7.1	
α-tocopherol	mg/kg		27.4	31.2	
Vitamin B6	mg/kg		8.73	9.93	
Alanine	%		0.77	0.87	
Arginine	%		1.09	1.24	
Aspartic acid	%		1.13	1.29	
Cystine	%		0.36	0.41	
Glutamic acid	%		2.87	3.27	
Glycine	%		0.89	1.01	
Histidine	%		0.45	0.51	
Isoleucine	%		0.47	0.53	
Leucine	%		0.93	1.06	
Lysine	%		0.60	0.69	
Methionine	%		0.21	0.24	
Phenylalanine	%		0.58	0.66	
Proline	%		0.87	0.99	
Serine	%		0.67	0.77	
Threonine	%		0.49	0.56	
Tyrosine	%		0.44	0.50	
Valine	%		0.64	0.73	

Wheat, hard red winter, flour, (4)

Ref No 4 08 110 United States

Feed Name or Analyses		Mean As Fed	Mean Dry	C.V. ± %
Dry matter	%	86.2	100.0	
Ash	%	0.4	0.5	
Ether extract	%	0.8	1.0	
Protein (N x 6.25)	%	10.7	12.5	
Cattle	dig prot %	* 6.4	7.5	
Goats	dig prot %	* 7.5	8.7	
Horses	dig prot %	* 7.5	8.7	
Sheep	dig prot %	* 7.5	8.7	
Starch	%	70.9	82.3	
Energy	GE Mcal/kg	3.78	4.39	
Aluminum	mg/kg	5.01	5.81	
Calcium	%	0.03	0.03	
Cobalt	mg/kg	0.060	0.070	
Copper	mg/kg	0.6	0.7	
Iron	%	0.000	0.001	
Magnesium	%	0.01	0.02	
Manganese	mg/kg	2.6	3.0	
Phosphorus	%	0.09	0.10	
Potassium	%	0.12	0.13	
Selenium	mg/kg	0.110	0.128	
Sodium	%	0.00	0.01	
Zinc	mg/kg	6.7	7.8	
Choline	mg/kg	766.	889.	
Folic acid	mg/kg	0.12	0.14	
Niacin	mg/kg	9.2	10.7	
Pantothenic acid	mg/kg	5.1	5.9	
Riboflavin	mg/kg	0.4	0.4	
Thiamine	mg/kg	1.1	1.3	
a-tocopherol	mg/kg	2.2	2.6	
Vitamin B6	mg/kg	0.71	0.83	
Alanine	%	0.50	0.58	
Arginine	%	0.42	0.49	
Aspartic acid	%	0.50	0.58	
Cystine	%	0.30	0.35	
Glutamic acid	%	4.46	5.17	
Glycine	%	0.43	0.50	
Histidine	%	0.24	0.28	
Isoleucine	%	0.44	0.51	
Leucine	%	0.94	1.09	
Lysine	%	0.23	0.27	
Methionine	%	0.18	0.21	
Phenylalanine	%	0.59	0.69	
Proline	%	1.44	1.67	
Serine	%	0.59	0.69	
Threonine	%	0.32	0.37	
Tyrosine	%	0.35	0.41	
Valine	%	0.50	0.58	

Wheat, hard red winter, flour by-product, coarse sift, mx 7% fiber, (4)

Wheat, hard red winter, shorts

Ref No 4 08 014 United States

Feed Name or Analyses		Mean As Fed	Mean Dry	C.V. ± %
Dry matter	%	87.8	100.0	
Ash	%	4.0	4.6	
Crude fiber	%	6.0	6.8	
Ether extract	%	4.8	5.5	
N-free extract	%	57.6	65.6	
Protein (N x 6.25)	%	15.4	17.5	
Cattle	dig prot %	* 10.6	12.1	
Goats	dig prot %	* 11.7	13.3	
Horses	dig prot %	* 11.7	13.3	
Sheep	dig prot %	* 11.7	13.3	
Starch	%	18.9	21.5	
Energy	GE Mcal/kg	4.35	4.95	
Cattle	DE Mcal/kg	* 2.99	3.41	
Sheep	DE Mcal/kg	* 3.06	3.49	
Swine	DE kcal/kg	*2981.	3395.	
Cattle	ME Mcal/kg	* 2.46	2.80	
Sheep	ME Mcal/kg	* 2.51	2.86	
Swine	ME kcal/kg	*2756.	3139.	
Cattle	TDN %	* 68.0	77.4	
Sheep	TDN %	* 69.5	79.2	
Swine	TDN %	* 67.6	77.0	
Aluminum	mg/kg	6.53	7.44	
Calcium	%	0.08	0.09	
Cobalt	mg/kg	0.112	0.128	
Copper	mg/kg	12.4	14.1	
Iron	%	0.006	0.007	
Magnesium	%	0.27	0.30	
Manganese	mg/kg	111.3	126.7	
Phosphorus	%	0.86	0.98	
Potassium	%	0.95	1.08	
Selenium	mg/kg	0.408	0.465	
Sodium	%	0.02	0.02	
Zinc	mg/kg	123.8	141.0	
Choline	mg/kg	2162.	2463.	
Folic acid	mg/kg	1.45	1.65	
Niacin	mg/kg	102.7	117.0	
Pantothenic acid	mg/kg	24.7	28.1	
Riboflavin	mg/kg	4.9	5.6	
Thiamine	mg/kg	20.3	23.1	
α-tocopherol	mg/kg	58.9	67.1	
Vitamin B6	mg/kg	8.08	9.20	
Alanine	%	0.98	1.12	
Arginine	%	1.31	1.49	
Aspartic acid	%	1.39	1.58	
Cystine	%	0.39	0.44	
Glutamic acid	%	3.20	3.64	
Glycine	%	1.00	1.14	
Histidine	%	0.48	0.55	
Isoleucine	%	0.58	0.66	
Leucine	%	1.12	1.28	
Lysine	%	0.87	0.99	
Methionine	%	0.29	0.33	
Phenylalanine	%	0.69	0.79	
Proline	%	0.99	1.13	
Serine	%	0.80	0.91	
Threonine	%	0.63	0.72	
Tyrosine	%	0.50	0.57	
Valine	%	0.86	0.98	

Wheat, hard red winter, flour by-product, fine sift, mx 4% fiber, (4)

Wheat, hard red winter, red dog

Ref No 4 08 018 United States

Feed Name or Analyses		Mean As Fed	Mean Dry	C.V. ± %
Dry matter	%	86.9	100.0	
Ash	%	2.0	2.3	
Crude fiber	%	2.1	2.4	
Ether extract	%	2.9	3.3	
N-free extract	%	66.6	76.6	
Protein (N x 6.25)	%	13.3	15.3	
Cattle	dig prot %	* 8.8	10.1	
Goats	dig prot %	* 9.8	11.3	
Horses	dig prot %	* 9.8	11.3	
Sheep	dig prot %	* 9.8	11.3	
Starch	%	45.4	52.3	
Energy	GE Mcal/kg	3.98	4.59	
Cattle	DE Mcal/kg	* 3.09	3.56	
Sheep	DE Mcal/kg	* 3.21	3.69	
Cattle	ME Mcal/kg	*3508.	4037.	
Cattle	ME Mcal/kg	* 2.54	2.92	
Sheep	ME Mcal/kg	* 2.63	3.03	
Swine	ME kcal/kg	*3259.	3750.	
Cattle	TDN %	* 70.2	80.8	
Sheep	TDN %	* 72.8	83.8	
Swine	TDN %	* 79.6	91.6	
Aluminum	mg/kg	5.05	5.81	
Calcium	%	0.04	0.05	
Cobalt	mg/kg	0.111	0.128	
Copper	mg/kg	6.1	7.0	
Iron	%	0.004	0.004	
Magnesium	%	0.12	0.14	
Manganese	mg/kg	52.8	60.8	
Phosphorus	%	0.47	0.55	
Potassium	%	0.45	0.52	
Selenium	mg/kg	0.333	0.384	
Sodium	%	0.01	0.01	
Zinc	mg/kg	72.8	83.7	
Choline	mg/kg	1523.	1752.	
Folic acid	mg/kg	0.60	0.69	
Niacin	mg/kg	35.9	41.3	
Pantothenic acid	mg/kg	12.4	14.3	
Riboflavin	mg/kg	2.3	2.7	
Thiamine	mg/kg	21.8	25.1	
a-tocopherol	mg/kg	30.9	35.6	
Vitamin B6	mg/kg	5.41	6.22	
Alanine	%	0.64	0.73	
Arginine	%	0.90	1.03	
Aspartic acid	%	0.89	1.02	
Cystine	%	0.35	0.41	
Glutamic acid	%	3.81	4.38	
Glycine	%	0.67	0.77	
Histidine	%	0.35	0.41	
Isoleucine	%	0.48	0.56	
Leucine	%	0.97	1.12	
Lysine	%	0.54	0.62	
Methionine	%	0.23	0.27	
Phenylalanine	%	0.62	0.71	
Proline	%	1.21	1.40	
Serine	%	0.68	0.78	
Threonine	%	0.45	0.52	
Tyrosine	%	0.42	0.49	
Valine	%	0.67	0.77	

Wheat, hard red winter, grain, (4)

Ref No 4 05 268 United States

Feed Name or Analyses		Mean As Fed	Mean Dry	C.V. ± %
Dry matter	%	88.5	100.0	1
Ash	%	1.7	1.9	
Crude fiber	%	2.3	2.6	
Sheep	dig coef %	* 33.	33.	
Ether extract	%	1.7	1.9	
Sheep	dig coef %	* 72.	72.	
N-free extract	%	70.7	79.9	
Sheep	dig coef %	* 92.	92.	
Protein (N x 6.25)	%	12.1	13.7	
Sheep	dig coef %	* 78.	78.	
Cattle	dig prot %	* 7.6	8.6	
Goats	dig prot %	* 8.7	9.8	
Horses	dig prot %	* 8.7	9.8	
Sheep	dig prot %	9.5	10.7	
Starch	%	58.0	65.5	
Energy	GE Mcal/kg	3.63	4.10	
Cattle	DE Mcal/kg	* 3.36	3.80	
Sheep	DE Mcal/kg	* 3.44	3.89	

(1) dry forages and roughages (2) pasture, range plants, and forages fed green (3) silages (4) energy feeds (5) protein supplements (6) minerals (7) vitamins (8) additives

Feed Name or Analyses			Mean As Fed	Mean Dry	C.V. ± %
Cattle	ME Mcal/kg	*	3558.	4020.	
Cattle	ME Mcal/kg	*	2.76	3.12	
Sheep	ME Mcal/kg	*	2.82	3.19	
Swine	ME kcal/kg	*	3317.	3748.	
Cattle	NE_m Mcal/kg	*	1.91	2.16	
Cattle	NE_{gain} Mcal/kg	*	1.26	1.42	
Cattle	$NE_{lactating\ cows}$ Mcal/kg	*	2.20	2.49	
Cattle	TDN %	*	76.4	86.3	
Sheep	TDN %		78.0	88.1	
Swine	TDN %	*	80.7	91.2	
Aluminum	mg/kg		5.06	5.71	
Calcium	%		0.04	0.05	28
Chlorine	%		0.03	0.03	
Cobalt	mg/kg		0.098	0.111	9
Copper	mg/kg		5.8	6.6	
Iron	%		0.003	0.004	34
Magnesium	%		0.10	0.12	
Manganese	mg/kg		33.2	37.5	35
Phosphorus	%		0.38	0.43	19
Potassium	%		0.42	0.48	
Selenium	mg/kg		0.283	0.320	
Sodium	%		0.01	0.01	
Sulphur	%		0.15	0.17	
Zinc	mg/kg		47.2	53.4	
Choline	mg/kg		1092.	1234.	
Folic acid	mg/kg		0.35	0.40	34
Niacin	mg/kg		48.6	54.9	
Pantothenic acid	mg/kg		9.9	11.2	
Riboflavin	mg/kg		1.4	1.6	
Thiamine	mg/kg		4.5	5.1	
α-tocopherol	mg/kg		14.3	16.1	
Vitamin B_6	mg/kg		2.24	2.53	
Alanine	%		0.45	0.50	
Arginine	%		0.51	0.58	
Aspartic acid	%		0.63	0.71	
Cystine	%		0.26	0.29	
Glutamic acid	%		4.06	4.58	
Glycine	%		0.52	0.58	
Histidine	%		0.19	0.21	
Isoleucine	%		0.43	0.49	
Leucine	%		0.85	0.96	
Lysine	%		0.34	0.39	
Methionine	%		0.24	0.27	
Phenylalanine	%		0.60	0.67	
Proline	%		1.32	1.50	
Serine	%		0.62	0.70	
Threonine	%		0.36	0.41	
Tryptophan	%		0.18	0.20	
Tyrosine	%		0.46	0.52	
Valine	%		0.53	0.59	

Wheat, hard red winter, grain, grnd, (4)

Ref No 4 08 785 — United States

Feed Name or Analyses			Mean As Fed	Mean Dry	C.V. ± %
Dry matter	%		86.0	100.0	
Ash	%		1.5	1.8	
Crude fiber	%		2.4	2.8	
Ether extract	%		1.4	1.7	
N-free extract	%		68.5	79.7	
Protein (N x 6.25)	%		12.1	14.1	
Cattle	dig prot %	*	7.7	9.0	
Goats	dig prot %	*	8.7	10.2	
Horses	dig prot %	*	8.7	10.2	
Sheep	dig prot %	*	8.7	10.2	
Starch	%		57.9	67.3	
Energy	GE Mcal/kg				
Cattle	DE Mcal/kg	*	3.12	3.63	
Sheep	DE Mcal/kg	*	3.18	3.70	
Swine	DE kcal/kg	*	3453.	4015.	
Cattle	ME Mcal/kg	*	2.56	2.98	
Sheep	ME Mcal/kg	*	2.61	3.03	
Swine	ME kcal/kg	*	3217.	3740.	
Cattle	TDN %	*	70.9	82.4	
Sheep	TDN %	*	72.1	83.8	
Swine	TDN %	*	78.3	91.1	
Calcium	%		0.04	0.05	
Cobalt	mg/kg		0.157	0.182	
Copper	mg/kg		4.3	5.0	
Iron	%		0.003	0.003	
Magnesium	%		0.10	0.12	
Manganese	mg/kg		29.7	34.5	
Phosphorus	%		0.38	0.44	
Potassium	%		0.42	0.49	
Selenium	mg/kg		0.567	0.659	
Sodium	%		0.01	0.02	
Zinc	mg/kg		38.7	45.0	
Choline	mg/kg		1108.	1288.	
Folic acid	mg/kg		0.39	0.45	
Niacin	mg/kg		53.4	62.1	
Pantothenic acid	mg/kg		8.8	10.2	
Riboflavin	mg/kg		1.6	1.9	
Thiamine	mg/kg		3.7	4.3	
Alanine	%		0.46	0.53	
Arginine	%		0.62	0.72	
Aspartic acid	%		0.65	0.75	
Cystine	%		0.34	0.39	
Glutamic acid	%		4.08	4.74	
Glycine	%		0.54	0.63	
Histidine	%		0.30	0.34	
Isoleucine	%		0.47	0.55	
Leucine	%		0.88	1.02	
Lysine	%		0.35	0.40	
Methionine	%		0.20	0.24	
Phenylalanine	%		0.62	0.72	
Proline	%		1.33	1.55	
Serine	%		0.60	0.70	
Threonine	%		0.37	0.43	
Tyrosine	%		0.40	0.46	
Valine	%		0.58	0.67	

Wheat, hard red winter, grain, gr 1 US mn wt 60 lb per bushel mx 0.5% foreign material, (4)

Ref No 4 05 261 — United States

Feed Name or Analyses			Mean As Fed	Mean Dry	C.V. ± %
Dry matter	%			100.0	
Starch	%			64.5	2
Sugars, total	%			3.0	6

Wheat, hard red winter, grain, gr 2 US mn wt 58 lb per bushel mx 1% foreign material, (4)

Ref No 4 05 263 — United States

Feed Name or Analyses			Mean As Fed	Mean Dry	C.V. ± %
Dry matter	%			100.0	
Ash	%			2.1	8
Ether extract	%			1.7	10
Protein (N x 6.25)	%			14.2	8
Cattle	dig prot %	*		9.1	
Goats	dig prot %	*		10.2	
Horses	dig prot %	*		10.2	
Sheep	dig prot %	*		10.2	
Starch	%			63.4	2
Sugars, total	%			2.9	5

Wheat, hard red winter, grain, gr 3 US mn wt 56 lb per bushel mx 2% foreign material, (4)

Ref No 4 05 265 — United States

Feed Name or Analyses			Mean As Fed	Mean Dry	C.V. ± %
Dry matter	%			100.0	
Ash	%			2.1	13
Ether extract	%			1.7	11
Protein (N x 6.25)	%			14.1	16
Cattle	dig prot %	*		9.0	
Goats	dig prot %	*		10.2	
Horses	dig prot %	*		10.2	
Sheep	dig prot %	*		10.2	
Starch	%			63.4	2
Sugars, total	%			2.8	5

Wheat, hard red winter, grain, gr 4 US mn wt 54 lb per bushel mx 3% foreign material, (4)

Ref No 4 05 266 — United States

Feed Name or Analyses			Mean As Fed	Mean Dry	C.V. ± %
Dry matter	%			100.0	
Ash	%			2.2	13
Ether extract	%			1.7	6
Protein (N x 6.25)	%			15.2	9
Cattle	dig prot %	*		10.0	
Goats	dig prot %	*		11.2	
Horses	dig prot %	*		11.2	
Sheep	dig prot %	*		11.2	
Starch	%			63.8	1
Sugars, total	%			2.7	4

Wheat, hard red winter, grain, gr sample US, (4)

Ref No 4 05 259 — United States

Feed Name or Analyses			Mean As Fed	Mean Dry	C.V. ± %
Dry matter	%			100.0	
Ash	%			2.0	
Ether extract	%			1.7	
Protein (N x 6.25)	%			12.6	
Cattle	dig prot %	*		7.6	
Goats	dig prot %	*		8.8	
Horses	dig prot %	*		8.8	
Sheep	dig prot %	*		8.8	
Starch	%			66.1	
Sugars, total	%			2.8	

Wheat, hard red winter, germ, (5)

Ref No 5 08 012 — United States

Feed Name or Analyses			Mean As Fed	Mean Dry	C.V. ± %
Dry matter	%		87.5	100.0	
Ash	%		4.2	4.8	
Crude fiber	%		3.3	3.8	
Ether extract	%		7.0	8.0	
N-free extract	%		50.9	58.2	
Protein (N x 6.25)	%		22.1	25.3	
Starch	%		17.0	19.4	
Energy	GE Mcal/kg		4.40	5.03	
Cattle	DE Mcal/kg	*	3.51	4.01	
Sheep	DE Mcal/kg	*	3.33	3.81	
Swine	DE kcal/kg	*	3765.	4303.	
Cattle	ME Mcal/kg	*	2.88	3.29	
Sheep	ME Mcal/kg	*	2.73	3.12	
Swine	ME kcal/kg	*	3422.	3911.	
Cattle	TDN %	*	79.5	90.9	
Sheep	TDN %	*	75.7	86.5	
Swine	TDN %	*	85.4	97.6	
Aluminum	mg/kg		5.09	5.81	

Continued

Feed Name or Analyses		Mean As Fed	Mean Dry	C.V. ± %
Calcium	%	0.06	0.07	
Cobalt	mg/kg	0.102	0.116	
Copper	mg/kg	9.3	10.6	
Iron	%	0.005	0.006	
Magnesium	%	0.24	0.28	
Manganese	mg/kg	113.6	129.9	
Phosphorus	%	0.94	1.07	
Potassium	%	1.02	1.16	
Selenium	mg/kg	0.346	0.395	
Sodium	%	0.02	0.02	
Zinc	mg/kg	122.8	140.3	
Choline	mg/kg	3097.	3540.	
Folic acid	mg/kg	1.64	1.87	
Niacin	mg/kg	78.8	90.0	
Pantothenic acid	mg/kg	24.4	27.9	
Riboflavin	mg/kg	5.7	6.5	
Thiamine	mg/kg	22.4	25.6	
α-tocopherol	mg/kg	156.3	178.6	
Vitamin B6	mg/kg	13.10	14.98	
Alanine	%	1.47	1.67	
Arginine	%	1.86	2.13	
Aspartic acid	%	2.07	2.36	
Cystine	%	0.47	0.53	
Glutamic acid	%	4.11	4.70	
Glycine	%	1.40	1.60	
Histidine	%	0.63	0.72	
Isoleucine	%	0.80	0.92	
Leucine	%	1.57	1.79	
Lysine	%	1.39	1.59	
Methionine	%	0.43	0.49	
Phenylalanine	%	0.95	1.08	
Proline	%	1.30	1.49	
Serine	%	1.12	1.28	
Threonine	%	0.96	1.09	
Tyrosine	%	0.70	0.80	
Valine	%	1.14	1.30	

WHEAT, HARD SPRING. Triticum aestivum

Wheat, hard spring, bran, (4)

Ref No 4 08 551 — United States

Feed Name or Analyses			Mean As Fed	Mean Dry	C.V. ± %
Dry matter		%	90.5	100.0	
Ash		%	6.3	7.0	
Crude fiber		%	10.5	11.6	
Ether extract		%	4.9	5.4	
N-free extract		%	51.4	56.8	
Protein (N x 6.25)		%	17.4	19.2	
Cattle	dig prot	%	* 12.4	13.7	
Goats	dig prot	%	* 13.4	14.9	
Horses	dig prot	%	* 13.4	14.9	
Sheep	dig prot	%	* 13.4	14.9	
Energy		GE Mcal/kg			
Cattle	DE Mcal/kg		* 2.90	3.21	
Sheep	DE Mcal/kg		* 2.94	3.25	
Swine	DE kcal/kg		*2781.	3073.	
Cattle	ME Mcal/kg		* 2.38	2.63	
Sheep	ME Mcal/kg		* 2.41	2.67	
Swine	ME kcal/kg		*2561.	2830.	
Cattle	TDN	%	* 65.9	72.8	
Sheep	TDN	%	* 66.8	73.8	
Swine	TDN	%	* 63.1	69.7	
Calcium		%	0.13	0.14	
Phosphorus		%	1.35	1.49	

(1) dry forages and roughages (2) pasture, range plants, and forages fed green (3) silages (4) energy feeds (5) protein supplements (6) minerals (7) vitamins (8) additives

Wheat, hard spring, grain, (1)

Ref No 1 08 317 — United States

Feed Name or Analyses			Mean As Fed	Mean Dry	C.V. ± %
Dry matter		%	90.1	100.0	
Ash		%	1.8	2.0	
Crude fiber		%	2.5	2.8	
Ether extract		%	2.2	2.4	
N-free extract		%	67.8	75.2	
Protein (N x 6.25)		%	15.8	17.5	
Cattle	dig prot	%	* 10.9	12.1	
Goats	dig prot	%	* 11.6	12.9	
Horses	dig prot	%	* 11.2	12.4	
Rabbits	dig coef	%	* 11.0	12.2	
Sheep	dig prot	%	* 11.1	12.3	
Energy		GE Mcal/kg			
Cattle	DE Mcal/kg		* 3.56	3.95	
Sheep	DE Mcal/kg		* 3.26	3.61	
Cattle	ME Mcal/kg		* 2.92	3.24	
Chickens	ME_n kcal/kg		3086.	3425.	
Sheep	ME Mcal/kg		* 2.67	2.96	
Cattle	TDN	%	* 80.7	89.6	
Sheep	TDN	%	* 73.9	82.0	
Calcium		%	0.04	0.04	
Phosphorus		%	0.40	0.44	
Choline		mg/kg	1014.	1126.	
Niacin		mg/kg	63.5	70.5	
Pantothenic acid		mg/kg	14.1	15.7	
Riboflavin		mg/kg	1.1	1.2	
Arginine		%	0.60	0.67	
Cystine		%	0.25	0.28	
Lysine		%	0.50	0.55	
Methionine		%	0.20	0.22	
Tryptophan		%	0.20	0.22	

Wheat, hard spring, germ, (5)

Ref No 5 05 269 — United States

Feed Name or Analyses			Mean As Fed	Mean Dry	C.V. ± %
Dry matter		%	90.8	100.0	
Ash		%	4.4	4.8	
Crude fiber		%	1.8	2.0	
Ether extract		%	11.6	12.8	
N-free extract		%	39.9	43.9	
Protein (N x 6.25)		%	33.1	36.5	
Energy		GE Mcal/kg			
Cattle	DE Mcal/kg		* 3.86	4.25	
Sheep	DE Mcal/kg		* 3.78	4.16	
Swine	DE kcal/kg		*3992.	4396.	
Cattle	ME Mcal/kg		* 3.17	3.49	
Sheep	ME Mcal/kg		* 3.10	3.41	
Swine	ME kcal/kg		*3537.	3895.	
Cattle	TDN	%	* 87.6	96.5	
Sheep	TDN	%	* 85.6	94.3	
Swine	TDN	%	* 90.5	99.7	
Calcium		%	0.05	0.06	
Chlorine		%	0.09	0.10	
Copper		mg/kg	15.0	16.5	
Iron		%	0.005	0.006	
Magnesium		%	0.33	0.36	
Manganese		mg/kg	155.2	170.9	
Phosphorus		%	1.13	1.24	
Potassium		%	0.92	1.01	
Sulphur		%	0.25	0.28	

WHEAT, HARD WINTER. Triticum aestivum

Wheat, hard winter, grain, (4)

Ref No 4 08 318 — United States

Feed Name or Analyses			Mean As Fed	Mean Dry	C.V. ± %
Dry matter		%	89.4	100.0	
Ash		%	2.1	2.3	
Crude fiber		%	2.8	3.1	
Ether extract		%	1.8	2.0	
N-free extract		%	69.2	77.4	
Protein (N x 6.25)		%	13.5	15.1	
Cattle	dig prot	%	* 8.8	9.9	
Goats	dig prot	%	* 9.9	11.1	
Horses	dig prot	%	* 9.9	11.1	
Sheep	dig prot	%	* 9.9	11.1	
Energy		GE Mcal/kg			
Cattle	DE Mcal/kg		* 3.30	3.69	
Sheep	DE Mcal/kg		* 3.27	3.66	
Swine	DE kcal/kg		*3518.	3936.	
Cattle	ME Mcal/kg		* 2.71	3.03	
Chickens	ME_n kcal/kg		3086.	3452.	
Sheep	ME Mcal/kg		* 2.68	3.00	
Swine	ME kcal/kg		*3270.	3658.	
Cattle	TDN	%	* 74.8	83.7	
Sheep	TDN	%	* 74.1	82.9	
Swine	TDN	%	* 79.8	89.3	
Calcium		%	0.05	0.06	
Phosphorus		%	0.42	0.47	
Choline		mg/kg	1014.	1134.	
Niacin		mg/kg	53.1	59.4	
Pantothenic acid		mg/kg	13.9	15.5	
Riboflavin		mg/kg	1.1	1.2	
Arginine		%	0.50	0.56	
Cystine		%	0.22	0.25	
Lysine		%	0.40	0.45	
Methionine		%	0.17	0.19	
Tryptophan		%	0.16	0.18	

Wheat, hard winter, grain, grnd, (4)

Ref No 4 08 783 — United States

Feed Name or Analyses			Mean As Fed	Mean Dry	C.V. ± %
Dry matter		%	86.0	100.0	
Ash		%	1.4	1.6	
Crude fiber		%	2.1	2.4	
Ether extract		%	1.5	1.7	
N-free extract		%	69.5	80.8	
Protein (N x 6.25)		%	11.5	13.4	
Cattle	dig prot	%	* 7.1	8.3	
Goats	dig prot	%	* 8.2	9.5	
Horses	dig prot	%	* 8.2	9.5	
Sheep	dig prot	%	* 8.2	9.5	
Starch		%	56.3	65.5	
Energy		GE Mcal/kg			
Cattle	DE Mcal/kg		* 3.26	3.79	
Sheep	DE Mcal/kg		* 3.20	3.72	
Swine	DE kcal/kg		*3486.	4054.	
Cattle	ME Mcal/kg		* 2.67	3.11	
Sheep	ME Mcal/kg		* 2.63	3.05	
Swine	ME kcal/kg		*3253.	3782.	
Cattle	TDN	%	* 73.9	85.9	
Sheep	TDN	%	* 72.6	84.5	
Swine	TDN	%	* 79.1	91.9	
Calcium		%	0.04	0.05	
Cobalt		mg/kg	0.070	0.081	
Copper		mg/kg	4.3	5.0	
Iron		%	0.002	0.003	
Magnesium		%	0.10	0.12	
Manganese		mg/kg	24.0	27.9	

Feed Name or Analyses				Mean As Fed	Mean Dry	C.V. ± %
Phosphorus		%		0.35	0.41	
Potassium		%		0.34	0.40	
Selenium		mg/kg		0.300	0.349	
Sodium		%		0.01	0.01	
Zinc		mg/kg		40.0	46.5	
Choline		mg/kg		970.	1128.	
Folic acid		mg/kg		0.41	0.48	
Niacin		mg/kg		56.0	65.1	
Pantothenic acid		mg/kg		10.1	11.7	
Riboflavin		mg/kg		1.6	1.9	
Thiamine		mg/kg		3.9	4.5	
Alanine		%		0.48	0.56	
Arginine		%		0.61	0.71	
Aspartic acid		%		0.67	0.78	
Cystine		%		0.32	0.37	
Glutamic acid		%		4.16	4.84	
Glycine		%		0.56	0.65	
Histidine		%		0.30	0.35	
Isoleucine		%		0.50	0.58	
Leucine		%		0.89	1.03	
Lysine		%		0.34	0.40	
Methionine		%		0.19	0.22	
Phenylalanine		%		0.61	0.71	
Proline		%		1.36	1.58	
Serine		%		0.54	0.63	
Threonine		%		0.35	0.41	
Tyrosine		%		0.39	0.45	
Valine		%		0.57	0.66	

Wheat, hard winter, germ, (5)

Ref No 5 05 270 United States

Feed Name or Analyses				Mean As Fed	Mean Dry	C.V. ± %
Dry matter		%		90.8	100.0	
Ash		%		4.3	4.7	
Crude fiber		%		2.2	2.4	
Ether extract		%		9.4	10.4	
N-free extract		%		43.7	48.1	
Protein (N x 6.25)		%		31.2	34.4	
Energy	GE	Mcal/kg				
Cattle	DE	Mcal/kg	*	3.76	4.14	
Sheep	DE	Mcal/kg	*	3.75	4.13	
Swine	DE	kcal/kg	*	3987.	4391.	
Cattle	ME	Mcal/kg	*	3.08	3.39	
Sheep	ME	Mcal/kg	*	3.08	3.39	
Swine	ME	kcal/kg	*	3551.	3912.	
Cattle	TDN	%	*	85.2	93.8	
Sheep	TDN	%	*	85.2	93.8	
Swine	TDN	%	*	90.4	99.6	
Calcium		%		0.06	0.07	
Chlorine		%		0.08	0.09	
Copper		mg/kg		20.1	22.1	
Iron		%		0.006	0.007	
Magnesium		%		0.35	0.39	
Manganese		mg/kg		227.3	250.3	
Phosphorus		%		1.12	1.23	
Potassium		%		0.98	1.08	
Sulphur		%		0.26	0.29	

WHEAT, LIGHT. Triticum aestivum

Wheat, light, grain, shrunken, (4)

Ref No 4 05 271 United States

Feed Name or Analyses				Mean As Fed	Mean Dry	C.V. ± %
Dry matter		%		86.2	100.0	
Ash		%		2.5	2.9	
Crude fiber		%		4.1	4.8	
Sheep	dig coef	%		33.	33.	
Ether extract		%		1.9	2.2	
Sheep	dig coef	%		74.	74.	
N-free extract		%		61.4	71.2	
Sheep	dig coef	%		89.	89.	
Protein (N x 6.25)		%		16.3	18.9	
Sheep	dig coef	%		80.	80.	
Cattle	dig prot	%	*	11.5	13.4	
Goats	dig prot	%	*	12.5	14.6	
Horses	dig prot	%	*	12.5	14.6	
Sheep	dig prot	%		13.0	15.1	
Energy	GE	Mcal/kg				
Cattle	DE	Mcal/kg	*	3.15	3.66	
Sheep	DE	Mcal/kg	*	3.18	3.69	
Swine	DE	kcal/kg	*	3244.	3764.	
Cattle	ME	Mcal/kg	*	2.59	3.00	
Sheep	ME	Mcal/kg	*	2.61	3.03	
Swine	ME	kcal/kg	*	2991.	3469.	
Cattle	TDN	%	*	71.6	83.1	
Sheep	TDN	%		72.2	83.7	
Swine	TDN	%	*	73.6	85.4	

WHEAT, MIXED. Triticum spp

Wheat, mixed, grain, (4)

Ref No 4 05 281 United States

Feed Name or Analyses				Mean As Fed	Mean Dry	C.V. ± %
Dry matter		%		90.5	100.0	2
Ash		%		4.9	5.4	
Crude fiber		%		7.9	8.8	
Sheep	dig coef	%		34.	34.	
Ether extract		%		4.3	4.8	
Sheep	dig coef	%		86.	86.	
N-free extract		%		56.8	62.8	
Sheep	dig coef	%		78.	78.	
Protein (N x 6.25)		%		16.6	18.3	
Sheep	dig coef	%		83.	83.	
Cattle	dig prot	%	*	11.6	12.8	
Goats	dig prot	%	*	12.7	14.0	
Horses	dig prot	%	*	12.7	14.0	
Sheep	dig prot	%		13.7	15.2	
Energy	GE	Mcal/kg				
Cattle	DE	Mcal/kg	*	3.01	3.32	
Sheep	DE	Mcal/kg	*	3.05	3.37	
Swine	DE	kcal/kg	*	2812.	3107.	
Cattle	ME	Mcal/kg	*	2.46	2.72	
Sheep	ME	Mcal/kg	*	2.50	2.76	
Swine	ME	kcal/kg	*	2596.	2868.	
Cattle	TDN	%	*	68.3	75.4	
Sheep	TDN	%		69.1	76.4	
Swine	TDN	%	*	63.8	70.5	

Wheat, mixed, grain, gr 5 US, (4)

Ref No 4 05 280 United States

Feed Name or Analyses				Mean As Fed	Mean Dry	C.V. ± %
Dry matter		%			100.0	
Starch		%			62.2	3
Sugars, total		%			2.8	6

WHEAT, PEMBINA. Triticum aestivum

Wheat, pembina, grain, grnd, (4)

Ref No 4 08 666 United States

Feed Name or Analyses				Mean As Fed	Mean Dry	C.V. ± %
Dry matter		%		86.0	100.0	
Ash		%		1.7	2.0	
Crude fiber		%		2.6	3.0	
Ether extract		%		1.9	2.2	
N-free extract		%		66.0	76.7	
Protein (N x 6.25)		%		13.8	16.0	
Cattle	dig prot	%	*	9.2	10.8	
Goats	dig prot	%	*	10.3	11.9	
Horses	dig prot	%	*	10.3	11.9	
Sheep	dig prot	%	*	10.3	11.9	
Starch		%		59.1	68.7	
Energy	GE	Mcal/kg				
Cattle	DE	Mcal/kg	*	3.07	3.57	
Sheep	DE	Mcal/kg	*	3.14	3.66	
Swine	DE	kcal/kg	*	3443.	4004.	
Cattle	ME	Mcal/kg	*	2.52	2.93	
Sheep	ME	Mcal/kg	*	2.58	3.00	
Swine	ME	kcal/kg	*	3194.	3714.	
Cattle	TDN	%	*	69.6	80.9	
Sheep	TDN	%	*	71.3	82.9	
Swine	TDN	%	*	78.1	90.8	
Calcium		%		0.02	0.02	
Cobalt		mg/kg		0.130	0.151	
Copper		mg/kg		6.2	7.2	
Iron		%		0.002	0.002	
Magnesium		%		0.10	0.12	
Manganese		mg/kg		37.0	43.0	
Phosphorus		%		0.34	0.40	
Potassium		%		0.32	0.37	
Selenium		mg/kg		0.300	0.349	
Sodium		%		0.01	0.01	
Zinc		mg/kg		49.0	57.0	
Choline		mg/kg		1258.	1463.	
Folic acid		mg/kg		0.45	0.52	
Niacin		mg/kg		62.5	72.7	
Pantothenic acid		mg/kg		10.1	11.7	
Riboflavin		mg/kg		1.6	1.9	
Thiamine		mg/kg		4.4	5.1	
Alanine		%		0.50	0.58	
Arginine		%		0.54	0.63	
Aspartic acid		%		0.69	0.80	
Cystine		%		0.34	0.40	
Glutamic acid		%		4.82	5.60	
Glycine		%		0.59	0.69	
Histidine		%		0.27	0.31	
Isoleucine		%		0.49	0.57	
Leucine		%		0.97	1.13	
Lysine		%		0.34	0.40	
Methionine		%		0.22	0.26	
Phenylalanine		%		0.67	0.78	
Proline		%		1.55	1.80	
Serine		%		0.69	0.80	
Threonine		%		0.40	0.47	
Tyrosine		%		0.42	0.49	
Valine		%		0.56	0.65	

WHEAT, RED BOBS. Triticum aestivum

Wheat, red bobs, grain, grnd, (4)

Wheat, triumph

Ref No 4 08 663 United States

Feed Name or Analyses				Mean As Fed	Mean Dry	C.V. ± %
Dry matter		%		86.0	100.0	
Ash		%		1.7	2.0	
Crude fiber		%		2.0	2.3	
Ether extract		%		1.6	1.9	
N-free extract		%		70.0	81.4	
Protein (N x 6.25)		%		10.7	12.4	
Cattle	dig prot	%	*	6.4	7.4	
Goats	dig prot	%	*	7.4	8.6	
Horses	dig prot	%	*	7.4	8.6	
Sheep	dig prot	%	*	7.4	8.6	

Continued

Feed Name or Analyses		Mean As Fed	Dry	C.V. ± %
Starch	%	54.1	62.9	
Energy	GE Mcal/kg			
Cattle	DE Mcal/kg	* 3.15	3.66	
Sheep	DE Mcal/kg	* 3.21	3.74	
Swine	DE kcal/kg	*3454.	4016.	
Cattle	ME Mcal/kg	* 2.58	3.00	
Sheep	ME Mcal/kg	* 2.63	3.06	
Swine	ME kcal/kg	*3229.	3755.	
Cattle	TDN %	* 71.5	83.1	
Sheep	TDN %	* 72.9	84.7	
Swine	TDN %	* 78.3	91.1	
Calcium	%	0.03	0.03	
Cobalt	mg/kg	0.120	0.140	
Copper	mg/kg	1.8	2.1	
Iron	%	0.002	0.003	
Magnesium	%	0.11	0.13	
Manganese	mg/kg	24.0	27.9	
Phosphorus	%	0.36	0.42	
Potassium	%	0.37	0.43	
Selenium	mg/kg	0.100	0.116	
Sodium	%	0.01	0.01	
Zinc	mg/kg	63.0	73.3	
Choline	mg/kg	970.	1128.	
Folic acid	mg/kg	0.23	0.27	
Niacin	mg/kg	51.4	59.8	
Pantothenic acid	mg/kg	10.3	12.0	
Riboflavin	mg/kg	1.5	1.7	
Thiamine	mg/kg	3.5	4.1	
Alanine	%	0.39	0.45	
Arginine	%	0.49	0.57	
Aspartic acid	%	0.56	0.65	
Cystine	%	0.30	0.35	
Glutamic acid	%	3.58	4.16	
Glycine	%	0.44	0.51	
Histidine	%	0.23	0.27	
Isoleucine	%	0.34	0.40	
Leucine	%	0.73	0.85	
Lysine	%	0.29	0.34	
Methionine	%	0.19	0.22	
Phenylalanine	%	0.50	0.58	
Proline	%	1.11	1.29	
Serine	%	0.59	0.69	
Threonine	%	0.34	0.40	
Tyrosine	%	0.34	0.40	
Valine	%	0.41	0.48	

WHEAT, RED SPRING. Triticum aestivum

Wheat, red spring, grain, Can 4 northern mn wt 56 lb per bushel mx 2.5% foreign material, (4)

Ref No 4 05 282 United States

Feed Name or Analyses		Mean As Fed	Dry	C.V. ± %
Dry matter	%	90.6	100.0	1
Crude fiber	%			
Sheep	dig coef %	* 33.	33.	
Swine	dig coef %	* 21.	21.	
Ether extract	%			
Sheep	dig coef %	* 72.	72.	
Swine	dig coef %	* 68.	68.	
N-free extract	%			
Sheep	dig coef %	* 92.	92.	
Swine	dig coef %	* 87.	87.	
Protein (N x 6.25)	%			
Sheep	dig coef %	* 78.	78.	

(1) dry forages and roughages (2) pasture, range plants, and forages fed green (3) silages (4) energy feeds (5) protein supplements (6) minerals (7) vitamins (8) additives

Feed Name or Analyses		Mean As Fed	Dry	C.V. ± %
Swine	dig coef %	* 82.	82.	
Energy	GE Mcal/kg	4.16	4.59	
Swine	GE dig coef %	77.	77.	
Sheep	DE Mcal/kg	* 3.52	3.88	
Swine	DE kcal/kg	3184.	3515.	
Chickens	ME_n kcal/kg	2709.	2990.	
Sheep	ME Mcal/kg	* 2.88	3.18	
Sheep	TDN %	79.7	88.0	
Swine	TDN %	76.6	84.6	
Calcium	%	0.05	0.06	34
Copper	mg/kg	10.8	11.9	
Iron	%	0.005	0.006	18
Manganese	mg/kg	72.7	80.2	9
Phosphorus	%	0.41	0.45	10

Ref No 4 05 282 Canada

Feed Name or Analyses		Mean As Fed	Dry	C.V. ± %
Dry matter	%	87.2	100.0	2
Swine	dig coef %	77.	77.	
Ash	%	1.5	1.8	11
Crude fiber	%	2.5	2.8	21
Swine	dig coef %	21.	21.	
Ether extract	%	1.9	2.2	31
Swine	dig coef %	68.	68.	
N-free extract	%	67.2	77.1	
Swine	dig coef %	87.	87.	
Protein (N x 6.25)	%	14.0	16.1	12
Swine	dig coef %	82.	82.	
Cattle	dig prot %	* 9.4	10.8	
Goats	dig prot %	* 10.4	12.0	
Horses	dig prot %	* 10.4	12.0	
Sheep	dig prot %	* 10.4	12.0	
Swine	dig prot %	11.5	13.2	
Energy	GE Mcal/kg	4.00	4.59	0
Swine	GE dig coef %	77.	77.	
Cattle	DE Mcal/kg	* 3.35	3.84	
Sheep	DE Mcal/kg	* 3.39	3.88	
Swine	DE kcal/kg	3069.	3518.	
Cattle	ME Mcal/kg	* 2.75	3.15	
Chickens	ME_n kcal/kg	2609.	2990.	6
Sheep	ME Mcal/kg	* 2.78	3.18	
Swine	ME kcal/kg	*2847.	3263.	
Cattle	TDN %	* 75.9	87.0	
Sheep	TDN %	76.8	88.0	
Swine	TDN %	73.8	84.6	
Calcium	%	0.05	0.06	64
Copper	mg/kg	6.2	7.1	
Phosphorus	%	0.36	0.42	19
Salt (NaCl)	%	0.12	0.14	
Niacin	mg/kg	32.9	37.7	
Pantothenic acid	mg/kg	8.5	9.7	
Riboflavin	mg/kg	1.0	1.2	
Thiamine	mg/kg	2.1	2.4	

WHEAT, SOFT. Triticum aestivum

Wheat, soft, bran, (4)

Ref No 4 08 554 United States

Feed Name or Analyses		Mean As Fed	Dry	C.V. ± %
Dry matter	%	90.5	100.0	
Ash	%	6.0	6.6	
Crude fiber	%	8.9	9.8	
Ether extract	%	3.9	4.3	
N-free extract	%	57.1	63.1	
Protein (N x 6.25)	%	14.6	16.1	
Cattle	dig prot %	* 9.8	10.8	
Goats	dig prot %	* 10.9	12.0	
Horses	dig prot %	* 10.9	12.0	
Sheep	dig prot %	* 10.9	12.0	

Feed Name or Analyses		Mean As Fed	Dry	C.V. ± %
Energy	GE Mcal/kg			
Cattle	DE Mcal/kg	* 2.75	3.03	
Sheep	DE Mcal/kg	* 3.03	3.35	
Swine	DE kcal/kg	*2554.	2822.	
Cattle	ME Mcal/kg	* 2.25	2.49	
Sheep	ME Mcal/kg	* 2.49	2.75	
Swine	ME kcal/kg	*2369.	2617.	
Cattle	TDN %	* 62.3	68.8	
Sheep	TDN %	* 68.8	76.0	
Swine	TDN %	* 57.9	64.0	

Wheat, soft, bran, dry milled, (4)

Ref No 4 05 283 United States

Feed Name or Analyses		Mean As Fed	Dry	C.V. ± %
Dry matter	%	88.2	100.0	0
Ash	%	6.5	7.3	16
Crude fiber	%	9.1	10.3	18
Ether extract	%	3.5	4.0	19
N-free extract	%	54.8	62.1	
Protein (N x 6.25)	%	14.3	16.2	1
Cattle	dig prot %	* 9.6	10.9	
Goats	dig prot %	* 10.7	12.1	
Horses	dig prot %	* 10.7	12.1	
Sheep	dig prot %	* 10.7	12.1	
Starch	%	5.8	6.6	
Energy	GE Mcal/kg			
Cattle	DE Mcal/kg	* 2.68	3.04	
Sheep	DE Mcal/kg	* 2.92	3.31	
Swine	DE kcal/kg	*2375.	2693.	
Cattle	ME Mcal/kg	* 2.20	2.49	
Sheep	ME Mcal/kg	* 2.40	2.72	
Swine	ME kcal/kg	*2203.	2497.	
Cattle	TDN %	* 60.8	68.9	
Sheep	TDN %	* 66.3	75.1	
Swine	TDN %	* 53.9	61.1	
Calcium	%	0.08	0.09	
Cobalt	mg/kg	0.082	0.093	
Copper	mg/kg	10.8	12.2	
Iron	%	0.008	0.009	
Magnesium	%	0.56	0.64	
Manganese	mg/kg	0.9	1.1	
Phosphorus	%	1.54	1.74	
Potassium	%	1.64	1.86	
Selenium	mg/kg	0.103	0.116	
Sodium	%	0.03	0.03	
Zinc	mg/kg	127.2	144.2	
Choline	mg/kg	2189.	2481.	
Folic acid	mg/kg	1.15	1.30	
Niacin	mg/kg	297.4	337.2	
Pantothenic acid	mg/kg	27.3	30.9	
Riboflavin	mg/kg	4.7	5.3	
Thiamine	mg/kg	6.1	6.9	
Alanine	%	0.82	0.93	
Arginine	%	1.09	1.23	
Aspartic acid	%	1.14	1.29	
Cystine	%	0.43	0.49	
Glutamic acid	%	3.29	3.73	
Glycine	%	0.91	1.03	
Histidine	%	0.40	0.45	
Isoleucine	%	0.47	0.53	
Leucine	%	0.98	1.12	
Lysine	%	0.59	0.67	
Methionine	%	0.24	0.27	
Phenylalanine	%	0.64	0.72	
Proline	%	1.03	1.16	
Serine	%	0.78	0.88	
Threonine	%	0.54	0.62	

Feed Name or Analyses		Mean As Fed	Mean Dry	C.V. ± %
Tyrosine	%	0.48	0.55	
Valine	%	0.70	0.79	

Wheat, soft, grain, (4)

Ref No 4 05 284 — United States

Feed Name or Analyses		Mean As Fed	Mean Dry	C.V. ± %
Dry matter	%	89.2	100.0	2
Ash	%	1.9	2.1	19
Crude fiber	%	2.4	2.7	28
Sheep	dig coef %	* 33.	33.	
Ether extract	%	1.8	2.0	22
Sheep	dig coef %	* 72.	72.	
N-free extract	%	72.6	81.4	
Sheep	dig coef %	* 92.	92.	
Protein (N x 6.25)	%	10.4	11.7	15
Sheep	dig coef %	* 78.	78.	
Cattle	dig prot %	* 6.0	6.8	
Goats	dig prot %	* 7.1	8.0	
Horses	dig prot %	* 7.1	8.0	
Sheep	dig prot %	8.1	9.1	
Linoleic	%	.599	.672	
Energy	GE Mcal/kg			
Cattle	DE Mcal/kg	* 3.35	3.75	
Sheep	DE Mcal/kg	* 3.47	3.89	
Swine	DE kcal/kg	*3522.	3951.	
Cattle	ME Mcal/kg	* 2.75	3.08	
Chickens	ME_n kcal/kg	3168.	3552.	
Sheep	ME Mcal/kg	* 2.84	3.19	
Swine	ME kcal/kg	*3298.	3699.	
Cattle	TDN %	* 75.9	85.1	
Sheep	TDN %	78.6	88.2	
Swine	TDN %	* 79.9	89.6	
Calcium	%	0.06	0.07	98
Copper	mg/kg	9.6	10.8	11
Iron	%	0.005	0.006	22
Magnesium	%	0.10	0.11	
Manganese	mg/kg	50.9	57.1	
Phosphorus	%	0.29	0.33	40
Potassium	%	0.39	0.44	
Choline	mg/kg	1014.	1137.	
Niacin	mg/kg	59.1	66.2	
Pantothenic acid	mg/kg	11.5	12.9	
Riboflavin	mg/kg	1.1	1.2	
α-tocopherol	mg/kg	11.0	12.3	
Arginine	%	0.40	0.45	
Cystine	%	0.17	0.19	
Lysine	%	0.30	0.34	
Methionine	%	0.13	0.15	
Tryptophan	%	0.12	0.13	

Wheat, soft, grain, Pacific coast, (4)

Ref No 4 08 555 — United States

Feed Name or Analyses		Mean As Fed	Mean Dry	C.V. ± %
Dry matter	%	89.1	100.0	
Ash	%	1.9	2.1	
Crude fiber	%	2.7	3.0	
Ether extract	%	2.0	2.2	
N-free extract	%	72.6	81.5	
Protein (N x 6.25)	%	9.9	11.1	
Cattle	dig prot %	* 5.5	6.2	
Goats	dig prot %	* 6.6	7.4	
Horses	dig prot %	* 6.6	7.4	
Sheep	dig prot %	* 6.6	7.4	
Energy	GE Mcal/kg			
Cattle	DE Mcal/kg	* 3.05	3.43	
Sheep	DE Mcal/kg	* 3.33	3.74	
Swine	DE kcal/kg	*3483.	3909.	
Cattle	ME Mcal/kg	* 2.50	2.81	
Sheep	ME Mcal/kg	* 2.73	3.07	
Swine	ME kcal/kg	*3266.	3665.	
Cattle	TDN %	* 69.2	77.7	
Sheep	TDN %	* 75.6	84.8	
Swine	TDN %	* 79.0	88.7	

WHEAT, SOFT RED WINTER. Triticum aestivum

Wheat, soft red winter, bran, (4)

Ref No 4 08 011 — United States

Feed Name or Analyses		Mean As Fed	Mean Dry	C.V. ± %
Dry matter	%	88.4	100.0	
Ash	%	7.0	7.9	
Crude fiber	%	9.2	10.4	
Ether extract	%	3.2	3.6	
N-free extract	%	55.1	62.3	
Protein (N x 6.25)	%	13.9	15.7	
Cattle	dig prot %	* 9.2	10.5	
Goats	dig prot %	* 10.3	11.6	
Horses	dig prot %	* 10.3	11.6	
Sheep	dig prot %	* 10.3	11.6	
Starch	%	5.7	6.4	
Energy	GE Mcal/kg			
Cattle	DE Mcal/kg	* 2.76	3.12	
Sheep	DE Mcal/kg	* 2.91	3.30	
Swine	DE kcal/kg	*2307.	2610.	
Cattle	ME Mcal/kg	* 2.26	2.56	
Sheep	ME Mcal/kg	* 2.39	2.70	
Swine	ME kcal/kg	*2142.	2423.	
Cattle	TDN %	* 62.5	70.8	
Sheep	TDN %	* 66.1	74.8	
Swine	TDN %	* 52.3	59.2	
Aluminum	mg/kg	5.14	5.81	
Calcium	%	0.08	0.09	
Cobalt	mg/kg	0.082	0.093	
Copper	mg/kg	10.8	12.2	
Iron	%	0.008	0.009	
Magnesium	%	0.57	0.64	
Manganese	mg/kg	93.5	105.8	
Phosphorus	%	1.54	1.74	
Potassium	%	1.64	1.86	
Selenium	mg/kg	0.103	0.116	
Sodium	%	0.03	0.03	
Zinc	mg/kg	127.5	144.2	
Choline	mg/kg	2194.	2481.	
Folic acid	mg/kg	1.15	1.30	
Niacin	mg/kg	298.1	337.2	
Pantothenic acid	mg/kg	27.3	30.9	
Riboflavin	mg/kg	4.7	5.3	
Thiamine	mg/kg	6.1	6.9	
α-tocopherol	mg/kg	20.4	23.0	
Vitamin B6	mg/kg	7.24	8.19	
Alanine	%	0.82	0.93	
Arginine	%	1.09	1.23	
Aspartic acid	%	1.14	1.29	
Cystine	%	0.43	0.49	
Glutamic acid	%	3.30	3.73	
Glycine	%	0.91	1.03	
Histidine	%	0.40	0.45	
Isoleucine	%	0.47	0.53	
Leucine	%	0.99	1.12	
Lysine	%	0.60	0.67	
Methionine	%	0.24	0.27	
Phenylalanine	%	0.64	0.72	
Proline	%	1.03	1.16	
Serine	%	0.78	0.88	
Threonine	%	0.54	0.62	
Tyrosine	%	0.48	0.55	
Valine	%	0.70	0.79	

Wheat, soft red winter, flour, (4)

Ref No 4 08 111 — United States

Feed Name or Analyses		Mean As Fed	Mean Dry	C.V. ± %
Dry matter	%	86.1	100.0	
Ash	%	0.3	0.4	
Ether extract	%	0.9	1.0	
Protein (N x 6.25)	%	10.8	12.5	
Cattle	dig prot %	* 6.5	7.5	
Goats	dig prot %	* 7.5	8.7	
Horses	dig prot %	* 7.5	8.7	
Sheep	dig prot %	* 7.5	8.7	
Starch	%	70.2	81.5	
Energy	GE Mcal/kg	3.78	4.39	
Aluminum	mg/kg	5.01	5.81	
Calcium	%	0.03	0.03	
Cobalt	mg/kg	0.070	0.081	
Copper	mg/kg	0.6	0.7	
Iron	%	0.000	0.001	
Magnesium	%	0.01	0.01	
Manganese	mg/kg	3.1	3.6	
Phosphorus	%	0.08	0.09	
Potassium	%	0.09	0.10	
Selenium	mg/kg	0.010	0.012	
Sodium	%	0.00	0.01	
Zinc	mg/kg	7.3	8.5	
Choline	mg/kg	698.	811.	
Folic acid	mg/kg	0.12	0.13	
Niacin	mg/kg	8.1	9.4	
Pantothenic acid	mg/kg	4.3	5.0	
Riboflavin	mg/kg	0.3	0.4	
Thiamine	mg/kg	1.4	1.6	
α-tocopherol	mg/kg	3.1	3.6	
Vitamin B6	mg/kg	1.36	1.58	
Alanine	%	0.38	0.44	
Arginine	%	0.43	0.50	
Aspartic acid	%	0.48	0.56	
Cystine	%	0.36	0.42	
Glutamic acid	%	4.31	5.00	
Glycine	%	0.43	0.50	
Histidine	%	0.26	0.30	
Isoleucine	%	0.42	0.49	
Leucine	%	0.84	0.98	
Lysine	%	0.23	0.27	
Methionine	%	0.21	0.24	
Phenylalanine	%	0.58	0.67	
Proline	%	1.44	1.67	
Serine	%	0.60	0.70	
Threonine	%	0.33	0.38	
Tyrosine	%	0.39	0.45	
Valine	%	0.48	0.56	

Wheat, soft red winter, flour by-product, coarse sift, mx 7% fiber, (4)

Wheat, soft red winter, shorts

Ref No 4 08 017 — United States

Feed Name or Analyses		Mean As Fed	Mean Dry	C.V. ± %
Dry matter	%	88.2	100.0	
Ash	%	3.9	4.4	
Crude fiber	%	6.0	6.8	
Ether extract	%	4.6	5.2	
N-free extract	%	59.8	67.8	
Protein (N x 6.25)	%	13.9	15.8	
Cattle	dig prot %	* 9.2	10.5	
Goats	dig prot %	* 10.3	11.7	

Continued

Feed Name or Analyses		Mean As Fed	Mean Dry	C.V. ± %
Horses	dig prot %	* 10.3	11.7	
Sheep	dig prot %	* 10.3	11.7	
Starch	%	21.3	24.2	
Energy	GE Mcal/kg	4.46	5.06	
Cattle	DE Mcal/kg	* 2.64	2.99	
Sheep	DE Mcal/kg	* 3.11	3.52	
Swine	DE kcal/kg	*2982.	3381.	
Cattle	ME Mcal/kg	* 2.16	2.45	
Sheep	ME Mcal/kg	* 2.55	2.89	
Swine	ME kcal/kg	*2768.	3138.	
Cattle	TDN %	* 59.8	67.8	
Sheep	TDN %	* 70.5	79.9	
Swine	TDN %	* 67.6	76.7	
Aluminum	mg/kg	5.13	5.81	
Calcium	%	0.07	0.08	
Cobalt	mg/kg	0.092	0.105	
Copper	mg/kg	13.7	15.6	
Iron	%	0.006	0.007	
Magnesium	%	0.25	0.28	
Manganese	mg/kg	112.8	127.9	
Phosphorus	%	0.88	1.00	
Potassium	%	0.93	1.06	
Selenium	mg/kg	0.041	0.047	
Sodium	%	0.02	0.02	
Zinc	mg/kg	117.9	133.7	
Choline	mg/kg	2069.	2345.	
Folic acid	mg/kg	1.76	2.00	
Niacin	mg/kg	94.4	107.0	
Pantothenic acid	mg/kg	23.9	27.1	
Riboflavin	mg/kg	4.8	5.5	
Thiamine	mg/kg	16.8	19.1	
α-tocopherol	mg/kg	56.8	64.4	
Vitamin B6	mg/kg	6.07	6.88	
Alanine	%	0.93	1.06	
Arginine	%	1.24	1.41	
Aspartic acid	%	1.31	1.49	
Cystine	%	0.43	0.49	
Glutamic acid	%	3.21	3.64	
Glycine	%	0.96	1.09	
Histidine	%	0.47	0.53	
Isoleucine	%	0.56	0.64	
Leucine	%	1.12	1.27	
Lysine	%	0.86	0.98	
Methionine	%	0.29	0.33	
Phenylalanine	%	0.70	0.79	
Proline	%	1.03	1.16	
Serine	%	0.80	0.91	
Threonine	%	0.62	0.70	
Tyrosine	%	0.52	0.59	
Valine	%	0.82	0.93	

Wheat, soft red winter, flour by-product, fine sift, mx 4% fiber, (4)

Wheat, soft red winter, red dog

Ref No 4 08 021 United States

Feed Name or Analyses		Mean As Fed	Mean Dry	C.V. ± %
Dry matter	%	87.8	100.0	
Ash	%	1.5	1.7	
Crude fiber	%	1.4	1.6	
Ether extract	%	2.9	3.3	
N-free extract	%	66.8	76.1	
Protein (N x 6.25)	%	15.2	17.3	
Cattle	dig prot %	* 10.5	11.9	
Goats	dig prot %	* 11.5	13.1	

Feed Name or Analyses		Mean As Fed	Mean Dry	C.V. ± %
Horses	dig prot %	* 11.5	13.1	
Sheep	dig prot %	* 11.5	13.1	
Starch	%	46.2	52.6	
Energy	GE Mcal/kg	4.06	4.63	
Cattle	DE Mcal/kg	* 3.35	3.82	
Sheep	DE Mcal/kg	* 3.25	3.70	
Swine	DE kcal/kg	*3703.	4217.	
Cattle	ME Mcal/kg	* 2.75	3.13	
Sheep	ME Mcal/kg	* 2.66	3.03	
Swine	ME kcal/kg	*3425.	3901.	
Cattle	TDN %	* 76.0	86.6	
Sheep	TDN %	* 73.6	83.9	
Swine	TDN %	* 84.0	95.6	
Aluminum	mg/kg	5.10	5.81	
Calcium	%	0.04	0.05	
Cobalt	mg/kg	0.112	0.128	
Copper	mg/kg	5.3	6.0	
Iron	%	0.003	0.003	
Magnesium	%	0.08	0.09	
Manganese	mg/kg	43.9	50.0	
Phosphorus	%	0.37	0.42	
Potassium	%	0.35	0.40	
Selenium	mg/kg	0.143	0.163	
Sodium	%	0.01	0.01	
Zinc	mg/kg	63.3	72.1	
Choline	mg/kg	1484.	1691.	
Folic acid	mg/kg	0.57	0.65	
Niacin	mg/kg	22.4	25.5	
Pantothenic acid	mg/kg	9.2	10.5	
Riboflavin	mg/kg	1.9	2.2	
Thiamine	mg/kg	19.4	22.1	
α-tocopherol	mg/kg	26.5	30.2	
Vitamin B6	mg/kg	2.72	3.09	
Alanine	%	0.70	0.80	
Arginine	%	0.94	1.07	
Aspartic acid	%	0.96	1.09	
Cystine	%	0.44	0.50	
Glutamic acid	%	4.90	5.58	
Glycine	%	0.75	0.85	
Histidine	%	0.42	0.48	
Isoleucine	%	0.61	0.70	
Leucine	%	1.15	1.31	
Lysine	%	0.56	0.64	
Methionine	%	0.26	0.29	
Phenylalanine	%	0.76	0.86	
Proline	%	1.52	1.73	
Serine	%	0.82	0.93	
Threonine	%	0.53	0.60	
Tyrosine	%	0.52	0.59	
Valine	%	0.80	0.91	

Wheat, soft red winter, grain, (4)

Ref No 4 05 294 United States

Feed Name or Analyses		Mean As Fed	Mean Dry	C.V. ± %
Dry matter	%	88.1	100.0	2
Ash	%	1.7	1.9	
Crude fiber	%	2.3	2.7	
Sheep	dig coef %	* 47.	47.	
Ether extract	%	1.9	2.2	
Sheep	dig coef %	* 75.	75.	
N-free extract	%	70.7	80.3	
Sheep	dig coef %	* 92.	92.	
Protein (N x 6.25)	%	11.5	13.0	
Sheep	dig coef %	* 75.	75.	
Cattle	dig prot %	* 7.0	8.0	
Goats	dig prot %	* 8.1	9.1	
Horses	dig prot %	* 8.1	9.1	
Sheep	dig prot %	8.6	9.8	
Starch	%	57.6	65.4	

Feed Name or Analyses		Mean As Fed	Mean Dry	C.V. ± %
Energy	GE Mcal/kg	3.61	4.10	
Cattle	DE Mcal/kg	* 3.32	3.77	
Sheep	DE Mcal/kg	* 3.44	3.90	
Swine	DE kcal/kg	*3523.	4000.	
Cattle	ME Mcal/kg	* 2.72	3.09	
Sheep	ME Mcal/kg	* 2.82	3.20	
Swine	ME kcal/kg	*3290.	3735.	
Cattle	NE_m Mcal/kg	* 1.90	2.16	
Cattle	NE_{gain} Mcal/kg	* 1.25	1.42	
Cattle	$NE_{lactating\ cows}$ Mcal/kg	* 2.19	2.49	
Cattle	TDN %	* 75.2	85.4	
Sheep	TDN %	77.9	88.5	
Swine	TDN %	* 79.9	90.7	
Aluminum	mg/kg	5.12	5.81	
Calcium	%	0.05	0.06	98
Chlorine	%	0.07	0.08	40
Cobalt	mg/kg	0.102	0.116	
Copper	mg/kg	7.0	8.0	
Iron	%	0.003	0.004	43
Magnesium	%	0.10	0.11	
Manganese	mg/kg	33.3	37.8	
Phosphorus	%	0.37	0.42	40
Potassium	%	0.40	0.45	
Selenium	mg/kg	0.041	0.047	
Sodium	%	0.01	0.01	
Sulphur	%	0.11	0.12	
Zinc	mg/kg	42.0	47.7	
Choline	mg/kg	1005.	1141.	
Folic acid	mg/kg	0.40	0.46	41
Niacin	mg/kg	43.2	49.1	
Pantothenic acid	mg/kg	8.8	10.0	
Riboflavin	mg/kg	1.3	1.5	
Thiamine	mg/kg	4.3	4.9	
α-tocopherol	mg/kg	15.6	17.7	
Vitamin B6	mg/kg	1.73	1.97	
Alanine	%	0.49	0.56	
Arginine	%	0.50	0.57	
Aspartic acid	%	0.67	0.76	
Cystine	%	0.27	0.30	
Glutamic acid	%	4.37	4.97	
Glycine	%	0.54	0.62	
Histidine	%	0.20	0.23	
Isoleucine	%	0.45	0.51	
Leucine	%	0.90	1.02	
Lysine	%	0.58	0.65	
Methionine	%	0.22	0.24	
Phenylalanine	%	0.63	0.72	
Proline	%	1.39	1.58	
Serine	%	0.65	0.73	
Threonine	%	0.39	0.44	
Tryptophan	%	0.26	0.30	
Tyrosine	%	0.37	0.42	
Valine	%	0.57	0.65	

Wheat, soft red winter, grain, grnd, (4)

Ref No 4 08 784 United States

Feed Name or Analyses		Mean As Fed	Mean Dry	C.V. ± %
Dry matter	%	86.0	100.0	
Ash	%	1.6	1.9	
Crude fiber	%	1.7	2.0	
Ether extract	%	1.7	2.0	
N-free extract	%	69.2	80.5	
Protein (N x 6.25)	%	11.8	13.7	
Cattle	dig prot %	* 7.4	8.6	
Goats	dig prot %	* 8.4	9.8	
Horses	dig prot %	* 8.4	9.8	
Sheep	dig prot %	* 8.4	9.8	
Starch	%	55.7	64.8	

(1) dry forages and roughages (2) pasture, range plants, and forages fed green (3) silages (4) energy feeds (5) protein supplements (6) minerals (7) vitamins (8) additives

Feed Name or Analyses			Mean As Fed	Mean Dry	C.V. ± %
Energy	GE Mcal/kg				
Cattle	DE Mcal/kg	*	3.11	3.62	
Sheep	DE Mcal/kg	*	3.21	3.73	
Swine	DE kcal/kg	*	3515.	4087.	
Cattle	ME Mcal/kg	*	2.55	2.97	
Sheep	ME Mcal/kg	*	2.63	3.06	
Swine	ME kcal/kg	*	3277.	3810.	
Cattle	TDN %	*	70.6	82.1	
Sheep	TDN %	*	72.7	84.5	
Swine	TDN %	*	79.7	92.7	
Calcium	%		0.02	0.02	
Cobalt	mg/kg		0.100	0.116	
Copper	mg/kg		4.2	4.9	
Iron	%		0.002	0.003	
Magnesium	%		0.10	0.12	
Manganese	mg/kg		28.0	32.6	
Phosphorus	%		0.41	0.48	
Potassium	%		0.41	0.48	
Selenium	mg/kg		0.040	0.047	
Sodium	%		0.01	0.01	
Zinc	mg/kg		41.0	47.7	
Choline	mg/kg		981.	1141.	
Folic acid	mg/kg		0.41	0.48	
Niacin	mg/kg		48.4	56.3	
Pantothenic acid	mg/kg		8.6	10.0	
Riboflavin	mg/kg		1.5	1.7	
Thiamine	mg/kg		4.1	4.8	
Alanine	%		0.48	0.56	
Arginine	%		0.63	0.73	
Aspartic acid	%		0.65	0.76	
Cystine	%		0.35	0.41	
Glutamic acid	%		4.27	4.97	
Glycine	%		0.53	0.62	
Histidine	%		0.31	0.36	
Isoleucine	%		0.44	0.51	
Leucine	%		0.88	1.02	
Lysine	%		0.35	0.41	
Methionine	%		0.21	0.24	
Phenylalanine	%		0.62	0.72	
Proline	%		1.36	1.58	
Serine	%		0.63	0.73	
Threonine	%		0.38	0.44	
Tyrosine	%		0.38	0.44	
Valine	%		0.56	0.65	

Wheat, soft red winter, grain, gr 2 US mn wt 58 lb per bushel mx 1% foreign material, (4)

Ref No 4 05 289 United States

Analyses			As Fed	Dry	C.V. ± %
Dry matter	%		90.7	100.0	1
Ash	%		1.8	2.0	9
Ether extract	%		1.6	1.8	11
Protein (N x 6.25)	%		10.4	11.5	11
Cattle	dig prot %	*	6.0	6.6	
Goats	dig prot %	*	7.1	7.8	
Horses	dig prot %	*	7.1	7.8	
Sheep	dig prot %	*	7.1	7.8	
Starch	%		59.3	65.4	2
Sugars, total	%		2.4	2.7	11

Wheat, soft red winter, grain, gr 3 US mn wt 56 lb per bushel mx 2% foreign material, (4)

Ref No 4 05 291 United States

Analyses			As Fed	Dry	C.V. ± %
Dry matter	%			100.0	
Ash	%			2.1	10
Ether extract	%			1.7	12
Protein (N x 6.25)	%			12.1	11
Cattle	dig prot %	*		7.1	
Goats	dig prot %	*		8.3	
Horses	dig prot %	*		8.3	
Sheep	dig prot %	*		8.3	
Starch	%			65.1	1
Sugars, total	%			2.7	5

Wheat, soft red winter, grain, gr 4 US mn wt 54 lb per bushel mx 3% foreign material, (4)

Ref No 4 05 292 United States

Analyses			As Fed	Dry	C.V. ± %
Dry matter	%		86.0	100.0	
Ash	%		1.8	2.1	7
Ether extract	%		1.5	1.7	9
Protein (N x 6.25)	%		10.8	12.5	11
Cattle	dig prot %	*	6.4	7.5	
Goats	dig prot %	*	7.5	8.7	
Horses	dig prot %	*	7.5	8.7	
Sheep	dig prot %	*	7.5	8.7	
Starch	%		54.4	63.3	
Sugars, total	%		2.2	2.6	

Wheat, soft red winter, grain, gr 5 US mn wt 51 lb per bushel mx 5% foreign material, (4)

Ref No 4 05 293 United States

Analyses			As Fed	Dry	C.V. ± %
Dry matter	%			100.0	
Ash	%			2.3	
Ether extract	%			1.6	
Protein (N x 6.25)	%			15.0	
Cattle	dig prot %	*		9.8	
Goats	dig prot %	*		11.0	
Horses	dig prot %	*		11.0	
Sheep	dig prot %	*		11.0	
Starch	%			62.8	
Sugars, total	%			2.8	

Wheat, soft red winter, germ, (5)

Ref No 5 09 077 United States

Analyses			As Fed	Dry	C.V. ± %
Dry matter	%		87.6	100.0	
Ash	%		4.0	4.6	
Crude fiber	%		3.0	3.5	
Ether extract	%		8.4	9.6	
N-free extract	%		50.4	57.6	
Protein (N x 6.25)	%		21.7	24.8	
Starch	%		16.0	18.2	
Energy	GE Mcal/kg		4.41	5.03	
Cattle	DE Mcal/kg	*	3.60	4.11	
Sheep	DE Mcal/kg	*	3.50	4.00	
Swine	DE kcal/kg	*	3854.	4400.	
Cattle	ME Mcal/kg	*	2.95	3.37	
Sheep	ME Mcal/kg	*	2.87	3.28	
Swine	ME kcal/kg	*	3507.	4004.	
Cattle	TDN %	*	81.6	93.2	
Sheep	TDN %	*	79.5	90.7	
Swine	TDN %	*	87.4	99.8	
Aluminum	mg/kg		5.09	5.81	
Calcium	%		0.04	0.05	
Cobalt	mg/kg		0.143	0.163	
Copper	mg/kg		7.3	8.4	
Iron	%		0.004	0.005	
Magnesium	%		0.20	0.23	
Manganese	mg/kg		102.9	117.4	
Phosphorus	%		0.78	0.90	
Potassium	%		0.88	1.00	
Selenium	mg/kg		0.010	0.012	
Sodium	%		0.02	0.02	
Zinc	mg/kg		127.3	145.3	
Choline	mg/kg		3045.	3477.	
Folic acid	mg/kg		2.08	2.37	
Niacin	mg/kg		66.4	75.8	
Pantothenic acid	mg/kg		19.6	22.3	
Riboflavin	mg/kg		5.9	6.8	
Thiamine	mg/kg		20.5	23.4	
α-tocopherol	mg/kg		138.7	158.4	
Vitamin B6	mg/kg		6.74	7.70	
Alanine	%		1.46	1.66	
Arginine	%		1.81	2.07	
Aspartic acid	%		1.96	2.23	
Cystine	%		0.47	0.53	
Glutamic acid	%		3.98	4.55	
Glycine	%		1.40	1.59	
Histidine	%		0.73	0.84	
Isoleucine	%		0.80	0.92	
Leucine	%		1.51	1.72	
Lysine	%		1.80	2.06	
Methionine	%		0.41	0.47	
Phenylalanine	%		0.88	1.00	
Proline	%		1.17	1.34	
Serine	%		1.07	1.22	
Threonine	%		0.92	1.05	
Tyrosine	%		0.68	0.78	
Valine	%		1.03	1.17	

WHEAT, SOFT WINTER. Triticum aestivum

Wheat, soft winter, grain, (4)

Ref No 4 08 556 United States

Analyses			As Fed	Dry	C.V. ± %
Dry matter	%		89.2	100.0	
Ash	%		1.8	2.0	
Crude fiber	%		2.1	2.4	
Ether extract	%		1.9	2.1	
N-free extract	%		73.2	82.1	
Protein (N x 6.25)	%		10.2	11.4	
Cattle	dig prot %	*	5.8	6.5	
Goats	dig prot %	*	6.9	7.7	
Horses	dig prot %	*	6.9	7.7	
Sheep	dig prot %	*	6.9	7.7	
Energy	GE Mcal/kg				
Cattle	DE Mcal/kg	*	3.37	3.78	
Sheep	DE Mcal/kg	*	3.35	3.76	
Swine	DE kcal/kg	*	3565.	3997.	
Cattle	ME Mcal/kg	*	2.77	3.10	
Sheep	ME Mcal/kg	*	2.75	3.08	
Swine	ME kcal/kg	*	3340.	3744.	
Cattle	TDN %	*	76.4	85.7	
Sheep	TDN %	*	76.0	85.2	
Swine	TDN %	*	80.9	90.6	
Phosphorus	%		0.29	0.33	

Wheat, soft winter, germ, (5)

Ref No 5 05 295 United States

Analyses			As Fed	Dry	C.V. ± %
Dry matter	%		89.9	100.0	
Ash	%		3.9	4.3	
Crude fiber	%		1.9	2.1	
Ether extract	%		7.6	8.5	
N-free extract	%		50.9	56.6	
Protein (N x 6.25)	%		25.6	28.5	
Energy	GE Mcal/kg				
Cattle	DE Mcal/kg	*	3.70	4.12	

Continued

Feed Name or Analyses		Mean As Fed	Dry	C.V. ± %
Sheep	DE Mcal/kg	* 3.79	4.22	
Swine	DE kcal/kg	*3908.	4347.	
Cattle	ME Mcal/kg	* 3.03	3.38	
Sheep	ME Mcal/kg	* 3.11	3.46	
Swine	ME kcal/kg	*3525.	3921.	
Cattle	TDN %	* 83.9	93.3	
Sheep	TDN %	* 86.0	95.7	
Swine	TDN %	* 88.6	98.6	
Calcium	%	0.05	0.06	
Chlorine	%	0.08	0.09	
Copper	mg/kg	12.6	14.0	
Iron	%	0.004	0.004	
Magnesium	%	0.29	0.32	
Manganese	mg/kg	237.2	263.9	
Phosphorus	%	1.04	1.16	
Potassium	%	0.93	1.03	
Sulphur	%	0.24	0.27	

WHEAT, SPRING. Triticum aestivum

Wheat, spring, bran, (4)

Ref No 4 05 296 United States

Feed Name or Analyses		Mean As Fed	Dry	C.V. ± %
Dry matter	%	91.6	100.0	
Ash	%	6.3	6.9	
Crude fiber	%	11.0	12.0	
Cattle	dig coef %	26.	26.	
Sheep	dig coef %	50.	50.	
Ether extract	%	4.9	5.4	
Cattle	dig coef %	78.	78.	
Sheep	dig coef %	55.	55.	
N-free extract	%	52.9	57.8	
Cattle	dig coef %	77.	77.	
Sheep	dig coef %	72.	72.	
Protein (N x 6.25)	%	16.4	17.9	
Cattle	dig coef %	74.	74.	
Sheep	dig coef %	77.	77.	
Cattle	dig prot %	12.1	13.2	
Goats	dig prot %	* 12.5	13.6	
Horses	dig prot %	* 12.5	13.6	
Sheep	dig prot %	12.6	13.8	
Energy	GE Mcal/kg			
Cattle	DE Mcal/kg	* 2.84	3.10	
Sheep	DE Mcal/kg	* 2.75	3.00	
Swine	DE kcal/kg	*2387.	2607.	
Cattle	ME Mcal/kg	* 2.33	2.54	
Sheep	ME Mcal/kg	* 2.25	2.46	
Swine	ME kcal/kg	*2205.	2408.	
Cattle	TDN %	64.4	70.3	
Sheep	TDN %	62.3	68.1	
Swine	TDN %	* 54.1	59.1	

WHEAT, TENMARQ. Triticum aestivum

Wheat, tenmarq, aerial part, fresh, (2)

Ref No 2 05 297 United States

Feed Name or Analyses		Mean As Fed	Dry	C.V. ± %
Dry matter	%	23.2	100.0	30
Chlorine	%	0.16	0.67	
Cobalt	mg/kg	0.017	0.075	77
Sodium	%	0.02	0.07	98
Sulphur	%	0.06	0.24	
Carotene	mg/kg	94.5	407.2	42
Niacin	mg/kg	12.1	52.2	34
Pantothenic acid	mg/kg	4.9	21.2	48
Riboflavin	mg/kg	5.6	24.3	38
Thiamine	mg/kg	0.8	3.5	
Vitamin A equivalent	IU/g	157.5	678.8	
Glutamic acid	%	0.49	2.10	9
Isoleucine	%	0.35	1.50	8
Leucine	%	0.53	2.30	8
Lysine	%	0.39	1.70	6
Methionine	%	0.07	0.30	
Phenylalanine	%	0.28	1.20	3
Threonine	%	0.46	2.00	9
Tryptophan	%	0.07	0.30	11
Valine	%	0.35	1.50	2

(1) dry forages and roughages (2) pasture, range plants, and forages fed green (3) silages (4) energy feeds (5) protein supplements (6) minerals (7) vitamins (8) additives

WHEAT, WHITE. Triticum aestivum

Wheat, white, bran, (4)

Ref No 4 08 010 United States

Feed Name or Analyses		Mean As Fed	Dry	C.V. ± %
Dry matter	%	89.2	100.0	
Ash	%	4.7	5.3	
Crude fiber	%	9.3	10.4	
Ether extract	%	3.1	3.5	
N-free extract	%	58.5	65.6	
Protein (N x 6.25)	%	13.6	15.2	
Cattle	dig prot %	* 8.9	10.0	
Goats	dig prot %	* 10.0	11.2	
Horses	dig prot %	* 10.0	11.2	
Sheep	dig prot %	* 10.0	11.2	
Starch	%	7.2	8.1	
Energy	GE Mcal/kg	4.05	4.54	
Cattle	DE Mcal/kg	* 2.78	3.11	
Sheep	DE Mcal/kg	* 3.01	3.38	
Swine	DE kcal/kg	*2568.	2878.	
Cattle	ME Mcal/kg	* 2.28	2.55	
Sheep	ME Mcal/kg	* 2.47	2.77	
Swine	ME kcal/kg	*2386.	2675.	
Cattle	TDN %	* 63.0	70.6	
Sheep	TDN %	* 68.4	76.6	
Swine	TDN %	* 58.2	65.3	
Aluminum	mg/kg	5.19	5.81	
Calcium	%	0.08	0.09	
Cobalt	mg/kg	0.124	0.140	
Copper	mg/kg	8.7	9.8	
Iron	%	0.010	0.011	
Magnesium	%	0.40	0.45	
Manganese	mg/kg	89.2	100.0	
Phosphorus	%	0.93	1.05	
Potassium	%	1.35	1.51	
Selenium	mg/kg	0.104	0.116	
Sodium	%	0.01	0.01	
Zinc	mg/kg	58.1	65.1	
Choline	mg/kg	2088.	2341.	
Folic acid	mg/kg	0.91	1.02	
Niacin	mg/kg	285.8	320.3	
Pantothenic acid	mg/kg	36.1	40.5	
Riboflavin	mg/kg	4.5	5.0	
Thiamine	mg/kg	6.0	6.7	
α-tocopherol	mg/kg	23.0	25.8	
Vitamin B6	mg/kg	7.36	8.26	
Alanine	%	0.76	0.85	
Arginine	%	1.05	1.17	
Aspartic acid	%	1.16	1.30	
Cystine	%	0.41	0.47	
Glutamic acid	%	2.91	3.27	
Glycine	%	0.92	1.03	
Histidine	%	0.41	0.47	
Isoleucine	%	0.44	0.49	
Leucine	%	0.90	1.01	
Lysine	%	0.58	0.65	
Methionine	%	0.27	0.30	
Phenylalanine	%	0.57	0.64	
Proline	%	0.89	1.00	
Serine	%	0.69	0.78	
Threonine	%	0.51	0.57	
Tyrosine	%	0.45	0.50	
Valine	%	0.66	0.74	

Wheat, white, flour, (4)

Ref No 4 08 112 United States

Feed Name or Analyses		Mean As Fed	Dry	C.V. ± %
Dry matter	%	86.7	100.0	
Ash	%	0.4	0.4	
Ether extract	%	0.9	1.0	
Protein (N x 6.25)	%	8.3	9.6	
Cattle	dig prot %	* 4.2	4.9	
Goats	dig prot %	* 5.3	6.1	
Horses	dig prot %	* 5.3	6.1	
Sheep	dig prot %	* 5.3	6.1	
Starch	%	64.3	74.2	
Energy	GE Mcal/kg	3.76	4.34	
Aluminum	mg/kg	5.04	5.81	
Calcium	%	0.03	0.03	
Cobalt	mg/kg	0.050	0.058	
Copper	mg/kg	0.6	0.7	
Iron	%	0.001	0.001	
Magnesium	%	0.02	0.02	
Manganese	mg/kg	3.5	4.1	
Phosphorus	%	0.08	0.09	
Potassium	%	0.13	0.15	
Selenium	mg/kg	0.020	0.023	
Sodium	%	0.01	0.01	
Zinc	mg/kg	3.4	4.0	
Choline	mg/kg	566.	652.	
Folic acid	mg/kg	0.12	0.14	
Niacin	mg/kg	11.5	13.3	
Pantothenic acid	mg/kg	3.8	4.3	
Riboflavin	mg/kg	0.3	0.4	
Thiamine	mg/kg	1.3	1.5	
α-tocopherol	mg/kg	2.2	2.6	
Vitamin B6	mg/kg	0.74	0.85	
Alanine	%	0.32	0.37	
Arginine	%	0.36	0.42	
Aspartic acid	%	0.41	0.48	
Cystine	%	0.23	0.27	
Glutamic acid	%	3.41	3.93	
Glycine	%	0.36	0.42	
Histidine	%	0.20	0.23	
Isoleucine	%	0.32	0.37	
Leucine	%	0.68	0.78	
Lysine	%	0.21	0.24	
Methionine	%	0.15	0.17	
Phenylalanine	%	0.49	0.57	
Proline	%	1.09	1.26	
Serine	%	0.50	0.58	
Threonine	%	0.28	0.33	
Tyrosine	%	0.29	0.34	
Valine	%	0.39	0.45	

Wheat, white, flour by-product, coarse sift, mx 7% fiber, (4)

Wheat, white, shorts

Ref No 4 08 016 United States

Feed Name or Analyses		Mean As Fed	Dry	C.V. ± %
Dry matter	%	88.6	100.0	
Ash	%	3.1	3.5	

Feed Name or Analyses			Mean As Fed	Mean Dry	C.V. ± %
Crude fiber	%		6.7	7.6	
Ether extract	%		3.7	4.2	
N-free extract	%		61.3	69.2	
Protein (N x 6.25)	%		13.8	15.6	
Cattle	dig prot %	*	9.1	10.3	
Goats	dig prot %	*	10.2	11.5	
Horses	dig prot %	*	10.2	11.5	
Sheep	dig prot %	*	10.2	11.5	
Starch	%		20.5	23.1	
Energy	GE Mcal/kg		4.34	4.90	
Cattle	DE Mcal/kg	*	2.75	3.35	
Sheep	DE Mcal/kg	*	3.12	3.52	
Swine	DE kcal/kg	*	2997.	3382.	
Cattle	ME Mcal/kg	*	2.44	2.75	
Sheep	ME Mcal/kg	*	2.55	2.88	
Swine	ME kcal/kg	*	2782.	3140.	
Cattle	TDN %	*	67.3	76.0	
Sheep	TDN %	*	70.7	79.8	
Swine	TDN %	*	68.0	76.7	
Aluminum	mg/kg		5.15	5.81	
Calcium	%		0.07	0.08	
Cobalt	mg/kg		0.113	0.128	
Copper	mg/kg		7.9	9.0	
Iron	%		0.008	0.009	
Magnesium	%		0.21	0.23	
Manganese	mg/kg		118.5	133.7	
Phosphorus	%		0.56	0.63	
Potassium	%		0.92	1.03	
Selenium	mg/kg		0.031	0.035	
Sodium	%		0.01	0.01	
Zinc	mg/kg		63.9	72.1	
Choline	mg/kg		1834.	2070.	
Folic acid	mg/kg		1.38	1.56	
Niacin	mg/kg		116.7	131.7	
Pantothenic acid	mg/kg		23.7	26.7	
Riboflavin	mg/kg		4.1	4.7	
Thiamine	mg/kg		18.3	20.7	
α-tocopherol	mg/kg		50.1	56.5	
Vitamin B6	mg/kg		6.55	7.40	
Alanine	%		0.83	0.94	
Arginine	%		1.05	1.19	
Aspartic acid	%		1.22	1.37	
Cystine	%		0.37	0.42	
Glutamic acid	%		2.83	3.20	
Glycine	%		0.86	0.97	
Histidine	%		0.38	0.43	
Isoleucine	%		0.47	0.53	
Leucine	%		0.96	1.08	
Lysine	%		0.70	0.79	
Methionine	%		0.24	0.27	
Phenylalanine	%		0.58	0.65	
Proline	%		0.92	1.03	
Serine	%		0.71	0.80	
Threonine	%		0.56	0.63	
Tyrosine	%		0.44	0.50	
Valine	%		0.65	0.73	

Wheat, white, flour by-product, fine sift, mx 4% fiber, (4)

Wheat, white, red dog

Ref No 4 08 020 United States

Feed Name or Analyses			Mean As Fed	Mean Dry	C.V. ± %
Dry matter	%		89.2	100.0	
Ash	%		2.0	2.2	
Crude fiber	%		3.2	3.6	
Ether extract	%		3.4	3.8	
N-free extract	%		68.0	76.2	
Protein (N x 6.25)	%		12.6	14.1	
Cattle	dig prot %	*	8.0	9.0	
Goats	dig prot %	*	9.1	10.2	
Horses	dig prot %	*	9.1	10.2	
Sheep	dig prot %	*	9.1	10.2	
Starch	%		40.0	44.8	
Energy	GE Mcal/kg		4.04	4.53	
Cattle	DE Mcal/kg	*	3.32	3.72	
Sheep	DE Mcal/kg	*	3.29	3.69	
Swine	DE kcal/kg	*	3485.	3907.	
Cattle	ME Mcal/kg	*	2.72	3.05	
Sheep	ME Mcal/kg	*	2.70	3.02	
Swine	ME kcal/kg	*	3246.	3639.	
Cattle	TDN %	*	75.3	84.4	
Sheep	TDN %	*	74.6	83.7	
Swine	TDN %	*	79.0	88.6	
Aluminum	mg/kg		5.19	5.81	
Calcium	%		0.04	0.05	
Cobalt	mg/kg		0.093	0.105	
Copper	mg/kg		6.3	7.1	
Iron	%		0.006	0.007	
Magnesium	%		0.11	0.13	
Manganese	mg/kg		71.6	80.2	
Phosphorus	%		0.37	0.42	
Potassium	%		0.54	0.60	
Selenium	mg/kg		0.311	0.349	
Sodium	%		0.01	0.01	
Zinc	mg/kg		50.8	57.0	
Choline	mg/kg		1696.	1901.	
Folic acid	mg/kg		1.05	1.17	
Niacin	mg/kg		48.0	53.8	
Pantothenic acid	mg/kg		13.1	14.7	
Riboflavin	mg/kg		2.7	3.0	
Thiamine	mg/kg		29.5	33.0	
α-tocopherol	mg/kg		33.3	37.3	
Vitamin B6	mg/kg		5.00	5.60	
Alanine	%		0.74	0.83	
Arginine	%		1.00	1.12	
Aspartic acid	%		1.05	1.17	
Cystine	%		0.40	0.45	
Glutamic acid	%		3.25	3.64	
Glycine	%		0.74	0.83	
Histidine	%		0.36	0.41	
Isoleucine	%		0.46	0.51	
Leucine	%		0.94	1.06	
Lysine	%		0.62	0.70	
Methionine	%		0.25	0.28	
Phenylalanine	%		0.58	0.65	
Proline	%		1.00	1.12	
Serine	%		0.69	0.78	
Threonine	%		0.51	0.57	
Tyrosine	%		0.41	0.47	
Valine	%		0.66	0.74	

Wheat, white, grain, (4)

Ref No 4 05 307 United States

Feed Name or Analyses			Mean As Fed	Mean Dry	C.V. ± %
Dry matter	%		88.0	100.0	1
Ash	%		1.5	1.7	
Crude fiber	%		2.0	2.3	
Sheep	dig coef %		17.	17.	
Ether extract	%		1.8	2.1	
Sheep	dig coef %		64.	64.	
N-free extract	%		72.5	82.4	
Sheep	dig coef %		93.	93.	
Protein (N x 6.25)	%		10.1	11.4	
Sheep	dig coef %		82.	82.	
Cattle	dig prot %	*	5.7	6.5	
Goats	dig prot %	*	6.8	7.7	
Horses	dig prot %	*	6.8	7.7	
Sheep	dig prot %		8.2	9.4	
Starch	%		59.9	68.1	
Energy	GE Mcal/kg		3.59	4.09	
Cattle	DE Mcal/kg	*	3.45	3.92	
Sheep	DE Mcal/kg	*	3.47	3.94	
Swine	DE kcal/kg	*	3546.	4030.	
Cattle	ME Mcal/kg	*	2.82	3.21	
Sheep	ME Mcal/kg	*	2.85	3.23	
Swine	ME kcal/kg	*	3322.	3776.	
Cattle	TDN %	*	78.1	88.8	
Sheep	TDN %		78.7	89.4	
Swine	TDN %	*	80.4	91.4	
Aluminum	mg/kg		5.12	5.81	
Calcium	%		0.05	0.05	98
Cobalt	mg/kg		0.133	0.151	
Copper	mg/kg		6.9	7.8	
Iron	%		0.003	0.003	36
Magnesium	%		0.09	0.10	
Manganese	mg/kg		46.1	52.4	
Phosphorus	%		0.32	0.37	43
Potassium	%		0.38	0.44	
Selenium	mg/kg		0.041	0.047	
Sodium	%		0.01	0.01	
Zinc	mg/kg		21.5	24.4	
Choline	mg/kg		1166.	1326.	
Folic acid	mg/kg		0.39	0.45	12
Niacin	mg/kg		50.2	57.0	
Pantothenic acid	mg/kg		8.6	9.8	
Riboflavin	mg/kg		1.3	1.4	
Thiamine	mg/kg		4.7	5.4	
α-tocopherol	mg/kg		14.8	16.9	
Vitamin B6	mg/kg		2.07	2.35	
Alanine	%		0.41	0.47	
Arginine	%		0.56	0.64	
Aspartic acid	%		0.58	0.66	
Cystine	%		0.30	0.34	
Glutamic acid	%		3.53	4.01	
Glycine	%		0.48	0.55	
Histidine	%		0.27	0.30	
Isoleucine	%		0.39	0.44	
Leucine	%		0.76	0.86	
Lysine	%		0.33	0.37	
Methionine	%		0.16	0.19	
Phenylalanine	%		0.50	0.57	
Proline	%		1.08	1.23	
Serine	%		0.53	0.60	
Threonine	%		0.33	0.37	
Tyrosine	%		0.33	0.37	
Valine	%		0.49	0.56	

Wheat, white, grain, gr 1 US mn wt 60 lb per bushel mx 0.5% foreign material, (4)

Ref No 4 05 300 United States

Feed Name or Analyses			Mean As Fed	Mean Dry	C.V. ± %
Dry matter	%			100.0	
Starch	%			67.5	
Sugars, total	%			3.8	

Wheat, white, grain, gr 2 US mn wt 58 lb per bushel mx 1% foreign material, (4)

Ref No 4 05 302 United States

Feed Name or Analyses			Mean As Fed	Mean Dry	C.V. ± %
Dry matter	%			100.0	
Sugars, total	%			3.6	9

Feed Name or Analyses		Mean As Fed	Mean Dry	C.V. ± %
Wheat, white, germ, (5)				
Ref No 5 05 308			United States	
Dry matter	%	88.5	100.0	
Ash	%	3.5	4.0	
Crude fiber	%	2.9	3.2	
Ether extract	%	7.9	8.9	
N-free extract	%	50.6	57.2	
Protein (N x 6.25)	%	23.6	26.6	
Starch	%	14.2	16.0	
Energy	GE Mcal/kg	4.32	4.88	
Cattle	DE Mcal/kg *	3.64	4.11	
Sheep	DE Mcal/kg *	3.49	3.94	
Swine	DE kcal/kg *	3797.	4290.	
Cattle	ME Mcal/kg *	2.98	3.37	
Sheep	ME Mcal/kg *	2.86	3.23	
Swine	ME kcal/kg *	3440.	3887.	
Cattle	TDN % *	82.5	93.3	
Sheep	TDN % *	79.0	89.3	
Swine	TDN % *	86.1	97.3	
Aluminum	mg/kg	5.15	5.81	
Calcium	%	0.04	0.05	
Chlorine	%	0.08	0.09	
Cobalt	mg/kg	0.093	0.105	
Copper	mg/kg	9.0	10.2	
Iron	%	0.006	0.007	
Magnesium	%	0.23	0.26	
Manganese	mg/kg	174.1	196.8	
Phosphorus	%	0.80	0.91	
Potassium	%	0.92	1.04	
Selenium	mg/kg	0.021	0.023	
Sodium	%	0.02	0.02	
Sulphur	%	0.15	0.18	
Zinc	mg/kg	103.9	117.4	
Choline	mg/kg	2667.	3014.	
Folic acid	mg/kg	2.67	3.01	
Niacin	mg/kg	73.6	83.1	
Pantothenic acid	mg/kg	18.4	20.8	
Riboflavin	mg/kg	6.4	7.3	
Thiamine	mg/kg	24.7	27.9	
α-tocopherol	mg/kg	164.1	185.5	
Vitamin B6	mg/kg	9.58	10.83	
Alanine	%	1.54	1.74	
Arginine	%	1.92	2.17	
Aspartic acid	%	2.32	2.62	
Cystine	%	0.48	0.55	
Glutamic acid	%	3.82	4.31	
Glycine	%	1.46	1.65	
Histidine	%	0.64	0.72	
Isoleucine	%	0.82	0.93	
Leucine	%	1.58	1.79	
Lysine	%	1.56	1.77	
Methionine	%	0.47	0.53	
Phenylalanine	%	0.93	1.05	
Proline	%	1.16	1.31	
Serine	%	1.10	1.24	
Threonine	%	0.99	1.12	
Tyrosine	%	0.70	0.79	
Valine	%	1.21	1.37	

(1) dry forages and roughages (2) pasture, range plants, and forages fed green (3) silages (4) energy feeds (5) protein supplements (6) minerals (7) vitamins (8) additives

Feed Name or Analyses		Mean As Fed	Mean Dry	C.V. ± %
WHEAT, WHITE CLUB. Triticum compactum				
Wheat, white club, grain, (4)				
Ref No 4 05 318			United States	
Dry matter	%		100.0	
Ash	%		1.8	12
Ether extract	%		1.9	10
Protein (N x 6.25)	%		9.8	12
Cattle	dig prot % *		5.0	
Goats	dig prot % *		6.2	
Horses	dig prot % *		6.2	
Sheep	dig prot % *		6.2	
Starch	%		66.9	6
Sugars, total	%		3.4	11
Niacin	mg/kg		62.2	
Pantothenic acid	mg/kg		12.6	
Vitamin B6	mg/kg		5.07	
Wheat, white club, grain, gr 1 US mn wt 60 lb per bushel mx 0.5% foreign material, (4)				
Ref No 4 05 311			United States	
Dry matter	%		100.0	
Ash	%		1.8	
Ether extract	%		1.9	
Protein (N x 6.25)	%		9.2	
Cattle	dig prot % *		4.5	
Goats	dig prot % *		5.7	
Horses	dig prot % *		5.7	
Sheep	dig prot % *		5.7	
Starch	%		68.7	2
Sugars, total	%		3.6	
Wheat, white club, grain, gr 2 US mn wt 58 lb per bushel mx 1% foreign material, (4)				
Ref No 4 05 313			United States	
Dry matter	%		100.0	
Ash	%		1.8	
Ether extract	%		2.0	
Protein (N x 6.25)	%		8.7	
Cattle	dig prot % *		4.0	
Goats	dig prot % *		5.2	
Horses	dig prot % *		5.2	
Sheep	dig prot % *		5.2	
Starch	%		69.3	
Sugars, total	%		3.5	
WHEAT, WHITE HARD. Triticum aestivum				
Wheat, white hard, grain, gr 1 US mn wt 60 lb per bushel mx 0.5% foreign material, (4)				
Ref No 4 05 321			United States	
Dry matter	%		100.0	
Ash	%		1.7	
Ether extract	%		2.0	
Protein (N x 6.25)	%		11.2	
Cattle	dig prot % *		6.3	
Goats	dig prot % *		7.5	
Horses	dig prot % *		7.5	
Sheep	dig prot % *		7.5	

Feed Name or Analyses		Mean As Fed	Mean Dry	C.V. ± %
Wheat, white hard, grain, gr 2 US mn wt 58 lb per bushel mx 1% foreign material, (4)				
Ref No 4 05 323			United States	
Dry matter	%		100.0	
Ash	%		2.2	
Ether extract	%		1.7	
Protein (N x 6.25)	%		14.2	
Cattle	dig prot % *		9.1	
Goats	dig prot % *		10.2	
Horses	dig prot % *		10.2	
Sheep	dig prot % *		10.2	
WHEAT, WHITE SOFT. Triticum aestivum				
Wheat, white soft, grain, (4)				
Ref No 4 05 336			United States	
Dry matter	%		100.0	
Niacin	mg/kg		63.3	3
Pantothenic acid	mg/kg		13.8	16
Riboflavin	mg/kg		1.3	52
Thiamine	mg/kg		5.5	17
Vitamin B6	mg/kg		5.51	12
Wheat, white soft, grain, gr 1 US mn wt 60 lb per bushel mx 0.5% foreign material, (4)				
Ref No 4 05 329			United States	
Dry matter	%		100.0	
Ash	%		1.8	8
Ether extract	%		2.0	9
Protein (N x 6.25)	%		10.0	9
Cattle	dig prot % *		5.2	
Goats	dig prot % *		6.4	
Horses	dig prot % *		6.4	
Sheep	dig prot % *		6.4	
Starch	%		66.8	0
Sugars, total	%		4.0	5
Wheat, white soft, grain, gr 2 US mn wt 58 lb per bushel mx 1% foreign material, (4)				
Ref No 4 05 331			United States	
Dry matter	%		100.0	
Ash	%		1.9	7
Ether extract	%		1.9	9
Protein (N x 6.25)	%		10.3	5
Cattle	dig prot % *		5.5	
Goats	dig prot % *		6.7	
Horses	dig prot % *		6.7	
Sheep	dig prot % *		6.7	
Starch	%		66.6	2
Sugars, total	%		3.9	9
WHEAT, WHITE SOFT WINTER. Triticum aestivum				
Wheat, white soft winter, grain, (4)				
Ref No 4 05 337			United States	
Dry matter	%		100.0	
Niacin	mg/kg		68.8	
Pantothenic acid	mg/kg		12.3	
Vitamin B6	mg/kg		5.29	

Feed Name or Analyses		Mean As Fed	Mean Dry	C.V. ± %

WHEAT, WINTER. Triticum aestivum

Wheat, winter, straw, (1)
Ref No 1 05 339 United States

Feed Name or Analyses		As Fed	Dry	C.V. ± %
Dry matter	%	84.3	100.0	
Ash	%	6.9	8.2	
Crude fiber	%	34.1	40.4	
Horses	dig coef %	16.	16.	
Sheep	dig coef %	39.	39.	
Ether extract	%	1.4	1.7	
Horses	dig coef %	82.	82.	
Sheep	dig coef %	52.	52.	
N-free extract	%	39.1	46.4	
Horses	dig coef %	26.	26.	
Sheep	dig coef %	46.	46.	
Protein (N x 6.25)	%	2.7	3.2	
Horses	dig coef %	30.	30.	
Sheep	dig coef %	6.	6.	
Cattle	dig prot %	* -0.1	-0.2	
Goats	dig prot %	* -0.3	-0.3	
Horses	dig prot %	0.8	1.0	
Rabbits	dig coef %	* 1.0	1.2	
Sheep	dig prot %	0.1	0.2	
Energy	GE Mcal/kg			
Cattle	DE Mcal/kg	* 1.62	1.92	
Horses	DE Mcal/kg	* 0.84	1.00	
Sheep	DE Mcal/kg	* 1.32	1.57	
Cattle	ME Mcal/kg	* 1.44	1.70	
Horses	ME Mcal/kg	* 36.7	43.5	
Sheep	ME Mcal/kg	* 1.18	1.40	
Cattle	TDN %	* 39.7	47.1	
Horses	TDN %	19.0	22.6	
Sheep	TDN %	32.7	38.8	

Wheat, winter, straw, steamed, (1)
Ref No 1 05 338 United States

Feed Name or Analyses		As Fed	Dry	C.V. ± %
Dry matter	%	81.7	100.0	
Ash	%	8.3	10.1	
Crude fiber	%	31.8	39.0	
Horses	dig coef %	46.	46.	
Sheep	dig coef %	60.	60.	
Ether extract	%	2.5	3.0	
Horses	dig coef %	70.	70.	
Sheep	dig coef %	69.	69.	
N-free extract	%	36.5	44.7	
Horses	dig coef %	36.	36.	
Sheep	dig coef %	64.	64.	
Protein (N x 6.25)	%	2.7	3.3	
Horses	dig coef %	0.	0.	
Sheep	dig coef %	8.	8.	
Cattle	dig prot %	* -0.1	-0.1	
Goats	dig prot %	* -0.2	-0.3	
Rabbits	dig coef %	* 1.0	1.2	
Sheep	dig prot %	0.2	0.3	
Energy	GE Mcal/kg			
Cattle	DE Mcal/kg	* 1.90	2.32	
Horses	DE Mcal/kg	* 1.39	1.71	
Sheep	DE Mcal/kg	* 2.05	2.51	
Cattle	ME Mcal/kg	* 1.55	1.90	
Horses	ME Mcal/kg	* 1.14	1.40	
Sheep	ME Mcal/kg	* 1.68	2.06	
Cattle	TDN %	* 43.0	52.6	
Horses	TDN %	31.6	38.7	
Sheep	TDN %	46.5	56.9	

Wheat, winter, straw, treated w sodium hydroxide dehy, (1)
Ref No 1 05 340 United States

Feed Name or Analyses		As Fed	Dry	C.V. ± %
Dry matter	%	79.7	100.0	
Ash	%	7.2	9.1	
Crude fiber	%	37.9	47.6	
Horses	dig coef %	56.	56.	
Sheep	dig coef %	66.	66.	
Ether extract	%	1.4	1.8	
Horses	dig coef %	67.	67.	
Sheep	dig coef %	34.	34.	
N-free extract	%	30.8	38.7	
Horses	dig coef %	41.	41.	
Sheep	dig coef %	62.	62.	
Protein (N x 6.25)	%	2.3	2.9	
Horses	dig coef %	0.	0.	
Sheep	dig coef %	-53.	-53.	
Cattle	dig prot %	* -0.4	-0.5	
Goats	dig prot %	* -0.5	-0.7	
Rabbits	dig coef %	* 0.7	0.9	
Sheep	dig prot %	-1.1	-1.4	
Energy	GE Mcal/kg			
Cattle	DE Mcal/kg	* 1.80	2.26	
Horses	DE Mcal/kg	* 1.59	1.99	
Sheep	DE Mcal/kg	* 1.47	1.85	
Cattle	ME Mcal/kg	* 1.05	1.31	
Horses	ME Mcal/kg	* 40.9	51.3	
Sheep	ME Mcal/kg	* 1.59	2.00	
Cattle	TDN %	* 28.9	36.3	
Horses	TDN %	36.0	45.2	
Sheep	TDN %	44.0	55.2	

Wheat, winter, bran, (4)
Ref No 4 05 341 United States

Feed Name or Analyses		As Fed	Dry	C.V. ± %
Dry matter	%	86.4	100.0	
Ash	%	6.2	7.2	
Crude fiber	%	8.8	10.2	
Sheep	dig coef %	28.	28.	
Ether extract	%	4.2	4.9	
Sheep	dig coef %	65.	65.	
N-free extract	%	53.2	61.6	
Sheep	dig coef %	71.	71.	
Protein (N x 6.25)	%	13.9	16.1	
Sheep	dig coef %	78.	78.	
Cattle	dig prot %	* 9.3	10.8	
Goats	dig prot %	* 10.4	12.0	
Horses	dig prot %	* 10.4	12.0	
Sheep	dig prot %	10.9	12.6	
Energy	GE Mcal/kg			
Cattle	DE Mcal/kg	* 2.68	3.10	
Sheep	DE Mcal/kg	* 2.53	2.92	
Swine	DE kcal/kg	*2359.	2730.	
Cattle	ME Mcal/kg	* 2.20	2.54	
Sheep	ME Mcal/kg	* 2.07	2.40	
Swine	ME kcal/kg	*2188.	2532.	
Cattle	TDN %	* 60.7	70.3	
Sheep	TDN %	57.3	66.3	
Swine	TDN %	* 53.5	61.9	

Wheat distillers dried grains (AAFCO) (CFA) –
see Wheat, distillers grains, dehy, (5)

Wheat distillers dried grains with solubles (AAFCO) –
see Wheat, distillers grains w solubles, dehy, mn 75% original solids, (5)

Wheat distillers dried solubles (AAFCO) –
see Wheat, distillers solubles, dehy, (5)

Wheat endosperm –
see Wheat, grits, cracked fine screened, (4)

Wheat feed flour, mx 1.5% fiber (AAFCO) –
see Wheat, flour, coarse bolt, feed gr mx 2% fiber, (4)

Wheat germ (CFA) –
see Wheat, germ, mn 75% embryo mx 2.5% fiber, (5)

Wheat germ meal (AAFCO) –
see Wheat, germ, grnd, mn 25% protein mn 7% fat, (5)

WHEATGRASS. Agropyron spp

Wheatgrass, hay, s-c, (1)
Ref No 1 05 351 United States

Feed Name or Analyses		As Fed	Dry	C.V. ± %
Dry matter	%	90.8	100.0	
Ash	%	7.3	8.0	
Crude fiber	%	31.3	34.5	
Sheep	dig coef %	62.	62.	
Ether extract	%	2.4	2.6	
Sheep	dig coef %	39.	39.	
N-free extract	%	42.5	46.8	
Sheep	dig coef %	58.	58.	
Protein (N x 6.25)	%	7.4	8.1	
Sheep	dig coef %	50.	50.	
Cattle	dig prot %	* 3.6	4.0	
Goats	dig prot %	* 3.7	4.1	
Horses	dig prot %	* 4.0	4.4	
Rabbits	dig coef %	* 4.5	4.9	
Sheep	dig prot %	3.7	4.1	
Energy	GE Mcal/kg			
Cattle	DE Mcal/kg	* 2.26	2.49	
Sheep	DE Mcal/kg	* 2.20	2.42	
Cattle	ME Mcal/kg	* 1.86	2.04	
Sheep	ME Mcal/kg	* 1.80	1.98	
Cattle	TDN %	* 51.4	56.6	
Sheep	TDN %	49.8	54.9	

Ref No 1 05 351 Canada

Feed Name or Analyses		As Fed	Dry	C.V. ± %
Dry matter	%	94.6	100.0	
Crude fiber	%	34.5	36.5	
Sheep	dig coef %	62.	62.	
Ether extract	%			
Sheep	dig coef %	39.	39.	
N-free extract	%			
Sheep	dig coef %	58.	58.	
Protein (N x 6.25)	%	7.3	7.7	
Sheep	dig coef %	50.	50.	
Cattle	dig prot %	* 3.4	3.6	
Goats	dig prot %	* 3.6	3.8	
Horses	dig prot %	* 3.9	4.1	
Rabbits	dig coef %	* 4.4	4.6	
Sheep	dig prot %	3.6	3.9	
Energy	GE Mcal/kg			
Sheep	DE Mcal/kg	* 2.29	2.42	
Sheep	ME Mcal/kg	* 1.88	1.99	
Sheep	TDN %	52.0	54.9	
Calcium	%	0.22	0.23	
Phosphorus	%	0.07	0.07	

Feed Name or Analyses			Mean As Fed	Mean Dry	C.V. ± %
Wheatgrass, hay, s-c, immature, (1)					
Ref No 1 05 344			United States		
Dry matter	%		89.1	100.0	2
Ash	%		8.4	9.4	
Crude fiber	%		19.9	22.3	
Sheep	dig coef %		60.	60.	
Ether extract	%		3.1	3.5	
Sheep	dig coef %		59.	59.	
N-free extract	%		44.0	49.4	
Sheep	dig coef %		72.	72.	
Protein (N x 6.25)	%		13.7	15.4	
Sheep	dig coef %		65.	65.	
Cattle	dig prot %	*	9.2	10.3	
Goats	dig prot %	*	9.7	10.9	
Horses	dig prot %	*	9.4	10.6	
Rabbits	dig coef %	*	9.4	10.6	
Sheep	dig prot %		8.9	10.0	
Energy	GE Mcal/kg				
Cattle	DE Mcal/kg	*	2.62	2.94	
Sheep	DE Mcal/kg	*	2.50	2.80	
Cattle	ME Mcal/kg	*	2.15	2.41	
Sheep	ME Mcal/kg	*	2.05	2.30	
Cattle	TDN %	*	59.4	66.7	
Sheep	TDN %		56.7	63.6	
Calcium	%		0.37	0.42	9
Magnesium	%		0.13	0.15	
Phosphorus	%		0.28	0.31	16

Feed Name or Analyses			Mean As Fed	Mean Dry	C.V. ± %
Wheatgrass, hay, s-c, early bloom, (1)					
Ref No 1 05 345			United States		
Dry matter	%		92.0	100.0	2
Ash	%		6.5	7.1	
Crude fiber	%		27.3	29.7	
Sheep	dig coef %		68.	68.	
Ether extract	%		2.2	2.4	
Sheep	dig coef %		57.	57.	
N-free extract	%		48.0	52.2	
Sheep	dig coef %		70.	70.	
Protein (N x 6.25)	%		7.9	8.6	
Sheep	dig coef %		48.	48.	
Cattle	dig prot %	*	4.0	4.4	
Goats	dig prot %	*	4.2	4.6	
Horses	dig prot %	*	4.4	4.8	
Rabbits	dig coef %	*	4.9	5.3	
Sheep	dig prot %		3.8	4.1	
Energy	GE Mcal/kg				
Cattle	DE Mcal/kg	*	2.57	2.80	
Sheep	DE Mcal/kg	*	2.59	2.82	
Cattle	ME Mcal/kg	*	2.11	2.29	
Sheep	ME Mcal/kg	*	2.13	2.31	
Cattle	TDN %	*	58.3	63.4	
Sheep	TDN %		58.8	63.9	
Calcium	%		0.29	0.31	16
Magnesium	%		0.22	0.24	
Phosphorus	%		0.25	0.27	14
Potassium	%		2.46	2.67	

(1) dry forages and roughages (2) pasture, range plants, and forages fed green (3) silages (4) energy feeds (5) protein supplements (6) minerals (7) vitamins (8) additives

Feed Name or Analyses			Mean As Fed	Mean Dry	C.V. ± %
Wheatgrass, hay, s-c, full bloom, (1)					
Ref No 1 05 346			United States		
Dry matter	%			100.0	
Crude fiber	%				
Cattle	dig coef %	*		64.	
Sheep	dig coef %	*		62.	
Ether extract	%				
Cattle	dig coef %	*		18.	
Sheep	dig coef %	*		39.	
N-free extract	%				
Sheep	dig coef %	*		58.	
Protein (N x 6.25)	%				
Cattle	dig coef %	*		63.	
Sheep	dig coef %	*		50.	
Energy	GE Mcal/kg	*		4.38	
Cattle	GE dig coef %	*		58.	
Cattle	DE Mcal/kg	*		2.54	
Sheep	DE Mcal/kg	*		2.40	
Cattle	ME Mcal/kg	*		2.08	
Sheep	ME Mcal/kg	*		1.97	
Sheep	TDN %	*		54.4	
Ref No 1 05 346			Canada		
Dry matter	%			100.0	
Ash	%			7.7	
Crude fiber	%			28.9	
Cattle	dig coef %			64.	
Ether extract	%			2.1	
Cattle	dig coef %			18.	
N-free extract	%			49.4	
Protein (N x 6.25)	%			11.9	
Cattle	dig coef %			63.	
Cattle	dig prot %			7.5	
Goats	dig prot %	*		7.7	
Horses	dig prot %	*		7.6	
Rabbits	dig coef %	*		7.9	
Sheep	dig prot %	*		7.2	
Cellulose (Matrone)	%			33.3	
Cattle	dig coef %			64.	
Energy	GE Mcal/kg			4.38	
Cattle	GE dig coef %			58.	
Cattle	DE Mcal/kg			2.54	
Sheep	DE Mcal/kg	*		2.40	
Cattle	ME Mcal/kg	*		2.08	
Sheep	ME Mcal/kg	*		1.97	
Cattle	TDN %	*		57.7	
Sheep	TDN %			54.4	

Feed Name or Analyses			Mean As Fed	Mean Dry	C.V. ± %
Wheatgrass, hay, s-c, late bloom, (1)					
Ref No 1 05 347			United States		
Dry matter	%		89.8	100.0	
Ash	%		5.7	6.3	
Crude fiber	%		34.0	37.9	
Sheep	dig coef %		51.	51.	
Ether extract	%		1.5	1.7	
Sheep	dig coef %		20.	20.	
N-free extract	%		43.8	48.8	
Sheep	dig coef %		53.	53.	
Protein (N x 6.25)	%		4.8	5.3	
Sheep	dig coef %		36.	36.	
Cattle	dig prot %	*	1.4	1.5	
Goats	dig prot %	*	1.4	1.5	
Horses	dig prot %	*	1.8	2.0	
Rabbits	dig coef %	*	2.5	2.8	
Sheep	dig prot %		1.7	1.9	
Energy	GE Mcal/kg				
Cattle	DE Mcal/kg	*	1.94	2.16	
Sheep	DE Mcal/kg	*	1.90	2.11	
Cattle	ME Mcal/kg	*	1.59	1.77	
Sheep	ME Mcal/kg	*	1.55	1.73	
Cattle	TDN %	*	44.1	49.1	
Sheep	TDN %		43.0	47.9	

Feed Name or Analyses			Mean As Fed	Mean Dry	C.V. ± %
Wheatgrass, hay, s-c, milk stage, (1)					
Ref No 1 05 348			United States		
Dry matter	%		91.4	100.0	4
Ash	%		5.5	6.0	6
Crude fiber	%		32.9	36.0	6
Ether extract	%		1.7	1.9	20
N-free extract	%		45.2	49.5	
Protein (N x 6.25)	%		6.0	6.6	12
Cattle	dig prot %	*	2.4	2.7	
Goats	dig prot %	*	2.5	2.7	
Horses	dig prot %	*	2.9	3.1	
Rabbits	dig coef %	*	3.4	3.8	
Sheep	dig prot %	*	2.3	2.5	
Energy	GE Mcal/kg				
Cattle	DE Mcal/kg	*	2.09	2.29	
Sheep	DE Mcal/kg	*	2.18	2.38	
Cattle	ME Mcal/kg	*	1.71	1.87	
Sheep	ME Mcal/kg	*	1.78	1.95	
Cattle	TDN %	*	47.4	51.8	
Sheep	TDN %	*	49.3	54.0	

Feed Name or Analyses			Mean As Fed	Mean Dry	C.V. ± %
Wheatgrass, hay, s-c, mature, (1)					
Ref No 1 05 349			United States		
Dry matter	%		90.4	100.0	2
Ash	%		7.9	8.7	16
Crude fiber	%		31.9	35.3	7
Ether extract	%		2.8	3.1	23
N-free extract	%		43.0	47.6	
Protein (N x 6.25)	%		4.8	5.3	13
Cattle	dig prot %	*	1.4	1.5	
Goats	dig prot %	*	1.4	1.5	
Horses	dig prot %	*	1.8	2.0	
Rabbits	dig coef %	*	2.5	2.8	
Sheep	dig prot %	*	1.2	1.3	
Energy	GE Mcal/kg				
Cattle	DE Mcal/kg	*	2.27	2.51	
Sheep	DE Mcal/kg	*	2.12	2.35	
Cattle	ME Mcal/kg	*	1.86	2.06	
Sheep	ME Mcal/kg	*	1.74	1.92	
Cattle	TDN %	*	51.5	57.0	
Sheep	TDN %	*	48.1	53.2	
Calcium	%		0.32	0.35	26
Iron	%		0.012	0.013	
Magnesium	%		0.08	0.09	
Manganese	mg/kg		6.0	6.6	80
Phosphorus	%		0.09	0.10	32

Feed Name or Analyses			Mean As Fed	Mean Dry	C.V. ± %
Wheatgrass, hay, s-c, over ripe, (1)					
Ref No 1 05 350			United States		
Dry matter	%		92.0	100.0	2
Ash	%		8.5	9.2	15
Crude fiber	%		33.2	36.1	5
Sheep	dig coef %	*	62.	62.	
Ether extract	%		2.6	2.8	32
Sheep	dig coef %	*	39.	39.	

Feed Name or Analyses			Mean As Fed	Mean Dry	C.V. ± %
N-free extract	%		44.0	47.8	
Sheep	dig coef %	*	58.	58.	
Protein (N x 6.25)	%		3.8	4.1	26
Sheep	dig coef %	*	50.	50.	
Cattle	dig prot %	*	0.5	0.5	
Goats	dig prot %	*	0.4	0.4	
Horses	dig prot %	*	0.9	1.0	
Rabbits	dig coef %	*	1.7	1.8	
Sheep	dig prot %		1.9	2.1	
Energy	GE Mcal/kg				
Cattle	DE Mcal/kg	*	2.28	2.48	
Sheep	DE Mcal/kg	*	2.22	2.41	
Cattle	ME Mcal/kg	*	1.87	2.03	
Sheep	ME Mcal/kg	*	1.82	1.97	
Cattle	TDN %	*	51.7	56.2	
Sheep	TDN %		50.2	54.6	
Calcium	%		0.23	0.25	10
Iron	%		0.014	0.015	
Manganese	mg/kg		8.7	9.5	
Phosphorus	%		0.06	0.06	31

Wheatgrass, aerial part, fresh, (2)
Ref No 2 05 365 United States

Feed Name or Analyses			Mean As Fed	Mean Dry	C.V. ± %
Dry matter	%		44.6	100.0	27
Ash	%		3.2	7.2	
Crude fiber	%		15.7	35.2	
Ether extract	%		1.2	2.8	
N-free extract	%		20.5	46.1	
Protein (N x 6.25)	%		3.9	8.7	
Cattle	dig prot %	*	2.4	5.3	
Goats	dig prot %	*	2.1	4.7	
Horses	dig prot %	*	2.2	5.0	
Rabbits	dig coef %	*	2.4	5.4	
Sheep	dig prot %	*	2.3	5.1	
Energy	GE Mcal/kg				
Cattle	DE Mcal/kg	*	1.30	2.91	
Sheep	DE Mcal/kg	*	1.27	2.85	
Cattle	ME Mcal/kg	*	1.06	2.38	
Sheep	ME Mcal/kg	*	1.04	2.34	
Cattle	TDN %	*	29.4	66.0	
Sheep	TDN %	*	28.9	64.7	
Carotene	mg/kg		91.4	205.0	
Choline	mg/kg		964.	2161.	
Riboflavin	mg/kg		4.4	9.9	20
Thiamine	mg/kg		2.0	4.4	27
Vitamin A equivalent	IU/g		152.4	341.8	

Wheatgrass, aerial part, fresh, early leaf, (2)
Ref No 2 08 841 United States

Feed Name or Analyses			Mean As Fed	Mean Dry	C.V. ± %
Dry matter	%			100.0	
Crude fiber	%			27.0	
Ether extract	%			3.0	
N-free extract	%			42.0	
Protein (N x 6.25)	%			14.0	
Cattle	dig prot %	*		9.8	
Goats	dig prot %	*		9.6	
Horses	dig prot %	*		9.4	
Rabbits	dig coef %	*		9.5	
Sheep	dig prot %	*		10.0	
Energy	GE Mcal/kg				
Cattle	DE Mcal/kg	*		2.64	
Sheep	DE Mcal/kg	*		2.76	
Cattle	ME Mcal/kg	*		2.17	
Sheep	ME Mcal/kg	*		2.26	

Feed Name or Analyses			Mean As Fed	Mean Dry	C.V. ± %
Cattle	TDN %	*		59.9	
Sheep	TDN %	*		62.5	

Wheatgrass, aerial part, fresh, immature, (2)
Ref No 2 05 354 United States

Feed Name or Analyses			Mean As Fed	Mean Dry	C.V. ± %
Dry matter	%		32.9	100.0	17
Ash	%		3.2	9.6	
Crude fiber	%		8.3	25.3	
Sheep	dig coef %		64.	64.	
Ether extract	%		0.9	2.8	
Sheep	dig coef %		66.	66.	
N-free extract	%		16.0	48.7	
Sheep	dig coef %		74.	74.	
Protein (N x 6.25)	%		4.5	13.6	
Sheep	dig coef %		74.	74.	
Cattle	dig prot %	*	3.1	9.5	
Goats	dig prot %	*	3.0	9.2	
Horses	dig prot %	*	3.0	9.1	
Rabbits	dig coef %	*	3.0	9.2	
Sheep	dig prot %		3.3	10.1	
Energy	GE Mcal/kg				
Cattle	DE Mcal/kg	*	0.96	2.91	
Sheep	DE Mcal/kg	*	0.96	2.93	
Cattle	ME Mcal/kg	*	0.78	2.38	
Sheep	ME Mcal/kg	*	0.79	2.40	
Cattle	TDN %	*	21.7	66.0	
Sheep	TDN %		21.9	66.5	
Carotene	mg/kg		120.5	366.2	48
Vitamin A equivalent	IU/g		200.8	610.4	

Wheatgrass, aerial part, fresh, immature, pure stand, (2)
Ref No 2 05 353 United States

Feed Name or Analyses			Mean As Fed	Mean Dry	C.V. ± %
Dry matter	%			100.0	
Carotene	mg/kg			529.5	26
Vitamin A equivalent	IU/g			882.8	

Wheatgrass, aerial part, fresh, pre-bloom, (2)
Ref No 2 05 356 United States

Feed Name or Analyses			Mean As Fed	Mean Dry	C.V. ± %
Dry matter	%		39.4	100.0	27
Crude fiber	%		12.6	32.0	
Ether extract	%		1.2	3.0	
N-free extract	%		16.5	42.0	
Protein (N x 6.25)	%		3.9	10.0	
Cattle	dig prot %	*	2.5	6.4	
Goats	dig prot %	*	2.3	5.9	
Horses	dig prot %	*	2.4	6.0	
Rabbits	dig coef %	*	2.5	6.4	
Sheep	dig prot %	*	2.5	6.3	
Energy	GE Mcal/kg				
Cattle	DE Mcal/kg	*	1.13	2.86	
Sheep	DE Mcal/kg	*	1.10	2.80	
Cattle	ME Mcal/kg	*	0.93	2.35	
Sheep	ME Mcal/kg	*	0.91	2.30	
Cattle	TDN %	*	25.5	64.8	
Sheep	TDN %	*	25.0	63.6	
Calcium	%		0.19	0.47	16
Phosphorus	%		0.09	0.23	37
Carotene	mg/kg		102.7	260.6	45
Vitamin A equivalent	IU/g		171.2	434.4	

Wheatgrass, aerial part, fresh, pre-bloom, pure stand, (2)
Ref No 2 05 355 United States

Feed Name or Analyses			Mean As Fed	Mean Dry	C.V. ± %
Dry matter	%			100.0	
Carotene	mg/kg			396.4	18
Vitamin A equivalent	IU/g			660.8	

Wheatgrass, aerial part, fresh, early bloom, (2)
Ref No 2 05 357 United States

Feed Name or Analyses			Mean As Fed	Mean Dry	C.V. ± %
Dry matter	%		36.6	100.0	
Ash	%		2.9	7.9	
Crude fiber	%		10.8	29.6	
Sheep	dig coef %		68.	68.	
Ether extract	%		1.1	3.1	
Sheep	dig coef %		53.	53.	
N-free extract	%		18.4	50.4	
Sheep	dig coef %		75.	75.	
Protein (N x 6.25)	%		3.3	9.1	
Sheep	dig coef %		60.	60.	
Cattle	dig prot %	*	2.1	5.6	
Goats	dig prot %	*	1.8	5.1	
Horses	dig prot %	*	1.9	5.3	
Rabbits	dig coef %	*	2.1	5.7	
Sheep	dig prot %		2.0	5.5	
Energy	GE Mcal/kg				
Cattle	DE Mcal/kg	*	1.03	2.81	
Sheep	DE Mcal/kg	*	1.08	2.95	
Cattle	ME Mcal/kg	*	0.84	2.31	
Sheep	ME Mcal/kg	*	0.89	2.42	
Cattle	TDN %	*	23.4	63.8	
Sheep	TDN %		24.5	67.0	

Wheatgrass, aerial part, fresh, mid-bloom, (2)
Ref No 2 05 358 United States

Feed Name or Analyses			Mean As Fed	Mean Dry	C.V. ± %
Dry matter	%		29.9	100.0	
Ash	%		2.2	7.2	
Crude fiber	%		9.5	31.8	9
Ether extract	%		0.7	2.4	11
N-free extract	%		14.5	48.4	
Protein (N x 6.25)	%		3.0	10.2	22
Cattle	dig prot %	*	2.0	6.6	
Goats	dig prot %	*	1.8	6.1	
Horses	dig prot %	*	1.9	6.2	
Rabbits	dig coef %	*	2.0	6.5	
Sheep	dig prot %	*	1.9	6.5	
Energy	GE Mcal/kg				
Cattle	DE Mcal/kg	*	0.89	2.98	
Sheep	DE Mcal/kg	*	0.86	2.88	
Cattle	ME Mcal/kg	*	0.73	2.44	
Sheep	ME Mcal/kg	*	0.71	2.36	
Cattle	TDN %	*	20.2	67.6	
Sheep	TDN %	*	19.5	65.3	
Calcium	%		0.13	0.42	5
Phosphorus	%		0.06	0.20	12

Wheatgrass, aerial part, fresh, full bloom, (2)
Ref No 2 05 360 United States

Feed Name or Analyses			Mean As Fed	Mean Dry	C.V. ± %
Dry matter	%		39.6	100.0	8
Carotene	mg/kg		58.1	146.6	40
Vitamin A equivalent	IU/g		96.8	244.4	

Feed Name or Analyses		Mean As Fed	Mean Dry	C.V. ± %

Wheatgrass, aerial part, fresh, full bloom, pure stand, (2)

Ref No 2 05 359 United States

Analyses		As Fed	Dry	C.V. ± %
Dry matter	%		100.0	
Carotene	mg/kg		182.3	25
Vitamin A equivalent	IU/g		303.9	

Wheatgrass, aerial part, fresh, milk stage, (2)

Ref No 2 05 361 United States

Analyses		As Fed	Dry	C.V. ± %
Dry matter	%	48.7	100.0	
Calcium	%	0.19	0.39	20
Phosphorus	%	0.12	0.24	37
Carotene	mg/kg	45.5	93.5	41
Vitamin A equivalent	IU/g	75.9	155.8	

Wheatgrass, aerial part, fresh, dough stage, (2)

Ref No 2 08 842 United States

Analyses		As Fed	Dry	C.V. ± %
Dry matter	%		100.0	
Crude fiber	%		29.0	
Ether extract	%		3.5	
N-free extract	%		44.0	
Protein (N x 6.25)	%		9.0	
Cattle	dig prot % *		5.5	
Goats	dig prot % *		5.0	
Horses	dig prot % *		5.2	
Rabbits	dig coef % *		5.6	
Sheep	dig prot % *		5.4	
Energy	GE Mcal/kg			
Cattle	DE Mcal/kg *		2.80	
Sheep	DE Mcal/kg *		2.88	
Cattle	ME Mcal/kg *		2.30	
Sheep	ME Mcal/kg *		2.36	
Cattle	TDN % *		63.4	
Sheep	TDN % *		65.3	

Wheatgrass, aerial part, fresh, mature, (2)

Ref No 2 05 363 United States

Analyses		As Fed	Dry	C.V. ± %
Dry matter	%	60.5	100.0	11
Crude fiber	%	17.5	29.0	
Ether extract	%	1.8	3.0	
N-free extract	%	26.0	43.0	
Protein (N x 6.25)	%	5.4	9.0	
Cattle	dig prot % *	3.4	5.5	
Goats	dig prot % *	3.0	5.0	
Horses	dig prot % *	3.1	5.2	
Rabbits	dig coef % *	3.4	5.6	
Sheep	dig prot % *	3.3	5.4	
Energy	GE Mcal/kg			
Cattle	DE Mcal/kg *	1.66	2.74	
Sheep	DE Mcal/kg *	1.73	2.87	
Cattle	ME Mcal/kg *	1.36	2.25	
Sheep	ME Mcal/kg *	1.42	2.35	
Cattle	TDN % *	37.6	62.1	
Sheep	TDN % *	39.3	65.0	
Carotene	mg/kg	41.3	68.3	33
Vitamin A equivalent	IU/g	68.9	113.9	

(1) dry forages and roughages (2) pasture, range plants, and forages fed green (3) silages (4) energy feeds (5) protein supplements (6) minerals (7) vitamins (8) additives

Wheatgrass, aerial part, fresh, mature, pure stand, (2)

Ref No 2 05 362 United States

Analyses		As Fed	Dry	C.V. ± %
Dry matter	%		100.0	
Carotene	mg/kg		74.7	26
Vitamin A equivalent	IU/g		124.6	

Wheatgrass, aerial part, fresh, over ripe, (2)

Ref No 2 05 364 United States

Analyses		As Fed	Dry	C.V. ± %
Dry matter	%	53.7	100.0	22
Ash	%	4.2	7.8	
Crude fiber	%	18.3	34.1	
Sheep	dig coef %	68.	68.	
Ether extract	%	1.8	3.4	
Sheep	dig coef %	48.	48.	
N-free extract	%	24.5	45.7	
Sheep	dig coef %	68.	68.	
Protein (N x 6.25)	%	4.8	9.0	
Sheep	dig coef %	64.	64.	
Cattle	dig prot % *	3.0	5.5	
Goats	dig prot % *	2.7	5.0	
Horses	dig prot % *	2.8	5.2	
Rabbits	dig coef % *	3.0	5.6	
Sheep	dig prot %	3.1	5.8	
Energy	GE Mcal/kg			
Cattle	DE Mcal/kg *	1.52	2.83	
Sheep	DE Mcal/kg *	1.51	2.81	
Cattle	ME Mcal/kg *	1.25	2.32	
Sheep	ME Mcal/kg *	1.24	2.30	
Cattle	TDN % *	34.5	64.3	
Sheep	TDN %	34.2	63.7	
Calcium	%	0.13	0.24	44
Copper	mg/kg	3.8	7.1	
Magnesium	%	0.03	0.06	
Manganese	mg/kg	23.9	44.5	
Phosphorus	%	0.05	0.09	90
Carotene	mg/kg	0.2	0.4	
Vitamin A equivalent	IU/g	0.4	0.7	

Wheatgrass, aerial part, fresh, pure stand, (2)

Ref No 2 05 366 United States

Analyses		As Fed	Dry	C.V. ± %
Dry matter	%		100.0	
Carotene	mg/kg		300.9	55
Vitamin A equivalent	IU/g		501.6	

WHEATGRASS, AMURENSE. Agropyron intermedium

Wheatgrass, amurense, hay, s-c, full bloom, (1)

Ref No 1 08 669 United States

Analyses		As Fed	Dry	C.V. ± %
Dry matter	%		100.0	
Sheep	dig coef %		51.	
Ash	%		6.3	
Crude fiber	%		39.1	
Ether extract	%		1.2	
N-free extract	%		48.9	
Protein (N x 6.25)	%		4.5	
Cattle	dig prot % *		0.8	
Goats	dig prot % *		0.8	
Horses	dig prot % *		1.4	
Rabbits	dig coef % *		2.1	
Sheep	dig prot % *		0.6	
Energy	GE Mcal/kg		4.32	
Sheep	GE dig coef %		49.	
Cattle	DE Mcal/kg *		2.06	
Sheep	DE Mcal/kg		2.11	
Cattle	ME Mcal/kg *		1.69	
Sheep	ME Mcal/kg *		1.73	
Cattle	TDN % *		46.8	
Sheep	TDN % *		51.8	

Wheatgrass, amurense, hay, s-c, chopped, early bloom, (1)

Ref No 1 08 671 United States

Analyses		As Fed	Dry	C.V. ± %
Dry matter	%		100.0	
Sheep	dig coef %		59.	
Ash	%		10.8	
Crude fiber	%		36.0	
Ether extract	%		2.0	
N-free extract	%		42.7	
Protein (N x 6.25)	%		8.5	
Cattle	dig prot % *		4.3	
Goats	dig prot % *		4.5	
Horses	dig prot % *		4.7	
Rabbits	dig coef % *		5.2	
Sheep	dig prot % *		4.2	
Lignin (Ellis)	%		7.5	
Energy	GE Mcal/kg		4.21	
Cattle	DE Mcal/kg *		2.47	
Sheep	DE Mcal/kg *		2.38	
Cattle	ME Mcal/kg *		2.03	
Sheep	ME Mcal/kg *		1.95	
Cattle	TDN % *		56.1	
Sheep	TDN % *		53.9	

Wheatgrass, amurense, hay, s-c chopped, full bloom, (1)

Ref No 1 08 670 United States

Analyses		As Fed	Dry	C.V. ± %
Dry matter	%		100.0	
Sheep	dig coef %		50.	
Ash	%		8.0	
Crude fiber	%		36.8	
Ether extract	%		2.2	
N-free extract	%		48.0	
Protein (N x 6.25)	%		5.0	
Cattle	dig prot % *		1.3	
Goats	dig prot % *		1.2	
Horses	dig prot % *		1.8	
Rabbits	dig coef % *		2.5	
Sheep	dig prot % *		1.1	
Lignin (Ellis)	%		8.5	
Energy	GE Mcal/kg		4.28	
Cattle	DE Mcal/kg *		2.34	
Sheep	DE Mcal/kg *		2.32	
Cattle	ME Mcal/kg *		1.92	
Sheep	ME Mcal/kg *		1.90	
Cattle	TDN % *		53.1	
Sheep	TDN % *		52.6	

WHEATGRASS, BEARDED. Agropyron subsecundum

Wheatgrass, bearded, hay, s-c, over ripe, (1)

Ref No 1 05 368 United States

Analyses		As Fed	Dry	C.V. ± %
Dry matter	%	88.9	100.0	
Ash	%	10.3	11.6	
Crude fiber	%	31.1	35.0	
Ether extract	%	3.6	4.1	
N-free extract	%	40.8	45.9	

Feed Name or Analyses			Mean As Fed	Mean Dry	C.V. ± %
Protein (N x 6.25)	%		3.0	3.4	
Cattle	dig prot %	*	0.0	0.0	
Goats	dig prot %	*	-0.1	-0.2	
Horses	dig prot %	*	0.4	0.4	
Rabbits	dig coef %	*	1.2	1.3	
Sheep	dig prot %	*	-0.2	-0.3	
Energy	GE Mcal/kg				
Cattle	DE Mcal/kg	*	2.04	2.30	
Sheep	DE Mcal/kg	*	2.04	2.29	
Cattle	ME Mcal/kg	*	1.68	1.89	
Sheep	ME Mcal/kg	*	1.67	1.88	
Cattle	TDN %	*	46.3	52.1	
Sheep	TDN %	*	46.3	52.0	

Ref No 1 05 368 Canada

Feed Name or Analyses			Mean As Fed	Mean Dry	C.V. ± %
Dry matter	%			100.0	
Protein (N x 6.25)	%			3.9	
Cattle	dig prot %	*		0.3	
Goats	dig prot %	*		0.2	
Horses	dig prot %	*		0.9	
Rabbits	dig coef %	*		1.7	
Sheep	dig prot %	*		0.1	
Phosphorus	%			0.10	
Potassium	%			0.09	

Wheatgrass, bearded, aerial part, fresh, (2)

Ref No 2 05 372 United States

Feed Name or Analyses			Mean As Fed	Mean Dry	C.V. ± %
Dry matter	%			100.0	
Ash	%			6.1	18
Crude fiber	%			34.9	15
Ether extract	%			2.2	26
N-free extract	%			47.7	
Protein (N x 6.25)	%			9.1	37
Cattle	dig prot %	*		5.6	
Goats	dig prot %	*		5.1	
Horses	dig prot %	*		5.3	
Rabbits	dig coef %	*		5.7	
Sheep	dig prot %	*		5.5	
Energy	GE Mcal/kg				
Cattle	DE Mcal/kg	*		3.04	
Sheep	DE Mcal/kg	*		2.87	
Cattle	ME Mcal/kg	*		2.49	
Sheep	ME Mcal/kg	*		2.35	
Cattle	TDN %	*		68.9	
Sheep	TDN %	*		65.1	

Wheatgrass, bearded, aerial part, fresh, immature, (2)

Ref No 2 05 369 United States

Feed Name or Analyses			Mean As Fed	Mean Dry	C.V. ± %
Dry matter	%			100.0	
Ash	%			6.9	
Crude fiber	%			26.6	
Ether extract	%			2.2	
N-free extract	%			50.2	
Protein (N x 6.25)	%			14.1	
Cattle	dig prot %	*		9.9	
Goats	dig prot %	*		9.7	
Horses	dig prot %	*		9.5	
Rabbits	dig coef %	*		9.6	
Sheep	dig prot %	*		10.1	
Calcium	%			0.57	11
Magnesium	%			0.14	26
Phosphorus	%			0.34	16
Sulphur	%			0.19	29

Wheat grass, bearded, aerial part, fresh, milk stage, (2)

Ref No 2 05 370 United States

Feed Name or Analyses			Mean As Fed	Mean Dry	C.V. ± %
Dry matter	%			100.0	
Ash	%			7.3	
Crude fiber	%			41.3	
Ether extract	%			1.6	
N-free extract	%			42.6	
Protein (N x 6.25)	%			7.2	
Cattle	dig prot %	*		4.0	
Goats	dig prot %	*		3.3	
Horses	dig prot %	*		3.6	
Rabbits	dig coef %	*		4.2	
Sheep	dig prot %	*		3.7	

Wheatgrass, bearded, aerial part, fresh, mature, (2)

Ref No 2 05 371 United States

Feed Name or Analyses			Mean As Fed	Mean Dry	C.V. ± %
Dry matter	%			100.0	
Ash	%			5.2	
Crude fiber	%			42.5	
Ether extract	%			2.2	
N-free extract	%			45.8	
Protein (N x 6.25)	%			4.3	
Cattle	dig prot %	*		1.5	
Goats	dig prot %	*		0.6	
Horses	dig prot %	*		1.2	
Rabbits	dig coef %	*		2.0	
Sheep	dig prot %	*		1.0	
Calcium	%			0.32	15
Magnesium	%			0.06	49
Phosphorus	%			0.10	49
Sulphur	%			0.09	32

WHEATGRASS, BEARDED BLUEBUNCH. Agropyron Spicatum

Wheatgrass, bearded bluebunch, hay, s-c, over ripe, (1)

Ref No 1 05 373 United States

Feed Name or Analyses			Mean As Fed	Mean Dry	C.V. ± %
Dry matter	%		88.9	100.0	
Ash	%		10.3	11.6	
Crude fiber	%		31.1	35.0	
Sheep	dig coef %		50.	50.	
Ether extract	%		3.6	4.1	
Sheep	dig coef %		35.	35.	
N-free extract	%		40.8	45.9	
Sheep	dig coef %		41.	41.	
Protein (N x 6.25)	%		3.0	3.4	
Sheep	dig coef %		6.	6.	
Cattle	dig prot %	*	0.0	0.0	
Goats	dig prot %	*	-0.1	-0.2	
Horses	dig prot %	*	0.4	0.4	
Rabbits	dig coef %	*	1.2	1.3	
Sheep	dig prot %		0.2	0.2	
Energy	GE Mcal/kg				
Sheep	DE Mcal/kg	*	1.56	1.75	
Sheep	ME Mcal/kg	*	1.28	1.44	
Sheep	TDN %		35.3	39.8	

WHEATGRASS, BEARDLESS. Agropyron inerme

Wheatgrass, beardless, aerial part, fresh, (2)

Ref No 2 05 381 United States

Feed Name or Analyses			Mean As Fed	Mean Dry	C.V. ± %
Dry matter	%		34.6	100.0	16
Ash	%		2.7	7.7	20
Crude fiber	%		10.3	29.7	8
Sheep	dig coef %	*	68.	68.	
Ether extract	%		1.3	3.7	19
Sheep	dig coef %	*	48.	48.	
N-free extract	%		16.5	47.6	
Sheep	dig coef %	*	68.	68.	
Protein (N x 6.25)	%		3.9	11.3	39
Sheep	dig coef %	*	64.	64.	
Cattle	dig prot %	*	2.6	7.5	
Goats	dig prot %	*	2.5	7.1	
Horses	dig prot %	*	2.5	7.1	
Rabbits	dig coef %	*	2.6	7.4	
Sheep	dig prot %		2.5	7.2	
Energy	GE Mcal/kg				
Cattle	DE Mcal/kg	*	1.00	2.90	
Sheep	DE Mcal/kg	*	0.97	2.81	
Cattle	ME Mcal/kg	*	0.82	2.38	
Sheep	ME Mcal/kg	*	0.80	2.31	
Cattle	TDN %	*	22.7	65.7	
Sheep	TDN %		22.1	63.8	
Sulphur	%		0.06	0.17	24
Carotene	mg/kg		107.9	311.7	69
Vitamin A equivalent	IU/g		179.8	519.7	

Wheatgrass, beardless, aerial part, fresh, immature, (2)

Ref No 2 05 375 United States

Feed Name or Analyses			Mean As Fed	Mean Dry	C.V. ± %
Dry matter	%		32.2	100.0	
Ash	%		2.4	7.6	13
Crude fiber	%		8.7	26.9	4
Sheep	dig coef %	*	74.	74.	
Ether extract	%		1.0	3.0	13
Sheep	dig coef %	*	66.	66.	
N-free extract	%		15.3	47.6	
Sheep	dig coef %	*	74.	74.	
Protein (N x 6.25)	%		4.8	14.9	14
Sheep	dig coef %	*	76.	76.	
Cattle	dig prot %	*	3.4	10.6	
Goats	dig prot %	*	3.4	10.5	
Horses	dig prot %	*	3.3	10.2	
Rabbits	dig coef %	*	3.3	10.2	
Sheep	dig prot %		3.6	11.3	
Energy	GE Mcal/kg				
Cattle	DE Mcal/kg	*	1.00	3.10	
Sheep	DE Mcal/kg	*	1.01	3.13	
Cattle	ME Mcal/kg	*	0.82	2.55	
Sheep	ME Mcal/kg	*	0.83	2.56	
Cattle	TDN %	*	22.7	70.4	
Sheep	TDN %		22.8	70.9	
Calcium	%		0.22	0.67	17
Phosphorus	%		0.09	0.28	37
Sulphur	%		0.06	0.20	19

Wheatgrass, beardless, aerial part, fresh, immature, pure stand, (2)

Ref No 2 05 374 United States

Feed Name or Analyses		Mean As Fed	Mean Dry	C.V. ± %
Dry matter	%		100.0	
Carotene	mg/kg		597.9	34
Vitamin A equivalent	IU/g		996.7	

Wheatgrass, beardless, aerial part, fresh, pre-bloom, pure stand, (2)

Ref No 2 05 376 United States

Feed Name or Analyses		Mean As Fed	Mean Dry	C.V. ± %
Dry matter	%		100.0	
Carotene	mg/kg		420.6	25
Vitamin A equivalent	IU/g		701.2	

Wheatgrass, beardless, aerial part, fresh, full bloom, (2)

Ref No 2 05 378 United States

Feed Name or Analyses		Mean As Fed	Mean Dry	C.V. ± %
Dry matter	%	45.5	100.0	
Ash	%	3.5	7.8	
Crude fiber	%	15.5	34.1	
Sheep	dig coef % *	68.	68.	
Ether extract	%	1.5	3.4	
Sheep	dig coef % *	48.	48.	
N-free extract	%	20.8	45.7	
Sheep	dig coef % *	68.	68.	
Protein (N x 6.25)	%	4.1	9.0	
Sheep	dig coef % *	64.	64.	
Cattle	dig prot % *	2.5	5.5	
Goats	dig prot % *	2.3	5.0	
Horses	dig prot % *	2.4	5.2	
Rabbits	dig coef % *	2.6	5.6	
Sheep	dig prot %	2.6	5.8	
Energy	GE Mcal/kg			
Cattle	DE Mcal/kg *	1.29	2.83	
Sheep	DE Mcal/kg *	1.28	2.81	
Cattle	ME Mcal/kg *	1.06	2.32	
Sheep	ME Mcal/kg *	1.05	2.30	
Cattle	TDN % *	29.2	64.3	
Sheep	TDN %	29.0	63.7	
Calcium	%	0.14	0.30	
Phosphorus	%	0.11	0.24	

Wheatgrass, beardless, aerial part, fresh, full bloom, pure stand, (2)

Ref No 2 05 377 United States

Feed Name or Analyses		Mean As Fed	Mean Dry	C.V. ± %
Dry matter	%		100.0	
Carotene	mg/kg		224.6	24
Vitamin A equivalent	IU/g		374.5	

Wheatgrass, beardless, aerial part, fresh, mature, (2)

Ref No 2 05 380 United States

Feed Name or Analyses		Mean As Fed	Mean Dry	C.V. ± %
Dry matter	%		100.0	
Ash	%		11.8	
Ether extract	%		4.6	
Protein (N x 6.25)	%		3.5	
Cattle	dig prot % *		0.9	
Goats	dig prot % *		0.1	
Horses	dig prot % *		0.5	
Rabbits	dig coef % *		1.4	
Sheep	dig prot % *		0.3	
Carotene	mg/kg		49.2	29
Vitamin A equivalent	IU/g		82.0	

(1) dry forages and roughages (2) pasture, range plants, and forages fed green (3) silages (4) energy feeds (5) protein supplements (6) minerals (7) vitamins (8) additives

Wheatgrass, beardless, aerial part, fresh, mature, pure stand, (2)

Ref No 2 05 379 United States

Feed Name or Analyses		Mean As Fed	Mean Dry	C.V. ± %
Dry matter	%		100.0	
Carotene	mg/kg		57.1	13
Vitamin A equivalent	IU/g		95.2	

Wheatgrass, beardless, aerial part, fresh, pure stand, (2)

Ref No 2 05 382 United States

Feed Name or Analyses		Mean As Fed	Mean Dry	C.V. ± %
Dry matter	%		100.0	
Carotene	mg/kg		322.1	64
Vitamin A equivalent	IU/g		536.9	

WHEATGRASS, BEARDLESS BLUEBUNCH. Agropyron inerme

Wheatgrass, beardless bluebunch, aerial part, fresh, immature, (2)

Ref No 2 05 383 United States

Feed Name or Analyses		Mean As Fed	Mean Dry	C.V. ± %
Dry matter	%	32.5	100.0	
Ash	%	3.3	10.2	
Crude fiber	%	8.1	24.8	
Sheep	dig coef %	78.	78.	
Ether extract	%	1.3	4.1	
Sheep	dig coef %	64.	64.	
N-free extract	%	13.0	40.1	
Sheep	dig coef %	76.	76.	
Protein (N x 6.25)	%	6.8	20.8	
Sheep	dig coef %	79.	79.	
Cattle	dig prot % *	5.1	15.6	
Goats	dig prot % *	5.2	16.0	
Horses	dig prot % *	4.9	15.2	
Rabbits	dig coef % *	4.8	14.7	
Sheep	dig prot %	5.3	16.4	
Energy	GE Mcal/kg			
Cattle	DE Mcal/kg *	1.02	3.14	
Sheep	DE Mcal/kg *	1.03	3.18	
Cattle	ME Mcal/kg *	0.84	2.57	
Sheep	ME Mcal/kg *	0.85	2.61	
Cattle	TDN % *	23.1	71.1	
Sheep	TDN %	23.5	72.2	

WHEATGRASS, BLUEBUNCH. Agropyron spicatum

Wheatgrass, bluebunch, aerial part, fresh, (2)

Ref No 2 05 394 United States

Feed Name or Analyses		Mean As Fed	Mean Dry	C.V. ± %
Dry matter	%		100.0	
Carotene	mg/kg		228.0	57
Vitamin A equivalent	IU/g		380.0	

Wheatgrass, bluebunch, aerial part, fresh, immature, (2)

Ref No 2 05 385 United States

Feed Name or Analyses		Mean As Fed	Mean Dry	C.V. ± %
Dry matter	%		100.0	
Carotene	mg/kg		411.6	32
Vitamin A equivalent	IU/g		686.1	

Wheatgrass, bluebunch, aerial part, fresh, immature, pure stand, (2)

Ref No 2 05 384 United States

Feed Name or Analyses		Mean As Fed	Mean Dry	C.V. ± %
Dry matter	%		100.0	
Carotene	mg/kg		539.7	16
Vitamin A equivalent	IU/g		899.7	

Wheatgrass, bluebunch, aerial part, fresh, pre-bloom, (2)

Ref No 2 05 387 United States

Feed Name or Analyses		Mean As Fed	Mean Dry	C.V. ± %
Dry matter	%		100.0	
Ash	%		6.3	11
Crude fiber	%		31.5	11
Ether extract	%		3.1	48
N-free extract	%		45.6	
Protein (N x 6.25)	%		13.5	13
Cattle	dig prot % *		9.4	
Goats	dig prot % *		9.2	
Horses	dig prot % *		9.0	
Rabbits	dig coef % *		9.1	
Sheep	dig prot % *		9.6	
Calcium	%		0.41	14
Phosphorus	%		0.30	20
Carotene	mg/kg		329.4	36
Vitamin A equivalent	IU/g		549.1	

Wheatgrass, bluebunch, aerial part, fresh, pre-bloom, pure stand, (2)

Ref No 2 05 386 United States

Feed Name or Analyses		Mean As Fed	Mean Dry	C.V. ± %
Dry matter	%		100.0	
Carotene	mg/kg		414.2	14
Vitamin A equivalent	IU/g		690.5	

Wheatgrass, bluebunch, aerial part, fresh, full bloom, (2)

Ref No 2 05 389 United States

Feed Name or Analyses		Mean As Fed	Mean Dry	C.V. ± %
Dry matter	%		100.0	
Ash	%		6.5	15
Crude fiber	%		32.3	9
Ether extract	%		3.2	30
N-free extract	%		48.8	
Protein (N x 6.25)	%		9.2	16
Cattle	dig prot % *		5.7	
Goats	dig prot % *		5.1	
Horses	dig prot % *		5.3	
Rabbits	dig coef % *		5.8	
Sheep	dig prot % *		5.6	
Carotene	mg/kg		131.6	33
Vitamin A equivalent	IU/g		219.4	

Feed Name or Analyses			Mean As Fed	Mean Dry	C.V. ± %

Wheatgrass, bluebunch, aerial part, fresh, full bloom, pure stand,(2)

Ref No 2 05 388 United States

Analyses	Unit		As Fed	Dry	C.V. ± %
Dry matter	%			100.0	
Carotene	mg/kg			177.9	12
Vitamin A equivalent	IU/g			296.6	

Wheatgrass, bluebunch, aerial part, fresh, milk stage, (2)

Ref No 2 05 390 United States

Analyses	Unit		As Fed	Dry	C.V. ± %
Dry matter	%			100.0	
Ash	%			7.4	25
Crude fiber	%			34.9	8
Ether extract	%			3.9	19
N-free extract	%			47.5	
Protein (N x 6.25)	%			6.3	10
Cattle	dig prot %	*		3.2	
Goats	dig prot %	*		2.4	
Horses	dig prot %	*		2.9	
Rabbits	dig coef %	*		3.5	
Sheep	dig prot %	*		2.9	
Calcium	%			0.37	19
Phosphorus	%			0.23	43

Wheatgrass, bluebunch, aerial part, fresh, mature, (2)

Ref No 2 05 392 United States

Analyses	Unit		As Fed	Dry	C.V. ± %
Dry matter	%			100.0	
Ash	%			9.2	34
Crude fiber	%			33.6	12
Ether extract	%			3.1	30
N-free extract	%			48.4	
Protein (N x 6.25)	%			5.7	35
Cattle	dig prot %	*		2.7	
Goats	dig prot %	*		1.9	
Horses	dig prot %	*		2.4	
Rabbits	dig coef %	*		3.1	
Sheep	dig prot %	*		2.3	
Calcium	%			0.32	26
Phosphorus	%			0.14	63
Carotene	mg/kg			76.9	35
Vitamin A equivalent	IU/g			128.3	

Wheatgrass, bluebunch, aerial part, fresh, mature, pure stand, (2)

Ref No 2 05 391 United States

Analyses	Unit		As Fed	Dry	C.V. ± %
Dry matter	%			100.0	
Carotene	mg/kg			86.0	15
Vitamin A equivalent	IU/g			143.3	

Wheatgrass, bluebunch, aerial part, fresh, over ripe, (2)

Ref No 2 05 393 United States

Analyses	Unit		As Fed	Dry	C.V. ± %
Dry matter	%			100.0	
Ash	%			9.9	33
Crude fiber	%			35.9	9
Ether extract	%			3.0	40
N-free extract	%			48.0	
Protein (N x 6.25)	%			3.2	33
Cattle	dig prot %	*		0.6	
Goats	dig prot %	*		-0.4	
Horses	dig prot %	*		0.2	
Rabbits	dig coef %	*		1.1	
Sheep	dig prot %	*		0.0	
Calcium	%			0.37	17
Phosphorus	%			0.05	39

Wheatgrass, bluebunch, aerial part, fresh, pure stand, (2)

Ref No 2 05 395 United States

Analyses	Unit		As Fed	Dry	C.V. ± %
Dry matter	%			100.0	
Carotene	mg/kg			311.3	42
Vitamin A equivalent	IU/g			518.9	

WHEATGRASS, BLUESTEM. Agropyron smithii

Wheatgrass, bluestem, hay, s-c, (1)

Wheatgrass hay, western

Ref No 1 05 400 United States

Analyses	Unit		As Fed	Dry	C.V. ± %
Dry matter	%		89.4	100.0	
Ash	%		7.5	8.4	
Crude fiber	%		28.8	32.2	
Sheep	dig coef %		69.	69.	
Ether extract	%		2.6	2.9	
Sheep	dig coef %		40.	40.	
N-free extract	%		42.7	47.8	
Sheep	dig coef %		63.	63.	
Protein (N x 6.25)	%		7.7	8.6	
Sheep	dig coef %		53.	53.	
Cattle	dig prot %	*	3.9	4.4	
Goats	dig prot %	*	4.1	4.6	
Horses	dig prot %	*	4.3	4.9	
Rabbits	dig coef %	*	4.8	5.3	
Sheep	dig prot %		4.1	4.6	
Fatty acids	%		2.6	3.0	
Energy	GE Mcal/kg				
Cattle	DE Mcal/kg	*	2.39	2.71	
Sheep	DE Mcal/kg	*	2.35	2.63	
Cattle	ME Mcal/kg	*	1.96	2.22	
Sheep	ME Mcal/kg	*	1.92	2.15	
Cattle	TDN %	*	54.2	61.5	
Sheep	TDN %		53.2	59.6	
Calcium	%		0.35	0.39	
Phosphorus	%		0.22	0.25	

Wheatgrass, bluestem, hay, s-c, immature, (1)

Wheatgrass hay, western immature

Ref No 1 05 396 United States

Analyses	Unit		As Fed	Dry	C.V. ± %
Dry matter	%		88.2	100.0	2
Ash	%		7.8	8.9	10
Crude fiber	%		26.4	29.9	9
Sheep	dig coef %	*	69.	69.	
Ether extract	%		2.8	3.2	16
Sheep	dig coef %	*	40.	40.	
N-free extract	%		40.6	46.0	
Sheep	dig coef %	*	63.	63.	
Protein (N x 6.25)	%		10.6	12.0	29
Sheep	dig coef %	*	53.	53.	
Cattle	dig prot %	*	6.5	7.3	
Goats	dig prot %	*	6.8	7.8	
Horses	dig prot %	*	6.8	7.7	
Rabbits	dig coef %	*	7.0	7.9	
Sheep	dig prot %		5.6	6.4	
Energy	GE Mcal/kg				
Cattle	DE Mcal/kg	*	2.52	2.85	
Sheep	DE Mcal/kg	*	2.29	2.59	
Cattle	ME Mcal/kg	*	2.06	2.34	
Sheep	ME Mcal/kg	*	1.88	2.13	
Cattle	TDN %	*	57.1	64.7	
Sheep	TDN %		51.9	58.9	
Calcium	%		0.42	0.48	
Magnesium	%		0.13	0.15	
Phosphorus	%		0.19	0.21	

Wheatgrass, bluestem, hay, s-c, full bloom, (1)

Wheatgrass hay, western, full bloom

Ref No 1 05 397 United States

Analyses	Unit		As Fed	Dry	C.V. ± %
Dry matter	%		93.4	100.0	1
Ash	%		6.6	7.1	6
Crude fiber	%		29.6	31.7	5
Sheep	dig coef %	*	69.	69.	
Ether extract	%		2.2	2.4	13
Sheep	dig coef %	*	40.	40.	
N-free extract	%		46.9	50.2	
Sheep	dig coef %	*	63.	63.	
Protein (N x 6.25)	%		8.0	8.6	8
Sheep	dig coef %	*	53.	53.	
Cattle	dig prot %	*	4.1	4.4	
Goats	dig prot %	*	4.3	4.6	
Horses	dig prot %	*	4.5	4.8	
Rabbits	dig coef %	*	5.0	5.3	
Sheep	dig prot %		4.3	4.6	
Energy	GE Mcal/kg				
Cattle	DE Mcal/kg	*	2.47	2.65	
Sheep	DE Mcal/kg	*	2.48	2.66	
Cattle	ME Mcal/kg	*	2.03	2.17	
Sheep	ME Mcal/kg	*	2.03	2.18	
Cattle	TDN %	*	56.1	60.1	
Sheep	TDN %	*	56.2	60.2	

Wheatgrass, bluestem, hay, s-c, mature, (1)

Wheatgrass hay, western, mature

Ref No 1 05 398 United States

Analyses	Unit		As Fed	Dry	C.V. ± %
Dry matter	%		90.2	100.0	2
Ash	%		9.3	10.3	13
Crude fiber	%		31.2	34.6	4
Sheep	dig coef %	*	64.	64.	
Ether extract	%		3.0	3.3	23
Sheep	dig coef %	*	46.	46.	
N-free extract	%		42.6	47.2	
Sheep	dig coef %	*	59.	59.	
Protein (N x 6.25)	%		4.2	4.6	13
Sheep	dig coef %	*	42.	42.	
Cattle	dig prot %	*	0.8	0.9	
Goats	dig prot %	*	0.8	0.9	
Horses	dig prot %	*	1.3	1.4	
Rabbits	dig coef %	*	2.0	2.2	
Sheep	dig prot %		1.7	1.9	
Lignin (Ellis)	%		7.3	8.0	
Fatty acids	%		2.8	3.1	
Energy	GE Mcal/kg				
Cattle	DE Mcal/kg	*	2.24	2.48	
Sheep	DE Mcal/kg	*	2.18	2.42	
Cattle	ME Mcal/kg	*	1.83	2.03	
Sheep	ME Mcal/kg	*	1.79	1.98	
Cattle	TDN %	*	50.8	56.3	
Sheep	TDN %		49.5	54.9	
Calcium	%		0.15	0.17	
Phosphorus	%		0.07	0.08	

Continued

Feed Name or Analyses			Mean As Fed	Mean Dry	C.V. ± %
Carotene	mg/kg		10.7	11.9	
Vitamin A equivalent	IU/g		17.9	19.8	

Wheatgrass, bluestem, hay, s-c, over ripe, (1)
Wheatgrass hay, western, over ripe
Ref No 1 05 399 United States

Feed Name or Analyses			Mean As Fed	Mean Dry	C.V. ± %
Dry matter	%		91.8	100.0	3
Ash	%		7.0	7.7	12
Crude fiber	%		30.9	33.7	4
Sheep	dig coef %		70.	70.	
Ether extract	%		2.7	2.9	5
Sheep	dig coef %		42.	42.	
N-free extract	%		45.9	50.0	
Sheep	dig coef %		68.	68.	
Protein (N x 6.25)	%		5.3	5.8	14
Sheep	dig coef %		56.	56.	
Cattle	dig prot %	*	1.8	2.0	
Goats	dig prot %	*	1.8	2.0	
Horses	dig prot %	*	2.3	2.5	
Rabbits	dig coef %	*	2.9	3.1	
Sheep	dig prot %		3.0	3.2	
Energy	GE Mcal/kg				
Cattle	DE Mcal/kg	*	2.36	2.57	
Sheep	DE Mcal/kg	*	2.57	2.80	
Cattle	ME Mcal/kg	*	1.94	2.11	
Sheep	ME Mcal/kg	*	2.11	2.30	
Cattle	TDN %	*	53.6	58.4	
Sheep	TDN %		58.3	63.5	
Calcium	%		0.26	0.28	
Iron	%		0.014	0.015	
Manganese	mg/kg		8.7	9.5	
Phosphorus	%		0.07	0.08	

Wheatgrass, bluestem, aerial part, fresh, (2)
Wheatgrass forage, western
Ref No 2 05 410 United States

Feed Name or Analyses			Mean As Fed	Mean Dry	C.V. ± %
Dry matter	%		41.3	100.0	21
Crude fiber	%				
Sheep	dig coef %	*	68.	68.	
Ether extract	%				
Sheep	dig coef %	*	53.	53.	
N-free extract	%				
Sheep	dig coef %	*	75.	75.	
Protein (N x 6.25)	%				
Sheep	dig coef %	*	60.	60.	
Cobalt	mg/kg		0.035	0.084	61
Sodium	%		0.21	0.51	
Sulphur	%		0.07	0.16	
Carotene	mg/kg		52.5	127.2	
Vitamin A equivalent	IU/g		87.6	212.1	

Wheatgrass, bluestem, aerial part, fresh, immature, (2)
Wheatgrass forage, western, immature
Ref No 2 05 401 United States

Feed Name or Analyses			Mean As Fed	Mean Dry	C.V. ± %
Dry matter	%		40.3	100.0	
Ash	%		4.8	12.0	
Crude fiber	%		10.5	26.0	
Sheep	dig coef %	*	64.	64.	
Ether extract	%		1.6	4.0	
Sheep	dig coef %	*	66.	66.	
N-free extract	%		16.4	40.9	
Sheep	dig coef %	*	74.	74.	
Protein (N x 6.25)	%		6.9	17.1	
Sheep	dig coef %	*	74.	74.	
Cattle	dig prot %	*	5.0	12.5	
Goats	dig prot %	*	5.1	12.6	
Horses	dig prot %	*	4.9	12.1	
Rabbits	dig coef %	*	4.8	11.9	
Sheep	dig prot %		5.1	12.7	
Energy	GE Mcal/kg				
Cattle	DE Mcal/kg	*	1.15	2.85	
Sheep	DE Mcal/kg	*	1.16	2.89	
Cattle	ME Mcal/kg	*	0.94	2.34	
Sheep	ME Mcal/kg	*	0.95	2.37	
Cattle	TDN %	*	26.0	64.7	
Sheep	TDN %		26.4	65.5	
Calcium	%		0.14	0.34	
Phosphorus	%		0.07	0.17	
Carotene	mg/kg		74.7	185.6	31
Vitamin A equivalent	IU/g		124.5	309.4	

(1) dry forages and roughages (2) pasture, range plants, and forages fed green (3) silages (4) energy feeds (5) protein supplements (6) minerals (7) vitamins (8) additives

Wheatgrass, bluestem, aerial part, fresh, early bloom, (2)
Wheatgrass forage, western, early bloom
Ref No 2 05 402 United States

Feed Name or Analyses			Mean As Fed	Mean Dry	C.V. ± %
Dry matter	%		34.8	100.0	
Crude fiber	%				
Sheep	dig coef %	*	64.	64.	
Ether extract	%				
Sheep	dig coef %	*	66.	66.	
N-free extract	%				
Sheep	dig coef %	*	74.	74.	
Protein (N x 6.25)	%				
Sheep	dig coef %	*	74.	74.	
Calcium	%		0.20	0.58	
Phosphorus	%		0.12	0.34	
Carotene	mg/kg		51.6	148.1	29
Vitamin A equivalent	IU/g		85.9	247.0	

Wheatgrass, bluestem, aerial part, fresh, mid-bloom, (2)
Wheatgrass forage, western, mid-bloom
Ref No 2 05 403 United States

Feed Name or Analyses			Mean As Fed	Mean Dry	C.V. ± %
Dry matter	%		29.9	100.0	
Ash	%		2.2	7.2	
Crude fiber	%		10.2	34.0	4
Sheep	dig coef %	*	68.	68.	
Ether extract	%		0.7	2.5	
Sheep	dig coef %	*	53.	53.	
N-free extract	%		14.2	47.6	
Sheep	dig coef %	*	75.	75.	
Protein (N x 6.25)	%		2.6	8.7	22
Sheep	dig coef %	*	60.	60.	
Cattle	dig prot %	*	1.6	5.3	
Goats	dig prot %	*	1.4	4.7	
Horses	dig prot %	*	1.5	4.9	
Rabbits	dig coef %	*	1.6	5.4	
Sheep	dig prot %		1.6	5.2	
Energy	GE Mcal/kg				
Cattle	DE Mcal/kg	*	0.87	2.92	
Sheep	DE Mcal/kg	*	0.88	2.96	
Cattle	ME Mcal/kg	*	0.72	2.39	
Sheep	ME Mcal/kg	*	0.72	2.42	
Cattle	TDN %	*	19.8	66.2	
Sheep	TDN %	*	20.0	67.0	

Wheatgrass, bluestem, aerial part, fresh, full bloom, (2)
Wheatgrass forage, western full bloom
Ref No 2 05 404 United States

Feed Name or Analyses			Mean As Fed	Mean Dry	C.V. ± %
Dry matter	%		38.9	100.0	9
Crude fiber	%				
Sheep	dig coef %	*	68.	68.	
Ether extract	%				
Sheep	dig coef %	*	53.	53.	
N-free extract	%				
Sheep	dig coef %	*	75.	75.	
Protein (N x 6.25)	%				
Sheep	dig coef %	*	60.	60.	
Calcium	%		0.18	0.45	
Magnesium	%		0.06	0.16	
Phosphorus	%		0.11	0.29	
Potassium	%		1.03	2.64	
Carotene	mg/kg		45.5	117.1	21
Vitamin A equivalent	IU/g		75.9	195.1	

Wheatgrass, bluestem, aerial part, fresh, milk stage, (2)
Wheatgrass forage, western, milk stage
Ref No 2 05 405 United States

Feed Name or Analyses			Mean As Fed	Mean Dry	C.V. ± %
Dry matter	%		48.7	100.0	
Carotene	mg/kg		55.3	113.5	31
Vitamin A equivalent	IU/g		92.2	189.3	

Wheatgrass, bluestem, aerial part, fresh, dough stage, (2)
Wheatgrass forage, western, dough stage
Ref No 2 05 406 United States

Feed Name or Analyses			Mean As Fed	Mean Dry	C.V. ± %
Dry matter	%			100.0	
Carotene	mg/kg			88.0	28
Vitamin A equivalent	IU/g			146.6	

Wheatgrass, bluestem, aerial part, fresh, mature, (2)
Wheatgrass forage, western, mature
Ref No 2 05 407 United States

Feed Name or Analyses			Mean As Fed	Mean Dry	C.V. ± %
Dry matter	%		62.6	100.0	
Ash	%		7.0	11.3	10
Crude fiber	%		20.9	33.3	12
Sheep	dig coef %	*	68.	68.	
Ether extract	%		2.1	3.4	29
Sheep	dig coef %	*	53.	53.	
N-free extract	%		29.3	46.8	
Sheep	dig coef %	*	75.	75.	
Protein (N x 6.25)	%		3.3	5.3	21
Sheep	dig coef %	*	60.	60.	
Cattle	dig prot %	*	1.5	2.4	
Goats	dig prot %	*	0.9	1.5	
Horses	dig prot %	*	1.2	2.0	
Rabbits	dig coef %	*	1.7	2.7	
Sheep	dig prot %		2.0	3.2	
Energy	GE Mcal/kg				
Cattle	DE Mcal/kg	*	1.51	2.42	
Sheep	DE Mcal/kg	*	1.79	2.86	
Cattle	ME Mcal/kg	*	1.24	1.98	
Sheep	ME Mcal/kg	*	1.47	2.35	

Feed Name or Analyses			Mean As Fed	Mean Dry	C.V. ± %
Cattle	TDN %	*	34.3	54.8	
Sheep	TDN %		40.6	64.9	
Calcium	%		0.21	0.34	18
Phosphorus	%		0.11	0.18	43
Carotene	mg/kg		38.9	62.2	27
Vitamin A equivalent	IU/g		64.9	103.6	

Wheatgrass, bluestem, aerial part, fresh, over ripe, (2)

Wheatgrass forage, western, over ripe

Ref No 2 05 408 United States

Feed Name or Analyses			Mean As Fed	Mean Dry	C.V. ± %
Dry matter	%		55.1	100.0	
Calcium	%		0.12	0.21	
Copper	mg/kg		3.2	5.7	
Magnesium	%		0.03	0.06	
Manganese	mg/kg		19.8	35.9	
Phosphorus	%		0.09	0.16	
Carotene	mg/kg		0.1	0.2	
Vitamin A equivalent	IU/g		0.2	0.4	

Wheatgrass, bluestem, aerial part, fresh, cut 1, (2)

Wheatgrass forage, western, cut 1

Ref No 2 05 409 United States

Feed Name or Analyses			Mean As Fed	Mean Dry	C.V. ± %
Dry matter	%			100.0	
Calcium	%			0.35	
Phosphorus	%			0.19	

Wheatgrass, bluestem, aerial part wo lower stems, fresh, early leaf, (2)

Ref No 2 08 698 United States

Feed Name or Analyses			Mean As Fed	Mean Dry	C.V. ± %
Dry matter	%			100.0	
Organic matter	%			89.3	2
Ash	%			10.7	17
Crude fiber	%			29.7	5
Ether extract	%			4.4	22
N-free extract	%			44.6	4
Protein (N x 6.25)	%			10.8	16
Cattle	dig prot %	*		7.0	
Goats	dig prot %	*		6.6	
Horses	dig prot %	*		6.7	
Rabbits	dig coef %	*		7.0	
Sheep	dig prot %	*		7.0	
Energy	GE Mcal/kg				
Cattle	DE Mcal/kg	*		2.63	
Sheep	DE Mcal/kg	*		2.84	
Cattle	ME Mcal/kg	*		2.16	
Sheep	ME Mcal/kg	*		2.33	
Cattle	TDN %	*		59.7	
Sheep	TDN %	*		64.4	
Calcium	%			0.35	30
Phosphorus	%			0.15	18
Carotene	mg/kg			71.5	
Vitamin A equivalent	IU/g			119.3	

Wheatgrass, bluestem, aerial part wo lower stems, fresh, mid-bloom, (2)

Ref No 2 08 699 United States

Feed Name or Analyses			Mean As Fed	Mean Dry	C.V. ± %
Dry matter	%			100.0	
Organic matter	%			88.7	4
Ash	%			11.3	32
Crude fiber	%			31.2	8
Ether extract	%			3.9	20
N-free extract	%			45.5	7
Protein (N x 6.25)	%			8.1	21
Cattle	dig prot %	*		4.8	
Goats	dig prot %	*		4.1	
Horses	dig prot %	*		4.4	
Rabbits	dig coef %	*		4.9	
Sheep	dig prot %	*		4.6	
Energy	GE Mcal/kg				
Cattle	DE Mcal/kg	*		2.70	
Sheep	DE Mcal/kg	*		2.90	
Cattle	ME Mcal/kg	*		2.21	
Sheep	ME Mcal/kg	*		2.38	
Cattle	TDN %	*		61.3	
Sheep	TDN %	*		65.7	
Calcium	%			0.37	28
Phosphorus	%			0.14	19

Wheatgrass, bluestem, aerial part wo lower stems, fresh, mature, (2)

Ref No 2 08 700 United States

Feed Name or Analyses			Mean As Fed	Mean Dry	C.V. ± %
Dry matter	%			100.0	
Organic matter	%			88.9	
Ash	%			11.1	
Crude fiber	%			29.4	
Ether extract	%			4.6	
N-free extract	%			46.8	
Protein (N x 6.25)	%			8.0	
Cattle	dig prot %	*		4.7	
Goats	dig prot %	*		4.0	
Horses	dig prot %	*		4.3	
Rabbits	dig coef %	*		4.8	
Sheep	dig prot %	*		4.4	
Calcium	%			0.52	
Phosphorus	%			0.12	

Wheatgrass, bluestem, aerial part wo lower stems, fresh, dormant, (2)

Ref No 2 08 697 United States

Feed Name or Analyses			Mean As Fed	Mean Dry	C.V. ± %
Dry matter	%			100.0	
Organic matter	%			86.3	3
Ash	%			13.7	21
Crude fiber	%			31.6	5
Ether extract	%			3.3	35
N-free extract	%			47.4	5
Protein (N x 6.25)	%			3.9	21
Cattle	dig prot %	*		1.2	
Goats	dig prot %	*		0.2	
Horses	dig prot %	*		0.9	
Rabbits	dig coef %	*		1.7	
Sheep	dig prot %	*		0.7	
Calcium	%			0.42	58
Phosphorus	%			0.07	25
Carotene	mg/kg			8.8	98
Vitamin A equivalent	IU/g			14.6	

WHEATGRASS, CRESTED. Agropyron cristatum

Wheatgrass, crested, hay, fan air dried, immature, (1)

Ref No 1 08 758 United States

Feed Name or Analyses			Mean As Fed	Mean Dry	C.V. ± %
Dry matter	%			100.0	
Lignin (Ellis)	%			6.6	

Wheatgrass, crested, hay, fan air dried, pre-bloom, (1)

Ref No 1 08 759 United States

Feed Name or Analyses			Mean As Fed	Mean Dry	C.V. ± %
Dry matter	%			100.0	
Lignin (Ellis)	%			4.9	

Wheatgrass, crested, hay, fan air dried, early bloom, (1)

Ref No 1 08 756 United States

Feed Name or Analyses			Mean As Fed	Mean Dry	C.V. ± %
Dry matter	%			100.0	
Lignin (Ellis)	%			7.5	

Wheatgrass, crested, hay, fan air dried, mid-bloom, (1)

Ref No 1 08 755 United States

Feed Name or Analyses			Mean As Fed	Mean Dry	C.V. ± %
Dry matter	%			100.0	
Lignin (Ellis)	%			8.1	

Wheatgrass, crested, hay, s-c, (1)

Ref No 1 05 418 United States

Feed Name or Analyses			Mean As Fed	Mean Dry	C.V. ± %
Dry matter	%		92.3	100.0	2
Ash	%		6.3	6.8	
Crude fiber	%		32.9	35.7	
Sheep	dig coef %	*	55.	55.	
Ether extract	%		2.3	2.4	
Sheep	dig coef %	*	28.	28.	
N-free extract	%		43.5	47.2	
Sheep	dig coef %	*	53.	53.	
Protein (N x 6.25)	%		7.3	7.9	
Sheep	dig coef %	*	41.	41.	
Cattle	dig prot %	*	3.5	3.8	
Goats	dig prot %	*	3.6	3.9	
Horses	dig prot %	*	3.9	4.2	
Rabbits	dig coef %	*	4.4	4.8	
Sheep	dig prot %		3.0	3.2	
Lignin (Ellis)	%		10.5	11.4	
Fatty acids	%		2.1	2.2	
Energy	GE Mcal/kg				
Cattle	DE Mcal/kg	*	2.17	2.35	
Sheep	DE Mcal/kg	*	2.00	2.17	
Cattle	ME Mcal/kg	*	1.78	1.93	
Sheep	ME Mcal/kg	*	1.64	1.78	
Cattle	NE_m Mcal/kg	*	1.14	1.24	
Cattle	NE_{gain} Mcal/kg	*	0.54	0.59	
Cattle	TDN %	*	49.2	53.3	
Sheep	TDN %		45.4	49.2	
Calcium	%		0.24	0.26	24
Cobalt	mg/kg		0.220	0.238	
Phosphorus	%		0.14	0.15	55
Carotene	mg/kg		20.6	22.3	
Vitamin A equivalent	IU/g		34.3	37.2	

Wheatgrass, crested, hay, s-c, immature, (1)

Ref No 1 05 411 United States

Feed Name or Analyses			Mean As Fed	Mean Dry	C.V. ± %
Dry matter	%		93.7	100.0	1
Ash	%		9.0	9.6	6
Crude fiber	%		19.6	20.9	5
Sheep	dig coef %		79.	79.	

Continued

Feed Name or Analyses				Mean As Fed	Mean Dry	C.V. ± %
Ether extract		%		2.5	2.7	5
Sheep	dig coef	%		62.	62.	
N-free extract		%		44.6	47.6	
Sheep	dig coef	%		82.	82.	
Protein (N x 6.25)		%		18.0	19.2	4
Sheep	dig coef	%		85.	85.	
Cattle	dig prot	%	*	12.7	13.5	
Goats	dig prot	%	*	13.5	14.4	
Horses	dig prot	%	*	12.9	13.8	
Rabbits	dig coef	%	*	12.6	13.5	
Sheep	dig prot	%		15.3	16.3	
Lignin (Ellis)		%		5.6	5.9	
Sheep	dig coef	%		84.	84.	
Energy	GE Mcal/kg					
Cattle	DE Mcal/kg		*	2.75	2.94	
Sheep	DE Mcal/kg		*	3.12	3.33	
Cattle	ME Mcal/kg		*	2.25	2.41	
Sheep	ME Mcal/kg		*	2.56	2.73	
Cattle	TDN	%	*	62.4	66.6	
Sheep	TDN	%		70.8	75.6	
Calcium		%		0.40	0.42	8
Phosphorus		%		0.25	0.26	13
Carotene		mg/kg		213.1	227.6	
Vitamin A equivalent		IU/g		355.3	379.3	

Wheatgrass, crested, hay, s-c, early bloom, (1)

Ref No 1 05 412 United States

Feed Name or Analyses				Mean As Fed	Mean Dry	C.V. ± %
Dry matter		%		90.9	100.0	1
Ash		%		7.5	8.2	
Crude fiber		%		33.9	37.3	
Cattle	dig coef	%		68.	68.	
Sheep	dig coef	%		64.	64.	
Ether extract		%		2.0	2.2	
Cattle	dig coef	%		25.	25.	
Sheep	dig coef	%		43.	43.	
N-free extract		%		33.5	36.8	
Sheep	dig coef	%		54.	54.	
Protein (N x 6.25)		%		14.1	15.5	
Cattle	dig coef	%		63.	63.	
Sheep	dig coef	%		71.	71.	
Cattle	dig prot	%		8.8	9.7	
Goats	dig prot	%	*	10.0	11.0	
Horses	dig prot	%	*	9.7	10.7	
Rabbits	dig coef	%	*	9.7	10.6	
Sheep	dig prot	%		10.0	11.0	
Cellulose (Matrone)		%		31.6	34.8	
Lignin (Ellis)		%		13.6	15.0	
Energy	GE Mcal/kg			4.02	4.42	
Cattle	GE dig coef	%		64.	64.	
Cattle	DE Mcal/kg			2.55	2.81	
Sheep	DE Mcal/kg		*	2.28	2.51	
Cattle	ME Mcal/kg		*	2.09	2.30	
Sheep	ME Mcal/kg		*	1.87	2.06	
Cattle	TDN	%	*	50.2	55.2	
Sheep	TDN	%		51.7	56.9	
Calcium		%		0.29	0.32	14
Phosphorus		%		0.25	0.27	12

Ref No 1 05 412 Canada

Feed Name or Analyses				Mean As Fed	Mean Dry	C.V. ± %
Dry matter		%		95.0	100.0	
Ash		%		6.7	7.1	

(1) dry forages and roughages (2) pasture, range plants, and forages fed green (3) silages (4) energy feeds (5) protein supplements (6) minerals (7) vitamins (8) additives

Feed Name or Analyses				Mean As Fed	Mean Dry	C.V. ± %
Crude fiber		%		30.0	31.6	
Cattle	dig coef	%		68.	68.	
Sheep	dig coef	%		64.	64.	
Ether extract		%		1.0	1.1	
Cattle	dig coef	%		25.	25.	
Sheep	dig coef	%		43.	43.	
N-free extract		%		47.4	49.9	
Sheep	dig coef	%		54.	54.	
Protein (N x 6.25)		%		9.9	10.4	
Cattle	dig coef	%		63.	63.	
Sheep	dig coef	%		71.	71.	
Cattle	dig prot	%		6.2	6.5	
Goats	dig prot	%	*	6.0	6.3	
Horses	dig prot	%	*	6.0	6.4	
Rabbits	dig coef	%	*	6.4	6.7	
Sheep	dig prot	%		7.0	7.4	
Cellulose (Matrone)		%		32.3	34.0	
Cattle	dig coef	%		70.	70.	
Energy	GE Mcal/kg			4.20	4.42	
Cattle	GE dig coef	%		64.	64.	
Cattle	DE Mcal/kg			2.67	2.81	
Sheep	DE Mcal/kg		*	2.33	2.45	
Cattle	ME Mcal/kg		*	2.19	2.30	
Sheep	ME Mcal/kg		*	1.91	2.01	
Cattle	TDN	%	*	55.7	58.6	
Sheep	TDN	%		52.8	55.5	

Wheatgrass, crested, hay, s-c, mid-bloom, (1)

Ref No 1 05 413 United States

Feed Name or Analyses				Mean As Fed	Mean Dry	C.V. ± %
Dry matter		%		96.4	100.0	
Ash		%		6.3	6.5	
Crude fiber		%		35.2	36.5	
Sheep	dig coef	%	*	51.	51.	
Ether extract		%		2.3	2.4	
Sheep	dig coef	%	*	20.	20.	
N-free extract		%		43.3	44.9	
Sheep	dig coef	%	*	53.	53.	
Protein (N x 6.25)		%		9.4	9.7	
Sheep	dig coef	%	*	36.	36.	
Cattle	dig prot	%	*	5.1	5.3	
Goats	dig prot	%	*	5.4	5.6	
Horses	dig prot	%	*	5.6	5.8	
Rabbits	dig coef	%	*	5.9	6.2	
Sheep	dig prot	%		3.4	3.5	
Energy	GE Mcal/kg					
Cattle	DE Mcal/kg		*	2.18	2.26	
Sheep	DE Mcal/kg		*	2.00	2.07	
Cattle	ME Mcal/kg		*	1.79	1.85	
Sheep	ME Mcal/kg		*	1.64	1.70	
Cattle	TDN	%	*	49.4	51.2	
Sheep	TDN	%	*	45.3	47.0	

Wheatgrass, crested, hay, s-c, full bloom, (1)

Ref No 1 05 414 United States

Feed Name or Analyses				Mean As Fed	Mean Dry	C.V. ± %
Dry matter		%		94.9	100.0	1
Ash		%		6.1	6.4	14
Crude fiber		%		31.3	33.0	19
Sheep	dig coef	%	*	51.	51.	
Ether extract		%		1.9	2.0	17
Sheep	dig coef	%	*	20.	20.	
N-free extract		%		47.4	49.9	
Sheep	dig coef	%	*	53.	53.	
Protein (N x 6.25)		%		8.3	8.7	13
Sheep	dig coef	%	*	36.	36.	
Cattle	dig prot	%	*	4.2	4.5	
Goats	dig prot	%	*	4.4	4.7	

Feed Name or Analyses				Mean As Fed	Mean Dry	C.V. ± %
Horses	dig prot	%	*	4.7	4.9	
Rabbits	dig coef	%	*	5.1	5.4	
Sheep	dig prot	%		3.0	3.1	
Energy	GE Mcal/kg			4.24	4.47	
Sheep	GE dig coef	%		55.	55.	
Cattle	DE Mcal/kg		*	2.38	2.51	
Sheep	DE Mcal/kg			2.33	2.46	
Cattle	ME Mcal/kg		*	1.95	2.05	
Sheep	ME Mcal/kg		*	1.91	2.02	
Cattle	TDN	%	*	53.9	56.8	
Sheep	TDN	%		44.9	47.3	
Calcium		%		0.25	0.26	11
Phosphorus		%		0.15	0.16	15

Ref No 1 05 414 Canada

Feed Name or Analyses				Mean As Fed	Mean Dry	C.V. ± %
Dry matter		%			100.0	
Sheep	dig coef	%			61.	
Organic matter		%			93.0	
Crude fiber		%			33.6	
Sheep	dig coef	%			55.	
Ether extract		%			1.5	
Sheep	dig coef	%			46.	
N-free extract		%			51.2	
Sheep	dig coef	%			70.	
Protein (N x 6.25)		%			6.5	
Sheep	dig coef	%			50.	
Cattle	dig prot	%	*		2.6	
Goats	dig prot	%	*		2.6	
Horses	dig prot	%	*		3.0	
Rabbits	dig coef	%	*		3.7	
Sheep	dig prot	%			3.3	
Lignin (Ellis)		%			8.7	
Sheep	dig coef	%			0.	
Energy	GE Mcal/kg				4.47	
Sheep	GE dig coef	%			55.	
Cattle	DE Mcal/kg		*		2.50	
Sheep	DE Mcal/kg				2.46	
Cattle	ME Mcal/kg		*		2.05	
Sheep	ME Mcal/kg		*		2.02	
Cattle	TDN	%	*		56.6	
Sheep	TDN	%			59.1	

Wheatgrass, crested, hay, s-c, milk stage, (1)

Ref No 1 05 415 United States

Feed Name or Analyses				Mean As Fed	Mean Dry	C.V. ± %
Dry matter		%		92.5	100.0	4
Ash		%		5.4	5.8	8
Crude fiber		%		31.8	34.4	7
Ether extract		%		1.9	2.0	23
N-free extract		%		46.5	50.3	
Protein (N x 6.25)		%		6.9	7.5	3
Cattle	dig prot	%	*	3.2	3.4	
Goats	dig prot	%	*	3.3	3.6	
Horses	dig prot	%	*	3.6	3.9	
Rabbits	dig coef	%	*	4.1	4.5	
Sheep	dig prot	%	*	3.1	3.3	
Cellulose (Matrone)		%		36.2	39.1	
Lignin (Ellis)		%		15.6	16.9	
Energy	GE Mcal/kg					
Cattle	DE Mcal/kg		*	2.21	2.39	
Sheep	DE Mcal/kg		*	2.25	2.43	
Cattle	ME Mcal/kg		*	1.81	1.96	
Sheep	ME Mcal/kg		*	1.84	1.99	
Cattle	TDN	%	*	50.1	54.2	
Sheep	TDN	%	*	51.0	55.1	
Calcium		%		0.24	0.26	11
Phosphorus		%		0.10	0.11	26

Feed Name or Analyses			Mean As Fed	Mean Dry	C.V. ± %

Wheatgrass, crested, hay, s-c, mature, (1)

Ref No 1 05 416 United States

Feed Name or Analyses			As Fed	Dry	C.V. ± %
Dry matter	%		94.4	100.0	3
Ash	%		5.5	5.8	4
Crude fiber	%		38.5	40.8	4
Sheep	dig coef %	*	51.	51.	
Ether extract	%		1.1	1.2	21
Sheep	dig coef %	*	20.	20.	
N-free extract	%		45.6	48.3	
Sheep	dig coef %	*	53.	53.	
Protein (N x 6.25)	%		3.7	3.9	19
Sheep	dig coef %	*	36.	36.	
Cattle	dig prot %	*	0.3	0.3	
Goats	dig prot %	*	0.2	0.2	
Horses	dig prot %	*	0.8	0.8	
Rabbits	dig coef %	*	1.6	1.7	
Sheep	dig prot %		1.3	1.4	
Energy	GE Mcal/kg	*	4.27	4.52	
Sheep	GE dig coef %	*	57.	57.	
Cattle	DE Mcal/kg	*	1.82	1.92	
Sheep	DE Mcal/kg	*	2.43	2.58	
Cattle	ME Mcal/kg	*	1.49	1.58	
Sheep	ME Mcal/kg	*	2.00	2.11	
Cattle	TDN %	*	41.2	43.6	
Sheep	TDN %	*	45.6	48.4	

Ref No 1 05 416 Canada

Feed Name or Analyses			As Fed	Dry	C.V. ± %
Dry matter	%		93.8	100.0	2
Sheep	dig coef %		58.	58.	
Ash	%		7.1	7.5	
Crude fiber	%		35.9	38.3	12
Protein (N x 6.25)	%		7.1	7.6	50
Sheep	dig coef %		62.	62.	
Cattle	dig prot %	*	3.3	3.5	
Goats	dig prot %	*	3.4	3.6	
Horses	dig prot %	*	3.7	4.0	
Rabbits	dig coef %	*	4.2	4.5	
Sheep	dig prot %		4.4	4.7	
Cellulose (Matrone)	%		29.5	31.5	
Sheep	dig coef %		58.	58.	
Energy	GE Mcal/kg		4.24	4.52	
Sheep	GE dig coef %		57.	57.	
Sheep	DE Mcal/kg		2.42	2.58	
Sheep	ME Mcal/kg	*	1.98	2.11	
Sheep	TDN %		45.5	48.5	
Calcium	%		0.19	0.20	
Phosphorus	%		0.11	0.12	
Carotene	mg/kg		28.9	30.8	
Vitamin A equivalent	IU/g		48.1	51.3	

Wheatgrass, crested, hay, s-c, over ripe, (1)

Ref No 1 05 417 United States

Feed Name or Analyses			As Fed	Dry	C.V. ± %
Dry matter	%		93.5	100.0	2
Calcium	%		0.22	0.23	11
Phosphorus	%		0.04	0.04	49

Wheatgrass, crested, aerial part, fresh, (2)

Ref No 2 05 429 United States

Feed Name or Analyses			As Fed	Dry	C.V. ± %
Dry matter	%		46.5	100.0	33
Carotene	mg/kg		116.3	250.0	
Choline	mg/kg		1005.	2161.	
Riboflavin	mg/kg		4.0	8.6	23
Thiamine	mg/kg		1.5	3.3	21
Vitamin A equivalent	IU/g		193.8	416.8	

Wheatgrass, crested, aerial part, fresh, immature, (2)

Ref No 2 05 420 United States

Feed Name or Analyses			As Fed	Dry	C.V. ± %
Dry matter	%		28.0	100.0	25
Ash	%		3.0	10.8	
Crude fiber	%		5.8	20.7	
Ether extract	%		1.0	3.6	
N-free extract	%		10.8	38.6	
Protein (N x 6.25)	%		7.3	26.3	
Cattle	dig prot %	*	5.7	20.2	
Goats	dig prot %	*	5.9	21.1	
Horses	dig prot %	*	5.5	19.9	
Rabbits	dig coef %	*	5.3	19.0	
Sheep	dig prot %	*	6.0	21.5	
Energy	GE Mcal/kg				
Cattle	DE Mcal/kg	*	0.94	3.37	
Sheep	DE Mcal/kg	*	0.90	3.20	
Cattle	ME Mcal/kg	*	0.77	2.76	
Sheep	ME Mcal/kg	*	0.73	2.62	
Cattle	NE_m Mcal/kg	*	0.46	1.65	
Cattle	NE_{gain} Mcal/kg	*	0.30	1.08	
Cattle	TDN %	*	21.4	76.4	
Sheep	TDN %	*	20.3	72.5	
Calcium	%		0.13	0.48	
Phosphorus	%		0.09	0.32	
Carotene	mg/kg		121.2	433.6	31
Vitamin A equivalent	IU/g		202.0	722.9	

Wheatgrass, crested, aerial part, fresh, immature, pure stand, (2)

Ref No 2 05 419 United States

Feed Name or Analyses			As Fed	Dry	C.V. ± %
Dry matter	%			100.0	
Carotene	mg/kg			468.3	13
Vitamin A equivalent	IU/g			780.6	

Wheatgrass, crested, aerial part, fresh, pre-bloom, (2)

Ref No 2 08 581 United States

Feed Name or Analyses			As Fed	Dry	C.V. ± %
Dry matter	%		34.3	100.0	
Ash	%		2.7	7.9	
Crude fiber	%		8.9	25.9	
Ether extract	%		0.9	2.6	
N-free extract	%		17.8	51.9	
Protein (N x 6.25)	%		4.0	11.7	
Cattle	dig prot %	*	2.7	7.8	
Goats	dig prot %	*	2.6	7.4	
Horses	dig prot %	*	2.5	7.4	
Rabbits	dig coef %	*	2.6	7.7	
Sheep	dig prot %	*	2.7	7.9	
Energy	GE Mcal/kg				
Cattle	DE Mcal/kg	*	1.01	2.94	
Sheep	DE Mcal/kg	*	1.01	2.94	
Cattle	ME Mcal/kg	*	0.83	2.41	
Sheep	ME Mcal/kg	*	0.83	2.41	
Cattle	TDN %	*	22.9	66.7	
Sheep	TDN %	*	22.9	66.8	
Calcium	%		0.14	0.41	
Phosphorus	%		0.07	0.20	

Wheatgrass, crested, aerial part, fresh, pre-bloom, pure stand, (2)

Ref No 2 05 421 United States

Feed Name or Analyses			As Fed	Dry	C.V. ± %
Dry matter	%			100.0	
Carotene	mg/kg			360.7	8
Vitamin A equivalent	IU/g			601.2	

Wheatgrass, crested, aerial part, fresh, early bloom, (2)

Ref No 2 05 422 United States

Feed Name or Analyses			As Fed	Dry	C.V. ± %
Dry matter	%		36.9	100.0	
Ash	%		2.7	7.3	
Crude fiber	%		12.2	33.1	
Ether extract	%		0.6	1.6	
N-free extract	%		18.7	50.7	
Protein (N x 6.25)	%		2.7	7.3	
Cattle	dig prot %	*	1.5	4.1	
Goats	dig prot %	*	1.2	3.4	
Horses	dig prot %	*	1.4	3.7	
Rabbits	dig coef %	*	1.6	4.3	
Sheep	dig prot %	*	1.4	3.8	
Energy	GE Mcal/kg				
Cattle	DE Mcal/kg	*	0.92	2.49	
Sheep	DE Mcal/kg	*	0.96	2.61	
Cattle	ME Mcal/kg	*	0.75	2.04	
Sheep	ME Mcal/kg	*	0.79	2.14	
Cattle	NE_m Mcal/kg	*	0.51	1.38	
Cattle	NE_{gain} Mcal/kg	*	0.29	0.79	
Cattle	TDN %	*	20.8	56.4	
Sheep	TDN %	*	21.8	59.1	

Wheatgrass, crested, aerial part, fresh, full bloom, (2)

Ref No 2 05 424 United States

Feed Name or Analyses			As Fed	Dry	C.V. ± %
Dry matter	%			100.0	
Energy	GE Mcal/kg				
Cattle	NE_m Mcal/kg	*		1.33	
Cattle	NE_{gain} Mcal/kg	*		0.72	
Carotene	mg/kg			153.4	7
Vitamin A equivalent	IU/g			255.8	

Wheatgrass, crested, aerial part, fresh, full bloom, pure stand, (2)

Ref No 2 05 423 United States

Feed Name or Analyses			As Fed	Dry	C.V. ± %
Dry matter	%			100.0	
Carotene	mg/kg			154.8	7
Vitamin A equivalent	IU/g			258.0	

Wheatgrass, crested, aerial part, fresh, milk stage, (2)

Ref No 2 05 425 United States

Feed Name or Analyses			As Fed	Dry	C.V. ± %
Dry matter	%			100.0	
Ash	%			16.9	
Crude fiber	%			30.9	
Ether extract	%			3.0	
N-free extract	%			41.2	
Protein (N x 6.25)	%			8.0	
Cattle	dig prot %	*		4.7	
Goats	dig prot %	*		4.0	
Horses	dig prot %	*		4.3	
Rabbits	dig coef %	*		4.8	
Sheep	dig prot %	*		4.4	

Continued

Feed Name or Analyses		Mean As Fed	Mean Dry	C.V. ± %
Energy	GE Mcal/kg			
Cattle	DE Mcal/kg *		2.63	
Sheep	DE Mcal/kg *		2.85	
Cattle	ME Mcal/kg *		2.16	
Sheep	ME Mcal/kg *		2.34	
Cattle	TDN % *		59.6	
Sheep	TDN % *		64.6	
Calcium	%		0.31	
Phosphorus	%		0.25	
Carotene	mg/kg		85.5	36
Vitamin A equivalent	IU/g		142.6	

Wheatgrass, crested, aerial part, fresh, mature, (2)
Ref No 2 05 427 United States

Feed Name or Analyses		Mean As Fed	Mean Dry	C.V. ± %
Dry matter	%	60.0	100.0	11
Ash	%	3.5	5.8	
Crude fiber	%	23.2	38.7	
Ether extract	%	1.2	2.0	
N-free extract	%	28.8	48.0	
Protein (N x 6.25)	%	3.3	5.5	
Cattle	dig prot % *	1.5	2.6	
Goats	dig prot % *	1.0	1.7	
Horses	dig prot % *	1.3	2.2	
Rabbits	dig coef % *	1.7	2.9	
Sheep	dig prot % *	1.3	2.1	
Energy	GE Mcal/kg			
Cattle	DE Mcal/kg *	1.43	2.39	
Sheep	DE Mcal/kg *	1.42	2.36	
Cattle	ME Mcal/kg *	1.18	1.96	
Sheep	ME Mcal/kg *	1.16	1.94	
Cattle	NE_m Mcal/kg *	0.82	1.36	
Cattle	NE_{gain} Mcal/kg *	0.46	0.76	
Cattle	TDN % *	32.5	54.1	
Sheep	TDN % *	32.2	53.6	
Calcium	%	0.16	0.27	27
Phosphorus	%	0.09	0.15	34
Carotene	mg/kg	45.2	75.4	23
Vitamin A equivalent	IU/g	75.4	125.7	

Wheatgrass, crested, aerial part, fresh, mature, pure stand, (2)
Ref No 2 05 426 United States

Feed Name or Analyses		Mean As Fed	Mean Dry	C.V. ± %
Dry matter	%		100.0	
Carotene	mg/kg		76.7	15
Vitamin A equivalent	IU/g		127.9	

Wheatgrass, crested, aerial part, fresh, over ripe, (2)
Ref No 2 05 428 United States

Feed Name or Analyses		Mean As Fed	Mean Dry	C.V. ± %
Dry matter	%	80.0	100.0	5
Energy	GE Mcal/kg			
Cattle	NE_m Mcal/kg *	1.19	1.49	
Cattle	NE_{gain} Mcal/kg *	0.74	0.92	
Calcium	%	0.22	0.27	
Cobalt	mg/kg	0.199	0.249	
Copper	mg/kg	6.7	8.4	
Manganese	mg/kg	42.3	52.9	
Phosphorus	%	0.06	0.07	
Carotene	mg/kg	0.2	0.2	
Vitamin A equivalent	IU/g	0.3	0.4	

(1) dry forages and roughages (2) pasture, range plants, and forages fed green (3) silages (4) energy feeds (5) protein supplements (6) minerals (7) vitamins (8) additives

Wheatgrass, crested, aerial part, fresh, pure stand, (2)
Ref No 2 05 430 United States

Feed Name or Analyses		Mean As Fed	Mean Dry	C.V. ± %
Dry matter	%		100.0	
Carotene	mg/kg		274.7	39
Vitamin A equivalent	IU/g		457.9	

Wheatgrass, crested, aerial part, fresh weathered, mature, (1)
Ref No 1 08 558 United States

Feed Name or Analyses		Mean As Fed	Mean Dry	C.V. ± %
Dry matter	%	81.0	100.0	
Ash	%	3.4	4.2	
Crude fiber	%	31.7	39.1	
Ether extract	%	1.2	1.5	
N-free extract	%	42.0	51.9	
Protein (N x 6.25)	%	2.7	3.3	
Cattle	dig prot % *	0.6	0.7	
Goats	dig prot % *	-0.2	-0.2	
Horses	dig prot % *	0.3	0.4	
Rabbits	dig coef % *	1.0	1.2	
Sheep	dig prot % *	0.1	0.1	
Energy	GE Mcal/kg			
Cattle	DE Mcal/kg *	1.77	2.19	
Sheep	DE Mcal/kg *	1.84	2.28	
Cattle	ME Mcal/kg *	1.46	1.80	
Sheep	ME Mcal/kg *	1.51	1.87	
Cattle	TDN % *	40.3	49.7	
Sheep	TDN % *	41.8	51.6	
Calcium	%	0.16	0.20	
Phosphorus	%	0.07	0.09	

WHEATGRASS, INTERMEDIATE. Agropyron intermedium

Wheatgrass, intermediate, hay, fan air dried, early leaf, (1)
Ref No 1 08 762 United States

Feed Name or Analyses		Mean As Fed	Mean Dry	C.V. ± %
Dry matter	%		100.0	
Lignin (Ellis)	%		3.9	

Wheatgrass, intermediate, hay, fan air dried, immature, (1)
Ref No 1 08 760 United States

Feed Name or Analyses		Mean As Fed	Mean Dry	C.V. ± %
Dry matter	%		100.0	
Lignin (Ellis)	%		6.4	

Wheatgrass, intermediate, hay, fan air dried, pre-bloom, (1)
Ref No 1 08 761 United States

Feed Name or Analyses		Mean As Fed	Mean Dry	C.V. ± %
Dry matter	%		100.0	
Lignin (Ellis)	%		5.6	

Wheatgrass, intermediate, hay, fan air dried, mid-bloom, (1)
Ref No 1 08 764 United States

Feed Name or Analyses		Mean As Fed	Mean Dry	C.V. ± %
Dry matter	%		100.0	
Lignin (Ellis)	%		7.8	

Wheatgrass, intermediate, hay, s-c, (1)
Ref No 1 05 431 United States

Feed Name or Analyses		Mean As Fed	Mean Dry	C.V. ± %
Dry matter	%	90.1	100.0	
Ash	%	8.3	9.2	
Crude fiber	%	31.6	35.0	
Sheep	dig coef %	57.	57.	
Ether extract	%	2.0	2.3	
Sheep	dig coef %	38.	38.	
N-free extract	%	40.9	45.3	
Sheep	dig coef %	68.	68.	
Protein (N x 6.25)	%	7.4	8.2	
Sheep	dig coef %	41.	41.	
Cattle	dig prot % *	3.6	4.0	
Goats	dig prot % *	3.8	4.2	
Horses	dig prot % *	4.0	4.5	
Rabbits	dig coef % *	4.5	5.0	
Sheep	dig prot %	3.0	3.4	
Fatty acids	%	2.0	2.2	
Energy	GE Mcal/kg	3.94	4.37	
Sheep	GE dig coef %	58.	58.	
Cattle	DE Mcal/kg *	2.25	2.49	
Sheep	DE Mcal/kg	2.28	2.53	
Cattle	ME Mcal/kg *	1.84	2.04	
Sheep	ME Mcal/kg *	1.87	2.08	
Cattle	TDN % *	50.9	56.5	
Sheep	TDN %	50.5	56.1	
Calcium	%	0.32	0.36	12

Ref No 1 05 431 Canada

Feed Name or Analyses		Mean As Fed	Mean Dry	C.V. ± %
Dry matter	%		100.0	
Sheep	dig coef %		58.	
Organic matter	%		91.0	
Crude fiber	%		35.2	
Sheep	dig coef %		57.	
Ether extract	%		1.5	
Sheep	dig coef %		38.	
N-free extract	%		48.6	
Sheep	dig coef %		68.	
Protein (N x 6.25)	%		5.8	
Sheep	dig coef %		41.	
Cattle	dig prot % *		2.0	
Goats	dig prot % *		2.0	
Horses	dig prot % *		2.5	
Rabbits	dig coef % *		3.1	
Sheep	dig prot %		2.4	
Lignin (Ellis)	%		8.9	
Sheep	dig coef %		3.	
Energy	GE Mcal/kg		4.37	
Sheep	GE dig coef %		58.	
Cattle	DE Mcal/kg *		2.46	
Sheep	DE Mcal/kg		2.53	
Cattle	ME Mcal/kg *		2.02	
Sheep	ME Mcal/kg *		2.08	
Cattle	TDN % *		55.7	
Sheep	TDN %		56.8	

Wheatgrass, intermediate, hay, s-c, immature, (1)
Ref No 1 05 636 United States

Feed Name or Analyses		Mean As Fed	Mean Dry	C.V. ± %
Dry matter	%		100.0	
Lignin (Ellis)	%		6.1	

Wheatgrass, intermediate, hay, s-c, pre-bloom, (1)

Ref No 1 05 635 United States

Feed Name or Analyses			Mean As Fed	Mean Dry	C.V. ± %
Dry matter	%			100.0	
Lignin (Ellis)	%			6.4	

Wheatgrass, intermediate, hay, s-c, early bloom, (1)

Ref No 1 05 634 United States

Feed Name or Analyses			Mean As Fed	Mean Dry	C.V. ± %
Dry matter	%			100.0	
Lignin (Ellis)	%			3.7	

Wheatgrass, intermediate, hay, s-c, mid-bloom, (1)

Ref No 1 05 638 United States

Feed Name or Analyses			Mean As Fed	Mean Dry	C.V. ± %
Dry matter	%			100.0	
Lignin (Ellis)	%			7.4	

Wheatgrass, intermediate, hay, fan air dried, early bloom, (1)

Ref No 1 08 763 United States

Feed Name or Analyses			Mean As Fed	Mean Dry	C.V. ± %
Dry matter	%			100.0	
Lignin (Ellis)	%			7.8	

WHEATGRASS, SLENDER. Agropyron tachycaulum

Wheatgrass, slender, hay, s-c, (1)

Ref No 1 05 436 United States

Feed Name or Analyses			Mean As Fed	Mean Dry	C.V. ± %
Dry matter	%		92.0	100.0	2
Ash	%		6.9	7.5	
Crude fiber	%		33.8	36.8	
Sheep	dig coef %		61.	61.	
Ether extract	%		2.1	2.3	
Sheep	dig coef %		47.	47.	
N-free extract	%		41.5	45.1	
Sheep	dig coef %		62.	62.	
Protein (N x 6.25)	%		7.7	8.4	
Sheep	dig coef %		61.	61.	
Cattle	dig prot %	*	3.9	4.2	
Goats	dig prot %	*	4.0	4.4	
Horses	dig prot %	*	4.3	4.7	
Rabbits	dig coef %	*	4.7	5.2	
Sheep	dig prot %		4.7	5.1	
Cellulose (Matrone)	%		32.3	35.1	
Lignin (Ellis)	%		10.8	11.7	
Fatty acids	%		2.1	2.3	
Energy	GE Mcal/kg				
Cattle	DE Mcal/kg	*	2.10	2.29	
Sheep	DE Mcal/kg	*	2.35	2.55	
Cattle	ME Mcal/kg	*	1.72	1.87	
Sheep	ME Mcal/kg	*	1.93	2.09	
Cattle	TDN %	*	47.7	51.9	
Sheep	TDN %		53.3	57.9	
Calcium	%		0.32	0.35	16
Magnesium	%		0.22	0.24	26
Phosphorus	%		0.23	0.25	28
Potassium	%		2.46	2.68	9
Sodium	%		0.76	0.83	10
Sulphur	%		0.11	0.12	15

Wheatgrass, slender, hay, s-c, immature, (1)

Ref No 1 05 432 United States

Feed Name or Analyses			Mean As Fed	Mean Dry	C.V. ± %
Dry matter	%		92.9	100.0	
Ash	%		8.2	8.8	21
Crude fiber	%		25.1	27.0	10
Ether extract	%		2.6	2.8	37
N-free extract	%		40.3	43.4	
Protein (N x 6.25)	%		16.7	18.0	22
Cattle	dig prot %	*	11.6	12.5	
Goats	dig prot %	*	12.4	13.4	
Horses	dig prot %	*	11.9	12.8	
Rabbits	dig coef %	*	11.7	12.6	
Sheep	dig prot %	*	11.8	12.7	
Energy	GE Mcal/kg				
Cattle	DE Mcal/kg	*	2.55	2.75	
Sheep	DE Mcal/kg	*	2.44	2.63	
Cattle	ME Mcal/kg	*	2.09	2.25	
Sheep	ME Mcal/kg	*	2.00	2.16	
Cattle	TDN %	*	57.9	62.3	
Sheep	TDN %	*	55.4	59.6	

Wheatgrass, slender, hay, s-c, early bloom, (1)

Ref No 1 05 433 United States

Feed Name or Analyses			Mean As Fed	Mean Dry	C.V. ± %
Dry matter	%		94.6	100.0	
Ash	%		9.6	10.1	48
Crude fiber	%		31.5	33.3	7
Ether extract	%		2.7	2.9	19
N-free extract	%		39.4	41.6	
Protein (N x 6.25)	%		11.4	12.1	12
Cattle	dig prot %	*	7.0	7.4	
Goats	dig prot %	*	7.4	7.8	
Horses	dig prot %	*	7.4	7.8	
Rabbits	dig coef %	*	7.6	8.0	
Sheep	dig prot %	*	7.0	7.4	
Energy	GE Mcal/kg				
Cattle	DE Mcal/kg	*	2.50	2.64	
Sheep	DE Mcal/kg	*	2.36	2.49	
Cattle	ME Mcal/kg	*	2.05	2.17	
Sheep	ME Mcal/kg	*	1.93	2.04	
Cattle	TDN %	*	56.7	59.9	
Sheep	TDN %	*	53.5	56.5	
Sodium	%		0.79	0.83	
Sulphur	%		0.13	0.14	

Wheatgrass, slender, hay, s-c, full bloom, (1)

Ref No 1 05 434 United States

Feed Name or Analyses			Mean As Fed	Mean Dry	C.V. ± %
Dry matter	%			100.0	
Ash	%			16.0	
Crude fiber	%			34.5	
Ether extract	%			1.9	
N-free extract	%			38.4	
Protein (N x 6.25)	%			9.2	
Cattle	dig prot %	*		4.9	
Goats	dig prot %	*		5.1	
Horses	dig prot %	*		5.3	
Rabbits	dig coef %	*		5.8	
Sheep	dig prot %	*		4.8	
Energy	GE Mcal/kg				
Cattle	DE Mcal/kg	*		2.52	
Sheep	DE Mcal/kg	*		2.38	
Cattle	ME Mcal/kg	*		2.07	
Sheep	ME Mcal/kg	*		1.96	
Cattle	TDN %	*		57.1	
Sheep	TDN %	*		54.1	

Wheatgrass, slender, hay, s-c, mature, (1)

Ref No 1 05 435 United States

Feed Name or Analyses			Mean As Fed	Mean Dry	C.V. ± %
Dry matter	%		95.0	100.0	1
Calcium	%		0.27	0.28	10
Phosphorus	%		0.09	0.09	22

Wheatgrass, slender, aerial part, fresh, (2)

Ref No 2 05 439 United States

Feed Name or Analyses			Mean As Fed	Mean Dry	C.V. ± %
Dry matter	%			100.0	
Ash	%			11.1	24
Crude fiber	%			32.1	12
Ether extract	%			4.2	31
N-free extract	%			44.8	
Protein (N x 6.25)	%			7.8	28
Cattle	dig prot %	*		4.5	
Goats	dig prot %	*		3.8	
Horses	dig prot %	*		4.2	
Rabbits	dig coef %	*		4.7	
Sheep	dig prot %	*		4.3	
Energy	GE Mcal/kg				
Cattle	DE Mcal/kg	*		2.64	
Sheep	DE Mcal/kg	*		2.89	
Cattle	ME Mcal/kg	*		2.17	
Sheep	ME Mcal/kg	*		2.37	
Cattle	TDN %	*		60.2	
Sheep	TDN %	*		65.5	
Calcium	%			0.47	22
Magnesium	%			0.36	
Phosphorus	%			0.14	36
Riboflavin	mg/kg			10.6	20
Thiamine	mg/kg			5.1	25

Wheatgrass, slender, aerial part, fresh, early bloom, (2)

Ref No 2 05 437 United States

Feed Name or Analyses			Mean As Fed	Mean Dry	C.V. ± %
Dry matter	%			100.0	
Ash	%			13.9	1
Crude fiber	%			24.2	0
Ether extract	%			6.2	23
N-free extract	%			43.4	
Protein (N x 6.25)	%			12.3	4
Cattle	dig prot %	*		8.3	
Goats	dig prot %	*		8.0	
Horses	dig prot %	*		8.0	
Rabbits	dig coef %	*		8.2	
Sheep	dig prot %	*		8.5	
Energy	GE Mcal/kg				
Cattle	DE Mcal/kg	*		2.64	
Cattle	DE Mcal/kg	*		2.33	
Sheep	DE Mcal/kg	*		2.84	
Cattle	ME Mcal/kg	*		2.16	
Sheep	ME Mcal/kg	*		2.33	
Cattle	TDN %	*		59.9	
Sheep	TDN %	*		64.4	
Calcium	%			0.57	13
Phosphorus	%			0.18	19

Feed Name or Analyses			Mean As Fed	Mean Dry	C.V. ± %

Wheatgrass, slender, aerial part, fresh, mature, (2)
Ref No 2 05 438 United States

Feed Name or Analyses			As Fed	Dry	C.V. ± %
Dry matter	%			100.0	
Ash	%			9.7	31
Crude fiber	%			36.4	8
Ether extract	%			3.2	3
N-free extract	%			45.8	
Protein (N x 6.25)	%			4.9	5
Cattle	dig prot %	*		2.1	
Goats	dig prot %	*		1.1	
Horses	dig prot %	*		1.7	
Rabbits	dig coef %	*		2.5	
Sheep	dig prot %	*		1.6	
Energy	GE Mcal/kg				
Cattle	DE Mcal/kg	*		2.75	
Sheep	DE Mcal/kg	*		2.93	
Cattle	ME Mcal/kg	*		2.26	
Sheep	ME Mcal/kg	*		2.40	
Cattle	TDN %	*		62.3	
Sheep	TDN %	*		66.4	
Calcium	%			0.31	17
Magnesium	%			0.36	
Phosphorus	%			0.08	43
Sulphur	%			0.12	

WHEATGRASS, STREAMBANK, Agropyron riparium

Wheatgrass, streambank, hay, s-c, (1)
Ref No 1 08 172 Canada

Feed Name or Analyses			As Fed	Dry	C.V. ± %
Dry matter	%			100.0	
Sheep	dig coef %			59.	
Organic matter	%			92.0	
Crude fiber	%			36.3	
Sheep	dig coef %			62.	
Ether extract	%			2.1	
Sheep	dig coef %			34.	
N-free extract	%			46.1	
Sheep	dig coef %			64.	
Protein (N x 6.25)	%			7.5	
Sheep	dig coef %			52.	
Cattle	dig prot %	*		3.4	
Goats	dig prot %	*		3.6	
Horses	dig prot %	*		3.9	
Rabbits	dig coef %	*		4.5	
Sheep	dig prot %			3.9	
Lignin (Ellis)	%			9.2	
Sheep	dig coef %			0.	
Energy	GE Mcal/kg			4.44	
Sheep	GE dig coef %			57.	
Cattle	DE Mcal/kg	*		2.35	
Sheep	DE Mcal/kg			2.53	
Cattle	ME Mcal/kg	*		1.92	
Sheep	ME Mcal/kg	*		2.08	
Cattle	TDN %	*		53.2	
Sheep	TDN %			57.5	

(1) dry forages and roughages (2) pasture, range plants, and forages fed green (3) silages (4) energy feeds (5) protein supplements (6) minerals (7) vitamins (8) additives

WHEATGRASS, TALL. Agropyron elongatum

Wheatgrass, tall, hay, s-c, (1)
Ref No 1 08 170 Canada

Feed Name or Analyses			As Fed	Dry	C.V. ± %
Dry matter	%			100.0	
Sheep	dig coef %			56.	
Organic matter	%			92.0	
Crude fiber	%			40.7	
Sheep	dig coef %			57.	
Ether extract	%			1.9	
Sheep	dig coef %			43.	
N-free extract	%			41.1	
Sheep	dig coef %			56.	
Protein (N x 6.25)	%			8.6	
Sheep	dig coef %			63.	
Cattle	dig prot %	*		4.4	
Goats	dig prot %	*		4.6	
Horses	dig prot %	*		4.8	
Rabbits	dig coef %	*		5.3	
Sheep	dig prot %			5.4	
Lignin (Ellis)	%			10.2	
Sheep	dig coef %			5.	
Energy	GE Mcal/kg			4.45	
Sheep	GE dig coef %			54.	
Cattle	DE Mcal/kg	*		2.23	
Sheep	DE Mcal/kg			2.40	
Cattle	ME Mcal/kg	*		1.83	
Sheep	ME Mcal/kg	*		1.97	
Cattle	TDN %	*		50.6	
Sheep	TDN %			53.5	

WHEATGRASS, BLUESTEM-BUFFALOGRASS.
Agropyron smithii, Buchloe dactyloides

Wheatgrass, bluestem-Buffalograss, aerial part, fresh, (2)
Ref No 2 05 444 United States

Feed Name or Analyses			As Fed	Dry	C.V. ± %
Dry matter	%			100.0	
Ash	%			9.3	16
Crude fiber	%			30.8	17
Ether extract	%			2.2	49
N-free extract	%			47.1	
Protein (N x 6.25)	%			10.6	33
Cattle	dig prot %	*		6.9	
Goats	dig prot %	*		6.4	
Horses	dig prot %	*		6.5	
Rabbits	dig coef %	*		6.9	
Sheep	dig prot %	*		6.9	
Energy	GE Mcal/kg				
Cattle	DE Mcal/kg	*		2.87	
Sheep	DE Mcal/kg	*		2.86	
Cattle	ME Mcal/kg	*		2.35	
Sheep	ME Mcal/kg	*		2.35	
Cattle	TDN %	*		65.0	
Sheep	TDN %	*		65.0	
Carotene	mg/kg			126.1	25
Vitamin A equivalent	IU/g			210.2	

Wheatgrass, bluestem-Buffalograss, aerial part, fresh, immature, (2)
Ref No 2 05 441 United States

Feed Name or Analyses			As Fed	Dry	C.V. ± %
Dry matter	%			100.0	
Ash	%			8.9	13
Crude fiber	%			25.8	9
Ether extract	%			3.7	35
N-free extract	%			45.3	
Protein (N x 6.25)	%			16.3	14
Cattle	dig prot %	*		11.7	
Goats	dig prot %	*		11.8	
Horses	dig prot %	*		11.4	
Rabbits	dig coef %	*		11.3	
Sheep	dig prot %	*		12.2	
Energy	GE Mcal/kg				
Cattle	DE Mcal/kg	*		3.03	
Sheep	DE Mcal/kg	*		2.76	
Cattle	ME Mcal/kg	*		2.49	
Sheep	ME Mcal/kg	*		2.26	
Cattle	TDN %	*		68.8	
Sheep	TDN %	*		62.5	

Wheatgrass, bluestem-Buffalograss, aerial part, fresh, mid-bloom, (2)
Ref No 2 05 442 United States

Feed Name or Analyses			As Fed	Dry	C.V. ± %
Dry matter	%			100.0	
Ash	%			9.2	15
Crude fiber	%			31.1	19
Ether extract	%			1.9	45
N-free extract	%			46.7	
Protein (N x 6.25)	%			11.1	16
Cattle	dig prot %	*		7.3	
Goats	dig prot %	*		6.9	
Horses	dig prot %	*		7.0	
Rabbits	dig coef %	*		7.2	
Sheep	dig prot %	*		7.3	
Energy	GE Mcal/kg				
Cattle	DE Mcal/kg	*		2.92	
Sheep	DE Mcal/kg	*		2.85	
Cattle	ME Mcal/kg	*		2.39	
Sheep	ME Mcal/kg	*		2.33	
Cattle	TDN %	*		66.2	
Sheep	TDN %	*		64.5	

Wheatgrass, bluestem-Buffalograss, aerial part, fresh, mature, (2)
Ref No 2 05 443 United States

Feed Name or Analyses			As Fed	Dry	C.V. ± %
Dry matter	%			100.0	
Ash	%			9.5	23
Crude fiber	%			33.7	11
Ether extract	%			1.7	37
N-free extract	%			48.1	
Protein (N x 6.25)	%			7.0	15
Cattle	dig prot %	*		3.8	
Goats	dig prot %	*		3.1	
Horses	dig prot %	*		3.5	
Rabbits	dig coef %	*		4.1	
Sheep	dig prot %	*		3.5	
Energy	GE Mcal/kg				
Cattle	DE Mcal/kg	*		2.73	
Sheep	DE Mcal/kg	*		2.93	
Cattle	ME Mcal/kg	*		2.24	
Sheep	ME Mcal/kg	*		2.41	
Cattle	TDN %	*		62.0	
Sheep	TDN %	*		66.5	

Feed Name or Analyses			Mean As Fed	Mean Dry	C.V. ± %

WHEATGRASS, BLUESTEM-GAMAGRASS. Agropyron smithii, Tripsacum spp

Wheatgrass, bluestem-Gamagrass, hay, s-c, (1)
Ref No 1 05 440 United States

			As Fed	Dry	C.V.
Dry matter	%			100.0	
Manganese	mg/kg			9.9	29

WHEATGRASS, BLUESTEM-GRAMA. Agropyron smithii, Bouteloua spp

Wheatgrass, bluestem-Grama, aerial part, fresh, full bloom, (2)
Ref No 2 05 445 United States

			As Fed	Dry	C.V.
Dry matter	%			100.0	
Ash	%			7.3	
Crude fiber	%			34.1	
Ether extract	%			2.1	
N-free extract	%			49.0	
Protein (N x 6.25)	%			7.5	
Cattle	dig prot %	*		4.3	
Goats	dig prot %	*		3.6	
Horses	dig prot %	*		3.9	
Rabbits	dig coef %	*		4.5	
Sheep	dig prot %	*		4.0	
Energy	GE Mcal/kg				
Cattle	DE Mcal/kg	*		2.88	
Sheep	DE Mcal/kg	*		2.93	
Cattle	ME Mcal/kg	*		2.36	
Sheep	ME Mcal/kg	*		2.40	
Cattle	TDN %	*		65.3	
Sheep	TDN %	*		66.5	
Carotene	mg/kg			120.8	
Vitamin A equivalent	IU/g			201.4	

Wheatgrass, bluestem-Grama, aerial part, fresh, milk stage, (2)
Ref No 2 05 446 United States

			As Fed	Dry	C.V.
Dry matter	%			100.0	
Ash	%			7.9	
Crude fiber	%			36.9	
Ether extract	%			1.4	
N-free extract	%			47.4	
Protein (N x 6.25)	%			6.4	
Cattle	dig prot %	*		3.3	
Goats	dig prot %	*		2.5	
Horses	dig prot %	*		3.0	
Rabbits	dig coef %	*		3.6	
Sheep	dig prot %	*		3.0	
Energy	GE Mcal/kg				
Cattle	DE Mcal/kg	*		2.85	
Sheep	DE Mcal/kg	*		2.91	
Cattle	ME Mcal/kg	*		2.34	
Sheep	ME Mcal/kg	*		2.39	
Cattle	TDN %	*		64.7	
Sheep	TDN %	*		66.0	
Carotene	mg/kg			73.9	
Vitamin A equivalent	IU/g			123.1	

Wheatgrass forage, western –
see Wheatgrass, bluestem, aerial part, fresh, (2)

Wheatgrass forage, western, cut 1 –
see Wheatgrass, bluestem, aerial part, fresh, cut 1, (2)

Wheatgrass, forage, western, dough stage –
see Wheatgrass, bluestem, aerial part, fresh, dough stage, (2)

Wheatgrass forage, western, early bloom –
see Wheatgrass, bluestem, aerial part, fresh, early bloom, (2)

Wheatgrass forage, western, full bloom –
see Wheatgrass, bluestem, aerial part, fresh, full bloom, (2)

Wheatgrass forage, western, immature –
see Wheatgrass, bluestem, aerial part, fresh, immature, (2)

Wheatgrass forage, western, mature –
see Wheatgrass, bluestem, aerial part, fresh, mature, (2)

Wheatgrass forage, western, mid-bloom –
see Wheatgrass,bluestem, aerial part, fresh, mid-bloom, (2)

Wheatgrass forage, western, milk stage –
see Wheatgrass, bluestem, aerial part, fresh, milk stage, (2)

Wheatgrass forage, western, over ripe –
see Wheatgrass, bluestem, aerial part, fresh, over ripe, (2)

Wheatgrass, hay, western –
see Wheatgrass, bluestem, hay, s-c, (1)

Wheatgrass, hay, western, full bloom –
see Wheatgrass, bluestem, hay, s-c, full bloom, (1)

Wheatgrass, hay, western, immature, –
see Wheatgrass, bluestem, hay, s-c, immature, (1)

Wheatgrass, hay, western, mature –
see Wheatgrass, bluestem, hay, s-c, mature, (1)

Wheatgrass, hay, western, over ripe –
see Wheatgrass, bluestem, hay, s-c, over ripe, (1)

Wheat, hard, mill run –
see Wheat, hard, flour by-product mill run, mx 9.5% fiber, (4)

Wheat, hard red winter, red dog –
see Wheat, hard red winter, flour by-product, fine sift, mx 4% fiber, (4)

Wheat, hard red winter, shorts –
see Wheat, hard red winter, flour by-product, coarse sift, mx 7% fiber, (4)

Wheat, hard, standard middlings –
see Wheat, hard flour by-product, mx 9.5% fiber, (4)

Wheat middlings (AAFCO) –
see Wheat, flour by-product, mx 9.5% fiber, (4)

Wheat mill run (AAFCO) –
see Wheat, flour by-product mill run, mx 9.5% fiber, (4)

WHEAT-PALM. Triticum spp, Elaeis spp

Wheat-Palm, flour by-product w palm oil added, mx 9.5% fiber, (4)
Palmo middlings
Ref No 4 08 557 United States

			As Fed	Dry	C.V.
Dry matter	%		93.4	100.0	
Ash	%		6.0	6.4	
Crude fiber	%		7.6	8.1	
Ether extract	%		8.3	8.9	
N-free extract	%		55.3	59.2	
Protein (N x 6.25)	%		16.2	17.3	
Cattle	dig prot %	*	11.2	11.9	
Goats	dig prot %	*	12.3	13.1	
Horses	dig prot %	*	12.3	13.1	
Sheep	dig prot %	*	12.3	13.1	
Energy	GE Mcal/kg				
Cattle	DE Mcal/kg	*	3.58	3.83	
Sheep	DE Mcal/kg	*	3.63	3.89	
Swine	DE kcal/kg	*	3306.	3540.	
Cattle	ME Mcal/kg	*	2.93	3.14	
Sheep	ME Mcal/kg	*	2.98	3.19	
Swine	ME kcal/kg	*	3059.	3275.	
Cattle	TDN %	*	81.1	86.8	
Sheep	TDN %	*	82.5	88.3	
Swine	TDN %	*	75.0	80.3	

Wheat red dog, mx 4% fiber (AAFCO) –
see Wheat, flour by-product, fine sift, mx 4% fiber, (4)

Wheat shorts, mx 7% fiber (AAFCO) –
see Wheat, flour by-product, coarse sift, mx 7% fiber, (4)

Wheat, soft red winter, red dog –
see Wheat, soft red winter, flour by-product, fine sift, mx 4% fiber, (4)

Wheat, soft red winter, shorts –
see Wheat, soft red winter, flour by-product, coarse sift, mx 7% fiber, (4)

Wheat standard middlings –
see Wheat, flour by-product, mx 9.5% fiber, (4)

WHEAT-TIMOTHY. Triticum aestivum, Phleum pratense

Wheat-Timothy, aerial part, ensiled, mature, (3)
Ref No 3 05 342 United States

			As Fed	Dry	C.V.
Dry matter	%			100.0	
Ash	%			5.2	
Crude fiber	%			43.7	
Ether extract	%			2.1	
N-free extract	%			44.6	
Protein (N x 6.25)	%			4.4	
Cattle	dig prot %	*		0.2	
Goats	dig prot %	*		0.2	
Horses	dig prot %	*		0.2	
Sheep	dig prot %	*		0.2	

Continued

Feed Name or Analyses	Mean As Fed	Mean Dry	C.V. ± %
Energy GE Mcal/kg			
Cattle DE Mcal/kg *		2.24	
Sheep DE Mcal/kg *		2.85	
Cattle ME Mcal/kg *		1.84	
Sheep ME Mcal/kg *		2.33	
Cattle TDN % *		50.9	
Sheep TDN % *		64.6	

WHEAT-VETCH. Triticum aestivum, Vicia spp

Wheat-Vetch, aerial part w molasses added, ensiled, (3)
Ref No 3 05 343 — United States

Analyses	As Fed	Dry	C.V. ± %
Dry matter %	29.2	100.0	6
Phosphorus %	0.09	0.31	15

Wheat, white, red dog –
see Wheat, white, flour by-product, fine sift, mx 4% fiber, (4)

Wheat, white, shorts –
see Wheat, white, flour by-product, coarse sift, mx 7% fiber, (4)

Whey, condensed –
see Cattle, whey, condensed, mn solids declared, (4)

Whey, dried –
see Cattle, whey, dehy, mn 65% lactose, (4)

Whey, evaporated –
see Cattle, whey, condensed, mn solids declared, (4)

Whey-product, condensed –
see Cattle, whey low lactose, condensed, mn solids declared, (4)

Whey product, dried –
see Cattle, whey low lactose, dehy, mn lactose declared, (4)

Whey-product, evaporated –
see Cattle, whey low lactose, condensed, mn solids declared, (4)

Whey, semisolid –
see Cattle, whey, condensed, mn solids declared, (4)

Whey solubles, condensed –
see Cattle, whey wo albumin low lactose, condensed, (4)

Whey solubles, dried –
see Cattle, whey wo albumin low lactose, dehy, (4)

Whey solubles, evaporated –
see Cattle, whey wo albumin low lactose, condensed, (4)

White fish meal (CFA) –
see Fish, white, whole or cuttings, cooked mech-extd dehy grnd, mx 4% oil, (5)

(1) dry forages and roughages (2) pasture, range plants, and forages fed green (3) silages (4) energy feeds (5) protein supplements (6) minerals (7) vitamins (8) additives

White hominy feed (AAFCO) (CFA) –
see Corn, white, grits by-product, mn 5% fat, (4)

Whitesage-see Winterfat, common, aerial part, fresh, mature, (2)

Whole eviscerated chicken (AAFCO) –
see Chicken, carcass, raw, (5)

Whole pressed sunflower meal (AAFCO) –
see Sunflower, seeds, extn unspecified grnd, (5)

WILDRYE. Elymus spp

Wildrye, aerial part, fresh, (2)
Ref No 2 05 451 — United States

Analyses	As Fed	Dry	C.V. ± %
Dry matter %		100.0	
Calcium %		0.44	35
Magnesium %		0.12	22
Phosphorus %		0.18	40
Potassium %		1.94	

Wildrye, aerial part, fresh, immature, (2)
Ref No 2 05 447 — United States

Analyses	As Fed	Dry	C.V. ± %
Dry matter %		100.0	
Ash %		9.6	48
Crude fiber %		29.2	12
Ether extract %		2.2	25
N-free extract %		43.1	
Protein (N x 6.25) %		15.9	13
Cattle dig prot % *		11.4	
Goats dig prot % *		11.4	
Horses dig prot % *		11.0	
Rabbits dig coef % *		10.9	
Sheep dig prot % *		11.8	
Calcium %		0.47	38
Phosphorus %		0.25	21

Wildrye, aerial part, fresh, mid-bloom, (2)
Ref No 2 05 448 — United States

Analyses	As Fed	Dry	C.V. ± %
Dry matter %		100.0	
Ash %		6.6	
Crude fiber %		29.5	
Ether extract %		2.0	
N-free extract %		50.2	
Protein (N x 6.25) %		11.7	
Cattle dig prot % *		7.8	
Goats dig prot % *		7.5	
Horses dig prot % *		7.5	
Rabbits dig coef % *		7.7	
Sheep dig prot % *		7.9	

Wildrye, aerial part, fresh, milk stage, (2)
Ref No 2 05 449 — United States

Analyses	As Fed	Dry	C.V. ± %
Dry matter %		100.0	
Ash %		7.1	
Crude fiber %		37.9	
Ether extract %		2.0	
N-free extract %		45.6	
Protein (N x 6.25) %		7.4	
Cattle dig prot % *		4.2	
Goats dig prot % *		3.5	
Horses dig prot % *		3.8	
Rabbits dig coef % *		4.4	
Sheep dig prot % *		3.9	

Wildrye, aerial part, fresh, mature, (2)
Ref No 2 05 450 — United States

Analyses	As Fed	Dry	C.V. ± %
Dry matter %		100.0	
Ash %		7.9	17
Crude fiber %		37.8	5
Ether extract %		2.8	10
N-free extract %		47.2	
Protein (N x 6.25) %		4.3	18
Cattle dig prot % *		1.5	
Goats dig prot % *		0.6	
Horses dig prot % *		1.2	
Rabbits dig coef % *		2.0	
Sheep dig prot % *		1.0	
Calcium %		0.37	17
Phosphorus %		0.14	31

WILDRYE, BLUE. Elymus glaucus

Wildrye, blue, aerial part, fresh, (2)
Ref No 2 05 452 — United States

Analyses	As Fed	Dry	C.V. ± %
Dry matter %		100.0	
Ash %		5.5	
Ether extract %		2.3	
Protein (N x 6.25) %		7.3	
Cattle dig prot % *		4.1	
Goats dig prot % *		3.4	
Horses dig prot % *		3.7	
Rabbits dig coef % *		4.3	
Sheep dig prot % *		3.8	
Calcium %		0.33	
Phosphorus %		0.23	

Wildrye, blue, leaves, fresh, (2)
Ref No 2 05 453 — United States

Analyses	As Fed	Dry	C.V. ± %
Dry matter %		100.0	
Ash %		9.5	
Ether extract %		6.3	
Protein (N x 6.25) %		14.1	
Cattle dig prot % *		9.9	
Goats dig prot % *		9.7	
Horses dig prot % *		9.5	
Rabbits dig coef % *		9.6	
Sheep dig prot % *		10.1	
Calcium %		0.73	
Phosphorus %		0.34	

Wildrye, blue, stems, fresh, (2)
Ref No 2 05 454 — United States

Analyses	As Fed	Dry	C.V. ± %
Dry matter %		100.0	
Ash %		4.6	
Ether extract %		1.4	
Protein (N x 6.25) %		4.8	
Cattle dig prot % *		2.0	
Goats dig prot % *		1.0	
Horses dig prot % *		1.6	
Rabbits dig coef % *		2.4	
Sheep dig prot % *		1.5	

Feed Name or Analyses			Mean As Fed	Mean Dry	C.V. ± %
Calcium		%		0.19	
Phosphorus		%		0.18	

WILDRYE, CANADA. Elymus canadensis

Wildrye, Canada, aerial part, fresh, (2)

Ref No 2 05 459 United States

Dry matter		%		100.0	
Ash		%		11.5	53
Crude fiber		%		30.6	27
Ether extract		%		2.3	36
N-free extract		%		47.2	
Protein (N x 6.25)		%		8.4	79
Cattle	dig prot	%	*	5.0	
Goats	dig prot	%	*	4.4	
Horses	dig prot	%	*	4.7	
Rabbits	dig coef	%	*	5.2	
Sheep	dig prot	%	*	4.8	

Wildrye, Canada, aerial part, fresh, immature, (2)

Ref No 2 05 455 United States

Dry matter		%		100.0	
Ash		%		22.7	
Crude fiber		%		19.2	
Ether extract		%		3.8	
N-free extract		%		34.3	
Protein (N x 6.25)		%		20.0	
Cattle	dig prot	%	*	14.9	
Goats	dig prot	%	*	15.2	
Horses	dig prot	%	*	14.5	
Rabbits	dig coef	%	*	14.1	
Sheep	dig prot	%	*	15.6	
Calcium		%		0.76	
Phosphorus		%		0.21	

Wildrye, Canada, aerial part, fresh, full bloom, (2)

Ref No 2 05 456 United States

Dry matter		%		100.0	
Ash		%		8.4	
Crude fiber		%		29.2	
Ether extract		%		2.0	
N-free extract		%		52.1	
Protein (N x 6.25)		%		8.3	
Cattle	dig prot	%	*	4.9	
Goats	dig prot	%	*	4.3	
Horses	dig prot	%	*	4.6	
Rabbits	dig coef	%	*	5.1	
Sheep	dig prot	%	*	4.7	
Calcium		%		0.44	
Phosphorus		%		0.13	

Wildrye, Canada, aerial part, fresh, milk stage, (2)

Ref No 2 05 457 United States

Dry matter		%		100.0	
Ash		%		7.8	
Crude fiber		%		40.4	
Ether extract		%		1.9	
N-free extract		%		42.6	
Protein (N x 6.25)		%		7.3	
Cattle	dig prot	%	*	4.1	
Goats	dig prot	%	*	3.4	
Horses	dig prot	%	*	3.7	
Rabbits	dig coef	%	*	4.3	
Sheep	dig prot	%	*	3.8	

Wildrye, Canada, aerial part, fresh, mature, (2)

Ref No 2 05 458 United States

Dry matter		%		100.0	
Ash		%		10.8	
Crude fiber		%		32.9	
Ether extract		%		2.1	
N-free extract		%		50.9	
Protein (N x 6.25)		%		3.3	
Cattle	dig prot	%	*	0.7	
Goats	dig prot	%	*	-0.3	
Horses	dig prot	%	*	0.3	
Rabbits	dig coef	%	*	1.2	
Sheep	dig prot	%	*	0.1	

WILDRYE, CANADA SMOOTHSCALE. Elymus Canadensis brachystachys

Wildrye, Canada smoothscale, aerial part, fresh, (2)

Ref No 2 05 460 United States

Dry matter		%		100.0	
Ash		%		9.5	
Crude fiber		%		35.3	
Ether extract		%		1.6	
N-free extract		%		40.8	
Protein (N x 6.25)		%		12.8	
Cattle	dig prot	%	*	8.8	
Goats	dig prot	%	*	8.5	
Horses	dig prot	%	*	8.4	
Rabbits	dig coef	%	*	8.6	
Sheep	dig prot	%	*	8.9	
Calcium		%		0.36	
Magnesium		%		0.13	
Phosphorus		%		0.21	
Potassium		%		1.94	

WILDRYE, COLORADO. Elymus ambiguus

Wildrye, Colorado, aerial part, fresh, (2)

Ref No 2 05 461 United States

Dry matter		%		100.0	
Ash		%		5.8	10
Crude fiber		%		36.5	15
Ether extract		%		1.8	49
N-free extract		%		48.0	
Protein (N x 6.25)		%		7.9	23
Cattle	dig prot	%	*	4.6	
Goats	dig prot	%	*	3.9	
Horses	dig prot	%	*	4.2	
Rabbits	dig coef	%	*	4.8	
Sheep	dig prot	%	*	4.4	

WILDRYE, GIANT. Elymus condensatus

Wildrye, giant, aerial part, fresh, (2)

Ref No 2 05 466 United States

Dry matter		%		100.0	
Ash		%		10.1	38
Crude fiber		%		35.9	12
Ether extract		%		3.2	36
N-free extract		%		41.5	
Protein (N x 6.25)		%		9.3	37
Cattle	dig prot	%	*	5.8	
Goats	dig prot	%	*	5.2	
Horses	dig prot	%	*	5.4	
Rabbits	dig coef	%	*	5.8	
Sheep	dig prot	%	*	5.7	
Calcium		%		0.39	42
Magnesium		%		0.12	22
Phosphorus		%		0.17	46

Wildrye, giant, aerial part, fresh, immature, (2)

Ref No 2 05 462 United States

Dry matter		%		100.0	
Ash		%		7.7	
Crude fiber		%		31.0	
Ether extract		%		2.0	
N-free extract		%		44.4	
Protein (N x 6.25)		%		14.9	
Cattle	dig prot	%	*	10.6	
Goats	dig prot	%	*	10.5	
Horses	dig prot	%	*	10.2	
Rabbits	dig coef	%	*	10.2	
Sheep	dig prot	%	*	10.9	
Calcium		%		0.43	44
Phosphorus		%		0.24	23

Wildrye, giant, aerial part, fresh, mid-bloom, (2)

Ref No 2 05 463 United States

Dry matter		%		100.0	
Ash		%		5.9	
Crude fiber		%		36.6	
Ether extract		%		1.8	
N-free extract		%		45.8	
Protein (N x 6.25)		%		9.9	
Cattle	dig prot	%	*	6.3	
Goats	dig prot	%	*	5.8	
Horses	dig prot	%	*	5.9	
Rabbits	dig coef	%	*	6.3	
Sheep	dig prot	%	*	6.2	

Wildrye, giant, aerial part, fresh, milk stage, (2)

Ref No 2 05 464 United States

Dry matter		%		100.0	
Ash		%		6.4	
Crude fiber		%		35.4	
Ether extract		%		2.1	
N-free extract		%		48.7	
Protein (N x 6.25)		%		7.4	
Cattle	dig prot	%	*	4.2	
Goats	dig prot	%	*	3.5	
Horses	dig prot	%	*	3.8	
Rabbits	dig coef	%	*	4.4	
Sheep	dig prot	%	*	3.9	

Wildrye, giant, aerial part, fresh, mature, (2)

Ref No 2 05 465 United States

Dry matter		%		100.0	
Ash		%		7.5	
Crude fiber		%		39.1	
Ether extract		%		3.0	

Continued

Feed Name or Analyses			Mean As Fed	Mean Dry	C.V. ± %
N-free extract	%			46.1	
Protein (N x 6.25)	%			4.3	
Cattle	dig prot %	*		1.5	
Goats	dig prot %	*		0.6	
Horses	dig prot %	*		1.2	
Rabbits	dig coef %	*		2.0	
Sheep	dig prot %	*		1.0	
Calcium	%			0.37	18
Phosphorus	%			0.13	25
Sulphur	%			0.11	

Wildrye, giant, aerial part, fresh, dormant, (2)
Ref No 2 07 995 United States

Feed Name or Analyses			Mean As Fed	Mean Dry	C.V. ± %
Dry matter	%			100.0	
Calcium	%			0.31	
Phosphorus	%			0.06	
Carotene	mg/kg			0.2	
Vitamin A equivalent	IU/g			0.4	

WILDRYE, HAIRY. Elymus villosus

Wildrye, hairy, aerial part, fresh, immature, (2)
Ref No 2 05 467 United States

Feed Name or Analyses			Mean As Fed	Mean Dry	C.V. ± %
Dry matter	%			100.0	
Ash	%			15.6	
Crude fiber	%			21.7	
Ether extract	%			3.0	
N-free extract	%			38.2	
Protein (N x 6.25)	%			21.5	
Cattle	dig prot %	*		16.2	
Goats	dig prot %	*		16.6	
Horses	dig prot %	*		15.8	
Rabbits	dig coef %	*		15.3	
Sheep	dig prot %	*		17.0	
Calcium	%			0.53	
Phosphorus	%			0.35	

WILDRYE, MEDUSAHEAD. Elymus caput-medusae

Wildrye, medusahead, aerial part, dehy grnd, early bloom, (1)
Ref No 1 08 285 United States

Feed Name or Analyses			Mean As Fed	Mean Dry	C.V. ± %
Dry matter	%		92.4	100.0	
Organic matter	%		85.8	92.9	
Ash	%		6.6	7.1	
Crude fiber	%		28.5	30.8	
Ether extract	%		2.1	2.3	
N-free extract	%		49.5	53.6	
Protein (N x 6.25)	%		5.7	6.2	
Cattle	dig prot %	*	2.1	2.3	
Goats	dig prot %	*	2.2	2.3	
Horses	dig prot %	*	2.6	2.8	
Rabbits	dig coef %	*	3.2	3.5	
Sheep	dig prot %	*	2.0	2.1	
Silicon	%		3.75	4.06	

Wildrye, medusahead, aerial part, dehy grnd, over ripe, (1)
Ref No 1 08 286 United States

Feed Name or Analyses			Mean As Fed	Mean Dry	C.V. ± %
Dry matter	%		92.7	100.0	
Organic matter	%		87.5	94.4	
Ash	%		5.2	5.6	
Crude fiber	%		28.0	30.2	
Ether extract	%		1.9	2.0	
N-free extract	%		53.6	57.8	
Protein (N x 6.25)	%		4.1	4.4	
Cattle	dig prot %	*	0.7	0.8	
Goats	dig prot %	*	0.6	0.7	
Horses	dig prot %	*	1.2	1.3	
Rabbits	dig coef %	*	1.9	2.1	
Sheep	dig prot %	*	0.5	0.5	
Silicon	%		3.24	3.50	

Wildrye, medusahead, aerial part, dehy grnd, dormant, (1)
Ref No 1 08 287 United States

Feed Name or Analyses			Mean As Fed	Mean Dry	C.V. ± %
Dry matter	%		92.5	100.0	0
Organic matter	%		86.7	93.8	1
Ash	%		5.8	6.2	15
Crude fiber	%		32.1	34.7	9
Ether extract	%		1.4	1.5	22
N-free extract	%		50.4	54.5	4
Protein (N x 6.25)	%		2.8	3.0	27
Cattle	dig prot %	*	-0.3	-0.3	
Goats	dig prot %	*	-0.5	-0.5	
Horses	dig prot %	*	0.1	0.1	
Rabbits	dig coef %	*	0.9	1.0	
Sheep	dig prot %	*	-0.6	-0.6	
Silicon	%		4.63	5.01	14

Wildrye, medusahead, leaves, dehy grnd, early leaf, (1)
Ref No 1 08 288 United States

Feed Name or Analyses			Mean As Fed	Mean Dry	C.V. ± %
Dry matter	%		93.6	100.0	
Organic matter	%		84.8	90.6	
Ash	%		8.8	9.4	
Crude fiber	%		14.3	15.3	
Ether extract	%		4.8	5.1	
N-free extract	%		43.7	46.7	
Protein (N x 6.25)	%		22.0	23.5	
Cattle	dig prot %	*	16.2	17.3	
Goats	dig prot %	*	17.3	18.5	
Horses	dig prot %	*	16.4	17.5	
Rabbits	dig coef %	*	15.7	16.8	
Sheep	dig prot %	*	16.5	17.6	
Silicon	%		1.55	1.66	

Wildrye, medusahead, leaves, dehy grnd, immature, (1)
Ref No 1 08 289 United States

Feed Name or Analyses			Mean As Fed	Mean Dry	C.V. ± %
Dry matter	%		93.3	100.0	
Organic matter	%		85.6	91.8	
Ash	%		7.6	8.2	
Crude fiber	%		19.9	21.3	
Ether extract	%		4.0	4.3	
N-free extract	%		46.3	49.6	
Protein (N x 6.25)	%		15.5	16.6	
Cattle	dig prot %	*	10.6	11.3	
Goats	dig prot %	*	11.2	12.0	
Horses	dig prot %	*	10.8	11.6	
Rabbits	dig coef %	*	10.7	11.5	
Sheep	dig prot %	*	10.7	11.5	
Silicon	%		2.84	3.05	

Wildrye, medusahead, leaves, dehy grnd, pre-bloom, (1)
Ref No 1 08 290 United States

Feed Name or Analyses			Mean As Fed	Mean Dry	C.V. ± %
Dry matter	%		92.1	100.0	
Organic matter	%		80.1	87.0	
Ash	%		12.0	13.0	
Protein (N x 6.25)	%		4.5	4.9	
Cattle	dig prot %	*	1.1	1.2	
Goats	dig prot %	*	1.0	1.1	
Horses	dig prot %	*	1.6	1.7	
Rabbits	dig coef %	*	2.3	2.5	
Sheep	dig prot %	*	0.9	1.0	
Silicon	%		8.80	9.55	

Wildrye, medusahead, leaves, dehy grnd, dormant, (1)
Ref No 1 08 291 United States

Feed Name or Analyses			Mean As Fed	Mean Dry	C.V. ± %
Dry matter	%		92.6	100.0	
Organic matter	%		80.7	87.2	
Ash	%		11.9	12.9	
Crude fiber	%		26.2	28.3	
Ether extract	%		2.6	2.8	
N-free extract	%		48.0	51.8	
Protein (N x 6.25)	%		3.9	4.2	
Cattle	dig prot %	*	0.5	0.6	
Goats	dig prot %	*	0.4	0.5	
Horses	dig prot %	*	1.0	1.1	
Rabbits	dig coef %	*	1.8	1.9	
Sheep	dig prot %	*	0.3	0.3	
Silicon	%		9.40	10.15	

Wildrye, medusahead, leaves w some stems, dehy grnd, pre-bloom, (1)
Ref No 1 08 292 United States

Feed Name or Analyses			Mean As Fed	Mean Dry	C.V. ± %
Dry matter	%		93.4	100.0	
Organic matter	%		87.7	93.9	
Ash	%		5.7	6.1	
Crude fiber	%		18.9	20.2	
Ether extract	%		2.5	2.7	
N-free extract	%		54.0	57.8	
Protein (N x 6.25)	%		12.3	13.2	
Cattle	dig prot %	*	7.8	8.4	
Goats	dig prot %	*	8.3	8.9	
Horses	dig prot %	*	8.2	8.7	
Rabbits	dig coef %	*	8.3	8.9	
Sheep	dig prot %	*	7.9	8.4	
Silicon	%		2.70	2.89	

WILDRYE, RUSSIAN. Elymus junceus

Wildrye, Russian, hay, s-c, (1)
Ref No 1 08 171 Canada

Feed Name or Analyses			Mean As Fed	Mean Dry	C.V. ± %
Dry matter	%			100.0	
Sheep	dig coef %			61.	
Organic matter	%			92.0	
Crude fiber	%			37.1	
Sheep	dig coef %			63.	
Ether extract	%			1.9	
Sheep	dig coef %			41.	

(1) dry forages and roughages (2) pasture, range plants, and forages fed green (3) silages (4) energy feeds (5) protein supplements (6) minerals (7) vitamins (8) additives

Feed Name or Analyses			Mean As Fed	Mean Dry	C.V. ± %
N-free extract	%			44.3	
Sheep	dig coef %			66.	
Protein (N x 6.25)	%			8.0	
Sheep	dig coef %			56.	
Cattle	dig prot %	*		3.9	
Goats	dig prot %	*		4.0	
Horses	dig prot %	*		4.3	
Rabbits	dig coef %	*		4.8	
Sheep	dig prot %			4.5	
Lignin (Ellis)	%			8.7	
Sheep	dig coef %			3.	
Energy	GE Mcal/kg			4.47	
Sheep	GE dig coef %			60.	
Cattle	DE Mcal/kg	*		2.51	
Sheep	DE Mcal/kg			2.68	
Cattle	ME Mcal/kg	*		2.06	
Sheep	ME Mcal/kg	*		2.20	
Cattle	TDN %	*		56.9	
Sheep	TDN %			58.8	

Wildrye, Russian, aerial part, fresh, (2)

Ref No 2 05 469 United States

Feed Name or Analyses			Mean As Fed	Mean Dry	C.V. ± %
Dry matter	%			100.0	
Ash	%			8.0	
Crude fiber	%			22.4	
Ether extract	%			2.6	
N-free extract	%			52.9	
Protein (N x 6.25)	%			14.1	
Cattle	dig prot %	*		9.9	
Goats	dig prot %	*		9.7	
Horses	dig prot %	*		9.5	
Rabbits	dig coef %	*		9.6	
Sheep	dig prot %	*		10.1	
Energy	GE Mcal/kg				
Cattle	DE Mcal/kg	*		3.03	
Sheep	DE Mcal/kg	*		2.93	
Cattle	ME Mcal/kg	*		2.48	
Sheep	ME Mcal/kg	*		2.41	
Cattle	TDN %	*		68.6	
Sheep	TDN %	*		66.5	

Wildrye, Russian, aerial part, fresh, mid-bloom, (2)

Ref No 2 05 468 United States

Feed Name or Analyses			Mean As Fed	Mean Dry	C.V. ± %
Dry matter	%			100.0	
Ash	%			7.2	
Crude fiber	%			22.4	
Ether extract	%			2.2	
N-free extract	%			54.8	
Protein (N x 6.25)	%			13.4	
Cattle	dig prot %	*		9.3	
Goats	dig prot %	*		9.1	
Horses	dig prot %	*		8.9	
Rabbits	dig coef %	*		9.0	
Sheep	dig prot %	*		9.5	
Energy	GE Mcal/kg				
Cattle	DE Mcal/kg	*		3.07	
Sheep	DE Mcal/kg	*		2.97	
Cattle	ME Mcal/kg	*		2.52	
Sheep	ME Mcal/kg	*		2.44	
Cattle	TDN %	*		69.6	
Sheep	TDN %	*		67.4	

WILDRYE, VIRGINIA. Elymus virginicus

Wildrye, Virginia, aerial part, fresh, mature, (2)

Ref No 2 05 470 United States

Feed Name or Analyses			Mean As Fed	Mean Dry	C.V. ± %
Dry matter	%			100.0	
Ash	%			6.3	
Crude fiber	%			34.4	
Ether extract	%			2.3	
N-free extract	%			51.1	
Protein (N x 6.25)	%			5.9	
Cattle	dig prot %	*		2.9	
Goats	dig prot %	*		2.1	
Horses	dig prot %	*		2.5	
Rabbits	dig coef %	*		3.2	
Sheep	dig prot %	*		2.5	
Energy	GE Mcal/kg				
Cattle	DE Mcal/kg	*		2.85	
Sheep	DE Mcal/kg	*		2.99	
Cattle	ME Mcal/kg	*		2.34	
Sheep	ME Mcal/kg	*		2.45	
Cattle	TDN %	*		64.7	
Sheep	TDN %	*		67.8	
Calcium	%			0.40	
Phosphorus	%			0.21	

WILLOW. Salix spp

Willow, leaves, s-c, (1)

Ref No 1 05 471 United States

Feed Name or Analyses			Mean As Fed	Mean Dry	C.V. ± %
Dry matter	%		79.6	100.0	
Ash	%		4.3	5.4	
Crude fiber	%		25.2	31.7	
Sheep	dig coef %		38.	38.	
Ether extract	%		1.8	2.3	
Sheep	dig coef %		41.	41.	
N-free extract	%		38.3	48.1	
Sheep	dig coef %		61.	61.	
Protein (N x 6.25)	%		10.0	12.5	
Sheep	dig coef %		39.	39.	
Cattle	dig prot %	*	6.2	7.8	
Goats	dig prot %	*	6.5	8.2	
Horses	dig prot %	*	6.5	8.1	
Rabbits	dig coef %	*	6.6	8.3	
Sheep	dig prot %		3.9	4.9	
Energy	GE Mcal/kg				
Cattle	DE Mcal/kg	*	1.69	2.12	
Sheep	DE Mcal/kg	*	1.70	2.13	
Cattle	ME Mcal/kg	*	1.38	1.74	
Sheep	ME Mcal/kg	*	1.39	1.75	
Cattle	TDN %	*	38.3	48.1	
Sheep	TDN %		38.5	48.4	

Willow, browse, fresh, (2)

Ref No 2 05 472 United States

Feed Name or Analyses			Mean As Fed	Mean Dry	C.V. ± %
Dry matter	%		41.0	100.0	
Ash	%		3.0	7.4	
Crude fiber	%		11.2	27.2	
Sheep	dig coef %		42.	42.	
Ether extract	%		2.0	4.9	
Sheep	dig coef %		70.	70.	
N-free extract	%		20.8	50.7	
Sheep	dig coef %		66.	66.	
Protein (N x 6.25)	%		4.0	9.8	
Sheep	dig coef %		29.	29.	
Cattle	dig prot %	*	2.6	6.2	
Goats	dig prot %	*	2.3	5.7	
Horses	dig prot %	*	2.4	5.9	
Rabbits	dig coef %	*	2.6	6.2	
Sheep	dig prot %		1.2	2.8	
Energy	GE Mcal/kg				
Cattle	DE Mcal/kg	*	1.02	2.48	
Sheep	DE Mcal/kg	*	1.00	2.44	
Cattle	ME Mcal/kg	*	0.83	2.03	
Sheep	ME Mcal/kg	*	0.82	2.00	
Cattle	TDN %	*	23.1	56.3	
Sheep	TDN %		22.7	55.4	

WILLOW, YELLOW. Salix lutea

Willow, yellow, browse, (2)

Ref No 2 05 475 United States

Feed Name or Analyses			Mean As Fed	Mean Dry	C.V. ± %
Dry matter	%			100.0	
Ash	%			6.8	7
Crude fiber	%			14.0	1
Ether extract	%			2.9	12
N-free extract	%			61.9	
Protein (N x 6.25)	%			14.4	12
Cattle	dig prot %	*		10.1	
Goats	dig prot %	*		10.0	
Horses	dig prot %	*		9.8	
Rabbits	dig coef %	*		9.8	
Sheep	dig prot %	*		10.4	

Willow, yellow, browse, immature, (2)

Ref No 2 05 473 United States

Feed Name or Analyses			Mean As Fed	Mean Dry	C.V. ± %
Dry matter	%			100.0	
Calcium	%			1.68	
Magnesium	%			0.34	
Phosphorus	%			0.30	
Sulphur	%			0.27	

Willow, yellow, browse, full bloom, (2)

Ref No 2 05 474 United States

Feed Name or Analyses			Mean As Fed	Mean Dry	C.V. ± %
Dry matter	%			100.0	
Calcium	%			0.25	
Magnesium	%			0.38	
Phosphorus	%			0.24	
Sulphur	%			0.18	

WINDMILLGRASS. Chloris spp

Windmillgrass, aerial part, fresh, (2)

Ref No 2 05 479 United States

Feed Name or Analyses			Mean As Fed	Mean Dry	C.V. ± %
Dry matter	%			100.0	
Ash	%			15.5	23
Crude fiber	%			29.2	6
Ether extract	%			2.1	13
N-free extract	%			44.1	
Protein (N x 6.25)	%			9.1	25
Cattle	dig prot %	*		5.6	
Goats	dig prot %	*		5.1	
Horses	dig prot %	*		5.3	
Rabbits	dig coef %	*		5.7	
Sheep	dig prot %	*		5.5	

Feed Name or Analyses		Mean As Fed	Mean Dry	C.V. ± %

Windmillgrass, aerial part, fresh, immature, (2)

Ref No 2 05 476 United States

Feed Name or Analyses		As Fed	Dry	C.V. ± %
Dry matter	%		100.0	
Ash	%		14.7	14
Crude fiber	%		28.3	4
Ether extract	%		2.4	7
N-free extract	%		42.2	
Protein (N x 6.25)	%		12.4	8
Cattle	dig prot % *		8.4	
Goats	dig prot % *		8.1	
Horses	dig prot % *		8.1	
Rabbits	dig coef % *		8.2	
Sheep	dig prot % *		8.5	
Calcium	%		0.68	8
Magnesium	%		0.28	
Phosphorus	%		0.22	15
Potassium	%		3.93	

Windmillgrass, aerial part, fresh, full bloom, (2)

Ref No 2 05 477 United States

Feed Name or Analyses		As Fed	Dry	C.V. ± %
Dry matter	%		100.0	
Ash	%		17.4	13
Crude fiber	%		30.0	1
Ether extract	%		1.6	5
N-free extract	%		43.4	
Protein (N x 6.25)	%		7.6	14
Cattle	dig prot % *		4.4	
Goats	dig prot % *		3.7	
Horses	dig prot % *		4.0	
Rabbits	dig coef % *		4.5	
Sheep	dig prot % *		4.1	

Windmillgrass, aerial part, fresh, mature, (2)

Ref No 2 05 478 United States

Feed Name or Analyses		As Fed	Dry	C.V. ± %
Dry matter	%		100.0	
Ash	%		17.7	32
Crude fiber	%		31.2	4
Ether extract	%		1.7	16
N-free extract	%		43.3	
Protein (N x 6.25)	%		6.1	12
Cattle	dig prot % *		3.1	
Goats	dig prot % *		2.3	
Horses	dig prot % *		2.7	
Rabbits	dig coef % *		3.4	
Sheep	dig prot % *		2.7	

WINDMILLGRASS, HOODED. Chloris cucullata

Windmillgrass, hooded, aerial part, fresh, (2)

Ref No 2 05 483 United States

Feed Name or Analyses		As Fed	Dry	C.V. ± %
Dry matter	%		100.0	
Ash	%		16.2	13
Crude fiber	%		28.6	2
Ether extract	%		2.2	14
N-free extract	%		43.1	
Protein (N x 6.25)	%		9.9	21
Cattle	dig prot % *		6.3	
Goats	dig prot % *		5.8	
Horses	dig prot % *		5.9	
Rabbits	dig coef % *		6.3	
Sheep	dig prot % *		6.2	

Windmillgrass, hooded, aerial part, fresh, immature, (2)

Ref No 2 05 480 United States

Feed Name or Analyses		As Fed	Dry	C.V. ± %
Dry matter	%		100.0	
Ash	%		14.5	
Crude fiber	%		28.1	
Ether extract	%		2.4	
N-free extract	%		43.1	
Protein (N x 6.25)	%		11.9	
Cattle	dig prot % *		8.0	
Goats	dig prot % *		7.7	
Horses	dig prot % *		7.6	
Rabbits	dig coef % *		7.9	
Sheep	dig prot % *		8.1	
Calcium	%		0.69	10
Phosphorus	%		0.23	15

Windmillgrass, hooded, aerial part, fresh, full bloom, (2)

Ref No 2 05 481 United States

Feed Name or Analyses		As Fed	Dry	C.V. ± %
Dry matter	%		100.0	
Calcium	%		0.50	
Phosphorus	%		0.17	

Windmillgrass, hooded, aerial part, fresh, mature, (2)

Ref No 2 05 482 United States

Feed Name or Analyses		As Fed	Dry	C.V. ± %
Dry matter	%		100.0	
Calcium	%		0.48	
Phosphorus	%		0.16	

WINDMILLGRASS, TUMBLE. Chloris verticillata

Windmillgrass, tumble, aerial part, fresh, (2)

Ref No 2 05 488 United States

Feed Name or Analyses		As Fed	Dry	C.V. ± %
Dry matter	%		100.0	
Ash	%		15.8	25
Crude fiber	%		29.2	7
Ether extract	%		2.0	13
N-free extract	%		44.8	
Protein (N x 6.25)	%		8.2	23
Cattle	dig prot % *		4.9	
Goats	dig prot % *		4.2	
Horses	dig prot % *		4.5	
Rabbits	dig coef % *		5.0	
Sheep	dig prot % *		4.6	
Calcium	%		0.64	25
Phosphorus	%		0.18	20

Windmillgrass, tumble, aerial part, fresh, immature, (2)

Ref No 2 05 484 United States

Feed Name or Analyses		As Fed	Dry	C.V. ± %
Dry matter	%		100.0	
Ash	%		16.6	
Crude fiber	%		28.0	
Ether extract	%		2.3	
N-free extract	%		40.7	
Protein (N x 6.25)	%		12.4	
Cattle	dig prot % *		8.4	
Goats	dig prot % *		8.1	
Horses	dig prot % *		8.1	
Rabbits	dig coef % *		8.2	
Sheep	dig prot % *		8.5	
Calcium	%		0.68	
Phosphorus	%		0.20	

Windmillgrass, tumble, aerial part, fresh, mid-bloom, (2)

Ref No 2 05 485 United States

Feed Name or Analyses		As Fed	Dry	C.V. ± %
Dry matter	%		100.0	
Calcium	%		0.58	11
Phosphorus	%		0.20	13

Windmillgrass, tumble, aerial part, fresh, full bloom, (2)

Ref No 2 05 486 United States

Feed Name or Analyses		As Fed	Dry	C.V. ± %
Dry matter	%		100.0	
Ash	%		15.3	
Crude fiber	%		30.2	
Ether extract	%		1.7	
N-free extract	%		44.2	
Protein (N x 6.25)	%		8.6	
Cattle	dig prot % *		5.2	
Goats	dig prot % *		4.6	
Horses	dig prot % *		4.8	
Rabbits	dig coef % *		5.3	
Sheep	dig prot % *		5.0	
Calcium	%		0.52	
Phosphorus	%		0.16	

Windmillgrass, tumble, aerial part, fresh, mature, (2)

Ref No 2 05 487 United States

Feed Name or Analyses		As Fed	Dry	C.V. ± %
Dry matter	%		100.0	
Ash	%		18.0	34
Crude fiber	%		31.1	5
Ether extract	%		1.6	19
N-free extract	%		43.5	
Protein (N x 6.25)	%		5.8	2
Cattle	dig prot % *		2.8	
Goats	dig prot % *		2.0	
Horses	dig prot % *		2.5	
Rabbits	dig coef % *		3.1	
Sheep	dig prot % *		2.4	
Calcium	%		0.31	
Phosphorus	%		0.12	

WINTERFAT, COMMON. Eurotia lanata

Winterfat, common, aerial part, s-c, (1)

Ref No 1 08 559 United States

Feed Name or Analyses		As Fed	Dry	C.V. ± %
Dry matter	%	92.6	100.0	
Ash	%	9.6	10.4	
Crude fiber	%	27.4	29.6	
Ether extract	%	1.9	2.1	
N-free extract	%	40.8	44.1	
Protein (N x 6.25)	%	12.9	13.9	
Cattle	dig prot % *	8.3	9.0	
Goats	dig prot % *	8.9	9.6	

(1) dry forages and roughages (2) pasture, range plants, and forages fed green (3) silages (4) energy feeds (5) protein supplements (6) minerals (7) vitamins (8) additives

Feed Name or Analyses			Mean As Fed	Mean Dry	C.V. ± %
Horses	dig prot %	*	8.7	9.4	
Rabbits	dig coef %	*	8.7	9.4	
Sheep	dig prot %	*	8.4	9.1	
Energy	GE Mcal/kg				
Cattle	DE Mcal/kg	*	2.21	2.39	
Sheep	DE Mcal/kg	*	1.98	2.14	
Cattle	ME Mcal/kg	*	1.81	1.96	
Sheep	ME Mcal/kg	*	1.62	1.75	
Cattle	TDN %	*	50.3	54.3	
Sheep	TDN %	*	45.0	48.6	

Winterfat, common, aerial part, fresh, (2)
Ref No 2 05 490 United States

Feed Name or Analyses			Mean As Fed	Mean Dry	C.V. ± %
Dry matter	%		65.0	100.0	
Ash	%		6.0	9.2	
Crude fiber	%		23.7	36.5	
Ether extract	%		1.3	2.0	
N-free extract	%		28.5	43.8	
Protein (N x 6.25)	%		5.5	8.5	
Cattle	dig prot %	*	3.3	5.1	
Goats	dig prot %	*	2.9	4.5	
Horses	dig prot %	*	3.1	4.7	
Rabbits	dig coef %	*	3.4	5.2	
Sheep	dig prot %	*	3.2	4.9	
Energy	GE Mcal/kg				
Cattle	DE Mcal/kg	*	1.55	2.39	
Sheep	DE Mcal/kg	*	1.51	2.33	
Cattle	ME Mcal/kg	*	1.27	1.96	
Sheep	ME Mcal/kg	*	1.24	1.91	
Cattle	TDN %	*	35.2	54.1	
Sheep	TDN %	*	43.3	52.8	
Calcium	%		1.04	1.60	
Phosphorus	%		0.08	0.12	
Sulphur	%		0.11	0.17	
Carotene	mg/kg		10.9	16.8	19
Vitamin A equivalent	IU/g		18.2	27.9	

Winterfat, common, aerial part, fresh, immature, (2)
Ref No 2 05 489 United States

Feed Name or Analyses			Mean As Fed	Mean Dry	C.V. ± %
Dry matter	%			100.0	
Ash	%			8.8	
Crude fiber	%			26.0	
Ether extract	%			1.6	
N-free extract	%			42.6	
Protein (N x 6.25)	%			21.0	
Cattle	dig prot %	*		15.7	
Goats	dig prot %	*		16.2	
Horses	dig prot %	*		15.4	
Rabbits	dig coef %	*		14.9	
Sheep	dig prot %	*		16.6	
Energy	GE Mcal/kg				
Cattle	DE Mcal/kg	*		2.35	
Sheep	DE Mcal/kg	*		2.28	
Cattle	ME Mcal/kg	*		1.93	
Sheep	ME Mcal/kg	*		1.87	
Cattle	TDN %	*		53.2	
Sheep	TDN %	*		51.7	

Winterfat, common, aerial part, fresh, mature, (2)
Whitesage
Ref No 2 08 828 United States

Feed Name or Analyses			Mean As Fed	Mean Dry	C.V. ± %
Dry matter	%			100.0	
Crude fiber	%			28.0	
Ether extract	%			3.0	

Feed Name or Analyses			Mean As Fed	Mean Dry	C.V. ± %
N-free extract	%			38.0	
Protein (N x 6.25)	%			11.0	
Cattle	dig prot %	*		7.2	
Goats	dig prot %	*		6.8	
Horses	dig prot %	*		6.9	
Rabbits	dig coef %	*		7.2	
Sheep	dig prot %	*		7.2	
Energy	GE Mcal/kg				
Cattle	DE Mcal/kg	*		2.41	
Sheep	DE Mcal/kg	*		2.30	
Cattle	ME Mcal/kg	*		1.98	
Sheep	ME Mcal/kg	*		1.89	
Cattle	TDN %	*		54.6	
Sheep	TDN %	*		52.1	

Wiregrass –
see Threeawn, aerial part, fresh, (2)

Wiregrass hay –
see Threeawn, hay, s-c, (1)

WITCHGRASS. Panicum spp

Witchgrass, aerial part, dehy, (1)
Ref No 1 05 491 United States

Feed Name or Analyses			Mean As Fed	Mean Dry	C.V. ± %
Dry matter	%		93.4	100.0	
Crude fiber	%		34.5	36.9	
Ether extract	%		2.7	2.9	
Protein (N x 6.25)	%		7.2	7.7	
Cattle	dig prot %	*	3.4	3.6	
Goats	dig prot %	*	3.5	3.7	
Horses	dig prot %	*	3.8	4.1	
Rabbits	dig coef %	*	4.3	4.6	
Sheep	dig prot %	*	3.2	3.5	
Carotene	mg/kg		98.0	104.9	
Vitamin A equivalent	IU/g		163.4	174.9	

Witchgrass, hay, s-c, (1)
Ref No 1 05 492 United States

Feed Name or Analyses			Mean As Fed	Mean Dry	C.V. ± %
Dry matter	%		93.4	100.0	
Ash	%		5.1	5.5	5
Crude fiber	%		33.3	35.7	3
Ether extract	%		2.7	2.9	8
N-free extract	%		45.4	48.6	
Protein (N x 6.25)	%		6.8	7.3	9
Cattle	dig prot %	*	3.0	3.3	
Goats	dig prot %	*	3.1	3.4	
Horses	dig prot %	*	3.5	3.7	
Rabbits	dig coef %	*	4.0	4.3	
Sheep	dig prot %	*	2.9	3.1	
Calcium	%		0.26	0.28	6
Phosphorus	%		0.21	0.22	14
Carotene	mg/kg		98.0	104.9	7
Riboflavin	mg/kg		6.8	7.3	15
Thiamine	mg/kg		2.5	2.6	12
Vitamin A equivalent	IU/g		163.4	174.9	

Witchgrass, aerial part, fresh, (2)
Ref No 2 05 494 United States

Feed Name or Analyses			Mean As Fed	Mean Dry	C.V. ± %
Dry matter	%			100.0	
Ash	%			9.8	26
Crude fiber	%			34.5	14
Ether extract	%			2.1	14
N-free extract	%			43.1	

Feed Name or Analyses			Mean As Fed	Mean Dry	C.V. ± %
Protein (N x 6.25)	%			10.5	36
Cattle	dig prot %	*		6.8	
Goats	dig prot %	*		6.4	
Horses	dig prot %	*		6.4	
Rabbits	dig coef %	*		6.8	
Sheep	dig prot %	*		6.8	
Calcium	%			0.58	27
Magnesium	%			0.32	
Phosphorus	%			0.21	46
Potassium	%			2.22	

Witchgrass, aerial part, fresh, immature, (2)
Ref No 2 05 493 United States

Feed Name or Analyses			Mean As Fed	Mean Dry	C.V. ± %
Dry matter	%			100.0	
Ash	%			11.5	
Crude fiber	%			31.3	
Ether extract	%			2.1	
N-free extract	%			43.2	
Protein (N x 6.25)	%			11.9	
Cattle	dig prot %	*		8.0	
Goats	dig prot %	*		7.7	
Horses	dig prot %	*		7.6	
Rabbits	dig coef %	*		7.9	
Sheep	dig prot %	*		8.1	

WITCHGRASS, COMMON. Panicum capillare

Witchgrass, common, aerial part, fresh, (2)
Ref No 2 05 496 United States

Feed Name or Analyses			Mean As Fed	Mean Dry	C.V. ± %
Dry matter	%			100.0	
Calcium	%			0.63	
Phosphorus	%			0.27	

Witchgrass, common, aerial part, fresh, immature, (2)
Ref No 2 05 495 United States

Feed Name or Analyses			Mean As Fed	Mean Dry	C.V. ± %
Dry matter	%			100.0	
Ash	%			13.2	
Crude fiber	%			27.8	
Ether extract	%			1.9	
N-free extract	%			40.5	
Protein (N x 6.25)	%			16.6	
Cattle	dig prot %	*		12.0	
Goats	dig prot %	*		12.0	
Horses	dig prot %	*		11.6	
Rabbits	dig coef %	*		11.5	
Sheep	dig prot %	*		12.5	
Calcium	%			0.84	
Phosphorus	%			0.36	

WITCHGRASS, CUSHION. Panicum capillare occidentale

Witchgrass, cushion, aerial part, fresh, (2)
Ref No 2 05 497 United States

Feed Name or Analyses			Mean As Fed	Mean Dry	C.V. ± %
Dry matter	%			100.0	
Ash	%			10.6	
Crude fiber	%			33.1	
Ether extract	%			2.2	
N-free extract	%			44.5	
Protein (N x 6.25)	%			9.6	
Cattle	dig prot %	*		6.1	

Continued

Feed Name or Analyses			Mean As Fed	Mean Dry	C.V. ± %
Goats	dig prot	% *		5.5	
Horses	dig prot	% *		5.7	
Rabbits	dig coef	% *		6.1	
Sheep	dig prot	% *		5.9	
Calcium		%		0.61	
Phosphorus		%		0.22	

WITCHGRASS, FALL. Leptoloma cognatum

Witchgrass, fall, aerial part, fresh, (2)
Ref No 2 05 500 United States

Feed Name or Analyses			Mean As Fed	Mean Dry	C.V. ± %
Dry matter		%		100.0	
Ash		%		8.1	
Crude fiber		%		37.1	
Ether extract		%		2.3	
N-free extract		%		42.8	
Protein (N x 6.25)		%		9.7	
Cattle	dig prot	% *		6.1	
Goats	dig prot	% *		5.6	
Horses	dig prot	% *		5.8	
Rabbits	dig coef	% *		6.2	
Sheep	dig prot	% *		6.0	
Calcium		%		0.54	
Magnesium		%		0.33	
Phosphorus		%		0.15	
Potassium		%		2.49	

Witchgrass, fall, aerial part, fresh, full bloom, (2)
Ref No 2 05 498 United States

Feed Name or Analyses			Mean As Fed	Mean Dry	C.V. ± %
Dry matter		%		100.0	
Ash		%		9.9	
Crude fiber		%		36.6	
Ether extract		%		1.9	
N-free extract		%		39.3	
Protein (N x 6.25)		%		12.3	
Cattle	dig prot	% *		8.3	
Goats	dig prot	% *		8.0	
Horses	dig prot	% *		8.0	
Rabbits	dig coef	% *		8.2	
Sheep	dig prot	% *		8.5	
Calcium		%		0.58	
Magnesium		%		0.37	
Phosphorus		%		0.19	
Potassium		%		1.75	

Witchgrass, fall, aerial part, fresh, mature, (2)
Ref No 2 05 499 United States

Feed Name or Analyses			Mean As Fed	Mean Dry	C.V. ± %
Dry matter		%		100.0	
Ash		%		6.3	
Crude fiber		%		37.5	
Ether extract		%		2.6	
N-free extract		%		46.5	
Protein (N x 6.25)		%		7.1	
Cattle	dig prot	% *		3.9	
Goats	dig prot	% *		3.2	
Horses	dig prot	% *		3.6	
Rabbits	dig coef	% *		4.2	
Sheep	dig prot	% *		3.6	

(1) dry forages and roughages (2) pasture, range plants, and forages fed green (3) silages (4) energy feeds (5) protein supplements (6) minerals (7) vitamins (8) additives

WITCHGRASS, ROUGHSTALK. Panicum hirticaule

Witchgrass, roughstalk, aerial part, fresh, (2)
Ref No 2 05 501 United States

Feed Name or Analyses			Mean As Fed	Mean Dry	C.V. ± %
Dry matter		%		100.0	
Ash		%		11.0	
Crude fiber		%		40.8	
Ether extract		%		1.8	
N-free extract		%		34.5	
Protein (N x 6.25)		%		11.9	
Cattle	dig prot	% *		8.0	
Goats	dig prot	% *		7.7	
Horses	dig prot	% *		7.6	
Rabbits	dig coef	% *		7.9	
Sheep	dig prot	% *		8.1	
Calcium		%		0.48	
Magnesium		%		0.30	
Phosphorus		%		0.18	
Potassium		%		1.69	

WOLFTAIL. Lycurus phleoides

Wolftail, aerial part wo lower stems, fresh, early leaf, (2)
Ref No 2 08 681 United States

Feed Name or Analyses			Mean As Fed	Mean Dry	C.V. ± %
Dry matter		%		100.0	
Organic matter		%		91.1	1
Ash		%		8.9	15
Crude fiber		%		32.3	3
Ether extract		%		1.7	25
N-free extract		%		52.0	2
Protein (N x 6.25)		%		5.2	41
Cattle	dig prot	% *		2.3	
Goats	dig prot	% *		1.4	
Horses	dig prot	% *		1.9	
Rabbits	dig coef	% *		2.7	
Sheep	dig prot	% *		1.8	
Calcium		%		0.23	20
Phosphorus		%		0.09	34
Carotene		mg/kg		52.8	
Vitamin A equivalent		IU/g		88.0	

Wolftail, aerial part wo lower stems, fresh, mid-bloom, (2)
Ref No 2 08 682 United States

Feed Name or Analyses			Mean As Fed	Mean Dry	C.V. ± %
Dry matter		%		100.0	
Organic matter		%		90.8	
Ash		%		9.2	
Crude fiber		%		32.3	
Ether extract		%		2.0	
N-free extract		%		50.3	
Protein (N x 6.25)		%		6.3	
Cattle	dig prot	% *		3.2	
Goats	dig prot	% *		2.4	
Horses	dig prot	% *		2.8	
Rabbits	dig coef	% *		3.5	
Sheep	dig prot	% *		2.8	
Calcium		%		0.32	
Phosphorus		%		0.13	

Wolftail, aerial part wo lower stems, fresh, mature, (2)
Ref No 2 08 683 United States

Feed Name or Analyses			Mean As Fed	Mean Dry	C.V. ± %
Dry matter		%		100.0	
Organic matter		%		88.4	
Ash		%		11.6	
Crude fiber		%		32.1	
Ether extract		%		1.9	
N-free extract		%		49.5	
Protein (N x 6.25)		%		4.9	
Cattle	dig prot	% *		2.1	
Goats	dig prot	% *		1.1	
Horses	dig prot	% *		1.7	
Rabbits	dig coef	% *		2.5	
Sheep	dig prot	% *		1.6	
Calcium		%		0.26	
Phosphorus		%		0.13	

Wolftail, aerial part wo lower stems, fresh, dormant, (2)
Ref No 2 08 680 United States

Feed Name or Analyses			Mean As Fed	Mean Dry	C.V. ± %
Dry matter		%		100.0	
Organic matter		%		92.3	2
Ash		%		7.7	18
Crude fiber		%		33.7	4
Ether extract		%		2.0	13
N-free extract		%		52.8	3
Protein (N x 6.25)		%		3.8	20
Cattle	dig prot	% *		1.1	
Goats	dig prot	% *		0.1	
Horses	dig prot	% *		0.7	
Rabbits	dig coef	% *		1.6	
Sheep	dig prot	% *		0.5	
Calcium		%		0.18	27
Phosphorus		%		0.09	27
Carotene		mg/kg		9.5	98
Vitamin A equivalent		IU/g		15.8	

WOOD. Scientific name not used

Wood, pulp, (1)
Ref No 1 07 714 United States

Feed Name or Analyses			Mean As Fed	Mean Dry	C.V. ± %
Dry matter		%		100.0	
Organic matter		%		100.0	
Crude fiber		%		84.0	
N-free extract		%		16.0	

Wood, molasses, (4)
Ref No 4 05 502 United States

Feed Name or Analyses			Mean As Fed	Mean Dry	C.V. ± %
Dry matter		%	62.4	100.0	8
Ash		%	3.1	5.0	26
Crude fiber		%	0.0	0.0	26
Ether extract		%	0.1	0.2	98
N-free extract		%	58.5	93.8	
Protein (N x 6.25)		%	0.6	1.0	65
Cattle	dig prot	% *	-1.8	-3.0	
Goats	dig prot	% *	-1.1	-1.8	
Horses	dig prot	% *	-1.1	-1.8	
Sheep	dig prot	% *	-1.1	-1.8	
Energy	GE Mcal/kg		2.49	4.00	1
Cattle	DE Mcal/kg	*	2.35	3.77	
Sheep	DE Mcal/kg	*	2.43	3.90	

Feed Name or Analyses		Mean As Fed	Mean Dry	C.V. ± %
Swine	DE kcal/kg	*2329.	3735.	
Cattle	ME Mcal/kg	* 1.93	3.09	
Sheep	ME Mcal/kg	* 1.99	3.20	
Swine	ME kcal/kg	*2231.	3579.	
Cattle	TDN %	* 53.4	85.6	
Sheep	TDN %	* 55.1	88.4	
Swine	TDN %	* 52.8	84.7	
Calcium	%	1.45	2.33	19
Chlorine	%	0.12	0.20	
Magnesium	%	0.07	0.11	
Manganese	mg/kg	12.6	20.3	
Phosphorus	%	0.03	0.05	98
Potassium	%	0.04	0.06	
Sodium	%	0.03	0.05	
Sulphur	%	0.03	0.05	

WOODWAXEN, BRIDALVEIL. Genista monosperma

Woodwaxen, bridalveil, seeds w hulls, (5)

Ref No 5 09 060 United States

Feed Name or Analyses		Mean As Fed	Mean Dry	C.V. ± %
Dry matter	%		100.0	
Protein (N x 6.25)	%		53.0	
Alanine	%		1.64	
Arginine	%		4.51	
Aspartic acid	%		4.77	
Glutamic acid	%		9.28	
Glycine	%		1.80	
Histidine	%		1.27	
Hydroxyproline	%		0.11	
Isoleucine	%		1.86	
Leucine	%		3.39	
Lysine	%		2.39	
Methionine	%		0.37	
Phenylalanine	%		1.75	
Proline	%		2.07	
Serine	%		2.01	
Threonine	%		1.54	
Tyrosine	%		1.70	
Valine	%		1.86	

WORMWOOD, MUGWORT. Artemisia vulgaris

Wormwood, mugwort, aerial part, (2)

Ref No 2 05 503 United States

Feed Name or Analyses		Mean As Fed	Mean Dry	C.V. ± %
Dry matter	%		100.0	
Ash	%		6.3	
Crude fiber	%		33.5	
Ether extract	%		6.0	
N-free extract	%		42.8	
Protein (N x 6.25)	%		11.4	
Cattle	dig prot %	*	7.6	
Goats	dig prot %	*	7.2	
Horses	dig prot %	*	7.2	
Rabbits	dig coef %	*	7.5	
Sheep	dig prot %	*	7.6	
Calcium	%		0.86	
Magnesium	%		0.27	
Phosphorus	%		0.19	
Potassium	%		2.20	

WYETHIA, WOOLY. Wyethia mollis

Wyethia, wooly, aerial part, fresh, over ripe, (2)

Ref No 2 05 504 United States

Feed Name or Analyses		Mean As Fed	Mean Dry	C.V. ± %
Dry matter	%	93.2	100.0	
Ash	%	10.2	10.9	
Crude fiber	%	15.9	17.1	
Sheep	dig coef %	54.	54.	
Ether extract	%	3.9	4.2	
Sheep	dig coef %	63.	63.	
N-free extract	%	47.0	50.4	
Sheep	dig coef %	61.	61.	
Protein (N x 6.25)	%	16.2	17.4	
Sheep	dig coef %	70.	70.	
Cattle	dig prot %	* 11.8	12.7	
Goats	dig prot %	* 11.9	12.8	
Horses	dig prot %	* 11.5	12.3	
Rabbits	dig coef %	* 11.3	12.1	
Sheep	dig prot %	11.4	12.2	
Energy	GE Mcal/kg			
Cattle	DE Mcal/kg	* 2.47	2.65	
Sheep	DE Mcal/kg	* 2.39	2.56	
Cattle	ME Mcal/kg	* 2.02	2.17	
Sheep	ME Mcal/kg	* 1.96	2.10	
Cattle	TDN %	* 56.0	60.1	
Sheep	TDN %	54.2	58.1	

YAMPA. Perideridia gairdneri

Yampa, aerial part, fresh, over ripe, (2)

Ref No 2 05 505 United States

Feed Name or Analyses		Mean As Fed	Mean Dry	C.V. ± %
Dry matter	%	93.2	100.0	
Ash	%	9.0	9.7	
Crude fiber	%	25.7	27.6	
Sheep	dig coef %	74.	74.	
Ether extract	%	4.8	5.1	
Sheep	dig coef %	77.	77.	
N-free extract	%	46.5	49.9	
Sheep	dig coef %	65.	65.	
Protein (N x 6.25)	%	7.2	7.7	
Sheep	dig coef %	57.	57.	
Cattle	dig prot %	* 4.1	4.4	
Goats	dig prot %	* 3.5	3.7	
Horses	dig prot %	* 3.8	4.1	
Rabbits	dig coef %	* 4.3	4.6	
Sheep	dig prot %	4.1	4.4	
Energy	GE Mcal/kg			
Cattle	DE Mcal/kg	* 2.77	2.97	
Sheep	DE Mcal/kg	* 2.72	2.91	
Cattle	ME Mcal/kg	* 2.27	2.44	
Sheep	ME Mcal/kg	* 2.23	2.39	
Cattle	TDN %	* 62.7	67.3	
Sheep	TDN %	61.6	66.1	

YARROW. Achillea spp

Yarrow, aerial part, fresh, (2)

Ref No 2 05 508 United States

Feed Name or Analyses		Mean As Fed	Mean Dry	C.V. ± %
Dry matter	%	44.6	100.0	11
Ash	%	4.8	10.8	17
Crude fiber	%	10.3	23.0	29
Ether extract	%	1.6	3.5	24
N-free extract	%	22.2	49.7	
Protein (N x 6.25)	%	5.8	13.0	27
Cattle	dig prot %	* 4.0	8.9	
Goats	dig prot %	* 3.9	8.7	
Horses	dig prot %	* 3.8	8.6	
Rabbits	dig coef %	* 3.9	8.7	
Sheep	dig prot %	* 4.1	9.1	

Yarrow, aerial part, fresh, immature, (2)

Ref No 2 05 506 United States

Feed Name or Analyses		Mean As Fed	Mean Dry	C.V. ± %
Dry matter	%		100.0	
Ash	%		11.9	10
Crude fiber	%		17.1	10
Ether extract	%		2.9	18
N-free extract	%		51.2	
Protein (N x 6.25)	%		16.9	13
Cattle	dig prot %	*	12.3	
Goats	dig prot %	*	12.3	
Horses	dig prot %	*	11.9	
Rabbits	dig coef %	*	11.7	
Sheep	dig prot %	*	12.7	

Yarrow, aerial part, fresh, full bloom, (2)

Ref No 2 05 507 United States

Feed Name or Analyses		Mean As Fed	Mean Dry	C.V. ± %
Dry matter	%		100.0	
Ash	%		10.9	4
Crude fiber	%		23.5	13
Ether extract	%		4.4	26
N-free extract	%		49.3	
Protein (N x 6.25)	%		11.9	5
Cattle	dig prot %	*	8.0	
Goats	dig prot %	*	7.7	
Horses	dig prot %	*	7.6	
Rabbits	dig coef %	*	7.9	
Sheep	dig prot %	*	8.1	

YARROW, COMMON. Achillea millefolium

Yarrow, common, aerial part, fresh, (2)

Ref No 2 05 511 United States

Feed Name or Analyses		Mean As Fed	Mean Dry	C.V. ± %
Dry matter	%		100.0	
Ash	%		12.5	14
Crude fiber	%		20.1	34
Ether extract	%		1.8	29
N-free extract	%		51.2	
Protein (N x 6.25)	%		14.4	40
Cattle	dig prot %	*	10.1	
Goats	dig prot %	*	10.0	
Horses	dig prot %	*	9.8	
Rabbits	dig coef %	*	9.8	
Sheep	dig prot %	*	10.4	

Yarrow, common, aerial part, fresh, immature, (2)

Ref No 2 05 509 United States

Feed Name or Analyses		Mean As Fed	Mean Dry	C.V. ± %
Dry matter	%		100.0	
Ash	%		14.1	
Crude fiber	%		14.7	
Ether extract	%		2.1	
N-free extract	%		49.9	
Protein (N x 6.25)	%		19.2	
Cattle	dig prot %	*	14.2	
Goats	dig prot %	*	14.5	

Continued

Feed Name or Analyses				Mean As Fed	Mean Dry	C.V. ± %
Horses	dig prot	%	*		13.8	
Rabbits	dig coef	%	*		13.5	
Sheep	dig prot	%	*		14.9	

Yarrow, common, aerial part, fresh, full bloom, (2)

Ref No 2 05 510 United States

Analyses				As Fed	Dry	C.V.
Dry matter		%			100.0	
Ash		%			10.9	
Crude fiber		%			23.8	
Ether extract		%			1.7	
N-free extract		%			50.8	
Protein (N x 6.25)		%			12.8	
Cattle	dig prot	%	*		8.8	
Goats	dig prot	%	*		8.5	
Horses	dig prot	%	*		8.4	
Rabbits	dig coef	%	*		8.6	
Sheep	dig prot	%	*		8.9	

YARROW, WESTERN. Achillea lanulosa

Yarrow, western, aerial part, fresh, immature, (2)

Ref No 2 05 512 United States

Analyses				As Fed	Dry	C.V.
Dry matter		%			100.0	
Ash		%			11.4	
Crude fiber		%			17.6	
Ether extract		%			3.0	
N-free extract		%			51.5	
Protein (N x 6.25)		%			16.5	
Cattle	dig prot	%	*		11.9	
Goats	dig prot	%	*		12.0	
Horses	dig prot	%	*		11.5	
Rabbits	dig coef	%	*		11.4	
Sheep	dig prot	%	*		12.4	
Sulphur		%			0.21	

Yarrow, western, aerial part, fresh, full bloom, (2)

Ref No 2 05 513 United States

Analyses				As Fed	Dry	C.V.
Dry matter		%			100.0	
Ash		%			10.9	
Crude fiber		%			23.4	
Ether extract		%			4.9	
N-free extract		%			49.1	
Protein (N x 6.25)		%			11.7	
Cattle	dig prot	%	*		7.8	
Goats	dig prot	%	*		7.5	
Horses	dig prot	%	*		7.5	
Rabbits	dig coef	%	*		7.7	
Sheep	dig prot	%	*		7.9	
Sulphur		%			0.16	

Yarrow, western, aerial part, fresh, mature, (2)

Ref No 2 05 514 United States

Analyses				As Fed	Dry	C.V.
Dry matter		%		44.6	100.0	11
Ash		%		4.6	10.4	17
Crude fiber		%		10.5	23.6	27
Ether extract		%		1.7	3.9	21
N-free extract		%		22.0	49.3	
Protein (N x 6.25)		%		5.7	12.8	21
Cattle	dig prot	%	*	3.9	8.8	
Goats	dig prot	%	*	3.8	8.5	
Horses	dig prot	%	*	3.7	8.4	
Rabbits	dig coef	%	*	3.8	8.6	
Sheep	dig prot	%	*	4.0	8.9	
Calcium		%		0.59	1.33	16
Magnesium		%		0.18	0.40	16
Phosphorus		%		0.16	0.36	16

Yarrow, western, heads, fresh, (2)

Ref No 2 05 515 United States

Analyses				As Fed	Dry	C.V.
Dry matter		%		35.7	100.0	24
Ash		%		2.7	7.6	10
Crude fiber		%		11.1	31.1	10
Ether extract		%		1.9	5.4	17
N-free extract		%		15.6	43.6	
Protein (N x 6.25)		%		4.4	12.3	17
Cattle	dig prot	%	*	3.0	8.3	
Goats	dig prot	%	*	2.9	8.0	
Horses	dig prot	%	*	2.8	8.0	
Rabbits	dig coef	%	*	2.9	8.2	
Sheep	dig prot	%	*	3.0	8.5	
Calcium		%		0.28	0.78	13
Phosphorus		%		0.14	0.39	25

Yarrow, western, leaves, fresh, (2)

Ref No 2 05 516 United States

Analyses				As Fed	Dry	C.V.
Dry matter		%		29.3	100.0	17
Ash		%		3.2	10.9	5
Crude fiber		%		6.0	20.6	4
Ether extract		%		1.6	5.4	7
N-free extract		%		14.6	49.9	
Protein (N x 6.25)		%		3.9	13.2	8
Cattle	dig prot	%	*	2.7	9.1	
Goats	dig prot	%	*	2.6	8.9	
Horses	dig prot	%	*	2.6	8.7	
Rabbits	dig coef	%	*	2.6	8.9	
Sheep	dig prot	%	*	2.7	9.3	
Calcium		%		0.39	1.33	8
Phosphorus		%		0.09	0.32	17

Yarrow, western, stems, fresh, (2)

Ref No 2 05 517 United States

Analyses				As Fed	Dry	C.V.
Dry matter		%		38.5	100.0	25
Ash		%		2.0	5.2	8
Crude fiber		%		19.4	50.5	4
Ether extract		%		0.5	1.3	11
N-free extract		%		15.2	39.5	
Protein (N x 6.25)		%		1.3	3.5	12
Cattle	dig prot	%	*	0.3	0.9	
Goats	dig prot	%	*	0.0	-0.1	
Horses	dig prot	%	*	0.2	0.5	
Rabbits	dig coef	%	*	0.5	1.4	
Sheep	dig prot	%	*	0.1	0.3	
Calcium		%		0.23	0.61	11
Phosphorus		%		0.08	0.20	19

YEAST. Saccharomyces cerevisiae

Yeast, medium w yeast, dehy, (7)
Yeast culture (AAFCO)

Ref No 7 05 520 United States

Analyses				As Fed	Dry	C.V.
Dry matter		%		91.0	100.0	1
Ash		%		3.5	3.8	28
Crude fiber		%		4.0	4.4	19
Ether extract		%		3.7	4.0	26
N-free extract		%		67.6	74.2	
Protein (N x 6.25)		%		12.3	13.5	19
Energy	GE	Mcal/kg				
Cattle	DE	Mcal/kg	*	2.92	3.21	
Sheep	DE	Mcal/kg	*	3.31	3.63	
Swine	DE	kcal/kg	*	3319.	3647.	
Cattle	ME	Mcal/kg	*	2.40	2.63	
Sheep	ME	Mcal/kg	*	2.71	2.98	
Swine	ME	kcal/kg	*	3095.	3401.	
Cattle	TDN	%	*	66.3	72.8	
Sheep	TDN	%	*	75.0	82.4	
Swine	TDN	%	*	75.3	82.7	
Calcium		%		2.53	2.78	
Iodine		mg/kg		0.271	0.298	
Phosphorus		%		0.61	0.67	
Choline		mg/kg		444.	488.	
Niacin		mg/kg		22.2	24.4	
Pantothenic acid		mg/kg		5.6	6.1	
Riboflavin		mg/kg		1.1	1.2	
Arginine		%		0.50	0.55	
Cystine		%		0.17	0.19	
Glycine		%		0.30	0.33	
Lysine		%		0.20	0.22	
Methionine		%		0.18	0.20	
Tryptophan		%		0.09	0.10	

Yeast, boiled pressed, (7)

Ref No 7 05 521 United States

Analyses				As Fed	Dry	C.V.
Dry matter		%		25.6	100.0	
Ash		%		2.8	11.0	
Ether extract		%		1.6	6.4	
Swine	dig coef	%		70.	70.	
N-free extract		%		4.9	19.2	
Swine	dig coef	%		66.	66.	
Protein (N x 6.25)		%		16.2	63.4	
Swine	dig coef	%		84.	84.	
Swine	dig prot	%		13.6	53.3	
Energy	GE	Mcal/kg				
Cattle	DE	Mcal/kg	*	0.82	3.20	
Sheep	DE	Mcal/kg	*	0.90	3.50	
Swine	DE	kcal/kg	*	858.	3351.	
Cattle	ME	Mcal/kg	*	0.67	2.62	
Sheep	ME	Mcal/kg	*	0.74	2.87	
Swine	ME	kcal/kg	*	714.	2788.	
Cattle	TDN	%	*	18.6	72.5	
Sheep	TDN	%	*	20.3	79.5	
Swine	TDN	%		19.5	76.0	

Yeast, dehy, human gr, (7)

Ref No 7 05 522 United States

Analyses				As Fed	Dry	C.V.
Dry matter		%		91.4	100.0	1
Ash		%		7.1	7.7	4
Crude fiber		%		2.5	2.7	23
Sheep	dig coef	%		-104.	-104.	
Swine	dig coef	%		3.	3.	

(1) dry forages and roughages (2) pasture, range plants, and forages fed green (3) silages (4) energy feeds (5) protein supplements (6) minerals (7) vitamins (8) additives

Feed Name or Analyses			Mean As Fed	Mean Dry	C.V. ± %
Ether extract		%	1.6	1.7	34
Cattle	dig coef	%	-88.	-88.	
Sheep	dig coef	%	7.	7.	
Swine	dig coef	%	-156.	-156.	
N-free extract		%	29.7	32.5	
Cattle	dig coef	%	89.	89.	
Sheep	dig coef	%	88.	88.	
Swine	dig coef	%	90.	90.	
Protein (N x 6.25)		%	50.7	55.4	7
Cattle	dig coef	%	92.	92.	
Sheep	dig coef	%	85.	85.	
Swine	dig coef	%	88.	88.	
Cattle	dig prot	%	46.6	51.0	
Sheep	dig prot	%	43.1	47.1	
Swine	dig prot	%	44.6	48.8	
Energy		GE Mcal/kg			
Cattle		DE Mcal/kg	* 3.30	3.61	
Sheep		DE Mcal/kg	* 2.95	3.22	
Swine		DE kcal/kg	*2903.	3175.	
Cattle		ME Mcal/kg	* 2.71	2.96	
Sheep		ME Mcal/kg	* 2.42	2.64	
Swine		ME kcal/kg	*2462.	2693.	
Cattle		TDN %	74.9	81.9	
Sheep		TDN %	66.8	73.1	
Swine		TDN %	65.8	72.0	
Calcium		%	0.14	0.15	39
Copper		mg/kg	25.0	27.4	94
Iron		%	0.006	0.006	14
Magnesium		%	0.16	0.18	24
Manganese		mg/kg	9.4	10.3	33
Phosphorus		%	1.11	1.21	2
Potassium		%	1.93	2.11	6
Sodium		%	0.08	0.08	80
Zinc		mg/kg	70.7	77.4	28
Biotin		mg/kg	0.73	0.79	17
Choline		mg/kg	2787.	3049.	3
Folic acid		mg/kg	46.51	50.87	68
Niacin		mg/kg	714.3	781.3	25
Pantothenic acid		mg/kg	195.6	213.9	6
Riboflavin		mg/kg	84.6	92.6	6
Thiamine		mg/kg	88.7	97.0	80
Vitamin B6		mg/kg	31.21	34.13	19
Arginine		%	2.23	2.44	16
Cystine		%	0.48	0.53	9
Glycine		%	1.94	2.12	11
Histidine		%	1.26	1.38	7
Isoleucine		%	1.94	2.12	11
Leucine		%	2.62	2.86	8
Lysine		%	3.10	3.39	11
Methionine		%	0.68	0.74	13
Threonine		%	2.42	2.65	5
Tryptophan		%	0.48	0.53	9
Tyrosine		%	1.65	1.80	11
Valine		%	2.23	2.44	14

Yeast, wet, (7)

Ref No 7 05 523 United States

Feed Name or Analyses			Mean As Fed	Mean Dry	C.V. ± %
Dry matter		%	19.1	100.0	
Ash		%	1.7	8.9	
Ether extract		%	0.6	3.3	
Swine	dig coef	%	30.	30.	
N-free extract		%	5.8	30.6	
Horses	dig coef	%	103.	103.	
Swine	dig coef	%	78.	78.	
Protein (N x 6.25)		%	11.2	58.9	
Horses	dig coef	%	91.	91.	
Swine	dig coef	%	89.	89.	
Horses	dig prot	%	10.2	53.6	
Swine	dig prot	%	10.0	52.4	
Energy		GE Mcal/kg			
Cattle		DE Mcal/kg	* 0.70	3.67	
Horses		DE Mcal/kg	* 0.74	3.87	
Sheep		DE Mcal/kg	* 0.70	3.68	
Swine		DE kcal/kg	* 647.	3386.	
Cattle		ME Mcal/kg	* 0.57	3.01	
Horses		ME Mcal/kg	* 0.61	3.17	
Sheep		ME Mcal/kg	* 0.58	3.02	
Swine		ME kcal/kg	* 544.	2848.	
Cattle		TDN %	* 15.9	83.2	
Horses		TDN %	16.8	87.7	
Sheep		TDN %	* 15.9	83.4	
Swine		TDN %	14.7	76.8	

YEAST, ACTIVE. Saccharomyces cerevisiae

Yeast, active, dehy, mn 15 billion live yeast cells per g, (7)

Active dry yeast (AAFCO)

Ref No 7 05 524 United States

Feed Name or Analyses			Mean As Fed	Mean Dry	C.V. ± %
Dry matter		%	89.9	100.0	1
Ash		%	2.9	3.2	6
Crude fiber		%	3.4	3.8	6
Cattle	dig coef	%	0.	0.	6
Sheep	dig coef	%	*-104.	-104.	
Swine	dig coef	%	* 3.	3.	
Ether extract		%	3.3	3.7	19
Cattle	dig coef	%	* -88.	-88.	
Sheep	dig coef	%	* 7.	7.	
Swine	dig coef	%	*-156.	-156.	
N-free extract		%	66.2	73.6	
Cattle	dig coef	%	* 89.	89.	
Sheep	dig coef	%	* 88.	88.	
Swine	dig coef	%	* 90.	90.	
Protein (N x 6.25)		%	14.1	15.7	12
Cattle	dig coef	%	* 92.	92.	
Sheep	dig coef	%	* 85.	85.	
Swine	dig coef	%	* 88.	88.	
Cattle	dig prot	%	13.0	14.4	
Sheep	dig prot	%	12.0	13.3	
Swine	dig prot	%	12.4	13.8	
Energy		GE Mcal/kg			
Cattle		DE Mcal/kg	* 2.88	3.20	
Sheep		DE Mcal/kg	* 2.96	3.30	
Swine		DE kcal/kg	*2663.	2963.	
Cattle		ME Mcal/kg	* 2.36	2.63	
Sheep		ME Mcal/kg	* 2.43	2.70	
Swine		ME kcal/kg	*2472.	2751.	
Cattle		TDN %	65.2	72.6	
Sheep		TDN %	67.2	74.7	
Swine		TDN %	60.4	67.2	
Calcium		%	0.14	0.16	
Phosphorus		%	0.88	0.98	

YEAST, BREWERS. Saccharomyces cerevisiae

Yeast, brewers, bitterness extd, (7)

Ref No 7 05 525 United States

Feed Name or Analyses		Mean As Fed	Mean Dry	C.V. ± %
Dry matter	%	90.0	100.0	
Sulphur	%	0.41	0.46	
Riboflavin	mg/kg	39.9	44.3	
Arginine	%	2.40	2.67	
Cystine	%	0.50	0.56	
Histidine	%	1.20	1.33	
Isoleucine	%	2.80	3.11	
Leucine	%	3.30	3.67	
Lysine	%	3.20	3.56	
Methionine	%	1.40	1.56	
Phenylalanine	%	2.20	2.44	
Threonine	%	2.50	2.78	
Tryptophan	%	0.70	0.78	
Tyrosine	%	1.60	1.78	
Valine	%	2.20	2.44	

YEAST, BREWERS SACCHAROMYCES. Saccharomyces cervisiae

Yeast, brewers Saccharomyces, dehy grnd, (7)

Brewers dried yeast (CFA)

Ref No 7 05 528 United States

Feed Name or Analyses		Mean As Fed	Mean Dry	C.V. ± %
Dry matter	%	93.0	100.0	
Ash	%	6.9	7.4	
Crude fiber	%	1.5	1.6	
Ether extract	%	1.0	1.1	
N-free extract	%	39.4	42.3	
Protein (N x 6.25)	%	44.3	47.6	
Energy	GE Mcal/kg	4.34	4.67	
Cattle	DE Mcal/kg	* 3.56	3.83	
Sheep	DE Mcal/kg	* 3.46	3.73	
Swine	DE kcal/kg	*3384.	3639.	
Cattle	ME Mcal/kg	* 2.92	3.14	
Chickens	MEn kcal/kg	2087.	2244.	
Sheep	ME Mcal/kg	* 2.84	3.05	
Swine	ME kcal/kg	*2923.	3143.	
Cattle	TDN %	* 80.8	86.9	
Sheep	TDN %	* 78.6	84.5	
Swine	TDN %	* 76.7	82.5	
Calcium	%	0.15	0.16	
Cobalt	mg/kg	1.476	1.587	
Copper	mg/kg	24.6	26.5	
Iron	%	0.005	0.005	
Magnesium	%	0.25	0.26	
Manganese	mg/kg	5.9	6.3	
Molybdenum	mg/kg	1.77	1.90	
Phosphorus	%	1.48	1.59	
Potassium	%	1.48	1.59	
Sodium	%	0.10	0.11	
Sulphur	%	0.49	0.53	
Zinc	mg/kg	37.4	40.2	
Betaine	%	.118 0	.126 9	
Biotin	mg/kg	1.08	1.16	
Choline	mg/kg	4183.	4497.	
Folic acid	mg/kg	14.76	15.87	
Niacin	mg/kg	467.5	502.6	
Pantothenic acid	mg/kg	110.7	119.0	
Riboflavin	mg/kg	32.0	34.4	
Thiamine	mg/kg	110.7	119.0	
Vitamin B6	mg/kg	46.75	50.26	

Ref No 7 05 528 Canada

Feed Name or Analyses		Mean As Fed	Mean Dry	C.V. ± %
Dry matter	%	93.5	100.0	1
Energy	GE Mcal/kg	4.36	4.67	
Chickens	MEn kcal/kg	2099.	2244.	
Niacin	mg/kg	378.0	404.3	
Pantothenic acid	mg/kg	61.0	65.3	14
Riboflavin	mg/kg	35.3	37.8	10
Thiamine	mg/kg	48.1	51.5	

Yeast, brewers Saccharomyces, dehy grnd, human gr, (7)

Ref No 7 05 526 United States

Feed Name or Analyses		As Fed	Dry	C.V. ± %
Dry matter	%	93.3	100.0	0
Ash	%	7.3	7.8	17
Crude fiber	%	5.7	6.1	23
Ether extract	%	1.3	1.4	13
N-free extract	%	27.3	29.3	
Protein (N x 6.25)	%	51.7	55.4	4
Energy	GE Mcal/kg			
Cattle	DE Mcal/kg	* 3.29	3.53	
Sheep	DE Mcal/kg	* 3.18	3.41	
Swine	DE kcal/kg	*3053.	3272.	
Cattle	ME Mcal/kg	* 2.70	2.89	
Sheep	ME Mcal/kg	* 2.61	2.80	
Swine	ME kcal/kg	*2589.	2775.	
Cattle	TDN %	* 74.7	80.1	
Sheep	TDN %	* 72.2	77.4	
Swine	TDN %	* 69.2	74.2	
Calcium	%	0.12	0.13	41
Copper	mg/kg	8.2	8.7	46
Iron	%	0.006	0.006	45
Magnesium	%	0.24	0.26	44
Manganese	mg/kg	6.4	6.9	29
Phosphorus	%	1.48	1.59	8
Potassium	%	1.62	1.74	41
Sodium	%	0.22	0.24	43
Zinc	mg/kg	45.6	48.9	36
Alanine	%	3.00	3.22	7
Cystine	%	0.50	0.54	11
Glycine	%	2.10	2.25	3
Histidine	%	1.30	1.39	13
Isoleucine	%	2.30	2.47	14
Leucine	%	3.30	3.54	18
Lysine	%	3.70	3.97	1
Methionine	%	0.70	0.75	16
Threonine	%	2.50	2.68	13
Tryptophan	%	0.60	0.64	
Tyrosine	%	1.70	1.82	19
Valine	%	2.60	2.79	2

Yeast, brewers Saccharomyces, dehy grnd, mn 40% protein, (7)

Brewers dried yeast (AAFCO)

Ref No 7 05 527 United States

Feed Name or Analyses		As Fed	Dry	C.V. ± %
Dry matter	%	93.7	100.0	2
Ash	%	6.6	7.1	12
Crude fiber	%	2.7	2.9	80
Cattle	dig coef %	0.	0.	80
Sheep	dig coef %	*-104.	-104.	
Swine	dig coef %	* 3.	3.	
Ether extract	%	0.9	1.0	61
Cattle	dig coef %	* -88.	-88.	
Sheep	dig coef %	* 7.	7.	
Swine	dig coef %	*-156.	-156.	
N-free extract	%	38.7	41.3	
Cattle	dig coef %	* 89.	89.	
Sheep	dig coef %	* 88.	88.	
Swine	dig coef %	* 90.	90.	
Protein (N x 6.25)	%	44.7	47.8	8
Cattle	dig coef %	* 92.	92.	
Sheep	dig coef %	* 85.	85.	
Swine	dig coef %	* 88.	88.	
Cattle	dig prot %	41.2	43.9	
Sheep	dig prot %	38.0	40.6	
Swine	dig prot %	39.4	42.0	
Linoleic	%	.050	.053	
Energy	GE Mcal/kg	3.98	4.25	10
Cattle	DE Mcal/kg	* 3.25	3.47	
Sheep	DE Mcal/kg	* 3.06	3.27	
Swine	DE kcal/kg	*3135.	3346.	
Cattle	ME Mcal/kg	* 2.67	2.85	
Chickens	ME_n kcal/kg	2210.	2412.	
Sheep	ME Mcal/kg	* 2.51	2.68	
Swine	ME kcal/kg	*2707.	2890.	
Cattle	NE_m Mcal/kg	* 1.66	1.77	
Cattle	NE_{gain} Mcal/kg	* 1.10	1.17	
Cattle	$NE_{lactating\ cows}$ Mcal/kg	* 1.99	2.12	
Cattle	TDN %	73.8	78.8	
Sheep	TDN %	69.4	74.1	
Swine	TDN %	71.1	75.9	
Calcium	%	0.13	0.14	39
Cobalt	mg/kg	0.184	0.196	29
Copper	mg/kg	33.1	35.3	41
Iron	%	0.013	0.014	91
Magnesium	%	0.23	0.25	41
Manganese	mg/kg	5.7	6.1	31
Phosphorus	%	1.46	1.56	14
Potassium	%	1.72	1.84	18
Sodium	%	0.07	0.07	54
Sulphur	%	0.38	0.41	
Zinc	mg/kg	38.9	41.5	61
Biotin	mg/kg	0.97	1.04	31
Carotene	mg/kg	0.0	0.0	31
Choline	mg/kg	3897.	4159.	2
Folic acid	mg/kg	9.73	10.39	33
Niacin	mg/kg	448.9	479.1	17
Pantothenic acid	mg/kg	110.1	117.5	33
Riboflavin	mg/kg	35.1	37.5	62
Thiamine	mg/kg	92.2	98.4	29
α-tocopherol	mg/kg	0.0	0.0	29
Vitamin B_6	mg/kg	43.57	46.50	41
Arginine	%	2.20	2.35	11
Cystine	%	0.50	0.53	39
Glycine	%	1.70	1.82	11
Histidine	%	1.10	1.18	18
Isoleucine	%	2.11	2.25	28
Leucine	%	3.21	3.43	39
Lysine	%	3.00	3.21	18
Methionine	%	0.70	0.75	29
Phenylalanine	%	1.81	1.93	12
Threonine	%	2.11	2.25	12
Tryptophan	%	0.50	0.53	10
Tyrosine	%	1.50	1.61	29
Valine	%	2.31	2.46	12

YEAST, IRRADIATED. Saccharomyces cerevisiae

Yeast, irradiated, dehy, (7)

Irradiated dried yeast (AAFCO)

Ref No 7 05 529 United States

Feed Name or Analyses		As Fed	Dry	C.V. ± %
Dry matter	%	93.9	100.0	2
Ash	%	6.2	6.6	31
Crude fiber	%	6.1	6.5	52
Cattle	dig coef %	0.	0.	52
Sheep	dig coef %	*-104.	-104.	
Swine	dig coef %	* 3.	3.	
Ether extract	%	1.1	1.2	12
Cattle	dig coef %	* -88.	-88.	
Sheep	dig coef %	* 7.	7.	
Swine	dig coef %	*-156.	-156.	
N-free extract	%	32.0	34.1	
Cattle	dig coef %	* 89.	89.	
Sheep	dig coef %	* 88.	88.	
Swine	dig coef %	* 90.	90.	
Protein (N x 6.25)	%	48.4	51.5	6
Cattle	dig coef %	* 92.	92.	
Sheep	dig coef %	* 85.	85.	
Swine	dig coef %	* 88.	88.	
Cattle	dig prot %	44.5	47.4	
Sheep	dig prot %	41.1	43.8	
Swine	dig prot %	42.6	45.4	
Energy	GE Mcal/kg	4.59	4.89	
Cattle	DE Mcal/kg	* 3.12	3.33	
Sheep	DE Mcal/kg	* 2.78	2.96	
Swine	DE kcal/kg	*2987.	3181.	
Cattle	ME Mcal/kg	* 2.56	2.73	
Sheep	ME Mcal/kg	* 2.28	2.43	
Swine	ME kcal/kg	*2556.	2722.	
Cattle	TDN %	70.8	75.5	
Sheep	TDN %	63.1	67.2	
Swine	TDN %	67.7	72.1	
Calcium	%	0.77	0.83	
Phosphorus	%	1.41	1.51	
Potassium	%	2.14	2.28	
Riboflavin	mg/kg	18.5	19.7	
Glutamic acid	%	6.30	6.71	13

YEAST, MINERAL. Torula utilis

Yeast, mineral, sulfite waste liquors, dehy, (7)

Ref No 7 05 530 United States

Feed Name or Analyses		As Fed	Dry	C.V. ± %
Dry matter	%	91.6	100.0	
Ash	%	9.1	9.9	
Crude fiber	%	2.1	2.3	
Sheep	dig coef %	-75.	-75.	
Ether extract	%	1.5	1.6	
Sheep	dig coef %	44.	44.	
N-free extract	%	36.1	39.4	
Sheep	dig coef %	62.	62.	
Protein (N x 6.25)	%	42.9	46.8	
Sheep	dig coef %	86.	86.	
Sheep	dig prot %	36.9	40.2	
Energy	GE Mcal/kg			
Cattle	DE Mcal/kg	* 2.67	2.92	
Sheep	DE Mcal/kg	* 2.61	2.84	
Swine	DE kcal/kg	*3254.	3552.	
Cattle	ME Mcal/kg	* 2.19	2.39	
Sheep	ME Mcal/kg	* 2.14	2.33	
Swine	ME kcal/kg	*2816.	3074.	
Cattle	TDN %	* 60.7	66.3	
Sheep	TDN %	59.1	64.5	
Swine	TDN %	* 73.8	80.6	

Yeast, mineral, (7)

Yeast, mineral

Ref No 7 05 531 United States

Feed Name or Analyses		As Fed	Dry	C.V. ± %
Dry matter	%	86.9	100.0	
Ash	%	9.7	11.2	
Ether extract	%	5.4	6.2	
Sheep	dig coef %	25.	25.	
N-free extract	%	28.2	32.4	
Sheep	dig coef %	73.	73.	
Protein (N x 6.25)	%	43.6	50.2	
Sheep	dig coef %	86.	86.	

(1) dry forages and roughages (2) pasture, range plants, and forages fed green (3) silages (4) energy feeds (5) protein supplements (6) minerals (7) vitamins (8) additives

Feed Name or Analyses		Mean As Fed	Mean Dry	C.V. ± %
Sheep	dig prot %	37.5	43.2	
Energy	GE Mcal/kg			
Cattle	DE Mcal/kg	* 2.77	3.19	
Sheep	DE Mcal/kg	* 2.69	3.10	
Swine	DE kcal/kg	*3343.	3846.	
Cattle	ME Mcal/kg	* 2.28	2.62	
Sheep	ME Mcal/kg	* 2.21	2.54	
Swine	ME kcal/kg	*2870.	3302.	
Cattle	TDN %	* 62.8	72.3	
Sheep	TDN %	61.1	70.3	
Swine	TDN %	* 75.8	87.2	

YEAST, MOLASSES DISTILLERS SACCHAROMYCES. Saccharomyces cerevisiae

Yeast, molasses distillers Saccharomyces, dehy, mn 40% protein, (7)
Molasses distillers dried yeast (AAFCO)
Ref No 7 05 532 United States

Feed Name or Analyses		Mean As Fed	Mean Dry	C.V. ± %
Dry matter	%	81.8	100.0	
Ash	%	11.7	14.3	
Crude fiber	%	5.8	7.1	
Cattle	dig coef %	0.	0.	
Sheep	dig coef %	*-104.	-104.	
Ether extract	%	1.3	1.6	
Cattle	dig coef %	* -88.	-88.	
Sheep	dig coef %	* 7.	7.	
N-free extract	%	35.2	43.1	
Cattle	dig coef %	* 89.	89.	
Sheep	dig coef %	* 88.	88.	
Swine	dig coef %	97.	97.	
Protein (N x 6.25)	%	27.8	33.9	
Cattle	dig coef %	* 92.	92.	
Sheep	dig coef %	* 85.	85.	
Swine	dig coef %	71.	71.	
Cattle	dig prot %	25.5	31.2	
Sheep	dig prot %	23.6	28.9	
Swine	dig prot %	19.7	24.1	
Energy	GE Mcal/kg			
Cattle	DE Mcal/kg	* 2.39	2.93	
Sheep	DE Mcal/kg	* 2.15	2.63	
Swine	DE kcal/kg	*2856.	3492.	
Cattle	ME Mcal/kg	* 1.96	2.40	
Sheep	ME Mcal/kg	* 1.76	2.15	
Swine	ME kcal/kg	*2546.	3113.	
Cattle	TDN %	54.3	66.4	
Sheep	TDN %	48.7	59.6	
Swine	TDN %	64.8	79.2	
Calcium	%	0.09	0.11	
Phosphorus	%	0.90	1.10	
Choline	mg/kg	1981.	2423.	
Folic acid	mg/kg	9.69	11.85	
Niacin	mg/kg	225.6	275.8	
Pantothenic acid	mg/kg	31.9	39.0	
Riboflavin	mg/kg	23.9	29.2	
Thiamine	mg/kg	66.4	81.2	
Vitamin B_{12}	µg/kg	8.4	10.3	
Arginine	%	1.80	2.20	
Cystine	%	0.45	0.55	
Glycine	%	1.80	2.20	
Lysine	%	2.70	3.30	
Methionine	%	0.63	0.77	
Tryptophan	%	0.45	0.55	

YEAST, PRIMARY SACCHAROMYCES. Saccharomyces cerevisiae

Yeast, primary Saccharomyces, dehy, mn 40% protein, (7)
Dried yeast (AAFCO)
Primary dried yeast (AAFCO)
Ref No 7 05 533 United States

Feed Name or Analyses		Mean As Fed	Mean Dry	C.V. ± %
Dry matter	%	92.6	100.0	3
Ash	%	8.0	8.6	16
Crude fiber	%	3.1	3.3	52
Cattle	dig coef %	0.	0.	52
Sheep	dig coef %	*-104.	-104.	
Ether extract	%	1.0	1.1	55
Cattle	dig coef %	* -88.	-88.	
Sheep	dig coef %	* 7.	7.	
N-free extract	%	32.5	35.1	
Cattle	dig coef %	* 89.	89.	
Sheep	dig coef %	* 88.	88.	
Protein (N x 6.25)	%	48.0	51.8	8
Cattle	dig coef %	* 92.	92.	
Sheep	dig coef %	* 85.	85.	
Cattle	dig prot %	44.2	47.7	
Sheep	dig prot %	40.8	44.1	
Energy	GE Mcal/kg	4.44	4.79	1
Cattle	DE Mcal/kg	* 3.13	3.38	
Sheep	DE Mcal/kg	* 2.92	3.16	
Swine	DE kcal/kg	*3162.	3415.	
Cattle	ME Mcal/kg	* 2.57	2.78	
Sheep	ME Mcal/kg	* 2.40	2.59	
Swine	ME kcal/kg	*2705.	2921.	
Cattle	TDN %	71.1	76.8	
Sheep	TDN %	66.3	71.6	
Swine	TDN %	* 71.7	77.4	
Calcium	%	0.36	0.39	98
Chlorine	%	0.02	0.02	
Iron	%	0.030	0.032	
Magnesium	%	0.36	0.39	
Manganese	mg/kg	3.7	4.0	23
Phosphorus	%	1.72	1.86	40
Sulphur	%	0.57	0.62	
Biotin	mg/kg	1.61	1.74	6
Folic acid	mg/kg	31.13	33.62	98
Niacin	mg/kg	300.7	324.7	98
Pantothenic acid	mg/kg	312.0	336.9	98
Riboflavin	mg/kg	38.8	41.9	56
Thiamine	mg/kg	6.4	6.9	42
Vitamin B_{12}	µg/kg	6.2	6.7	
Arginine	%	2.60	2.81	25
Cystine	%	0.50	0.54	
Histidine	%	5.60	6.05	
Isoleucine	%	3.60	3.89	5
Leucine	%	3.70	4.00	2
Lysine	%	3.80	4.10	18
Methionine	%	1.00	1.08	
Phenylalanine	%	2.50	2.70	
Threonine	%	2.50	2.70	3
Tryptophan	%	0.40	0.43	34
Valine	%	3.20	3.46	8

YEAST, TORULOPSIS. Torulopsis utilis

Yeast, Torulopsis, dehy, mn 40% protein, (7)
Torula dried yeast (AAFCO)
Ref No 7 05 534 United States

Feed Name or Analyses		Mean As Fed	Mean Dry	C.V. ± %
Dry matter	%	92.6	100.0	1
Ash	%	8.1	8.7	11
Crude fiber	%	2.5	2.7	49
Cattle	dig coef %	0.	0.	49
Sheep	dig coef %	* -75.	-75.	
Swine	dig coef %	* 101.	101.	
Ether extract	%	1.6	1.8	94
Cattle	dig coef %	* 40.	40.	
Sheep	dig coef %	* 44.	44.	
Swine	dig coef %	* -31.	-31.	
N-free extract	%	33.4	36.1	
Cattle	dig coef %	* 89.	89.	
Sheep	dig coef %	* 62.	62.	
Swine	dig coef %	* 76.	76.	
Protein (N x 6.25)	%	47.0	50.8	8
Cattle	dig coef %	* 91.	91.	
Sheep	dig coef %	* 86.	86.	
Swine	dig coef %	* 82.	82.	
Cattle	dig prot %	42.8	46.2	
Sheep	dig prot %	40.4	43.7	
Swine	dig prot %	38.6	41.6	
Linoleic	%	.050	.054	
Energy	GE Mcal/kg	4.41	4.76	0
Cattle	DE Mcal/kg	* 3.26	3.52	
Sheep	DE Mcal/kg	* 2.69	2.90	
Swine	DE kcal/kg	*2879.	3108.	
Cattle	ME Mcal/kg	* 2.68	2.89	
Chickens	ME_n kcal/kg	2141.	2312.	
Sheep	ME Mcal/kg	* 2.20	2.38	
Swine	ME kcal/kg	*2469.	2665.	
Cattle	NE_m Mcal/kg	* 1.72	1.86	
Cattle	NE_{gain} Mcal/kg	* 1.15	1.24	
Cattle	$NE_{lactating\ cows}$ Mcal/kg	* 2.04	2.20	
Cattle	TDN %	74.0	79.9	
Sheep	TDN %	60.9	65.8	
Swine	TDN %	65.3	70.5	
Calcium	%	0.57	0.62	42
Copper	mg/kg	13.4	14.5	98
Iron	%	0.009	0.010	18
Magnesium	%	0.13	0.14	19
Manganese	mg/kg	12.8	13.8	33
Phosphorus	%	1.68	1.81	13
Potassium	%	1.88	2.03	11
Sodium	%	0.01	0.01	
Zinc	mg/kg	98.7	106.6	
Biotin	mg/kg	1.07	1.16	81
Choline	mg/kg	2886.	3116.	34
Folic acid	mg/kg	23.16	25.00	61
Niacin	mg/kg	500.4	540.2	67
Pantothenic acid	mg/kg	67.8	73.2	98
Riboflavin	mg/kg	44.5	48.0	39
Thiamine	mg/kg	6.1	6.6	37
Vitamin B_6	mg/kg	29.33	31.66	42
Arginine	%	2.60	2.80	20
Cystine	%	0.60	0.65	83
Glycine	%	2.54	2.75	47
Histidine	%	1.39	1.50	17
Isoleucine	%	2.88	3.11	37
Leucine	%	3.47	3.75	23
Lysine	%	3.79	4.09	17
Methionine	%	0.80	0.86	17
Phenylalanine	%	2.98	3.22	33

Continued

Feed Name or Analyses		Mean As Fed	Mean Dry	C.V. ± %
Threonine	%	2.58	2.79	10
Tryptophan	%	0.50	0.54	40
Tyrosine	%	2.08	2.25	29
Valine	%	2.88	3.11	17

Yeast culture (AAFCO) –
see Yeast, medium w yeast, dehy, (7)

Yeast dried grains (AAFCO) –
see Grains, yeast fermentation grains, dehy, (7)

Yellow hominy feed (AAFCO) (CFA) –
see Corn, yellow, grits by-product, mn 5% fat, (4)

YUCCA. Yucca spp

Yucca, flowers, fresh, (2)
Ref No 2 05 536 United States

Analyses		As Fed	Dry	C.V. ± %
Dry matter	%		100.0	
Ash	%		8.7	
Crude fiber	%		13.3	
Ether extract	%		4.4	
N-free extract	%		53.3	
Protein (N x 6.25)	%		20.3	
Cattle	dig prot % *		15.1	
Goats	dig prot % *		15.5	
Horses	dig prot % *		14.8	
Rabbits	dig coef % *		14.3	
Sheep	dig prot % *		15.9	

Yucca, flowers, fresh, pre-bloom, (2)
Ref No 2 05 535 United States

Analyses		As Fed	Dry	C.V. ± %
Dry matter	%		100.0	
Calcium	%		1.15	
Magnesium	%		0.32	
Phosphorus	%		0.51	
Potassium	%		3.03	

Yucca, leaves, fresh, (2)
Ref No 2 05 537 United States

Analyses		As Fed	Dry	C.V. ± %
Dry matter	%	33.1	100.0	
Ash	%	2.3	7.0	46
Crude fiber	%	12.3	37.2	
Ether extract	%	0.6	1.9	
N-free extract	%	15.5	46.7	
Protein (N x 6.25)	%	2.4	7.2	56
Cattle	dig prot % *	1.3	4.0	
Goats	dig prot % *	1.1	3.3	
Horses	dig prot % *	1.2	3.6	
Rabbits	dig coef % *	1.4	4.2	
Sheep	dig prot % *	1.2	3.7	
Calcium	%	0.66	1.99	
Iron	%	0.004	0.012	
Magnesium	%	0.15	0.46	
Phosphorus	%	0.04	0.13	
Potassium	%	0.37	1.12	

Yucca, shoots, fresh, (2)
Ref No 2 05 538 United States

Analyses		As Fed	Dry	C.V. ± %
Dry matter	%		100.0	
Ash	%		7.1	
Crude fiber	%		25.2	
Ether extract	%		1.5	
N-free extract	%		50.3	
Protein (N x 6.25)	%		15.9	
Cattle	dig prot % *		11.4	
Goats	dig prot % *		11.4	
Horses	dig prot % *		11.0	
Rabbits	dig coef % *		10.9	
Sheep	dig prot % *		11.8	
Calcium	%		0.12	
Magnesium	%		0.22	
Phosphorus	%		0.38	
Potassium	%		2.30	

Yucca, stems, fresh, (2)
Ref No 2 05 539 United States

Analyses		As Fed	Dry	C.V. ± %
Dry matter	%	28.3	100.0	
Ash	%	2.3	8.1	31
Crude fiber	%	9.0	31.8	
Ether extract	%	0.4	1.4	
N-free extract	%	15.5	54.9	
Protein (N x 6.25)	%	1.1	3.8	36
Cattle	dig prot % *	0.3	1.1	
Goats	dig prot % *	0.0	0.1	
Horses	dig prot % *	0.2	0.8	
Rabbits	dig coef % *	0.5	1.6	
Sheep	dig prot % *	0.2	0.5	

YUCCA, ARIZONICA. Yucca arizonica

Yucca, Arizonica, seeds w hulls, (4)
Ref No 4 09 292 United States

Analyses		As Fed	Dry	C.V. ± %
Dry matter	%		100.0	
Protein (N x 6.25)	%		14.0	
Cattle	dig prot % *		8.9	
Goats	dig prot % *		10.1	
Horses	dig prot % *		10.1	
Sheep	dig prot % *		10.1	
Alanine	%		0.53	
Arginine	%		1.50	
Aspartic acid	%		1.05	
Glutamic acid	%		2.10	
Glycine	%		0.56	
Histidine	%		0.31	
Hydroxyproline	%		0.43	
Isoleucine	%		0.46	
Leucine	%		0.71	
Lysine	%		0.60	
Methionine	%		0.27	
Phenylalanine	%		0.60	
Proline	%		0.67	
Serine	%		0.60	
Threonine	%		0.45	
Tyrosine	%		0.67	
Valine	%		0.76	

YUCCA, BEARGRASS. Yucca spp

Yucca, beargrass, aerial part, s-c, (1)
Beargrass, yucca, dried
Ref No 1 08 560 United States

Analyses		As Fed	Dry	C.V. ± %
Dry matter	%	92.6	100.0	
Ash	%	6.9	7.5	
Crude fiber	%	38.6	41.7	
Ether extract	%	2.2	2.4	
N-free extract	%	38.3	41.4	
Protein (N x 6.25)	%	6.6	7.1	
Cattle	dig prot % *	2.9	3.1	
Goats	dig prot % *	3.0	3.2	
Horses	dig prot % *	3.3	3.6	
Rabbits	dig coef % *	3.9	4.2	
Sheep	dig prot % *	2.7	3.0	
Energy	GE Mcal/kg			
Cattle	DE Mcal/kg *	2.17	2.34	
Sheep	DE Mcal/kg *	2.34	2.53	
Cattle	ME Mcal/kg *	1.78	1.92	
Sheep	ME Mcal/kg *	1.92	2.07	
Cattle	TDN % *	47.2	53.1	
Sheep	TDN % *	53.1	57.3	

Yucca, beargrass, aerial part, fresh, (2)
Beargrass, yucca, fresh
Ref No 2 08 561 United States

Analyses		As Fed	Dry	C.V. ± %
Dry matter	%	49.4	100.0	
Ash	%	3.5	7.1	
Crude fiber	%	21.1	42.7	
Ether extract	%	1.0	2.0	
N-free extract	%	20.0	40.5	
Protein (N x 6.25)	%	3.8	7.7	
Cattle	dig prot % *	2.2	4.4	
Goats	dig prot % *	1.8	3.7	
Horses	dig prot % *	2.0	4.1	
Rabbits	dig coef % *	2.3	4.6	
Sheep	dig prot % *	2.1	4.2	
Energy	GE Mcal/kg			
Cattle	DE Mcal/kg *	1.24	2.50	
Sheep	DE Mcal/kg *	1.18	2.39	
Cattle	ME Mcal/kg *	1.01	2.05	
Sheep	ME Mcal/kg *	0.97	1.96	
Cattle	TDN % *	28.1	56.8	
Sheep	TDN % *	26.7	54.1	

YUCCA, SOAPTREE. Yucca elata

Yucca, soaptree, stems, fresh, (2)
Ref No 2 05 540 United States

Analyses		As Fed	Dry	C.V. ± %
Dry matter	%		100.0	
Ash	%		6.5	
Crude fiber	%		29.6	
Ether extract	%		1.4	
N-free extract	%		58.0	
Protein (N x 6.25)	%		4.5	
Cattle	dig prot % *		1.7	
Goats	dig prot % *		0.8	
Horses	dig prot % *		1.4	
Rabbits	dig coef % *		2.1	
Sheep	dig prot % *		1.2	

(1) dry forages and roughages (2) pasture, range plants, and forages fed green (3) silages (4) energy feeds (5) protein supplements (6) minerals (7) vitamins (8) additives

Feed Name or Analyses				Mean As Fed	Mean Dry	C.V. ± %

YUCCA, SOAPWEED. Yucca spp

Yucca, soapweed, aerial part, fresh, (2)

Ref No 2 08 562 United States

Analyses				As Fed	Dry	C.V. ± %
Dry matter		%		44.6	100.0	
Ash		%		3.3	7.4	
Crude fiber		%		14.1	31.6	
Ether extract		%		0.8	1.8	
N-free extract		%		24.0	53.8	
Protein (N x 6.25)		%		2.4	5.4	
Cattle	dig prot	%	*	1.1	2.5	
Goats	dig prot	%	*	0.7	1.6	
Horses	dig prot	%	*	0.9	2.1	
Rabbits	dig coef	%	*	1.3	2.8	
Sheep	dig prot	%	*	0.9	2.0	
Energy		GE Mcal/kg				
Cattle		DE Mcal/kg	*	1.01	2.25	
Sheep		DE Mcal/kg	*	1.07	2.41	
Cattle		ME Mcal/kg	*	0.82	1.84	
Sheep		ME Mcal/kg	*	0.88	1.98	
Cattle		TDN %	*	22.7	51.0	
Sheep		TDN %	*	24.4	54.6	
Calcium		%		0.41	0.92	
Phosphorus		%		0.08	0.18	

Yucca, soapweed, leaves, fresh, (2)

Ref No 2 08 563 United States

Analyses				As Fed	Dry	C.V. ± %
Dry matter		%		56.2	100.0	
Ash		%		3.7	6.6	
Crude fiber		%		18.7	33.3	
Ether extract		%		1.7	3.0	
N-free extract		%		27.7	49.3	
Protein (N x 6.25)		%		4.4	7.8	
Cattle	dig prot	%	*	2.6	4.5	
Goats	dig prot	%	*	2.2	3.9	
Horses	dig prot	%	*	2.3	4.2	
Rabbits	dig coef	%	*	2.6	4.7	
Sheep	dig prot	%	*	2.4	4.3	
Energy		GE Mcal/kg				
Cattle		DE Mcal/kg	*	1.22	2.17	
Sheep		DE Mcal/kg	*	1.28	2.28	
Cattle		ME Mcal/kg	*	1.00	1.78	
Sheep		ME Mcal/kg	*	1.05	1.87	
Cattle		TDN %	*	27.7	49.2	
Sheep		TDN %	*	29.1	51.7	

YUCCA, TORREY. Yucca torreyi

Yucca, torrey, leaves, fresh, (2)

Ref No 2 05 541 United States

Analyses				As Fed	Dry	C.V. ± %
Dry matter		%		33.1	100.0	
Ash		%		2.3	6.9	
Crude fiber		%		9.9	30.0	
Ether extract		%		0.7	2.1	
N-free extract		%		18.8	56.8	
Protein (N x 6.25)		%		1.4	4.2	
Cattle	dig prot	%	*	0.5	1.5	
Goats	dig prot	%	*	0.2	0.5	
Horses	dig prot	%	*	0.4	1.1	
Rabbits	dig coef	%	*	0.6	1.9	
Sheep	dig prot	%	*	0.3	0.9	
Calcium		%		0.28	0.86	
Chlorine		%		0.13	0.40	
Iron		%		0.004	0.012	
Magnesium		%		0.27	0.83	
Phosphorus		%		0.05	0.14	
Potassium		%		0.28	0.86	
Sodium		%		0.04	0.13	
Sulphur		%		0.07	0.22	

Yucca, torrey, stems, fresh, (2)

Ref No 2 05 542 United States

Analyses				As Fed	Dry	C.V. ± %
Dry matter		%		28.3	100.0	
Ash		%		2.7	9.7	
Crude fiber		%		10.3	36.4	
Ether extract		%		0.4	1.4	
N-free extract		%		14.0	49.5	
Protein (N x 6.25)		%		0.8	3.0	
Cattle	dig prot	%	*	0.1	0.4	
Goats	dig prot	%	*	-0.1	-0.5	
Horses	dig prot	%	*	0.0	0.1	
Rabbits	dig coef	%	*	0.3	1.0	
Sheep	dig prot	%	*	0.0	-0.1	
Calcium		%		0.52	1.84	
Chlorine		%		0.08	0.30	
Iron		%		0.005	0.018	
Magnesium		%		0.25	0.89	
Phosphorus		%		0.03	0.09	
Potassium		%		0.17	0.61	
Sodium		%		0.19	0.68	
Sulphur		%		0.05	0.16	

YUCCA, TRECUL. Yucca treculeana

Yucca, trecul, buds, fresh, (2)

Ref No 2 05 543 United States

Analyses				As Fed	Dry	C.V. ± %
Dry matter		%			100.0	
Ash		%			8.8	
Crude fiber		%			14.4	
Ether extract		%			1.3	
N-free extract		%			50.7	
Protein (N x 6.25)		%			24.8	
Cattle	dig prot	%	*		19.0	
Goats	dig prot	%	*		19.7	
Horses	dig prot	%	*		18.6	
Rabbits	dig coef	%	*		17.8	
Sheep	dig prot	%	*		20.1	
Calcium		%			1.52	
Magnesium		%			0.28	
Phosphorus		%			0.64	
Potassium		%			2.75	

Yucca, trecul, leaves, fresh, immature, (2)

Ref No 2 05 544 United States

Analyses				As Fed	Dry	C.V. ± %
Dry matter		%			100.0	
Ash		%			10.5	
Crude fiber		%			36.1	
Ether extract		%			1.9	
N-free extract		%			39.5	
Protein (N x 6.25)		%			12.0	
Cattle	dig prot	%	*		8.1	
Goats	dig prot	%	*		7.8	
Horses	dig prot	%	*		7.7	
Rabbits	dig coef	%	*		7.9	
Sheep	dig prot	%	*		8.2	
Calcium		%			3.82	
Magnesium		%			0.36	
Phosphorus		%			0.17	
Potassium		%			1.78	

ZEXMENIA, ORANGE. Zexmenia hispida

Zexmenia, orange, aerial part, fresh, pre-bloom, (2)

Ref No 2 05 545 United States

Analyses				As Fed	Dry	C.V. ± %
Dry matter		%			100.0	
Ash		%			15.5	
Crude fiber		%			21.3	
Ether extract		%			3.2	
N-free extract		%			42.5	
Protein (N x 6.25)		%			17.5	
Cattle	dig prot	%	*		12.8	
Goats	dig prot	%	*		12.9	
Horses	dig prot	%	*		12.4	
Rabbits	dig coef	%	*		12.2	
Sheep	dig prot	%	*		13.3	
Calcium		%			3.51	
Magnesium		%			0.66	
Phosphorus		%			0.20	
Potassium		%			2.48	

Zexmenia, orange, aerial part, fresh, mid-bloom, (2)

Ref No 2 05 546 United States

Analyses				As Fed	Dry	C.V. ± %
Dry matter		%			100.0	
Ash		%			16.5	
Crude fiber		%			25.7	
Ether extract		%			1.3	
N-free extract		%			44.3	
Protein (N x 6.25)		%			12.2	
Cattle	dig prot	%	*		8.3	
Goats	dig prot	%	*		7.9	
Horses	dig prot	%	*		7.9	
Rabbits	dig coef	%	*		8.1	
Sheep	dig prot	%	*		8.4	
Calcium		%			2.93	
Magnesium		%			0.46	
Phosphorus		%			0.18	
Potassium		%			3.15	

ZINC ACETATE

Zinc acetate, $Zn(C_2H_3O_2)_2 \cdot 2H_2O$, cp, (6)

Ref No 6 05 548 United States

Analyses				As Fed	Dry	C.V. ± %
Dry matter		%		83.6	100.0	
Iron		%		0.001	0.001	
Sulphur		%		0.07	0.08	
Zinc		%		29.8	35.6	

ZINC CARBONATE

Zinc carbonate, cp, (6)

Ref No 6 05 550 United States

Analyses				As Fed	Dry	C.V. ± %
Dry matter		%			100.0	
Chlorine		%			0.00	
Iron		%			0.002	
Sulphur		%			0.14	
Zinc		%			56.0	

Feed Name or Analyses		Mean As Fed	Mean Dry	C.V. ± %

ZINC CHLORIDE

Zinc chloride, $ZnCl_2$, cp, (6)
Ref No 6 05 552 United States

		As Fed	Dry	C.V. ± %
Dry matter	%		100.0	
Chlorine	%		52.03	
Iron	%		0.001	
Sulphur	%		0.07	
Zinc	%		48.0	

ZINC OXIDE

Zinc oxide, ZnO, cp, (6)
Ref No 6 05 554 United States

		As Fed	Dry	C.V. ± %
Dry matter	%		100.0	
Iron	%		0.001	
Manganese	%		0.0	
Sulphur	%		0.03	
Zinc	%		80.3	

ZINC SULFATE

Zinc sulfate, $ZnSO_4 \cdot 7H_2O$, cp, (6)
Ref No 6 05 556 United States

		As Fed	Dry	C.V. ± %
Dry matter	%	56.2	100.0	
Iron	%	0.001	0.001	
Manganese	%	0.0	0.0	
Sulphur	%	11.15	19.84	
Zinc	%	22.7	40.4	

(1) dry forages and roughages (2) pasture, range plants, and forages fed green (3) silages (4) energy feeds (5) protein supplements (6) minerals (7) vitamins (8) additives

GLOSSARY OF FEED TERMS

acid hydrolyzed See **hydrolyzed.**

acidified Treated with acid to lower the pH.

acid precipitated Separated from a solution by the action of acid

additive An ingredient or combination of ingredients added, usually in minute quantities, to the basic feed mix or parts thereof to fulfill a specific need. See **feed additive concentrate, feed additive supplement, feed additive premix, food additive.**

aerial part The above-ground part of a plant.

air-ashed Reduced by combustion in air to a mineral residue.

AIV preservative A preservative for silage consisting of a mixture of hydrochloric and sulfuric acids.

alcohol-extracted Treated with alcohol to remove all alcohol-soluble substances.

ammoniated Combined or impregnated with ammonia or an ammonium compound.

antibiotic A drug synthesized by a microorganism and having the power (in proper concentration) to inhibit the growth of other microorganisms.

apparent digestible energy (DE) Food-intake gross energy minus fecal energy. Syn: **apparent absorbed energy, energy of apparently digested food.** See **gross energy digestion coefficient.**

$$\text{DE} = (\text{GE of food per unit dry wt} \times \text{dry wt of food}) - (\text{GE of feces per unit dry wt} \times \text{dry wt of feces})$$

artificially dried Moisture removed by other than natural means. See **fan air-dried with heat.**

as fed See **dry-matter content of feed samples.**

aspirated Removal of light materials from heavier materials by use of air. Refers to chaff, dust, or other light materials.

bagasse Pulp from sugarcane. NRC term: **pulp.**

balanced Containing nutrients in amounts and proportions that fulfill physiological needs of animals as specified by recognized authorities in animal nutrition.*

barn-cured Dried with forced ventilation in an enclosure. Refers to forage. See **fan air-dried** (NRC term).

biscuits Shaped and baked dough.

bisulfite preservative An acid sulfite used to prevent decomposition of stored products.

bitterness-extracted Treated to remove bitter taste.

blended Mingled or combined. Refers to ingredients of a mixed feed; does not imply uniformity of dispersion.

block Agglomerated feed compressed into a solid mass (usually weighing 30 to 50 pounts) cohesive enough to hold its form. See **brick, pellets.**

blocked Compressed into a large, solid mass.

blood albumin One of the blood proteins.

blowings See **mill dust.**

bolls The pod or capsules of certain plants (e.g., flax and cotton).

*The species for which the feed or ration is intended, and its functions, such as maintenance or maintenance plus production (growth, fetus, fat, milk, eggs, wool, feathers, or work) should be specified.

bolls process residue The residue from immature and unopened cotton bolls after removal of fiber and seed.

bolted Separated from parent material by means of a bolting cloth. Refers to two ingredients (e.g., bran separated from flour).

bone ash The white, porous residue that remains after burning bones in air.

bone charcoal The product obtained by charring bones in a closed retort.

bone glue residue Part of the bone remaining (chiefly calcium phosphate) after removal of the part used in manufacturing bone glue.

bone phosphate The residue of bones that have been treated first in caustic solution, then in hydrochloric acid solution, and thereafter precipitated with lime and dried.

bone protein colloids The gelatinous material extracted from bones by moist heat treatment.

bran Pericarp of grain.

brand name Defined by the Association of American Feed Control Officials as "any word, name, symbol, or device, or any combination thereof, identifying the commercial feed of a distributor and distinguishing it from that of others."

brewers' grains The coarse, insoluble residue from brewed malt.

brick Agglomerated feed compressed into a solid mass (weighing less than 2 pounds) cohesive enough to hold its form. See **block, pellets.**

browse Small stems, leaves, flowers, and fruits of shrubs, trees, or woody vines.

cake The mass that results from pressing seeds, meat, or fish to remove oils, fats, or other liquids.

calcined Heated to high temperature in the presence of air.

calorie (cal) The unit for measuring chemical energy. It is defined as the amount of heat required to raise the temperature of 1 gram of water from 14.5° to 15.5°C at standard pressure. One thousand calories is designated as 1 kilocalorie (kcal),* and 1 million calories as 1 megacalorie (Mcal). One calorie is equivalent to 4.184 joules (J), which is the unit of electrical energy defined as 10^7 ergs or, practically, the energy expanded in 1 second by an electric current of 1 ampere in a resistance of 1 ohm. The standard calorie used for expressing the chemical energy in feeds and metabolic processes is based on the heat of combustion of benzoic acid, which has been precisely determined to be 771.36 ± 0.03 kcal/mole.

canned Processed, packaged, sealed, and sterilized in cans or similar containers.

cannery residue Edible residue that remains after a product is prepared for canning.

capsule chaff The light, fibrous material obtained by aspiration of flaxseed or flaxseed capsule.

carcass The body of an animal exclusive of the intestinal tract and lung tissue. (If head and skin are included, the

*The term "kilocalorie" is preferred to "Calorie" (capitalized) because "Calorie" and "calorie" are often confused.

term *carcass with head and skin* is used; if head and skin are not included, the term is *carcass without head and skin*.)

carcass meat trimmings Soft tissues obtained from slaughtered animals. The tissues consist chiefly of striate, skeletal, and cardiac muscles, but they may include the accompanying fat, skin, sinews, nerves, and blood vessels.

carcass residue, mammals Residue from carcasses exclusive of hair, hoofs, horns, and contents of the digestive tract. (If bones are included, the term *carcass residue with bones* is used.)

carrier An edible material (e.g., soybean meal) to which ingredients (e.g., vitamin A or riboflavin) are added. The added ingredients are absorbed, impregnated, or coated into or onto the edible material.

casein The protein precipitate that results from treating skim milk with acid or rennet.

centrifuged Separated by a force moving away from a center.

cereal by-product Secondary product resulting from the manufacture of a table cereal.

chaff Glumes, hulls, joints, and small fragments of straw that are separated from seed in threshing or processing.

chaff and dust Defined by the Association of American Feed Control Officials as material

> separated from grains or seeds in the usual commercial cleaning processes. It may include hulls, joints, straw, mill or elevator dust, sweepings, sand, dirt, grains, seeds. It must be labeled "chaff and/or dust." If it contains more than 15% ash the words "sand" and "dirt" must appear on the label.

See **screenings.**

charcoal Dark, porous forms of carbon made by incomplete combustion of plant or animal material.

chipped Cut or broken into fragments or cut into small, thin slices.

chopped Reduced in particle size by cutting.

cleaned Subjected to any process (e.g., scalping, screening, aspiration, or magnetic separation) by which unwanted material is removed.

cleanings Chaff, weed seeds, dust, and other foreign matter removed from cereal grains.

clipped Refers to removal of ends of whole grain.

close-planted Planted with less than normal distance between rows.

coagulated Curdled, clotted, or congealed, usually by the action of a coagulant.

coarse-bolted Separated from parent material by means of a coarsely woven bolting cloth.

coarse-sifted Separated from parent material by means of coarsely woven wire sieves.

cob fractions A mixture containing rings, or disks, cut from corn (maize) cobs and all or some of the following: glumes, lemmas, paleae, and sterile florets.

cobs furfural residue The residue from extraction of furfurals from corn (maize) cobs.

cobs with grain See **ears** (NRC term).

cobs with husks Corn (maize) cobs with the enveloping husks but without the grain.

commercial feed Defined in the Uniform Feed Bill (an Act) of the Association of American Feed Control Officials as follows:

> The term "commercial feed" means all materials which are distributed for use as feed or for mixing in feed, for animals other than man except:
>
> (1) Option A—Unmixed seed, whole or processed, made directly from the entire seed which are not adulterated within the meaning of Section 7 of this Act.
> Option B—Unmixed or unprocessed whole seeds which are not adulterated within the meaning of Section 7 of this Act.
> (2) Hay, straw, stover, silage, cobs, husks and hulls
> (i) when unground, and
> (ii) when unmixed with other materials.
>
> (3) Individual chemical compounds when not mixed with other materials which are by regulation exempted.

complete feed A nutritionally adequate feed for a specific animal in a specific physiological state. It is compounded to be fed as the sole diet and is capable of maintaining life or promoting production (or both) without the consumption of any additional substance except water.

concentrate A feed used with another to improve the nutritive balance of the total and intended to be diluted and mixed to produce a supplement or a complete feed.

condensed Reduced in volume by removal of moisture.

conditioned See **tempered** (NRC term).

cooked Heated in the presence of moisture to alter chemical or physical characteristics (or both) or to sterilize.

cracked Reduced in size by a combined breaking and crushing action. Refers to particles of grain.

cracklings The residue that remains after removal (by dry heat) of fat from adipose tissue or skin of animals.

crimped Rolled with corrugated rollers. The grain to which this term refers may be tempered or conditioned before it is crimped and may be cooled afterward.

crown On a seed plant, the point (usually at ground level) at which stems and roots merge.

crumbled Broken with corrugated rollers. Refers to pellets.

crumbles Pelleted feed reduced to granular form with corrugated rollers.

crushed See **rolled.**

cubes See **pellets, range cubes.**

cull Material rejected, in grading or separating, as inferior.

culture Nutrient medium bearing a colony of specific microorganisms.

cultured Produced in a culture. Refers to biological material.

cured Prepared for keeping or use (e.g., by drying, smoking, or salting or by using a chemical preservative).

customer-formula feed A commercial feed whose components are mixed according to the specific instructions of the final purchaser or contract feeder.

cuttings Parts or sections of a plant or animal.

D-activated Activated with vitamin D (e.g., by ultraviolet light). Refers to plant or animal sterol fractions.

debittered Having had bitter substances removed.

defluorinated Having had fluorine partially removed.

degermed Having had the embryos wholly or partially separated from the starch endosperms. Refers to seeds. See **without germ** (NRC term).

dehulled See **without hulls** (NRC term).

dehydrated Having had most of the moisture removed by heat.

deribbed Having had the primary veins removed. Refers to leaves.

dextrose equivalent A measurement of the reducing power of sugars and starch hydrolyzates calculated as dextrose. The equivalent is expressed as a percentage of the dry substance.

diet The feed and water regularly offered to or consumed by an animal.

digested Subjected to prolonged heat and moisture, or to chemicals or enzymes with a resultant change or decomposition of the physical or chemical nature.

digestible energy See **apparent digestible energy.**

diluent An edible substance that is mixed with a nutrient or additive to reduce its concentration and thereby make it more acceptable to animals, safer to use, or more amenable to being mixed uniformly in a feed. A diluent may also be a carrier.

distillers' grains Grains from which alcohol or alcoholic beverages have been distilled.

distillers' residue See **stillage.**

distillers' solubles Stillage filtrate.

distillers' stillage See **stillage.**

dressed Made uniform in texture by breaking or screening lumps or by applying liquid(s). Refers to feed.

dried See **dehydrated** (NRC term).

drug Defined by the U.S. Food and Drug Administration as follows:

> A substance (a) intended for use in the diagnosis, cure, mitigation, treatment or prevention of disease in man or other animals or (b) a substance other than food intended to affect the structure or any function of the body of man or other animals.

dry See **dry-matter content of feed samples.**

dry-matter content of feed samples Three terms are used to express the dry-matter content of feed samples and other materials: *as fed, partially dry,* and *dry.* Definitions of these terms follow.*

as fed . . . As fed refers to the feed as it is consumed by the animal; the term "as collected" is used for materials which are not usually fed to the animal, i.e., urine, feces, etc. If the analyses on a sample are affected by partial drying, the analyses are made on the "as fed" or "as collected" sample. Similar terms: air dry, i.e., hay; as received; fresh; green; wet.

partially dry . . . Partially dry refers to a sample of "as fed" or "as collected" material that has been dried in an oven (usually with forced air) at a temperature usually at 60°C or freeze dried and has been equilibrated with the air; the sample after these processes would usually contain more than 88% dry matter (12% moisture); some materials are prepared in this way so they may be sampled, chemically analyzed and stored. This analysis is referred to as "partial dry matter % of 'as fed' or 'as collected' sample." The partially dry sample must be analyzed for dry matter (determined in an oven at 105°C) to correct subsequent chemical analyses of the sample to a "dry" basis. This analysis is referred to as "dry matter % of partial dry sample." Similar term: air dry (sometimes air dry is used for "as fed"; see as fed).

dry . . . Dry refers to a sample of material that has been dried at 105°C until all the moisture has been removed. Similar terms: 100% dry matter; moisture free. If dry matter (in an oven at 105°C) is determined on an "as fed" sample it is referred to as "dry matter" of as fed sample. If dry matter is determined on a partial dry sample it is referred to as "dry matter of partial dry sample." It is recommended that analyses be reported on the "dry" basis (100% dry matter or moisture free), and in addition the "as fed dry matter" should be reported.

*Source: L. E. Harris. 1970. Nutrition research techniques for domestic and wild animals. Vol. I. An international record system and procedures for analyzing samples. Lorin E. Harris, 1408 Highland Drive, Logan, Utah.

dry-milled Milled by tempering with a small amount of water or steam to facilitate separation into component parts. Refers to kernels of grain.

dry-rendered Having undergone (1) cooking in open steam-jacketed vessels until the water has evaporated and (2) removal of fats by draining and pressing. Refers to residues of animal tissues.

dust Fine, dry particles of matter usually resulting from the cleaning or grinding of grain or other feedstuff.

ears Fruiting heads of corn, including cobs and grain but not the husks. Syn: **cobs with grain.**

egg albumen Whites of birds' eggs.

emulsifier A material that lowers the surface tension of the system to which it is added.

endosperm Starchy part of a seed.

endosperm oil Oil obtained from endosperms.

ensiled Preserved by ensiling, a process in which finely cut parts of plants, packed in an airtight chamber (e.g., a silo), undergo an acid fermentation that retards spoilage.

entire plant The whole plant, including the roots.

etiolated Grown in reduced light. Refers to plants.

eviscerated Subjected to removal of all organs. Refers to the great cavity of an animal's body.

expanded Increased in volume as a result of abrupt reduction in pressure. Refers to a feed or feed mixture that is extruded after being subjected to moisture, pressure, and temperature to gelatinize the starchy part.

expeller-extracted See **mechanically extracted**

extruded Pushed through orifices of a die under pressure. Refers to feed.

fan air-dried Dried with a device producing a current of air. See **barn-cured.**

fan air-dried with heat Dried with a device producing a current of heated air. See **artificially dried.**

fat A substance, solid or plastic at room temperature, composed chiefly of triglycerides of fatty acids.

fatty acids Aliphatic monobasic organic acids containing only the elements carbon, hydrogen, and oxygen.

fatty acids ethyl Saturated aliphatic monocarboxylic acids occurring naturally in fats, waxes, and essential oils in the form of ethyl ester, which is a class of compounds that yield ethyl alcohol on hydrolysis.

fatty acids methyl ester Saturated aliphatic monocarboxylic acids occurring naturally in fats, waxes, and essential oils in the form of methyl ester, which is a class of compounds that yield methyl alcohol on hydrolysis.

fatty acids nonglyceride ester Saturated aliphatic monocarboxylic acids occurring naturally in fats, waxes, and essential oils in the form of esters other than those of glycerol.

feed(s) Material(s) consumed by animals that contribute energy and nutrients (or both) to the diet.

feed additive concentrate Defined by the U.S. Food and Drug Administration as follows:

> An article intended to be further diluted to produce a complete feed or a feed additive supplement and is not suitable for offering as a supplement or for offering free choice without dilution. It contains, among other things, one or more additives in amounts in a suitable feed base such that from 100 to 1000 pounds of concentrate must be diluted to produce 1 ton of a complete feed. A "feed additive concentrate" is unsafe if fed free choice or as a supplement because of danger to the health of the animal or because of the production of residues in the edible products from food producing animals in excess of the safe levels established.

feed additive premix Defined by the U.S. Food and Drug Administration as follows:

> An article that must be diluted for safe use in a feed additive concentrate, a feed additive supplement or a complete feed. It contains, among other things, one or more additives in high concentration in a suitable feed base such that up to 100 pounds must be diluted to produce 1 ton of a complete feed. A feed additive premix contains additives at levels for which safety to the animal has not been demonstrated and/or which may result when fed undiluted in residues in the edible products from food producing animals in excess of the safe levels established.

feed additive supplement Defined by the U.S. Food and Drug Administration as follows:

> An article for the diet of an animal which contains one or more food additives and is intended to be:
> (1) Further diluted and mixed to produce a complete feed; or
> (2) Fed undiluted as a supplement to other feeds; or
> (3) Offered free choice with other parts of the rations separately available.
>
> A "feed additive supplement" is safe for the animal and will not produce unsafe residues in the edible products from food producing animals if fed according to directions.

feed grade suitable for animal, but not human, consumption.

feed mixture See **formula feed.**

fermentation product Product formed by enzymatic transformation of organic substrates.

fermentation solubles Parts of stillage that pass through screens, consisting chiefly of water, water-soluble substances, and fine particles from the fermentation process.

fermented Acted upon by yeasts, filamentous fungi, or bacteria in a controlled aerobic or anaerobic process. Refers to products (e.g., grains and molasses) used in the manufacture of alcohols, acids, vitamins of the B-complex group, and antibiotics.

fiber An elongate tapering plant cell that has at maturity no protoplasm. It is found chiefly in the vascular tissues of plants but may occur in other sites.

fiber by-product A secondary product obtained during the manufacture of a fiber product (e.g., flax fiber by-product, which is obtained during the manufacture of flax).

finely ground Reduced to very small particles by impact, shearing, or attrition.

finely screened Separated according to particle size by passage over or through wire screens.

finely sifted Separated according to particle size by passage through a finely woven meshed material.

fines Material that passes through a screen whose openings are smaller than the specified minimum size of crumbles, pellets, or substances such as citrus pulp.

fish stickwater An aqueous, oil-free extract of cooked fish. It contains the aqueous cell solutions of the fish and any water used in processing.

flaked 1. Prepared by a method involving the use of high heat, tempering, and rollers set close together. 2. Cut into flat pieces (e.g., potato flakes).

flakes An ingredient rolled or cut into flat pieces with or without prior steam conditioning. See **flaked.**

flour Soft, finely ground, bolted meal obtained by milling cereal grains and other seeds. It consists essentially of the starch and gluten of the endosperm.

flour by-product A secondary product obtained during the milling of grain for preparation of bread flour.

flower extract Material removed from flowers by leaching with a liquid.

fodder Green or cured plants (e.g., corn and sorghum) that are fed in their entirety, except for the roots, as forage. See **aerial part** (NRC term).

food additive Defined by the U.S. Food and Drug Administration as

> any substance which becomes a component of or affects the characteristics of a feed or food if such substance is not generally recognized among experts qualified by scientific training and experience to evaluate its safety as having been adequately shown through scientific procedures to be safe under the conditions of its intended use. Excepted are substances having "prior sanction" and pesticide chemicals under certain conditions.

forage Aerial plant material, primarily grasses and legumes containing more than 18 percent crude fiber on a dry basis, used as animal feed. The term usually refers only to such plant materials as pasture, hay, silage, and green chopped feeds.

formula feed Feed consisting of two or more ingredients proportioned, mixed, and processed according to the manufacturer's specifications.

free choice A feeding system in which animals are given unlimited access to the separate feeds or mixtures of feeds constituting the diet.

fresh Recently produced or gathered; not stored, cured, or preserved.

fungal amylase process distillers' grains with solubles The solid residue resulting from combining distillers' grains and solubles and drying after hydrolysis of the starch by fungal amylase.

fused Blended by melting.

gelatinized Ruptured by a combination of moisture, heat, and pressure. Refers to starch granules of a feed.

germ Embryo of a seed.

germ oil Oil extracted from the germ of cereal grains or other seeds.

gin by-product Material that remains after cotton fibers and seeds of cotton bolls are separated in ginning.

gland tissue An aggregate of cells of various special secreting organs with their intercellular contents.

glue by-product A secondary product obtained in manufacturing glue.

gluten The tough, viscid, nitrogenous substance that remains after the flour of wheat or other grain has been washed to remove the starch.

gluten-low glutamic acid Gluten from which some of the glutamic acid has been removed.

gossypol A phenolic pigment in cottonseed that is toxic to some animals.

graham flour Whole wheat flour; often a mixture of flour and bran.

grain clippings The hulls, fragments of groats, immature grains, and chaffy material obtained during the dehulling of oats and other cereal grains.

grain distillers' saccharomyces A genus of unicellular yeasts, which are fungi having little or no mycelial growth, reproducing asexually by budding, and typically produce alcoholic fermentations on carbohydrate substrates.

grain fines Small particles screened from cracked grain.

grain screenings Defined by the Association of American Feed Control Officials as materials obtained from screening grains and consisting "of 70% or more of grains, including light and broken grains, wild buckwheat, and wild oats. It must contain not more than 6.5% ash." See **screenings.**

grease Animal fats with a titer below 40°C.

grits Coarsely ground grain from which the bran and germ have been removed.

groats Grain from which the hulls have been removed.

gross energy (GE) The amount of heat that is released when a substance is completely oxidized in a bomb calorimeter containing 25 to 30 atmospheres of oxygen. Syn: **heat of combustion.**

gross energy digestion coefficient The percentage of gross energy apparently absorbed. GE digestion coefficient equals:

$$\frac{\text{(GE of food per unit dry wt} \times \text{dry wt of food)} - \text{(GE of feces per unit dry wt} \times \text{dry wt of feces)}}{\text{GE of food per unit dry wt} \times \text{dry wt of food}} \times 100$$

ground Reduced in particle size by impact shearing or attrition.

grounds with chicory residue Sediment (e.g., coffee grounds) that contains chicory residue.

hatchery by-product A mixture of eggshells, unhatched eggs, and culled chicks that has been cooked, dehydrated, and ground, with or without partial removal of fat.

hay The aerial parts of grass or herbage cut and cured for animal feeding.

heads The parts of a plant that contain the seeds (e.g., sorghum heads).

heads without seeds Heads from which seeds have been removed.

heat- and acid-precipitated Separated from a suspension or solution by action of heat and acid.

heat-hydrolyzed See **hydrolyzed.**

heat-processed Prepared by a method involving the use of elevated temperatures, with or without pressure.

heat-processed flaked See **flaked.**

heat-rendered Melted, extracted, or clarified by heating. (Water and fat are usually removed.)

homogenized Broken down into evenly distributed globules small enough to remain as an emulsion for long periods. Refers to particles of fat.

hulls Outer covering of seeds.

husks 1. Leaves enveloping an ear of corn. 2. Outer coverings of kernels or seeds, especially when dry and membranous (e.g., almond husks).

hydraulically extracted See **mechanically extracted.**

hydrolyzed Subjected to hydrolysis, a process by which complex molecules (e.g., those in proteins) are split into simpler units by chemical reaction with water molecules. (The reaction may be produced by an enzyme, catalyst, or acid or by heat and pressure.)

iodinated Treated with iodine.

irradiated Treated, prepared, or altered by exposure to radiated energy.

joints Nodes of plant stems.

joule (J) The International Organization for Standardization defines 1 joule as "the work done when the point of application of a force of one newton (N) is displaced through a distance of one meter (m) in the direction of the force." One calorie is equal to 4.184 joules. See **calorie.**

juice The aqueous substance obtained from biological tissue by pressing or filtering, with or without addition of water.

katabolizable energy See **nitrogen equilibrium metabolizable energy (ME_n).**

kernel In cereals, a whole grain; in other species, a dehulled seed.

kibbled Cracked or crushed. Refers to baked dough or to extruded feed that was cooked before or during the extrusion process.

kilocalorie (kcal) One thousand small calories. This term is preferable to Calorie (capitalized). See **calorie.**

lactase-hydrolyzed See **hydrolyzed.**

lactic acid bacteria Any of various bacteria (chiefly of the genera *Lactobacillus* and *Streptococcus*) that produce predominantly lactic fermentation on suitable media.

lactose A white crystalline disaccharide found in milk.

leached Affected by the action of percolating water or other liquid.

lecithin A specific phospholipid; the principal constituent of crude phosphatides derived from oil-bearing seeds.

lint A fibrous coat of thickened convoluted hairs on the seeds of cotton plants.

litter Fibrous material used on the floor of poultry houses, with the poultry excreta.

low oil Containing very little oil (usually 5 percent or less).

magnetic separation Removal of ferrous material by magnets (e.g., removal of iron objects from mixed feed).

malt Sprouted and steamed whole grain from which the radicle has been removed.

malted Converted into malt or treated with malt or malt extract.

malt hulls Product, consisting almost entirely of hulls, that is obtained in cleaning malted barley.

maltose processed Treated with the enzyme maltase.

malt sprout cleanings with hulls Product obtained in cleaning malted barley or in recleaning malt. (Contains less protein than malt sprouts with hulls.)

malt sprouts with hulls Sprouts from malted barley combined with malt hulls.

marc Pulp, seeds, and skins from grapes. See **pulp** (NRC term).

mash A mixture of ingredients in meal form. Syn: **mash feed.**

meal An ingredient(s) that has been ground or otherwise reduced to a particle size somewhat larger than flour, unbolted.

meat Flesh obtained from slaughtered mammals. (The term includes skeletal muscle, cardiac muscle, and the tongue, diaphragm, and esophagus; it sometimes includes the accompanying fat, skin, sinews, nerves, and blood vessels; it does not include the lips, snout, and ears.)

meat stickwater An aqueous, fat-free extract of meat. (It is obtained in wet-rendering meat products and contains the aqueous cell solutions, the soluble glue proteins, and water condensed from the steam used in wet-rendering.)

meats See **nut meats.**

meats with hulls Certain nut meats (e.g., those of peanuts) combined with their hulls.

mechanically extracted Extracted by heat and mechanical pressure. Refers to removal of fat or oil from the seeds. Syn: **expeller-extracted, hydraulically extracted, old process.**

mechanically extracted caked Extracted from seeds by heat and mechanical pressure in such a way that the remaining product (e.g., cottonseed oil) is caked. Refers to fat or oil.

media Nutrient substrate for culturing bacteria (or other organisms) or cells.

medicated feed 1. A feed that contains drug ingredients intended (a) to cure, mitigate, treat, or prevent diseases of animals other than man or (b) to affect the structure or functioning of the bodies of animals other than man. 2. A feed that contains an antibiotic intended to promote growth or increase feed efficiency.

medium with yeast Cells of yeast combined with (1) the liquor containing the medium in which the cells grow and (2) the by-products of the cells' metabolism.

megacalorie (Mcal) One million small calories. See **calorie.**

metabolizable energy (ME) Food intake gross energy, minus fecal energy, minus energy in the gaseous products of digestion, minus urinary energy.

micro-ingredients Vitamins, minerals, antibiotics, drugs, and other materials normally required in small amounts and measured in milligrams, micrograms, or parts per million.

middlings A by-product of flour milling that contains varying proportions of endosperm, bran, and germ.

milk albumin The coagulated protein fraction from whey.

mill dust Fine feed particles resulting from handling and processing feed and feed ingredients.

mill residue Part of a feed or feed ingredient that remains after a milling process.

mill run A product as it comes from the mill, ungraded and usually uninspected.

mixed Two or more materials combined by agitation to a specific degree of dispersion.

mixed screenings Defined by the Association of American Feed Control Officials as a mixture of material obtained from screening grains and of the screenings that are "excluded from the preceding definition (grain screenings). It must contain not more than 27% crude fiber and not more than 15% ash." See **screenings, grain screenings.**

molasses The thick, viscous by-product resulting from the manufacture of refined sugar.

molasses distillers' solubles Liquid containing dissolved substances obtained from molasses stillage.

molded Overgrown or otherwise acted upon by fungi.

needles Slender, pointed leaves, as of pine, spruce, and larch.

net energy The difference between metabolizable energy and heat increment. It includes the amount of energy used for maintenance only or for maintenance plus production. Net energy can also be defined as the gross energy of the gain in tissue or of the products synthesized plus the energy required for maintenance. Below the critical temperature, net energy includes the heat increment.*

net energy for maintenance (NE_m) The part of net energy expended to keep an animal in energy equilibrium. When

*Reports on net energy should clearly state which functions are included. Subscripts are suggested. For example, there may be values for net energy for maintenance plus production (NE_{m+p}), net energy for maintenance only (NE_m), or net energy for production only (NE_p).

an animal is in this state, there is no net gain or loss of energy in the body tissues. The net energy for maintenance for a producing animal may be different from that for a nonproducing animal of the same weight. The difference is due to changes in amounts of hormones produced and to differences in voluntary activity. This difference may be charged to maintenance, but in practice it is usually charged to the production requirement.

net energy for production (NE_p) The part of net energy, in addition to that needed for body maintenance, that is used for work or for tissue gain (growth or fat production or both), or for the synthesis of, for example, a fetus, milk, eggs or wool.*

new process See **solvent-extracted.**

nitrogen equilibrium metabolizable energy (ME_n) Food intake gross energy minus fecal energy, minus energy in the gaseous products of digestion, minus urinary energy, corrected for nitrogen retained in or lost from the body. For birds and monogastric mammals, the gaseous products of digestion need not be considered.†

nuts with shells Dry indehiscent fruit having a hard, bony ovary wall.

nuts with shells with husks Dry indehiscent fruit having a hard, bony wall enclosed by a dry outer covering (e.g., almond).

nut meats Nuts from which the shells have been removed.

offal Low-grade residue left from the milling of some product.

oil A substance that consists chiefly of triglycerides of fatty acids and is liquid at room temperature.

oil refinery lipid By-product obtained in refining an edible oil.

old process See **mechanically extracted.**

partially dry See **dry-matter content of feed samples.**

partially extracted Partially removed from a feed by a chemical or mechanical process. Refers to fat or oil.

paunch contents See **rument contents.**

pasture Grass or other plants grown for grazing animals; herbage.

pearl by-product By-product obtained in pearling barley. See **pearled.**

pearled Reduced by machine brushing to smaller, smooth particles. Refers to dehulled grains (e.g., pearled barley).

pectin Any of the group of colorless amorphous methylated pectic substances that occur in plant tissues or are obtained by restricted treatment of protopectin obtained from fruits or succulent vegetables and that yield viscous solutions with water and when combined with acid and sugar yield a gel.

peel See **skin.**

peelings Outer layers of fruits or vegetables that have been removed.

pellets Agglomerated feed formed by compacting and forcing feed through die openings mechanically. Syn: **pelleted feed, hard pellet.** *Soft pellets* are those containing sufficient liquid to require immediate dusting and cooling. Syn: **high molasses pellets.** See **block, brick.**

pH The negative logarithm (base 10) of the hydrogen ion concentration.

pith Continuous central strand of parenchymatous tissue occurring in the stems of many vascular plants.

pits Stones of drupacous fruits.

pit silo A below-ground bin sealed when full to exclude air and used for storing silage.

pod A dehiscent seed vessel or fruit (e.g., pea or bean pod).

polished Smoothed by a mechanical process. Refers to grain (e.g., polished rice).

polishing A by-product of rice consisting of the fine residue that accumulates as the rice kernels are polished (after hulls and bran have been removed).

pollen A mass of microspores (usually resembling a fine dust) in a seed plant.

pomace Pulp, seeds, and stems from fruit. See **pulp** (NRC term).

potassium salts A mineral compound containing potassium.

precipitated Separated from suspension or solution as a result of a chemical or physical change.

premix A uniform mixture consisting of one or more micro-ingredients and a diluent or carrier (or both) and used to facilitate uniform distribution of the micro-ingredients within a larger mixture.

premixed Mixed with a diluent or carrier (or both) preliminary to final mixing with other ingredients. Refers to micro-ingredients.

prepressed solvent-extracted Removed from materials partly by heat and mechanical pressure and (later) partly by organic solvents. Refers to fat or oil.

pressed Compacted or molded by pressure, or extracted under pressure. Refers to fat, oil, or juice.

presswater Aqueous extract obtained from fish or meat by hydraulic pressing of the fish or meat followed by removal of fat or oil (or both), usualy by centrifuging.

process residue Material remaining after some of the constituents of the original material (e.g., pineapple slices for canning) have been removed in a manufacturing process.

protein Any of a large class of naturally occurring complex combinations of amino acids.

pulp The solid residue (including seeds and skins, if present) remaining after extraction of juices from fruits, roots, or stems. See **bagassee, pomace, marc.**

*It should always be clearly stated which production fractions are included. For example, there could be NE_{egg}, NE_{gain}, NE_{milk}, NE_{preg}, NE_{wool}, or NE_{work}.

†For mammals the correction is made as follows: For each gram of nitrogen lost from the body (equal to negative nitrogen balance), 7.45 kcal are added to the metabolizable energy, and for each gram of nitrogen retained in the body (equal to positive nitrogen balance), 7.45 kcal are subtracted from the metabolizable energy. Since this value was obtained with dogs, it may not be entirely correct for other animals. In the case of animals synthesizing products such as milk or eggs, no correction is made for the nitrogen in these products. Similar term:

$$ME_n = GE_i - FE - GPD \pm \text{correction}$$

For birds, the preferable factor is 8.73 kcal because it represents the average gross energy of urine not contaminated with feces.

pulp fines See **fines.**

pulverized See **ground.**

raisin syrup by-product Residue from the manufacture of raisin syrup.

range cake Cake fed on the range, usually on the ground (e.g., cottonseed cake). See **cake.**

range cubes Large pellets intended to be fed on the ground. Syn: **range wafers.**

ration The total amount of feed (diet) allotted to one animal for a 24-hr period.

refuse Damaged, defective, or excess edible material produced during or left over from a manufacturing or industrial process.

retort-charred Partly burned in a closed retort. Refers to bone black.

rolled Compressed between rollers. Rolling may entail tempering or conditioning.

roughage Plant material, primarily by-products of crop production, high in crude fiber, low in digestibility, and low in protein. Examples are straw, stover, bagasse, peanut and oat hulls, and corn (maize) cobs.

rumen contents Contents of the first two compartments of the stomach of a ruminant. Syn: **paunch contents.**

scalped Removal by screening. Refers to large particles.

scoured Cleansed by impact or friction. Refers to removal of the beard from the wheat kernel.

scratch grain Whole, cracked, or coarsely cut grain. Syn: **scratch feed.**

screened Separated into different sizes by being passed over or through screens.

screenings Defined by the Association of American Feed Control Officials as material

> obtained in the cleaning of grains which are included in the United States Grain Standard Act and other agricultural seeds. It may include light and broken grains and agricultural seeds, weed seeds, hulls, chaff, joints, straw, elevator or mill dust, sand, and dirt. It must be designated as Grain Screenings, Mixed Screenings and Chaff and/or Dust. No grade of screenings must contain any seeds or other material in amount that is either injurious to animals or will impart an objectionable odor or flavor to their milk or flesh. The screenings must contain not more than four whole prohibited noxious weed seeds per pound and must contain not more than 100 whole restricted noxious weed seeds per pound. The prohibited and restricted noxious weed seeds must be those named as such by the seed control law of the states in which the screenings is sold or used.

See **chaff and dust, grain screenings, mixed screenings.**

seedballs Rounded and usually dry or capsular fruits (e.g., potato seed).

seed skins Outer layers of some seeds (e.g., beans and peas).

self-fed Provided on a continuous basis. Refers to a component of a diet or to mixed components. Self-feeding enables animals to eat at will.

separation Classification of particles by size, shape, or density.

separation, magnetic See **magnetic separation.**

shoots The immature aerial parts of plants.

shorts A by-product of flour milling that consists of germ, offal, fine particles of bran, and small amounts of flour.

shredded Cut into long, thin pieces.

sifted Separated into different sizes by being passed through wire or nylon sieves.

silk The styles on an ear of corn.

sizing See **screenings.**

skimmed Removed by settling, flotation, or centrifuging. Refers to removal of the lighter part of a liquid from the heavier part (e.g., removal of cream from milk).

skin 1. The outer covering of a fruit or seed. Syn: **rind, husk, peel.** 2. The dermal tissue of animals.

skin scrapings Scrapings from hides of slaughtered animals.

solubles Dissolved substances (and possibly fine solids) in liquids obtained in processing animal or plant materials.

solubles with low potassium salts and glutamic acid The residue from manufacturing monosodium glutamate from Steffen's filtrate.

solvent-extracted Removed from materials (e.g., soybean seeds) by organic solvents. Refers to fat or oil. Syn: **new process.**

solvent-extracted flakes with reduced protein and carbohydrates The product remaining after some of the protein and nitrogen-free extract have been removed from dehulled, solvent-extracted soybean flakes.

spent Exhausted of absorbing properties (e.g., spent bone black).

spent residue liquid The liquid residue that remains after extracting starch from potatoes.

spice Any of various aromatic vegetable products used to season foods.

spine A specialized, stiff, sharp-pointed leaf form.

split pea by-product The residue from the manufacture of split peas, consisting primarily of skins and broken and rejected peas.

stabilized Made more resistant to chemical change by an added substance.

stack-ensiled Ensiled while in a pile above ground.

stalk The main stem of a herbaceous plant.

starch by-product The residue from the manufacture of starch (e.g., starch from potatoes).

steamed Treated with steam, as in steam cooking. Syn: **steam-cooked, steam-rendered, tanked.**

steep-extracted Soaked in water or other liquid to remove soluble materials. Refers to grain (e.g., corn that is being wet-milled).

steepwater Water containing soluble materials removed by steep-extraction. See **steep-extracted.**

Steffen process A process for treating beet molasses to recover additional sugar through precipitation of calcium sucrate.

Steffen's filtrate The filtrate obtained from the precipitation of calcium sucrate in the Steffen process and used chiefly as a source of amino acids.

stem butts Proximal ends of stems.

sterols Solid cyclic alcohols that are the major constituents of the unsaponifiable part of animal and vegetable fats and oils.

stick Condensed stickwater or presswater. See **fish stickwater, meat stickwater, presswater.**

stickwater See **fish stickwater, meat stickwater.**

stickwater solubles Water-soluble fraction from fish from which the liquid, originally obtained by steam-cooking and pressing the fish, has been removed.

stickwater solubles precipitated Precipitated water-soluble fraction from fish stickwater.

stillage The mash from fermentation of grains or molasses after removal of alcohol by distillation.

stover Stalks and leaves of corn or sorghum after the ears of corn or heads of sorghum have been harvested. NRC terms: **aerial part without ears without husks, aerial part without heads.**

straw Plant residue remaining after separation of the seeds (grain, peas, or beans) by threshing. See **threshed.**

straw pulp A moist, slightly cohering mass consisting of ground straw treated with water.

stubble The lower parts of plant stems that remain standing in the field after harvest.

sulfite waste liquors Residues from products (e.g., wood pulp) treated with sulfite.

sulfur-fertilized Supplied with a form of elemental sulfur as a nutrient. Refers to plants.

sun-cured Dried by exposure to the sun.

sun-cured brown Partially dried by exposure to the direct rays of the sun, then put in a stack or bale, where heat from microbial action causes browning.

sun-cured on riders Stacked on tripods made of poles, then sun-cured. Refers to forage.

supplement A feed used with another to improve nutritive balance or performance. It may be fed undiluted as a supplement to other feeds, offered free choice with other parts of the ration separately available, or mixed with other feed ingredients to produce a complete feed.

syrup Concentrated juice of a fruit or plant.

syrup by-product A secondary product consisting chiefly of the fatty fractic of corn starch together with protein and residual carbohydrates. Syn: **corn syrup refinery insolubles.**

tallow Animal fats with titer above 40°C.

tankage See **carcass residue with blood** (NRC term).

tanked See **steamed.**

tassels Male inflorescences of some plants (e.g., the tassels at the end of a stalk of corn).

tempered Brought to predetermined moisture characteristics or temperature (or both) before further processing. Syn: **conditioned.**

threshed Separated from straw by impaction and subsequent screening. Refers to grain, peas, or beans. See **straw.**

titer The solidification point (determined by heating or cooling) of the fatty acids liberated from a fat by hydrolysis.

toasted Browned, dried, or parched by exposure to a wood fire or to gas or electric heat.

tops The uppermost parts of plants (e.g., sugarcane tops). See **aerial parts.**

toxicity-extracted Treated to remove a poisonous substance.

trace mineral Mineral nutrient required by animals in very small amounts.

trench silo A trench that is filled with fresh forage and then sealed to exclude air and permit the formation of silage.

tubers Short, thickened fleshy stems, or rhizomes, that usually form underground and bear minute scaled leaves, each with a bud capable of developing into a new plant (e.g., potato).

unsaponifiable matter Ether-soluble material extractable after complete reaction with strong alkali.

urea A highly soluble, crystalline, white compound used as a source of nonprotein nitrogen for ruminants. It is produced by mammals during nitrogen metabolism and is also produced synthetically.

vacuum-dehydrated Dehydrated under vacuum. See **dehydrated**

vinegar fermentation grains Grains used as the substrate to provide a source of carbohydrate that is transformed into vinegar.

viscera All organs in the great cavity of the body. The viscera of fish include the gills, heart, liver, spleen, stomach, and intestines and their contents. The viscera of mammals include the esophagus, heart, lungs, liver, spleen, stomach, and intestines (but not their contents). The viscera of poultry include the esophagus, heart, liver, spleen, stomach, crop, gizzard, undeveloped eggs, and intestines and their contents.

vitamins Organic compounds that function as parts of enzyme systems that are essential for transmitting energy and regulating metabolism.

wafered Agglomerated by compressing into a form that usually measures more in diameter or cross-section than in length. Refers to fibrous feeds (e.g., wafered alfalfa hay).

waste See **refuse** (NRC term).

water-extracted Removed with water. Refers to a product from which soluble substances have been removed.

weathered Exposed to air, sunlight, and precipitation.

wet-milled Steeped in water, which may contain sulfur dioxide, to facilitate separation of the parts. Refers to kernels of corn.

wet-rendered Cooked with steam under pressure in closed tanks.

whey The watery part of milk separated from the coagulated curd.

whey albumin One of the whey proteins.

whey fermentation solubles Whey together with the water-soluble substances produced during the fermentation of whey.

whey low lactose The product resulting from removal of some of the milk sugar from whey.

whole plant See **entire plant** (NRC term).

whole-pressed Pressed to remove oil. Refers to seeds with hulls (e.g., cotton seeds).

wilted A product that has lost turgor as a result of desiccation.

without germ See **degermed.**

without hulls See **dehulled.**

wort The liquid portion of malted grain. It is a solution of malt sugar and other water-soluble extracts from malted mash.

yeast fermentation grains Residue of grains after being used as a source of carbohydrate for yeast fermentation.

yeast with medium See **medium with yeast.**

These definitions were prepared by the Subcommittee on Feed Composition of the Committee on Animal Nutrition of the National Research Council and the Definitions Committee of the American Feed Manufacturers Association (formerly the Midwest Feed Production School's Definitions Committee) and were reviewed by representatives of the Association of American Feed Control Officials.

Certain definitions in this glossary are adaptations of those to be found in standard dictionaries, and some are verbatim reproductions.